CHILTON'S™

AUTO REPAIR MANUAL
1995–99

President	Dean F. Morgantini, S.A.E.
Vice President–Finance	Barry L. Beck
Vice President–Sales	Glenn D. Potere
Executive Editor	Kevin M. G. Maher, A.S.E.
Production Manager	Ben Greisler, S.A.E.
Production Assistant	Melinda Possinger
Project Managers	George B. Heinrich III, A.S.E., S.A.E., Will Kessler, A.S.E., S.A.E., James R. Marotta, A.S.E., S.T.S., Richard Schwartz, A.S.E., Todd W. Stidham
Schematics Editor	Christopher G. Ritchie
Editors	Leonard Davis, A.S.E., S.T.S. Frank Keytanjian, A.S.E., S.A.E.

CHILTON™ *Automotive Books*

PUBLISHED BY **W. G. NICHOLS, INC.**

Manufactured in USA
© 1998 W. G. Nichols
1020 Andrew Drive
West Chester, PA 19380
ISBN 0-8019-7922-6
Library of Congress Catalog Card No. 98-71351
1234567890 7654321098

Table of Contents

Car Sections

Specifications	1	Specifications, Scheduled Maintenance Interval and Scheduled Maintenance Labor Time Charts
MIL Resetting and DTC Retrieval	2	Maintenance Light (MIL) Resetting and Diagnostic Code (DTC) Retrieval
Firing Orders	3	Firing Order Diagrams
Accessory Drive Belts	4	Accessory Serpentine and V-Belt Service Procedures
Timing Belts	5	Timing Belt Removal and Installation Procedures
Brakes	6	Brake System Component Service Procedures
Driveshafts, U-Joints and CV-Joint Boots	7	Driveshaft, U-Joint and CV-Joint Boot Service Procedures
Oxygen (O2) Sensors	8	O2 Sensor General Information, Testing, Removal and Installation Service Procedures
Electric Cooling Fans	9	Electric Cooling Fan Testing, Removal and Installation Procedures, Fan System Wiring Schematics
Starting and Charging Systems	10	Starting and Charging System Description, Testing and Service Procedures
Piston, Piston Ring and Connecting Rod Positioning	11	Piston, Piston Ring and Connecting Rod Positioning Diagrams
Chrysler Corp.	12	300M, Concord, LHS, New Yorker, Intrepid, Vision
	13	Sebring Coupe, Avenger
	14	Talon
	15	Neon
	16	Cirrus, Sebring Convertible, Stratus, Breeze
Ford Motor Co.	17	Aspire
	18	Probe

Table of Contents

Car Sections

Ford Motor Co. (cont.)	19	Contour, Mystique, Cougar (1999)
	20	Taurus, Sable
	21	Continental
	22	Escort, Escort ZX2, Tracer
	23	Mustang
	24	Mark VIII
	25	Thunderbird, Cougar (1995-97)
	26	Crown Victoria, Town Car, Grand Marquis
General Motors	27	A Body—Century, Cutlass Ciera, Cutlass Cruiser
	28	B Body—Roadmaster, Fleetwood, Caprice, Impala SS
	29	C & H Bodies—Le Sabre, Park Ave., Eighty Eight, Eighty Eight LS, LSS, Regency, Bonneville
	30	E & K Bodies—DeVille, Concours, Eldorado, Seville
	31	F Body—Camaro, Z28, Firebird, Trans Am
	32	G Body—Riviera, Aurora
	33	J Body—Cavalier, Sunfire
	34	L Body—Beretta, Corsica
	35	L/N Body—Malibu, Cutlass
	36	N Body—Skylark, Achieva, Grand Am
	37	V Body—Catera
	38	W Body—Century, Regal, Lumina, Monte Carlo, Cutlass Supreme, Intrigue, Grand Prix
	39	Y Body—Corvette
Geo/Chevrolet	40	Metro, Prizm
Saturn	41	SC1, SC2, SL, SL1, SL2, SW1, SW2

Model Index

Model	Section No.	Model	Section No.	Model	Section No.
3		**D**		**N**	
300M	12-1	DeVille	30-1	Neon	15-1
A		**E**		New Yorker	12-1
Achieva	36-1	Eighty Eight	29-1	**P**	
Aspire	17-1	Eighty Eight LS	29-1	Park Ave.	29-1
Aurora	32-1	Eldorado	30-1	Prizm	40-1
Avenger	13-1	Escort	22-1	Probe	18-1
B		Escort ZX2	22-1	**R**	
Beretta	34-1	**F**		Regal	38-1
Bonneville	29-1	Firebird	31-1	Regency	29-1
Breeze	16-1	Fleetwood	28-1	Riviera	32-1
C		**G**		Roadmaster	28-1
Camaro	31-1	Grand Am	36-1	**S**	
Caprice	28-1	Grand Marquis	26-1	Sable	20-1
Catera	37-1	Grand Prix	38-1	Saturn Coupe	41-1
Cavalier	33-1	**I**		Saturn Sedan	41-1
Century	27-1	Impala SS	28-1	Saturn Wagon	41-1
Century	38-1	Intrepid	12-1	Sebring Convertible	16-1
Cirrus	16-1	Intrigue	38-1	Sebring Coupe	13-1
Concord	12-1	**L**		Seville	30-1
Concours	30-1	Le Sabre	29-1	Skylark	36-1
Continental	21-1	LHS	12-1	Stratus	16-1
Contour	19-1	LSS	29-1	Sunfire	33-1
Corsica	34-1	Lumina	38-1	**T**	
Corvette	39-1	**M**		Talon	14-1
Cougar (1995-97)	25-1	Malibu	35-1	Taurus	20-1
Cougar (1999)	19-1	Mark VIII	24-1	Thunderbird	25-1
Crown Victoria	26-1	Metro	40-1	Town Car	26-1
Cutlass	35-1	Monte Carlo	38-1	Tracer	22-1
Cutlass Ciera	27-1	Mustang	23-1	Trans Am	31-1
Cutlass Cruiser	27-1	Mystique	19-1	**V**	
Cutlass Supreme	38-1			Vision	12-1

HOW TO USE THIS MANUAL

Model Specific Sections

The model specific sections are grouped by manufacturer and arranged in alphabetical order. The text and illustrations that comprise the service procedures in each model specific section are arranged in the following order of systems and components: Engine Repair (Gasoline, then Diesel if applicable), Fuel System (Gasoline, then Diesel if applicable), Drive Train, Steering and Suspension.

All illustrations are located as close as possible to the applicable procedure. Procedures are for all models in the particular section unless specifically noted otherwise.

Unit Repair Sections

The Unit Repair Sections (URS's) are written to cover all applicable 1995-99 models for the specific URS system or component, unless specifically noted otherwise. The procedures covered in the 10 URS's are not repeated in the model specific sections; therefore, refer to the URS's for the service procedures for the applicable systems or components. Refer to the Table of Contents for URS coverage.

Locating Information

The Table of Contents, located at the front of the book, lists each Unit Repair Section (URS) and model specific section in this manual.

To find where a particular model specific section is located in the book, you need only look in the Table of Contents. Once you have found the proper section, you may wish to find where specific procedures located in that section. Turn to the Index at the front of the model specific section. At the upper left-hand side is a listing of the main topics within that section and the page number on which they may be found. Following the main topics is an alphabetical listing of all of the procedures within the section and their page numbers.

The Model Index, located just after the Table of Contents in the beginning of this manual, may also be used to locate the specific section for any vehicle model covered in this manual.

Safety Notice

Proper service and repair procedures are vital to the safe, reliable operation of all motor vehicles, as well as the personal safety of those performing the repairs. This manual outlines procedures for servicing and repairing vehicles using safe effective methods. The procedures contain many NOTES, WARNINGS and CAUTIONS which should be followed along with standard safety procedures to eliminate the possibility of personal injury or improper service which could damage the vehicle or compromise its safety.

It is important to note that repair procedures and techniques, tools and parts for servicing vehicles, as well as the skill and experience of the individual performing the work vary widely. It is not possible to anticipate all of the conceivable ways or conditions under which vehicles may be serviced, or to provide cautions as to all of the possible hazards that may result. Standard and accepted safety precautions and equipment should be used when handling toxic or flammable fluids, and safety goggles or other protection should be used during cutting, grinding, chiseling, prying, or any other process that can cause material removal or projectiles.

Some procedures require the use of tools specially designed for a specific purpose. Before substituting another tool or procedure, you must be completely satisfied that neither your personal safety, nor the performance of the vehicle will be endangered.

Although information in this manual is based on industry sources and is as complete as possible at the time of publication, the possibility exists that some vehicle manufacturers made later changes which could not be included here. Information on very late models may not be available in some circumstances. While striving for total accuracy, NP/Chilton cannot assume responsibility for any errors, changes, or omissions that may occur in the compilation of this data.

Part Numbers

Part numbers listed in this book are not recommendations by NP/Chilton for any product by brand name. They are references that can be used with interchanges manuals and aftermarket supplier catalogs to locate each brand supplier's discrete part number.

Special Tools

Special tools are recommended by the vehicle manufacturer to perform their specific job. Use has been kept to a minimum, but where absolutely necessary, they are referred to in the text by the part number of the tool manufacturer. These tools may be purchased, under the appropriate part number, from your local dealer or regional distributor, or an equivalent tool can be purchased locally from a tool supplier or parts outlet. Before substituting any tool for the one recommended, read the previous Safety Notice.

Copyright Notice

NP/Chilton would like to thank all manufacturer's involved for their generous assistance.

Get ready for ASE testing with Motor Age Self Study Guides.

Each training unit contains a complete description of the ASE Task Analysis and Test Specifications, and covers the subject areas of the corresponding ASE test question group. Also included are sample ASE test questions. In addition, each book includes a special glossary, sample questions and an expanded answer analysis to increase your knowledge of the subject.

AA Car & Light Truck
A1 Engine Repair
A2 Automatic
Transmission/Transaxle
A3 Manual Drive Train & Axles
A4 Suspension & Steering
A5 Brakes
A6 Electrical/Electronic Systems
A7 Heating & A/C
A8 Engine Performance

Parts Specialist
P1 Medium/Heavy Parts Specialist
P2 Automobile Parts Specialist

Advanced Level
L1 Advanced Engine Performance Specialist
L2 Med/Hvy Vehicle Electronic Diesel Engine Diagnosis Specialist
F1 Light Vehicle Compressed Natural Gas

ALSO AVAILABLE:

TT Medium/Heavy Truck Service: T1 Gasoline Engines, T2 Diesel Engines, T3 Drive Train, T4 Brakes, T5 Suspension & Steering, T6 Electrical/Electronic Systems, T7 Heating, Ventilation & A/C, T8

Preventive Maintenance Inspection (PMI), & MM Engine Machinist (M1, M2, M3),

BB Collision Repair/Paint & Refinish: B2 Paint & Refinishing, B3 Non-Structural Analysis & Damage Repair, B4 Structural Analysis & Damage Repair, B5 Mechanical & Electrical Components, B6 Damage Analysis & Estimating

Name: _____
        first   middle   last

Company _____

Address: _____ Apt. # _____

City: _____ State: _____ Zip: _____

Phone (DAYTIME): _____ Fax _____

For pricing and shipping information, fax this form to Trudy Kolb, 610-964-4251

SPECIFICATIONS

1

CHRYSLER CORP.
Chrysler 300M • Concorde • LHS • New Yorker • Dodge Intrepid • Eagle Vision 1-2
Chrysler Cirrus • Sebring Convertible • Dodge Stratus • Plymouth Breeze 1-9
Chrysler Lebaron • Dodge Spirit • Plymouth Acclaim 1-16
Chrysler Sebring Coupe • Dodge Avenger 1-23
Dodge/Plymouth Neon 1-28
Dodge Stealth 1-33
Eagle Talon 1-38

FORD MOTOR CO.
Ford Aspire...................................1-50
Ford Contour • Mercury Mystique • Cougar (1999)............................1-44
Ford Crown Victoria • Lincoln Town Car • Mercury Grand Marquis....1-55
Ford Escort • Escort ZX2 • Mercury Tracer 1-60
Ford Mustang • Thunderbird • Lincoln Mark III • Mercury Cougar 1-68

Ford Probe 1-78
Ford Taurus • Taurus SHO • Lincoln Continental • Mercury Sable 1-83

GENERAL MOTORS A BODY
Buick Century • Oldsmobile Cutlass • Cutlass Ciera..........................1-92

GENERAL MOTORS B BODY
Buick Roadmaster • Cadillac Fleetwood Brougham • Chevrolet Caprice • Impala SS 1-98

GENERAL MOTORS C & H BODIES
Buick LeSabre • Park Avenue • Oldsmobile Eighty-Eight • Ninety-Eight • LSS • Regency • Pontiac Bonneville 1-103

GENERAL MOTORS E & K BODIES
Cadillac Deville • Deville Concours • Eldorado • Seville 1-110

GENERAL MOTORS F BODY
Chevrolet Camaro • Pontiac Firebird.................................1-117

GENERAL MOTORS G BODY
Buick Riviera • Oldsmobile Aurora 1-123

GENERAL MOTORS J BODY
Chevrolet Cavalier • Pontiac Sunfire....................................1-129

GENERAL MOTORS L BODY
Chevrolet Corsica • Beretta 1-134

GENERAL MOTORS L/N BODY
Chevrolet Malibu • Oldsmobile Cutlass 1-139

GENERAL MOTORS N BODY
Buick Skylark • Oldsmobile Achieva • Pontiac Grand Am 1-144

GENERAL MOTORS V BODY
Cadillac Catera 1-150

GENERAL MOTORS W BODY
Buick Regal • Century (1997-99) • Chevrolet Lumina • Monte Carlo • Oldsmobile Cutlass Supreme • Intrigue • Pontiac Grand Prix ...1-152

GENERAL MOTORS Y BODY
Chevrolet Corvette.......................1-160

GEO/CHEVROLET
Metro • Prizm............................1-165

SATURN
SC1 • SC2 • SL • SL1 • SL2 • SW1 • SW2..................1-173

CHRYSLER CORP.
Chrysler 300M • Concorde • LHS • New Yorker
Dodge Intrepid • Eagle Vision

VEHICLE IDENTIFICATION CHART

Engine Code						Model Year	
Code	Liters	Cu. In. (cc)	Cyl.	Fuel Sys.	Eng. Mfg.	Code	Year
F	3.5	215 (3518)	V6	MFI	Chrysler	W	1998
G	3.5	215 (3518)	V6	MFI	Chrysler	X	1999
J	3.2	195 (3231)	V6	MFI	Chrysler	T	1996
R	2.7	167 (2736)	V6	MFI	Chrysler	S	1995
T	3.3	201 (3301)	V6	MFI	Chrysler	V	1997

MFI - Multi-point Fuel Injection

79221C01

ENGINE IDENTIFICATION
All measurements are given in inches.

Year	Model	Engine Displacement Liters (cc)	Engine Series (ID/VIN)	Fuel System	No. of Cylinders	Engine Type
1995	Concorde	3.5 (3518)	F	MFI	6	SOHC
	Concorde	3.3 (3301)	T	MFI	6	OHV
	Intrepid	3.5 (3518)	F	MFI	6	SOHC
	Intrepid	3.3 (3300)	T	MFI	6	OHV
	LHS	3.5 (3518)	F	MFI	6	SOHC
	New Yorker	3.5 (3518)	F	MFI	6	SOHC
	Vision	3.5 (3518)	F	MFI	6	SOHC
	Vision	3.3 (3301)	T	MFI	6	OHV
1996	Concorde	3.5 (3518)	F	MFI	6	SOHC
	Concorde	3.3 (3301)	T	MFI	6	OHV
	Intrepid	3.5 (3518)	F	MFI	6	SOHC
	Intrepid	3.3 (3300)	T	MFI	6	OHV
	LHS	3.5 (3518)	F	MFI	6	SOHC
	New Yorker	3.5 (3518)	F	MFI	6	SOHC
	Vision	3.5 (3518)	F	MFI	6	SOHC
	Vision	3.3 (3301)	T	MFI	6	OHV
1997	Concorde	3.5 (3518)	F	MFI	6	SOHC
	Intrepid	3.5 (3518)	F	MFI	6	SOHC
	Intrepid	3.3 (3300)	T	MFI	6	OHV
	LHS	3.5 (3518)	F	MFI	6	SOHC
	Vision	3.5 (3518)	F	MFI	6	SOHC
1998-99	300M	3.5 (3518)	G	MFI	6	SOHC
	Concorde	3.2 (3231)	J	MFI	6	SOHC
	Concorde	2.7 (2736)	R	MFI	6	DOHC
	Intrepid	3.2 (3231)	J	MFI	6	SOHC
	Intrepid	2.7 (2736)	R	MFI	6	DOHC
	LHS	3.5 (3518)	G	MFI	6	SOHC
	Vision	3.2 (3231)	J	MFI	6	SOHC
	Vision	2.7 (2736)	R	MFI	6	DOHC

MFI - Multi-point Fuel Injection
OHV - Overhead Valve
SOHC - Single Overhead Camshaft
DOHC - Double Overhead Camshafts

79221C02

GENERAL ENGINE SPECIFICATIONS

Year	Engine ID/VIN	Engine Displacement Liters (cc)	Fuel System Type	Net Horsepower @ rpm	Net Torque @ rpm (ft. lbs.)	Bore x Stroke (in.)	Compression Ratio	Oil Pressure @ rpm
1995	F	3.5 (3518)	MFI	214@5800	221@3100	3.78x3.19	9.6:1	27-70@3000
	T	3.3 (3301)	MFI	161@5300	181@3200	3.66x3.19	8.9:1	30-80@3000
1996	F	3.5 (3518)	MFI	214@5800	221@3100	3.78x3.19	9.6:1	25-80@3000
	T	3.3 (3301)	MFI	161@5300	181@3200	3.66x3.19	8.9:1	30-80@3000
1997	F	3.5 (3518)	MFI	214@5800	221@3100	3.78x3.19	9.6:1	25-80@3000
	T	3.3 (3301)	MFI	161@5300	181@3200	3.66x3.19	8.9:1	30-80@3000
1998-99	G	3.5 (3518)	MFI	214@5800	221@3100	3.78x3.19	9.6:1	25-80@3000
	J	3.2 (3231)	MFI	225@6300	225@3800	3.62x3.19	9.5:1	25-80@3000
	R	2.7 (2736)	MFI	200@5800	190@4850	3.39x3.09	9.6:1	25-80@3000

MFI - Multi-point Fuel Injection

79221C03

GASOLINE ENGINE TUNE-UP SPECIFICATIONS

Year	Engine ID/VIN	Engine Displacement Liters (cc)	Spark Plugs Gap (in.)	Ignition Timing (deg.) MT	Ignition Timing (deg.) AT	Fuel Pump (psi)		Idle Speed (rpm) MT	Idle Speed (rpm) AT	Valve Clearance In.	Valve Clearance Ex.
1995	F	3.5 (3518)	0.048-0.053	—	①	48	②	—	750-1100	HYD	HYD
	T	3.3 (3301)	0.048-0.053	—	①	55	②	—	600-840	HYD	HYD
1996	F	3.5 (3518)	0.033-0.038	—	①	48	②	—	750-1100	HYD	HYD
	T	3.3 (3301)	0.048-0.053	—	①	55	②	—	600-840	HYD	HYD
1997	F	3.5 (3518)	0.033-0.038	—	①	48	②	—	750-1100	HYD	HYD
	T	3.3 (3301)	0.048-0.053	—	①	55	②	—	600-840	HYD	HYD
1998-99	G	3.5 (3518)	0.048-0.053	—	①	49	②	—	750-1100	HYD	HYD
	J	3.2 (3231)	0.048-0.053	—	①	49	②	—	600-840	HYD	HYD
	R	2.7 (2736)	0.048-0.053	—	①	49	②	—	600-840	HYD	HYD

NOTE: The Vehicle Emission Control Information label often reflects specification changes made during production. The label figures must be used if they differ from those in this chart.

HYD - Hydraulic

① Basic ignition timing not adjustable

② This reading measured with vacuum hose disconnected from fuel pressure regulator

79221C04

CAPACITIES

Year	Model	Engine ID/VIN	Engine Displacement Liters (cc)	Engine Oil with Filter (qts.)	Transmission (pts.) 4-Spd	5-Spd	Auto.	Transfer Case (pts.)	Drive Axle Front (pts.)	Rear (pts.)	Fuel Tank (gal.)	Cooling System (qts.)
1995	Concorde	F	3.5 (3518)	5.5	—	—	19.8 ①	—	2.0	—	18.0	11.8
	Concorde	T	3.3 (3301)	5.0	—	—	19.8 ①	—	2.0	—	18.0	10.2
	Intrepid	F	3.5 (3518)	5.5	—	—	19.8 ①	—	2.0	—	18.0	11.8
	Intrepid	T	3.3 (3300)	5.0	—	—	19.8 ①	—	2.0	—	18.0	10.2
	LHS	F	3.5 (3518)	5.5	—	—	19.8 ①	—	2.0	—	18.0	11.8
	New Yorker	F	3.5 (3518)	5.5	—	—	19.8 ①	—	2.0	—	18.0	11.8
	Vision	F	3.5 (3518)	5.5	—	—	19.8 ①	—	2.0	—	18.0	11.8
	Vision	T	3.3 (3301)	5.0	—	—	19.8 ①	—	2.0	—	18.0	10.1
1996	Concorde	F	3.5 (3518)	5.5	—	—	19.7 ①	—	2.0	—	18.0	11.8
	Concorde	T	3.3 (3301)	5.0	—	—	19.8 ①	—	2.0	—	18.0	10.2
	Intrepid	F	3.5 (3518)	5.5	—	—	19.8 ①	—	2.0	—	18.0	11.8
	Intrepid	T	3.3 (3300)	5.0	—	—	19.8 ①	—	2.0	—	18.0	10.2
	LHS	F	3.5 (3518)	5.5	—	—	19.8 ①	—	2.0	—	18.0	11.8
	New Yorker	F	3.5 (3518)	5.5	—	—	19.8 ①	—	2.0	—	18.0	11.8
	Vision	F	3.5 (3518)	5.5	—	—	19.8 ①	—	2.0	—	18.0	11.8
	Vision	T	3.3 (3301)	5.0	—	—	19.8 ①	—	2.0	—	18.0	10.2
1997	Concorde	F	3.5 (3518)	5.5	—	—	19.7 ①	—	2.0	—	18.0	11.8
	Intrepid	F	3.5 (3518)	5.5	—	—	19.8 ①	—	2.0	—	18.0	11.8
	Intrepid	T	3.3 (3300)	5.0	—	—	19.8 ①	—	2.0	—	18.0	10.2
	LHS	F	3.5 (3518)	5.5	—	—	19.8 ①	—	2.0	—	18.0	11.8
	Vision	T	3.3 (3301)	5.0	—	—	19.8 ①	—	2.0	—	18.0	10.2
1998-99	300M	G	3.5 (3518)	5.5	—	—	19.8 ①	—	2.0	—	18.0	11.8
	Concorde	J	3.2 (3231)	5.0	—	—	19.8 ①	—	2.0	—	18.0	8.0
	Concorde	R	2.7 (2736)	5.0	—	—	19.8 ①	—	2.0	—	18.0	8.0
	Intrepid	J	3.2 (3231)	5.0	—	—	19.8 ①	—	2.0	—	18.0	8.0
	Intrepid	R	2.7 (2736)	5.0	—	—	19.8 ①	—	2.0	—	18.0	8.0
	LHS	G	3.5 (3518)	5.5	—	—	19.8 ①	—	2.0	—	18.0	11.8
	Vision	J	3.2 (3231)	5.0	—	—	19.8 ①	—	2.0	—	18.0	8.0
	Vision	R	2.7 (2736)	5.0	—	—	19.8 ①	—	2.0	—	18.0	8.0

NOTE: All capacities are approximate. Add fluid gradually and ensure a proper fluid level is obtained.

① Overhaul fill capacity with torque converter empty
Estimated service fill 9 pts.

79221C05

VALVE SPECIFICATIONS

Year	Engine ID/VIN	Engine Displacement Liters (cc)	Seat Angle (deg.)	Face Angle (deg.)	Spring Test Pressure (lbs. @ in.)	Spring Installed Height (in.)	Stem-to-Guide Clearance (in.)		Stem Diameter (in.)	
							Intake	Exhaust	Intake	Exhaust
1995	F	3.5 (3518)	45-45.5	44.5-45	①	1.496	0.0009-0.0026	0.0020-0.0037	0.2730-0.2737	0.2719-0.2726
	T	3.3 (3301)	45-45.5	44.5-45	②	1.622-1.681	0.0010-0.0100	0.0020-0.0060	0.3120-0.3130	0.3112-0.3119
1996	F	3.5 (3518)	45-45.5	44.5-45	①	1.496	0.0009-0.0026	0.0020-0.0037	0.2730-0.2737	0.2719-0.2726
	T	3.3 (3301)	45-45.5	44.5-45	②	1.622-1.681	0.0010-0.0100	0.0020-0.0060	0.3120-0.3130	0.3112-0.3119
1997	F	3.5 (3518)	45-45.5	44.5-45	①	1.496	0.0009-0.0026	0.0020-0.0037	0.2730-0.2737	0.2719-0.2726
	T	3.3 (3301)	45-45.5	44.5-45	②	1.622-1.681	0.0010-0.0100	0.0020-0.0060	0.3120-0.3130	0.3112-0.3119
1998-99	G	3.5 (3518)	45-45.5	44.5-45	①	1.496	0.0009-0.0026	0.0020-0.0037	0.2730-0.2737	0.2719-0.2726
	J	3.2 (3231)	45-45.5	44.5-45	③	1.496	0.0009-0.0026	0.0020-0.0037	0.2730-0.2737	0.2719-0.2726
	R	2.5 (2736)	45-45.5	44.5-45	④	1.496	0.0009-0.0026	0.0020-0.0033	0.2337-0.2344	0.2326-0.2333

① Intake: 201.7-218.3 lbs.@1.1752 in.
   Exhaust: 158.5-171.5 lbs.@1.239 in.
② Intake: 95-100 lbs.@1.570 in. valve closed
   Exhaust: 207-229 lbs.@1.169 in. valve closed
③ Intake: 69.5-80.5 lbs.@1.496 in. valve closed
   Intake: 188.0-204.0 lbs.@1.1594 in. valve opened
   Exhaust: 56-64 lbs.@1.4961 in. valve closed
   Exhaust: 124-136 lbs.@1.239 in. valve opened
④ 56-64 lbs.@ 1.496 in. valve closed
   Intake: 147.9-162.1 lbs.@1.1417 in. valve opened
   Exhaust: 138.0-150.8 lbs.@1.811 in. valve opened

79221C06

TORQUE SPECIFICATIONS
All readings in ft. lbs.

Year	Engine ID/VIN	Engine Displacement Liters (cc)	Cylinder Head Bolts	Main Bearing Bolts	Rod Bearing Bolts	Crankshaft Damper Bolts	Flywheel Bolts	Manifold		Spark Plugs	Lug Nut
								Intake	Exhaust		
1995	F	3.5 (3518)	①	②	③	85	75	21	17	20	95
	T	3.3 (3301)	④	⑤	③	40	75	17	17	20	95
1996	F	3.5 (3518)	①	②	③	85	75	21	17	20	100
	T	3.3 (3301)	④	⑤	③	40	75	17	17	20	100
1997	F	3.5 (3518)	①	②	③	85	75	21	17	20	100
	T	3.3 (3301)	④	⑤	③	40	75	17	17	20	100
1998-99	G	3.5 (3518)	①	②	③	85	75	⑥	⑦	20	100
	J	3.2 (3231)	①	⑧	③	75	75	17	17	20	100
	R	2.7 (2736)	⑨	⑩	⑧	125	75	⑫	⑦	15	100

① Step 1: 45 ft. lbs.
  Step 2: 65 ft. lbs.
  Step 3: 65 ft. lbs.
  Step 4: Plus 1/4 turn
  Final torque should be over 90 ft. lbs.
② Main cap bolts: 30 ft. lbs. plus 1/4 turn
  Main cap tie bolts: 40 ft. lbs.
③ Step 1: 40 ft. lbs.
  Step 2: Plus 1/4 turn
④ Step 1: 45 ft. lbs.
  Step 2: 65 ft. lbs.
  Step 3: 65 ft. lbs.
  Step 4: Plus 1/4 turn
  Torque small bolt in rear of cylinder head to 25 ft. lbs.
⑤ Step 1: 30 ft. lbs
  Step 2: Plus 1/4 turn
⑥ M8 bolts: 250 in. lbs.
  M6 bolts: 105 in. lbs.
⑦ 200 in. lbs.
⑧ Main cap inside bolts: 15 ft. lbs. plus 1/4 turn
  Main cap outside bolts: 20 ft. lbs. plus 1/4 turn
  Main cap tie bolts: 250 in. lbs.
⑨ Step 1: 35 ft. lbs.
  Step 2: 55 ft. lbs.
  Step 3: 55 ft. lbs.
  Step 4: Plus 90 degrees
  Step 5: M8 bolts (3 front) 250 in. lbs.
⑩ Main cap tie bolts: 250 in. lbs.
  Main cap inner bolts: 15 ft. lbs. plus 1/4 turn
  Main cap outer bolts: 20 ft. lbs. plus 1/4 turn
⑪ Step 1: 20 ft. lbs.
  Step 2: Plus 1/4 turn
⑫ 105 in. lbs.

79221C07

SCHEDULED MAINTENANCE INTERVALS
(CHRYSLER CONCORDE, LHS, NEW YORKER, DODGE INTREPID & EAGLE VISION)

TO BE SERVICED	TYPE OF SERVICE	VEHICLE MILEAGE INTERVAL (x1000)												
		7.5	15	22.5	30	37.5	45	52.5	60	67.5	75	82.5	90	97.5
Engine oil & filter	R	✓	✓	✓	✓	✓	✓	✓	✓	✓	✓	✓	✓	✓
Exhaust system	S/I	✓	✓	✓	✓	✓	✓	✓	✓	✓	✓	✓	✓	✓
Brake hoses	S/I	✓	✓	✓	✓	✓	✓	✓	✓	✓	✓	✓	✓	✓
CV joints & front suspension components	S/I	✓	✓	✓	✓	✓	✓	✓	✓	✓	✓	✓	✓	✓
Rotate tires	S/I	✓	✓	✓	✓	✓	✓	✓	✓	✓	✓	✓	✓	✓
Coolant level, hoses & clamps	S/I	✓	✓	✓	✓	✓	✓	✓	✓	✓	✓	✓	✓	✓
Accessory drive belts	S/I		✓		✓		✓		✓		✓		✓	
Brake linings	S/I		✓	✓				✓	✓				✓	
Spark plugs	R				✓				✓				✓	
Air filter element	R				✓				✓				✓	
Lubricate steering linkage & tie rod ends	S/I				✓				✓				✓	
Engine Coolant	R						✓				✓			
PCV valve	S/I								✓				✓	
Ignition cables	R								✓					
Camshaft timing belt	R								✓					

R – Replace   S/I – Service or Inspect

FREQUENT OPERATION MAINTENANCE (SEVERE SERVICE)

If a vehicle is operated under any of the following conditions it is considered severe service:
- Extremely dusty areas.
- 50% or more of the vehicle operation is in 32°C (90°F) or higher temperatures, or constant operation in temperatures below 0°C (32°F).
- Prolonged idling (vehicle operation in stop and go traffic).
- Frequent short running periods (engine does not warm to normal operating temperatures).
- Police, taxi, delivery usage or trailer towing usage.

CV joints & front suspension components – check every 3000 miles.
Oil & oil filter change – change every 3000 miles.
Rotate tires every 3000 miles.
Brake linings – check every 9000 miles.
Air filter element – change every 15,000 miles.
Automatic transaxle fluid – change every 15,000 miles.
Differential fluid – change every 15,000 miles.
Tie rod ends & steering linkage – lubricate every 15,000 miles.
PCV valve – check every 30,000 miles.

79221C08

SCHEDULED MAINTENANCE
CHRYSLER CORPORATION
CHRYSLER CONCORDE, LHS, NEW YORKER, 300M
DODGE INTREPID, EAGLE VISION

The following should be used as a guide when determining the amount of work required for a particular service if taken to a repair shop. In estimating how long a particular Scheduled Maintenance Service should take, please observe the following:

- **Chilton Time** is time based on field research and data supplied by the vehicle manufacturer.
- Labor time operations are given in hours and tenths of an hour.
- All labor operations, are to be used as a guide.

Mechanic Skill Level Codes:
    (G) GENERAL: Normally skilled with certification.
    (M) MAINTENANCE: Semi-skilled working on certication.
    (P) PRECISION: Really skilled with multiple certification.

	Chilton Time
(M) 7500 Mile Service	
1995-99	1.2
(M) 15000 Mile Service	
1995-99	1.8
(M) 22500 Mile Service	
1995-99	1.7
(G) 30000 Mile Service	
1995-99	2.7
(M) 37500 Mile Service	
1995-99	1.2

	Chilton Time
(G) 45000 Mile Service	
1995-99	2.3
(M) 52500 Mile Service	
1995-99	1.2
(G) 60000 Mile Service	
1995-99	3.6
Renew timing belt, add	3.5

	Chilton Time
(M) 67500 Mile Service	
1995-99	1.7
(M) 75000 Mile Service	
1995-99	1.8
(M) 82500 Mile Service	
1995-99	1.2
(G) 90000 Mile Service	
1995-99	3.4
(G) 97500 Mile Service	
1995-99	1.2

79221C09

CHRYSLER CORP.
Chrysler Cirrus • Sebring Convertible
Dodge Stratus • Plymouth Breeze

VEHICLE IDENTIFICATION CHART

Engine Code						Model Year	
Code	Liters	Cu. In. (cc)	Cyl.	Fuel Sys.	Eng. Mfg.	Code	Year
C	2.0	122 (1996)	I4	MFI	Chrysler	S	1995
H	2.5	152 (2497)	V6	MFI	Mitsubishi	T	1996
X	2.4	148 (2429)	I4	MFI	Chrysler	V	1997
						W	1998
						X	1999

MFI - Multi-point Fuel Injection

79221C31

ENGINE IDENTIFICATION

All measurements are given in inches.

Year	Model	Engine Displacement Liters (cc)	Engine Series (ID/VIN)	Fuel System	No. of Cylinders	Engine Type
1995	Cirrus	2.5 (2497)	H	MFI	6	SOHC
	Stratus	2.0 (1996)	C	MFI	4	SOHC
	Stratus	2.5 (2497)	H	MFI	6	SOHC
	Stratus	2.4 (2429)	X	MFI	4	DOHC
1996	Breeze	2.0 (1996)	C	MFI	4	SOHC
	Cirrus	2.5 (2497)	H	MFI	6	SOHC
	Cirrus	2.4 (2429)	X	MFI	4	DOHC
	Sebring Convertible	2.5 (2497)	H	MFI	6	SOHC
	Sebring Convertible	2.4 (2429)	X	MFI	4	DOHC
	Stratus	2.0 (1996)	C	MFI	4	SOHC
	Stratus	2.5 (2497)	H	MFI	6	SOHC
	Stratus	2.4 (2429)	X	MFI	4	DOHC
1997	Breeze	2.0 (1996)	C	MFI	4	SOHC
	Cirrus	2.5 (2497)	H	MFI	6	SOHC
	Cirrus	2.4 (2429)	X	MFI	4	DOHC
	Sebring Convertible	2.5 (2497)	H	MFI	6	SOHC
	Sebring Convertible	2.4 (2429)	X	MFI	4	DOHC
	Stratus	2.0 (1996)	C	MFI	4	SOHC
	Stratus	2.5 (2497)	H	MFI	6	SOHC
	Stratus	2.4 (2429)	X	MFI	4	DOHC
1998-99	Breeze	2.0 (1996)	C	MFI	4	SOHC
	Cirrus	2.5 (2497)	H	MFI	6	SOHC
	Cirrus	2.4 (2429)	X	MFI	4	DOHC
	Sebring Convertible	2.5 (2497)	H	MFI	6	SOHC
	Sebring Convertible	2.4 (2429)	X	MFI	4	DOHC
	Stratus	2.0 (1996)	C	MFI	4	SOHC
	Stratus	2.5 (2497)	H	MFI	6	SOHC
	Stratus	2.4 (2429)	X	MFI	4	DOHC

MFI - Multi-point Fuel Injectiom
SOHC - Single Overhead Camshaft
DOHC - Double Overhead Camshafts

79221C32

GENERAL ENGINE SPECIFICATIONS

Year	Engine ID/VIN	Engine Displacement Liters (cc)	Fuel System Type	Net Horsepower @ rpm	Net Torque @ rpm (ft. lbs.)	Bore x Stroke (in.)	Compression Ratio	Oil Pressure @ rpm
1995	C	2.0 (1996)	MFI	132@6000	129@5000	3.44x3.26	9.8:1	25-80@3000
	H	2.5 (2497)	MFI	164@5900	163@4350	3.29x2.99	9.4:1	35-75@3000
	X	2.4 (2429)	MFI	140@5200	160@4000	3.44x3.98	9.4:1	25-80@3000
1996	C	2.0 (1996)	MFI	132@6000	129@5000	3.44x3.26	9.8:1	25-80@3000
	H	2.5 (2497)	MFI	164@5900	163@4350	3.29x2.99	9.4:1	35-75@3000
	X	2.4 (2429)	MFI	140@5200	160@4000	3.44x3.98	9.4:1	25-80@3000
1997	C	2.0 (1996)	MFI	132@6000	129@5000	3.44x3.26	9.8:1	25-80@3000
	H	2.5 (2497)	MFI	164@5900	163@4350	3.29x2.99	9.4:1	35-75@3000
	X	2.4 (2429)	MFI	140@5200	160@4000	3.44x3.98	9.4:1	25-80@3000
1998-99	C	2.0 (1996)	MFI	132@6000	129@5000	3.44x3.26	9.8:1	25-80@3000
	H	2.5 (2497)	MFI	164@5900	163@4350	3.29x2.99	9.4:1	35-75@3000
	X	2.4 (2429)	MFI	140@5200	160@4000	3.44x3.98	9.4:1	25-80@3000

MFI - Multi-point Fuel Injection

79221C33

GASOLINE ENGINE TUNE-UP SPECIFICATIONS

Year	Engine ID/VIN	Engine Displacement Liters (cc)	Spark Plugs Gap (in.)	Ignition Timing (deg.) MT	Ignition Timing (deg.) AT	Fuel Pump (psi)	Idle Speed (rpm) MT	Idle Speed (rpm) AT	Valve Clearance In.	Valve Clearance Ex.
1995	C	2.0 (1996)	0.035	①	①	48	②	②	HYD	HYD
	H	2.5 (2497)	0.038-0.043	—	①	49 ③	—	500-1100	HYD	HYD
	X	2.4 (2429)	0.050	—	①	49	—	②	HYD	HYD
1996	C	2.0 (1996)	0.035	①	①	48	②	②	HYD	HYD
	H	2.5 (2497)	0.038-0.043	—	①	49 ③	—	500-1100	HYD	HYD
	X	2.4 (2429)	0.050	—	①	49	—	②	HYD	HYD
1997	C	2.0 (1996)	0.035	①	①	48	②	②	HYD	HYD
	H	2.5 (2497)	0.038-0.043	—	①	49 ③	—	500-1100	HYD	HYD
	X	2.4 (2429)	0.050	—	①	49	—	②	HYD	HYD
1998-99	C	2.0 (1996)	0.035	①	①	48	②	②	HYD	HYD
	H	2.5 (2497)	0.038-0.043	—	①	49 ③	—	500-1100	HYD	HYD
	X	2.4 (2429)	0.050	—	①	49	—	②	HYD	HYD

NOTE: The Vehicle Emission Control Information label often reflects specification changes made during production. The label figures must be used if they differ from those in this chart.

HYD - Hydraulic

① Ignition timing cannot be adjusted. Base engine timing is set at TDC during assembly.
② Refer to the Vehicle Emission Control Information label for correct timing specifications with a range of +/- 2 degrees
③ This reading measured with vacuum hose disconnected from fuel pressure regulator

79221C34

CAPACITIES

Year	Model	Engine ID/VIN	Engine Displacement Liters (cc)	Engine Oil with Filter (qts.)	Transmission (pts.)			Transfer Case (pts.)	Drive Axle		Fuel Tank (gal.)	Cooling System (qts.)
					4-Spd	5-Spd	Auto.		Front (pts.)	Rear (pts.)		
1995	Cirrus	H	2.5 (2497)	4.5	—	—	18.2 ②	—	—	—	16.0	10.5
	Stratus	C	2.0 (1996)	4.5	—	4.4 ①	—	—	—	—	16.0	7.4
	Stratus	H	2.5 (2497)	4.5	—	—	8.0	—	—	—	16.0	10.5
1996	Breeze	C	2.0 (1996)	4.5	—	①	8.0	—	—	—	16.0	7.4
	Cirrus	H	2.5 (2497)	4.5	—	—	18.2 ②	—	—	—	16.0	10.5
	Cirrus	X	2.4 (2429)	5.0	—	—	18.2 ②	—	—	—	16.0	9.0
	Sebring Convertible	H	2.5 (2497)	4.5	—	—	18.2 ②	—	—	—	16.0	10.5
	Sebring Convertible	X	2.4 (2429)	4.5	—	—	18.2 ②	—	—	—	16.0	9.0
	Stratus	C	2.0 (1996)	4.5	—	4.4 ①	—	—	—	—	16.0	7.4
	Stratus	H	2.5 (2497)	4.5	—	—	8.0	—	—	—	16.0	10.5
	Stratus	X	2.4 (2429)	4.5	—	—	8.0	—	—	—	16.0	9.0
	Stratus	X	2.4 (2429)	4.5	—	—	8.0	—	—	—	16.0	9.0
1997	Breeze	C	2.0 (1996)	4.5	—	①	8.0	—	—	—	16.0	7.4
	Cirrus	H	2.5 (2497)	4.5	—	—	18.2 ②	—	—	—	16.0	10.5
	Cirrus	X	2.4 (2429)	5.0	—	—	18.2 ②	—	—	—	16.0	9.0
	Sebring Convertible	H	2.5 (2497)	4.5	—	—	18.2 ②	—	—	—	16.0	10.5
	Sebring Convertible	X	2.4 (2429)	4.5	—	—	18.2 ②	—	—	—	16.0	9.0
	Stratus	C	2.0 (1996)	4.5	—	4.4 ①	—	—	—	—	16.0	7.4
	Stratus	H	2.5 (2497)	4.5	—	—	8.0	—	—	—	16.0	10.5
	Stratus	X	2.4 (2429)	4.5	—	—	8.0	—	—	—	16.0	9.0
1998-99	Breeze	C	2.0 (1996)	4.5	—	①	8.0	—	—	—	16.0	7.4
	Cirrus	H	2.5 (2497)	4.5	—	—	18.2 ②	—	—	—	16.0	10.5
	Cirrus	X	2.4 (2429)	5.0	—	—	18.2 ②	—	—	—	16.0	9.0
	Sebring Convertible	H	2.5 (2497)	4.5	—	—	18.2 ②	—	—	—	16.0	10.5
	Sebring Convertible	X	2.4 (2429)	4.5	—	—	18.2 ②	—	—	—	16.0	9.0
	Stratus	C	2.0 (1996)	4.5	—	4.4 ①	—	—	—	—	16.0	7.4
	Stratus	H	2.5 (2497)	4.5	—	—	8.0	—	—	—	16.0	10.5
	Stratus	X	2.4 (2429)	4.5	—	—	8.0	—	—	—	16.0	9.0

NOTE: All capacities are approximate. Add fluid gradually and ensure a proper fluid level is obtained.

① Fill to bottom of fill hole

② Overhaul fill capacity with torque converter empty

79221C35

VALVE SPECIFICATIONS

Year	Engine ID/VIN	Engine Displacement Liters (cc)	Seat Angle (deg.)	Face Angle (deg.)	Spring Test Pressure (lbs. @ in.)	Spring Installed Height (in.)	Stem-to-Guide Clearance (in.)		Stem Diameter (in.)	
							Intake	Exhaust	Intake	Exhaust
1995	C	2.0 (1996)	44.5-45	45-45.5	75@1.54	1.540	0.0018-0.0025	0.0029-0.0037	0.2340	0.2330
	H	2.5 (2497)	44-44.5	45-45.5	60@1.74	1.740	0.0008-0.0020	0.0016-0.0028	0.2360	0.2360
	X	2.4 (2429)	45-45.5	44.5-45	76@1.50	1.496	0.0018-0.0025	0.0029-0.0037	0.2340	0.2330
1996	C	2.0 (1996)	44.5-45	45-45.5	75@1.54	1.540	0.0018-0.0025	0.0029-0.0037	0.2340	0.2330
	H	2.5 (2497)	45-44.5	45-45.5	60@1.74	1.740	0.0008-0.0020	0.0016-0.0028	0.2360	0.2360
	X	2.4 (2429)	45-44.5	45-45.5	76@1.50	1.496	0.0018-0.0025	0.0029-0.0037	0.2340	0.2330
1997	C	2.0 (1996)	44.5-45	45-45.5	75@1.54	1.540	0.0018-0.0025	0.0029-0.0037	0.2340	0.2330
	H	2.5 (2497)	45-44.5	45-45.5	60@1.74	1.740	0.0008-0.0020	0.0016-0.0028	0.2360	0.2360
	X	2.4 (2429)	45-44.5	45-45.5	76@1.50	1.496	0.0018-0.0025	0.0029-0.0037	0.2340	0.2330
1998-99	C	2.0 (1996)	44.5-45	45-45.5	75@1.54	1.540	0.0018-0.0025	0.0029-0.0037	0.2340	0.2330
	H	2.5 (2497)	45-44.5	45-45.5	60@1.74	1.740	0.0008-0.0020	0.0016-0.0028	0.2360	0.2360
	X	2.4 (2429)	45-44.5	45-45.5	76@1.50	1.496	0.0018-0.0025	0.0029-0.0037	0.2340	0.2330

79221C36

TORQUE SPECIFICATIONS
All readings in ft. lbs.

Year	Engine ID/VIN	Engine Displacement Liters (cc)	Cylinder Head Bolts	Main Bearing Bolts	Rod Bearing Bolts	Crankshaft Damper Bolts	Flywheel Bolts	Manifold		Spark Plugs	Lug Nut
								Intake	Exhaust		
1995	C	2.0 (1996)	①	②	③	105	70	17	17	20	95
	H	2.5 (2497)	80	60	37	134	70	16	22	18	95
	X	2.4 (2429)	①	②	③	100	70	17	17	20	95
1996	C	2.0 (1996)	①	②	③	105	70	17	17	20	95
	H	2.5 (2497)	80	60	37	134	70	16	22	18	95
	X	2.4 (2429)	①	②	③	100	70	17	17	20	95
1997	C	2.0 (1996)	①	②	③	105	70	17	17	20	95
	H	2.5 (2497)	80	60	37	134	70	16	22	18	95
	X	2.4 (2429)	①	②	③	100	70	17	17	20	95
1998-99	C	2.0 (1996)	①	②	③	105	70	17	17	20	95
	H	2.5 (2497)	80	60	37	134	70	16	22	18	95
	X	2.4 (2429)	①	②	③	100	70	17	17	20	95

① Step 1: 25 ft. lbs.
   Step 2: 50 ft. lbs.
   Step 3: 50 ft. lbs.
   Step 4: Plus 1/4 turn
② Step 1: 30 ft. lbs
   Step 2: Plus 1/4 turn
③ Step 1: 20 ft. lbs.
   Step 2: +90 degrees

79221C37

SCHEDULED MAINTENANCE INTERVALS
(CHRYSLER STRATUS & SEBRING CONVERTIBLE, DODGE CIRRUS, PLYMOUTH BREEZE)

TO BE SERVICED	TYPE OF SERVICE	VEHICLE MILEAGE INTERVAL (x1000)												
		7.5	15	22.5	30	37.5	45	52.5	60	67.5	75	82.5	90	97.5
Engine oil & filter	R	✓	✓	✓	✓	✓	✓	✓	✓	✓	✓	✓	✓	✓
Brake hoses	S/I	✓	✓	✓	✓	✓	✓	✓	✓	✓	✓	✓	✓	✓
Coolant level, hoses & clamps	S/I	✓	✓	✓	✓	✓	✓	✓	✓	✓	✓	✓	✓	✓
CV joints & front suspension components	S/I	✓	✓	✓	✓	✓	✓	✓	✓	✓	✓	✓	✓	✓
Exhaust system	S/I	✓	✓	✓	✓	✓	✓	✓	✓	✓	✓	✓	✓	✓
Rotate tires	S/I	✓	✓	✓	✓	✓	✓	✓	✓	✓	✓	✓	✓	✓
Accessory drive belts	S/I		✓			✓			✓		✓		✓	
Brake linings	S/I			✓			✓			✓			✓	
Air filter element	R				✓				✓				✓	
Spark plugs①	R				✓				✓				✓	
Lubricate front & rear ball joints	S/I				✓				✓				✓	
Engine Coolant	R					✓					✓			
PCV valve	S/I								✓				✓	
Ignition cables②	R								✓					
Camshaft timing belt③	R													

① 4-cylinder shown; 6-cylinder = 100,000 miles.
② 1995 shown. 1996-99: 4-cylinder 60,000 miles, 6-cylinder 100,000 miles.
③ Replace at 105,000 miles for normal service; replace at 102,000 miles for severe service.
R – Replace   S/I – Service or Inspect

FREQUENT OPERATION MAINTENANCE (SEVERE SERVICE)
If a vehicle is operated under any of the following conditions it is considered severe service:
- Extremely dusty areas.
- 50% or more of the vehicle operation is in 32°C (90°F) or higher temperatures, or constant operation in temperatures below 0°C (32°F).
- Prolonged idling (vehicle operation in stop and go traffic).
- Frequent short running periods (engine does not warm to normal operating temperatures).
- Police, taxi, delivery usage or trailer towing usage.
Oil & filter change – change every 3000 miles.
Rotate tires every 6000 miles.
Brake linings – check every 12,000 miles.
Air filter element – service or inspect every 15,000 miles.
Automatic transaxle – change fluid & adjust bands (if equipped) every 15,000 miles.
PCV valve – check every 30,000 miles.
Engine coolant, replace at 36,000, 51,000 & 81,000 miles.

79221C38

SCHEDULED MAINTENANCE
CHRYSLER CORPORATION
CHRYSLER CIRRUS, SEBRING CONVERTIBLE
DODGE STRATUS, PLYMOUTH BREEZE

The following should be used as a guide when determining the amount of work required for a particular service if taken to a repair shop. In estimating how long a particular Scheduled Maintenance Service should take, please observe the following:

- **Chilton Time** is time based on field research and data supplied by the vehicle manufacturer.
- Labor time operations are given in hours and tenths of an hour.
- All labor operations, are to be used as a guide.

Mechanic Skill Level Codes:
>   **(G) GENERAL:** Normally skilled with certification.
>   **(M) MAINTENANCE:** Semi-skilled working on certication.
>   **(P) PRECISION:** Really skilled with multiple certification.

	Chilton Time
(M) 7500 Mile Service 1995-99	1.2
(M) 15000 Mile Service 1995-99	1.3
(M) 22500 Mile Service 1995-99	1.5
(G) 30000 Mile Service 1995-99	2.5
(M) 37500 Mile Service 1995-99	1.2

	Chilton Time
(G) 45000 Mile Service 1995-99	2.0
(M) 52500 Mile Service 1995-99	1.2
(G) 60000 Mile Service 1995-99	3.5

	Chilton Time
(M) 67500 Mile Service 1995-99	1.5
(M) 75000 Mile Service 1995-99	1.8
(M) 82500 Mile Service 1995-99	1.2
(G) 90000 Mile Service 1995-99	3.0
(M) 97500 Mile Service 1995-99	1.2

79221C39

CHRYSLER CORP.
Chrysler Lebaron • Dodge Spirit • Plymouth Acclaim

VEHICLE IDENTIFICATION CHART

		Engine Code				Model Year	
Code	Liters	Cu. In. (cc)	Cyl.	Fuel Sys.	Eng. Mfg.	Code	Year
3	3.0	181 (2972)	V6	MFI	Mitsubishi	S	1995
K	2.5	153 (2507)	I4	TFI	Chrysler		
V	2.5	153 (2507)	I4	MFI-M85	Chrysler		

M85-85% Methanol, Flexible Fuel Engine
MFI - Multi-point Fuel Injection
TFI - Throttle body Fuel Injection

79221C10

ENGINE IDENTIFICATION
All measurements are given in inches.

Year	Model	Engine Displacement Liters (cc)	Engine Series (ID/VIN)	Fuel System	No. of Cylinders	Engine Type
1995	Acclaim	3.0 (2972)	3	MFI	6	SOHC
	Acclaim	2.5 (2507)	K	TFI	4	SOHC
	Acclaim	2.5 (2507)	V	MFI-M85	4	SOHC
	LeBaron	3.0 (2972)	3	MFI	6	SOHC
	Spirit	3.0 (2972)	3	MFI	6	SOHC
	Spirit	2.5 (2507)	K	TFI	4	SOHC
	Spirit	2.5 (2507)	V	MFI-M85	4	SOHC

TFI - Throttle body Fuel Injection
SOHC - Single Overhead Camshaft
MFI - Multi-point Fuel Injection
M85 - 85% Methanol Flexible Fuel Engine

79221C11

GENERAL ENGINE SPECIFICATIONS

Year	Engine ID/VIN	Engine Displacement Liters (cc)	Fuel System Type	Net Horsepower @ rpm	Net Torque @ rpm (ft. lbs.)	Bore x Stroke (in.)	Compression Ratio	Oil Pressure @ rpm
1995	3	3.0 (2972)	MFI	141@5000	170@2800	3.59x2.99	8.9:1	35-75@3000
	K	2.5 (2507)	TFI	100@4800	135@2800	3.44x4.09	8.9:1	25-80@3000
	V	2.5 (2507)	MFI-M85	106@4400	145@2400	3.44x4.09	9.0:1	25-80@3000

MFI - Multi-point Fuel Injection
TFI - Throttle body Fuel Injection
M85 - 85% Methanol Flexible Fuel Engine

79221C12

GASOLINE ENGINE TUNE-UP SPECIFICATIONS

Year	Engine ID/VIN	Engine Displacement Liters (cc)	Spark Plugs Gap (in.)	Ignition Timing (deg.) MT	Ignition Timing (deg.) AT	Fuel Pump (psi)	Idle Speed (rpm) MT	Idle Speed (rpm) AT	Valve Clearance In.	Valve Clearance Ex.
1995	3	3.0 (2972)	0.035	—	①	48	—	700	HYD	HYD
	K	2.5 (2507)	0.035	—	①	39	—	850	HYD	HYD
	V	2.5 (2507)	0.035	—	①	55	—	850	HYD	HYD

NOTE: The Vehicle Emission Control Information label often reflects specification changes made during production. The label figures must be used if they differ from those in this chart.

HYD - Hydraulic

① Refer to the Vehicle Emission Control Information label for proper timing specifications with a range of +/- 2 degrees

79221C13

CAPACITIES

Year	Model	Engine ID/VIN	Engine Displacement Liters (cc)	Engine Oil with Filter (qts.)	Transmission (pts.) 4-Spd	Transmission (pts.) 5-Spd	Transmission (pts.) Auto.	Transfer Case (pts.)	Drive Axle Front (pts.)	Drive Axle Rear (pts.)	Fuel Tank (gal.)	Cooling System (qts.)
1995	Acclaim	3	3.0 (2972)	4.5	—	—	①	—	—	—	16.0	9.5
	Acclaim	K	2.5 (2507)	4.5	—	4.8	②	—	—	—	16.0	9.0
	Acclaim	V	2.5 (2507)	4.5	—	—	①	—	—	—	18.0	9.0
	Lebaron	3	3.0 (2972)	4.5	—	—	19.8 ②	—	—	—	14.0	9.5
	Spirit	3	3.0 (2972)	4.5	—	—	③	—	—	—	16.0	9.5
	Spirit	K	2.5 (2507)	4.5	—	4.8	③	—	—	—	16.0	9.0
	Spirit	V	2.5 (2507)	4.5	—	—	③	—	—	—	18.0	9.0

NOTE: All capacities are approximate. Add fluid gradually and ensure a proper fluid level.

① Non-fleet models: 8.9 qts.
Fleet models: 9.2 qts.
A413 with lock-up converter: 8.5 qts.
A604 transaxle: 9.1 qts.

② Overhaul fill capacity with torque converter empty

③ A413 - 17.8 pts.
A413 (fleet) - 18.4 pts.
A413 (lock-up) - 17.0 pts.
A604 (electronic) - 18.2 pts.

79221C14

VALVE SPECIFICATIONS

Year	Engine ID/VIN	Engine Displacement Liters (cc)	Seat Angle (deg.)	Face Angle (deg.)	Spring Test Pressure (lbs. @ in.)	Spring Installed Height (in.)	Stem-to-Guide Clearance (in.) Intake	Stem-to-Guide Clearance (in.) Exhaust	Stem Diameter (in.) Intake	Stem Diameter (in.) Exhaust
1995	3	3.0 (2972)	44.0-44.3	45-45.5	73@1.59	1.590	0.0010-0.0020	0.0019-0.0030	0.3140	0.3125
	K	2.5 (2507)	45	45	114@1.65	1.650	0.0009-0.0026	0.0030-0.0047	0.3124	0.3103
	V	2.5 (2507)	45	45	114@1.65	1.650	0.0010-0.0030	0.0030-0.0047	0.3124	0.3103

79221C15

TORQUE SPECIFICATIONS
All readings in ft. lbs.

Year	Engine ID/VIN	Engine Displacement Liters (cc)	Cylinder Head Bolts	Main Bearing Bolts	Rod Bearing Bolts	Crankshaft Damper Bolts	Flywheel Bolts	Manifold Intake	Manifold Exhaust	Spark Plugs	Lug Nut
1995	3	3.0 (2972)	80	60	38	112	70	15	16	20	95
	K	2.5 (2507)	①	②	③	85	70	17	17	20	95
	V	2.5 (2507)	①	②	③	85	70	17	17	26	95

① Step 1: 45 ft. lbs.
  Step 2: 65 ft. lbs.
  Step 3: 65 ft. lbs.
  Step 4: Plus 1/4 turn
② Step 1: 30 ft. lbs
  Step 2: Plus 1/4 turn
③ Step 1: 40 ft. lbs.
  Step 2: Plus 1/4 turn

79221C16

SCHEDULED MAINTENANCE INTERVALS
(CHRYSLER LEBARON, DODGE SPIRIT & PLYMOUTH ACCLAIM)

TO BE SERVICED	TYPE OF SERVICE	VEHICLE MILEAGE INTERVAL (x1000)												
		7.5	15	22.5	30	37.5	45	52.5	60	67.5	75	82.5	90	97.5
Engine oil & filter	R	✓	✓	✓	✓	✓	✓	✓	✓	✓	✓	✓	✓	✓
Exhaust system	S/I	✓	✓	✓	✓	✓	✓	✓	✓	✓	✓	✓	✓	✓
Brake hoses	S/I	✓	✓	✓	✓	✓	✓	✓	✓	✓	✓	✓	✓	✓
CV joints & front suspension components	S/I	✓	✓	✓	✓	✓	✓	✓	✓	✓	✓	✓	✓	✓
Rotate tires	S/I	✓	✓	✓	✓	✓	✓	✓	✓	✓	✓	✓	✓	✓
Coolant level, hoses & clamps	S/I	✓	✓	✓	✓	✓	✓	✓	✓	✓	✓	✓	✓	✓
Accessory drive belts	S/I		✓		✓			✓	✓		✓		✓	
Brake linings & rear wheel bearings	S/I			✓			✓			✓			✓	
Spark plugs	R				✓				✓				✓	
Air filter element	R				✓				✓				✓	
Lubricate steering linkage, tie rod ends & ball joints	S/I				✓				✓				✓	
Engine Coolant	R							✓				✓		
Automatic transaxle fluid & filter	R				✓				✓					
Ignition cables	R								✓					
Camshaft timing belt (2.2L & 2.5L)	R												✓	
Camshaft timing belt (3.0L)	R								✓					
PCV valve	S/I								✓					

R – Replace   S/I – Service or Inspect

FREQUENT OPERATION MAINTENANCE (SEVERE SERVICE)

If a vehicle is operated under any of the following conditions it is considered severe service:
- Extremely dusty areas.
- 50% or more of the vehicle operation is in 32°C (90°F) or higher temperatures, or constant operation in temperatures below 0°C (32°F).
- Prolonged idling (vehicle operation in stop and go traffic).
- Frequent short running periods (engine does not warm to normal operating temperatures).
- Police, taxi, delivery usage or trailer towing usage.

CV joints & front suspension components – check every 3000 miles.
Oil & oil filter change – change every 3000 miles.
Rotate tires every 6000 miles.
Brake linings & rear wheel bearings– check every 9000 miles.
Air filter element – change every 15,000 miles.
Automatic transaxle fluid – change every 15,000 miles.
Tie rod ends, steering linkage & ball joints – lubricate every 15,000 miles.
PCV valve – check every 30,000 miles.

79221C17

FREQUENT MAINTENANCE LABOR
CHRYSLER LEBARON

The following should be used as a guide when determining the amount of work required for a particular service if taken to a repair shop. In estimating how long a particular Frequent Maintenance Service item should take, please observe the following:
- **Factory Time** is time that is generated by the vehicle manufacturer.
- **Chilton Time** is time that is based on field research and data supplied by the vehicle manufacturer.
- All labor time operations are given in hours and tenths of an hour.
- All labor operations, are to be used as a **guide**.

	(Factory Time)	Chilton Time

COOLING

(G) Winterize Cooling System

Includes: Run engine to check for leaks, tighten all hose connections. Test radiator and pressure cap, drain radiator and engine block. Add antifreeze and refill system.

1995		.5

(G) Belt, Drive, Renew

1995
V belt		
Fan & Alternator (.2)		.3
Power Steering (.4)		.6
w/AC add		.1
Air Conditioning (.3)		.4
Serpentine belt (.2)		.4

(G) Belt, Drive, Adjust

1995
one		.2
each adtnl.		.1

(G) Hoses, Radiator, Renew

Includes: Drain and refill cooling system as required.

1995
upper (.3)		.4
lower (.4)		.6

(G) Thermostat, Coolant, Renew

1995 (.4)		.6

FUEL

(M) Air Cleaner, Service

1995		.3

(G) Filter, Fuel, Renew

1995 in line (.3)		.4
1995 in tank (1.0)		1.4

BRAKES

(G) Bleed Brakes (Four Wheels)

Includes: Add fluid.

1995 (.4)		.6
Bleed modulator, add		.6
1995 (.4)		.6
w/Antilock 4 add		.4
Bleed modulator, add		1.5

(G) Brakes, Adjust (Minor)

Includes: Adjust brakes, fill master cylinder.

1995, two wheels		.4

(M) Parking Brake, Adjust

1995 (.3)		.4

LUBRICATION SERVICES

(M) Engine Oil & Filter, Renew

Includes: Inspect and correct all fluid levels.

1995 (.2)		.3

(M) Lubricate Chassis, Change Oil & Filter

Includes: Inspect and correct all fluid levels.

1995		.6
Install grease fittings, add		.1

(M) Lubricate Chassis

Includes: Inspect and correct all fluid levels.

1995		.4
Install grease fittings, add		.1

WHEELS

(G) Wheel, Renew (One)

1995		.5

(G) Wheel, Balance

1995
one		.3
each adtnl.		.2

(G) Wheels, Rotate (All)

1995		.5

ELECTRICAL

(G) Headlamps, Aim

1995
two		.4
four		.6

79221C18

FREQUENT MAINTENANCE LABOR (cont.)
CHRYSLER LEBARON

	(Factory Time)	Chilton Time
(G) Halogen Headlamp Bulb, Renew		
1995, each (.2)		.3
(G) High Mount Stop Lamp and/or Lens, Renew		
1995 (.7)		1.0

	(Factory Time)	Chilton Time
(G) License Lamp Assy., Renew		
1995 (.2)		.3
(G) Tail Lamp Assy., Renew		
1995		
Daytona, Laser (.5)		.7
LeBaron (.3)		.4

	(Factory Time)	Chilton Time
(G) Turn Signal & Parking Lamp Assy., Renew		
1995 (.2)		.4
(G) Horn, Renew		
1995, one electric (.2)		.4
(M) Terminals, Battery, Clean		
1995		.3

FREQUENT MAINTENANCE LABOR
DODGE SPIRIT, PLYMOUTH ACCLAIM

The following should be used as a guide when determining the amount of work required for a particular service if taken to a repair shop. In estimating how long a particular Frequent Maintenance Service item should take, please observe the following:
- **Factory Time** is time that is generated by the vehicle manufacturer.
- **Chilton Time** is time that is based on field research and data supplied by the vehicle manufacturer.
- All labor time operations are given in hours and tenths of an hour.
- All labor operations, are to be used as a **guide**.

	(Factory Time)	Chilton Time
COOLING		
(G) Winterize Cooling System		
Includes: Run engine to check for leaks, tighten all hose connections. Test radiator and pressure cap, drain radiator and engine block. Add antifreeze and refill system.		
1995		.5
(G) Belt, Drive, Renew		
1995		
V belt		
Fan & alternator		.3
Power steering		.6
w/AC add		.1
Air conditioner		.3
Serpentine		.4
(G) Belt, Drive, Adjust		
1995		
one		.2
each adtnl.		.1
(G) Hoses, Radiator, Renew		
Includes: Drain and refill cooling system as required.		
1995		
upper		.4
lower		.6
(G) Thermostat, Coolant, Renew		
1995		.6

	(Factory Time)	Chilton Time
FUEL		
(M) Air Cleaner, Service		
1995		.2
(G) Filter, Fuel, Renew		
1995		
in line		.8
in tank		1.4
BRAKES		
(G) Bleed Brakes (Four Wheels)		
Includes: Add fluid.		
1995 (.4)		.6
Bleed modulator, add		.6
1995 (.4)		.6
w/antilock 4 add		.6
Bleed modulator, add		1.5
(G) Brakes, Adjust (Minor)		
Includes: Adjust brakes, fill master cylinder.		
1995, two wheels		.4
(M) Parking Brake, Adjust		
1995 (.3)		.4
LUBRICATION SERVICES		
(M) Engine Oil & Filter, Renew		
Includes: Inspect and correct all fluid levels.		
1995		.3

	(Factory Time)	Chilton Time
(M) Lubricate Chassis, Change Oil & Filter		
Includes: Inspect and correct all fluid levels.		
1995		.6
Install grease fittings, add		.1
(M) Lubricate Chassis		
Includes: Inspect and correct all fluid levels.		
1995		.4
Install grease fittings, add		.1
WHEELS		
(G) Wheel, Renew (One)		
1995, one		.5
(G) Wheel, Balance		
1995		
one		.3
each adtnl.		.2
(G) Wheels, Rotate (All)		
1995		.5
ELECTRICAL		
(G) Headlamps, Aim		
1995		
two		.4
four		.6
(G) Halogen Headlamp Bulb, Renew		
1995, each		.3

FREQUENT MAINTENANCE LABOR (cont.)
DODGE SPIRIT, PLYMOUTH ACCLAIM

(G) High Mount Stop Lamp and/or Lens, Renew
1995 4

(G) License Lamp Assy., Renew
1995 3

(G) Tail Lamp Assy., Renew
1995 4

(G) Tail Lamp Lens or Bulb, Renew
1995 5

(G) Turn Signal & Parking Lamp Assy., Renew
1995 3

(G) Horn, Renew
1995, one 4

(M) Terminals, Battery, Clean
1995 3

79221C20

CHRYSLER CORP.
Chrysler Sebring Coupe • Dodge Avenger

VEHICLE IDENTIFICATION CHART

		Engine Code						Model Year	
Code	Liters	Cu. In. (cc)	Cyl.	Fuel Sys.	Eng. Mfg.			Code	Year
N	2.5	152 (2497)	V6	MFI	Mitsubishi			S	1995
Y	2.0	122 (1996)	I4	MFI	Chrysler			T	1996
								V	1997
								W	1998
								Z	1999

MFI - Multi-port Fuel Injection

79221C21

ENGINE IDENTIFICATION
All measurements are given in inches.

Year	Model	Engine Displacement Liters (cc)	Engine Series (ID/VIN)	Fuel System	No. of Cylinders	Engine Type
1995	Avenger	2.5 (2497)	N	MFI	6	SOHC
	Avenger	2.0 (1996)	Y	MFI	4	DOHC
	Sebring	2.5 (2497)	N	MFI	6	SOHC
	Sebring	2.0 (1996)	Y	MFI	4	DOHC
1996	Avenger	2.5 (2497)	N	MFI	6	SOHC
	Avenger	2.0 (1996)	Y	MFI	4	DOHC
	Sebring Coupe	2.5 (2497)	N	MFI	6	SOHC
	Sebring Coupe	2.0 (1996)	Y	MFI	4	DOHC
1997	Avenger	2.5 (2497)	N	MFI	6	SOHC
	Avenger	2.0 (1996)	Y	MFI	4	DOHC
	Sebring Coupe	2.5 (2497)	N	MFI	6	SOHC
	Sebring Coupe	2.0 (1996)	Y	MFI	4	DOHC
1998-99	Avenger	2.5 (2497)	N	MFI	6	SOHC
	Avenger	2.0 (1996)	Y	MFI	4	DOHC
	Sebring Coupe	2.5 (2497)	N	MFI	6	SOHC
	Sebring Coupe	2.0 (1996)	Y	MFI	4	DOHC

MFI - Multi-port Fuel Injection
DOHC - Double Overhead Camshaft
SOHC - Single Overhead Camshaft

79221C22

GENERAL ENGINE SPECIFICATIONS

Year	Engine ID/VIN	Engine Displacement Liters (cc)	Fuel System Type	Net Horsepower @ rpm	Net Torque @ rpm (ft. lbs.)	Bore x Stroke (in.)	Compression Ratio	Oil Pressure @ rpm
1995	N	2.5 (2497)	MFI	155@5500	161@4400	3.29 x 2.99	9.4:1	35-75@3000
	Y	2.0 (1996)	MFI	140@6000	130@4800	3.44 x 3.27	9.6:1	25-80@3000
1996	N	2.5 (2497)	MFI	163@5500	170@4400	3.29 x 2.99	9.4:1	35-75@3000
	Y	2.0 (1996)	MFI	140@6000	130@4800	3.44 x 3.27	9.6:1	25-80@3000
1997	N	2.5 (2497)	MFI	163@5500	170@4400	3.29 x 2.99	9.4:1	35-75@3000
	Y	2.0 (1996)	MFI	140@6000	130@4800	3.44 x 3.27	9.6:1	25-80@3000
1998-99	N	2.5 (2497)	MFI	163@5500	170@4400	3.29 x 2.99	9.4:1	35-75@3000
	Y	2.0 (1996)	MFI	140@6000	130@4800	3.44 x 3.27	9.6:1	25-80@3000

MFI - Multi-port Fuel Injection

79221C23

GASOLINE ENGINE TUNE-UP SPECIFICATIONS

Year	Engine ID/VIN	Engine Displacement Liters (cc)	Spark Plugs Gap (in.)	Ignition Timing (deg.) MT	Ignition Timing (deg.) AT	Fuel Pump (psi)	Idle Speed (rpm) MT	Idle Speed (rpm) AT	Valve Clearance In.	Valve Clearance Ex.
1995	N	2.5 (2497)	0.039-0.043	—	①	47-50 ②	—	700	HYD	HYD
	Y	2.0 (1996)	0.033-0.038	①	①	47-50	800	800	HYD	HYD
1996	N	2.5 (2497)	0.039-0.043	—	①	47-50 ②	—	750	HYD	HYD
	Y	2.0 (1996)	0.033-0.038	①	①	47-50	800	800	HYD	HYD
1997	N	2.5 (2497)	0.039-0.043	—	①	47-50 ②	—	750	HYD	HYD
	Y	2.0 (1996)	0.033-0.038	①	①	47-50	800	800	HYD	HYD
1998-99	N	2.5 (2497)	0.039-0.043	—	①	47-50 ②	—	750	HYD	HYD
	Y	2.0 (1996)	0.033-0.038	①	①	47-50	800	800	HYD	HYD

NOTE: The Vehicle Emission Control Information label often reflects specification changes made during production. The label figures must be used if they differ from those in this chart.

HYD - Hydraulic

① Basic ignition timing not adjustable

② This reading measured with vacuum hose disconnected from fuel pressure regulator

79221C24

CAPACITIES

Year	Model	Engine ID/VIN	Engine Displacement Liters (cc)	Engine Oil with Filter (qts.)	Transmission (pts.) 4-Spd	Transmission (pts.) 5-Spd	Transmission (pts.) Auto.	Transfer Case (pts.)	Drive Axle Front (pts.)	Drive Axle Rear (pts.)	Fuel Tank (gal.)	Cooling System (qts.)
1995	Avenger	N	2.5 (2497)	4.5	—	—	18.0 ①	—	—	—	16.9	7.4
	Avenger	Y	2.0 (1996)	4.5	—	4.2	18.0 ①	—	—	—	16.9	7.4
	Sebring	N	2.5 (2497)	4.5	—	—	18.2 ①	—	—	—	16.9	7.4
	Sebring	Y	2.0 (1996)	4.5	—	4.2	18.2 ①	—	—	—	16.9	7.4
1996	Avenger	N	2.5 (2497)	4.5	—	—	18.0 ①	—	—	—	16.9	7.4
	Avenger	Y	2.0 (1996)	4.5	—	4.2	18.0 ①	—	—	—	16.9	7.4
	Sebring Coupe	N	2.5 (2497)	4.5	—	—	18.2 ①	—	—	—	16.9	7.4
	Sebring Coupe	Y	2.0 (1996)	4.5	—	4.2	18.2 ①	—	—	—	16.9	7.4
1997	Avenger	N	2.5 (2497)	4.5	—	—	18.0 ①	—	—	—	16.9	7.4
	Avenger	Y	2.0 (1996)	4.5	—	4.2	18.0 ①	—	—	—	16.9	7.4
	Sebring Coupe	N	2.5 (2497)	4.5	—	—	18.2 ①	—	—	—	16.9	7.4
	Sebring Coupe	Y	2.0 (1996)	4.5	—	4.2	18.2 ①	—	—	—	16.9	7.4
1998-99	Avenger	N	2.5 (2497)	4.5	—	—	18.0 ①	—	—	—	16.9	7.4
	Avenger	Y	2.0 (1996)	4.5	—	4.2	18.0 ①	—	—	—	16.9	7.4
	Sebring Coupe	N	2.5 (2497)	4.5	—	—	18.2 ①	—	—	—	16.9	7.4
	Sebring Coupe	Y	2.0 (1996)	4.5	—	4.2	18.2 ①	—	—	—	16.9	7.4

NOTE: All capacities are approximate. Add fluid gradually and ensure a proper fluid level is obtained.

① Overhaul fill capacity with torque converter empty

79221C25

VALVE SPECIFICATIONS

Year	Engine ID/VIN	Engine Displacement Liters (cc)	Seat Angle (deg.)	Face Angle (deg.)	Spring Test Pressure (lbs. @ in.)	Spring Installed Height (in.)	Stem-to-Guide Clearance (in.)		Stem Diameter (in.)	
							Intake	Exhaust	Intake	Exhaust
1995	N	2.5 (2497)	44-44.5	45-45.5	60@1.740	1.740	0.0008-0.0040	0.0016-0.0060	0.2360	0.2360
	Y	2.0 (1996)	44.5-45	45-45.5	123-137@1.153 ①	1.496	0.0019-0.0030	0.0029-0.0040	0.2336-0.2343	0.2325-0.2332
1996	N	2.5 (2497)	44-44.5	45-45.5	60@1.740	1.740	0.0008-0.0040	0.0016-0.0060	0.2360	0.2360
	Y	2.0 (1996)	44.5-45	45-45.5	123-137@1.153 ①	1.496	0.0019-0.0030	0.0029-0.0040	0.2336-0.2343	0.2325-0.2332
1997	N	2.5 (2497)	44-44.5	45-45.5	60@1.740	1.740	0.0008-0.0040	0.0016-0.0060	0.2360	0.2360
	Y	2.0 (1996)	44.5-45	45-45.5	123-137@1.153 ①	1.496	0.0019-0.0030	0.0029-0.0040	0.2336-0.2343	0.2325-0.2332
1998-99	N	2.5 (2497)	44-44.5	45-45.5	60@1.740	1.740	0.0008-0.0040	0.0016-0.0060	0.2360	0.2360
	Y	2.0 (1996)	44.5-45	45-45.5	123-137@1.153 ①	1.496	0.0019-0.0030	0.0029-0.0040	0.2336-0.2343	0.2325-0.2332

① With valves open

79221C26

TORQUE SPECIFICATIONS
All readings in ft. lbs.

Year	Engine ID/VIN	Engine Displacement Liters (cc)	Cylinder Head Bolts	Main Bearing Bolts	Rod Bearing Bolts	Crankshaft Damper Bolts	Flywheel Bolts	Manifold		Spark Plugs	Lug Nut
								Intake	Exhaust		
1995	N	2.5 (2497)	80	69	37	134	70	16	22	18	65-80
	Y	2.0 (1996)	①	55	①	105	70	17	17	20	65-80
1996	N	2.5 (2497)	80	69	37	134	70	16	22	18	88-103
	Y	2.0 (1996)	①	55	①	105	70	17	17	20	88-103
1997	N	2.5 (2497)	80	69	37	134	70	16	22	18	88-103
	Y	2.0 (1996)	①	55	①	105	70	17	17	20	88-103
1998-99	N	2.5 (2497)	80	69	37	134	70	16	22	18	88-103
	Y	2.0 (1996)	①	55	①	105	70	17	17	20	88-103

① Step 1: 20 ft. lbs.
   Step 2: Plus 1/4 turn

79221C27

SCHEDULED MAINTENANCE INTERVALS
(CHRYSLER SEBRING, DODGE AVENGER)

TO BE SERVICED	TYPE OF SERVICE	VEHICLE MILEAGE INTERVAL (x1000)												
		7.5	15	22.5	30	37.5	45	52.5	60	67.5	75	82.5	90	97.5
Engine oil & filter	R	✓	✓	✓	✓	✓	✓	✓	✓	✓	✓	✓	✓	✓
Coolant level, hoses & clamps	S/I	✓	✓	✓	✓	✓	✓	✓	✓	✓	✓	✓	✓	✓
Rotate tires	S/I	✓	✓	✓	✓	✓	✓	✓	✓	✓	✓	✓	✓	✓
Automatic transaxle fluid level	S/I		✓		✓		✓		✓		✓		✓	

79221C28

SCHEDULED MAINTENANCE INTERVALS
(CHRYSLER SEBRING, DODGE AVENGER) (Cont.)

TO BE SERVICED	TYPE OF SERVICE	VEHICLE MILEAGE INTERVAL (x1000)												
		7.5	15	22.5	30	37.5	45	52.5	60	67.5	75	82.5	90	97.5
Brake hoses & disc brake pads	S/I		✓		✓		✓		✓		✓		✓	
Drive shaft boots & front suspension components	S/I		✓		✓		✓		✓		✓		✓	
Air filter element	R				✓				✓				✓	
Engine Coolant	R				✓				✓				✓	
Spark plugs (DOHC)	R				✓				✓				✓	
Spark plugs (SOHC 1995)①	R								✓					
Accessory drive belts	S/I				✓				✓				✓	
Ball joints & steering linkage seals	S/I				✓				✓				✓	
Exhaust system	S/I				✓				✓				✓	
Fuel hoses	S/I				✓				✓				✓	
Manual transaxle oil	S/I				✓				✓				✓	
PCV valve	S/I				✓				✓				✓	
Rear drum brake lining & rear wheel cylinders	S/I				✓				✓				✓	
Camshaft timing belt	R								✓					
Ignition cables	R								✓					
Distributor cap & rotor	S/I								✓					
EVAP system	S/I								✓					
Fuel system	S/I								✓					

① Spark plugs (SOHC 1996) - replace every 100,000 miles.
R – Replace   S/I – Service or Inspect

FREQUENT OPERATION MAINTENANCE (SEVERE SERVICE)

If a vehicle is operated under any of the following conditions it is considered severe service:
- Extremely dusty areas.
- 50% or more of the vehicle operation is in 32°C (90°F) or higher temperatures, or constant operation in temperatures below 0°C (32°F).
- Prolonged idling (vehicle operation in stop and go traffic).
- Frequent short running periods (engine does not warm to normal operating temperatures).
- Police, taxi, delivery usage or trailer towing usage.

Oil & filter change – change every 3000 miles.
Disc brake pads - check every 6000 miles.
Air filter element – change every 15,000 miles.
Automatic transaxle fluid – change every 15,000 miles.
Rear drum brake linings & rear wheel cylinders - check every 15,000 miles.
Spark plugs - change every 15,000 miles.

79221C29

SCHEDULED MAINTENANCE
CHRYSLER CORPORATION
CHRYSLER SEBRING COUPE
DODGE AVENGER

The following should be used as a guide when determining the amount of work required for a particular service if taken to a repair shop. In estimating how long a particular Scheduled Maintenance Service should take, please observe the following:

- **Chilton Time** is time based on field research and data supplied by the vehicle manufacturer.
- Labor time operations are given in hours and tenths of an hour.
- All labor operations, are to be used as a guide.

Mechanic Skill Level Codes:
> **(G)** GENERAL: Normally skilled with certification.
> **(M)** MAINTENANCE: Semi-skilled working on certicication.
> **(P)** PRECISION: Really skilled with multiple certification.

	Chilton Time
(M) 7500 Mile Service	
1995-99	1.2
(M) 15000 Mile Service	
1995-99.	1.5
(M) 22500 Mile Service	
1995-99.	1.2
(G) 30000 Mile Service	
1995-99.	3.2
(M) 37500 Mile Service	
1995-99	1.2

	Chilton Time
(M) 45000 Mile Service	
1995-99	1.6
(M) 52500 Mile Service	
1995-99	1.2
(G) 60000 Mile Service	
1995-99.	6.0
(M) 67500 Mile Service	
1995-99.	1.2

	Chilton Time
(M) 75000 Mile Service	
1995-99	1.6
(M) 82500 Mile Service	
1995-99	1.2
(G) 90000 Mile Service	
1995-99.	3.2
(M) 97500 Mile Service	
1995-99	1.2

79221C30

CHRYSLER CORP.
Dodge/Plymouth Neon

VEHICLE IDENTIFICATION CHART

Code	Liters	Cu. In. (cc)	Cyl.	Fuel Sys.	Eng. Mfg.
		Engine Code			
C	2.0	122 (1996)	I4	MFI	Chrysler
Y	2.0	122 (1996)	I4	MFI	Chrysler

Code	Year
	Model Year
S	1995
T	1996
V	1997
W	1998
X	1999

MFI - Multi-point Fuel Injection

79221C40

ENGINE IDENTIFICATION
All measurements are given in inches.

Year	Model	Engine Displacement Liters (cc)	Engine Series (ID/VIN)	Fuel System	No. of Cylinders	Engine Type
1995	Neon	2.0 (1996)	C	MFI	4	SOHC
	Neon	2.0 (1996)	Y	MFI	4	DOHC
1996	Neon	2.0 (1996)	C	MFI	4	SOHC
	Neon	2.0 (1996)	Y	MFI	4	DOHC
1997	Neon	2.0 (1996)	C	MFI	4	SOHC
	Neon	2.0 (1996)	Y	MFI	4	DOHC
1998-99	Neon	2.0 (1996)	C	MFI	4	SOHC
	Neon	2.0 (1996)	Y	MFI	4	DOHC

MFI - Multi-point Fuel Injection
SOHC - Single Overhead Camshaft
DOHC - Double Overhead Camshaft

79221C41

GENERAL ENGINE SPECIFICATIONS

Year	Engine ID/VIN	Engine Displacement Liters (cc)	Fuel System Type	Net Horsepower @ rpm	Net Torque @ rpm (ft. lbs.)	Bore x Stroke (in.)	Compression Ratio	Oil Pressure @ rpm
1995	C	2.0 (1996)	MFI	132@6000	129@5000	3.44x3.26	9.8:1	25-80@3000
	Y	2.0 (1996)	MFI	150@4400	NA	3.44x3.26	9.6:1	25-80@3000
1996	C	2.0 (1996)	MFI	132@6000	129@5000	3.44x3.26	9.8:1	25-80@3000
	Y	2.0 (1996)	MFI	150@4400	NA	3.44x3.26	9.6:1	25-80@3000
1997	C	2.0 (1996)	MFI	132@6000	129@5000	3.44x3.26	9.8:1	25-80@3000
	Y	2.0 (1996)	MFI	150@4400	NA	3.44x3.26	9.6:1	25-80@3000
1998-99	C	2.0 (1996)	MFI	132@6000	129@5000	3.44x3.26	9.8:1	25-80@3000
	Y	2.0 (1996)	MFI	150@4400	NA	3.44x3.26	9.6:1	25-80@3000

MFI - Multi-point Fuel Injection

79221C42

GASOLINE ENGINE TUNE-UP SPECIFICATIONS

Year	Engine ID/VIN	Engine Displacement Liters (cc)	Spark Plugs Gap (in.)	Ignition Timing (deg.) MT	Ignition Timing (deg.) AT	Fuel Pump (psi)	Idle Speed (rpm) MT	Idle Speed (rpm) AT	Valve Clearance In.	Valve Clearance Ex.
1995	C	2.0 (1996)	0.035	①	①	48	②	②	HYD	HYD
	Y	2.0 (1996)	0.035	①	①	48	②	②	HYD	HYD
1996	C	2.0 (1996)	0.035	①	①	48	②	②	HYD	HYD
	Y	2.0 (1996)	0.035	①	①	48	②	②	HYD	HYD
1997	C	2.0 (1996)	0.035	①	①	48	②	②	HYD	HYD
	Y	2.0 (1996)	0.035	①	①	48	②	②	HYD	HYD
1998-99	C	2.0 (1996)	0.035	①	①	48	②	②	HYD	HYD
	Y	2.0 (1996)	0.035	①	①	48	②	②	HYD	HYD

NOTE: The Vehicle Emission Control Information label often reflects specification changes made during production. The label figures must be used if they differ from those in this chart.

HYD - Hydraulic

NA - Not Available

① Refer to the Vehicle Emission Control Information label for correct timing specifications with a range of +/- 2 degrees

   Ignition timing cannot be adjusted. Base engine timing is set at TDC during assembly.

② Refer to Vehicle Emissions Control Information label for proper specification

79221C43

CAPACITIES

Year	Model	Engine ID/VIN	Engine Displacement Liters (cc)	Engine Oil with Filter (qts.)	Transmission (pts.) 4-Spd	Transmission (pts.) 5-Spd	Transmission (pts.) Auto.	Transfer Case (pts.)	Drive Axle Front (pts.)	Drive Axle Rear (pts.)	Fuel Tank (gal.)	Cooling System (qts.)
1995	Neon	C	2.0 (1996)	4.5	—	①	8.0	—	—	—	11.0	7.4
	Neon	Y	2.0 (1996)	4.5	—	①	8.0	—	—	—	11.0	7.4
1996	Neon	C	2.0 (1996)	4.5	—	①	8.0	—	—	—	11.0	7.4
	Neon	Y	2.0 (1996)	4.5	—	①	8.0	—	—	—	11.0	7.4
1997	Neon	C	2.0 (1996)	4.5	—	①	8.0	—	—	—	11.0	7.4
	Neon	Y	2.0 (1996)	4.5	—	①	8.0	—	—	—	11.0	7.4
1998-99	Neon	C	2.0 (1996)	4.5	—	①	8.0	—	—	—	11.0	7.4
	Neon	Y	2.0 (1996)	4.5	—	①	8.0	—	—	—	11.0	7.4

NOTE: All capacities are approximate. Add fluid gradually and ensure a proper fluid level is obtained.

① Fill to bottom of fill hole

79221C44

VALVE SPECIFICATIONS

Year	Engine ID/VIN	Engine Displacement Liters (cc)	Seat Angle (deg.)	Face Angle (deg.)	Spring Test Pressure (lbs. @ in.)	Spring Installed Height (in.)	Stem-to-Guide Clearance (in.) Intake	Stem-to-Guide Clearance (in.) Exhaust	Stem Diameter (in.) Intake	Stem Diameter (in.) Exhaust
1995	C	2.0 (1996)	44.5-45	45-45.5	75@1.54	1.540	0.0018-0.0025	0.0029-0.0037	0.2340	0.2330
	Y	2.0 (1996)	44.5-45	45-45.5	55-60@1.49	1.490	0.0018-0.0025	0.0029-0.0037	0.2340	0.2330
1996	C	2.0 (1996)	44.5-45	45-45.5	75@1.54	1.540	0.0018-0.0025	0.0029-0.0037	0.2340	0.2330
	Y	2.0 (1996)	44.5-45	45-45.5	55-60@1.49	1.490	0.0018-0.0025	0.0029-0.0037	0.2340	0.2330
1997	C	2.0 (1996)	44.5-45	45-45.5	75@1.54	1.540	0.0018-0.0025	0.0029-0.0037	0.2340	0.2330
	Y	2.0 (1996)	44.5-45	45-45.5	55-60@1.49	1.490	0.0018-0.0025	0.0029-0.0037	0.2340	0.2330
1998-99	C	2.0 (1996)	44.5-45	45-45.5	75@1.54	1.540	0.0018-0.0025	0.0029-0.0037	0.2340	0.2330
	Y	2.0 (1996)	44.5-45	45-45.5	55-60@1.49	1.490	0.0018-0.0025	0.0029-0.0037	0.2340	0.2330

79221C45

TORQUE SPECIFICATIONS
All readings in ft. lbs.

Year	Engine ID/VIN	Engine Displacement Liters (cc)	Cylinder Head Bolts	Main Bearing Bolts	Rod Bearing Bolts	Crankshaft Damper Bolts	Flywheel Bolts	Manifold		Spark Plugs	Lug Nut
								Intake	Exhaust		
1995	C	2.0 (1996)	①	②	③	105	70	17	17	20	95
	Y	2.0 (1996)	④	②	③	105	70	17	17	20	95
1996	C	2.0 (1996)	①	②	③	105	70	17	17	20	95
	Y	2.0 (1996)	④	②	③	105	70	17	17	20	95
1997	C	2.0 (1996)	①	②	③	105	70	17	17	20	95
	Y	2.0 (1996)	④	②	③	105	70	17	17	20	95
1998-99	C	2.0 (1996)	①	②	③	105	70	17	17	20	95
	Y	2.0 (1996)	④	②	③	105	70	17	17	20	95

① Step 1: 25 ft. lbs.
  Step 2: 50 ft. lbs.
  Step 3: 50 ft. lbs.
  Step 4: +90 degrees
② Step 1: 30 ft. lbs.
  Step 2: +90 degrees
③ Step 1: 20 ft. lbs.
  Step 2: +90 degrees
④ Step 1:
    Bolts 1-6: 25 ft. lbs.
    Bolts 7-10: 20 ft. lbs.
  Step 2:
    Bolts 1-6: 50 ft. lbs.
    Bolts 7-10: 20 ft. lbs.
  Step 3:
    Bolts 1-6: 50 ft. lbs.
    Bolts 7-10: 20 ft. lbs.
  Step 4: +90 degrees

79221C46

SCHEDULED MAINTENANCE INTERVALS
(DODGE NEON, PLYMOUTH NEON)

TO BE SERVICED	TYPE OF SERVICE	VEHICLE MILEAGE INTERVAL (x1000)												
		7.5	15	22.5	30	37.5	45	52.5	60	67.5	75	82.5	90	97.5
Engine oil & filter	R	✓	✓	✓	✓	✓	✓	✓	✓	✓	✓	✓	✓	✓
Brake hoses	S/I	✓	✓	✓	✓	✓	✓	✓	✓	✓	✓	✓	✓	✓
Coolant level, hoses & clamps	S/I	✓	✓	✓	✓	✓	✓	✓	✓	✓	✓	✓	✓	✓
CV joints & front suspension components	S/I	✓	✓	✓	✓	✓	✓	✓	✓	✓	✓	✓	✓	✓
Exhaust system	S/I	✓	✓	✓	✓	✓	✓	✓	✓	✓	✓	✓	✓	✓
Manual transaxle oil	S/I	✓	✓	✓	✓	✓	✓	✓	✓	✓	✓	✓	✓	✓
Rotate tires	S/I	✓	✓	✓	✓	✓	✓	✓	✓	✓	✓	✓	✓	✓
Accessory drive belts	S/I		✓		✓		✓		✓		✓		✓	
Brake linings	S/I			✓			✓			✓			✓	
Air filter element	R				✓				✓				✓	
Spark plugs	R				✓				✓				✓	
Lubricate ball joints	S/I				✓				✓				✓	
Engine Coolant	R						✓				✓			
PCV valve	S/I								✓				✓	
Ignition cables	R								✓					
Camshaft timing belt①	R													

① Camshaft timing belt - replace at 105,000 miles for normal service, or 102,000 miles for severe service (1995) or 105,000 miles (1996-99).

R – Replace   S/I – Service or Inspect

FREQUENT OPERATION MAINTENANCE (SEVERE SERVICE)

If a vehicle is operated under any of the following conditions it is considered severe service:
- Extremely dusty areas.
- 50% or more of the vehicle operation is in 32°C (90°F) or higher temperatures, or constant operation in temperatures below 0°C (32°F).
- Prolonged idling (vehicle operation in stop and go traffic).
- Frequent short running periods (engine does not warm to normal operating temperatures).
- Police, taxi, delivery usage or trailer towing usage.
Oil & filter change – change every 3000 miles.
Rotate tires every 6000 miles.
Brake linings - inspect every 12,000 miles.
Air filter element - service or inspect every 15,000 miles.
Automatic transaxle – change fluid & adjust bands every 15,000 miles.
Manual transaxle fluid - replace every 15,000 miles.
Engine coolant - replace at 36,000 miles, and every 30,000 miles thereafter.

79221C47

SCHEDULED MAINTENANCE
CHRYSLER CORPORATION
DODGE NEON
PLYMOUTH NEON

The following should be used as a guide when determining the amount of work required for a particular service if taken to a repair shop.
In estimating how long a particular Scheduled Maintenance Service should take, please observe the following:

- **Chilton Time** is time based on field research and data supplied by the vehicle manufacturer.
- Labor time operations are given in hours and tenths of an hour.
- All labor operations, are to be used as a guide.

Mechanic Skill Level Codes:
- **(G)** GENERAL: Normally skilled with certification.
- **(M)** MAINTENANCE: Semi-skilled working on certicication.
- **(P)** PRECISION: Really skilled with multiple certification.

	Chilton Time
(M) 7500 Mile Service	
1995-99	1.3
(M) 15000 Mile Service	
1995-99	1.4
(M) 22500 Mile Service	
1995-99	1.6
(G) 30000 Mile Service	
1995-99	2.2

	Chilton Time
(M) 37500 Mile Service	
1995-99	1.3
(G) 45000 Mile Service	
1995-99	2.2
(M) 52500 Mile Service	
1995-99	1.3
(G) 60000 Mile Service	
1995-99	2.8
(M) 67500 Mile Service	
1995-99	1.6

	Chilton Time
(M) 75000 Mile Service	
1995-99	1.9
(M) 82500 Mile Service	
1995-99	1.3
(G) 90000 Mile Service	
1995-99	2.7
(M) 97500 Mile Service	
1995-99	1.3

79221C48

CHRYSLER CORP.
Dodge Stealth

VEHICLE IDENTIFICATION CHART

Code	Liters	Cu. In. (cc)	Cyl.	Fuel Sys.	Eng. Mfg.
		Engine Code			
H	3.0	181 (2972)	V6	MFI	Mitsubishi
J	3.0	181 (2972)	V6	MFI	Mitsubishi
K	3.0	181 (2972)	V6	MFI-TT	Mitsubishi

Code	Year
	Model Year
S	1995
T	1996

MFI - Multi-point Fuel Injection
TT - Twin Turbocharged

79221C49

ENGINE IDENTIFICATION
All measurements are given in inches.

Year	Model	Engine Displacement Liters (cc)	Engine Series (ID/VIN)	Fuel System	No. of Cylinders	Engine Type
1995	Stealth	3.0 (2972)	H	MFI	6	SOHC
	Stealth	3.0 (2972)	J	MFI	6	DOHC
	Stealth	3.0 (2972)	K	MFI-TT	6	DOHC
1996	Stealth	3.0 (2972)	H	MFI	6	SOHC
	Stealth	3.0 (2972)	J	MFI	6	DOHC
	Stealth	3.0 (2972)	K	MFI-TT	6	DOHC

MFI - Multi-point Fuel Injection
TT - Twin Turbocharged
SOHC - Single Overhead Camshaft
DOHC - Double Overhead Camshaft

79221C50

GENERAL ENGINE SPECIFICATIONS

Year	Engine ID/VIN	Engine Displacement Liters (cc)	Fuel System Type	Net Horsepower @ rpm	Net Torque @ rpm (ft. lbs.)	Bore x Stroke (in.)	Compression Ratio	Oil Pressure @ rpm
1995	H	3.0 (2972)	MFI	164@5500	185@4000	3.59 x 2.99	8.9:1	35-100@2000
	J	3.0 (2972)	MFI	222@6000	201@4500	3.59 x 2.99	10.0:1	35-100@2000
	K	3.0 (2972)	MFI-TT	320@6000	315@2500	3.59 x 2.99	8.0:1	35-100@2000
1996	H	3.0 (2972)	MFI	164@5500	185@4000	3.59 x 2.99	8.9:1	35-100@2000
	J	3.0 (2972)	MFI	222@6000	201@4500	3.59 x 2.99	10.0:1	35-100@2000
	K	3.0 (2972)	MFI-TT	320@6000	315@2500	3.59 x 2.99	8.0:1	35-100@2000

MFI - Multi-point Fuel Injection
TT - Twin Turbocharged

79221C51

GASOLINE ENGINE TUNE-UP SPECIFICATIONS

Year	Engine ID/VIN	Engine Displacement Liters (cc)	Spark Plugs Gap (in.)	Ignition Timing (deg.)		Fuel Pump (psi)	Idle Speed (rpm)		Valve Clearance	
				MT	AT		MT	AT	In.	Ex.
1995	H	3.0 (2972)	0.039-0.043	5B	5B	38	750	750	HYD	HYD
	J	3.0 (2972)	0.039-0.043	5B	5B	38	750	750	HYD	HYD
	K	3.0 (2972)	0.039-0.043	5B	5B	34	750	750	HYD	HYD
1996	H	3.0 (2972)	0.039-0.043	5B	5B	38	750	750	HYD	HYD
	J	3.0 (2972)	0.039-0.043	5B	5B	38	750	750	HYD	HYD
	K	3.0 (2972)	0.039-0.043	5B	5B	34	750	750	HYD	HYD

NOTE: The Vehicle Emission Control Information label often reflects specification changes made during production. The label figures must be used if they differ from those in this chart.
B - Before top dead center
HYD - Hydraulic

79221C52

CAPACITIES

Year	Model	Engine ID/VIN	Engine Displacement Liters (cc)	Engine Oil with Filter (qts.)	Transmission (pts.)			Transfer Case (pts.)	Drive Axle		Fuel Tank (gal.)
					4-Spd	5-Spd	Auto.		Front (pts.)	Rear (pts.)	
1995	Stealth	H	3.0 (2972)	4.7	—	4.8	15.8	—	—	—	19.8
	Stealth	J	3.0 (2972)	4.7	—	4.8	15.8	—	—	—	19.8
	Stealth	K	3.0 (2972)	5.2	—	5.0 ①	15.8	0.58	—	2.3	19.8
1996	Stealth	H	3.0 (2972)	4.7	—	4.8	15.8	—	—	—	19.8
	Stealth	J	3.0 (2972)	4.7	—	4.8	15.8	—	—	—	19.8
	Stealth	K	3.0 (2972)	5.2	—	5.0 ①	15.8	0.58	—	2.3	19.8

NOTE: All capacities are approximate. Be sure to add fluid gradually and check often to ensure a proper fluid level has been obtained.
① 6 speed manual transaxle

79221C53

VALVE SPECIFICATIONS

Year	Engine ID/VIN	Engine Displacement Liters (cc)	Seat Angle (deg.)	Face Angle (deg.)	Spring Test Pressure (lbs. @ in.)	Spring Installed Height (in.)	Stem-to-Guide Clearance (in.)		Stem Diameter (in.)	
							Intake	Exhaust	Intake	Exhaust
1995	H	3.0 (2972)	44-44.5	45-45.5	①	1.591-1.630	0.0012-0.0024 ②	0.0020-0.0035 ③	0.3140	0.3140
	J	3.0 (2972)	44-44.5	45-45.5	④	1.492	0.0008-0.0020 ②	0.0020-0.0035 ⑤	0.2600	0.2600
	K	3.0 (2972)	44-44.5	45-45.5	④	1.492	0.0008-0.0020 ②	0.0020-0.0035 ⑤	0.2600	0.2600
1996	H	3.0 (2972)	45-44.5	45-45.5	①	1.591-1.630	0.0012-0.0024 ②	0.0020-0.0035 ③	0.3140	0.3140
	J	3.0 (2972)	45-44.5	45-45.5	④	1.492	0.0008-0.0020 ②	0.0020-0.0035 ⑤	0.2600	0.2600
	K	3.0 (2972)	45-44.5	45-45.5	④	1.492	0.0008-0.0020 ②	0.0020-0.0035 ⑤	0.2600	0.2600

① 74 @ installed height
② Wear limit: 0.0039
③ Wear limit: 0.0059
④ 53 @ installed height
⑤ Wear limit: 0.0047

79221C54

TORQUE SPECIFICATIONS
All readings in ft. lbs.

Year	Engine ID/VIN	Engine Displacement Liters (cc)	Cylinder Head Bolts	Main Bearing Bolts	Rod Bearing Bolts	Crankshaft Damper Bolts	Flywheel Bolts	Manifold		Spark Plugs	Lug Nut
								Intake	Exhaust		
1995	H	3.0 (2972)	76-83	58	38	108-116	55	13	13	15	87-101
	J	3.0 (2972)	①	67	38	130-137	55	13	22	15	87-101
	K	3.0 (2972)	①	54	38	130-137	55	13	22	15	87-101
1996	H	3.0 (2972)	76-83	58	38	108-116	55	13	13	15	87-101
	J	3.0 (2972)	①	67	38	130-137	55	13	22	15	87-101
	K	3.0 (2972)	①	54	38	130-137	55	13	22	15	87-101

① Step 1: 87-94 ft. lbs.
  Step 2: Fully loosen
  Step 3: 87-94 ft. lbs.

79221C55

SCHEDULED MAINTENANCE INTERVALS
(DODGE STEALTH)

TO BE SERVICED	TYPE OF SERVICE	VEHICLE MILEAGE INTERVAL (x1000)												
		7.5	15	22.5	30	37.5	45	52.5	60	67.5	75	82.5	90	97.5
Engine oil & filter (Non-turbo) ①	R	✓	✓	✓	✓	✓	✓	✓	✓	✓	✓	✓	✓	✓
Coolant level, hoses & clamps	S/I	✓	✓	✓	✓	✓		✓	✓	✓	✓	✓	✓	✓
Rotate tires	S/I	✓	✓	✓	✓	✓	✓	✓	✓	✓	✓	✓	✓	✓
Brake hoses	S/I		✓		✓		✓		✓		✓		✓	
Drive shaft boots & front suspension components	S/I		✓		✓		✓		✓		✓		✓	
Brake linings	S/I		✓		✓		✓		✓		✓		✓	
Air filter element	R				✓				✓				✓	
Automatic transaxle fluid & filter	R				✓				✓				✓	
Differential fluid (AWD)	R				✓				✓				✓	
Engine Coolant	R				✓				✓				✓	
Spark plugs (Non-platinum)	R				✓				✓				✓	
Accessory drive belts	S/I				✓				✓				✓	
Ball joints & steering linkage seals	S/I				✓				✓				✓	
Exhaust system	S/I				✓				✓				✓	
Fuel hoses	S/I				✓				✓				✓	
Manual transaxle oil (including transfer)	S/I				✓				✓				✓	
PCV valve	S/I				✓				✓				✓	
Spark plugs (Platinum)	R								✓					
Camshaft timing belt	R								✓					
Ignition cables	R								✓					
EVAP system	S/I								✓					
Distributor cap & rotor	S/I								✓					
Fuel system	S/I								✓					

① Engine oil & filter (Turbo) - change every 5000 miles
R – Replace   S/I – Service or Inspect

FREQUENT OPERATION MAINTENANCE (SEVERE SERVICE)

If a vehicle is operated under any of the following conditions it is considered severe service:
- Extremely dusty areas.
- 50% or more of the vehicle operation is in 32°C (90°F) or higher temperatures, or constant operation in temperatures below 0°C (32°F).
- Prolonged idling (vehicle operation in stop and go traffic).
- Frequent short running periods (engine does not warm to normal operating temperatures).
- Police, taxi, delivery usage or trailer towing usage.

CV joints & front suspension components - check every 3000 miles.
Oil & oil filter change – change every 3000 miles.
Brake linings - check every 7500 miles (1993-95) or 6000 miles (1996).
Air filter element – service or inspect every 7500 miles.
Automatic transaxle fluid – change every 15,000 miles.
Differential fluid – change every 15,000 miles.
Manual transaxle (including transfer) - change every 15,000 miles.
Spark plugs - change every 15,000 miles.
Tie rod ends & steering linkage – lubricate every 15,000 miles.

79221C56

SCHEDULED MAINTENANCE
CHRYSLER CORPORATION
DODGE STEALTH

The following should be used as a guide when determining the amount of work required for a particular service if taken to a repair shop.
In estimating how long a particular Scheduled Maintenance Service should take, please observe the following:

- **Chilton Time** is time based on field research and data supplied by the vehicle manufacturer.
- Labor time operations are given in hours and tenths of an hour.
- All labor operations, are to be used as a guide.

Mechanic Skill Level Codes:
  (G) **GENERAL:** Normally skilled with certification.
  (M) **MAINTENANCE:** Semi-skilled working on certication.
  (P) **PRECISION:** Really skilled with multiple certification.

	Chilton Time		Chilton Time		Chilton Time
(M) 7500 Mile Service		**(G) 45000 Mile Service**		**(M) 67500 Mile Service**	
1995-99	.9	1995-99	1.5	1995-99	.9
(G) 15000 Mile Service		**(M) 52500 Mile Service**		**(G) 75000 Mile Service**	
1995-99	1.6	1995-99	.9	1995-99	1.6
(M) 22500 Mile Service		**(G) 60000 Mile Service**		**(M) 82500 Mile Service**	
1995-99	.9	1995-99	6.7	1995-99	.9
(G) 30000 Mile Service				**(G) 90000 Mile Service**	
1995-99	3.2			1995-99	3.2
(M) 37500 Mile Service				**(M) 97500 Mile Service**	
1995-99	.9			1995-99	.9

79221C57

CHRYSLER CORP.
Eagle Talon

VEHICLE IDENTIFICATION CHART

	Engine Code						Model Year	
Code	Liters	Cu. In. (cc)	Cyl.	Fuel Sys.	Eng. Mfg.		Code	Year
F	2.0	122 (1999)	4	MFI Turbo	Mitsubsihi		S	1995
Y	2.0	122 (1996)	4	MFI	Chrysler		T	1996
							V	1997
							W	1998

MFI - Multi-point Fuel Injection

79221C58

ENGINE IDENTIFICATION
All measurements are given in inches.

Year	Model	Engine Displacement Liters (cc)	Engine Series (ID/VIN)	Fuel System	No. of Cylinders	Engine Type
1995	Talon	2.0 (1997)	F	MFI-Turbo	4	DOHC
	Talon	2.0 (1996)	Y	MFI	4	DOHC
1996	Talon	2.0 (1997)	F	MFI-Turbo	4	DOHC
	Talon	2.0 (1996)	Y	MFI	4	DOHC
1997	Talon	2.0 (1997)	F	MFI-Turbo	4	DOHC
	Talon	2.0 (1996)	Y	MFI	4	DOHC
1998	Talon	2.0 (1997)	F	MFI-Turbo	4	DOHC
	Talon	2.0 (1996)	Y	MFI	4	DOHC

MFI - Multi-point Fuel Injection
DOHC - Double Overhead Camshaft

79221C59

GENERAL ENGINE SPECIFICATIONS

Year	Engine ID/VIN	Engine Displacement Liters (cc)	Fuel System Type	Net Horsepower @ rpm	Net Torque @ rpm (ft. lbs.)	Bore x Stroke (in.)	Compression Ratio	Oil Pressure @ rpm
1995	F	2.0 (1997)	MFI	210@6000 ①	214@3000 ②	3.35x3.46	8.5:1	③
	Y	2.0 (1996)	MFI	140@6000	131@4800	3.44x3.27	9.6:1	④
1996	F	2.0 (1997)	MFI	210@6000 ①	214@3000 ②	3.35x3.46	8.5:1	③
	Y	2.0 (1996)	MFI	140@6000	131@4800	3.44x3.27	9.6:1	④
1997	F	2.0 (1997)	MFI	210@6000 ①	214@3000 ②	3.35x3.46	8.5:1	③
	Y	2.0 (1996)	MFI	140@6000	131@4800	3.44x3.27	9.6:1	④
1998	F	2.0 (1997)	MFI	210@6000 ①	214@3000 ②	3.35x3.46	8.5:1	③
	Y	2.0 (1996)	MFI	140@6000	131@4800	3.44x3.27	9.6:1	④

① Automatic: 205@6000
② Automatic: 220@3000
③ 11.4 psi or more at curb idle speed
④ 4 psi or more at curb idle speed

79221C60

GASOLINE ENGINE TUNE-UP SPECIFICATIONS

Year	Engine ID/VIN	Engine Displacement Liters (cc)	Spark Plugs Gap (in.)	Ignition Timing (deg.) MT	Ignition Timing (deg.) AT	Fuel Pump (psi)	Idle Speed (rpm) MT	Idle Speed (rpm) AT	Valve Clearance In.	Valve Clearance Ex.
1995	F	2.0 (1997)	0.028-0.030	5B	5B	33 ①	750	750	HYD	HYD
	Y	2.0 (1996)	0.033-0.038	②	②	38 ①	700	700	HYD	HYD
1996	F	2.0 (1997)	0.028-0.030	5B	5B	33 ①	750	750	HYD	HYD
	Y	2.0 (1996)	0.033-0.038	②	②	38 ①	800	800	HYD	HYD
1997	F	2.0 (1997)	0.028-0.030	5B	5B	33 ①	750	750	HYD	HYD
	Y	2.0 (1996)	0.033-0.038	②	②	38 ①	800	800	HYD	HYD
1998	F	2.0 (1997)	0.028-0.030	5B	5B	33 ①	750	750	HYD	HYD
	Y	2.0 (1996)	0.033-0.038	②	②	38 ①	800	800	HYD	HYD

NOTE: The Vehicle Emission Control Information label often reflects specification changes made during production. The label figures must be used if they differ from those in this chart.

HYD - Hydraulic

① Pressure at idle with vacuum applied to fuel pressure regulator

② Basic ignition timing is not adjustable

79221C61

CAPACITIES

Year	Model	Engine ID/VIN	Engine Displacement Liters (cc)	Engine Oil with Filter (qts.)	Transmission (pts.) 4-Spd	Transmission (pts.) 5-Spd	Transmission (pts.) Auto.	Transfer Case (pts.)	Drive Axle Front (pts.)	Drive Axle Rear (pts.)	Fuel Tank (gal.)	Cooling System (qts.)
1995	Talon	F	2.0 (1997)	4.6	—	①	14.2	1.06	—	1.8	16.0	7.4
	Talon	Y	2.0 (1996)	4.5	—	4.2	18.2	—	—	—	16.0	7.4
1996	Talon	F	2.0 (1997)	4.6	—	①	14.2	1.06	—	1.8	16.0	7.4
	Talon	Y	2.0 (1996)	4.5	—	4.2	18.2	—	—	—	16.0	7.4
1997	Talon	F	2.0 (1997)	4.6	—	①	14.2	1.06	—	1.8	16.0	7.4
	Talon	Y	2.0 (1996)	4.5	—	4.2	18.2	—	—	—	16.0	7.4
1998	Talon	F	2.0 (1997)	4.6	—	①	14.2	1.06	—	1.8	16.0	7.4
	Talon	Y	2.0 (1996)	4.5	—	4.2	18.2	—	—	—	16.0	7.4

NOTE: All capacities are approximate. Add fluid gradually and ensure a proper fluid level is obtained.

① 2WD: 4.6
  4WD: 4.8

79221C62

VALVE SPECIFICATIONS

Year	Engine ID/VIN	Engine Displacement Liters (cc)	Seat Angle (deg.)	Face Angle (deg.)	Spring Test Pressure (lbs. @ in.)	Spring Installed Height (in.)	Stem-to-Guide Clearance (in.)		Stem Diameter (in.)	
							Intake	Exhaust	Intake	Exhaust
1995	F	2.0 (1997)	44-44.5	45-45.5	54 ①	1.570	0.0008-0.0040	0.0020-0.0060	0.2600	0.2560
	Y	2.0 (1996)	45	45-45.5	110-120@ ② 1.173	1.496	0.0019-0.0030	0.0029-0.0040	0.2336-0.2343	0.2325-0.2332
1996	F	2.0 (1997)	44-44.5	45-45.5	54 ①	1.570	0.0008-0.0040	0.0020-0.0060	0.2600	0.2560
	Y	2.0 (1996)	45	44.5-45	123-137@ ② 1.153	1.496	0.0019-0.0030	0.0029-0.0040	0.2336-0.2343	0.2325-0.2332
1997	F	2.0 (1997)	44-44.5	45-45.5	54 ①	1.570	0.0008-0.0040	0.0020-0.0060	0.2600	0.2560
	Y	2.0 (1996)	45	44.5-45	123-137@ ② 1.153	1.496	0.0019-0.0030	0.0029-0.0040	0.2336-0.2343	0.2325-0.2332
1998	F	2.0 (1997)	44-44.5	45-45.5	54 ①	1.570	0.0008-0.0040	0.0020-0.0060	0.2600	0.2560
	Y	2.0 (1996)	45	44.5-45	123-137@ ② 1.153	1.496	0.0019-0.0030	0.0029-0.0040	0.2336-0.2343	0.2325-0.2332

① At installed height
② With valves open

79221C63

TORQUE SPECIFICATIONS
All readings in ft. lbs.

Year	Engine ID/VIN	Engine Displacement Liters (cc)	Cylinder Head Bolts	Main Bearing Bolts	Rod Bearing Bolts	Crankshaft Damper Bolts	Flywheel Bolts	Manifold		Spark Plugs	Lug Nut
								Intake	Exhaust		
1995	F	2.0 (1997)	①	②	14.5 ③	94	94-101	14	18-22	18	65-80
	Y	2.0 (1996)	④	55	⑤	105	—	17	17	20	65-80
1996	F	2.0 (1997)	①	②	14.5 ③	94	94-101	14	18-22	18	65-80
	Y	2.0 (1996)	④	55	⑤	105	—	17	17	20	65-80
1997	F	2.0 (1997)	①	②	14.5 ③	94	94-101	14	18-22	18	65-80
	Y	2.0 (1996)	④	55	⑤	105	—	17	17	20	65-80
1998	F	2.0 (1997)	①	②	14.5 ③	94	94-101	14	18-22	18	65-80
	Y	2.0 (1996)	④	55	⑤	105	—	17	17	20	65-80

① Step 1: 58 ft. lbs.
  Step 2: Fully loosen
  Step 3: 15 ft. lbs.
  Step 4: Plus 90 degrees
  Step 5: Repeat Step 4
② Step 1: 18 ft. lbs.
  Step 2: Plus 90 degrees
③ Plus 90 degrees
④ Step 1:
  Bolts 1-6: 24 ft. lbs.
  Bolts 7-10: 20 ft. lbs.
  Step 2:
  Bolts 1-6: 49 ft. lbs.
  Bolts 7-10: 20 ft. lbs.
  Step 3: Plus 90 degrees
⑤ 20 ft. lbs. plus 90 degrees

79221C64

SCHEDULED MAINTENANCE INTERVALS
(EAGLE TALON)

TO BE SERVICED	TYPE OF SERVICE	VEHICLE MILEAGE INTERVAL (x1000)												
		7.5	15	22.5	30	37.5	45	52.5	60	67.5	75	82.5	90	97.5
Engine oil & filter (Non-turbo)①	R	✓	✓	✓	✓	✓	✓	✓	✓	✓	✓	✓	✓	✓
Coolant level, hoses & clamps	S/I	✓	✓	✓	✓	✓	✓	✓	✓	✓	✓	✓	✓	✓
Rotate tires	S/I	✓	✓	✓	✓	✓	✓	✓	✓	✓	✓	✓	✓	✓
Automatic transaxle fluid level	S/I		✓		✓		✓		✓		✓		✓	
Brake hoses & disc brake pads	S/I		✓		✓		✓		✓		✓		✓	
Drive shaft boots & front suspension components	S/I		✓		✓		✓		✓		✓		✓	
Air filter element	R				✓				✓				✓	
Automatic transaxle fluid & filter②	R				✓				✓				✓	
Engine Coolant	R				✓				✓				✓	
Spark plugs	R				✓				✓				✓	
Accessory drive belts	S/I				✓				✓				✓	
Ball joints & steering linkage seals	S/I				✓				✓				✓	
Exhaust system	S/I				✓				✓				✓	
Fuel hoses	S/I				✓				✓				✓	

79221C65

SCHEDULED MAINTENANCE INTERVALS
(EAGLE TALON Cont.)

TO BE SERVICED	TYPE OF SERVICE	VEHICLE MILEAGE INTERVAL (x1000)												
		7.5	15	22.5	30	37.5	45	52.5	60	67.5	75	82.5	90	97.5
Manual transaxle oil (including transfer)	S/I				✓				✓				✓	
Rear axle oil (AWD)	S/I				✓				✓				✓	
Camshaft timing belt	R								✓					
Ignition cables	R								✓					
EVAP & fuel system	S/I								✓					

① Engine oil & filter (Turbo) - change every 5000 miles.   ② 1995-98: Turbo w/A/T only.

R – Replace   S/I – Service or Inspect

FREQUENT OPERATION MAINTENANCE (SEVERE SERVICE)

If a vehicle is operated under any of the following conditions it is considered severe service:
- Extremely dusty areas.
- 50% or more of the vehicle operation is in 32°C (90°F) or higher temperatures, or constant operation in temperatures below 0°C (32°F).
- Prolonged idling (vehicle operation in stop and go traffic).
- Frequent short running periods (engine does not warm to normal operating temperatures).
- Police, taxi, delivery usage or trailer towing usage.

Oil & filter change – change every 3000 miles.
Air filter element – service or inspect every 7500 miles.
Automatic transaxle fluid – change every 15,000 miles.
Spark plugs - change every 15,000 miles.
Disc brake pads - check more frequently than every 7500 miles (1996-98) or 6000 miles (1995).

79221C66

SCHEDULED MAINTENANCE
CHRYSLER CORPORATION
EAGLE TALON

The following should be used as a guide when determining the amount of work required for a particular service if taken to a repair shop.
In estimating how long a particular Scheduled Maintenance Service should take, please observe the following:

- **Chilton Time** is time based on field research and data supplied by the vehicle manufacturer.
- Labor time operations are given in hours and tenths of an hour.
- All labor operations, are to be used as a guide.

Mechanic Skill Level Codes:
- **(G) GENERAL:** Normally skilled with certification.
- **(M) MAINTENANCE:** Semi-skilled working on certication.
- **(P) PRECISION:** Really skilled with multiple certification.

	Chilton Time
(M) 7500 Mile Service	
1995-98	.9
(G) 15000 Mile Service	
1995-98	1.2
(M) 22500 Mile Service	
1995-98	.9
(G) 30000 Mile Service	
1995-98	3.9
(M) 37500 Mile Service	
1995-98	.9

	Chilton Time
(G) 45000 Mile Service	
1995-98	1.2
(M) 52500 Mile Service	
1995-98	.9
(G) 60000 Mile Service	
1995-98	7.0

	Chilton Time
(M) 67500 Mile Service	
1995-98	.9
(G) 75000 Mile Service	
1995-98	1.2
(M) 82500 Mile Service	
1995-98	.9
(G) 90000 Mile Service	
1995-98	3.9
(M) 97500 Mile Service	
1995-98	.9

79221C67

FORD MOTOR CO.
Ford Contour • Mercury Mystique • Cougar (1999)

VEHICLE IDENTIFICATION CHART

		Engine Code				Model Year	
Code	Liters	Cu. In. (cc)	Cyl.	Fuel Sys.	Eng. Mfg.	Code	Year
3	2.0	122 (1999)	4	SFI	Ford	S	1995
L	2.5	153 (2507)	6	SFI	Ford	T	1996
						V	1997
						W	1998
						X	1999

SFI - Sequential Fuel Injection

79221C68

ENGINE IDENTIFICATION
All measurements are given in inches.

Year	Model	Engine Displacement Liters (cc)	Engine Series (ID/VIN)	Fuel System	No. of Cylinders	Engine Type
1995	Contour	2.0 (1999)	3	SFI	4	DOHC
	Contour	2.5 (2507)	L	SFI	6	DOHC
	Mystique	2.0 (1999)	3	SFI	4	DOHC
	Mystique	2.5 (2507)	L	SFI	6	DOHC
1996	Contour	2.0 (1999)	3	SFI	4	DOHC
	Contour	2.5 (2507)	L	SFI	6	DOHC
	Mystique	2.0 (1999)	3	SFI	4	DOHC
	Mystique	2.5 (2507)	L	SFI	6	DOHC
1997	Contour	2.0 (1999)	3	SFI	4	DOHC
	Contour	2.5 (2507)	L	SFI	6	DOHC
	Mystique	2.0 (1999)	3	SFI	4	DOHC
	Mystique	2.5 (2507)	L	SFI	6	DOHC
1998-99	Contour	2.0 (1999)	3	SFI	4	DOHC
	Contour	2.5 (2507)	L	SFI	6	DOHC
	Cougar	2.0 (1999)	3	SFI	4	DOHC
	Cougar	2.5 (2507)	L	SFI	6	DOHC
	Mystique	2.0 (1999)	3	SFI	4	DOHC
	Mystique	2.5 (2507)	L	SFI	6	DOHC

SFI - Sequential Fuel Injection
DOHC - Double Overhead Camshafts

79221C69

GENERAL ENGINE SPECIFICATIONS

Year	Engine ID/VIN	Engine Displacement Liters (cc)	Fuel System Type	Net Horsepower @ rpm	Net Torque @ rpm (ft. lbs.)	Bore x Stroke (in.)	Compression Ratio	Oil Pressure @ rpm
1995	3	2.0 (1999)	SFI	125@6000	130@4500	3.39x3.46	9.6:1	20-45@1500
	L	2.5 (2507)	SFI	170@6200	165@4200	3.25x3.13	9.7:1	25-45@1500
1996	3	2.0 (1999)	SFI	125@5500	130@4000	3.39x3.46	9.6:1	20-45@1500
	L	2.5 (2507)	SFI	170@6200	165@4200	3.25x3.13	9.7:1	20-45@1500
1997	3	2.0 (1999)	SFI	125@5500	130@4000	3.39x3.46	9.6:1	20-45@1500
	L	2.5 (2507)	SFI	170@6200	165@4200	3.25x3.13	9.7:1	20-45@1500
1998-99	3	2.0 (1999)	SFI	125@5500	130@4000	3.39x3.46	9.6:1	20-45@1500
	L	2.5 (2507)	SFI	170@6200	165@4200	3.25x3.13	9.7:1	20-45@1500

SFI - Sequential Fuel Injection

79221C70

GASOLINE ENGINE TUNE-UP SPECIFICATIONS

Year	Engine ID/VIN	Engine Displacement Liters (cc)	Spark Plugs Gap (in.)	Ignition Timing (deg.) MT	Ignition Timing (deg.) AT	Fuel Pump (psi)	Idle Speed (rpm) MT	Idle Speed (rpm) AT	Valve Clearance In.	Valve Clearance Ex.
1995	3	2.0 (1999)	0.050	10B	10B	30-38 ①	880	800	HYD	HYD
	L	2.5 (2507)	0.054	10B	10B	30-36 ①	②	②	HYD	HYD
1996	3	2.0 (1999)	0.050	10B	10B	37-41 ①	②	②	HYD	HYD
	L	2.5 (2507)	0.054	10B	10B	37-41 ①	②	②	HYD	HYD
1997	3	2.0 (1999)	0.050	10B	10B	37-41 ①	②	②	HYD	HYD
	L	2.5 (2507)	0.054	10B	10B	37-41 ①	②	②	HYD	HYD
1998-99	3	2.0 (1999)	0.050	10B	10B	37-41 ①	②	②	HYD	HYD
	L	2.5 (2507)	0.054	10B	10B	37-41 ①	②	②	HYD	HYD

NOTE: The Vehicle Emission Control Information label often reflects specification changes made during production. The label figures must be used if they differ from those in this chart.

B - Before Top Dead Center

HYD - Hydraulic

① Fuel pressure with engine running, pressure regulator vacuum hose connected

② Refer to Vehicle Emission Control Information label

79221C71

CAPACITIES

Year	Model	Engine ID/VIN	Engine Displacement Liters (cc)	Engine Oil with Filter (qts.)	Transmission (pts.) 4-Spd	5-Spd	Auto.	Drive Axle Front (pts.)	Rear (pts.)	Fuel Tank (gal.)	Cooling System (qts.)
1995	Contour	3	2.0 (1999)	4.5	—	5.5	18.0 ①	②	—	14.5	③
	Contour	L	2.5 (2507)	5.5	—	5.5	20.6 ①	②	—	14.5	④
	Mystique	3	2.0 (1999)	4.5	—	5.5	18.0 ①	②	—	14.5	③
	Mystique	L	2.5 (2507)	5.5	—	5.5	20.6 ①	②	—	14.5	④
1996	Contour	3	2.0 (1999)	4.5	—	5.5	18.0 ①	②	—	14.5	③
	Contour	L	2.5 (2507)	5.8	—	5.5	20.6 ①	②	—	14.5	④
	Mystique	3	2.0 (1999)	4.5	—	5.5	18.0 ①	②	—	14.5	③
	Mystique	L	2.5 (2507)	5.8	—	5.5	20.6 ①	②	—	14.5	④
1997	Contour	3	2.0 (1999)	4.5	—	5.5	18.0 ①	②	—	14.5	③
	Contour	L	2.5 (2507)	5.8	—	5.5	20.6 ①	②	—	14.5	④
	Mystique	3	2.0 (1999)	4.5	—	5.5	18.0 ①	②	—	14.5	③
	Mystique	L	2.5 (2507)	5.8	—	5.5	20.6 ①	②	—	14.5	④
1998-99	Contour	3	2.0 (1999)	4.5	—	5.5	18.0 ①	②	—	14.5	③
	Contour	L	2.5 (2507)	5.8	—	5.5	20.6 ①	②	—	14.5	④
	Cougar	3	2.0 (1999)	4.5	—	5.5	18.0 ①	②	—	14.5	④
	Cougar	L	2.5 (2507)	5.8	—	5.5	20.6 ①	②	—	14.5	④
	Mystique	3	2.0 (1999)	4.5	—	5.5	18.0 ①	②	—	14.5	③
	Mystique	L	2.5 (2507)	5.8	—	5.5	20.6 ①	②	—	14.5	④

NOTE: All capacities are approximate. Add fluid gradually and ensure a proper fluid level is obtained.
① Includes torque converter
② Included in transaxle capacity
③ Automatic transaxle: 7.5 qts.
  Manual transaxle: 7.0 qts.
④ Automatic transaxle: 9.1 qts.
  Manual transaxle: 8.9 qts.

79221C72

VALVE SPECIFICATIONS

Year	Engine ID/VIN	Engine Displacement Liters (cc)	Seat Angle (deg.)	Face Angle (deg.)	Spring Test Pressure (lbs. @ in.)	Spring Installed Height (in.)	Stem-to-Guide Clearance (in.) Intake	Exhaust	Stem Diameter (in.) Intake	Exhaust
1995	3	2.0 (1999)	45	45	NA	1.346	0.0007-0.0025	0.0014-0.0032	0.2373-0.2379	0.2366-0.2372
	L	2.5 (2507)	44.75	45.5	153@1.18	1.570	0.0007-0.0027	0.0017-0.0037	0.2350-0.2358	0.2343-0.2350
1996	3	2.0 (1999)	45	45	NA	1.346	0.0007-0.0025	0.0014-0.0032	0.2373-0.2379	0.2366-0.2372
	L	2.5 (2507)	44.75	45.5	153@1.18	1.570	0.0007-0.0027	0.0017-0.0037	0.2350-0.2358	0.2343-0.2350
1997	3	2.0 (1999)	45	45	NA	1.346	0.0007-0.0025	0.0014-0.0032	0.2373-0.2379	0.2366-0.2372
	L	2.5 (2507)	44.75	45.5	153@1.18	1.570	0.0007-0.0027	0.0017-0.0037	0.2350-0.2358	0.2343-0.2350
1998-99	3	2.0 (1999)	45	45	NA	1.346	0.0007-0.0025	0.0014-0.0032	0.2373-0.2379	0.2366-0.2372
	L	2.5 (2507)	44.75	45.5	153@1.18	1.570	0.0007-0.0027	0.0017-0.0037	0.2350-0.2358	0.2343-0.2350

79221C73

TORQUE SPECIFICATIONS
All readings in ft. lbs.

Year	Engine ID/VIN	Engine Displacement Liters (cc)	Cylinder Head Bolts	Main Bearing Bolts	Rod Bearing Bolts	Crankshaft Damper Bolts	Flywheel Bolts	Manifold		Spark Plugs	Lug Nut
								Intake	Exhaust		
1995	3	2.0 (1999)	①	55-66	②	81-89	80-87	12-15	13-16	9-13	63
	L	2.5 (2507)	③	④	⑤	⑥	54-64	6-9	13-16	7-15	63
1996	3	2.0 (1999)	①	55-66	②	81-89	80-87	12-15	13-16	9-13	63
	L	2.5 (2507)	③	④	⑤	⑥	54-64	6-9	13-16	7-15	63
1997	3	2.0 (1999)	①	55-66	②	81-89	80-87	12-15	13-16	9-13	63
	L	2.5 (2507)	③	④	⑤	⑥	54-64	6-9	13-16	7-15	63
1998-99	3	2.0 (1999)	①	55-66	②	81-89	80-87	12-15	13-16	9-13	63
	L	2.5 (2507)	③	④	⑤	⑥	54-64	6-9	13-16	7-15	63

NOTE: Always follow proper torque patterns

NOTE: Stretch bolts are used in all procedures that require rotating the fastener a certain number of degrees. The bolts stretch and cannot be reused. For reassembly, replace with new fastners.

① Step 1: 15-22 ft. lbs.
  Step 2: 30-37 ft. lbs.
  Step 3: Rotate 90-120 degrees
② Step 1: 22-25 ft. lbs.
  Step 2: Rotate each bolt 85-95 degrees
③ Step 1: 27-32 ft. lbs.
  Step 2: Rotate 85-95 degrees
  Step 3: Loosen bolts then repeat Step 1
  Step 4: Rotate 85-95 degrees
  Step 5: Repeat Step 5
④ Step 1: 2.0-3.6 ft. lbs.
  Step 2: Push crankshaft rearward.
  Lightly seat crankshaft washer forward
  Step 3: Outer cap bolts: 16-21 ft. lbs.
  Step 4: Inner cap bolts: 27-32 ft. lbs.
  Step 5: Rotate inner and outer cap bolts 85-95 degrees
  Step 6: Remaining bolts: 15-22 ft. lbs.
⑤ 26-33 ft. lbs. plus 90-120 degrees
⑥ Step 1: 89 ft. lbs.
  Step 2: Loosen bolt
  Step 3: 35-39 ft. lbs.
  Step 4: Rotate 85-95 degrees

79221C74

SCHEDULED MAINTENANCE INTERVALS
(FORD CONTOUR, MERCURY MYSTIQUE)

TO BE SERVICED	TYPE OF SERVICE	VEHICLE MILEAGE INTERVAL (x1000)												
		5	10	15	20	25	30	35	40	45	50	55	60	65
Engine oil & filter	R	✓	✓	✓	✓	✓	✓	✓	✓	✓	✓	✓	✓	✓
Rotate tires	S/I	✓		✓		✓		✓		✓		✓		✓
Front & rear brakes	S/I		✓		✓		✓		✓		✓		✓	
Cooling system, hoses, clamps & coolant strength	S/I			✓			✓			✓			✓	
Passenger compartment air filter	R				✓				✓				✓	
Air cleaner element	R						✓						✓	
Automatic transaxle fluid & filter (1995)	R						✓						✓	
Automatic transaxle fluid & filter (1996-99)	S/I												✓	
Exhaust heat shields	S/I						✓						✓	
Accessory drive belt(s)	S/I						✓						✓	
Fuel lines & hoses	S/I						✓						✓	
Crankcase emission filter (2.0L)	R						✓							
Engine coolant①	R										✓			

79221C75

SCHEDULED MAINTENANCE INTERVALS
(FORD CONTOUR, MERCURY MYSTIQUE) (Cont.)

TO BE SERVICED	TYPE OF SERVICE	VEHICLE MILEAGE INTERVAL (x1000)												
		5	10	15	20	25	30	35	40	45	50	55	60	65
Spark plugs②	R												✓	
PCV valve	R												✓	

① Change initially at 50,000 miles, & every 30,000 miles thereafter.
② 2.0L shown; 2.5L - replace every 100,000 miles.
R – Replace   S/I – Service or Inspect

FREQUENT OPERATION MAINTENANCE (SEVERE SERVICE)

If a vehicle is operated under any of the following conditions it is considered severe service:
- Extremely dusty areas.
- 50% or more of the vehicle operation is in 32°C (90°F) or higher temperatures, or constant operation in temperatures below 0°C (32°F).
- Prolonged idling (vehicle operation in stop and go traffic).
- Frequent short running periods (engine does not warm to normal operating temperatures).
- Police, taxi, delivery usage or trailer towing usage.

Oil & oil filter – change every 3000 miles.
Front & rear brakes - check every 9000 miles.
Rotate tires at 6000 miles & every 9000 miles thereafter.
Air cleaner element - check every 15,000 miles.
Passenger compartment air filter - change every 18,000 miles.
Automatic transaxle fluid & filter - change every 21,000 miles (1995), 30,000 miles (1996-99).
Spark plugs - replace every 60,000 miles.

79221C76

SCHEDULED MAINTENANCE
FORD MOTOR COMPANY
FORD CONTOUR
MERCURY MYSTIQUE, COUGAR (1999)

The following should be used as a guide when determining the amount of work required for a particular service if taken to a repair shop.
In estimating how long a particular Scheduled Maintenance Service should take, please observe the following:

- **Chilton Time** is time based on field research and data supplied by the vehicle manufacturer.
- Labor time operations are given in hours and tenths of an hour.
- All labor operations, are to be used as a guide.

Mechanic Skill Level Codes:
  (G) GENERAL: Normally skilled with certification.
  (M) MAINTENANCE: Semi-skilled working on certicication.
  (P) PRECISION: Really skilled with multiple certification.

	Chilton Time
(M) 5000 Mile Service	
1995-99	.8
(M) 10000 Mile Service	
1995-99	.6
(M) 15000 Mile Service	
1995-99	.9
(M) 20000 Mile Service	
1995-99	.9
(M) 25000 Mile Service	
1995-99	.8

	Chilton Time
(G) 30000 Mile Service	
1995-99	1.6
(M) 35000 Mile Service	
1995-99	.8
(M) 40000 Mile Service	
1995-99	.9
(M) 45000 Mile Service	
1995-99	.9

	Chilton Time
(G) 50000 Mile Service	
1995-99	1.1
(M) 55000 Mile Service	
1995-99	.6
(G) 60000 Mile Service	
1995-99	3.3
Renew auto. trans. fluid & filter add	.5
(G) 65000 Mile Service	
1995-99	.8

79221C77

FORD MOTOR CO.
Ford Aspire

VEHICLE IDENTIFICATION CHART

Engine Code						Model Year	
Code	Liters	Cu. In. (cc)	Cyl.	Fuel Sys.	Eng. Mfg.	Code	Year
H	1.3	81 (1319)	4	SFI	Kia Motors	S	1995
						T	1996
						V	1997

SFI - Sequential Fuel Injection

79221C00

ENGINE IDENTIFICATION
All measurements are given in inches.

Year	Model	Engine Displacement Liters (cc)	Engine Series (ID/VIN)	Fuel System	No. of Cylinders	Engine Type
1995	Aspire	1.3 (1319)	H	SFI	4	SOHC
1996	Aspire	1.3 (1319)	H	SFI	4	SOHC
1997	Aspire	1.3 (1319)	H	SFI	4	SOHC

SFI - Sequential Fuel Injection
SOHC - Single Overhead Camshaft

79221CA1

GENERAL ENGINE SPECIFICATIONS

Year	Engine ID/VIN	Engine Displacement Liters (cc)	Fuel System Type	Net Horsepower @ rpm	Net Torque @ rpm (ft. lbs.)	Bore x Stroke (in.)	Compression Ratio	Oil Pressure @ rpm
1995	H	1.3 (1319)	SFI	63@5000	73@3000	2.79x3.29	9.7:1	50-64@3000
1996	H	1.3 (1319)	SFI	63@5000	73@3000	2.79x3.29	9.7:1	50-64@3000
1997	H	1.3 (1319)	SFI	63@5000	73@3000	2.79x3.29	9.7:1	50-64@3000

SFI - Sequential Fuel Injection

79221CA2

GASOLINE ENGINE TUNE-UP SPECIFICATIONS

Year	Engine ID/VIN	Engine Displacement Liters (cc)	Spark Plugs Gap (in.)	Ignition Timing (deg.) MT	Ignition Timing (deg.) AT	Fuel Pump (psi)	Idle Speed (rpm) MT	Idle Speed (rpm) AT	Valve Clearance In.	Valve Clearance Ex.
1995	H	1.3 (1319)	0.040	10B	10B	30-38 ①	700	750	HYD	HYD
1996	H	1.3 (1319)	0.040	10B	10B	30-38 ①	700	750	HYD	HYD
1997	H	1.3 (1319)	0.040	10B	10B	30-38 ①	700	750	HYD	HYD

NOTE: The Vehicle Emission Control Information label often reflects specification changes made during production. The label figures must be used if they differ from those in this chart.

HYD - Hydraulic

① Fuel pressure with engine running, pressure regulator vacuum hose connected

79221CA3

CAPACITIES

Year	Model	Engine ID/VIN	Engine Displacement Liters (cc)	Engine Oil with Filter (qts.)	Transmission (pts.) 4-Spd	Transmission (pts.) 5-Spd	Transmission (pts.) Auto.	Drive Axle Front (pts.)	Drive Axle Rear (pts.)	Fuel Tank (gal.)	Cooling System (qts.)
1995	Aspire	H	1.3 (1319)	3.6	—	5.2	12.0 ①	②	—	10.0	6.3
1996	Aspire	H	1.3 (1319)	3.6	—	5.2	12.0 ①	②	—	10.0	6.3
1997	Aspire	H	1.3 (1319)	3.6	—	5.2	12.0 ①	②	—	10.0	6.3

NOTE: All capacities are approximate. Add fluid gradually and ensure a proper fluid level is obtained.

① Includes torque converter

② Included in transaxle capacity

79221CA4

VALVE SPECIFICATIONS

Year	Engine ID/VIN	Engine Displacement Liters (cc)	Seat Angle (deg.)	Face Angle (deg.)	Spring Test Pressure (lbs. @ in.)	Spring Installed Height (in.)	Stem-to-Guide Clearance (in.) Intake	Stem-to-Guide Clearance (in.) Exhaust	Stem Diameter (in.) Intake	Stem Diameter (in.) Exhaust
1995	H	1.3 (1319)	45	45	NA	1.717 ①	0.0010-0.0024	0.0012-0.0026	0.2744-0.2750	0.2742-0.2748
1996	H	1.3 (1319)	45	45	NA	1.717 ①	0.0010-0.0024	0.0012-0.0026	0.2744-0.2750	0.2742-0.2748
1997	H	1.3 (1319)	45	45	NA	1.717 ①	0.0010-0.0024	0.0012-0.0026	0.2744-0.2750	0.2742-0.2748

① Spring height measured unloaded

79221CA5

TORQUE SPECIFICATIONS
All readings in ft. lbs.

Year	Engine ID/VIN	Engine Displacement Liters (cc)	Cylinder Head Bolts	Main Bearing Bolts	Rod Bearing Bolts	Crankshaft Damper Bolts	Flywheel Bolts	Manifold Intake	Manifold Exhaust	Spark Plugs	Lug Nut
1995	H	1.3 (1319)	①	40-43	②	③	71-76	14-20	12-17	15-22	85
1996	H	1.3 (1319)	①	40-43	②	③	71-76	14-20	12-17	15-22	76
1997	H	1.3 (1319)	①	40-43	②	③	71-76	14-20	12-17	15-22	76

NOTE: Always follow proper torque patterns

NOTE: Stretch bolts are used in all procedures that require rotating the fastener a certain number of degrees. The bolts stretch and cannot be reused. For reassembly, replace with new fastners.

① Step 1: 35-40 ft. lbs.
   Step 2: 56-60 ft. lbs.
② Step 1: 11-13 ft. lbs.
   Step 2: 22-25 ft. lbs.
③ Pulley bolts: 9-13 ft. lbs.
   Sprocket bolt: 80-87 ft. lbs.

79221CA6

SCHEDULED MAINTENANCE INTERVALS
(FORD ASPIRE)

TO BE SERVICED	TYPE OF SERVICE	VEHICLE MILEAGE INTERVAL (x1000)												
		5	10	15	20	25	30	35	40	45	50	55	60	65
Engine oil & filter	R	✓	✓	✓	✓	✓	✓	✓	✓	✓	✓	✓	✓	✓
Rotate tires	S/I	✓		✓		✓		✓		✓		✓		✓
Air cleaner element & engine coolant	R						✓						✓	
Spark plugs	R						✓						✓	
Automatic transaxle fluid & filter	R						✓						✓	
Exhaust heat shields	S/I						✓						✓	
Disc brake pads & rotors, brake linings, drum, brake lines, hoses & connections	S/I						✓						✓	
Accessory drive belt(s)	S/I						✓						✓	
Fuel lines, hoses & idle speed	S/I						✓						✓	
Cooling system, hoses, clamps & coolant strength	S/I						✓						✓	
Clutch pedal operation	S/I						✓						✓	
Front wheel driveshaft joint boots	S/I						✓						✓	
Front suspension ball joints, steering operation & linkage	S/I						✓						✓	
Timing belt/chain & fuel filter	R												✓	
Fuel lines & tubes (emission)	S/I												✓	
Ignition timing	S/I												✓	
Repack front & rear wheel bearings	S/I												✓	

R – Replace   S/I – Service or Inspect

FREQUENT OPERATION MAINTENANCE (SEVERE SERVICE)
If a vehicle is operated under any of the following conditions it is considered severe service:
- Extremely dusty areas.
- 50% or more of the vehicle operation is in 32°C (90°F) or higher temperatures, or constant operation in temperatures below 0°C (32°F).
- Prolonged idling (vehicle operation in stop and go traffic).
- Frequent short running periods (engine does not warm to normal operating temperatures).
- Police, taxi, delivery usage or trailer towing usage.

Oil & oil filter – change every 3000 miles.
Rotate tires at 6000 miles & every 9000 miles thereafter.
Air cleaner element - service or inspect every 15,000 miles.
Automatic transaxle fluid & filter - change every 21,000 miles.

79221CA7

SCHEDULED MAINTENANCE
FORD MOTOR COMPANY
FORD ASPIRE

The following should be used as a guide when determining the amount of work required for a particular service if taken to a repair shop. In estimating how long a particular Scheduled Maintenance Service should take, please observe the following:

- **Chilton Time** is time based on field research and data supplied by the vehicle manufacturer.
- Labor time operations are given in hours and tenths of an hour.
- All labor operations, are to be used as a guide.

Mechanic Skill Level Codes:
  - **(G)** GENERAL: Normally skilled with certification.
  - **(M)** MAINTENANCE: Semi-skilled working on certicication.
  - **(P)** PRECISION: Really skilled with multiple certification.

	Chilton Time			Chilton Time			Chilton Time
(M) 5000 Mile Service			**(G) 30000 Mile Service**			**(M) 45000 Mile Service**	
1995-97	.8		1995-97	2.9		1995-97	.8
(M) 10000 Mile Service			**(M) 35000 Mile Service**			**(M) 50000 Mile Service**	
1995-97	.3		1995-97	.8		1995-97	.3
(M) 15000 Mile Service			**(M) 40000 Mile Service**			**(M) 55000 Mile Service**	
1995-97	.8		1995-97	.3		1995-97	.8
(M) 20000 Mile Service						**(G) 60000 Mile Service**	
1995-97	.3					1995-97	6.0
(M) 25000 Mile Service						**(M) 65000 Mile Service**	
1995-97	.8					1995-97	.8

79221CA8

FORD MOTOR CO.
Ford Crown Victoria • Lincoln Town Car
Mercury Grand Marquis

VEHICLE IDENTIFICATION CHART

		Engine Code					Model Year	
Code	Liters	Cu. In. (cc)	Cyl.	Fuel Sys.	Eng. Mfg.		Code	Year
W	4.6	281 (4593)	8	SFI	Ford		S	1995
							T	1996
SFI - Sequential Fuel Injection							V	1997
							W	1998
							X	1999

79221C78

ENGINE IDENTIFICATION
All measurements are given in inches.

Year	Model	Engine Displacement Liters (cc)	Engine Series (ID/VIN)	Fuel System	No. of Cylinders	Engine Type
1995	Crown Victoria	4.6 (4593)	W	SFI	8	SOHC
	Grand Marquis	4.6 (4593)	W	SFI	8	SOHC
	Town Car	4.6 (4593)	W	SFI	8	SOHC
1996	Crown Victoria	4.6 (4593)	W	SFI	8	SOHC
	Grand Marquis	4.6 (4593)	W	SFI	8	SOHC
	Town Car	4.6 (4593)	W	SFI	8	SOHC
1997	Crown Victoria	4.6 (4593)	W	SFI	8	SOHC
	Grand Marquis	4.6 (4593)	W	SFI	8	SOHC
	Town Car	4.6 (4593)	W	SFI	8	SOHC
1998-99	Crown Victoria	4.6 (4593)	W	SFI	8	SOHC
	Grand Marquis	4.6 (4593)	W	SFI	8	SOHC
	Town Car	4.6 (4593)	W	SFI	8	SOHC

SFI - Sequential Fuel Injection
SOHC - Single Overhead Camshaft

79221C79

GENERAL ENGINE SPECIFICATIONS

Year	Engine ID/VIN	Engine Displacement Liters (cc)	Fuel System Type	Net Horsepower @ rpm	Net Torque @ rpm (ft. lbs.)	Bore x Stroke (in.)	Compression Ratio	Oil Pressure @ rpm
1995	W	4.6 (4593)	SFI	①	②	3.55x3.54	9.0:1	20-45@2000
1996	W	4.6 (4593)	SFI	③	④	3.55x3.54	⑤	20-45@1500
1997	W	4.6 (4593)	SFI	③	④	3.55x3.54	⑤	20-45@1500
1998-99	W	4.6 (4593)	SFI	③	④	3.55x3.54	⑤	20-45@1500

MFI - Mulyi-point Fuel Injection
SFI - Sequential Fuel Injection
① Single exhaust: 190@4250
   Dual exhaust:: 210@4250
② Single exhaust: 260@3250
   Dual exhaust:: 270@3250
③ Single exhaust: 190@4250
   Dual exhaust: 210@4250
   Crown Victoria with natural gas: 178@4500
④ Single exhaust: 265@3250
   Dual exhaust: 275@3250
   Crown Victoria with natural gas: 237@3500
⑤ Base engine: 9.0:1
   Crown Victoria with natural gas: 10.0:1

79221C80

GASOLINE ENGINE TUNE-UP SPECIFICATIONS

Year	Engine ID/VIN	Engine Displacement Liters (cc)	Spark Plugs Gap (in.)	Ignition Timing (deg.) MT	Ignition Timing (deg.) AT	Fuel Pump (psi)	Idle Speed (rpm) MT	Idle Speed (rpm) AT	Valve Clearance In.	Valve Clearance Ex.
1995	W	4.6 (4593)	0.054	—	10B	30-45 ①	—	②	HYD	HYD
1996	W	4.6 (4593)	0.054	—	10B	35-45 ①	—	②	HYD	HYD
1997	W	4.6 (4593)	0.054	—	10B	35-45 ①	—	②	HYD	HYD
1998-99	W	4.6 (4593)	0.054	—	10B	35-45 ①	—	②	HYD	HYD

NOTE: The Vehicle Emission Control Information label often reflects specification changes made during production. The label figures must be used if they differ from those in this chart.

B - Before Top Dead Center
HYD - Hydraulic
① Fuel pressure with engine running, pressure regulator vacuum hose connected
② Refer to Vehicle Emission Control Information label

79221C81

CAPACITIES

Year	Model	Engine ID/VIN	Engine Displacement Liters (cc)	Engine Oil with Filter (qts.)	Transmission (pts.) 4-Spd	Transmission (pts.) 5-Spd	Transmission (pts.) Auto.	Drive Axle Front (pts.)	Drive Axle Rear (pts.)	Fuel Tank (gal.)	Cooling System (qts.)
1995	Crown Victoria	W	4.6 (4593)	5.0	—	—	27.2 ①	—	3.0	20.0	14.1
	Grand Marquis	W	4.6 (4593)	5.0	—	—	28.2 ①	—	3.1	20.0	14.1
	Town Car	W	4.6 (4593)	5.0	—	—	28.2 ①	—	3.1	20.0	14.1
1996	Crown Victoria	W	4.6 (4593)	5.0	—	—	27.2 ①	—	3.75 ②	20.0	14.1
	Grand Marquis	W	4.6 (4593)	5.0	—	—	28.2 ①	—	3.75 ②	20.0	15.1
	Town Car	W	4.6 (4593)	5.0	—	—	28.2 ①	—	3.75 ②	20.0	15.1
1997	Crown Victoria	W	4.6 (4593)	5.0	—	—	27.2 ①	—	3.75 ②	20.0	14.1
	Grand Marquis	W	4.6 (4593)	5.0	—	—	28.2 ①	—	3.75 ②	20.0	15.1
	Town Car	W	4.6 (4593)	5.0	—	—	28.2 ①	—	3.75 ②	20.0	15.1
1998-99	Crown Victoria	W	4.6 (4593)	5.0	—	—	27.2 ①	—	3.75 ②	20.0	14.1
	Grand Marquis	W	4.6 (4593)	5.0	—	—	28.2 ①	—	3.75 ②	20.0	15.1
	Town Car	W	4.6 (4593)	5.0	—	—	28.2 ①	—	3.75 ②	20.0	15.1

NOTE: All capacities are approximate. Add fluid gradually and ensure a proper fluid level is obtained.
① Includes torque converter
② 7.50" axle: 3.0 pts.
   8.80" axle: 3.25 pts.

79221C82

VALVE SPECIFICATIONS

Year	Engine ID/VIN	Engine Displacement Liters (cc)	Seat Angle (deg.)	Face Angle (deg.)	Spring Test Pressure (lbs. @ in.)	Spring Installed Height (in.)	Stem-to-Guide Clearance (in.) Intake	Stem-to-Guide Clearance (in.) Exhaust	Stem Diameter (in.) Intake	Stem Diameter (in.) Exhaust
1995	W	4.6 (4593)	45	45.5	132@1.10	1..570	0.0008-0.0027	0.0018-0.0037	0.2746-0.2754	0.2736-0.2744
1996	W	4.6 (4593)	45	45.5	132@1.10	1.570	0.0008-0.0027	0.0018-0.0037	0.2746-0.2754	0.2736-0.2744
1997	W	4.6 (4593)	45	45.5	132@1.10	1.570	0.0008-0.0027	0.0018-0.0037	0.2746-0.2754	0.2736-0.2744
1998-99	W	4.6 (4593)	45	45.5	132@1.10	1.570	0.0008-0.0027	0.0018-0.0037	0.2746-0.2754	0.2736-0.2744

79221C83

TORQUE SPECIFICATIONS
All readings in ft. lbs.

Year	Engine ID/VIN	Engine Displacement Liters (cc)	Cylinder Head Bolts	Main Bearing Bolts	Rod Bearing Bolts	Crankshaft Damper Bolts	Flywheel Bolts	Manifold Intake	Manifold Exhaust	Spark Plugs	Lug Nut
1995	W	4.6 (4593)	①	②	③	114-121	54-64	15-22	13-16	7-15	95
1996	W	4.6 (4593)	①	②	③	114-121	54-64	15-22	13-16	7-15	95
1997	W	4.6 (4593)	①	②	③	114-121	54-64	15-22	13-16	7-15	95
1998-99	W	4.6 (4593)	①	②	③	114-121	54-64	15-22	13-16	7-15	95

NOTE: Always follow proper torque patterns
NOTE: Stretch bolts are used in all procedures that require rotating the fastener a certain number of degrees. The bolts stretch and cannot be reused. For reassembly, replace with new fasteners.

① Step 1: 22-30 ft. lbs.
  Step 2: Rotate each bolt 85-95 degrees
  Step 3: Repeat Step 2
② Step 1: Main bearing cap bolts: 22-25 ft. lbs.
  Step 2: Rotate each bolt 85-95 degrees
  Step 3: Main bearing cap adjusting screws:
  44 inch lbs. then 80-90 inch lbs.
  Step 4: Main bearing cap side bolts:
  7 ft. lbs. then 14-17 ft. lbs.
③ Step 1: 8 ft. lbs.
  Step 2: 12 ft. lbs.
  Step 3: 25-34 ft. lbs.
  Step 4: Rotate 85-95 degrees

79221C84

SCHEDULED MAINTENANCE INTERVALS
(FORD CROWN VICTORIA, LINCOLN TOWN CAR, MERCURY GRAND MARQUIS)

TO BE SERVICED	TYPE OF SERVICE	VEHICLE MILEAGE INTERVAL (x1000) 5	10	15	20	25	30	35	40	45	50	55	60	65
Engine oil & filter	R	✓	✓	✓	✓	✓	✓	✓	✓	✓	✓	✓	✓	✓
Rotate tires	S/I	✓		✓		✓		✓		✓		✓		✓
Cooling system, hoses, clamps & coolant strength	S/I			✓			✓			✓			✓	
Lubricate steering linkage (Crown Victoria, Grand Marquis)	S/I			✓			✓			✓			✓	
Air cleaner element	R						✓						✓	
Automatic transmission fluid & filter	R						✓						✓	
Spark plugs (1995)	R						✓						✓	
Spark plugs (1996-99)②	R													
Exhaust heat shields	S/I						✓						✓	

79221C85

SCHEDULED MAINTENANCE INTERVALS
(FORD CROWN VICTORIA, LINCOLN TOWN CAR, MERCURY GRAND MARQUIS)
(Cont.)

TO BE SERVICED	TYPE OF SERVICE	VEHICLE MILEAGE INTERVAL (x1000)												
		5	10	15	20	25	30	35	40	45	50	55	60	65
Fuel filter (NGV Crown Victoria)③	R					✓					✓			
Lubricate steering linkage (Town Car)	S/I						✓						✓	
Front & rear brakes	S/I						✓						✓	
Lubricate suspension (Town Car)	S/I						✓						✓	
Engine coolant①	R										✓			
PCV valve	R												✓	
Accessory drive belt(s)	S/I												✓	

① Change initially at 50,000 miles, & thereafter every 30,000 miles.
② Replace every 100,000 miles.
③ Also drain coalescer. Perform every 24,000 miles for severe service.
R – Replace   S/I – Service or Inspect

FREQUENT OPERATION MAINTENANCE (SEVERE SERVICE)

If a vehicle is operated under any of the following conditions it is considered severe service:
- Extremely dusty areas.
- 50% or more of the vehicle operation is in 32°C (90°F) or higher temperatures, or constant operation in temperatures below 0°C (32°F).
- Prolonged idling (vehicle operation in stop and go traffic).
- Frequent short running periods (engine does not warm to normal operating temperatures).
- Police, taxi, delivery usage or trailer towing usage.

Oil & oil filter – change every 3000 miles.
Rotate tires at 6000 miles & every 9000 miles thereafter.
Automatic transmission fluid & filter - change every 21,000 miles

79221C86

SCHEDULED MAINTENANCE
FORD MOTOR COMPANY
FORD CROWN VICTORIA, MERCURY GRAND MARQUIS
LINCOLN TOWN CAR

The following should be used as a guide when determining the amount of work required for a particular service if taken to a repair shop. In estimating how long a particular Scheduled Maintenance Service should take, please observe the following:

- **Chilton Time** is time based on field research and data supplied by the vehicle manufacturer.
- Labor time operations are given in hours and tenths of an hour.
- All labor operations, are to be used as a guide.

Mechanic Skill Level Codes:
- (G) GENERAL: Normally skilled with certification.
- (M) MAINTENANCE: Semi-skilled working on certicication.
- (P) PRECISION: Really skilled with multiple certification.

(M) 5000 Mile Service
1995-99 8

(M) 10000 Mile Service
1995-99 3

(M) 15000 Mile Service
1995-99 9

(M) 20000 Mile Service
1995-99 3

(M) 25000 Mile Service
1995-99 8

(G) 30000 Mile Service
1995-99 Crown Victoria 1.5
1995-99 Grand Marquis 1.5
Renew spark plugs (1995)
  add 1.1
1995-99 Town Car 1.7
Renew auto. trans. fluid &
  filter add 5

(M) 35000 Mile Service
1995-99 8

(M) 40000 Mile Service
1995-99 3

(M) 45000 Mile Service
1995-99 9

(G) 50000 Mile Service
1995-99 . . . . 1.4
Renew engine coolant add 5

(M) 55000 Mile Service
1995-99 8

(G) 60000 Mile Service
1995-99 Crown Victoria 1.5
1995-99 Grand Marquis 1.5
  Renew spark plugs (1995)
  add 1.1
1995-99 Town Car 1.7
Renew auto. trans. fluid &
  filter add 5

(M) 65000 Mile Service
1995-99 8

79221C87

FORD MOTOR CO.
Ford Escort • Escort ZX2 • Mercury Tracer

VEHICLE IDENTIFICATION CHART

		Engine Code					Model Year	
Code	Liters	Cu. In. (cc)	Cyl.	Fuel Sys.	Eng. Mfg.		Code	Year
3	2.0	121 (1999)	4	SFI	Ford		S	1995
8	1.8	112 (1844)	4	MFI	Mazda		T	1996
J	1.9	116 (1901)	4	SFI	Ford		V	1997
P	2.0	121 (1999)	4	SFI	Ford		W	1998
							X	1999

MFI - Multi-point Fuel Injection

79221C88

ENGINE IDENTIFICATION
All measurements are given in inches.

Year	Model	Engine Displacement Liters (cc)	Engine Series (ID/VIN)	Fuel System	No. of Cylinders	Engine Type
1995	Escort	1.8 (1844)	8	MFI	4	DOHC
	Escort	1.9 (1901)	J	SFI	4	SOHC
	Tracer	1.8 (1844)	8	MFI	4	DOHC
	Tracer	1.9 (1901)	J	SFI	4	SOHC
1996	Escort	1.8 (1844)	8	MFI	4	DOHC
	Escort	1.9 (1901)	J	SFI	4	SOHC
	Tracer	1.8 (1844)	8	MFI	4	DOHC
	Tracer	1.9 (1901)	J	SFI	4	SOHC
1997	Escort	2.0 (1999)	P	SFI	4	DOHC
	Tracer	2.0 (1999)	P	SFI	4	SOHC
1998-99	Escort	2.0 (1999)	P	SFI	4	SOHC
	Escort ZX2	2.0 (1999)	3	SFI	4	DOHC
	Tracer	2.0 (1999)	P	SFI	4	SOHC

MFI - Multi-point Fuel Injection
SFI - Sequential Fuel Injection
DOHC - Double Overhead Camshafts
SOHC - Single Overhead Camshaft

79221C89

GENERAL ENGINE SPECIFICATIONS

Year	Engine ID/VIN	Engine Displacement Liters (cc)	Fuel System Type	Net Horsepower @ rpm	Net Torque @ rpm (ft. lbs.)	Bore x Stroke (in.)	Compression Ratio	Oil Pressure @ rpm
1995	8	1.8 (1844)	MFI	127@6500	114@4500	3.27x3.35	9.0:1	35-65@2000
	J	1.9 (1901)	SFI	88@4400	108@4000	3.23x3.46	9.0:1	35-65@2000
1996	8	1.8 (1844)	MFI	127@6500	114@4500	3.27x3.35	9.0:1	28-43@1000
	J	1.9 (1901)	SFI	88@4400	108@3800	3.23x3.35	9.0:1	35-65@2000
1997	P	2.0 (1999)	SFI	110@5000	125@3750	3.34x3.46	9.2:1	35-65@2000
1998-99	3	2.0 (1999)	SFI	125@5500	130@4000	3.34x3.46	10.0:1	35-65@2000
	P	2.0 (1999)	SFI	110@5000	125@3750	3.34x3.46	9.2:1	35-65@2000

MFI - Multi-point Fuel Injection
SFI - Sequential Fuel Injection

79221C90

GASOLINE ENGINE TUNE-UP SPECIFICATIONS

Year	Engine ID/VIN	Engine Displacement Liters (cc)	Spark Plugs Gap (in.)	Ignition Timing (deg.) MT	Ignition Timing (deg.) AT	Fuel Pump (psi)	Idle Speed (rpm) MT	Idle Speed (rpm) AT	Valve Clearance In.	Valve Clearance Ex.
1995	8	1.8 (1844)	0.041	10B	10B	31-38 ①	750	750	HYD	HYD
	J	1.9 (1901)	0.054	10B	10B	30-34 ①	780	780	HYD	HYD
1996	8	1.8 (1844)	0.041	10B	10B	31-38 ①	750	750	HYD	HYD
	J	1.9 (1901)	0.054	10B	10B	38-45 ①	②	②	HYD	HYD
1997	P	2.0 (1999)	0.054	10B	10B	38-45 ①	②	②	HYD	HYD
1998-99	3	2.0 (1999)	0.050	10B	10B	31-38 ①	②	②	HYD	HYD
	P	2.0 (1999)	0.054	10B	10B	38-45 ①	②	②	HYD	HYD

NOTE: The Vehicle Emission Control Information label often reflects specification changes made during production. The label figures must be used if they differ from those in this chart.

B - Before Top Dead Center
HYD - Hydraulic
① Fuel pressure with engine running, pressure regulator vacuum hose connected
② Refer to Vehicle Emission Control Information label

79221C91

CAPACITIES

Year	Model	Engine ID/VIN	Engine Displacement Liters (cc)	Oil with Filter (qts.)	Transmission (pts.) 4-Spd	Transmission (pts.) 5-Spd	Transmission (pts.) Auto.	Axle Front (pts.)	Axle Rear (pts.)	Fuel Tank (gal.)	Cooling System (qts.)
1995	Escort	8	1.8 (1844)	4.0	—	7.2	13.4 ①	②	—	13.2	③
	Escort	J	1.9 (1901)	4.0	—	5.7	13.4 ①	②	—	11.9	③
	Tracer	8	1.8 (1844)	4.0	—	7.1	13.4 ①	②	—	13.2	③
	Tracer	J	1.9 (1901)	4.0	—	5.7	13.4 ①	②	—	11.9	③
1996	Escort	8	1.8 (1944)	4.0	—	6.7	13.4 ①	②	—	13.2	6.3
	Escort	J	1.9 (1901)	4.0	—	5.7	13.4 ①	②	—	11.9	③
	Tracer	8	1.8 (1944)	4.0	—	6.7	13.4 ①	②	—	13.2	6.3
	Tracer	J	1.9 (1901)	4.0	—	5.7	13.4 ①	②	—	11.9	③
1997	Escort	P	2.0 (1999)	4.0	—	6.7	13.4 ①	②	—	13.2	④
	Tracer	P	2.0 (1999)	4.0	—	5.7	13.4 ①	②	—	11.9	④
1998-99	Escort	P	2.0 (1999)	4.0	—	5.7	13.4 ①	②	—	11.9	④
	Escort ZX2	3	2.0 (1999)	4.0	—	6.7	13.4 ①	②	—	13.2	⑤
	Tracer	P	2.0 (1999)	4.0	—	5.7	13.4 ①	②	—	11.9	④

Note: All capacities are approximates. Add fluid gradually and ensure a proper fluid level is obtained.
① Includes torque converter
② Included in transaxle capacity
③ Manual transaxle: 5.3 qts.
 Automatic transaxle: 6.3 qts.
④ Manual transaxle 7.9
 Automatic transaxle 5.8
⑤ Manual transaxle 7.0
 Automatic transaxle 7.5

79221C92

VALVE SPECIFICATIONS

Year	Engine ID/VIN	Engine Displacement Liters (cc)	Seat Angle (deg.)	Face Angle (deg.)	Spring Test Pressure (lbs. @ in.)	Spring Installed Height (in.)	Stem-to-Guide Clearance (in.)		Stem Diameter (in.)	
							Intake	Exhaust	Intake	Exhaust
1995	8	1.8 (1844)	45	45	NA	①	0.0010-0.0024	0.0012-0.0026	0.2350-0.2356	0.2348-0.2354
	J	1.9 (1901)	45	45.6	200@1.09	1.440-1.480	0.0008-0.0027	0.0018-0.0037	0.3159-0.3167	0.3149-0.3156
1996	8	1.8 (1844)	45	45	NA	①	0.0010-0.0024	0.0012-0.0026	0.2350-0.2356	0.2348-0.2354
	J	1.9 (1901)	45	45.6	200@1.09	1.440-1.480	0.0008-0.0027	0.0018-0.0037	0.3159-0.3167	0.3149-0.3156
1997	P	2.0 (1999)	45	45.6	200@1.09	1.420-1.540	0.0008-0.0027	0.0018-0.0037	0.3159-0.3167	0.3149-0.3156
1998-99	3	2.0 (1999)	45	45	NA	1.346	0.0007-0.0025	0.0014-0.0032	0.2373-0.2379	0.2366-0.2372
	P	2.0 (1999)	45	45.6	200@1.09	1.420-1.540	0.0008-0.0027	0.0018-0.0037	0.3159-0.3167	0.3149-0.3156

① Spring height measured unloaded
Minimum length: 1.821

79221C93

TORQUE SPECIFICATIONS
All readings in ft. lbs.

Year	Engine ID/VIN	Engine Displacement Liters (cc)	Cylinder Head Bolts	Main Bearing Bolts	Rod Bearing Bolts	Crankshaft Damper Bolts	Flywheel Bolts	Manifold Intake	Manifold Exhaust	Spark Plugs	Lug Nut
1995	8	1.8 (1844)	56-60	40-43	35-37	80-87	71-76	14-19	28-34	11-17	85
	J	1.9 (1901)	①	67-80	26-30	81-96	54-67	12-15	15-20	8-15	95
1996	8	1.8 (1844)	50-60	40-43	35-37	80-87	71-76	14-19	28-34	11-17	76
	J	1.9 (1901)	①	67-80	26-30	81-96	54-67	12-15	15-20	8-15	76
1997	P	2.0 (1999)	①	66-79	26-30	80-87	71-76	15-22	15-17	12-15	76
1998-99	3	2.0 (1999)	②	55-65	③	80-87	71-76	11-12	10-12	10-12	76
	P	2.0 (1999)	①	66-79	26-30	80-87	71-76	15-22	15-17	12-15	76

NOTE: Always follow proper torque patterns
NOTE: Stretch bolts are used in all procedures that require rotating the fastener a certain number of degrees. The bolts stretch and cannot be reused. For reassembly, replace with new fastners.

① Do not reuse cylinder head bolts.
  Step 1: Tighten bolts, in sequence, to 44 ft. lbs.
  Step 2: Loosen bolts approx. two turns,
      retighten in sequence to 44 ft. lbs.
  Step 3: Turn all bolts, in sequence, +90 degrees
  Step 4: Repeat Step 3
② Step 1: 15-22 ft. lbs.
  Step 2: 30-37 ft. lbs.
  Step 3: Rotate 90-120 degrees
③ Step 1: 22-25 ft. lbs.
  Step 2: Rotate each bolt 85-95 degrees

79221C94

SCHEDULED MAINTENANCE INTERVALS
(FORD ESCORT, MERCURY TRACER (1994-96))

TO BE SERVICED	TYPE OF SERVICE	VEHICLE MILEAGE INTERVAL (x1000)												
		5	10	15	20	25	30	35	40	45	50	55	60	65
Engine oil & filter	R	✓	✓	✓	✓	✓	✓	✓	✓	✓	✓	✓	✓	✓
Rotate tires	S/I	✓		✓		✓		✓		✓		✓		✓
Cooling system, hoses, clamps & coolant strength (1.9L)	S/I			✓			✓			✓			✓	
Air cleaner element, automatic transmission fluid & filter	R						✓						✓	

79221C95

SCHEDULED MAINTENANCE INTERVALS
(FORD ESCORT, MERCURY TRACER (1994-96)) (Cont.)

TO BE SERVICED	TYPE OF SERVICE	VEHICLE MILEAGE INTERVAL (x1000)												
		5	10	15	20	25	30	35	40	45	50	55	60	65
Crankcase emission filter (1.9L)	R						✓						✓	
Engine coolant (1.8L)	R						✓						✓	
Engine coolant (1.9L 1994)	R						✓						✓	
Engine coolant (1.9L 1995-97)①	R										✓			
Accessory drive belt(s) (1.8L)	S/I						✓						✓	
Bolts & nuts on chassis & body	S/I						✓						✓	
Idle speed (1.8L)	S/I						✓						✓	
Clutch pedal operation, brake lines, hoses & connections	S/I						✓						✓	
Cooling system, hoses, clamps & coolant strength (1.8L)	S/I						✓						✓	
Exhaust heat shields, front & rear brakes	S/I						✓						✓	
Front suspension lower arm ball joints, steering operation & linkage, & front wheel driveshaft joint boots	S/I						✓						✓	
Fuel lines & hoses (1.8L)	S/I						✓						✓	
Fuel filter (1.8L)	R												✓	
PCV valve (1.9L)	R												✓	
Spark plugs (1.8L)	R						✓						✓	

79221C96

SCHEDULED MAINTENANCE INTERVALS
(FORD ESCORT, MERCURY TRACER (1994-96)) (Cont.)

TO BE SERVICED	TYPE OF SERVICE	VEHICLE MILEAGE INTERVAL (x1000)												
		5	10	15	20	25	30	35	40	45	50	55	60	65
Spark plugs (1.9L)	R												✓	
Timing belt (1.8L)	R												✓	
Accessory drive belt(s) (1.9L)	S/I												✓	
Evaporative emission hose for emissions (1.8L)	S/I												✓	

① Change initially at 50,000 miles & every 30,000 miles thereafter.

R – Replace   S/I – Service or Inspect

FREQUENT OPERATION MAINTENANCE (SEVERE SERVICE)

If a vehicle is operated under any of the following conditions it is considered severe service:
- Extremely dusty areas.
- 50% or more of the vehicle operation is in 32°C (90°F) or higher temperatures, or constant operation in temperatures below 0°C (32°F).
- Prolonged idling (vehicle operation in stop and go traffic).
- Frequent short running periods (engine does not warm to normal operating temperatures).
- Police, taxi, delivery usage or trailer towing usage.

Oil & oil filter – change every 3000 miles.
Air cleaner element - check every 15,000 miles.
Front & rear brakes - check every 15,000 miles.
Nuts & bolts on chassis & body - check every 15,000 miles.
Automatic transaxle fluid & filter - change every 21,000 miles.

79221C97

1997-99 SCHEDULED MAINTENANCE INTERVALS
(FORD ESCORT & MERCURY TRACER)

TO BE SERVICED	TYPE OF SERVICE	VEHICLE MILEAGE INTERVAL (x1000)																			
		5	10	15	20	25	30	35	40	45	50	55	60	65	70	75	80	85	90	95	100
Engine oil & filter	R	✓	✓	✓	✓	✓	✓	✓	✓	✓	✓	✓	✓	✓	✓	✓	✓	✓	✓	✓	✓
Tires ①	S/I	✓		✓		✓		✓		✓		✓		✓		✓		✓		✓	
Air Cleaner	S/I						✓						✓						✓		
	R						✓						✓						✓		
Spark Plugs	R	Every 100,000 miles																			
Drive Belts	S/I												✓								
Cooling system	S/I			✓			✓			✓			✓			✓			✓		
Engine coolant	R										✓					✓					
PCV valve	R												✓								
Exhaust heat shields	S/I						✓						✓						✓		
Brake linings & drums	S/I						✓						✓						✓		
Brake line hoses & connections	S/I						✓						✓						✓		
Front ball joints	S/I												✓								
Bolts & nuts on chassis body	S/I						✓						✓						✓		
Steering linkage operation	S/I						✓						✓						✓		
Brake pads & rotor	S/I						✓						✓						✓		
Clutch pedal operation	S/I						✓						✓						✓		
Halfshaft dust boots	S/I						✓						✓						✓		

① Rotate, inspect the tire tread for wear, and adjust air pressure.

R - Replace   S/I - Inspect and service, if needed   L - Lubricate   A - Adjust   C - Clean

79221C98

SCHEDULED MAINTENANCE
FORD MOTOR COMPANY
FORD ESCORT
MERCURY TRACER

The following should be used as a guide when determining the amount of work required for a particular service if taken to a repair shop. In estimating how long a particular Scheduled Maintenance Service should take, please observe the following:

- **Chilton Time** is time based on field research and data supplied by the vehicle manufacturer.
- Labor time operations are given in hours and tenths of an hour.
- All labor operations, are to be used as a guide.

Mechanic Skill Level Codes:
   (G) GENERAL: Normally skilled with certification.
   (M) MAINTENANCE: Semi-skilled working on certicication.
   (P) PRECISION: Really skilled with multiple certification.

	Chilton Time
(M) 5000 Mile Service	
1995-99	.8
(M) 10000 Mile Service	
1995-99	.3
(M) 15000 Mile Service	
1995-99	
1.8L	.8
1.9L	.9
2.0L	1.0
(M) 20000 Mile Service	
1995-99	.3
(M) 25000 Mile Service	
1995-99	.8
(G) 30000 Mile Service	
1995-99	
1.8L	1.9
1.9L	2.5
2.0L	1.2

	Chilton Time
(M) 35000 Mile Service	
1995-99	.8
(M) 40000 Mile Service	
1995-99	.3
(M) 45000 Mile Service	
1995-99	
1.8L	.8
1.9L	.9
2.0L	1.0
(M) 50000 Mile Service	
1995-99	
1.8L	.3
1.9L	.8
2.0L	.6
(M) 55000 Mile Service	
1995-99	.8
(G) 60000 Mile Service	
1995-99	
1.8L	5.8
1.9L	2.7
2.0L	1.5

	Chilton Time
(M) 65000 Mile Service	
1995-99	.8
(M) 70000 Mile Service	
1997-99	.3
(M) 75000 Mile Service	
1997-99	1.0
(M) 80000 Mile Service	
1997-99	.6
(M) 85000 Mile Service	
1997-99	.8
(G) 90000 Mile Service	
1997-99	1.2
(M) 95000 Mile Service	
1997-99	.8
(M) 100000 Mile Service	
1995-99	.3

79221C99

FORD MOTOR CO.
Ford Mustang • Thunderbird
Lincoln Mark III • Mercury Cougar

VEHICLE IDENTIFICATION CHART

Engine Code							Model Year	
Code	Liters	Cu. In. (cc)	Cyl.	Fuel Sys.	Eng. Mfg.		Code	Year
4	3.8	232 (3802)	6	MFI	Ford		S	1995
D ①	5.0	302 (4949)	8	SFI	Ford		T	1996
R ②	3.8	232 (3802)	6	MFI	Ford		V	1997
T ③	5.0	302 (4949)	8	SFI	Ford		W	1998
V	4.6	281 (4593)	8	SFI	Ford		X	1999
W	4.6	281 (4593)	8	SFI	Ford			

MFI - Multi-point Fuel Injection
SFI - Sequential Fuel Injection
① Special High Performance
② Supercharged
③ High Output

79221CA9

ENGINE IDENTIFICATION

All measurements are given in inches.

Year	Model	Engine Displacement Liters (cc)	Engine Series (ID/VIN)	Fuel System	No. of Cylinders	Engine Type
1995	Cougar	3.8 (3802)	4	SFI	6	OHV
	Cougar	4.5 (4593)	W	SFI	8	SOHC
	Mark VIII	4.6 (4593)	V	SFI	8	DOHC
	Mustang	3.8 (3802)	4	SFI	6	OHV
	Mustang	5.0 (4949)	D	SFI	8	OHV
	Mustang	5.0 (4949)	T	SFI	8	OHV
	Thunderbird	3.8 (3802)	4	SFI	6	OHV
	Thunderbird	3.8 (3802)	R	SFI	6	OHV
	Thunderbird	4.6 (4593)	W	SFI	8	SOHC
1996	Cougar	3.8 (3802)	4	SFI	6	OHV
	Cougar	4.6 (4593)	W	SFI	8	SOHC
	Mark VIII	4.6 (4593)	V	SFI	8	DOHC
	Mustang	3.8 (3802)	4	SFI	6	OHV
	Mustang	4.6 (4593)	V	SFI	8	DOHC
	Mustang	4.6 (4593)	W	SFI	8	SOHC
	Thunderbird	3.8 (3802)	4	SFI	6	OHV
	Thunderbird	4.6 (4593)	W	SFI	8	SOHC
1997	Cougar	3.8 (3802)	4	SFI	6	OHV
	Cougar	4.6 (4593)	W	SFI	8	SOHC
	Mark VIII	4.6 (4593)	V	SFI	8	DOHC
	Mustang	3.8 (3802)	4	SFI	6	OHV
	Mustang	4.6 (4593)	V	SFI	8	DOHC
	Mustang	4.6 (4593)	W	SFI	8	SOHC
	Thunderbird	3.8 (3802)	4	SFI	6	OHV
	Thunderbird	4.6 (4593)	W	SFI	8	SOHC
1998-99	Mark VIII	4.6 (4593)	V	SFI	8	DOHC
	Mustang	3.8 (3802)	4	SFI	6	OHV
	Mustang	4.6 (4593)	V	SFI	8	DOHC
	Mustang	4.6 (4593)	W	SFI	8	SOHC

MFI - Multi-point Fuel Injection
SFI - Sequential Fuel Injection
OHV - Overhead Valves
SOHC - Single Overhead Camshaft
DOHC - Double Overhead Camshafts

79221CA0

GENERAL ENGINE SPECIFICATIONS

Year	Engine ID/VIN	Engine Displacement Liters (cc)	Fuel System Type	Net Horsepower @ rpm	Net Torque @ rpm (ft. lbs.)	Bore x Stroke (in.)	Compression Ratio	Oil Pressure @ rpm
1995	4	3.8 (3802)	SFI	140@3800	215@2400	3.81x3.39	9.0:1	40-60@2500
	D ①	5.0 (4949)	SFI	225@4200	315@2600	4.00x3.00	9.0:1	40-60@2000
	R ②	3.8 (3802)	SFI	210@2000	315@2600	3.81x3.39	8.2:1	40-60@2500
	T ③	5.0 (4949)	SFI	200@4000	275@3000	4.00x3.00	9.0:1	40-60@2000
	V	4.6 (4593)	SFI	280@4500	285@4500	3.55x3.54	9.8:1	33@1500
	W	4.6 (4593)	SFI	④	⑤	3.55x3.54	9.0:1	20-45@2000
1996	4	3.8 (3802)	SFI	145@4000	215@2750	3.81x3.39	9.0:1	40-60@2500
	V	4.6 (4593)	SFI	⑥	⑦	3.55x3.54	9.5:1	20-45@1500
	W	4.6 (4593)	SFI	⑧	⑨	3.55x3.54	⑩	20-45@1500
1997	4	3.8 (3802)	SFI	145@4000	215@2750	3.81x3.39	9.0:1	40-60@2500
	V	4.6 (4593)	SFI	⑥	⑦	3.55x3.54	9.5:1	20-45@1500
	W	4.6 (4593)	SFI	⑧	⑨	3.55x3.54	⑩	20-45@1500
1998-99	4	3.8 (3802)	SFI	145@4000	215@2750	3.81x3.39	9.0:1	40-60@2500
	V	4.6 (4593)	SFI	⑥	⑦	3.55x3.54	9.5:1	20-45@1500
	W	4.6 (4593)	SFI	⑧	⑨	3.55x3.54	⑩	20-45@1500

SFI - Sequential Fuel Injection
① Special high performance
② Supercharged
③ High output
④ Single exhaust: 190@4250
 Dual exhaust:
 Thunderbird: 205@4500
⑤ Single exhaust: 260@3250
 Dual exhaust:
 Thunderbird: 265@3200
⑥ Mark VIII without LSC package: 280@5500
 Mark VIII with LSC package: 290@5750
 Mustang: 305@5800
⑦ Mark VIII without LSC package: 285@4500
 Mark VIII with LSC package: 292@4500
 Mustang: 300@4800
⑧ Thunderbird: 205@4250
 Mustang: 215@4400
⑨ Thunderbird: 280@3000
 Mustang: 285@3500
⑩ Base engine: 9.0:1

79221CB1

GASOLINE ENGINE TUNE-UP SPECIFICATIONS

Year	Engine ID/VIN	Engine Displacement Liters (cc)	Spark Plugs Gap (in.)	Ignition Timing (deg.)		Fuel Pump (psi)	Idle Speed (rpm)		Valve Clearance	
				MT	AT		MT	AT	In.	Ex.
1995	4	3.8 (3802)	0.054	—	10B	30-45 ①	—	②	HYD	HYD
	D	5.0 (4949)	0.054	10B	10B	30-45 ①	②	②	HYD	HYD
	R	3.8 (3802)	0.054	10B	10B	30-40 ①	②	②	HYD	HYD
	T	5.0 (4949)	0.054	10B	10B	30-45 ①	②	②	HYD	HYD
	V	4.6 (4593)	0.054	—	10B	30-45 ①	—	②	HYD	HYD
	W	4.6 (4593)	0.054	—	10B	30-45 ①	—	②	HYD	HYD
1996	4	3.8 (3802)	0.054	②	②	28-54 ①	②	②	HYD	HYD
	V	4.6 (4593)	0.054	10B	—	35-45 ①	②	—	HYD	HYD
	W	4.6 (4593)	0.054	10B	10B	35-45 ①	②	②	HYD	HYD
1997	4	3.8 (3802)	0.054	②	②	28-54 ①	②	②	HYD	HYD
	V	4.6 (4593)	0.054	10B	—	35-45 ①	②	—	HYD	HYD
	W	4.6 (4593)	0.054	10B	10B	35-45 ①	②	②	HYD	HYD
1998-99	4	3.8 (3802)	0.054	②	②	28-54 ①	②	②	HYD	HYD
	V	4.6 (4593)	0.054	10B	—	35-45 ①	②	—	HYD	HYD
	W	4.6 (4593)	0.054	10B	10B	35-45 ①	②	②	HYD	HYD

NOTE: The Vehicle Emission Control Information label often reflects specification changes made during production. The label figures must be used if they differ from those in this chart.

B - Before Top Dead Center

HYD - Hydraulic

① Fuel pressure with engine running, pressure regulator vacuum hose connected

② Refer to Vehicle Emission Control Information label

79221CB2

CAPACITIES

Year	Model	Engine ID/VIN	Engine Displacement Liters (cc)	Engine Oil with Filter (qts.)	Transmission (pts.) 4-Spd	Transmission (pts.) 5-Spd	Transmission (pts.) Auto.	Drive Axle Front (pts.)	Drive Axle Rear (pts.)	Fuel Tank (gal.)	Cooling System (qts.)
1995	Cougar	4	3.8 (3802)	5.0	—	—	27.2 ①	—	②	18.0	12.6
	Cougar	W	4.6 (4593)	5.0	—	—	27.2 ①	—	②	18.0	14.1
	Mark VIII	V	4.6 (4593)	6.0	—	—	25.0 ①	—	3.00	18.0	16.0
	Mustang	4	3.8 (3802)	5.0	—	5.6	27.2 ①	—	③	15.4	11.8
	Mustang	D	5.0 (4949)	5.0	—	5.6	27.2 ①	—	③	15.4	14.1
	Mustang	T	5.0 (4949)	5.0	—	5.6	27.2 ①	—	③	15.4	14.1
	Thunderbird	4	3.8 (3802)	5.0	—	—	27.2 ①	—	②	18.0	12.6
	Thunderbird	R	3.8 (3802) ④	5.0	—	6.3	27.2 ①	—	②	18.0	12.5
	Thunderbird	W	4.6 (4593)	5.0	—	—	27.2 ①	—	②	18.0	14.1
1996	Cougar	4	3.8 (3802)	5.0	—	—	27.8 ①	—	②	18.0	12.6
	Cougar	W	4.6 (4593)	5.3	—	—	27.8 ①	—	②	18.0	14.1
	Mark VIII	V	4.6 (4593)	6.0	—	—	25.6 ①	—	3.00	18.0	16.0
	Mustang	4	3.8 (3802)	5.0	—	5.6	27.8 ①	—	3.50	15.4	11.8
	Mustang	V	4.6 (4593)	6.0	—	6.5	27.8 ①	—	3.75	15.4	14.1
	Mustang	W	4.6 (4593)	⑤	—	6.5	25.6 ①	—	3.75	15.4	14.1
	Thunderbird	4	3.8 (3802)	5.0	—	—	27.8 ①	—	⑥	18.0	12.6
	Thunderbird	W	4.6 (4593)	5.3	—	—	27.8 ①	—	⑥	18.0	14.1
1997	Cougar	4	3.8 (3802)	5.0	—	—	27.8 ①	—	②	18.0	12.6
	Cougar	W	4.6 (4593)	5.3	—	—	27.8 ①	—	②	18.0	14.1
	Mark VIII	V	4.6 (4593)	6.0	—	—	25.6 ①	—	3.00	18.0	16.0
	Mustang	4	3.8 (3802)	5.0	—	5.6	27.8 ①	—	3.50	15.4	11.8
	Mustang	V	4.6 (4593)	6.0	—	6.5	27.8 ①	—	3.75	15.4	14.1
	Mustang	W	4.6 (4593)	⑤	—	6.5	25.6 ①	—	3.75	15.4	14.1
	Thunderbird	4	3.8 (3802)	5.0	—	—	27.8 ①	—	⑥	18.0	12.6
	Thunderbird	W	4.6 (4593)	5.3	—	—	27.8 ①	—	⑥	18.0	14.1
1998-99	Mark VIII	V	4.6 (4593)	6.0	—	—	25.6 ①	—	3.00	18.0	16.0
	Mustang	4	3.8 (3802)	5.0	—	5.6	27.8 ①	—	3.50	15.4	11.8
	Mustang	V	4.6 (4593)	6.0	—	6.5	27.8 ①	—	3.75	15.4	14.1
	Mustang	W	4.6 (4593)	⑤	—	6.5	25.6 ①	—	3.75	15.4	14.1

NOTE: All capacities are approximate. Add fluid gradually and ensure a proper fluid level is obtained.

① Includes torque converter
② 7.50" limited slip axle: 2.75 pts.
   7.50" axle: 3.0 pts.
   8.80" axle: 3.25 pts.
③ 7.50" axle: 3.5 pts.
   8.80" axle: 3.75 pts.
④ Supercharged
⑤ Automatic transmission: 6.7 qts.
   Manual transmission: 6.4 qts.
⑥ 7.50" axle: 3.0 pts.
   8.80" axle: 3.25 pts.

79221CB3

VALVE SPECIFICATIONS

Year	Engine ID/VIN	Engine Displacement Liters (cc)	Seat Angle (deg.)	Face Angle (deg.)	Spring Test Pressure (lbs. @ in.)	Spring Installed Height (in.)	Stem-to-Guide Clearance (in.)		Stem Diameter (in.)	
							Intake	Exhaust	Intake	Exhaust
1995	4	3.8 (3802)	44.5	45.8	220@1.18	1.650	0.0010-0.0027	0.0015-0.0032	0.3415-0.3423	0.3410-0.3418
	D	5.0 (4949) ①	45	44	②	③	0.0010-0.0027	0.0015-0.0032	0.3416-0.3423	0.3411-0.3418
	R	3.8 (3802)	44.5	45.8	220@1.18	1.970	0.0010-0.0027	0.0015-0.0032	0.3415-0.3423	0.3410-0.3418
	T	5.0 (4949)	45	44	④	⑤	0.0010-0.0027	0.0015-0.0032	0.3416-0.3423	0.3411-0.3418
	V	4.6 (4593)	45	45.5	160@1.103	1.425	0.0008-0.0027	0.0018-0.0037	0.2746-0.2754	0.2736-0.2744
	W	4.6 (4593)	45	45.5	132@1.10	1.570	0.0008-0.0027	0.0018-0.0037	0.2746-0.2754	0.2736-0.2744
1996	4	3.8 (3802)	44.5	45.8	220@1.18	1.650	0.0010-0.0027	0.0015-0.0032	0.3415-0.3423	0.3410-0.3418
	V	4.6 (4593)	45	45.5	160@1.10	1.425	0.0008-0.0027	0.0018-0.0037	0.2746-0.2754	0.2736-0.2744
	W	4.6 (4593)	45	45.5	132@1.10	1.570	0.0008-0.0027	0.0018-0.0037	0.2746-0.2754	0.2736-0.2744
1997	4	3.8 (3802)	44.5	45.8	220@1.18	1.650	0.0010-0.0027	0.0015-0.0032	0.3415-0.3423	0.3410-0.3418
	V	4.6 (4593)	45	45.5	160@1.10	1.425	0.0008-0.0027	0.0018-0.0037	0.2746-0.2754	0.2736-0.2744
	W	4.6 (4593)	45	45.5	132@1.10	1.570	0.0008-0.0027	0.0018-0.0037	0.2746-0.2754	0.2736-0.2744
1998-99	4	3.8 (3802)	44.5	45.8	220@1.18	1.650	0.0010-0.0027	0.0015-0.0032	0.3415-0.3423	0.3410-0.3418
	V	4.6 (4593)	45	45.5	160@1.10	1.425	0.0008-0.0027	0.0018-0.0037	0.2746-0.2754	0.2736-0.2744
	W	4.6 (4593)	45	45.5	132@1.10	1.570	0.0008-0.0027	0.0018-0.0037	0.2746-0.2754	0.2736-0.2744

① Cobra
② Intake: 280@1.30
   Exhaust: 264@1.12
③ Intake: 1.80
   Exhaust: 1.62
④ Exhaust: 200-226@1.15
⑤ Intake: 1.75-1.80
   Exhaust: 1.58-1.64

79221CB4

TORQUE SPECIFICATIONS
All readings in ft. lbs.

Year	Engine ID/VIN	Engine Displacement Liters (cc)	Cylinder Head Bolts	Main Bearing Bolts	Rod Bearing Bolts	Crankshaft Damper Bolts	Flywheel Bolts	Manifold		Spark Plugs	Lug Nut
								Intake	Exhaust		
1995	4	3.8 (3802)	①	65-81	31-36	103-132	54-64	②	15-22	7-15	95
	D	5.0 (4949)	⑩	60-70	19-24	110-130	75-85	⑪	26-32	10-15	95
	R	3.8 (3802)	①	65-81	31-36	103-132	54-64	③	15-22	7-15	95
	T	5.0 (4949)	⑩	60-70	19-24	110-130	75-85	⑪	26-32	10-15	95
	V	4.6 (4593)	④	⑦	⑧	114-121	54-64	⑨	13-16	7-15	95
	W	4.6 (4593)	④	⑤	⑥	114-121	54-64	15-22	15-22	7-15	95
1996	4	3.8 (3802)	①	65-81	31-36	103-132	54-64	②	15-22	7-15	95
	V	4.6 (4593)	④	⑦	⑧	114-121	54-64	⑨	13-16	7-15	95
	W	4.6 (4593)	④	⑤	⑥	114-121	54-64	15-22	15-22	7-15	95
1997	4	3.8 (3802)	①	65-81	31-36	103-132	54-64	②	15-22	7-15	95
	V	4.6 (4593)	④	⑦	⑧	114-121	54-64	⑨	13-16	7-15	95
	W	4.6 (4593)	④	⑤	⑥	114-121	54-64	15-22	15-22	7-15	95
1998-99	4	3.8 (3802)	①	65-81	31-36	103-132	54-64	②	15-22	7-15	95
	V	4.6 (4593)	④	⑦	⑧	114-121	54-64	⑨	13-16	7-15	95
	W	4.6 (4593)	④	⑤	⑥	114-121	54-64	15-22	15-22	7-15	95

① Do not reuse cylinder head bolts
Step 1: 15 ft. lbs.
Step 2: 29 ft. lbs.
Step 3: 37 ft. lbs.
Step 4: Loosen bolts one at a time and retorque as follows:
Long bolts: 11-18 ft. lbs.
Short bolts: 7-15 ft. lbs.
Step 5: Rotate 85-95 degrees

② Upper intake manifold bolts
Step 1: 8 ft. lbs.
Step 2: 15 ft. lbs.
Step 3: 24 ft. lbs.
Lower intake manifold bolts
Step 1: 13 ft. lbs.
Step 2: 16 ft. lbs.

③ Supercharger to lower intake manifold bolts
M8 x 43mm bolts: 20-28 ft.lbs
M8 x 108mm bolts: 15-22 ft. lbs.
M12 bolt: 52-70 ft. lbs.
Lower intake manifold bolts
Step 1: 13 ft. lbs.
Step 2: 16 ft. lbs.

④ Do not reuse cylinder head bolts
Step 1: 27-32 ft. lbs.
Step 2: Rotate each bolt 85-95 degrees
Step 3: Repeat Step 2

⑤ Do not reuse main cap bolts
Step 1: Main bearing cap bolts: 22-25 ft. lbs.
Step 2: Rotate each bolt 85-95 degrees
Step 3: Main bearing cap adjust screws: 4 ft. lbs. then 6-8 ft. lbs.
Step 4: Main bearing cap side bolts: 7 ft. lbs. then 14-17 ft. lbs.

⑥ Do not reuse rod bolts
Step 1: 12 ft. lbs.
Step 2: Rotate 85-95 degrees

⑦ Step 1: Main bearing cap bolts: 6-9 ft. lbs.
Step 2: Main bearing cap bolts, outer: 16-21 ft. lbs.
Step 3: Main bearing cap bolts, inner: 27-32 ft. lbs.
Step 4: Rotate main bearing cap bolts 85-95 degrees
Step 5: Main cap adjusting screws: 4 ft. lbs. then 7.5 ft. lbs.
Step 6: Main cap side bolts: 7 ft. lbs. then 14-17 ft. lbs.

⑧ Step 1: 5 ft. lbs.
Step 2: 10 ft. lbs.
Step 3: 18-25 ft. lbs.
Step 4: Rotate 85-95 degrees

⑨ Step 1: Four inside short bolts: 9-11 ft. lbs.
Step 2: All other bolts: 13-16 ft. lbs.
Step 3: Rotate 85-95 degrees

⑩ Do not reuse cylinder head bolts
Step 1: 22-35 ft. lbs.
Step 2: 44-55 ft. lbs.
Step 3: Rotate 85-95 degrees

⑪ Step 1: 8 ft. lbs.
Step 2: 16 ft. lbs.
Step 3: 23-25 ft. lbs.

79221CB5

SCHEDULED MAINTENANCE INTERVALS
(FORD MUSTANG, THUNDERBIRD, LINCOLN MARK VIII, MERCURY COUGAR)

TO BE SERVICED	TYPE OF SERVICE	VEHICLE MILEAGE INTERVAL (x1000)												
		5	10	15	20	25	30	35	40	45	50	55	60	65
Engine oil & filter	R	✓	✓	✓	✓	✓	✓	✓	✓	✓	✓	✓	✓	✓
Adjust clutch pedal by lifting pedal	S/I	✓	✓	✓	✓	✓	✓	✓	✓	✓	✓	✓	✓	✓
Rotate tires	S/I	✓		✓		✓		✓		✓		✓		✓
Cooling system, hoses, clamps & coolant strength	S/I			✓			✓			✓			✓	
Lubricate steering linkage (T-Bird/Cougar)	S/I			✓			✓			✓			✓	
Air cleaner element	R						✓						✓	
Automatic transmission fluid & filter	R						✓						✓	

79221CB6

SCHEDULED MAINTENANCE INTERVALS
(FORD MUSTANG, THUNDERBIRD, LINCOLN MARK VIII, MERCURY COUGAR)
(Cont.)

TO BE SERVICED	TYPE OF SERVICE	VEHICLE MILEAGE INTERVAL (x1000)												
		5	10	15	20	25	30	35	40	45	50	55	60	65
Engine coolant①	R						✓						✓	
Spark plugs (Mark VIII)②	R													
Spark plugs (T-Bird & Cougar Exc. 3.8L SC, 3.8L Calif. 1995)	R						✓						✓	
Spark plugs (T-Bird & Cougar 1996-99)	R												✓	
Spark plugs (T-Bird & Cougar w/3.8L SC, 3.8L Calif. 1995)	R												✓	
Spark plugs (Mustang 3.8L 1995)	R												✓	
Spark plugs (Mustang 5.0L 1995)	R						✓						✓	
Spark plugs (Mustang 1996-99)②	R													
Accessory drive belt(s)	S/I						✓						✓	
Brake lines, hoses & connections	S/I						✓						✓	
Clutch fluid level (T-Bird)	S/I						✓						✓	
Exhaust heat shields	S/I						✓						✓	
Front & rear brakes	S/I						✓						✓	
PCV filter (Mustang 5.0L)	R						✓						✓	
PCV valve	R												✓	
Rear axle lubricant②	R													
Supercharger fluid level	S/I												✓	

① Engine coolant (1996-99 models) - change engine coolant at 48,000 to 50,000 miles and thereafter every 30,000 miles.
② Replace every 100,000 miles.

R – Replace   S/I – Service or Inspect

79221CB7

SCHEDULED MAINTENANCE INTERVALS
(FORD MUSTANG, THUNDERBIRD, LINCOLN MARK VIII, MERCURY COUGAR)
(Cont.)

FREQUENT OPERATION MAINTENANCE (SEVERE SERVICE)
If a vehicle is operated under any of the following conditions it is considered severe service:
- Extremely dusty areas.
- 50% or more of the vehicle operation is in 32°C (90°F) or higher temperatures, or constant operation in temperatures below 0°C (32°F).
- Prolonged idling (vehicle operation in stop and go traffic).
- Frequent short running periods (engine does not warm to normal operating temperatures).
- Police, taxi, delivery usage or trailer towing usage.

Oil & oil filter – change every 3000 miles.
Adjust clutch by lifting pedal every 3000 miles.
Rotate tires at 6000 miles & every 9000 miles thereafter.
Front & rear brakes - check every 15,000 miles.
Automatic transmission fluid & filter - change every 21,000 miles.

79221CB8

SCHEDULED MAINTENANCE
FORD MOTOR COMPANY
FORD MUSTANG, THUNDERBIRD
LINCOLN MARK VIII, MERCURY COUGAR (1995-97)

The following should be used as a guide when determining the amount of work required for a particular service if taken to a repair shop.
In estimating how long a particular Scheduled Maintenance Service should take, please observe the following:

- **Chilton Time** is time based on field research and data supplied by the vehicle manufacturer.

- Labor time operations are given in hours and tenths of an hour.

- All labor operations, are to be used as a guide.

Mechanic Skill Level Codes:
- **(G)** GENERAL: Normally skilled with certification.
- **(M)** MAINTENANCE: Semi-skilled working on certicication.
- **(P)** PRECISION: Really skilled with multiple certification.

	Chilton Time
(G) 5000 Mile Service	
1995-99	.9
(G) 10000 Mile Service	
1995-99	.4
(G) 15000 Mile Service	
1995-99	1.1
(G) 20000 Mile Service	
1995-99	.4
(G) 25000 Mile Service	
1995-99	.9
(G) 30000 Mile Service	
1995-99 Mustang	3.2

	Chilton Time
1995-99 Mark VIII	3.4
1995-99 Thunderbird	3.6
1995-97 Cougar	3.4
(G) 35000 Mile Service	
1995-99	.9
(G) 40000 Mile Service	
1995-99	.4
(G) 45000 Mile Service	
1995-99 Mustang	1.1
(G) 50000 Mile Service	
1995-99	.4

	Chilton Time
(G) 55000 Mile Service	
1995-99 Mustang	.9
(G) 60000 Mile Service	
1995 Mustang	
3.8L	3.7
5.0L	3.9
1996-99 Mustang	3.7
Renew PCV filter (5.0L) add . .	.3
1995-99 Mark VIII	3.9
1995-99 Thunderbird	3.6
1995-97 Cougar	3.4
(G) 65000 Mile Service	
1995-99	.9

79221CB9

FORD MOTOR CO.
Ford Probe

VEHICLE IDENTIFICATION CHART

Engine Code						Model Year	
Code	Liters	Cu. In. (cc)	Cyl.	Fuel Sys.	Eng. Mfg.	Code	Year
A	2.0	122 (1993)	4	MFI	Mazda	S	1995
B	2.5	153 (2501)	6	MFI	Mazda	T	1996
						V	1997

MFI - Multi-point Fuel Injection

79221CB0

ENGINE IDENTIFICATION
All measurements are given in inches.

Year	Model	Engine Displacement Liters (cc)	Engine Series (ID/VIN)	Fuel System	No. of Cylinders	Engine Type
1995	Probe	2.0 (1993)	A	SFI	4	DOHC
	Probe	2.5 (2501)	B	SFI	6	DOHC
1996	Probe	2.0 (1993)	A	SFI	4	DOHC
	Probe	2.5 (2501)	B	SFI	6	DOHC
1997	Probe	2.0 (1993)	A	SFI	4	DOHC
	Probe	2.5 (2501)	B	SFI	6	DOHC

SFI - Sequential Fuel Injection
DOHC - Double Overhead Camshafts

79221CC1

GENERAL ENGINE SPECIFICATIONS

Year	Engine ID/VIN	Engine Displacement Liters (cc)	Fuel System Type	Net Horsepower @ rpm	Net Torque @ rpm (ft. lbs.)	Bore x Stroke (in.)	Compression Ratio	Oil Pressure @ rpm
1995	A	2.0 (1993)	SFI	115@5500	124@3500	3.27x3.62	9.0:1	57-71@2000
	B	2.5 (2501)	SFI	164@6000	156@4000	3.33x2.92	9.2:1	49-71@3000
1996	A	2.0 (1993)	SFI	118@5500	127@4500	3.27x3.62	9.0:1	57-71@2000
	B	2.5 (2501)	SFI	①	②	3.33x2.92	9.2:1	49-71@3000
1997	A	2.0 (1993)	SFI	118@5500	127@4500	3.27x3.62	9.0:1	57-71@2000
	B	2.5 (2501)	SFI	①	②	3.33x2.92	9.2:1	49-71@3000

SFI - Sequential Fuel Injection
① California: 160@5500
   Except California: 164@5600
② California: 156@5000
   Except California: 160@4800

79221CC2

GASOLINE ENGINE TUNE-UP SPECIFICATIONS

Year	Engine ID/VIN	Engine Displacement Liters (cc)	Spark Plugs Gap (in.)	Ignition Timing (deg.)		Fuel Pump (psi)	Idle Speed (rpm)		Valve Clearance	
				MT	AT		MT	AT	In.	Ex.
1995	A	2.0 (1993)	0.040	10B	12B	30-38 ①	700	700	HYD	HYD
	B	2.5 (2501)	0.040	10B	10B	30-36 ①	650	650	HYD	HYD
1996	A	2.0 (1993)	0.041	10B	12B	30-38 ①	700	700	HYD	HYD
	B	2.5 (2501)	0.041	10B	10B	30-36 ①	700	700	HYD	HYD
1997	A	2.0 (1993)	0.041	10B	12B	30-38 ①	700	700	HYD	HYD
	B	2.5 (2501)	0.041	10B	10B	30-36 ①	700	700	HYD	HYD

NOTE: The Vehicle Emission Control Information label often reflects specification changes made during production. The label figures must be used if they differ from those in this chart.

HYD - Hydraulic

① Fuel pressure with engine running, pressure regulator vacuum hose connected

79221CC3

CAPACITIES

Year	Model	Engine ID/VIN	Engine Displacement Liters (cc)	Engine Oil with Filter (qts.)	Transmission (pts.)			Drive Axle		Fuel Tank (gal.)	Cooling System (qts.)
					4-Spd	5-Spd	Auto.	Front (pts.)	Rear (pts.)		
1995	Probe	A	2.0 (1993)	3.7	—	5.8	17.6 ①	②	—	15.5	7.4
	Probe	B	2.5 (2501)	4.2	—	5.8	14.4 ①	②	—	15.5	7.9
1996	Probe	A	2.0 (1993)	3.7	—	5.8	17.6 ①	②	—	15.5	7.4
	Probe	B	2.5 (2501)	4.2	—	5.8	14.4 ①	②	—	15.5	7.9
1997	Probe	A	2.0 (1993)	3.7	—	5.8	17.6 ①	②	—	15.5	7.4
	Probe	B	2.5 (2501)	4.2	—	5.8	14.4 ①	②	—	15.5	7.9

NOTE: All capacities are approximate. Add fluid gradually and ensure a proper fluid level is obtained.

① Includes torque converter

② Included in transaxle capacity

79221CC4

VALVE SPECIFICATIONS

Year	Engine ID/VIN	Engine Displacement Liters (cc)	Seat Angle (deg.)	Face Angle (deg.)	Spring Test Pressure (lbs. @ in.)	Spring Installed Height (in.)	Stem-to-Guide Clearance (in.)		Stem Diameter (in.)	
							Intake	Exhaust	Intake	Exhaust
1995	A	2.0 (1993)	45	45	①	①	0.0010-0.0024	0.0012-0.0026	0.2350-0.2356	0.2348-0.2354
	B	2.5 (2501)	45	45	②	②	0.0010-0.0023	0.0012-0.0026	0.2351-0.2356	0.2349-0.2354
1996	A	2.0 (1993)	45	45	①	①	0.0010-0.0024	0.0012-0.0026	0.2350-0.2356	0.2348-0.2354
	B	2.5 (2501)	45	45	②	②	0.0010-0.0023	0.0012-0.0026	0.2351-0.2356	0.2349-0.2354
1997	A	2.0 (1993)	45	45	①	①	0.0010-0.0024	0.0012-0.0026	0.2350-0.2356	0.2348-0.2354
	B	2.5 (2501)	45	45	②	②	0.0010-0.0023	0.0012-0.0026	0.2351-0.2356	0.2349-0.2354

① Measure spring free length and out of square.
Maximum allowable out-of-square: 0.061
Spring free length: 1.732

② Measure spring free length and out of square.
Maximum allowable out-of-square: 0.642
Spring free length: Intake: 1.729, Exhaust: 1.847

79221CC5

TORQUE SPECIFICATIONS
All readings in ft. lbs.

Year	Engine ID/VIN	Engine Displacement Liters (cc)	Cylinder Head Bolts	Main Bearing Bolts	Rod Bearing Bolts	Crankshaft Damper Bolts	Flywheel Bolts	Manifold		Spark Plugs	Lug Nut
								Intake	Exhaust		
1995	A	2.0 (1993)	①	②	③	116-123	70-75	14-19	14-21	11-17	85
	B	2.5 (2501)	①	④	③	116-123	45-49	14-18	14-18	11-16	85
1996	A	2.0 (1993)	①	②	⑤	116-123	70-75	14-19	14-21	11-17	85
	B	2.5 (2501)	①	④	⑤	116-123	45-49	14-18	14-18	11-16	85
1997	A	2.0 (1993)	①	②	⑤	116-123	70-75	14-19	14-21	11-17	85
	B	2.5 (2501)	①	④	⑤	116-123	45-49	14-18	14-18	11-16	85

NOTE: Always follow proper torque patterns

NOTE: Stretch bolts are used in all procedures that require rotating the fastener a certain number of degrees. The bolts stretch and cannot be reused. For reassembly, replace with new fastners.

① Step 1: 8-10 ft. lbs.
  Step 2: 13-16 ft. lbs.
  Step 3: Rotate 90 degrees
  Step 4: Repeat Step 3
② Step 1: 12 ft. lbs.
  Step 2: Rotate each bolt 85-95 degrees
③ Step 1: 16-19 ft. lbs.
  Step 2: Rotate each bolt 90 degrees
  Step 3: Repeat Step 2
④ Step 1: Inner main bolts: 10-12 ft. lbs.
  Step 2: Inner main bolts: 17-19 ft. lbs.
  Step 3: Outer main bolts: 6-8 ft. lbs.
  Step 4: Outer main bolts: 13-15 ft. lbs.
  Step 5: Rotate inner bolts 75 degrees
  Step 6: Rotate outer bolts 60 degrees
  Step 7: Repeat Steps 5 and 6
  Step 8: Outer cylinder block bolts: 14-15 ft. lbs.
⑤ 16-19 ft. lbs. plus 90 degrees

79221CC6

SCHEDULED MAINTENANCE INTERVALS
(FORD PROBE)

TO BE SERVICED	TYPE OF SERVICE	VEHICLE MILEAGE INTERVAL (x1000)												
		5	10	15	20	25	30	35	40	45	50	55	60	65
Engine oil & filter	R	✓	✓	✓	✓	✓	✓	✓	✓	✓	✓	✓	✓	✓
Rotate tires	S/I	✓		✓		✓		✓		✓		✓		✓
Air cleaner element	R						✓						✓	
Spark plugs	R						✓						✓	
Automatic transmission fluid & filter	R						✓						✓	
Exhaust heat shields	S/I						✓						✓	
Front & rear brakes	S/I						✓						✓	
Accessory drive belt(s)	S/I						✓						✓	
Fuel lines & hoses	S/I						✓						✓	
Cooling system, hoses, clamps & coolant strength	S/I						✓						✓	
Front wheel driveshaft joint boots	S/I						✓						✓	
Brake lines, hoses & connections	S/I						✓						✓	
Front suspension ball joints, steering operation & linkage	S/I						✓						✓	
Idle speed	S/I						✓						✓	
Bolts & nuts on chassis & body	S/I						✓						✓	
Engine coolant	R										✓			
Timing belt/chain & fuel filter	R												✓	
Fuel lines & tubes (emission)	S/I												✓	

R – Replace   S/I – Service or Inspect

FREQUENT OPERATION MAINTENANCE (SEVERE SERVICE)

If a vehicle is operated under any of the following conditions it is considered severe service:
- **Extremely dusty areas.**
- 50% or more of the vehicle operation is in 32°C (90°F) or higher temperatures, or constant operation in temperatures below 0°C (32°F).
- Prolonged idling (vehicle operation in stop and go traffic).
- Frequent short running periods (engine does not warm to normal operating temperatures).
- Police, taxi, delivery usage or trailer towing usage.

Oil & oil filter – change every 3000 miles.
Air cleaner element - check every 15,000 miles.
Bolts & nuts on chassis & body - check every 15,000 miles.
Front & rear brakes - check every 15,000 miles.
Automatic transaxle fluid & filter - change every 21,000 miles.

SCHEDULED MAINTENANCE
FORD MOTOR COMPANY
FORD PROBE

The following should be used as a guide when determining the amount of work required for a particular service if taken to a repair shop. In estimating how long a particular Scheduled Maintenance Service should take, please observe the following:

- **Chilton Time** is time based on field research and data supplied by the vehicle manufacturer.
- Labor time operations are given in hours and tenths of an hour.
- All labor operations, are to be used as a guide.

Mechanic Skill Level Codes:
  (G) GENERAL: Normally skilled with certification.
  (M) MAINTENANCE: Semi-skilled working on certicication.
  (P) PRECISION: Really skilled with multiple certification.

	Chilton Time
(M) 5000 Mile Service	
1995-97	.8
(M) 10000 Mile Service	
1995-97	.3
(M) 15000 Mile Service	
1995-97	.8
(M) 20000 Mile Service	
1995-97	.3

	Chilton Time
(M) 25000 Mile Service	
1995-97	.8
(G) 30000 Mile Service	
1995-97	3.2
(M) 35000 Mile Service	
1995-97	.8
(M) 40000 Mile Service	
1995-97	.3
(M) 45000 Mile Service	
1995-97	.8

	Chilton Time
(G) 50000 Mile Service	
1995-97	1.0
(M) 55000 Mile Service	
1995-97	.8
(G) 60000 Mile Service	
1995-97	5.3
(M) 65000 Mile Service	
1995-97	.8

79221CC8

FORD MOTOR CO.
Ford Taurus • Taurus SHO • Lincoln Continental • Mercury Sable

VEHICLE IDENTIFICATION CHART

Engine Code						Model Year	
Code	Liters	Cu. In. (cc)	Cyl.	Fuel Sys.	Eng. Mfg.	Code	Year
4	3.8	232 (3802)	6	SFI	Ford	S	1995
N	3.4	207 (3393)	8	SFI	Ford	T	1996
S	3.0	183 (3049)	6	SFI	Ford	V	1997
U	3.0	181 (2971)	6	SFI	Ford	W	1998
V	4.6	281 (4593)	8	SFI	Ford	X	1999
Y	3.0	182 (2980)	6	SFI	Yamaha		

SFI - Sequential Fuel Injection

79221CC9

ENGINE IDENTIFICATION
All measurements are given in inches.

Year	Model	Engine Displacement Liters (cc)	Engine Series (ID/VIN)	Fuel System	No. of Cylinders	Engine Type
1995	Continental	4.6 (4593)	V	SFI	8	DOHC
	Sable	3.8 (3802)	4	SFI	6	OHV
	Sable	3.0 (2980)	U	SFI	6	OHV
	Taurus	3.8 (3802)	4	SFI	6	OHV
	Taurus	3.0 (2980)	U	SFI	6	OHV
	Taurus SHO	3.2 (3191)	P	SFI	6	DOHC
	Taurus SHO	3.0 (2980)	Y	SFI	6	DOHC
1996	Continental	4.6 (4593)	V	SFI	8	DOHC
	Sable	3.0 (2998)	S	SFI	6	DOHC
	Sable	3.0 (2982)	U	SFI	6	OHV
	Taurus	3.0 (3049)	S	SFI	6	DOHC
	Taurus	3.0 (2982)	U	SFI	6	OHV
	Taurus SHO	3.4 (3393)	N	SFI	8	DOHC
1997	Continental	4.6 (4593)	V	SFI	8	DOHC
	Sable	3.0 (2998)	S	SFI	6	DOHC
	Sable	3.0 (2982)	U	SFI	6	OHV
	Taurus	3.0 (3049)	S	SFI	6	DOHC
	Taurus	3.0 (2982)	U	SFI	6	OHV
	Taurus SHO	3.4 (3393)	N	SFI	8	DOHC
1998-99	Continental	4.6 (4593)	V	SFI	8	DOHC
	Sable	3.0 (2998)	S	SFI	6	DOHC
	Sable	3.0 (2982)	U	SFI	6	OHV
	Taurus	3.0 (3049)	S	SFI	6	DOHC
	Taurus	3.0 (2982)	U	SFI	6	OHV
	Taurus SHO	3.4 (3393)	N	SFI	8	DOHC

SFI - Sequential Fuel Injection
OHV - Overhead Valves
DOHC - Double Overhead Camshafts

79221CC0

GENERAL ENGINE SPECIFICATIONS

Year	Engine ID/VIN	Engine Displacement Liters (cc)	Fuel System Type	Net Horsepower @ rpm	Net Torque @ rpm (ft. lbs.)	Bore x Stroke (in.)	Compression Ratio	Oil Pressure @ rpm
1995	4	3.8 (3802)	SFI	140@3800	215@2400	3.81x3.39	9.0:1	40-60@2500
	P	3.2 (3191)	SFI	220@6200	215@4800	3.62x3.15	9.8:1	40-60@2000
	U	3.0 (2980)	SFI	140@4800	160@3000	3.50x3.15	9.3:1	55-70@2500
	V	4.6 (4593)	SFI	260@5750	265@4750	3.55x3.54	9.8:1	33@1500
	Y	3.0 (2980)	SFI	220@6200	200@4800	3.50x3.15	9.8:1	40-65@2000
1996	N	3.4 (3393)	SFI	235@6100	230@4800	3.25x3.13	10.0:1	20-45@1500
	S	3.0 (2998)	SFI	200@5750	200@4500	3.50x3.13	10.0:1	20-45@1515
	U	3.0 (2982)	SFI	145@5250	170@3250	3.50x3.15	9.3:1	55-70@2500
	V	4.6 (4593)	SFI	260@5750	265@4750	3.55x3.54	9.8:1	33@1500
1997	N	3.4 (3393)	SFI	235@6100	230@4800	3.25x3.13	10.0:1	20-45@1500
	S	3.0 (2998)	SFI	200@5750	200@4500	3.50x3.13	10.0:1	20-45@1515
	U	3.0 (2982)	SFI	145@5250	170@3250	3.50x3.15	9.3:1	55-70@2500
	V	4.6 (4593)	SFI	260@5750	265@4750	3.55x3.54	9.8:1	33@1500
1998-99	N	3.4 (3393)	SFI	235@6100	230@4800	3.25x3.13	10.0:1	20-45@1500
	S	3.0 (2998)	SFI	200@5750	200@4500	3.50x3.13	10.0:1	20-45@1515
	U	3.0 (2982)	SFI	145@5250	170@3250	3.50x3.15	9.3:1	55-70@2500
	V	4.6 (4593)	SFI	260@5750	265@4750	3.55x3.54	9.8:1	33@1500

SFI - Sequential Fuel Injection

79221CD1

GASOLINE ENGINE TUNE-UP SPECIFICATIONS

Year	Engine ID/VIN	Engine Displacement Liters (cc)	Spark Plugs Gap (in.)	Ignition Timing (deg.) MT	Ignition Timing (deg.) AT	Fuel Pump (psi)	Idle Speed (rpm) MT	Idle Speed (rpm) AT	Valve Clearance In.	Valve Clearance Ex.
1995	4	3.8 (3802)	0.054	—	10B	30-45 ①	—	②	HYD	HYD
	P	3.2 (3191)	0.044	—	10B	30-45 ①	—	800	0.006-0.010	0.010-0.014
	U	3.0 (2980)	0.044	10B	10B	30-45 ①	—	②	HYD	HYD
	V	4.6 (4593)	0.054	—	10B	30-45 ①	—	②	HYD	HYD
	Y	3.0 (2980)	0.044	10B	—	28-33 ①	②	—	0.006-0.010	0.010-0.014
1996	N	3.4 (3393)	0.042-0.046	—	10B	35-45 ①	—	②	HYD	HYD
	S	3.0 (2998)	0.054	—	10B	30-45 ①	—	②	HYD	HYD
	U	3.0 (2982)	0.044	—	10B	30-45 ①	—	②	HYD	HYD
	V	4.6 (4593)	0.054	—	10B	30-45 ①	—	②	HYD	HYD
1997	N	3.4 (3393)	0.042-0.046	—	10B	35-45 ①	—	②	HYD	HYD
	S	3.0 (2998)	0.054	—	10B	30-45 ①	—	②	HYD	HYD
	U	3.0 (2982)	0.044	—	10B	30-45 ①	—	②	HYD	HYD
	V	4.6 (4593)	0.054	—	10B	30-45 ①	—	②	HYD	HYD
1998-99	N	3.4 (3393)	0.042-0.046	—	10B	35-45 ①	—	②	HYD	HYD
	S	3.0 (2998)	0.054	—	10B	30-45 ①	—	②	HYD	HYD
	U	3.0 (2982)	0.044	—	10B	30-45 ①	—	②	HYD	HYD
	V	4.6 (4593)	0.054	—	10B	30-45 ①	—	②	HYD	HYD

NOTE: The Vehicle Emission Control Information label often reflects specification changes made during production. The label figures must be used if they differ from those in this chart.

B - Before Top Dead Center

HYD - Hydraulic

① Fuel pressure with engine running, pressure regulator vacuum hose connected

② Refer to Vehicle Emission Control Information label

79221CD2

CAPACITIES

Year	Model	Engine ID/VIN	Engine Displacement Liters (cc)	Engine Oil with Filter (qts.)	Transmission (pts.) 4-Spd	5-Spd	Auto.	Drive Axle Front (pts.)	Rear (pts.)	Fuel Tank (gal.)	Cooling System (qts.)
1995	Continental	V	4.6 (4593)	6.0	—	—	26.6 ①	②	—	18.4	14.3
	Sable	4	3.8 (3802)	4.5	—	—	24.5 ①	②	—	③	12.1
	Sable	U	3.0 (2980)	4.5	—	—	24.5 ①	②	—	③	11.0
	Taurus	4	3.8 (3802)	4.5	—	—	24.5 ①	②	—	③	12.1
	Taurus	U	3.0 (2980)	4.5	—	—	24.5 ①	②	—	③	11.0
	Taurus SHO	P	3.2 (3191)	5.0	—	—	24.5 ①	②	—	18.6	11.4
	Taurus SHO	Y	3.0 (2980)	5.0	—	6.2	24.5 ①	②	—	18.6	11.6
1996	Continental	V	4.6 (4593)	6.0	—	—	27.4 ①	②	—	17.8	14.3
	Sable	S	3.0 (2998)	5.8	—	—	27.0 ①	②	—	③	16.0
	Sable	U	3.0 (2982)	4.5	—	—	24.5 ①	②	—	③	16.0
	Taurus	S	3.0 (2998)	5.8	—	—	27.0 ①	②	—	③	16.0
	Taurus	U	3.0 (2982)	4.5	—	—	24.5 ①	②	—	③	16.0
	Taurus SHO	N	3.4 (3393)	5.8	—	—	27.0 ①	②	—	③	16.0
1997	Continental	V	4.6 (4593)	6.0	—	—	27.4 ①	②	—	17.8	14.3
	Sable	S	3.0 (2998)	5.8	—	—	27.0 ①	②	—	③	16.0
	Sable	U	3.0 (2982)	4.5	—	—	24.5 ①	②	—	③	16.0
	Taurus	S	3.0 (2998)	5.8	—	—	27.0 ①	②	—	③	16.0
	Taurus	U	3.0 (2982)	4.5	—	—	24.5 ①	②	—	③	16.0
	Taurus SHO	N	3.4 (3393)	5.8	—	—	27.0 ①	②	—	③	16.0
1998-99	Continental	V	4.6 (4593)	6.0	—	—	27.4 ①	②	—	17.8	14.3
	Sable	S	3.0 (2998)	5.8	—	—	27.0 ①	②	—	③	16.0
	Sable	U	3.0 (2982)	4.5	—	—	24.5 ①	②	—	③	16.0
	Taurus	S	3.0 (2998)	5.8	—	—	27.0 ①	②	—	③	16.0
	Taurus	U	3.0 (2982)	4.5	—	—	24.5 ①	②	—	③	16.0
	Taurus SHO	N	3.4 (3393)	5.8	—	—	27.0 ①	②	—	③	16.0

NOTE: All capacities are approximate. Add fluid gradually and ensure a proper fluid level is obtained.

① Includes torque converter
② Included in transaxle capacity
③ Standard tank: 16.0 gals.
Optional extended range tank: 18.6 gals.

79221CD3

VALVE SPECIFICATIONS

Year	Engine ID/VIN	Engine Displacement Liters (cc)	Seat Angle (deg.)	Face Angle (deg.)	Spring Test Pressure (lbs. @ in.)	Spring Installed Height (in.)	Stem-to-Guide Clearance (in.)		Stem Diameter (in.)	
							Intake	Exhaust	Intake	Exhaust
1995	4	3.8 (3802)	44.5	45.8	220@1.18	1.970	0.0010-0.0027	0.0015-0.0032	0.3415-0.3423	0.3410-0.3418
	P	3.2 (3191)	45	45.5	121@1.19	1.520	0.0010-0.0023	0.0012-0.0025	0.2346-0.2352	0.2344-0.2350
	U	3.0 (2980)	45	44	180@1.06	1.580	0.0001-0.0027	0.0015-0.0032	0.3126-0.3129	0.3121-0.3134
	V	4.6 (4593)	45	45.5	160@1.103	1.425	0.0008-0.0027	0.0018-0.0037	0.2746-0.2754	0.2736-0.2744
	Y	3.0 (2980)	45	45.5	121@1.19	1.520	0.0010-0.0023	0.0012-0.0025	0.2346-0.2352	0.2344-0.2350
1996	N	3.4 (3393)	45	45.5	89@1.00	1.360	0.0010-0.0023	0.0012-0.0025	0.2346-0.2352	0.2344-0.2350
	S	3.0 (2998)	44.75	45.5	153@1.18	1.570	0.0007-0.0027	0.0017-0.0037	0.2350-0.2358	0.2343-0.2350
	U	3.0 (2982)	45	44	180@1.06	1.580	0.0001-0.0027	0.0015-0.0032	0.3126-0.3134	0.3121-0.3129
	V	4.6 (4593)	45	45.5	160@1.103	1.425	0.0008-0.0027	0.0018-0.0037	0.2746-0.2754	0.2736-0.2744
1997	N	3.4 (3393)	45	45.5	89@1.00	1.360	0.0010-0.0023	0.0012-0.0025	0.2346-0.2352	0.2344-0.2350
	S	3.0 (2998)	44.75	45.5	153@1.18	1.570	0.0007-0.0027	0.0017-0.0037	0.2350-0.2358	0.2343-0.2350
	U	3.0 (2982)	45	44	180@1.06	1.580	0.0001-0.0027	0.0015-0.0032	0.3126-0.3134	0.3121-0.3129
	V	4.6 (4593)	45	45.5	160@1.103	1.425	0.0008-0.0027	0.0018-0.0037	0.2746-0.2754	0.2736-0.2744
1998-99	N	3.4 (3393)	45	45.5	89@1.00	1.360	0.0010-0.0023	0.0012-0.0025	0.2346-0.2352	0.2344-0.2350
	S	3.0 (2998)	44.75	45.5	153@1.18	1.570	0.0007-0.0027	0.0017-0.0037	0.2350-0.2358	0.2343-0.2350
	U	3.0 (2982)	45	44	180@1.06	1.580	0.0001-0.0027	0.0015-0.0032	0.3126-0.3134	0.3121-0.3129
	V	4.6 (4593)	45	45.5	160@1.103	1.425	0.0008-0.0027	0.0018-0.0037	0.2746-0.2754	0.2736-0.2744

79221CD4

TORQUE SPECIFICATIONS
All readings in ft. lbs.

Year	Engine ID/VIN	Engine Displacement Liters (cc)	Cylinder Head Bolts	Main Bearing Bolts	Rod Bearing Bolts	Crankshaft Damper Bolts	Flywheel Bolts	Manifold		Spark Plugs	Lug Nut
								Intake	Exhaust		
1995	4	3.8 (3802)	⑥	65-81	31-36	103-132	54-64	⑦	15-22	7-15	95
	P	3.2 (3191)	③	④	⑤	112-127	51-58	11-17	26-38	15-22	95
	U	3.0 (2980)	①	55-63	26	93-121	54-64	②	15-22	7-15	95
	V	4.6 (4593)	⑧	⑨	⑩	114-121	54-64	⑪	13-16	7-15	95
	Y	3.0 (2980)	③	④	⑤	113-126	51-58	11-17	26-38	15-22	95
1996	N	3.4 (3393)	⑫	⑬	⑭	⑮	54-64	14-20	11-19	11-15	95
	S	3.0 (2998)	⑯	⑰	⑱	⑮	54-64	⑲	13-16	7-15	95
	U	3.0 (2982)	①	55-63	23-29	93-121	54-64	②	15-18	7-15	95
	V	4.6 (4593)	⑧	⑨	⑩	114-121	54-64	⑪	13-16	7-15	95
1997	N	3.4 (3393)	⑫	⑬	⑭	⑮	54-64	14-20	11-19	11-15	95
	S	3.0 (2998)	⑯	⑰	⑱	⑮	54-64	⑲	13-16	7-15	95
	U	3.0 (2982)	①	55-63	23-29	93-121	54-64	②	15-18	7-15	95
	V	4.6 (4593)	⑧	⑨	⑩	114-121	54-64	⑪	13-16	7-15	95
1998-99	N	3.4 (3393)	⑫	⑬	⑭	⑮	54-64	14-20	11-19	11-15	95
	S	3.0 (2998)	⑯	⑰	⑱	⑮	54-64	⑲	13-16	7-15	95
	U	3.0 (2982)	①	55-63	23-29	93-121	54-64	②	15-18	7-15	95
	V	4.6 (4593)	⑧	⑨	⑩	114-121	54-64	⑪	13-16	7-15	95

① Step 1: 33-41 ft.lbs
Step 2: 63-73 ft. lbs.

② Step 1: 15-22 ft. lbs.
Step 2: 19-24 ft. lbs.

③ Step 1: 37-50 ft. lbs.
Step 2: 58-64 ft. lbs.

④ Step 1: 22-26 ft. lbs.
Step 2: 33-36 ft. lbs.
Step 1: Inner bolts - 17-19 ft. lbs. in two to three steps
Step 2: Outer bolts - 13-15 ft. lbs. in two to three steps
Step 3: Inner bolts - 1-3 70 degrees; Inner bolt 4; 80 degrees
Step 4: Tighten outer bolts 60 degrees
Step 5: Repeat Step 4

⑤ Step 1: 16-19 ft. lbs.
Step 2: Rotate each bolt 90 degrees
Step 3: Repeat Step 2

⑥ Do not reuse cylinder head bolts
Step 1: 15 ft. lbs.
Step 2: 29 ft. lbs.
Step 3: 37 ft. lbs.
Step 4: Loosen bolts one at a time and retorque as follows;
Long bolts: 11-18 ft. lbs.
Short bolts: 7-15 ft. lbs
Step 5: Rotate 85-95 degrees

⑦ Step 1: 13 ft. lbs.
Step 2: 16 ft. lbs.
NOTE: Always follow the proper torque patterns

⑧ Step 1: 27-32 ft. lbs.
Step 2: Rotate 85-95 degrees
Step 3: Repeat Step 2

⑨ Step 1: Main bearing cap bolts: 6-9 ft. lbs.
Step 2: Main bearing cap bolts, outer: 16-21 ft. lbs.
Step 3: Main bearing cap bolts, inner: 27-32 ft. lbs.
Step 4: Rotate main bearing cap bolts 85-95 degrees
Step 5: Main cap adjusting screws 4 ft. lbs. then 7.5 ft. lbs.
Step 6: Main cap side bolts: 7 ft. lbs. then 14-17 ft. lbs.

⑩ Step 1: 5 ft. lbs.
Step 2: 10 ft. lbs.
Step 3: 18-25 ft. lbs.
Step 4: Rotate 85-95 degrees

⑪ Step 1: Four inside short bolts: 9-11 ft. lbs.
Step 2: All other bolts: 13-16 ft. lbs.
Step 3: Rotate 85-95 degrees

⑫ Step 1: 20-23 ft. lbs.
Step 2: Rotate 85-95 degrees

⑬ Step 1: Cap bolts 1-10 (outer) 17-20 ft. lbs.
Step 2: Cap bolts 11-20 (inner) 28-31 ft. lbs.
Step 3: Rotate bolts 1-20, 85-95 degrees
Step 4: Bolts 21-31, 15-22 ft. lbs.

⑭ Step 1: 30-33 ft. lbs.
Step 2: Rotate 90-120 degrees

⑮ Step 1: 77-99 ft. lbs.
Step 2: Loosen 360 degrees
Step 3: Tighten to 35-39 ft. lbs.
Step 4: Rotate 85-95 degrees

⑯ Step 1: 28-31 ft. lbs.
Step 2: Rotate 85-95 ft. lbs.
Step 3: Loosen one turn
Step 4: 28-31 ft. lbs.
Step 5: Rotate 85-95 degrees
Step 6: Repeat Step 5

⑰ Step 1: Cap bolts 1-8 (outer) 17-20 ft. lbs.
Step 2: Cap bolts 9-16 (inner) 28-31 ft. lbs.
Step 3: Rotate bolts 1-16, 85-95 degrees
Step 4: Bolts 17-22; 15-22 ft. lbs.

⑱ Step 1: 30-33 ft. lbs.
Step 2: Rotate 90-120 degrees

⑲ 71-106 inch lbs.

79221CD5

SCHEDULED MAINTENANCE INTERVALS
(FORD TAURUS, LINCOLN CONTINENTAL, MERCURY SABLE)

TO BE SERVICED	TYPE OF SERVICE	VEHICLE MILEAGE INTERVAL (x1000)												
		5	10	15	20	25	30	35	40	45	50	55	60	65
Engine oil & filter	R	✓	✓	✓	✓	✓	✓	✓	✓	✓	✓	✓	✓	✓
Rotate tires	S/I	✓		✓		✓		✓		✓		✓		✓
Engine coolant protection, hoses & clamps	S/I			✓			✓			✓			✓	
Pass. compartment air filter (Continental)	R			✓			✓			✓			✓	
Pass. compartment air filter (Taurus, Sable)	R				✓				✓				✓	
Air cleaner filter	R						✓						✓	
Automatic transaxle fluid & filter	R						✓						✓	
Brake lines & connections	S/I						✓						✓	
Exhaust heat shields	S/I						✓						✓	
Front and rear disc brake pads & rotors	S/I						✓						✓	

79221CD6

SCHEDULED MAINTENANCE INTERVALS
(FORD TAURUS, LINCOLN CONTINENTAL, MERCURY SABLE Cont.)

TO BE SERVICED	TYPE OF SERVICE	VEHICLE MILEAGE INTERVAL (x1000)												
		5	10	15	20	25	30	35	40	45	50	55	60	65
Accessory drive belt(s)	S/I												✓	
Engine coolant (1996-99)①	R										✓			
Engine coolant (1995)	R						✓						✓	
Spark plugs (1995 exc. 3.8L Calif.)	R						✓							
Spark plugs (1995 3.8L Calif.)	R												✓	
Spark plugs (1996-99 exc. 3.0L FF)②	R													
Spark plugs (1996-99 3.0L FF)	R						✓						✓	
PCV valve (except 3.0L 4-valve)	R												✓	
PCV valve (3.0L 4-valve)③	R													

① Engine coolant - change initially at 50,000 miles & thereafter every 30,000 miles.
② Platinum tip spark plugs - change every 100,000 miles.
③ Replace every 100,000 miles.
R – Replace   S/I – Service or Inspect

FREQUENT OPERATION MAINTENANCE (SEVERE SERVICE)
If a vehicle is operated under any of the following conditions it is considered severe service:
- Extremely dusty areas.
- 50% or more of the vehicle operation is in 32°C (90°F) or higher temperatures, or constant operation in temperatures below 0°C (32°F).
- Prolonged idling (vehicle operation in stop and go traffic).
- Frequent short running periods (engine does not warm to normal operating temperatures).
- Police, taxi, delivery usage or trailer towing usage.
Oil & oil filter - change every 3000 miles.
Rotate tires at 6000 miles & every 9000 miles thereafter.
Air cleaner element - service or inspect every 15,000 miles.
Automatic transaxle fluid & filter - change every 21,000 miles (1995-97) or every 30,000 miles (1994).

79221CD7

SCHEDULED MAINTENANCE
FORD MOTOR COMPANY
FORD TAURUS, MERCURY SABLE
LINCOLN CONTINENTAL

The following should be used as a guide when determining the amount of work required for a particular service if taken to a repair shop. In estimating how long a particular Scheduled Maintenance Service should take, please observe the following:

- **Chilton Time** is time based on field research and data supplied by the vehicle manufacturer.
- Labor time operations are given in hours and tenths of an hour.
- All labor operations, are to be used as a guide.

Mechanic Skill Level Codes:
(G) GENERAL: Normally skilled with certification.
(M) MAINTENANCE: Semi-skilled working on certicication.
(P) PRECISION: Really skilled with multiple certification.

	Chilton Time
(M) 5000 Mile Service	
1995-99	.9
(M) 10000 Mile Service	
1995-99	.4
(G) 15000 Mile Service	
1995-99	1.2
(M) 20000 Mile Service	
1995-99	.6
(M) 25000 Mile Service	
1995-99	.9

	Chilton Time
(G) 30000 Mile Service	
1995-99	3.1
Renew auto. trans. fluid & filter add	.5
Renew engine coolant add	.5
(M) 35000 Mile Service	
1995-99	.9
(M) 40000 Mile Service	
1995-99	.6
(G) 45000 Mile Service	
1995-99	1.0

	Chilton Time
(M) 50000 Mile Service	
1995-99	.9
(M) 55000 Mile Service	
1995-99	.9
(G) 60000 Mile Service	
1995-99	3.3
Renew auto. trans. fluid & filter add	.5
Renew engine coolant add	.5
(M) 65000 Mile Service	
1995-99	.9

79221CD8

GENERAL MOTORS A BODY
Buick Century • Oldsmobile Cutlass • Cutlass Ciera

VEHICLE IDENTIFICATION CHART

Engine Code						Model Year	
Code	Liters	Cu. In. (cc)	Cyl.	Fuel Sys.	Eng. Mfg.	Code	Year
4	2.2	134 (2195)	4	MFI	BOC	S	1995
M	3.1	191 (3130)	6	MFI	BOC	T	1996

MFI - Multi-point Fuel Injection
BOC - Buick/Oldsmobile/Cadillac

79221CD9

ENGINE IDENTIFICATION

Year	Model	Engine Displacement Liters (cc)	Engine Series (ID/VIN)	Fuel System	No. of Cylinders	Engine Type
1995	Century	3.1 (3130)	M	MFI	6	OHV
	Cutlass Ciera	2.2 (2195)	4	MFI	4	OHV
	Cutlass Ciera	3.1 (3130)	M	MFI	6	OHV
	Cutlass Cruiser	2.2 (2195)	4	MFI	4	OHV
	Cutlass Cruiser	3.1 (3130)	M	MFI	6	OHV
1996	Century	2.2 (2195)	4	MFI	4	OHV
	Century	3.1 (3130)	M	MFI	6	OHV
	Cutlass Ciera	2.2 (2195)	4	MFI	4	OHV
	Cutlass Ciera	3.1 (3130)	M	MFI	6	OHV
	Cutlass Cruiser	3.1 (3130)	M	MFI	6	OHV

MFI - Multi-point Fuel Injection
OHV - Overhead Valves

79221CD0

GENERAL ENGINE SPECIFICATIONS

Year	Engine ID/VIN	Engine Displacement Liters (cc)	Fuel System Type	Net Horsepower @ rpm	Net Torque @ rpm (ft. lbs.)	Bore x Stroke (in.)	Compression Ratio	Oil Pressure @ rpm
1995	4	2.2 (2195)	MFI	120@5200	130@4000	3.50x3.46	8.85:1	56@3000
	M	3.1 (3130)	MFI	160@5200	185@4000	3.50x3.31	9.5:1	15@1100
1996	4	2.2 (2195)	MFI	120@5200	130@4000	3.50x3.46	8.85:1	56@3000
	M	3.1 (3130)	MFI	160@5200	185@4000	3.50x3.31	9.5:1	15@1100

MFI - Multi-pont Fuel Injection

79221CE1

GASOLINE ENGINE TUNE-UP SPECIFICATIONS

Year	Engine ID/VIN	Engine Displacement Liters (cc)	Spark Plugs Gap (in.)	Ignition Timing (deg.) MT	Ignition Timing (deg.) AT	Fuel Pump (psi)	Idle Speed (rpm) MT	Idle Speed (rpm) AT	Valve Clearance In.	Valve Clearance Ex.
1995	4	2.2 (2195)	0.060	①	①	41-47 ②	③	③	HYD	HYD
	M	3.1 (3130)	0.060	①	①	41-47 ②	③	③	HYD	HYD
1996	4	2.2 (2195)	0.060	①	①	41-47	③	③	HYD	HYD
	M	3.1 (3130)	0.060	①	①	41-47	③	③	HYD	HYD

NOTE: The Vehicle Emission Control Information label often reflects specification changes made during production. The label figures must be used if they differ from those in this chart.

HYD - Hydraulic

① DIS Ignition System timing not adjustable
② Pressure at fuel pump
③ Idle speed maintained by ECM. There is no recommended adjustment procedure

79221CE2

CAPACITIES

Year	Model	Engine ID/VIN	Engine Displacement Liters (cc)	Engine Oil with Filter (qts.)	Transmission (pts.) 4-Spd	Transmission (pts.) 5-Spd	Transmission (pts.) Auto.	Drive Axle Front (pts.)	Drive Axle Rear (pts.)	Fuel Tank (gal.)	Cooling System (qts.)
1995	Century	4	2.2 (2195)	3.8 ①	—	—	②	—	—	16.5	8.7
	Century	M	3.1 (3130)	3.8 ①	—	—	②	—	—	16.5	11.6
	Cutlass Ciera	4	2.2 (2195)	3.8 ①	—	—	②	—	—	16.5	8.7
	Cutlass Ciera	M	3.1 (3130)	3.8 ①	—	—	②	—	—	16.5	11.6
	Cutlass Cruiser	4	2.2 (2195)	3.8 ①	—	—	②	—	—	16.5	8.7
	Cutlass Cruiser	M	3.1 (3130)	3.8 ①	—	—	②	—	—	16.5	11.6
1996	Century	4	2.2 (2195)	4.0 ①	—	—	②	—	—	16.5	8.7
	Century	M	3.1 (3130)	4.0 ①	—	—	②	—	—	16.5	11.6
	Cutlass Ciera	4	2.2 (2195)	4.0 ①	—	—	②	—	—	16.5	8.7
	Cutlass Ciera	M	3.1 (3130)	3.8 ①	—	—	②	—	—	16.5	11.6
	Cutlass Cruiser	M	3.1 (3130)	4.0 ①	—	—	②	—	—	16.5	11.6

NOTE: All capacities are approximate. Add fluid gradually and ensure a proper fluid level is obtained.

① Specification is without filter replacement; Additional oil may be required
② 3 speed: 8.0
   4 speed: 14.8

79221CE3

VALVE SPECIFICATIONS

Year	Engine ID/VIN	Engine Displacement Liters (cc)	Seat Angle (deg.)	Face Angle (deg.)	Spring Test Pressure (lbs. @ in.)	Spring Installed Height (in.)	Stem-to-Guide Clearance (in.) Intake	Stem-to-Guide Clearance (in.) Exhaust	Stem Diameter (in.) Intake	Stem Diameter (in.) Exhaust
1995	4	2.2 (2195)	46	45	220-236@ 1.278	1.710	0.0010-0.0027	0.0014-0.0031	NA	NA
	M	3.1 (3130)	45	45	250@1.239	1.710	0.0010-0.0027	0.0010-0.0027	NA	NA
1996	4	2.2 (2195)	46	45	220-236@ 1.278	1.710	0.0010-0.0027	0.0014-0.0031	NA	NA
	M	3.1 (3130)	45	45	230@1.260	1.701	0.0010-0.0027	0.0010-0.0027	NA	NA

NA - Not Available

79221CE4

TORQUE SPECIFICATIONS
All readings in ft. lbs.

Year	Engine ID/VIN	Engine Displacement Liters (cc)	Cylinder Head Bolts	Main Bearing Bolts	Rod Bearing Bolts	Crankshaft Damper Bolts	Flywheel Bolts	Manifold Intake	Manifold Exhaust	Spark Plugs	Lug Nut
1995	4	2.2 (2195)	①	70	38	77	55	24	10	②	100
	M	3.1 (3130)	③	④	⑤	76	59	⑥	12	②	100
1996	4	2.2 (2195)	①	70	38	77	55	24	10	②	100
	M	3.1 (3130)	③	④	⑤	76	59	⑥	12	②	100

① Short bolts: 43 ft. lbs. plus 90 degrees
  Long bolts: 46 ft. lbs. plus 90 degrees
② New cylinder first-time installation: 21 ft. lbs.
  All others: 11 ft. lbs.
③ Coat threads with sealer torque to 37 ft. lbs.,
  then turn 1/4 turn (90 degrees)
④ 37 ft. lbs. plus 75 degrees
⑤ 15 ft. lbs. plus 75 degrees
⑥ Torque all bolts to 15 ft. lbs.,
  Retorque to 24 ft. lbs.

79221CE5

SCHEDULED MAINTENANCE INTERVALS
(GM A BODY- BUICK CENTURY, OLDSMOBILE CUTLASS CIERA & CUTLASS CRUISER)

TO BE SERVICED	TYPE OF SERVICE	VEHICLE MILEAGE INTERVAL (x1000)												
		7.5	15	22.5	30	37.5	45	52.5	60	67.5	75	82.5	90	97.5
Engine oil & filter	R	✓	✓	✓	✓	✓	✓	✓	✓	✓	✓	✓	✓	✓
Brake hoses	S/I	✓	✓	✓	✓	✓	✓	✓	✓	✓	✓	✓	✓	✓
Chassis lubrication	S/I	✓	✓	✓	✓	✓	✓	✓	✓	✓	✓	✓	✓	✓
Coolant level, hoses & clamps	S/I	✓	✓	✓	✓	✓	✓	✓	✓	✓	✓	✓	✓	✓
Drive shaft boots & front suspension components	S/I	✓	✓	✓	✓	✓	✓	✓	✓	✓	✓	✓	✓	✓
Exhaust system	S/I	✓	✓	✓	✓	✓	✓	✓	✓	✓	✓	✓	✓	✓
Lubricate transaxle shift linkage, parking brake cable guides, underbody contact points & linkage	S/I	✓	✓	✓	✓	✓	✓	✓	✓	✓	✓	✓	✓	✓
Rotate tires	S/I	✓		✓		✓		✓		✓		✓		✓
Brake linings	S/I	✓		✓		✓		✓		✓		✓		✓
Air filter element	R				✓				✓				✓	
Engine Coolant②	R				✓				✓				✓	
Spark plugs①	R				✓				✓				✓	
Accessory drive belt(s)	S/I				✓				✓				✓	

79221CE6

SCHEDULED MAINTENANCE INTERVALS

(GM A BODY- BUICK CENTURY, OLDSMOBILE CUTLASS CIERA & CUTLASS CRUISER Cont.)

TO BE SERVICED	TYPE OF SERVICE	VEHICLE MILEAGE INTERVAL (x1000)												
		7.5	15	22.5	30	37.5	45	52.5	60	67.5	75	82.5	90	97.5
EGR & fuel systems	S/I				✓				✓				✓	
PCV valve & filter	S/I				✓				✓				✓	
Ignition Cables	S/I				✓				✓				✓	
Automatic transaxle fluid & filter③	R													
Throttle body mount bolt torque	S/I	✓												

① Platinum tip spark plugs - replace every 100,000 miles.
② Engine coolant (1996) - replace every 100,000 miles. Use O.E. specified (DEX-COOL™) coolant only. If any silicate coolant is used, the service interval is every 30,000 miles.
③ Replace fluid & filter every 100,000 miles (1995).
R – Replace   S/I – Service or Inspect

FREQUENT OPERATION MAINTENANCE (SEVERE SERVICE)

If a vehicle is operated under any of the following conditions it is considered severe service:
- Extremely dusty areas.
- 50% or more of the vehicle operation is in 32°C (90°F) or higher temperatures, or constant operation in temperatures below 0°C (32°F).
- Prolonged idling (vehicle operation in stop and go traffic).
- Frequent short running periods (engine does not warm to normal operating temperatures).
- Police, taxi, delivery usage or trailer towing usage.
Engine oil & filter change – change every 3000 miles.
Inspect CV joints & front suspension components - check every 3000 miles.
Throttle body mount bolt torque - tighten at 6000 miles.
Rotate tires at 6000 miles, then every 12,000 miles.
Brake linings – check every 9000 miles.
Air filter element – service or inspect every 15,000 miles.
Automatic transaxle fluid & filter - change every 15,000 miles (1995) or every 50,000 miles (1996).

79221CE7

SCHEDULED MAINTENANCE
GENERAL MOTORS CORPORATION
A BODY
BUICK CENTURY, OLDSMOBILE CUTLASS CIERA, CUTLASS CRUISER

The following should be used as a guide when determining the amount of work required for a particular service if taken to a repair shop.
In estimating how long a particular Scheduled Maintenance Service should take, please observe the following:

- **Chilton Time** is time based on field research and data supplied by the vehicle manufacturer.

- Labor time operations are given in hours and tenths of an hour.

- All labor operations, are to be used as a guide.

Mechanic Skill Level Codes:
> **(G)** GENERAL: Normally skilled with certification.
> **(M)** MAINTENANCE: Semi-skilled working on certicication.
> **(P)** PRECISION: Really skilled with multiple certification.

	Chilton Time		Chilton Time		Chilton Time
(G) 7500 Mile Service		**(M) 45000 Mile Service**		**(G) 67500 Mile Service**	
1995-96	1.6	1995-96	.9	1995-96	1.6
(M) 15000 Mile Service		**(G) 52500 Mile Service**		**(M) 75000 Mile Service**	
1995-96	.9	1995-96	1.6	1995-96	.9
(G) 22500 Mile Service		**(G) 60000 Mile Service**		**(G) 82500 Mile Service**	
1995-96	1.6	1995-96	4.1	1995-96	1.6
(G) 30000 Mile Service		**(G) 37500 Mile Service**		**(G) 90000 Mile Service**	
1995-96	4.1			1995-96	4.1
(G) 37500 Mile Service				**(G) 97500 Mile Service**	
1995-96	1.6			1995-96	1.6

79221CE8

GENERAL MOTORS B BODY
Buick Roadmaster • Cadillac Fleetwood Brougham
Chevrolet Caprice • Impala SS

VEHICLE IDENTIFICATION CHART

Code	Liters	Cu. In. (cc)	Cyl.	Fuel Sys.	Eng. Mfg.
Engine Code					
P	5.7	350 (5737)	8	MFI	CPC
W	4.3	265 (4294)	8	MFI	CPC

Code	Year
Model Year	
S	1995
T	1996

CPC - Chevrolet/Pontiac/Canada

MFI - Multi-point Fuel Injection

79221CE9

ENGINE IDENTIFICATION

Year	Model	Engine Displacement Liters (cc)	Engine Series (ID/VIN)	Fuel System	No. of Cylinders	Engine Type
1995	Caprice	5.7 (5737)	P	MFI	8	OHV
	Caprice	4.3 (4294)	W	MFI	8	OHV
	Fleetwood	5.7 (5733)	P	MFI	8	OHV
	Impala SS	5.7 (5737)	P	MFI	8	OHV
	Roadmaster	5.7 (5737)	P	MFI	8	OHV
1996	Caprice	5.7 (5737)	P	MFI	8	OHV
	Caprice	4.3 (4294)	W	MFI	8	OHV
	Fleetwood	5.7 (5733)	P	MFI	8	OHV
	Impala SS	5.7 (5737)	P	MFI	8	OHV
	Roadmaster	5.7 (5737)	P	MFI	8	OHV

MFI - Multi-point Fuel Injection

OHV - Overhead Valves

79221CE0

GENERAL ENGINE SPECIFICATIONS

Year	Engine ID/VIN	Engine Displacement Liters (cc)	Fuel System Type	Net Horsepower @ rpm	Net Torque @ rpm (ft. lbs.)	Bore x Stroke (in.)	Compression Ratio	Oil Pressure @ rpm
1995	P	5.7 (5737)	MFI	260@5000	330@3200	4.00x3.48	10.25:1	18@2000
	W	4.3 (4294)	MFI	200@5200	245@2400	3.74x3.48	9.93:1	18@2000
1996	P	5.7 (5737)	MFI	260@5000	330@3200	4.00x3.48	10.25:1	18@2000
	W	4.3 (4294)	MFI	200@5200	245@2400	3.74x3.48	9.93:1	18@2000

MFI - Multi-point Fuel Injection

79221CF1

GASOLINE ENGINE TUNE-UP SPECIFICATIONS

Year	Engine ID/VIN	Engine Displacement Liters (cc)	Spark Plugs Gap (in.)	Ignition Timing (deg.) MT	Ignition Timing (deg.) AT	Fuel Pump (psi)	Idle Speed (rpm) MT	Idle Speed (rpm) AT	Valve Clearance In.	Valve Clearance Ex.
1995	P	5.7 (5737)	0.035	—	①	41-47	—	①	HYD	HYD
	W	4.3 (4294)	0.050	—	①	6-24	—	①	HYD	HYD
1996	P	5.7 (5737)	0.035	—	①	41-47	—	①	HYD	HYD
	W	4.3 (4294)	0.050	—	①	6-24	—	①	HYD	HYD

NOTE: The Vehicle Emission Control Information label often reflects specification changes made during production. The label figures must be used if they differ from those in this chart.

HYD - Hydraulic

① Refer to Vehicle Emission Control Information label

79221CF2

CAPACITIES

Year	Model	Engine ID/VIN	Engine Displacement Liters (cc)	Engine Oil with Filter (qts.)	Transmission (pts.) 4-Spd	Transmission (pts.) 5-Spd	Transmission (pts.) Auto.	Transfer Case (pts.)	Drive Axle Front (pts.)	Drive Axle Rear (pts.)	Fuel Tank (gal.)	Cooling System (qts.)
1995	Caprice	P	5.7 (5737)	5.0	—	—	7.0 ①	—	—	②	22.0	14.6 ③
	Caprice	W	4.3 (4294)	4.5	—	—	7.0 ①	—	—	②	23.0	12.6 ③
	Fleetwood	P	5.7 (5737)	5.0	—	—	10.0 ④	—	—	4.2	23.0	14.6
	Impala SS	P	5.7 (5737)	5.0	—	—	7.0 ①	—	—	②	22.0	14.6 ③
	Roadmaster	P	5.7 (5737)	5.0	—	—	10.0	—	—	②	22.0	⑤
1996	Caprice	P	5.7 (5737)	5.0	—	—	7.0 ⑥	—	—	⑦	22.0	14.6 ⑧
	Caprice	W	4.3 (4294)	4.5	—	—	7.0 ⑥	—	—	⑦	23.0	12.6 ⑧
	Fleetwood	P	5.7 (5733)	5.0	—	—	10.0 ④	—	—	4.2	23.0	14.6
	Impala SS	P	5.7 (5737)	5.0	—	—	7.0 ⑥	—	—	⑦	22.0	14.6 ⑧
	Roadmaster	P	5.7 (5733)	5.0	—	—	10.0	—	—	—	22.0	⑤

NOTE: All capacities are approximate. Add fluid gradually and ensure a proper fluid level is obtained.

① 4L60 trans.: 10.0 pts.

② With 7 5/8" ring gear: 3.50 pts.
   With 8.5" ring gear: 4.25 pts.
   With 8.75" ring gear: 5.4 pts.

③ Add 0.6 qts. for HD radiator

④ Fluid change with filter

⑤ With standard cooling: 16.4
   With heavy duty cooling: 16.9

⑥ With std. cooling: 14.40
   With heavy duty cooling: 15.10

⑦ With std. cooling: 14.30
   With heavy duty cooling: 14.60

79221CF3

VALVE SPECIFICATIONS

Year	Engine ID/VIN	Engine Displacement Liters (cc)	Seat Angle (deg.)	Face Angle (deg.)	Spring Test Pressure (lbs. @ in.)	Spring Installed Height (in.)	Stem-to-Guide Clearance (in.)		Stem Diameter (in.)	
							Intake	Exhaust	Intake	Exhaust
1995	P	5.7 (5737)	46	45	187-203@ 1.27	1.70	0.0009- 0.0027	0.0009- 0.0027	NA	NA
	W	4.3 (4294)	46	45	187-203@ 1.27	1.70	0.0009- 0.0027	0.0009- 0.0027	NA	NA
1996	P	5.7 (5737)	46	45	187-203@ 1.27	1.70	0.0009- 0.0027	0.0009- 0.0027	NA	NA
	W	4.3 (4294)	46	45	187-203@ 1.27	1.70	0.0009- 0.0027	0.0009- 0.0027	NA	NA

NA - Not Available

79221CF4

TORQUE SPECIFICATIONS
All readings in ft. lbs.

Year	Engine ID/VIN	Engine Displacement Liters (cc)	Cylinder Head Bolts	Main Bearing Bolts	Rod Bearing Bolts	Crankshaft Damper Bolts	Flywheel Bolts	Manifold		Spark Plugs	Lug Nut
								Intake	Exhaust		
1995	P	5.7 (5737)	65	78	47	60	74	①	35	11	100
	W	4.3 (4294)	65	78	47	60	74	①	35	11	100
1996	P	5.7 (5737)	65	78	47	60	74	①	35	11	100
	W	4.3 (4294)	65	78	47	60	74	①	35	11	100

① Step 1: 71 inch lbs.
  Step 2: 35 ft. lbs.

79221CF5

SCHEDULED MAINTENANCE INTERVALS
(GM B BODY – BUICK ROADMASTER , CADILLAC FLEETWOOD, CHEVROLET CAPRICE & IMPALA SS)

TO BE SERVICED	TYPE OF SERVICE	VEHICLE MILEAGE INTERVAL (x1000)												
		7.5	15	22.5	30	37.5	45	52.5	60	67.5	75	82.5	90	97.5
Engine oil & filter	R	✓	✓	✓	✓	✓	✓	✓	✓	✓	✓	✓	✓	✓
Automatic transmission fluid & filter	S/I	✓	✓	✓	✓	✓	✓	✓	✓	✓	✓	✓	✓	✓
Brake hoses	S/I	✓	✓	✓	✓	✓	✓	✓	✓	✓	✓	✓	✓	✓
Engine coolant level, hoses & clamps	S/I	✓	✓	✓	✓	✓	✓	✓	✓	✓	✓	✓	✓	✓
Exhaust system & throttle linkage	S/I	✓	✓	✓	✓	✓	✓	✓	✓	✓	✓	✓	✓	✓
Front suspension components	S/I	✓	✓	✓	✓	✓	✓	✓	✓	✓	✓	✓	✓	✓
Lubricate chassis, suspension, steering linkage, transmission shift linkage, parking brake cable guides, underbody contact points & linkage	S/I	✓	✓	✓	✓	✓	✓	✓	✓	✓	✓	✓	✓	✓
Brake linings & rotate tires	S/I	✓		✓		✓		✓		✓		✓		✓
Air filter element	R				✓				✓				✓	
Engine coolant②	R				✓				✓				✓	
Spark plugs①	R				✓				✓				✓	
Front wheel bearings	S/I				✓				✓				✓	
Ignition cables, EGR & fuel systems	S/I				✓				✓				✓	
Serpentine drive belt	S/I				✓				✓				✓	
Thermostatically controlled air cleaner	S/I				✓				✓				✓	
Rear axle oil (Limited slip)	R	✓												
Throttle body mount bolt torque	S/I	✓												

① Platinum tip spark plugs - replace every 100,000 miles.
② Engine coolant (1996-99) - replace every 100,000 miles. Use O.E. specified (DEX-COOL™) coolant only. If any silicate coolant is used, the service interval is every 30,000 miles.

R – Replace   S/I – Service or Inspect

79221CF6

SCHEDULED MAINTENANCE INTERVALS
(GM B BODY – BUICK ROADMASTER , CADILLAC FLEETWOOD, CHEVROLET CAPRICE & IMPALA SS Cont.)

FREQUENT OPERATION MAINTENANCE (SEVERE SERVICE)

If a vehicle is operated under any of the following conditions it is considered severe service:
- Extremely dusty areas.
- 50% or more of the vehicle operation is in 32°C (90°F) or higher temperatures, or constant operation in temperatures below 0°C (32°F).
- Prolonged idling (vehicle operation in stop and go traffic).
- Frequent short running periods (engine does not warm to normal operating temperatures).
- Police, taxi, delivery usage or trailer towing usage.

Oil & oil filter – change every 3000 miles.
Chassis lubrication - lubricate every 6000 miles.
Rear axle oil (Limited slip) - replace every 6000 miles. If vehicle is used for towing, police or taxi service, replace every 6000 miles in either type of axle.
Throttle body mount bolt torque - tighten at 6000 miles.
Air filter element - service or inspect every 15,000 miles.
Automatic transmission fluid & filter - change 50,000 miles.
Front wheel bearings - repack every 15,000 miles.
Rotate tires at 6000 miles, then every 15,000 miles.

79221CF6A

SCHEDULED MAINTENANCE
GENERAL MOTORS CORPORATION
B BODY
BUICK ROADMASTER, CADILLAC FLEETWOOD BROUGHAM
CHEVROLET CAPRICE, IMPALA SS

The following should be used as a guide when determining the amount of work required for a particular service if taken to a repair shop. In estimating how long a particular Scheduled Maintenance Service should take, please observe the following:

- **Chilton Time** is time based on field research and data supplied by the vehicle manufacturer.
- Labor time operations are given in hours and tenths of an hour.
- All labor operations, are to be used as a guide.

Mechanic Skill Level Codes:
   (G) GENERAL: Normally skilled with certification.
   (M) MAINTENANCE: Semi-skilled working on certicication.
   (P) PRECISION: Really skilled with multiple certification.

	Chilton Time		Chilton Time		Chilton Time
(G) 7500 Mile Service		**(G) 37500 Mile Service**		**(M) 75000 Mile Service**	
1995-96	2.2	1995-96	1.7	1995-96	1.0
(M) 15000 Mile Service		**(M) 45000 Mile Service**		**(G) 82500 Mile Service**	
1995-96	1.0	1995-96	1.0	1995-96	1.7
(G) 22500 Mile Service		**(G) 52500 Mile Service**		**(G) 90000 Mile Service**	
1995-96	1.7	1995-96	1.7	1995-96	3.2
(G) 30000 Mile Service		**(G) 60000 Mile Service**		**(G) 97500 Mile Service**	
1995-96	3.2	1995-96	3.2	1995-96	1.7
		(G) 67500 Mile Service			
		1995-96	1.7		

79221CF7

GENERAL MOTORS C & H BODIES
Buick LeSabre • Park Avenue • Oldsmobile Eighty-Eight
Ninety-Eight • LSS • Regency • Pontiac Bonneville

VEHICLE IDENTIFICATION CHART

Engine Code							Model Year	
Code	Liters	Cu. In. (cc)	Cyl.	Fuel Sys.	Eng. Mfg.		Code	Year
1	3.8 ①	231 (3785)	6	MFI	BOC		S	1995
K	3.8	231 (3785)	6	MFI	BOC		T	1996
L	3.8	231 (3785)	6	MFI	BOC		V	1997
							W	1998
							X	1999

MFI - Multi-point Fuel Injection

BOC - Buick/Oldsmobile/Cadillac

① Supercharged Engine

79221CF8

ENGINE IDENTIFICATION

Year	Model	Engine Displacement Liters (cc)	Engine Series (ID/VIN)	Fuel System		No. of Cylinders	Engine Type
1995	Bonneville	3.8 (3785)	1	MFI	①	6	OHV
	Bonneville	3.8 (3785)	K	MFI		6	OHV
	Eighty-Eight/LSS	3.8 (3785)	1	MFI	①	6	OHV
	Eighty-Eight/Royale	3.8 (3785)	K	MFI		6	OHV
	LeSabre	3.8 (3785)	L	MFI		6	OHV
	Ninety-Eight	3.8 (3785)	1	MFI	①	6	OHV
	Ninety-Eight	3.8 (3785)	K	MFI		6	OHV
	Park Avenue	3.8 (3785)	K	MFI		6	OHV
	Park Avenue Ultra	3.8 (3785)	1	MFI	①	6	OHV
1996	Bonneville	3.8 (3785)	1	MFI	①	6	OHV
	Bonneville	3.8 (3785)	K	MFI		6	OHV
	Eighty-Eight	3.8 (3786)	K	MFI		6	OHV
	Eighty-Eight/LSS	3.8 (3785)	1	MFI	①	6	OHV
	LeSabre	3.8 (3785)	K	MFI		6	OHV
	Ninety-Eight	3.8 (3786)	K	MFI		6	OHV
	Park Avenue	3.8 (3785)	K	MFI		6	OHV
	Park Avenue Ultra	3.8 (3785)	1	MFI	①	6	OHV
1997	Bonneville	3.8 (3785)	1	MFI	①	6	OHV
	Bonneville	3.8 (3785)	K	MFI		6	OHV
	Eighty-Eight	3.8 (3786)	K	MFI		6	OHV
	LeSabre	3.8 (3785)	K	MFI		6	OHV
	LSS	3.8 (3785)	1	MFI	①	6	OHV
	LSS	3.8 (3786)	K	MFI		6	OHV
	Park Avenue	3.8 (3785)	1	MFI	①	6	OHV
	Park Avenue	3.8 (3785)	K	MFI		6	OHV
	Regency	3.8 (3786)	K	MFI		6	OHV
1998-99	Bonneville	3.8 (3785)	1	MFI	①	6	OHV
	Bonneville	3.8 (3785)	K	MFI		6	OHV
	Eighty-Eight	3.8 (3786)	K	MFI		6	OHV
	LeSabre	3.8 (3785)	K	MFI		6	OHV
	LSS	3.8 (3786)	K	MFI		6	OHV
	Park Avenue	3.8 (3785)	1	MFI	①	6	OHV
	Park Avenue	3.8 (3785)	K	MFI		6	OHV
	Regency	3.8 (3786)	K	MFI		6	OHV

MFI - Multi-point Fuel Injection
OHV - Overhead Valves
① Supercharged Engine

79221CF9

GENERAL ENGINE SPECIFICATIONS

Year	Engine ID/VIN	Engine Displacement Liters (cc)	Fuel System Type	Net Horsepower @ rpm	Net Torque @ rpm (ft. lbs.)	Bore x Stroke (in.)	Compression Ratio	Oil Pressure @ rpm
1995	1	3.8 (3875)	MFI	225@5000	275@3200	3.80x3.40	9.0:1	60@1850 ①
	K	3.8 (3875)	MFI	205@5200	230@4000	3.80x3.40	9.4:1	60@1850
	L	3.8 (3875)	MFI	170@4800	225@3200	3.80x3.40	9.0:1	60@1850
1996	1	3.8 (3786)	MFI	240@5200	280@3200	3.80x3.40	9.0:1	60@1850 ①
	K	3.8 (3785)	MFI	205@5200	230@4000	3.80x3.40	9.4:1	60@1850
1997	1	3.8 (3786)	MFI	240@5200	280@3200	3.80x3.40	9.0:1	60@1850 ①
	K	3.8 (3785)	MFI	205@5200	230@4000	3.80x3.40	9.4:1	60@1850
1998-99	1	3.8 (3786)	MFI	240@5200	280@3200	3.80x3.40	9.0:1	60@1850 ①
	K	3.8 (3785)	MFI	205@5200	230@4000	3.80x3.40	9.4:1	60@1850

MFI - Multi-point Fuel Injection
① Supercharged

79221CF0

GASOLINE ENGINE TUNE-UP SPECIFICATIONS

Year	Engine ID/VIN	Engine Displacement Liters (cc)	Spark Plugs Gap (in.)	Ignition Timing (deg.) MT	Ignition Timing (deg.) AT	Fuel Pump (psi)	Idle Speed (rpm) MT	Idle Speed (rpm) AT	Valve Clearance In.	Valve Clearance Ex.
1995	1	3.8 (3785)	0.060	—	①	40-47 ②	—	③	HYD	HYD
	K	3.8 (3785)	0.060	—	①	40-47 ②	—	③	HYD	HYD
	L	3.8 (3785)	0.060	—	①	40-47 ②	—	③	HYD	HYD
1996	1	3.8 (3786)	0.060	—	①	41-47	—	③	HYD	HYD
	K	3.8 (3785)	0.060	—	①	41-47	—	③	HYD	HYD
1997	1	3.8 (3786)	0.060	—	①	41-47	—	③	HYD	HYD
	K	3.8 (3785)	0.060	—	①	41-47	—	③	HYD	HYD
1998-99	1	3.8 (3786)	0.060	—	①	41-47	—	③	HYD	HYD
	K	3.8 (3785)	0.060	—	①	41-47	—	③	HYD	HYD

NOTE: The Vehicle Emission Control Information label often reflects specification changes made during production. The label figures must be used if they differ from those in this chart.

HYD - Hydraulic
① DIS Ignition System timing not adjustable
② Pressure at fuel pump
③ Idle speed maintained by ECM. There is no recommended adjustment procedure

79221CG1

CAPACITIES

Year	Model	Engine ID/VIN	Engine Displacement Liters (cc)	Engine Oil with Filter (qts.)		Transmission (pts.)			Drive Axle		Fuel Tank (gal.)	Cooling System (qts.)
						4-Spd	5-Spd	Auto.	Front (pts.)	Rear (pts.)		
1995	Bonneville	1	3.8 (3785)	3.8	①	—	—	13.0	—	—	18.0	13.0
	Bonneville	K	3.8 (3785)	3.8	①	—	—	13.0	—	—	18.0	13.0
	Eighty-Eight	1	3.8 (3785)	3.8	①	—	—	13.0	—	—	18.0	13.0
	Eighty-Eight	K	3.8 (3785)	3.8	①	—	—	13.0	—	—	18.0	13.0
	LeSabre	L	3.8 (3785)	4.0	①	—	—	13.0	—	—	18.0	13.0
	Ninety-Eight	1	3.8 (3785)	3.8	①	—	—	13.0	—	—	18.0	13.0
	Ninety-Eight	K	3.8 (3785)	3.8	①	—	—	13.0	—	—	18.0	13.0
	Park Avenue	L	3.8 (3785)	4.0	①	—	—	13.0	—	—	18.0	13.0
	Park Avenue Ultra	1	3.8 (3785)	4.0	①	—	—	13.0	—	—	18.0	13.0
1996	Bonneville	1	3.8 (3785)	5.0	②	—	—	12.0	—	—	18.0	13.0
	Bonneville	K	3.8 (3785)	5.0	②	—	—	12.0	—	—	18.0	13.0
	Eighty-Eight	K	3.8 (3785)	5.0	②	—	—	12.0	—	—	18.0	13.0
	Eighty-Eight/LSS	1	3.8 (3785)	5.0	②	—	—	12.0	—	—	18.0	13.0
	LeSabre	K	3.8 (3785)	5.0	②	—	—	12.0	—	—	18.0	13.0
	Ninety-Eight	K	3.8 (3785)	5.0	②	—	—	12.0	—	—	18.0	13.0
	Park Avenue	K	3.8 (3785)	5.0	②	—	—	12.0	—	—	18.0	13.0
	Park Avenue Ultra	1	3.8 (3785)	5.0	②	—	—	12.0	—	—	18.0	13.0
1997	Bonneville	1	3.8 (3785)	5.0	②	—	—	12.0	—	—	18.0	13.0
	Bonneville	K	3.8 (3785)	5.0	②	—	—	12.0	—	—	18.0	13.0
	Eighty-Eight	K	3.8 (3785)	5.0	②	—	—	12.0	—	—	18.0	13.0
	LeSabre	K	3.8 (3785)	5.0	②	—	—	12.0	—	—	18.0	13.0
	LSS	1	3.8 (3785)	5.0	②	—	—	12.0	—	—	18.0	13.0
	LSS	K	3.8 (3785)	5.0	②	—	—	12.0	—	—	18.0	13.0
	Park Avenue	1	3.8 (3785)	5.0	②	—	—	12.0	—	—	18.0	13.0
	Park Avenue	K	3.8 (3785)	5.0	②	—	—	12.0	—	—	18.0	13.0
	Regency	K	3.8 (3785)	5.0	②	—	—	12.0	—	—	18.0	13.0
1998-99	Bonneville	1	3.8 (3785)	5.0	②	—	—	12.0	—	—	18.0	13.0
	Bonneville	K	3.8 (3785)	5.0	②	—	—	12.0	—	—	18.0	13.0
	Eighty-Eight	K	3.8 (3785)	5.0	②	—	—	12.0	—	—	18.0	13.0
	LeSabre	K	3.8 (3785)	5.0	②	—	—	12.0	—	—	18.0	13.0
	LSS	K	3.8 (3785)	5.0	②	—	—	12.0	—	—	18.0	13.0
	Park Avenue	1	3.8 (3785)	5.0	②	—	—	12.0	—	—	18.0	13.0
	Park Avenue	K	3.8 (3785)	5.0	②	—	—	12.0	—	—	18.0	13.0
	Regency	K	3.8 (3785)	5.0	②	—	—	12.0	—	—	18.0	13.0

NOTE: All capacities are approximate. Add fluid gradually and ensure a proper fluid level is obtained.
① Specification is without filter replacement; Additional oil may be required
② Fluid change with filter

79221CG2

VALVE SPECIFICATIONS

Year	Engine ID/VIN	Engine Displacement Liters (cc)	Seat Angle (deg.)	Face Angle (deg.)	Spring Test Pressure (lbs. @ in.)	Spring Installed Height (in.)	Stem-to-Guide Clearance (in.)		Stem Diameter (in.)	
							Intake	Exhaust	Intake	Exhaust
1995	1	3.8 (3785)	46	45	210@1.315	1.690-1.720	0.0015-0.0035	0.0015-0.0032	NA	NA
	K	3.8 (3785)	46	45	210@1.315	1.690-1.720	0.0015-0.0035	0.0015-0.0032	NA	NA
	L	3.8 (3785)	45	45	210@1.315	1.690-1.720	0.0015-0.0035	0.0015-0.0032	NA	NA
1996	1	3.8 (3785)	45	45	80@1.750	1.690-1.720	0.0015-0.0032	0.0015-0.0032	NA	NA
	K	3.8 (3785)	45	45	80@1.750	1.690-1.720	0.0015-0.0032	0.0015-0.0032	NA	NA
1997	1	3.8 (3785)	45	45	80@1.750	1.690-1.720	0.0015-0.0032	0.0015-0.0032	NA	NA
	K	3.8 (3785)	45	45	80@1.750	1.690-1.720	0.0015-0.0032	0.0015-0.0032	NA	NA
1998-99	1	3.8 (3785)	45	45	80@1.750	1.690-1.720	0.0015-0.0032	0.0015-0.0032	NA	NA
	K	3.8 (3785)	45	45	80@1.750	1.690-1.720	0.0015-0.0032	0.0015-0.0032	NA	NA

79221CG3

TORQUE SPECIFICATIONS
All readings in ft. lbs.

Year	Engine ID/VIN	Engine Displacement Liters (cc)	Cylinder Head Bolts	Main Bearing Bolts	Rod Bearing Bolts	Crankshaft Damper Bolts	Flywheel Bolts	Manifold		Spark Plugs	Lug Nut
								Intake	Exhaust		
1995	1	3.7 (3785)	①	②	③	④	⑤	11	38	11	100
	K	3.7 (3785)	①	②	③	④	⑤	11	38	11	100
	L	3.8 (3785)	①	②	③	④	⑤	11	38	11	100
1996	1	3.8 (3786)	①	⑥	③	④	⑤	⑦	22	11	100
	K	3.8 (3785)	①	②	③	④	⑤	11	38	11	100
1997	1	3.8 (3786)	①	⑥	③	④	⑤	⑦	22	11	100
	K	3.8 (3785)	①	②	③	④	⑤	11	38	11	100
1998-99	1	3.8 (3786)	①	⑥	③	④	⑤	⑦	22	11	100
	K	3.8 (3785)	①	②	③	④	⑤	11	38	11	100

① Step 1: Tighten all bolts to 35 ft. lbs.
   Step 2: Turn all bolts 130 degrees
   Step 3: Rotate four center bolts an additional 30 degrees
② 26 ft. lbs. plus 50 degrees
③ 20 ft. lbs. plus 50 degrees
④ 110 ft. lbs. plus 76 degrees
⑤ 11 ft. lbs. plus 50 degrees
⑥ Step 1: Tighten caps in equal increments to 52 ft. lbs.
   Step 2: Loosen 360 degrees
   Step 3: 15 ft. lbs.
   Step 4: 54 ft. lbs.
   Step 5: Plus three turns of 35 degrees for a total of 105 degrees
⑦ Upper manifold: 8 ft. lbs.
   Lower manifold: 11 ft. lbs.

79221CG4

SCHEDULED MAINTENANCE INTERVALS
(GM C & H BODIES — BUICK LESABRE, PARK AVENUE, OLDSMOBILE LSS, REGENCY & EIGHTY-EIGHT, NINETY-EIGHT, PONTIAC BONNEVILLE)

TO BE SERVICED	TYPE OF SERVICE	VEHICLE MILEAGE INTERVAL (x1000)												
		7.5	15	22.5	30	37.5	45	52.5	60	67.5	75	82.5	90	97.5
Engine oil & filter	R	✓	✓	✓	✓	✓	✓	✓	✓	✓	✓	✓	✓	✓
Exhaust system & brake hoses	S/I	✓	✓	✓	✓	✓	✓	✓	✓	✓	✓	✓	✓	✓
Drive shaft boots & front suspension components	S/I	✓	✓	✓	✓	✓	✓	✓	✓	✓	✓	✓	✓	✓
Lubricate chassis, suspension, steering linkage, transaxle shift linkage, parking brake cable guides, underbody contact points & linkage	S/I	✓	✓	✓	✓	✓	✓	✓	✓	✓	✓	✓	✓	✓
Coolant level, hoses & clamps	S/I	✓	✓	✓	✓	✓	✓	✓	✓	✓	✓	✓	✓	✓
Throttle linkage	S/I	✓	✓	✓	✓	✓	✓	✓	✓	✓	✓	✓	✓	✓
Brake linings & rotate tires	S/I	✓		✓		✓		✓		✓		✓		✓
Accessory drive belts supercharger oil	S/I				✓				✓				✓	
Engine coolant (1995)	R				✓				✓				✓	
Spark plugs①	R				✓				✓				✓	
Air filter element	R				✓				✓				✓	
PCV filter	R				✓				✓				✓	
Ignition cables	S/I				✓				✓				✓	
EGR & fuel systems	S/I				✓				✓				✓	
Automatic transaxle fluid & filter	R													✓
Throttle body mount bolt torque	S/I	✓												

① Platinum tip spark plugs - replace every 100,000 miles.
② Engine coolant (1996-99) - replace every 100,000 miles. Use O.E. specified (DEX-COOL™) only. If any silicate coolant is used, the service interval is every 30,000 miles.

R – Replace   S/I – Service or Inspect

79221CG5

SCHEDULED MAINTENANCE INTERVALS
(GM C & H BODIES — BUICK LESABRE, PARK AVENUE, OLDSMOBILE LSS, REGENCY & EIGHTY-EIGHT, NINETY-EIGHT, PONTIAC BONNEVILLE Cont.)

FREQUENT OPERATION MAINTENANCE (SEVERE SERVICE)

If a vehicle is operated under any of the following conditions it is considered severe service:
- Extremely dusty areas.
- 50% or more of the vehicle operation is in 32°C (90°F) or higher temperatures, or constant operation in temperatures below 0°C (32°F).
- Prolonged idling (vehicle operation in stop and go traffic).
- Frequent short running periods (engine does not warm to normal operating temperatures).
- Police, taxi, delivery usage or trailer towing usage.

Engine oil & filter change – change every 3000 miles.
Inspect CV joints & front suspension components - check every 3000 miles.
Brake linings – check every 6000 miles.
Chassis lubrication - lubricate every 6000 miles.
Throttle body mount bolt torque - tighten at 6000 miles.
Air filter element – service or inspect every 15,000 miles.
Automatic transaxle fluid – change every 50,000 miles.
Rotate tires at 6000 miles, then every 15,000 miles.

79221CG6

SCHEDULED MAINTENANCE
GENERAL MOTORS CORPORATION
C & H BODIES
BUICK LESABRE, PARK AVENUE, OLDSMOBILE EIGHTY-EIGHT, NINETY-EIGHT, LSS, REGENCY, PONTIAC BONNEVILLE

The following should be used as a guide when determining the amount of work required for a particular service if taken to a repair shop. In estimating how long a particular Scheduled Maintenance Service should take, please observe the following:

- **Chilton Time** is time based on field research and data supplied by the vehicle manufacturer.
- Labor time operations are given in hours and tenths of an hour.
- All labor operations, are to be used as a guide.

Mechanic Skill Level Codes:
(G) GENERAL: Normally skilled with certification.
(M) MAINTENANCE: Semi-skilled working on certicication.
(P) PRECISION: Really skilled with multiple certification.

	Chilton Time		Chilton Time		Chilton Time
(M) 7500 Mile Service		**(M) 45000 Mile Service**		**(M) 67500 Mile Service**	
1995-99	1.4	1995-99	.8	1995-99	1.4
(M) 15000 Mile Service		**(M) 52500 Mile Service**		**(M) 75000 Mile Service**	
1995-99	.8	1995-99	1.4	1995-99	.8
(M) 22500 Mile Service		**(G) 60000 Mile Service**		**(M) 82500 Mile Service**	
1995-99	1.4	1995-99	3.7	1995-99	1.4
(G) 30000 Mile Service		Renew engine coolant (1995) add	.5	**(G) 90000 Mile Service**	
1995-99	3.7			1995-99	3.7
Renew engine coolant (1995) add	.5			Renew engine coolant (1995) add	.5
(M) 37500 Mile Service				**(M) 97500 Mile Service**	
1995-99	1.4			1995-99	1.9

79221CG7

GENERAL MOTORS E & K BODIES
Cadillac Deville • Deville Concours • Eldorado • Seville

VEHICLE IDENTIFICATION CHART

Engine Code						Model Year	
Code	Liters	Cu. In. (cc)	Cyl.	Fuel Sys.	Eng. Mfg.	Code	Year
9	4.6	279 (4573)	8	MFI	Cadillac	S	1995
B	4.9	300 (4917)	8	MFI	Cadillac	T	1996
Y	4.6	279 (4573)	8	MFI	Cadillac	V	1997
						W	1998
						X	1999

MFI - Multi-point Fuel Injection

79221CG8

ENGINE IDENTIFICATION

Year	Model	Engine Displacement Liters (cc)	Engine Series (ID/VIN)	Fuel System	No. of Cylinders	Engine Type
1995	DeVille	4.9 (4917)	B	MFI	8	OHV
	DeVille Concours	4.6 (4573)	Y	MFI	8	DOHC
	Eldorado	4.6 (4573)	9	MFI	8	DOHC
	Eldorado ETC	4.6 (4573)	Y	MFI	8	DOHC
	Seville SLS	4.6 (4573)	Y	MFI	8	DOHC
	Seville STS	4.6 (4573)	9	MFI	8	DOHC
1996	DeVille	4.6 (4573)	Y	MFI	8	DOHC
	DeVille Concours	4.6 (4573)	9	MFI	8	DOHC
	Eldorado	4.6 (4573)	Y	MFI	8	DOHC
	Eldorado ETC	4.6 (4573)	9	MFI	8	DOHC
	Seville SLS	4.6 (4573)	Y	MFI	8	DOHC
	Seville STS	4.6 (4573)	9	MFI	8	DOHC
1997	DeVille	4.6 (4573)	Y	MFI	8	DOHC
	DeVille Concours	4.6 (4573)	9	MFI	8	DOHC
	Eldorado	4.6 (4573)	Y	MFI	8	DOHC
	Eldorado ETC	4.6 (4573)	9	MFI	8	DOHC
	Seville SLS	4.6 (4573)	Y	MFI	8	DOHC
	Seville STS	4.6 (4573)	9	MFI	8	DOHC
1998-99	DeVille	4.6 (4573)	Y	MFI	8	DOHC
	DeVille Concours	4.6 (4573)	9	MFI	8	DOHC
	Eldorado	4.6 (4573)	Y	MFI	8	DOHC
	Eldorado ETC	4.6 (4573)	9	MFI	8	DOHC
	Seville SLS	4.6 (4573)	Y	MFI	8	DOHC
	Seville STS	4.6 (4573)	9	MFI	8	DOHC

MFI - Multi-point Fuel Injection
OHV - Overhead Valves
DOHC - Double Overhead Camshafts

79221CG9

GENERAL ENGINE SPECIFICATIONS

Year	Engine ID/VIN	Engine Displacement Liters (cc)	Fuel System Type	Net Horsepower @ rpm	Net Torque @ rpm (ft. lbs.)	Bore x Stroke (in.)	Com- pression Ratio	Oil Pressure @ rpm
1995	9	4.6 (4573)	MFI	300@6000	290@4400	3.66x3.31	10.3:1	35@2000
	B	4.9 (4917)	MFI	200@4400	275@3000	3.62x3.62	9.5:1	53@2000
	Y	4.6 (4573)	MFI	275@5600	300@4000	3.66x3.31	10.3:1	35@2000
1996	9	4.6 (4573)	MFI	300@6000	290@4400	3.66x3.31	10.3:1	35@2000
	Y	4.6 (4573)	MFI	275@5600	300@4000	3.66x3.31	10.3:1	35@2000
1997	9	4.6 (4573)	MFI	300@6000	290@4400	3.66x3.31	10.3:1	35@2000
	Y	4.6 (4573)	MFI	275@5600	300@4000	3.66x3.31	10.3:1	35@2000
1998-99	9	4.6 (4573)	MFI	300@6000	290@4400	3.66x3.31	10.3:1	35@2000
	Y	4.6 (4573)	MFI	275@5600	300@4000	3.66x3.31	10.3:1	35@2000

MFI - Multi-point Fuel Injection

79221CG0

GASOLINE ENGINE TUNE-UP SPECIFICATIONS

Year	Engine ID/VIN	Engine Displacement Liters (cc)	Spark Plugs Gap (in.)	Ignition Timing (deg.) MT	Ignition Timing (deg.) AT	Fuel Pump (psi)	Idle Speed (rpm) MT	Idle Speed (rpm) AT	Valve Clearance In.	Valve Clearance Ex.
1995	9	4.6 (4573)	0.050	—	①	40-50	—	①	HYD	HYD
	B	4.9 (4917)	0.060	—	①	40-50	—	①	HYD	HYD
	Y	4.6 (4573)	0.050	—	①	40-50	—	①	HYD	HYD
1996	9	4.6 (4573)	0.050	—	①	40-50	—	①	HYD	HYD
	Y	4.6 (4573)	0.050	—	①	40-50	—	①	HYD	HYD
1997	9	4.6 (4573)	0.050	—	①	40-50	—	①	HYD	HYD
	Y	4.6 (4573)	0.050	—	①	40-50	—	①	HYD	HYD
1998-99	9	4.6 (4573)	0.050	—	①	40-50	—	①	HYD	HYD
	Y	4.6 (4573)	0.050	—	①	40-50	—	①	HYD	HYD

NOTE: The Vehicle Emission Control Information label often reflects specification changes made during production. The label figures must be used if they differ from those in this chart.

HYD - Hydraulic

① Refer to Vehicle Emission Control Information label

79221CH1

CAPACITIES

Year	Model	Engine ID/VIN	Engine Displacement Liters (cc)	Engine Oil with Filter (qts.)	Transmission (pts.)			Drive Axle		Fuel Tank (gal.)	Cooling System (qts.)
					4-Spd	5-Spd	Auto.	Front (pts.)	Rear (pts.)		
1995	DeVille	B	4.9 (4917)	5.5	—	—	13.0 ①	—	—	18.0	12.1
	DeVille Concours	Y	4.6 (4573)	7.5	—	—	16.0 ①	—	—	18.0	12.1
	Eldorado	Y	4.6 (4573)	7.5	—	—	16.0 ①	—	—	20.0	12.3
	Eldorado ETC	9	4.6 (4573)	7.5	—	—	16.0 ①	—	—	20.0	12.3
	Seville SLS	Y	4.6 (4573)	7.5	—	—	16.0 ①	—	—	20.0	12.3
	Seville STS	9	4.6 (4573)	7.5	—	—	16.0 ①	—	—	20.0	12.3
1996	DeVille	Y	4.6 (4573)	7.5	—	—	16.0 ①	—	—	20.0	12.3 ②
	DeVille Concours	9	4.6 (4573)	7.5	—	—	16.0 ①	—	—	18.0	12.1 ②
	Eldorado	Y	4.6 (4573)	7.5	—	—	16.0 ①	—	—	20.0	12.3 ②
	Eldorado ETC	9	4.6 (4573)	7.5	—	—	16.0 ①	—	—	20.0	12.3 ②
	Seville SLS	Y	4.6 (4573)	7.5	—	—	16.0 ①	—	—	20.0	12.3 ②
	Seville STS	9	4.6 (4573)	7.5	—	—	16.0 ①	—	—	20.0	12.3 ②
1997	DeVille	Y	4.6 (4573)	7.5	—	—	16.0 ①	—	—	20.0	12.3 ②
	DeVille Concours	9	4.6 (4573)	7.5	—	—	16.0 ①	—	—	18.0	12.1 ②
	Eldorado	Y	4.6 (4573)	7.5	—	—	16.0 ①	—	—	20.0	12.3 ②
	Eldorado ETC	9	4.6 (4573)	7.5	—	—	16.0 ①	—	—	20.0	12.3 ②
	Seville SLS	Y	4.6 (4573)	7.5	—	—	16.0 ①	—	—	20.0	12.3 ②
	Seville STS	9	4.6 (4573)	7.5	—	—	16.0 ①	—	—	20.0	12.3 ②
1998-99	DeVille	Y	4.6 (4573)	7.5	—	—	16.0 ①	—	—	20.0	12.3 ②
	DeVille Concours	9	4.6 (4573)	7.5	—	—	16.0 ①	—	—	18.0	12.1 ②
	Eldorado	Y	4.6 (4573)	7.5	—	—	16.0 ①	—	—	20.0	12.3 ②
	Eldorado ETC	9	4.6 (4573)	7.5	—	—	16.0 ①	—	—	20.0	12.3 ②
	Seville SLS	Y	4.6 (4573)	7.5	—	—	16.0 ①	—	—	20.0	12.3 ②
	Seville STS	9	4.6 (4573)	7.5	—	—	16.0 ①	—	—	20.0	12.3 ②

NOTE: All capacities are approximate. Add fluid gradually and ensure a proper fluid level is obtained.
① Bottom pan and side cover
② Dex-Cool engine coolant and three pellets of P/N 1052753 or equivalent
 (Do not mix with ethylene glycol base)

79221CH2

VALVE SPECIFICATIONS

Year	Engine ID/VIN	Engine Displacement Liters (cc)	Seat Angle (deg.)	Face Angle (deg.)	Spring Test Pressure (lbs. @ in.)	Spring Installed Height (in.)	Stem-to-Guide Clearance (in.)		Stem Diameter (in.)	
							Intake	Exhaust	Intake	Exhaust
1995	9	4.6 (4573)	46	45	53@1.190	1.190	0.0010-0.0030	0.0020-0.0040	0.2331-0.2339	0.2331-0.2339
	B	4.9 (4917)	45	44	68-76@1.730	1.730	0.0010-0.0030	0.0020-0.0040	0.3413-0.3420	0.3401-0.3408
	Y	4.6 (4573)	46	45	46@1.190	1.190	0.0010-0.0030	0.0020-0.0040	0.2331-0.2339	0.2331-0.2339
1996	9	4.6 (4573)	46	45	53@1.190	1.190	0.0010-0.0030	0.0020-0.0040	0.2331-0.2339	0.2331-0.2339
	Y	4.6 (4573)	46	45	46@1.190	1.190	0.0010-0.0030	0.0020-0.0040	0.2331-0.2339	0.2331-0.2339
1997	9	4.6 (4573)	46	45	53@1.190	1.190	0.0010-0.0030	0.0020-0.0040	0.2331-0.2339	0.2331-0.2339
	Y	4.6 (4573)	46	45	46@1.190	1.190	0.0010-0.0030	0.0020-0.0040	0.2331-0.2339	0.2331-0.2339
1998-99	9	4.6 (4573)	46	45	53@1.190	1.190	0.0010-0.0030	0.0020-0.0040	0.2331-0.2339	0.2331-0.2339
	Y	4.6 (4573)	46	45	46@1.190	1.190	0.0010-0.0030	0.0020-0.0040	0.2331-0.2339	0.2331-0.2339

79221CH3

TORQUE SPECIFICATIONS
All readings in ft. lbs.

Year	Engine ID/VIN	Displacement Liters (cc)	Head Bolts	Bearing Bolts	Bearing Bolts	Damper Bolts	Flywheel Bolts	Manifold		Spark Plugs	Lug Nut
								Intake	Exhaust		
1995	9	4.6 (4573)	①	②	③	④	⑤	⑥	18	11	100
	B	4.9 (4917)	⑦	85	25	118	70	⑧	16	11	100
	Y	4.6 (4573)	①	②	③	④	⑤	⑥	18	11	100
1996	9	4.6 (4573)	①	②	⑨	⑩	⑤	⑥	18	11	100
	Y	4.6 (4573)	①	②	⑨	⑩	⑤	⑥	18	11	100
1997	9	4.6 (4573)	①	②	⑨	⑩	⑤	⑥	18	11	100
	Y	4.6 (4573)	①	②	⑨	⑩	⑤	⑥	18	11	100
1998-99	9	4.6 (4573)	①	②	⑨	⑩	⑤	⑥	18	11	100
	Y	4.6 (4573)	①	②	⑨	⑩	⑤	⑥	18	11	100

① Step 1: 22 ft. lbs.
  Step 2: Plus two turns of 90 degrees
② Step 1: 15 ft. lbs.
  Step 2: Plus 65 degrees
③ Step 1: 18 ft. lbs.
  Step 2: Plus 90 degrees
④ Step 1: 44 ft. lbs.
  Step 2: Plus 120 degrees
⑤ Step 1: 11 ft. lbs.
  Step 2: Plus 50 degrees
⑥ 89 in. lbs.
⑦ Short Bolts (39–41mm)
  (must be taper ground to prevent bottoming out)
  Step 1: 29 ft. lbs.
  Step 2: 51 ft. lbs.
  Step 3: 85 ft. lbs.
  Step 4: Tighten 3 center inboard
      studs 1,3 and 5 to 88 ft. lbs.
  Long Bolts (47–50mm) - should not be modified
⑧ Step 1: 8 ft. lbs.
  Step 2: 12 ft. lbs.
⑨ Step 1: 18 ft. lbs.
  Step 2: Plus 110 degrees
⑩ Step 1: 37 ft. lbs.
  Step 2: Plus 120 degrees

79221CH4

SCHEDULED MAINTENANCE INTERVALS
(GM E & K BODIES - CADILLAC DEVILLE, DEVILLE CONCOURS, ELDORADO, SEVILLE)

TO BE SERVICED	TYPE OF SERVICE	VEHICLE MILEAGE INTERVAL (x1000)												
		7.5	15	22.5	30	37.5	45	52.5	60	67.5	75	82.5	90	97.5
Engine oil & filter	R	✓	✓	✓	✓	✓	✓	✓	✓	✓	✓	✓	✓	✓
Coolant level, hoses & clamps	S/I	✓	✓	✓	✓	✓	✓	✓	✓	✓	✓	✓	✓	✓
Drive shaft boots & front suspension components	S/I	✓	✓	✓	✓	✓	✓	✓	✓	✓	✓	✓	✓	✓
Exhaust system, brake hoses & throttle linkage	S/I	✓	✓	✓	✓	✓	✓	✓	✓	✓	✓	✓	✓	✓
Lubricate chassis, suspension, steering linkage, transaxle shift linkage, parking brake cable guides, underbody contact points & linkage	S/I	✓	✓	✓	✓	✓	✓	✓	✓	✓	✓	✓	✓	✓
Throttle body mount bolt torque	S/I	✓												
Brake linings	S/I	✓		✓		✓		✓		✓		✓		✓
Rotate tires	S/I	✓		✓		✓		✓		✓		✓		✓
Inspect throttle body bore & throttle plate for deposits	S/I		✓				✓				✓		✓	
Air filter element	R				✓				✓				✓	
Engine coolant②	R				✓				✓				✓	
PCV valve	R				✓				✓				✓	
Spark plugs①	R				✓				✓				✓	
Accessory drive belt(s)	S/I				✓				✓				✓	
Automatic transaxle fluid & filter	S/I				✓				✓				✓	
EGR & fuel systems	S/I				✓				✓				✓	
Engine timing (4.9L)	S/I				✓				✓				✓	
Ignition cables	S/I				✓				✓				✓	

① Platinum tip spark plugs - replace every 100,000 miles.
② Engine coolant (1996-99) - replace every 100,000 miles. Use O.E. specified (DEX-COOL™) coolant only. If any silicate coolant is used, the service interval is every 30,000 miles.

R – Replace   S/I – Service or Inspect

79221CH5

SCHEDULED MAINTENANCE INTERVALS
(GM E & K BODIES—CADILLAC DEVILLE, CADILLAC CONCOURS, ELDORADO, SEVILLE Cont.)

FREQUENT OPERATION MAINTENANCE (SEVERE SERVICE)

If a vehicle is operated under any of the following conditions it is considered severe service:
- Extremely dusty areas.
- 50% or more of the vehicle operation is in 32°C (90°F) or higher temperatures, or constant operation in temperatures below 0°C (32°F).
- Prolonged idling (vehicle operation in stop and go traffic).
- Frequent short running periods (engine does not warm to normal operating temperatures).
- Police, taxi, delivery usage or trailer towing usage.

CV joints & front suspension components - service or inspect every 3000 miles.
Engine oil & filter change – change every 3000 miles
Brake linings - check every 6000 miles.
Chassis lubrication - lubricate every 6000 miles.
Suspension, steering linkage, transaxle shift linkage, parking cable guides, underbody contact points - lubricate every 6000 miles.
Throttle body mount bolt torque - tighten at 6000 miles.
Air filter element – service or inspect every 15,000 miles.
Automatic transaxle fluid – change every 50,000 miles (1995-97).
Inspect throttle body bore & throttle plate for deposits - clean as required every 15,000 miles.
Rotate tires at 6000 miles, then every 15,000 miles.

79221CH6

SCHEDULED MAINTENANCE
GENERAL MOTORS CORPORATION
E & K BODIES
CADILLAC DEVILLE, DEVILLE CONCOURS
ELDORADO, SEVILLE

The following should be used as a guide when determining the amount of work required for a particular service if taken to a repair shop. In estimating how long a particular Scheduled Maintenance Service should take, please observe the following:

- **Chilton Time** is time based on field research and data supplied by the vehicle manufacturer.
- Labor time operations are given in hours and tenths of an hour.
- All labor operations, are to be used as a guide.

Mechanic Skill Level Codes:
- **(G) GENERAL:** Normally skilled with certification.
- **(M) MAINTENANCE:** Semi-skilled working on certicication.
- **(P) PRECISION:** Really skilled with multiple certification.

	Chilton Time		Chilton Time		Chilton Time
(M) 7500 Mile Service		**(M) 45000 Mile Service**		**(M) 67500 Mile Service**	
1995-99	1.5	1995-99	.9	1995-99	1.4
(M) 15000 Mile Service		**(M) 52500 Mile Service**		**(M) 75000 Mile Service**	
1995-99	.8	1995-99	1.4	1995-99	.9
(M) 22500 Mile Service		**(G) 60000 Mile Service**		**(M) 82500 Mile Service**	
1995-99	1.4	1995-99	3.8	1995-99	1.4
(G) 30000 Mile Service		Inspect engine timing (4.9L) add	.2	**(G) 90000 Mile Service**	
1995-99	3.8			1995-99	4.0
Inspect ignition timing (4.9L) add	.2			Inspect engine timing (4.9L) add	.2
(M) 37500 Mile Service				**(M) 97500 Mile Service**	
1995-99	1.4			1995-99	1.4

79221CH7

GENERAL MOTORS F BODY
Chevrolet Camaro • Pontiac Firebird

VEHICLE IDENTIFICATION CHART

Engine Code						Model Year	
Code	Liters	Cu. In. (cc)	Cyl.	Fuel Sys.	Eng. Mfg.	Code	Year
G	5.7	350 (5665)	8	MFI	CPC	S	1995
K	3.8	231 (3785)	6	MFI	CPC	T	1996
P	5.7	350 (5737)	8	MFI	CPC	V	1997
S	3.4	207 (3393)	6	MFI	CPC	W	1998
						X	1999

MFI - Multi-point Fuel Injection
CPC - Chevrolet/Pontiac/Canada

79221CH8

ENGINE IDENTIFICATION

Year	Model	Engine Displacement Liters (cc)	Engine Series (ID/VIN)	Fuel System	No. of Cylinders	Engine Type
1995	Camaro	5.7 (5737)	P	MFI	8	OHV
	Camaro	3.4 (3393)	S	MFI	6	OHV
	Firebird	5.7 (5737)	P	MFI	8	OHV
	Firebird	3.4 (3393)	S	MFI	6	OHV
1996	Camaro	3.8 (3785)	K	MFI	6	OHV
	Camaro	5.7 (5737)	P	MFI	8	OHV
	Firebird	3.8 (3785)	K	MFI	6	OHV
	Firebird	5.7 (5737)	P	MFI	8	OHV
1997	Camaro	3.8 (3785)	K	MFI	6	OHV
	Camaro	5.7 (5737)	P	MFI	8	OHV
	Firebird	3.8 (3785)	K	MFI	6	OHV
	Firebird	5.7 (5737)	P	MFI	8	OHV
1998-99	Camaro	5.7 (5665)	G	MFI	8	OHV
	Camaro	3.8 (3785)	K	MFI	6	OHV
	Firebird	5.7 (5665)	G	MFI	8	OHV
	Firebird	3.8 (3785)	K	MFI	6	OHV

MFI - Multi-point Fuel Injection
OHV - Overhead Valves

79221CH9

GENERAL ENGINE SPECIFICATIONS

Year	Engine ID/VIN	Engine Displacement Liters (cc)	Fuel System Type	Net Horsepower @ rpm	Net Torque @ rpm (ft. lbs.)	Bore x Stroke (in.)	Compression Ratio	Oil Pressure @ rpm
1995	P	5.7 (5737)	MFI	275@5000	325@2400	4.00x3.48	10.25:1	18@2000
	S	3.4 (3393)	MFI	160@4600	200@3600	3.62x3.31	9.0:1	15@1100
1996	K	3.8 (3785)	MFI	160@4600	200@3600	3.80x3.40	9.4:1	60@1850
	P	5.7 (5737)	MFI	275@5000	325@2400	4.00x3.48	10.25:1	18@2000
1997	K	3.8 (3785)	MFI	160@4600	200@3600	3.80x3.40	9.4:1	60@1850
	P	5.7 (5737)	MFI	275@5000	325@2400	4.00x3.48	10.25:1	18@2000
1998-99	G	5.7 (5665)	MFI	305@5200	335@4000	3.89x3.62	10.0:1	18@2000
	K	3.8 (3785)	MFI	200@5200	225@4000	3.80x3.40	9.4:1	60@1850

MFI - Multi-point Fuel Injection

79221CH0

GASOLINE ENGINE TUNE-UP SPECIFICATIONS

Year	Engine ID/VIN	Engine Displacement Liters (cc)	Spark Plugs Gap (in.)	Ignition Timing (deg.) MT	AT	Fuel Pump (psi)	Idle Speed (rpm) MT	AT	Valve Clearance In.	Ex.
1995	P	5.7 (5737)	0.035	—	①	41-47	—	①	HYD	HYD
	S	3.4 (3393)	0.045	①	①	41-47	①	①	HYD	HYD
1996	K	3.8 (3785)	0.045	①	①	41-47	①	①	HYD	HYD
	P	5.7 (5737)	0.035	—	①	41-47	—	①	HYD	HYD
1997	K	3.8 (3785)	0.045	①	①	41-47	①	①	HYD	HYD
	P	5.7 (5737)	0.035	—	①	41-47	—	①	HYD	HYD
1998-99	G	5.7 (5665)	0.060	①	①	48-55	①	①	HYD	HYD
	K	3.8 (3785)	0.045	①	①	41-47	①	①	HYD	HYD

NOTE: The Vehicle Emission Control Information label often reflects specification changes made during production. The label figures must be used if they differ from those in this chart.

HYD - Hydraulic

① Refer to Vehicle Emission Control Information label

79221CJ1

CAPACITIES

Year	Model	Engine ID/VIN	Engine Displacement Liters (cc)	Engine Oil with Filter (qts.)	Transmission (pts.) 4-Spd	5-Spd	Auto.	Drive Axle Front (pts.)	Rear (pts.)	Fuel Tank (gal.)	Cooling System (qts.)
1995	Camaro	P	5.7 (5737)	5.0 ①	—	5.9 ②	10.0	—	3.5	15.5	15.1 ③
	Camaro	S	3.4 (3393)	4.5 ①	—	5.9	10.0	—	3.5	15.5	12.3 ④
	Firebird	P	5.7 (5737)	5.0 ①	—	5.9 ②	10.0	—	3.5	15.5	15.1 ③
	Firebird	S	3.4 (3393)	4.5 ①	—	5.9	10.0	—	3.5	15.5	12.3 ④
1996	Camaro	K	3.8 (3785)	4.5 ①	—	5.9	10.0	—	3.5	15.5	12.5
	Camaro	P	5.7 (5737)	4.5 ①	—	⑤	10.0	—	3.5	15.5	15.2
	Firebird	K	3.8 (3785)	4.5 ①	—	5.9	10.0	—	3.5	15.5	12.5
	Firebird	P	5.7 (5737)	4.5 ①	—	⑤	10.0	—	3.5	15.5	15.2
1997	Camaro	K	3.8 (3785)	4.5 ①	—	5.9	10.0	—	3.5	15.5	12.5
	Camaro	P	5.7 (5737)	4.5 ①	—	⑤	10.0	—	3.5	15.5	15.2
	Firebird	K	3.8 (3785)	4.5 ①	—	5.9	10.0	—	3.5	15.5	12.5
	Firebird	P	5.7 (5737)	4.5 ①	—	⑤	10.0	—	3.5	15.5	15.2
1998-99	Camaro	G	5.7 (5665)	4.5 ①	—	⑤	10.0	—	3.5	15.5	⑥
	Camaro	K	3.8 (3785)	4.5 ①	—	5.9	10.0	—	3.5	15.5	⑦
	Firebird	G	5.7 (5665)	4.5 ①	—	⑤	10.0	—	3.5	15.5	⑥
	Firebird	K	3.8 (3785)	4.5 ①	—	5.9	10.0	—	3.5	15.5	⑦

NOTE: All capacities are approximate. Add fluid gradually and ensure a proper fluid level is obtained.

① With vehicle on level surface, check oil level. Add as required to fill

② With 6 speed transmission: 8.0 pts.

③ With manual transmission: 15.3 qts.

④ With manual transmission: 12.5 qts.

⑤ ZF 6 speed trans.: 4.4 pts.

⑥ With mauual transmission: 12.5 qts.

   With automatic transmission: 12.3 qts.

⑦ With manual transmission: 15.3 qts.

   With automatic transmission: 15.1 qts.

79221CJ2

VALVE SPECIFICATIONS

Year	Engine ID/VIN	Engine Displacement Liters (cc)	Seat Angle (deg.)	Face Angle (deg.)	Spring Test Pressure (lbs. @ in.)	Spring Installed Height (in.)	Stem-to-Guide Clearance (in.)		Stem Diameter (in.)	
							Intake	Exhaust	Intake	Exhaust
1995	P	5.7 (5737)	46	45	187-203@ 1.27	1.70	0.0009- 0.0027	0.0009- 0.0027	NA	NA
	S	3.4 (3393)	46	45	190@1.20	1.61	0.0014- 0.0025	0.0015- 0.0029	NA	NA
1996	K	3.8 (3785)	45	45	210@1.32	1.69- 1.72	0.0015- 0.0035	0.0015- 0.0032	NA	NA
	P	5.7 (5737)	46	45	245-265@1.33 ①	1.78	0.0009- 0.0027	0.0009- 0.0027	NA	NA
1997	K	3.8 (3785)	45	45	210@1.32	1.69- 1.72	0.0015- 0.0035	0.0015- 0.0032	NA	NA
	P	5.7 (5737)	46	45	245-265@1.33 ①	1.78	0.0009- 0.0027	0.0009- 0.0027	NA	NA
1998-99	G	5.7 (5665)	46	45	220@1.32	1.80	0.0010- 0.0026	0.0010- 0.0026	0.0313- 0.0314	0.0313- 0.0314
	K	3.8 (3785)	45	45	210@1.32	1.69- 1.72	0.0015- 0.0035	0.0015- 0.0032	NA	NA

NA - Not Available
① With valve open

79221CJ3

TORQUE SPECIFICATIONS
All readings in ft. lbs.

Year	Engine ID/VIN	Engine Displacement Liters (cc)	Cylinder Head Bolts	Main Bearing Bolts	Rod Bearing Bolts	Crankshaft Damper Bolts	Flywheel Bolts	Manifold Intake	Manifold Exhaust	Spark Plugs	Lug Nut
1995	P	5.7 (5737)	65	78	47	60	74	①	35	11	100
	S	3.4 (3393)	②	③	37	58	61	④	18	23	100
1996	K	3.8 (3785)	⑤	⑥	20	⑦	⑧	④	18	23	100
	P	5.7 (5737)	65	78	47	60	74	35	22	11	100
1997	K	3.8 (3785)	⑤	⑥	20	⑦	⑧	④	18	23	100
	P	5.7 (5737)	65	78	47	60	74	35	22	11	100
1998-99	G	5.7 (5665)	⑨	⑩	⑪	⑫	⑬	⑭	⑮	12	100
	K	3.8 (3785)	⑤	⑥	20	⑦	⑧	④	18	23	100

NA - Not Available

① Step 1: 71 inch lbs.
  Step 2: 35 ft. lbs.
② 40 ft. lbs. plus 90 degrees
③ 37 ft. lbs. plus 75 degrees
④ Upper manifold: 18 ft. lbs.
  Lower manifold bolt/nut: 22 ft. lbs.
  Upper manifold studs: 89 inch lbs.
⑤ Step 1: 37 ft. lbs. plus 130 degrees
  Step 2: Turn center bolts an additional 30 degrees
⑥ Step 1: Tighten to 52 ft. lbs. to fully seat caps
  Step 2: Loosen bearing cap 360 degrees counter-clockwise
  Step 3: Tighten caps 15 ft. lbs., then 30 ft. lbs., then 35 degrees,
   then an additional 35 degrees plus 40 degrees -- for a total of 110 degrees
⑦ 111 ft. lbs. plus 76 degrees
⑧ 11 ft. lbs. plus 50 degrees
⑨ Step 1: 22 ft. lbs.
  Step 2: Rotate 76 degrees
  Step 3: Repeat Step 2
  Step 3: Rotate 34 degrees
  Step 4: Tighten inner (M8) bolts to 22 ft. lbs
⑩ Step 1: Inner bolts: 15 ft. lbs.
  Step 2: Inner bolts: Rotate 80 degrees
  Step 3: Outer side studs: 18 ft. lbs.
  Step 4: Outer side studs: Rotate 53 degrees
⑪ Step 1: 15 ft. lbs.
  Step 2: Rotate 60 degrees
⑫ Step 1: Ensure damper is instll fully, tighten old bolt 240 ft. lbs.
  Step 2: Install new bolt, tighten 37 ft. lbs.
  Step 3: Rotate 120 degrees
⑬ Step 1: 15 ft. lbs.
  Step 2: 37 ft. lbs.
  Step 3: 74 ft. lbs.
⑭ Step 1: 44 inch lbs.
  Step 2: 74 inch lbs
⑮ Step 1: 11 ft. lbs.
  Step 2: 18 ft. lbs

79221CJ4

SCHEDULED MAINTENANCE INTERVALS
(GM F BODY - CHEVROLET CAMARO, PONTIAC FIREBIRD)

TO BE SERVICED	TYPE OF SERVICE	VEHICLE MILEAGE INTERVAL (x1000)												
		7.5	15	22.5	30	37.5	45	52.5	60	67.5	75	82.5	90	97.5
Engine oil & filter	R	✓	✓	✓	✓	✓	✓	✓	✓	✓	✓	✓	✓	✓
Coolant level, hoses & clamps	S/I	✓	✓	✓	✓	✓	✓	✓	✓	✓	✓	✓	✓	✓
Exhaust system & throttle linkage	S/I	✓	✓	✓	✓	✓	✓	✓	✓	✓	✓	✓	✓	✓
Lubricate chassis, suspension, steering linkage, transmission shift linkage, parking brake cable guides, underbody contact points & linkage	S/I	✓	✓	✓	✓	✓	✓	✓	✓	✓	✓	✓	✓	✓
Brake hoses & brake lining	S/I	✓		✓		✓		✓		✓		✓		✓
Rotate tires③	S/I	✓		✓		✓		✓		✓		✓		✓
Automatic transmission fluid & filter⑤	S/I		✓		✓		✓		✓		✓		✓	
Air filter element & PCV filter	R				✓				✓				✓	
Engine coolant (1995)	R				✓				✓				✓	
Spark plugs①	R				✓				✓				✓	
Ignition cables, EGR & fuel systems	S/I				✓				✓				✓	
Serpentine drive belt	S/I				✓				✓				✓	
Rear axle oil (Limited slip)④	R	✓												

① Platinum tip spark plugs - replace every 100,000 miles.
② Engine coolant (1996-99) - replace every 100,000 miles. Use O.E. specified (DEX-COOL™) coolant only. If any silicate coolant is used, the service interval is every 30,000 miles.
③ For models with P245/50ZR16 tires, rotate front-to-rear only, & be sure that the tires roll in the direction indicated by the arrows on the side walls.
④ If the vehicle is used to tow a trailer, change the rear axle fluid every 7500 miles in either type of differential.
⑤ Automatic transmission fluid & filter (1995) - replace every 100,000 miles.
R – Replace   S/I – Service or Inspect

FREQUENT OPERATION MAINTENANCE (SEVERE SERVICE)

If a vehicle is operated under any of the following conditions it is considered severe service:
- Extremely dusty areas.
- 50% or more of the vehicle operation is in 32°C (90°F) or higher temperatures, or constant operation in temperatures below 0°C (32°F).
- Prolonged idling (vehicle operation in stop and go traffic).
- Frequent short running periods (engine does not warm to normal operating temperatures).
- Police, taxi, delivery usage or trailer towing usage.

Oil & oil filter – change every 3000 miles.
Chassis lubrication - lubricate every 6000 miles.
Automatic transmission fluid & filter - change every 15,000 miles.
Air filter element - service or inspect every 15,000 miles.
Rotate tires at 6000 miles, then every 15,000 miles.③

79221CJ5

SCHEDULED MAINTENANCE
GENERAL MOTORS CORPORATION
F BODY
CHEVROLET CAMARO
PONTIAC FIREBIRD, TRANS AM

The following should be used as a guide when determining the amount of work required for a particular service if taken to a repair shop. In estimating how long a particular Scheduled Maintenance Service should take, please observe the following:

- **Chilton Time** is time based on field research and data supplied by the vehicle manufacturer.
- Labor time operations are given in hours and tenths of an hour.
- All labor operations, are to be used as a guide.

Mechanic Skill Level Codes:
- **(G)** GENERAL: Normally skilled with certification.
- **(M)** MAINTENANCE: Semi-skilled working on certicication.
- **(P)** PRECISION: Really skilled with multiple certification.

	Chilton Time
(G) 7500 Mile Service	
1995-99	1.7
(M) 15000 Mile Service	
1995-99	.8
(G) 22500 Mile Service	
1995-99	1.4
(G) 30000 Mile Service	
1995-99	2.8
Renew engine coolant (1995) add	.5

	Chilton Time
(G) 37500 Mile Service	
1995-99	1.4
(M) 45000 Mile Service	
1995-99	.8
(G) 52500 Mile Service	
1995-99	1.4
(G) 60000 Mile Service	
1995-99	2.8
Renew engine coolant (1995) add	.5
(G) 67500 Mile Service	
1995-99	1.4

	Chilton Time
(M) 75000 Mile Service	
1995-99	.8
(G) 82500 Mile Service	
1995-99	1.4
(G) 90000 Mile Service	
1995-99	2.8
Renew engine coolant (1995) add	.5
(G) 97500 Mile Service	
1995-99	1.4

79221CJ6

GENERAL MOTORS G BODY
Buick Riviera • Oldsmobile Aurora

VEHICLE IDENTIFICATION CHART

Code	Liters		Cu. In. (cc)	Cyl.	Fuel Sys.	Eng. Mfg.
1	3.8	①	231 (3785)	6	MFI	BOC
C	4.0		244 (3995)	8	MFI	BOC
K	3.8		231 (3785)	6	MFI	BOC

Code	Year
S	1995
T	1996
V	1997
W	1998
X	1999

Model Year

MFI - Multi-point Fuel Injection

BOC - Buick/Oldsmobile/Cadillac

① Supercharged Engine

79221CJ7

ENGINE IDENTIFICATION

Year	Model	Engine Displacement Liters (cc)	Engine Series (ID/VIN)	Fuel System		No. of Cylinders	Engine Type
1995	Aurora	4.0 (3995)	C	MFI		8	DOHC
	Riviera	3.8 (3785)	1	MFI	①	6	OHV
	Riviera	3.8 (3785)	K	MFI		6	OHV
1996	Aurora	4.0 (3995)	C	MFI		8	DOHC
	Riviera	3.8 (3785)	1	MFI	①	6	OHV
	Riviera	3.8 (3785)	K	MFI		6	OHV
1997	Aurora	4.0 (3995)	C	MFI		8	DOHC
	Riviera	3.8 (3785)	1	MFI	①	6	OHV
	Riviera	3.8 (3785)	K	MFI		6	OHV
1998-99	Aurora	4.0 (3995)	C	MFI		8	DOHC
	Riviera	3.8 (3785)	1	MFI	①	6	OHV
	Riviera	3.8 (3785)	K	MFI		6	OHV

MFI - Multi-point Fuel Injection

DOHC - Double Overhead Camshafts

OHV - Overhead Valves

① Supercharged Engine

79221CJ8

GENERAL ENGINE SPECIFICATIONS

Year	Engine ID/VIN		Engine Displacement Liters (cc)	Fuel System Type	Net Horsepower @ rpm	Net Torque @ rpm (ft. lbs.)	Bore x Stroke (in.)	Compression Ratio	Oil Pressure @ rpm
1995	1	①	3.8 (3785)	MFI	225@5000	275@3200	3.80x3.40	9.0:1	60@1850
	C		4.0 (3995)	MFI	250@5600	245@4400	3.43x3.31	10.2:1	30@2000
	K		3.8 (3785)	MFI	205@5200	230@4000	3.80x3.40	9.4:1	60@1850
1996	1	①	3.8 (3785)	MFI	225@5000	275@3200	3.80x3.40	8.5:1	60@1850
	C		4.0 (3995)	MFI	250@5600	245@4400	3.43x3.31	10.2:1	20@2000
	K		3.8 (3785)	MFI	205@5200	230@4000	3.80x3.40	9.4:1	60@1850
1997	1	①	3.8 (3785)	MFI	225@5000	275@3200	3.80x3.40	8.5:1	60@1850
	C		4.0 (3995)	MFI	250@5600	245@4400	3.43x3.31	10.2:1	20@2000
	K		3.8 (3785)	MFI	205@5200	230@4000	3.80x3.40	9.4:1	60@1850
1998-99	1	①	3.8 (3785)	MFI	225@5000	275@3200	3.80x3.40	8.5:1	60@1850
	C		4.0 (3995)	MFI	250@5600	245@4400	3.43x3.31	10.2:1	20@2000
	K		3.8 (3785)	MFI	205@5200	230@4000	3.80x3.40	9.4:1	60@1850

MFI - Multi-point Fuel Injection

① Supercharged Engine

79221CJ9

GASOLINE ENGINE TUNE-UP SPECIFICATIONS

Year	Engine ID/VIN	Engine Displacement Liters (cc)	Spark Plugs Gap (in.)	Ignition Timing (deg.) MT	Ignition Timing (deg.) AT	Fuel Pump (psi)	Idle Speed (rpm) MT	Idle Speed (rpm) AT	Valve Clearance In.	Valve Clearance Ex.
1995	1	3.8 (3785)	0.060	—	①	40-47	—	③	HYD	HYD
	C	4.0 (3995)	0.050	—	②	41-47	—	③	HYD	HYD
	K	3.8 (3785)	0.060	—	①	40-47	—	③	HYD	HYD
1996	1	3.8 (3785)	0.060	—	②	41-47	—	③	HYD	HYD
	C	4.0 (3995)	0.050	—	②	41-47	—	③	HYD	HYD
	K	3.8 (3785)	0.060	—	②	41-47	—	③	HYD	HYD
1997	1	3.8 (3785)	0.060	—	②	41-47	—	③	HYD	HYD
	C	4.0 (3995)	0.050	—	②	41-47	—	③	HYD	HYD
	K	3.8 (3785)	0.060	—	②	41-47	—	③	HYD	HYD
1998-99	1	3.8 (3785)	0.060	—	②	41-47	—	③	HYD	HYD
	C	4.0 (3995)	0.050	—	②	41-47	—	③	HYD	HYD
	K	3.8 (3785)	0.060	—	②	41-47	—	③	HYD	HYD

HYD - Hydraulic
① Refer to underhood sticker
② DIS Ignition System timing is not adjustable
③ Idle speed is maintained by the ECM. There is no recommended adjustment procedure

79221CJ0

CAPACITIES

Year	Model	Engine ID/VIN	Engine Displacement Liters (cc)	Engine Oil with Filter (qts.)	Transmission (pts.) 4-Spd	Transmission (pts.) 5-Spd	Transmission (pts.) Auto.	Drive Axle Front (pts.)	Drive Axle Rear (pts.)	Fuel Tank (gal.)	Cooling System (qts.)
1995	Aurora	C	4.0 (3995)	7.0 ①	—	—	6.5	—	—	20.0	13.0
	Riviera	1	3.8 (3785)	5.0	—	—	13.2	—	—	20.0	13.0
	Riviera	K	3.8 (3785)	5.0	—	—	13.2	—	—	20.0	13.0
1996	Aurora	C	4.0 (3995)	7.5	—	—	6.5	—	—	20.0	13.0
	Riviera	1	3.8 (3785)	5.0	—	—	12.0	—	—	20.0	13.0
	Riviera	K	3.8 (3785)	4.5	—	—	12.0	—	—	20.0	13.0
1997	Aurora	C	4.0 (3995)	7.5	—	—	6.5	—	—	20.0	13.0
	Riviera	1	3.8 (3785)	5.0	—	—	12.0	—	—	20.0	13.0
	Riviera	K	3.8 (3785)	4.5	—	—	12.0	—	—	20.0	13.0
1998-99	Aurora	C	4.0 (3995)	7.5	—	—	6.5	—	—	20.0	13.0
	Riviera	1	3.8 (3785)	5.0	—	—	12.0	—	—	20.0	13.0
	Riviera	K	3.8 (3785)	4.5	—	—	12.0	—	—	20.0	13.0

NOTE: All capacities are approximate. Add fluid gradually and ensure a proper fluid level is obtained.
① Specification is without filter replacement; Additional oil may be required

79221CK1

VALVE SPECIFICATIONS

Year	Engine ID/VIN	Engine Displacement Liters (cc)	Seat Angle (deg.)	Face Angle (deg.)	Spring Test Pressure (lbs. @ in.)	Spring Installed Height (in.)	Stem-to-Guide Clearance (in.)		Stem Diameter (in.)	
							Intake	Exhaust	Intake	Exhaust
1995	1	3.8 (3785)	45	45	210@1.315	1.690-1.720	0.0015-0.0035	0.0015-0.0032	NA	NA
	C	4.0 (3995)	46	45	92@0.854	1.190	0.0010-0.0030	0.0020-0.0040	NA	NA
	K	3.8 (3785)	45	45	210@1.315	1.690-1.720	0.0015-0.0035	0.0015-0.0032	NA	NA
1996	1	3.8 (3785)	45	45	210@1.315	1.690-1.720	0.0015-0.0035	0.0015-0.0032	NA	NA
	C	4.0 (3995)	46	45	92@0.854	1.190	0.0010-0.0030	0.0020-0.0040	NA	NA
	K	3.8 (3785)	45	45	210@1.315	1.690-1.720	0.0015-0.0035	0.0015-0.0032	NA	NA
1997	1	3.8 (3785)	45	45	210@1.315	1.690-1.720	0.0015-0.0035	0.0015-0.0032	NA	NA
	C	4.0 (3995)	46	45	92@0.854	1.190	0.0010-0.0030	0.0020-0.0040	NA	NA
	K	3.8 (3785)	45	45	210@1.315	1.690-1.720	0.0015-0.0035	0.0015-0.0032	NA	NA
1998-99	1	3.8 (3785)	45	45	210@1.315	1.690-1.720	0.0015-0.0035	0.0015-0.0032	NA	NA
	C	4.0 (3995)	46	45	92@0.854	1.190	0.0010-0.0030	0.0020-0.0040	NA	NA
	K	3.8 (3785)	45	45	210@1.315	1.690-1.720	0.0015-0.0035	0.0015-0.0032	NA	NA

79221CK2

TORQUE SPECIFICATIONS
All readings in ft. lbs.

Year	Engine ID/VIN	Engine Displacement Liters (cc)	Cylinder Head Bolts	Main Bearing Bolts	Rod Bearing Bolts	Crankshaft Damper Bolts	Flywheel Bolts	Manifold Intake	Manifold Exhaust	Spark Plugs	Lug Nut
1995	1	3.8 (3785)	①	②	③	④	⑤	7	38	11	100
	C	4.0 (3995)	⑦	⑧	⑨	⑩	⑤	89	18	⑥	100
	K	3.8 (3785)	①	②	③	④	⑤	7	38	11	100
1996	1	3.8 (3786)	①	⑪	③	④	⑤	⑬	22	11	100
	C	4.0 (3995)	⑦	⑧	⑨	⑩	⑤	89	18	⑥	100
	K	3.8 (3786)	①	⑫	③	④	⑤	11	38	11	100
1997	1	3.8 (3786)	①	⑪	③	④	⑤	⑬	22	11	100
	C	4.0 (3995)	⑦	⑧	⑨	⑩	⑤	89	18	⑥	100
	K	3.8 (3786)	①	⑫	③	④	⑤	11	38	11	100
1998-99	1	3.8 (3786)	①	⑪	③	④	⑤	⑬	22	11	100
	C	4.0 (3995)	⑦	⑧	⑨	⑩	⑤	89	18	⑥	100
	K	3.8 (3786)	①	⑫	③	④	⑤	11	38	11	100

① Step 1: 36 ft. lbs.
　Step 2: 130 degrees
　Step 3: Plus 30 degrees additional on four center bolts
　(NOTE: Must use new bolts)
② 26 ft. lbs. plus 50 degrees
③ 20 ft. lbs. plus 50 degrees
④ 110 ft. lbs. plus 76 degrees
⑤ Step 1: 11 ft. lbs.
　Step 2: Plus 50 degrees
⑥ New cylinder head 1st-time installation: 20 ft. lbs.
　All others: 11 ft. lbs.
⑦ Step 1: Torque to 22 ft. lbs. then 90 degrees.
　Step 2: Retorque an additional 75 degrees.
⑧ 15 ft. lbs. plus 65 degrees
⑨ 18 ft. lbs. plus 90 degrees
⑩ 44 ft. lbs. plus 120 degrees
⑪ Step 1: Tighten caps in equal increments to 52 ft. lbs.
　Step 2: Loosen 360 degrees
　Step 3: 15 ft. lbs.
　Step 4: 54 ft. lbs.
　Step 5: Plus three turns of 35 degrees-total of 105 degrees
⑫ 26 ft. lbs. plus 90 degrees
⑬ Upper manifold: 8 ft. lbs.
　Lower manifold: 11 ft. lbs.

79221CK3

SCHEDULED MAINTENANCE INTERVALS

(GM G BODY - BUICK RIVIERA, OLDSMOBILE AURORA)

TO BE SERVICED	TYPE OF SERVICE	VEHICLE MILEAGE INTERVAL (x1000)												
		7.5	15	22.5	30	37.5	45	52.5	60	67.5	75	82.5	90	97.5
Engine oil & filter	R	✓	✓	✓	✓	✓	✓	✓	✓	✓	✓	✓	✓	✓
Coolant level, hoses & clamps	S/I	✓	✓	✓	✓	✓	✓	✓	✓	✓	✓	✓	✓	✓
Drive shaft boots & front suspension components	S/I	✓	✓	✓	✓	✓	✓	✓	✓	✓	✓	✓	✓	✓
Exhaust system & brake hoses	S/I	✓	✓	✓	✓	✓	✓	✓	✓	✓	✓	✓	✓	✓
Lubricate, chassis, suspension, steering linkage, transaxle shift linkage, parking brake cable guides, underbody contact points & linkage	S/I	✓	✓	✓	✓	✓	✓	✓	✓	✓	✓	✓	✓	✓
Throttle linkage	S/I	✓	✓	✓	✓	✓	✓	✓	✓	✓	✓	✓	✓	✓
Brake linings	S/I	✓		✓		✓		✓		✓		✓		
Rotate tires	S/I	✓		✓		✓		✓		✓		✓		✓
Air filter element	R				✓				✓				✓	
Engine coolant (1995)②	R				✓				✓				✓	
Spark plugs①	R				✓				✓				✓	
Accessory drive belt(s)	S/I				✓				✓				✓	
Automatic transaxle fluid & filter	S/I				✓				✓				✓	
Fuel system	S/I				✓				✓				✓	
Ignition cables	S/I				✓				✓				✓	
Inspect throttle body bore & throttle plate for deposits (Oldsmobile)	S/I		✓				✓				✓			
Supercharger oil	S/I				✓				✓				✓	
Throttle body mounting torque	S/I	✓												

① Platinum tip spark plugs - replace every 100,000 miles.

② Engine coolant (1996-99) - replace every 100,000 miles. Use O.E. specified (DEX-COOL™) coolant only. If any silicate coolant is used, the service interval is every 30,000 miles.

R – Replace   S/I – Service or Inspect

79221CK4

SCHEDULED MAINTENANCE INTERVALS
(GM G BODY - BUICK RIVIERA, OLDSMOBILE AURORA Cont.)

FREQUENT OPERATION MAINTENANCE (SEVERE SERVICE)

If a vehicle is operated under any of the following conditions it is considered severe service:
- Extremely dusty areas.
- 50% or more of the vehicle operation is in 32°C (90°F) or higher temperatures, or constant operation in temperatures below 0°C (32°F).
- Prolonged idling (vehicle operation in stop and go traffic).
- Frequent short running periods (engine does not warm to normal operating temperatures).
- Police, taxi, delivery usage or trailer towing usage.

CV joints & front suspension components - check every 3000 miles.
Oil & oil filter – change every 3000 miles.
Brake linings – check every 6000 miles.
Chassis lubrication - lubricate every 6000 miles.
Suspension, steering linkage, transaxle shift linkage, parking cable guides, underbody contact points, & linkage - lubricate every 6000 miles.
Throttle body mounting bolt torque - check at 6000 miles.
Air filter element – service or inspect every 15,000 miles.
Rotate tires initially at 6000 miles, then every 15,000 miles.
Automatic transaxle fluid – change every 50,000 miles.

79221CK5

SCHEDULED MAINTENANCE
GENERAL MOTORS CORPORATION
G BODY
BUICK RIVIERA, OLDSMOBILE AURORA

The following should be used as a guide when determining the amount of work required for a particular service if taken to a repair shop. In estimating how long a particular Scheduled Maintenance Service should take, please observe the following:

- **Chilton Time** is time based on field research and data supplied by the vehicle manufacturer.
- Labor time operations are given in hours and tenths of an hour.
- All labor operations, are to be used as a guide.

Mechanic Skill Level Codes:
- **(G) GENERAL:** Normally skilled with certification.
- **(M) MAINTENANCE:** Semi-skilled working on certicication.
- **(P) PRECISION:** Really skilled with multiple certification.

	Chilton Time		Chilton Time		Chilton Time
(G) 7500 Mile Service		**(G) 45000 Mile Service**		**(G) 67500 Mile Service**	
1995-99	1.7	1995-99	.8	1995-99	1.5
(G) 15000 Mile Service		Inspect throttle body for deposits (Aurora) add	.1	**(G) 75000 Mile Service**	
1995-99	.8	**(G) 52500 Mile Service**		1995-99	.8
Inspect throttle body for deposits (Aurora) add	.1	1995-99	1.5	Inspect throttle body for deposits (Aurora) add	.1
(G) 22500 Mile Service		**(G) 60000 Mile Service**		**(G) 82500 Mile Service**	
1995-99	1.5	1995-99	2.7	1995-99	1.5
(G) 30000 Mile Service		Renew engine coolant (1995) add	.5	**(G) 90000 Mile Service**	
1995-99	2.7			1995-99	2.7
Renew engine coolant (1995) add	.5			Renew engine coolant (1995) add	.5
(G) 37500 Mile Service				**(G) 97500 Mile Service**	
1995-99	1.5			1995-99	1.5

79221CK6

GENERAL MOTORS J BODY
Chevrolet Cavalier • Pontiac Sunfire

VEHICLE IDENTIFICATION CHART

Engine Code						Model Year	
Code	Liters	Cu. In. (cc)	Cyl.	Fuel Sys.	Eng. Mfg.	Code	Year
4	2.2	133 (2180)	4	MFI	CUS	S	1995
D	2.3	138 (2262)	4	MFI	CUS	T	1996
T	2.4	146 (2392)	4	MFI	CUS	V	1997
T	3.1	191 (3130)	6	MFI	CPC	W	1998
						X	1999

CUS - Chevrolet/United States
CPC - Chevrolet/Pontiac/Canada
MFI - Multi-point Fuel Injection

79221CK7

ENGINE IDENTIFICATION

Year	Model	Engine Displacement Liters (cc)	Engine Series (ID/VIN)	Fuel System	No. of Cylinders	Engine Type
1995	Cavalier	2.2 (2180)	4	MFI	4	OHV
	Cavalier	2.3 (2262)	D	MFI	4	DOHC
	Sunfire	2.2 (2180)	4	MFI	4	OHV
	Sunfire	2.3 (2262)	D	MFI	4	DOHC
1996	Cavalier	2.2 (2180)	4	MFI	4	OHV
	Cavalier	2.4 (2392)	T	MFI	4	DOHC
	Sunfire	2.2 (2180)	4	MFI	4	OHV
	Sunfire	2.4 (2392)	T	MFI	4	DOHC
1997	Cavalier	2.2 (2180)	4	MFI	4	OHV
	Cavalier	2.4 (2392)	T	MFI	4	DOHC
	Sunfire	2.2 (2180)	4	MFI	4	OHV
	Sunfire	2.4 (2392)	T	MFI	4	DOHC
1998-99	Cavalier	2.2 (2180)	4	MFI	4	OHV
	Cavalier	2.4 (2392)	T	MFI	4	DOHC
	Sunfire	2.2 (2180)	4	MFI	4	OHV
	Sunfire	2.4 (2392)	T	MFI	4	DOHC

MFI - Multi-point Fuel Injection
OHV - Overhead Valves
DOHC - Double Overhead Camshafts

79221CK8

GENERAL ENGINE SPECIFICATIONS

Year	Engine ID/VIN	Engine Displacement Liters (cc)	Fuel System Type	Net Horsepower @ rpm	Net Torque @ rpm (ft. lbs.)	Bore x Stroke (in.)	Compression Ratio	Oil Pressure @ rpm
1995	4	2.2 (2180)	MFI	120@5200	130@3200	3.50x3.46	9.0:1	63-77@1200
	D	2.3 (2262)	MFI	150@6100	145@4800	3.63x3.35	9.5:1	30@2000
1996	4	2.2 (2180)	MFI	120@5200	130@3200	3.50x3.46	9.0:1	56@3000
	T	2.4 (2392)	MFI	150@6000	155@4400	3.54x3.70	9.5:1	30@3000
1997	4	2.2 (2180)	MFI	120@5200	130@3200	3.50x3.46	9.0:1	56@3000
	T	2.4 (2392)	MFI	150@6000	155@4400	3.54x3.70	9.5:1	30@3000
1998-99	4	2.2 (2180)	MFI	120@5200	130@3200	3.50x3.46	9.0:1	56@3000
	T	2.4 (2392)	MFI	150@6000	155@4400	3.54x3.70	9.5:1	30@3000

MFI - Multi-point Fuel Injection

79221CK9

GASOLINE ENGINE TUNE-UP SPECIFICATIONS

Year	Engine ID/VIN	Engine Displacement Liters (cc)	Spark Plugs Gap (in.)	Ignition Timing (deg.) MT	AT	Fuel Pump (psi)	Idle Speed (rpm) MT	AT	Valve Clearance In.	Ex.
1995	4	2.2 (2180)	0.045	①	①	41-47	①	①	HYD	HYD
	D	2.3 (2262)	0.035	①	①	41-47	①	①	HYD	HYD
1996	4	2.2 (2180)	0.045	①	①	41-47	①	①	HYD	HYD
	T	2.4 (2392)	0.035	①	①	41-47	①	①	HYD	HYD
1997	4	2.2 (2180)	0.045	①	①	41-47	①	①	HYD	HYD
	T	2.4 (2392)	0.035	①	①	41-47	①	①	HYD	HYD
1998-99	4	2.2 (2180)	0.045	①	①	41-47	①	①	HYD	HYD
	T	2.4 (2392)	0.035	①	①	41-47	①	①	HYD	HYD

NOTE: The Vehicle Emission Control Information label often reflects specification changes made during production. The label figures must be used if they differ from those in this chart.

HYD - Hydraulic

① Refer to Vehicle Emission Control Information label

79221CK0

CAPACITIES

Year	Model	Engine ID/VIN	Engine Displacement Liters (cc)	Engine Oil with Filter (qts.)	Transmission (pts.) 4-Spd	5-Spd	Auto.	Transfer Case (pts.)	Drive Axle Front (pts.)	Rear (pts.)	Fuel Tank (gal.)	Cooling System (qts.)
1995	Cavalier	4	2.2 (2180)	4.5	—	4.0	8.0 ①	—	—	—	13.6	8.5
	Cavalier	D	2.3 (2262)	4.5	—	4.0	14.0	—	—	—	13.6	10.4
	Sunfire	4	2.2 (2180)	4.5	—	4.0	8.0 ①	—	—	—	15.2	10.7
	Sunfire	D	2.3 (2262)	4.5	—	4.0	22.0	—	—	—	15.2	10.4
1996	Cavalier	4	2.2 (2180)	4.5	—	4.0	8.0 ①	—	—	—	13.6	8.5
	Cavalier	T	2.4 (2392)	4.5	—	4.0	14.0	—	—	—	13.6	10.4
	Sunfire	4	2.2 (2180)	4.5	—	4.0	8.0 ①	—	—	—	15.2	10.7
	Sunfire	T	2.4 (2392)	4.5	—	4.0	22.0	—	—	—	15.2	10.4
1997	Cavalier	4	2.2 (2180)	4.5	—	4.0	8.0 ①	—	—	—	13.6	8.5
	Cavalier	T	2.4 (2392)	4.5	—	4.0	14.0	—	—	—	13.6	10.4
	Sunfire	4	2.2 (2180)	4.5	—	4.0	8.0 ①	—	—	—	15.2	10.7
	Sunfire	T	2.4 (2392)	4.5	—	4.0	22.0	—	—	—	15.2	10.4
1998-99	Cavalier	4	2.2 (2180)	4.5	—	4.0	8.0 ①	—	—	—	13.6	8.5
	Cavalier	T	2.4 (2392)	4.5	—	4.0	14.0	—	—	—	13.6	10.4
	Sunfire	4	2.2 (2180)	4.5	—	4.0	8.0 ①	—	—	—	15.2	10.7
	Sunfire	T	2.4 (2392)	4.5	—	4.0	22.0	—	—	—	15.2	10.4

NOTE: All capacities are approximate. Add fluid gradually and ensure a proper fluid level is obtained.

① 10.0 pts. if equipped with O/D

79221CL1

VALVE SPECIFICATIONS

Year	Engine ID/VIN	Engine Displacement Liters (cc)	Seat Angle (deg.)	Face Angle (deg.)	Spring Test Pressure (lbs. @ in.)	Spring Installed Height (in.)	Stem-to-Guide Clearance (in.)		Stem Diameter (in.)	
							Intake	Exhaust	Intake	Exhaust
1995	4	2.2 (2180)	46	45	225-233@ ① 1.25	1.640 ②	0.0011-0.0026	0.0014-0.0031	NA	NA
	D	2.3 (2262)	45	44	193-207@ 1.04	1.440 ②	0.0010-0.0027	0.0015-0.0032	0.2740-0.2750	0.2740-0.2750
1996	4	2.2 (2180)	46	45	75-81@1.71	1.710	0.0010-0.0027	0.0014-0.0031	NA	NA
	T	2.4 (2392)	45	46	50-55@1.44	1.437	0.0009-0.0025	0.0016-0.0032	0.2331-0.2339	0.2326-0.2334
1997	4	2.2 (2180)	46	45	75-81@1.71	1.710	0.0010-0.0027	0.0014-0.0031	NA	NA
	T	2.4 (2392)	45	46	50-55@1.44	1.437	0.0009-0.0025	0.0016-0.0032	0.2331-0.2339	0.2326-0.2334
1998-99	4	2.2 (2180)	46	45	75-81@1.71	1.710	0.0010-0.0027	0.0014-0.0031	NA	NA
	T	2.4 (2392)	45	46	50-55@1.44	1.437	0.0009-0.0025	0.0016-0.0032	0.2331-0.2339	0.2326-0.2334

NA - Not Available
① With valve open
② With valve closed

79221CL2

TORQUE SPECIFICATIONS
All readings in ft. lbs.

Year	Engine ID/VIN	Engine Displacement Liters (cc)	Cylinder Head Bolts	Main Bearing Bolts	Rod Bearing Bolts	Crankshaft Damper Bolts	Flywheel Bolts	Manifold		Spark Plugs	Lug Nut
								Intake	Exhaust		
1995	4	2.2 (2180)	①	77	38	85 ②	52-55	18	6-13	20	100
	D	2.3 (2262)	26 ③	④	⑤	⑥	⑦	⑧	⑨	17	100
1996	4	2.2 (2180)	①	70	38	77 ②	52-55	24	18	11	100
	T	2.4 (2392)	⑩	④	⑤	⑥	⑦	⑧	⑨	11	100
1997	4	2.2 (2180)	①	70	38	77 ②	52-55	24	18	11	100
	T	2.4 (2392)	⑩	④	⑤	⑥	⑦	⑧	⑨	11	100
1998-99	4	2.2 (2180)	①	70	38	77 ②	52-55	24	18	11	100
	T	2.4 (2392)	⑩	④	⑤	⑥	⑦	⑧	⑨	11	100

NA - Not Available
① Step 1: 41 ft. lbs.
Step 2: Tighten an additional 45 degrees
Step 3: Tighten an additional 45 degrees
Step 4: Long bolts 1, 4-5, 8-9 an additional 20 degrees
Step 5: Short bolts 2-3, 6-7, 10 an additional 10 degrees
② Center bolt spec shown; Pulley to hub bolts: 37 ft. lbs.
③ Cylinder head bolts should be torqued 26 ft. lbs.
Long bolts: 100 degrees
Short bolts: 120 degrees
④ 15 ft. lbs. plus 90 degrees
⑤ 18 ft. lbs. plus 80 degrees
⑥ 74 ft. lbs. plus 90 degrees
⑦ 22 ft. lbs. plus 45 degrees
⑧ Nuts: 18 ft. lbs.
Studs: 96 inch lbs.
⑨ Bolts: 27 ft. lbs.
Studs: 106 inch lbs.
⑩ Cylinder head bolts: 40 ft. lbs.
plus 90 degrees

79221CL3

SCHEDULED MAINTENANCE INTERVALS
(GM J BODY - CHEVROLET CAVALIER, PONTIAC SUNFIRE)

TO BE SERVICED	TYPE OF SERVICE	VEHICLE MILEAGE INTERVAL (x1000)												
		7.5	15	22.5	30	37.5	45	52.5	60	67.5	75	82.5	90	97.5
Engine oil & filter	R	✓	✓	✓	✓	✓	✓	✓	✓	✓	✓	✓	✓	✓
Exhaust system & brake hoses	S/I	✓	✓	✓	✓	✓	✓	✓	✓	✓	✓	✓	✓	✓
Drive shaft boots & front suspension components	S/I	✓	✓	✓	✓	✓	✓	✓	✓	✓	✓	✓	✓	✓
Coolant level, hoses & clamps	S/I	✓	✓	✓	✓	✓	✓	✓	✓	✓	✓	✓	✓	✓
Throttle linkage	S/I	✓	✓	✓	✓	✓	✓	✓	✓	✓	✓	✓	✓	✓
Lubricate, chassis, suspension, steering linkage, transaxle shift linkage, parking brake cable guides, underbody contact points & linkage	S/I	✓	✓	✓	✓	✓	✓	✓	✓	✓	✓	✓	✓	✓
Brake linings & rotate tires	S/I	✓		✓		✓		✓		✓		✓		✓
Automatic transaxle fluid & filter③	S/I	✓		✓		✓		✓		✓		✓		✓
Air filter element & PCV filter	R				✓				✓				✓	
Engine coolant (1995)②	R				✓				✓				✓	
Spark plugs①	R				✓				✓				✓	
Accessory drive belt(s)	S/I				✓				✓				✓	
EGR & fuel systems	S/I				✓				✓				✓	
Ignition cables	S/I				✓				✓				✓	
Throttle body mount bolt torque	S/I	✓												

① Platinum tip spark plugs - replace every 100,000 miles.
② Engine coolant (1996-99) - replace every 100,000 miles. Use O.E. specified (DEX-COOL™) coolant only. If any silicate coolant is used, the service interval is every 30,000 miles.
③ Automatic transaxle fluid & filter - replace at 100,000 miles (if not changed previously).
R – Replace   S/I – Service or Inspect

79221CL4

SCHEDULED MAINTENANCE INTERVALS
(GM J BODY - CHEVROLET CAVALIER, PONTIAC SUNFIRE Cont.)

FREQUENT OPERATION MAINTENANCE (SEVERE SERVICE)

If a vehicle is operated under any of the following conditions it is considered severe service:
- Extremely dusty areas.
- 50% or more of the vehicle operation is in 32°C (90°F) or higher temperatures, or constant operation in temperatures below 0°C (32°F).
- Prolonged idling (vehicle operation in stop and go traffic).
- Frequent short running periods (engine does not warm to normal operating temperatures).
- Police, taxi, delivery usage or trailer towing usage.

CV joints & front suspension components - check every 3000 miles.
Oil & oil filter – change every 3000 miles.
Brake linings – check every 6000 miles.
Chassis lubrication - lubricate every 6000 miles.
Lubricate suspension, steering linkage, transaxle shift linkage, parking cable guides, underbody contact points, & linkage - lubricate every 6000 miles.
Throttle body mount bolt torque- check at 6000 miles.
Rotate tires at 6000 miles, then every 12,000 miles.
Air filter element – service or inspect every 15,000 miles.
Automatic transaxle fluid – change every 15,000 miles.

79221CL5

SCHEDULED MAINTENANCE
GENERAL MOTORS CORPORATION
J BODY
CHEVROLET CAVALIER, PONTIAC SUNFIRE

The following should be used as a guide when determining the amount of work required for a particular service if taken to a repair shop.
In estimating how long a particular Scheduled Maintenance Service should take, please observe the following:

- **Chilton Time** is time based on field research and data supplied by the vehicle manufacturer.

- Labor time operations are given in hours and tenths of an hour.

- All labor operations, are to be used as a guide.

Mechanic Skill Level Codes:
    (G) GENERAL: Normally skilled with certification.
    (M) MAINTENANCE: Semi-skilled working on certication.
    (P) PRECISION: Really skilled with multiple certification.

	Chilton Time
(M) 7500 Mile Service	
1995-99	1.9
(M) 15000 Mile Service	
1995-99	.8
(M) 22500 Mile Service	
1995-99	1.6
(G) 30000 Mile Service	
1995-99	2.4
Renew engine coolant (1995) add	.5

	Chilton Time
(M) 37500 Mile Service	
1995-99	1.6
(M) 45000 Mile Service	
1995-99	.8
(M) 52500 Mile Service	
1995-99	1.6
(G) 60000 Mile Service	
1995-99	2.4
Renew engine coolant (1995) add	.5
(M) 67500 Mile Service	
1995-99	1.6

	Chilton Time
(M) 75000 Mile Service	
1995-99	.8
(M) 82500 Mile Service	
1995-99	1.6
(G) 90000 Mile Service	
1995-99	2.4
Renew engine coolant (1995) add	.5
(M) 97500 Mile Service	
1994-95	1.6

79221CL6

GENERAL MOTORS L BODY
Chevrolet Corsica • Beretta

VEHICLE IDENTIFICATION CHART

Engine Code						Model Year	
Code	Liters	Cu. In. (cc)	Cyl.	Fuel Sys.	Eng. Mfg.	Code	Year
4	2.2	133 (2180)	4	MFI	CUS	S	1995
M	3.1	191 (3130)	6	MFI	CPC	T	1996

CUS - Chevrolet/United States
CPC - Chevrolet/Pontiac/Canada
MFI - Multi-point Fuel Injection

79221CL7

ENGINE IDENTIFICATION

Year	Model	Engine Displacement Liters (cc)	Engine Series (ID/VIN)	Fuel System	No. of Cylinders	Engine Type
1995	Beretta	2.2 (2180)	4	MFI	4	OHV
	Beretta	3.1 (3130)	M	MFI	6	OHV
	Corsica	2.2 (2180)	4	MFI	4	OHV
	Corsica	3.1 (3130)	M	MFI	6	OHV
1996	Beretta	2.2 (2180)	4	MFI	4	OHV
	Beretta	3.1 (3130)	M	MFI	6	OHV
	Corsica	2.2 (2180)	4	MFI	4	OHV
	Corsica	3.1 (3130)	M	MFI	6	OHV

MFI - Multi-point Fuel Injection
OHV - Overhead Valves

79221CL8

GENERAL ENGINE SPECIFICATIONS

Year	Engine ID/VIN	Engine Displacement Liters (cc)	Fuel System Type	Net Horsepower @ rpm	Net Torque @ rpm (ft. lbs.)	Bore x Stroke (in.)	Compression Ratio	Oil Pressure @ rpm
1995	4	2.2 (2180)	MFI	120@5200	130@3200	3.50x3.46	9.0:1	63-77@1200
	M	3.1 (3130)	MFI	155@5200	185@4000	3.50x3.31	9.6:1	15@1100
1996	4	2.2 (2180)	MFI	120@5200	130@3200	3.50x3.46	9.0:1	56@3000
	M	3.1 (3130)	MFI	155@5200	185@4000	3.50x3.31	9.5:1	15@1100

MFI - Multi-point Fuel Injection

79221CL9

GASOLINE ENGINE TUNE-UP SPECIFICATIONS

Year	Engine ID/VIN	Engine Displacement Liters (cc)	Spark Plugs Gap (in.)	Ignition Timing (deg.) MT	Ignition Timing (deg.) AT	Fuel Pump (psi)	Idle Speed (rpm) MT	Idle Speed (rpm) AT	Valve Clearance In.	Valve Clearance Ex.
1995	4	2.2 (2180)	0.045	①	①	41-47	①	①	HYD	HYD
	M	3.1 (3130)	0.045	①	①	41-47	①	①	HYD	HYD
1996	4	2.2 (2180)	0.045	①	①	41-47	①	①	HYD	HYD
	M	3.1 (3130)	0.045	①	①	41-47	①	①	HYD	HYD

NOTE: The Vehicle Emission Control Information label often reflects specification changes made during production. The label figures must be used if they differ from those in this chart.

HYD - Hydraulic

① Refer to Vehicle Emission Control Information label

79221CL0

CAPACITIES

Year	Model	Engine ID/VIN	Engine Displacement Liters (cc)	Engine Oil with Filter (qts.)	Transmission (pts.) 4-Spd	Transmission (pts.) 5-Spd	Transmission (pts.) Auto.	Drive Axle Front (pts.)	Drive Axle Rear (pts.)	Fuel Tank (gal.)	Cooling System (qts.)
1995	Beretta	4	2.2 (2180)	4.5	—	4.0	14.0 ①	—	—	15.6	9.5
	Beretta	M	3.1 (3130)	4.5	—	4.0	14.0 ①	—	—	15.6	②
	Corsica	4	2.2 (2180)	4.5	—	4.0	14.0 ①	—	—	15.6	9.5
	Corsica	M	3.1 (3130)	4.5	—	4.0	14.0 ①	—	—	15.6	②
1996	Beretta	4	2.2 (2180)	4.5	—	4.0	14.0 ①	—	—	15.6	9.5
	Beretta	M	3.1 (3130)	4.5	—	4.0	14.0 ①	—	—	15.6	②
	Corsica	4	2.2 (2180)	4.5	—	4.0	14.0 ①	—	—	15.6	9.5
	Corsica	M	3.1 (3130)	4.5	—	4.0	14.0 ①	—	—	15.6	②

NOTE: All capacities are approximate. Add fluid gradually and ensure a proper fluid level is obtained.

① Drain and refill figure, overhaul: 16.0 pts.

② Automatic transmission: 12.4 qts.
Manual transmission: 11.8 qts.

79221CM1

VALVE SPECIFICATIONS

Year	Engine ID/VIN	Engine Displacement Liters (cc)	Seat Angle (deg.)	Face Angle (deg.)	Spring Test Pressure (lbs. @ in.)	Spring Installed Height (in.)	Stem-to-Guide Clearance (in.) Intake	Stem-to-Guide Clearance (in.) Exhaust	Stem Diameter (in.) Intake	Stem Diameter (in.) Exhaust
1995	4	2.2 (2180)	46	45	225-233@ ① 1.25	1.64 ②	0.0011-0.0026	0.0014-0.0031	NA	NA
	M	3.1 (3130)	45	45	80@1.71	1.71	0.0010-0.0027	0.0010-0.0027	NA	NA
1996	4	2.2 (2195)	46	45	75-81@1.71	1.71	0.0010-0.0027	0.0014-0.0031	NA	NA
	M	3.1 (3136)	45	45	80@1.71	1.71	0.0010-0.0027	0.0010-0.0027	NA	NA

NA - Not Available

① With valve open

② With valve closed

79221CM2

TORQUE SPECIFICATIONS
All readings in ft. lbs.

Year	Engine ID/VIN	Engine Displacement Liters (cc)	Cylinder Head Bolts	Main Bearing Bolts	Rod Bearing Bolts	Crankshaft Damper Bolts	Flywheel Bolts	Manifold		Spark Plugs	Lug Nut
								Intake	Exhaust		
1995	4	2.2 (2180)	①	77	38	85 ②	52-55	18	6-13	20	100
	M	3.1 (3130)	③	④	37	76	61	⑤	12	11	100
1996	4	2.2 (2180)	①	70	38	77 ②	52-55	24	18	11	100
	M	3.1 (3130)	③	④	37	76	61	⑤	12	11	100

① Step 1: 41 ft. lbs.
   Step 2: Tighten an additional 45 degrees
   Step 3: Tighten an additional 45 degrees
   Step 4: Long bolts 1, 4-5, 8-9 an additional 20 degrees
   Step 4: Short bolts 2-3, 6-7, 10 an additional 10 degrees
② Center bolt spec shown; Pulley to hub bolts: 37 ft. lbs.
③ 33 ft. lbs. plus 90 degrees
④ 37 ft. lbs. plus 75 degrees
⑤ 115 inch lbs.

79221CM3

SCHEDULED MAINTENANCE INTERVALS
(GM L BODY— CHEVROLET CORSICA & BERETTA)

TO BE SERVICED	TYPE OF SERVICE	VEHICLE MILEAGE INTERVAL (x1000)												
		7.5	15	22.5	30	37.5	45	52.5	60	67.5	75	82.5	90	97.5
Engine oil & filter	R	✓	✓	✓	✓	✓	✓	✓	✓	✓	✓	✓	✓	✓
Automatic transaxle fluid & filter④	S/I	✓	✓	✓	✓	✓	✓	✓	✓	✓	✓	✓	✓	✓
Brake hoses	S/I	✓	✓	✓	✓	✓	✓	✓	✓	✓	✓	✓	✓	✓
Chassis lubrication	S/I	✓	✓	✓	✓	✓	✓	✓	✓	✓	✓	✓	✓	✓
Coolant level, hoses & clamps	S/I	✓	✓	✓	✓	✓	✓	✓	✓	✓	✓	✓	✓	✓
Drive shaft boots & front suspension components	S/I	✓	✓	✓	✓	✓	✓	✓	✓	✓	✓	✓	✓	✓
Exhaust system	S/I	✓	✓	✓	✓	✓	✓	✓	✓	✓	✓	✓	✓	✓
Lubricate suspension, steering linkage, transaxle shift linkage, parking brake cable guides, underbody contact points & linkage	S/I		✓	✓	✓	✓	✓	✓	✓	✓	✓	✓	✓	✓
Manual transaxle oil	S/I	✓	✓	✓	✓	✓	✓	✓	✓	✓	✓	✓	✓	✓
Throttle linkage	S/I	✓	✓	✓	✓	✓	✓	✓	✓	✓	✓	✓	✓	✓
Brake linings	S/I	✓		✓		✓		✓		✓		✓		✓
Rotate tires③	S/I	✓		✓		✓		✓		✓		✓		✓
Air filter element	R				✓				✓				✓	
Engine coolant (1995)②	R				✓				✓				✓	
PCV filter	R				✓				✓				✓	
Spark plugs①	R				✓				✓				✓	
Accessory drive belt(s)	S/I				✓				✓				✓	
EGR & fuel systems	S/I				✓				✓				✓	
Ignition cables	S/I				✓				✓				✓	
Throttle body mount bolt torque	S/I	✓												

① Platinum tip spark plugs - replace every 100,000 miles.
② Engine coolant (1996) - replace every 100,000 miles. Use O.E. specified (DEX-COOL™) coolant only. If any silicate coolant is used, the service interval is every 30,000 miles.
③ Rotate tires front-to-rear only on Beretta GTZ and Z26.
④ Automatic transaxle fluid & filter - change at 100,000 miles (unless changed previously).
R – Replace   S/I – Service or Inspect

79221CM4

SCHEDULED MAINTENANCE INTERVALS
(GM L BODY - CHEVROLET CORSICA & BERETTA Cont.)

FREQUENT OPERATION MAINTENANCE (SEVERE SERVICE)

If a vehicle is operated under any of the following conditions it is considered severe service:
- Extremely dusty areas.
- 50% or more of the vehicle operation is in 32°C (90°F) or higher temperatures, or constant operation in temperatures below 0°C (32°F).
- Prolonged idling (vehicle operation in stop and go traffic).
- Frequent short running periods (engine does not warm to normal operating temperatures).
- Police, taxi, delivery usage or trailer towing usage.

Oil & oil filter – change every 3000 miles.
Chassis lubrication - lubricate every 6000 miles.
Throttle body mount bolt torque - tighten at 6000 miles.
Rotate tires at 6000 miles, then every 12,000 miles. ③
Air filter element - service or inspect every 15,000 miles.
Automatic transaxle fluid & filter - change every 15,000 miles.

79221CM5

SCHEDULED MAINTENANCE
GENERAL MOTORS CORPORATION
L BODY
CHEVROLET CORSICA, BERETTA

The following should be used as a guide when determining the amount of work required for a particular service if taken to a repair shop.
In estimating how long a particular Scheduled Maintenance Service should take, please observe the following:

- **Chilton Time** is time based on field research and data supplied by the vehicle manufacturer.
- Labor time operations are given in hours and tenths of an hour.
- All labor operations, are to be used as a guide.

Mechanic Skill Level Codes:
- **(G)** GENERAL: Normally skilled with certification.
- **(M)** MAINTENANCE: Semi-skilled working on certicication.
- **(P)** PRECISION: Really skilled with multiple certification.

	Chilton Time			Chilton Time			Chilton Time
(G) 7500 Mile Service			**(M) 37500 Mile Service**			**(M) 75000 Mile Service**	
1995-96	2.0		1995-96	1.9		1995-96	1.2
(M) 15000 Mile Service			**(M) 45000 Mile Service**			**(G) 82500 Mile Service**	
1995-96	1.1		1995-96	1.2		1995-96	1.9
(G) 22500 Mile Service			**(G) 52500 Mile Service**			**(G) 90000 Mile Service**	
1995-96	1.9		1995-96	1.9		1995-96	2.9
(G) 30000 Mile Service			**(G) 60000 Mile Service**			Renew engine coolant (1996) add	.5
1995-96	2.9		1995-96	2.9		**(G) 97500 Mile Service**	
Renew engine coolant (1995) add	.5		Renew engine coolant (1995) add	.5		1995-96	1.9
			(G) 67500 Mile Service				
			1995-96	1.9			

79221CM6

GENERAL MOTORS L/N BODY
Chevrolet Malibu • Oldsmobile Cutlass

VEHICLE IDENTIFICATION CHART

Engine Code						Model Year	
Code	Liters	Cu. In. (cc)	Cyl.	Fuel Sys.	Eng. Mfg.	Code	Year
M	3.1	191 (3130)	6	MFI	BOC	V	1997
T	2.4	146 (2392)	4	MFI	CUS	W	1998
						X	1999

BOC - Buick/Oldsmobile/Cadillac

CUS - Chevrolet/United States

MFI - Multi-point Fuel Injection

79221CM7

ENGINE IDENTIFICATION

Year	Model	Engine Displacement Liters (cc)	Engine Series (ID/VIN)	Fuel System	No. of Cylinders	Engine Type
1997	Cutlass	3.1 (3130)	M	MFI	6	OHV
	Malibu	3.1 (3130)	M	MFI	6	OHV
	Malibu	2.4 (2392)	T	MFI	4	DOHC
1998-99	Cutlass	3.1 (3130)	M	MFI	6	OHV
	Malibu	3.1 (3130)	M	MFI	6	OHV
	Malibu	2.4 (2392)	T	MFI	4	DOHC

MFI - Multi-point Fuel Injection

DOHC - Double Overhead Camshafts

OHV - Overhead Valves

79221CM8

GENERAL ENGINE SPECIFICATIONS

Year	Engine ID/VIN	Engine Displacement Liters (cc)	Fuel System Type	Net Horsepower @ rpm	Net Torque @ rpm (ft. lbs.)	Bore x Stroke (in.)	Compression Ratio	Oil Pressure @ rpm
1997	M	3.1 (3130)	MFI	160@5200	185@4000	3.50x3.31	9.5:1	15@1100
	T	2.4 (2392)	MFI	150@6000	150@5600	3.54x3.70	9.5:1	30@3000
1998-99	M	3.1 (3130)	MFI	160@5200	185@4000	3.50x3.31	9.5:1	15@1100
	T	2.4 (2392)	MFI	150@6000	150@5600	3.54x3.70	9.5:1	30@3000

MFI - Multi-point Fuel Injection

79221CM9

GASOLINE ENGINE TUNE-UP SPECIFICATIONS

Year	Engine ID/VIN	Engine Displacement Liters (cc)	Spark Plugs Gap (in.)	Ignition Timing (deg.) MT	Ignition Timing (deg.) AT	Fuel Pump (psi)	Idle Speed (rpm) MT	Idle Speed (rpm) AT	Valve Clearance In.	Valve Clearance Ex.
1997	M	3.1 (3130)	0.060	①	①	41-47	①	①	HYD	HYD
	T	2.4 (2392)	0.035	①	①	41-47	①	①	HYD	HYD
1998-99	M	3.1 (3130)	0.060	①	①	41-47	①	①	HYD	HYD
	T	2.4 (2392)	0.035	①	①	41-47	①	①	HYD	HYD

NOTE: The Vehicle Emission Control Information label often reflects specification changes made during production. The label figures must be used if they differ from those in this chart.

HYD - Hydraulic

① Refer to Vehicle Emission Control Information label

79221CM0

CAPACITIES

Year	Model	Engine ID/VIN	Engine Displacement Liters (cc)	Engine Oil with Filter (qts.)	Transmission (pts.) 4-Spd	Transmission (pts.) 5-Spd	Transmission (pts.) Auto.	Drive Axle Front (pts.)	Drive Axle Rear (pts.)	Fuel Tank (gal.)	Cooling System (qts.)
1997	Cutlass	M	3.1 (3130)	4.0 ①	—	—	②	—	—	15.2	13.1
	Malibu	M	3.1 (3130)	4.0 ①	—	—	②	—	—	15.2	10.8
	Malibu	T	2.4 (2392)	4.0 ①	—	—	②	—	—	15.2	10.4
1998-99	Cutlass	M	3.1 (3130)	4.0 ①	—	—	②	—	—	15.2	13.1
	Malibu	M	3.1 (3130)	4.0 ①	—	—	②	—	—	15.2	10.8
	Malibu	T	2.4 (2392)	4.0 ①	—	—	②	—	—	15.2	10.4

NOTE: All capacities are approximate. Add fluid gradually and ensure a proper fluid level is obtained.

① Capacity is without filter replacement; Additional oil may be required

② 4 speed: 12

79221CN1

VALVE SPECIFICATIONS

Year	Engine ID/VIN	Engine Displacement Liters (cc)	Seat Angle (deg.)	Face Angle (deg.)	Spring Test Pressure (lbs. @ in.)	Spring Installed Height (in.)	Stem-to-Guide Clearance (in.) Intake	Stem-to-Guide Clearance (in.) Exhaust	Stem Diameter (in.) Intake	Stem Diameter (in.) Exhaust
1997	M	3.1 (3130)	45	45	250@1.239	1.710	0.0010-0.0027	0.0010-0.0027	NA	NA
	T	2.4 (2392)	45	46	50-55@1.437	1.437	0.0009-0.0025	0.0016-0.0032	0.2331-0.2339	0.2326-0.2334
1998-99	M	3.1 (3130)	45	45	250@1.239	1.710	0.0010-0.0027	0.0010-0.0027	NA	NA
	T	2.4 (2392)	45	46	50-55@1.437	1.437	0.0009-0.0025	0.0016-0.0032	0.2331-0.2339	0.2326-0.2334

NA - Not Available

79221CN2

TORQUE SPECIFICATIONS
All readings in ft. lbs.

Year	Engine ID/VIN	Engine Displacement Liters (cc)	Cylinder Head Bolts	Main Bearing Bolts	Rod Bearing Bolts	Crankshaft Damper Bolts	Flywheel Bolts	Manifold Intake	Manifold Exhaust	Spark Plugs	Lug Nut
1997	M	3.1 (3130)	⑥	⑦	⑧	76	59	⑨	12	⑩	100
	T	2.4 (2392)	①	②	③	④	⑤	19	31	16	100
1998-99	M	3.1 (3130)	⑥	⑦	⑧	76	59	⑨	12	⑩	100
	T	2.4 (2392)	①	②	③	④	⑤	19	31	16	100

NA - Not Available

① Nos. 1-8: 30 ft. lbs.
  Nos. 9-10: 26 ft. lbs.
  Tighten all bolts an additional 90 degrees
② 15 ft. lbs. plus 90 degrees
③ 18 ft. lbs. plus 80 degrees
④ 129 ft. lbs. plus 90 degrees
⑤ 22 ft. lbs. plus 45 degrees
⑥ Coat threads with sealer torque to 37 ft. lbs.,
  then turn 1/4 turn (90 degrees)
⑦ 37 ft. lbs. plus 75 degrees
⑧ 15 ft. lbs. plus 75 degrees
⑨ Torque all bolts to 15 ft. lbs.,
  Retorque to 24 ft. lbs.
⑩ New cylinder first-time installation: 21 ft. lbs.
  All others: 11 ft. lbs.

79221CN3

SCHEDULED MAINTENANCE INTERVALS
(GM L/N BODY—CHEVROLET MALIBU, OLDSMOBILE CUTLASS)

TO BE SERVICED	TYPE OF SERVICE	VEHICLE MILEAGE INTERVAL (x1000)																	
		7.5	15	22.5	30	37.5	45	52.5	60	67.5	75	82.5	90	97.5	100	105	112.5	120	150
Accessory drive belts	S/I								✓									✓	
Air cleaner element	R				✓				✓				✓					✓	
Cooling system	S/I ①																		✓
Cooling system hoses	S/I																		✓
Engine coolant	R																		✓
Engine oil and filter	R	✓	✓	✓	✓	✓	✓	✓	✓	✓	✓	✓	✓	✓		✓	✓	✓	
Fuel tank, cap and lines	S/I				✓				✓				✓					✓	
Radiator, condenser, pressure cap and neck	C																		✓
Spark plug wires	S/I														✓				
Spark plugs	R														✓				
Tires and wheels	S/I ②	✓	✓	✓	✓	✓	✓	✓	✓	✓	✓	✓	✓	✓		✓	✓	✓	

① Pressure test the cooling system and pressure cap.

② This includes rotating the tires.

R - Replace　　S/I - Inspect and service, if needed　　L - Lubricate　　A - Adjust　　C - Clean

FREQUENT OPERATION MAINTENANCE (SEVERE SERVICE) ADDITIONS

If a vehicle is operated under any of the following conditions it is considered severe service:

- Towing a trailer or using a camper or car-top carrier.
- Repeated short trips of less than 5 miles in temperatures below freezing, or trips of less than 10 miles in any temperature.
- Extensive idling or low-speed driving for long distances as in heavy commercial use, such as delivery, taxi or police cars.
- Operating on rough, muddy or salt-covered roads.
- Operating on unpaved or dusty roads.
- Driving in extremely hot (over 90°) conditions.

Engine oil and filter - change every 3,000 miles or 3 months, whichever occurs first.

Tires and wheels - inspect and rotate every 6,000 miles.

Air cleaner element - inspect every 15,000 miles, and replace or clean as needed. Replace it at least every 30,000 miles.

Transaxle fluid and filter - replace every 50,000 miles.

79221CN4

SCHEDULED MAINTENANCE
GENERAL MOTORS CORPORATION
L/N BODY
CHEVROLET MALIBU, OLDSMOBILE CUTLASS

The following should be used as a guide when determining the amount of work required for a particular service if taken to a repair shop.
In estimating how long a particular Scheduled Maintenance Service should take, please observe the following:

- **Chilton Time** is time based on field research and data supplied by the vehicle manufacturer.
- Labor time operations are given in hours and tenths of an hour.
- All labor operations, are to be used as a guide.

Mechanic Skill Level Codes:
- **(G)** GENERAL: Normally skilled with certification.
- **(M)** MAINTENANCE: Semi-skilled working on certication.
- **(P)** PRECISION: Really skilled with multiple certification.

	Chilton Time
(G) 7500 Mile Service	
1997-99	.9
(M) 15000 Mile Service	
1997-99	.9
(M) 22500 Mile Service	
1997-99	.9
(G) 30000 Mile Service	
1997-99	1.3
(G) 37500 Mile Service	
1997-99	.9
(M) 45000 Mile Service	
1997-99	.9

	Chilton Time
(M) 52500 Mile Service	
1997-99	.9
(G) 60000 Mile Service	
1997-99	1.4
(M) 67500 Mile Service	
1997-99	.9
(M) 75000 Mile Service	
1997-99	.9
(M) 82500 Mile Service	
1997-99	.9
(G) 90000 Mile Service	
1997-99	1.3

	Chilton Time
(M) 97500 Mile Service	
1997-99	.9
(G) 100000 Mile Service	
1997-99	
Replace Spark Plugs	1.0
Replace Spark Plug Wires	1.1
(M) 105000 Mile Service	
1997-99	.9
(M) 112500 Mile Service	
1997-99	.9
(G) 120000 Mile Service	
1997-99	1.4

79221CN5

GENERAL MOTORS N BODY
Buick Skylark • Oldsmobile Achieva • Pontiac Grand Am

VEHICLE IDENTIFICATION CHART

Engine Code						Model Year	
Code	Liters	Cu. In. (cc)	Cyl.	Fuel Sys.	Eng. Mfg.	Code	Year
D	2.3	138 (2261)	4	MFI	BOC	S	1995
M	3.1	191 (3130)	6	MFI	BOC	T	1996
T	2.4	146 (2392)	4	MFI	CUS	V	1997
						W	1998
						X	1999

BOC - Buick/Oldsmobile/Cadillac

CUS - Chevrolet/United States

MFI - Multi-point Fuel Injection

79221CN6

ENGINE IDENTIFICATION

Year	Model	Engine Displacement Liters (cc)	Engine Series (ID/VIN)	Fuel System	No. of Cylinders	Engine Type
1995	Achieva	2.3 (2261)	D	MFI	4	DOHC
	Achieva	3.1 (3130)	M	MFI	6	OHV
	Grand Am	2.3 (2261)	D	MFI	4	DOHC
	Grand Am	3.1 (3130)	M	MFI	6	OHV
	Skylark	2.3 (2261)	D	MFI	4	DOHC
	Skylark	3.1 (3130)	M	MFI	6	OHV
1996	Achieva	3.1 (3130)	M	MFI	6	OHV
	Achieva	2.4 (2392)	T	MFI	4	DOHC
	Grand Am	3.1 (3130)	M	MFI	6	OHV
	Grand Am	2.4 (2392)	T	MFI	4	DOHC
	Skylark	3.1 (3130)	M	MFI	6	OHV
	Skylark	2.4 (2392)	T	MFI	4	DOHC
1997	Achieva	3.1 (3130)	M	MFI	6	OHV
	Achieva	2.4 (2392)	T	MFI	4	DOHC
	Grand Am	3.1 (3130)	M	MFI	6	OHV
	Grand Am	2.4 (2392)	T	MFI	4	DOHC
	Skylark	3.1 (3130)	M	MFI	6	OHV
	Skylark	2.4 (2392)	T	MFI	4	DOHC
1998-99	Achieva	3.1 (3130)	M	MFI	6	OHV
	Achieva	2.4 (2392)	T	MFI	4	DOHC
	Grand Am	3.1 (3130)	M	MFI	6	OHV
	Grand Am	2.4 (2392)	T	MFI	4	DOHC

MFI - Multi-point Fuel Injection

DOHC - Double Overhead Camshafts

OHV - Overhead Valves

79221CN7

GENERAL ENGINE SPECIFICATIONS

Year	Engine ID/VIN	Engine Displacement Liters (cc)	Fuel System Type	Net Horsepower @ rpm	Net Torque @ rpm (ft. lbs.)	Bore x Stroke (in.)	Compression Ratio	Oil Pressure @ rpm
1995	D	2.3 (2261)	MFI	155@6000	145@4800	3.62x3.35	9.5:1	30@2000
	M	3.1 (3130)	MFI	160@5200	185@4000	3.50x3.31	9.5:1	15@1100
1996	M	3.1 (3130)	MFI	160@5200	185@4000	3.50x3.31	9.5:1	15@1100
	T	2.4 (2392)	MFI	150@6000	150@4400	3.54x3.70	9.5:1	30@3000
1997	M	3.1 (3130)	MFI	160@5200	185@4000	3.50x3.31	9.5:1	15@1100
	T	2.4 (2392)	MFI	150@6000	150@4400	3.54x3.70	9.5:1	30@3000
1998-99	M	3.1 (3130)	MFI	160@5200	185@4000	3.50x3.31	9.5:1	15@1100
	T	2.4 (2392)	MFI	150@6000	150@4400	3.54x3.70	9.5:1	30@3000

MFI - Multi-point Fuel Injection

79221CN8

GASOLINE ENGINE TUNE-UP SPECIFICATIONS

Year	Engine ID/VIN	Engine Displacement Liters (cc)	Spark Plugs Gap (in.)	Ignition Timing (deg.) MT	Ignition Timing (deg.) AT	Fuel Pump (psi)	Idle Speed (rpm) MT	Idle Speed (rpm) AT	Valve Clearance In.	Valve Clearance Ex.
1995	D	2.3 (2262)	0.035	①	①	41-47	②	②	HYD	HYD
	M	3.1 (3130)	0.060	①	①	41-47	②	②	HYD	HYD
1996	M	3.1 (3130)	0.060	③	③	41-47	③	③	HYD	HYD
	T	2.4 (2392)	0.035	③	③	41-47	③	③	HYD	HYD
1997	M	3.1 (3130)	0.060	③	③	41-47	③	③	HYD	HYD
	T	2.4 (2392)	0.035	③	③	41-47	③	③	HYD	HYD
1998-99	M	3.1 (3130)	0.060	③	③	41-47	③	③	HYD	HYD
	T	2.4 (2392)	0.035	③	③	41-47	③	③	HYD	HYD

NOTE: The Vehicle Emission Control Information label often reflects specification changes made during production. The label figures must be used if they differ from those in this chart.

HYD - Hydraulic

① DIS Ignition System timing is not adjustable

② Idle speed is maintained by the ECM. There is no recommended adjustment procedure

③ Refer to Vehicle Emission Control Information label

79221CN9

CAPACITIES

Year	Model	Engine ID/VIN	Engine Displacement Liters (cc)	Engine Oil with Filter (qts.)		Transmission (pts.)			Drive Axle		Fuel Tank (gal.)	Cooling System (qts.)
						4-Spd	5-Spd	Auto.	Front (pts.)	Rear (pts.)		
1995	Achieva	D	2.3 (2261)	4.0	①	—	4.0	②	—	—	15.2	10.4
	Achieva	M	3.1 (3130)	3.8	①	—	4.0	②	—	—	15.2	10.8
	Grand Am	D	2.3 (2261)	4.0	①	—	4.0	8.0 ③	—	—	15.2	10.4
	Grand Am	M	3.1 (3130)	3.8	①	—	4.0	8.0 ③	—	—	15.2	13.1
	Skylark	D	2.3 (2261)	4.0	①	—	—	8.0 ③	—	—	15.2	10.4
	Skylark	M	3.1 (3130)	4.0	①	—	—	8.0 ③	—	—	15.2	13.1
1996	Achieva	M	3.1 (3130)	4.0	①	—	4.0	②	—	—	15.2	10.8
	Achieva	T	2.4 (2392)	4.0	①	—	4.0	②	—	—	15.2	10.4
	Grand Am	M	3.1 (3130)	4.5		—	4.0	8.0 ③	—	—	15.2	13.1
	Grand Am	T	2.4 (2392)	4.5		—	4.0	— ③	—	—	15.2	10.4
	Skylark	M	3.1 (3130)	4.0	①	—	12.0	—	—	—	15.2	13.1
	Skylark	T	2.4 (2392)	4.0	①	—	12.0	—	—	—	15.2	10.4
1997	Achieva	M	3.1 (3130)	4.0	①	—	4.0	②	—	—	15.2	10.8
	Achieva	T	2.4 (2392)	4.0	①	—	4.0	②	—	—	15.2	10.4
	Grand Am	M	3.1 (3130)	4.5		—	4.0	8.0 ③	—	—	15.2	13.1
	Grand Am	T	2.4 (2392)	4.5		—	4.0	— ③	—	—	15.2	10.4
	Skylark	M	3.1 (3130)	4.0	①	—	12.0	—	—	—	15.2	13.1
	Skylark	T	2.4 (2392)	4.0	①	—	12.0	—	—	—	15.2	10.4
1998-99	Achieva	M	3.1 (3130)	4.0	①	—	4.0	④	—	—	15.2	10.8
	Achieva	T	2.4 (2392)	4.0	①	—	4.0	②	—	—	15.2	10.4
	Grand Am	M	3.1 (3130)	4.5		—	4.0	8.0 ③	—	—	15.2	13.1
	Grand Am	T	2.4 (2392)	4.5		—	4.0	— ③	—	—	15.2	10.4

NOTE: All capacities are approximate. Add fluid gradually and ensure a proper fluid level is obtained.

① Capacity is without filter replacement; Additional oil may be required

② 3 speed: 8.0
4 speed: 12

③ With 4T60E transaxle: 12.0 pts.

④ 3 speed: 8.0
4 speed: 12

79221CN0

VALVE SPECIFICATIONS

Year	Engine ID/VIN	Engine Displacement Liters (cc)	Seat Angle (deg.)	Face Angle (deg.)	Spring Test Pressure (lbs. @ in.)	Spring Installed Height (in.)	Stem-to-Guide Clearance (in.)		Stem Diameter (in.)	
							Intake	Exhaust	Intake	Exhaust
1995	D	2.3 (2261)	45	①	193-207@ 1.043	0.984- 1.004 ②	0.0010- 0.0027	0.0015- 0.0032	0.2751- 0.2745	0.2740- 0.2747
	M	3.1 (3130)	45	45	250@ 1.239	1.710	0.0010- 0.0027	0.0010- 0.0027	NA	NA
1996	M	3.1 (3130)	45	45	250@1.239	1.710	0.0010- 0.0027	0.0010- 0.0027	NA	NA
	T	2.4 (2392)	45	46	50-55@ 1.437	1.437	0.0009- 0.0025	0.0016- 0.0032	0.2331- 0.2339	0.2326- 0.2334
1997	M	3.1 (3130)	45	45	250@1.239	1.710	0.0010- 0.0027	0.0010- 0.0027	NA	NA
	T	2.4 (2392)	45	46	50-55@ 1.437	1.437	0.0009- 0.0025	0.0016- 0.0032	0.2331- 0.2339	0.2326- 0.2334
1998-99	M	3.1 (3130)	45	45	250@1.239	1.710	0.0010- 0.0027	0.0010- 0.0027	NA	NA
	T	2.4 (2392)	45	46	50-55@ 1.437	1.437	0.0009- 0.0025	0.0016- 0.0032	0.2331- 0.2339	0.2326- 0.2334

NA - Not Available
① Intake face angle: 44 degrees
  Exhaust face angle: 44.5 degrees
② Measured from top of valve stem to top of camshaft housing

79221CP1

TORQUE SPECIFICATIONS
All readings in ft. lbs.

Year	Engine ID/VIN	Engine Displacement Liters (cc)	Cylinder Head Bolts	Main Bearing Bolts	Rod Bearing Bolts	Crankshaft Damper Bolts	Flywheel Bolts	Manifold		Spark Plugs	Lug Nut
								Intake	Exhaust		
1995	D	2.3 (2261)	①	②	③	④	⑤	19	31	16	100
	M	3.1 (3130)	⑥	⑦	⑧	76	59	⑨	12	⑩	100
1996	M	3.1 (3130)	⑥	⑦	⑧	76	59	⑨	12	⑩	100
	T	2.4 (2392)	①	②	③	④	⑤	19	31	16	100
1997	M	3.1 (3130)	⑥	⑦	⑧	76	59	⑨	12	⑩	100
	T	2.4 (2392)	①	②	③	④	⑤	19	31	16	100
1998-99	M	3.1 (3130)	⑥	⑦	⑧	76	59	⑨	12	⑩	100
	T	2.4 (2392)	①	⑩	③	④	⑤	19	31	16	100

① Nos. 1-8: 30 ft. lbs.
  Nos. 9-10: 26 ft. lbs.
  Tighten all bolts an additional 90 degrees
② 15 ft. lbs. plus 90 degrees
③ 18 ft. lbs. plus 80 degrees
④ 129 ft. lbs. plus 90 degrees
⑤ 22 ft. lbs. plus 45 degrees
⑥ Coat threads with sealer torque to 37 ft. lbs.,
  then turn 1/4 turn (90 degrees)
⑦ 37 ft. lbs. plus 75 degrees
⑧ 15 ft. lbs. plus 75 degrees
⑨ Torque all bolts to 15 ft. lbs.,
  Retorque to 24 ft. lbs.
⑩ New cylinder first-time installation: 21 ft. lbs.
  All others: 11 ft. lbs.

79221CP2

SCHEDULED MAINTENANCE INTERVALS
(GM N BODY - BUICK SKYLARK, OLDSMOBILE ACHIEVA, PONTIAC GRAND AM)

TO BE SERVICED	TYPE OF SERVICE	VEHICLE MILEAGE INTERVAL (x1000)												
		7.5	15	22.5	30	37.5	45	52.5	60	67.5	75	82.5	90	97.5
Engine oil & filter	R	✓	✓	✓	✓	✓	✓	✓	✓	✓	✓	✓	✓	✓
Automatic transaxle fluid & filter③	S/I	✓	✓	✓	✓	✓	✓	✓	✓	✓	✓	✓	✓	✓
Brake hoses	S/I	✓	✓	✓	✓	✓	✓	✓	✓	✓	✓	✓	✓	✓
Chassis lubrication	S/I	✓	✓	✓	✓	✓	✓	✓	✓	✓	✓	✓	✓	✓
Coolant level, hoses & clamps	S/I	✓	✓	✓	✓	✓	✓	✓	✓	✓	✓	✓	✓	✓
Drive shaft boots & front suspension components	S/I	✓	✓	✓	✓	✓	✓	✓	✓	✓	✓	✓	✓	✓
Exhaust system	S/I	✓	✓	✓	✓	✓	✓	✓	✓	✓	✓	✓	✓	✓
Lubricate suspension, steering linkage, transaxle shift linkage, parking brake cable guides, underbody contact points & linkage	S/I	✓	✓	✓	✓	✓	✓	✓	✓	✓	✓	✓	✓	✓
Manual transaxle oil	S/I	✓	✓	✓	✓	✓	✓	✓	✓	✓	✓	✓	✓	✓
Throttle linkage	S/I	✓	✓	✓	✓	✓	✓	✓	✓	✓	✓	✓	✓	✓
Brake linings	S/I	✓		✓		✓		✓		✓		✓		✓
Rotate tires	S/I	✓		✓		✓		✓		✓		✓		✓
Air filter element & PCV filter	R				✓				✓				✓	
Engine coolant (1995)②	R				✓				✓				✓	
Spark plugs①	R				✓				✓				✓	
Accessory drive belt(s)	S/I				✓				✓				✓	
EGR & fuel systems	S/I				✓				✓				✓	
Ignition cables	S/I				✓				✓				✓	
Throttle body mount bolt torque	S/I	✓												

① Platinum tip spark plugs - replace every 100,000 miles.
② Engine coolant (1996-99) - replace every 100,000 miles. Use O.E. specified (DEX-COOL™) coolant only. If any silicate coolant is used, the service interval is every 30,000 miles.
③ Automatic transaxle fluid & filter - replace every 100,000 miles (if not changed previously).

R – Replace   S/I – Service or Inspect

79221CP3

SCHEDULED MAINTENANCE INTERVALS
(GM N BODY - BUICK SKYLARK, OLDSMOBILE ACHIEVA, PONTIAC GRAND AM Cont.)

FREQUENT OPERATION MAINTENANCE (SEVERE SERVICE)

If a vehicle is operated under any of the following conditions it is considered severe service:

- Extremely dusty areas.
- 50% or more of the vehicle operation is in 32°C (90°F) or higher temperatures, or constant operation in temperatures below 0°C (32°F).
- Prolonged idling (vehicle operation in stop and go traffic).
- Frequent short running periods (engine does not warm to normal operating temperatures).
- Police, taxi, delivery usage or trailer towing usage.

Oil & oil filter – change every 3000 miles.
Throttle body mount bolt torque - tighten at 6000 miles
Rotate tires at 6000 miles, & then every 12,000 miles.
Chassis lubrication - lubricate every 6000 miles.
Automatic transaxle fluid - change every 15,000 miles.
Air filter element - service or inspect every 15,000 miles.

79221CP4

SCHEDULED MAINTENANCE
GENERAL MOTORS CORPORATION
N BODY
BUICK SKYLARK, OLDSMOBILE ACHIEVA
PONTIAC GRAND AM

The following should be used as a guide when determining the amount of work required for a particular service if taken to a repair shop.
In estimating how long a particular Scheduled Maintenance Service should take, please observe the following:

- **Chilton Time** is time based on field research and data supplied by the vehicle manufacturer.
- Labor time operations are given in hours and tenths of an hour.
- All labor operations, are to be used as a guide.

Mechanic Skill Level Codes:
  (G) GENERAL: Normally skilled with certification.
  (M) MAINTENANCE: Semi-skilled working on certication.
  (P) PRECISION: Really skilled with multiple certification.

	Chilton Time		Chilton Time		Chilton Time
(G) 7500 Mile Service		**(G) 37500 Mile Service**		**(M) 75000 Mile Service**	
1995-99	2.1	1995-99	1.9	1995-99	1.2
(M) 15000 Mile Service		**(M) 45000 Mile Service**		**(G) 82500 Mile Service**	
1995-99	1.2	1995-99	1.3	1995-99	1.9
(G) 22500 Mile Service		**(G) 52500 Mile Service**		**(G) 90000 Mile Service**	
1995-99	1.9	1995-99	1.9	1995-99	2.9
(G) 30000 Mile Service		**(G) 60000 Mile Service**		Renew engine coolant (1995) add	.5
1995-99	2.9	1995-99	2.9	**(G) 97500 Mile Service**	
Renew engine coolant (1995) add	.5	Renew engine coolant (1995) add	.5	1995-99	1.9
		(G) 67500 Mile Service			
		1995-99	1.9		

79221CP5

GENERAL MOTORS V BODY
Cadillac Catera

VEHICLE IDENTIFICATION CHART

Engine Code						Model Year	
Code	Liters	Cu. In. (cc)	Cyl.	Fuel Sys.	Eng. Mfg.	Code	Year
R	3.0	181 (2972)	6	MFI	Opel	V	1997
						W	1998
						X	1999

MFI - Multi-point Fuel Injection

79221CP6

ENGINE IDENTIFICATION

Year	Model	Engine Displacement Liters (cc)	Engine Series (ID/VIN)	Fuel System	No. of Cylinders	Engine Type
1997	Catera	3.0 (2972)	R	MFI	6	DOHC
1998-99	Catera	3.0 (2972)	R	MFI	6	DOHC

MFI - Multi-point Fuel Injection
DOHC - Double Overhead Camshafts

79221CP7

GENERAL ENGINE SPECIFICATIONS

Year	Engine ID/VIN	Engine Displacement Liters (cc)	Fuel System Type	Net Horsepower @ rpm	Net Torque @ rpm (ft. lbs.)	Bore x Stroke (in.)	Compression Ratio	Oil Pressure @ rpm
1997	R	3.0 (2972)	MFI	200@6000	192@3600	3.38x3.34	10.0:1	22@900
1998-99	R	3.0 (2972)	MFI	200@6000	192@3600	3.38x3.34	10.0:1	22@900

MFI - Multi-point Fuel Injection

79221CP8

GASOLINE ENGINE TUNE-UP SPECIFICATIONS

Year	Engine ID/VIN	Engine Displacement Liters (cc)	Spark Plugs Gap (in.)	Ignition Timing (deg.) MT	Ignition Timing (deg.) AT	Fuel Pump (psi)	Idle Speed (rpm) MT	Idle Speed (rpm) AT	Valve Clearance In.	Valve Clearance Ex.
1997	R	3.0 (2972)	0.034-0.043	—	①	41-47	—	①	HYD	HYD
1998-99	R	3.0 (2972)	0.034-0.043	—	①	41-47	—	①	HYD	HYD

NOTE: The Vehicle Emission Control Information label often reflects specification changes made during production. The label figures must be used if they differ from those in this chart.
HYD - Hydraulic
① Refer to Vehicle Emission Control Information label

79221CP9

CAPACITIES

Year	Model	Engine ID/VIN	Engine Displacement Liters (cc)	Engine Oil with Filter (qts.)	Transmission (pts.) 4-Spd	Transmission (pts.) 5-Spd	Transmission (pts.) Auto.	Drive Axle Front (pts.)	Drive Axle Rear (pts.)	Fuel Tank (gal.)	Cooling System (qts.)
1997	Catera	R	3.0 (2972)	6.0	—	—	14	—	3.5	18	10
1998-99	Catera	R	3.0 (2972)	6.0	—	—	14	—	3.5	18	10

NOTE: All capacities are approximate. Add fluid gradually and ensure a proper fluid level is obtained.

79221CP0

VALVE SPECIFICATIONS

Year	Engine ID/VIN	Engine Displacement Liters (cc)	Seat Angle (deg.)	Face Angle (deg.)	Spring Test Pressure (lbs. @ in.)	Spring Installed Height (in.)	Stem-to-Guide Clearance (in.) Intake	Stem-to-Guide Clearance (in.) Exhaust	Stem Diameter (in.) Intake	Stem Diameter (in.) Exhaust
1997	R	3.0 (2972)	①	45	56.6@1.338	1.543	0.0012-0.0022	0.0016-0.0026	0.2344-0.2350	0.2341-0.2346
1998-99	R	3.0 (2972)	①	45	56.6@1.338	1.543	0.0012-0.0022	0.0016-0.0026	0.2344-0.2350	0.2341-0.2346

① 45 degrees 20 minutes

79221CQ1

TORQUE SPECIFICATIONS
All readings in ft. lbs.

Year	Engine ID/VIN	Engine Displacement Liters (cc)	Cylinder Head Bolts	Main Bearing Bolts	Rod Bearing Bolts	Crankshaft Damper Bolts	Flywheel Bolts	Manifold Intake	Manifold Exhaust	Spark Plugs	Lug Nut
1997	R	3.0 (2972)	①	②	③	④	⑤	15	15	19	100
1998-99	R	3.0 (2972)	①	②	③	④	⑤	15	15	19	100

① Step 1: 18 ft. lbs.
  Step 2: Rotate 90 degrees
  Step 3: Rotate 90 degrees
  Step 4: Rotate 90 degrees
  Step 5: Rotate 15 degrees
② Step 1: 37 ft. lbs.
  Step 2: Rotate 60 degrees
  Step 3: Rotate 15 degrees
③ Step 1: 26 ft. lbs.
  Step 2: Rotate 45 degrees
  Step 3: Rotate 15 degrees
④ Step 1: 184 ft. lbs.
  Step 2: Rotate 45 degrees
  Step 3: Rotate 15 degrees
  Step 4: Harmonic balancer 15 ft. lbs.
⑤ Step 1: 48 ft. lbs.
  Step 2: Rotate 30 degrees
  Step 3: Rotate 15 degrees

79221CQ2

GENERAL MOTORS W BODY
Buick Regal • Century (1997-99) • Chevrolet Lumina • Monte Carlo
Oldsmobile Cutlass Supreme • Intrigue • Pontiac Grand Prix

VEHICLE IDENTIFICATION CHART

Engine Code							Model Year	
Code	Liters	Cu. In. (cc)	Cyl.	Fuel Sys.	Eng. Mfg.		Code	Year
1	3.8	231 (3785)	6	MFI	BOC		S	1995
K	3.8	231 (3785)	6	MFI	CPC		T	1996
L	3.8	231 (3785)	6	MFI	BOC		V	1997
M	3.1	191 (3130)	6	MFI	BOC		W	1998
X	3.4	207 (3393)	6	MFI	CPC		X	1999

MFI - Multi-point Fuel Injection
BOC - Buick/Oldsmobile/Cadillac
CPC - Chevrolet/Pontiac/Canada

79221C03

ENGINE IDENTIFICATION

Year	Model	Engine Displacement Liters (cc)	Engine Series (ID/VIN)	Fuel System	No. of Cylinders	Engine Type
1995	Cutlass Supreme	3.1 (3130)	M	MFI	6	OHV
	Cutlass Supreme	3.4 (3393)	X	MFI	6	DOHC
	Grand Prix	3.1 (3130)	M	MFI	6	OHV
	Grand Prix	3.4 (3393)	X	MFI	6	DOHC
	Lumina	3.1 (3130)	M	MFI	6	OHV
	Lumina	3.4 (3393)	X	MFI	6	DOHC
	Monte Carlo	3.1 (3130)	M	MFI	6	OHV
	Monte Carlo	3.4 (3393)	X	MFI	6	DOHC
	Regal	3.8 (3785)	L	MFI	6	OHV
	Regal	3.1 (3130)	M	MFI	6	OHV
1996	Cutlass Supreme	3.1 (3130)	M	MFI	6	OHV
	Cutlass Supreme	3.4 (3393)	X	MFI	6	DOHC
	Grand Prix	3.1 (3130)	M	MFI	6	OHV
	Grand Prix	3.4 (3393)	X	MFI	6	DOHC
	Lumina	3.1 (3130)	M	MFI	6	OHV
	Lumina	3.4 (3393)	X	MFI	6	DOHC
	Monte Carlo	3.1 (3130)	M	MFI	6	OHV
	Monte Carlo	3.4 (3393)	X	MFI	6	DOHC
	Regal	3.8 (3785)	K	MFI	6	OHV
	Regal	3.1 (3130)	M	MFI	6	OHV
1997	Cutlass Supreme	3.1 (3130)	M	MFI	6	OHV
	Grand Prix	3.8 (3785)	1	MFI	6	OHV
	Grand Prix	3.8 (3785)	K	MFI	6	OHV
	Grand Prix	3.1 (3130)	M	MFI	6	OHV
	Lumina	3.1 (3130)	M	MFI	6	OHV
	Lumina	3.4 (3393)	X	MFI	6	DOHC
	Monte Carlo	3.1 (3130)	M	MFI	6	OHV
	Monte Carlo	3.4 (3393)	X	MFI	6	DOHC
	Regal	3.8 (3785)	1	MFI	6	OHV
	Regal	3.8 (3785)	K	MFI	6	OHV
1998-99	Century	3.1 (3130)	M	MFI	6	OHV
	Grand Prix	3.8 (3785)	1	MFI	6	OHV
	Grand Prix	3.8 (3785)	K	MFI	6	OHV
	Grand Prix	3.1 (3130)	M	MFI	6	OHV
	Intrigue	3.8 (3785)	K	MFI	6	OHV
	Lumina	3.8 (3785)	K	MFI	6	OHV
	Lumina	3.1 (3130)	M	MFI	6	OHV
	Monte Carlo	3.8 (3785)	K	MFI	6	OHV
	Monte Carlo	3.1 (3130)	M	MFI	6	OHV
	Regal	3.8 (3785)	1	MFI	6	OHV
	Regal	3.8 (3785)	K	MFI	6	OHV
	Regal	3.1 (3130)	M	MFI	6	OHV

MFI - Multi-point Fuel Injection
OHV - Overhead Valves
DOHC - Double Overhead Camshafts

79221CQ4

GENERAL ENGINE SPECIFICATIONS

Year	Engine ID/VIN	Engine Displacement Liters (cc)	Fuel System Type	Net Horsepower @ rpm	Net Torque @ rpm (ft. lbs.)	Bore x Stroke (in.)	Compression Ratio	Oil Pressure @ rpm
1995	L	3.8 (3785)	MFI	170@4800	225@3200	3.80x3.40	9.0:1	60@1850
	M	3.1 (3130)	MFI	160@5200	185@4000	3.50x3.31	9.5:1	15@1100
	X	3.4 (3393)	MFI	210@5200	215@4000	3.62x3.31	9.25:1	15@1100
1996	K	3.8 (3785)	MFI	205@5200	230@4000	3.80x3.40	9.4:1	60@1850
	M	3.1 (3130)	MFI	160@5200	185@4000	3.50x3.31	9.5:1	15@1100
	X	3.4 (3393)	MFI	210@5200	215@4000	3.62x3.31	9.25:1	15@1100
1997	1	3.8 (3785)	MFI	240@5200	280@3200	3.80x3.40	9.0:1	60@1850
	K	3.8 (3785)	MFI	205@5200	230@4000	3.80x3.40	9.4:1	60@1850
	M	3.1 (3130)	MFI	160@5200	185@4000	3.50x3.31	9.5:1	15@1100
	X	3.4 (3393)	MFI	210@5200	215@4000	3.62x3.31	9.25:1	15@1100
1998-99	1	3.8 (3785)	MFI	240@5200	280@3200	3.80x3.40	9.0:1	60@1850
	K	3.8 (3785)	MFI	205@5200	230@4000	3.80x3.40	9.4:1	60@1850
	M	3.1 (3130)	MFI	160@5200	185@4000	3.50x3.31	9.5:1	15@1100

MFI - Multi-point Fuel Injection

79221CQ5

GASOLINE ENGINE TUNE-UP SPECIFICATIONS

Year	Engine ID/VIN	Engine Displacement Liters (cc)	Spark Plugs Gap (in.)	Ignition Timing (deg.) MT	Ignition Timing (deg.) AT	Fuel Pump (psi)	Idle Speed (rpm) MT	Idle Speed (rpm) AT	Valve Clearance In.	Valve Clearance Ex.
1995	L	3.8 (3785)	0.060	—	①	40-47	—	②	HYD	HYD
	M	3.1 (3130)	0.060	—	①	41-47	—	②	HYD	HYD
	X	3.4 (3393)	0.045	—	①	41-47	—	②	HYD	HYD
1996	K	3.8 (3785)	0.060	—	①	41-47	—	②	HYD	HYD
	M	3.1 (3130)	0.060	—	①	41-47	—	②	HYD	HYD
	X	3.4 (3393)	0.045	—	①	41-47	—	②	HYD	HYD
1997	1	3.8 (3785)	0.060	—	①	41-47	—	②	HYD	HYD
	K	3.8 (3785)	0.060	—	①	41-47	—	②	HYD	HYD
	M	3.1 (3130)	0.060	—	①	41-47	—	②	HYD	HYD
	X	3.4 (3393)	0.045	—	①	41-47	—	②	HYD	HYD
1998-99	1	3.8 (3785)	0.060	—	①	41-47	—	②	HYD	HYD
	K	3.8 (3785)	0.060	—	①	41-47	—	②	HYD	HYD
	M	3.1 (3130)	0.060	—	①	41-47	—	②	HYD	HYD

NOTE: The Vehicle Emission Control Information label often reflects specification changes made during production. The label figures must be used if they differ from those in this chart.
HYD - Hydraulic
① DIS Ignition System timing is not adjustable
② Idle speed is maintained by the ECM. There is no recommended adjustment procedure

79221CQ6

CAPACITIES

Year	Model	Engine ID/VIN	Engine Displacement Liters (cc)	Engine Oil with Filter (qts.)	Transmission (pts.)			Drive Axle		Fuel Tank (gal.)	Cooling System (qts.)
					4-Spd	5-Spd	Auto.	Front (pts.)	Rear (pts.)		
1995	Cutlass Supreme	M	3.1 (3130)	4.5	—	—	②	—	—	16.0	12.6
	Cutlass Supreme	X	3.4 (3393)	5.0	—	—	②	—	—	16.5	12.7
	Grand Prix	M	3.1 (3130)	4.5	—	—	②	—	—	16.0	12.6
	Grand Prix	X	3.4 (3393)	5.0	—	—	②	—	—	16.5	12.7
	Lumina	M	3.1 (3130)	4.5	—	—	②	—	—	16.0	12.6
	Lumina	X	3.4 (3393)	5.0	—	—	②	—	—	16.5	12.7
	Monte Carlo	M	3.1 (3130)	4.5	—	—	②	—	—	16.5	12.6
	Monte Carlo	X	3.4 (3393)	5.0	—	—	②	—	—	16.5	12.7
	Regal	L	3.8 (3785)	4.0 ①	—	—	12.0	—	—	16.5	11.1
	Regal	M	3.1 (3130)	4.0 ①	—	—	12.0	—	—	16.5	11.8
1996	Cutlass Supreme	M	3.1 (3130)	4.5	—	—	②	—	—	16.0	12.6
	Cutlass Supreme	X	3.4 (3393)	5.0	—	—	②	—	—	16.5	12.7
	Grand Prix	M	3.1 (3130)	4.5	—	—	②	—	—	16.0	12.6
	Grand Prix	X	3.4 (3393)	5.0	—	—	②	—	—	16.5	12.7
	Lumina	M	3.1 (3130)	4.5	—	—	②	—	—	16.0	12.6
	Lumina	X	3.4 (3393)	5.0	—	—	②	—	—	16.5	12.7
	Monte Carlo	M	3.1 (3130)	4.5	—	—	②	—	—	16.5	12.6
	Monte Carlo	X	3.4 (3393)	5.0	—	—	②	—	—	16.5	12.7
	Regal	K	3.8 (3785)	4.0 ①	—	—	②	—	—	17.1	11.1
	Regal	M	3.1 (3130)	4.0 ①	—	—	②	—	—	17.1	12.5
1997	Cutlass Supreme	M	3.1 (3130)	4.5	—	—	②	—	—	16.0	12.6
	Grand Prix	1	3.8 (3785)	5.0	—	—	②	—	—	16.5	12.7
	Grand Prix	K	3.8 (3785)	5.0	—	—	②	—	—	16.5	12.7
	Grand Prix	M	3.1 (3130)	4.5	—	—	②	—	—	16.0	12.6
	Lumina	M	3.1 (3130)	4.5	—	—	②	—	—	16.0	12.6
	Lumina	X	3.4 (3393)	5.0	—	—	②	—	—	16.5	12.7
	Monte Carlo	M	3.1 (3130)	4.5	—	—	②	—	—	16.5	12.6
	Monte Carlo	X	3.4 (3393)	5.0	—	—	②	—	—	16.5	12.7
	Regal	1	3.8 (3785)	5.0	—	—	②	—	—	16.5	12.7
	Regal	K	3.8 (3785)	5.0	—	—	②	—	—	16.5	12.7
1998-99	Century	M	3.1 (3130)	4.5	—	—	②	—	—	16.0	12.6
	Grand Prix	1	3.8 (3785)	5.0	—	—	②	—	—	16.5	12.7
	Grand Prix	K	3.8 (3785)	5.0	—	—	②	—	—	16.5	12.7
	Grand Prix	M	3.1 (3130)	4.5	—	—	②	—	—	16.0	12.6
	Intrigue	K	3.8 (3785)	5.0	—	—	②	—	—	16.5	12.7
	Lumina	K	3.8 (3785)	5.0	—	—	②	—	—	16.5	12.7
	Lumina	M	3.1 (3130)	4.5	—	—	②	—	—	16.0	12.6
	Monte Carlo	K	3.8 (3785)	5.0	—	—	②	—	—	16.5	12.7
	Monte Carlo	M	3.1 (3130)	4.5	—	—	②	—	—	16.5	12.6
	Regal	1	3.8 (3785)	5.0	—	—	②	—	—	16.5	12.7
	Regal	K	3.8 (3785)	5.0	—	—	②	—	—	17.1	11.1
	Regal	M	3.1 (3130)	4.5	—	—	②	—	—	17.1	12.5

NOTE: All capacities are approximate. Add fluid gradually and ensure a proper fluid is obtained.

① Capacity is without filter replacement; Additional oil may be required

② 3T40 trans.: 8.0 pts.
4T60 trans.: 12.0 pts.
4T60E trans.: 14.8 pts.

79221CQ7

VALVE SPECIFICATIONS

Year	Engine ID/VIN	Engine Displacement Liters (cc)	Seat Angle (deg.)	Face Angle (deg.)	Spring Test Pressure (lbs. @ in.)	Spring Installed Height (in.)	Stem-to-Guide Clearance (in.)		Stem Diameter (in.)	
							Intake	Exhaust	Intake	Exhaust
1995	L	3.8 (3785)	45	45	210@1.315	1.690-1.720	0.0015-0.0035	0.0015-0.0032	NA	NA
	M	3.1 (3130)	45	45	250@1.239	1.710	0.0001-0.0027	0.0010-0.0027	NA	NA
	X	3.4 (3393)	46	45	75@1.40	1.40	0.0011-0.0026	0.0014-0.0031	NA	NA
1996	K	3.8 (3785)	45	45	210@1.32	1.69-1.72	0.0015-0.0035	0.0015-0.0032	NA	NA
	M	3.1 (3130)	45	45	250@1.239	1.710	0.0001-0.0027	0.0010-0.0027	NA	NA
	X	3.4 (3393)	46	45	75@1.40	1.40	0.0011-0.0026	0.0014-0.0031	NA	NA
1997	1	3.8 (3785)	45	45	80@1.750	1.690-1.720	0.0015-0.0032	0.0015-0.0032	NA	NA
	K	3.8 (3785)	45	45	210@1.32	1.69-1.72	0.0015-0.0035	0.0015-0.0032	NA	NA
	M	3.1 (3130)	45	45	250@1.239	1.710	0.0001-0.0027	0.0010-0.0027	NA	NA
	X	3.4 (3393)	46	45	75@1.40	1.40	0.0011-0.0026	0.0014-0.0031	NA	NA
1998-99	1	3.8 (3785)	45	45	80@1.750	1.690-1.720	0.0015-0.0032	0.0015-0.0032	NA	NA
	K	3.8 (3785)	45	45	210@1.32	1.69-1.72	0.0015-0.0035	0.0015-0.0032	NA	NA
	M	3.1 (3130)	45	45	250@1.239	1.710	0.0001-0.0027	0.0010-0.0027	NA	NA

NA - Not Available

79221CQ8

TORQUE SPECIFICATIONS
All readings in ft. lbs.

Year	Engine ID/VIN	Engine Displacement Liters (cc)	Cylinder Head Bolts	Main Bearing Bolts	Rod Bearing Bolts	Crankshaft Damper Bolts	Flywheel Bolts	Manifold Intake	Manifold Exhaust	Spark Plugs	Lug Nut
1995	L	3.8 (3785)	⑧	⑨	⑩	⑪	⑫	⑩	38	11	100
	M	3.1 (3130)	①	②	③	76	61	④	10	⑤	100
	X	3.4 (3393)	⑥	②	39	78	61	18	⑦	11	100
1996	K	3.8 (3785)	⑧	⑬	20	⑪	⑫	⑭	18	11	100
	M	3.1 (3130)	①	②	③	76	61	④	10	⑤	100
	X	3.4 (3393)	⑥	②	39	78	61	18	⑦	11	100
1997	1	3.8 (3785)	⑧	⑬	20	⑪	⑫	⑯	22	11	100
	K	3.8 (3785)	⑧	⑬	20	⑪	⑫	⑭	38	11	100
	M	3.1 (3130)	①	②	③	76	61	④	10	⑤	100
	X	3.4 (3393)	⑥	②	39	78	61	18	⑦	11	100
1998-99	1	3.8 (3785)	⑧	⑬	⑮	⑪	⑫	⑯	22	11	100
	K	3.8 (3785)	⑧	⑬	⑮	⑪	⑫	⑭	38	11	100
	M	3.1 (3130)	①	②	③	76	61	④	10	⑩	100

① Coat threads with sealer torque to 33 ft. lbs., then turn 1/4 turn (90 degrees)

② 37 ft. lbs. plus 77 degrees

③ 15 ft. lbs. plus 75 degrees

④ Torque all bolts to 15 ft. lbs. Retorque to 24 ft. lbs.

⑤ New cylinder head:
1st-time installation: 20 ft. lbs.
All other installations: 11 ft. lbs.

⑥ 37 ft. lbs. plus 90 degrees

⑦ 115 inch lbs.

⑧ Step 1: 35 ft. lbs.
Step 2: 130 degrees
Step 3: Rotate four center bolts an additional 30 degrees

⑨ 26 ft. lbs. plus 50 degrees

⑩ 20 ft. lbs. plus 50 degrees

⑪ 110 ft. lbs. plus 76 degrees

⑫ 11 ft. lbs. plus 50 degrees

⑬ 30 ft. lbs. plus 110 degrees
Side bolts: 11 ft. lbs. plus 45 degrees

⑭ Upper manifold: 18 ft. lbs.
Lower manifold bolt/nut: 22 ft. lbs.
Upper manifold studs: 89 in. lbs.

⑮ 20 ft. lbs. plus 50 degrees

⑯ Upper manifold: 8 ft. lbs.
Lower manifold: 11 ft. lbs.

79221CQ9

SCHEDULED MAINTENANCE INTERVALS
(GM W BODY - BUICK CENTURY & REGAL, CHEVROLET LUMINA & MONTE CARLO, OLDSMOBILE INTRIGUE & CUTLASS SUPREME, PONTIAC GRAND PRIX)

TO BE SERVICED	TYPE OF SERVICE	VEHICLE MILEAGE INTERVAL (x1000)												
		7.5	15	22.5	30	37.5	45	52.5	60	67.5	75	82.5	90	97.5
Engine oil & filter	R	✓	✓	✓	✓	✓	✓	✓	✓	✓	✓	✓	✓	✓
Automatic transaxle fluid & filter③	S/I	✓	✓	✓	✓	✓	✓	✓	✓	✓	✓	✓	✓	✓
Brake hoses	S/I	✓	✓	✓	✓	✓	✓	✓	✓	✓	✓	✓	✓	✓
Coolant level, hoses & clamps	S/I	✓	✓	✓	✓	✓	✓	✓	✓	✓	✓	✓	✓	✓
Drive shaft boots & front suspension components	S/I	✓	✓	✓	✓	✓	✓	✓	✓	✓	✓	✓	✓	✓
Exhaust system & throttle linkage	S/I	✓	✓	✓	✓	✓	✓	✓	✓	✓	✓	✓	✓	✓
Lubricate chassis, suspension, steering linkage, transaxle shift linkage, parking brake cable guides, underbody contact points & linkage	S/I	✓	✓	✓	✓	✓	✓	✓	✓	✓	✓	✓	✓	✓
Rotate tires	S/I	✓		✓		✓		✓		✓		✓		✓
Air filter element	R				✓				✓				✓	
Engine coolant②	R				✓				✓				✓	
PCV filter	R				✓				✓				✓	
Spark plugs①	R				✓				✓				✓	
Accessory drive belt(s)	S/I				✓				✓				✓	
Ignition cables, EGR & fuel systems	S/I				✓				✓				✓	
Camshaft timing belt	R								✓					
Throttle body mount bolt torque	S/I	✓												

① Platinum tip spark plugs - replace every 100,000 miles.
② Engine coolant (1996-99) - replace every 100,000 miles. Use O.E. specified (DEX-COOL™) coolant only. If any silicate coolant is used, the service interval is every 30,000 miles.
③ Automatic transaxle fluid & filter (1995) - replace every 100,000 miles.
R – Replace   S/I – Service or Inspect

79221CQ0

SCHEDULED MAINTENANCE INTERVALS
(GM W BODY - BUICK CENTURY & REGAL, CHEVROLET LUMINA & MONTE CARLO, OLDSMOBILE INTRIGUE & CUTLASS SUPREME, PONTIAC GRAND PRIX Cont.)

FREQUENT OPERATION MAINTENANCE (SEVERE SERVICE)

If a vehicle is operated under any of the following conditions it is considered severe service:
- Extremely dusty areas.
- 50% or more of the vehicle operation is in 32°C (90°F) or higher temperatures, or constant operation in temperatures below 0°C (32°F).
- Prolonged idling (vehicle operation in stop and go traffic).
- Frequent short running periods (engine does not warm to normal operating temperatures).
- Police, taxi, delivery usage or trailer towing usage.

Oil & oil filter – change every 3000 miles.
Chassis lubrication - lubricate every 6000 miles.
Rotate tires at 6000 miles, then every 15,000 miles (1993-94) or every 12,000 miles (1995-97).
Throttle body mount bolt torque - tighten at 6000 miles
Air filter element - service or inspect every 15,000 miles.
Automatic transaxle fluid - change every 15,000 miles (1993-95) or every 50,000 miles (1996-97).
Camshaft timing belt - change every 60,000 miles.

79221CR1

SCHEDULED MAINTENANCE
GENERAL MOTORS CORPORATION
W BODY
BUICK REGAL, CENTURY (1997-99), CHEVROLET LUMINA, MONTE CARLO
OLDSMOBILE CUTLASS SUPREME, OLDSMOBILE INTRIGUE, PONTIAC GRAND PRIX

The following should be used as a guide when determining the amount of work required for a particular service if taken to a repair shop.
In estimating how long a particular Scheduled Maintenance Service should take, please observe the following:

- **Chilton Time** is time based on field research and data supplied by the vehicle manufacturer.
- Labor time operations are given in hours and tenths of an hour.
- All labor operations, are to be used as a guide.

Mechanic Skill Level Codes:
  (G) GENERAL: Normally skilled with certification.
  (M) MAINTENANCE: Semi-skilled working on certication.
  (P) PRECISION: Really skilled with multiple certification.

	Chilton Time			Chilton Time			Chilton Time
(G) 7500 Mile Service			**(M) 45000 Mile Service**			**(G) 67500 Mile Service**	
1995-99	1.6		1995-99	.9		1995-99	1.4
(M) 15000 Mile Service			**(G) 52500 Mile Service**			**(M) 75000 Mile Service**	
1995-99	.9		1995-99	1.4		1995-99	.9
(G) 22500 Mile Service			**(G) 60000 Mile Service**			**(G) 82500 Mile Service**	
1995-99	1.4		1995-99	3.2		1995-99	1.4
(G) 30000 Mile Service			Renew timing belt			**(G) 90000 Mile Service**	
1995-99	3.2		add	5.0		1995-99	3.2
(G) 37500 Mile Service						**(G) 97500 Mile Service**	
1995-99	1.4					1995-99	1.4

79221CR2

GENERAL MOTORS Y BODY
Chevrolet Corvette

VEHICLE IDENTIFICATION CHART

Code	Liters	Cu. In. (cc)	Cyl.	Fuel Sys.	Eng. Mfg.
		Engine Code			
5	5.7	350 (5737)	8	MFI	CPC
G	5.7	350 (5665)	8	MFI	CPC
J	5.7	350 (5737)	8	MFI	①
P	5.7	350 (5737)	8	MFI	CPC

Code	Year
	Model Year
S	1995
T	1996
V	1997
W	1998
X	1999

CPC - Chevrolet/Pontiac/Canada

MFI - Multi-point Fuel Injection

① Manufactured by Mercury Marine

79221CR3

ENGINE IDENTIFICATION

Year	Model	Engine Displacement Liters (cc)	Engine Series (ID/VIN)	Fuel System	No. of Cylinders	Engine Type
1995	Corvette	5.7 (5737)	J	MFI	8	DOHC
	Corvette	5.7 (5737)	P	MFI	8	OHV
1996	Corvette	5.7 (5737)	5	MFI	8	DOHC
	Corvette	5.7 (5737)	P	MFI	8	OHV
1997	Corvette	5.7 (5665)	G	MFI	8	OHV
1998-99	Corvette	5.7 (5665)	G	MFI	8	OHV

MFI - Multi-point Fuel Injection

OHV - Overhead Valves

DOHC - Double Overhead Camshafts

79221CR4

GENERAL ENGINE SPECIFICATIONS

Year	Engine ID/VIN	Engine Displacement Liters (cc)	Fuel System Type	Net Horsepower @ rpm	Net Torque @ rpm (ft. lbs.)	Bore x Stroke (in.)	Compression Ratio	Oil Pressure @ rpm
1995	J	5.7 (5737)	MFI	375@5800	370@4800	3.90x3.66	11.0:1	40@2000
	P	5.7 (5737)	MFI	300@5000	340@3600	4.00x3.48	10.25:1	18@2000
1996	5	5.7 (5737)	MFI	375@5800	370@4800	3.90x3.66	11.0:1	40@2000
	P	5.7 (5737)	MFI	300@5000	340@3600	4.00x3.48	10.25:1	18@2000
1997	G	5.7 (5665)	MFI	345@5600	350@4400	3.89x3.62	10.1:1	18@2000
1998-99	G	5.7 (5665)	MFI	345@5600	350@4400	3.89x3.62	10.1:1	18@2000

MFI - Multi-point Fuel Injection

79221CR5

GASOLINE ENGINE TUNE-UP SPECIFICATIONS

Year	Engine ID/VIN	Engine Displacement Liters (cc)	Spark Plugs Gap (in.)	Ignition Timing (deg.) MT	Ignition Timing (deg.) AT	Fuel Pump (psi)	Idle Speed (rpm) MT	Idle Speed (rpm) AT	Valve Clearance In.	Valve Clearance Ex.
1995	J	5.7 (5737)	0.035	①	①	48-55	①	①	HYD	HYD
	P	5.7 (5737)	0.035	—	①	41-47	—	①	HYD	HYD
1996	5	5.7 (5737)	0.035	①	①	48-55	①	①	HYD	HYD
	P	5.7 (5737)	0.035	—	①	41-47	—	①	HYD	HYD
1997	G	5.7 (5665)	0.060	①	①	48-55	①	①	HYD	HYD
1998-99	G	5.7 (5665)	0.060	①	①	48-55	①	①	HYD	HYD

NOTE: The Vehicle Emission Control Information label often reflects specification changes made during production. The label figures must be used if they differ from those in this chart.

HYD - Hydraulic

① Refer to Vehicle Emission Control Information label

79221CR6

CAPACITIES

Year	Model	Engine ID/VIN	Engine Displacement Liters (cc)	Engine Oil with Filter (qts.)	Transmission (pts.) 5-Spd	Transmission (pts.) 6-Spd	Transmission (pts.) Auto.	Drive Axle (pts.)	Fuel Tank (gal.)	Cooling System (qts.)
1995	Corvette	J	5.7 (5737)	8.6	—	①	—	3.75	20.0	14.7
	Corvette	P	5.7 (5737)	4.5	—	①	②	3.75	20.0	17.8
1996	Corvette	5	5.7 (5737)	7.6	—	①	②	3.8	20.0	14.7
	Corvette	P	5.7 (5737)	5.0	—	①	②	3.8	20.0	17.8
1997	Corvette	G	5.7 (5737)	7.6	—	①	②	3.8	20.0	14.7
1998-99	Corvette	G	5.7 (5737)	7.6	—	①	②	3.8	20.0	14.7

NOTE: All capacities are approximate. Add fluid gradually and ensure a proper fluid level is obtained.

① ZF 6 speed trans.: 4.4 pts.
   MM 6 speed trans.: 4.1 pts.
② 440T4 trans.: 13.0 pts.
   125C trans.: 8.0 pts.
   4L60E trans.: 10.0 pts.

79221CR7

VALVE SPECIFICATIONS

Year	Engine ID/VIN	Engine Displacement Liters (cc)	Seat Angle (deg.)	Face Angle (deg.)	Spring Test Pressure (lbs. @ in.)	Spring Installed Height (in.)	Stem-to-Guide Clearance (in.) Intake	Stem-to-Guide Clearance (in.) Exhaust	Stem Diameter (in.) Intake	Stem Diameter (in.) Exhaust
1995	J	5.7 (5737)	44	45	147-166@0.95	1.34 ①	0.0012-0.0026	0.0014-0.0030	NA	NA
	P	5.7 (5737)	46	45	187-203@1.27	1.70	0.0009-0.0027	0.0009-0.0027	NA	NA
1996	5	5.7 (5737)	44	45	147-166@0.95	1.34	0.0012-0.0026	0.0014-0.0030	NA	NA
	P	5.7 (5737)	46	45	81-89@1.78	1.78	0.0009-0.0027	0.0009-0.0027	NA	NA
1997	G	5.7 (5665)	46	45	76@1.80	1.80	0.0010-0.0026	0.0010-0.0026	NA	NA
1998-99	G	5.7 (5665)	46	45	76@1.80	1.80	0.0010-0.0026	0.0010-0.0026	NA	NA

NA - Not Available

① Inner spring: 1.18 in.

79221CR8

TORQUE SPECIFICATIONS
All readings in ft. lbs.

Year	Engine ID/VIN	Engine Displacement Liters (cc)	Cylinder Head Bolts	Main Bearing Bolts	Rod Bearing Bolts	Crankshaft Damper Bolts	Flywheel Bolts	Manifold Intake	Manifold Exhaust	Spark Plugs	Lug Nut
1995	J	5.7 (5737)	①	②	③	148	66	④	⑤	15	100
	P	5.7 (5737)	65	78	47	60	74	⑥	35	11	100
1996	P	5.7 (5737)	65	78	47	60	74	35	30	11	100
	5	5.7 (5737)	①	⑦	③	148	66	④	⑤	15	100
1997	G	5.7 (5665)	⑧	⑨	⑩	⑪	⑫	⑬	⑭	11	100
1998-99	G	5.7 (5665)	⑧	⑨	⑩	⑪	⑫	⑬	⑭	11	100

NA - Not Available

① Step 1: 45 ft. lbs.
   Step 2: 74 ft. lbs.
   Step 3: 118 ft. lbs.
② Step 1: 15 ft. lbs.
   Step 2: Inner: 65-70 degrees
   Step 3: Outer: 50-55 degrees
③ 22 ft. lbs. plus 80-85 degrees
④ Injector housing and fuel rail bolts: 20 ft. lbs.
⑤ Studs: 22 ft. lbs.
   Bolts: 18 ft. lbs.
⑥ Step 1: 71 inch lbs.
   Step 2: 35 ft. lbs.
⑦ Step 1: 15 ft. lbs.
   Step 2: Inner: 65-70 degrees
   Step 3: Outer: 80-85 degrees
⑧ Step 1: 22 ft. lbs.
   Step 2: Rotate 76 degrees
   Step 3: Repeat Step 2 except for the medium
           lenght bolts at the front and rear.
   Step 4: Rotate the medium bolts at the front
           and rear of each head 34 degrees
⑨ Step 1: Inner bolts 15 ft. lbs.
   Step 2: Inner bolts rotate 80 degrees
   Step 3: Side bolts 18 ft. lbs.

   Step 4: Outer studs 15 ft. lbs.
   Step 5: Outer studs rotate 53 degrees
⑩ Step 1: 15 ft. lbs.
   Step 2: Rotate 60 degrees
⑪ Step 1: 240 ft. lbs.
   Step 2: Loosen bolt and retighten to 37 ft. lbs.
   Step 3: Rotate 120 degrees
⑫ Step 1: 15 ft. lbs.
   Step 2: 37 ft. lbs.
   Step 3: 74 ft. lbs.
⑬ Step 1: 44 inch lbs.
   Step 2: 74 inch lbs.
⑭ Step 1: 11 ft. lbs.
   Step 2: 18 ft. lbs.

79221CR9

SCHEDULED MAINTENANCE INTERVALS
(GM Y BODY - CHEVROLET CORVETTE)

TO BE SERVICED	TYPE OF SERVICE	VEHICLE MILEAGE INTERVAL (x1000)												
		7.5	15	22.5	30	37.5	45	52.5	60	67.5	75	82.5	90	97.5
Engine oil & filter③	R	✓	✓	✓	✓	✓	✓	✓	✓	✓	✓	✓	✓	✓
Brake hoses & brake lining	S/I	✓	✓	✓	✓	✓	✓	✓	✓	✓	✓	✓	✓	✓
Coolant level, hoses & clamps	S/I	✓	✓	✓	✓	✓	✓	✓	✓	✓	✓	✓	✓	✓
Exhaust system & throttle linkage	S/I	✓	✓	✓	✓	✓	✓	✓	✓	✓	✓	✓	✓	✓
Lubricate chassis, suspension, steering linkage, transmission shift linkage, parking brake cable guides, underbody contact points & linkage	S/I	✓	✓	✓	✓	✓	✓	✓	✓	✓	✓	✓	✓	✓
Rear axle fluid level	S/I	✓	✓	✓	✓	✓	✓	✓	✓	✓	✓	✓	✓	✓
Automatic transmission fluid & filter④	S/I		✓		✓		✓		✓		✓		✓	
Air filter element	R				✓				✓				✓	
Engine coolant (1995)②	R				✓				✓				✓	
Ignition cables, EGR & fuel systems	S/I				✓				✓				✓	
Serpentine drive belt	S/I				✓				✓				✓	
Spark plugs①	S/I													

① Platinum tip spark plugs - replace every 100,000 miles.
② Engine coolant (1996-99) - replace every 100,000 miles. Use O.E. specified (DEX-COOL™) coolant only. If any silicate coolant is used, the service interval is every 30,000 miles.
③ Corvette engines require a special oil meeting GM Standard 4718M.
④ Automatic transmission fluid & filter - replace every 100,000 miles (unless previously replaced).
R – Replace   S/I – Service or Inspect

FREQUENT OPERATION MAINTENANCE (SEVERE SERVICE)

If a vehicle is operated under any of the following conditions it is considered severe service:
- Extremely dusty areas.
- 50% or more of the vehicle operation is in 32°C (90°F) or higher temperatures, or constant operation in temperatures below 0°C (32°F).
- Prolonged idling (vehicle operation in stop and go traffic).
- Frequent short running periods (engine does not warm to normal operating temperatures).
- Police, taxi, delivery usage or trailer towing usage.
Engine oil & oil filter – change every 3000 miles. ③
Chassis lubrication - lubricate every 6000 miles.
Lubricate suspension, parking brake cable guides, underbody contact points & linkage - lubricate every 6000 miles.
Air filter element - service or inspect every 15,000 miles.
Automatic transmission fluid & filter - change every 15,000 miles.

79221CR0

SCHEDULED MAINTENANCE
GENERAL MOTORS CORPORATION
Y BODY
CHEVROLET CORVETTE

The following should be used as a guide when determining the amount of work required for a particular service if taken to a repair shop. In estimating how long a particular Scheduled Maintenance Service should take, please observe the following:

- **Chilton Time** is time based on field research and data supplied by the vehicle manufacturer.
- Labor time operations are given in hours and tenths of an hour.
- All labor operations, are to be used as a guide.

Mechanic Skill Level Codes:
- **(G)** GENERAL: Normally skilled with certification.
- **(M)** MAINTENANCE: Semi-skilled working on certication.
- **(P)** PRECISION: Really skilled with multiple certification.

	Chilton Time
(G) 7500 Mile Service	
1995-99	1.1
(G) 15000 Mile Service	
1995-99	1.2
(G) 22500 Mile Service	
1995-99	1.1
(G) 30000 Mile Service	
1995-99	1.9
Renew engine coolant (1995) add	.5

	Chilton Time
(G) 37500 Mile Service	
1995-99	1.1
(G) 45000 Mile Service	
1995-99	1.2
(G) 52500 Mile Service	
1995-99	1.1
(G) 60000 Mile Service	
1995-99	1.9
Renew engine coolant (1995) add	.5
(G) 67500 Mile Service	
1995-99	1.1

	Chilton Time
(G) 75000 Mile Service	
1995-99	1.2
(G) 82500 Mile Service	
1995-99	1.1
(G) 90000 Mile Service	
1995-99	1.9
Renew engine coolant (1995) add	.5
(G) 97500 Mile Service	
1995-99	1.1

79221CS1

GEO/CHEVROLET
Metro • Prizm

VEHICLE IDENTIFICATION CHART

		Engine Code				Model Year	
Code	Liters	Cu. In. (cc)	Cyl.	Fuel Sys.	Eng. Mfg.	Code	Year
6	1.0	(993)	3	TFI	Suzuki	S	1995
6	1.6	(1590)	4	MFI	Suzuki	T	1996
8	1.8	(1803)	4	MFI	Toyota	V	1997
9	1.3	(1300)	4	TFI	Suzuki	W	1998
						X	1999

TBI - Throttle body Fuel Injection

MFI - Multi-point Fuel Injection

79221CS2

ENGINE IDENTIFICATION

Year	Model	Engine Displacement Liters (cc)	Engine Series (ID/VIN)	Fuel System	No. of Cylinders	Engine Type
1995	Metro	1.0 (993)	6	TFI	3	SOHC
	Metro	1.3 (1300)	9	TFI	4	SOHC
	Prism	1.6 (1590)	6	MFI	4	DOHC
	Prizm	1.8 (1803)	8	MFI	4	DOHC
1996	Metro	1.0 (993)	6	TFI	3	SOHC
	Metro	1.3 (1300)	9	TFI	4	SOHC
	Prism	1.6 (1590)	6	MFI	4	DOHC
	Prizm	1.8 (1803)	8	MFI	4	DOHC
1997	Metro	1.0 (993)	6	TFI	3	SOHC
	Metro	1.3 (1300)	9	TFI	4	SOHC
	Prism	1.6 (1590)	6	MFI	4	DOHC
	Prizm	1.8 (1803)	8	MFI	4	DOHC
1998-99	Metro	1.0 (993)	6	TFI	3	SOHC
	Metro	1.3 (1300)	9	TFI	4	SOHC
	Prizm	1.8 (1803)	8	MFI	4	DOHC

TFI - Throttle body Fuel Injection

MFI - Multi-point Fuel Injection

SOHC - Single Overhead Camshaft

DOHC - Double Overhead Camshafts

79221CS3

GENERAL ENGINE SPECIFICATIONS

Year	Engine ID/VIN	Engine Displacement Liters (cc)	Fuel System Type	Net Horsepower @ rpm	Net Torque @ rpm (ft. lbs.)	Bore x Stroke (in.)	Compression Ratio	Oil Pressure @ rpm
1995	6	1.0 (1993)	TFI	55@5700	58@3300	2.91x3.03	9.5:1	54@3000
	6	1.6 (1590)	MFI	105@5800	100@4800	3.20x3.00	9.5:1	36-71@3000
	8	1.8 (1803)	MFI	115@5600	115@2800	3.20x3.40	9.5:1	36-71@3000
	9	1.3 (1300)	TFI	70@5500	74@3000	2.91x3.03	9.5:1	54@3000
1996	6	1.0 (993)	TFI	55@5700	58@3300	2.91x3.03	9.5:1	54@3000
	6	1.6 (1590)	MFI	105@5800	100@4800	3.20x3.00	9.5:1	36-71@3000
	8	1.8 (1803)	MFI	115@5200	117@2800	3.20x3.40	9.5:1	36-71@3000
	9	1.3 (1300)	TFI	70@5500	74@3500	2.91x3.03	9.5:1	54@3000
1997	6	1.0 (993)	TFI	55@5700	58@3300	2.91x3.03	9.5:1	54@3000
	6	1.6 (1590)	MFI	105@5800	100@4800	3.20x3.00	9.5:1	36-71@3000
	8	1.8 (1803)	MFI	115@5200	117@2800	3.20x3.40	9.5:1	36-71@3000
	9	1.3 (1300)	TFI	70@5500	74@3500	2.91x3.03	9.5:1	54@3000
1998-99	6	1.0 (993)	TFI	55@5700	58@3300	2.91x3.03	9.5:1	54@3000
	8	1.8 (1803)	MFI	120@5600	127@4400	3.20x3.40	9.5:1	36-71@3000
	9	1.3 (1300)	TFI	70@5500	74@3500	2.91x3.03	9.5:1	54@3000

TFI - Throttle body Fuel Injection
MFI - Multi-point Fuel Injection

79221CS4

GASOLINE ENGINE TUNE-UP SPECIFICATIONS

Year	Engine ID/VIN	Engine Displacement Liters (cc)	Spark Plugs Gap (in.)	Ignition Timing (deg.) MT		Ignition Timing (deg.) AT		Fuel Pump (psi)	Idle Speed (rpm) MT	Idle Speed (rpm) AT	Valve Clearance In.	Valve Clearance Ex.
1995	6	1.0 (993)	0.041	5B	①	B	①	23-30	800	850	HYD	HYD
	6	1.6 (1590)	0.031	10B	②	10B	②	31-37	700-750	700-750	0.0060-0.0100	0.0100-0.0140
	8	1.8 (1803)	0.031	10B	①	10B	②	31-37	700-750	700-750	0.0060-0.010	0.0100-0.0140
	9	1.3 (1300)	0.041	5B	①	5B	①	23-30	800	850	HYD	HYD
1996	6	1.0 (993)	0.041	5B	①	5B	①	23-30	800	850	HYD	HYD
	6	1.6 (1590)	0.031	10B	②	10B	②	31-37	700-750	700-750	0.0060-0.010	0.0100-0.0140
	8	1.8 (1803)	0.031	10B	②	10B	②	31-37	700-750	700-750	0.0060-0.010	0.0100-0.0140
	9	1.3 (1300)	0.041	5B	①	5B	①	23-30	800	850	HYD	HYD
1997	6	1.0 (993)	0.041	5B	①	5B	①	23-30	800	850	HYD	HYD
	6	1.6 (1590)	0.031	10B	②	10B	②	31-37	700-750	700-750	0.0060-0.010	0.0100-0.0140
	8	1.8 (1803)	0.031	10B	②	10B	②	31-37	700-750	700-750	0.0060-0.010	0.0100-0.0140
	9	1.3 (1300)	0.041	5B	①	5B	①	23-30	800	850	HYD	HYD
1998-99	6	1.0 (993)	0.041	5B	①	5B	①	23-30	800	850	HYD	HYD
	8	1.8 (1803)	0.031	10B	②	10B	②	31-37	700-750	700-750	0.0060-0.010	0.0100-0.0140
	9	1.3 (1300)	0.041	5B	①	5B	①	23-30	800	850	HYD	HYD

NOTE: The Vehicle Emission Control Information label often reflects specification changes made during production. The label figures must be used if they differ from those in this chart.
B - E Before Top Dead Center
HYD - Hydraulic
① Connect a fused jumper from Duty Check cavity 4 to cavity 5 for fixed timing (DLC connector located at left strut tower)
② Insert jumper wire between terminals in DLC connector E1 and TE1

79221CS5

CAPACITIES

Year	Model	Engine ID/VIN	Engine Displacement Liters (cc)	Engine Oil with Filter	Transmission (pts.)			Drive Axle		Fuel Tank (gal.)	Cooling System (qts.)
					4-Spd	5-Spd	Auto.	Front (pts.)	Rear (pts.)		
1995	Metro	6	1.0 (993)	3.7	—	5.0	10.1 ①	—	—	10.6	③
	Metro	9	1.3 (1300)	3.7	—	5.0	10.4 ①	—	—	10.6	4.9
	Prizm	6	1.6 (1590)	3.2	—	3.0	6.6	—	3.0 ②	13.2	6.7
	Prizm GSi	8	1.8 (1803)	3.9	—	3.0	6.6	—	3.0 ②	13.2	6.7
1996	Metro	6	1.0 (993)	3.7	—	5.0	10.1 ①	—	—	10.6	③
	Metro	9	1.3 (1300)	3.7	—	5.0	10.1 ①	—	—	10.6	4.9
	Prizm	6	1.6 (1590)	3.2	—	4.0	6.6	—	3.0 ②	13.2	6.7
	Prizm GSi	8	1.8 (1803)	3.9	—	4.0	6.6	—	3.0 ②	13.2	6.7
1997	Metro	6	1.0 (993)	3.7	—	5.0	10.1 ①	—	—	10.6	③
	Metro	9	1.3 (1300)	3.7	—	5.0	10.1 ①	—	—	10.6	4.9
	Prizm	6	1.6 (1590)	3.2	—	4.0	6.6	—	3.0 ②	13.2	6.7
	Prizm GSi	8	1.8 (1803)	3.9	—	4.0	6.6	—	3.0 ②	13.2	6.7
1998-99	Metro	6	1.0 (993)	3.7	—	5.0	10.1 ①	—	—	10.6	③
	Metro	9	1.3 (1300)	3.7	—	5.0	10.1 ①	—	—	10.6	4.9
	Prizm	8	1.8 (1803)	3.9	—	4.0	6.6	—	3.0 ②	13.2	6.7

NOTE: All capacities are approximate. Add fluid gradually and ensure a propwer fluid level is obtained.

① Automatic transmission - Specification is after complete overhaul. Drain and fill will be less

② 3 speed automatic only

③ Manual transaxle: 4.1 qts.
 Automatic transaxle: 4.2 qts.

79221CS6

VALVE SPECIFICATIONS

Year	Engine ID/VIN	Engine Displacement Liters (cc)	Seat Angle (deg.)	Face Angle (deg.)	Spring Test Pressure (lbs. @ in.)	Spring Installed Height (in.)	Stem-to-Guide Clearance (in.)		Stem Diameter (in.)	
							Intake	Exhaust	Intake	Exhaust
1995	6	1.0 (993)	45	45	46.1-51.8@ 1.28	1.28	0.0008- 0.0022	0.0018- 0.0028	0.2148- 0.2157	0.2142- 0.2148
	6	1.6 (1590)	45	45.5	37.3@1.25	1.25	0.0010- 0.0024	0.0012- 0.0026	0.2350- 0.2356	0.2348- 0.2354
	8	1.8 (1803)	45	45.5	37.3@1.25	1.25	0.0010- 0.0024	0.0012- 0.0026	0.2350- 0.2356	0.2348- 0.2354
	9	1.3 (1300)	45	45	54.7-64.3@ 1.63	1.63	0.0008- 0.0019	0.0014- 0.0025	0.2742- 0.2748	0.2737- 0.2742
1996	6	1.0 (993)	45	45	46.1-51.8@ 1.28	1.28	0.0008- 0.0022	0.0018- 0.0028	0.2148- 0.2157	0.2142- 0.2148
	6	1.6 (1590)	45	45.5	37.3@1.25	1.25	0.0010- 0.0024	0.0012- 0.0026	0.2350- 0.2356	0.2348- 0.2354
	8	1.8 (1803)	45	45.5	37.3@1.25	1.25	0.0010- 0.0024	0.0012- 0.0026	0.2350- 0.2356	0.2348- 0.2354
	9	1.3 (1300)	45	45	54.7-64.3@ 1.63	1.63	0.0008- 0.0019	0.0014- 0.0025	0.2742- 0.2748	0.2737- 0.2742
1997	6	1.0 (993)	45	45	46.1-51.8@ 1.28	1.28	0.0008- 0.0022	0.0018- 0.0028	0.2148- 0.2157	0.2142- 0.2148
	6	1.6 (1590)	45	45.5	37.3@1.25	1.25	0.0010- 0.0024	0.0012- 0.0026	0.2350- 0.2356	0.2348- 0.2354
	8	1.8 (1803)	45	45.5	37.3@1.25	1.25	0.0010- 0.0024	0.0012- 0.0026	0.2350- 0.2356	0.2348- 0.2354
	9	1.3 (1300)	45	45	54.7-64.3@ 1.63	1.63	0.0008- 0.0019	0.0014- 0.0025	0.2742- 0.2748	0.2737- 0.2742
1998-99	6	1.0 (993)	45	45	46.1-51.8@ 1.28	1.28	0.0008- 0.0022	0.0018- 0.0028	0.2148- 0.2157	0.2142- 0.2148
	8	1.8 (1803)	45	45.5	37.3@1.25	1.25	0.0010- 0.0024	0.0012- 0.0026	0.2350- 0.2356	0.2348- 0.2354
	9	1.3 (1300)	45	45	54.7-64.3@ 1.63	1.63	0.0008- 0.0019	0.0014- 0.0025	0.2742- 0.2748	0.2737- 0.2742

79221CS7

TORQUE SPECIFICATIONS
All readings in ft. lbs.

Year	Engine ID/VIN	Engine Displacement Liters (cc)	Cylinder Head Bolts	Main Bearing Bolts	Rod Bearing Bolts	Crankshaft Damper Bolts	Flywheel Bolts	Manifold		Spark Plugs	Lug Nut
								Intake	Exhaust		
1995	6	1.0 (993)	54	40	26	81 ①	52	17	17	21	44
	6	1.6 (1590)	②	44	26	87 ①	③	14	29	21	76
	8	1.8 (1590)	④	40	26	81 ①	58	17	17	21	70
	9	1.3 (1300)	54	40	26	79 ①	45	17	17	21	44
1996	6	1.0 (993)	54	40	26	81 ①	52	17	17	21	44
	6	1.6 (1590)	②	44	②	87 ①	③	14	25	21	76
	8	1.8 (1803)	②	44	18 ⑤	87 ①	③	14	25	21	76
	9	1.3 (1300)	54	40	26	79 ①	45	17	17	21	44
1997	6	1.0 (993)	54	40	26	81 ①	52	17	17	21	44
	6	1.6 (1590)	②	44	②	87 ①	③	14	25	21	76
	8	1.8 (1803)	②	44	18 ⑤	87 ①	③	14	25	21	76
	9	1.3 (1300)	54	40	26	79 ①	45	17	17	21	44
1998-99	6	1.0 (993)	54	40	26	81 ①	52	17	17	21	44
	8	1.8 (1803)	②	44	18 ⑤	87 ①	③	14	25	21	76
	9	1.3 (1300)	54	40	26	79 ①	45	17	17	21	44

① Crankshaft timing belt sprocket
② Step 1: 22 ft. lbs.
   Step 2: Plus 90 degrees
③ Manual transaxle: 58 ft. lbs.
   Automatic transaxle: 47 ft. lbs.
④ Step 1: 26 ft. lbs.
   Step 2: 41 ft. lbs.
   Step 3: 52 ft. lbs.
⑤ Tighten an additional 90 degrees

79221CS8

SCHEDULED MAINTENANCE INTERVALS
(GEO METRO, PRIZM & STORM)

TO BE SERVICED	TYPE OF SERVICE	VEHICLE MILEAGE INTERVAL (x1000)												
		7.5	15	22.5	30	37.5	45	52.5	60	67.5	75	82.5	90	97.5
Engine oil & filter	R	✓	✓	✓	✓	✓	✓	✓	✓	✓	✓	✓	✓	✓
Chassis lubrication (Metro & Storm)	S/I	✓	✓	✓	✓	✓	✓	✓	✓	✓	✓	✓	✓	✓
Locking front hubs (1996-99 Metro)	S/I	✓	✓	✓	✓	✓	✓	✓	✓	✓	✓	✓	✓	✓
Lubricate parking brake cable guides, underbody contact points & linkage	S/I	✓	✓	✓	✓	✓	✓	✓	✓	✓	✓	✓	✓	✓
Rotate tires	S/I	✓	✓	✓	✓	✓	✓	✓	✓	✓	✓	✓	✓	✓
Brake system	S/I	✓		✓		✓		✓		✓		✓		✓
Engine idle speed (Prizm)	S/I	✓		✓		✓		✓		✓		✓		✓
Exhaust system	S/I	✓		✓		✓		✓		✓		✓		✓
Chassis lubrication (Prizm)	S/I		✓		✓		✓		✓		✓		✓	
Fuel tank, cap & lines (Metro & Storm)	S/I		✓		✓		✓		✓		✓		✓	
Fuel tank, cap & lines (Prizm)	S/I				✓				✓				✓	
Valve clearance (Storm 1.6L)	S/I		✓		✓		✓		✓		✓		✓	
Valve clearance (Prizm)	S/I								✓					
Air cleaner filter	R				✓				✓				✓	
Engine coolant	R				✓				✓				✓	
Fuel filter (Metro)	R				✓				✓				✓	
Fuel tank cap gasket (Prizm)	R				✓				✓				✓	
Manual transaxle oil	R				✓				✓				✓	
Spark plugs	R				✓				✓				✓	
Accessory drive belt(s) (1996-99 Prizm)	S/I								✓		✓		✓	✓
Accessory drive belt(s) (except 1996-99 Prizm)	S/I				✓				✓				✓	

79221CS9

SCHEDULED MAINTENANCE INTERVALS
(GEO METRO, PRIZM & STORM Cont.)

TO BE SERVICED	TYPE OF SERVICE	VEHICLE MILEAGE INTERVAL (x1000)												
		7.5	15	22.5	30	37.5	45	52.5	60	67.5	75	82.5	90	97.5
Cooling system (Metro & Storm)	S/I				✓				✓				✓	
Cooling system (Prizm)	S/I						✓				✓			
Automatic transaxle fluid & filter	S/I				✓				✓				✓	
EGR system	S/I				✓				✓				✓	
Engine timing (1995 Metro)	S/I				✓				✓				✓	
Engine timing (1996-99 Metro)	S/I								✓					
Ignition cables (Storm)	S/I				✓				✓				✓	
Ignition cables (Metro & Prizm)	S/I								✓					
PCV system	S/I				✓				✓				✓	
Brake fluid (1996-99 Metro)	R								✓					
PCV valve (1996-99 Metro)①	R													
Timing belt	R								✓					
EVAP canister (1996-99 Prizm)	S/I								✓					
Throttle body unit mount bolt torque	S/I	✓												

① PCV valve (1996-99 Metro) - replace at 50,000 miles.
R – Replace   S/I – Service or Inspect

FREQUENT OPERATION MAINTENANCE (SEVERE SERVICE)

If a vehicle is operated under any of the following conditions it is considered severe service:
- Extremely dusty areas.
- 50% or more of the vehicle operation is in 32°C (90°F) or higher temperatures, or constant operation in temperatures below 0°C (32°F).
- Prolonged idling (vehicle operation in stop and go traffic).
- Frequent short running periods (engine does not warm to normal operating temperatures).
- Police, taxi, delivery usage or trailer towing usage.

Oil & oil filter (Metro, Storm & Prizm) – change every 3000 miles.
Chassis lubrication (Metro, Storm & Prizm) - lubricate every 6000 miles.
Throttle body mount bolt torque - torque at 6000 miles.
Air cleaner filter - service or inspect every 15,000 miles.
Differential fluid (1996-99 Prizm) - replace every 15,000 miles.
Rotate tires - rotate at 6000 miles and then every 15,000 miles thereafter.
Automatic transaxle fluid & filter (Prizm & Storm) - replace every 15,000 miles.
Automatic transaxle fluid & filter (Metro) - replace every 50,000 miles.

79221CS0

SCHEDULED MAINTENANCE
GENERAL MOTORS CORPORATION
GEO/CHEVROLET
METRO & PRIZM

The following should be used as a guide when determining the amount of work required for a particular service if taken to a repair shop. In estimating how long a particular Scheduled Maintenance Service should take, please observe the following:

- **Chilton Time** is time based on field research and data supplied by the vehicle manufacturer.
- Labor time operations are given in hours and tenths of an hour.
- All labor operations, are to be used as a guide.

Mechanic Skill Level Codes:
- **(G) GENERAL:** Normally skilled with certification.
- **(M) MAINTENANCE:** Semi-skilled working on certication.
- **(P) PRECISION:** Really skilled with multiple certification.

	Chilton Time
(M) 7500 Mile Service	
1995-99	1.7
Inspect locking hubs (Metro) add	.1
(M) 15000 Mile Service	
1995-99	.6
Inspect locking hubs (Metro) add	.1
Rotate tires add	.5
(M) 22500 Mile Service	
1995-99	1.5
Inspect locking hubs (Metro) add	.1
(G) 30000 Mile Service	
1995-99	2.6
Inspect engine timing (1995 Metro) add	.2
Inspect locking hubs (Metro) add	.1
Rotate tires add	.5
(M) 37500 Mile Service	
1995-99	1.5
Inspect locking hubs (Metro) add	.1

	Chilton Time
(M) 45000 Mile Service	
1995-99	.6
Inspect locking hubs (Metro) add	.1
Rotate tires add	.5
(M) 52500 Mile Service	
1995-99 Metro	1.5
Inspect locking hubs (Metro) add	.1
(G) 60000 Mile Service	
1995-99 Metro	5.1
Inspect locking hubs (Metro) add	.1
Replace brake fluid (Metro) add	.5
1995-99 Prizm	4.8

	Chilton Time
(M) 67500 Mile Service	
1995-99	1.5
Inspect locking hubs (Metro) add	.1
(M) 75000 Mile Service	
1995-99	.6
Inspect locking hubs (Metro) add	.1
(M) 82500 Mile Service	
1995-99	1.5
Inspect locking hubs (Metro) add	.1
(G) 90000 Mile Service	
1995-99	2.7
Inspect locking hubs (Metro) add	.1
Inspect engine timing (1995 Metro) add	.2
(M) 97500 Mile Service	
1995-99	1.5
Inspect locking hubs (Metro) add	.1

79221CT1

SATURN
SC1 • SC2 • SL • SL1 • SL2 • SW1 • SW2

VEHICLE IDENTIFICATION CHART

Code	Liters	Cu. In. (cc)	Cyl.	Fuel Sys.	Eng. Mfg.
7	1.9	116 (1901)	4	MFI	Saturn
8	1.9	116 (1901)	4	MFI	Saturn

MFI - Multi-point Fuel Injection

Code	Year
S	1995
T	1996
V	1997
W	1998
X	1999

79221CT2

ENGINE IDENTIFICATION

Year	Model	Engine Displacement Liters (cc)	Engine Series (ID/VIN)	Fuel System	No. of Cylinders	Engine Type
1995	Sedan	1.9 (1901)	7	MFI	4	DOHC
	Sedan	1.9 (1901)	8	MFI	4	SOHC
	Coupe	1.9 (1901)	7	MFI	4	DOHC
	Coupe	1.9 (1901)	8	MFI	4	SOHC
	Wagon	1.9 (1901)	7	MFI	4	DOHC
	Wagon	1.9 (1901)	8	MFI	4	SOHC
1996	Sedan	1.9 (1901)	7	MFI	4	DOHC
	Sedan	1.9 (1901)	8	MFI	4	SOHC
	Coupe	1.9 (1901)	7	MFI	4	DOHC
	Coupe	1.9 (1901)	8	MFI	4	SOHC
	Wagon	1.9 (1901)	7	MFI	4	DOHC
	Wagon	1.9 (1901)	8	MFI	4	SOHC
1997	Sedan	1.9 (1901)	7	MFI	4	DOHC
	Sedan	1.9 (1901)	8	MFI	4	SOHC
	Coupe	1.9 (1901)	7	MFI	4	DOHC
	Coupe	1.9 (1901)	8	MFI	4	SOHC
	Wagon	1.9 (1901)	7	MFI	4	DOHC
	Wagon	1.9 (1901)	8	MFI	4	SOHC
1998-99	Sedan	1.9 (1901)	7	MFI	4	DOHC
	Sedan	1.9 (1901)	8	MFI	4	SOHC
	Coupe	1.9 (1901)	7	MFI	4	DOHC
	Coupe	1.9 (1901)	8	MFI	4	SOHC
	Wagon	1.9 (1901)	7	MFI	4	DOHC
	Wagon	1.9 (1901)	8	MFI	4	SOHC

MFI - Multi-point Fuel Injection
DOHC - Double Overhead Camshafts
SOHC - Single Overhead Camshaft

79221CT3

GENERAL ENGINE SPECIFICATIONS

Year	Engine ID/VIN	Engine Displacement Liters (cc)	Fuel System Type	Net Horsepower @ rpm	Net Torque @ rpm (ft. lbs.)	Bore x Stroke (in.)	Compression Ratio	Oil Pressure @ rpm
1995	7	1.9 (1901)	MFI	124@5600	122@4800	3.23x3.54	9.5:1	29@2000
	8	1.9 (1901)	MFI	85@5000	107@2400	3.23x3.54	9.3:1	36@2000
1996	7	1.9 (1901)	MFI	124@5600	122@4800	3.23x3.54	9.5:1	29@2000
	8	1.9 (1901)	MFI	100@5000	114@2400	3.23x3.54	9.3:1	36@2000
1997	7	1.9 (1901)	MFI	124@5600	122@4800	3.23x3.54	9.5:1	29@2000
	8	1.9 (1901)	MFI	100@5000	114@2400	3.23x3.54	9.3:1	36@2000
1998-99	7	1.9 (1901)	MFI	124@5600	122@4800	3.23x3.54	9.5:1	29@2000
	8	1.9 (1901)	MFI	100@5000	114@2400	3.23x3.54	9.3:1	36@2000

MFI - Multi-point Fuel Injection

79221CT4

GASOLINE ENGINE TUNE-UP SPECIFICATIONS

Year	Engine ID/VIN	Engine Displacement Liters (cc)	Spark Plugs Gap (in.)	Ignition Timing (deg.) MT	Ignition Timing (deg.) AT	Fuel Pump (psi)	Idle Speed (rpm) MT	Idle Speed (rpm) AT	Valve Clearance In.	Valve Clearance Ex.
1995	7	1.9 (1901)	0.040	①	①	31-36 ②	850 ③	750 ③	HYD	HYD
	8	1.9 (1901)	0.040	①	①	26-31 ②	750 ③	650 ③	HYD	HYD
1996	7	1.9 (1901)	0.040	①	①	31-36 ②	850 ③	750 ③	HYD	HYD
	8	1.9 (1901)	0.040	①	①	31-36 ②	750 ③	650 ③	HYD	HYD
1997	7	1.9 (1901)	0.040	①	①	31-36 ②	850 ③	750 ③	HYD	HYD
	8	1.9 (1901)	0.040	①	①	31-36 ②	750 ③	650 ③	HYD	HYD
1998-99	7	1.9 (1901)	0.040	①	①	31-36 ②	850 ③	750 ③	HYD	HYD
	8	1.9 (1901)	0.040	①	①	31-36 ②	750 ③	650 ③	HYD	HYD

NOTE: The Vehicle Emission Control Information label often reflects specification changes made during production. The label figures must be used if they differ from those in this chart.

HYD - Hydraulic

① Engines equipped with Distributorless Ignition System (DIS). Ignition timing is not adjustable

② Pressure measured at idle

③ Idle speed measured with manual transmission in neutral; automatic transmission in drive

79221CT5

CAPACITIES

Year	Model	Engine ID/VIN	Engine Displacement Liters (cc)	Engine Oil with Filter (qts.)	Transmission (pts.)			Drive Axle		Fuel Tank (gal.)	Cooling System (qts.)
					4-Spd	5-Spd	Auto.	Front (pts.)	Rear (pts.)		
1995	Sedan	7	1.9 (1901)	4.0	—	5.2	7.5 ①	—	—	12.8	7.0
	Sedan	8	1.9 (1901)	4.0	—	5.2	7.5 ①	—	—	12.8	7.0
	Coupe	7	1.9 (1901)	4.0	—	5.2	7.5 ①	—	—	12.8	7.0
	Coupe	8	1.9 (1901)	4.0	—	5.2	7.5 ①	—	—	12.8	7.0
	Wagon	7	1.9 (1901)	4.0	—	5.2	7.5 ①	—	—	12.8	7.0
	Wagon	8	1.9 (1901)	4.0	—	5.2	7.5 ①	—	—	12.8	7.0
1996	Sedan	7	1.9 (1901)	4.0	—	5.2	7.5 ①	—	—	12.8	7.0
	Sedan	8	1.9 (1901)	4.0	—	5.2	7.5 ①	—	—	12.8	7.0
	Coupe	7	1.9 (1901)	4.0	—	5.2	7.5 ①	—	—	12.8	7.0
	Wagon	7	1.9 (1901)	4.0	—	5.2	7.5 ①	—	—	12.8	7.0
	Wagon	8	1.9 (1901)	4.0	—	5.2	7.5 ①	—	—	12.8	7.0
1997	Sedan	7	1.9 (1901)	4.0	—	5.2	7.5 ①	—	—	12.8	7.0
	Sedan	8	1.9 (1901)	4.0	—	5.2	7.5 ①	—	—	12.8	7.0
	Coupe	7	1.9 (1901)	4.0	—	5.2	7.5 ①	—	—	12.8	7.0
	Coupe	8	1.9 (1901)	4.0	—	5.2	7.5 ①	—	—	12.8	7.0
	Wagon	7	1.9 (1901)	4.0	—	5.2	7.5 ①	—	—	12.8	7.0
	Wagon	8	1.9 (1901)	4.0	—	5.2	7.5 ①	—	—	12.8	7.0
1998-99	Sedan	7	1.9 (1901)	4.0	—	5.2	7.5 ①	—	—	12.1	7.0
	Sedan	8	1.9 (1901)	4.0	—	5.2	7.5 ①	—	—	12.1	7.0
	Coupe	7	1.9 (1901)	4.0	—	5.2	7.5 ①	—	—	12.1	7.0
	Coupe	8	1.9 (1901)	4.0	—	5.2	7.5 ①	—	—	12.1	7.0
	Wagon	7	1.9 (1901)	4.0	—	5.2	7.5 ①	—	—	12.1	7.0
	Wagon	8	1.9 (1901)	4.0	—	5.2	7.5 ①	—	—	12.1	7.0

NOTE: All capacities are approximate. Add fluid gradually and ensure a proper fluid level is obtained.

① Specification is for overhaul. 4.2 qts. with fluid and filter change

79221CT6

VALVE SPECIFICATIONS

Year	Engine ID/VIN	Engine Displacement Liters (cc)	Seat Angle (deg.)	Face Angle (deg.)	Spring Test Pressure (lbs. @ in.)	Spring Installed Height (in.)	Stem-to-Guide Clearance (in.)		Stem Diameter (in.)	
							Intake	Exhaust	Intake	Exhaust
1995	7	1.9 (1901)	44.5-45.4	45-45.5	163-180@ 0.984	①	0.0010-0.0025	0.0015-0.0032	0.2736-0.2740	0.2729-0.2736
	8	1.9 (1901)	44.5-45.4	45-45.5	202-211@ 1.280	①	0.0010-0.0025	0.0015-0.0032	0.2736-0.2741	0.2736-0.2740
1996	7	1.9 (1901)	44.5-45.4	45-45.5	163-180@ 0.984	①	0.0010-0.0025	0.0015-0.0032	0.2736-0.2740	0.2729-0.2736
	8	1.9 (1901)	44.5-45.4	45-45.25	202-211@ 1.280	①	0.0010-0.0025	0.0015-0.0032	0.2736-0.2741	0.2736-0.2740
1997	7	1.9 (1901)	44.5-45.4	45-45.5	163-180@ 0.984	①	0.0010-0.0025	0.0015-0.0032	0.2736-0.2740	0.2729-0.2736
	8	1.9 (1901)	44.5-45.4	45-45.25	202-211@ 1.280	①	0.0010-0.0025	0.0015-0.0032	0.2736-0.2741	0.2736-0.2740
1998-99	7	1.9 (1901)	44.5-45.4	45-45.5	163-180@ 0.984	①	0.0010-0.0025	0.0015-0.0032	0.2736-0.2740	0.2729-0.2736
	8	1.9 (1901)	44.5-45.4	45-45.25	202-211@ 1.280	①	0.0010-0.0025	0.0015-0.0032	0.2736-0.2741	0.2736-0.2740

① Installed height not available
  Free length SOHC: 1.8898-1.9134
  Free length DOHC: 1.6100

79221CT7

TORQUE SPECIFICATIONS
All readings in ft. lbs.

Year	Engine ID/VIN	Engine Displacement Liters (cc)	Cylinder Head Bolts	Main Bearing Bolts	Rod Bearing Bolts	Crankshaft Damper Bolts	Flywheel Bolts	Manifold		Spark Plugs	Lug Nut
								Intake	Exhaust		
1995	7	1.9 (1901)	①	37	33	159	59 ④	22 ③	22 ③	20	103
	8	1.9 (1901)	②	37	33	159	59 ④	22 ③	22 ③	20	103
1996	7	1.9 (1901)	①	37	33	159	59 ④	22 ③	19 ③	20	103
	8	1.9 (1901)	②	37	33	159	59 ④	22 ③	16 ③	20	103
1997	7	1.9 (1901)	①	37	33	159	59 ④	22 ③	19 ③	20	103
	8	1.9 (1901)	②	37	33	159	59 ④	22 ③	16 ③	20	103
1998-99	7	1.9 (1901)	①	37	33	159	59 ④	22 ③	19 ③	20	103
	8	1.9 (1901)	②	37	33	159	59 ④	22 ③	16 ③	20	103

① Step 1: 22 ft. lbs.
  Step 2: 37 ft. lbs.
  Step 3: 90 degrees
② Step 1: 22 ft. lbs.
  Step 2: 33 ft. lbs.
  Step 3: 90 degrees
③ Studs: 106 in. lbs.
④ Flexplate specification: 44 ft. lbs.

79221CT8

SCHEDULED MAINTENANCE INTERVALS
(SATURN SC, SC1, SC2, SL, SL1, SL2, SW1, & SW2)

TO BE SERVICED	TYPE OF SERVICE	VEHICLE MILEAGE INTERVAL (x1000)												
		3	6	9	12	15	18	21	24	27	30	33	36	39
Engine oil & filter	R		✓		✓		✓		✓		✓		✓	
Lubricate chassis, suspension, steering linkage, transmission shift linkage, parking brake cable guides, underbody contact points & linkage	S/I		✓		✓		✓		✓		✓		✓	
Drive shaft boots, suspension bushings & ball joint seals	S/I		✓		✓		✓		✓		✓		✓	
Exhaust system & throttle linkage	S/I		✓		✓		✓		✓		✓		✓	
Rotate tires	S/I		✓				✓				✓			
Brake hoses & brake lining	S/I		✓				✓				✓			
Accessory drive belt(s)	S/I						✓						✓	
Engine coolant level, hoses & clamps	S/I						✓						✓	
Air filter element	R										✓			
Engine coolant	R												✓	
Manual transaxle oil	R		✓											
Spark plugs①	R										✓			
Automatic transaxle fluid & filter	R										✓			
Ignition cables & fuel systems	S/I										✓			
Vacuum line/hose	S/I										✓			
Fuel filter②	R													

① Platinum tip spark plugs - replace every 100,000 miles.
② Replace every 60,000 miles.
R – Replace   S/I – Service or Inspect

FREQUENT OPERATION MAINTENANCE (SEVERE SERVICE)

If a vehicle is operated under any of the following conditions it is considered severe service:
- Extremely dusty areas.
- 50% or more of the vehicle operation is in 32°C (90°F) or higher temperatures, or constant operation in temperatures below 0°C (32°F).
- Prolonged idling (vehicle operation in stop and go traffic).
- Frequent short running periods (engine does not warm to normal operating temperatures).
- Police, taxi, delivery usage or trailer towing usage.

Engine oil & oil filter - change every 3000 miles.

79221CT9

SCHEDULED MAINTENANCE
GENERAL MOTORS CORPORATION
SATURN
SC1,SC2, SL, SL1, SL2, SW1, SW2

The following should be used as a guide when determining the amount of work required for a particular service if taken to a repair shop. In estimating how long a particular Scheduled Maintenance Service should take, please observe the following:

- **Chilton Time** is time based on field research and data supplied by the vehicle manufacturer.
- Labor time operations are given in hours and tenths of an hour.
- All labor operations, are to be used as a guide.

Mechanic Skill Level Codes:
  (G) GENERAL: Normally skilled with certification.
  (M) MAINTENANCE: Semi-skilled working on certicication.
  (P) PRECISION: Really skilled with multiple certification.

	Chilton Time		Chilton Time		Chilton Time
(G) 6000 Mile Service		**(G) 18000 Mile Service**		**(G) 30000 Mile Service**	
1995-99	1.7	1995-99	1.6	1995-99	3.0
(M) 12000 Mile Service		**(M) 24000 Mile Service**		**(G) 36000 Mile Service**	
1995-99	.8	1995-99	.8	1995-99	1.5

79221CT0

MAINTENANCE LIGHT RESETTING AND DTC RETRIEVAL

2

**MAINTENANCE INDICATOR
LIGHT (MIL) RESETTING 2-2**
**OBD I DIAGNOSTIC TROUBLE
CODES.........................2-4**
**OBD II DIAGNOSTIC TROUBLE
CODES 2-43**
Chrysler Corp. (OBD I Diagnostic
Trouble Codes) 2-4
  Domestic Vehicle
    Self-Diagnostics.........................2-9
  Reading Codes...........................2-5
  Scan Tool Functions.....................2-7
  Self-Diagnostics.........................2-4
Chrysler Corp. (OBD II Diagnostic
Trouble Codes) 2-44
  Clearing Codes 2-44
  OBD II Trouble Code
    Equivalents...............................2-49
  OBD II Trouble Codes..................2-45
  Reading Codes 2-44
Ford Motor Co. (Maintenance
Indicator Light Resetting) 2-2
  Resetting 2-2

Ford Motor Co. (OBD I Diagnostic
Trouble Codes) 2-12
  Diagnostic Trouble Codes
    (DTC's)......................................2-26
  EEC-IV and EEC-V Diagnostic
    Systems.....................................2-12
  EEC-IV and EEC-V Reading
    Codes..2-15
  EEC-IV and EEC-V Scan Tool
    Functions 2-13
  EEC-IV and EEC-V Self-
    Diagnostics 2-14
  Introduction to Ford
    Self-Diagnostics.........................2-12
  MCU Carbureted Diagnostic
    System 2-23
  MCU Carbureted Reading
    Codes..2-24
  MCU Carbureted
    Self-Diagnostics 2-23
  Non-NAAO Diagnostic System....2-19
  Non-NAAO Reading Codes 2-20
  Non-NAAO Scan Tool Functions....2-19

  Non-NAAO Self-Diagnostics 2-20
Ford Motor Co. (OBD II Diagnostic
Trouble Codes) 2-50
  Clearing Codes 2-51
  OBD II Trouble Codes.................2-51
  Reading Codes 2-50
General Information 2-4
General Motors (Maintenance
Indicator Light Resetting) 2-2
  Resetting 2-2
General Motors (OBD I Diagnostic
Trouble Codes) 2-35
  Diagnostic Trouble Codes 2-41
  Reading and Clearing Codes.......2-37
  Self-Diagnostics.........................2-35
General Motors Corporation
(OBD II Diagnostic Trouble
Codes)...2-60
  Clearing Codes 2-60
  OBD II Trouble Codes.................2-61
  Reading Codes 2-60
Introduction 2-43
  Trouble Code Description...........2-43

MAINTENANCE INDICATOR LIGHT (MIL) RESETTING

This section describes reset procedures for maintenance indicator lights. Maintenance indicator lights are used to indicate to the operator that some type of routine maintenance should be performed. Unlike a Check Engine light that will be displayed when there is a fault with the engine management system, the maintenance light will be displayed when an engine or transmission oil change is recommended according to driving conditions. Also, the light will be displayed to indicate when the emission control system is in need of maintenance service.

Ford Motor Co.

RESETTING

1995 Continental

The 1995 Continental is equipped with a SERVICE INTERVAL REMINDER light. During the SYSTEM CHECK sequence, the SERVICE symbol activates and displays the miles (kilometers) left until the next normal service is due. After the necessary service is complete, reset the reminder as follows:
1. Press the **SYSTEM CHECK** button.
2. Press the **RESET** button.
3. Press the **SYSTEM CHECK** and the **RESET** button at the same time. The display should now show 7,200 miles (11,580 km). The mileage until the next service will count down from this point.

1995–97 Mark VIII

The 1995–97 Mark VIII continues to use the CHANGE OIL SOON or OIL CHANGE REQUIRED light. When the oil life left is between five percent and zero percent, CHANGE OIL SOON will be displayed on the message center. When oil life reaches zero percent, the OIL CHANGE REQUIRED message will be displayed. The message center indicator will indicate the percentage of oil life left during the System Check (during start up). This percentage is based on the driver's driving history and the amount of time since the last oil change. In order to ensure oil life left indications, the driver should only perform the OIL CHANGE RESET procedure after every oil change. Reset the system by pressing the **OIL CHANGE RESET** switch and hold for 5 seconds.

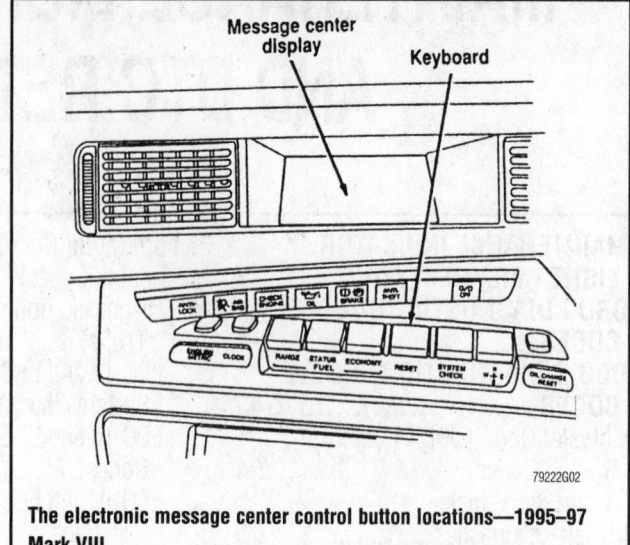

The electronic message center control button locations—1995–97 Mark VIII

After a successful reset the message center will display oil life indicators. The CHANGE OIL SOON or OIL CHANGE REQUIRED message will disappear after the 5 second interval.

General Motors

RESETTING

B Body Vehicles

The General Motors B body class designation includes the Caprice, the Impala, the Roadmaster and the Fleetwood.
After changing the engine oil, reset the engine oil life indicator whether the CHANGE OIL warning/indicator lamp illuminated or not.
1. Reset the engine oil life monitor as follows:
   a. Turn the ignition switch to the **ON** position, but don't start the engine.

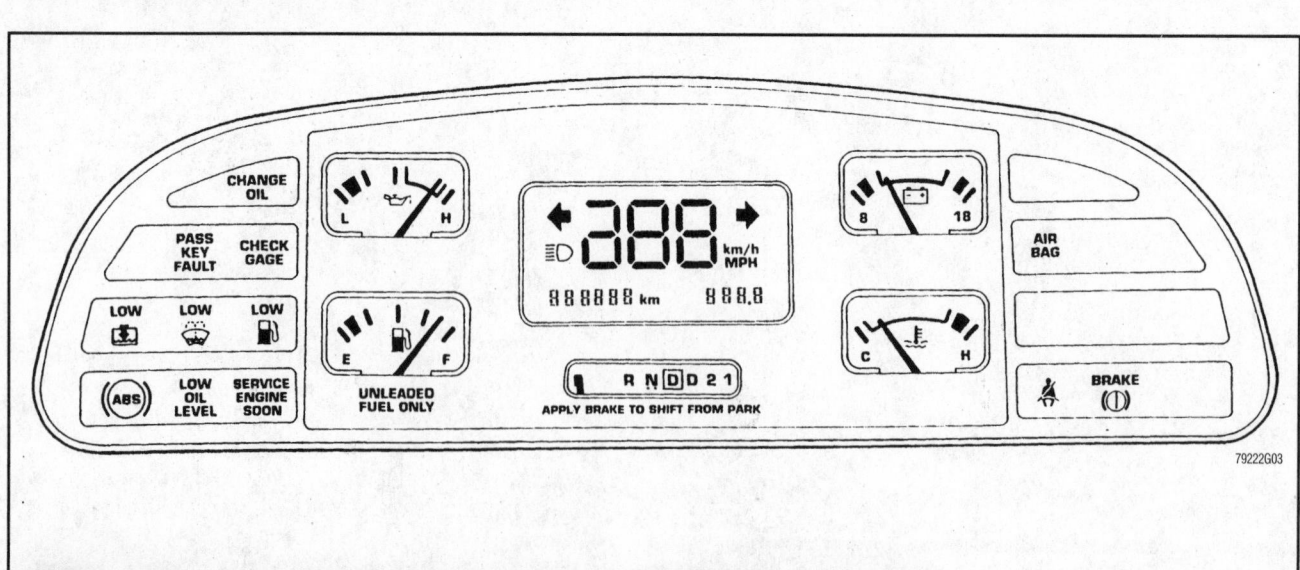

Instrument cluster warning/indicator lights, showing the location of the CHANGE OIL lamp—B Body vehicles

b. Depress the accelerator pedal to the Wide Open Throttle (WOT) position and release it three times within five seconds.

c. If the CHANGE OIL warning/indicator lamp blinks twice, then goes out, the system has been reset. If the CHANGE OIL warning/indicator lamp does not reset, turn the ignition switch to the **OFF** position and repeat the procedure.

C & H and G Body Vehicles

The General Motors C & H body class designations include the LeSabre, the Park Ave., the Eighty Eight, the Ninety Eight, the Bonneville, the Regency and the LSS. The G body designation refers to the Riviera and the Aurora.

These vehicles may be equipped with an ENGINE OIL LIFE INDEX (EOLI) in the display located on the Drivers Information Center (DIC). The Powertrain Control Module (PCM) determines approximately when the engine oil should be changed by calculating information based on vehicle speed, coolant temperature and engine RPM. Once the PCM determines it is time to change the engine oil, it will illuminate the CHANGE OIL SOON light on the DIC. This indicates that the remaining oil life is below 10 percent, (not to be confused with oil level). On a new vehicle, or one that has just been reset, the oil life is 100 percent. This percentage will slowly decrease based on inputs the PCM receives. When oil life reaches the 0 percent mark, the PCM will illuminate the CHANGE OIL NOW light on the DIC. At this time both messages will be displayed accompanied by a slow 5 second audible chime. The reset button must be pressed to acknowledge each of these messages. If a steady CHANGE OIL SOON light persists, Diagnostic Trouble Code (DTC) 61 has been set, indicating a possible shorted switch. Remaining oil life percentage can be displayed by pressing the **OIL** button on the DIC and advancing through the messages until the oil life index is displayed. The oil life index will not detect abnormal conditions, such as excessively dusty conditions or engine malfunctions that could otherwise affect engine oil life.

Reset the EOLI as follows:

1. Acknowledge all diagnostic messages in the Drivers Information Center by pressing the **RESET** button.

2. Press the **SEL** button on the left to select OIL. Press the **SEL** button on the right, if necessary, to display oil life.

3. Press and hold the **RESET** button for about 5 seconds. Once the oil life index has been reset, a RESET message will be displayed, then oil life will change to 100 percent. Be careful not reset the oil life accidentally at any time other than when the oil has just been changed. It can not be reset accurately until the next oil change.

E & K Body Vehicles

ENGINE OIL

The General Motors E & K body class designations include the DeVille, the Eldorado, the Seville, and the DeVille Concours.

These vehicles are equipped with an Engine Oil Life Index (EOLI) feature as part of the Driver Information Center (DIC) display. Engine oil life is displayed through engine data as the EOLI and as a CHANGE ENGINE OIL message. The EOLI is displayed following a number between 0 (zero) and 100. This is the percentage of oil life remaining, based on driving conditions, engine oil temperature, and mileage driven since the last time the oil life indicator was reset. When the oil life index reaches 10 percent or less a CHANGE OIL SOON message will appear as a reminder to schedule an oil change. When the oil life index reaches 0, the CHANGE ENGINE OIL message will appear indicating that the oil should be changed within the next 200 miles (320 km). After the oil has been changed, display the EOLI message by pressing the **INFORMATION** button several times. Press and hold the **RESET** button until the display shows 100. This will reset the oil life index. The CHANGE ENGINE OIL message will remain off until the next oil change is needed. The percentage of oil life remaining may be checked at any time by pressing the **INFORMATION** button several times until the EOLI appears.

TRANSAXLE FLUID

➡**This procedure is only applicable for 1995–96 models. Also, on 1996 models, a scan tool is necessary to reset the CHANGE TRANS FLUID indicator.**

All Cadillacs with 4.6L Northstar engines and 4T80-E transaxles are equipped with a transaxle fluid change indicator. A CHANGE TRANS FLUID message will display on the Information Center when the Powertrain Control Module (PCM) monitors actual operating conditions and displays the CHANGE TRANS FLUID message due either to calculations based on those conditions, or at the 100,000 mile (160,000 km) mark. Change the fluid by removing the lower pan and the sidecover drain plug. Change both the transmission fluid and the filter every 15,000 miles (25,000 km), if the vehicle is mainly driven under one or more of these conditions:

• In heavy city traffic where outside temperatures regularly reach 90° F (32° C) or higher.
• On high or mountain terrain.
• Frequent trailer pulling.
• Used for delivery service.

If the vehicle is not used under any of the preceding conditions, change both the fluid and filter (or service the screen) every 100,000 miles (160,000 km).

1. Reset the light on 1995 models as follows:

a. When the CHANGE TRANS FLUID message appears, change the fluid in both the pan and sidecover.

b. Turn the key **ON** with the engine stopped. Press and hold the **OFF** and **REAR DEFOG** buttons on the climate control simultaneously until the TRANS FLUID RESET message appears in the Information Center (between 5–20 seconds) The system is now reset.

2. Reset the indicator on 1996 models as follows:

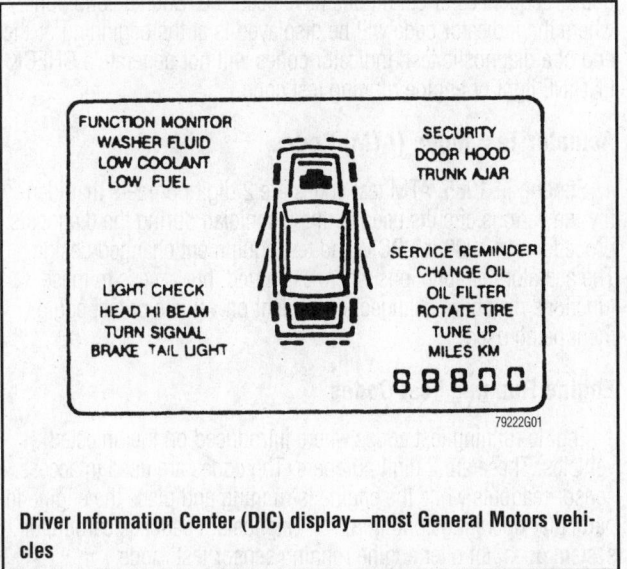

Driver Information Center (DIC) display—most General Motors vehicles

a. When the CHANGE TRANS FLUID message appears, change the fluid in both the pan and side cover.

b. To reset the transaxle oil life indicator it requires the use of a scan tool. Set the scan tool to the MISC TEST, then perform the OIL LIFE RESET followed by the TRANS OIL ST. The system is now reset.

Y Body Vehicles

The General Motors Y body class designation refers to the Corvette.

1995–96 VEHICLES

1. Turn the key to the **ON** position, but do not start the engine.
2. Press the **ENG MET** button on the trip monitor and release. Press the button again within 5 seconds.
3. Within 5 seconds of pressing the button the second time, press and hold the **GAUGES** button on the trip monitor. The CHANGE OIL light will flash.

4. Hold the **GAUGES** button until the CHANGE OIL light stops flashing and extinguishes. The monitor is reset at this point. Repeat this procedure if the light does not go out.

1997–99 VEHICLES

The 1997–99 Corvette is equipped with an Oil Life Monitor. The Oil Life Monitor is used to inform the driver when the next service is required. A scan tool can be used to monitor and display the amount of oil life left (shown as a percentage).

➡**Repair any Throttle Position (TP) or Accelerator Pedal Position (APP) sensor diagnostic trouble codes before resetting the oil life monitor.**

1. Turn the ignition **ON** , but do not start the engine.
2. Depress the accelerator pedal to the Wide Open Throttle (WOT) position and release it three times within five seconds.
3. If the Oil Life Monitor is not reset, perform this procedure again.
4. Change the oil and the filter.

OBD I DIAGNOSTIC TROUBLE CODES

General Information

For years, vehicles have been capable of storing diagnostic trouble codes. Codes prior to the 1996 OBD II legislation have been proprietary to the vehicle manufacturer. In some cases, the codes are specific to the individual make and model.

Furthermore, some manufacturers have developed specialized devices to read their codes. This complicates code reading and clearing.

Chrysler Corp.

SELF-DIAGNOSTICS

The Chrysler fuel injection systems combine electronic spark advance and fuel control. At the center of these systems is a digital, pre-programmed computer, known as an Powertrain Control Module (PCM). The PCM can also be referred to as the Single Module Engine Controller (SMEC) or as the Single Board Engine Controller (SBEC). The PCM regulates ignition timing, air-fuel ratio, emission control devices, cooling fan, charging system idle speed and speed control. It has the ability to update and revise its commands to meet changing operating conditions.

Various sensors provide the input necessary for PCM to correctly regulate fuel flow at the injectors. These include the Manifold Absolute Pressure (MAP), Throttle Position Sensor (TPS), oxygen sensor, coolant temperature sensor, charge temperature sensor, and vehicle speed sensors.

In addition to the sensors, various switches are used to provide important information to the PCM. These include the neutral safety switch, air conditioning clutch switch, brake switch and speed control switch. These signals cause the PCM to change either the fuel flow at the injectors or the ignition timing or both.

The PCM is designed to test it's own input and output circuits, If a fault is found in a major system, this information is stored in the PCM for eventual display to the technician. Information on this fault can be displayed to the technician by means of the instrument panel CHECK ENGINE light or by connecting a diagnostic read-out tester and reading a numbered display code, which directly relates to a general fault. Some inputs and outputs are checked continuously and others are checked under certain conditions. If the problem is repaired or no longer exists, the PCM cancels the fault code after approximately 50 key **ON/OFF** cycles.

When a fault code is detected, it appears as either a flash of the CHECK ENGINE light on the instrument panel or by watching the Diagnostic Readout Box version II (ORB-II). This indicates that an abnormal signal in the system has been recognized by the PCM. Fault codes do indicate the presence of a failure but they don't identify the failed component directly.

Fault Codes

Fault codes are 2 digit numbers that tell the technician which circuit is bad. Fault codes do indicate the presence of a failure but they don't identify the failed component directly. Therefore a fault code a result and not always the reason for the problem.

Indicator Codes

Indicator codes are 2 digit numbers that tell the technician if particular sequences or conditions have occurred. Such a condition where the indicator code will be displayed is at the beginning or the end of a diagnostic test. Indicator codes will not generate a CHECK ENGINE light or engine running test code.

Actuator Test Mode (ATM) Codes

Starting in 1985, ATM test codes are 2 digit numbers that identify the various circuits used by the technician during the diagnosis procedure. In 1989 the PCM and test equipment changed design. The actuator test functions where expanded, but access to these functions may have changed, dependent on vehicle or test equipment being used.

Engine Running Test Codes

Engine running test codes where introduced on fuel injected vehicles. These are 2 digit numbers. The codes are used to access sensor readouts while the engine is running and place the engine in particular operating conditions for diagnosis. Feedback carburetor system does not offer engine running sensor test mode.

Check Engine Light

This is possibly the most critical step of diagnosis. A detailed examination of connectors, wiring and vacuum hoses can often lead to a repair without further diagnosis. A careful inspector will check the undersides of hoses as well as the integrity of hard-to-reach hoses blocked by the air cleaner or other component. Wiring should be checked carefully for any sign of strain, burning, crimping, or terminals pulled-out from a connector. Checking connectors at components or in harnesses is required; usually, pushing them together will reveal a loose fit.

The CHECK ENGINE or Maintenance Indicator Lamp (MIL) light has 2 modes of operation: diagnostic mode and switch test mode.

If a ORB-II diagnostic tester is not available, the PCM can show the technician fault codes by flashing the CHECK ENGINE light on the instrument panel in the diagnostic mode. In the switch test mode, after all codes are displayed, switch function can be confirmed. The light will turn on and off when a switch is turned ON and OFF.

Even though the light can be used as a diagnostic tool, it cannot do the following:

Once the light starts to display fault codes, it cannot be stopped. If the technician loses count, he must start the test procedure again. The light cannot display all of the codes or any blank displays.

The light cannot tell the technician if the oxygen feed-back system is lean or rich and if the idle motor and detonation systems are operational. The light cannot perform the actuation test mode, sensor test mode or engine running test mode.

➡**Be advised that the CHECK ENGINE light can only perform a limited amount of functions and is not to be used as a substitute for a diagnostic tester. All diagnostic procedure described herein are intended for use with a Diagnostic Readout Box II (DRB-II) or equivalent tool.**

Limp-In Mode

The limp-in mode is the attempt by the PCM to compensate for the failure of certain components by substituting information from other sources. If the PCM senses incorrect data or no data at all from the MAP sensor, throttle position sensor or coolant temperature sensor, the system is placed into limp-in mode and the CHECK ENGINE light on the instrument panel is activated. This mode will keep the vehicle drive able until the customer can get it to a service facility.

Test Modes

There are 5 modes of testing required for the proper diagnosis of the system. They are as follows:

Diagnostic Test Mode This mode is used to access the fault codes from the PCM's memory.

Circuit Actuation Test Mode (ATM Test) This mode is used to turn a certain circuit on and off in order to test it. ATM test codes are used in this mode.

Switch Test Mode This mode is used to determine if specific switch inputs are being received by the PCM.

Sensor Test Mode This mode looks at the output signals of certain sensors as they are received by the PCM when the engine is not running. Sensor access codes are read in this mode. Also this mode is used to clear the PCM memory of stored codes.

Engine Running Test Mode This mode looks at sensor output signals as seen by the PCM when the engine is running. Also this mode is used to determine some specific running conditions necessary for diagnosis.

READING CODES

Obtaining Trouble Codes

Entering the Jeep or Eagle self-diagnostic system requires the use of a special adapter that connects with the Diagnostic Readout Box II (DRB-II). These systems require the adapter because all of the system diagnosis is done Off-Board instead of On-Board like most vehicles. The adapter, which is a computer module itself, measures signals at the diagnostic connector and converts the signals into a form which the ORBII can use to perform tests. On vehicles other than Jeep and Eagle the following procedures will obtain stored Diagnostic Trouble Codes (DTC).

USING THE CHECK ENGINE LAMP

Codes display on vehicles built before 1989 are displayed in numerical order, after 1989 codes are displayed in order of occurrence.

1. Connect the readout box to the diagnostic connector located in the engine compartment near PCM.

2. Start the engine, if possible, cycle the transmission selector and the A/C switch if applicable. Shut off the engine.

3. Turn the ignition switch ON—OFF, ON—OFF, ON—OFF, ON within 5 seconds.

4. Observe the CHECK ENGINE light on the instrument panel.

5. Just after the last ON cycle, the dash warning (MIL) lamp will begin flashing the stored codes.

6. The codes are transmitted as two digit flashes.

7. Example would be Code 21 will be displayed as a FLASH FLASH pause FLASH.

8. Be ready to write down the codes as they appear; the only way to repeat the codes is to start over at the beginning.

SCAN TOOL

The scan tool is the preferred choice for fault recovery and system diagnosis. Some hints on using the ORB-II include:

• To use the HELP screen, press and hold F3 at any time.

• To restart the ORB-II at any time, hold the MODE button and press ATM at the same time.

• Pressing the up or down arrows will move forward or backward one item within a menu.

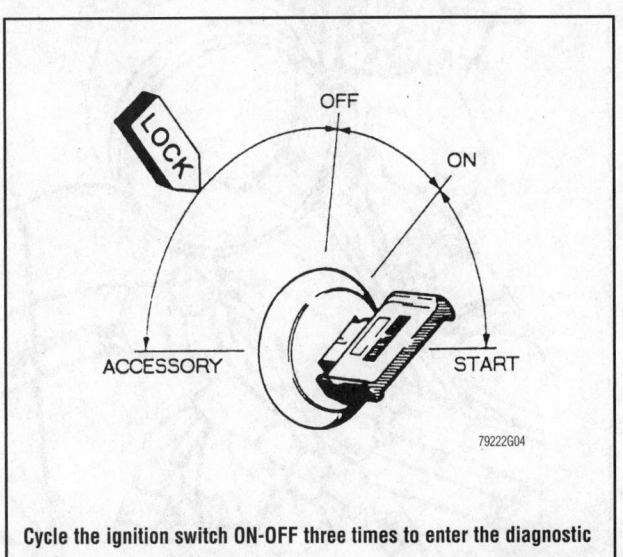

79222G04

Cycle the ignition switch ON-OFF three times to enter the diagnostic mode

• To select an item, either press the number of the item or move the cursor arrow to the selection, then press ENTER.

• To return to the previous display (screen), press ATM.

• Some test screens display multiple items. To view only one, move the cursor arrow to the desired item, then press ENTER.

To read stored faults with the ORB-II:

1. With the ignition switch OFF, connect the tool to the diagnostic connector near the engine controller under the hood. On some 1988 and earlier models, cycling the ignition key ON-OFF three times may be necessary to enter the diagnostics. On 1989 and newer models, simply turn the ignition switch ON to access the read fault code data.

2. Start the engine if possible. Cycle the transmission from Park to a forward gear, then back to Park. Cycle the air conditioning ON and OFF. Turn the ignition switch OFF.

3. Turn the ignition switch ON but do not start the engine. The ORB-II will begin its power-up sequence; do not touch any keys on the scan tool during this sequence.

4. Reading faults must be selected from the FUEL/IGN MENU. To reach this menu on the ORB-II:

a. When the initial menu is displayed after the power-up sequence, use the down arrow to display choice 4) SELECT SYSTEM and select this choice.

b. Once on the—SELECT SYSTEM—screen, choose 1) ENGINE. This will enter the engine diagnostics section of the program.

c. The screen will momentarily display the engine family and SBEC identification numbers. After a few seconds the screen displays the choices 1) With A/C and 2) Without A/C. Select and enter the correct choice for the vehicle.

d. When the—ENGINE SYSTEM—screen appears, select 1) FUEL/IGNITION from the menu.

e. On the next screen, select 2) READ FAULTS.

5. If any faults are stored, the display will show how many are stored (1 of 4 faults, etc.) and issue a text description of the problem, such as COOLANT SENSOR VOLTAGE TOO LOW. The last line of the display shows the number of engine starts since the code was set. If the number displayed is 0 starts, this indicates a hard or current fault. Faults are displayed in reverse order of occurrence; the

```
----- FUEL/IGN FAULTS -----
NO FAULTS DETECTED

X STARTS SINCE ERS
```

```
1 OF X FAULTS
[message
appears here]
X STARTS SINCE SET
```

79222G06

Example of the DRB-II display screen while reading the trouble codes

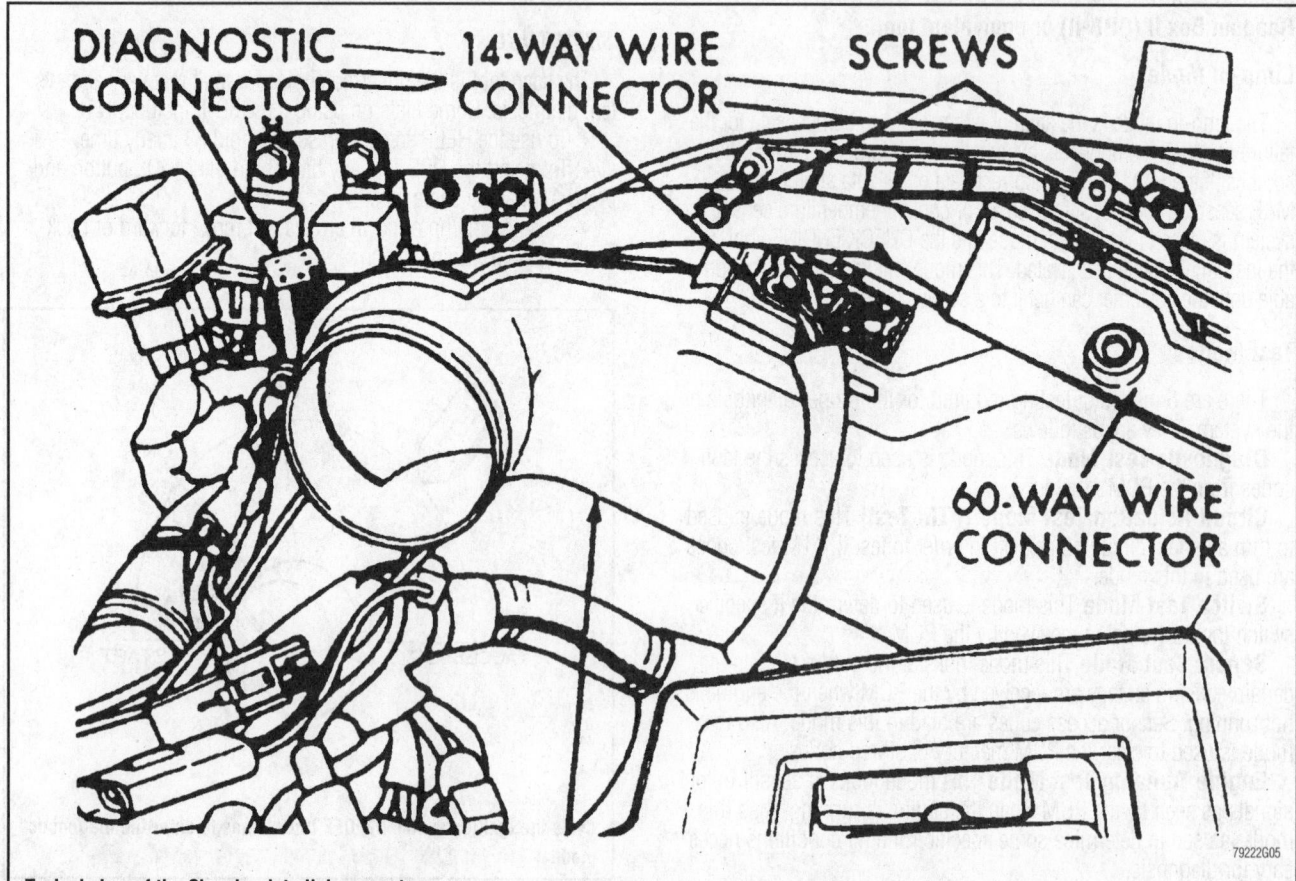

Typical view of the Chrysler data link connector

DIAGNOSTIC CONNECTOR — 14-WAY WIRE CONNECTOR   SCREWS

60-WAY WIRE CONNECTOR

79222G05

first fault shown is the most current and the last fault shown is the oldest.

6. Press the down arrow to read each fault after the first. Record the screen data carefully for easy reference.

7. If no faults are stored in the controller, the display will state NO FAULTS DETECTED and show the number of starts since the system memory was last erased.

8. After all faults have been read and recorded, press ATM.

9. Refer to the appropriate diagnostic chart for a diagnostic path. Remember that the fault message identifies a circuit problem, not a component. Use of the charts is required to sequentially test a circuit and identify the fault.

SWITCH TEST

The PCM only recognizes 2 switch input states—HI and LOW. For this reason the PCM cannot tell the difference between a selected switch position and an open circuit, short circuit or an open switch. However, if one of the switches is toggled, the controller does have the ability to respond to the change of state in the switch. If the change is displayed, it can be assumed that entire switch circuit to the PCM is operational.

1988 AND EARLIER MODELS:

After all codes have been shown and has indicated Code 55 end of message, actuate the following component switches. The digital display must change its numbers between 00 and 88 and the CHECK ENGINE light will blink when the following switches are activated and released:

- Brake pedal
- Gear shift selector
- A/C switch
- Electric defogger switch (1984)

1989 AND NEWER MODELS:

To enter the switch test mode, activate read input states or equivalent function on the readout box for the following switch tests:

- Z1 Voltage Sense
- Speed Control Set
- Speed Control ON/OFF
- Speed Control Resume
- A/C Switch Sense
- Brake Switch
- Park/neutral Switch

SCAN TOOL FUNCTIONS

✳✳ CAUTION

Always apply the parking brake and block the wheels before performing any diagnostic procedures with the engine running. Failure to do so may result in personal injury and/or property damage.

After stored faults have been read and recorded, the scan tool may be used to investigate states and functions of various components. This ability compliments but does not replace the use of diagnostic charts. The ORB-II functions are useful in identifying circuits which are or are not operating correctly as well as checking component function or signal.

When diagnosing an emissions-related problem, keep in mind that the SBEC system only enters closed loop mode under certain conditions. The single most important criteria for entry into closed loop operation is that the engine be at normal operating tempera-

ture; i.e., fully warmed up. The engine is considered to be at normal operating temperature if any of the following are true: the electric cooling fan cycles on at least once or the upper radiator hose is hot to the touch or the heater is able to deliver hot air.

In open loop operation, the signal from the oxygen sensor is ignored by the engine controller and the fuel injection is controlled by pre-programmed values within the computer. Once closed loop operation is begun, the signal from the oxygen sensor is used by the engine controller to constantly adjust the fuel injection to maintain the proper air/fuel ratio. The system will switch in and out of closed loop operation depending on sensor signals and driver input. In most cases, the system will be in closed loop operation during normal driving, acceleration or deceleration and idle. Wide open throttle will cause the system to momentarily switch to open loop operation. Additionally, some engine control systems will momentarily switch to open loop under hard acceleration or deceleration until the MAP sensor signal stabilizes.

The ORB-II may be operated in the following diagnostic modes from the FUEL/IGN MENU screen.

Sensors

This function displays current data being transmitted from the fuel and ignition sensors to the engine controller. Examples of sensor data available include MAP voltage, throttle position sensor voltage and percentage, RPM, coolant temperature, voltage sensor, total spark advance, and vehicle speed. Many other sensors may be monitored depending on engine/transmission combinations.

Data for each sensor is displayed in the appropriate units, such as volts, mph, in. Hg, degrees F, etc.

1988 AND EARLIER MODELS

1. Put the system into the diagnostic test mode and wait for Code 55 to appear on the display screen.

2. Press the ATM button on the diagnostic tool to activate the display. If a specific sensor read test is desired, hold the ATM button down until the desired test code appears.

3. Slide the READ/HOLD switch to the HOLD position to display the corresponding sensor output.

Sensor Read Test Display codes:

Code 01 Battery temperature sensor; display voltage divided by 10 equals sensor temperature

Code 02 Oxygen sensor voltage; display number divided by 10 equals sensor voltage

Code 03 Charge temperature sensor voltage; display number divided by 10 equals sensor voltage

Code 04 Engine coolant temperature sensor; display number multiplied by 10 equals degrees of engine coolant sensor

Code 05 Throttle position sensor voltage; display number divided by 10 equals sensor voltage or temperature

Code 06 Peak knock sensor voltage; display number is sensor voltage

Code 07 Battery voltage; display number is battery voltage

Code 08 Map sensor voltage; display number divided by 10 equals sensor voltage

Code 09 Speed control switches:
- Display is blank—Cruise OFF
- Display shows 00—Cruise ON
- Display shows 10—Cruise SET
- Display shows 01—Cruise RESUME

Code 10 Fault code erase routine; display will flash 0's for 4 seconds

The State Display programs allow the operator to view the present conditions in the SBEC system. These choices are displayed on the FUEL IGN STATE screen and offer the choices of MODULE INFO, SENSORS, INPUTS/OUTPUTS or MONITORS. Viewing system data through these windows can be helpful in observing the effects of repairs or to compare the problem vehicle to a known-good vehicle.

1989 AND NEWER MODELS

To enter the sensor test mode, activate read sensor voltage or read sensor values or equivalent functions on the readout box for the following sensor displays:

Read Sensor Voltage
- Battery temperature sensor
- Oxygen sensor input
- Throttle body temperature sensor
- Coolant temperature sensor
- Throttle position
- Minimum throttle
- Battery voltage
- MAP sensor voltage

Read Sensor Values
- Throttle body temperature
- Coolant temperature
- MAP gauge reading
- AIS motor position
- Added adaptive fuel
- Adaptive fuel factor
- Barometric pressure
- Engine speed
- Module spark advance
- Vehicle speed
- Oxygen sensor state

Engine Running Test Mode

1988 AND EARLIER MODELS

The Engine Running Test Mode monitors the sensors on the vehicle which check operating conditions while the engine is running. The engine running test mode can be performed with the engine idling in NEUTRAL and with parking brake set or under actual driving conditions. With the diagnostic readout box READ/HOLD switch in the READ position, the engine running test mode is initiated after the engine is started.

Select a test code by switching the READ/HOLD switch to the READ position and pressing the actuator button until the desired code appears. Release actuator button and switch the READ/HOLD switch to the HOLD position. The logic module will monitor that system test and results will be displayed.

Only fuel injected engines offer this function. The Feedback carburetor system does not offer engine running sensor test mode.

ENGINE RUNNING TEST DISPLAY CODES:

Code 61 Battery temperature sensor; display number divided by 10 equals voltage

Code 62 Oxygen sensor; display number divided by 10 equals voltage

Code 63 Fuel injector temperature sensor; display number divided by 10 equals voltage

Code 64 Engine coolant temperature sensor; display number multiplied by 10 equals degrees F

Code 65 Throttle position sensor; display number divided by 10 equals voltage

Code 67 Battery voltage sensor; display is voltage

Code 68 Manifold vacuum sensor; display is in. Hg code 69—Minimum throttle position sensor; display number divided by 10 equals voltage

Code 70 Minimum airflow idle speed sensor; display number multiplied by 10 equals rpm (see minimum air flow check procedure)

Code 71 Vehicle speed sensor; display is mph Code 72—Engine speed sensor; display number multiplied by 10 equals rpm

FUEL/IGNITION INPUT/OUTPUT:

The engine controller recognizes only two states of electrical signals, voltage high or low. In some cases this corresponds to a switch or circuit being on or off; in other circuits a voltage signal may change from low voltage to higher voltage as a sensor opens. The controller cannot recognize the difference between a selected switch position and an open or shorted circuit.

In this test mode, the change in the circuit may be viewed as the switch is operated. For example, if the BRAKE SWITCH state is selected, the display should change from Low to High as the brake pedal is pressed. If a change in a circuit is displayed as the switch is used, it may be reasonably assumed that the entire switch circuit into the engine controller is operating correctly.

Depending on the engine/transmission in the vehicle, some of the switch states which may be checked include the air conditioning switch, brake switch, park/neutral switch, fuel flow signal, air conditioning clutch relay, radiator fan relay, CHECK ENGINE lamp, overdrive solenoid(s), lock-up solenoid and the speed control vent or vacuum solenoids. The scan tool will recognize the correct choices for each vehicle and only offer the appropriate systems on the screen.

MONITORS

On vehicles built before 1991, this display is called ENGINE PARAMETERS. 1991 and newer vehicles name the screen MONITORS. This display allows close observation of groups of related signals.

For example, if RPM is chosen, the screen will display data for many of the factors affecting the rpm such as throttle position sensor, advance, air conditioning status, park/neutral status, AIS status and coolant temperature.

One of the screens within this test is NO START. When this display is selected, the screen shows the initial data sent to the engine controller during cranking. Using this screen to identify missing or unusual signals can shorten diagnostic time.

Actuator Tests

The purpose of the circuit actuation mode test is to check for proper operation of the output circuits that the PCM cannot internally recognize. The PCM can attempt to activate these outputs and allow the technician to affirm proper operation. Most of the tests performed in this mode issue an audible click or visual indication of component operation (click of relay contacts, injector spray, etc.). Except for intermittent conditions, if a component functions properly when it is tested, it can be assumed that the component, attendant wiring and driving circuit are functioning properly.

1988 AND EARLIER MODELS

The Actuator Test Mode 10 Code number was introduced in 1985. In 1983–84 ATM function only provided 3 ignition sparks, 2 AIS motor cycles and 1 injector pulse.

1. Put the system into the diagnostic test mode and wait for Code 55 to appear on the display screen.

2. Press ATM button on the tool to activate the display. If a specific ATM test is desired, hold the ATM button down until the desired test code appears.

3. The computer will continue to turn the selected circuit on and off for as long as 5 minutes or until the ATM button is pressed again or the ignition switch is turned to the OFF position.

4. If the ATM button is not pressed again, the computer will continue to cycle the selected circuit for 5 minutes and then shut the system off. Turning the ignition to the OFF position will also turn the test mode off.

ACTUATOR TEST DISPLAY CODES:

Code 01 Spark activation—once every 2 seconds

Code 02 Injector activation—once every 2 seconds

Code 03 AIS activation—one step open, one step closed every 4 seconds

Code 04 Radiator fan relay—once every 2 seconds code 05-A/C WOT cutout relay—once every 2 seconds Code 06—ASD relay activation—once every 2 seconds code 07—Purge solenoid activation—one toggle every 2 seconds (The A/C fan will run continuously and the A/C switch must be in the ON position to allow for actuation)

Code 08 Speed control activation—speed control vent and vacuum every 2 seconds (Speed control switch must be in the ON position to allow for activation)

Code 09 Alternator control field activation—one toggle every 2 seconds

Code 10 Shift indicator activation—one toggle every 2 seconds

Code 11 EGR diagnosis solenoid activation—one toggle every 2 seconds

1989 AND NEWER MODELS

This family of tests is chosen from the FUEL/IGN MENU screen. The actuator tests allow the operation of the output circuits not recognized by the engine controller to be checked by energizing them on command. Testing in this fashion is necessary because the controller does not recognize the function of all the external components. If an output to a relay is triggered, and the relay is heard to click, it may be reasonably assumed that both the output circuit and the relay are operating properly. In this mode, most of the tests cause a response that may be seen or heard, although close attention may be necessary to notice the change.

Once selected, the ACTUATOR TEST screen offers a choice of items to be activated. Depending on engine and fuel system, some of the choices include:

- Stop all tests
- Engine rpm
- Ignition coil
- Fuel injector
- Fuel system
- Solenoid/relay
- AIS motor

The engine speed may be set to a desired level through the ENGINE RPM screen. Once a system is chosen, related screens will appear allowing detailed selection of which relay, injector or component is to be operated.

Exiting Diagnostic Test

By turning the ignition switch to the OFF position, the test mode system is exited. With a Diagnostic Readout Box attached to the system and the ATM control button not pressed, the computer will continue to cycle the selected circuits for 5 minutes and then automatically shut the system down.

Clearing Codes

Stored faults should only be cleared by use of the ORB-II or similar scan tool. Disconnecting the battery will clear codes but is not recommended as doing so will also clear all other memories on the vehicle and may affect drive ability. Disconnecting the PCM connector will also clear codes, but on newer models it may store a power loss code and will affect driveability until the vehicle is driven and the PCM can relearn it's drive ability memory.

The—ERASE—screen will appear when ATM is pressed at the end of the stored faults. Select the desired action from ERASE or DON'T ERASE. If ERASE is chosen, the display asks ARE YOU SURE? Pressing ENTER erases stored faults and displays the message FAULTS ERASED. After the faults are erased, press ATM to return the FUEL/IGN MENU.

DOMESTIC VEHICLE SELF-DIAGNOSTICS

Chrysler Domestic Built Fuel Injection System

Code 88 Display used for start of test

Code 11 Camshaft signal or Ignition signal—no reference signal detected during engine cranking

Code 12 Memory to controller has been cleared within 50-100 engine starts

Code 13 MAP sensor pneumatic signal—no variation in MAP sensor signal is detected or no difference is recognized between the engine MAP reading and the stored barometric pressure reading

Code 14 MAP voltage too high or too low

Code 15 Vehicle speed sensor signal—no distance sensor signal detected during road load conditions

Code 16 Knock sensor circuit—Open or short has been detected in the knock sensor circuit

Code 16 Battery input sensor—battery voltage sensor input below 4 volts with engine running

Code 17 Low engine temperature—engine coolant temperature remains below normal operating temperature during vehicle travel; possible thermostat problem

Code 21 Oxygen sensor signal—neither rich or lean condition is detected from the oxygen sensor input

Code 22 Coolant voltage low—coolant temperature sensor input below the minimum acceptable voltage/Coolant voltage high—coolant temperature sensor input above the maximum acceptable voltage

Code 23 Air Charge or Throttle Body temperature voltage HIGH/LOW—charge air temperature sensor input is above or below the acceptable voltage limits

Code 24 Throttle Position sensor voltage high or low. Code 25—Automatic Idle Speed (AIS) motor driver circuit—short or open detected in 1 or more of the AIS control circuits

Code 26 Injectors No. 1, 2, or 3 peak current not reached, high resistance in circuit

Code 27 Injector control circuit—bank output driver stage does not respond properly to the control signal

Code 27 Injectors No. 1, 2, or 3 control circuit and peak current not reached

Code 31 Purge solenoid circuit—open or short detected in the purge solenoid circuit

Code 32 Exhaust Gas Recirculation (EGR) solenoid circuit—open or short detected in the EGR solenoid circuit EGR system failure—required change in fuel/air ratio not detected during diagnostic test

Code 32 Surge valve solenoid—open or short in turbocharger surge valve circuit—some 1993 vehicles

Code 33 Air conditioner clutch relay circuit—open or short detected in the air conditioner clutch relay circuit. If vehicle doesn't have air conditioning ignore this code

Code 34 Speed control servo solenoids or MUX speed control circuit HIGH/LOW—open or short detected in the vacuum or vent solenoid circuits or speed control switch input above or below allowable voltage

Code 35 Radiator fan control relay circuit—open or short detected in the radiator fan relay circuit

Code 35 Idle switch shorted—switch input shorted to ground—some 1993 vehicles

Code 36 Wastegate solenoid—open or short detected in the turbocharger wastegate control solenoid circuit

Code 37 Part Throttle Unlock (PTU) circuit for torque converter clutch—open or short detected in the torque converter part throttle unlock solenoid circuit

Code 37 Baro Reed Solenoid—solenoid does not turn off when it should

Code 37 Shift indicator circuit (manual transaxle)

Code 37 Transaxle temperature out of range—some 1993 models

Code 41 Charging system circuit—output driver stage for generator field does not respond properly to the voltage regulator control signal

Code 42 Fuel pump or no Auto shut-down (ASD) relay voltage sense at controller

Code 43 Ignition control circuit—peak primary circuit current not respond properly with maximum dwell time

Code 43 Ignition coil #1, 2, or 3 primary circuits—peak primary was not achieved within the maximum allowable dwell time

Code 44 Battery temperature voltage—problem exists in the PCM battery temperature circuit or there is an open or short in the engine coolant temperature circuit

Code 44 Fused J2 circuit in not present in the logic board; used on the single engine module controller system

Code 45 Turbo boost limit exceeded—MAP sensor detects overboost

Code 44 Overdrive solenoid circuit—open or short in overdrive solenoid circuit

Code 46 Battery voltage too high—battery voltage sense input above target charging voltage during engine operation

Code 47 Battery voltage too low—battery voltage sense input below target charging voltage

Code 51 Air/fuel at limit—oxygen sensor signal input indicates LEAN air/fuel ratio condition during engine operation

Code 52 Air/fuel at limit—oxygen sensor signal input indicates RICH air/fuel ratio condition during engine operation

Code 52 Logic module fault—1984 vehicles. Code 53—Internal controller failure—internal engine controller fault condition detected during self test

Code 54 Camshaft or (distributor sync.) reference circuit—No camshaft position sensor signal detected during engine rotation

Code 55 End of message

Code 61 Baro read solenoid—open or short detected in the baro read solenoid circuit

Code 62 EMR mileage not stored—unsuccessful attempt to update EMR mileage in the controller EEPROM

Code 63 EEPROM write denied—unsuccessful attempt to write to an EEPROM location by the controller

Code 64 Flex fuel sensor—Flex fuel sensor signal out of range—(new in 1993)- CNG Temperature voltage out of range—CN gas pressure out of range

Code 65 Manifold tuning valve—an open or short has been detected in the manifold tuning valve solenoid circuit (3.3L and 3.SL LH-Platform)

Code 66 No CCO messages or no BODY CCO messages or no EATX CCO messages—messages from the CCO bus or the BODY CCD or the EATX CCO were not received by the PCM

Code 76 Ballast bypass relay—open or short in fuel pump relay circuit

Code 77 Speed control relay—an open or short has been detected in the speed control relay

Code 88 Display used for start of test

Code Error Fault code error—Unrecognized fault 10 received by ORBII

➡**This list is for reference and does not mean that a component is defective. The code identifies the circuit and component that require further testing.**

Chrysler Domestic Built Feedback Carburetor System

Code 88 Display used for start of test—must appear or other codes aren't valid

Code 11 Carburetor oxygen solenoid

Code 12 Transmission unlock relay—3.7L and 5.2L

Code 13 Air switching solenoid—3.7L and 5.2L—or Vacuum operated secondary control solenoid—2.2L

Code 14 Battery feed to computer disconnected with 20–40 engine starts

Code 16 Ignore

Code 17 Electronic throttle control solenoid

Code 18 EGR or Purge control solenoid

Code 21 Distributor pick-up signal

Code 22 Oxygen feedback stays rich or lean too long -3.7L and 5.2L—or Oxygen feedback is LEAN too long—2.2L

Code 23 Oxygen feedback is RICH too long—2.2L

Code 24 Vacuum transducer signal problem

Code 25 Charge temperature switch signal—3.7L and 5.2L engine—or Radiator fan temperature switch signal—2.2L engine

Code 26 Charge temperature sensor signal—3.7L and 5.2L engine—or Engine temperature sensor signal—2.2L

Code 28 Speed sensor circuit (if equipped)

Code 31 Battery feed to computer

Code 32 Computer can't enter diagnostics

Code 33 Computer can't enter diagnostics

Code 55 End of message

Code 88 Display used for start of test

Code 00 Diagnostic readout box is powered up and waiting for codes

➡**This list is for reference and does not mean that a component is defective. The code identifies the circuit and component that require further testing.**

Chrysler Import Built Fuel Systems

1984–1988 COLT, VISTA, SUMMIT AND D50

Code 1 Oxygen sensor
Code 2 Crank angle sensor—or Ignition signal
Code 3 Air flow sensor
Code 4 Barometric pressure sensor
Code 5 Throttle Position Sensor (TPS)
Code 6 Motor Position Sensor (MPS)—or Idle Speed Control (ISC) position sensor
Code 7 Engine Coolant Temperature Sensor
Code 8 No. 1 cylinder TDC Sensor—or Vehicle speed sensor

➡Some 1988 Multi-Point injected vehicles use 1989 2-digit codes.

1988–93 COLT, SUMMIT, VISTA, LASER, TALON, STEALTH AND D50

Code 11 Oxygen sensor
Code 12 Air flow sensor
Code 13 Intake Air Temperature Sensor
Code 14 Throttle Position Sensor (TPS)
Code 15 SC Motor Position Sensor (MPS)
Code 21 Engine Coolant Temperature Sensor
Code 22 Crank angle sensor
Code 23 No. 1 cylinder TOC (Camshaft position) Sensor
Code 24 Vehicle speed sensor
Code 25 Barometric pressure sensor
Code 31 Knock (KS) sensor
Code 32 Manifold pressure sensor
Code 36 Ignition timing adjustment signal
Code 39 Oxygen sensor (rear—turbocharged)
Code 41 Injector
Code 42 Fuel pump
Code 43 EGR-California
Code 44 Ignition Coil—power transistor unit (No. 1 and No. 4 cylinders) on 3.0L Stealth
Code 52 Ignition Coil—power transistor unit (No. 2 and No. 5 cylinders) on 3.0L Stealth
Code 53 Ignition coil, power transistor unit (No. 3 and No. 6 cylinders)
Code 55 IAC valve position sensor
Code 59 Heated oxygen sensor
Code 61 Transaxle control unit cable (automatic transmission)
Code 62 Warm up control valve position sensor (non-turbo)

Jeep and Eagle Built Fuel Systems

1988–90 2.5L, 3.0L AND 4.0L ENGINE

Code 1000 Ignition line low
Code 1001 Ignition line high
Code 1002 Oxygen heater line
Code 1004 Battery voltage low
Code 1005 Sensor ground line out of limits
Code 1010 Diagnostic enable line low
Code 1011 Diagnostic enable line high
Code 1012 MAP line low
Code 1013 MAP line high
Code 1014 Fuel pump line low
Code 1015 Fuel pump line high

Code 1016 Charge air temperature sensor low
Code 1017 Charge air temperature sensor high
Code 1018 No serial data from the ECU
Code 1021 Engine failed to start due to mechanical, fuel, or ignition problem
Code 1022 Start line low
Code 1024 ECU does not see start signal
Code 1025 Wide open throttle circuit low
Code 1027 ECU sees wide open throttle
Code 1028 ECU does not see wide open throttle
Code 1031 ECU sees closed throttle
Code 1032 ECU does not see closed throttle
Code 1033 Idle speed increase line low
Code 1034 Idle speed increase line high
Code 1035 Idle speed decrease line low
Code 1036 Idle speed decrease line high
Code 1037 Throttle position sensor reads low
Code 1038 Park/Neutral line high
Code 1040 Latched B+ line low
Code 1041 Latched B+ line high
Code 1042 No Latched B+ 1/2 volt drop
Code 1047 Wrong ECU
Code 1048 Manual vehicle equipped with automatic ECU
Code 1949 Automatic vehicle equipped with manual ECU
Code 1050 Idle RPM less than 500
Code 1051 Idle RPM greater than 2000
Code 1052 MAP sensor out of limits
Code 1053 Change in MAP reading out of limits
Code 1054 Coolant temperature sensor line low
Code 1055 Coolant temperature sensor line high
Code 1056 Inactive coolant temperature sensor
Code 1057 Knock circuit shorted
Code 1058 Knock value out of limits
Code 1059 A/C request line low
Code 1060 A/C request line high
Code 1061 A/C select line low
Code 1062 A/C select line high
Code 1063 A/C clutch line low
Code 1064 A/C clutch line high
Code 1065 Oxygen reads rich
Code 1066 Oxygen reads lean
Code 1067 Latch relay line low
Code 1068 Latch relay line high
Code 1070 A/C cutout line low
Code 1071 A/C cutout line high
Code 1073 ECU does not see speed sensor signal
Code 1200 ECU defective
Code 1202 Injector shorted to ground
Code 1209 Injector open
Code 1218 No voltage at ECU from power latch relay
Code 1220 No voltage at ECU from EGR solenoid
Code 1221 No injector voltage
Code 1222 MAP not grounded
Code 1223 No ECU tests run

➡Prior to 1988 vehicles used an Off-Board Diagnostic system which required special diagnostic equipment to read codes. After 1991 Jeep and Eagle vehicles used the Chrysler Domestic Built Engine Control system. The code list for Chrysler Built Domestic Fuel injection System also covers 1991 and newer Jeep and Eagle vehicles.

Ford Motor Co.

INTRODUCTION TO FORD SELF-DIAGNOSTICS

The engine control systems are used in conjunction with either a throttle body (CFI) injection or multi-point (EFI and SEFI) injection fuel delivery system or feedback carburetor systems depending on the year, model and powertrain. Although the individual system components vary slightly, the electronic control system operation is basically the same. The major difference is the number and type of output devices being controlled by the ECA.

Automotive manufacturers have developed on-board computers to control engines, transmissions and many other components. These on-board computers with dozens of sensors and actuators have become almost impossible to test without the help of electronic test equipment.

One of these electronic test devices has become the on-board computer itself. The Powertrain Control Modules (PCM), sometimes called the Electronic Control Assembly (ECA), used on toadies vehicles has a built in self testing system. This self test ability is called self-diagnosis. The self-diagnosis system will test many or all of the sensors and controlled devices for proper function. When a malfunction is detected this system will store a fault code in memory that's related to that specific circuit. You can access the computer to obtain fault codes recorded in memory by using an analog voltmeter or special diagnostic scan tool. This will help narrow down what area to begin testing.

Fault code meanings can vary from year to year even on the same model. It is extremely important after retrieving a fault code to verify its meaning with a proper manual. Servicing a fault code incorrectly will not only lead to the wrong conclusion but could also cause damage if tested or serviced incorrectly. There is a list of general code descriptions provide later in this manual.

What System Is On My Car?

There are 3 electronic fuel control systems used by Ford Motor Company. These systems all operate using similar components and on-board computers. Self-Diagnostic on these systems will vary, but, the basic fuel control operation is the same. Ford uses the following systems:

- **EEC-IV and EEC-V** engine control system: used on most domestic built Ford vehicles since 1984.
- **Non-NAAO EEC** engine control system: used on import built Ford vehicles, referred to as Non-NAAO cars.
- **MCU** feedback carburetor system: used on most Ford vehicles before 1984 and some later model vehicles equipped with a V8 engine and feedback carburetor.

Most Ford vehicles made after 1983 except for Capri, Festiva, Probe, Escort and Tracer use the 4th generation Electronic Engine Control system, commonly designated EEC-IV.

If you own a Capri, Festiva, Probe 2.0L, 2.2L, or 2.SL, an Escort or Tracer with a 1 .6L or 1 .8L engine, then the fuel control system is referred to as NON-NAAO (Not North American Automotive Operations produced vehicles) system. The fuel system used on these vehicles is called Electronic Engine Control (EEC). This Non-NAAO EEC system components and operation are basically the same as the EEC-IV system. The self-diagnostic function on the EEC system differs from the EEC-VI system and is covered under NON-NAAO vehicle.

Most 1984–94 Ford domestic built vehicles employ the 4th generation Electronic Engine Control system, commonly called EEC-IV, to manage fuel, ignition and emissions on vehicle engines. In 1994 the EEC-V system was introduced on some models. The diagnostic system on EEC-V provides 3 digit codes in place of 2 digit codes, and it is capable of monitoring more inputs and outputs.

If your vehicle was made before 1984, or has a feedback carburetor equipped V8 engine, then it probably uses the Microprocessor Control Unit (MCU). The MCU system was used on most 1981-83 carburetor equipped vehicles, and 1984 and newer V8 engines with feedback carburetors. The MCU system uses a large six sided connector, identical to the one used with EEC-IV systems. The MCU system does NOT use the small single wire connector, like the EEC-IV system. The MCU system is covered in greater detail later in this manual.

EEC-IV & EEC-V DIAGNOSTIC SYSTEMS

This system includes all Ford Motor Company vehicles with the exception of imported vehicles like the Capri, Festiva, Probe 2.0L, 2.2L and 2.SL engine and the Escort and Tracer equipped with the 1 .8L engine.

Most 1984–94 Ford domestic built vehicles employ the 4th generation Electronic Engine Control system, commonly designated EEC-IV, to manage fuel, ignition and emissions on vehicle engines. In 1994 the EEC-V system was introduced on some models. The diagnostic system on EEC-V provides 3 digit codes in place of 2 digit codes and monitors more components.

Engine Control System

The Powertrain Control Modules (PCM), usually referred to as the Electronic Control Assembly (ECA) by Ford, is given responsibility for the operation of the emission control devices, cooling fans, ignition and advance and in some cases, automatic transmission functions. Because the EEC-IV oversees both the ignition timing and the fuel injector operation, a precise air/fuel ratio will be maintained under all operating conditions. The ECA is a microprocessor or small computer which receives electrical inputs from several sensors, switches and relays on and around the engine.

Based on combinations of these inputs, the ECA controls outputs to various devices concerned with engine operation and emissions. The engine control assembly relies on the signals to form a correct picture of current vehicle operation. If any of the input signals is incorrect, the ECA reacts to what ever picture is painted for it. For example, if the coolant temperature sensor is inaccurate and reads too low, the ECA may see a picture of the engine never warming up. Consequently, the engine settings will be maintained as if the engine were cold. Because so many inputs can affect one output, correct diagnostic procedures are essential on these systems.

One part of the ECA is devoted to monitoring both input and output functions within the system. This ability forms the core of the self-diagnostic system. If a problem is detected within a circuit, the controller will recognize the fault, assign it an identification code, and store the code in a memory section. Depending on the year and model, the fault code(s) may be represented by two or three digit numbers. The stored code(s) may be retrieved during diagnosis.

When the term Powertrain Control Module (PCM) is used in this manual it will refer to the engine control computer regardless that it may also be called an Electronic Control Assembly (ECA).

While the EEC-IV system is capable of recognizing many internal faults, certain faults will not be recognized. Because the computer system sees only electrical signals, it cannot sense or react to mechanical or vacuum faults affecting engine operation. Some of these faults may affect another component which will set a code. For example, the ECA monitors the output signal to the fuel injectors, but cannot detect a partially clogged injector. As long as the output driver responds correctly, the computer will read the system as functioning correctly. However, the improper flow of fuel may result in a lean mixture. This would, in turn, be detected by the oxygen sensor and noticed as a constantly lean signal by the ECA. Once the signal falls outside the pre-programmed limits, the engine control assembly would notice the fault and set an identification code.

Additionally, the EEC-IV system employs adaptive fuel logic. This process is used to compensate for normal wear and variability within the fuel system. Once the engine enters steady-state operation, the engine control assembly watches the oxygen sensor signal for a bias or tendency to run slightly rich or lean. If such a bias is detected, the adaptive logic corrects the fuel delivery to bring the air/fuel mixture towards a centered or 14.7:1 ratio. This compensating shift is stored in a non-volatile memory which is retained by battery power even with the ignition switched off. The correction factor is then available the next time the vehicle is operated.

➡**If the battery is disconnected for longer than 5 minutes, the adaptive fuel factor will be lost. After repair it will be necessary to drive the car at least 10 miles to allow the processor to relearn the correct factors. The driving period should include steady-throttle open road driving if possible. During the drive, the vehicle may exhibit driveability symptoms not noticed before. These symptoms should clear as the ECA computes the correction factor. The ECA will also store Code 19 indicating loss of power to the controller.**

FAILURE MODE EFFECTS MANAGEMENT (FMEM)

The engine controller assembly contains back-up programs which allow the engine to operate if a sensor signal is lost. If a sensor input is seen to be out of range—either high or low—the FMEM program is used. The processor substitutes a fixed value for the missing sensor signal. The engine will continue to operate, although performance and driveability may be noticeably reduced. This function of the controller is sometimes referred to as the limp-in or fail-safe mode. If the missing sensor signal is restored, the FMEM system immediately returns the system to normal operation. The dashboard warning lamp will be lit when FMEM is in effect.

HARDWARE LIMITED OPERATION STRATEGY (HLOS)

This mode is only used if the fault is too extreme for the FMEM circuit to handle. In this mode, the processor has ceased all computation and control; the entire system is run on fixed values. The vehicle may be operated but performance and driveability will be greatly reduced. The fixed or default settings provide minimal calibration, allowing the vehicle to be carefully driven in for service. The dashboard warning lamp will be lit when HLOS is engaged. Codes cannot be read while the system is operating in this mode.

Dashboard Warning Lamp

The CHECK ENGINE or SERVICE ENGINE SOON dashboard warning lamp is referred to as the Malfunction Indicator Lamp (MIL). The lamp is connected to the engine control assembly and will alert the driver to certain malfunctions within the EEC-IV system. When the lamp is lit, the ECA has detected a fault and stored an identity code in memory. The engine control system will usually enter either FMEM or HLOS mode and driveability will be impaired.

The light will stay on as long as the fault causing it is present. Should the fault self-correct, the MIL will extinguish but the stored code will remain in memory.

Under normal operating conditions, the MIL should light briefly when the ignition key is turned ON. As soon as the ECA receives a signal that the engine is cranking, the lamp will be extinguished. The dash warning lamp should remain out during the entire operating cycle.

➡**On Continental, the CHECK ENGINE message is displayed on the message center. When a fault is detected, the message is accompanied by a 1 second tone every 5 seconds. The tone stops after 1 minute. When the Continental system enters HLOS, the additional message CHECK DCL is displayed. DCL refers to the Data Communications Link running between the engine controller and the message center.**

EEC-IV & EEC-V SCAN TOOL FUNCTIONS

Although stored codes may be read by using a analog voltmeter, the use of hand-held scan tools such as Ford's Self-Test Automatic Readout (STAR) tester or the second generation SUPER STAR II tester or their equivalent is recommended. There are many manufacturers of these tools; the purchaser must be certain that the tool is proper for the intended use.

Both the STAR and SUPER STAR testers are designed to communicate directly with the EEC-IV system and interpret the electrical signals. The SUPER STAR tester may be used to read either 2 or 3 digit codes; the original STAR tester will not read the 3 digit codes used on many 1990 and newer vehicles.

The scan tool allows any stored faults to be read from the engine controller memory. Use of the scan tool provides additional data during troubleshooting but does not eliminate the use of the charts. The scan tool makes collecting information easier; the data must be correctly interpreted by an operator familiar with the system.

Electrical Tools

The most commonly required electrical diagnostic tool is the Digital Multimeter, allowing voltage, resistance and amperage to be read by one instrument. Many of the diagnostic charts require the use of a volt or ohmmeter during diagnosis.

The multimeter must be a high impedance unit, with 10 megohms of impedance in the voltmeter. This type of meter will not place an additional load on the circuit it is testing; this is extremely important in low voltage circuits. The multimeter must be of high quality in all respects. It should be handled carefully and protected from impact or damage. Replace the batteries frequently in the unit.

Additionally, an analog (needle type) voltmeter may be used to read stored fault codes if the STAR tester is not available. The codes are transmitted as visible needle sweeps on the face of the instrument. Almost all diagnostic procedures will require the use of the Breakout Box, a device which connects into the EEC-IV harness and provides testing ports for the 60 wires in the harness. Direct testing of the harness connectors at the terminals or by back-probing is not recommended; damage to the wiring and terminals is almost certain to occur.

Other necessary tools include a quality tachometer with inductive (clip-on) pickup, a fuel pressure gauge with system adapters and a vacuum gauge with an auxiliary source of vacuum.

EEC-IV & EEC-V SELF-DIAGNOSTICS

Diagnosis of a driveability problem requires attention to detail and following the diagnostic procedures in the correct order. Resist the temptation to begin extensive testing before completing the preliminary diagnostic steps. The preliminary or visual inspection must be completed in detail before diagnosis begins. In many cases this will shorten diagnostic time and often cure the problem without electronic testing.

Visual Inspection

This is possibly the most critical step of diagnosis. A detailed examination of all connectors, wiring and vacuum hoses can often lead to a repair without further diagnosis. Performance of this step relies on the skill of the technician performing it; a careful inspector will check the undersides of hoses as well as the integrity of hard-to-reach hoses blocked by the air cleaner or other components. Wiring should be checked carefully for any sign of strain , burning, crimping or terminal pull-out from a connector.

Checking connectors at components or in harnesses is required; usually, pushing them together will reveal a loose fit. Pay particular attention to ground circuits, making sure they are not loose or corroded. Remember to inspect connectors and hose fittings at components not mounted on the engine, such as the evaporative canister or relays mounted on the fender aprons. Any component or wiring in the vicinity of a fluid leak or spillage should be given extra attention during inspection.

Additionally, inspect maintenance items such as belt condition and tension, battery charge and condition and the radiator cap carefully. Any of these very simple items may affect the system enough to set a fault.

Diagnostic Connector Location

The Diagnostic Link Connectors (DLC) are located a 6 basic locations:
- Near the bulkhead (right or left side of vehicle)
- Near the wheel well (right or left side of vehicle)
- Near the front corner of the engine compartment (right or left side of vehicle)

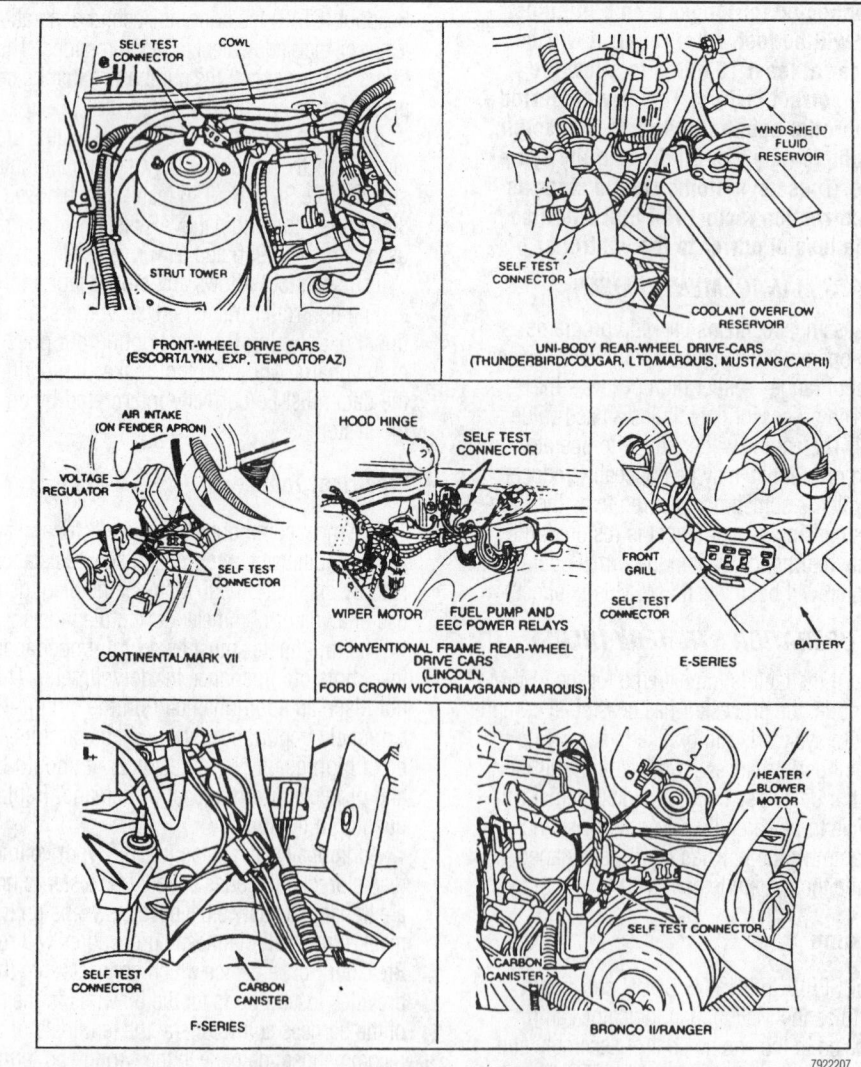

7922207

Typical diagnostic link connector locations. Locations will vary depending on the year or model

EEC-IV & EEC-V READING CODES

The EEC-IV system may be interrogated for stored codes using the Quick Test procedures. These tests will reveal faults immediately present during the test as well as any intermittent codes set within the previous 80 warm up cycles. If a code was set before a problem self-corrected (such as a momentarily loose connector), the code will be erased if the problem does not reoccur within 80 warm-up cycles.

The Quick Test procedure is divided into 2 sections, Key On Engine Off (KOEO) and Key On Engine Running (KOER). These 2 procedures must be performed correctly if the system is to run the internal self-checks and provide accurate fault codes. Codes will be output and displayed as numbers on the hand scan tool, i.e. 23. Code 23 would be displayed as 2 needle sweeps and pause and 3 more needle sweeps. For codes being read on an analog voltmeter, the needle sweeps indicate the code digits in the same manner as the lamp flashes on other systems.

In all cases, the codes 11 or 111 are used to indicate PASS during testing. Note that the PASS code may appear, followed by other stored codes. These are codes from the continuous memory and may indicate intermittent faults, even though the system does not presently contain the fault. The PASS designation only indicates the system passes all internal tests at the moment.

Once the Quick Test has been performed and all fault codes recorded, refer to the code charts. The charts direct the use of specific pinpoint tests for the appropriate circuit and will allow complete circuit testing.

✳✳ CAUTION

To prevent injury and/or property damage, always block the drive wheels, firmly apply the parking brake, place the transmission in Park or Neutral and turn all electrical loads off before performing the Quick Test procedures.

Reading Codes With Analog Voltmeter

➡**There are inexpensive tools available at auto parts stores that make reading and clear Ford engine codes very easy. Reading the voltmeter needle sweeps is sometimes difficult. Always check the code more than once to make certain it was read correctly.**

In the absence of a scan tool, an analog voltmeter may be used to retrieve stored fault codes. Set the meter range to read DC 0–15 volts. Connect the positive (+) lead of the meter to the battery positive terminal and connect the negative (-) lead of the meter to the self-test output pin of the diagnostic connector.

Follow the directions given for performing the KOEO and KOER tests. To activate the tests, use a jumper wire to connect the signal return pin on the diagnostic connector to the self-test input connector. The self-test input line is the separate wire and connector with or near the diagnostic connector.

The codes will be transmitted as groups of needle sweeps. This method may be used to read either 2 or 3 digit codes. The Continuous Memory codes are separated from the KOEO codes by 6 seconds, a single sweep and another 6 second delay.

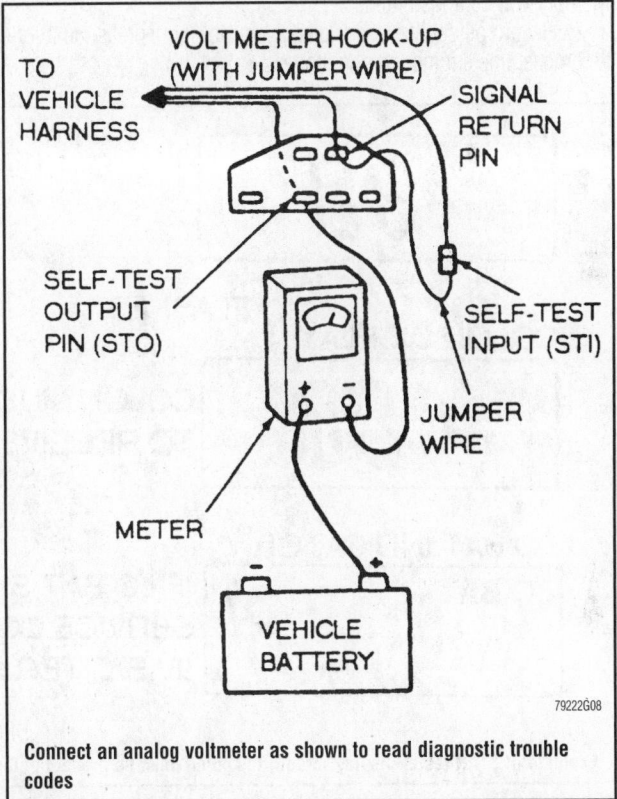

Connect an analog voltmeter as shown to read diagnostic trouble codes

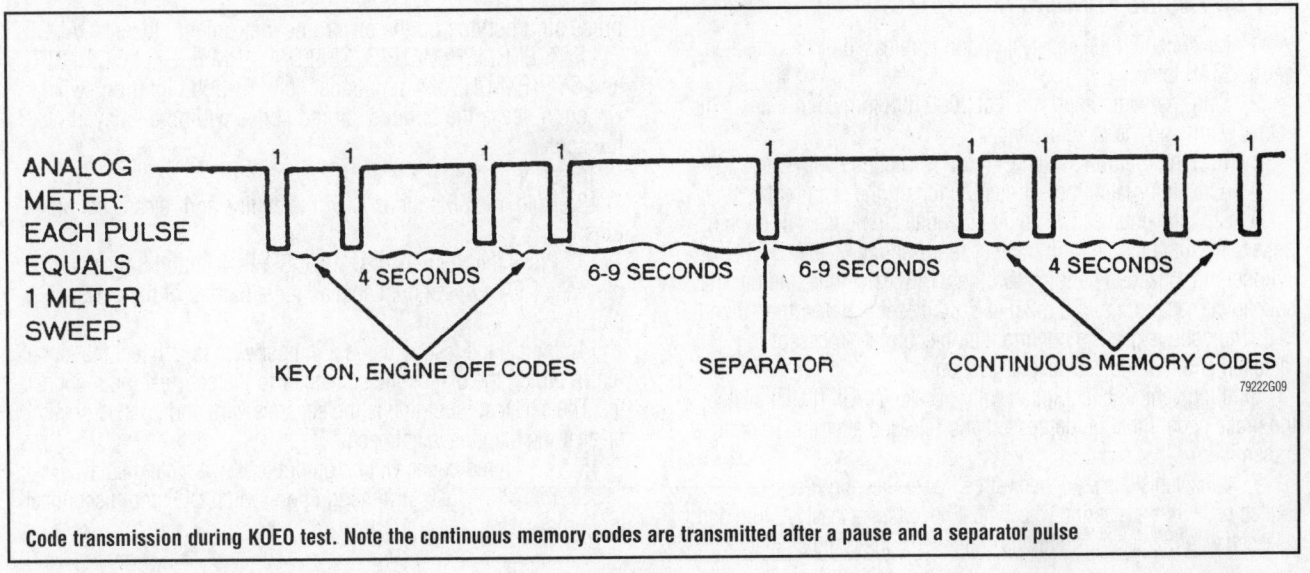

Code transmission during KOEO test. Note the continuous memory codes are transmitted after a pause and a separator pulse

KEY ON ENGINE OFF (KOEO) TEST

1. Connect the scan tool to the self-test connectors. Make certain the test button is unlatched or up.

2. Start the engine and run it until normal operating temperature is reached.

3. Turn the engine OFF for 10 seconds.

4. Activate the test button on the STAR tester.

5. Turn the ignition switch ON but do not start the engine. For vehicles with 4.9L engines, depress the clutch during the entire test. For vehicles with the 7.3L diesel engine, hold the accelerator to the floor during the test.

6. The KOEO codes will be transmitted. Six to nine seconds after the last KOEO code, a single separator pulse will be transmitted. Six to nine seconds after this pulse, the codes from the Continuous Memory will be transmitted.

7. Record all service codes displayed. Do not depress the throttle on gasoline engines during the test.

8. If the vehicle is equipped with the E400 transmission, the Overdrive Cancel Switch (OCS) must be cycled after the engine 10 code is transmitted.

9. Certain Ford vehicles will display a Dynamic Response code 6–20 seconds after the engine 10 code. This will appear as one pulse on a meter or as a 10 on the STAR tester. When this code appears, briefly take the engine to wide open throttle. This allows the system to test the throttle position, MAF and MAP sensors.

10. All relevant codes will be displayed and should be recorded. Remember that the codes refer only to faults present during this test cycle. Codes stored in Continuous Memory are not displayed in this test mode.

11. Do not depress the throttle during testing unless a dynamic response code is displayed.

TESTING WITH CONTINENTAL MESSAGE CENTER:

The stored fault codes may be displayed on the electronic message screen in Continentals so equipped. To perform the KOEO test,

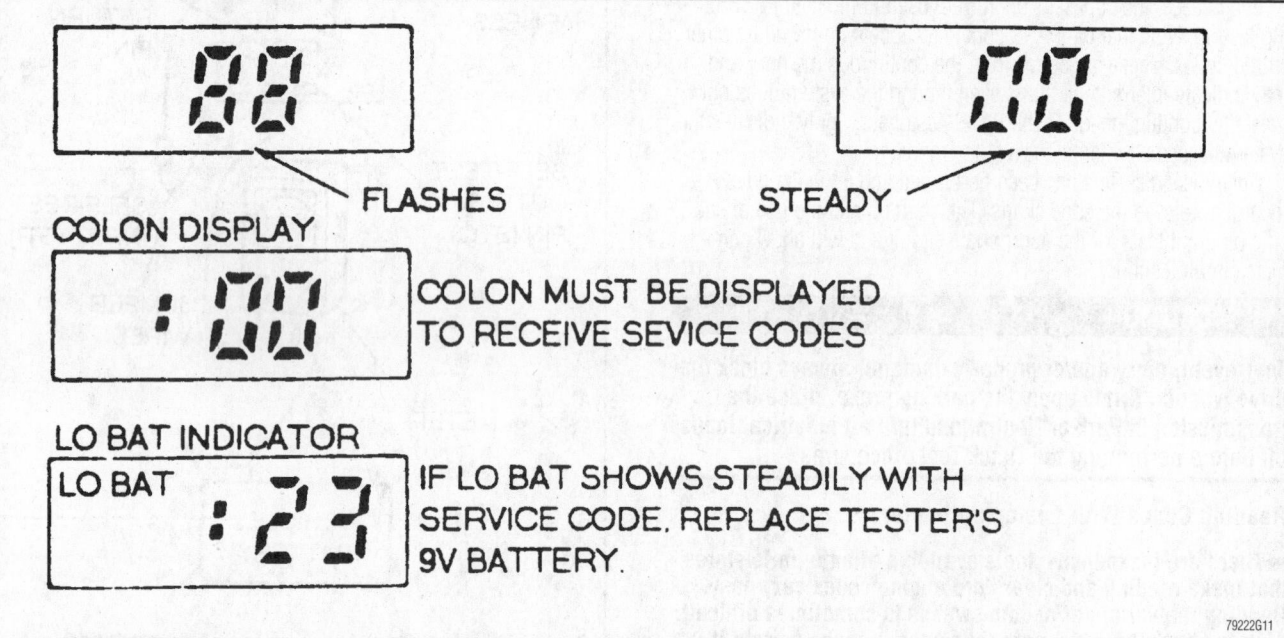

Example of STAR tester display screen. the colon must be present before the codes can be retrieved

KEY ON ENGINE RUNNING (KOER) TEST

1. Make certain the self-test button is released or de-activated on the STAR tester.

2. Start the engine and run it at 2000 rpm for two minutes. This action warms up the oxygen sensor.

3. Turn the ignition switch OFF for 10 seconds.

4. Activate or latch the self-test button on the scan tool.

5. Start the engine. The engine identification code will be transmitted. This is a single digit number representing ½ the number of cylinders in a gasoline engine. On the STAR tester, this number may appear with a zero, i.e., 20 = 2. For 7.3L diesel engines, the 10 code is 5. The code is used to confirm that the correct processor is installed and that the self-test has begun.

6. If the vehicle is equipped with a Brake On/Off (BOO) switch, the brake pedal must be depressed and released after the 10 code is transmitted.

7. If the vehicle is equipped with a Power Steering Pressure Switch (PSPS), the steering wheel must be turned at least `/2 turn and released within 2 seconds after the engine 10 code is transmitted.

press all 3 buttons on the electronic instrument cluster (GAUGE SELECT, ENGLISH/METRIC, SPEED ALARM or SELECT, RESET and SYSTEM CHECK) simultaneously. Turn the ignition switch ON and release the buttons; stored codes will be displayed on the screen.

To perform the KOER test:

12. Hold in all 3 buttons, start the engine and release the buttons.

13. Press the SELECT or GAUGE SELECT button 3 times. The message DEALER 4 should appear at the bottom of the message panel.

14. Initiate the test by using a jumper wire to connect the signal return pin on the diagnostic connector to the self-test input connector. The self-test input line is the separate wire and connector with or near the diagnostic connector.

15. The stored codes will be output to the vehicle display.

16. To exit the test, turn the ignition switch OFF and disconnect the jumper wire.

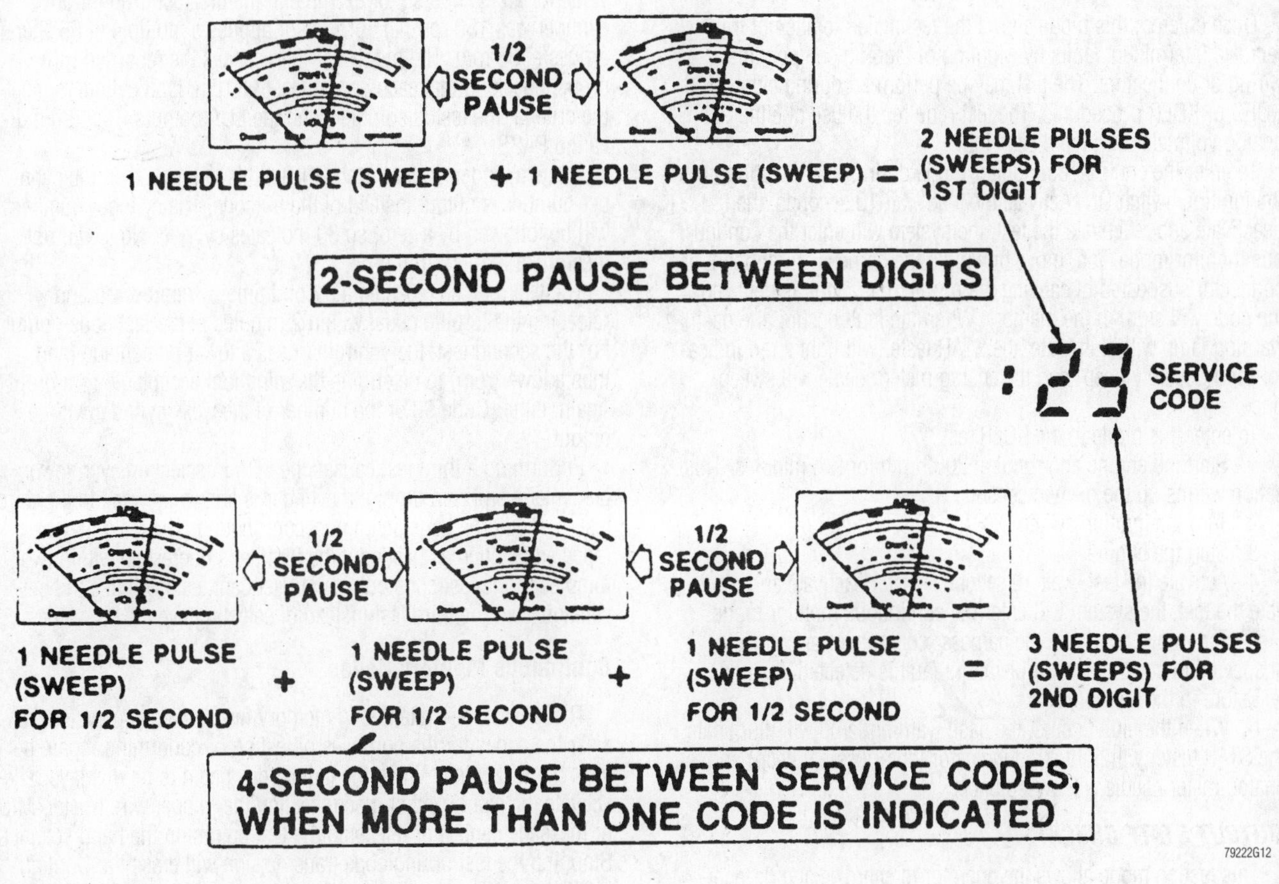

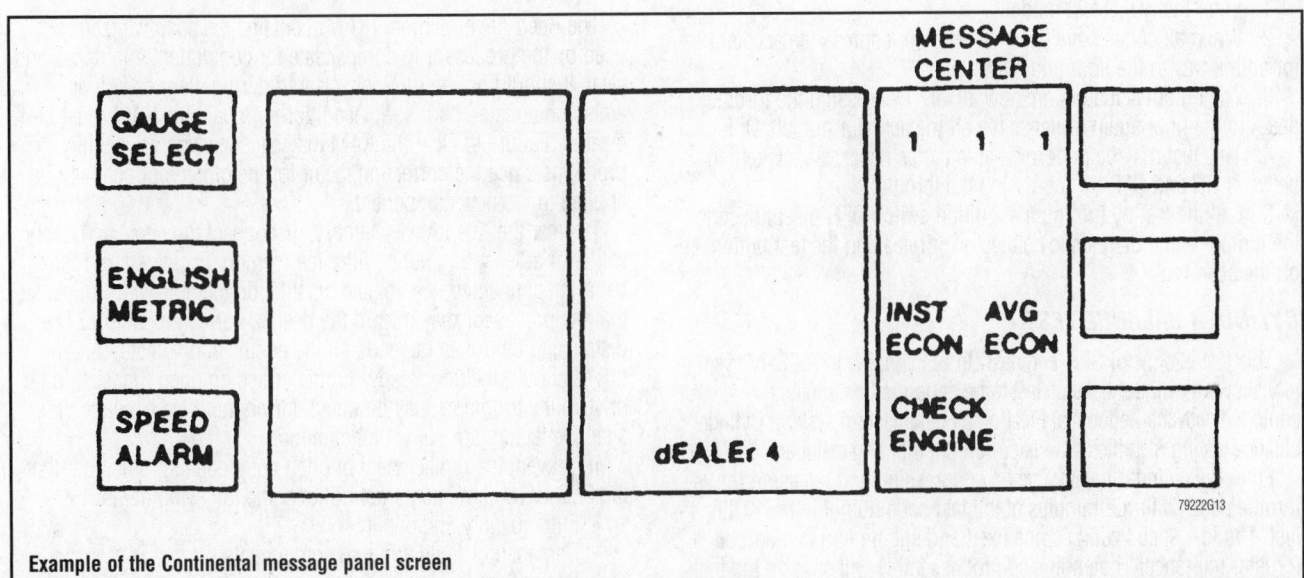

2-SECOND PAUSE BETWEEN DIGITS

1 NEEDLE PULSE (SWEEP) + 1 NEEDLE PULSE (SWEEP) = 2 NEEDLE PULSES (SWEEPS) FOR 1ST DIGIT

1/2 SECOND PAUSE

:23 SERVICE CODE

1 NEEDLE PULSE (SWEEP) FOR 1/2 SECOND + 1 NEEDLE PULSE (SWEEP) FOR 1/2 SECOND + 1 NEEDLE PULSE (SWEEP) FOR 1/2 SECOND = 3 NEEDLE PULSES (SWEEPS) FOR 2ND DIGIT

1/2 SECOND PAUSE   1/2 SECOND PAUSE

4-SECOND PAUSE BETWEEN SERVICE CODES, WHEN MORE THAN ONE CODE IS INDICATED

79222G12

Example of code display pattern using an analog voltmeter

MESSAGE CENTER

GAUGE SELECT

ENGLISH METRIC

SPEED ALARM

1 1 1

INST AVG
ECON ECON

CHECK ENGINE

dEALEr 4

79222G13

Example of the Continental message panel screen

Advanced Test Modes

CONTINUOUS MONITOR OR WIGGLE TEST MODE

Once entered, this mode allows the technician to attempt to recreate intermittent faults by wiggling or tapping components, wiring or connectors. The test may be performed during either KOEO or KOER procedures. The test requires the use of either an analog voltmeter or a hand scan tool.

To enter the continuous monitor mode during KOEO testing, turn the ignition switch ON. Activate the test, wait 10 seconds, then deactivate and reactivate the test; the system will enter the continuous monitor mode. Tap, move or wiggle the harness, component or connector suspected of causing the problem; if a fault is detected, the code will store in the memory. When the fault occurs, the dash warning lamp will illuminate, the STAR tester will light a red indicator (and possibly beep) and the analog meter needle will sweep once.

To enter this mode in the KOER test:

1. Start the engine and run it at 2000 rpm for two minutes. This action warms up the oxygen sensor.
2. Turn the ignition switch OFF for 10 seconds.
3. Start the engine.
4. Activate the test, wait 10 seconds, then deactivate and reactivate the test; the system will enter the continuous monitor mode.
5. Tap, move or wiggle the harness, component or connector suspected of causing the problem; if a fault is detected, the code will store in the memory.
6. When the fault occurs, the dash warning lamp will illuminate, the STAR tester will light a red indicator (and possibly beep) and the analog meter needle will sweep once.

OUTPUT STATE CHECK

This testing mode allows the operator to energize and de-energize most of the outputs controlled by the EEC-IV system. Many of the outputs may be checked at the component by listening for a click or feeling the item move or engage by a hand placed on the case. To enter this check:

1. Enter the KOEO test mode.
2. When all codes have been transmitted, depress the accelerator all the way to the floor and release it.
3. The output actuators are now all ON. Depressing the throttle pedal to the floor again switches the all the actuator outputs OFF.
4. This test may be performed as often as necessary, switching between ON and OFF by depressing the throttle.
5. Exit the test by turning the ignition switch OFF, disconnecting the jumper at the diagnostic connector or releasing the test button on the scan tool.

CYLINDER BALANCE TEST

This test is only for SEFI engines. On SEFI engine the EEC-IV system allows a cylinder balance test to be performed on engines equipped with the Sequential Electronic Fuel Injection system. Cylinder balance testing identifies a weak or non-contributing cylinder.

Enter the cylinder balance test by depressing and releasing the throttle pedal within 2 minutes of the last code output in the KOER test. The idle speed will become fixed and engine mm is recorded for later reference. The engine control assembly will shut off the fuel to the highest numbered cylinder (4, 6 or 8), allow the engine to stabilize and then record the rpm. The injector is turned back on and the next one shut off and the process continues through cylinder No. 1.

The controller selects the highest rpm drop from all the cylinders tested, multiplies it by a percentage and arrives at an rpm drop value for all cylinders. For example, if the greatest drop for any cylinder was 150 rpm, the processor applies a multiple of 65% and arrives at 98 mm. The processor then checks the recorded rpm drops, checking that each was at least 98 rpm. If all cylinders meet the criteria, the test is complete and the ECA outputs Code 90 indicating PASS.

If one cylinder did not drop at least this amount, then the cylinder number is output instead of the 90 code. The cylinder number will be followed by a zero, so 30 indicates cylinder No. 3 did not meet the minimum rpm drop.

The test may be repeated a second time by depressing and releasing the throttle pedal within 2 minutes of the last code output. For the second test, the controller uses a lower percentage (and thus a lower rpm) to determine the minimum acceptable rpm drop. Again, either Code 90 or the number of the weak cylinder will be output.

Performing a third test causes the ECA to select an even lower percentage and rpm drop. If a cylinder is shown as weak in the third test, it should be considered non-contributing. The tests may be repeated as often as needed if the throttle is depressed within two minutes of the last code output. Subsequent tests will use the percentage from the third test instead of selecting even lower values.

Continuous Memory Codes

These codes are retained in memory for 80 warm-up cycles. To clear the codes for the purposes of testing or confirming repair, perform the KOEO test. When the fault codes begin to be displayed, deactivate the test by either disconnecting the jumper wire (meter, MIL or message center) or releasing the test button on the hand scanner. Stopping the test during code transmission will erase the Continuous Memory. Do not disconnect the negative battery cable to clear these codes; the Keep Alive memory will be cleared and a new code, 19, will be stored for loss of ECA power.

KEEP ALIVE MEMORY

The Keep Alive Memory (KAM) contains the adaptive factors used by the processor to compensate for component tolerances and wear. It should not be routinely cleared during diagnosis. If an emissions related part is replaced during repair, the KAM must be cleared. Failure to clear the KAM may cause severe driveability problems since the correction factor for the old component will be applied to the new component.

To clear the Keep Alive Memory, disconnect the negative battery cable for at least 5 minutes. After the memory is cleared and the battery reconnected, the vehicle must be driven at least 10 miles so that the processor may relearn the needed correction factors. The distance to be driven depends on the engine and vehicle, but all drives should include steady-throttle cruise on open roads. Certain driveability problems may be noted during the drive because the adaptive factors are not yet functioning.

To prevent the replacement of good components, remember that the EEC-IV system has no control over the following items:
- Fuel quantity and quality
- Damaged or faulty ignition components
- Internal engine condition—rings, valves, timing belt, etc.
- Starter and battery circuit
- Dual Hall sensor
- TFI or DIS module
- Distributor condition or function

- Camshaft sensor
- Crankshaft sensor
- Ignition or DIS coil
- Engine governor module.

Any of these systems can cause erratic engine behavior easily mistaken for an EEC-IV problem.

NON-NAAO DIAGNOSTIC SYSTEM

The Capri, Festiva, Probe 2.0L, 2.2L and 2.5L engine, Escort and Tracer with 1.8L engine diagnostic system vehicles are referred to by Ford Motor Company as NON-NAAO, indicating the vehicles and/or their engines originate outside North American Automotive Operations. Note that some of the models also contain North American engines, such as the Probe with 3.0L engine, Escort or Tracer with 1.9L engine.

Although these vehicles share many similarities in their engine control systems, differences must also be considered. While the fault codes are almost standardized (i.e., Code 14 indicates the barometric pressure sensor), not all engines use the same components so a code may be unique to a particular engine or family. These procedures encompass both turbocharged and non-turbocharged engines.

Beside the engine diagnostic function, these procedures will also display codes related to the 4-speed Electronically-controlled Automatic Transaxle (4EAT) used in these vehicles. Note that the 4EAT codes are displayed by these procedures even though retrieving the engine fault codes may require the North American procedures described at the beginning of this section. The Probe with 3.0L engine, Escort and Tracer with 1.9L engine are examples of this situation.

Engine Control System

These vehicles employ the Electronic Engine Control system, commonly designated EEC, to manage fuel, ignition and emissions on vehicle engines. This system is not EEC-IV, but does share some similarities.

The engine control assembly (ECA) is given responsibility for the operation of the emission control devices, cooling fans, ignition and advance and in some cases, automatic transmission functions. Because the EEC oversees both the ignition timing and the fuel injector operation, a precise air/fuel ratio will be maintained under all operating conditions. The ECA is a microprocessor or small computer which receives electrical in-puts from several sensors, switches and relays on and around the engine.

Based on combinations of these inputs, the ECA controls outputs to various devices concerned with engine operation and emissions. The engine control assembly relies on the signals to form a correct picture of current vehicle operation. If any of the input signals is incorrect, the ECA reacts to what ever picture is painted for it. For example, if the coolant temperature sensor is inaccurate and reads too low, the ECA may see a picture of the engine never warming up. Consequently, the engine settings will be maintained as if the engine were cold. Because so many inputs can affect one output, correct diagnostic procedures are essential on these systems.

One part of the ECA is devoted to monitoring both input and output functions within the system. This ability forms the core of the self-diagnostic system. If a problem is detected within a circuit, the controller will recognize the fault, assign it an identification code, and store the code in a memory section. Most NON-NAAO vehicles use two-digit codes for both engine and 4EAT transaxle faults. The stored code(s) may be retrieved during diagnosis.

➡**When the term Powertrain Control Module (PCM) is used in this manual it will refer to the engine control computer regardless that it may also be called an Electronic Control Assembly (ECA).**

While the EEC system is capable of recognizing many internal faults, certain faults will not be recognized. Because the computer system sees only electrical signals, it cannot sense or react to mechanical or vacuum faults affecting engine operation. Some of these faults may affect another component which will set a code. For example, the ECA monitors the output signal to the fuel injectors, but cannot detect a partially clogged injector. As long as the output driver responds correctly, the computer will read the system as functioning correctly. However, the improper flow of fuel may result in a lean mixture. This would, in turn, be detected by the oxygen sensor and noticed as a constantly lean signal by the ECA. Once the signal falls outside the pre-programmed limits, the engine control assembly would notice the fault and set an identification code.

Dashboard Warning Lamp

The CHECK ENGINE dashboard warning lamp is referred to as the Malfunction Indicator Lamp (MIL). The lamp is connected to the engine control assembly and will alert the driver to certain malfunctions within the EEC system. When the lamp is lit, the ECA has detected a fault and stored an identity code in memory.

The light will stay on as long as the fault causing it is present. Should the fault self-correct, the MIL will extinguish but the stored code will remain in memory.

Under normal operating conditions, the MIL should light briefly when the ignition key is turned ON. As soon as the ECA receives a signal that the engine is running, the lamp will be extinguished. The dash warning lamp should remain out during the entire operating cycle.

Vehicles with a 4EAT transaxle (except 1.8L and 1.9L engines) also provide a manual shift light, indicating when the transmission is in manual shift mode. On 2.2L turbocharged engines, this lamp will light to advise the driver of certain electronic malfunctions.

NON-NAAO SCAN TOOL FUNCTIONS

Although stored codes may be read by using an analog voltmeter by counting the needle sweeps, the use of hand-held scan tools such as Ford's second generation SUPER STAR II tester or equivalent is recommended. There are many manufacturers of these tools; the purchaser must be certain that the tool is proper for the intended use.

➡**The engine and 4EAT fault codes on NON-NAAO vehicles may only be read with the SUPER STAR II or its equivalent. The regular STAR tester or voltmeter may be capable not retrieve the stored codes.**

The SUPER STAR II tester is designed to communicate directly with the EEC system and interpret the electrical signals. The scan tool allows any stored faults to be read from the engine controller memory. Use of the scan tool provides additional data during troubleshooting but does not eliminate the use of the charts. The scan tool makes collecting information easier; the data must be correctly interpreted by an operator familiar with the system.

An adapter cable will be required to connect the scan tool to the vehicle; the adapter(s) may differ depending on the vehicle being tested.

Electrical Tools

The most commonly required electrical diagnostic tool is the Digital Multimeter, allowing voltage, resistance and amperage to be read by one instrument. Many of the diagnostic charts require the use of a voltmeter or ohmmeter during diagnosis.

The multimeter must be a high impedance unit, with 10 megohms of impedance in the voltmeter. This type of meter will not place an additional load on the circuit it is testing; this is extremely important in low voltage circuits. The multimeter must be of high quality in all respects. It should be handled carefully and protected from impact or damage. Replace the batteries frequently in the unit.

Additionally, an analog (needle type) voltmeter may be used to read stored fault codes if the SUPER STAR II tester is not available. The codes are transmitted as visible needle sweeps on the face of the instrument.

Almost all diagnostic procedures will require the use of the Break-out Box, a device which connects into the EEC harness and provides testing ports for the 60 wires in the harness. Direct testing of the harness connectors at the terminals or by backprobing is not recommended; damage to the wiring and terminals is almost certain to occur.

Other necessary tools include a quality tachometer with inductive (clip-on) pickup, a fuel pressure gauge with system adapters and a vacuum gauge with an auxiliary source of vacuum.

NON-NAAO SELF-DIAGNOSTICS

Diagnosis of a driveability problem requires attention to detail and following the diagnostic procedures in the correct order. Resist the temptation to begin extensive testing before completing the preliminary diagnostic steps. The preliminary or visual inspection must be completed in detail before diagnosis begins. In many cases this will shorten diagnostic time and often cure the problem without electronic testing.

Keep in mind that all the things that previously went wrong with vehicles, before the age of electronics, can still go wrong and are still the cause of the majority of the driveability problems. The best diagnosis starts with a list of symptoms and possible causes, followed by careful checking of those causes in the most likely order. Eliminate all the possible mechanical causes before considering electrical faults.

Visual Inspection

This is possibly the most critical step of diagnosis. A detailed examination of all connectors, wiring and vacuum hoses can often lead to a repair without further diagnosis. Performance of this step relies on the skill of the technician performing it; a careful inspector will check the undersides of hoses as well as the integrity of hard-to-reach hoses blocked by the air cleaner or other components. Wiring should be checked carefully for any sign of strain , burning, crimping or terminal pull-out from a connector.

Checking connectors at components or in harnesses is required; usually, pushing them together will reveal a loose fit. Pay particular attention to ground circuits, making sure they are not loose or corroded. Remember to inspect connectors and hose fittings at components not mounted on the engine, such as the evaporative canister or relays mounted on the fender aprons. Any component or wiring in the vicinity of a fluid leak or spillage should be given extra attention during inspection.

Additionally, inspect maintenance items such as belt condition and tension, battery charge and condition and the radiator cap carefully. Any of these very simple items may affect the system enough to set a fault.

NON-NAAO READING CODES

The EEC system may be interrogated for stored codes using the Quick Test procedures. If a code was set before a problem self-corrected (such as a momentarily loose connector), the code will remain in memory until cleared.

The Quick Test procedure is divided into 3 sections, Key On Engine Off (KOEO), Key On Engine Running (KOER) and the Switch Monitor test. These 3 procedures must be performed correctly if the system is to run the internal self-checks and provide accurate fault codes. Codes will be output and displayed as numbers on the hand scan tool, i.e. 23. If the codes are being read by an analog voltmeter, the codes will be displayed as groups of needle sweeps separated by pauses.

Code 23 would be shown as two sweeps, a pause and three more sweeps. A longer pause will occur between codes. Unlike the EEC-IV system, the EEC system does not broadcast a PASS designator or code. If no fault codes are stored, the display screen of the hand scanner will remain blank. Additionally, the EEC system does not operate switches or sensors during KOEO or KOER testing.

Once the Quick Test has been performed and all fault codes recorded, refer to the service code charts. The charts direct the use of specific pinpoint tests for the appropriate circuit and will allow complete circuit testing.

The EEC diagnostic connector is located at the left rear corner of the engine compartment on most vehicles except when equipped with the 1.6L engine. The 1.6L diagnostic connector is in the right rear corner of the engine compartment. When connecting the test equipment and adapters, note that the Self-Test Input (STI) connector is separate from the main diagnostic connector on all NON-NAAO engines except for the 1.8L engine. The Self-Test Output (STO) connector is contained within the main diagnostic connector.

✳✳ CAUTION

To prevent injury and/or property damage, always block the drive wheels, firmly apply the parking brake, place the transmission in Park or Neutral and turn all electrical loads off before performing the Quick Test procedures.

Reading Codes With Analog Voltmeter

In the absence of a scan tool, an analog voltmeter may be used to retrieve stored fault codes. Set the meter range to read DC 0–20 volts. Connect the + lead of the meter to the STO pin in the diagnostic connector and connect the—lead of the meter to the negative battery terminal or a good engine ground.

Follow the directions given for performing the KOEO and KOER tests. To activate the tests, use a jumper wire to connect the STI connector to ground. The codes will be transmitted as groups of needle sweeps.

1 NEEDLE PULSE (SWEEP) + 1 NEEDLE PULSE (SWEEP) = 2 NEEDLE PULSES (SWEEPS) FOR 1ST DIGIT

1.6-SECOND PULSE BETWEEN DIGITS

:23 SERVICE CODE

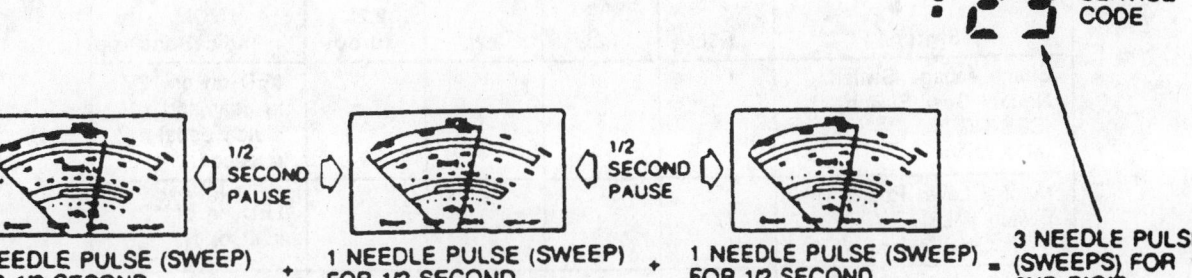

1 NEEDLE PULSE (SWEEP) FOR 1/2 SECOND + 1/2 SECOND PAUSE + 1 NEEDLE PULSE (SWEEP) FOR 1/2 SECOND + 1/2 SECOND PAUSE + 1 NEEDLE PULSE (SWEEP) FOR 1/2 SECOND = 3 NEEDLE PULSES (SWEEPS) FOR 2ND DIGIT

4-SECOND PAUSE BETWEEN SERVICE CODES, WHEN MORE THAN ONE CODE IS INDICATED

79222G14

Code display patterns on an analog voltmeter

KEY ON ENGINE OFF (KOEO) TEST

1. Make certain the scan tool is OFF; connect it to the self-test connectors. Switch the scan tool to the MECS position. Except on 1.8L engines, make certain the adapter ground cable is connected to the negative battery terminal. On the 1.8L engine, make certain the switch on the adapter is set to EEC or ECA if engine codes are to be retrieved. The other switch position will retrieve codes from the 4EAT.

2. Make certain the scan tool test button is ON or latched down.

3. For all engine or 4EAT codes except 1.8L and 1.9L engines, turn the ignition switch ON but do not start the engine, then turn the scan tool ON. On 1.8L and 1.9L engines, turn the scan tool ON first, then turn the ignition switch ON.

4. Once energized, the tester should display 888 and beep for 2 seconds. Release the test button; 00 should appear, signifying the tool is ready to read codes.

5. Re-engage the test button.

6. The KOEO codes will be transmitted.

7. Record all service codes displayed.

8. After all codes are received, release the test button to review all the codes retained in tester memory.

9. Make sure all codes displayed are recorded. Clear the ECA memory and perform the KOEO test again. This will isolate hard faults from intermittent ones. Any hard faults will cause the code(s) to be repeated in the 2nd test. An intermittent which is not now present will not set a new code.

10. Record all codes from the 2nd test. After repairs are made on hard fault items, the intermittent ones must be recreated by tapping suspect sensors, wiggling wires or connectors or reproducing circumstances on a test drive.

➡ **For both KOEO and KOER tests, the message STO LO always displayed on the screen indicates that the system cannot initiate the Self-Test. The message STI LO displayed with an otherwise blank screen indicates Pass or No Codes Stored.**

KEY ON ENGINE RUNNING (KOER) TEST

1. Make certain the self-test button is released or de-activated on the SUPER STAR II tester and that the tester is properly connected.

2. Start the engine and run it at 2000 rpm for 2 minutes. This action warms up the oxygen sensor.

3. Turn the ignition switch OFF.

4. Turn the ignition switch ON for 10 seconds but do not start the engine.

5. Start the engine and run it at idle.

6. Activate or latch the self-test button on the scan tool.

7. All relevant codes will be displayed and should be recorded.

SWITCH MONITOR TESTS

This test mode allows the operator to check the input signal from individual switches to the ECA. All switches to be tested must be OFF at the time the test begins; if one switch is on, it will affect the testing of another. The test must begin with the engine cool. The tests may be performed with either the SUPER STAR II tester or an analog voltmeter. When using the scan tool, the small LED on the adapter cable will light to show that the ECA has received the switch signal. If the voltmeter is used, the voltage will change when the switch is engaged or disengaged.

1. The engine must be off and cooled. Place the transmission in Park or Neutral.

2. Turn all accessories OFF.

3. If using the SUPER STAR II, connect it properly. If using an analog voltmeter, use a jumper to ground the STI terminal. Connect the positive (+) voltmeter lead to the SML terminal of the diagnostic connector and connect the negative (-) lead to a good engine ground.

4. Turn the ignition switch **ON**. Engage the center button on the SUPER STAR II. Most switches can be exercised without starting the engine.

5. Operate each switch according to the test chart and note the response either on the LED or the volt scale. Remember that an improper response means the ECA did not see the switch operation; check circuitry and connectors before assuming the switch is faulty.

6. Turn the ignition switch **OFF** when testing is complete.

Switch	1.3L	1.8L	2.2L	2.2L Turbo	SUPER STAR II Tester LED or Analog VOM Indications
Clutch engage Switch/ Neutral Gear Switch (CES/NGS) (MTX only)	X	X	X	X	LED on or 12V in gear and clutch pedal released
Manual Lever Position Switch (MLP) (ATX Only)	X	X	X		LED on or 12V in P or N
Idle Switch (IDL)	X	X	X	X	LED on or 12V with accelerator pedal depressed
Brake On-Off Switch (BOO)	X	X MTX	X	X	LED on or 12V with brake pedal depressed
Headlamps Switch (HLDT)	X	X	X	X	LED on or 12V with headlamp switch on
Blower Motor Switch (BLMT)	X	X	X	X	LED on or 12V with blower switch at 2nd or above position
A/C Switch (ACS)	X	X			LED on or 12V with A/C switch on and blower on
Defrost Switch (DEF)	X	X	X	X	LED on or 12V with defrost switch on
Coolant Temperature Switch (CTS)	X	X		X	LED on or 12V with cooling fan on
Wide Open Throttle Switch (WOT)	X	X MTX			LED off or 0V with accelerator pedal fully depressed

Switch tests for 1990 and early Ford Non-NAAO vehicles

79222G15

Switch	1.3L	1.6L	1.8L	2.2L	2.2L Turbo	SUPER STAR II Tester LED or Analog VOM Indications
Clutch engage Switch/ Neutral Gear Switch (CES/NGS) (MTX only)	X	X	X	X	X	LED on or less than 1.5V in gear and clutch pedal released
Manual Lever Position Switch (MLP) (ATX Only)	X	X	X	X	X	LED on or less than 1.5V in P or N
Idle Switch (IDL)	X	X	X	X	X	LED on or less than 1.5V with accelerator pedal depressed
Brake On-Off Switch (BOO)	X	X	X MTX	X	X	LED on or less than 1.5V with brake pedal depressed (not fully)
Headlamps Switch (HLDT)	X	X	X	X	X	LED on or less than 1.5V with headlamp switch on
Blower Motor Switch (BLMT)	X	X	X	X	X	LED on or less than 1.5V with blower switch at 2nd or above position
A/C Switch (ACS)	X	X	X	X	X	LED on or less than 1.5V with A/C switch on and blower on
Defrost Switch (DEF)	X	X	X	X	X	LED on or less than 1.5V with defrost switch on
Coolant Temperature Switch (CTS)	X	X	X	X	X	LED on or less than 1.5V with cooling fan on
Wide Open Throttle Switch (WOT)	X		X			LED off or 0V with accelerator pedal fully depressed
Knock Control (KC)					X	LED on or less than 1.5V while tapping on engine

79222G16

Switch tests for 1991 and newer Ford Non-NAAO vehicles

Clearing Codes

Codes stored within the memory must be erased when repairs are completed. Additionally, erasing codes during diagnosis can separate hard faults from intermittent ones.

To erase stored codes, disconnect the negative battery cable, then depress the brake pedal for at least 10 seconds. Reconnect the battery cable and recheck the system for any remaining or newly-set codes.

MCU CARBURETED DIAGNOSTIC SYSTEM

The Microprocessor Control Unit (MCU) system was used on most 1981–83 carburetor equipped vehicles, and 1984 and newer V8 engines with feedback carburetors. The MCU system uses a large six sided connector, identical to the one used with EEC-IV systems. The MCU system does NOT use the small single wire connector, like the EEC-IV system.

This system has limited ability to diagnose a malfunction within itself. Through the use of trouble codes, the system will indicate where to test. When an analog voltmeter or special tester is con-nected to the diagnostic link connector and the system is triggered, the self-test simulates a variety of engine operating conditions and evaluates all the responses received from the various MCU compo-nents, so any abnormal operating conditions can be detected.

MCU CARBURETED SELF-DIAGNOSTICS

Diagnosis of a driveability problem requires attention to detail and following the diagnostic procedures in the correct order. Resist the temptation to begin extensive testing before completing the pre-liminary diagnostic steps. The preliminary or visual inspection must be completed in detail before diagnosis begins. In many cases this will shorten diagnostic time and often cure the problem without electronic testing.

Visual Inspection

This is possibly the most critical step of diagnosis. A detailed examination of all connectors, wiring and vacuum hoses can often lead to a repair without further diagnosis. Performance of this step relies on the skill of the technician performing it; a careful inspector will check the undersides of hoses as well as the integrity of hard-to-reach hoses blocked by the air cleaner or other components. Wiring should be checked carefully for any sign of strain, burning, crimping or terminal pull-out from a connector.

Checking connectors at components or in harnesses is required; usually, pushing them together will reveal a loose fit. Pay particular attention to ground circuits, making sure they are not loose or corroded. Remember to inspect connectors and hose fittings at components not mounted on the engine, such as the evaporative canister or relays mounted on the fender aprons. Any component or wiring in the vicinity of a fluid leak or spillage should be given extra attention during inspection.

Additionally, inspect maintenance items such as belt condition and tension, battery charge and condition and the radiator cap carefully. Any of these very simple items may affect the system enough to set a fault.

MCU CARBURETED READING CODES

Preparation For Reading Codes

1. Turn OFF all electrical equipment and accessories in vehicle.
2. Follow all safety precautions during testing.
3. Make sure all fluids are at proper levels.
4. Perform 'Visual Inspection' as detailed in EEC-IV system testing earlier in this section.
5. Start the engine and let it idle, until the engine reaches normal operating temperature. This is when the upper radiator hose is Hot and engine RPM has dropped to its normal warm idle speed.
6. Turn ignition switch OFF.

✳✳ CAUTION

Always operate the vehicle in a well ventilated area. Exhaust gases are very poisonous.

INLINE 4 AND 6 CYLINDER ENGINES

On Inline 4- and 6-cylinder engines with canister control valves, remove the hose that goes to the carbon canister (this simulates a clean carbon canister). Do NOT plug this hose for the remainder of

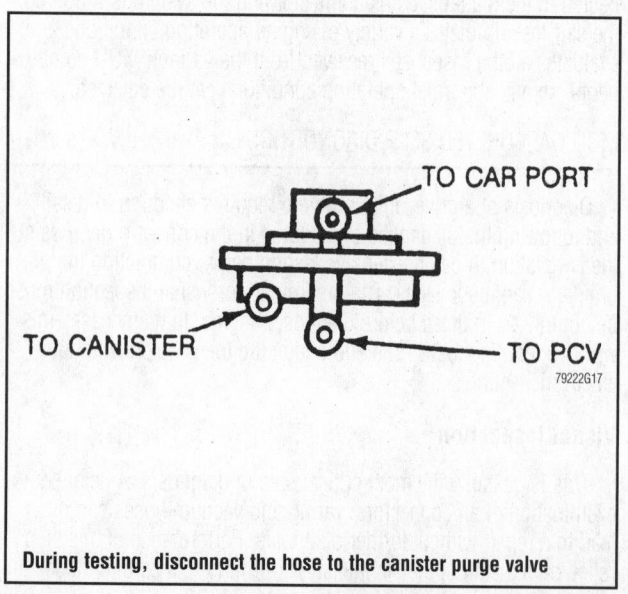

During testing, disconnect the hose to the canister purge valve

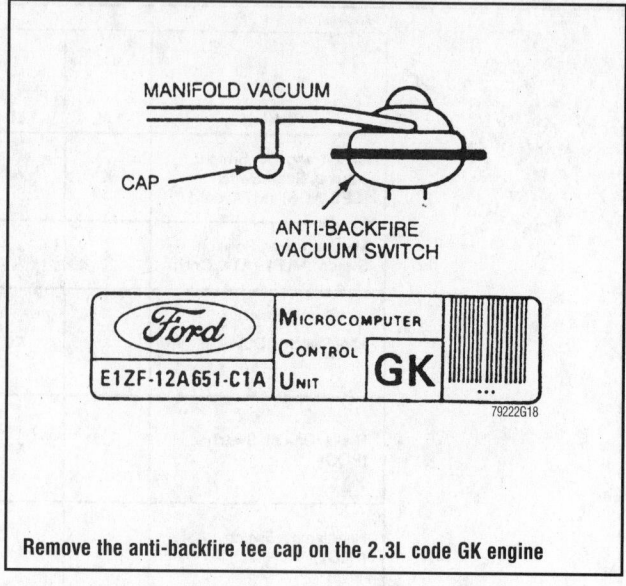

Remove the anti-backfire tee cap on the 2.3L code GK engine

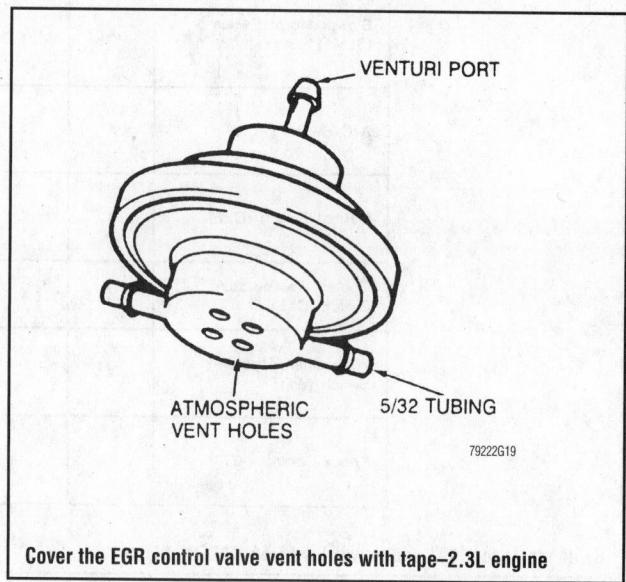

Cover the EGR control valve vent holes with tape—2.3L engine

the test procedure. Make certain the throttle linkage is off of the high choke cam setting.

The 2.3L engines with GK code, you must remove the cap from the anti-backfire vacuum switch tee during testing. The switch is near the rear of the MCU module. On 2.3L engines with an EGR vacuum load control (wide open throttle) valve, you must cover the atmospheric vent holes with a piece of tape.

V6 AND V8 ENGINES

On V6 and V8 engines, remove the PCV valve from breather cap on valve cover. On the 4.2L and 5.8L engines with vacuum delay valves, uncap the restrictor on the thermactor vacuum control line. On the V6 4.2L engine the vacuum cap is on the TAD line on the 5.8L engine the vacuum cap is on the TAB line.

➡**Remember to replace vacuum lines, tee caps and return all components to original condition after testing is complete.**

After you have performed any special procedures for your vehicle, have a pencil and paper nearby to write down codes. Now you are ready to perform the KOEO test.

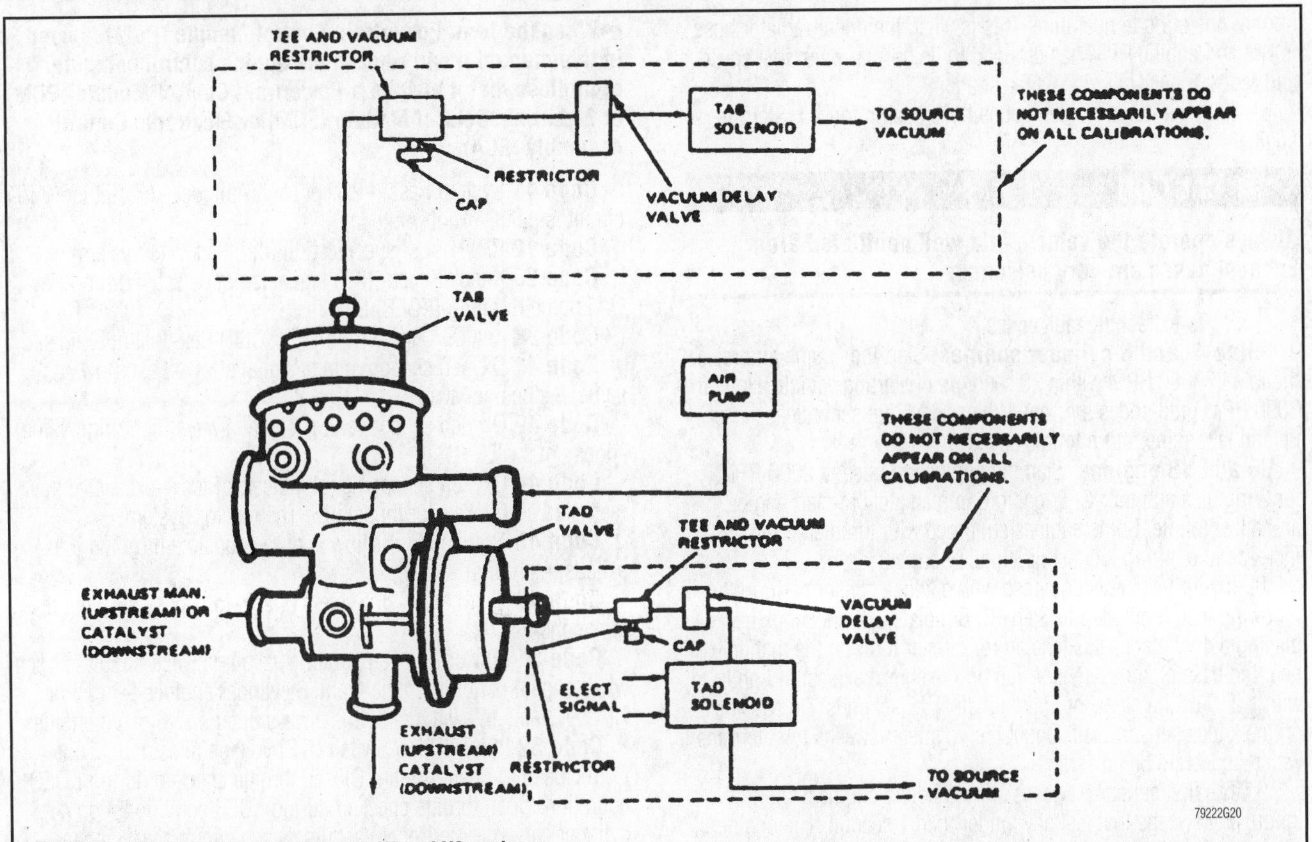

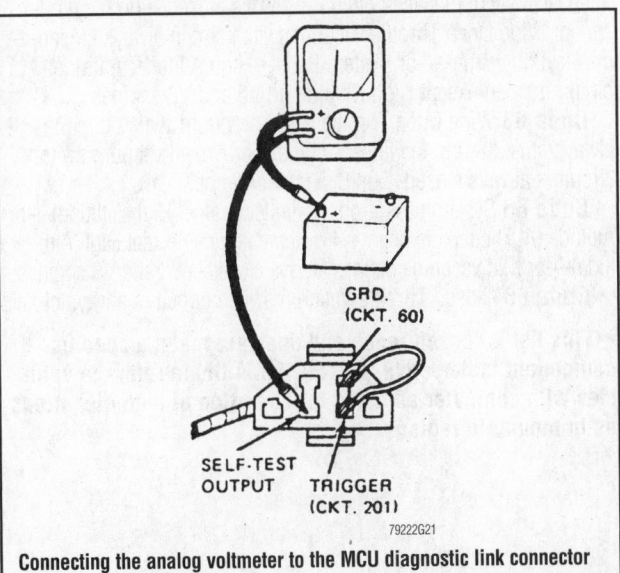

TEE AND VACUUM RESTRICTOR

THESE COMPONENTS DO NOT NECESSARILY APPEAR ON ALL CALIBRATIONS.

CAP

RESTRICTOR

VACUUM DELAY VALVE

TAB SOLENOID

TO SOURCE VACUUM

TAB VALVE

AIR PUMP

THESE COMPONENTS DO NOT NECESSARILY APPEAR ON ALL CALIBRATIONS.

TAD VALVE

TEE AND VACUUM RESTRICTOR

VACUUM DELAY VALVE

CAP

EXHAUST MAN. (UPSTREAM) OR CATALYST (DOWNSTREAM)

ELECT SIGNAL

TAD SOLENOID

EXHAUST (UPSTREAM) CATALYST (DOWNSTREAM)

RESTRICTOR

TO SOURCE VACUUM

79222G20

Remove the TAB or TAD tee vacuum cap—V6 and V8 engines

Using an Analog Voltmeter

KEY ON ENGINE OFF (KOEO) TEST

1. With the ignition switch in the **OFF** position, connect a jumper wire between circuits 60 and 201 on the self-test connector.

2. Connect the analog voltmeter from the battery positive post to the self-test output connector.

3. Self the voltmeter scale to 0-15 volt range.

4. Turn the ignition switch **ON**, but do NOT start the engine. One quick initialization pulse may occur. The output codes will follow in about 5 seconds.

GRD. (CKT. 60)

SELF-TEST OUTPUT

TRIGGER (CKT. 201)

79222G21

Connecting the analog voltmeter to the MCU diagnostic link connector

5. Count the voltmeter sweeps, to determine which codes are being transmitted.

6. The MCU system uses 2-digit codes with the pause between each digit being about 2 seconds long. The pause between the two different codes is about 4 seconds long. The code group is sent twice. This allows you to check the accuracy of the codes as you record them.

7. Once this test has been performed and all fault codes recorded, you can refer to the 'Code Descriptions' in this manual for the meaning of the fault code(s). For more detailed code retrieval information and to repair faults refer to your specific vehicle service manual.

KEY ON ENGINE RUNNING (KOER) TEST

1. Turn OFF all electrical equipment and accessories in vehicle.
2. Follow all safety precautions during testing.
3. Make sure all fluids are at the proper levels.
4. Perform 'Visual Inspection' as detailed earlier in this manual.

✲✲ CAUTION

The following steps involve servicing the engine with the engine running. Observe safety precautions.

- Apply the parking brake.
- Put the shift lever in P (automatic transmission) or NEUTRAL (manual transmission).
- Block the drive wheels.
- Always operate vehicle in well ventilated area. Exhaust gases are very poisonous.
- Stay clear of hot and moving engine parts.

5. The engine should be at normal operating temperature for this test. If not, start the engine and let it idle, until the engine reaches

normal operating temperature. This is when the upper radiator hose is Hot and engine RPM has dropped to its normal warm idle speed and repeat Key ON Engine OFF test again.

6. If engine is warm, after codes have been retrieved, Start the engine.

✳✳ CAUTION

Always operate the vehicle in a well ventilated area. Exhaust gases are very poisonous.

7. To extract the fault codes:

Inline 4- and 6-cylinder engines: Start the engine and raise the idle to 3000 RPM within 20 seconds of starting vehicle. Hold at 3000 RPM until codes are sent. When codes are sent release throttle and let engine return to idle speed.

V6 and V8 engines: Start the engine are raise to 2000 RPM for 2 minutes and turn OFF ignition. Immediately re-start engine and allow to idle. Some engines equipped with throttle kicker will increase idle during the testing, this is normal.

8. If your vehicle is equipped with a knock sensor perform the following test, if not skip to Step 10. Simulate a spark knock by placing a 4 inch socket extension (or similar tool) on the manifold near the base of knock sensor. Tap on the end of extension lightly with a 2–6 oz. hammer for approximately 15 seconds. Do NOT hit on the knock sensor itself. Count the voltmeter sweeps to determine which codes are being sent.

9. The first series of sweeps should be the engine ID code, ignore any sweeps that last any longer than 1 second. The engine ID code will be ½ the number of cylinders. For example, a 4 cylinder would appear as 2 sweeps and a 6 cylinder as 3 sweeps and an 8 cylinder as 4 sweeps.

10. If no sweeps occur repeat KOER test procedures, starting with Step 1. If the meter still does not sweep, you have a problem which must be repaired before proceeding. Refer to your specific vehicle service manual.

11. Count the sweeps on the meter to find out which codes are being sent. All codes are 2-digits long and will appear the same way as in KOEO Self-Test. Ignore any sweeps lasting more than 1 second. Write codes down on a piece of paper, codes will be sent twice so you can check your list for accuracy. Write codes down in the order they appear. Turn the ignition switch OFF when codes are finished and remove jumper wire.

DIAGNOSTIC TROUBLE CODES (DTC'S)

MCU System

The code definitions listed are general for Ford Vehicles using the Microprocessor Control Unit (MCU) engine control system. Most Ford vehicles up to 1983 and feedback carburetor equipped V8 engines into the 1990's use the MCU engine control system. For a specific code definition or component test procedure consult service manual for your vehicle. A diagnostic code does not mean the component is defective. For example a Code 44 is an oxygen sensor code (rich oxygen sensor signal). This code may set if a carburetor is flooding or has a very restricted air cleaner. Replacing the oxygen sensor would not fix the problem.

➡**When the term Powertrain Control Module (PCM) is used in this manual it will refer to the engine control computer regardless that it may be a Powertrain Control Module (PCM) or Electronic Control Module (ECM) or Electronic Control Assembly (ECA).**

Code 11 System Pass—Except High Altitude—or Altitude (ALT) circuit is open—High Altitude

Code 12 RPM out of specification (throttle kicker system)

Code 25 Knock Sensor (KS) signal is not detected during Key On Engine Running (KOER) Self-Test

Code 33 Key On Engine Running (KOER) Self-Test not initiated

Code 41 Oxygen sensor voltage signal always Lean (low value)—does not switch

Code 42 Oxygen sensor voltage signal always Rich (high value)-does not switch

Code 44 Oxygen sensor signal indicates Rich—excessive fuel, restricted air intake—or Inoperative Thermactor System

Code 45 Thermactor Air flow is always upstream (going into exhaust manifold)

Code 46 Thermactor Air System unable to bypass air (vent to atmosphere)

Code 51 Low or Mid Temperature vacuum switch circuit is open when engine is hot on Inline 4 and 6 cylinder engines—or HI or HI/LOW vacuum switch circuit is always open on V6 or V8 engines

Code 52 Idle Tracking Switch (ITS) voltage does not change from closed to open throttle (Closed throttle checked during KOEO condition. Open throttle checked during KOER conditions) on 4 cylinder car—or Idle/Decel Vacuum switch circuit always open—on 4 cylinder truck—or Wide Open Throttle vacuum switch circuit always open—on Inline 6 cylinder engine

Code 53 Wide Open Throttle vacuum switch circuit always open on 4 cylinder engine—or Crowd vacuum switch circuit is always open—on Inline 6 cylinder engine—or Dual temperature switch circuit is always open -on V6 and V8 engines

Code 54 Mid temperature switch circuit is always open

Code 55 Road load vacuum switch circuit is always open—on 4 cylinder engine—or Mid vacuum switch circuit is always open—on V6 and V8 engines

Code 56 Closed throttle vacuum switch circuit is always open

Code 61 Hi/Low Vacuum switch circuit is always closed

Code 62 Idle Tracking Switch (ITS) circuit is closed at idle—or Idle/Decel vacuum switch circuit is always closed—on 4 cylinder car -or Wide Open Throttle vacuum switch circuit always closed—on 4 cylinder truck—or System Pass—High Altitude; Altitude (ALT) circuit is open except High Altitude on V6 and V8 engines

Code 63 Wide Open Throttle (WOT) vacuum switch circuit is always closed—on 4 cylinder engine—or Crowd vacuum switch circuit is always closed—on 6 cylinder engine

Code 65 System pass—on 4 cylinder engine (High Altitude)—or Altitude (ALT) circuit is open—4 cylinder engine (except High Altitude)—or Mid vacuum circuit is always closed—V6 and V8 engines

Code 66 Closed Throttle Vacuum switch circuit is always closed

➡**This list is for reference and does not mean a specific component is defective. NOTE: High Altitude refers to vehicles with computer adjusted for operation at high elevations as in mountain regions.**

EEC-IV System

The code definitions listed general 2-digit codes for Ford Vehicles using the Ford EEC-IV engine control system. In 1991 Ford started introducing vehicles that use 3-digit codes. The code definitions for both the 2 and 3-digit codes are found in this section. For a specific code definition or component test procedure consult your 'Chilton Total Car Care' manual for your vehicle. A diagnostic code does not mean the component is defective. For example a Code 29 is a vehicle speed sensor code. This does not mean the sensor is defective, but to check the sensor and related components. A defective speedometer cable or transmission problem will also set this code.

➡ **When the term Powertrain Control Module (PCM) is used in this manual it will refer to the engine control computer regardless that it may be a PCM or Electronic Control Module (ECM) or Electronic Control Assembly (ECA).**

2-DIGIT DTC'S

1981–94 PASSENGER CARS AND 1984–94 LIGHT TRUCKS:

Code 11 System Pass
Code 12 (R) Idle control fault—RPM Unable To Reach Upper Limit Self-Test
Code 13 (C) DC Motor Did Follow Dashpot
Code 13 (O) DC Motor Did Not Move
Code 13 (R) Idle control fault—Cannot control RPM during Self-Test low RPM check
Code 14(C)—Engine RPM signal fault—Profile Ignition Pickup (PIP) circuit failure or RPM sensor.
Code 15 (C) EEC Processor, power to Keep Alive Memory (KAM) interrupted or test failed
Code 15 (O) Power Interrupted To Processor or EEC Processor ROM Test failure
Code 16 (O,R) RPM too low to perform Exhaust Gas Oxygen (EGO) sensor test or fuel control error.
Code 1 (O)7 CFI Fuel Control System fault—Rich/Lean condition indicated; 3.8L V-6/5.0LV-8 (1984).
Code 17 (R) RPM Below Self-Test Limit, Set Too Low. Code 18 (C)—Ignition diagnostic monitor (1DM) circuit failure, loss of RPM signal or SPOUT circuit grounded
Code 18 (O)—Ignition Diagnostic Monitor (1DM) circuit
Code 18 (R) SPOUT or SAW circuit open
Code 19 (C) Cylinder Identification (CID) Sensor Input failure
Code 19 (O) Failure in EEC Processor internal voltage. Code 19 (R)—Erratic RPM During EGR Test or RPM Too Low During ISC Off Test
Code 21 Engine Coolant Temperature (ECT) out of Self-Test range
Code 22 (O, R) Manifold Absolute Pressure (MAP)/Barometric Pressure (BP/BARO) Sensor circuit out of Self-Test range
Code 23 Throttle Position (TP) Sensor out of Self-Test range
Code 24 (O, R) Air Charge (ACT) or Intake Air (IAT) Temperature out of Self-Test range
Code 25 (R) Knock not sensed during dynamic response test
Code 26 (O, R) Transmission Fluid Temp (TFT) out of Self-Test range
Code 26 (O, R) Vane Air (VAF) or Mass Air (MAF) sensor out of self-test range
Code 28 (C) Loss Of Primary Tach, Right Side. Code 29 (C)—Insufficient input from Vehicle Speed Sensor (VSS) or Programmable Speedometer/Odometer Module (PSOM)

Code 31 EGR valve position sensor circuit below minimum voltage
Code 32 EGR Valve Position (EVP) sensor circuit voltage below closed limit
Code 33 (C) Throttle Position (TP) sensor noisy/harsh on line
Code 33 (R, C) EGR valve position sensor circuit, EGR valve opening not detected
Code 34 EGR valve circuit out of self-test range or valve not closing
Code 35 EGR valve circuit above maximum voltage—except 2.3L HSC with Feedback Carburetor System—or—Throttle Kicker on 2.3L HSC with Feedback Carburetor System.
Code 38 (C) Idle Track Switch Circuit Open. Code 39 (C)—AXOD Torque Converter or Bypass Clutch Not Applying Properly
Code 41 (R,C) Oxygen Sensor circuit indicates system always lean
Code 42 (R,C) Oxygen Sensor circuit indicates system always rich, right side if 2 sensors used
Code 43 (C) Oxygen Sensor Out Of Test Range—on 1992 and earlier vehicles—or—Throttle Position Sensor failure—on 1993 and newer vehicles
Code 43 (R) Exhaust Gas Oxygen (EGO) sensor cool down has occurred during testing—2.3L HSC and 2.8L FBC truck
Code 44 (R) Air injection control system failure (right side cylinders, if a split system)
Code 45 (C) Coil 1 primary circuit failure
Code 45 (R) Air injection control system air flow misdirected
Code 46 (C) Coil Primary Circuit failure
Code 46 (R) Thermactor air not bypassed during Self-Test
Code 47 (C) 4x4 switch is closed—on Truck.
Code 47 (R) Airflow low at idle—on fuel injected engines—or—4 x 4 switch is closed—on Truck—or—Fuel control system/Exhaust Gas Oxygen (EGO) Sensor fault—on 2.3L HSC and 2.8L FBC truck
Code 48 (C) Coil Primary Circuit failure; Except 2.3L Truck—or—Loss Of Secondary Tach, Left Side—with 2.3L Truck engine
Code 48 Airflow high at base idle
Code 49 (C) El electronic Transmission Shift Error—on Truck and 1992 and later cars—or—SPOUT Signal Defaulted To 10 Degrees BTDC or SPOUT Open—Up to 1991 passenger cars
Code 51 (O, C) Engine Coolant Temperature (ECT) circuit open or out of range during self-test
Code 52 (O) Power Steering Pressure Switch (PSPS) circuit open
Code 52 (R) Power Steering Pressure Switch (PSPS) circuit did not change states
Code 53 (O, C) Throttle Position (TP) circuit above maximum voltage
Code 54 (O, C) Air Charge (ACT) or Intake Air (IAT) Temperature circuit open
Code 55 (R) Key Power Input To Processor—open circuit
Code 56 (O, C) Mass Air (MAF) or Vane Air (VAF) Flow circuit above maximum voltage—Port fuel injected engines—or—Transmission oil temperature (TOT) circuit open—on vehicles with automatic transaxle
Code 57 (C) AXOD Circuit failure—on vehicles with automatic overdrive transaxle—or—Octane Adjust Circuit failure—on some 1992 and newer cars
Code 58 (R) Idle Tracking Switch circuit fault.
Code 59 (C) Automatic Transmission Shift Error—on 1991 and newer—or—AXOD 4/3 or Neutral Pressure Switch Failed Open—on 3.0L EFI and 3.8L AXOD—vehicles with automatic overdrive transaxle

Code 59 (O) AXOD 4/3 Pressure Switch Failed Closed -on 3.8L engine AXOD—vehicles with automatic transaxle—or—Idle Adjust Service Pin In Use—on 2.9L EFI engine—or—Low Speed Fuel Pump Circuit failure—on 3.0L SHO engine

Code 61 (O, C) Engine Coolant Temperature (ECT) circuit grounded

Code 62 (C) Converter clutch error

Code 62 (O) Electronic Transmission Shift Error. Code 63 (O, C)—Throttle Position (TP) circuit below minimum voltage

Code 64 (O, C) Air Charge (ACT) or Intake Air (IAT) Temperature circuit grounded

Code 65 (C) Fuel System Failed To Enter Closed Loop Mode or key power

Code 65 (O) Key Power Check—Possible Charging System overvoltage condition

Code 65 (R) Overdrive Cancel Switch (OCS) circuit did not switch

Code 66 (C) Mass Air (MAF) or Vane Air (VAF) Flow circuit below minimum voltage—engine with Port fuel injection—or—Transmission Oil Temperature (TOT) circuit grounded—vehicles with automatic transaxle

Code 67 (O, C) Manual Lever Position (MLP) sensor out of range and A/C ON

Code 67 (O, C) Neutral/Drive Switch (NDS) circuit open/A/C on during Self-Test

Code 67 (O, R) Neutral Drive Circuit Failed or A/C Input High -or—Clutch Switch Circuit failed—on vehicles with manual transaxle—or—Manual Lever Position Sensor out of range—on vehicles with automatic transaxle

Code 68 (C) Transmission Fluid Temp (TFT) transmission over temp (over heated)

Code 68 (O) Idle Tracking Switch circuit—on 2.8L FBC truck only—or—Air temperature sensor—except FBC truck.

Code 68 (R,C) Air Temperature Sensor Circuit failure -on 1.9L EFI engine—or—Idle Tracking Switch Circuit failure—on CFI engine—or—Transmission Temperature Circuit

Code 69 (O, C) Transmission Shift Error

Code 70 (C) Data Communications Link Circuit failure

Code 71(C) Software Re-Initialization Detected—on 1.9L EFI and 2.3L Turbo—or—Idle Tracking Switch failure—on CFI engine—or -Message Center Control Circuit failure—on vehicles with Message Center Control Center—or—Power Interrupt Detected—except vehicles with 3.8L AXOD (automatic overdrive transaxle)

Code 72 (R) Insufficient Manifold Absolute Pressure (MAP) change during Dynamic Response Test

Code 73 (R) Insufficient Throttle Position (TP) change during Dynamic Response Test

Code 74 (R,C) Brake On/Off (BOO) circuit open/not actuated during Self-Test

Code 75 (R) Brake On/Off (BOO) circuit closed/EEC processor input open

Code 76 (R) Insufficient Airflow Output Change During Test

Code 77 (R) Brief Wide Open Throttle (WOT) not sensed during Self-Test/operator error (Dynamic Response/Cylinder Balance Tests)

Code 78 (C) Power Interrupt Detected

Code 79 (O) A/C on/Defrost on during Self-Test

Code 81(C) MAP Sensor Has Not Changing Normally

Code 81(O) Air Management Circuit failure

Code 82 (O) Supercharger Bypass Circuit failure, 3.8L SC engine—or—Air Management Circuit failure, Except 3.8L SC engine—or -EGR Solenoid Circuit failure, 2.3L OHC engine

Code 83 OIC—Low speed fuel pump relay circuit failure

Code 83 (O) High Speed Electro Drive Fan Circuit failure, Except 2.3L OHC and 3.0L SHO engine—or—Low Speed Fuel Pump Relay Circuit failure, 3.0L SHO engine

Code 84 (O) EGR Vacuum Regulator (EVR) circuit failure

Code 84 (R) EGR Solenoid Circuit failure

Code 85 (C) Adaptive Lean Limit Reached

Code 85 (O) Canister Purge (CANP) circuit failure

Code 86 (C) Adaptive Rich Limit Reached

Code 86 (O) Shift Solenoid (SS) circuit failure—or—Wide Open Throttle (WOT) A/C Cutoff Solenoid circuit—on Carbureted engine

Code 87 Fuel Pump circuit fault

Code 88 (C) Loss Of Dual Plug Input control

Code 88 (O) Electro Drive Fan Circuit failure—fuel injected engine—or—Throttle Kicker, feedback carburetor system

Code 89 (O) Transmission solenoid circuit failure. **Code 89 (O)** Clutch Converter Override (CCO) circuit failure—or—Exhaust Heat Control (EHC) Solenoid circuit—3.8L CFI engine

Code 91(C) No Heated Exhaust Gas Oxygen (HEGO) sensor switching detected—left HEGO

Code 91(O) Shift Solenoid 1 (SS1) circuit failure.

Code 91(R) Heated Exhaust Gas Oxygen (H EGO) sensor circuit indicates system lean—left HEGO

Code 92 (O) Shift Solenoid Circuit failure

Code 92 (R) Oxygen Sensor Circuit failure

Code 93 (O) Throttle Position Sensor (TPS) input low at maximum DC motor extension—OR—Shift solenoid circuit failure

Code 93 (O) Coast Clutch Solenoid (CCS) circuit failure

Code 94 (O) Torque Converter Clutch (TCC) solenoid circuit failure

Code 94 (O) Converter Clutch Control (CCC) Solenoid circuit failure

Code 94 (R) Thermactor Air System inoperative, left side

Code 95 (O, C) Fuel Pump secondary circuit failure/Fuel Pump circuit open—EEC processor to motor ground

Code 96 (O, C) Fuel Pump secondary circuit failure/Fuel Pump circuit open—battery to EEC processor

Code 97 (O) Overdrive Cancel Indicator Light (OCIL) circuit failure

Code 98 (R) Electronic control assembly failure

Code 98 (O)—Electronic Pressure Control (EPC) Driver open in EEC processor

Code 98 (R) Hard fault is present—FMEM mode

Code 99 (O,C) Electronic Pressure Control (EPC) circuit failure

Code 92 (O) Shift Solenoid 2 (SS2) circuit failure

Code 92 (R) Heated Exhaust Gas Oxygen (HEGO) sensor circuit indicates system rich—left HEGO

Code 93 (O) Throttle Position Sensor Input Low At Max DC Motor Extension, CFI engine—or—Shift Solenoid Circuit failure—Except CFI engine

Code 94 (O) Converter Clutch Solenoid Circuit failure

Code 94 (R) Thermactor Air System Inoperative

Code 95 (O, C) Fuel Pump Circuit failure, ECA To ground

Code 96 (O, C) Fuel Pump Circuit failure

Code 97 (O) Transmission Indicator Circuit failure

Code 98 (O) Electronic Pressure Control Circuit failure

Code 98 (R) Electronic Control Assembly failure

Code 99 (O, C) Electronic Pressure Control Circuit or Transmission Shift failure

Code 99 (R) EEC System Has Not Learned To Control Idle: Ignore Codes 12 & 13

No Code—Unable to Run Self Test or Output Codes, or list does not apply to vehicle tested, refer to service manual.

➡**This list is to be used as a reference for testing and does not mean a specific component Is defective.**

(O)—Key On, Engine Off
(R)—Engine running
(C)—Continuous Memory

3-DIGIT DTC'S

1991–95 VEHICLES:

Code 111 System pass
Code 112 Intake Air Temperature (IAT) Sensor circuit below minimum voltage
Code 113 Intake Air Temperature (IAT) Sensor circuit above maximum voltage
Code 114 Intake Air Temperature (IAT) higher or lower than expected
Code 116 Engine Coolant Temperature (ECT) higher or lower than expected
Code 117 Engine Coolant Temperature (ECT) Sensor circuit below minimum voltage
Code 118 Engine Coolant Temperature (ECT) Sensor circuit above maximum voltage
Code 121 Closed throttle voltage higher or lower than expected
Code 121 Indicates Throttle Position voltage inconsistent with Mass Air Flow (MAF) Sensor
Code 122 Throttle Position (TP) Sensor circuit below minimum voltage
Code 123 Throttle Position (TP) Sensor circuit above maximum voltage
Code 124 Throttle Position (TP) Sensor circuit voltage higher than expected
Code 125 Throttle Position (TP) Sensor circuit voltage lower than expected
Code 126 Manifold Absolute Pressure/Barometric Pressure (MAP/BARO) Sensor higher or lower than expected
Code 128 Manifold Absolute Pressure (MAP) Sensor vacuum hose damaged/disconnected
Code 129 Insufficient Manifold Absolute Pressure (MAP)/Mass Air Flow (MAF) change during Dynamic Response Test-KOER
Code 136 Lack of Heated Oxygen Sensor (HO2S-2) switches during KOER, indicates lean—Bank # 2
Code 137 Lack of Heated Oxygen Sensor (HO2S-2) switches during KOER, indicates rich—Bank # 2
Code 138 Cold Start Injector (CSI) flow insufficient—KOER
Code 139 No Heated Oxygen Sensor (HO2S-2) switches detected—Bank # 2
Code 141 Fuel system indicates lean
Code 144 No Heated Oxygen Sensor (HO2S-1) switches detected—Bank # 1
Code 157 Mass Air Flow (MAF) Sensor circuit below minimum voltage
Code 158 Mass Air Flow (MAF) Sensor circuit above maximum voltage
Code 159 Mass Air Flow (MAF) higher or lower than expected
Code 167 Insufficient Throttle Position (TP) change during Dynamic Response Test—KOER
Code 171 Fuel system at adaptive limits, Heated Oxygen Sensor (HO2S-I) unable to switch—Bank # 1
Code 172 Lack of Heated Oxygen Sensor (HO2S-1) switches, indicates lean—Bank # 1

Code 173 Lack of Heated Oxygen Sensor (HO2S-1) switches, indicates rich—Bank # 1
Code 174 Heated Oxygen Sensor (HO2S) switching time is slow—Right side—1992 vehicles only
Code 175 Fuel system at adaptive limits, Heated Oxygen Sensor (HO2S-2) unable to switch—Bank # 2
Code 176 Lack of Heated Oxygen Sensor (HO2S-2) switches, indicates lean—Bank # 2
Code 177 Lack of Heated Oxygen Sensor (HO2S-2) switches, indicates rich—Bank # 2
Code 178 Heated Oxygen Sensor (HO2S) switching time is slow—Left side—1992 vehicles only
Code 179 Fuel system at lean adaptive limit at part throttle, system rich—Bank # 1
Code 181 Fuel system at rich adaptive limit at part throttle, system lean—Bank # 1
Code 182 Fuel system at lean adaptive limit at idle, system rich—Right side—1992 vehicles only
Code 183 Fuel system at rich adaptive limit at idle, system lean—Right side—1992 vehicles only
Code 184 Mass Air Flow (MAF) higher than expected
Code 185 Mass Air Flow (MAF) lower than expected
Code 186 Injector pulse width higher or Mass Air Flow (MAF) lower than expected (without BARO Sensor)
Code 187 Injector pulse width lower than expected (with BARO Sensor)
Code 187 Injector pulse width lower or Mass Air Flow (MAF) higher than expected (without BARO Sensor)
Code 188 Fuel system at lean adaptive limit at part throttle, system rich—Bank # 2
Code 189 Fuel system at rich adaptive limit at part throttle, system lean—Bank # 2
Code 191 Adaptive fuel lean limit is reached at idle—Left side—1992 vehicles only
Code 192 Adaptive fuel rich limit is reached at idle—Left side—1992 vehicles only
Code 193 Flexible Fuel (FF) Sensor circuit failure
Code 211 Profile Ignition Pickup (PIP) circuit failure
Code 212 Loss of Ignition Diagnostic monitor (1DM) input to Powertrain Control Module (PCM)/SPOUT circuit grounded
Code 213 SPOUT circuit open
Code 214 Cylinder Identification (CID) circuit failure
Code 215 Powertrain Control Module (PCM) detected Coil 1 Primary circuit failure (EI)
Code 216 Powertrain Control Module (PCM) detected Coil 2 Primary circuit failure (EI)
Code 217 Powertrain Control Module (PCM) detected Coil 3 Primary circuit failure (EI)
Code 218 Loss of Ignition Diagnostic Monitor (1DM) signal left side (dual plug EI)
Code 219 Spark Timing defaulted to 10 degrees—SPOUT circuit open (EI)
Code 221 Spark Timing error (EI)
Code 222 Loss of Ignition Diagnostic Monitor (1DM) signal—right side (dual plug EI)
Code 223 Loss of Dual Plug Inhibit (DPI) control (Dual Plug EI)
Code 224 Powertrain Control Module (PCM) detected Coil 1,2,3,or 4 Primary circuit failure (Dual Plug EI)
Code 225 Knock not sensed during Dynamic Response Test—KOER
Code 226 Ignition Diagnostic Monitor (1DM) signal not received (EI)

Code 232 Powertrain Control Module (PCM) detected Coil 1,2,3,or 4 Primary circuit failure (EI)

Code 238 Powertrain Control Module (PCM) detected Coil 4 Primary circuit failure (EI)

Code 241 Ignition Control Module (1CM) to Powertrain Control Module (PCM) Ignition Diagnostic Monitor (1DM) Pulse Width Transmission error (EI)

Code 244 Cylinder Identification (CID) circuit fault present when Cylinder Balance Test requested

Code 311 Secondary Air Injection (AIR) system inoperative during KOER Bank # 1 with dual HO_2S

Code 312 Secondary Air Injection (AIR) misdirected during KOER

Code 313 Secondary Air Injection (AIR) not bypassed during KOER

Code 314 Secondary Air Injection (AIR) system inoperative during KOER—Bank # 2 with dual HO_2S

Code 326 EGR (PFE/DPFE) circuit voltage lower than expected

Code 327 EGR (EVP/PFE/DPFE) circuit below minimum voltage

Code 328 EGR (EVP) closed valve voltage lower than expected

Code 332 Insufficient EGR flow detected/EGR Valve opening not detected (EVP/PFE/DPFE)

Code 334 EGR (EVP) closed valve voltage higher than expected

Code 335 EGR (PFE/DPFE) Sensor voltage higher or lower than expected during KOEO

Code 336 Exhaust pressure high/EGR (PFE/DPFE) circuit voltage higher than expected

Code 337 EGR (EVP/PFE/DPFE) circuit above maximum voltage

Code 338 Engine Coolant Temperature (ECT) lower than expected (thermostat test)

Code 339 Engine Coolant Temperature (ECT) higher than expected (thermostat test)

Code 341 Octane Adjust service pin open

Code 411 Cannot control RPM during KOER low rpm check

Code 412 Cannot control RPM during KOER high rpm check

Code 415 Idle Air Control (IAC) system at maximum adaptive lower limit

Code 416 Idle Air Control (IAC) system at upper adaptive learning limit

Code 452 Insufficient input from Vehicle Speed Sensor (VSS) to PCM

Code 453 Servo leaking down (KOER IVSC test)

Code 454 Servo leaking up (KOER IVSC test)

Code 455 Insufficient RPM increase (KOER IVSC test)

Code 456 Insufficient RPM decrease (KOER IVSC test)

Code 457 Speed Control Command Switch(s) circuit not functioning (KOEO IVSC test)

Code 458 Speed Control Command Switch(s) stuck/circuit grounded (KOEO IVSC test)

Code 459 Speed Control ground circuit open (KOEO IVSC test)

Code 511 Powertrain Control Module (PCM) Read Only Memory (ROM) test failure (KOEO)

Code 512 Powertrain Control Module (PCM) Keep Alive Memory (KAM) test failure

Code 513 Powertrain Control Module (PCM) internal voltage failure (KOEO)

Code 519 Power Steering Pressure (PSP) Switch circuit open—KOEO

Code 521 Power Steering Pressure (PSP) Switch circuit did not change states—KOER

Code 522 Vehicle not in park or neutral during KOEO/Park/Neutral Position (PNP) Switch circuit open

Code 524 Low speed Fuel Pump circuit open—battery to PCM

Code 525 Indicates vehicle in gear/A/C on

Code 526 Neutral Pressure Switch (NPS) circuit closed; A/C on -1992 vehicles only

Code 527 Park/Neutral Position (PNP) Switch open—A/C on, KOEO

Code 528 Clutch Pedal Position (CPP) switch circuit failure

Code 529 Data Communications Link (DCL) or PCM circuit failure

Code 532 Cluster Control Assembly (CCA) circuit failure

Code 533 Data Communications Link (DCL) or Electronic Instrument Cluster (EIC) circuit failure

Code 536 Brake On/Off (BOO) circuit failure/not actuated during KOER

Code 538 Insufficient RPM change during KOER Dynamic Response Test

Code 538 Invalid Cylinder Balance Test due to throttle movement during test—SFI only

Code 538 Invalid Cylinder Balance test due to Cylinder Identification (CID) circuit failure

Code 539 A/C on/Defrost on during Self-Test

Code 542 Fuel Pump secondary circuit failure

Code 543 Fuel Pump secondary circuit failure

Code 551 Idle Air Control (IAC) circuit failure—KOEO

Code 552 Secondary Air Injection Bypass (AIRB) circuit failure -KOEO

Code 553 Secondary Air Injection Diverter (AIRD) circuit failure—KOEO

Code 554 Fuel Pressure Regulator Control (FPRC) circuit failure

Code 556 Fuel Pump Relay primary circuit failure

Code 557 Low speed Fuel Pump primary circuit failure

Code 558 EGR Vacuum Regulator (EVR) circuit failure—KOEO

Code 559 Air Conditioning On (ACON) Relay circuit failure—KOEO

Code 563 High Fan Control (HFC) circuit failure—KOEO

Code 564 Fan Control (FC) circuit failure—KOEO

Code 565 Canister Purge (CANP) circuit failure—KOEO

Code 566 3-4 Shift Solenoid circuit failure, A4LD transmission -KOEO

Code 567 Speed Control Vent (SCVNT) circuit failure—KOEO IVSC test

Code 568 Speed Control Vacuum (SCVAC) circuit failure—KOEO IVSC test

Code 569 Auxiliary Canister Purge (CANP2) circuit failure—KOEO

Code 571 EGRA solenoid circuit failure KOEO

Code 572 EGRV solenoid circuit failure KOEO

Code 578 A/C Pressure Sensor circuit shorted (VCRM) mode

Code 579 Insufficient A/C pressure change (VCRM) mode

Code 581 Power to fan circuit over current (VCRM) mode

Code 582 Fan circuit open (VCRM) mode

Code 583 Power to Fuel Pump over current (VCRM) mode

Code 584 Power ground circuit open (Pin 1) (VCRM) mode

Code 585 Power to A/C Clutch over current (VCRM) mode

Code 586 A/C Clutch circuit open (VCRM) mode

Code 587 Variable Control Relay Module (VCRM) communication failure

Code 593 Heated Oxygen Sensor Heater (HO_2S HTR)

Code 617 1-2 Shift error

Code 618 2-3 Shift error

Code 619 3-4 Shift error

Code 621 Shift Solenoid 1 (SS1) circuit failure—KOEO

Code 622 Shift Solenoid 2 (SS2) circuit failure—KOEO

Code 623 Transmission Control Indicator Lamp (TCIL) circuit failure

Code 624 Electronic Pressure Control (EPC) circuit failure

Code 625 Electronic Pressure Control (EPC) driver open in PCM

Code 626 Coast Clutch Solenoid (CCS) circuit failure—KOEO

Code 627 Torque Converter Clutch (TCC) solenoid circuit failure

Code 628 Excessive Converter Clutch slippage

Code 629 Torque Converter Clutch (TCC) solenoid circuit failure

Code 631 Transmission Control Indicator Lamp (TCIL) circuit failure—KOEO

Code 632 Transmission Control Switch (TCS) circuit did not change states during KOER

Code 633 4 x 4L Switch closed during KOEO

Code 634 Manual Lever Position (MLP) voltage higher or lower than expected/ error in Transmission Select Switch (TSS) circuit(s)

Code 636 Transmission Oil Temperature (TOT) higher or lower than expected

Code 637 Transmission Oil Temperature (TOT) Sensor circuit above maximum voltage/circuit open

Code 638 Transmission Oil Temperature (TOT) Sensor circuit below minimum voltage/circuit shorted

Code 639 Insufficient input from Transmission Speed Sensor (TSS)

Code 641 Shift Solenoid 3 (SS3) circuit failure

Code 643 Torque Converter Clutch (TCC) circuit failure

Code 645 Incorrect gear ratio obtained for first gear

Code 646 Incorrect gear ratio obtained for second gear

Code 647 Incorrect gear ratio obtained for third gear

Code 648 Incorrect gear ratio obtained for fourth gear

Code 649 Electronic Pressure Control (EPC) higher or lower than expected

Code 651 Electronic Pressure Control (EPC) circuit failure

Code 652 Torque Converter Clutch (TCC) Solenoid circuit failure

Code 654 Manual Lever Position (MLP) Sensor not indicating park during KOEO

Code 655 Manual Lever Position (MLP) Sensor indicating not in neutral during Self-Test

Code 656 Torque Converter Clutch (TCC) continuous slip error

Code 657 Transmission Over Temperature condition occurred

Code 659 High vehicle speed in park indicated

Code 667 Transmission Range sensor circuit voltage below minimum voltage

Code 668 Transmission Range sensor circuit voltage above maximum voltage

Code 675 Transmission Range sensor circuit voltage out of range

Code 691 4x4 Low switch open or short circuit

Code 692 Transmission state does not match calculated ratio

Code 998 Hard fault present—FMEM Mode

➡**If specific cylinder banks or sides are referred to in any of the above codes, but the vehicle code is being obtained from has a 4 cylinder engine, or only one Oxygen Sensor, disregard the bank side reference, but the code definition and components it pertains to is always the same.**

EEC-V System

1994 PASSENGER CARS AND LIGHT TRUCKS

DTC P0102 Mass Air Flow (MAF) Sensor circuit low input

DTC P0103 Mass Air Flow (MAF) Sensor circuit high input

DTC P0112 Intake Air Temperature (IAT) Sensor circuit low input

DTC P0113 Intake Air Temperature (IAT) Sensor high input

DTC P0117 Engine Coolant Temperature (ECT) low input

DTC P0118 Engine Coolant Temperature (ECT) Sensor circuit high input

DTC P0122 Throttle Position (TP) Sensor circuit low input

DTC P0123 Throttle Position (TP) Sensor high input

DTC P0125 Insufficient coolant temperature to enter closed loop fuel control

DTC P0132 Upstream Heated Oxygen Sensor (HO2S 11) circuit high voltage (Bank #1)

DTC P0135 Heated Oxygen Sensor Heater (HTR 11) circuit malfunction

DTC P0138 Downstream Heated Oxygen Sensor (HO2S 12) circuit high voltage (Bank #1)

DTC P0140 Heated Oxygen Sensor (HO2S 12) circuit no activity detected (Bank #1)

DTC P0141 Heated Oxygen Sensor Heater (HTR 12) circuit malfunction

DTC P0152 Upstream Heated Oxygen Sensor (HO2S 21) circuit high voltage (Bank #2)

DTC P0155 Heated Oxygen Sensor Heater (HTR 21) circuit malfunction

DTC P0158 Downstream Heated Oxygen Sensor (HO2S 22) circuit high voltage (Bank #2)

DTC P0160 Heated Oxygen Sensor (HO2S 12) circuit no activity detected (Bank #2)

DTC P0161 Heated Oxygen Sensor Heater (HTR 22) circuit malfunction

DTC P0171 System (adaptive fuel) too lean (Bank #1)

DTC P0172 System (adaptive fuel) too lean (Bank #1)

DTC P0174 System (adaptive fuel) too lean (Bank #1)

DTC P0175 System (adaptive fuel) too lean (Bank #1)

DTC P0300 Random misfire detected

DTC P0301 Cylinder #1 misfire detected

DTC P0302 Cylinder #2 misfire detected

DTC P0303 Cylinder #3 misfire detected

DTC P0304 Cylinder #4 misfire detected

DTC P0305 Cylinder #5 misfire detected

DTC P0306 Cylinder #6 misfire detected

DTC P0307 Cylinder #7 misfire detected

DTC P0308 Cylinder #8 misfire detected

DTC P0320 Ignition engine speed (Profile Ignition Pickup) input circuit malfunction

DTC P0340 Camshaft Position (CMP) sensor circuit malfunction (CID)

DTC P0402 Exhaust Gas Recirculation (EGR) excess flow detected (valve open at idle)

DTC P0420 Catalyst system efficiency below threshold (Bank #1)

DTC P0430 Catalyst system efficiency below threshold (Bank #2)

DTC P0443 Evaporative emission control system Canister Purge (CANP) Control Valve circuit malfunction

DTC P0500 Vehicle Speed Sensor (VSS) malfunction

DTC P0505 Idle Air Control (IAC) system malfunction

DTC P0605 Powertrain Control Module (PCM)—Read Only Memory (ROM) test error

DTC P0703 Brake On/Off (BOO) switch input malfunction

DTC P0707 Manual Lever Position (MLP) sensor circuit low input

DTC P0708 Manual Lever Position (MLP) sensor circuit high input

DTC P0720 Output Shaft Speed (OSS) sensor circuit malfunction

DTC P0741 Torque Converter Clutch (TCC) system incorrect mechanical performance

DTC P0743 Torque Converter Clutch (TCC) system electrical failure

DTC P0750 Shift Solenoid #1(SS1) circuit malfunction

DTC P0751 Shift Solenoid #1(SS1) performance

DTC P0755 Shift Solenoid #2 (SS2) circuit malfunction

DTC P0756 Shift Solenoid #2 (SS2) performance

DTC P1000 OBD II Monitor Testing not complete

DTC P1100 Mass Air Flow (MAF) sensor intermittent

DTC P1101 Mass Air Flow (MAF) sensor out of Self-Test range

DTC P1112 Intake Air Temperature (IAT) sensor intermittent

DTC P1116 Engine Coolant Temperature (ECT) sensor out of Self-Test range

DTC P1117 Engine Coolant Temperature (ECT) sensor intermittent

DTC P1120 Throttle Position (TP) sensor out of range low

DTC P1121 Throttle Position (TP) sensor inconsistent with MAF sensor

DTC P1124 Throttle Position (TP) sensor out of Self-Test range

DTC P1125 Throttle Position (TP) sensor circuit intermittent

DTC P1130 Lack of HO_2S 11 switch, adaptive fuel at limit

DTC P1131 Lack of HO_2S 11 switch, sensor indicates lean (Bank #1)

DTC P1132 Lack of HO_2S 11 switch, sensor indicates rich (Bank #1)

DTC P1137 Lack of HO_2S 12 switch, sensor indicates lean (Bank #1)

DTC P1138 Lack of HO_2S 12 switch, sensor indicates rich (Bank #1)

DTC P1150 Lack of HO_2S 21 switch, adaptive fuel at limit

DTC P1151 Lack of HO_2S 21 switch, sensor indicates lean (Bank #2)

DTC P1152 Lack of HO_2S 21 switch, sensor indicates rich (Bank #2)

DTC P1157 Lack of HO_2S 22 switch, sensor indicates lean (Bank #2)

DTC P1158 Lack of HO_2S 22 switch, sensor indicates rich (Bank #2)

DTC P1351 Ignition Diagnostic Monitor (1DM) circuit input malfunction

DTC P1352 Ignition coil A primary circuit malfunction

DTC P1353 Ignition coil B primary circuit malfunction

DTC P1354 Ignition coil C primary circuit malfunction

DTC P1355 Ignition coil D primary circuit malfunction

DTC P1364 Ignition coil primary circuit malfunction

DTC P1390 Octane Adjust (OCT ADJ) out of Self-Test range

DTC P1400 Differential Pressure Feedback Electronic (DPFE) sensor circuit low voltage detected

DTC P1401 Differential Pressure Feedback Electronic (DPFE) sensor circuit high voltage detected

DTC P1403 Differential Pressure Feedback Electronic (DPFE) sensor hoses reversed

DTC P1405 Differential Pressure Feedback Electronic (DPFE) sensor upstream hose off or plugged

DTC P1406 Differential Pressure Feedback Electronic (DPFE) sensor downstream hose off or plugged

DTC P1407 Exhaust Gas Recirculation (EGR) no flow detected (valve stuck closed or inoperative)

DTC P1408 Exhaust Gas Recirculation (EGR) flow out of Self-Test range

DTC P1473 Fan Secondary High with fan(s) off

DTC P1474 Low Fan Control primary circuit malfunction

DTC P1479 High Fan Control primary circuit malfunction

DTC P1480 Fan Secondary low with low fan on

DTC P1481 Fan Secondary low with high fan on

DTC P1500 Vehicle Speed Sensor (VSS) circuit intermittent

DTC P1505 Idle Air Control (IAC) system at adaptive clip

DTC P1605 Powertrain Control Module (PCM)—Keep Alive Memory (KAM) test error

DTC P1703 Brake On/Off (BOO) switch out of Self-Test range

DTC P1705 Manual Lever Position (MLP) sensor out of Self-Test range

DTC P1711 Transmission Fluid Temperature (TFT) sensor out of Self-Test range

DTC P1742 Torque Converter Clutch (TCC) solenoid mechanically failed (turns MIL on)

DTC P1743 Torque Converter Clutch (TCC) solenoid mechanically failed (turns TCIL on)

DTC P1744 Torque Converter Clutch (TCC) system mechanically stuck in off position

DTC P1746 Electronic Pressure Control (EPC) solenoid circuit low input (open circuit)

DTC P1747 Electronic Pressure Control (EPC) solenoid circuit high input (short circuit)

DTC P1751 Shift Solenoid #1(SS1) performance

DTC P1756 Shift Solenoid #2 (SS2) performance

DTC P1780 Transmission Control Switch (TCS) circuit out of Self-Test range

1995 PASSENGER CARS AND LIGHT TRUCKS

DTC P0102 Mass Air Flow (MAF) Sensor circuit low input

DTC P0103 Mass Air Flow (MAF) Sensor circuit high input

DTC P0112 Intake Air Temperature (IAT) Sensor circuit low input

DTC P0113 Intake Air Temperature (IAT) Sensor high input

DTC P0117 Engine Coolant Temperature (ECT) low input

DTC P0118 Engine Coolant Temperature (ECT) Sensor circuit high input

DTC P0121 In range operating Throttle Position (TP) sensor circuit failure

DTC P0122 Throttle Position (TP) Sensor circuit low input

DTC P0123 Throttle Position (TP) Sensor high input

DTC P0125 Insufficient coolant temperature to enter closed loop fuel control

DTC P0126 Insufficient coolant temperature for stable operation

DTC P0131 Upstream Heated Oxygen Sensor (HO2S 11) circuit out of range low voltage (bank #1)

DTC P0132 Upstream Heated Oxygen Sensor (HO2S 11) circuit high voltage (Bank #1)

DTC P0133 Upstream Heated Oxygen Sensor (HO2S 11) circuit slow response (Bank #1)

DTC P0135 Heated Oxygen Sensor Heater (HTR 11) circuit malfunction

DTC P0136 Downstream Heated Oxygen Sensor (HO2S 12) circuit malfunction (Bank #1

DTC P0138 Downstream Heated Oxygen Sensor (HO2S 12) circuit high voltage (Bank #1)

DTC P0140 Heated Oxygen Sensor (HO2S 12) circuit no activity detected (Bank #1)

DTC P0141 Heated Oxygen Sensor Heater (HTR 12) circuit malfunction

DTC P0151 Upstream Heated Oxygen Sensor (HO2S 21) circuit out of range low voltage (Bank #2)

DTC P0152 Upstream Heated Oxygen Sensor (HO2S 21) circuit high voltage (Bank #2)

DTC P0153 Upstream Heated Oxygen Sensor (HO2S 21) circuit slow response (Bank #2)

DTC P0155 Heated Oxygen Sensor Heater (HTR 21) circuit malfunction

DTC P0156 Downstream Heated Oxygen Sensor (HO2S 22) circuit malfunction (Bank #2)

DTC P0158 Downstream Heated Oxygen Sensor (HO2S 22) circuit high voltage (Bank #2)

DTC P0160 Heated Oxygen Sensor (HO2S 12) circuit no activity detected (Bank #2)

DTC P0161 Heated Oxygen Sensor Heater (HTR 22) circuit malfunction

DTC P0171 System (adaptive fuel) too lean (Bank #1)

DTC P0172 System (adaptive fuel) too rich (Bank #1)

DTC P0174 System (adaptive fuel) too lean (Bank #2)

DTC P0175 System (adaptive fuel) too rich (Bank #2)

DTC P0222 Throttle Position Sensor B (TP-B) circuit low input

DTC P0223 Throttle Position Sensor B (TP-B) circuit high input

DTC P0230 Fuel pump primary circuit malfunction

DTC P0231 Fuel pump secondary circuit low

DTC P0232 Fuel pump secondary circuit high

DTC P0300 Random misfire detected

DTC P0301 Cylinder #1 misfire detected

DTC P0302 Cylinder #2 misfire detected

DTC P0303 Cylinder #3 misfire detected

DTC P0304 Cylinder #4 misfire detected

DTC P0305 Cylinder #5 misfire detected

DTC P0306 Cylinder #6 misfire detected

DTC P0307 Cylinder #7 misfire detected

DTC P0308 Cylinder #8 misfire detected

DTC P0320 Ignition engine speed (Profile Ignition Pickup) input circuit malfunction

DTC P0340 Camshaft Position (CMP) sensor circuit malfunction (CID)

DTC P0350 Ignition Coil primary circuit malfunction

DTC P0351 Ignition Coil A primary circuit malfunction

DTC P0352 Ignition Coil B primary circuit malfunction

DTC P0353 Ignition Coil C primary circuit malfunction

DTC P0354 Ignition Coil D primary circuit malfunction

DTC P0400 Exhaust Gas Recirculation (EGR) flow malfunction

DTC P0401 Exhaust Gas Recirculation (EGR) flow insufficient detected

DTC P0402 Exhaust Gas Recirculation (EGR) excess flow detected (valve open at idle)

DTC P0411 Secondary Air Injection system incorrect flow detected

DTC P0412 Secondary Air Injection system control valve malfunction

DTC P0420 Catalyst system efficiency below threshold (Bank #1)

DTC P0430 Catalyst system efficiency below threshold (Bank #2)

DTC P0443 Evaporative emission control system Canister Purge (CANP) Control Valve circuit malfunction

DTC P0500 Vehicle Speed Sensor (VSS) malfunction

DTC P0505 Idle Air Control (IAC) system malfunction

DTC P0603 Powertrain Control Module (PCM)—Keep Alive Memory (KAM) test error

DTC P0605 Powertrain Control Module (PCM)—Read Only Memory (ROM) test error

DTC P0704 Clutch Pedal Position (CPP) switch input circuit malfunction

DTC P0703 Brake On/Off (BOO) switch input malfunction

DTC P0707 Manual Lever Position (MLP) sensor circuit low input

DTC P0708 Manual Lever Position (MLP) sensor circuit high input

DTC P0712 Transmission Fluid Temperature (TFT) sensor circuit low input

DTC P0713 Transmission Fluid Temperature (TFT) sensor circuit high input

DTC P0715 Turbine Shaft Speed (TSS) sensor circuit malfunction

DTC P0720 Output Shaft Speed (OSS) sensor circuit malfunction

DTC P0731 Incorrect ratio for first gear

DTC P0732 Incorrect ratio for second gear

DTC P0733 Incorrect ratio for third gear

DTC P0734 Incorrect ratio for fourth gear

DTC P0736 Reverse incorrect gear

DTC P0741 Torque Converter Clutch (TCC) system incorrect mechanical performance

DTC P0746 Electronic Pressure Control (EPC) solenoid performance

DTC P0743 Torque Converter Clutch (TCC) system electrical failure

DTC P0750 Shift Solenoid #1(SS1) circuit malfunction

DTC P0751 Shift Solenoid #1(SS1) performance

DTC P0755 Shift Solenoid #2 (SS2) circuit malfunction

DTC P0756 Shift Solenoid #2 (SS2) performance

DTC P0760 Shift Solenoid #3 (SS3) circuit malfunction

DTC P0761 Shift Solenoid #3 (SS3) performance

DTC P0781 1 to 2 shift error

DTC P0782 2 to 3 shift error

DTC P0783 3 to 4 shift error

DTC P0784 4 to 5 shift error

DTC P1000 OBD II Monitor Testing not complete

DTC U1039 OBD II Monitor not complete

DTC UIOS1 Brake switch signal missing or incorrect

DTC P1100 Mass Air Flow (MAF) sensor intermittent

DTC P1101 Mass Air Flow (MAF) sensor out of Self-Test range

DTC P1112 Intake Air Temperature (IAT) sensor intermittent

DTC P1116 Engine Coolant Temperature (ECT) sensor out of Self-Test range

DTC P1117 Engine Coolant Temperature (ECT) sensor intermittent

DTC P1120 Throttle Position (TP) sensor out of range low

DTC P1121 Throttle Position (TP) sensor inconsistent with MAF sensor

DTC P1124 Throttle Position (TP) sensor out of Self-Test range

DTC P1125 Throttle Position (TP) sensor circuit intermittent

DTC P1130 Lack of HO2S 11 switch, adaptive fuel at limit

DTC P1131 Lack of HO2S 11 switch, sensor indicates lean (Bank #1)

DTC P1132 Lack of HO2S 11 switch, sensor indicates rich (Bank #1)

DTC U1135 Ignition switch signal missing or incorrect

DTC P1137 Lack of HO2S 12 switch, sensor indicates lean (Bank #1)

DTC P1138 Lack of HO2S 12 switch, sensor indicates rich (Bank #1)

DTC P1150 Lack of HO2S 21 switch, adaptive fuel at limit

DTC P1151 Lack of HO2S 21 switch, sensor indicates lean (Bank #2)

DTC P1152 Lack of HO2S 21 switch, sensor indicates rich (Bank #2)

DTC P1157 Lack of HO2S 22 switch, sensor indicates lean (Bank #2)

DTC P1158 Lack of HO2S 22 switch, sensor indicates rich (Bank #2)

DTC P1220 Series Throttle Control malfunction

DTC P1224 Throttle Position Sensor (TP-B) out of Self-test range

DTC P1233 Fuel Pump driver Module off-line

DTC P1234 Fuel Pump driver Module off-line

DTC P1235 Fuel Pump control out of range

DTC P1236 Fuel Pump control out of range

DTC P1237 Fuel Pump secondary circuit malfunction

DTC P1238 Fuel Pump secondary circuit malfunction

DTC P1260 THEFT detected—engine disabled

DTC P1270 Engine RPM or vehicle speed limiter reached

DTC P1351 Ignition Diagnostic Monitor (1DM) circuit input malfunction

DTC P1352 Ignition coil A primary circuit malfunction

DTC P1353 Ignition coil B primary circuit malfunction

DTC P1354 Ignition coil C primary circuit malfunction

DTC P1355 Ignition coil D primary circuit malfunction

DTC P1358 Ignition Diagnostic Monitor (1DM) signal out of Self-Test range

DTC P1359 Spark output circuit malfunction

DTC P1364 Ignition coil primary circuit malfunction

DTC P1390 Octane Adjust (OCT ADJ) out of Self-Test range

DTC P1400 Differential Pressure Feedback Electronic (DPFE) sensor circuit low voltage detected

DTC P1401 Differential Pressure Feedback Electronic (DPFE) sensor circuit high voltage detected

DTC P1403 Differential Pressure Feedback Electronic (DPFE) sensor hoses reversed

DTC P1405 Differential Pressure Feedback Electronic (DPFE) sensor upstream hose off or plugged

DTC P1406 Differential Pressure Feedback Electronic (DPFE) sensor downstream hose off or plugged

DTC P1407 Exhaust Gas Recirculation (EGR) no flow detected (valve stuck closed or inoperative)

DTC P1408 Exhaust Gas Recirculation (EGR) flow out of Self-Test range

DTC P1409 Electronic Vacuum Regulator (EVR) control circuit malfunction

DTC P1414 Secondary Air Injection system monitor circuit high voltage

DTC P1443 Evaporative emission control system—vacuum system purge control solenoid or purge control valve malfunction

DTC P1444 4- Purge Flow Sensor (PFS) circuit low input

DTC P1445 Purge Flow Sensor (PFS) circuit high input

DTC U1451 Lack of response from Passive Anti-Theft system (PATS) module—engine disabled

DTC P1460 Wide Open Throttle Air Conditioning Cut-off (WAC) circuit malfunction

DTC P1461 Air Conditioning Pressure (ACP) sensor circuit low input

DTC P1462 Air Conditioning Pressure (ACP) sensor circuit high input

DTC P1463 Air Conditioning Pressure (ACP) sensor insufficient pressure change

DTC P1469 Low air conditioning cycling period

DTC P1473 Fan Secondary High with fan(s) off

DTC P1474 Low Fan Control primary circuit malfunction

DTC P1479 High Fan Control primary circuit malfunction

DTC P1480 Fan Secondary low with low fan on

DTC P1481 Fan Secondary low with high fan on

DTC P1500 Vehicle Speed Sensor (VSS) circuit intermittent

DTC P1505 Idle Air Control (IAC) system at adaptive clip

DTC P1506 Idle Air control (IAC) over speed error

DTC P1518 Intake Manifold Runner Control (IMRC) malfunction (stuck open)

DTC P1519 Intake Manifold Runner Control (IMRC) malfunction (stuck closed)

DTC P1520 Intake Manifold Runner Control (IMRC) circuit malfunction

DTC P1507 Idle Air control (IAC) under speed error

DTC P1605 Powertrain Control Module (PCM)—Keep Alive Memory (KAM) test error

DTC P1650 Power steering Pressure (PSP) switch out of Self-Test range

DTC P1651 Power steering Pressure (PSP) switch input malfunction

DTC P1701 Reverse engagement error

DTC P1703 Brake On/Off (BOO) switch out of Self-Test range

DTC P1705 Manual Lever Position (MLP) sensor out of Self-Test range

DTC P1709 Park or Neutral Position (PNP) switch out of Self-test range

DTC P1729 4X4 Low switch error

DTC P1711 Transmission Fluid Temperature (TFT) sensor out of Self-Test range

DTC P1741 Torque Converter Clutch (TOG) control error

DTC P1742 Torque Converter Clutch (TOO) solenoid mechanically failed (turns MIL on)

DTC P1743 Torque Converter Clutch (TOO) solenoid mechanically failed (turns TOIL on)

DTC P1744 Torque Converter Clutch (TOO) system mechanically stuck in off position

DTC P1748 Electronic Pressure Control (EPC) solenoid circuit low input (open circuit)

DTO P1747 Electronic Pressure Control (EPC) solenoid circuit high input (short circuit)

DTC P1749 Electric Pressure Control (EPC) solenoid failed low

DTC P1751 Shift Solenoid #1(SS1) performance

DTC P1756 Shift Solenoid #2 (SS2) performance

DTC P1780 Transmission Control Switch (TCS) circuit out of Self-Test range

EEC NON-NAAO System

The Capri, Festiva, Probe, Escort and Tracer are referred to as NON-NAAO (Not North American Automotive Operations) produced vehicles. The fuel system used on these vehicles is called Electronic Engine Control (EEC). This EEC system components and operation are basically the same as the EEC-IV system. The self-diagnostic function on the EEC system differs from the EEC-VI system and is covered under NON-NAAO vehicle.

The Non-NAAO EEC system self-diagnostics and code retrieval many be different than domestic built Ford engines. The code descriptions are mostly the same for all of the NON NAAO vehicles. Take note that not all vehicles use all codes. The transaxle codes were introduced in 1992 some Capri, Escort, Probe and Tracer models. Late model Probe and Escort may be equipped with EEC-IV fuel control system, the diagnostic link connector (DLC) will be of the EEC-IV design.

The code definitions listed general 2-digit codes for Ford Vehicles using the Ford EEC engine control system. For a specific code definition or component test procedure consult a vehicle service manual for your vehicle. A diagnostic code does not mean the component is defective. For example a Code 6 is a vehicle speed sensor code. This does not mean the sensor is defective, but to check the sensor and related components. A defective speedometer cable or transmission problem will also set this code.

➡ **When the term Powertrain Control Module (PCM) is used in this manual it will refer to the engine control computer regardless that it may be a Powertrain Control Module (PCM) or Electronic Control Module (ECM) or Electronic Control Assembly (ECA).**

Code 02 Crankshaft position sensor
Code 03 Cylinder identification sensor #1, 1.6L engine
Code 06 Vehicle speed sensor, 1.6L engine
Code 08 Air flow signal
Code 09 Engine coolant temperature sensor
Code 10 Air temperature sensor
Code 12 Throttle position sensor, 1.6L engine
Code 14 Barometric pressure sensor
Code 15 Oxygen sensor, signal LEAN
Code 16 EGR position valve, 2.2L engine
Code 17 Oxygen sensor, signal RICH
Code 25 Fuel pressure regulator solenoid, 1.6L engine
Code 26 Canister purge solenoid
Code 34 Idle speed control solenoid
Code 41 High speed inlet air control solenoid
Code 55 Pulse generator, transaxle code
Code 57 Down shift signal, transaxle code
Code 60 1-2 Shift solenoid, transaxle code
Code 61 2-3 Shift solenoid, transaxle code
Code 62 3-4 Shift solenoid, transaxle code
Code 63 Lock-up control, transaxle code

➡ **This list is to be used as a reference for testing and does not mean a specific component is defective.**

No Code Unable to Run Self Test or Output Codes, or list does not apply to vehicle tested, refer to specific vehicle service manual

General Motors

SELF-DIAGNOSTICS

Automotive manufacturers have developed on-board computers to control engines, transmissions and many other components. These on-board computers with dozens of sensors and actuators have become almost impossible to test without the help of electronic test equipment.

One of these electronic test devices has become the on-board computer itself. The Powertrain Control Modules (PCM), sometimes called the Electronic Control Module (ECM), used on toadies vehicles has a built in self testing system. This self test ability is called self-diagnosis. The self-diagnosis system will test many or all of the sensors and controlled devices for proper function. When a malfunction is detected this system will store a code in memory that's related to that specific circuit. The computer can later be accessed to obtain fault codes recorded in memory using the procedures for Reading Codes. This helps narrow down what area to begin testing.

Fault code meanings can vary from year to year even on the same model. It is extremely important after retrieving a fault code to verify its meaning with a proper manual. Servicing a code incorrectly will not only lead to the wrong conclusion but could also cause damage if tested or serviced incorrectly.

Since the control module is programmed to recognize the presence and value of electrical inputs, it will also note the lack of a signal or a radical change in values. It will, for example, react to the loss of signal from the vehicle speed sensor or note that engine coolant temperature has risen beyond acceptable (programmed) limits. Once a fault is recognized, a numeric code is assigned and held in memory. The dashboard warning lamp—CHECK ENGINE or SERVICE ENGINE SOON—will illuminate to advise the operator that the system has detected a fault.

More than one code may be stored. Although not every engine uses every code and the same code may carry different meanings relative to each engine or engine family. For example, on the 3.3L (VIN N), Code 46 indicates a fault found in the power steering pressure switch circuit. The same code on the 5.7L (VIN F) engine indicates a fault in the VATS anti-theft system. The list of codes and descriptions can be found in the 'Code Descriptions' section of the manual.

In the event of an PCM failure, the system will default to a pre-programmed set of values. These are compromise values which allow the engine to operate, although possibly at reduced efficiency. This is also known as the default, limp-in or back-up mode. Driveability is almost always affected when the PCM enters this mode.

Service Precautions

• Protect the on-board solid-state components from rough handling or extremes of temperature.

• Always turn the ignition OFF when connecting or disconnecting battery cables, jumper cables, or a battery charger. Failure to do this can result in PCM or other electronic component damage.

• Remove the PCM before any arc welding is performed to the vehicle

• Electronic components are very susceptible to damage caused by electrostatic discharge (static electricity). To prevent electronic component damage, do not touch the control module connector pins or soldered components on the control module circuit board.

Visual Inspection

This is possibly the most critical step of diagnosis. A detailed examination of all connectors, wiring and vacuum hoses can often lead to a repair without further diagnosis. Also, take into consideration if the vehicle has been serviced recently? Sometimes things get reconnected in the wrong place, or not at all. A careful inspector will check the undersides of hoses as well as the integrity of hard-to-reach hoses blocked by the air cleaner or other components. Correct routing for vacuum hoses can be obtained from your specific vehicle service manual or Vehicle Emission Control Information (VECI) label in the engine compartment of the vehicle. Wiring should be checked carefully for any sign of strain, burning, crimping or terminals pulled-out from a connector.

Checking connectors at components or in harnesses is required; usually, pushing them together will reveal a loose fit. Also, check electrical connectors for corroded, bent, damaged, improperly seated pins, and bad wire crimps to terminals. Pay particular attention to ground circuits, making sure they are not loose or corroded. Remember to inspect connectors and hose fittings at components not mounted on the engine, such as the evaporative canister or relays mounted on the fender aprons. Any component or wiring in the vicinity of a fluid leak or spillage should be given extra attention during inspection.

➡There are many problems with connectors on electronic engine control systems. Due to the low voltage signals that these systems use any dirt, corrosion or damage will affect their operation. Note that some connectors use a special grease on the contacts to prevent corrosion. Do not wipe this grease off, it is a special type for this purpose. You can obtain this grease from your vehicle dealer.

Additionally, inspect maintenance items such as belt condition and tension, battery charge and condition and the radiator cap carefully. Any of these very simple items may affect the system enough to set a fault.

Dashboard Warning Lamp

The primary function of the dash warning lamp is to advise the operator that a fault has been detected, and, in most cases, a code stored. Under normal conditions, the dash warning lamp will illuminate when the ignition is turned ON. Once the engine is started and running, the PCM will perform a system check and extinguish the warning lamp if no fault is found.

Additionally, the dash warning lamp can be used to retrieve stored codes after the system is placed in the Diagnostic Mode. Codes are transmitted as a series of flashes with short or long pauses. When the system is placed in the Field Service Mode (available on fuel injected model), the dash lamp will indicate open loop or closed loop function.

SERVICE ENGINE SOON

"SERVICE ENGINE SOON"

79222G22

The Check Engine or Service Engine Soon (MIL) lamp is used for reading trouble codes

Intermittent Problems

If a fault occurs intermittently, such as a loose connector pin breaking contact as the vehicle hits a bump, the PCM will note the fault as it occurs and energize the dash warning lamp. If the problem self-corrects, as with the terminal pin again making contact, the dash lamp will extinguish after 10 seconds but a code will remain stored in the PCM memory. When an unexpected code appears during an intermittent failure that self-corrected; the codes are still useful in diagnosis and should not be discounted.

Diagnostic Connector Location

The Assembly Line Communication Link (ALCL) or Assembly Line Diagnostic Link (ALDL) is a Diagnostic Link Connector (DLC) located in the passenger compartment. It has terminals which are

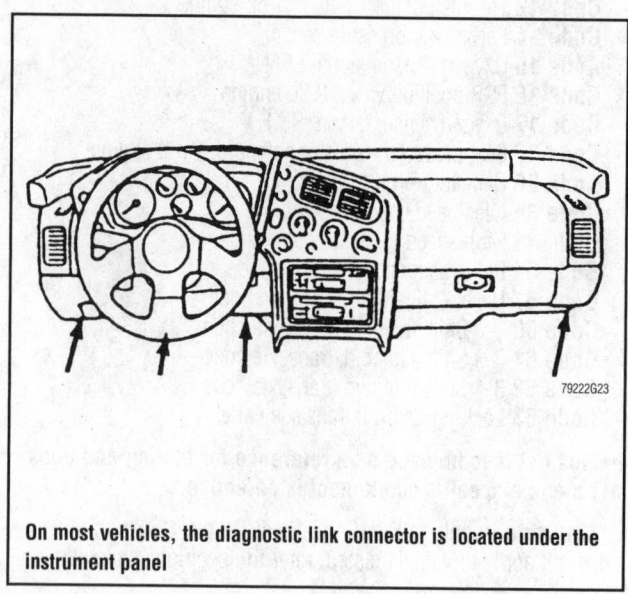

79222G23

On most vehicles, the diagnostic link connector is located under the instrument panel

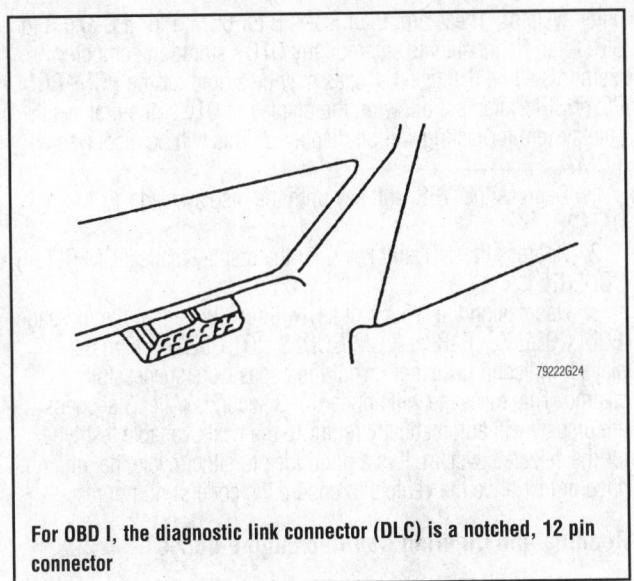

For OBD I, the diagnostic link connector (DLC) is a notched, 12 pin connector

used in the assembly plant to check that the engine is operating properly before it leaves the plant.

This DLC is where you connect you jumper the terminals to place the engine control computer into self-diagnostic mode. The standard term DLC is sometimes referred to as the ALCL or the ALDL in different manuals. Either way it is referred to, they all still perform the same function.

READING & CLEARING CODES

Reading Codes (Except Cadillac)

Since the inception of electronic engine management systems on General Motors vehicles, there has been a variety of connectors provided to the technician for retrieving Diagnostic Trouble Codes (DTC)s. Additionally, there have been a number of different names given to these connectors over the years; Assembly Line Communication Link (ALCL), Assembly Line Diagnostic Link (ALDL), Data Link Connector (DLC). Actually when the system was initially introduced to the 49 states in 1979, early 1980, there was no connector used at all. On these early vehicles there was a green spade terminal taped to the ECM harness and connected to the diagnostic enable line at the computer. When this terminal was grounded with the key ON, the system would flash any stored diagnostic trouble codes. The introduction of the ALOL was found to be a much more convenient way of retrieving fault codes. This connector was located underneath the instrument panel on most GM vehicles, however on some models it will not be found there. On early Corvettes the ALOL is located underneath the ashtray, it can be found in the glove compartment of some early FWD Oldsmobiles, and between the seats in the Pontiac Fiero. The connector was first introduced as a square connector with four terminals, then progressed to a flat five terminal connector, and finally to what is still used in 1993, a 12 terminal double row connector. To access stored Diagnostic Trouble Codes (DTC) from the square connector, turn the ignition ON and identify the diagnostic enable terminal (usually a white wire with a black tracer) and ground it. The flat five terminal connector is identified from left to right as A, B, C, D, and E. There is a space between terminal D and E which permits a spade to be inserted for the purposes of diagnostics when the ignition key is ON. On this connector terminal D is the diagnostic enable line, and E is a ground. The 12 terminal double row connector has been continually expanded through the years as vehicles acquired more on-board electronic systems such as Anti-lock Brakes. Despite this the terminals used for engine code retrieval have remained the same. The 12 terminal connector is identified from right-to-left on the top row A-F, and on the bottom row from left-to-right, G-L. To access engine codes turn the ignition ON and insert a jumper between terminals A and B. Terminal A is a ground, and terminal B is the diagnostic request line. Stored trouble codes can be read through the flashing of the Check Engine Light or on later vehicles the Service Engine Soon lamp. Trouble codes are identified by the timed flash of the indicator light. When diagnostics are first entered the light will flash once, pause; then two quick flashes.

This reads as DTC 12 which indicates that the diagnostic system is working. This code will flash indefinitely if there are no stored trouble codes. If codes are stored in memory, Code 12 will flash three times before the next code appears. Codes are displayed in the next highest numerical sequence. For example, Code 13 would be displayed next if it was stored in memory and would read as follow: flash, pause, flash, flash, flash, long pause, repeat twice. This sequence will continue until all codes have been displayed, and then start all over again with Code 12.

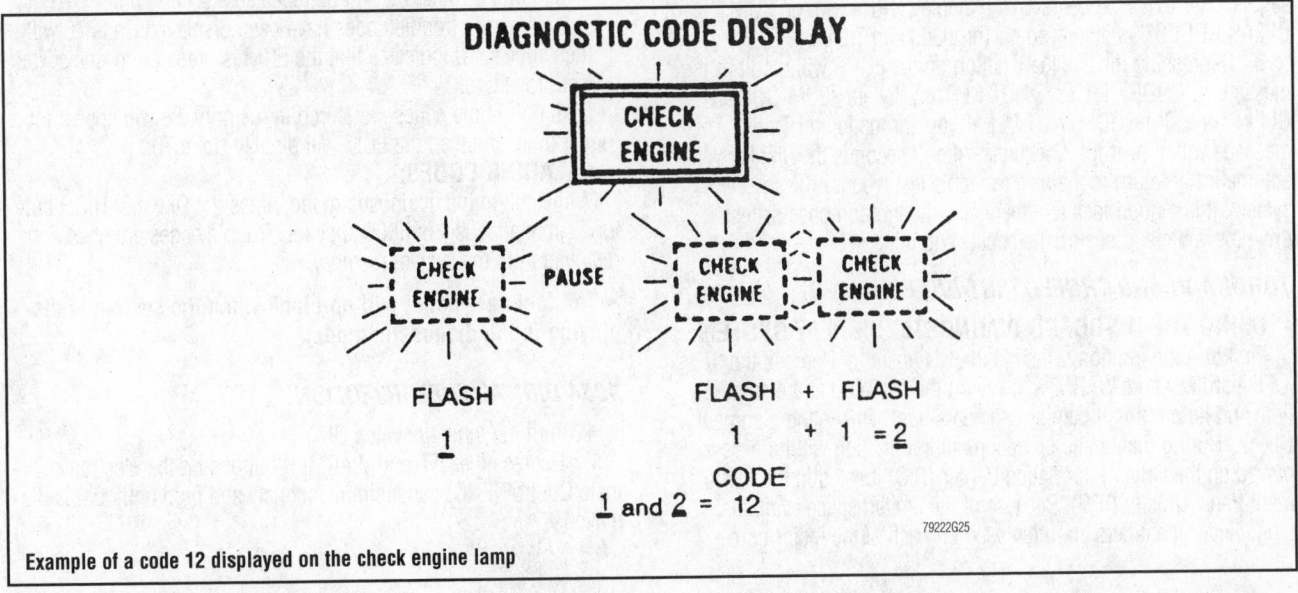

Example of a code 12 displayed on the check engine lamp

Clearing Codes (Except Cadillac)

EXCEPT RIVIERA, TORONADO AND TROFEO

To clear any Diagnostic Trouble Codes (DTC's) from the PCM memory, either to determine if the malfunction will occur again or because repair has been completed, power feed must be disconnected for at least 30 seconds. Depending on how the vehicle is equipped, the system power feed can be disconnected at the positive battery terminal pigtail, the inline fuseholder that originates at the positive connection at the battery, or the ECM/PCM fuse in the fuse block. The negative battery terminal may be disconnected but other on-board memory data such as preset radio tuning will also be lost. To prevent system damage, the ignition switch must be in the OFF position when disconnecting or reconnecting power.

When using a Diagnostic Computer such as Tech 1, or equivalent scan tool to read the diagnostic trouble codes, clearing the codes is done in the same manner. On some systems, OTC's may be cleared through the Tech 1, or equivalent scan tool.

On Riviera, Toronado and Trofeo, clearing codes is part of the dashboard display menu or diagnostic routine. Because of the amount of electronic equipment on these vehicles, clearing codes by disconnecting the battery is not recommended.

RIVIERA, TORONADO AND TROFEO (NON-CRT/DID VEHICLES)

USING THE ON-BOARD DIAGNOSTIC DISPLAY SYSTEM:

First turn the ignition to the ON position. On Riviera depress the OFF and TEMP buttons on the ECCP at the same time and hold until all display segments light. This is known as the Segment Check. On Toronado and Trofeo follow the same procedure, however, depress the OFF and WARMER buttons on the ECOP instead. After diagnostics is entered, any OTC's stored in computer memory will be displayed. Codes may be stored for the PCM, BCM, PC or SIR systems. Following the display of OTC's, the first available system for testing will be displayed. For example, 'EC?' would be displayed on Riviera for EOM testing, while on Toronado and Trofeo the message 'ECM?' will appear. The message is more clear on these vehicles due to increased character space in the IPO display area.

1. Depress the 'FAN UP' button on the ECCP until the message 'DATA EC?' appears on the display for Riviera, or 'ECM DATA?' is displayed on Toronado and Trofeo.

2. Depress the 'FAN DOWN' button on the ECCP until the message 'CLR E CODE' appears on the display for Riviera, or 'ECM CLEAR CODES?' is displayed on Toronado and Trofeo.

3. Depressing the 'FAN UP' button on the ECCP will result in the message 'E CODE CLR' or 'E NOT CLR' on Riviera, 'EOM CODES CLEAR' or 'ECM CODES NOT CLEAR' on Toronado and Trofeo. This message will appear for 3 seconds. After 3 seconds the display will automatically return to the next available test type for the selected system. It is a good idea to either cycle the ignition once or test drive the vehicle to ensure the code(s) do not reset.

TORONADO AND TROFEO (CRT/DID EQUIPPED)

USING THE ON-BOARD DIAGNOSTIC DISPLAY SYSTEM:

First turn the ignition switch to the ON position. Depress the 'OFF' hard key and 'WARM' soft key on the CRT/DID at the same time and hold until all display segments light. This is the 'Segment Check.' During diagnostic operation, all information will be displayed on the Driver Information Center (DIC) located in the Instrument Panel Cluster (IPC). Because of the limited space available single letter identiffers are often used for each of the major com-

puter systems. These are: E for ECM, B for 6CM, I for IPC and R for SIR. After diagnostics is entered, any OTC's stored in computer memory will be displayed. Codes may be stored for the PCM, BCM, PC or SIR systems. Following the display of OTC's, the first available system for testing will be displayed. This will be displayed as 'ECM?'.

1. Depress the 'YES' soft key until the display reads 'ECM DATA?'.

2. Depress the 'NO' soft key until the display reads 'ECM CLEAR CODES?'.

3. Depressing the 'YES' soft key will result in either the message 'ECM CODES CLEAR' or 'ECM CODES NOT CLEAR' being displayed, indicating whether or not the codes were successfully cleared. This message will appear for 3 seconds. After 3 seconds the display will automatically return to the next available test type for the selected system. It is a good idea to either cycle the ignition once or test drive the vehicle to ensure the code(s) do not reset.

Reading and Clearing Cadillac Engine Codes

➡ **The Cadillac Cimmaron used the 12 terminal DLC and codes can be accessed in the conventional manner as all other General Motors vehicles. The rear wheel drive Cadillac equipped with either the 4.1L V6, 5.0L V8, or the 5.7L V8 all can also be accessed in the conventional manner using the DLC.**

1980–1983 DIGITAL FUEL INJECTION

1. Turn the ignition switch **ON**.

2. Depress the OFF and WARMER buttons on the Electronic Climate Control (ECO) panel simultaneously and hold until .. is displayed.

3. The numerals 88 should then appear indicating that all display segments are functional. Diagnosis should not be attempted unless the entire 88 is displayed or misdiagnosis will result.

4. If trouble codes are present they will appear on the digital ECO panel as follows:

    a. The lowest numbered code will be displayed for approximately two seconds.

    b. Progressively higher numbered codes, if present, will be displayed consecutively for two second intervals until the highest code has been displayed.

    c. 88 is again displayed.

    d. Parts A, B, and C will be repeated a second time.

    e. After the trouble codes have been displayed Code 70 will then appear. 70 indicates that the ECM is prepared for the switch test procedure.

5. If no trouble codes are stored in memory, 88 will appear for a longer time, and then the ECM will display Code 70.

CLEARING CODES:

While still in the diagnostic mode, press the OFF and HIGH buttons simultaneously until 00 appears. Trouble codes are now removed from the system memory.

➡ **The fuel data panel will go blank when the system is displaying in the diagnostic mode.**

1984 DIGITAL FUEL INJECTION

1. Turn the ignition switch ON.

2. Depress the OFF and WARMER buttons on the Electronic Climate Control (ECC) panel simultaneously and hold until .0.0 is displayed.

3. —1.8.8 should then appear indicating that all display segments are functional. Diagnosis should not be attempted unless the entire —1.8.8 is displayed or misdiagnosis will result.

4. If trouble codes are present they will appear on the digital ECO panel as follows:

a. The lowest numbered code will be displayed for approximately two seconds.

b. Progressively higher numbered codes, if present, will be displayed consecutively for two second intervals until the highest code has been displayed.

c. 88 is again displayed.

d. Parts A, B, and C will be repeated a second time.

e. After the trouble codes have been displayed Code 70 will then appear. 70 indicates that the ECM is prepared for the switch test procedure.

5. If no trouble codes are stored in memory, —1.8.8 will appear for a longer time, and then the ECM will display Code 70.

CLEARING CODES:

While still in the diagnostic mode, press the OFF and HIGH buttons simultaneously until .0.0 appears. Trouble codes are now removed from the system memory.

➡ **The fuel data panel will go blank when the system is displaying in the diagnostic mode.**

1985–1986 DIGITAL FUEL INJECTION

1. Turn the ignition switch **ON**.

2. Depress the OFF and WARMER buttons on the Climate Control Panel (CCP) simultaneously and hold until −188 is displayed.

3. -188 should then appear indicating that all display segments are functional. Diagnosis should not be attempted unless the entire -188 is displayed or misdiagnosis will result.

4. If trouble codes are present they will appear on the Fuel Data Center (FDC) panel as follows:

a. Display of trouble codes will begin with an 8.8.8 on the FDC panel for approximately one second. . . . E will then be displayed which indicates beginning of the ECM stored trouble codes. The initial pass of ECM codes includes all the detected malfunctions whether or not they are currently present. If no ECM codes are stored the . . . E display will be bypassed.

b. Following the display of . . . E the lowest numbered ECM code will be displayed for approximately two seconds. All ECM codes will be prefixed with an E (i.e. E12, E13, etc.).

c. Progressively higher numbered codes, if present, will be displayed consecutively for two second intervals until the highest code has been displayed.

d. .E.E is again displayed which indicates the start of the second pass of ECM trouble codes. On the second pass only current faults (hard codes) will be displayed. Codes displayed on the first pass are history failures or (soft codes). If all displayed codes were history codes the .E.E will be bypassed.

e. When all ECM codes have been displayed, BCM codes will appear with the prefix F in the same manner as the ECM did.

f. After the display of all codes, or if no codes were stored, Code .7.0 will appear indicating the start of the switch tests.

CLEARING CODES:

While still in the diagnostic mode, press the OFF and HIGH buttons simultaneously until E.0.0 appears. Trouble codes are now removed from the ECM memory.

1987–1993 DEVILLE AND FLEETWOOD

1. Turn the ignition switch **ON**.

2. Depress the OFF and WARMER buttons on the Climate Control Panel (CCP) simultaneously and hold until -188 is displayed.

3. -188 should then appear indicating that all display segments are functional. Diagnosis should not be attempted unless the entire -188 is displayed or misdiagnosis will result.

4. If trouble codes are present they will appear on the Fuel Data Center (FDC) panel as follows:

a. Display of trouble codes will begin with an 8.8.8 on the FDC panel for approximately one second. . . . E will then be displayed which indicates beginning of the ECM stored trouble codes. The initial pass of ECM codes includes all the detected malfunctions whether or not they are currently present. If no ECM codes are stored the . . . E display will be bypassed.

b. Following the display of . . . E the lowest numbered ECM code will be displayed for approximately two seconds. All ECM codes will be prefixed with an E (i.e. E12, E13, etc.).

c. Progressively higher numbered codes, if present, will be displayed consecutively for two second intervals until the highest code has been displayed.

d. .E.E is again displayed which indicates the start of the second pass of ECM trouble codes. On the second pass only current faults (hard codes) will be displayed. Codes displayed on the first pass are history failures or (soft codes). If all displayed codes were history codes the .E.E will be bypassed.

e. When all ECM codes have been displayed, BCM codes will appear with the prefix F in the same manner as the ECM did.

5. After the display of all codes, or if no codes were stored, Code .7.0 will appear indicating the start of the switch tests.

CLEARING CODES:

While still in the diagnostic mode, press the OFF and HIGH buttons simultaneously until E.0.0 appears. Trouble codes are now removed from the ECM memory.

1987–1993 ALLANTE, ELDORADO AND SEVILLE

1. Turn the ignition switch **ON**.

2. Depress the OFF and WARMER buttons on the Climate Control Panel (CCP) simultaneously and hold until the segment check appears on the Instrument Panel Cluster (IPC) and the Climate Control Driver Information Center (CODIC).

3. Diagnosis should not be attempted unless all of the segments of the vacuum fluorescent display are working as this could lead to misdiagnosis. On the PC however the turn signal indicators do not light during this check.

4. After the service mode is entered, any trouble codes stored in the computer memory will be displayed, starting with ECM codes prefixed with an E.

5. If no trouble codes are present, the message NO ECM CODES will be displayed. Some later systems will display NO X CODES present, with X representing the system selected such as ECM, BOM, SIR, etc.

CLEARING CODES:

6. While still in the service mode, and the ECM diagnostic code display has been completed press the HI button on the CCP.

7. This action should cause the display to read 'ECM DATA?'.

8. Press the LO button on the COP until the display reads 'ECM CLEAR CODES?'.

9. Press the HI button on the CCP and the display should read 'ECM CODES CLEAR'.

10. After approximately 3 seconds, all stored ECM codes will be erased.

➡The Cadillac Cimmaron used the 12 terminal DLC and codes can be accessed in the conventional manner. The rear wheel drive Cadillacs equipped with either the 4.1 L V6, 5.0L V8, or the 5.7L V8 all can be accessed in the conventional manner using the DLC.

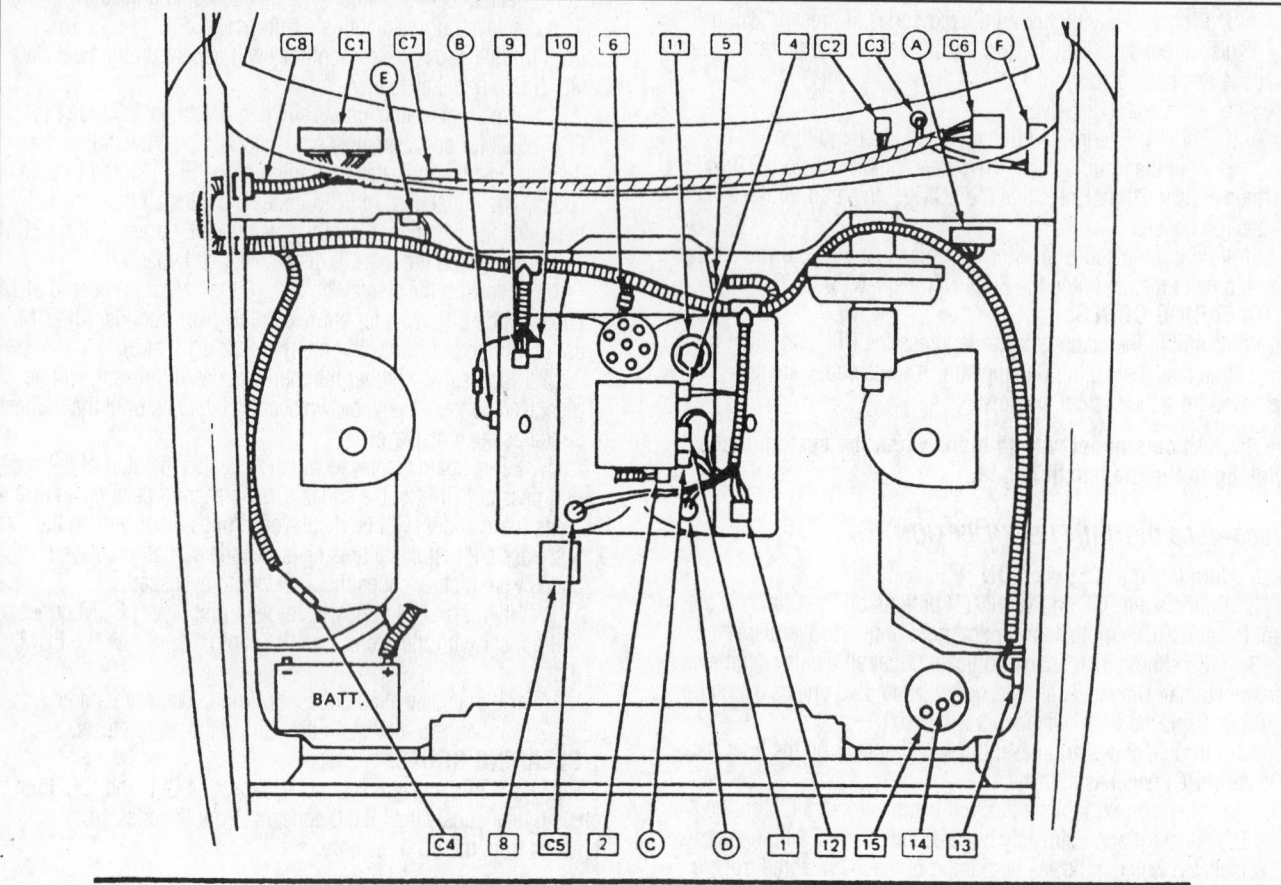

"F" SERIES
2.8L (173 CID) V6   RPO: LC1/LL1 H.O.   V.I.N. CODE: 1/L

COMPUTER SYSTEM

C1	Electronic Control Module (ECM)
C2	ALCL Connector
C3	"CHECK ENGINE" Light
C4	System Power
C5	System Ground
C6	Fuse Panel
C7	Lamp Driver
C8	Computer Control Harness

AIR/FUEL SYSTEM

1	Mixture Control
2	Idle Speed Solenoid
4	Heated Grid EFE

TRANSMISSION CONVERTER CLUTCH CONTROL SYSTEM

5	Trans. Conv. Clutch Connector

IGNITION SYSTEM

6	Electronic Spark Timing Connector

AIR INJECTION SYSTEM

8	Air Injection Pump
9	Air Control Solenoid Valve (Divert)
10	Air Switching Solenoid Valve

EXHAUST GAS RECIRCULATION CONTROL SYSTEM

11	Exhaust Gas Recirculation Valve
12	Exhaust Gas Recirculation Solenoid Valve

FUEL VAPOR CONTROL SYSTEM

13	Canister Purge Solenoid Valve
14	From Fuel Tank
15	Vapor Canister

SEM   S/SWITCHES

A	Differential Pressure Sensor
B	Exhaust Oxygen Sensor
C	Throttle Position Sensor
D	Coolant Sensor
E	Barometric Pressure Sensor
F	Vehicle Speed Sensor

79222G26

Example of component locator on a vehicle with electronic engine controls—2.8L Firebird shown

DIAGNOSTIC TROUBLE CODES

Except Front Wheel Drive Cadillac

Code 12 No engine RPM reference pulses—System Normal

Code 13 Oxygen Sensor (02S) circuit open—left side on 2 sensor system

Code 14 Engine Coolant Temperature (ECT) sensor -possible circuit high or shorted sensor

Code 15 Engine Coolant Temperature (ECT) sensor -circuit low or open circuit

Code 16 Direct ignition system (DIS), fault line circuit or Distributor ignition system (low resolution pulse) or Missing 2x reference circuit or OPTI-Spark ignition timing system (low resolution pulse) or System voltage out of range

Code 17 Camshaft Position Sensor (OPS) or spark reference circuit error

Code 18 Crank/Cam error

Code 19 Crankshaft Position Sensor (CPS) circuit

Code 21 Throttle Position (TP) sensor circuit—signal voltage out of range, probably high

Code 22 Throttle Position (TP) sensor circuit—signal voltage low

Code 23 Intake Air Temperature (IAT or MAT) sensor circuit temperature out of range, low or Open or grounded M/C solenoid Feedback Carburetor system

Code 24 Vehicle Speed Sensor (VSS) circuit

Code 25 Intake Air Temperature (IAT or MAT) sensor circuit temperature out of range, high

Code 26 Quad-Driver Module #1 circuit or Transaxle gear switch circuit

Code 27 Quad-Driver Module circuit or Transaxle gear switch, probably 2nd gear switch circuit

Code 28 Quad-Driver Module (QDM) #2 circuit or Transaxle gear switch, probably 3rd gear switch circuit

Code 29 Transaxle gear switch, probably 4th gear switch circuit

Code 31 Camshaft sensor circuit fault or Park/Neutral Position (PNP) switch circuit or Wastegate circuit signal

Code 32 Exhaust Gas Recirculation (EGR) circuit fault or Barometric Pressure Sensor circuit low Feedback Carburetor system

Code 33 Manifold Absolute Pressure (MAP) sensor—signal voltage out of range, high or Mass Air Flow (MAE) sensor—signal voltage out of range, probably high

Code 34—Manifold Absolute Pressure (MAP) sensor—circuit out of range voltage, low or Mass Air Flow (MAF) sensor circuit (gm/sec low)

Code 35—Idle Air Control (IAC) or idle speed error or Idle Speed Control (ISO) circuit throttle switch shorted Feedback Carburetor system

Code 36 Ignition system circuit error or Transaxle shift problem—4T60E Transaxle

Code 38 Brake input circuit fault—Torque converter clutch signal

Code 39 Clutch input circuit fault—Torque converter clutch signal

Code 41 Cam sensor or cylinder select circuit fault ignition control (IC) reference pulse system fault or Electronic Spark Timing (EST) circuit open or shorted

Code 42 Electronic Spark Timing (EST) circuit grounded or Ignition Control (IC) circuit grounded or faulty bypass line

Code 43 Knock Sensor (KS) or Electronic Spark Control (ESC) circuit fault

Code 44 Oxygen Sensor (02S), left side on 2 sensor system lean exhaust indicated

Code 45 Oxygen Sensor (02S), left side on 2 sensor system rich exhaust indicated

Code 46 Personal Automotive Security System (PASSKey II) circuit or Power Steering Pressure Switch (PSPS) circuit

Code 47 PCM-BCM data circuit

Code 48 Misfire diagnosis

Code 51 Calibration error, faulty MEM-CAL, ECM or EEPROM failure

Code 52 Engine oil temperature sensor circuit, low temperature indicated or Fuel Calpac missing or Over voltage condition or EGR Circuit fault

Code 53 Battery voltage error or EGR problem or Personal Automotive Security System (PASS-Key) circuit

Code 54 EGR #2 problem or Fuel pump circuit (low voltage) or Shorted mixture control solenoid circuit Feedback Carburetor system

Code 55 A/D Converter error, PCM error or not grounded, EGR #3 problem, Fuel lean monitor, Grounded voltage reference, faulty oxygen sensor or fuel lean Feedback Carburetor system

Code 56 Quad-Driver Module (QDM) #2 circuit or Secondary air inlet valve actuator vacuum sensor circuit signal high 5.7L (VIN J)

Code 57 Boost control problem

Code 58 Vehicle Anti-theft System fuel enable circuit

Code 61 A/C system performance or Cruise vent solenoid circuit fault or Oxygen Sensor (02S) degraded signal or Secondary port throttle valve system fault 5.7L (VIN J) or Transaxle gear switch signal

Code 62 Cruise vacuum solenoid circuit fault or Engine oil temperature sensor, high temperature indicated or Transaxle gear switch signal circuit fault

Code 63 Oxygen Sensor (02S), right side circuit open or Cruise system problem (speed error) or Manifold Absolute Pressure (MAP) sensor circuit out of range

Code 64 Oxygen Sensor (02S), right side—lean exhaust indicated

Code 65 Oxygen Sensor (02S), right side—rich exhaust indicated or Cruise servo position circuit or Fuel injector circuit low current

Code 66 A/C pressure sensor circuit fault, probably low pressure or Engine power switch, voltage high or low or PCM fault 5.7L (VIN J)

Code 67 A/C pressure sensor circuit, sensor or A/C clutch circuit failure or Cruise switch circuit fault

Code 68 A/C compressor relay (shorted circuit) or Cruise system fault

Code 69 A/C clutch circuit or head pressure high

Code 70 A/C refrigerant pressure sensor circuit (high pressure)

Code 71 A/C evaporator temperature sensor circuit (low temperature)

Code 72 Gear selector switch circuit

Code 73 A/C evaporator temperature sensor circuit (high temperature)

Code 75 Digital EGR #1 solenoid error

Code 76 Digital EGR #2 solenoid error

Code 77 Digital EGR #3 solenoid error

Code 79 Vehicle Speed Sensor (VSS) circuit signal high

Code 80 Vehicle Speed Sensor (VSS) circuit signal low
Code 81 Brake input circuit fault—Torque converter clutch signal
Code 82 Ignition Control (IC) 3X signal error
Code 85 PROM error
Code 86 Analog/Digital ECM error
Code 87 EEPROM error
Code 99 Power management

➡**This list is for reference and does not mean a specific component is defective.**

Front Wheel Drive Cadillac

Cadillac Codes may start with an 'E', 'EO', 'P or 'P0' dependent on model or type of code display. This prefix has been left off the following code description list.
Code 12 No spark reference from ignition control module or distributor
Code 13 Oxygen sensor No.1 not ready
Code 14 Engine Coolant Temperature (ECT) sensor circuit shorted
Code 15 Engine Coolant Temperature (ECT) sensor circuit open
Code 16 System voltage out of range
Code 17 Oxygen sensor No.2 not ready
Code 19 Fuel pump circuit shorted
Code 20 Fuel pump circuit open
Code 21 Throttle Position Sensor (TPS) circuit shorted
Code 22 Throttle Position Sensor (TPS) circuit open
Code 23 Electronic Spark Timing (EST) circuit fault or Ignition Control (IC) circuit problem
Code 24 Vehicle Speed Sensor (VSS) circuit problem
Code 26 Throttle Position (TP) switch circuit shorted
Code 27 Throttle Position (TP) switch circuit open
Code 28 Transaxle pressure switch problem
Code 29 Transaxle shift 'B' solenoid problem
Code 30 Idle Speed Control (ISO) RPM out of range
Code 31 Manifold Absolute Pressure (MAP) sensor circuit shorted
Code 32 Manifold Absolute Pressure (MAP) sensor circuit open
Code 33 Extended travel brake switch input circuit problem
Code 34 Manifold Absolute Pressure (MAP) sensor signal too high
Code 35 Ignition ground voltage out of range
Code 36 EGR valve pintle position out of range
Code 37 Intake Air Temperature (IAT) Manifold Air Temperature (MAT) circuit shorted
Code 38 Intake Air Temperature (IAT) sensor, Manifold Air Temperature (MAT) circuit open
Code 39 Torque Converter Clutch (TCC) engagement problem
Code 40 Power Steering Pressure Switch (PSPS) open
Code 41 Cam sensor circuit fault
Code 42 Oxygen sensor No.1 LEAN exhaust signal
Code 43 Oxygen sensor No.1 RICH exhaust signal
Code 44 Oxygen sensor No.2 LEAN exhaust signal
Code 45 Oxygen sensor No.2 RICH exhaust signal
Code 46 Bank-to-bank fueling difference
Code 47 ECM Body Control Module (BCM) or IPC/PCM data fault
Code 48 EGR control system fault
Code 50 2nd gear pressure circuit fault
Code 51 MEM-CAL error or PROM checksum mismatch
Code 52 ECM memory reset indicator or PCM keep alive memory reset

Code 53 Spark reference signal interrupt from Ignition Control (IC) module
Code 55 Closed throttle angle out of range or Throttle Position Sensor (TPS) misadjusted
Code 56 Transaxle input speed sensor circuit problem
Code 57 Shorted transaxle temperature sensor circuit
Code 58 Personal Automotive Security System (PASS) control fault
Code 59 Open transaxle temperature sensor circuit
Code 60 Cruise transaxle not in drive
Code 61 Cruise vent solenoid circuit fault
Code 62 Cruise vacuum solenoid circuit fault
Code 63 Cruise vehicle speed and set speed difference
Code 64 Cruise vehicle acceleration too high
Code 65 Cruise servo position sensor failure
Code 66 Cruise engine RPM too high
Code 67 Cruise set/coast or resume/accel input shorted
Code 68 Cruise Control Command (CCC) fault or servo position out of range
Code 69 Traction control active in cruise
Code 70 Intermittent Throttle Position (TP) sensor signal
Code 71 Intermittent Manifold Absolute Pressure (MAP) sensor signal
Code 73 Intermittent Engine Coolant Temperature (ECT) sensor signal
Code 74 Intermittent Intake Air Temperature (IAT) sensor signal
Code 75 Vehicle Speed Sensor (VSS) signal intermittent
Code 76 Transaxle pressure control solenoid circuit malfunction
Code 80 Fuel system rich or TP Sensor/idle learn not complete
Code 81 Cam to 4X reference correlation problem
Code 83 24X Reference signal high
Code 85 Idle throttle angle too high, Throttle body service required
Code 86 Undefined gear ratio
Code 88 Torque Converter Clutch (TOO) not disengaging
Code 89 Long shift and maximum adapt
Code 90 Viscous Converter Clutch (VOC) brake switch input fault
Code 91 Park/neutral switch fault
Code 92 Heated windshield fault
Code 93 Traction control system PWM link failure
Code 94 Transaxle shift 'A' solenoid problem
Code 95 Engine stall detected
Code 96 Torque converter overstress
Code 97 Park/neutral to drive/reverse at high throttle angle
Code 98 High RPM P/N to D/R shift under Idle Speed Control (ISO)
Code 99 Cruise control servo not applied in cruise
Code P102 Shorted Brake Booster Vacuum (BBV) sensor
Code P103 Open Brake Booster Vacuum (BBV) sensor
Code P105 Brake Booster Vacuum (BBV) too low
Code P106 Stop lamp switch input circuit problem
Code P107 PCM/BCM data link problem
Code P108 PROM checksum mismatch
Code P109 POM keep alive memory reset
Code P110 Generator L-terminal circuit problem
Code P112 Total EEPROM failure
Code P117 Shift 'A/Shift 'B' circuit output open or shorted
Code P131 Active Knock Sensor (KS) failure
Code P132 Knock Sensor (KS) circuit failure
Code P137 Loss of ABS/TCS data
This list is for reference and does not mean a specific component is defective.

OBD II DIAGNOSTIC TROUBLE CODES

Introduction

To comply with On Board Diagnostics Second Generation (OBD II) regulations, the Control Module is equipped with software designed to allow it to monitor vehicle emission control systems and components. Once the ignition is turned on or the engine is started, and certain test conditions are met, the PCM runs a series of monitors to test the emission control systems and components. Test conditions include different inputs such as time since startup, run-time, engine speed and temperature, transaxle gear position, and the engine open or closed loop status. Once the monitor is started, the control module attempts to run it to completion. If a particular monitor fails a test, a code is set and operating conditions at that time are recorded in memory. If the same component or system fails twice in succession, the Malfunction Indicator Lamp (MIL) is activated.

Monitors are divided into two types: Main Monitors and the Comprehensive Component Monitors.

- Catalyst Monitor
- EGR Monitor
- EVAP Monitor
- Fuel System Monitor
- Misfire Monitor
- Oxygen Sensor Monitor
- Oxygen Sensor Heater Monitor

Certain monitors, in particular the fuel system and misfire monitors, have limitations that are different from any of the other monitors. The first time either of these monitors fail, the MIL is activated, and engine conditions at the time of the fault are recorded. In order for the control module to turn off an MIL related to these two monitors, it must determine that no faults are present with engine operating conditions similar to when it detected the fault. To qualify, the engine must be operated within a specified speed range, engine load range and temperature range.

A warm-up cycle is considered to be vehicle operation after the engine has been turned off for a period of time, with the ECT input rising a specified amount and reaching normal operating temperature. When the MIL is turned off because a fault is no longer present, most OBD II codes will be erased after a minimum of 40 warm-up cycles. Misfire and fuel system codes require a minimum of 80 warm-up cycles before they clear.

OBD II Systems use a standardized test connector, called the Data Link Connector (DLC). It is located beneath the left side of the instrument panel. The DLC is located out of the line of sight of vehicle passengers, but is easily viewable from a kneeling position outside the vehicle. The connector is rectangular in design and contains up to 16 terminals. It has keying features to allow for easy connection. Both the DLC and Scan Tool connectors have latching features that ensure the scan tool will remain properly connected.

Some common uses of the Scan Tool are to identify and clear Diagnostic Trouble Codes (OTC°s) and to read control module freeze frame.

The Malfunction Indicator Lamp (MIL) looks similar to the "Check Engine" lamp. However, on OBD II Systems, it is controlled under a strict set of guidelines that dictate when the MIL is illuminated. If any of the control module monitors detects a fault that could impact vehicle emissions, a fault code is set. A One-Trip Monitor requires that a test fail once, a Two-Trip Monitor requires a test fail twice in succession, and a Three-Trip Monitor requires that a test fail three times in succession to activate the MIL.

The MIL is mounted in the instrument panel and has two functions: To act as a bulb check at key On and to inform the driver that an emissions fault has occurred.

Once the engine is started, if no faults are detected, the control module should extinguish the MIL after a few seconds. If the MIL remains On or flashes with the engine running a driveability symptom is present.

Federal law required all vehicle manufacturers to meet On Board Diagnostics, Second Generation or OBD II standards by 1996. In order to meet this standard, the automobile's on-board computer must monitor and perform diagnostic tests on vehicle emissions to ensure that the vehicle is operating at an acceptable (legal) emission level. The maximum allowable emission level is set by the Federal Test Procedure (FTP).

Some 1995 and all 1996–99 vehicles are OBD II compliant. All OBD II vehicles have the same 16 pin diagnostic connector or DLC. This eliminates the need to have a manufacturer specific connector to plug a scan tool into your vehicle.

➡**Many 1995 vehicles have a 16 pin OBD II connector, however, this does not mean that the vehicle is OBD II compliant.**

TROUBLE CODE DESCRIPTION

In the past, trouble code numbers varied between manufacturers, years, makes and models. OBD II requires that all vehicle manufacturers use a common Diagnostic Trouble Code (DTC) numbering system. Since the generic listing was not specific enough, most manufacturers came up with their own DTC listings which are called manufacturer specific codes. Both generic and manufacturer specific codes are 5 digits. The numbers can be decoded as follows:

The first digit is a letter which identifies the function of the device or circuit which has the fault. This digit can be either:

- P—Powertrain
- B—Body
- C—Chassis
- U—Network or data link code

The second digit is either a 0 or 1 and indicates whether the code is generic or manufacturer specific.

- 0—Generic
- 1—Manufacturer Specific

The third digit represents the specific vehicle circuit or system that has the fault. Listed below are the number identifiers for the powertrain system.

- 1—Fuel and Air Metering
- 2—Fuel and Air Metering (Injector Circuit Malfunctions Only)
- 3—Ignition System or Misfire
- 4—Auxiliary Emission Control
- 5—Vehicle Speed Control and Idle Control System
- 6—Computer and Auxiliary Outputs
- 7—Transmission
- 8—Transmission

The last two digits indicate the specific trouble code.

On OBD II vehicles there are two different types of OTC°s: Stored and Pending. For a DTC to become Stored, certain malfunction conditions must occur. The condition(s) required to Store codes are different for every DTC and vary by vehicle manufacturer.

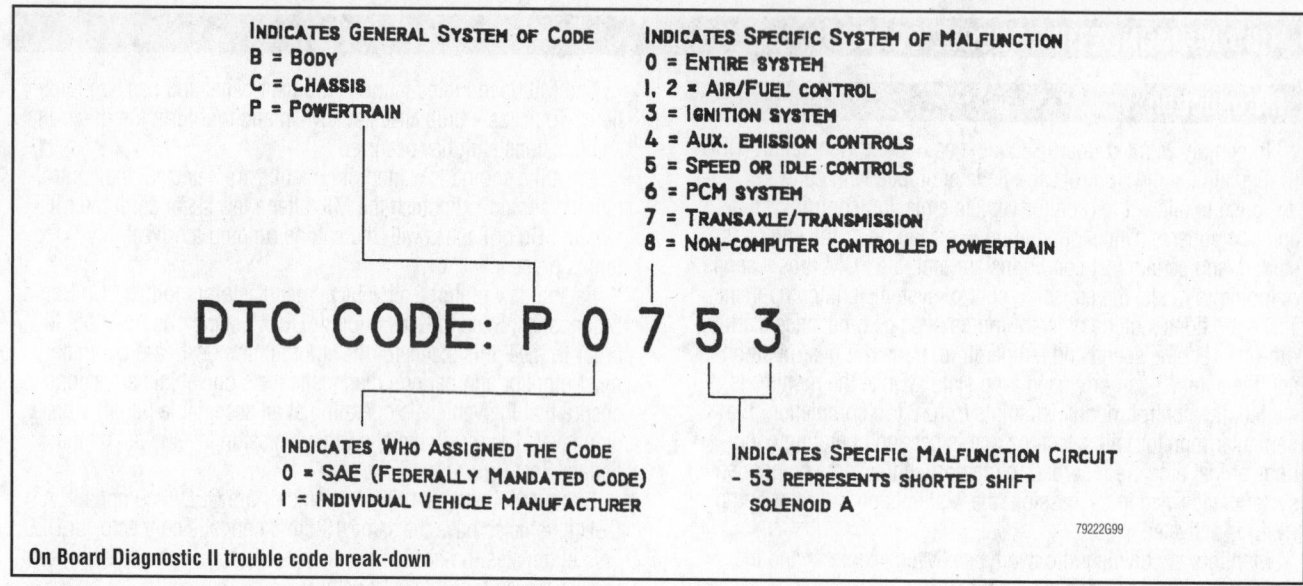

On Board Diagnostic II trouble code break-down

In order for some OTCs to become Stored, a malfunction condition has to happen more than once. If the malfunction conditions are required to occur more than once, the potential malfunction is called a Pending DTC. The DTC remains pending until the malfunction condition occurs the required number of times to make the code stored. If the malfunction condition does not occur again after a set time the pending DTC will be cleared.

Chrysler Corp.

READING CODES

With Scan Tool

Reading the control module memory is on of the first steps in OBD II system diagnostics. This step should be initially performed to determine the general nature of the fault. Subsequent readings will determine if the fault has been cleared.

Reading codes can be performed by any of the methods below:
• Read the control module memory with the Generic Scan Tool (GST)
• Read the control module memory with the vehicle manufacturer's specific tester

To read the fault codes, connect the scan tool or tester according to the manufacturer's instructions. Follow the manufacturer's specified procedure for reading the codes.

Without Scan Tool

On all 1995–97 vehicles, as well as the 1998 Sebring coupe and Avenger models with a 2.0L engine, DTC's can be accessed by observing the 2-digit number (referred to as an OBD II Equivalent code) displayed by the Malfunction Indicator Lamp (MIL). The MIL is shown on the instrument panel as the Check Engine lamp. This method should be used as a "quick test" only. You should always use a scan tool to get the most detailed information.

➡Be advised that the MIL can only perform a limited number of functions, and it is a good idea to have the system checked with a scan tool to double check the circuit function.

Within a period of 5 seconds, cycle the ignition key **ON—OFF—ON—OFF—ON**.

1. Count the number of times the MIL (check engine lamp) on the instrument panel flashes on and off.

The number of flashes represents the trouble code. There is a short pause between the flashes representing the 1st and 2nd digits of the code. Longer pauses are used to separate individual 2-digit trouble codes.

An example of a flashed DTC is as follows:
• Lamp flashes 4 times, pauses, then flashes 6 more times. This denotes a DTC number 46.
• Lamp flashes 5 times, pauses, then flashes 5 more times. This indicates a DTC number 55.

DTC 55 will always be the last code to be displayed.

CLEARING CODES

With Scan Tool

Control module reset procedures are a very important part of OBD II System diagnostics. This step should be done at the end of any fault code repair and at the end of any driveability repair.

Clearing codes can be performed by any of the methods below:
• Clear the control module memory with the Generic Scan Tool (GST)
• Clear the control module memory with the vehicle manufacturer's specific tester
• Turn the ignition off and remove the negative battery cable for at least 1 minute.

Removing the negative battery cable may cause other systems in the vehicle to loose their memory. Prior to removing the cable, ensure you have the proper reset codes for radios and alarms.

➡The MIL will may also be de-activated for some codes if the vehicle completes three consecutive trips without a fault detected with vehicle conditions similar to those present during the fault.

Without Scan Tool

Control module reset procedures are a very important part of OBD II System diagnostics. This step should be done at the end of any fault code repair and at the end of any driveability repair.

Clearing codes can be performed by turning the ignition off and removing the negative battery cable for at least 1 minute. Removing the negative battery cable may cause other systems in the vehicle to loose their memory. Prior to removing the cable, ensure you have the proper reset codes for radios and alarms.

➡**The MIL will may also be de-activated for some codes if the vehicle completes three consecutive trips without a fault detected with vehicle conditions similar to those present during the fault.**

OBD II TROUBLE CODES

P0100 Mass or Volume Air Flow Circuit Malfunction
P0101 Mass or Volume Air Flow Circuit Range/Performance Problem
P0102 Mass or Volume Air Flow Circuit Low Input
P0103 Mass or Volume Air Flow Circuit High Input
P0104 Mass or Volume Air Flow Circuit Intermittent
P0105 Manifold Absolute Pressure/Barometric Pressure Circuit Malfunction
P0106 Manifold Absolute Pressure/Barometric Pressure Circuit Range/Performance Problem
P0107 Manifold Absolute Pressure/Barometric Pressure Circuit Low Input
P0108 Manifold Absolute Pressure/Barometric Pressure Circuit High Input
P0109 Manifold Absolute Pressure/Barometric Pressure Circuit Intermittent
P0110 Intake Air Temperature Circuit Malfunction
P0111 Intake Air Temperature Circuit Range/Performance Problem
P0112 Intake Air Temperature Circuit Low Input
P0113 Intake Air Temperature Circuit High Input
P0114 Intake Air Temperature Circuit Intermittent
P0115 Engine Coolant Temperature Circuit Malfunction
P0116 Engine Coolant Temperature Circuit Range/Performance Problem
P0117 Engine Coolant Temperature Circuit Low Input
P0118 Engine Coolant Temperature Circuit High Input
P0119 Engine Coolant Temperature Circuit Intermittent
P0120 Throttle/Pedal Position Sensor/Switch "A" Circuit Malfunction
P0121 Throttle/Pedal Position Sensor/Switch "A" Circuit Range/Performance Problem
P0122 Throttle/Pedal Position Sensor/Switch "A" Circuit Low Input
P0123 Throttle/Pedal Position Sensor/Switch "A" Circuit High Input
P0124 Throttle/Pedal Position Sensor/Switch "A" Circuit Intermittent
P0125 Insufficient Coolant Temperature For Closed Loop Fuel Control
P0126 Insufficient Coolant Temperature For Stable Operation
P0130 O2 Circuit Malfunction (Bank no. 1 Sensor no. 1)

P0131 O2 Sensor Circuit Low Voltage (Bank no. 1 Sensor no. 1)
P0132 O2 Sensor Circuit High Voltage (Bank no. 1 Sensor no. 1)
P0133 O2 Sensor Circuit Slow Response (Bank no. 1 Sensor no. 1)
P0134 O2 Sensor Circuit No Activity Detected (Bank no. 1 Sensor no. 1)
P0135 O2 Sensor Heater Circuit Malfunction (Bank no. 1 Sensor no. 1)
P0136 O2 Sensor Circuit Malfunction (Bank no. 1 Sensor no. 2)
P0137 O2 Sensor Circuit Low Voltage (Bank no. 1 Sensor no. 2)
P0138 O2 Sensor Circuit High Voltage (Bank no. 1 Sensor no. 2)
P0139 O2 Sensor Circuit Slow Response (Bank no. 1 Sensor no. 2)
P0140 O2 Sensor Circuit No Activity Detected (Bank no. 1 Sensor no. 2)
P0141 O2 Sensor Heater Circuit Malfunction (Bank no. 1 Sensor no. 2)
P0142 O2 Sensor Circuit Malfunction (Bank no. 1 Sensor no. 3)
P0143 O2 Sensor Circuit Low Voltage (Bank no. 1 Sensor no. 3)
P0144 O2 Sensor Circuit High Voltage (Bank no. 1 Sensor no. 3)
P0145 O2 Sensor Circuit Slow Response (Bank no. 1 Sensor no. 3)
P0146 O2 Sensor Circuit No Activity Detected (Bank no. 1 Sensor no. 3)
P0147 O2 Sensor Heater Circuit Malfunction (Bank no. 1 Sensor no. 3)
P0150 O2 Sensor Circuit Malfunction (Bank no. 2 Sensor no. 1)
P0151 O2 Sensor Circuit Low Voltage (Bank no. 2 Sensor no. 1)
P0152 O2 Sensor Circuit High Voltage (Bank no. 2 Sensor no. 1)
P0153 O2 Sensor Circuit Slow Response (Bank no. 2 Sensor no. 1)
P0154 O2 Sensor Circuit No Activity Detected (Bank no. 2 Sensor no. 1)
P0155 O2 Sensor Heater Circuit Malfunction (Bank no. 2 Sensor no. 1)
P0156 O2 Sensor Circuit Malfunction (Bank no. 2 Sensor no. 2)
P0157 O2 Sensor Circuit Low Voltage (Bank no. 2 Sensor no. 2)
P0158 O2 Sensor Circuit High Voltage (Bank no. 2 Sensor no. 2)
P0159 O2 Sensor Circuit Slow Response (Bank no. 2 Sensor no. 2)
P0160 O2 Sensor Circuit No Activity Detected (Bank no. 2 Sensor no. 2)
P0161 O2 Sensor Heater Circuit Malfunction (Bank no. 2 Sensor no. 2)
P0162 O2 Sensor Circuit Malfunction (Bank no. 2 Sensor no. 3)
P0163 O2 Sensor Circuit Low Voltage (Bank no. 2 Sensor no. 3)
P0164 O2 Sensor Circuit High Voltage (Bank no. 2 Sensor no. 3)
P0165 O2 Sensor Circuit Slow Response (Bank no. 2 Sensor no. 3)
P0166 O2 Sensor Circuit No Activity Detected (Bank no. 2 Sensor no. 3)
P0167 O2 Sensor Heater Circuit Malfunction (Bank no. 2 Sensor no. 3)
P0170 Fuel Trim Malfunction (Bank no. 1)
P0171 System Too Lean (Bank no. 1)

P0172 System Too Rich (Bank no. 1)
P0173 Fuel Trim Malfunction (Bank no. 2)
P0174 System Too Lean (Bank no. 2)
P0175 System Too Rich (Bank no. 2)
P0176 Fuel Composition Sensor Circuit Malfunction
P0177 Fuel Composition Sensor Circuit Range/Performance
P0178 Fuel Composition Sensor Circuit Low Input
P0179 Fuel Composition Sensor Circuit High Input
P0180 Fuel Temperature Sensor "A" Circuit Malfunction
P0181 Fuel Temperature Sensor "A" Circuit Range/Performance
P0182 Fuel Temperature Sensor "A" Circuit Low Input
P0183 Fuel Temperature Sensor "A" Circuit High Input
P0184 Fuel Temperature Sensor "A" Circuit Intermittent
P0185 Fuel Temperature Sensor "B" Circuit Malfunction
P0186 Fuel Temperature Sensor "B" Circuit Range/Performance
P0187 Fuel Temperature Sensor "B" Circuit Low Input
P0188 Fuel Temperature Sensor "B" Circuit High Input
P0189 Fuel Temperature Sensor "B" Circuit Intermittent
P0190 Fuel Rail Pressure Sensor Circuit Malfunction
P0191 Fuel Rail Pressure Sensor Circuit Range/Performance
P0192 Fuel Rail Pressure Sensor Circuit Low Input
P0193 Fuel Rail Pressure Sensor Circuit High Input
P0194 Fuel Rail Pressure Sensor Circuit Intermittent
P0195 Engine Oil Temperature Sensor Malfunction
P0196 Engine Oil Temperature Sensor Range/Performance
P0197 Engine Oil Temperature Sensor Low
P0198 Engine Oil Temperature Sensor High
P0199 Engine Oil Temperature Sensor Intermittent
P0200 Injector Circuit Malfunction
P0201 Injector Circuit Malfunction—Cylinder no. 1
P0202 Injector Circuit Malfunction—Cylinder no. 2
P0203 Injector Circuit Malfunction—Cylinder no. 3
P0204 Injector Circuit Malfunction—Cylinder no. 4
P0205 Injector Circuit Malfunction—Cylinder no. 5
P0206 Injector Circuit Malfunction—Cylinder no. 6
P0207 Injector Circuit Malfunction—Cylinder no. 7
P0208 Injector Circuit Malfunction—Cylinder no. 8
P0209 Injector Circuit Malfunction—Cylinder no. 9
P0210 Injector Circuit Malfunction—Cylinder no. 10
P0211 Injector Circuit Malfunction—Cylinder no. 11
P0212 Injector Circuit Malfunction—Cylinder no. 12
P0213 Cold Start Injector no. 1 Malfunction
P0214 Cold Start Injector no. 2 Malfunction
P0215 Engine Shutoff Solenoid Malfunction
P0216 Injection Timing Control Circuit Malfunction
P0217 Engine Over Temperature Condition
P0218 Transmission Over Temperature Condition
P0219 Engine Over Speed Condition
P0220 Throttle/Pedal Position Sensor/Switch "B" Circuit Malfunction
P0221 Throttle/Pedal Position Sensor/Switch "B" Circuit Range/Performance Problem
P0222 Throttle/Pedal Position Sensor/Switch "B" Circuit Low Input
P0223 Throttle/Pedal Position Sensor/Switch "B" Circuit High Input
P0224 Throttle/Pedal Position Sensor/Switch "B" Circuit Intermittent
P0225 Throttle/Pedal Position Sensor/Switch "C" Circuit Malfunction

P0226 Throttle/Pedal Position Sensor/Switch "C" Circuit Range/Performance Problem
P0227 Throttle/Pedal Position Sensor/Switch "C" Circuit Low Input
P0228 Throttle/Pedal Position Sensor/Switch "C" Circuit High Input
P0229 Throttle/Pedal Position Sensor/Switch "C" Circuit Intermittent
P0230 Fuel Pump Primary Circuit Malfunction
P0231 Fuel Pump Secondary Circuit Low
P0232 Fuel Pump Secondary Circuit High
P0233 Fuel Pump Secondary Circuit Intermittent
P0234 Engine Over Boost Condition
P0261 Cylinder no. 1 Injector Circuit Low
P0262 Cylinder no. 1 Injector Circuit High
P0263 Cylinder no. 1 Contribution/Balance Fault
P0264 Cylinder no. 2 Injector Circuit Low
P0265 Cylinder no. 2 Injector Circuit High
P0266 Cylinder no. 2 Contribution/Balance Fault
P0267 Cylinder no. 3 Injector Circuit Low
P0268 Cylinder no. 3 Injector Circuit High
P0269 Cylinder no. 3 Contribution/Balance Fault
P0270 Cylinder no. 4 Injector Circuit Low
P0271 Cylinder no. 4 Injector Circuit High
P0272 Cylinder no. 4 Contribution/Balance Fault
P0273 Cylinder no. 5 Injector Circuit Low
P0274 Cylinder no. 5 Injector Circuit High
P0275 Cylinder no. 5 Contribution/Balance Fault
P0276 Cylinder no. 6 Injector Circuit Low
P0277 Cylinder no. 6 Injector Circuit High
P0278 Cylinder no. 6 Contribution/Balance Fault
P0279 Cylinder no. 7 Injector Circuit Low
P0280 Cylinder no. 7 Injector Circuit High
P0281 Cylinder no. 7 Contribution/Balance Fault
P0282 Cylinder no. 8 Injector Circuit Low
P0283 Cylinder no. 8 Injector Circuit High
P0284 Cylinder no. 8 Contribution/Balance Fault
P0285 Cylinder no. 9 Injector Circuit Low
P0286 Cylinder no. 9 Injector Circuit High
P0287 Cylinder no. 9 Contribution/Balance Fault
P0288 Cylinder no. 10 Injector Circuit Low
P0289 Cylinder no. 10 Injector Circuit High
P0290 Cylinder no. 10 Contribution/Balance Fault
P0291 Cylinder no. 11 Injector Circuit Low
P0292 Cylinder no. 11 Injector Circuit High
P0293 Cylinder no. 11 Contribution/Balance Fault
P0294 Cylinder no. 12 Injector Circuit Low
P0295 Cylinder no. 12 Injector Circuit High
P0296 Cylinder no. 12 Contribution/Balance Fault
P0300 Random/Multiple Cylinder Misfire Detected
P0301 Cylinder no. 1—Misfire Detected
P0302 Cylinder no. 2—Misfire Detected
P0303 Cylinder no. 3—Misfire Detected
P0304 Cylinder no. 4—Misfire Detected
P0305 Cylinder no. 5—Misfire Detected
P0306 Cylinder no. 6—Misfire Detected
P0307 Cylinder no. 7—Misfire Detected
P0308 Cylinder no. 8—Misfire Detected
P0309 Cylinder no. 9—Misfire Detected
P0310 Cylinder no. 10—Misfire Detected

P0311 Cylinder no. 11—Misfire Detected

P0312 Cylinder no. 12—Misfire Detected

P0320 Ignition/Distributor Engine Speed Input Circuit Malfunction

P0321 Ignition/Distributor Engine Speed Input Circuit Range/Performance

P0322 Ignition/Distributor Engine Speed Input Circuit No Signal

P0323 Ignition/Distributor Engine Speed Input Circuit Intermittent

P0325 Knock Sensor no. 1—Circuit Malfunction (Bank no. 1 or Single Sensor)

P0326 Knock Sensor no. 1—Circuit Range/Performance (Bank no. 1 or Single Sensor)

P0327 Knock Sensor no. 1—Circuit Low Input (Bank no. 1 or Single Sensor)

P0328 Knock Sensor no. 1—Circuit High Input (Bank no. 1 or Single Sensor)

P0329 Knock Sensor no. 1—Circuit Input Intermittent (Bank no. 1 or Single Sensor)

P0330 Knock Sensor no. 2—Circuit Malfunction (Bank no. 2)

P0331 Knock Sensor no. 2—Circuit Range/Performance (Bank no. 2)

P0332 Knock Sensor no. 2—Circuit Low Input (Bank no. 2)

P0333 Knock Sensor no. 2—Circuit High Input (Bank no. 2)

P0334 Knock Sensor no. 2—Circuit Input Intermittent (Bank no. 2)

P0335 Crankshaft Position Sensor "A" Circuit Malfunction

P0336 Crankshaft Position Sensor "A" Circuit Range/Performance

P0337 Crankshaft Position Sensor "A" Circuit Low Input

P0338 Crankshaft Position Sensor "A" Circuit High Input

P0339 Crankshaft Position Sensor "A" Circuit Intermittent

P0340 Camshaft Position Sensor Circuit Malfunction

P0341 Camshaft Position Sensor Circuit Range/Performance

P0342 Camshaft Position Sensor Circuit Low Input

P0343 Camshaft Position Sensor Circuit High Input

P0344 Camshaft Position Sensor Circuit Intermittent

P0350 Ignition Coil Primary/Secondary Circuit Malfunction

P0351 Ignition Coil "A" Primary/Secondary Circuit Malfunction

P0352 Ignition Coil "B" Primary/Secondary Circuit Malfunction

P0353 Ignition Coil "C" Primary/Secondary Circuit Malfunction

P0354 Ignition Coil "D" Primary/Secondary Circuit Malfunction

P0355 Ignition Coil "E" Primary/Secondary Circuit Malfunction

P0356 Ignition Coil "F" Primary/Secondary Circuit Malfunction

P0357 Ignition Coil "G" Primary/Secondary Circuit Malfunction

P0358 Ignition Coil "H" Primary/Secondary Circuit Malfunction

P0359 Ignition Coil "I" Primary/Secondary Circuit Malfunction

P0360 Ignition Coil "J" Primary/Secondary Circuit Malfunction

P0361 Ignition Coil "K" Primary/Secondary Circuit Malfunction

P0362 Ignition Coil "L" Primary/Secondary Circuit Malfunction

P0370 Timing Reference High Resolution Signal "A" Malfunction

P0371 Timing Reference High Resolution Signal "A" Too Many Pulses

P0372 Timing Reference High Resolution Signal "A" Too Few Pulses

P0373 Timing Reference High Resolution Signal "A" Intermittent/Erratic Pulses

P0374 Timing Reference High Resolution Signal "A" No Pulses

P0375 Timing Reference High Resolution Signal "B" Malfunction

P0376 Timing Reference High Resolution Signal "B" Too Many Pulses

P0377 Timing Reference High Resolution Signal "B" Too Few Pulses

P0378 Timing Reference High Resolution Signal "B" Intermittent/Erratic Pulses

P0379 Timing Reference High Resolution Signal "B" No Pulses

P0380 Glow Plug/Heater Circuit "A" Malfunction

P0381 Glow Plug/Heater Indicator Circuit Malfunction

P0382 Glow Plug/Heater Circuit "B" Malfunction

P0385 Crankshaft Position Sensor "B" Circuit Malfunction

P0386 Crankshaft Position Sensor "B" Circuit Range/Performance

P0387 Crankshaft Position Sensor "B" Circuit Low Input

P0388 Crankshaft Position Sensor "B" Circuit High Input

P0389 Crankshaft Position Sensor "B" Circuit Intermittent

P0400 Exhaust Gas Recirculation Flow Malfunction

P0401 Exhaust Gas Recirculation Flow Insufficient Detected

P0402 Exhaust Gas Recirculation Flow Excessive Detected

P0403 Exhaust Gas Recirculation Circuit Malfunction

P0404 Exhaust Gas Recirculation Circuit Range/Performance

P0405 Exhaust Gas Recirculation Sensor "A" Circuit Low

P0406 Exhaust Gas Recirculation Sensor "A" Circuit High

P0407 Exhaust Gas Recirculation Sensor "B" Circuit Low

P0408 Exhaust Gas Recirculation Sensor "B" Circuit High

P0410 Secondary Air Injection System Malfunction

P0411 Secondary Air Injection System Incorrect Flow Detected

P0412 Secondary Air Injection System Switching Valve "A" Circuit Malfunction

P0413 Secondary Air Injection System Switching Valve "A" Circuit Open

P0414 Secondary Air Injection System Switching Valve "A" Circuit Shorted

P0415 Secondary Air Injection System Switching Valve "B" Circuit Malfunction

P0416 Secondary Air Injection System Switching Valve "B" Circuit Open

P0417 Secondary Air Injection System Switching Valve "B" Circuit Shorted

P0418 Secondary Air Injection System Relay "A" Circuit Malfunction

P0419 Secondary Air Injection System Relay "B" Circuit Malfunction

P0420 Catalyst System Efficiency Below Threshold (Bank no. 1)

P0421 Warm Up Catalyst Efficiency Below Threshold (Bank no. 1)

P0422 Main Catalyst Efficiency Below Threshold (Bank no. 1)

P0423 Heated Catalyst Efficiency Below Threshold (Bank no. 1)

P0424 Heated Catalyst Temperature Below Threshold (Bank no. 1)

P0430 Catalyst System Efficiency Below Threshold (Bank no. 2)

P0431 Warm Up Catalyst Efficiency Below Threshold (Bank no. 2)

P0432 Main Catalyst Efficiency Below Threshold (Bank no. 2)

P0433 Heated Catalyst Efficiency Below Threshold (Bank no. 2)

P0434 Heated Catalyst Temperature Below Threshold (Bank no. 2)

P0440 Evaporative Emission Control System Malfunction

P0441 Evaporative Emission Control System Incorrect Purge Flow

P0442 Evaporative Emission Control System Leak Detected (Small Leak)

P0443 Evaporative Emission Control System Purge Control Valve Circuit Malfunction

P0444 Evaporative Emission Control System Purge Control Valve Circuit Open

P0445 Evaporative Emission Control System Purge Control Valve Circuit Shorted

P0446 Evaporative Emission Control System Vent Control Circuit Malfunction

P0447 Evaporative Emission Control System Vent Control Circuit Open

P0448 Evaporative Emission Control System Vent Control Circuit Shorted

P0449 Evaporative Emission Control System Vent Valve/Solenoid Circuit Malfunction

P0450 Evaporative Emission Control System Pressure Sensor Malfunction

P0451 Evaporative Emission Control System Pressure Sensor Range/Performance

P0452 Evaporative Emission Control System Pressure Sensor Low Input

P0453 Evaporative Emission Control System Pressure Sensor High Input

P0454 Evaporative Emission Control System Pressure Sensor Intermittent

P0455 Evaporative Emission Control System Leak Detected (Gross Leak)

P0460 Fuel Level Sensor Circuit Malfunction

P0461 Fuel Level Sensor Circuit Range/Performance

P0462 Fuel Level Sensor Circuit Low Input

P0463 Fuel Level Sensor Circuit High Input

P0464 Fuel Level Sensor Circuit Intermittent

P0465 Purge Flow Sensor Circuit Malfunction

P0466 Purge Flow Sensor Circuit Range/Performance

P0467 Purge Flow Sensor Circuit Low Input

P0468 Purge Flow Sensor Circuit High Input

P0469 Purge Flow Sensor Circuit Intermittent

P0470 Exhaust Pressure Sensor Malfunction

P0471 Exhaust Pressure Sensor Range/Performance

P0472 Exhaust Pressure Sensor Low

P0473 Exhaust Pressure Sensor High

P0474 Exhaust Pressure Sensor Intermittent

P0475 Exhaust Pressure Control Valve Malfunction

P0476 Exhaust Pressure Control Valve Range/Performance

P0477 Exhaust Pressure Control Valve Low

P0478 Exhaust Pressure Control Valve High

P0479 Exhaust Pressure Control Valve Intermittent

P0480 Cooling Fan no. 1 Control Circuit Malfunction

P0481 Cooling Fan no. 2 Control Circuit Malfunction

P0482 Cooling Fan no. 3 Control Circuit Malfunction

P0483 Cooling Fan Rationality Check Malfunction

P0484 Cooling Fan Circuit Over Current

P0485 Cooling Fan Power/Ground Circuit Malfunction

P0500 Vehicle Speed Sensor Malfunction

P0501 Vehicle Speed Sensor Range/Performance

P0502 Vehicle Speed Sensor Circuit Low Input

P0503 Vehicle Speed Sensor Intermittent/Erratic/High

P0505 Idle Control System Malfunction

P0506 Idle Control System RPM Lower Than Expected

P0507 Idle Control System RPM Higher Than Expected

P0510 Closed Throttle Position Switch Malfunction

P0520 Engine Oil Pressure Sensor/Switch Circuit Malfunction

P0521 Engine Oil Pressure Sensor/Switch Range/Performance

P0522 Engine Oil Pressure Sensor/Switch Low Voltage

P0523 Engine Oil Pressure Sensor/Switch High Voltage

P0530 A/C Refrigerant Pressure Sensor Circuit Malfunction

P0531 A/C Refrigerant Pressure Sensor Circuit Range/Performance

P0532 A/C Refrigerant Pressure Sensor Circuit Low Input

P0533 A/C Refrigerant Pressure Sensor Circuit High Input

P0534 A/C Refrigerant Charge Loss

P0550 Power Steering Pressure Sensor Circuit Malfunction

P0551 Power Steering Pressure Sensor Circuit Range/Performance

P0552 Power Steering Pressure Sensor Circuit Low Input

P0553 Power Steering Pressure Sensor Circuit High Input

P0554 Power Steering Pressure Sensor Circuit Intermittent

P0560 System Voltage Malfunction

P0561 System Voltage Unstable

P0562 System Voltage Low

P0563 System Voltage High

P0565 Cruise Control On Signal Malfunction

P0566 Cruise Control Off Signal Malfunction

P0567 Cruise Control Resume Signal Malfunction

P0568 Cruise Control Set Signal Malfunction

P0569 Cruise Control Coast Signal Malfunction

P0570 Cruise Control Accel Signal Malfunction

P0571 Cruise Control/Brake Switch "A" Circuit Malfunction

P0572 Cruise Control/Brake Switch "A" Circuit Low

P0573 Cruise Control/Brake Switch "A" Circuit High

P0574 Through P0580 Reserved for Cruise Codes

P0600 Serial Communication Link Malfunction

P0601 Internal Control Module Memory Check Sum Error

P0602 Control Module Programming Error

P0603 Internal Control Module Keep Alive Memory (KAM) Error

P0604 Internal Control Module Random Access Memory (RAM) Error

P0605 Internal Control Module Read Only Memory (ROM) Error

P0606 PCM Processor Fault

P0608 Control Module VSS Output "A" Malfunction

P0609 Control Module VSS Output "B" Malfunction

P0620 Generator Control Circuit Malfunction

P0621 Generator Lamp "L" Control Circuit Malfunction

P0622 Generator Field "F" Control Circuit Malfunction

P0650 Malfunction Indicator Lamp (MIL) Control Circuit Malfunction

P0654 Engine RPM Output Circuit Malfunction

P0655 Engine Hot Lamp Output Control Circuit Malfunction

P0656 Fuel Level Output Circuit Malfunction

P0700 Transmission Control System Malfunction

P0701 Transmission Control System Range/Performance

P0702 Transmission Control System Electrical

P0703 Torque Converter/Brake Switch "B" Circuit Malfunction

P0704 Clutch Switch Input Circuit Malfunction

P0705 Transmission Range Sensor Circuit Malfunction (PRNDL Input)

P0706 Transmission Range Sensor Circuit Range/Performance

P0707 Transmission Range Sensor Circuit Low Input

P0708 Transmission Range Sensor Circuit High Input

P0709 Transmission Range Sensor Circuit Intermittent

P0710 Transmission Fluid Temperature Sensor Circuit Malfunction

P0711 Transmission Fluid Temperature Sensor Circuit Range/Performance

P0712 Transmission Fluid Temperature Sensor Circuit Low Input

P0713 Transmission Fluid Temperature Sensor Circuit High Input

P0714 Transmission Fluid Temperature Sensor Circuit Intermittent

P0715 Input/Turbine Speed Sensor Circuit Malfunction

P0716 Input/Turbine Speed Sensor Circuit Range/Performance

P0717 Input/Turbine Speed Sensor Circuit No Signal

P0718 Input/Turbine Speed Sensor Circuit Intermittent

P0719 Torque Converter/Brake Switch "B" Circuit Low

P0720 Output Speed Sensor Circuit Malfunction

P0721 Output Speed Sensor Circuit Range/Performance

P0722 Output Speed Sensor Circuit No Signal

P0723 Output Speed Sensor Circuit Intermittent

P0724 Torque Converter/Brake Switch "B" Circuit High

P0725 Engine Speed Input Circuit Malfunction

P0726 Engine Speed Input Circuit Range/Performance

P0727 Engine Speed Input Circuit No Signal

P0728 Engine Speed Input Circuit Intermittent

P0730 Incorrect Gear Ratio

P0731 Gear no. 1 Incorrect Ratio

P0732 Gear no. 2 Incorrect Ratio

P0733 Gear no. 3 Incorrect Ratio

P0734 Gear no. 4 Incorrect Ratio

P0735 Gear no. 5 Incorrect Ratio

P0736 Reverse Incorrect Ratio

P0740 Torque Converter Clutch Circuit Malfunction

P0741 Torque Converter Clutch Circuit Performance or Stuck Off

P0742 Torque Converter Clutch Circuit Stuck On

P0743 Torque Converter Clutch Circuit Electrical

P0744 Torque Converter Clutch Circuit Intermittent

P0745 Pressure Control Solenoid Malfunction

P0746 Pressure Control Solenoid Performance or Stuck Off

P0747 Pressure Control Solenoid Stuck On

P0748 Pressure Control Solenoid Electrical

P0749 Pressure Control Solenoid Intermittent

P0750 Shift Solenoid "A" Malfunction

P0751 Shift Solenoid "A" Performance or Stuck Off

P0752 Shift Solenoid "A" Stuck On

P0753 Shift Solenoid "A" Electrical

P0754 Shift Solenoid "A" Intermittent

P0755 Shift Solenoid "B" Malfunction

P0756 Shift Solenoid "B" Performance or Stuck Off

P0757 Shift Solenoid "B" Stuck On

P0758 Shift Solenoid "B" Electrical

P0759 Shift Solenoid "B" Intermittent

P0760 Shift Solenoid "C" Malfunction

P0761 Shift Solenoid "C" Performance Or Stuck Off

P0762 Shift Solenoid "C" Stuck On

P0763 Shift Solenoid "C" Electrical

P0764 Shift Solenoid "C" Intermittent

P0765 Shift Solenoid "D" Malfunction

P0766 Shift Solenoid "D" Performance Or Stuck Off

P0767 Shift Solenoid "D" Stuck On

P0768 Shift Solenoid "D" Electrical

P0769 Shift Solenoid "D" Intermittent

P0770 Shift Solenoid "E" Malfunction

P0771 Shift Solenoid "E" Performance Or Stuck Off

P0772 Shift Solenoid "E" Stuck On

P0773 Shift Solenoid "E" Electrical

P0774 Shift Solenoid "E" Intermittent

P0780 Shift Malfunction

P0781 1–2 Shift Malfunction

P0782 2–3 Shift Malfunction

P0783 3–4 Shift Malfunction

P0784 4–5 Shift Malfunction

P0785 Shift/Timing Solenoid Malfunction

P0786 Shift/Timing Solenoid Range/Performance

P0787 Shift/Timing Solenoid Low

P0788 Shift/Timing Solenoid High

P0789 Shift/Timing Solenoid Intermittent

P0790 Normal/Performance Switch Circuit Malfunction

P0801 Reverse Inhibit Control Circuit Malfunction

P0803 1–4 Upshift (Skip Shift) Solenoid Control Circuit Malfunction

P0804 1–4 Upshift (Skip Shift) Lamp Control Circuit Malfunction

P1290 CNG Fuel System Pressure Too High—3.3L CNG vehicles only

P1291 No Temp Rise Seen From Intake Air Heaters

P1292 CNG Pressure Sensor Voltage Too High—3.3L CNG vehicles only

P1293 CNG Pressure Sensor Voltage Too Low—3.3L CNG vehicles only

P1294 Target Idle Not Reached

P1296 No 5-Volts To MAP Sensor

P1297 No Change In MAP From Start To Run

P1391 Intermittent Loss Of CMP Or CKP

P1398 Misfire Adaptive Numerator At Limit

P1486 EVAP Leak Monitor Pinched Hose Or Obstruction Found

P1491 Radiator Fan Control Relay Circuit

P1492 Battery Temp Sensor Voltage Too High

P1493 Battery Temp Sensor Voltage Too Low

P1494 Leak Detection Pump Pressure Switch Or Mechanical Fault

P1495 Leak Detection Pump Solenoid Circuit

P1498 Auxiliary 5-Volt Supply Output Too Low

P1697 PCM Failure SRI Mile Not Stored

P1698 PCM Failure EEPROM Write Denied

P1756 Governor Pressure Not Equal To Target @ 15–20 PSI

P1757 Governor Pressure Above 3 PSI In Gear With 0 MPH

P1762 Governor Pressure Sensor Offset Volts Too Low Or High

P1763 Governor Pressure Sensor Volts Too High

P1764 Governor Pressure Sensor Volts Too Low

P1765 Trans 12-Volt Supply Relay Control Circuit

P1899 P/N Switch Stuck In Park Or In Gear

OBD II TROUBLE CODE EQUIVALENTS

11 No crank reference signal at PCM

11 Timing belt skipped t tooth or more

11 Intermittent loss of CMP or CKP

11 Misfire adaptive numerator at limit

12 Battery disconnect

13 Slow change in idle MAP sensor signal (VIN N engine)

13 No change in MAP from start to run

14 MAP sensor voltage too low

14 MAP sensor voltage too high

14 No 5 volts to MAP sensor
15 5 volt supply output too low
16 No vehicle speed sensor signal
16 Knock sensor signal
17 Engine cold too long
17 Closed loop temperature not reached
21 Front 02S shorted to voltage
21 Front 02S stays at center
21 Rear 02S shorted to voltage
21 Rear 02S stays at center
21 Upstream 02S shorted to ground
21 Upstream 02S shorted to voltage
21 Upstream 02S response
21 Upstream 02S stays at center
21 Upstream 02S heater failure
21 Downstream 02S shorted to ground
21 Downstream 02S shorted to voltage
21 Downstream 025 response
21 Downstream 02S signal inactive
21 Downstream 02S heater failure
21 Front bank upstream 02S shorted to ground (6 cylinder)
21 Front bank upstream 02S shorted to voltage (6 cylinder)
21 Front bank upstream 02S slow response 6 cylinder)
21 Front bank upstream 02S stays at center (6 cylinder)
21 Front bank upstream 025 heater failure (6 cylinder)
21 Front bank downstream 025 shorted to ground (6 cylinder)
21 Front bank downstream 02S shorted to voltage (6 cylinder)
21 Front bank downstream 02S stays at center (6 cylinder)
21 Front bank downstream 02S heater failure (6 cylinder)
22 ECT sensor voltage too low
22 ECT sensor voltage too high
23 Intake air temperature voltage low
23 Intake air temperature voltage high
24 TPS voltage does not agree with MAP
24 Throttle position sensor voltage low
24 Throttle position sensor voltage high
24 No 5 volts to TPS
25 Idle air control motor circuits
25 Target idle not reached
25 Vacuum leak found (IAC fully seated)
27 Injector #l control circuit
27 Injector #2 control circuit
27 Injector#3control circuit
27 Injector #4 control circuit
27 Injector #5 control circuit (6 cylinder)
27 Injector #6 control circuit (6 cylinder)
31 EVAP purge flow monitor failure
31 EVAP system small leak
31 EVAP solenoid circuit
31 EVAP system large leak
31 EVAP leak monitor pinched hose
31 Leak detection pump pressure switch
31 EVAP emission vent solenoid switch or mechanical failure
31 Leak detection pump solenoid circuit
31 EVAP emission vent solenoid circuit
31 High speed radiator fan ground control relay circuit
32 EGR system failure
32 EGR solenoid circuit
33 A/C pressure sensor volts too high
33 A/C pressure sensor volts too low
33 A/C clutch relay circuit
34 Speed control switch always low

34 Speed control switch always high
31 Speed control solenoid circuit
35 High speed condenser fan control relay circuit
35 High fan and high fan ground control relay circuit
35 High speed radiator fan control relay circuit
35 High speed fan control relay circuit
35 Low speed fan control relay circuit
37 Park/Neutral switch failure
41 Alternator field not switching properly
42 Auto shutdown relay circuit
42 No ASD relay output voltage at PCM
42 Fuel level sending unit volts too low
42 Fuel level sending unit volts too high
42 Fuel level unit no change over miles
42 Fuel pump relay control circuit
43 Multiple cylinder misfire
43 Cylinder #l misfire
43 Cylinder #2 misfire
43 Cylinder #3mislire
43 Cylinder #4 misfire
43 Cylinder #5 misfire
43 Cylinder #6 misfire
43 Ignition coil #1 primary circuit
43 Ignition coil #2 primary circuit
44 Ambient temperature sensor
44 Battery temperature sensor volts out of limit
44 Battery temperature sensor voltage too high
44 Battery temperature sensor voltage too low
45 Transaxle fault present
46 Charging system voltage too high
47 Charging system voltage too low
51 Fuel system lean (4 cylinder)
51 Rear bank fuel system lean (6 cylinder)
51 Front bank fuel system lean (6 cylinder)
52 Fuel system rich (4 cylinder)
52 Rear bank fuel system rich (6 cylinder)
52 Front bank fuel system rich (6 cylinder)
53 Internal controller failure
53 PCM failure SPI communications
53 Internal controller failure
53 PCM failure SPI communications
54 No cam signal at PCM
55 Completion or fault code display on Check Engine Lamp
62 PCM failure SRI mile not stared
63 PCM failure EEPROM write denied
64 Catalytic converter efficiency failure
64 Rear bank catalytic converter efficiency failure
65 Power steering switch failure
65 Brake switch performance circuit
66 No CCD message from body controller
66 No CCD message from TCM
71 5 volt output low speed control power circuit
72 Catalytic Converter efficiency failure
72 Front bank catalytic converter efficiency failure
77 Malfunction detected with power feed to speed control servo

Ford Motor Co.

READING CODES

Reading the control module memory is on of the first steps in OBD II system diagnostics. This step should be initially performed to determine the general nature of the fault. Subsequent readings will determine if the fault has been cleared.

Reading codes can be performed by any of the methods below:

• Read the control module memory with the Generic Scan Tool (GST)

• Read the control module memory with the vehicle manufacturer's specific tester

To read the fault codes, connect the scan tool or tester according to the manufacturer's instructions. Follow the manufacturer's specified procedure for reading the codes.

CLEARING CODES

Control module reset procedures are a very important part of OBD II System diagnostics. This step should be done at the end of any fault code repair and at the end of any driveability repair.

Clearing codes can be performed by any of the methods below:

• Clear the control module memory with the Generic Scan Tool (GST)

• Clear the control module memory with the vehicle manufacturer's specific tester

• Turn the ignition off and remove the negative battery cable for at least 1 minute.

Removing the negative battery cable may cause other systems in the vehicle to loose their memory. Prior to removing the cable, ensure you have the proper reset codes for radios and alarms.

➡ **The MIL will may also be de-activated for some codes if the vehicle completes three consecutive trips without a fault detected with vehicle conditions similar to those present during the fault.**

OBD II TROUBLE CODES

1995 Models

P0102 Mass Air Flow (MAF) Sensor circuit low input
P0103 Mass Air Flow (MAF) Sensor circuit high input
P0112 Intake Air Temperature (IAT) Sensor circuit low input
P0113 Intake Air Temperature (IAT) Sensor high input
P0117 Engine Coolant Temperature (ECT) low input
P0118 Engine Coolant Temperature (ECT) Sensor circuit high input
P0121 In range operating Throttle Position (TP) sensor circuit failure
P0122 Throttle Position (TP) Sensor circuit low input
P0123 Throttle Position (TP) Sensor high input
P0125 Insufficient coolant temperature to enter closed loop fuel control
P0126 Insufficient coolant temperature for stable operation
P0131 Upstream Heated Oxygen Sensor (HO$_2$S 11) circuit out of range low voltage (bank #1)
P0132 Upstream Heated Oxygen Sensor (HO$_2$S 11) circuit high voltage (Bank #1)
P0133 Upstream Heated Oxygen Sensor (HO$_2$S 11) circuit slow response (Bank #1)
P0135 Heated Oxygen Sensor Heater (HTR 11) circuit malfunction
P0136 Downstream Heated Oxygen Sensor (HO$_2$S 12) circuit malfunction (Bank #1

P0138 Downstream Heated Oxygen Sensor (HO$_2$S 12) circuit high voltage (Bank #1)
P0140 Heated Oxygen Sensor (HO$_2$S 12) circuit no activity detected (Bank #1)
P0141 Heated Oxygen Sensor Heater (HTR 12) circuit malfunction
P0151 Upstream Heated Oxygen Sensor (HO$_2$S 21) circuit out of range low voltage (Bank #2)
P0152 Upstream Heated Oxygen Sensor (HO$_2$S 21) circuit high voltage (Bank #2)
P0153 Upstream Heated Oxygen Sensor (HO$_2$S 21) circuit slow response (Bank #2)
P0155 Heated Oxygen Sensor Heater (HTR 21) circuit malfunction
P0156 Downstream Heated Oxygen Sensor (HO$_2$S 22) circuit malfunction (Bank #2)
P0158 Downstream Heated Oxygen Sensor (HO$_2$S 22) circuit high voltage (Bank #2)
P0160 Heated Oxygen Sensor (HO$_2$S 12) circuit no activity detected (Bank #2)
P0161 Heated Oxygen Sensor Heater (HTR 22) circuit malfunction
P0171 System (adaptive fuel) too lean (Bank #1)
P0172 System (adaptive fuel) too rich (Bank #1)
P0174 System (adaptive fuel) too lean (Bank #2)
P0175 System (adaptive fuel) too rich (Bank #2)
P0222 Throttle Position Sensor B (TP-B) circuit low input
P0223 Throttle Position Sensor B (TP-B) circuit high input
P0230 Fuel pump primary circuit malfunction
P0231 Fuel pump secondary circuit low
P0232 Fuel pump secondary circuit high
P0300 Random misfire detected
P0301 Cylinder #1 misfire detected
P0302 Cylinder #2 misfire detected
P0303 Cylinder #3 misfire detected
P0304 Cylinder #4 misfire detected
P0305 Cylinder #5 misfire detected
P0306 Cylinder #6 misfire detected
P0307 Cylinder #7 misfire detected
P0308 Cylinder #8 misfire detected
P0320 Ignition engine speed (Profile Ignition Pick-up) input circuit malfunction
P0340 Camshaft Position (CMP) sensor circuit malfunction (CID)
P0350 Ignition Coil primary circuit malfunction
P0351 Ignition Coil A primary circuit malfunction
P0352 Ignition Coil B primary circuit malfunction
P0353 Ignition Coil C primary circuit malfunction
P0354 Ignition Coil D primary circuit malfunction
P0400 Exhaust Gas Recirculation (EGR) flow malfunction
P0401 Exhaust Gas Recirculation (EGR) flow insufficient detected
P0402 Exhaust Gas Recirculation (EGR) excess flow detected (valve open at idle)
P0411 Secondary Air Injection system incorrect flow detected
P0412 Secondary Air Injection system control valve malfunction
P0420 Catalyst system efficiency below threshold (Bank #1)
P0430 Catalyst system efficiency below threshold (Bank #2)
P0443 Evaporative emission control system Canister Purge (CANP) Control Valve circuit malfunction
P0500 Vehicle Speed Sensor (VSS) malfunction
P0505 Idle Air Control (IAC) system malfunction

P0603 Powertrain Control Module (PCM)—Keep Alive Memory (KAM) test error

P0605 Powertrain Control Module (PCM)—Read Only Memory (ROM) test error

P0704 Clutch Pedal Position (CPP) switch input circuit malfunction

P0703 Brake On/Off (BOO) switch input malfunction

P0707 Manual Lever Position (MLP) sensor circuit low input

P0708 Manual Lever Position (MLP) sensor circuit high input

P0712 Transmission Fluid Temperature (TFT) sensor circuit low input

P0713 Transmission Fluid Temperature (TFT) sensor circuit high input

P0715 Turbine Shaft Speed (TSS) sensor circuit malfunction

P0720 Output Shaft Speed (OSS) sensor circuit malfunction

P0731 Incorrect ratio for first gear

P0732 Incorrect ratio for second gear

P0733 Incorrect ratio for third gear

P0734 Incorrect ratio for fourth gear

P0736 Reverse incorrect gear

P0741 Torque Converter Clutch (TCC) system incorrect mechanical performance

P0746 Electronic Pressure Control (EPC) solenoid performance

P0743 Torque Converter Clutch (TCC) system electrical failure

P0750 Shift Solenoid #1 (SS1) circuit malfunction

P0751 Shift Solenoid #1 (SS1) performance

P0755 Shift Solenoid #2 (SS2) circuit malfunction

P0756 Shift Solenoid #2 (SS2) performance

P0760 Shift Solenoid #3 (SS3) circuit malfunction

P0761 Shift Solenoid #3 (SS3) performance

P0781 1 to 2 shift error

P0782 2 to 3 shift error

P0783 3 to 4 shift error

P0784 4 to 5 shift error

P1000 OBD II Monitor Testing not complete

U1039 OBD II Monitor not complete

UI051 Brake switch signal missing or incorrect

P1100 Mass Air Flow (MAF) sensor intermittent

P1101 Mass Air Flow (MAF) sensor out of Self-Test range

P1112 Intake Air Temperature (IAT) sensor intermittent

P1116 Engine Coolant Temperature (ECT) sensor out of Self-Test range

P1117 Engine Coolant Temperature (ECT) sensor intermittent

P1120 Throttle Position (TP) sensor out of range low

P1121 Throttle Position (TP) sensor inconsistent with MAF sensor

P1124 Throttle Position (TP) sensor out of Self-Test range

P1125 Throttle Position (TP) sensor circuit intermittent

P1130 Lack of HO_2S 11 switch, adaptive fuel at limit

P1131 Lack of HO_2S 11 switch, sensor indicates lean (Bank #1)

P1132 Lack of HO_2S 11 switch, sensor indicates rich (Bank #1)

U1135 Ignition switch signal missing or incorrect

P1137 Lack of HO_2S 12 switch, sensor indicates lean (Bank #1)

P1138 Lack of HO_2S 12 switch, sensor indicates rich (Bank #1)

P1150 Lack of HO_2S 21 switch, adaptive fuel at limit

P1151 Lack of HO_2S 21 switch, sensor indicates lean (Bank #2)

P1152 Lack of HO_2S 21 switch, sensor indicates rich (Bank #2)

P1157 Lack of HO_2S 22 switch, sensor indicates lean (Bank #2)

P1158 Lack of HO_2S 22 switch, sensor indicates rich (Bank #2)

P1220 Series Throttle Control malfunction

P1224 Throttle Position Sensor (TP-B) out of Self-test range

P1233 Fuel Pump driver Module off-line

P1234 Fuel Pump driver Module off-line

P1235 Fuel Pump control out of range

P1236 Fuel Pump control out of range

P1237 Fuel Pump secondary circuit malfunction

P1238 Fuel Pump secondary circuit malfunction

P1260 THEFT detected—engine disabled

P1270 Engine RPM or vehicle speed limiter reached

P1351 Ignition Diagnostic Monitor (IDM) circuit input malfunction

P1352 Ignition coil A primary circuit malfunction

P1353 Ignition coil B primary circuit malfunction

P1354 Ignition coil C primary circuit malfunction

P1355 Ignition coil D primary circuit malfunction

P1358 Ignition Diagnostic Monitor (IDM) signal out of Self-Test range

P1359 Spark output circuit malfunction

P1364 Ignition coil primary circuit malfunction

P1390 Octane Adjust (OCT ADJ) out of Self-Test range

P1400 Differential Pressure Feedback Electronic (DPFE) sensor circuit low voltage detected

P1401 Differential Pressure Feedback Electronic (DPFE) sensor circuit high voltage detected

P1403 Differential Pressure Feedback Electronic (DPFE) sensor hoses reversed

P1405 Differential Pressure Feedback Electronic (DPFE) sensor upstream hose off or plugged

P1406 Differential Pressure Feedback Electronic (DPFE) sensor downstream hose off or plugged

P1407 Exhaust Gas Recirculation (EGR) no flow detected (valve stuck closed or inoperative)

P1408 Exhaust Gas Recirculation (EGR) flow out of Self-Test range

P1409 Electronic Vacuum Regulator (EVR) control circuit malfunction

P1414 Secondary Air Injection system monitor circuit high voltage

P1443 Evaporative emission control system—vacuum system purge control solenoid or purge control valve malfunction

P1444 Purge Flow Sensor (PFS) circuit low input

P1445 Purge Flow Sensor (PFS) circuit high input

U1451 Lack of response from Passive Anti-Theft system (PATS) module—engine disabled

P1460 Wide Open Throttle Air Conditioning Cut-off (WAC) circuit malfunction

P1461 Air Conditioning Pressure (ACP) sensor circuit low input

P1462 Air Conditioning Pressure (ACP) sensor circuit high input

P1463 Air Conditioning Pressure (ACP) sensor insufficient pressure change

P1469 Low air conditioning cycling period

P1473 Fan Secondary High with fan(s) off

P1474 Low Fan Control primary circuit malfunction

P1479 High Fan Control primary circuit malfunction

P1480 Fan Secondary low with low fan on

P1481 Fan Secondary low with high fan on

P1500 Vehicle Speed Sensor (VSS) circuit intermittent

P1505 Idle Air Control (IAC) system at adaptive clip

P1506 Idle Air control (IAC) overspeed error

P1518 Intake Manifold Runner Control (IMRC) malfunction (stuck open)

P1519 Intake Manifold Runner Control (IMRC) malfunction (stuck closed)

P1520 Intake Manifold Runner Control (IMRC) circuit malfunction

P1507 Idle Air control (IAC) under speed error

P1605 Powertrain Control Module (POM)—Keep Alive Memory (KAM) test error

P1650 Power steering Pressure (PSP) switch out of Self-Test range

P1651 Power steering Pressure (PSP) switch input malfunction

P1701 Reverse engagement error

P1703 Brake On/Off (BOO) switch out of Self-Test range

P1705 Manual Lever Position (MLP) sensor out of Self-Test range

P1709 Park or Neutral Position (PNP) switch out of Self-test range

P1729 4X4 Low switch error

P1711 Transmission Fluid Temperature (TFT) sensor out of Self-Test range

P1741 Torque Converter Clutch (TCC) control error

P1742 Torque Converter Clutch (TCC) solenoid mechanically failed (turns MIL on)

P1743 Torque Converter Clutch (TCC) solenoid mechanically failed (turns TOIL on)

P1744 Torque Converter Clutch (TCC) system mechanically stuck in off position

P1748 Electronic Pressure Control (EPC) solenoid circuit low input (open circuit)

P1747 Electronic Pressure Control (EPC) solenoid circuit high input (short circuit)

P1749 Electric Pressure Control (EPC) solenoid failed low

P1751 Shift Solenoid #1(SS1) performance

P1756 Shift Solenoid #2 (SS2) performance

P1780 Transmission Control Switch (TCS) circuit out of Self-Test range

1996–99 Models

P0000 No Failures

P0100 Mass or Volume Air Flow Circuit Malfunction

P0101 Mass or Volume Air Flow Circuit Range/Performance Problem

P0102 Mass or Volume Air Flow Circuit Low Input

P0103 Mass or Volume Air Flow Circuit High Input

P0104 Mass or Volume Air Flow Circuit Intermittent

P0105 Manifold Absolute Pressure/Barometric Pressure Circuit Malfunction

P0106 Manifold Absolute Pressure/Barometric Pressure Circuit Range/Performance Problem

P0107 Manifold Absolute Pressure/Barometric Pressure Circuit Low Input

P0108 Manifold Absolute Pressure/Barometric Pressure Circuit High Input

P0109 Manifold Absolute Pressure/Barometric Pressure Circuit Intermittent

P0110 Intake Air Temperature Circuit Malfunction

P0111 Intake Air Temperature Circuit Range/Performance Problem

P0112 Intake Air Temperature Circuit Low Input

P0113 Intake Air Temperature Circuit High Input

P0114 Intake Air Temperature Circuit Intermittent

P0115 Engine Coolant Temperature Circuit Malfunction

P0116 Engine Coolant Temperature Circuit Range/Performance Problem

P0117 Engine Coolant Temperature Circuit Low Input

P0118 Engine Coolant Temperature Circuit High Input

P0119 Engine Coolant Temperature Circuit Intermittent

P0120 Throttle/Pedal Position Sensor/Switch "A" Circuit Malfunction

P0121 Throttle/Pedal Position Sensor/Switch "A" Circuit Range/Performance Problem

P0122 Throttle/Pedal Position Sensor/Switch "A" Circuit Low Input

P0123 Throttle/Pedal Position Sensor/Switch "A" Circuit High Input

P0124 Throttle/Pedal Position Sensor/Switch "A" Circuit Intermittent

P0125 Insufficient Coolant Temperature For Closed Loop Fuel Control

P0126 Insufficient Coolant Temperature For Stable Operation

P0130 O2 Circuit Malfunction (Bank no. 1 Sensor no. 1)

P0131 O2 Sensor Circuit Low Voltage (Bank no. 1 Sensor no. 1)

P0132 O2 Sensor Circuit High Voltage (Bank no. 1 Sensor no. 1)

P0133 O2 Sensor Circuit Slow Response (Bank no. 1 Sensor no. 1)

P0134 O2 Sensor Circuit No Activity Detected (Bank no. 1 Sensor no. 1)

P0135 O2 Sensor Heater Circuit Malfunction (Bank no. 1 Sensor no. 1)

P0136 O2 Sensor Circuit Malfunction (Bank no. 1 Sensor no. 2)

P0137 O2 Sensor Circuit Low Voltage (Bank no. 1 Sensor no. 2)

P0138 O2 Sensor Circuit High Voltage (Bank no. 1 Sensor no. 2)

P0139 O2 Sensor Circuit Slow Response (Bank no. 1 Sensor no. 2)

P0140 O2 Sensor Circuit No Activity Detected (Bank no. 1 Sensor no. 2)

P0141 O2 Sensor Heater Circuit Malfunction (Bank no. 1 Sensor no. 2)

P0142 O2 Sensor Circuit Malfunction (Bank no. 1 Sensor no. 3)

P0143 O2 Sensor Circuit Low Voltage (Bank no. 1 Sensor no. 3)

P0144 O2 Sensor Circuit High Voltage (Bank no. 1 Sensor no. 3)

P0145 O2 Sensor Circuit Slow Response (Bank no. 1 Sensor no. 3)

P0146 O2 Sensor Circuit No Activity Detected (Bank no. 1 Sensor no. 3)

P0147 O2 Sensor Heater Circuit Malfunction (Bank no. 1 Sensor no. 3)

P0150 O2 Sensor Circuit Malfunction (Bank no. 2 Sensor no. 1)

P0151 O2 Sensor Circuit Low Voltage (Bank no. 2 Sensor no. 1)

P0152 O2 Sensor Circuit High Voltage (Bank no. 2 Sensor no. 1)

P0153 O2 Sensor Circuit Slow Response (Bank no. 2 Sensor no. 1)

P0154 O2 Sensor Circuit No Activity Detected (Bank no. 2 Sensor no. 1)

P0155 O2 Sensor Heater Circuit Malfunction (Bank no. 2 Sensor no. 1)

P0156 O2 Sensor Circuit Malfunction (Bank no. 2 Sensor no. 2)

P0157 O2 Sensor Circuit Low Voltage (Bank no. 2 Sensor no. 2)

P0158 O2 Sensor Circuit High Voltage (Bank no. 2 Sensor no. 2)

P0159 O2 Sensor Circuit Slow Response (Bank no. 2 Sensor no. 2)

P0160 O2 Sensor Circuit No Activity Detected (Bank no. 2 Sensor no. 2)

P0161 O2 Sensor Heater Circuit Malfunction (Bank no. 2 Sensor no. 2)

P0162 O2 Sensor Circuit Malfunction (Bank no. 2 Sensor no. 3)

P0163 O2 Sensor Circuit Low Voltage (Bank no. 2 Sensor no. 3)

P0164 O2 Sensor Circuit High Voltage (Bank no. 2 Sensor no. 3)

P0165 O2 Sensor Circuit Slow Response (Bank no. 2 Sensor no. 3)

P0166 O2 Sensor Circuit No Activity Detected (Bank no. 2 Sensor no. 3)

P0167 O2 Sensor Heater Circuit Malfunction (Bank no. 2 Sensor no. 3)

P0170 Fuel Trim Malfunction (Bank no. 1)

P0171 System Too Lean (Bank no. 1)

P0172 System Too Rich (Bank no. 1)

P0173 Fuel Trim Malfunction (Bank no. 2)

P0174 System Too Lean (Bank no. 2)

P0175 System Too Rich (Bank no. 2)

P0176 Fuel Composition Sensor Circuit Malfunction

P0177 Fuel Composition Sensor Circuit Range/Performance

P0178 Fuel Composition Sensor Circuit Low Input

P0179 Fuel Composition Sensor Circuit High Input

P0180 Fuel Temperature Sensor "A" Circuit Malfunction

P0181 Fuel Temperature Sensor "A" Circuit Range/Performance

P0182 Fuel Temperature Sensor "A" Circuit Low Input

P0183 Fuel Temperature Sensor "A" Circuit High Input

P0184 Fuel Temperature Sensor "A" Circuit Intermittent

P0185 Fuel Temperature Sensor "B" Circuit Malfunction

P0186 Fuel Temperature Sensor "B" Circuit Range/Performance

P0187 Fuel Temperature Sensor "B" Circuit Low Input

P0188 Fuel Temperature Sensor "B" Circuit High Input

P0189 Fuel Temperature Sensor "B" Circuit Intermittent

P0190 Fuel Rail Pressure Sensor Circuit Malfunction

P0191 Fuel Rail Pressure Sensor Circuit Range/Performance

P0192 Fuel Rail Pressure Sensor Circuit Low Input

P0193 Fuel Rail Pressure Sensor Circuit High Input

P0194 Fuel Rail Pressure Sensor Circuit Intermittent

P0195 Engine Oil Temperature Sensor Malfunction

P0196 Engine Oil Temperature Sensor Range/Performance

P0197 Engine Oil Temperature Sensor Low

P0198 Engine Oil Temperature Sensor High

P0199 Engine Oil Temperature Sensor Intermittent

P0200 Injector Circuit Malfunction

P0201 Injector Circuit Malfunction—Cylinder no. 1

P0202 Injector Circuit Malfunction—Cylinder no. 2

P0203 Injector Circuit Malfunction—Cylinder no. 3

P0204 Injector Circuit Malfunction—Cylinder no. 4

P0205 Injector Circuit Malfunction—Cylinder no. 5

P0206 Injector Circuit Malfunction—Cylinder no. 6

P0207 Injector Circuit Malfunction—Cylinder no. 7

P0208 Injector Circuit Malfunction—Cylinder no. 8

P0209 Injector Circuit Malfunction—Cylinder no. 9

P0210 Injector Circuit Malfunction—Cylinder no. 10

P0211 Injector Circuit Malfunction—Cylinder no. 11

P0212 Injector Circuit Malfunction—Cylinder no. 12

P0213 Cold Start Injector no. 1 Malfunction

P0214 Cold Start Injector no. 2 Malfunction

P0215 Engine Shutoff Solenoid Malfunction

P0216 Injection Timing Control Circuit Malfunction

P0217 Engine Over Temperature Condition

P0218 Transmission Over Temperature Condition

P0219 Engine Over Speed Condition

P0220 Throttle/Pedal Position Sensor/Switch "B" Circuit Malfunction

P0221 Throttle/Pedal Position Sensor/Switch "B" Circuit Range/Performance Problem

P0222 Throttle/Pedal Position Sensor/Switch "B" Circuit Low Input

P0223 Throttle/Pedal Position Sensor/Switch "B" Circuit High Input

P0224 Throttle/Pedal Position Sensor/Switch "B" Circuit Intermittent

P0225 Throttle/Pedal Position Sensor/Switch "C" Circuit Malfunction

P0226 Throttle/Pedal Position Sensor/Switch "C" Circuit Range/Performance Problem

P0227 Throttle/Pedal Position Sensor/Switch "C" Circuit Low Input

P0228 Throttle/Pedal Position Sensor/Switch "C" Circuit High Input

P0229 Throttle/Pedal Position Sensor/Switch "C" Circuit Intermittent

P0230 Fuel Pump Primary Circuit Malfunction

P0231 Fuel Pump Secondary Circuit Low

P0232 Fuel Pump Secondary Circuit High

P0233 Fuel Pump Secondary Circuit Intermittent

P0234 Engine Over Boost Condition

P0261 Cylinder no. 1 Injector Circuit Low

P0262 Cylinder no. 1 Injector Circuit High

P0263 Cylinder no. 1 Contribution/Balance Fault

P0264 Cylinder no. 2 Injector Circuit Low

P0265 Cylinder no. 2 Injector Circuit High

P0266 Cylinder no. 2 Contribution/Balance Fault

P0267 Cylinder no. 3 Injector Circuit Low

P0268 Cylinder no. 3 Injector Circuit High

P0269 Cylinder no. 3 Contribution/Balance Fault

P0270 Cylinder no. 4 Injector Circuit Low

P0271 Cylinder no. 4 Injector Circuit High

P0272 Cylinder no. 4 Contribution/Balance Fault

P0273 Cylinder no. 5 Injector Circuit Low

P0274 Cylinder no. 5 Injector Circuit High

P0275 Cylinder no. 5 Contribution/Balance Fault

P0276 Cylinder no. 6 Injector Circuit Low

P0277 Cylinder no. 6 Injector Circuit High

P0278 Cylinder no. 6 Contribution/Balance Fault

P0279 Cylinder no. 7 Injector Circuit Low

P0280 Cylinder no. 7 Injector Circuit High

P0281 Cylinder no. 7 Contribution/Balance Fault

P0282 Cylinder no. 8 Injector Circuit Low

P0283 Cylinder no. 8 Injector Circuit High

P0284 Cylinder no. 8 Contribution/Balance Fault

P0285 Cylinder no. 9 Injector Circuit Low

P0286 Cylinder no. 9 Injector Circuit High

P0287 Cylinder no. 9 Contribution/Balance Fault

P0288 Cylinder no. 10 Injector Circuit Low

P0289 Cylinder no. 10 Injector Circuit High

P0290 Cylinder no. 10 Contribution/Balance Fault
P0291 Cylinder no. 11 Injector Circuit Low
P0292 Cylinder no. 11 Injector Circuit High
P0293 Cylinder no. 11 Contribution/Balance Fault
P0294 Cylinder no. 12 Injector Circuit Low
P0295 Cylinder no. 12 Injector Circuit High
P0296 Cylinder no. 12 Contribution/Balance Fault
P0300 Random/Multiple Cylinder Misfire Detected
P0301 Cylinder no. 1—Misfire Detected
P0302 Cylinder no. 2—Misfire Detected
P0303 Cylinder no. 3—Misfire Detected
P0304 Cylinder no. 4—Misfire Detected
P0305 Cylinder no. 5—Misfire Detected
P0306 Cylinder no. 6—Misfire Detected
P0307 Cylinder no. 7—Misfire Detected
P0308 Cylinder no. 8—Misfire Detected
P0309 Cylinder no. 9—Misfire Detected
P0310 Cylinder no. 10—Misfire Detected
P0311 Cylinder no. 11—Misfire Detected
P0312 Cylinder no. 12—Misfire Detected
P0320 Ignition/Distributor Engine Speed Input Circuit Malfunction
P0321 Ignition/Distributor Engine Speed Input Circuit Range/Performance
P0322 Ignition/Distributor Engine Speed Input Circuit No Signal
P0323 Ignition/Distributor Engine Speed Input Circuit Intermittent
P0325 Knock Sensor no. 1—Circuit Malfunction (Bank no. 1 or Single Sensor)
P0326 Knock Sensor no. 1—Circuit Range/Performance (Bank no. 1 or Single Sensor)
P0327 Knock Sensor no. 1—Circuit Low Input (Bank no. 1 or Single Sensor)
P0328 Knock Sensor no. 1—Circuit High Input (Bank no. 1 or Single Sensor)
P0329 Knock Sensor no. 1—Circuit Input Intermittent (Bank no. 1 or Single Sensor)
P0330 Knock Sensor no. 2—Circuit Malfunction (Bank no. 2)
P0331 Knock Sensor no. 2—Circuit Range/Performance (Bank no. 2)
P0332 Knock Sensor no. 2—Circuit Low Input (Bank no. 2)
P0333 Knock Sensor no. 2—Circuit High Input (Bank no. 2)
P0334 Knock Sensor no. 2—Circuit Input Intermittent (Bank no. 2)
P0335 Crankshaft Position Sensor "A" Circuit Malfunction
P0336 Crankshaft Position Sensor "A" Circuit Range/Performance
P0337 Crankshaft Position Sensor "A" Circuit Low Input
P0338 Crankshaft Position Sensor "A" Circuit High Input
P0339 Crankshaft Position Sensor "A" Circuit Intermittent
P0340 Camshaft Position Sensor Circuit Malfunction
P0341 Camshaft Position Sensor Circuit Range/Performance
P0342 Camshaft Position Sensor Circuit Low Input
P0343 Camshaft Position Sensor Circuit High Input
P0344 Camshaft Position Sensor Circuit Intermittent
P0350 Ignition Coil Primary/Secondary Circuit Malfunction
P0351 Ignition Coil "A" Primary/Secondary Circuit Malfunction
P0352 Ignition Coil "B" Primary/Secondary Circuit Malfunction
P0353 Ignition Coil "C" Primary/Secondary Circuit Malfunction
P0354 Ignition Coil "D" Primary/Secondary Circuit Malfunction

P0355 Ignition Coil "E" Primary/Secondary Circuit Malfunction
P0356 Ignition Coil "F" Primary/Secondary Circuit Malfunction
P0357 Ignition Coil "G" Primary/Secondary Circuit Malfunction
P0358 Ignition Coil "H" Primary/Secondary Circuit Malfunction
P0359 Ignition Coil "I" Primary/Secondary Circuit Malfunction
P0360 Ignition Coil "J" Primary/Secondary Circuit Malfunction
P0361 Ignition Coil "K" Primary/Secondary Circuit Malfunction
P0362 Ignition Coil "L" Primary/Secondary Circuit Malfunction
P0370 Timing Reference High Resolution Signal "A" Malfunction
P0371 Timing Reference High Resolution Signal "A" Too Many Pulses
P0372 Timing Reference High Resolution Signal "A" Too Few Pulses
P0373 Timing Reference High Resolution Signal "A" Intermittent/Erratic Pulses
P0374 Timing Reference High Resolution Signal "A" No Pulses
P0375 Timing Reference High Resolution Signal "B" Malfunction
P0376 Timing Reference High Resolution Signal "B" Too Many Pulses
P0377 Timing Reference High Resolution Signal "B" Too Few Pulses
P0378 Timing Reference High Resolution Signal "B" Intermittent/Erratic Pulses
P0379 Timing Reference High Resolution Signal "B" No Pulses
P0380 Glow Plug/Heater Circuit "A" Malfunction
P0381 Glow Plug/Heater Indicator Circuit Malfunction
P0382 Glow Plug/Heater Circuit "B" Malfunction
P0385 Crankshaft Position Sensor "B" Circuit Malfunction
P0386 Crankshaft Position Sensor "B" Circuit Range/Performance
P0387 Crankshaft Position Sensor "B" Circuit Low Input
P0388 Crankshaft Position Sensor "B" Circuit High Input
P0389 Crankshaft Position Sensor "B" Circuit Intermittent
P0400 Exhaust Gas Recirculation Flow Malfunction
P0401 Exhaust Gas Recirculation Flow Insufficient Detected
P0402 Exhaust Gas Recirculation Flow Excessive Detected
P0403 Exhaust Gas Recirculation Circuit Malfunction
P0404 Exhaust Gas Recirculation Circuit Range/Performance
P0405 Exhaust Gas Recirculation Sensor "A" Circuit Low
P0406 Exhaust Gas Recirculation Sensor "A" Circuit High
P0407 Exhaust Gas Recirculation Sensor "B" Circuit Low
P0408 Exhaust Gas Recirculation Sensor "B" Circuit High
P0410 Secondary Air Injection System Malfunction
P0411 Secondary Air Injection System Incorrect Flow Detected
P0412 Secondary Air Injection System Switching Valve "A" Circuit Malfunction
P0413 Secondary Air Injection System Switching Valve "A" Circuit Open
P0414 Secondary Air Injection System Switching Valve "A" Circuit Shorted
P0415 Secondary Air Injection System Switching Valve "B" Circuit Malfunction
P0416 Secondary Air Injection System Switching Valve "B" Circuit Open
P0417 Secondary Air Injection System Switching Valve "B" Circuit Shorted
P0418 Secondary Air Injection System Relay "A" Circuit Malfunction

P0419 Secondary Air Injection System Relay "B" Circuit Malfunction

P0420 Catalyst System Efficiency Below Threshold (Bank no. 1)

P0421 Warm Up Catalyst Efficiency Below Threshold (Bank no. 1)

P0422 Main Catalyst Efficiency Below Threshold (Bank no. 1)

P0423 Heated Catalyst Efficiency Below Threshold (Bank no. 1)

P0424 Heated Catalyst Temperature Below Threshold (Bank no. 1)

P0430 Catalyst System Efficiency Below Threshold (Bank no. 2)

P0431 Warm Up Catalyst Efficiency Below Threshold (Bank no. 2)

P0432 Main Catalyst Efficiency Below Threshold (Bank no. 2)

P0433 Heated Catalyst Efficiency Below Threshold (Bank no. 2)

P0434 Heated Catalyst Temperature Below Threshold (Bank no. 2)

P0440 Evaporative Emission Control System Malfunction

P0441 Evaporative Emission Control System Incorrect Purge Flow

P0442 Evaporative Emission Control System Leak Detected (Small Leak)

P0443 Evaporative Emission Control System Purge Control Valve Circuit Malfunction

P0444 Evaporative Emission Control System Purge Control Valve Circuit Open

P0445 Evaporative Emission Control System Purge Control Valve Circuit Shorted

P0446 Evaporative Emission Control System Vent Control Circuit Malfunction

P0447 Evaporative Emission Control System Vent Control Circuit Open

P0448 Evaporative Emission Control System Vent Control Circuit Shorted

P0449 Evaporative Emission Control System Vent Valve/Solenoid Circuit Malfunction

P0450 Evaporative Emission Control System Pressure Sensor Malfunction

P0451 Evaporative Emission Control System Pressure Sensor Range/Performance

P0452 Evaporative Emission Control System Pressure Sensor Low Input

P0453 Evaporative Emission Control System Pressure Sensor High Input

P0454 Evaporative Emission Control System Pressure Sensor Intermittent

P0455 Evaporative Emission Control System Leak Detected (Gross Leak)

P0460 Fuel Level Sensor Circuit Malfunction

P0461 Fuel Level Sensor Circuit Range/Performance

P0462 Fuel Level Sensor Circuit Low Input

P0463 Fuel Level Sensor Circuit High Input

P0464 Fuel Level Sensor Circuit Intermittent

P0465 Purge Flow Sensor Circuit Malfunction

P0466 Purge Flow Sensor Circuit Range/Performance

P0467 Purge Flow Sensor Circuit Low Input

P0468 Purge Flow Sensor Circuit High Input

P0469 Purge Flow Sensor Circuit Intermittent

P0470 Exhaust Pressure Sensor Malfunction

P0471 Exhaust Pressure Sensor Range/Performance

P0472 Exhaust Pressure Sensor Low

P0473 Exhaust Pressure Sensor High

P0474 Exhaust Pressure Sensor Intermittent

P0475 Exhaust Pressure Control Valve Malfunction

P0476 Exhaust Pressure Control Valve Range/Performance

P0477 Exhaust Pressure Control Valve Low

P0478 Exhaust Pressure Control Valve High

P0479 Exhaust Pressure Control Valve Intermittent

P0480 Cooling Fan no. 1 Control Circuit Malfunction

P0481 Cooling Fan no. 2 Control Circuit Malfunction

P0482 Cooling Fan no. 3 Control Circuit Malfunction

P0483 Cooling Fan Rationality Check Malfunction

P0484 Cooling Fan Circuit Over Current

P0485 Cooling Fan Power/Ground Circuit Malfunction

P0500 Vehicle Speed Sensor Malfunction

P0501 Vehicle Speed Sensor Range/Performance

P0502 Vehicle Speed Sensor Circuit Low Input

P0503 Vehicle Speed Sensor Intermittent/Erratic/High

P0505 Idle Control System Malfunction

P0506 Idle Control System RPM Lower Than Expected

P0507 Idle Control System RPM Higher Than Expected

P0510 Closed Throttle Position Switch Malfunction

P0520 Engine Oil Pressure Sensor/Switch Circuit Malfunction

P0521 Engine Oil Pressure Sensor/Switch Range/Performance

P0522 Engine Oil Pressure Sensor/Switch Low Voltage

P0523 Engine Oil Pressure Sensor/Switch High Voltage

P0530 A/C Refrigerant Pressure Sensor Circuit Malfunction

P0531 A/C Refrigerant Pressure Sensor Circuit Range/Performance

P0532 A/C Refrigerant Pressure Sensor Circuit Low Input

P0533 A/C Refrigerant Pressure Sensor Circuit High Input

P0534 A/C Refrigerant Charge Loss

P0550 Power Steering Pressure Sensor Circuit Malfunction

P0551 Power Steering Pressure Sensor Circuit Range/Performance

P0552 Power Steering Pressure Sensor Circuit Low Input

P0553 Power Steering Pressure Sensor Circuit High Input

P0554 Power Steering Pressure Sensor Circuit Intermittent

P0560 System Voltage Malfunction

P0561 System Voltage Unstable

P0562 System Voltage Low

P0563 System Voltage High

P0565 Cruise Control On Signal Malfunction

P0566 Cruise Control Off Signal Malfunction

P0567 Cruise Control Resume Signal Malfunction

P0568 Cruise Control Set Signal Malfunction

P0569 Cruise Control Coast Signal Malfunction

P0570 Cruise Control Accel Signal Malfunction

P0571 Cruise Control/Brake Switch "A" Circuit Malfunction

P0572 Cruise Control/Brake Switch "A" Circuit Low

P0573 Cruise Control/Brake Switch "A" Circuit High

P0574 Through P0580 Reserved for Cruise Codes

P0600 Serial Communication Link Malfunction

P0601 Internal Control Module Memory Check Sum Error

P0602 Control Module Programming Error

P0603 Internal Control Module Keep Alive Memory (KAM) Error

P0604 Internal Control Module Random Access Memory (RAM) Error

P0605 Internal Control Module Read Only Memory (ROM) Error

P0606 PCM Processor Fault

P0608 Control Module VSS Output "A" Malfunction
P0609 Control Module VSS Output "B" Malfunction
P0620 Generator Control Circuit Malfunction
P0621 Generator Lamp "L" Control Circuit Malfunction
P0622 Generator Field "F" Control Circuit Malfunction
P0650 Malfunction Indicator Lamp (MIL) Control Circuit Malfunction
P0654 Engine RPM Output Circuit Malfunction
P0655 Engine Hot Lamp Output Control Circuit Malfunction
P0656 Fuel Level Output Circuit Malfunction
P0700 Transmission Control System Malfunction
P0701 Transmission Control System Range/Performance
P0702 Transmission Control System Electrical
P0703 Torque Converter/Brake Switch "B" Circuit Malfunction
P0704 Clutch Switch Input Circuit Malfunction
P0705 Transmission Range Sensor Circuit Malfunction (PRNDL Input)
P0706 Transmission Range Sensor Circuit Range/Performance
P0707 Transmission Range Sensor Circuit Low Input
P0708 Transmission Range Sensor Circuit High Input
P0709 Transmission Range Sensor Circuit Intermittent
P0710 Transmission Fluid Temperature Sensor Circuit Malfunction
P0711 Transmission Fluid Temperature Sensor Circuit Range/Performance
P0712 Transmission Fluid Temperature Sensor Circuit Low Input
P0713 Transmission Fluid Temperature Sensor Circuit High Input
P0714 Transmission Fluid Temperature Sensor Circuit Intermittent
P0715 Input/Turbine Speed Sensor Circuit Malfunction
P0716 Input/Turbine Speed Sensor Circuit Range/Performance
P0717 Input/Turbine Speed Sensor Circuit No Signal
P0718 Input/Turbine Speed Sensor Circuit Intermittent
P0719 Torque Converter/Brake Switch "B" Circuit Low
P0720 Output Speed Sensor Circuit Malfunction
P0721 Output Speed Sensor Circuit Range/Performance
P0722 Output Speed Sensor Circuit No Signal
P0723 Output Speed Sensor Circuit Intermittent
P0724 Torque Converter/Brake Switch "B" Circuit High
P0725 Engine Speed Input Circuit Malfunction
P0726 Engine Speed Input Circuit Range/Performance
P0727 Engine Speed Input Circuit No Signal
P0728 Engine Speed Input Circuit Intermittent
P0730 Incorrect Gear Ratio
P0731 Gear no. 1 Incorrect Ratio
P0732 Gear no. 2 Incorrect Ratio
P0733 Gear no. 3 Incorrect Ratio
P0734 Gear no. 4 Incorrect Ratio
P0735 Gear no. 5 Incorrect Ratio
P0736 Reverse Incorrect Ratio
P0740 Torque Converter Clutch Circuit Malfunction
P0741 Torque Converter Clutch Circuit Performance or Stuck Off
P0742 Torque Converter Clutch Circuit Stuck On
P0743 Torque Converter Clutch Circuit Electrical
P0744 Torque Converter Clutch Circuit Intermittent
P0745 Pressure Control Solenoid Malfunction
P0746 Pressure Control Solenoid Performance or Stuck Off
P0747 Pressure Control Solenoid Stuck On

P0748 Pressure Control Solenoid Electrical
P0749 Pressure Control Solenoid Intermittent
P0750 Shift Solenoid "A" Malfunction
P0751 Shift Solenoid "A" Performance or Stuck Off
P0752 Shift Solenoid "A" Stuck On
P0753 Shift Solenoid "A" Electrical
P0754 Shift Solenoid "A" Intermittent
P0755 Shift Solenoid "B" Malfunction
P0756 Shift Solenoid "B" Performance or Stuck Off
P0757 Shift Solenoid "B" Stuck On
P0758 Shift Solenoid "B" Electrical
P0759 Shift Solenoid "B" Intermittent
P0760 Shift Solenoid "C" Malfunction
P0761 Shift Solenoid "C" Performance Or Stuck Off
P0762 Shift Solenoid "C" Stuck On
P0763 Shift Solenoid "C" Electrical
P0764 Shift Solenoid "C" Intermittent
P0765 Shift Solenoid "D" Malfunction
P0766 Shift Solenoid "D" Performance Or Stuck Off
P0767 Shift Solenoid "D" Stuck On
P0768 Shift Solenoid "D" Electrical
P0769 Shift Solenoid "D" Intermittent
P0770 Shift Solenoid "E" Malfunction
P0771 Shift Solenoid "E" Performance Or Stuck Off
P0772 Shift Solenoid "E" Stuck On
P0773 Shift Solenoid "E" Electrical
P0774 Shift Solenoid "E" Intermittent
P0780 Shift Malfunction
P0781 1–2 Shift Malfunction
P0782 2–3 Shift Malfunction
P0783 3–4 Shift Malfunction
P0784 4–5 Shift Malfunction
P0785 Shift/Timing Solenoid Malfunction
P0786 Shift/Timing Solenoid Range/Performance
P0787 Shift/Timing Solenoid Low
P0788 Shift/Timing Solenoid High
P0789 Shift/Timing Solenoid Intermittent
P0790 Normal/Performance Switch Circuit Malfunction
P0801 Reverse Inhibit Control Circuit Malfunction
P0803 1–4 Upshift (Skip Shift) Solenoid Control Circuit Malfunction
P0804 1–4 Upshift (Skip Shift) Lamp Control Circuit Malfunction
P1000 OBD II Monitor Testing Not Complete More Driving Required
P1001 Key On Engine Running (KOER) Self-Test Not Able To Complete, KOER Aborted
P1100 Mass Air Flow (MAF) Sensor Intermittent
P1101 Mass Air Flow (MAF) Sensor Out Of Self-Test Range
P1111 System Pass 49 State Except Econoline
P1112 Intake Air Temperature (IAT) Sensor Intermittent
P1116 Engine Coolant Temperature (ECT) Sensor Out Of Self-Test Range
P1117 Engine Coolant Temperature (ECT) Sensor Intermittent
P1120 Throttle Position (TP) Sensor Out Of Range (Low)
P1121 Throttle Position (TP) Sensor Inconsistent With MAF Sensor
P1124 Throttle Position (TP) Sensor Out Of Self-Test Range
P1125 Throttle Position (TP) Sensor Circuit Intermittent
P1127 Exhaust Not Warm Enough, Downstream Heated Oxygen Sensors (HO2S) Not Tested

P1128 Upstream Heated Oxygen Sensors (HO2S) Swapped From Bank To Bank

P1129 Downstream Heated Oxygen Sensors (HO2S) Swapped From Bank To Bank

P1130 Lack Of Upstream Heated Oxygen Sensor (HO2S 11) Switch, Adaptive Fuel At Limit (Bank #1)

P1131 Lack Of Upstream Heated Oxygen Sensor (HO2S 11) Switch, Sensor Indicates Lean (Bank #1)

P1132 Lack Of Upstream Heated Oxygen Sensor (HO2S 11) Switch, Sensor Indicates Rich (Bank#1)

P1137 Lack Of Downstream Heated Oxygen Sensor (HO2S 12) Switch, Sensor Indicates Lean (Bank#1)

P1138 Lack Of Downstream Heated Oxygen Sensor (HO2S 12) Switch, Sensor Indicates Rich (Bank#1)

P1150 Lack Of Upstream Heated Oxygen Sensor (HO2S 21) Switch, Adaptive Fuel At Limit (Bank #2)

P1151 Lack Of Upstream Heated Oxygen Sensor (HO2S 21) Switch, Sensor Indicates Lean (Bank#2)

P1152 Lack Of Upstream Heated Oxygen Sensor (HO2S 21) Switch, Sensor Indicates Rich (Bank #2)

P1157 Lack Of Downstream Heated Oxygen Sensor (HO2S 22) Switch, Sensor Indicates Lean (Bank #2)

P1158 Lack Of Downstream Heated Oxygen Sensor (HO2S 22) Switch, Sensor Indicates Rich (Bank#2)

P1169 (HO2S 12) Signal Remained Unchanged For More Than 20 Seconds After Closed Loop

P1170 (HO2S 11) Signal Remained Unchanged For More Than 20 Seconds After Closed Loop

P1173 Feedback A/F Mixture Control (HO2S 21) Signal Remained Unchanged For More Than 20 Seconds After Closed Loop

P1184 Engine Oil Temp Sensor Circuit Performance

P1195 Barometric (BARO) Pressure Sensor Circuit Malfunction (Signal Is From EGR Boost Sensor)

P1196 Starter Switch Circuit Malfunction

P1209 Injection Control Pressure (ICP) Peak Fault

P1210 Injection Control Pressure (ICP) Above Expected Level

P1211 Injection Control Pressure (ICP) Not Controllable—Pressure Above/Below Desired

P1212 Injection Control Pressure (ICP) Voltage Not At Expected Level

P1218 Cylinder Identification (CID) Stuck High

P1219 Cylinder Identification (CID) Stuck Low

P1220 Series Throttle Control Malfunction (Traction Control System)

P1224 Throttle Position Sensor "B" (TP-B) Out Of Self-Test Range (Traction Control System)

P1230 Fuel Pump Low Speed Malfunction

P1231 Fuel Pump Secondary Circuit Low With High Speed Pump On

P1232 Low Speed Fuel Pump Primary Circuit Malfunction

P1233 Fuel Pump Driver Module Off-line (MIL DTC)

P1234 Fuel Pump Driver Module Disabled Or Off-line (No MIL)

P1235 Fuel Pump Control Out Of Range (MIL DTC)

P1236 Fuel Pump Control Out Of Range (No MIL)

P1237 Fuel Pump Secondary Circuit Malfunction (MIL DTC)

P1238 Fuel Pump Secondary Circuit Malfunction (No DMIL)

P1250 Fuel Pressure Regulator Control (FPRC) Solenoid Malfunction

P1260 THEFT Detected—Engine Disabled

P1261 High To Low Side Short—Cylinder #1 (Indicates Low side Circuit Is Shorted To B+ Or To The High Side Between The IDM And The Injector)

P1262 High To Low Side Short—Cylinder #2 (Indicates Low side Circuit Is Shorted To B+ Or To The High Side Between The IDM And The Injector)

P1263 High To Low Side Short—Cylinder #3 (Indicates Low side Circuit Is Shorted To B+ Or To The High Side Between The IDM And The Injector)

P1264 High To Low Side Short—Cylinder #4 (Indicates Low side Circuit Is Shorted To B+ Or To The High Side Between The IDM And The Injector)

P1265 High To Low Side Short—Cylinder #5 (Indicates Low side Circuit Is Shorted To B+ Or To The High Side Between The IDM And The Injector)

P1266 High To Low Side Short—Cylinder #6 (Indicates Low side Circuit Is Shorted To B+ Or To The High Side Between The IDM And The Injector)

P1267 High To Low Side Short—Cylinder #7 (Indicates Low side Circuit Is Shorted To B+ Or To The High Side Between The IDM And The Injector)

P1268 High To Low Side Short—Cylinder #8 (Indicates Low side Circuit Is Shorted To B+ Or To The High Side Between The IDM And The Injector)

P1270 Engine RPM Or Vehicle Speed Limiter Reached

P1271 High To Low Side Open—Cylinder #1 (Indicates A High To Low Side Open Between The Injector And The IDM)

P1272 High To Low Side Open—Cylinder #2 (Indicates A High To Low Side Open Between The Injector And The IDM)

P1273 High To Low Side Open—Cylinder #3 (Indicates A High To Low Side Open Between The Injector And The IDM)

P1274 High To Low Side Open—Cylinder #4 (Indicates A High To Low Side Open Between The Injector And The IDM)

P1275 High To Low Side Open—Cylinder #5 (Indicates A High To Low Side Open Between The Injector And The IDM)

P1276 High To Low Side Open—Cylinder #6 (Indicates A High To Low Side Open Between The Injector And The IDM)

P1277 High To Low Side Open—Cylinder #7 (Indicates A High To Low Side Open Between The Injector And The IDM)

P1278 High To Low Side Open—Cylinder #8 (Indicates A High To Low Side Open Between The Injector And The IDM)

P1280 Injection Control Pressure (ICP) Circuit Out Of Range Low

P1281 Injection Control Pressure (ICP) Circuit Out Of Range High

P1282 Injection Control Pressure (ICP) Excessive

P1283 Injection Pressure Regulator (IPR) Circuit Failure

P1284 Injection Control Pressure (ICP) Failure—Aborts KOER Or CCT Test

P1285 Cylinder Head Temperature (CHT) Over Temperature Sensed

P1288 Cylinder Head Temperature (CHT) Sensor Out Of Self-Test Range

P1289 Cylinder Head Temperature (CHT) Sensor Circuit Low Input

P1290 Cylinder Head Temperature (CHT) Sensor Circuit High Input

P1291 IDM To Injector High Side Circuit #1 (Right Bank) Short To GND Or B+

P1292 IDM To Injector High Side Circuit #2 (Right Bank) Short To GND Or B+

P1293 IDM To Injector High Side Circuit Open Bank #1 (Right Bank)

P1294 IDM To Injector High Side Circuit Open Bank #2 (Left Bank)

P1295 Multiple IDM/Injector Circuit Faults On Bank #1 (Right)

P1296 Multiple IDM/Injector Circuit Faults On Bank#2 (Left)

P1297 High Sides Shorted Together

P1298 IDM Failure

P1299 Engine Over Temperature Condition

P1309 Misfire Detection Monitor Is Not Enabled

P1316 Injector Circuit/IDM Codes Detected

P1320 Distributor Signal Interrupt

P1336 Crankshaft Position Sensor (Gear)

P1345 No Camshaft Position Sensor Signal

P1351 Ignition Diagnostic Monitor (IDM) Circuit Input Malfunction

P1351 Indicates Ignition System Malfunction

P1352 Indicates Ignition System Malfunction

P1353 Indicates Ignition System Malfunction

P1354 Indicates Ignition System Malfunction

P1355 Indicates Ignition System Malfunction

P1356 PIPs Occurred While IDM Pulse width Indicates Engine Not Turning

P1357 Ignition Diagnostic Monitor (IDM) Pulse width Not Defined

P1358 Ignition Diagnostic Monitor (IDM) Signal Out Of Self-Test Range

P1359 Spark Output Circuit Malfunction

P1364 Spark Output Circuit Malfunction

P1390 Octane Adjust (OCT ADJ) Out Of Self-Test Range

P1391 Glow Plug Circuit Low Input Bank #1 (Right)

P1392 Glow Plug Circuit High Input Bank #1 (Right)

P1393 Glow Plug Circuit Low Input Bank #2 (Left)

P1394 Glow Plug Circuit High Input Bank #2 (Left)

P1395 Glow Plug Monitor Fault Bank #1

P1396 Glow Plug Monitor Fault Bank #2

P1397 System Voltage Out Of Self Test Range

P1400 Differential Pressure Feedback EGR (DPFE) Sensor Circuit Low Voltage Detected

P1401 Differential Pressure Feedback EGR (DPFE) Sensor Circuit High Voltage Detected/EGR Temperature Sensor

P1402 EGR Valve Position Sensor Open Or Short

P1403 Differential Pressure Feedback EGR (DPFE) Sensor Hoses Reversed

P1405 Differential Pressure Feedback EGR (DPFE) Sensor Upstream Hose Off Or Plugged

P1406 Differential Pressure Feedback EGR (DPFE) Sensor Downstream Hose Off Or Plugged

P1407 Exhaust Gas Recirculation (EGR) No Flow Detected (Valve Stuck Closed Or Inoperative)

P1408 Exhaust Gas Recirculation (EGR) Flow Out Of Self-Test Range

P1409 Electronic Vacuum Regulator (EVR) Control Circuit Malfunction

P1410 Check That Fuel Pressure Regulator Control Solenoid And The EGR Check Solenoid Connectors Are Not Swapped

P1411 Secondary Air Injection System Incorrect Downstream Flow Detected

P1413 Secondary Air Injection System Monitor Circuit Low Voltage

P1414 Secondary Air Injection System Monitor Circuit High Voltage

P1442 Evaporative Emission Control System Small Leak Detected

P1443 Evaporative Emission Control System—Vacuum System, Purge Control Solenoid Or Purge Control Valve Malfunction

P1444 Purge Flow Sensor (PFS) Circuit Low Input

P1445 Purge Flow Sensor (PFS) Circuit High Input

P1449 Evaporative Emission Control System Unable To Hold Vacuum

P1450 Unable To Bleed Up Fuel Tank Vacuum

P1455 Evaporative Emission Control System Control Leak Detected (Gross Leak)

P1460 Wide Open Throttle Air Conditioning Cut-Off Circuit Malfunction

P1461 Air Conditioning Pressure (ACP) Sensor Circuit Low Input

P1462 Air Conditioning Pressure (ACP) Sensor Circuit High Input

P1463 Air Conditioning Pressure (ACP) Sensor Insufficient Pressure Change

P1464 Air Conditioning (A/C) Demand Out Of Self-Test Range/A/C On During KOER Or CCT Test

P1469 Low Air Conditioning Cycling Period

P1473 Fan Secondary High, With Fan(s) Off

P1474 Low Fan Control Primary Circuit Malfunction

P1479 High Fan Control Primary Circuit Malfunction

P1480 Fan Secondary Low, With Low Fan On

P1481 Fan Secondary Low, With High Fan On

P1483 Power To Fan Circuit Over current

P1484 Open Power/Ground To Variable Load Control Module (VLCM)

P1485 EGR Control Solenoid Open Or Short

P1486 EGR Vent Solenoid Open Or Short

P1487 EGR Boost Check Solenoid Open Or Short

P1500 Vehicle Speed Sensor (VSS) Circuit Intermittent

P1501 Vehicle Speed Sensor (VSS) Out Of Self-Test Range/Vehicle Moved During Test

P1502 Invalid Self Test—Auxiliary Powertrain Control Module (APCM) Functioning

P1504 Idle Air Control (IAC) Circuit Malfunction

P1505 Idle Air Control (IAC) System At Adaptive Clip

P1506 Idle Air Control (IAC) Overspeed Error

P1507 Idle Air Control (IAC) Underspeed Error

P1512 Intake Manifold Runner Control (IMRC) Malfunction (Bank#1 Stuck Closed)

P1513 Intake Manifold Runner Control (IMRC) Malfunction (Bank#2 Stuck Closed)

P1516 Intake Manifold Runner Control (IMRC) Input Error (Bank #1)

P1517 Intake Manifold Runner Control (IMRC) Input Error (Bank #2)

P1518 Intake Manifold Runner Control (IMRC) Malfunction (Stuck Open)

P1519 Intake Manifold Runner Control (IMRC) Malfunction (Stuck Closed)

P1520 Intake Manifold Runner Control (IMRC) Circuit Malfunction

P1521 Variable Resonance Induction System (VRIS) Solenoid #1 Open Or Short

P1522 Variable Resonance Induction System (VRIS) Solenoid#2 Open Or Short

P1523 High Speed Inlet Air (HSIA) Solenoid Open Or Short

P1530 Air Condition (A/C) Clutch Circuit Malfunction

P1531 Invalid Test—Accelerator Pedal Movement

P1536 Parking Brake Applied Failure

P1537 Intake Manifold Runner Control (IMRC) Malfunction (Bank#1 Stuck Open)

P1538 Intake Manifold Runner Control (IMRC) Malfunction (Bank#2 Stuck Open)

P1539 Power To Air Condition (A/C) Clutch Circuit Overcurrent

P1549 Problem In Intake Manifold Tuning (IMT) Valve System

P1550 Power Steering Pressure (PSP) Sensor Out Of Self-Test Range

P1601 Serial Communication Error

P1605 Powertrain Control Module (PCM)—Keep Alive Memory (KAM) Test Error

P1608 PCM Internal Circuit Malfunction

P1609 PCM Internal Circuit Malfunction (2.5L Only)

P1625 B+ Supply To Variable Load Control Module (VLCM) Fan Circuit Malfunction

P1626 B+ Supply To Variable Load Control Module (VLCM) Air Conditioning (A/C) Circuit

P1650 Power Steering Pressure (PSP) Switch Out Of Self-Test Range

P1651 Power Steering Pressure (PSP) Switch Input Malfunction

P1660 Output Circuit Check Signal High

P1661 Output Circuit Check Signal Low

P1662 Injection Driver Module Enable (IDM EN) Circuit Failure

P1663 Fuel Delivery Command Signal (FDCS) Circuit Failure

P1667 Cylinder Identification (CID) Circuit Failure

P1668 PCM—IDM Diagnostic Communication Error

P1670 EF Feedback Signal Not Detected

P1701 Reverse Engagement Error

P1701 Fuel Trim Malfunction (Villager)

P1703 Brake On/Off (BOO) Switch Out Of Self-Test Range

P1704 Digital Transmission Range (TR) Sensor Failed To Transition State

P1705 Transmission Range (TR) Sensor Out Of Self-Test Range

P1705 TP Sensor (AT) Villager

P1705 Clutch Pedal Position (CPP) Or Park Neutral Position (PNP) Problem

P1706 High Vehicle Speed In Park

P1709 Park Or Neutral Position (PNP) Or Clutch Pedal Position (CPP) Switch Out Of Self-Test Range

P1709 Throttle Position (TP) Sensor Malfunction (Aspire 1.3L, Escort/ Tracer 1.8L, Probe 2.5L)

P1711 Transmission Fluid Temperature (TFT) Sensor Out Of Self-Test Range

P1714 Shift Solenoid "A" Inductive Signature Malfunction

P1715 Shift Solenoid "B" Inductive Signature Malfunction

P1716 Transmission Malfunction

P1717 Transmission Malfunction

P1719 Transmission Malfunction

P1720 Vehicle Speed Sensor (VSS) Circuit Malfunction

P1727 Coast Clutch Solenoid Inductive Signature Malfunction

P1728 Transmission Slip Error—Converter Clutch Failed

P1729 4x4 Low Switch Error

P1731 Improper 1–2 Shift

P1732 Improper 2–3 Shift

P1733 Improper 3–4 Shift

P1734 Improper 4–5 Shift

P1740 Torque Converter Clutch (TCC) Inductive Signature Malfunction

P1741 Torque Converter Clutch (TCC) Control Error

P1742 Torque Converter Clutch (TCC) Solenoid Failed On (Turns On MIL)

P1743 Torque Converter Clutch (TCC) Solenoid Failed On (Turns On TCIL)

P1744 Torque Converter Clutch (TCC) System Mechanically Stuck In Off Position

P1744 Torque Converter Clutch (TCC) Solenoid Malfunction (2.5L Only)

P1746 Electronic Pressure Control (EPC) Solenoid Open Circuit (Low Input)

P1747 Electronic Pressure Control (EPC) Solenoid Short Circuit (High Input)

P1748 Electronic Pressure Control (EPC) Malfunction

P1749 Electronic Pressure Control (EPC) Solenoid Failed Low

P1751 Shift Solenoid#1 (SS1) Performance

P1754 Coast Clutch Solenoid (CCS) Circuit Malfunction

P1756 Shift Solenoid#2 (SS2) Performance

P1760 Overrun Clutch SN

P1761 Shift Solenoid #(SS2) Performance

P1762 Transmission Malfunction

P1765 3–2 Timing Solenoid Malfunction (2.5L Only)

P1779 TCIL Circuit Malfunction

P1780 Transmission Control Switch (TCS) Circuit Out Of Self-Test Range

P1781 4x4 Low Switch, Out Of Self-Test Range

P1783 Transmission Over Temperature Condition

P1784 Transmission Malfunction

P1785 Transmission Malfunction

P1786 Transmission Malfunction

P1787 Transmission Malfunction

P1788 3–2 Timing/Coast Clutch Solenoid (3–2/CCS) Circuit Open

P1789 3–2 Timing/Coast Clutch Solenoid (3–2/CCS) Circuit Shorted

P1792 Idle (IDL) Switch (Closed Throttle Position Switch) Malfunction

P1794 Loss Of Battery Voltage Input

P1795 EGR Boost Sensor Malfunction

P1797 Clutch Pedal Position (CPP) Switch Or Neutral Switch Circuit Malfunction

P1900 Cooling Fan

U1021 SCP Indicating The Lack Of Air Conditioning (A/C) Clutch Status Response

U1039 Vehicle Speed Signal (VSS) Missing Or Incorrect

U1051 Brake Switch Signal Missing Or Incorrect

U1073 SCP Indicating The Lack Of Engine Coolant Fan Status Response

U1131 SCP Indicating The Lack Of Fuel Pump Status Response

U1135 SCP Indicating The Ignition Switch Signal Missing Or Incorrect

U1256 SCP Indicating A Communications Error

U1451 Lack Of Response From Passive Anti-Theft System (PATS) Module—Engine Disabled

General Motors Corporation

READING CODES

Reading the control module memory is on of the first steps in OBD II system diagnostics. This step should be initially performed to determine the general nature of the fault. Subsequent readings will determine if the fault has been cleared.

Reading codes can be performed by any of the methods below:
- Read the control module memory with the Generic Scan Tool (GST)
- Read the control module memory with the vehicle manufacturer's specific tester

To read the fault codes, connect the scan tool or tester according to the manufacturer's instructions. Follow the manufacturer's specified procedure for reading the codes.

CLEARING CODES

Control module reset procedures are a very important part of OBD II System diagnostics. This step should be done at the end of any fault code repair and at the end of any driveability repair.

Clearing codes can be performed by any of the methods below:
- Clear the control module memory with the Generic Scan Tool (GST)
- Clear the control module memory with the vehicle manufacturer's specific tester
- Turn the ignition off and remove the negative battery cable for at least 1 minute.

Removing the negative battery cable may cause other systems in the vehicle to loose their memory. Prior to removing the cable, ensure you have the proper reset codes for radios and alarms.

➡**The MIL will may also be de-activated for some codes if the vehicle completes three consecutive trips without a fault detected with vehicle conditions similar to those present during the fault.**

OBD II TROUBLE CODES

P0100 Mass or Volume Air Flow Circuit Malfunction
P0101 Mass or Volume Air Flow Circuit Range/Performance Problem
P0102 Mass or Volume Air Flow Circuit Low Input
P0103 Mass or Volume Air Flow Circuit High Input
P0104 Mass or Volume Air Flow Circuit Intermittent
P0105 Manifold Absolute Pressure/Barometric Pressure Circuit Malfunction
P0106 Manifold Absolute Pressure/Barometric Pressure Circuit Range/Performance Problem
P0107 Manifold Absolute Pressure/Barometric Pressure Circuit Low Input
P0108 Manifold Absolute Pressure/Barometric Pressure Circuit High Input
P0109 Manifold Absolute Pressure/Barometric Pressure Circuit Intermittent
P0110 Intake Air Temperature Circuit Malfunction
P0111 Intake Air Temperature Circuit Range/Performance Problem
P0112 Intake Air Temperature Circuit Low Input
P0113 Intake Air Temperature Circuit High Input
P0114 Intake Air Temperature Circuit Intermittent
P0115 Engine Coolant Temperature Circuit Malfunction
P0116 Engine Coolant Temperature Circuit Range/Performance Problem
P0117 Engine Coolant Temperature Circuit Low Input
P0118 Engine Coolant Temperature Circuit High Input
P0119 Engine Coolant Temperature Circuit Intermittent
P0120 Throttle/Pedal Position Sensor/Switch "A" Circuit Malfunction
P0121 Throttle/Pedal Position Sensor/Switch "A" Circuit Range/Performance Problem
P0122 Throttle/Pedal Position Sensor/Switch "A" Circuit Low Input
P0123 Throttle/Pedal Position Sensor/Switch "A" Circuit High Input
P0124 Throttle/Pedal Position Sensor/Switch "A" Circuit Intermittent
P0125 Insufficient Coolant Temperature For Closed Loop Fuel Control
P0126 Insufficient Coolant Temperature For Stable Operation
P0130 O2 Circuit Malfunction (Bank no. 1 Sensor no. 1)
P0131 O2 Sensor Circuit Low Voltage (Bank no. 1 Sensor no. 1)
P0132 O2 Sensor Circuit High Voltage (Bank no. 1 Sensor no. 1)
P0133 O2 Sensor Circuit Slow Response (Bank no. 1 Sensor no. 1)
P0134 O2 Sensor Circuit No Activity Detected (Bank no. 1 Sensor no. 1)
P0135 O2 Sensor Heater Circuit Malfunction (Bank no. 1 Sensor no. 1)
P0136 O2 Sensor Circuit Malfunction (Bank no. 1 Sensor no. 2)
P0137 O2 Sensor Circuit Low Voltage (Bank no. 1 Sensor no. 2)
P0138 O2 Sensor Circuit High Voltage (Bank no. 1 Sensor no. 2)
P0139 O2 Sensor Circuit Slow Response (Bank no. 1 Sensor no. 2)
P0140 O2 Sensor Circuit No Activity Detected (Bank no. 1 Sensor no. 2)
P0141 O2 Sensor Heater Circuit Malfunction (Bank no. 1 Sensor no. 2)
P0142 O2 Sensor Circuit Malfunction (Bank no. 1 Sensor no. 3)
P0143 O2 Sensor Circuit Low Voltage (Bank no. 1 Sensor no. 3)
P0144 O2 Sensor Circuit High Voltage (Bank no. 1 Sensor no. 3)
P0145 O2 Sensor Circuit Slow Response (Bank no. 1 Sensor no. 3)
P0146 O2 Sensor Circuit No Activity Detected (Bank no. 1 Sensor no. 3)
P0147 O2 Sensor Heater Circuit Malfunction (Bank no. 1 Sensor no. 3)
P0150 O2 Sensor Circuit Malfunction (Bank no. 2 Sensor no. 1)
P0151 O2 Sensor Circuit Low Voltage (Bank no. 2 Sensor no. 1)
P0152 O2 Sensor Circuit High Voltage (Bank no. 2 Sensor no. 1)
P0153 O2 Sensor Circuit Slow Response (Bank no. 2 Sensor no. 1)
P0154 O2 Sensor Circuit No Activity Detected (Bank no. 2 Sensor no. 1)

P0155 O2 Sensor Heater Circuit Malfunction (Bank no. 2 Sensor no. 1)

P0156 O2 Sensor Circuit Malfunction (Bank no. 2 Sensor no. 2)

P0157 O2 Sensor Circuit Low Voltage (Bank no. 2 Sensor no. 2)

P0158 O2 Sensor Circuit High Voltage (Bank no. 2 Sensor no. 2)

P0159 O2 Sensor Circuit Slow Response (Bank no. 2 Sensor no. 2)

P0160 O2 Sensor Circuit No Activity Detected (Bank no. 2 Sensor no. 2)

P0161 O2 Sensor Heater Circuit Malfunction (Bank no. 2 Sensor no. 2)

P0162 O2 Sensor Circuit Malfunction (Bank no. 2 Sensor no. 3)

P0163 O2 Sensor Circuit Low Voltage (Bank no. 2 Sensor no. 3)

P0164 O2 Sensor Circuit High Voltage (Bank no. 2 Sensor no. 3)

P0165 O2 Sensor Circuit Slow Response (Bank no. 2 Sensor no. 3)

P0166 O2 Sensor Circuit No Activity Detected (Bank no. 2 Sensor no. 3)

P0167 O2 Sensor Heater Circuit Malfunction (Bank no. 2 Sensor no. 3)

P0170 Fuel Trim Malfunction (Bank no. 1)

P0171 System Too Lean (Bank no. 1)

P0172 System Too Rich (Bank no. 1)

P0173 Fuel Trim Malfunction (Bank no. 2)

P0174 System Too Lean (Bank no. 2)

P0175 System Too Rich (Bank no. 2)

P0176 Fuel Composition Sensor Circuit Malfunction

P0177 Fuel Composition Sensor Circuit Range/Performance

P0178 Fuel Composition Sensor Circuit Low Input

P0179 Fuel Composition Sensor Circuit High Input

P0180 Fuel Temperature Sensor "A" Circuit Malfunction

P0181 Fuel Temperature Sensor "A" Circuit Range/Performance

P0182 Fuel Temperature Sensor "A" Circuit Low Input

P0183 Fuel Temperature Sensor "A" Circuit High Input

P0184 Fuel Temperature Sensor "A" Circuit Intermittent

P0185 Fuel Temperature Sensor "B" Circuit Malfunction

P0186 Fuel Temperature Sensor "B" Circuit Range/Performance

P0187 Fuel Temperature Sensor "B" Circuit Low Input

P0188 Fuel Temperature Sensor "B" Circuit High Input

P0189 Fuel Temperature Sensor "B" Circuit Intermittent

P0190 Fuel Rail Pressure Sensor Circuit Malfunction

P0191 Fuel Rail Pressure Sensor Circuit Range/Performance

P0192 Fuel Rail Pressure Sensor Circuit Low Input

P0193 Fuel Rail Pressure Sensor Circuit High Input

P0194 Fuel Rail Pressure Sensor Circuit Intermittent

P0195 Engine Oil Temperature Sensor Malfunction

P0196 Engine Oil Temperature Sensor Range/Performance

P0197 Engine Oil Temperature Sensor Low

P0198 Engine Oil Temperature Sensor High

P0199 Engine Oil Temperature Sensor Intermittent

P0200 Injector Circuit Malfunction

P0201 Injector Circuit Malfunction—Cylinder no. 1

P0202 Injector Circuit Malfunction—Cylinder no. 2

P0203 Injector Circuit Malfunction—Cylinder no. 3

P0204 Injector Circuit Malfunction—Cylinder no. 4

P0205 Injector Circuit Malfunction—Cylinder no. 5

P0206 Injector Circuit Malfunction—Cylinder no. 6

P0207 Injector Circuit Malfunction—Cylinder no. 7

P0208 Injector Circuit Malfunction—Cylinder no. 8

P0209 Injector Circuit Malfunction—Cylinder no. 9

P0210 Injector Circuit Malfunction—Cylinder no. 10

P0211 Injector Circuit Malfunction—Cylinder no. 11

P0212 Injector Circuit Malfunction—Cylinder no. 12

P0213 Cold Start Injector no. 1 Malfunction

P0214 Cold Start Injector no. 2 Malfunction

P0215 Engine Shutoff Solenoid Malfunction

P0216 Injection Timing Control Circuit Malfunction

P0217 Engine Over Temperature Condition

P0218 Transmission Over Temperature Condition

P0219 Engine Over Speed Condition

P0220 Throttle/Pedal Position Sensor/Switch "B" Circuit Malfunction

P0221 Throttle/Pedal Position Sensor/Switch "B" Circuit Range/Performance Problem

P0222 Throttle/Pedal Position Sensor/Switch "B" Circuit Low Input

P0223 Throttle/Pedal Position Sensor/Switch "B" Circuit High Input

P0224 Throttle/Pedal Position Sensor/Switch "B" Circuit Intermittent

P0225 Throttle/Pedal Position Sensor/Switch "C" Circuit Malfunction

P0226 Throttle/Pedal Position Sensor/Switch "C" Circuit Range/Performance Problem

P0227 Throttle/Pedal Position Sensor/Switch "C" Circuit Low Input

P0228 Throttle/Pedal Position Sensor/Switch "C" Circuit High Input

P0229 Throttle/Pedal Position Sensor/Switch "C" Circuit Intermittent

P0230 Fuel Pump Primary Circuit Malfunction

P0231 Fuel Pump Secondary Circuit Low

P0232 Fuel Pump Secondary Circuit High

P0233 Fuel Pump Secondary Circuit Intermittent

P0234 Engine Over Boost Condition

P0261 Cylinder no. 1 Injector Circuit Low

P0262 Cylinder no. 1 Injector Circuit High

P0263 Cylinder no. 1 Contribution/Balance Fault

P0264 Cylinder no. 2 Injector Circuit Low

P0265 Cylinder no. 2 Injector Circuit High

P0266 Cylinder no. 2 Contribution/Balance Fault

P0267 Cylinder no. 3 Injector Circuit Low

P0268 Cylinder no. 3 Injector Circuit High

P0269 Cylinder no. 3 Contribution/Balance Fault

P0270 Cylinder no. 4 Injector Circuit Low

P0271 Cylinder no. 4 Injector Circuit High

P0272 Cylinder no. 4 Contribution/Balance Fault

P0273 Cylinder no. 5 Injector Circuit Low

P0274 Cylinder no. 5 Injector Circuit High

P0275 Cylinder no. 5 Contribution/Balance Fault

P0276 Cylinder no. 6 Injector Circuit Low

P0277 Cylinder no. 6 Injector Circuit High

P0278 Cylinder no. 6 Contribution/Balance Fault

P0279 Cylinder no. 7 Injector Circuit Low

P0280 Cylinder no. 7 Injector Circuit High

P0281 Cylinder no. 7 Contribution/Balance Fault

P0282 Cylinder no. 8 Injector Circuit Low

P0283 Cylinder no. 8 Injector Circuit High

P0284 Cylinder no. 8 Contribution/Balance Fault

P0285 Cylinder no. 9 Injector Circuit Low
P0286 Cylinder no. 9 Injector Circuit High
P0287 Cylinder no. 9 Contribution/Balance Fault
P0288 Cylinder no. 10 Injector Circuit Low
P0289 Cylinder no. 10 Injector Circuit High
P0290 Cylinder no. 10 Contribution/Balance Fault
P0291 Cylinder no. 11 Injector Circuit Low
P0292 Cylinder no. 11 Injector Circuit High
P0293 Cylinder no. 11 Contribution/Balance Fault
P0294 Cylinder no. 12 Injector Circuit Low
P0295 Cylinder no. 12 Injector Circuit High
P0296 Cylinder no. 12 Contribution/Balance Fault
P0300 Random/Multiple Cylinder Misfire Detected
P0301 Cylinder no. 1—Misfire Detected
P0302 Cylinder no. 2—Misfire Detected
P0303 Cylinder no. 3—Misfire Detected
P0304 Cylinder no. 4—Misfire Detected
P0305 Cylinder no. 5—Misfire Detected
P0306 Cylinder no. 6—Misfire Detected
P0307 Cylinder no. 7—Misfire Detected
P0308 Cylinder no. 8—Misfire Detected
P0309 Cylinder no. 9—Misfire Detected
P0310 Cylinder no. 10—Misfire Detected
P0311 Cylinder no. 11—Misfire Detected
P0312 Cylinder no. 12—Misfire Detected
P0320 Ignition/Distributor Engine Speed Input Circuit Malfunction
P0321 Ignition/Distributor Engine Speed Input Circuit Range/Performance
P0322 Ignition/Distributor Engine Speed Input Circuit No Signal
P0323 Ignition/Distributor Engine Speed Input Circuit Intermittent
P0325 Knock Sensor no. 1—Circuit Malfunction (Bank no. 1 or Single Sensor)
P0326 Knock Sensor no. 1—Circuit Range/Performance (Bank no. 1 or Single Sensor)
P0327 Knock Sensor no. 1—Circuit Low Input (Bank no. 1 or Single Sensor)
P0328 Knock Sensor no. 1—Circuit High Input (Bank no. 1 or Single Sensor)
P0329 Knock Sensor no. 1—Circuit Input Intermittent (Bank no. 1 or Single Sensor)
P0330 Knock Sensor no. 2—Circuit Malfunction (Bank no. 2)
P0331 Knock Sensor no. 2—Circuit Range/Performance (Bank no. 2)
P0332 Knock Sensor no. 2—Circuit Low Input (Bank no. 2)
P0333 Knock Sensor no. 2—Circuit High Input (Bank no. 2)
P0334 Knock Sensor no. 2—Circuit Input Intermittent (Bank no. 2)
P0335 Crankshaft Position Sensor "A" Circuit Malfunction
P0336 Crankshaft Position Sensor "A" Circuit Range/Performance
P0337 Crankshaft Position Sensor "A" Circuit Low Input
P0338 Crankshaft Position Sensor "A" Circuit High Input
P0339 Crankshaft Position Sensor "A" Circuit Intermittent
P0340 Camshaft Position Sensor Circuit Malfunction
P0341 Camshaft Position Sensor Circuit Range/Performance
P0342 Camshaft Position Sensor Circuit Low Input
P0343 Camshaft Position Sensor Circuit High Input
P0344 Camshaft Position Sensor Circuit Intermittent
P0350 Ignition Coil Primary/Secondary Circuit Malfunction

P0351 Ignition Coil "A" Primary/Secondary Circuit Malfunction
P0352 Ignition Coil "B" Primary/Secondary Circuit Malfunction
P0353 Ignition Coil "C" Primary/Secondary Circuit Malfunction
P0354 Ignition Coil "D" Primary/Secondary Circuit Malfunction
P0355 Ignition Coil "E" Primary/Secondary Circuit Malfunction
P0356 Ignition Coil "F" Primary/Secondary Circuit Malfunction
P0357 Ignition Coil "G" Primary/Secondary Circuit Malfunction
P0358 Ignition Coil "H" Primary/Secondary Circuit Malfunction
P0359 Ignition Coil "I" Primary/Secondary Circuit Malfunction
P0360 Ignition Coil "J" Primary/Secondary Circuit Malfunction
P0361 Ignition Coil "K" Primary/Secondary Circuit Malfunction
P0362 Ignition Coil "L" Primary/Secondary Circuit Malfunction
P0370 Timing Reference High Resolution Signal "A" Malfunction
P0371 Timing Reference High Resolution Signal "A" Too Many Pulses
P0372 Timing Reference High Resolution Signal "A" Too Few Pulses
P0373 Timing Reference High Resolution Signal "A" Intermittent/Erratic Pulses
P0374 Timing Reference High Resolution Signal "A" No Pulses
P0375 Timing Reference High Resolution Signal "B" Malfunction
P0376 Timing Reference High Resolution Signal "B" Too Many Pulses
P0377 Timing Reference High Resolution Signal "B" Too Few Pulses
P0378 Timing Reference High Resolution Signal "B" Intermittent/Erratic Pulses
P0379 Timing Reference High Resolution Signal "B" No Pulses
P0380 Glow Plug/Heater Circuit "A" Malfunction
P0381 Glow Plug/Heater Indicator Circuit Malfunction
P0382 Glow Plug/Heater Circuit "B" Malfunction
P0385 Crankshaft Position Sensor "B" Circuit Malfunction
P0386 Crankshaft Position Sensor "B" Circuit Range/Performance
P0387 Crankshaft Position Sensor "B" Circuit Low Input
P0388 Crankshaft Position Sensor "B" Circuit High Input
P0389 Crankshaft Position Sensor "B" Circuit Intermittent
P0400 Exhaust Gas Recirculation Flow Malfunction
P0401 Exhaust Gas Recirculation Flow Insufficient Detected
P0402 Exhaust Gas Recirculation Flow Excessive Detected
P0403 Exhaust Gas Recirculation Circuit Malfunction
P0404 Exhaust Gas Recirculation Circuit Range/Performance
P0405 Exhaust Gas Recirculation Sensor "A" Circuit Low
P0406 Exhaust Gas Recirculation Sensor "A" Circuit High
P0407 Exhaust Gas Recirculation Sensor "B" Circuit Low
P0408 Exhaust Gas Recirculation Sensor "B" Circuit High
P0410 Secondary Air Injection System Malfunction
P0411 Secondary Air Injection System Incorrect Flow Detected
P0412 Secondary Air Injection System Switching Valve "A" Circuit Malfunction
P0413 Secondary Air Injection System Switching Valve "A" Circuit Open
P0414 Secondary Air Injection System Switching Valve "A" Circuit Shorted
P0415 Secondary Air Injection System Switching Valve "B" Circuit Malfunction
P0416 Secondary Air Injection System Switching Valve "B" Circuit Open

P0417 Secondary Air Injection System Switching Valve "B" Circuit Shorted

P0418 Secondary Air Injection System Relay "A" Circuit Malfunction

P0419 Secondary Air Injection System Relay "B" Circuit Malfunction

P0420 Catalyst System Efficiency Below Threshold (Bank no. 1)

P0421 Warm Up Catalyst Efficiency Below Threshold (Bank no. 1)

P0422 Main Catalyst Efficiency Below Threshold (Bank no. 1)

P0423 Heated Catalyst Efficiency Below Threshold (Bank no. 1)

P0424 Heated Catalyst Temperature Below Threshold (Bank no. 1)

P0430 Catalyst System Efficiency Below Threshold (Bank no. 2)

P0431 Warm Up Catalyst Efficiency Below Threshold (Bank no. 2)

P0432 Main Catalyst Efficiency Below Threshold (Bank no. 2)

P0433 Heated Catalyst Efficiency Below Threshold (Bank no. 2)

P0434 Heated Catalyst Temperature Below Threshold (Bank no. 2)

P0440 Evaporative Emission Control System Malfunction

P0441 Evaporative Emission Control System Incorrect Purge Flow

P0442 Evaporative Emission Control System Leak Detected (Small Leak)

P0443 Evaporative Emission Control System Purge Control Valve Circuit Malfunction

P0444 Evaporative Emission Control System Purge Control Valve Circuit Open

P0445 Evaporative Emission Control System Purge Control Valve Circuit Shorted

P0446 Evaporative Emission Control System Vent Control Circuit Malfunction

P0447 Evaporative Emission Control System Vent Control Circuit Open

P0448 Evaporative Emission Control System Vent Control Circuit Shorted

P0449 Evaporative Emission Control System Vent Valve/Solenoid Circuit Malfunction

P0450 Evaporative Emission Control System Pressure Sensor Malfunction

P0451 Evaporative Emission Control System Pressure Sensor Range/Performance

P0452 Evaporative Emission Control System Pressure Sensor Low Input

P0453 Evaporative Emission Control System Pressure Sensor High Input

P0454 Evaporative Emission Control System Pressure Sensor Intermittent

P0455 Evaporative Emission Control System Leak Detected (Gross Leak)

P0460 Fuel Level Sensor Circuit Malfunction

P0461 Fuel Level Sensor Circuit Range/Performance

P0462 Fuel Level Sensor Circuit Low Input

P0463 Fuel Level Sensor Circuit High Input

P0464 Fuel Level Sensor Circuit Intermittent

P0465 Purge Flow Sensor Circuit Malfunction

P0466 Purge Flow Sensor Circuit Range/Performance

P0467 Purge Flow Sensor Circuit Low Input

P0468 Purge Flow Sensor Circuit High Input

P0469 Purge Flow Sensor Circuit Intermittent

P0470 Exhaust Pressure Sensor Malfunction

P0471 Exhaust Pressure Sensor Range/Performance

P0472 Exhaust Pressure Sensor Low

P0473 Exhaust Pressure Sensor High

P0474 Exhaust Pressure Sensor Intermittent

P0475 Exhaust Pressure Control Valve Malfunction

P0476 Exhaust Pressure Control Valve Range/Performance

P0477 Exhaust Pressure Control Valve Low

P0478 Exhaust Pressure Control Valve High

P0479 Exhaust Pressure Control Valve Intermittent

P0480 Cooling Fan no. 1 Control Circuit Malfunction

P0481 Cooling Fan no. 2 Control Circuit Malfunction

P0482 Cooling Fan no. 3 Control Circuit Malfunction

P0483 Cooling Fan Rationality Check Malfunction

P0484 Cooling Fan Circuit Over Current

P0485 Cooling Fan Power/Ground Circuit Malfunction

P0500 Vehicle Speed Sensor Malfunction

P0501 Vehicle Speed Sensor Range/Performance

P0502 Vehicle Speed Sensor Circuit Low Input

P0503 Vehicle Speed Sensor Intermittent/Erratic/High

P0505 Idle Control System Malfunction

P0506 Idle Control System RPM Lower Than Expected

P0507 Idle Control System RPM Higher Than Expected

P0510 Closed Throttle Position Switch Malfunction

P0520 Engine Oil Pressure Sensor/Switch Circuit Malfunction

P0521 Engine Oil Pressure Sensor/Switch Range/Performance

P0522 Engine Oil Pressure Sensor/Switch Low Voltage

P0523 Engine Oil Pressure Sensor/Switch High Voltage

P0530 A/C Refrigerant Pressure Sensor Circuit Malfunction

P0531 A/C Refrigerant Pressure Sensor Circuit Range/Performance

P0532 A/C Refrigerant Pressure Sensor Circuit Low Input

P0533 A/C Refrigerant Pressure Sensor Circuit High Input

P0534 A/C Refrigerant Charge Loss

P0550 Power Steering Pressure Sensor Circuit Malfunction

P0551 Power Steering Pressure Sensor Circuit Range/Performance

P0552 Power Steering Pressure Sensor Circuit Low Input

P0553 Power Steering Pressure Sensor Circuit High Input

P0554 Power Steering Pressure Sensor Circuit Intermittent

P0560 System Voltage Malfunction

P0561 System Voltage Unstable

P0562 System Voltage Low

P0563 System Voltage High

P0565 Cruise Control On Signal Malfunction

P0566 Cruise Control Off Signal Malfunction

P0567 Cruise Control Resume Signal Malfunction

P0568 Cruise Control Set Signal Malfunction

P0569 Cruise Control Coast Signal Malfunction

P0570 Cruise Control Accel Signal Malfunction

P0571 Cruise Control/Brake Switch "A" Circuit Malfunction

P0572 Cruise Control/Brake Switch "A" Circuit Low

P0573 Cruise Control/Brake Switch "A" Circuit High

P0574 **Through P0580** Reserved for Cruise Codes

P0600 Serial Communication Link Malfunction

P0601 Internal Control Module Memory Check Sum Error

P0602 Control Module Programming Error

P0603 Internal Control Module Keep Alive Memory (KAM) Error

P0604 Internal Control Module Random Access Memory (RAM) Error

P0605 Internal Control Module Read Only Memory (ROM) Error

P0606 PCM Processor Fault

P0608 Control Module VSS Output "A" Malfunction

P0609 Control Module VSS Output "B" Malfunction

P0620 Generator Control Circuit Malfunction

P0621 Generator Lamp "L" Control Circuit Malfunction

P0622 Generator Field "F" Control Circuit Malfunction

P0650 Malfunction Indicator Lamp (MIL) Control Circuit Malfunction

P0654 Engine RPM Output Circuit Malfunction

P0655 Engine Hot Lamp Output Control Circuit Malfunction

P0656 Fuel Level Output Circuit Malfunction

P0700 Transmission Control System Malfunction

P0701 Transmission Control System Range/Performance

P0702 Transmission Control System Electrical

P0703 Torque Converter/Brake Switch "B" Circuit Malfunction

P0704 Clutch Switch Input Circuit Malfunction

P0705 Transmission Range Sensor Circuit Malfunction (PRNDL Input)

P0706 Transmission Range Sensor Circuit Range/Performance

P0707 Transmission Range Sensor Circuit Low Input

P0708 Transmission Range Sensor Circuit High Input

P0709 Transmission Range Sensor Circuit Intermittent

P0710 Transmission Fluid Temperature Sensor Circuit Malfunction

P0711 Transmission Fluid Temperature Sensor Circuit Range/Performance

P0712 Transmission Fluid Temperature Sensor Circuit Low Input

P0713 Transmission Fluid Temperature Sensor Circuit High Input

P0714 Transmission Fluid Temperature Sensor Circuit Intermittent

P0715 Input/Turbine Speed Sensor Circuit Malfunction

P0716 Input/Turbine Speed Sensor Circuit Range/Performance

P0717 Input/Turbine Speed Sensor Circuit No Signal

P0718 Input/Turbine Speed Sensor Circuit Intermittent

P0719 Torque Converter/Brake Switch "B" Circuit Low

P0720 Output Speed Sensor Circuit Malfunction

P0721 Output Speed Sensor Circuit Range/Performance

P0722 Output Speed Sensor Circuit No Signal

P0723 Output Speed Sensor Circuit Intermittent

P0724 Torque Converter/Brake Switch "B" Circuit High

P0725 Engine Speed Input Circuit Malfunction

P0726 Engine Speed Input Circuit Range/Performance

P0727 Engine Speed Input Circuit No Signal

P0728 Engine Speed Input Circuit Intermittent

P0730 Incorrect Gear Ratio

P0731 Gear no. 1 Incorrect Ratio

P0732 Gear no. 2 Incorrect Ratio

P0733 Gear no. 3 Incorrect Ratio

P0734 Gear no. 4 Incorrect Ratio

P0735 Gear no. 5 Incorrect Ratio

P0736 Reverse Incorrect Ratio

P0740 Torque Converter Clutch Circuit Malfunction

P0741 Torque Converter Clutch Circuit Performance or Stuck Off

P0742 Torque Converter Clutch Circuit Stuck On

P0743 Torque Converter Clutch Circuit Electrical

P0744 Torque Converter Clutch Circuit Intermittent

P0745 Pressure Control Solenoid Malfunction

P0746 Pressure Control Solenoid Performance or Stuck Off

P0747 Pressure Control Solenoid Stuck On

P0748 Pressure Control Solenoid Electrical

P0749 Pressure Control Solenoid Intermittent

P0750 Shift Solenoid "A" Malfunction

P0751 Shift Solenoid "A" Performance or Stuck Off

P0752 Shift Solenoid "A" Stuck On

P0753 Shift Solenoid "A" Electrical

P0754 Shift Solenoid "A" Intermittent

P0755 Shift Solenoid "B" Malfunction

P0756 Shift Solenoid "B" Performance or Stuck Off

P0757 Shift Solenoid "B" Stuck On

P0758 Shift Solenoid "B" Electrical

P0759 Shift Solenoid "B" Intermittent

P0760 Shift Solenoid "C" Malfunction

P0761 Shift Solenoid "C" Performance Or Stuck Off

P0762 Shift Solenoid "C" Stuck On

P0763 Shift Solenoid "C" Electrical

P0764 Shift Solenoid "C" Intermittent

P0765 Shift Solenoid "D" Malfunction

P0766 Shift Solenoid "D" Performance Or Stuck Off

P0767 Shift Solenoid "D" Stuck On

P0768 Shift Solenoid "D" Electrical

P0769 Shift Solenoid "D" Intermittent

P0770 Shift Solenoid "E" Malfunction

P0771 Shift Solenoid "E" Performance Or Stuck Off

P0772 Shift Solenoid "E" Stuck On

P0773 Shift Solenoid "E" Electrical

P0774 Shift Solenoid "E" Intermittent

P0780 Shift Malfunction

P0781 1–2 Shift Malfunction

P0782 2–3 Shift Malfunction

P0783 3–4 Shift Malfunction

P0784 4–5 Shift Malfunction

P0785 Shift/Timing Solenoid Malfunction

P0786 Shift/Timing Solenoid Range/Performance

P0787 Shift/Timing Solenoid Low

P0788 Shift/Timing Solenoid High

P0789 Shift/Timing Solenoid Intermittent

P0790 Normal/Performance Switch Circuit Malfunction

P0801 Reverse Inhibit Control Circuit Malfunction

P0803 1–4 Upshift (Skip Shift) Solenoid Control Circuit Malfunction

P0804 1–4 Upshift (Skip Shift) Lamp Control Circuit Malfunction

P1106 MAP Sensor Voltage Intermittently High

P1107 MAP Sensor Voltage Intermittently Low

P1111 IAT Sensor Circuit Intermittent High Voltage

P1112 IAT Sensor Circuit Intermittent Low Voltage

P1114 ECT Sensor Circuit Intermittent Low Voltage

P1115 ECT Sensor Circuit Intermittent High Voltage

P1121 TP Sensor Voltage Intermittently High

P1122 TP Sensor Voltage Intermittently Low

P1133 HO$_2$S Insufficient Switching Sensor

P1133 HO$_2$S Insufficient Switching Bank #1, Sensor #1

P1134 HO$_2$S #1 Transition Time Ratio

P1134 HO$_2$S Transition Time Ratio Bank #1, Sensor #1

P1153 HO$_2$S Insufficient Switching Sensor Bank #2, Sensor #1

P1154 HO$_2$S Transition Time Ratio Bank #2, Sensor #1
P1345 Crankshaft/Camshaft (CKP/CMP) Correlation
P1350 Ignition Control (IC) Circuit Malfunction
P1351 Ignition Control (IC) Circuit High Voltage
P1361 Ignition Control (IC) Circuit Not Toggling
P1361 Ignition Control (IC) Circuit Low Voltage
P1380 Electronic Brake Control Module (EBCM) DTC Detected Rough Road Data Unusable
P1381 Misfire Detected, No EBCM/PCM/VCM Serial Data
P1406 EGR Pintle Position Circuit Fault
P1415 AIR System Bank #1
P1416 AIR System Bank #2
P1441 EVAP Control System Flow During Non-Purge
P1442 EVAP Vacuum Switch Circuit
P1450 Barometric Pressure Sensor Circuit Fault
P1451 Barometric Pressure Sensor Performance

P1460 Cooling Fan Control System Fault
P1500 Starter Signal Circuit Fault
P1510 Back-up Power Supply Fault
P1508 IAC System Low RPM
P1509 IAC System High RPM
P1520 PNP Circuit
P1530 Ignition Timing Adjustment Switch Circuit
P1600 PCM Battery Circuit Fault
P1635 5-Volt Reference "A" Circuit
P1639 5-Volt Reference "B" Circuit
P1641 MIL Control Circuit
P1651 Fan #1 Relay Control Circuit
P1652 Fan #2 Relay Control Circuit
P1654 A/C Relay Control
P1655 EVAP Purge Solenoid Control Circuit
P1672 Low Engine Oil Level Light Control Circuit

FIRING ORDERS

3

FIRING ORDERS 3-1
FIRING ORDER INDEX 3-2

FIRING ORDERS

On every vehicle manufactured between 1995 and 1999, there are essentially only two basic methods for distributing the ignition system spark to the spark plugs: distributor system and Distributorless Ignition System (DIS). The distributor system uses a rotating rotor within a distributor cap to dispense the system's spark to the applicable spark plug. DIS systems use one of three general set-ups for spark distribution: remote coil pack(s), waste spark system, and direct ignition system (also often referred to as DIS). All DIS systems are controlled by the engine control computer, which computes the proper ignition timing based upon incoming reference signals from engine sensors.

The remote coil pack set-up uses one or more coil packs connected to the spark plugs via plug wires. The waste spark system is actually a sub-type of the remote coil pack system. The only difference being that two spark plugs are fired simultaneously because they share one coil. Many waste spark systems are designed as a hybrid of a direct ignition system and a remote coil pack system, because a coil is mounted directly on top of one spark plug and attached to another spark plug via a plug wire. Direct ignition does away with the spark plug wires completely and uses a single coil pack mounted directly on top of the spark plug for each cylinder.

Firing orders are most important for vehicles equipped with distributor ignition systems because the distributor can be rotated (which can lead to confusion as to which plug tower is what). If the distributor is rotated and the spark plug wires are installed on the original cap towers, the ignition timing will be adversely affected. DIS systems are not adjustable in the same manner as distributor systems. Therefore, if you connect the wires (when applicable) to the proper coil pack towers, the ignition timing will always be correct. Thus, if your vehicle is equipped with a DIS system and the firing order illustration does not contain a specific firing order, simply attach the wires to the proper coil pack towers and the ignition system will function properly.

➡ **The coil packs used on DIS systems are often labeled with the cylinder number of their corresponding spark plugs.**

If your vehicle is equipped with a distributor which is not keyed for installation with only one orientation, it could have been removed previously and rewired. The resultant wiring would hold the correct firing order, but could change the relative placement of the plug towers in relation to the engine. For this reason it is imperative that you label all wires before disconnecting any of them. Also, before removal, compare the current wiring with the accompanying illustrations. If the current wiring does not match, make notes in your book to reflect how your engine is wired.

➡ **To avoid confusion, remove and tag the spark plug wires one at a time, for replacement.**

FIRING ORDER INDEX

MANUFACTURER ENGINE	FIGURE
Chrysler Corp.	
2.0L (VIN C) Engines	1
2.0L (VIN F) Engines	2
2.0L (VIN Y) Engines	3
2.5L (VIN H) Engines	4
2.5L (VIN N) Engines	5
2.7L Engines	6
3.2L Engines	6
3.3L Engines	7
3.5L Engines	7
Ford Motor Co.	
1.3L Engines	8
1.8L Engines	9
1.9L Engines	10
2.0L (VIN A) Engines	11
2.0L (VIN P and 3) Engines	12
2.0L (VIN 3) Engines	13
2.5L (VIN B) Engines	14
2.5L (VIN L) Engines	15
3.0L (VIN U and 1) Engines	
1995 Models	16
1996-97 Models	17
3.0L (VIN Y) SHO Engines	17
3.2L (VIN P) SHO Engines	17
3.0L (VIN S) Engines	18
3.4L Engines	19
3.8L (VIN 4) Engines	20
3.8L (VIN R) Engines	21
4.6L Engines	22
5.0L Engines	23
General Motors	
2.2L Engines	24
2.3L Engines	25
2.4L Engines	25
3.0L Engines	26
3.1L Engines	27
3.4L (VIN X) Engines	28
3.8L Engines	29
4.0L Engines	30
4.3L (VIN W) Engines	31
4.6L Engines	
1995-97 Models	32
1998-99 Models	33
4.9L Engines	34
5.7L (VIN G) Engines	35
5.7L (VIN J) Engines	36
5.7L (VIN P) Engines	31

FIRING ORDER INDEX

MANUFACTURER ENGINE	FIGURE
Geo/Chevrolet	
1.0L Engines	37
1.3L Engines	38
1.6L Engines	39
1.8L Engines	39
Saturn	
1.9L Engines	40

79223C02

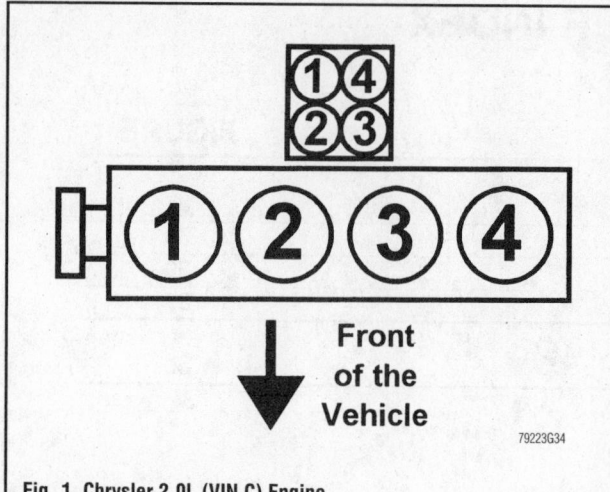

Fig. 1 Chrysler 2.0L (VIN C) Engine
Firing order: 1–3–4–2
Distributorless ignition system

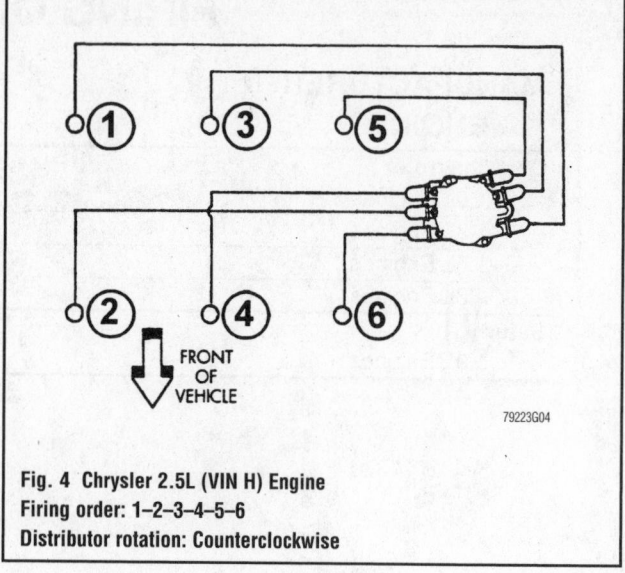

Fig. 4 Chrysler 2.5L (VIN H) Engine
Firing order: 1–2–3–4–5–6
Distributor rotation: Counterclockwise

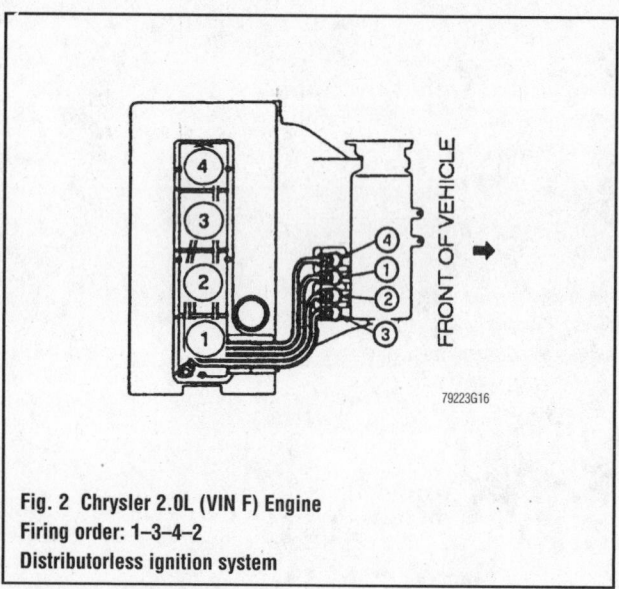

Fig. 2 Chrysler 2.0L (VIN F) Engine
Firing order: 1–3–4–2
Distributorless ignition system

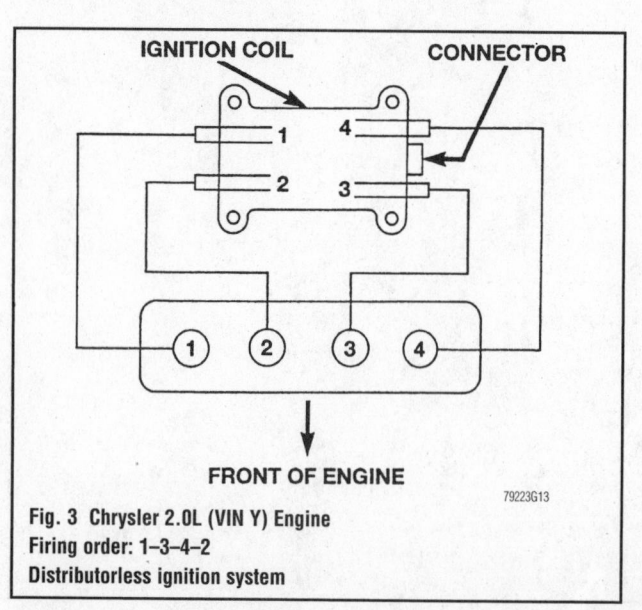

Fig. 3 Chrysler 2.0L (VIN Y) Engine
Firing order: 1–3–4–2
Distributorless ignition system

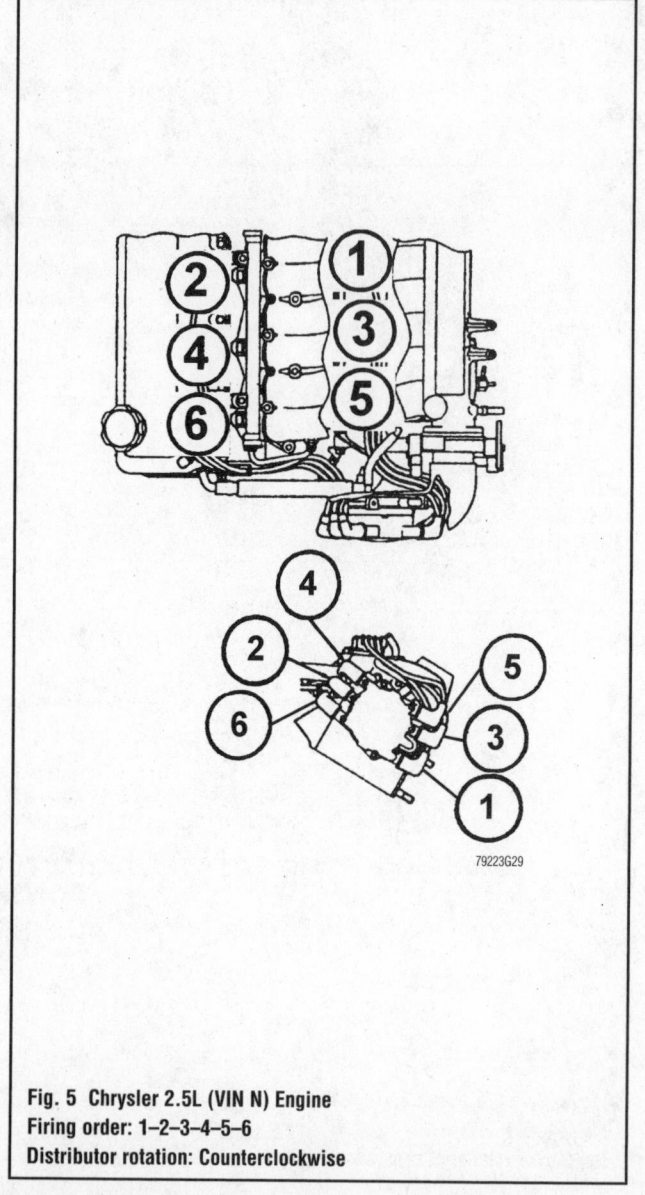

Fig. 5 Chrysler 2.5L (VIN N) Engine
Firing order: 1–2–3–4–5–6
Distributor rotation: Counterclockwise

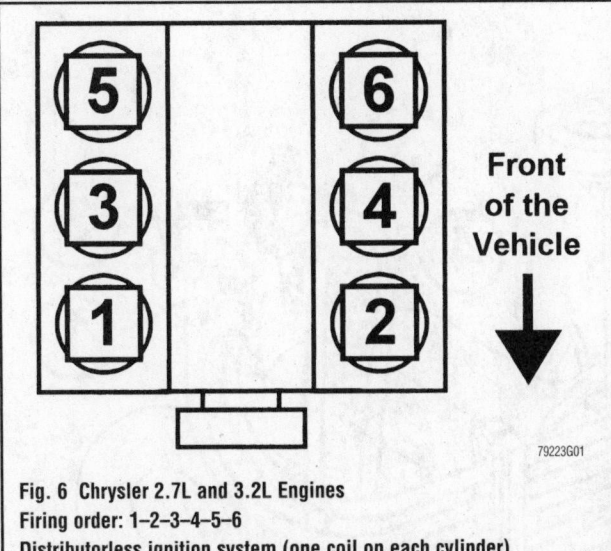

Fig. 6 Chrysler 2.7L and 3.2L Engines
Firing order: 1–2–3–4–5–6
Distributorless ignition system (one coil on each cylinder)

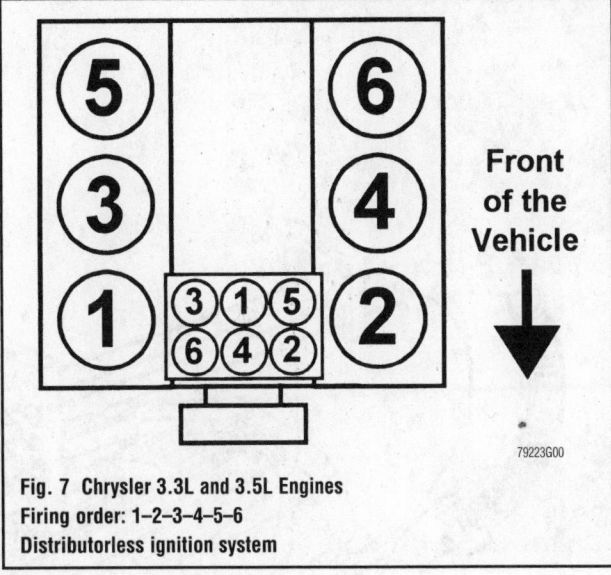

Fig. 7 Chrysler 3.3L and 3.5L Engines
Firing order: 1–2–3–4–5–6
Distributorless ignition system

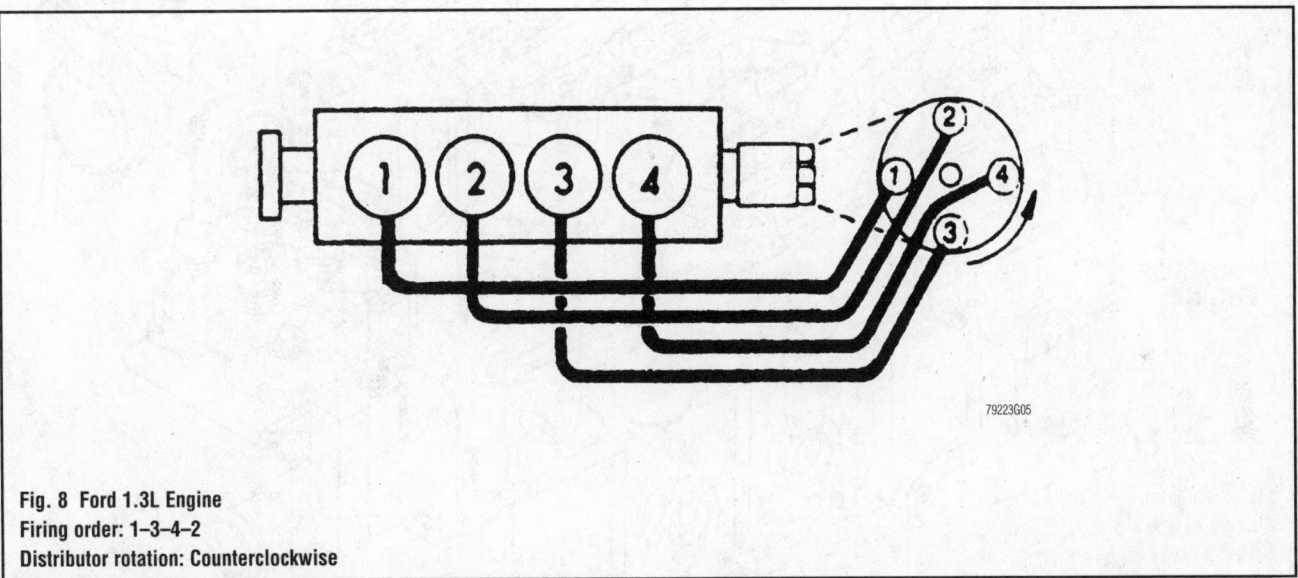

Fig. 8 Ford 1.3L Engine
Firing order: 1–3–4–2
Distributor rotation: Counterclockwise

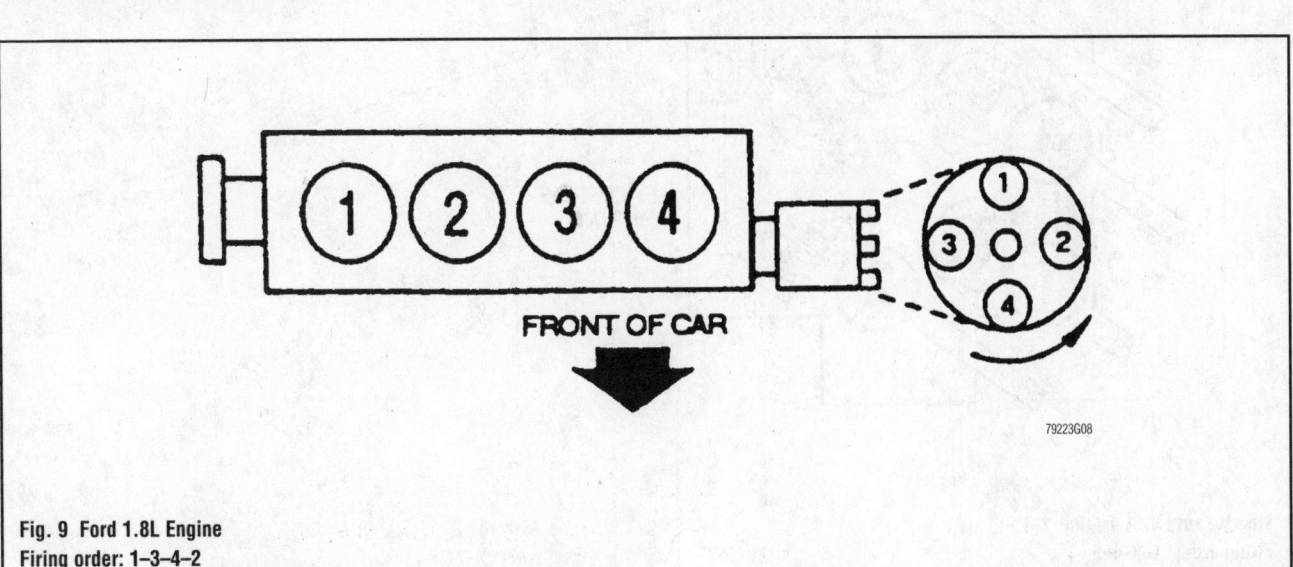

Fig. 9 Ford 1.8L Engine
Firing order: 1–3–4–2
Distributor rotation: Counterclockwise

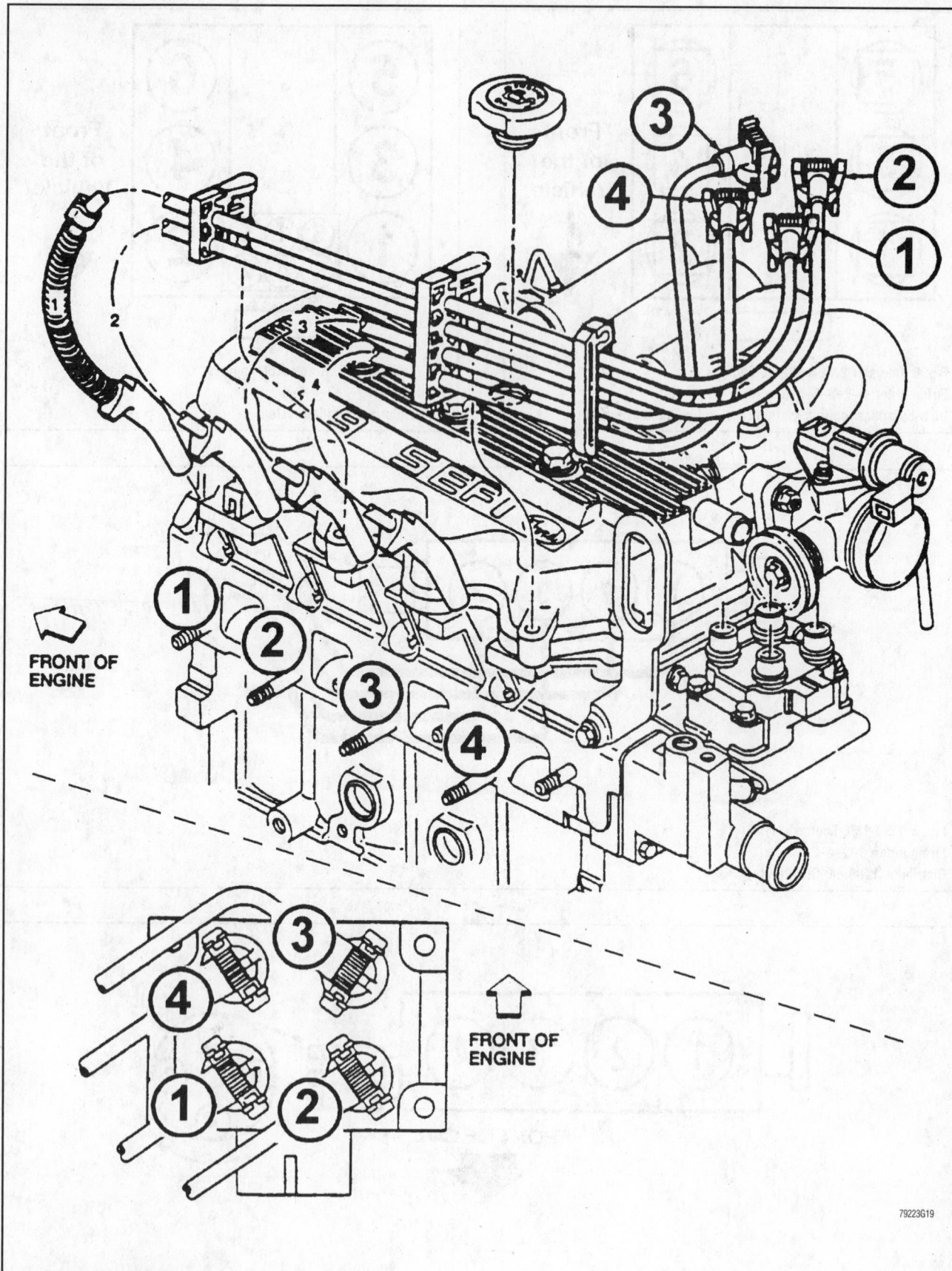

FRONT OF ENGINE

FRONT OF ENGINE

79223G19

Fig. 10 Ford 1.9L Engine
Firing order: 1–3–4–2
Distributorless ignition system

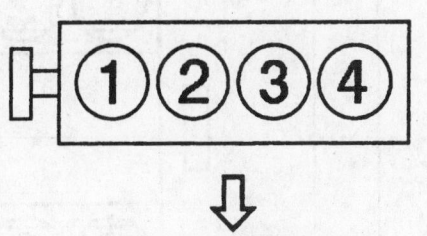
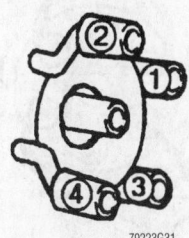

Fig. 11 Ford 2.0L (VIN A) Engine
Firing order: 1–3–4–2
Distributor rotation: Clockwise

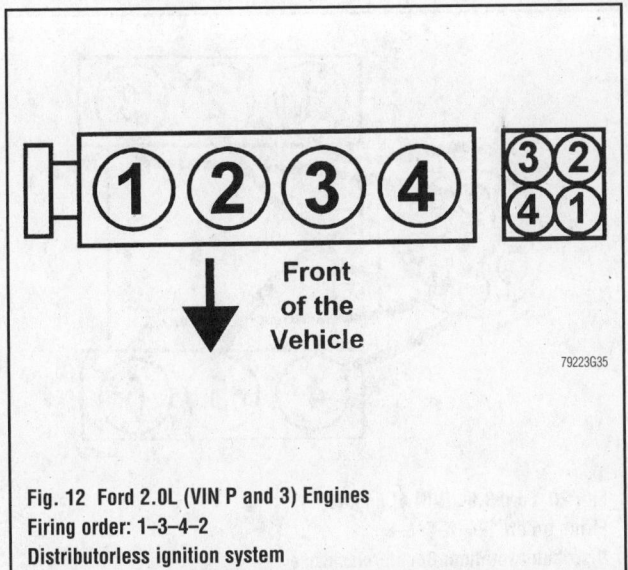

Fig. 12 Ford 2.0L (VIN P and 3) Engines
Firing order: 1–3–4–2
Distributorless ignition system

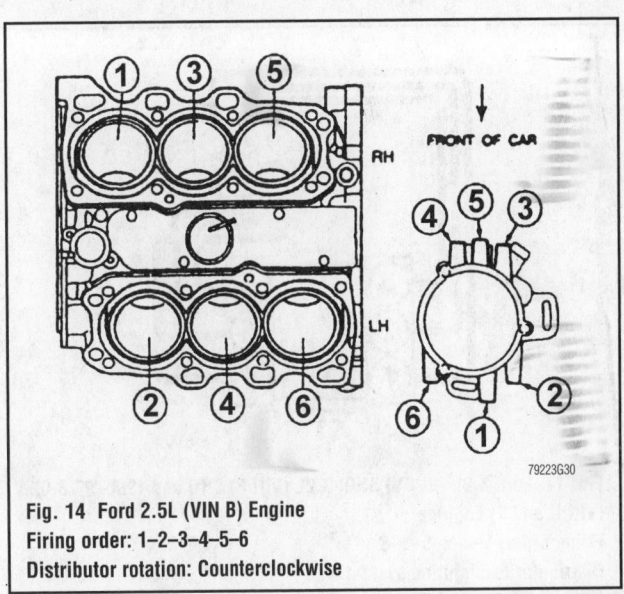

Fig. 14 Ford 2.5L (VIN B) Engine
Firing order: 1–2–3–4–5–6
Distributor rotation: Counterclockwise

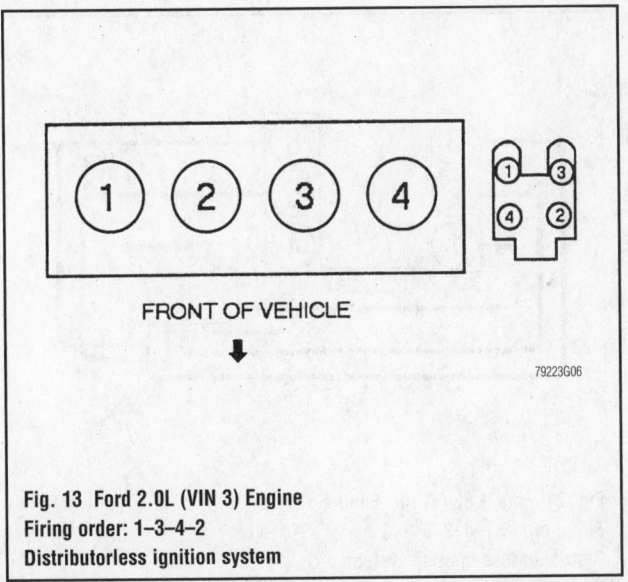

Fig. 13 Ford 2.0L (VIN 3) Engine
Firing order: 1–3–4–2
Distributorless ignition system

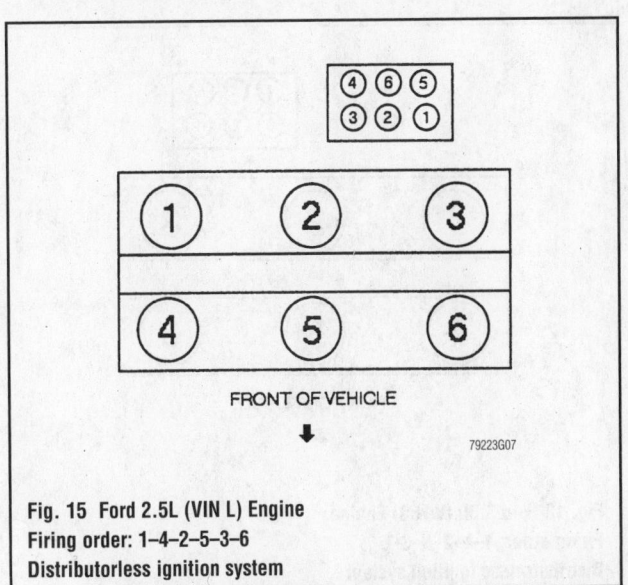

Fig. 15 Ford 2.5L (VIN L) Engine
Firing order: 1–4–2–5–3–6
Distributorless ignition system

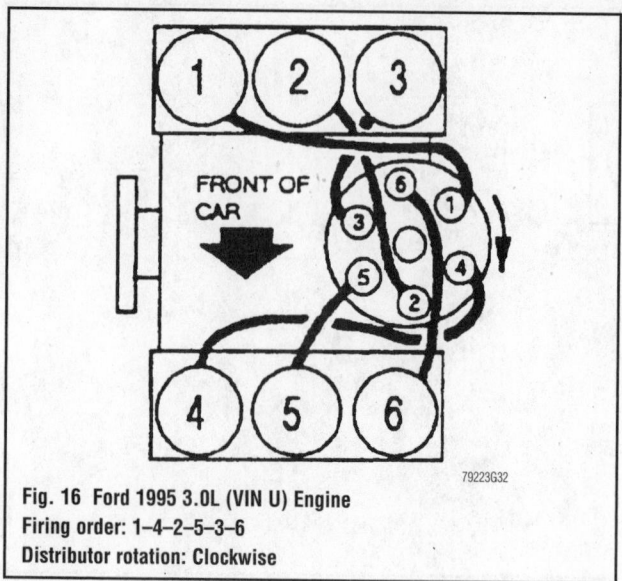

Fig. 16 Ford 1995 3.0L (VIN U) Engine
Firing order: 1–4–2–5–3–6
Distributor rotation: Clockwise

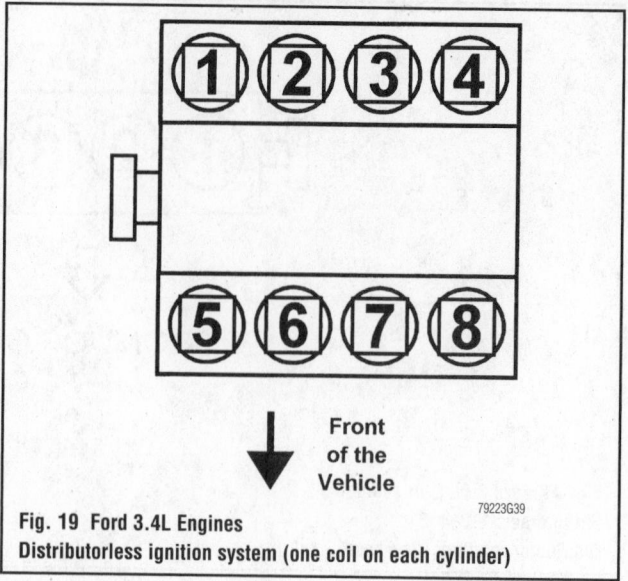

Fig. 19 Ford 3.4L Engines
Distributorless ignition system (one coil on each cylinder)

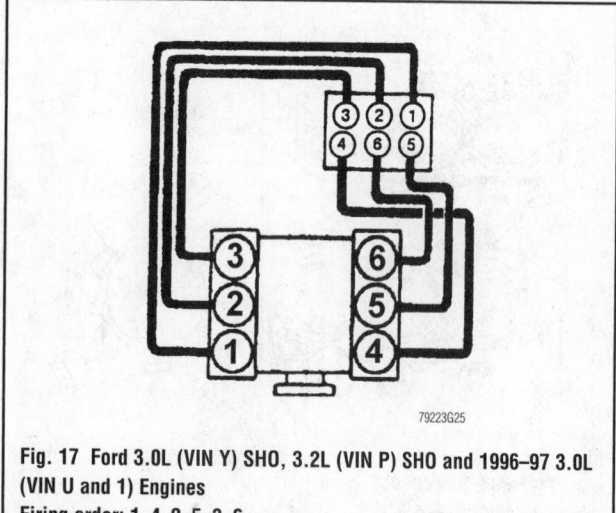

Fig. 17 Ford 3.0L (VIN Y) SHO, 3.2L (VIN P) SHO and 1996–97 3.0L (VIN U and 1) Engines
Firing order: 1–4–2–5–3–6
Distributorless ignition system

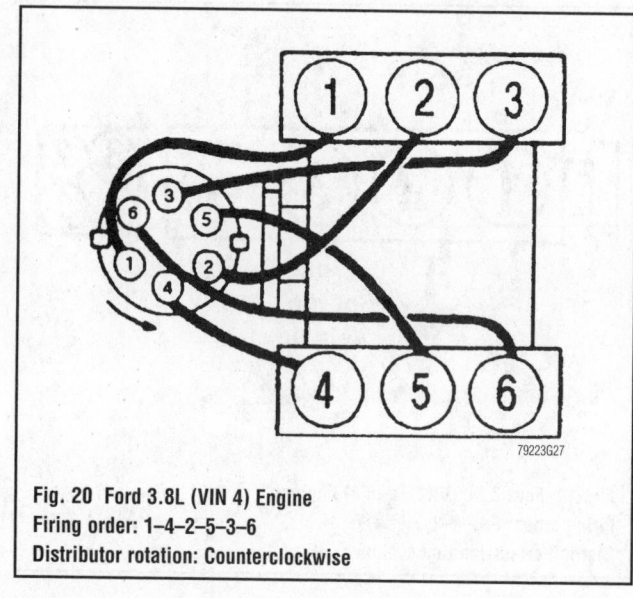

Fig. 20 Ford 3.8L (VIN 4) Engine
Firing order: 1–4–2–5–3–6
Distributor rotation: Counterclockwise

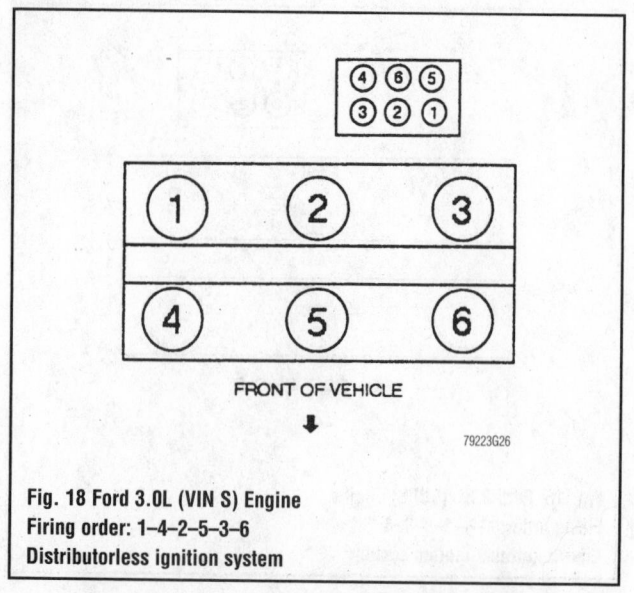

Fig. 18 Ford 3.0L (VIN S) Engine
Firing order: 1–4–2–5–3–6
Distributorless ignition system

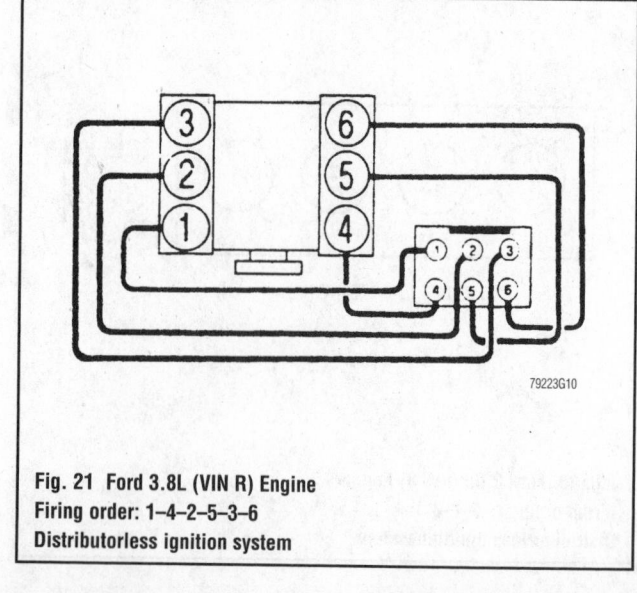

Fig. 21 Ford 3.8L (VIN R) Engine
Firing order: 1–4–2–5–3–6
Distributorless ignition system

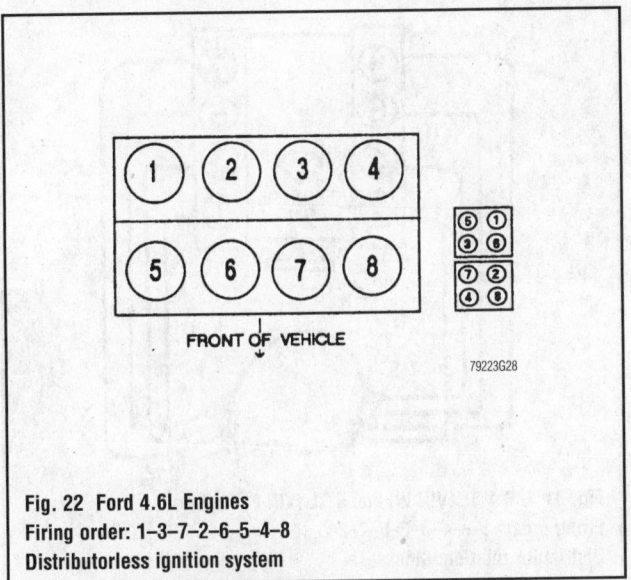

Fig. 22 Ford 4.6L Engines
Firing order: 1–3–7–2–6–5–4–8
Distributorless ignition system

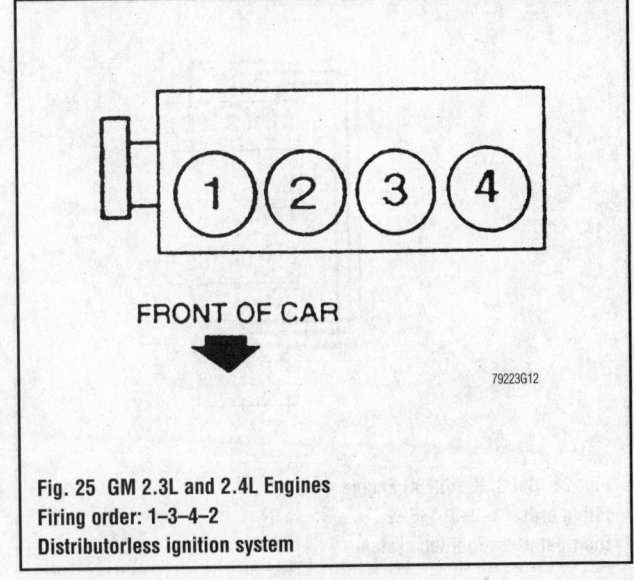

FRONT OF CAR

Fig. 25 GM 2.3L and 2.4L Engines
Firing order: 1–3–4–2
Distributorless ignition system

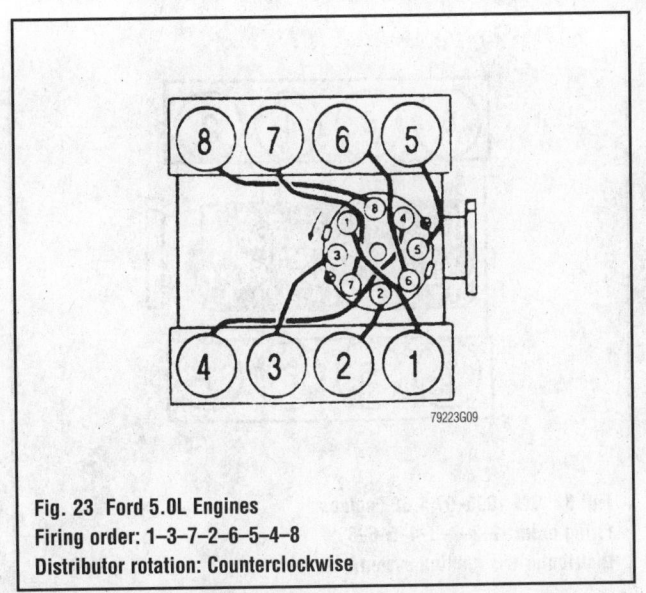

Fig. 23 Ford 5.0L Engines
Firing order: 1–3–7–2–6–5–4–8
Distributor rotation: Counterclockwise

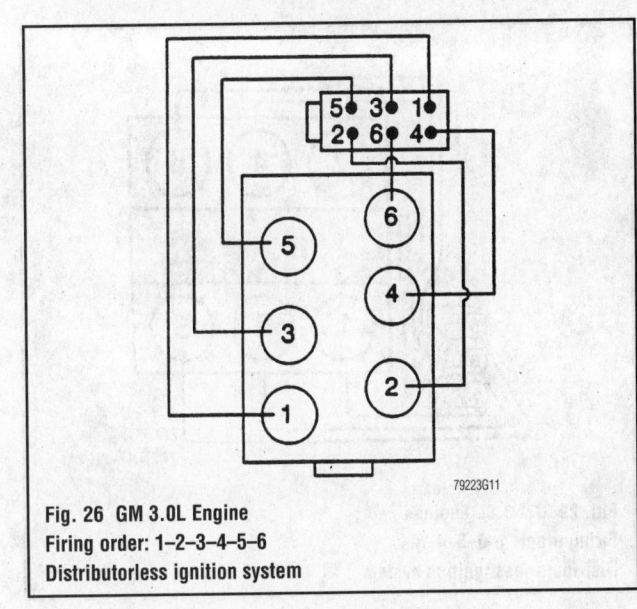

Fig. 26 GM 3.0L Engine
Firing order: 1–2–3–4–5–6
Distributorless ignition system

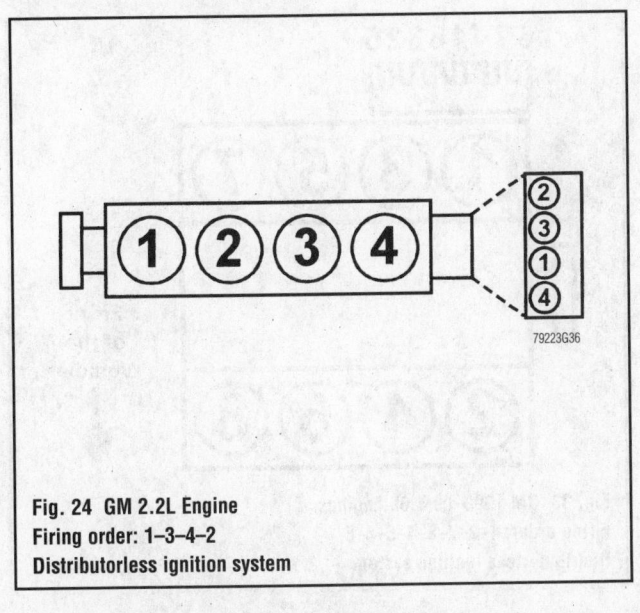

Fig. 24 GM 2.2L Engine
Firing order: 1–3–4–2
Distributorless ignition system

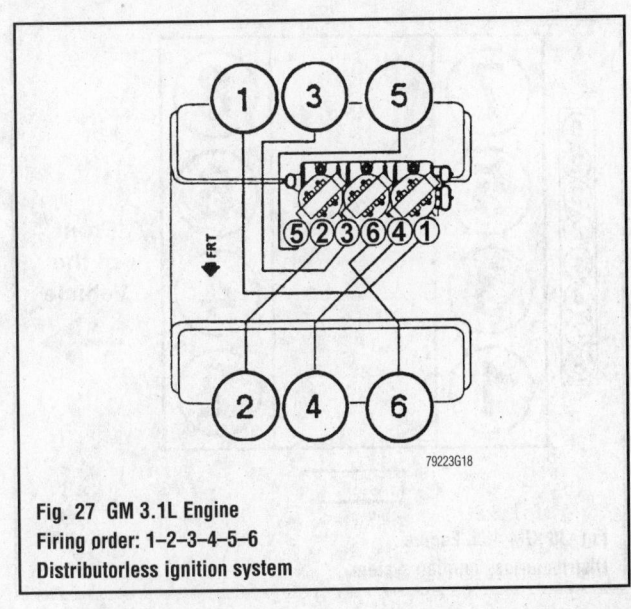

Fig. 27 GM 3.1L Engine
Firing order: 1–2–3–4–5–6
Distributorless ignition system

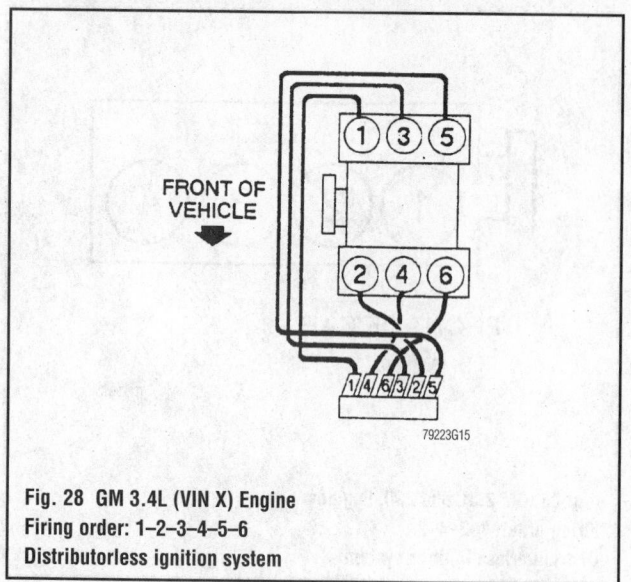

Fig. 28 GM 3.4L (VIN X) Engine
Firing order: 1–2–3–4–5–6
Distributorless ignition system

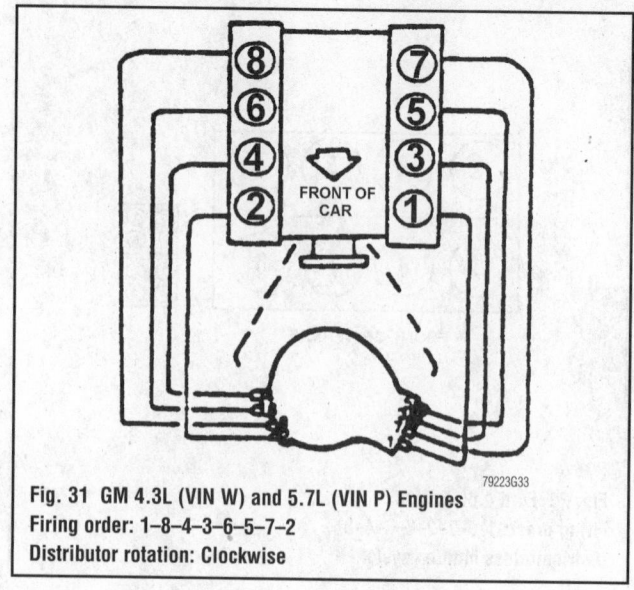

Fig. 31 GM 4.3L (VIN W) and 5.7L (VIN P) Engines
Firing order: 1–8–4–3–6–5–7–2
Distributor rotation: Clockwise

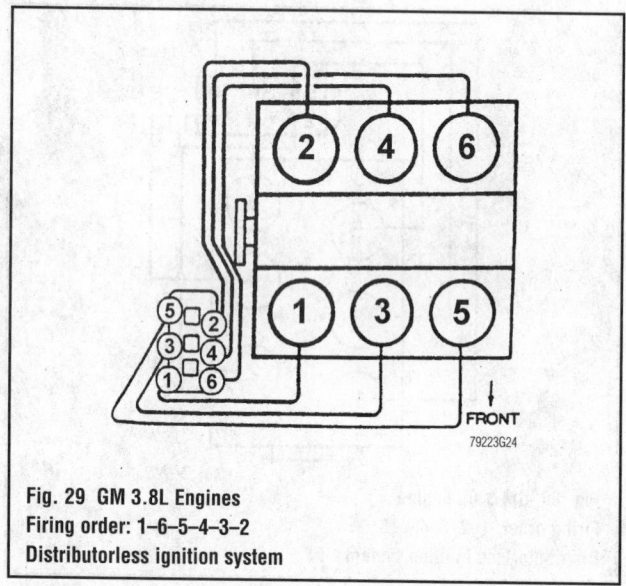

Fig. 29 GM 3.8L Engines
Firing order: 1–6–5–4–3–2
Distributorless ignition system

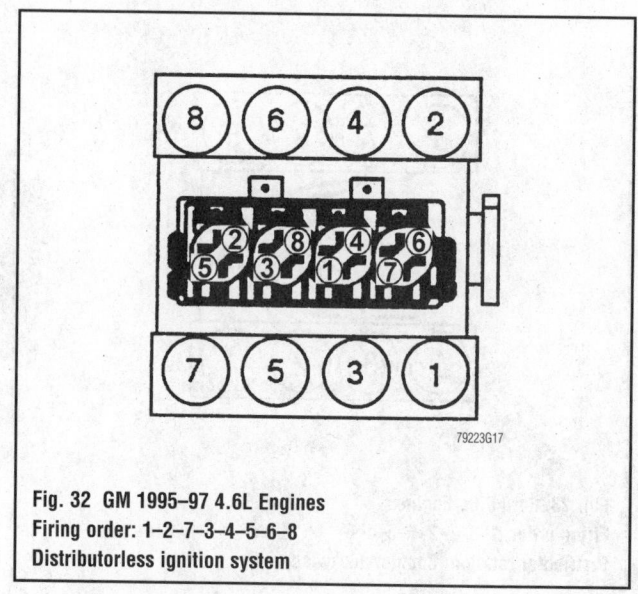

Fig. 32 GM 1995–97 4.6L Engines
Firing order: 1–2–7–3–4–5–6–8
Distributorless ignition system

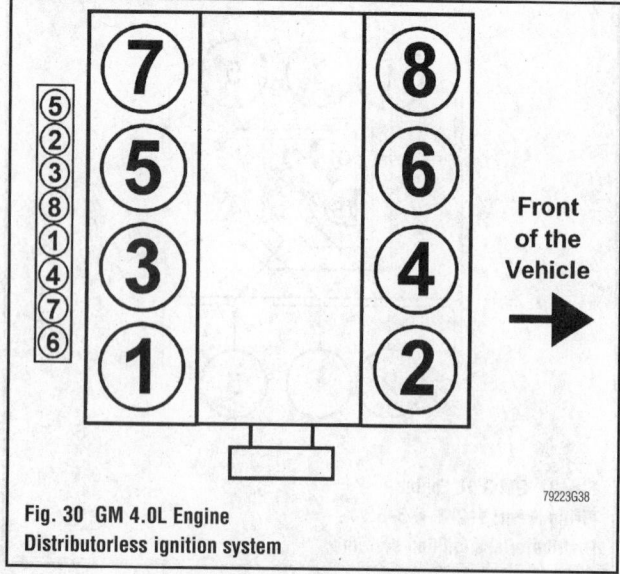

Fig. 30 GM 4.0L Engine
Distributorless ignition system

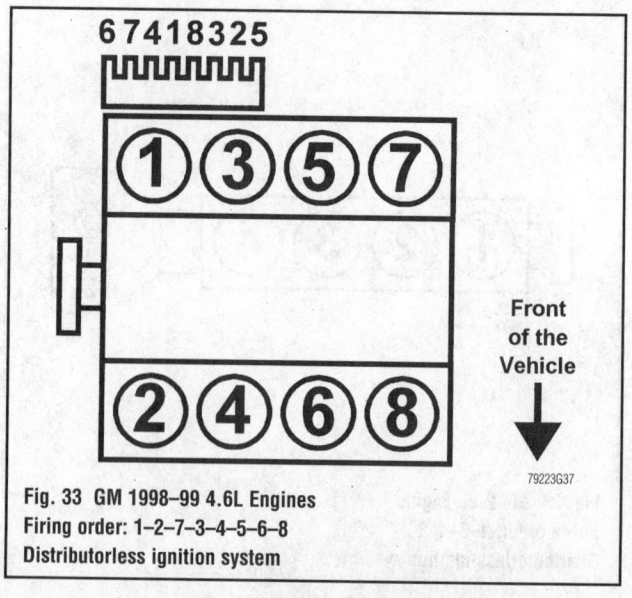

Fig. 33 GM 1998–99 4.6L Engines
Firing order: 1–2–7–3–4–5–6–8
Distributorless ignition system

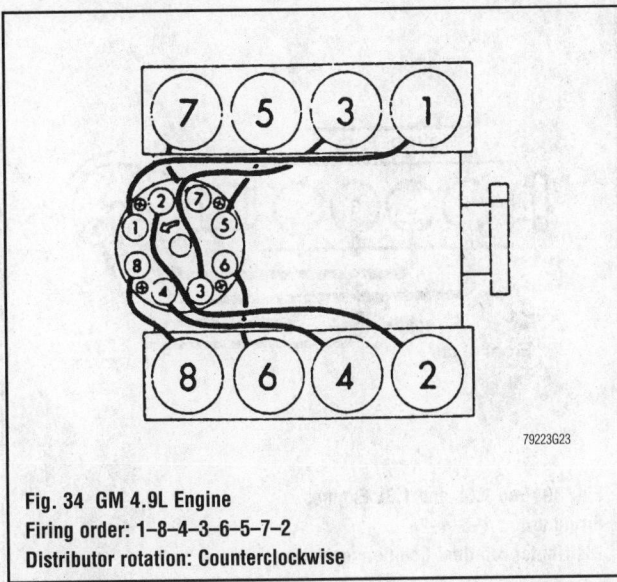

Fig. 34 GM 4.9L Engine
Firing order: 1–8–4–3–6–5–7–2
Distributor rotation: Counterclockwise

79223G23

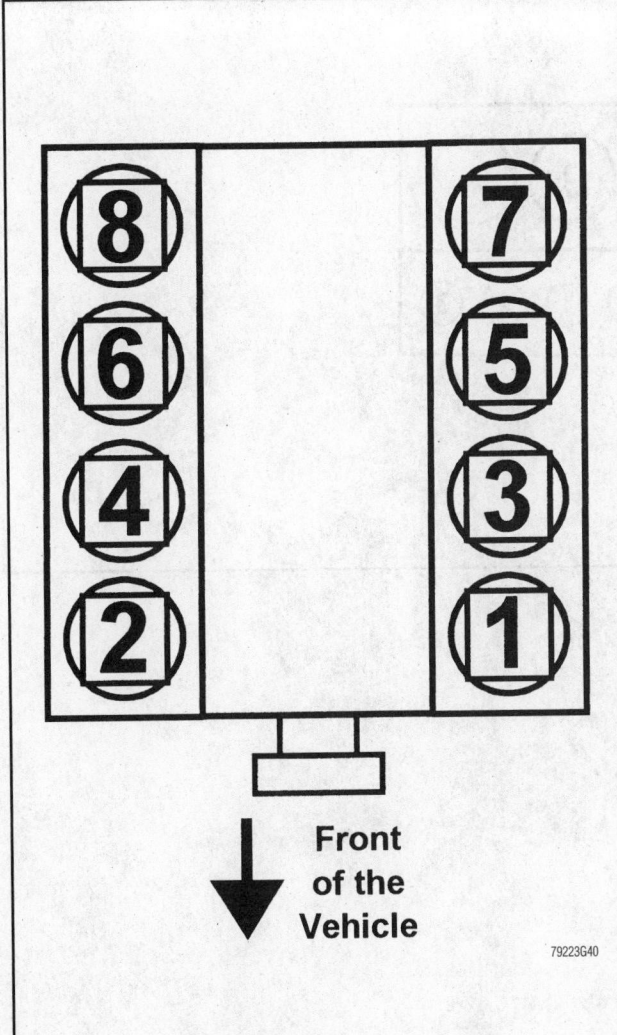

Front of the Vehicle

Fig. 35 GM 5.7L (VIN G) Engines
Firing order: 1–8–4–3–6–5–7–2
Distributorless ignition system (one coil on each cylinder)

79223G40

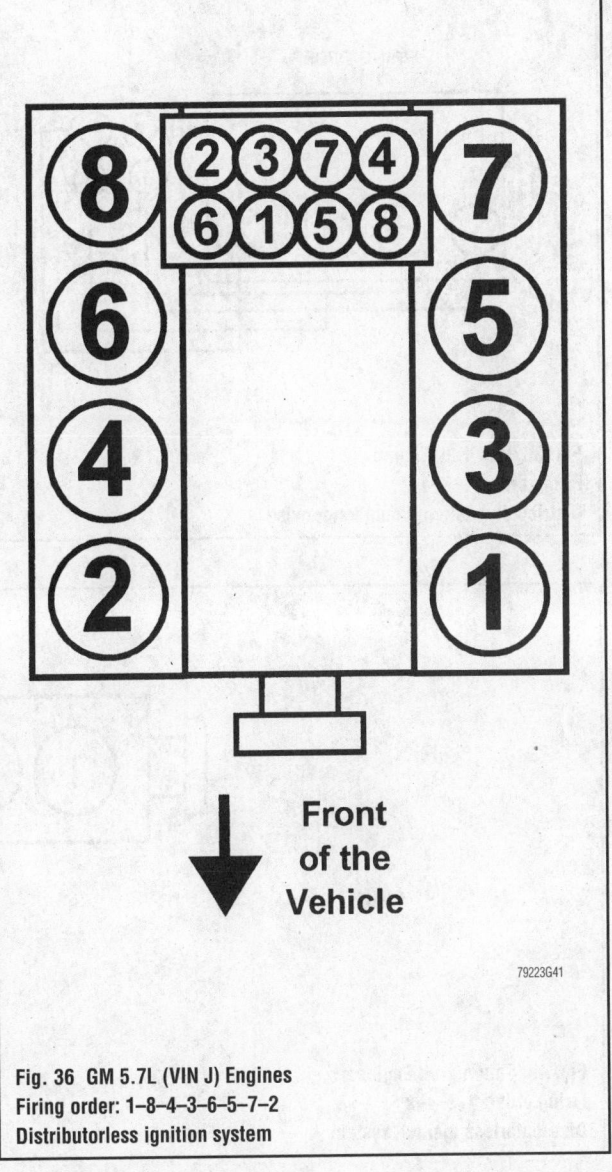

Front of the Vehicle

Fig. 36 GM 5.7L (VIN J) Engines
Firing order: 1–8–4–3–6–5–7–2
Distributorless ignition system

79223G41

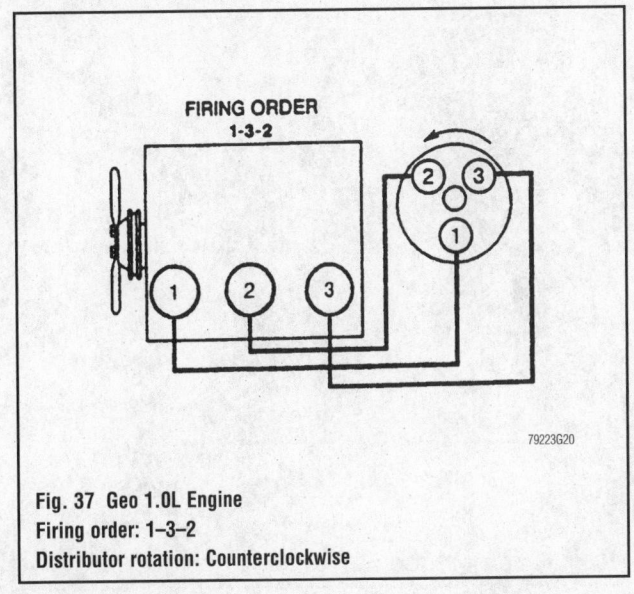

FIRING ORDER
1-3-2

Fig. 37 Geo 1.0L Engine
Firing order: 1–3–2
Distributor rotation: Counterclockwise

79223G20

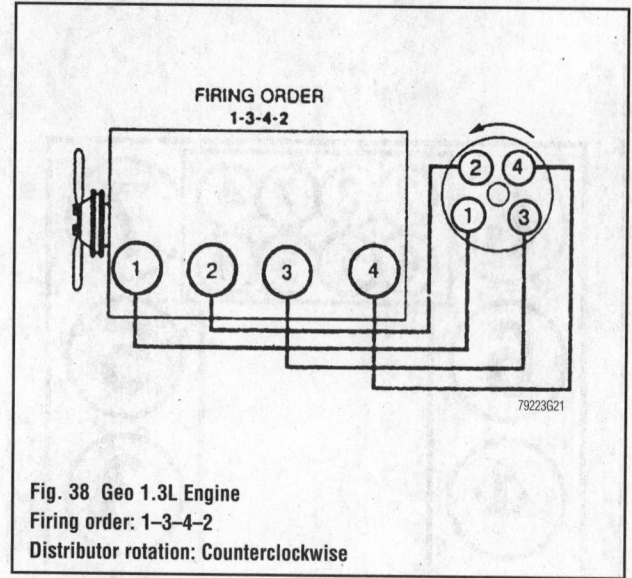

Fig. 38 Geo 1.3L Engine
Firing order: 1–3–4–2
Distributor rotation: Counterclockwise

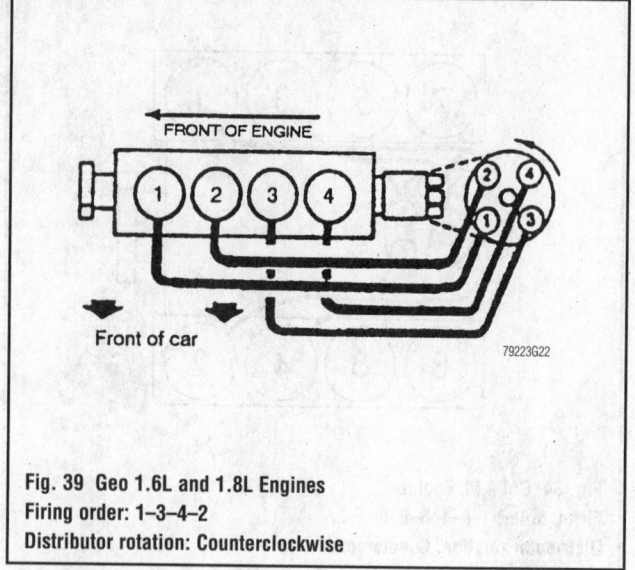

Fig. 39 Geo 1.6L and 1.8L Engines
Firing order: 1–3–4–2
Distributor rotation: Counterclockwise

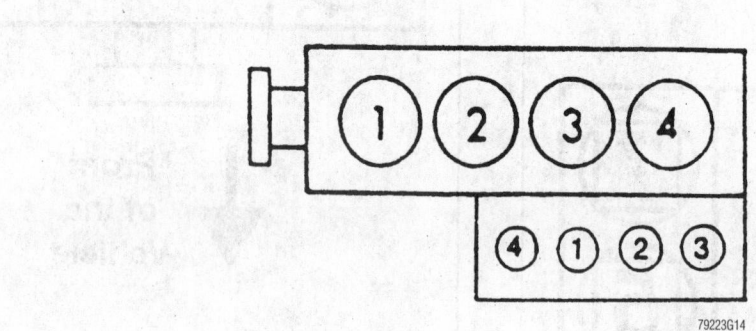

Fig. 40 Saturn 1.9L Engines
Firing order: 1–3–4–2
Distributorless ignition system

ACCESSORY DRIVE BELTS

4

ACCESSORY DRIVE BELTS **4-1**
ACCESSORY DRIVE BELT
 ROUTING INDEX **4-8**
TROUBLESHOOTING THE
 SERPENTINE DRIVE BELT **4-5**

Serpentine Belts.............................4-4
 Adjustment 4-6
 Inspection...................................4-4
 Removal & Installation 4-7

V-Belts..4-1
 Adjustment 4-2
 Inspection...................................4-1
 Removal & Installation 4-4

ACCESSORY DRIVE BELTS

Accessory drive belts are usually divided into two basic types: V-belts (conventional, cogged, and flat multi-ribbed) and serpentine (multi-ribbed) belts. The flat multi-ribbed V-belt actually resembles a serpentine belt, however, unlike a serpentine belt, only the inner surface of the belt makes contact with the components' pulleys. (Rarely, the back of multi-ribbed V-belts may ride against an idler or tensioner pulley, however.) V-belts ride in pulleys with V-shaped groove(s) to rotate various accessories, such as the power steering pump, air conditioner compressor, alternator/generator, water pump, and air pump. Only the inside of a V-belt is used, unlike a serpentine belt which utilizes both sides. V-belts typically operate one or two accessories per belt, whereas a single serpentine belt can drive all of the accessories. V-belts and a few serpentine belts require periodic adjustment because the belts are under tension and stretch over time. Most serpentine belts utilize an automatic belt tensioner that constantly provides the proper tension to the belt.

V-Belts

INSPECTION

Although different maintenance intervals are given by each manufacturer, it is a good rule of thumb to inspect the drive belts every 15,000 miles (24,000 km) or 12 months (whichever occurs first). Determine the belt tension at a point half-way between the pulleys by pressing on the belt with moderate thumb pressure. The belt should deflect about ¼–½ in. (6)–13mm) at this point. Note that "deflection" is not play, but the ability of the belt, under actual tension, to stretch slightly and give.

Inspect the belts for the following signs of damage or wear: glazing, cracking, fraying, crumbling or missing chunks. A glazed belt will be perfectly smooth from slippage, while a good belt will have a slight texture of fabric visible. Cracks will usually start at the inner edge of the belt and run outward. A belt that is fraying will have the fabric backing de-laminating its self from the belt. A belt that is crumbling or missing chunks will have voids in the cross-section of the belt, some times the section missing chunks will be in the pulley groove and not easily seen. All worn or damaged drive belts

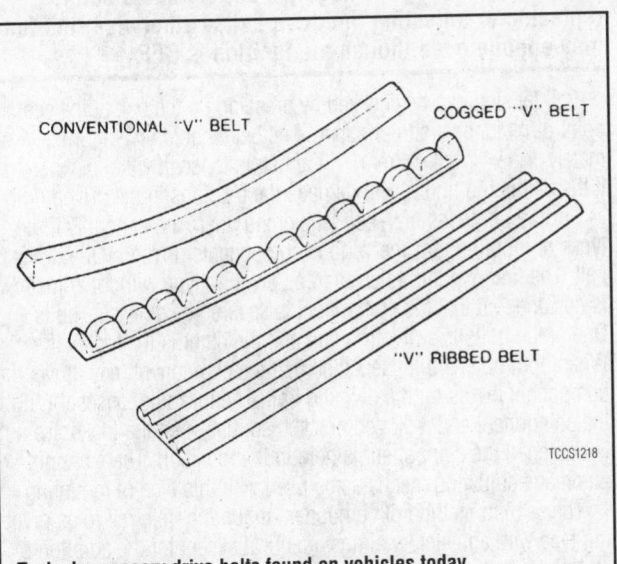

CONVENTIONAL "V" BELT

COGGED "V" BELT

"V" RIBBED BELT

TCCS1218

Typical accessory drive belts found on vehicles today

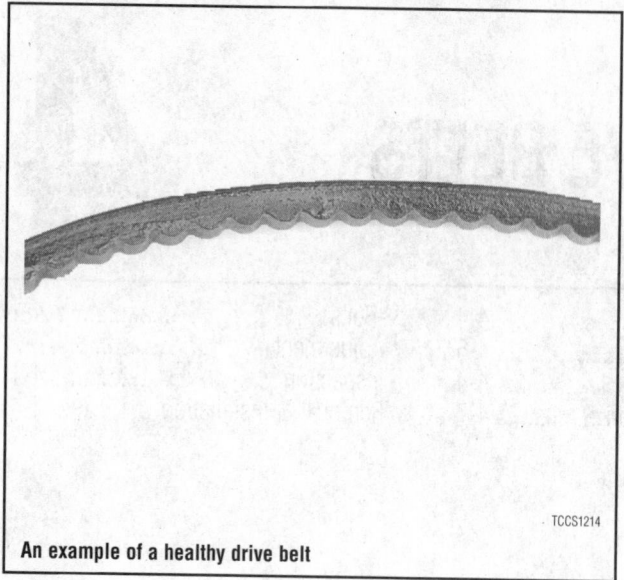

TCCS1214

An example of a healthy drive belt

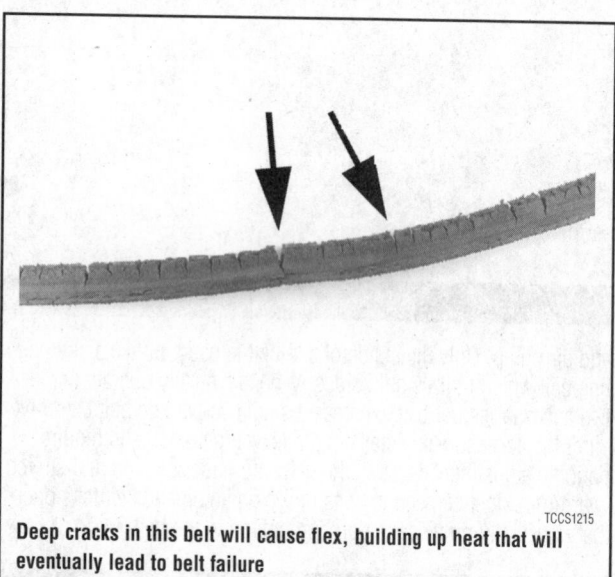

TCCS1215

Deep cracks in this belt will cause flex, building up heat that will eventually lead to belt failure

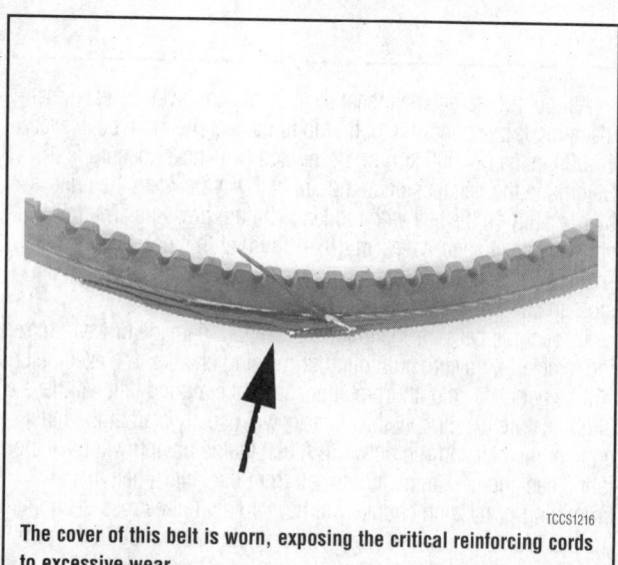

TCCS1216

The cover of this belt is worn, exposing the critical reinforcing cords to excessive wear

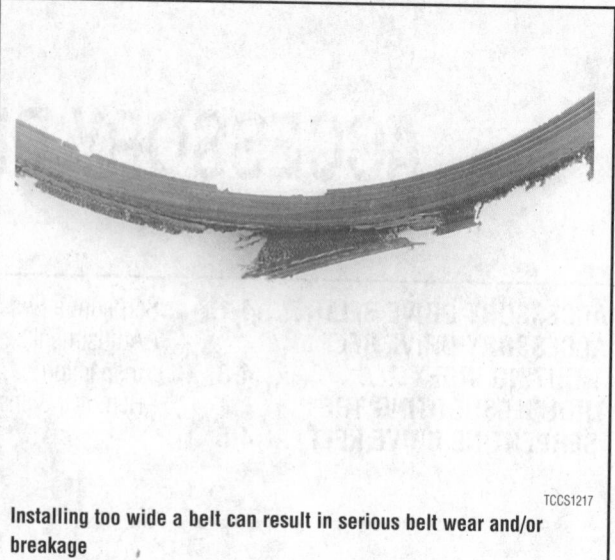

TCCS1217

Installing too wide a belt can result in serious belt wear and/or breakage

should be replaced immediately. It is best to replace all drive belts at one time, as a preventive maintenance measure.

Although it is generally easier on the component to have the belt too loose than too tight, a very loose belt may place a high impact load on a bearing due to the whipping or snapping action of the belt. A belt that is slightly loose may slip, especially when component loads are high. This slippage may be hard to identify. For example, the generator belt may run okay during the day,, then slip at night when headlights are turned on. Slipping belts wear quickly not only due to the direct effect of slippage but also because of the heat the slippage generates. Extreme slippage may even cause a belt to burn. A very smooth, glazed appearance on the belt's sides, as opposed to the obvious pattern of a fabric cover, indicates that the belt has been slipping.

ADJUSTMENT

❈❈ CAUTION

On vehicles with an electric cooling fan, disable the power to the fan by disengaging the fan motor wiring connector or removing the negative battery cable before replacing or adjusting the drive belts. Otherwise, the fan may engage even though the ignition is OFF.

Belt tension can be checked by pressing on the belt at the center point of its longest straight span. The belt should give approximately ¼–½ in. (6–13mm). If the belt is loose it will slip, whereas if the belt is too tight it will damage the bearings in the driven unit.

For the purposes of V-belt tensioning, there are generally three types of mounting for the various components driven by the drive belt. The first method, referred to as pivoting type without adjuster, is designed so that the component is secured by at least 2 bolts. One of the bolts is a pivoting bolt and the other is the lockbolt. When both bolts are loosened so that the component may move, the component pivots on the pivoting bolt. The lockbolt passes through the component and a slotted bracket, so that when the lockbolt's nut is tightened the component is held in that position. There are not automatic adjusting mechanisms used with this type of mounting.

The second method of component mounting, referred to as pivoting type with adjuster, is almost identical except for the addition of an adjuster of some sort. Usually the adjuster is composed of a

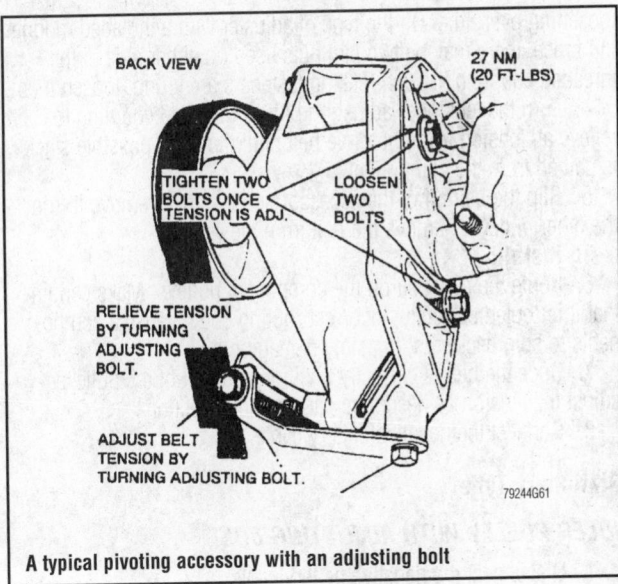

A typical pivoting accessory with an adjusting bolt

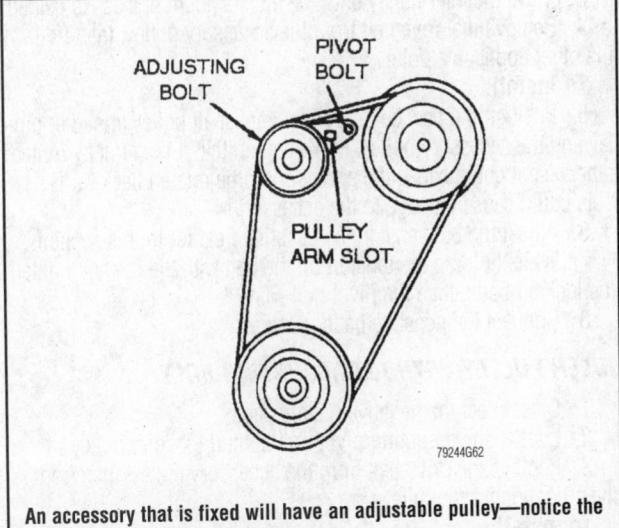

An accessory that is fixed will have an adjustable pulley—notice the square slot to aid the adjustment

bracket attached to the component and a threaded adjusting bolt. After loosening the pivoting and lockbolts, the adjusting bolt can be tightened or loosened to increase or decrease the drive belt's tension. With this type of mounting, you do not have to hold the component in a tensioned position and tighten the pivoting and lockbolts; the adjusting bolt does the job for you.

Some versions of this method of mounting use an adjuster which is built into one of the components mounting braces. The brace attaches the component to the engine and incorporates a threaded adjuster in its mid-span, so that when the threaded adjuster is turned the brace shortens or lengthens. This in turn increases or decreases the amount of tension on the component.

The third type of mounting, referred to as stationary type, is designed so that the component is mounted on its brackets. There are no pivoting or lockbolts, and the component is not designed to be moved. Rather, this type of mounting uses an extra tensioner idler pulley assembly. The drive belt is tensioned by adjusting the position of the idler pulley, usually accomplished by turning the adjuster bolt on the idler mechanism.

Pivoting Type

WITHOUT ADJUSTER

1. Disconnect the negative battery cable.
2. Loosen the component's lockbolt and pivoting bolt only enough for the component to move.
3. Using a strong wooden, plastic or metal prytool, move the component either closer to, or farther away from, the engine to provide the correct tension on the belt.

✳✳ WARNING

If using a metal prytool, always wrap the end with a rag or towel to prevent accidentally damaging the component from undue stress.

4. Once the proper amount of tension is applied to the drive belt, hold the prytool with one hand while tightening the lockbolt securely with the other hand.
5. Release the pressure from the prytool and tighten the pivoting bolt securely.
6. Double check the drive belt's tension, in case the component moved slightly while tightening the bolts.
7. Connect the negative battery cable.

WITH ADJUSTER

This type of drive belt is tensioned by a tensioner, which makes precise tension adjustment easy.
1. Disconnect the negative battery cable.
2. Loosen the component's pivot and lockbolts.
3. Inspect the tensioner assembly on the component; the tensioner adjusting bolt may use a locknut or screw to prevent it from loosening over time. On the type of adjuster with a threaded mounting brace, there may be two jam nuts used on either side of the threaded coupling. If such locking fasteners are found, loosen them.
4. Turn the tensioner adjusting bolt or threaded coupling to increase or decrease the amount of tension on the drive belt, as necessary.
5. When the belt tension is correct, tighten the lockbolt and the pivot bolt.
6. If equipped, tighten the tension adjusting bolt locknut or screw to prevent the adjuster from slowly loosening over time. If equipped, tighten the two jam nuts.
7. Connect the negative battery cable.

Stationary Type

IDLER PULLEY WITH ADJUSTING BOLT

1. Loosen the idler bracket pivot bolt and locking bolts.
2. Adjust the belt tension by inserting the proper size ratchet in the square slot of the idler bracket and rotating the bracket until tension is applied.
3. While holding the tension on the belt with the ratchet, tighten the locking bolts, then the pivot bolt.

IDLER PULLEY WITHOUT ADJUSTING BOLT

1. Loosen the mounting/pivot bolt behind the idler pulley.
2. Swivel the idler pulley with a pair of pliers or a wrench on the bearing mounting until the proper tension is achieved.
3. While holding the idler pulley, at the proper tension, tighten the mounting/pivot bolt.

REMOVAL & INSTALLATION

If a belt must be replaced, the driven unit or idler pulley must be loosened and moved to its extreme loosest position, generally by moving it toward the center of the motor. After removing the old belt, check the pulleys for dirt or built-up material which could affect belt contact. Carefully install the new belt, remembering that it is new and unused; it may appear to be just a little too small to fit over the pulley flanges. Fit the belt over the largest pulley (usually the crankshaft pulley at the bottom center of the motor) first, then work on the smaller one(s). Gentle pressure in the direction of rotation is helpful. Some belts run around a third, or idler pulley, which acts as an additional pivot in the belt's path. It may be possible to loosen the idler pulley as well as the main component, making your job much easier. Depending on which belt(s) you are changing, it may be necessary to loosen or remove other interfering belts to get at the one(s) you want.

When buying replacement belts, remember that the fit is critical according to the length of the belt ("diameter"), the width of the belt, the depth of the belt and the angle or profile of the V shape or the ribs. The belt shape should match the shape of the pulley exactly; belts that are not an exact match can cause noise, slippage and premature failure.

After the new belt is installed, draw tension on it by moving the driven unit or idler pulley away from the motor and tighten its mounting bolts. This is sometimes a three or four-handed job; you may find an assistant helpful. Be sure that all the bolts you loosened get retightened and that any other loosened belts also have the correct tension. A new belt can be expected to stretch a bit after installation so be prepared to readjust your new belt, if needed, within the first two hundred miles of use.

Pivoting Type

> **❋❋ CAUTION**
>
> **On vehicles with an electric cooling fan, disable the power to the fan by disengaging the fan motor wiring connector or removing the negative battery cable before replacing or adjusting the drive belts. Otherwise, the fan may engage even though the ignition is OFF.**

WITHOUT ADJUSTER

1. Disconnect the negative battery cable.
2. Loosen the accessory's slotted adjusting bracket bolt. If the hinge bolt is excessively tight, it too will have to be loosened.
3. Push the component toward the engine to provide enough slack in the belt so that it will slide over one of the accessory drive pulleys. Remove the drive belt from the accessory drive pulleys and from the vehicle.

To install:

4. Position the new drive belt over the component pulleys. Be sure that it is routed correctly.
5. Adjust the tension of the belt, as described earlier in this section.
6. Connect the negative battery cable.

WITH ADJUSTER

1. Disconnect the negative battery cable.
2. Loosen the component's pivot and lockbolts.
3. Inspect the tensioner assembly on the component; the tensioner adjusting bolt may use a locknut or screw to prevent it from

loosening over time. On the type of adjuster with a threaded mounting brace, there may be two jam nuts used on either side of the threaded coupling. If such locking fasteners are found, loosen them.

4. Turn the tensioner adjusting bolt or threaded coupling to relieve all tension from the drive belt until the most possible slack is gained from the component.
5. Slip the belt off of the accessory pulley, then remove it from the other pulleys. Remove the belt from the vehicle.

To install:

6. Route the new belt on the component pulleys. Make certain that it is routed correctly; incorrect routing could cause a components to spin backward, possibly damaging it.
7. Once the belt is correctly positioned on all of the pulleys, adjust the tension as described earlier in this section.
8. Connect the negative battery cable.

Stationary Type

IDLER PULLEY WITH ADJUSTING BOLT

1. Disconnect the negative battery cable.
2. Loosen the idler bracket pivot bolt and locking bolts.
3. Move the idler pulley until the most amount of slack is gained.
4. Remove the drive belt from the accessory pulley, then from the other applicable pulleys.

To install:

5. Position the new belt over the crankshaft pulley, the idler pulley and the accessory pulley. Make certain that it is correctly routed, otherwise it could cause the accessory to be rotated backwards. This could cause damage to the accessory.
6. Adjust the belt tension, as described earlier in this section.
7. While holding the tension on the belt with the ratchet, tighten the locking bolts, then the pivot bolt.
8. Connect the negative battery cable.

IDLER PULLEY WITHOUT ADJUSTING BOLT

1. Disconnect the negative battery cable.
2. Loosen the mounting/pivot bolt behind the idler pulley.
3. Remove the drive belt from the accessory pulley, then from the other applicable pulleys.

To install:

4. Position the new belt over the crankshaft pulley, the idler pulley and the accessory pulley. Make certain that it is correctly routed, otherwise it could cause the accessory to be rotated backwards. This could cause damage to the accessory.
5. Swivel the idler pulley with a pair of pliers or a wrench on the bearing mounting until the proper tension is achieved.
6. While holding the idler pulley, at the proper tension, tighten the mounting/pivot bolt.
7. Connect the negative battery cable.

Serpentine Belts

INSPECTION

Although many manufacturers recommend that the drive belt(s) be inspected every 30,000 miles (48,000 km) or more, it is really a good idea to check them at least once a year, or at every major fluid change. Whichever interval you choose, the belts should be checked for wear or damage. Obviously, a damaged drive belt can cause problems should it give way while the vehicle is in operation. But, improper length belts (too short or long), as well as excessively

Troubleshooting the Serpentine Drive Belt

Problem	Cause	Solution
Tension sheeting fabric failure (woven fabric on outside circumference of belt has cracked or separated from body of belt)	• Grooved or backside idler pulley diameters are less than minimum recommended • Tension sheeting contacting (rubbing) stationary object • Excessive heat causing woven fabric to age • Tension sheeting splice has fractured	• Replace pulley(s) not conforming to specification • Correct rubbing condition • Replace belt • Replace belt
Noise (objectional squeal, squeak, or rumble is heard or felt while drive belt is in operation)	• Belt slippage • Bearing noise • Belt misalignment • Belt-to-pulley mismatch • Driven component inducing vibration • System resonant frequency inducing vibration	• Adjust belt • Locate and repair • Align belt/pulley(s) • Install correct belt • Locate defective driven component and repair • Vary belt tension within specifications. Replace belt.
Rib chunking (one or more ribs has separated from belt body)	• Foreign objects imbedded in pulley grooves • Installation damage • Drive loads in excess of design specifications • Insufficient internal belt adhesion	• Remove foreign objects from pulley grooves • Replace belt • Adjust belt tension • Replace belt
Rib or belt wear (belt ribs contact bottom of pulley grooves)	• Pulley(s) misaligned • Mismatch of belt and pulley groove widths • Abrasive environment • Rusted pulley(s) • Sharp or jagged pulley groove tips • Rubber deteriorated	• Align pulley(s) • Replace belt • Replace belt • Clean rust from pulley(s) • Replace pulley • Replace belt
Longitudinal belt cracking (cracks between two ribs)	• Belt has mistracked from pulley groove • Pulley groove tip has worn away rubber-to-tensile member	• Replace belt • Replace belt
Belt slips	• Belt slipping because of insufficient tension • Belt or pulley subjected to substance (belt dressing, oil, ethylene glycol) that has reduced friction • Driven component bearing failure • Belt glazed and hardened from heat and excessive slippage	• Adjust tension • Replace belt and clean pulleys • Replace faulty component bearing • Replace belt
"Groove jumping" (belt does not maintain correct position on pulley, or turns over and/or runs off pulleys)	• Insufficient belt tension • Pulley(s) not within design tolerance • Foreign object(s) in grooves	• Adjust belt tension • Replace pulley(s) • Remove foreign objects from grooves

Troubleshooting the Serpentine Drive Belt

Problem	Cause	Solution
"Groove jumping" (belt does not maintain correct position on pulley, or turns over and/or runs off pulleys)	• Excessive belt speed • Pulley misalignment • Belt-to-pulley profile mismatched • Belt cordline is distorted	• Avoid excessive engine acceleration • Align pulley(s) • Install correct belt • Replace belt
Belt broken (Note: identify and correct problem before replacement belt is installed)	• Excessive tension • Tensile members damaged during belt installation • Belt turnover • Severe pulley misalignment • Bracket, pulley, or bearing failure	• Replace belt and adjust tension to specification • Replace belt • Replace belt • Align pulley(s) • Replace defective component and belt
Cord edge failure (tensile member exposed at edges of belt or separated from belt body)	• Excessive tension • Drive pulley misalignment • Belt contacting stationary object • Pulley irregularities • Improper pulley construction • Insufficient adhesion between tensile member and rubber matrix	• Adjust belt tension • Align pulley • Correct as necessary • Replace pulley • Replace pulley • Replace belt and adjust tension to specifications
Sporadic rib cracking (multiple cracks in belt ribs at random intervals)	• Ribbed pulley(s) diameter less than minimum specification • Backside bend flat pulley(s) diameter less than minimum • Excessive heat condition causing rubber to harden • Excessive belt thickness • Belt overcured • Excessive tension	• Replace pulley(s) • Replace pulley(s) • Correct heat condition as necessary • Replace belt • Replace belt • Adjust belt tension

TCCS3C10

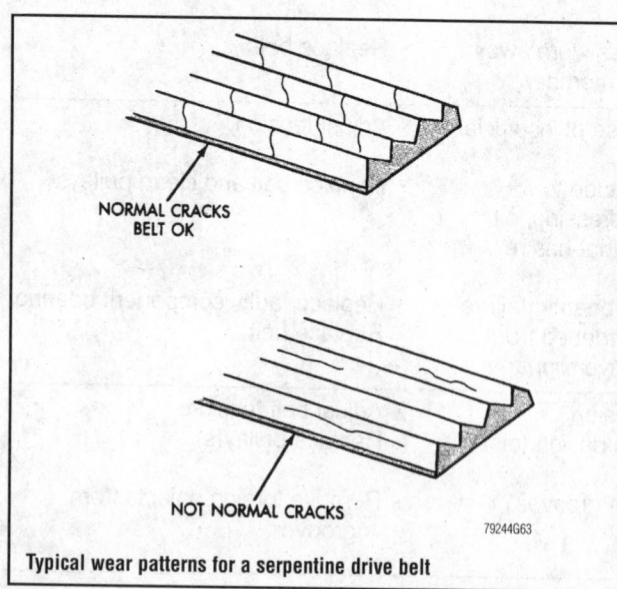

NORMAL CRACKS
BELT OK

NOT NORMAL CRACKS

79244G63

Typical wear patterns for a serpentine drive belt

worn belts, can also cause problems. Loose accessory drive belts can lead to poor engine cooling and diminished output from the alternator, air conditioning compressor or power steering pump. A belt that is too tight places a severe strain on the driven unit and can wear out bearings quickly.

Serpentine drive belts should be inspected for rib chunking (pieces of the ribs breaking off), severe glazing, frayed cords or other visible damage. Any belt which is missing sections of 2 or more adjacent ribs which are ½ in. (13mm) or longer must be replaced. You might want to note that serpentine belts do tend to form small cracks across the backing. If the only wear you find is in the form of one or more cracks are across the backing and NOT parallel to the ribs, the belt is still good and does not need to be replaced.

ADJUSTMENT

Periodic drive belt tensioning is not necessary, because an automatic spring-loaded tensioner is used with these belts to maintain proper adjustment at all times. The tensioner is also useful as a

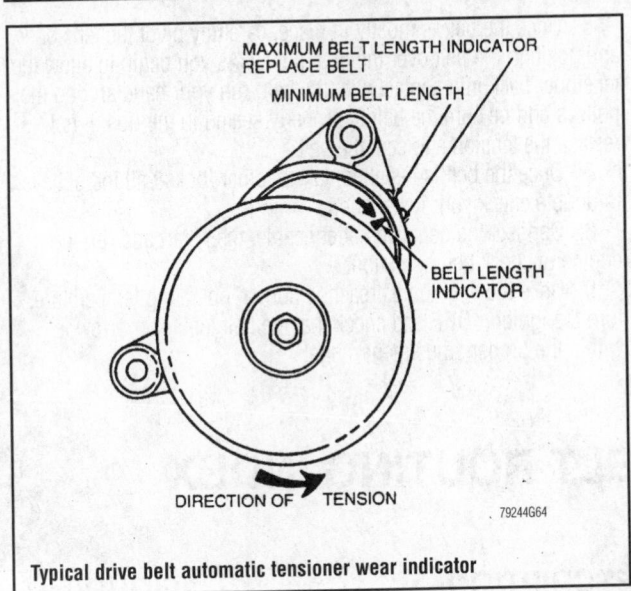

Typical drive belt automatic tensioner wear indicator

Often the underhood label will display the serpentine drive belt routing

wear indicator. When the belt is properly installed, the arrow on the tensioner housing must point within the acceptable range lines on the tensioner's face. If the arrow falls outside the range, either an improper belt has been installed or the belt is worn beyond its useful life span. In either case, a new belt must be installed immediately to assure proper engine operation and to prevent possible accessory damage.

REMOVAL & INSTALLATION

Because serpentine belts use a spring loaded tensioner for adjustment, belt replacement tends to be somewhat easier than it used to be on engines where accessories were pivoted and bolted in place for tension adjustment. Basically, all belt replacement involves is to pivot the tensioner to loosen the belt, then slide the belt off of the pulleys. The two most important points are to pay CLOSE attention to the proper belt routing (since serpentine belts tend to be "snaked" all different ways through the pulleys) and to be sure the V-ribs are properly seated in all the pulleys.

Although belt routing diagrams have been included in this section, the first places you should check for proper belt routing are the labels in your engine compartment. These should include a belt routing diagram which may reflect changes made during a production run.

1. Disconnect the negative battery cable for safety. This will help assure that no one mistakenly cranks the engine over with your hands between the pulleys, and that the cooling fan cannot activate while servicing the belt(s).

➡️**Take a good look at the installed belt and make a note of the routing. Before removing the belt, be sure the routing matches that of the belt routing label or one of the diagrams in this book. If for some reason a diagram does not match (you may not have the original engine or it may have been modified), carefully note the changes on a piece of paper.**

2. For tensioners equipped with a ½ in. (13mm) square hole, insert the drive end of a large breaker bar into the hole. Use the breaker bar to pivot the tensioner away from the drive belt. For tensioners not equipped with this hole, use the proper-sized socket and breaker bar (or a large handled wrench) on the tensioner idler pulley center bolt to pivot the tensioner away from the belt. This will

Relieve the belt tension by pivoting the automatic tensioner away from the belt, then remove the belt

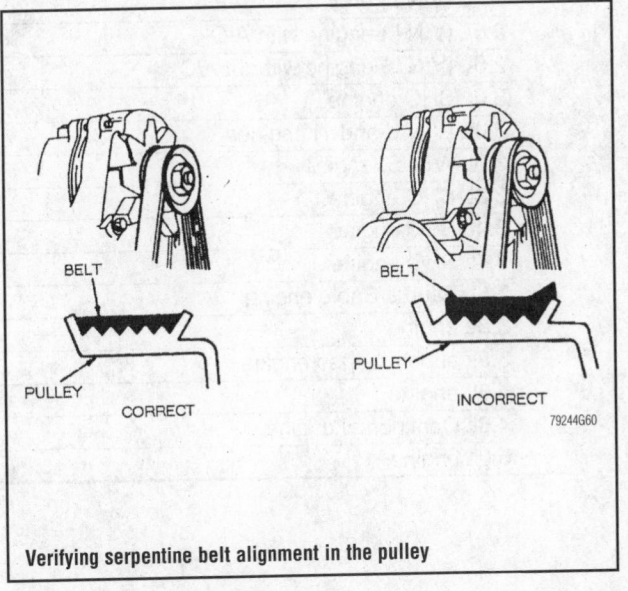

Verifying serpentine belt alignment in the pulley

loosen the belt sufficiently that it can be pulled off of one or more of the pulleys. It is usually easiest to carefully pull the belt out from underneath the tensioner pulley itself.

3. Once the belt is off one of the pulleys, gently pivot the tensioner back into position. DO NOT allow the tensioner to snap back, as this could damage the tensioner's internal parts.

4. Now finish removing the belt from the other pulleys and remove it from the engine.

To install:

5. While referring to the proper routing diagram (which you identified earlier), begin to route the belt over the pulleys, leaving whichever pulley you first released it from for last.

6. Once the belt is mostly in place, carefully pivot the tensioner and position the belt over the final pulley. As you begin to allow the tensioner back into contact with the belt, run your hand around the pulleys and be sure the belt is properly seated in the ribs. If not, release the tension and seat the belt.

7. Once the belt is installed, take another look at all the pulleys to double check your installation.

8. Connect the negative battery cable, then start and run the engine to check belt operation.

9. Once the engine has reached normal operating temperature, turn the ignition **OFF** and check that the belt tensioner arrow is within the proper adjustment range.

ACCESSORY DRIVE BELT ROUTING INDEX

MANUFACTURER

ENGINES	DESCRIPTION	FIGURE
Chrysler Corp.		
2.0L (VIN C, Y) Neon engine	Accessory drive belt routing	1
2.0L (VIN Y) except Neon engine	Accessory drive belt routing	2
2.0L (VIN F) turbo engine	Accessory drive belt routing	3
2.4L engine	Accessory drive belt routing	4
2.5L engine	Accessory drive belt routing	5
2.7L engine	Accessory drive belt routing	6
3.2 L engine	Accessory drive belt routing	7
3.3L engine	Accessory drive belt routing	8
3.5L engine	Accessory drive belt routing	9
Ford Motor Co.		
1.3L Aspire engine	Accessory drive belt routing	10
1.8L engine	Accessory drive belt routing	11
1.9L engine	Serpentine accessory drive belt routing	12
2.0L Probe engine	Accessory drive belt routing	13
2.0L (VIN 3) engine	Accessory drive belt routing	14
2.5L (VIN B) Probe engine without A/C	Alternator drive belt routing	15
2.5L (VIN B) Probe engine with A/C	Alternator drive belt routing	16
2.5L (VIN B) Probe engine	Power steering and water pump drive belts	17
2.5L (VIN L) engine	Water pump drive belt routing	18
2.5L (VIN L) engine with A/C	Accessory drive belt routing	19
2.5L (VIN L) engine without A/C	Accessory drive belt routing	20
3.0L SHO engine	Serpentine drive belt routing	21
3.0L (VIN U and 1) engines	Serpentine drive belt routing	22
3.0L (VIN S) engine	Serpentine drive belt routing	23
3.2L SHO engine	Serpentine drive belt routing	24
3.4L SHO engine	Serpentine drive belt routing	25
3.4L SHO engine	Water pump drive belt routing	26
3.8L Taurus/Sable engine	Serpentine drive belt routing	27
3.8L engine	Accessory drive belt routing	28
3.8L supercharged engine	Accessory drive belt routing	29
4.6L engine	Accessory drive belt routing	30
4.6L Continental engine	Serpentine drive belt routing	31
5.0L engine	Accessory drive belt routing	32

ACCESSORY DRIVE BELT ROUTING INDEX

MANUFACTURER ENGINES	DESCRIPTION	FIGURE
General Motors		
2.2L engines		
A body models	Serpentine drive belt routing	33
Except A body models		
1995-97 engines	Serpentine drive belt routing	34
1998-99 engines with A/C	Serpentine drive belt routing	35
1998-99 engines without A/C	Serpentine drive belt routing	36
2.3L engines	Serpentine drive belt routing	37
2.4L engines	Serpentine drive belt routing	37
3.0L engine	Serpentine drive belt routing	38
3.1L engines		
A body models	Serpentine drive belt routing	39
Except A and L/N body models	Serpentine drive belt routing	40
L/N body models	Serpentine drive belt routing	41
3.4L engines		
Except F body models	Serpentine drive belt routing	42
F body models	Serpentine drive belt routing	43
3.8L engines		
1995 C & H bodies 3.8L (VIN L/K) engines	Serpentine drive belt routing	44
1996-99 C & H bodies 3.8L (VIN K) engines	Serpentine drive belt routing	45
C & H bodies 3.8L (VIN 1) engines	Serpentine drive belt routing	46
F body models	Serpentine drive belt routing	47
G body 3.8L (VIN 1) engines	Serpentine drive belt routing	48
G body 3.8L (VIN K) engines	Serpentine drive belt routing	49
W body models	Serpentine drive belt routing	50
4.0L engines	Serpentine drive belt routing	51
4.3L engines	Serpentine drive belt routing	52
4.6L engines	Serpentine drive belt routing	53
4.9L engines	Serpentine drive belt routing	54
5.7L engines		
5.7L (VIN P) except B or F body engines	Serpentine drive belt routing	55
5.7L (VIN J) except B or F body engines	Serpentine drive belt routing	56
5.7L (VIN G) except B or F body engines	Serpentine drive belt routing	57
5.7L B body engines	Serpentine drive belt routing	52
5.7L 1995-97 F body engines	Serpentine drive belt routing	58
5.7L 1998-99 F body engines	Serpentine drive belt routing	59
5.7L 1998-99 F body engines	A/C drive belt routing	60
Geo/Chevrolet		
All engines	Accessory drive belt	61
Saturn		
All engines	Serpentine drive belt routing	62

79224C02

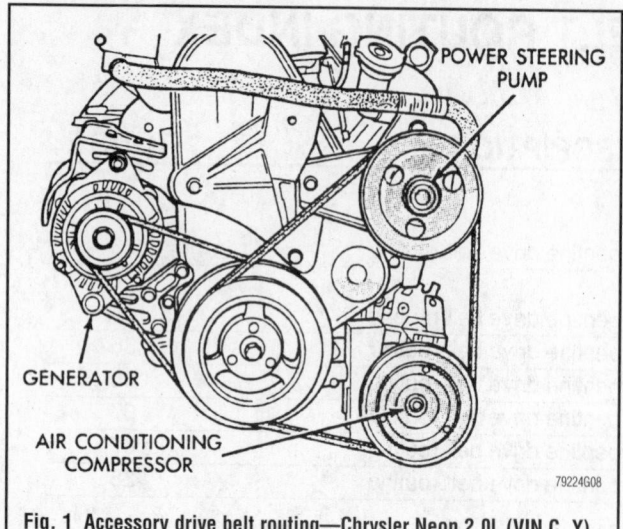

Fig. 1 Accessory drive belt routing—Chrysler Neon 2.0L (VIN C, Y) engine

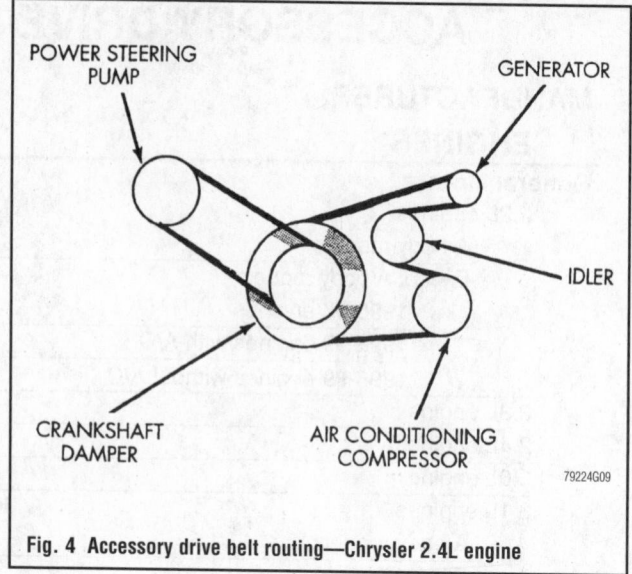

Fig. 4 Accessory drive belt routing—Chrysler 2.4L engine

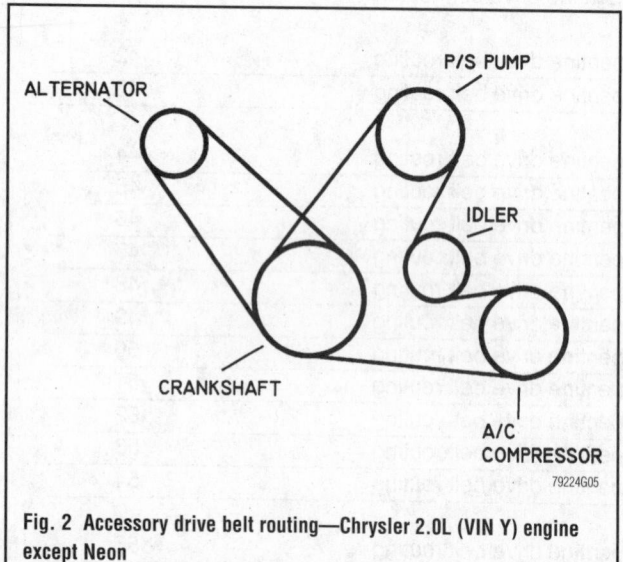

Fig. 2 Accessory drive belt routing—Chrysler 2.0L (VIN Y) engine except Neon

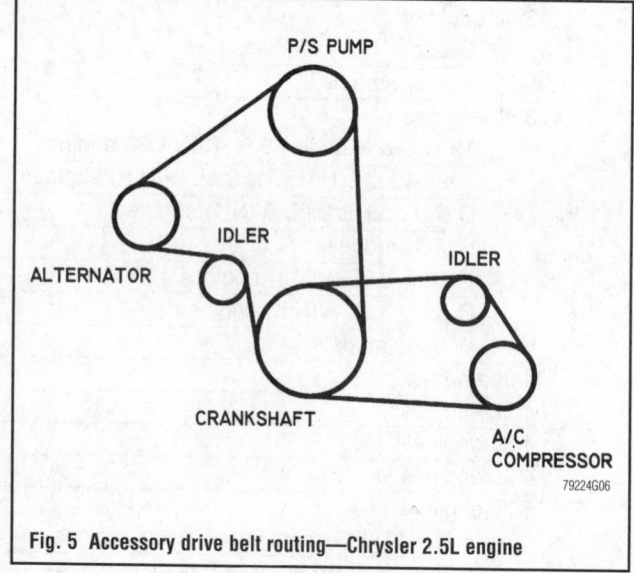

Fig. 5 Accessory drive belt routing—Chrysler 2.5L engine

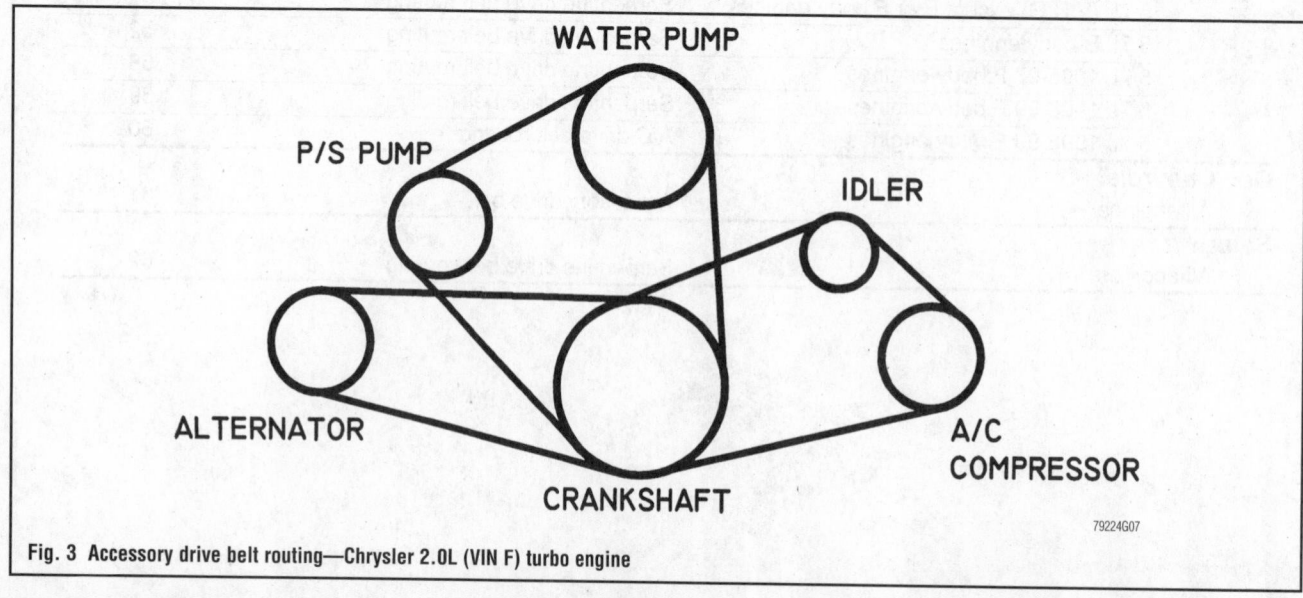

Fig. 3 Accessory drive belt routing—Chrysler 2.0L (VIN F) turbo engine

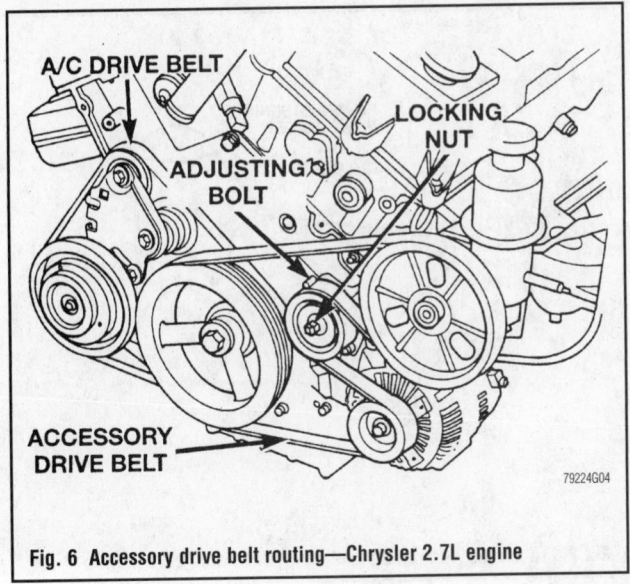

Fig. 6 Accessory drive belt routing—Chrysler 2.7L engine

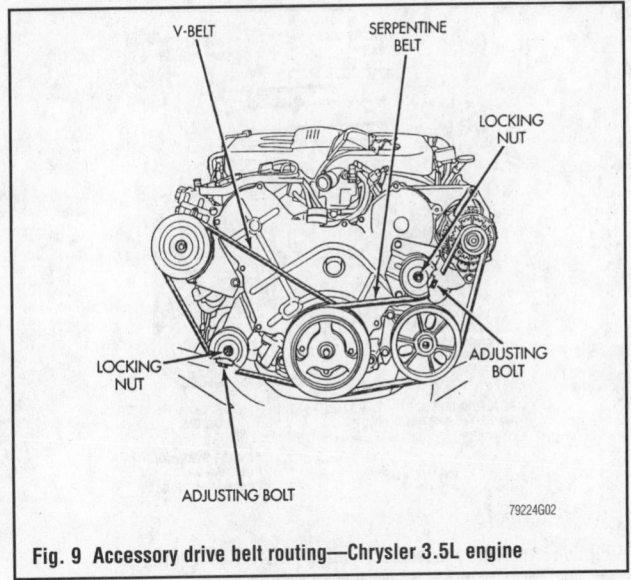

Fig. 9 Accessory drive belt routing—Chrysler 3.5L engine

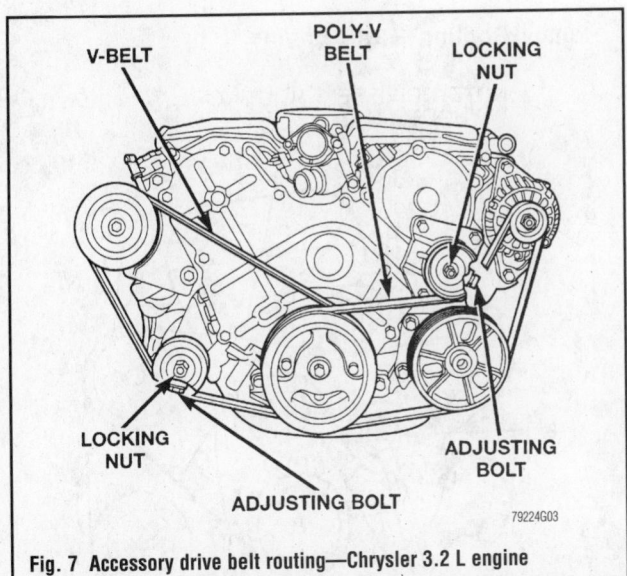

Fig. 7 Accessory drive belt routing—Chrysler 3.2 L engine

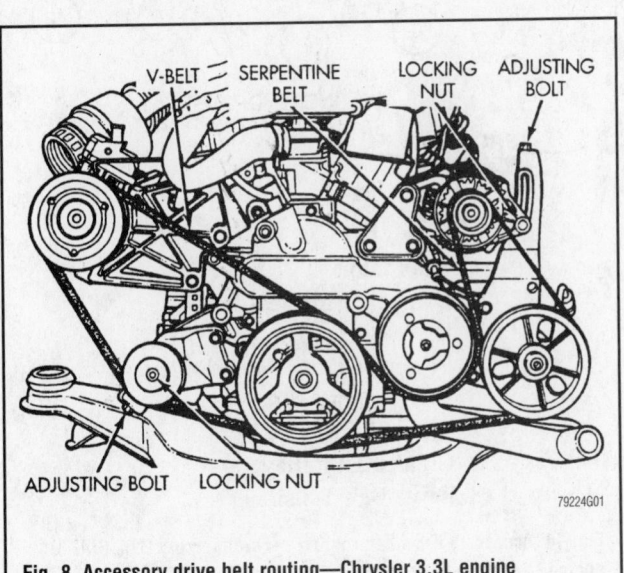

Fig. 8 Accessory drive belt routing—Chrysler 3.3L engine

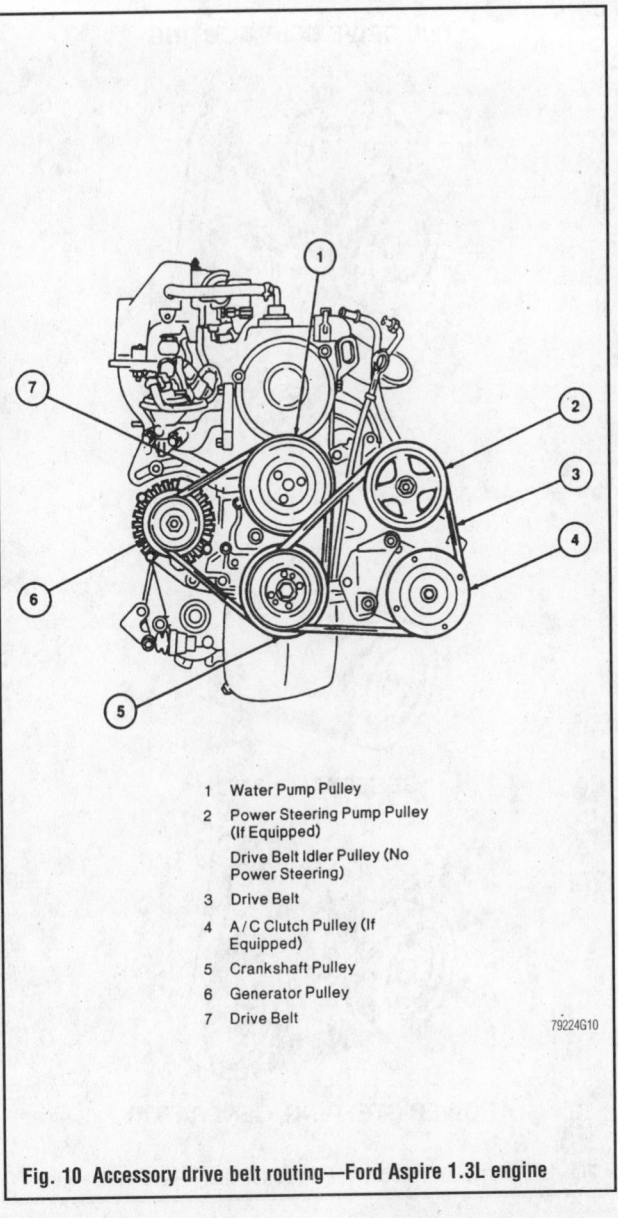

1   Water Pump Pulley

2   Power Steering Pump Pulley
    (If Equipped)

    Drive Belt Idler Pulley (No
    Power Steering)

3   Drive Belt

4   A / C Clutch Pulley (If
    Equipped)

5   Crankshaft Pulley

6   Generator Pulley

7   Drive Belt

Fig. 10 Accessory drive belt routing—Ford Aspire 1.3L engine

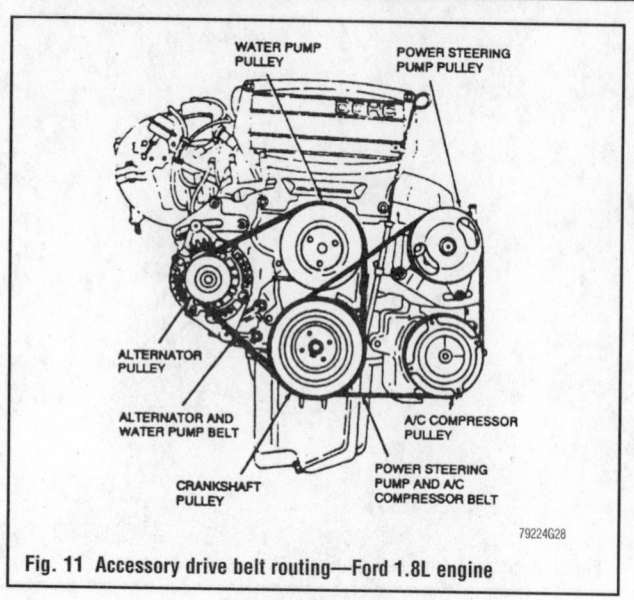

Fig. 11 Accessory drive belt routing—Ford 1.8L engine

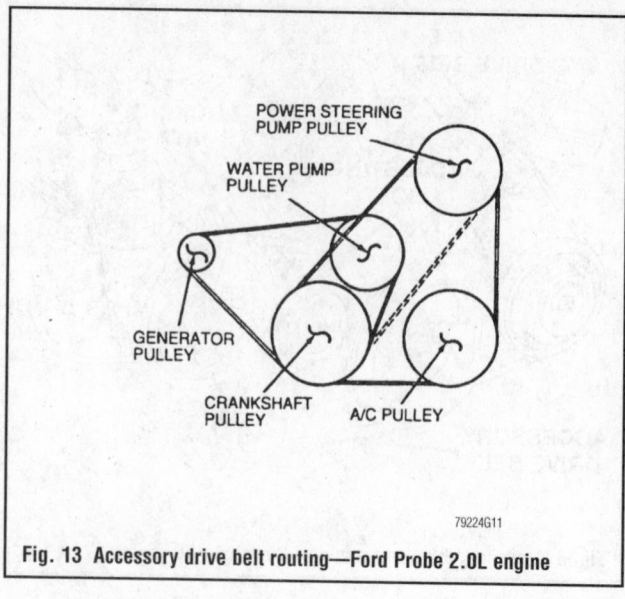

Fig. 13 Accessory drive belt routing—Ford Probe 2.0L engine

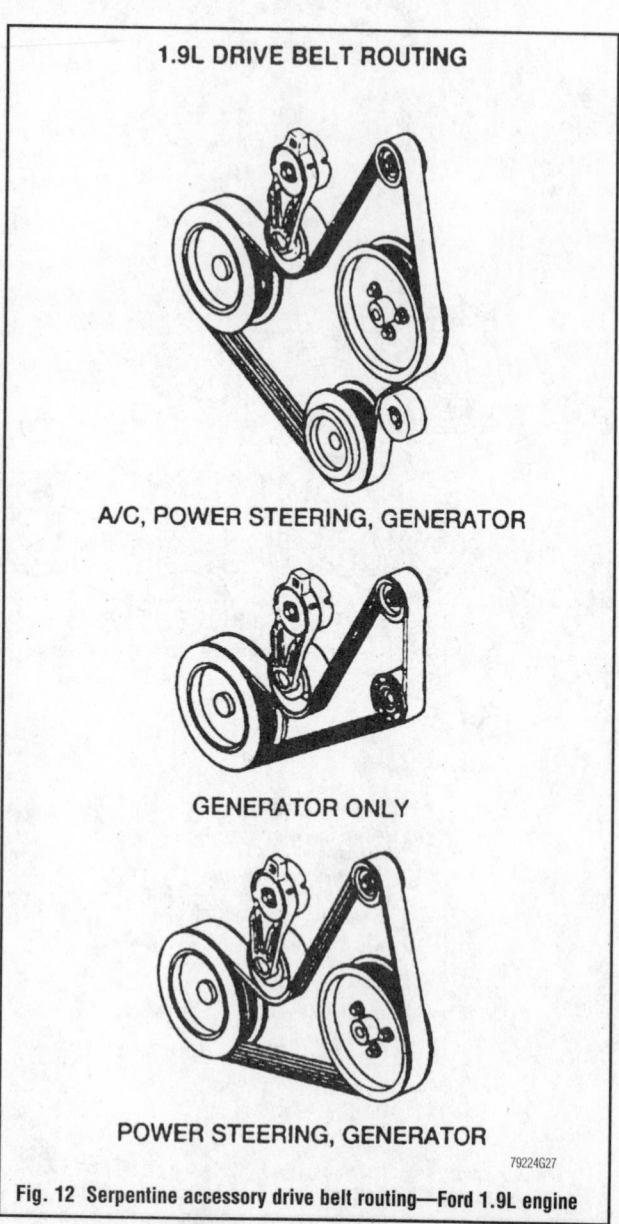

1.9L DRIVE BELT ROUTING

A/C, POWER STEERING, GENERATOR

GENERATOR ONLY

POWER STEERING, GENERATOR

Fig. 12 Serpentine accessory drive belt routing—Ford 1.9L engine

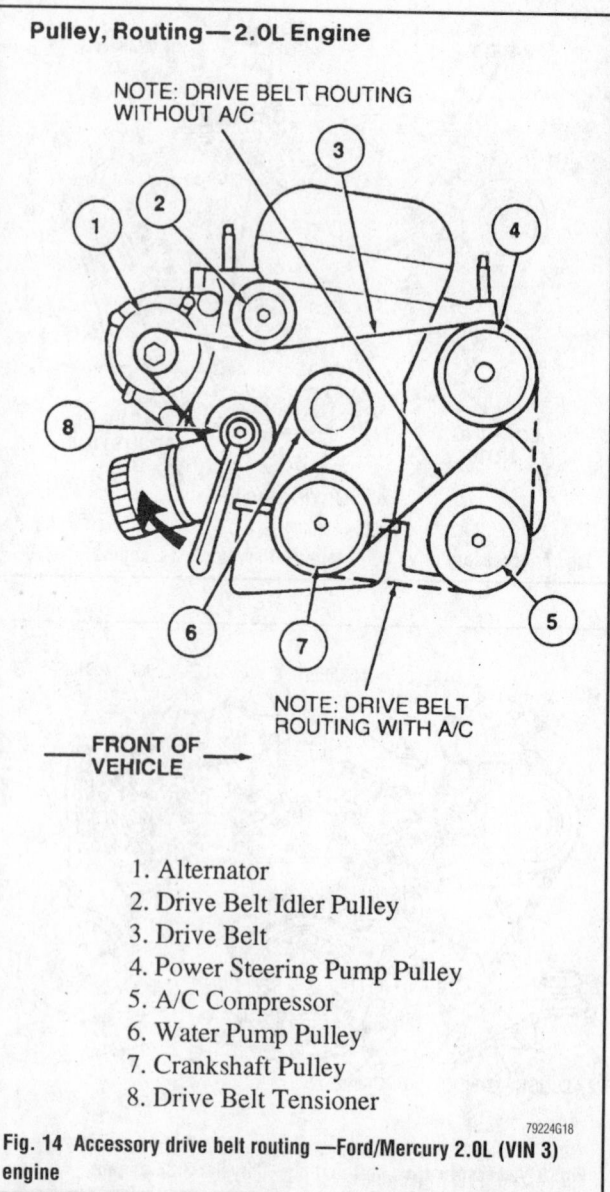

Pulley, Routing—2.0L Engine

NOTE: DRIVE BELT ROUTING WITHOUT A/C

NOTE: DRIVE BELT ROUTING WITH A/C

FRONT OF VEHICLE

1. Alternator
2. Drive Belt Idler Pulley
3. Drive Belt
4. Power Steering Pump Pulley
5. A/C Compressor
6. Water Pump Pulley
7. Crankshaft Pulley
8. Drive Belt Tensioner

Fig. 14 Accessory drive belt routing —Ford/Mercury 2.0L (VIN 3) engine

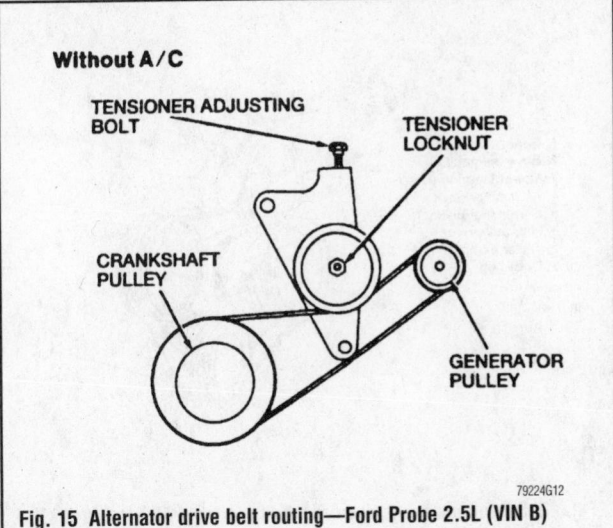

Fig. 15 Alternator drive belt routing—Ford Probe 2.5L (VIN B) engine without A/C

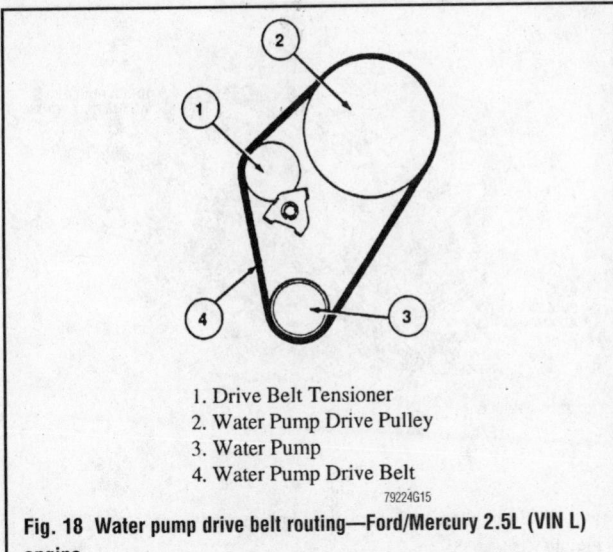

1. Drive Belt Tensioner
2. Water Pump Drive Pulley
3. Water Pump
4. Water Pump Drive Belt

Fig. 18 Water pump drive belt routing—Ford/Mercury 2.5L (VIN L) engine

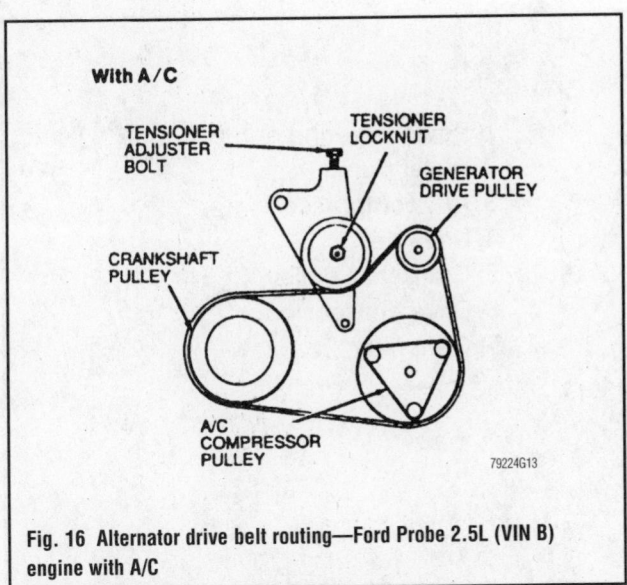

Fig. 16 Alternator drive belt routing—Ford Probe 2.5L (VIN B) engine with A/C

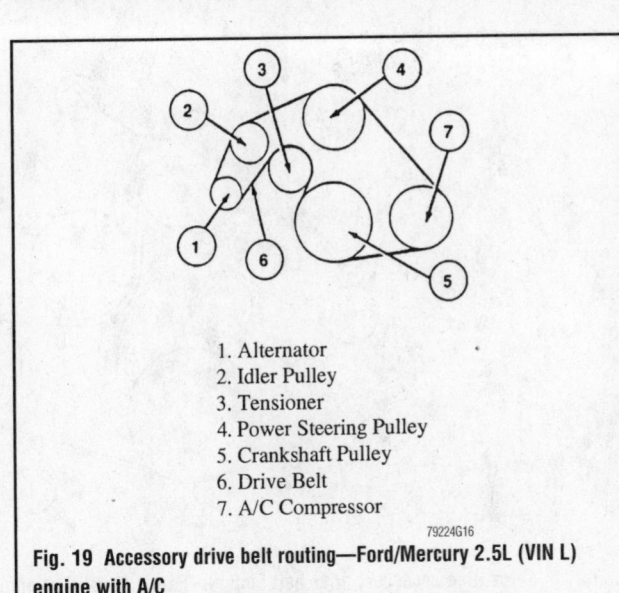

1. Alternator
2. Idler Pulley
3. Tensioner
4. Power Steering Pulley
5. Crankshaft Pulley
6. Drive Belt
7. A/C Compressor

Fig. 19 Accessory drive belt routing—Ford/Mercury 2.5L (VIN L) engine with A/C

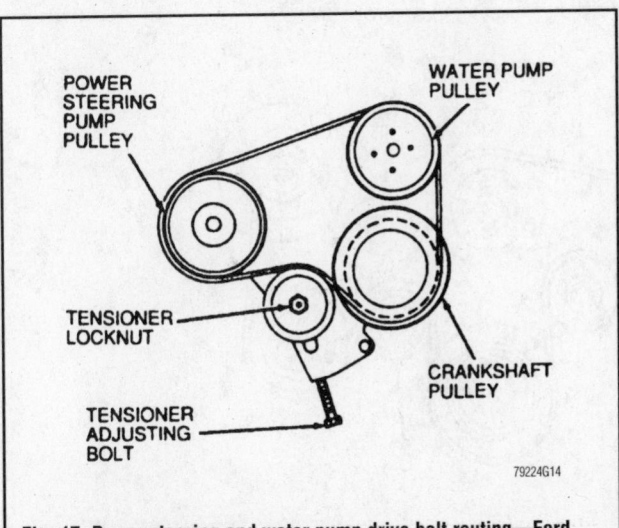

Fig. 17 Power steering and water pump drive belt routing—Ford Probe 2.5L (VIN B) engine

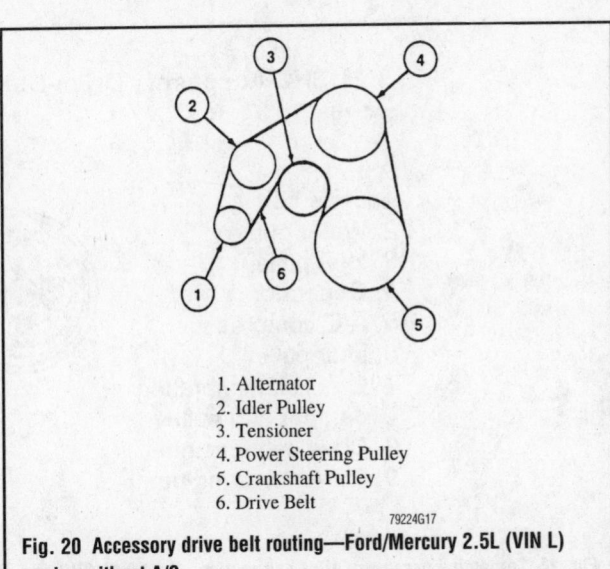

1. Alternator
2. Idler Pulley
3. Tensioner
4. Power Steering Pulley
5. Crankshaft Pulley
6. Drive Belt

Fig. 20 Accessory drive belt routing—Ford/Mercury 2.5L (VIN L) engine without A/C

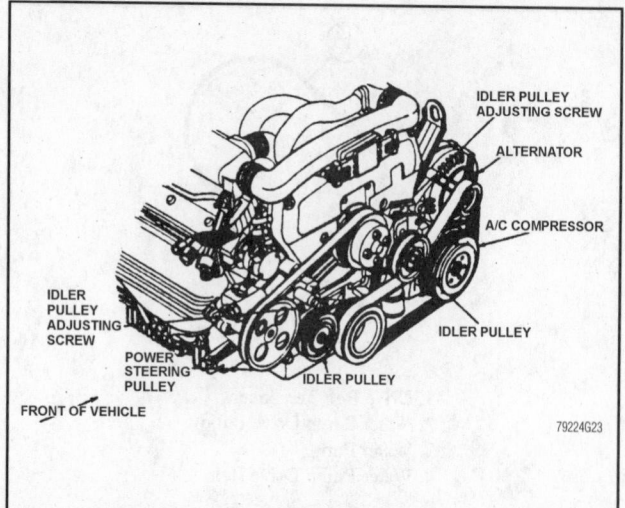

Fig. 21 Serpentine accessory drive belt routing—Ford 3.0L SHO engine

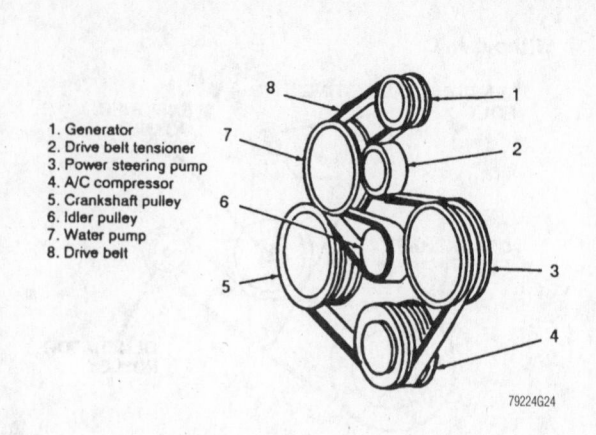

1. Generator
2. Drive belt tensioner
3. Power steering pump
4. A/C compressor
5. Crankshaft pulley
6. Idler pulley
7. Water pump
8. Drive belt

Fig. 22 Serpentine accessory drive belt routing—Ford 3.0L (VIN U and 1) engines

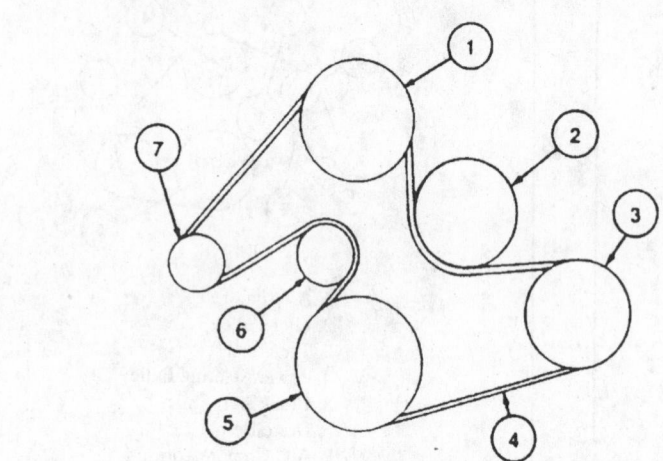

1. Power steering pump
2. Water pump
3. A/C compressor
4. Drive belt
5. Crankshaft pulley
6. Drive belt tensioner
7. Generator

Fig. 23 Serpentine accessory drive belt routing—Ford 3.0L (VIN S) engine

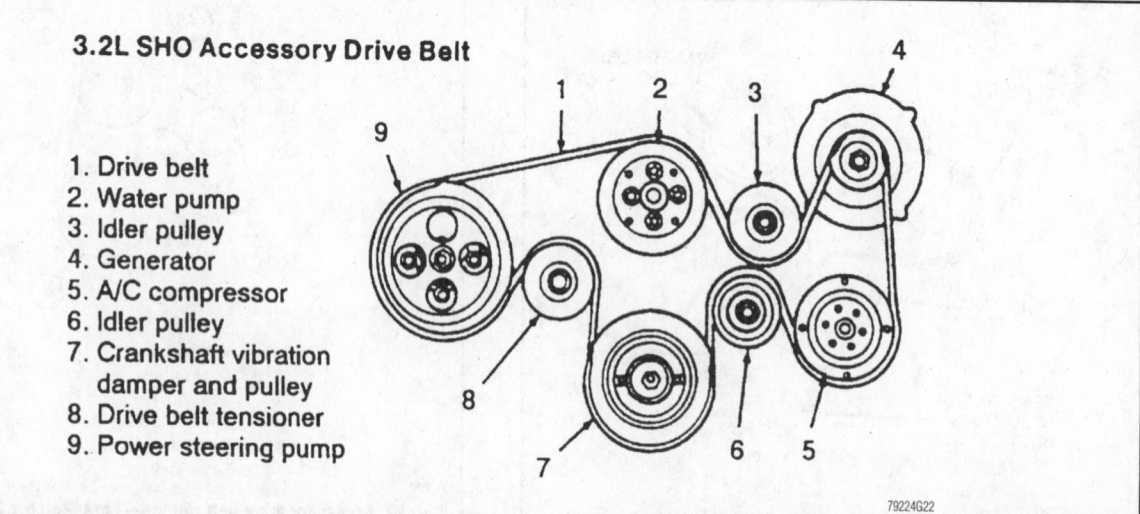

3.2L SHO Accessory Drive Belt

1. Drive belt
2. Water pump
3. Idler pulley
4. Generator
5. A/C compressor
6. Idler pulley
7. Crankshaft vibration damper and pulley
8. Drive belt tensioner
9. Power steering pump

Fig. 24 Serpentine accessory drive belt routing—Ford 3.2L SHO engine

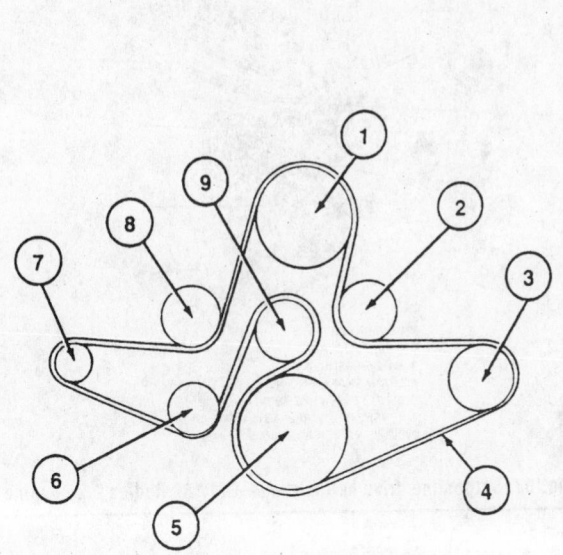

1. Power Steering Pump Pulley
2. Drive Belt Idler Pulley (RH)
3. A/C Compressor
4. Drive Belt
5. Crankshaft Pulley
6. Drive Belt Tensioner
7. Alternator
8. Drive Belt Idler Pulley (LH)
9. Drive Belt Idler Pulley (Center)

79224G20

Fig. 25 Serpentine accessory drive belt routing—Ford 3.4L SHO engine

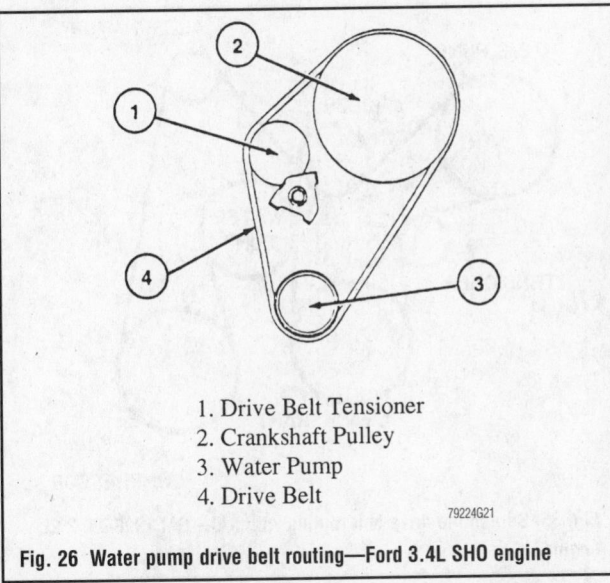

1. Drive Belt Tensioner
2. Crankshaft Pulley
3. Water Pump
4. Drive Belt

79224G21

Fig. 26 Water pump drive belt routing—Ford 3.4L SHO engine

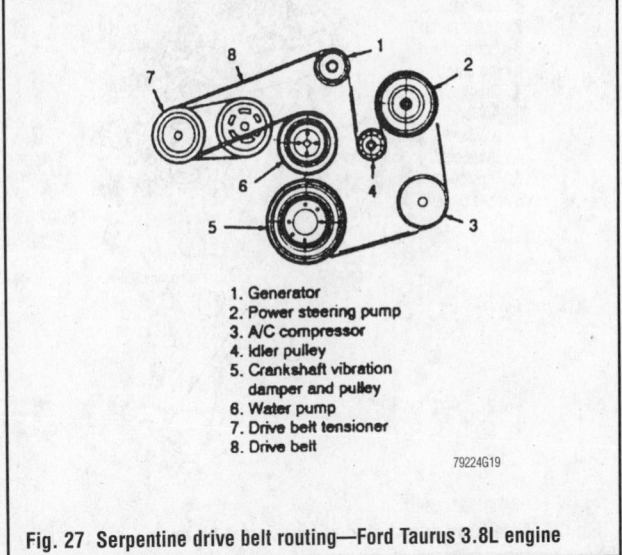

1. Generator
2. Power steering pump
3. A/C compressor
4. Idler pulley
5. Crankshaft vibration damper and pulley
6. Water pump
7. Drive belt tensioner
8. Drive belt

79224G19

Fig. 27 Serpentine drive belt routing—Ford Taurus 3.8L engine

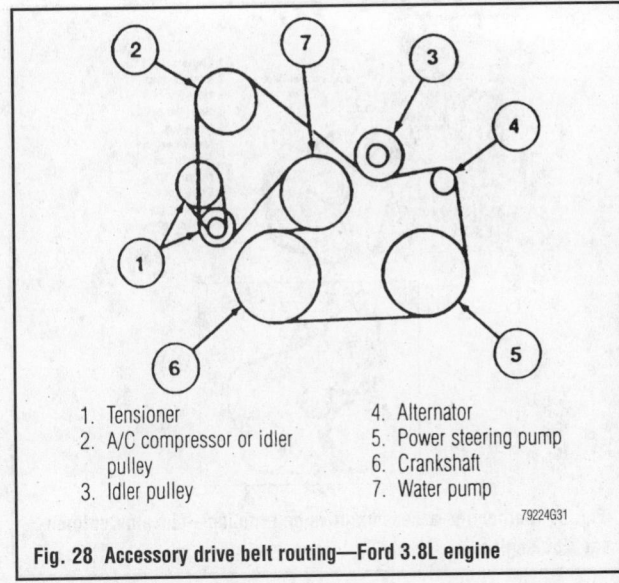

1. Tensioner
2. A/C compressor or idler pulley
3. Idler pulley
4. Alternator
5. Power steering pump
6. Crankshaft
7. Water pump

79224G31

Fig. 28 Accessory drive belt routing—Ford 3.8L engine

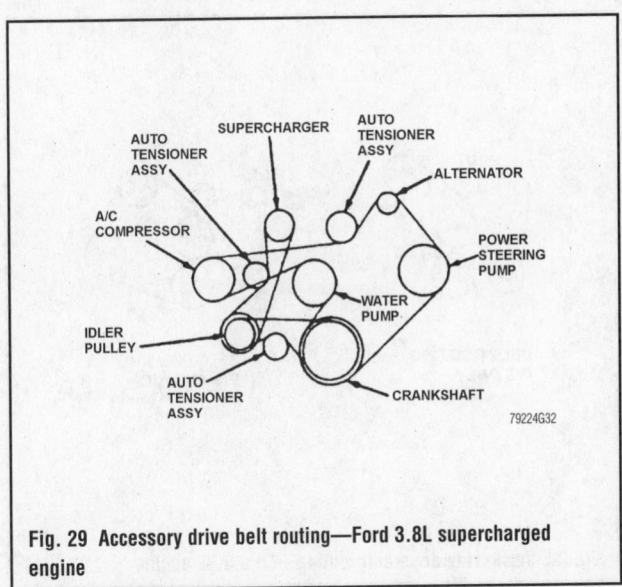

79224G32

Fig. 29 Accessory drive belt routing—Ford 3.8L supercharged engine

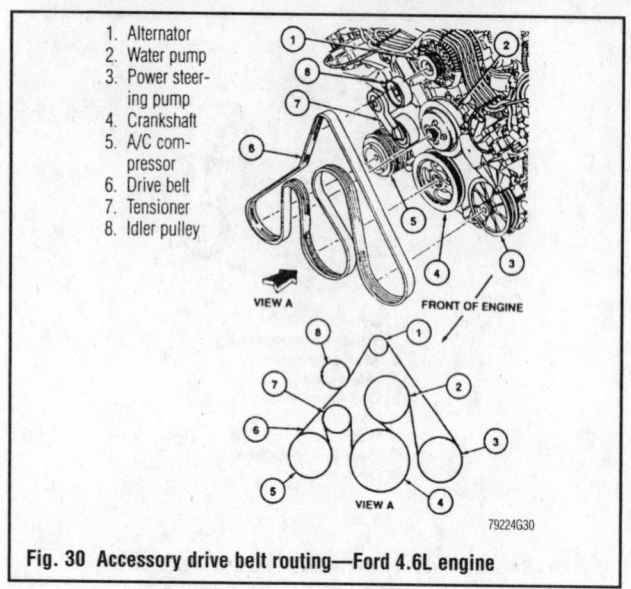

1. Alternator
2. Water pump
3. Power steering pump
4. Crankshaft
5. A/C compressor
6. Drive belt
7. Tensioner
8. Idler pulley

VIEW A

FRONT OF ENGINE

VIEW A

79224G30

Fig. 30 Accessory drive belt routing—Ford 4.6L engine

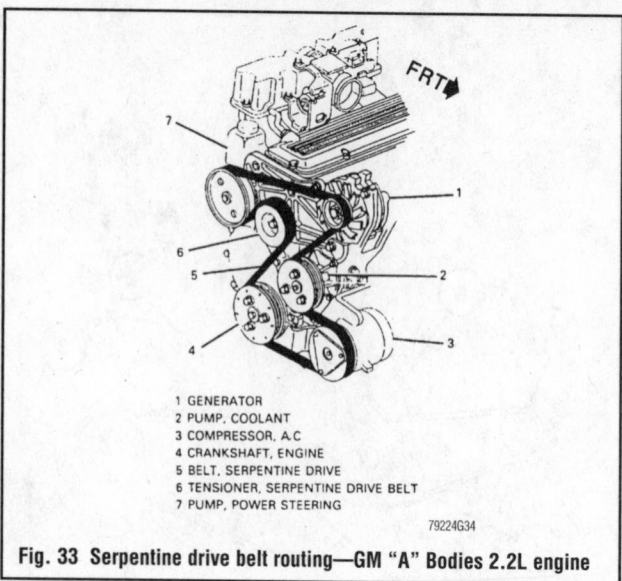

FRT

1 GENERATOR
2 PUMP, COOLANT
3 COMPRESSOR, A.C
4 CRANKSHAFT, ENGINE
5 BELT, SERPENTINE DRIVE
6 TENSIONER, SERPENTINE DRIVE BELT
7 PUMP, POWER STEERING

79224G34

Fig. 33 Serpentine drive belt routing—GM "A" Bodies 2.2L engine

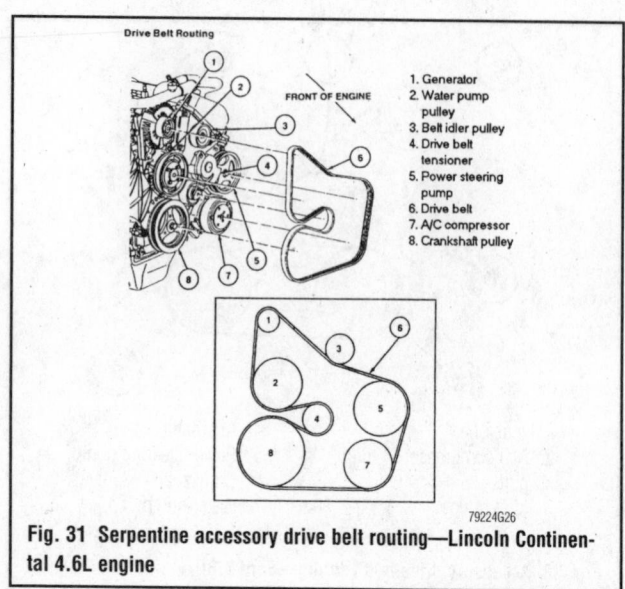

Drive Belt Routing

FRONT OF ENGINE

1. Generator
2. Water pump pulley
3. Belt idler pulley
4. Drive belt tensioner
5. Power steering pump
6. Drive belt
7. A/C compressor
8. Crankshaft pulley

79224G26

Fig. 31 Serpentine accessory drive belt routing—Lincoln Continental 4.6L engine

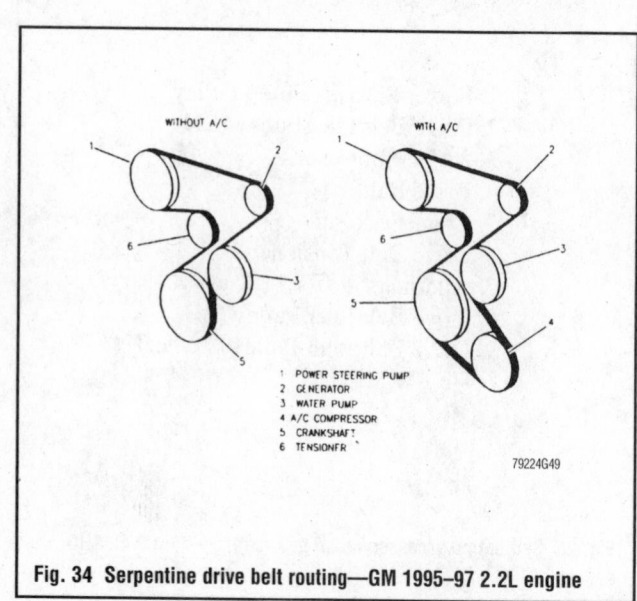

WITHOUT A/C

WITH A/C

1 POWER STEERING PUMP
2 GENERATOR
3 WATER PUMP
4 A/C COMPRESSOR
5 CRANKSHAFT
6 TENSIONER

79224G49

Fig. 34 Serpentine drive belt routing—GM 1995–97 2.2L engine

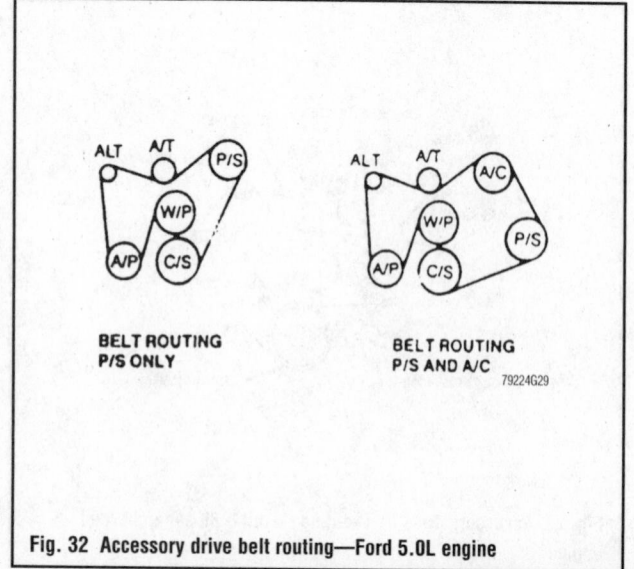

ALT   A/T   P/S

W/P

A/P   C/S

BELT ROUTING
P/S ONLY

ALT   A/T   A/C

W/P

A/P   C/S   P/S

BELT ROUTING
P/S AND A/C

79224G29

Fig. 32 Accessory drive belt routing—Ford 5.0L engine

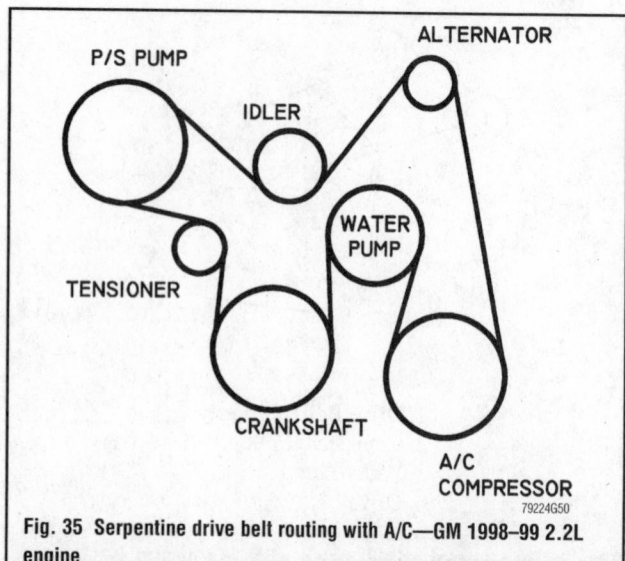

P/S PUMP

ALTERNATOR

IDLER

WATER
PUMP

TENSIONER

CRANKSHAFT

A/C
COMPRESSOR

79224G50

Fig. 35 Serpentine drive belt routing with A/C—GM 1998–99 2.2L engine

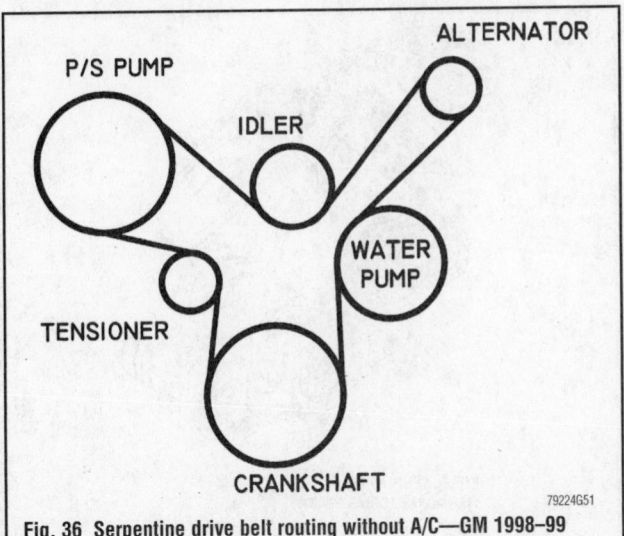

Fig. 36 Serpentine drive belt routing without A/C—GM 1998–99 2.2L engine

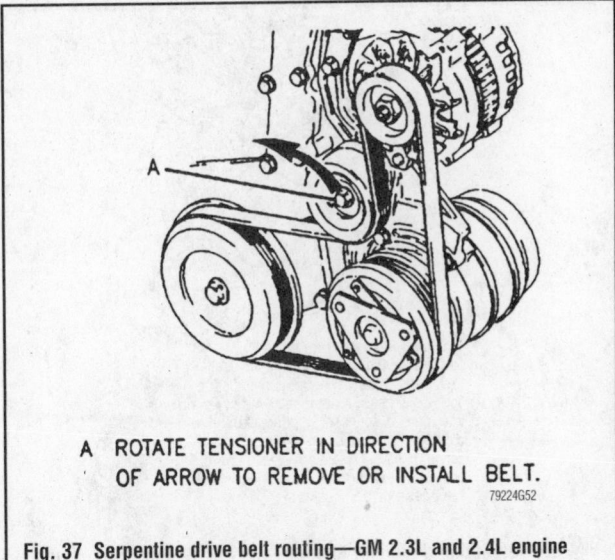

A ROTATE TENSIONER IN DIRECTION OF ARROW TO REMOVE OR INSTALL BELT.

Fig. 37 Serpentine drive belt routing—GM 2.3L and 2.4L engine

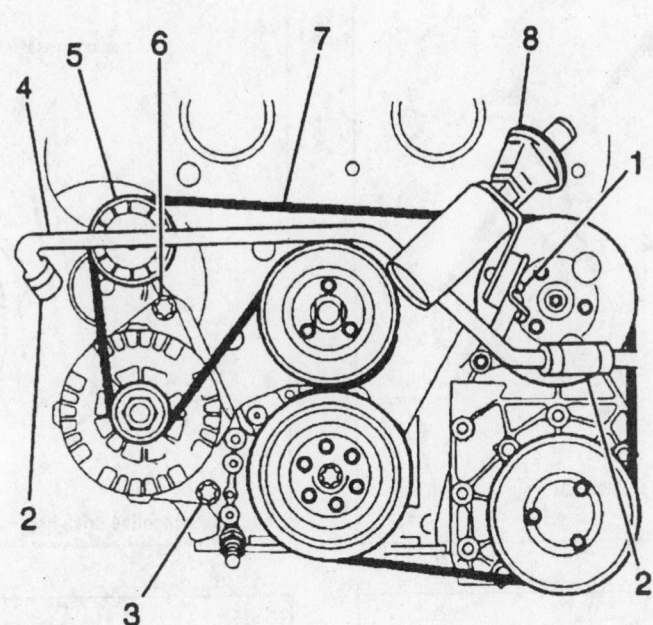

Legend

(1) AIR Injection Crossover Pipe Bushing Nut
(2) AIR Injection Rubber Hose Connection
(3) Lower Generator Bolt
(4) AIR Injection Crossover Pipe
(5) Serpentine Drive Belt Tensioner
(6) AIR Injection Crossover Pipe Support (Generator) Bracket Nut
(7) Serpentine Drive Belt
(8) AIR Injection Diverter Valve

Fig. 38 Serpentine drive belt routing—GM 3.0L engine

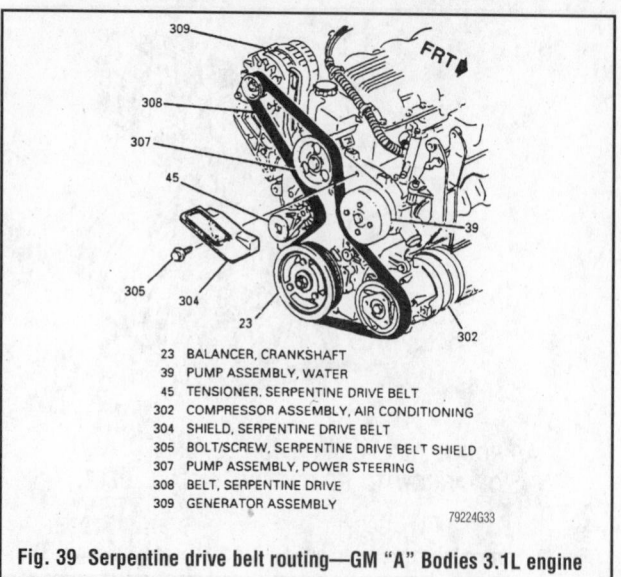

23 BALANCER, CRANKSHAFT
39 PUMP ASSEMBLY, WATER
45 TENSIONER, SERPENTINE DRIVE BELT
302 COMPRESSOR ASSEMBLY, AIR CONDITIONING
304 SHIELD, SERPENTINE DRIVE BELT
305 BOLT/SCREW, SERPENTINE DRIVE BELT SHIELD
307 PUMP ASSEMBLY, POWER STEERING
308 BELT, SERPENTINE DRIVE
309 GENERATOR ASSEMBLY

79224G33

Fig. 39 Serpentine drive belt routing—GM "A" Bodies 3.1L engine

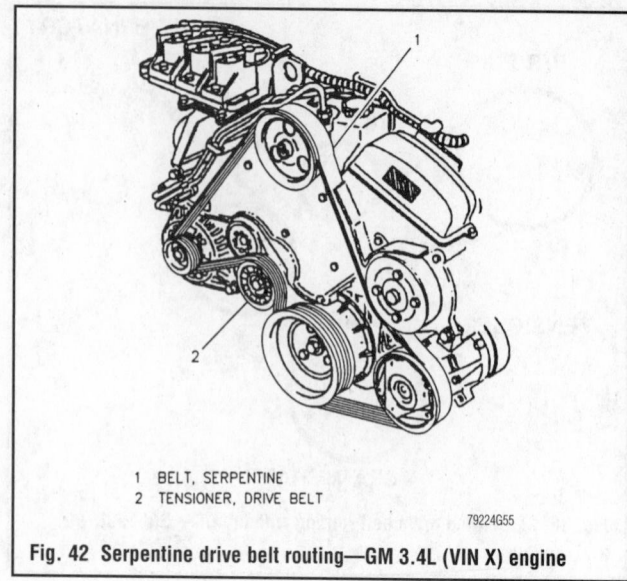

1 BELT, SERPENTINE
2 TENSIONER, DRIVE BELT

79224G55

Fig. 42 Serpentine drive belt routing—GM 3.4L (VIN X) engine

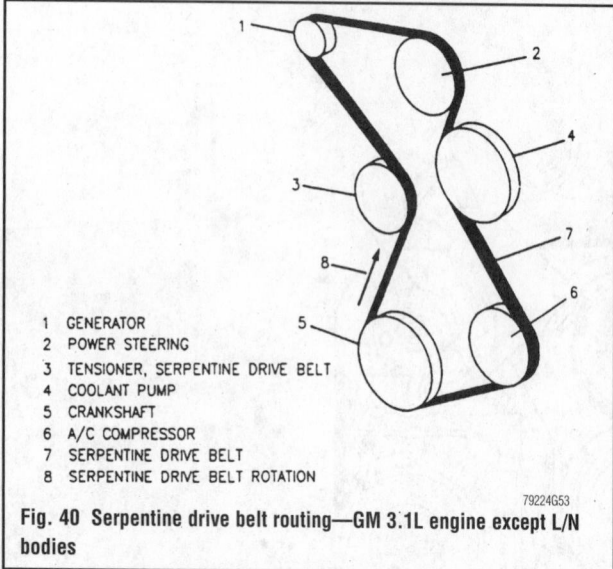

1 GENERATOR
2 POWER STEERING
3 TENSIONER, SERPENTINE DRIVE BELT
4 COOLANT PUMP
5 CRANKSHAFT
6 A/C COMPRESSOR
7 SERPENTINE DRIVE BELT
8 SERPENTINE DRIVE BELT ROTATION

79224G53

Fig. 40 Serpentine drive belt routing—GM 3.1L engine except L/N bodies

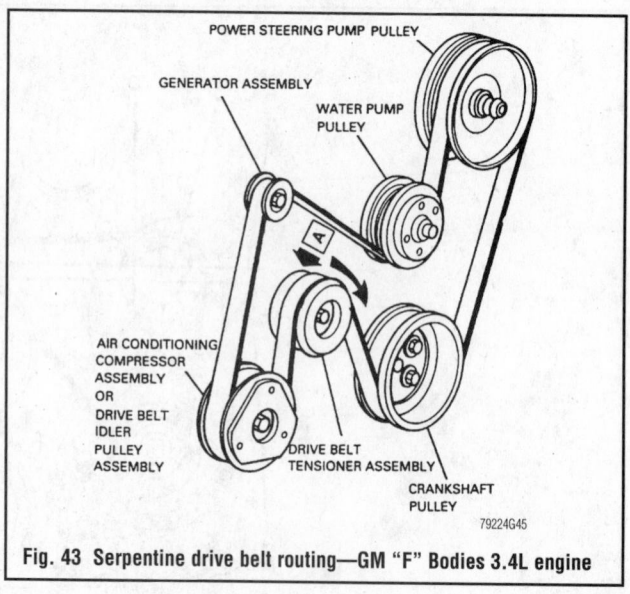

POWER STEERING PUMP PULLEY

GENERATOR ASSEMBLY

WATER PUMP PULLEY

AIR CONDITIONING COMPRESSOR ASSEMBLY OR DRIVE BELT IDLER PULLEY ASSEMBLY

DRIVE BELT TENSIONER ASSEMBLY

CRANKSHAFT PULLEY

79224G45

Fig. 43 Serpentine drive belt routing—GM "F" Bodies 3.4L engine

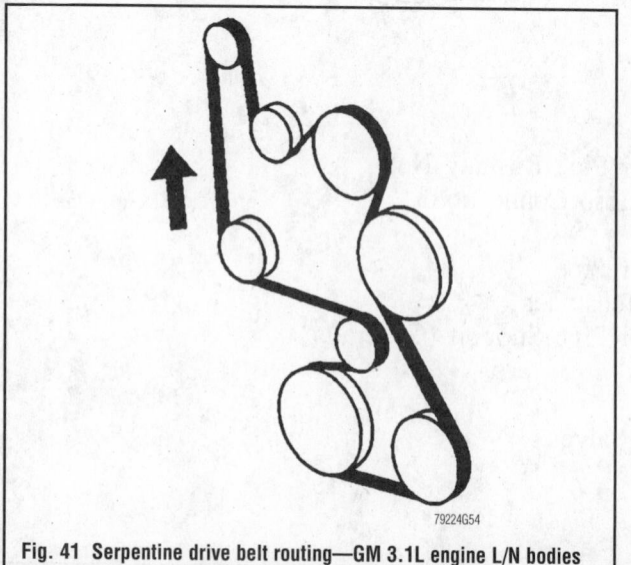

79224G54

Fig. 41 Serpentine drive belt routing—GM 3.1L engine L/N bodies

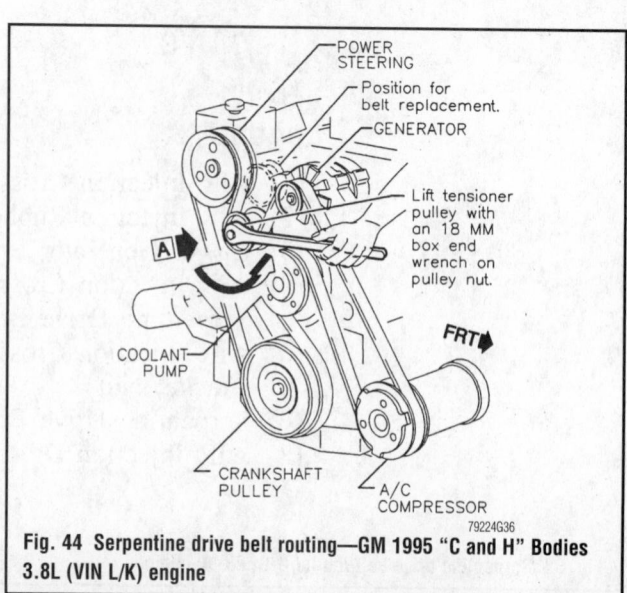

POWER STEERING

Position for belt replacement.

GENERATOR

Lift tensioner pulley with an 18 MM box end wrench on pulley nut.

FRT

COOLANT PUMP

CRANKSHAFT PULLEY

A/C COMPRESSOR

79224G36

Fig. 44 Serpentine drive belt routing—GM 1995 "C and H" Bodies 3.8L (VIN L/K) engine

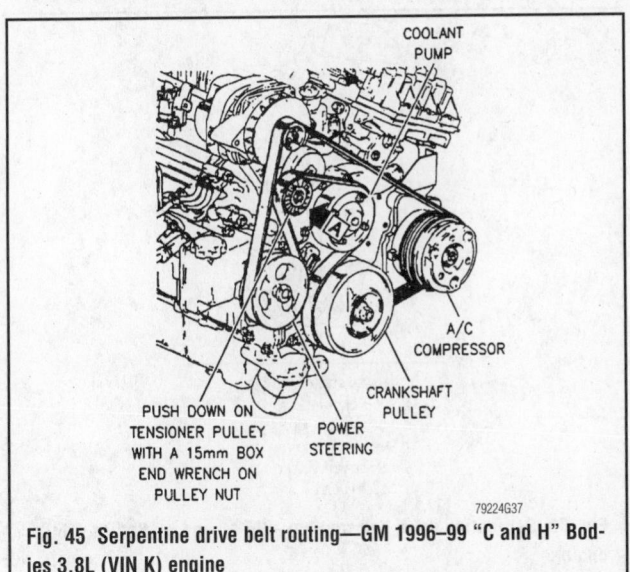

Fig. 45 Serpentine drive belt routing—GM 1996–99 "C and H" Bodies 3.8L (VIN K) engine

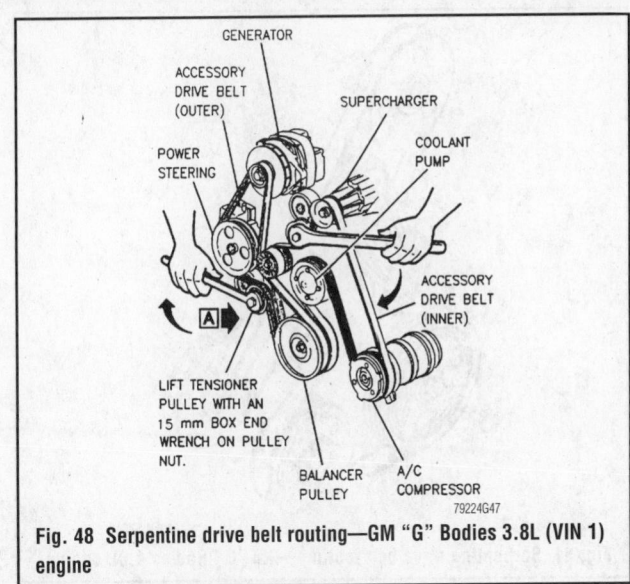

Fig. 48 Serpentine drive belt routing—GM "G" Bodies 3.8L (VIN 1) engine

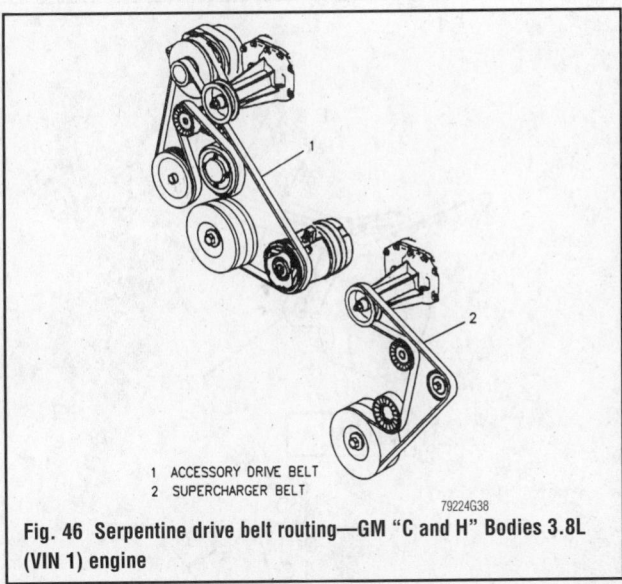

1 ACCESSORY DRIVE BELT
2 SUPERCHARGER BELT

Fig. 46 Serpentine drive belt routing—GM "C and H" Bodies 3.8L (VIN 1) engine

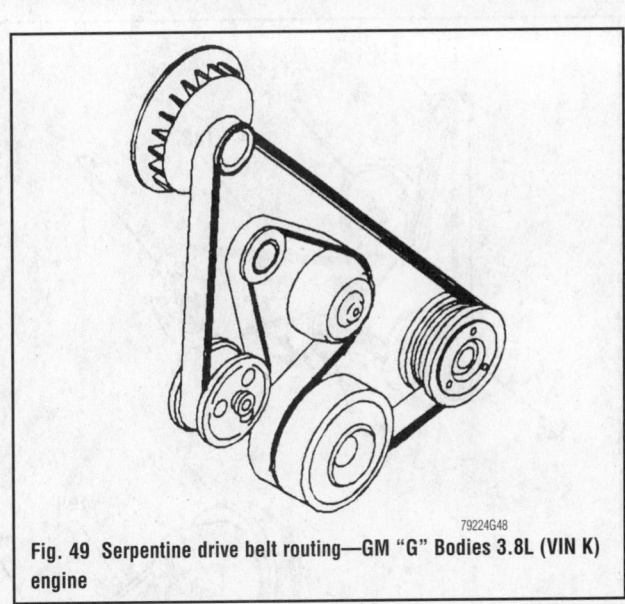

Fig. 49 Serpentine drive belt routing—GM "G" Bodies 3.8L (VIN K) engine

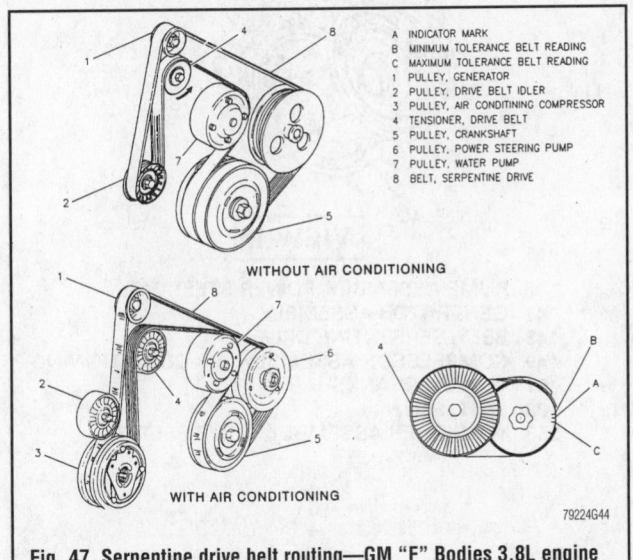

A INDICATOR MARK
B MINIMUM TOLERANCE BELT READING
C MAXIMUM TOLERANCE BELT READING
1 PULLEY, GENERATOR
2 PULLEY, DRIVE BELT IDLER
3 PULLEY, AIR CONDITIONING COMPRESSOR
4 TENSIONER, DRIVE BELT
5 PULLEY, CRANKSHAFT
6 PULLEY, POWER STEERING PUMP
7 PULLEY, WATER PUMP
8 BELT, SERPENTINE DRIVE

WITHOUT AIR CONDITIONING

WITH AIR CONDITIONING

Fig. 47 Serpentine drive belt routing—GM "F" Bodies 3.8L engine

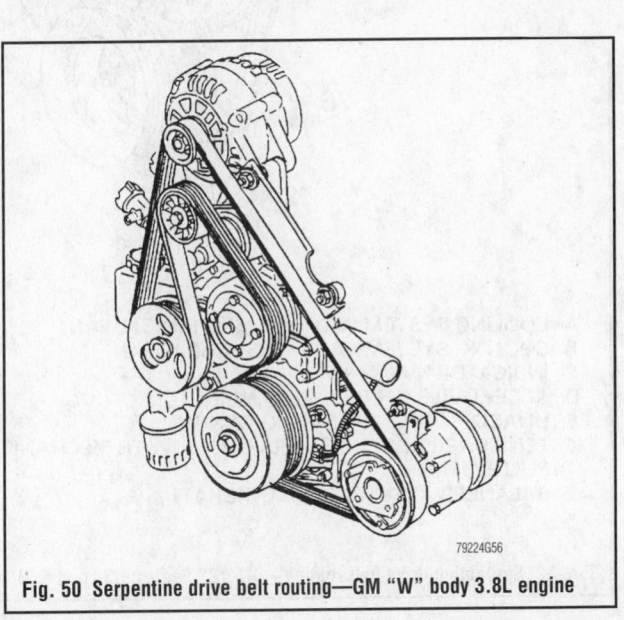

Fig. 50 Serpentine drive belt routing—GM "W" body 3.8L engine

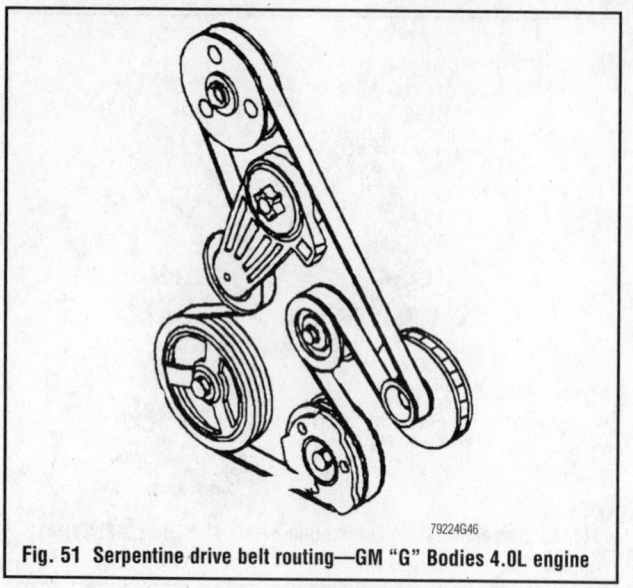

Fig. 51 Serpentine drive belt routing—GM "G" Bodies 4.0L engine

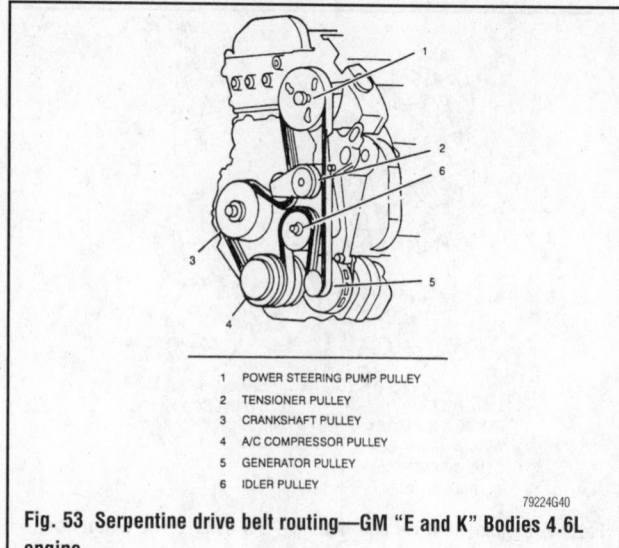

1	POWER STEERING PUMP PULLEY
2	TENSIONER PULLEY
3	CRANKSHAFT PULLEY
4	A/C COMPRESSOR PULLEY
5	GENERATOR PULLEY
6	IDLER PULLEY

Fig. 53 Serpentine drive belt routing—GM "E and K" Bodies 4.6L engine

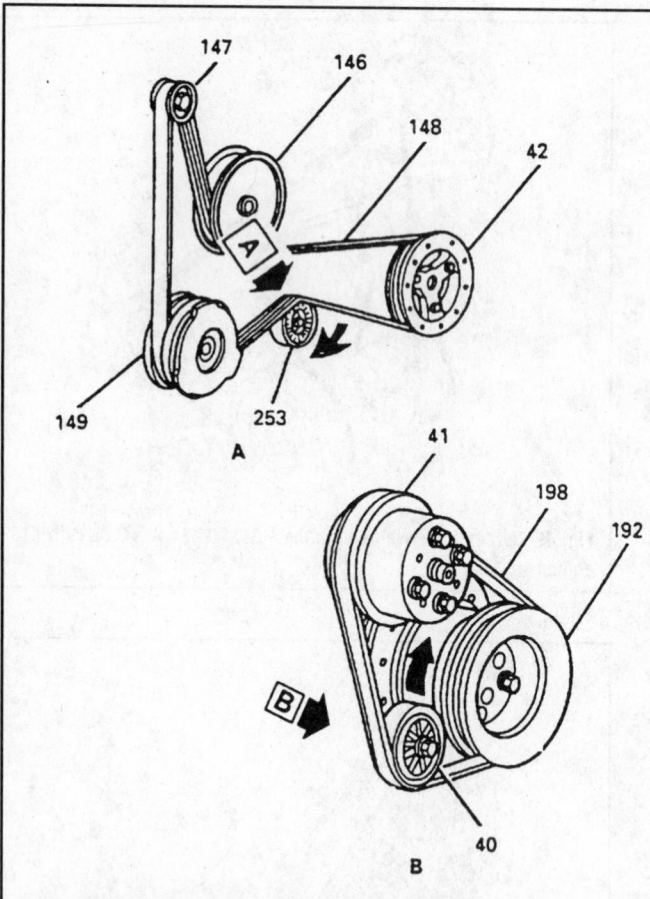

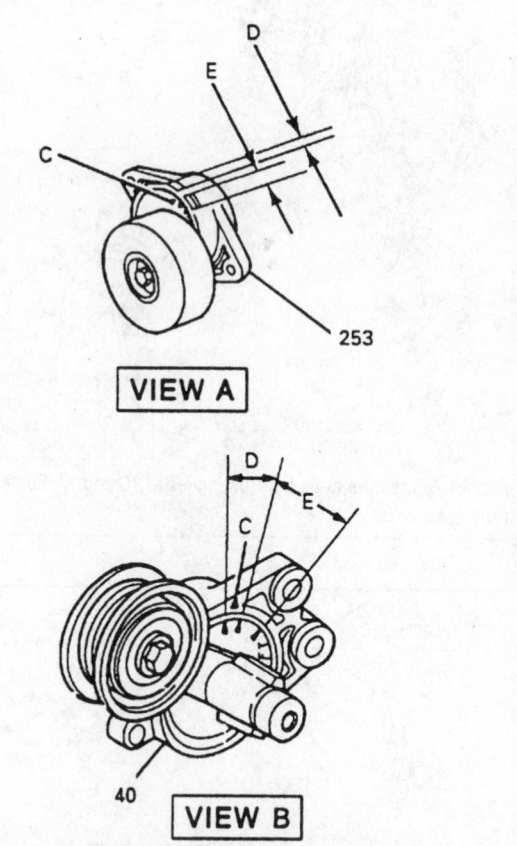

A   COOLING SYSTEM WITHOUT MECHANICAL FAN
B   COOLING SYSTEM WITH MECHANICAL FAN
C   INDICATOR MARK
D   ACCEPTABLE OPERATING RANGE
E   UNACCEPTABLE OPERATING RANGE
40 TENSIONER ASSEMBLY, DRIVE BELT (WITH MECHANICAL FAN)
41 PULLEY, FAN
42 BALANCER ASSEMBLY, CRANKSHAFT

46 PUMP ASSEMBLY, POWER STEERING
47 GENERATOR ASSEMBLY
48 BELT, SERPENTINE DRIVE
49 COMPRESSOR ASSEMBLY, AIR CONDITIONING
92 PULLEY, CRANKSHAFT
98 BELT, FAN
253 TENSIONER ASSEMBLY, DRIVE BELT

Fig. 52 Serpentine drive belt routing—GM "B" Bodies 4.3L and 5.7L engine

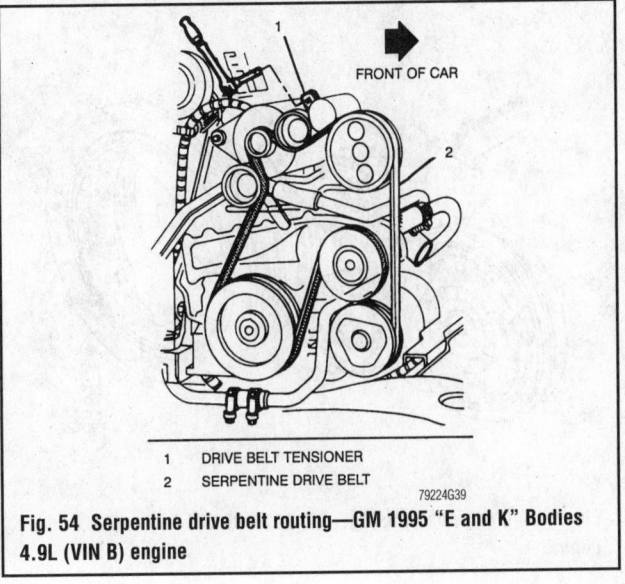

1 DRIVE BELT TENSIONER
2 SERPENTINE DRIVE BELT

79224G39

Fig. 54 Serpentine drive belt routing—GM 1995 "E and K" Bodies 4.9L (VIN B) engine

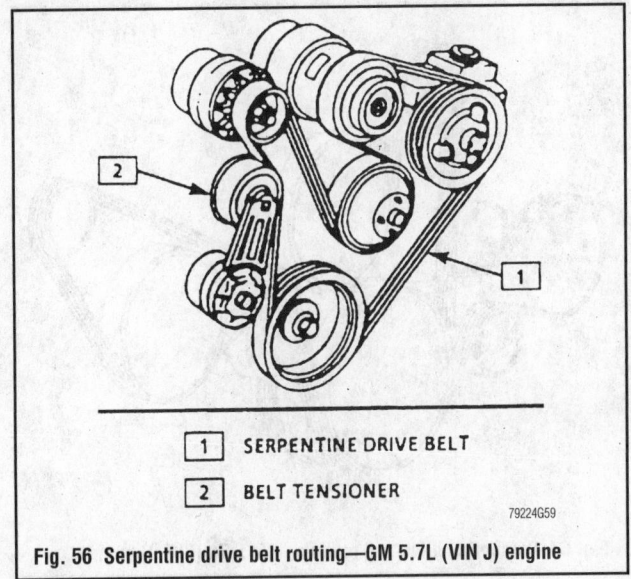

1 SERPENTINE DRIVE BELT

2 BELT TENSIONER

79224G59

Fig. 56 Serpentine drive belt routing—GM 5.7L (VIN J) engine

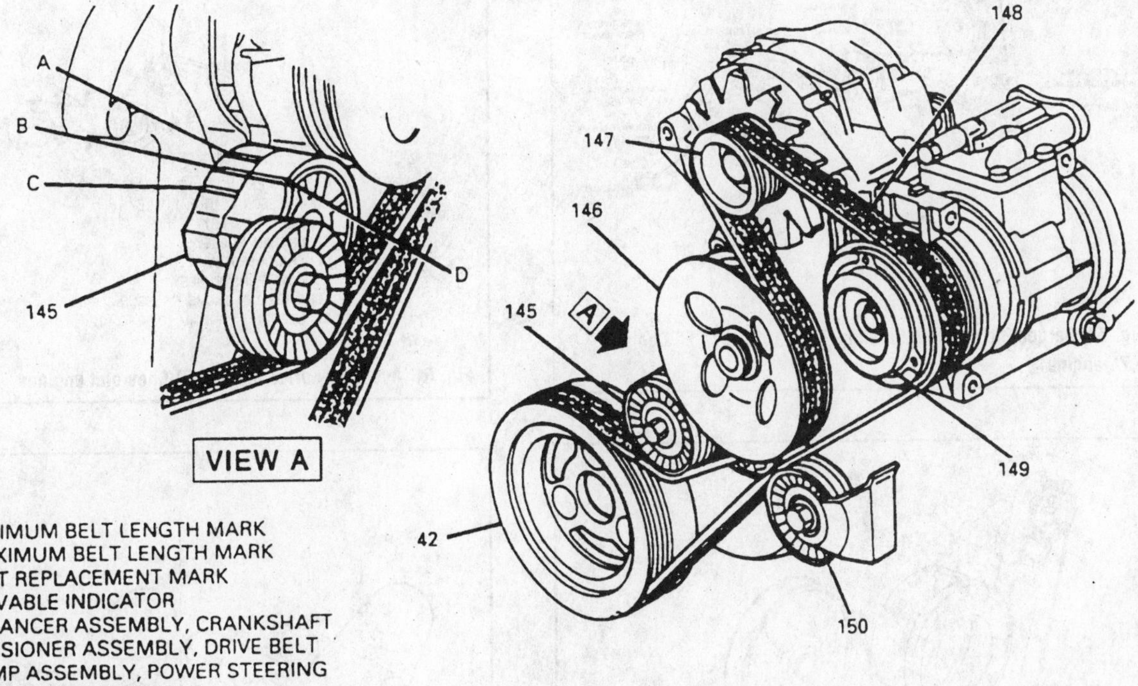

VIEW A

A MINIMUM BELT LENGTH MARK
B MAXIMUM BELT LENGTH MARK
C BELT REPLACEMENT MARK
D MOVABLE INDICATOR
42 BALANCER ASSEMBLY, CRANKSHAFT
145 TENSIONER ASSEMBLY, DRIVE BELT
146 PUMP ASSEMBLY, POWER STEERING
147 GENERATOR ASSEMBLY
148 BELT, SERPENTINE DRIVE
149 COMPRESSOR ASSEMBLY, AIR CONDITIONING
150 PULLEY, DRIVE BELT IDLER

79224G58

Fig. 55 Serpentine drive belt routing—GM 5.7L (VIN P) engine

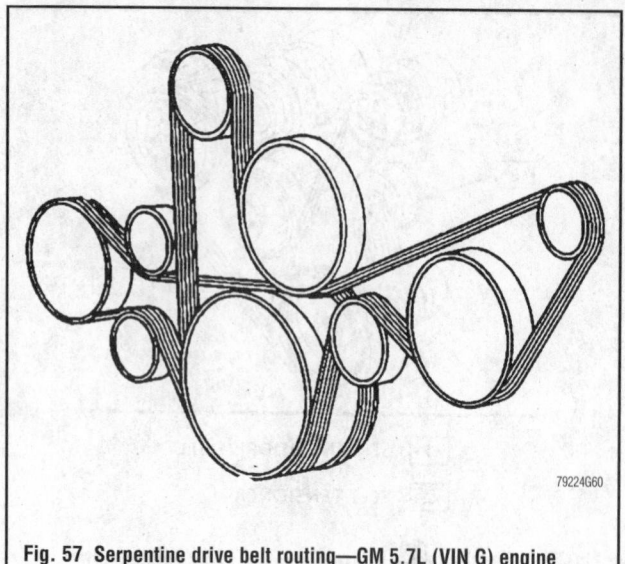

Fig. 57 Serpentine drive belt routing—GM 5.7L (VIN G) engine

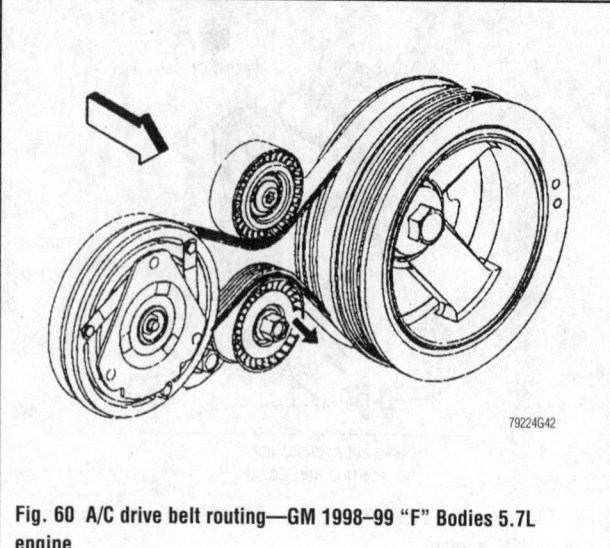

Fig. 60 A/C drive belt routing—GM 1998–99 "F" Bodies 5.7L engine

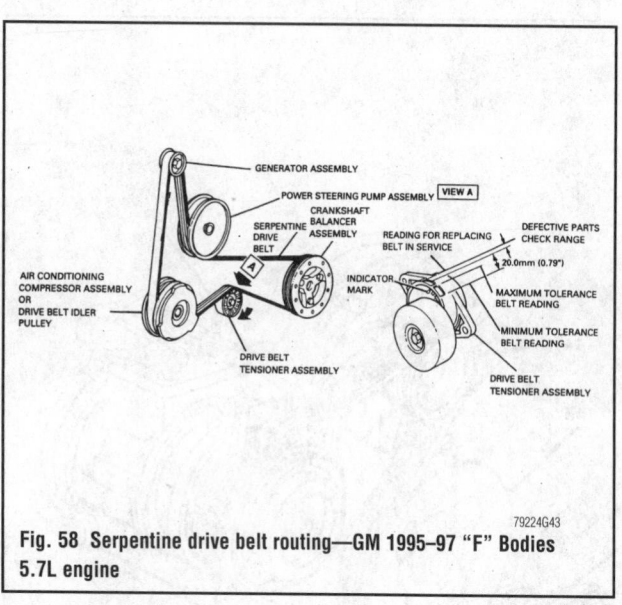

Fig. 58 Serpentine drive belt routing—GM 1995–97 "F" Bodies 5.7L engine

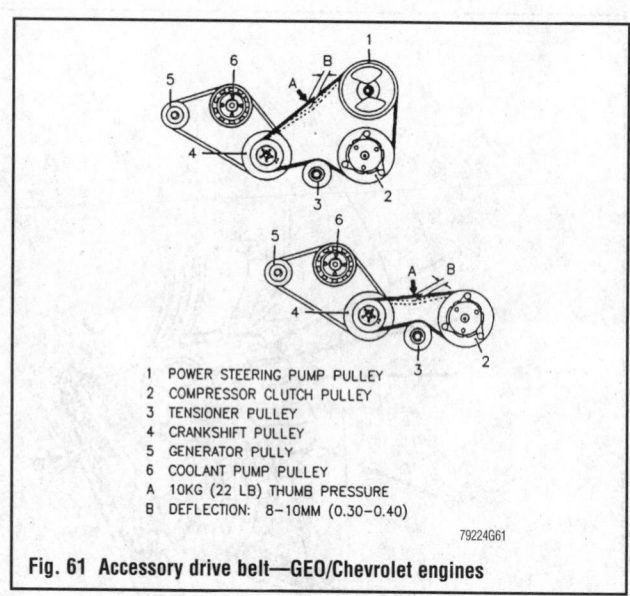

1 POWER STEERING PUMP PULLEY
2 COMPRESSOR CLUTCH PULLEY
3 TENSIONER PULLEY
4 CRANKSHIFT PULLEY
5 GENERATOR PULLY
6 COOLANT PUMP PULLEY
A 10KG (22 LB) THUMB PRESSURE
B DEFLECTION: 8–10MM (0.30–0.40)

Fig. 61 Accessory drive belt—GEO/Chevrolet engines

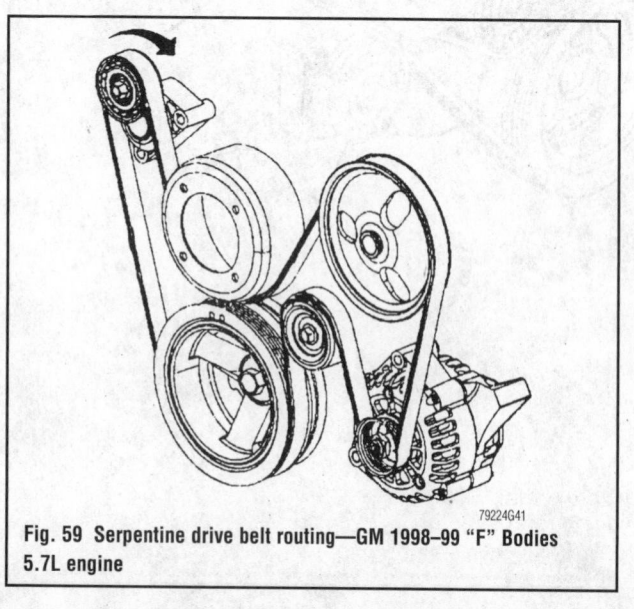

Fig. 59 Serpentine drive belt routing—GM 1998–99 "F" Bodies 5.7L engine

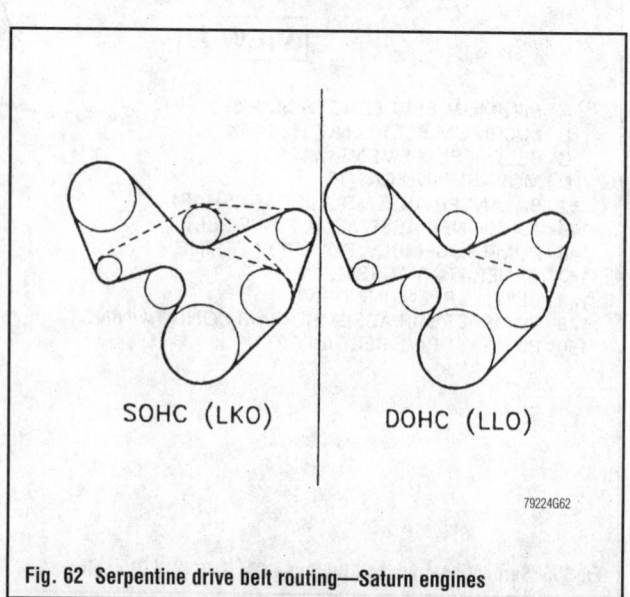

SOHC (LKO)   DOHC (LLO)

Fig. 62 Serpentine drive belt routing—Saturn engines

TIMING BELTS

5

GENERAL INFORMATION 5-1
TIMING BELT SERVICE..........5-1
Interference Engines 5-1
Inspection.......................................5-1

Removal & Installation.....................5-3
  Chrysler Corporation.....................5-3
  Ford Motor Company....................5-9
  General Motors 5-19

GENERAL INFORMATION

Timing belts are typically only used on overhead camshaft engines. Timing belts are used to synchronize the crankshaft with the camshaft, similar to a timing chain on a overhead valve (pushrod) engine. Unlike a timing belt, a timing chain will normally last the life of the engine without needing service or replacement. Timing belts use raised teeth to mesh with sprockets to operate the valvetrain of an overhead camshaft engine.

Whenever a vehicle with an unknown service history comes into your repair facility or is recently purchased, here are some points that should be asked to help prevent costly engine damage:

• Does the owner know if, or when the belt was replaced?

• If the vehicle purchased is used, or the condition and mileage of the last timing belt replacement are unknown, it is recommended to inspect, replace, or at least inform the owner that the vehicle is equipped with a timing belt.

• Note the mileage of the vehicle. The average replacement interval for a timing belt is approximately 60,000 miles (96,000 km).

Interference Engines

Engines, chain- or belt-driven, can be classified as either free-running or interference, depending on what would happen if the piston-to-valve timing is disrupted. A free-running engine is designed with enough clearance between the pistons and valves to allow the crankshaft to rotate (pistons still moving) while the camshaft stays in one position (several valves fully open). If this condition occurs normally, no internal engine damage will result. In an interference engine, there is not enough clearance between the pistons and valves to allow the crankshaft to turn without the camshaft being in time.

An interference engine can suffer extensive internal damage if a timing belt fails. The piston design does not allow clearance for the valve to be fully open and the piston to be at the top of its stroke. If the belt fails, the piston will collide with the valve and will bend or break the valve, damage the piston, and/or bend a connecting rod. When this type of failure occurs, the engine will need to be replaced or disassembled for further internal inspection; either choice costing many times that of replacing the timing belt.

TIMING BELT SERVICE

Inspection

➡For manufacturer's recommended service interval, refer to the maintenance interval chart located in this manual.

The average replacement interval for a timing belt is approximately 60,000 miles (96,000km). If, however, the timing belt is inspected earlier or more frequently than suggested, and shows signs of wear or defects, the belt should be replaced at that time.

✷✷ WARNING

Never allow antifreeze, oil or solvents to come into with a timing belt. If this occurs immediately wash the solution from the timing belt. Also, never excessive bend or twist the timing belt; this can damage the belt so that its lifetime is severely shortened.

Never bend or twist a timing belt excessively, and do not allow solvents, antifreeze, gasoline, acid or oil to come into contact with the belt

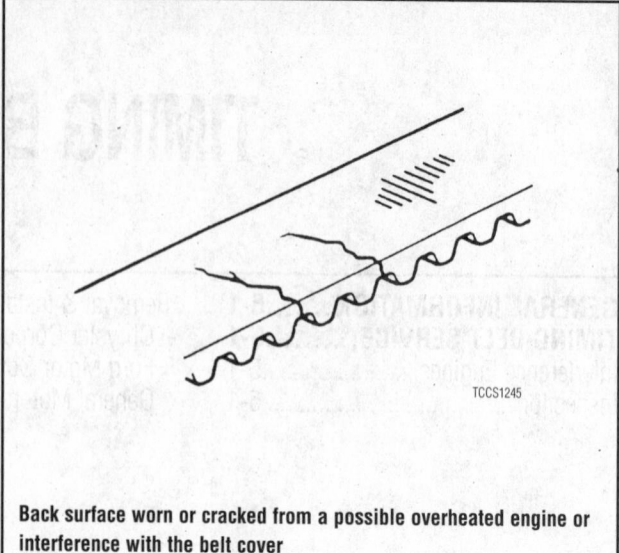

Back surface worn or cracked from a possible overheated engine or interference with the belt cover

Inspect both sides of the timing belt. Replace the belt with a new one if any of the following conditions exist:
- Hardening of the rubber—back side is glossy without resilience and leaves no indentation when pressed with a fingernail
- Cracks on the rubber backing
- Cracks or peeling of the canvas backing
- Cracks on rib root
- Cracks on belt sides
- Missing teeth or chunks of teeth
- Abnormal wear of belt sides—the sides are normal if they are sharp, as if cut by a knife.

If none of these conditions exist, the belt does not need replacement unless it is at the recommended interval. The belt MUST be replaced at the recommended interval.

✷✷ WARNING

On interference engines, it is very important to replace the timing belt at the recommended intervals, otherwise expensive engine damage will likely result if the belt fails.

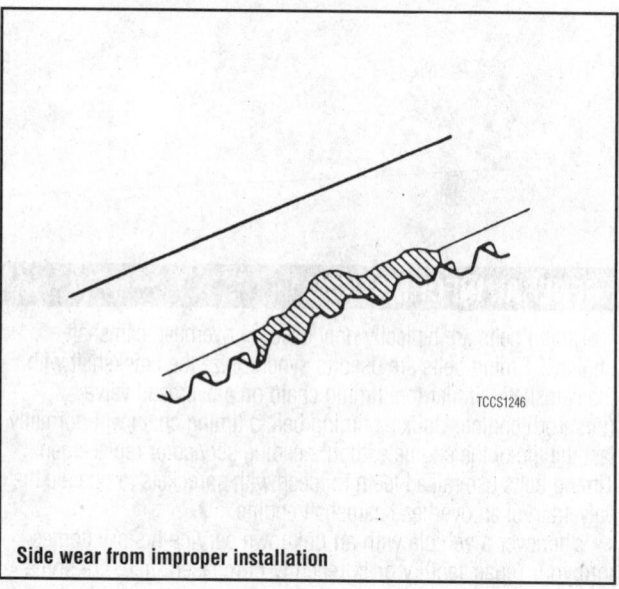

Side wear from improper installation

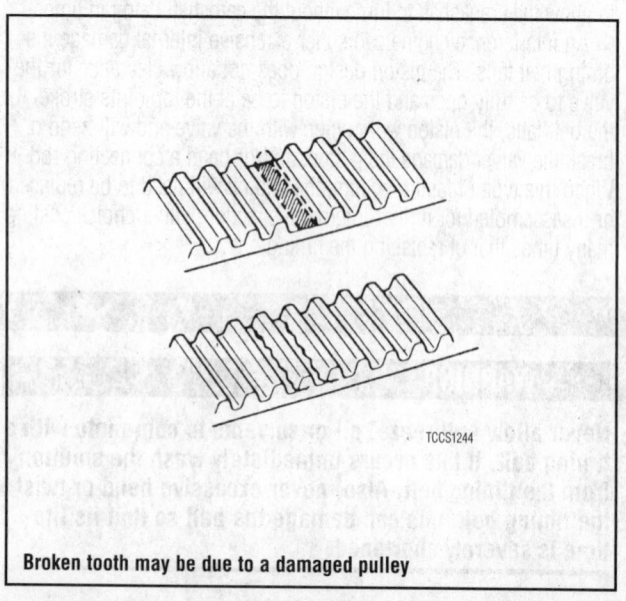

Broken tooth may be due to a damaged pulley

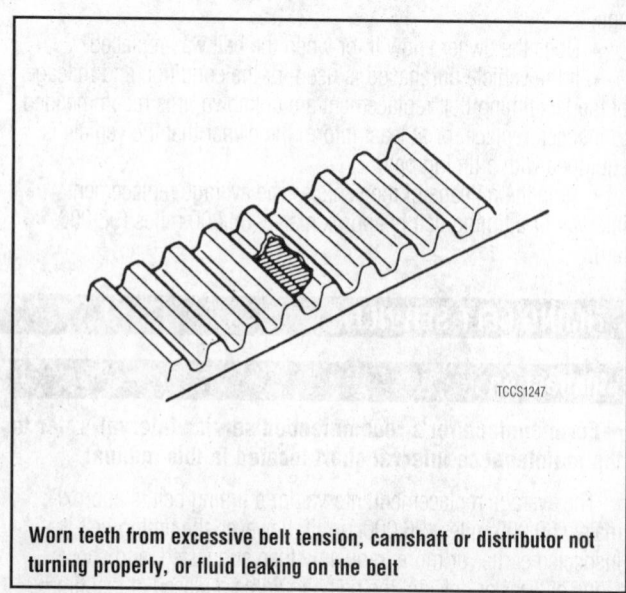

Worn teeth from excessive belt tension, camshaft or distributor not turning properly, or fluid leaking on the belt

Removal & Installation

CHRYSLER CORPORATION

2.0L (VIN C) SOHC Engine

1. On Cirrus, Stratus, Breeze, and Sebring Convertible models, disconnect the negative battery cable from the left strut tower. The ground cable is equipped with a insulator grommet which should be placed on the stud to prevent the negative battery cable from accidentally grounding.
2. On Neon models, disconnect the negative battery cable.
3. Remove the drive belts and accessories.
4. Raise and safely support the vehicle.
5. Remove the right inner splash-shield.
6. Remove the crankshaft damper.
7. Remove the right engine mount
8. Place a support under the engine.
9. Remove the engine mount bracket
10. Remove the timing belt cover.
11. Loosen the timing belt tensioner bolts.
12. Remove the timing belt and the tensioner.

➡ **When tensioner is removed from the engine it is necessary to compress the plunger into the tensioner body.**

13. Place the tensioner into a soft-jawed vise to compress the tensioner.
14. After compressing the tensioner place a pin (5/64 in. Allen wrench will work) into the plunger side hole to retain the plunger until installation.

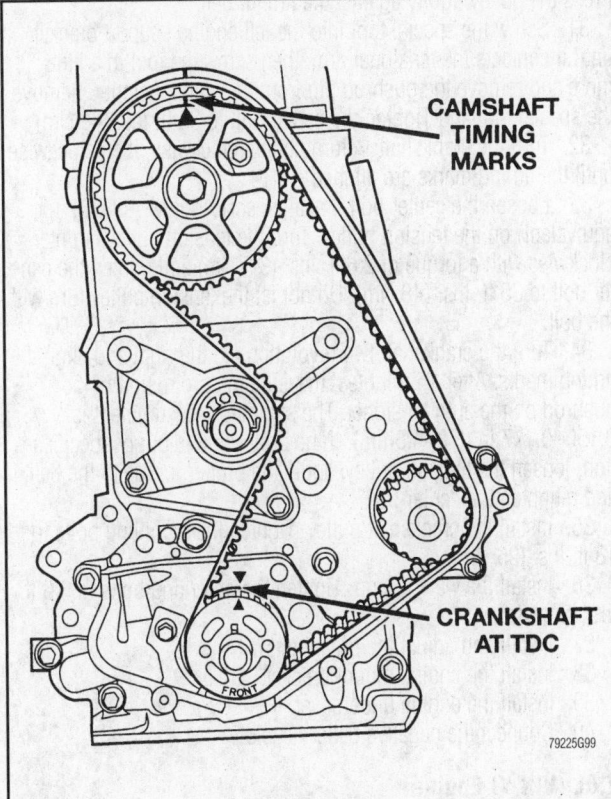

TDC alignment for timing belt installation—2.0L (VIN C) SOHC Engine

CAMSHAFT TIMING MARKS

CRANKSHAFT AT TDC

79225G99

To install:

15. Set the crankshaft sprocket to Top Dead Center (TDC) by aligning the notch on the sprocket with the arrow on the oil pump housing, then back off the sprocket three notches before TDC.
16. Set the camshaft to align the timing marks.
17. Move the crankshaft to ½ notch before TDC.
18. Install the timing belt starting at the crankshaft, around the water pump, then around the camshaft last.
19. Move the crankshaft to TDC to take up the belt slack.
20. Reinstall the tensioner to the block but do not tighten it.
21. Using a torque wrench apply 250 inch lbs. (28 Nm) of torque to the tensioner pulley.
22. With torque being applied to the tensioner pulley, move the tensioner up against the tensioner bracket and tighten the fasteners to 275 inch lbs. (31 Nm).
23. Remove the tensioner plunger pin, the tension is correct when the plunger pin can be removed and replaced easily.
24. Rotate the crankshaft two revolutions and recheck the timing marks.
25. Reinstall the timing belt cover.
26. Reinstall the engine mount bracket.
27. Reinstall the right engine mount.
28. Remove the engine support.
29. Reinstall the crankshaft damper and tighten to 105 ft. lbs. (142 Nm).
30. Reinstall the drive belts and accessories.
31. Reinstall the right inner splash-shield.
32. Perform the crankshaft and camshaft relearn alignment procedure using the DRB scan tool or equivalent.

2.0L (VIN F) Engine

1. Disconnect the negative battery cable.
2. Remove the engine undercover.
3. Remove the engine mount bracket.
4. Remove the drive belts.
5. Remove the belt tensioner pulley.
6. Remove the water pump pulleys.
7. Remove the crankshaft pulley.
8. Remove the stud bolt from the engine support bracket and remove the timing belt covers.
9. Rotate the crankshaft clockwise to line up the camshaft timing marks. Always turn the crankshaft in the forward direction only.
10. Loosen the tension pulley center bolt.

➡ **If the timing belt is to be reused, mark the direction of rotation on the flat side of the belt with an arrow.**

11. Move the tension pulley towards the water pump and remove the timing belt.
12. Remove the crankshaft sprocket center bolt using special tool MB990767 to hold the crankshaft sprocket while removing the center bolt. Then, use MB998778 or equivalent puller to remove the sprocket.
13. Mark the direction of rotation on the timing belt B with a arrow.
14. Loosen the center bolt on the tensioner and remove the belt.
15. To remove the camshaft sprocket, remove the cylinder head cover. Use a wrench to hold the hexagonal part of the camshaft and remove the sprocket mounting bolt.

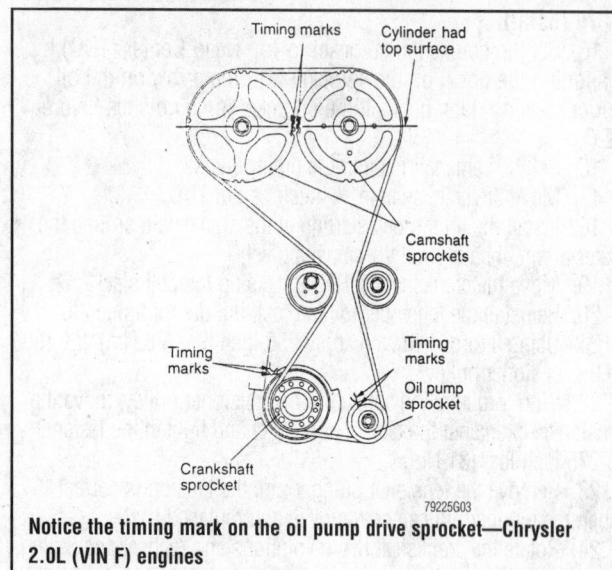

Notice the timing mark on the oil pump drive sprocket—Chrysler 2.0L (VIN F) engines

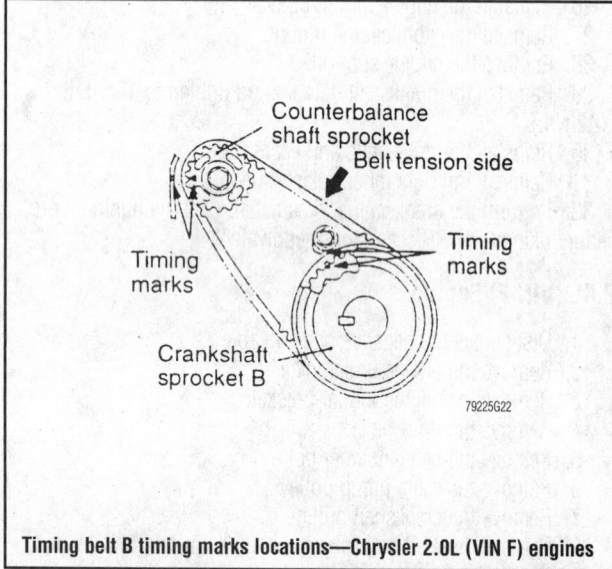

Timing belt B timing marks locations—Chrysler 2.0L (VIN F) engines

Do not rotate the camshafts or the crankshaft while the timing belt is removed.

To install:

16. Use a wrench to hold the camshaft, and install the sprocket and mounting bolt. Tighten the bolt (s) to 65 ft. lbs. (88 Nm).

17. Install the cylinder head cover.

18. Place the crankshaft sprocket on the crankshaft. Use tool MB990767 (or equivalent) to hold the crankshaft sprocket while tightening the center bolt. Tighten the center bolt to 80–94 ft. lbs. (108–127 Nm).

19. Align the timing marks on the crankshaft sprocket B and the balance shaft.

20. Install timing belt B on the sprockets. Position the center of the tensioner pulley to the left and above the center of the mounting bolt.

21. Push the pulley clockwise toward the crankshaft to apply tension to the belt and tighten the mounting bolt to 14 ft. lbs. (19

Nm). Do not let the pulley turn when tightening the bolt because it will cause excessive tension on the belt. The belt should deflect 0.20–0.28 in. (5–7mm) when finger pressure is applied between the pulleys.

22. Install the crankshaft sensing blade and the crankshaft sprocket. Apply engine oil to the mounting bolt and tighten the bolt to 80–94 ft. lbs. (108–127 Nm).

23. Use a press or vise to compress the auto tensioner pushrod. Insert a set pin when the holed are lined up.

Do not compress the pushrod too quickly, damage to the pushrod can occur.

24. Install the auto tensioner on the engine.

25. Align the timing marks on the camshaft sprocket, crankshaft sprocket and the oil pump sprocket.

26. After aligning the mark on the oil pump sprocket, remove the cylinder block plug and insert a Phillips screwdriver in the hole to check the position of the counter balance shaft. The screwdriver should go in at least 2.36 in. or more. If not, rotate the oil pump sprocket once and realign the timing mark so the screwdriver goes in. Do not remove the screwdriver until the timing belt is installed.

27. Install the timing belt on the intake camshaft and secure it with a clip.

28. Install the timing belt on the exhaust camshaft. Align the timing marks with the cylinder head top surface using two wrenches. Secure the belt with another clip.

29. Install the belt around the idler pulley, oil pump sprocket, crankshaft sprocket and the tensioner pulley.

30. Turn the tension pulley so the pinholes are at the bottom. Press the pulley lightly against the timing belt.

31. Screw the special tool into the left engine support bracket until it contacts the tensioner arm, then screw the tool in a little more and remove the pushrod pin from the auto tensioner. Remove the special tool and Tighten the center bolt to 35 ft. lbs. (48 Nm).

32. Turn the crankshaft ¼ turn counterclockwise, then clockwise until the timing marks are aligned.

33. Loosen the center bolt. Install special tool MD998767 (or equivalent) on the tension pulley. Turn the tension pulley counterclockwise with a torque of 2.6 ft. lbs. (3.5 Nm) and tighten the center bolt to 35 ft. lbs. (48 Nm). Do not let the tension pulley turn with the bolt.

34. Turn the crankshaft two revolutions to the right and align the timing marks. After 15 minutes, measure the protrusion of the pushrod on the auto tensioner. The standard measurement is 0.150–0.177 in (3.8–4.5mm). If the protrusion is out of specification, loosen the tension pulley, apply the proper torque to the belt and retighten the center bolt.

35. Install the crankshaft pulley. Tighten the mounting bolts to 18 ft. lbs. (25 Nm).

36. Install the water pump. Tighten the mounting bolts to 6.5 ft. lbs. (8.8 Nm).

37. Install and adjust the drive belts.

38. Install the engine mount bracket.

39. Install the engine undercover.

40. Connect the negative battery cable.

2.0L (VIN Y) Engine

Valve timing is critical to engine operation. Use care when servicing the timing belt. There are a number of timing marks that must be properly aligned or engine damage will result. If the timing belt has not broken, or jumped teeth, it is recommended that the crankshaft be turned by hand (clockwise) to TDC No.1 cylinder compression stroke (firing position) before beginning work. This should align all the timing marks and serve as a reference for later work. Some technicians will apply a small amount of white paint to all timing marks. This helps make them more visible under the low-light conditions found underhood.

1. Disconnect the negative battery cable.
2. Remove the accessory drive belts.
3. Using C 3281, or equivalent crankshaft holding tool, remove the crankshaft pulley center retaining bolt.
4. Using puller tools 1026 and 6827 or equivalent, remove the crankshaft pulley.
5. Remove the power steering pump from the bracket and position it out of the way. Do not disconnect the hoses.
6. Remove the power steering pump bracket from the engine.
7. Use a floor jack with a piece of wood on it and jack up the engine to take the weight off of the engine mount.
8. Remove the engine mount and bracket.
9. Remove the front timing belt cover.
10. If not done so previously, align the timing marks. Loosen the timing belt tensioner and remove the belt.

❊❊ WARNING

Do not rotate the crankshaft or camshafts after removing the timing belt or valvetrain components may be damaged. Always align the timing marks before removing the timing belt.

11. For 1995–97 models:
　a. If the timing belt tensioner is to be replaced, remove the retaining bolts and remove the timing belt tensioner. When the timing belt tensioner is removed from the engine it is necessary to compress the plunger into the tensioner body.
　b. Place the tensioner in a vise and slowly compress the plunger.

➡**Position the tensioner in the vise the same way it will be installed on the engine. This is to ensure proper pin orientation for when the tensioner is installed on the engine.**

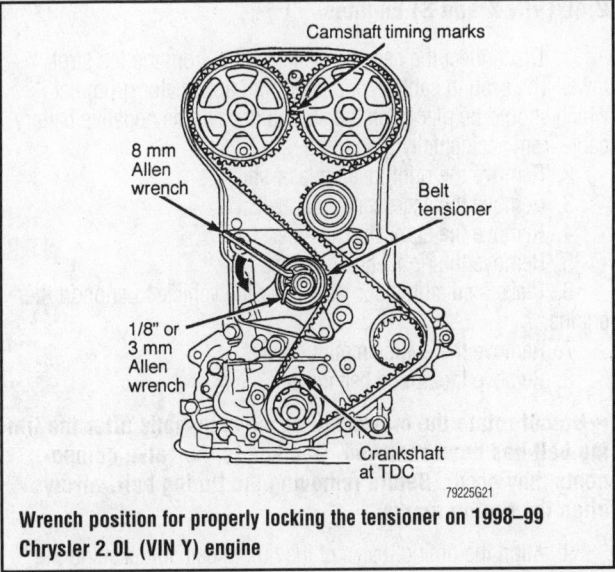

Wrench position for properly locking the tensioner on 1998–99 Chrysler 2.0L (VIN Y) engine

　c. When the plunger is compressed into the tensioner body, install a pin through the body and plunger to hold the plunger in place until the tensioner is installed.
12. For 1998–99 models:
　a. Place an 8mm Allen wrench into the belt tensioner, then using the long end of a ⅛ in. (3mm) Allen wrench, rotate the tensioner counterclockwise until it slides into the locking hole.
13. Remove the timing belt.

To install:

14. Check that all timing marks are still aligned. Bring the crankshaft sprocket to ½ a notch before TDC.
15. Install the timing belt. Starting at the crankshaft, route the belt around the water pump sprocket, idler pulley, camshaft sprockets, then around the tensioner pulley.
16. Move the crankshaft to TDC to take up the slack in the belt. Install the tensioner to the block but do not tighten the retaining bolts.
17. For 1995–97 models:
　a. Using a torque wrench on the tensioner pulley, apply 21 ft. lbs. (28 Nm) of torque to the pulley.
　b. With the torque being applied to the tensioner pulley, move the tensioner up against the tensioner pulley bracket and tighten the retaining bolts to 23 ft. lbs. (31 Nm).
　c. Remove the tensioner plunger pin. Pretension is correct when the pin can be removed and installed.
18. For 1998–99 models, remove the wrenches from the tensioner.
19. Rotate the crankshaft two revolutions and check the timing marks. If the timing marks are not properly aligned, remove the belt and reinstall it as described.
20. Install the front timing belt cover.
21. Lower the engine enough to install the engine mount bracket.
22. Install the bracket and remove the floor jack.
23. Install the power steering pump bracket and pump.
24. Install the crankshaft pulley using C-4685-C or an equivalent pulley installer.
25. Tighten the mounting bolt to 105 ft. lbs. (142 Nm).
26. Install the accessory drive belts.
27. Connect the negative battery cable.
28. Start the engine. Check for leaks and proper engine operation.

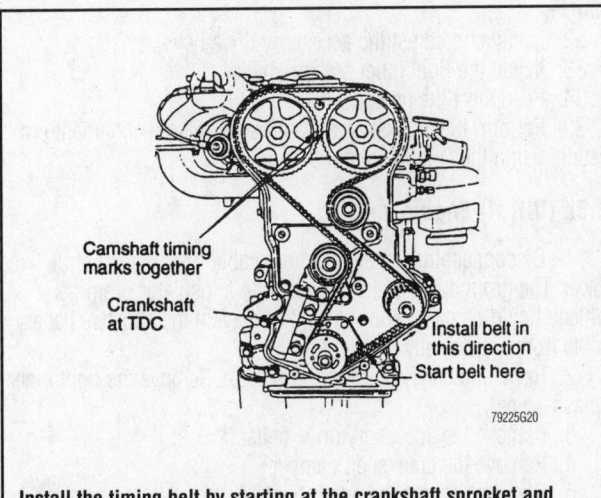

Install the timing belt by starting at the crankshaft sprocket and working around the other pulleys in a counterclockwise direction— Chrysler 2.0L (VIN Y) engine

2.4L (VIN X and S) Engines

1. Disconnect the negative battery cable from the left strut tower. The ground cable is equipped with a insulator grommet which should be placed on the stud to prevent the negative battery cable from accidentally grounding.
2. Remove the right inner splash-shield.
3. Remove the accessory drive belts.
4. Remove the crankshaft damper.
5. Remove the right engine mount.
6. Place a suitable floor jack under the vehicle to support the engine.
7. Remove the engine mount bracket
8. Remove the timing belt cover.

➡**Do not rotate the crankshaft or the camshafts after the timing belt has been removed. Damage to the valve components may occur. Before removing the timing belt, always align the timing marks.**

9. Align the timing marks of the timing belt sprockets to the

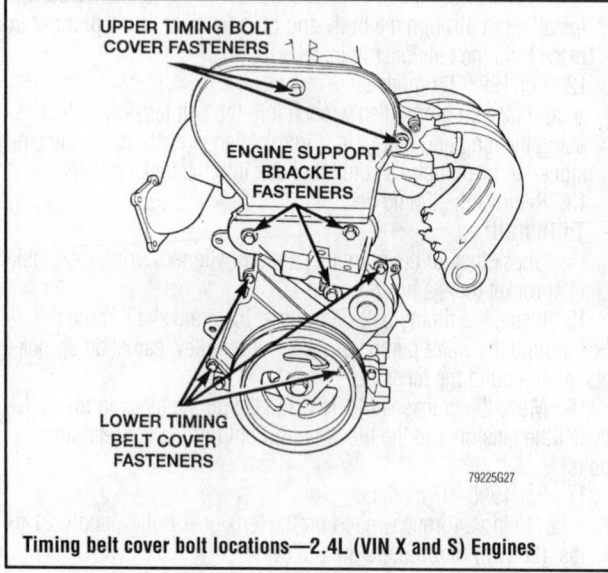

Timing belt cover bolt locations—2.4L (VIN X and S) Engines

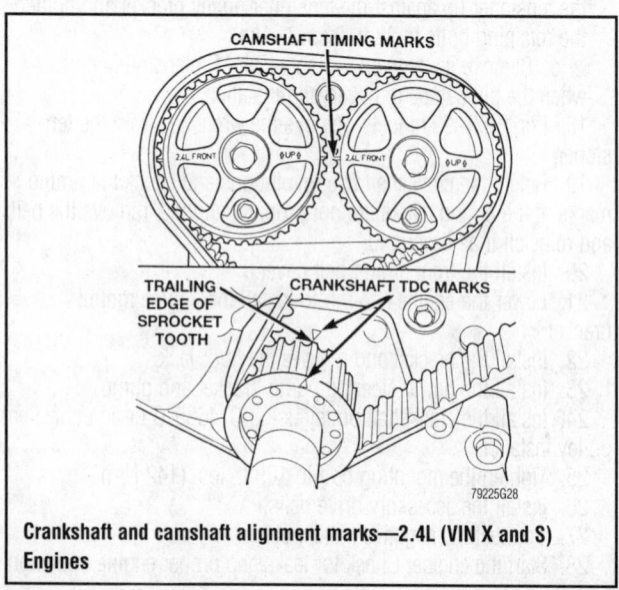

Crankshaft and camshaft alignment marks—2.4L (VIN X and S) Engines

timing marks on the rear timing belt cover and oil pump cover. Loosen the timing belt tensioner bolts.
10. Remove the timing belt and the tensioner.
11. Remove the camshaft timing belt sprockets.
12. Remove the crankshaft timing belt sprocket using special removal tool No. 6793 or equivalent.
13. Place the tensioner into a soft-jawed vise to compress the tensioner.
14. After compressing the tensioner, place a pin (⁵⁄₆₄ in. Allen wrench will work) into the plunger side hole to retain the plunger until installation.

To install:

15. Using special tool No. 6792, or its equivalent, install the crankshaft timing belt sprocket onto the crankshaft.
16. Install the camshaft sprockets onto the camshafts. Install and tighten the camshaft sprocket bolts to 75 ft. lbs. (101 Nm).
17. Set the crankshaft sprocket to Top Dead Center (TDC) by aligning the notch on the sprocket with the arrow on the oil pump housing.
18. Set the camshafts to align the timing marks on the sprockets.
19. Move the crankshaft to ½ notch before TDC.
20. Install the timing belt starting at the crankshaft, then around the water pump sprocket, idler pulley, camshaft sprockets, then around the tensioner pulley.
21. Move the crankshaft sprocket to TDC to take up the belt slack.
22. Reinstall the tensioner to the block, but do not tighten it at this time.
23. Using a torque wrench on the tensioner pulley, apply 250 inch. lbs. (28 Nm) of torque to the tensioner pulley.
24. With torque being applied to the tensioner pulley, move the tensioner up against the tensioner pulley bracket and tighten the fasteners to 275 inch. lbs. (31 Nm).
25. Remove the tensioner plunger pin, the tension is correct when the plunger pin can be removed and replaced easily.
26. Rotate the crankshaft two revolutions and recheck the timing marks. Wait several minutes, then recheck that the plunger pin can easily be removed and installed.
27. Reinstall the front timing belt cover.
28. Reinstall the engine mount bracket.
29. Reinstall the right engine mount.
30. Remove the floor jack from under the vehicle.
31. Install the crankshaft damper and tighten to 105 ft. lbs. (142 Nm).
32. Install and adjust the accessory drive belts.
33. Install the right inner splash-shield.
34. Reconnect the negative battery cable.
35. Perform the crankshaft and camshaft relearn alignment procedure using the DRB scan tool or equivalent.

2.5L (VIN H) Engine

1. Disconnect the negative battery cable from the left strut tower. The ground cable is equipped with a insulator grommet which should be placed on the stud to prevent the negative battery cable from accidentally grounding.
2. Raise and safely support the vehicle. Remove the right inner splash-shield.
3. Remove the accessory drive belts.
4. Remove the crankshaft damper.
5. Remove the right engine mount
6. Place a suitable floor jack under the vehicle to support the engine.

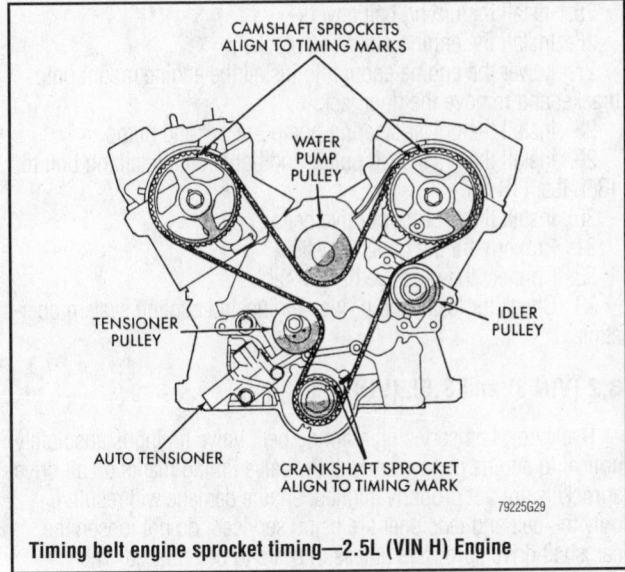

CAMSHAFT SPROCKETS
ALIGN TO TIMING MARKS

WATER
PUMP
PULLEY

TENSIONER
PULLEY

IDLER
PULLEY

AUTO TENSIONER

CRANKSHAFT SPROCKET
ALIGN TO TIMING MARK

79225G29

Timing belt engine sprocket timing—2.5L (VIN H) Engine

7. Remove the right engine mount bracket

8. Remove the timing belt upper left cover, upper right cover and lower cover.

9. Loosen the timing belt tensioner bolts.

➡**Before removing timing belt, be sure to align the sprocket timing marks to the timing marks on the rear timing belt cover.**

10. If the present timing belt is going to be reused, mark the running direction of the timing belt for installation. Remove the timing belt and the tensioner.

11. Remove the camshaft timing belt sprockets from the camshaft, if necessary.

12. Remove the crankshaft timing belt sprocket and key.

13. Place the tensioner into a soft-jawed vise to compress the tensioner.

14. After compressing the tensioner place a pin into the plunger side hole to retain the plunger until installation.

To install:

15. If removed, reinstall the camshaft sprockets onto the camshaft. Install the camshaft sprocket bolt and tighten to 65 ft. lbs. (88 Nm).

16. If removed, reinstall the crankshaft timing belt sprocket and key onto the crankshaft.

17. Set the crankshaft sprocket to Top Dead Center (TDC) by aligning the notch on the sprocket with the arrow on the oil pump housing, then back off the sprocket three notches before TDC.

18. Set the camshafts to align the timing marks on the sprockets with the marks on the rear timing belt cover.

19. Install the belt on the rear camshaft sprocket first.

20. Install a binder clip on the belt to the sprocket so it won't slip out of position.

21. Keeping the belt taut, install it under the water pump pulley and around the front camshaft sprocket.

22. Install a binder on the front sprocket and belt.

23. Rotate the crankshaft to TDC.

24. Continue routing the belt by the idler pulley and around the crankshaft sprocket to the tensioner pulley.

25. Move the crankshaft sprocket clockwise to TDC to take up the belt slack. Check that all timing marks are in alignment.

26. Reinstall the tensioner to the block but do not tighten it at this time.

27. Using special tool No. MD998767 (or equivalent) and a torque wrench on the tensioner pulley, apply 39 inch lbs. (4.4 Nm) of torque to the tensioner. Tighten the tensioner pulley bolt to 35 ft. lbs. (48 Nm).

28. With torque being applied to the tensioner pulley, move the tensioner up against the tensioner bracket and tighten the fasteners to 17 ft. lbs. (23 Nm).

29. Remove the tensioner plunger pin, the tension is correct when the plunger pin can be removed and replaced easily.

30. Rotate the crankshaft two revolutions clockwise and recheck the timing marks. Check to be sure the tensioner plunger pin can be easily installed and removed. If the pin does not remove and install easily, perform the procedure again.

31. Reinstall the timing belt cover.

32. Reinstall the engine mount bracket.

33. Reinstall the right engine mount.

34. Remove the engine support.

35. Install the crankshaft damper and tighten to 134 ft. lbs. (182 Nm).

36. Reinstall the accessory drive belts and adjust them.

37. Reinstall the right inner splash-shield.

38. Perform the crankshaft and camshaft relearn alignment procedure using the DRB scan tool or equivalent.

2.5L (VIN N) Engine

1. Disconnect the negative battery cable.

2. Remove the accessory drive belts.

3. Using Crankshaft Holding Tools MB990767 and MB998754, or their equivalents, remove the crankshaft bolt and remove the pulley.

4. Remove the heated oxygen sensor connection.

5. Remove the power steering pump with the hose attached and position it aside.

6. Remove the power steering pump bracket.

7. Place a floor jack under the engine oil pan, with a block of wood in between, and jack up the engine so that the weight of the engine is no longer being applied to the engine support bracket.

8. Remove the upper engine mount. Spraying lubricant, slowly remove the reamer (alignment) bolt and remaining bolts and remove the engine support bracket.

➡**The reamer bolt is sometimes heat-seized on the engine support bracket**

9. Remove the front timing belt covers.

10. If the timing belt is to be reused, draw an arrow indicating the direction of rotation on the back of the belt for reinstallation.

11. Align the timing marks by turning the crankshaft with MD998769 Crankshaft Turning tool or its equivalent. Loosen the center bolt on the timing belt tensioner pulley and remove the belt.

✲✲ WARNING

Do not rotate the crankshaft or camshaft after removing the timing belt, or valvetrain components may be damaged. Always align the timing marks before removing the timing belt.

12. Check the belt tensioner for leaks and check the pushrod for cracks.

13. If the timing belt tensioner is to be replaced, remove the retaining bolts and remove the timing belt tensioner. When the tim-

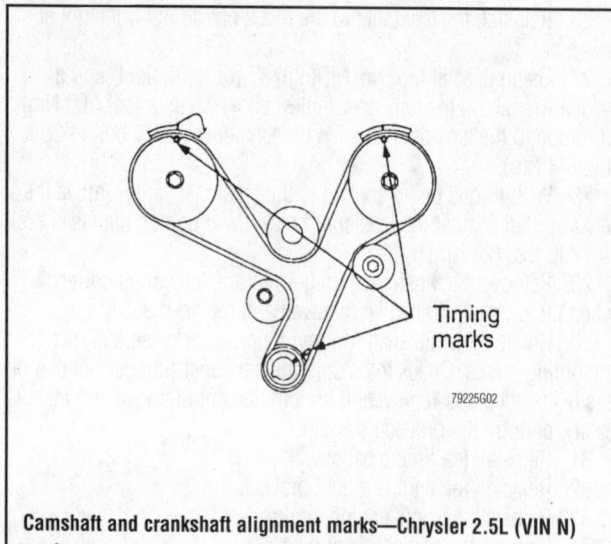

Camshaft and crankshaft alignment marks—Chrysler 2.5L (VIN N) engines

ing belt tensioner is removed from the engine it is necessary to compress the plunger into the tensioner body.

14. Place the tensioner in a vise and slowly compress the plunger. Take care not to damage the pushrod.

➡ **Position the tensioner in the vise the same way it will be installed on the engine. This is to ensure proper pin orientation for when the tensioner is installed on the engine.**

15. When the plunger is compressed into the tensioner body, install a pin through the body and plunger to hold the plunger in place until the tensioner is installed.

To install:

16. Install the timing belt tensioner and tighten the retaining bolts to 17 ft. lbs. (24 Nm), but do not remove the pin at this time.

17. Check that all timing marks are still aligned.

18. Use bulldog clips (large paper binder clips) or other suitable tool to secure the timing belt and to prevent it from slacking. Install the timing belt. Starting at the crankshaft, go around the idler pulley, then the front camshaft sprocket, the water pump pulley, the rear camshaft sprocket, then around the tensioner pulley.

19. Be sure the belt is tight between the crankshaft and front camshaft sprocket, between the camshaft sprockets and the water pump. Gently raise the tensioner pulley, so that the belt does not sag, and temporarily tighten the center bolt.

20. Move the crankshaft ¼ turn counterclockwise,, then turn it clockwise to the position where the timing marks are aligned.

21. Loosen the center bolt of the tensioner pulley. Using MD998767, or equivalent tensioner tool, and a torque wrench apply 3.3 ft. lbs. (4.4 Nm) tensional torque to the timing belt and tighten the center bolt to 35 ft. lbs. (48 Nm). When tightening the bolt, be sure that the tensioner pulley shaft does not rotate with the bolt.

22. Remove the tensioner plunger pin. Pretension is correct when the pin can be removed and installed easily. If the pin cannot be easily removed and installed it is still satisfactory as long as it is within its standard value.

23. Check that the tensioner pushrod is within the standard value. When the tensioner is engaged the pushrod should measure 0.149–0.177 in. (3.8–4.5mm).

24. Rotate the crankshaft two revolutions and check the timing marks. If the timing marks are not properly aligned remove the belt and repeat Steps 17 through 23.

25. Install the timing belt covers.

26. Install the engine mounting bracket.

27. Lower the engine enough to install the engine mount onto bracket and remove the floor jack.

28. Install the power steering pump bracket and pump.

29. Install the crankshaft pulley and tighten the retaining bolt to 13 ft. lbs. (18 Nm).

30. Install the accessory drive belts.

31. Properly fill the cooling system.

32. Connect the negative battery cable.

33. Check for leaks and proper engine and cooling system operation.

3.2 (VIN J) and 3.5L (VIN F) Engines

Use care when servicing a timing belt. Valve timing is absolutely critical to engine performance. If the valve timing marks on all drive sprockets are not properly aligned, engine damage will result. If only the belt and tensioner are being serviced, do not loosen the camshaft drive sprockets unless they are to be replaced. The sprockets have oversized openings and can be rotated several degrees in each direction on their shafts. This means the sprockets must be retimed, requiring some special tools.

✳✳ CAUTION

Fuel injection systems remain under pressure, even after the engine has been turned off. The fuel system pressure must be relieved before disconnecting any fuel lines. Failure to do so may result in fire and/or personal injury.

1. Disconnect the negative battery cable.

2. Rotate the engine to Top Dead Center (TDC) on the compression stroke for cylinder No. 1.

3. Release the fuel system pressure using the recommended procedure.

4. Place a pan under the radiator and drain the coolant.

5. Remove the radiator and cooling fan assemblies.

6. Remove the accessory drive belts.

7. Remove the upper radiator hose.

8. Remove the crankshaft damper with a quality puller tool gripping the inside of the pulley.

9. Remove the stamped steel cover. Do not remove the sealer on the cover; it may be reusable.

10. Remove the left side cast cover. If necessary, remove the lower belt cover, located behind the crankshaft damper.

11. If the timing belt is to be reused, mark the timing belt with the running direction for installation.

12. Align the camshaft sprockets with the marks on the rear covers.

13. Remove the timing belt and tensioner.

14. If it is necessary to service the camshaft sprockets, use the following procedure:

   a. Hold the camshaft sprocket with a 36mm box end wrench, loosen and remove the sprocket retaining bolt and washer.

➡ **To remove the camshaft sprocket retainer bolt while the engine is in the vehicle, it may be necessary to raise that side of the engine due to the length of the retainer bolt. The right bolt is 8 ⅜-in. (213mm) long, while the left bolt is 10in. (254mm) long. These bolts are not interchangeable and their original location during removal should be noted.**

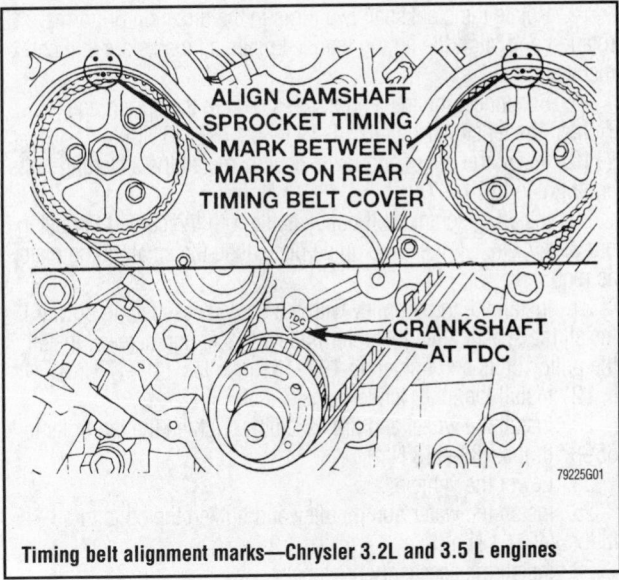

ALIGN CAMSHAFT SPROCKET TIMING MARK BETWEEN MARKS ON REAR TIMING BELT COVER

CRANKSHAFT AT TDC

79225G01

Timing belt alignment marks—Chrysler 3.2L and 3.5 L engines

b. Remove the camshaft sprocket from the camshaft. The camshaft sprockets are not interchangeable from side-to-side.

c. Remove the crankshaft sprocket using Puller L-4407A, or equivalent.

To install:

15. If it was necessary to remove the camshaft sprockets, use the following procedure:

➡ **This procedure can only be used when the camshaft sprockets have been loosened or removed from the camshafts. Each sprocket has a D -shaped hole that allows it to be rotated several degrees in each direction on its shaft. This design must be timed with the engine to ensure proper performance.**

a. Install the crankshaft sprocket, using tool C-4685-C1, thrust bearing, washer and 12mm bolt.

b. When the camshaft sprockets are loosened or removed, the camshafts must be timed to the engine. Install the Camshaft Alignment tools 6642-A, or their exact equivalents, to the rear of the cylinder heads. These tools lock the camshafts in the proper position.

16. Preload the belt tensioner as follows:

a. Place the tensioner in a vise the same way it is mounted on the engine.

b. Slowly compress the plunger into the tensioner body.

c. When the plunger is compressed into the tensioner body, install a pin through the body and plunger to retain the plunger in place until the tensioner is installed.

17. Install both camshaft sprockets to the appropriate shafts. The left camshaft sprocket has the DIS pick-up as part of the sprocket.

➡ **The right bolt is 8 ⅜ in. (213mm) long, while the left bolt is 10in. (254mm) long. These bolts are not interchangeable.**

18. Apply Loctite®271 or equivalent, to the threads of the camshaft sprocket retainer bolts and install to the appropriate shafts. Do not tighten the bolts at this time.

19. Align the camshaft sprockets between the marks on the rear belt covers.

20. Align the crankshaft sprocket with the TDC mark on the oil pump cover.

21. Install the timing belt, starting at the crankshaft sprocket and going in a counterclockwise direction. After the belt is installed on the right sprocket, keep tension on the belt until it is past the tensioner pulley.

22. Holding the tensioner pulley against the belt, install the timing belt tensioner into the housing and tighten to 21 ft. lbs. (28 Nm).

23. When the tensioner is in place, pull the retainer pin out to allow tensioner to extend to the pulley bracket.

➡ **Be sure that the timing marks on the cam sprockets are still between the marks on the rear cover.**

24. Remove the spark plug in the No.1 cylinder and install a dial indicator to check for Top Dead Center (TDC) of the piston. Rotate the crankshaft until the piston is exactly at TDC.

25. Hold the camshaft sprocket hex with a 36mm wrench and tighten the right camshaft sprocket bolt to 75 ft. lbs. (102 Nm) plus an additional 90 degree turn. Tighten the left camshaft sprocket bolt to 85 ft. lbs. (115 Nm) plus an additional 90 degree turn.

26. Remove the dial indicator. Install the spark plug and tighten to 20 ft. lbs. (28 Nm).

27. Remove the camshaft alignment tools from the back of the cylinder heads and install the cam covers with new O-rings.

28. Tighten the fasteners to 20 ft. lbs. (27 Nm). Repeat this procedure on the other camshaft.

29. Rotate the crankshaft sprocket two revolutions and check for proper alignment of the timing marks on the camshaft and the crankshaft. If the timing marks do not line up, repeat the procedure.

30. Before installing, inspect the sealer on the stamped steel cover. If some sealer is missing, use MOPAR Silicone Rubber Adhesive sealant or equivalent to replace the missing sealer.

31. Install the lower belt cover behind the crankshaft damper, if necessary.

32. Install the stamped steel cover and the left side cast cover. Tighten the 6mm bolts to 105 in.. lbs. (12 Nm), the 8mm bolts to 250 inch lbs. (28 Nm) and the 10mm bolts to 40 ft. lbs. (54 Nm).

33. Install the crankshaft damper using special tool L-4524, a 5.9 in. long bolt, thrust bearing and washer or equivalent damper installation tools. Tighten the center bolt to 85 ft. lbs. (115 Nm).

34. Install the upper radiator hose.

35. Install the accessory drive belts and adjust them to the proper tension.

36. Install the radiator and cooling fan assemblies.

37. Refill and bleed the cooling system.

38. Connect the negative battery cable.

39. With the radiator cap off so coolant can be added, run the engine. Watch for leaks and listen for unusual engine noises.

FORD MOTOR COMPANY

1.3L (VIN H) Engine

1. Disconnect the negative battery cable.

2. Remove the accessory drive belts.

3. Remove the three water pump pulley attaching bolts and remove the water pump pulley.

4. Raise and safely support the vehicle on jack stands.

5. Remove the right front wheel and tire assembly and the right inner fender panel.

6. Remove the four attaching bolts and the screws from the crankshaft pulley. Remove the spacer and outer pulley, if equipped.

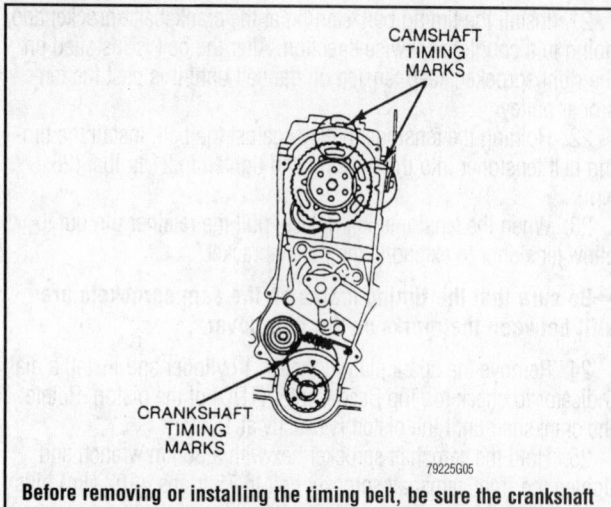

CAMSHAFT
TIMING
MARKS

CRANKSHAFT
TIMING
MARKS

79225G05

Before removing or installing the timing belt, be sure the crankshaft and both camshaft matchmarks are aligned as shown—Ford 1.3L (VIN H) engine

Remove the inner spacer, inner pulley and the baffle or guide plates, as required.

7. Remove the attaching bolts and the upper and lower covers.

8. Rotate the crankshaft until the sprocket timing marks are aligned.

9. Remove the timing belt tensioner spring, spring cover and timing belt tensioner bolt. Remove the timing belt.

➡**If the timing belt is to be reused, mark the direction of rotation on the belt, using a crayon, so the belt can be reinstalled in the same direction.**

10. If the camshaft sprocket requires removal, proceed as follows:

   a. Hold the camshaft stationary with an open end wrench and remove the camshaft sprocket retaining bolt.

   b. Pull the camshaft sprocket with the dowel pin off of the camshaft. Use care not to drop the dowel pin.

11. If the crankshaft sprocket requires removal, proceed as follows:

   a. Remove the crankshaft pulley retaining bolt.

   b. Pull the crankshaft pulley hub, sprocket and key from the crankshaft. Be sure not to drop the crankshaft key.

To install:

12. If removed, install the crankshaft sprocket as follows:

   a. Install the sprocket with the key onto the crankshaft.

   b. Install the crankshaft pulley hub.

   c. Clean the threads of the crankshaft pulley bolt and coat with a non-hardening sealer.

   d. Install the bolt and tighten to 80–85 ft. lbs. (108–118 Nm).

13. If removed, install the camshaft sprocket as follows:

   a. Position the sprocket and dowel pin to the camshaft and install the retaining bolt.

   b. Hold the camshaft stationary with an open end wrench and tighten the retaining bolt to 36–45 ft. lbs. (49–61 Nm).

14. Align the camshaft and crankshaft timing marks with the marks located on the cylinder head and oil pump housing.

15. If reusing the original timing belt, install the timing belt with the mark made indicating the direction of rotation.

16. Install the timing belt tensioner spring and cover on the tensioner. Position the tensioner and spring assembly on the engine and install the attaching bolt. Do not tighten the bolt at this time.

17. Rotate the crankshaft two turns in the direction of normal rotation and align the timing marks. Ensure all marks are still correctly aligned.

18. Reconnect the free end of the spring to the spring anchor. Tighten the tensioner bolt to 14–19 ft. lbs. (19–26 Nm).

19. Install the upper and lower covers. Install the attaching bolts and tighten to 71–97 inch lbs. (8–11 Nm).

20. Install the crankshaft pulley baffle with the curved lip facing outward, or install the large guide plate, then the small guide plate, as required.

21. Install the inner pulley with the deep recess facing outward. Install the spacer, then the outer pulley, spacer and screws. Install the pulley bolts and tighten to 109–152 inch lbs. (12–17 Nm).

22. Install the inner fender panel.

23. Install the wheel and tire assembly. Tighten the lug bolts to 65–87 ft. lbs. (88–118 Nm).

24. Lower the vehicle.

25. Install the water pump pulley and tighten the bolts to 36–45 ft. lbs. (49–61 Nm).

26. Install the accessory drive belts.

27. Connect the negative battery cable.

28. Run the engine and check for proper operation.

1.8L (VIN 8) Engine

1. Disconnect the negative battery cable.

2. Remove the timing belt upper cover and gasket.

3. Remove the accessory drive belts.

4. Remove the water pump pulley bolts and remove the pulley.

5. Raise and safely support the vehicle on jack stands.

6. Remove the right front wheel and tire assembly.

7. Remove the right upper and lower splash-shields.

8. Remove the timing belt middle and lower covers along with the gaskets.

9. Remove the crankshaft pulley hub bolt and hub.

10. Rotate the crankshaft and align the timing marks located on the camshaft sprockets and seal plate.

11. Check that the crankshaft sprocket and the oil pump are aligned.

➡**If the timing belt is to be reused, mark an arrow on the belt to indicate it's rotational direction for installation reference.**

12. Loosen the timing belt tensioner bolt.

13. Turn the timing belt tensioner counterclockwise and hand-tighten the tensioner bolt to relieve the tension on the timing belt.

14. Remove the timing belt.

15. If the camshaft sprockets are to be removed, continue as follows:

   a. Disconnect and tag the ignition wires and vacuum lines blocking the removal of the cylinder head cover.

   b. Remove the cylinder head cover retaining bolts and remove the cover and gasket.

   c. While holding the camshaft with a wrench, remove the camshaft sprocket retaining bolt.

   d. Remove the camshaft sprocket.

   e. If removing both camshaft sprockets, tag the sprockets for identification at reassembly.

16. If removing the crankshaft sprocket, remove the crankshaft pulley bolt and hub, if not already done. Slide the crankshaft sprocket off the crankshaft.

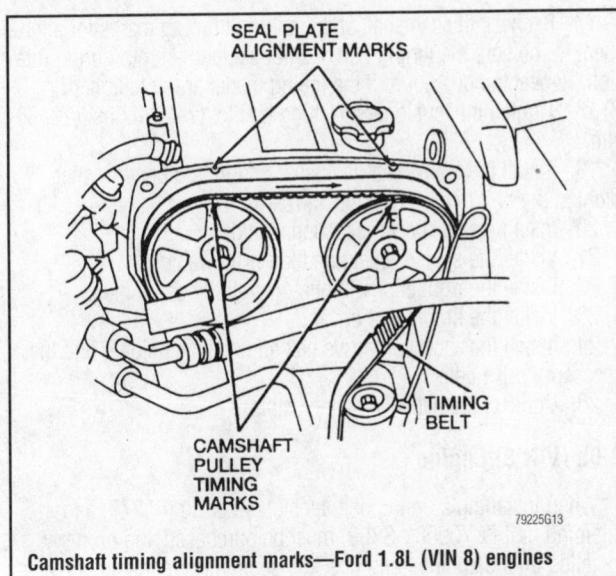

Camshaft timing alignment marks—Ford 1.8L (VIN 8) engines

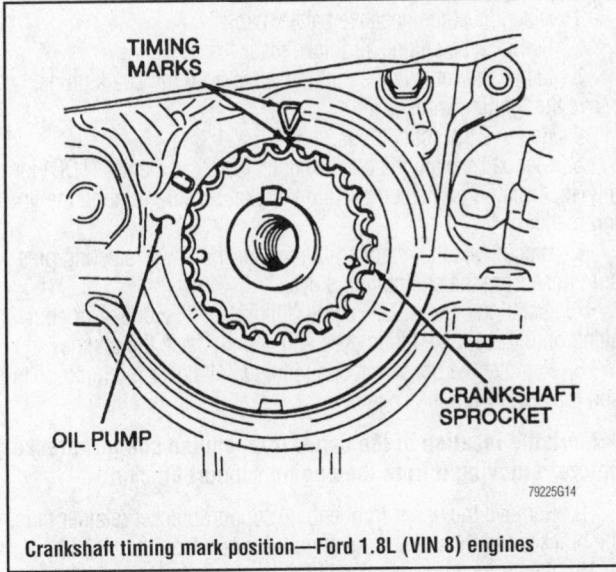

Crankshaft timing mark position—Ford 1.8L (VIN 8) engines

17. Inspect the timing belt tensioner and spring, replace if necessary.

To install:

18. If the crankshaft sprocket was removed, install the crankshaft key with the tapered end facing the oil pump. Install the crankshaft sprocket onto the crankshaft while making sure to match the alignment grooves.

19. If removed, install the camshaft sprockets as follows

a. Turn the camshaft until the dowel pins face straight up.

b. Install the camshaft sprocket with the **I** mark straight up for the intake camshaft or with the **E** mark straight up for the exhaust camshaft.

c. Align the camshaft sprockets with the timing marks on the seal plate.

d. While holding each camshaft with a wrench, install the camshaft sprocket retaining bolts. Tighten the bolts to 36–45 ft. lbs. (49–61 Nm).

e. Install a new cylinder head cover gasket onto the cylinder head.

f. Place the cylinder head cover into its mounting position and install the retaining bolts. Tighten the cylinder head cover bolts to 43–78 inch lbs. (4.9–8.8 Nm).

g. Install the ignition wires to the spark plugs and connect the vacuum hoses to the cylinder head cover.

20. Temporarily secure the timing belt tensioner in the far left position.

21. Verify that the timing marks on the crankshaft sprocket and the oil pump are aligned.

22. Verify that the timing marks on the camshaft sprockets and the seal plate are aligned.

23. Install the timing belt in a counterclockwise motion. Be sure there is no looseness on the idler side of the timing belt or between the camshaft sprockets.

➡**If using the old timing belt, be sure to install the belt in the same direction of travel as it was removed.**

24. Loosen the timing belt tensioner bolt. Allow the tensioner spring to apply tension to the timing belt.

25. Rotate the crankshaft 1 ⅚ turns clockwise and align the timing belt pulley mark with the tension set mark which is located at approximately the 10 o'clock position.

26. Turn the crankshaft two turns clockwise and align the crankshaft sprocket with the tension set mark on the oil pump.

27. Verify that all timing marks are aligned. If not, remove the timing belt and repeat the installation procedures.

28. Apply tension to the timing belt tensioner and tighten the tensioner lockbolt to 27–38 ft. lbs. (37–52 Nm).

29. Rotate the crankshaft 2 ⅙ (780 degrees) turns clockwise and verify that the camshaft and crankshaft timing marks are aligned.

30. Measure the timing belt deflection by applying 22 lbs. (98 N) of pressure on the timing belt between the camshaft sprockets. The timing belt deflection should be 0.35–0.45 in. (9–11.5mm). If necessary to adjust the timing belt deflection, rotate the crankshaft two turns clockwise and ensure that the timing marks are still aligned. If the timing marks are not aligned, repeat the installation procedure.

31. Install the crankshaft pulley hub and tighten the retaining bolt to 80–87 ft. lbs. (108–118 Nm).

32. Install the crankshaft pulley and washer. Install the retaining bolts and tighten to 109–152 inch lbs. (12–17 Nm).

33. Install the timing belt middle and lower covers with the gaskets. Tighten the middle and lower timing belt cover retaining bolts to 65–95 inch lbs. (7.8–11 Nm,).

34. Install the power steering drive belt.

35. Install the water pump pulley and retaining bolts. Tighten the bolts to 69–95 inch lbs. (7.8–11.0 Nm).

36. Install the alternator/water pump drive belt.

37. Install the splash-shields. Tighten the bolts to 69–95 inch lbs. (7.8–11.0 Nm).

38. Install the right wheel and tire assembly. Tighten the lug nuts to 65–87 ft. lbs. (88–118 Nm).

39. Lower the vehicle.

40. Install the timing belt upper cover and gasket. Tighten the bolts to 69–95 inch lbs. (7.8–11.0 Nm).

41. Connect the negative battery cable.

42. Run the engine and check for leaks.

43. Road test the vehicle and check for proper engine operation.

1.9L (VIN J) Engine

1. Disconnect the negative battery cable.
2. Remove the accessory drive belt automatic tensioner and the accessory drive belt.
3. Remove the timing belt cover.
4. Align the timing mark on the camshaft sprocket with the timing mark on the cylinder head.
5. Confirm that the timing mark on the crankshaft sprocket is aligned with the timing mark on the oil pump housing.
6. Loosen the belt tensioner attaching bolt, pry the tensioner away from the timing belt and retighten the bolt.
7. Remove the spark plugs. Remove the right engine mount.
8. Raise and safely support the vehicle on jack stands.
9. Remove the right side splash-shield.
10. Remove the flywheel inspection shield.
11. Use a suitable tool to hold the flywheel in place.
12. Remove the crankshaft damper bolt and washer and remove the bolt.
13. Remove the timing belt.

➡ **With the timing belt removed and the No. 1 piston at TDC, do not rotate the camshaft. If the camshaft must be rotated, align the crankshaft damper 90 degrees BTDC.**

To install:

14. Install the timing belt over the sprockets in a counterclockwise direction starting at the crankshaft. Keep the belt span from the crankshaft to the camshaft tight while installing over the remaining sprocket.
15. Loosen the belt tensioner attaching bolt, allowing the tensioner to snap against the belt.
16. Rotate the crankshaft clockwise two complete revolutions, stopping at TDC. This will allow the tensioner spring to load the timing belt.

➡ **Do not turn the engine counterclockwise to align the timing marks. Do not rotate the crankshaft with the spark plugs installed.**

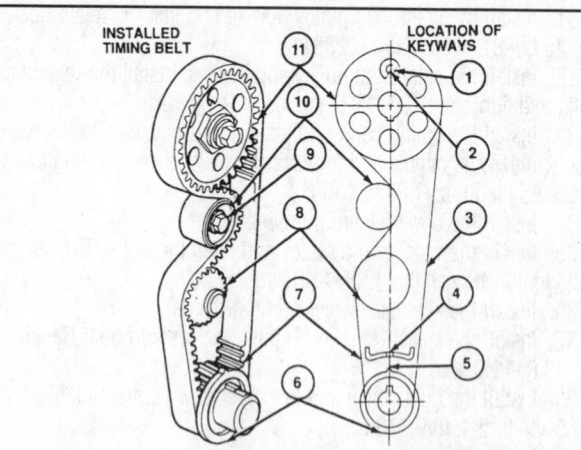

INSTALLED TIMING BELT — LOCATION OF KEYWAYS

1. Cylinder head timing mark
2. Camshaft sprocket timing pointer
3. Tension check point
4. Oil pump timing mark
5. Crankshaft timing pointer
6. Crankshaft sprocket
7. Timing belt
8. Water pump sprocket
9. Timing belt tensioner bolt
10. Timing belt tensioner
11. Camshaft sprocket

View of timing belt and alignment positions—Ford 1.9L (VIN J) Engine

79225G98

17. Recheck the camshaft and crankshaft timing marks for alignment, to be sure the timing belt has not skipped a tooth during rotation. Repeat the procedure if the timing marks are not aligned.
18. Tighten the tensioner attaching bolt to 17–22 ft. lbs. (23–30 Nm).
19. Install the crankshaft dampener and the bolt and washer. Tighten the bolt to 81–96 ft. lbs. (110–130 Nm).
20. Install the flywheel inspection shield.
21. Install the splash-shield and lower the vehicle.
22. Install the right engine mount. Install the spark plugs.
23. Install the timing belt cover.
24. Install the accessory drive belt automatic tensioner and the accessory drive belt.
25. Connect the negative battery cable.

2.0L (VIN 3) Engine

When installing a timing belt, tensioner spring (6L277) and retaining bolt (W700001-S309) must be purchased and properly installed on the engine. First check to see if these parts are already installed. The tensioner spring will adjust the timing belts tension and should not require further adjustments.

1. Disconnect the negative battery cable.
2. Remove the engine air intake resonators.
3. Label and remove the ignition wires from the spark plugs. Move the ignition wires aside.
4. Remove the spark plugs.
5. Manually rotate the crankshaft to Top Dead Center (TDC) for the No. 1 piston on its compression stroke. Be sure to align the timing marks.
6. Disconnect the retaining bracket for the power steering pressure hose from the engine lifting eye.
7. Install the Three Bar Engine Support D88L-6000-A or equivalent, onto the engine lifting eyes and slightly raise the engine.
8. Remove the upper camshaft timing belt cover retaining bolts and the cover from the engine.

➡ **Mark the location of the upper front engine support bracket before removing it from the engine support bracket.**

9. Remove the upper front engine support bracket retainer nuts, the bracket and the upper front engine support insulator.
10. If equipped, remove the wiring harness connector from the low coolant level sensor at the radiator coolant recovery reservoir.
11. Remove the radiator coolant recovery reservoir retainers and move the reservoir aside.
12. Remove the upper front engine support insulator.
13. Set the coolant recovery reservoir back into position temporarily.
14. Loosen the water pump pulley retaining bolts. Do not remove the bolts completely.
15. Remove the accessory drive belt.
16. Remove the drive belt idler pulley retaining bolt and pulley from the alternator mounting bracket.
17. Finish removing the water pump retaining bolts and remove the water pump pulley.
18. Remove the center camshaft timing belt cover retaining bolts and the cover from the engine.
19. Raise and safely support the vehicle on jack stands.
20. Remove the crankshaft pulley.
21. Remove the lower camshaft timing belt cover bolts and the cover from the engine.
22. Remove the valve cover as follows:

a. Disconnect the crankcase ventilation tube from the valve cover.

b. Remove the retaining bolt and nut for the power steering pressure hose retaining bracket and move the hose aside.

c. Remove the valve cover retaining bolts in a standard removal sequence starting from the outside of the valve cover and working toward the inside of the valve cover.

d. Remove the valve cover and gasket from the engine.

23. Place Camshaft Alignment Timing Tool T94P-6256-CH or equivalent, into the slots of both camshafts at the rear of the cylinder head to lock the camshafts into position.

24. Loosen the camshaft timing belt tensioner pulley retaining bolt and move the tensioner pulley to relieve the tension on the timing belt.

25. Temporarily tighten the tensioner in this position.

➡ **If the timing belt is to be reused, mark the belt for the direction of rotation before removing to prevent premature wear or failure.**

26. Remove the timing belt.

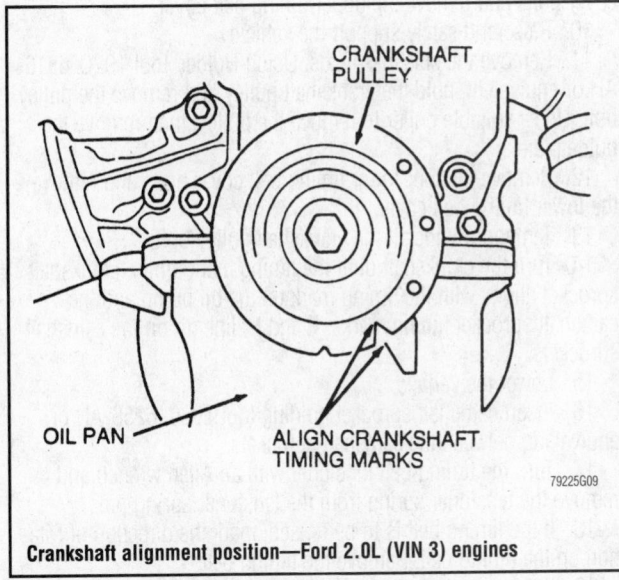

Crankshaft alignment position—Ford 2.0L (VIN 3) engines

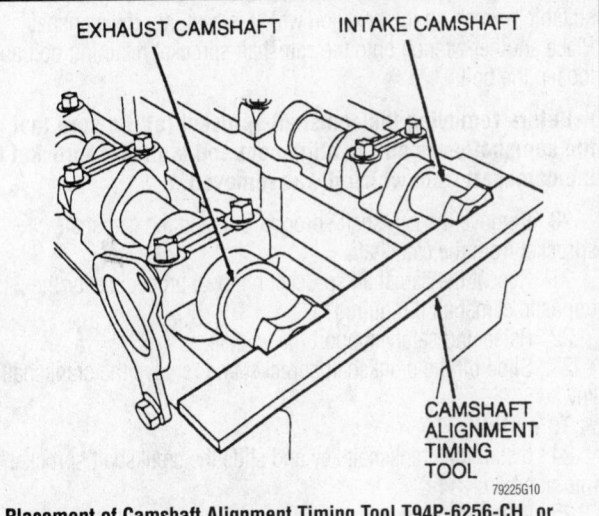

Placement of Camshaft Alignment Timing Tool T94P-6256-CH, or equivalent—Ford 2.0L (VIN 3) engines

27. If required, remove the sprockets as follows:

a. Hold the camshaft with the Camshaft Sprocket Holding Tool T74P-6256-B or equivalent.

b. Loosen and remove the camshaft sprocket retaining bolt.

c. Remove the sprocket from the camshaft.

d. Repeat the procedure for the 2nd camshaft sprocket.

e. Remove the crankshaft sprocket.

To install:

28. Slide the crankshaft sprocket onto the crankshaft aligning the keyway.

29. Align the camshafts using the Camshaft Alignment Timing Tool T94P-6256-CH.

30. Reinstall the sprockets onto the camshafts and loosely install the camshaft retaining bolts.

31. Tighten the camshaft sprocket retaining bolts to 47–53 ft. lbs. (64–72 Nm).

32. Loosely install the crankshaft pulley to verify that the engine is at TDC. Realign the marks if they have moved.

33. Verify that the camshafts are aligned.

➡ **It is recommended to purchase a tensioner spring and retaining bolt through the dealer parts to apply the proper tension for used or new belt installations. The spring is bolted to the tensioner assembly and becomes a part of the engine. Ignore this notice if the tensioner spring is already installed.**

34. Reinstall the retaining bolt (W700001-S309) into the hole provided in the cylinder block and place the tensioner spring (6L277) between the bolt and the camshaft timing belt tensioner pulley.

35. Tighten the retainer bolt to 71–97 inch lbs. (8–11 Nm).

36. Remove the crankshaft pulley and install the timing belt onto the crankshaft sprocket, then onto the camshaft sprockets working in a counterclockwise direction.

37. Tighten the camshaft sprocket retaining bolts to 47–53 ft. lbs. (64–72 Nm).

38. Be sure that the span of the camshaft timing belt between the crankshaft sprocket and the exhaust camshaft sprocket is not loose.

39. Be sure that the camshaft timing belt is securely aligned on all sprockets.

40. Reinstall the lower timing belt cover and tighten the retaining bolts to 53–71 inch lbs. (6–8 Nm).

41. Apply silicone sealer to the keyway of the crankshaft pulley and install. Tighten the retaining bolt to 81–89 ft. lbs. (110–120 Nm).

42. Inspect the timing mark on the crankshaft pulley to verify that the engine is still at TDC.

43. Loosen the camshaft timing belt tensioner pulley retaining bolt and allow the tensioner spring attached to the pulley to draw the tensioner pulley against the camshaft timing belt.

44. Remove the camshaft alignment timing tool from the camshafts at the rear of the engine.

45. Turn the crankshaft two revolutions in a clockwise direction.

46. Tighten the camshaft timing belt tensioner pulley retaining bolt to 26–30 ft. lbs. (35–40 Nm).

47. Recheck that the crankshaft timing mark is at TDC for the No. 1 piston, and that both camshafts are in alignment using the camshaft alignment timing tool.

➡ **A slight adjustment of the camshafts to allow the insertion of the camshaft alignment timing tool is permissible as long as the crankshaft stays at the TDC location.**

48. Camshaft Sprocket Holding Tool T74P-6256-b can be used to move the camshaft sprocket (s) if a slight adjustment is required.

49. If a camshaft is not properly aligned, perform the following procedure:

a. Loosen the retaining bolt securing the sprocket to the camshaft while holding the camshaft sprocket from turning with the sprocket holding tool.

b. Turn the camshaft until the camshaft alignment timing tool can be installed.

c. Verify that the crankshaft timing mark is at TDC for the No. 1 cylinder.

d. While holding the camshaft sprocket with the camshaft sprocket holding tool, tighten the retaining bolt to 47–53 ft. lbs. (64–72 Nm).

e. Remove the tool and rotate the crankshaft two revolutions (clockwise).

f. Verify that the camshafts are aligned and that the crankshaft is at TDC for the No. 1 cylinder.

50. Reinstall the valve cover as follows:

a. Clean the gasket sealing surfaces.

b. Inspect the valve cover gasket and O-rings; replace as required.

c. Reinstall the valve cover retaining bolts and tighten them in a standard sequence starting from the center and working towards the outside of the valve cover to 53–71 inch lbs. (6–8 Nm).

d. Reinstall the power steering hose retaining bracket and the power steering hose.

e. Reinstall the crankcase ventilation tube to the valve cover.

51. Position the center camshaft timing belt cover.

52. Reinstall the center camshaft timing belt cover retaining bolts and tighten them to 53–71 inch lbs. (6–8 Nm).

53. Reinstall the water pump pulley and the retaining bolts. Reinstall the bolts, finger-tight.

54. Reinstall the drive belt idler pulley.

55. Reinstall the drive belt idler pulley retaining bolt and tighten it to 35 ft. lbs. (48 Nm).

56. Reinstall the accessory drive belt.

57. Tighten the water pump pulley retaining bolts to 89–124 inch lbs. (10–14 Nm).

58. Move the radiator coolant recovery reservoir aside.

59. Reinstall the upper front engine support insulator.

60. Position the radiator coolant recovery reservoir and install the retainers.

61. If equipped, install the wiring harness to the low coolant level sensor on the coolant recovery reservoir.

62. Reinstall the upper front engine support bracket to the engine and the upper front engine support insulator using the mark made during the removal procedure for reference.

63. Install the upper camshaft timing belt cover.

64. Reinstall the upper camshaft timing belt cover retaining bolts and tighten to 27–44 inch lbs. (3–5 Nm).

65. Remove the engine support.

66. Reinstall the retaining bracket for the power steering pressure hose to the engine lifting eye.

67. Reinstall the spark plugs and the ignition wires.

68. Reinstall the engine air intake resonators.

69. Replace the engine oil.

70. Reconnect the negative battery cable.

71. Run the engine and check for leaks and proper operation.

2.0L (VIN A) Engine

1. Disconnect the negative battery cable.

2. Label and disconnect the spark plug wires and clips from the cylinder head cover. Remove the ignition distributor with wiring and set it aside.

3. Remove the power steering hose brackets from the cylinder head cover. If necessary disconnect the crankshaft position sensor.

4. Disconnect the breather tube and PCV valve from the cylinder head cover.

5. Loosen the cylinder head cover bolts in 2–3 steps. Remove the cylinder head cover.

6. Remove the power steering belt shield. Loosen the power steering adjusting bolt, lockbolt and through-bolt and remove the power steering belt.

7. Loosen the alternator adjusting bolt and upper mounting bolt. Remove the alternator belt.

8. Support the engine with Engine Support Tool 014–00750 or equivalent. Raise the engine slightly with a jack and remove the right side engine support insulator (mount).

9. Remove the oil level indicator bolt and four upper timing belt cover bolts and remove the upper timing belt cover.

10. Raise and safely support the vehicle.

11. Remove the splash-shields. Using Holder Tool T92C-6316-AH or equivalent, hold the crankshaft pulley and remove the pulley bolt. Use a suitable puller to remove the pulley, then remove the guide plate.

12. Remove the four lower timing belt cover bolts and remove the lower timing belt cover.

13. Temporarily install the crankshaft pulley bolt.

14. Turn the crankshaft until the timing mark on the crankshaft sprocket aligns with the timing mark on the oil pump and the camshaft sprocket timing marks, **E** and **I**, line up on the camshaft sprockets.

15. Lower the vehicle.

16. Insert camshaft sprocket holding tool T92C-6256-AH or equivalent, between the camshaft sprockets.

17. Turn the timing belt tensioner with an Allen wrench and remove the tensioner spring from the tensioner spring pin.

18. If the timing belt is to be reused, mark the direction of rotation on the timing belt. Remove the timing belt.

19. If it is necessary to remove the sprockets, remove the camshaft sprocket holding tool. Hold the camshaft by placing a suitable wrench on the hexagon which is cast into the camshaft. Place another wrench onto the camshaft sprocket retaining bolt and loosen the bolt.

➡ **Before removing the camshaft sprocket (s), be sure that the camshafts are still in alignment and tag each sprocket to the camshaft from which it was removed.**

20. Remove the camshaft sprocket bolt and the camshaft sprocket from the camshaft.

21. Repeat the camshaft sprocket removal procedure for the opposite camshaft if required.

22. Raise and safely support the vehicle.

23. Slide off the crankshaft sprocket and remove the crankshaft key.

To install:

24. Install the crankshaft key and slide the crankshaft sprocket into position.

25. Lower the vehicle.

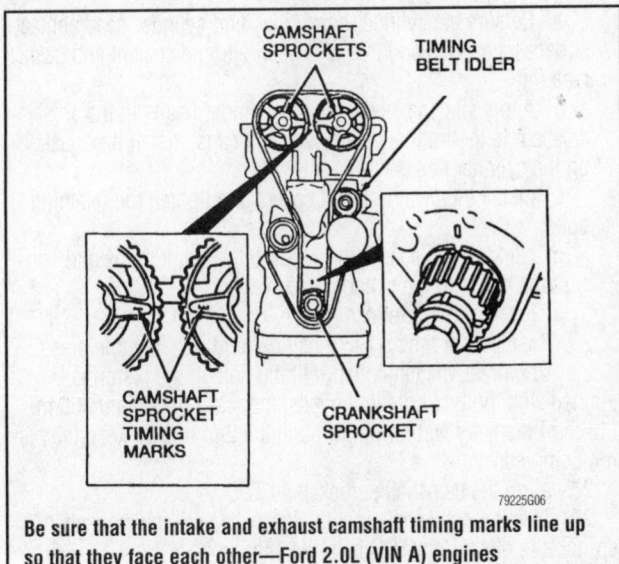

Be sure that the intake and exhaust camshaft timing marks line up so that they face each other—Ford 2.0L (VIN A) engines

26. Install the camshaft sprocket onto the proper camshaft, making sure to align the dowel pin.

27. Be sure that the **I** and **E** are in alignment.

28. Install the camshaft sprocket bolt. Hold the hexagon on the camshaft with a suitable wrench and tighten the sprocket bolt to 35–48 ft. lbs. (47–65 Nm). Be sure the camshaft sprockets are still properly aligned and reinstall the sprocket holding tool.

29. Be sure the timing marks on the camshaft and crankshaft sprockets are still aligned.

30. Install the timing belt. If reusing the original timing belt, be sure it is installed in the same direction of rotation.

31. Turn the tensioner clockwise with an Allen wrench and install the tensioner spring. Remove the holding tool from between the camshaft sprockets.

32. Rotate the crankshaft clockwise two turns and align the timing marks. Be sure all marks are still correctly aligned.

➡ **The timing chain tensioner automatically adjusts the tension on the timing belt.**

33. Raise and safely support vehicle on jack stands.

34. Install the timing belt lower cover and tighten the four bolts to 71–88 inch lbs. (8–10 Nm).

35. Install the guide plate, crankshaft pulley and pulley bolt. Secure the pulley with the holder tool and tighten the bolt to 116–123 ft. lbs. (157–167 Nm).

36. Install the splash-shields and lower the vehicle.

37. Raise the engine slightly with the jack and install the right side engine mount. Tighten the mount through-bolt to 63–86 ft. lbs. (86–116 Nm) and the mount attaching nuts to 54–75 ft. lbs. (74–103 Nm). Remove the engine support tool.

38. Install the upper timing belt cover and tighten the bolts to 71–88 inch lbs. (8–10 Nm).

39. Clean the cylinder head and valve cover mating surfaces thoroughly.

40. Apply silicone sealant to the cylinder head surface in the area adjacent to the front camshaft bearing caps. Apply sealant to a new gasket and install it on the cylinder head cover.

41. Install the cylinder head cover and tighten the bolts.

42. Install the power steering hose brackets and tighten the bolts to 71–88 inch lbs. (8–10 Nm). Connect the spark plug wires and

wire clips. Connect the breather tube and PCV valve. If necessary, connect the crankshaft position sensor.

43. Install the alternator belt and adjust the tension. Tighten the upper mounting bolt to 14–18 ft. lbs. (19–25 Nm) and the lower through-bolt to 27–38 ft. lbs. (37–52 Nm).

44. Install the power steering belt and adjust the tension. Tighten the through-bolt to 32–45 ft. lbs. (43–61 Nm) and the lockbolt to 23–34 ft. lbs. (31–46 Nm). Install the power steering belt shield and tighten the bolts to 61–86 inch lbs. (7–9 Nm).

45. Connect the negative battery cable.

46. Run the engine and check for leaks and proper engine operation.

2.5L (VIN B) Engine

1. Disconnect the negative battery cable.

2. Label and disengage the electrical connectors from the coolant elbow. Label and remove the electrical connectors from the knock sensor (if required) and crankshaft position sensor.

3. Loosen the drive belt tensioner locknuts and adjusting bolts. Remove the accessory drive belts.

4. Raise and safely support the vehicle on jack stands.

5. Remove the lower bolt from the A/C and alternator tensioner bracket.

6. Remove the right wheel and splash-shields.

7. Hold the crankshaft pulley (damper) with Holder Tool T92C-6316-AH or equivalent, and remove the crankshaft pulley bolt. Remove the crankshaft pulley, using a puller if needed.

8. Remove the 5 front timing belt cover bolts.

9. Hold the water pump pulley with Holder Tool T92C-6312-AH or equivalent, remove the four bolts and the water pump pulley.

10. Hold the power steering pump pulley with Strap Wrench D85L-6000-A or equivalent and remove the power steering pump pulley nut and pulley.

11. Lower the vehicle.

12. Remove the upper bolt from the belt idler bracket and remove the bracket.

13. Remove the engine oil dipstick tube retaining bolt and the tube.

14. Remove the 8 rear timing belt cover retaining bolts and remove the timing belt covers.

15. Temporarily reinstall the crankshaft pulley bolt.

16. Remove the three nuts and through-bolt from the right-hand engine support insulator and remove the support insulator. Remove the support insulator bracket.

17. Raise and safely support the vehicle on jack stand.

18. Turn the crankshaft to TDC No. 1 cylinder in the direction of normal rotation. Be sure that the timing mark on the crankshaft sprocket aligns with the timing mark on the oil pump.

19. Remove the two bolts from the timing belt tensioner arm, removing the lower bolt first.

20. Remove the timing belt tensioner arm.

21. If the timing belt is to be reused, mark the direction of rotation on the timing belt.

22. Loosen the Allen bolt on the timing belt tensioner.

23. Remove the timing belt.

24. If the timing belt sprockets are to be removed, proceed as follows:

    a. Remove the intake manifold.

    b. Label and disconnect the necessary hoses from the cylinder head covers.

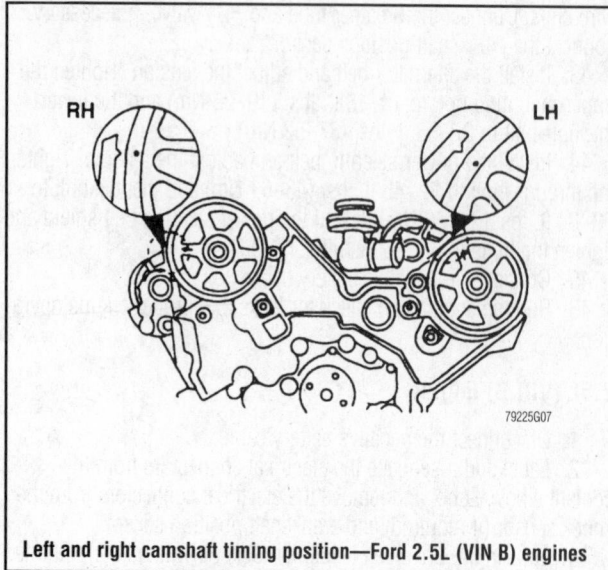

Left and right camshaft timing position—Ford 2.5L (VIN B) engines

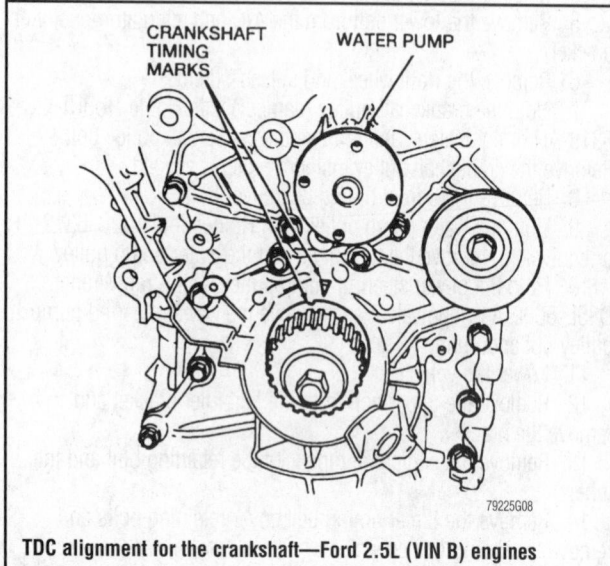

TDC alignment for the crankshaft—Ford 2.5L (VIN B) engines

   c. Label and disconnect the spark plug wires from the spark plugs.

   d. Remove the cylinder head cover retaining bolts and remove the cylinder head covers.

   e. Hold the camshaft using a suitable wrench on the hexagon cast into the camshaft. Remove the camshaft sprocket bolts and the camshaft sprockets.

   f. Use Crankshaft Damper Puller T74P-6316-A or equivalent, to remove the crankshaft sprocket. Remove the crankshaft sprocket key.

To install:

25. If the timing belt sprockets were removed, proceed as follows:

   a. Install the crankshaft sprocket key and crankshaft sprocket.

   b. Install the camshaft sprockets on the camshafts with the retaining bolts.

   c. Hold the camshaft using a suitable wrench on the hexagon cast into the camshaft. Tighten the camshaft sprocket bolts to 90–103 ft. lbs. (123–140 Nm).

   d. Be sure the cylinder head cover and cylinder head contact surfaces are clean and free of dirt, oil and old sealant and gasket material.

   e. Apply silicone sealant to the cylinder heads in the area adjacent to the front and rear camshaft caps. Install new gaskets on the cylinder heads.

   f. Install the cylinder head covers and tighten the retaining bolts.

   g. Connect the spark plug wires to the spark plugs and connect the hoses to the cylinder head covers.

   h. Install the intake manifold.

26. Position the timing belt tensioner arm in a suitable press.

27. Compress the tensioner until the hole in the piston is aligned with the 2nd hole in the tensioner case. Insert a 0.060 in. (1.6mm) diameter wire or pin through the 2nd hole to keep the piston compressed.

28. Align the camshaft sprockets to TDC.

29. Turn the crankshaft counterclockwise until the crankshaft sprocket is offset from TDC by one tooth.

30. Install the timing belt.

31. If the original belt is being reused, be sure it is installed in the same direction of rotation.

32. Turn the crankshaft in the direction of normal engine rotation until the crankshaft sprocket timing mark is at TDC. This should place all of the belt slack in the timing belt tensioner portion of the timing belt.

33. Install the timing belt tensioner arm and two bolts. Tighten the bolts to 14–18 ft. lbs. (19–25 Nm).

34. Remove the wire or pin from the tensioner.

➡ **When properly timed, the crankshaft timing marks will line up and the crankshaft sprocket timing mark will no longer be one tooth off.**

35. Rotate the crankshaft two complete turns in the direction of normal rotation and align the timing marks. Be sure all marks are still correctly aligned. This will also set the timing belt tension.

➡ **The timing belt tensioner will automatically adjust the timing belt tension.**

36. Tighten the timing belt tensioner Allen bolt to 28–32 ft. lbs. (35–51 Nm).

37. Install the right-hand engine support insulator. Tighten the three nuts to 54–76 ft. lbs. (74–103 Nm) and the through-bolt to 50–68 ft. lbs. (67–93 Nm).

38. Remove the crankshaft damper bolt.

39. Install the timing belt covers with the rear 8 bolts. Tighten to 71–88 inch lbs. (8–10 Nm).

40. Install the engine oil dipstick tube and retaining nut.

41. Install the belt idler bracket and the upper retaining bolt.

42. Raise and safely support the vehicle on jack stands.

43. Install the power steering pump pulley and nut. Tighten the nut to 36–43 ft. lbs. (49–59 Nm) while holding the pulley with a strap wrench.

44. Install the water pump pulley and four bolts. Secure the pulley with the holder tool and tighten the bolts to 71–88 inch lbs. (8–10 Nm).

45. Install the 5 front timing belt cover bolts and tighten to 71–88 inch lbs. (8–10 Nm).

46. Install the crankshaft pulley (damper) with the bolt. Hold the crankshaft pulley with the holding tool and tighten to 116–122 ft. lbs. (157–166 Nm).

47. Install the splash-shields.

48. Install the wheel and tighten the lug nuts to 65–87 ft. lbs. (88–118 Nm).

49. Install the lower bolt into the A/C and alternator tensioner bracket.

50. Lower the vehicle.

51. Install the accessory drive belts and adjust the tension.

52. Engage the electrical connectors to the sensors at the coolant elbow and the knock sensor and crankshaft position sensor.

53. Connect the negative battery cable.

54. Run the engine and check for leaks and proper engine operation.

3.0L (VIN Y) SHO Engine

1. Disconnect both battery cables, negative cable first. Remove the battery.

2. Remove the right engine roll damper. Disconnect the wiring to the ignition module.

3. Remove the intake manifold crossover tube bolts. Loosen the intake manifold tube hose clamps. Remove the intake manifold crossover tube.

4. Loosen the alternator/air conditioning belt tensioner pulley and remove the drive belt.

5. Loosen the water pump/power steering belt tensioner pulley and remove the drive belt.

6. Remove the alternator/air conditioning belt tensioner pulley and bracket assembly.

7. Remove the water pump/power steering belt tensioner pulley only. Remove the upper timing belt cover.

8. Unplug the Crankshaft Position (CKP) sensor electrical connector. Place the transaxle gear selector in **N** (Neutral).

9. Rotate the crankshaft until the piston for the No. 1 cylinder is at TDC on its compression stroke. Be sure the white mark on the crankshaft damper aligns with the **0** degree index mark on the lower timing belt cover and the marks on the intake camshaft sprockets align with the index marks on the metal timing belt cover.

10. Raise and safely support the vehicle on jack stands. Remove the right front wheel and tire assembly. Loosen the fender splash-shield and place it aside.

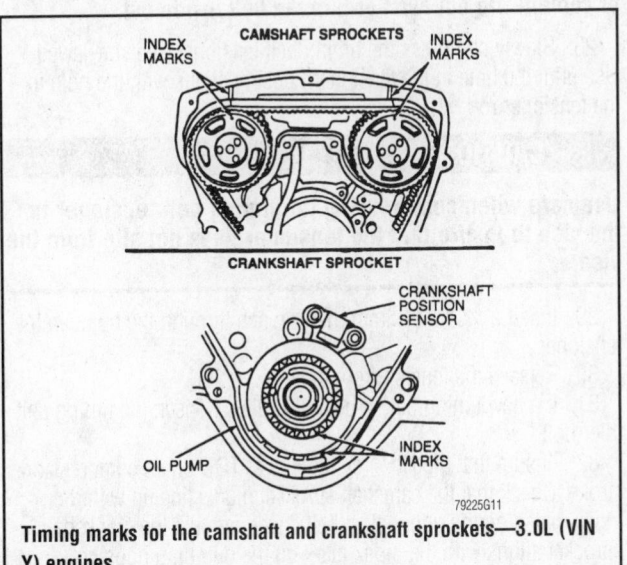

Timing marks for the camshaft and crankshaft sprockets—3.0L (VIN Y) engines

11. Remove the crankshaft damper and pulley retaining bolt. Using Puller T67L-3600-A or equivalent, remove the crankshaft damper and pulley.

12. Remove the lower timing belt cover.

13. Remove the center timing belt cover and disconnect the CKP sensor wire and grommet from the slot in the cover and the stud on the water pump.

14. Loosen the timing belt tensioner idler pulley. Rotate the idler pulley 180 degrees clockwise and tighten the tensioner nut to hold the pulley in an unloaded position.

15. Lower the vehicle.

16. Remove the timing belt. If the belt is to be reused, use crayon to mark an arrow on the belt to indicate the direction of rotation, for installation reference.

17. If removing one or more camshaft sprockets, remove two retaining bolts securing each camshaft sprocket and remove the sprocket, noting the location of the dowel pin.

18. If removing the crankshaft sprocket, install Puller T67L-3600-A, or equivalent and pull the crankshaft sprocket off the crankshaft using care not to damage the pulse wheel.

To install:

19. If the crankshaft sprocket was removed, install the crankshaft sprocket by aligning the keyway and pushing the sprocket on by hand.

20. If one or more camshaft sprockets was removed, install each camshaft sprocket by aligning the timing marks on the sprockets with the camshaft using the dowel pin as a guide. Install two retaining bolts and tighten to 10–13 ft. lbs. (14–18 Nm).

➡ **Before installing the timing belt, inspect it for cracks, wear or other damage and replace, if necessary. Do not allow the timing belt to come into contact with gasoline, oil or coolant. Do not twist or turn the belt inside out.**

21. Be sure the engine is at TDC for the No. 1 cylinder. Check that the camshaft sprocket marks line up with the index marks on the upper steel belt cover and that the crankshaft sprocket aligns with the index mark on the oil pump housing.

➡ **The timing belt has three yellow lines. Each line aligns with the index marks.**

22. Install the timing belt over the crankshaft and camshaft sprockets. The lettering on the belt **KOA** should be readable from the rear of the engine (top of the lettering to the front of the engine). Be sure the yellow lines are aligned with the index marks on the sprockets.

23. Release the timing belt tensioner idler pulley locknut. Leave the nut loose. Raise and safely support the vehicle on jack stands.

24. Install the center timing belt cover. Be sure the CKP sensor wiring and grommet are installed and routed properly. Tighten the mounting bolts to 60–90 inch lbs. (7–11 Nm).

25. Install the lower timing belt cover. Tighten the bolts to 60–90 inch lbs. (7–11 Nm).

26. Install the crankshaft damper and pulley using Installer T88T-6701-A, or equivalent. Install the retaining bolt and tighten to 113–126 ft. lbs. (152–172 Nm).

27. Rotate the crankshaft two revolutions in the clockwise direction until the yellow mark on the damper aligns with the **0** degree mark on the lower timing belt cover.

28. Remove the plastic door in the lower timing belt cover. Tighten the tensioner locknut to 25–37 ft. lbs. (33–51 Nm) and install the plastic door.

29. Rotate the crankshaft 60 degrees more in the clockwise direction until the white mark on the damper aligns with the **0** degree mark on the lower timing belt cover.

30. Lower the vehicle. Be sure the index marks on the camshaft sprockets align with the marks on the rear metal timing belt cover.

31. Route the CKP sensor wiring and connect it to the engine wiring harness.

32. Install the upper timing belt cover. Tighten the bolts to 60–90 inch lbs. (7–11 Nm).

33. Install the water pump/power steering tensioner pulley. Tighten the nut to 11–17 ft. lbs. (15–23 Nm).

34. Install the alternator/air conditioning tensioner pulley and bracket assembly. Tighten the bolts to 11–17 ft. lbs. (15–23 Nm).

35. Install the water pump/power steering and alternator/air conditioning drive belts, and set the tension. Tighten the idler pulley nut to 25–36 ft. lbs. (34–50 Nm).

36. Install the intake manifold crossover tube. Tighten the bolts to 11–17 ft. lbs. (15–23 Nm).

37. Install the engine roll damper. Install the battery. Connect the wiring to the ignition module.

38. Connect both battery cables, negative cable last. Raise and safely support the vehicle on jack stands.

39. Install the splash-shield. Install the right front wheel and tire assembly.

40. Lower the vehicle. Run the engine and check for proper operation.

3.2L (VIN P) SHO Engine

1. Disconnect the battery cables, negative cable first.
2. Remove the battery.
3. Remove the right engine roll damper.
4. Disconnect the wiring to the ignition module.
5. Remove the intake manifold crossover tube bolts. Loosen the intake manifold tube hose clamps. Remove the intake manifold crossover tube.
6. Rotate the accessory drive belt tensioner clockwise to relieve belt tension. Remove the accessory drive belt.
7. Disconnect the surge tank fitting.
8. Remove the bolts retaining the upper and lower idler pulleys to the engine and remove the pulleys.
9. Using Strap Wrench D85L-6000-A or equivalent, hold the power steering pump pulley. Remove the retaining nut and washer. Remove the power steering pulley.
10. Remove the retaining bolt from the belt tensioner and remove the tensioner.
11. Remove the upper and center timing belt covers.
12. Disengage the Crankshaft Position (CKP) sensor electrical connector.
13. Place the transaxle selector in **N** (Neutral).
14. Rotate the crankshaft until the piston for No. 1 cylinder is at TDC on its compression stroke. Be sure the white mark on the crankshaft damper aligns with the **0** degree index mark on the lower timing belt cover and the marks on the intake camshaft sprockets align with the index marks on the metal timing belt cover.
15. Raise and safely support the vehicle on jack stands.
16. Remove the right front wheel and tire assembly.
17. Loosen the fender splash-shield and place it aside.
18. Remove the crankshaft pulley and damper using Puller T67L-3600-A with the appropriate adapters, or equivalent.
19. Remove the lower timing belt cover and belt guide.
20. Remove the upper timing belt tensioner bolt.
21. Slowly loosen the lower timing belt tension bolt and remove the tensioner.
22. Lower the vehicle.

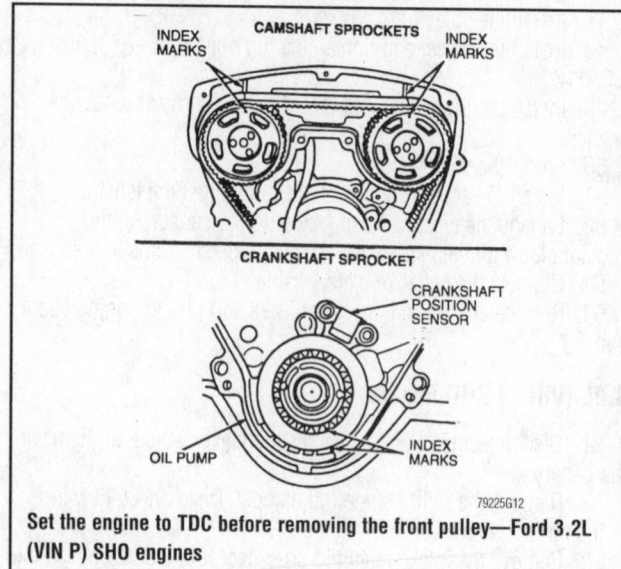

Set the engine to TDC before removing the front pulley—Ford 3.2L (VIN P) SHO engines

23. Remove the timing belt. If the belt is to be reused, use crayon to mark an arrow on the belt to indicate the direction of rotation, for installation reference.

24. If removing one or more camshaft sprockets, remove two retaining bolts securing each camshaft sprocket and remove the sprocket noting the location of the dowel pin.

25. If removing the crankshaft sprocket, install Puller T67L-3600-A, or equivalent and pull the crankshaft sprocket off the crankshaft using care not to damage the pulse wheel.

To install:

26. If the crankshaft sprocket was removed, install the crankshaft sprocket by aligning the keyway and pushing the sprocket on by hand.

27. If one or more camshaft sprockets was removed, install each camshaft sprocket by aligning the timing marks on the camshaft sprocket with the camshaft using the dowel pin as a guide. Install two retaining bolts and tighten to 10–13 ft. lbs. (14–18 Nm).

➡Before installing the timing belt, inspect it for cracks, wear or other damage and replace as necessary. Do not allow the timing belt to come into contact with gasoline, oil or coolant. Do not twist or turn the belt inside out.

28. Slowly compress the timing belt tensioner in a soft-jawed vise until the hole in the tensioner housing aligns with the hole in the tensioner rod.

❄❄ CAUTION

Use care when compressing the timing belt tensioner in the vise to insure that the tensioner does not slip from the vise.

29. Insert a 1/20 in. (1.5mm) hex wrench through the holes in the tensioner.

30. Release the tension from the vise.

31. If a new timing belt is being installed, loosen the timing belt idler bolt.

32. Ensure that the No. 1 cylinder is at TDC on its compression stroke. Check that the camshaft sprocket marks line up with the index marks on the upper steel belt cover and that the crankshaft sprocket aligns with the index mark on the oil pump housing.

➡ **The timing belt has three yellow lines. Each line aligns with the index marks.**

33. Install the timing belt over the crankshaft and camshaft sprockets. The lettering on the belt **KOB** should be readable from the rear of the engine (top of the lettering to the front of the engine). Be sure the yellow lines are aligned with the index marks on the sprockets.

❄❄ WARNING

Do not install the timing belt tensioner with the tensioner rod extended.

34. Install the timing belt tensioner on the cylinder block while pushing the timing belt idler toward the timing belt. Install and tighten the tensioner bolts to 12–17 ft. lbs. (16–23 Nm).

35. Install the grommets between the timing belt tensioner and the oil pump.

36. Remove the hex wrench from the timing belt tensioner.

37. If a new timing belt is being installed, perform the following steps:

  a. Remove the hex wrench from the timing belt tensioner, if installed.

  b. Mount Timing Belt Tensioner tool T93P-6254-B or equivalent, using the holes in the power steering pump support.

  c. Hand-tighten the timing belt idler pulley bolt.

  d. Using an in. pound torque wrench with attachment T93P-6254-A or equivalent, rotate the timing belt tensioner clockwise to 4.3 inch lbs. (0.5 Nm).

  e. Tighten the timing belt idler pulley bolt to 27–37 ft. lbs. (36–50 Nm). Remove both timing belt tensioning tools.

38. Raise and safely support the vehicle securely on jackstands.

39. Install the belt guide and lower timing belt cover. Tighten the retaining bolts to 12–17 ft. lbs. (16–23 Nm).

40. Using a suitable tool, install the crankshaft damper. Tighten the damper attaching bolt to 113–126 ft. lbs. (152–172 Nm).

41. Rotate the crankshaft two revolutions clockwise until the yellow mark on the damper aligns with the **0** degree mark on the lower timing belt cover.

42. Lower the vehicle.

43. Ensure that the index marks on the camshaft sprockets align with the marks on the rear metal timing belt cover.

44. Route the CKP sensor wiring and connect with the engine wiring harness.

45. Install the center and upper timing belt covers. Tighten the bolts to 12–17 ft. lbs. (16–23 Nm).

46. Install the steering pump pulley. Tighten the nut to 12–17 ft. lbs. (16–23 Nm).

47. Install the accessory drive belt while rotating the accessory drive belt tensioner clockwise.

48. Install the surge tank fitting.

49. Install the intake manifold crossover tube. Tighten the bolts to 11–17 ft. lbs. (15–23 Nm).

50. Install the engine roll damper

51. Install the battery.

52. Connect the wiring to the ignition module.

53. Connect both battery cables, negative cable last.

54. Raise and safely support the vehicle.

55. Install the splash-shield.

56. Install the right front wheel and tire assembly.

57. Lower the vehicle.

58. Run the engine and check for leaks and proper engine operation.

GENERAL MOTORS

1.0L (VIN 6) and 1.3 (VIN 9) Engines

➡ **Timing belts must always be completely free of dirt, grease, fluids and lubricants. This includes the sprockets and contact surfaces on which the belt rides. The belt must never be crimped, twisted or bent. Never use tools to pry or wedge the belt.**

1. Disconnect the negative battery cable.

2. Raise and safely support the vehicle.

3. Remove the clips and right side splash-shield.

4. Remove the lower alternator cover plate.

5. Remove the alternator drive belt, if equipped, the A/C drive belt.

6. Remove the water pump pulley.

➡ **It is not necessary to remove the crankshaft timing sprocket bolt (center bolt) to remove the crankshaft pulley.**

7. Remove the crankshaft pulley bolts and crankshaft pulley.

8. Remove the retaining bolts and nut from the timing belt outside cover.

9. Remove the timing belt outside cover.

10. Turn the crankshaft to align the timing marks. The mark on the crankshaft sprocket should align with the arrow mark on the oil pump housing. The mark on the camshaft sprocket should align with the **V** mark on the timing belt inner cover or cylinder head cover.

11. If the timing belt is to be reused, mark the direction of rotation on the belt.

12. Remove the timing belt tensioner, tensioner plate, tensioner spring, spring damper and timing belt.

➡ **Never turn the camshaft or crankshaft independently after the timing belt has been removed. Interference may occur between the pistons and valves, and parts may be damaged.**

13. Inspect the timing belt for wear or cracks, and replace as necessary. Check the tensioner for smooth rotation.

14. If the timing belt sprockets are to be removed, proceed as follows:

  a. Using a 0.39 in. (10mm) rod inserted into the camshaft, hold the camshaft and remove the retaining bolt and camshaft sprocket.

❄❄ WARNING

Be careful not to damage the cylinder head or cylinder head cover mating surfaces. Place a clean shop cloth between the rod and cylinder head. Do not bump the rod hard against the cylinder head when loosening the bolt.

  b. Raise and safely support the vehicle on jack stands.

  c. If equipped with a manual transaxle, lock the crankshaft in position by inserting a suitable flat-bladed tool into the hole in the bottom of the bell housing to engage the flywheel teeth.

  d. If equipped with an automatic transaxle, lock the crankshaft in position by inserting a suitable flat-bladed tool between the flywheel teeth and against the engine block.

  e. Remove the crankshaft sprocket bolt and crankshaft sprocket.

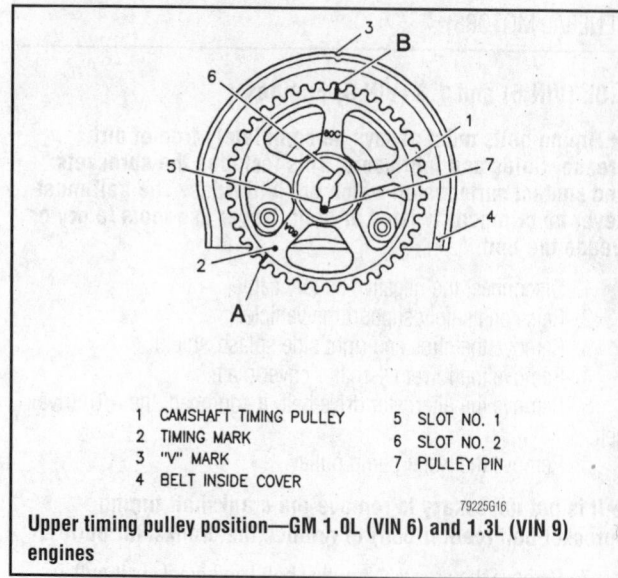

1	CAMSHAFT TIMING PULLEY	5	SLOT NO. 1
2	TIMING MARK	6	SLOT NO. 2
3	"V" MARK	7	PULLEY PIN
4	BELT INSIDE COVER		

79225G16

Upper timing pulley position—GM 1.0L (VIN 6) and 1.3L (VIN 9) engines

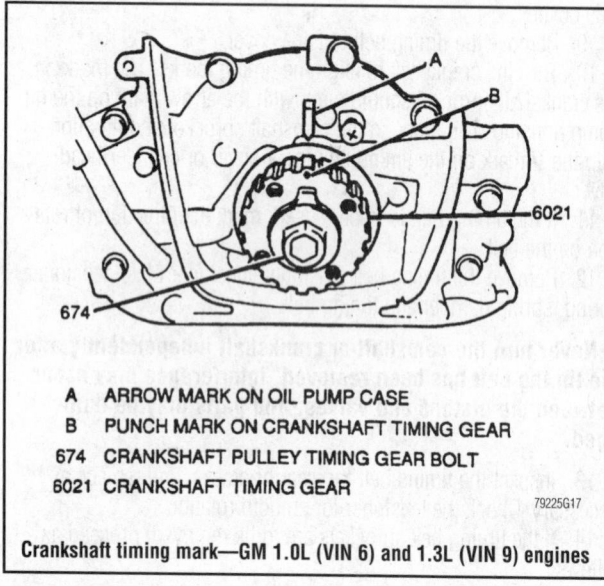

A	ARROW MARK ON OIL PUMP CASE
B	PUNCH MARK ON CRANKSHAFT TIMING GEAR
674	CRANKSHAFT PULLEY TIMING GEAR BOLT
6021	CRANKSHAFT TIMING GEAR

79225G17

Crankshaft timing mark—GM 1.0L (VIN 6) and 1.3L (VIN 9) engines

To install:

15. If the timing belt sprockets were removed, proceed as follows:

  a. Install the crankshaft sprocket, aligning the keyway. Lock the crankshaft in place and tighten the crankshaft sprocket bolt to 81 ft. lbs. (110 Nm).

  b. Lower the vehicle.

  c. Install the camshaft sprocket and retaining bolt. Lock the camshaft in place using the rod, and tighten the bolt to 44 ft. lbs. (60 Nm). Remove the locking rod.

16. Install the tensioner plate to the tensioner.

17. Insert the lug of the tensioner plate into the hole of the tensioner.

18. Install the tensioner, tensioner plate and spring. Do not fully tighten the tensioner bolt and stud at this time.

19. Move the tensioner plate in a counterclockwise direction. This should cause the tensioner to move in the same direction. If it does not, remove the tensioner and tensioner plate, and reinsert the tensioner plate lug in the timing plate tensioner hole.

20. Check that the camshaft timing marks are aligned. If not, align the two marks by turning the camshaft.

21. Check that the punch mark on the crankshaft timing belt sprocket is aligned with the arrow mark on the oil pump case. If not, align the two marks by turning the crankshaft.

22. With the timing marks aligned, install the timing belt on the two sprockets. If the old belt is being reused, be sure to install it running in the same direction of original rotation.

23. Install the tensioner spring and spring damper. Turn the timing belt two rotations clockwise after installing the tensioner spring and damper to remove any belt slack. Tighten the tensioner stud to 8 ft. lbs. (11 Nm), then the tensioner bolt to 20 ft. lbs. (27 Nm).

➡**Confirm that both sets of timing marks are aligned properly.**

24. Using a new seal, install the timing belt cover and tighten the bolts and nut to 97 inch lbs. (11 Nm).

25. Install the crankshaft pulley. Fit the keyway on the pulley to the crankshaft timing belt sprocket and tighten the bolts to 8–12 ft. lbs. (11–16 Nm).

26. Install the alternator drive belt, if equipped, A/C drive belt.

27. Install the lower alternator cover plate. Tighten the bolts to 89 inch lbs. (10 Nm).

28. Install the right side splash-shield and clips.

29. Lower the vehicle and connect the negative battery cable.

30. Run the engine. Check for leaks.

1.6L (VIN 6) and 1.8L (VIN 8) Engines

1. Disconnect the negative battery cable.

2. Remove the windshield washer reservoir from the engine compartment.

3. If equipped with cruise control, proceed as follows:

  a. Remove the cruise control actuator cover.

  b. Disconnect the cruise control harnesses.

  c. Disconnect the control cable.

  d. Remove the bolts and actuator from the vehicle.

4. Raise and safely support the vehicle on jack stands.

5. Remove the right front wheel.

6. Remove the bolts and plastic clips and the right front wheel housing.

7. Remove the alternator/water pump drive belt.

8. Lower the vehicle.

9. If equipped with A/C, proceed as follows:

  a. Remove the A/C compressor drive belt.

  b. Disconnect the compressor harness.

  c. Remove the bolts and compressor, without disconnecting the refrigerant lines. Suspend the compressor out of the way.

  d. Remove the compressor mounting bracket.

10. Remove the power steering pump drive belt.

11. Disconnect the wiring from the alternator and oil pressure switch.

12. Remove the engine wiring harness cover.

13. Remove the wiring harness from the cylinder head cover.

14. Disconnect the ignition wires from the spark plugs, then remove the spark plugs.

15. Remove the PCV hoses from the valve cover.

16. Remove the cap nuts, the seal washers and the cylinder head cover with the gasket.

17. Turn the crankshaft to align the timing mark on the crankshaft pulley at **0**, setting the piston in the No. 1 cylinder at Top Dead Center (TDC) on the compression stroke. Check that the valve

lash adjusters on the No. 1 cylinder are loose. If not, turn the crankshaft pulley one complete revolution (360 degrees).

18. Remove the engine ground wire from the right fender apron.

19. Install a suitable support under the engine and remove the engine mount.

20. Remove the water pump pulley.

21. Remove the crankshaft pulley using a suitable puller.

22. Remove the 9 retaining bolts and the timing belt covers.

23. Slide the timing belt guide from the crankshaft.

24. Be sure the timing belt sprockets are properly aligned.

✳✳ WARNING

Do not turn the crankshaft or camshaft independently after removal of the timing belt; binding or damage to engine components could result. If the timing belt is to be reused, mark the belt with an arrow showing the direction of engine revolution.

25. Remove the timing belt tensioner bolt, tensioner and tension spring.

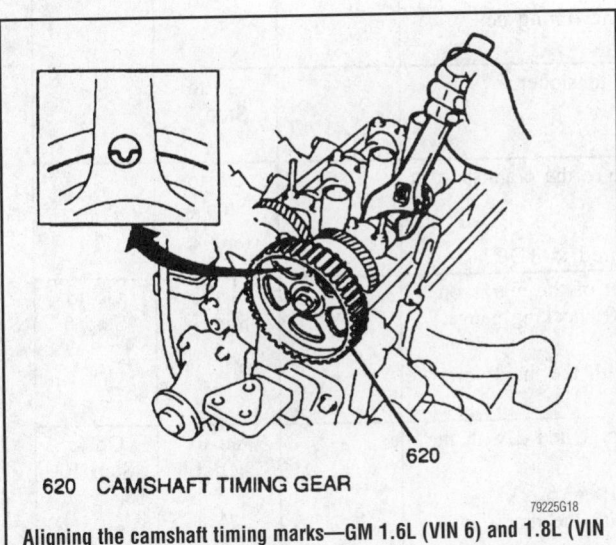

620 CAMSHAFT TIMING GEAR

79225G18

Aligning the camshaft timing marks—GM 1.6L (VIN 6) and 1.8L (VIN 8) engines

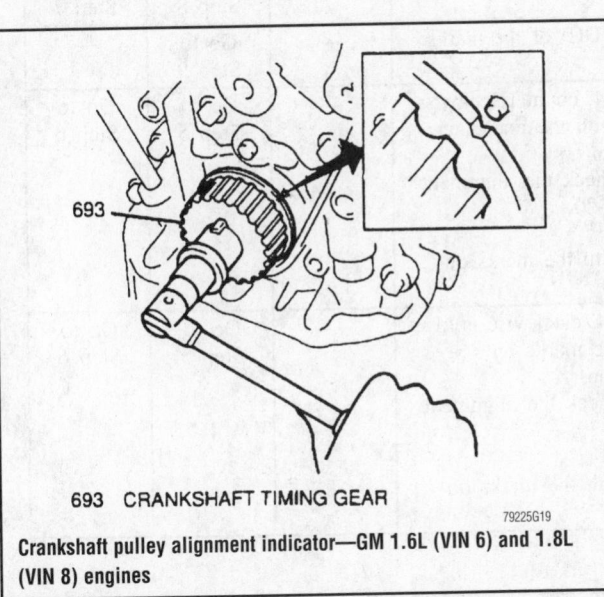

693 CRANKSHAFT TIMING GEAR

79225G19

Crankshaft pulley alignment indicator—GM 1.6L (VIN 6) and 1.8L (VIN 8) engines

26. Remove the timing belt from the sprockets. Inspect the timing belt for cracked or damaged teeth. Replace as necessary.

✳✳ WARNING

Do not bend, twist or turn the timing belt.

27. If the camshaft sprocket is to be removed, hold the camshaft stationary using a wrench positioned on the hexagon cast into the camshaft, and remove the sprocket retaining bolt and sprocket.

✳✳ WARNING

Be careful not to damage the cylinder head when holding the camshaft in place.

28. If the crankshaft sprocket is to be removed, pry it from the crankshaft using two flat-bladed prybars.

To install:

29. Align the camshaft key with the groove on the sprocket and slide the sprocket on. Hold the camshaft with the wrench at the hexagonal portion of the camshaft, and tighten the camshaft timing sprocket bolt to 43 ft. lbs. (59 Nm).

30. Be sure the sprocket is still properly aligned.

31. Install the crankshaft timing sprocket. Align the crankshaft key with the groove on the sprocket and slide it on.

32. Reinstall the timing belt tensioner and the tension spring. Pry the tensioner to the left as far as it will go and temporarily tighten the retaining bolt.

33. Install the timing belt. If installing the old belt, observe the matchmarks made during removal.

34. Loosen the retaining bolt for the timing belt tensioner and allow it to tension the belt.

35. Temporarily install the crankshaft pulley bolt and turn the crankshaft clockwise two full revolutions. Be sure each timing mark realigns exactly.

36. Tighten the timing belt tensioner bolt to 27 ft. lbs. (37 Nm).

37. Measure the timing belt deflection. Correct deflection should be 0.20–0.24 in. (5–6mm) at 4 lbs. (20 Nm) of pressure. If the deflection is not correct, adjust it with the timing belt tensioner.

38. Install the timing belt guide, with the cup side facing outward.

39. Install the timing belt covers, installing the bottom one first. Tighten the 9 cover bolts to 62 inch lbs. (7 Nm).

40. Install the crankshaft pulley after aligning the pulley key with the slot on the pulley. Hold the pulley with tool J-8614–01 or equivalent, and tighten the pulley bolt to 87 ft. lbs. (118 Nm).

41. Temporarily install the water pump pulley.

42. Raise and safely support the vehicle on jack stands.

43. Install the engine mount.

44. Install or connect the remaining components.

45. If equipped with A/C, proceed as follows:

   a. Install the compressor mounting bracket and tighten the bolts to 35 ft. lbs. (47 Nm).

   b. Install the compressor and tighten the bolts to 18 ft. lbs. (25 Nm).

   c. Engage the compressor wiring connector.

   d. Install the compressor drive belt and adjust the tension.

46. If equipped with cruise control, proceed as follows:

   a. Install the cruise control actuator and tighten the bolts to 89 inch lbs. (10 Nm).

   b. Connect the cruise control cable.

c. Install the cruise control actuator cover.
47. Connect the negative battery cable.
48. Start the engine and check vehicle operation.

3.0L (VIN R) Engine

➡ **The steps in this procedure are critical in preventing catastrophic engine damage, adhering to this sequence is imperative. There are special tools needed to perform this procedure. It is a good idea to read this procedure several times before attempting to perform this job. This is an interference engine.**

Always turn the crankshaft in the direction of rotation (clockwise), never against engine rotation. Never remove the timing belt without first setting the camshaft gears and crankshaft drive gear to TDC and locking them in place with tool J42069 (or equivalent).

The "Timing Belt Installation and Adjustment Table" provides an overview of the steps needed to properly install and adjust the timing belt. Use this table as a reference, not as a substitution, for the steps in this procedure.

1. Disconnect the negative battery cable.
2. Remove the resonance chamber.

TIMING BELT INSTALLATION AND ADJUSTMENT TABLE

Step	Action	Value	Yes	No
1	Install the timing belt and align marks on the belt with the marks on the camshaft gears and the crankshaft gear. Check the timing belt deflection between the idler pulley for camshafts 3 & 4 and camshaft number 4. Is the timing belt installed, the marks aligned and the timing belt deflection adjusted?	1 cm (0.4 in) maximum	Go to Step 2	—
2	Set the initial timing belt tension at the timing belt tensioner. Is the initial timing belt tension set?	—	Go to Step 3	—
3	Rotate the engine two complete revolutions and secure the crankshaft at Top Dead Center (TDC) with the J 42069-10. Has the engine been rotated and the crankshaft secured to TDC?	—	Go to Step 4	—
4	Starting with camshafts 3 and 4, check the alignment of the marks on the camshaft gears with the marks on the J 42069-20 checking gauge. Do the marks on the camshaft gears align exactly with the marks on J 42069-20?	—	Go to Step 5	Go to Step 6
5	Check the alignment of the marks on camshafts gears 1 and 2 with the marks on the J 42069-20 checking gauge. Do the marks on the camshaft gears align exactly with the marks on J 42069-20?	—	Go to Step 14	Go to Step 10
6	Do the camshaft gear marks line up to the left (BTDC) of the marks on the J 42069-20 checking gauge?	—	Go to Step 8	Go to Step 7
7	Do the camshaft gear marks line up to the right (ATDC) of the marks on the J 42069-20 checking gauge?	—	Go to Step 9	—
8	Turn the idler pulley eccentric, for camshafts 3 and 4, counterclockwise until the marks on the camshaft gear align exactly with the marks on J 42069-20. Rotate the engine two complete revolutions, lock the crankshaft at TDC with J 42069-10 and recheck the alignment of the camshaft gear marks to the marks on J 42069-20. Do the marks on the camshaft gears align exactly with the marks on J 42069-20?	—	Go to Step 5	Go to Step 6
9	Turn the idler pulley eccentric, for camshafts 3 and 4, clockwise until the marks on the camshaft gear align exactly with the marks on J 42069-20. Rotate the engine two complete revolutions, lock the crankshaft at TDC with J 42069-10 and recheck the alignment of the camshaft gear marks to the marks on J 42069-20. Do the marks on the camshaft gears align exactly with the marks on J 42069-20?	—	Go to Step 5	Go to Step 6

Timing belt installation and adjustment table—GM 3.0L (VIN R) engine

79225G35

Step	Action	Value	Yes	No
10	Do the camshaft gear marks line up to the left (BTDC) of the marks on the J 42069-20 checking gauge?	—	Go to Step 12	Go to Step 11
11	Do the camshaft gear marks line up to the right (ATDC) of the marks on the J 42069-20 checking gauge?	—	Go to Step 13	—
12	Turn the idler pulley eccentric, for camshafts 1 and 2, counterclockwise until the marks on the camshaft gear align exactly with the marks on J 42069-20. Rotate the engine two complete revolutions, lock the crankshaft at TDC with J 42069-10 and recheck the alignment of the camshaft gear marks to the marks on J 42069-20. Do the marks on the camshaft gears align exactly with the marks on J 42069-20?	—	Go to Step 14	Go to Step 10
13	Turn the idler pulley eccentric, for camshafts 1 and 2, clockwise until the marks on the camshaft gear align exactly with the marks on J 42069-20. Rotate the engine two complete revolutions, lock the crankshaft at TDC with J 42069-10 and recheck the alignment of the camshaft gear marks to the marks on J 42069-20. Do the marks on the camshaft gears align exactly with the marks on J 42069-20?	—	Go to Step 14	Go to Step 10
14	Set the final timing belt tension at the timing belt tensioner. Is the final timing belt tension set?	—	Go to Step 15	—
15	Again, rotate the engine two complete revolutions and lock the crankshaft at TDC. Do a final inspection of the camshaft gear marks' relationship to the J 42069-20 marks. The marks must align exactly. Do the marks on the camshaft gears align exactly with the marks on the J 42069-20?	—	Go to Step 16	Go to Step 2
16	Remove all checking tools and ensure all idler pulleys and the tensioner locking nut are tightened to specifications. Continue with re-assembly of the engine.	—	—	—

79225G36

Timing belt installation and adjustment table (continued)—GM 3.0L (VIN R) engine

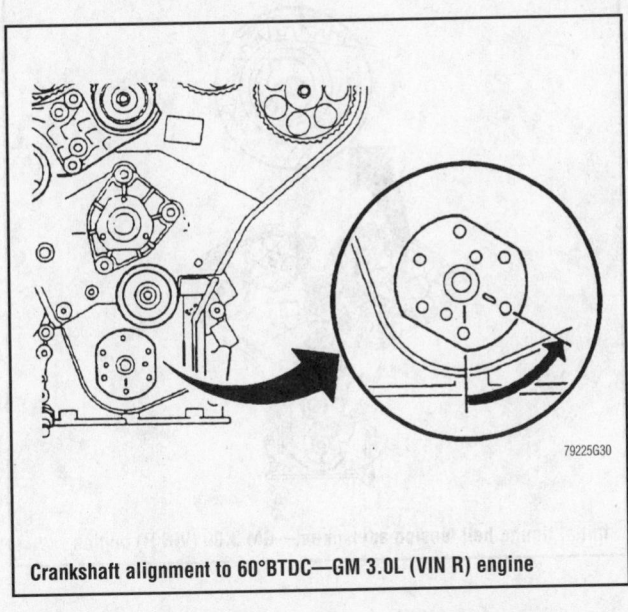

79225G30

Crankshaft alignment to 60°BTDC—GM 3.0L (VIN R) engine

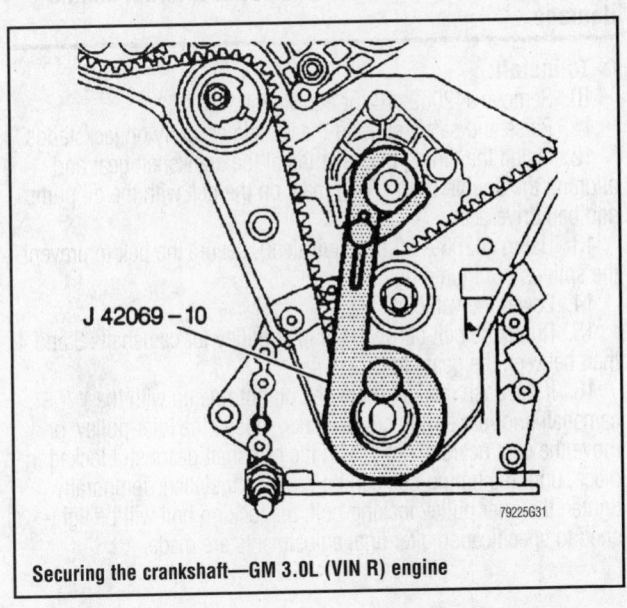

J 42069-10

79225G31

Securing the crankshaft—GM 3.0L (VIN R) engine

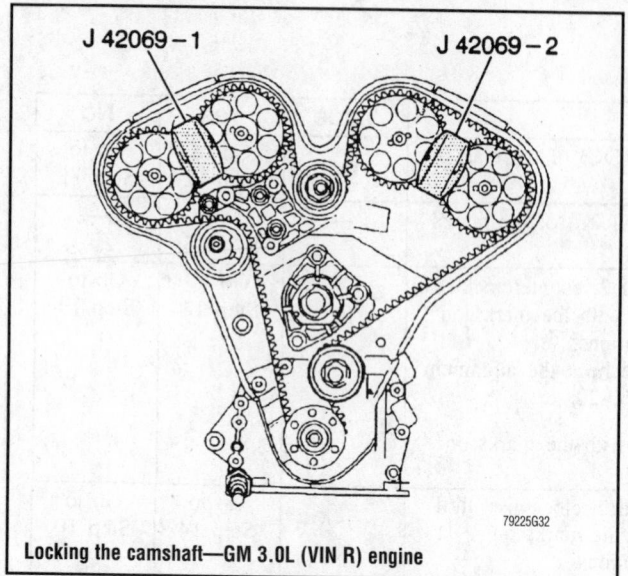

Locking the camshaft—GM 3.0L (VIN R) engine

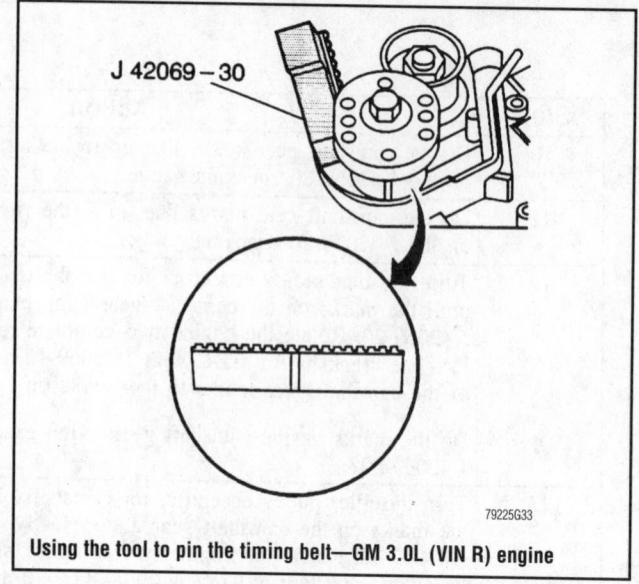

Using the tool to pin the timing belt—GM 3.0L (VIN R) engine

3. Remove the front timing belt cover.

4. Remove the harmonic balancer from the crankshaft.

5. Rotate the crankshaft clockwise to 60°BTDC.

6. Install J42069–10 (or equivalent) to the crankshaft drive gear with knurled bolt.

7. Turn the engine clockwise, with J42098 (or equivalent), until the lever of J42069–10 (or equivalent) firmly contacts the water pump pulley flange. Secure the lever to the water pump.

➡**Be sure the engine is not 180°off. The camshaft marks must line up with the rear timing cover.**

8. Lock the camshaft gears, using J42069–1 (or equivalent) and J42069–2 (or equivalent). It may be necessary to loosen the relevant idler pulley to lock the gears, then tighten the idler pulley bolt to 30 ft. lb. (40 Nm).

9. Loosen the timing belt tensioner and remove the belt.

✸✸ WARNING

With the belt removed, do not rotate the camshaft or crankshaft, or remove the locking tools, because the pistons may contact the valves and cause internal engine damage.

To install:

10. Remove J42069–10 (or equivalent).

11. Raise and safely support the vehicle securely on jackstands.

12. Install the timing belt, starting at the crankshaft gear and aligning the double dash (TDC) mark on the belt with the oil pump and belt drive gear.

13. Using J42069–30 (or equivalent), secure the belt to prevent the splines from jumping.

14. Lower the vehicle.

15. Route the belt between the idler pulley for camshafts 3 and 4, then between the gears for 3 and 4.

16. If the dash marks on the belt do not line up with the camshaft and rear timing cover marks, loosen the idler pulley, or move the cam gears slightly, with the camshaft gears still locked in place, until the timing belt can be properly installed. Temporally tighten the idler pulley locking bolt; the locking bolt will be tightened to specification after final adjustments are made.

➡**The timing belt deflection must be no more than 0.4 in (10mm) between camshaft gear 4 and the idler pulley. To adjust the deflection, rotate the timing belt idler pulley for camshafts 3 and 4 counterclockwise with tool J42069–40 (or equivalent).**

17. Route the belt between the idler pulley for camshafts 1 and 2, then between the gears for 1and 2.

18. If the dash marks on the belt do not line up with the camshaft and rear timing cover marks, loosen the idler pulley, or move the cam gears slightly, with the camshaft gears still locked in place, until the timing belt can be installed. Temporally tighten the idler pulley; the locking bolt will be tightened to specification after final adjustments are made.

19. Apply tension to the belt to keep it from slipping off the gears by turning the timing belt idler pulley for camshafts 1and 2 counterclockwise with J42069–40 (or equivalent). Temporally tighten the idler pulley; the locking bolt will be tightened to specification after final adjustments are made.

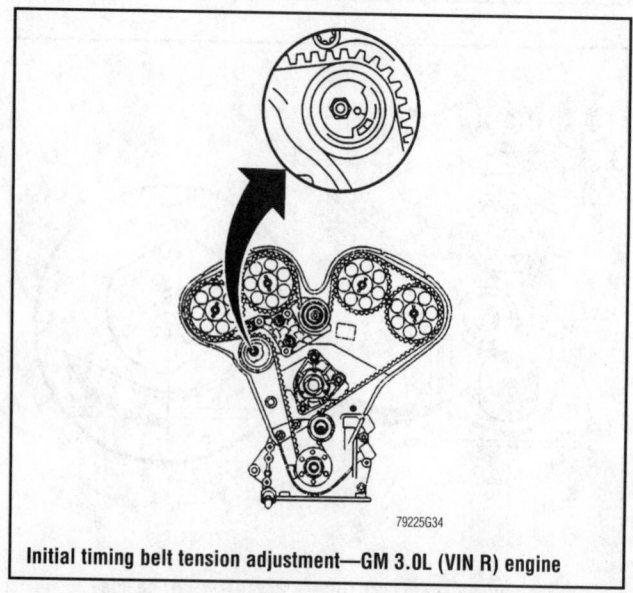

Initial timing belt tension adjustment—GM 3.0L (VIN R) engine

20. Complete the routing of the timing belt through the belt tensioner.

21. Apply initial tension by turning the timing belt tensioner counterclockwise, with a .5mm Allen wrench, until the marks are set as shown in the initial timing belt adjustment illustration.

22. Tighten the tensioner locking nut to 15 ft. lbs. (20 Nm).

23. Ensure that the alignment marks are at their specific reference points.

24. Remove J42069–30, J42069–1 and J42069–2.

25. Rotate the engine two revolutions clockwise, stopping at 60°BTDC.

26. Install J42069–10 to the crankshaft gear with the knurled bolt.

27. Turn the engine clockwise, with J42098 or equivalent, until the lever of J42069–10 or equivalent firmly contacts the water pump pulley flange. Secure the lever to the water pump.

➡ **The alignment marks on the timing belt will no longer match the marks on the camshaft gears after one or more revolutions. The marks on the camshaft gears must line up with the notches on the rear timing cover, and the crankshaft drive gear and the oil pump housing should match up to their mark.**

28. If timing belt adjustment is necessary, first adjust camshafts 3 and 4.

29. Check the alignment of camshafts 3 and 4, 1 and 2 with gauge J42069–20.

30. If alignment is OK, then set final belt tension as follows:

   a. Loosen the timing belt tensioner locking nut.

   b. Adjust the timing belt tensioner by turning the eccentric cam, with a 5mm hex wrench, until the marks are set.

31. Tighten the tensioner locking nut to 15 ft. lb. (20 Nm).

32. Remove J42069–10 and J42069–20.

33. Rotate the engine clockwise two revolutions, stopping at 60°BTDC.

34. If further adjustment is needed, then repeat the applicable steps.

35. Tighten the idler pulley bolts for the camshafts to 30 ft lb. (40 Nm).

36. Be sure all tools are removed from the engine.

37. Install the harmonic balancer and tighten the bolts to 15 ft. lb. (20 Nm).

38. Install the timing belt covers.

39. Install the resonance chamber.

40. Connect the negative battery cable, and reprogram applicable accessories.

3.4L (VIN X) Engine

The 3.4L (VIN X) engine uses a timing chain and camshaft timing belts.

1. Disconnect the negative battery cable.

2. Disconnect and remove the power steering pump from the pump mounting bracket.

3. Remove the left, right and center timing belt covers.

4. Rotate the engine clockwise to align the timing marks, TDC on the No.1 exhaust stroke, on the camshaft sprockets and intermediate shaft.

5. Loosely clamp the two camshaft sprockets on each side of the engine together using clamping pliers or the equivalent. Secure the belt to the right side cam sprocket with a C-clamp and a wide pad on the belt.

➡ **When clamping the sprockets no deflection should be noticed. If any deflection is noticed, loosen the clamping devices. DO NOT mar the camshaft sprockets with the clamping device.**

6. Remove the tensioner side plate retaining bolts from the tensioner and remove the side plate from the actuator and base.

7. Rotate the actuator assembly around the arm pivot and out of the base. Removal of the tensioner from the base allows it to extend to its maximum travel.

8. Set the actuator aside on a table in a vertical position to allow the oil to drain into the boot end. The tensioner should be allowed to sit for 5 minutes prior to refilling with oil.

9. Reset the timing belt actuator as follows:

   a. Straighten out a paper clip or a piece of stiff wire 0.032 in. (.75mm) diameter to a minimum straight length of 1.85 in. (47mm). Form a double loop in the remaining end.

   b. Remove the rubber end plug from the rear of the tensioner assembly. This will aid in allowing the oil in the tensioner to escape.

   c. Hold the tensioner in your hand with the rubber boot end of the tensioner pointing down.

   d. DO NOT remove the vent plug. Push the paper clip through the center hole in the vent plug and into the pilot hole.

   e. Insert a small screwdriver into the screw slot inside the end of the tensioner.

   f. Retract the tensioner by rotating the tensioner plunger in a clockwise direction while pushing the rod tip against a table top.

   g. Align the screw slot to align with the vent hole, and push the straight section of the wire into the screw slot to retain the plunger in the retracted position.

   h. If tensioner oil has been lost, fill the tensioner with SAE 5W30 Mobil 1®. Fill the tensioner to the bottom of the plug. The tensioner **MUST** be fully retracted before being filled with oil.

10. If the belt is being reused, mark the direction of rotation on the belt.

11. Remove the timing belt tensioner pulley mounting bolt and pulley.

12. Remove the timing belt after first removing the C-clamp retaining the belt to the right side camshaft sprocket.

13. Remove the Torx®head bolts securing the idler pulleys, if the idlers need to be replaced.

14. Remove the intermediate shaft sprocket using the following procedure:

   a. Use a suitable tool to hold the engine from turning.

   b. Remove the intermediate shaft sprocket mounting bolt and washer.

   c. Using J-38616, or an equivalent sprocket puller, remove the sprocket from the intermediate shaft.

15. If the camshaft sprockets need to be removed, proceed as follows:

a. Remove the camshaft carrier cover (s).

b. Remove the camshaft sprocket clamping pliers.

c. Rotate the camshaft being serviced so the flats on the camshaft are face up.

d. Install a camshaft hold-down tool, J-38613 or the equivalent, and tighten to 22 ft. lbs. (30 Nm).

e. Remove the camshaft sprocket mounting bolt and washer while holding the camshaft from turning with J-38613 and J-38614, or the equivalent

f. Using J-38616, or an equivalent sprocket puller remove the sprocket from the camshaft.

g. Repeat for each camshaft sprocket as necessary.

16. Drain the cooling system.

17. Disconnect the lower radiator hose from water pump inlet pipe.

18. If equipped with a manual transaxle, disconnect the front AIR hose at the AIR pipe.

19. Disconnect the heater hose at the front cover.

20. Remove the heater pipe bracket mounting bolts at the frame.

21. Raise and safely support the vehicle on jack stands. Remove the right front tire and wheel assembly and the right inner fender splash-shield.

22. Remove the crankshaft pulley mounting bolts and remove the pulley from the damper.

23. Remove the crankshaft damper as follows:

a. While holding the crankshaft from turning using a suitable tool, remove the damper mounting bolt and washer.

b. Install tool J-24420-B, or the equivalent, and remove the damper from the crankshaft.

24. Place an oil catch pan under the oil filter and remove the oil filter.

25. Remove the A/C compressor mounting bracket bolts.

26. Remove the lower front cover bolts.

27. On automatic transaxle vehicles, remove the halfshaft following the recommended procedure.

28. Remove the rear alternator bracket.

29. Disconnect and remove the starter following the recommended procedure.

30. Lower the vehicle.

31. Remove the intermediate shaft drive belt sprocket retaining bolt and remove the intermediate shaft drive belt sprocket using J38616, or an equivalent puller.

32. Remove the upper alternator mounting bolts.

33. Remove the forward light relay center screws and position the relay center aside.

34. Disconnect the oil cooler hose from the front cover.

35. Remove the water pump pulley.

36. Remove the upper front cover bolts and remove the front cover.

37. Mark the intermediate shaft sprocket, chain link, crankshaft sprocket and cylinder block for assembly reference. The marks should be made with paint so they won't be lost when the components are removed.

38. Retract the timing chain tensioner shoe as follows:

a. Insert J-33875 or an equivalent, on both sides of the tensioner.

b. Pull on the through-pin in the tensioner arm to retract the spring located in the tensioner arm.

c. While compressing the spring, use a suitable tool, a cotter pin or nail, and insert the pin in the hole in the tensioner assembly to hold the tensioner compressed. The tool used must be strong enough to hold the tensioner compressed.

➡**The timing chain, crankshaft sprocket and intermediate shaft sprocket will be removed at the same time. If, when removing the assembly, the intermediate shaft sprocket does not easily come off the intermediate shaft, rotate the crankshaft back and forth to loosen the intermediate shaft sprocket.**

39. Install a suitable puller, J-38611 and J-8433 or their equivalents.

40. Tighten the bolt on the puller and slowly pull the crankshaft sprocket off the crankshaft. Be sure the intermediate shaft sprocket is moving along with the crankshaft sprocket.

41. Remove the timing chain and sprockets.

42. Remove the tensioner mounting bolts and remove the tensioner assembly.

To install:

43. Install the tensioner assembly and tensioner assembly mounting bolts finger-tight first. Tighten the bolt in the slotted hole first to 18 ft. lbs. (25 Nm), then tighten the remainder of the bolts to 18 ft. lbs. (25 Nm).

44. Check to ensure that the crankshaft key is fully seated in the crankshaft cutout and the tensioner assembly is fully retracted.

45. Assemble the timing chain, intermediate shaft sprocket and crankshaft sprocket on a work bench. The timing marks made should be in alignment. The large chamfer and counterbore of the crankshaft sprocket are installed facing toward the engine and the intermediate shaft spline sockets are installed facing away from the engine.

46. Install the sprocket and chain assembly onto the engine. As the sprockets are installed, parallel alignment must be maintained.

47. The crankshaft sprocket will have to be pressed on the final 0.31 in. (8mm). This can be done using J-38612, or an equivalent puller.

48. Ensure timing was maintained.

49. Remove the retaining pin from tensioner. Clean all gasket surfaces completely.

50. Apply GM Sealer 1052080 or equivalent, to the lower edges of the sealing surface of the front cover. Install a new gasket on the front cover.

51. Install the front cover on the engine. Apply thread sealant to the large bolts and tighten the bolts enough to pull the front cover against the engine block.

52. Install the water pump pulley.

53. Connect the oil cooler hose to the front cover.

54. Position the forward light relay center and install the mounting screws.

55. Install the upper alternator mounting bolt and tighten it to 22 ft. lbs. (30 Nm).

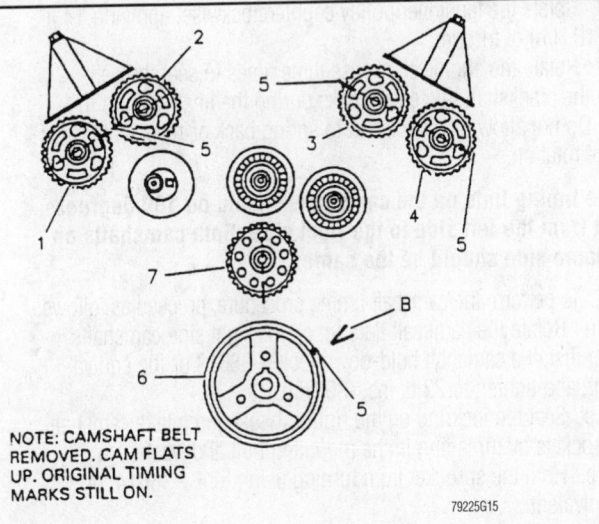

A   LOCATION OF TIMING MARKS WITH CAM HOLD
    DOWN TOOLS J 38613 INSTALLED
    (#4 TDC COMPRESSION STROKE)
B   FRONT COVER TIMING MARK
C   LOCATION OF TIMING MARKS WITH DRIVE
    BELT INSTALLED
D   LOCATION WHERE CAM HOLD DOWN TOOLS
    ARE INSTALLED
1   RH EXHAUST CAMSHAFT SPROCKET
2   RH INTAKE CAMSHAFT SPROCKET
3   LH INTAKE CAMSHAFT SPROCKET
4   LH EXHAUST CAMSHAFT SPROCKET
5   PERMANENT MARKS PAINTED DOTS REMOVE
    PREVIOUS MARKS IF TIMING IS BEING CHANGED
    AND MARKS AGAIN IN THESE LOCATIONS
6   CRANKSHAFT BALANCER
7   INTERMEDIATE SHAFT SPROCKET

NOTE: CAMSHAFT BELT
REMOVED. CAM FLATS
UP. ORIGINAL TIMING
MARKS STILL ON.

79225G15

Timing marks with hold-down tool in place—GM 3.4L (VIN X) engines

56. Install the intermediate shaft drive belt sprocket. The sprocket must lock into the intermediate shaft timing chain sprocket. Install the mounting bolt and washer. While holding the engine from turning, tighten the bolt to 95 ft. lbs. (130 Nm).

57. Raise and safely support vehicle on jack stands.

58. Connect and install the starter. Tighten the starter mounting bolts to 32 ft. lbs. (43 Nm).

59. Install the halfshaft following the recommended procedure.

60. Install the rear alternator bracket. Tighten the mounting bolt to 22 ft. lbs. (30 Nm) and the mounting stud to 41 ft. lbs. (55 Nm) and the lower bolt to 61 ft. lbs. (83 Nm).

61. Install the lower front cover bolts and tighten the small bolts to 18 ft. lbs. (25 Nm).

62. Install the A/C compressor mounting bolts and tighten the mounting bolts to 37 ft. lbs. (50 Nm).

63. Install a new oil filter.

64. Install the crankshaft damper as follows:
   a. Coat the seal contact area on the damper with clean engine oil.
   b. Line up the notch inside the damper with the crankshaft key and slide the damper on until the key is started into the notch.
   c. Using J-29113, or an equivalent puller, press the damper into position on the crankshaft.
   d. Install the crankshaft pulley and pulley mounting bolts. Tighten the pulley mounting bolts to 37 ft. lbs. (50 Nm).
   e. Install the crankshaft damper mounting bolt and washer, and tighten to 78 ft. lbs. (105 Nm).

65. Install the right side inner fender splash-shield.

66. Install the tire and wheel assembly.

67. Lower the vehicle.

68. Tighten the upper front cover small bolts to 18 ft. lbs. (25 Nm) and the large bolts to 35 ft. lbs. (47 Nm).

69. Connect the heater hose to the front cover.

70. Install the heater pipe bracket mounting bolts.

71. If equipped with a manual transaxle, connect the front AIR hose to the AIR pipe.

72. Connect the lower radiator hose to the water pump.

73. Install the right side cooling fan. Install the upper radiator support.

74. Position the torque strut mounting bracket and install the mounting bolts and tighten to 52 ft. lbs. (70 Nm).

75. Install the front engine lift hook and tighten the mounting bolt to 52 ft. lbs. (70 Nm).

76. To install the camshaft sprockets, proceed as follows:
   a. Wipe the camshaft noses with clean engine oil.
   b. Install the camshaft sprocket onto the nose of the camshaft.
   c. Install the lockring and shim ring.
   d. Install, but DO NOT tighten, the camshaft sprocket mounting bolts at this time.

77. Install the intermediate shaft sprocket as follows:
   a. Lubricate the seal contact area on the intermediate shaft sprocket with clean engine oil.
   b. Slide the sprocket through the intermediate shaft sprocket seal and engage the locking tangs into the sockets of the chain sprocket.
   c. Lightly lubricate the shaft seal and place it in position on the end of the intermediate shaft.
   d. Install the intermediate shaft sprocket mounting bolt and washer. Tighten the bolt to 96 ft. lbs. (130 Nm) while holding the crankshaft from turning.

78. Install the timing belt idler pulleys and tighten the Torx® bolts to 37 ft. lbs. (50 Nm).

79. Install the actuator assembly and side plate. Tighten the actuator mounting bracket bolts to 37 ft. lbs. (50 Nm).

80. Install the belt, taking note of direction of rotation if the old belt was used.

81. Install the tensioner pulley to the mounting base. Tighten the bolt to 37 ft. lbs. (50 Nm).

82. Rotate the tensioner pulley counterclockwise into the belt using the cast square lug on the body and engage the ball end of the actuator into the socket on the pulley arm.

83. Remove the tensioner lockpin allowing the tensioner shaft to extend and the pulley to move into the belt.

84. Rotate the tensioner pulley counterclockwise, applying 14 ft. lbs. (18 Nm) of torque.

85. Rotate the engine clockwise three times to seat the belt. Align the crankshaft reference marks during the final rotation to TDC. Do not allow the crankshaft to spring back or reverse its direction of rotation.

➡**The timing flats on the camshafts should be 180 degrees apart from the left side to the right side. Both camshafts on the same side should be the same.**

86. To perform the camshaft timing procedure, proceed as follows:

  a. Rotate the camshaft flats up on the right side camshafts and install a camshaft hold-down tool, J-38613 or the equivalent, and tighten to 22 ft. lbs. (30 Nm).

  b. Seat the lockring on the right exhaust and intake camshaft sprockets by threading in the mounting bolt and washer.

  c. Hold the sprocket from turning using tool J-38614, or equivalent.

➡**Running torque of the bolts before seating should be 55 ft. lbs. (75 Nm).**

  d. If less torque is required, replace the shim ring and lockring.

  e. If more torque is required, replace the shim ring and lockring and inspect the bolts for burring.

  f. Seating of the lockring is accomplished when the edge is flush with the sprocket hub.

  g. With the lockring seated tighten the bolt to final torque of 81 ft. lbs. (110 Nm).

  h. Remove J-38613, or equivalent.

  i. Rotate the engine clockwise one full revolution, or any number of odd revolutions. DO NOT rotate the engine backward.

  j. Be sure the timing mark on the damper lines up with the mark on the front cover.

  k. Repeat Substeps a through i for the left side.

87. Install the camshaft carrier covers.

88. Install timing the belt left, right and center covers and retaining bolts.

89. Connect the negative battery cable.

90. Start the engine and verify proper operation and engine performance.

BRAKES

6

ANTI-LOCK BRAKE
  SYSTEMS.......................6-51
BRAKE OPERATING
  SYSTEM.........................6-1
BRAKE SPECIFICATIONS......6-53
DISC BRAKES 6-9
DRUM BRAKES 6-26
Basic Operating Principles.........6-1
  Disc Brakes 6-2
  Drum Brakes 6-2
  Power Boosters 6-2
Brake Calipers 6-12
  Removal & Installation 6-12

Brake Drums.................................6-26
  Inspection.................................6-26
  Removal & Installation 6-26
Brake Pads.....................................6-9
  Inspection...................................6-9
  Removal & Installation 6-9
Brake Rotors 6-22
  Inspection.................................6-22
  Removal & Installation 6-23
Brake Shoes 6-30
  Adjustment 6-46
  General Information.................6-30
  Inspection................................6-30

Removal & Installation 6-30
Brake System Bleeding 6-6
  Models With ABS........................6-8
  Models Without ABS....................6-6
General Information 6-51
  ABS Depressurizing 6-52
  Diagnostic Trouble Codes 6-52
  Precautions 6-51
Master Cylinder 6-4
  Removal & Installation 6-4
Wheel Cylinders 6-47
  Overhaul.................................6-49
  Removal & Installation 6-47

BRAKE OPERATING SYSTEM

Basic Operating Principles

Hydraulic systems are used to actuate the brakes of all modern automobiles. The system transports the power required to force the frictional surfaces of the braking system together from the pedal to the individual brake units at each wheel. A hydraulic system is used for two reasons.

First, fluid under pressure can be carried to all parts of an automobile by small pipes and flexible hoses without taking up a significant amount of room or posing routing problems.

Second, a great mechanical advantage can be given to the brake pedal end of the system, and the foot pressure required to actuate the brakes can be reduced by making the surface area of the master cylinder pistons smaller than that of any of the pistons in the wheel cylinders or calipers.

The master cylinder consists of a fluid reservoir along with a double cylinder and piston assembly. Double type master cylinders are designed to separate the front and rear braking systems hydraulically in case of a leak. The master cylinder coverts mechanical motion from the pedal into hydraulic pressure within the lines. This pressure is translated back into mechanical motion at the wheels by either the wheel cylinder (drum brakes) or the caliper (disc brakes).

Steel lines carry the brake fluid to a point on the vehicle's frame near each of the vehicle's wheels. The fluid is, then carried to the calipers and wheel cylinders by flexible tubes in order to allow for suspension and steering movements.

In drum brake systems, each wheel cylinder contains two pistons, one at either end, which push outward in opposite directions and force the brake shoe into contact with the drum.

In disc brake systems, the cylinders are part of the calipers. At least one cylinder in each caliper is used to force the brake pads against the disc.

All pistons employ some type of seal, usually made of rubber, to minimize fluid leakage. A rubber dust boot seals the outer end of the cylinder against dust and dirt. The boot fits around the outer end of the piston on disc brake calipers, and around the brake actuating rod on wheel cylinders.

The hydraulic system operates as follows: When at rest, the entire system, from the piston(s) in the master cylinder to those in the wheel cylinders or calipers, is full of brake fluid. Upon application of the brake pedal, fluid trapped in front of the master cylinder piston(s) is forced through the lines to the wheel cylinders. Here, it forces the pistons outward, in the case of drum brakes, and inward toward the disc, in the case of disc brakes. The motion of the pistons is opposed by return springs mounted outside the cylinders in drum brakes, and by spring seals, in disc brakes.

Upon release of the brake pedal, a spring located inside the master cylinder immediately returns the master cylinder pistons to the normal position. The pistons contain check valves and the master cylinder has compensating ports drilled in it. These are uncovered as the pistons reach their normal position. The piston check valves allow fluid to flow toward the wheel cylinders or calipers as the pistons withdraw. Then, as the return springs force the brake pads or shoes into the released position, the excess fluid reservoir through the compensating ports. It is during the time the pedal is in the released position that any fluid that has leaked out of the system will be replaced through the compensating ports.

Dual circuit master cylinders employ two pistons, located one behind the other, in the same cylinder. The primary piston is actuated directly by mechanical linkage from the brake pedal through the power booster. The secondary piston is actuated by fluid trapped between the two pistons. If a leak develops in front of the secondary piston, it moves forward until it bottoms against the front of the master cylinder, and the fluid trapped between the pistons will operate the rear brakes. If the rear brakes develop a leak, the primary piston will move forward until direct contact with the secondary piston takes place, and it will force the secondary piston to actuate the front brakes. In either case, the brake pedal moves farther when the brakes are applied, and less braking power is available.

All dual circuit systems use a switch to warn the driver when only half of the brake system is operational. This switch is usually located in a valve body which is mounted on the firewall or the frame below the master cylinder. A hydraulic piston receives pressure from both circuits, each circuit's pressure being applied to one end of the piston. When the pressures are in balance, the piston remains stationary. When one circuit has a leak, however, the greater pressure in that circuit during application of the brakes will push the piston to one side, closing the switch and activating the brake warning light.

In disc brake systems, this valve body also contains a metering valve, in some cases, a proportioning valve. The metering valve keeps pressure from traveling to the disc brakes on the front wheels until the brake shoes on the rear wheels have contacted the drums, ensuring that the front brakes will never be used alone. The proportioning valve controls the pressure to the rear brakes to lessen the chance of rear wheel lock-up during very hard braking.

Warning lights may be tested by depressing the brake pedal and holding it while opening one of the wheel cylinder bleeder screws. If this does not cause the light to go on, substitute a new lamp, make continuity checks, finally, replace the switch as necessary.

The hydraulic system may be checked for leaks by applying pressure to the pedal gradually and steadily. If the pedal sinks very slowly to the floor, the system has a leak. This is not to be confused with a springy or spongy feel due to the compression of air within the lines. If the system leaks, there will be a gradual change in the position of the pedal with a constant pressure.

Check for leaks along all lines and at wheel cylinders. If no external leaks are apparent, the problem is inside the master cylinder.

DISC BRAKES

Instead of the traditional expanding brakes that press outward against a circular drum, disc brake systems utilize a disc (rotor) with brake pads positioned on either side of it. An easily-seen analogy is the hand brake arrangement on a bicycle. The pads squeeze onto the rim of the bike wheel, slowing its motion. Automobile disc brakes use the identical principle but apply the braking effort to a separate disc instead of the wheel.

The disc (rotor) is a casting, usually equipped with cooling fins between the two braking surfaces. This enables air to circulate between the braking surfaces making them less sensitive to heat buildup and more resistant to fade. Dirt and water do not drastically affect braking action since contaminants are thrown off by the centrifugal action of the rotor or scraped off the by the pads. Also, the equal clamping action of the two brake pads tends to ensure uniform, straight line stops. Disc brakes are inherently self-adjusting. There are three general types of disc brake:

1. Fixed calipers.
2. Floating calipers.
3. Sliding calipers.

The fixed caliper design uses one or two pistons mounted on each side of the rotor (in each side of the caliper). The caliper is mounted rigidly and does not move.

The sliding and floating designs are quite similar. In fact, these two types are often lumped together. In both designs, the pad on the inside of the rotor is moved into contact with the rotor by hydraulic force. The caliper, which is not held in a fixed position, moves slightly, bringing the outside pad into contact with the rotor.

Floating calipers use threaded guide pins and bushings, or sleeves to allow the caliper to slide and apply the brake pads.

There are typically three methods of securing a sliding caliper to its mounting bracket: with a retaining pin, with a key and bolt, or with a wedge and pin. On calipers which use the retaining pin method, you will find pins driven into the slot between the caliper and the caliper mount. On calipers which use the bolt and key method, a key is used between the caliper and the mounting bracket to allow the caliper to slide. The key is held in position by a lockbolt. On calipers which use the pin and wedge method, a wedge, retained by a pin, is used between the caliper and the mounting bracket.

For pad removal purposes, fixed calipers are usually not removed, floating calipers are either removed or flipped (hinged up or down on one pin), and sliding calipers are removed.

DRUM BRAKES

Drum brakes employ two brake shoes mounted on a stationary backing plate. These shoes are positioned inside a circular drum which rotates with the wheel assembly. The shoes are held in place by springs. This allows them to slide toward the drums (when they are applied) while keeping the linings and drums in alignment. The shoes are actuated by a wheel cylinder which is mounted at the top of the backing plate. When the brakes are applied, hydraulic pressure forces the wheel cylinder's actuating links outward. Since these links bear directly against the top of the brake shoes, the tops of the shoes are, then forced against the inner side of the drum. This action forces the bottoms of the two shoes to contact the brake drum by rotating the entire assembly slightly (known as servo action). When pressure within the wheel cylinder is relaxed, return springs pull the shoes back away from the drum.

Most modern drum brakes are designed to self-adjust themselves during application when the vehicle is moving in reverse. This motion causes both shoes to rotate very slightly with the drum, rocking an adjusting lever, thereby causing rotation of the adjusting screw. Some drum brake systems are designed to self-adjust during application whenever the brakes are applied. This on-board adjustment system reduces the need for maintenance adjustments and keeps both the brake function and pedal feel satisfactory.

POWER BOOSTERS

Virtually all modern vehicles use a power assisted brake system to multiply the braking force and reduce pedal effort. There are two types of power assist used. The most widely used, by far, is the vacuum assist booster. The other is the hydraulically assisted booster.

Vacuum-Assisted Boosters

Most modern vehicles use a vacuum assisted power brake. This system was likely developed, since on all internal combustion engines, except diesels, vacuum is always available when the engine is operating, making the system is simple and efficient.

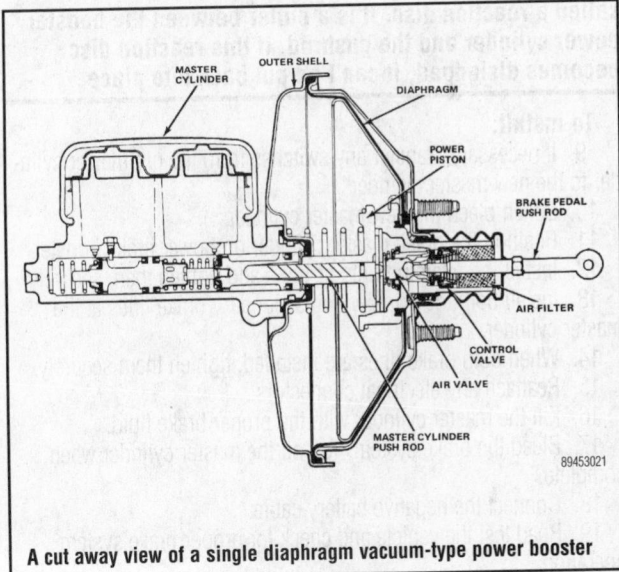

A cut away view of a single diaphragm vacuum-type power booster

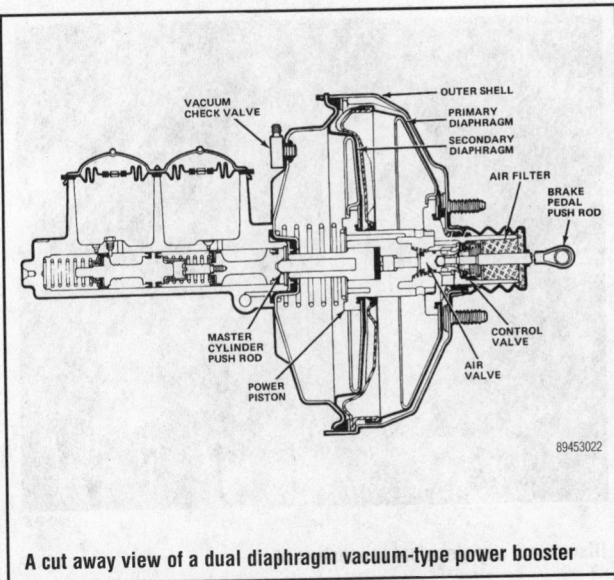

A cut away view of a dual diaphragm vacuum-type power booster

With diesel engines, vacuum is created and stored by way of a belt-driven vacuum pump and reservoir. In either case, the operation of the vacuum assist is the same.

A vacuum diaphragm is located on the front of the master cylinder and assists the driver in applying the brakes, reducing both the effort and travel one must put into moving the brake pedal. The vacuum diaphragm housing is normally connected to the intake manifold by a vacuum hose. A check valve is placed at the point where the hose enters the diaphragm housing, so that during periods of low manifold vacuum brake assist will not be lost.

Depressing the brake pedal closes off the vacuum source and allows atmospheric pressure to enter on one side of the diaphragm. This causes the master cylinder pistons to move and apply the brakes. When the brake pedal is released, vacuum is applied to both sides of the diaphragm and springs return the diaphragm and master cylinder pistons to the released position.

If the vacuum supply fails, the brake pedal rod will contact the end of the master cylinder actuator rod and the system will apply the brakes without any power assistance. The driver will notice that much higher pedal effort is needed to stop the vehicle and that the pedal feels harder than usual.

If you think this is the case you can check it as follows:

VACUUM LEAK TEST

1. Operate the engine at idle without touching the brake pedal for at least one minute.
2. Turn off the engine and wait one minute.
3. Test for the presence of assist vacuum by depressing the brake pedal and releasing it several times. If vacuum is present in the system, light application will produce less and less pedal travel. If there is no vacuum, air is leaking into the system.

SYSTEM OPERATION TEST

1. With the engine **OFF**, pump the brake pedal until the supply vacuum is entirely gone.
2. Apply light, steady pressure to the brake pedal.
3. Start the engine and let it idle. If the system is operating correctly, the brake pedal should fall toward the floor if constant pressure is maintained.

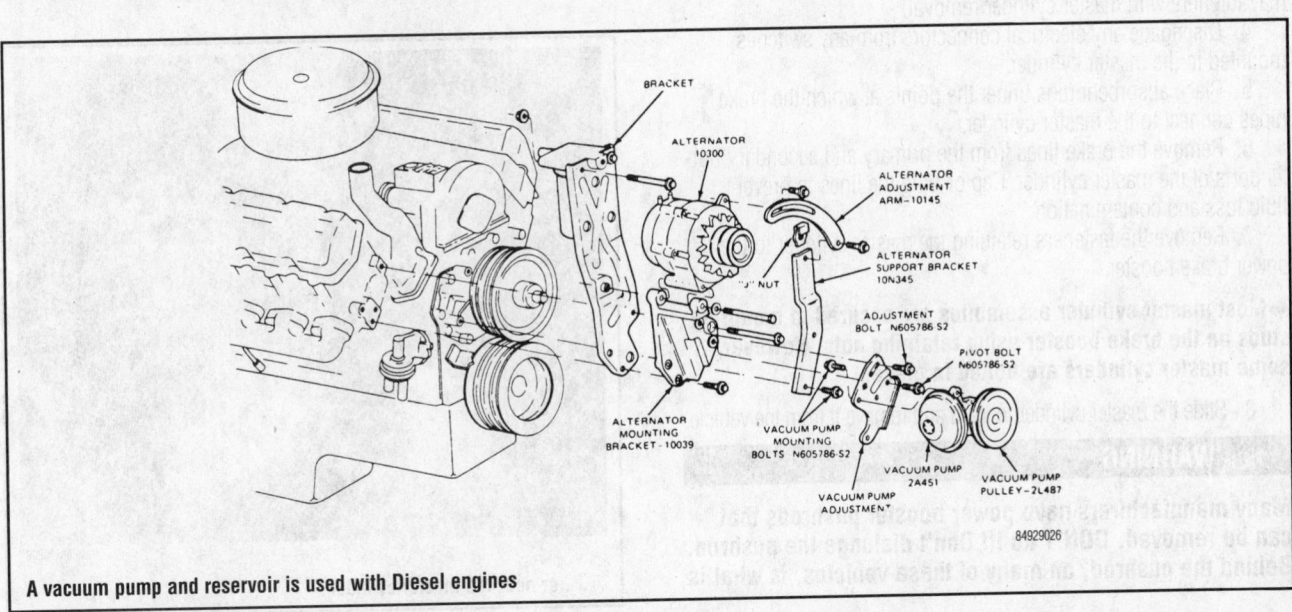

A vacuum pump and reservoir is used with Diesel engines

Power brake systems may be tested for hydraulic leaks just as ordinary systems are tested.

Hydraulically-Assisted Boosters

Used on some light vehicles, the unit is fed hydraulic fluid through the power steering system. The booster assembly, sometimes known generically by the brand name Hydro-Boost, contains a valve which controls pump pressure while braking, a lever to control the position of the valve and a boost piston to provide the force to operate the master cylinder attached to the front of the booster. The unit has a reserve system designed to store pressurized fluid to provide at least 2 brake applications in the event of hydraulic supply system failure, such as a broken power steering belt. The brakes can also be applied unassisted in the event of system depletion.

Master Cylinder

➡The following procedures apply to non-ABS systems and ABS system master cylinders that are separate from other ABS system components. ABS systems with integral master cylinder components often require special tools and model-specific procedures.

REMOVAL & INSTALLATION

With Power-Assisted Brakes

1. Disconnect the negative battery cable.
2. If applicable, apply the brake pedal several times to exhaust all vacuum from the power boost system.
3. Remove any components in the engine compartment which may interfere with master cylinder removal.
4. Disengage any electrical connectors from any switches mounted in the master cylinder.
5. Place absorbent rags under the points at which the brake pipes connect to the master cylinder.
6. Remove the brake lines from the primary and secondary outlet ports of the master cylinder. Cap or plug the lines to prevent fluid loss and contamination.
7. Remove the fasteners retaining the master cylinder to the power brake booster.

➡Most master cylinder assemblies are secured to mounting studs on the brake booster using retaining nuts. However, some master cylinders are bolted in place.

8. Slide the master cylinder forward and remove it from the vehicle.

✳✳ WARNING

Many manufacturers have power booster pushrods that can be removed. DON'T do it! Don't dislodge the pushrod. Behind the pushrod, on many of these vehicles, is what is called a reaction disc. It is a buffer between the booster power cylinder and the pushrod. If this reaction disc becomes dislodged, it can't be put back into place.

To install:

9. If necessary, transfer any switches from the old master cylinder to the new master cylinder.
10. Bench bleed the new master cylinder.
11. Position the brake master cylinder on power brake booster.
12. Install the retaining nuts or bolts and tighten them securely.
13. Install both the primary and secondary brake lines at the master cylinder.
14. When both brake lines are installed, tighten them securely.
15. Reattach any electrical connectors.
16. Fill the master cylinder with the proper brake fluid.
17. Bleed the brake system. Top off the master cylinder when complete.
18. Connect the negative battery cable.
19. Road test the vehicle and check for proper brake system operation.

Disconnect any electrical connectors at . . .

88489P65

. . . or near the master cylinder

88279P29

Use an open-end wrench (a line wrench is preferable) to loosen the brake pipe fittings . . .

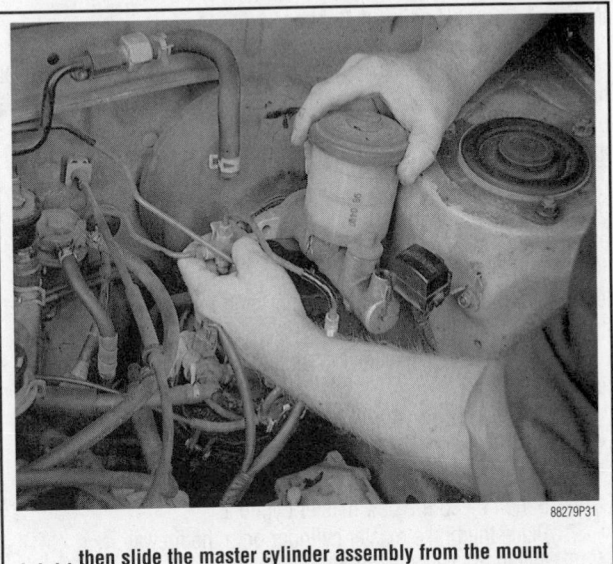

. . . , then slide the master cylinder assembly from the mount

. . . the disconnect the pipes from the master cylinder assembly

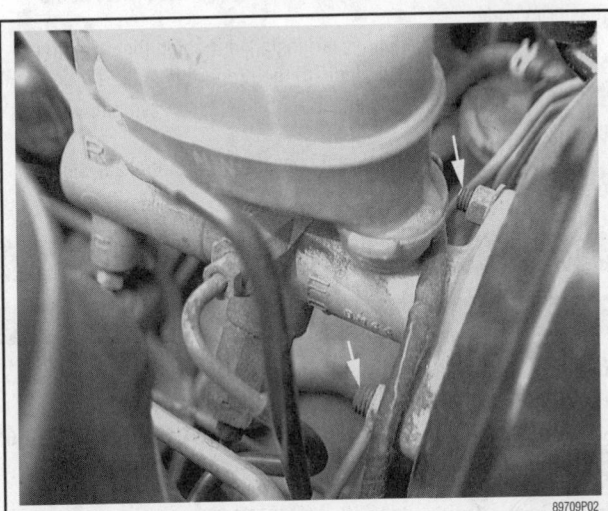

Most master cylinders are secured to the brake booster using 2 retaining nuts

Loosen the master cylinder retainers . . .

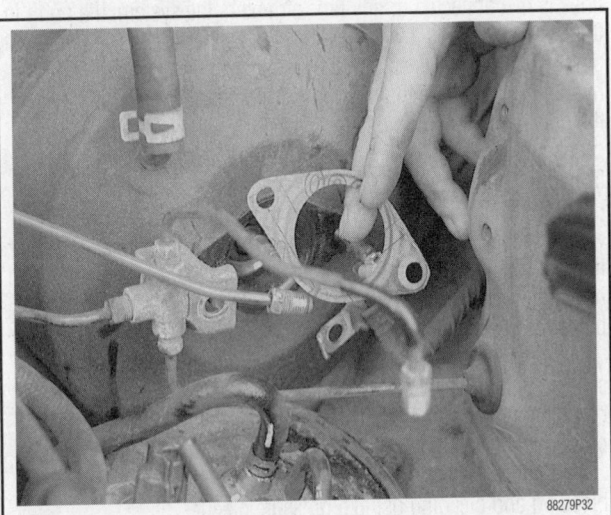

Some vehicles have gaskets between the master cylinder and booster

Without Power-Assisted Brakes

1. Disconnect the master cylinder pushrod from the brake pedal linkage. This connection can be either inside the passenger compartment or inside the engine compartment. The connection is usually by way of a rod fitting over a stud on the pedal arm, retained by washers and a cotter pin or clip.

2. Place absorbent rags under the points at which the brake pipes connect to the master cylinder.

3. Remove the brake lines from the primary and secondary outlet ports of the master cylinder. Cap or plug the lines to prevent fluid loss and contamination.

4. Remove the fasteners retaining the master cylinder to the firewall.

5. Slide the master cylinder forward and remove it from the vehicle.

To install:

6. Bench bleed the new master cylinder.

7. Place the brake master cylinder onto the firewall.

8. Install the fasteners and tighten them securely.

9. Install both the primary and secondary brake lines at the master cylinder.

10. When both brake lines are installed, tighten them securely.

11. Fill the master cylinder with the proper brake fluid.

12. Bleed the brake system. Top off the master cylinder when complete.

13. Road test the vehicle and check for proper brake system operation.

Brake System Bleeding

✳✳ CAUTION

Brake fluid contains polyglycol ethers and polyglycols. Avoid contact with the eyes and wash your hands thoroughly after handling brake fluid. If you do get brake fluid in your eyes, flush your eyes with clean, running water for 15 minutes. If eye irritation persists, or if you have taken brake fluid internally, IMMEDIATELY seek medical assistance.

The hydraulic brake system must be bled any time any of the lines is disconnected or any time air enters the system. If a point in the system, such as a wheel cylinder or caliper brake line is the only point which was opened, the bleeder screws down stream in the hydraulic system are the only ones which must be bled. If however, the master cylinder fittings are opened, or if the reservoir level drops sufficiently that air is drawn into the system, air must be bled from the entire hydraulic system. If the brake pedal feels spongy upon application and travels almost to the floor but regains height when pumped, air has entered the system. It must be bled out. If no fittings were recently opened for service, check for leaks that would have allowed the entry of air and repair them before attempting to bleed the system.

As a general rule, once the master cylinder (and the brake pressure modulator valve or combination valve on ABS systems) is bled, the remainder of the hydraulic system should be bled in the proper sequence.

The hydraulic system can be bled in one of two ways: manual bleeding and bleeding using a pressure bleeder.

MODELS WITHOUT ABS

Manual Bleeding

MASTER CYLINDER

If the unit is removed from the vehicle, there are 2 ways to "bench-bleed" a master cylinder.

One method is with a large, clear plastic syringe made for the purpose. They are usually available at auto parts stores. In this procedure, the master cylinder is clamped in a soft-jawed vise and filled with fluid. The outlet ports are capped or plugged. Then, uncap each port, place the syringe securely in the outlet port and draw fluid into the syringe until no air is left in the master cylinder, capping the ports when done.

The other is with 2 lengths of hose or pipe (to use as bleeder tubes). Plastic hoses, made for the purpose, are available at most auto parts stores. These hoses have threaded ends for attachment to

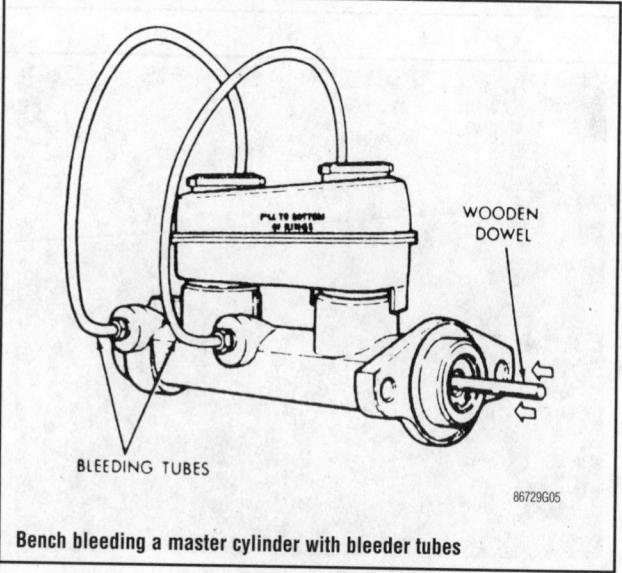

Bench bleeding a master cylinder with bleeder tubes

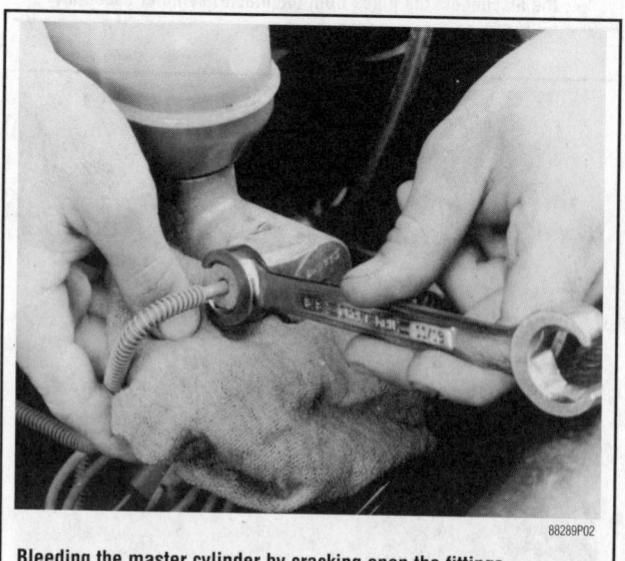

Bleeding the master cylinder by cracking open the fittings

the outlet ports. Otherwise, you'll have to make your own bleeder pipes from 2 lengths of brake pipe equipped with threaded ends. Try to get the plastic ones. In this procedure, clamp the master cylinder in a soft-jawed vise. Connect the pieces of brake pipe or the plastic hoses to the outlet fittings, bend them until the free end is the master cylinder reservoir. Fill the reservoir with fresh DOT 3, or equivalent, brake fluid from a closed container, completely covering the tube ends. Pump the piston slowly until no more air bubbles appear in the reservoir. Remove the tubes, refill the brake master cylinder and securely install the caps or plugs in the ports.

If the brake master cylinder is on the vehicle, place a large, absorbent rag under the fittings. Open the brake lines slightly with the flare nut wrench while pressure is applied to the brake pedal by a helper inside the vehicle. Be sure to tighten the line before the brake pedal is released. Repeat the process with both lines until no air bubbles come out.

In both cases, the rest of the brake system must be bled to assure that all trapped has been removed and that the system will operate properly.

CALIPERS AND WHEEL CYLINDERS

We recommend that the brake system be bled using the jar and tube method. We know some people just let the fluid spray all over the place from the nipple. This is not only unprofessional, but it's messy and potentially dangerous. Brake fluid damages paint, concrete, your clothes, your skin, most importantly, your eyes.

➡**Hydraulic brake systems must be totally flushed if the fluid becomes contaminated with water, dirt or other corrosive chemicals. Also, many manufacturers recommend that the system be flushed routinely, every 2 years or so. To flush, bleed the entire system until all fluid has been replaced and the new brake fluid runs clear.**

The hydraulic system on vehicles with a split system—a 2-chambered master cylinder—can be split either into front/rear or diagonally. In the diagonally split system there is one front and one rear component in each circuit. If you are in doubt as to the design of your vehicle's system, you can check the brake lines. Follow them to each wheel and see which are paired.

➡**If, during the bleeding procedure, you can't get a good flow of fluid from the front brakes, the problem is with the**

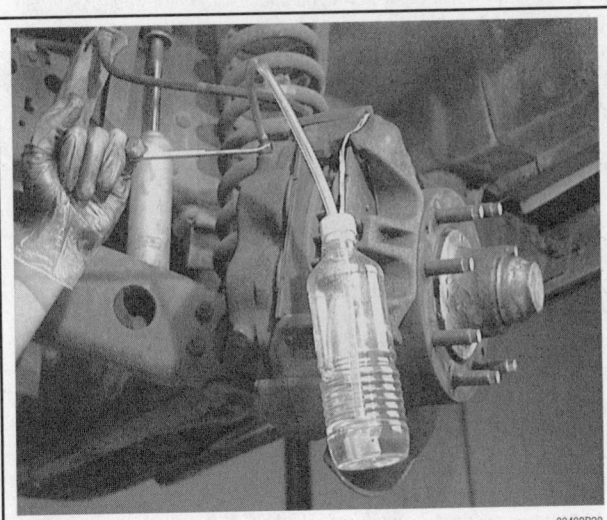

Bleeding the calipers

88489P48

Bleeding the wheel cylinders

metering part of the combination valve. Check the valve and you'll see a small stem sticking out of one end. You'll have to fabricate a little clip to hold the stem out as far as it will go. This will allow a full flow to the front brakes. Also, when using this clip on vehicles with power brakes, try bleeding with the engine running. The greater pressure allowed by the power booster will aid in purging the system.

1. Fill the brake master cylinder with the fluid recommended for your vehicle. Check the level often during the procedure. Never let the master cylinder go dry or the procedure will have to be performed again.

2. Raise and support the vehicle safely.

3. If necessary for better access, remove the wheels.

4. On vehicles with a single chamber system or dual chambered systems split front/rear, you can bleed the system in the following order:
   - Right rear
   - Left rear
   - Right front
   - Left front

5. On vehicles with a dual chambered system split diagonally, the usual bleeding order is:
   - Right rear
   - Left front
   - Left rear
   - Right front

6. Find a wrench, a box wrench if possible, of the right size for the bleeder screw and place it on the nipple of the first cylinder to be bled.

7. Connect a clear, vinyl tube to the bleeder nipple. Place the other end of the tube in a clear glass jar of at least 8 oz. (237mL) capacity. The jar should be about ½ full of clean brake fluid. Submerge the end of the tube in the brake fluid.

8. Have an assistant pump the brake pedal, then hold it down. Slowly open the bleeder screw. When the brake pedal reaches the floor, close the bleeder and have the helper slowly release the pedal. Wait 15 seconds, then repeat the procedure until no more air comes out of the bleeder.

9. Repeat the procedure on the remaining calipers or wheel cylinders in the appropriate order.

10. If the brake pedal has a spongy feel, the brake system must be bled again to remove air still trapped in the system.

11. Install the bleeder caps to keep dirt out.

12. If removed for access, install the wheels.

13. Lower the vehicle.

14. Road test the vehicle and check for proper brake system operation.

Pressure Bleeding

A pressure bleeder is a device that uses compressed air and a series of adapters to forcibly expel air from the hydraulic system. When using a pressure bleeder, always follow the manufacturer's instructions. What we've given you here are general instructions.

When using pressure bleeding equipment, it's best to use a bladder-type bleeder tank. In this type of bleeder, the brake fluid is separated from the air by a rubber diaphragm. The bleeder tank must contain enough brake fluid to complete the bleeding operation and should be charged with only 10–30 psi (69–207 kPa). Never exceed 50 psi (345 kPa).

1. Clean all dirt from the master cylinder fluid reservoir filler cap.

➡**The reservoir must be at least ¾ -full during the bleeding procedure. Fill the reservoir as necessary. Use only clean, fresh brake fluid from a sealed container. Fill to the MAX level line on the reservoir.**

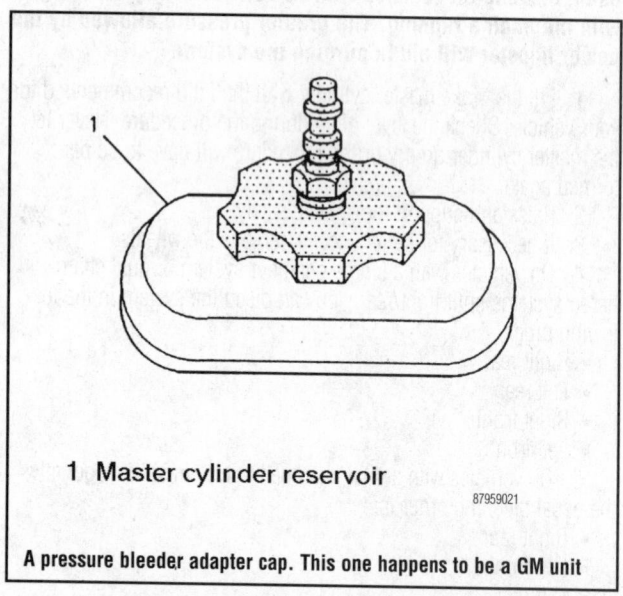

1 **Master cylinder reservoir**

87959021

A pressure bleeder adapter cap. This one happens to be a GM unit

2. Install the bleeder adapter tool on the master cylinder and attach the hose from the bleeder tank to the fitting on the adapter. Follow the manufacturer's instructions when installing and connecting the master cylinder adapter.

3. Open the valve on the bleeder tank.

MASTER CYLINDER

1. If the master cylinder is known or suspected to contain air, it must be bled before the wheel cylinders or calipers. Place a large, absorbent rag under the pipe fittings. Beginning at the front of the master cylinder, alternately loosen and tighten the brake line flare nuts. Allow the fluid to flow for several seconds before tightening the flare nut. Repeat this operation several times to be sure all air has been removed from the master cylinder.

CALIPERS AND WHEEL CYLINDERS

Pressure system bleeding must be performed in the correct order. Refer to the manual bleeding procedure for proper bleeding sequences.

1. Raise and safely support the vehicle.

2. Remove the protective bleeder screw cap from the caliper or wheel cylinder and clean the nipple.

3. Place a wrench, preferably a box end wrench, on the bleeder screw.

4. Attach a length of clear vinyl hose onto the bleeder nipple. The hose must fit tightly around the bleeder screw.

5. Submerge the free end of the hose in a large (approximately 16 oz./475mL) clean glass jar about half filled with clean brake fluid.

6. Loosen the bleeder screw approximately ¾ of a turn. When the fluid entering the jar is completely free of bubbles, tighten the bleeder screw.

7. Remove the bleeder hose and attach the protective screw cap.

8. Repeat the bleeding procedure at each brake.

9. Close the valve at the bleeder tank, disconnect the hose from the master cylinder adapter and remove the master cylinder adapter.

10. Check the fluid level in the remote reservoir, refilling with clean, fresh brake fluid, as necessary.

11. Check the brake pedal feel. If spongy, repeat the bleeding process and/or look for defective system components.

MODELS WITH ABS

There are 2 potential problems with attempting to bleed an ABS system. The first is that many use control valves and pressure modulators which might trap air if they are not opened and closed during the procedure using a scan tool. The second potential problem is that some ABS systems operate under extremely high pressure (making bleeding dangerous at worst or messy at best).

With this said, there are still many systems which can be bled with common tools. Many of the control valves have pressure relief knobs at one end of the valve which can be held open using a small tool (or pair of locking pliers)., just about all systems can be bled at the wheels provided that the openings are capped immediately during service. The caps keep enough fluid in the lines to prevent air from working its way back to the control or modulator valves.

Before starting, remember that many manufacturers require the use of special scan tools to bleed any part of the system other than the caliper or wheel cylinders. Some manufacturers recommend the scan tool be used when bleeding any part of the system on some of their models. All manufacturers recommend the use of pressure bleeding equipment for ABS systems, especially when bleeding the rear brakes even though manual bleeding can be done successfully in most cases.

If you decide to attempt bleeding the calipers or wheel cylinders, and you are sure that any residual high pressure is depleted, use the same procedure for bleeding as described for non-ABS systems. During the bleeding procedure, wait 10–15 seconds after closing the bleeder screw before reopening it each time. This is recommended by most manufacturers due to the number of valved components in the system.

Once the procedure is complete, start the engine and allow it to run for 15–30 seconds. Depress the brake pedal. The ABS light should not be **ON**. If the light is **ON**, there is a system problem, probably air still trapped somewhere. At this point, you can try the bleeding procedure again or have the vehicle towed to a dealer or repair shop for system bleeding.

As in all bleeding procedures, DO NOT attempt to move the vehicle unless a firm brake pedal feel has been obtained.

DISC BRAKES

Brake Pads

INSPECTION

To inspect the brake pads, remove the wheel. It is usually possible to view the pad thickness through a large hole in the caliper, or by looking at the side of the pad. However, on a few models, it may be necessary to remove the pads for inspection.

As a rule of thumb, the brake pad lining material should be worn no more than 1/8 in. (3mm). On brake pads glued to the backing material, the pad material can be measured from the edge of the backing material. However, on pads which are riveted to the backing material, the lining should be measured from the rivet heads (in the holes in the lining material)

The brake lining material should not exhibit any dampness, crumbling or cracking. If any such damage is evident the pads must be replaced. If the pads showed evidence of dampness, locate the source of the fluid leak and repair it before installing the new pads. If the brake pads exhibit uneven wear, (such as, one pair of pads is worn more on one side of the vehicle than the other pair of pads on the other side; the inner pad is worn more than the outer pad, or vice versa, on one wheel; the pad lining material is worn more on the front edge of a pad, or more on the rear edge of a pad) the disc brake caliper is either defective or mounted improperly.

✳✳ WARNING

Never polish the pad lining with sandpaper, because hard particles from the sandpaper will become imbedded in the lining, which will damage the brake rotor. If the pad lining is damaged or worn excessively or unevenly, replace the pads with new ones.

REMOVAL & INSTALLATION

✳✳ CAUTION

Brake dust may contain asbestos! Asbestos is harmful to your health. Never use compressed air to clean any brake component. A filtering mask should be worn during any brake repair.

Brake pad replacement should always be performed on both front or rear wheels at the same time. Never replace pads on only one wheel. When servicing any brakes use only OEM or better quality pads and parts. When the caliper is removed some brake pads stay with the caliper, others remain on the caliper mounting bracket. Use new pad mounting hardware (springs, anti-rattle clips, or shims) whenever possible to ensure a better repair.

Sliding and Floating Calipers

➡**On certain floating calipers it may be possible to remove one of the guide pins and pivot the caliper up or down to gain access to the brake pads. If you decide to do this, be sure that pivoting the caliper will not damage the flexible brake hose.**

1. Open the hood and locate the master brake cylinder fluid reservoir. Clean the area surrounding the reservoir cap, then remove the cap. Remove some of the brake fluid from the reservoir.

2. Loosen the lug nuts on the applicable wheels.
3. Raise and safely support the vehicle.
4. Remove the wheels.
5. Disconnect any electrical brake pad wear sensors.

➡**It is not necessary, and actually discouraged, to detach the brake hose from the caliper during this procedure. If you decide to detach the hose, it will be necessary for you to bleed your brake system.**

6. Remove and suspend the caliper with a piece of wire, cord or strong string. Be sure that it is not placing any stress on the brake hose.
7. For caliper bracket-mounted pads, perform the following:
   a. If present, remove any anti-squeal shims, noting their positions.
   b. Also, remove any anti-rattle springs that may be present. If these springs don't provide good tension, then replace them.

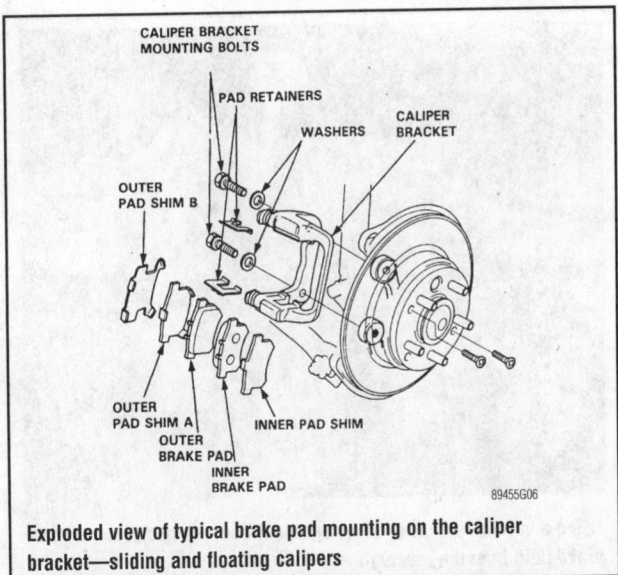

Exploded view of typical brake pad mounting on the caliper bracket—sliding and floating calipers

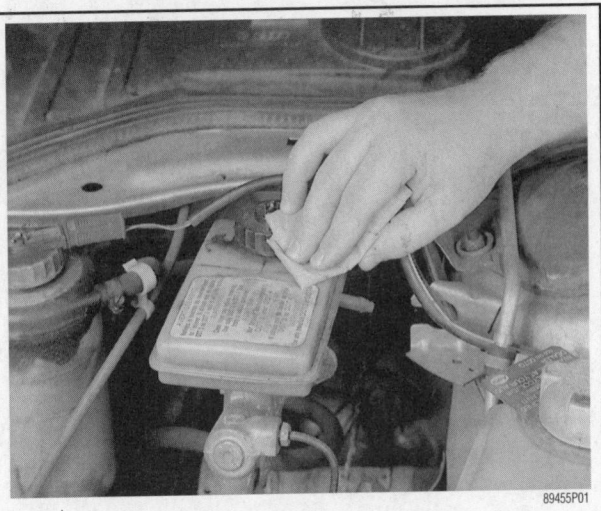

To remove the brake pads, first clean the brake master cylinder reservoir cap . . .

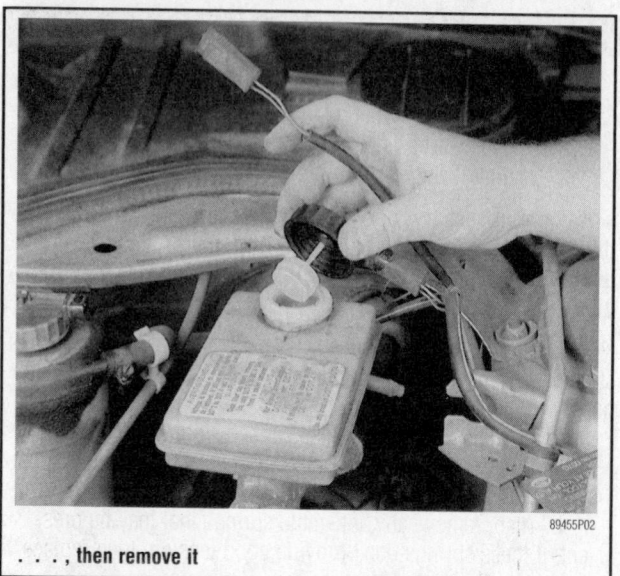

. . . , then remove it

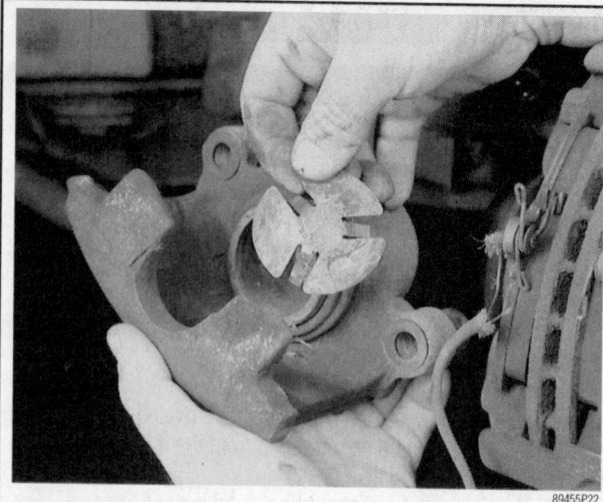

Be sure to note the positions of any clips or springs on the caliper—sliding and floating calipers

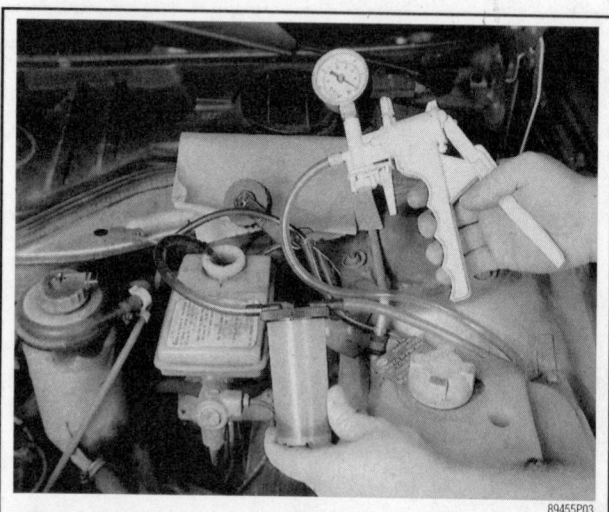

Using a vacuum pump, or some other method, remove some of the brake fluid from the reservoir

Remove the outboard pad from the mounting bracket . . .

Remove the disc brake caliper from the rotor—sliding and floating calipers

. . . , then remove the inboard pad—sliding and floating calipers

c. Remove the brake pads from the caliper bracket by lifting the pad out by hand or with a slight tap of a hammer to help.

8. For caliper mounted pads, perform the following:

a. Some outer pads have tabs that are bent over the edge of the caliper, which hold the pads tight in the caliper. Straighten the tabs with pliers before trying to remove the brake pad from the caliper.

b. Then, remove the outer brake pad with a slight tap to the back of the pad with a hammer.

c. Other outer pads use a spring-clip to mount to the caliper. To remove this type of pad, press the pad towards the center of the caliper and slide it off. It may be helpful to use a small pry-bar.

d. Remove the inner pad by pulling it out of the piston.

To install:

9. Clean the caliper sliding area using a wire brush and spray brake cleaner.

10. Lubricate the sliding area of the caliper and the pins with high temperature brake grease.

11. Apply anti-squeal compound to the back side of both brake pads. Allow the compound to set-up according to the instructions on the package.

12. Install one of the old brake pads against the caliper piston, then use a large C-clamp to press the piston back into its bore.

13. Install any new hardware provided with the new pads.

14. For bracket-mounted pads, perform the following steps:

a. Install the pads onto the caliper bracket. Some pads are marked for position.

b. Be sure that the notches or ears of the brake pads are properly engaged on the bracket.

c. Place the caliper over the pads and onto the caliper mounting bracket.

d. Install the caliper mounting hardware and anti-rattle clips. Tighten the guide pins or lockbolt to the proper specification.

➡ **It is a good idea to use some thread-locking compound (removable type) to the threaded fasteners of the caliper.**

15. For caliper mounted pads, perform the following:

a. Install the inner pad by pushing the retaining fingers of the pad into the piston of the caliper.

Apply a thin coat of high-temperature brake grease to the sliding surfaces of the bracket and caliper

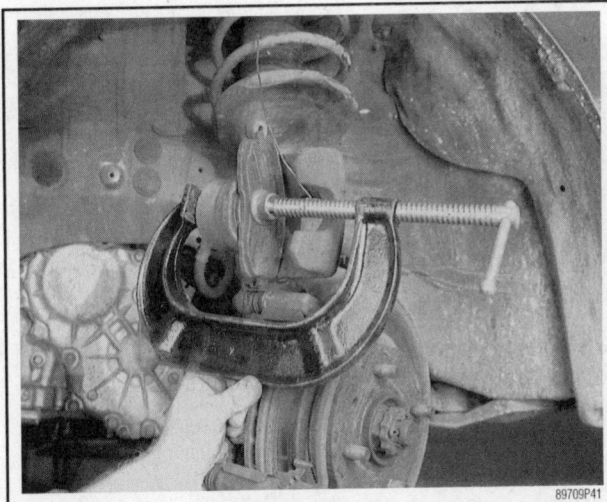

On calipers without integral parking brake mechanisms, a C-clamp can seat the piston in the caliper bore

Clean the caliper and mounting bracket with spray brake solvent and a wire brush

Install all of the springs and clips in their original positions—sliding and floating calipers

When installing the caliper and pads, be sure not to pinch the sensor wire (if equipped)—sliding and floating calipers

b. If the outer pad has a spring-clip, slide the pad over the edge of the caliper into the caliper frame.

c. If you have the bent-tab style outer brake pad, then test fit the pad; it should fit tight. If the tabs do not secure the pad snugly in the caliper, place the pad on a piece of wood and tap the tab with a hammer to adjust it. It may take a few tries to get it right.

d. Place the caliper with the pads onto the rotor, if equipped, caliper bracket.

e. Install the caliper mounting hardware and anti-rattle clips. Tighten the guide pins or lockbolt(s) securely.

➡**It is a good idea to use some thread-locking compound (removable type) on the threaded fasteners of the caliper.**

16. Connect any electrical brake pad wear sensors.

17. Seat the brake pads, otherwise the vehicle may coast out of the work area and into traffic before the brakes become effective. It will take several pumps of the brake pedal to seat the pads against the rotor.

18. If a firm pedal is not achieved, it may be necessary to bleed the brakes.

19. Check the brake fluid level in the reservoir and top off as needed.

20. Install the wheels and tighten the lug nuts.

21. Road test the vehicle.

Fixed Calipers

➡**It is usually not necessary to remove the caliper to replace the brake pads on a fixed caliper.**

1. Loosen the lug nuts on the applicable wheels.
2. Raise and safely support the vehicle.
3. Remove the wheels.
4. Disconnect any electrical brake pad ware sensors.
5. Remove the pad retaining pins by pulling out the spring-clip or cotter pin, then use a punch and hammer to drive the pin out. Pins without a spring-clip or cotter pin, may be equipped with a spring steel collar on the head of the pin. To remove this style pin, just drive the pin out with a punch and hammer.
6. On calipers with hold-down clips, remove the bolt that holds the clip down.

7. Remove the pads from the caliper with a pair of pliers.
8. To seat the pistons of a fixed caliper, use a piece of wood or a prybar with a rag wrapped around the end, then wedge it between the rotor and the piston and slide the piston into its seat.

➡**It is helpful to replace one pad at a time, to reduce the risk of a piston coming out of its bore, which would lead to the caliper needing to be rebuilt.**

9. Lubricate the sliding area of the caliper and the brake pads with high temperature brake grease.
10. Apply anti-squeal compound to the back side of both brake pads. Allow the compound to set-up according to the instructions on the product.
11. Insert the new pads into the caliper.
12. If equipped, install the anti-rattle clip or retaining pin spring-clip or cotter pin. On pins with a spring steel collar, you must knock them in until seated against the shoulder in the caliper.

➡**It is a good idea to use some thread-locking compound (removable type) to the threaded fasteners of the caliper.**

13. Connect any electrical brake pad wear sensors.
14. Seat the brake pads, otherwise the vehicle may coast out of the work area and into traffic before the brakes become effective. It will take several pumps of the brake pedal to seat the pads against the rotor.

➡**If a firm pedal is not achieved, it may be necessary to bleed the brakes.**

15. Check the brake fluid level in the reservoir and top off as needed.
16. Install the wheels and tighten the lug nuts.
17. Road test the vehicle.

Brake Calipers

REMOVAL & INSTALLATION

Calipers without Integral Parking Brake Mechanisms

SLIDING CALIPERS

✳✳ CAUTION

Brake dust may contain asbestos! Asbestos is harmful to your health. Never use compressed air to clean any brake component. A filtering mask should be worn during any brake repair.

There are typically three methods of securing a sliding caliper to its mounting bracket: with a retaining pin, with a key and bolt, or with a wedge and pin. On calipers which use the retaining pin method, you will find pins driven into the slot between the caliper and the caliper mount. On calipers which use the bolt and key method, a key (small piece of metal) is used between the caliper and the mounting bracket to allow the caliper to slide. The key is held in position by a lockbolt. On calipers which use the pin and wedge method, a wedge, retained by a pin, is used between the caliper and the mounting bracket in much the same manner as with the key and bolt method.

1. Loosen the lug nuts on the applicable wheels.
2. Raise and safely support the vehicle.
3. Remove the wheels.

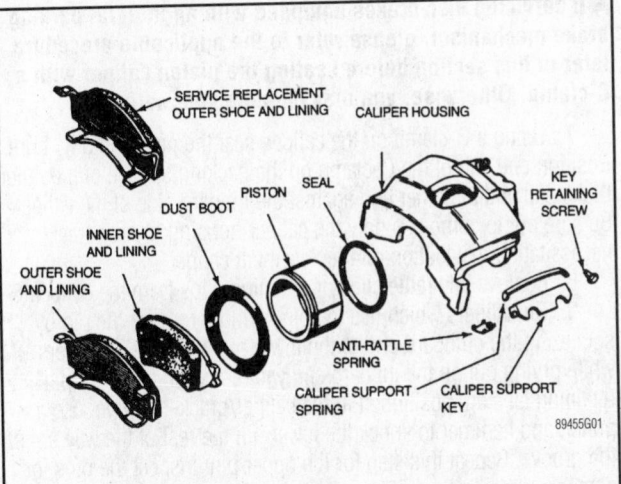

Exploded view of a typical sliding caliper, showing the key and bolt (retaining screw)

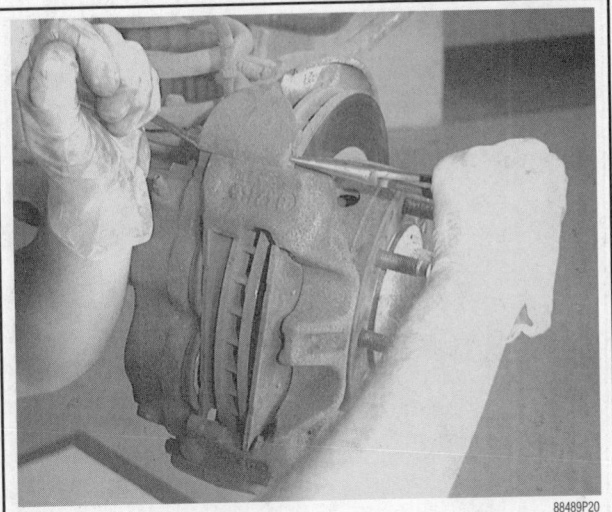

Compress the upper pin tabs and pry on the inner end of the pin . . .

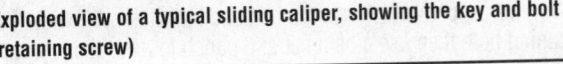

※ CAUTION

Any brake fluid that is removed from the system should be discarded. Also, do not allow any brake fluid to come in contact with a painted surface; it will damage the paint. Also, brake fluid contains polyglycol ethers and polyglycols. Avoid contact with the eyes and wash your hands thoroughly after handling brake fluid. If you do get brake fluid in your eyes, flush your eyes with clean, running water for 15 minutes. If eye irritation persists, or if you have taken brake fluid internally, IMMEDIATELY seek medical assistance.

4. Remove some brake fluid from the brake fluid reservoir. Use a clean suction pump, a turkey baster, or an absorbent pad to do so. Never reuse any brake fluid.

5. Place a drain pan under the work area. Clean the brake pad and rotor area with spray brake cleaner.

6. Disconnect any electrical brake pad wear sensor.

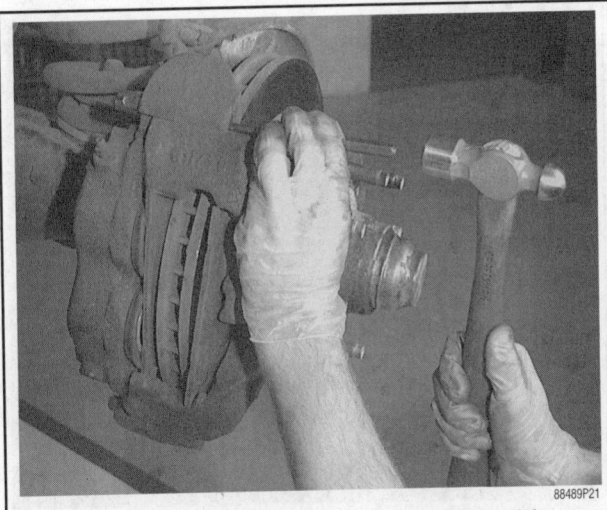

. . ., then use a hammer and punch to drive the pin out of the groove . . .

To remove a typical sliding caliper, remove the anti-rattle clips (if equipped)

. . . until it can be removed by hand

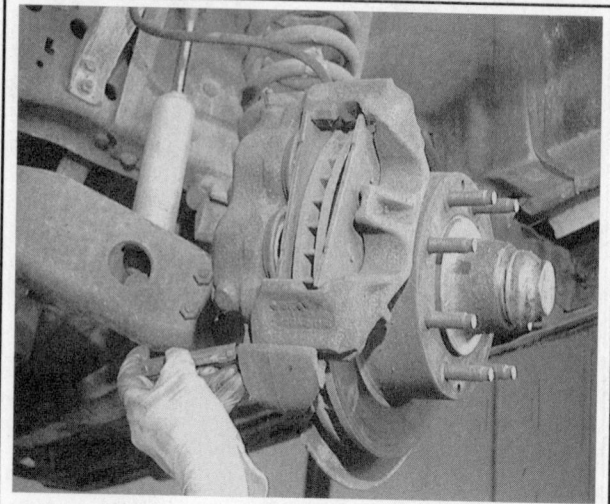

Perform the same for the lower pin as well . . . .

88489P25

→If servicing disc brakes equipped with an integral parking brake mechanism, please refer to the applicable procedure later in this section before seating the piston caliper with a C-clamp. Otherwise, you may damage your caliper.

7. Using a C-clamp on the caliper, seat the piston into its bore. Position one end of the C-clamp on the backing surface of the outer brake pad and the other end against the inboard side of the caliper. Be sure not to compress only the caliper housing; it may crack, necessitating installation of a replacement caliper.

8. Remove any rattle clips or retaining clips from the caliper.

9. On calipers which use the pin method, remove the pin by squeezing the outboard end of the lower pin with a pair of pliers while prying out on the inboard end with a prybar. Once the pin retaining tabs are positioned in the caliper/bracket groove, use a punch and hammer to knock the lower pin the rest of the way out of the groove. Repeat this step for the upper pin. Inspect the pins for damage, wear, and rust. Replace as needed in pairs.

10. On calipers which use the bolt and key method, remove the retaining bolt, then use a hammer and punch to drive the key out.

. . ., then pull the caliper off of the rotor and bracket

88489P26

A vacuum pump setup can be used to draw brake fluid from the reservoir

89709P10

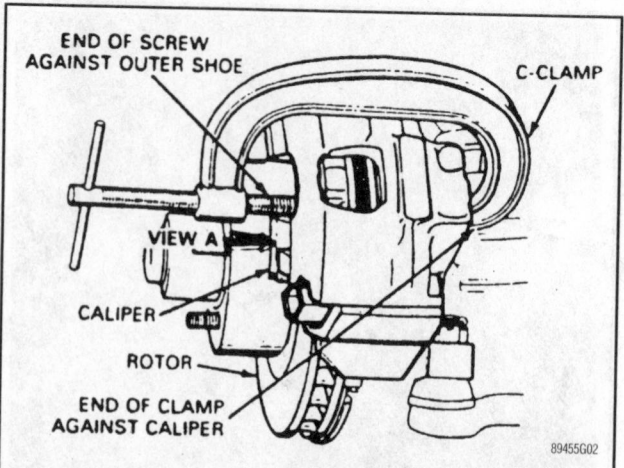

Use a large C-clamp to seat the caliper piston, be sure that one end of the clamp is positioned against the outer shoe—calipers without integral parking brake mechanisms

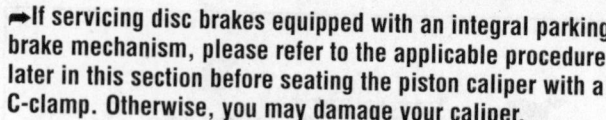

Once the caliper is removed, the brake pads can be removed

88489P27

(Be careful not to lose the caliper support spring, if equipped.) Check parts for wear and replace as necessary.

11. On calipers which use the wedge and pin method, remove the retaining pin from the guide plate, then use a punch and hammer to tap out the guide plate. Inspect parts for wear and replace as necessary.

12. If the caliper is going to be removed for overhaul or replacement, loosen the brake hose, lift off the caliper and remove the brake hose completely. Immediately plug the open end of the rubber brake hose to prevent contamination of the brake fluid. If the brake hose was attached to the caliper with a banjo connection, be sure to remove and discard the two copper washers.

13. If the caliper does not require overhaul or replacement, prepare a length of wire (a coat hanger works well), cord, or a length of strong string to support the caliper. DO NOT let the caliper hang from the brake hose; it may be damaged.

14. Remove the caliper and suspend it from the wire.

15. If the brake pads came off the rotor with the caliper, remove them by prying the pads out of the caliper piston.

16. Inspect the caliper for fluid leakage, torn dust boots, or missing parts. Rebuild or replace the caliper if a problem is found.

17. Inspect the rubber brake hose for cracks or signs of rubbing against the body or steering components. Also, it is a good idea to replace them if they are over 10 years old to maintain proper brake operation.

18. Inspect metal lines for corrosion and kinks from road debris kicked up under the vehicle. If a problem is found, replace the line.

19. Inspect the rotor for non-machine grooves, heat stress cracks, glazing, minimum wear thickness, and disk run-out. Replace the rotor or have it machined to repair the damage.

20. Inspect the brake pads for minimum thickness, loose rivets, or glazing. Install new brake pads if any such problems exist.

To install:

21. Clean the sliding surfaces of the caliper and mounting bracket with spray brake cleaner and a small wire brush, then lubricate them with high temperature brake grease.

22. If necessary, place the pad(s) back onto the caliper or mounting bracket.

23. If the brake hose was removed, reattach it to the caliper. If so equipped, use two new copper washers for the banjo fitting.

24. Install the caliper onto its mounting bracket.

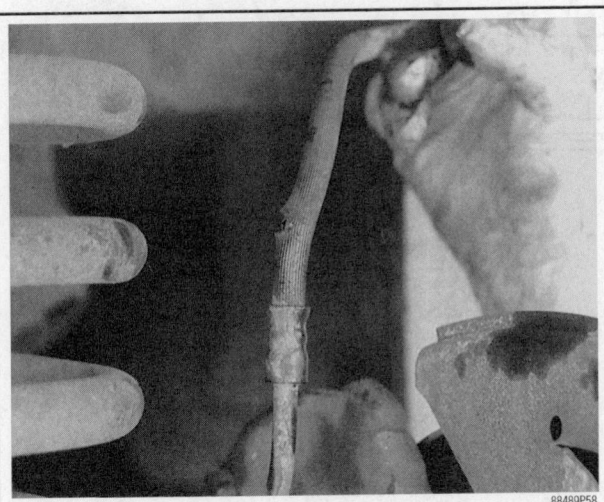

When inspecting the flexible brake hoses, check for rips (as shown), tears and cracks

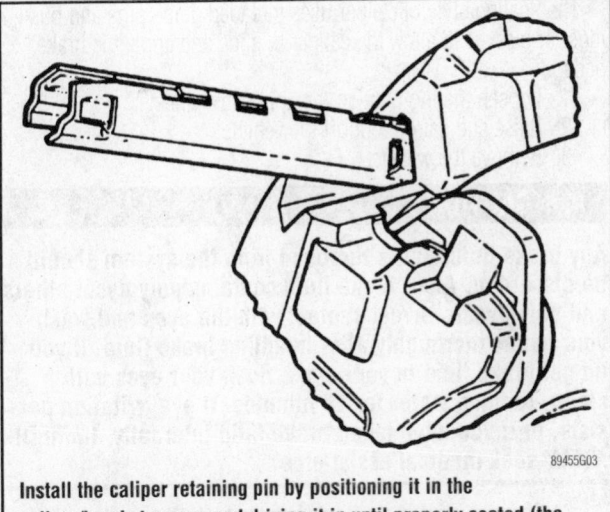

Install the caliper retaining pin by positioning it in the caliper/bracket groove and driving it in until properly seated (the retaining tabs on both ends should protrude from the groove)

25. For calipers which use the pin retaining method, use a hammer to tap the pins back into position, then install any anti-rattle clips.

26. For calipers which use the bolt and key method, use a prybar to lift the caliper up to create a gap into which the key and spring can slide. Tap the key and spring into position, then install the locking bolt and any anti-rattle clips. Tighten the locking bolt securely.

27. For calipers which use the wedge and pin method, slide the guide plates (wedge) between the gaps of the caliper and mounting bracket, then install the retaining pin. Tighten the retaining pin securely.

28. Reattach any electrical brake pad sensors.

❊❊ WARNING

Clean, high quality brake fluid is essential to the safe and proper operation of the brake system. You should always buy the highest quality brake fluid that is available. If the brake fluid becomes contaminated, drain and flush the system, then refill the master cylinder with new fluid. Never reuse any brake fluid. Any brake fluid that is removed from the system should be discarded. Also, do not allow any brake fluid to come in contact with a painted surface; it will damage the paint.

29. Bleed the brakes if a brake line was replaced, or the caliper was detached from a brake line.

30. Seat the brake pads, otherwise the vehicle may coast out of the work area and into traffic before the brakes become effective. It will take several pumps of the brake pedal to seat the pads against the rotor.

31. Check the brake fluid level in the reservoir and top off as needed.

32. Install the wheels and tighten the lug nuts.

33. Road test the vehicle.

FLOATING CALIPERS

❊❊ CAUTION

Brake dust may contain asbestos! Asbestos is harmful to your health. Never use compressed air to clean any brake component. A filtering mask should be worn during any brake repair.

The floating style of caliper uses threaded guide pins and bushings, or sleeves to allow the caliper to slide and apply the brake pads.

1. Loosen the lug nuts on the applicable wheels.
2. Raise and safely support the vehicle.
3. Remove the wheels.

✳✳ CAUTION

Any brake fluid that is removed from the system should be discarded. Also, brake fluid contains polyglycol ethers and polyglycols. Avoid contact with the eyes and wash your hands thoroughly after handling brake fluid. If you do get brake fluid in your eyes, flush your eyes with clean, running water for 15 minutes. If eye irritation persists, or if you have taken brake fluid internally, IMMEDIATELY seek medical assistance.

4. Remove some brake fluid from the brake fluid reservoir. Use a clean suction pump, a turkey baster (not to be returned to the kitchen), or an absorbent pad to do so. Never reuse any brake fluid.
5. Place a drain pan under the work area. Clean the brake pad and rotor area with spray brake cleaner.
6. Disconnect any electrical brake pad wear sensor.
7. If an anti-rattle spring is used and is not part of the brake pad, it can usually be pried off or pulled out.

➡**If servicing disc brakes equipped with an integral parking brake mechanism, please refer to the applicable procedure later in this section before seating the piston caliper with a C-clamp. Otherwise, you may damage the caliper.**

8. Using a C-clamp on the caliper, seat the piston into its bore. Position one end of the C-clamp on the backing surface of the outer brake pad and the other end against the inboard side of the caliper. Be sure not to compress only the caliper; it may crack, necessitating installation of a replacement caliper.
9. Loosen and remove the guide pins from the caliper.

10. If the caliper is going to be removed for overhaul or replacement, loosen the brake hose, lift off the caliper and remove the brake hose completely. Immediately plug the open end of the rubber brake hose to prevent contamination of the brake fluid. If the brake hose was attached to the caliper via a banjo connection, be sure to remove and discard the two copper washers.
11. If the caliper does not require overhaul or replacement, prepare a length of wire (a coat hanger works well), cord, or a length of strong string to support the caliper. DO NOT let the caliper hang from the brake hose; it may be damaged.
12. Remove the caliper from the rotor, if equipped, the mounting bracket.

➡**The pads may or may not come off with the caliper; this is normal.**

13. If the brake pads stay on the caliper, they can usually be tapped off with a hammer, or pried out by hand or with prytool.
14. If the brake pads remain on the bracket, when applicable, they can be removed from the bracket by hand.
15. Inspect the caliper for fluid leakage, torn dust boot, or missing parts. Rebuild or replace if a problem is found.
16. Inspect the rubber brake hose for cracks or signs of rubbing against the body or steering components. Install a new rubber hose if any such conditions exist. Also, it is a good idea to replace them if over 10 years old to maintain proper brake operation.
17. Inspect the metal lines for corrosion and kinks from road debris kicked up under the vehicle. If a problem is found, replace the line.
18. Inspect the rotor for non-machine grooves, heat stress cracks, glazing, minimum wear thickness, and disk run-out. Replace the rotor or have it machined to repair the damage.
19. Inspect the brake pads for minimum thickness, loose rivets, or glazing. If such a problem is found, new pads must be installed.
To install:
20. If equipped with a mounting bracket, clean the sliding surfaces of the caliper and mounting bracket with spray brake cleaner

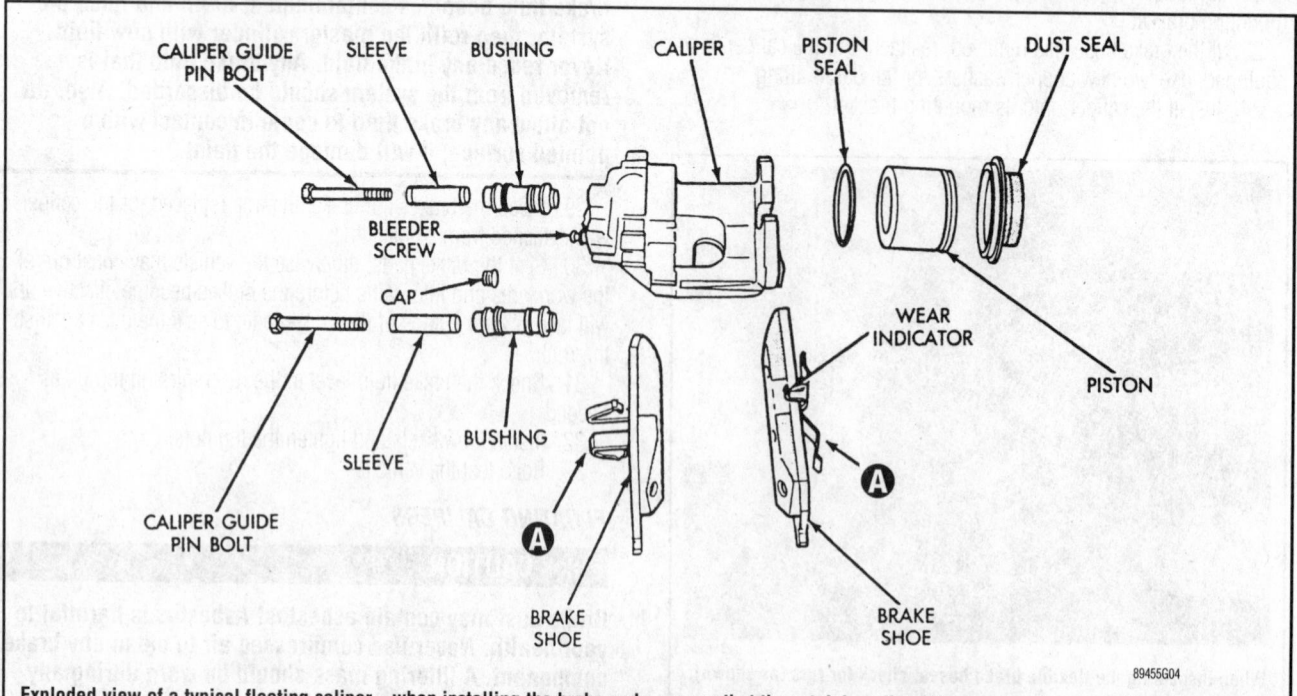

Exploded view of a typical floating caliper—when installing the brake pads, ensure that the retaining clips (A) are properly engaged in the caliper

89455G04

and a small wire brush, then lubricate them with high temperature brake grease.

21. If the brake hose was removed, reattach it to the caliper. If so equipped, use two new copper washers for the banjo fitting.

22. Transfer old pad hardware to the new pads, or install new hardware.

23. Clean and inspect the caliper guide pins, if they are okay, then lubricate them with high temperature brake grease.

24. On caliper mounted pads, position the pads on the caliper, then install the caliper on the rotor.

25. On bracket mounted pads, install the pads on the mounting bracket. Install the caliper on the rotor.

26. Tighten the caliper guide pins securely, and replace any anti rattle clips.

27. Connect any electrical brake pad sensors.

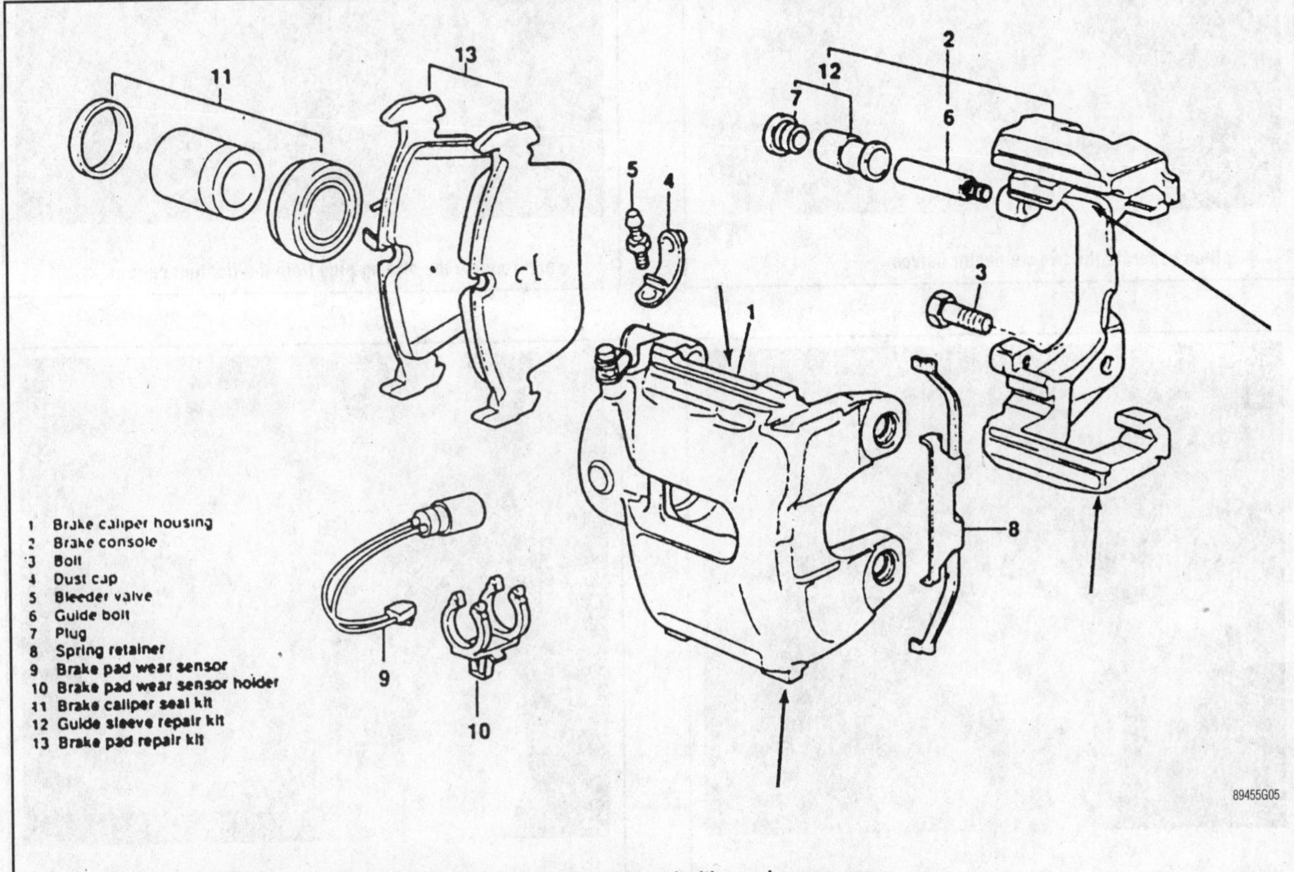

1. Brake caliper housing
2. Brake console
3. Bolt
4. Dust cap
5. Bleeder valve
6. Guide bolt
7. Plug
8. Spring retainer
9. Brake pad wear sensor
10. Brake pad wear sensor holder
11. Brake caliper seal kit
12. Guide sleeve repair kit
13. Brake pad repair kit

89455G05

Exploded view of another floating caliper—note that this vehicle is equipped with a pad wear sensor

89455P04

Some vehicles are equipped with brake pad wear sensors, indicated by the wire leading to the pad

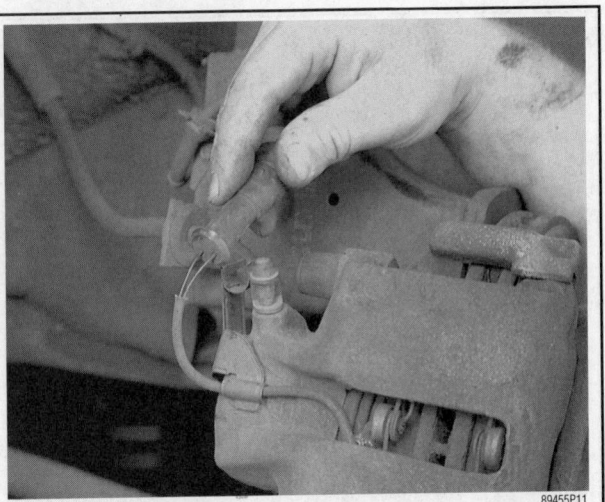

89455P11

To remove a typical floating caliper, disengage the sensor wire connector from its mounting clip (if equipped) . . .

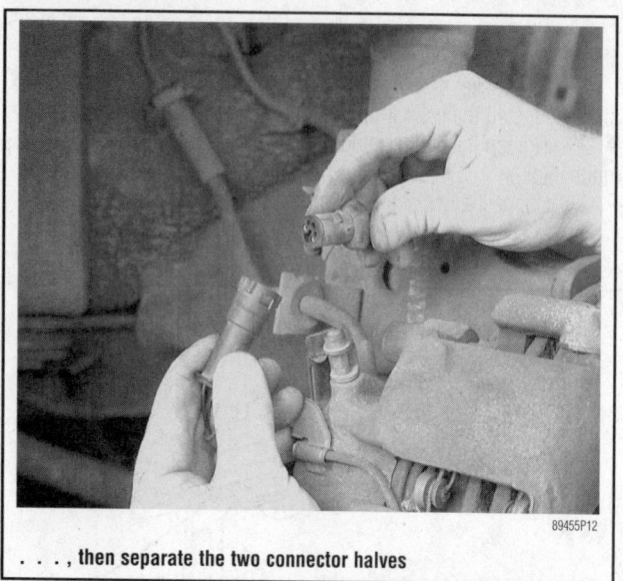

. . ., then separate the two connector halves

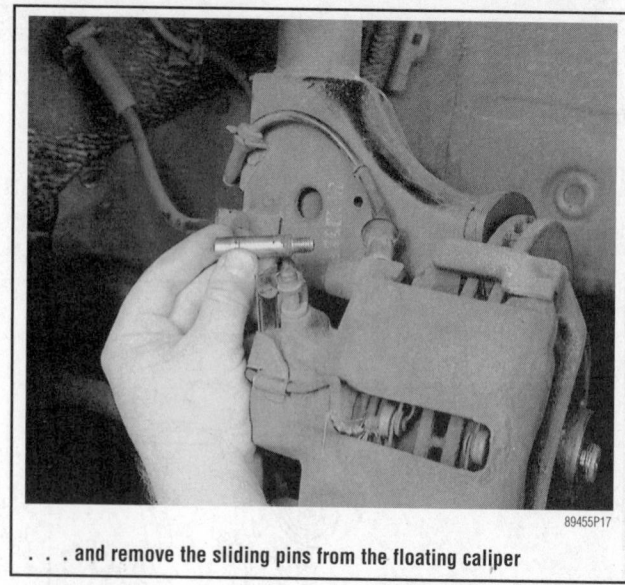

. . . and remove the sliding pins from the floating caliper

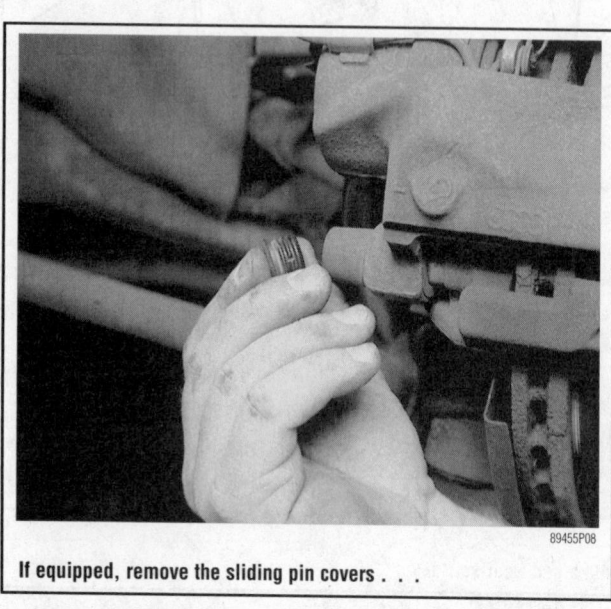

If equipped, remove the sliding pin covers . . .

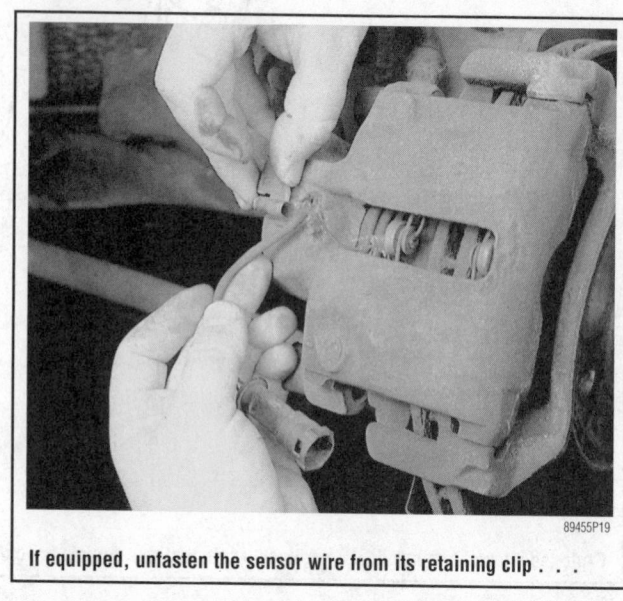

If equipped, unfasten the sensor wire from its retaining clip . . .

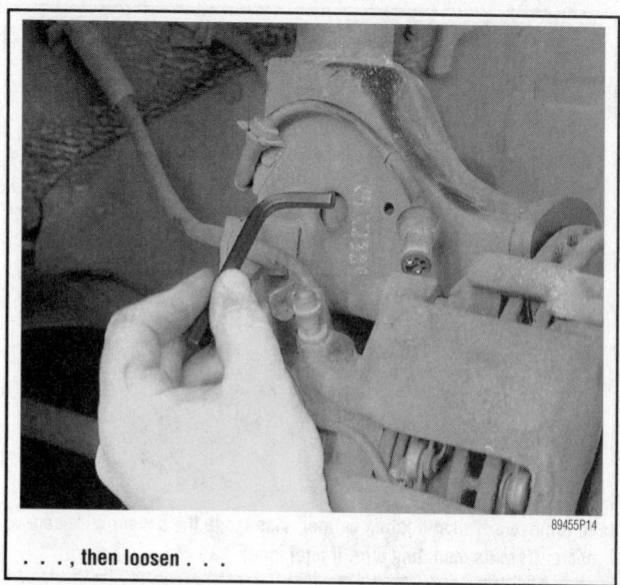

. . ., then loosen . . .

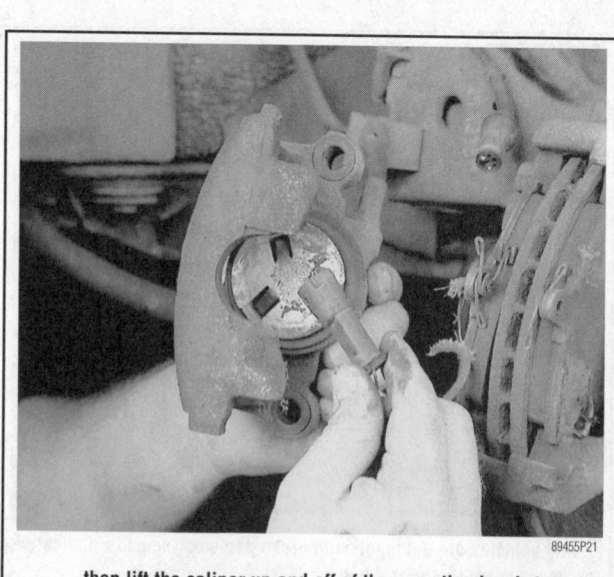

. . ., then lift the caliper up and off of the mounting bracket

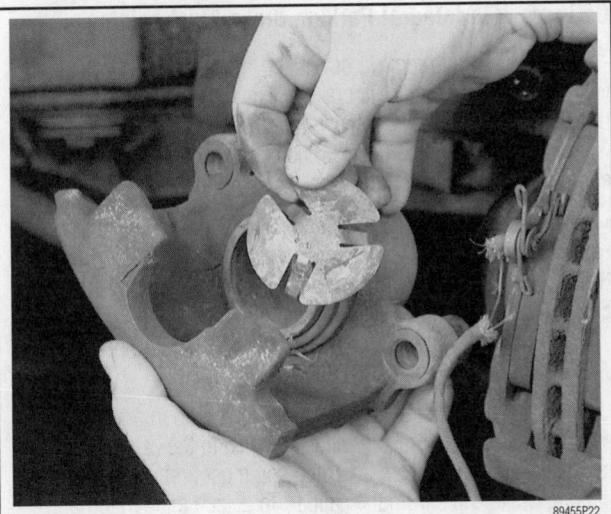

Be sure to note the positions of any clips or springs on the caliper

✳✳ WARNING

Clean, high quality brake fluid is essential to the safe and proper operation of the brake system. You should always buy the highest quality brake fluid that is available. If the brake fluid becomes contaminated, drain and flush the system, then refill the master cylinder with new fluid. Never reuse any brake fluid. Any brake fluid that is removed from the system should be discarded. Also, do not allow any brake fluid to come in contact with a painted surface; it will damage the paint.

28. Bleed the brakes, if a brake line was replaced or the caliper was detached from a brake line.

29. Seat the brake pads, otherwise the vehicle may coast out of the work area and into traffic before the brakes become effective. It will take several pumps of the brake pedal to seat the pads against the rotor.

30. Check the brake fluid level in the reservoir and top off as needed.

31. Install the wheels and tighten the lug nuts.

32. Road test the vehicle.

FIXED CALIPERS

✳✳ CAUTION

Brake dust may contain asbestos! Asbestos is harmful to your health. Never use compressed air to clean any brake component. A filtering mask should be worn during any brake repair.

The fixed type caliper is bolted to the steering knuckle. The brake pads on this style of caliper are typically held in place by one or two retaining pins. Some other pads use hold down clips. It may not be necessary to remove the brake pads in order to remove the caliper.

1. Loosen the lug nuts on the applicable wheels.

2. Raise and safely support the vehicle.

✳✳ WARNING

Any brake fluid that is removed from the system should be discarded. Also, do not allow any brake fluid to come in contact with a painted surface; it will damage the paint.

3. Remove the wheels.

✳✳ CAUTION

Brake fluid contains polyglycol ethers and polyglycols. Avoid contact with the eyes and wash your hands thoroughly after handling brake fluid. If you do get brake fluid in your eyes, flush your eyes with clean, running water for 15 minutes. If eye irritation persists, or if you have taken brake fluid internally, IMMEDIATELY seek medical assistance.

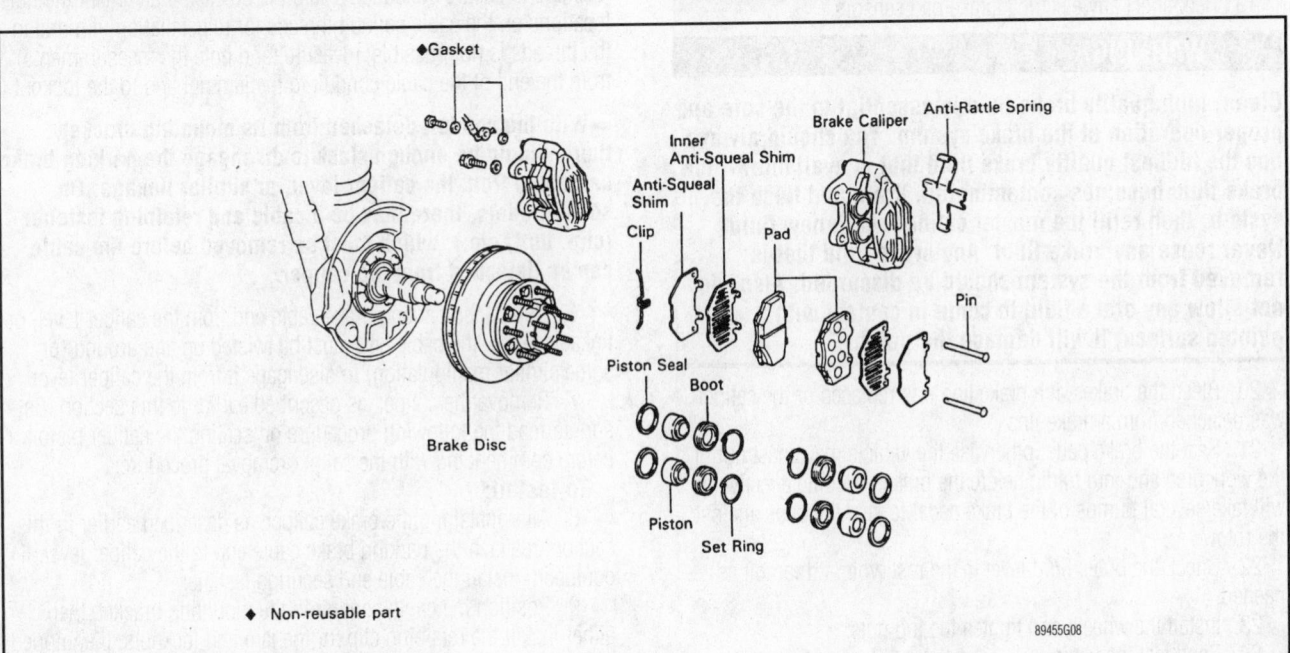

◆ Non-reusable part

Exploded view of a common four piston fixed caliper

4. Remove some brake fluid from the brake fluid reservoir. Use a clean suction pump, a turkey baster, or an absorbent pad to do so. Never reuse any brake fluid.

5. Place a drain pan under the work area. Clean the brake pad and rotor area with spray brake cleaner.

6. If equipped, disconnect any electrical brake pad wear sensor.

7. Although not necessary for caliper removal, the brake pads can now be removed from the caliper.

8. Loosen the caliper mounting bolts.

9. If the caliper is going to be removed for overhaul or replacement purposes, loosen the brake hose, remove the caliper bolts, and disconnect the brake line.

10. If the caliper does not require overhaul or replacement (in other words, you only need to remove it for access to some other component), prepare a length of wire (coat hanger), cord, or a length of strong string from which the caliper can be hung. DO NOT let the caliper hang from the brake hose; it may be damaged and need to be replaced. Remove the caliper and hang it from the wire.

11. Inspect the caliper for fluid leakage, torn dust boot, or missing parts. Rebuild or replace if a problem is found.

12. Inspect the rubber brake hose for cracks or signs of rubbing against the body or steering components. Install a new brake hose if any such damage is evident. Also, it is a good idea to replace them if over 10 years old to maintain proper brake operation.

13. Inspect the metal brake lines for corrosion and kinks from road debris kicked up under the vehicle. If a problem is found replace the line.

14. Inspect the rotor for non-machine grooves, heat stress cracks, glazing, minimum wear thickness, and disk run-out. Replace the rotor or have it machined to repair the damage.

15. Inspect the brake pads for minimum thickness, loose rivets, or glazing. If any such problem is found, new pads must be installed.

To install:

16. Install the caliper and tighten the mounting bolts securely.

17. If the brake hose was removed, reattach it to the caliper. If so equipped, use two new copper washers for the banjo fitting.

18. If removed, install the brake pads.

19. Reconnect any electrical brake pad sensors.

✳✳ WARNING

Clean, high quality brake fluid is essential to the safe and proper operation of the brake system. You should always buy the highest quality brake fluid that is available. If the brake fluid becomes contaminated, drain and flush the system, then refill the master cylinder with new fluid. Never reuse any brake fluid. Any brake fluid that is removed from the system should be discarded. Also, do not allow any brake fluid to come in contact with a painted surface; it will damage the paint.

20. Bleed the brakes, if a brake line was replaced or the caliper was detached from a brake line.

21. Seat the brake pads, otherwise the vehicle may coast out of the work area and into traffic before the brakes become effective. It will take several pumps of the brake pedal to seat the pads against the rotor.

22. Check the brake fluid level in the reservoir and top off as needed.

23. Install the wheels and tighten the lug nuts.

24. Road test the vehicle.

Calipers with Integral Parking Brake Mechanisms

The procedure to remove or replace the caliper and/or pads on vehicles equipped with disc brakes designed with integral parking brake mechanisms is essentially the same as disc brake calipers without integral parking brakes. There are usually two major differences between these two disc brake caliper designs.

➡ **For the actual caliper removal and installation process, refer to the applicable procedure earlier in this section. Read the following two procedures, and perform them in conjunction with the caliper procedures.**

REMOVING THE PARKING BRAKE CABLE

The first, and most obvious, difference is that, in one fashion or another, the parking brake cable is attached to the caliper. Before removing the caliper from the rotor, you must first disengage the parking brake cable from the caliper. To detach the parking brake cable from the caliper, perform the following:

➡ **This is a general procedure and may need slight alteration to apply fully to your specific vehicle. The most important thing to remember is to carefully inspect your caliper to identify the applicable parking brake cable components before disconnecting anything.**

1. Loosen the lug nuts on the applicable wheels.
2. Raise and safely support the vehicle.

➡ **Some vehicles, in fact, may be designed with front parking brake assemblies.**

3. Remove the wheels for easier access to the brake assembly.
4. Relieve the parking brake cable tension.
5. Carefully inspect the parking brake cable mounting and attaching (to the caliper) points. Most parking brake cable conduits are retained to a mounting bracket either by a jam nut and locknut setup, or by a retaining clip. Either remove the jam and locknuts, or pull the retaining clip off of the bracket, then disengage the cable conduit from the mounting bracket. If your vehicle utilizes jam and locknuts to secure the conduit onto the bracket, matchmark the nuts' locations on the cable conduit threads for reinstallation; if marking the threads is not possible, measure (and note the measurements) from the end of the cable conduit to the jam nut and to the locknut.

➡ **With the conduit detached from its mounting bracket, there should be enough slack to disengage the parking brake cable end from the caliper lever, or similar linkage. On some models, there may be a cable end retaining fastener (clip, bolt, etc.), which must be removed before the cable can be detached from the caliper.**

6. Detach the parking brake cable end from the caliper lever, or linkage. Often, the cable end must be twisted up and around (or some similar manipulation) to disengage it from the caliper lever.

7. Remove the caliper, as described earlier in this section. Be sure to read the following procedure on seating the caliper piston before commencing with the caliper removal procedure.

To install:

8. After installing the brake caliper, as described earlier in this section, reattach the parking brake cable end to the caliper lever. If equipped, install the cable end securing fastener.

9. Position the cable conduit in the mounting bracket, then either install the retaining clip, or the jam and locknuts. If equipped with jam and locknuts, position the nuts on the cable conduit so

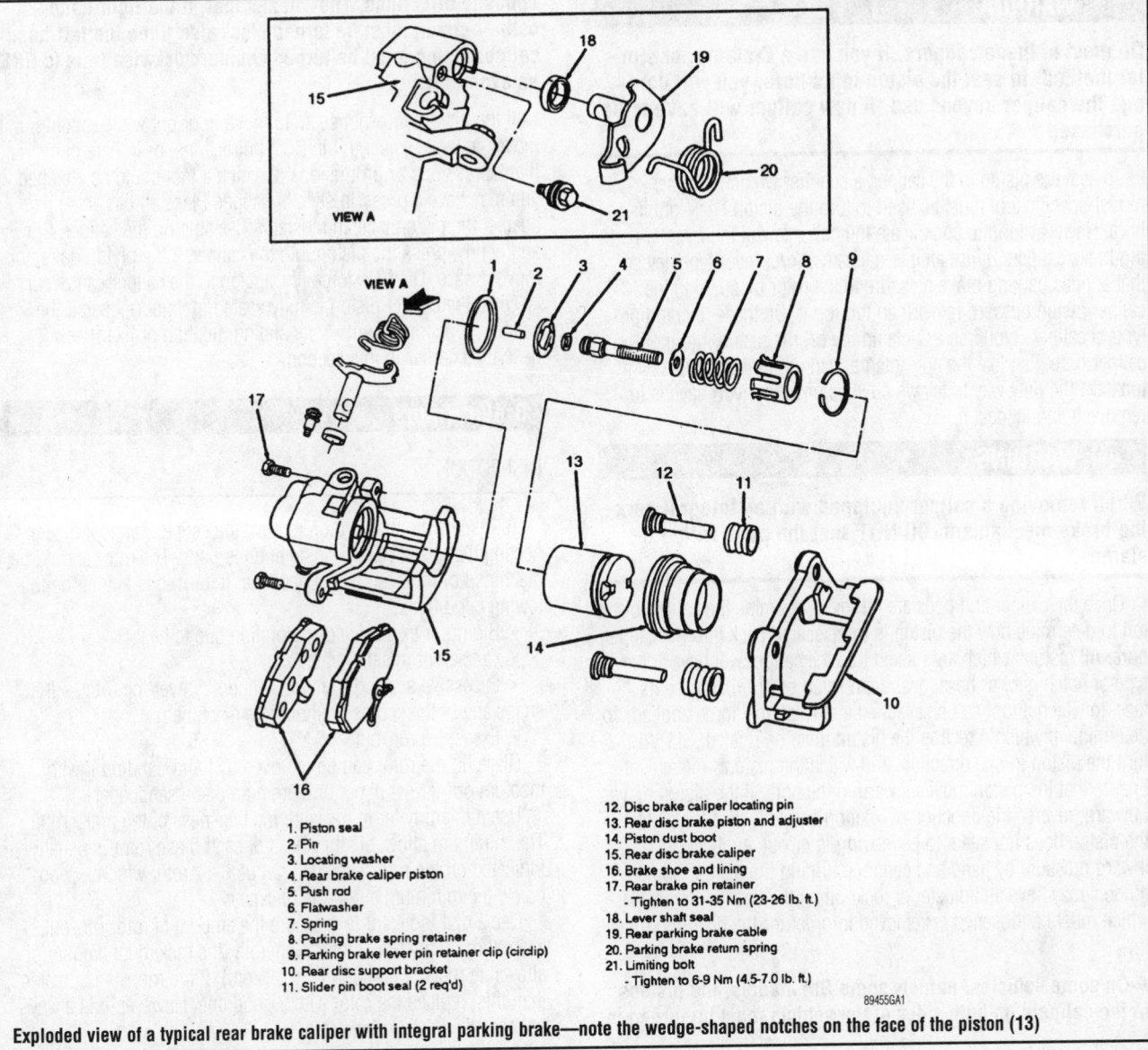

1. Piston seal
2. Pin
3. Locating washer
4. Rear brake caliper piston
5. Push rod
6. Flatwasher
7. Spring
8. Parking brake spring retainer
9. Parking brake lever pin retainer clip (circlip)
10. Rear disc support bracket
11. Slider pin boot seal (2 req'd)

12. Disc brake caliper locating pin
13. Rear disc brake piston and adjuster
14. Piston dust boot
15. Rear disc brake caliper
16. Brake shoe and lining
17. Rear brake pin retainer
    - Tighten to 31-35 Nm (23-26 lb. ft.)
18. Lever shaft seal
19. Rear parking brake cable
20. Parking brake return spring
21. Limiting bolt
    - Tighten to 6-9 Nm (4.5-7.0 lb. ft.)

89455GA1

Exploded view of a typical rear brake caliper with integral parking brake—note the wedge-shaped notches on the face of the piston (13)

that the nuts are positioned as before (using the marks on the threads or a ruler).

10. Adjust the parking brake cable tension, as described in Section 3.

11. Install the wheels and snug the lug nuts.

12. Lower the vehicle.

13. Tighten the lug nuts fully.

14. Depress the brake pedal a few times to ensure that the brake pads are fully seated.

✳✳ CAUTION

If you do not seat the pads before driving the vehicle, the first few times you apply the brake pedal the vehicle may not stop as anticipated; this could lead to an accident with a telephone pole or one of your neighbors' cars.

SEATING THE CALIPER PISTON

➡Be sure to read this entirely before commencing with caliper service.

The second difference between brake calipers with and without integral parking brake mechanisms is in how the caliper pistons should be seated into their bores.

Whereas most pistons on calipers which are not equipped with integral parking brake mechanisms can be seated by using a large C-clamp, this is USUALLY not the case with calipers designed with integral parking brake mechanisms. Most integral parking brake calipers apply parking brake pressure to the rotor as follows: when the parking brake is applied, the cable pulls on the caliper lever. The lever, in turn, applies a rotational (spinning) movement to the caliper piston. The piston is designed much like an ordinary screw, so that when a rotational movement is applied to the piston, it slowly presses in against the rotor. To prevent having to constantly adjust the parking brake cable tension as the brake pads slowly wear down, the internal parking brake mechanism is designed with a ratcheting apparatus, which automatically readjusts the parking brake tension.

Since the caliper is designed to protrude from its bore when turned, usually, it cannot be seated in its bore in a conventional manner (with a large C-clamp).

⁂ WARNING

On most of these calipers, if you use a C-clamp, or similar method, to seat the piston in its bore, you will damage the caliper beyond use. A new caliper will have to be purchased.

To seat the piston in the caliper, a spanner wrench or other model-specific tool must be used to turn the piston back into its bore. However (and to complicate things), a few of the integral parking brake calipers utilize an internal cam and/or lever type device that applies parking brake pressure to the rotor by pushing the caliper piston outward rather than turning it. On these uncommon type of calipers, you use a C-clamp to seat the piston into the caliper bore, just like the non-integral parking brake calipers. Unfortunately, the only way to tell which style of caliper you have is to remove it and inspect it.

⁂ WARNING

When removing a caliper equipped with an integral parking brake mechanism, DO NOT seat the pads with a C-clamp.

Once the caliper and pads are removed, examine the caliper piston to determine how the piston is to be seated back into the caliper bore. All pistons which are rotated into the caliper will have some type of notch, slot or hexagonal depression or protrusion on its face, to which a tool can be attached and rotational force applied. To determine in which direction the piston must be rotated, SLOWLY turn the piston in one direction, and watch the piston's movement. Ensure that the piston moves inward in the bore. If the piston moves outward, reverse the direction of rotation and fully seat the piston. If the piston does not seem to be moving in or out, apply slight inward pressure by hand and continue turning the piston. Some models may have an adjuster or lockbolt on the back of the caliper which must be loosened or removed in order for the piston to rotate in.

➡On some vehicles, namely some GM models, the pistons in the calipers on both sides of the vehicle must be turned in opposite directions. That means that, if the right-hand caliper piston must be turned clockwise, then the left-hand caliper piston must be turned counterclockwise (this is ONLY an example).

If the piston does not seem to move in or out while rotating, if it moves in while rotating it in BOTH directions, or if there is no visible depressions or protrusions to which a tool could be attached, you may have a press-in style of caliper. Place an old brake pad against the piston face and install a C-clamp on the caliper. Slowly, and gently, press the piston into the caliper. If the piston does not move inward, DO NOT force it! Damage to the caliper can occur.

Once the caliper piston is fully seated in its bore, install the caliper (depending on its type: sliding, floating or fixed) as described earlier in this section.

Brake Rotors

INSPECTION

To inspect the brake rotor, remove the caliper (without disconnecting the flexible brake hose) and the pads. The rotor should be machined or replaced with a new one, if it exhibits any of the following conditions:
- Bluing or excessive discoloration due to heat
- Cracks, or missing chunks
- Excessive scoring (run your fingernail over the rotor—if it snags any of the scores, it should be machined)
- Excessive run-out

Glaze on the rotor can be removed by hand-sanding it with medium grit garnet paper or aluminum oxide sandpaper.

Use a micrometer to measure the thickness of the brake rotor. The minimum allowable thickness of each brake rotor is usually indicated on the rotor itself. Do not utilize a rotor which is worn below the minimum allowable thickness

Use a dial indicator to measure the amount of rotor run-out, while turning the brake rotor. Generally, the maximum amount of allowable run-out is 0.006 in. (0.15mm); if the run-out is greater than this, replace the rotor with a good one. However, it is always better to have less rotor run-out.

PARKING BRAKE LEVER
PARKING BRAKE CABLE
ADJUSTING NUT
BRACKET

89455GA3

If it is necessary to remove the parking brake cable, carefully inspect it to determine how it is attached and adjusted

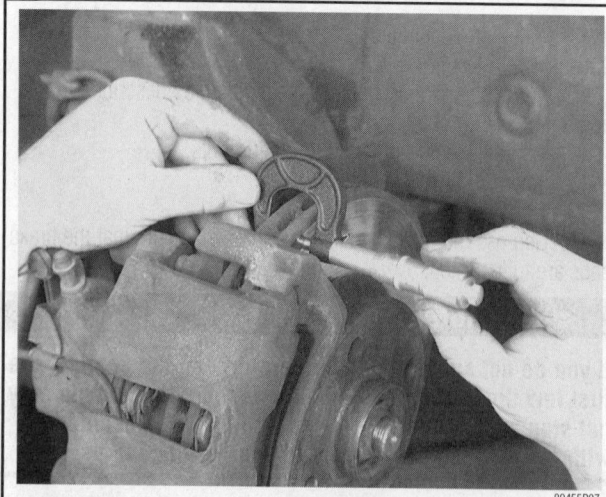

89455P07

Use a micrometer to measure the rotor thickness, and replace it if it is below specifications

REMOVAL & INSTALLATION

Rotors mount in one of 2 ways: either directly on the hub (held in place by the wheels or small fasteners), which are referred to as non-integral (they are not one piece with the hub), or are integral with the hub.

➡**On some vehicles, the manufacturer installs retaining clips over one or two of the wheel lugs to hold the rotor in place during assembly. Although it is generally thought that these retainers are not necessary and may be discarded, it is a good idea to reinstall them anyway (better safe than sorry). Other manufacturers use one or two small machine screws to hold the rotor in place on the hub; these screws MUST be reinstalled.**

Non-Integral Rotors

1. Loosen the lug nuts on the applicable wheels.
2. Raise and safely support the vehicle.
3. Remove the wheels.
4. Clean the brake assembly thoroughly with spray brake cleaner.

If the rotor is equipped with holes and is difficult to remove, it can be loosened using two small bolts . . .

To remove the rotor, remove the disc brake caliper . . .

. . . , then removed by hand—non-integral rotors

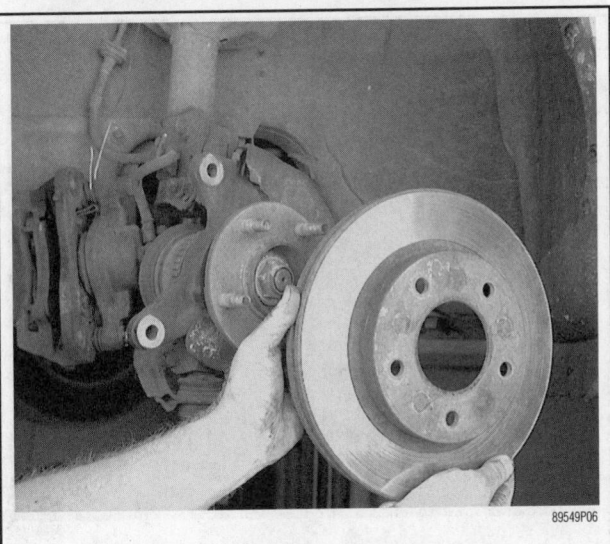

. . . , then pull the rotor off by hand—non-integral rotors

5. Remove the caliper.
6. If any rotor retainers are present, remove them. The push-nut type of retainer is usually damaged during removal; discard the old ones and purchase new ones.
7. Remove the rotor. On some vehicles, the rotor simply slides off the wheel studs. However, some rotors are pressed into place and must be removed by screwing bolts in the threaded holes provided, thereby forcing the rotor off the hub. Other rotors, not equipped with the threaded holes for press-off bolts, may require the use of a puller to dislodge them from the hub.

➡**The rotor may be rusted in place. Spray the area liberally with WD-40®, Liquid Wrench® or equivalent and tap the rotor loose.**

To install:

➡**New rotors come with an oily, rust-preventive coating on the braking surface. This coating can be removed with brake parts cleaner or most cleaners which are good for oil removal. Be sure that all traces of the coating are removed. Allow the rotor to dry before installation.**

8. Position the rotor on the hub and install any retainers.
9. Install the caliper.
10. Install the wheels.
11. Lower the vehicle.
12. Seat the brake pads, otherwise the vehicle may coast out of the work area and into traffic before the brakes become effective. It will take several pumps of the brake pedal to seat the pads against the rotor.
13. Check the brake system for proper operation.

Integral Rotor/Hub Assemblies

NON-SEALED HUB/BEARING ASSEMBLIES

1. Loosen the lug nuts on the applicable wheels.
2. Raise and safely support the vehicle.
3. Remove the wheels.
4. Clean the brake assembly thoroughly with spray brake cleaner.
5. Remove the caliper and suspend it out of the way with wire.
6. Remove the hub grease cap.

Remove the cotter pin . . .

Pry loose the grease cap . . .

. . . , then pull the locking cap off of the end of the axle shaft

. . . , then remove it from the rotor hub

Remove the adjusting nut . . .

. . . , then pull the outer bearing out of the hub

Remove the hub/rotor assembly—integral non-sealed hub/bearing

7. Remove the cotter pin and wheel bearing nut locking cap. Discard the cotter pin.

8. Remove the wheel bearing nut.

➡**On some vehicles, a left-hand threaded nut is used on the right wheel spindle. Turn this locknut clockwise to loosen.**

9. Remove the brake rotor/hub, washer and bearings as an assembly. Be careful not to let the outer wheel bearing fall out of the hub during removal.

10. If the brake rotor is to be machined or replaced, remove the wheel bearings and grease seal.

To install:

➡**New rotors come with an oily, rust-preventive coating on the braking surface. This coating can be removed with brake parts cleaner or most cleaners which are good for oil removal. Be sure that all traces of the coating are removed. Allow the rotor to dry before installation.**

11. If removed, install the inner wheel bearing and a new grease seal.

12. Be sure the bearings and hub contain an adequate amount of clean wheel bearing grease.

13. Position the rotor/hub assembly on the spindle. Keep the hub centered on the spindle to prevent damage to the grease seal and spindle threads.

14. Install the outer wheel bearing, washer and wheel bearing nut.

15. Properly adjust the wheel bearing. On most vehicles (those with tapered roller bearing) this is done by tightening the adjusting nut until drag is felt on the bearing while rotating the rotor;, then, back off the nut about ¼ turn (90°). The rotor/hub should turn freely with no end-play. If you are in any doubt about the proper adjustment procedure, refer to the appropriate model-specific section in this manual.

➡**On some vehicles (those with ball bearings), the nut is not so much an adjuster as a locknut. Tighten this nut to the manufacturer's specifications.**

16. Install the wheel bearing nut cover and a new cotter pin.

17. Install the caliper.

18. Install the wheels and snug the lug nuts.

19. Lower the vehicle.

20. Tighten the lug nuts fully.

21. Seat the brake pads, otherwise the vehicle may coast out of the work area and into traffic before the brakes become effective. It will take several pumps of the brake pedal to seat the pads against the rotor.

22. Check the brake system for proper operation.

SEALED HUB/BEARING ASSEMBLIES

These are unitized hubs that contain the bearing assembly. The hub/bearing unit is replaced as an assembly.

1. Loosen the lug nuts on the applicable wheels.

2. Raise and safely support the vehicle.

3. Remove the wheels.

4. Clean the brake assembly thoroughly with spray brake cleaner.

5. Remove the caliper and suspend it out of the way with wire.

6. On models so equipped, disconnect the ABS sensor wire.

7. Working through the hole provided in the rotor, or working from behind the rotor, remove the hub retaining bolts or nuts.

8. Remove the hub assembly.

To install:

➡**New rotors come with an oily, rust-preventive coating on the braking surface. This coating can be removed with brake parts cleaner or most cleaners which are good for oil removal. Be sure that all traces of the coating are removed. Allow the rotor to dry before installation.**

9. Clean the mounting surfaces of the hub and spindle.

10. Install the hub assembly and tighten the bolts/nuts securely.

11. Connect the ABS wire on models so equipped.

12. Install the caliper.

13. Install the wheels and snug the lug nuts.

14. Lower the vehicle.

15. Tighten the lug nuts fully.

16. Seat the brake pads, otherwise the vehicle may coast out of the work area and into traffic before the brakes become effective. It will take several pumps of the brake pedal to seat the pads against the rotor.

17. Check the brake system for proper operation.

DRUM BRAKES

Brake Drums

→ Most vehicles have rubber plugs in the backing plates that are removed to access the brake adjusters. However, some vehicles are built with what are called knock-out plugs. These are areas in the backing plate that are made to be knocked out with a hammer and punch. Once the drum is off, the knock-out plug is removed and a rubber plug used in its place.

INSPECTION

→ While the brake drum is removed from the vehicle, inspect the wheel cylinder for damage and leakage.

1. Remove the brake drum from the vehicle.

❊❊ CAUTION

Older brake pads or shoes may contain asbestos, which has been determined to be a cancer causing agent. Never clean the brake surfaces with compressed air! Avoid inhaling any dust from any brake surface! When cleaning brake surfaces, use a commercially available brake cleaning fluid.

2. Thoroughly clean the brake drum.
3. Inspect the brake drum for cracks, scores deep grooves, etc. A damaged drum is unsafe for use, and should be replaced immediately. Do not attempt to weld a cracked drum. If the drum exhibits scoring, and there is enough metal left on the inside diameter of the drum, have the drum cut by a qualified automotive machine shop. Slight scoring can be smoothed using emery cloth.
4. Inspect the drum for excessive wear by measuring the inside diameter of the brake drum with a caliper gauge. The maximum inside drum diameter allowable should be imprinted in the drum itself.
5. If the brake drum exhibits damage, or if the inside diameter is larger than specified, replace it with a new one.

REMOVAL & INSTALLATION

Brake drums are either separate components or an integral part of the hub assembly. Non-integral brake drums are held onto the axle flange or hub by the wheel and lug nuts; once the wheel is removed, the brake drum can be pulled off of the axle flange. Integral (with the hub assembly) brake drums are combined with the bearing hub to comprise one piece, which means that the wheel bearings must be disturbed (loosened or removed) in one way or another to remove the drum/hub assembly.

❊❊ WARNING

If the drum is excessively difficult to remove, loosen the brake pads by adjusting their position with a brake spoon. Access for adjusting the brake pads is often gained through a small hole in the backing plate. If a brake drum is forced off of an axle flange without loosening the brake pads, damage can occur to the brake or axle components.

Non-integral drums (those that are not part of the hub) are usually fairly easy to remove. There are always exceptions to the rule, how-

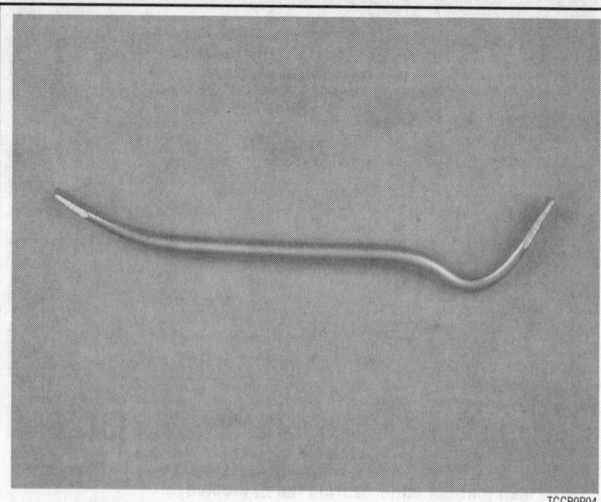

TCCB9P04

A brake spoon can be used to back off the shoe adjustment to allow drum removal

ever. There are drums that are retained to the hub with one or two small bolts. Some drums can be drawn off of the hub by installing two small bolts into threaded holes in the drum; as these bolts are tightened, they slowly press the drum off of the hub. Occasionally a drum is difficult to remove because it binds on the hub flange; these must be worked off by prying gently between the drum and backing plate while applying penetrating oil to the drum/flange contact point. Some older vehicles have a drum assembly that fits over splines on the end of the axle shaft. Others just rust in place. If this occurs, just spray the area around each lug stud and the hub flange with a penetrant such as WD-40®, Liquid Wrench® or equivalent. Let the stuff work for a while, then try pulling or prying the drum off.

Non-Integral Drums

❊❊ CAUTION

It is always a good idea to wear eye protection when working on brake components, especially drum brakes. Drum brakes often use powerful springs which could cause severe eye injury if they accidentally break.

FREE-MOUNTED TYPE

❊❊ CAUTION

Brake shoes may contain asbestos, which is a known cancer-causing agent. As soon as the drum is removed, generously spray the entire brake assembly with brake parts cleaner. Let it dry before proceeding. It's a good idea to wear a filter mask when doing brake work.

→ Some vehicles are built with retainers threaded over 2 or more lug studs to hold the drum in place during assembly. Although these retainers may not be necessary (according to the manufacturer), it may be a good idea to reinstall new retainers anyhow.

1. Loosen the lug nuts on the applicable wheels.
2. Raise and safely support the vehicle.

To remove a free-mounted brake drum, first safely raise the vehicle and remove the wheel . . .

. . . , then grasp hold of the drum and pull it from the axle flange and brake shoes

3. Remove the wheels.

4. If necessary, remove and discard the retainers holding the drum to the hub.

5. If applicable, back off the parking brake adjustment.

6. Back off the brake adjustment until the wheels rotate freely, as follows:

   a. On vehicles with a starwheel-type adjuster: Remove the plug on the backing plate, then insert a thin prytool and a brake spoon into the slot. Hold the adjuster lever away from the adjuster wheel with the thin prytool and back of the adjuster wheel with the brake spoon.

   b. On vehicles with an expanding-type adjuster, remove the plug and rotate the adjuster screw (usually in an upward motion).

   c. On vehicles with ratcheting-type adjusters, remove the plug and insert a thin punch in the hole until it contacts the adjuster assembly pivot. Apply side pressure on this pivot point to allow the adjuster quadrant to ratchet and release the brake adjustment.

   d. Some vehicles, notably with manual adjusters, use adjusting cams. On these vehicles, the cam can be turned back from behind the backing plate.

7. Grasp the drum and pull it off the hub.

➡On some vehicles, the drum won't come off even with the shoes completely backed off. This is due to the drum binding on the hub boss. The safest way to remove the drum when this happens, is to spray the binding point with lubricant and to carefully pry between the hub and backing plate. Use a small prybar and pry at various points while rotating the drum. It helps to occasionally tap the hub with a deadblow, or brass mallet.

8. Spray the brake shoe assembly thoroughly with brake parts cleaner and let it dry. Similarly, spray the inside of the drum.

9. Inspect the drum for wear and/or damage, such as deep grooves, excessive thinness, cracks, etc. Machine or replace the drum as necessary. When machining, observe the maximum diameter specification. The maximum machining diameter is stamped into the drum. If the drum braking surface shows signs of blue discoloration, overheating is indicated. If the bluing is extensive the drum must be replaced. Extensive bluing indicates a weakening of the metal.

To install:

➡New brake drums come with an oily, rust-preventive coating on the braking surface. This coating can be removed with brake parts cleaner or most cleaners which are good for oil removal. Be sure that all traces of the coating are removed. Allow the drum to dry before installation.

10. If a new brake drum is being installed, remove the protective coating from the inner braking surface.

11. Adjust the brake shoes to just smaller than the inside diameter of the brake drum.

12. Slide the brake drum onto the hub. Be sure that the brake shoes are not dragging on the brake drum. Install new brake drum retainers.

13. Install the wheels and tighten the lug nuts in a star pattern until tight.

14. Adjust the brakes shoes.

15. Adjust the parking brake.

16. Install the rubber plug in the access hole.

17. Lower the vehicle. To activate the adjusters, some vehicles require you to make several quick pulls on the parking brake lever. On most, however, several short back-ups, about 10 ft. (3m) each, should do it.

18. Road test the vehicle and check for proper brake operation.

FORCE-FIT TYPE

✳✳ CAUTION

Brake shoes may contain asbestos, which is a known cancer-causing agent. As soon as the drum is removed, generously spray the entire brake assembly with brake parts cleaner. Let it dry before proceeding. It's a good idea to wear a filter mask when doing brake work.

1. Loosen the lug nuts on the applicable wheels.

2. Raise and safely support the vehicle.

3. Remove the wheels.

4. If necessary, remove and discard the retainers holding the drum to the hub.

5. If applicable, back off the parking brake adjustment.

6. Back off the brake adjustment until the wheels rotate freely, as follows:

88279P13

If equipped with threaded holes, it is possible to press a force-fit type drum off of the hub using bolts, as shown

a. On vehicles with a starwheel-type adjuster: Remove the plug on the backing plate, then insert a thin prytool and a brake spoon into the slot. Hold the adjuster lever away from the adjuster wheel with the thin prytool and back of the adjuster wheel with the brake spoon.

b. On vehicles with an expanding-type adjuster, remove the plug and rotate the adjuster screw (usually in an upward motion).

c. On vehicles with ratcheting-type adjusters, remove the plug and insert a thin punch in the hole until it contacts the adjuster assembly pivot. Apply side pressure on this pivot point to allow the adjuster quadrant to ratchet and release the brake adjustment.

d. Some vehicles, notably with manual adjusters, use adjusting cams. On these vehicles, the cam can be turned back from behind the backing plate.

7. Thread the proper size bolts into the holes provided in the drum until each contacts the hub. Turn the bolts evenly, a little at a time, until the drum slides free.

8. Grasp the drum and remove it from the axle flange or hub assembly. Remove the forcing bolts.

9. Spray the brake assembly thoroughly with brake parts cleaner and let it dry. Similarly, spray the inside of the drum.

10. Inspect the drum for wear and/or damage, such as deep grooves, excessive thinness, cracks, etc. Machine or replace the drum as necessary. When machining, observe the maximum diameter specification. The maximum machining diameter is stamped into the drum. If the drum braking surface shows signs of blue discoloration, overheating is indicated. If the bluing is extensive the drum must be replaced. Extensive bluing indicates a weakening of the metal.

To install:

➡New brake drums come with an oily, rust-preventive coating on the braking surface. This coating can be removed with brake parts cleaner or most cleaners which are good for oil removal. Be sure that all traces of the coating are removed. Allow the drum to dry before installation.

11. If a new brake drum is being installed, remove the protective coating from the inner braking surface.

12. Adjust the brake shoes to match the inside diameter of the brake drum.

13. Slide the brake drum onto the hub. Install 2 wheel lug nuts and tighten them, forcing the drum into place on the hub. Remove the lug nuts, then, if equipped, install new drum retainers.

14. Install the wheels.

15. Adjust the brake shoes.

16. Adjust the parking brake.

17. Install the rubber plug in the access hole.

18. Lower the vehicle. To activate the adjusters, some vehicles require you to make several quick pulls on the parking brake lever. On most, however, several short back-ups, about 10 ft. (3m) each, should do it.

19. Road test the vehicle and check for proper brake operation.

BOLTED-IN-PLACE TYPE

✳✳ CAUTION

Brake shoes may contain asbestos, which is a known cancer-causing agent. As soon as the drum is removed, generously spray the entire brake assembly with brake parts cleaner. Let it dry before proceeding. It's a good idea to wear a filter mask when doing brake work.

1. Loosen the lug nuts on the applicable wheels.

2. Raise and safely support the vehicle.

3. Remove the wheels.

4. If applicable, back off the parking brake adjustment.

5. Back off the brake adjustment until the wheels rotate freely, as follows:

a. On vehicles with a starwheel-type adjuster: Remove the plug on the backing plate, then insert a thin prytool and a brake spoon into the slot. Hold the adjuster lever away from the adjuster wheel with the thin prytool and back of the adjuster wheel with the brake spoon.

b. On vehicles with an expanding-type adjuster, remove the plug and rotate the adjuster screw (usually in an upward motion).

c. On vehicles with ratcheting-type adjusters, remove the plug and insert a thin punch in the hole until it contacts the adjuster assembly pivot. Apply side pressure on this pivot point to allow the adjuster quadrant to ratchet and release the brake adjustment.

d. Some vehicles, notably with manual adjusters, use adjusting cams. On these vehicles, the cam can be turned back from behind the backing plate.

6. Remove the drum-to-hub attaching bolts.

7. Grasp the drum and remove it from the axle flange or hub assembly.

8. Spray the brake assembly thoroughly with brake parts cleaner and let it dry. Similarly, spray the inside of the drum.

➡On some vehicles, the drum won't come off even with the shoes completely backed off. This is due to the drum binding on the hub boss. The safest way to remove the drum when this happens, is to spray the binding point with lubricant and pry, carefully between the hub and backing plate. Use a small prybar and pry at various points while rotating the drum. It helps to occasionally rap the hub with a deadblow, or brass mallet.

9. Inspect the drum for wear and/or damage, such as deep grooves, excessive thinness, cracks, etc. Machine or replace the drum as necessary. When machining, observe the maximum diameter specification. The maximum machining diameter is stamped into the drum. If the drum braking surface shows signs of blue discol-

oration, overheating is indicated. If the bluing is extensive the drum must be replaced. Extensive bluing indicates a weakening of the metal.

To install:

➡**New brake drums come with an oily, rust-preventive coating on the braking surface. This coating can be removed with brake parts cleaner or most cleaners which are good for oil removal. Be sure that all traces of the coating are removed. Allow the drum to dry before installation.**

10. If a new brake drum is being installed, remove the protective coating from the inner braking surface.

11. Adjust the brake shoes to match the inside diameter of the brake drum.

12. Slide the brake drum onto the hub. Be sure that the brake shoes are not dragging on the brake drum.

13. Install the drum-to-hub attaching bolts and tighten them securely.

14. Install the wheels.

15. Adjust the brakes as follows:

   a. Adjust the brake shoes so that you can feel a slight drag on, or hear a scraping noise coming from the wheel when you spin it.

   b. Back the shoes off just until the drag is no longer felt, or the rasping noise is no longer heard.

16. Adjust the parking brake.

17. Install the rubber plug in the access hole.

18. Lower the vehicle. To activate the adjusters, some vehicles require you to make several quick pulls on the parking brake lever, On most, however, several short back-ups, about 10 ft. (3m) each, should do it.

19. Road test the vehicle and check for proper brake operation.

Integral Drum/Hub Assemblies

> ❊❊ **CAUTION**
>
> **It is always a good idea to wear eye protection when working on brake components, especially drum brakes. Drum brakes often use powerful springs which could cause severe eye injury if they accidentally break.**

Some Rear Wheel Drive (RWD) front drums and some Front Wheel Drive (FWD) rear drums are designed with the bearing hub as an integral assembly with the drum.

> ❊❊ **CAUTION**
>
> **Brake shoes may contain asbestos, which is a known cancer-causing agent. As soon as the drum is removed, generously spray the entire brake assembly with brake parts cleaner. Let it dry before proceeding. It's a good idea to wear a filter mask when doing brake work.**

1. Raise and safely support the vehicle.

2. Remove the wheels.

3. If applicable, back off the parking brake adjustment.

4. Back off the brake adjustment until the wheels rotate freely.

   a. On vehicles with a starwheel-type adjuster: Remove the plug on the backing plate, then insert a thin prytool and a brake spoon into the slot. Hold the adjuster lever away from the adjuster wheel with the thin prytool and back of the adjuster wheel with the brake spoon.

   b. On vehicles with an expanding-type adjuster, remove the plug and rotate the adjuster screw (usually in an upward motion).

   c. On vehicles with ratcheting-type adjusters, remove the plug and insert a thin punch in the hole until it contacts the adjuster assembly pivot. Apply side pressure on this pivot point to allow the adjuster quadrant to ratchet and release the brake adjustment.

   d. Some vehicles, notably with manual adjusters, use adjusting cams. On these vehicles, the cam can be turned back from behind the backing plate.

5. Remove the hub grease cap.

6. Remove the cotter pin and wheel bearing adjusting nut cover. Discard the cotter pin.

7. Remove the wheel bearing nut.

> ❊❊ **WARNING**
>
> **On some vehicles, a left-hand threaded nut is used on the right wheel spindle. Turn this locknut clockwise to loosen, otherwise damage to the spindle threads will occur.**

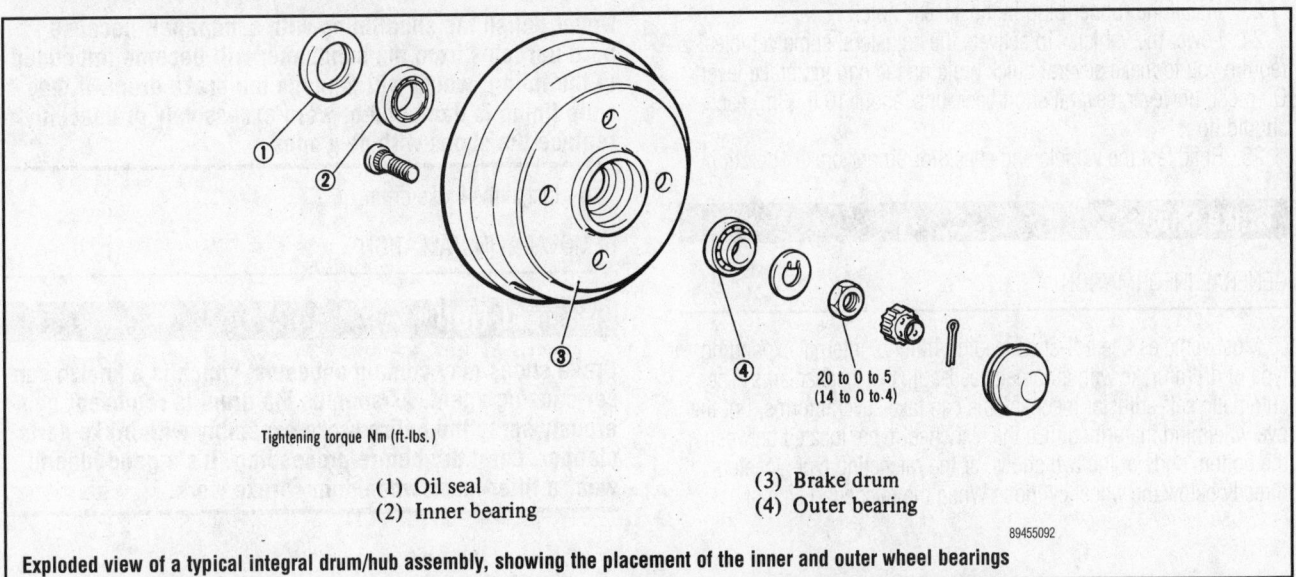

Tightening torque Nm (ft-lbs.)

20 to 0 to 5
(14 to 0 to 4)

(1) Oil seal
(2) Inner bearing

(3) Brake drum
(4) Outer bearing

89455092

Exploded view of a typical integral drum/hub assembly, showing the placement of the inner and outer wheel bearings

8. Remove the brake drum, washer and bearings as an assembly. Be careful not to let the outer wheel bearing fall out of the hub during removal.

9. Spray the brake assembly thoroughly with brake parts cleaner and let it dry. Similarly, spray the inside of the drum.

10. Remove the brake drum/hub assembly. Inspect the drum for wear and/or damage. Machine or replace as necessary. When machining, observe the maximum diameter specification. The maximum machining diameter is stamped into the drum. If the drum braking surface shows signs of blue discoloration, overheating is indicated. If the bluing is extensive the drum/hub assembly must be replaced. Extensive bluing indicates a weakening of the metal.

➡**If the brake drum is to be machined or replaced, remove the inner wheel bearing and grease seal.**

To install:

➡**New brake drums come with an oily, rust-preventive coating on the braking surface. This coating can be removed with brake parts cleaner or most cleaners which are good for oil removal. Be sure that all traces of the coating are removed. Allow the drum to dry before installation.**

11. If a new brake drum is being installed, remove the protective coating from the inner braking surface.

12. If removed, install the inner wheel bearing and a new grease seal.

13. Be sure the bearings and hub contain an adequate amount of clean wheel bearing grease.

14. Adjust the distance between the brake shoes to match the inner diameter of the brake drum.

15. Position the brake drum on the spindle. Keep the drum centered on the spindle to prevent damage to the grease seal and spindle threads.

16. Install the outer wheel bearing, washer and wheel bearing nut.

17. Properly adjust the wheel bearing; refer to the appropriate model-specific section in this manual.

18. Install the wheel bearing nut cover and a new cotter pin.

19. Install the hub grease cap.

20. Install the wheels.

21. Adjust the brake shoes.

22. Adjust the parking brake.

23. Install the rubber plug in the access hole.

24. Lower the vehicle. To activate the adjusters, some vehicles require you to make several quick pulls on the parking brake lever. On most, however, several short back-ups, about 10 ft. (3m) each, should do it.

25. Road test the vehicle and check for proper brake operation.

Brake Shoes

GENERAL INFORMATION

Most vehicles use a 2-shoe leading/trailing, internal expanding type of drum brake with automatic self-adjuster mechanisms. The automatic self-adjuster mechanisms can take several forms, but the overwhelming majority utilize the starwheel-type, located between the bottom ends of the two shoes, or the ratcheting type, located directly below the wheel cylinder. When the ratcheting type of adjuster is used, the lower ends of the brake shoes usually rest on an anchor plate.

➡**On some vehicles, notably those with unitized rear hubs, and some vehicles with full-floating axles, not only does the brake drum have to be removed, but the hub assembly must be removed as well.**

❋❋ CAUTION

Brake shoes must always be replaced as an axle set. That is, do not just replace the shoes on one side of the vehicle. Replace them on both sides. Replacing shoes on only one side will result in poor braking performance. Besides, if the shoes wore out on one side faster than the other side, there is a malfunction in the brake system. Inspect the brake system, if necessary, repair the problem before proceeding.

➡**It is not a good idea to disassemble the brakes on both sides at the same time. There are a lot of parts involved which must be replaced in a certain way. Work on one side at a time, only. If you become confused as to the particular position of the various brake parts during the brake shoe replacement, refer to the other side. Remember, however, the other side is a mirror image (everything is reversed).**

INSPECTION

1. Remove the brake drum.

2. Inspect the brake shoe lining material for cracks, crumbling or evidence of wetness. Replace the shoes with new ones if any such damage is found. If evidence of wetness is evident, repair the leaking component prior to installing the new shoes.

3. Measure the thickness of the brake shoe lining (not including the shoe backing). Generally, the minimum allowable lining thickness is either $\frac{1}{16}$ in. (1.6mm) above the head of the rivet (for rivet mounted linings), or $\frac{3}{32}$ in. (2.4mm) from the shoe backing (for glued linings).

4. If one of the brake linings is worn to or beyond the allowable limit, all four of the rear brake shoes must be replaced.

❋❋ WARNING

Never polish the shoe lining with sandpaper, because hard particles from the sandpaper will become imbedded in the lining, which will damage the brake drum. If the shoe lining is damaged or worn excessively or unevenly, replace the shoes with new ones.

5. Install the brake drum.

REMOVAL & INSTALLATION

❋❋ CAUTION

Brake shoes may contain asbestos, which is a known cancer-causing agent. As soon as the drum is removed, generously spray the entire brake assembly with brake parts cleaner. Let it dry before proceeding. It's a good idea to wear a filter mask when doing brake work.

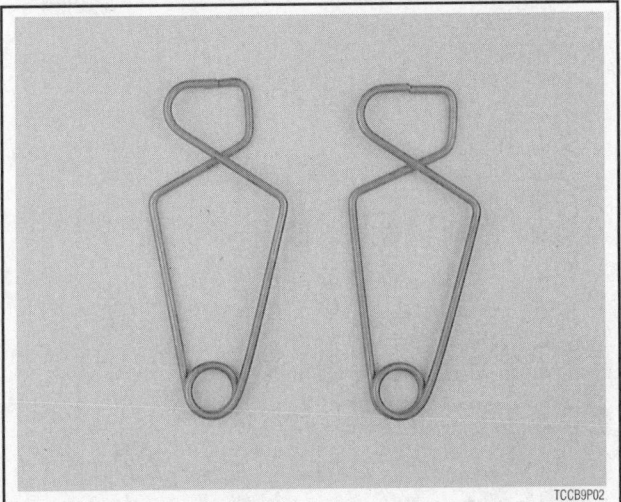

TCCB9P02

Spring clamp tools, such as those shown, can hold the wheel cylinder pistons in while servicing the shoes

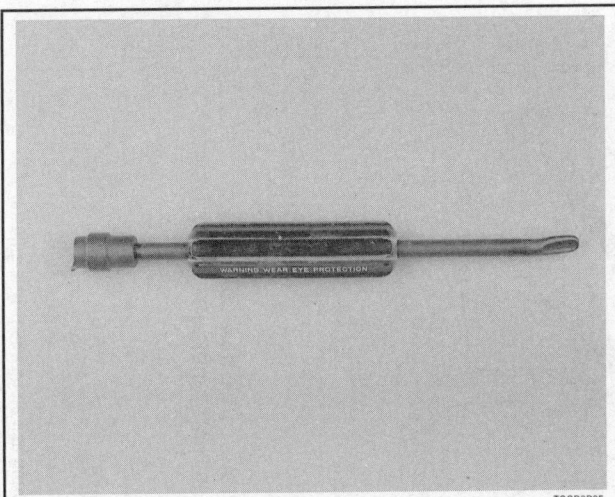

TCCB9P05

There are several varieties of spring removal and installation tools available, such as this straight one . . .

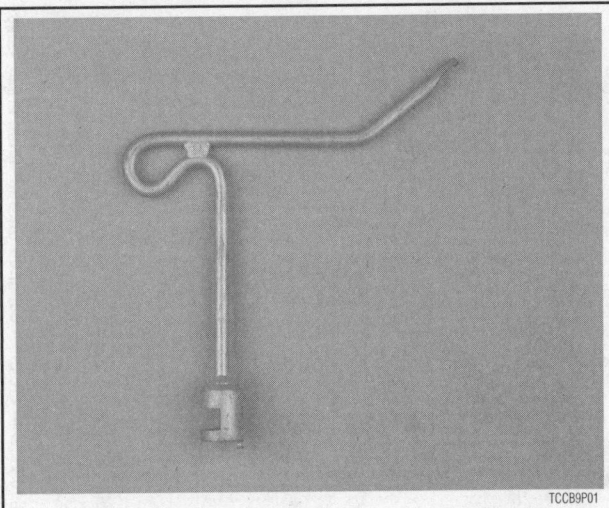

TCCB9P01

. . . and this curved one—The shape of this tool is designed to provide more leverage during use

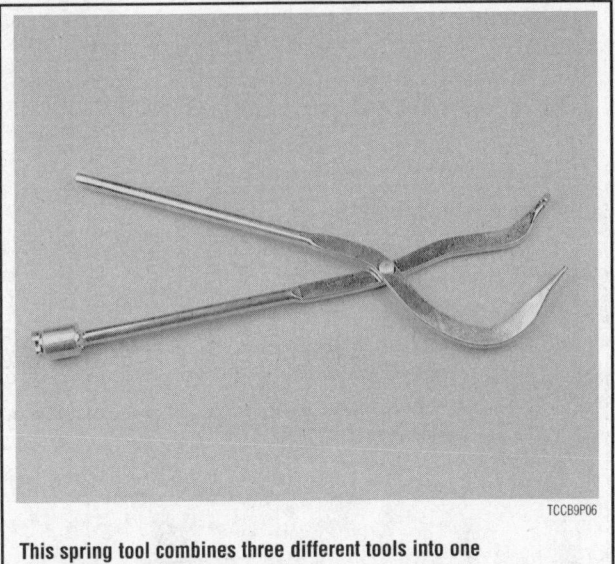

TCCB9P06

This spring tool combines three different tools into one

Models with Dual Return Springs and a Starwheel-type Adjuster

> **✷✷ CAUTION**
>
> It is always a good idea to wear eye protection when working on brake components, especially drum brakes. Drum brakes often use powerful springs which could cause severe eye injury if they accidentally break.

1. Remove the brake drum.
2. Spray the brake assembly thoroughly with brake parts cleaner and let it dry. Similarly, spray the inside of the drum.
3. Inspect the drum for wear and/or damage. Machine or replace as necessary. When machining, observe the maximum diameter specification. The maximum machining diameter is stamped into the drum. If the drum braking surface shows signs of blue discoloration, overheating is indicated. If the bluing is extensive the drum/hub assembly must be replaced. Extensive bluing indicates a weakening of the metal.

➡ **Note the location of all springs and clips for proper assembly. If an instant camera is handy, it may be a good idea to take a picture of the brake assembly with the brake drum removed. This will make reassembly much easier.**

4. Completely retract the adjuster by rotating the starwheel to relieve tension on the lower spring.
5. Remove the starwheel assembly and adjuster lever from between the two brake shoes.
6. Using a brake spring tool, remove the 2 upper return springs.
7. Remove the adjuster cable and cable guide.
8. Remove the anchor block plate.
9. Using a hold-down spring tool or pliers, while holding the back of the spring mounting pin with one hand, press inward on the hold-down spring plate, turn it slightly to align the notches and pin ears, then remove the hold-down spring assembly with your other hand. Remove the other hold-down spring in the same manner.
10. Lift the shoes off the pins and remove the pins from the backing plate.
11. Remove the parking brake link.

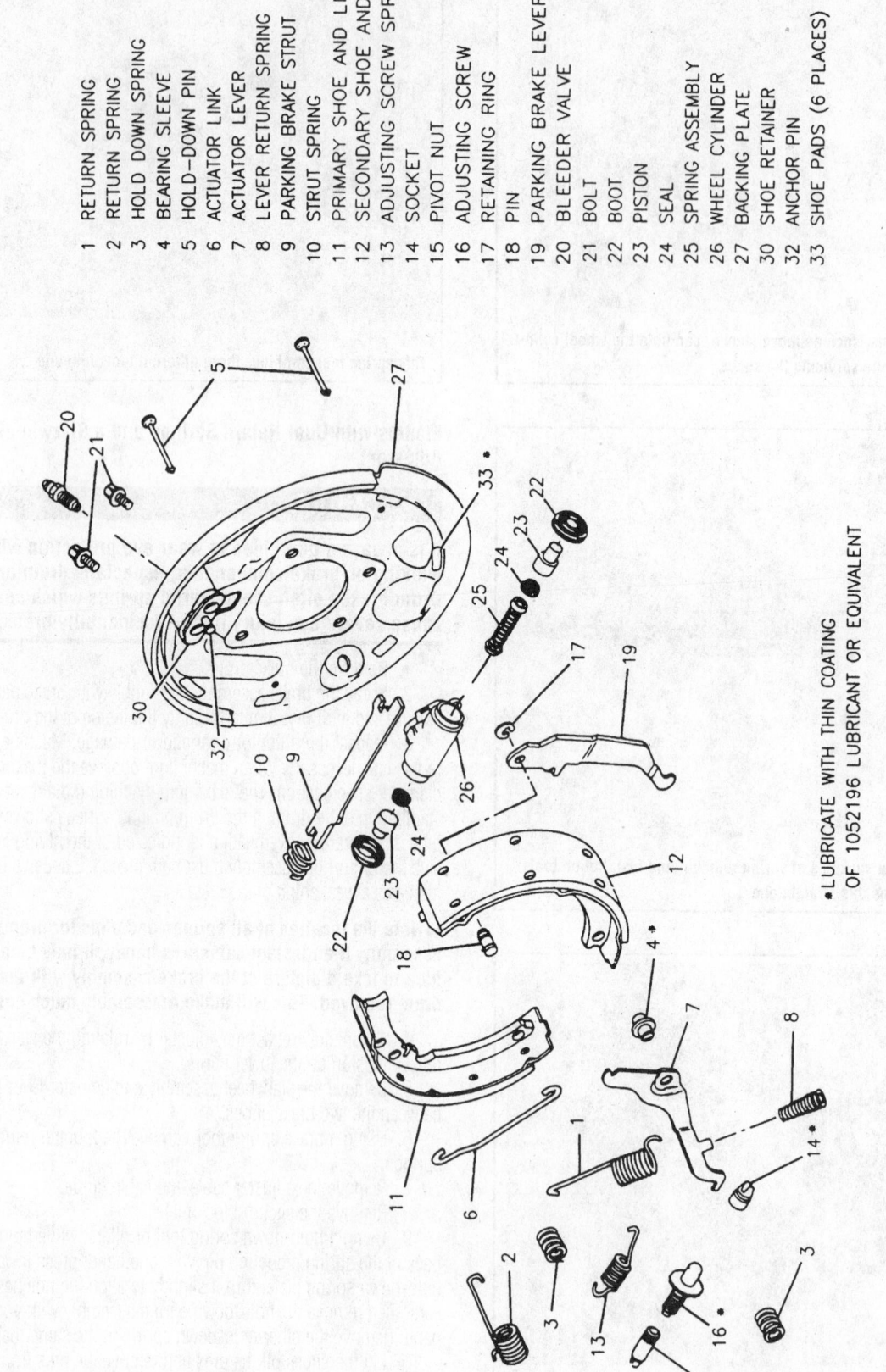

1 RETURN SPRING
2 RETURN SPRING
3 HOLD DOWN SPRING
4 BEARING SLEEVE
5 HOLD-DOWN PIN
6 ACTUATOR LINK
7 ACTUATOR LEVER
8 LEVER RETURN SPRING
9 PARKING BRAKE STRUT
10 STRUT SPRING
11 PRIMARY SHOE AND LINING
12 SECONDARY SHOE AND LINING
13 ADJUSTING SCREW SPRING
14 SOCKET
15 PIVOT NUT
16 ADJUSTING SCREW
17 RETAINING RING
18 PIN
19 PARKING BRAKE LEVER
20 BLEEDER VALVE
21 BOLT
22 BOOT
23 PISTON
24 SEAL
25 SPRING ASSEMBLY
26 WHEEL CYLINDER
27 BACKING PLATE
30 SHOE RETAINER
32 ANCHOR PIN
33 SHOE PADS (6 PLACES)

*LUBRICATE WITH THIN COATING
OF 1052196 LUBRICANT OR EQUIVALENT

Exploded view of the most common GM rear drum brake setup—dual return spring and a starwheel-type adjuster type

REAR DRUM BRAKE COMPONENTS

1. Secondary shoe
2. Adjusting screw assembly
3. Primary shoe
4. Adjuster spring
5. Adjuster lever
6. Hold-down pin
7. Hold-down spring
8. Hold-down assembly
9. Adjuster cable guide
10. Parking brake lever
11. Parking brake link
12. Link spring
13. Primary shoe return spring
14. Anchor pin plate
15. Secondary shoe return spring
16. Adjuster cable

Typical Ford dual return spring drum brake setup component identification—dual return spring and a starwheel-type adjuster type

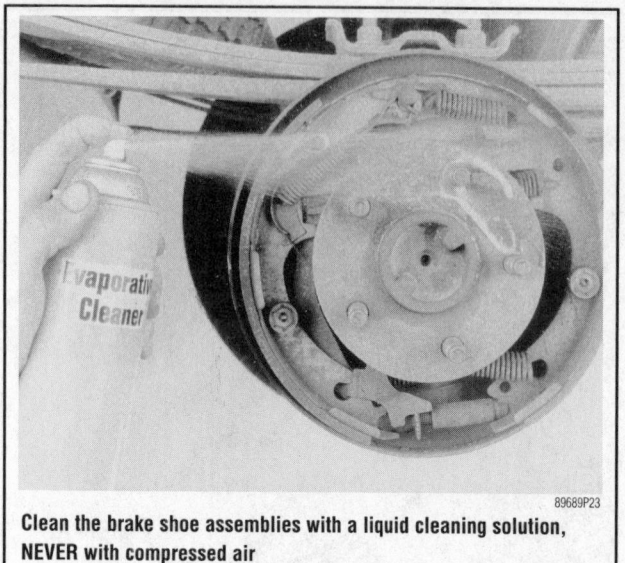

Clean the brake shoe assemblies with a liquid cleaning solution, NEVER with compressed air

A specially-designed brake tool can make disconnecting the upper return springs much easier—dual return spring and a starwheel-type adjuster type

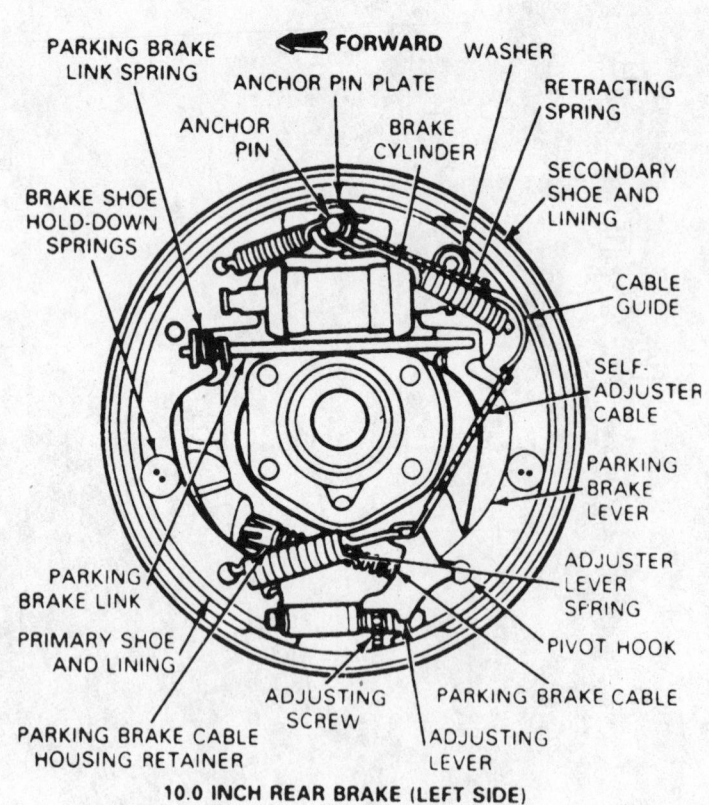

← FORWARD

PARKING BRAKE LINK SPRING

ANCHOR PIN PLATE

WASHER

ANCHOR PIN

BRAKE CYLINDER

RETRACTING SPRING

BRAKE SHOE HOLD-DOWN SPRINGS

SECONDARY SHOE AND LINING

CABLE GUIDE

SELF-ADJUSTER CABLE

PARKING BRAKE LEVER

PARKING BRAKE LINK

ADJUSTER LEVER SPRING

PRIMARY SHOE AND LINING

PIVOT HOOK

PARKING BRAKE CABLE

ADJUSTING SCREW

ADJUSTING LEVER

PARKING BRAKE CABLE HOUSING RETAINER

10.0 INCH REAR BRAKE (LEFT SIDE)

Identify the brake components and note their locations prior to disassembling the brake assembly

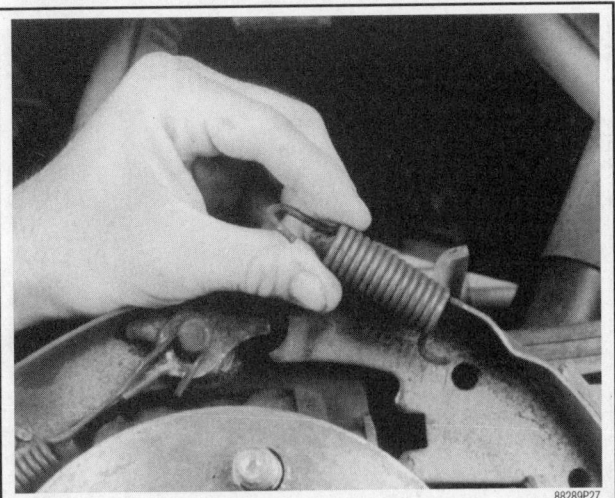

Detach the upper return springs first from the anchor bolt, then from the brake shoes . . .

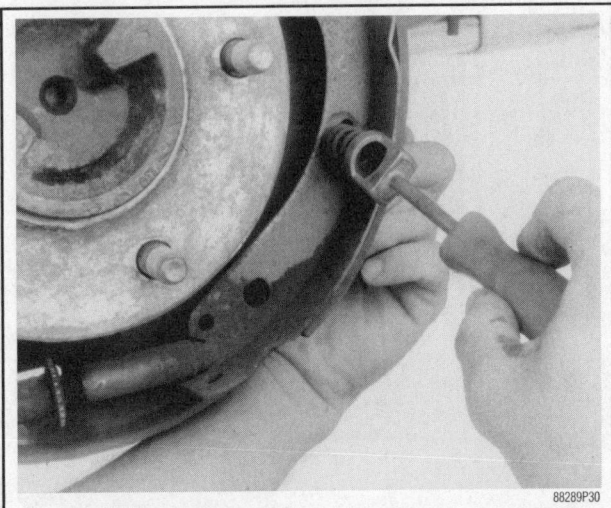

. . . , then remove the hold-down springs, retainers and pins from both shoes—dual return spring and a starwheel-type adjuster type

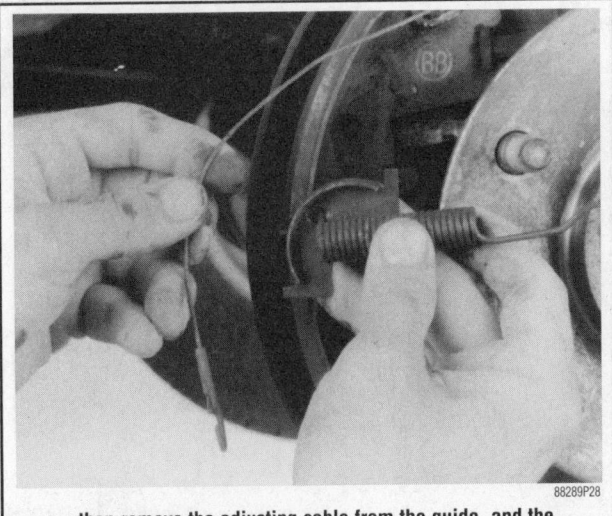

. . . , then remove the adjusting cable from the guide, and the guide from the brake shoe

Lift the brake shoes off of the backing plate . . .

Remove the anchor block plate . . .

. . . , then detach the parking brake cable from the lever—dual return spring and a starwheel-type adjuster type

Another way to remove the shoes for a dual spring setup is to pull the adjuster cable toward the shoe . . .

. . . and remove the spring and the lever—dual return spring and a starwheel-type adjuster type

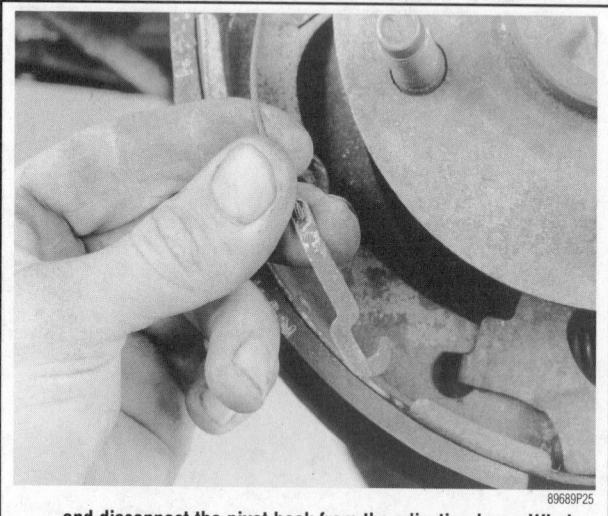

. . . and disconnect the pivot hook from the adjusting lever. Wind the starwheel all the way in

Next, using a brake spring removal tool . . .

Disconnect the adjuster lever return spring from the lever . . .

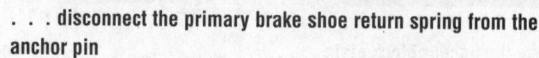

. . . disconnect the primary brake shoe return spring from the anchor pin

Repeat the procedure and remove the secondary return spring, adjuster cable and its guide

Press in the hold-down springs while holding in on the nail from behind, then turn the cup 90° . . .

Also remove the anchor pin plate—dual return spring and a star-wheel-type adjuster type

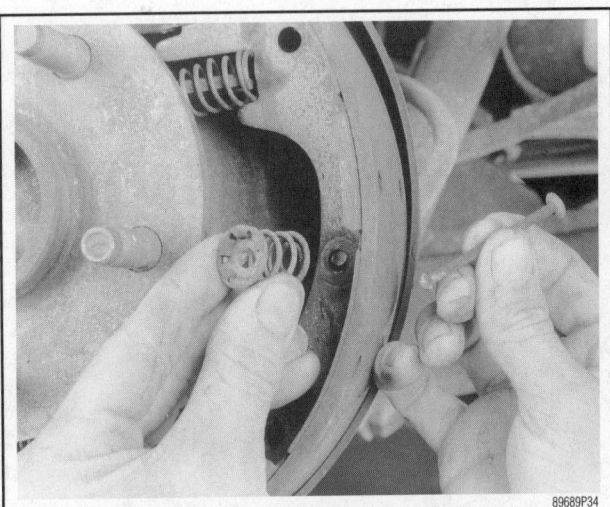

. . . and release to remove the hold-down spring. Pull the nail out from the backing plate

Pull the bottoms of the shoes apart and remove the adjuster screw assembly

Remove the primary (front) brake shoe from the backing plate . . .

. . . and the parking brake strut as well—dual return spring and a starwheel-type adjuster type

Remove the secondary shoe hold-down, pull the shoe out, then press up on the cable spring . . .

. . . and disconnect the parking brake cable from its lever by pulling it from the slot

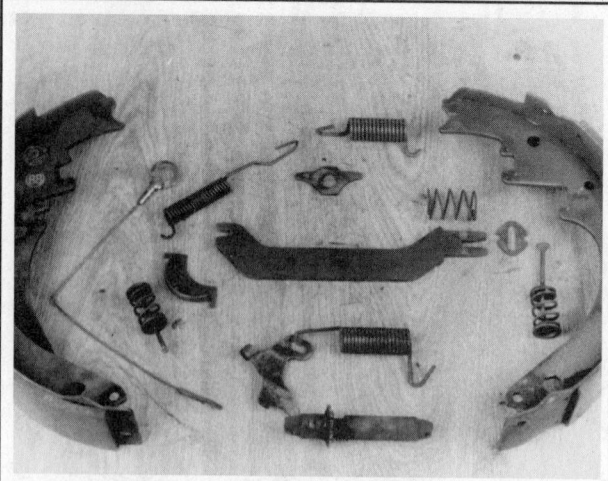

It's a good idea to arrange all the parts in their approximate installed positions on a clean work surface

12. Pull back on the parking brake cable spring and twist the cable out of the parking brake lever.

13. The parking brake lever is held onto the rear shoe with a horseshoe clip. Spread the clip and remove the lever and washer.

To install:

14. Thoroughly clean and dry the backing plate and starwheel assembly.

15. Lubricate the backing plate bosses, anchor plate surfaces, and starwheel threads and contact points with silicone grease. High-temperature wheel bearing grease or synthetic brake grease also work well for this application.

✲✲ CAUTION

When applying lubricant to the backing plate and other components, do not use' so much grease that it may get spread onto the new brake shoes' friction material; this can adversely affect the performance of the new brake shoes, therefore, increase vehicle stopping distance.

Thoroughly clean the backing plate, then be sure to lubricate the brake shoe bosses on the backing plate

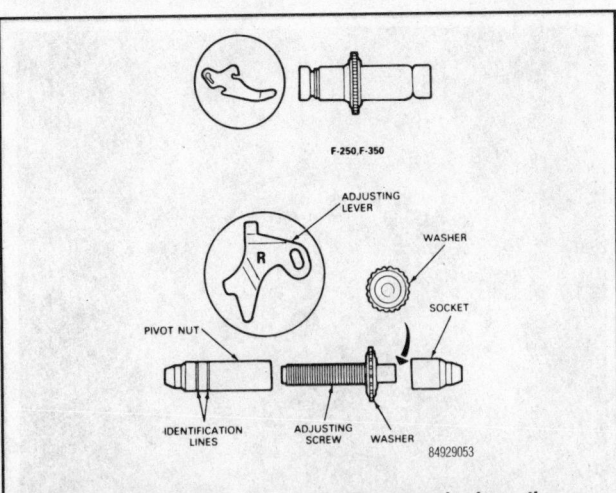

Exploded view of a typical starwheel adjuster mechanism—the adjusting levers may be stamped for left side and right side applications

16. Insert the parking brake lever pivot stud through the applicable hole in the rear shoe, then install a new wave washer and horseshoe clip. Squeeze the clip ends until the clip cannot be pulled from the lever pivot stud.

17. Connect the parking brake cable to the lever.

18. Position the rear shoe assembly on the backing plate and install the hold-down pin and spring assembly.

19. Install the front shoe and secure it with the hold-down spring assembly.

20. Position the parking brake link and spring between the front shoe and parking brake lever.

21. Position the adjuster cable on the anchor plate pin, install the cable guide and lay the cable across the guide.

22. Be sure that the notch in the upper end of the shoe is engaging the wheel cylinder piston or piston pin.

23. Position the rear shoe return spring into the guide and shoe hole, using a brake spring tool, stretch the spring onto the anchor plate pin. Be sure that the cable guide remained in place.

24. Position the front shoe return spring in its hole in the shoe.

25. Be sure that the parking brake link is properly positioned and that the upper end of the shoe will enter the wheel cylinder or engage the wheel cylinder piston.

26. Using the spring tool, stretch the spring into position on the anchor plate pin.

➡**If the shoe doesn't properly engage the link or wheel cylinder piston, try again by removing the spring.**

27. Position the adjuster lever in its hole in the rear shoe and hook the cable to it.

28. Position the lower spring in its hole in the front shoe. Now comes the hard part. Clamp a pair of locking pliers, like Vise Grips® on the spring and stretch it to engage the hole in the adjuster lever. Be sure that the cable stays in place on the guide.

29. Check that the shoes are evenly positioned on the backing plate.

30. Turn the starwheel to spread the shoes to the point at which the drum can be installed with very slight drag.

31. Install the drum and adjust the starwheel until the drum can't be turned. Then, back off the adjustment until the drum can just be turned without drag.

32. Install the wheels, lower the vehicle and check brake action. A firm pedal should be felt.

33. To activate the adjusters, some vehicles require you to make several quick pulls on the parking brake lever. On most, however, several short back-ups, about 10 ft. (3m) each, should do it.

Models with a Single Upper Shoe-to-Shoe Return Spring

✳✳ CAUTION

It is always a good idea to wear eye protection when working on brake components, especially drum brakes. Drum brakes often use powerful springs which could cause severe eye injury if they accidentally break. Also, Brake shoes may contain asbestos, which is a known cancer-causing agent. As soon as the drum is removed, generously spray the entire brake assembly with brake parts cleaner. Let it dry before proceeding. It's a good idea to wear a filter mask when doing brake work.

WITH LOWER ANCHOR PLATE

1. Remove the brake drum.

2. Clean the brake assembly and drum thoroughly with brake parts cleaner and let it dry.

Inspect the drum for wear and/or damage. Machine or replace as necessary. When machining, observe the maximum diameter specification. The maximum machining diameter is stamped into the drum. If the drum braking surface shows signs of blue discoloration, overheating is indicated. If the bluing is extensive the drum/hub assembly must be replaced. Extensive bluing indicates a weakening of the metal.

➡**Note the location of all springs and clips for proper assembly. If you own an instant camera, to make installation easier it may be a good idea to take a picture of your brake assembly with the brake drum removed.**

3. Remove the shoe-to-lever spring and remove the adjuster lever.

4. Remove the auto-adjuster assembly.

5. Remove the retainer spring.

6. Using a hold-down spring tool or pliers, while holding the back of the spring mounting pin with one hand, press inward on the

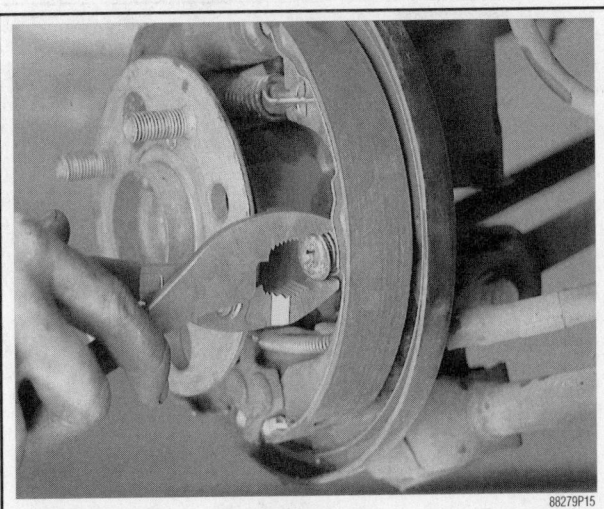

Pliers can be used to disengage the hold-down spring retainer by rotating it until aligned with the pin tabs . . .

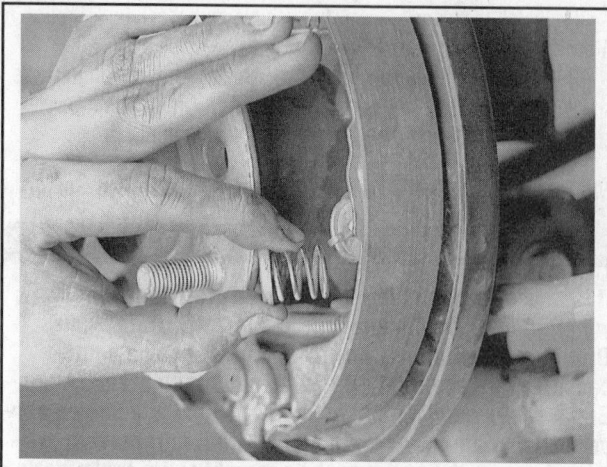

. . . , then remove the retainer, spring and pin from the shoe and backing plate—models with a single upper shoe-to-shoe return spring and lower anchor plate

. . . , then remove the brake shoes from the backing plate . . .

hold-down spring plate, turn it slightly to align the notches and pin ears, then remove the hold-down spring assemblies with your other hand.

7. Remove the shoe-to-shoe spring.

8. Remove the brake shoes from the backing plate.

9. Using a flat-tipped tool, pry open the parking brake lever retaining clip. Remove the clip and washer from the pin on the shoe assembly and remove the shoe from the lever assembly.

➡On some vehicles, the parking brake actuating lever is permanently attached to the trailing brake shoe assembly. Do not attempt to remove it from the original brake shoe assembly or reuse the original actuating lever on a replacement brake shoe assembly. All replacement brake shoe assemblies for these vehicles must come with the actuating lever as part of the trailing brake shoe assembly.

To install:

10. Thoroughly clean all parts.

11. On vehicles with the ratcheting upper mounted adjuster, clean and inspect the brake support plate and the automatic adjuster

. . . and detach the parking brake cable from the applicable brake shoe—models with a single upper shoe-to-shoe return spring and lower anchor plate

mechanism. Be sure the quadrant (toothed part) of the adjuster is free to rotate throughout its entire tooth contact range and is free to slide the full length of its mounting slot. Check the knurled pin. It should be securely attached to the adjuster mechanism and its teeth should be in good condition. If the adjuster is worn or damaged, replace it. If the adjuster is serviceable, lubricate lightly with high-temperature grease between the strut and the quadrant.

✳✳ CAUTION

The trailing brake shoe assemblies used on the rear brakes of these vehicles are different for the left and right side of the vehicle. Care must be taken to ensure the brake shoes are properly installed in their correct side of the vehicle. Otherwise the brakes will probably malfunction, thereby creating a very dangerous condition. When the trailing shoes are properly installed on their correct side of the vehicle, the park brake actuating lever will be positioned under the brake shoe web.

Use a pair of needlenose pliers, or similar tool, to detach the upper return spring from both shoes . . .

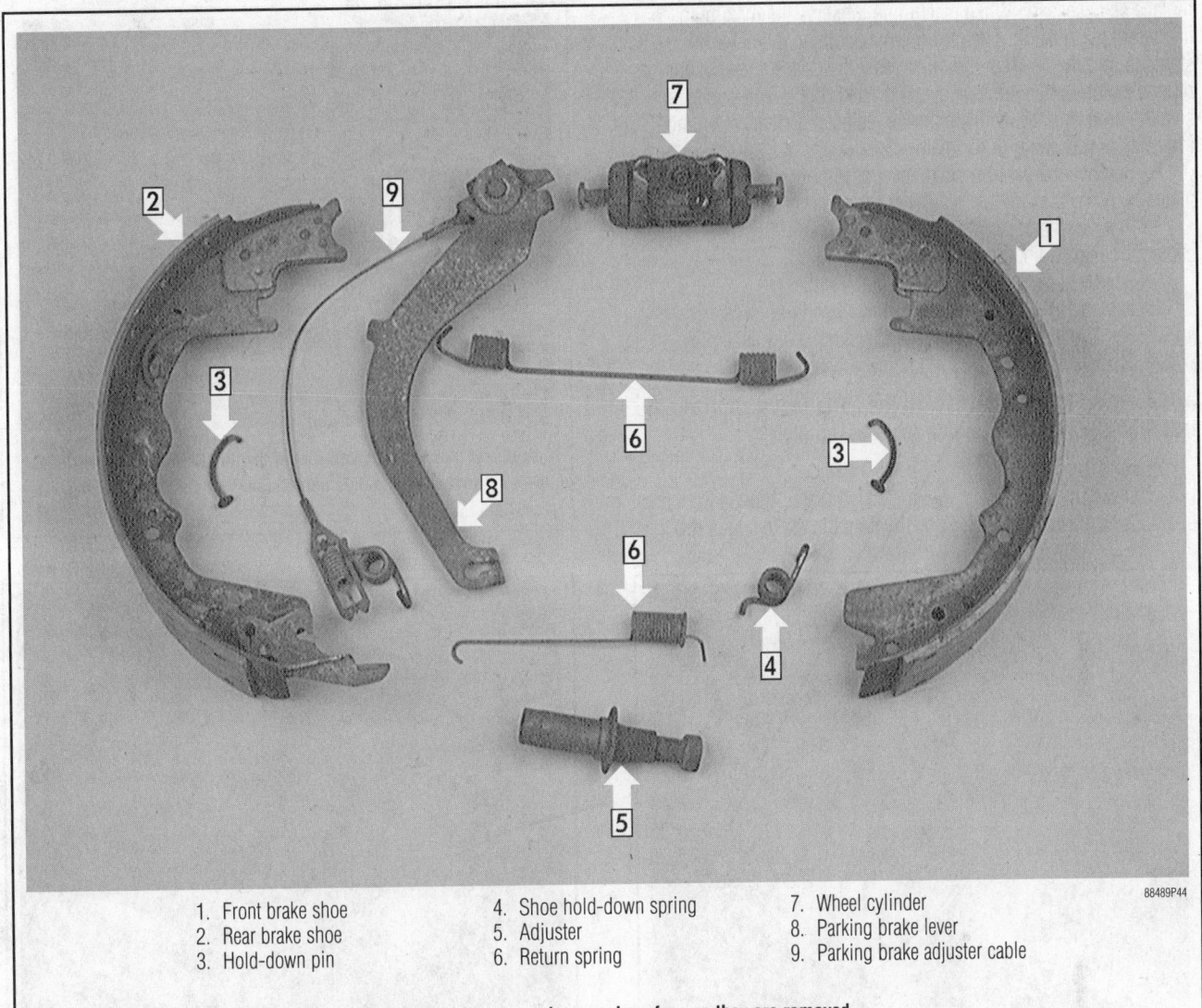

1. Front brake shoe
2. Rear brake shoe
3. Hold-down pin
4. Shoe hold-down spring
5. Adjuster
6. Return spring
7. Wheel cylinder
8. Parking brake lever
9. Parking brake adjuster cable

It is a good idea to lay the brake parts out in their positions on a clean work surface as they are removed

12. Thoroughly clean and dry the backing plate. Lubricate the backing plate at the brake shoe contact points. Also, lubricate backing plate bosses, anchor pin, and parking brake actuating mechanism with silicone grease. High-temperature wheel bearing grease or synthetic brake grease also work well for this application.

13. Install the parking brake lever assembly on the lever pin. Install the wave washer and a new retaining clip. Use pliers, or the like, to install the retainer on the pin. If removed, connect the parking brake lever to the parking brake cable and verify that the cable is properly routed.

14. Clean and lubricate the adjuster assembly. Be sure the nut-adjuster is drawn all the way to the stop, but the nut must NOT lock firmly at the end of the assembly.

15. Install the brake shoes on the backing plate with the hold-down springs, washers and pins.

16. Install the shoe-to-shoe spring.

17. Install the retainer spring.

18. Install the auto-adjuster assembly and install the adjuster lever and the shoe-to-lever spring.

19. Pre-adjust the shoes so the drum slides on with a light drag and install the brake drum.

20. Adjust the brake shoes.

21. Install the rear wheels.

22. To activate the adjusters, some vehicles require you to make several quick pulls on the parking brake lever. On most, however, several short back-ups, about 10 ft. (3m) each, should do it.

23. Adjust the parking brake cable.

24. Lower the vehicle and check for proper brake operation.

WITH LOWER STARWHEEL-TYPE ADJUSTER

1. Loosen the lug nuts on the applicable wheels.

2. If servicing the front brakes, apply the parking brake, block the rear wheels, then raise and safely support the front of the vehicle securely.

3. If servicing the rear brakes, block the front wheels, then raise and safely support the rear of the vehicle securely.

4. Remove the wheels.

5. Remove the drums.

6. Spray the brake assembly thoroughly with brake parts cleaner and let it dry. Similarly, spray the inside of the drum.

7. Inspect the drum for wear and/or damage. Machine or

replace as necessary. When machining, observe the maximum diameter specification. The maximum machining diameter is stamped into the drum. If the drum braking surface shows signs of blue discoloration, overheating is indicated. If the bluing is extensive the drum/hub assembly must be replaced. Extensive bluing indicates a weakening of the metal.

8. Remove the parking brake lever assembly from the backing plate.

9. Remove the adjusting cable assembly from the anchor pin, cable guide and adjusting lever.

10. Remove the brake shoe retracting springs.

11. Remove the brake shoe hold-down spring from each shoe.

12. Remove the brake shoes and adjusting screw assembly.

13. Disassemble the adjusting screw assembly.

➡️**It's a good idea to arrange all the parts in the approximate installed positions as a guide for reassembly.**

To install:

14. Clean the ledge pads on the backing plate. Apply a light coat of silicone grease to the ledge pads (where the brake shoes rub the

Disconnect the adjusting cable from the anchor pin, guide and lever—models with a single upper shoe-to-shoe return spring and starwheel adjuster

Remove the brake drum from the rear axle

Slide the parking brake lever out from its mounting—models with a single upper shoe-to-shoe return spring and starwheel adjuster

Remove the parking brake lever retaining nut which is located behind the backing plate

Disconnect the parking brake cable from the lever

Use an appropriate tool to disconnect the return springs from their retaining holes

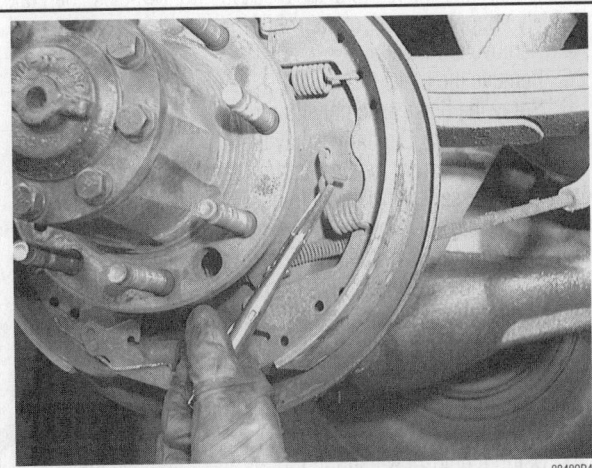

Disengage the hold-down springs from the retaining clips on the backing plate—models with a single upper shoe-to-shoe return spring and starwheel adjuster

Back off the adjusting screw and remove it from the brake assembly

Spread the shoes apart and remove them from the backing plate

backing plate). High-temperature wheel bearing grease or synthetic brake grease (designed specifically for this) also work well. Also, apply grease to the adjusting screw assembly and the hold-down and retracting spring contacts on the brake shoes.

15. Install the upper retracting spring on the primary and secondary shoes, then position the shoe assembly on the backing plate with the wheel cylinder pistons engaged with the shoes.

16. Install the brake shoe hold-down springs.

17. Install the brake shoe adjustment screw assembly so that the slot in the head of the adjusting screw is toward the primary (leading) shoe, along with the lower retracting spring, adjusting lever spring, adjusting lever assembly and connect the adjusting cable to the adjusting lever. Position the cable in the cable guide and install the cable anchor fitting on the anchor pin.

18. Install the adjusting screw assemblies in the same locations from which they were removed.

✳✳ CAUTION

Interchanging the brake shoe adjusting screws from one side of the vehicle to the other will cause the brake shoes

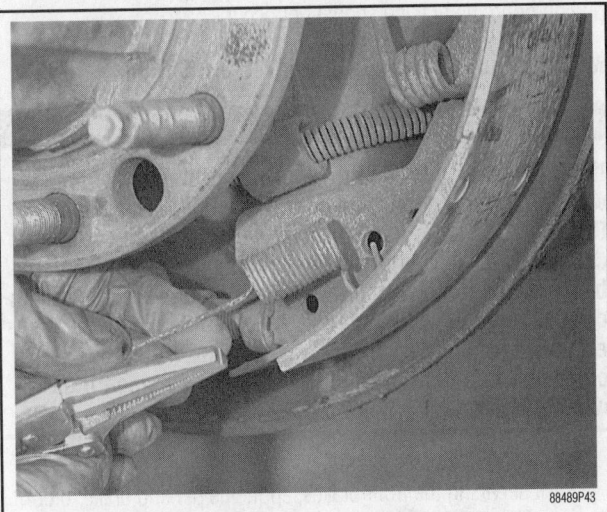

Connecting the lower retracting spring can often be difficult—be careful and have patience

This is how everything should look after assembly—models with a single upper shoe-to-shoe return spring and starwheel adjuster

Before removing any parts, make a note of their positions

to retract rather than expand each time the automatic adjusting mechanism is operated; this will create an extremely dangerous condition when driving the vehicle. To prevent incorrect installation, the socket end of each adjusting screw is usually stamped with an R or an L to indicate their installation on the right or left side of the vehicle. In some cases, the adjusting pivot nuts can be distinguished by the number of lines machined around the body of the nut. Two lines indicate a nut which should be installed on the right side of the vehicle; one line indicates a nut that must be installed on the left side of the vehicle.

19. Install the parking brake assembly in the anchor pin and secure with the retaining nut behind the backing plate.

20. Adjust the brakes before installing the brake drums and wheels. Install the brake drums and wheels.

21. To activate the adjusters, some vehicles require you to make several quick pulls on the parking brake lever. On most, however, several short back-ups, about 10 ft. (3m) each, should do it.

22. Lower the vehicle and road test the brakes. New brakes may pull to one side or the other before they are seated. Continued pulling or erratic braking should not occur.

Models with a Single U-Shaped Return Spring

✳✳ CAUTION

It is always a good idea to wear eye protection when working on brake components, especially drum brakes. Drum brakes often use powerful springs which could cause severe eye injury if they accidentally break. Also, brake shoes may contain asbestos, which is a known cancer-causing agent. As soon as the drum is removed, generously spray the entire brake assembly with brake parts cleaner. Let it dry before proceeding. It's a good idea to wear a filter mask when doing brake work.

1. Loosen the lug nuts on the applicable wheels.
2. If servicing the front brakes, apply the parking brake, block the rear wheels, then raise and safely support the front of the vehicle securely.

3. If servicing the rear brakes, block the front wheels, then raise and safely support the rear of the vehicle securely.

4. Remove the wheels.

5. Remove the brake drum.

6. Spray the brake assembly thoroughly with brake parts cleaner and let it dry. Similarly, spray the inside of the drum.

7. Inspect the drum for wear and/or damage. Machine or replace as necessary. When machining, observe the maximum diameter specification. The maximum machining diameter is stamped into the drum. If the drum braking surface shows signs of blue discoloration, overheating is indicated. If the bluing is extensive the drum/hub assembly must be replaced. Extensive bluing indicates a weakening of the metal.

8. Remove the return spring clip from the lower anchor block.

9. Squeeze the upper ends of the return spring slightly and remove it from the shoes.

10. Using a hold-down spring tool or pliers, remove the hold-down springs. While holding the back of the spring mounting pin with one hand, press inward on the hold-down spring plate, turn it slightly to align the notches and pin ears, then remove the hold-down spring assemblies with your other hand.

For models with a single U-shaped return spring, depress and rotate the hold-down spring retainer . . .

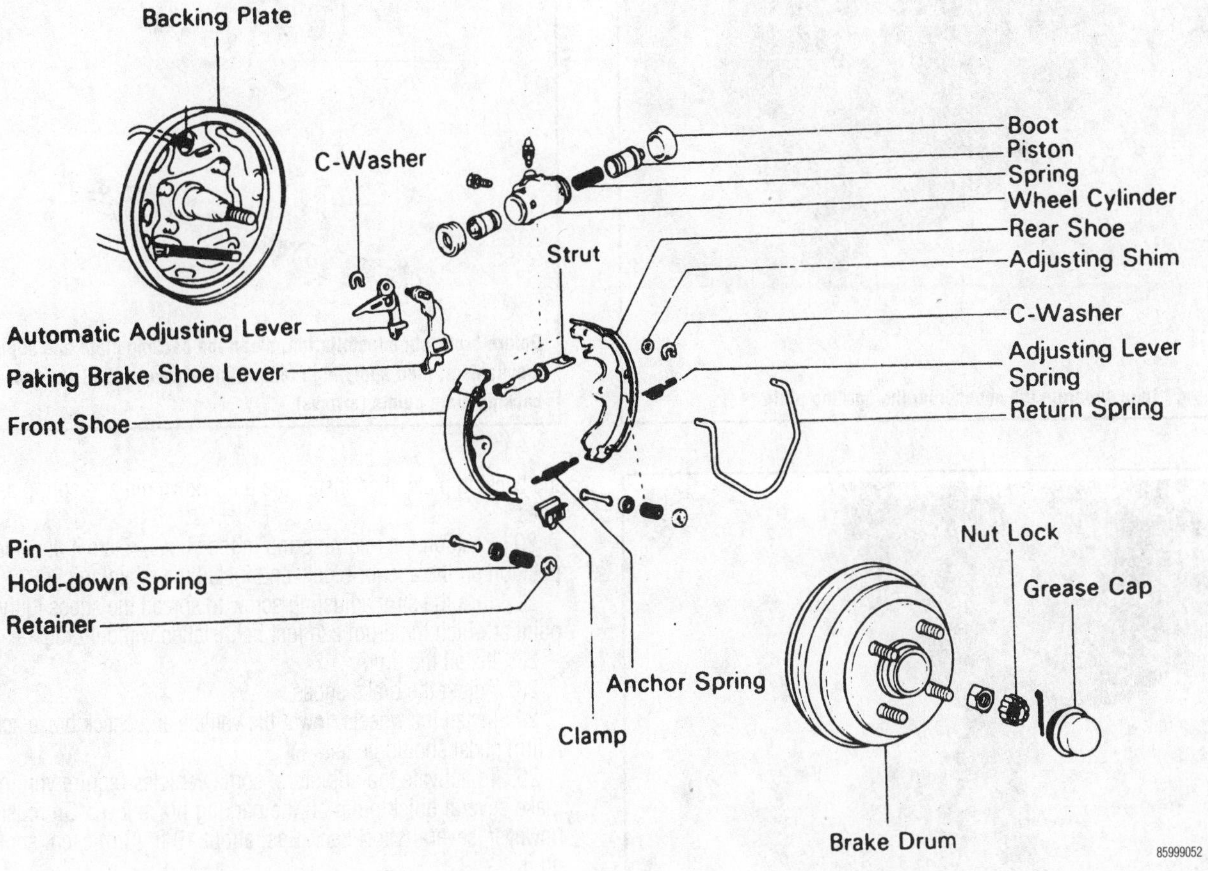

Backing Plate

C-Washer

Boot
Piston
Spring
Wheel Cylinder
Rear Shoe
Adjusting Shim

Strut

C-Washer

Automatic Adjusting Lever

Adjusting Lever
Spring

Paking Brake Shoe Lever

Return Spring

Front Shoe

Pin
Hold-down Spring

Retainer

Nut Lock

Grease Cap

Anchor Spring

Clamp

Brake Drum

85999052

Exploded view of a typical single U-shaped return spring drum brake setup

. . . , then remove the spring, retainer and pin from the backing plate and shoes

85999057

Remove the return spring from both brake shoes . . .

85999058

. . . , then separate the shoes from the backing plate

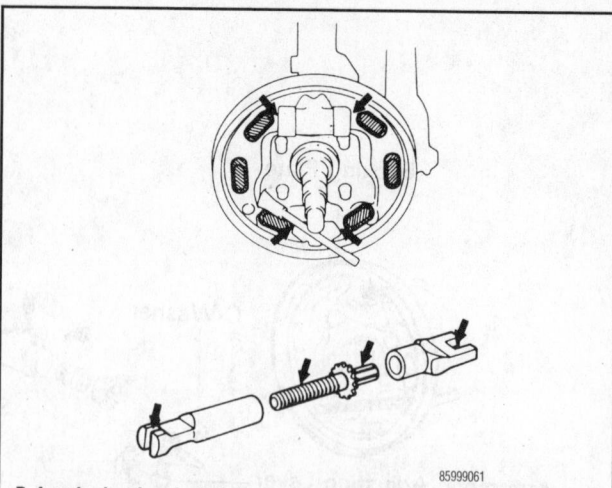

Before brake shoe installation, clean the backing plate and adjuster mechanism, then apply high temperature grease at all shoe-to-backing plate points (arrows)

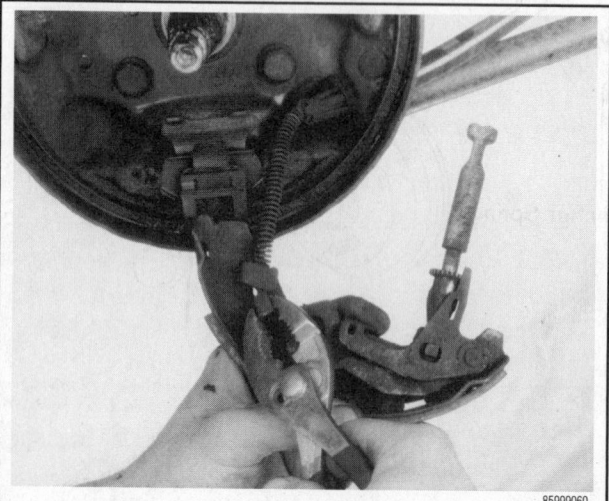

A large pair of pliers can be used to disconnect the parking brake cable from the lever

11. Lift the shoes off of the pins, then remove the pins from the backing plate.

12. Remove the shoes and adjuster as an assembly.

13. Pull back on the parking brake cable spring and twist the cable out of the parking brake lever.

14. The parking brake lever is held onto the rear shoe with a horseshoe clip. Spread the clip and detach the lever and washer from the shoe.

To install:

15. Thoroughly clean and dry the backing plate assembly.

16. Lubricate the backing plate bosses, anchor plate surfaces, and all contact points with silicone grease. High-temperature wheel bearing grease or synthetic brake grease (designed specifically for this) also work well.

17. Lubricate the parking brake lever pivot stud, then insert the pivot stud through the applicable hole in the rear shoe, then install a new wave washer and horseshoe clip. Squeeze the clip ends until the clip cannot be pulled from the lever pivot stud.

18. Connect the parking brake cable to the lever.

19. Position the front and rear shoe assemblies and adjuster on

the backing plate, then install the hold-down pin and spring assemblies.

20. Position the return spring in the shoes, rotate it down into position on the anchor block, and install the retaining clip.

21. Turn the strut adjusting screw to spread the shoes to the point at which the drum can just be installed without drag.

22. Install the drum.

23. Adjust the brake shoes.

24. Install the wheels, lower the vehicle and check brake action. A firm pedal should be felt.

25. To activate the adjusters, some vehicles require you to make several quick pulls on the parking brake lever. On most, however, several short back-ups, about 10 ft. (3m) each, should do it.

ADJUSTMENT

Drum brakes on all modern vehicles are self-adjusting, however, when the shoes are replaced, a preliminary adjustment makes the job easier.

On most vehicles, the adjustment is made with an expanding adjuster that is a threaded sleeve/stud assembly. Turning the knurled nut or starwheel expands or contracts the spring-loaded brake shoes. On most vehicles, this adjuster can be accessed without removing the drum, or, for that matter, the wheel.

Raise the vehicle and support it safely. Release the parking brake. Put the transmission in neutral. All this allows the wheels to turn freely. Remove the rubber plug in the brake backing plate and insert a brake adjusting tool. If you're applying brake pressure, that is, expanding the brakes, just turn the starwheel or knurled adjuster until the brake shoes lock the drum; meaning you can't turn it. Then, back off the adjustment until the drum can JUST turn freely without any drag. Some manufacturers even say it's okay to have a SLIGHT amount of drag. If the vehicle at hand is equipped with self-adjusters, you'll find that the adjuster can't be backed off. That's because the adjusting lever is holding it in place. You'll have to insert a thin punch or similar device in the hole with the brake adjusting tool. Just push slightly on the adjusting lever. That'll free the adjuster.

There are a few vehicle models that use cam-type adjusters. With these, a hex or square headed stud protrudes through the backing plate. Turning this stud rotates an eccentric cam that contacts the brake shoe. Turning it one way pushes the shoe outward; turning it the other way rotates the cam away from the shoe allowing the springs to pull the shoe away from the drum.

Wheel Cylinders

REMOVAL & INSTALLATION

Wheel cylinders are held in place on the backing plate with either bolts or spring clips. A first glance, this looks like a fairly easy job, and it can be. However, a lot can go wrong. If the wheel cylinder has been there a long time, the bolts or clips can be rusted in place. Worse, the brake line flare nut may be rusted in place. The flats on the nut are easily rounded off. Also, the flare nut can be rusted to the line, meaning the line will twist when the nut is turned. So, before starting, it's best to thoroughly soak the area with penetrating oil where the brake line threads into the wheel cylinder. Also, apply penetrating oil to the mounting bolts or clips.

If you run into problems, here are some general tips:

• Use a flare nut wrench on the flare nuts. Sounds logical, doesn't it? Flare nut wrenches are designed to reduce the possibility of rounding-off.

• Use a box end wrench, or, if room permits, a socket on the bolts. The better grip of a box end wrench or socket will help prevent rounding off the bolt head(s).

• If you round off a bolt head, you'll have to try using Vise-Grips® (or equivalent), one of those wrenches designed for rounded-off bolts (space permitting), a nut splitter (again, space permitting), or grind off the bolt head.

• If the brake line won't budge, you fear kinking or twisting the line, or you rounded off the flare nut, try this: remove the wheel cylinder bolts or clips and pull the wheel cylinder, line attached, away from the backing plate. Usually, there is enough play in the brake line. Hold the flare nut with Vise-Grips® or equivalent, and try turning the wheel cylinder. The wheel cylinder gives you greater mechanical advantage than the flare nut. If nothing works, disconnect the line at the junction box. You'll have to install a new line.

Bolt-on Type

✳✳ CAUTION

It is always a good idea to wear eye protection when working on brake components, especially drum brakes. Drum brakes often use powerful springs which could cause severe eye injury if they accidentally break. Also, brake shoes may contain asbestos, which is a known cancer-causing agent. As soon as the drum is removed, generously spray the entire brake assembly with brake parts cleaner. Let it dry before proceeding. It's a good idea to wear a filter mask when doing brake work.

1. Loosen the lug nuts on the applicable wheels.
2. Raise and safely support the vehicle.
3. Remove the wheels.
4. Remove the drum.
5. Remove the brake shoes.

➡**On some vehicles, it may be possible to just remove the return springs and pull the shoes apart far enough for wheel cylinder removal. We do not recommend this for two reasons: wheel cylinder removal involves spilling some brake fluid—brake fluid can contaminate brake shoe friction material—and leaving the brake shoes on the backing plate can reduce working space and interfere with the job.**

6. Loosen the brake fluid line fitting, then separate the line from the wheel cylinder.

✳✳ CAUTION

Plug the line immediately to prevent contamination of the brake fluid, because brake fluid absorbs water from the atmosphere very quickly. Water reduces the effectiveness of brake fluid, leading to increased brake fade.

7. Remove the wheel cylinder bolts, and separate the cylinder from the backing plate.

To install:

8. Clean the backing plate thoroughly.

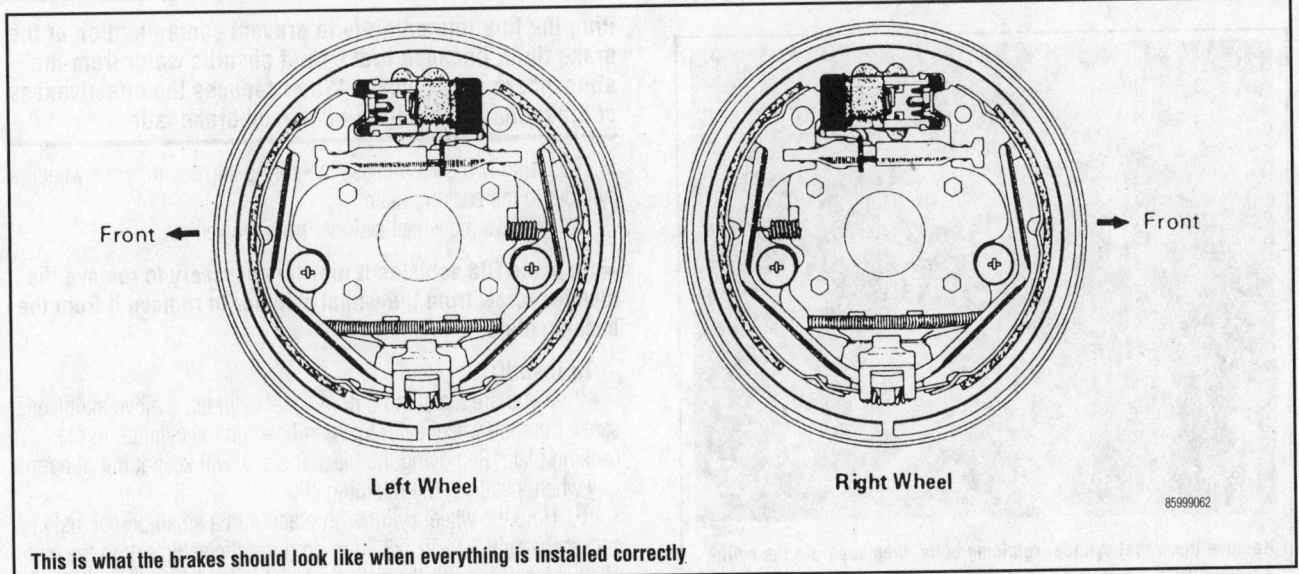

Front ← **Left Wheel**   → Front **Right Wheel**

85999062

This is what the brakes should look like when everything is installed correctly

Use a flare nut wrench to loosen the brake line fitting from the inboard side of the wheel cylinder

When the brake line is disconnected there will be some fluid leakage—plug the line to avoid contamination

Remove the wheel cylinder retaining bolts, then separate the cylinder from the backing plate—bolt-on type

9. Apply a very thin coating of RTV silicone sealer to the cylinder mounting surface. This will aid in keeping moisture and dirt out of the brakes.

10. Position the cylinder on the backing plate, then install the retaining bolts.

11. Reattach the brake line to the wheel cylinder.

12. Install the brake shoes.

13. Install the drum.

14. Bleed the brake system.

15. Adjust the brake shoes.

16. Install the wheels and tighten the lug nuts.

Spring Clip Type

✳✳ CAUTION

It is always a good idea to wear eye protection when working on brake components, especially drum brakes. Drum brakes often use powerful springs which could cause severe eye injury if they accidentally break. Also, brake shoes may contain asbestos, which is a known cancer-causing agent. As soon as the drum is removed, generously spray the entire brake assembly with brake parts cleaner. Let it dry before proceeding. It's a good idea to wear a filter mask when doing brake work.

1. Loosen the lug nuts on the applicable wheels.
2. Raise and safely support the vehicle.
3. Remove the wheels.
4. Remove the brake drum.
5. Remove the brake shoes.

➡On some vehicles, it may be possible to just remove the return springs and pull the shoes apart far enough for wheel cylinder removal. We do not recommend this for two reasons: wheel cylinder removal involves spilling some brake fluid—brake fluid can contaminate brake shoe friction material—and leaving the brake shoes on the backing plate can reduce working space and interfere with the job.

6. Disconnect and cap the brake line at the wheel cylinder.

✳✳ CAUTION

Plug the line immediately to prevent contamination of the brake fluid, because brake fluid absorbs water from the atmosphere very quickly. Water reduces the effectiveness of brake fluid, leading to increased brake fade.

7. Using two awls, release the spring clip securing the wheel cylinder to the backing plate.

8. Remove the wheel cylinder from the vehicle.

➡On some GM vehicles it may be necessary to remove the bleeder screw from the wheel cylinder to remove it from the backing plate.

To install:

9. If you are installing a new wheel cylinder, remove the bleeder screw from the wheel cylinder, then position the cylinder in the backing plate. Removing the bleeder screw will keep it out of harm's way when installing the retaining clip.

10. Hold the wheel cylinder in place with a small prybar, using a socket (usually 1 ⅛ in./28.5mm on domestic vehicles) on the end of an extension, push the spring clip into place. Be sure both spring clip ears are seated correctly.

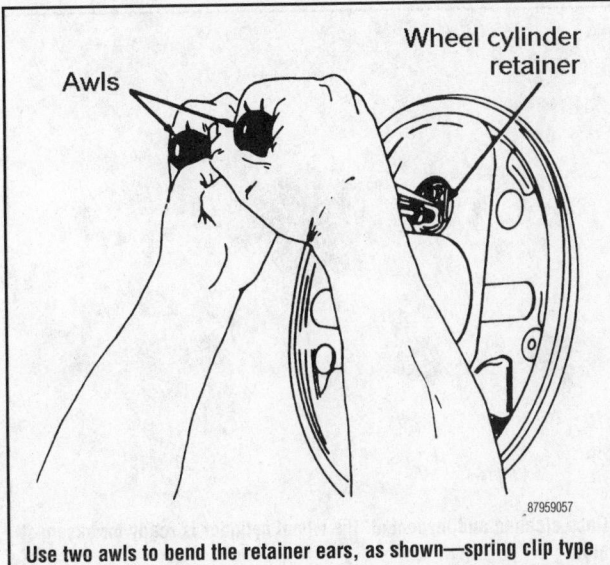

Use two awls to bend the retainer ears, as shown—spring clip type

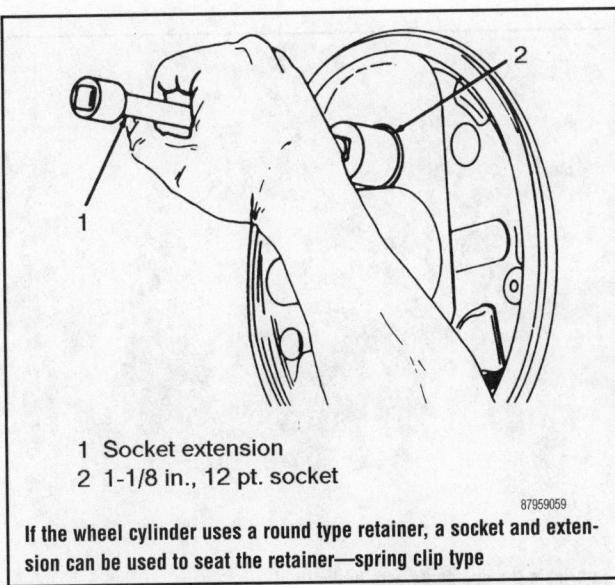

1 Socket extension
2 1-1/8 in., 12 pt. socket

If the wheel cylinder uses a round type retainer, a socket and extension can be used to seat the retainer—spring clip type

11. Connect the brake line to the wheel cylinder.
12. Install the bleeder screw and temporarily tighten it.
13. Install the brake shoes.
14. Install the brake drum.
15. Bleed the brake system.
16. Adjust the brake shoes.
17. Install the wheels and tighten the lug nuts.

OVERHAUL

Wheel cylinders can be overhauled, although most people do not bother. Replacing the wheel cylinder is much easier and requires no special tools or experience. If the cost difference between a rebuilding kit and new cylinder is not great, it's much safer to install the new cylinder.

If you decide to overhaul your wheel cylinder(s), you will need a wheel cylinder hone and a rebuild parts kit.

➡It is possible to rebuild the wheel cylinder while still in place on the backing plate. There is no good reason to do so other than that, for some reason, you can't remove the cylin-

der. If you choose to do this, it is of the UTMOST importance that all material be flushed out of the bore before installing new parts. We DO NOT recommend rebuilding a wheel cylinder while it is installed on the backing plate.

1. Remove the old wheel cylinder.
2. Thoroughly clean the outside of the unit with brake parts cleaner.
3. Place the cylinder on a clean work surface.
4. Remove the boots, then use a finger to push the pistons, cups and spring out of the bore.
5. Inspect the inner bore surface. If it is not badly pitted, rusted or scored, it can be rebuilt.
6. Remove the bleeder screw.
7. Install a wheel cylinder hone into a low-speed drill, and coat the inside of the cylinder with clean brake fluid.
8. Make several passes through the cylinder bore with the hone, never stopping in one place or passing completely through the bore.
9. Remove just enough material to establish a clean, cross-hatched inner surface.
10. Thoroughly clean the wheel cylinder bore with alcohol and

To disassemble the wheel cylinder, first remove the outer boots . . .

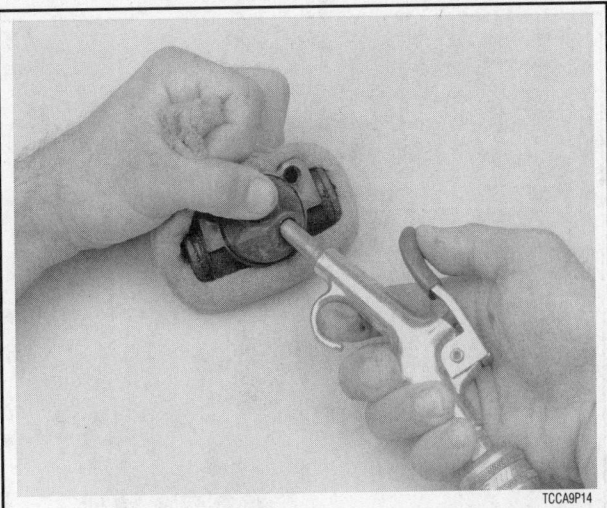

. . . , then carefully apply compressed air to the bleeder valve hole to extract the pistons and seals

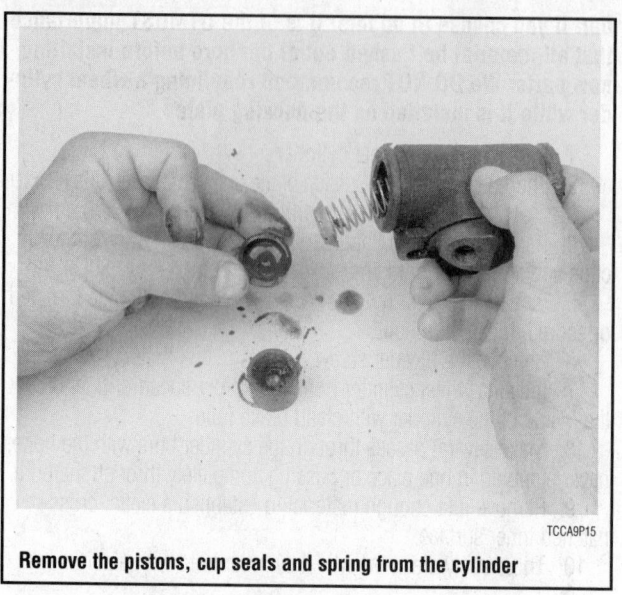

TCCA9P15

Remove the pistons, cup seals and spring from the cylinder

TCCA9P18

Once cleaned and inspected, the wheel cylinder is ready for assembly

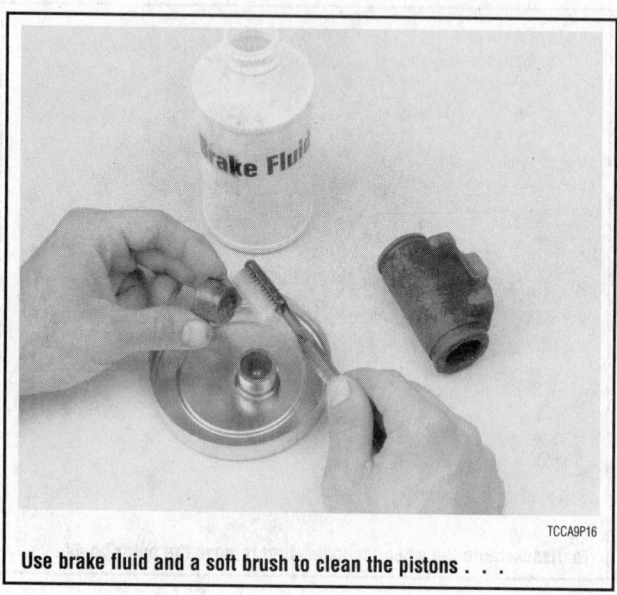

TCCA9P16

Use brake fluid and a soft brush to clean the pistons . . .

TCCA9P19

Lubricate the cup seals with brake fluid . . .

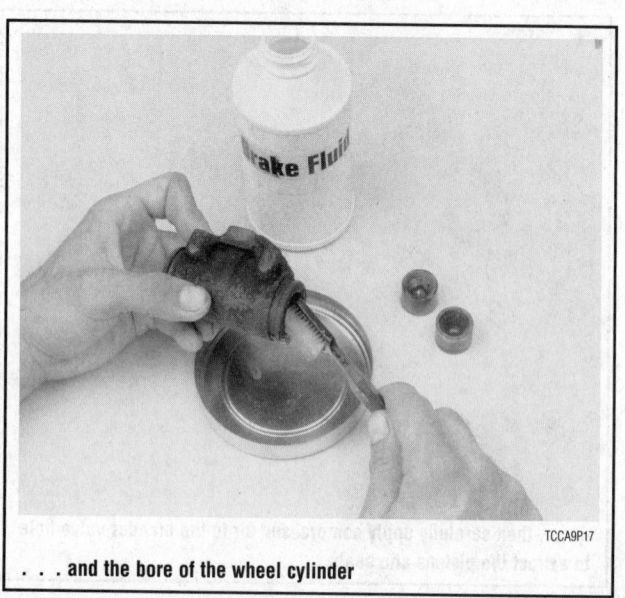

TCCA9P17

. . . and the bore of the wheel cylinder

TCCA9P20

. . . , then install the spring and cup seals in the bore

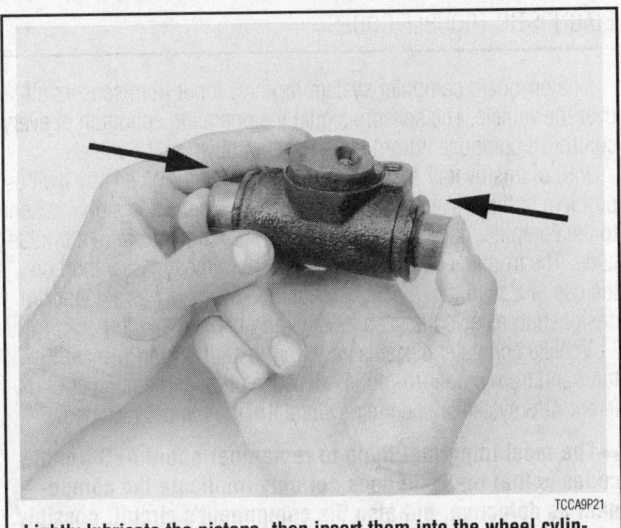

Lightly lubricate the pistons, then insert them into the wheel cylinder bore

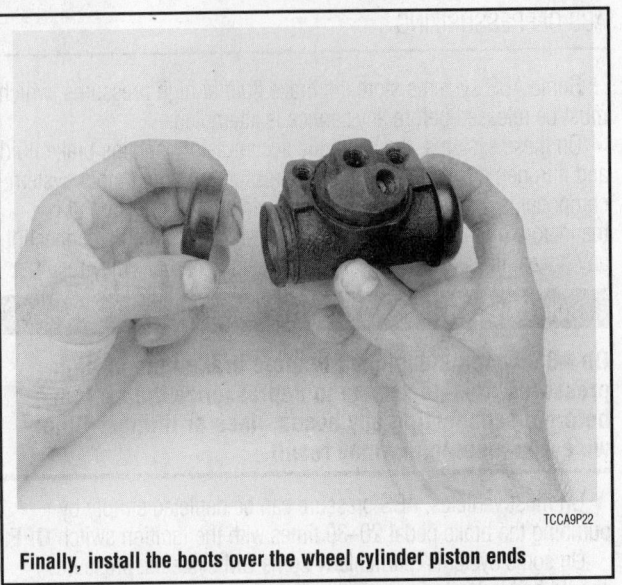

Finally, install the boots over the wheel cylinder piston ends

let it dry. Blow out all passages with compressed air, including the bleeder screw area.

11. Coat the bore with clean brake fluid.

✳✳ WARNING

Be sure to use all of the replacement parts which come with the rebuild kit you purchased, otherwise the rebuilt wheel cylinder may not function properly.

ANTI-LOCK BRAKE SYSTEMS

General Information

The purpose of the Anti-lock Brake System (ABS) is to prevent wheel lock-up under hard braking conditions. This is especially critical on wet or slippery surfaces. ABS is desirable because a vehicle that is stopped without locking one or more wheels, can stop with more control and in a shorter distance than a vehicle with locked wheels.

Under normal braking conditions, the ABS system operates just like a standard system. When one or more wheels shows a tendency to lock during braking, the ABS computer detects this and puts the system into the anti-lock mode. In this mode, hydraulic pressure is modulated to each wheel, preventing any one wheel from locking. The system can hold or reduce pressure at each wheel as necessary, depending on the signal received by the computer.

The effect is sort of like pumping your brakes, although it's done hundreds of time faster. In fact, when driving an ABS vehicle on ice or snow, a driver must overcome the urge to pump the brake during a stop. Let the ABS system work. Pumping the pedal on an ABS equipped vehicle will defeat the system.

PRECAUTIONS

• Do not use rubber hoses or other parts not specifically designed for the ABS system used by your vehicle. When using repair kits, replace all parts included in the kit. Partial or incorrect repair may lead to functional problems and require the replacement of components. NEVER fabricate your own replacement parts!

12. Coat all replacement parts with clean brake fluid.

13. Install a cup and piston in one side, place the spring into the other side, followed by the other cup and piston. Push the pistons in until both are within the bore.

14. Install the end caps.

15. Loosely install the bleeder screw.

16. Install the rebuilt wheel cylinder.

• Lubricate rubber parts with clean, fresh brake fluid to ease assembly. Do not use lubricated shop air to clean parts; damage to rubber components may result.

• Use only specified brake fluid from an unopened container.

• If any hydraulic component or line is removed or replaced, it may be necessary to bleed the entire system. This is always true when any upper end component (master cylinder, accumulator, control unit, etc.) is opened. It is also true when any lower end component (caliper or wheel cylinder) is opened and too much brake fluid has been lost; this does not happen often. If simply servicing a brake caliper, wheel cylinder, etc. and the line was adequately plugged after it was disconnected, the entire system will not need bleeding; only the component which was serviced. However, when in doubt, play it safe and bleed the entire system.

• A clean repair area is essential. Always clean the reservoir and cap thoroughly before removing the cap. The slightest amount of dirt in the fluid may plug an orifice and impair system function. Perform repairs after components have been thoroughly cleaned; use only denatured alcohol to clean components. Do not allow ABS components to come into contact with any substance containing mineral oil; this includes used shop rags.

• The anti-lock control unit is a microprocessor similar to other computer units in the vehicle. Ensure that the ignition switch is **OFF** before removing or installing controller wiring harnesses. Avoid static electricity discharge at or near the controller.

• If any arc welding is to be done on the vehicle, the control unit should be unplugged before welding operations begin.

ABS DEPRESSURIZING

Some ABS systems store the brake fluid at high pressures, which must be released before any service is attempted.

On these systems, the hydraulic accumulator contains brake fluid and nitrogen gas at extremely high pressures. Certain other system components may also contain brake fluid at high pressure. It is mandatory that the system pressure is relieved before disconnecting any hoses, lines or fittings, otherwise personal injury may result.

✵ CAUTION

On ABS systems designed to store brake fluid at high pressures, it is necessary to depressurize the system before disconnecting any hoses, lines or fittings. Otherwise, personal injury may result.

On most vehicles, ABS pressure can be depleted simply by pumping the brake pedal 20–30 times with the ignition switch **OFF**.

On some systems, particularly some GM systems, pressure should be bled using a specific, expensive scan tool. For this reason, we recommend that when in doubt, all ABS system service be referred to a professional, qualified technician.

DIAGNOSTIC TROUBLE CODES

The on-board computer system receives input from sensors all over the vehicle. The sensors signal the operating condition of every controlled component from the engine on down to the wheels.

Part of this overall system is the brake system. When any fault or problem in the brake system is detected by a sensor, a signal is sent to the computer and recorded in its memory in the form of a trouble code. The trouble codes can be accessed, in most cases, through the use of a scan tool. Each ABS equipped vehicle has a connector designed to receive the scan tools wiring harness plug(s).

Vehicle computer systems vary from manufacturer-to-manufacturer and from model-to-model. Because of the large number of different ABS systems, code retrieval information is not included.

➡**The most important thing to remember about ABS trouble codes is that the code does not only implicate the component as defective, but also the component's circuit, possibly, the diagnostic monitor computer. Always check the circuit for faults when diagnosing the brake system based on a trouble code.**

BRAKE SPECIFICATIONS
CHRYSLER 300M, CONCORD, LHS, NEW YORKER, DODGE INTREPID, EAGLE VISION
All measurements in inches unless noted

Year	Model		Master Cylinder Bore	Brake Disc Original Thickness	Brake Disc Minimum Thickness	Brake Disc Maximum Runout	Brake Drum Diameter Original Inside Diameter	Brake Drum Diameter Max. Wear Limit	Brake Drum Diameter Maximum Machine Diameter	Minimum Lining Thickness Front	Minimum Lining Thickness Rear	Brake Caliper Bracket Bolts (ft. lbs.)	Brake Caliper Mounting Bolts (ft. lbs.)
1995	Concorde	F	0.937	0.945	0.882	0.003	—	—	—	0.310	0.280	—	17
		R	—	0.468	0.409	0.003	8.00	①	①	—	0.280	—	17
	Intrepid	F	0.937	0.945	0.882	0.005	—	—	—	0.310	—	—	17
		R	—	0.468	0.409	0.003	8.00	①	①	—	0.280	—	17
	LHS	F	0.937	0.945	0.882	0.003	—	—	—	0.310	—	—	17
		R	—	0.468	0.409	0.003	—	—	—	—	0.280	—	17
	New Yorker	F	0.937	0.945	0.882	0.003	—	—	—	0.300	—	—	17
		R	—	0.468	0.409	0.003	—	—	—	—	0.280	—	17
	Vision	F	0.937	0.945	0.882	0.003	—	—	—	0.312	—	—	17
		R	—	0.468	0.409	0.003	8.00	①	—	—	0.281	—	17
1996	Concorde	F	0.874	0.945	0.882	0.003	—	—	—	0.310	—	—	17
		R	—	0.468	0.409	0.003	—	—	—	—	0.280	—	17
	Intrepid	F	0.937	0.945	0.882	0.005	—	—	—	0.310	—	—	17
		R	—	0.468	0.409	0.003	8.00	①	①	—	0.280	—	17
	LHS	F	0.874	0.945	0.882	0.003	—	—	—	0.310	—	—	17
		R	—	0.468	0.409	0.003	—	—	—	—	0.280	—	17
	Vision	F	0.937	0.945	0.882	0.003	—	—	—	0.312	—	—	17
		R	—	0.468	0.409	0.003	8.00	①	—	—	0.281	—	17
1997	Concorde	F	0.874	0.945	0.882	0.003	—	—	—	0.310	—	—	17
		R	—	0.468	0.409	0.003	—	—	—	—	0.280	—	17
	Intrepid	F	0.937	0.945	0.882	0.005	—	—	—	0.310	—	—	17
		R	—	0.468	0.409	0.003	8.00	①	①	—	0.280	—	17
	LHS	F	0.874	0.945	0.882	0.003	—	—	—	0.310	—	—	17
		R	—	0.468	0.409	0.003	—	—	—	—	0.280	—	17
	Vision	F	0.937	0.945	0.882	0.003	—	—	—	0.312	—	—	17
		R	—	0.468	0.409	0.003	8.00	①	—	—	0.281	—	17
1998-99	300M	F	0.937	1.019	0.882	0.003	—	—	—	0.310	—	—	17
		R	—	0.468	0.409	0.003	—	—	—	—	0.281	—	17
	Concorde	F	0.937	1.019	0.882	0.003	—	—	—	0.310	—	—	17
		R	—	0.468	0.409	0.003	—	—	—	—	0.280	—	17
	Intrepid	F	0.937	1.019	0.882	0.005	—	—	—	0.310	—	—	17
		R	—	0.468	0.409	0.003	—	—	—	—	0.280	—	17
	LHS	F	0.937	1.019	0.882	0.003	—	—	—	0.310	—	—	17
		R	—	0.468	0.409	0.003	—	—	—	—	0.280	—	17
	Vision	F	0.937	1.019	0.882	0.003	—	—	—	0.310	—	—	17
		R	—	0.468	0.409	0.003	—	—	—	—	0.281	—	17

F -Front
R - Rear
① Maximum diameter is stamped on drum

79226C50

BRAKE SPECIFICATIONS
CHRYSLER CIRRUS, SEBRING CONVERTIBLE, DODGE STRATUS, PLYMOUTH BREEZE
All measurements in inches unless noted

Year	Model	Master Cylinder Bore	Brake Disc Original Thickness	Brake Disc Minimum Thickness	Brake Disc Maximum Runout	Brake Drum Diameter Original Inside Diameter	Brake Drum Diameter Max. Wear Limit	Brake Drum Diameter Maximum Machine Diameter	Minimum Lining Thickness Front	Minimum Lining Thickness Rear	Brake Caliper Bracket Bolts (ft. lbs.)	Brake Caliper Mounting Bolts (ft. lbs.)
1995	Cirrus	0.874	0.911	0.843	0.005	7.88	①	①	0.035	0.062	—	16
	Stratus	0.875	0.911	0.843	0.005	7.88	①	①	0.035	0.062	—	16
1996	Breeze	0.875	0.911	0.843	0.003	7.88	①	①	0.035	0.062	—	16
	Cirrus	0.874	0.911	0.843	0.005	7.88	①	①	0.035	0.062	—	16
	Sebring Convertible	0.874	0.911	0.843	0.005	8.66	①	①	0.035	0.062	—	16
	Stratus	0.875	0.911	0.843	0.003	7.88	①	①	0.035	0.062	—	16
1997	Breeze	0.875	0.911	0.843	0.003	7.88	①	①	0.035	0.062	—	16
	Cirrus	0.874	0.911	0.843	0.005	7.88	①	①	0.035	0.062	—	16
	Sebring Convertible	0.874	0.911	0.843	0.005	8.66	①	①	0.035	0.062	—	16
	Stratus	0.875	0.911	0.843	0.003	7.88	①	①	0.035	0.062	—	16
1998-99	Breeze	0.875	0.911	0.843	0.003	7.88	①	①	0.035	0.062	—	16
	Cirrus	0.874	0.911	0.843	0.005	7.88	①	①	0.035	0.062	—	16
	Sebring Convertible	0.874	0.911	0.843	0.005	8.66	①	①	0.035	0.062	—	16
	Stratus	0.875	0.911	0.843	0.003	7.88	①	①	0.035	0.062	—	16

① Maximum diameter is stamped on drum

79226C51

BRAKE SPECIFICATIONS
CHRYSLER LEBARON, DODGE SPIRIRT, PLYMOUTH ACCLAIM
All measurements in inches unless noted

Year	Model		Master Cylinder Bore	Brake Disc Original Thickness	Brake Disc Minimum Thickness	Brake Disc Maximum Runout	Brake Drum Diameter Original Inside Diameter	Brake Drum Diameter Max. Wear Limit	Brake Drum Diameter Maximum Machine Diameter	Minimum Lining Thickness Front	Minimum Lining Thickness Rear	Brake Caliper Bracket Bolts (ft. lbs.)	Brake Caliper Mounting Bolts (ft. lbs.)
1995	Acclaim	F	0.827	0.935	0.882	0.005	10.24	NA	NA	0.300	0.300	165	25
	①	R	—	0.468	0.409	0.005	—	—	—	—	0.280	—	16
	LeBaron	F	0.827	0.930	0.882	0.005	—	—	—	0.300	0.062	165	25
	②	R	—	0.467	0.409	0.005	8.66	③	③	—	0.280	—	16
	④	R	—	0.856	0.797	0.005	—	—	—	—	0.280	—	16
	Spirit	F	0.827	0.935	0.882	0.005	—	—	—	0.300	0.300	165	25
		R	①	0.468	0.409	0.005	10.24	③	③	—	0.280	—	16

F - Front
R - Rear
① Optional vented rear disc brakes:
   Original thickness: 0.856
   Minimum thickness: 0.797
② Solid rear disc
③ Maximum diameter is stamped on drum
④ Vented rear disc

79226C52

BRAKE SPECIFICATIONS
CHRYSLER SEBRING COUPE, DODGE AVENGER
All measurements in inches unless noted

| Year | Model | | Master Cylinder Bore | Brake Disc | | | Brake Drum Diameter | | | Minimum Lining Thickness | | Brake Caliper | |
				Original Thickness	Minimum Thickness	Maximum Runout	Original Inside Diameter	Max. Wear Limit	Maximum Machine Diameter	Front	Rear	Bracket Bolts (ft. lbs.)	Mounting Bolts (ft. lbs.)
1995	Avenger	F	0.937	0.940	0.880	0.003	—	—	—	0.080	—	65	54
		R	—	0.390	0.330	0.003	9.00	①	①	—	0.080	—	38
	Sebring	F	0.937 ②	0.940	0.880	0.003	—	—	—	0.080	—	65	54
		R ③	—	0.390	0.330	0.003	④	①	①	—	⑤	—	38
1996	Avenger	F	0.937	0.940	0.880	0.003	—	—	—	0.080	—	65	54
		R	—	0.390	0.330	0.003	9.00	①	①	—	0.080	—	38
	Sebring Coupe	F	1.000 ⑥	0.940	0.880	0.003	—	—	—	0.080	—	65	54
		R ③	—	0.390 ⑦	0.330 ⑧	0.003	④	①	①	—	⑤	—	38
1997	Avenger	F	0.937	0.940	0.880	0.003	—	—	—	0.080	—	65	54
		R	—	0.390	0.330	0.003	9.00	①	①	—	0.080	—	38
	Sebring Coupe	F	1.000 ⑥	0.940	0.880	0.003	—	—	—	0.080	—	65	54
		R ③	—	0.390 ⑦	0.330 ⑧	0.003	④	①	①	—	⑤	—	38
1998-99	Avenger	F	0.937	0.940	0.880	0.003	—	—	—	0.080	—	65	54
		R	—	0.390	0.330	0.003	9.00	①	①	—	0.080	—	38
	Sebring Coupe	F	1.000 ⑥	0.940	0.880	0.003	—	—	—	0.080	—	66	54
		R ③	—	0.390 ⑦	0.330 ⑧	0.003	④	①	①	—	⑤	—	38

F - Front
R - Rear
① Maximum diameter is stamped on drum
② Equipped with ABS: 1.000
③ Solid rear disc
④ Minimum diameter: 9.00 in.
⑤ .039 in. with drum brakes
　.080 in. with disc brakes
⑥ Equipped with standard ABS
⑦ Vented rear disc: 0.790
⑧ Vented rear disc: 0.720

79226C53

BRAKE SPECIFICATIONS
DODGE/PLYMOUTH NEON
All measurements in inches unless noted

| Year | Model | | Master Cylinder Bore | Brake Disc | | | Brake Drum Diameter | | | Minimum Lining Thickness | | Brake Caliper | |
				Original Thickness	Minimum Thickness	Maximum Runout	Original Inside Diameter	Max. Wear Limit	Maximum Machine Diameter	Front	Rear	Bracket Bolts (ft. lbs.)	Mounting Bolts (ft. lbs.)
1995	Neon	F	①	0.792	0.724	0.005	—	—	—	0.300	—	—	16
		R	—	NA	NA	NA	7.88	NA	NA	—	②	55	16
1996	Neon	F	①	0.792	0.724	0.005	—	—	—	0.300	—	—	16
		R	—	NA	NA	NA	7.88	NA	NA	—	②	55	16
1997	Neon	F	①	0.792	0.724	0.005	—	—	—	0.300	—	—	16
		R	—	NA	NA	NA	7.88	NA	NA	—	②	55	16
1998-99	Neon	F	①	0.792	0.724	0.005	—	—	—	0.300	—	—	16
		R	—	NA	NA	NA	7.88	NA	NA	—	②	55	16

NA - Not Available
① If equipped with rear drum brakes: 0.827
　If equipped with rear disc brakes: 0.875
② Rear disc: NA
　Rear drum: 0.280

79226C54

BRAKE SPECIFICATIONS
DODGE STEALTH
All measurements in inches unless noted

Year	Model		Master Cylinder Bore	Brake Disc Original Thickness	Brake Disc Minimum Thickness	Brake Disc Maximum Runout	Minimum Lining Thickness Front	Minimum Lining Thickness Rear	Brake Caliper-to-Bracket (ft. lbs.)	Brake Caliper-to-Knuckle (ft. lbs.)
1995	Stealth	F	①	0.940	0.880	0.003	0.080	—	54	65
		R	—	0.710	0.650	0.003	—	0.080	—	36-43
	Stealth AWD	F	1.063	1.180	1.120	0.003	0.080	—	—	NA
		R	—	0.790	0.720	0.003	—	0.080	—	36-43
1996	Stealth	F	①	0.940	0.880	0.003	0.080	—	54	65
		R	—	0.710	0.650	0.003	—	0.080	—	36-43
	Stealth AWD	F	1.063	1.180	1.120	0.003	0.080	—	—	NA
		R	—	0.790	0.720	0.003	—	0.080	—	36-43

① With ABS: 1.000
Without ABS: 1.063

79226C90

BRAKE SPECIFICATIONS
EAGLE TALON
All measurements in inches unless noted

Year	Model		Master Cylinder Bore	Brake Disc Original Thickness	Brake Disc Minimum Thickness	Brake Disc Maximum Runout	Brake Drum Diameter Original Inside Diameter	Brake Drum Diameter Max. Wear Limit	Brake Drum Diameter Maximum Machine Diameter	Minimum Lining Thickness Front	Minimum Lining Thickness Rear	Bake Caliper Bracket Bolts (ft.bs.)	Bake Caliper Mounting Bolts (ft.bs.)
1995	Talon	F	①	②	③	0.003	—	—	—	0.080	0.080	65	54
		R	—	—	—	0.003	—	—	—	0.080	0.080	—	38
1996	Talon	F	①	0.940	0.882	0.003	—	—	—	0.080	—	65	54
		R	①	0.390	0.331	0.003	—	—	—	—	0.080	—	38
1997	Talon	F	①	0.940	0.882	0.003	—	—	—	0.080	—	65	54
		R	①	0.390	0.331	0.003	—	—	—	—	0.080	—	38
1998	Talon	F	①	0.940	0.882	0.003	—	—	—	0.080	—	65	54
		R	①	0.390	0.331	0.003	—	—	—	—	0.080	—	38

① Non-turbocharged: 0.938 ② Front: 0.940 ③ Front: 0.882
FWD turbocharged: 0.938   Rear: 0.390   Rear: 0.331
AWD turbocharged: 1.000   AWD: 0.790   AWD: 0.721

79226C55

BRAKE SPECIFICATIONS
FORD ASPIRE
All measurements in inches unless noted

Year	Model			Master Cylinder Bore	Brake Disc Original Thickness	Brake Disc Minimum Thickness	Brake Disc Maximum Runout	Brake Drum Diameter Original Inside Diameter	Brake Drum Diameter Max. Wear Limit	Brake Drum Diameter Maximum Machine Diameter	Minimum Lining Thickness Front	Minimum Lining Thickness Rear	Brake Caliper Bracket Bolts (ft. lbs.)	Brake Caliper Mounting Bolts (ft. lbs.)
1995	Aspire	①	F	②	0.710	0.630	0.004	7.87	7.93	—	0.080	0.040	—	29-36
		③	R	②	0.860	0.780	0.004	7.87	7.93	—	0.080	0.040	—	29-36
1996	Aspire	①	F	②	0.710	0.630	0.004	7.87	7.93	—	0.080	0.040	—	29-36
		③	R	②	0.860	0.780	0.004	7.87	7.93	—	0.080	0.040	—	29-36
1997	Aspire	①	F	②	0.710	0.630	0.004	7.87	7.93	—	0.080	0.040	—	29-36
		③	R	②	0.860	0.780	0.004	7.87	7.93	—	0.080	0.040	—	29-36

NOTE: Follow specifications stamped on rotor or drum if figures differ from those in this chart.
NA - Not Available
F - Front
R - Rear

① Manual transaxle
② Without ABS: 0.810
   With ABS: 0.870
③ Automatic transaxle

79226C56

BRAKE SPECIFICATIONS
FORD CONTOUR, MERCURY MYSTIQUE, COUGAR (1999)
All measurements in inches unless noted

Year	Model		Master Cylinder Bore	Brake Disc Original Thickness	Brake Disc Minimum Thickness	Brake Disc Maximum Runout	Brake Drum Diameter Original Inside Diameter	Brake Drum Diameter Max. Wear Limit	Brake Drum Diameter Maximum Machine Diameter	Minimum Lining Thickness Front	Minimum Lining Thickness Rear	Brake Caliper Bracket Bolts (ft. lbs.)	Brake Caliper Mounting Bolts (ft. lbs.)
1995	Contour	F	NA	0.950	0.870	0.006	—	—	—	0.125	—	—	20
		R	—	0.790	0.710	0.006	8.00	—	—	—	0.125	—	30
	Mystique	F	NA	0.950	0.870	0.006	—	—	—	0.125	—	—	20
		R	—	0.790	0.710	0.006	8.00	—	—	—	0.125	—	30
1996	Contour	F	NA	0.950	0.870	0.006	—	—	—	0.125	—	—	20
		R	—	0.790	0.710	0.006	8.00	NA	8.04	—	0.125	—	30
	Mystique	F	NA	0.950	0.870	0.006	—	—	—	0.125	—	—	20
		R	—	0.790	0.710	0.006	8.00	NA	8.04	—	0.125	—	30
1997	Contour	F	NA	0.950	0.870	0.006	—	—	—	0.125	—	—	20
		R	—	0.790	0.710	0.006	8.00	NA	8.04	—	0.125	—	30
	Mystique	F	NA	0.950	0.870	0.006	—	—	—	0.125	—	—	20
		R	—	0.790	0.710	0.006	8.00	NA	8.04	—	0.125	—	30
1998-99	Contour	F	NA	0.950	0.870	0.006	—	—	—	0.125	—	—	20
		R	—	0.790	0.710	0.006	8.00	NA	8.04	—	0.125	—	30
	Cougar	F	NA	0.950	0.870	0.006	—	—	—	0.125	—	—	20
		R	—	0.790	0.710	0.006	8.00	NA	8.04	—	0.125	—	30
	Mystique	F	NA	0.950	0.870	0.006	—	—	—	0.125	—	—	20
		R	—	0.790	0.710	0.006	8.00	NA	8.04	—	0.125	—	30

NOTE: Follow specifications stamped on rotor or drum if figures differ from those in this chart.

NA - Not Available

F - Front

R - Rear

79226C57

BRAKE SPECIFICATIONS
FORD CROWN VICTORIA, LINCOLN TOWN CAR, MERCURY GRAND MARQUIS
All measurements in inches unless noted

Year	Model		Master Cylinder Bore	Brake Disc Original Thickness	Brake Disc Minimum Thickness	Brake Disc Maximum Runout	Brake Drum Diameter Original Inside Diameter	Brake Drum Diameter Max. Wear Limit	Brake Drum Diameter Maximum Machine Diameter	Minimum Lining Thickness Front	Minimum Lining Thickness Rear	Brake Caliper Bracket Bolts (ft. lbs.)	Brake Caliper Mounting Bolts (ft. lbs.)
1995	Crown Victoria	F	1.000	1.030	0.974	0.003	—	—	—	0.125	—	170-230	26
		R	—	0.500	0.440	0.003	—	—	—	—	0.125	—	26
	Grand Marquis	F	1.000	1.030	0.974	0.003	—	—	—	0.125	—	170-230	26
		R	—	0.500	0.440	0.003	—	—	—	—	0.125	—	26
	Town Car	F	1.000	1.030	0.974	0.003	—	—	—	0.125	—	170-230	26
		R	—	0.500	0.440	0.003	—	—	—	—	0.125	—	26
1996	Crown Victoria	F	1.000	1.024	0.974	0.002	—	—	—	0.125	—	170-230	21-26
		R	—	0.550	0.510	0.002	—	—	—	—	0.125	—	20
	Grand Marquis	F	1.000	1.024	0.974	0.002	—	—	—	0.125	—	170-230	21-26
		R	—	0.550	0.510	0.002	—	—	—	—	0.125	—	20
	Town Car	F	1.000	1.024	0.974	0.002	—	—	—	0.125	—	170-230	21-26
		R	—	0.550	0.510	0.002	—	—	—	—	0.125	—	20
1997	Crown Victoria	F	1.000	1.024	0.974	0.002	—	—	—	0.125	—	170-230	21-26
		R	—	0.550	0.510	0.002	—	—	—	—	0.125	—	20
	Grand Marquis	F	1.000	1.024	0.974	0.002	—	—	—	0.125	—	170-230	21-26
		R	—	0.550	0.510	0.002	—	—	—	—	0.125	—	20
	Town Car	F	1.000	1.024	0.974	0.002	—	—	—	0.125	—	170-230	21-26
		R	—	0.550	0.510	0.002	—	—	—	-	0.125	—	20
1998-99	Crown Victoria	F	1.000	1.024	0.974	0.002	—	—	—	0.125	—	170-230	21-26
		R	—	0.550	0.510	0.002	—	—	—	—	0.125	—	20
	Grand Marquis	F	1.000	1.024	0.974	0.002	—	—	—	0.125	—	170-230	21-26
		R	—	0.550	0.510	0.002	—	—	—	—	0.125	—	20
	Town Car	F	1.000	1.024	0.974	0.002	—	—	—	0.125	—	170-230	21-26
		R	—	0.550	0.510	0.002	—	—	—	—	0.125	—	20

NOTE: Follow specifications stamped on rotor or drum if figures differ from those in this chart.

F - Front

R - Rear

79226C58

BRAKE SPECIFICATIONS
FORD ESCORT, ESCORT ZX2, MERCURY TRACER
All measurements in inches unless noted

Year	Model		Master Cylinder Bore	Brake Disc Original Thickness	Brake Disc Minimum Thickness	Brake Disc Maximum Runout	Brake Drum Diameter Original Inside Diameter	Brake Drum Diameter Max. Wear Limit	Brake Drum Diameter Maximum Machine Diameter	Minimum Lining Thickness Front	Minimum Lining Thickness Rear	Brake Caliper Bracket Bolts (ft. lbs.)	Brake Caliper Mounting Bolts (ft. lbs.
1995	Escort	F	0.875	0.870	0.790	0.004	7.87	7.95	7.91	0.080	0.040	—	35
		R	—	0.350	0.280	0.004	—	—	—	—	0.040	—	—
	Tracer	F	0.875	0.870	0.790	0.004	—	—	—	0.080	—	—	35
		R	—	0.350	0.280	0.004	7.87	7.95	7.91	—	0.040	—	—
1996	Escort	F	0.875	0.870	0.790	0.004	—	—	—	0.080	—	—	35
		R	—	0.350	0.280	0.004	7.87	7.95	7.91	—	0.040	—	—
	Tracer	F	0.875	0.870	0.790	0.004	—	—	—	0.080	—	—	35
		R	—	0.350	0.280	0.004	7.87	7.95	7.91	—	0.040	—	—
1997	Escort	F	0.875	0.870	0.790	0.004	—	—	—	0.080	—	—	36-43
		R	—	0.350	0.280	0.004	7.87	7.95	7.91	—	0.040	—	—
	Tracer	F	0.875	0.870	0.790	0.004	—	—	—	0.080	—	—	36-43
		R	—	0.350	0.280	0.004	7.87	7.95	7.91	—	0.040	—	—
1998-99	Escort	F	0.875	0.870	0.790	0.004	—	—	—	0.080	—	—	36-43
		R	—	0.350	0.280	0.004	7.87	7.95	7.91	—	0.040	—	—
	Escort ZX2	F	0.875	0.870	0.790	0.004	—	—	—	0.080	-	—	36-43
		R	—	0.350	0.280	0.004	7.87	7.95	7.91	—	0.040	—	—
	Tracer	F	0.875	0.870	0.790	0.004	—	—	—	0.080	—	—	36-43
		R	—	0.350	0.280	0.004	7.87	7.95	7.91	—	0.040	—	—

NOTE: Follow specifications stamped on rotor or drum if figures differ from those in this chart.
NA - Not Available
F - Front
R - Rear

79226C59

BRAKE SPECIFICATIONS
FORD MARK VIII, MUSTANG, THUNDERBIRD, MERCURY COUGAR
All measurements in inches unless noted

Year	Model		Master Cylinder Bore	Brake Disc Original Thickness	Brake Disc Minimum Thickness	Brake Disc Maximum Runout	Brake Drum Diameter Original Inside Diameter	Brake Drum Diameter Max. Wear Limit	Brake Drum Diameter Maximum Machine Diameter	Minimum Lining Thickness Front	Minimum Lining Thickness Rear	Brake Caliper Bracket Bolts (ft. lbs.)	Brake Caliper Mounting Bolts (ft. lbs.)
1995	Cougar	F	0.938 ①	1.025	0.974	0.003	9.84	9.89	9.86	0.040	0.030	—	60
		R	—	0.709	0.657	0.003	—	—	—	—	0.123	—	26
	Mark VIII	F	NA	1.024	0.974	0.002	—	—	—	0.040	—	—	60
		R	—	0.709	0.657	0.002	—	—	—	—	0.123	—	26
	Mustang	F	1.060	1.030	0.970	0.002	—	—	—	0.040	—	—	64
	②	F	1.000	1.100	1.040	0.002	—	—	—	0.040	—	—	64
		R	—	0.550	0.500	0.002	—	—	—	—	0.123	—	26
	Thunderbird ③	F	0.938 ①	1.025	0.974	0.003	9.84	9.89	9.86	0.040	0.030	—	60
	④	R	—	0.709	0.657	0.003	—	—	—	—	0.123	—	26
1996	Cougar	F	0.938	1.025	0.974	0.002	—	—	—	0.125	—	—	60
		R	—	0.710	0.657	—	9.80	NA	9.90	—	0.125	—	26
	Mark VIII	F	1.000	1.024	0.974	0.003	—	—	—	0.125	—	—	60
		R	—	0.709	0.657	0.002	—	—	—	—	0.125	—	26
	Mustang ⑤	F	1.060	1.030	0.970	0.001	—	—	—	0.125	—	—	64
	⑥	F	1.000	1.030	0.970	0.002	—	—	—	0.125	—	—	64
	②	F	1.000	1.100	1.040	0.001	—	—	—	0.125	—	—	64
		R	—	0.550	0.500	0.002	—	—	—	—	0.123	—	26
	②	R	—	0.710	0.660	0.002	—	—	—	—	0.123	—	26
	Thunderbird	F	0.938	1.025	0.974	0.002	—	—	—	0.125	—	—	60
		R	—	0.710	0.657	—	9.80	NA	9.90	—	0.125	—	26
1997	Cougar	F	0.938	1.025	0.974	0.002	—	—	—	0.125	—	—	60
		R	—	0.710	0.657	—	9.80	NA	9.90	—	0.125	—	26
	Mark VIII	F	1.000	1.024	0.974	0.003	—	—	—	0.125	—	—	60
		R	—	0.709	0.657	0.002	—	—	—	—	0.125	—	26
	Mustang ⑤	F	1.060	1.030	0.970	0.001	—	—	—	0.125	—	—	64
	⑥	F	1.000	1.030	0.970	0.002	—	—	—	0.125	—	—	64
	②	F	1.000	1.100	1.040	0.001	—	—	—	0.125	—	—	64
		R	—	0.550	0.500	0.002	—	—	—	—	0.123	—	26
	②	R	—	0.710	0.660	0.002	—	—	—	—	0.123	—	26
	Thunderbird	F	0.938	1.025	0.974	0.002	—	—	—	0.125	—	—	60
		R	—	0.710	0.657	—	9.80	NA	9.90	—	0.125	—	26
1998-99	Mark VIII	F	1.000	1.024	0.974	0.003	—	—	—	0.125	—	—	60
		R	—	0.709	0.657	0.002	—	—	—	—	0.125	—	26
	Mustang ⑤	F	1.060	1.030	0.970	0.001	—	—	—	0.125	—	—	64
	⑥	F	1.000	1.030	0.970	0.002	—	—	—	0.125	—	—	64
	②	F	1.000	1.100	1.040	0.001	—	—	—	0.125	—	—	64
		R	—	0.550	0.500	0.002	—	—	—	—	0.123	—	26
	②	R	—	0.710	0.660	0.002	—	—	—	—	0.123	—	26

NOTE: Follow specifications stamped on rotor or drum if figures differ from those in this chart.

NA - Not Available

F - Front

R - Rear

① Except ABS
   With drum brakes: 0.030

② Cobra

③ Except rear disc

④ With rear disc

⑤ 3.8L engine

⑥ 4.6L except Cobra

79226C60

BRAKE SPECIFICATIONS
FORD PROBE
All measurements in inches unless noted

Year	Model		Master Cylinder Bore	Brake Disc Original Thickness	Brake Disc Minimum Thickness	Brake Disc Maximum Runout	Brake Drum Diameter Original Inside Diameter	Brake Drum Diameter Max. Wear Limit	Brake Drum Diameter Maximum Machine Diameter	Minimum Lining Thickness Front	Minimum Lining Thickness Rear	Brake Caliper Bracket Bolts (ft. lbs.)	Brake Caliper Mounting Bolts (ft. lbs.)
1995	Probe	F	0.937	0.890	0.860	0.004	9.00	9.06	9.86	0.040	0.040	58-74	33-36
		R	—	0.390	0.315	0.004	—	—	—	—	0.040	33-49	25-29
1996	Probe	F	0.937	0.890	0.860	0.004	—	—	—	0.040	—	58-74	33-36
		R	—	0.345	0.315	0.004	—	—	—	—	0.040	33-49	25-29
1997	Probe	F	0.937	0.890	0.860	0.004	—	—	—	0.040	—	58-74	33-36
		R	—	0.345	0.315	0.004	—	—	—	—	0.040	33-49	25-29

NOTE: Follow specifications stamped on rotor or drum if figures differ from those in this chart.

NA - Not Available

F - Front

R - Rear

79226C61

BRAKE SPECIFICATIONS
FORD TAURUS, LINCOLN CONTINENTAL, MERCURY SABLE
All measurements in inches unless noted

Year	Model		Master Cylinder Bore	Brake Disc Original Thickness	Brake Disc Minimum Thickness	Brake Disc Maximum Runout	Brake Drum Diameter Original Inside Diameter	Brake Drum Diameter Max. Wear Limit	Brake Drum Diameter Maximum Machine Diameter	Minimum Lining Thickness Front	Minimum Lining Thickness Rear	Brake Caliper Bracket Bolts (ft. lbs.)	Brake Caliper Mounting Bolts (ft. lbs.)
1995	Continental	F	NA	1.020	0.974	0.003	—	—	—	0.060	—	—	25
		R	—	0.550	0.502	0.001	—	—	—	—	0.123	—	25
	Sable		1.000	①	②	③	④	NA	⑤	0.040	⑥	65-85	25
	Taurus		1.000	①	②	③	④	NA	⑤	0.040	⑥	65-85	25
	Taurus SHO		1.000	①	②	③	—	—	—	0.040	0.123	65-85	25
1996	Continental	F	NA	1.020	0.974	0.003	—	—	—	0.060	—	—	25
		R	—	0.550	0.502	0.001	—	—	—	—	0.130	—	25
	Sable	F	1.000	1.020	0.974	0.002	—	—	—	0.040	—	65-85	23-28
		R	—	0.940	0.500	0.002	④	NA	⑤	—	⑦	65-87	23-25
	Taurus	F	1.000	1.020	0.974	0.002	—	—	—	0.040	—	65-85	23-28
		R	—	0.940	0.500	0.002	④	NA	⑤	—	⑦	65-87	23-25
1997	Continental	F	NA	1.020	0.974	0.003	—	—	—	0.060	—	—	25
		R	—	0.550	0.502	0.001	—	—	—	—	0.130	—	25
	Sable	F	1.000	1.020	0.974	0.002	—	—	—	0.040	—	65-85	23-28
		R	—	0.940	0.500	0.002	④	NA	⑤	—	⑦	65-87	23-25
	Taurus	F	1.000	1.020	0.974	0.002	—	—	—	0.040	—	65-85	23-28
		R	—	0.940	0.500	0.002	④	NA	⑤	—	⑦	65-87	23-25
1998-99	Continental	F	NA	1.020	0.974	0.003	—	—	—	0.060	—	—	25
		R	—	0.550	0.502	0.001	—	—	—	—	0.130	—	25
	Sable	F	1.000	1.020	0.974	0.002	—	—	—	0.040	—	65-85	23-28
		R	—	0.940	0.500	0.002	④	NA	⑤	—	⑦	65-87	23-25
	Taurus	F	1.000	1.020	0.974	0.002	—	—	—	0.040	—	65-85	23-28
		R	—	0.940	0.500	0.002	④	NA	⑤	—	⑦	65-87	23-25

NOTE: Follow specifications stamped on rotor or drum if figures differ from those in this chart.

NA - Not Available

F - Front

R - Rear

① Front: 0.003
　Rear: 0.002

② Sedan: 8.85
　Wagon: 9.84

③ Sedan: 8.91
　Wagon: 9.90

④ Front: 1.020
　Rear: 0.940

⑤ Front: 0.974
　Rear: 0.500

⑥ With disc brakes: 0.123
　With drum brakes: 0.030

⑦ Riveted lining: 0.031
　Bonded lining: 0.125

79226C62

BRAKE SPECIFICATIONS
GM A BODY
All measurements in inches unless noted

| Year | Model | Master Cylinder Bore | Brake Disc | | | Brake Drum Diameter | | | Minimum Lining Thickness | | Brake Caliper | |
			Original Thickness	Minimum Thickness	Maximum Runout	Original Inside Diameter	Max. Wear Limit	Maximum Machine Diameter	Front	Rear	Bracket Bolts (ft. lbs.)	Mounting Bolts (ft. lbs.)
1995	Century	0.944	1.028	0.957	0.002	8.863	8.909	8.880	0.030	②	—	③
	Cutlass Ciera	0.944	1.028	0.957	0.002	8.863	8.909	8.880	0.030	②	—	③
	Cutlass Cruiser	0.944	1.028	0.957	0.002	8.863	①	8.880	0.030	②	—	③
1996	Century	0.944	1.028	0.957	0.002	8.863	8.909	8.920	0.030	②	—	③
	Cutlass Ciera	0.944	1.028	0.957	0.002	8.863	8.909	8.920	0.030	②	—	③
	Cutlass Cruiser	0.944	1.028	0.957	0.002	8.863	8.909	8.920	0.030	②	—	③

NA - Not Available

F - Front

R - Rear

① 0.030 over rivet head; If bonded lining, use 0.062 from shoe

② Use discard diameter cast into drum

③ Front: 38 ft. lbs.

  Rear: 74 ft. lbs.

79226C63

BRAKE SPECIFICATIONS
GM B BODY
All measurements in inches unless noted

| Year | Model | Master Cylinder Bore | Brake Disc | | | Brake Drum Diameter | | | Minimum Lining Thickness | | Brake Caliper | |
			Original Thickness	Minimum Thickness	Maximum Runout	Original Inside Diameter	Max. Wear Limit	Maximum Machine Diameter	Front	Rear	Bracket Bolts (ft. lbs.)	Mounting Bolts (ft. lbs.)
1995	Caprice	1.125	1.043	0.980	0.004	11.00	11.09	11.06	0.030	0.030	—	38
	Fleetwood	1.125	1.043	0.965	0.004	11.00	11.09	11.06	0.030	0.030	—	38
	Impala SS	1.125	1.043	0.980	0.004	11.00	11.09	11.06	0.030	0.030	—	38
	Roadmaster	1.125	1.043	0.965	0.003	11.00	11.09	11.06	0.030	①	—	38
1996	Caprice	1.125	1.043	0.980	0.004	11.00	11.09	11.06	0.030	0.030	—	38
	Impala SS	1.125	1.043	0.980	0.004	11.00	11.09	11.06	0.030	0.030	—	38
	Fleetwood	1.125	1.125	1.043	0.004	11.00	11.09	11.06	0.030	0.030	—	38
	Roadmaster	1.125	1.043	0.965	0.004	9.500	9.590	9.560	0.030	①	—	38

F - Front

R - Rear

① 0.030 over rivet head; If bonded lining, use 0.062 from shoe

79226C64

BRAKE SPECIFICATIONS
GM C & H BODIES
All measurements in inches unless noted

Year	Model	Master Cylinder Bore	Brake Disc Original Thickness	Brake Disc Minimum Thickness	Brake Disc Maximum Runout	Brake Drum Diameter Original Inside Diameter	Brake Drum Diameter Max. Wear Limit	Brake Drum Diameter Maximum Machine Diameter	Minimum Lining Thickness Front	Minimum Lining Thickness Rear	Brake Caliper Bracket Bolts (ft. lbs.)	Brake Caliper Mounting Bolts (ft. lbs.)
1995	Bonneville	1.000	1.276	1.209	0.004	8.860	8.909	8.800	0.030	①	—	38
	Eighty-Eight	1.000	1.276	1.209	0.004	8.860	8.909	8.800	0.030	①	—	38
	LeSabre	1.000	1.276	1.209	0.004	8.860	8.909	8.800	0.030	①	—	38
	Ninety-Eight	1.000	1.276	1.209	0.004	8.860	8.909	8.800	0.030	①	—	38
	Park Avenue	1.000	1.276	1.209	0.004	8.860	8.909	8.800	0.030	①	—	38
1996	Bonneville	1.000	1.260	1.209	0.002	8.860	8.909	8.920	0.030	0.030	—	38
	Eighty-Eight	1.000	1.276	1.209	0.004	8.860	8.909	8.800	0.030	①	—	38
	LeSabre	1.000	1.260	1.209	0.002	8.863	8.909	8.920	0.030	①	—	38
	Ninety-Eight	1.000	1.276	1.209	0.004	8.860	8.909	8.800	0.030	①	—	38
	Park Avenue	1.000	1.260	1.209	0.002	8.863	8.909	8.920	0.030	①	—	38
1997	Bonneville	1.000	1.260	1.209	0.002	8.860	8.909	8.920	0.030	0.030	—	38
	Eighty-Eight	1.000	1.276	1.209	0.004	8.860	8.909	8.800	0.030	①	—	38
	LeSabre	1.000	1.260	1.209	0.002	8.863	8.909	8.920	0.030	①	—	38
	LSS	1.000	1.276	1.209	0.004	8.860	8.909	8.800	0.030	①	—	38
	Park Avenue	1.000	1.260	1.209	0.002	8.863	8.909	8.920	0.030	①	—	38
	Regency	1.000	1.276	1.209	0.004	8.860	8.909	8.800	0.030	①	—	38
1998-99	Bonneville	1.000	1.260	1.209	0.002	8.860	8.909	8.920	0.030	0.030	—	38
	Eighty-Eight	1.000	1.276	1.209	0.004	8.860	8.909	8.800	0.030	①	—	38
	LeSabre	1.000	1.260	1.209	0.002	8.863	8.909	8.920	0.030	①	—	38
	LSS	1.000	1.276	1.209	0.004	8.860	8.909	8.800	0.030	①	—	38
	Park Avenue	1.000	1.260	1.209	0.002	8.863	8.909	8.920	0.030	①	—	38
	Regency	1.000	1.276	1.209	0.004	8.860	8.909	8.800	0.030	①	—	38

NA - Not Available

F - Front

R - Rear

① 0.030 over rivet head; If bonded lining, use 0.062 from shoe

79226C65

BRAKE SPECIFICATIONS
GM E & K BODIES
All measurements in inches unless noted

Year	Model		Master Cylinder Bore	Brake Disc Original Thickness	Brake Disc Minimum Thickness	Brake Disc Maximum Runout	Brake Drum Diameter Original Inside Diameter	Brake Drum Diameter Max. Wear Limit	Brake Drum Diameter Maximum Machine Diameter	Minimum Lining Thickness Front	Minimum Lining Thickness Rear	Brake Caliper Bracket Bolts (ft. lbs.)	Brake Caliper Mounting Bolts (ft. lbs.)
1995	DeVille	F	1.000	1.268	1.209	0.002	—	—	—	0.030	—	—	38
		R	—	0.433	0.374	0.002	—	—	—	—	0.030	—	20
	DeVille Concours	F	1.000	1.268	1.209	0.002	—	—	—	0.030	—	—	38
		R	—	0.433	0.374	0.002	—	—	—	—	0.030	—	20
	Eldorado	F	1.000	1.268	1.209	0.002	—	—	—	0.030	—	—	38
		R	—	0.433	0.374	0.002	—	—	—	—	0.030	—	20
	Eldorado ETC	F	1.000	1.268	1.209	0.002	—	—	—	0.030	—	—	38
		R	—	0.433	0.374	0.002	—	—	—	—	0.030	—	20
	Seville SLS	F	1.000	1.268	1.209	0.002	—	—	—	0.030	—	—	38
		R	—	0.433	0.374	0.002	—	—	—	-	0.030	—	20
	Seville STS	F	1.000	1.268	1.209	0.002	—	—	—	0.030	—	—	38
		R	—	0.433	0.374	0.002	—	—	—	—	0.030	—	20
1996	DeVille	F	1.000	1.268	1.209	0.002	—	—	—	0.030	—	—	38
		R	—	0.433	0.374	0.002	—	—	—	—	0.030	—	20
	DeVille Concours	F	1.000	1.268	1.209	0.002	—	—	—	0.030	—	—	38
		R	—	0.433	0.374	0.002	—	—	—	—	0.030	—	20
	Eldorado	F	1.000	1.268	1.209	0.002	—	—	—	0.030	—	—	38
		R	—	0.433	0.374	0.002	—	—	—	—	0.030	—	20
	Eldorado ETC	F	1.000	1.268	1.209	0.002	—	—	—	0.030	—	—	38
		R	—	0.433	0.374	0.002	—	—	—	—	0.030	—	20
	Seville SLS	F	1.000	1.268	1.209	0.002	—	—	—	0.030	—	—	38
		R	—	0.433	0.374	0.002	—	—	—	—	0.030	—	20
	Seville STS	F	1.000	1.268	1.209	0.002	—	—	—	0.030	—	—	38
		R	—	0.433	0.374	0.002	—	—	—	—	0.030	—	20
1997	DeVille	F	1.000	1.268	1.209	0.002	—	—	—	0.030	—	—	38
		R	—	0.433	0.374	0.002	—	—	—	—	0.030	—	20
	DeVille Concours	F	1.000	1.268	1.209	0.002	—	—	—	0.030	—	—	38
		R	—	0.433	0.374	0.002	—	—	—	—	0.030	—	20
	Eldorado	F	1.000	1.268	1.209	0.002	—	—	—	0.030	—	—	38
		R	—	0.433	0.374	0.002	—	—	—	—	0.030	—	20
	Eldorado ETC	F	1.000	1.268	1.209	0.002	—	—	—	0.030	—	—	38
		R	—	0.433	0.374	0.002	—	—	—	—	0.030	—	20
	Seville SLS	F	1.000	1.268	1.209	0.002	—	—	—	0.030	—	—	38
		R	—	0.433	0.374	0.002	—	—	—	—	0.030	—	20
	Seville STS	F	1.000	1.268	1.209	0.002	—	—	—	0.030	—	—	38
		R	—	0.433	0.374	0.002	—	—	—	—	0.030	—	20
1998-99	DeVille	F	1.000	1.268	1.209	0.002	—	—	—	0.030	—	—	38
		R	—	0.433	0.374	0.002	—	—	—	—	0.030	—	20
	DeVille Concours	F	1.000	1.268	1.209	0.002	—	—	—	0.030	—	—	38
		R	—	0.433	0.374	0.002	—	—	—	—	0.030	—	20
	Eldorado	F	1.000	1.268	1.209	0.002	—	—	—	0.030	—	—	38
		R	—	0.433	0.374	0.002	—	—	—	—	0.030	—	20
	Eldorado ETC	F	1.000	1.268	1.209	0.002	—	—	—	0.030	—	—	38
		R	—	0.433	0.374	0.002	—	—	—	—	0.030	—	20
	Seville SLS	F	1.000	1.268	1.209	0.002	—	—	—	0.030	—	—	38
		R	—	0.433	0.374	0.002	—	—	—	—	0.030	—	20
	Seville STS	F	1.000	1.268	1.209	0.002	—	—	—	0.030	—	—	38
		R	—	0.433	0.374	0.002	—	—	—	—	0.030	—	20

79226C66

BRAKE SPECIFICATIONS
GM F BODY
All measurements in inches unless noted

Year	Model		Master Cylinder Bore	Brake Disc Original Thickness	Brake Disc Minimum Thickness	Brake Disc Maximum Runout	Brake Drum Diameter Original Inside Diameter	Brake Drum Diameter Max. Wear Limit	Brake Drum Diameter Maximum Machine Diameter	Minimum Lining Thickness Front	Minimum Lining Thickness Rear	Brake Caliper Bracket Bolt (ft. lbs.)	Brake Caliper Mounting Bolt (ft. lbs.)
1995	Camaro		①	②	③	0.005	9.50	9.59	9.56	0.030	0.030	160	38
	Firebird		①	②	③	0.005	9.50	9.59	9.56	0.030	0.030	160	38
1996	Camaro	F	①	1.043	0.980	0.005	—	—	—	0.030	—	160	38
		R		0.795	0.744	0.005	9.50	9.59	9.56	—	0.030	160	27
	Firebird	F	①	1.043	0.980	0.005	—	—	—	0.030	—	160	38
		R		0.795	0.744	0.005	9.50	9.59	9.56	—	0.030	160	27
1997	Camaro	F	①	1.043	0.980	0.005	—	—	—	0.030	—	160	38
		R		0.795	0.744	0.005	9.50	9.59	9.56	—	0.030	160	27
	Firebird	F	①	1.043	0.980	0.005	—	—	—	0.030	—	160	38
		R		0.795	0.744	0.005	9.50	9.59	9.56	—	0.030	160	27
1998-99	Camaro	F	①	1.260	1.223	0.005	—	—	—	0.030	—	160	23
		R		—	0.985	0.005	—	—	—	—	0.030	160	27
	Firebird	F	①	1.260	1.223	0.005	—	—	—	0.030	—	160	23
		R		—	0.985	0.005	—	—	—	—	0.030	160	27

NA - Not Available

F - Front

R - Rear

① Rear drum: 0.945; Rear disc: 1.00

② Front: 1.043; Rear: 0.795

③ Front: 0.980; Rear: 0.744

79226C67

BRAKE SPECIFICATIONS
GM G BODY
All measurements in inches unless noted

Year	Model		Master Cylinder Bore	Brake Disc Original Thickness	Brake Disc Minimum Thickness	Brake Disc Maximum Runout	Brake Drum Diameter Original Inside Diameter	Brake Drum Diameter Max. Wear Limit	Brake Drum Diameter Maximum Machine Diameter	Minimum Lining Thickness Front	Minimum Lining Thickness Rear	Brake Caliper Bracket Bolt (ft. lbs.)	Brake Caliper Mounting Bolt (ft. lbs.)
1995	Aurora	F	1.000	1.260	1.209	0.002	—	—	—	0.030	—	—	38
		R	1.000	0.433	0.374	0.002	—	—	—	—	0.030	—	20
	Riviera	F	1.000	1.260	1.209	0.002	—	—	—	0.030	—	—	38
		R	1.000	0.433	0.374	0.002	—	—	—	0.030	—	—	20
1996	Aurora	F	1.000	1.260	1.209	0.002	—	—	—	0.030	—	—	38
		R	1.000	0.433	0.374	0.002	—	—	—	—	0.030	—	20
	Riviera	F	1.000	1.260	1.209	0.002	—	—	—	0.030	—	—	38
		R	1.000	0.433	0.374	0.002	—	—	—	0.030	—	—	20
1997	Aurora	F	1.000	1.260	1.209	0.002	—	—	—	0.030	—	—	38
		R	1.000	0.433	0.374	0.002	—	—	—	—	0.030	—	20
	Riviera	F	1.000	1.260	1.209	0.002	—	—	—	0.030	—	—	38
		R	1.000	0.433	0.374	0.002	—	—	—	0.030	—	—	20
1998-99	Aurora	F	1.000	1.260	1.209	0.002	—	—	—	0.030	—	—	38
		R	1.000	0.433	0.374	0.002	—	—	—	—	0.030	—	20
	Riviera	F	1.000	1.260	1.209	0.002	—	—	—	0.030	—	—	38
		R	1.000	0.433	0.374	0.002	—	—	—	0.030	—	—	20

F - Front

R - Rear

79226C68

BRAKE SPECIFICATIONS
GM J BODY
All measurements in inches unless noted

Year	Model	Master Cylinder Bore	Brake Disc Original Thickness	Brake Disc Minimum Thickness	Brake Disc Maximum Runout	Brake Drum Diameter Original Inside Diameter	Brake Drum Diameter Max. Wear Limit	Brake Drum Diameter Maximum Machine Diameter	Minimum Lining Thickness Front	Minimum Lining Thickness Rear	Brake Caliper Bracket Bolts (ft. lbs.)	Brake Caliper Mounting Bolts (ft. lbs.)
1995	Cavalier	0.875	0.786	0.736	0.003	7.879	7.929	7.899	0.125	0.125	—	40
	Sunfire	0.874	0.786	0.736	0.003	7.870	7.930	7.900	0.030	0.030	—	40
1996	Cavalier	0.874	0.806	0.736	0.003	7.880	7.930	7.900	0.030	0.030	—	40
	Sunfire	0.874	0.786	0.736	0.003	7.870	7.930	7.900	0.030	0.030	—	40
1997	Cavalier	0.874	0.806	0.736	0.003	7.880	7.930	7.900	0.030	0.030	—	40
	Sunfire	0.874	0.786	0.736	0.003	7.870	7.930	7.900	0.030	0.030	—	40
1998-99	Cavalier	0.874	0.806	0.736	0.003	7.880	7.930	7.900	0.030	0.030	—	40
	Sunfire	0.874	0.786	0.736	0.003	7.870	7.930	7.900	0.030	0.030	—	40

79226C69

BRAKE SPECIFICATIONS
GM L BODY
All measurements in inches unless noted

Year	Model	Master Cylinder Bore	Brake Disc Original Thickness	Brake Disc Minimum Thickness	Brake Disc Maximum Runout	Brake Drum Diameter Original Inside Diameter	Brake Drum Diameter Max. Wear Limit	Brake Drum Diameter Maximum Machine Diameter	Minimum Lining Thickness Front	Minimum Lining Thickness Rear	Brake Caliper Bracket Bolts (ft. lbs.)	Brake Caliper Mounting Bolts (ft. lbs.)
1995	Beretta	0.945	0.885	0.830	0.004	7.879	7.929	7.899	0.030	0.030	—	38
	Corsica	0.945	0.885	0.830	0.004	7.879	7.929	7.899	0.030	0.030	—	38
1996	Beretta	0.945	0.885	0.830	0.004	7.879	7.929	7.899	0.030	0.030	—	38
	Corsica	0.945	0.885	0.830	0.004	7.879	7.929	7.899	0.030	0.030	—	38

79226C70

BRAKE SPECIFICATIONS
GM L/N BODY
All measurements in inches unless noted

Year	Model	Master Cylinder Bore	Brake Disc Original Thickness	Brake Disc Minimum Thickness	Brake Disc Maximum Runout	Brake Drum Diameter Original Inside Diameter	Brake Drum Diameter Max. Wear Limit	Brake Drum Diameter Maximum Machine Diameter	Minimum Lining Thickness Front	Minimum Lining Thickness Rear	Brake Caliper Bracket Bolts (ft. lbs.)	Brake Caliper Mounting Bolts (ft. lbs.)
1997	Cutlass	0.874	0.806	0.736	0.003	7.874-7.890	7.929	7.899	0.030	①	—	40
	Malibu	0.874	0.806	0.736	0.003	7.874-7.890	7.930	7.899	0.030	①	—	40
1998-99	Cutlass	0.874	0.806	0.736	0.003	7.874-7.890	7.929	7.899	0.030	①	—	40
	Malibu	0.874	0.806	0.736	0.003	7.874-7.890	7.930	7.899	0.030	①	—	40

① 0.030 over rivet head; If bonded lining, use 0.062 from shoe

79226C71

BRAKE SPECIFICATIONS
GM N BODY
All measurements in inches unless noted

Year	Model	Master Cylinder Bore	Brake Disc			Brake Drum Diameter			Minimum Lining Thickness		Brake Caliper	
			Original Thickness	Minimum Thickness	Maximum Runout	Original Inside Diameter	Max. Wear Limit	Maximum Machine Diameter	Front	Rear	Bracket Bolts (ft. lbs.)	Mounting Bolts (ft. lbs.)
1995	Achieva	0.874	0.806	0.736	0.003	7.874-7.890	7.929	7.899	0.030	①	—	40
	Grand Am	0.874	0.806	0.736	0.003	7.874	7.929	7.899	0.030	①	—	40
	Skylark	0.874	0.806	0.736	0.003	7.874-7.890	7.929	7.899	0.030	①	—	40
1996	Achieva	0.874	0.806	0.736	0.003	7.874-7.890	7.930	7.899	0.030	①	—	40
	Grand Am	0.874	0.806	0.736	0.003	7.874	7.930	7.900	0.030	0.030	—	40
	Skylark	0.874	0.806	0.736	0.003	7.874-7.890	7.929	7.899	0.030	①	—	40
1997	Achieva	0.874	0.806	0.736	0.003	7.874-7.890	7.930	7.899	0.030	①	—	40
	Grand Am	0.874	0.806	0.736	0.003	7.874	7.930	7.900	0.030	0.030	—	40
	Skylark	0.874	0.806	0.736	0.003	7.874-7.890	7.929	7.899	0.030	①	—	40
1998-99	Achieva	0.874	0.806	0.736	0.003	7.874-7.890	7.930	7.899	0.030	①	—	40
	Grand Am	0.874	0.806	0.736	0.003	7.874	7.930	7.900	0.030	0.030	—	40

① 0.030 over rivet head; If bonded lining, use 0.062 from shoe

79226C72

BRAKE SPECIFICATIONS
GM V BODY
All measurements in inches unless noted

Year	Model		Master Cylinder Bore	Brake Disc			Brake Drum Diameter			Minimum Lining Thickness		Brake Caliper	
				Original Thickness	Minimum Thickness	Maximum Runout	Original Inside Diameter	Max. Wear Limit	Maximum Machine Diameter	Front	Rear	Bracket Bolts (ft. lbs.)	Mounting Bolts (ft. lbs.)
1997	Catera	F	1.000	1.270	1.250	0.003	—	—	—	—	—	137	63
		R	—	0.430	0.410	0.003	—	—	—	—	—	92	32
1998-99	Catera	F	1.000	1.270	1.250	0.003	—	—	—	—	—	137	63
		R	—	0.430	0.410	0.003	—	—	—	—	—	92	32

F - Front
R - Rear

79226C73

BRAKE SPECIFICATIONS
GM W BODY
All measurements in inches unless noted

Year	Model		Master Cylinder Bore	Brake Disc Original Thickness	Minimum Thickness	Maximum Runout	Brake Drum Diameter Original Inside Diameter	Max. Wear Limit	Maximum Machine Diameter	Min Lining Front	Thickness Rear	Bracket Bolts (ft. lbs.)	Mounting Bolts (ft. lbs.)
1995	Cutlass Supreme	F	1.000	1.039	0.972	0.003	—	—	—	0.030	0.030	—	80
		R	1.000	0.492	0.429	0.003	—	—	—	0.030	0.030	—	32
	Grand Prix	F	0.944	1.039	0.972	0.004	NA	NA	NA	0.030	NA	—	80
		R	NA	0.492	0.429	0.004	NA	NA	NA	NA	0.030	—	32
	Lumina	F	0.945	1.040	0.972	0.004	NA	NA	NA	0.030	—	—	80
		R	0.945	0.492	0.429	0.004	NA	NA	NA	—	0.030	—	32
	Monte Carlo	F	0.945	1.040	0.972	0.004	NA	NA	NA	0.030	—	—	80
		R	0.945	0.492	0.429	0.004	NA	NA	NA	—	0.030	—	32
	Regal	F	0.945	1.039	0.972	0.003	—	—	—	0.030	0.030	—	80
		R	0.945	0.492	0.429	0.003	—	—	—	0.030	0.030	—	32
1996	Cutlass Supreme	F	1.000	1.039	0.972	0.003	—	—	—	0.030	0.030	—	80
		R	1.000	0.492	0.429	0.003	—	—	—	0.030	0.030	—	32
	Grand Prix	F	0.944	1.039	0.972	0.004	NA	NA	NA	0.030	NA	—	80
		R	NA	0.492	0.429	0.004	NA	NA	NA	NA	0.030	—	32
	Lumina	F	0.945	1.040	0.972	0.004	NA	NA	NA	0.030	—	—	80
		R	0.945	0.492	0.429	0.004	NA	NA	NA	—	0.030	—	32
	Monte Carlo	F	0.945	1.040	0.972	0.004	NA	NA	NA	0.030	—	—	80
		R	0.945	0.492	0.429	0.004	NA	NA	NA	—	0.030	—	32
	Regal	F	0.945	1.039	0.972	0.003	—	—	—	0.030	0.030	—	80
		R	0.945	0.492	0.429	0.003	—	—	—	0.030	0.030	—	32
1997	Cutlass Supreme	F	1.000	1.039	0.972	0.003	—	—	—	0.030	0.030	—	80
		R	1.000	0.492	0.429	0.003	—	—	—	0.030	0.030	—	32
	Grand Prix	F	0.944	1.039	0.972	0.004	NA	NA	NA	0.030	NA	—	80
		R	NA	0.492	0.429	0.004	NA	NA	NA	NA	0.030	—	32
	Lumina	F	0.945	1.040	0.972	0.004	NA	NA	NA	0.030	—	—	80
		R	0.945	0.492	0.429	0.004	NA	NA	NA	—	0.030	—	32
	Monte Carlo	F	0.945	1.040	0.972	0.004	NA	NA	NA	0.030	—	—	80
		R	0.945	0.492	0.429	0.004	NA	NA	NA	—	0.030	—	32
	Regal	F	0.945	1.039	0.972	0.003	—	—	—	0.030	0.030	—	80
		R	0.945	0.492	0.429	0.003	—	—	—	0.030	0.030	—	32
1998-99	Century	F	1.000	1.039	0.972	0.003	—	—	—	0.030	0.030	—	80
		R	1.000	0.492	0.429	0.003	—	—	—	0.030	0.030	—	32
	Grand Prix	F	1.000	1.039	0.972	0.003	—	—	—	0.030	0.030	—	80
		R	1.000	0.492	0.429	0.003	—	—	—	0.030	0.030	—	32
	Intrigue	F	1.000	1.039	0.972	0.003	—	—	—	0.030	0.030	—	80
		R	1.000	0.492	0.429	0.003	—	—	—	0.030	0.030	—	32
	Lumina	F	0.945	1.040	0.972	0.004	NA	NA	NA	0.030	—	—	80
		R	0.945	0.492	0.429	0.004	NA	NA	NA	—	0.030	—	32
	Monte Carlo	F	0.945	1.040	0.972	0.004	NA	NA	NA	0.030	—	—	80
		R	0.945	0.492	0.429	0.004	NA	NA	NA	—	0.030	—	32
	Regal	F	1.000	1.039	0.972	0.003	—	—	—	0.030	0.030	—	80
		R	1.000	0.492	0.429	0.003	—	—	—	0.030	0.030	—	32

F - Front
R - Rear

79226C74

BRAKE SPECIFICATIONS
GM Y BODY
All measurements in inches unless noted

| Year | Model | Master Cylinder Bore | Brake Disc | | | Brake Drum Diameter | | | Minimum Lining Thickness | | Brake Caliper | |
			Original Thickness	Minimum Thickness	Maximum Runout	Original Inside Diameter	Max. Wear Limit	Maximum Machine Diameter	Front	Rear	Bracket Bolts (ft. lbs.)	Mounting Bolts (ft. lbs.)
1995	Corvette	NA	①	②	0.006	NA	NA	NA	0.030	0.030	—	③
1996	Corvette	NA	①	②	0.006	NA	NA	NA	0.030	0.030	—	③
1997	Corvette	NA	①	②	0.006	NA	NA	NA	0.030	0.030	—	③
1998-99	Corvette	NA	①	②	0.006	NA	NA	NA	0.030	0.030	—	③

NA - Not Available
① Heavy duty: 1.110; Std.: 0.795
② Heavy duty: 1.059; Std.: 0.744
③ Front: NA, Rear: upper 26 ft. lbs., lower 16 ft. lbs

79226C75

BRAKE SPECIFICATIONS
GEO/CHEVROLET
All measurements in inches unless noted

| Year | Model | Master Cylinder Bore | Brake Disc | | | Brake Drum Diameter | | | Minimum Lining Thickness | | Brake Caliper | |
			Original Thickness	Minimum Thickness	Maximum Runout	Original Inside Diameter	Max. Wear Limit	Maximum Machine Diameter	Front	Rear	Bracket Bolts (ft. lbs.)	Mounting Bolts (ft. lbs.)
1995	Metro	NA	0.670	0.590	0.004	①	①	①	0.236 ②	0.111 ②	—	18
	Prizm	NA	0.866	0.787	0.003	7.87	7.91	7.91	0.039	0.039	—	④
1996	Metro	NA	0.670	0.590	0.004	①	①	①	0.236 ③	0.111 ②	—	18
	Prizm	NA	0.866	0.787	0.003	7.87	7.91	7.91	0.390	0.390	—	④
1997	Metro	NA	0.670	0.590	0.004	①	①	①	0.236 ③	0.111 ②	—	18
	Prizm	NA	0.866	0.787	0.003	7.87	7.91	7.91	0.390	0.390	—	④
1998-99	Metro	NA	0.670	0.590	0.004	①	①	①	0.236 ③	0.111 ②	—	18
	Prizm	NA	0.866	0.787	0.003	7.87	7.91	7.91	0.390	0.390	—	④

NA - Not Available
① 2 door: 7.09
  4 door: 7.87
② Minimum lining thickness includes pad/shoe backing
③ 2 door: 8.66 (service limit 0.874)
④ 4 door: 10.00 (service limit 10.07)
④ Front: 18 ft. lbs.
  Rear: 14 ft. lbs.

79226C76

BRAKE SPECIFICATIONS
SATURN
All measurements in inches unless noted

| Year | Model | Master Cylinder Bore | Brake Disc | | | Brake Drum Diameter | | | Minimum Lining Thickness | | Brake Caliper | |
			Original Thickness	Minimum Thickness	Maximum Runout	Original Inside Diameter	Max. Wear Limit	Maximum Machine Diameter	Front	Rear	Bracket Bolts (ft. lbs.)	Mounting Bolts (ft. lbs.)
1995	Sedan	NA	①	②	0.0024	7.87	7.93	7.91	0.080	0.040	③	27
	Coupe	NA	①	②	0.0024	7.87	7.93	7.91	0.080	0.040	③	27
	Wagon	NA	①	②	0.0024	7.87	7.93	7.91	0.080	0.040	③	27
1996	Sedan	NA	①	②	0.0024	7.87	7.93	7.91	0.080	0.040	③	27
	Coupe	NA	①	②	0.0024	7.87	7.93	7.91	0.080	0.040	③	27
	Wagon	NA	①	②	0.0024	7.87	7.93	7.91	0.080	0.040	③	27
1997	Sedan	NA	①	②	0.0024	7.87	7.93	7.91	0.080	0.040	③	27
	Coupe	NA	①	②	0.0024	7.87	7.93	7.91	0.080	0.040	③	27
	Wagon	NA	①	②	0.0024	7.87	7.93	7.91	0.080	0.040	③	27
1998-99	Sedan	NA	①	②	0.0024	7.87	7.93	7.91	0.080	0.040	③	27
	Coupe	NA	①	②	0.0024	7.87	7.93	7.91	0.080	0.040	③	27
	Wagon	NA	①	②	0.0024	7.87	7.93	7.91	0.080	0.040	③	27

NA - Not Available
① Front: 0.710
  Rear: 0.430
② Front: 0.633
  Rear: 0.370
③ Front: 81 ft. lbs. Rear: 63 ft. lbs

79226C77

DRIVESHAFTS, U-JOINTS AND CV-JOINT BOOTS

7

**DRIVESHAFTS, U-JOINTS AND
   CV-JOINT BOOTS** 7-1
**TROUBLESHOOTING BASIC
   DRIVESHAFT PROBLEMS** 7-4
Constant Velocity Joint (CV-Joint)
Boots...7-7

Inspection.....................................7-7
Replacement.................................7-8
Driveshafts...................................7-1
Balancing 7-3
General Information.....................7-1
Removal & Installation 7-2

Universal Joints (U-Joints)..............7-5
General Information.......................7-5
Inspection.....................................7-5
Overhaul.......................................7-6

DRIVESHAFTS, U-JOINTS AND CV-JOINT BOOTS

Driveshafts

➡The term driveshafts does not refer to halfshafts (often termed driveshafts by various manufacturers), which are used on front wheel drive vehicles.

GENERAL INFORMATION

The driveshaft is a long steel tube used to transmit power from the transmission to the rear differential. Located at either end of the

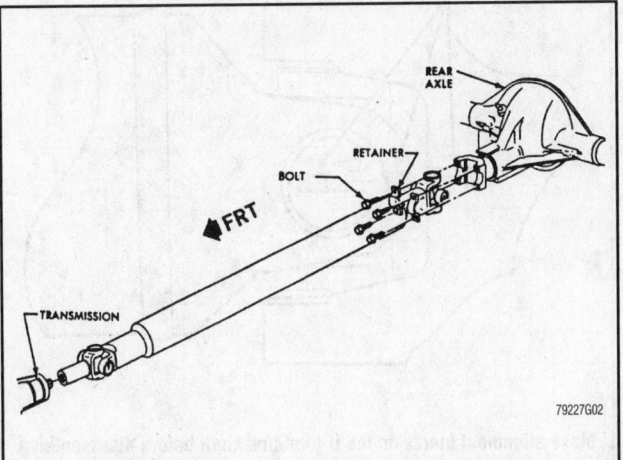

Exploded view of a typical driveshaft assembly and attachment points

driveshaft is a universal joint (U-joint), which allows the driveshaft to move up and down (within designed limits) in order to match the motion of the rear axle. A slip joint is often attached to the U-joint closest to the transmission. The shaft is designed with yokes at each end that are inline with each other in order to produce the smoothest possible running shaft.

➡ALWAYS matchmark the shaft ends to the yoke or flange before removal.

At the front of the driveshaft, the U-joint usually connects the driveshaft to a slip-jointed yoke. This yoke is internally splined and allows the driveshaft to move in and out on the transmission output shaft, which is externally splined. At the rear of the driveshaft, the U-joint is clamped to the yoke attached to the rear axle pinion flange. The rear yoke may also be a flange that mates to the pinion flange on the differential. It is usually secured in the yoke by a bracket with two small bolts, one at either end.

Some rear U-joints are pressed into the yoke which is then bolted to the pinion flange on the differential.

On some production U-joints, nylon (plastic) is injected through a small hole in the yoke during manufacture and flows along a circular groove between the U-joint and the yoke, creating a non-metallic snapring.

➡ Since plastic retaining rings must be sheared for removal and no snapring grooves are supplied, the production joints must be replaced with service U-joints with a snapring groove whenever they are removed from the shaft.

Bad U-joints, requiring replacement, will produce a clunking sound when the vehicle is put into gear and when the transmission

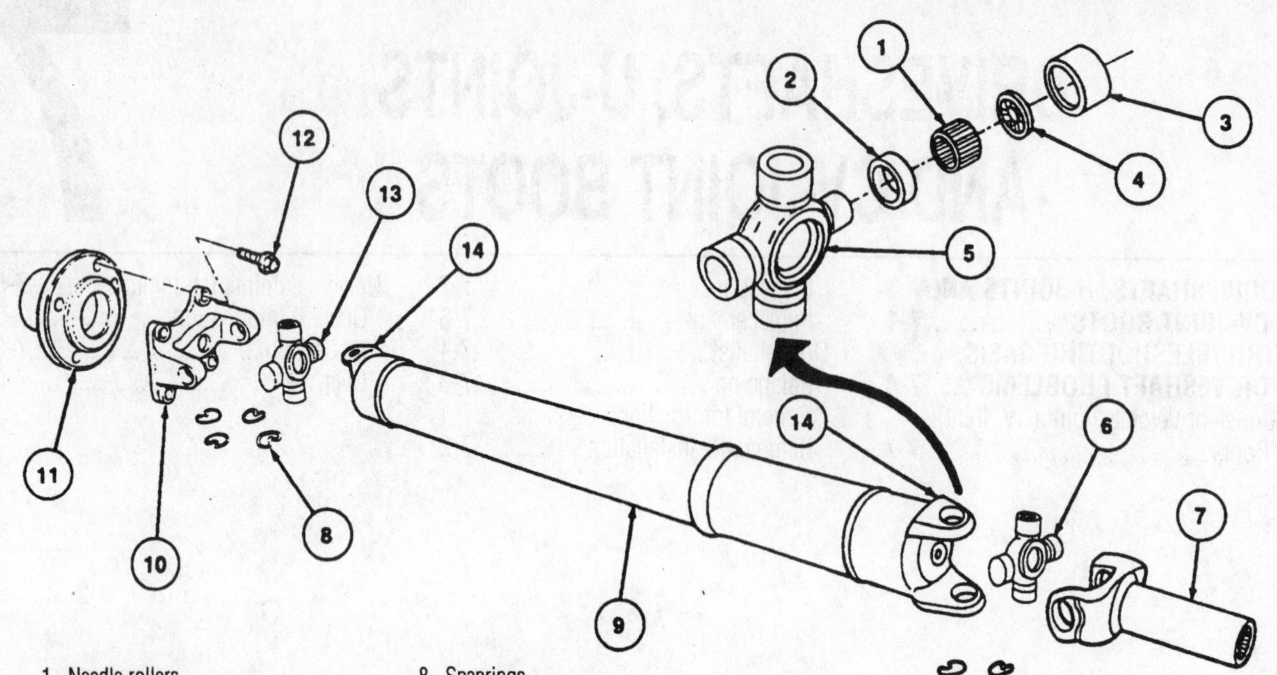

1. Needle rollers
2. Grease seal
3. Bearing cup
4. Thrust washer
5. Spider
6. Universal joint
7. Driveshaft slip yoke
8. Snaprings
9. Driveshaft
10. Driveshaft centering socket yoke
11. Rear axle universal joint flange
12. Attaching bolt
13. Universal joint
14. End yoke

79227G26

Exploded view of a typical driveshaft where the U-joint is pressed into the rear yoke, rather than bolted to it

shifts from gear-to-gear. This is due to worn needle bearings or scored trunnion ends. Most U-joints are permanently lubricated at the factory and require no periodic maintenance. Those that do have grease fittings should be lubricated at every oil change. Clean the fitting with a shop rag before applying grease to avoid forcing dirt into the joint.

REMOVAL & INSTALLATION

1. Raise and safely support the vehicle.
2. Mark the relationship of the driveshaft-to-pinion flange or yoke and disconnect the rear universal joint by removing the bolts. If the bearing cups are loose, tape them together to prevent dropping them and losing the bearing rollers.
3. Slide the driveshaft forward to disengage it from the rear axle flange or yoke.
4. Drop the rear end of the driveshaft below the differential, then move the driveshaft rearward to disengage it from the transmission slip-joint, passing it under the differential housing.

✳✳ WARNING

DO NOT allow the driveshaft to hang by the U-joint or bend to extreme angles, as damage to the U-joint may occur. Support the driveshaft shaft during removal.

To install:

5. Inspect the slip-joint and transmission output shaft for damage, burrs or wear, for this will damage the transmission seal or make installation difficult. Apply engine oil to all splined driveshaft yokes.

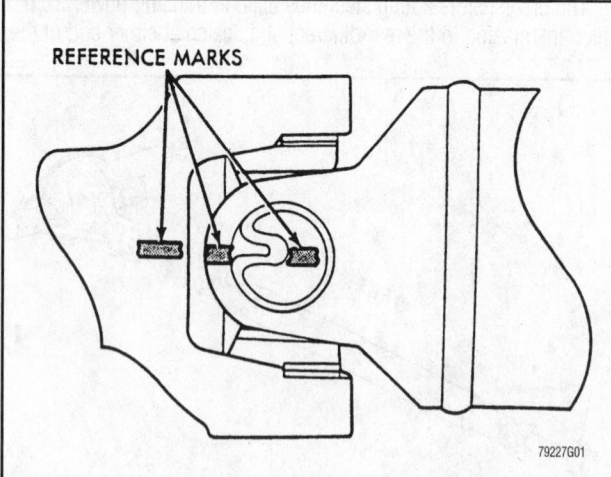

REFERENCE MARKS

79227G01

Make alignment marks on the U-joint and shaft before disassembly to prevent possible vibration when assembled

Make alignment marks on the U-joint and shaft before disassembly to prevent possible vibration when assembled

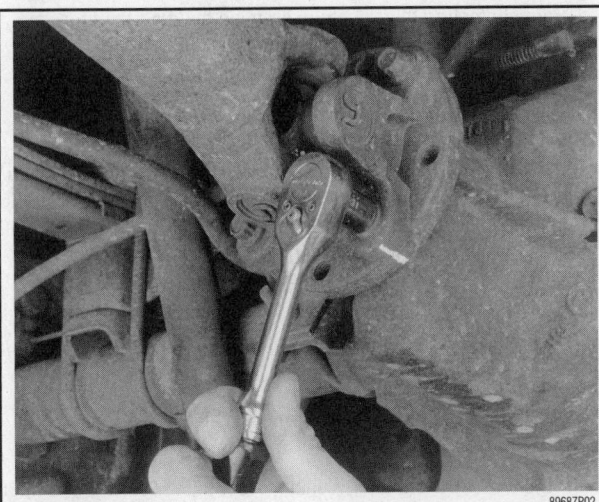

This type of driveshaft is attached to the pinion flange with four bolts and nuts

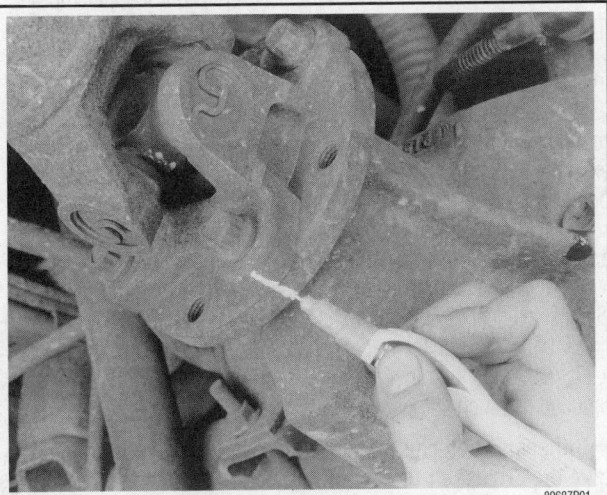

On this type of driveshaft, make alignment marks across the two flanges so it can be installed in the same position

The driveshaft can be detached from the pinion shaft yoke after removing the fasteners and brackets

The U-joint on this driveshaft is attached to the pinion yoke by two small brackets and four bolts

** WARNING

DO NOT use a hammer to force the driveshaft into place. Check for burrs on the transmission output shaft spline, twisted slip yoke splines or possibly the wrong U-joint. Make sure the splines agree in number and fit. To prevent trunnion seal damage, DO NOT place any tool between the yoke and splines.

6. Slide the driveshaft into the transmission.
7. Align the rear universal joint to the rear axle pinion flange, ensuring the bearings are properly seated in the pinion flange yoke, if so equipped.
8. Install the retaining bolts and tighten them bolts securely.
9. Road test the vehicle.

BALANCING

The following procedure is used to help eliminate minor driveshaft vibration of an otherwise good driveshaft.

Before attempting this, carefully examine the driveshaft for damage such as dents and deformations. Driveshafts are subjected to large amounts of twisting force which can literally twist the driveshaft. Also check for missing weights that may have been knocked off of the shaft. If the driveshaft is deformed, replace it. If any weights appear to be missing, take the driveshaft to a machine shop that is equipped to balance the shaft and have it repaired. Since driveshafts typically turn at speeds 2½ to 4 (or more) times faster than the rear axle, do not use a damaged driveshaft.

This type of balancing is performed by installing one or two hose clamps near the end of the driveshaft closest to the drive axle. The trial and error method is used to determine the best position of the clamp(s).

➡**Removing and turning the driveshaft 180° relative to the yoke may reduce some vibration. This should be done prior to the hose clamp method.**

1. Mark the rear of the driveshaft in four equal sections. Number the marks 1 through 4.

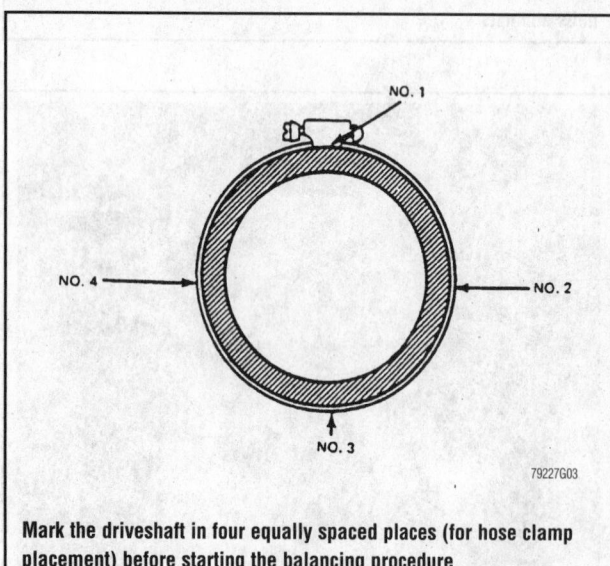

Mark the driveshaft in four equally spaced places (for hose clamp placement) before starting the balancing procedure

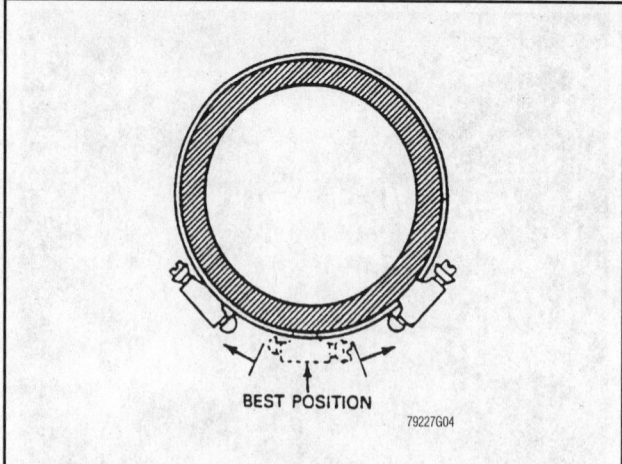

BEST POSITION

79227G04

Move the hose clamp heads an equal distance from the best position a little at a time until the vibration is reduced to an acceptable level

2. Install a hose clamp with the screw portion of the clamp on the No. 1 mark.

3. Test drive the vehicle to see if the vibration condition has improved.

4. Recheck the vibration with the clamp positioned at the remaining three positions. If the vibration is equally reduced at, for example, position number two and position number three, then position the screw portion of the clamp halfway between the marks.

5. Test drive the vehicle. If the vibration is still apparent, install another clamp in the same position as the first.

6. Test drive the vehicle. If the vibration is the same, move both clamps an equal distance from the point determined to be the best position. At first, position the clamps approximately ½ in. (12mm) apart.

7. Continue the moving the clamp(s) until the vibration is reduced as much as possible.

➡**If the vibration cannot be reduced to an acceptable level, take the driveshaft to a qualified machine shop for balancing.**

Troubleshooting Basic Driveshaft Problems

When abnormal vibrations or noises are detected in the driveshaft area, this chart can be used to help diagnose possible causes. Remember that other components such as wheels, tires, rear axle and suspension can also produce similar conditions.

BASIC DRIVESHAFT PROBLEMS

Problem	Cause	Solution
Shudder as car accelerates from stop or low speed	• Loose U-joint • Defective center bearing	• Replace U-joint • Replace center bearing
Loud clunk in driveshaft when shifting gears	• Worn U-joints	• Replace U-joints
Roughness or vibration at any speed	• Out-of-balance, bent or dented driveshaft • Worn U-joints • U-joint clamp bolts loose	• Balance or replace driveshaft • Replace U-joints • Tighten U-joint clamp bolts
Squeaking noise at low speeds	• Lack of U-joint lubrication	• Lubricate U-joint; if problem persists, replace U-joint
Knock or clicking noise	• U-joint or driveshaft hitting frame tunnel • Worn CV joint	• Correct overloaded condition • Replace CV joint

79227C01

Universal Joints (U-Joints)

GENERAL INFORMATION

The universal joint (U-Joint) is used to provide a strong and flexible connection between the driveshaft and the axle (differential) assembly. A flexible joint is necessary because of the constant movement of the axle assembly relative to the body of the vehicle. A U-Joint consists of the spider (trunnion), needle (roller) bearings, bearing cups, seals and snaprings. In most cases, U-Joints will last the life of the vehicle. The life of the U-Joint may decrease significantly if the operating angle has been changed or exceeded. This occurs when the vehicle ride height is changed. Vehicles that have been lifted will benefit by using a Double Cardon type joint. The Double Cardon type joint has a greater operating angle than the single U-joint.

When two components are connected by a conventional U-joint, the bend that is formed is called the operating angle. The larger the angle, the larger the amount of angular acceleration and deceleration of the joint. In other words, when the driveshaft is turning at a steady speed,

PROPELLER SHAFT R.P.M.	MAX. NORMAL OPERATING ANGLES
5000	3°
4500	3°
4000	4°
3500	5°
3000	5°
2500	7°
2000	8°
1500	11°

79227G07

Maximum normal operating angle between the driveshaft and transmission and/or axle assembly

the pinion gear in the differential will actually speed up and slow down. This takes place as long as the driveshaft and pinion gear shaft are at different angles (not in the same plane). The speeding up and slowing down must be canceled out to ensure a smooth flow of power. This is why both yokes on the driveshaft are in line with each other. For example, whereas the transmission output is at a steady speed, the angle at the U-joint causes the driveshaft speed to vary. In such a case, the rear U-joint cancels the fluctuations caused by the front U-joint.

➡**The operating angle is the difference in degrees between the centerline of the driveshaft and the centerline of the transmission and/or axle assembly. The maximum allowable operating angle is determined by engine speed.**

INSPECTION

Remove and replace the U-joint if any of the following conditions are present:
- Knocking or clunking noise from the driveshaft when the vehicle is put into gear, or when coasting at 10 mph (16 km/h) in neutral.

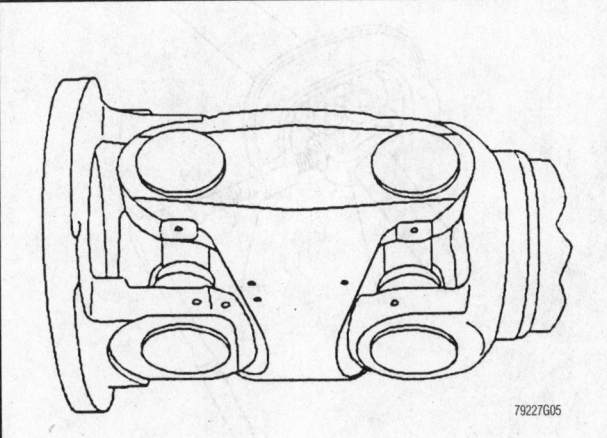

79227G05

A Double Cardon universal joint has a greater operating angle than a single joint. This joint has been punch marked before disassembly so the components can be reassembled in their original positions

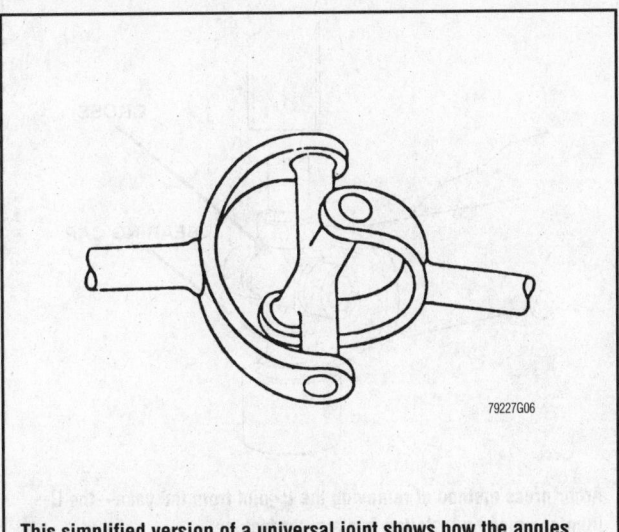

79227G06

This simplified version of a universal joint shows how the angles can change while still transmitting power

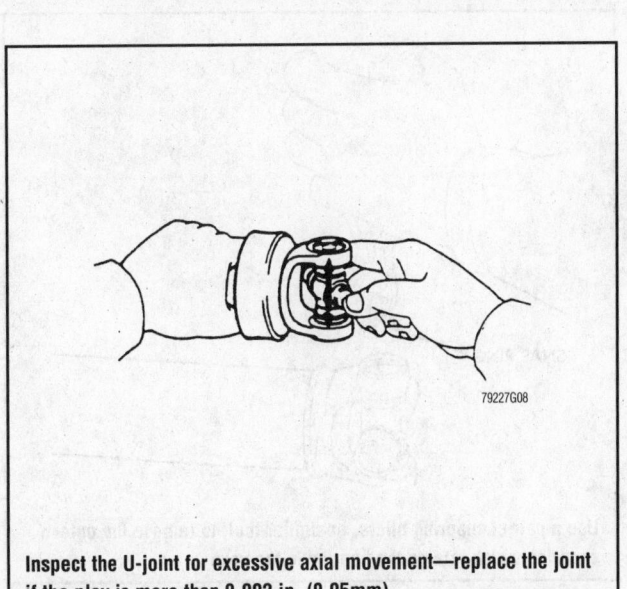

79227G08

Inspect the U-joint for excessive axial movement—replace the joint if the play is more than 0.002 in. (0.05mm).

• Squeaking noise from the U-joint that increases in frequency as the speed of the vehicle increases.
• Roughness in the U-joint bearing when felt by hand. The U-joint should turn smoothly.
• Axial play (up and down movement). Replace the U-joint if the axial play is more than 0.002 in. (0.05mm).

OVERHAUL

1. Position the driveshaft assembly in a sturdy soft-jawed vise, BUT DO NOT place a significant clamp load on the shaft or you will risk deforming and ruining it.

➡**Some original equipment U-joints are secured in the yoke by nylon (plastic) that has been injected at the factory. To remove this type of U-joint from the yoke, press the bearing cup until the plastic retaining ring breaks. The replacement U-joint will have a snapring groove like a conventional joint.**

2. Remove the snaprings which retain the bearings in the yoke.

➡**A U-joint removal and installation tool (which looks like a large C-clamp) is available to significantly ease the task, but it is very possible to replace the U-joints using an arbor press or a large vise and a variety of sockets.**

3. Using a large C-clamp, vise or an arbor press, along with a socket smaller than the bearing cap (on one side) and a socket larger than the bearing cap (on the other side), drive one of the bearings in toward the center of the universal joint, which will force the opposite bearing out.

➡**The smaller socket is used as a driver here, as it can pass through the opening of the U-joint or slip yoke flange. The larger socket is used to support the other side of the flange so that the bearing cap has room to exit the flange (into the socket).**

4. As each bearing is forced far enough out of the universal joint to be accessible, grip it with a pair of pliers and pull it from the driveshaft yoke. Drive the spider in the opposite direction in order to

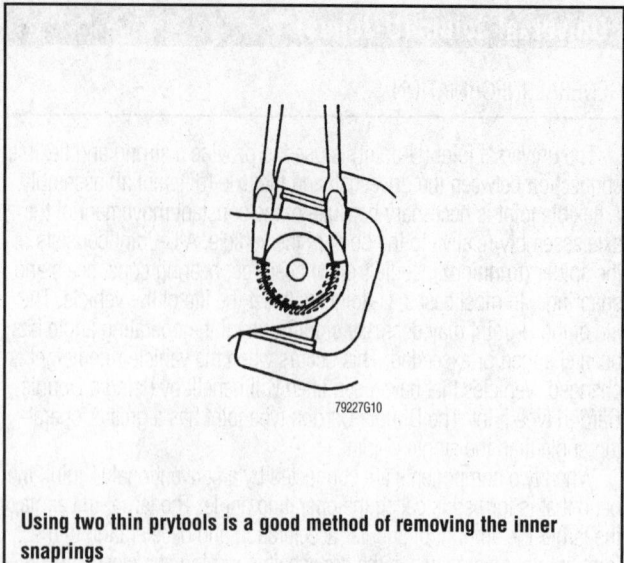

Using two thin prytools is a good method of removing the inner snaprings

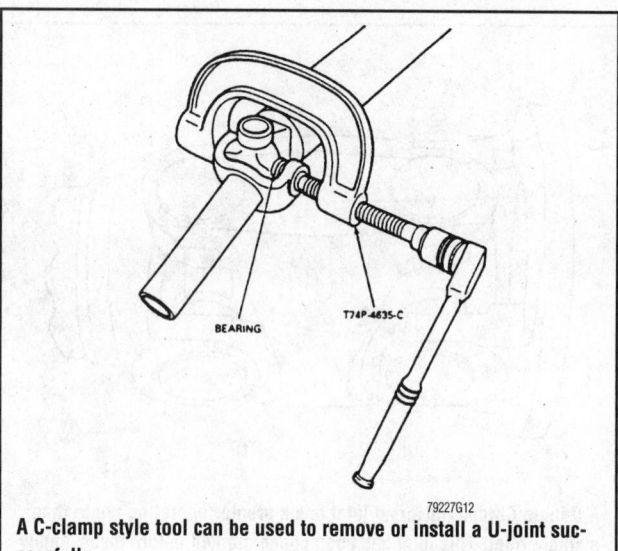

A C-clamp style tool can be used to remove or install a U-joint successfully

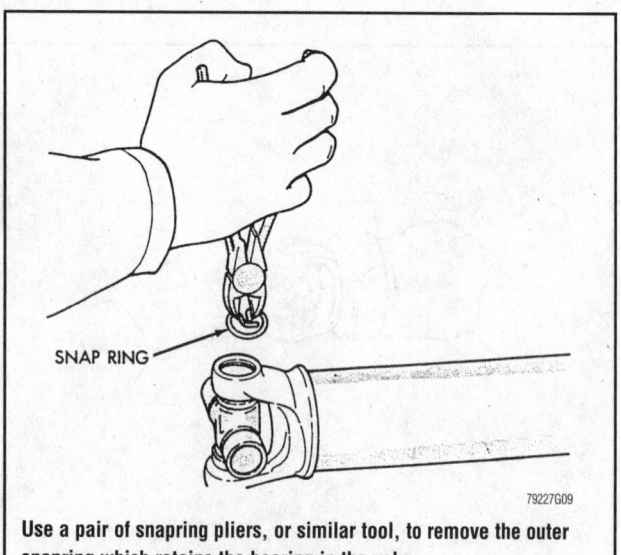

Use a pair of snapring pliers, or similar tool, to remove the outer snapring which retains the bearing in the yoke

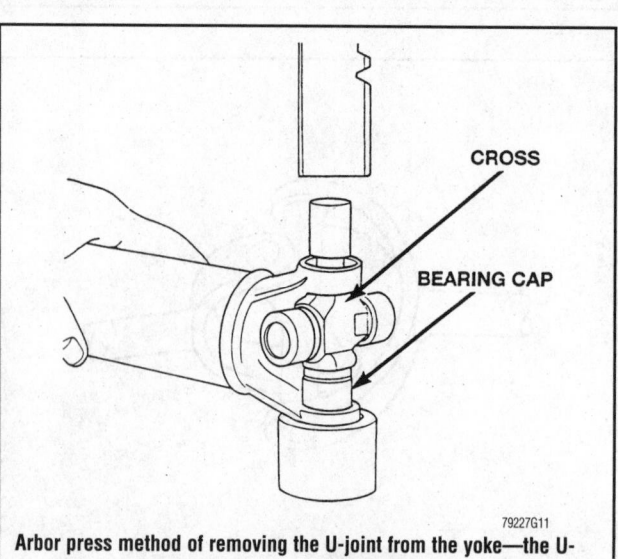

Arbor press method of removing the U-joint from the yoke—the U-joint can also be installed in a similar fashion

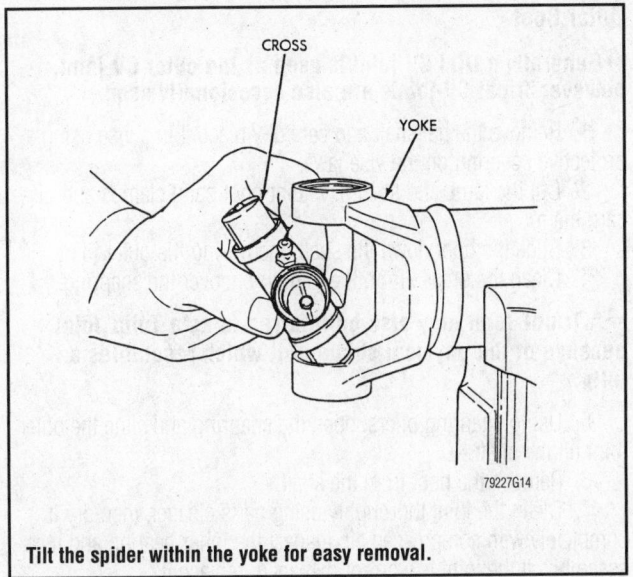

Tilt the spider within the yoke for easy removal.

make the opposite bearing accessible and pull it free with a pair of pliers. Use this procedure to remove all the bearings from both universal joints.

5. After removing the bearings, lift the spider from the yoke.

6. Thoroughly clean all dirt and foreign matter from the yokes on both ends of the driveshaft.

To assemble:

✳✳ WARNING

When installing new bearings in the yokes, it is advisable to use an arbor press or the special C-clamp tool. If this tool is not available, the bearings should be pressed into position with extreme care, as a heavy jolt on the needle bearings can easily damage or misalign them. This will greatly shorten their life and hamper their efficiency.

7. Start a new bearing into the yoke at the rear of the driveshaft.

8. Position a new spider in the rear yoke and press the new bearing ¼ in. (6mm) below the outer surface of the yoke.

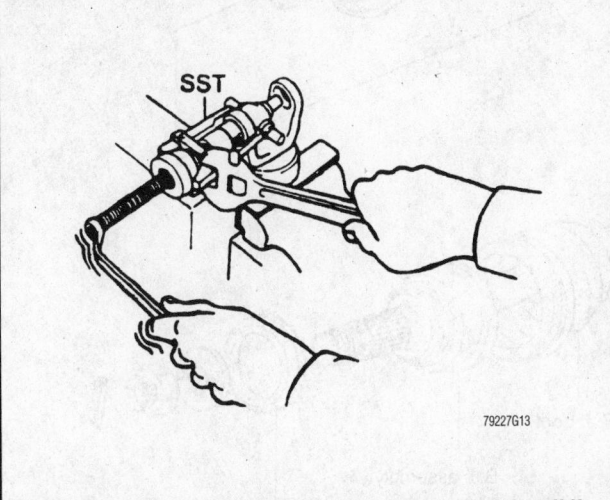

The 2-jawed puller method can also be used to remove or install U-joints

9. With the bearing in position, install a new snapring.

10. Start a new bearing into the opposite side of the yoke. Press the bearing until the opposite bearing, which you have just installed, contacts the inner surface of the snapring.

11. Install a new snapring on the second bearing. It may be necessary to grind the surface of the second snapring.

12. Reposition the driveshaft in the vise, so that the front universal joint is accessible.

13. Install the new bearings, new spider and new snaprings in the same manner as for the previously assembled rear joint.

14. Position the slip yoke on the spider. Install new bearings, nylon thrust bearings (if applicable) and snaprings.

15. Check both reassembled joints for freedom of movement, If misalignment of any part is causing a bind, a sharp rap on the side of the yoke with a brass hammer should seat the needle bearings and provide the desired freedom of movement. Care should be exercised to firmly support the shaft end during this operation, as well as to prevent blows to the bearings themselves. Under no circumstance should the driveshaft be installed in a car if there is any binding in the universal joints.

16. Apply grease to the fittings, if equipped.

Constant Velocity Joint (CV-Joint) Boots

INSPECTION

Whenever undercarriage work is performed such as brakes, exhaust or suspension work, the Constant Velocity Joint (CV-Joint) boots should be inspected for breaks and tears. The first sign of boot damage will be dark spots (grease) on the inside of the tire and wheel. If boot damage is caught early enough, the joint can be saved by cleaning, regreasing and replacing the boot. If the boot is left unrepaired, damage to the bearing will occur and replacement of the CV-joint is required. In many cases, it may be more economical to replace the entire halfshaft with a remanufactured one (already assembled with new CV-joints and boots).

➡**Check with your parts supplier for price and availability to determine whether you should replace the entire halfshaft or separate components.**

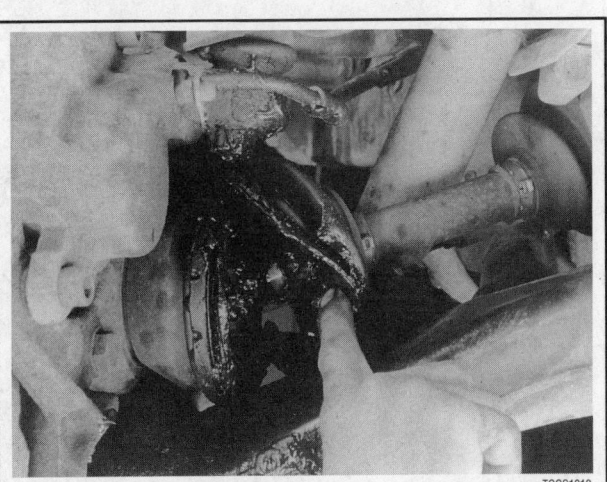

Remove, clean and inspect this CV joint-it may be possible to save the joint by replacing the boot as long as the joint is not beyond repair

Push apart the bellows to inspect for tears or cracks that may be developing

TCCS1011

REPLACEMENT

➡**Birfield outer CV-joints, should not be disassembled. To replace the outer boot, disassemble the inner joint, then slide the outer boot off the inner end of the shaft.**

Outer Boot

➡**Generally a DOJ CV-joint is used as the outer CV-joint, however Tripot CV-joints are also occasionally used.**

1. Remove the halfshaft and carefully place it in a vise using a protective covering on the vise jaws.
2. Cut the large and small CV-joint boot band clamps and discard them.
3. Slide the boot down the shaft uncovering the outer joint.
4. Clean the grease from the joint to uncover the snapring.

➡**A Tripot Joint may also be referred to as a Tulip Joint because of the physical shape of it which resembles a tulip.**

5. Using snapring pliers, open the snapring and slide the outer joint off the shaft.
6. Remove the boot from the shaft.
7. Clean the joint thoroughly using parts cleaner, then dry it completely with compressed air. Inspect the inner bearing and race assembly. If the joint is worn or damaged, replace it.

To install:

8. Wrap the splines on the end of the halfshaft with tape to prevent damage to the boot during installation.
9. Slide the small CV-Joint boot clamp onto the halfshaft and push the boot down several inches past the seal mounting area. Remove the tape from the halfshaft splines.

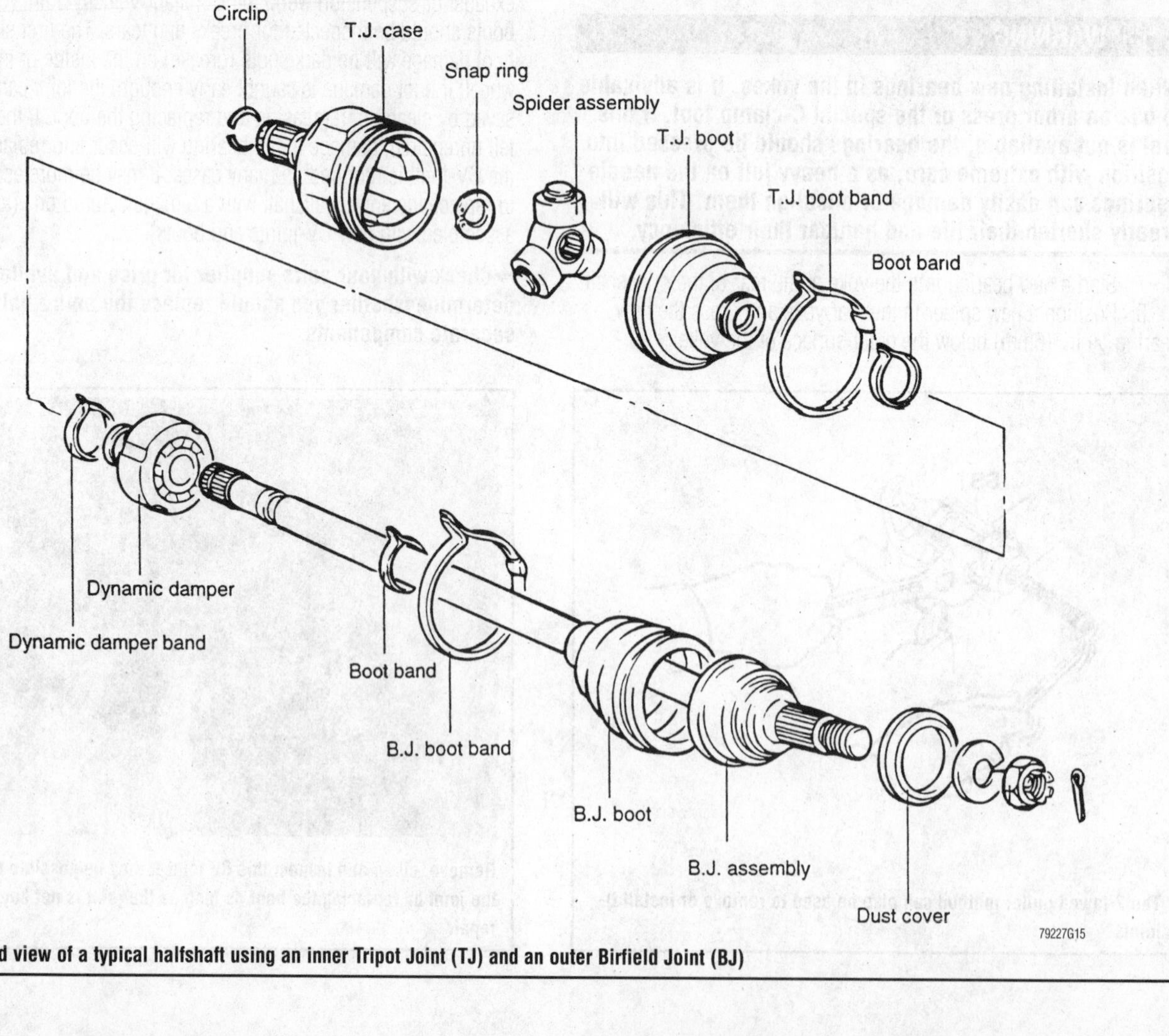

Circlip
T.J. case
Snap ring
Spider assembly
T.J. boot
T.J. boot band
Boot band

Dynamic damper
Dynamic damper band
Boot band
B.J. boot band
B.J. boot
B.J. assembly
Dust cover

79227G15

Exploded view of a typical halfshaft using an inner Tripot Joint (TJ) and an outer Birfield Joint (BJ)

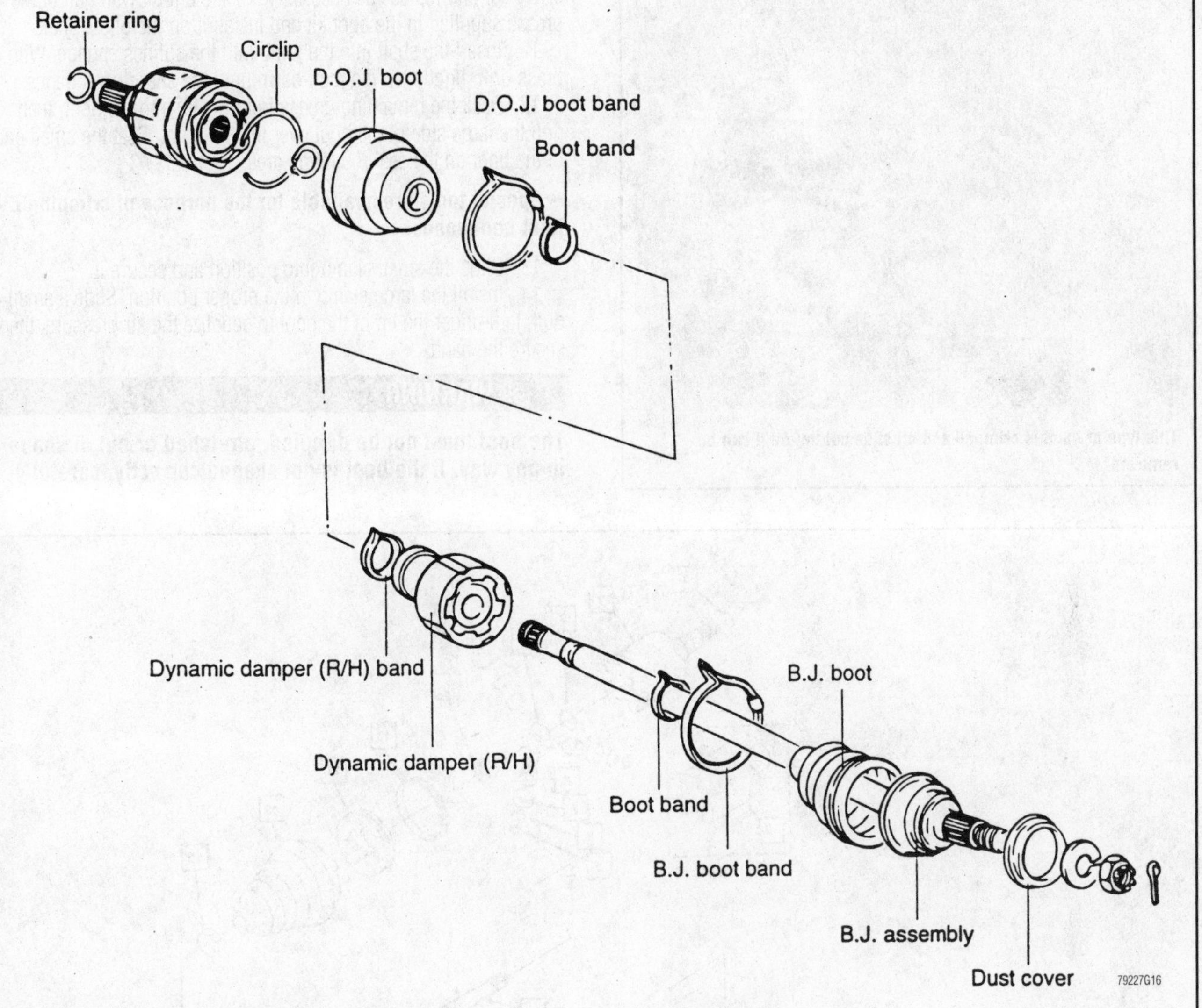

Retainer ring

Circlip

D.O.J. boot

D.O.J. boot band

Boot band

Dynamic damper (R/H) band

Dynamic damper (R/H)

Boot band

B.J. boot band

B.J. boot

B.J. assembly

Dust cover

79227G16

Exploded view of a typical halfshaft using an inner Double Offset Joint (DOJ) and an outer Birfield Joint (BJ)

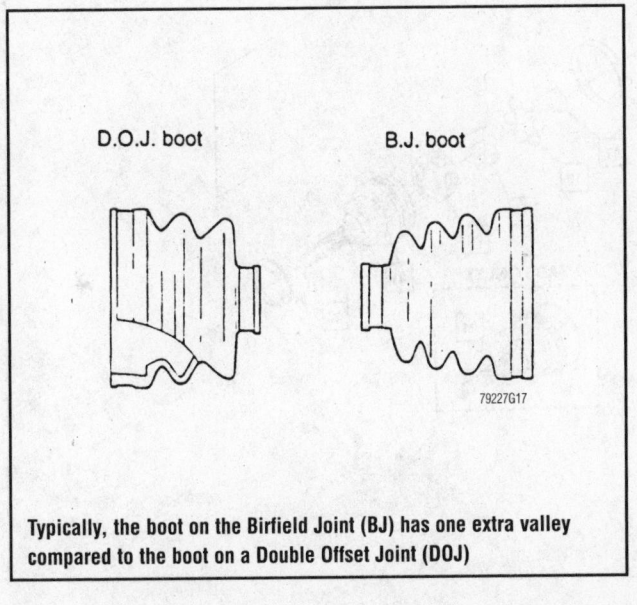

D.O.J. boot

B.J. boot

79227G17

Typically, the boot on the Birfield Joint (BJ) has one extra valley compared to the boot on a Double Offset Joint (DOJ)

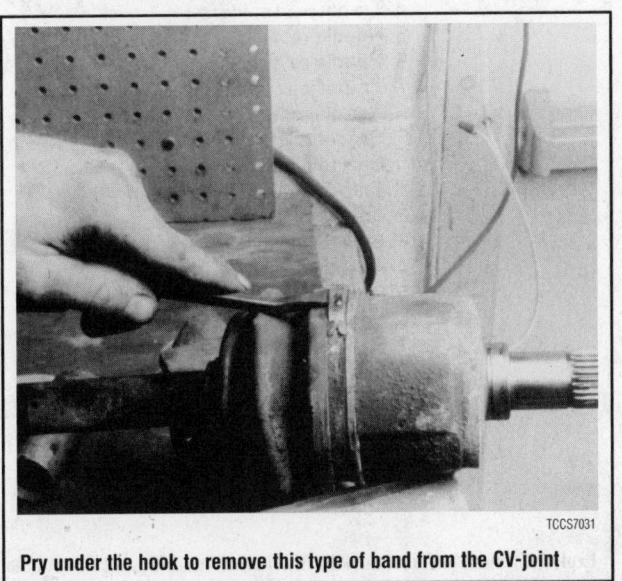

TCCS7031

Pry under the hook to remove this type of band from the CV-joint

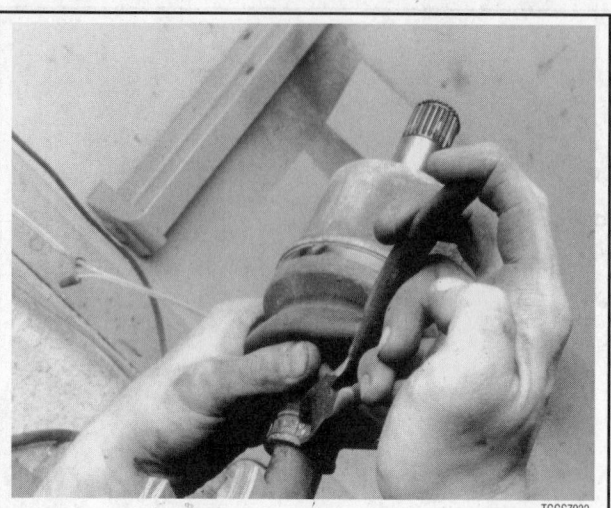

This type of band is crimped and must be cut before it can be removed

TCCS7032

10. Check the snapring in the outer joint for damage or excessive wear and replace as necessary. Pack the joint with half of the grease supplied in the boot kit and install it on the of the shaft.

11. Insert the shaft into the joint until the splines engage. With a brass drift, lightly tap the joint down until the snapring engages.

12. Pack the remaining grease from the kit into the boot, then pull the large side of the boot over the CV-Joint. Seat the small end of the boot on the seal mounting area.

➡**Special tools are available for the purpose of crimping CV-joint boot bands.**

13. Slide the small clamp into position and secure it.

14. Install the large clamp in the proper position. Slide a small dull, tool under the lip of the boot to equalize the air pressure, then secure the band.

✳✳ WARNING

The boot must not be dimpled, stretched or out of shape in any way. If the boot is not shaped correctly, carefully

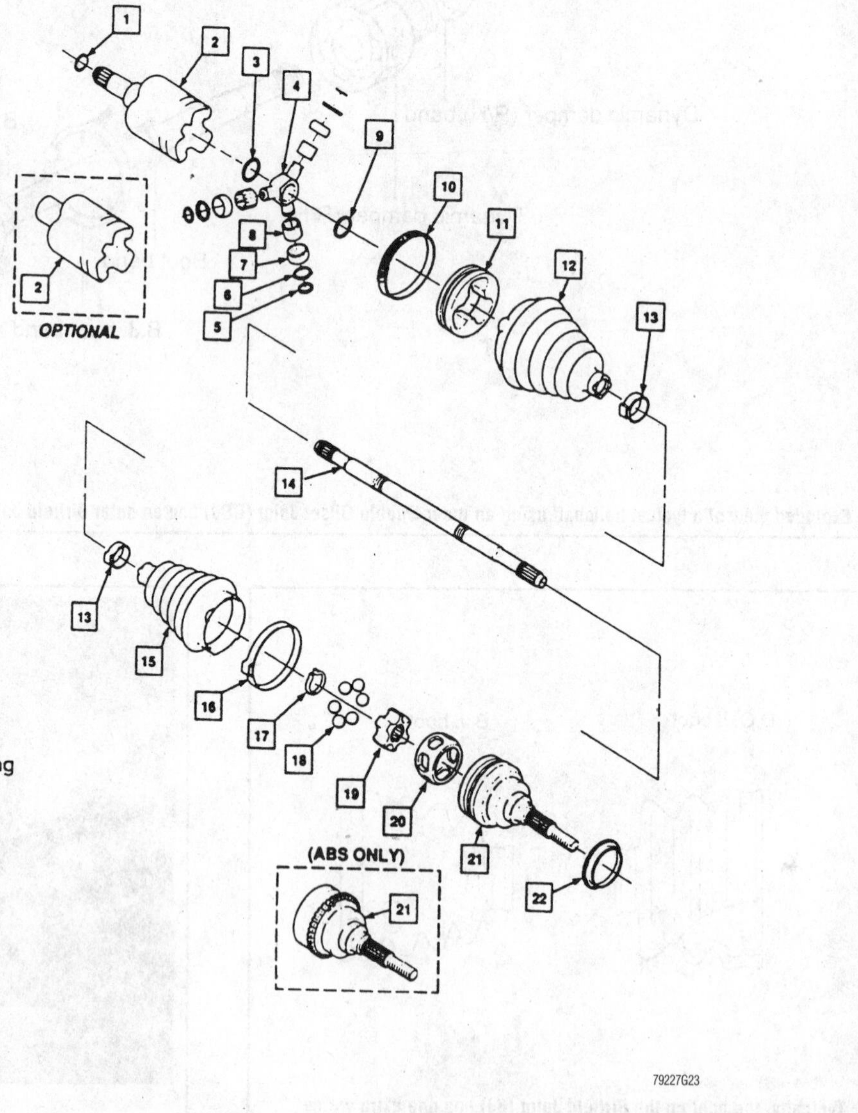

---

1 Retaining ring
2 Tri-pot housing asm.
3 Shaft retaining ring
4 Tri-pot joint spider
5 Needle retainer ring
6 Needle retainer
7 Tri-pot joint ball
8 Needle roller
9 Spacer ring
10 Seal retaining clamp
11 Trilobal tri-pot bushing
12 Tri-pot joint seal
13 Seal retaining clamp
14 Axle shaft
15 C/V joint seal
16 Seal retaining clamp
17 Race retaining ring
18 Ball
19 C/V joint inner race
20 C/V joint cage
21 C/V joint outer race
22 Deflector ring

OPTIONAL

(ABS ONLY)

79227G23

Exploded view of a halfshaft with Tripot inner and a Double Offset outer joints

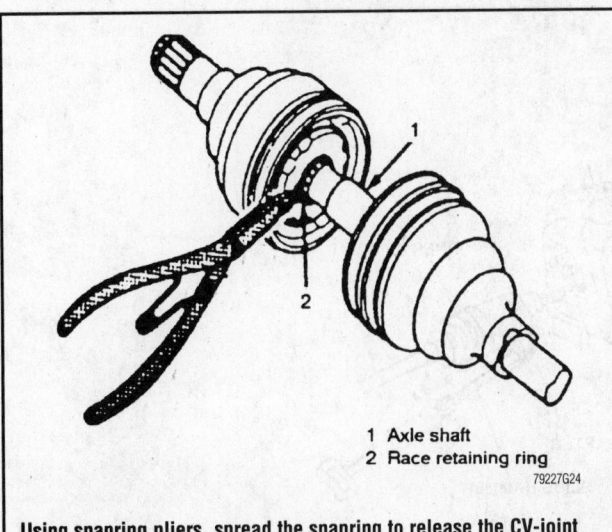

1 Axle shaft
2 Race retaining ring

79227G24

Using snapring pliers, spread the snapring to release the CV-joint from the shaft—outer CV-joint shown

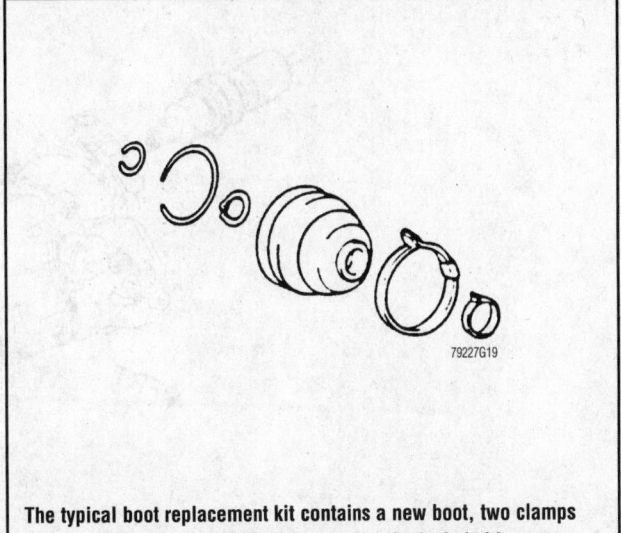

79227G19

The typical boot replacement kit contains a new boot, two clamps and special grease—new circlips may also be included is some

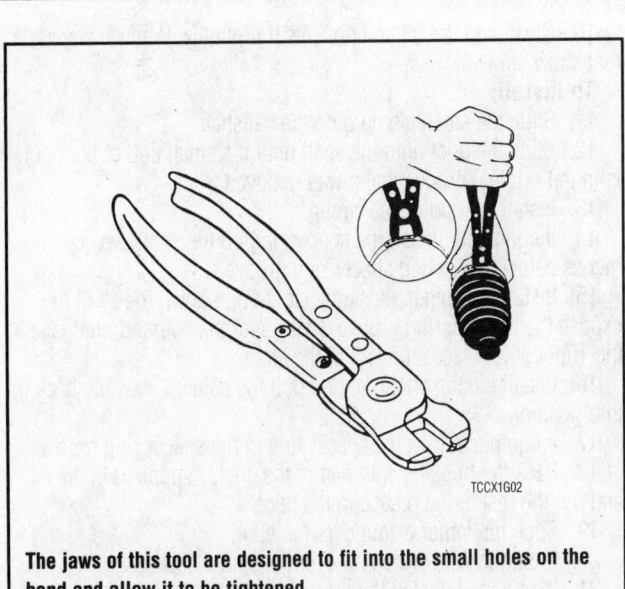

TCCX1G02

The jaws of this tool are designed to fit into the small holes on the band and allow it to be tightened

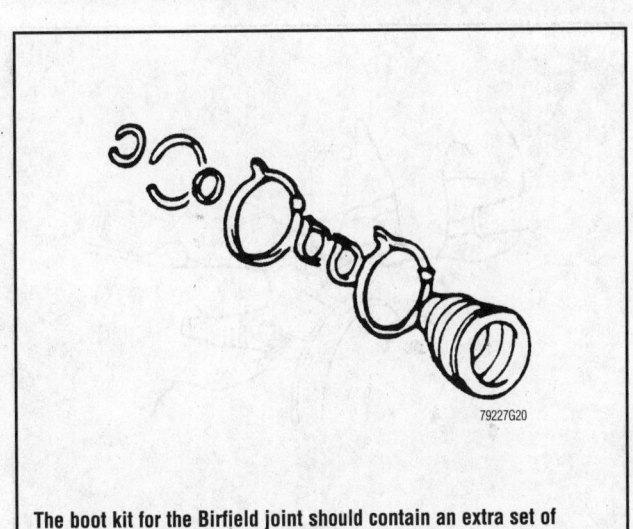

79227G20

The boot kit for the Birfield joint should contain an extra set of bands, because the inner joint must be removed in order to install a new boot on the outer (Birfield) joint

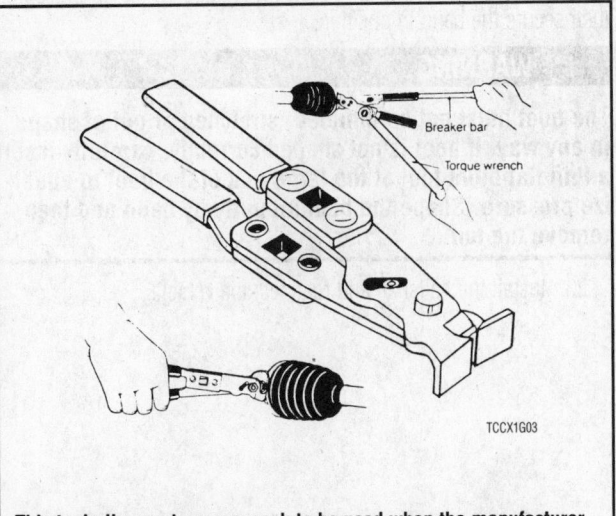

Breaker bar

Torque wrench

TCCX1G03

This tool allows a torque wrench to be used when the manufacturer specifies that a certain pressure is required to crimp the band

insert a thin, blunt tool under the large end of the boot to equalize air pressure. Shape the boot properly by hand, then remove the tool.

15. Install the halfshaft. Road test the vehicle to check for abnormal noise or vibration.

Inner Boot

1. Remove the halfshaft and carefully place it in a vise using a protective covering on the vise jaws.
2. Cut the large and small boot clamps and discard.

✳✳ WARNING

Do not cut through the boot and damage the sealing surface of the Tripot outer housing.

3. Pull the boot down the shaft to expose the joint.
4. If equipped, remove the large circlip from the inner edge of the outer bearing race.

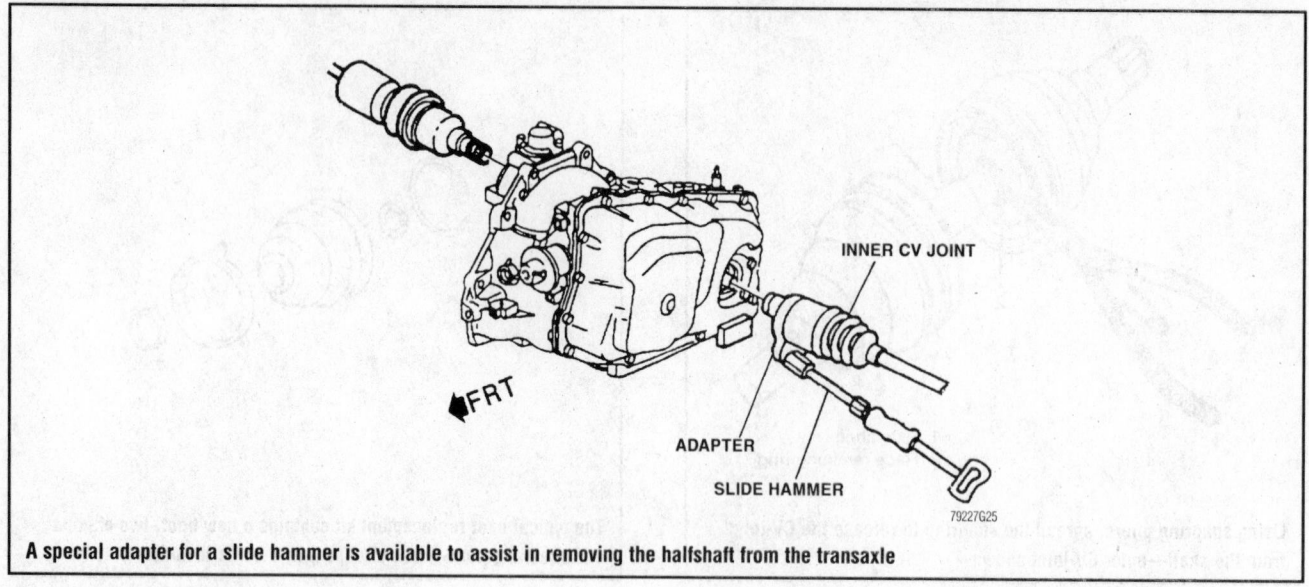

A special adapter for a slide hammer is available to assist in removing the halfshaft from the transaxle

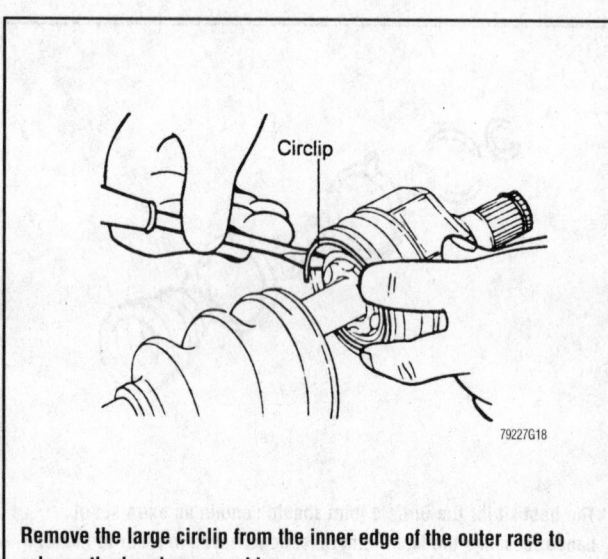

Remove the large circlip from the inner edge of the outer race to release the bearing assembly

5. Matchmark the bearing and outer case so they can be installed in their original positions.

6. Remove the Tripot housing from the spider and axle. Clean and dry all components thoroughly. Replace any parts that show signs of wear.

7. Push the spider assembly down the shaft to uncover the snapring on the end of the shaft. Remove the snapring and slide the spider assembly off the end of the shaft.

8. If equipped, remove the spacer ring from the shaft.

9. Remove the remaining circlip and slide the boot off the shaft.

10. Clean and dry all components thoroughly. Replace any parts that show signs of wear.

To install:

11. Slide the small clamp onto the halfshaft.

12. Slide the boot onto the shaft until the small end of the boot is in the original groove that it was removed from.

13. Install the applicable circlip.

14. If equipped, install the spacer ring on the shaft, several inches below the second spacer ring groove.

15. Install the spider assembly far enough down the shaft to expose the top snapring groove. Make sure the counterbored face of the Tripot spider faces the end of the shaft.

16. Install the top snapring and pull the spider assembly back up into position.

17. If equipped, lock the spacer ring in the spacer ring groove.

18. Pack the housing with half of the grease supplied in the kit and put the rest of the grease in the boot.

19. Slide the larger clamp over the boot.

20. Push the Tripot housing over the spider assembly.

21. If equipped, install the large circlip in the outer race.

22. Slide the larger diameter of the boot into position. Slide a small dull tool under the lip of the boot to equalize the air pressure, then secure the band in position.

✳✳ WARNING

The boot must not be dimpled, stretched or out of shape in any way. If boot is not shaped correctly, carefully insert a thin flat blunt tool at the large end of the boot to equalize pressure. Shape the boot properly by hand and then remove the tool.

23. Install the halfshaft and road test the vehicle.

OXYGEN (O$_2$) SENSORS

8

OXYGEN (O2) SENSORS........8-1
OXYGEN SENSOR LOCATION
  CHART 8-13

General Information 8-1
O$_2$ (Oxygen) Sensor Service 8-2
  Locations.................................8-11

Precautions 8-2
Removal & Installation 8-9
Testing..8-2

OXYGEN (O$_2$) SENSORS

General Information

An Oxygen (O$_2$) sensor is an input device used by the engine control computer to monitor the amount of oxygen in the exhaust gas stream. This information is used by the computer, along with other inputs, to fine-tune the air/fuel mixture so that the engine can run with the greatest efficiency in all conditions. The O$_2$ sensor sends this information to the computer in the form of a 100–900 millivolt (mV) reference signal, which is actually created by the O$_2$ sensor itself through chemical interactions between the sensor tip material (zirconium dioxide in almost all cases) and the oxygen levels in the exhaust gas stream and ambient atmosphere gas. At operating temperatures, approximately 1100°F (600°C), the element becomes a semiconductor. Essentially, through the differing levels of oxygen in the exhaust gas stream and in the surrounding atmosphere, the sensor creates a voltage signal which is directly and consistently related to the concentration of oxygen in the exhaust stream. Typically, a higher than normal amount of oxygen in the exhaust stream indicates that not all of the available oxygen was used in the combustion process, because there was not enough fuel (lean condition) present. Inversely, a lower than normal concentration of oxygen in the exhaust stream indicates that a large amount was used in the combustion process, because a larger than necessary amount of fuel was present (rich condition). Thus, the engine control computer can correct the amount of fuel introduced into the combustion chambers.

Since the control computer uses the O$_2$ sensor output voltage as an indication of the oxygen concentration, and the oxygen concentration directly affects O$_2$ sensor output, the signal voltage from the sensor to the computer fluctuates constantly. This fluctuation is caused by the nature of the interaction between the computer and the O$_2$ sensor, which follows a general pattern: detect, compare, compensate, detect, compare, compensate, etc. This means that when the computer detects a lean signal from the O$_2$ sensor, it compares the reading with known parameters stored within its memory. It calculates that there is too much oxygen present in the exhaust gases, so it compensates by adding more fuel to the air/fuel mixture. This, in turn, causes the O$_2$ sensor to send a rich signal to the computer, which then compares this new signal, and adjusts the air/fuel mixture again. This pattern constantly repeats itself: detect rich, compare, compensate lean, detect lean, compare, compensate rich, etc. Since the O$_2$ sensor fluctuates between rich and lean, and because the lean limit for sensor output is 100 mV and the rich limit is 900 mV, the proper voltage signal from a normally functioning O$_2$ sensor consistently fluctuates between 100–300 and 700–900 mV.

➡ The sensor voltage may never quite reach 100 or 900 mV, but it should fluctuate from at least below 300 mV to above 700 mV, and the mid-point of the fluctuations should be centered around 500 mV.

To improve O$_2$ sensor efficiency, newer O$_2$ sensors were designed with a built-in heating element, and were called Heated O$_2$ (HO$_2$) sensors. This heating element was incorporated into the sensor so that the sensor would reach optimal operating temperature quicker, meaning that the O$_2$ sensor output signal could be used by the engine control computer sooner. Because the sensor reaches optimal temperature quicker, modern vehicles enjoy improved driveability and fuel economy even before the engine reaches normal operating temperature.

Although a few manufacturers changed earlier, in 1995 all vehicles were required to implement a new set of engine control param-

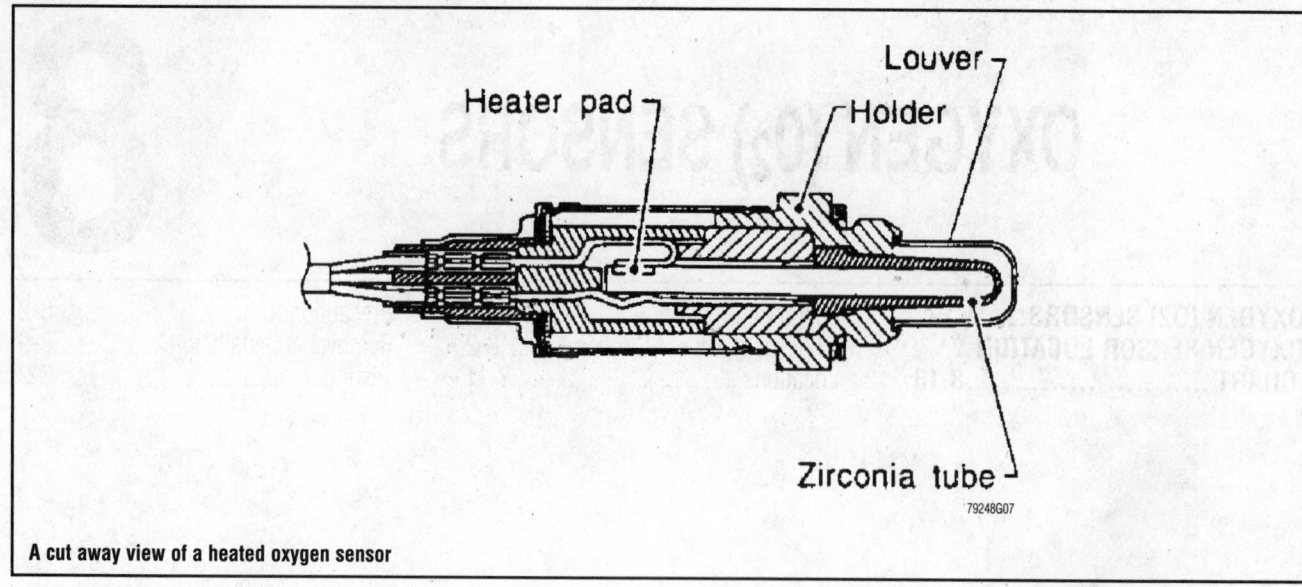

A cut away view of a heated oxygen sensor

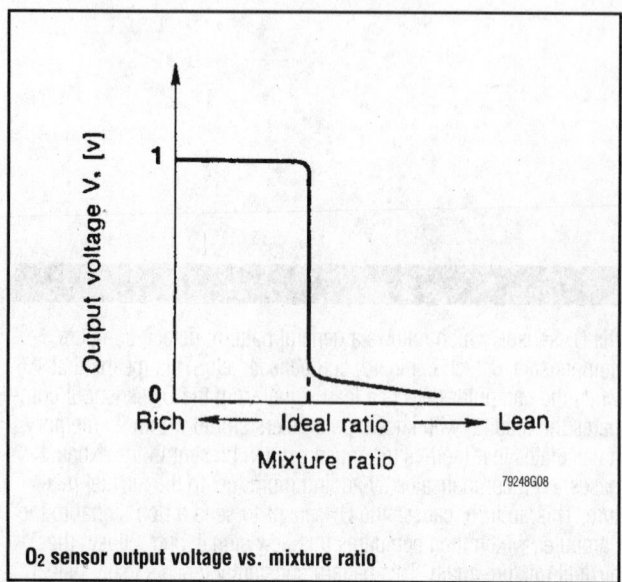

O₂ sensor output voltage vs. mixture ratio

eters, referred to as On-Board Diagnostics second generation (OBD-II). This updated system (based on the former OBD-I), called for additional O₂ sensors to be used after the catalytic converter, so that catalytic converter efficiency could be measured by the vehicle's engine control computer. The O₂ sensors mounted in the exhaust system after the catalytic converters are not used to affect air/fuel mixture; they are used solely to monitor catalytic converter efficiency.

O2 (Oxygen) Sensor Service

PRECAUTIONS

When testing or servicing an O₂ sensor you will need to start and warm the engine to operating temperature in order to either perform the necessary testing procedures or to easily remove the sensor from its fitting. This will create a situation in which you will be working around a **HOT** exhaust system. The following is a list of precautions to consider during this service:

- Do not pierce any wires when testing an O₂ sensor, as this can lead to wiring harness damage. Backprobe the connector, when necessary.
- While testing the sensor, be sure to keep out of the way of moving engine components, such as the cooling fan. Refrain from wearing loose clothing which may become tangled in moving engine components.
- Safety glasses must be worn at all times when working on, or near, the exhaust system. Older exhaust systems may be covered with loose rust particles which can shower you when disturbed. These particles are more than a nuisance and can injure your eye.
- Be cautious when working on and around the hot exhaust system. Painful burns will result if skin is exposed to the exhaust system pipes or manifolds.
- The O₂ sensor may be difficult to remove when the engine temperature is below 120°F (48°C). Excessive force may damage the threads in the exhaust manifold or pipe, therefore always start the engine and allow it to reach normal operating temperature prior to removal.
- Since O₂ sensors are usually designed with a permanently-attached wiring pigtail (this allows the wiring harness and sensor connectors to be positioned away from the hot exhaust system), it may be necessary to use a socket or wrench that is designed specifically for this purpose. Before purchasing such a socket, be sure that you can't save some money by using a box end wrench for sensor removal.

TESTING

The best, and most accurate method to test the operation of an O₂ sensor is with the use of either an oscilloscope or a Diagnostic Scan Tool (DST), following their specific instructions for testing. It is possible, however, to test whether the O₂ sensor is functioning properly within general parameters using a Digital Volt-Ohmmeter (DVOM), also referred to as a Digital Multi-Meter (DMM). Newer DMM's are often designed to perform many advanced diagnostic functions, and some are even constructed to be used as an oscilloscope. Two in-vehicle testing procedures, and one bench test procedure, will be provided for the common zirconium dioxide oxygen sensor. The first in-vehicle test makes use of a standard DVOM with a 10 megohm impedance, whereas the second in-vehicle test presented necessitates the usage of an advanced DMM with MIN/MAX/Average functions. Both of these in-vehicle test procedures are likely to set Diagnostic Trouble Codes (DTC's) in the engine control computer. Therefore, after testing, be sure to clear all DTC's before retesting the sensor, if necessary.

These are some of the common DTC's which may be set during testing:

- Open in the O_2 sensor circuit
- Constant low voltage in the O_2 sensor circuit
- Constant high voltage in the O_2 sensor circuit
- Other fuel system problems could set a O_2 sensor code

➡**Because an improperly functioning fuel delivery and/or control system can adversely affect the O_2 sensor voltage output signal, testing only the O_2 sensor is an inaccurate method for diagnosing an engine driveability problem.**

If after testing the sensor, the sensor is thought to be defective because of high or low readings, be sure to check that the fuel delivery and engine management system is working properly before condemning the O_2 sensor. Otherwise, the new O_2 sensor may continue to register the same high or low readings.

Often, by testing the O_2 sensor, another problem in the engine control management system can be diagnosed. If the sensor appears to be defective while installed in the vehicle, perform the bench test. If the sensor functions properly during the bench test, chances are that there may be a larger problem in the vehicle's fuel delivery and/or control system.

Many things can cause an O_2 sensor to fail, including old age, antifreeze contamination, physical damage, prolonged exposure to overly-rich exhaust gases, and exposure to silicone sealant fumes. Be sure to remedy any such condition prior to installing a new sensor, otherwise the new sensor may be damaged as well.

➡**Perform a visual inspection of the sensor. Black sooty deposits may indicate a rich air/fuel mixture, brown deposits may indicate an oil consumption problem, and white gritty deposits may indicate an internal coolant leak. All of these conditions can destroy a new sensor if not corrected before installation.**

O₂ Sensor Terminal Identification

The easiest method for determining sensor terminal identification is to use a wiring diagram for the vehicle and engine in question. However, if a wiring diagram is not available there is a method for determining terminal identification. Throughout the testing procedures, the following terms will be used for clarity:

- Vehicle harness connector—this refers to the connector on the wires which are attached to the vehicle; NOT the connector at the end of the sensor pigtail.

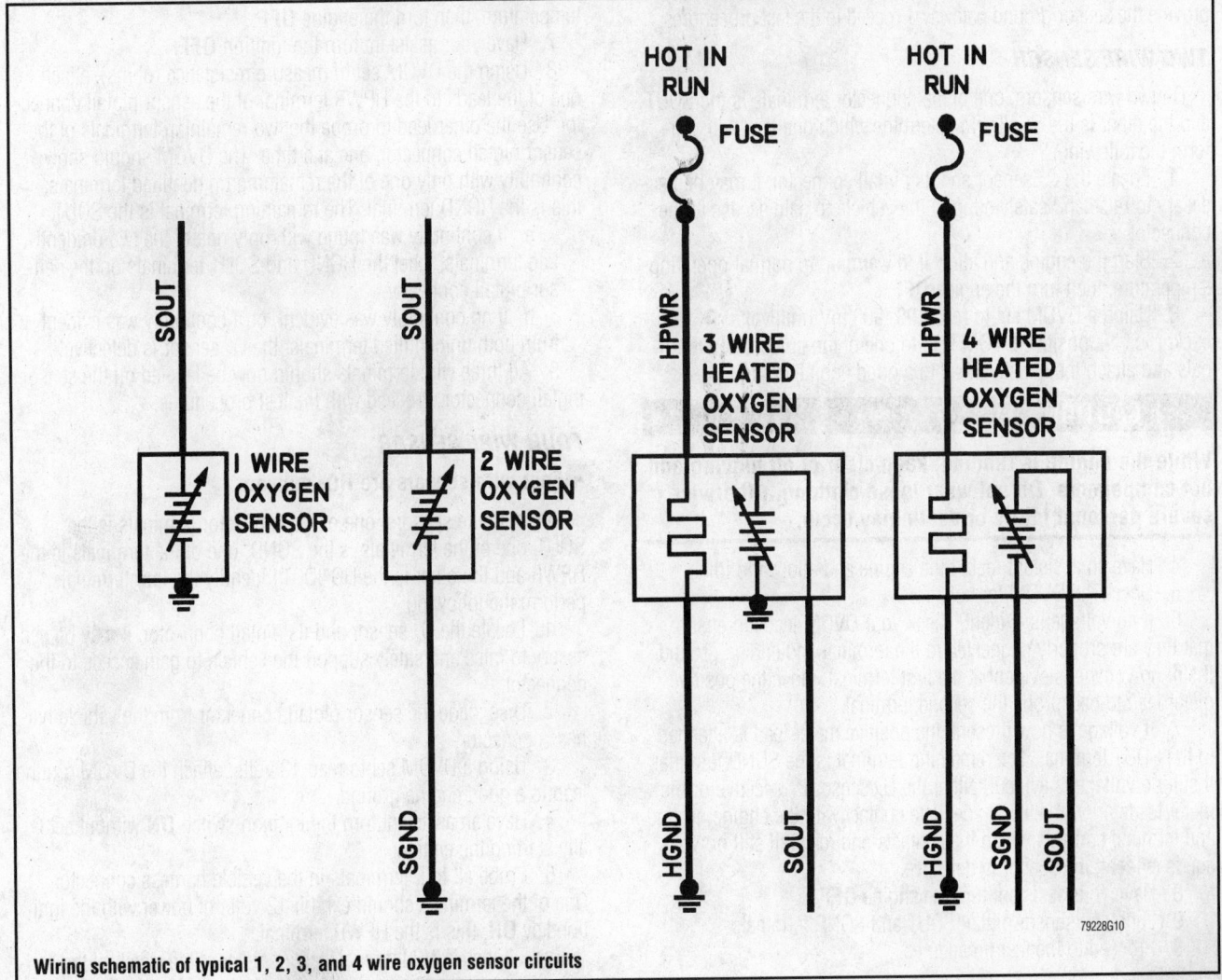

Wiring schematic of typical 1, 2, 3, and 4 wire oxygen sensor circuits

79228G10

- Sensor pigtail connector—this refers to the connector attached to the sensor itself.
- O_2 circuit—this refers to the circuit in a Heated O_2 (HO_2) sensor which corresponds to the oxygen-sensing function of the sensor; NOT the heating element circuit.
- Heating circuit—this refers to the circuit in a HO_2 sensor which is designed to warm the HO_2 sensor quickly to improve driveability.
- Sensor Output (SOUT) terminal—this is the terminal which corresponds to the O_2 circuit output. This is the terminal which will register the millivolt signals created by the sensor based upon the amount of oxygen in the exhaust gas stream.
- Sensor Ground (SGND) terminal—when a sensor is so equipped, this refers to the O_2 circuit ground terminal. Many O_2 sensors are not equipped with a ground wire, rather they utilize the exhaust system for the ground circuit.
- Heating Power (HPWR) terminal—this terminal corresponds to the circuit which provides the O_2 sensor heating circuit with power when the ignition key is turned to the **ON** or **RUN** positions.
- Heating Ground (HGND) terminal—this is the terminal connected to the heating circuit ground wire.

ONE WIRE SENSOR

One wire sensors are by far the easiest to determine sensor terminal identification, but this is self-evident. On one wire O_2 sensors, the single wire terminal is the SOUT and the exhaust system is used to provide the sensor ground pathway. Proceed to the test procedures.

TWO WIRE SENSOR

On two wire sensors, one of the connector terminals is the SOUT and the other is the SGND. To determine which one is which, perform the following:

1. Locate the O_2 sensor and its pigtail connector. It may be necessary to raise and safely support the vehicle to gain access to the connector.
2. Start the engine and allow it to warm up to normal operating temperature, then turn the engine **OFF**.
3. Using a DVOM set to read 100–900 mV (millivolts) DC, backprobe the positive DVOM lead to one of the unidentified terminals and attach the negative lead to a good engine ground.

※※ CAUTION

While the engine is running, keep clear of all moving and hot components. Do not wear loose clothing. Otherwise severe personal injury or death may occur.

4. Have an assistant restart the engine and allow it to idle.
5. Check the DVOM for voltage.
6. If no voltage is evident, check your DVOM leads to ensure that they are properly connected to the terminal and engine ground. If still no voltage is evident at the first terminal, move the positive meter lead to backprobe the second terminal.
7. If voltage is now present, the positive meter lead is attached to the SOUT terminal. The remaining terminal is the SGND terminal. If still no voltage is evident, either the O_2 sensor is defective or the meter leads are not making adequate contact with the engine ground and terminal contacts; clean the contacts and retest. If still no voltage is evident, the sensor is defective.
8. Have your assistant turn the engine **OFF**.
9. Label the sensor pigtail SOUT and SGND terminals.
10. Proceed to the test procedures.

THREE WIRE SENSOR

➡**Three wire sensors are HO_2 sensors.**

On three wire sensors, one of the connector terminals is the SOUT, one of the terminals is the HPWR and the other is the HGND. The SGND is achieved through the exhaust system, as with the one wire O_2 sensor. To identify the three terminals, perform the following:

1. Locate the O_2 sensor and its pigtail connector. It may be necessary to raise and safely support the vehicle to gain access to the connector.
2. Disengage the sensor pigtail connector from the vehicle harness connector.
3. Using a DVOM set to read 12 volts, attach the DVOM ground lead to a good engine ground.
4. Have an assistant turn the ignition switch **ON** without actually starting the engine.
5. Probe all three terminals in the vehicle harness connector. One of the terminals should exhibit 12 volts of power with the ignition key **ON**; this is the HPWR terminal.
   a. If the HPWR terminal was identified, note which of the sensor harness connector terminals is the HPWR, then match the vehicle harness connector to the sensor pigtail connector. Label the corresponding sensor pigtail connector terminal with HPWR.
   b. If none of the terminals showed 12 volts of power, locate and test the heater relay or fuse. Then, perform Steps 3–6 again.
6. Start the engine and allow it to warm up to normal operating temperature, then turn the engine **OFF**.
7. Have your assistant turn the ignition **OFF**.
8. Using the DVOM set to measure resistance (ohms), attach one of the leads to the HPWR terminal of the sensor pigtail connector. Use the other lead to probe the two remaining terminals of the sensor pigtail connector, one at a time. The DVOM should show continuity with only one of the remaining unidentified terminals; this is the HGND terminal. The remaining terminal is the SOUT.
   a. If continuity was found with only one of the two unidentified terminals, label the HGND and SOUT terminals on the sensor pigtail connector.
   b. If no continuity was evident, or if continuity was evident from both unidentified terminals, the O_2 sensor is defective.
9. All three wire terminals should now be labeled on the sensor pigtail connector. Proceed with the test procedures.

FOUR WIRE SENSOR

➡**Four wire sensors are HO_2 sensors.**

On four wire sensors, one of the connector terminals is the SOUT, one of the terminals is the SGND, one of the terminals is the HPWR and the other is the HGND. To identify the four terminals, perform the following:

1. Locate the O_2 sensor and its pigtail connector. It may be necessary to raise and safely support the vehicle to gain access to the connector.
2. Disengage the sensor pigtail connector from the vehicle harness connector.
3. Using a DVOM set to read 12 volts, attach the DVOM ground lead to a good engine ground.
4. Have an assistant turn the ignition switch **ON** without actually starting the engine.
5. Probe all four terminals in the vehicle harness connector. One of the terminals should exhibit 12 volts of power with the ignition key **ON**; this is the HPWR terminal.
   a. If the HPWR terminal was identified, note which of the sen-

sor harness connector terminals is the HPWR, then match the vehicle harness connector to the sensor pigtail connector. Label the corresponding sensor pigtail connector terminal with HPWR.

b. If none of the terminals showed 12 volts of power, locate and test the heater relay or fuse. Then, perform Steps 2–6 again.

6. Have your assistant turn the ignition **OFF**.

7. Using the DVOM set to measure resistance (ohms), attach one of the leads to the HPWR terminal of the sensor pigtail connector. Use the other lead to probe the three remaining terminals of the sensor pigtail connector, one at a time. The DVOM should show continuity with only one of the remaining unidentified terminals; this is the HGND terminal.

a. If continuity was found with only one of the two unidentified terminals, label the HGND terminal on the sensor pigtail connector.

b. If no continuity was evident, or if continuity was evident from all unidentified terminals, the O₂ sensor is defective.

c. If continuity was found at two of the other terminals, the sensor is probably defective. However, the sensor may not necessarily be defective, because it may have been designed with the two ground wires joined inside the sensor in case one of the ground wires is damaged; the other circuit could still function properly. Though, this is highly unlikely. A wiring diagram is necessary in this particular case to know whether the sensor was so designed.

8. Reattach the sensor pigtail connector to the vehicle harness connector.

9. Start the engine and allow it to warm up to normal operating temperature, then turn the engine **OFF**.

10. Using a DVOM set to read 100–900 mV (millivolts) DC, backprobe the negative DVOM lead to one of the unidentified terminals and the positive lead to the other unidentified terminal.

✷✷ CAUTION

While the engine is running, keep clear of all moving and hot components. Do not wear loose clothing. Otherwise severe personal injury or death may occur.

11. Have an assistant restart the engine and allow it to idle.

12. Check the DVOM for voltage.

a. If no voltage is evident, check your DVOM leads to ensure that they are properly connected to the terminals. If still no voltage is evident at either of the terminals, either the terminals were accidentally marked incorrectly or the sensor is defective.

b. If voltage is present, but the polarity is reversed (the DVOM will show a negative voltage amount), turn the engine **OFF** and swap the two DVOM leads on the terminals. Start the engine and ensure that the voltage now shows the proper polarity.

c. If voltage is evident and is the proper polarity, the positive DVOM lead is attached to the SOUT and the negative lead to the SGND terminals.

13. Have your assistant turn the engine **OFF**.

14. Label the sensor pigtail SOUT and SGND terminals.

In-Vehicle Tests

✷✷ WARNING

Never apply voltage to the O₂ circuit of the sensor, otherwise it may be damaged. Also, never connect an ohmmeter (or a DVOM set on the ohm function) to both of the O₂ circuit terminals (SOUT and SGND) of the sensor pigtail connector; it may damage the sensor.

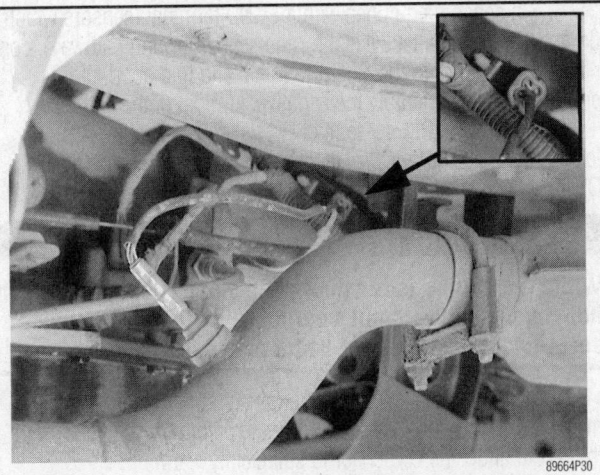

89664P30

To test the O₂ sensor, locate it and its connector (inset), which should be positioned away from the exhaust system to prevent heat damage

Test 1 makes use of a standard DVOM with a 10 megohm impedance, whereas Test 2 necessitates the usage of an advanced Digital Multi-Meter (DMM) with MIN/MAX/Average functions or a sliding bar graph function. Both of these in-vehicle test procedures are likely to set Diagnostic Trouble Codes (DTC's) in the engine control computer. Therefore, after testing, be sure to clear all DTC's before retesting the sensor, if necessary. The third in-vehicle test is designed for the use of a scan tool or oscilloscope. The fourth test (Heating Circuit Test) is designed to check the function of the heating circuit in a HO₂ sensor.

➡ **If the O₂ sensor being tested is designed to use the exhaust system for the SGND, excessive corrosion between the exhaust and the O₂ sensor may affect sensor functioning.**

The in-vehicle tests may be performed for O₂ sensors located in the exhaust system after the catalytic converter. However, the O₂ sensors located behind the catalytic converter will not fluctuate like the sensors mounted before the converter, because the converter, when functioning properly, emits a steady amount of oxygen. If the O₂ sensor mounted after the catalytic converter exhibits a fluctuating signal (like other O₂ sensors), the catalytic converter is most likely defective.

TEST 1—DIGITAL VOLT-OHMMETER

This test will not only verify proper sensor functioning, but is also designed to ensure the engine control computer and associated wiring is functioning properly as well.

1. Start the engine and allow it to warm up to normal operating temperature.

➡ **If you are using the opening of the thermostat to gauge normal operating temperature, be forewarned: a defective thermostat can open too early and prevent the engine from reaching normal operating temperature. This can cause a slightly rich condition in the exhaust, which can throw the O₂ sensor readings off slightly.**

2. Turn the ignition switch **OFF**, then locate the O₂ sensor pigtail connector.

3. Perform a visual inspection of the connector to ensure it is properly engaged and all terminals are straight, tight and free from corrosion or damage.

4. Disengage the sensor pigtail connector from the vehicle harness connector.

5. On sensors equipped with a SGND terminal (sensors which do not use the exhaust system for the sensor ground pathway), connect a jumper wire to the SGND terminal and to a good, clean engine ground (preferably the negative terminal of the battery).

6. Using a DVOM set to read DC voltage, attach the positive lead to the SOUT terminal of the sensor pigtail connector, and the DVOM negative lead to a good engine ground.

✳✳ CAUTION

While the engine is running, keep clear of all moving and hot components. Do not wear loose clothing. Otherwise severe personal injury or death may occur.

7. Have an assistant start the engine and hold it at approximately 2,000 rpm. Wait at least 1 minute before commencing with the test to allow the O₂ sensor to sufficiently warm up.

➡ **Some carbureted Asian models may not switch into closed loop operation until engine speed is above 2,500 rpm.**

8. Using a jumper wire, connect the SOUT terminal of the **vehicle harness connector** to a good engine ground. This will fool the engine control computer into thinking it is receiving a lean signal from the O₂ sensor, and, therefore, the computer will enrich the air/fuel ratio. With the SOUT terminal so grounded, the DVOM should register at least 800 mV, as the control computer adds additional fuel to the air/fuel ratio.

9. While observing the DVOM, disconnect the vehicle harness connector SOUT jumper wire from the engine ground. Use the jumper wire to apply slightly less than 1 volt to the SOUT terminal of the vehicle harness connector. One method to do this is by grasping and squeezing the end of the jumper between your forefinger and thumb of one hand while touching the positive terminal of the battery post with your other hand. This allows your body to act as a resistor for the battery positive voltage, and fools the engine control computer into thinking it is receiving a rich signal. Or, use a mostly-drained AA battery by connecting the positive terminal of the AA battery to the jumper wire and the negative terminal of the battery to a good engine ground. (Another jumper wire may be necessary to do this.) The computer should lean the air/fuel mixture out. This lean mixture should register as 150 mV or less on the DVOM.

10. If the DVOM did not register millivoltages as indicated, the problem may be either the sensor, the engine control computer or the associated wiring. Perform the following to determine which is the defective component:

 a. Remove the vehicle harness connector SOUT jumper wire.

 b. While observing the DVOM, artificially enrich the air/fuel charge using propane. The DVOM reading should register higher than normal millivoltages. (Normal voltage for an ideal air/fuel mixture is approximately 450–550 mV DC). Then, lean the air/fuel intake charger by either disconnecting one of the fuel injector wiring harness connectors (to prevent the injector from delivering fuel) or by detaching one or two vacuum lines (to add additional non-metered air into the engine). The DVOM should now register lower than normal millivoltages. If the DVOM functioned as indicated, the problem lies elsewhere in the fuel delivery and control system. If the DVOM readings were still unresponsive, the O₂ sensor is defective; replace the sensor and retest.

➡ **Poor wire connections and/or ground circuits may shift a normal O₂ sensor's millivoltage readings up into the rich range or down into the lean range. It is a good idea to check**

the wire condition and continuity before replacing a component which will not fix the problem. A voltage drop test between the sensor case and ground which reveals 14–16 mV, or more, indicates a probable bad ground.

11. Turn the engine **OFF**, remove the DVOM and all associated jumper wires. Reattach the vehicle harness connector to the sensor pigtail connector. If applicable, reattach the fuel injector wiring connector and/or the vacuum line(s).

12. Clear any DTC's present in the engine control computer memory, as necessary.

TEST 2—DIGITAL MULTI-METER

This test method is a more straight forward O₂ sensor test, and does not test the engine control computer's response to the O₂ sensor signal. The use of a DMM with the MIN/MAX/Average function or sliding bar graph/wave function is necessary for this test. Don't forget that the O₂ sensor mounted after the catalytic converter (if equipped) will not fluctuate like the other O₂ sensor(s) will.

1. Start the engine and allow it to warm up to normal operating temperature.

➡ **If you are using the opening of the thermostat to gauge normal operating temperature, be forewarned: a defective thermostat can open too early and prevent the engine from reaching normal operating temperature. This can cause a slightly rich condition in the exhaust, which can throw the O₂ sensor readings off slightly.**

2. Turn the ignition switch **OFF**, then locate the O₂ sensor pigtail connector.

3. Perform a visual inspection of the connector to ensure it is properly engaged and all terminals are straight, tight and free from corrosion or damage.

4. Backprobe the O₂ sensor connector terminals. Attach the DMM positive test lead to the SOUT terminal of the sensor pigtail connector and the negative lead to either the SGND terminal of the sensor pigtail connector (if equipped—refer to the terminal identification procedures earlier in this section for clarification) or to a good, clean engine ground.

5. Activate the MIN/MAX/Average or sliding bar graph/wave function on the DMM.

✳✳ CAUTION

While the engine is running, keep clear of all moving and hot components. Do not wear loose clothing. Otherwise severe personal injury or death may occur.

6. Have an assistant start the engine and wait a few minutes before commencing with the test to allow the O₂ sensor to sufficiently warm up.

7. Read the minimum, maximum and average readings exhibited by the O₂ sensor, or observe the bar graph/wave form. The average reading for a properly functioning O₂ sensor is be approximately 450–550 mV DC. The minimum and maximum readings should vary more than 300–600 mV. A typical O₂ sensor can fluctuate from as low as 100 mV to as high as 900 mV; if the sensor range of fluctuation is not large enough, the sensor is defective. Also, if the fluctuation range is biased up or down in the scale. For example, if the fluctuation range is 400 mV to 900 mV the sensor is defective, because the readings are pushed up into the rich range (as long as the fuel delivery system is functioning properly). The same goes for a fluctuation range pushed down into the lean range.

The mid-point of the fluctuation range should be around 400–500 mV. Finally, if the O₂ sensor voltage fluctuates too slowly (usually the voltage wave should oscillate past the mid-way point of 500 mV several times per second) the sensor is defective. (Technician's refer to this state as "lazy.")

➡️ **Poor wire connections and/or ground circuits may shift a normal O₂ sensor's millivoltage readings up into the rich range or down into the lean range. It is a good idea to check the wire condition and continuity before replacing a component which will not fix the problem. A voltage drop test between the sensor case and ground which reveals 14–16 mV, or more, indicates a probable bad ground.**

8. Using the propane method, enrichen the air/fuel mixture and observe the DMM readings. The average O₂ sensor output signal voltage should rise into the rich range.

9. Lean the air/fuel mixture by either disconnecting a fuel injector wiring harness connector or by disconnecting a vacuum line. The O₂ sensor average output signal voltage should drop into the lean range.

10. If the O₂ sensor did not react as indicated, the sensor is defective and should be replaced.

11. Turn the engine **OFF**, remove the DMM and all associated jumper wires. Reattach the vehicle harness connector to the sensor pigtail connector. If applicable, reattach the fuel injector wiring connector and/or the vacuum line(s).

12. Clear any DTC's present in the engine control computer memory, as necessary.

TEST 3—OSCILLOSCOPE

This test is designed for the use of an oscilloscope to test the functioning of an O₂ sensor.

➡️ **This test is only applicable for O₂ sensors mounted in the exhaust system before the catalytic converter.**

1. Start the engine and allow it to reach normal operating temperature.

2. Turn the engine **OFF**, and locate the O₂ sensor connector. Backprobe the scope lead to the O₂ sensor connector SOUT terminal. Refer to the manufacturer's instructions for more information on attaching the scope to the vehicle.

3. Turn the scope ON.

4. Set the oscilloscope amplitude to 200 mV per division, and the time to 1 second per division. Use the 1:1 setting of the probe, and be sure to connect the scope's ground lead to a good, clean engine ground. Set the signal function to automatic or internal triggering.

5. Start the engine and run it at 2,000 rpm.

6. The oscilloscope should display a wave form, representative of the O₂ sensor switching between lean (100–300 mV) and rich (700–900 mV). The sensor should switch between rich and lean, or lean and rich (crossing the mid-point of 500 mV) several times per second. Also, the range of each wave should reach at least above 700 mV and below 300 mV. However, an occasional low peak is acceptable.

7. Force the air/fuel mixture rich by introducing propane into the engine, then observe the oscilloscope readings. The fluctuating range of the O₂ sensor should climb into the rich range.

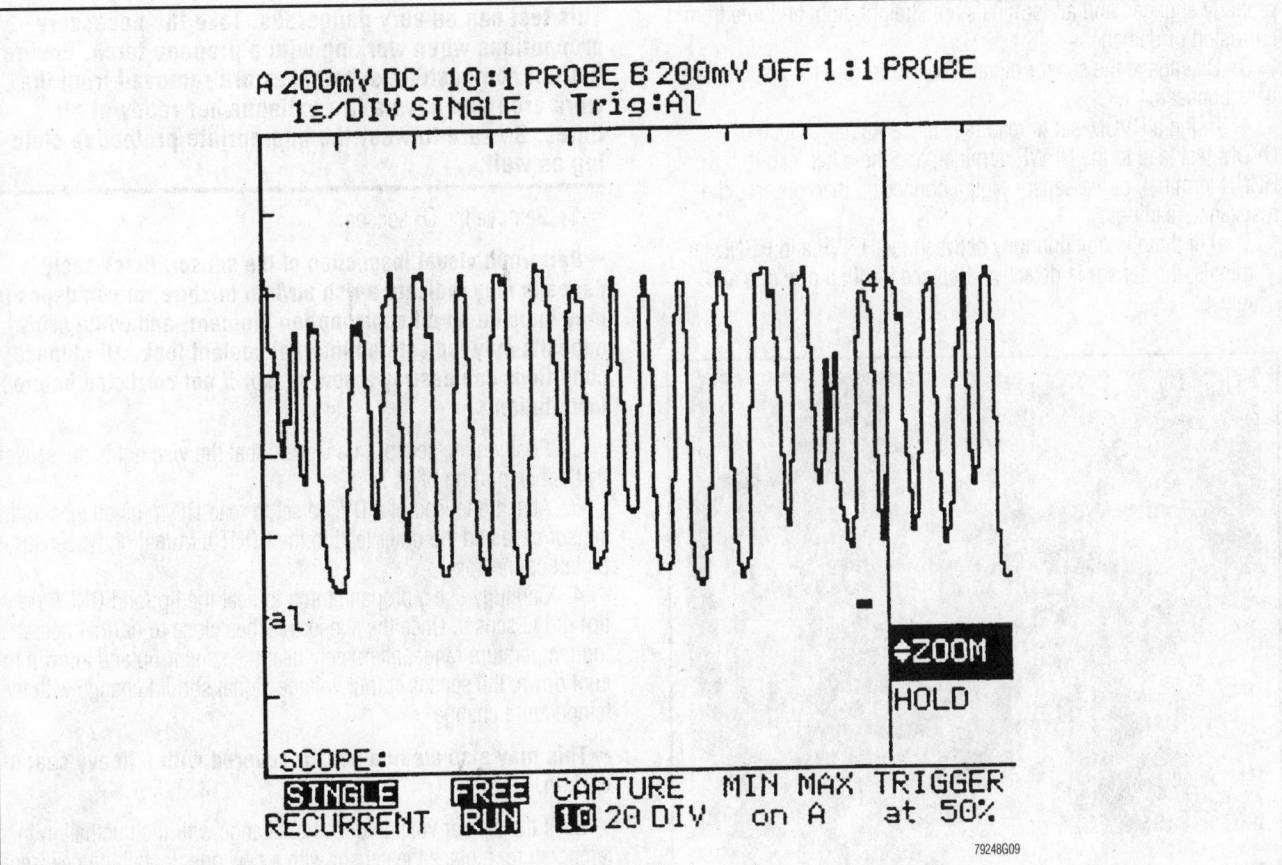

An oscilloscope wave form of a typical good O₂ sensor as it fluctuates from rich to lean

8. Lean the air/fuel mixture out by either detaching a vacuum line or by disengaging one of the fuel injector's wiring connectors. Watch the scope readings; the O₂ sensor wave form should drop toward the lean range.

9. If the O₂ sensor's wave form does not fluctuate adequately, is not centered around 500 mV during normal engine operation, does not climb toward the rich range when propane is added to the engine, or does not drop toward the lean range when a vacuum hose or fuel injector connector is detached, the sensor is defective.

10. Reattach the fuel injector connector or vacuum hose.

11. Disconnect the oscilloscope from the vehicle.

HEATING CIRCUIT TEST

The heating circuit in an O₂ sensor is designed only to heat the sensor quicker than a non-heated sensor. This provides an advantage of increased engine driveability and fuel economy while the engine temperature is still below normal operating temperature, because the fuel management system can enter closed loop operation (more efficient than open loop operation) sooner.

Therefore, if the heating element goes bad, the O₂ sensor may still function properly once the sensor warms up to its normal temperature. This will take longer than normal and may cause mild driveability-related problems while the engine has not reached normal operating temperature.

If the heating element is found to be defective, replace the O₂ sensor without wasting your time testing the O₂ circuit; if necessary, you can perform the O₂ circuit test with the new O₂ sensor and save yourself some time.

1. Locate the O₂ sensor pigtail connector.

2. Perform a visual inspection of the connector to ensure it is properly engaged and all terminals are straight, tight and free from corrosion or damage.

3. Disengage the sensor pigtail connector from the vehicle harness connector.

4. Using a DVOM set to read resistance (ohms), attach one DVOM test lead to the HPWR terminal, and the other lead to the HGND terminal, of the sensor pigtail connector, then observe the resistance readings.

   a. If there is no continuity between the HPWR and HGND terminals, the sensor is defective. Replace it with a new one and retest.

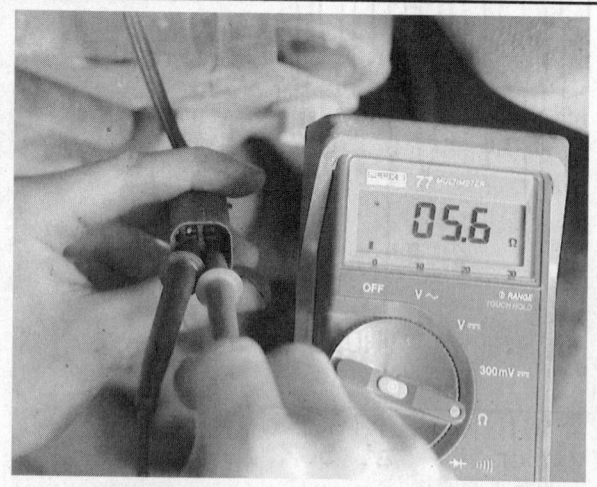

The heating circuit of the O₂ sensor can be tested with a DMM set to measure resistance

89714P27

b. If there is continuity between the two terminals, but the resistance is greater than approximately 20 ohms, the sensor is defective. Replace it with a new one and retest.

➡**For the following step, the HO₂ sensor should be approximately 75°F (23°C) for the proper resistance values.**

   c. If there is continuity between the two terminals and it is less than 20 ohms, the sensor is probably not defective. Because of the large diversity of engine control systems used in vehicles today, O₂ sensor heating circuit resistance specifications change often. Generally, the amount of resistance an O₂ sensor heating circuit should exhibit is between 2–9 ohms. However, some manufacturer's O₂ sensors may show resistance as high as 15–20 ohms. As a rule of thumb, 20 ohms of resistance is the upper limit allowable.

5. Turn the engine **OFF**, remove the DVOM and all associated jumper wires. Reattach the vehicle harness connector to the sensor pigtail connector.

6. Clear any DTC's present in the engine control computer memory, as necessary.

Bench Test

➡**Utilize one of the in-vehicle tests before performing this test.**

This test is designed to test an O₂ sensor which does not seem to fluctuate fully beyond 400–700 mV. The sensor is to be secured in a table-mounted vise.

✷✷ CAUTION

This test can be very dangerous. Take the necessary precautions when working with a propane torch. Ensure that all combustible substances are removed from the work area and have a fire extinguisher ready at all times. Be sure to wear the appropriate protective clothing as well.

1. Remove the O₂ sensor.

➡**Perform a visual inspection of the sensor. Black sooty deposits may indicate a rich air/fuel mixture, brown deposits may indicate an oil consumption problem, and white gritty deposits may indicate an internal coolant leak. All of these conditions can destroy a new sensor if not corrected before installation.**

2. Position the sensor in a vise so that the vise holds the sensor by the hex portion of its case.

3. Attach one lead of a DVOM set to read DC millivoltages to the sensor case and the other lead to the SOUT terminal of the sensor pigtail connector.

4. Carefully use a propane torch to heat the tip (and ONLY the tip) of the sensor. Once the sensor reaches close to normal operating temperature range, alternately heat the sensor up and allow it to cool down; the sensor output voltage signal should change with the temperature change.

➡**This may also clean a sensor covered with a heavy coat of carbon.**

5. If the sensor voltage does not change with the fluctuation in temperature, replace the sensor with a new one. Install the new sensor and perform one of the in-vehicle tests to rule out additional fuel management system faults.

REMOVAL & INSTALLATION

1. Start the engine and allow it to reach normal operating temperature, then turn the ignition switch **OFF**.
2. Disconnect the negative battery cable.
3. Open the hood and locate the O₂ sensor connector. It may be necessary to raise and safely support the vehicle for access to the sensor and its connector.

➡ **On a few models, it may be necessary to remove the passenger seat and lift the carpeting in order to access the connector for a downstream O₂ sensor.**

4. Disengage the O₂ sensor pigtail connector from the vehicle harness connector.

➡ **There are generally two methods used to mount an O₂ sensor in the exhaust system: either the O₂ sensor is threaded directly into the exhaust component (screw-in type), or the O₂ sensor is retained by a flange and two nuts or bolts (flange type).**

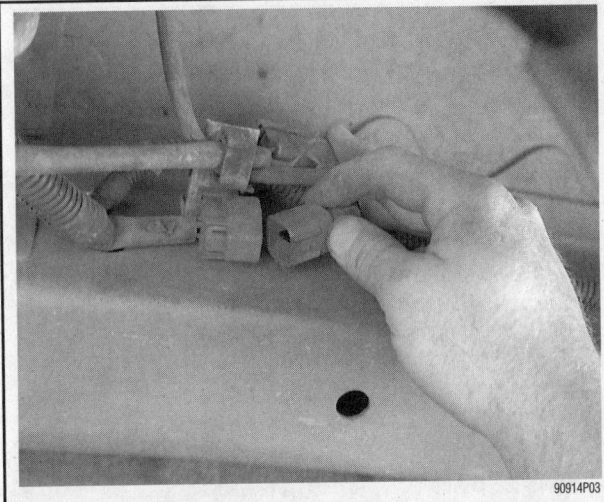

Disengage the sensor pigtail connector half from the vehicle harness connector half

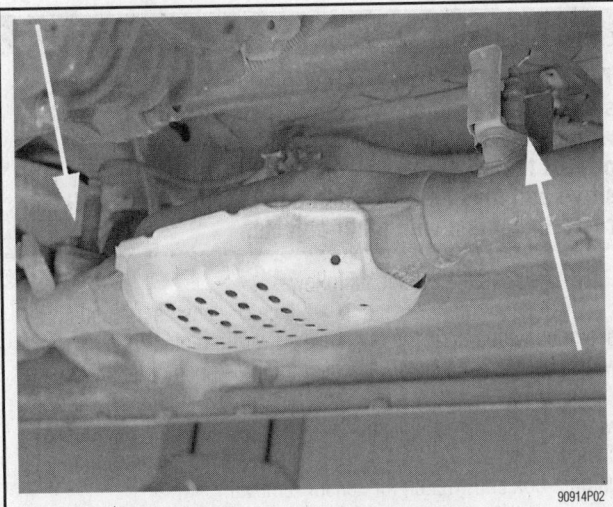

Since sensor locations vary between vehicles, the first step in removal is to locate the O₂ sensors (arrows) . . .

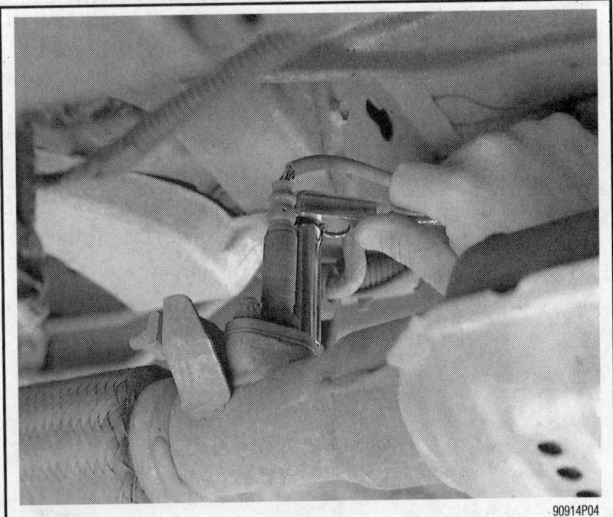

For flange type sensors, loosen the hold-down fasteners . . .

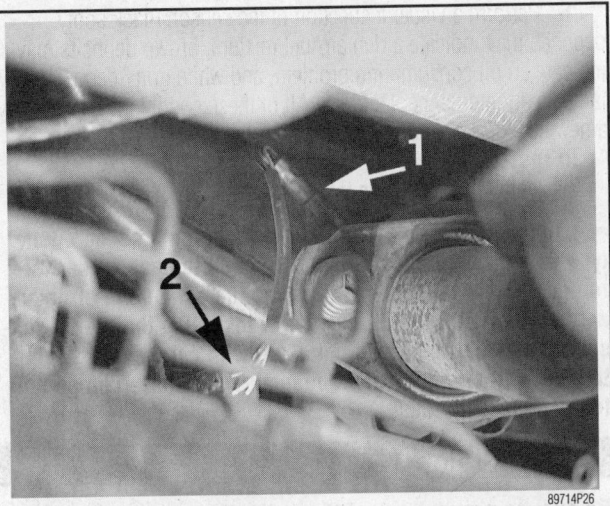

. . . and the sensor connector (2), which is usually near the O₂ sensor (1), but removed enough from the heat of the exhaust system

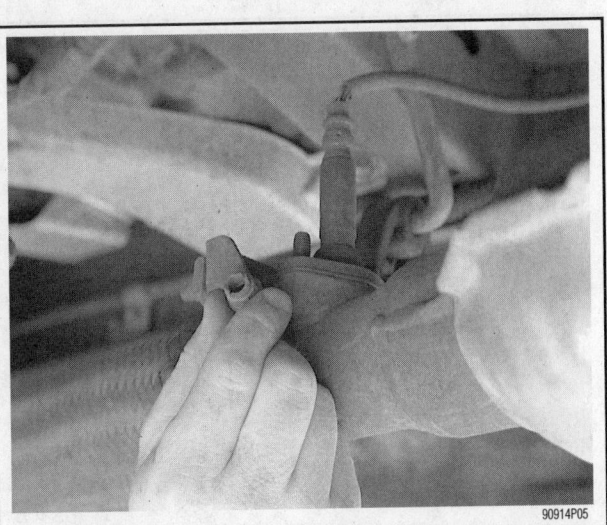

. . . which happen to be nuts in this particular case—some models may use bolts rather than nuts

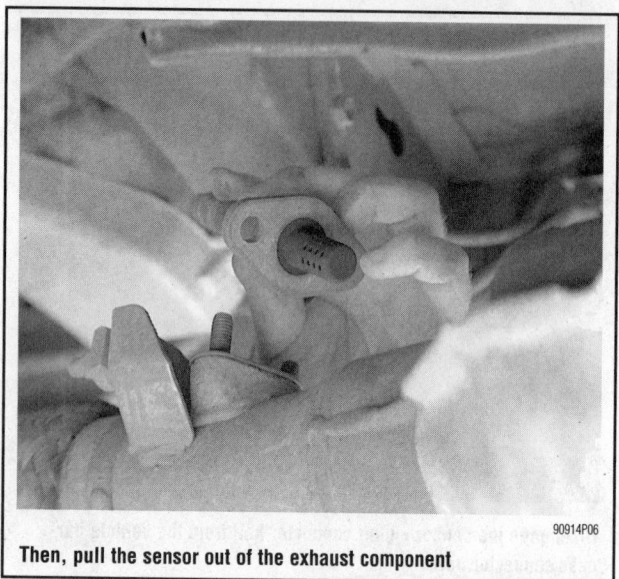

Then, pull the sensor out of the exhaust component

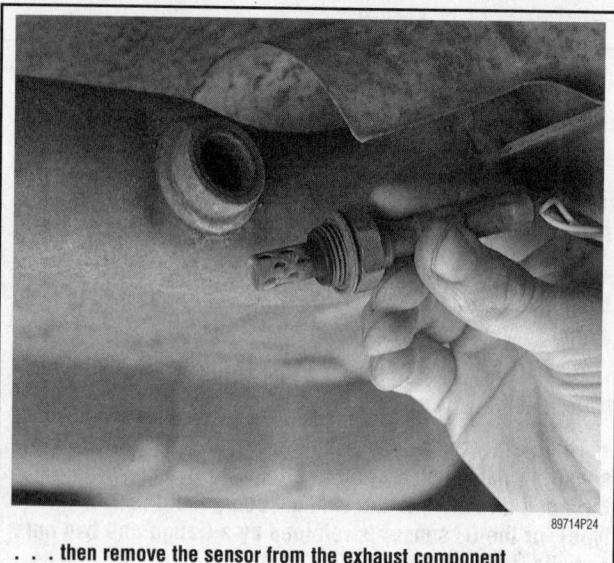

. . . then remove the sensor from the exhaust component

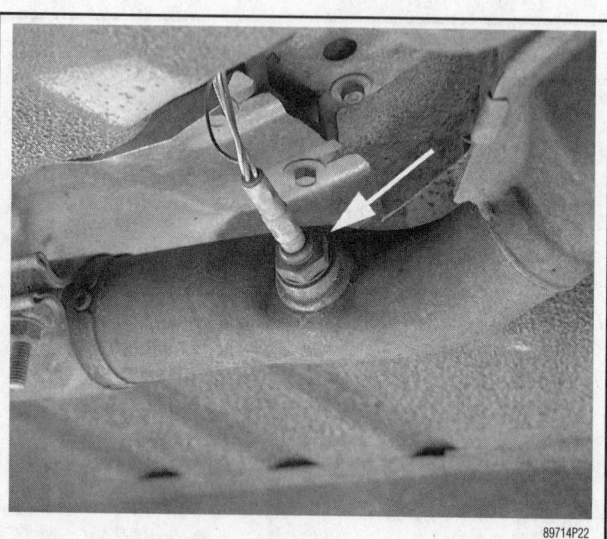

For screw-in type sensors (arrow) . . .

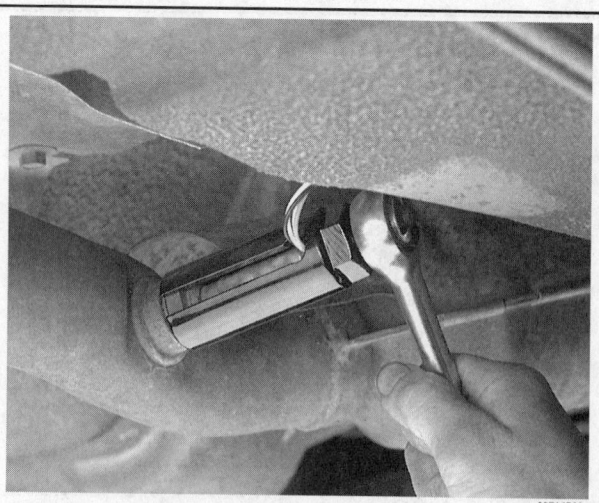

. . . either use a box end wrench to loosen the sensor or a socket designed expressly for this purpose . . .

✴ WARNING

To prevent damaging a screw-in type O₂ sensor, if excessive force is needed to remove the sensor lubricate it with penetrating oil prior to removal. Also, be sure to protect the tip of the sensor; O₂ sensor tips are very sensitive and may be easily damaged if allowed to strike or come in contact with other objects.

  5. Remove the sensor, as follows:
  • For screw-in type sensors—Since O₂ sensors are usually designed with a permanently-attached wiring pigtail (this allows the wiring harness and sensor connectors to be positioned away from the hot exhaust system), it may be necessary to use a socket or wrench that is designed specifically for this purpose. Before purchasing such a socket, be sure that you can't save some money by using a box end wrench for sensor removal.
  • For flange type sensors—Loosen the hold-down nuts or bolts and pull the sensor out of the exhaust component. Be sure to remove and discard the old sensor gasket, if equipped. You will need a new gasket for installation.
  6. Perform a visual inspection of the sensor. Black sooty deposits may indicate a rich air/fuel mixture, brown deposits may indicate an oil consumption problem, and white gritty deposits may indicate an internal coolant leak. All of these conditions can destroy a new sensor if not corrected before installation.
  To install:
  7. Install the sensor, as follows:

➡**A special anti-seize compound is used on most screw-in type O₂ sensor threads, and is designed to ease O₂ sensor removal. New sensors usually have the compound already applied to the threads. However, if installing the old O₂ sensor or the new sensor did not come with compound, apply a thin coating of electrically-conductive anti-seize compound to the sensor threads.**

✴ WARNING

Be sure to prevent any of the anti-seize compound from coming in contact with the O₂ sensor tip. Also, take pre-

cautions to protect the sensor tip from physical damage during installation.

• For screw-in type sensors—Install the sensor in the mounting boss, then tighten it securely.

• For flange type sensors—Position a new sensor gasket on the exhaust component and insert the sensor. Tighten the hold-down fasteners securely and evenly.

8. Reattach the sensor pigtail connector to the vehicle harness connector.

9. Lower the vehicle.

10. Connect the negative battery cable.

11. Start the engine and ensure no Diagnostic Trouble Codes (DTC's) are set.

LOCATIONS

Generally, there are only five different locations in the exhaust system where O₂ sensors are positioned. The five locations have been given numbers and will be used in the accompanying charts to identify the positions of O₂ sensors in most 1995–99 vehicles.

Due to mid-year production changes or factory inconsistencies, all models may not be covered. If a vehicle you are servicing is not covered in the charts, inspect the exhaust system (while cold!) in the five general locations to find the applicable O₂ sensors.

➡ **On models equipped with dual exhaust systems, there may be up to 4 or 5 O₂ sensors in the exhaust system. Be sure to locate all of them before commencing with any testing or service.**

The five locations are as follows:

• Location No.1—exhaust manifold or down pipe.

• Location No.2—both exhaust manifolds or down pipes of a V-type engine.

• Location No.3—exhaust collector.

• Location No.4—outlet of the catalytic converter.

• Location No.5—both the inlet and outlet of catalytic converter. This location is used to monitor the efficiency of the catalytic converter.

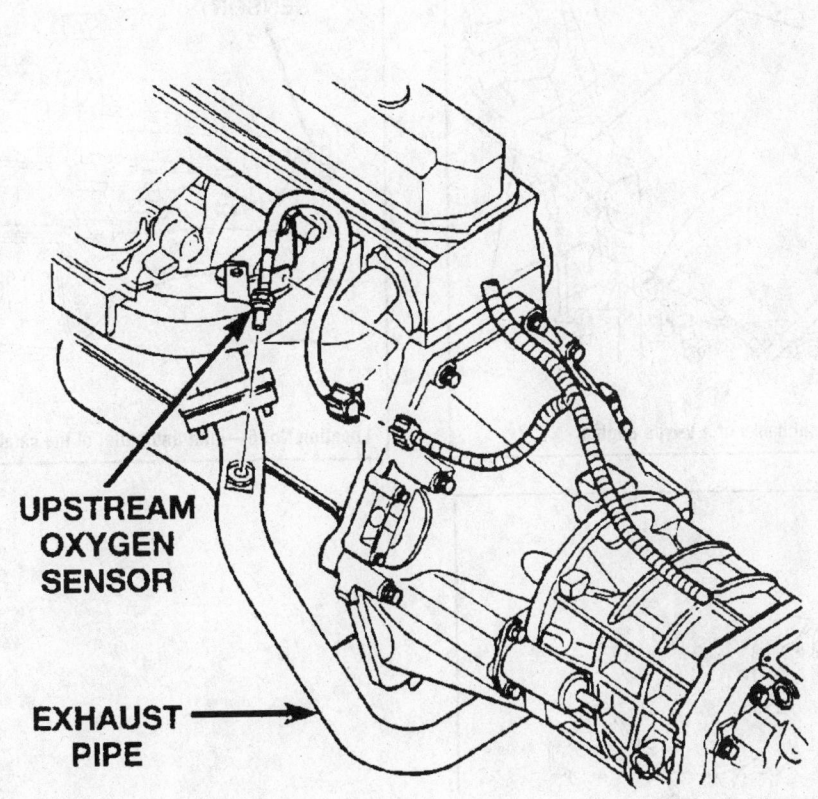

UPSTREAM OXYGEN SENSOR

EXHAUST PIPE

79248G01

Location No. 1—down pipe or exhaust manifold

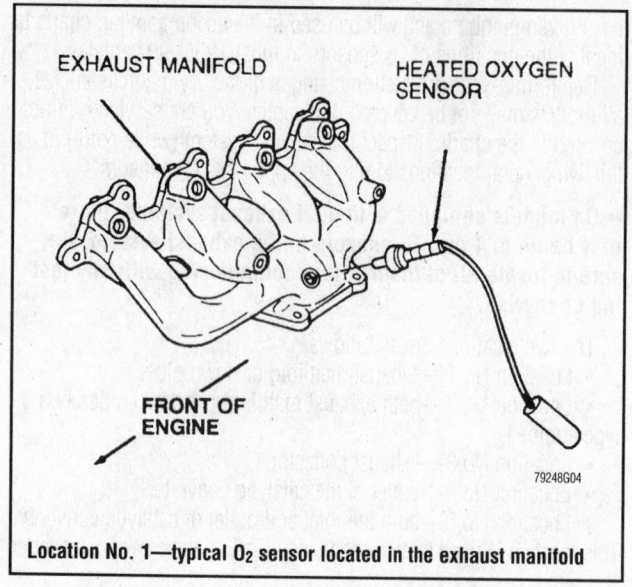

Location No. 1—typical O₂ sensor located in the exhaust manifold

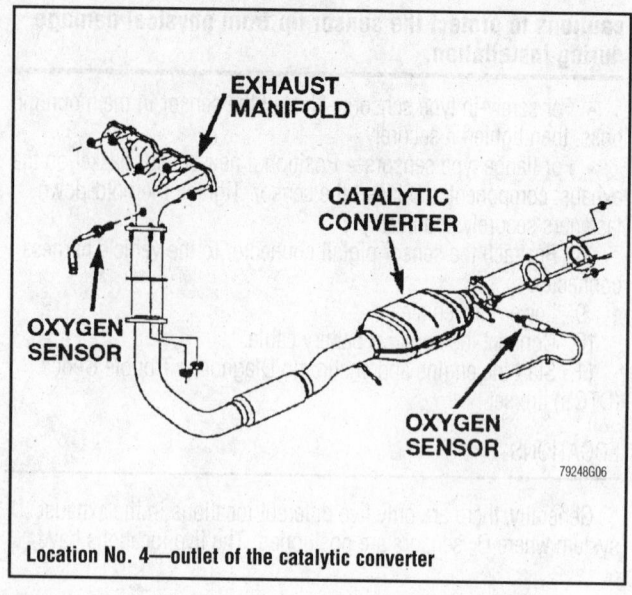

Location No. 4—outlet of the catalytic converter

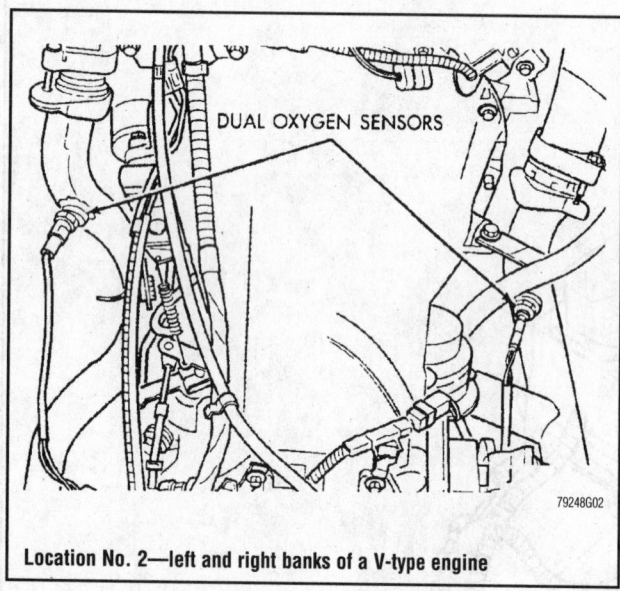

Location No. 2—left and right banks of a V-type engine

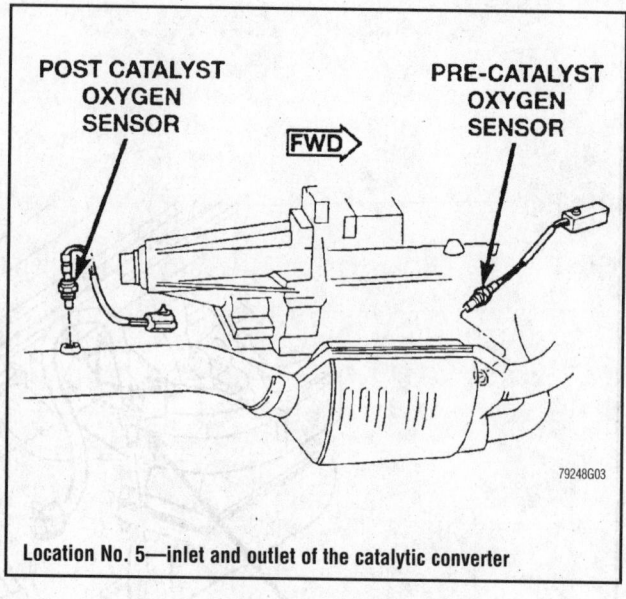

Location No. 5—inlet and outlet of the catalytic converter

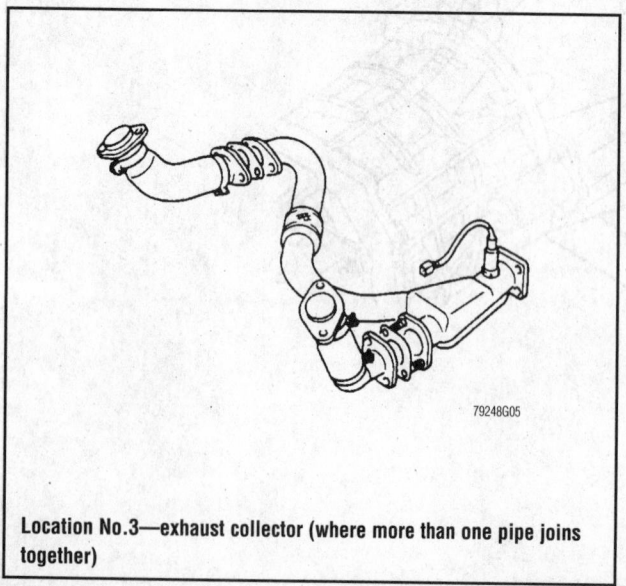

Location No.3—exhaust collector (where more than one pipe joins together)

OXYGEN SENSOR LOCATIONS

Manufacturer Years	Engines	No. of Sensors	Location
Chrysler LH Vehicles ①			
1995	3.3L	1	1
	3.5L	1	1
1996-99	ALL	4	2, 4
Chrysler Avenger and Sebring Coupe			
1995-99	2.0L	2	1, 4
	2.5L	3	2, 4
Eagle Talon			
1995-99	ALL	2	1, 4
Plymouth Neon			
1995-99	ALL	2	1, 4
Chrysler JA/JX Vehicles ②			
1995-99	ALL	2	1, 4
Ford Aspire			
1995	1.3L	1	1
1996-97	1.3L	2	1, 4
Ford Probe			
1995	2.0L	1	1
	2.5L	2	2
1996-97	2.0L	2	1, 4
	2.5L	4	2, 4
Ford Contour, Mystique and 1999 Cougar			
1995-99	2.0L	2	1, 4
	2.5L	3	1, 4
Ford Taurus and Sable			
1995	ALL	2	1, 4
1996-99	ALL	4	2, 4
Lincoln Continental			
1995-99	4.6L	4	5
Ford Escort, Tracer and ZX2			
1995-97	1.8L	2	1, 4
	1.9L	2	1, 4
1998-99	2.0L	2	1, 4
Ford Mustang			
1995	3.8L	2	1, 4
1996-99	ALL	4	2, 4
Lincoln Mark VIII			
1995-99	4.6L	4	5
Ford Cougar (1995-98)			
1995-98	3.8L	2	1, 4
	4.6L	4	5
Ford Full-size			
1995-99	4.6L	4	5

Manufacturer Years	Engines	No. of Sensors	Location
General Motors A Body			
1995	2.2L	1	1
	3.1L	1	1
1996	2.2L	2	1, 4
	3.1L	2	1, 4
General Motors B Body			
1995	4.3L	2	1, 4
	5.7L	2	1, 4
1996	4.3L	4	2, 4
	5.7L	4	2, 4
General Motors C & H Bodies			
1995	3.8L	1	1
1996-99	3.8L	2	1, 4
General Motors E & K Bodies			
1995	4.6L	2	1, 4
	4.9L	2	1, 4
1996-99	4.6L	4	5
General Motors F Body			
1995	3.4L	2	1, 4
	5.7L	2	1, 4
1996-99	3.8L	4	5
	5.7L	4	5
General Motors G Body			
1995	3.8L	1	1
	4.0L	1	1
1996-99	3.8L	2	1, 4
	4.0L	2	1, 4
General Motors J Body			
1995	2.2L	1	1
	2.3L	1	1
1996-99	2.2L	2	5
	2.4L	2	5
General Motors L Body			
1995	2.2L	1	1
	3.1L	1	1
1996	2.2L	2	5
	3.1L	2	5
General Motors L/N Body			
1997-99	2.4L	2	1, 4
	3.1L	2	1, 4
General Motors N Body			
1995	2.3L	1	1
	3.1L	1	1

79228C01

OXYGEN SENSOR LOCATIONS

Manufacturer Years	Engines	No. of Sensors	Location	Manufacturer Years	Engines	No. of Sensors	Location
General Motors N Body (cont.)				**General Motors Y Body**			
1996-99	2.4L	2	1, 4	1995	5.7L	2	2
	3.1L	2	1, 4	1996-99	5.7L	4	5
General Motors V Body				**GEO/Chevrolet**			
1997-99	3.0L	4	5	1995	ALL	1	1
General Motors W Body				1996-99	ALL	2	1, 4
1995	3.1L	1	1	**Saturn**			
	3.4L	1	1	1995	1.9L	1	1
1996-99	3.1L	2	5	1996-99	1.9L	2	1, 4
	3.4L	2	5				
	3.8L	2	5				

① Chrysler LH class designation refers to the Chrysler Concorde, LHS, New Yorker, Dodge Intrepid, and Eagle Vision.
② Chrysler JA class designation refers to the Chrysler Cirrus, Plymouth Breeze, and Dodge Stratus.
Chrysler JX class designation refers to the Chrysler Sebring Convertible.

79228C02

ELECTRIC COOLING FANS

9

COOLING FAN DIAGRAM
    INDEX..........................**9-8**
ELECTRIC COOLING FANS **9-1**
Electric Cooling Fan Service............9-2
    Removal & Installation 9-2
    Troubleshooting 9-6
General Information 9-1

ELECTRIC COOLING FANS

General Information

A basic vehicle cooling system consists of a radiator, water pump, thermostat, electric or engine-driven cooling fan, and hoses. Electric cooling fans are common on today's vehicles due to engine compartment space limitations or engine layout. Electric cooling fans operate in either a pusher or a puller capacity. A pusher type fan is typically mounted on the front of the radiator assembly and forces air through the radiator, whereas a puller type fan is mounted on the engine side of the radiator and draws air through the grill and radiator assembly. Vehicles that utilize a transversely-mounted engine will always be equipped with at least one electric cooling fan (most having two), because none of the engine pulleys are inline with the radiator air-flow.

There are generally two types of electric cooling fans: primary cooling fans and secondary cooling fans. Primary cooling fans are typically of the puller style. Vehicles that do not incorporate an engine-driven mechanical cooling fan will utilize a primary cooling fan. The secondary cooling fan, also known as a A/C condenser fan or auxiliary cooling fan by certain manufacturers, could be of either a pusher or a puller style. Vehicles equipped with A/C will either utilize the radiator cooling fan or a separate fan as the A/C condenser cooling fan (which performs the same function as an auxiliary cooling fan on vehicles with a primary mechanical fan). The engine control computer that receives inputs from various sensors in the engine compartment commonly controls electric cooling fans. The engine control computer receives inputs from the engine coolant temperature sensors and A/C system pressure switches, then actuates the necessary cooling fan relays to engage the applicable cooling fan for the condition. On models equipped with only one electric primary cooling fan, the fan can operate at two speeds: low speed and high speed. The low speed condition is enabled when the engine begins to heat up or when the A/C is engaged. As the engine demands more cooling, the cooling fan will be stepped-up to high speed.

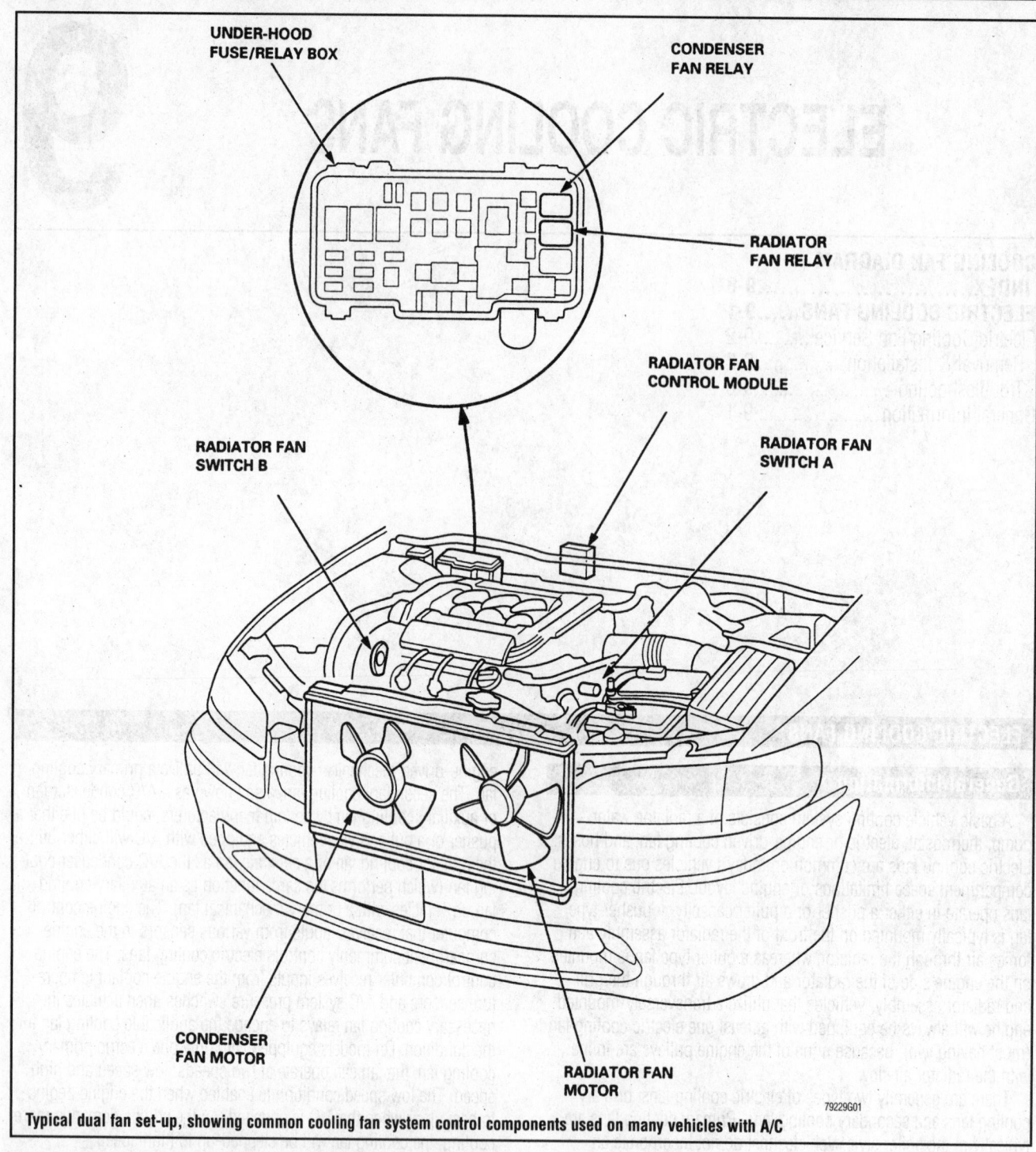

UNDER-HOOD FUSE/RELAY BOX

CONDENSER FAN RELAY

RADIATOR FAN RELAY

RADIATOR FAN CONTROL MODULE

RADIATOR FAN SWITCH B

RADIATOR FAN SWITCH A

CONDENSER FAN MOTOR

RADIATOR FAN MOTOR

79229G01

Typical dual fan set-up, showing common cooling fan system control components used on many vehicles with A/C

Electric Cooling Fan Service

Due to the wide variety of vehicle manufacturers and suppliers of electric cooling fans it is almost impossible to cover every specific combination of cooling fan and model. The following procedures will cover the most common types of mountings and troubleshooting techniques.

REMOVAL & INSTALLATION

Puller Type

➡**It may be simpler to remove the cooling fan(s) with the radiator as an assembly.**

1. Disconnect the negative battery cable.
2. Inspect the cooling fan and take note of any wires, hoses or A/C lines which may hamper fan removal. Also at this time, decide whether it is necessary to remove the fan along with the radiator or not.
3. Position aside all wires, hoses and A/C lines for fan removal. It may not always be possible to create enough clearance for fan removal by simply moving these obstructions aside; often they must be disconnected. If any cooling system lines must be disconnected, drain and recycle the engine coolant. If any of the A/C lines must be disconnected, the A/C system will need to be discharged and evacuated by a MVAC-trained technician using an approved recovery machine.
4. Disengage the cooling fan wiring harness connector.
5. If the fan can be removed without the radiator, perform the following:

   a. Loosen the mounting fasteners. Usually there are two nuts or bolts along the top edge of the cooling fan shroud and either two retaining clips or bolts along the bottom edge.

   b. Carefully lift the fan up and out of the engine compartment, making sure that no wires or hoses get hung up on it.
6. If it is necessary to remove the radiator for fan removal, perform the following:

   c. Disconnect all cooling system hoses from it after draining the cooling system.

   d. Locate all of the radiator mounting fasteners (usually two or more nuts or bolts along the top, possibly two along the bottom).

➡**Quite a few radiators are secured along the bottom by two posts which fit into rubber grommets. The rubber grommets help isolate the radiator from harsh vibrations in the frame. If no nuts or bolts can be located along the bottom of the radiator, chances are that the radiator is secured with the posts and grommets.**

   e. Lift the radiator and cooling fan up and out of the engine compartment together.

   f. Separate the cooling fan from the radiator by removing the attaching fasteners.

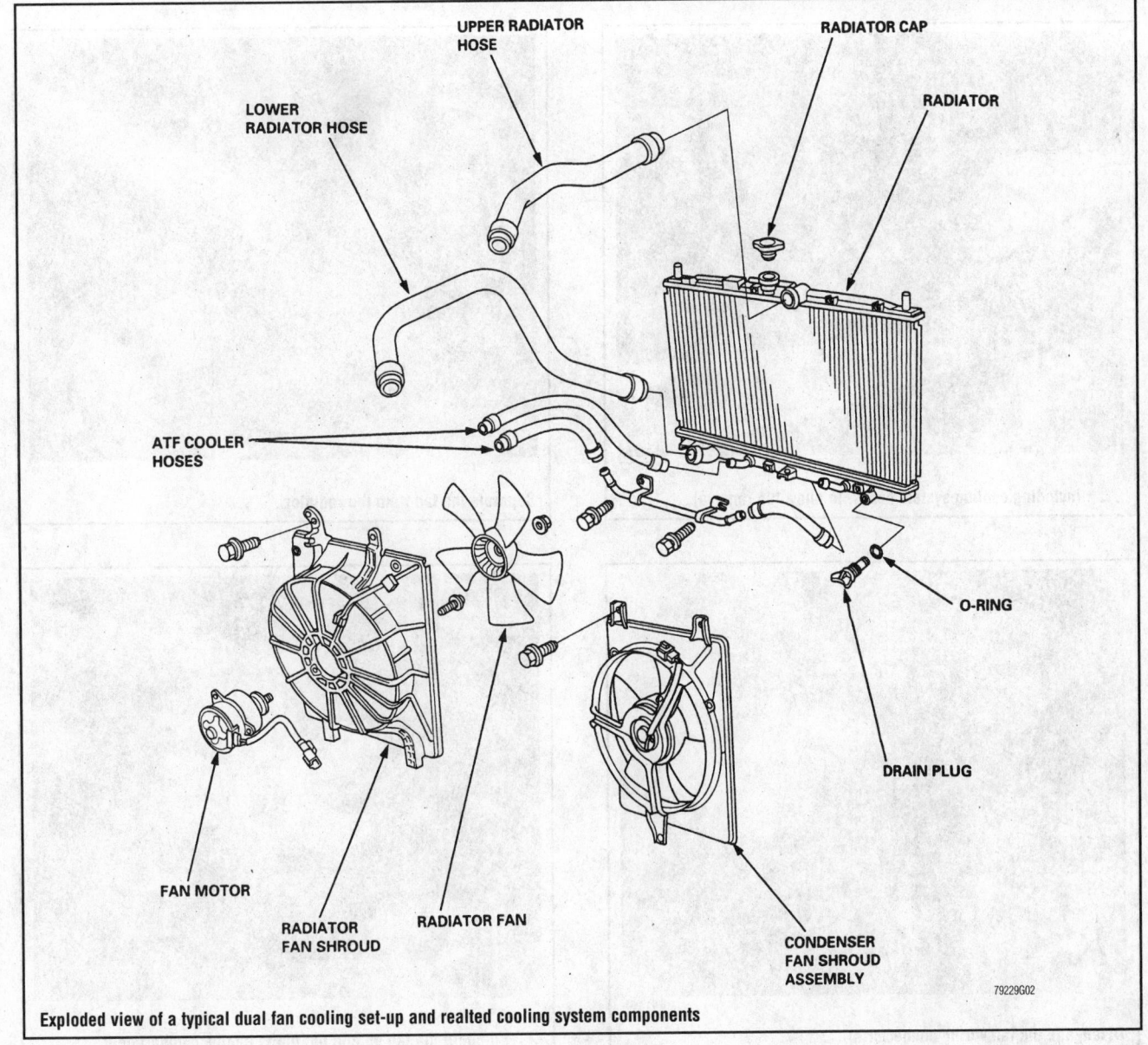

UPPER RADIATOR HOSE — RADIATOR CAP — RADIATOR — LOWER RADIATOR HOSE — ATF COOLER HOSES — O-RING — DRAIN PLUG — FAN MOTOR — RADIATOR FAN SHROUD — RADIATOR FAN — CONDENSER FAN SHROUD ASSEMBLY

79229G02

Exploded view of a typical dual fan cooling set-up and realted cooling system components

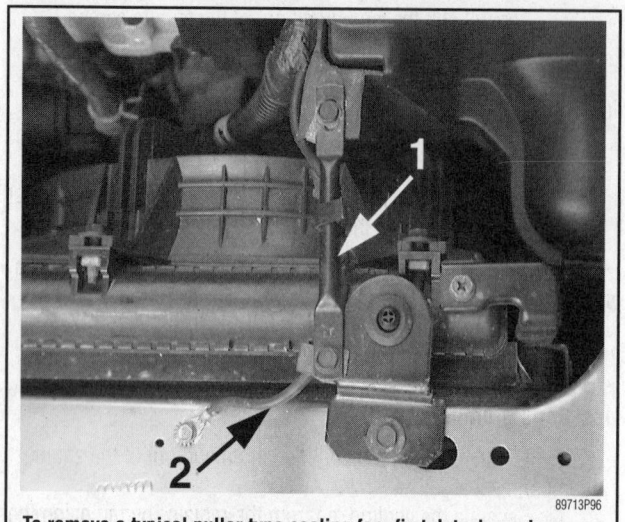

To remove a typical puller type cooling fan, first detach any braces (1), wires (2) or other obstructions . . .

. . . and loosen all fan mounting fasteners

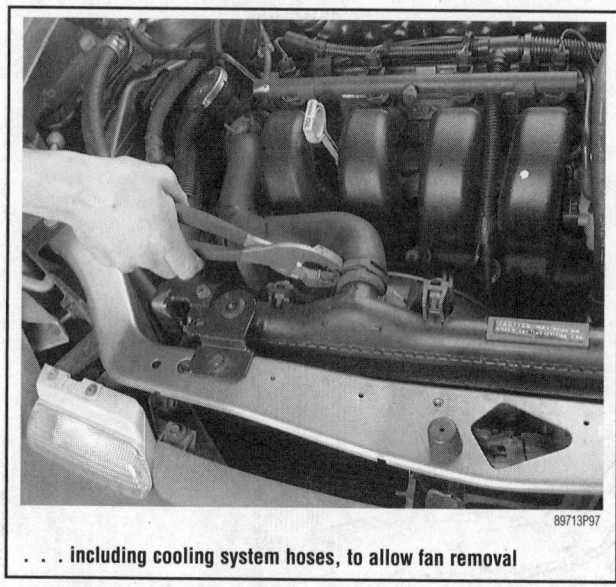

. . . including cooling system hoses, to allow fan removal

Separate the fan from the radiator . . .

Disengage the fan wiring connector(s) . . .

. . . then lift the fan up and out of the engine compartment

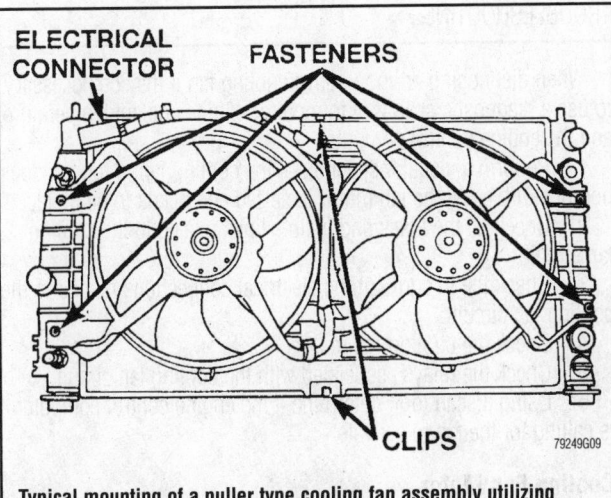

Typical mounting of a puller type cooling fan assembly utilizing retaining clips and screws—note that this particular model uses a dual puller fan setup

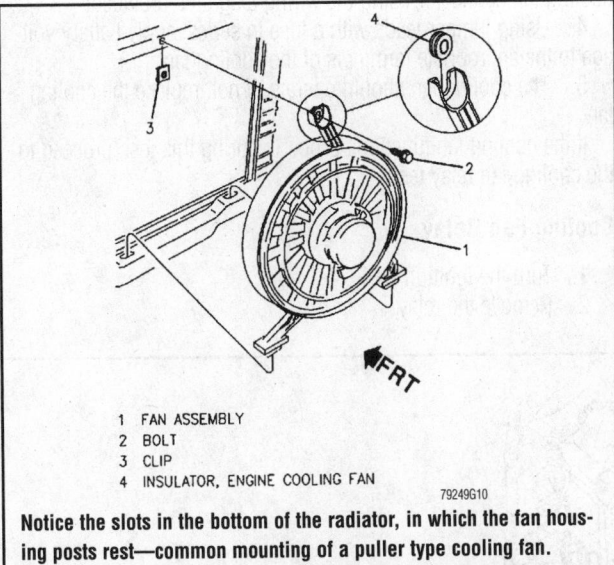

1 FAN ASSEMBLY
2 BOLT
3 CLIP
4 INSULATOR, ENGINE COOLING FAN

Notice the slots in the bottom of the radiator, in which the fan housing posts rest—common mounting of a puller type cooling fan.

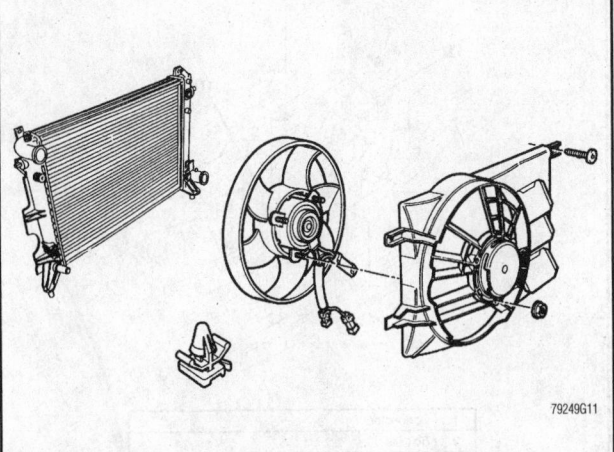

This fan mounts to the fan shroud, then the shroud mounts to the radiator—molded clips in the radiator hold the bottom in place and screws at the top.

To install:

7. If applicable, install the cooling fan on the radiator.

8. Install the cooling fan and shroud assembly (also the radiator if necessary). Tighten the fan shroud mounting bolts.

9. Reattach all wires, hoses and A/C lines as applicable. If the A/C lines were detached, the system must be evacuated and recharged by a MVAC-trained technician.

10. If drained, refill and bleed the cooling system.

11. Reattach the cooling fan electrical harness connector.

12. Connect the negative battery cable.

13. Start the engine and check for leaks.

14. Verify the operation of the cooling fan(s).

Pusher Type

Vehicles that utilize the pusher type of electric cooling fan, may require the removal of the grilles and/or upper radiator shroud in order to gain access the fasteners that mount the fan assembly in the vehicle.

1. Disconnect the negative battery cable.

2. Access the cooling fan.

3. Label and disconnect the cooling fan electrical harness.

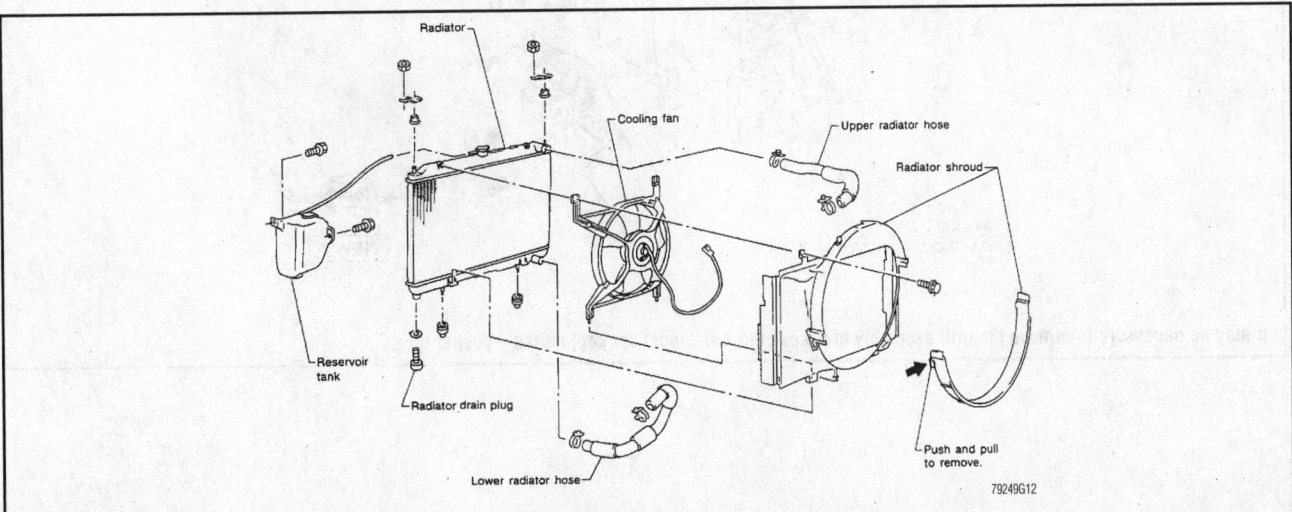

Typically the cooling fan is rubber mounted to isolate vibration and noise—usually the rubber grommets are located at the mount, verify their position before installation

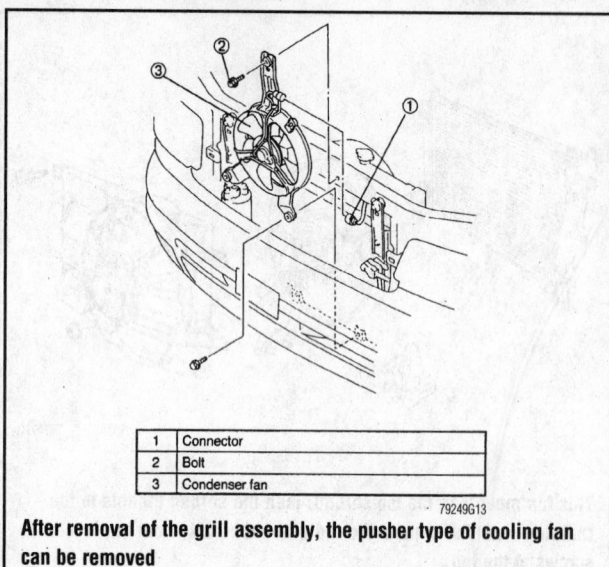

1	Connector
2	Bolt
3	Condenser fan

79249G13

After removal of the grill assembly, the pusher type of cooling fan can be removed

➡**It may be necessary to loosen the mounting bolts for the A/C condenser to the body**

4. Remove the fasteners that mount the cooling fan to the A/C condenser or radiator.
5. Lift the cooling fan out of the vehicle.

To install:

6. Insert the cooling fan into the vehicle.
7. Mount the cooling fan to the A/C condenser or radiator
8. Connect the cooling fan electrical harness.
9. If removed, install any shrouding or grills.
10. Connect the negative battery cable.

TROUBLESHOOTING

When diagnosing an inoperative cooling fan it may be necessary to use a diagnostic scan tool to monitor engine coolant temperature and the engine control computer.

1. Perform a visual inspection of the cooling fan. If the fan does not turn with ease, the fan motor is seized and needs to be replaced.
2. Check all the fuses and fusible links related to the cooling fan circuit.
3. Check the integrity of the electrical connections related to the cooling fan circuit.
4. Check the cooling fan motor.
5. Check the relays associated with the cooling fan circuit.
6. Using a scan tool, determine if the engine control computer is calling for the fan to activate.

Cooling Fan Motor

1. Disconnect the negative battery cable.
2. Disengage the cooling fan motor connector.
3. Identify and label the ground and the power terminals of the cooling fan connector using the wiring diagrams provided.
4. Using jumper leads with a fuse in series, apply battery voltage to the appropriate terminals of the cooling fan.
5. The cooling fan should operate. If not, replace the cooling fan.

If the cooling fan functions properly during this test, proceed to the cooling fan relay test.

Cooling Fan Relay

1. Turn the ignition **OFF**.
2. Remove the relay.

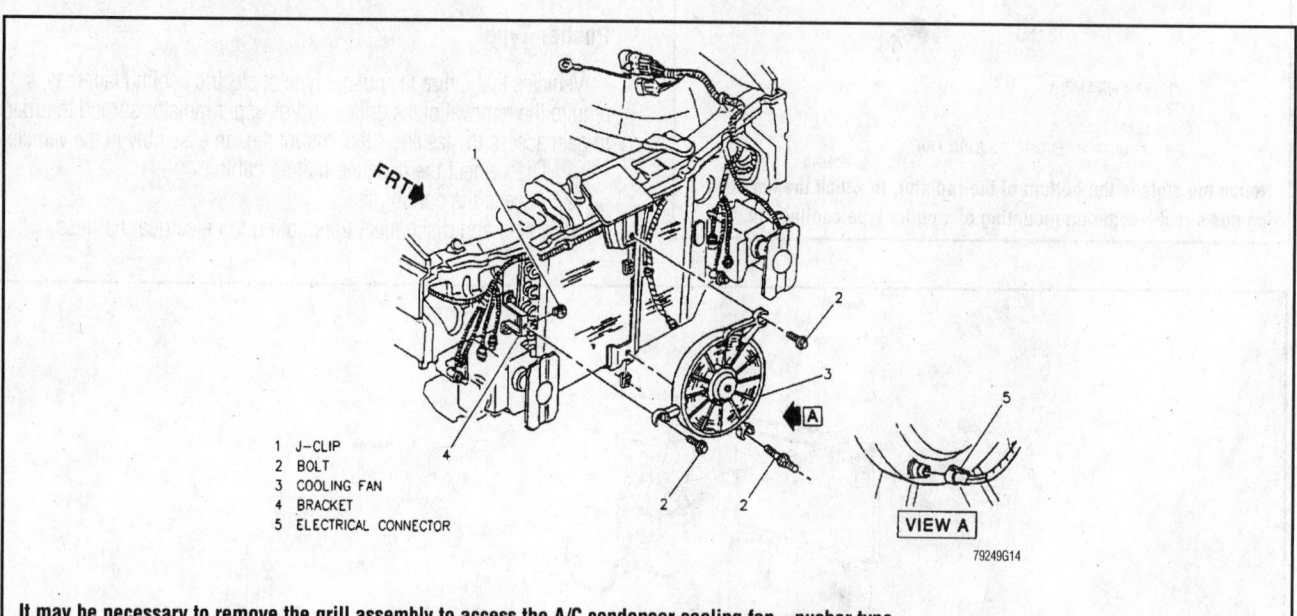

1	J–CLIP
2	BOLT
3	COOLING FAN
4	BRACKET
5	ELECTRICAL CONNECTOR

VIEW A

79249G14

It may be necessary to remove the grill assembly to access the A/C condenser cooling fan—pusher type

3. Locate the two terminals on the relay, which are connected to the coil windings. Check the relay coil for continuity. Connect the common meter lead to terminal 85 and positive meter lead to terminal 86. There should be continuity. If not, replace the relay.

4. Check the operation of the internal relay contacts.

    a. Connect the meter leads to terminals 30 and 87. Meter polarity does not matter for this step.

    b. Apply positive battery voltage to terminal 86 and ground to terminal 85. The relay should click as the contacts are drawn toward the coil and the meter should indicate continuity. Replace the relay if your results are different.

If the relay functions properly during this test, inspect the coolant temperature sensor and the cooling fan system wiring for defects.

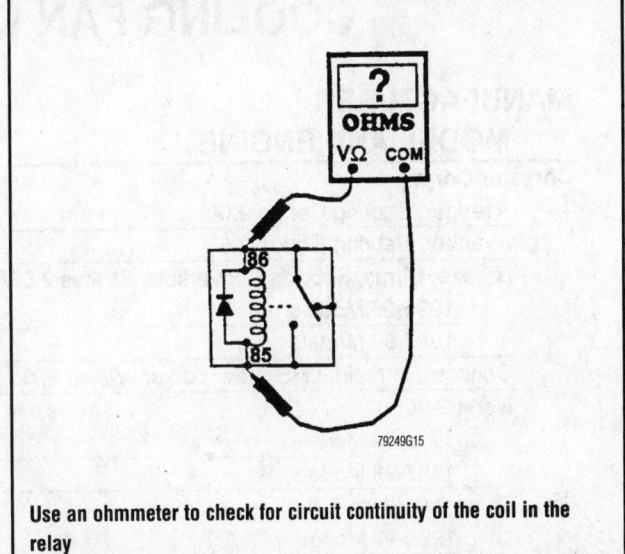

Use an ohmmeter to check for circuit continuity of the coil in the relay

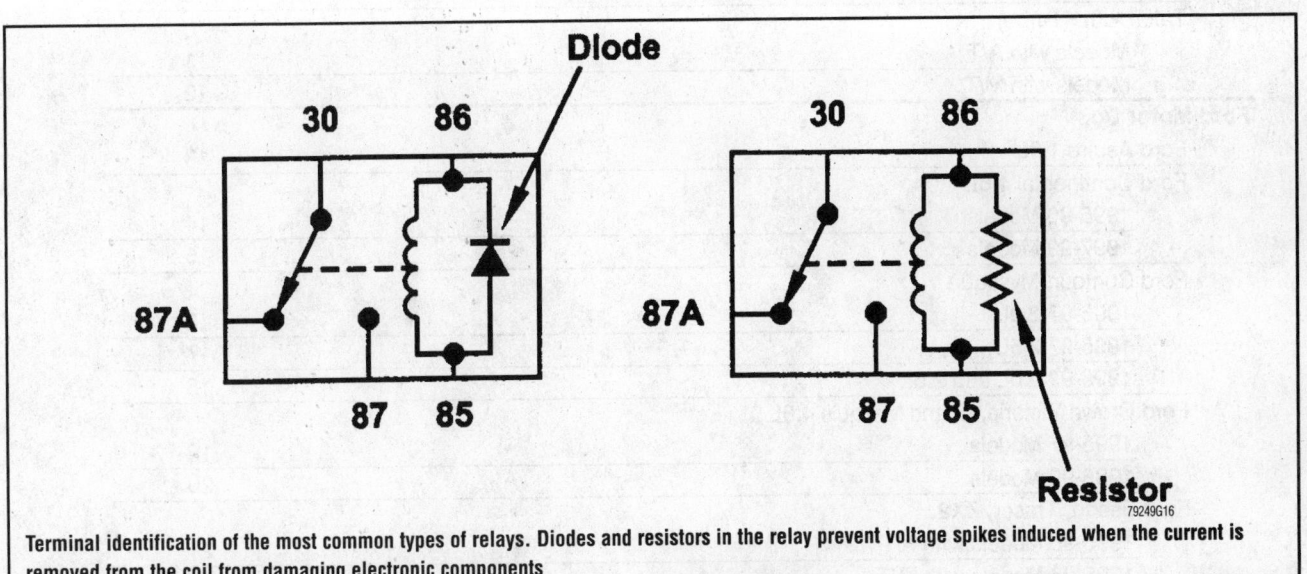

Terminal identification of the most common types of relays. Diodes and resistors in the relay prevent voltage spikes induced when the current is removed from the coil from damaging electronic components

COOLING FAN DIAGRAM INDEX

MANUFACTURER MODEL AND ENGINE	DIAGRAM
Chrysler Corp.	
Avenger, Sebring Coupe 2.0L	1
Avenger, Sebring Coupe 2.5L	2
Breeze, Cirrus, Sebring Convertible, Stratus 2.0L/ 2.4L/ 2.5L	
1995-96 Models	3
1997-99 Models	4
Concord, Intrepid, LHS, New Yorker, Vision 2.7L/ 3.2L/ 3.3L 3.5L	5
Neon 2.0L	
1995-96 Models	6
1997-99 Models	7
Talon 2.0L (Non-turbo)	
1995-97 Models with A/T	8
1995-97 Models with M/T	9
1998 Models	10
Talon 2.0L (Turbo)	
Models with A/T	11
Models with M/T	12
Ford Motor Co.	
Ford Aspire 1.3L	13
Ford Continental 4.6L	
1995-96 Models	14
1997-99 Models	15
Ford Contour, Mystique	
1995-97 2.0L	16
1995-97 2.5L	17
1998-99 2.0L and 2.5L	18
Ford Crown Victoria, Grand Marquis 4.6L	
1995-97 Models	19
1998-99 Models	20
Ford Escort, Tracer, ZX2	
1995-96 Models with A/T	21
1995-96 Models with M/T	22
1997-99 Models	23
Ford Mark VIII 4.6L	
1995-96 Models	24
1997-99 Models	25
Ford Mustang	
1995 Models	26
1996-99 Models with 3.8L	27
1996-99 Models with 4.6L	28
Ford Probe 2.0L/ 2.5L	29
Ford Taurus, Sable	
1995 Models with 3.0L, 3.0L SHO and 3.8L	30
1995 Models with 3.2L SHO	31
1996-97 Models with 3.0L and 3.4L	32
1998-99 Models with 3.0L and 3.4L	33

COOLING FAN DIAGRAM INDEX

MANUFACTURER MODEL AND ENGINE	DIAGRAM
Ford Motor Co. (cont.)	
Ford Thunderbird, Cougar XR7 3.8L/ 4.6L	34
1995 Models	35
1996-99 Models	
Lincoln Town Car 4.6L	36
1995-97 Models	37
1998-99 Models	
General Motors	
A Body (Century, Cutlass Ciera, Cruiser) 2.2L/ 3.1L	38
B Body (Caprice, Impala SS, Roadmaster) 4.3L/ 5.7L (w/o Mechanical Fan)	39
B Body (Caprice, Impala SS, Roadmaster) 4.3L/ 5.7L (w/ Mechanical Fan)	40
B Body (Fleetwood) 5.7L (w/o Mechanical Fan)	41
B Body (Fleetwood) 5.7L (w/ Mechanical Fan)	42
C & H Bodies (Bonneville, Eighty-Eight, Ninety-Eight, Park Ave., Le Sabre, LSS, Regency) 3.8L	43
E & K Bodies (DeVille, ElDarado, Seville) 4.6L/ 4.9L	
1995-96 Models	44
1997 Models	45
1998-99 Models	46
F Body (Camaro, Firebird) 3.4L (w/ C41)	47
F Body (Camaro, Firebird) 3.4L (w/ C60)	48
F Body (Camaro, Firebird) 3.8L	49
F Body (Camaro, Firebird) 5.7L	
1995 Models with C41 option	47
1995 Models with C60 option	48
1996-97 Models	49
1998-99 Models	50
G Body (Aurora, Riviera) 3.8L	
1995 Models with 3.8L	51
1996 Models with 3.8L	52
1997-99 Models with 3.8L	53
1995 Models with 4.0L	51
1996-99 Models with 4.0L	54
J Body (Cavalier, Sunfire) 2.2L/ 2.3L/ 2.4L	55
L Body (Beretta, Corsica) 2.2L/ 3.1L	56
L/N Bodies (Cutlass, Malibu) 2.4L/ 3.1L	57
N Body (Acheiva, Grand Am, Skylark) 2.3L/ 2.4L/ 3.1L	58
V Body (Catera) 3.0L	59
W Body (Lumina, Monte Carlo, Grand Prix, Cutlass Supreme, Regal, Intrigue) 3.1L/ 3.4L	
1995-96 Models (except Cutlass Supreme)	60
1995-97 Cutlass Supreme	60
1997-99 Models (except Cutlass Supreme)	61
Y Body (Corvette) 5.7L	
1995-96 Models	62
1997-99 Models	63

79229C02

COOLING FAN DIAGRAM INDEX

MANUFACTURER MODEL AND ENGINE	DIAGRAM
Geo/Chevrolet	
Metro 1.0L/ 1.3L (w/o A/C)	64
Metro 1.0L/ 1.3L (w/ A/C)	65
Prism 1.6L/ 1.8L (w/ A/C)	66
Prism 1.6L/ 1.8L (w/o A/C)	67
Saturn	
1.9L	68

79229C03

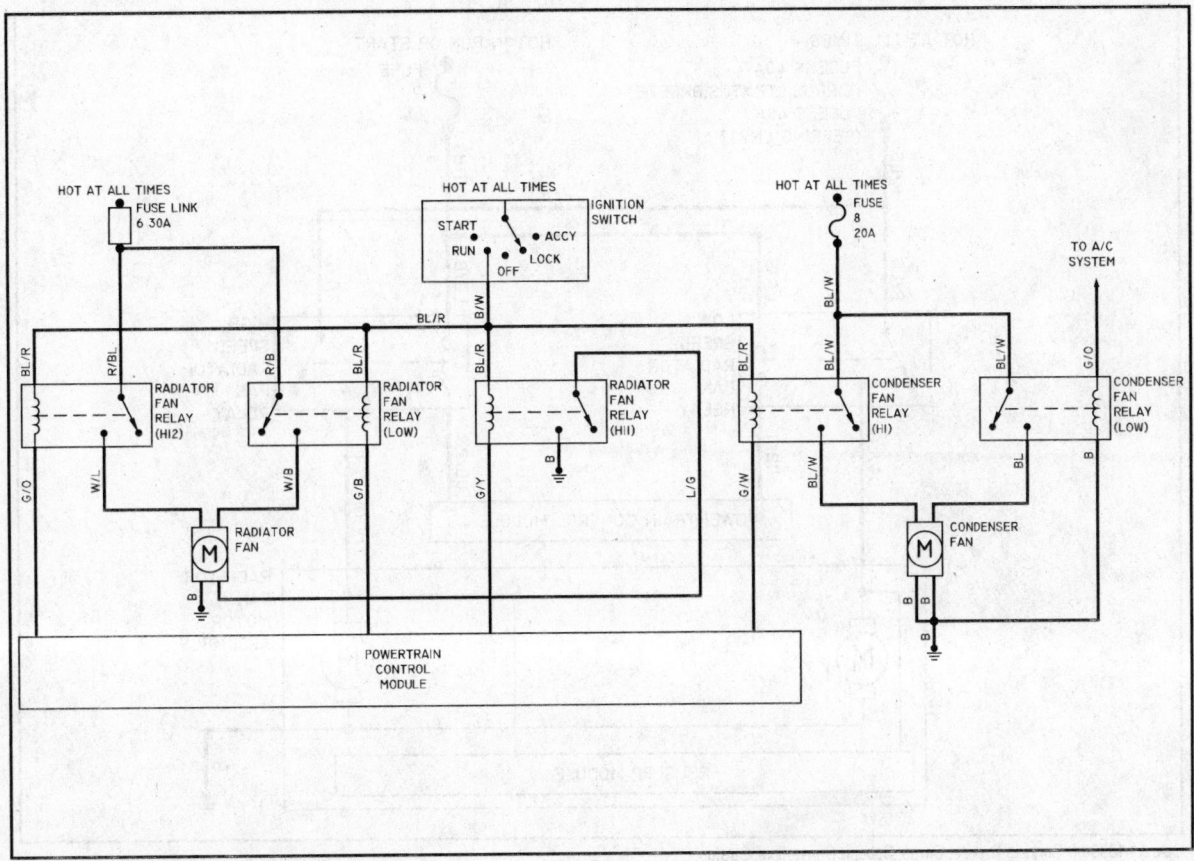

DIA. 1- 1995-99 Chrysler Avenger, Sebring Coupe 2.0L

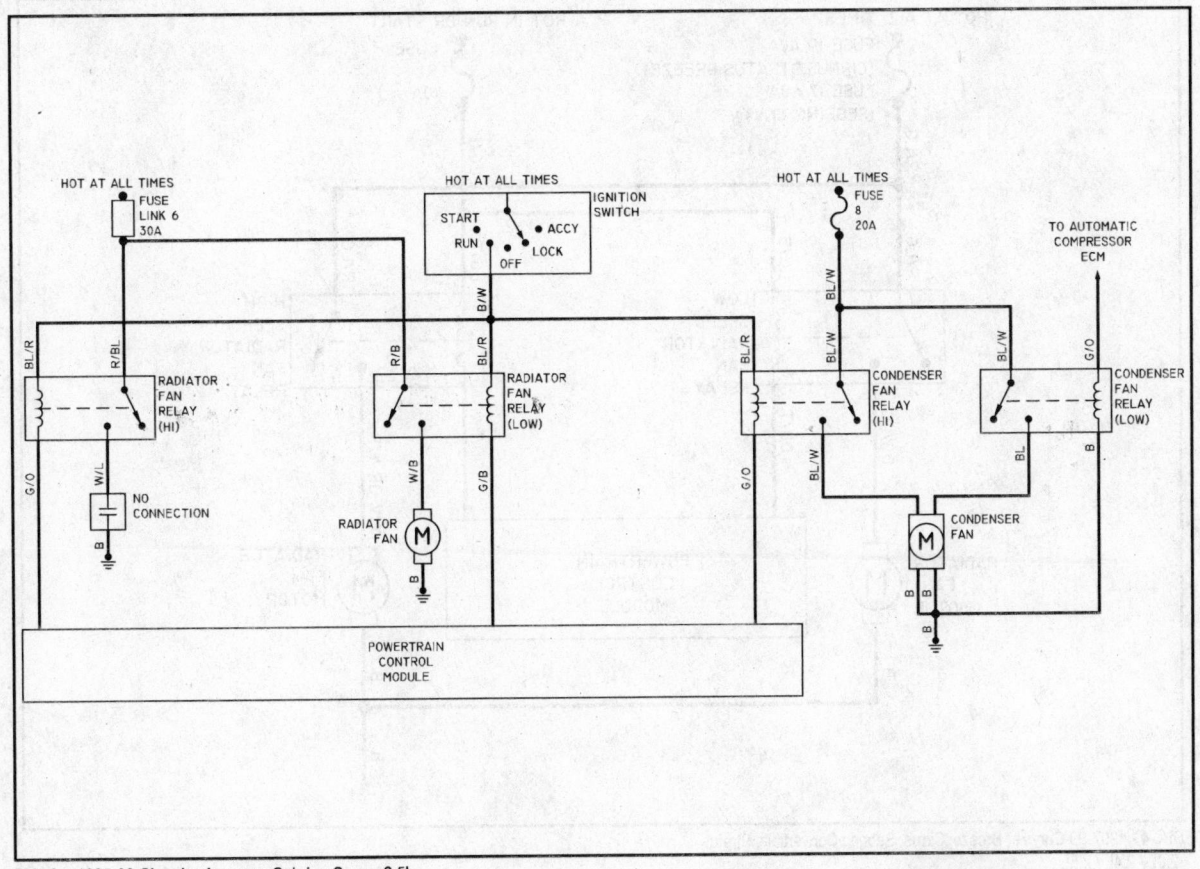

DIA. 2 - 1995-99 Chrysler Avenger, Sebring Coupe 2.5L

79229W01

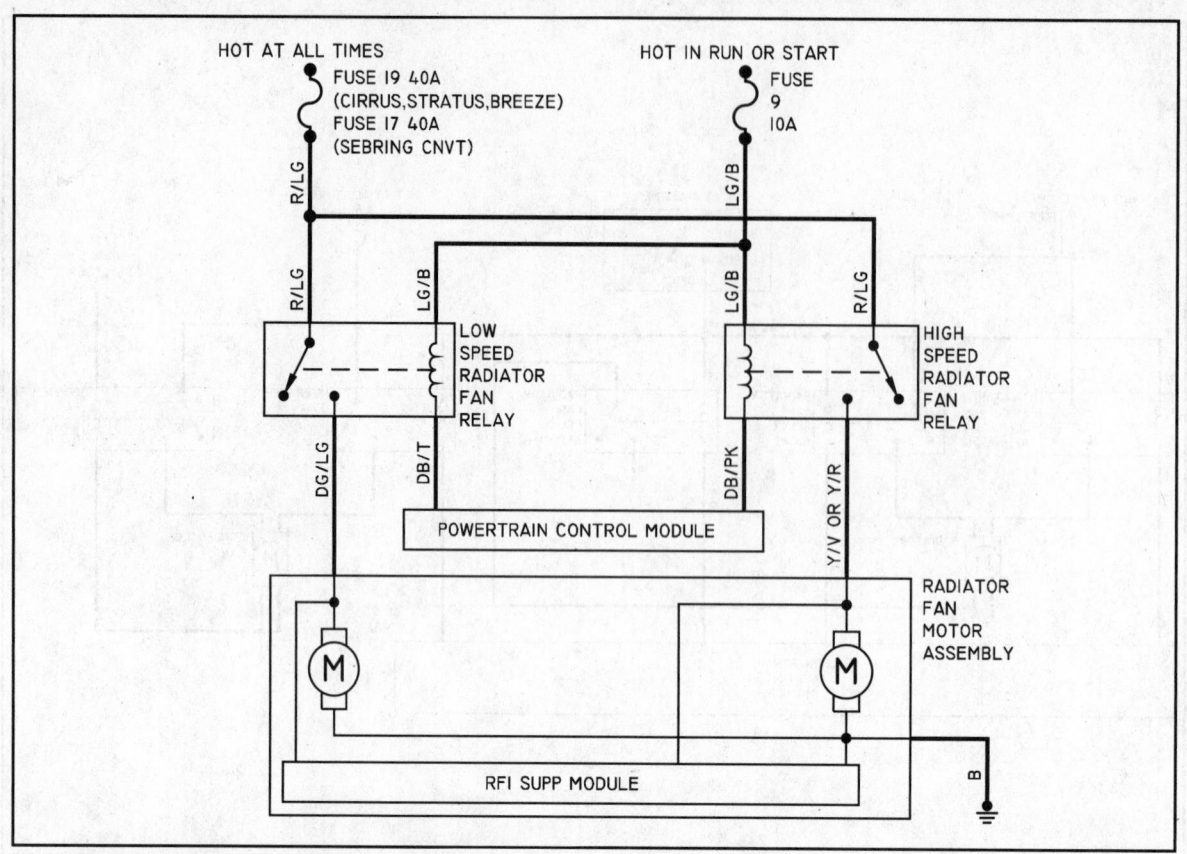

DIA. 3 - 1995-96 Chrysler Breeze, Cirrus, Sebring Convertible, Stratus
2.0L / 2.4L / 2.5L

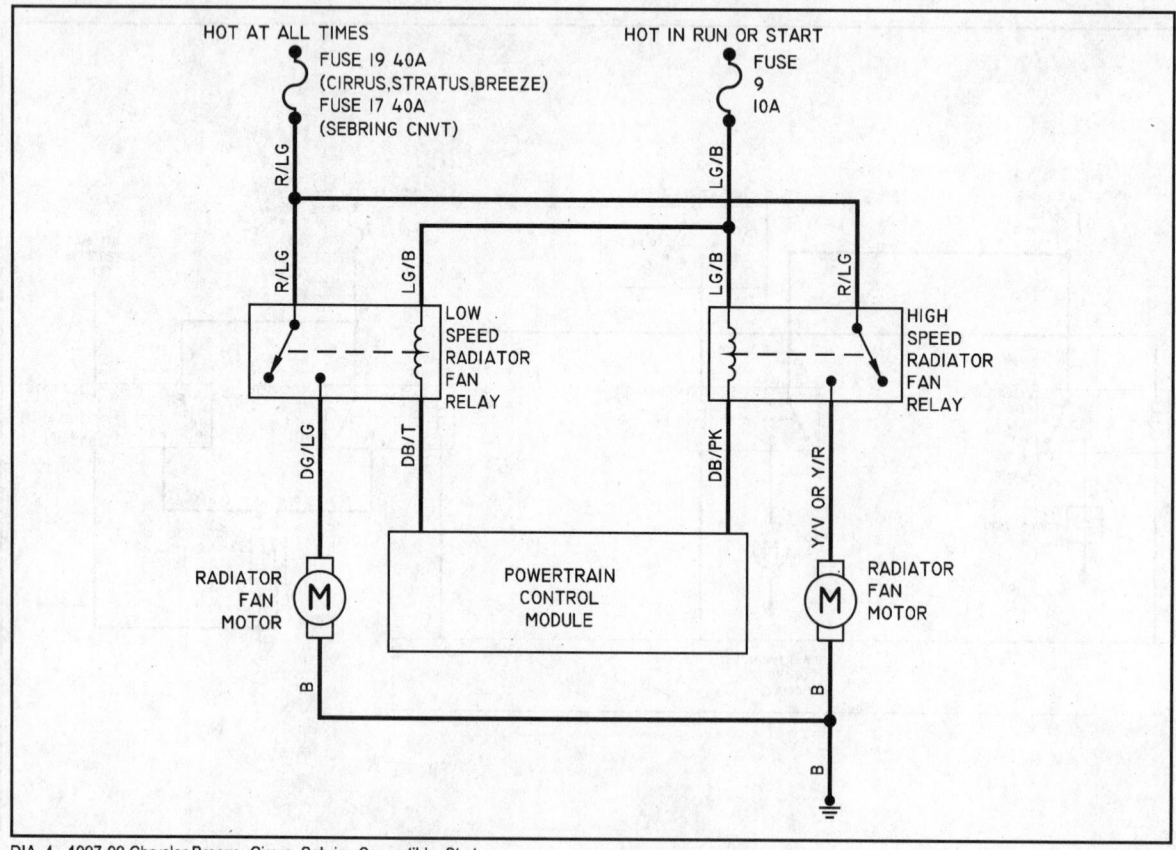

DIA. 4 - 1997-99 Chrysler Breeze, Cirrus, Sebring Convertible, Stratus
2.0L / 2.4L / 2.5L

79229W02

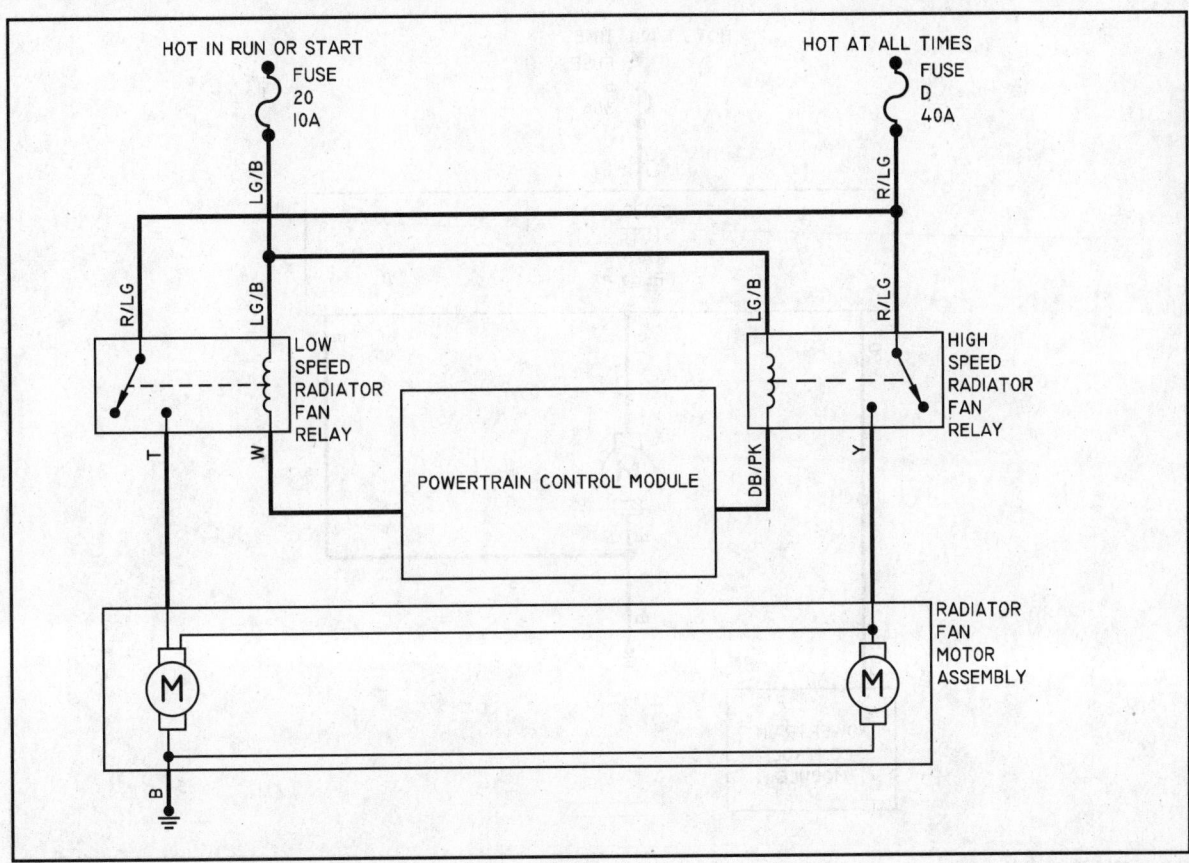

DIA. 5 - 1995-99 Chrysler Concorde, Intrepid, LHS, New Yorker, Vision 2.7L/3.2L/3.3L/3.5L

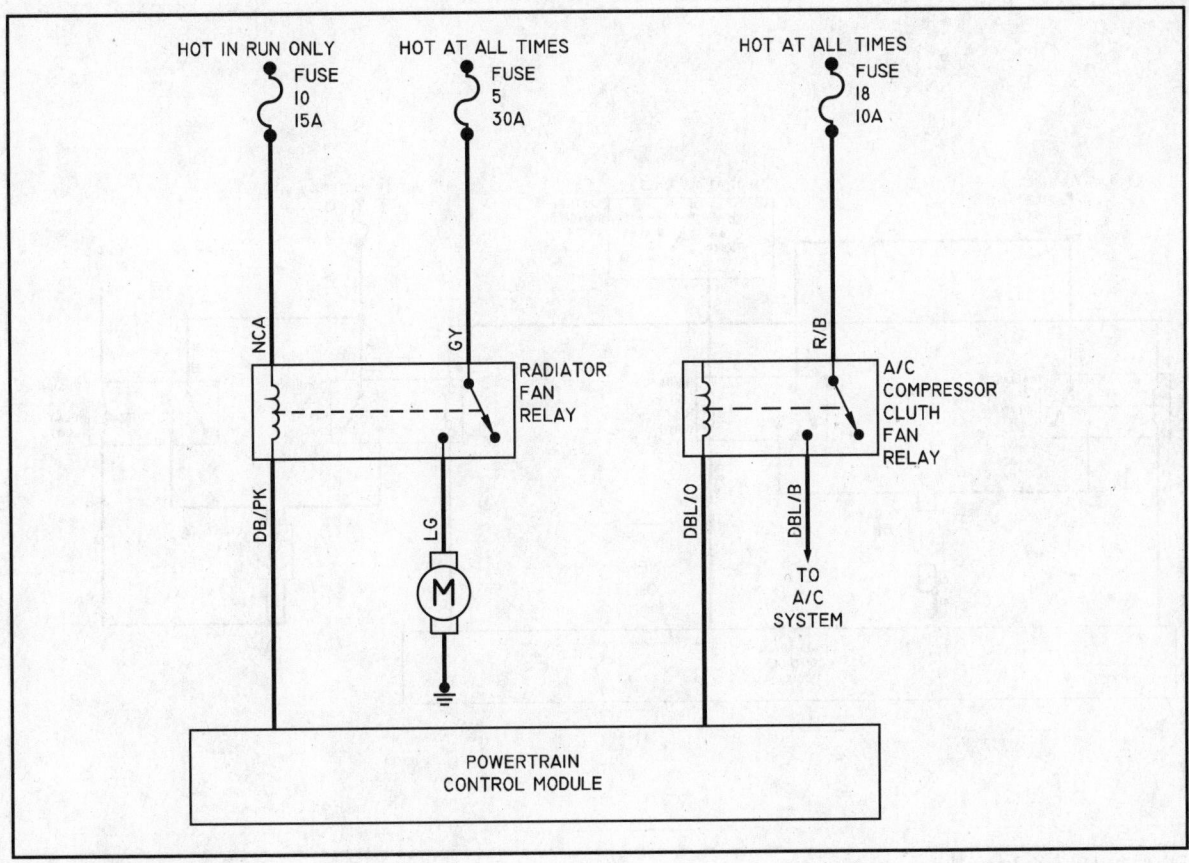

DIA. 6 - 1995-96 Dodge Neon 2.0L

79229W03

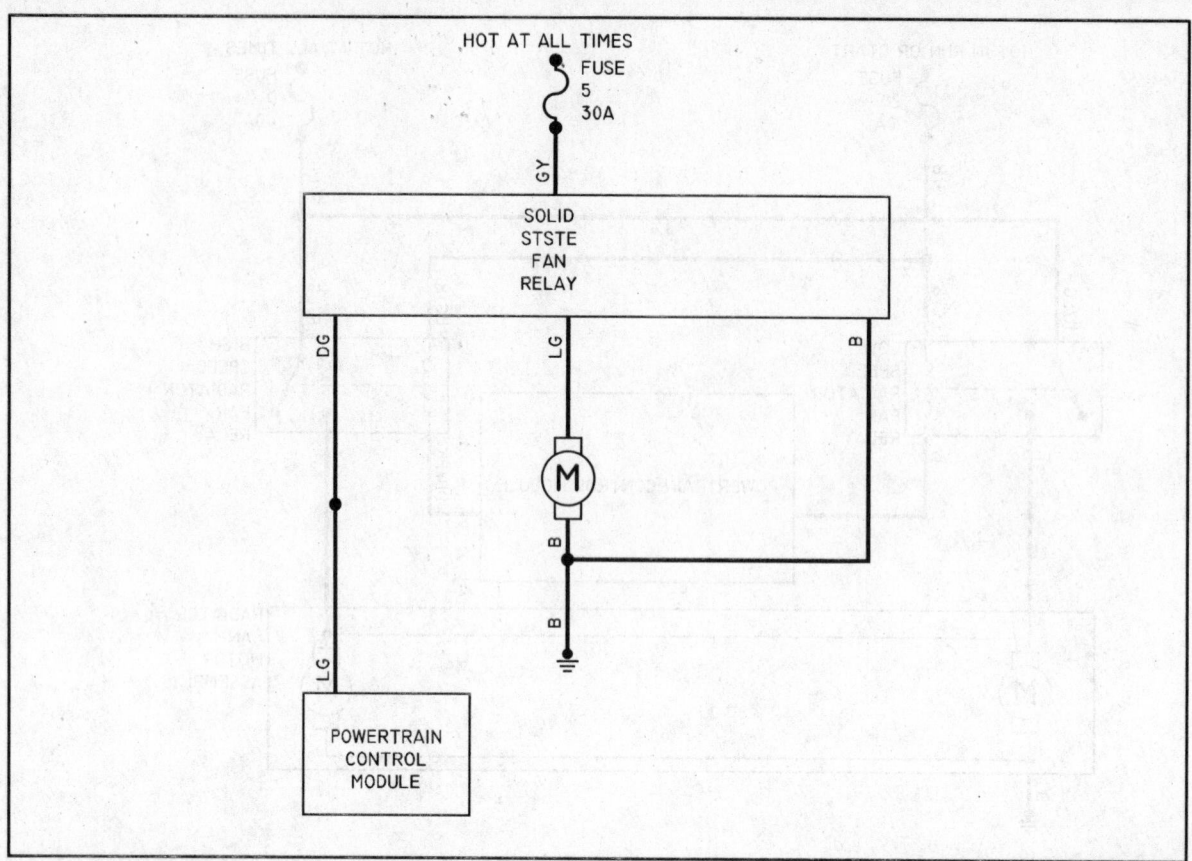

DIA. 7 - 1997-99 Dodge Neon 2.0L

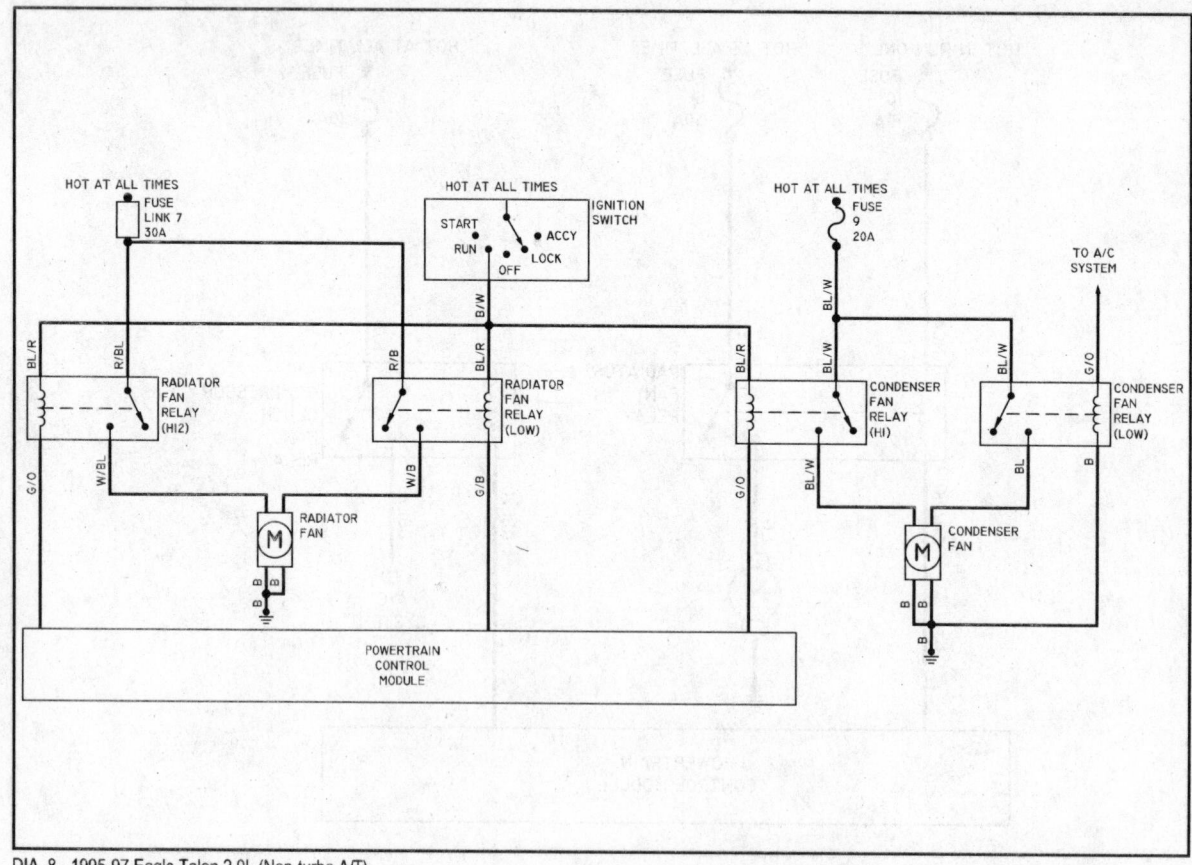

DIA. 8 - 1995-97 Eagle Talon 2.0L (Non-turbo A/T)

79229W04

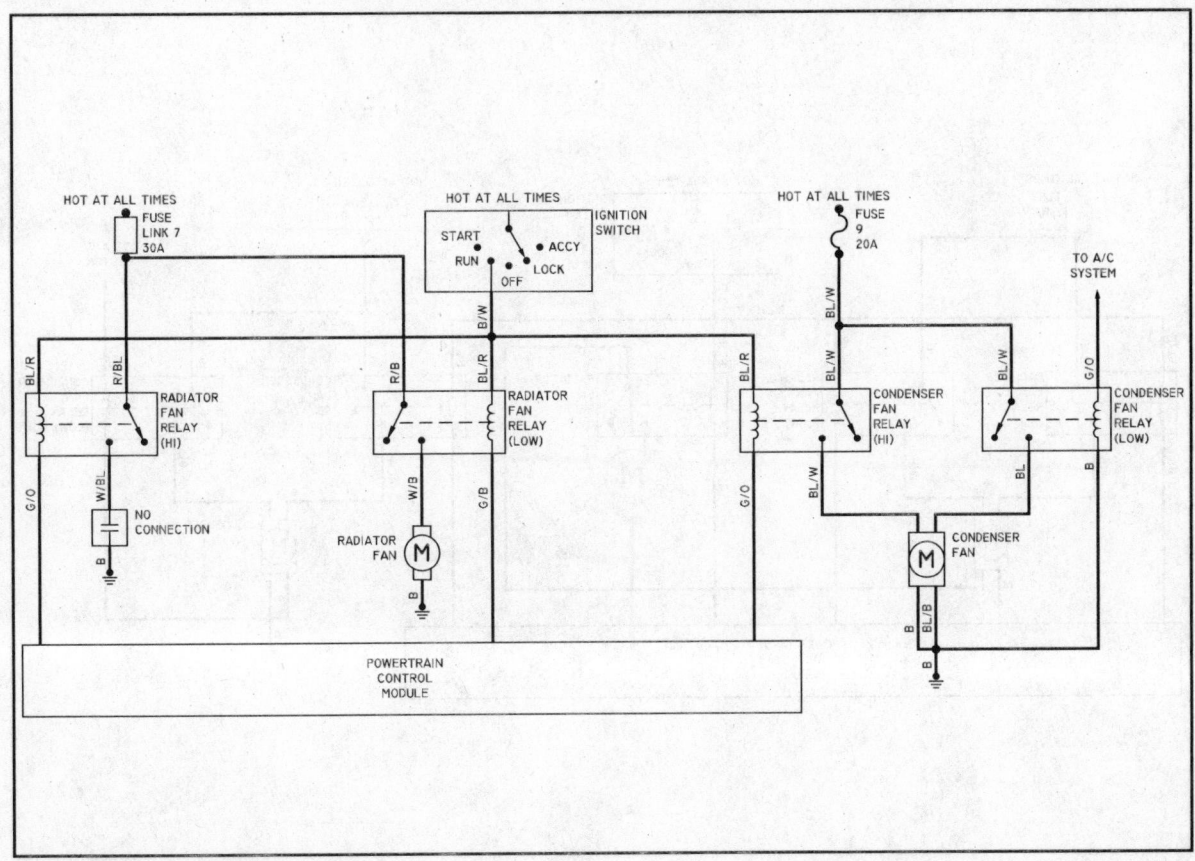

DIA. 9 - 1995-97 Eagle Talon 2.0 (Non-turbo M/T)

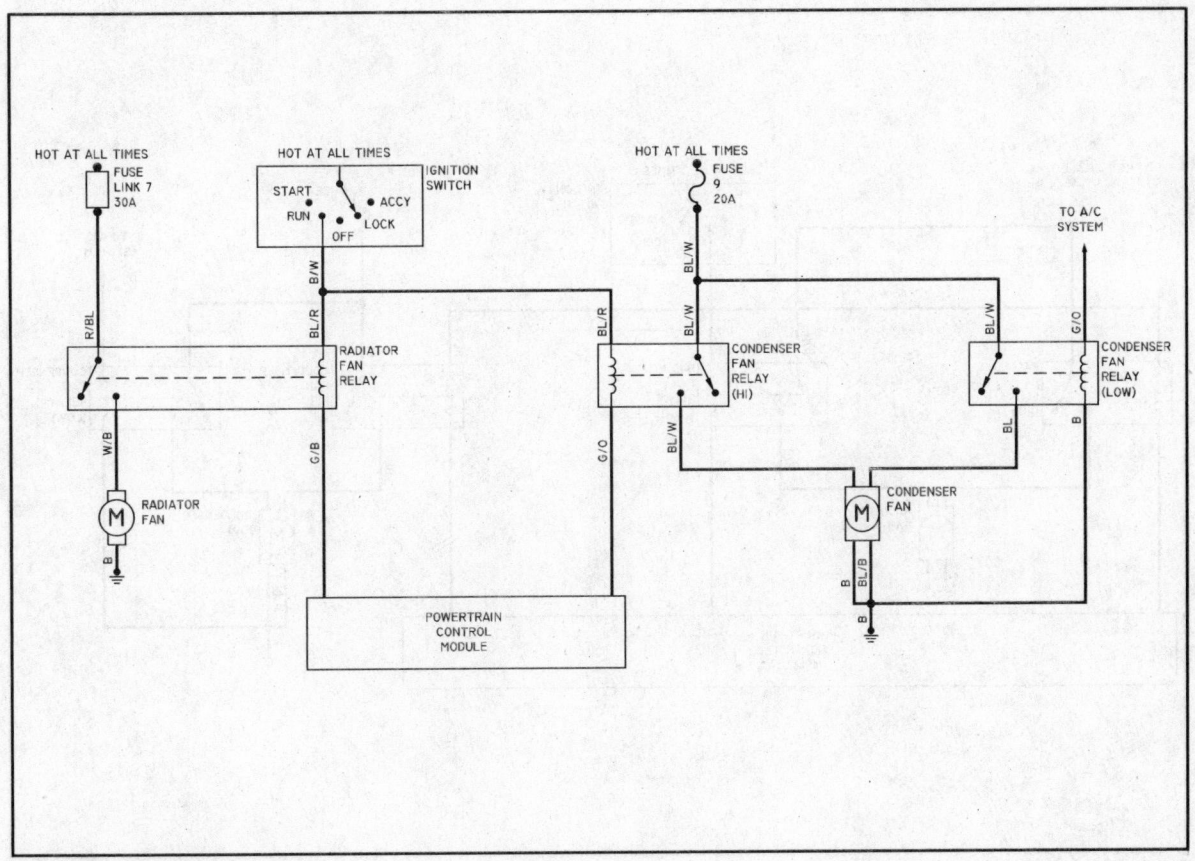

DIA. 10 - 1998 Eagle Talon 2.0L (Non-turbo)

79229W05

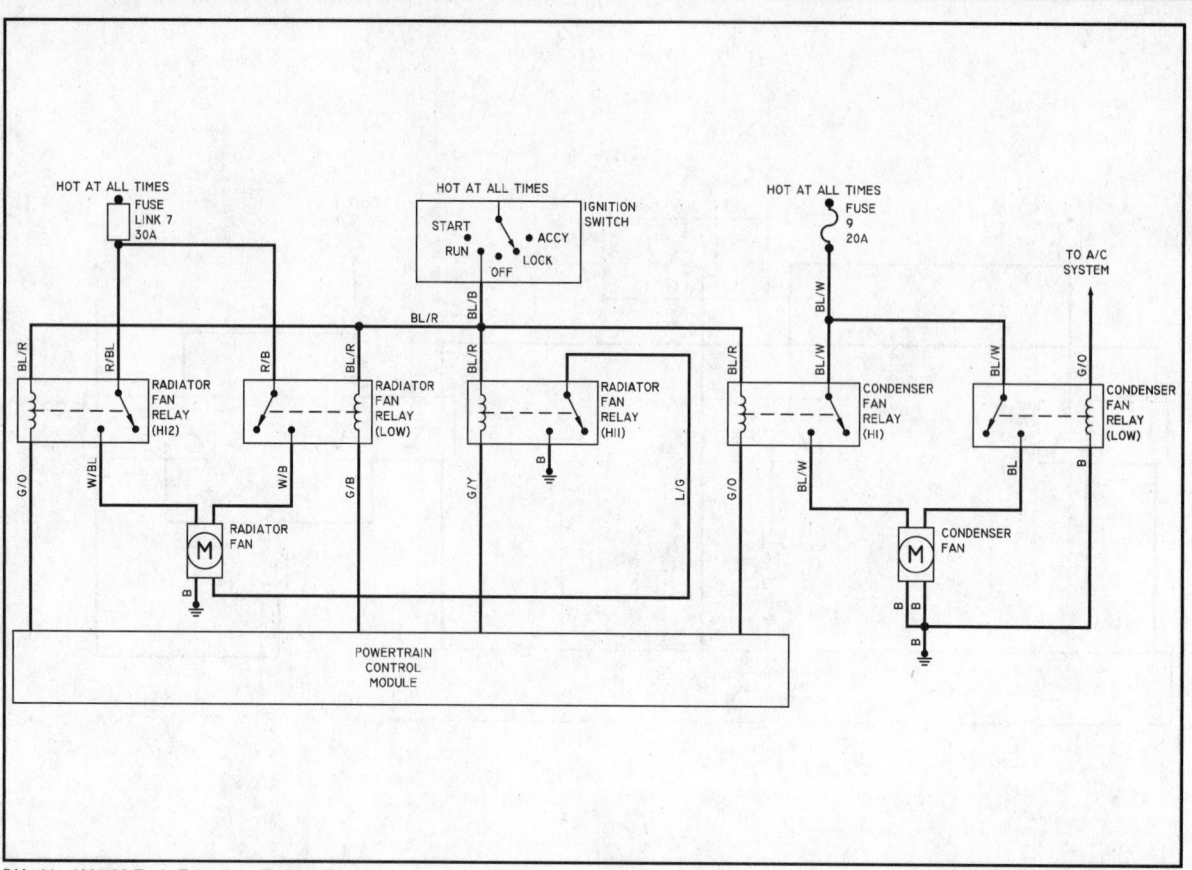

DIA. 11 - 1995-98 Eagle Talon 2.0L (Turbo A/T)

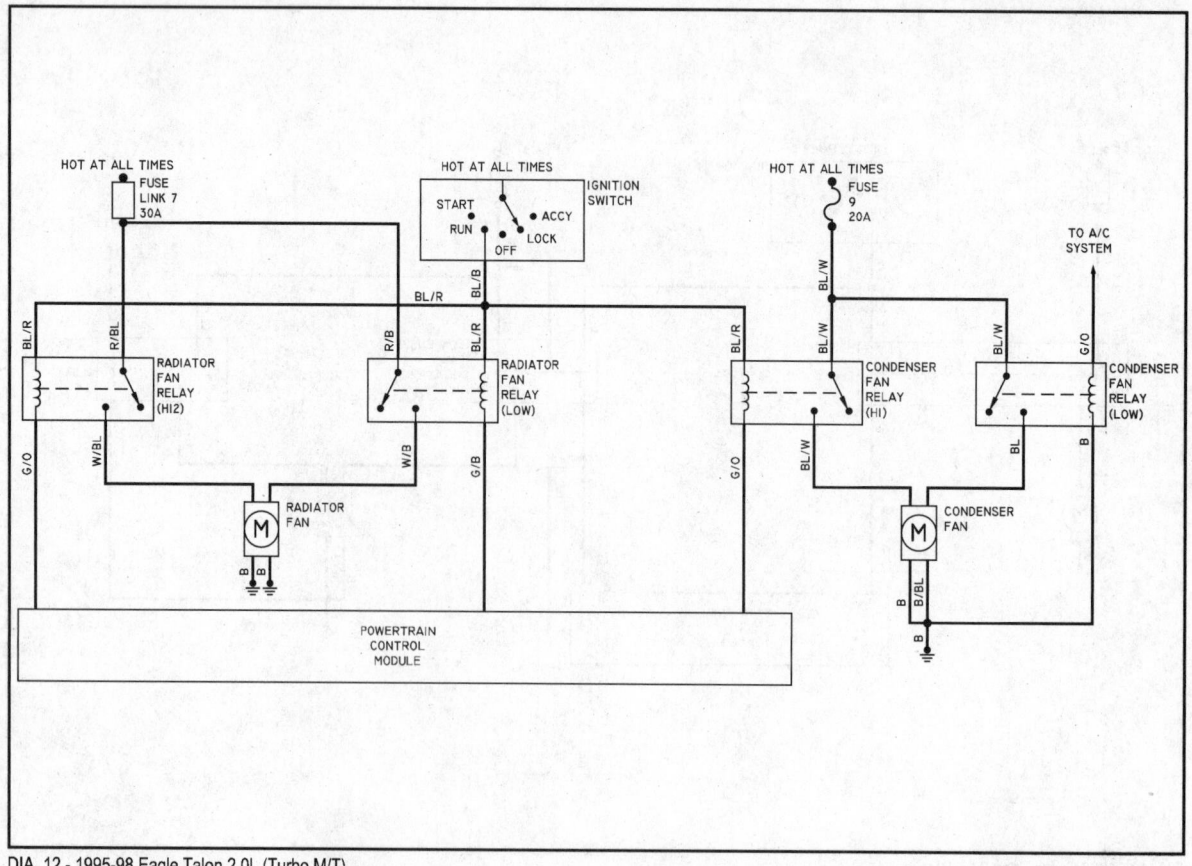

DIA. 12 - 1995-98 Eagle Talon 2.0L (Turbo M/T)

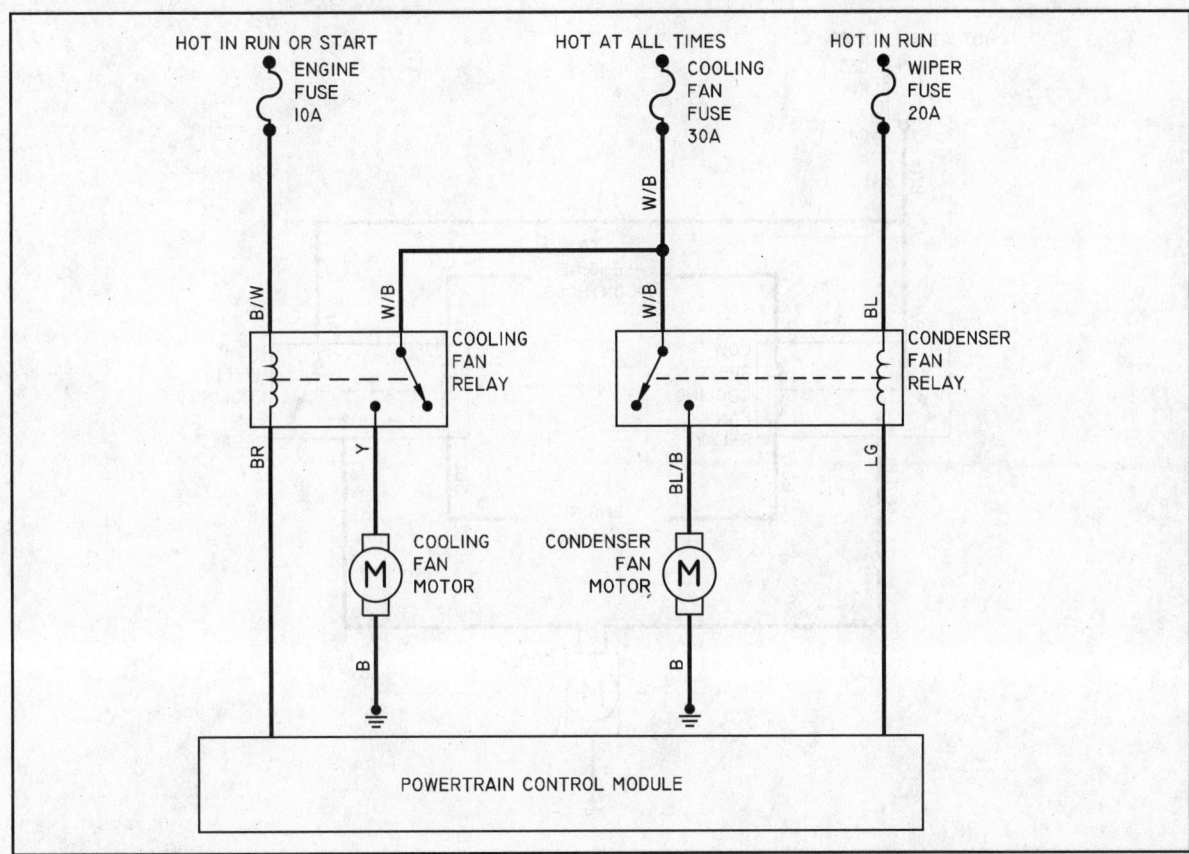

DIA. 13 - 1995-97 Ford Aspire 1.3L

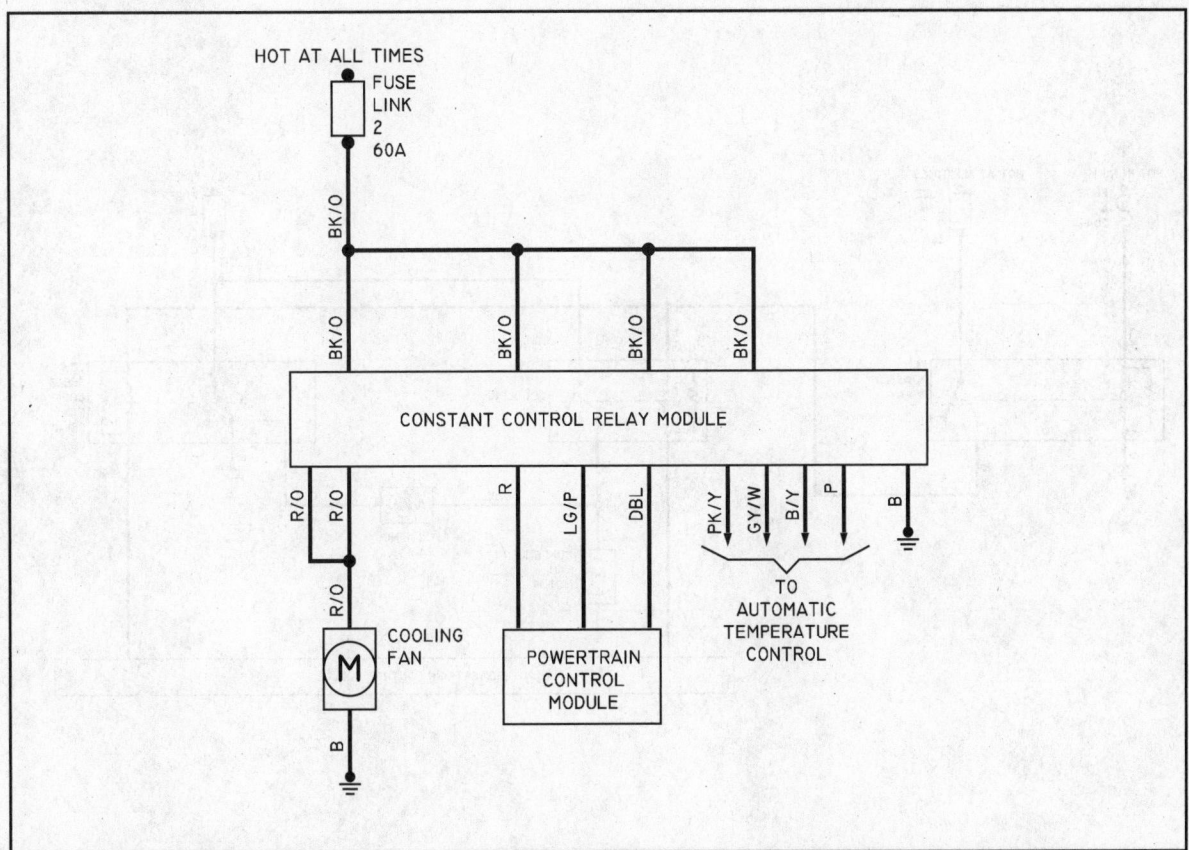

DIA. 14 - 1995-96 Ford Continental 4.6L

79229W07

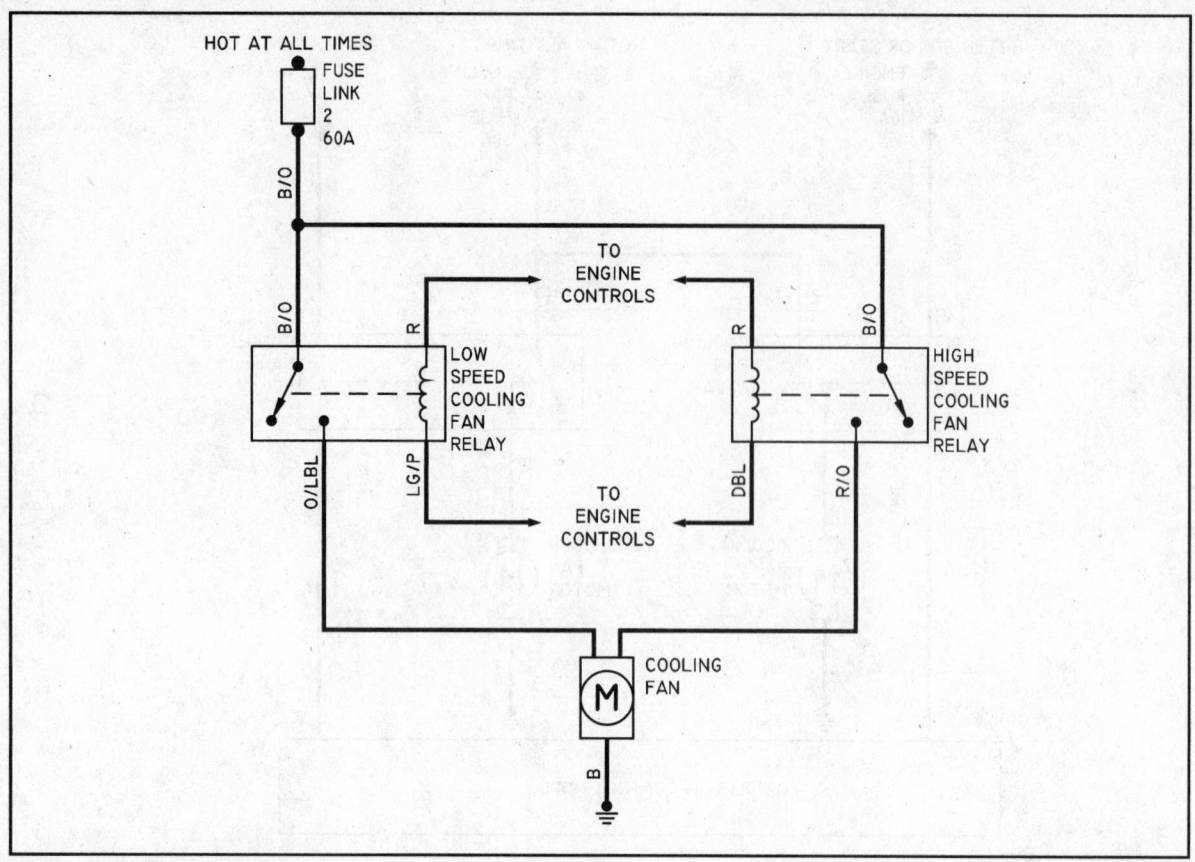

DIA. 15 - 1997-99 Ford Continental 4.6L

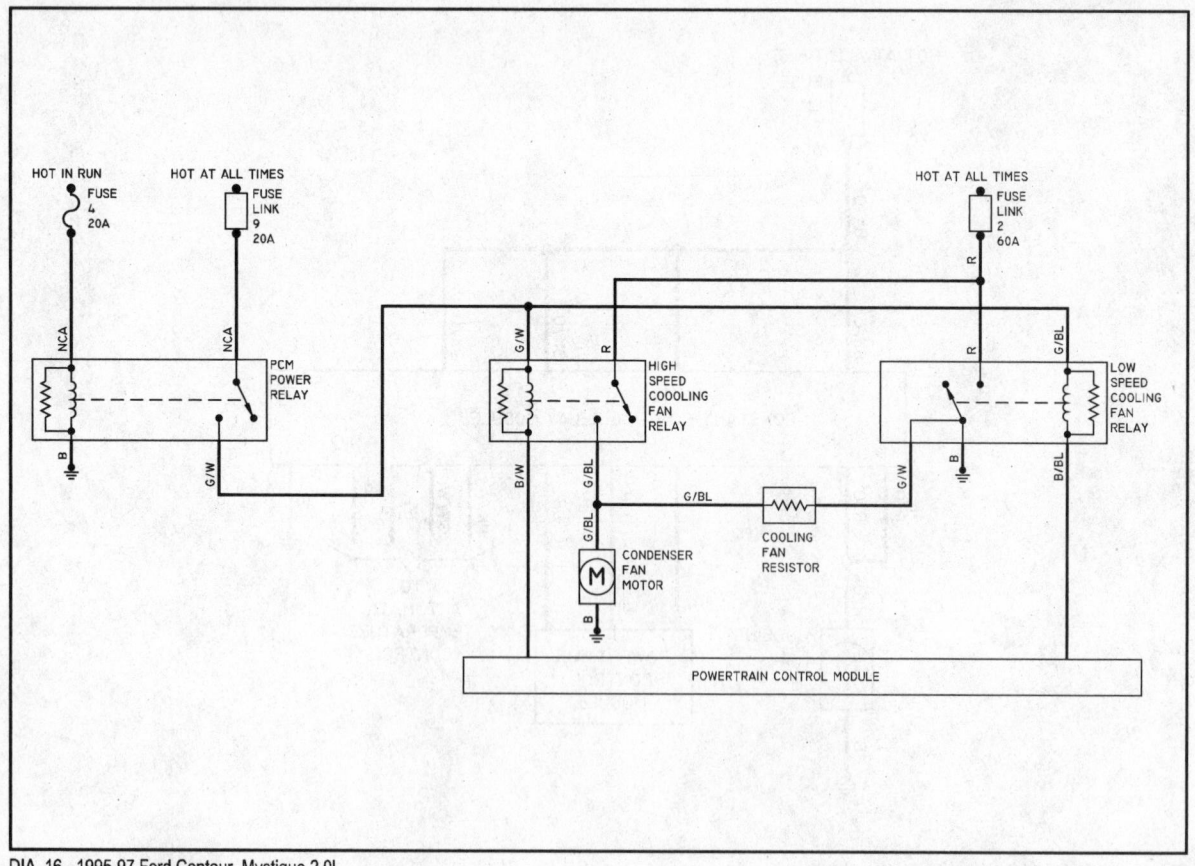

DIA. 16 - 1995-97 Ford Contour, Mystique 2.0L

79229W08

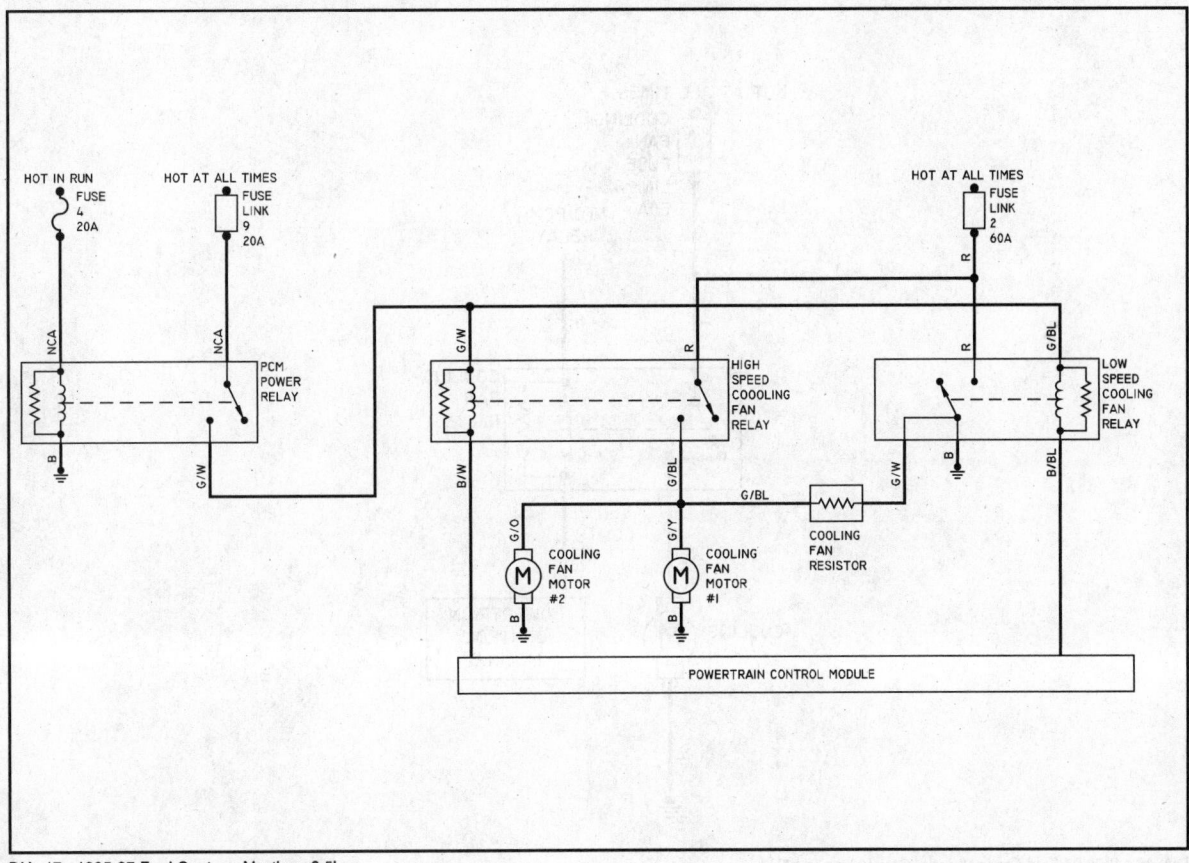

DIA. 17 - 1995-97 Ford Contour, Mystique 2.5L

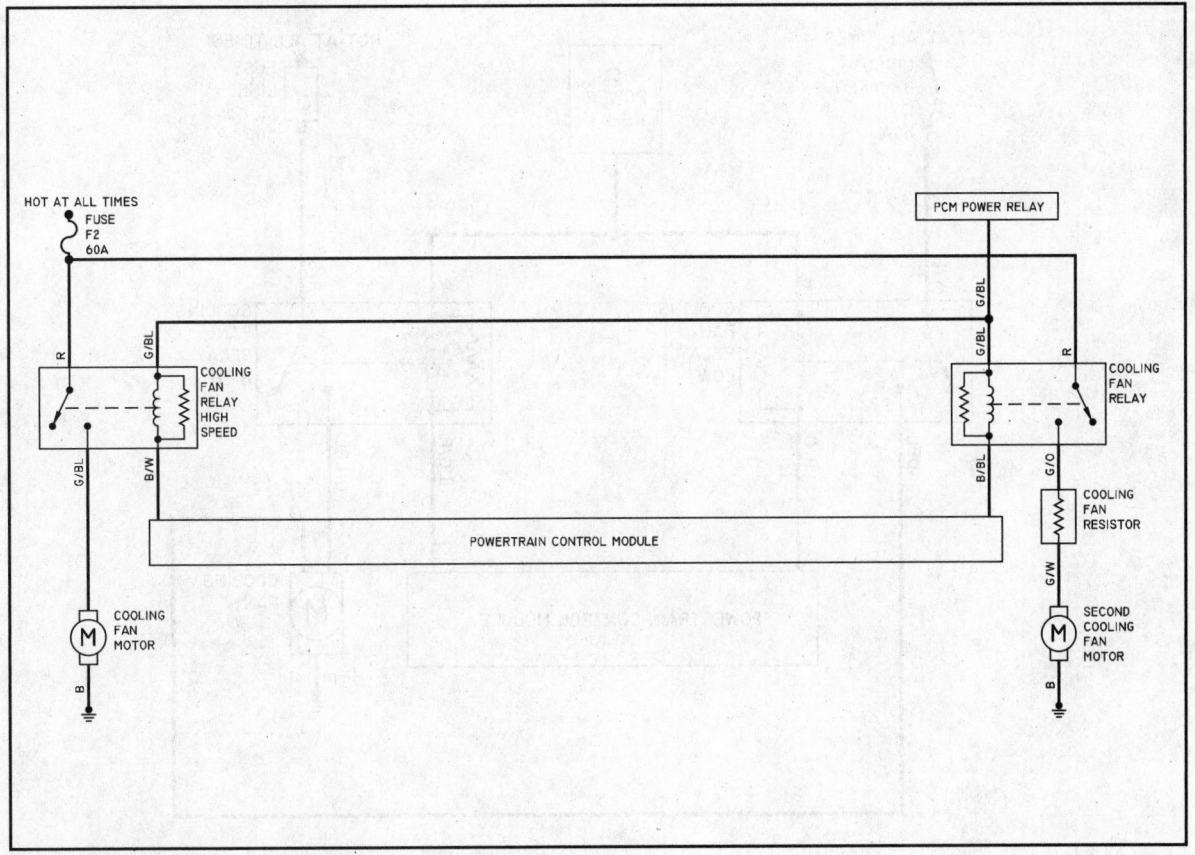

DIA. 18 - 1998-99 Ford Contour, Mystique 2.0L / 2.5L

79229W09

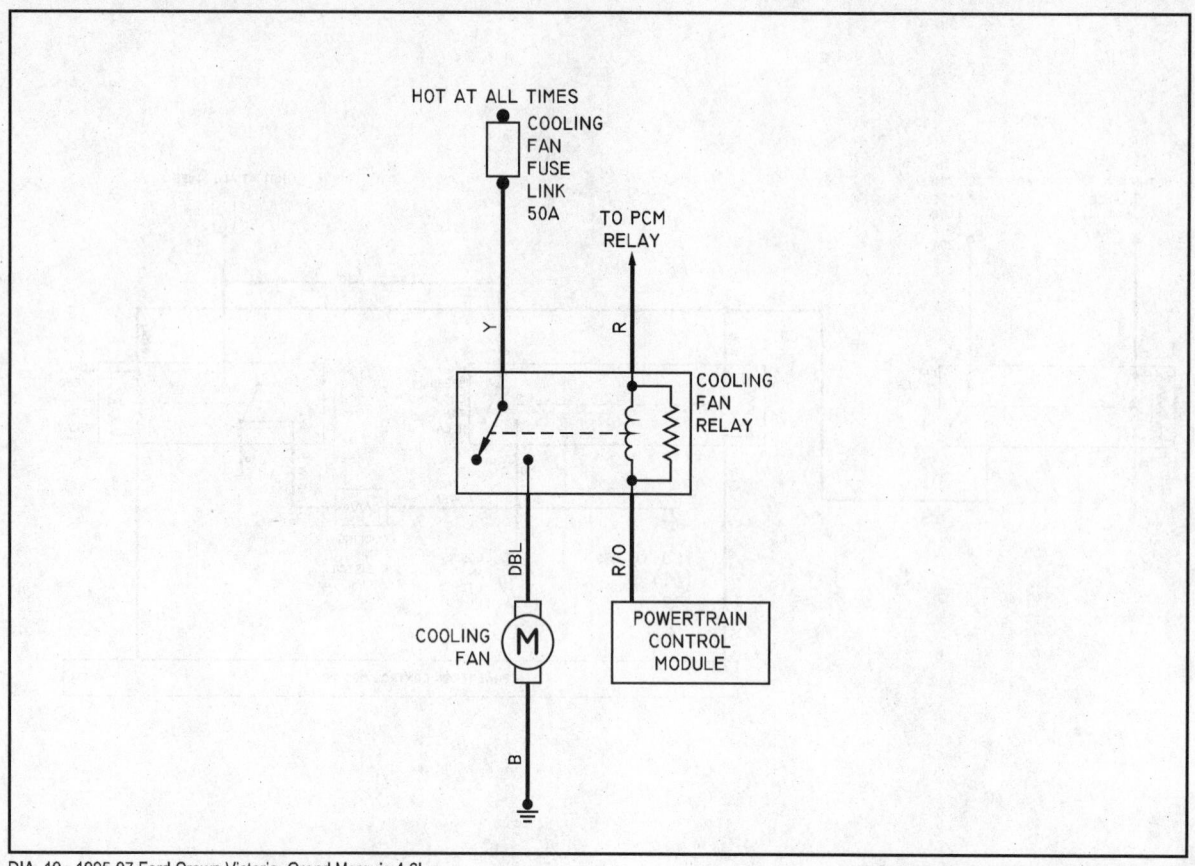

DIA. 19 - 1995-97 Ford Crown Victoria, Grand Marquis 4.6L

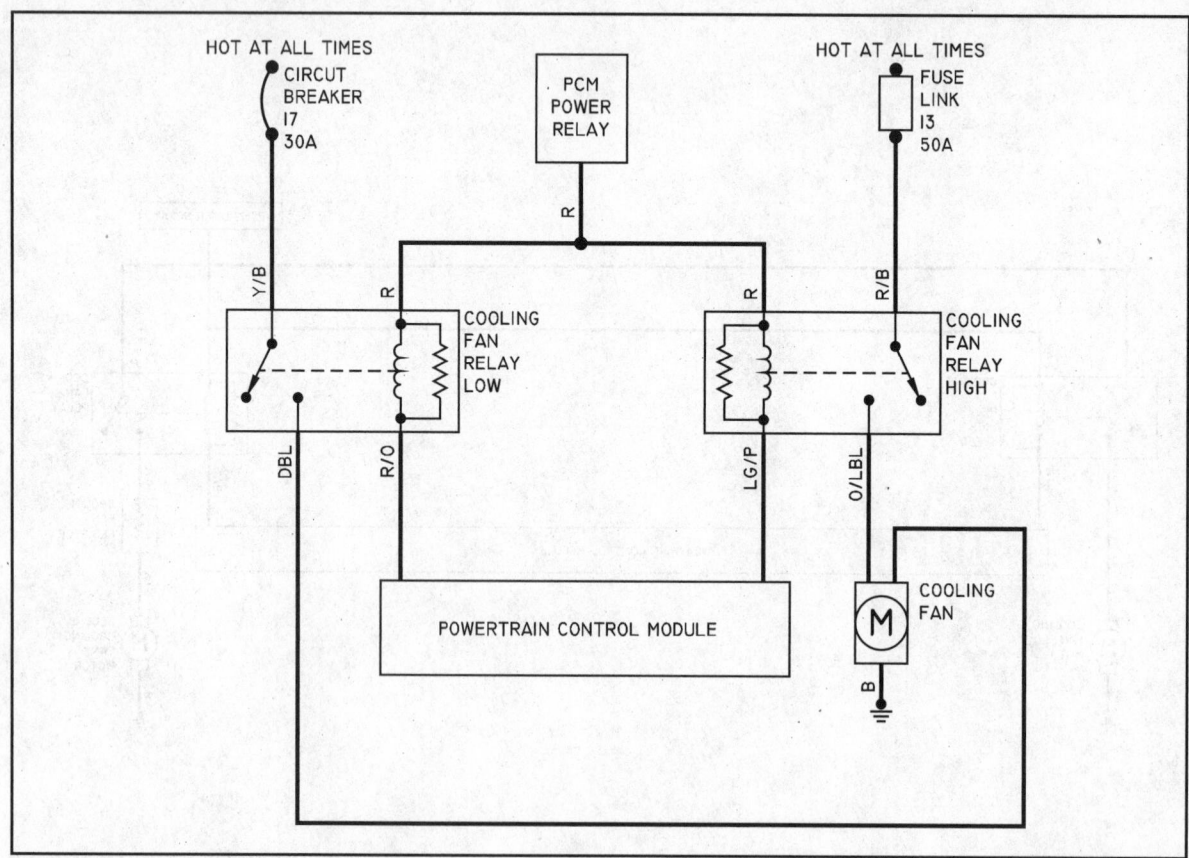

DIA. 20 - 1998-99 Ford Crown Victoria, Grand Marquis 4.6L

79229W10

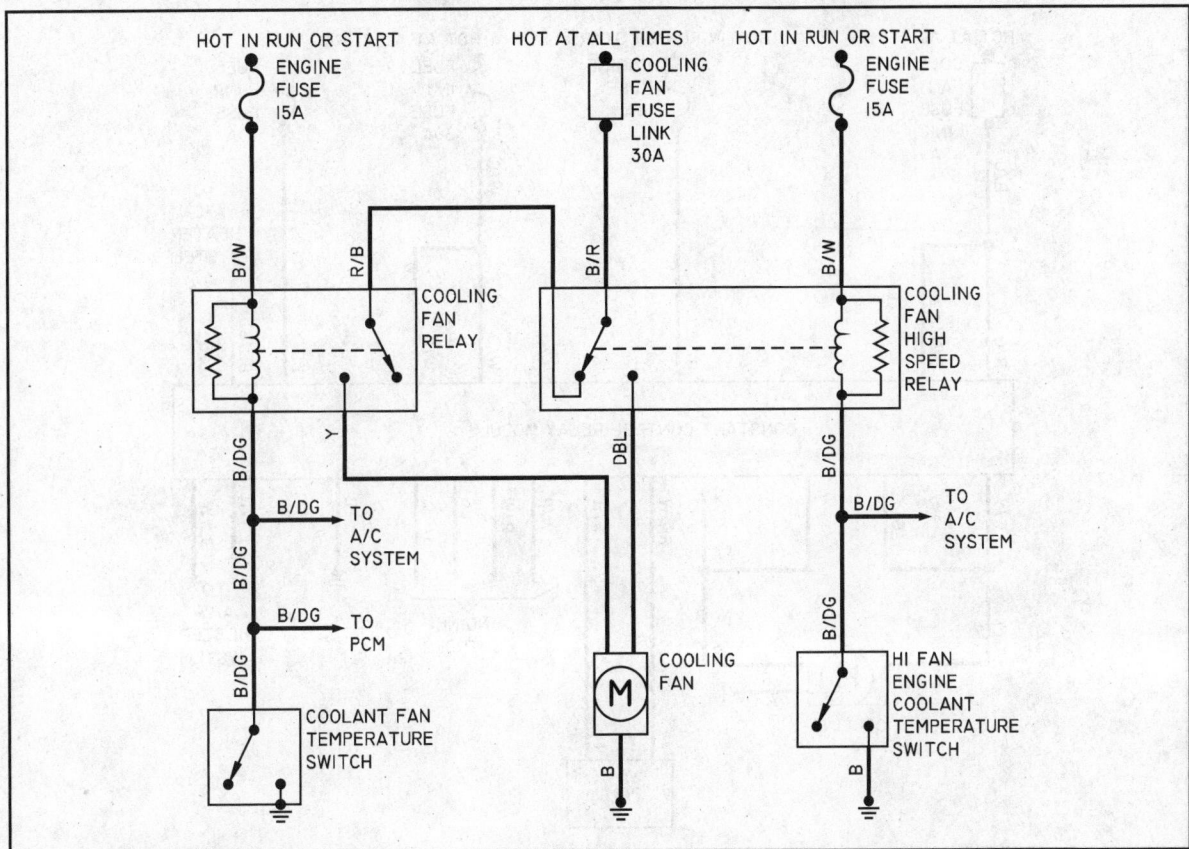

DIA. 21 - 1995-96 Ford Escort, Tracer 1.8L A/T

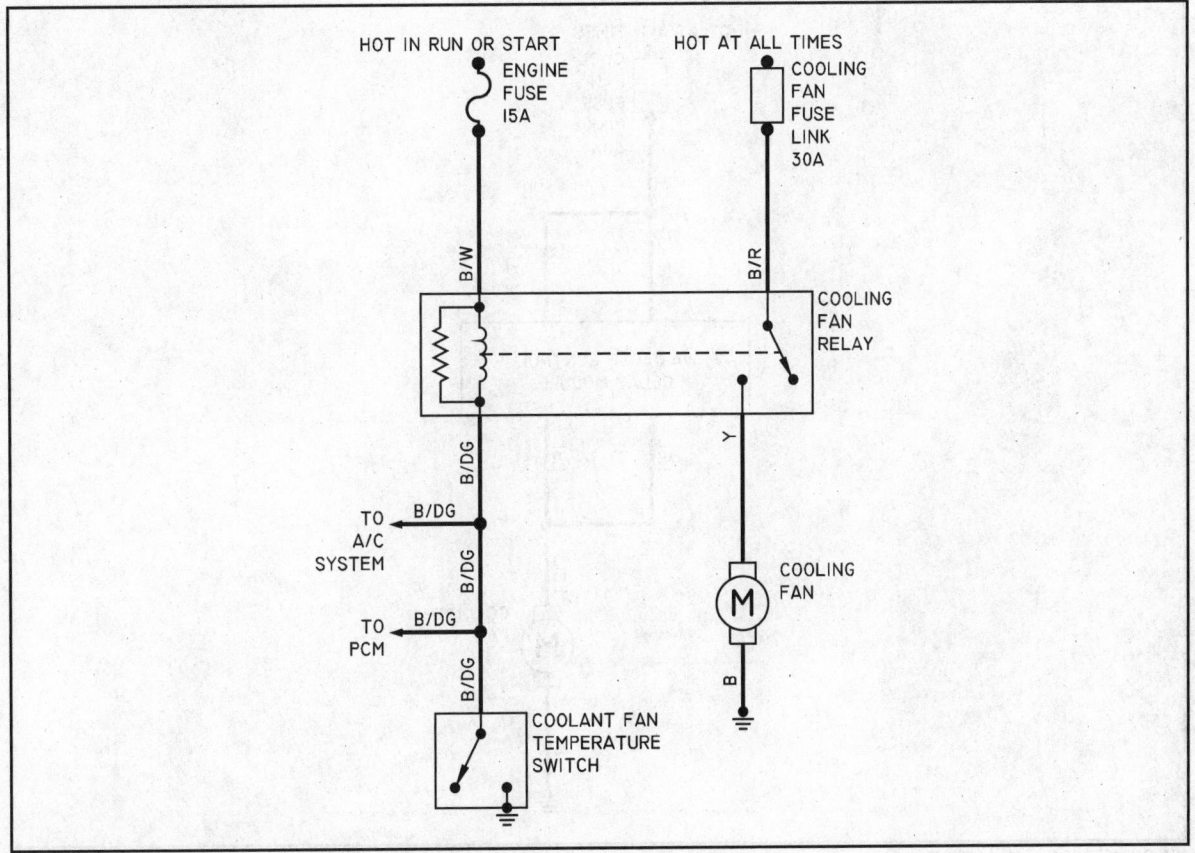

DIA. 22 - 1995-96 Ford Escort, Tracer 1.8L M/T

79229W11

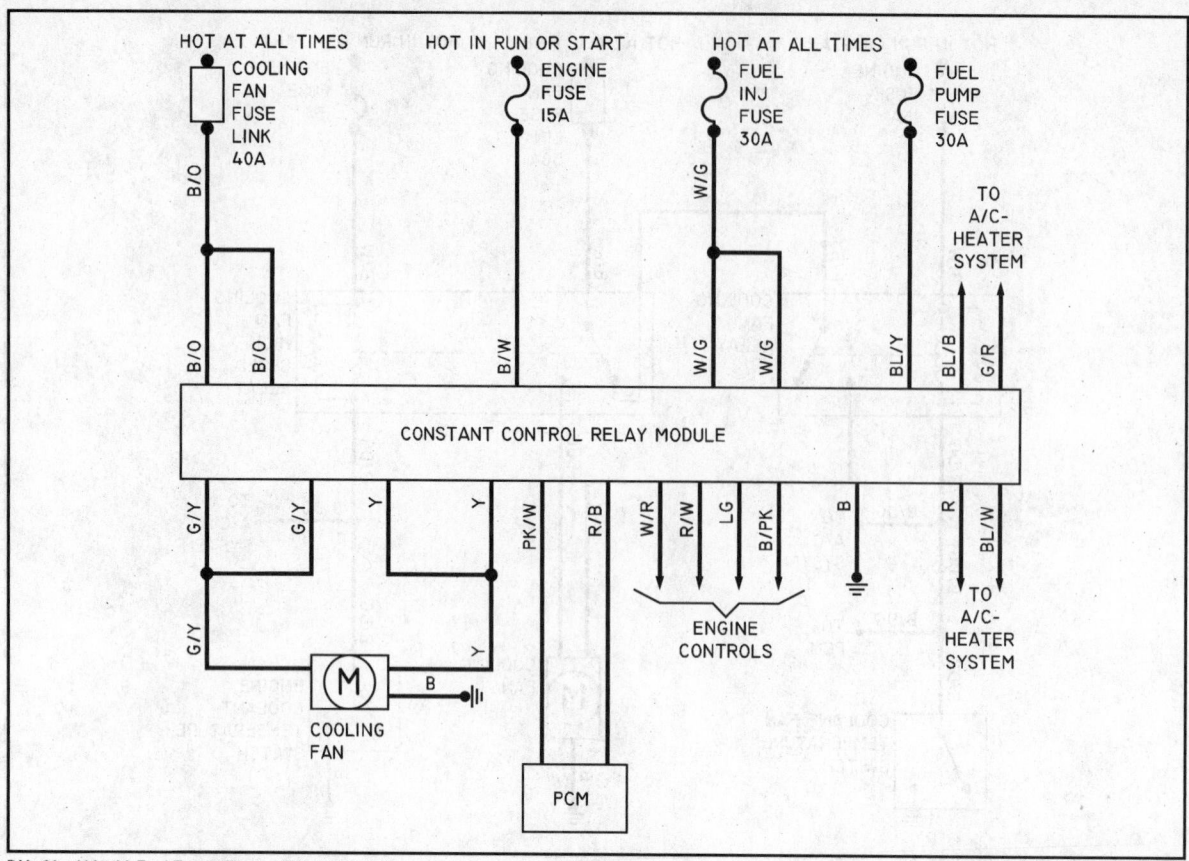

DIA. 23 - 1997-99 Ford Escort, Tracer, ZX2 1.8L / 1.9L

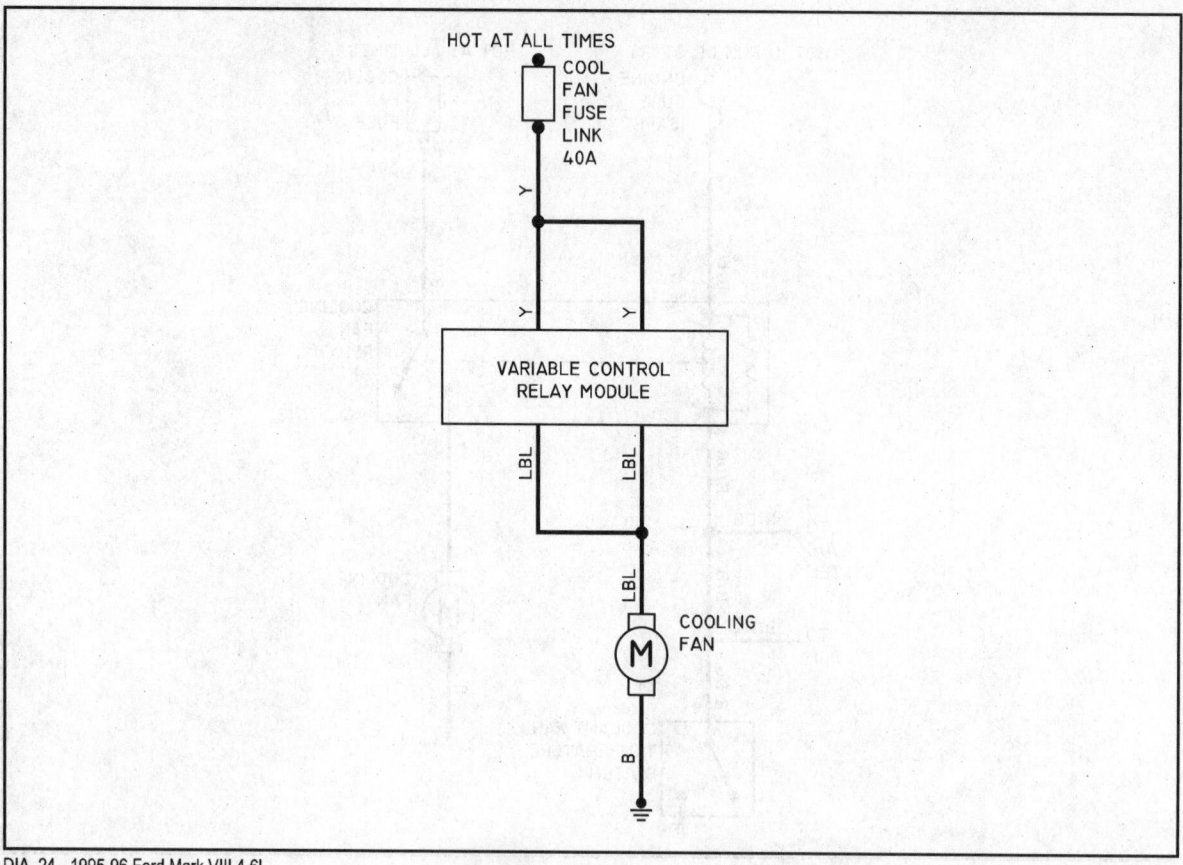

DIA. 24 - 1995-96 Ford Mark VIII 4.6L

79229W12

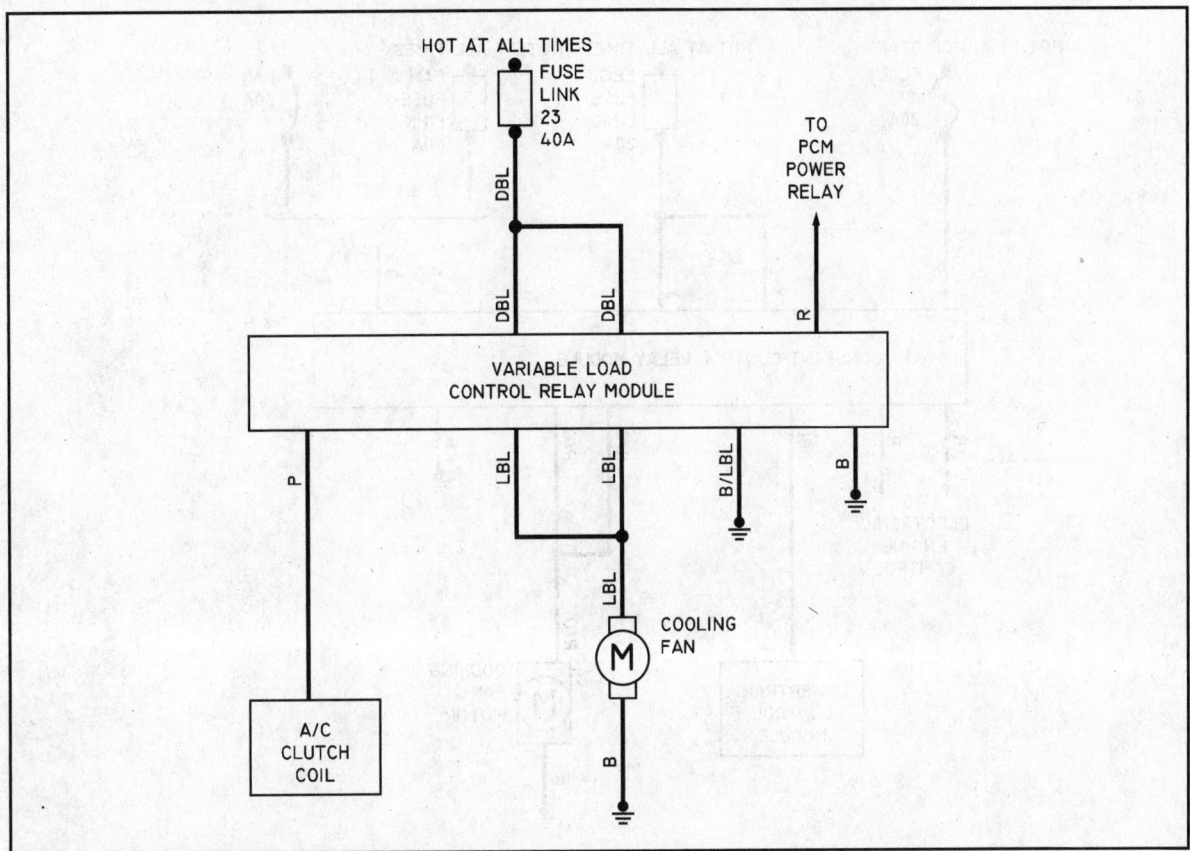

DIA. 25 - 1997-99 Ford Mark VIII 4.6L

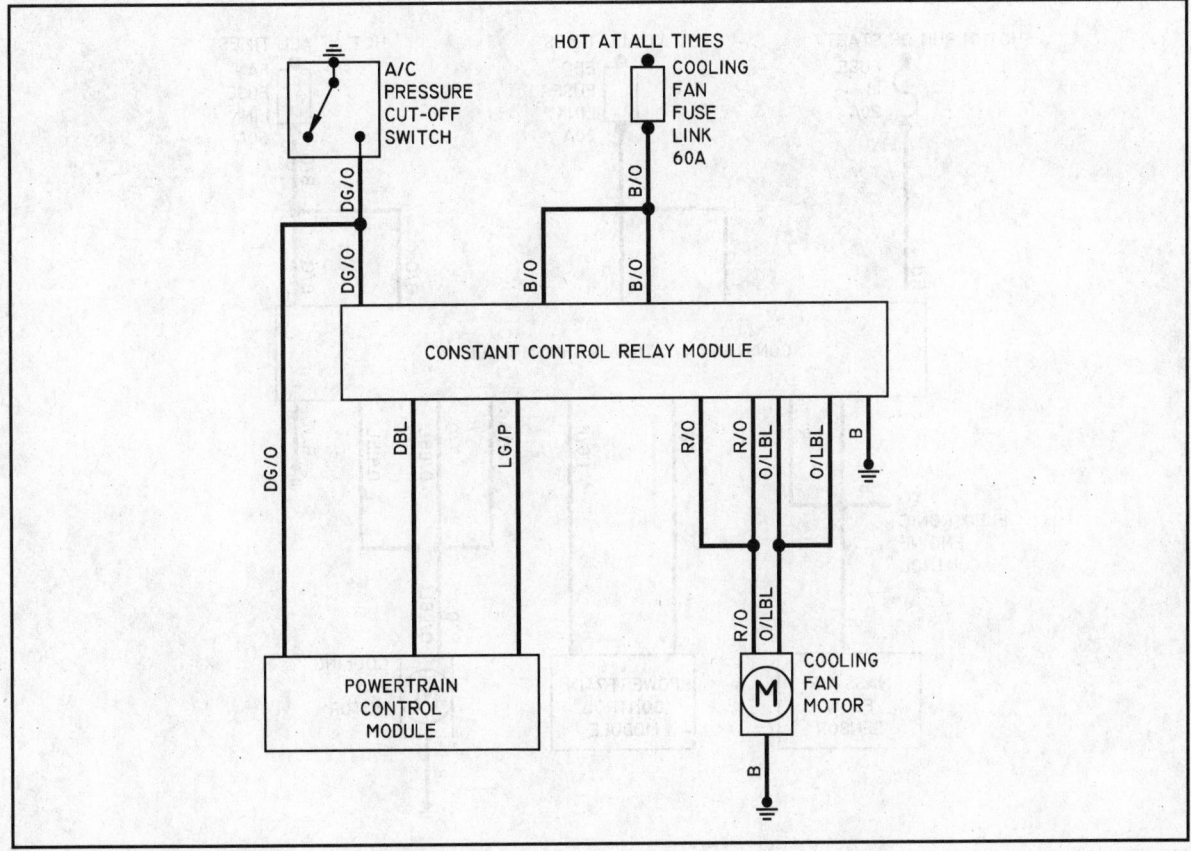

DIA. 26 - 1995 Ford Mustang 3.8L / 5.0L

79229W13

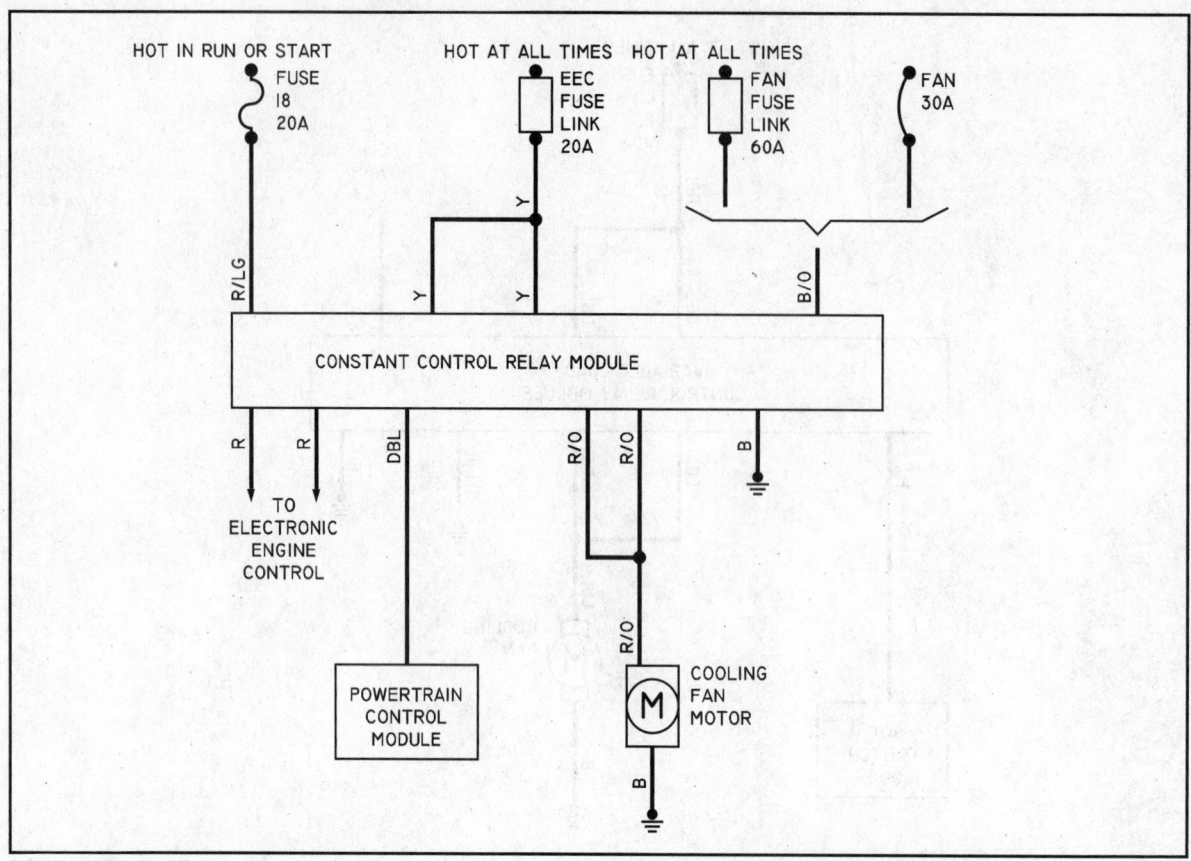

DIA. 27 - 1996-99 Ford Mustang 3.8L

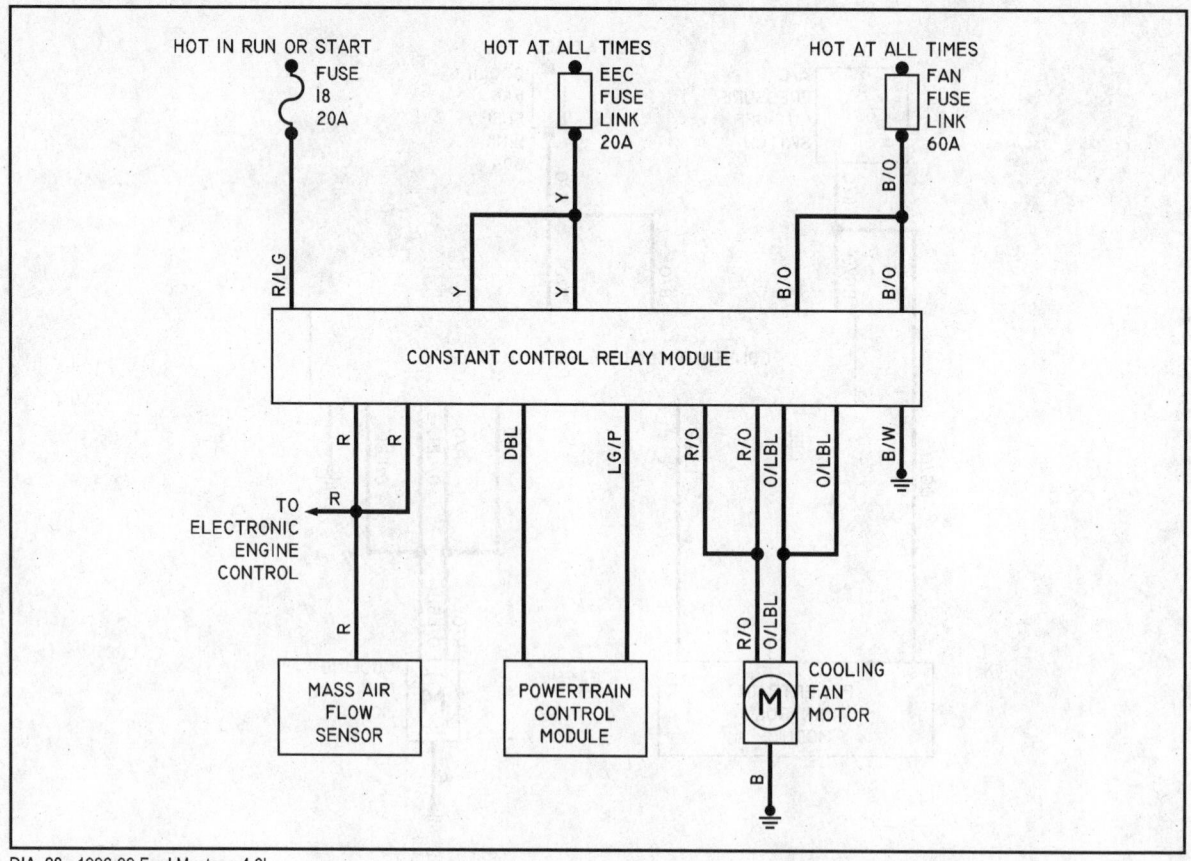

DIA. 28 - 1996-99 Ford Mustang 4.6L

79229W14

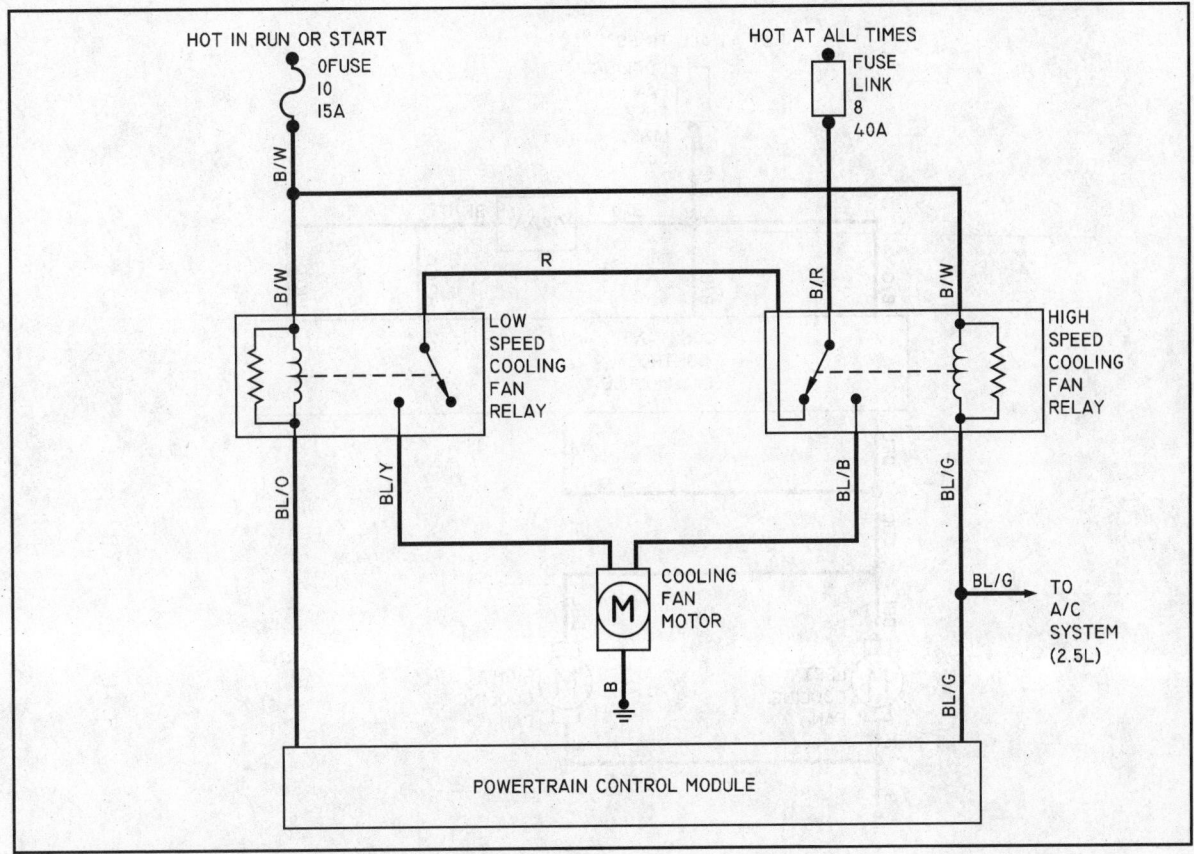

DIA. 29 - 1995-97 Ford Probe 2.0L / 2.5L

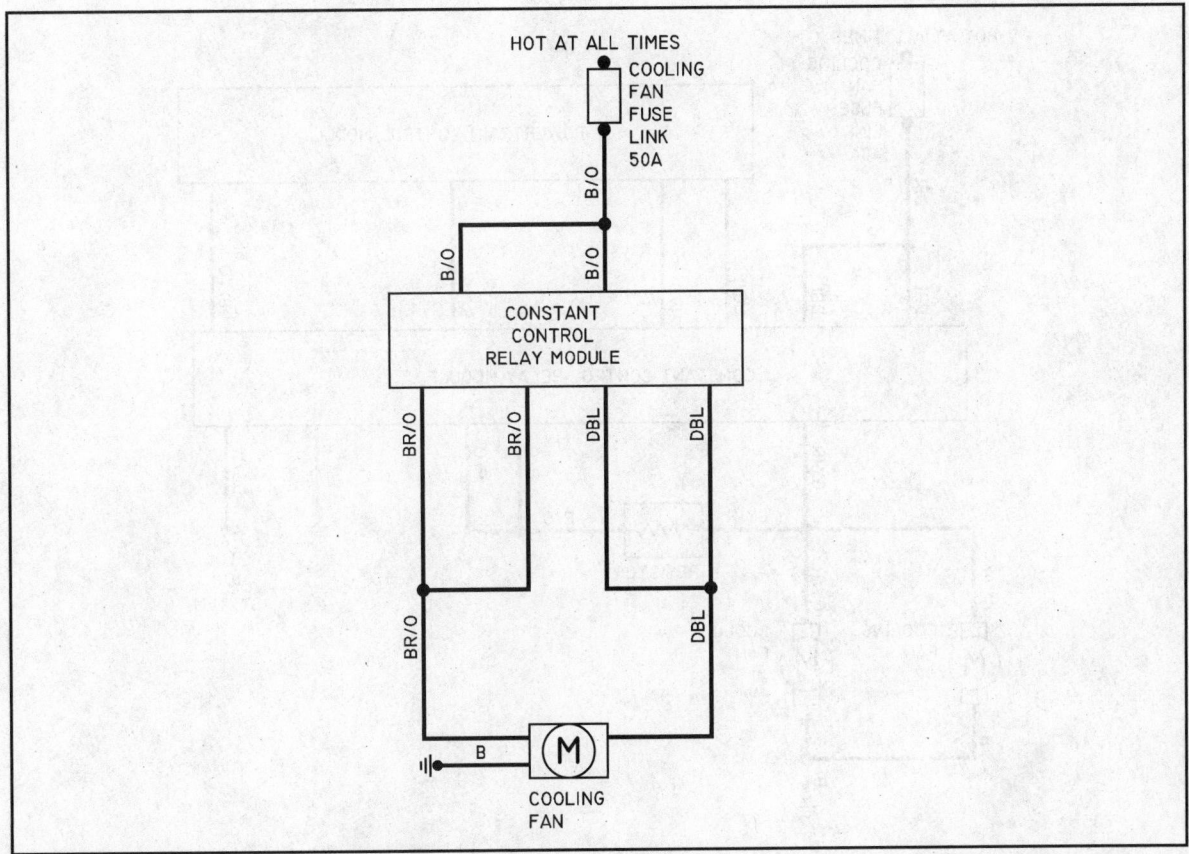

DIA. 30 - 1995 Ford Taurus, Sable 3.0L / 3.0L SHO / 3.8L

79229W15

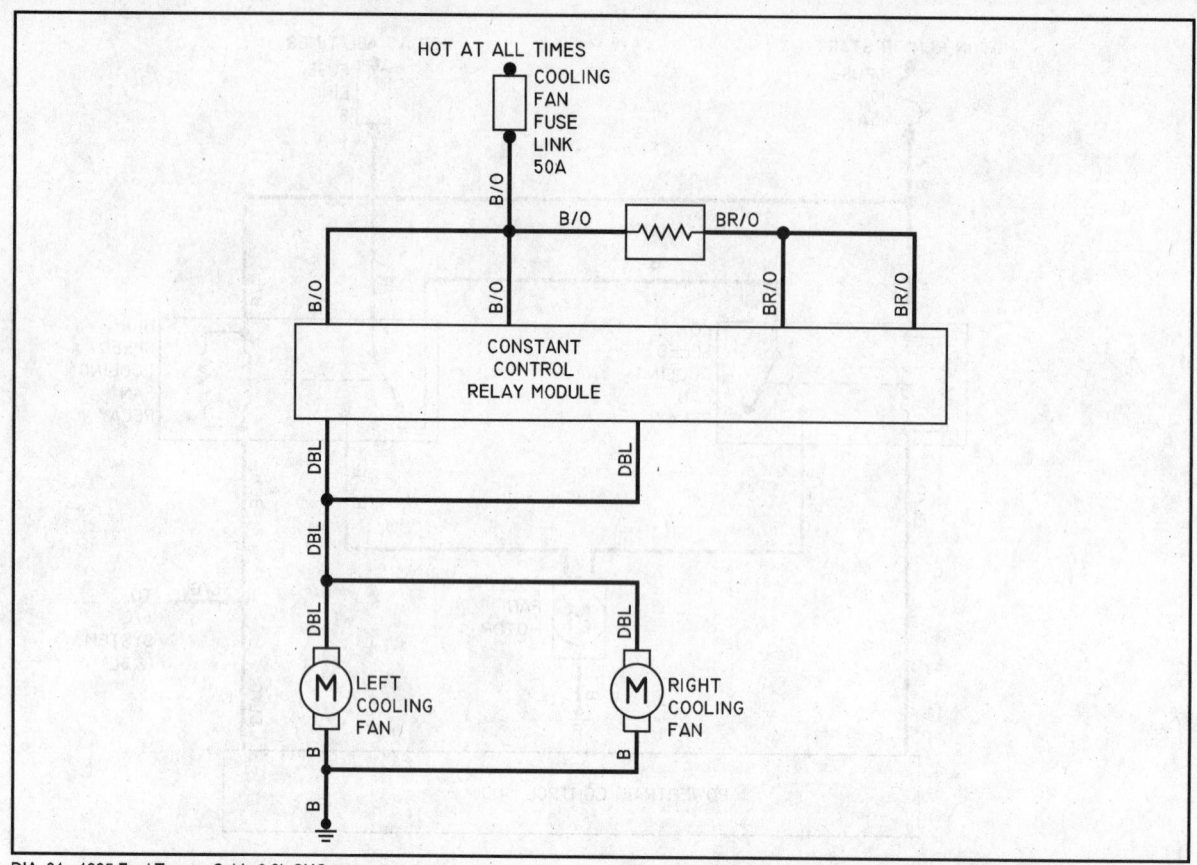

DIA. 31 - 1995 Ford Taurus, Sable 3.2L SHO

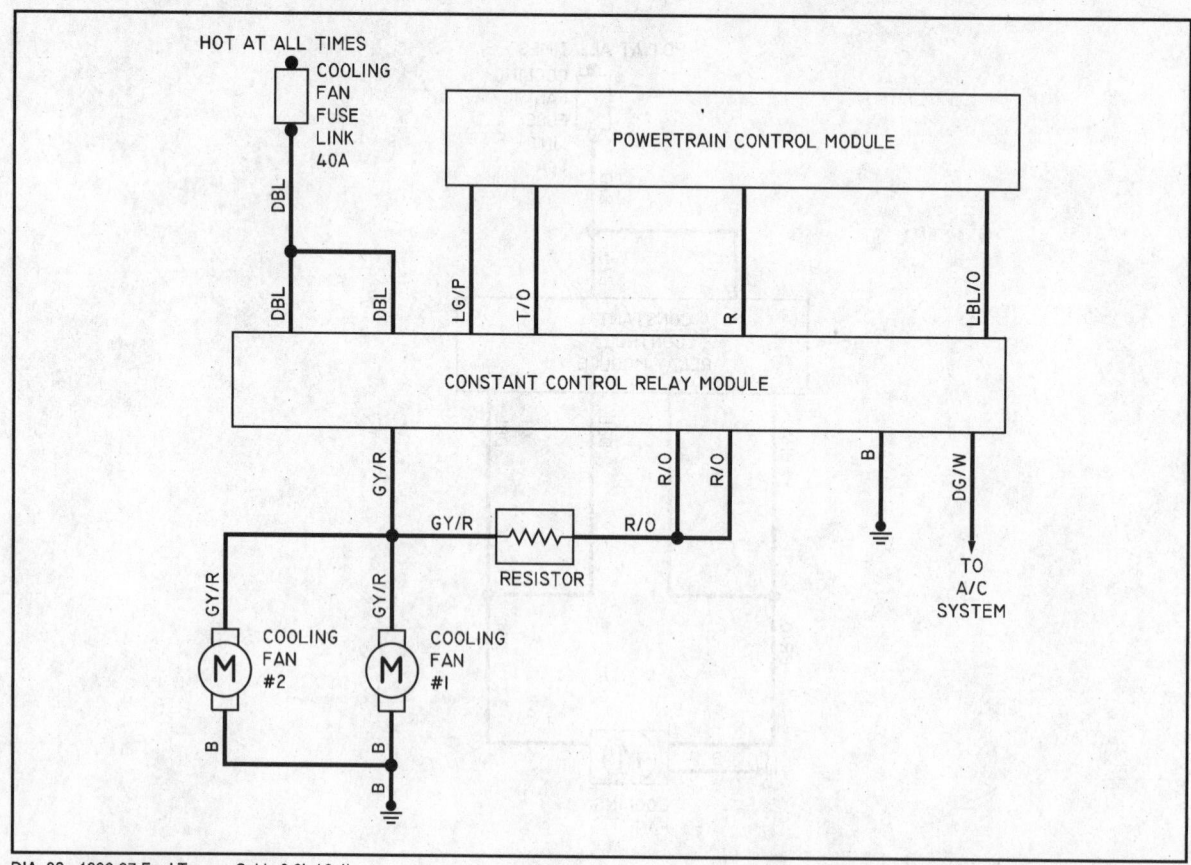

DIA. 32 - 1996-97 Ford Taurus, Sable 3.0L / 3.4L

79229W16

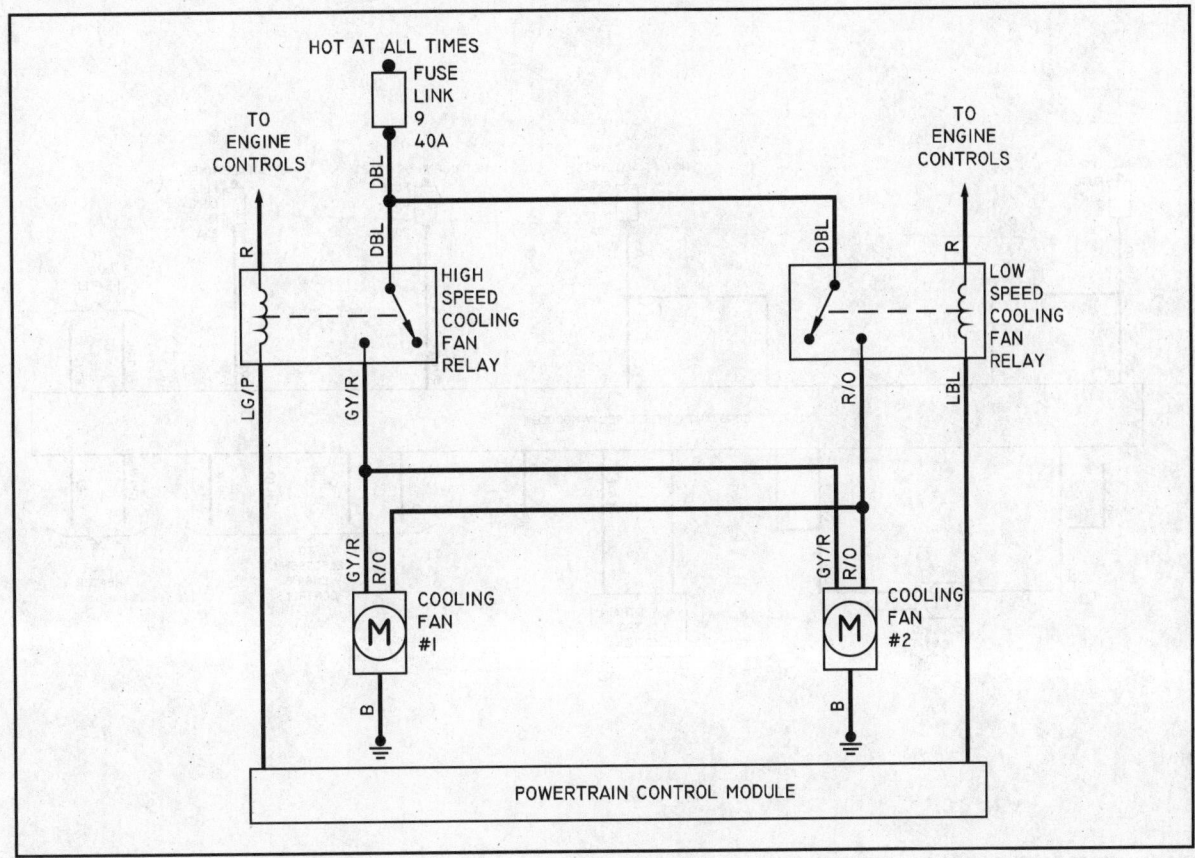

DIA. 33 - 1998-99 Ford Taurus, Sable 3.0L / 3.4L

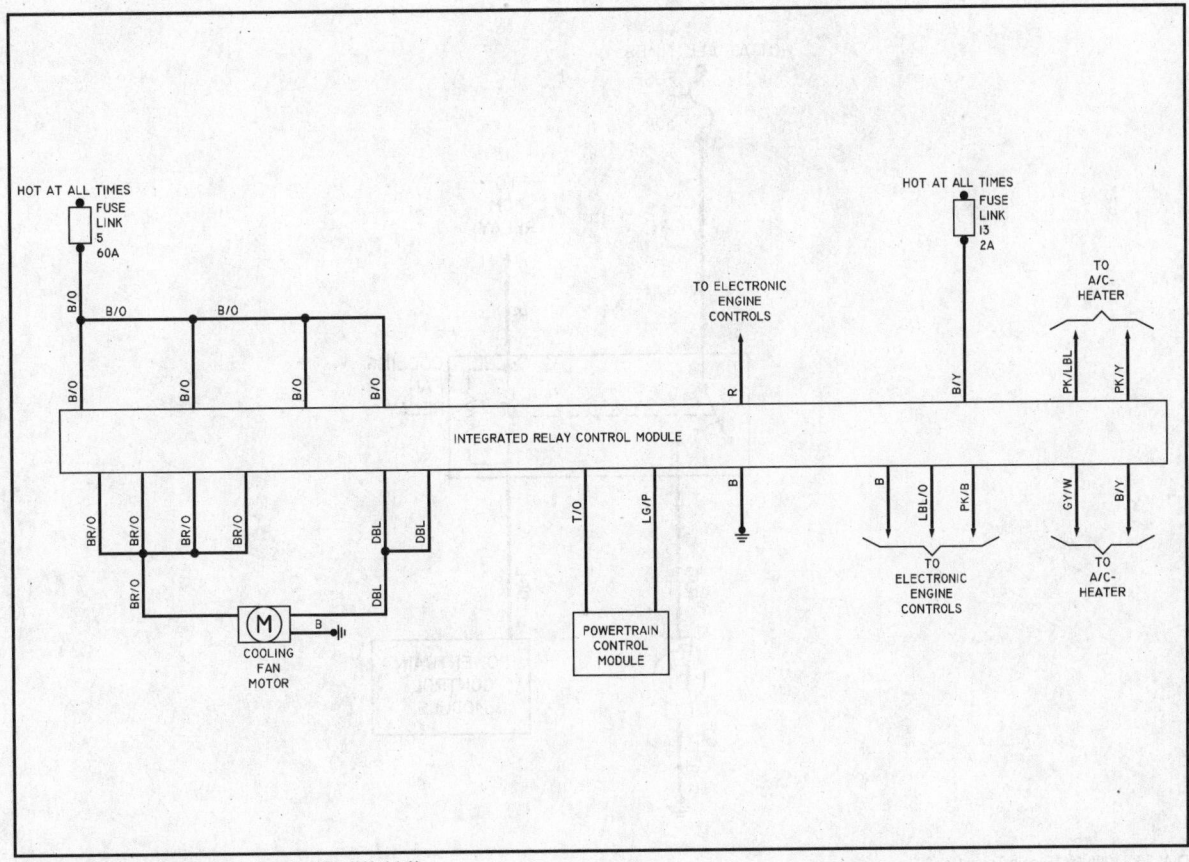

DIA. 34 - 1995 Ford Thunderbird, Cougar, XR7 3.8L / 4.6L

79229W17

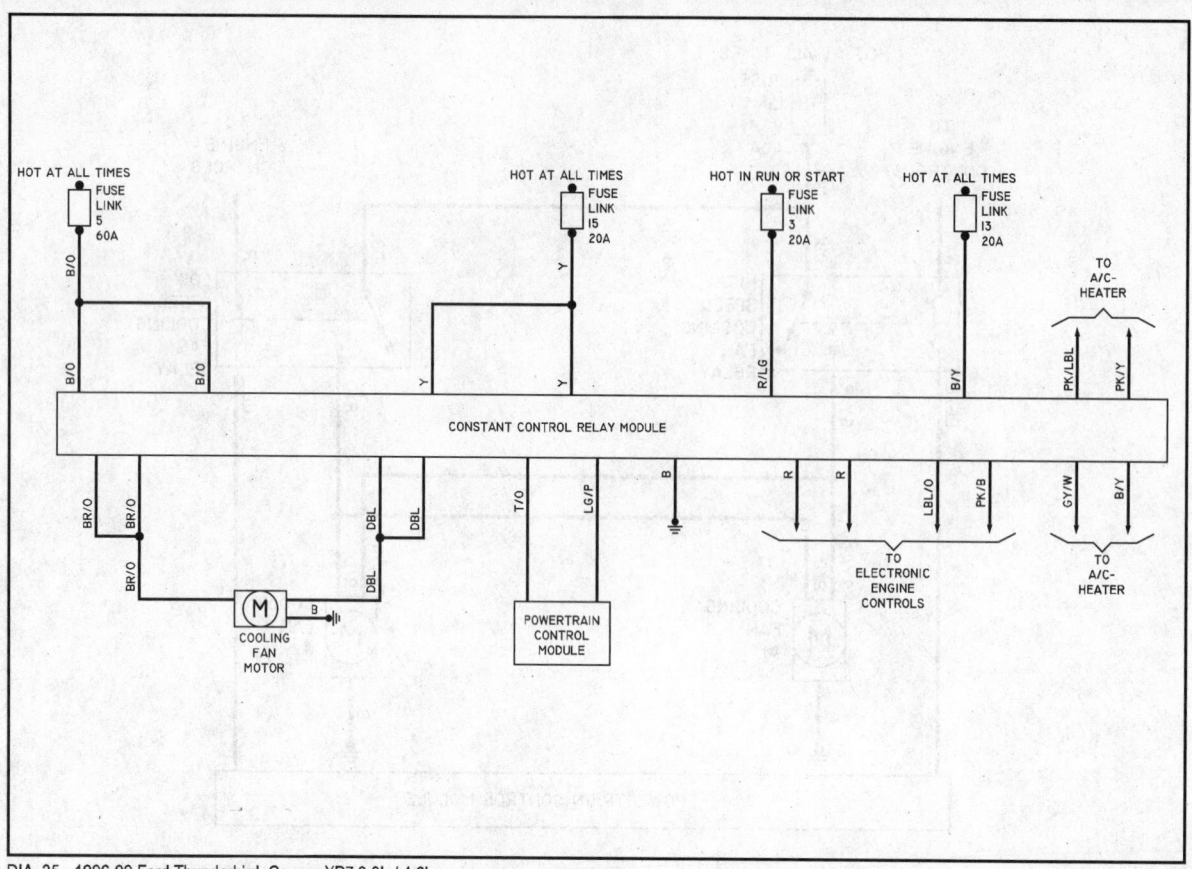

DIA. 35 - 1996-99 Ford Thunderbird, Cougar XR7 3.8L / 4.6L

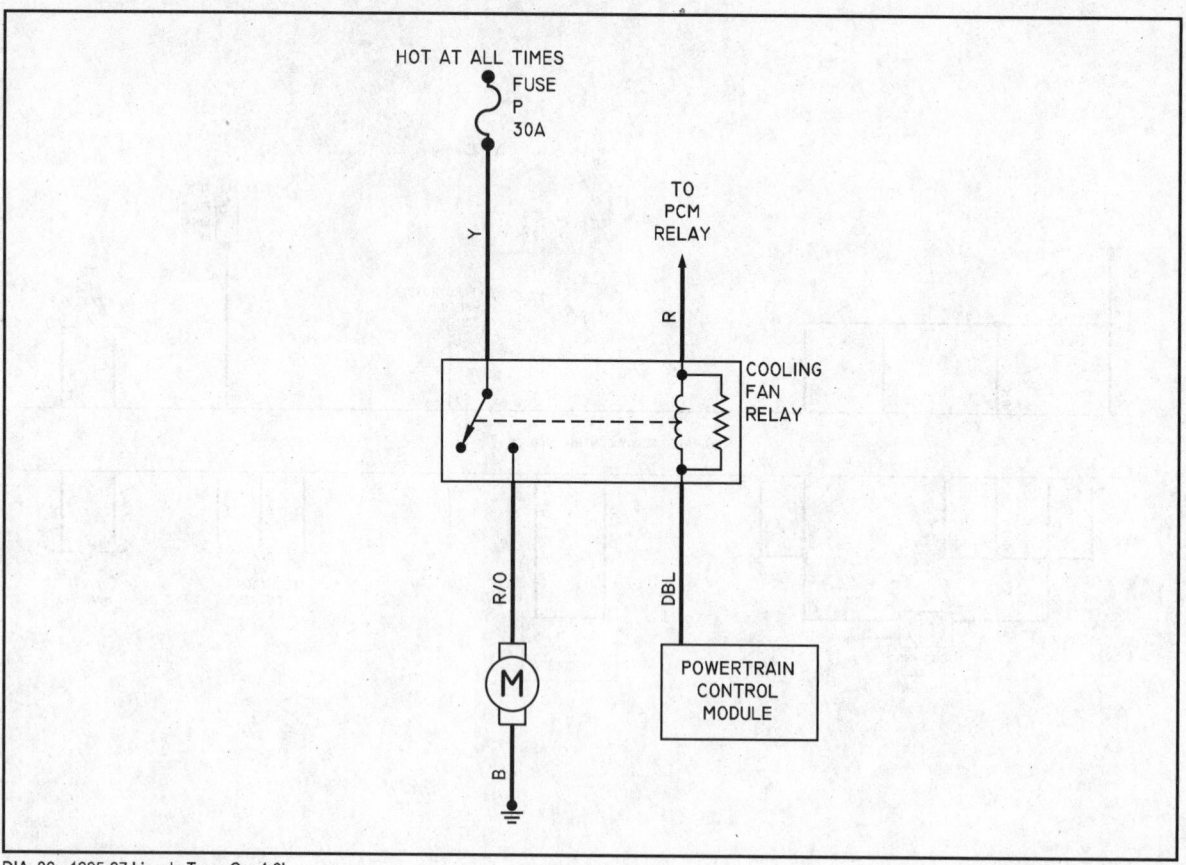

DIA. 36 - 1995-97 Lincoln Town Car 4.6L

79229W18

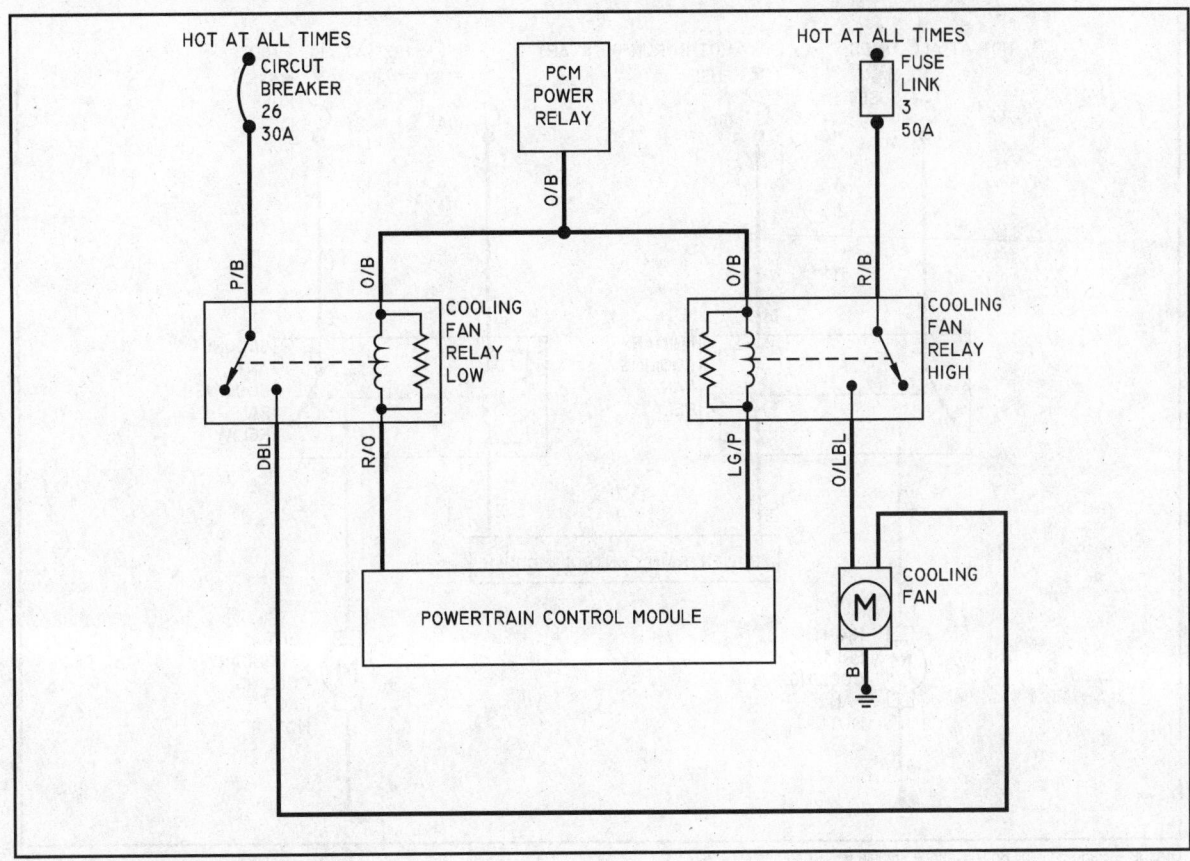

DIA. 37 - 1998-99 Lincoln Town Car 4.6L

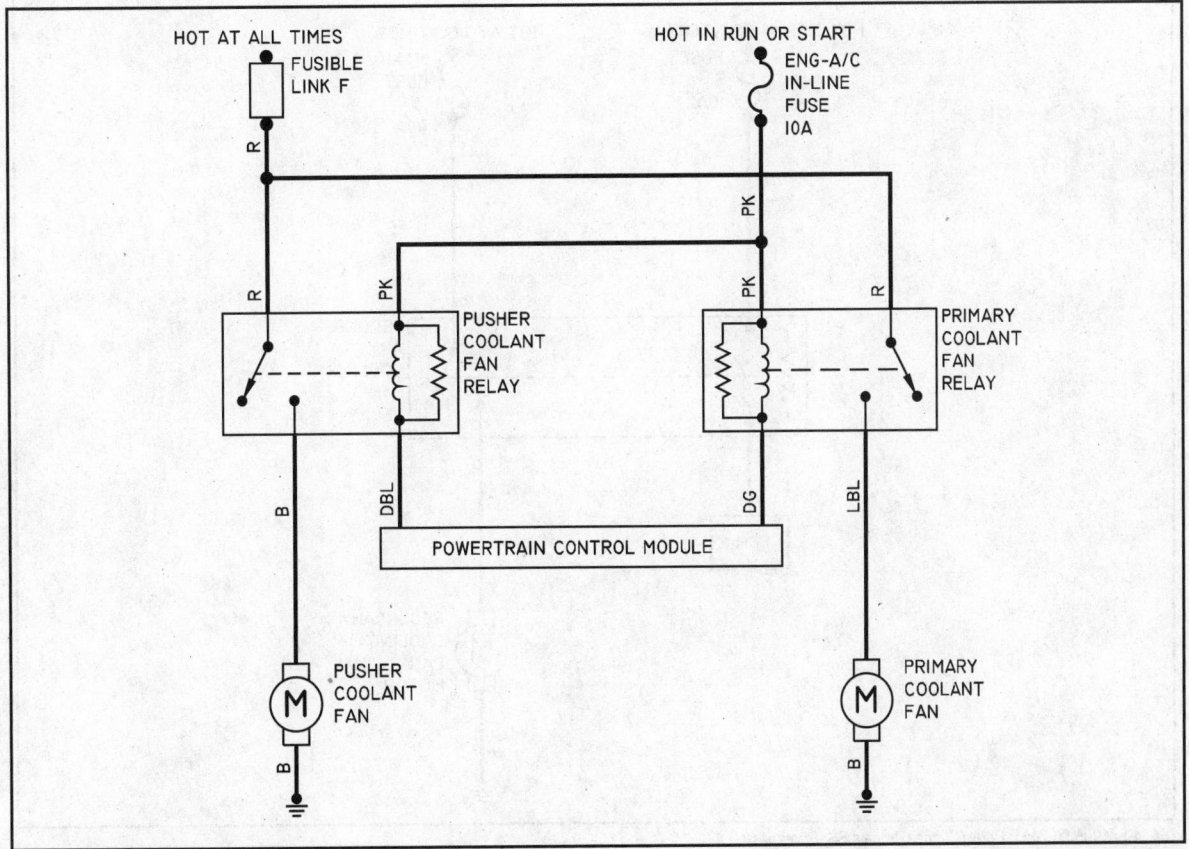

DIA. 38 - 1995-96 GM A Body (Century, Cutlass Ciera, Cutlass Cruiser) 2.2L / 3.1L

79229W19

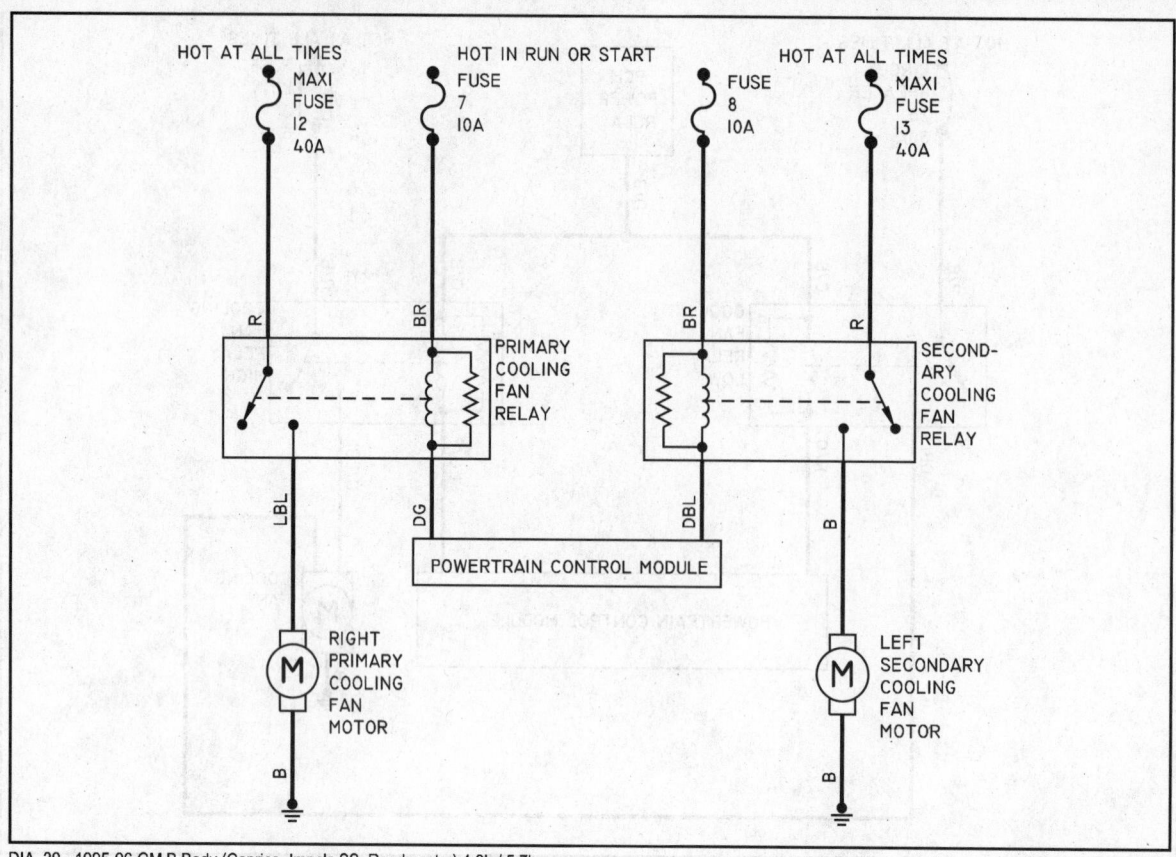

DIA. 39 - 1995-96 GM B Body (Caprice, Impala SS, Roadmaster) 4.3L / 5.7L
(W/O Mech Fan)

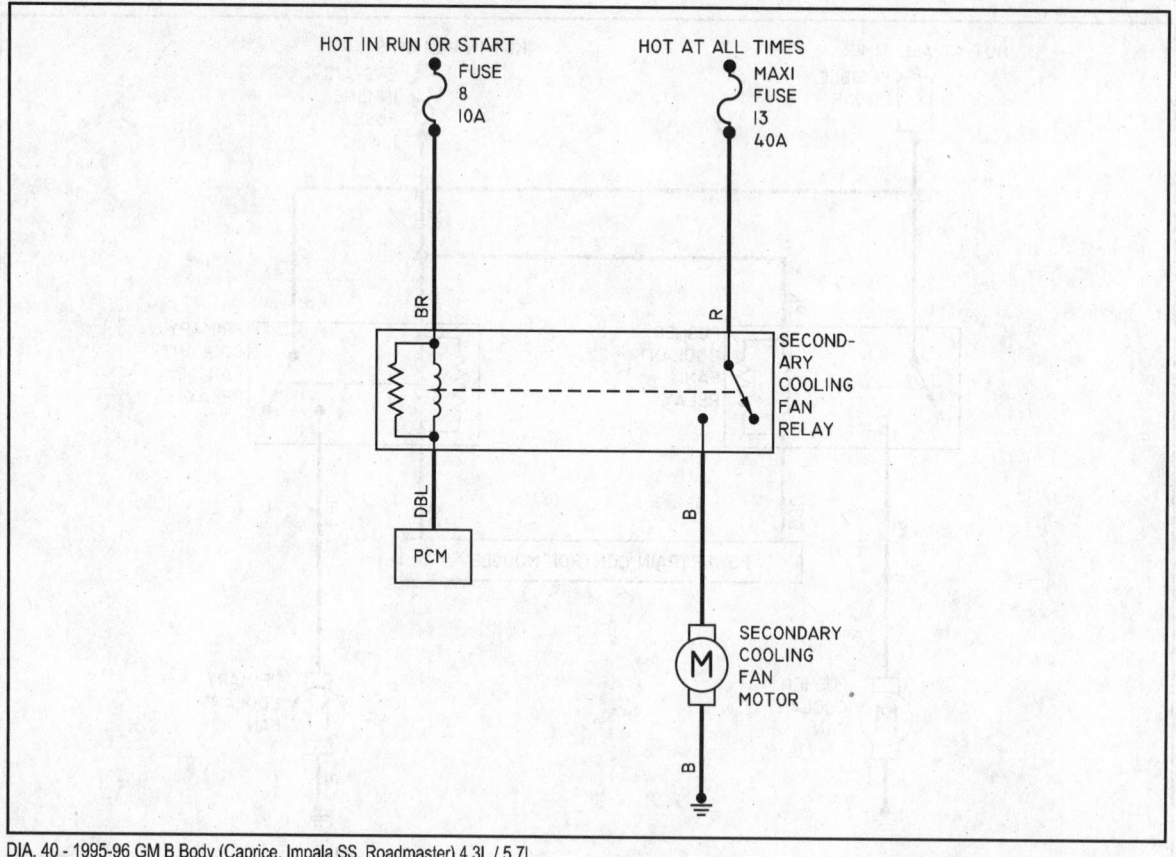

DIA. 40 - 1995-96 GM B Body (Caprice, Impala SS, Roadmaster) 4.3L / 5.7L
(W/ Mech Fan)

79229W20

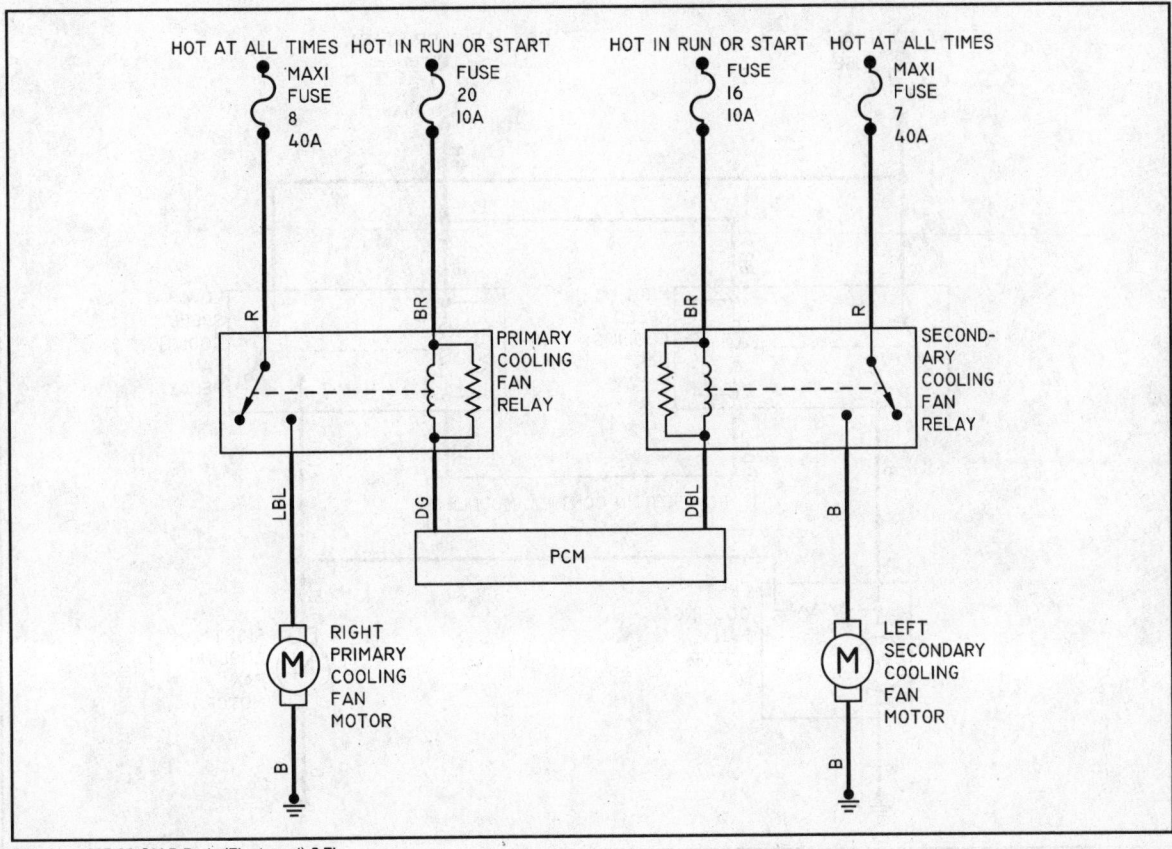

DIA. 41 - 1995-96 GM B Body (Fleetwood) 5.7L
(W/O Mech Fan)

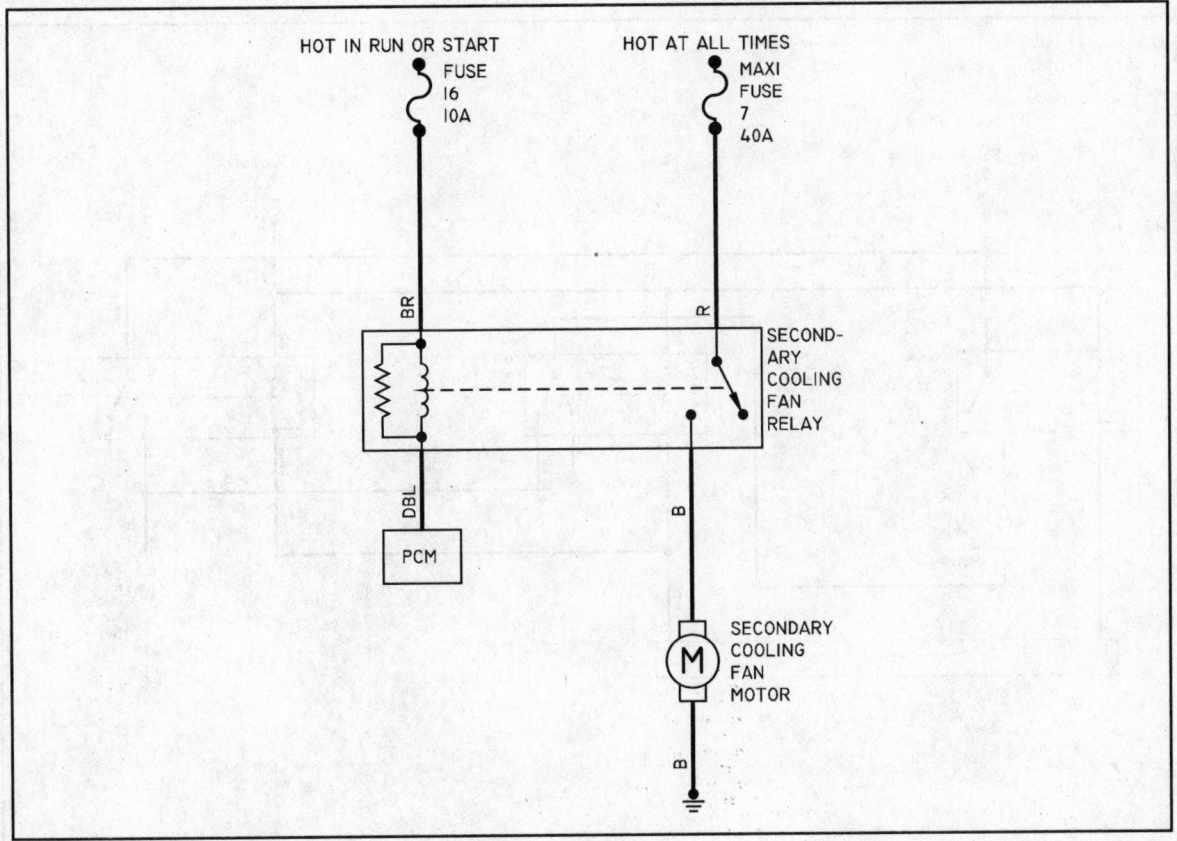

DIA. 42 - 1995-96 GM B Body (Fleetwood) 5.7L
(W/ Mech Fan)

79229W21

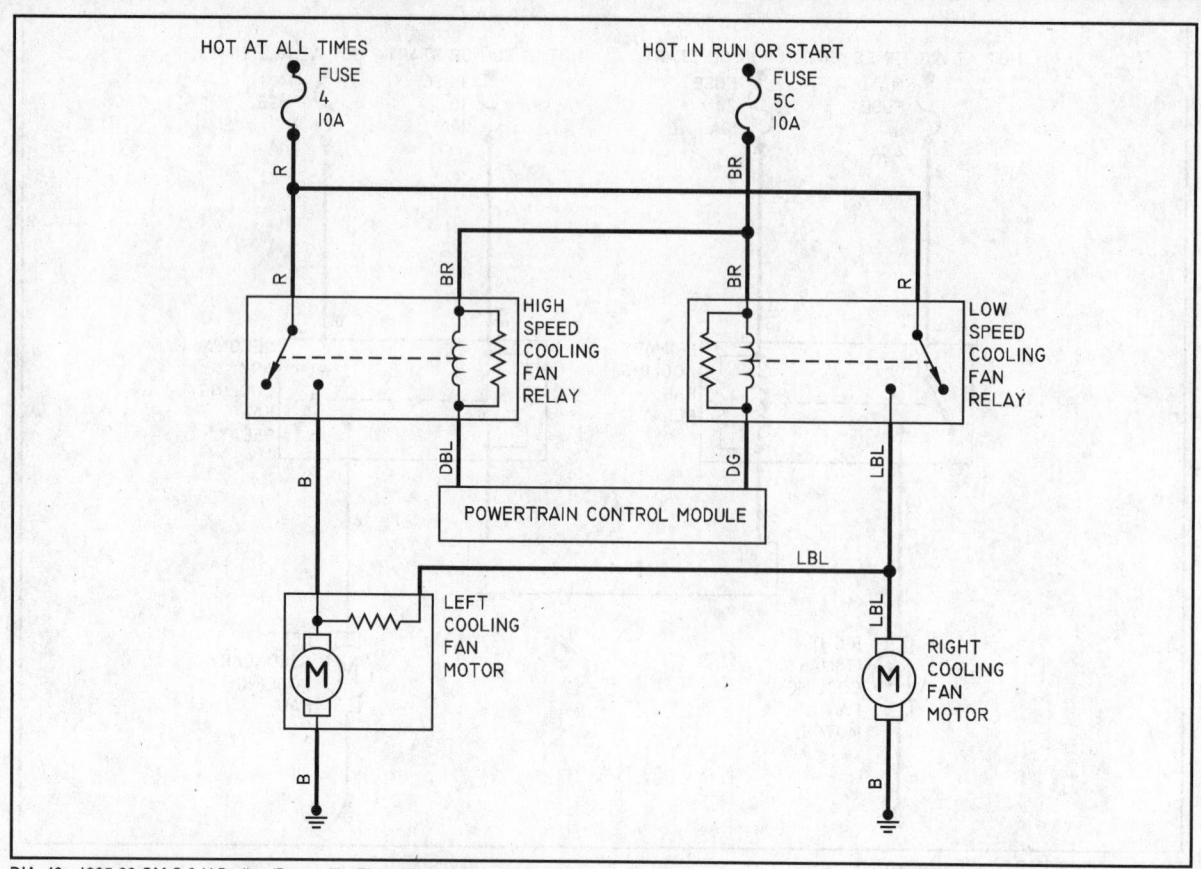

DIA. 43 - 1995-99 GM C & H Bodies (Bonneville, Eighty-Eight, Ninety-Eight, Park Ave, Le Sabre, LSS, Regency) 3.8L

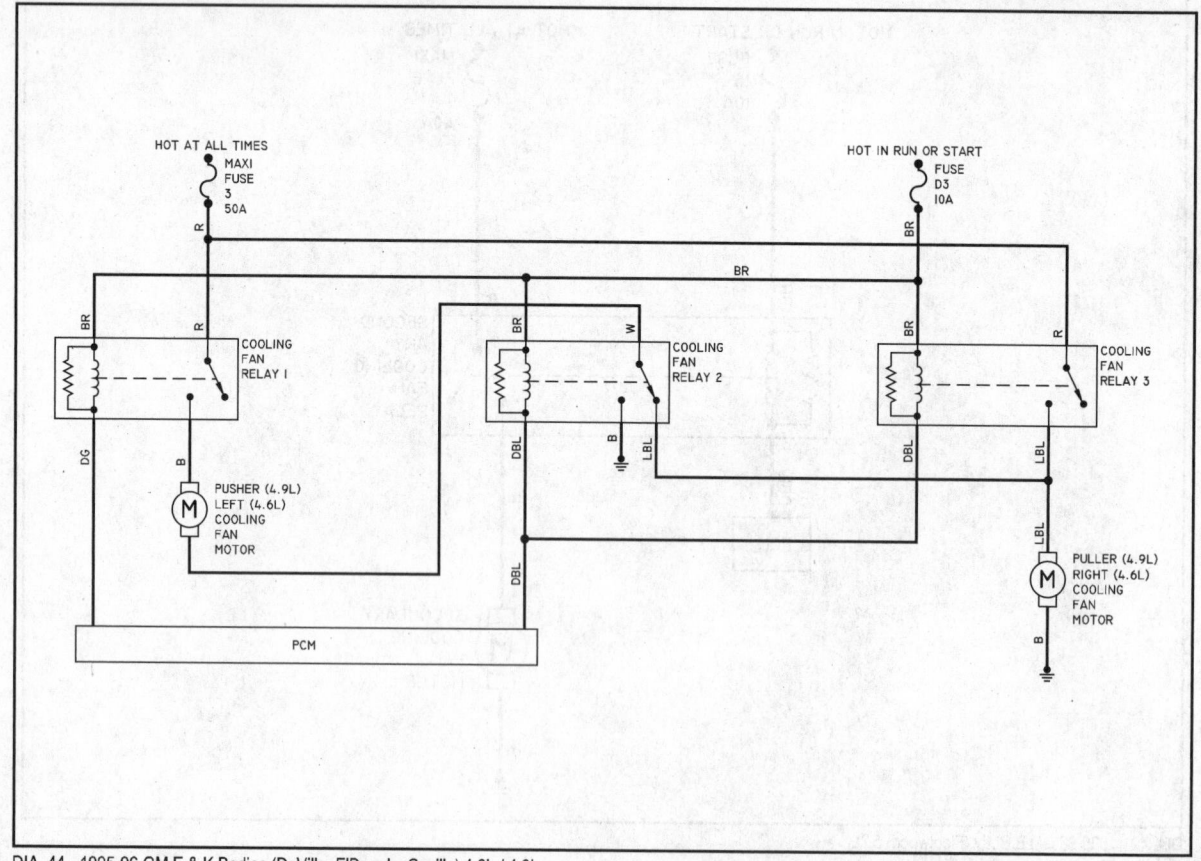

DIA. 44 - 1995-96 GM E & K Bodies (DeVille, ElDarado, Seville) 4.6L / 4.9L

79229W22

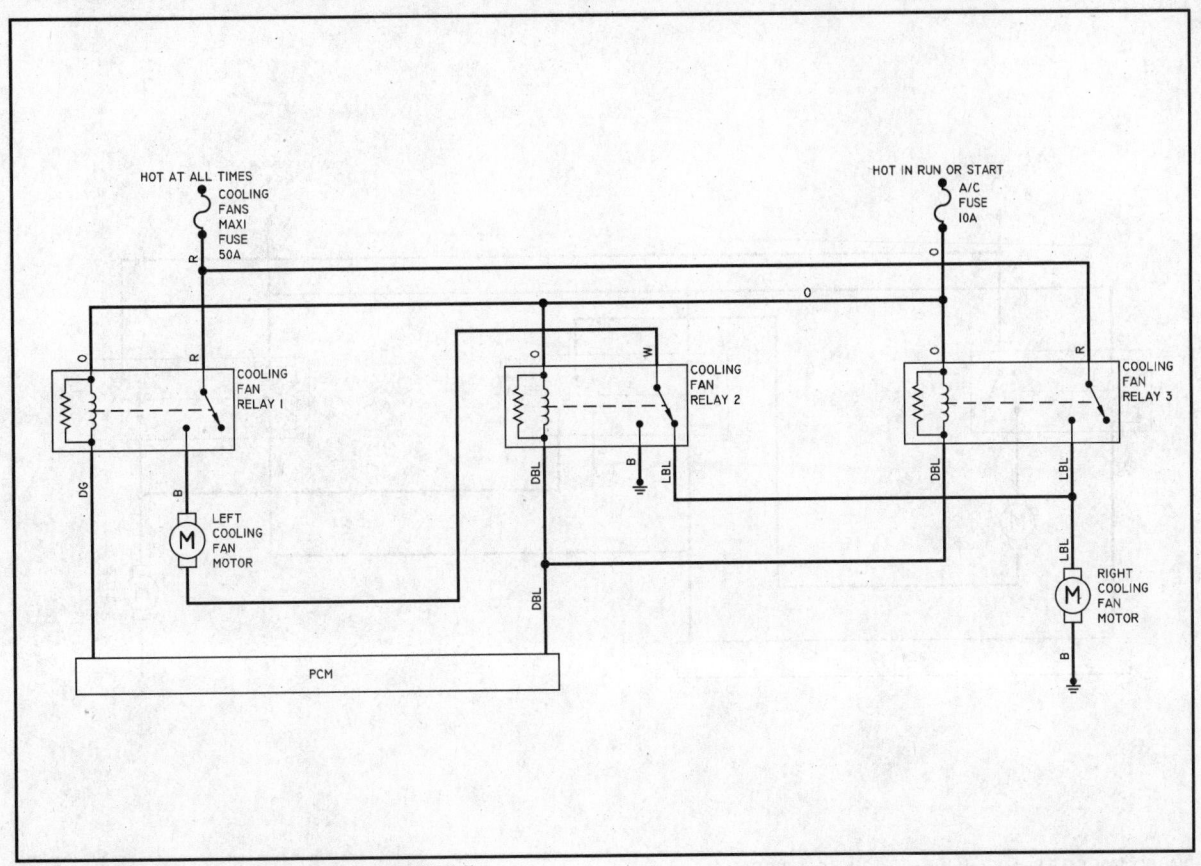

DIA. 45 - 1997 GM E & K Bodies (DeVille, ElDarado, Seville) 4.6L

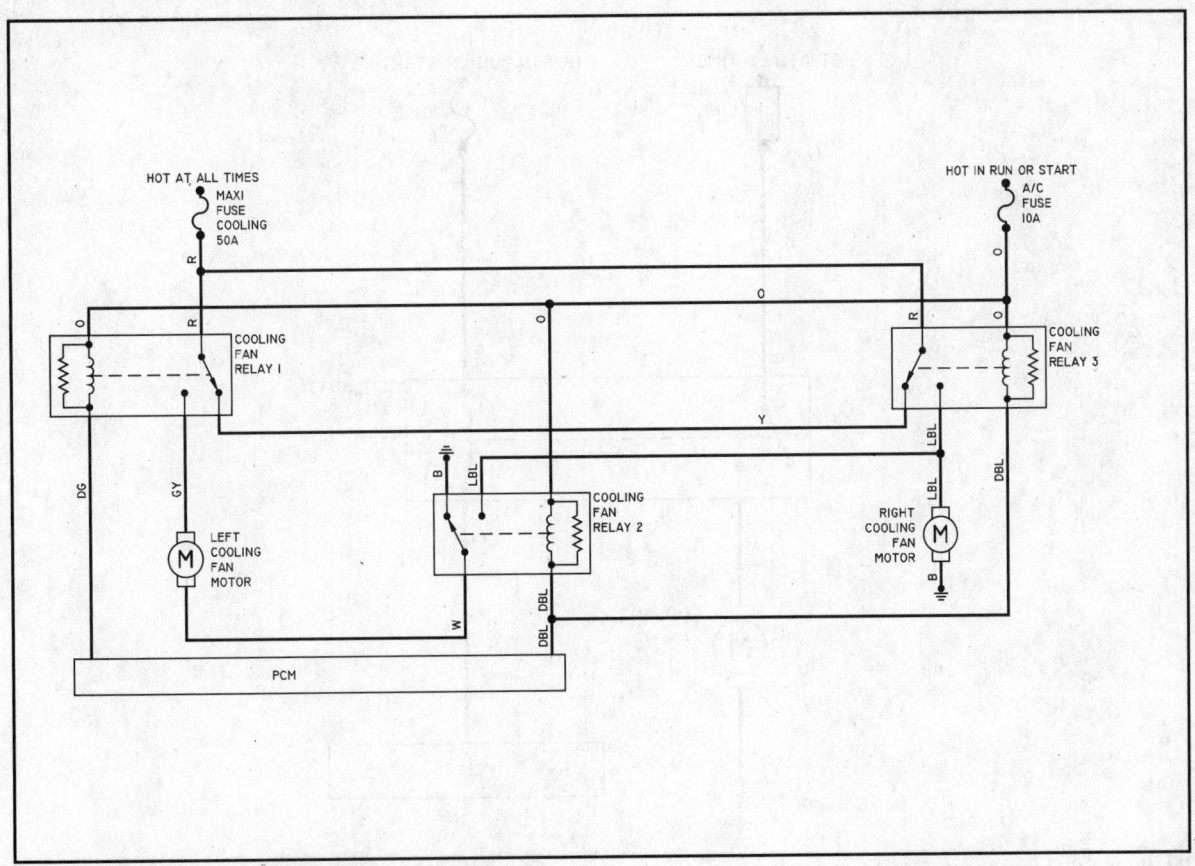

DIA. 46 - 1998-99 GM E & K Bodies (DeVille, ElDarado, Seville) 4.6L

79229W23

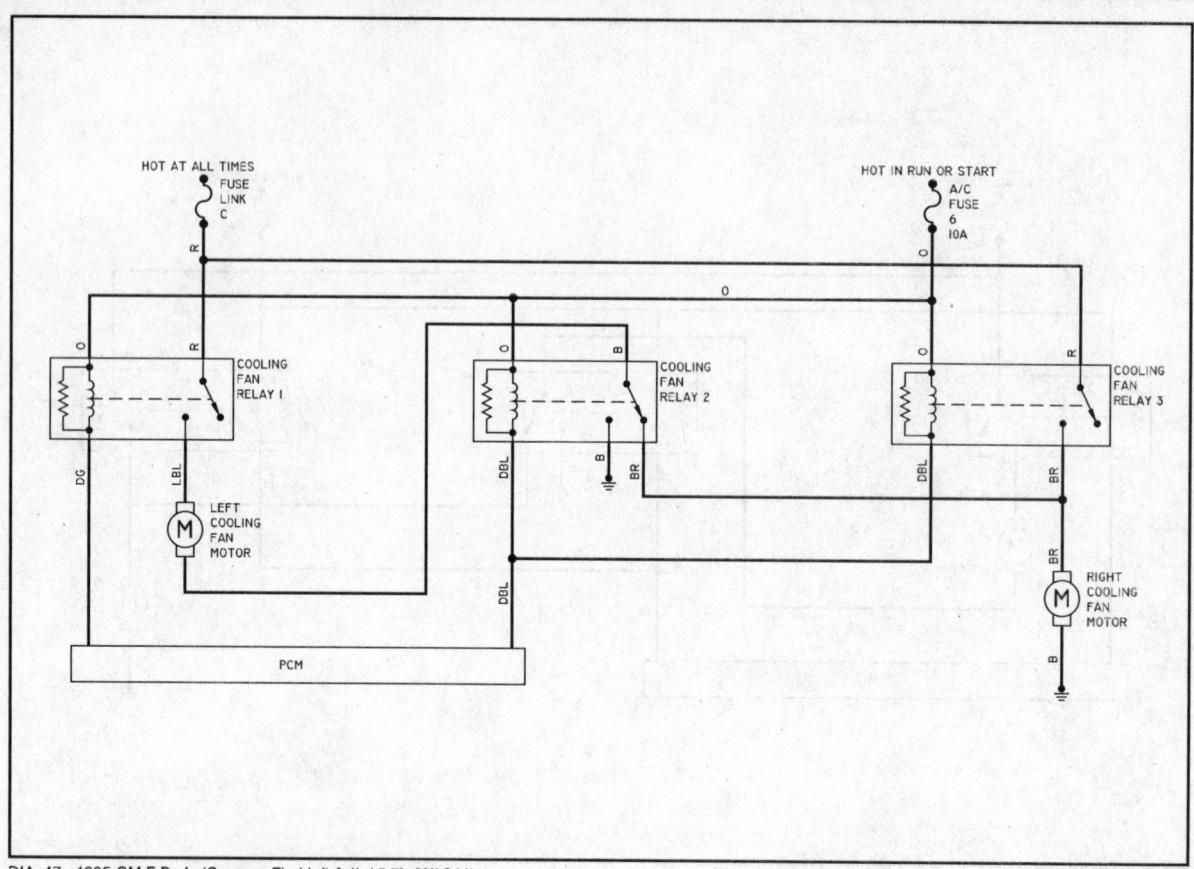

DIA. 47 - 1995 GM F Body (Camaro, Firebird) 3.4L / 5.7L (W/ C41)

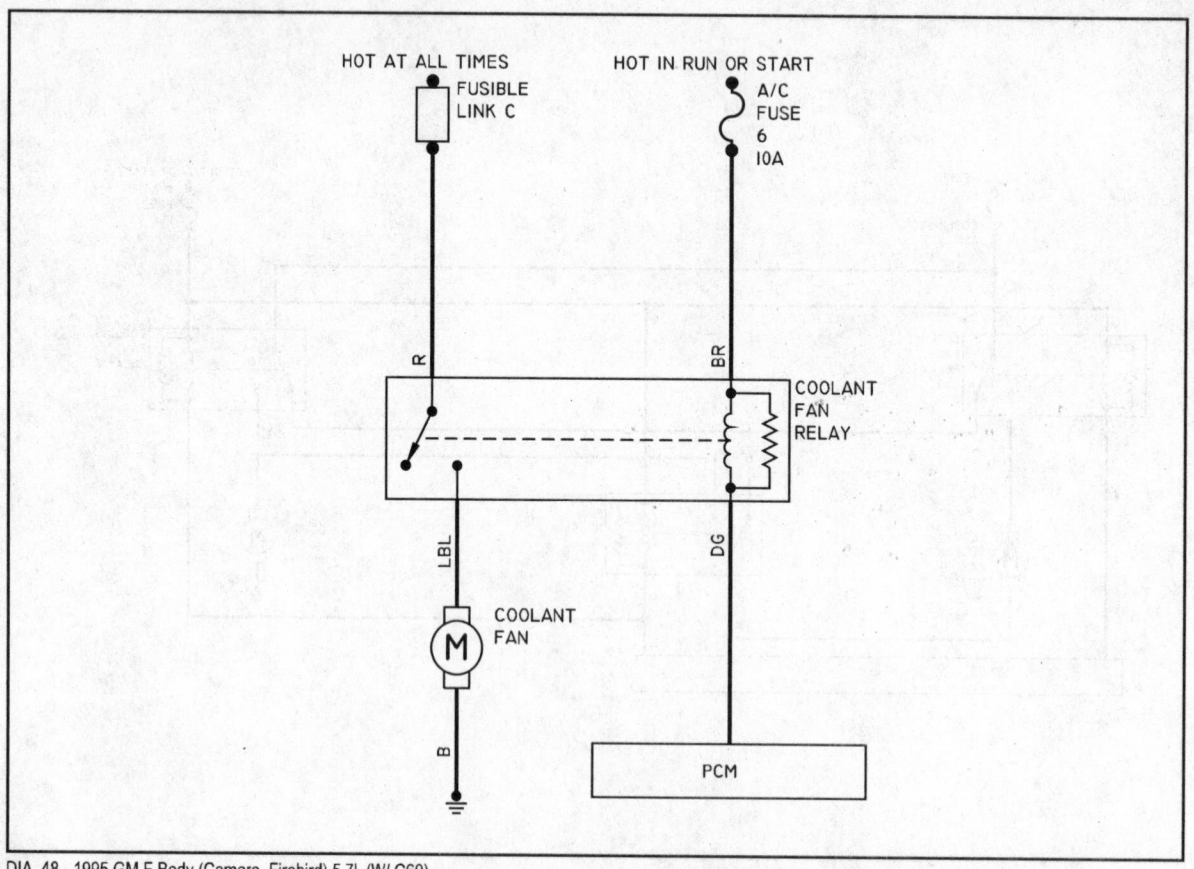

DIA. 48 - 1995 GM F Body (Camaro, Firebird) 5.7L (W/ C60)

79229W24

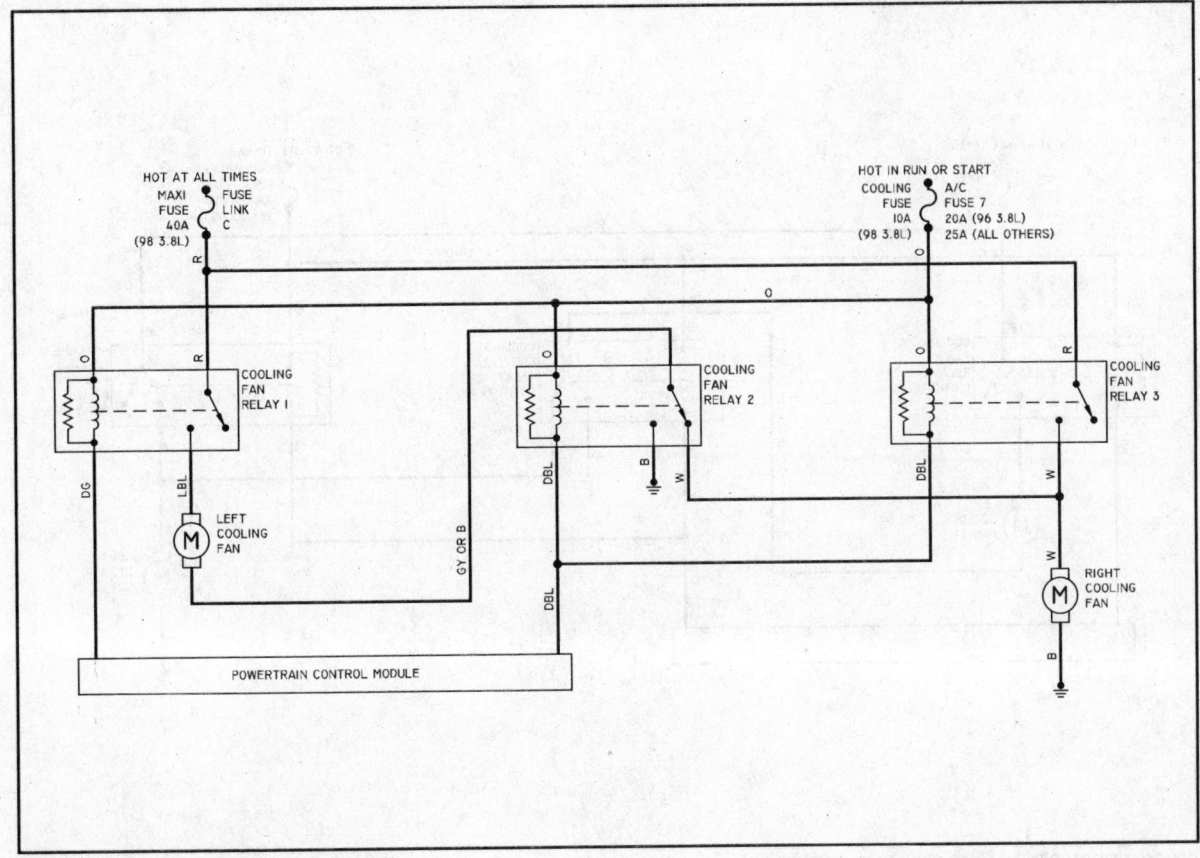

DIA. 49 - 1996-99 GM F Body (Camaro, Firebird) 3.8L
1996-97 GM F Body (Camaro, Firebird) 5.7L

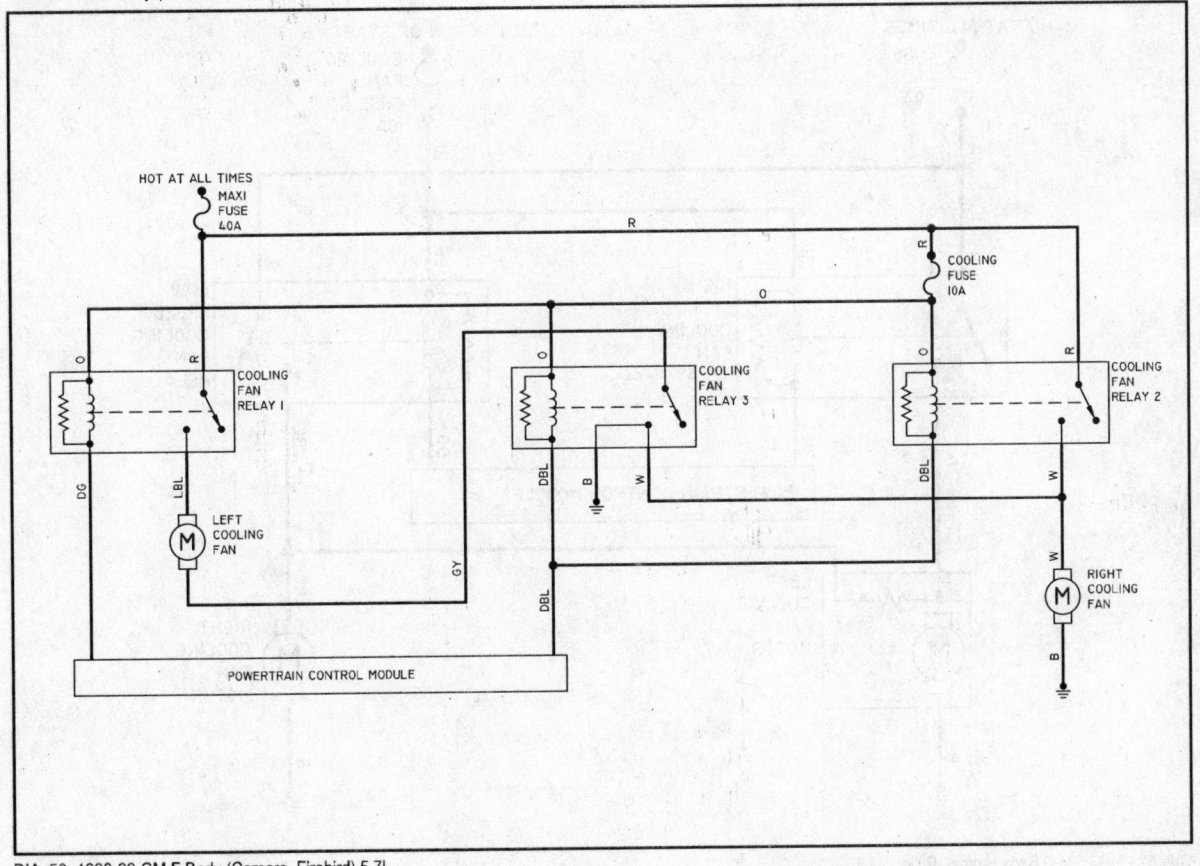

DIA. 50- 1998-99 GM F Body (Camaro, Firebird) 5.7L

79229W25

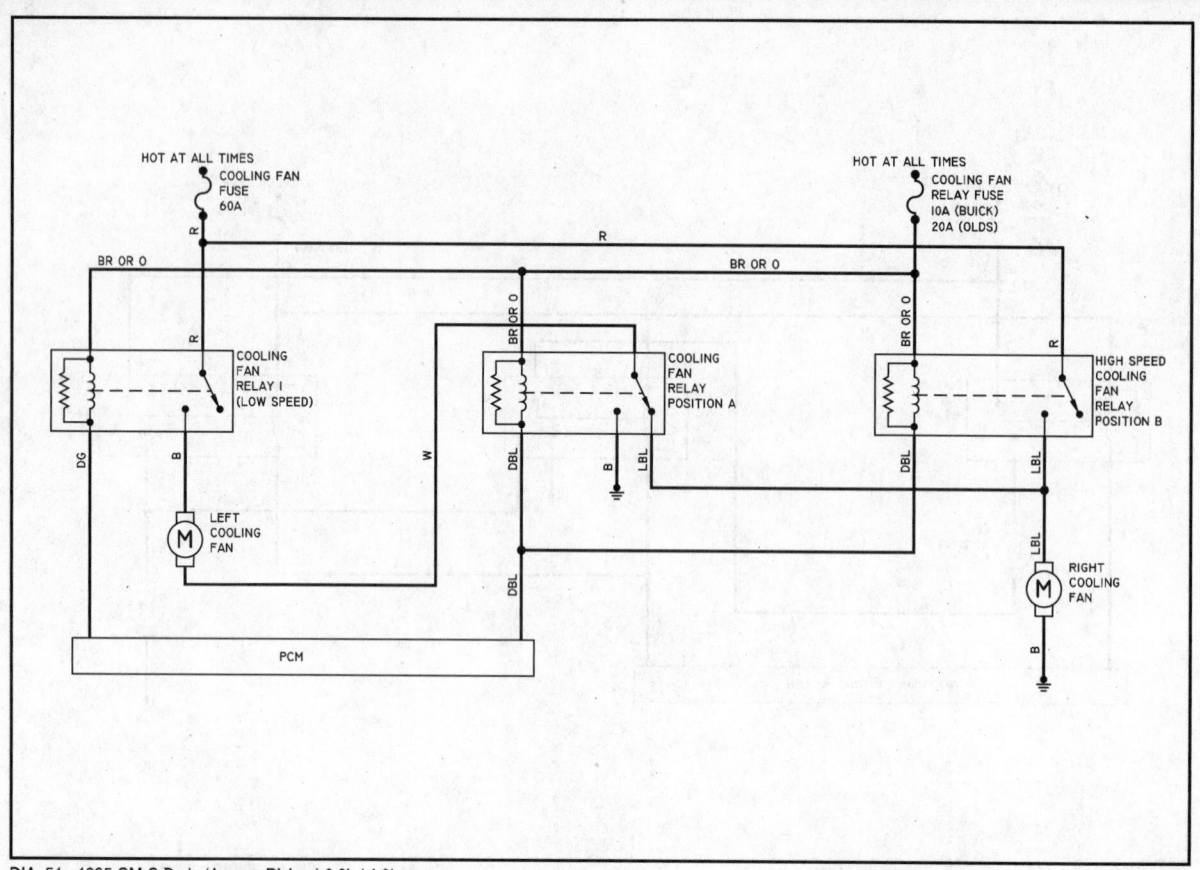

DIA. 51 - 1995 GM G Body (Aurora, Riviera) 3.8L / 4.0L

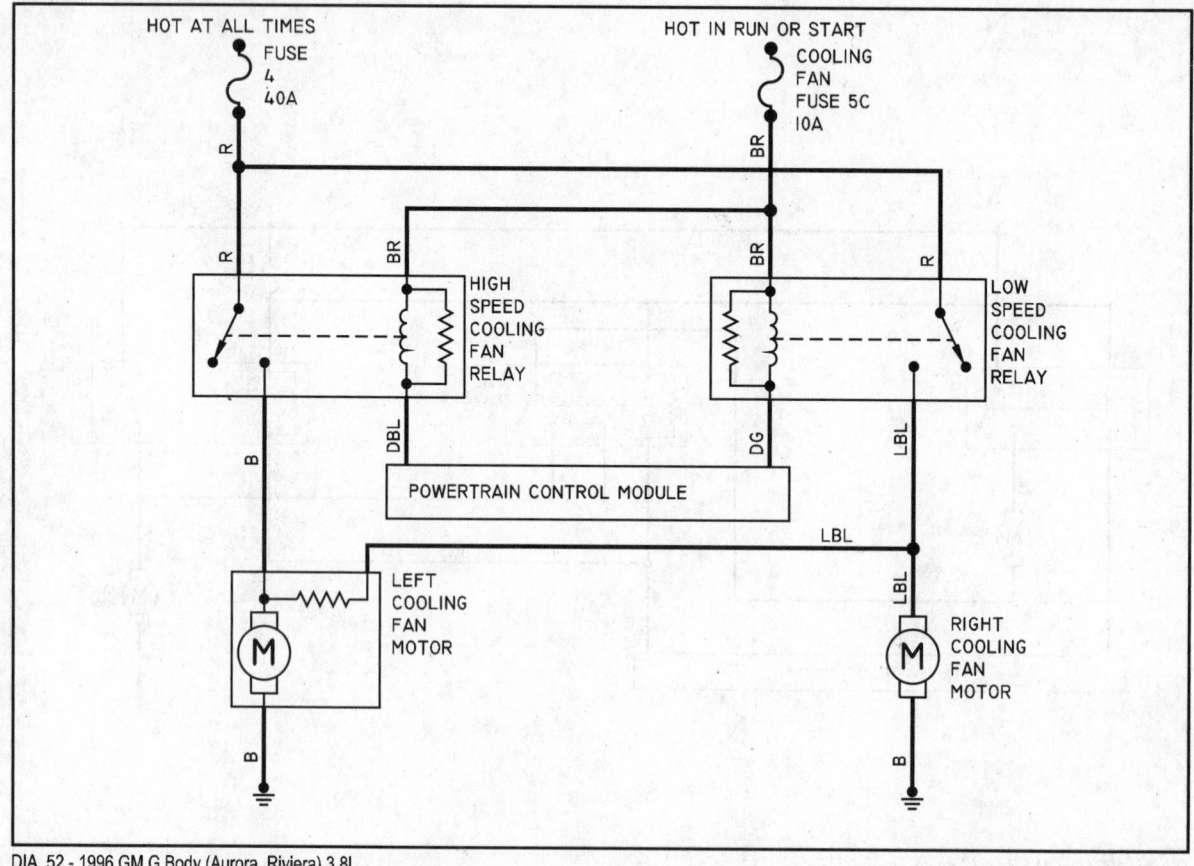

DIA. 52 - 1996 GM G Body (Aurora, Riviera) 3.8L

79229W26

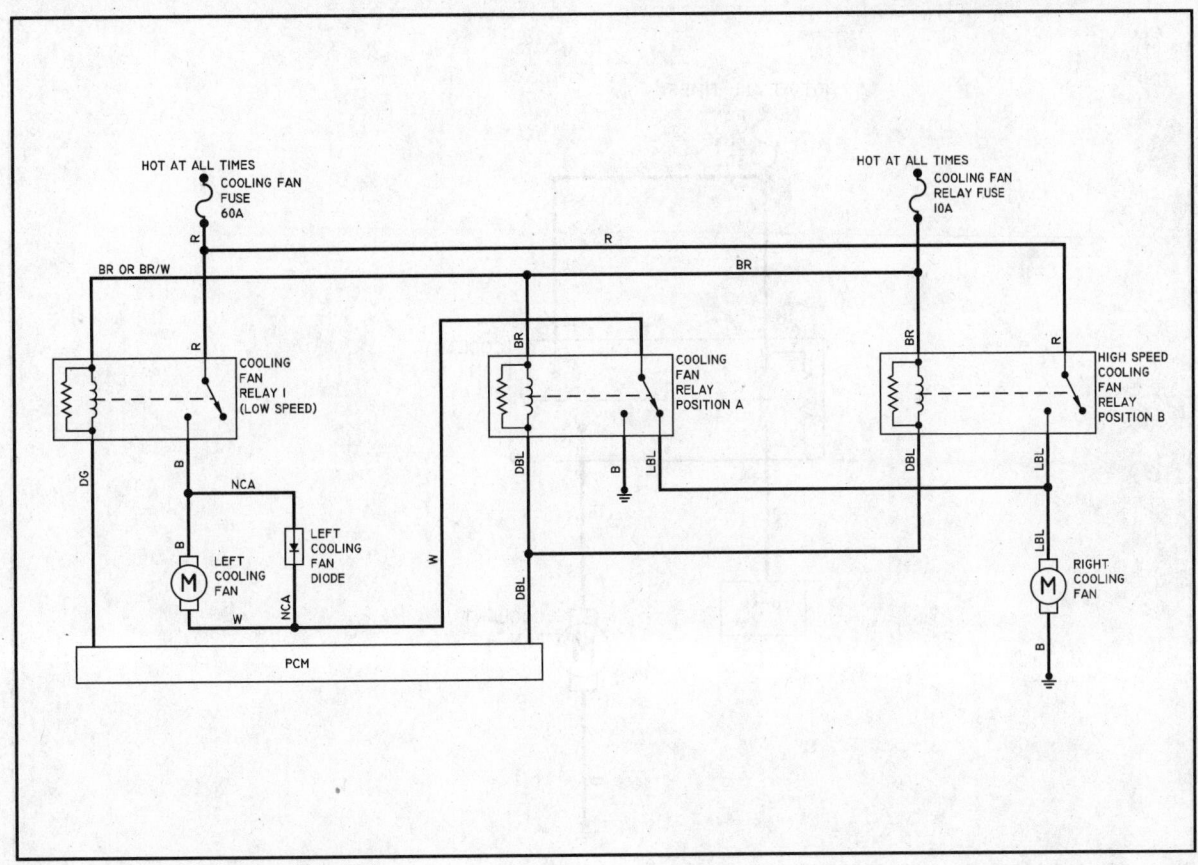

DIA. 53 - 1997-99 GM G Body (Aurora, Riviera) 3.8L

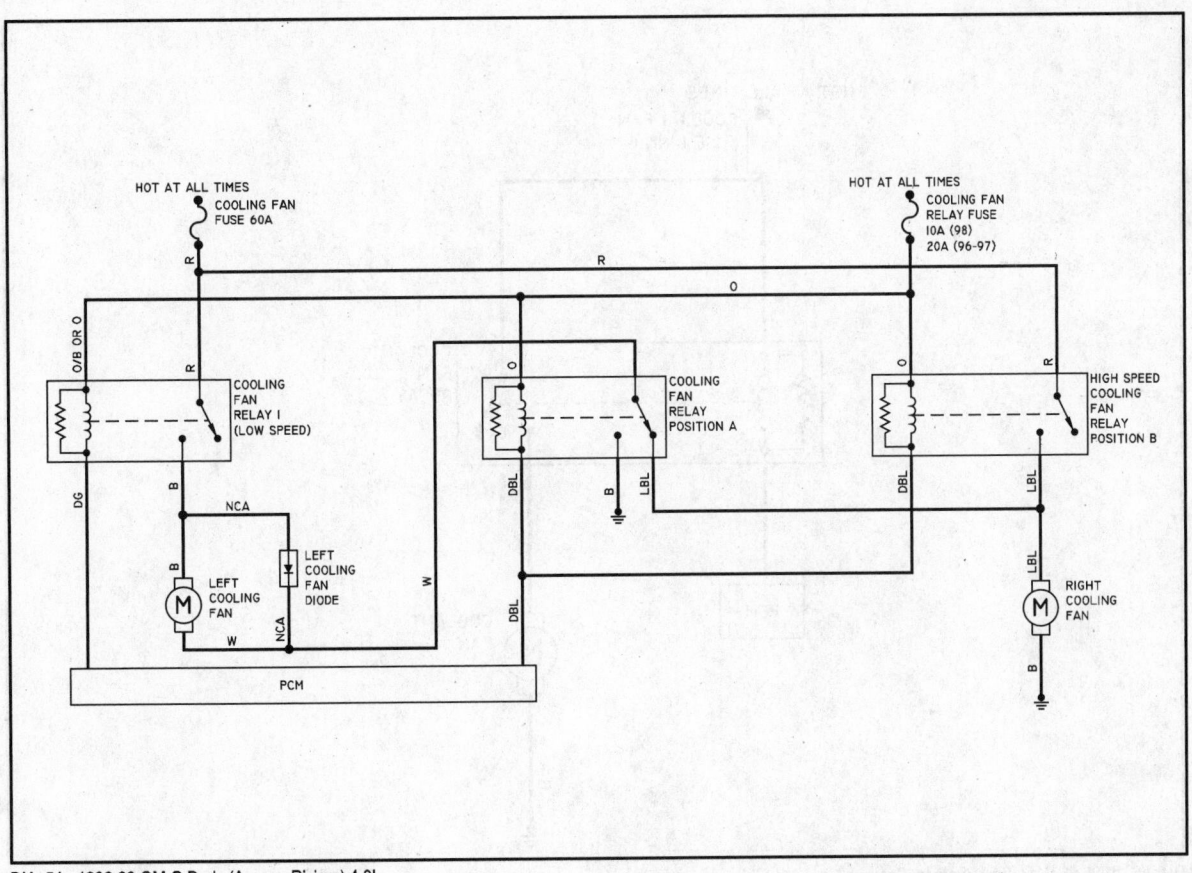

DIA. 54 - 1996-99 GM G Body (Aurora, Riviera) 4.0L

79229W27

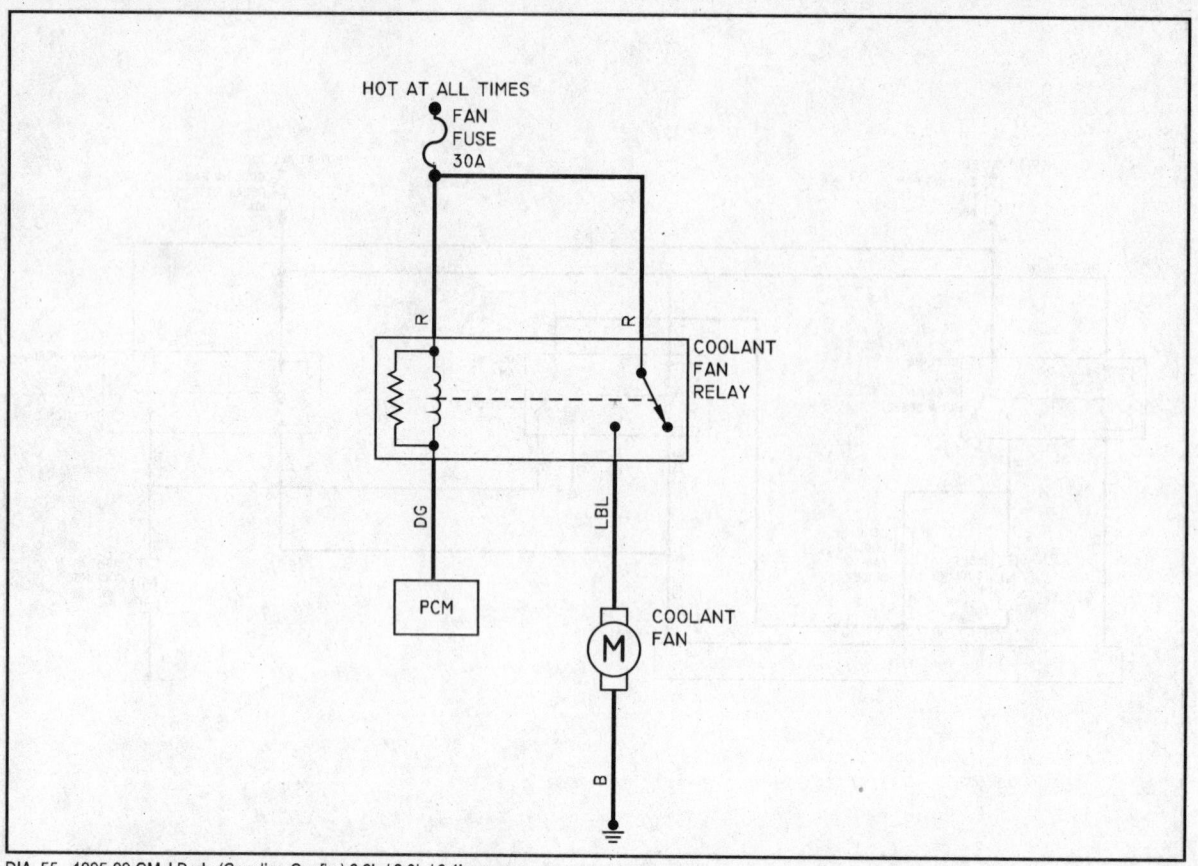

DIA. 55 - 1995-99 GM J Body (Cavalier, Sunfire) 2.2L / 2.3L / 2.4L

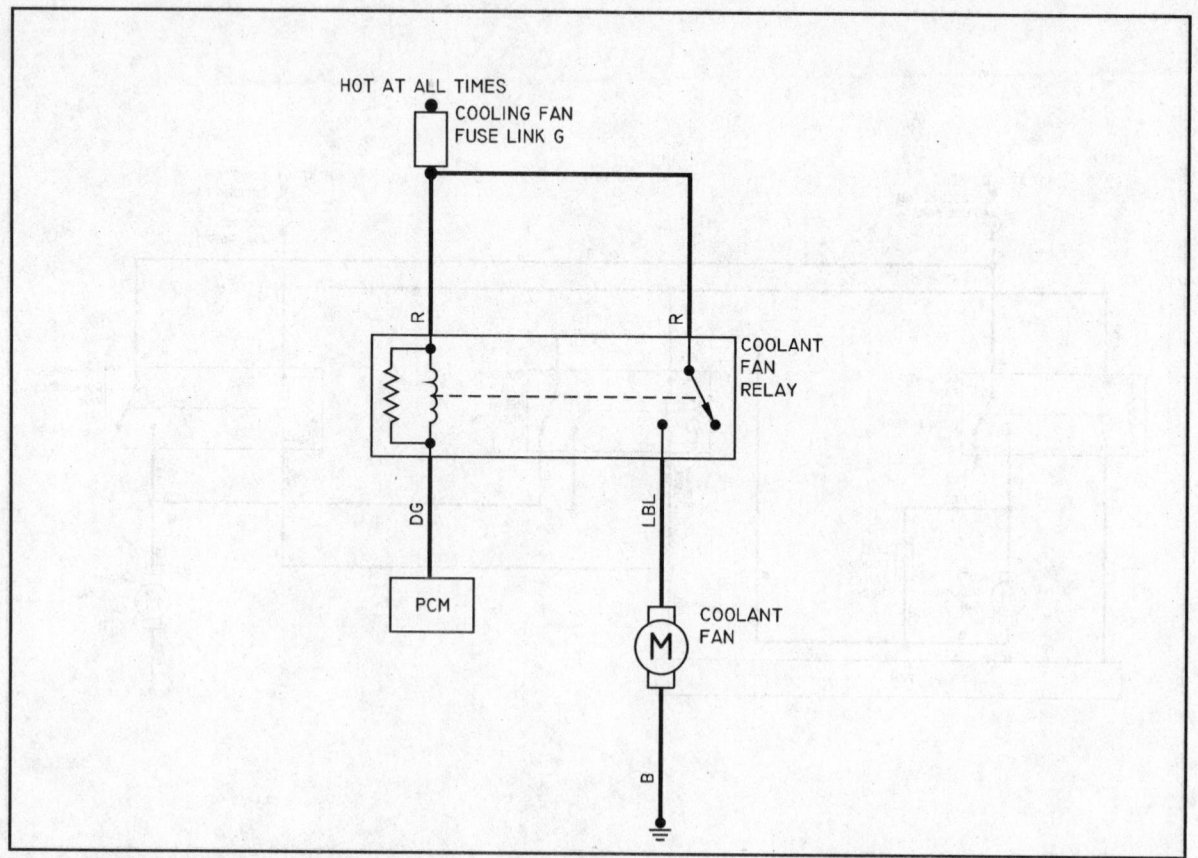

DIA. 56 - 1995-96 GM L Body (Beretta, Corsica) 2.2L / 3.1L

79229W28

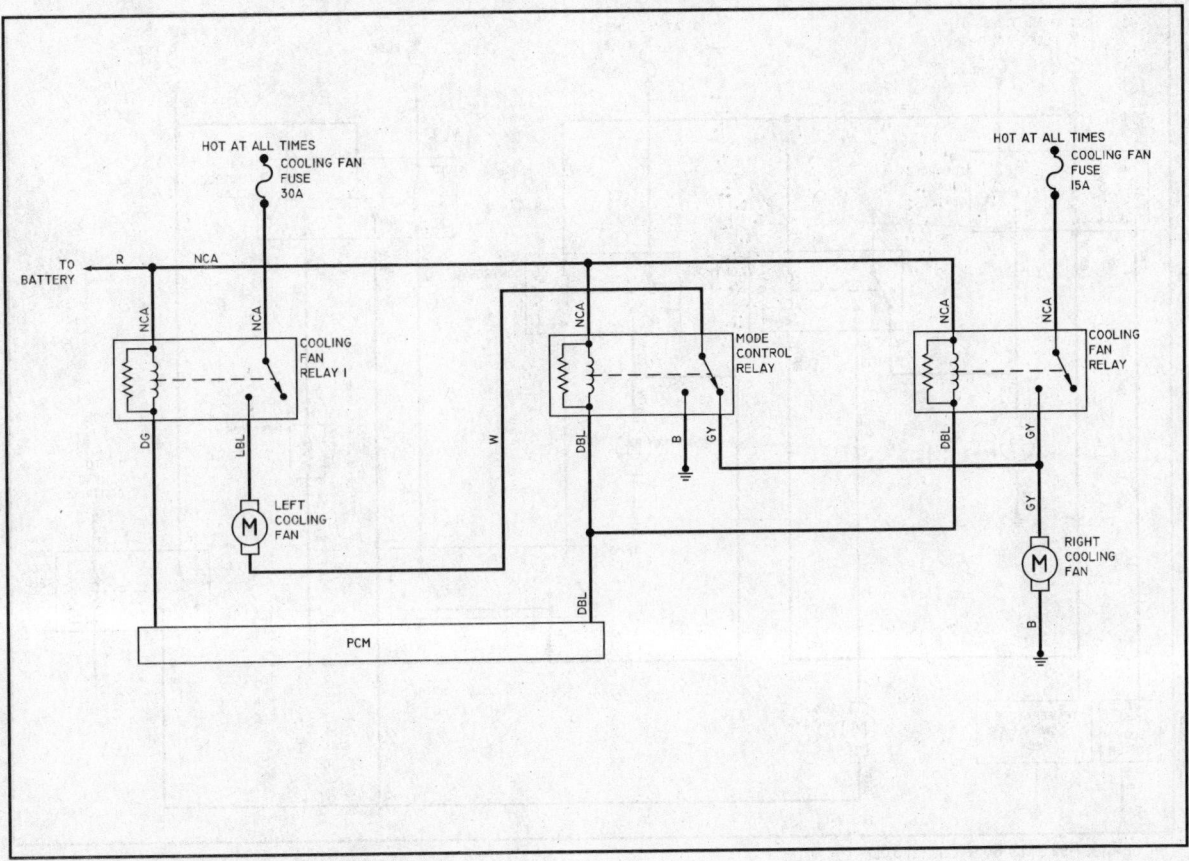

DIA. 57 - 1997-99 GM L/N Bodies (Cutlass, Malibu) 2.4L / 3.1L

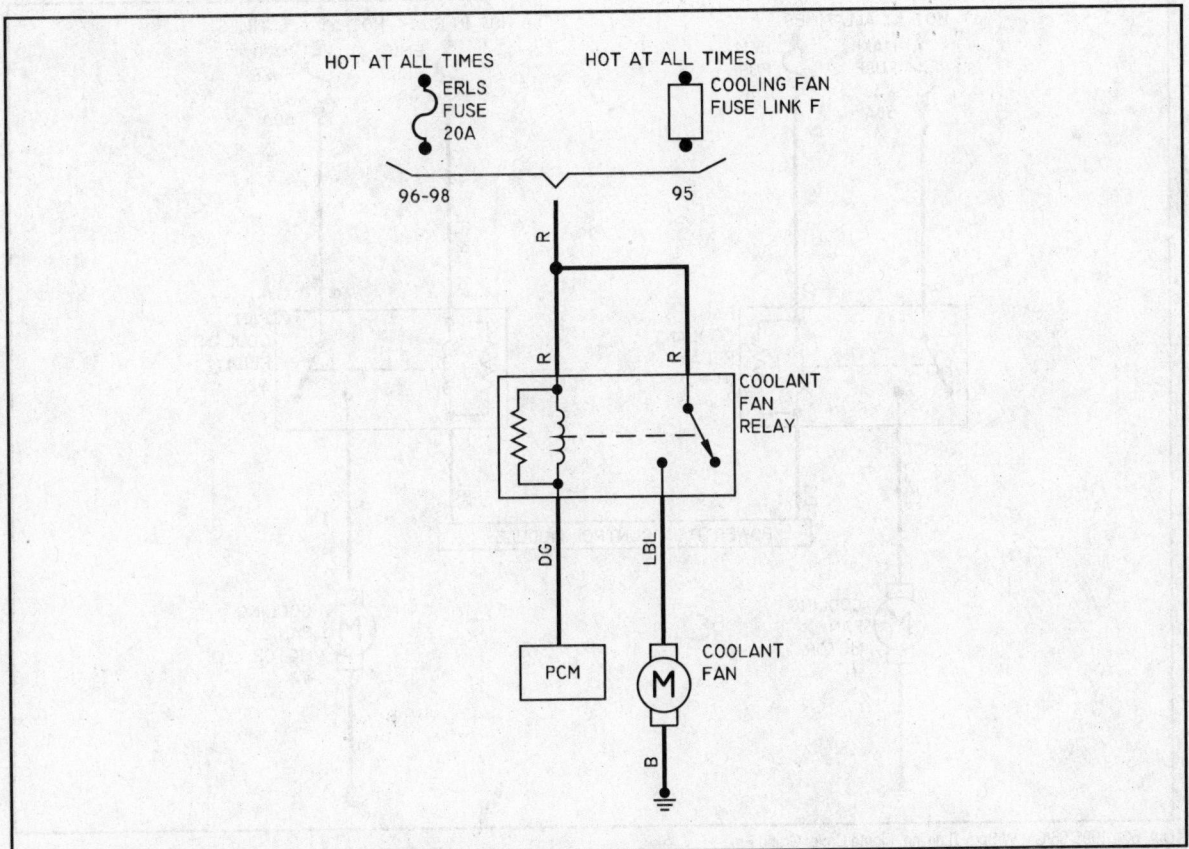

DIA. 58 - 1995-99 GM N Body (Acheiva, Grand Am, Skylark) 2.3L / 2.4L / 3.1L

79229W29

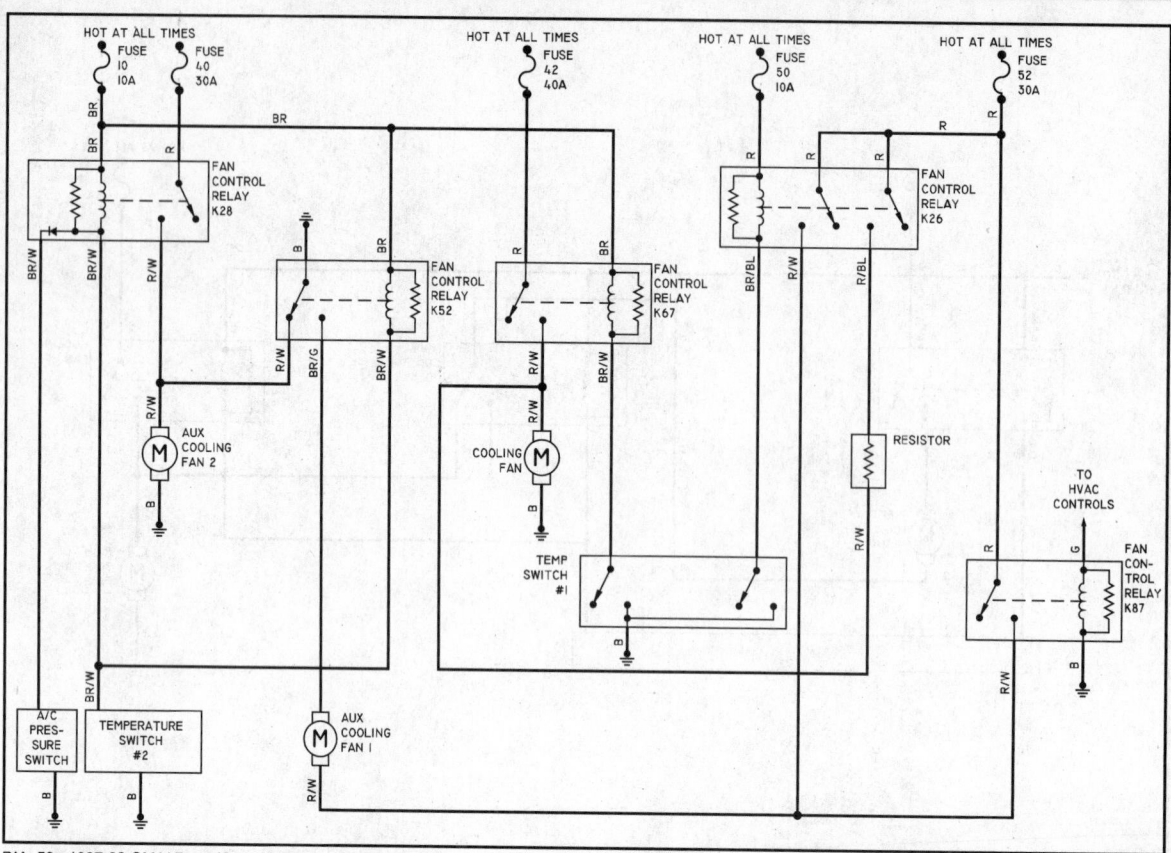

DIA. 59 - 1997-99 GM V Body (Catera) 3.0L

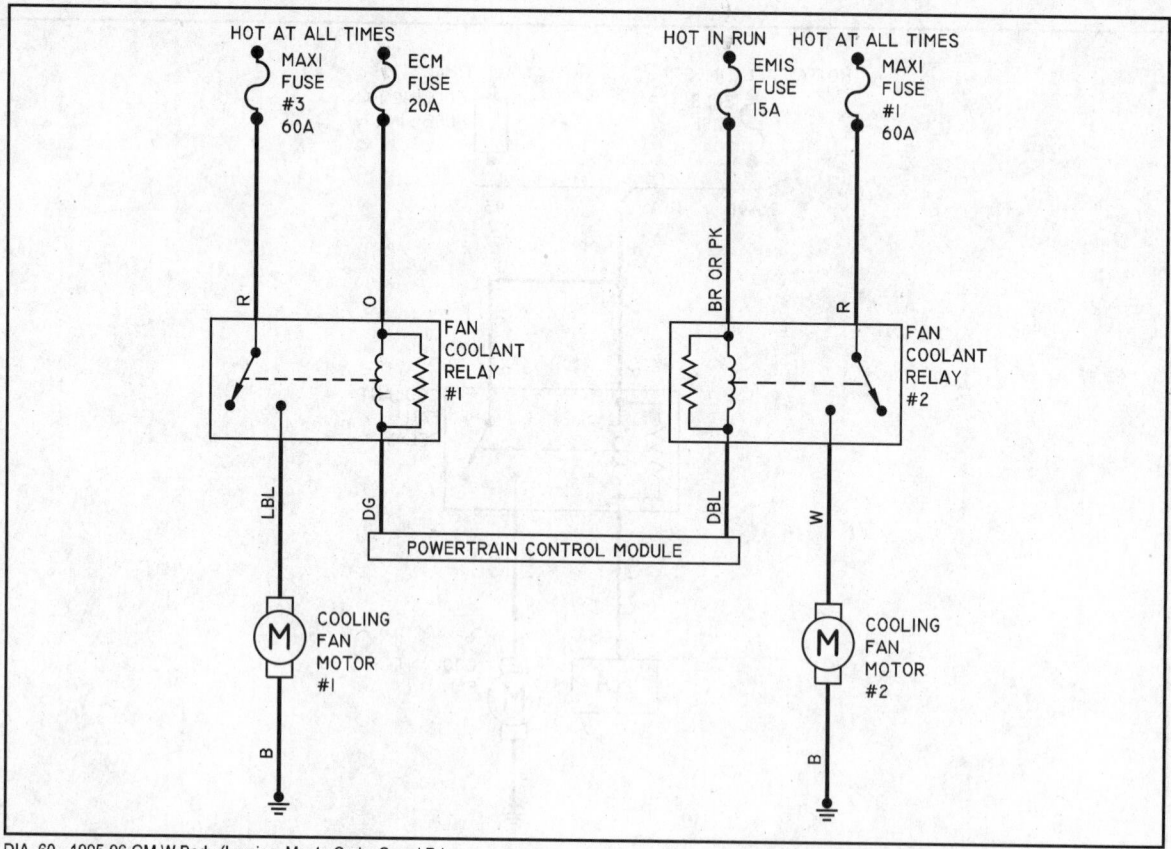

DIA. 60 - 1995-96 GM W Body (Lumina, Monte Carlo, Grand Prix,
Cutlass Supreme, Regal & 1997 Cutlass Supreme) 3.1L / 3.4L

79229W30

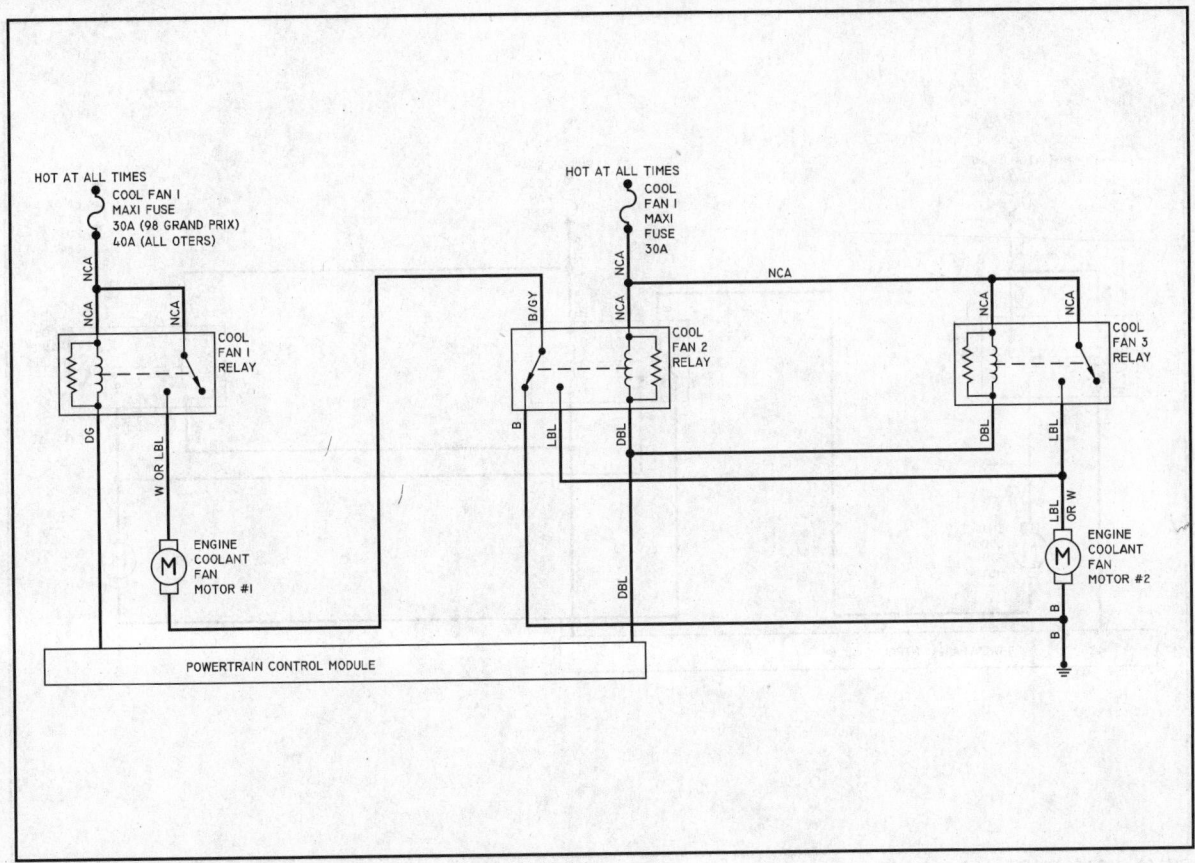

DIA. 61 - 1997-99 GM W Body (Century, Regal, Grand Prix, Intrigue) 3.1L / 3.8L

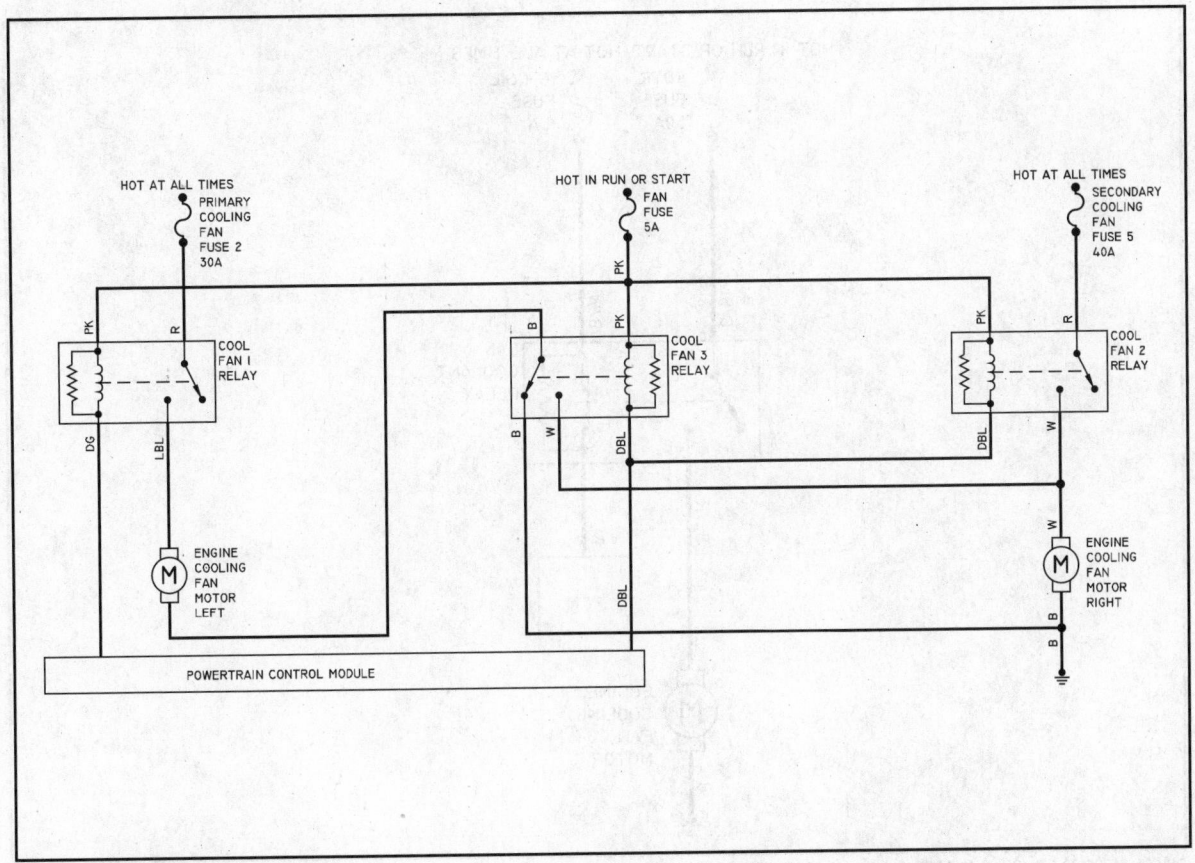

DIA. 62 - 1995-96 GM Y Body (Corvette) 5.7L

79229W31

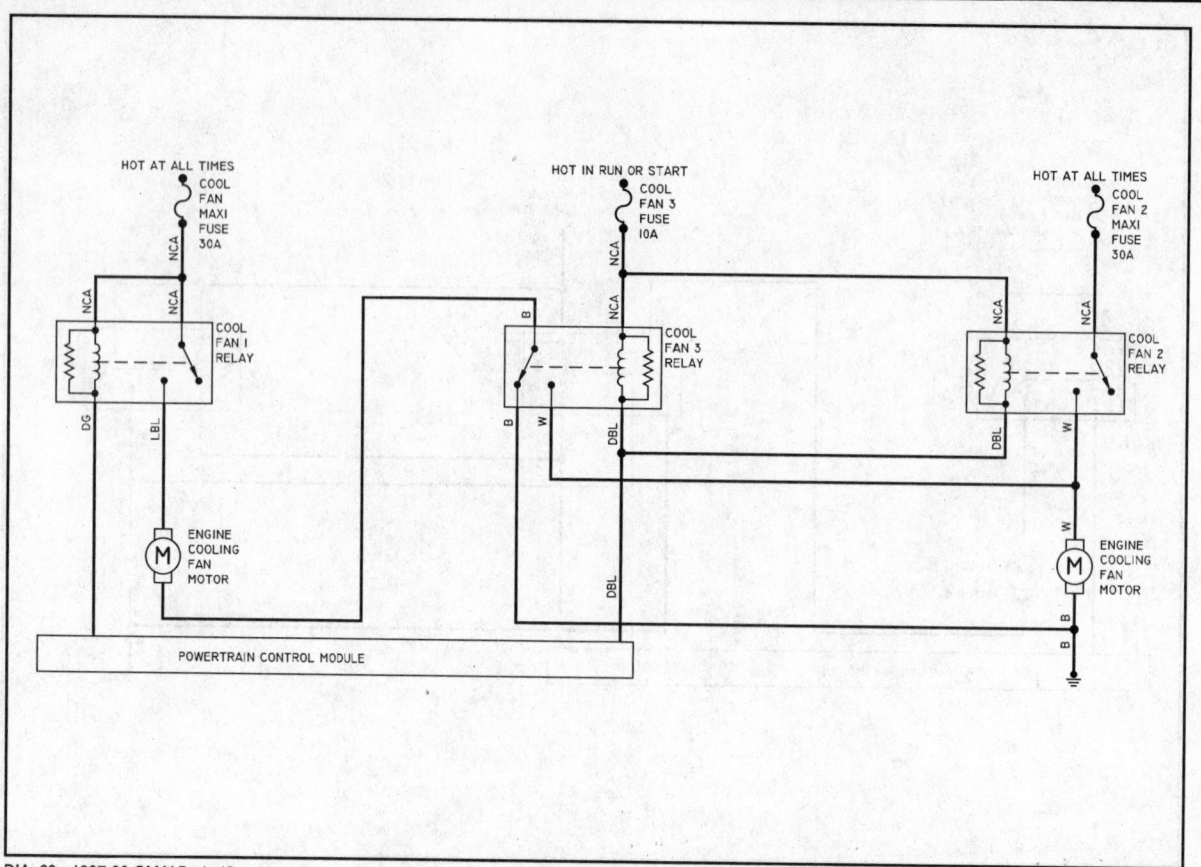

DIA. 63 - 1997-99 GM Y Body (Corvette) 5.7L

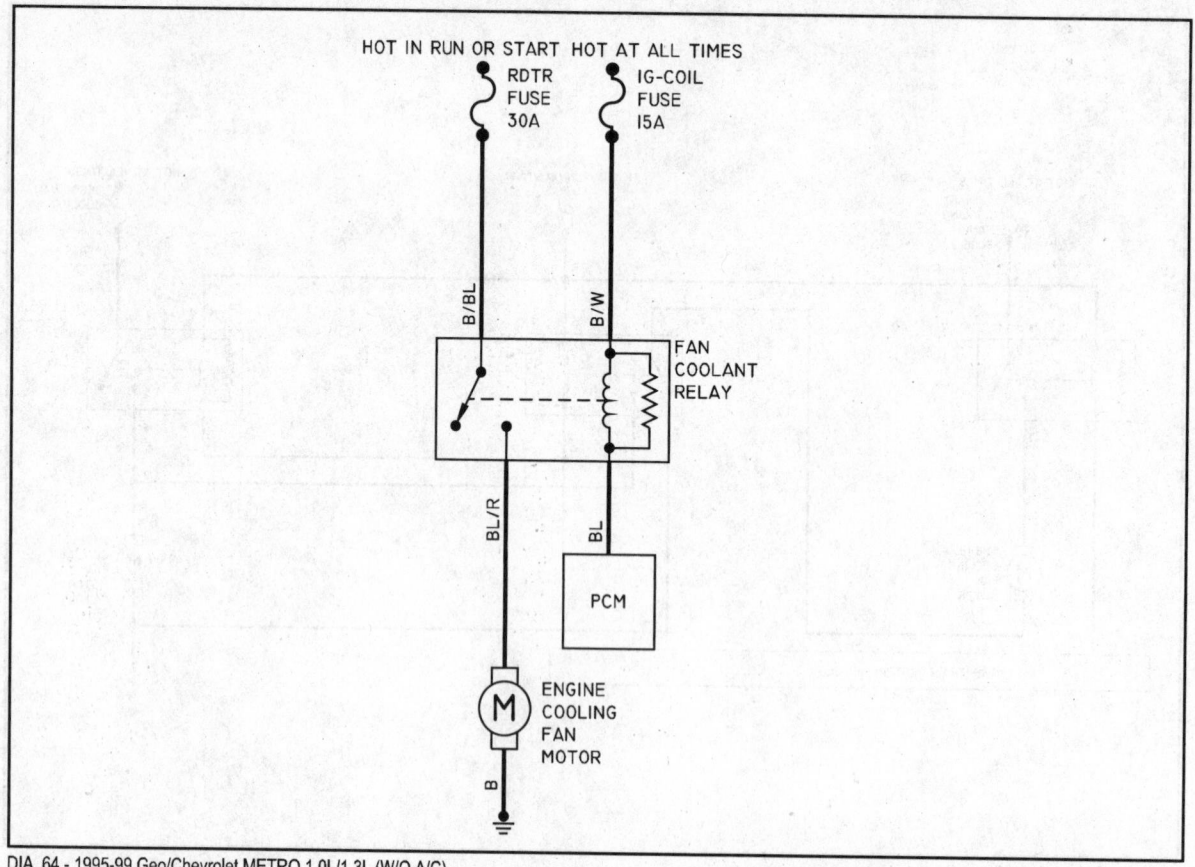

DIA. 64 - 1995-99 Geo/Chevrolet METRO 1.0L/1.3L (W/O A/C)

79229W32

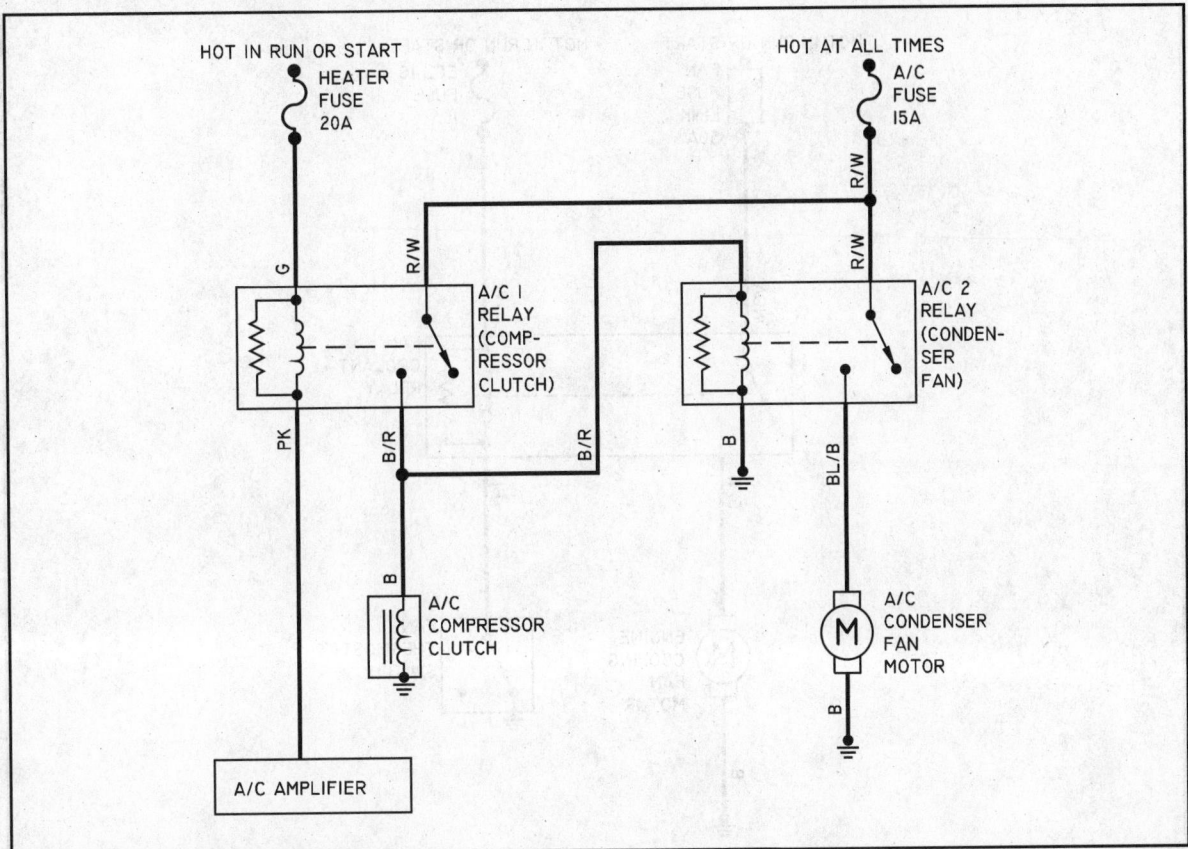

DIA. 65 - 1995-99 Geo/Chevrolet METRO 1.0L/1.3L (W/ A/C)

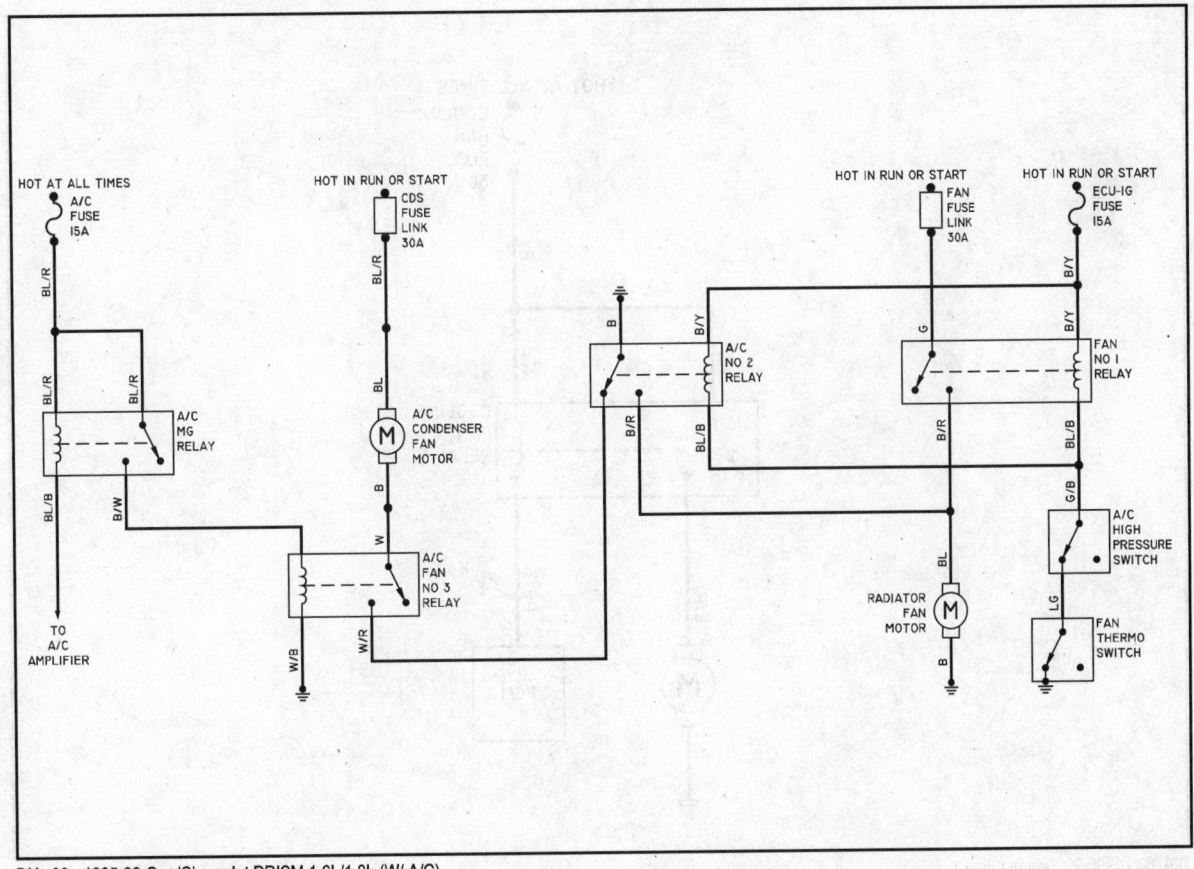

DIA. 66 - 1995-99 Geo/Chevrolet PRISM 1.6L/1.8L (W/ A/C)

79229W33

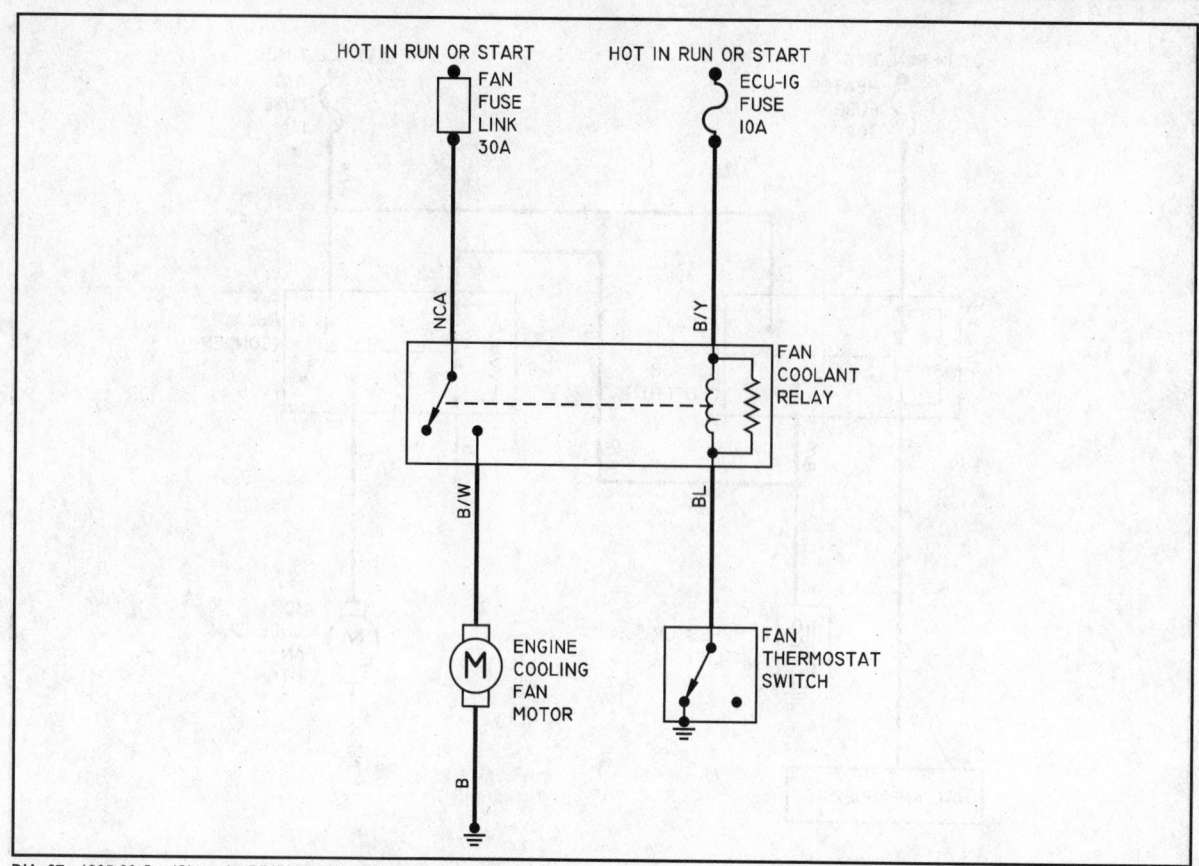

DIA. 67 - 1995-99 Geo/Chevrolet PRISM 1.6L/1.8L (W/O A/C)

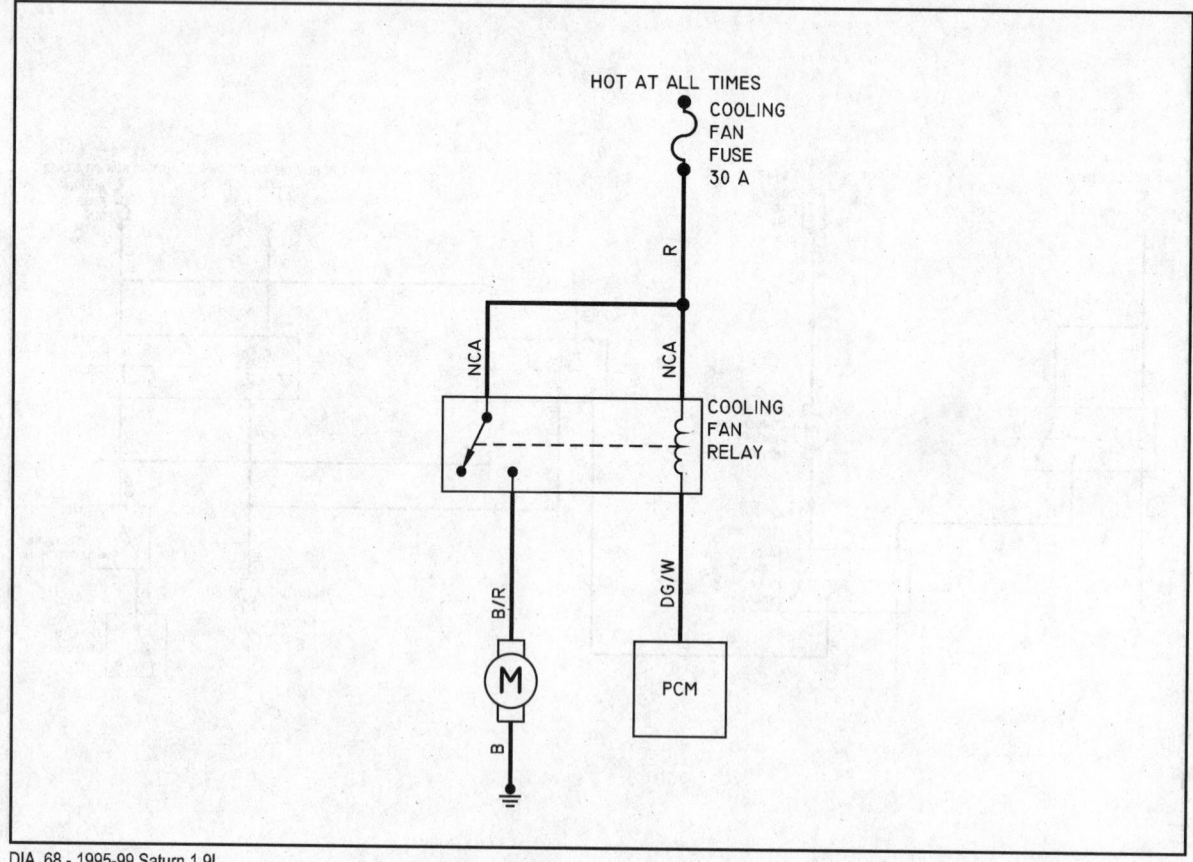

DIA. 68 - 1995-99 Saturn 1.9L

79229W34

STARTING AND CHARGING SYSTEMS

10

CHARGING SYSTEM **10-7**
STARTING SYSTEM **10-1**
Alternator 10-8
  Removal & Installation 10-10
  Testing...................................10-8
Battery 10-12
  Jump Starting a Dead Battery 10-14
  Removal & Installation 10-14
  Testing...................................10-12
General Information (Charging) 10-7

Precautions 10-8
System Testing 10-8
General Information (Starting) 10-1
  Precautions 10-1
Solenoid 10-7
  Removal & Installation 10-7
  Testing...................................10-7
Starter Motor 10-4
  Adjustments 10-5
  Removal & Installation 10-4

Starter Relay 10-6
  Testing..10-6
System Testing 10-2
  With External Solenoid...............10-3
  With Starter Mounted Solenoid ...10-2
Voltage Regulator 10-15
  Removal & Installation 10-15
  Testing...................................10-15

STARTING SYSTEM

General Information

The starting system includes the battery, starter motor, solenoid, ignition switch, and in some cases, a starter relay. An inhibitor (neutral safety) switch is included in the starting system circuit to prevent the vehicle from being started while in gear.

When the ignition key is turned to the **START** position, current flows and energizes the starter's solenoid coil. The energized coil becomes an electromagnet which pulls the plunger into the coil, the plunger closes a set of contacts which allow high current to reach the starter motor. On models where the solenoid is mounted on the starter, the plunger also serves to push the starter pinion to mesh with the teeth on the flywheel/flexplate.

To prevent damage to the starter motor when the engine starts, the pinion gear incorporates an over-running (one-way) clutch which is splined to the starter armature shaft. The rotation of the running engine may speed the rotation of the pinion but not the starter motor itself.

Some starting systems employ a starter relay in addition to the solenoid. This relay may be located under the instrument panel, in the kickpanel or in the fuse/relay center under the hood. This relay is used to reduce the amount of current the starting (ignition) switch must carry.

PRECAUTIONS

To prevent damage to the on-board computer, alternator and regulator, the following precautionary measures must be taken when working with the electrical system.

• Always disconnect the negative battery cable before servicing the starter motor. Battery voltage is always present at the large (**B**) terminal on the solenoid. When removing the starter motor, be prepared to support its weight after the last bolt is removed because the starter motor is a fairly heavy component.

• Never operate the starter motor for more than 30 seconds at a time. Too much cranking will cause the starter motor to overheat, causing permanent damage. Allow the starter motor to cool for at least two minutes between starting attempts.

• Wear safety glasses when working on or near the battery.

• Don't wear a watch with a metal band when servicing the battery. Serious burns can result if the band completes the circuit between the positive battery terminal and ground.

TCCA1P02

Before servicing the electrical system always disconnect the negative battery cable to prevent system damage

• Be absolutely sure of the polarity of a booster battery before making connections. Connect the cables positive to positive, and negative to negative. Connect positive cables first and then make the last connection to ground on the body of the booster vehicle so that arcing cannot ignite hydrogen gas that may have accumulated near the battery. Even momentary connection of a booster battery with the polarity reversed will damage the alternator diodes.

• Disconnect both vehicle battery cables before attempting to charge a battery.

• Be cautious when using metal tools around a battery to avoid creating a short circuit between the terminals.

• When installing a battery, make sure that the positive and negative cables are not reversed.

• When jump-starting the car, be sure that like terminals are connected. This also applies to using a battery charger. Reversed polarity will burn out the alternator and regulator in a matter of seconds.

• Always disconnect the battery (negative cable first) when charging it.

System Testing

➡**A good quality digital multimeter with at least 10 megohm/volt impedance should be used when testing modern automotive circuits. These meters can accurately detect very small amounts of voltage, current and resistance. This type of meter also has a high internal resistance that will not load the circuit being tested. Loading the circuit causes inaccurate readings, and may cause damage to sensitive computer circuits. Although we are not testing computer circuits in this section, accuracy is very important.**

WITH STARTER MOUNTED SOLENOID

1. Check the battery and clean the connections as follows:
   a. If the battery cells have removable caps, check the water level. Add distilled water if low. Load test the battery and charge if necessary. See Battery Testing in this section for the procedure.
   b. Remove the cables and clean them with a wire brush. Reconnect the cables.
2. Check the starter motor ground circuit with a voltage drop test as follows:
   a. Set the meter to read DC voltage on the lowest possible scale.
   b. Connect the negative lead of your multimeter to the negative terminal of the battery.
   c. Connect the positive lead to the body of the starter. Make sure the starter mounting bolts are tight. The meter should read 0.2 volts or less. If the voltage reading is greater, remove and clean the negative battery connection on the engine block. The voltage reading should now be within specification: if not, replace the negative battery cable.
3. Check the motor feed circuit with a voltage drop test as follows:
   a. Disconnect the coil wire or the fuel injector harness to prevent the engine from starting.
   b. Connect the positive lead of your meter to the positive terminal of the battery.
   c. Connect the negative meter lead to the motor feed terminal. The motor feed terminal comes out of the body of the starter motor and connects to the solenoid.
   d. Turn the ignition key to the **START** position. The meter should read 0.2 volts or less. If the voltage reading is greater,

remove and clean the positive battery connection on the starter solenoid. The voltage reading should now be within specification, if not replace the positive battery cable.
   e. Connect the coil wire or fuel injector harness.
4. Check for battery voltage at the **S** terminal on the starter solenoid as follows:
   a. Disconnect the coil wire or the fuel injector harness to prevent the engine from starting.
   b. Set the meter to read battery voltage. Move it to the next higher range if set on the 2 volt scale.
   c. Connect the positive lead to the **S** terminal on the starter solenoid and the negative lead to a good ground.
   d. Turn the ignition key to the **START** position and crank the engine. The meter should read battery voltage. If battery voltage is not present, check the inhibitor (neutral safety) switch, fuse(s) and wiring between the ignition switch and starter solenoid. If

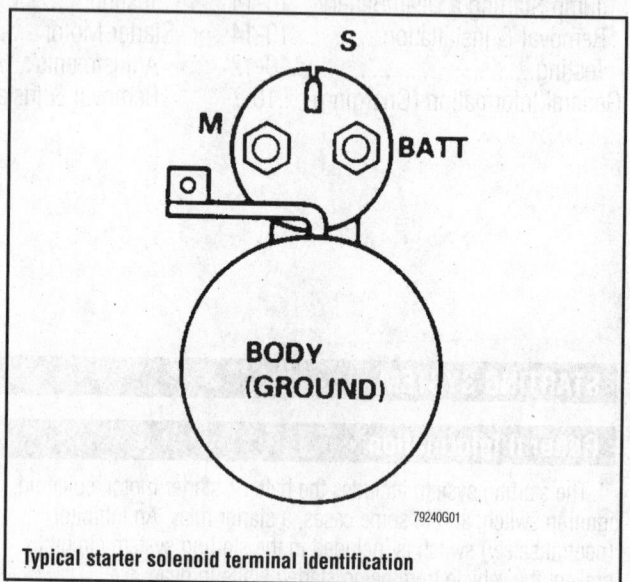

Typical starter solenoid terminal identification

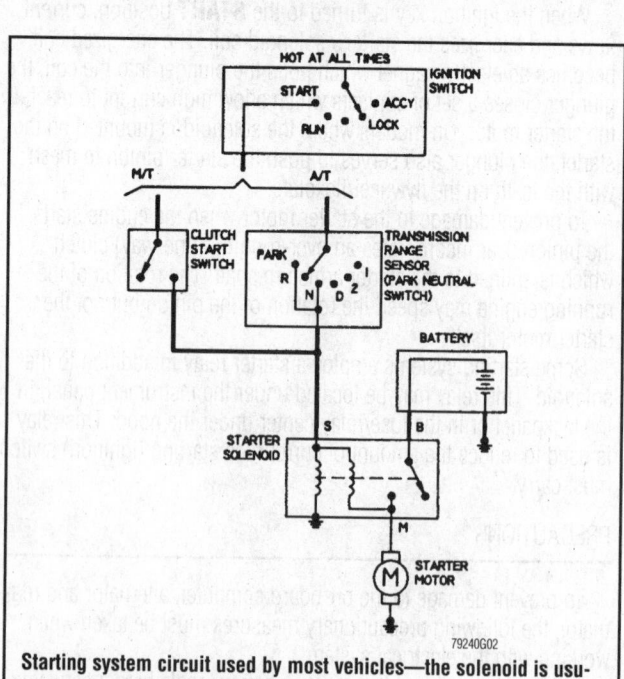

Starting system circuit used by most vehicles—the solenoid is usually mounted on the starter as indicated

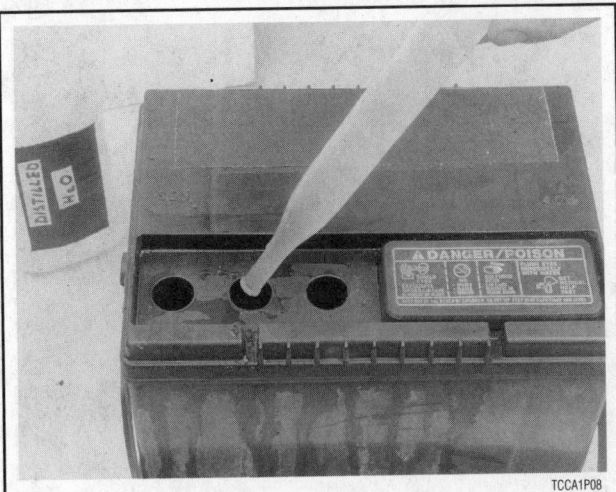

Before testing the system, be sure that the battery is in good shape, which includes ensuring the cells are full on serviceable batteries

Also, disconnect both battery cables (negative first) . . .

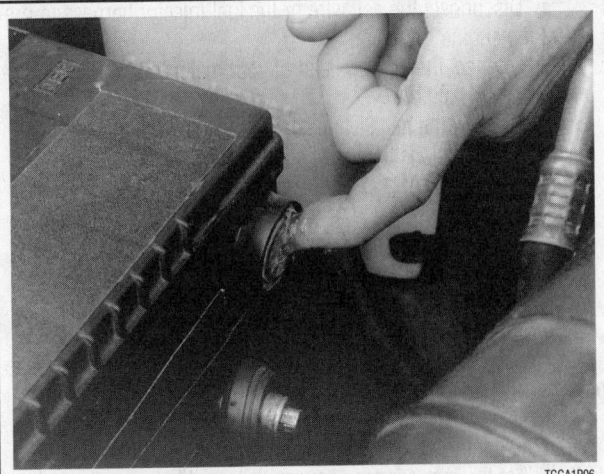

. . . and apply petroleum jelly or multi-purpose grease to the terminals before reattaching the cables

. . . clean the cable terminals of all dirt and corrosion with a wire brush . . .

battery voltage is present at the **S** terminal on the solenoid and the starter does not operate, replace the starter and solenoid assembly.

    e. Connect the coil wire or fuel injector harness.

WITH EXTERNAL SOLENOID

➡ Not all solenoids are mounted on the starter motor. Some models use a solenoid (relay) mounted on the inner fender or firewall. Both types of solenoids serve to make the connection between the battery and starter motor. Trace the wires for positive identification. The small wire comes from the ignition switch, one large cable from the battery and the other large cable to the starter. The terminals are S, B and M respectively.

  1. Check the battery and clean the connections as follows:

    a. If the battery cells have removable caps, check the water level. Add distilled water if low. Load test the battery and charge if necessary. See Battery Testing in this section for the procedure.

✳✳ CAUTION

Always remove the negative battery cable first, and install it last.

    b. Remove the cables and clean them with a wire brush. Disconnect and clean the cables on the solenoid in the same manner. Reconnect the cables on the solenoid, then the battery.

  2. Check the starter motor ground circuit with a voltage drop test as follows:

    a. Set the meter to read DC voltage on the lowest possible scale.

    b. Connect the negative lead of your multimeter to the negative terminal of the battery.

    c. Connect the positive lead to the body of the starter. Make sure the starter mounting bolts are tight. The meter should read 0.2 volts or less. If the voltage reading is greater, remove and clean the negative battery connection on the engine block. The voltage reading should now be within specification; if not, replace the negative battery cable.

  3. Check the motor feed circuit with a voltage drop test as follows:

a. Disconnect the coil wire or the fuel injector harness to prevent the engine from starting.

b. Connect the positive lead of your meter to the positive terminal of the battery.

c. Connect the negative meter lead to the motor feed terminal at the starter. This is the heavy cable on the starter. Turn the ignition key to the **START** position and crank the engine. The meter should read 0.2 volts or less. If the voltage reading is greater, remove and clean the positive battery connections on the starter and solenoid. The voltage reading should now be within specification; if not, replace the positive battery cable.

d. Connect the coil wire or fuel injector harness.

4. Check for battery voltage at the **S** terminal on the starter solenoid as follows:

a. Disconnect the coil wire or the fuel injector harness to prevent the engine from starting.

b. Set the meter to read battery voltage. Move it to next higher range, if previously set on the 2 volt scale.

c. Connect the positive lead to the **S** terminal on the starter solenoid and the negative lead to a good ground.

d. Turn the ignition key to the **START** position. The meter should read battery voltage. If battery voltage is not present, check the inhibitor (neutral safety) switch, fuse(s) and wiring between the ignition switch and starter solenoid. If battery voltage is present at the **S** and **B** terminals but not at the motor feed terminal, replace the solenoid. If battery voltage is present at all three terminals and the starter does not operate, replace the starter motor.

e. Connect the coil wire or fuel injector harness.

Starter Motor

REMOVAL & INSTALLATION

1. Disconnect the negative battery cable.

2. Remove all components necessary to gain access to the starter motor (such as exhaust pipes, air intake ducts, hoses, brackets and heat shields).

3. Disconnect the wiring from the starter. In some cases, the wiring may be more accessible after removing the mounting bolts and moving the starter.

. . . then pull the starter out of the transmission bell housing . . .

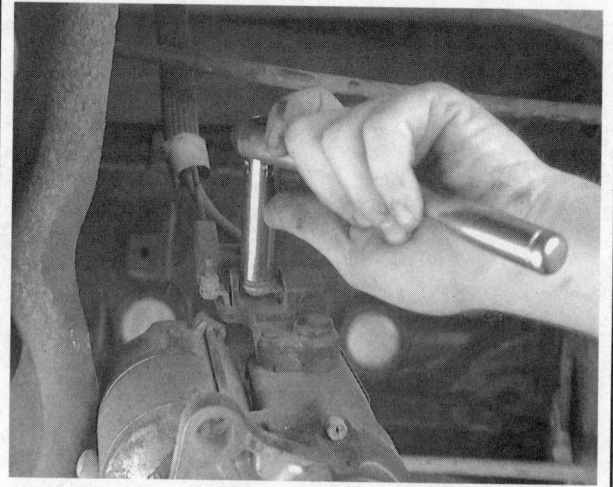

. . . and disconnect the starter motor wires, if not already done

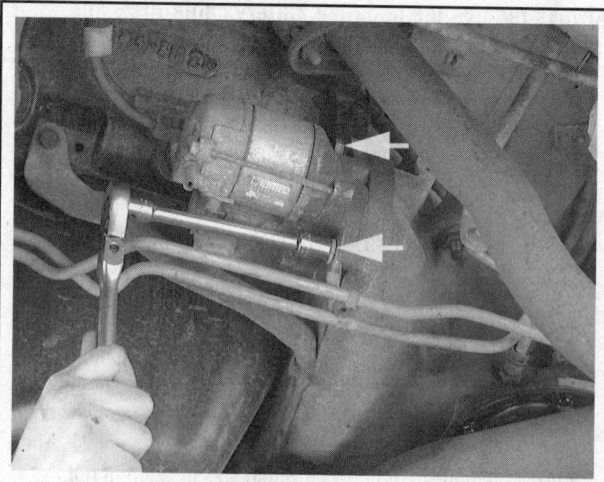

To remove a common starter, raise the vehicle if needed and loosen the starter mounting fasteners (arrows) . . .

Before installing the starter motor, be sure to inspect the gear teeth (arrow) . . .

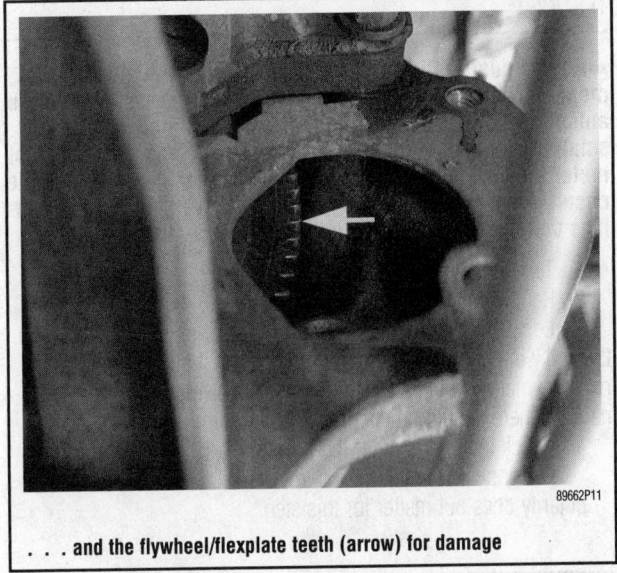

. . . and the flywheel/flexplate teeth (arrow) for damage

4. Remove the starter mounting bolts, if not already done.
5. Remove the starter assembly from the vehicle. In some cases, the starter will have to be turned to a different angle to clear obstructions. Don't loose any shims that may fall out from between the starter and the mounting boss, they will need to be returned to their original position when installing the starter. The shims are used to adjust the clearance between the starter pinion and flywheel/flexplate teeth.

To install:
6. If necessary, measure and adjust the pinion-to-ring gear clearance.
7. Position the shim (if any) and the starter motor on the mounting boss. Tighten the mounting bolts securely.
8. Connect the wiring, if not already done.
9. Install any components that were removed to gain access to the starter.
10. Connect the negative battery cable.

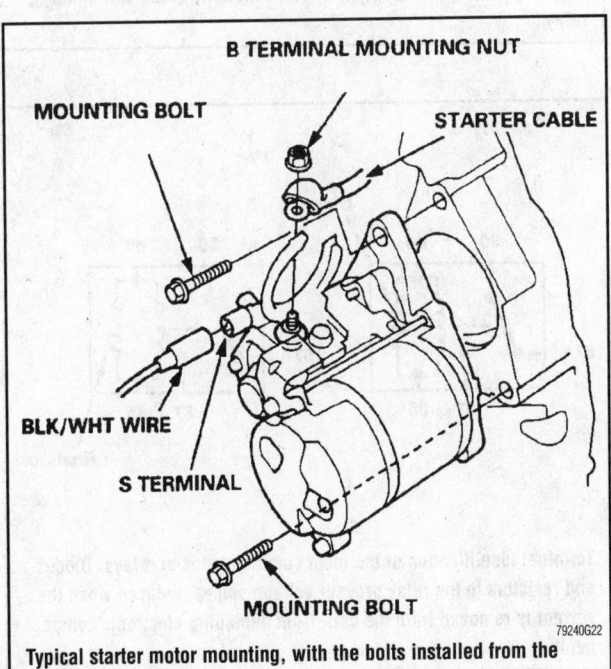

Typical starter motor mounting, with the bolts installed from the starter motor side

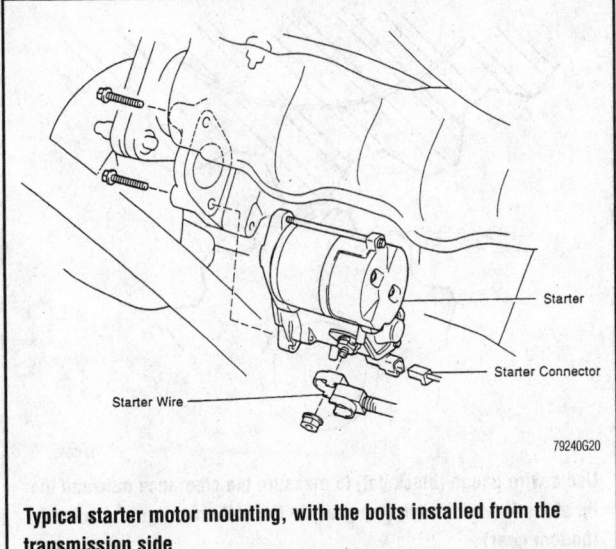

Typical starter motor mounting, with the bolts installed from the transmission side

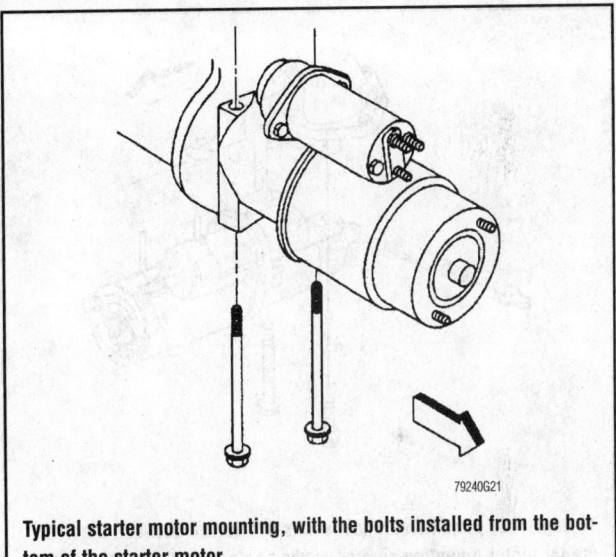

Typical starter motor mounting, with the bolts installed from the bottom of the starter motor

ADJUSTMENTS

Starter Pinion Depth

➡This procedure is used to diagnose starter noise caused by incorrect clearance between the starter pinion and flywheel while the starter is engaged.

1. Raise and safely support the front of the vehicle securely.
2. Remove the flywheel cover.
3. Inspect the flywheel for chipped or missing teeth, abnormal wear, cracks and warpage. Replace the damaged component, if any, and continue with the procedure.
4. Make sure the vehicle is in Park or Neutral. Apply the parking brake.
5. Have an assistant slowly and smoothly rotate the crankshaft in the normal direction of rotation.
6. Slowly move a piece of chalk toward the edge of the flywheel until it just touches, which will highlight the high spot of the ring gear.

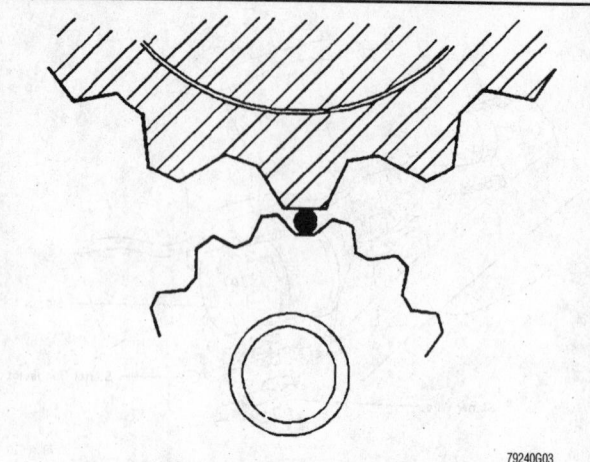

Use a wire gauge (black dot) to measure the clearance between the tip of the flywheel tooth (top gear) to the bottom of the pinion teeth (bottom gear)

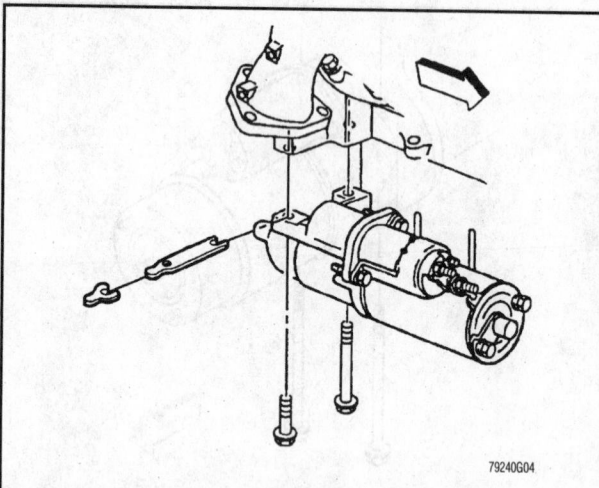

To adjust the pinion clearance, shims are placed between the starter motor mounting surface on the engine and the starter motor

7. Disconnect the negative battery cable.
8. Turn the high spot of the flywheel to the area of the starter drive pinion.
9. Using a wire gauge, measure the clearance between the tip of the ring gear tooth and the bottom of the pinion gear teeth. Clearance should generally be 0.02–0.06 in. (0.5–1.5mm).
10. Add or remove shims to adjust the clearance, if needed.
11. Install the flywheel cover.
12. Lower the vehicle to the floor.
13. Connect the negative battery cable.
Generally, add shims if the starter whines after the engine starts, and remove shims if the starter whines only during cranking.

Starter Relay

➥The starter relay is usually located in the fuse/relay panel. Depending on the manufacturer, it may be in the engine compartment, under the dash or behind a kickpanel. Refer to the owner's manual for the location of the fuse/relay box.

TESTING

➥A good quality Digital Multimeter (DMM) with at least 10 megohm/volt impedance should be used when testing modern automotive circuits. These meters can accurately detect very small amounts of voltage, current and resistance. This type of meter also has a high internal resistance that will not load the circuit being tested. Loading the circuit gives inaccurate readings and may cause damage to sensitive computer circuits.

1. Turn the ignition **OFF**.
2. Remove the relay.
3. Locate the two terminals on the relay which are connected to the coil windings. Check the relay coil for continuity. Connect the negative meter lead to terminal **85** and positive meter lead to terminal **86**. There should be continuity. If not, replace the relay.
4. Check the operation of the internal relay contacts, as follows:
   a. Connect the meter leads to terminals **30** and **87**. Meter polarity does not matter for this step.

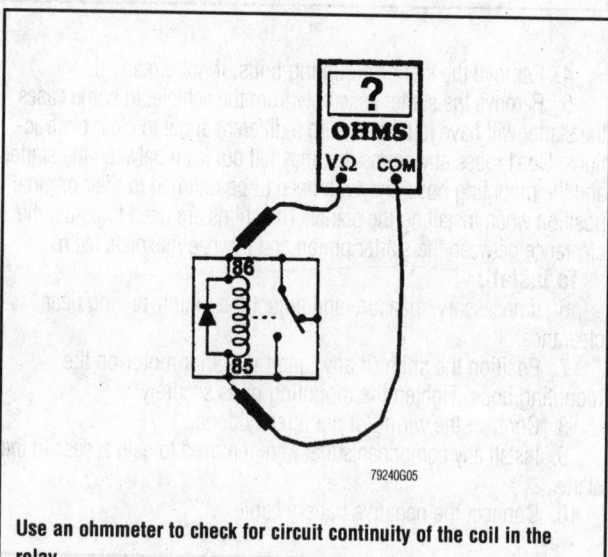

Use an ohmmeter to check for circuit continuity of the coil in the relay

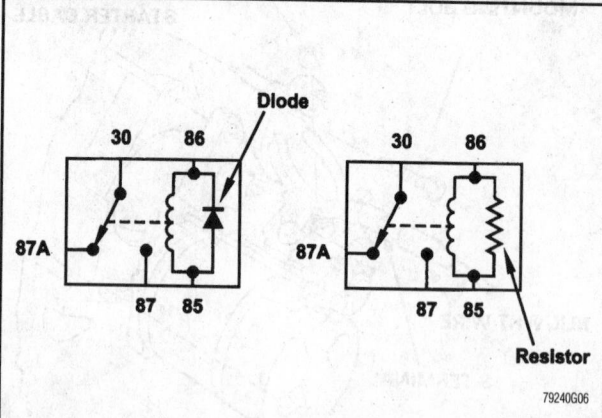

Terminal identification of the most common types of relays. Diodes and resistors in the relay prevent voltage spikes, induced when the current is removed from the coil, from damaging electronic components

b. Apply positive battery voltage to terminal **86** and ground to terminal **85**. The relay should click as the contacts are drawn toward the coil and the meter should indicate continuity. Replace the relay if your results are different.

Solenoid

TESTING

1. Disconnect the negative battery cable.
2. Remove the wire connections from the starter solenoid.

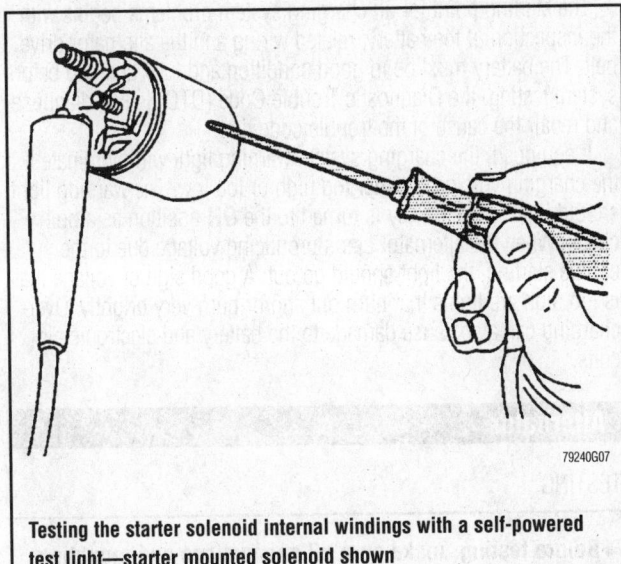

Testing the starter solenoid internal windings with a self-powered test light—starter mounted solenoid shown

3. Using a self-powered test light or ohmmeter, check for continuity between the following:
- Solenoid **B** terminal and solenoid case or ground terminal—no continuity
- **S** terminal and solenoid case or ground terminal—continuity
- **S** terminal and **M** terminal—continuity
- **M** terminal and solenoid case or ground terminal—continuity

4. If the actual results of the test are different than indicated, replace the starter solenoid.

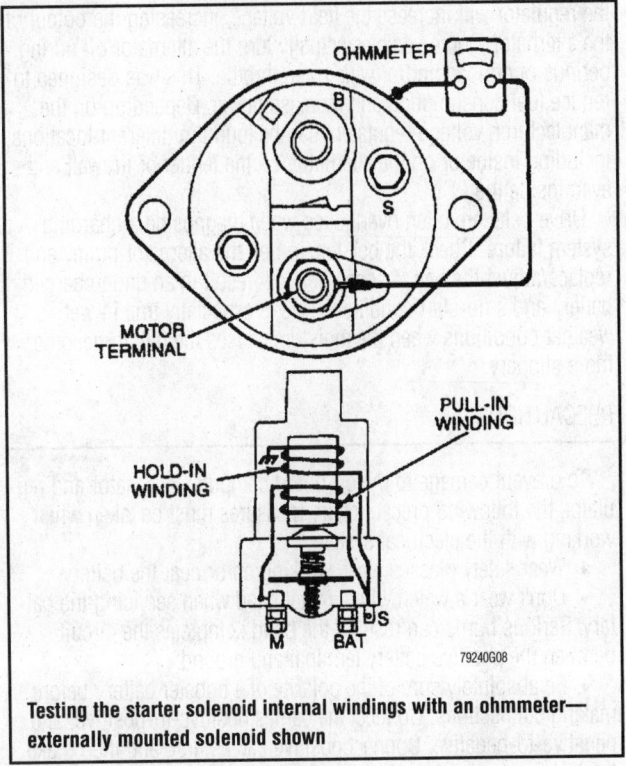

Testing the starter solenoid internal windings with an ohmmeter— externally mounted solenoid shown

REMOVAL & INSTALLATION

➡**This procedure is for externally mounted starter solenoids only. For solenoids mounted on the starter, we recommend replacing the complete assembly.**

1. Disconnect the negative battery cable.
2. Remove the wiring from the starter solenoid. Label the wires and the corresponding terminals if necessary for installation.
3. Remove the fasteners securing the solenoid to the fender or firewall.
4. Remove the solenoid.

To install:

5. Clean the solenoid mounting and the solenoid to ensure good electrical contact.
6. Install the solenoid.
7. Connect the wiring to the proper terminals.
8. Connect the negative battery cable.

CHARGING SYSTEM

General Information

A typical charging system contains an alternator (generator), drive belt, battery, voltage regulator and the associated wiring. The charging system, like the starting system is a series circuit with the battery wired in parallel. After the engine is started and running, the alternator takes over as the source of power and the battery then becomes part of the load on the charging system.

Some vehicle manufacturers use the term generator instead of alternator. Many years ago there used to be a difference, now they are one and the same. The alternator which is driven by the belt, consists of a rotating coil of laminated wire called the rotor. Surrounding the rotor are more coils of laminated wire that remain stationary just inside the alternator case. This is how we get the name of stator.

When current is passed through the rotor via the slip rings and brushes, the rotor becomes a rotating magnet with, of course, a magnetic field. When a magnetic field passes through a conductor (the stator), alternating current (A/C) is generated. This A/C current is rectified, turned into direct current (D/C), by the diodes located within the alternator.

The voltage regulator controls the alternator's field voltage by grounding one end of the field windings very rapidly. The frequency varies according to current demand. The more the field is grounded, the more voltage and current the alternator produces. Voltage is maintained at about 13.5–15 volts. During high engine speeds and low current demands, the regulator will adjust the voltage of the alternator field to lower the alternator output voltage. Conversely, when the vehicle is idling and the current demands may be high, the regulator will increase the field voltage, increasing the output of the alternator. Some vehicles actually turn the alternator off during periods of no load and/or wide open throttle. This was designed to reduce fuel consumption and increase power. Depending on the manufacturer, voltage regulators can be found in different locations, including inside or on the alternator, on the fender or firewall and even inside the PCM.

Drive belts are often overlooked when diagnosing a charging system failure. Check the belt tension on the alternator pulley and replace/adjust the belt. A loose belt will result in an undercharged battery and a no-start condition. This is especially true in wet weather conditions when the moisture causes the belt to become more slippery.

PRECAUTIONS

To prevent damage to the on-board computer, alternator and regulator, the following precautionary measures must be taken when working with the electrical system:
- Wear safety glasses when working on or near the battery.
- Don't wear a watch with a metal band when servicing the battery. Serious burns can result if the band completes the circuit between the positive battery terminal and ground.
- Be absolutely sure of the polarity of a booster battery before making connections. Connect the cables positive-to-positive, and negative-to-negative. Connect positive cables first, and then make the last connection to ground on the body of the booster vehicle so that arcing cannot ignite hydrogen gas that may have accumulated near the battery. Even momentary connection of a booster battery with the polarity reversed will damage alternator diodes.
- Disconnect both vehicle battery cables before attempting to charge a battery.
- Never ground the alternator or generator output or battery terminal. Be cautious when using metal tools around a battery to avoid creating a short circuit between the terminals.
- Never ground the field circuit between the alternator and regulator.
- Never run an alternator or generator without load unless the field circuit is disconnected.
- Never attempt to polarize an alternator.
- When installing a battery, make sure that the positive and negative cables are not reversed.
- When jump-starting the car, be sure that like terminals are connected. This also applies to using a battery charger. Reversed polarity will burn out the alternator and regulator in a matter of seconds.
- Never operate the alternator with the battery disconnected or on an otherwise uncontrolled open circuit.
- Do not short across or ground any alternator or regulator terminals.
- Do not try to polarize the alternator.
- Do not apply full battery voltage to the field (brown) connector.
- Always disconnect the battery ground cable before disconnecting the alternator lead.
- Always disconnect the battery (negative cable first) when charging it.
- Never subject the alternator to excessive heat or dampness. If you are steam cleaning the engine, cover the alternator.
- Never use arc-welding equipment on the car with the alternator connected.

SYSTEM TESTING

The charging system should be inspected if:
- A Diagnostic Trouble Code (DTC) is set relating to the charging system
- The charging system warning light is illuminated
- The voltmeter on the instrument panel indicates improper charging (either high or low) voltage
- The battery is overcharged (electrolyte level is low and/or boiling out)
- The battery is undercharged (insufficient power to crank the starter)

The starting point for all charging system problems begins with the inspection of the battery, related wiring and the alternator drive belt. The battery must be in good condition and fully charged before system testing. If a Diagnostic Trouble Code (DTC) is set, diagnose and repair the cause of the trouble code first.

If equipped, the charging system warning light will illuminate if the charging voltage is either too high or too low. The warning light should light when the key is turned to the **ON** position as a bulb check. When the alternator starts producing voltage due to the engine starting, the light should go out. A good sign of voltage that is too high are lights that burn out and/or burn very brightly. Overcharging can also cause damage to the battery and electronic circuits.

Alternator

TESTING

➡**Before testing, make sure all connections and mounting bolts are clean and tight. Many charging system problems are related to loose and corroded terminals or bad grounds. Don't overlook the engine ground connection to the body, or the tension of the alternator drive belt.**

Voltage Drop Test

➡**A good quality Digital Multimeter (DMM) with at least 10 megohm/volt impedance should be used when testing modern automotive circuits. These meters can accurately detect very small amounts of voltage, current and resistance. This type of meter also has a high internal resistance that will not load the circuit being tested. Loading the circuit gives inaccurate readings and may cause damage to sensitive computer circuits.**

1. Make sure the battery is in good condition and fully charged.
2. Perform a voltage drop test of the positive side of the circuit as follows:
- Start the engine and allow it to reach normal operating temperature.
- Turn the headlamps, heater blower motor and interior lights on.
- Bring the engine to about 2,500 rpm and hold it there.
- Connect the negative (-) voltmeter lead directly to the battery positive (+) terminal.
- Touch the positive (+) voltmeter lead directly to the alternator **B+** output stud, not the nut. The meter should read no higher than about 0.5 volts. If it does, then there is higher than normal resistance between the positive side of the battery and the **B+** output at the alternator.

• Move the positive (+) meter lead to the nut and compare the voltage reading with the previous measurement. If the voltage reading drops substantially, then there is resistance between the stud and the nut.

➡The theory is to keep moving closer to the battery terminal one connection at a time in order to find the area of high resistance (bad connection).

3. Perform a voltage drop test of the negative side of the circuit as follows:

　　a. Start the engine and allow it to reach normal operating temperature.

　　b. Turn the headlamps, heater blower motor and interior lights ON.

　　c. Bring the engine to about 2,500 rpm and hold it there.

　　d. Connect the negative (-) voltmeter lead directly to the negative battery terminal.

　　e. Touch the positive (+) voltmeter lead directly to the alternator case or ground connection. The meter should read no higher than about 0.3 volts. If it does, then there is higher than normal resistance between the battery ground terminal and the alternator ground.

　　f. Move the positive (+) meter lead to the alternator mounting bracket, if the voltage reading drops substantially then you know that there is a bad electrical connection between the alternator and the mounting bracket.

➡The theory is to keep moving closer to the battery terminal one connection at a time in order to find the area of high resistance (bad connection).

Current Output Test

➡A good quality Digital Multimeter (DMM) with at least 10 megohm/volt impedance should be used when testing modern automotive circuits. These meters can accurately detect very small amounts of voltage, current and resistance. This type of meter also has a high internal resistance that will not load the circuit being tested. Loading the circuit gives inaccurate readings and may cause damage to sensitive computer circuits.

1. Perform a current output test as follows:

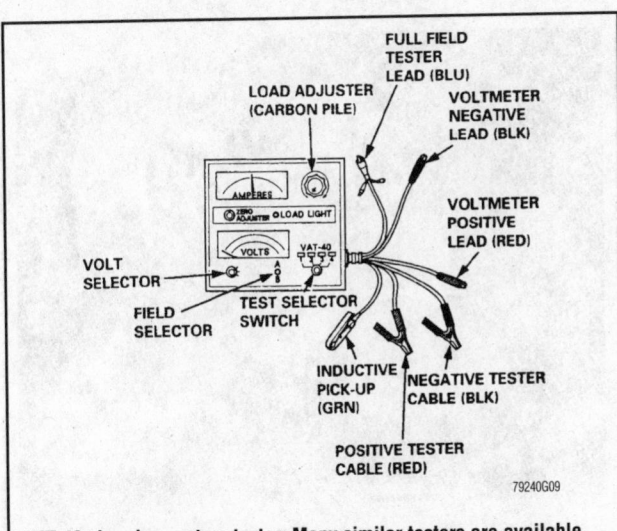

VAT-40 charging system tester. Many similar testers are available that perform equally as well

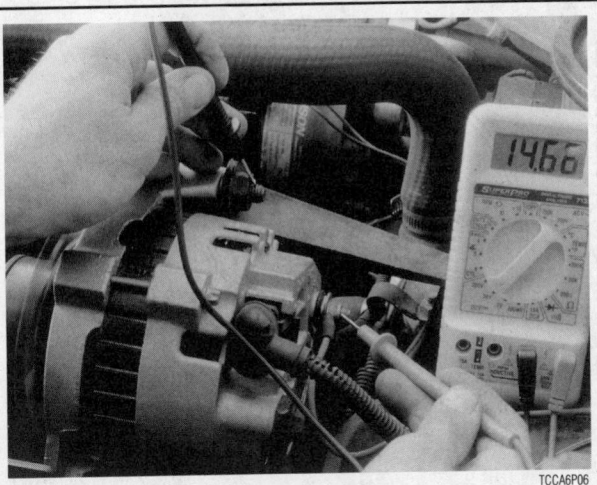

The output voltage of the alternator can be quickly measured by probing between the output terminal and a good ground

➡The current output test requires the use of a volt/amp tester with battery load control and an inductive amperage pick-up. Follow the manufacturer's instructions on the use of the equipment.

　　a. Start the engine and allow it to reach normal operating temperature.

　　b. Apply the parking brake and turn OFF all electrical accessories.

　　c. Connect the tester to the battery terminals and cable according to the instructions.

　　d. Bring the engine to about 2,500 rpm and hold it there.

　　e. Apply a load to the charging system with the rheostat on the tester. Do not let the voltage drop below 12 volts.

　　f. The alternator should deliver to within 10% of the rated output. If the amperage is not within 10% and all other components test good, replace the alternator.

Alternator Isolation Test

➡A good quality Digital Multimeter (DMM) with at least 10 megohm/volt impedance should be used when testing modern automotive circuits. These meters can accurately detect very small amounts of voltage, current and resistance. This type of meter also has a high internal resistance that will not load the circuit being tested. Loading the circuit gives inaccurate readings and may cause damage to sensitive computer circuits.

On some models it is possible to isolate the alternator from the regulator by grounding the **F** (field) terminal. Grounding the **F** terminal removes the regulator from the circuit and forces full alternator output. On alternators equipped with internal regulators, we recommend replacing the complete assembly if either the alternator or regulator is defective.

✳✳ WARNING

Do not allow the voltage to rise above 18 volts. Damage to electrical circuits may occur.

1. Connect a voltmeter across the battery terminals so the voltage can be monitored.

2. Start the engine and allow it reach normal operating temperature.

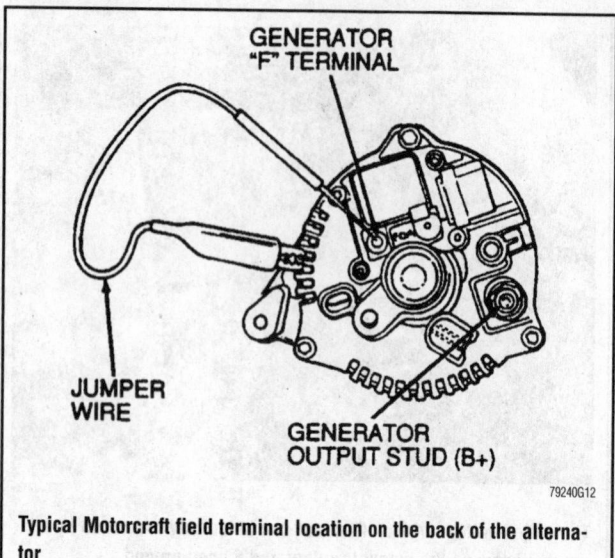

Typical Motorcraft field terminal location on the back of the alternator

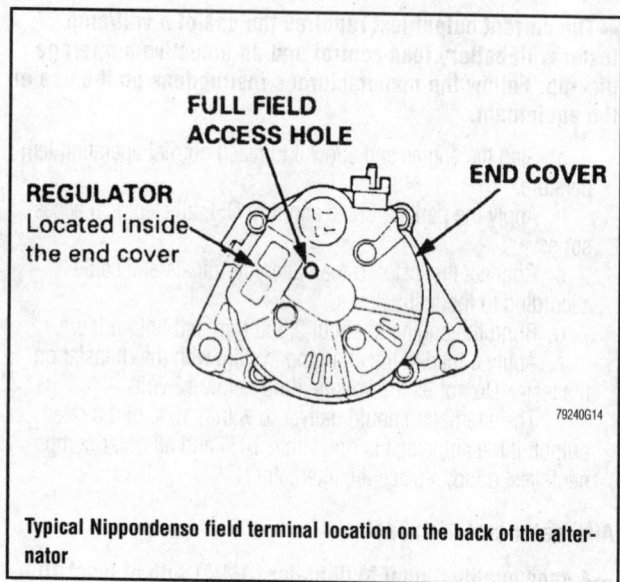

Typical Nippondenso field terminal location on the back of the alternator

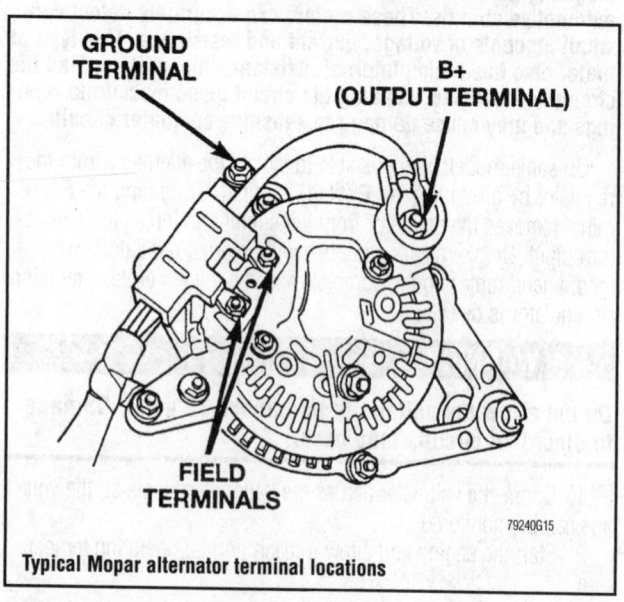

Typical Mopar alternator terminal locations

3. Connect a jumper lead to a good ground.
4. Locate the field terminal (negative) on the back of the alternator.
5. Momentarily connect the grounded jumper to the field terminal. If the alternator is OK, the voltage will climb rapidly. Disconnect the jumper before the output reaches 18 volts. If the voltage does not rise, replace the alternator. If the voltage rises, then the regulator is bad.

➡Chrysler models have two field terminals, one positive and one negative. The positive (+) terminal will have battery voltage present and the negative (-) terminal will have 3–5 volts less. Ground the negative (-) terminal when testing this type of alternator.

REMOVAL & INSTALLATION

1. Disconnect the negative battery cable.
2. Remove the drive belt from the alternator pulley.

➡In some cases, it may be easier to disconnect the wiring after the alternator has been removed. Be sure to support the alternator by hand while removing the wiring.

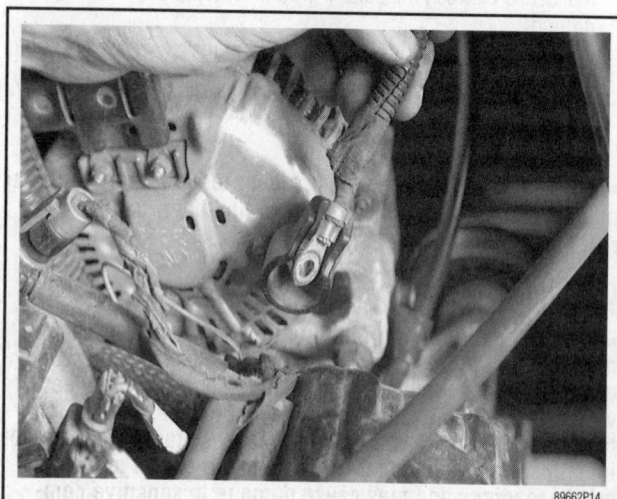

To remove a typical alternator, first detach the wiring terminals from it (if possible) . . .

. . . then loosen the alternator mounting fasteners—this alternator uses mounting bolts (arrows)

When removing alternator fasteners, be sure to retain any washers, spacers or nuts for reassembly

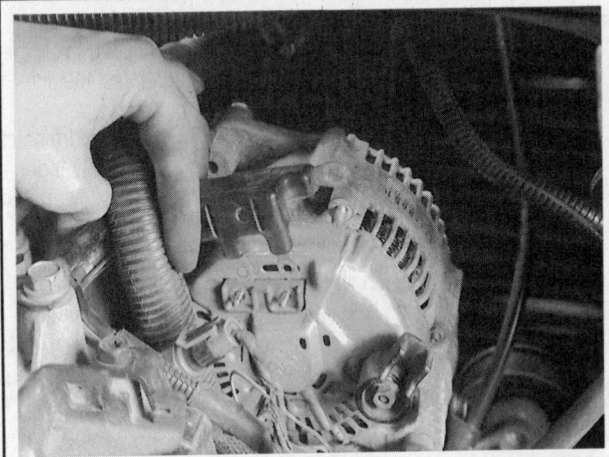

If not possible earlier, now disconnect any applicable wiring from the alternator

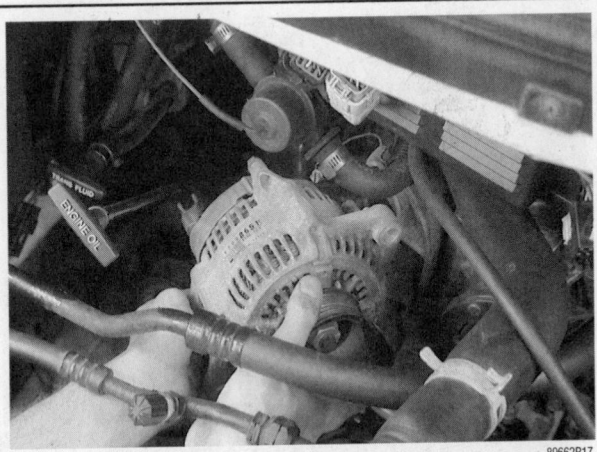

Finally, carefully remove the alternator from the engine compartment—although not shown here some alternators must be dropped out the bottom of the vehicle

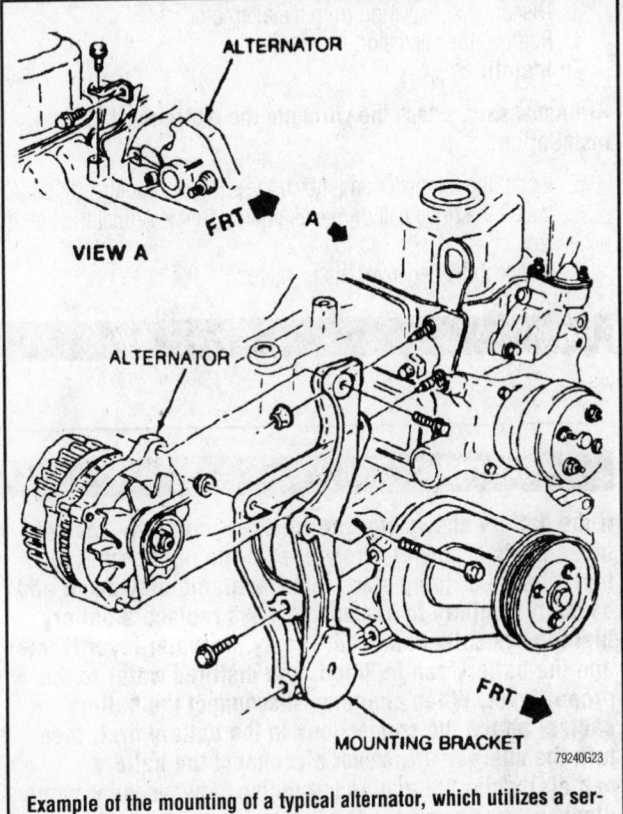

Example of the mounting of a typical alternator, which utilizes a serpentine belt

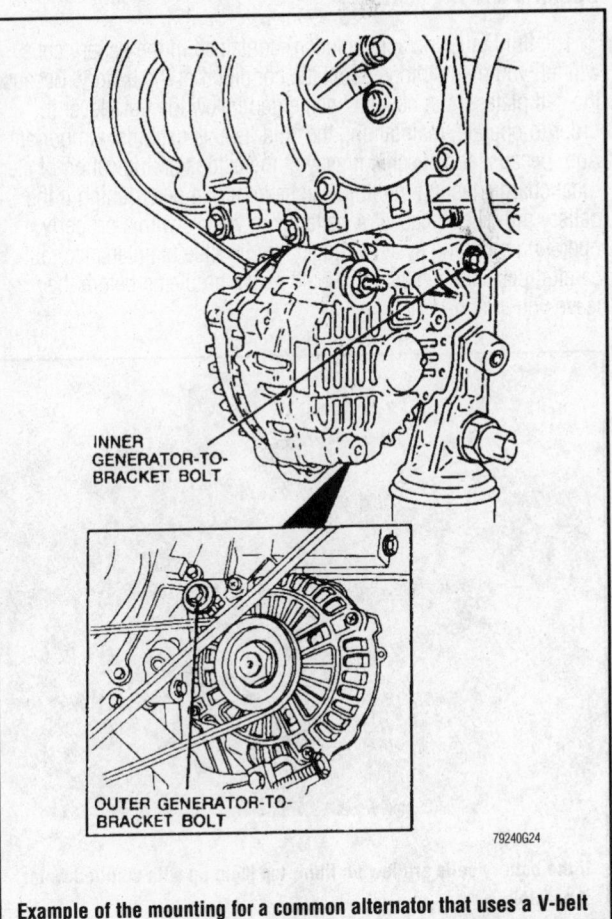

Example of the mounting for a common alternator that uses a V-belt

3. Disconnect the wiring from the alternator.
4. Remove the alternator.

To install:

➡**If necessary, attach the wiring to the alternator before installation.**

5. Install the alternator and attach the wiring if not already done.
6. Install the drive belt on the alternator pulley. Adjust the belt if necessary.
7. Connect the negative battery cable.

Battery

TESTING

✳✳ CAUTION

If the battery shows signs of freezing, cracking, leaking, loose posts or low electrolyte level, do not attempt to test, charge or jump start. Internal arcing may occur and cause the battery to explode. Always replace a battery that is physically damaged. If only the water level is low and the battery can be filled, add distilled water to the proper level. When charging, disconnect the battery cables, attach the connections to the battery first, then turn the charger ON. Never disconnect the battery cable(s) while the engine is running. Always wear safety glasses when servicing the battery.

Specific Gravity Test

The fluid (sulfuric acid solution) contained in the battery cells will tell you many things about the condition of the battery. Because the cell plates must be kept submerged below the fluid level in order to operate, maintaining the fluid level is extremely important. And, because the specific gravity of the acid is an indication of electrical charge, testing the fluid can be an aid in determining if the battery must be replaced. A battery in a vehicle with a properly operating charging system should require little maintenance, but careful, periodic inspection should reveal problems before they leave you stranded.

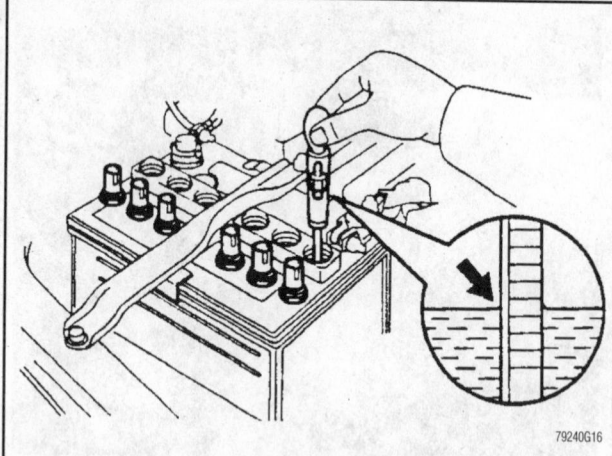

On serviceable batteries, draw some battery fluid into the hydrometer and read the specific gravity indicated by the float inside the tester

At least once a year, check the specific gravity of the battery. It should be between 1.20 and 1.26 on the gravity scale. Most auto supply stores carry a variety of inexpensive battery testing hydrometers. These can be used on any non-sealed battery to test the specific gravity in each cell.

Draw some of the electrolyte from the battery into the hydrometer until the float is lifted from its seat. Read the specific gravity indicated by the position of the float. If the specific gravity is low in one or more cells, the battery should be slowly charged and checked again to see if the gravity has come up. Generally, if after charging, the specific gravity between any two cells varies more than 50 points (0.50), replace the battery, as it can no longer produce sufficient voltage to guarantee proper operation.

No Load Voltage Test

➡**A good quality Digital Multimeter (DMM) with at least 10 megohm/volt impedance should be used when testing modern automotive circuits. These meters can accurately detect very small amounts of voltage, current and resistance. This type of meter also has a high internal resistance that will not**

If the battery cells are low on fluid, top them up with distilled water if possible

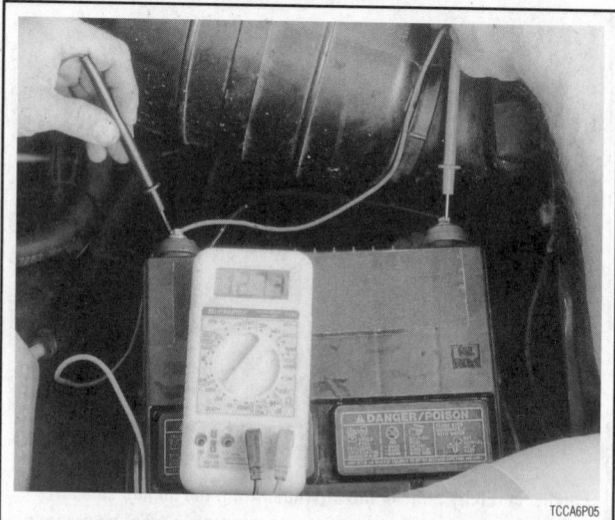

Use a high quality DMM to measure the battery voltage

Open Circuit Voltage	
Open Circuit Volts	**Charge Percentage**
11.7 volts or less	0%
12.0 volts	25%
12.2 volts	50%
12.4 volts	75%
12.6 volts or more	100%

79240G10

Compare the actual voltage measured with these values to determine the percent of charge based on no load test results

load the circuit being tested. Loading the circuit gives inaccurate readings and may cause damage to sensitive computer circuits.

1. Perform a no load voltage test to determine the state of charge by doing the following:

a. If the battery has just been charged, remove the surface charge by turning on the headlamps for 15 seconds, then let the voltage stabilize for about 5 minutes before making any measurements.

b. Disconnect the negative battery cable.

c. Measure the battery voltage with a DMM.

d. Compare the readings to the chart to determine the state of charge.

High Capacity Discharge Test

1. Perform a high capacity discharge test to determine the cranking capacity as follows:

a. Fully charge the battery.

b. Connect a VAT-40 or equivalent load tester to the battery.

Load Test Temperature		
Minimum Voltage	**Temperature**	
	°F	°C
9.6 volts	70° and above	21° and above
9.5 volts	60°	16°
9.4 volts	50°	10°
9.3 volts	40°	4°
9.1 volts	30°	-1°
8.9 volts	20°	-7°
8.7 volts	10°	-12°
8.5 volts	0°	-18°

79240G11

High capacity discharge test minimum voltage/temperature chart

c. Apply a load equal to $FR1/2 of the Cold Cranking Amp (CCA) rating of the battery for 15 seconds. The CCA is usually found on the battery label, if not, apply a load equal to 200 amps.

d. If the voltmeter reading falls below 9.6 volts at 70°F (21°C) or more, the battery should be replaced. The minimum battery voltage will be lower depending on the ambient temperature. Refer to the chart for testing in temperatures lower than 70°F (21°C).

Parasitic Draw Test

→**A good quality Digital Multimeter (DMM) with at least 10 megohm/volt impedance should be used when testing modern automotive circuits. These meters can accurately detect very small amounts of voltage, current and resistance. This type of meter also has a high internal resistance that will not load the circuit being tested. Loading the circuit gives inaccurate readings and may cause damage to sensitive computer circuits.**

This test measures the amount of current that the vehicle draws while it is parked and not in use. A small amount of current should be flowing for such things as the on-board computer memory, automatic climate control, clock, and radio station presets. If there is a short in the vehicle electrical system or something has been left on, the excess current draw will eventually drain the battery and cause a no-start condition.

1. Be sure all accessories are turned **OFF**. Disconnect the negative battery cable.

2. Install a battery quick-disconnect switch (such as GM Parasitic Draw Test Switch J 38758) between the negative cable and the negative battery terminal. A battery disconnect switch will work in most cases.

3. Road test the vehicle while activating all accessories including the radio and air conditioning. Then, turn all accessories **OFF**.

4. Turn the vehicle **OFF** and open the hood.

5. If equipped, disable the underhood light.

6. Allow approximately 20 minutes for the vehicle computer system(s) to power down.

7. Connect one end of a jumper with a 10 amp fuse to the side of the quick-disconnect switch closest to the negative battery terminal. Be sure the jumper is on the metal part of the switch.

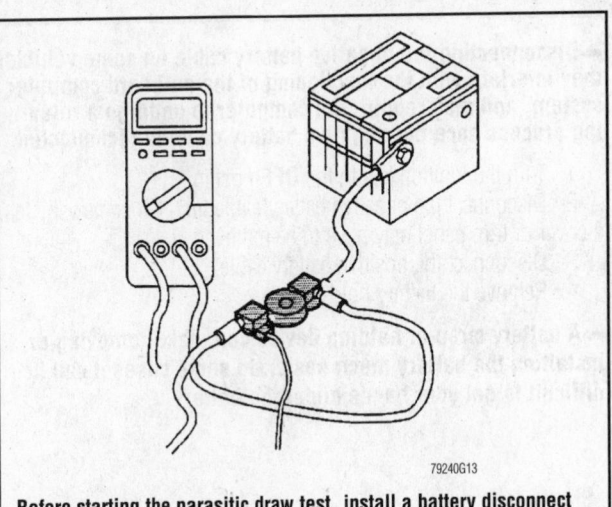

79240G13

Before starting the parasitic draw test, install a battery disconnect switch between the negative battery cable and the battery terminal, as shown

8. Connect the remaining end of the jumper to the other side of the switch closest to the negative battery cable.

✳✳ WARNING

Do not connect the multimeter to the circuit if more than 10 amps are flowing. Damage to the meter may occur.

9. Open the switch so all current flows through the jumper with the 10 amp fuse. If the fuse blows, there is more than 10 amps flowing in the circuit. This indicates that a component was left on (glove box light or other accessory) or there is a short in the electrical system. Find and correct the cause of the large current flow, then continue with this test.

10. If the fuse does not blow, close the disconnect switch and remove the jumper.

11. Set the multimeter to read 10 amps.

12. Connect the multimeter leads in place of the jumper used previously. When the switch is opened, current will flow through the meter.

13. The current draw should now be below 2 amps. If not, then something in the vehicle has been left on. Find the cause and correct it. When the current is less than 2 amps, set the meter to the 2 amp range. This will allow you to measure small amounts of current.

✳✳ WARNING

Do not open the door of the vehicle. The interior lights coming on will blow the fuse of the meter while on the 2 amp range.

14. Normal current draw should be less than ¼ of the reserve capacity of the battery. If the reserve capacity is unknown, normal current draw should be somewhere in the range of 0.005–0.040 amps depending on the type and amount of equipment on the vehicle.

➡ **The reserve capacity is the amount of time, in minutes, it takes for the battery voltage to fall below 10.5 volts at a discharge rate of 25 amps at 80°F (26.7°C). In most cases, this number can be found on the battery label.**

15. If the current draw is higher than specified, pull fuses and/or disconnect components until the problem is found. Don't overlook the alternator connection.

REMOVAL & INSTALLATION

➡ **Disconnecting the negative battery cable on some vehicles may interfere with the functioning of the on-board computer system, and may require the computer to undergo a relearning process once the negative battery cable is reconnected.**

1. Turn the ignition key to the **OFF** position.

2. Disconnect the negative battery cable first. On some vehicles, a cover or trim panel may have to be removed first.

3. Disconnect the positive battery cable.

4. Remove the battery hold-down.

➡ **A battery strap or holding device can make removing or installing the battery much easier. In some cases it can be difficult to get your hands under the battery.**

To install:

5. Position the battery in the vehicle. Pay attention to the location of the terminals.

6. Install the battery hold-down. A loose battery may cause a vehicle fire or severe damage to the electrical system.

7. Clean the terminals and connect the positive battery cable first, then the negative cable.

8. If equipped, install the cover or trim panel.

JUMP STARTING A DEAD BATTERY

Whenever a vehicle is jump started, precautions must be followed in order to prevent the possibility of personal injury. Remember that batteries contain a small amount of explosive hydrogen gas which is a by-product of battery charging. Sparks should always be avoided when working around batteries, especially when attaching jumper cables. To minimize the possibility of accidental sparks, follow the procedure carefully.

✳✳ CAUTION

NEVER hook the batteries up in a series circuit or the entire electrical system will go up in smoke, including the starter!

Vehicles equipped with a diesel engine may utilize two 12 volt batteries. If so, the batteries are connected in a parallel circuit (positive terminal-to-positive terminal, negative terminal-to-negative terminal). Hooking the batteries up in parallel circuit increases battery cranking power without increasing total battery voltage output. Output remains at 12 volts. On the other hand, hooking two 12 volt batteries up in a series circuit (positive terminal-to-negative terminal, positive terminal-to-negative terminal) increases total battery output to 24 volts (12 volts plus 12 volts).

Jump Starting Precautions

To avoid personal injury and/or vehicle damage, please read all of the following precautions prior to jump starting a discharged battery:

• NEVER hook the batteries up in a series circuit or the entire electrical system will go up in smoke, including the starter!

• Be sure that both batteries are of the same voltage. Vehicles covered by this manual and most vehicles on the road today utilize a 12 volt charging system.

• Be sure that both batteries are of the same polarity (have the same terminal, in most cases NEGATIVE grounded).

• Be sure that the vehicles are not touching, otherwise a short could occur.

• On serviceable batteries, be sure the vent cap holes are not obstructed.

• Do not smoke or allow sparks anywhere near the batteries.

• In cold weather, make sure the battery electrolyte is not frozen. This can occur more readily in a battery that has been in a state of discharge.

• Do not allow electrolyte to contact your skin or clothing.

Jump Starting Procedure

1. Make sure that the voltages of the two batteries are the same. Most batteries and charging systems are of the 12 volt variety.

2. Pull the vehicle with the good battery into a position so the jumper cables can reach the dead battery and that vehicle's engine compartment. Make sure that the vehicles DO NOT touch.

➡**Remote power terminals are usually provided on vehicles where the battery is located in the fender or other location that makes connecting jumper cables difficult. These power terminals are located in the engine compartment. If this is the situation, use the remote terminals instead of the terminals on the battery.**

3. Place the transmissions/transaxles of both vehicles in Neutral, manual transmissions, or P (park), automatic transmissions, as applicable, then firmly set their parking brakes.

➡**If necessary for safety reasons, the hazard lights on both vehicles may be operated throughout the entire procedure without significantly increasing the difficulty of jumping the dead battery.**

4. Turn all lights and accessories OFF on both vehicles. Be sure the ignition switches on both vehicles are turned to the **OFF** position.

5. Cover the battery cell caps with a rag, but do not cover the terminals.

6. Make sure the terminals on both batteries are clean and free of corrosion, otherwise proper electrical connection will be impeded. If necessary, clean the battery terminals before proceeding.

7. Identify the positive (+) and negative (-) terminals on both batteries.

8. Connect the first jumper cable to the positive (+) terminal of the dead battery, then attach the other end of that cable to the positive (+) terminal of the booster (good) battery.

9. Connect the clamp of the negative jumper cable to the negative (-) terminal on the good battery and the final cable clamp to an engine bolt head, alternator bracket or other solid, metallic point on the engine with the dead battery. Try to pick a ground on the engine that is positioned away from the battery in order to minimize the possibility of explosion due to the sparks created when the last connection is made. DO NOT connect this clamp to the negative terminal of the bad battery.

✳✳ WARNING

Be very careful to keep the jumper cables away from moving parts (cooling fan, belts, etc.) on both engines.

10. Ensure the cables are routed away from any moving parts, then start the donor vehicle's engine. Run the engine at moderate speed for several minutes to allow the dead battery a chance to receive some initial charge.

11. With the donor vehicle's engine still running slightly above idle, try to start the vehicle with the dead battery. Crank the engine for no more than 10 seconds at a time and let the starter cool for at least 20 seconds between tries. If the vehicle does not start in 3 tries, it is likely that something else is also wrong, or that the battery needs additional time to charge.

12. Once the vehicle is started, allow it to run at idle for a few seconds to make sure that it is operating properly.

13. Turn ON the headlights, heater blower and, if equipped, the rear defroster of both vehicles in order to reduce the severity of voltage spikes and subsequent risk of damage to the vehicles' electrical systems when the cables are disconnected. This step is especially important to any vehicle equipped with computer control modules.

14. Carefully remove the cables in the reverse order of connection. Start with the negative cable that is attached to the engine ground, then the negative cable on the donor battery. Disconnect the positive cable from the donor battery, and finally disconnect the positive cable from the formerly dead battery. Be careful when disconnecting the cables from the positive terminals not to allow the alligator clips to touch any metal on either vehicle or a short and sparks will occur.

Voltage Regulator

TESTING

➡**Most regulators are integral (built in) to the alternator or Powertrain Control Module (PCM). If the regulator is found to be defective on these models, the alternator or PCM should be replaced.**

For voltage regulator testing, refer to the Alternator Isolation test.

REMOVAL & INSTALLATION

➡**The following procedure is only for voltage regulators mounted on the back (outside) of the alternator or elsewhere in the engine compartment.**

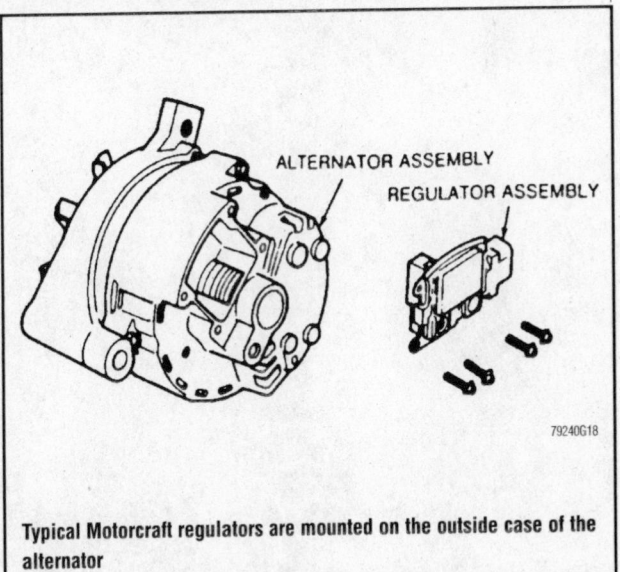

Typical Motorcraft regulators are mounted on the outside case of the alternator

1. Disconnect the negative battery cable.
2. If equipped, remove the exterior alternator cover to expose the regulator. Do not disassemble the alternator case that houses the rotor and stator.
3. If equipped, disengage the electrical connector from the regulator.
4. Remove the regulator mounting screws and remove the regulator.

To install:

5. Position the regulator in its original position and install the mounting screws.
6. Connect any wiring that was removed from the regulator.
7. If equipped, install the cover.
8. Connect the negative battery cable.

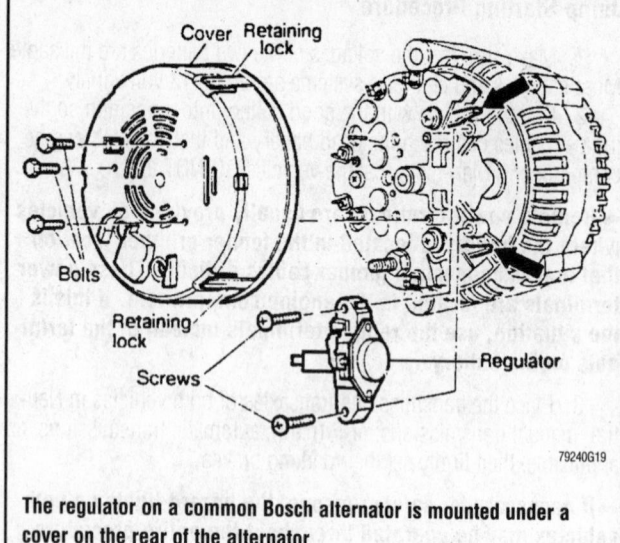

The regulator on a common Bosch alternator is mounted under a cover on the rear of the alternator

PISTON, PISTON RING & CONNECTING ROD POSITIONING

PISTON, PISTON RING &
CONNECTING ROD
POSITIONING 11-1
PISTON, PISTON RING AND
CONNECTING ROD POSITIONING
INDEX 11-2

PISTON, PISTON RING AND CONNECTING ROD POSITIONING

When assembling the pistons, piston rings and connecting rods, and when installing these assemblies into the engine block, it is vitally important to ensure that these three components are properly positioned with respect to each other. Often times the engine block is designed so that if a connecting rod or piston is installed backwards, or in the wrong bank of cylinders, internal engine damage may occur once the engine is started. The piston ring end-gap spacing that is recommended by the engine manufacturer is often with the purpose of increased compression pressures during the engine break-in period. Failure to properly space the piston ring end-gaps may lead to increased oil consumption and extended break-in time. Therefore, always be sure to position the pistons, rings and connecting rods as shown in the accompanying illustrations.

✳✳ WARNING

Always be sure to matchmark the connecting rods and caps prior to disassembly so that they may be reassembled with their original counterparts. If the caps are not installed on their original connecting rods, the assemblies will most likely need machining to avoid bearing, connecting rod and/or crankshaft damage.

PISTON, PISTON RING AND CONNECTING ROD POSITIONING INDEX

MANUFACTURER ENGINE	DESCRIPTION	FIGURE
Chrysler Corp.		
All engines	Connecting Rod And Cap Installation	1
	Piston Ring Identification Mark Locations	2
2.0L engines	Piston Ring Orientation	3
	Piston Ring End-Gap Spacing	5
	Piston Positioning	6
	Piston Positioning	8
	Piston Ring End-Gap Spacing	7
	Piston Positioning	9
2.4L engines	Piston Positioning	6
	Piston Ring End-Gap Spacing	5
	Piston Ring Orientation	3
2.5L engines	Piston Ring End-Gap Spacing	7
	Piston Ring Orientation	3
	Piston Positioning Mark Locations	10
	Piston Positioning	9
2.7L engines	Piston Positioning	11
	Piston Ring End-Gap Spacing	7
	Piston Ring Orientation	4
3.2L engines	Piston Positioning	9
	Piston Ring End-Gap Spacing	7
	Piston Ring Orientation	4
3.3L engines	Piston Positioning	12
	Piston Ring End-Gap Spacing	5
	Piston Ring Orientation	4
3.5L engines	Piston Positioning Mark Locations	13
	Piston Ring End-Gap Spacing	7
	Piston Ring Orientation	4
Ford Motor Co.		
All engines	Connecting Rod And Cap Installation	14
1.3L engines	Piston And Connecting Rod Assembly	17
	Piston Ring End-Gap Spacing	16
	Piston Ring Positioning	15
1.8L engines	Piston Ring End-Gap Spacing	18
	Piston Positioning	20
1.9L engines	Piston Ring End-Gap Spacing	19
	Piston Positioning	20
2.0L (VIN 3) engines	Piston Ring Positioning, End-Gap Spacing And Piston Positioning	21
2.0L (VIN A) engines	Piston And Connecting Rod Assembly	17
	Piston Ring End-Gap Spacing	16
	Piston Ring Positioning	15
2.5L (VIN B) engines	Piston And Connecting Rod Assembly	23
	Piston Ring End-Gap Spacing	22
2.5L (VIN L) engines	Piston Ring Positioning, End-Gap Spacing And Piston Positioning	21

7922AC01

PISTON, PISTON RING AND CONNECTING ROD POSITIONING INDEX

MANUFACTURER ENGINE	DESCRIPTION	FIGURE
Ford Motor Co. (cont.)		
3.0L (VIN N, S, U and 1) engines	Piston And Connecting Rod Positioning	25
	Piston Ring End-Gap Spacing	24
3.0L (VIN Y) SHO engines	Piston And Connecting Rod Positioning	27
	Piston Positioning Mark Locations	28
	Piston Ring Positioning And End-Gap Spacing	26
3.2L (VIN P) SHO engines	Piston And Connecting Rod Positioning	27
	Piston Positioning Mark Locations	28
	Piston Ring Positioning And End-Gap Spacing	26
3.4L engines	Piston And Connecting Rod Positioning	31
	Piston Ring End-Gap Spacing	30
	Piston Ring Positioning	29
3.8L engines	Piston And Connecting Rod Positioning	25
	Piston Ring End-Gap Spacing	24
4.6L (VIN V) engines	Piston Ring End-Gap Spacing And Piston Positioning	33
	Piston Ring Positioning	32
4.6L (VIN W and 9) engines	Piston Ring End-Gap Spacing And Piston Positioning	35
	Piston Ring Positioning	34
5.0L engines	Piston Assembly Positioning	37
	Piston Ring End-Gap Spacing	36
General Motors		
All engines	Connecting Rod And Cap Installation	38
2.2L engines	Piston Positioning	43
	Piston Ring End-Gap Spacing	42
	Piston Ring Positioning	41
2.3L engines	Piston And Connecting Rod Assembly Positioning	40
	Piston Ring End-Gap Spacing	39
2.4L engines	Piston And Connecting Rod Assembly Positioning	40
	Piston Ring End-Gap Spacing	39
3.0L engines	Piston Ring End-Gap Spacing	44
3.1L engines	Piston Positioning	43
	Piston Ring End-Gap Spacing	42
	Piston Ring Positioning	41
3.4L engines	Piston Positioning	43
	Piston Ring End-Gap Spacing	42
	Piston Ring Positioning	41
3.8L engines	Piston Positioning	43
	Piston Ring End-Gap Spacing	42
	Piston Ring Positioning	41
4.0L engines	Piston And Connecting Rod Assembly Positioning	46
	Piston Ring And Ring End-Gap Positioning	45

7922AC02

PISTON, PISTON RING AND CONNECTING ROD POSITIONING INDEX

MANUFACTURER

ENGINE	DESCRIPTION	FIGURE
General Motors (cont.)		
4.3L engines	Piston And Connecting Rod Assembly Positioning	49
	Piston Ring End-Gap Spacing	48
	Piston Ring Positioning	47
4.6L engines	Piston And Connecting Rod Assembly Positioning	46
	Piston Ring And Ring End-Gap Positioning	45
4.9L engines	Piston Ring End-Gap Spacing	51
5.7L engines	Piston Ring End-Gap Spacing	48
	Piston Ring Positioning	47
	Piston And Connecting Rod Assembly Positioning	49
	Piston And Connecting Rod Assembly Positioning	50
Geo/Chevrolet		
All engines	Piston Ring Positioning	52
	Piston Ring End-Gap Spacing	53
	Piston And Connecting Rod Assembly Positioning	54
Saturn		
1.9L engines	Piston Ring End-Gap Spacing	55
	Upper Ring Identification	56
	Piston And Connecting Rod Assembly Positioning Mark Locations	57

7922AC03

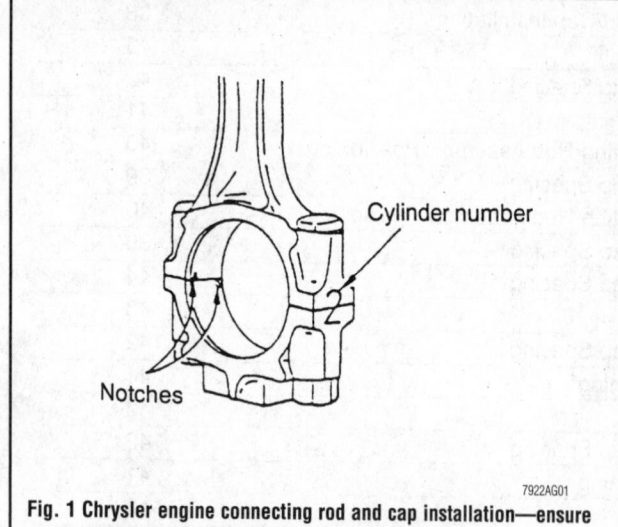

Fig. 1 Chrysler engine connecting rod and cap installation—ensure to matchmark the cap and rod prior to disassembly

7922AG01

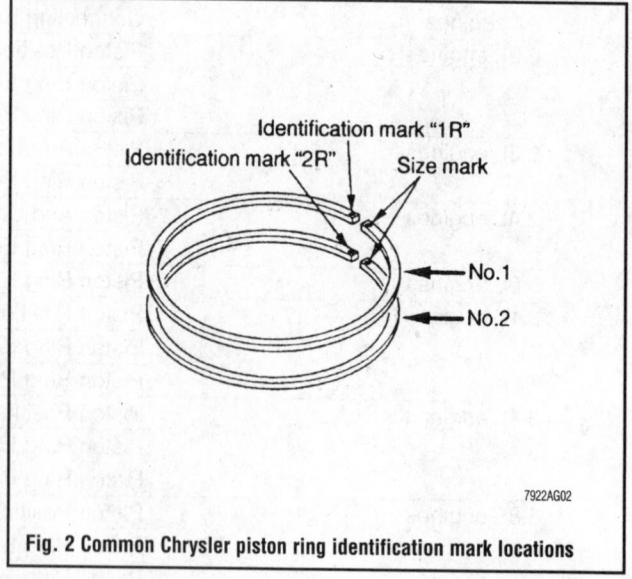

Fig. 2 Common Chrysler piston ring identification mark locations

7922AG02

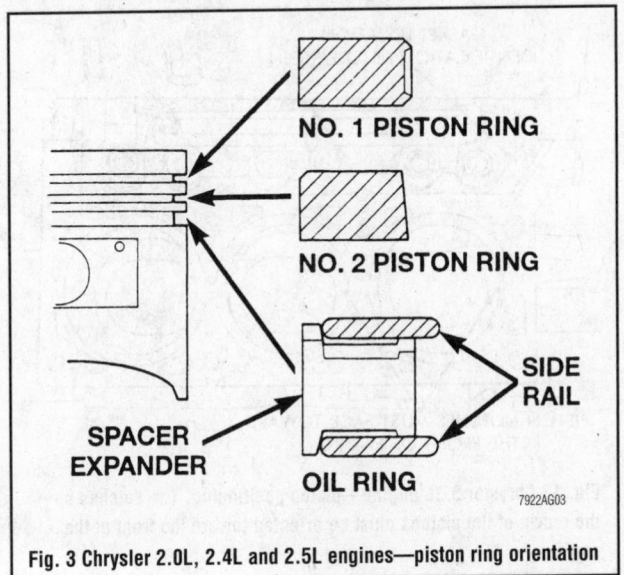

Fig. 3 Chrysler 2.0L, 2.4L and 2.5L engines—piston ring orientation

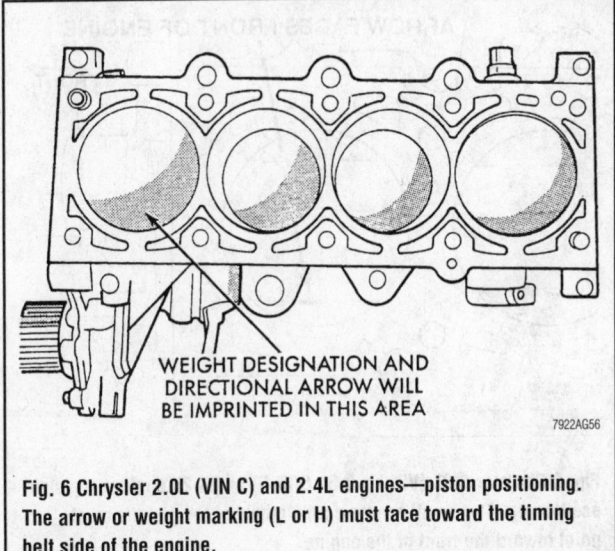

Fig. 6 Chrysler 2.0L (VIN C) and 2.4L engines—piston positioning. The arrow or weight marking (L or H) must face toward the timing belt side of the engine.

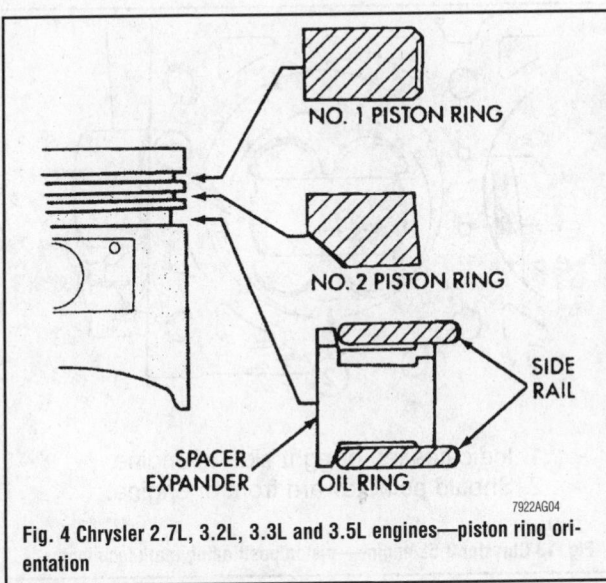

Fig. 4 Chrysler 2.7L, 3.2L, 3.3L and 3.5L engines—piston ring orientation

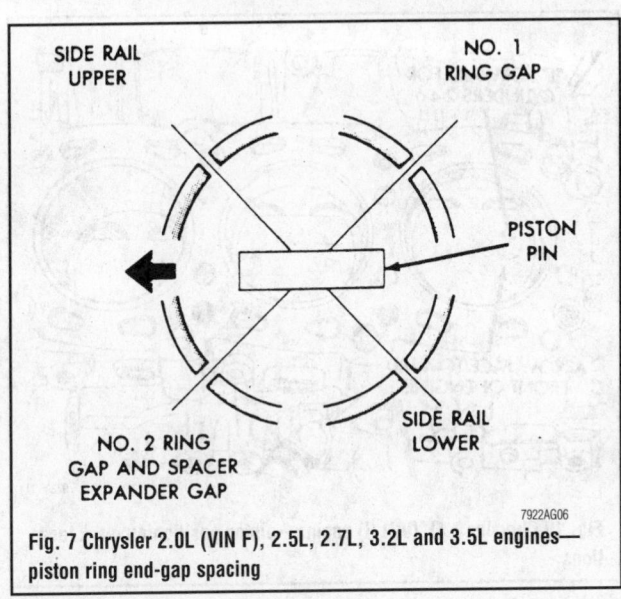

Fig. 7 Chrysler 2.0L (VIN F), 2.5L, 2.7L, 3.2L and 3.5L engines—piston ring end-gap spacing

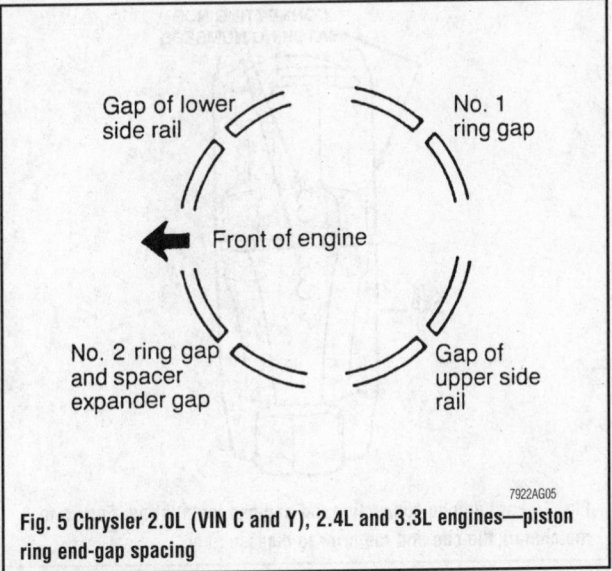

Fig. 5 Chrysler 2.0L (VIN C and Y), 2.4L and 3.3L engines—piston ring end-gap spacing

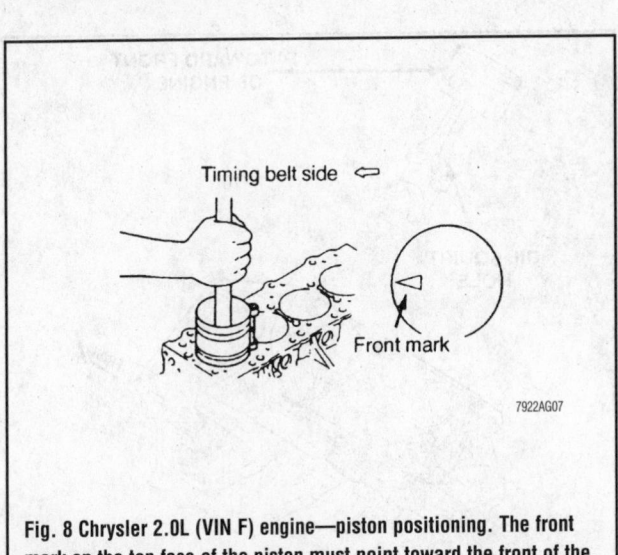

Fig. 8 Chrysler 2.0L (VIN F) engine—piston positioning. The front mark on the top face of the piston must point toward the front of the engine

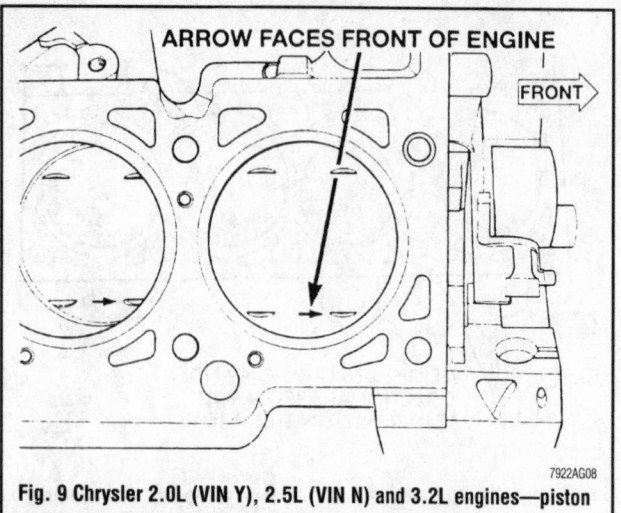

Fig. 9 Chrysler 2.0L (VIN Y), 2.5L (VIN N) and 3.2L engines—piston positioning. The small arrows on the crown of the pistons must point toward the front of the engine.

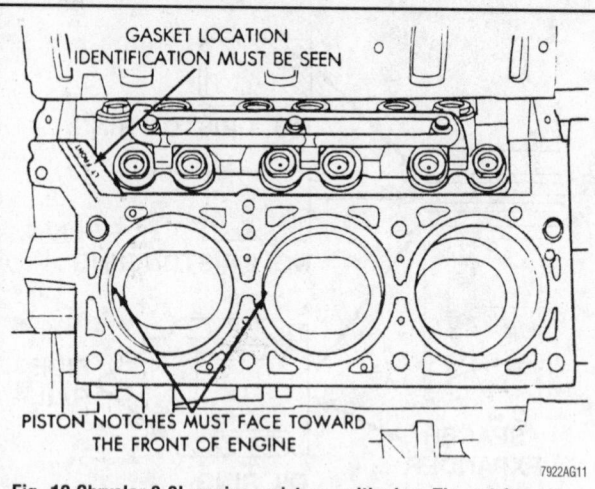

Fig. 12 Chrysler 3.3L engine—piston positioning. The notches on the crown of the pistons must be oriented toward the front of the engine.

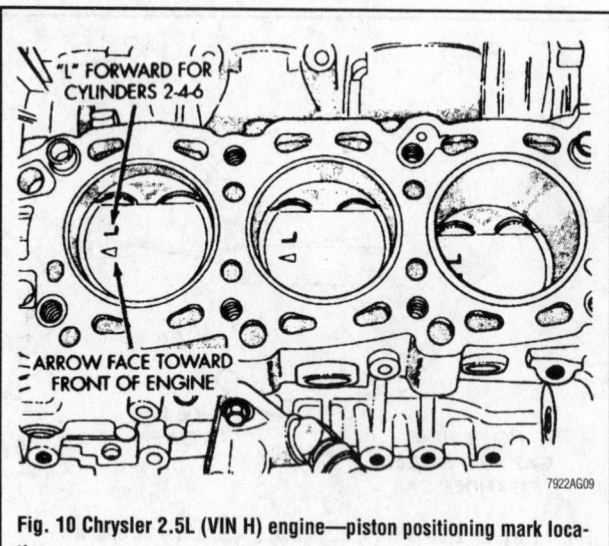

Fig. 10 Chrysler 2.5L (VIN H) engine—piston positioning mark locations

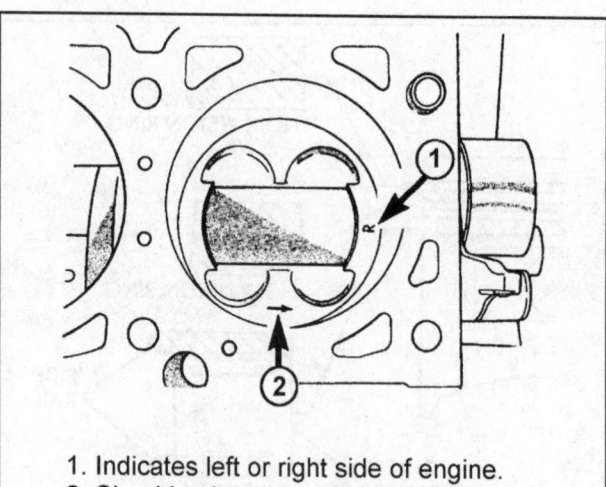

1. Indicates left or right side of engine.
2. Should point toward front of engine.

Fig. 13 Chrysler 3.5L engine—piston positioning mark locations

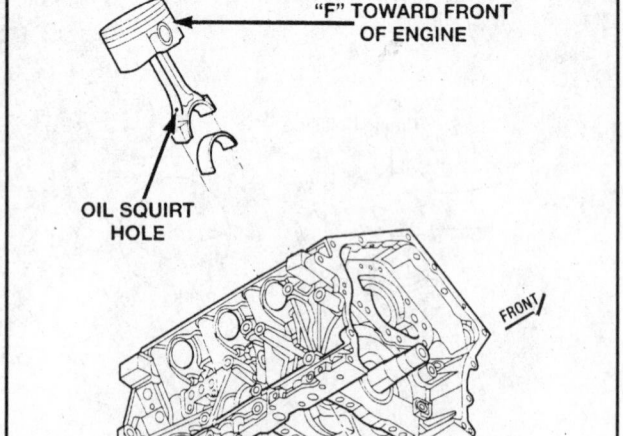

Fig. 11 Chrysler 2.7L engine—piston positioning

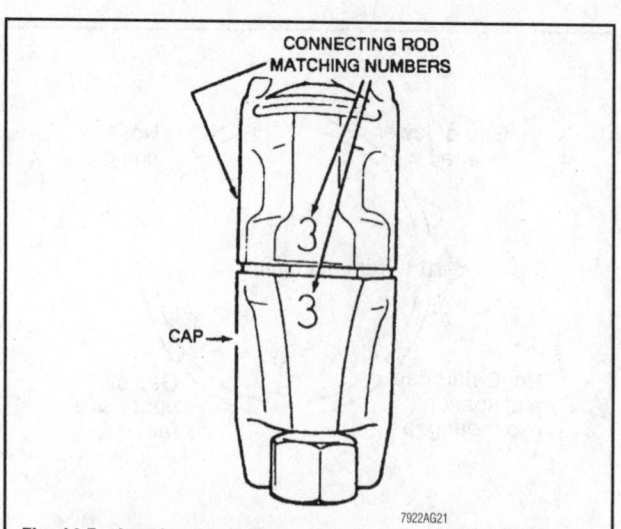

Fig. 14 Ford engine connecting rod and cap installation. Ensure to matchmark the cap and rod prior to disassembly.

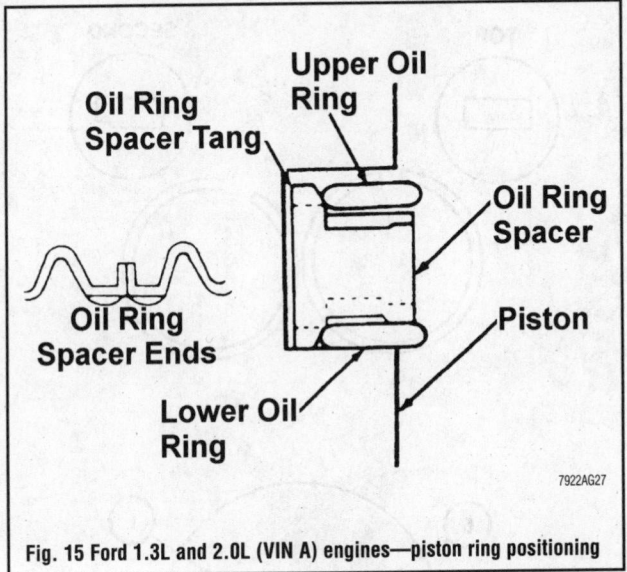

Fig. 15 Ford 1.3L and 2.0L (VIN A) engines—piston ring positioning

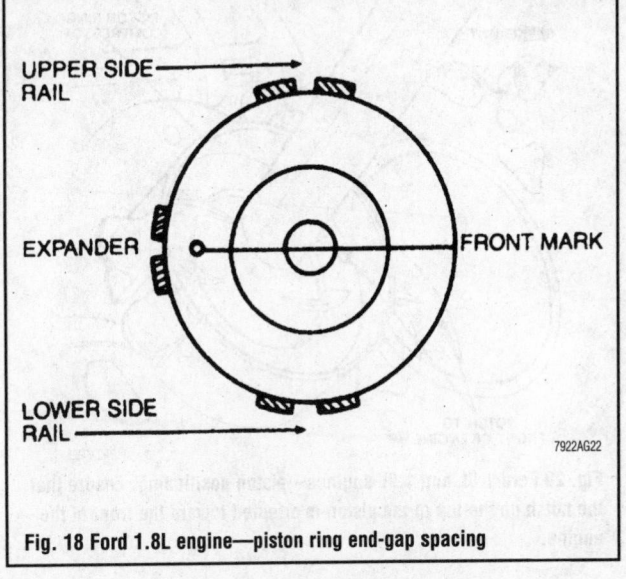

Fig. 18 Ford 1.8L engine—piston ring end-gap spacing

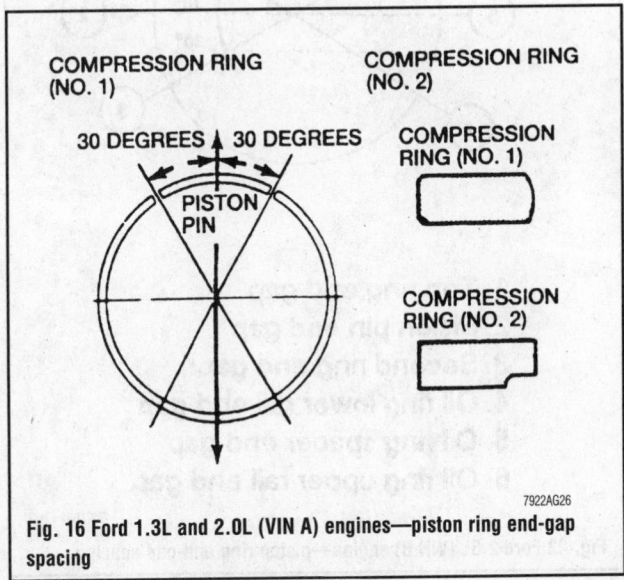

Fig. 16 Ford 1.3L and 2.0L (VIN A) engines—piston ring end-gap spacing

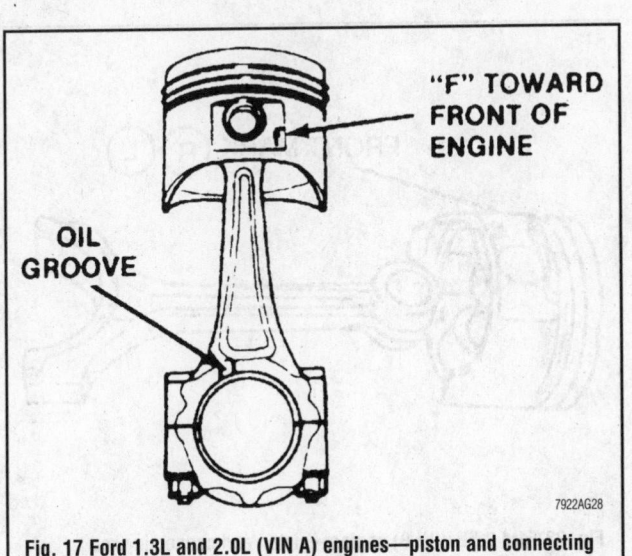

Fig. 17 Ford 1.3L and 2.0L (VIN A) engines—piston and connecting rod assembly

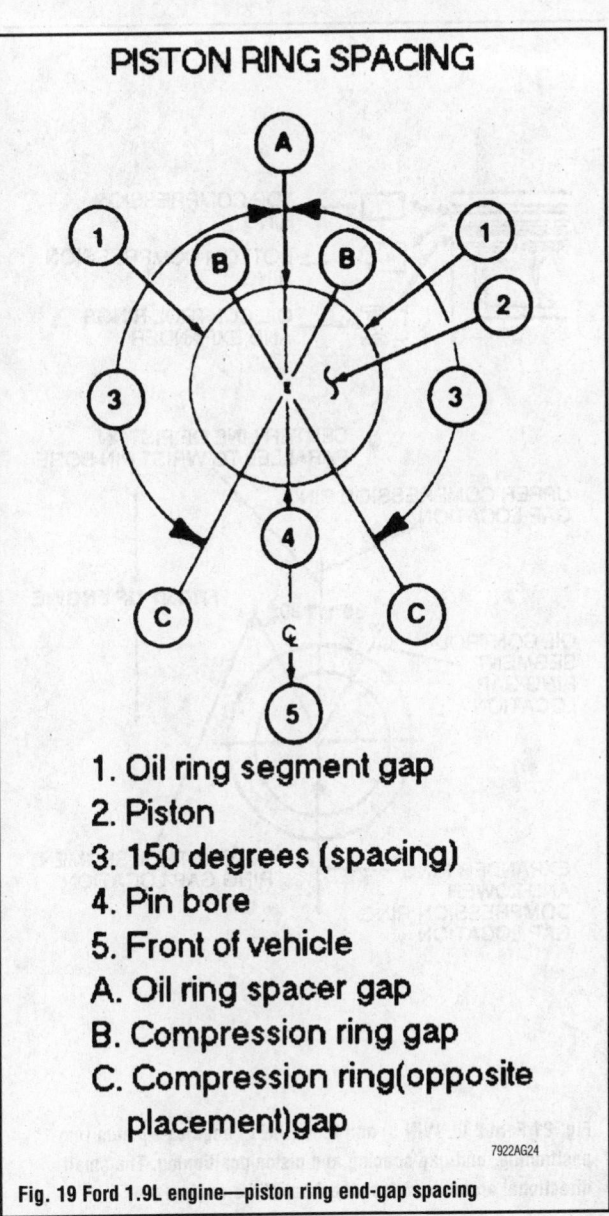

PISTON RING SPACING

1. Oil ring segment gap
2. Piston
3. 150 degrees (spacing)
4. Pin bore
5. Front of vehicle
A. Oil ring spacer gap
B. Compression ring gap
C. Compression ring(opposite placement)gap

Fig. 19 Ford 1.9L engine—piston ring end-gap spacing

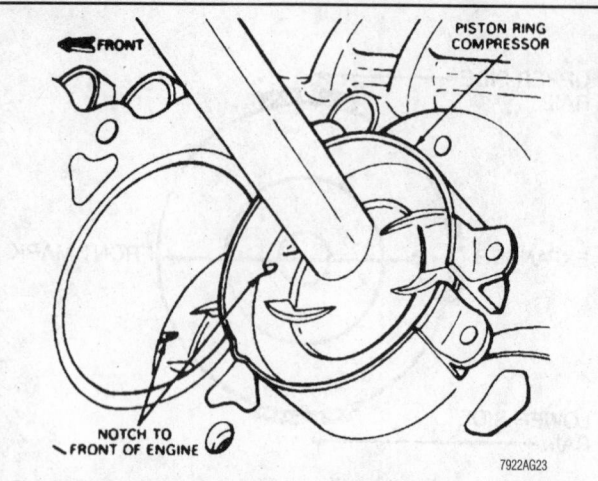

Fig. 20 Ford 1.8L and 1.9L engines—piston positioning. Ensure that the notch on the top of the piston is oriented toward the front of the engine.

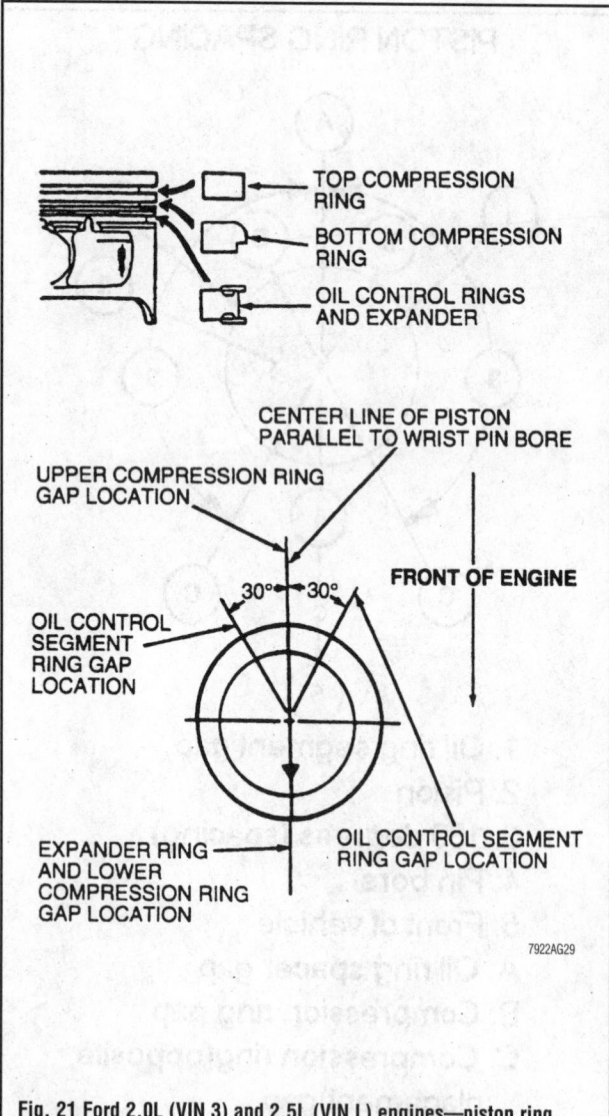

Fig. 21 Ford 2.0L (VIN 3) and 2.5L (VIN L) engines—piston ring positioning, end-gap spacing and piston positioning. The small directional arrow must face the front of the engine.

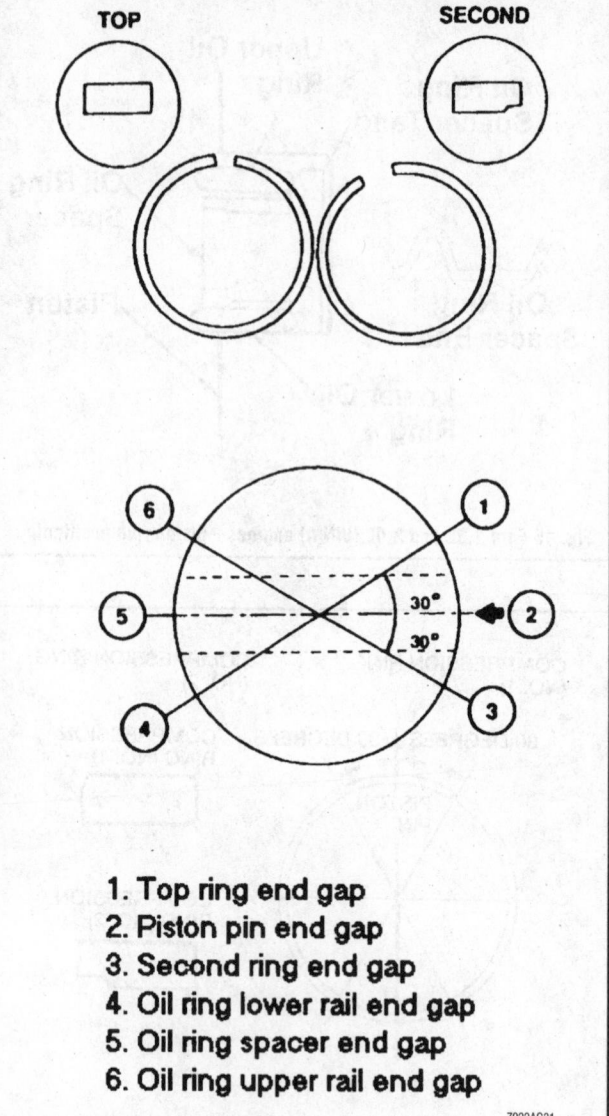

1. Top ring end gap
2. Piston pin end gap
3. Second ring end gap
4. Oil ring lower rail end gap
5. Oil ring spacer end gap
6. Oil ring upper rail end gap

Fig. 22 Ford 2.5L (VIN B) engine—piston ring end-gap spacing

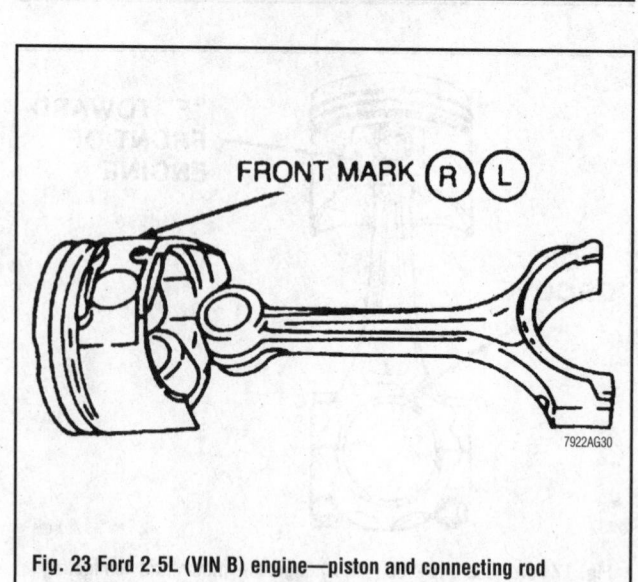

Fig. 23 Ford 2.5L (VIN B) engine—piston and connecting rod assembly

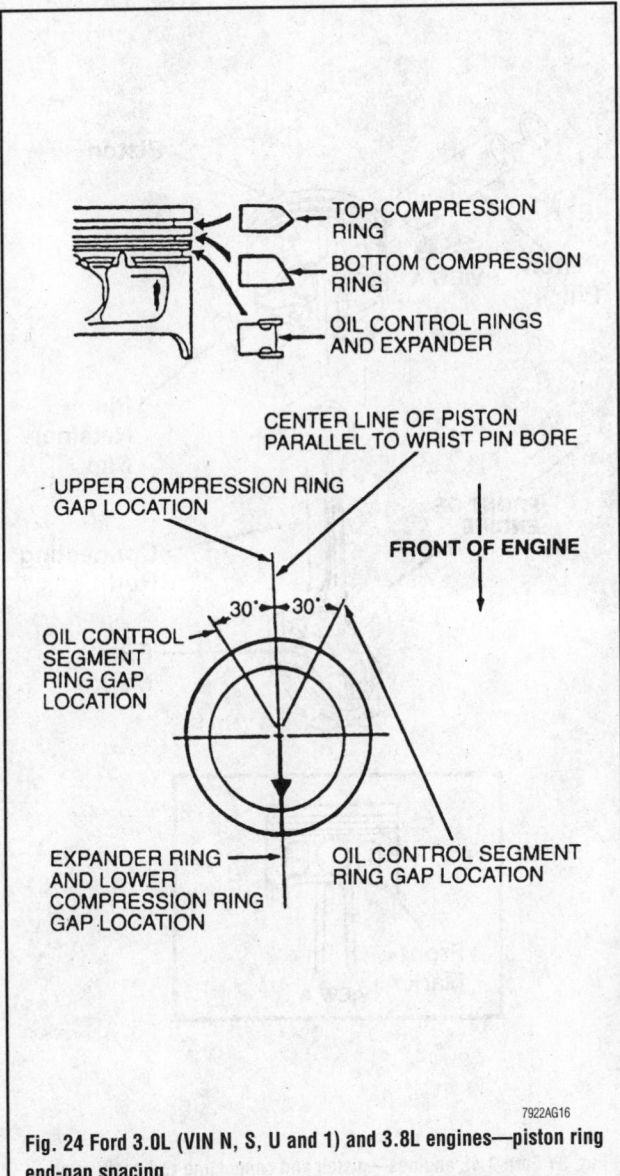

Fig. 24 Ford 3.0L (VIN N, S, U and 1) and 3.8L engines—piston ring end-gap spacing

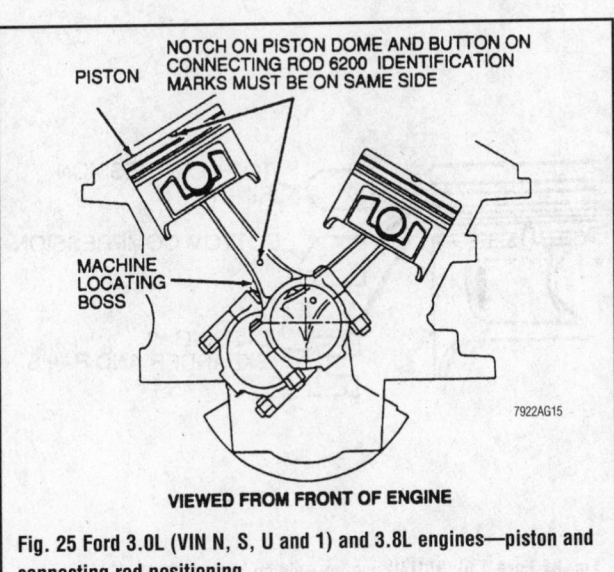

Fig. 25 Ford 3.0L (VIN N, S, U and 1) and 3.8L engines—piston and connecting rod positioning

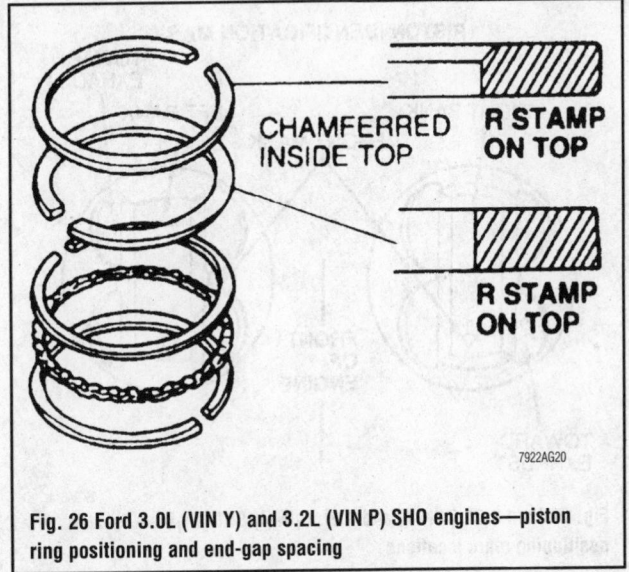

Fig. 26 Ford 3.0L (VIN Y) and 3.2L (VIN P) SHO engines—piston ring positioning and end-gap spacing

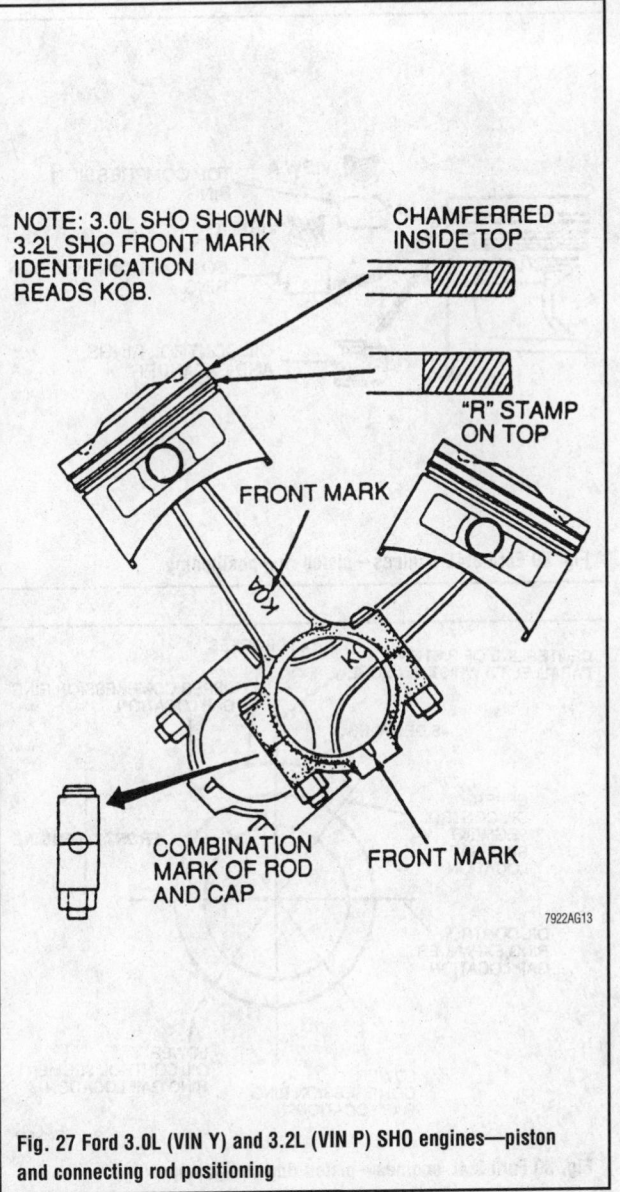

Fig. 27 Ford 3.0L (VIN Y) and 3.2L (VIN P) SHO engines—piston and connecting rod positioning

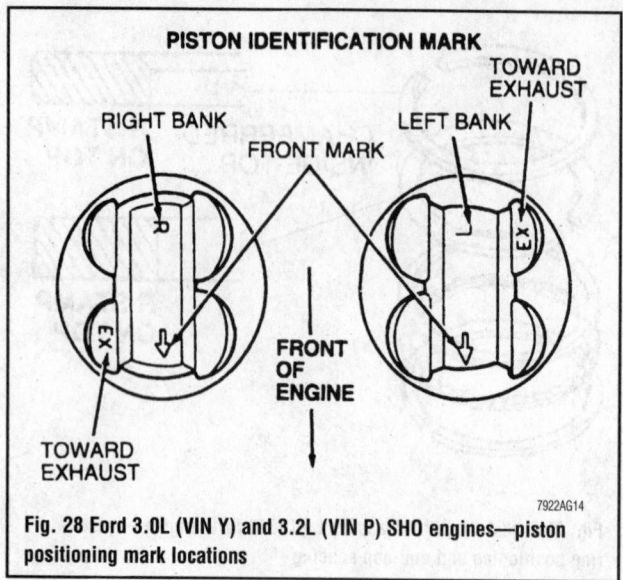

PISTON IDENTIFICATION MARK

Fig. 28 Ford 3.0L (VIN Y) and 3.2L (VIN P) SHO engines—piston positioning mark locations

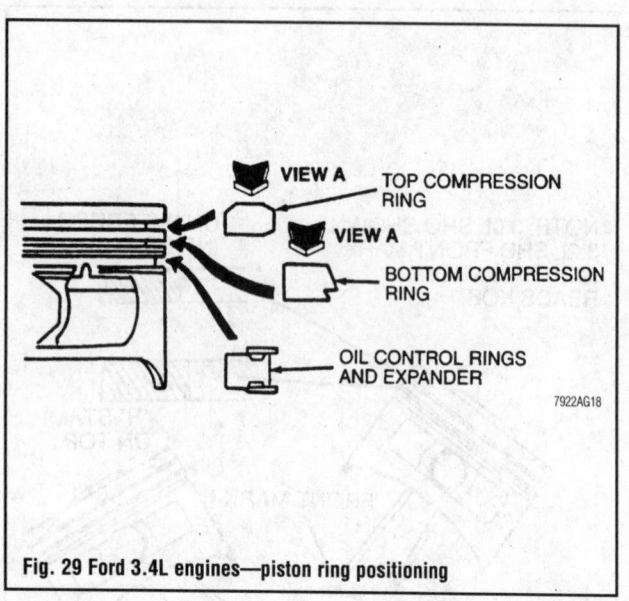

Fig. 29 Ford 3.4L engines—piston ring positioning

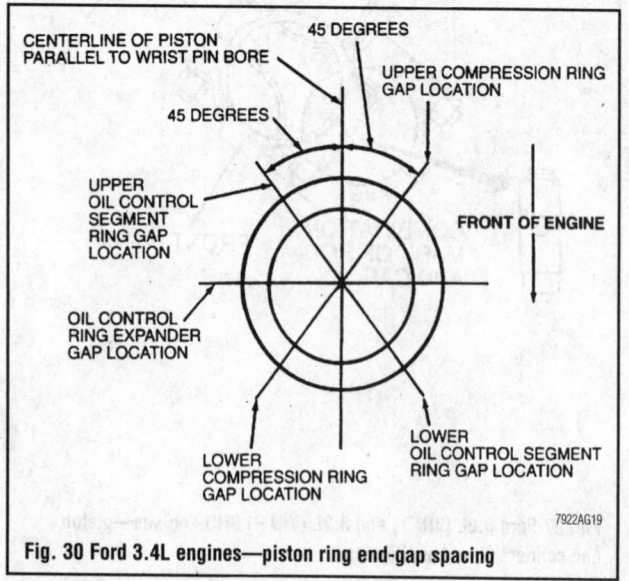

Fig. 30 Ford 3.4L engines—piston ring end-gap spacing

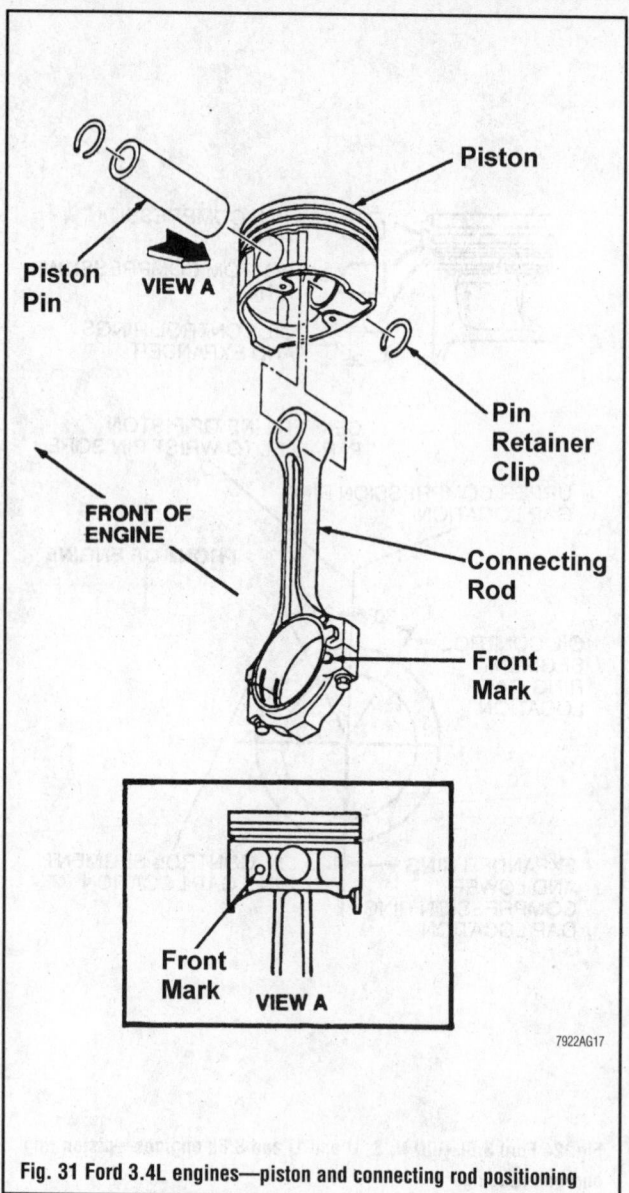

Fig. 31 Ford 3.4L engines—piston and connecting rod positioning

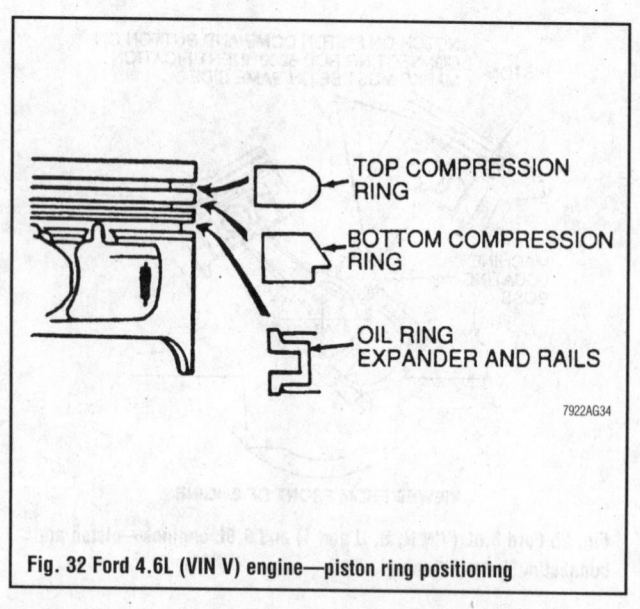

Fig. 32 Ford 4.6L (VIN V) engine—piston ring positioning

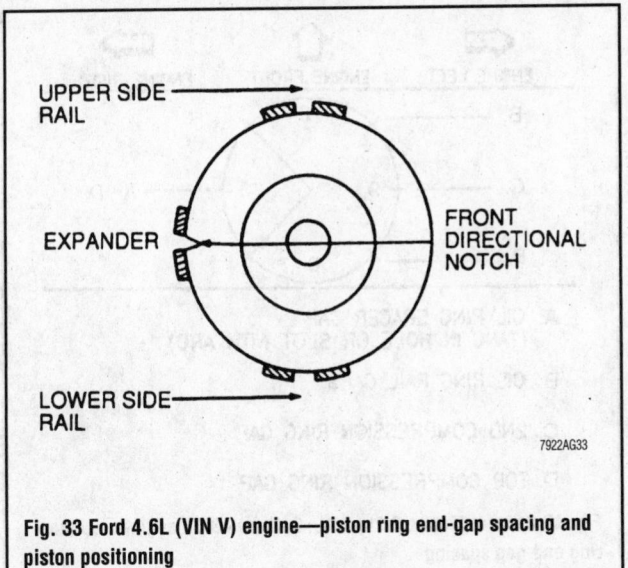

Fig. 33 Ford 4.6L (VIN V) engine—piston ring end-gap spacing and piston positioning

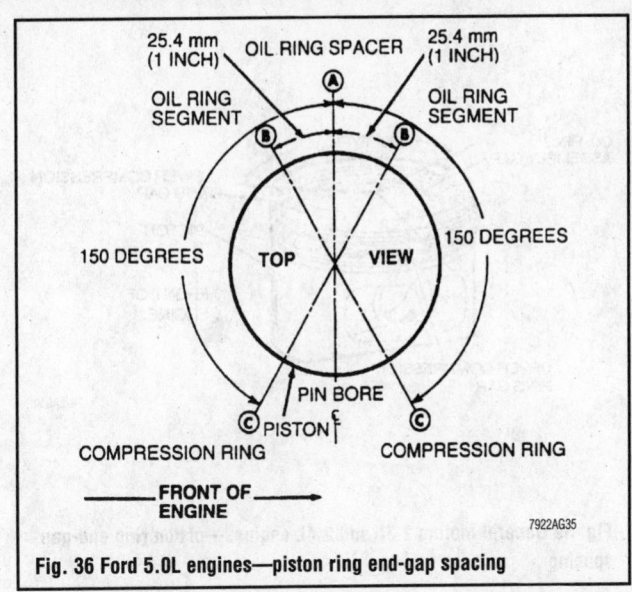

Fig. 36 Ford 5.0L engines—piston ring end-gap spacing

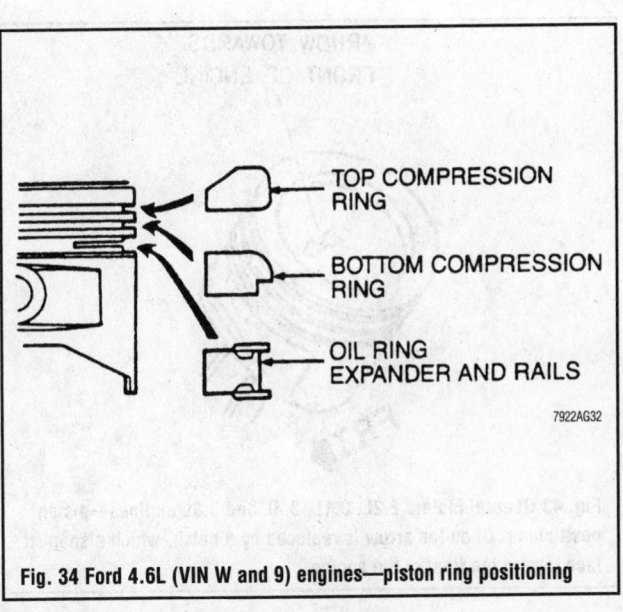

Fig. 34 Ford 4.6L (VIN W and 9) engines—piston ring positioning

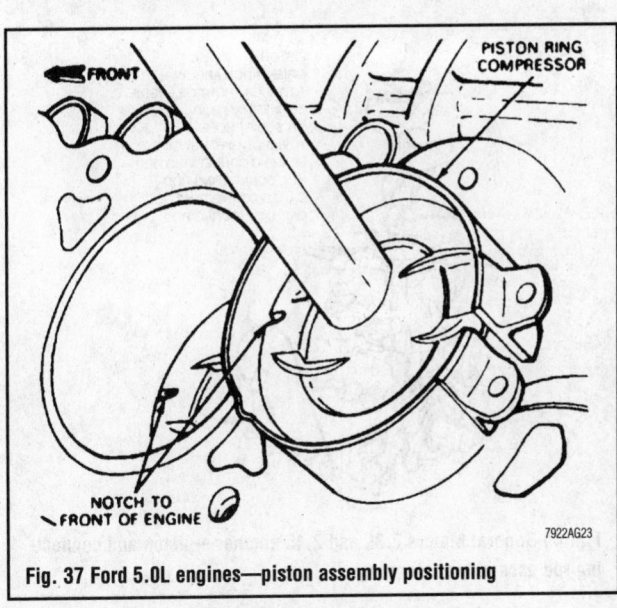

Fig. 37 Ford 5.0L engines—piston assembly positioning

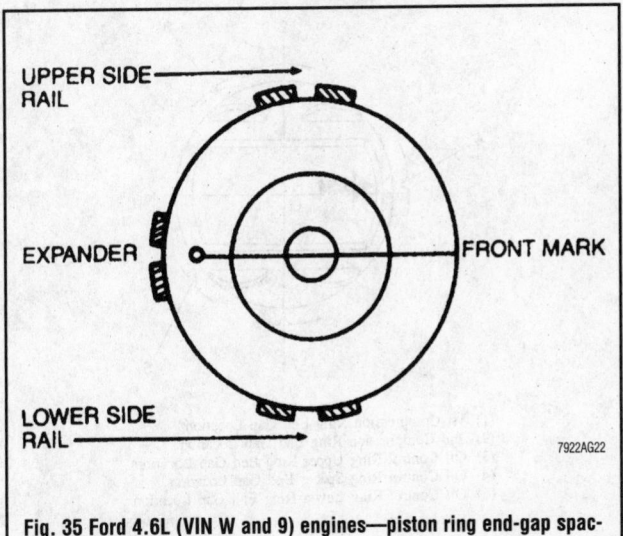

Fig. 35 Ford 4.6L (VIN W and 9) engines—piston ring end-gap spacing and piston positioning

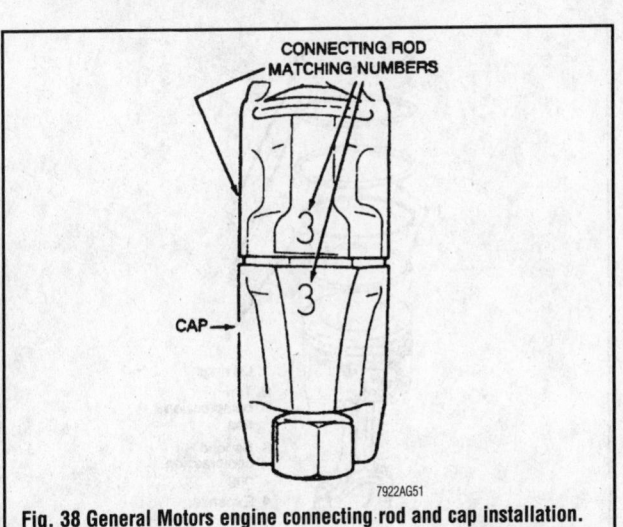

Fig. 38 General Motors engine connecting rod and cap installation. Be sure to matchmark the cap and rod prior to disassembly, as shown.

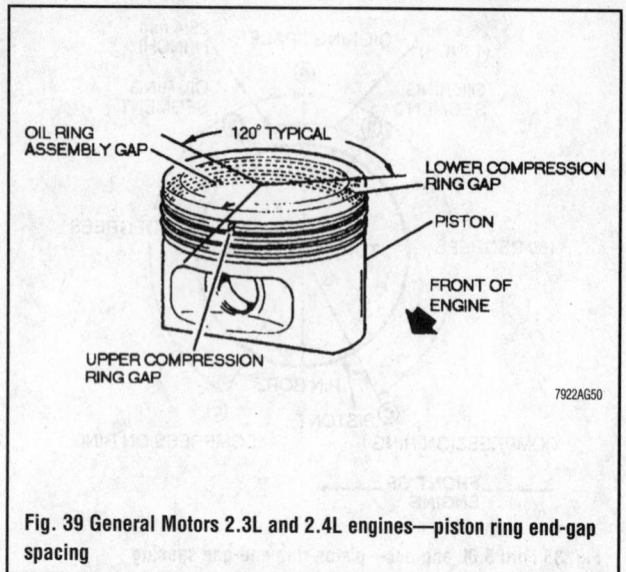

Fig. 39 General Motors 2.3L and 2.4L engines—piston ring end-gap spacing

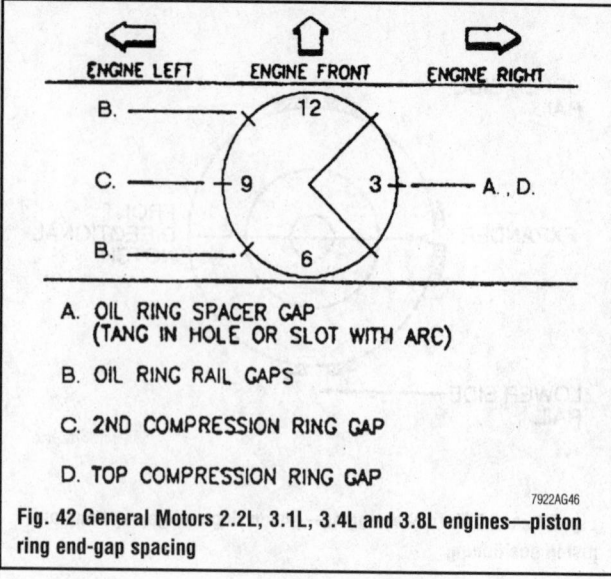

A. OIL RING SPACER GAP
   (TANG IN HOLE OR SLOT WITH ARC)

B. OIL RING RAIL GAPS

C. 2ND COMPRESSION RING GAP

D. TOP COMPRESSION RING GAP

Fig. 42 General Motors 2.2L, 3.1L, 3.4L and 3.8L engines—piston ring end-gap spacing

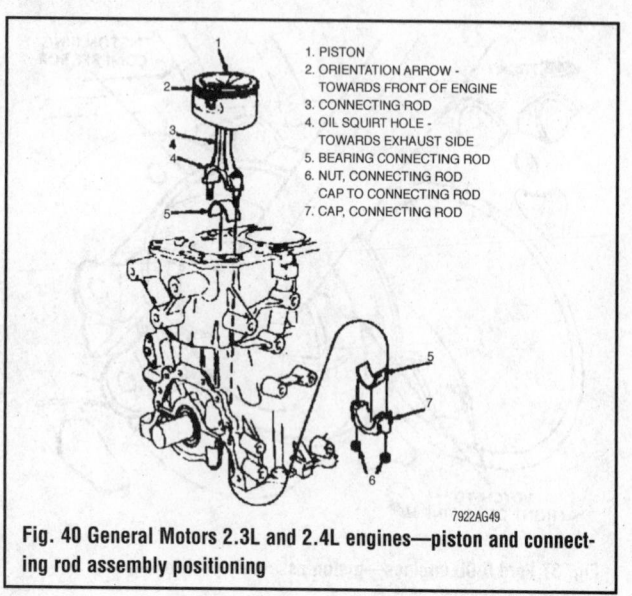

1. PISTON
2. ORIENTATION ARROW - TOWARDS FRONT OF ENGINE
3. CONNECTING ROD
4. OIL SQUIRT HOLE - TOWARDS EXHAUST SIDE
5. BEARING CONNECTING ROD
6. NUT, CONNECTING ROD CAP TO CONNECTING ROD
7. CAP, CONNECTING ROD

Fig. 40 General Motors 2.3L and 2.4L engines—piston and connecting rod assembly positioning

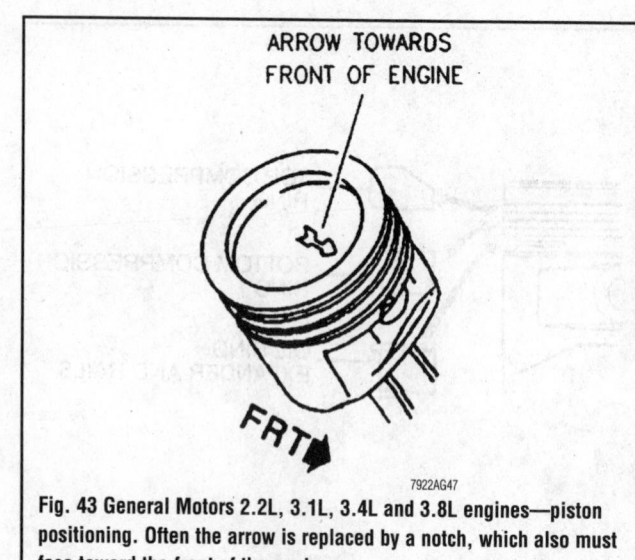

ARROW TOWARDS FRONT OF ENGINE

FRT

Fig. 43 General Motors 2.2L, 3.1L, 3.4L and 3.8L engines—piston positioning. Often the arrow is replaced by a notch, which also must face toward the front of the engine.

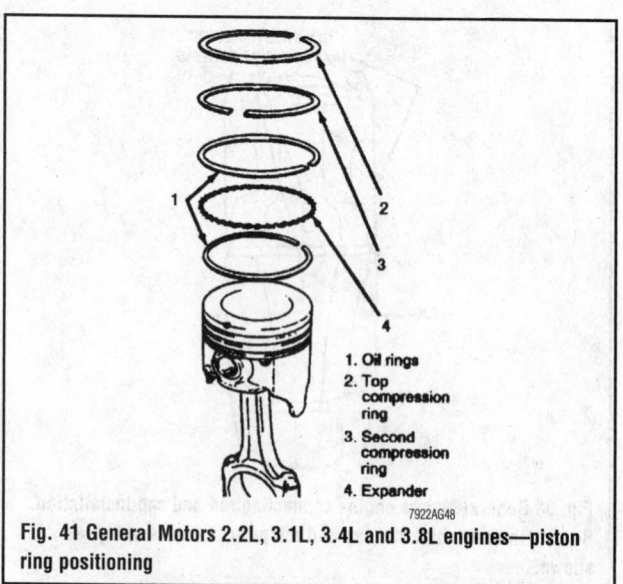

1. Oil rings
2. Top compression ring
3. Second compression ring
4. Expander

Fig. 41 General Motors 2.2L, 3.1L, 3.4L and 3.8L engines—piston ring positioning

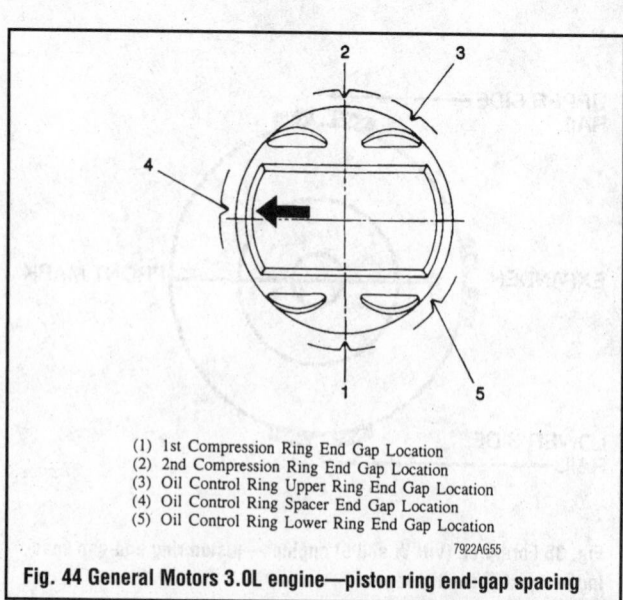

(1) 1st Compression Ring End Gap Location
(2) 2nd Compression Ring End Gap Location
(3) Oil Control Ring Upper Ring End Gap Location
(4) Oil Control Ring Spacer End Gap Location
(5) Oil Control Ring Lower Ring End Gap Location

Fig. 44 General Motors 3.0L engine—piston ring end-gap spacing

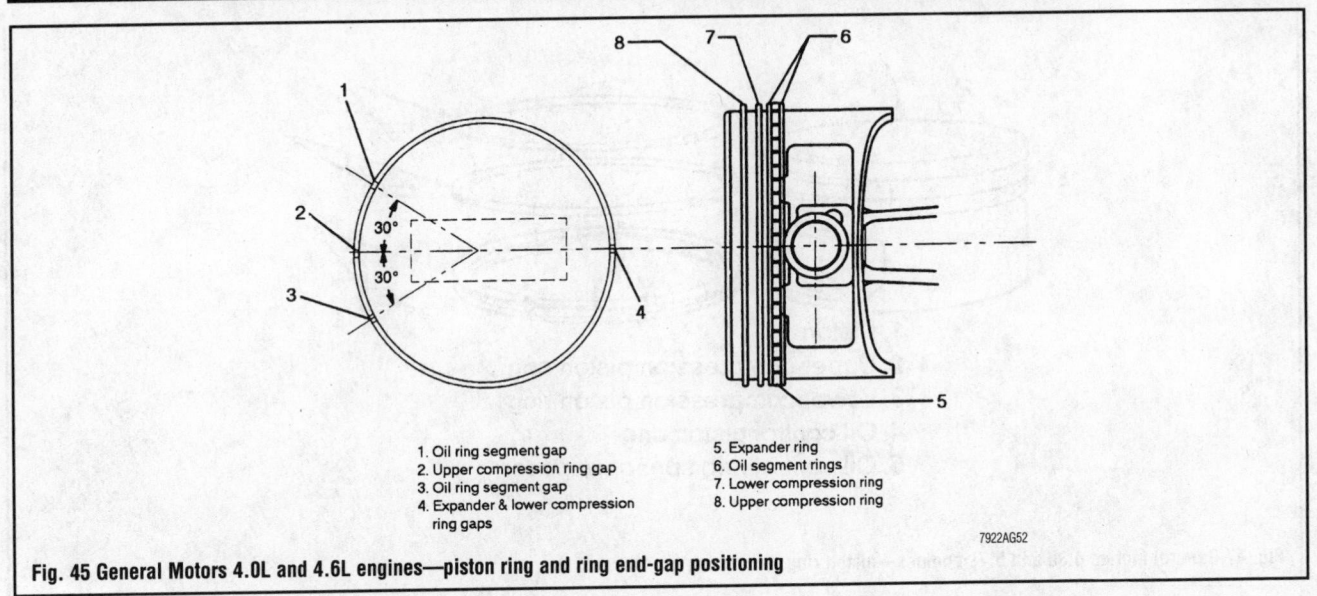

1. Oil ring segment gap
2. Upper compression ring gap
3. Oil ring segment gap
4. Expander & lower compression ring gaps

5. Expander ring
6. Oil segment rings
7. Lower compression ring
8. Upper compression ring

7922AG52

Fig. 45 General Motors 4.0L and 4.6L engines—piston ring and ring end-gap positioning

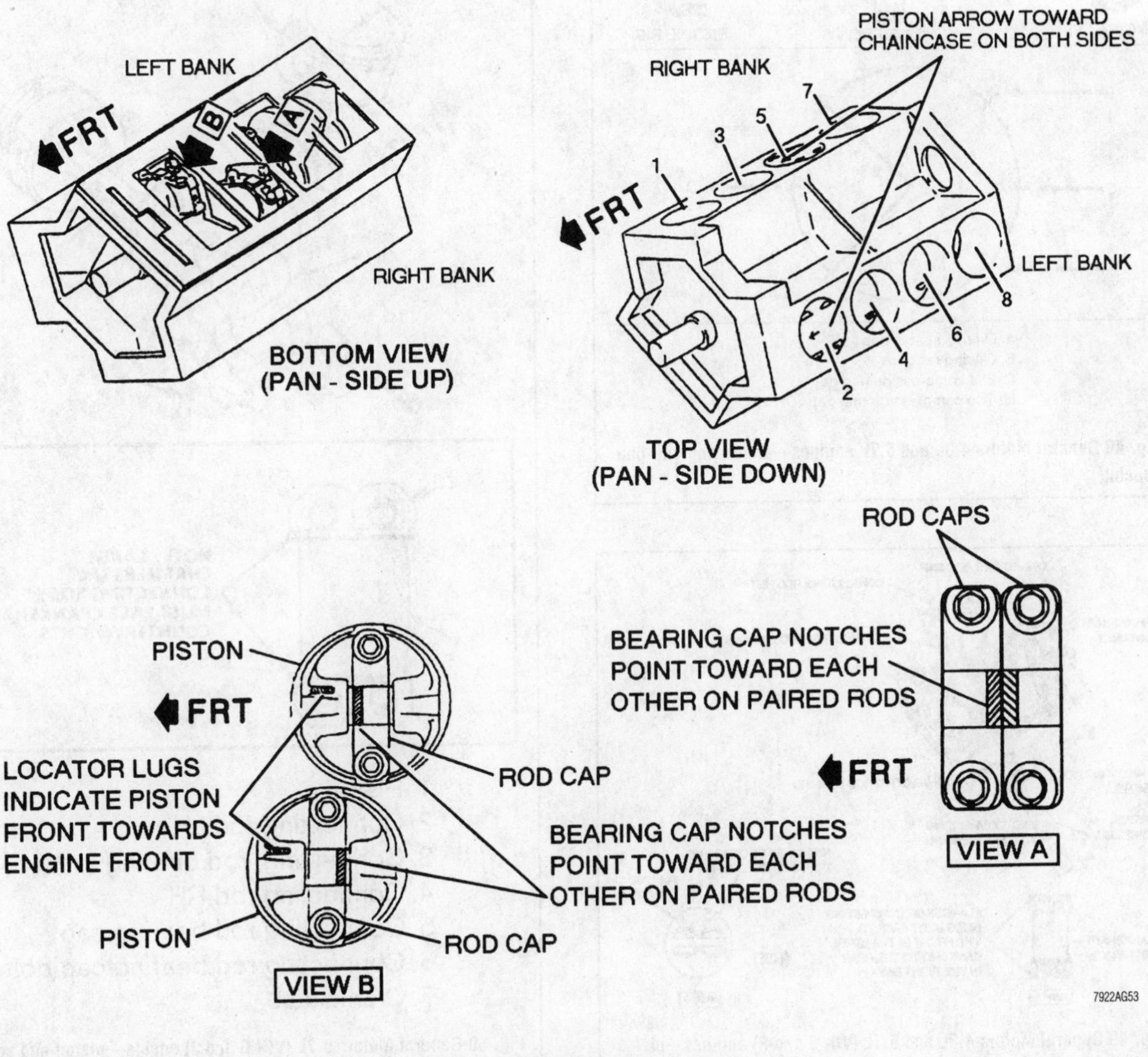

LEFT BANK

FRT

B   A

RIGHT BANK

**BOTTOM VIEW
(PAN - SIDE UP)**

PISTON ARROW TOWARD
CHAINCASE ON BOTH SIDES

RIGHT BANK

FRT

1 3 5 7

2 4 6 8

LEFT BANK

**TOP VIEW
(PAN - SIDE DOWN)**

PISTON

FRT

LOCATOR LUGS
INDICATE PISTON
FRONT TOWARDS
ENGINE FRONT

ROD CAP

PISTON

ROD CAP

VIEW B

ROD CAPS

BEARING CAP NOTCHES
POINT TOWARD EACH
OTHER ON PAIRED RODS

BEARING CAP NOTCHES
POINT TOWARD EACH
OTHER ON PAIRED RODS

FRT

VIEW A

7922AG53

Fig. 46 General Motors 4.0L and 4.6L engines—piston and connecting rod assembly positioning

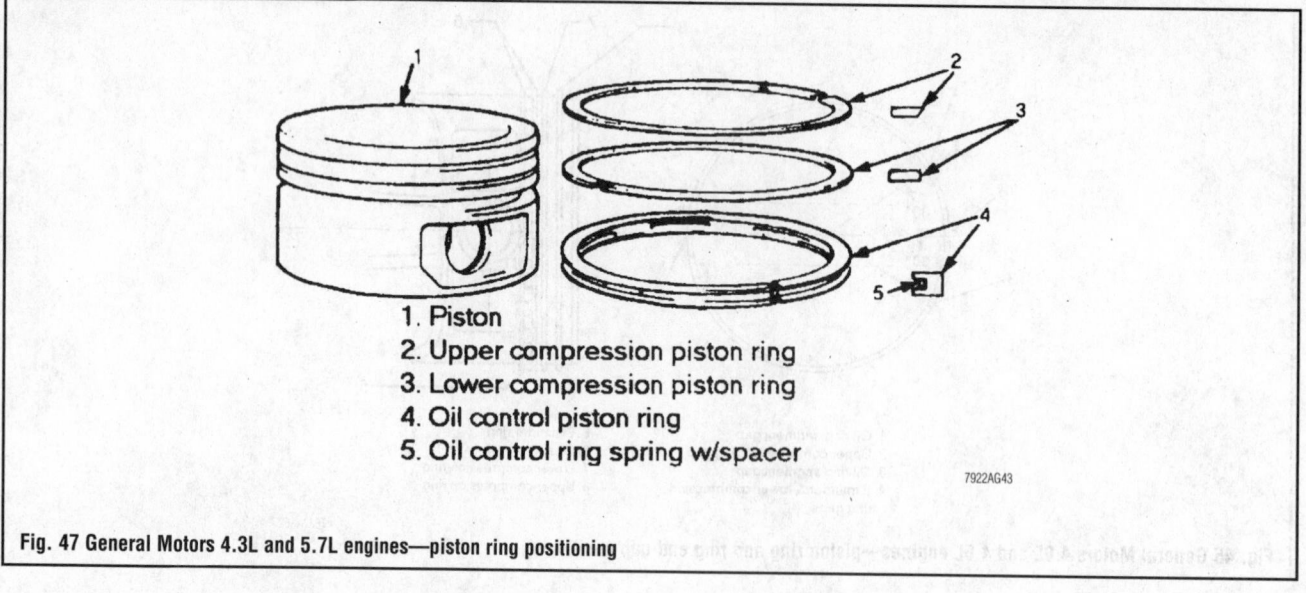

1. Piston
2. Upper compression piston ring
3. Lower compression piston ring
4. Oil control piston ring
5. Oil control ring spring w/spacer

7922AG43

Fig. 47 General Motors 4.3L and 5.7L engines—piston ring positioning

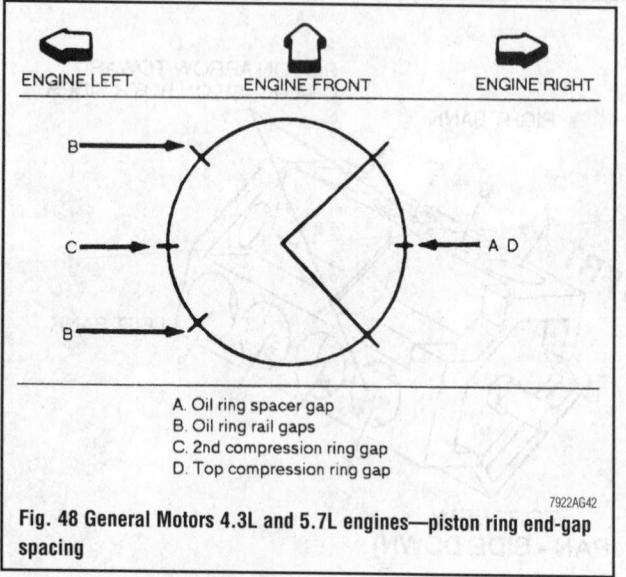

ENGINE LEFT   ENGINE FRONT   ENGINE RIGHT

A. Oil ring spacer gap
B. Oil ring rail gaps
C. 2nd compression ring gap
D. Top compression ring gap

7922AG42

Fig. 48 General Motors 4.3L and 5.7L engines—piston ring end-gap spacing

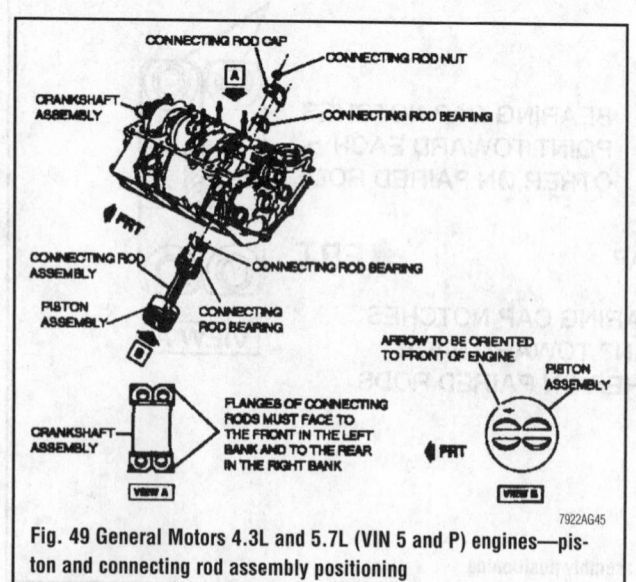

7922AG45

Fig. 49 General Motors 4.3L and 5.7L (VIN 5 and P) engines—piston and connecting rod assembly positioning

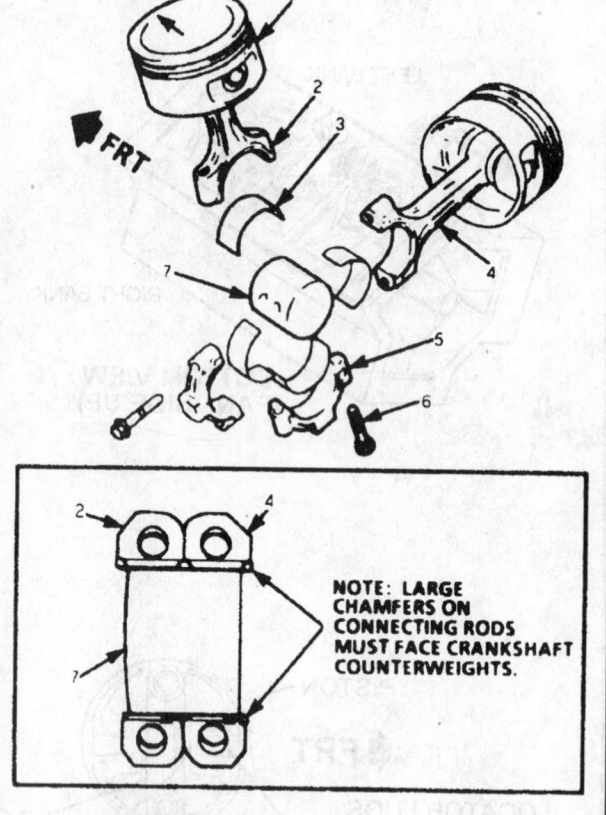

NOTE: LARGE CHAMFERS ON CONNECTING RODS MUST FACE CRANKSHAFT COUNTERWEIGHTS.

1. Piston
2. Connecting rod LH
3. Connecting rod bearing
4. Connecting rod RH
5. Connecting rod bearing cap
6. Connecting rod bearing cap bolt
7. Crankshaft

7922AG44

Fig. 50 General Motors 5.7L (VIN G and J) engine—piston and connecting rod assembly positioning

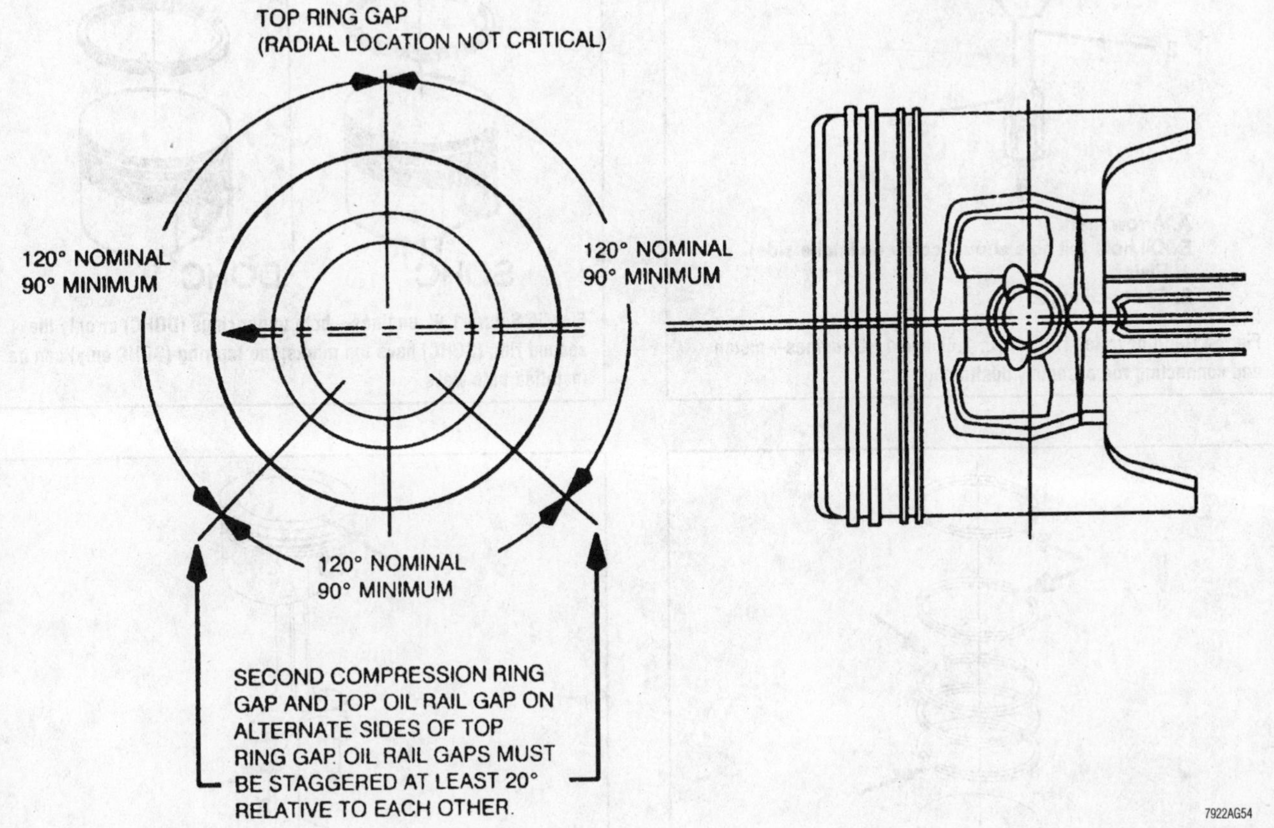

TOP RING GAP
(RADIAL LOCATION NOT CRITICAL)

120° NOMINAL
90° MINIMUM

120° NOMINAL
90° MINIMUM

120° NOMINAL
90° MINIMUM

SECOND COMPRESSION RING
GAP AND TOP OIL RAIL GAP ON
ALTERNATE SIDES OF TOP
RING GAP. OIL RAIL GAPS MUST
BE STAGGERED AT LEAST 20°
RELATIVE TO EACH OTHER.

7922AG54

Fig. 51 General Motors 4.9L engines—piston ring end-gap spacing

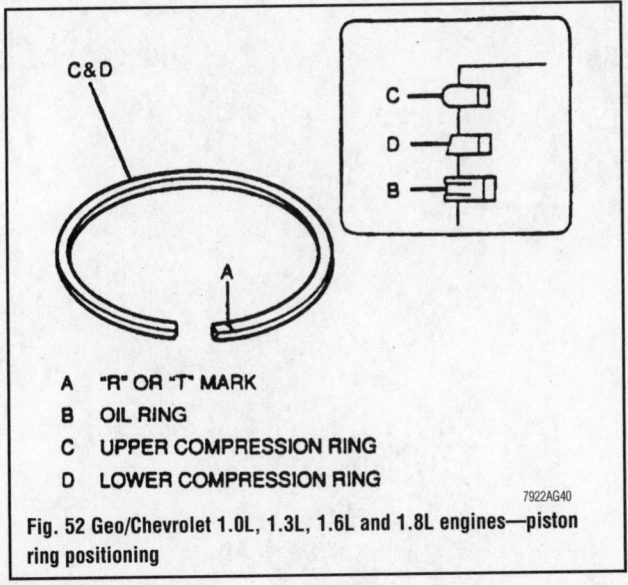

C & D

C
D
B

A   "R" OR "T" MARK
B   OIL RING
C   UPPER COMPRESSION RING
D   LOWER COMPRESSION RING

7922AG40

Fig. 52 Geo/Chevrolet 1.0L, 1.3L, 1.6L and 1.8L engines—piston ring positioning

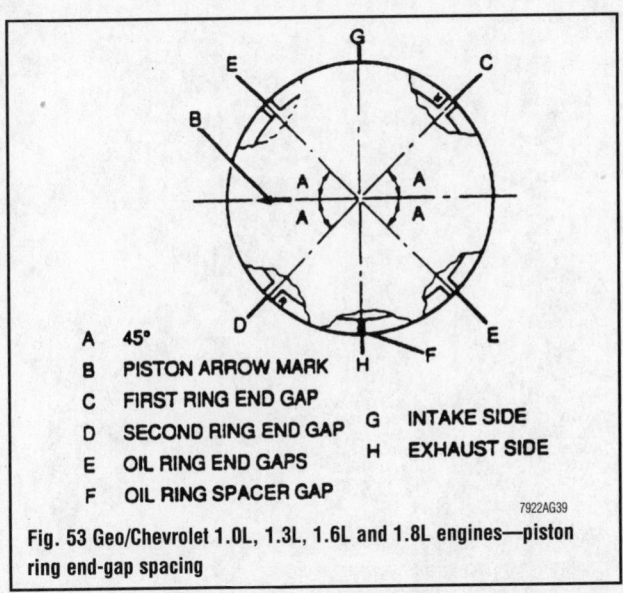

E   G   C
B

A   A
A   A

D   E
H   F

A   45°
B   PISTON ARROW MARK
C   FIRST RING END GAP
D   SECOND RING END GAP
E   OIL RING END GAPS
F   OIL RING SPACER GAP

G   INTAKE SIDE
H   EXHAUST SIDE

7922AG39

Fig. 53 Geo/Chevrolet 1.0L, 1.3L, 1.6L and 1.8L engines—piston ring end-gap spacing

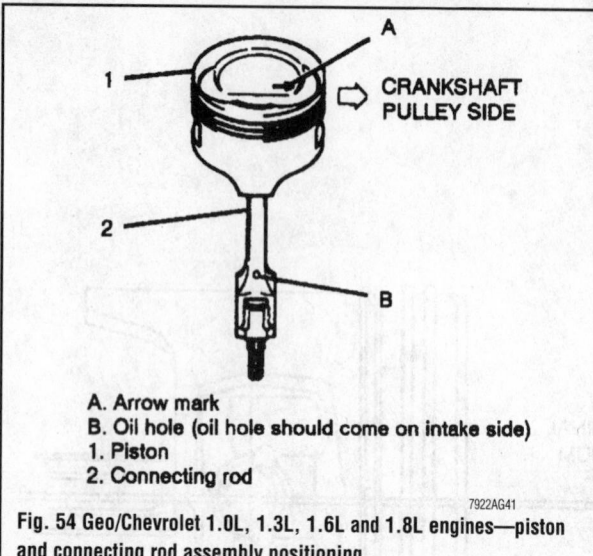

A. Arrow mark
B. Oil hole (oil hole should come on intake side)
1. Piston
2. Connecting rod

7922AG41

Fig. 54 Geo/Chevrolet 1.0L, 1.3L, 1.6L and 1.8L engines—piston and connecting rod assembly positioning

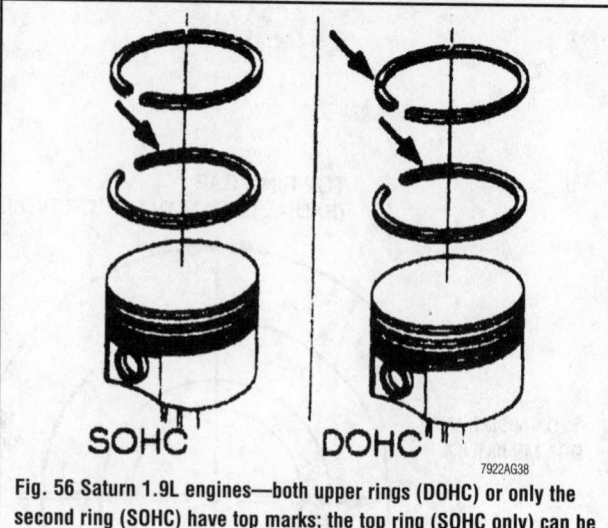

7922AG38

Fig. 56 Saturn 1.9L engines—both upper rings (DOHC) or only the second ring (SOHC) have top marks; the top ring (SOHC only) can be installed both ways.

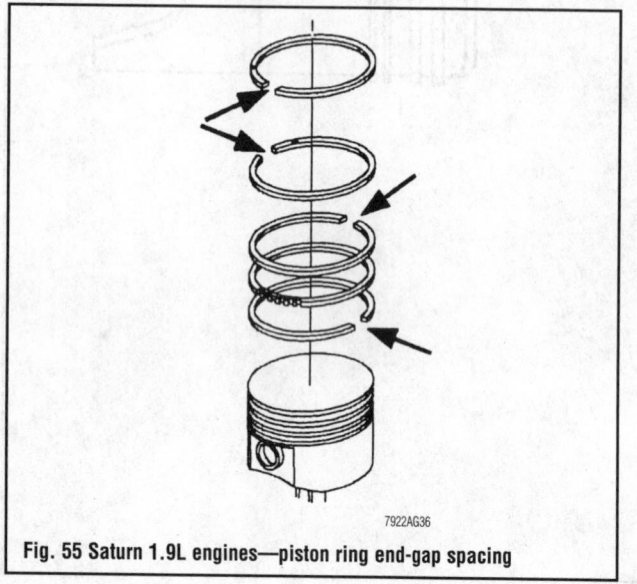

7922AG36

Fig. 55 Saturn 1.9L engines—piston ring end-gap spacing

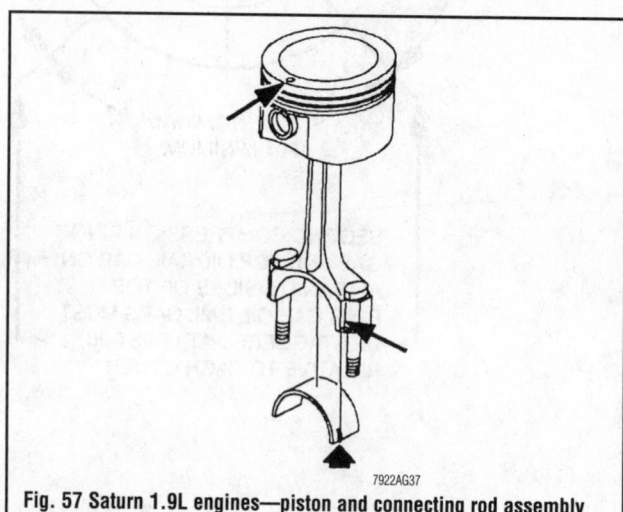

7922AG37

Fig. 57 Saturn 1.9L engines—piston and connecting rod assembly positioning mark locations. The mark on the piston and the rod bearing tang slots must face the front of the engine.

CHRYSLER CORPORATION

12

Chrysler-300M • Concord • LHS • New Yorker • Dodge-Intrepid • Eagle-Vision

DRITE TRAIN 12-27
DRIVE TRAIN 12-27
ENGINE REPAIR 12-2
FUEL SYSTEM 12-24
STEERING AND
  SUSPENSION 12-29

A

Air Bag.........................12-29
  Disarming....................12-29
  Precautions 12-29

C

Camshaft and Valve Lifters 12-15
  Removal & Installation 12-15
Coil Springs 12-33
  Removal & Installation 12-33
Cylinder Head....................12-5
  Removal & Installation 12-5

E

Engine Assembly 12-2
  Removal & Installation 12-2
Exhaust Manifold..................12-13
  Removal & Installation 12-13

F

Front Crankshaft Seal 12-14
  Removal & Installation 12-14
Fuel Filter 12-25
  Removal & Installation 12-25
Fuel Pump 12-26
  Removal & Installation 12-26

Fuel System Pressure 12-25
  Relieving 12-25
Fuel System Service
  Precautions......................12-24

H

Halfshaft.........................12-28
  Removal & Installation 12-28

I

Ignition Timing 12-2
  Adjustment......................12-2
Intake Manifold..................12-10
  Removal & Installation 12-10

L

Lower Ball Joint..................12-34
Lower Control Arm 12-34
  Removal & Installation 12-34

O

Oil Pan..........................12-17
  Removal & Installation 12-17
Oil Pump 12-19
  Removal & Installation 12-19

P

Power Rack and Pinion Steering
  Gear............................12-29
  Removal & Installation 12-29

R

Rear Main Seal 12-20
  Removal & Installation 12-20

Rocker Arm/Shafts...................12-9
  Removal & Installation 12-9

S

Strut.............................12-31
  Removal & Installation 12-31

T

Timing Chain, Sprockets,
  Front Cover and Seal.................12-21
  Removal & Installation 12-21
Transaxle Assembly 12-27
  Removal & Installation 12-27

V

Valve Lash 12-17
  Adjustment......................12-17

W

Water Pump......................12-4
  Removal & Installation 12-4
Wheel Bearings....................12-35
  Adjustment......................12-35
  Removal & Installation 12-35

■ ENGINE REPAIR

➡ **Disconnecting the negative battery cable on some vehicles may interfere with the functions of the on board computer systems and may require the computer to undergo a relearning process, once the negative battery cable is reconnected.**

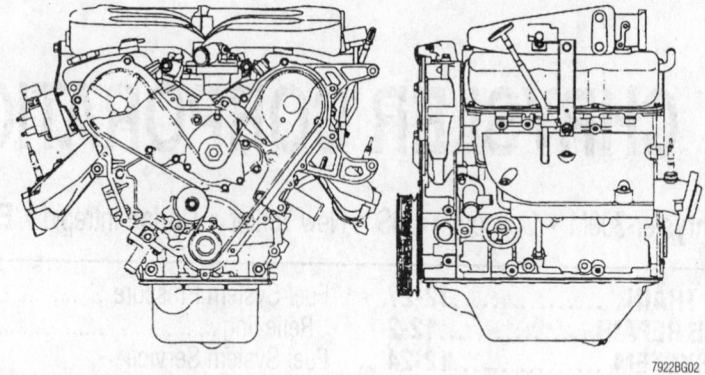

Engine front and side views—1995–97 3.5L engines

Ignition Timing

ADJUSTMENT

All models utilize a distributorless ignition system referred to as Direct Ignition System (DIS). It is a fixed ignition timing system which means that basic ignition timing cannot be adjusted. All spark advance is permanently set by the Powertrain Control Module (PCM).

Engine Assembly

REMOVAL & INSTALLATION

1995–97 Models

❄❄ CAUTION

The fuel injection system remains under pressure, even after the engine has been turned OFF. The fuel system pressure must be relieved before disconnecting any fuel lines. Failure to do so may result in fire and/or personal injury.

1. Disconnect negative battery cable.
2. Properly relieve the fuel system pressure.
3. Matchmark the hood and hinges and remove the hood assembly.
4. Drain the cooling system.

5. Remove the radiator and cooling fan assemblies.
6. Label and unplug all electrical connections.
7. Remove the coolant hoses from the engine.
8. Disconnect the fuel lines using the following procedure:

   a. At the fuel rail, push the quick-connect fitting toward the fuel tube while depressing the built-in disconnect tool with Quick-Connect Fitting Tool 6751 or equivalent.

   b. To disconnect the fitting from the fuel rail, slightly twist the fitting while maintaining downward pressure on tool 6751.

   c. Wrap shop towels around the fuel hoses to absorb any fuel spillage and be sure to cover the openings to prevent system contamination.

9. Disconnect the accelerator and cruise control cables from the throttle body.
10. Remove the air cleaner assembly. Raise and safely support the vehicle.
11. Place a drain pan under the vehicle and drain the engine oil.
12. Remove the air conditioning compressor mounting bolts and position the compressor aside. It should not be neces-

sary to disconnect the refrigerant lines from the compressor.
13. Disconnect the exhaust pipe from the exhaust manifold.
14. Remove the transaxle inspection cover and mark the flexplate for reference during installation.
15. Remove the screws holding the torque converter to the flexplate. Attach a C-clamp or some other restraint on the converter housing to prevent the torque converter from falling out during removal of the engine.
16. Remove the power steering pump mounting bolts and position the pump aside.
17. Remove the 2 lower transaxle-to-blockscrews. Remove the starter motor from the transaxle housing.
18. Lower the vehicle and disconnect the vacuum lines and ground strap. Support the transaxle with a floor jack.
19. Attach an engine lifting hoist to the engine and support.
20. Remove the upper transaxle mounting bolts.
21. Remove the insulator mounting nuts from the engine mounts.
22. Lift the engine from the vehicle.
To install:
23. Lower the engine into the engine compartment. Align the engine mounts and install all nuts. Once all mount bolts are installed, tighten the fasteners to 45 ft. lbs. (61 Nm).
24. Install the transaxle to the engine block and tighten the bolts to 75 ft. lbs. (102 Nm).
25. Remove the engine hoist and the transaxle holding fixture.
26. Remove the C-clamp from the converter housing, if installed.
27. Align the flexplate to the converter using the marks made during the removal procedure. Install the converter mounting screws and tighten to 55 ft. lbs. (75 Nm).

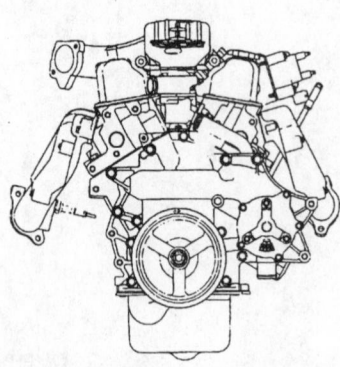
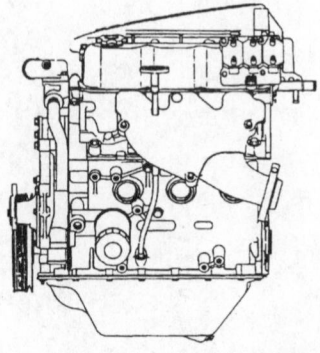

Engine front and side views—3.3L engines

28. Install the transaxle inspection cover.

29. Reconnect the exhaust system to the engine exhaust manifold and install the starter. Tighten the exhaust pipe-to-manifold bolts to 25 ft. lbs. (34 Nm).

30. Install the power steering pump and the air conditioning compressor to their mounting brackets and secure. Tighten the mounting bracket bolts to 30 ft. lbs. (41 Nm).

31. Lower the vehicle and reconnect all vacuum lines.

32. Attach all electrical connectors including the ground strap.

33. Reconnect the fuel lines to the fuel rail as follows:

   a. Be sure that the black plastic release ring to the quick-connect fitting is in the OUT position. Place Special Tool 6751 under the largest diameter of the quick-connect fitting.

   b. Pull Tool 6751 toward the fuel rail until the quick-connect fitting clicks into place.

   c. Place the special tool between the shoulder of the built-in disconnect tool and the top of the quick-connect fitting, then inspect the security of the fitting by applying a slight downward force against the fitting. It should be locked in place.

34. Reconnect the accelerator and cruise control cables to the throttle lever.

35. Install the radiator and cooling fan assemblies.

36. Refill the crankcase with the proper amount of engine oil and replace the oil filter, if necessary.

✷✷ WARNING

Operating the engine without the proper amount and type of engine oil will result in severe engine damage.

37. Refill and bleed the cooling system.

38. Install the hood onto the vehicle aligning the marks made during removal.

39. Check to be sure that all hoses, wiring connectors, cables, vacuum and fluid lines are reconnected.

40. Reconnect the negative battery cable. Check all fluid levels.

41. Start the engine and allow to idle until normal operating temperature is reached. Inspect all fluid systems for leaks and correct level.

42. Road test the vehicle. Adjust the transaxle linkage, as necessary.

1998–99 Models

✷✷ CAUTION

The fuel injection system remains under pressure, even after the engine has been turned OFF. The fuel system pressure must be relieved before disconnecting any fuel lines. Failure to do so may result in fire and/or personal injury.

1. Disconnect negative battery cable.

2. Properly relieve the fuel system pressure.

3. Matchmark the hood and hinges and remove the hood assembly.

4. Remove the wiper arms.

5. Remove the cowl covers and supports.

6. Disconnect and remove the air cleaner assembly and the air inlet duct.

7. Remove the upper radiator support and the hood release cable.

8. Remove the fan module.

9. Drain and recycle the engine coolant.

10. Remove the accessory drive belts.

11. Remove the upper and lower radiator hoses.

12. Disconnect the engine oil and transmission cooler lines at the radiator.

13. If necessary, label and disconnect the generator wiring harness, then remove the generator.

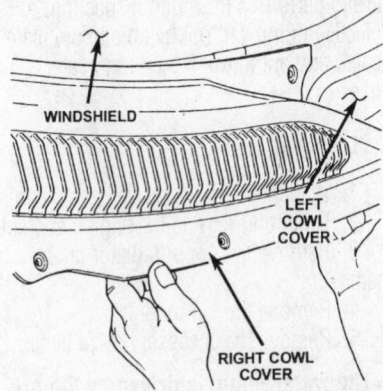

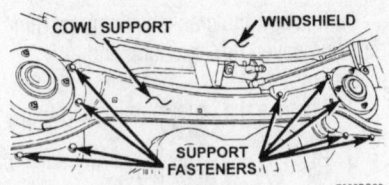

Be careful not to damage the windshield when removing the cowl covers and supports—1998–99 models

14. Remove the air conditioning compressor mounting bolts and position the compressor aside. It should not be necessary to disconnect the refrigerant lines from the compressor.

15. Remove the power steering pump mounting bolts and position the pump aside.

16. Loosen, then remove the V-band clamp at the right exhaust manifold.

17. Remove the right side bracket fasteners attaching the right catalytic converter down pipe.

18. Label and disconnect the fuel lines.

19. Disconnect the throttle and cruise control cables.

20. Label and disconnect all vacuum hoses.

21. Disconnect the ground straps at both cylinder heads.

✷✷ CAUTION

The intake manifold is a composite design, and should be removed before lifting the engine or it will be damaged and have to be replaced.

22. Remove the upper intake manifold, and cover the openings to prevent debris from entering the engine.

23. Disconnect the heater hoses.

24. Remove the throttle body support brackets.

25. Remove the upper transaxle-to-engine mounting bolts.

26. Label and detach all electrical and vacuum hose connections.

27. Raise and safely support the vehicle.

28. Drain the engine oil into a suitable container.

29. Remove the structural collar, then matchmark the flexplate-to-torque converter position.

30. Remove the flexplate-to-torque converter mounting bolts.

31. Loosen and remove the left exhaust manifold V-band clamp, then the left catalytic converter support brackets.

32. Remove the starter.

33. Remove the left and right engine mounting bolts.

34. Remove the Crankshaft Position Sensor (CPS).

35. Remove the lower engine-to-transaxle mounting bolts.

36. Lower the vehicle.

37. Attach a suitable lifting device to the engine.

38. Support the transaxle using a floor jack with a small block of wood in between.

39. Slowly lift the engine from the vehicle.

To install:

40. Carefully lower the engine into the vehicle.

41. Align the engine mounts and install the fasteners, but do not tighten them until all of the mounting bolts have been installed.

42. Install the engine-to-transaxle mounting bolts and tighten to 75 ft. lbs. (102 Nm).

43. Remove the engine lifting device.

44. Raise and safely support the vehicle.

45. Tighten the engine mount nuts/bolts 45 ft. lbs. (61 Nm).

46. Align the flexplate-to-torque converter matchmarks and install the mounting bolts, then tighten to 55 ft. lbs. (75 Nm).

47. Install the CPS and the starter.

48. Install the left-side exhaust manifold V-band clamp and tighten to 100 inch lbs. (11 Nm).

49. Install the mounting bracket fasteners for the left-side catalytic converter.

50. Install the structural collar utilizing the following procedure:

   a. Install the vertical collar to the oil pan mounting bolts and tighten, temporarily to 10 inch lbs. (1.1 Nm).

   b. Install the collar-to-transaxle bolts and tighten to 40 ft. lbs. (55 Nm).

   c. Starting with the center vertical bolt and working outward, tighten the bolts to 40 ft. lbs. (55 Nm).

51. Lower the vehicle.

52. Install the throttle body support bracket.

53. Connect the heater hoses, all ground straps, and vacuum hoses.

54. Install the upper intake manifold.

55. Attach all engine wiring harnesses.

56. Connect and adjust the throttle and cruise control cables.

57. Connect the fuel lines.

58. Install the V-band clamp and on the right exhaust manifold and tighten to 100 inch lbs. (11 Nm).

59. Install the mounting bracket fasteners for the right-side catalytic converter.

60. Install the air conditioning compressor, and tighten the mounting bolts to 250 inch lbs. (28 Nm).

61. If removed, install the generator, and attach its wiring harness.

62. Install the radiator.

63. Install the power steering pump.

64. Connect the engine oil and transaxle cooler lines.

65. Connect the radiator hoses and install the accessory drive belt.

66. Fill the cooling system and the engine oil pan to the proper level.

** WARNING

Operating the engine without the proper amount and type of engine oil will result in severe engine damage.

67. Install the hood release cable and the upper radiator support.

68. Install the air cleaner and inlet hose.

69. Install the cowl covers and supports, then the wiper arms.

70. Line up the matchmarks and install the hood.

71. Connect the negative battery cable.

Water Pump

REMOVAL & INSTALLATION

The water pump has a die cast aluminum body and a stamped steel impeller. It bolts directly to the chain case cover using an O-ring for sealing. It is driven by the back side of the serpentine belt.

It is normal for a small amount of coolant to drip from the weep hole located on the water pump body (small black spot). If this condition exists, do **NOT** replace the water pump. Only replace the water pump if a heavy deposit or steady flow of brown/green coolant is visible on the water pump body from the weep hole, which would indicate shaft seal failure. Before replacing the water pump, be sure to perform a thorough analysis. Before replacing the water pump, be sure to perform a thorough inspection. A defective pump will not be able to circulate heated coolant through the long heater hose.

2.7L Engine

1. Disconnect the negative battery cable.
2. Drain and recycle the engine coolant.
3. Remove the upper radiator crossmember.
4. Remove the fan module.
5. Remove the accessory drive belts.

➡**The water pump is driven by the primary timing chain.**

6. Remove the crankshaft damper, timing chain cover, timing chain, and all guides.

7. Remove the water pump mounting bolts.

8. Remove the water pump, then clean the mounting surface.

To install:

9. Install the water pump and gasket, then tighten the mounting bolts to 105 inch lbs. (12 Nm).

10. Install the all guides, timing chain, and timing chain cover.

11. Install the crankshaft damper and tighten the center bolt to 125 ft. lbs. (170 Nm).

12. Install the accessory drive belts.

13. Install the fan module and the upper radiator crossmember.

14. Fill the cooling system.

15. Connect the negative battery cable.

3.2L and 3.5L Engines

1. Disconnect the negative battery cable.

** CAUTION

Do not open the radiator drain or the coolant pressure bottle cap with the system hot and under pressure or serious burns from coolant may occur.

2. Place a drain pan under the radiator. Open the radiator drain located at the lower right side of the radiator. Do **NOT** use pliers to open the plastic drain.

3. Remove the coolant pressure bottle cap and open the thermostat bleed valve.

4. Remove the timing belt using the recommended procedure. It is good practice to turn the crankshaft until the engine is at TDC No. 1 cylinder, compression stroke (firing position). This aligns all the timing marks and serves as a reference point for all the work that follows.

5. Remove the water pump mounting bolts and pump. Discard the O-ring seal.

6. Clean the gasket sealing surfaces. Do not scratch the aluminum surfaces.

To install:

7. Install a new O-ring and wet with clean coolant. Be sure to keep the new O-ring free of any oil or grease.

8. Install the water pump and O-ring to the engine block.

9. Install the retaining bolts and tighten to 105 inch lbs. (12 Nm).

10. Rotate the pump and check for freedom of movement.

11. Install the timing belt using the recommended procedure. Verify that all valve timing marks align. This is most important. An engine out-of-time will be seriously damaged when first started.

12. Be sure that the radiator drain is closed. Open the thermostat bleed valve. Install a ¼ in. (6mm) clear hose about 48 in. (1.2m) long to the end of the bleed valve and the other end into a clean container. The intent is to keep coolant off of the drive belt(s).

13. Slowly refill the coolant pressure bottle until a steady stream of coolant flows out of the thermostat bleed valve. Gently squeeze the upper radiator hose until all of the air is removed from the system.

14. Close the bleed valve and continue to fill up the coolant pressure bottle to the proper level. Install the cap back on the bottle and remove the hose from the bleed valve.

15. Reconnect the negative battery cable. Start the engine and allow to run until normal operating temperature is reached.

16. Check the cooling system for leaks and correct coolant level. Be sure that the thermostat bleed valve is closed once the cooling system has been bled of any trapped air.

3.3L Engines

1. Disconnect the negative battery cable.

✱✱ CAUTION

Do not remove the radiator drain with the system hot and under pressure or serious burns from coolant may occur.

2. Place a drain pan under the radiator. Open the radiator drain located at the lower right side of the radiator. Do **NOT** use pliers to open the plastic drain.

3. Remove the coolant pressure bottle cap and open the thermostat bleed valve.

4. Remove the serpentine belt. If necessary, remove the right front lower fender shield.

5. Remove the water pump pulley bolts and pulley.

6. Remove the water pump mounting bolts and pump. Discard the O-ring seal.

7. Clean the gasket sealing surfaces. Do not scratch the aluminum surfaces.

To install:

8. Install a new O-ring into the O-ring groove and the water pump to the timing chain case. Be sure to keep the O-ring free of any oil or grease.

9. Install the retaining bolts and tighten to 105 inch lbs. (12 Nm).

10. Rotate the pump and check for freedom of movement.

11. Install the pump pulley and tighten the bolts to 250 inch lbs. (30 Nm).

12. Install the serpentine belt and right lower fender shield.

13. Be sure that the radiator drain is closed. Open the thermostat bleed valve. Install a ¼ in. (6mm) diameter clear hose about 48 in. (1.2m) long, to the end of the

bleed valve and the other end into a clean container. The intent is to keep coolant off the drive belt(s).

14. Slowly refill the coolant pressure bottle until a steady stream of coolant flows out of the thermostat bleed valve. Gently squeeze the upper radiator hose until all of the air is removed from the system.

15. Close the bleed valve and continue to fill up the coolant pressure bottle to the proper level. Install the cap back on the bottle and remove the hose from the bleed valve.

16. Reconnect the negative battery cable. Start the engine and allow to run until normal operating temperature is reached.

17. Check the cooling system for leaks and correct coolant level. Be sure that the thermostat bleed valve is closed once the cooling system has been bled of any trapped air.

Cylinder Head

REMOVAL & INSTALLATION

✱✱ CAUTION

The EPA warns that prolonged contact with used engine oil may cause a number of skin disorders, including cancer! You should make every effort to minimize your exposure to used engine oil. Protective gloves should be worn when changing the oil. Wash your hands and any other exposed skin areas as soon as possible after exposure to used engine oil. Soap and water, or waterless hand cleaner should be used.

2.7L Engine

1. Disconnect the negative battery cable.

2. Properly relieve the fuel system pressure.

3. Drain and recycle the engine coolant.

4. Remove the accessory drive belts.

5. Remove the crankshaft damper.

6. Remove the intake plenum, lower intake manifold, and exhaust manifold. Place shop rags in the openings to prevent debris from entering the engine.

7. Remove the valve and timing chain covers.

8. Remove the coolant connections for the cylinder heads.

9. Rotate the crankshaft until the crankshaft timing mark aligns with the timing mark on the oil pump.

10. Remove the primary timing chain.

11. Remove the camshaft bearing caps, gradually, in reverse order of installation.

12. Remove the camshafts from the cylinder heads.

✱✱ CAUTION

Be sure the head bolts 11–9 are removed before attempting to remove the cylinder head, the head and/or block may be damaged.

13. Remove the cylinder head bolts in reverse order of installation starting with bolts 11–9, then 8–1.

14. Remove the cylinder head(s).

To install:

15. Thoroughly clean and dry the mating surfaces of the head and block. Check the cylinder head for cracks, damage or engine coolant leakage. Remove scale, sealing compound and carbon. Clean the oil passages thoroughly.

16. Place a new head gasket on the cylinder block over the locating dowels.

17. Inspect the cylinder head bolts for necking (stretching) by holding a straight-edge against the threads of each bolt. If all of the threads are not contacting the scale, the bolt should be replaced. New head bolts are recommended.

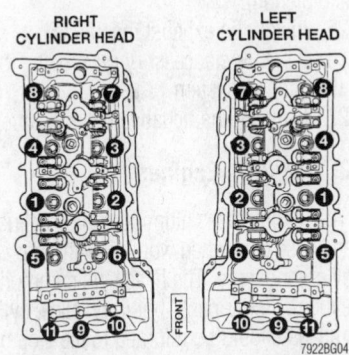

Cylinder head tightening sequence—2.7L engines

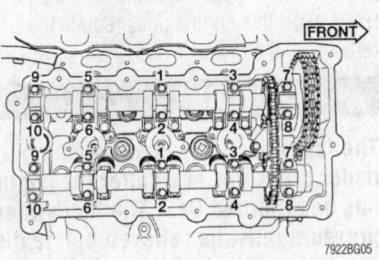

Camshaft tightening sequence—2.7L engines

❋❋ WARNING

Due to the cylinder head bolt torque method used, it is imperative that the threads of the bolts be inspected for necking (stretching) prior to installation. If the threads are necked down, the bolt should be replaced. Failure to do so may result in parts failure or damage.

18. Lubricate the bolt threads with clean engine oil, then install them.

19. Tighten the head bolts in sequence as shown, utilizing the following steps and tightening values:
- Step 1. Bolts 1–8 to 35 ft. lbs. (48 Nm)
- Step 2. Bolts 1–8 to 55 ft. lbs. (75 Nm)
- Step 3. Bolts 1–8 to 55 ft. lbs. (75 Nm)
- Step 4. Bolts 1–8 to +90° turn. Do not use a torque wrench for this step.
- Step 5. Bolts 9–11 to 250 inch lbs. (28 Nm).

20. Install the camshafts, timing chain, and sprockets.

21. Install the water connections to the cylinder head.

22. Install the valve and timing chain covers.

23. Install the crankshaft damper and tighten the center bolt to 125 ft. lbs. (170 Nm)

24. Install the lower intake manifold and intake plenum.

25. Install the exhaust manifolds.

26. Install the accessory drive belts, then fill the cooling system.

27. Connect the negative battery cable.

3.2L and 3.5L Engines

This engine uses aluminum alloy cylinder heads. Use care when working with light alloy components. The heads are common to either cylinder bank, but in practice, cylinder heads should be returned to the side of the engine from which it was removed. Removal of a cylinder head involves removal of the timing belt. Great care is required to install the belt, paying attention to all valve timing marks. Please note that camshaft removal on this engine does require the removal of the cylinder head.

❋❋ CAUTION

The fuel injection system remains under pressure, even after the engine has been turned OFF. The fuel system pressure must be relieved before disconnecting any fuel lines. Failure to do so may result in fire and/or personal injury.

1. Release the fuel system pressure using the recommended procedure.

2. Disconnect negative battery cable.

3. Drain the cooling system.

4. Remove the radiator and cooling fan assemblies.

5. Remove the air cleaner assembly and the intake manifold plenum. Cover the lower intake manifold during service.

6. Remove the accessory drive belts.

7. Remove the crankshaft damper using the proper puller.

8. Remove the engine valve covers.

9. Remove the timing belt covers.

10. Mark the timing belt running direction for installation. Align the camshaft sprockets with the marks on the rear covers. Remove the timing belt and tensioner.

11. Pre-load the timing belt tensioner as follows:

    a. Place tensioner in a vise the same way it is mounted on the engine.

    b. Slowly compress the plunger into the tensioner body.

    c. When the plunger is compressed into the tensioner body, install a pin through the body and plunger to retain plunger in place until the tensioner is installed.

12. Hold the camshaft sprocket with 36mm box wrench, loosen and remove the sprocket retaining bolt and washer.

➡**To remove the camshaft sprocket retainer bolt while the engine is in the vehicle, it may be necessary to raise that side of the engine due to the length of the retainer bolt. The right bolt is 8.370 in. (212.6mm) long, while the left bolt is 10.0 in. (253mm) long. These bolts are not interchangeable and their original location during removal should be noted.**

13. Remove the camshaft sprocket from the camshaft. The camshaft sprockets are not interchangeable from side to side.

14. Remove the intake manifold assembly using the recommended procedure.

15. Remove the rear timing belt cover to cylinder head fasteners. If the right timing belt cover is to be removed, there are O-rings located behind it for the water pump passages.

16. Remove the cylinder head mounting bolts in the reverse order of the tightening sequence. Remove the cylinder head.

To install:

17. Thoroughly clean and dry the mating surfaces of the head and block.

➡**When cleaning the cylinder head and block mating surfaces, do not use a**

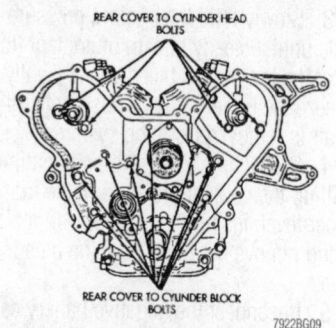

Remove the rear timing belt cover-to-cylinder head bolts, note the bolt locations—3.2L and 3.5L engines

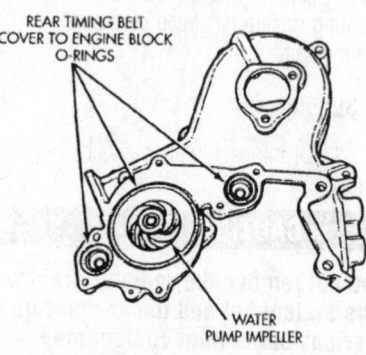

Right side belt cover, water pump and O-rings—3.2L and 3.5L engines

metal scraper because the soft aluminum surfaces could be cut or damaged. Instead, use a scraper made of wood or plastic.

18. Check the cylinder head for cracks, damage or engine coolant leakage. Check the head for flatness. End to end, the head should be within 0.002 in. (0.051mm) normally with 0.008 in. (0.203mm) the maximum allowed out of true. The resurface limit is 0.008 in. (0.203mm) maximum, the combined total dimension of stock removal from the cylinder head if any and block top surface.

19. Place a new head gaskets on the cylinder block locating dowels being sure the gasket is on the correct side.

20. Inspect the cylinder head bolts for necking by holding a straightedge against the threads of each bolt. If all of the threads are not contacting the scale, the bolt should be replaced.

❋❋ WARNING

Due to the cylinder head bolt torque method used, it is imperative that the threads of the bolts be inspected for necking prior to installation. If the threads are necked down, the bolt should be replaced. Failure to do so

may result in parts failure or damage. New bolts are always recommended.

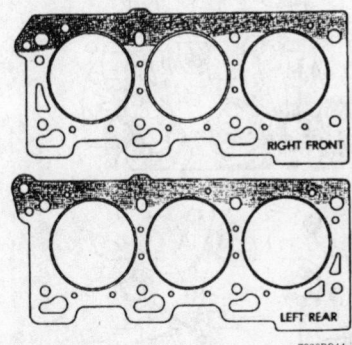

Correct positioning for the head gaskets—3.2L and 3.5L engines

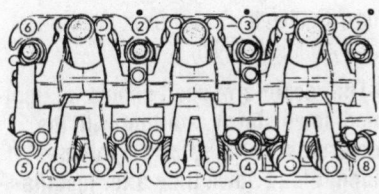

Cylinder head bolt tightening sequence—3.2L and 3.5L engines

21. Install the cylinder head into position on the engine block and over the dowels. Install the cylinder head bolts, lubricating the threads with clean engine oil prior to installation. Tighten bolts following the proper sequence as listed below:
- Step 1—45 ft. lbs. (61 Nm)
- Step 2—65 ft. lbs. (88 Nm)
- Step 3—again, 65 ft. lbs. (88 Nm)
- Step 4—additional ¼ -turn

➡**Do not use a torque wrench for Step 4. Inspect the bolt torque after tightening. The torque should be over 90 ft. lbs. (122 Nm). If not, replace the cylinder head bolt.**

22. Install the rear timing belt cover bolts and tighten as follows:
- M6 bolts—105 inch lbs. (12 Nm)
- M8 bolts—21 ft. lbs. (28 Nm)
- M10 bolts—40 ft. lbs. (54 Nm)

23. Install the intake manifold assembly and tighten bolts following the proper sequence to 21 ft. lbs. (28 Nm).

➡**The following procedure can only be used when the camshaft sprockets have been loosened or removed from the shafts.**

24. When the camshaft sprockets are loosened or removed, the camshafts must be timed to the engine. Install the Camshaft Alignment Tools 6642-A or exact equivalent, to the rear of the cylinder heads.

25. Install both camshaft sprockets to the appropriate shafts. The left camshaft sprocket has the DIS pick-up as part of the sprocket.

26. Apply thread locking compound to the threads of the camshaft sprocket retainer bolts and install to the appropriate shafts. The right bolt is 8.380 in. (21.3cm) long, while the left bolt is 10.0 in. (25.4cm) long. These bolts are not interchangeable. Do not tighten the bolts at this time. The camshaft marks should be between the marks on the cover.

27. Place the crankshaft sprocket to the TDC mark on the oil pump housing. Install the timing belt starting at the crankshaft sprocket and working in a counterclockwise direction.

28. After the belt is installed around the last sprocket keep tension on the belt until it is past the tensioner pulley.

29. Holding the tensioner pulley against the belt, install the tensioner housing and tighten to 250 inch lbs. (28 Nm)

30. When the tensioner is in place pull the retainer pin to allow tensioner to extend to the pulley bracket.

31. Install a dial indicator in No. 1 cylinder to check Top Dead Center (TDC) of the piston. Rotate the crankshaft until the piston is exactly at TDC.

32. Hold the right camshaft sprocket hex with a 36mm box wrench and tighten the right camshaft sprocket bolt to 75 ft. lbs. (102 Nm). Turn the sprocket bolt an additional 90 degrees.

33. Hold the left camshaft sprocket hex with a 36mm box wrench and tighten the left camshaft sprocket bolt to 85 ft. lbs. (115 Nm). Turn the sprocket bolt an additional 90 degrees.

34. Remove the dial indicator.

35. Remove the camshaft alignment tools from the back of the cylinder heads and install the cam covers and new O-rings. Tighten the fasteners to 20 ft. lbs. (27 Nm). Repeat this procedure on the other camshaft.

36. Install the timing belt covers and crankshaft damper. Tighten the crankshaft damper bolt to 85 ft. lbs. (115 Nm).

37. Install the valve covers and tighten bolts to 105 inch lbs. (12 Nm).

38. Install the spark plug tube nut and O-ring. Tighten the nut to 60 inch lbs. (7 Nm). Install spark plugs and tighten to 20 ft. lbs. (28 Nm).

39. Install the air conditioning compressor to the mounting bracket. Tighten the mounting bracket bolts to 30 ft. lbs. (41 Nm).

40. Reconnect the spark plug wires to the correct spark plugs.

41. Install the accessory drive belts. Adjust to the proper tension.

42. Install the intake manifold plenum using the recommended procedure.

43. Install the air cleaner assembly.

44. Install the radiator and cooling fan assemblies.

45. Check to be sure that all hoses, wiring connectors, cables, fluid and vacuum lines are reconnected.

46. Change the engine oil and oil filter.

47. Refill and bleed the cooling system.

48. Reconnect the negative battery cable, run the vehicle with the radiator cap off so coolant can be added as required until the thermostat opens. Watch for leaks and for unusual engine noises. Fill the radiator completely as required.

49. Once the vehicle has cooled, recheck the coolant and oil level.

3.3L Engines

The cylinder heads on this engine are aluminum alloy, retained by nine bolts. Valve seats and guides are inserts. Use care when handling alloy parts. In addition, the cylinder head bolts are torque-to-yield type which stretch during the torque process. Head bolts must be checked carefully before reuse. New head bolts are recommended.

1. Release the fuel system pressure using the recommended procedure.

2. Disconnect negative battery cable.

3. Drain the cooling system.

4. Remove the intake manifold and throttle body assemblies using the recommended procedure. Use care when handling the intake manifold gasket. It is made of very thin metal and could cause cuts if carelessly handled.

5. Disconnect the coil wires, sending unit wire, heater hoses and bypass hose.

6. Remove the evaporation control system, closed ventilation system and the cylinder head covers.

7. Remove the exhaust manifold(s) from the engine.

8. Remove rocker arm and shaft assemblies. Remove pushrods and identify to assure installation in original location.

9. Loosen and remove the nine head bolts in the reverse order of the tightening sequence from the cylinder head.

10. Lift the head off the cylinder block.

To install:

11. Thoroughly clean and dry the mating surfaces of the head and block. Check the cylinder head for cracks, damage or engine coolant leakage. Remove scale, sealing compound and carbon. Clean the oil passages thoroughly. Check the head for flatness. End to end, the head should be within 0.002 in. (0.051mm) normally with 0.008 in. (0.203mm) the maximum allowed out of true. The total thickness allowed to be removed from the head and block is 0.008 in. (0.203mm) maximum.

12. Place a new head gasket on the cylinder block with the identification marks facing upward. Do not use sealer on factory type gasket.

13. Inspect the cylinder head bolts for necking (stretching) by holding a straight-edge against the threads of each bolt. If all of the threads are not contacting the scale, the bolt should be replaced. New head bolts are recommended.

✳✳ WARNING

Due to the cylinder head bolt torque method used, it is imperative that the threads of the bolts be inspected for necking (stretching) prior to installa-

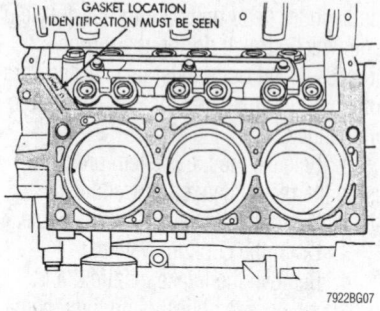

When installing the end gasket, the markings on the gasket must be seen—3.3L engine

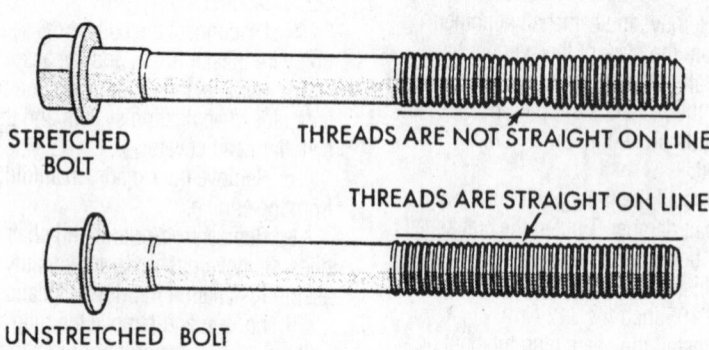

When reusing them, inspect the cylinder head bolts for necking (stretching). You can see a difference in these bolts—all models

Cylinder head bolt tightening sequence—3.3L engine

tion. If the threads are necked down, the bolt should be replaced. Failure to do so may result in parts failure or damage.

14. Install the cylinder head bolts. Tighten bolts Nos. 1 through 8 following the proper sequence as listed below:

- Step 1—45 ft. lbs. (61 Nm)
- Step 2—65 ft. lbs. (88 Nm)
- Step 3— (again) 65 ft. lbs. (88 Nm)
- Step 4—additional ¼ -turn

➡**Do not use a torque wrench for Step 4. Inspect the bolt torque after tightening. The torque should be over 90 ft. lbs. (122 Nm). If not, replace the cylinder head bolt.**

15. Tighten bolt No. 9 to 25 ft. lbs. (33 Nm) only after bolts 1–8 have been tightened to specification.

16. Inspect the pushrods and replace worn or bent rods. Install the pushrods, rocker arm and shaft assemblies with the stamped steel retainers in the forward posi-

tions. Tighten the rocker shaft retainers to 250 inch lbs. (28 Nm).

➡**The rocker arm shaft should be tightened down slowly, starting with the centermost bolts. Allow 20 minutes tappet bleed down time after installation of the rocker shafts before engine operation.**

17. Install the cylinder head covers with new gaskets in place. Tighten the retainers to 105 inch lbs. (12 Nm).

➡**The factory type intake manifold gasket is made of very thin metal and is very sharp. Handle with care or personal injury may occur.**

18. Using all new gaskets, install the intake manifold, throttle body, air intake plenum and exhaust manifold, using the recommended procedure and following the proper torque sequences.

19. Connect all exhaust and fuel connections.

20. Reconnect the coil wires, sending unit wire, heater hoses and bypass hose.

21. Install the air intake hose.

22. Change the engine oil and oil filter.

23. Refill and bleed the cooling system.

24. Reconnect the negative battery cable and run the vehicle with the radiator cap off so coolant can be added as required until the thermostat opens. Watch for leaks and unusual engine noises that might indicate a problem. Fill the radiator completely.

25. Once the vehicle has cooled, recheck the coolant and oil level.

Rocker Arm/Shafts

REMOVAL & INSTALLATION

2.7L Engines

1. Disconnect the negative battery cable.
2. Remove the valve covers.
3. Position the camshaft so that the base circle (heel) is facing the rocker arm being serviced.

❋❋ CAUTION

Depress the valve spring only enough to remove the rocker arm.

4. Using Valve Spring tool 8215 and Adapter 8216, or their equivalent, depress the valve spring enough to release the tension on the rocker arm.
5. If the rocker arms are to be reused, identify their positions for reassembly in their original positions.
6. Repeat this procedure for each rocker arm being removed.

To install:
7. Lubricate the rocker arms with clean engine oil, prior to installation.
8. Position the camshaft so that the base circle (heel) is facing the rocker arm being installed.

❋❋ CAUTION

Depress the valve spring only enough to install the rocker arm.

9. Using Valve Spring tool 8215 and Adapter 8216, or their equivalent, depress the valve spring enough to install the rocker arm.

10. Install the rocker arm in the original position (if reused) over the valve and lash adjuster.

➡**Inspect the rocker arm for proper engagement into the lash adjuster and valve tip.**

11. Release the tension on the valve spring and remove the tools.
12. Install the valve covers.
13. Connect the negative battery cable.

3.2L and 3.5L Engines

1. Disconnect negative battery cable.

❋❋ CAUTION

The fuel injection system remains under pressure even after the engine has been turned OFF. The fuel system pressure must be relieved before disconnecting any fuel lines. Failure to do so may result in fire and/or personal injury.

2. Relieve the fuel system pressure.
3. Remove the air cleaner assembly and the intake manifold plenum. Cover the lower intake manifold during service.
4. Remove the cylinder head covers.
5. Remove the rocker arm assembly mounting bolts and remove the assembly from the engine.
6. Inspect the rocker arms for wear or damage. Inspect the roller for scuffing or wear. Replace assembly as necessary.

➡**Do not remove the lash adjusters from the rocker arm assembly. The rocker arm and the adjuster are serviced as an assembly.**

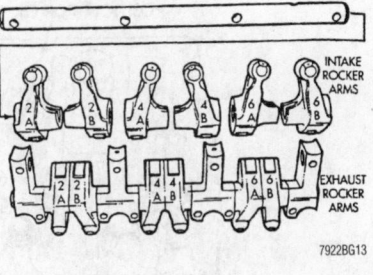

Left bank rocker arm and shaft identification—3.2L and 3.5L engines

7. Identify the rocker arm assemblies and rocker arms and disassemble the shaft as follows:
a. Thread a nut, washer and spacer onto a 4mm screw.
b. Insert and tighten a 4mm screw into the dowel pin on the shaft.
c. Loosen the nut on the screw. This will pull the dowel pin from the shaft support.
d. Remove the rocker arms and pedestals keeping in order.
e. Check the oil holes for restrictions with a small wire and clean as required.

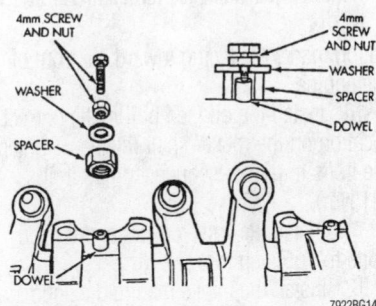

Remove the dowel pin using a 4mm screw, nut, spacer and washer installed into the pin—3.2L and 3.5L engines

To install:
8. Assemble the rocker shaft as follows:
a. Install the rocker arms and pedestals onto the shaft keeping in original order.
b. Press the dowel pins into the pedestals until they bottom out in the pedestals.
9. Position the camshaft so that the timing mark on the right camshaft timing belt sprocket aligns with the timing mark on the rear timing belt cover and the timing mark on the left sprocket is 45 degrees from the mark on the rear timing belt cover. There will be no load on the shaft during installation. Install the rocker shafts so the identifi-

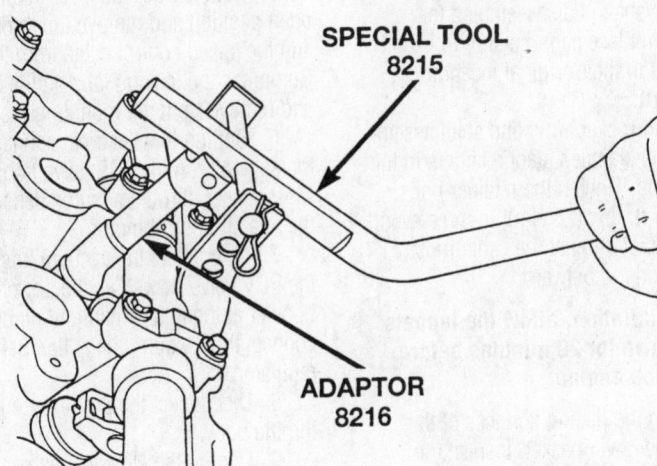

Only depress the valve spring enough to remove the rocker arm—2.7L engine

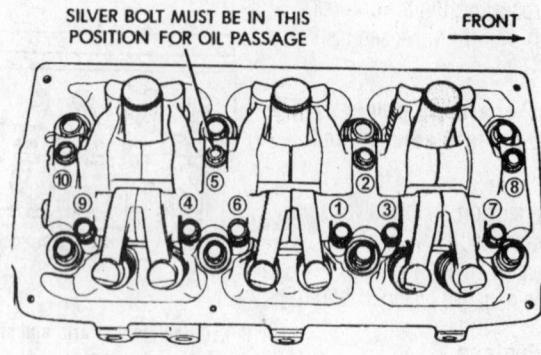

SILVER BOLT MUST BE IN THIS
POSITION FOR OIL PASSAGE

FRONT →

RIGHT SIDE SHOWN

← FRONT

SILVER BOLT MUST BE IN THIS
POSITION FOR OIL PASSAGE

LEFT SIDE SHOWN

7922BG15

Proper torque sequence for the rocker arm and shaft assemblies—3.2L and 3.5L engines

cation marks are facing toward the front of the engine.

10. Install the oil feed bolt in the correct location on the rocker shaft retainer. Tighten the bolts in proper sequence to 23 ft. lbs. (31 Nm).

11. Install the valve covers and tighten bolts to 105 inch lbs. (12 Nm).

12. Install the intake manifold plenum and the air cleaner assembly.

13. Reconnect the negative battery cable.

3.3L Engines

1. Disconnect negative battery cable.

> ✳✳ **CAUTION**
>
> **The fuel injection system remains under pressure, even after the engine has been turned OFF. The fuel system pressure must be relieved before disconnecting any fuel lines. Failure to do so may result in fire and/or personal injury.**

2. Relieve the fuel system pressure.

3. Remove the upper intake manifold assembly.

4. Disconnect and label the spark plug wires from the plugs. Remove by pulling on the boot in a straight out in line with the spark plug.

5. Disconnect the closed crankcase ventilation system and the evaporative control system from the cylinder head cover.

6. Remove the cylinder head cover and gasket from the engine.

7. Remove the four rocker shaft bolts and retainers. Remove the rocker arms and shafts from the engine.

8. Inspect rocker arm and components for wear or damage and replace as required. If the rocker shaft is disassembled for cleaning or replacement, be sure to install components in their original location.

To install:

9. Install rocker arms and shaft assemblies with the stamped steel retainers in the four positions. Tighten the retainer bolts slowly to 21 ft. lbs. (28 Nm), in three even progressions starting at the centermost bolts and working outward.

➡ **After installation, allow the tappets to bleed down for 20 minutes before operating the engine.**

10. Clean the mating surfaces of the cylinder head cover gasket. Inspect the cylinder head cover and straighten out if distorted.

11. Install the cylinder head cover with new gasket in place. Tighten fasteners to 105 inch lbs. (12 Nm).

12. Reconnect the closed crankcase ventilation system and the evaporative control system.

13. Reconnect the spark plug wires making sure that each wire is connected to the correct spark plug.

14. Install the upper intake manifold assembly and reconnect the negative battery cable.

Intake Manifold

REMOVAL & INSTALLATION

2.7L and 3.3L Engines

> ✳✳ **CAUTION**
>
> **The fuel injection system remains under pressure, even after the engine has been turned OFF. The fuel system pressure must be relieved before disconnecting any fuel lines. Failure to do so may result in fire and/or personal injury.**

1. Disconnect negative battery cable.

2. Remove the fuel filler cap. Release the fuel system pressure using the recommended procedure.

> ✳✳ **CAUTION**
>
> **Do not open the radiator drain with the system hot and under pressure or serious burns from coolant may occur.**

3. Drain the cooling system.

4. Disconnect the air tube from the air cleaner and the throttle body.

5. Hold the throttle lever in the wide-open position and remove the throttle cable and the speed control cable from the lever. Compress the locking tabs on the cables and remove from the mounting brackets.

6. Unplug the electrical connector from the solenoid on the EGR valve transducer, MAP sensor, throttle position sensor and the idle air control motor.

7. Disconnect the vacuum hose from the PCV valve as well as the power brake booster at the intake manifold nipple. Disconnect the vacuum line at the fuel pressure regulator.

8. Disconnect the purge hose from the throttle body.

9. Unplug the electrical harnesses from the throttle position sensor and the idle air control motor.

10. Remove the EGR tube mounting screws at the intake manifold plenum.

11. Remove the intake manifold plenum (upper part of the manifold) mounting bolts and remove the plenum from the engine. Cover the lower part of the intake manifold to prevent foreign material from entering the engine.

12. Disconnect the fuel supply and return tubes to the fuel rail at the rear of the intake manifold by pushing the quick-connect fitting toward the fuel tube while depressing the built-in disconnect tool with Quick-Connect Fitting Tool 6751. To disconnect the fitting from the fuel rail, slightly twist the fitting while maintaining downward pressure on tool 6751. Wrap shop towels around the fuel hoses to absorb any fuel spillage.

13. Cover the fuel line openings to prevent system contamination.

14. Remove the screw from the fuel clamp and separate the fuel tubes from the bracket.

15. Tag each connector for identification, then disconnect the electrical harness from the injectors and turn toward the center of the engine.

16. Remove the fuel rail mounting bolts and lift fuel rail with the injectors attached straight up and off the engine. Cover the injector openings.

17. On the 3.3L engine, remove the upper radiator hose, heater hose and the rear intake manifold hose.

18. Remove the intake manifold bolts and the manifold from the engine.

19. Remove the intake manifold seal retainers screws and remove the intake manifold gasket. Clean all mating surfaces.

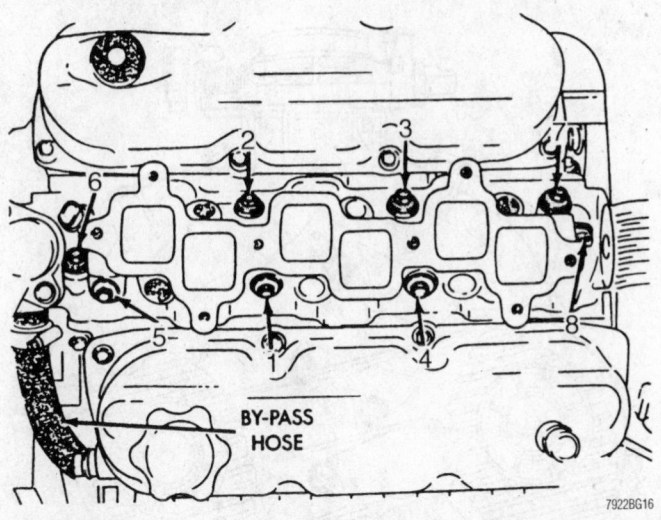

Loosening and tightening sequence for the intake manifold bolts—3.3L engine

20. Inspect the manifold for damage, cracks or clogged passages. Repair, clean or replace the manifold as required.

To install:

21. Verify that all intake manifold and cylinder head sealing surfaces are clean. Place a drop of sealant onto each of the four corners of the intake manifold gasket, where the cylinder head meets the engine block. On the 3.3L engine, carefully install the intake manifold gasket and tighten the end seal retainers to 105 inch lbs. (12 Nm).

➡**The intake manifold gasket is made of very thin metal and can cause cuts if handled carelessly.**

22. Install the intake manifold and eight mounting bolts. Snug down evenly to just 10 inch lbs. (1.1 Nm).

23. On the 3.3L engine, tighten the lower intake manifold bolts in the proper sequence to 16 ft. lbs. (22 Nm). Once all bolts are tightened, repeat the sequence again tightening the bolts to 16 ft. lbs. (22 Nm). Inspect to be sure all seals are still in place.

24. On the 2.7L engine, tighten the lower intake manifold bolts in the proper sequence to 105 inch lbs. (12 Nm).

25. Apply a light coat of clean engine oil to the O-ring on the nozzle end of each injector.

26. Insert the fuel injector nozzles into the openings in the intake manifold. Seat the injectors in place and install the fuel rail mounting bolts, tightening to 16 ft. lbs. (22 Nm).

27. Attach the electrical connectors to each fuel injector. Rotate the injectors toward the cylinder head covers.

28. Reconnect the fuel supply and return tubes to the fuel rail. Be sure that the black plastic release ring to the quick-connect fitting is in the OUT position. Place special tool 6751 under the largest diameter of the quick-connect fitting.

29. Pull Tool 6751 toward the fuel rail until the quick-connect fitting clicks into place. Place the special tool between the shoulder of the built-in disconnect tool and top of the quick-connect fitting, then inspect the security of the fitting by applying a slight downward force against the fitting. It should be locked in place.

30. Install the intake plenum with new gasket onto the intake manifold. Loosely install the mounting bolts.

31. Install the EGR tube to the manifold with new gasket in place. Loosely install the mounting screws.

32. On the 3.3L engine, tighten the intake manifold plenum mounting bolts to 21 ft. lbs. (28 Nm) following the outlined sequence.

33. On the 2.7L engine, tighten the intake manifold plenum mounting bolts to 105 inch lbs. (12 Nm) following the outlined sequence.

34. On the 2.7L engine, install the left and right supports brackets to the manifold. Tighten the lower fasteners to 50 inch lbs. (6 Nm) and the upper fasteners to 105 inch lbs. (12 Nm).

35. Tighten the EGR tube mounting bolts.

36. Reconnect the PCV valve hose and power brake booster hose.

37. Attach the electrical connectors to the EGR transducer solenoid, idle air control motor, map and throttle position sensors.

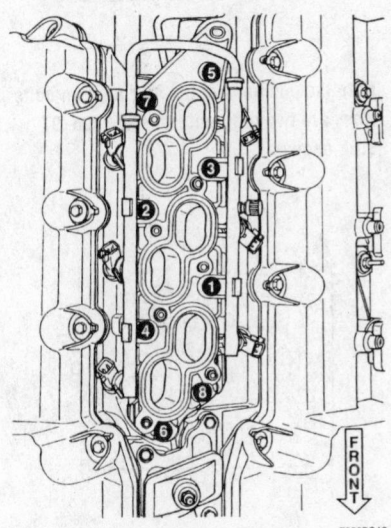

Loosening and tightening sequence for the intake manifold bolts—2.7L engine

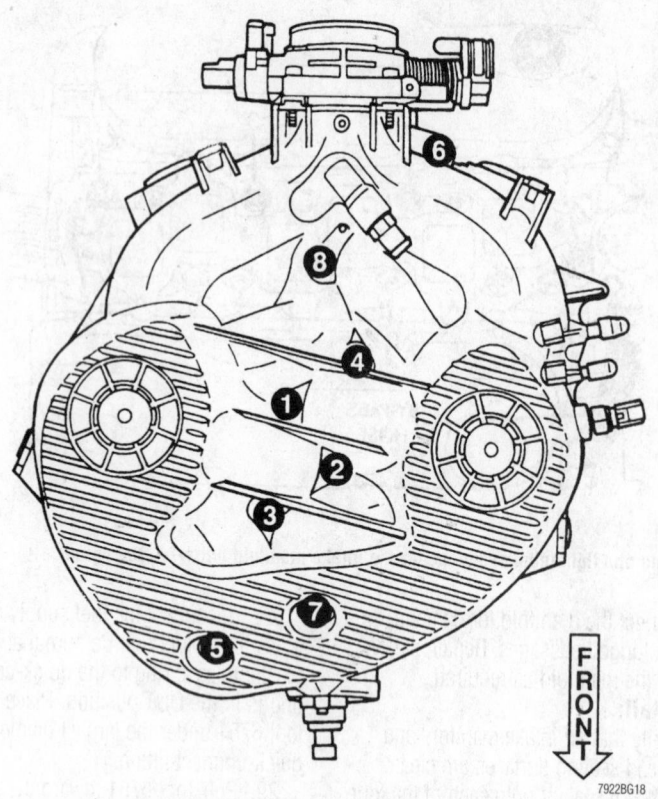

Tightening sequence for the intake plenum—2.7L engine

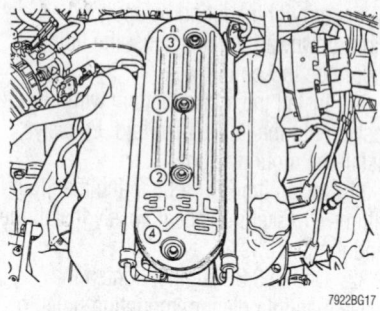

Tightening sequence for the intake plenum—3.3L engine

38. Install the throttle cable and speed control cable to the mounting bracket and connect to the throttle body lever while holding lever in the wide-open position.

39. Reconnect the purge hose to the throttle body. Reconnect the air tube to the air cleaner and the throttle body.

40. Drain and replace the engine oil and oil filter.

41. Reconnect the negative battery cable. Refill the cooling system. Run the vehicle with the radiator cap removed until the thermostat opens, adding coolant as required. Watch for fuel and coolant leaks and for correct engine operation.

42. Once the vehicle has cooled, recheck the coolant level and add, if necessary.

3.2L and 3.5L Engines

✳✳ **CAUTION**

The fuel injection system remains under pressure, even after the engine has been turned OFF. The fuel system pressure must be relieved before disconnecting any fuel lines. Failure to do so may result in fire and/or personal injury.

1. Disconnect negative battery cable.
2. Remove the fuel filler cap. Release the fuel system pressure using the recommended procedure.

✳✳ **CAUTION**

Do not open the radiator drain with the system hot and under pressure or serious burns from coolant may occur.

3. Drain the cooling system.
4. Remove the engine cover from the top of the intake manifold.
5. Remove the accelerator and the speed control cable from the throttle lever.

6. Unplug the electrical connector from the idle air control motor, intake air temperature sensor and the Manifold Absolute Pressure (MAP) sensor.

7. Remove the ground screw from the intake manifold. Unplug the electrical connector to the throttle position sensor.

8. Disconnect the vacuum hoses from the manifold tuning valve, PCV make-up air hose, idle air control motor supply hose and the purge hose from the throttle bodies.

9. Disconnect the brake booster hose, PCV hose and the remaining vacuum hoses from the intake manifold. If required, label for proper installation.

10. Remove the mounting bolts for the EGR tube at the intake manifold plenum.

11. Remove the plenum support bracket mounting bolts on each side of the plenum. Remove the intake plenum mounting bolts.

➡The intake manifold plenum (upper half of the intake manifold assembly) uses two different length bolts. Take note of their position and be sure they are installed in the same location during installation.

12. Remove the intake manifold plenum from the intake manifold. Discard the old gasket. Cover the intake manifold openings with tape to keep debris from entering the engine.

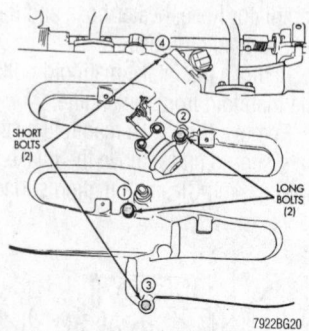

Note the positions of the four plenum bolts (they are two different sizes)—1995–97 3.5L engines

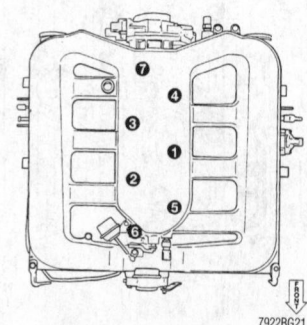

Upper intake manifold tightening and loosening sequence—1998–99 3.2L and 3.5L engines

13. Remove the upper radiator hose from the thermostat housing and the heater hose from the rear of the intake manifold.

14. Remove the lower intake manifold retaining bolts and the manifold from the engine. Clean all gasket mating surfaces and inspect for distortion with a good straightedge.

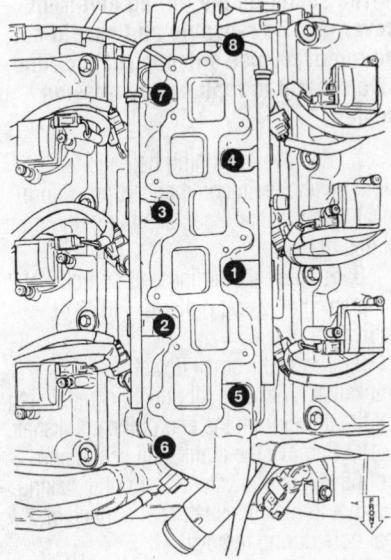

Loosening and tightening sequence for the lower manifold bolts—3.2L and 3.5L engines

To install:

15. Verify that all intake manifold and cylinder head sealing surfaces are clean. Carefully install the intake manifold gasket, then the lower manifold. Tighten bolts in proper sequence to 250 inch lbs. (28 Nm).

16. Reconnect the upper radiator hose to the thermostat housing and the heater hose to the rear of the intake manifold.

17. Ensure the ignition cables are routed out of the way of the intake plenum. Install the intake manifold plenum with new gasket in place and tighten in sequence to 250 inch lbs. (28 Nm).

18. Install the intake manifold plenum with a new gasket in place. Tighten mounting bolts working from the center outward, to 250 inch lbs. (28 Nm). Do not over-tighten bolts when working with light alloys.

19. Install and tighten the support bracket bolts.

20. Reconnect the electrical connectors to the manifold absolute pressure sensor, throttle position sensor, idle air control motor and intake air temperature sensor.

21. Reconnect the vacuum hose to the manifold tuning valve.

22. Install the EGR tube. Tighten the EGR tube to intake manifold plenum screws to 200 inch lbs. (22 Nm). Be sure that the insulation on the EGR tube aligns with and contacts the insulation on the vacuum harness at the rear of the engine. Rotate the throttle lever to the wide-open position and reconnect the speed control and throttle cables.

23. Reconnect the PCV valve hose. Install the air cleaner plenum and reconnect plenum hose.

24. Reconnect the ground wire the intake manifold plenum. Reconnect the brake booster hose to the fitting on the intake manifold plenum.

25. Reconnect the throttle body purge tubes.

26. Install the cover on the intake manifold plenum.

27. Reconnect the negative battery cable. Fill and bleed the cooling system.

28. Raise and safely support the vehicle. Change the engine oil and filter.

29. Test run the engine, check for fuel and coolant leaks and verify correct engine operation.

Exhaust Manifold

REMOVAL & INSTALLATION

2.7L Engine

RIGHT MANIFOLD

1. Disconnect the negative battery cable.

2. Remove the air intake plenum and the air filter housing.

3. Remove the bolt holding the battery cable housing tube to the transaxle.

4. Remove the EGR valve and tube from the vehicle.

5. Disconnect the oxygen sensor wiring harness and remove.

6. Loosen and remove the V-band clamp from the manifold.

➡**Do not reuse the V-band clamps.**

7. Remove the heat shield, then the manifold retaining bolts and remove the manifold.

8. Remove all traces of the old manifold gasket and clean both gasket mating surfaces.

To install:

9. Install the exhaust manifold and new gasket. Tighten the bolts to 200 inch lbs. (23 Nm) working from the center outward.

10. Install the heat shields and tighten the mounting bolts to 105 inch lbs. (12 Nm).

11. Install the new V-band clamp and tighten to 100 inch lbs. (11.3 Nm).

12. Install the oxygen sensor and connect its wiring harness.

13. Install the EGR valve and tube, using new gaskets, then tighten to 95 inch lbs. (11 Nm).

14. Attach the battery cable tube to the transaxle and tighten the bolt to 75 ft. lbs. (101 Nm).

15. Connect the air inlet plenum and install the air filter housing.

16. Connect the negative battery cable.

LEFT MANIFOLD

1. Disconnect the negative battery cable.

2. Raise and safely support the vehicle.

3. Remove the exhaust system, then the V-band clamps and left catalytic converter.

➡**Do not reuse the V-band clamps.**

4. Loosen and rotate out of the way the transaxle dipstick tube.

5. Lower the vehicle.

6. Remove the engine wiring harness support bracket from the cylinder head.

7. Disconnect the oxygen sensor wiring harness, then remove it from the manifold.

8. Remove the engine oil dipstick tube and the manifold heat shield.

9. Remove the manifold mounting bolts, then remove the manifold from the vehicle.

To install:

10. Remove all traces of the old manifold gasket and clean both gasket mating surfaces.

11. Install the exhaust manifold and new gasket. Tighten the bolts to 200 inch lbs. (23 Nm) working from the center outward.

12. Install the heat shields and tighten the mounting bolts to 105 inch lbs. (12 Nm).

13. Install the oxygen sensor and connect its wiring harness.

14. Raise and safely support the vehicle.

15. Reposition the transaxle dipstick tube.

16. Replace the o-ring for the engine oil dipstick tube and install the tube.

17. Install the catalytic converter, then the new V-band clamp and tighten to 100 inch lbs. (10 Nm).

18. Install the engine wiring harness support bracket to the cylinder head.

19. Install the exhaust system, then lower the vehicle.

20. Connect the negative battery cable.

3.2L and 3.5L Engines

1. Disconnect the negative battery cable. Raise and safely support the vehicle.

2. Disconnect the exhaust pipes from the exhaust manifold.

3. Disconnect the heated oxygen sensor electrical wiring.

4. Lower the vehicle and remove the screws attaching the heat shield to the exhaust manifold.

5. Remove the manifold attaching bolts, then remove the manifold from the cylinder head.

6. Inspect the manifold for damage or cracks. Check for distortion against a straightedge or thickness gauge. Replace manifold if required.

7. Remove all traces of the old manifold gasket and clean both gasket mating surfaces.

To install:

8. Install the new manifold gasket and exhaust manifold to the cylinder head. Install the retainer bolts and tighten to 15 ft. lbs. (20 Nm).

9. Raise and safely support the vehicle.

10. Reconnect the exhaust pipe to the exhaust manifold and tighten nuts to 21 ft. lbs. (28 Nm).

11. Lower the vehicle. Install the heat shield to the manifold and tighten the retaining screws to 11 ft. lbs. (15 Nm).

12. Reattach the electrical wiring on the oxygen sensor.

13. Connect the negative battery cable. Operate the vehicle and inspect for exhaust leaks.

3.3L Engines

1. Disconnect the negative battery cable.

2. Raise and safely support the vehicle. Disconnect the exhaust pipe from the exhaust manifold.

3. Disconnect the heated oxygen sensor lead wire and remove the EGR tube from the exhaust manifold.

4. Remove the screws attaching the heat shield to the exhaust manifold. Lower the vehicle, if necessary.

5. Remove the manifold attaching bolts and remove the manifold from the cylinder head.

6. These manifolds are thin-wall designs to save weight. Inspect the manifold carefully for cracks or other damage. Check for distortion against a straightedge or thickness gauge. Replace manifold if required.

7. Remove all traces of the old manifold gasket and clean both gasket mating surfaces.

To install:

8. Install the exhaust manifold and a new manifold gasket to the cylinder head. Install the retainer bolts and tighten to 17 ft. lbs. (23 Nm).

9. Reconnect the exhaust pipe to the exhaust manifold and tighten the nuts to 21 ft. lbs. (28 Nm).

10. Install the EGR tube back onto the manifold.

11. Install the heat shield to the manifold. Lower the vehicle, if necessary.

12. Attach the electrical connector at the oxygen sensor.

13. Reconnect the negative battery cable. Operate the vehicle and inspect for exhaust leaks.

Front Crankshaft Seal

➡ **The front crankshaft seal procedures are for timing belt equipped engines only. For engines which utilize timing chains, please refer to the applicable procedure later in this section.**

REMOVAL & INSTALLATION

3.2L and 3.5L Engines

Note that the timing belt must be removed from the vehicle to perform this service. Use care to be sure all valve timing marks are carefully aligned both before removing the belt and after belt installation and all service has been completed. It may be good practice to set the engine to TDC No. 1 cylinder compression stroke (firing position) and aligning all timing marks before removing the timing belt. This serves as a reference for all work that follows.

1. Disconnect negative battery cable.

✳✳ CAUTION

The fuel injection system remains under pressure, even after the engine has been turned OFF. The fuel system pressure must be relieved before disconnecting any fuel lines. Failure to do so may result in fire and/or personal injury.

2. Release the fuel system pressure using the recommended procedure.

3. Remove the radiator and cooling fan module assembly.

4. Remove the accessory drive belts.

5. Hold the crankshaft from turning and remove the crankshaft damper bolt. Use a balancer puller and remove the crankshaft damper.

6. Remove the stamped steel timing belt front cover.

➡ **The sealer on the timing belt front cover may be reusable and should not be removed. Use silicone rubber adhesive sealant to replace any missing sealer.**

7. Remove the timing belt and tensioner using the recommended procedure, located in the unit repair section in the front of this manual.

8. Remove the timing belt sprocket at the crankshaft using puller L-4407A or equivalent puller.

9. Locate the small dowel pin in the crankshaft. With a small punch, carefully tap out the dowel from the end of the crankshaft.

10. Remove the crankshaft seal using tool 6341A or equivalent seal puller, taking care not to nick the shaft seal surface or seal bore during removal.

To install:

11. Inspect the crankshaft seal lip surface for varnish and dirt. Polish area using 400 grit paper to remove varnish as necessary.

12. Install crankshaft seal using seal installer tool 6342 or equivalent seal driver.

13. Install the rear lower timing belt cover.

14. Install the dowel into the crankshaft to 0.047 in. (1.2mm) protrusion.

15. Install the timing belt sprocket at the crankshaft using tool C-4685C1, thrust bearing, washer and 12mm bolt or equivalent setup to pull the sprocket onto crankshaft. Do not hammer on the sprocket.

16. Verify that all valve timing marks are aligned.

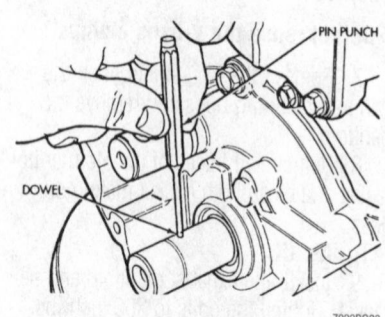

Removing the timing belt sprocket dowel pin on the crankshaft—3.2L and 3.5L engines

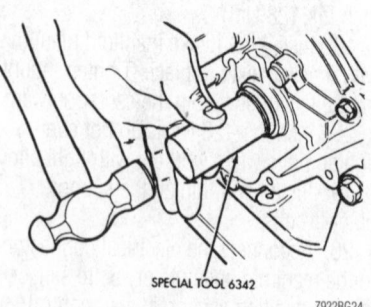

Installing the crankshaft oil seal—3.2L and 3.5L engines

17. Install the timing belt and tensioner using the recommended procedure.

18. Rotate the crankshaft two complete turns and recheck the timing marks on the camshafts and crankshaft. The marks must line up with their respective locations. If the marks do not line up, repeat the timing belt installation procedure. When correct valve timing has been verified, install the timing belt covers.

19. Install the crankshaft damper. Hold the crankshaft damper using tool L-3281 or equivalent and tighten the crankshaft bolt to 85 ft. lbs. (115 Nm).

20. Install the accessory drive belts and adjust to the proper tension.

21. Install the radiator and cooling fan assemblies.

22. Refill and bleed the cooling system.

23. Connect the negative battery cable.

Camshaft and Valve Lifters

REMOVAL & INSTALLATION

2.7L Engine

※※ CAUTION

When the timing chain is removed and the cylinder heads are installed, DO NOT turn the crankshaft or camshaft without first locating the proper crankshaft position. Failure to do so will result in piston-to-valve contact.

1. Remove the primary timing chain.

2. Remove the second chain tensioner mounting bolts.

3. Slowly loosen the camshaft bearing cap retaining bolts in the reverse order of installation.

4. Remove the bearing caps.

5. Remove the camshafts, secondary chain, and tensioner as an assembly.

6. Remove the tensioner and chain form the crankshaft.

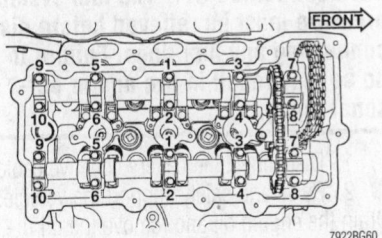

Camshaft bearing cap tightening sequence—2.7L engine

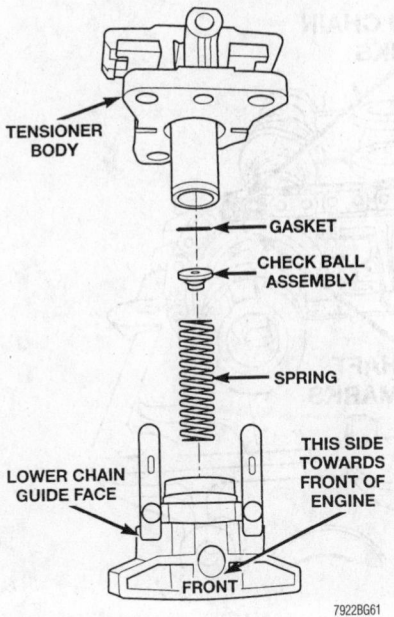

TENSIONER BODY

GASKET

CHECK BALL ASSEMBLY

SPRING

THIS SIDE TOWARDS FRONT OF ENGINE

LOWER CHAIN GUIDE FACE

FRONT

7922BG61

Exploded view of the camshaft (secondary) chain tensioner, early build—2.7L engine

To install:

7. Assemble the chain on the camshafts. Ensure the plated links are facing toward the front. Align the plated links to the dot on the camshaft sprocket.

➡**There are two different styles of camshaft (secondary) chain tensioners will be used. The Early Build tensioners will separate into subcomponents, the Later Build tensioners will not.**

8. Compress the camshaft (secondary) chain as follows:

9. For Early Build vehicles, perform the following:

   a. Separate the tensioner cylinder from the tensioner housing.

   b. Carefully drain the oil from the housing using care not to remove the internal tensioner components.

   c. Assemble the tensioner housing as shown.

   d. Using hand pressure, compress and lock the tensioner using a fabricated lockpin.

10. For Late Build vehicles, perform the following:

   a. Place the tensioner in a soft jawed vise.

   b. Slowly compress the tensioner until the fabricated lockpin can be installed.

   c. Remove the compressed and locked tensioner from the vise.

11. Insert the compressed and locked camshaft chain tensioner in-between the camshafts and chain.

12. Position the camshafts so that the plated links and dots are facing 12:00 O'clock.

13. Install the cams to the cylinder head. Ensure that the rocker arms are correctly seated and in proper positions.

14. Install the camshaft bearing caps in their original positions.

15. Tighten the camshaft bearing cap retaining bolts gradually in sequence, as shown, to 105 inch lbs. (12 Nm).

16. Install the secondary chain tensioner bolts and tighten to 105 inch lbs. (12 Nm).

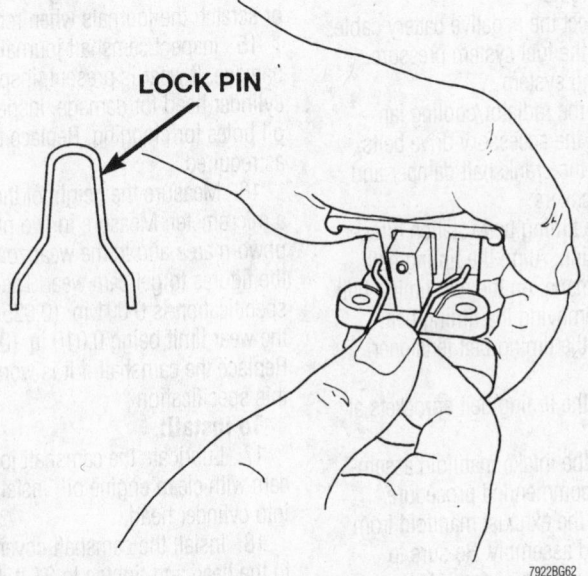

LOCK PIN

7922BG62

Fabricate a lockpin as shown, to keep the tensioner compressed—2.7L engine

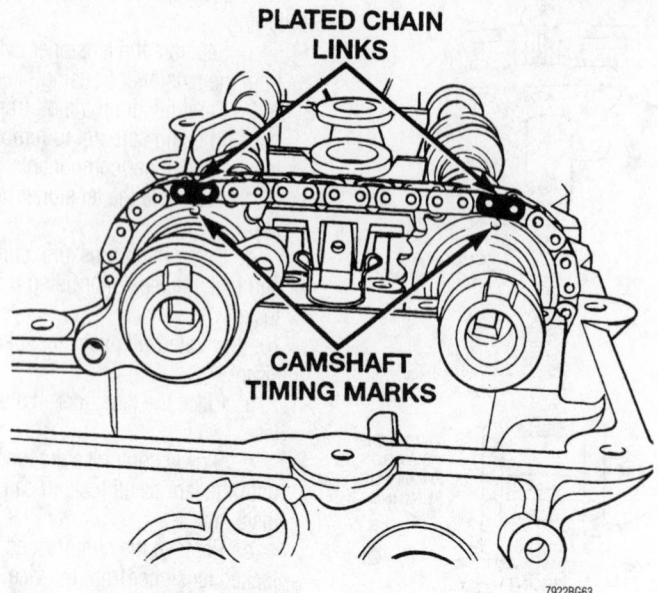

Proper camshaft (secondary) chain alignment—2.7L engine

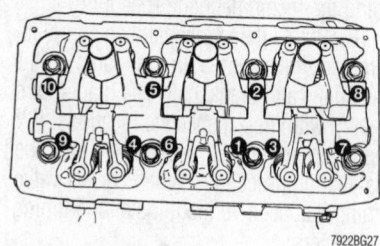

Rocker arm/shaft tightening sequence—3.2L and 3.5L engines

17. Remove the lockpin from the secondary chain tensioner.

18. Install the primary timing chain.

19. Connect the negative battery cable.

3.2L and 3.5L Engines

Camshafts are serviced from the rear of the cylinder head. Although the engine does not need to be removed for camshaft service, the cylinder head must be removed from the vehicle. Note too, that the camshaft sprockets have a D-shaped hole that allows it to rotate several degrees in each direction on its shaft. When cam sprockets are loosened or removed, the camshafts must be timed to the engine to ensure proper performance of the engine.

1. Disconnect the negative battery cable.

2. Release the fuel system pressure. Drain the cooling system.

3. Remove the radiator/cooling fan assemblies and the accessory drive belts.

4. Remove the crankshaft damper and the timing belt covers.

5. Mark the timing belt rotation direction for installation. Align the timing belt sprockets with marks on the rear timing belt covers before removing the timing belt.

6. Remove the timing belt tensioner and timing belt.

7. Remove the timing belt sprockets at each camshaft.

8. Remove the intake manifold assembly using the recommended procedure.

9. Separate the exhaust manifold from the cylinder head assembly. Be sure to clean the gasket mating surfaces between the exhaust manifold and the cylinder head.

10. The rear timing belt cover must be removed to remove the cylinder heads. Remove the rear timing belt cover-to-cylinder head bolts. Remove the rear timing belt covers. The right-hand side timing belt cover has O-rings located behind it for the water pump passages.

11. Remove the cylinder head bolts and remove the cylinder head from the engine.

12. Mark the rocker arm assembly to note component locations prior to disassembly. Remove the rocker arm and shaft assemblies from the cylinder head.

13. Remove the rear camshaft cover and O-ring from the head.

14. Carefully remove the camshaft from the rear of the head taking care not to nick or scratch the journals when removing.

15. Inspect camshaft journals for wear or damage. If wear is present, inspect cylinder head for damage. Inspect the head oil holes for clogging. Replace the camshaft as required.

16. Measure the height of the cam using a micrometer. Measure in two places; the unworn area and in the wear zone. Subtract the figures to get cam wear. The standard specification is 0.001 in. (0.0254mm) with the wear limit being 0.010 in. (0.254mm). Replace the camshaft if it is worn beyond this specification

To install:

17. Lubricate the camshaft journals and cam with clean engine oil. Install camshaft into cylinder head.

18. Install the camshaft cover and O-ring to the head and tighten to 21 ft. lbs. (28 Nm).

19. Install the rocker arm assemblies in their original location.

20. Install the cylinder head assembly to the engine block. New head bolts are recommended.

21. Install the rear timing belt covers and tighten the rear cover-to-cylinder head bolts.

22. Install the timing belt sprocket, timing belt and timing belt tensioner using the recommended procedure. Be sure that once they are all installed, the timing of the camshaft(s) is accurate.

23. Install the exhaust manifold, with a new manifold gasket, to the cylinder head.

24. Install the intake manifold assembly using the recommended procedure.

25. Install the timing belt covers and crankshaft damper. Install the accessory drive belts and set them to the proper tension.

26. Install the radiator and cooling fan assembly.

27. Refill and bleed the cooling system. An oil and filter change is recommended.

28. Reconnect the negative battery cable.

3.3L Engines

To remove and replace the camshaft on this engine and in this vehicle, the engine assembly must be removed from the vehicle.

※※ CAUTION

The fuel injection system remains under pressure, even after the engine has been turned OFF. The fuel system pressure must be relieved before disconnecting any fuel lines. Failure to do so may result in fire and/or personal injury.

1. Disconnect the battery negative cable.

2. Raise and safely support the vehicle. Drain the engine oil and remove the oil filter.

3. Relieve the fuel system pressure using the recommended procedure.

Do not open the radiator drain cock with the system hot and under pressure or serious burns from coolant may occur.

4. Remove the radiator and cooling fan assemblies.

5. Remove the engine from the vehicle using the recommended procedure.

6. With the engine removed from the vehicle, remove the cylinder head covers, rocker arm and rocker arm shaft assemblies.

7. Remove the intake manifold assembly and cylinder heads from the engine.

8. Remove the harmonic balancer. Remove the timing chain case cover and timing chain from the engine.

9. Remove the pushrods and tappets. If any valvetrain components are to be reused, identify each part and its location so each can be installed in its original location.

➡ **If the camshaft is being removed to replace with a new one, new valve lifters MUST be installed. Installing used lifters on a new camshaft will quickly fail the camshaft.**

10. Remove the camshaft thrust plate. Install a long bolt into the front of the camshaft to act as a handle and aid in removal. Remove the camshaft being careful not to damage the cam bearings with the cam lobes.

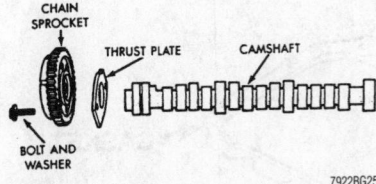

Sprocket, thrust plate and camshaft in the order of removal—3.3L engine

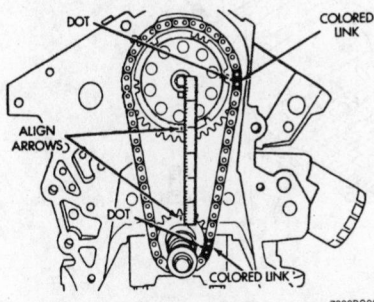

Using a straightedge, check the alignment of the timing arrows during timing chain installation—3.3L engine

11. Inspect the bearing journals and the lobes on the shaft for damage and replace the camshaft is required. New lifters must be installed on a new camshaft.

To install:

12. Lubricate the camshaft lobes and the camshaft bearing journals. Inspect the bearing journals on the camshaft and install the shaft within 2 in. (5cm) of its final position in the cylinder block.

➡ **Chrysler recommends the addition of 1 pint (0.473L) of Chrysler Crankcase Conditioner or equivalent, be added to the crankcase when the camshaft has been replaced. This will aid in break-in. Leave the oil mixture in the engine for a minimum of 500 miles (804 km) and drain at the next normal oil change.**

13. Install the camshaft thrust plate with the 2 screws and tighten to 105 inch lbs. (12 Nm).

14. Rotate the crankshaft so the timing arrow is in the 12 o'clock position.

15. Position the camshaft sprocket so the timing arrow is at the 6 o'clock position. Place the timing chain around the camshaft sprocket aligning the dark colored link of the chain with the dot on the camshaft sprocket.

16. Place the timing chain around the crankshaft sprocket aligning the dot on the crankshaft sprocket with the dark colored link on the chain. Install the camshaft sprocket in position on the shaft.

17. Using a straightedge, check the alignment of the timing arrows. Install the camshaft bolt and washer and tighten to 40 ft. lbs. (54 Nm).

18. Rotate the crankshaft 2 revolutions in the direction of engine rotation. Check the alignment of the timing arrows, which should line up with each other.

   a. If they do not align, remove the camshaft sprocket and re-time the engine.

   b. Again, rotate the crankshaft 2 revolutions in the direction of rotation, and confirm alignment of the timing marks.

19. Check the camshaft end-play. With new thrust plate the specification is 0.005–0.012 in. (0.0127–0.3040mm) or 0.012 in. (0.3040mm) for old thrust plate. If not within specifications, replace thrust plate.

20. Lubricate and install the valve lifters (tappets) in their original position. If the camshaft was replaced, all lifters must be replaced with new parts.

21. Install the timing chain front cover using new seals and O-rings.

22. Install the cylinder heads and intake manifold assemblies onto the engine using the recommended procedures.

23. Lubricate and install the pushrods in their original position.

24. Install the rocker arm and rocker arm shaft assemblies. Install the cylinder head covers.

25. Tighten the oil pan drain plug and install a new oil filter.

26. Install the engine into the vehicle using the recommended procedure.

27. Install the radiator and cooling fan assemblies.

28. Be sure that all fluid lines, cables, hoses, and electrical connectors are attached and secured.

29. Refill the engine with the correct amount of clean SAE 5W-30 **OR** SAE 10W-30 engine oil only. Do not mix the 2 grades of oil.

30. Refill and bleed the cooling system.

31. Reconnect the negative battery cable. Start the engine and inspect for leaks. Test drive the vehicle.

32. Check engine fluid levels and top off if necessary.

Valve Lash

ADJUSTMENT

These engines use hydraulic roller lifters to take up the free play in the valve train system, therefore no lash adjustments are necessary.

Oil Pan

REMOVAL & INSTALLATION

The EPA warns that prolonged contact with used engine oil may cause a number of skin disorders, including cancer! You should make every effort to minimize your exposure to used engine oil. Protective gloves should be worn when changing the oil. Wash your hands and any other exposed skin areas as soon as possible after exposure to used engine oil. Soap and water, or waterless hand cleaner should be used.

2.7L, 3.2L, and 1998–99 3.5L Engines

1. Disconnect the negative battery cable.

2. Remove the dipstick and housing.

3. Raise and safely support the vehicle, then drain the engine oil and remove the oil filter.

4. Remove the structural collar from the rear of the oil pan and transmission housing.

5. If equipped, remove the engine oil cooler lines from the oil pan.

6. If necessary, remove the transmission oil cooler line clips.

7. Remove the oil pan mounting bolts, then remove the oil pan and gasket.

8. Clean the oil pan and all gasket surfaces.

To install:

9. Apply a ⅛ in. (3mm) bead of sealer at the parting line of the oil pump body and the rear seal retainer.

10. Install the oil pan to the engine block and tighten the M8 nuts/bolts to 250 inch lbs. (28 Nm) and the M6 nuts/bolts to 105 inch lbs. (12 Nm).

11. Install the structural collar utilizing the following procedure:

a. Install the vertical collar to the oil pan mounting bolts and tighten, temporarily to 10 inch lbs. (1.1 Nm).

b. Install the collar-to-transaxle bolts and tighten to 40 ft. lbs. (55 Nm).

c. Starting with the center vertical bolt and working outward, tighten the bolts to 40 ft. lbs. (55 Nm).

12. Lower the vehicle and install the dipstick and housing.

13. Refill the engine with the proper amount of clean SAE 5W-30 **OR** SAE 10W-30 engine oil only. Do not mix the 2 grades of oil.

14. Connect the negative battery cable.

15. Start the engine and check for leaks.

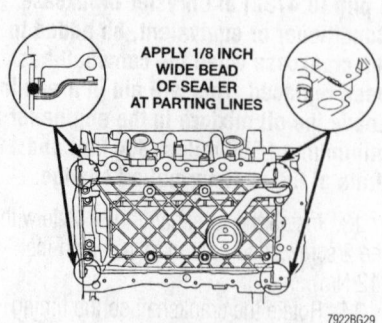

APPLY 1/8 INCH WIDE BEAD OF SEALER AT PARTING LINES

7922BG29

To ensure a proper seal, apply sealer as shown—2.7L and 1998–99 3.5L engines

3.3L Engines

1. Disconnect negative battery cable.
2. Remove the engine oil dipstick.
3. Raise and safely support the vehicle.
4. Place a drain pan underneath the vehicle and drain the engine oil. Remove the oil filter.
5. Disconnect the sway bar and move to the rear of the vehicle, if necessary.
6. Remove the transaxle support bracket and inspection cover.
7. Remove the oil pan bolts, then the oil pan.
8. Remove the oil pick-up tube, if necessary. Discard the old oil pick-up tube O-ring.

To install:

9. Thoroughly clean and dry the oil pan, cylinder block bolts and bolt holes. Inspect the oil pan flange for bends or distortion. Straighten the flange if necessary. Clean the oil screen and pipe in clean solvent. Inspect the condition of the screen and replace if necessary.

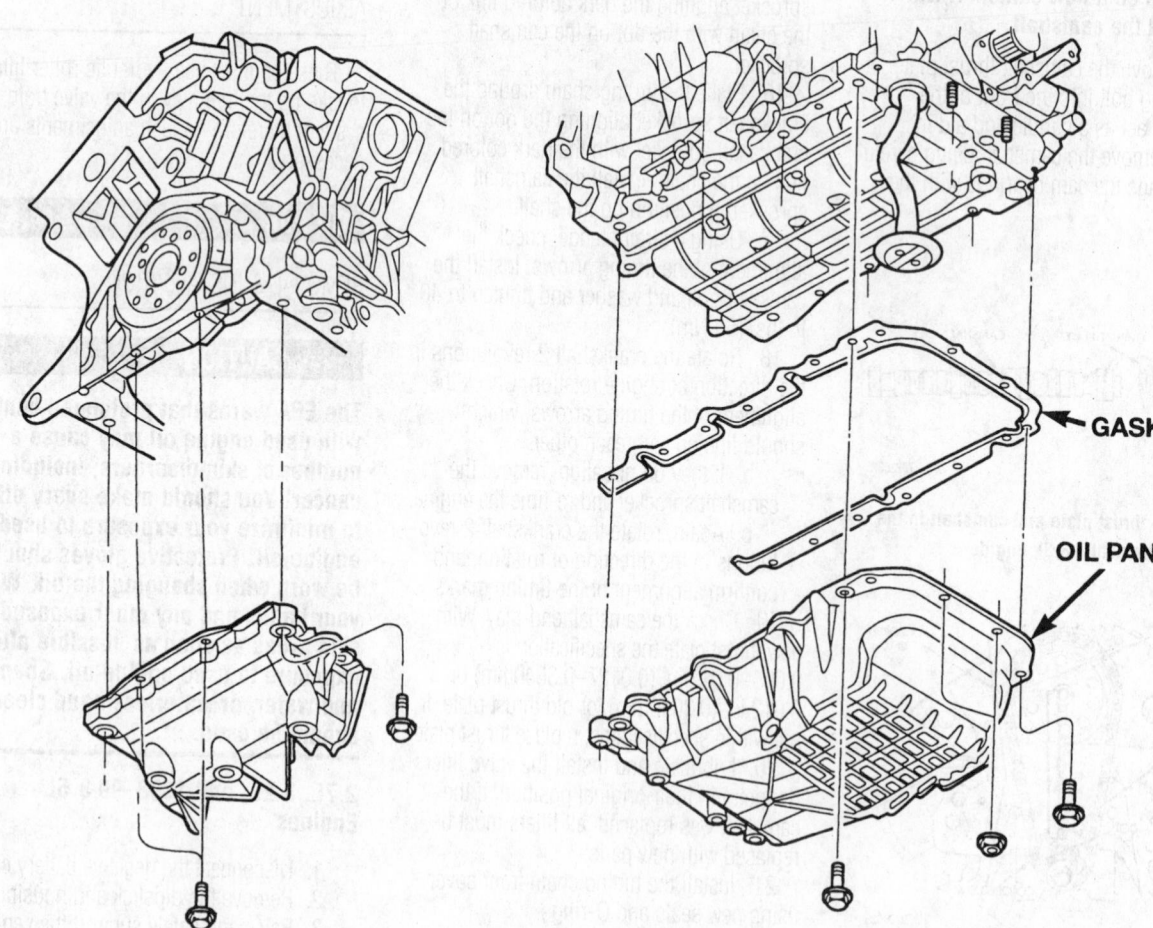

GASKET

OIL PAN

7922BG28

Exploded view of the oil pan removal and installation—2.7L and 1998–99 3.5L engines

10. Apply a ⅛ in. (3mm) bead of sealer at the parting line of the chain case cover and the rear seal retainer.

11. Install a new O-ring on the oil pick-up tube and install it into the pump body. Tighten the screws to 20 ft. lbs. (28 Nm), if removed.

12. Install a new oil pan gasket.

13. Install the oil pan and retaining bolts and tighten to 9 ft. lbs. (12 Nm).

14. Install the transaxle support bracket and inspection cover.

15. Reconnect the sway bar.

16. Tighten the oil pan drain plug and install a new oil filter.

17. Lower the vehicle and install the oil dipstick.

18. Refill the engine with the proper amount of clean SAE 5W-30 **OR** SAE 10W-30 engine oil only. Do not mix the 2 grades of oil.

19. Reconnect the negative battery cable.

20. Start the engine and check for leaks.

1995–97 3.5L Engine

1. Disconnect negative battery cable.

2. Remove the engine oil dipstick.

3. Raise and safely support the vehicle.

4. Place a drain pan underneath the vehicle and drain the engine oil. Remove the oil filter.

5. Disconnect the sway bar and move to the rear of the vehicle, if necessary.

6. Remove the transaxle support bracket and inspection cover.

7. Remove the oil pan screws and remove the oil pan.

8. Remove the oil pick-up tube, if necessary and the windage tray/oil pan gaskets. The windage tray and oil pan gasket are integral. The silicone rubber gaskets are bonded directly to both sides of the windage tray. This assembly is reusable if it is not damaged upon removal. Discard the old oil pick-up tube O-ring.

➡Any old sealant must be carefully removed if the gasket is going to be used again.

To install:

9. Thoroughly clean and dry the oil pan, cylinder block bolts and bolt holes. Inspect the oil pan flange for bends or distortion. Straighten flange if necessary. Clean the oil screen and pipe in clean solvent. Inspect the condition of the screen and replace if necessary.

10. Apply a ⅛ in. (3mm) bead of sealer at the parting line of the oil pump body and the rear seal retainer.

11. Install a new O-ring on the oil pick-up tube and install into the pump body. Tighten the screws to 20 ft. lbs. (28 Nm), if removed.

12. Install the windage tray/oil pan gasket.

13. Install the oil pan and retaining bolts. Tighten screws to 9 ft. lbs. (12 Nm).

14. Install the transaxle support bracket and inspection cover.

15. Reconnect the sway bar.

16. Tighten the oil pan drain plug and install a new oil filter.

17. Lower the vehicle and install the oil dipstick.

18. Refill the engine with the proper amount of clean SAE 5W-30 **OR** SAE 10W-30 engine oil only. Do not mix the 2 grades of oil.

19. Reconnect the negative battery cable.

20. Start the engine and check for leaks.

Oil Pump

REMOVAL & INSTALLATION

✳✳ CAUTION

The EPA warns that prolonged contact with used engine oil may cause a number of skin disorders, including cancer! You should make every effort to minimize your exposure to used engine oil. Protective gloves should be worn when changing the oil. Wash your hands and any other exposed skin areas as soon as possible after exposure to used engine oil. Soap and water, or waterless hand cleaner should be used.

2.7L Engine

1. Remove the crankshaft damper, timing chain cover, timing chain, crankshaft sprocket, and oil pan.

2. Unbolt the oil pick-up tube and remove the O-ring.

3. Remove the oil pump mounting bolts, then remove the pump.

To install:

4. Fill the oil pump rotor cavity with clean engine oil.

5. Carefully install the oil pump over the crankshaft and into position.

6. Install the oil pump mounting bolts and tighten to 250 inch lbs. (28 Nm).

7. Lubricate the new pick-up tube O-ring with clean engine oil and tighten the pick-up tube mounting bolts to 250 inch lbs. (28 Nm).

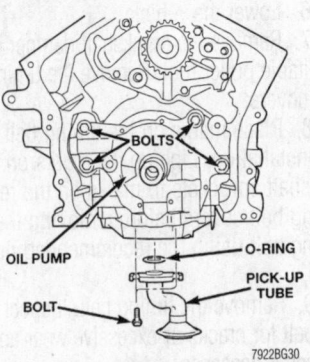

Oil pump mounting bolt locations—2.7L engine

8. Install the oil pan, crankshaft sprocket, timing chain, timing chain cover, and crankshaft damper.

9. Fill the crankcase with clean engine oil to the proper level.

3.2L and 3.5L Engines

Please note that the timing belt must be removed to access the oil pump behind the crankshaft drive sprocket. It is good practice to turn the crankshaft to TDC No. 1 cylinder compression stroke (firing position) before starting disassembly. This should align all timing marks and be a good point of reference for all work to follow.

1. Disconnect the negative battery cable.

✳✳ CAUTION

Do not open the radiator drain or the coolant pressure bottle cap with the system hot and under pressure or serious burns from coolant may occur.

2. The cooling system must be drained and the radiator removed for access to the timing belt covers and vibration damper. Use the following procedure.

a. Place a drain pan under the radiator. Open the radiator drain fitting located at the lower right side of the radiator. Do **NOT** use pliers to open the plastic drain fitting.

b. Remove the coolant pressure bottle cap and open the thermostat bleed valve.

c. Remove the radiator hoses and radiator for access to the timing belt covers and crankshaft damper.

3. Remove the accessory drive belts.

4. Raise and safely support the vehicle. Place a drain pan under the vehicle and drain the engine oil.

5. Remove the oil filter. Remove the oil pan, oil pump pick-up tube and windage tray/oil pan gasket.

6. Lower the vehicle.

7. Remove the crankshaft damper using a suitable puller tool. Remove the timing belt covers.

8. Place marks on the timing belt to aid installation. Line up the marks on the camshaft sprockets to marks on the rear timing belt covers before removing the timing belt using the recommended procedure.

9. Remove the timing belt. Inspect timing belt for cracks or excessive wear and replace if necessary.

10. Remove the crankshaft sprocket using a suitable puller tool.

11. Remove the oil pump mounting screws and the pump from the engine.

12. Remove the oil pump cover retaining screws and lift off the oil pump cover.

13. Remove the pump rotors. Wash all parts in solvent and inspect carefully for damage or wear.

To install:

14. Clean all parts well. There should be no traces of old gasket/sealer on any components.

15. Assemble the oil pump with new parts as required.

16. Install the pump cover and tighten the fasteners to 9 ft. lbs. (12 Nm).

17. Prime the oil pump prior to installation by filling the rotor cavity with clean engine oil.

18. Install the oil pump over the crankshaft and carefully into position. Tighten the retaining screws as follows:
- M8 screws—21 ft. lbs. (28 Nm)
- M10 screws—40 ft. lbs. (55 Nm)

19. Raise and safely support the vehicle. Tighten the oil pan drain plug and install a new oil filter.

20. Install the oil pump pick-up tube, windage tray/oil pan gasket and the oil pan. Tighten the oil pan fasteners to 9 ft. lbs. (12 Nm). Pay attention to sealing the oil pan gasket and its integral windage tray.

21. Lower the vehicle.

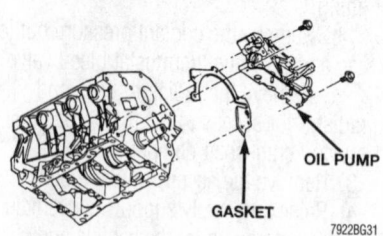

7922BG31

Be sure to prime the oil pump before installation, because a dry pump will wear prematurely and cause low oil pressure— 3.2L and 3.5L engines

22. Install the crankshaft sprocket using tool C-4685C1, thrust bearing, washer and 12mm bolt to draw the sprocket onto the crankshaft.

23. Install the timing belt and set to the correct tension using the recommended procedure.

24. Install the timing belt covers and install the vibration damper using tool L-4524, thrust bearing and washer plate or equivalent damper installation tools.

25. Install the accessory drive belts and adjust to the correct tension.

26. Install the radiator into the vehicle. Install the radiator hoses.

27. Refill and bleed the cooling system.

28. Refill the engine with the correct amount of clean SAE 5W-30 **OR** SAE 10W-30 engine oil only. Do not mix the 2 grades of oil.

29. Reconnect the negative battery cable.

30. Run the engine. Check for leaks and proper oil pressure.

3.3L Engines

1. Disconnect negative battery cable and drain the cooling system.

2. Remove the radiator assembly.

3. Raise and safely support the vehicle.

4. Drain the engine oil. Remove the oil filter.

5. Disconnect the sway bar and place it to the rear of the vehicle to gain access to the oil pan.

6. Remove the transmission support brackets and inspection cover.

7. Remove the oil pan and the oil pump pick-up tube.

8. Lower the vehicle.

9. Remove the accessory drive belts and tensioner pulley bracket.

10. Remove the power steering pump and set aside. Remove the air compressor mounting bolts and set compressor aside. Remove the compressor mounting bracket. It is not necessary to disconnect the refrigerant lines or evacuate the refrigerant from the air conditioning system.

11. Using an appropriate puller, remove the crankshaft pulley.

12. Remove the tensioner pulley bracket.

13. Remove the camshaft sensor from the chain case cover.

14. Remove the timing chain case cover mounting bolts and the cover from the front of the engine.

15. Clean the gasket material from the mating surfaces of the cover and the block.

16. Remove the oil pump cover retaining screws from the timing cover. Lift off the oil pump cover.

17. Remove the pump rotors. Wash all parts in solvent and inspect carefully for damage or wear.

To install:

18. Clean all parts well. Assemble the oil pump with new parts as required. Install the inner rotor with chamfer facing the cast iron oil pump cover.

19. Install the pump cover and tighten the fasteners to 9 ft. lbs. (12 Nm).

20. Prime the oil pump prior to installation by filling the rotor cavity with clean engine oil.

21. Remove the crankshaft oil seal from the front cover.

22. Install a new cover gasket and O-ring onto the cover.

23. Rotate the crankshaft so the oil pump drive flats are vertical. Position the oil pump inner rotor so the mating flats are in the same position as the crankshaft drive flats.

24. Install the front cover making sure the pump is correctly engaged on the crankshaft, or severe damage may result.

25. Install the chain case cover screws and snug the two bottom screws and the top center screw. Ensure the cover is seated to the block, then tighten all screws to 20 ft. lbs. (27 Nm).

26. Install the oil seal and the crankshaft damper.

27. Install the tensioner pulley bracket and the cam sensor.

28. Install the air conditioning compressor to the mounting bracket.

29. Install the accessory drive belt.

30. Raise and safely support the vehicle. Tighten the oil pan drain plug and install a new oil filter.

31. Install the oil pump pick-up tube and oil pan. Install the transaxle inspection cover, if removed.

32. Fill the crankcase with clean engine oil to the proper level. Install a new oil filter.

33. Install radiator assembly. Check condition of radiator hoses. Fill and bleed the cooling system.

34. Reconnect the negative battery cable. Run engine and check for leaks. Verify correct oil pressure with a gauge.

Rear Main Seal

REMOVAL & INSTALLATION

1. Disconnect the negative battery cable.

2. Raise and support the vehicle.

3. Remove the transaxle inspection cover and flywheel/flexplate.

4. Using a small prytool, carefully pry out the rear oil seal. Be careful not to nick

or damage the crankshaft flange seal surface or the retainer bore.

To install:

5. Place the Seal Pilot Tool C-4681 (or equivalent) onto the crankshaft.

6. Lightly coat the oil seal outside diameter with Loctite Stud N' Bearing Mount® or the equivalent.

7. Apply a light coating of engine oil to the entire circumference of the oil seal lip.

8. Place the seal over the special tool and tap the seal in place with a plastic mallet.

9. Install the flexplate or flywheel. Install the transaxle.

10. Lower the vehicle and connect the battery cable.

Timing Chain, Sprockets, Front Cover and Seal

REMOVAL & INSTALLATION

2.7L Engine

✳✴ CAUTION

When aligning the timing marks, rotate the engine using the crankshaft, not the camshafts. DO NOT rotate the camshafts or crankshaft with the tim-

ing chain removed without locating the crankshaft position, piston and/or valve damage may occur.

1. Remove the upper intake manifold and valve covers.

2. Remove the upper radiator cross-member.

3. Remove the fan module.

4. Remove the accessory drive belts.

5. Using Crankshaft Damper Holder tool 8191, or equivalent, hold the crankshaft and remove the center bolt.

6. Remove the damper using a three jaw puller.

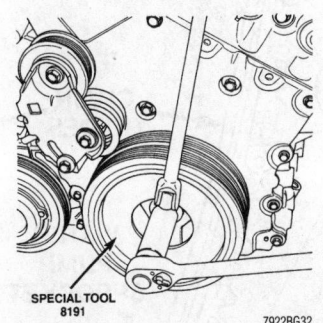

SPECIAL TOOL
8191
7922BG32

Removing the crankshaft center bolt using the Crankshaft Damper Holder tool—2.7L engine

7. Remove the power steering pump, and position it aside without disconnecting the hydraulic lines.

8. Remove the accessory drive belt tensioner pulley.

9. Remove the timing chain cover mounting bolts.

10. Clean and inspect the sealing surfaces.

11. Align the crankshaft sprocket timing mark to the mark on the oil pump housing.

➡**The mark on the oil pump housing is 60° After Top Dead Center (ATDC).**

12. Remove the primary timing chain tensioner from the right cylinder head.

13. Remove the camshaft position sensor and the timing chain access plug from the left cylinder head.

➡**The camshafts will rotate clockwise, when the camshaft sprocket bolts are removed.**

14. Starting with the right camshaft, remove the sprocket mounting bolts, camshaft damper, and the sprocket.

15. Remove the left camshaft sprocket attaching bolts and sprocket.

16. Remove the lower timing chain guide and tensioner arm, then the primary timing chain.

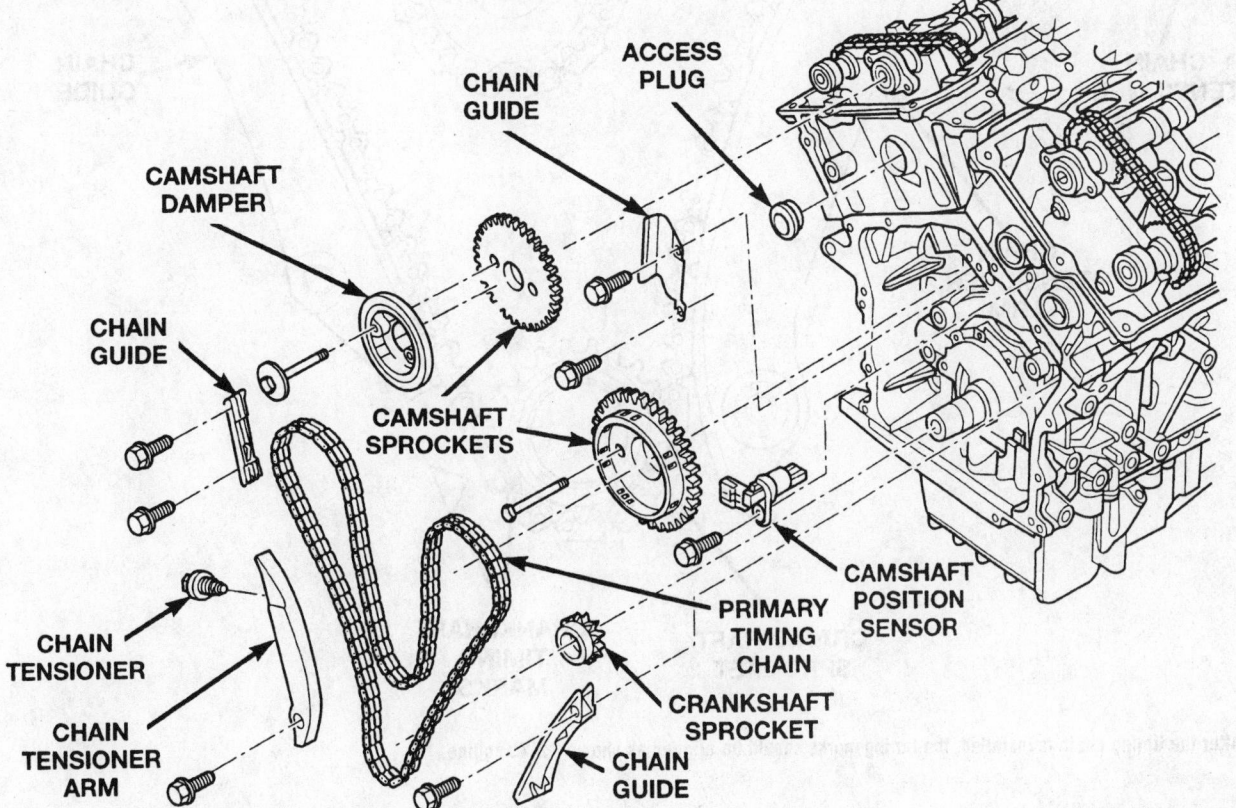

7922BG33

Exploded view of the timing chain drive assembly—2.7L engine

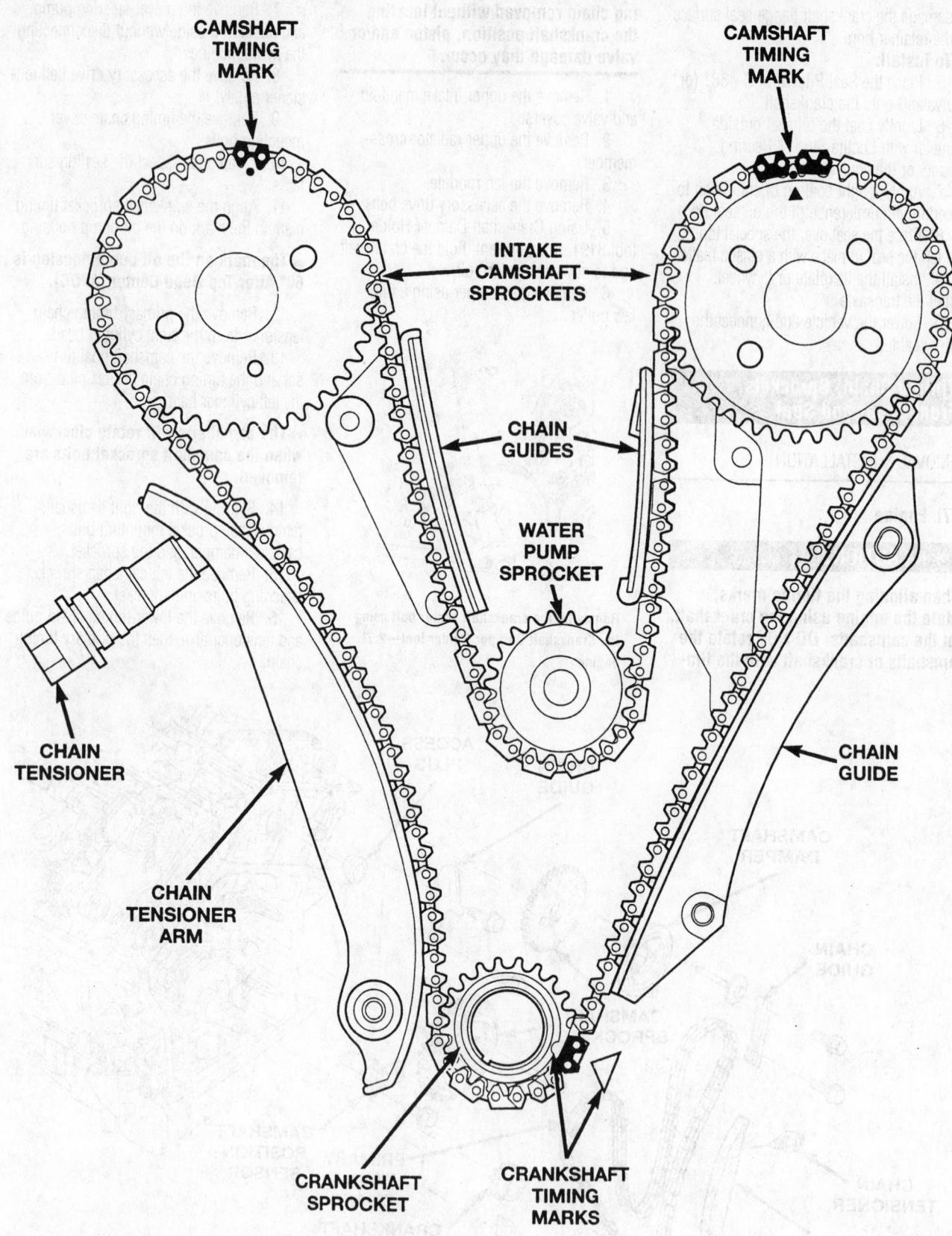

CAMSHAFT TIMING MARK

CAMSHAFT TIMING MARK

INTAKE CAMSHAFT SPROCKETS

CHAIN GUIDES

WATER PUMP SPROCKET

CHAIN GUIDE

CHAIN TENSIONER

CHAIN TENSIONER ARM

CRANKSHAFT SPROCKET

CRANKSHAFT TIMING MARKS

7922BG34

After the timing chain is installed, the timing marks should be aligned as shown—2.7L engine

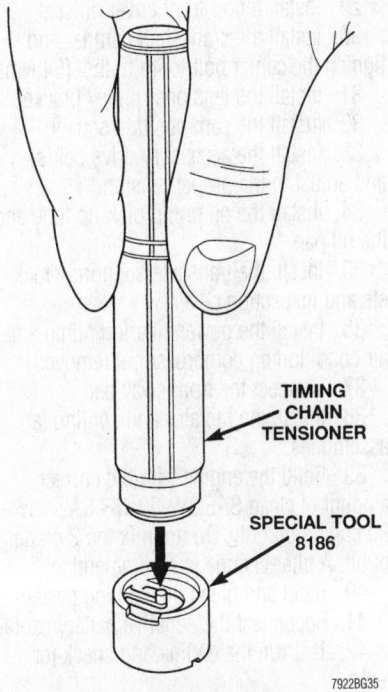

Using the Tensioner Resetting Special tool, or equivalent, to purge the oil from the tensioner—2.7L engine

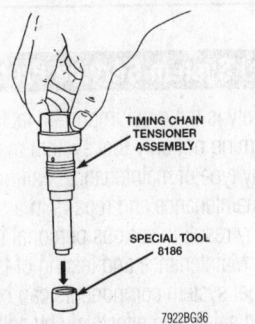

Resetting the timing chain tensioner using the Tensioner Resetting Special tool, or equivalent—2.7L engine

To install:

➡**Lubricate the timing chain and guides with clean engine oil before installation.**

17. Verify that the crankshaft sprocket timing mark is aligned with the mark on the oil pump housing.

18. Place the left side primary timing chain sprocket onto the chain, while aligning the timing mark on the sprocket to the two plated links on the chain.

19. Lower the chain with the left sprocket through the left cylinder head opening.

20. Loosely position the left camshaft sprocket over the camshaft hub.

21. Align the plated link to the crankshaft sprocket timing mark.

22. Position the timing chain around the water pump drive sprocket.

23. Align the right camshaft sprocket timing mark to the plated link on the timing chain and loosely position the sprocket over the camshaft hub.

24. Verify that all the plated links are aligned to their proper timing marks.

25. Install the left lower timing chain guide and tensioner, then tighten the bolts to 250 inch lbs. (28 Nm).

➡**Inspect the timing chain guide access plug O-rings before installing. Replace damaged O-rings as necessary.**

26. Install the timing chain guide access plug to the left cylinder head and tighten to 15 ft. lbs. (20 Nm).

➡**To reset the timing chain tensioner, oil will first need to be purged from the tensioner.**

27. Purge oil from the timing chain tensioner using the following procedure:

a. Remove the tensioner from the tensioner housing.

b. Place the check ball end of the tensioner into the shallow end of the Tensioner Resetting Special tool 8186 or equivalent.

c. Using hand pressure, slowly depress the tensioner until oil is purged from the cylinder.

d. Reinstall the tensioner into the tensioner housing.

28. Reset the timing chain tensioner using the following procedure:

a. Position the cylinder plunger into the deeper side of the Tensioner Resetting Special tool 8186 or equivalent.

b. Apply a downward force until the tensioner is reset.

❋❋ CAUTION

Ensure that the tensioner is properly reset. The tensioner body must be bottom against the top edge of the Tensioner Resetting Special tool 8186 or equivalent. Failure to properly perform the resetting procedure may cause tensioner jamming.

29. Install the chain tensioner into the right cylinder head.

30. Starting at the right cylinder head, insert a ⅜ in. (9.5mm) square drive extension with a breaker bar into the intake camshaft drive hub. Rotate the camshaft until the camshaft hub aligns with the camshaft sprocket and damper attaching holes. Install the sprocket attaching bolts and tighten to 250 inch lbs. (28 Nm).

31. Turn the left camshaft by inserting a ⅜ in. (9.5mm) square drive extension with a breaker bar into the intake camshaft drive hub. Rotate the camshaft until the camshaft hub aligns with the camshaft sprocket and damper attaching holes. Install the sprocket attaching bolts and tighten to 250 inch lbs. (28 Nm).

32. If necessary, rotate the engine slightly clockwise to remove any slack in the timing chain.

33. To arm the timing chain tensioner: Use a flat bladed prytool to gently pry the tensioner arm towards the tensioner slightly. Then, release the tensioner arm. Verify the tensioner extends.

34. Inspect and replace the timing chain cover gasket and oil seal.

35. Apply a ⅛ in. (3mm) bead of sealer at the parting line of the oil pan and engine block.

➡**When installing the timing cover, guide the seal over the crankshaft to prevent damage to the lip of the seal.**

36. Install the timing cover and gasket, then tighten the M10 bolts to 40 ft. lbs. (54 Nm) and the M6 bolts to 105 inch lbs. (12 Nm).

37. Install the crankshaft damper.

38. Install the accessory drive belt tensioner pulley.

39. Install the power steering pump.

40. Install the crankshaft damper onto the crankshaft.

41. Utilizing tool 8191, tighten the crankshaft center bolt to 125 ft. lbs. (170 Nm).

42. Install the accessory drive belts.

43. Install the fan module and engage its electrical wiring harness.

44. Install the upper radiator crossmember.

45. Connect the negative battery cable.

3.3L Engines

1. Relieve the fuel system pressure using the recommended procedure.

2. Rotate the engine to Top Dead Center (TDC) on No. 1 cylinder. This provides a reference point.

3. Disconnect negative battery cable.

❋❋ CAUTION

The fuel injection system remains under pressure even after the engine has been turned OFF. The fuel system pressure must re-relieved before disconnecting any fuel lines. Failure to do so may result in fire and/or personal injury.

4. Drain the cooling system. Remove the radiator and cooling fan assemblies.

5. Raise and safely support the vehicle. Place a drain pan under the engine oil pan and drain the engine oil.

6. Disconnect the sway bar and place it to the rear of the vehicle to gain access to the oil pan.

7. Remove the transaxle supports brackets and inspection cover.

8. Remove the engine oil pan and oil pump pick-up.

9. Remove the accessory drive belt(s).

10. Remove the power steering pump and set aside. It may be necessary to also remove the air conditioning compressor and set aside.

11. Remove the crankshaft damper using a puller to draw the damper from the crankshaft.

12. Remove the tensioner pulley bracket.

13. Remove the cam sensor from the chain case cover.

14. Remove the timing chain front cover and remove the front cover oil seal.

15. If the chain is found to be out of specification, remove the camshaft sprocket attaching bolt, and remove the timing chain with the camshaft sprocket.

16. Using a gear puller, remove the crankshaft sprocket. Be careful not to damage the crankshaft surface.

To install:

17. Position a new crankshaft sprocket onto the shaft, then install the sprocket with a suitably-sized socket and a rubber or plastic mallet. Be sure that the sprocket is seated in position.

18. Rotate the crankshaft, if needed, until the timing mark is in the 12 o'clock position.

19. Situate the timing chain on the camshaft sprocket and hold the camshaft sprocket so that the timing mark is in the 6 o'clock position.

20. Align the dark colored links with the dot on the camshaft sprocket, place the timing chain around the crankshaft sprocket with the dark colored link lined up with the

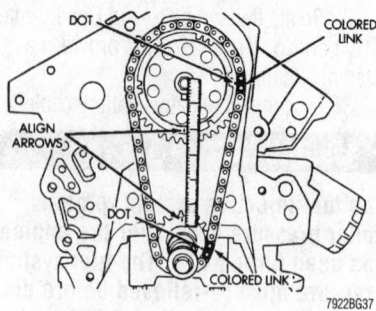

Using a straightedge to align the timing marks—3.3L engine

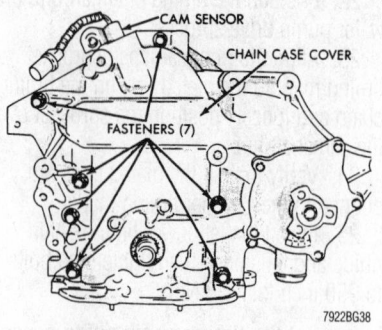

Timing cover mounting bolt locations—3.3L engine

dot on the sprocket and install the camshaft sprocket onto the end of the camshaft.

21. Using a straightedge, check the alignment of the crankshaft and camshaft timing marks.

22. Install the camshaft bolt and washer. Tighten the bolt to 40 ft. lbs. (54 Nm).

23. Rotate the crankshaft 2 full revolutions. The timing marks should line up. If the timing marks do not line up, remove the camshaft sprocket and realign it.

24. Check the camshaft end-play. With a new thrust plate the specification is 0.005–0.012 in. (0.0127–0.304mm). The old thrust plate specification is 0.012 in. (0.31mm) maximum. If not within these limits install a new thrust plate.

25. Install the timing chain snubbers. Tighten the retaining screws to 9 ft. lbs. (12 Nm). Each model year of these engines may use different length bolts, therefore do not use bolts for this component from any other model year engine.

26. Clean all parts well. Use care to remove all old sealer and gasket material from the timing chain cover.

➡ **The crankshaft oil seal must be removed to insure correct oil pump engagement.**

27. Remove the old oil seal from the timing case cover. Be sure that the mating surfaces for the timing chain cover gasket are clean and free of any burrs. Rotate the crankshaft so that the oil pump drive flats are vertical. Position the oil pump inner rotor so the mating flats are in the same position as the crankshaft drive flats. Install the timing chain front cover using new gasket and O-rings. Be sure the oil pump is engaged on the crankshaft correctly or severe damage may result.

28. Install the chain case cover screws. Snug down the two bottom screws and the top center screw. Be sure the cover is seated on the block, then tighten all the other screws to 20 ft. lbs. (27 Nm).

29. Install a new front cover oil seal.

30. Install the crankshaft damper and tighten the center bolt to 40 ft. lbs. (54 Nm).

31. Install the tensioner pulley bracket.

32. Install the cam position sensor.

33. Install the accessory drive belt(s) and adjust to the proper tension.

34. Install the oil pump pick-up tube and the oil pan.

35. Install the transaxle supports brackets and inspection cover.

36. Install the power steering pump and air conditioning compressor, if removed.

37. Connect the front sway bar.

38. Install the radiator and cooling fan assemblies.

39. Refill the engine with the correct amount of clean SAE 5W-30 **OR** SAE 10W-30 engine oil only. Do not mix the 2 grades of oil. A filter change is recommended.

40. Refill and bleed the cooling system.

41. Reconnect the negative battery cable.

42. Test run the engine and check for leaks.

FUEL SYSTEM

Fuel System Service Precautions

Safety is the most important factor when performing not only fuel system maintenance but any type of maintenance. Failure to conduct maintenance and repairs in a safe manner may result in serious personal injury or death. Maintenance and testing of the vehicle's fuel system components can be accomplished safely and effectively by adhering to the following rules and guidelines.

• To avoid the possibility of fire and personal injury, always disconnect the negative battery cable unless the repair or test procedure requires that battery voltage be applied.

• Always relieve the fuel system pressure prior to disconnecting any fuel system component (injector, fuel rail, pressure regulator, etc.), fitting or fuel line connection. Exercise extreme caution whenever relieving fuel system pressure to avoid exposing skin, face and eyes to fuel spray. Please be advised that fuel under pressure may penetrate the skin or any part of the body that it contacts.

• Always place a shop towel or cloth around the fitting or connection prior to loosening to absorb any excess fuel due to spillage. Ensure that all fuel spillage (should it occur) is quickly removed from engine surfaces. Ensure that all fuel soaked cloths or towels are deposited into a suitable waste container.

• Always keep a dry chemical (Class B) fire extinguisher near the work area.

• Do not allow fuel spray or fuel vapors to come into contact with a spark or open flame.

• Always use a back-up wrench when loosening and tightening fuel line connection fittings. This will prevent unnecessary stress and torsion to fuel line piping. Always follow the proper torque specifications.

• Always replace worn fuel fitting O-rings with new. Do not substitute fuel hose or equivalent, where fuel pipe is installed.

Fuel System Pressure

RELIEVING

✳✳ CAUTION

The fuel injection system remains under pressure, even after the engine has been turned OFF. The fuel system pressure must be relieved before disconnecting any fuel lines. Failure to do so may result in fire and/or personal injury.

1. Disconnect the negative battery cable.
2. Remove the fuel filler cap.
3. Remove the safety cap from the fuel pressure test port located on the fuel rail.
4. Place the open end of the fuel pressure release hose tool C-4799–1 or equivalent, into a proper gasoline container.
5. Connect the other end of the hose to the fuel pressure test port.
6. The fuel pressure will bleed off through the hose into the container.

Fuel Filter

✳✳ CAUTION

Never smoke when working around gasoline! Avoid all sources of sparks or ignition. Gasoline vapors are EXTREMELY volatile!

The fuel filter mounts to the frame rail in front of the fuel tank. The inlet and outlet ends of the filter are marked for installation purposes.

REMOVAL & INSTALLATION

1995–97 Vehicles

1. Relieve the fuel system pressure.
2. Raise and safely support the vehicle, then locate the filter.

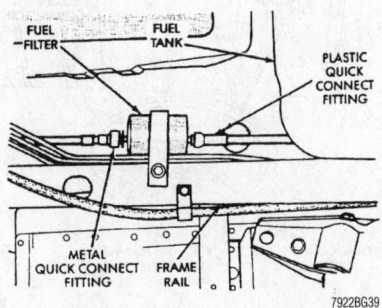

The fuel filter mounts to the frame rail in front of the fuel tank—1995–97 vehicles

3. Disengage the quick connect fittings from the filter.
4. Remove the filter mounting bracket, then the filter.

To install:

5. The inlet and outlet sides of the filter are marked, install the filter with the inlet side to the fuel tank.
6. Place the filter into the bracket. Place the bracket against the frame rail, tighten the mounting screw to 110 inch lbs. (12 Nm).
7. Apply a light coat of clean 30 weight engine oil to the fuel filter nipples. Install the fuel lines.
8. Lower the vehicle.
9. Start the engine and check for leaks.

1998–99 Vehicles

➡The fuel delivery system uses quick connect fittings. The fuel filter mounts to the top of the fuel tank.

1. Properly relieve the fuel system pressure.
2. Disconnect the negative battery cable.
3. Raise and safely support the vehicle.
4. Locate the fuel filter in its mounting on top of the fuel tank. Disconnect the quick-connect fittings from the chassis fuel supply tube and fuel pump module by squeezing the quick-connect fitting retainer tabs together and pulling the fitting assembly away from the fuel line nipple. The retainer will remain on the fuel tube.
5. Remove the fuel filter mounting bolt and remove the fuel filter from the fuel tank.

To install:

6. Install the fuel filter to the top of the fuel tank and tighten the mounting bolt.
7. The fuel supply (to chassis fuel line), return tube (to pump module) and fuel supply (to fuel filter) tube are permanently attached to the fuel filter. The quick-connect fitting ends of the fuel supply and return tubes are of different sizes.
8. Apply a light coating of 30W engine oil to the nipples of the fuel filter.
9. Push the quick-connect fitting over the fuel line until the retainer seats and

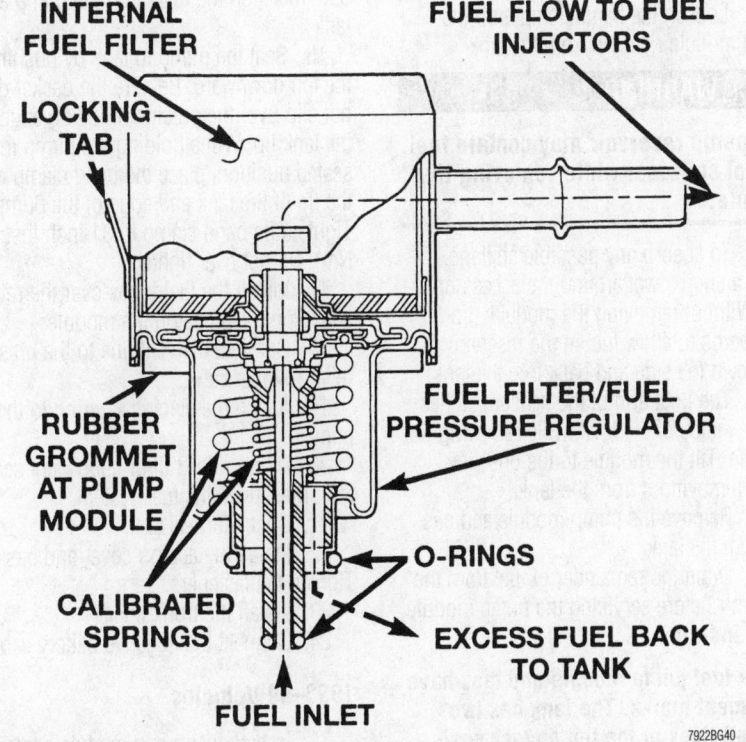

Cut away view of the fuel filter/pressure regulator—1998–99 vehicles

clicks into place. Be sure the retainer tabs have locked into the case of the quick-connect fitting.

10. Lower the vehicle. Start the engine and check for leaks.

Fuel Pump

REMOVAL & INSTALLATION

1995–97 Vehicles

An electric fuel pump is used with fuel injection systems in order to provide higher and more uniform fuel pressures. It is located in the tank. To perform the testing or servicing, you will need a DRB III or equivalent scan tool.

1. Release the fuel system pressure.
2. Disconnect the negative battery cable.
3. Open the trunk lid, then remove the trunk liner.
4. Remove the access panel fasteners.
5. Remove the access panel and gasket. Inspect the gasket for damage. If necessary, replace the gasket.
6. Disconnect the electrical wiring from the top of the pump module.
7. Disconnect the fuel supply and return tubes from the fuel pump module.
8. Disconnect the hose from the pressure relief/rollover valve.
9. A band clamp fastens the pump module to the tank. The module rises up from the tank after loosening the clamp.
10. Loosen the band clamp until the pump module rises up from the tank.

❄❄ WARNING

The pump reservoir may contain fuel. Do not spill fuel while removing the module.

11. To absorb any possible spillage, place a shop towel around the access opening. Without removing the modular, tip it backwards to allow fuel in the reservoir to run down the side and back into the tank.
12. The float arm of the sensor catches on the inside of the tank while removing the module. Tilt the module to the one side when removing it from the tank.
13. Remove the pump module and gasket from the tank.
14. Drain the remainder of fuel from the reservoir before servicing the pump module.

To install:

➡The fuel pump module and tank have alignment marks. The tank has two molded lines at the ten o'clock position. The fuel pump has a triangular

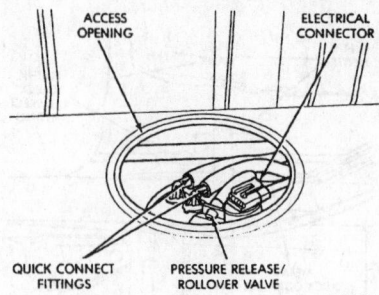

ACCESS OPENING — ELECTRICAL CONNECTOR — QUICK CONNECT FITTINGS — PRESSURE RELEASE/ ROLLOVER VALVE

7922BG41

Once the cover is removed, access to the pump is obtained—1995–97 vehicles

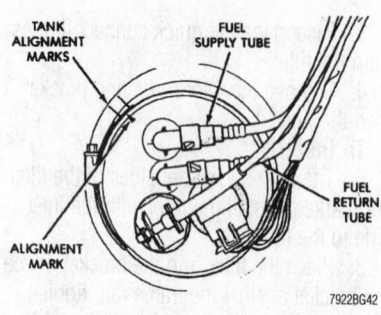

TANK ALIGNMENT MARKS — FUEL SUPPLY TUBE — FUEL RETURN TUBE — ALIGNMENT MARK

7922BG42

Note that the alignment marks must be aligned to install the pump correctly—1995–97 vehicles

alignment mark. Align the marks to correctly position the pump.

15. With a new gasket, insert the pump straight into the tank to set the float on the bottom. Align the marks on the pump and tank.
16. Seat the pump to tank by pushing the top downward. Be sure the gasket does not slip over the outside or inside edge of the tank lip. While holding the pump in the seated position, place the band clamp over the lip of the tank and edge of the pump. Tighten the band clamp to 31 inch lbs. (4 Nm). Do not over tighten.
17. Install the fuel tubes over the return and supply nipples on the module.
18. Connect the vent line to the pressure relief/rollover valve.
19. Attach the electrical wiring to the pump.
20. Using a DRB III or equivalent scan tool, pressurize the fuel system.
21. Check for leaks.
22. Install the access cover and gasket. Tighten the fasteners.
23. Install the trunk liner.
24. Connect the negative battery cable.

1998–99 Vehicles

The in-tank fuel pump module contains the fuel pump and pressure regulator which

adjusts fuel system pressure to approximately 49 psi. Voltage to the fuel pump is supplied through the fuel pump relay.

The fuel pump is serviced as part of the fuel pump module. The fuel pump module is installed in the top of the fuel tank and contains the electric fuel pump, fuel pump reservoir, inlet strainer fuel gauge sending unit, fuel supply and return line connections and the pressure regulator. The inlet strainer, fuel pressure regulator and level sensor are the only serviceable items. If the fuel pump requires service, replace the fuel pump module. Use the following procedure.

❄❄ CAUTION

The fuel injection system remains under pressure, even after the engine has been turned OFF . The fuel system pressure must be relieved before disconnecting any fuel lines. Failure to do so may result in fire and/or personal injury.

1. Remove the fuel filler cap and properly relieve the fuel system pressure.
2. Disconnect and isolate the negative battery cable.
3. Drain and remove the fuel tank.

❄❄ CAUTION

Observe all applicable safety precautions when working around fuel. Do not allow fuel spray or fuel vapors to come in contact with a spark or open flame. Keep a dry chemical (Class B) fire extinguisher near the work area. Never drain or store fuel in an open container due to the possibility of fire or explosion.

4. Clean the top of the tank to remove any loose dirt.
5. Disconnect fuel lines from the fuel pump module by squeezing the quick-connect fitting with thumb and fore finger.

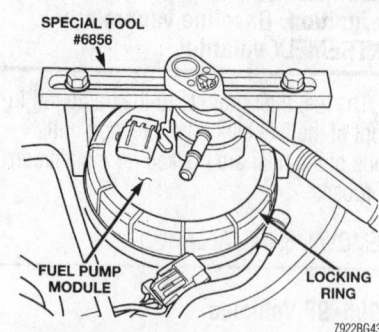

SPECIAL TOOL #6856 — FUEL PUMP MODULE — LOCKING RING

7922BG43

Using special tool 6856 or equivalent, remove the fuel pump module locknut—1998–99 vehicles

6. Disconnect the fuel pump module electrical connector from the top of the fuel pump module.

7. Using special tool 6856 or equivalent, remove the fuel pump locknut by turning counterclockwise.

✳✳ CAUTION

The fuel reservoir of the fuel pump module does not empty out when the tank is drained. The fuel in the reservoir may spill out when the module is removed.

8. Remove the fuel pump and O-ring from the tank. Discard the O-ring.

To install:

9. Thoroughly clean all parts. Wipe the seal area of the tank clean. Place a new O-ring on the ledge between the tank threads and the pump module opening.

10. Position the fuel pump module in the tank. Be sure the alignment tab on the underside of the pump module flange sits in the corresponding notch in the fuel tank.

11. While holding the fuel pump module in place install the locking ring and tighten to 40 inch lbs. (5 Nm) using special tool 6856 or equivalent spanner-type tool.

12. Install the fuel tank assembly.

13. Connect fuel pump module electrical connector.

14. Reconnect the negative battery cable.

15. Fill the fuel tank with fuel. Install the fuel filler cap. Turn the ignition switch to the **ON** position to pressurize the system. Check the fuel system for leaks.

DRIVE TRAIN

Transaxle Assembly

REMOVAL & INSTALLATION

The 42LE four speed transaxle uses fully-adaptive controls. Adaptive controls are those which perform their functions based on real-time feedback sensor information. The transaxle is conventional in the use of hydraulically applied clutches to shift a planetary gear train. However, it uses electronics to control virtually all other functions. The following components are serviceable in the vehicle: valve body assembly, solenoid pack, manual valve lever position sensor, input and output speed sensors, transfer chain and sprockets, short (right side) stub shaft seal and the long

(left side) stub shaft and ball bearing. Note that the factory recommends that before attempting any repair on the 42LE four-speed automatic transaxle, always check for proper shift linkage adjustment. Also check for diagnostic trouble codes with the Chrysler DRB scan tool (or equivalent).

Use MOPAR Type 7176 Automatic Transmission Fluid only. Do not substitute transaxle fluid. If the differential sump requires fluid, use 80W-90 petroleum based Hypoid gear lubricant. The transaxle can be removed without removing the engine. Use the following procedure.

1. Disconnect negative battery cable.

2. Remove the engine air inlet tube.

3. The crankshaft position sensor is located on the upper right side of the transaxle bell housing. Unplug the crankshaft position sensor connector and remove sensor.

4. Unplug the transaxle wiring connector block located on the right shock tower. To free the connector from the harness, remove the wire ties.

5. Raise and safely support the vehicle.

6. Remove the front wheels.

7. Remove the strut to steering knuckle bolts on both sides of the vehicle. Disconnect the tie-rod ends if required.

8. Remove the Anti-lock Brake System (ABS) wheel speed sensor, if equipped.

9. Remove the halfshafts from the transfer case by inserting a prybar between the halfshaft and the transaxle case and pry the shaft from the transaxle housing. Swing the shafts out of the way keeping the joints straight and suspend using wire. Be careful not to damage the halfshaft seals.

➡**Do not let the halfshafts or CV-joints hang unsupported. Internal joint damage may result if allowed to hang free.**

10. Remove the engine-to-transaxle brackets and the transaxle bell housing cover.

11. Mark the driveplate to the torque converter and remove the torque converter bolts. The driveplate-to-torque converter bolts are not to be reused.

12. Unbolt and remove the starter assembly from the bell housing and allow the starter motor to sit between the engine and the frame.

13. Disconnect the oil cooler lines from the transaxle and plug to prevent excess fluid leakage.

14. Remove the transaxle dipstick.

15. Disconnect the gear selector cable from the transaxle.

16. Disconnect the exhaust pipe from the exhaust manifold and position out of the

way. If the clearance will not allow for transaxle removal, remove the exhaust system from the vehicle.

17. Support the transaxle using a transmission jack. Raise the transaxle slightly to relieve the weight off the rear transaxle mount.

18. Remove the engine-to-transaxle brackets and the transaxle mount through-bolt.

19. Remove the rear crossmember mounting bolts. Pry the transaxle mount rearward to separate the mount from the transaxle. Remove the rear crossmember.

20. Lower the rear of the transaxle to gain access to the bell housing bolts. Remove the bell housing bolts.

21. Place a drain pan under the dipstick in the transaxle to catch transaxle fluid that will drain out of the case. Remove the dipstick tube from the transaxle and plug hole.

22. Remove the engine-to-transaxle bolts and lower the transaxle from the vehicle.

➡**The driveplate-to-torque converter bolts and the driveplate-to-crankshaft bolts must not be reused. Install new bolts when ever these bolts are removed.**

23. Inspect the driveplate for cracks. If cracks are present, replace the driveplate.

To install:

➡**Apply a light coating of grease to the pilot hole of the crankshaft if the torque converter is being replaced.**

✳✳ WARNING

When installing the transaxle unit into the vehicle, be careful that the fuel tubes at the rear of the engine do not contact the following:

- Tie rod attachment plate at the power steering rack
- Exhaust Gas Recirculation (EGR) tube
- Transaxle wiring harness

24. Install the driveplate to the engine and secure using new fasteners. Tighten the fastener to 75 ft. lbs. (101 Nm).

25. Install the transaxle into the vehicle and install the engine-to-transaxle case mounting bolts. Tighten the bolts to 75 ft. lbs. (101 Nm).

26. Install the rear transaxle case mount and the rear crossmember in position and secure all fasteners.

27. Install the dipstick tube.

28. Reconnect the exhaust pipe to the engine exhaust manifold.

29. Reconnect the gear selector cable to the transaxle. Reconnect the transaxle oil cooler lines.

30. Install the starter assembly and secure with the mounting bolts tightened to 40 ft. lbs. (54 Nm). Be sure that the starter ground strap is installed correctly.

31. Position the torque converter so matchmarks made during disassembly are in alignment. Install new torque converter to driveplate bolts and tighten to 60 ft. lbs. (81 Nm).

32. Install the transaxle bell housing cover. Install the engine to transaxle brackets.

33. While pulling the top of the steering knuckle outward, install the inner CV-joint with new retainer clip in place, into the transaxle.

34. Install the ABS wheel sensor, if removed. Install the strut-to-steering knuckle bolts and secure.

35. Install the front wheels and lug nuts. Tighten the lug nuts, in a star pattern sequence, to 95–100 ft. lbs. (129–135 Nm).

36. Lower the vehicle to the ground. Install the transaxle dipstick.

37. Engage the transaxle wiring harness connector on the right shock tower.

38. Install and reconnect the crankshaft position sensor.

39. Install the air inlet tube and reconnect the negative battery cable.

40. Start the engine and allow to idle for two minutes. Apply parking brake and move selector through each gear position, ending in **N**. Recheck fluid level and add if necessary. Be sure the vehicle is level when refilling the transaxle. Use Mopar Type 7176 Automatic Transmission Fluid only. Do not substitute transaxle fluid. If the differential sump requires fluid, use 80W-90 petroleum based Hypoid gear lubricant.

41. Check the transaxle or proper operation. Adjust the shift linkage, if necessary. Be sure the reverse lamps come on when in reverse.

Halfshaft

REMOVAL & INSTALLATION

✳✳ WARNING

Allowing the CV-joint assemblies to dangle unsupported or pulling or pushing the ends can damage boots or CV-joints. Always support both ends of the halfshaft to prevent damage or disengagement of the Tripod joint.

1. Disconnect the negative battery cable.

2. Raise and support the vehicle safely.

3. Remove the front wheels.

4. Remove the front caliper assembly from the steering knuckle.

5. Remove the front brake rotor from the hub by pulling it straight off wheel mounting studs.

6. Remove the speed sensor cable routing bracket from the strut assembly.

7. Remove the hub and bearing-to-stub axle retainer nut.

8. Install a puller tool onto the hub and bearing assembly and secure it into place using the wheel lug nuts.

9. Protect wheel stud threads by installing a wheel lug nut onto a wheel stud. Install a flat blade prying tool to prevent the hub from turning. Using the puller tool,

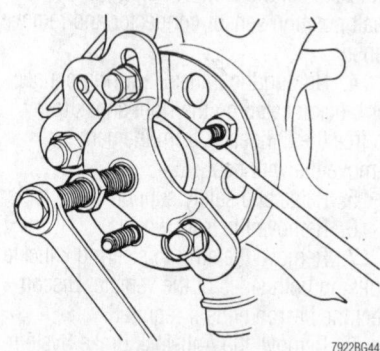

Removing the stub axle from the front hub/bearing assembly—halfshaft service

force the halfshaft outer stub axle from the hub and bearing assembly.

10. Dislodge the inner Tripod joint from the stub shaft retaining snapring on the transaxle. To do this, insert a prybar between the transaxle case and the inner Tripod joint and pry on Tripod joint.

➡**Do not try to remove the inner Tripod joint from the transaxle stub shaft at this time. Only disengage the inner Tripod joint from the retainer snapring.**

11. Remove the strut assembly-to-steering knuckle attaching bolts from the strut assembly.

✳✳ WARNING

The strut assembly to steering knuckle bolts are serrated (toothed) where they go through the strut assembly and steering knuckle. When removing the bolts, turn the nuts off the bolt. Do not turn the bolts in the steering knuckle or damage to the steering knuckle will result.

12. Separate the top of the steering knuckle from the lower end of the strut.

13. Hold the outer joint assembly with one hand. Grasp the steering knuckle with the other hand and rotate it out and to the rear of the vehicle, until the outer CV-joint clears the hub and bearing assembly.

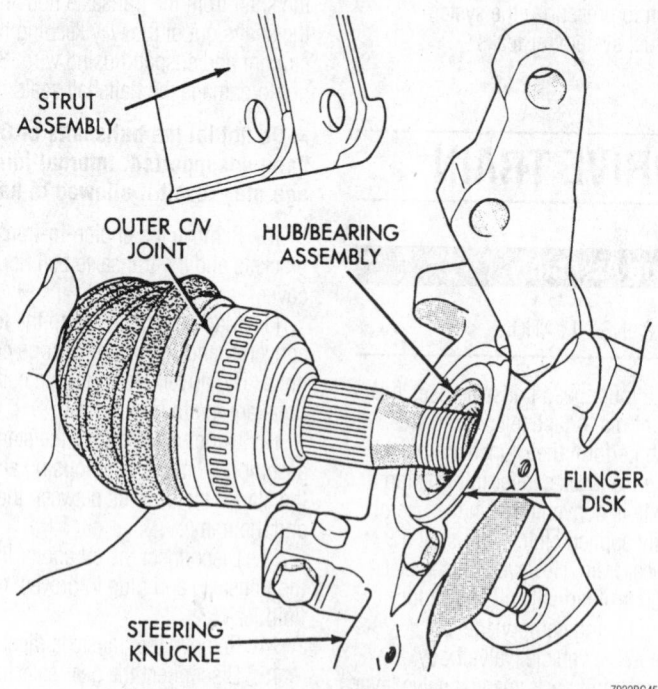

Be careful not to damage the threads for the axle nut when removing the outer CV joint from the steering knuckle—halfshaft service

✳✳ WARNING

When removing the outer CV-joint from the hub and bearing assembly, do not allow the flange disc on the hub and bearing assembly to become damaged. If this happens, dirt and water can enter the bearing which will cause premature bearing failure.

14. Remove the halfshaft inner joint from the transaxle stub shaft by grasping the inner Tripod joint and the interconnecting shaft and pulling both pieces at the same time. Take care not to pull on the interconnecting shaft to remove or separation of the spider assembly will occur.

To install:

15. Replace the inner Tripod joint retaining circlip and O-ring seal on the transaxle stub shaft. These components are not reusable and must be replaced whenever the halfshaft is removed.

16. Apply an even coat of grease on the splines of the inner Tripod joint, where the O-ring seats against the Tripod joint.

17. Install the halfshaft through the hole in the splash shield. Grasp the inner joint in 1 hand and interconnecting shaft in the other. Align the inner Tripod joint spline with the stub shaft spline on the transaxle. Use a rocking motion with the inner Tripod joint to get it past the circlip on the transaxle stub shaft.

18. Continue pushing Tripod joint onto transaxle stub shaft until it stops moving. The O-ring on the stub shaft should not be visible when the inner Tripod joint is fully installed. Check that the inner Tripod joint is locked in position by grasping the inner joint and pulling. If locked in position, the joint will not move on the stub shaft.

19. Hold the outer CV-joint assembly with one hand. Grasp the steering knuckle with the other and rotate it out and to the rear of the vehicle. Install the outer CV-joint into the hub and bearing assembly.

20. Install the top of the steering knuckle into the strut assembly. Align the steering knuckle to strut assembly mounting holes.

21. Install the strut assembly-to-steering knuckle attaching bolts. Install the nuts to the attaching bolts and while holding the bolt heads, tighten nuts to 125 ft. lbs. (170 Nm). Turn the nuts on the bolts. Do **NOT** turn the bolts.

22. Install a new hub and bearing assembly-to-stub shaft retainer nut. Tighten but do not torque the nut at this time.

23. Install the speed sensor cable routing bracket and secure attaching screw.

24. Install the brake rotor and the caliper assembly. Install the caliper guide pin bolts to steering knuckle and tighten to 30 ft. lbs. (41 Nm).

25. Install the front wheels and lug nuts. Lower the vehicle to the ground. Pump the brakes until a firm pedal is obtained.

26. Apply the brakes and tighten the new stub shaft-to-hub and bearing assembly retainer nut to 120 ft. lbs. (163 Nm).

✳✳ WARNING

When tightening the stub shaft retaining nut, be careful not to exceed the maximum torque specification of 120 ft. lbs. (163 Nm). If this specification is exceeded, failure of the halfshaft could result.

27. Reconnect the negative battery cable. Road test vehicle to check for noise or vibration.

STEERING AND SUSPENSION

Air Bag

✳✳ CAUTION

Some vehicles are equipped with an air bag system, also known as the Supplemental Inflatable Restraint (SIR) or Supplemental Restraint System (SRS). The system must be disabled before performing service on or around system components, steering column, instrument panel components, wiring and sensors. Failure to follow safety and disabling procedures could result in accidental air bag deployment, possible personal injury and unnecessary system repairs.

PRECAUTIONS

Several precautions must be observed when handling the inflator module to avoid accidental deployment and possible personal injury.

• Never carry the inflator module by the wires or connector on the underside of the module.

• When carrying a live inflator module, hold securely with both hands, and ensure that the bag and trim cover are pointed away.

• Place the inflator module on a bench or other surface with the bag and trim cover facing up.

• With the inflator module on the bench, never place anything on or close to the module which may be thrown in the event of an accidental deployment.

DISARMING

✳✳ CAUTION

The Air Bag system must be disarmed before repair and/or removal of any component in its immediate area including the air bag itself. Failure to do so may cause accidental deployment of the air bag, resulting in unnecessary system repairs and/or personal injury.

1. Disconnect the negative battery cable and isolate the cable using an appropriate insulator (wrap with quality electrical tape).

2. Allow the system capacitor to discharge for 2 minutes before starting any repair on any air bag system or related components. This will disable the air bag system.

✳✳ CAUTION

Always wear safety goggles when working with, or around, the air bag system. When carrying a live air bag, be sure the bag and trim cover are pointed away from the body. In the unlikely event of an accidental deployment, the bag will, then deploy with minimal chance of injury. When placing a live air bag on a bench or other surface, always face the bag and trim cover up, away from the surface. This will reduce the motion of the module if it is accidentally deployed.

Power Rack and Pinion Steering Gear

REMOVAL & INSTALLATION

✳✳ CAUTION

The Supplemental Inflatable Restraint (SIR) system must be disarmed before removing the rack and pinion steering gear. Failure to do so may cause accidental deployment of the air bag, resulting in unnecessary SIR system repairs and/or personal injury.

1. Disconnect negative battery cable. Disarm the air bag system.

2. Raise and safely support the vehicle.

3. Remove the gear shift cable from the shifter lever on the transaxle.

4. Loosen the bolt at the gear shift cable to transaxle mount. Remove the cable from the transaxle.

5. Lower the vehicle.

6. If necessary, disconnect the throttle cable from the throttle body and remove throttle cable bracket.

7. Remove both wiper arm assemblies from the wiper arm pivots. Remove the cowl closure panel and weatherstripping as an assembly from the cowl.

8. Disconnect the wiper module wiring harness from the vehicle wiring harness. Remove the wiper module assembly from the vehicle cowl panel.

9. Disconnect the air plenum from the throttle body, PCV make up air tube and the idle air control motor. Remove he plenum from the right side of the vehicle through the wiper module area.

10. Remove the vacuum connector from the power brake booster at the intake manifold.

11. Turn the front wheels to the full left position. Then, turn the wheels back in the other direction until the roll pin in the lower steering coupler is accessible. Turn the ignition key switch to the **LOCK** position to keep the steering column from rotating after the coupler is removed from the steering gear. If the steering column shaft rotates beyond the normal number of turns in either direction, the air bag clock spring will be damaged.

12. Using paint, mark the steering coupling and steering gear shaft for orientation. Using the correct size punch, remove the roll pin from the steering coupling.

13. If equipped with a brake pedal travel sensor, remove the pedal travel sensor from the brake booster as follows:

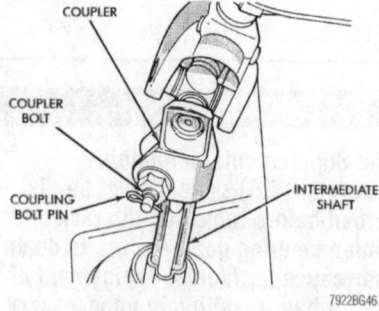

Matchmark the steering coupler to the intermediate shaft for proper alignment

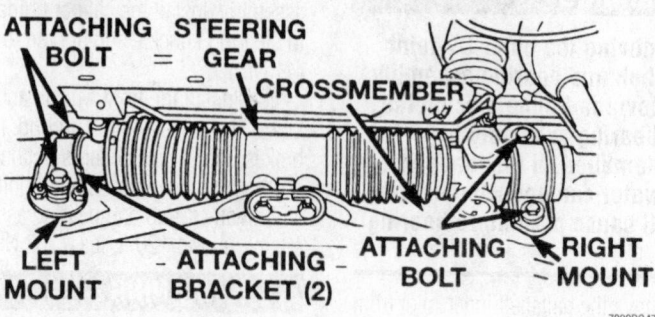

Steering rack-to-subframe mounting bolt locations—replace mounts if worn

a. Pump the brake pedal approximately 20 times. This will bleed the vacuum stored in the booster.

b. Remove the wiring harness connector from the sensor.

c. Using a small flat tipped tool, lift the retainer ring from the notch. Then, remove the retaining ring from the grommet.

d. Remove the pedal travel sensor from the brake booster by carefully pulling it straight out of its mounting grommet. Do not twist the sensor.

14. Loosen and remove the two nuts attaching the master cylinder to the brake booster. Remove the master cylinder with the brake lines connected, and position aside.

15. Remove the power steering pressure hose and return hose from the power steering gear.

16. Bend back the retaining tabs on bolt attaching the tie rods to the steering gear and remove the bolts.

17. Lay the tie rods, bolts and plate as an assembly on the bell housing of the transaxle.

18. If the rack and pinion steering gear unit being removed is a speed proportional steering gear, disconnect the vehicle wiring harness from the solenoid control module.

19. Remove the four bolts attaching the steering gear assembly to the crossmember. Slide the steering gear forward in the vehicle to disengage steering coupler from the steering gear shaft. After gear is disengaged, do not rotate the steering gear shaft.

20. Remove the steering gear assembly from the vehicle through the area in the cowl from which the windshield wiper module was previously removed.

To install:

21. If a replacement rack is being installed, grasp the shaft of the steering gear and rotate until steering gear center take off

is in a full left turn position. Install the steering gear into the vehicle through the wiper module opening in cowl.

22. If the original gear is being installed, align the paint mark on the steering coupler with the mark on the steering gear shaft and install the steering gear shaft into the steering gear coupler.

23. If a replacement rack is being installed, the steering gear shaft and steering coupler must be aligned. Rotate the steering gear shaft back from the full left turn position until the master spline on the steering gear shaft is aligned with the master spline on the steering coupler. At this point, install the steering gear into the coupler.

24. Align the steering gear with the mounting holes in the crossmember and install bolts. Be sure the brake line routing clip is installed under the left steering gear mounting bracket. Tighten the mounting bolts to 50 ft. lbs. (68 Nm).

25. Install the steering coupler to steering gear shaft retaining roll pin until it is flush with the top edge of the steering coupler.

26. If equipped with 3.5L engine, correct orientation of the power steering pressure hose at the power steering pump must be maintained. Be sure the power steering hose is installed in orientation clip at the power steering pump prior to tightening tube fitting. Attach the power steering pressure and return lines onto the proper ports of the power steering gear. Tighten both fittings to 23 ft. lbs. (31 Nm).

27. Align the center take off on the steering gear with the tie rod assemblies. Install the tie rod attaching bolts and washers into the steering gear assembly. Be sure the washers are installed between the tie rods and the steering gear. Tighten the tie rod to steering gear attaching bolts to 55 ft. lbs. (75 Nm). Bend the retaining tabs against the heads of the bolts.

28. Install the pedal travel sensor retainer ring on the travel sensor grommet in the vacuum booster. The tab on the retaining ring should be located in top notch of the mounting grommet.

29. Sparingly lubricate pedal travel sensor O-ring with fresh brake fluid. Install the pedal travel O-ring into the pedal travel sensor mounting grommet. Coat the end of the sensor with fresh brake fluid and install by pushing straight into the mounting grommet on the brake booster until the tab on the sensor is past the retaining ring on grommet.

30. Install the wiring harness connector to the pedal travel sensor.

31. Install the master cylinder and tighten nuts to 250 inch lbs. (28 Nm).

32. Install the power booster vacuum hose to the intake manifold. Install the windshield wiper module to the cowl panel. Reconnect the electrical harness to the module.

33. If removed, install the air intake plenum and reconnect to the idle air control motor, PCV make up air tube and throttle body.

34. Install the windshield wiper module assembly into the vehicle cowl area and reconnect the wiring harness from the wiper module to the vehicle wiring harness.

35. Install the cowl closure panel and tighten the six mounting screws. Install the weather strip on shock towers.

36. Install the windshield washer hoses on the wiper arms, then install arms on the windshield wiper pivots.

37. If removed, install the throttle cable to the bracket and install to the throttle body.

38. Reconnect the wiring harness from the solenoid control valve onto the solenoid control module. Be sure that the harness connector seal is in good condition before installation.

39. Raise and safely support the vehicle.

40. Install the gear shift cable onto the shift lever on transaxle. Install gear shift cable on cable mounting bracket of transaxle and securely tighten bolt.

41. Lower the vehicle. Connect the negative battery cable.

42. Refill the pump reservoir to the correct lever with Mopar Power Steering Fluid or equivalent. Do not use any type of automatic transmission fluid. Start the engine and turn the steering wheel several times from stop to stop to bleed the air from the fluid in the system. Check and add fluid as required.

43. Adjust the front suspension toe setting.

Strut

REMOVAL & INSTALLATION

Front

➡ **Service of the coil spring requires the use of a coil spring compressor tool. It is required that 5 coils be captured within the jaws of the compressor tool.**

1. Disconnect the negative battery cable.
2. Raise and safely support vehicle. Do not support vehicle by placing supports under the suspension arms. The suspension arms must hang freely.
3. Remove the front wheel(s).
4. Remove the stabilizer bar attaching link at the strut assembly.
5. Loosen but do not remove the outer tie rod end to strut assembly steering arm attaching nut. Then, remove the outer tie rod end from the steering arm using puller MB-990635 or equivalent.
6. If equipped with ABS, remove the speed sensor wiring harness mounting bracket from the strut.
7. Remove the brake caliper assembly. Support the caliper assembly from the vehicle frame with a strong piece of wire. Do not allow the assembly to hang by the brake hose. Remove the front brake rotor disc.

➡ **The strut assembly to steering knuckle bolts are serrated where they go through the strut and steering knuckle. Do not turn the bolts during removal. If bolts are turned, damage to the steering knuckle will result.**

8. The strut assembly to steering knuckle bolts must not be turned during strut removal. Hold the bolt head with a wrench and turn the nuts off the bolts.
9. Remove the three strut assembly upper mount to shock tower mounting nuts and washers. Remove the strut from the vehicle.
10. Securely mount the strut assembly into a vise or use the aide of a helper to hold the unit. Using paint, mark the strut unit, lower spring isolator, spring and upper strut mount for indexing of the parts at assembly.
11. Position the spring compressor tool onto the strut. Compress the coil spring until all load is off the upper strut mount assembly.
12. Install Strut Rod Socket tool L-4558A on the strut shaft nut and a 10mm socket on the end of the strut shaft to prevent it from turning. Remove the strut shaft nut.

13. Remove the upper mount assembly, jounce bumper and seat bearing and dust shield as an assembly.
14. Remove the coil spring and compressor as an assembly from the strut. Remove the lower spring isolator from the strut assembly lower spring seat.
15. Inspect all components for abnormal wear, oil leakage or failure. Replace parts as required.

To install:

16. Inspect the strut assembly for signs of leakage. Actual leakage will be a stream of fluid running down the side and dripping off the lower end of the strut. A slight amount of seepage between the strut rod and strut shaft seal is not unusual and does not affect performance of the strut assembly.
17. Install the lower spring isolator on the strut unit. Install the compressed coil spring onto the strut assembly aligning the paint marks made during removal.
18. Install the strut bearing into the bearing seat. Bearing must be installed into the seat with notches on the bearings facing down.
19. Lower the seat bearing and dust shield onto the strut and spring assembly. Align the paint marks made during removal.
20. Install jounce bumper and upper mount on the strut shaft aligning the paint marks.
21. Install the strut mount-to-shaft retainer nut. Inspect all alignment marks made during removal and align as required. While holding the strut shaft from turning with a 10mm socket, tighten the strut shaft nut to 70 ft. lbs. (94 Nm).
22. Equally loosen the spring compressor tool until all tension is released. Remove the spring compressor tool.
23. Install the front strut into the shock tower and install the three upper mount nuts and washers. Tighten mounting nuts to 25 ft. lbs. (33 Nm).
24. Position the steering knuckle neck into the strut assembly. Install the strut assembly to steering knuckle bolts. Install the nuts onto the attaching bolts and tighten to 125 ft. lbs. (169 Nm). Do not turn the serrated bolt heads during installation. Turn only the nuts.

➡ **The strut assembly to steering knuckle bolts are serrated (toothed) where they go through the strut and steering knuckle. Do not turn the bolts during removal. If bolts are turned, damage to the steering knuckle will result.**

25. Install the brake rotor and the caliper assembly to the adapter. Tighten the caliper mounting bolts 14 ft. lbs. (19 Nm).

26. Install the front speed sensor cable routing bracket onto the front strut, if equipped.

27. Install the outer tie rod on steering arm and tighten attaching nut to 27 ft. lbs. (37 Nm).

28. Install the stabilizer link assembly onto the strut assembly and tighten the attaching nut to 70 ft. lbs. (95 Nm).

29. Install the front wheel and lug nuts. Tighten the lug nuts, in sequence, to 95–100 ft. lbs. (129–135 Nm).

30. Lower the vehicle to the ground. Connect the negative battery cable.

Rear

The rear strut assemblies support the weight of the vehicle using coil springs positioned around the struts. The coil springs are contained between the upper mount of the strut assembly and a lower spring seat on the body of the strut assembly. The strut is attached to the spindle by a split collar on the rear spindle with a pinch bolt to hold the spindle to the strut.

➡**Service of the coil spring requires the use of a coil spring compressor tool. It is required that 5 coils be captured within the jaws of the compressor tool.**

1. Raise and safely support the vehicle.
2. Remove the rear wheel and tire assembly.
3. If equipped with rear disc, remove the caliper assembly and rotor from the hub. If equipped with rear drum brakes, disconnect the brake flex hose from the support bracket and wheel cylinder. Plug the brake flex hose to prevent system contamination. Do not allow the rear caliper to hang down by the brake hose. Support the caliper off of the frame with a strong piece of wire.
4. If equipped with ABS, remove the speed sensor cable routing bracket and tube.
5. Remove the bolts attaching the lateral links to the rear spindle assembly.
6. Remove the rear strut assembly to stabilizer bar attaching link at the stabilizer bar. Hold the hex on the attaching link stud while breaking nut loose. The attaching link does not have to be removed from the strut.
7. Remove the rear spindle to strut assembly pinch bolt. Install a center punch in hole on spindle and tap punch into hole until jammed. This will spread spindle casting allowing it to be removed from strut.
8. Using a hammer, tap on the top surface of the spindle, driving spindle down and off the end of the strut assembly. Let the

spindle and assembled components hang from the trailing arm while the strut is being serviced.

9. Lower the vehicle. From inside the trunk of the vehicle, remove the three upper strut mounting bolts and remove the strut from the vehicle.

10. Securely mount the strut assembly into a vise. Using paint, mark the strut assembly, lower spring isolator, spring and upper strut mount for indexing of the parts at reassembly.

11. Position a spring compressor tool onto the coil spring. Compress the coil spring until all load is off of the upper strut mount assembly.

12. Install strut rod socket tool L-4558 on the strut shaft nut and an 8mm Allen wrench on the end of the strut shaft to prevent it from turning. Remove the strut shaft nut.

13. Remove the upper strut mount assembly off of the strut shaft. Remove the coil spring and compressor tool as an assembly from the strut.

14. Remove the plate, dust shield and jounce bumper off of the strut unit.

15. Inspect all components for abnormal wear, oil leakage or failure. Replace parts as required.

To install:

16. Install the lower spring isolator on the strut unit.

17. Install the lower spring isolator on the strut unit. If it is the original isolator, align the paint marks.

18. Install the jounce bumper into the dust shield. Install the plate on top of the dust shield and into the jounce bumper.

19. Install the dust shield, jounce bumper and the top plate onto the strut unit as an assembly.

20. Install the coil spring and compressor tool onto the strut unit and align the paint marks on the spring to that of the strut unit.

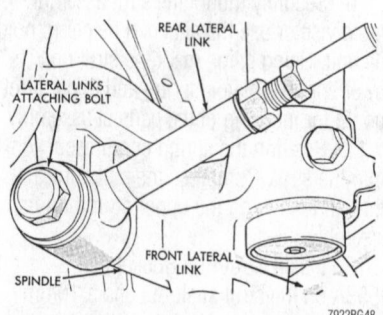

Separate the lateral links from the spindle—rear strut service

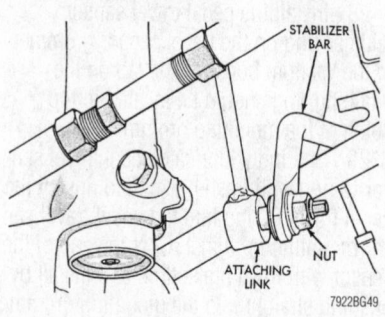

Remove the nut from the stabilizer-to-strut attaching link stud at the bar—rear strut service

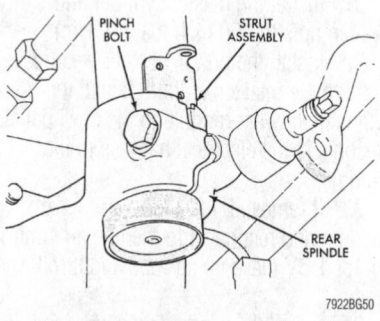

Loosen, then remove the rear spindle-to-strut pinch bolt—rear strut service

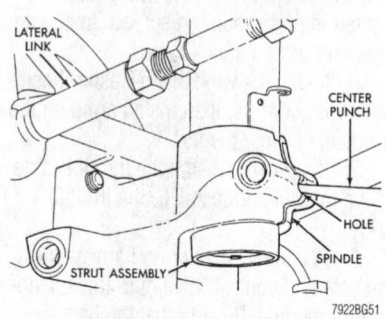

Insert a center punch into the hole on the spindle and tap until the casting is spread—rear strut service

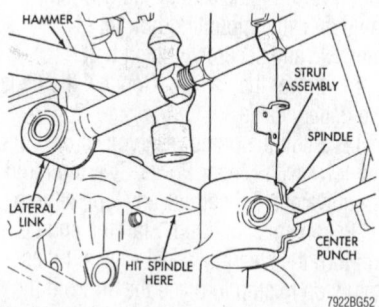

Tap with a hammer on the surface of the spindle driving it down and off the end of the strut—rear strut service

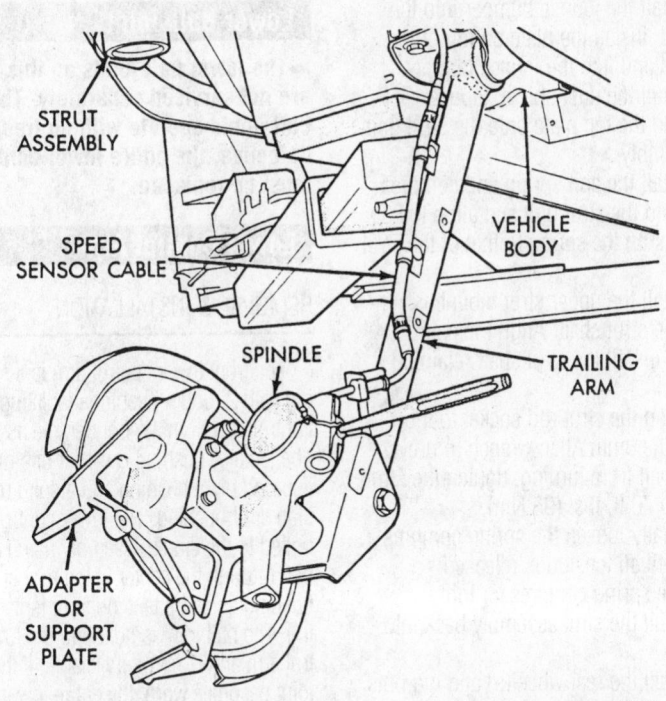

Allow the components to hang from the trailing arm as shown—rear strut service

7922BG53

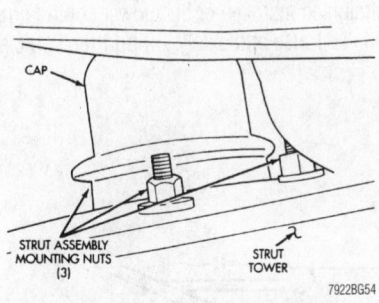

7922BG54

Access the 3 nuts through the trunk—rear strut service

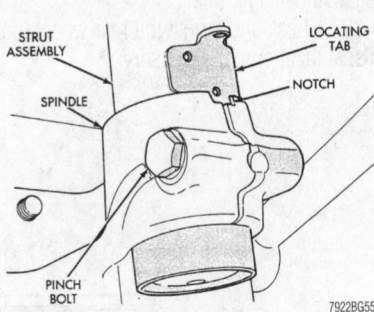

7922BG55

Tap the spindle onto the strut until the notch in the spindle is tightly seated against the tab—rear strut service

21. Install the upper strut mount assembly onto the strut shaft. Align the paint marks and install the strut shaft retaining nut.

22. Using the strut rod socket tool L-4558 and the 8mm Allen wrench to prevent the strut shaft from turning, tighten the strut shaft nut to 70 ft. lbs. (95 Nm).

23. Equally loosen the spring compressor tool until all tension is released. Remove the spring compressor tool.

24. Position the strut in vehicle and install the three upper mounting nuts tightening to 20 ft. lbs. (28 Nm).

25. Install the spindle assembly onto bottom of strut. Push or tap spindle assembly onto strut until notch in spindle is tightly seated against locating tap on strut assembly. Remove the center punch from the hole in the spindle.

26. Install the strut to spindle pinch bolt and tighten to 40 ft. lbs. (55 Nm).

27. Install the lateral link to the spindle attaching bolt and tighten to 105 ft. lbs. (140 Nm).

28. Install the stabilizer bar attaching link onto the stabilizer bar and install stabilizer link to stabilizer bar attaching nut. Tighten attaching nut to 70 ft. lbs. (95 Nm), while holding the stabilizer link stud at hex with wrench.

29. If equipped with ABS, mount rear speed sensor cable routing tube and bracket in position. Install rear disc brake rotor and caliper assembly to the adapter plate. Tighten caliper mounting bolts to 16 ft. lbs. (22 Nm).

30. If equipped with rear drum brakes, install rear brake flex hose to the wheel

cylinder and support plate. The brake system will require bleeding.

31. Install the rear wheel(s) and lug nuts. Tighten the lug nuts, in sequence, to 95 ft. lbs. (129 Nm). Lower the vehicle. Bleed the brake system, if equipped with rear drum brakes.

32. Have the rear wheel toe set to specifications.

Coil Springs

REMOVAL & INSTALLATION

Front

The front strut and suspension of the vehicle is supported by coil springs positioned around the struts, contained between an upper seat, located just below the top strut mount and a lower spring seat on the strut lower housing.

➡ Service of the coil spring requires the use of a coil spring compressor tool. It is required that 5 coils be captured within the jaws of the compressor tool.

1. Disconnect the negative battery cable.

2. Raise and safely support the vehicle.

3. Remove the appropriate wheel(s).

4. Remove the strut assembly from the vehicle.

5. Securely mount the strut assembly into a vice. Using paint, mark the strut unit, lower spring isolator, spring and upper strut mount for indexing of the parts at assembly.

6. Position the spring compressor tool, such as C-4838 or equivalent onto the strut. Compress the coil spring until all load is off the upper strut mount assembly.

7. Install strut rod socket tool L-4558A on the strut shaft nut and a 10mm socket on the end of the strut shaft to prevent it from turning. Remove the strut shaft nut.

8. Remove the upper mount assembly, jounce bumper and seat bearing and dust shield as an assembly.

9. Remove the coil spring and compressor as an assembly from the strut. Remove the lower spring isolator from the strut assembly lower spring seat.

10. Inspect all components for abnormal wear, oil leakage or failure. Replace parts as required.

To install:

11. Install the lower spring isolator on the strut unit. Install the compressed coil spring onto the strut assembly aligning the paint marks made during removal.

12. Install the strut bearing into the bearing seat. Bearing must be installed into the seat with notches on the bearings facing down.

13. Lower the seat bearing and dust shield onto the strut and spring assembly. Align the paint marks made during removal.

14. Install jounce bumper and upper mount on the strut shaft aligning the paint marks.

15. Install the strut mount-to-shaft retainer nut. Inspect all alignment marks made during removal and align as required. While holding the strut shaft from turning with a 10mm socket, tighten the strut shaft nut to 70 ft. lbs. (94 Nm).

16. Equally loosen the spring compressor tool until all tension is released. Remove the spring compressor tool.

17. Install the strut assembly back into the vehicle.

18. Install the wheels and lug nuts. Tighten the lug nuts, in sequence, to 95 ft. lbs. (129 Nm).

19. Lower the vehicle. Connect the negative battery cable.

Rear

1. Disconnect the negative battery cable.

2. Raise and safely support the vehicle. Remove the rear wheel(s).

3. Remove the rear strut assembly from the vehicle.

4. Securely mount the strut assembly into a vice. Using paint, mark the strut assembly, lower spring isolator, spring and upper strut mount for indexing of the parts at reassembly.

5. Position a spring compressor tool such as C-4838 onto the coil spring. Compress the coil spring until all load is off of the upper strut mount assembly.

6. Install strut rod socket tool L-4558 on the strut shaft nut and an 8mm Allen® wrench on the end of the strut shaft to prevent it from turning. Remove the strut shaft nut.

7. Remove the upper strut mount assembly off of the strut shaft. Remove the coil spring and compressor tool as an assembly from the strut.

8. Remove the plate, dust shield and jounce bumper off of the strut unit.

9. Inspect all components for abnormal wear, oil leakage or failure. Replace parts as required.

To install:

10. Install the lower spring isolator on the strut unit. if it is the original isolator, align the paint marks.

11. Install the jounce bumper into the dust shield. Install the plate on top of the dust shield and into the jounce bumper.

12. Install the dust shield, jounce bumper and the top plate onto the strut unit as an assembly.

13. Install the coil spring and compressor tool onto the strut unit and align the paint marks on the spring to that of the strut unit.

14. Install the upper strut mount assembly onto the strut shaft. Align the paint marks and install the strut shaft retaining nut.

15. Using the strut rod socket tool L-4558 and the 8mm Allen wrench to prevent the strut shaft from turning, tighten the strut shaft nut to 70 ft. lbs. (95 Nm).

16. Equally loosen the spring compressor tool until all tension is released. Remove the spring compressor tool.

17. Install the strut assembly back into the vehicle.

18. Install the rear wheel(s) and lug nuts. Tighten the lug nuts, in sequence, to 95 ft. lbs. (129 Nm).

19. Lower the vehicle. Reconnect the negative battery cable.

20. Check and adjust the rear wheel toe to specifications, if necessary.

Lower Ball Joint

➡The lower ball joints on this vehicle are not serviced separately. The lower ball joints operate with no free-play. If defective, the entire lower control arm must be replaced.

Lower Control Arm

REMOVAL & INSTALLATION

The front lower control arm is a steel forging with 2 rubber bushings isolating the lower control arm from the front cradle assembly. The isolator bushings consist of a metal encased pivot bushing and a solid rubber tension strut bushing. The lower control arm is bolted to the cradle assembly using a pivot bolt through the center of the rubber pivot bushing an at the tension strut isolator bushing. The ball joint is built into the lower control arm and is non-serviceable. If the ball joint becomes worn, the entire lower control arm must be replaced. The ball joint seal, however, is replaceable as well as the lower control arm inner bushing. If the lower control arm is damaged, do not attempt to repair or straighten a broken or bent lower control arm.

1. Raise and safely support the vehicle.

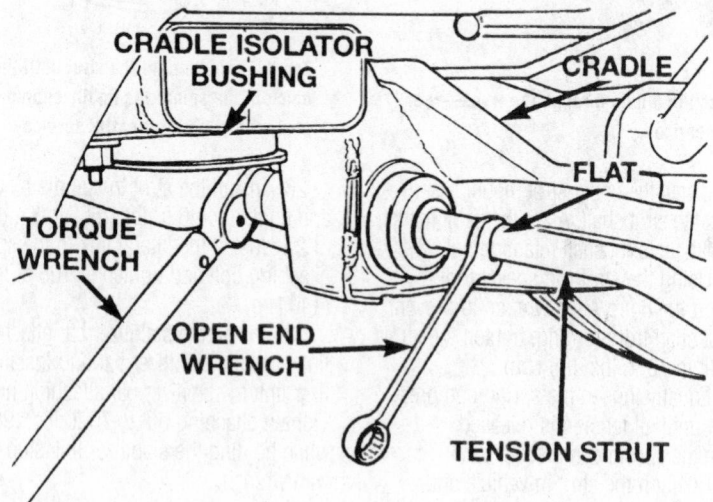

Remove the tension strut-to-cradle nut and washer—lower control arm service

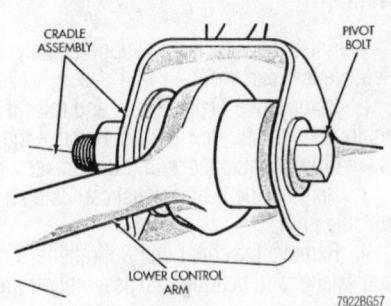

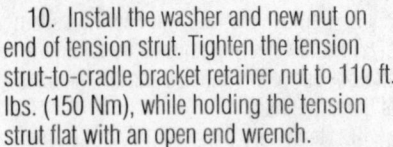

Loosen the pivot bolt and remove—lower control arm service

2. Remove the front wheel(s).

3. Remove the ball joint stud to steering knuckle clamp nut and bolt.

4. Carefully insert a prybar between the lower control arm and the steering knuckle and separate ball joint from knuckle. Be sure ball joint seal does not get damaged during separation.

✳✳ WARNING

Pulling the steering knuckle out from the vehicle after releasing from the ball joint can separate the inner CV-joint. Do not separate the inner CV-joint or it can be damaged.

5. Remove the tension strut-to-cradle attaching nut and washer from the end of the tension strut. When removing the nut, keep the strut from turning by holding the tension strut at the flats using an open end wrench. Discard the tension strut-to-cradle retainer nut. A new nut must be used during installation.

➡ **A new tension strut-to-cradle attaching nut must be used when installing the tension strut.**

6. Loosen and remove the lower control arm pivot bushing-to-cradle assembly pivot bolt.

7. Separate the lower control arm and tension strut from the cradle as an assembly by first removing the pivot bushing from the cradle, then sliding tension strut out of isolator bushing. Inspect control arm and tension strut for distortion, check the rubber bushings for excessive wear or deterioration and replace these components, if necessary.

To install:

8. Install the tension strut and isolator bushing into the cradle first, then install lower control arm pivot bushing into bracket on the cradle.

9. Installing the lower control arm-to-cradle bracket attaching bolt. Do not tighten the bolt at this time.

10. Install the washer and new nut on end of tension strut. Tighten the tension strut-to-cradle bracket retainer nut to 110 ft. lbs. (150 Nm), while holding the tension strut flat with an open end wrench.

11. Inspect the ball joint seal and replace if damaged. Install the lower ball joint stud into the steering knuckle and install the clamp bolt and nut. Tighten the bolt to 40 ft. lbs. (55 Nm).

12. Install the front wheel and lug nuts. Tighten the lug nuts, in sequence, to 95–100 ft. lbs. (129–135 Nm). Lower the vehicle so the suspension is supporting the weight of the vehicle.

13. Tighten the lower control arm pivot bushing-to-cradle bracket attaching bolt to 90 ft. lbs. (123 Nm).

Wheel Bearings

ADJUSTMENT

These front wheel drive vehicles are equipped with permanently sealed front and rear wheel bearings. There is no periodic lubrication or maintenance recommended for these units.

REMOVAL & INSTALLATION

Front

1. Raise and support the vehicle safely.

2. Remove the front wheel.

3. Remove the front caliper assembly from the steering knuckle using the procedure found in the brake section.

4. Remove the front brake rotor from the hub by pulling it straight off of the wheel mounting stud.

5. Remove the hub and bearing to stub axle retainer nut.

➡ **This hub nut is a torque prevailing retaining nut and can not be reused. A NEW retaining nut MUST be used when assembling the hub.**

6. Remove the three attaching bolts that mount the hub and bearing assembly to the steering knuckle assembly.

➡ **If the metal seal on the hub and bearing assembly is seized to the steering knuckle and becomes dislodged on the hub and bearing during removal, the hub and bearing must be replaced. If the flinger disc becomes damaged during the removal procedure, the hub and bearing assembly must be replaced.**

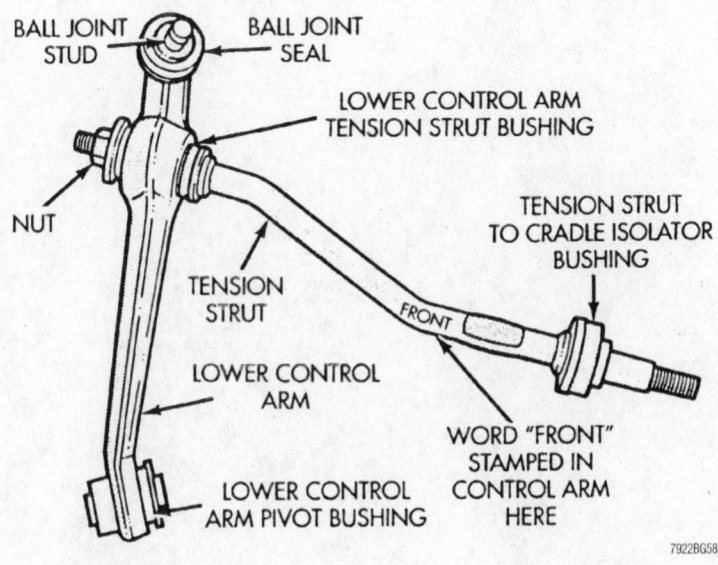

Inspect the control arm and tension strut for distortion—lower control arm service

7. Remove the hub and bearing assembly from the steering knuckle by sliding it straight out of the knuckle and off the ends of the stub shaft.

8. Gently pry the assembly out with a prybar or tap it out with a soft face hammer, if necessary. Be very careful not to damage the hub and bearing assembly.

To install:

9. Clean the hub and bearing mounting surfaces of dirt and be sure there are no nicks present.

10. Install the hub and bearing squarely onto the stub shaft and the steering knuckle.

11. Install the bearing assembly mounting bolts and tighten equally until the bearing assembly is seated squarely against the front of the steering knuckle. Tighten the mounting bolts to 80 ft. lbs. (110 Nm).

12. Install a new hub and bearing assembly-to-stub shaft retainer nut. A NEW retaining nut MUST be used when assembling the hub. Tighten but do not torque the nut at this time.

13. Install the brake rotor and the caliper assembly following the procedure in the brake section. Install the caliper-to-steering knuckle retainer bolts and tighten to 14 ft. lbs. (19 Nm).

14. Install the wheel and lug nuts. Tighten the lug nuts in sequence to 95–100 ft. lbs. (129–135 Nm).

15. Lower the vehicle.

16. Pump the brakes until a firm pedal is obtained.

17. With the weight on the vehicle on its wheels, apply the brakes to keep the vehicle from moving. Tighten the hub and bearing assembly to stub shaft retaining nut to 120 ft. lbs. (163 Nm).

✳✳ WARNING

When tightening the hub and bearing assembly to stub shaft retaining nut, do not exceed the maximum torque of 120 ft. lbs. (163 Nm). If the maximum torque is exceeded this may result in a failure of the halfshaft.

18. Inspect the toe setting on the vehicle and adjust, if necessary.

Rear

1. Raise the vehicle and support safely. Remove the rear wheel.

2. Remove the brake caliper and rotor if equipped with rear disc brakes. Remove the brake drum if equipped with drum brakes.

3. Remove the bearing dust cap using a suitable prybar.

4. Remove the cotter pin, nut retainer, nut, washer and bearing/hub assembly from the spindle.

To install:

5. Install the bearing/hub assembly. Install the bearing/hub assembly washer and retaining nut. Tighten the nut to 124 ft. lbs. (168 Nm). Install the nut retainer and new cotter pin. Bend over the cotter pin.

6. Install the dust cap.

7. Install the brake drum or rotor and caliper assembly.

8. Install the rear wheel and tighten the lug nuts, in sequence, to 95–100 ft. lbs. (129–135 Nm). Lower the vehicle safely.

9. Road test vehicle to verify no excessive noise from rear wheel bearing area.

CHRYSLER CORPORATION

Chrysler-Sebring Coupe • **Dodge-**Avenger

13

DRIVE TRAIN **13-20**
ENGINE REPAIR **13-2**
FUEL SYSTEM **13-19**
STEERING AND
SUSPENSION **13-24**

A
Air Bag.......................................13-24
  Disarming................................13-25
  Precautions 13-25

C
Camshaft and Valve Lifters 13-14
  Removal & Installation 13-14
Clutch.......................................13-22
  Adjustment 13-22
  Removal & Installation 13-23
Coil Spring 13-28
  Removal & Installation 13-28
Cylinder Head 13-5
  Removal & Installation 13-5

D
Distributor..................................13-2
  Installation...............................13-2
  Removal 13-2

E
Engine Assembly 13-2
  Removal & Installation 13-2
Exhaust Manifold........................13-11
  Removal & Installation 13-11

F
Front Crankshaft Seal 13-14
  Removal & Installation 13-14

Fuel Filter 13-20
  Removal & Installation 13-20
Fuel Pump 13-20
  Removal & Installation 13-20
Fuel System Pressure 13-19
  Relieving 13-19
Fuel System Service
  Precautions.............................13-19

H
Halfshaft....................................13-23
  Removal & Installation 13-23
Hydraulic Clutch System 13-23
  Bleeding 13-23

I
Ignition Timing 13-2
Intake Manifold...........................13-9
  Removal & Installation 13-9

L
Lower Ball Joint..........................13-29
  Removal & Installation 13-29
Lower Control Arm 13-31
  Removal & Installation 13-31

O
Oil Pan......................................13-16
  Removal & Installation 13-16
Oil Pump 13-17
  Removal & Installation 13-17

P
Power Rack and Pinion Steering
  Gear.......................................13-25
  Removal & Installation 13-25

R
Rear Main Seal 13-19
  Removal & Installation 13-19
Rocker Arm/Shafts......................13-7
  Removal & Installation 13-7

S
Shock Absorber 13-27
  Removal & Installation 13-27

T
Transaxle Assembly 13-20
  Removal & Installation 13-20

U
Upper Ball Joint..........................13-29
  Removal & Installation 13-29
Upper Control Arm 13-29
  Removal & Installation 13-29

V
Valve Lash 13-16
  Adjustment 13-16

W
Water Pump 13-4
  Removal & Installation 13-4
Wheel Bearings..........................13-31
  Adjustment 13-31
  Removal & Installation 13-31

ENGINE REPAIR

➡ **Disconnecting the negative battery cable on some vehicles may interfere with the functions of the on board computer systems and may require the computer to undergo a relearning process, once the negative battery cable is reconnected.**

Distributor

REMOVAL

2.5L Engine

The 2.5L engine is equipped with a camshaft driven mechanical distributor. This engine uses a fixed ignition timing system, in which the basic ignition timing is not adjustable. The Powertrain Control Module (PCM) determines spark advance. The crankshaft position sensor and camshaft position sensor are Hall effect devices. The crankshaft sensor is mounted remotely from the distributor, while the camshaft position sensor is mounted inside the distributor housing. Both sensors generate pulses which serve as inputs to the PCM; the PCM determines crankshaft position from these sensors, then calculates injector sequence and ignition timing, based on the data.

1. Disconnect the negative battery cable.
2. If necessary for access, perform the following:

   a. Remove the bolt attaching the air inlet resonator to the intake manifold.

   b. Loosen the clamps holding the air cleaner cover to the air cleaner housing.

   c. Remove the PCV make-up air hose from the air inlet tube.

   d. Loosen the hose clamp at the throttle body.

   e. Remove the air cleaner cover, resonator and inlet tube.

   f. Remove the EGR tube.

3. Mark for identification, if necessary, and remove the spark plug wires from the distributor cap.
4. Remove the distributor cap.
5. Mark the rotor position with a scribe mark to indicate where to position the rotor when reinstalling the distributor. Remove the rotor.
6. Unfasten the two electrical harness connections from the distributor.
7. Remove the two distributor hold-down nuts and washers.
8. If necessary, remove the spark plug cable mounting bracket.
9. Remove the transaxle dipstick tube.
10. Carefully remove the distributor from the engine.

INSTALLATION

2.5L Engine

TIMING NOT DISTURBED

1. Inspect the rotor for cracks or burned electrodes, and replace if defective. Install the rotor onto the distributor.
2. Inspect the O-ring seal. If nicked or cracked, replace with a new one. Be sure the O-ring is properly seated on the distributor.
3. Carefully engage the distributor drive with the slotted end of the camshaft. When the distributor is installed properly, the rotor will be in line with the previously made mark.
4. Verify proper rotor alignment with the mark made at disassembly.
5. Reinstall the distributor hold-down nuts and washers. Tighten the nuts to 9 ft. lbs. (13 Nm).
6. Reinstall the spark plug cable bracket.
7. Reconnect the two distributor wiring connectors.
8. Reinstall the distributor cap.
9. Reinstall the spark plug cables, following the identification marks made at disassembly.
10. Reinstall the transaxle dipstick tube.
11. If removed earlier, install the following:

   a. Install the EGR tube and tighten the mounting bolts to 95 inch lbs. (11 Nm).

   b. Install the air cleaner cover, resonator and inlet tube.

   c. Tighten the hose clamp at the throttle body.

   d. Install the PCV hose.

   e. Tighten the clamps holding the air cleaner cover to the air cleaner housing.

   f. Install the bolt attaching the air inlet resonator to the intake manifold.

12. Reconnect the negative battery cable.

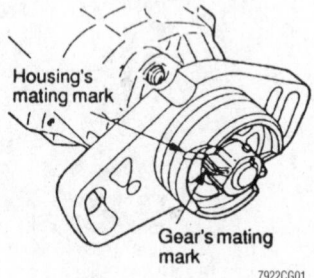

Housing's mating mark

Gear's mating mark

7922CG01

Prior to installation, align the distributor shaft with the distributor housing—2.5L engine

TIMING DISTURBED

1. Rotate the crankshaft until the No. 1 piston is at Top Dead Center (TDC) of the compression stroke.
2. Rotate the rotor to the No. 1 terminal position on the distributor cap.
3. Lower the distributor into place, engaging the distributor drive with the drive on the camshaft. With the distributor fully seated on the engine, the rotor should be under the No. 1 terminal.
4. Verify proper rotor alignment with the mark made at disassembly.
5. Reinstall the distributor hold-down nuts and washers. Tighten the nuts to 9 ft. lbs. (13 Nm).
6. Reinstall the spark plug cable bracket.
7. Reconnect the two distributor wiring connectors.
8. Reinstall the distributor cap.
9. Reinstall the spark plug cables, following the identification marks made at disassembly.
10. Reinstall the transaxle dipstick tube.
11. If removed earlier, install the following:

   a. Install the EGR tube and tighten the mounting bolts to 95 inch lbs. (11 Nm).

   b. Install the air cleaner cover, resonator and inlet tube.

   c. Tighten the hose clamp at the throttle body.

   d. Install the PCV hose.

   e. Tighten the clamps holding the air cleaner cover to the air cleaner housing.

   f. Install the bolt attaching the air inlet resonator to the intake manifold.

12. Reconnect the negative battery cable.

Ignition Timing

All engines in the vehicles covered by this manual are equipped with a fixed ignition system. Accordingly, ignition timing is controlled by the Powertrain Control Module (PCM) and is not adjustable.

Engine Assembly

REMOVAL & INSTALLATION

✳✳ CAUTION

The fuel injection system remains under pressure, even after the engine has been turned OFF. The fuel system pressure must be relieved before disconnecting any fuel lines. Failure to do so may result in fire and/or personal injury.

The transaxle must be removed before removing the engine. They will not come out as a unit.

1. Disconnect the negative battery cable.
2. Drain the engine coolant.
3. Drain the engine oil and the transmission oil.
4. Safely relieve the pressure within the fuel injection system.
5. Matchmark the hood to the hinges and remove the hood.
6. Remove the engine under covering.
7. Remove the transaxle assembly using the recommended procedure.
8. Remove the radiator, disconnecting the hoses at the engine.
9. Disconnect the accelerator cable and remove the bracket.
10. Disconnect the heater hoses.
11. Disconnect the brake booster vacuum hose at the engine.
12. Label and disconnect the vacuum hoses running to the bulkhead.
13. Disconnect the high pressure fuel line and discard the O-ring. It is not reusable.
14. Remove the fuel return hose.
15. Label and disengage the electrical connectors to the engine components. All wires and connectors should be labeled at the time of engine removal. This should save much time at assembly.
16. Remove the accessory drive belts. Remove the bolts holding the power steering pump to its bracket and hang the pump out of the way. Do not disconnect the hoses and do not allow the pump to hang by the hoses. Remove the power steering pump bracket.
17. Remove the air conditioning compressor from its mount and hang it from a stiff wire out of the way. Note that the hoses should be left still attached. Do not loosen them or discharge the system.
18. Remove the bolts at the exhaust system joint just below the manifold. Separate the exhaust pipes. Discard the gasket and the two nuts.
19. Raise and safely support the vehicle. Install the engine hoist equipment and make certain the attaching points on the engine are secure. Draw tension on the hoist just enough to support the engine's weight but no more. Do not disturb the placement of the vehicle on the stands.
20. Remove the through-bolt from the rear (bulkhead side) roll stopper. Remove the through-bolt from the front engine roll stopper.
21. Remove the nuts and bolts holding the upper (right-side) engine mount to the engine. Remove the through-bolt and remove the mount assembly. Also remove the support bracket below the mount.

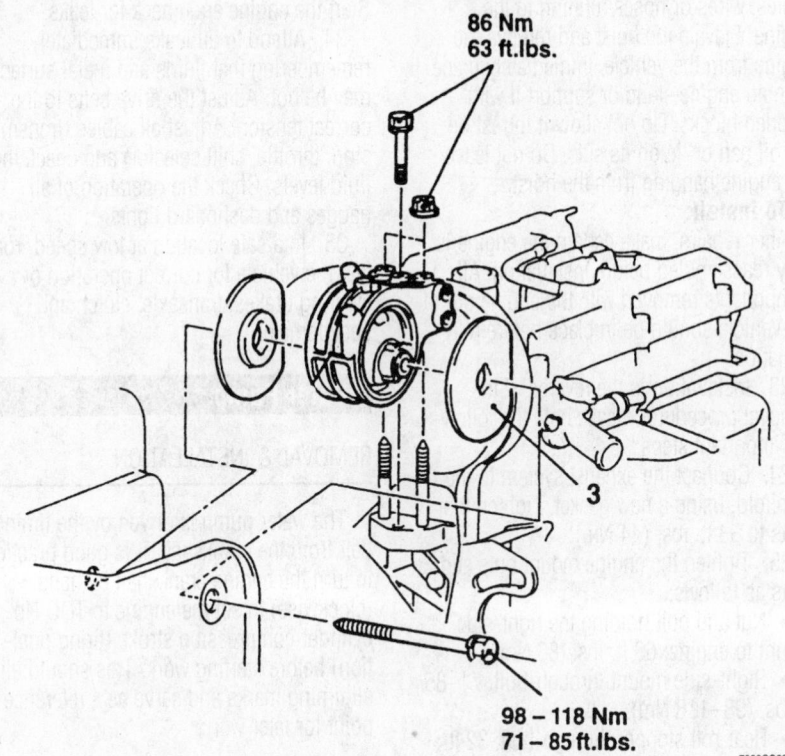

86 Nm
63 ft.lbs.

98 – 118 Nm
71 – 85 ft.lbs.

7922CG02

Exploded view of the right-side engine mount—2.0L engine

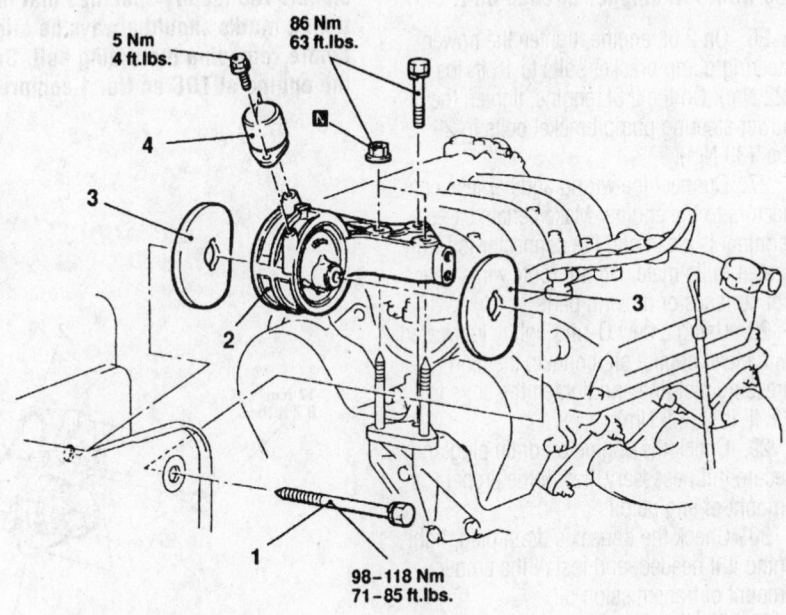

5 Nm
4 ft.lbs.

86 Nm
63 ft.lbs.

98–118 Nm
71–85 ft.lbs.

7922CG03

1. Engine mount insulator mounting bolt
2. Engine mount bracket
3. Engine mount stopper
4. Dynamic damper

Exploded view of the right-side engine mount—2.5L engine

22. Double check for any remaining cables, wires or hoses running to the engine. Elevate the hoist and remove the engine from the vehicle. Immediately place it on an engine stand or support it with wooden blocks. Do not allow it to rest on the oil pan or lie on its side. Do not leave the engine hanging from the hoist.

To install:

After repairs, make certain the engine is fully reassembled before installation. All components removed with the engine out of the vehicle should be in place before reinstallation.

23. Installation is the reverse of the removal procedure. Please note the following important steps.

24. Connect the exhaust system to the manifold, using a new gasket. Tighten the bolts to 33 ft. lbs. (44 Nm).

25. Tighten the engine mount nuts and bolts as follows:
- Nut and bolt holding the right-side mount to engine: 63 ft. lbs. (86 Nm)
- Right-side mount through-bolt: 71–85 ft. lbs. (98–118 Nm)
- Rear roll stopper through-bolt: 32 ft. lbs. (44 Nm)
- Front roll stopper through-bolt: 41 ft. lbs. (56 Nm)

➡**Allow the mounts to support the engine weight before final tightening the front roll stopper through-bolt.**

26. On 2.0L engine, tighten the power steering pump bracket bolts to 16 ft. lbs. (22 Nm). On the 2.5L engine, tighten the power steering pump bracket bolts to 29 ft. lbs. (39 Nm).

27. Connect the wiring and harness connectors to the engine. Make certain each terminal is clean and the connector is firmly seated to its mate. Do not route wires near hot surfaces or moving parts.

28. Using a new O-ring lightly lubricated with clean engine oil, connect the high pressure fuel line and tighten the bolts to 1.8 ft. lbs. (2.5 Nm).

29. Check the engine oil drain plug and secure it if necessary. Install the proper amount of engine oil.

30. Check the transaxle drain plug, tightening it if needed, and install the proper amount of transmission oil.

31. Check the radiator and engine drain cocks, closing them if necessary and refill the coolant system.

32. Double check all installation items, paying particular attention to loose hoses or hanging wires, loosened nuts, poor routing of hoses and wires (too tight or rubbing) and tools left in the engine area.

33. Connect the negative battery cable. Start the engine and check for leaks.

34. Attend to all leaks immediately, remembering that fluids and metal surfaces may be hot. Adjust the drive belts to the correct tension. Adjust all cables (transmission, throttle, shift selector) and check the fluid levels. Check the operation of all gauges and dashboard lights.

35. In a safe location at low speed, road test the vehicle for correct operation of steering brakes, transaxle, clutch and speedometer.

Water Pump

REMOVAL & INSTALLATION

The water pump is driven by the timing belt from the crankshaft. It is good practice to turn the engine crankshaft by hand (clockwise) to set the engine to TDC No. 1 cylinder compression stroke (firing position) before starting work. This should align all timing marks and serve as a reference point for later work.

2.0L Engine

1. Disconnect the negative battery cable.

➡**This procedure requires removing the engine timing belt and the auto tensioner. The factory specifies that the timing marks should always be aligned before removing the timing belt. Set the engine at TDC on No. 1 compression stroke. This should align all timing marks on the crankshaft sprocket and both camshaft sprockets.**

2. Raise and safely support the vehicle to a level that allows access from above and underneath.

3. Remove the right inner splash shield.

4. Remove the accessory drive belts.

5. Place a drain pan under the radiator drain plug. Drain and properly contain the cooling system.

6. Support the engine using a floor jack and block of wood, then remove the right motor mount.

7. Remove the timing belt, tensioner and camshaft sprockets.

✳✳ WARNING

With the timing belt removed, DO NOT rotate the camshaft or crankshaft or damage to the engine could occur.

8. Remove the rear timing belt cover to access the water pump.

9. Remove the water pump attaching bolts.

10. Remove the water pump.

To install:

11. Thoroughly clean all sealing surfaces. Replace the water pump if there are any cracks, signs of coolant leakage from the shaft seal, loose or rough turning bearings, damaged impeller or sprocket or sprocket flange loose or damaged.

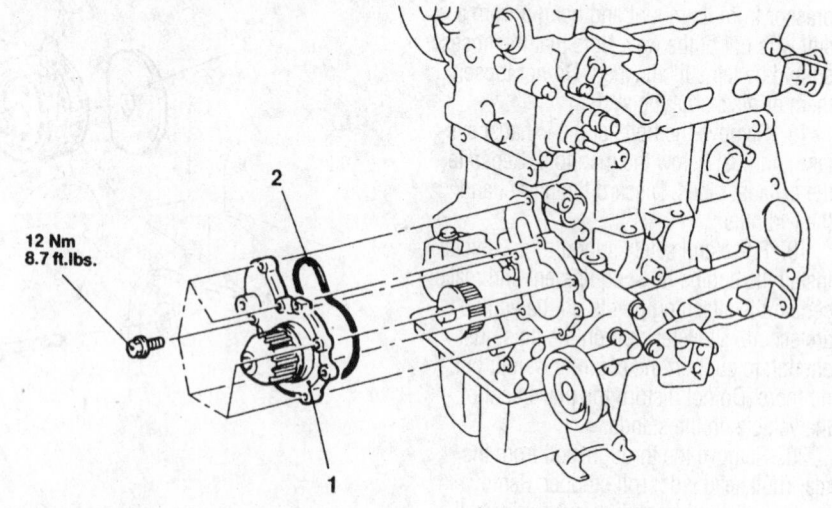

1. Water pump
2. O-ring

Exploded view of the water pump mounting—2.0L engine

7922CG04

12. Install a new rubber O-ring into the water pump.

➡**Be sure the O-ring is properly seated in the water pump groove before tightening the screws. An improperly located O-ring may cause damage to the O-ring and cause a coolant leak.**

13. Install the water pump and tighten the bolts to 105 inch lbs. (12 Nm).

14. Using a cooling system pressure tester, pressurize the cooling system to 15 psi and check for leaks. If okay, release the pressure and continue the engine assembly process.

15. Rotate the water pump by hand to check for freedom of movement.

16. Install the rear timing belt cover.

17. Install the camshaft sprocket(s), timing belt and tensioner. DO NOT allow the camshafts to turn while the sprockets bolts are being tightened to maintain timing mark alignment.

※ WARNING

Do not attempt to compress the tensioner plunger with the tensioner assembly installed in the engine. This will cause damage to the tensioner and other related components. The tensioner MUST be compressed in a vise.

18. Install the timing belt covers.

19. Install the right engine mount bracket and engine mount.

20. Remove the floor jack and wood block from underneath the engine.

21. Install the crankshaft damper.

22. Install the right inner splash shield.

23. Lower the vehicle.

24. Install and tension the accessory drive belts.

25. Refill the cooling system using the correct quantity and type of coolant. Bleed the cooling system.

26. Start the engine and check for proper operation.

27. Check and top off cooling system, if necessary.

2.5L Engine

1. Disconnect the negative battery cable.

2. Place a large drain pan under the radiator drain plug. Drain and properly contain the engine coolant.

➡**This procedure requires removing the engine timing belt and the auto tensioner. To help assure proper alignment at assembly, it may be helpful to**

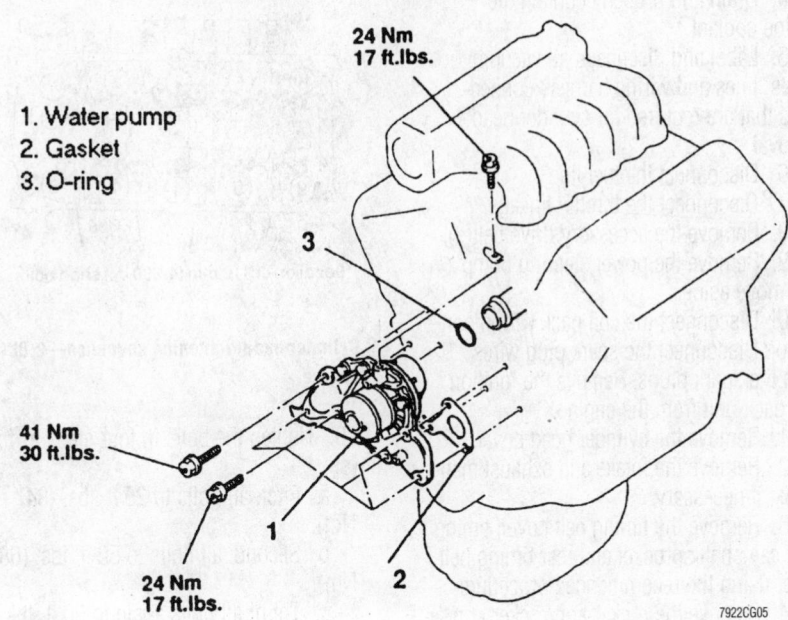

1. Water pump
2. Gasket
3. O-ring

24 Nm
17 ft.lbs.

41 Nm
30 ft.lbs.

24 Nm
17 ft.lbs.

7922CG05

Exploded view of the water pump mounting—2.5L engine

set the engine at TDC on No. 1 compression stroke. This should align all timing marks on the crankshaft sprocket and both camshaft sprockets.

3. Remove the accessory drive belts and crankshaft damper.

4. Remove the right engine mount. This requires safely supporting the engine with a floor jack and wood block so the mount can be removed.

5. Remove the timing belt covers.

6. Remove the timing belt and tensioner.

7. Remove the water pump mounting bolts.

8. Separate the water pump from the water inlet pipe and remove the pump.

To install:

9. Thoroughly clean all sealing surfaces. Inspect the pump for damage or cracks, signs of coolant leakage at the vent and excessive looseness or rough turning bearing. Any problems require a new pump.

10. Install a new O-ring on the water inlet pipe. Wet the O-ring with water to make installation easier. DO NOT use oil or grease on the O-ring.

11. Install a new gasket on the water pump and fit the pump inlet opening over the water pipe. Press the assembly together to force the pipe into the water pump.

12. Install the water pump-to-engine bolts and tighten to 20 ft. lbs. (27 Nm).

13. Install the timing belt and timing belt tensioner. Set the timing belt tension.

14. Install the timing belt covers. Install the right engine mount. Remove the floor

jack and engine block from underneath the engine.

15. Install the crankshaft damper.

16. Install the accessory drive belts and set to proper tension.

17. Connect the negative battery cable.

18. Fill and bleed the engine cooling system.

19. Start the engine and verify proper operation.

Cylinder Head

REMOVAL & INSTALLATION

2.0L Engine

※ CAUTION

The fuel injection system remains under pressure, even after the engine has been turned OFF. The fuel system pressure MUST BE relieved before disconnecting any fuel lines. Failure to do so may result in fire and/or personal injury.

1. Disconnect the negative battery cable from the left shock absorber tower. The ground cable is equipped with a insulator grommet which should be placed on the stud to prevent the negative battery cable from accidentally grounding.

2. Properly relieve the fuel system pressure.

3. Remove the air cleaner assembly.

4. Drain and properly contain the engine coolant.

5. Label and disengage all vacuum hoses, lines and wiring harness connections that are required for cylinder head removal.

6. Disconnect the fuel line.

7. Disconnect the throttle linkage.

8. Remove the accessory drive belt(s).

9. Remove the power steering pump and move aside.

10. Disconnect the coil pack wiring connector. Disconnect the spark plug wires from the spark plugs. Remove the ignition coil pack unit from the engine.

11. Remove the cylinder head cover.

12. Remove the intake and exhaust manifolds, if necessary.

13. Remove the timing belt cover, timing belt, camshaft sprocket and rear timing belt cover using the recommended procedure.

14. Remove the rocker arm/rocker arm shaft assemblies.

15. Remove the cylinder head bolts and remove the cylinder head.

To install:

➡**The cylinder head bolts should be checked for stretching before reuse. If the thread area of the bolt is necked down the bolts must be replaced with new. New head bolts are recommended.**

16. Thoroughly clean all parts. Clean all sealing surfaces. Use care not to scratch the aluminum cylinder head sealing surface. Check the cylinder head for flatness using a feeler gauge and a straight-edge. The cylinder head must be flat within 0.004 in. (0.1mm).

17. Check the cylinder head for cracks or other damage.

18. Install a new gasket and the cylinder head to the engine block.

19. Be sure to oil the cylinder head bolt threads with clean engine oil. Install the four cylinder head bolts, the 4.330 in. (110mm) short bolts are to be installed in positions 7, 8, 9 and 10. Tighten the bolts in proper sequence.

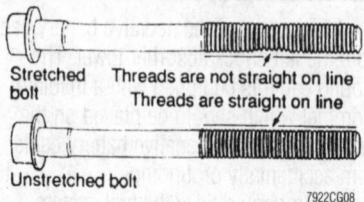

Checking the cylinder head bolts for necking (stretching)—2.0L engine

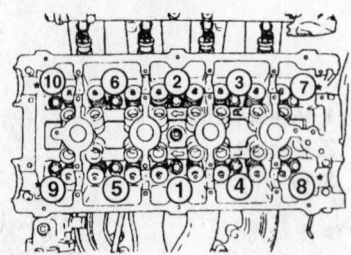

*Location of 110 mm (4.330 in.) short bolts.

7922CG09

Cylinder head tightening sequence—2.0L engine

20. Tighten the bolts in four steps as follows:

    a. First: all bolts to 25 ft. lbs. (34 Nm).

    b. Second: all bolts to 50 ft. lbs. (68 Nm).

    c. Third: all bolts again to 50 ft. lbs. (68 Nm).

    d. Fourth: all bolts an additional ¼ turn.

➡**Do not use a torque wrench for the fourth step.**

21. Install the rocker arm/rocker arm shaft assemblies.

22. Install the cylinder head cover.

23. Install the timing belt rear cover and camshaft sprocket. Install the timing belt.

24. Install the timing belt cover.

25. Install the intake and exhaust manifolds, if removed.

26. Install the ignition coil pack onto the engine. Reconnect the coil pack wiring connector and the spark plug wires to the correct spark plugs.

27. Install the power steering pump.

28. Install and adjust the accessory drive belts.

29. Connect the throttle linkage.

30. Check to be sure all ducts, hoses, fuel lines and wiring harness connectors have been properly engaged.

31. Install the air cleaner assembly.

32. Fill the cooling system with a ⁵⁰⁄₅₀ mixture of clean antifreeze and water. A complete oil and filter change is recommended.

33. Connect the negative battery cable.

34. Start the engine and check for leaks. Run the engine with the radiator cap off so as the engine warms and the thermostat opens, coolant can be added to the radiator. When satisfied that the cooling system is full, shut the engine **OFF**, install the radiator cap and allow the engine to cool.

35. With the engine cool, check all fluid levels. Add coolant and oil as required.

Restart the engine and test drive vehicle to check for proper operation.

2.5L Engine

1. Release the fuel system pressure.

2. Disconnect the negative battery cable.

3. Drain the engine cooling system.

4. Remove the timing belt and camshaft sprockets.

5. Label and disengage any wiring harnesses, vacuum hoses and lines that would in some way inhibit removal of the cylinder head.

6. Remove the intake manifold assembly.

7. Remove the water pump inlet pipe retaining bolt on the inner rear part of the front cylinder head.

8. Remove the valve covers and rocker arm assemblies.

9. Remove the distributor assembly.

10. On the front cylinder head, remove the ground strap on the left-end of the head.

11. Remove the exhaust manifolds and crossover pipe.

12. Remove the cylinder head mounting bolts and place them in numbered order through holes made in piece of cardboard.

13. Remove the cylinder head.

To install:

14. Thoroughly clean and dry the mating surfaces of the head and block. Check the cylinder head for cracks, damage or engine coolant leakage. Remove scale, sealing compound and carbon. Clean the oil passages thoroughly. Check the head for flatness. End to end, the head should

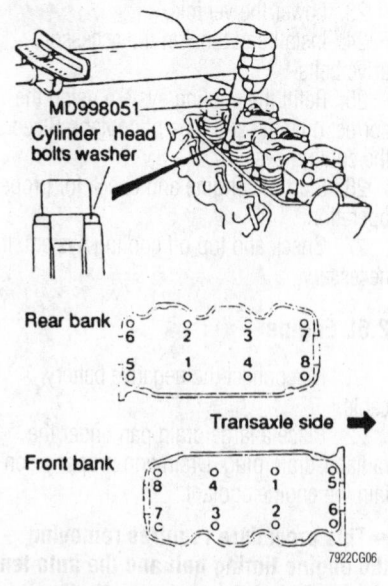

Cylinder head bolt tightening sequence—2.5L engine

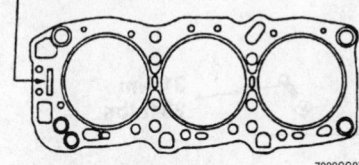

Identification mark

7922CG07

Install the head gasket with the identification mark at the front facing upward—2.5L engine

be no more than 0.008 in. (0.2mm) out-of-true. If the service limit is exceeded, correct to meet specifications. Note that the maximum amount from stock allowed to be removed from the cylinder head and mating cylinder block is 0.0079 in. (0.2mm). If the cylinder head cannot be made serviceable by removing this amount, replace the head.

15. Check that the head gaskets have the proper identification marks for the engine. Lay the head gasket with the identification mark at the front top.

➡**Do not apply sealant to the head gasket or mating surfaces.**

16. Inspect the cylinder head bolts prior to installation. If the threads are necked down (stretched), the bolts should be replaced. Necking can be checked by holding a straight edge against the threads. If all of the threads do not contact the straightedge, the bolt should be replaced. All new head bolts are recommended.

17. Install the head straight down onto the block. Try to eliminate most of the side-to-side adjustments, as this may move the gasket out of position or damage the gasket. Before installing the bolts, the threads should be oiled with clean engine oil. Install the bolts and the special washers by hand and just start each bolt one or two turns on the threads.

➡**The washers must be installed correctly. The rounded shoulder of the washer denotes the face in contact with the bolt. The flat face contacts the head.**

18. Correct tightening of the head bolts requires three steps:
   a. Following the tightening sequence, draw each bolt to 62 ft. lbs. (84 Nm).
   b. Follow the tightening sequence and tighten each bolt to 70 ft. lbs. (95 Nm).

   c. Follow the tightening sequence and tighten each bolt to 80 ft. lbs. (108 Nm).

19. Install the valve cover and gasket.

20. Install the exhaust manifolds and crossover pipe.

21. Install the distributor assembly.

22. Install the intake manifold assembly.

23. Check to be sure that all wiring harnesses, vacuum hoses and lines are all properly connected.

➡**Before proceeding, double check all installation items, paying particular attention to loose hoses or hanging wires, nuts not properly tightened, poor routing of hoses and wires (too tight or rubbing) and tools left in the engine area.**

24. Fill the cooling system with coolant. Changing the oil and filter is recommended to eliminate pollutants such as coolant in the oil.

25. Connect the negative battery cable. With the radiator cap off, start the engine and check for leaks of fuel, vacuum, oil or coolant. Check the operation of all engine electrical systems as well as dashboard gauges and lights. Add coolant as the engine warms.

26. Perform necessary adjustments to the accelerator cable and drive belts. Allow the engine to cool and once again check and adjust the coolant level.

Rocker Arm/Shafts

REMOVAL & INSTALLATION

This procedure applies to 2.5L engine only. On 2.0L engine, the valves are actuated directly by the camshafts.

2.5L Engine

✳✳ CAUTION

The fuel injection system remains under pressure, even after the engine has been turned OFF. The fuel system pressure MUST BE relieved before disconnecting any fuel lines. Failure to do so may result in fire and/or personal injury.

1. Disconnect the negative battery cable.

2. Properly relieve the fuel system pressure.

3. If removing the right (firewall) side rocker arm/shaft assembly, Remove the

upper intake manifold (air intake plenum), which is a 2-piece unit of aluminum alloy.

4. Remove the valve cover(s).

5. Identify the rocker arm shaft assemblies before removal.

6. Install the auto lash adjuster retainers Special Tool MD 998443 or equivalent to keep the auto lash adjusters from falling out of the rocker arms when the rocker arm assembly is removed.

7. Loosen the attaching fasteners and remove the rocker arm shaft assemblies from the cylinder head.

➡**The hydraulic automatic lash adjusters are precision units installed in the machined openings in the rocker arm units. Do not disassemble the auto lash adjusters from the rocker arms.**

To install:

8. The rocker arm shafts are hollow and used as a lubrication oil duct. Ensure all valvetrain parts are clean. Check the rocker arm mounting portion of the shafts for wear or damage. Replace if necessary. Check all oil holes for clogging with a small wire and clean as required. If any rockers were removed, lubricate and install on the shafts in their original positions.

9. Install the rocker arm and shaft assemblies with the FLAT in the rocker arm shafts facing toward the timing belt side of the engine for the right cylinder head. For the left cylinder head install the rocker arm and shaft assembly with the FLAT in the rocker arm shaft facing toward the transaxle side of the engine. Install the retainers and spring clips in their original positions on the exhaust and intake shafts. Tighten the retainer bolts to 276 inch lbs. (31 Nm) working from the center, outward. Remove the valve lash retainer tools that should have been installed at disassembly.

10. Inspect the spark plug tube seals located on the ends of each tube. These seals slide onto each tube to seal the cylinder head cover to the spark plug tube. If these seals show signs of hardness and/or cracks, they should be replaced.

11. Install the valve cover(s).

12. Install the upper intake manifold (plenum), if necessary.

13. Check to be sure all remaining electrical connectors have been reconnected. Tighten the air tube connections.

14. An oil and filter change is recommended.

15. Connect the negative battery cable. Start the engine and check for leaks, abnormal noises and vibrations.

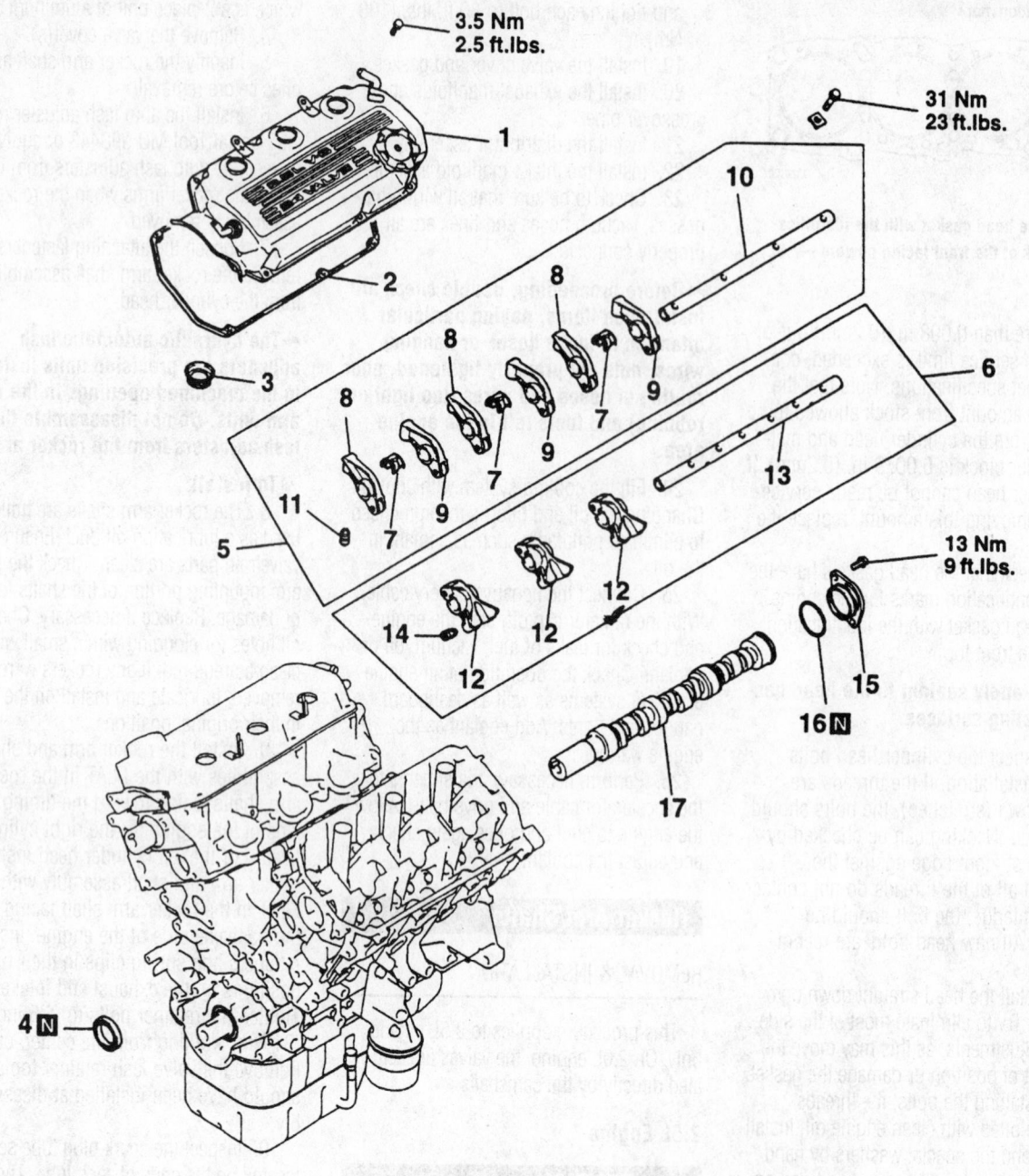

1. Rocker cover
2. Rocker cover gasket
3. Oil seal
4. Oil seal
5. Rocker arm and rocker arm shaft
6. Rocker arm and rocker arm shaft
7. Rocker shaft spring
8. Rocker arm A
9. Rocker arm B

10. Rocker arm shaft
11. Lash adjuster
12. Rocker arm C
13. Rocker arm shaft
14. Lash adjuster
15. Thrust case
16. O-ring
17. Camshaft

7922CG10

Exploded view of the cylinder head valvetrain assembly—2.5L engine

Intake Manifold

REMOVAL & INSTALLATION

2.0L Engine

This engine uses a two-piece aluminum intake manifold. The upper half of the manifold (also called a plenum) mounts the throttle body. The lower half of the manifold contains the fuel rail and injectors. A non-reusable gasket joins the two halves. Use care when working with light alloy parts.

✴✴ CAUTION

The fuel injection system remains under pressure, even after the engine has been turned OFF. The fuel system pressure must be relieved before disconnecting any fuel lines. Failure to do so may result in fire and/or personal injury.

1. Relieve the fuel system pressure.
2. Disconnect the negative battery cable and drain the cooling system.
3. Disconnect the accelerator cable, breather hose and air intake hose.
4. Disengage the vacuum connection at the power brake booster and the PCV valve. Disconnect all remaining vacuum hoses and pipes, as necessary. Tag for identification, if necessary, to save time at assembly.
5. Disconnect the fuel line(s) remove the throttle control cable and brackets.

✴✴ CAUTION

Do not use conventional fuel filters, hoses or clamps when servicing fuel injection systems. They are not compatible with the injection system and could fail, causing personal injury or damage to the vehicle. Use only hoses and clamps specifically designed for fuel injection.

6. Unplug the alternator wiring harness connection.
7. Disengage the MAP sensor and the intake air temperature sensor connectors.
8. Disengage the TPS connector and position the engine wiring harness aside.
9. Disengage the EGR pipe connection.

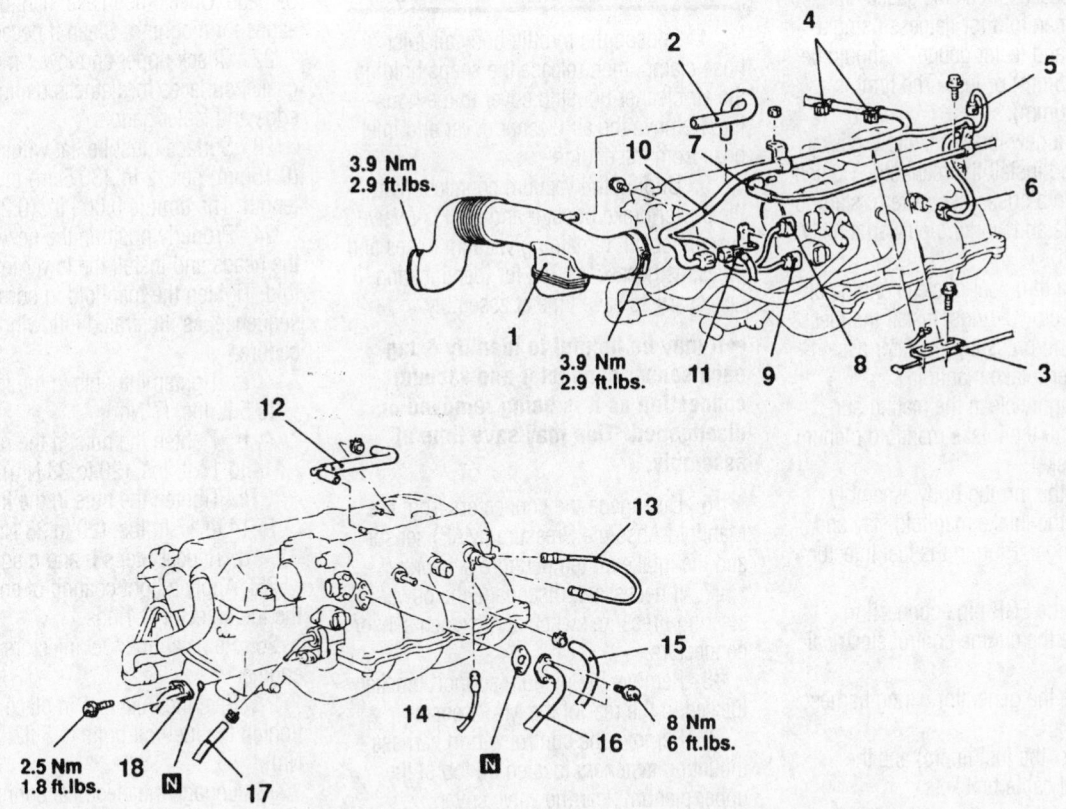

1. Air intake hose
2. Breather hose
3. Accelerator cable connection
4. Clip
5. MAP sensor connector
6. Intake air temperature sensor connector
7. Vacuum hose connection
8. TPS connector
9. Idle air control motor connector
10. Control wiring harness
11. Generator wiring harness connection
12. PCV hose assembly
13. Vacuum hose
14. Vacuum hose connection
15. Brake booster vacuum hose connection
16. EGR pipe connection
17. Fuel return hose connection
18. High-pressure fuel hose connection
N. use new components

Intake manifold hose, cable and wire attachment identification—2.0L engine

7922CG11

10. Remove the intake manifold stay and the engine hanger. Disengage the fuel injector connectors.

11. Remove the throttle body assembly.

12. Remove the mounting bolts and remove the intake manifold plenum and gasket.

13. Remove the complete fuel rail assembly. Use care since the fuel injectors can drop out of the fuel rail as it is being removed.

14. Remove the mounting bolts, then remove the intake manifold and gasket from the engine.

To install:

15. Clean all gasket material from the cylinder head and intake manifold assembly. Check both surfaces for cracks or other damage. Check the intake manifold water passages and air passages for clogging. Clean if necessary. Check the gasket surface of the intake manifold for flatness using a straight edge and feeler gauge. It should be 0.006 in. (0.15mm) or less. The limit is 0.008 in. (0.20mm).

16. Install a new intake manifold gasket to the head and install the manifold. Tighten the manifold in a crisscross pattern, starting from the inside and working outwards to 17 ft. lbs. (23 Nm).

17. Apply a thin coat of clean engine oil to the fuel injector O-rings. Install the fuel rail, injector and pressure regulator assembly to the lower intake manifold.

18. Thoroughly clean the mating surfaces and install the intake manifold plenum with a new gasket.

19. Install the throttle body assembly.

20. Install the intake manifold stay and the engine hanger. Plug in the fuel injector connectors.

21. Attach the EGR pipe connection.

22. Engage the engine control electrical connectors.

23. Engage the generator wiring harness connection.

24. Connect the fuel line(s) and the throttle control cable brackets.

25. Connect the vacuum hose at the power brake booster and the PCV valve. Connect all remaining vacuum hoses and pipes.

26. Connect the accelerator cable, breather hose and air intake hose.

27. Connect the negative battery cable.

28. Start the engine and check for proper operation.

2.5L Engine

1. Disconnect the negative battery cable.

2. Properly relieve the fuel system pressure.

✳✳ CAUTION

The fuel injection system remains under pressure, even after the engine has been turned OFF. The fuel system pressure MUST BE relieved before disconnecting any fuel lines. Failure to do so may result in fire and/or personal injury.

3. Disconnect the fuel line(s) from the fuel rail assembly. On quick-connect fittings, squeeze the fitting retainer tabs together and separate the connection.

✳✳ CAUTION

Wrap shop towels around the connection to catch any gasoline spillage.

4. Loosen the throttle body air inlet hose clamp, then release the snaps holding the air cleaner housing cover to the housing. Remove the air cleaner cover and inlet hose from the engine.

5. Unplug the vacuum connection at the power brake booster and the PCV valve. Disconnect all remaining vacuum hoses and pipes, as necessary. Tag for identification, if necessary, to save time at assembly.

➡ **It may be helpful to identify & tag each sensor connector and vacuum connection as it is being removed or disengaged. This may save time at assembly.**

6. Disengage the connectors from the Manifold Absolute Pressure (MAP) sensor and the intake air temperature sensors.

7. If necessary, disengage the power steering pressure switch and oxygen sensor connectors.

8. Remove the plenum support bracket located to the rear of the MAP sensor.

9. Remove the control wiring harness mounting fasteners located on top of the upper plenum near the valve cover.

10. Disconnect the Throttle Position Sensor (TPS) and the Idle Air Control (IAC) motor electrical connections.

11. Remove the throttle body assembly.

12. Remove the throttle cable bracket.

13. Remove the EGR tube from the engine intake manifold.

14. Remove the EGR valve and transducer assembly.

15. Remove the plenum support bracket located to the rear of the EGR tube.

16. Remove the seven bolts attaching the upper intake plenum to the lower manifold

and remove plenum. Remove the intake plenum-to-lower manifold gasket.

17. Disconnect the fuel injector electrical connectors.

18. Remove the four bolts attaching the fuel rail to the intake manifold. Use care. There are spacers under each fuel rail bolt. Remove the fuel rail.

➡ **It may be necessary to remove the power steering fluid reservoir and mounting bracket to access all of the lower intake manifold fasteners.**

19. Remove the lower intake manifold attaching bolts.

20. Remove the intake manifold and discard the old gaskets.

To install:

21. Clean all gasket sealing surfaces. Check both surfaces for cracks or other damage. Check the intake manifold air passages for clogging. Clean if necessary.

22. Check upper and lower manifold gasket surfaces for flatness using a straight-edge and feeler gauge.

23. Surface must be flat within 0.006 in. (0.15mm) per 12 in. (30.5cm) of manifold length. The limit is 0.008 in. (0.20mm).

24. Properly position the new gaskets to the heads and install the lower intake manifold. Tighten the manifold in correct sequence, as illustrated following this procedure.

   a. Tighten the nuts in the front bank to 5 ft. lbs. (7 Nm).

   b. Tighten the nuts in the rear bank to 14 to 17 ft. lbs. (20 to 23 Nm).

   c. Tighten the nuts in the front bank to 14 to 17 ft. lbs. (20 to 23 Nm).

   d. Repeat Steps **b** and **c** again.

25. Apply a light coating of engine oil to the fuel injector O-rings.

26. Reinstall the fuel injectors into the engine.

27. Seat the injectors in place and tighten the fuel rail bolts to 8 ft. lbs. (12 Nm).

28. Engage the electrical connectors to the fuel injectors.

29. Reconnect the fuel line(s) to the fuel rail assembly. Exert a slight tug on the fuel line away from the fuel rail to verify positive engagement.

30. Install the upper intake plenum with new gaskets.

31. Tighten the plenum bolts to 13 ft. lbs. (18 Nm).

32. Reinstall the plenum support brackets and tighten to 13 ft. lbs. (18 Nm).

33. Install the EGR valve and transducer assembly.

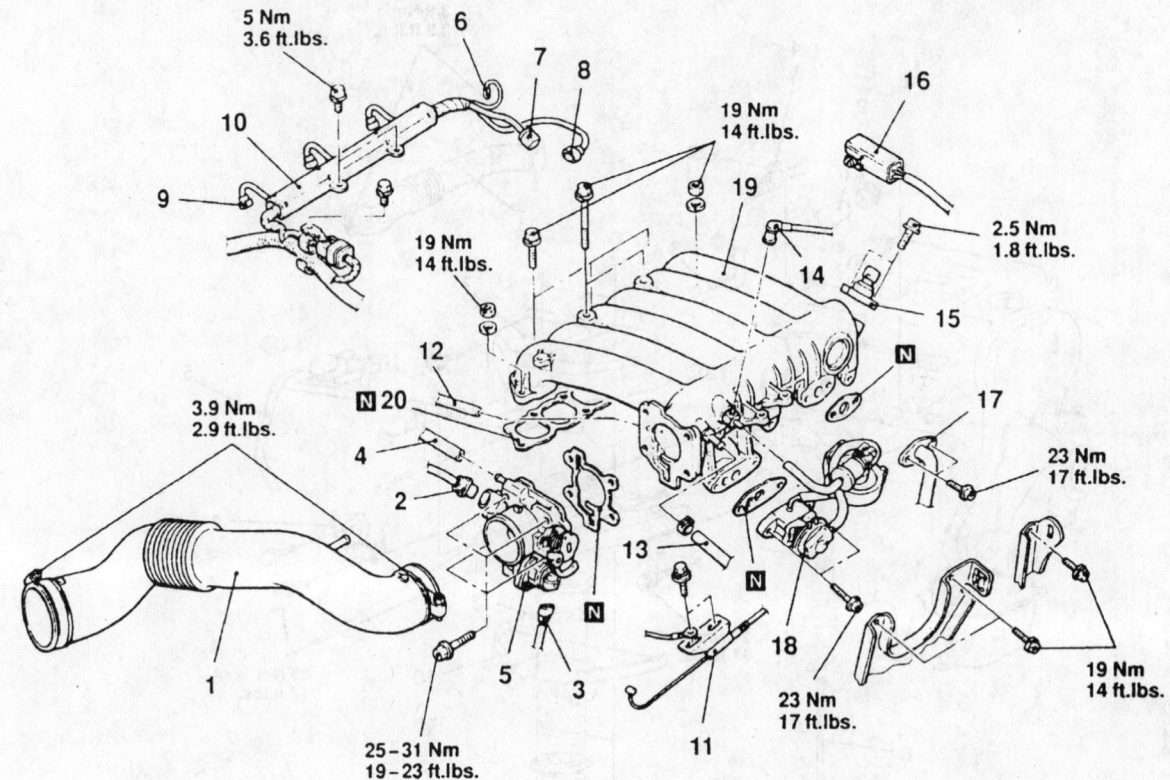

5 Nm
3.6 ft.lbs.

19 Nm
14 ft.lbs.

2.5 Nm
1.8 ft.lbs.

19 Nm
14 ft.lbs.

3.9 Nm
2.9 ft.lbs.

23 Nm
17 ft.lbs.

23 Nm
17 ft.lbs.

19 Nm
14 ft.lbs.

25–31 Nm
19–23 ft.lbs.

1. Air intake hose
2. TPS connector
3. Idle air control motor connector
4. Vacuum hose connection
5. Throttle body assembly
6. Power steering oil pressure switch connector
7. Heated oxygen sensor connector
8. Intake air temperature sensor connector
9. Injector connector
10. Control wiring harness
11. Accelerator cable connection
12. Vacuum hose connection
13. Brake booster vacuum hose connection
14. Vacuum hose connection
15. MAP sensor
16. Heated oxygen sensor harness
17. EGR pipe connection
18. EGR valve and EGR transducer assembly
19. Intake manifold plenum
20. Intake manifold plenum gasket

7922CG12

Exploded view of the plenum and intake manifold—2.5L engine

34. Install the EGR tube and tighten the screws to 95 inch lbs. (11 Nm).

35. Install the throttle cable bracket.

36. Install the throttle body assembly.

37. Reconnect the TPS and IAC electrical connections.

38. Place the control wiring harness into correct position on the engine and tighten the mounting fasteners.

39. Engage the power steering pressure switch and oxygen sensor connectors, if previously disconnected.

40. Engage the MAP sensor and the intake air temperature sensor connectors.

41. Connect the vacuum hose at the power brake booster and the PCV valve. Connect all remaining vacuum hoses and pipes.

42. Connect the remaining engine control system electrical connectors.

43. Reinstall the air cleaner cover and air inlet hose. Tighten the intake hose-to-throttle body hose clamp.

44. Reconnect the negative battery cable.

45. Start the engine and check for leaks.

Exhaust Manifold

REMOVAL & INSTALLATION

2.0L Engine

1. Disconnect the negative battery cable.

2. Remove the air intake hose and the small air hose connection.

3. Properly drain the engine coolant.

4. Disconnect the upper radiator hose from the thermostat housing.

5. Disengage the control wiring harness connection.

6. Remove the water pipe assembly and the engine oil level dipstick.

7. Remove the heat shield and the engine hanger.

8. Remove the pulsed secondary air injection valve, if equipped.

9. Raise and safely support the vehicle.

10. Remove the exhaust pipe to exhaust manifold lock-nuts and separate the exhaust pipe. Discard the gasket.

11. Lower the vehicle.

12. Loosen the mounting fasteners, and remove the exhaust manifold.

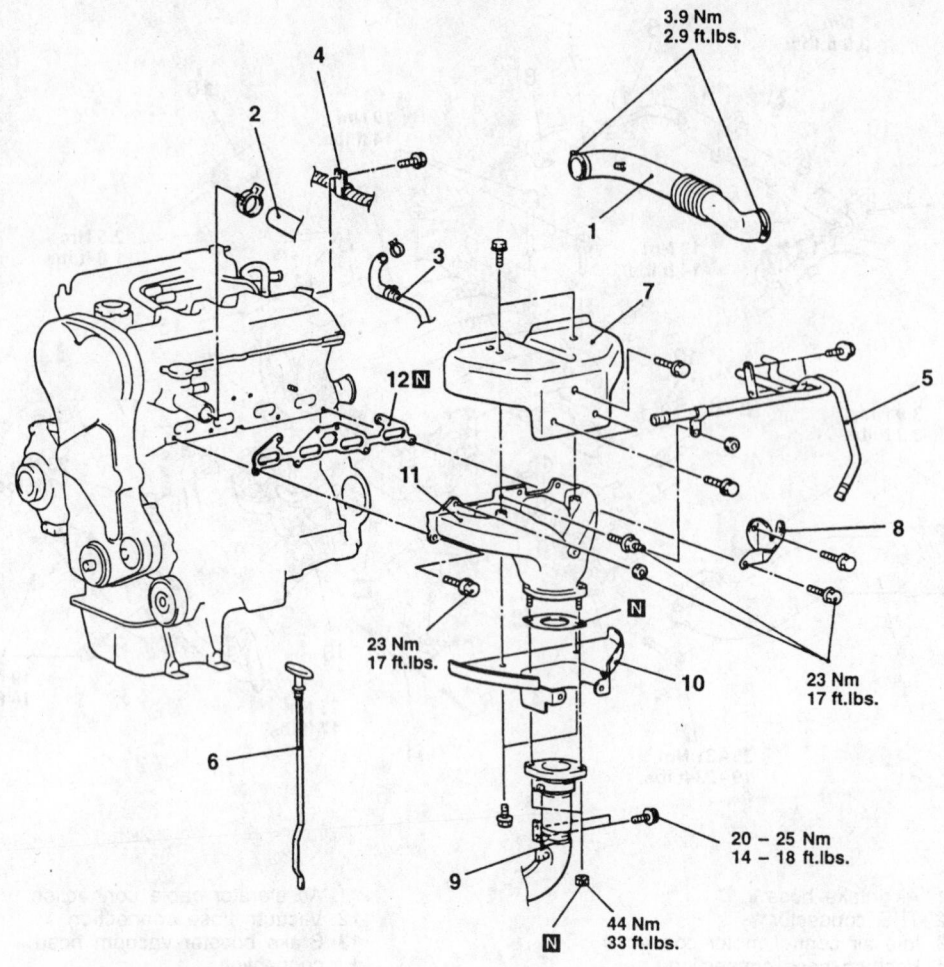

1. Air intake hose
2. Radiator upper hose connection
3. Air hose connection
4. Control wiring harness connection
5. Water pipe assembly
6. Engine oil level gauge
7. Heat protector
8. Engine hanger
9. Front exhaust pipe connection
10. Heat protector
11. Exhaust manifold
12. Exhaust manifold gasket

7922CG13

Exploded view of the exhaust manifold and related components—2.0L engine

To install:

13. Clean all gasket material from the mating surfaces and check the manifold for cracks or warpage.

14. Install a new gasket and install the manifold. Tighten the fasteners, in a criss-cross pattern to 17 ft. lbs. (23 Nm).

15. Raise and safely support the vehicle.

16. Install the exhaust pipe to the exhaust manifold with a new gasket and new lock-nuts. Tighten the nuts to 33 ft. lbs. (44 Nm).

17. Lower the vehicle.

18. Install the pulsed secondary air injection valve, if equipped.

19. Install the heat shield and the engine hanger.

20. Engage the control wiring harness connection.

21. Connect the upper radiator hose to the thermostat housing.

22. Properly fill the engine cooling system.

23. Install the air intake hose and the small air hose connection.

24. Connect the negative battery cable and check for exhaust leaks.

2.5L Engine

FRONT BANK

1. Disconnect the negative battery cable.

2. Remove the cooling fan motor assembly.

3. Remove the engine oil level dipstick and tube.

4. Remove the engine hanger or lower heat shield, if equipped.

5. Raise and safely support the vehicle.

6. Remove the exhaust pipe to exhaust manifold lock-nuts and separate the exhaust pipe. Discard the gasket.

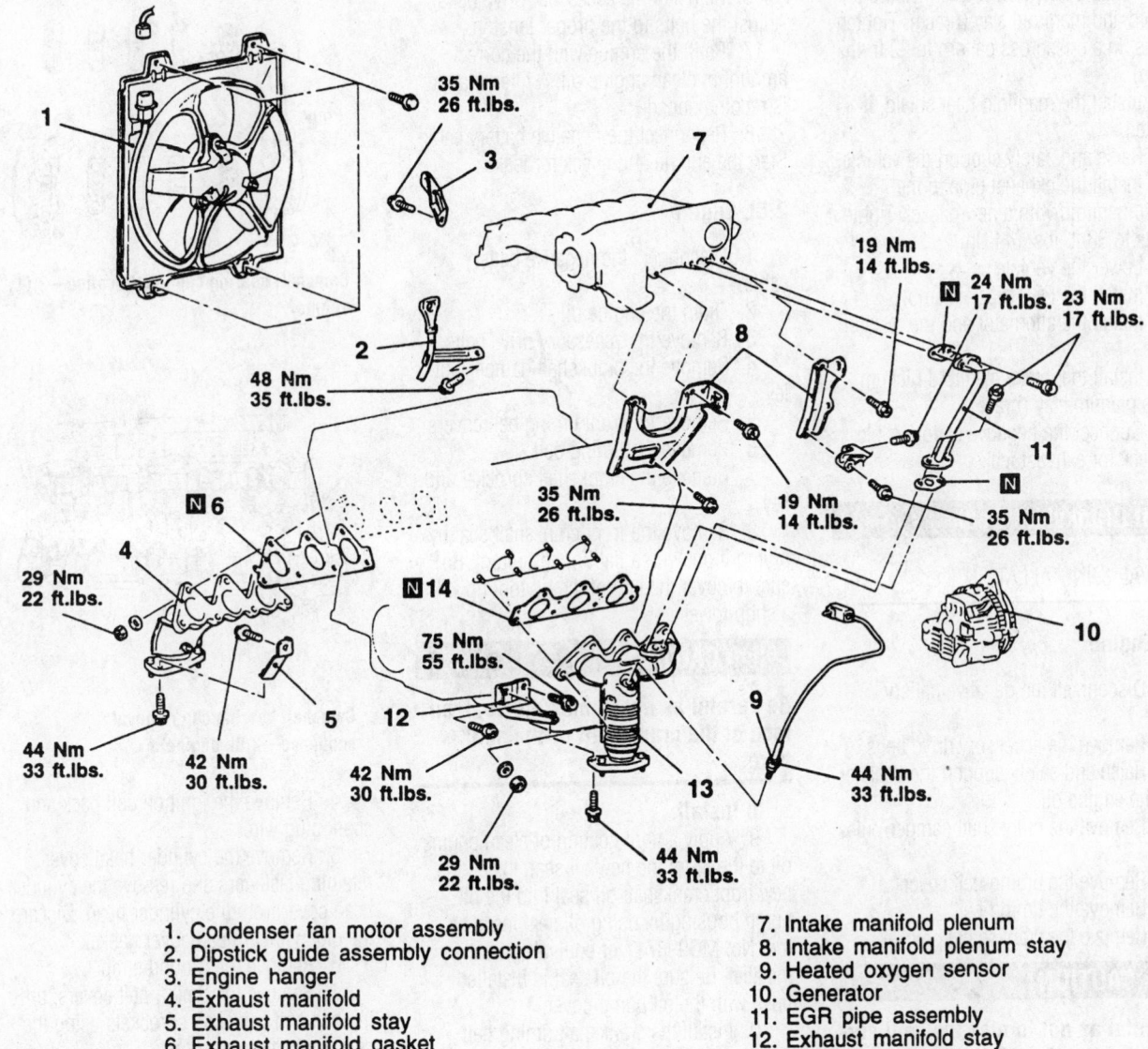

1. Condenser fan motor assembly
2. Dipstick guide assembly connection
3. Engine hanger
4. Exhaust manifold
5. Exhaust manifold stay
6. Exhaust manifold gasket
7. Intake manifold plenum
8. Intake manifold plenum stay
9. Heated oxygen sensor
10. Generator
11. EGR pipe assembly
12. Exhaust manifold stay
13. Exhaust manifold
14. Exhaust manifold gasket

7922CG14

Exploded view of the exhaust manifold assembly and related components—2.5L engine

7. Lower the vehicle.

8. Remove the exhaust manifold mounting fasteners, the exhaust manifold stay (brace), the exhaust manifold and the gasket.

To install:

9. Clean all gasket material from the mating surfaces and check the manifold for cracks or warpage.

10. Install a new gasket and install the manifold and manifold stay (brace). Tighten the fasteners, in a crisscross pattern to 22 ft. lbs. (29 Nm).

11. Raise and safely support the vehicle.

12. Install the exhaust pipe to the exhaust manifold with a new gasket. Tighten the nuts to 33 ft. lbs. (44 Nm).

13. Lower the vehicle.

14. Install the engine hanger or lower heat shield, if equipped. Tighten the lower heat shield fasteners to 10 ft. lbs. (13 Nm).

15. Install the engine oil level dipstick and tube.

16. Install the cooling fan motor assembly.

17. Connect the negative battery cable and check for exhaust leaks.

REAR BANK

1. Disconnect the negative battery cable.

2. Remove the intake manifold plenum and the plenum stay.

3. Remove the heated oxygen sensor.

4. Remove the alternator.

5. Remove the EGR pipe assembly.

6. Raise and safely support the vehicle.

7. Remove the exhaust pipe to exhaust manifold lock-nuts and separate the exhaust pipe. Discard the gasket.

8. Lower the vehicle.

9. If equipped, remove the manifold heat shield.

10. Remove the exhaust manifold mounting bolts, the exhaust manifold stay (brace), the exhaust manifold and the gasket.

To install:

11. Clean all gasket material from the mating surfaces and check the manifold for cracks or warpage.

12. Install a new gasket and install the manifold and manifold stay (brace). Tighten the nuts, in a crisscross pattern to 22 ft. lbs. (30 Nm).

13. Install the manifold heat shield, if equipped.

14. Raise and safely support the vehicle.

15. Install the exhaust pipe to the exhaust manifold with a new gasket. Tighten the nuts to 33 ft. lbs. (44 Nm).

16. Lower the vehicle.

17. Install the EGR pipe assembly.

18. Install the alternator and the oxygen sensor.

19. Install the intake manifold plenum and the plenum stay (brace).

20. Connect the negative battery cable and check for exhaust leaks.

Front Crankshaft Seal

REMOVAL & INSTALLATION

2.0L Engine

1. Disconnect the negative battery cable.

2. Remove the accessory drive belts.

3. Raise and safely support the vehicle. Drain the engine oil.

4. Remove the crankshaft damper/pulley.

5. Remove the timing belt cover.

6. Remove the timing belt.

7. Remove the crankshaft sprocket.

✳✳ CAUTION

Be careful as not to nick the seal surface of the crankshaft or the seal bore.

8. Remove the front crankshaft seal using a seal puller tool. Be careful not to damage the seal contact area of the crankshaft.

To install

9. Apply a light coating of clean engine oil to the lip of the new oil seal. Install the new front crankshaft oil seal by using Oil Seal Installer tool No. 6780–1 or equivalent seal tool.

10. Place new oil seal into the opening with the seal spring facing the inside of the engine. Be sure the oil seal is installed flush with the front cover.

11. Install the crankshaft timing belt sprocket.

12. Install the timing belt.

13. Install the timing belt cover.

14. Install the crankshaft damper/pulley.

15. Lower the vehicle.

16. Reinstall the accessory drive belts. Adjust the belts to the proper tension.

17. Refill the engine with the correct amount of clean engine oil. A filter change is recommended.

18. Reconnect the negative battery cable. Start the engine and check for leaks.

2.5L Engine

1. Disconnect the negative battery cable.

2. Drain the engine oil.

3. Remove the accessory drive belts.

4. Remove the crankshaft damper/pulley.

5. Remove the front timing belt covers.

6. Remove the timing belt.

7. Remove the crankshaft sprocket and key.

8. Remove the front crankshaft seal by prying it out with a flat tipped prytool. Be sure to cover the end of the prytool tip with a shop towel.

✳✳ CAUTION

Be careful as not to nick the seal surface of the crankshaft or the seal bore.

To install:

9. Apply a light coating of clean engine oil to the lip of the new oil seal. Install the new front crankshaft oil seal into the oil pump housing by using oil seal installer tool No. MD998717 or equivalent seal installer. Be sure the oil seal is installed flush with the oil pump cover.

10. Install the crankshaft timing belt sprocket and key.

11. Install the timing belt.

12. Install the timing belt covers.

13. Install the crankshaft damper/pulley onto the crankshaft.

14. Install the accessory drive belts.

15. Fill the engine with the correct amount of clean engine oil.

16. Connect the negative battery cable. Start the engine and check for leaks.

Camshaft and Valve Lifters

REMOVAL & INSTALLATION

2.0L Engine

1. Disconnect the negative battery cable.

2. Properly relieve the fuel system pressure.

3. Label and disconnect the spark plug wires from the spark plugs.

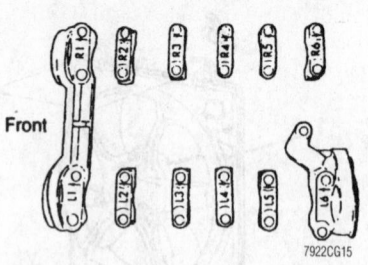

Camshaft bearing cap identification—2.0L engine

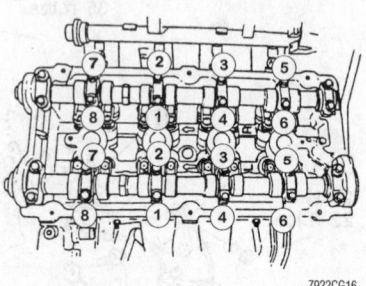

Camshaft bearing cap removal sequence—2.0L engine

4. Remove the ignition coil pack and spark plug wires.

5. Remove the cylinder head cover retaining fasteners and remove the cylinder head cover from the cylinder head. Discard the old cylinder head cover gasket.

6. Remove the ground strap.

7. Remove the timing belt covers, timing belt and camshaft sprockets using the recommended procedure.

8. Take note that the camshaft bearing caps are numbered for correct location during installation. Remove the outer bearing caps first.

9. Loosen, but do not remove, the camshaft bearing cap retaining fasteners in the correct sequence, inside working outward. Perform this step on one camshaft at a time.

10. Identify the camshafts, if they are to be reused, for later installation. The camshafts are not interchangeable. Remove the camshaft bearing caps and remove the camshafts.

11. Remove the camshaft followers. Any components that are to be reused must be installed in their original locations. Use care to identify and mark the positions of any removed valvetrain components so they may be reinstalled correctly.

12. Inspect the camshaft bearing oil feed holes in the cylinder head for clogging. Inspect the camshaft bearing journals for

wear or scoring. Check the cam surface for abnormal wear and damage. A visible worn groove in the roller path or on the cam lobes is cause for replacement.

To install:

13. Thoroughly clean all camshaft and related parts.

14. If the fit and condition of the cam-shafts are acceptable, remove the camshafts for installation of the cam followers.

15. The hydraulic valve lash adjusters are inside the roller cam followers. Be sure they are clean, well-lubricated with clean engine oil and properly positioned. Install the cam followers in their original positions on the hydraulic adjuster and valve stem.

✸✸ WARNING

Be sure NONE of the pistons are at Top Dead Center when installing the camshafts.

16. Lubricate the camshaft bearing journals and cam followers with clean engine oil and install the camshafts. Install right and left camshaft bearing caps No. 2 through No. 5 and right side No. 6. Tighten the M6 fasteners to 105 inch lbs. (12 Nm) in correct sequence.

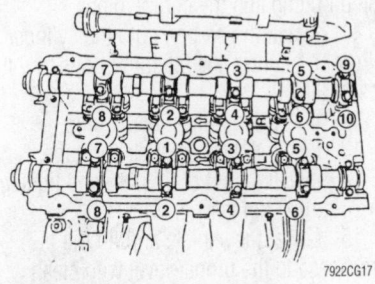

Camshaft bearing cap tightening sequence—2.0L engine

17. Apply Mopar® Gasket Maker or equivalent sealer to the No. 1 and left-side No. 6 bearing caps. Install the bearing caps and tighten the M8 fasteners to 250 inch lbs. (28 Nm). The end caps must be installed before the seals may be installed.

18. Install the camshaft end seals.

19. Reinstall the camshaft sprockets, if removed. Install the timing belt using care to be sure all timing marks are properly aligned, using the recommended procedure. Install the timing belt covers.

✸✸ WARNING

Verify that all timing marks are correct. If the timing belt or sprockets are incorrectly installed, engine damage will occur. Take time to be sure all timing marks are correctly aligned.

20. Clean all sealing surfaces. Make certain the rails are flat.

21. Install new cylinder head cover gaskets. Use care. DO NOT allow oil or solvents to contact the timing belt as they can deteriorate the rubber and cause tooth skipping. Apply Mopar Silicone Rubber Adhesive Sealant, or equivalent, at the camshaft cap corners and at the top edge of the ½ round seal.

➡Inspect the spark plug well seals for cracking and/or swelling and replace if necessary.

22. Install the cylinder head cover assembly to the head and tighten the fasteners in sequence using the following three Steps:

  a. First: tighten all cylinder head cover fasteners to 40 inch lbs. (4.5 Nm).

  b. Second: tighten all fasteners to 80 inch lbs. (9 Nm).

  c. Third: tighten all fasteners to 105 inch lbs. (12 Nm).

23. Install the ignition coil pack and connect the spark plug wiring to the correct spark plugs. Tighten the coil pack retaining fasteners to 105 inch lbs. (12 Nm).

24. Reconnect the ground strap.

25. Check to be sure all vacuum lines and remaining wiring have been reconnected.

26. An oil and filter change is recommended.

27. Reconnect the negative battery cable and test run vehicle. Check for leaks and for proper operation.

2.5L Engine

➡For camshaft service, the cylinder head must be removed.

1. Disconnect the negative battery cable.

2. Relieve the fuel system pressure.

3. Place a large drain pan under the radiator drain plug. Drain the cooling system.

4. Remove the timing belt covers, timing belt and camshaft sprockets using the recommended procedure.

5. The intake manifold is a two-piece unit. The upper part is a large air intake plenum of aluminum alloy. Use care working with light alloy parts. Remove the air intake plenum first, then remove the lower intake manifold.

6. Remove the cylinder head bolts and remove the cylinder head from the vehicle.

7. Remove the thrust case from the left head assembly and remove the camshaft from the rear of the head. If not already done, remove the distributor from the right cylinder head and remove the camshaft from the rear of the head.

To install:

8. Lubricate the camshaft journals and carefully install the camshaft into the cylinder head. Install the thrust case and tighten the fasteners to 9 ft. lbs. (13 Nm).

9. Apply a light coating of engine oil to the camshaft oil seal lip and install the camshaft seal. The camshaft must be installed before installing the seal. Be sure the seal is installed flush with the cylinder head surface. Install the camshaft sprocket and tighten to 65 ft. lbs. (88 Nm).

10. Install the cylinder head.

11. Install the lower intake manifold using new gaskets.

12. Install the rocker arm and shaft assemblies.

13. Install the timing belt.

14. Inspect the spark plug tube seals located on the ends of each tube. These seals slide onto each tube to seal the cylinder head cover to the spark plug tube. If these seals show signs of hardness and/or cracks, they should be replaced.

15. Install the cylinder head cover. Reconnect the spark plug wires.

16. Install the intake manifold plenum.

17. Connect the throttle and speed control cables.

18. Install the air inlet resonator, air inlet hose and the air cleaner housing cover.

19. Check to be sure all remaining electrical connectors have been reconnected. Tighten the air tube connections.

20. Refill the cooling system. An oil and filter change is recommended whenever a cylinder head has been removed since coolant can get into the oil system.

21. Reconnect the negative battery cable. Start the engine and check for leaks, abnormal noises and vibrations. Bleed the cooling system.

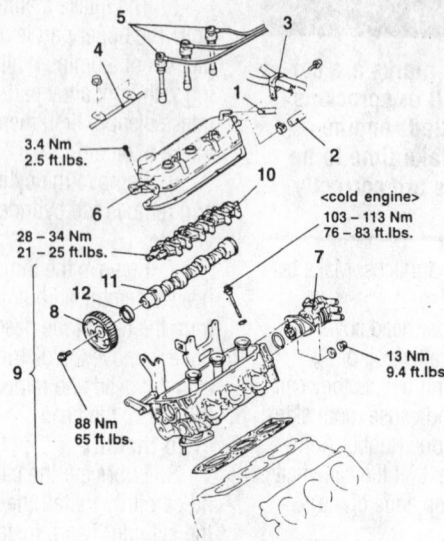

3.4 Nm
2.5 ft.lbs.

103 – 113 Nm
76 – 83 ft.lbs.
<cold engine>

28 – 34 Nm
21 – 25 ft.lbs.

88 Nm
65 ft.lbs.

13 Nm
9.4 ft.lbs.

1. Breather hose connection
2. Blow-by hose
3. Fuel hose assembly connection
4. Vacuum pipe connection
5. Spark plug cable
6. Rocker cover
7. Distributor
8. Camshaft sprocket
9. Cylinder head assembly
10. Rocker arm and rocker shaft
    assembly
11. Camshaft
12. Camshaft oil seal

7922CG18

Exploded view of the rear cylinder head and camshaft—2.5L engine

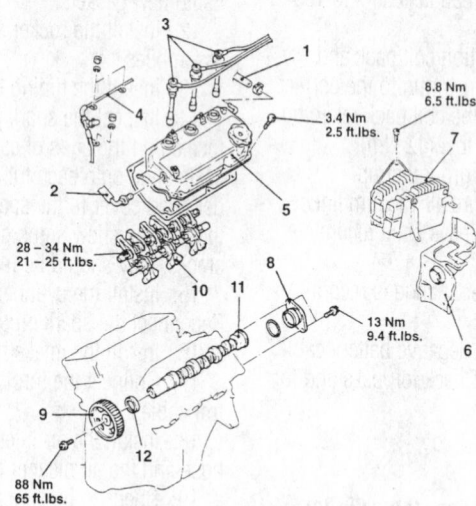

8.8 Nm
6.5 ft.lbs.

3.4 Nm
2.5 ft.lbs.

28 – 34 Nm
21 – 25 ft.lbs.

13 Nm
9.4 ft.lbs.

88 Nm
65 ft.lbs.

1. Blow-by hose
2. PCV valve and hose assembly
   connection
3. Spark plug cable
4. Vacuum pipe and hose asembly
5. Rocker cover
6. Relay box bracket assembly
7. Control module and bracket
   assembly
8. Thrust case
9. Camshaft sprocket
10. Rocker arm and rocker shaft
    assembly
11. Camshaft
12. Camshaft oil seal

7922CG19

Exploded view of the front cylinder head and camshaft—2.5L engine

Valve Lash

ADJUSTMENT

The engines in these vehicles do not require periodic valve lash adjustment.

Oil Pan

REMOVAL & INSTALLATION

2.0L Engine

1. Disconnect the negative battery cable.
2. Raise and safely support the vehicle.
3. Remove the oil pan drain plug and drain the engine oil.
4. Remove the oil dipstick and tube.
5. Remove the front plate.
6. Remove the front exhaust pipe.
7. Remove the oil pan retaining bolts and carefully remove the oil pan.

To install:

8. Inspect the oil pan for damage and cracks. Replace if faulty. While the pan is removed, inspect the oil screen for clogging, damage and cracks. Clean and/or replace if faulty.
9. Thoroughly clean the mating surfaces of the cylinder block and the oil pan.
10. Apply sealant to the seams between the oil pump and the engine block.
11. Install the oil pan onto the cylinder block and tighten the retaining bolts to 9 ft. lbs. (12 Nm).
12. Install the front exhaust pipe.
13. Install the oil dipstick and tube.
14. Install the oil drain plug and tighten to 25 ft. lbs. (34 Nm).
15. Lower the vehicle and fill the crankcase to the proper level with clean engine oil.
16. Connect the negative battery cable. Start the engine and check for leaks.

2.5L Engine

1. Disconnect the negative battery cable.
2. Raise and safely support the vehicle.
3. Place a large drain pan under the oil pan drain plug and drain the oil from the engine.
4. Disconnect and lower the front exhaust pipe.
5. Remove the center member.
6. Remove the engine oil dipstick tube and dipstick.
7. Remove the starter motor.
8. Remove the front and rear plates.

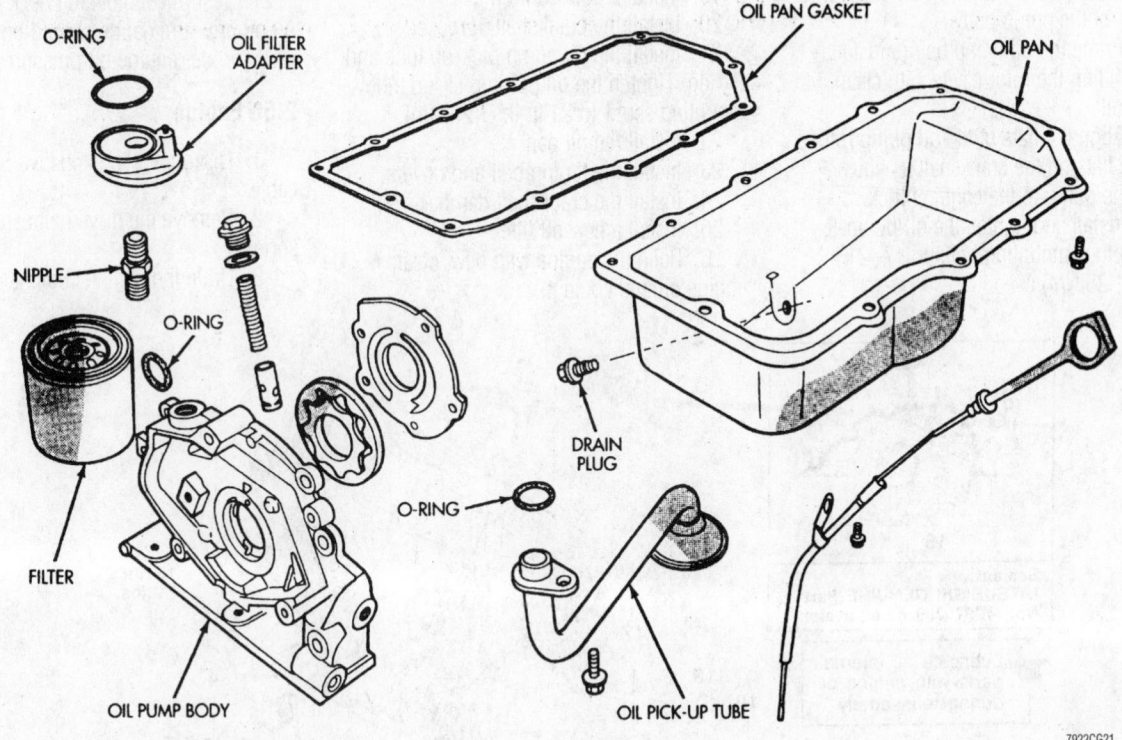

Exploded view of the engine lubricating components—2.0L engine

9. Remove the transaxle inspection cover.

10. Remove the oil pan attaching bolts.

11. Remove the oil pan. If necessary, use a rubber faced mallet, or a hammer with a block of wood to separate the oil pan from the engine block.

To install:

12. Thoroughly clean and dry the oil pan, cylinder block and cylinder block bolts and bolt holes.

13. Apply a continuous 0.157 in. (4mm) bead of MOPAR® Silicone Adhesive Sealant or equivalent, to the oil pan gasket surface. Be sure to circle all mounting bolt holes as well. Install the oil pan within a 10–15 minute period of applying the gasket material to ensure proper sealing.

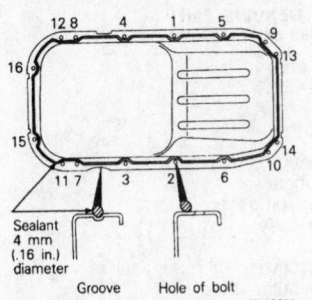

To ensure a leak-free seal, apply sealer as shown and tighten the bolts following the specified sequence—2.5L engine

14. Install the oil pan to the engine.

15. Tighten the oil pan attaching bolts to 53 inch lbs. (6 Nm).

16. Install transaxle inspection cover.

17. Install the front and rear plates and tighten the bolts to 80 ft. lbs. (108 Nm).

18. Install the starter motor.

19. Install the engine oil dipstick tube and dipstick.

20. Install the center member. Tighten the mounting bolts to 65 ft. lbs. (88 Nm).

21. Connect the front exhaust pipe.

22. Reinstall the oil pan drain plug and gasket. Tighten the drain plug to 29 ft. lbs. (40 Nm).

23. Lower the vehicle.

24. Refill the engine with fresh oil to the proper level. An oil filter change is recommended.

25. Reconnect the negative battery cable. Start the engine and check for leaks.

Oil Pump

REMOVAL & INSTALLATION

2.0L Engine

1. Disconnect the negative battery cable.

2. Remove the timing belt.

3. Remove the oil pan.

4. Using a suitable puller, draw the crankshaft sprocket from the front of the crankshaft.

5. Remove the oil pump pick-up tube and O-ring.

6. Remove the oil pump and front crankshaft seal. The front cover/oil pump mounting bolts may be different sizes and must be reinstalled in their original locations. Remove and tag the front cover mounting bolts.

7. Inspect the oil pump case for damage and remove the rear cover.

8. Remove the pump rotors and inspect the inside of the case for excessive wear.

9. Check that the oil relief plunger slides smoothly and check for a broken spring.

To install:

10. Clean all parts well. Be sure the block and pump surfaces are clean and free of old sealer.

11. Assemble the pump using new parts as required with clean oil. Align the marks on the inner and outer rotors when assembling.

12. Install the pump back cover and tighten the screws to 88 inch lbs. (10 Nm).

13. Reinstall the pump relief valve, spring, gasket and valve cap. Tighten the valve cap to 30–33 ft. lbs. (41–44 Nm).

14. Apply gasket maker to the engine block mounting surface of the oil pump body.

15. Install the oil ring into the discharge passage of the pump body.

16. Prime the oil pump before installation by filling the rotor cavity with clean engine oil.

17. Align the flats of the oil pump rotor with the flats on the crankshaft as you install the pump to the engine block.

18. Install and tighten the oil pump-to-engine block mounting bolts to 17–21 ft. lbs. (23–28 Nm).

19. Install a new front oil seal.

20. Install the crankshaft sprocket.

21. Install the oil pump pick-up tube and O-ring. Tighten the oil pump pick-up tube mounting screw to 21 ft. lbs. (28 Nm).

22. Install the oil pan.

23. Install the timing belt and covers.

24. Install the crankshaft damper.

25. Install a new oil filter.

26. Refill the engine with new, clean engine oil and coolant.

27. Test run vehicle to check for leaks. An oil pressure gauge should be installed to verify proper engine oil pressure.

2.5L Engine

1. Disconnect the negative battery cable.

2. Remove the drive belts and accessories.

3. Drain the engine coolant.

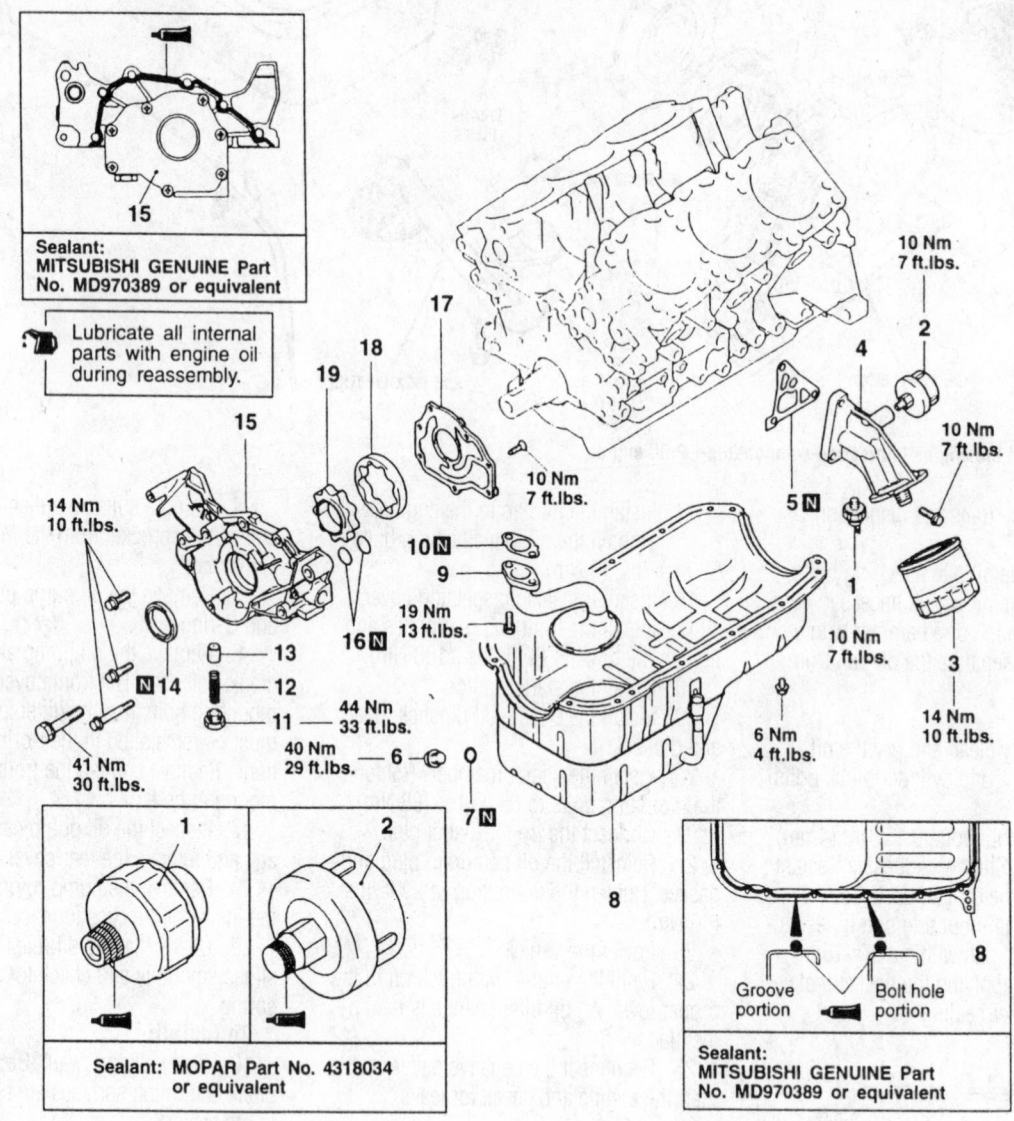

1. Oil pressure switch
2. Oil pressure gauge unit
3. Oil filter
4. Oil filter bracket
5. Oil filter bracket gasket
6. Drain plug
7. Drain plug gasket
8. Oil pan
9. Oil screen
10. Oil screen gasket
11. Plug
12. Relief spring
13. Relief plunger
14. Crankshaft oil seal
15. Oil pump case
16. O-ring
17. Oil pump cover
18. Oil pump outer rotor
19. Oil pump inner rotor

Exploded view of the oil pan and pump assembly—2.5L engine

7922CG20

4. Raise and safely support the vehicle. Drain the engine oil.

5. Remove the crankshaft damper.

6. Remove the timing belt upper and lower covers.

7. Loosen the timing belt and crankshaft sprocket from the crankshaft.

8. Remove the five bolts that attach the oil pump to the block and remove the oil pump.

9. Inspect the oil pump case for damage and remove the rear cover.

10. Remove the pump rotors and inspect the inside of the case for excessive wear.

11. Check that the oil relief plunger slides smoothly and check for a broken spring.

To install:

12. Clean all parts well. Be sure the block and pump surfaces are clean and free of old sealer.

13. Assemble the pump using new parts as required with clean oil. Align the marks on the inner and outer rotors when assembling.

14. Install the pump back cover and tighten the screws to 88 inch lbs. (10 Nm).

15. Reinstall the pump relief valve, spring, gasket and valve cap. Tighten the valve cap to 30–33 ft. lbs. (41–44 Nm).

16. Prime the pump before installation by filling the rotor cavity with clean engine oil.

17. Apply gasket maker or equivalent sealer to the pump. Install the O-ring into the counterbore on the pump body discharge passage. Position the pump onto the crankshaft until seated on the block. Tighten the size M8 fasteners to 10 ft. lbs. (14 Nm) and size M10 fasteners to 30 ft. lbs. (41 Nm).

18. Install the timing belt and crankshaft sprocket.

19. Install the timing belt cover.

20. Install the crankshaft damper.

21. Install the drive belts and accessories.

22. Refill the cooling system. Install a new oil filter and refill the engine with oil.

23. Road test the vehicle. Check for proper operation as well as leaks.

Rear Main Seal

REMOVAL & INSTALLATION

2.0L Engine

1. Remove the transaxle and flexplate/flywheel.

2. Remove the rear crankshaft oil seal from the oil seal housing using a suitable flat bladed prying tool.

To Install:

➡ **When installing seal there is no need to lubricate sealing surface.**

3. Install seal into housing using a suitable installation tool.

4. Install the flexplate/flywheel, and tighten the bolts to 68 ft. lbs. (92 Nm).

5. Install the transaxle.

2.5L Engine

1. Remove the transaxle and flexplate/flywheel.

2. Remove the five rear seal housing mounting bolts.

3. Remove the housing from the engine.

4. Remove the rear crankshaft oil seal from the oil seal housing using a suitable flat bladed prying tool.

To Install:

➡ **When installing seal there is no need to lubricate sealing surface.**

5. Install seal into housing using a suitable installation tool.

6. Apply silicone rubber adhesive sealant to the mating surface of the seal housing.

7. Apply a light coating of engine oil to the entire oil seal lip circumference.

8. Install the oil seal and housing to the engine cylinder block. Install and tighten the mounting bolts to 96 inch lbs. (11 Nm).

9. Install the flexplate/flywheel, and tighten the bolts to 68 ft. lbs. (92 Nm).

10. Install the transaxle.

FUEL SYSTEM

Fuel System Service Precautions

Safety is the most important factor when performing not only fuel system maintenance but any type of maintenance. Failure to conduct maintenance and repairs in a safe manner may result in serious personal injury or death. Maintenance and testing of the vehicle's fuel system components can be accomplished safely and effectively by adhering to the following rules and guidelines.

• To avoid the possibility of fire and personal injury, always disconnect the negative battery cable unless the repair or test procedure requires that battery voltage be applied.

• Always relieve the fuel system pressure prior to disconnecting any fuel system component (injector, fuel rail, pressure regulator, etc.), fitting or fuel line connection. Exercise extreme caution whenever relieving fuel system pressure to avoid exposing skin, face and eyes to fuel spray. Please be advised that fuel under pressure may penetrate the skin or any part of the body that it contacts.

• Always place a shop towel or cloth around the fitting or connection prior to loosening to absorb any excess fuel due to spillage. Ensure that all fuel spillage (should it occur) is quickly removed from engine surfaces. Ensure that all fuel soaked cloths or towels are deposited into a suitable waste container.

• Always keep a dry chemical (Class B) fire extinguisher near the work area.

• Do not allow fuel spray or fuel vapors to come into contact with a spark or open flame.

• Always use a back-up wrench when loosening and tightening fuel line connection fittings. This will prevent unnecessary stress and torsion to fuel line piping. Always follow the proper torque specifications.

• Always replace worn fuel fitting O-rings with new. Do not substitute fuel hose or equivalent, where fuel pipe is installed.

Fuel System Pressure

RELIEVING

✳✳ CAUTION

The fuel injection system remains under pressure even after the engine has been turned OFF. The fuel system pressure must be relieved before disconnecting any fuel lines. Failure to do so may result in fire and/or personal injury.

1. Remove the fuel filler cap to release fuel tank pressure.

2. Remove the rear seat cushion.

3. At the fuel tank, disconnect the fuel pump harness connector.

4. Start the vehicle and allow it to run until it stalls from lack of fuel. Turn the key to the **OFF** position.

5. Disconnect the negative battery cable, then reconnect the fuel pump connector.

6. Install the rear seat cushion and the fuel filler cap.

✳✳ CAUTION

Always wrap shop towels around a fitting that is being disconnected to absorb residual fuel in the lines.

Fuel Filter

REMOVAL & INSTALLATION

A replaceable fuel filter is located in the engine compartment, on the bulkhead, next to the brake booster.

✳✳ CAUTION

The fuel injection system remains under pressure, even after the engine has been turned OFF. The fuel system pressure must be relieved before disconnecting any fuel lines. Failure to do so may result in fire and/or personal injury.

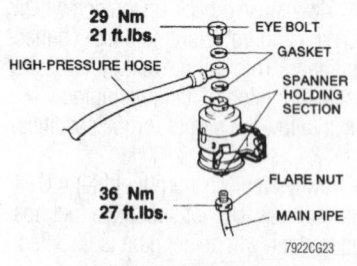

29 Nm
21 ft.lbs. EYE BOLT
GASKET
HIGH-PRESSURE HOSE
SPANNER HOLDING SECTION
FLARE NUT
36 Nm
27 ft.lbs. MAIN PIPE

7922CG23

Exploded view of the fuel line-to-filter connection

1. Following proper procedures, relieve the fuel system residual pressure.

➡**Wrap shop towels around the fitting that is being disconnected to absorb residual fuel in the lines.**

2. Disconnect the negative battery cable.
3. Remove the air intake hose for access.
4. Hold the fuel filter housing securely with a wrench. Cover the hoses with shop towels and remove the eye bolt. Discard the gaskets.
5. Separate the flare nut connection at the bottom of the filter.
6. Remove the mounting bolts and the fuel filter from the vehicle.

➡**Do not use conventional fuel filters, hoses or clamps when servicing fuel injection systems. They are not compatible with the injection system and the high pressures in fuel injection systems and could cause substandard parts to fail, causing personal injury or damage to the vehicle. Use only hoses and clamps specifically designed for fuel injection.**

To install:

7. Tighten the flare nut fitting by hand before mounting the filter to the bracket.
8. Install the filter to its bracket only finger-tight. Movement of the filter will ease attachment of the fuel lines.
9. Using new gaskets, connect the high pressure hose and eye bolt. While holding the fuel filter housing, tighten the eye bolt to 21 ft. lbs. (29 Nm). Tighten the flare nut to 27 ft. lbs. (36 Nm).
10. Tighten the filter mounting bolts fully.
11. Install the intake air hose.
12. Connect the negative battery cable, turn the key to the **ON** position to pressurize the fuel system and check for leaks.
13. If necessary, release the fuel pressure and repair leaks.

Fuel Pump

REMOVAL & INSTALLATION

Do not use conventional fuel filters, hoses or clamps when servicing fuel injection systems. They are not compatible with the injection system and could fail, causing personal injury or damage to the vehicle. Use only hoses and clamps specifically designed for fuel injection.

1. Relieve the fuel system pressure, using proper procedures. Disconnect negative battery cable.

➡ **The rear seat cushion must be removed in order to gain access to the fuel pump.**

2. Remove the rear seat cushion by pulling the seat stopper outward and lifting the lower cushion upward. There are two access covers underneath the seat. The panel on the far right-side is for the fuel pump.
3. Remove the access cover.
4. Disconnect the fuel pump wiring.

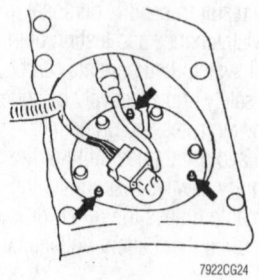

7922CG24

Aligning the fuel pump for installation

5. Disconnect the return hose and the high pressure fuel hose.

✳✳ CAUTION

Observe all applicable safety precautions when working around fuel. Do not allow fuel spray or fuel vapors to come into contact with a spark or open flame. Keep a dry chemical (Class B) fire extinguisher near the work area. Never drain or store fuel in an open container due to the possibility of fire or explosion. Cover all fuel hose connections with a shop towel, prior to disconnecting, to prevent splash of fuel that could be caused by residual pressure remaining in the fuel line.

6. Remove the pump mounting nuts and remove the pump assembly.

To install:

7. Align the seal position projections with the holes in the fuel pump assembly and install the assembly in the tank. Tighten the retaining nuts to 22 inch lbs. (2.5 Nm).
8. Connect the high pressure hose, return hose and the fuel pump wiring.
9. Connect the negative battery cable.
10. Check the fuel pump for proper pressure and inspect the entire system for leaks.
11. Apply sealant to the access cover and install the cover.
12. Install the rear seat cushion.
13. Pressurize the fuel system by turning the ignition key to the **ON** position. Check for leaks. Start the engine to verify proper fuel pump performance.

DRIVE TRAIN

Transaxle Assembly

REMOVAL & INSTALLATION

Manual

1. Disconnect both battery cables, negative side first. Remove the battery and battery tray.
2. Remove the battery stay (brace).
3. Remove the air cleaner and intake hoses.
4. Drain the transaxle into a suitable waste container.
5. Remove the cotter pin securing the select and shift cables and remove the cable ends from the transaxle.

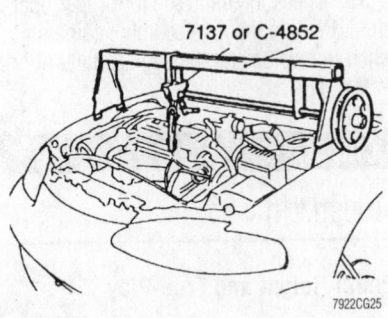

For transaxle removal, properly support
the engine assembly as shown

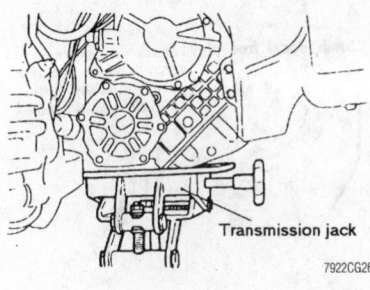

Also, use a transmission jack to support
the transaxle assembly

6. Disconnect the back-up light switch
harness and position it aside.

7. Disconnect the speedometer electrical connector, from the transaxle assembly.

8. Remove the starter motor and position it aside.

9. Using special tool 7137 or C-4852
or equivalent, support the engine assembly.

10. Remove the rear roll stopper mounting bracket.

11. Remove the transaxle mount bracket.

12. Remove the upper transaxle mounting bolts.

13. Raise and safely support the vehicle.

14. Remove the front wheel assemblies.

15. Remove the under cover.

16. Remove the cotter pin and disconnect
the tie rod end, from the steering knuckle.

17. Disconnect the stabilizer bar link,
from the damper fork.

18. Disconnect the damper fork, from the
lateral lower control arm.

19. Disconnect the later lower arm, and
the compression lower arm, lower ball
joints, from the steering knuckle.

20. Pry the halfshafts from the transaxle,
and secure aside.

21. Remove the connection for the clutch
release cylinder and without disconnecting
the hydraulic line, secure aside.

22. Remove the cover from the transaxle
bell housing.

23. Remove the engine front roll stopper
through-bolt.

24. Remove the centermember.

25. Support the transaxle, using a transmission jack, and remove the transaxle
lower coupling bolt.

➡ The coupling bolt threads from the
engine side, into the transaxle, and is
located just above the halfshaft opening.

26. Slide the transaxle rearward and
carefully lower it from the vehicle.

To install:

27. Install the transaxle to the engine
and install the mounting bolts and tighten to
70 ft. lbs. (95 Nm).

28. Install the cover to the transaxle bell
housing and tighten the mounting bolts to 7
ft. lbs. (9 Nm).

29. Install the centermember and tighten
the front mounting bolts to 65 ft. lbs. (88
Nm) and the rear bolt to 54 ft. lbs. (73 Nm).
Install the front engine roll stopper through-
bolt and lightly tighten. Once the full weight
of the engine is on the mounts, tighten the
bolt to 42 ft. lbs. (57 Nm).

30. Connect the clutch release cylinder.

31. Install the halfshafts, using new circlips on the axle ends.

✳✳ WARNING

When installing the halfshaft, keep
the inboard joint straight in relation
to the axle, to avoid damaging the oil
seal lip of the transaxle, with the serrated part of the halfshaft.

32. Connect the tie rod and ball joints to
the steering knuckle. Tighten the ball joint
self-locking nuts to 48 ft. lbs. (65 Nm).
Tighten the tie rod end nut to 21 ft. lbs. (28
Nm) and secure with a new cotter pin.

33. Connect the damper fork to the lower
control arm and tighten the through-bolt to
65 ft. lbs. (88 Nm).

34. Connect the stabilizer link to the
damper fork, and tighten the self-locking
nut to 29 ft. lbs. (39 Nm).

35. Install the under cover.

36. Install wheels and lower vehicle.

37. Install the transaxle mount bracket,
to the transaxle, and tighten the mounting
nuts to 32 ft. lbs. (43 Nm).

38. Install the rear roll stopper mounting
bracket.

39. Remove the engine support. Tighten
the transaxle mount through-bolt to 51 ft.
lbs. (69 Nm) and tighten the front engine
roll stopper through-bolt.

40. Install the upper transaxle mounting
bolts and tighten to 35 ft. lbs. (48 Nm).

41. Install the starter motor.

42. Connect the back-up light switch
and the speedometer connector.

43. Connect the select and shift cables
and install new cotter pins.

44. Install the air cleaner and the air
intake hose.

45. Install the battery tray and battery.

46. Install the battery stay.

47. Be sure the vehicle is level, and refill
the transaxle.

48. Check the transaxle for proper operation. Be sure the reverse lights come on
when in reverse.

Automatic

1. Disconnect both battery cables, negative side first. Remove the battery and battery tray.

2. Remove the battery stay (brace).

3. Remove the air cleaner and intake
hoses.

4. Drain the transaxle fluid into a suitable waste container.

5. Remove the nut securing the shifter
lever to the transaxle. Remove the cable
retaining clip and remove the cable from the
transaxle.

6. Remove the shifter cable mounting
bracket.

7. Disconnect and tag the electrical
connectors for the speedometer, solenoid,
neutral safety switch (inhibitor switch), the
pulse generator, kickdown servo switch, oil
temperature sensor.

8. Disconnect and tag the oil cooler
lines, at the transaxle.

9. Remove the bolt securing the fluid
dipstick tube, to the transaxle. Remove the
dipstick and tube from the transaxle.

10. Remove the starter motor and position aside.

11. Using special tool 7137 or C-4852
or equivalent, support the engine assembly.

12. Remove the rear roll stopper mounting bracket.

13. Remove the transaxle mount bracket.

14. Remove the upper transaxle mounting bolts.

15. Raise and safely support the vehicle.

16. Remove the front wheels.

17. Remove the left-side undercover.

18. Remove the cotter pin and disconnect the tie rod end, from the steering
knuckle.

19. Disconnect the stabilizer bar link,
from the damper fork.

20. Disconnect the damper fork, from the
lateral lower control arm.

21. Disconnect the later lower arm, and the compression lower arm, lower ball joints, from the steering knuckle.

22. Pry the halfshafts from the transaxle, and secure aside.

23. Remove the cover from the transaxle bell housing.

24. Remove the engine front roll stopper through-bolt.

25. Remove the centermember.

26. Remove the bolts holding the flex-plate to the torque converter with a box wrench. Rotate the crankshaft to bring the bolts into a position for removal, one at a time.

27. To make installation easier, use chalk or paint to make matchmarks on the torque converter and flexplate. These marks will be used at assembly to realign the assembly, keeping these parts in balance. After removing the bolts, push the torque converter toward the transaxle. This will prevent the converter from remaining in contact with the engine, possibly damaging the converter.

28. Support the transaxle using a transmission jack (at the side of the case, NOT at the pan), and remove the transaxle lower coupling bolt.

➡ **The coupling bolt threads from the engine side, into the transaxle, and is located just above the halfshaft opening.**

29. Slide the transaxle rearward and carefully lower it from the vehicle.

To install:

30. After the torque converter has been mounted on the transaxle, install the transaxle assembly to the engine. Install the mounting bolts and tighten to 70 ft. lbs. (95 Nm).

31. Align the balance matchmarks made at disassembly, connect the torque converter to the flexplate and tighten the bolts to 55 ft. lbs. (75 Nm).

32. Install the cover to the transaxle bell housing and tighten the mounting bolts to 9 ft. lbs. (12 Nm).

33. Install the centermember and tighten the front mounting bolts to 65 ft. lbs. (88 Nm) and the rear bolts to 51–58 ft. lbs. (69–78 Nm). Install the front engine roll stopper through-bolt and lightly tighten. Once the full weight of the engine is on the mounts, tighten the bolt to 42 ft. lbs. (56 Nm).

34. Install the halfshafts, using new circlips on the axle ends.

✳✳ WARNING

When installing the halfshaft, keep the inboard joint straight in relation to the axle, to avoid damaging the oil seal lip of the transaxle, with the serrated part of the halfshaft.

35. Connect the tie rod and ball joints to the steering knuckle. Tighten the ball joint self-locking nuts to 48 ft. lbs. (65 Nm). Tighten the tie rod end nut to 21 ft. lbs. (28 Nm) and secure with a new cotter pin.

36. Connect the damper fork to the lower control arm and tighten the through-bolt to 65 ft. lbs. (88 Nm).

37. Connect the stabilizer link to the damper fork, and tighten the self-locking nut to 29 ft. lbs. (39 Nm).

38. Install the left-side undercover.

39. Install the wheels and lower the vehicle.

40. Install the transaxle mount bracket, to the transaxle, and tighten the mounting nuts to 32 ft. lbs. (43 Nm).

41. Install the rear roll stopper mounting bracket.

42. Remove the engine support. Tighten the transaxle mount through-bolt to 51 ft. lbs. (69 Nm) and tighten the front engine roll stopper through-bolt.

43. Install the upper transaxle mounting bolts and tighten to 35 ft. lbs. (48 Nm).

44. Install the starter motor.

45. Install the dipstick tube and the dipstick.

46. Install the shifter cable mounting bracket.

47. Connect the shifter lever and tighten the retaining nut to 14 ft. lbs. (19 Nm).

48. Connect the oil cooler lines and secure with clamps.

49. Connect the electrical connectors for the speedometer, solenoid, neutral safety switch (inhibitor switch), the pulse generator, kickdown servo switch and oil temperature sensor.

50. Install the air cleaner and the air intake hose.

51. Install the battery tray and battery.

52. Be sure the vehicle is level, and refill the transaxle with MOPAR® ATF PLUS or equivalent transmission fluid. Start the engine and allow it to idle for two minutes. Apply the parking brake and move the selector through each gear position, ending in **N**. Recheck fluid level and add if necessary. Fluid level should be between the marks in the **HOT** range.

53. Check the transaxle for proper operation. Be sure the reverse lights come on when in reverse and the engine starts only in **P** or **N**.

Clutch

ADJUSTMENT

Pedal Height and Free-Play

1. Measure the clutch pedal height from the face of the pedal pad to the bulkhead. Compare the measured value with the desired distance of 7.0–7.09 in. (175–180mm).

Clutch pedal free play	Distance between the clutch pedal and the firewall when the clutch is disengaged

Clutch pedal free-play adjustment measurements

2. Measure the clutch pedal clevis pin play at the face of the pedal pad. Press the pedal lightly until resistance is met, and measure this distance. The clutch pedal clevis pin play should be within 0.040–0.120 in. (1–3mm).

3. If the clutch pedal height or clevis pin play are not within the standard values, adjust as follows:

    a. For vehicles without cruise control, turn and adjust the stop bolt so the pedal height is the standard value, then tighten the locknut.

    b. Vehicles with auto-cruise control system, disconnect the clutch switch connector and turn the switch to obtain the standard clutch pedal height. Then, lock by tightening the locknut.

    c. Turn the pushrod to adjust the clutch pedal clevis pin play to agree with the standard value and secure the pushrod with the locknut.

➡ **When adjusting the clutch pedal height or the clutch pedal clevis pin play, be careful not to push the pushrod toward the master cylinder.**

d. Check that when the clutch pedal is depressed all the way, the interlock switch switches over from **ON** to **OFF**.

4. Move the clutch pedal until the resistance begins to increase; measure between this point and the pedal resting point, to determine the clutch pedal free-play. The clutch pedal free-play measurement should be between 0.240–0.510 in. (6–13mm). With the pedal fully disengaged, check the distance between the bulkhead and the top of the pedal pad. The measurement should be 2.760 in. (70mm) or more.

5. If the measurements are not within specification, bleed the clutch hydraulic system. If after bleeding the measurements are still not within specified range, there is a faulty component in the system, which must be replaced.

REMOVAL & INSTALLATION

1. Disconnect the negative battery cable.

2. Raise and safely support the vehicle.

3. Remove the transaxle assembly from the vehicle using the recommended procedure.

4. Remove the pressure plate attaching bolts, pressure plate and clutch disc. If the pressure plate is to be reused, loosen the bolts in a diagonal pattern, one or two turns at a time. This will prevent warping the clutch cover assembly.

5. Remove the return clip and the pressure plate release bearing. Do not use solvent to clean the bearing.

6. Inspect the clutch release fork and fulcrum for damage or wear. If necessary,

remove the release fork and the fulcrum from the transaxle.

7. Carefully inspect the condition of the clutch components and replace any worn or damaged parts.

To install:

8. Inspect the flywheel for heat damage or cracks. Resurface or replace the flywheel as required. Install the flywheel using new bolts.

9. Install the fulcrum, if removed, and tighten. Install the release fork. Apply a coating of multi-purpose grease to the point of contact with the fulcrum and the point of contact with the release bearing. Apply a coating of multi-purpose grease to the end of the release cylinder's pushrod and the pushrod hole in the release fork.

➡**When installing the clutch, apply grease to each part, but be careful not to apply excessive grease. Excessive grease will cause clutch slippage and shudder.**

10. Apply multi-purpose grease to the clutch release bearing. Pack the bearing inner surface and the groove with grease. Do not apply grease to the resin portion of the bearing. Place the bearing in position and install the return clip.

11. Apply a coating of grease to the clutch disc splines, then use a brush to rub it in the grooves. Using a universal clutch disc alignment tool, position the clutch disc on the flywheel. Install the retainer bolts and tighten a little at a time, in a diagonal sequence.

12. Install the transaxle assembly using the recommended procedure and check the fluid level.

13. Verify proper clutch operation.

Hydraulic Clutch System

BLEEDING

✳✳ WARNING

The clutch hydraulic system uses DOT 3 or DOT 4 brake fluid. Use care. Brake fluid is harmful to painted surfaces.

1. Fill the reservoir with clean DOT 3 or DOT 4 brake fluid.

2. Loosen the bleed screw, have the clutch pedal pressed to the floor.

3. Tighten the bleed screw and release the clutch pedal.

4. Repeat the procedure until the fluid is free of air bubbles.

➡**It is suggested that a hose be attached to the bleeder with the other end immersed in a container at least half full of brake fluid during the bleeding operation. Do not allow the reservoir to run out of fluid during bleeding.**

5. Refill the reservoir with clean brake fluid.

6. Check the clutch for proper operation.

Halfshaft

REMOVAL & INSTALLATION

➡**If the vehicle is going to be rolled while the halfshafts are out of the vehicle, obtain two outer CV-joints or proper equivalent tools and install to the hubs. If the vehicle is rolled without the proper torque applied to the front wheel bearings, the bearings will no longer be usable.**

1. Disconnect the negative battery cable.

2. Remove the cotter pin, halfshaft nut and washer.

3. Raise and safely support the vehicle.

4. Remove the wheel.

5. Using Joint Separation tool MB991113 or equivalent, disconnect the tie rod end from the steering knuckle.

✳✳ CAUTION

Use of improper methods of joint separation can result in damage to the joint, leading to possible failure.

6. Disconnect the sway bar link from the damper fork.

7. Remove the damper fork lower through-bolts and upper pinch bolt. Remove the damper fork assembly.

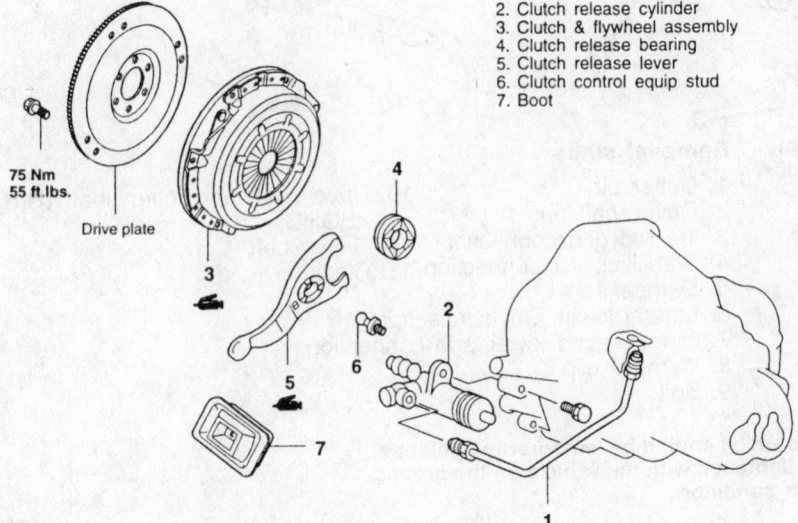

1. Oil tube
2. Clutch release cylinder
3. Clutch & flywheel assembly
4. Clutch release bearing
5. Clutch release lever
6. Clutch control equip stud
7. Boot

75 Nm
55 ft.lbs.

Drive plate

7922CG28

Exploded view of the clutch assembly

8. Using a joint separation tool, disconnect the lateral arm and the compression arm from the steering knuckle.

9. Remove the halfshaft from the hub/knuckle by setting up a puller on the outside wheel hub and pushing the halfshaft from the front hub. After pressing the outer shaft, insert a prybar between the transaxle case and the halfshaft and pry the shaft from the transaxle.

➡**Do not pull on the shaft. Doing so damages the inboard joint. Do not insert the prybar too far or the oil seal in the case may be damaged.**

To install:

10. Inspect the halfshaft boot for damage or deterioration. Check the ball joints and splines for wear.

11. Replace the circlips on the ends of the halfshaft(s).

12. Insert the halfshaft into the transaxle. Be sure it is fully seated.

13. Pull the knuckle assembly outward and install the other end of the halfshaft into the hub.

14. Install the washer so the chamfered edge faces outward. Install the halfshaft nut and tighten temporarily.

15. Connect the lateral arm and the compression arm to the steering knuckle. Tighten the self-locking nuts to 43–52 ft. lbs. (59–71 Nm).

16. Install the damper fork. Tighten the lower through-bolt/nut to 65 ft. lbs. (88 Nm) and the upper pinch bolt to 76 ft. lbs. (103 Nm).

17. Connect the tie rod end to the steering knuckle. Tighten the retaining nut to 17–25 ft. lbs. (24–33 Nm) and install a new cotter pin.

18. Connect the sway bar link to the damper fork and tighten the link nut to 29 ft. lbs. (39 Nm).

19. Install the lock-washer and axle nut. Tighten the axle nut with the special tool MB990767 to hold the hub from turning. Tighten the nut to 145–188 ft. lbs. (200–260 Nm).

➡**Before securely tightening the axle nut be sure there is no load on the wheel bearings.**

20. Install a new cotter pin and bend to secure.

21. Install the wheel.

22. Check the transaxle fluid level and top off, if necessary.

23. Connect the negative battery cable.

24. Test drive the vehicle and check for proper operation.

STEERING AND SUSPENSION

Air Bag

✳✳ CAUTION

Some vehicles are equipped with an air bag system, also known as the Supplemental Inflatable Restraint (SIR) system or Supplemental Restraint System (SRS). The system must be disabled before performing service on or around system components, steering column, instrument panel components, wiring and sensors. Failure to follow safety and disabling procedures could result in accidental air bag deployment, possible personal injury and unnecessary system repairs.

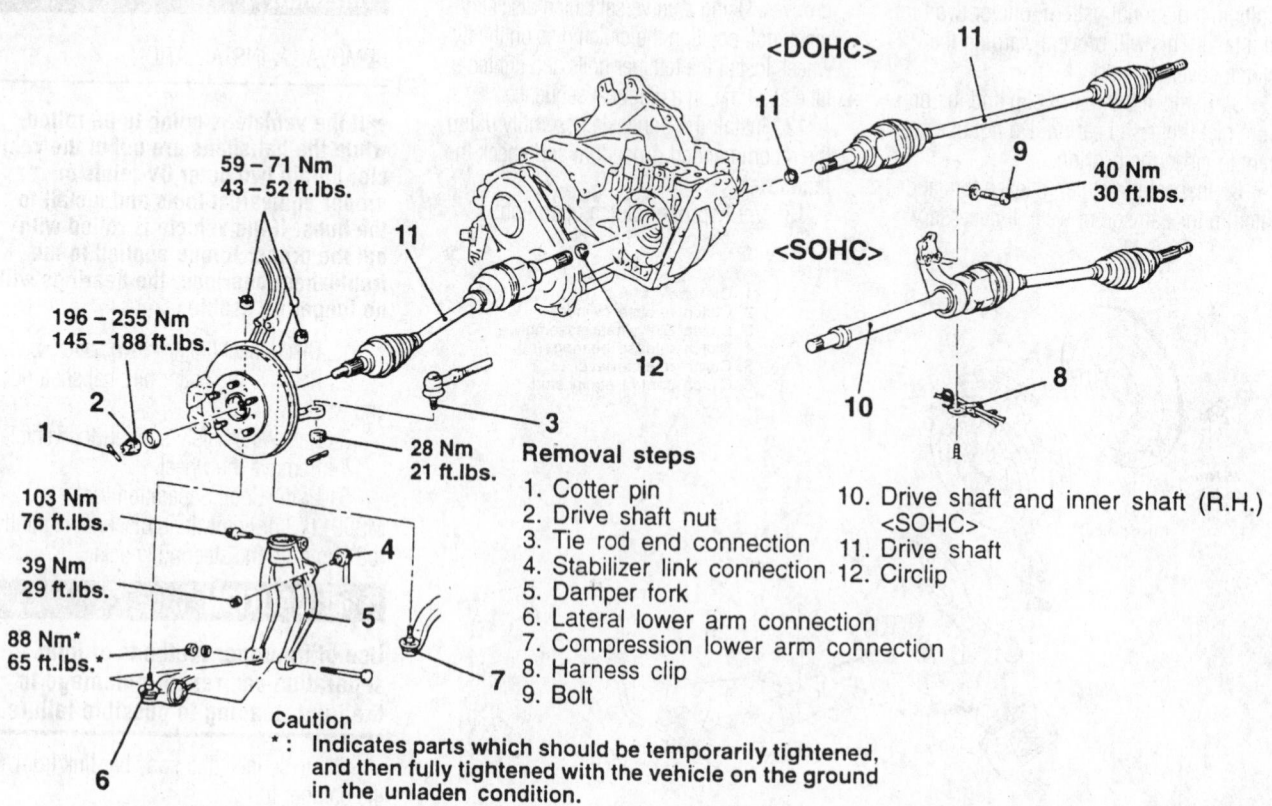

Removal steps
1. Cotter pin
2. Drive shaft nut
3. Tie rod end connection
4. Stabilizer link connection
5. Damper fork
6. Lateral lower arm connection
7. Compression lower arm connection
8. Harness clip
9. Bolt
10. Drive shaft and inner shaft (R.H.) <SOHC>
11. Drive shaft
12. Circlip

Caution
* : Indicates parts which should be temporarily tightened, and then fully tightened with the vehicle on the ground in the unladen condition.

7922CG29

Exploded view of the left and right halfshaft assemblies' mounting

PRECAUTIONS

Several precautions must be observed when handling the inflator module to avoid accidental deployment and possible personal injury.

• Never carry the inflator module by the wires or connector on the underside of the module.

• When carrying a live inflator module, hold securely with both hands, and ensure that the bag and trim cover are pointed away.

• Place the inflator module on a bench or other surface with the bag and trim cover facing up.

• With the inflator module on the bench, never place anything on or close to the module which may be thrown in the event of an accidental deployment.

1. DO NOT use any electrical test equipment on or near any SRS components except those specified by Chrysler corporation:

a. Use a digital multi-meter for which the maximum test current is 2mA or less at the minimum range of resistance measurement for use with the Chrysler SRS Check Harness when checking the SRS electrical circuitry.

b. Chrysler special tool MB991613 SRS Check Harness acts like a "break-out box" for checking SRS wiring. There are other factory special tool wiring adapters that are available and may be used.

c. DRB III or equivalent scan tool for reading and erasing air bag diagnostic codes.

2. NEVER ATTEMPT TO REPAIR THE FOLLOWING COMPONENTS:

a. Air Bag Control Unit (SRS-ECU)

b. Clock Spring under steering wheel.

c. Air Bag Modules

d. If any of these components are diagnosed as faulty, they should only be replaced.

3. Do not attempt to repair any of the air bag system wiring harness connectors. If any of the connectors or wires are faulty, replace that harness.

4. Air bag components should not be subjected to heat over 200° F. Remove the SRS-ECU, the air bag modules themselves and the clock spring before drying or baking the vehicle after painting.

5. After air bag system service, check the SRS warning light operation to be sure that the system functions properly.

6. Make certain that the ignition switch is in the **OFF** position when a scan tool is connected or disconnected.

DISARMING

The Air Bag system (;BSChrysler also calls it the Supplemental Restraint System or SRS) is designed to supplement the driver's and passenger's seat belts to help reduce the risk or severity of injury to the driver and front passenger by activating and deploying both air bags in certain frontal collisions. The system consists of two air bag modules, one located in the center of the steering wheel and another located above the glove box, which contains the folded air bag and an inflator unit. The air bag electronic control unit (SRS-ECU) located under the floor console assembly monitors the system and which contains a safing G sensor and analog G sensor. An SRS warning light is located on the instrument panel which indicates the status of the air bag system. A clock spring interconnection is located within the steering column.

To deploy the air bags, the SRS-ECU must respond to the output signal from the analog G sensor and the safing G sensor must be ON. The SRS-ECU, then causes the air bag modules to ignite and deploy.

Service technicians should use care when working around any vehicle equipped with an air bag system, to avoid injury to the technician by inadvertent deployment of the air bag or to the driver by rendering the air bag system inoperative.

The SRS-ECU not only controls the air bag system, it can provide diagnostic information. The SRS-ECU monitors the air bag system and stores data concerning any detected faults in the system. When the ignition key is turned to the **ON** or **START** position, the SRS warning light should illuminate for about 7 seconds, then turn off. That indicates that the SRS system is in operating condition. If the SRS warning light does not illuminate as described or stays on for more than 7 seconds or if the SRS light illuminates while driving, immediate inspection is required. If the vehicle's SRS warning light is in any of these three conditions, the SRS system must be inspected, diagnosed and serviced.

To avoid injury from accidental deployment of the air bag during vehicle servicing, the following service precautions must be observed.

✻✻ CAUTION

The Air Bag system must be disarmed before removing many components. Failure to do so may cause accidental deployment of the air bag, resulting in unnecessary system repairs and/or personal injury.

1. Disarm the air bag system using the following procedure:

a. Position the front wheels in the straight-ahead position and place the key in the **LOCK** position. Remove the key from the ignition lock cylinder.

b. Disconnect the negative battery cable and insulate the cable end with high-quality electrical tape or similar non-conductive wrapping.

c. Wait at least one minute before working on the vehicle. The air bag system is designed to retain enough voltage to deploy the air bag for a short period of time even after the battery has been disconnected.

Power Rack and Pinion Steering Gear

REMOVAL & INSTALLATION

✻✻ CAUTION

Prior to removal of the steering rack and pinion unit, center the front wheels and remove the ignition key. Failure to do so may damage the SRS (air bag system) clock spring under the steering wheel and render SRS system inoperative, risking serious driver injury.

1. Drain the power steering fluid using the following procedure:

a. Disconnect the power steering return (low side) hose.

b. Connect a suitable container to the hose.

c. Properly disable the ignition system.

d. While cranking the engine, turn the wheels, several times, from side to side, until the fluid is removed.

✻✻ CAUTION

The Air Bag system must be disarmed before working around the steering column. Failure to do so may cause accidental deployment of the air bag, resulting in unnecessary system repairs and/or personal injury.

2. Disarm the SRS system as follows:

a. Position the front wheels in the straight-ahead position and place the key in the **LOCK** position. Remove the key from the ignition lock cylinder.

b. Disconnect the negative battery cable and insulate the cable end with high-quality electrical tape or similar non-conductive wrapping.

c. Wait at least one minute before working on the vehicle. The air bag system is designed to retain enough voltage to deploy the air bag for a short period of time even after the battery has been disconnected.

3. Raise and safely support the vehicle.

4. Remove both front wheels.

5. Remove the bolt holding the lower steering column joint to the rack and pinion input shaft.

6. Remove the stabilizer bar.

7. Remove the cotter pins and using Joint Separator MB991113, disconnect the tie rod ends, from the steering knuckles.

8. On vehicles equipped with Electronic Control Power steering (EPS), disconnect the wiring harness, from the solenoid connector.

9. Locate the two triangular braces near the crossmember and remove both.

10. Support the center crossmember. Remove the through-bolt from the front round roll stopper and remove the three bolts securing the center crossmember.

11. Remove the center crossmember.

12. Properly support the engine and remove the rear roll stopper through-bolt. Lower the engine slightly.

✳✳ WARNING

In order to prevent damage to the engine, when supporting and jacking the engine, place a block of wood between the jack and the oil pan.

13. Disconnect the power steering fluid pressure pipe and return hose from the rack fittings. Plug the fittings to prevent excessive fluid leakage.

14. Remove the clamp bolts and the two bolts securing the rack assembly to the chassis.

15. Remove the rack and pinion steering assembly and its rubber mounts.

➡ **When removing the rack and pinion assembly, tilt the assembly to the inner side of the compression lower arm, and remove from the left-side of the vehicle. Use caution to avoid damaging the boots.**

To install:

16. Align the rack assembly so the splines are inserted into the steering column shaft.

17. Install the rack and with the mounting bolts. Tighten the mounting bolts to 51 ft. lbs. (69 Nm).

18. Install the pinch bolt and tighten the bolt to 13 ft. lbs. (18 Nm).

19. Connect the power steering fluid lines to the rack and tighten to high side fitting to 11 ft. lbs. (15 Nm). Secure the low side hose with the clamp.

20. Raise the engine into position. Install the rear roll stopper through-bolt and tighten to 32 ft. lbs. (43 Nm).

21. Raise the crossmember into position. Install the center member mounting bolts and tighten the front bolts to 58–65 ft. lbs. (78–88 Nm) and the rear bolt to 51–58 ft. lbs. (69–78 Nm).

22. Install the front roll stopper bolt and tighten the nut to 32 ft. lbs. (43 Nm).

23. Install the two triangular braces and tighten the mounting bolts to 50–56 ft. lbs. (69–78 Nm).

24. Install the stabilizer bar.

25. Connect the tie rod ends and tighten the nuts to 20 ft. lbs. (27 Nm).

26. On vehicles equipped with EPS, connect the wiring harness to the solenoid connector.

27. Install the wheels and lower the vehicle.

28. Refill the reservoir with power steering fluid and properly bleed the power steering system.

29. Perform a front end alignment.

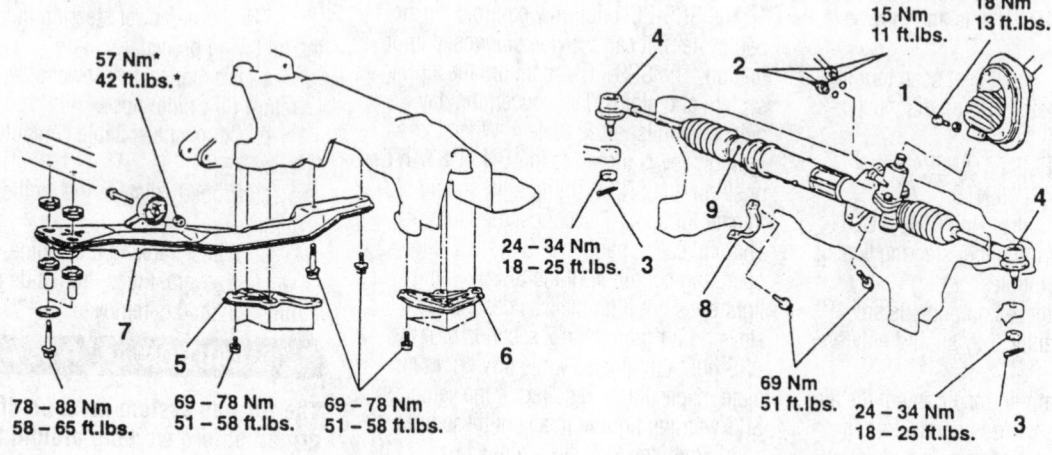

1. Joint assembly and gear box connecting bolt
2. Power steering pipe connection
3. Cotter pin
4. Tie rod end and knuckle connection
5. Stay (L.H.)
6. Stay (R.H.)
7. Center member assembly
8. Clamp
9. Gear box assembly

Caution
The fasteners marked * should be temporarily tightened before they are finally tightened once the total weight of the engine has been placed on the vehicle body.

Exploded view of the power rack and pinion steering gear mounting

7922CG30

Shock Absorber

REMOVAL & INSTALLATION

Front

1. Disconnect the negative battery cable.
2. Raise and safely support the vehicle.
3. Remove the appropriate wheel assembly.
4. Disconnect the sway bar link from the damper fork.
5. Remove the damper fork lower through-bolt and upper pinch bolt. Remove the damper fork assembly.

6. Remove the shock absorber upper nuts and remove the shock absorber assembly from the vehicle. Do NOT remove the large center nut.

To install:

7. Install the shock to the vehicle and tighten the upper mounting nuts to 32 ft. lbs. (44 Nm).
8. Align the shock to the damper fork and install the damper fork. Tighten the lower through-bolt/nut to 65 ft. lbs. (88 Nm) and the upper pinch bolt to 76 ft. lbs. (103 Nm).
9. Connect the sway bar link to the damper fork and tighten the link nut to 29 ft. lbs. (39 Nm).

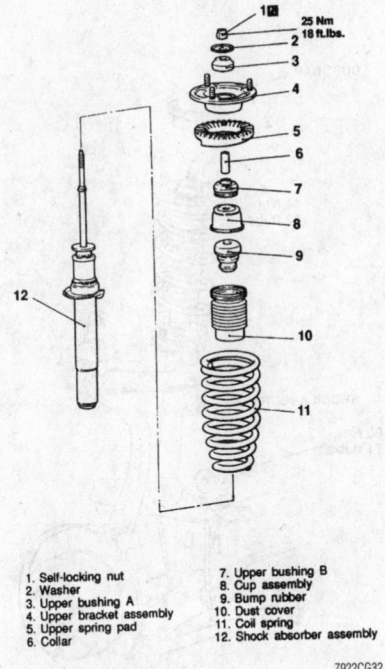

1. Self-locking nut	7. Upper bushing B
2. Washer	8. Cup assembly
3. Upper bushing A	9. Bump rubber
4. Upper bracket assembly	10. Dust cover
5. Upper spring pad	11. Coil spring
6. Collar	12. Shock absorber assembly

7922CG32

Exploded view of the front shock absorber

10. Install the wheel and tire assembly.
11. Perform a front end alignment.

Rear

➡The shock absorber assembly is a load bearing component, therefore the vehicle chassis and axle weight must be supported separately, requiring the use of two separate lifting devices.

1. The rear package shelf front cover(s) must be removed to access the top mounting nuts. Most connections are plastic clips. Use care when removing these components to avoid unnecessary damage.
   a. Remove the rear shelf speaker covers.
   b. Remove the rear shelf top assembly.
   c. Remove the front cover(s) to access the shock absorber top mounting nuts.
2. Raise and support vehicle chassis.
3. Raise and support lower control arm assembly slightly.
4. Remove the shock absorber upper mounting nuts.
5. Remove the shock absorber lower mounting bolt and remove the assembly from the vehicle.

To install

6. Position the shock absorber assem-

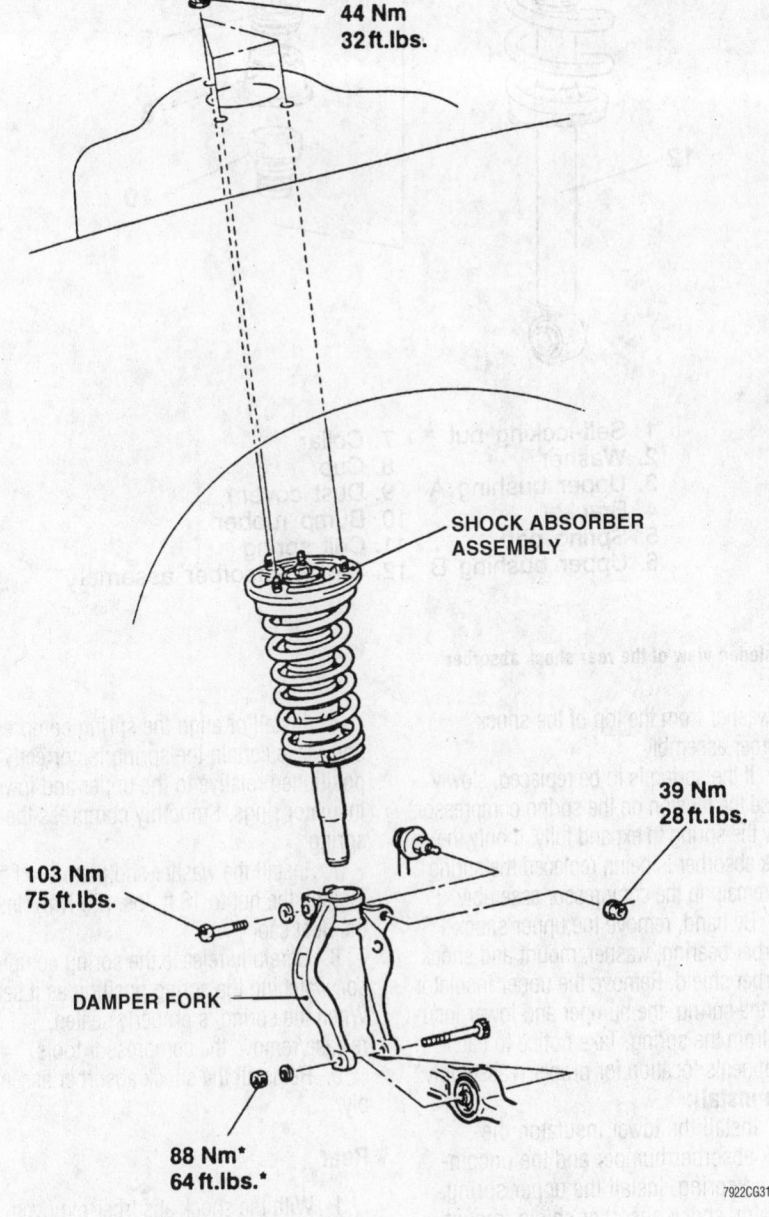

44 Nm
32 ft.lbs.

SHOCK ABSORBER ASSEMBLY

39 Nm
28 ft.lbs.

103 Nm
75 ft.lbs.

DAMPER FORK

88 Nm*
64 ft.lbs.*

7922CG31

Exploded view of the front shock absorber assembly mounting

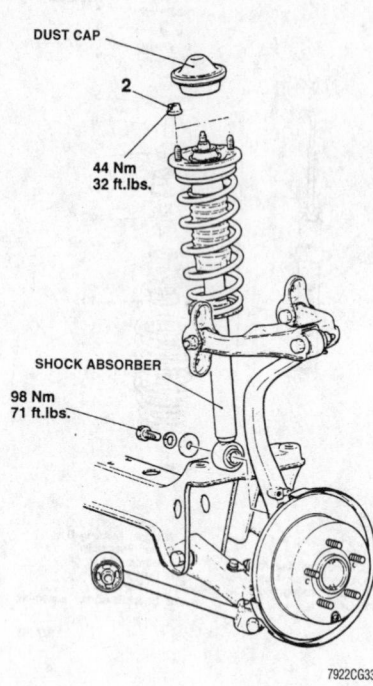

DUST CAP

2

44 Nm
32 ft.lbs.

SHOCK ABSORBER

98 Nm
71 ft.lbs.

7922CG33

Exploded view of the rear shock absorber assembly mounting

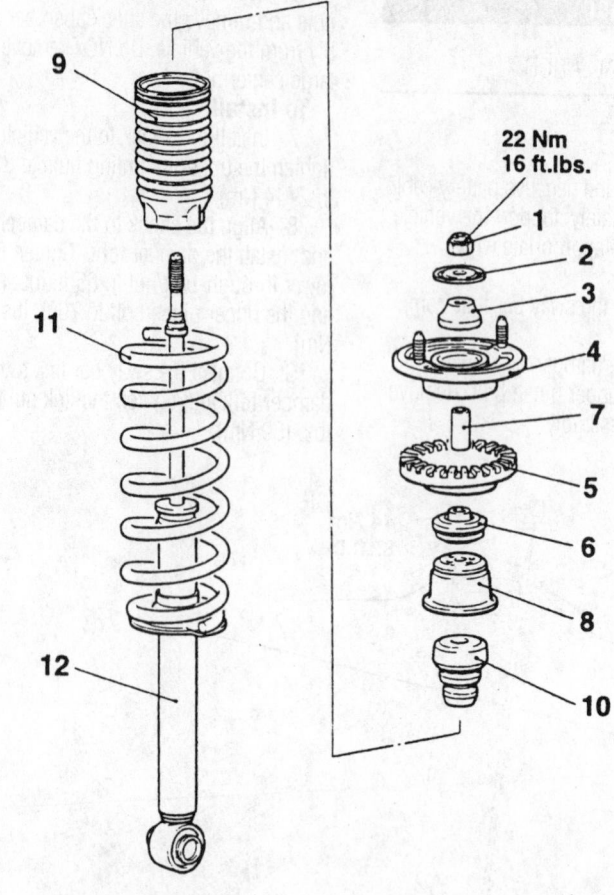

7922CG34

1. Self-locking nut
2. Washer
3. Upper bushing A
4. Bracket
5. Spring pad
6. Upper bushing B
7. Collar
8. Cup
9. Dust cover
10. Bump rubber
11. Coil spring
12. Shock absorber assembly

Exploded view of the rear shock absorber

bly so that the lower mounting bolt can be installed and lightly tightened.

7. Use a jack to raise or lower the lower control arm, so that top shock absorber plate studs aligns through the body. Raise the jack to hold the shock absorber assembly in position.

8. Install top plate nuts on studs and tighten the mounting nuts to 32 ft. lbs. (44 Nm).

9. Tighten the lower mounting bolt to 71 ft. lbs. (98 Nm).

10. Install the interior trim pieces to complete shock absorber installation.

Coil Spring

REMOVAL & INSTALLATION

Front

1. Place the shock absorber assembly in a MB991237 and MB991238 spring compressor assembly or equivalent. Tighten the compressor and compress the spring slowly. Make certain the compressor is properly engaged before tightening.

2. After tension has been removed from the shock absorber assembly and shock absorber plate, remove the piston rod nut

and washer from the top of the shock absorber assembly.

3. If the spring is to be replaced, slowly release the tension on the spring compressor. Allow the spring to expand fully. If only the shock absorber is being replaced the spring may remain in the compressor assembly.

4. By hand, remove the upper shock absorber bearing, washer, mount and shock absorber shield. Remove the upper insulator ring, the spring, the bumper and lower insulator from the spring. Take notice to each components location for proper reassembly.

To install:

5. Install the lower insulator, the shock absorber bumper and the uncompressed spring. Install the upper spring insulator, shock absorber shield, mount and washer.

6. Install or align the spring compressor. Make certain the spring is correctly positioned relative to the upper and lower insulator rings. Smoothly compress the spring.

7. Install the washer and piston rod nut. Tighten the nut to 18 ft. lbs. (25 Nm). Install the dust cap.

8. Carefully release the spring compressor, watching the spring position as it seats. When the spring is properly seated, release/remove the compressor tools.

9. Reinstall the shock absorber assembly.

Rear

1. With the shock absorber removed from the vehicle, place the shock absorber assembly in a MB991237 and MB991239

spring compressor assembly or equivalent shock absorber service tool. Tighten the compressor and compress the spring slowly. Make certain the compressor is properly engaged before tightening.

2. After tension has been removed from the shock absorber assembly and shock absorber plate, remove the piston rod nut and washer from the top of the shock absorber assembly.

3. If the spring is to be replaced, slowly release the tension on the spring compressor. Allow the spring to expand fully. If only the shock absorber is being replaced the spring may remain in the compressor assembly.

4. By hand, remove the upper shock absorber bearing, washer, mount and shock absorber shield. Remove the upper insulator ring, the spring, the bumper and lower insulator from the spring. Take notice to each components location for proper reassembly.

To install:

5. Install the lower insulator, the shock absorber bumper and the uncompressed spring. Install the upper spring

insulator, shock absorber shield, mount and washer.

6. Install or align the spring compressor. Make certain the spring is correctly positioned relative to the upper and lower insulator rings. Smoothly compress the spring.

7. Install the washer and piston rod nut. Tighten the nut to 16 ft. lbs. (22 Nm). Install the dust cap.

8. Carefully release the spring compressor, watching the spring position as it seats. When the spring is properly seated, release/remove the compressor tools.

9. Reinstall the shock absorber assembly.

Upper Ball Joint

REMOVAL & INSTALLATION

The upper ball joint is an integrated part of the upper control arm assembly, and can not be serviced separately. A worn or damaged ball joint requires replacement of upper control arm assembly.

Lower Ball Joint

REMOVAL & INSTALLATION

The front suspension is called a Multi-Link Suspension. There are two lower arms used in this front suspension; a curved arm called the Compression Lower Arm and also a straight arm called the Lateral Lower Arm. Both arms contain lower ball joints since there are two sockets in the steering knuckle. Ball joints and lower arms are removed and replaced as an assembly. A front end alignment is required after these procedures.

Upper Control Arm

REMOVAL & INSTALLATION

1. Raise and safely support the vehicle.

2. Remove the appropriate wheel.

3. Using the Joint Separation tool, MB991113 or equivalent, disconnect the ball joint stud from the steering knuckle.

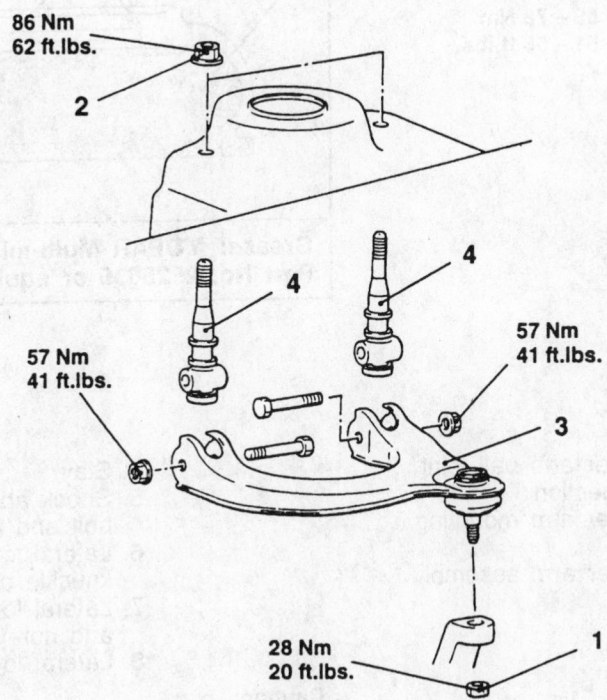

1. Upper arm ball joint and knuckle connection
2. Self-locking nut for upper arm installation
3. Upper arm assembly
4. Upper arm shaft assembly

7922CG35

Exploded view of the upper control arm mounting

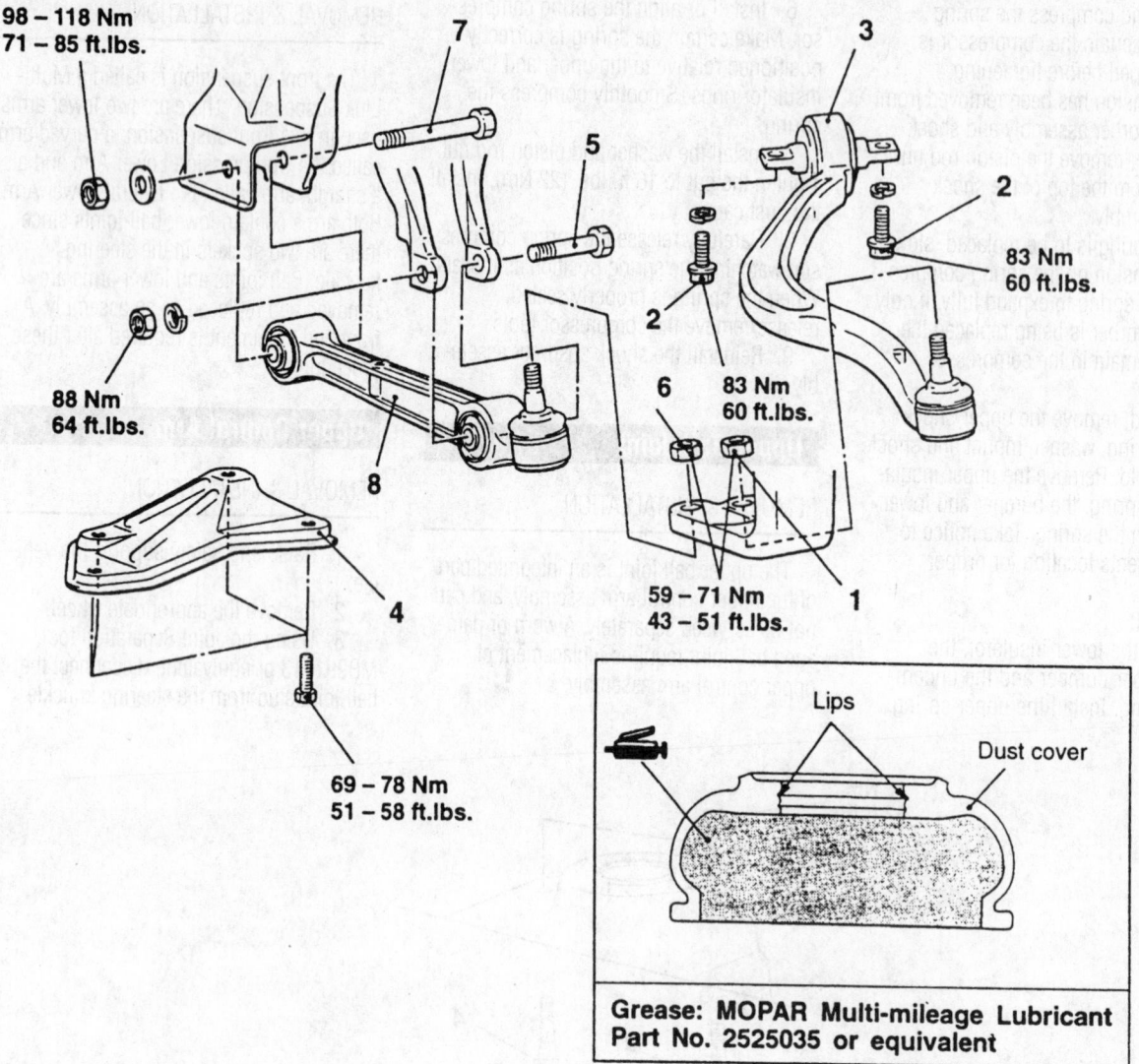

98 – 118 Nm
71 – 85 ft.lbs.

88 Nm
64 ft.lbs.

69 – 78 Nm
51 – 58 ft.lbs.

83 Nm
60 ft.lbs.

83 Nm
60 ft.lbs.

59 – 71 Nm
43 – 51 ft.lbs.

Lips

Dust cover

**Grease: MOPAR Multi-mileage Lubricant
Part No. 2525035 or equivalent**

1. Compression lower arm ball joint and knuckle connection
2. Compression lower arm mounting bolt
3. Compression lower arm assembly

4. Stay
5. Shock absorber lower mounting bolt and nut
6. Lateral lower arm ball joint and knuckle connection
7. Lateral lower arm mounting bolt and nut
8. Lateral lower arm assembly

Caution
*: Indicates parts which should be temporarily tightened, and then fully tightened with the vehicle on the ground in the unladen condition.

7922CG36•

Exploded view of the lateral lower arm and the compression lower arm mounting

4. The ball joint can be checked using the following procedure.

   a. An adapter (MB 990326) is available that fits onto the ball joint stud and adapts to an inch-pound torque wrench. If this tool is not available, a shop-made substitute can be fabricated.

   b. Turn the ball joint stud with the torque wrench. The factory standard for breakaway torque is 3–13 inch lbs. (0.34–1.45 Nm).

   c. If the ball joint stud is out of specification (turns too easily or is too stiff), continue with the ball joint/control arm assembly replacement. Note that the boot over the ball joint can be removed, and the joint greased. A new replacement ball joint boot is recommended.

5. Inside the engine compartment, at the shock absorber tower, locate the upper control arm mounting nuts. Remove the nuts and using the joint separation tool, separate the upper arm shafts from the shock absorber tower.

6. Remove the control arm assembly.

To install:

7. Align the upper control arm shafts to the shock absorber tower and secure it with the mounting nuts. Tighten the mounting nuts to 62 ft. lbs. (86 Nm).

8. Connect the ball joint to the knuckle and tighten the locking nut to 20 ft. lbs. (28 Nm).

9. Install the wheel and lower the vehicle.

10. Check the wheel alignment and adjust if necessary.

Lower Control Arm

REMOVAL & INSTALLATION

Lateral Lower Arm

1. Raise and safely support the vehicle.
2. Remove the appropriate wheel assembly.
3. Remove the stay bracket from the crossmember.
4. Using joint separator MB991113, disconnect the ball joint stud from the steering knuckle.
5. Remove the through-bolt, connecting the damper fork to the lower control arm.
6. Remove the mounting bolt connecting the lower control arm to the suspension crossmember.
7. Remove the lower control arm from the vehicle.

To install:

8. When installing the control arm, temporarily tighten the nuts and/or bolts securing the control arm to the suspension crossmember. Tighten them fully only after the vehicle is sitting on its wheels.

9. Connect the damper fork to the lower control arm and tighten the through-bolt to 64 ft. lbs. (88 Nm).

10. Connect the ball joint stud to the knuckle and tighten the nut to 65–80 ft. lbs. (88–108 Nm). Install a new cotter pin.

11. Connect the tension rod to the control arm and tighten the nuts to 80–94 ft. lbs. (108–127 Nm).

12. Connect the stay bracket to the crossmember and tighten the mounting bolts to 50–56 ft. lbs. (69–78 Nm).

13. Install the wheels and lower the vehicle to the floor.

14. Once the full weight of the vehicle is on the suspension, tighten the inner lower arm mounting bolt nut to 71–85 ft. lbs. (98–118 Nm).

15. Check the front end alignment and adjust as required.

Compression Lower Arm

1. Raise and support the vehicle safely.
2. Remove the appropriate wheel assembly.
3. Using joint separator MB991113, disconnect the ball joint stud from the steering knuckle.
4. Remove the mounting bolts connecting the lower control arm to the suspension crossmember.

To install:

5. Connect the control arm to the suspension crossmember, and tighten the bolts to 60 ft. lbs. (83 Nm).

6. Connect the ball joint stud to the knuckle and tighten the nut to 43–51 ft. lbs. (59–71 Nm).

7. Install the wheels and lower the vehicle to the floor.

8. Check the front end alignment and adjust as required.

Wheel Bearings

ADJUSTMENT

Front

To check hub and bearing assembly end-play, remove the caliper and rotor. Rig a dial indicator to bear against the hub flange near the center ridge. Wiggle the hub back and forth. If end-play exceeds 0.002 in. (0.05mm), replace the front hub and bearing assembly.

Rear

The rear hub and wheel bearing assembly is designed for the life of the vehicle and requires no type of adjustment or periodic maintenance. The bearing is a sealed unit with the wheel hub and can only be removed and/or replaced as one unit.

REMOVAL & INSTALLATION

Front

1. Remove the cotter pin, halfshaft nut and washer.
2. Raise and safely support the vehicle.
3. Remove the appropriate wheel assembly.
4. If equipped with ABS, remove the vehicle speed sensor.
5. Remove the caliper and brake pads. Support the caliper out of the way using wire.
6. Remove the brake rotor from the hub assembly.
7. Disconnect the upper ball joint from the steering knuckle using a press type tool and pull the knuckle outward.

✳✳ CAUTION

Use of improper methods of joint separation can result in damage to joint, leading to possible failure. Never use wedge-type tools or the ball joint can be damaged.

8. From the back of the knuckle, remove the four bolts securing the hub to the knuckle.

9. Remove the hub and bearing assembly from the knuckle.

➡**The hub and wheel bearing assembly is not serviceable and should not be disassembled.**

To install:

10. Install the hub to the steering knuckle and tighten the mounting bolts to 65 ft. lbs. (88 Nm).

11. Connect the upper ball joint to the steering knuckle and tighten the self-locking nut to 21 ft. lbs. (28 Nm).

12. Position the rotor on the hub. Install a couple of lug nuts and lightly tighten to hold rotor on hub.

13. Install the caliper holder and place brake pads in holder. Slide caliper over brake pads and install guide pins. Once caliper is secured, lug nuts can be removed.

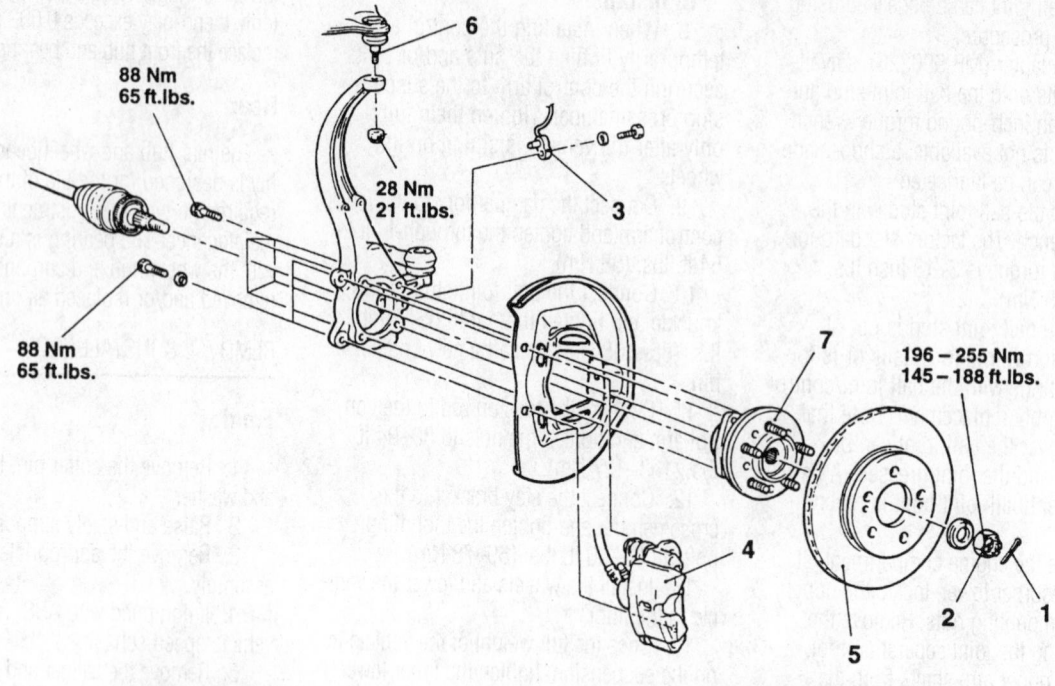

88 Nm
65 ft.lbs.

6

28 Nm
21 ft.lbs.

3

88 Nm
65 ft.lbs.

7

196 – 255 Nm
145 – 188 ft.lbs.

4

2   1

5

1. Cotter pin
2. Drive shaft nut
3. Front speed sensor <Vehicles with ABS>
4. Caliper assembly

5. Brake disc
6. Upper arm connection
7. Front hub assebly

Caution
Do not disassemle the front hub assembly.

7922CG37

Exploded view of the front hub assembly mounting and related components

14. If equipped with ABS, install the vehicle speed sensor.

15. Install the wheel assembly and lower the vehicle.

16. Install the wheel and lower the vehicle to the floor. Examine the drive shaft (half-shaft) hub washer. Locate the chamfered side. This side is installed outward, away from the hub. Install the washer and hub nut. Tighten the axle nut with the brakes applied. Tighten the nut to 145–188 ft. lbs. (200–260 Nm).

17. Install a new cotter pin and bend to secure.

✳✳ CAUTION

Pump the brake pedal until hard, before attempting to move the vehicle.

Rear

WITH DRUM BRAKES

1. Raise and safely support the vehicle.
2. Remove the appropriate wheel assembly.
3. If equipped with ABS, remove the vehicle speed sensor.

4. Remove the brake drum from the hub assembly.

5. From the back of the knuckle, remove the four bolts securing the hub to the knuckle.

6. Remove the hub and bearing assembly from the knuckle.

➡**The hub assembly is not serviceable and should not be disassembled.**

7. If replacing the hub, use special socket MB991248 and a press, to remove the wheel sensor rotor from the hub.

To install:

8. Press the wheel sensor rotor onto the hub.

9. Install the hub to the knuckle and tighten the mounting bolts to 54–65 ft. lbs. (74–88 Nm).

10. Install the brake drum on the hub.

11. If equipped with ABS, install the vehicle speed sensor.

12. Install the wheel assembly and lower the vehicle.

WITH DISC BRAKES

1. Raise and safely support the vehicle.

2. Remove the appropriate wheel assembly.

3. If equipped with ABS, remove the vehicle speed sensor.

4. Remove the caliper and brake pads. Support the caliper out of the way using wire.

5. Remove the brake rotor from the hub assembly.

6. Remove the parking brake shoes as follows:

   a. Remove the upper shoe to anchor springs.

   b. Remove the lower shoe to shoe spring.

   c. Remove the brake shoe hold-down springs.

   d. Disconnect the parking brake cable from the actuating lever.

7. From the back of the knuckle, remove the four bolts securing the hub to the knuckle.

8. Remove the hub and bearing assembly from the knuckle.

➡**The hub assembly is not serviceable and should not be disassembled.**

9. If replacing the hub, use special socket MB991248 and a press, to remove the wheel sensor rotor from the hub.

<Vehicles with drum brakes>

<Vehicles with disc brakes>

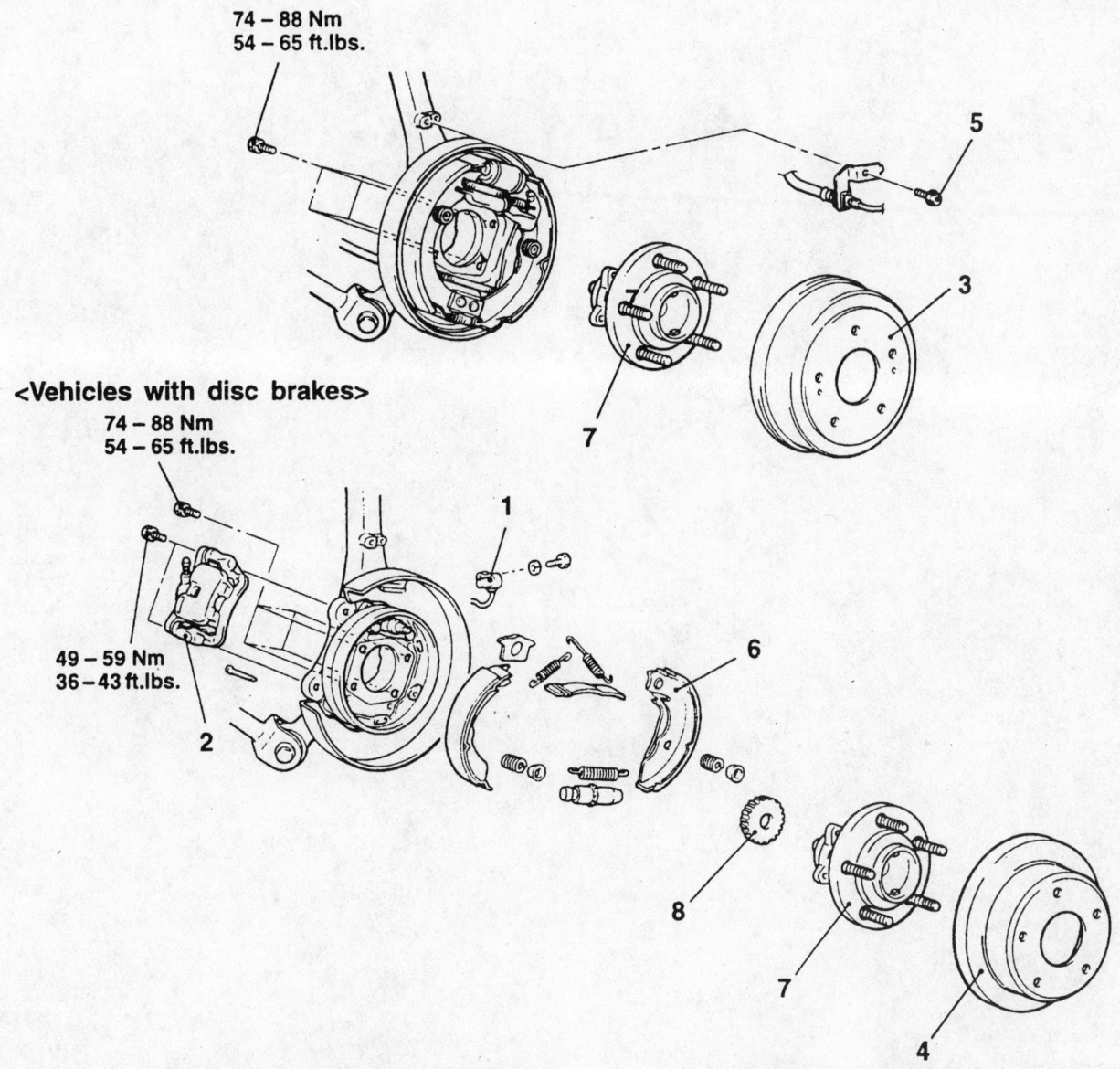

1. Rear speed sensor <Vehicles with ABS>
2. Caliper assembly
3. Brake drum
4. Brake disc
5. Clip mounting bolt

6. Shoe and lining assembly <Drum in disc brake>
7. Rear hub assembly
8. ABS-rotor <Vehicles with ABS>

Caution
Do not disassemble the rear hub assembly.

7922CG38

Exploded view of the rear hub assembly mounting

To install:

10. Press the wheel sensor rotor onto the hub.

11. Install the hub to the knuckle and tighten the mounting bolts to 54–65 ft. lbs. (74–88 Nm).

12. Install the parking brake shoes.

13. Position the rotor on the hub. Install a couple of lug nuts and lightly tighten to hold rotor on hub.

14. Install the caliper holder and place brake pads in holder. Slide the caliper over brake pads and install guide pins. Once caliper is secured, lug nuts can be removed.

15. If equipped with ABS, install the vehicle speed sensor.

16. Install the wheel assembly and lower the vehicle.

CHRYSLER CORPORATION

14

Eagle-Talon

DRIVE TRAIN **14-12**
ENGINE REPAIR **14-2**
FUEL SYSTEM **14-11**
STEERING AND
SUSPENSION **14-16**

A
Air Bag 14-16
  Disarming 14-16
  Precautions 14-16

C
Camshaft and Valve Lifters 14-7
  Removal & Installation 14-7
Clutch 14-13
  Adjustments 14-13
  Removal & Installation 14-13
Cylinder Head 14-3
  Removal & Installation 14-3

E
Engine Assembly 14-2
  Removal & Installation 14-2
Exhaust Manifold 14-6
  Removal & Installation 14-6

F
Front Crankshaft Seal 14-7
  Removal & Installation 14-7
Fuel Filter 14-11
  Removal & Installation 14-11
Fuel Pump 14-12
  Removal & Installation 14-12

Fuel System Pressure 14-11
  Relieving 14-11
Fuel System Service
  Precautions 14-11

H
Halfshaft 14-15
  Removal & Installation 14-15
Hydraulic Clutch System 14-14
  Bleeding 14-14

I
Ignition Timing 14-2
Intake Manifold 14-5
  Removal & Installation 14-5

L
Lower Ball Joint 14-18
  Removal & Installation 14-18

O
Oil Pan 14-8
  Removal & Installation 14-8
Oil Pump 14-9
  Removal & Installation 14-9

P
Power Rack and Pinion
  Steering Gear 14-16
  Removal & Installation 14-16

R
Rear Main Seal 14-11
  Removal & Installation 14-11

Rocker Arm/Shafts 14-4
  Removal & Installation 14-4

S
Strut 14-17
  Overhaul 14-17
  Removal & Installation 14-17

T
Transaxle Assembly 14-12
  Removal & Installation 14-12
Transfer Case Assembly 14-14
  Removal & Installation 14-14
Turbocharger 14-4
  Removal & Installation 14-4

U
Upper Ball Joint 14-18
  Removal & Installation 14-18

V
Valve Lash 14-8
  Adjustment 14-8

W
Water Pump 14-2
  Removal & Installation 14-2
Wheel Bearings 14-19
  Adjustment 14-19
  Removal & Installation 14-19

ENGINE REPAIR

➡ **Disconnecting the negative battery cable on some vehicles may interfere with the functioning of the on-board computer system, and may require the computer to undergo a relearning process once the negative battery cable is reconnected.**

Ignition Timing

It is not necessary to check the ignition timing, because the crankshaft position is detected directly and the ignition timing is controlled electronically.

Engine Assembly

REMOVAL & INSTALLATION

The following procedure can be used on all vehicles. Slight variations may occur due to extra connections, etc., but the basic procedure should cover all models.

1. Relieve the fuel system pressure.
2. Disconnect the negative battery cable.
3. If equipped, remove the engine undercover.
4. Matchmark the hood and hinges and remove the hood assembly. Remove the air cleaner assembly and all adjoining air intake duct work.
5. Drain the engine coolant into a suitable container, then remove the radiator assembly, coolant reservoir and intercooler.
6. Drain the engine oil.
7. If equipped with AWD, remove the transaxle and transfer case.
8. Tag and detach the following components: accelerator cable, heater hoses, brake vacuum hose, connection for vacuum hoses, high pressure fuel line, fuel return line, oxygen sensor connection, coolant temperature gauge connection, coolant temperature sensor connector, connection for thermo switch sensor, if equipped with automatic transaxle, the connection for the idle speed control, motor position sensor connector, throttle position sensor connector, EGR temperature sensor connection (California vehicles), fuel injector connectors, power transistor connector, ignition coil connector, condenser and noise filter connector, distributor and control harness, connections for the alternator and oil pressure switch wires.
9. If equipped with A/C, remove the A/C drive belt and the compressor. Leave the A/C lines attached. Do NOT discharge the system. Wire the compressor aside.

10. Remove the power steering pump and wire aside.
11. Remove the exhaust manifold-to-head pipe nuts. Discard the gasket and replace with a new one during installation.
12. Attach a hoist to the engine and take up the engine weight. Remove the engine mount bracket. Remove any torque control brackets (roll stoppers). Note that some engine mount pieces have arrows on them for proper assembly. Double check that all cables, hoses, harness connectors, etc., are disconnected from the engine. Lift the engine slowly from the engine compartment.

To install:

13. Install the engine and secure in position. The front lower mount through-bolt nut should not be tightened until the full weight of the engine is on the mount.
   a. Tighten the engine mount bolts to 50 ft. lbs. (69 Nm).
14. Install the exhaust pipe, power steering pump and A/C compressor.
15. Checking the tags installed during removal, attach all electrical and vacuum connections.
16. Install the transaxle in the vehicle and tighten the upper mounting bolts to 65 ft. lbs. (90 Nm). Install the starter assembly and tighten both mounting bolts to 54–65 ft. lbs. (75–90 Nm).
17. Install the radiator assembly and intercooler.

18. Install the air cleaner assembly. Install all control brackets, if not already done.
19. Fill the engine with the proper amount of engine oil. Connect the negative battery cable.

✳✳ WARNING

Operating the engine without the proper amount and type of engine oil will result in severe engine damage.

20. Refill the cooling system. Start the engine, and allow it to reach normal operating temperature. Check for leaks.
21. Check the ignition timing and adjust, if necessary.
22. Install the hood, making sure to align the matchmarks made during disassembly.
23. Road test the vehicle and check all functions for proper operation.

Water Pump

REMOVAL & INSTALLATION

1. Drain the engine coolant into a suitable container.
2. Remove the timing belt, as described in the timing belt unit repair section in the beginning of this manual.

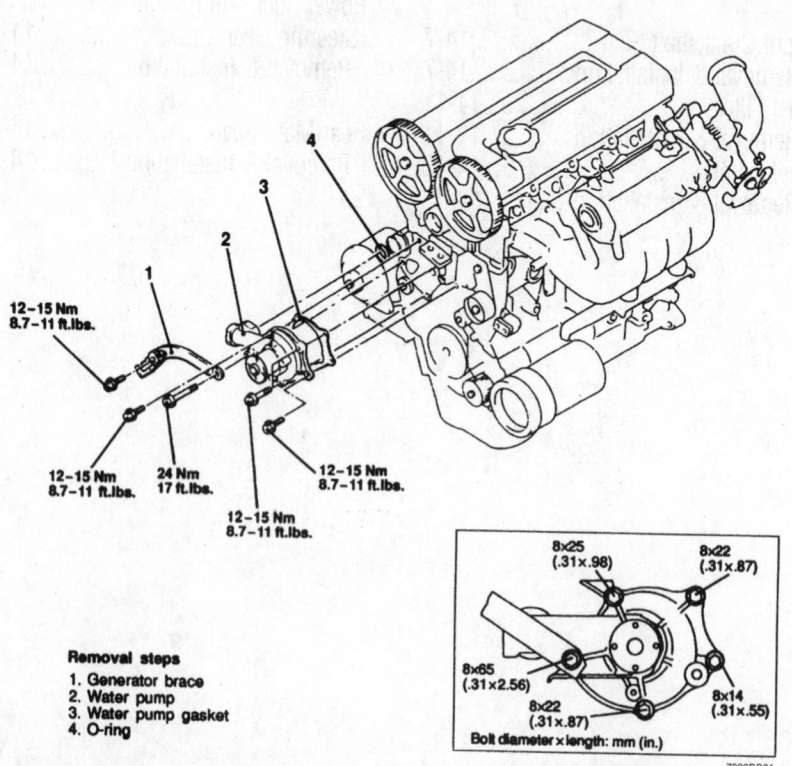

12–15 Nm
8.7–11 ft.lbs.

12–15 Nm
8.7–11 ft.lbs.

24 Nm
17 ft.lbs.

12–15 Nm
8.7–11 ft.lbs.

12–15 Nm
8.7–11 ft.lbs.

Removal steps
1. Generator brace
2. Water pump
3. Water pump gasket
4. O-ring

8x25 (.31x.98)
8x22 (.31x.87)
8x65 (.31x2.56)
8x22 (.31x.87)
8x14 (.31x.55)
Bolt diameter x length: mm (in.)

7922DG01

Water pump mounting and bolt locations—2.0L (VIN F) engine

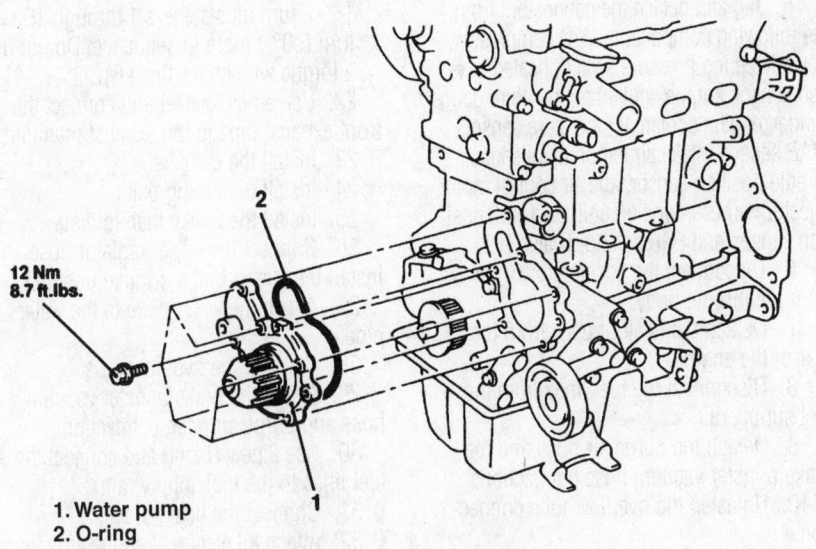

12 Nm
8.7 ft.lbs.

2

1

1. Water pump
2. O-ring

7922DG02

Exploded view of the water pump and O-ring mounting—2.0L (VIN Y) engine

3. If necessary, remove the alternator brace from the water pump.

4. Unfasten the retainers, then remove any brackets for access to the rear cover.

5. If necessary, remove the timing belt rear cover.

6. Remove the water pump mounting bolts.

7. Remove the water pump, gasket and O-ring. Discard the gasket and O-ring and replace with new ones during installation.

To install:

8. Install a new O-ring on the water inlet pipe. Coat the O-ring with water or coolant. Do not allow oil or other grease to contact the O-ring.

9. Use a new gasket and install the water pump on the engine block. Tighten the mounting bolts to 8.7–11 ft. lbs. (12–15 Nm). Install the alternator brace on the water pump. Tighten the brace pivot bolt to 17 ft. lbs. (24 Nm).

10. If removed, install the timing belt rear cover.

11. Install the timing belt.

12. Install the remaining components.

13. Refill the engine with coolant.

14. Connect the negative battery cable, start the engine and check for leaks.

Cylinder Head

REMOVAL & INSTALLATION

2.0L (VIN F) Engine

1. Properly relieve the fuel system pressure.

2. If not already done, disconnect the negative battery cable.

3. Drain the engine coolant and engine oil into suitable containers.

4. Disconnect the accelerator cable and remove the mounting bracket.

5. Detach the intake air duct (hose) from the throttle body.

6. Tag and detach the electrical connectors from the following components: Idle Air Control (IAC) motor, knock sensor, heated oxygen sensor, engine coolant temperature gauge sender, engine coolant temperature sensor, ignition module (power transistor), throttle position sensor, condenser, manifold differential pressure sensor, fuel injectors, ignition coil, camshaft position sensor, crankshaft position sensor, A/C compressor, and engine control wiring harness

7. Remove the engine center cover. Tag and disconnect the spark plug wires.

8. Disconnect the brake booster vacuum hose.

9. Detach the fuel lines from the fuel supply rail.

10. Disconnect the bypass hose and the water hose connections.

11. Disconnect the vacuum hoses, breather hose and the PCV hose.

12. Remove the timing belt, as described in the timing belt unit repair section in the beginning of this manual.

13. Remove the power steering pump.

14. Remove the cylinder head cover and the semi-circular packing.

15. Remove the heat protector.

16. Mark the position of the hose clamps on the hoses and disconnect the water hoses and the radiator hoses.

17. Remove the thermostat housing and the O-ring.

18. Remove the intake manifold stay.

19. Remove the turbocharger assembly from the exhaust manifold.

20. Gradually loosen the cylinder head bolts in two or three steps using the specified sequence and remove the bolts.

21. Remove the cylinder head and the gasket. Discard the gasket and replace with a new one during installation.

To install:

22. Thoroughly clean the deck surface of the engine block and the sealing surface of the cylinder head. Check the cylinder head for warpage.

23. Measure the length of the cylinder head bolts from below the head to the end, if the bolt measures more than 3.913 in. (9.94cm), replace the bolt.

24. Install a new gasket on the engine block with the identification mark facing upwards.

25. Carefully place the cylinder head on the engine. Apply clean engine oil to the bolts and install the bolts finger-tight.

26. Tighten the bolts in the proper sequence using the following procedure:

   a. Tighten the bolts in sequence to 58 ft. lbs. (78 Nm).

   b. Loosen the bolts completely in the reverse order.

   c. Tighten the bolts in sequence to 15 ft. lbs. (20 Nm).

   d. Make a paint mark on the head of the bolt and the cylinder head at the same spot. Tighten the bolt 90° or ¼ turn from the mark.

   e. Tighten the bolt an additional 90° or ¼ turn so that the mark on the head of the bolt is opposite the original mark on the cylinder head.

27. Use a new gasket and install the turbocharger on the exhaust manifold.

28. Install the intake manifold stay.

29. Install the thermostat housing and connect the hoses. Align the matchmarks and install the clamps.

30. Install the heat protector.

31. Apply sealant to the semi-circular packing and install it on the cylinder head.

32. Apply sealant at the front of the cylinder head where the camshaft oil seal retainer and the cylinder head come together and install the cylinder head cover using a new gasket.

33. Install the power steering pump.

34. Install the timing belt.

35. Connect the PCV, breather and vacuum hoses.

36. Connect the water hose and the bypass hose.

37. Use a new O-ring and connect the lines to the fuel supply rail. Apply a small amount of engine oil to the new O-ring.

38. Connect the brake booster vacuum hose.

39. Connect the spark plug wires and install the center cover.

40. Attach all of the connectors, as tagged during removal.

41. Connect the intake air hose to the throttle body.

42. Connect and adjust the accelerator cable.

43. Refill the engine with coolant.

44. Replace the oil filter and refill the engine with the proper amount of oil.

45. Connect the negative battery cable, start the engine and check for fuel, coolant and oil leaks.

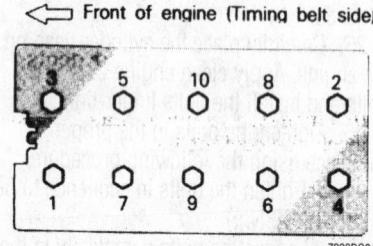

⟵ Front of engine (Timing belt side)

7922DG04

Cylinder head bolt removal sequence— 2.0L (VIN F and Y) engines

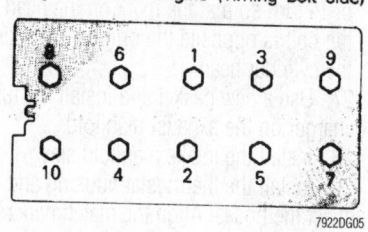

⟵ Front of engine (Timing belt side)

7922DG05

Cylinder head bolt tightening sequence— 2.0L (VIN F and Y) engines

2.0L (VIN Y) Engine

1. Relieve the fuel system pressure.

2. Disconnect the negative battery cable.

3. Drain the engine coolant and oil into suitable containers.

4. Remove the air cleaner and air intake duct.

5. Tag and detach the connectors from the following components: A/C compressor, power steering pressure switch, heated oxygen sensor, engine coolant temperature gauge sender, engine coolant temperature sensor, MAP sensor, intake air temperature sensor, throttle position sensor, idle air control motor, injector harness, ignition coil, camshaft position sensor, and EGR solenoid valve.

6. Disconnect the accelerator cable from the throttle body.

7. Detach the heater hoses from the rear of the engine.

8. Disconnect the fuel lines from the fuel supply rail.

9. Detach the purge air hose and the brake booster vacuum hose connections.

10. Unfasten the overflow tube connection.

11. Mark the position of the clamp on the hose and disconnect the upper radiator hose and the water hose connections.

12. Remove the timing belt, as described in the timing belt unit repair section in the beginning of this manual.

13. Remove the intake manifold stay.

14. Remove the intake and exhaust camshafts.

15. Disconnect the exhaust pipe connection from the exhaust manifold.

16. Unfasten the cylinder head mounting bolts, in the proper sequence, then remove the cylinder head from the vehicle.

To install:

17. Thoroughly clean the cylinder head and engine block sealing surfaces. Check the deck and the cylinder head for warpage.

18. Clean the cylinder head bolts and inspect them for stretching. If a bolt appears to by stretched, replace it.

19. Place a new head gasket on the engine block and carefully place the cylinder head on the engine.

20. Coat the threads of the bolts with clean engine oil and install the bolts finger-tight in the engine block. The short bolts go in the corners.

21. Tighten the cylinder head bolts in the proper sequence and in the following the three steps:

    a. Tighten the center bolts 1 through 6 to 25 ft. lbs. (33 Nm), then tighten the outer bolts 7 through 10 to 20 ft. lbs. (27 Nm).

    b. Tighten the center bolts 1 through 6 to 50 ft. lbs. (67 Nm), then tighten the outer bolts 7 through 10 to 20 ft. lbs. (27 Nm).

    c. Tighten the center bolts 1 through 6 to 50 ft. lbs. (67 Nm), then tighten the outer bolts 7 through 10 to 20 ft. lbs. (27 Nm).

• Turn all fasteners 1 through 10 ¼ turn (90°) more in sequence. Do not use a torque wrench for this step.

22. Use a new gasket and connect the front exhaust pipe to the exhaust manifold.

23. Install the camshafts.

24. Install the timing belts.

25. Install the intake manifold stay.

26. Connect the upper radiator hose. Install the clamp in the original position.

27. Attach the water hose to the water pipe.

28. Connect the overflow tube.

29. Attach the brake booster vacuum hose and purge air hose connection.

30. Use a new O-ring and connect the fuel lines to the fuel supply rail.

31. Connect the heater hose.

32. Attach all electrical connectors, as tagged during removal.

33. Install and adjust the accelerator cable.

34. Install the air intake duct and the air cleaner assembly.

35. Refill the engine with oil and coolant. Replace the oil filter.

36. Turn the ignition to the **ON** position and check for fuel leaks. Then, start the engine and check for coolant leaks and proper operation.

Rocker Arm/Shafts

REMOVAL & INSTALLATION

➡**The DOHC engines do not use rocker arm shafts; the valves are directly actuated by rocker arms. To remove the arms, the camshaft must first be removed. It is recommended that all rocker arms and lash adjusters are replaced together.**

Turbocharger

REMOVAL & INSTALLATION

1. Disconnect the negative battery cable.

2. Drain the engine oil, cooling system and remove the radiator. On vehicles equipped with A/C, remove the condenser fan assembly with the radiator.

3. Disconnect the oxygen sensor connector and remove the sensor.

4. Remove the oil dipstick and tube.

5. Remove the air intake bellows hose, the wastegate vacuum hose, the connections for the air outlet hose, and the upper and lower heat shield.

6. Unbolt the power steering pump and bracket assembly and leaving the hoses connected, wire it aside.

7. Remove the self-locking exhaust manifold nuts, the triangular engine hanger bracket, the eyebolt and gaskets that connect the oil feed line to the turbo center section and the water cooling lines. The water line under the turbo has a threaded connection.

8. Remove the exhaust pipe nuts and gasket and lift off the exhaust manifold. Discard the gasket.

9. Remove the two through-bolts and two nuts that hold the exhaust manifold to the turbocharger.

10. Remove the two capscrews from the oil return line (under the turbo). Discard the gasket. Separate the turbo from the exhaust manifold. The two water pipes and oil feed line can still be attached.

11. Visually check the turbine wheel (hot side) and compressor wheel (cold side) for cracking or other damage. Check whether the turbine wheel and the compressor wheel can be easily turned by hand. Check for oil leakage. Check whether or not the wastegate valve remains open. If any problem is found, replace the part. Inspect oil passages for restriction or deposits and clean as required.

12. The wastegate can be checked with a pressure tester. Apply approximately 9 psi (62 kPa) to the actuator and be sure the rod moves. Do not apply more than 10.3 psi (71 kPa) or the diaphragm in the wastegate may be damaged. Vacuum applied to the wastegate actuator should be maintained, replace if leaks vacuum. Do not attempt to adjust the wastegate valve.

To install:

13. Prime the oil return line with clean engine oil. Replace all locking nuts. Before installing the threaded connection for the water inlet pipe, apply light oil to the inner surface of the pipe flange. Assemble the turbocharger and exhaust manifold.

14. Install the exhaust manifold using a new gasket.

15. Connect the water cooling lines, oil feed line and engine hanger.

16. If removed, install the power steering pump and bracket.

17. Install the heat shields, air outlet hose, wastegate hose and air intake bellows.

18. Install the oil dipstick tube and dipstick. Install the oxygen sensor.

19. Install the radiator assembly.

20. Fill the engine with fresh oil, fill the cooling system and reconnect the negative battery cable.

✱✱ WARNING

Operating the engine without the proper amount and type of engine oil will result in severe engine damage.

Intake Manifold

REMOVAL & INSTALLATION

2.0L (VIN F) Engine

1. Relieve the fuel system pressure.
2. Disconnect the negative battery cable.
3. Drain the cooling system.
4. Disconnect the accelerator cable, breather hose and air intake hose.
5. Disconnect the upper radiator hose, heater hose and water bypass hose.
6. Remove all vacuum hoses and pipes as necessary, including the brake booster vacuum line.
7. Disconnect the high pressure fuel line and the fuel return hose.
8. Tag and detach the electrical connectors from the oxygen sensor, coolant temperature sensor, thermo switch, idle speed control, EGR temperature sensor and spark plug wires.
9. Remove the fuel rail, fuel injectors, pressure regulator and insulators from the engine.
10. Remove the intake manifold bracket.
11. Disconnect the water hose at the throttle body, and the water inlet and heater connections.
12. Remove the thermostat housing, if necessary.
13. Disconnect the vacuum line from the power brake booster and PCV valve if connected.
14. Remove the intake manifold mounting bolts and remove the manifold.

To install:

15. Clean the gasket material from the intake mounting surface and manifold assembly. Check both surfaces for cracks or other damage. Check the intake manifold water passages and air passages for clogging. Clean if necessary.

16. Install a new gasket to the head, and install the manifold. Tighten the manifold in a crisscross pattern, starting from the inside and working outwards to 11–14 ft. lbs. (15–19 Nm).

17. Reinstall the fuel delivery pipe, injectors and pressure regulator, lubricating all seals lightly with oil. Tighten the retaining bolts to 84–108 inch lbs. (10–13 Nm).

18. Reinstall the thermostat housing, if removed, intake manifold brace bracket, and throttle body bracket.

19. Attach or install all hoses, cables and electrical connectors that were removed.

20. Fill the system with coolant.

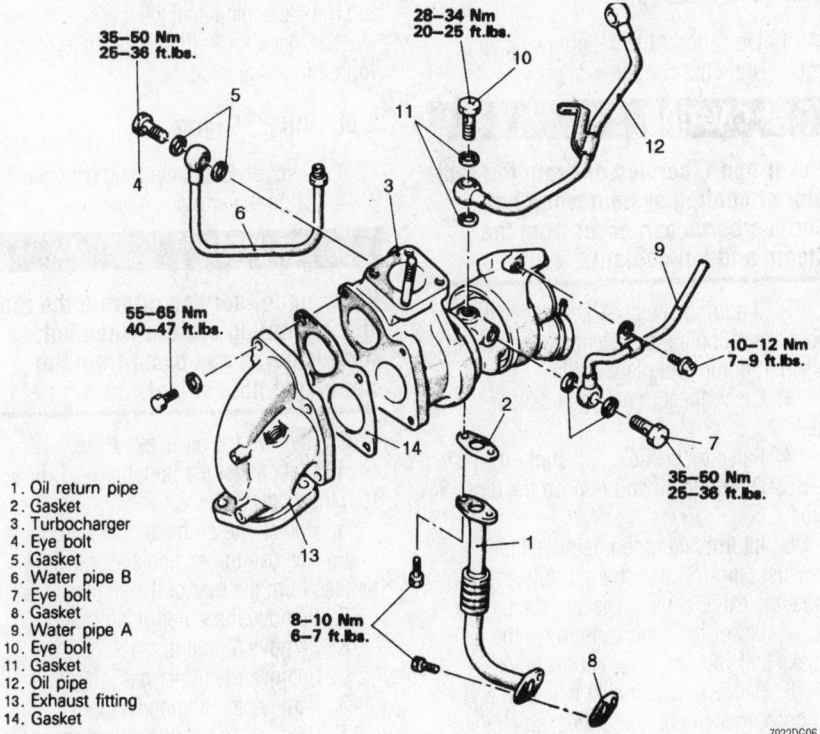

35–50 Nm
25–36 ft.lbs.

28–34 Nm
20–25 ft.lbs.

55–65 Nm
40–47 ft.lbs.

10–12 Nm
7–9 ft.lbs.

35–50 Nm
25–36 ft.lbs.

8–10 Nm
6–7 ft.lbs.

1. Oil return pipe
2. Gasket
3. Turbocharger
4. Eye bolt
5. Gasket
6. Water pipe B
7. Eye bolt
8. Gasket
9. Water pipe A
10. Eye bolt
11. Gasket
12. Oil pipe
13. Exhaust fitting
14. Gasket

7922DG06

Exploded view of the turbocharger assembly—2.0L (VIN F) engine

21. Reconnect the negative battery cable. Run the vehicle until the thermostat opens, fill the radiator completely. Check for leaks.

22. Adjust the accelerator cable. Check and adjust the ignition timing. Once the vehicle has cooled, recheck the coolant level.

2.0L (VIN Y) Engine

✳✳ CAUTION

The fuel injection system remains under pressure after the engine has been turned OFF. Properly relieve fuel pressure before disconnecting any fuel lines. Failure to do so may result in fire or personal injury.

1. Properly relieve the fuel system pressure.

2. Disconnect the negative battery cable.

3. Drain the engine coolant.

4. Remove the vacuum reservoir if equipped with cruise control.

5. Remove the air intake and breather hoses.

6. Remove the accelerator cable from the bracket.

7. Disconnect the engine harness retaining clips.

8. Disconnect the Manifold Absolute Pressure (MAP) sensor.

9. Disconnect the charge temperature sensor connector.

10. Disconnect the Throttle Position Sensor (TPS) and the Auto Idle Speed (AIS) motor connectors.

11. Position the engine control wiring harness out of the way.

12. Disconnect the alternator wiring harness.

13. Remove the PCV hose assembly.

14. Label and disconnect the vacuum hoses.

15. Detach the EGR pipe connection.

16. Disconnect the fuel lines from the fuel rail.

17. Remove the intake manifold stay and the engine hanger.

18. Remove the throttle body.

19. Remove the intake manifold plenum and gasket.

20. Disconnect the injector connectors.

21. Remove the fuel rail with the injectors.

22. Remove the intake manifold.

To install:

23. Use a new gasket and install the intake manifold.

24. Reinstall the fuel rail assembly and connect the injectors.

25. Use a new gasket and install the intake plenum.

26. Use a new gasket and install the throttle body.

27. Reinstall the intake manifold stay and the engine hanger.

28. Use a new O-ring and connect the fuel lines to the fuel rail.

29. Reconnect the EGR pipe.

30. Reconnect the vacuum hoses and the PCV hose assembly.

31. Reconnect the alternator wiring harness.

32. Reposition the engine control wiring harness and install the brackets and clips.

33. Reconnect the AIS motor and the TPS sensor.

34. Reconnect the vacuum hose to the throttle body.

35. Reconnect the MAP and the charge temperature sensor.

36. Reinstall the accelerator cable in the bracket and connect it to the throttle body.

37. Reconnect the breather hose and the air intake hose.

38. Reinstall the vacuum reservoir, if equipped.

39. Reconnect the negative battery cable.

40. Refill the engine with coolant.

41. Adjust the accelerator cable.

Exhaust Manifold

REMOVAL & INSTALLATION

2.0L (VIN F) Engine

1. Disconnect the battery negative cable. Drain the cooling system.

✳✳ CAUTION

Never open, service or drain the radiator or cooling system when hot; serious burns can occur from the steam and hot coolant.

2. If equipped with A/C, remove the condenser cooling fan. Remove the power steering pump and place aside.

3. Disconnect the oxygen sensor harness.

4. Raise the vehicle and support safely.

5. Drain the oil and remove the dipstick tube.

6. If turbo equipped, remove the exhaust pipe-to-turbocharger nuts and separate the exhaust pipe. Discard the gasket.

7. Lower the vehicle. Remove the air intake and vacuum hose connections.

8. Remove the exhaust manifold and turbocharger (if equipped) heat shields.

9. Remove the exhaust manifold-to-turbocharger attaching bolts and nut.

10. If applicable, remove the engine hanger, water and oil lines from the turbo.

11. Remove the exhaust manifold mounting nuts. Remove the exhaust manifold and gasket from the engine.

To install:

12. Clean all gasket material from the mating surfaces and check the manifold for damage.

13. Install new gaskets and install the manifold. Tighten the manifold-to-head nuts in a crisscross pattern to 18–22 ft. lbs. (25–30 Nm). Tighten the manifold-to-turbo nut and bolts to 40–47 ft. lbs. (55–65 Nm).

14. Reinstall the engine hanger, water and oil lines to the turbocharger.

15. Reinstall the heat shields.

16. Install the new gasket and connect the exhaust pipe.

17. Reinstall the condenser cooling fan and power steering pump. Reconnect the oxygen sensor harness.

18. Reinstall the oil level indicator and tube replacing the O-ring as required.

19. Fill the crankcase with clean oil and refill the cooling system.

✳✳ WARNING

Operating the engine without the proper amount and type of engine oil will result in severe engine damage.

20. Reconnect the negative battery cable and run the engine until the thermostat opens.

21. Check for fluid and exhaust leaks. Top off the engine coolant.

2.0L (VIN Y) Engine

1. Disconnect the negative battery cable.

2. Drain the engine coolant.

✳✳ CAUTION

Never open, service or drain the radiator or cooling system when hot; serious burns can occur from the steam and hot coolant.

3. Remove the air intake hose.

4. Disconnect the upper radiator hose from the water outlet.

5. Detach the air hose connection.

6. Remove the engine control wiring harness from the rear of the engine.

7. Remove the water pipe assembly.

8. Remove the oil dipstick.

9. Remove the upper heat shield.

10. Remove the engine hanger.

11. Disconnect the pulsed secondary air injection (check valve) valve from the exhaust pipe (manual transaxle only).

12. Disconnect the front exhaust pipe from the manifold.

13. Remove the lower heat shield.

14. Remove the exhaust manifold and gasket.

To install:

15. Use a new gasket and install the exhaust manifold. Tighten the nuts and bolts to 17 ft. lbs. (23 Nm).

16. Reinstall the lower heat shield.

17. Use a new gasket and connect the front exhaust pipe to the manifold.

18. On vehicles with manual transaxles, connect the pulsed secondary air injection valve to the exhaust pipe.

19. Reinstall the engine hanger.

20. Reinstall the upper heat shield.

21. Reinstall the dipstick and the water pipe.

22. Attach the engine wiring harness to the rear of the engine.

23. Reconnect the air hose and the upper radiator hose.

24. Reinstall the air intake hose.

25. Reconnect the negative battery cable.

26. Refill the engine with coolant, start the engine and check for leaks.

Front Crankshaft Seal

REMOVAL & INSTALLATION

2.0L (VIN F) Engine

➡Timing belt service is covered in the applicable unit repair section in the front of this manual.

1. Disconnect the negative battery cable.
2. Remove the timing belts.

➡When reusing the timing belts, be sure to mark the direction of rotation on the belt. This will extend belt life, ensuring the same direction of rotation.

3. Remove the crankshaft pulley.

4. Remove the timing belt crankshaft sprocket. If the sprocket is difficult to remove, an appropriate puller may be used.

5. Pry the seal from the bore using the proper tools.

To install:

6. Using a proper-sized driver, install a new seal.

7. Reinstall the crankshaft sprocket.

8. Reinstall the timing belt and remaining components.

9. Reinstall the engine undercover. Reconnect the negative battery cable, start the engine and check for leaks.

2.0L (VIN Y) Engine

➡Timing belt service is covered in the applicable unit repair section in the front of this manual.

1. Remove the timing belt.

2. Using Crankshaft Sprocket Removal tool MB995027 or equivalent, remove the crankshaft sprocket.

3. Using Crankshaft Oil Seal Removal tool MB995020 or equivalent, remove the oil seal. Be careful not the scratch the oil seal bore or the crankshaft sealing surface.

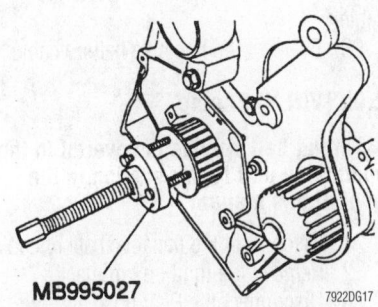

MB995027

To prevent damaging the end of the crankshaft, use the proper sprocket removal tool—2.0L (VIN Y) engine

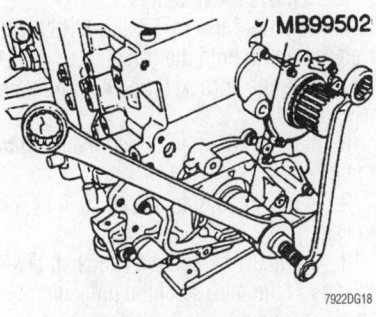

MB99502

Using the special tool to remove the front crankshaft seal—2.0L (VIN Y) engine

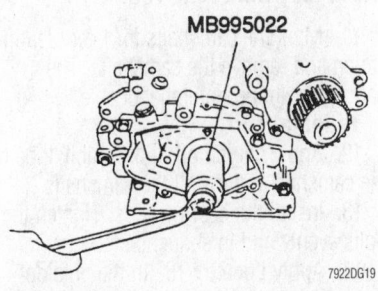

MB995022

To avoid damaging the front seal, use the Front Oil Seal Installer tool MB995022 or equivalent, as shown—2.0L (VIN Y) engine

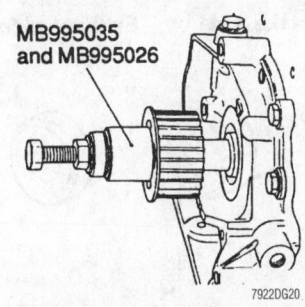

MB995035 and MB995026

Installing the front crankshaft sprocket using the **Crankshaft Sprocket Installer tools MB995035 and MB995026 or equivalents—2.0L (VIN Y) engine**

To install:

4. Apply clean engine oil to the oil seal. Using the Front Oil Seal Installer tool MB995022 or equivalent, install the oil seal.

5. Using Crankshaft Sprocket Installer tools MB995035 and MB995026 or equivalent, install the crankshaft sprocket.

6. Reinstall the timing belt using the recommended procedure.

Camshaft and Valve Lifters

REMOVAL & INSTALLATION

2.0L (VIN F) Engine

➡Timing belt service is covered in the applicable unit repair section in the front of this manual.

1. Disconnect the negative battery cable.

2. Disconnect the accelerator cable from the throttle body and remove the cable bracket from the intake plenum.

3. Remove the engine center cover.

4. Disconnect the spark plug cables from the spark plugs. Label them if necessary.

5. Disconnect the breather hose and the PCV hose from the rocker cover.

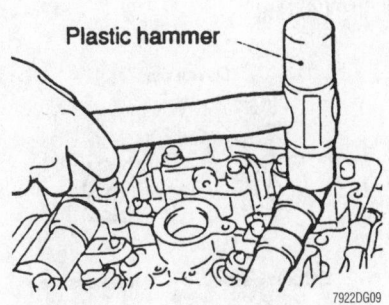

Plastic hammer

Tap the camshaft with a plastic hammer to loosen the bearing caps—2.0L (VIN F) engine

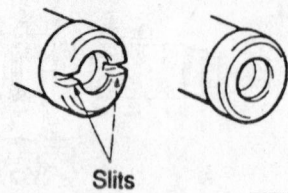

Intake side   **Exhaust side**

Slits

7922DG16

Identifying the rocker arm shafts—notice the slits in the intake side

6. Remove the rocker cover.

7. Position the No. 1 cylinder at TDC on the compression stroke.

8. Remove the timing belt.

9. Use a wrench on the hex shaped part of the camshaft to hold the cam and remove the sprockets.

10. Loosen the bearing cap bolts in two or three steps and remove the bearing caps. If the bearing caps are hard to remove, tap the rear of the camshaft with a plastic hammer.

11. Remove the camshaft(s) and the seals.

To install:

12. Apply engine oil or assembly lube to the camshafts and install them on the cylinder head.

✳✳ WARNING

If new camshaft(s) are being installed, remove the rocker arms and install the camshaft(s) and the bearing caps. Be sure the camshaft(s) can be turned by hand. After checking, remove the camshafts and install the rocker arms.

➥Bearing caps and rocker arms must be installed in the same location from which they were remove.

13. Install the bearing caps and tighten the bolts evenly in two or three steps to 14 ft. lbs. (20 Nm).

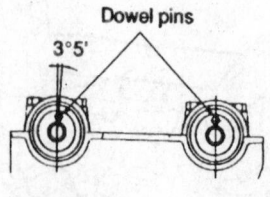

Dowel pins

3°5'

Exhaust side   **Intake side**

7922DG08

For installation, position the camshafts with the dowels facing up, as shown— 2.0L (VIN F) engine

14. Apply engine oil to the lip of the seal. Using MB998713 or a similar seal installer, install the front oil seal.

15. Install the camshaft sprockets.

16. Install the timing belt.

17. Apply sealant to the semi-circular packing and install it in the cylinder head.

18. Apply sealant to the lower part of the front and rear bearing caps where they meet the cylinder head. Use a new gasket and install the rocker cover.

19. Connect the PCV hose and the breather hose.

20. Connect the spark plug wires.

21. Install the center cover.

22. Install and adjust the accelerator cable.

23. Connect the negative battery cable.

2.0L (VIN Y) Engine

➥Timing belt service is covered in the applicable unit repair section in the front of this manual.

1. Disconnect the negative battery cable.

2. Remove the ignition coil pack.

3. Disconnect the PCV hose and the breather hose from the cylinder head cover.

4. Remove the semi-circular packing from the rear of the head.

5. Remove the camshaft position sensor.

6. Remove the timing belt.

7. Use tool MB990767 and MB998719 or equivalent to hold the camshaft sprockets and remove the sprocket mounting bolt and the sprocket.

8. Remove the bracket and the rear timing belt cover.

9. Remove the outside camshaft bearing cap.

10. Gradually loosen the camshaft bearing caps in the reverse of the tightening sequence, one camshaft at a time, and remove the bearing caps.

➥Keep the bearing caps in order. They must be installed in the location from which they were removed.

11. Mark the camshafts for later identification and remove the camshafts. The camshafts are not interchangeable.

To install:

12. Apply engine oil or assembly lube to the camshaft and install the camshafts.

13. Install the bearing caps. Tighten the bolts evenly and in sequence.

14. Apply Loctite 518® to the outside camshaft bearing caps and install them.

15. Install the camshaft oil seal.

16. Install the rear timing belt cover and the bracket.

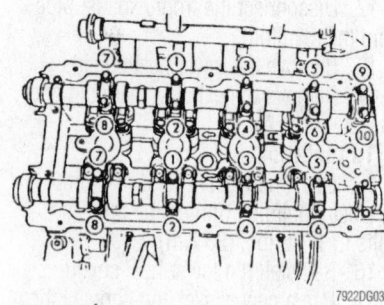

7922DG03

Camshaft bearing cap bolt tightening sequence—2.0L (VIN Y) engine

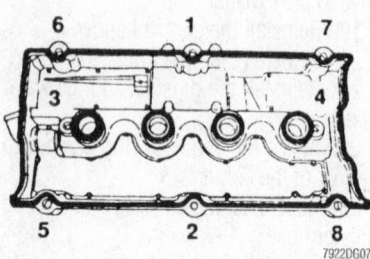

7922DG07

Cylinder head cover bolt tightening sequence—2.0L (VIN Y) engine

17. Use the special tools and install the camshaft sprockets.

18. Install the timing belt.

19. Apply Loctite 5699® or equivalent to the semi-circular packing and install it in the rear of the cylinder head.

20. Install the camshaft position sensor.

21. Install the cylinder head cover. Tighten the bolts in the proper sequence, evenly as follows:

    a. Step 1: 40 inch lbs. (4.5 Nm).

    b. Step 2: 80 inch lbs. (9.0 Nm).

    c. Step 3: 106 inch lbs. (12 Nm).

22. Install the air, breather and PCV hoses.

23. Install the coil pack.

24. Connect the negative battery cable.

Valve Lash

ADJUSTMENT

Valve clearance is not adjustable. Proper valve clearance is maintained by the hydraulic lash adjusters.

Oil Pan

REMOVAL & INSTALLATION

2.0L (VIN F) Engine

1. Disconnect the negative battery cable.

2. Raise the vehicle and support safely.

3. Remove the oil pan drain plug and drain the engine oil.

✳✳ CAUTION

The EPA warns that prolonged contact with used engine oil may cause a number of skin disorders, including cancer! You should make every effort to minimize your exposure to used engine oil. Protective gloves should be worn when changing the oil. Wash your hands and any other exposed skin areas as soon as possible after exposure to used engine oil. Soap and water, or waterless hand cleaner should be used.

4. Disconnect and lower the exhaust pipe from the engine manifold.
5. On AWD models, remove the transfer assembly and right driveshaft.
6. Using the appropriate equipment, support the weight of the engine and remove the crossmember.
7. Disconnect the return pipe for the turbocharger from the side of the oil pan.
8. Remove the oil pan bolts. Tap a thin prybar between the engine block and the oil pan.

➡**Do not use a chisel, screwdriver or similar tool when removing the oil pan. Damage to engine components may occur.**

9. Inspect the oil pan for damage and cracks. Replace if faulty. While the pan is removed, inspect the oil screen for clogging, damage and cracks. Replace if faulty.

To install:
10. Using a wire brush or other tool, scrape clean all gasket surfaces of the cylinder block and the oil pan so that all loose material is removed. Clean sealing surfaces of all dirt and oil.
11. Apply sealant around the gasket surfaces of the oil pan in such a manner that all bolt holes are circled and there is a continuous bead of sealer around the entire perimeter of the oil pan.

➡**The continuous bead of sealer should be applied in a bead approximately 0.16 in. (4mm) in diameter.**

12. Install the oil pan onto the cylinder block within 15 minutes after applying sealant. Reinstall the fasteners and tighten to 48–72 inch lbs. (6–8 Nm).
13. Reinstall the oil return pipe using a new gasket, if removed. Tighten the retainers to 62–88 inch lbs. (7–10 Nm).

14. On FWD models, reinstall the crossmember and tighten the mounting bolts to 72 ft. lbs. (100 Nm).
15. Reinstall the left member and tighten the forward retainer bolts to 72 ft. lbs. (100 Nm). Tighten the rearward left member bolts to 58 ft. lbs. (80 Nm).
16. Reinstall the transfer assembly and right driveshaft.
17. Reconnect the exhaust pipe from the engine manifold with a new gasket in place. Tighten the exhaust pipe-to-manifold flange nuts to 43 ft. lbs. (60 Nm).
18. Reinstall the oil drain plug and tighten to 33 ft. lbs. (42 Nm).
19. Lower the vehicle and fill the crankcase to the proper level with clean engine oil.

✳✳ WARNING

Operating the engine without the proper amount and type of engine oil will result in severe engine damage.

20. Reconnect the negative battery cable. Start the engine and check for leaks.

2.0L (VIN Y) Engine

1. Disconnect the negative battery cable.
2. Raise and safely support the vehicle.
3. Drain the engine oil.

✳✳ CAUTION

The EPA warns that prolonged contact with used engine oil may cause a number of skin disorders, including cancer! You should make every effort to minimize your exposure to used engine oil. Protective gloves should be worn when changing the oil. Wash your hands and any other exposed skin areas as soon as possible after exposure to used engine oil. Soap and water, or waterless hand cleaner should be used.

4. Remove the front exhaust pipe.
5. Remove the dipstick and tube assembly.
6. Remove the front plate.
7. Remove the oil pan mounting bolts.
8. Remove the oil pan and gasket.

To install:
9. Clean all traces of old gasket or sealer material from the oil pan and engine block mating surfaces.
10. Apply sealant at the point where the engine block meets the oil pump.
11. Use a new gasket and install the oil pan. Tighten the mounting bolts to 107 inch lbs. (12 Nm).

12. Reinstall the front plate.
13. Reinstall the front exhaust pipe.
14. Reinstall the dipstick and tube assembly.
15. Lower the vehicle.
16. Refill the crankcase with oil to the proper level.

✳✳ WARNING

Operating the engine without the proper amount and type of engine oil will result in severe engine damage.

17. Reconnect the negative battery cable.
18. Start the engine and check for leaks.

Oil Pump

REMOVAL & INSTALLATION

2.0L (VIN F) Engine

➡**Whenever the oil pump is disassembled or the cover removed, the gear cavity must be filled with petroleum jelly. This seals the pump and acts like a primer so the oil pump draws oil as soon as the engine turns. Do not use grease.**

1. Disconnect the negative battery cable. Rotate the engine so that No. 1 cylinder is at Top Dead Center (TDC) of its compression stroke. The timing marks should be aligned at this point.
2. Raise and safely support the vehicle.

✳✳ CAUTION

The EPA warns that prolonged contact with used engine oil may cause a number of skin disorders, including cancer! You should make every effort to minimize your exposure to used engine oil. Protective gloves should be worn when changing the oil. Wash your hands and any other exposed skin areas as soon as possible after exposure to used engine oil. Soap and water, or waterless hand cleaner should be used.

3. Drain the engine oil. Lower the vehicle.
4. Using the proper equipment, support the weight of the engine. Remove the front engine mount bracket and accessory drive belts.

➡**Timing belt service is covered in the applicable unit repair section in the front of this manual.**

5. Remove timing belt upper and lower covers.

6. Remove the timing belt and crankshaft sprocket.

7. Detach the electrical connector from the oil pressure sending unit and remove the oil pressure sensor. Remove the oil filter and bracket.

8. Remove the oil pan, oil screen and gasket.

9. Using Special Plug Removal tool MD998162, remove the plug cap in the engine front cover.

10. Remove the plug on the side of the engine block. Insert a suitable tool with a shaft diameter of 0.32 in. (8mm) into the plug hole. This will hold the silent shaft in position.

11. Remove the driven gear bolt that secures the oil pump driven gear to the silent shaft.

12. Remove and tag the front cover mounting bolts. Note the lengths of the mounting bolts as they are removed for proper installation.

13. Remove the front case cover and oil pump assembly. If necessary, the silent shaft can come out with the cover assembly.

14. Remove the oil pump cover, located on the back of the engine front cover. Remove the oil pump drive and driven gears.

15. After disassembling the oil pump, clean all components and remove all residual gasket material from the mating surfaces.

16. Assemble the oil pump gears into the front case and rotate it to ensure smooth rotation and no looseness. Be sure there is no ridge wear on the contact surface between the front case and the gear surface of the oil pump front cover.

To install:

17. Align the timing mark on the oil pump drive gear with that on the driven gear and install them into the engine front case. Apply engine oil to the gears.

18. Reinstall the oil pump cover and tighten the retainer bolts to 13 ft. lbs. (18 Nm).

19. Using the appropriate driver, install a new crankshaft seal into the front case.

20. Position a new front case gasket in place. Set Seal Guide tool MD998285 or equivalent on the front end of the crankshaft to protect the seal from damage. Apply a thin coat of oil to the outer circumference of the seal pilot tool.

21. Reinstall the front case assembly through a new front case gasket and temporarily tighten the flange bolts.

22. Mount the oil filter on the bracket with a new oil filter bracket gasket in place. Reinstall the four bolts with washers and tighten to 25 ft. lbs. (34 Nm).

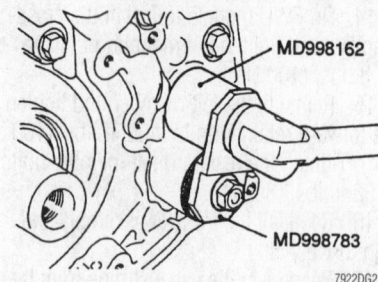

Using the special tool to tighten the plug cap—2.0L (VIN F) engine

23. Insert a suitable tool into the hole in the left side of the engine block to lock the silent shaft in place.

24. Secure the oil pump drive gear onto the left silent shaft by installing and tightening the driven gear bolt to 29 ft. lbs. (40 Nm).

25. Install a new O-ring in the groove in the front case and install the plug cap. Using the special tool MD998162 or equivalent, tighten the cap to 20 ft. lbs. (27 Nm).

26. Reinstall the oil screen in position with new gasket in place.

27. Clean both mating surfaces of the oil pan and the cylinder block. Apply sealant in the groove in the oil pan flange, keeping towards the inside of the bolt holes. The width of the sealant bead applied is to be about 0.16 in. (4mm) wide.

➡**After applying sealant to the oil pan, do not exceed 15 minutes before installing the oil pan.**

28. Install the oil pan to the engine and secure with the retainers. Tighten the bolts to 108 inch lbs. (12 Nm).

29. Reinstall the oil pressure gauge unit and the oil pressure switch. Reconnect the electrical harness.

30. Reinstall the oil cooler. Secure it with the oil cooler bolt tightened to 33 ft. lbs. (45 Nm).

31. Refill the crankcase. Reinstall new oil filter.

✳✳ WARNING

Operating the engine without the proper amount and type of engine oil will result in severe engine damage.

32. Reconnect the negative battery cable and start the engine. Verify oil pressure and inspect for leaks.

2.0L (VIN Y) Engine

1. Disconnect the negative battery cable.

2. Raise and safely support the vehicle.

✳✳ CAUTION

The EPA warns that prolonged contact with used engine oil may cause a number of skin disorders, including cancer! You should make every effort to minimize your exposure to used engine oil. Protective gloves should be worn when changing the oil. Wash your hands and any other exposed skin areas as soon as possible after exposure to used engine oil. Soap and water, or waterless hand cleaner should be used.

3. Drain the engine oil.

4. Remove the rear plate.

5. Remove the oil filter and adapter.

6. Remove the oil pan.

7. Remove the oil pick-up tube.

➡**Timing belt service is covered in the applicable unit repair section in the front of this manual.**

8. Remove the timing belt.

9. Using the Crankshaft Sprocket Removal tool MB995027 or equivalent, remove the crankshaft sprocket.

✳✳ WARNING

Do not nick the crankshaft sealing surface or the seal bore.

10. Using Crankshaft Oil Seal tool MB995020 or equivalent, remove the crankshaft oil seal.

11. Remove the oil pump mounting bolts.

12. Remove the oil pump.

To install:

13. Apply a bead of sealant to the sealing surface of the oil pump and install a new O-ring into the counterbore on the oil pump discharge passage.

14. Carefully install the oil pump on the crankshaft until seated to the engine block. Tighten the bolts to 17 ft. lbs. (23 Nm).

15. Install a new crankshaft oil seal in the oil pump.

16. Reinstall the crankshaft sprocket using the proper installation tools.

17. Reinstall the timing belt and related components.

18. Reinstall the oil pick-up tube.

19. Apply Loctite® 18718 or equivalent at the point where the oil pump meets the engine block.

20. Reinstall the oil pan using a new gasket. Tighten the mounting bolts to 108 inch lbs. (12 Nm).

21. Use a new O-ring and install the oil filter adapter onto the engine. Be sure the

roll pin aligns with the hole. Tighten the assembly to 40 ft. lbs. (55 Nm).

22. Reinstall a new oil filter.
23. Reinstall the rear plate.
24. Safely lower the vehicle to the floor.
25. Refill the engine with the proper amount of oil.

✳✳ WARNING

Operating the engine without the proper amount and type of engine oil will result in severe engine damage.

26. Start the engine and check for leaks.

Rear Main Seal

REMOVAL & INSTALLATION

1. Disconnect the negative battery cable.
2. Remove the transaxle and transfer case, if equipped, from the vehicle.
3. If equipped with an automatic transaxle, remove the flywheel/ring gear assembly from the crankshaft.
4. If equipped with a manual transaxle, remove the rear engine plate and the bell housing cover.
5. If the crankshaft rear oil seal case is leaking, remove it. Otherwise, just remove the oil seal. Some engines have a separator that should also be removed.

To install:
6. Lubricate the inner diameter of the new seal with clean engine oil.
7. Install the oil seal in the crankshaft rear oil seal case using tool MD998376 or an equivalent seal installer. Press the seal all the way in without tilting it. Force the oil separator into the oil seal case so the oil hole in the separator is downward.
8. Install the seal case on the crankshaft with a new gasket, if removed.
9. Install the flywheel or drive plate, transfer case and transaxle.
10. Connect the negative battery cable and check for leaks.

✳✳ WARNING

Operating the engine without the proper amount and type of engine oil will result in severe engine damage.

11. Fill the crankcase with clean oil to the proper level. Start the engine and check for leaks.

FUEL SYSTEM

Fuel System Service Precautions

Safety is an important factor when servicing the fuel system. Failure to conduct maintenance and repairs in a safe manner may result in serious personal injury. Maintenance and testing of the vehicle's fuel system components can be accomplished safely and effectively by adhering to the following rules and guidelines.

• To avoid the possibility of fire and personal injury, always disconnect the negative battery cable unless the repair or test procedure requires that battery voltage be applied.

• Always relieve the fuel system pressure prior to disconnecting any fuel system component (injector, fuel rail, pressure regulator, etc.), fitting or fuel line connection. Exercise extreme caution whenever relieving fuel system pressure to avoid exposing skin, face and eyes to fuel spray. Please be advised that fuel under pressure may penetrate the skin or any part of the body that it contacts.

• Always place a shop towel or cloth around the fitting or connection prior to loosening to absorb any excess fuel due to spillage. Ensure that all fuel spillage is quickly removed from engine surfaces. Ensure that all fuel soaked cloths or towels are deposited into a suitable waste container.

• Always keep a dry chemical (Class B) fire extinguisher near the work area.

• Do not allow fuel spray or fuel vapors to come into contact with a spark or open flame.

• Always use a back-up wrench when loosening and tightening fuel line connection fittings. This will prevent unnecessary stress and torsion to fuel line piping. Always follow the proper torque specifications.

• Always replace worn fuel fitting O-rings. Do not substitute fuel hose where fuel pipe is installed.

Fuel System Pressure

RELIEVING

✳✳ CAUTION

Observe all applicable safety precautions when working around fuel. Whenever servicing the fuel system,

always work in a well ventilated area. Do not allow fuel spray or vapors to come in contact with a spark or open flame. Keep a dry chemical fire extinguisher near the work area. Always keep fuel in a container specifically designed for fuel storage; also, always properly seal fuel containers to avoid the possibility of fire or explosion.

1. Remove the rear seat cushion.
2. Remove the protector, then detach the fuel pump connector.
3. Start the engine and allow it to run until it stops, due to lack of fuel. Turn the ignition switch to the **OFF** position.
4. Disconnect the negative battery cable.
5. Attach the fuel pump connector and install the protector.
6. Install the rear seat cushion.

Fuel Filter

REMOVAL & INSTALLATION

✳✳ CAUTION

Observe all applicable safety precautions when working around fuel. Whenever servicing the fuel system, always work in a well ventilated area. Do not allow fuel spray or vapors to come in contact with a spark or open flame. Keep a dry chemical fire extinguisher near the work area. Always keep fuel in a container specifically designed for fuel storage; also, always properly seal fuel containers to avoid the possibility of fire or explosion.

On most vehicles covered by this manual, the fuel filter is located in the engine compartment, mounted on the firewall. On some 1996–98 2.0L non-turbo engines, the fuel filter is mounted under the vehicle near the fuel tank.

✳✳ CAUTION

Do not use conventional fuel filters, hoses or clamps when servicing fuel injection systems. They are not compatible with the injection system and could fail, causing personal injury or damage to the vehicle. Use only hoses and clamps specifically designed for fuel injection systems.

1. Properly relieve the fuel system pressure.

2. If not already done, disconnect the negative battery cable.

3. On non-turbo engines, raise and safely support the vehicle to gain access to the filter.

➡**Wrap shop towels around the fitting that is being disconnected to absorb residual fuel in the lines.**

4. Cover the hose connection with shop towels to prevent any splash of fuel that could be caused by residual pressure in the fuel pipe line. Hold the fuel filter nut securely with a back-up wrench, then remove the eye bolt. Disconnect the high-pressure fuel line from the filter. Remove and discard the gaskets.

5. While holding the fuel filter nut securely with a back-up wrench, loosen the main pipe flare nut. Separate the flare nut connection from the filter. Remove and discard the gaskets.

6. If equipped with a fuel tank mounted filter, perform the following:

   a. Remove the eye bolt, gasket and connector.

   b. Remove the pressure regulator.

7. Remove the mounting bolts and remove the fuel filter. If necessary, remove the fuel filter bracket.

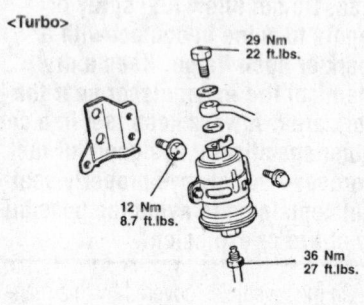

Exploded view of the fuel filter mounting—2.0L (VIN F) engine

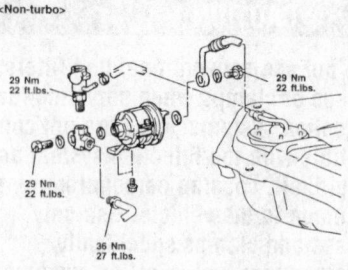

Exploded view of the fuel filter mounting—2.0L (VIN Y) engine

To install:

8. Install the filter in its bracket only finger-tight. Movement of the filter will ease attachment of the fuel lines.

➡**Be sure new O-rings are installed prior to assembly.**

9. Insert the main pipe at the connector part of the filter and manually screw in the main pipe's flare nut.

10. While holding the fuel filter nut with a back-up wrench, tighten the eye bolts to 22 ft. lbs. (30 Nm). Tighten the flare nut to 25 ft. lbs. (35 Nm), with a back-up wrench on the nut.

11. Tighten the filter mounting bolts to 10 ft. lbs. (14 Nm).

12. Connect the negative battery cable. Turn the key to the **ON** position to pressurize the fuel system and check for leaks.

13. If repairs of a leak are required, remember to release the fuel pressure before opening the fuel system.

Fuel Pump

REMOVAL & INSTALLATION

✳✳ CAUTION

The fuel injection system remains under pressure after the engine has been turned OFF. Properly relieve fuel pressure before disconnecting any fuel lines. Failure to do so may result in fire or personal injury.

1. Relieve the fuel system pressure.
2. Disconnect the negative battery cable.
3. Remove the rear seat cushion by pulling the seat stopper near the floor and lifting the cushion up.
4. Remove the inspection cover on the right side of the vehicle.
5. Disconnect the harness connector and the fuel lines.
6. Remove the fuel pump assembly from the tank. Use Fuel Pump Locking Ring

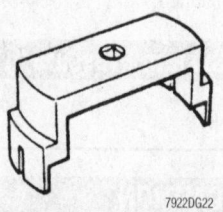

To service the fuel pump on AWD vehicles, a Fuel Pump Locking Ring Removal tool MB991480 or equivalent is needed

Removal tool MB991480 or equivalent to remove the locking ring on the AWD model.

To install:

7. Install the fuel pump in the tank.
8. Attach the hoses and the harness connector.
9. Reinstall the inspection cover.
10. Reinstall the rear seat.
11. Reconnect the negative battery cable.

DRIVE TRAIN

Transaxle Assembly

REMOVAL & INSTALLATION

Manual

1. Remove the battery and the air intake hoses.
2. Remove the battery tray and support.
3. If equipped with cruise control, remove the auto-cruise actuator and bracket.
4. Drain the transaxle and transfer case fluid, if equipped, into a suitable container.
5. Remove the charcoal canister and bracket.
6. Disconnect the shift and select cables from the transaxle.
7. Disconnect the back-up light switch and the vehicle speed sensor.
8. Remove the starter assembly.
9. Attach a support fixture to the engine and remove the transaxle mounting bolts.
10. Remove the rear roll stopper bracket mounting bolts.
11. Remove the transaxle mounting bracket nuts.
12. Raise and safely support the vehicle and remove the engine undercover.
13. If equipped with All Wheel Drive (AWD), remove the transfer case assembly.

✳✳ WARNING

Do not remove or install the axle shaft nut when the vehicle is on the floor or damage to the bearings will occur.

14. Remove the axle shafts.
15. Remove the slave cylinder from the bell housing but do not disconnect the fluid line. Position it out of the way.
16. Remove the bell housing cover and the right-hand center member stay (support).

17. Remove the center member.

18. Place a transmission jack under the transaxle and remove the transaxle mounting bolt.

19. Remove the transaxle mounting and lower the transaxle.

To install:

20. Raise the transaxle into position and install the transaxle mounting. Tighten the through-bolt to 50 ft. lbs. (69 Nm).

21. Reinstall the transaxle assembly mounting bolt. Tighten the bolt to 22–25 ft. lbs. (30–34 Nm).

22. Reinstall the center member assembly and the right-hand stay.

23. Reinstall the bell housing cover and the slave cylinder.

24. Reinstall the axle shafts. Be sure to install the washer in the proper direction.

25. Reinstall the engine undercover and lower the vehicle.

26. Reinstall the transfer case assembly, if removed.

27. Reinstall the transaxle mounting bracket mounting nuts.

28. Reinstall the rear roll stopper bracket mounting bolts.

29. Reinstall the transaxle assembly mounting bolts. Tighten the mounting bolts to 35 ft. lbs. (48 Nm).

30. Remove the engine support fixture.

31. Reinstall the starter assembly.

32. Attach the vehicle speed sensor and the back-up light connectors.

33. Reinstall the cruise control actuator, if removed.

34. Reinstall the battery tray support and the tray.

35. Reinstall the charcoal canister bracket and the canister.

36. Reinstall the air duct and the air cleaner assembly.

37. Refill the transaxle. On turbo models, fill with gear oil of classification GL-4 or higher, SAE 75W-85W or 75W-90. On non-turbo models, fill with Mopar MS9417 MTX Fluid P/N 4773167 or equivalent.

38. Refill the transfer case with gear oil of classification GL-4 or higher, SAE 75W-85W or 75W-90.

Automatic

1. Disconnect the negative battery cable. Remove the battery and battery tray.

2. On vehicles equipped with auto-cruise, remove the control actuator and bracket.

3. Drain the transaxle fluid.

4. Remove the air cleaner assembly, intercooler and air hose, as required.

5. Mark the shift cable. Remove the adjusting nut and disconnect the shift cable.

6. Remove the dipstick and tube assembly.

7. Detach the electrical connectors for the solenoid, neutral safety switch (inhibitor switch), the pulse generator kickdown servo switch and oil temperature sensor.

8. Disconnect the speedometer cable and oil cooler lines.

9. Disconnect the wires to the starter motor and remove the starter.

10. Remove the upper transaxle to engine bolts.

11. Support the transaxle and remove the transaxle mounting bracket.

12. Raise the vehicle and support safely. Remove the sheet metal undercover.

13. Disconnect the tie rod ends and ball joints from the steering knuckle.

14. Remove the halfshafts.

15. On AWD vehicles, disconnect the exhaust pipe and remove the transfer case.

16. Remove the lower bell housing cover and remove the special bolts holding the flexplate to the torque converter. To remove, turn the engine crankshaft with a box wrench and bring the bolts into a position appropriate for removal, one at a time. After removing the bolts, push the torque converter toward the transaxle so it doesn't stay on the engine allowing oil to pour out the converter hub or cause damage to the converter.

17. Remove the lower transaxle-to-engine bolts and remove the transaxle assembly.

To install:

18. After the torque converter has been mounted on the transaxle, install the transaxle assembly on the engine. Tighten the driveplate bolts to 34–38 ft. lbs. (46–53 Nm). Reinstall the bell housing cover.

19. On AWD models, install the transfer case and frame pieces. Reconnect the exhaust pipe using a new gasket.

20. Replace the circlips and install the halfshafts.

21. Reinstall the tie rods and ball joint.

22. Reinstall the transaxle mounting bracket.

23. Reinstall the underguard.

24. Reinstall the starter.

25. Reconnect the speedometer cable and oil cooler lines.

26. Reconnect the solenoid, neutral safety switch (inhibitor switch), the pulse generator kickdown servo switch and oil temperature sensor.

27. Reinstall the remaining components.

28. Refill with Dexron II, Mopar ATF Plus type 7176, Mitsubishi Plus ATF, or equivalent, automatic transaxle fluid. If the vehicle is equipped with AWD, check and fill the transfer case.

29. Start the engine and allow it to idle for 2 minutes. Apply the parking brake and move the selector through each gear position, ending in N (neutral). Recheck the fluid level and add if necessary. Fluid level should be between the marks in the HOT range.

Clutch

ADJUSTMENTS

Clutch Pedal Free-Play

1. Measure the clutch pedal height from the face of the pedal pad to the firewall.

2. Compare the measured value with the proper distance of 6.93–7.17 in. (176–182mm).

3. Measure the clutch pedal clevis pin play at the face of the pedal pad. Press the pedal lightly until resistance is met, and measure this distance. The clutch pedal clevis pin play should be within 0.04–0.12 in. (1–3mm).

4. If the clutch pedal height or clevis pin play are not within the standard values, adjust as follows:

   a. For vehicles without cruise control, turn and adjust the stop bolt so the pedal height is the standard value, then tighten the locknut.

   b. For vehicles with cruise control, detach the clutch switch connector and turn the switch to obtain the standard clutch pedal height. Then, lock by tightening the locknut.

   c. Turn the pushrod to adjust the clutch pedal clevis pin play to agree with the standard value and secure the pushrod with the locknut.

➡**When adjusting the clutch pedal height or the clutch pedal clevis pin play, be careful not to push the pushrod toward the master cylinder.**

   d. Check that when the clutch pedal is depressed all the way, the interlock switch changes from ON to OFF.

REMOVAL & INSTALLATION

2.0L (VIN F) Engine

1. Remove the transaxle from the vehicle.

2. Unscrew the fittings, then remove the clutch oil tubes.

3. Unfasten the retaining bolt, then remove the clutch oil fluid chamber.

4. Remove the clutch release (slave) cylinder union bolt, the gaskets and the union.

5. Remove the valve plate and valve plate spring.

6. Remove the clutch release (slave) cylinder.

7. Unfasten the clutch cover retaining bolts, slowly and evenly in a crisscross pattern, then remove the clutch cover. Remove the clutch disc.

8. Unfasten the return clip, then remove the clutch release bearing.

9. Slide the release fork in the direction of the arrow shown in the accompanying figure, then detach the fulcrum from the clip to remove the release fork. Be careful not to cause damage to the clip by pushing the fork in any other direction or removing it with force.

10. Remove the release fork boot and fulcrum.

To install:

11. Install the fulcrum and the release fork boot.

12. Apply grease to the areas shown on the clutch release fork, then install the fork.

13. Before installation, apply a suitable grease to the areas shown on the clutch release bearing.

14. Install the return clip.

15. Apply a suitable grease to the clutch disc splines, squeezing it in plate with a brush. Install the clutch disc, using a suitable guide to position the disc on the flywheel.

16. Install the clutch cover.

17. Install the clutch release (slave) cylinder.

18. Install the valve plate spring and valve plate.

19. Install the gasket, union, gasket and union bolt on the release cylinder.

20. Install the clutch oil fluid chamber. Uncap or unplug the clutch oil tubes, then attach them and tighten the fittings to 11 ft. lbs. (15 Nm).

21. Install the transaxle assembly.

2.0L (VIN Y) Engine

1. Remove the transaxle from the vehicle.

2. Unscrew the fittings, then remove the clutch oil tube. Plug or cap the ends to prevent contamination from entering the line.

3. Remove the release (slave) cylinder.

4. Unfasten the retaining bolts, then remove the clutch and flywheel assembly. Remove the drive plate.

5. Remove the clutch release bearing.

6. Remove the clutch release lever.

7. Remove the clutch control equip stud.

8. Remove the boot.

To install:

9. Install the boot assembly.

10. Install the clutch control equip stud.

11. Apply a suitable grease to the clutch release lever, as shown in the accompanying figure. Install the clutch release lever.

12. Install the clutch release bearing.

13. Apply a suitable grease to the clutch disc splines, squeezing it in place with a brush. Install the driven plate, and the clutch and flywheel assembly.

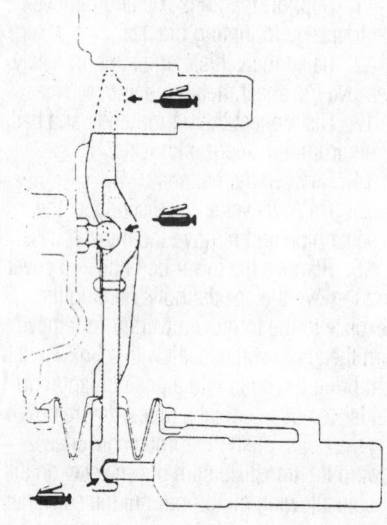

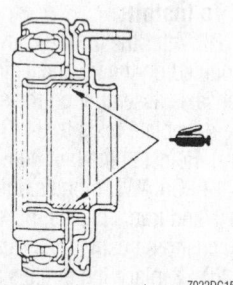

Lubrication points for the clutch release lever and bearing

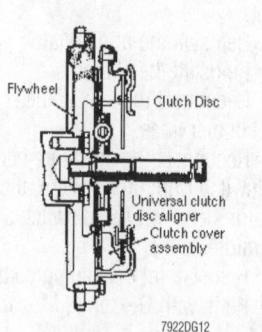

Cross-sectional view of proper clutch disk and pressure plate alignment, showing the aligner tool

14. Install the clutch release (slave) cylinder.

15. Install the clutch oil tube.

16. Install the transaxle in the vehicle.

Hydraulic Clutch System

BLEEDING

➡**Do not allow the reservoir to run out of fluid during bleeding.**

1. Fill the reservoir with clean brake fluid meeting DOT 3 specifications.

2. Attach a hose to the bleeder valve on the slave cylinder with the other end of the hose submerged in a container at least half full of fresh DOT 3 brake fluid from a sealed container.

3. Press the clutch pedal to the floor, then loosen the bleed screw on the slave cylinder.

4. Tighten the bleed screw and release the clutch pedal.

5. Repeat the procedure until the fluid is free of air bubbles.

Transfer Case Assembly

REMOVAL & INSTALLATION

1. Disconnect the battery negative cable.

2. Raise the vehicle and support safely. Drain the transfer oil.

3. Disconnect the front exhaust pipe.

4. Unbolt the transfer case assembly and remove by sliding it off the rear driveshaft. Be careful not to damage the oil seal in the transfer case output housing. Do not let the rear driveshaft hang; suspend it from a frame piece. Cover the opening in the transaxle and transfer case to keep oil from dripping and to keep dirt out.

To install:

5. Lubricate the driveshaft sleeve yoke and oil seal lip on the transfer extension housing with clean engine oil. Install the transfer case assembly onto the transaxle. Use care when installing the rear driveshaft to the transfer case output shaft.

6. Tighten the transfer case-to-transaxle bolts to 40–43 ft. lbs. (55–60 Nm) on manual transaxle vehicles, or 43–58 ft. lbs. (60–80 Nm) on automatic transaxle vehicles.

7. Install the exhaust pipe, using a new gasket.

8. Refill the transfer case with gear oil of classification GL-4 or higher, SAE 75W-85W or 75W-90. Check fluid level in transaxle and add as required.

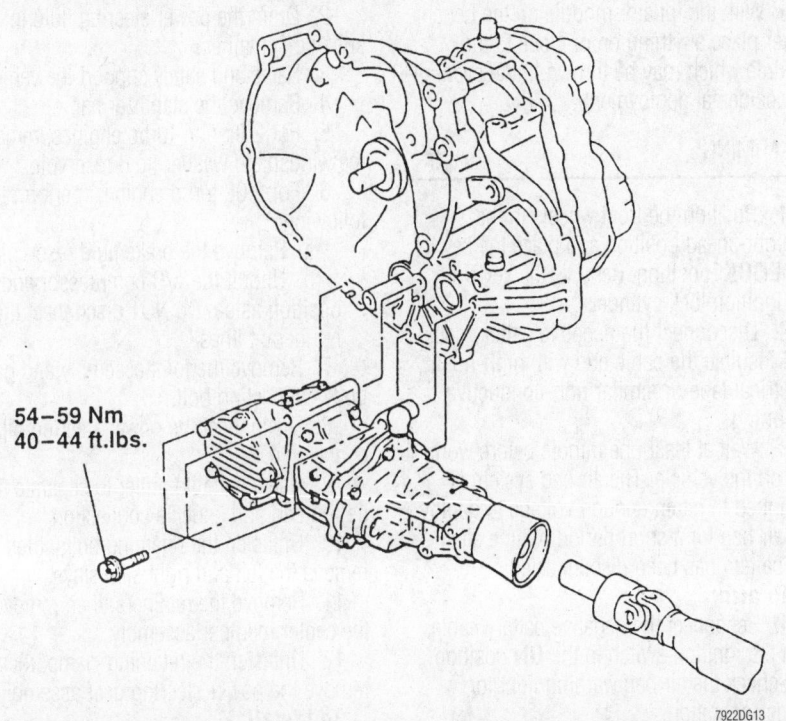

54–59 Nm
40–44 ft.lbs.

7922DG13

Exploded view of the transfer case assembly mounting—vehicles with manual transaxle shown

Halfshaft

REMOVAL & INSTALLATION

Front

✳✳ WARNING

If the vehicle is going to be rolled while the halfshafts are out of the vehicle, obtain two outer CV-joints or proper equivalent tools and install to the hubs. If the vehicle is rolled without the proper torque applied to the front wheel bearings, the bearings will no longer be usable.

1. Disconnect the negative battery cable.
2. Remove the cotter pin, halfshaft nut and washer.
3. Raise the vehicle and support safely. If removing the right halfshaft, remove the retainer bolt and the speedometer drive from the right extension housing.
4. Using the proper tool, disconnect the tie rod from the knuckle.
5. Disconnect the stabilizer link from the damper fork.
6. Disconnect the damper fork from the lateral lower arm.
7. Remove the lateral lower arm from the knuckle.

✳✳ WARNING

Use of improper methods of joint separation can result in damage to the joint, leading to possible failure.

8. On halfshafts with an inner shaft (AWD vehicles), remove the center support bearing bracket bolts and washers. Then, remove the halfshaft by setting up a puller on the outside wheel hub and pushing the halfshaft from the front hub. Then, tap the shaft union at the joint case with a plastic hammer to remove the halfshaft and inner shaft from the transaxle.
9. On one piece halfshafts (FWD vehicles), remove the halfshaft from the hub/knuckle by setting up a puller on the outside wheel hub and pushing the halfshaft from the front hub. After pressing the outer shaft, insert a prybar between the transaxle case and the halfshaft and pry the shaft from the transaxle.

➡**Do not pull on the shaft. Doing so damages the inboard joint. Do not insert the prybar too far or the oil seal in the case may be damaged.**

To install:

10. Inspect the halfshaft boot for damage or deterioration. Check the ball joints and splines for wear.
11. Replace the circlips on the ends of the halfshafts.

12. Insert the halfshaft into the transaxle. Be sure it is fully seated.
13. Pull the strut assembly outward and install the other end of the halfshaft into the hub.
14. Reinstall the center bearing bracket bolts and tighten to 33 ft. lbs. (45 Nm), if equipped.
15. Reinstall the washer so the chamfered edge faces outward. Reinstall the halfshaft nut and tighten temporarily.
16. Reinstall the tie rod end and ball joint to the steering knuckle.
17. Reinstall the wheel and lower the vehicle to the floor. Tighten the axle nut with the brakes applied. Tighten the nut to 145–188 ft. lbs. (200–260 Nm).
18. Reinstall a new cotter pin and bend to secure.

Rear

1. Raise and safely support the vehicle.
2. Remove the rear wheel(s).
3. If equipped with ABS, remove the wheel speed sensor.
4. Remove the brake caliper and rotor or brake drum.
5. Remove the parking brake shoes or the shoe and lever assembly.
6. Remove the parking brake cable from the backing plate.
7. If equipped with drum brakes, disconnect the brake line from the wheel cylinder.
8. Disconnect the shock absorber from the knuckle.
9. Disconnect the trailing arm and the lower arm from the knuckle.
10. Disconnect the toe control arm from the knuckle.
11. Remove the cotter pin, nut and washer from the axle shaft.
12. Remove the differential mount support.
13. Pull the lower end of the knuckle outward and pry the axle shaft out of the differential housing.
14. Remove the axle shaft from the hub assembly.

To install:

15. Install the axle shaft in the hub assembly.
16. Install a new circlip on the inner shaft, and install the shaft in the differential.
17. Reinstall the differential mount support.
18. Reinstall the washer on the axle shaft in the correct direction.
19. Reinstall the nut and tighten to 145–188 ft. lbs. (196–255 Nm). If the hole

for the cotter pin does not line up, tighten the nut up to 188 ft. lbs. (255 Nm) and install the cotter pin in the first hole that lines up.

20. Reconnect the toe control arm, trailing arm and the lower arm to the knuckle.

21. Reinstall the lower shock mount to the knuckle.

22. Reconnect the brake line to the wheel cylinder, if removed.

23. Reinstall the parking brake cable through the backing plate.

24. Assemble the remaining brake components.

25. Reinstall the ABS wheel speed sensor, if removed.

26. Reinstall the wheel.

27. If equipped with drum brakes, bleed the brake system.

STEERING AND SUSPENSION

Air Bag

✳✳ CAUTION

Some vehicles are equipped with an air bag system, also known as the Supplemental Inflatable Restraint (SIR) or Supplemental Restraint System (SRS). The system must be disabled before performing service on or around system components, steering column, instrument panel components, wiring and sensors. Failure to follow safety and disabling procedures could result in accidental air bag deployment, possible personal injury and unnecessary system repairs.

PRECAUTIONS

Several precautions must be observed when handling the inflator module to avoid accidental deployment and possible personal injury.

• Never carry the inflator module by the wires or connector on the underside of the module.

• When carrying a live inflator module, hold securely with both hands, and ensure that the bag and trim cover are pointed away from you.

• Place the inflator module on a bench or other surface with the bag and trim cover facing up.

• With the inflator module on the bench, never place anything on or close to the module which may be thrown in the event of an accidental deployment.

DISARMING

1. Position the front wheels in the straight-ahead position and place the key in the **LOCK** position. Remove the key from the ignition lock cylinder.

2. Disconnect the negative battery cable and insulate the cable end with high-quality electrical tape or similar non-conductive wrapping.

3. Wait at least one minute before working on the vehicle. The air bag system is designed to retain enough voltage to deploy the air bag for a short period of time after the battery has been disconnected.

To arm:

4. Reconnect the negative battery cable, turn the ignition switch to the **ON** position and check the air bag warning light for proper operation.

Power Rack and Pinion Steering Gear

REMOVAL & INSTALLATION

1. Disconnect the negative battery cable.

2. Drain the power steering fluid into a suitable container.

3. Raise and safely support the vehicle.

4. Remove the stabilizer bar.

5. For 2.0L non-turbo engines, remove the windshield washer fluid reservoir.

6. For 2.0L turbo engines, perform the following:

a. Remove the brake fluid reservoir.

b. Unbolt the A/C compressor and position aside. Do NOT disconnect the refrigerant lines.

7. Remove the joint assembly and gear body connecting bolt.

8. Disconnect the power steering pipe connection.

9. Use a suitable puller to separate the tie rod end and knuckle connection.

10. Unfasten the retaining bolts, then remove the left and right side stays.

11. Remove the retainers, then remove the center member assembly.

12. Unfasten the retaining clamp, then remove the power steering gear assembly.

To install:

13. Install the power steering gear assembly onto the vehicle, and tighten the mounting bolts to specification as shown.

14. Connect the tie rod ends to the knuckle.

15. Installation is the reverse of the removal procedure.

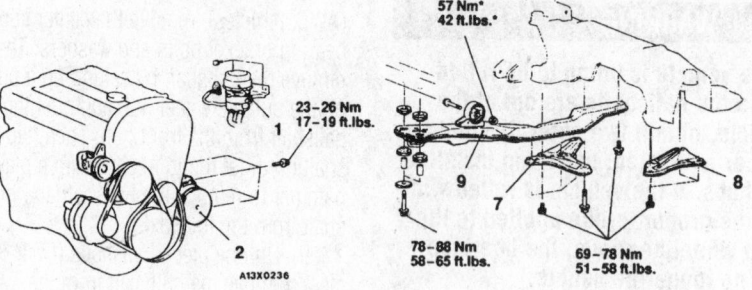

1. Brake fluid reservoir assembly
2. A/C compressor
3. Joint assembly and gear box connecting bolt
4. Power steering pipe connection
5. Cotter pin
6. Tie-rod end and knuckle connection
7. Stay (L.H.)
8. Stay (R.H.)
9. Centermember assembly
10. Clamp
11. Gear box assembly
12. Return tube

NOTE
The fasteners marked * should be temporarily tightened before they are finally tightened once the total weight of the engine has been placed on the vehicle body.

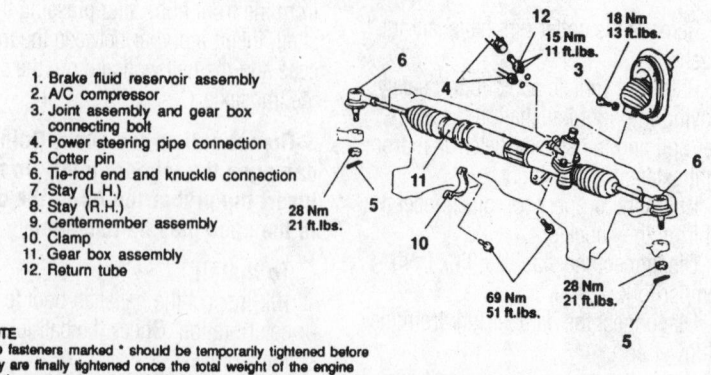

Exploded view of the power steering gear assembly mounting, showing related components

7922DG14

16. Refill the reservoir with power steering fluid and bleed the system.

17. Check the front end alignment, and adjust if necessary.

Strut

REMOVAL & INSTALLATION

Front

1. Remove the top strut assembly mounting nuts from inside the engine compartment.

2. Raise and safely support the vehicle. Remove the wheel and tire assembly.

3. Remove the stabilizer link mounting nut.

4. Unfasten the strut assembly lower mounting bolts.

5. Remove the damper fork mounting bolt, then remove the damper fork from the vehicle.

6. Remove the strut assembly by maneuvering it out of the vehicle.

7. Installation is the reverse of the removal procedure. Be sure to tighten the retainers to the specifications shown in the accompanying figure.

Rear

1. Remove the service lid in the luggage compartment.

2. Remove the cap and flange nuts securing the upper mounting bracket to the

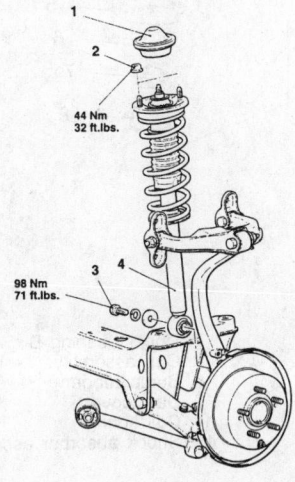

44 Nm
32 ft.lbs.

98 Nm
71 ft.lbs.

1. Cap
2. Flange nuts
3. Bolt
4. Shock absorber

7922DG26

Exploded view of the rear strut mounting

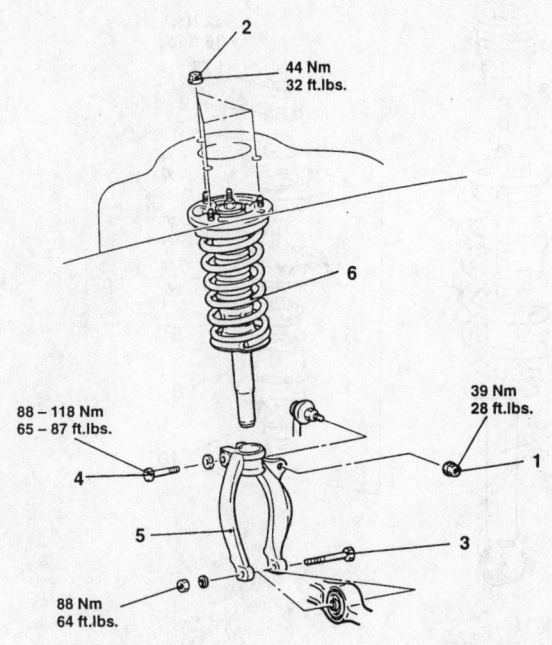

2 — 44 Nm 32 ft.lbs.

6

88 – 118 Nm
65 – 87 ft.lbs.

39 Nm
28 ft.lbs.

88 Nm
64 ft.lbs.

1. Stabilizer link mounting nut
2. Shock absorber upper mounting nuts
3. Shock absorber lower mounting bolt
4. Damper fork mounting bolt
5. Damper fork
6. Shock absorber assembly

7922DG23

Exploded view of the front strut mounting

body. Do not remove the larger nut in the center of the strut assembly.

3. Raise and safely support the vehicle.

4. Remove the bolt attaching the lower end of the strut to the knuckle and remove the strut from the vehicle.

To install:

5. Reinstall the upper bracket of the shock to the vehicle. Tighten the mounting nuts to 32 ft. lbs. (44 Nm).

6. Raise the suspension up with a jack or adjustable stand to align the strut lower mounting holes.

7. Reinstall the lower mounting bolt. Tighten the bolt to 71 ft. lbs.

8. Remove the jack or stand and safely lower the vehicle to the floor.

9. Reinstall the cap and service lid.

OVERHAUL

Front

1. Remove the shock and strut assembly from the vehicle.

❊❊ CAUTION

Do not use air tools to tighten the compressor tool bolt.

2. Use the proper spring compressing tools (MB991237 and MB991239) to compress the coil spring. Be sure to install the tools evenly so the maximum length will be attained within the installation range.

3. While holding the piston rod, remove the self-locking nut.

4. Remove the washer.

5. Remove the upper bushing A, the upper bracket assembly and the upper spring pad.

6. Remove the collar, upper bushing B, cup assembly, rubber bumper and dust cover.

7. Separate the coil spring and strut assemblies.

To assemble:

8. Use the compressor tools to compress the coil spring, then install it to the strut. Align the edge of the coil spring to the stepped portion of the strut spring seat.

9. Install the dust cover, rubber bumper, cup, upper bushing B, collar and upper spring pad.

10. Install the upper bracket assembly, installing it so the position of the three bolts are in the proper orientation with the damper fork, as shown in the accompanying figure.

11. Install the other upper bushing.

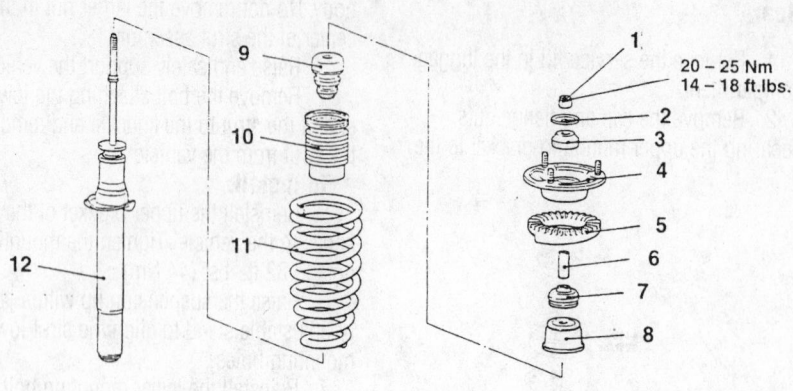

Disassembly steps
1. Self-locking nut
2. Washer
3. Upper bushing A
4. Upper bracket assembly
5. Upper spring pad
6. Collar

7. Upper bushing B
8. Cup assembly
9. Bump rubber
10. Dust cover
11. Coil spring
12. Shock absorber assembly

7922DG28

Exploded view of the front strut assembly

12. Install the washer and the self-locking nut. Temporarily tighten the self-locking nut, then remove the spring compressor tools and tighten the nut to 14–18 ft. lbs. (20–25 Nm) using a torque wrench. Do not use air tools to tighten the self-locking nut!

13. Install the strut assembly in the vehicle.

Rear

1. Remove the strut from the vehicle.
2. Use a coil spring compressor to compress the spring.
3. While holding the piston rod, remove the self locking nut.
4. Remove the upper bracket assembly and spring pad.
5. Remove the collar, upper bushing, cup assembly, bump rubber and dust cover.
6. Remove the coil spring from the strut.

To assemble:

7. Align the end of the coil spring with the stepped part of the spring seat and install the compressed coil spring on the strut.
8. Reinstall the dust cover, bump rubber, cup assembly, upper bushing, collar, upper spring pad and bracket assembly on the strut.
9. Reinstall the upper bushing and washer on the piston rod.
10. Install a new self locking nut on the piston rod. Temporarily tighten the nut.

11. Carefully remove the spring compressor from the spring. Tighten the self locking nut to 16 ft. lbs. (25 Nm).

12. Install the strut assembly in the vehicle.

Upper Ball Joint

REMOVAL & INSTALLATION

The upper ball joint is an integral part of the upper control arm. If the upper ball joint is to be serviced, the upper control arm will have to be replaced.

Lower Ball Joint

REMOVAL & INSTALLATION

The lower ball joint is an integral part of the lower control arm assembly, and cannot be serviced separately. A worn or damaged ball joint, requires replacement of lower control arm assembly.

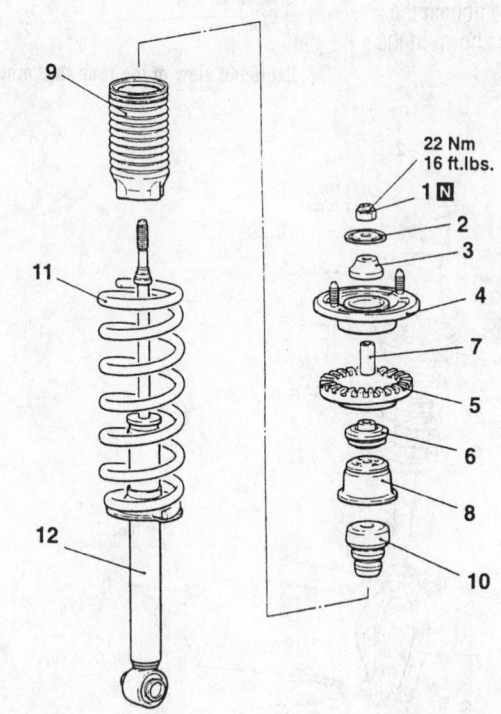

Disassembly steps
1. Self-locking nut
2. Washer
3. Upper bushing A
4. Upper bracket assembly
5. Upper spring pad
6. Upper bushing B

7. Collar
8. Cup assembly
9. Dust cover
10. Bump rubber
11. Coil spring
12. Shock absorber assembly

7922DG27

Exploded view of the rear strut assembly

Wheel Bearings

ADJUSTMENT

Front

The front wheel bearing is a sealed and not adjustable. If wheel bearing shows signs of play, the hub assembly must be replaced.

Rear

The wheel bearing play is not adjustable. If the wheel bearing play is not within specifications, the hub assembly must be replaced.

FWD VEHICLES

1. Release the parking brake and remove the brake drum.
2. If equipped with rear disc brakes, remove the caliper assembly and the brake disc (rotor).
3. Place a dial gauge against the hub surface, then move the hub in the axial direction and check whether or not there is end-play. Specification is 0.002 in (0.05mm).
4. To check for rotary sliding resistance: turn the hub a few times to seat the bearing. Wind a rope around the hub bolts and turn the hub by pulling at a 90° angle with a spring balance. Measure to determine whether or not the rotary sliding resistance of the rear hub is at the limit value. Specification is: 3.9 lbs. (1.8 kg) or less.

AWD VEHICLES

1. Remove the caliper assembly and the brake disc (rotor).
2. Place a dial gauge against the hub surface, then move the hub in the axial direction and check whether or not there is end-play. Specification is 0.002 in (0.05mm).

REMOVAL & INSTALLATION

Front

➡ **The front hub assembly is a sealed unit and should not be disassembled.**

1. Raise and safely support the vehicle.
2. Remove the wheel and tire assembly.
3. Remove and discard the halfshaft nut cotter pin.
4. Remove the halfshaft nut, holding the rotor in place with a suitable tool.
5. If equipped with ABS, remove the front wheel speed sensor.
6. Unbolt the caliper assembly and suspend out of the way with a piece of wire. Do not disconnect the fluid line.

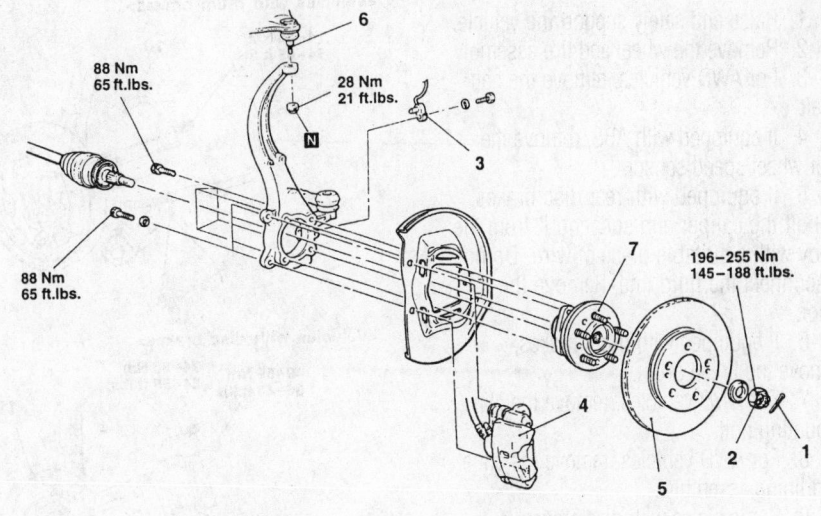

88 Nm
65 ft.lbs.

28 Nm
21 ft.lbs.

88 Nm
65 ft.lbs.

196–255 Nm
145–188 ft.lbs.

Removal steps
1. Cotter pin
2. Drive shaft nut
3. Front wheel speed sensor
   <Vehicles with ABS>
4. Caliper assembly
5. Brake disc
6. Upper arm ball joint and knuckle connection
7. Front hub assembly

7922DG25

Exploded view of the front hub assembly

7. Use a suitable puller to separate the upper control arm ball joint from the steering knuckle.
8. Unfasten the retaining bolts, then remove the front hub assembly. Shift the knuckle to the outside in order to keep the clearance between the front hub mounting bolts and halfshaft. Be careful not to damage the ball joint boot. If equipped with ABS, be careful not to damage the rotor.
9. If steering knuckle removal is necessary proceed with the following steps:
   a. Unfasten the retaining bolts, then remove the dust shield.
   b. Remove and discard the cotter pin, then separate the tie rod end from the steering knuckle using a puller.
   c. Use a suitable puller to separate the compression and lateral lower control arm ball joints from the steering knuckle.
   d. Unfasten the connecting nut and bolt of the damper fork and lateral lower control arm.
   e. Remove the steering knuckle from the vehicle.

To install:
10. If the steering knuckle was removed, perform the following:
   a. Position the steering knuckle in the vehicle. Install the connecting bolt and nut and secure hand-tight.

   b. Attach the lateral and compression lower control arm ball joints into the steering knuckle.
   c. Fasten the tie rod end ball joint to the steering knuckle, then install the retaining nut and tighten to 17–25 ft. lbs. (24–33 Nm). Install a new cotter pin and bend over securely.
   d. Install the dust shield and secure with the retaining bolt. 6.5 ft. lbs. (8.5 Nm).
11. Position the front hub assembly, then secure with the retaining bolts. Tighten to 65 ft. lbs. (88 Nm).
12. Fasten the upper control arm ball joint to the steering knuckle.
13. Install the brake disc (rotor) and caliper.
14. If equipped, install the front wheel speed sensor.
15. Install the halfshaft nut. Hold the rotor in position while tightening the halfshaft nut to 145–188 ft. lbs. (196–255 Nm). If the cotter pin holes does not match, tighten the nut up to 188 ft. lbs. (225 Nm) maximum. Install a new cotter pin in the first matching holes, then bend it over securely.
16. Install the wheel and tire assembly, then carefully lower the vehicle.

Rear

1. Raise and safely support the vehicle.
2. Remove the wheel and tire assembly.
3. For AWD vehicles, remove the half-shaft.
4. If equipped with ABS, remove the rear wheel speed sensor.
5. If equipped with rear disc brakes, unbolt the caliper and suspend it from the body with a suitable piece of wire. Do not disconnect the fluid line. Remove the rotor.
6. If equipped with drum brakes, remove the brake drum.
7. For FWD vehicles, remove the clip mounting bolt.
8. For AWD vehicles, remove the shoe and lining assembly.
9. For vehicles with disc brakes, remove the parking brake shoe and lining assembly.
10. For AWD vehicles, remove the parking brake cable clip, then detach the cable.
11. Remove the retaining bolts, then remove the rear hub and bearing assembly.
12. For FWD vehicles equipped with ABS, use a suitable inner shaft remover to press off the ABS rotor.

To install:

13. If removed, install the ABS rotor.
14. Install the rear hub and bearing and secure with the mounting bolts.
15. For AWD vehicles, attach the parking brake cable, then install the retaining clip.
16. If equipped with disc brakes, install the parking brake shoe and lining assembly.
17. For AWD vehicles, install the shoe and lining assembly.
18. For FWD vehicles, install the clip mounting bolt.
19. If equipped with drum brakes, install the brake drum.

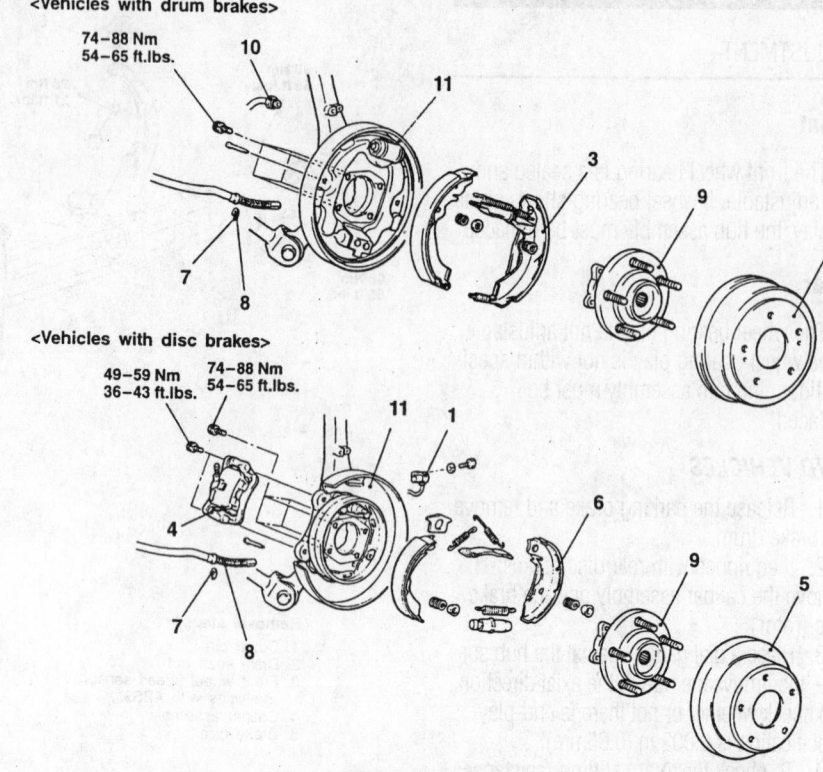

<Vehicles with drum brakes>
74–88 Nm
54–65 ft.lbs.

<Vehicles with disc brakes>
49–59 Nm / 74–88 Nm
36–43 ft.lbs. / 54–65 ft.lbs.

1. Rear wheel speed sensor <Vehicles with ABS>
2. Brake drum
3. Shoe and lever assembly
4. Caliper assembly
5. Brake disc
6. Shoe and lining assembly
7. Clip
8. Parking brake cable
9. Rear hub assembly
10. Brake pipe connection
11. Dust seal

7922DG24

Exploded view of the rear hub assembly

20. If equipped with disc brakes, install the rotor, then position the caliper and secure with the retaining bolts.
21. If equipped, install the rear wheel speed sensor.

22. For AWD vehicles, install the half-shaft.
23. Install the front wheel and tire assembly, then carefully lower the vehicle.

CHRYSLER CORPORATION

Dodge/Plymouth-Neon

DRIVE TRAIN 15-16
ENGINE REPAIR 15-2
FUEL SYSTEM 15-14
STEERING AND
SUSPENSION 15-20

A
Air Bag.........................15-20
  Disarming.....................15-21
  Precautions 15-20

C
Camshaft and Lifters.............15-8
  Removal & Installation 15-8
Clutch..........................15-18
  Adjustment....................15-18
  Removal & Installation 15-18
Coil Spring.....................15-24
  Removal & Installation 15-24
Cylinder Head...................15-3
  Removal & Installation 15-3

E
Engine Assembly 15-2
  Removal & Installation 15-2
Exhaust Manifold.................15-7
  Removal & Installation 15-7

F
Front Crankshaft Seal 15-8
  Removal & Installation 15-8
Fuel Filter 15-14
  Removal & Installation 15-14
Fuel Pump 15-15
  Removal & Installation 15-15
Fuel System Pressure 15-14
  Relieving 15-14
Fuel System Service
  Precaution 15-14

H
Halfshafts......................15-19
  Removal & Installation 15-19

I
Ignition Timing 15-2
  Adjustment....................15-2
Intake Manifold 15-6
  Removal & Installation 15-6

L
Lower Ball Joint.................15-25
  Removal & Installation 15-26

O
Oil Pan.........................15-11
  Removal & Installation 15-11

Oil Pump 15-12
  Removal & Installation 15-12

R
Rack and Pinion Steering Gear 15-21
  Removal & Installation 15-21
Rear Main Seal 15-13
  Removal & Installation 15-13
Rocker Arm/Shafts 15-5
  Removal & Installation 15-5

S
Strut...........................15-22
  Removal & Installation 15-22

T
Transaxle Assembly 15-16
  Removal & Installation 15-16

V
Valve Lash 15-11
  Adjustment....................15-11

W
Water Pump 15-3
  Removal & Installation 15-3
Wheel Bearings...................15-26
  Adjustment....................15-26
  Removal & Installation 15-26

ENGINE REPAIR

➥Disconnecting the negative battery cable on some vehicles may interfere with the functions of the on board computer systems and may require the computer to undergo a relearning process, once the negative battery cable is reconnected.

Ignition Timing

ADJUSTMENT

Ignition timing is controlled by the Powertrain Control Module (PCM). No adjustment is necessary or possible.

Engine Assembly

REMOVAL & INSTALLATION

➥After all components are installed on the engine, a DRB scan tool is necessary to perform the camshaft and crankshaft timing relearn procedure.

1. Properly relieve the fuel system pressure.
2. Disconnect the negative, then the positive battery cables. Remove the battery and the battery tray. Set the PCM aside.
3. Drain the cooling system into a suitable container.
4. Remove the upper radiator hose, radiator and fan module assembly. Remove the lower radiator hose.
5. If equipped with an automatic transaxle, disconnect and plug the transaxle cooler lines.
6. Disconnect the clutch cable (manual transaxles) and transmission shift linkage.
7. Disconnect the throttle body linkage.
8. Detach the engine wiring harness.
9. Disconnect the heater hoses.
10. Raise and safely support the vehicle, then remove the right inner splash shield.
11. Drain the engine oil.
12. Remove the accessory drive belts.
13. Remove the halfshafts.
14. Disconnect the exhaust pipe from the manifold.
15. Support the engine and transaxle assembly with a suitable jack, then remove the front engine mount.
16. On 1996–99 vehicles equipped with manual transaxle, remove the power hop damper.
17. Carefully lower the vehicle.
18. Remove the air cleaner assembly.

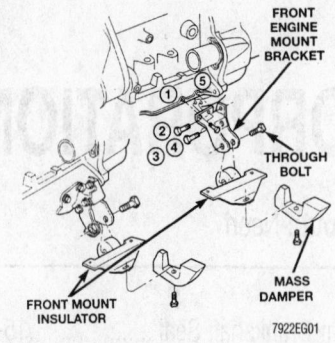

Front engine mount location and bolt identification

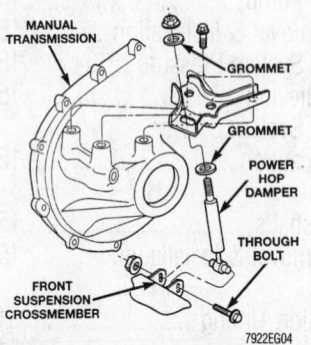

On vehicles with manual transaxles, you must remove the power hop damper

19. Unbolt the power steering pump and reservoir and position them aside.
20. If equipped, remove the A/C compressor.
21. Remove the ground straps to body.
22. Raise the vehicle enough to allow a suitable engine dolly and cradle to be placed under the engine.
23. Loosen the engine support posts in order to allow movement for positioning onto the engine locating holes and flange on the engine bedplate. Carefully lower the vehicle and position the cradle until the engine is resting on the support posts. Tighten the mounts to the cradle frame. This will keep the support posts from moving when removing or installing the engine and transmission.
24. Install safety straps around the engine to the cradle; tighten straps and lock them into position.
25. Raise the vehicle enough to see if the straps are tight enough to hold the cradle assembly to the engine.
26. Lower the vehicle so the weight of the engine and transaxle ONLY is on the cradle.
27. Remove the engine and transaxle mount through-bolts.
28. Raise the vehicle slowly, it might be necessary to move the engine/transaxle

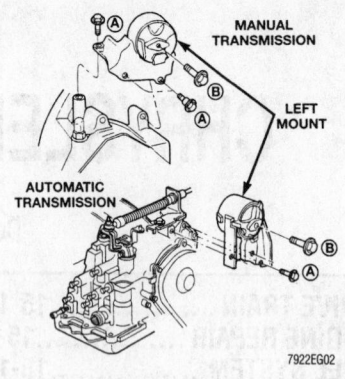

Exploded view of the left engine mount

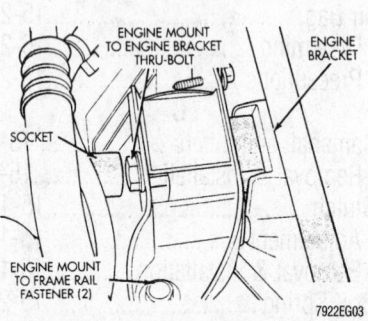

Location of the right engine mount

assembly with the cradle to allow to remove around the body flanges.

To install:
29. Installation is the reverse of the removal procedure. Please note the following important steps.
30. Tighten the front engine mount retainers as follows (bolt specifications are on accompanying figure:
   a. If the engine mount bracket was removed, tighten bolt 1 to 20 inch lbs. (3 Nm) and bolts 2, 3 and 4 to 80 ft. lbs. (108 Nm).
   b. If the engine mount bracket was removed, tighten bolts 5 and 1 to 40 ft. lbs. (54 Nm).
   c. Tighten the engine mount bracket-to-insulator assembly through-bolt to 40 ft. lbs. (54 Nm).
   d. Tighten the insulator assembly nuts to the lower radiator crossmember to 40 ft. lbs. (54 Nm).
   e. Tighten the mass damper bolt to 40 ft. lbs. (54 Nm).
31. Tighten the left engine mount as follows:
   a. To 40 ft. lbs. (54 Nm) for (A) fasteners
   b. To 80 ft. lbs. (108 Nm) for (B) fasteners.

32. Tighten the right engine mount as follows:

    a. Engine mount-to-rail fasteners to 40 ft. lbs. (54 Nm).

    b. Engine mount-to-engine bracket to 80 ft. lbs. (108 Nm).

33. Tighten the power hop damper retainers to 40 ft. lbs. (54 Nm).

34. Fill the engine with fresh engine oil.

✳✳ WARNING

Operating the engine without the proper amount and type of engine oil will result in severe engine damage.

35. After all components are installed, perform the camshaft and crankshaft timing relearn procedure as follows:

    a. Connect a DRB or equivalent scan tool to the Data Link Connector (located under the instrument panel, near the steering column).

    b. Turn the ignition switch **ON**, and access the "miscellaneous" screen.

    c. Select "re-learn cam/crank" option and follow the directions on the scan tool screen.

36. If equipped with A/C, take your vehicle to a reputable repair shop to have the A/C system recharged.

Water Pump

REMOVAL & INSTALLATION

➡**After all components are installed on the engine, a DRB scan tool is necessary to perform the camshaft and crankshaft timing relearn procedure.**

1. Disconnect the negative battery cable.

2. Raise and safely support the vehicle. Remove the right inner splash shield.

3. Remove the accessory drive belts and power steering pump.

4. Drain the cooling system into a suitable container.

5. Securely support the engine from the bottom, then remove the right engine mount.

6. Remove the power steering pump bracket bolts, then set the pump and bracket assembly aside, but the power steering lines do not need to be disconnected.

7. Remove the right engine mount bracket.

8. Remove the timing belt tensioner and timing belt.

9. Remove the camshaft sprocket(s) and inner timing belt cover.

10. Unfasten the water pump-to-engine attaching screws, then remove the water pump from the engine.

11. Remove and discard the water pump O-ring, and thoroughly clean the mating surfaces.

    To install:

12. Install a new O-ring gasket in the water pump O-ring groove. Hold the O-ring in place with a few small dabs of suitable silicone sealant.

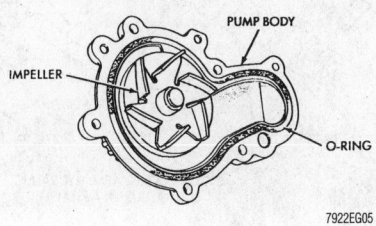

7922EG05

Be sure the O-ring is seated into its groove before installing the new pump

✳✳ WARNING

Before proceeding, be sure the O-ring gasket is properly seated in the water pump groove before tightening the screws. A improperly installed O-ring could cause a coolant leak.

13. Position the water pump to the block and install the retainers. Tighten the retainers to 9 ft. lbs. (12 Nm). Use a pressure tester to pressurize the cooling system to 15 psi and check the water pump shaft seal and O-ring for leaks.

14. Rotate the pump by hand to check for freedom of movement.

15. Install the inner timing belt cover, timing belt and tensioner.

16. Install the right engine mount bracket and engine mount.

17. Refill the cooling system with the proper type and amount of coolant.

18. Install the power steering pump and accessory drive belts.

19. Connect the negative battery cable.

20. Use a DRB or equivalent scan tool to perform the camshaft and crankshaft timing relearn procedure, as follows:

    a. Connect the scan tool to the Data Link Connector (located under the instrument panel, near the steering column).

    b. Turn the ignition switch **ON**, and access the "miscellaneous" screen.

    c. Select the "re-learn cam/crank" option, then follow the instructions on the scan tool screen.

Cylinder Head

REMOVAL & INSTALLATION

➡**After all components are installed on the engine, a DRB scan tool is necessary to perform the camshaft and crankshaft timing relearn procedure.**

1. Disconnect the negative battery cable.

2. Properly relieve the fuel system pressure.

3. Drain the cooling system into a suitable container.

4. Remove the air cleaner inlet duct and air cleaner. Tag and disconnect all vacuum lines, electrical wiring and fuel lines from the throttle body.

5. Remove the throttle linkage.

6. Remove the accessory drive belts.

7. Disconnect the power brake vacuum hose from the intake manifold.

8. Raise and safely support the vehicle, then separate the exhaust pipe from the manifold. Carefully lower the vehicle.

9. Unbolt the power steering pump and set aside. Do NOT disconnect the fluid lines!

10. Detach the ignition coil pack wiring connector, then remove the coil pack and bracket from the engine.

11. Detach the cam sensor and fuel injector electrical connectors.

12. For 1996–99 vehicles, remove the intake manifold.

13. For SOHC engines, remove the timing belt and camshaft sprocket.

14. For DOHC engines, remove the timing belt, timing belt tensioner and camshaft sprocket. Remove the inner timing belt cover.

15. Remove the rocker arm (valve) cover.

16. For SOHC engine, remove the rocker arm shaft assemblies.

17. For DOHC engines, remove the camshaft and cam follower assemblies.

18. For SOHC engines, if necessary, remove the oil separator assembly bolts, then remove the oil separator.

19. For SOHC engines, unfasten the hose clamps, then disconnect the heater hoses. Remove any remaining hoses or lines.

20. Use a ratchet to unfasten the cylinder head bolts, working from the center outward, then remove the cylinder head from the engine block.

21. The cylinder head bolts must be inspected before they can be reused. If the threads of bolts are stretched, they must be

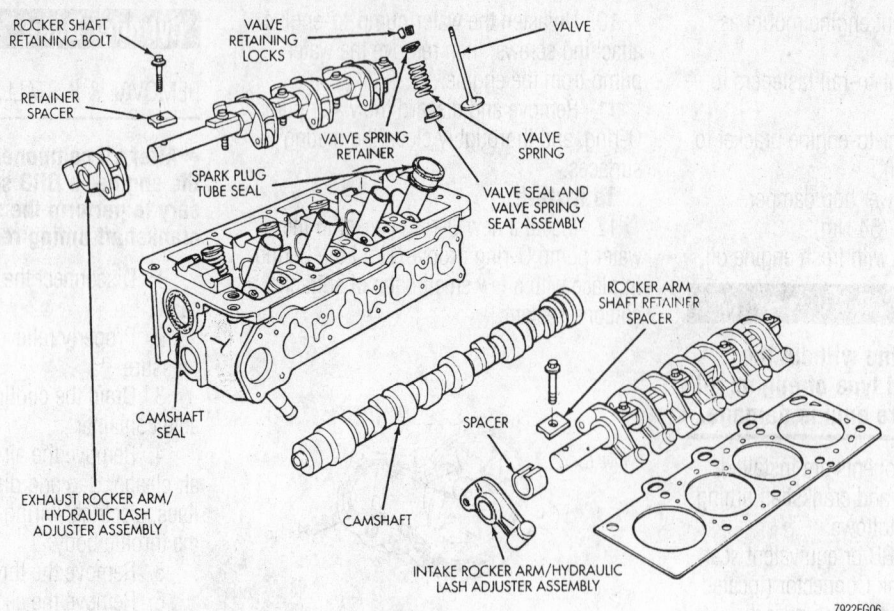

Exploded view of the cylinder head and valvetrain components—SOHC engine

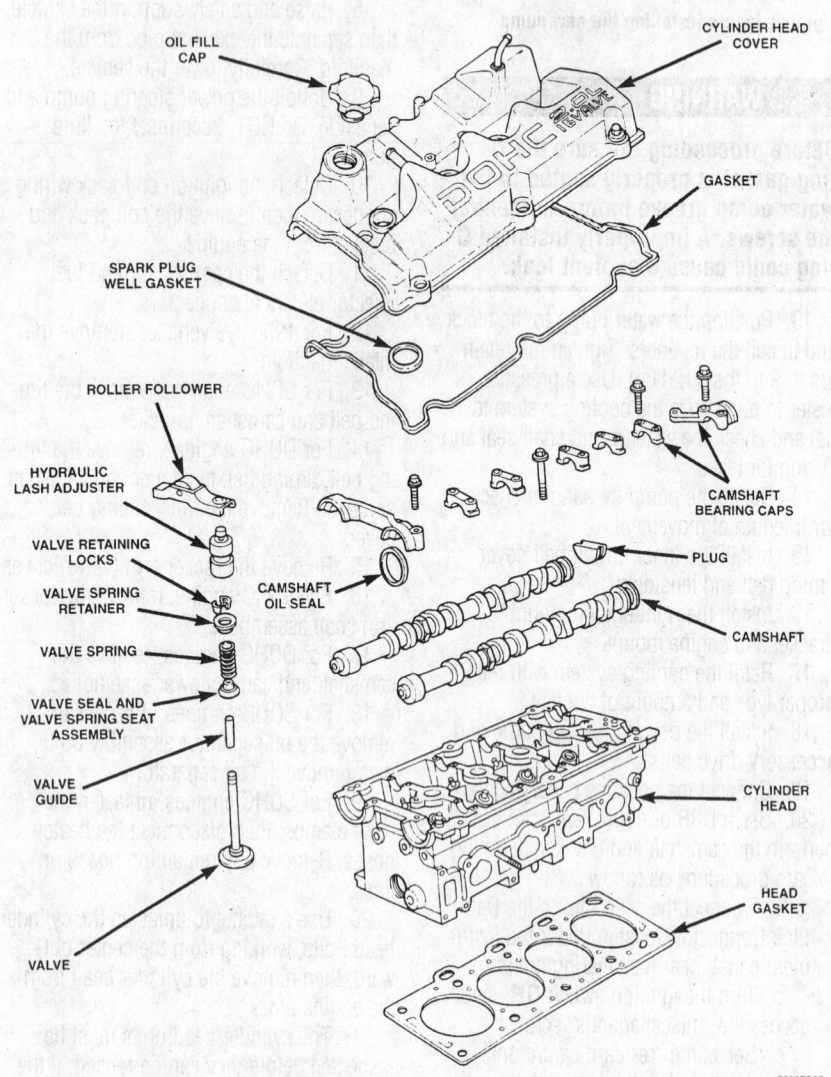

Exploded view of the cylinder head and valvetrain components—DOHC engine

replaced. Check for thread stretching by holding a scale or other straight edge against the threads. If all the threads do not contact the scale, the bolts must be replaced.

✳✳ WARNING

Use only a plastic scraper to clean the mating surfaces. NEVER use metal, as this may gouge the metal surfaces and cause leaks!

22. Cover the combustion chambers, then use a plastic scraper to thoroughly and carefully clean the engine block and cylinder head mating surfaces.

To install:

23. Position a new gasket on the engine block, then place the cylinder head over the gasket.

24. Apply a thin coat of engine oil to the cylinder head bolt threads. The four short bolts 4.33 in. (110mm) are installed in

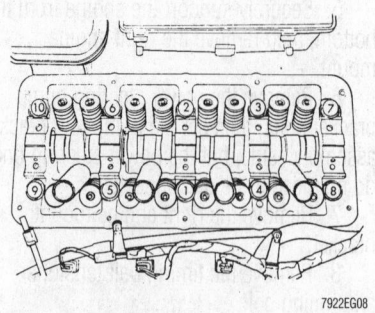

Cylinder head tightening sequence—SOHC engine

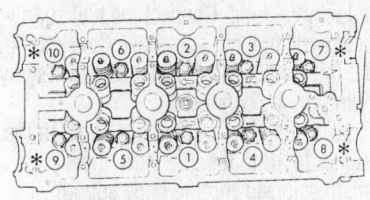

✴ LOCATION OF 110 mm (4.330 in.) BOLTS

7922EG09

Cylinder head tightening sequence—DOHC engine

positions 7, 8, 9 and 10, as shown in the accompanying figure.

25. For SOHC engines, tighten the cylinder head bolts, in the sequence shown, to the following specifications:

a. Step 1: Tighten the bolts to 25 ft. lbs. (34 Nm).

b. Step 2: Tighten the bolts to 50 ft. lbs. (68 Nm).

c. Step 3: Loosen, then re-tighten the bolts to 50 ft. lbs. (68 Nm).

d. Step 4: Tighten all bolts an additional ¼ turn. Do NOT use a torque wrench for this step.

26. For DOHC engines, tighten the cylinder head bolts, in the sequence shown, to the following specifications:

a. Step 1: Tighten bolts 1–6 to 25 ft. lbs. (34 Nm) and bolts 7–10 to 20 ft. lbs. (28 Nm).

b. Step 2: Tighten bolts 1–6 to 50 ft. lbs. (68 Nm) and bolts 7–10 to 20 ft. lbs. (28 Nm).

c. Step 3: Loosen, then re-tighten bolts 1–6 to 50 ft. lbs. (68 Nm) and bolts 7–10 to 20 ft. lbs. (28 Nm).

d. Step 4: Tighten all bolts an additional ¼ turn. Do NOT use a torque wrench for this step.

27. The remainder of installation is the reverse of the removal procedure.

28. Connect the negative battery cable.

29. Fill the cooling system with the proper amount and type of coolant.

30. Use a DRB or equivalent scan tool to perform the camshaft and crankshaft timing relearn procedure, as follows:

a. Connect the scan tool to the Data Link Connector (located under the instrument panel, near the steering column).

b. Turn the ignition switch **ON**, and access the "miscellaneous" screen.

c. Select the "re-learn cam/crank" option, then follow the instructions on the scan tool screen.

Rocker Arm/Shafts

REMOVAL & INSTALLATION

This procedure applies to SOHC engines only. On DOHC engines, the valves are actuated directly by the camshafts.

1. Disconnect the negative battery cable.
2. Remove the cylinder head cover.

➡**Be sure to note the installed positions of the rocker arm shaft assemblies before removal.**

3. Loosen the rocker arm shaft attaching fasteners.

4. Remove the rocker arm shaft assembly from the cylinder head.

5. If necessary, disassemble the rocker arm assemblies by removing the attaching bolts from the shaft.

6. Slide the rocker arms and spacers off the shaft. Be sure to keep the spacers and rocker arms in their original locations for installation.

7. Inspect the rocker arm for scoring, wear on the roller or damage to the rocker arm and replace if necessary. Check the location where the rocker arms mount to the shafts for war or damage. Replace if dam-

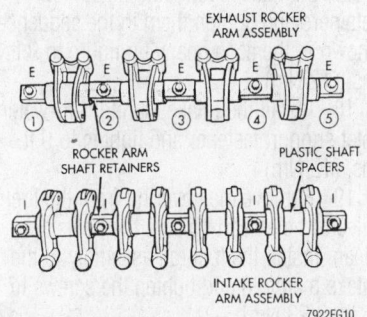

7922EG10

Be sure to note the rocker arm shaft component positions before disassembling the rocker arm shafts

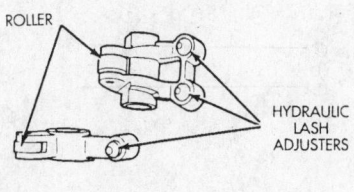

7922EG11

Disassembled view of an intake and exhaust rocker arm

aged or worn. The rocker arm shaft is hollow and is used a lubrication oil duct. Check the oil holes for clogs with a small piece of wire, and clean as required. Lubricate the rocker arms and spacers. Be sure to install in their original locations.

To install:

✴✴ WARNING

You MUST set the crankshaft to three notches before TDC before installing the rocker arm shafts.

8. Set the crankshaft sprocket to TDC by aligning the mark on the sprocket with the arrow on the oil pump housing, then back off to three notches before TDC, as shown in the accompanying figure.

9. Install the rocker arm/hydraulic lash adjuster assembly making sure that the adjusters are at least partially full of oil. This is indicated by little or no plunger travel when the lash adjuster is depressed. If there is excessive plunger travel, place the rocker arm assembly into clean engine oil and pump the plunger until the lash adjuster travel is taken up. If travel is not reduced, replace the assembly. The hydraulic lash adjuster and rocker arm are serviced as an assembly.

10. Install the rocker arm and shaft assemblies with the NOTCH in the rocker arm shafts pointing up and toward the timing belt side of the engine. Install the retainers in their original positions on the exhaust and intake shafts.

✴✴ WARNING

When installing the intake rocker arm shaft assembly, be sure the plastic spacers do not interfere with the spark plug tubes. If the spacers do interfere, rotate until they are at the proper angle. To avoid damaging the spark plug tubes, do not try to rotate the spacers by forcing the shaft down.

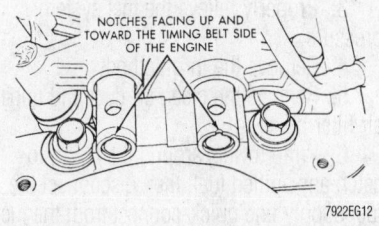

7922EG12

Be sure the notches in the rocker arm shafts points up and toward the timing belt side of the engine

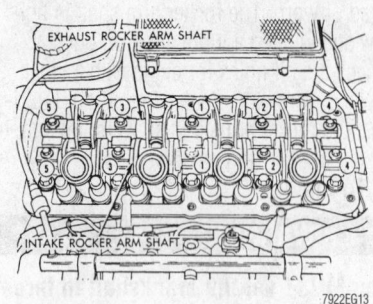

Rocker arm shaft assembly bolt tightening sequence

11. Tighten the bolts to 17 ft. lbs. (23 Nm) for 1995 vehicles or to 21 ft. lbs. (28 Nm) in the sequence shown in the accompanying figure.

12. Install the rocker arm (valve) cover.

13. Connect the negative battery cable.

Intake Manifold

REMOVAL & INSTALLATION

SOHC Engine

1. Disconnect the negative battery cable.

2. Remove the fresh air inlet duct from the air cleaner.

✳✳ CAUTION

Observe all applicable safety precautions when working around fuel. Whenever servicing the fuel system, always work in a well ventilated area. Do not allow fuel spray or vapors to come in contact with a spark or open flame. Keep a dry chemical fire extinguisher near the work area. Always keep fuel in a container specifically designed for fuel storage; also, always properly seal fuel containers to avoid the possibility of fire or explosion.

3. Properly relieve the fuel system pressure.

4. Remove the throttle body.

5. Remove the clean air duct and upper air filter housing.

6. Wrap towels around the fitting to catch any spilled fuel, then disconnect the fuel supply line quick-connect from the fuel tube assembly.

7. Unfasten the fuel rail attaching screws, then remove the fuel rail from the engine. Be sure to cover the injector openings.

✳✳ WARNING

Do NOT set the fuel injectors on their tips, as this may damage them.

8. Detach the electrical connector(s) from the MAP and IAT sensors. On early models there are two sensors, on later models they are combined into one sensor.

9. Unplug the knock sensor electrical connector.

10. Disconnect the wiring from the starter, then unclip the wiring harness and position it out of the way.

11. Remove the EGR tube bolts at the intake manifold, then remove the tube from the manifold. Remove the gasket.

12. Disconnect the brake booster vacuum hose.

13. Disconnect the PCV vapor hose.

14. Unfasten the intake manifold-to-inlet water tube support fastener.

15. Remove and discard the intake manifold retainers. Remove the intake manifold from the vehicle.

16. Remove and discard the gaskets and seals. Thoroughly clean the gasket mating surfaces.

To install:

17. Position new gaskets and seals, then install the intake manifold. Install new retainers and tighten them in the sequence shown in the accompanying figure to 9 ft. lbs. (12 Nm).

18. Install the intake manifold-to-water inlet support fastener and tighten to 9 ft. lbs. (12 Nm).

19. Remove the covering from the fuel injector holes and be sure the holes are clean. Install the fuel rail assembly to the intake manifold and tighten the screws to 17 ft. lbs. (23 Nm).

20. Connect the PCV and brake booster vacuum hoses.

21. Inspect the fuel line quick-connect fittings for damage and replace if necessary.

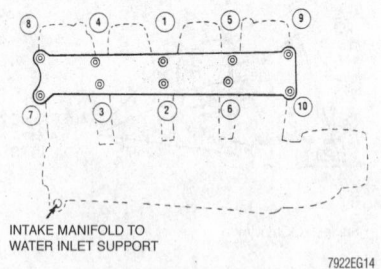

Intake manifold retainer tightening sequence—SOHC engines

Apply a small amount of clean engine oil to the fuel inlet tube. Connect the fuel supply hose to the fuel rail assembly. Check to be sure the connection is fastened securely by pulling on the connector.

22. Install the throttle body. Tighten the fastener to 16 ft. lbs. (22 Nm). Install the transmission-to-throttle body support bracket and tighten to 9 ft. lbs. (12 Nm) at the throttle body first. Next, tighten the bracket at the transmission.

23. Attach the MAP and IAT electrical connector(s).

24. Connect the knock sensor wiring and the wiring at the starter. Fasten the wiring harness to the intake manifold tab.

25. Attach the Idle Air Control (IAC) motor and Throttle Position Sensor (TPS) wiring connectors.

26. Connect the vacuum hoses to the throttle body.

27. Install accelerator, kickdown and speed control cables to their bracket and connect them to the throttle lever.

28. Loosely assemble the EGR tube to the intake manifold finger-tight, then tighten the retainers to 95 inch lbs. (11 Nm).

29. Install the clean air duct to the air filter housing, then tighten the clamp to 30 inch lbs. (3 Nm).

30. Connect the negative battery cable.

31. Attach the fresh air duct to the air cleaner and tighten the wing nut securely.

DOHC Engine

1. Disconnect the negative battery cable.

2. Loosen the wing nut on the intake, then remove the fresh air inlet duct.

3. Properly relieve the fuel system pressure.

✳✳ CAUTION

Observe all applicable safety precautions when working around fuel. Whenever servicing the fuel system, always work in a well ventilated area. Do not allow fuel spray or vapors to come in contact with a spark or open flame. Keep a dry chemical fire extinguisher near the work area. Always keep fuel in a container specifically designed for fuel storage; also, always properly seal fuel containers to avoid the possibility of fire or explosion.

4. Wrap towels around the fitting to catch any spilled fuel, then disconnect the fuel supply line quick-connect from the fuel tube assembly.

5. Remove the clean air inlet duct.

6. Detach the coolant temperature sensor electrical connector.

7. Disconnect the heater hose from the intake manifold and the heater tube from the bottom of the intake manifold.

8. Disconnect the upper radiator and coolant recovery hoses.

❋❋ WARNING

Do not allow the injectors to rest on their tips, as this may cause damage.

9. Unfasten the attaching screws, then remove the fuel rail from the engine. Cover the injector openings with a suitable covering to prevent debris from entering the ports.

10. Remove the accelerator, kickdown and speed control (if equipped) cables from the throttle lever and bracket.

11. Detach the Idle Air Control (IAC) motor and Throttle Position Sensor (TPS) wiring connectors.

12. Tag and disconnect the vacuum hoses from the throttle body, then remove the throttle body from the vehicle.

13. Detach the Manifold Absolute Pressure (MAP) and Intake Air Temperature (IAT) sensor electrical connector. Disconnect the vapor and brake booster hoses.

14. Detach the knock sensor electrical connector and detach the wiring harness from the tab located on the heater tube.

15. Disconnect the wiring from the starter.

16. Unfasten the EGR tube bolts at the valve and at the intake manifold. Remove the tube from the engine.

17. Remove the intake manifold fasteners, then remove the upper and lower intake manifold assemblies. If necessary, you can separate the upper and lower manifolds.

18. Remove and discard the gaskets, then thoroughly clean all of the gasket mating surfaces.

To install:

19. If separated, position a new gasket, then assemble the lower manifold to the

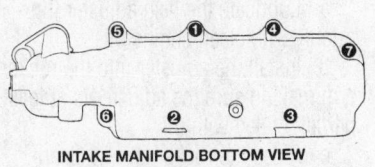

INTAKE MANIFOLD BOTTOM VIEW

7922EG15

Lower-to-upper intake manifold bolt tightening sequence—DOHC engines

---

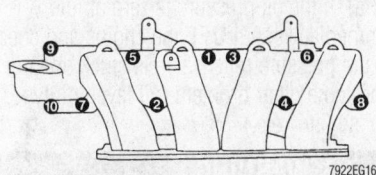

7922EG16

Be sure to follow the proper sequence when tightening the intake manifold-to-cylinder head retainers

upper, then tighten the retaining bolts to 21 ft. lbs. (28 Nm) in the sequence shown in the accompanying figure.

20. Position a new gasket on the cylinder head, then install the intake manifold on the head. Tighten the fasteners to 21 ft. lbs. (28 Nm).

21. Remove the covering from the fuel injector holes and be sure the holes are clean. Install the fuel rail assembly to the intake manifold and tighten the screws to 17 ft. lbs. (23 Nm).

22. Connect the PCV and brake booster vacuum hoses.

23. Inspect the fuel line quick-connect fittings for damage and replace if necessary. Apply a small amount of clean engine oil to the fuel inlet tube. Connect the fuel supply hose to the fuel rail assembly. Check to be sure the connection is fastened securely by pulling on the connector.

24. Connect the heater tube and hose to the intake manifold.

25. Attach the upper radiator hose and coolant recovery reservoir hose.

26. Attach the coolant temperature sensor wiring.

27. Install the throttle body. Tighten the fastener to 16 ft. lbs. (22 Nm).

28. Attach the MAP and IAT electrical connector.

29. Connect the knock sensor wiring and the wiring at the starter. Fasten the wiring harness to the heater tube tab.

30. Attach the Idle Air Control (IAC) motor and Throttle Position Sensor (TPS) wiring connectors.

31. Connect the vacuum hoses to the throttle body.

32. Install accelerator, kickdown and speed control cables to their bracket and connect them to the throttle lever.

33. Loosely assemble the EGR tube to valve and the intake manifold finger-tight. Tighten the tube fasteners to the at the EGR valve first to 95 inch lbs. (11 Nm), then tighten the intake manifold side retainers to 95 inch lbs. (11 Nm).

---

34. Install the clean air duct to the air filter housing, then tighten the clamp to 25 inch lbs. (3 Nm).

35. Connect the negative battery cable.

36. Attach the fresh air duct to the air cleaner and tighten the wing nut securely.

Exhaust Manifold

REMOVAL & INSTALLATION

1. Disconnect the negative battery cable.

2. Remove the air cleaner assembly and bracket.

3. Unfasten the retainers, then separate the exhaust pipe from the manifold. If Low Emission Vehicle (LEV) equipped, discard the manifold-to-flex joint gasket.

4. Unbolt the power steering pump reservoir and position it aside. Do NOT disconnect the fluid lines.

5. Unfasten the retainers, then remove the exhaust manifold heat shield.

6. For 1996–99 vehicles, detach the upstream heated oxygen sensor connector.

➡**It may be necessary to loosen the alternator bracket bolt to remove the outer exhaust manifold bolt.**

7. Remove the eight exhaust manifold retaining fasteners, then remove the exhaust manifold and gasket.

8. Discard the gasket, then thoroughly clean the mating surfaces.

To install:

9. Position a new gasket and the exhaust manifold it their proper positions. Apply Mopar stud and bearing mount, or equivalent, to the fasteners. Install the fasteners and tighten to 17 ft. lbs. (23 Nm), starting at the center and working outward in both directions. Repeat this procedure until all fasteners are tightened to specifications.

10. If loosened, tighten the alternator bracket bolt.

11. Install the exhaust manifold heat shield.

12. Place the power steering pump reservoir in position and secure with the mounting bolts.

13. If necessary, attach the upstream heated oxygen sensor.

14. Install the air cleaner bracket and assembly.

15. If Low Emission Vehicle (LEV) equipped, replace the manifold-to-flex fold gasket.

16. Attach the exhaust pipe to the manifold and tighten to 21 ft. lbs. (28 Nm).

17. Connect the negative battery cable.

Front Crankshaft Seal

REMOVAL & INSTALLATION

1. Disconnect the negative battery cable.
2. Turn the crankshaft clockwise until the engine is at TDC No. 1 cylinder compression stroke (firing position). Remove the timing belt using the recommended procedure.
3. Using a suitable puller, remove the crankshaft pulley and sprocket.
4. Remove the crankshaft oil seal using a suitable puller.
5. Using Seal Puller tool 6771 or equivalent, remove the front crankshaft oil seal. Take care not to damage the seal surface of the cover.

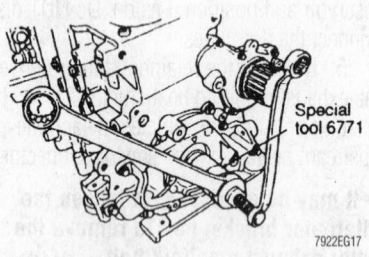

Removing the front crankshaft oil seal

To install:

6. Using tool 6780–1 or equivalent, install a new front crankshaft oil seal.
7. Reinstall the crankshaft timing belt drive pulley and sprocket. Install the timing belt following the recommended procedure. Take care that all timing marks are properly aligned or engine damage will result.
8. Reconnect the negative battery cable.
9. Remove the engine oil pressure sending unit and install an oil pressure gauge. Start and run the engine until the thermostat opens. Curb idle oil pressure should be 4 psi minimum. At 3000 rpm, the oil pressure

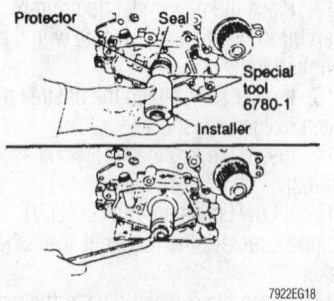

Installing a new seal using Seal Installer 6780–1—use care if using substitute tools

should be between 25–80 psi (172–551 kPa). If the oil pressure is zero at idle, immediately shut **OFF** the engine and check if the pressure relief valve is stuck open or for some other problem (oil level, oil type, loose filter, etc.).

✳✳ WARNING

If the oil pressure is zero at idle DO NOT run the engine at 3000 rpm looking for a pressure increase or the engine could be damaged.

Camshaft and Lifters

REMOVAL & INSTALLATION

SOHC Engine

This engine uses a Single Over Head Camshaft (SOHC) running in an aluminum cylinder head. Rocker arm shafts mount directly to the cylinder head. Care must be taken to be sure all valve timing marks align after cylinder head and valvetrain service. Please note that the cylinder head must be removed from the vehicle to service the camshaft.

➡**After all components are installed on the engine, a DRB scan tool is necessary to perform the camshaft and crankshaft timing relearn procedure.**

✳✳ CAUTION

Fuel injection systems remain under pressure, even after the engine has been turned OFF. The fuel system pressure MUST BE relieved before disconnecting any fuel lines. Failure to do so may result in fire and/or personal injury.

1. Disconnect the negative battery cable.
2. Relieve the fuel system pressure using the recommended procedure.
3. Remove the cylinder head cover using the recommended procedure.
4. Mark the rocker arm shaft assemblies to identify them for later installation.
5. Remove the rocker arm shaft bolts and remove the rocker arm assemblies from the cylinder head.
6. Remove the timing belt and camshaft sprocket using the recommended procedure.
7. Remove the cylinder head using the recommended procedure.
8. Remove the camshaft sensor and remove the camshaft from the rear of the cylinder head.

To install:

➡**The cylinder head bolts should be checked for stretching before reuse. If the thread area of the bolt is necked down the bolts must be replaced with new. New head bolts are recommended.**

9. Clean all parts well. Inspect the camshaft journals for scoring. Check the oil feed holes in the head for blockage. Check the camshaft bearing journals for scoring. If light scratches are present, they may be removed with 400 grit abrasive paper. If deep scratches are present, replace the camshaft and check the cylinder head for damage. Replace the cylinder head if worn or damaged.
10. If the camshaft lobes show signs of wear, check the corresponding rocker arm roller for wear or damage. Replace rocker arms/hydraulic lash adjuster if worn or damaged. If the camshaft lobes show signs of pitting on the nose, flank or base circle, replace the camshaft.
11. If the rocker arms and shaft are to be serviced, mark the rocker arms so any that are to be returned to service will be installed in their original locations. Slide the rocker arms off the shaft. Keep the spacers and rocker arms in the same location for reassembly.
   a. Inspect the rocker arms for scoring, wear on the roller or damage to the shaft. Replace parts as necessary.
   b. The rocker arm shaft is hollow and used as a lubrication oil duct. Check that the shaft is clean inside and out.
   c. Check all oil holes for clogging with a small wire and clean and required.
   d. To assemble, thoroughly lubricate all rocker arm components and spacers and install on the rocker arm shaft in the original locations.
12. If the vehicle exhibited a tappet-like noise, the valve lash adjusters built into the rocker arms should be cleaned and checked. Lash adjusters removed from a rocker arm should be returned to their original locations. Replace worn or defective lash adjusters. To install a lash adjuster, use the following procedure.
   a. Lubricate the lash adjuster thoroughly with clean engine oil.
   b. Install the adjuster into the rocker arm making sure the adjuster is at least partially filled with oil.
   c. Place the rocker arm in clean engine oil and pump the plunger until the lash adjuster travel is taken up. If travel is not reduced, replace the adjuster.

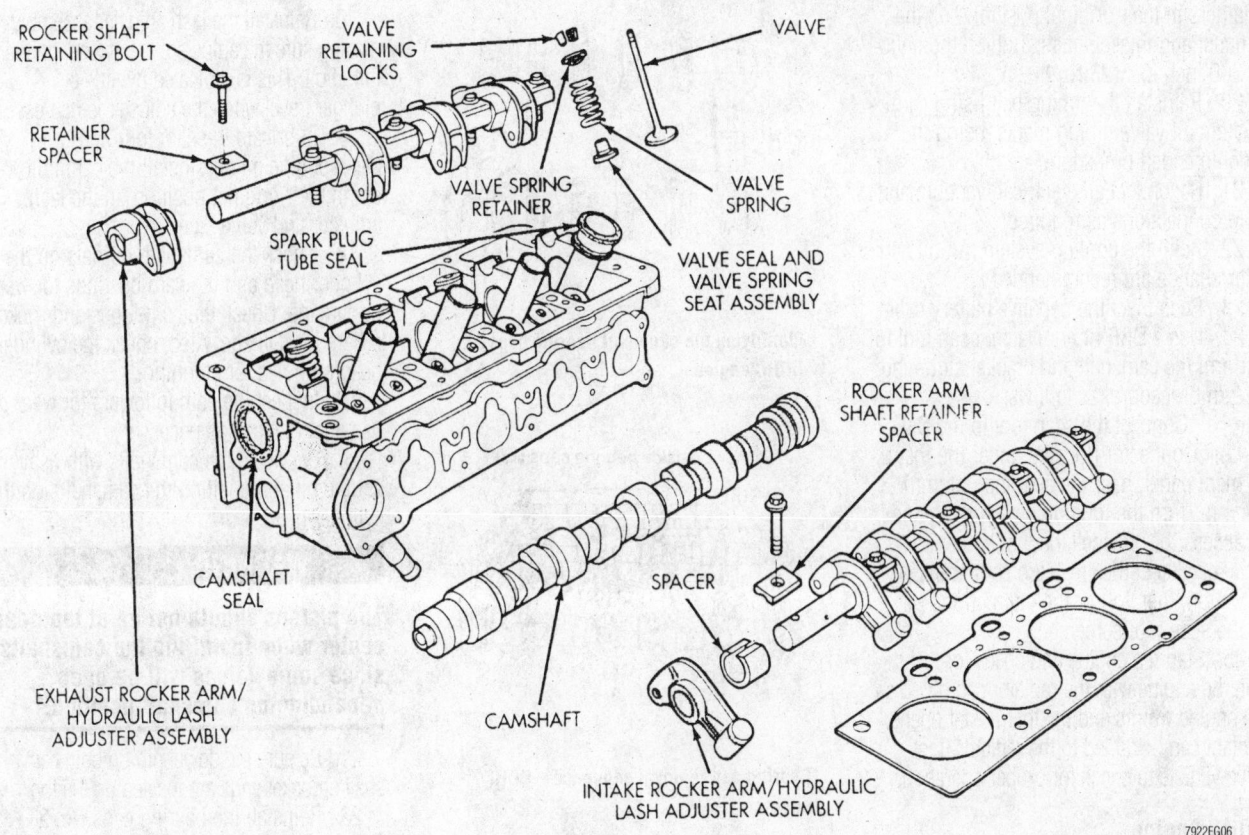

Exploded view of the cylinder head and valve assembly—SOHC engine

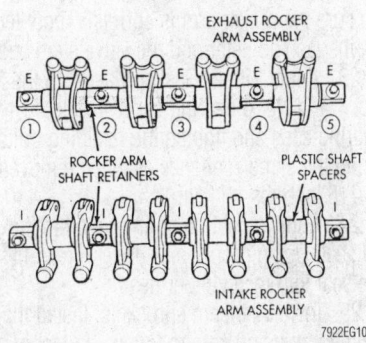

Rocker arm shaft identification—SOHC engine

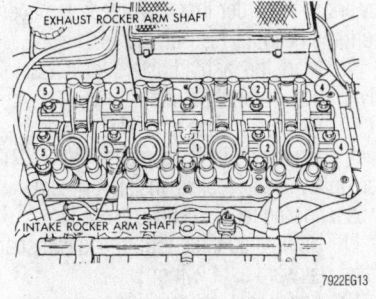

Rocker arm shaft tightening sequence—SOHC engine

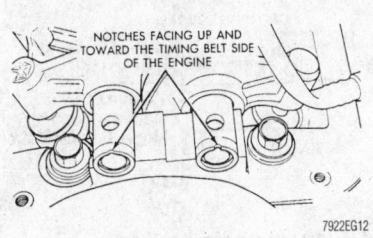

Rocker arm shaft notch location—SOHC engine

d. Install the rocker arm back on the rocker arm shaft.

13. Camshaft end-play can be checked. Use the following procedure.

a. Oil the camshaft journals and install the camshaft without the rocker arm assemblies. Install the cam sensor and tighten the screws to 85 inch lbs. (9.6 Nm).

b. Setup a dial indicator to touch on the nose of the camshaft.

c. Using a suitable tool, move the camshaft as far rearward as it will go. Be sure the dial indicator probe is in contact with the camshaft.

d. Zero the dial indicator.

e. Move the camshaft as far forward as it will go.

f. Read the end-play on the dial indicator. Specification is 0.005–0.013 inch (0.13–0.33mm).

14. To install the camshaft, lubricate the bearing journals thoroughly. Install the camshaft into the cylinder head carefully. Be sure it turns freely. If the camshaft installation is satisfactory, install the cam sensor and tighten the screws to 85 inch lbs. (9.6 Nm).

15. Reinstall the camshaft seal. The camshaft must be installed before the camshaft seal is installed. The seal should be flush with the cylinder head after installation.

16. Reinstall the camshaft sprocket and tighten to the bolt to 85 ft. lbs. (115 Nm).

17. Reinstall the cylinder head using the recommended procedure.

18. Before installing the rocker arm and shaft assemblies, set the crankshaft to three notches before TDC on the crankshaft sprocket.

19. Reinstall the rocker arm and shaft assemblies with the small notches in the rocker shafts pointing up and toward the timing belt side of the engine. Install the

retainers in their original positions on the exhaust and intake shafts. Tighten the bolts to 200 inch lbs. (23 Nm).

20. Reinstall the timing belt using care to align all valve timing marks, using the recommended procedure.

21. Reconnect all electrical, vacuum and fluid connections as required.

22. Refill the cooling system. An oil and filter change are recommended.

23. Reconnect the negative battery cable.

24. Use a DRB or equivalent scan tool to perform the camshaft and crankshaft timing relearn procedure, as follows:

a. Connect the scan tool to the Data Link Connector (located under the instrument panel, near the steering column).

b. Turn the ignition switch **ON**, and access the "miscellaneous" screen.

c. Select the "re-learn cam/crank" option, then follow the instructions on the scan tool screen.

25. Start the engine and check for leaks. Run the engine with the radiator cap off so as the engine warms and the thermostat opens, coolant can be added to the radiator. Test drive vehicle to check for proper operation.

DOHC Engine

This engine is called a Dual Over Head Camshaft (DOHC) engine since it uses two camshafts. Care must be taken to be sure all valve timing marks align after any cylinder head and valvetrain service.

➡️**After all components are installed on the engine, a DRB scan tool is necessary to perform the camshaft and crankshaft timing relearn procedure.**

1. Disconnect the negative battery cable.
2. Remove the cam cover.

➡️**Always rotate the crankshaft in a clockwise direction. Make a mark on the back of the timing belt indicating the direction of rotation so it may be reassembled in the same direction if it is to be reused.**

3. Rotate the crankshaft clockwise and align the timing marks so the No. 1 piston will be at TDC of the compression stroke.

4. Remove the timing belt cover and the timing belt.

5. Remove both camshaft sprockets.

6. Loosen the bearing caps in sequence, 1 camshaft at a time. The bearing caps are identified for location. Remove the outside bearing caps first.

➡️**If the bearing caps are difficult to remove, use a plastic hammer to gently tap the rear part of the camshaft.**

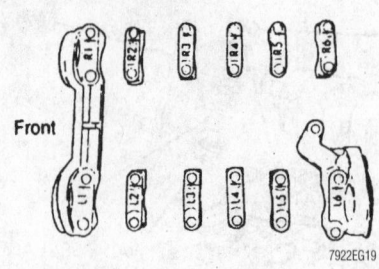

Identifying the camshaft bearing caps—DOHC engine

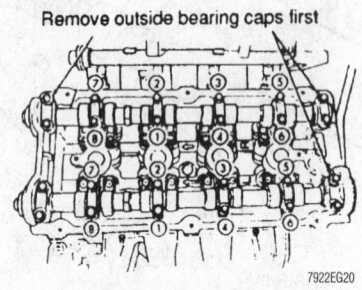

Bearing cap removal sequence—DOHC engine

7. Remove the intake and exhaust camshafts.

8. Remove the cam follower assemblies from the cylinder head. Keep the cam followers in the order they have been removed from the head for reassembly.

9. Mark the lash adjusters for reassembly in their original positions.

10. Check camshaft end-play. Oil camshaft journals and install camshaft **WITHOUT** cam follower assemblies. Install rear cam caps following recommended procedure.

11. Using a suitable tool, move camshaft as far rearward as it will go.

12. Attach dial indicator and zero.

13. Move camshaft as far forward as it will go.

14. Measure and record end-play. End play specs: 0.002–0.06 inches (0.05–0.15mm)

❊❊ WARNING

The camshafts and their components are NOT interchangeable. Place an identifying mark on each component. Be sure to keep all parts organized for proper reassembly in their original positions.

To install:

15. Before installation, clean the cylinder head and cover mating surfaces. Make certain that the rails are flat.

16. Reinstall the lash adjuster assembly making sure the adjusters are at least partially full of oil. This is indicated by little or no plunger travel when the adjuster is depressed.

17. Lubricate the cam followers with clean engine oil and install the cam followers in their original position on the lash adjuster and valve stem.

18. Check the camshaft journals on the cylinder head and the cam bearings for wear or damage. Check the cam lobes and rocker rollers for damage. Also, check the cylinder head oil holes for clogging.

19. Inspect the cam followers for wear or damage. Replace as necessary.

20. Lubricate the camshafts with heavy engine oil and position the camshafts on the cylinder head.

❊❊ WARNING

The pistons should not be at top dead center when installing the camshafts since some valves will be open depending on camshaft position.

21. Be sure the dowel pin on both camshaft sprocket ends are located on the top.

22. Reinstall the bearing caps No. 2 through No. 5 and right No. 6 and tighten the caps in sequence to 9 ft. lbs. (12 Nm). Check the markings on the caps to identify the cap number and intake/exhaust symbol. Be sure the rocker arm is correctly mounted on the lash adjuster and the valve stem end.

23. Apply Mopar Gasket Maker® to the No. 1 and No. 6 bearing caps. Install the bearing caps and tighten the retaining bolts, using the same sequence as when removed, to 215 inch lbs. (24 Nm).

24. Apply a coating of engine oil to the oil seal. Using proper size driver, press-fit the seal into the cylinder head.

25. Install the cam sprockets. Install the timing belt using care to follow the recommended procedure. Cam timing is critical or engine damage will result. Install the timing belt cover and related components.

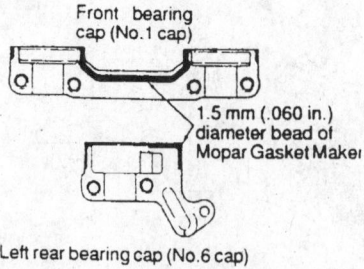

To prevent leaking from the bearing cap ends, apply sealant as shown—DOHC engine

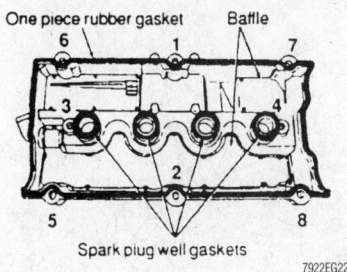

Cam cover tightening sequence—DOHC engine

26. Apply Mopar® silicone rubber adhesive sealant at the camshaft cap corners and at the top edge of the half round seal.

27. Reinstall the cam cover using a new gasket. Tighten the cam cover retaining bolts in sequence, using a three step torque method:

    a. Tighten all bolts, in sequence to 3.3 ft. lbs. (4.5 Nm)

    b. Tighten all bolts, in sequence to 6.5 ft. lbs. (9 Nm)

    c. Tighten all bolts, in sequence to 9 ft. lbs. (12 Nm)

28. Check engine oil level. After this type of service, an oil and filter change is recommended. If a camshaft had failed, metal particles may be spread throughout the engine so an oil and filter change should be mandatory.

29. Reconnect the negative battery cable.

30. Use a DRB or equivalent scan tool to perform the camshaft and crankshaft timing relearn procedure, as follows:

    a. Connect the scan tool to the Data Link Connector (located under the instrument panel, near the steering column).

    b. Turn the ignition switch **ON**, and access the "miscellaneous" screen.

    c. Select the "re-learn cam/crank" option, then follow the instructions on the scan tool screen.

31. Start the engine and check for proper operation and leaks.

Valve Lash

ADJUSTMENT

The engines in these vehicles do not require periodic valve lash adjustment.

Oil Pan

REMOVAL & INSTALLATION

1995 Vehicles

1. Raise and safely support the vehicle.
2. Drain the engine oil into a suitable container.

3. Unfasten the pan retaining bolts, then lower the oil pan from the engine.

4. Thoroughly clean all of the gasket mating surfaces.

To install:

5. Apply suitable silicone sealer to the oil pump-to-engine block parting line, as shown in the accompanying figure.

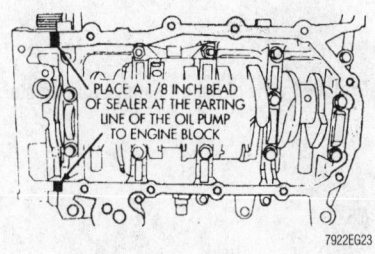

Silicone sealer application locations—oil pan service

6. Position the oil pan gasket to the block, using silicone sealer to hold the gasket in place.

7. Install the oil pan and secure with the retaining bolts. Tighten the bolts to 85 inch lbs. (9.5 Nm).

8. Carefully lower the vehicle. Fill the crankcase with the proper type and amount of engine oil.

9. Start the engine and check for leaks, then recheck the fluid level and add as necessary.

1996–97 Vehicles

1. Raise and safely support the vehicle.
2. Drain the engine oil into a suitable container.
3. Remove the transmission bending bracket.
4. Support the engine and transaxle assembly, then remove the front engine mount and bracket.
5. Remove the transmission inspection cover.
6. If equipped with A/C, remove the oil filter and adapter.
7. Unfasten the retainers, then remove the oil pan.
8. Thoroughly clean the gasket mating surfaces.

To install:

9. Apply suitable silicone sealer to the oil pump-to-engine block parting line, as shown in the accompanying figure.

10. Position a new oil pan gasket on the pan.

11. Install the pan, then tighten the retainers to 105 inch lbs. (12 Nm).

12. If removed, install the oil filter and adapter.

13. Install the transmission inspection cover.

14. Install the front engine mount and bracket.

15. Install the transmission bending bracket.

16. Carefully lower the vehicle. Fill the crankcase with the proper type and amount of engine oil.

17. Start the engine and check for leaks, then recheck the fluid level and add as necessary.

1998–99 Vehicles

1. Raise and safely support the vehicle.
2. Drain the engine oil into a suitable container.
3. Properly support the engine and transaxle assembly, then remove the front engine mount bracket.
4. Remove the powertrain bending strut.
5. Remove the structural collar from the oil pan-to-transaxle.
6. Remove the transaxle lower dust cover.
7. If equipped with A/C, remove the oil filter and adapter.
8. Unfasten the retaining bolts, then remove the oil pan.
9. Thoroughly clean the gasket mating surfaces.

To install:

10. Apply suitable silicone sealer to the oil pump-to-engine block parting line, as shown in the accompanying figure.

11. Position a new oil pan gasket on the pan.

12. Install the pan, then tighten the retainers to 105 inch lbs. (12 Nm).

13. If removed, install the oil filter and adapter.

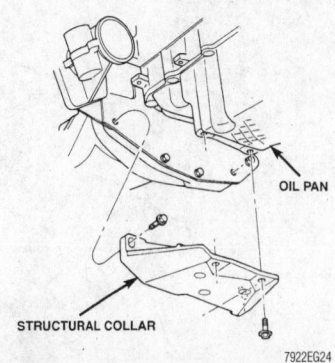

Exploded view of the structural collar mounting—1998–99 vehicles

14. Install the transaxle lower dust cover.
15. Install the powertrain bending strut.
16. Install the front engine mount and bracket.

✳✳ WARNING

You MUST follow the proper tightening sequence for the structural collar or damage to the collar or oil pan may occur!

17. Install the structural collar, then tighten the retainers as follows:

a. Step 1: Install the collar-to-oil pan bolts and tighten to 30 inch lbs. (3 Nm).

b. Step 2: Install the collar-to-transaxle bolts and tighten to 80 ft. lbs. (108 Nm).

c. Step 3: Final-tighten the collar-to-oil pan bolts to 40 ft. lbs. (54 Nm).

18. Carefully lower the vehicle. Fill the crankcase with the proper type and amount of engine oil.

19. Start the engine and check for leaks, then recheck the fluid level and add as necessary.

Oil Pump

REMOVAL & INSTALLATION

1. Disconnect the negative battery cable.
2. Remove the timing belt.
3. Raise and safely support the vehicle.
4. Drain the engine oil into a suitable container.
5. Remove the oil pan.
6. Remove the crankshaft sprocket using a suitable puller.
7. Remove the oil pick-up tube.
8. Remove the oil pump and the front crankshaft seal.
9. Remove the oil pump cover screws, then lift the cover off.
10. Remove the oil pump rotors.

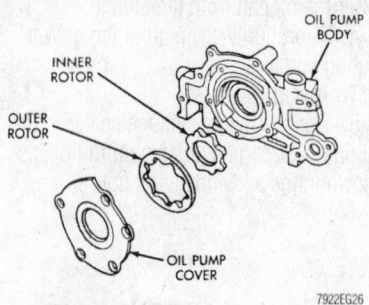

Exploded view of the oil pump components

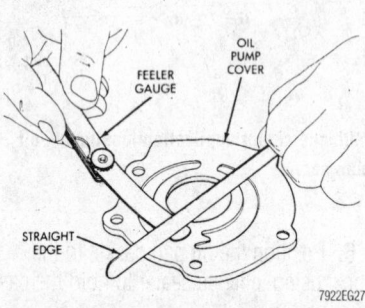

Checking the oil pump cover flatness

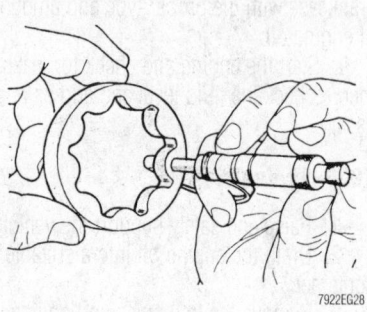

Use calipers to measure the outer rotor thickness . . .

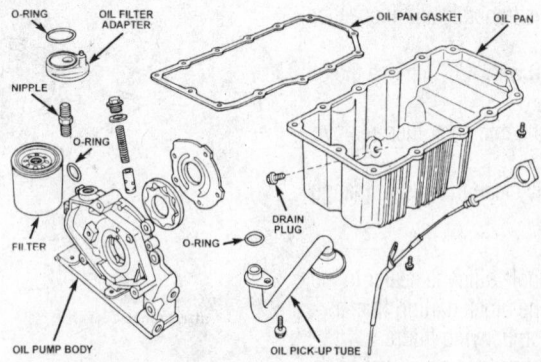

Exploded view of the oil pump and related components' mounting

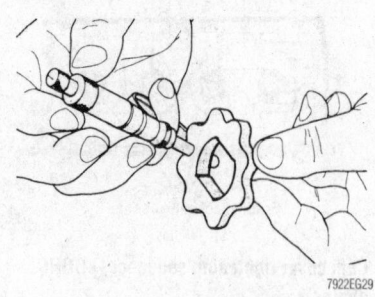

. . . and the inner rotor thickness

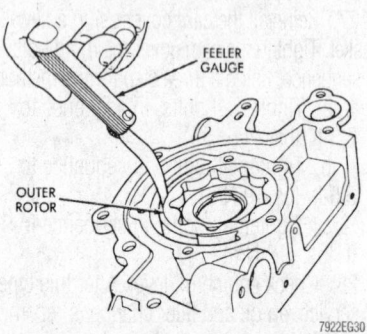

Measure the outer rotor clearance in the housing

11. Wash all parts in a suitable solvent, then inspect carefully for damage or wear, as follows:

a. Clean all parts thoroughly. The mating surface of the oil pump should be smooth. Replace the pump cover if it is scratched or grooved.

b. Lay a straightedge across the pump cover surface. If a 0.003 in. (0.076mm) feeler gauge can be inserted between the cover and the straight edge, the cover should be replaced.

c. Measure the thickness and diameter of the outer rotor. If the outer rotor thickness measures 0.301 in. (7.64mm) or less, or if the diameter is 3.148 in. (79.95mm) or less, replace the outer rotor.

d. If the inner rotor measures 0.301 in. (7.64mm) or less, replace the inner rotor.

e. Slide the outer rotor into the pump housing, press to one side with your fingers and measure the clearance between the rotor and the housing. If the measurement is 0.015 in. (0.39mm) or more, replace the housing only f the outer rotor is within specification.

f. Install the inner rotor into the pump housing. If the clearance between the inner and outer rotors is 0.008 in. (0.203mm) or more, replace both rotors.

g. Place a straightedge across the face of the pump housing, between the bolt holes. If a feeler gauge of 0.004 in. (0.102mm) or more can be inserted between the rotors and the straightedge. replace the pump assembly ONLY if the rotors are within specifications.

h. Inspect the oil pressure relief valve plunger for scoring and free operation in its bore. Small marks may be removed with 400-grit wet or dry sandpaper.

i. The relief valve spring has a free length of about 2.39 in. (60.7mm) and should test between 18–19 lbs. (8.1–8.6 kg) when compressed to 1.60 in. (4.05mm). Replace the spring if it falls outside of specifications.

j. If the oil pressure is low and the pump is within specifications, inspect for worn engine bearings or other reasons for oil pressure loss.

To install:

12. Assemble the pump, using new parts as required. Install the inner rotor with the chamfer facing the cast iron oil pump cover.

13. Apply Mopar® or equivalent gasket maker to the oil pump as shown in the accompanying figure. Install the oil ring into the oil pump body discharge passage.

14. Prime the oil pump before installation by filling the rotor cavity with engine oil.

15. Align the oil pump rotor flats with the flats on the crankshaft as your install the oil pump to the block.

✳✳ WARNING

The front crankshaft seal MUST be out of the pump to align, or damage may result.

16. Tighten all of the pump attaching bolts to 21 ft. lbs. (28 Nm).

17. Install a new front crankshaft seal using seal driver tool 6780 or equivalent.

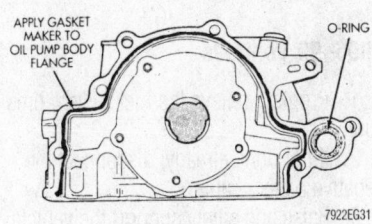

Apply a small amount of gasket maker to the pump body cover mounting surface

18. Install the crankshaft sprocket using a suitable crankshaft damper installation tool.

19. Install the oil pump pick-up tube and oil pan.

20. Install the timing belt.

21. Carefully lower the vehicle. Fill the crankcase with the proper type and amount of engine oil.

22. Start the engine and check for leaks, then recheck the fluid level and add as necessary.

Rear Main Seal

REMOVAL & INSTALLATION

1. Insert a ³⁄₁₆ in. flat-bladed screwdriver between the dust lip and the metal case of the crankshaft seal. Angle the screwdriver through the dust lip against the metal case of the seal. Pry out the seal.

✳✳ WARNING

Do NOT let the screwdriver blade contact the crankshaft seal surface. Contact of the screwdriver blade against the crankshaft edge (chamfer) is permitted.

To install:

✳✳ WARNING

If the crankshaft edge (chamfer) has any burrs or scratches on the, you

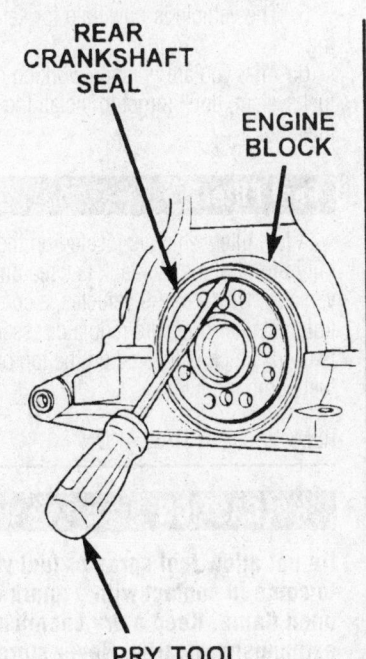

When prying the seal out, be sure to use the prytool at the proper angle

can clean it up with 400 grit sand paper to prevent seal damage during installation of the new seal.

➡ **No lubrication is necessary when installing the seal.**

2. Place special tool 6926–1, or equivalent on the crankshaft. This is a pilot tool with a magnetic base.

3. Position the seal over the pilot tool. Be sure you can read the words THIS SIDE OUT on the seal. The pilot tool should stay on the crankshaft during installation of the seal. Be sure that the lip of the seal if facing towards the crankcase during installation.

✳✳ WARNING

If the seal is driven in the block past flush, this may cause an oil leak.

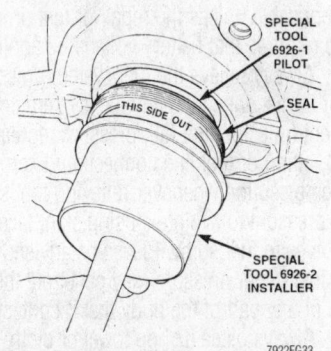

Place a proper size pilot tool with a magnetic base on the crankshaft

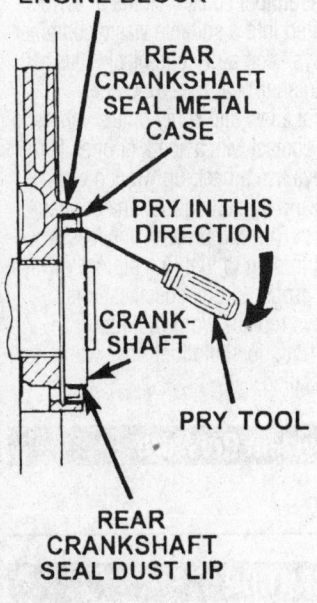

4. Drive the seal into the block using special seal installation tools 6926–2 and handle C-4171 until the tool bottoms out against the block.

FUEL SYSTEM

Fuel System Service Precaution

Safety is an important factor when servicing the fuel system. Failure to conduct maintenance and repairs in a safe manner may result in serious personal injury. Maintenance and testing of the vehicle's fuel system components can be accomplished safely and effectively by adhering to the following rules and guidelines.

• To avoid the possibility of fire and personal injury, always disconnect the negative battery cable unless the repair or test procedure requires that battery voltage be applied.

• Always relieve the fuel system pressure prior to disconnecting any fuel system component (injector, fuel rail, pressure regulator, etc.), fitting or fuel line connection. Exercise extreme caution whenever relieving fuel system pressure to avoid exposing skin, face and eyes to fuel spray. Please be advised that fuel under pressure may penetrate the skin or any part of the body that it contacts.

• Always place a shop towel or cloth around the fitting or connection prior to loosening to absorb any excess fuel due to spillage. Ensure that all fuel spillage is quickly removed from engine surfaces. Ensure that all fuel soaked cloths or towels are deposited into a suitable waste container.

• Always keep a dry chemical (Class B) fire extinguisher near the work area.

• Do not allow fuel spray or fuel vapors to come into contact with a spark or open flame.

• Always use a back-up wrench when loosening and tightening fuel line connection fittings. This will prevent unnecessary stress and torsion to fuel line piping. Always follow the proper torque specifications.

• Always replace worn fuel fitting O-rings. Do not substitute fuel hose where fuel pipe is installed.

Fuel System Pressure

RELIEVING

✳✳ CAUTION

You MUST relieve the fuel system pressure before servicing any components of the fuel system. Service

vehicles in well ventilated areas and avoid ignition sources. NEVER smoke while servicing the vehicle!

1. Disconnect the negative battery cable.
2. Remove the fuel filler cap.
3. Remove the protective cap from the fuel pressure port on the fuel rail.
4. Place the open end of a suitable fuel pressure release hose (tool C-4799–1 or equivalent) into an approved gasoline container. Connect the other end of the hose to the fuel pressure test port. The fuel pressure will bleed off through the hose into the gasoline container.

➡ **Fuel pressure gauge kit C-4799-B contains hose C-4799–1.**

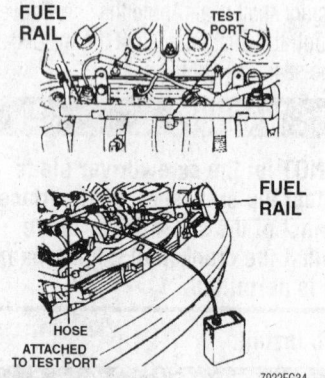

Relieve the fuel system, allowing the pressure to bleed off through the hose into the container

5. The vehicle is now safe for servicing.
6. After you are finished working on the fuel system, don't forget to install the fuel filler cap.

Fuel Filter

A fuel filter, which is located on the frame rail in front of the fuel tank is used on 1995 vehicles. On 1996–99 vehicles, a combination fuel filter/pressure regulator assembly is used, which is located on the top of the fuel pump module.

REMOVAL & INSTALLATION

✳✳ CAUTION

Do not allow fuel spray or fuel vapors to come in contact with a spark or open flame. Keep a dry chemical fire extinguisher nearby. Never store fuel in an open container due to risk of fire or explosion.

1995 Vehicles

1. Properly relieve the fuel system pressure.
2. If not done already, disconnect the negative battery cable.
3. Raise and safely support the vehicle.
4. If necessary, unfasten the retainers, then remove the shield covering the fuel filter and pump assembly.
5. Unfasten the quick-connect fittings from the fuel pump module and chassis fuel supply tube.

➡ **The fuel lines are permanently attached to the fuel filter. The ends of the fuel supply and return lines have different size quick-connect fittings. The larger quick-connect fittings attach to the large nipple on the fuel pump module. The smaller fitting connects to the small nipple on the fuel pump module.**

6. Remove the fuel filter mounting screw, then remove the fuel filter from the vehicle.

To install:

7. Position the fuel filter on the frame rail and secure with the retaining screw. Tighten the screw to 85 inch lbs. (9.5 Nm).
8. Apply a thin coating of clean engine oil to the fuel filter nipples, then attach the quick-connect fuel lines.
9. Carefully lower the vehicle, then connect the negative battery cable.

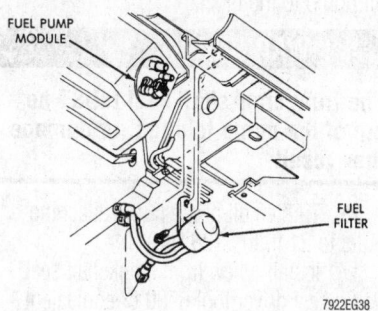

View of the fuel filter mounting—1995 models

1996–99 Vehicles

1. Properly relieve the fuel system pressure.
2. If not done already, disconnect the negative battery cable.
3. Raise and safely support the vehicle.
4. Unfasten the quick-connect fuel supply line from the filter/regulator nipple.
5. Depress the locking spring tab, located on the side of the fuel filter/regula-

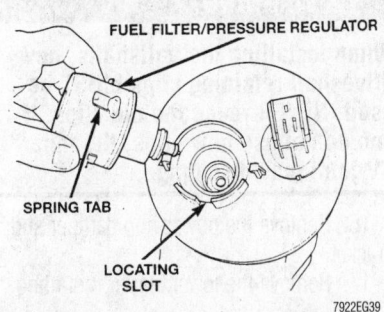

Depress the spring tab, then rotate and pull the fuel filter assembly out

tor, then rotate 90° and pull out. Be sure the upper and lower O-rings are still on the filter assembly.

To install:

6. Lightly coat the filter O-rings with clean engine oil. Insert the filter into the opening in the fuel pump module, then align the two hold-down tabs with the flange.

7. While applying downward pressure, rotate the filter clockwise until the spring tab catches in the locating slot.

8. Attach the fuel line to the filter/regulator assembly.

9. Carefully lower the vehicle, then connect the negative battery cable.

Fuel Pump

The fuel pump is integral with the pump module, which also contains the fuel reservoir, level sensor, inlet strainer and fuel pressure regulator. The inlet strainer, fuel pressure regulator and level sensor are the only serviceable items. If the fuel pump requires service, replace the entire fuel pump module.

REMOVAL & INSTALLATION

1995 Vehicles

1. Disconnect the negative battery cable.
2. Properly relieve the fuel system pressure.
3. Raise and safely support the vehicle.
4. Drain the fuel tank, as outlined under the fuel tank removal and installation procedure.

✳✳ WARNING

The fuel reservoir of the fuel pump module does not empty out when the tank is drained. The fuel in the reservoir will spill out when the module is removed.

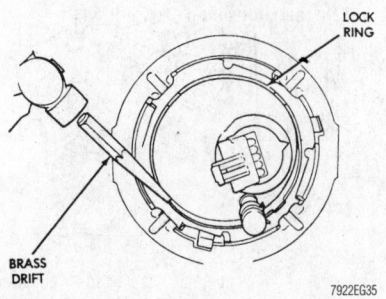

Carefully tap the lockring counterclockwise to release the fuel pump

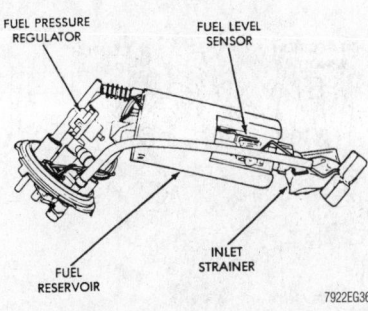

The fuel module assembly contains the pump, pressure regulator, reservoir, inlet strainer and level sensor

5. Disconnect the fuel lines from the fuel pump module by depressing the quick-connect retainers with your thumb and fore-finger.

6. Using a hammer and brass drift punch, carefully tap the lockring counter-clockwise to release the pump.

7. Remove the fuel pump and O-ring seal from the tank. Discard the old seal.

To install:

8. Wipe the area of the tank clean, then place a new O-ring seal in proper position on the pump.

9. Position the fuel pump in the tank with the locking ring.

10. Using a hammer and brass drift, drive the ring around in a clockwise direction to lock the pump in place.

✳✳ CAUTION

Do not overtighten the pump lockring, as this may cause a fuel leak.

11. Carefully lower the vehicle.
12. Fill the fuel tank, the check for leaks.
13. Connect the negative battery cable.

1996–99 Vehicles

1. Disconnect the negative battery cable.
2. Properly relieve the fuel system pressure.

3. Raise and safely support the vehicle.
4. Drain the fuel tank, as outlined under the fuel tank removal and installation procedure.

✳✳ WARNING

The fuel reservoir of the fuel pump module does not empty out when the tank is drained. The fuel in the reservoir will spill out when the module is removed.

5. Disconnect the fuel lines from the fuel pump module by depressing the quick-connect retainers with your thumb and fore-finger.

6. Slide the fuel pump module electrical lock to unlock it.

7. Detach the electrical connection from the fuel pump module, by pushing down on the connector retainer and pulling the connector off of the module.

8. Use a transmission jack to safely support the fuel tank, then remove the bolts from the fuel tank straps.

9. Carefully lower the tank slightly for access to the module.

10. Use a ratchet and spanner wrench to remove the fuel pump module locknut.

11. Remove the fuel pump and O-ring seal from the tank. Discard the old seal.

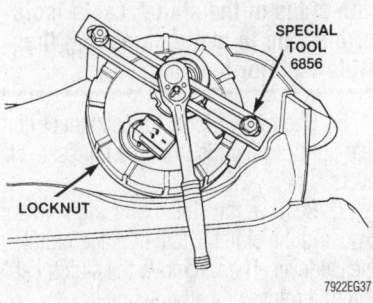

To loosen the fuel pump module locknut, use a ratchet and spanner wrench

To install:

12. Wipe the area of the tank clean, then place a new O-ring seal in position in the tank opening.

13. Position the fuel pump in the tank. Be sure the alignment tab on the underside of the fuel pump module flange sits in the notch on the fuel tank.

14. Position the locknut over the fuel pump module.

15. Using the ratchet and spanner wrench, tighten the locknut to 41 ft. lbs. (55 Nm).

16. Carefully lower the vehicle, then check for leaks.

17. Connect the negative battery cable.

DRIVE TRAIN

Transaxle Assembly

REMOVAL & INSTALLATION

Manual

1. Disconnect the negative, then the positive battery cables.

2. Pull the Power Distribution Center (PDC) up and out of its holding bracket. Position the PDC aside for working clearance.

3. Remove the battery heat shield, then remove the battery from the vehicle. Remove the battery tray from the engine compartment. If equipped, disconnect the cruise control.

4. Remove the vehicle speed sensor wire.

5. Detach the back-up lamp switch wiring from the transaxle.

✳✳ WARNING

Pry with equal amounts of force on both sides of the shifter cable isolator bushing to avoid damaging the cable isolator bushing.

6. Use two prytools to disconnect both gear shift cable ends from the transaxle shift levers.

7. Remove the clutch housing vent cap, exposing the clutch cable end and clutch release lever. Then, remove the clutch cable from the transaxle bell housing.

8. Unfasten the retaining bolts, then remove the shift cable (linkage) mounting bracket.

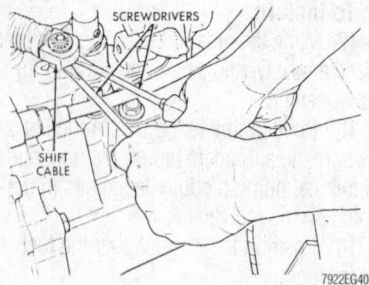

Use two prytools to disconnect the gear shift cable ends from the shift levers—manual transaxle

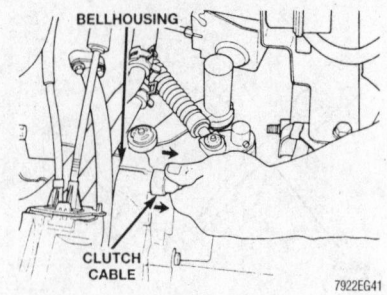

Pull the clutch cable backward, to disconnect the clutch cable from the bell housing—manual transaxle

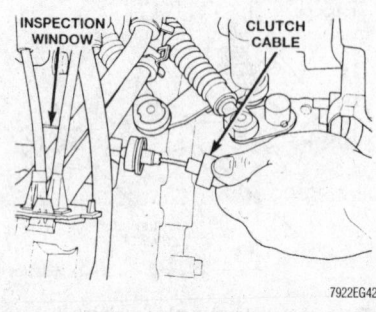

Remove the clutch cable from the lever—manual transaxle

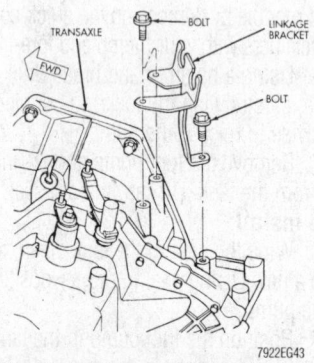

Exploded view of the linkage mounting bracket—manual transaxle

9. If equipped, remove the accelerator cable shield.

10. Remove the intake manifold support bracket and the upper starter bolt.

11. Remove the upper bell housing bolt.

12. Install a suitable engine bridge fixture and support the engine securely.

13. Raise and safely support the vehicle, then remove the front wheel and tire assemblies.

14. Position a suitable drain pan under the vehicle, then drain the transaxle.

15. Remove both front halfshafts.

✳✳ WARNING

When installing the halfshafts, new driveshaft retaining clips MUST be used. NEVER reuse the old clips. If you do not use new clips, the inner CV-joint may disengage.

16. Remove the power hop damper and bracket.

17. Remove the lower starter mounting bolt.

18. Remove the transaxle-to-rear lateral bending strut from the engine and transaxle.

19. Support the transaxle with a suitable jack.

20. Remove the front motor mount through-bolt. Remove the front motor mount bolts from the engine and transaxle.

21. Remove the lower dust shield screw and dust shield.

22. Rotate the crankshaft clockwise in order to access the driveplate-to-modular clutch bolts.

➡**For installation purposes, matchmark the driveplate and pressure plate before removing any bolts.**

23. Unfasten the four driveplate-to-modular clutch bolts in order to separate the driveplate from the clutch.

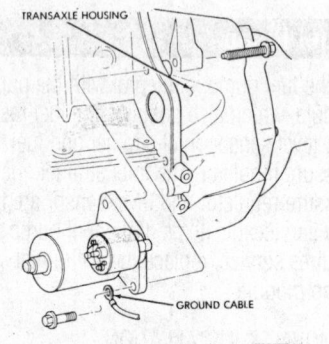

Removing the lower starter mounting bolts—manual transaxle

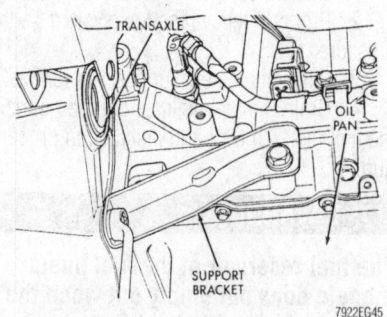

Unfasten the two bolts, then remove the bracket (strut) from the engine and transaxle—manual transaxle

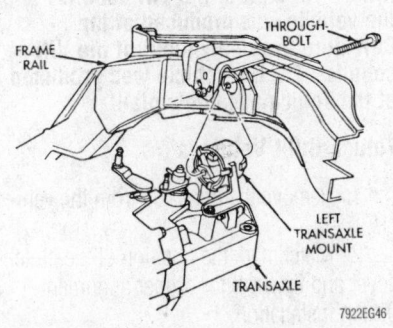

Exploded view of the left manual transaxle mount through-bolt

24. Push the modular clutch assembly into the transaxle bell housing for easier transaxle removal.

25. Remove the frame rail-to-left transaxle mount through-bolt.

26. Remove the left transaxle mount from the transaxle. Then, push the mount up to get clearance for transaxle removal.

27. Remove the transaxle from the vehicle.

28. Remove the modular clutch assembly from the transaxle input shaft.

To install:

29. Installation is the reverse of the removal procedure. Please note the following important steps.

30. The following items must be tightened to the specifications listed.
- Dust shield: 9 ft. lbs. (12 Nm)
- Front engine mount-to-transaxle: 80 ft. lbs. (108 Nm)
- Front mount through-bolt: 45 ft. lbs. (61 Nm)
- Front mount-to-engine bolt: 40 ft. lbs. (54 Nm)
- Lateral bending strut bolts: 40 ft. lbs. (54 Nm)
- Left mount through-bolt: 80 ft. lbs. (108 Nm)
- Left mount-to-transaxle bolt: 40 ft. lbs. (54 Nm)
- Power hop damper bolts: 40 ft. lbs. (54 Nm)
- Transaxle-to-engine bolt: 70 ft. lbs. (95 Nm)
- Transaxle-to-engine intake bracket bolts: 70 ft. lbs. (95 Nm)

31. After installing the transaxle, before lowering the vehicle to the floor, fill the transaxle to the bottom of the fill plug hole with Mopar® type M.S. 9417 or equivalent manual transaxle fluid.

32. Be sure the vehicle's back-up lights and speedometer are functioning properly.

33. After installing the transaxle in the vehicle, you must adjust the crossover cable

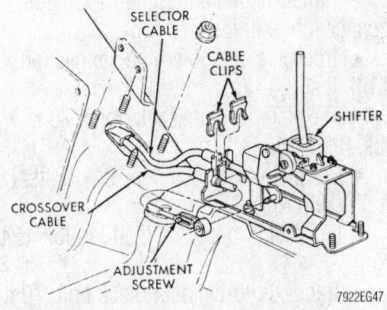

Loosen the adjustment screw on the crossover cable at the shifter—manual transaxle

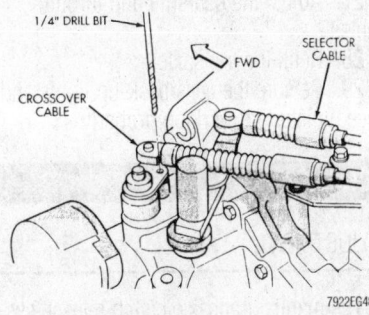

Insert a ¼ in. diameter drill bit to pin the crossover lever into the 3–4 neutral position—manual transaxle

in order to ensure proper shifter adjustment. Adjust as follows:

a. Remove the floor shift console from the vehicle.

b. Loosen the adjusting screw on the crossover cable at the shifter.

c. Pin the transaxle crossover cable in the 3–4 neutral position using a ¼ in. drill bit. Align the hole in the crossover lever with the hole in the boss on the transaxle case. Be sure the drill bit goes into the transaxle case at least ½ in. (12mm).

d. The shifter is spring loaded and self centering. Allow the shifter to rest in its neutral position. Tighten the adjustment screw to 70 inch lbs. (8 Nm). You must be careful to avoid moving the shift mechanism off-center during screw tightening.

e. Remove the drill bit from the transaxle case and perform a functional check by shifting the transaxle into all gears.

f. Reinstall the center shift console. Blouse the boot out around the console. Seat the boot lip on the top of the console.

34. Road test the vehicle to be sure the transaxle is operating properly.

Automatic

The transaxle and torque converter must be removed as an assembly; otherwise the torque converter drive plate, pump bushing or oil seal may be damaged. The drive plate will not support a load; therefore, none of the weight of the transaxle should be allowed to rest on the plate during removal.

1. Disconnect the negative, then the positive battery cables.

2. Pull the Power Distribution Center (PDC) up and out of its holding bracket. Set the PDC aside to gain clearance.

3. Remove the battery heat shield, then remove the battery from the engine compartment. Remove the battery tray from the engine compartment. If equipped, disconnect the cruise control.

4. Remove the vehicle speed sensor wiring.

5. Disconnect the neutral safety switch and torque converter control wiring from the transaxle.

✷✷ WARNING

Pry up on both sides of the shift cable isolator bushing, evenly, to avoid damaging the cable isolator bushing.

6. Disconnect the gear shift cable end from the transaxle shift lever. Remove the bracket bolt from the transaxle.

7. Remove the throttle pressure control cable from the lever. Then, remove the bracket bolts from the transaxle.

8. Remove the transaxle dipstick tube.

9. Disconnect the transaxle oil cooler lines, and plug them to prevent contamination from entering.

10. Remove the throttle pressure control cable support bracket bolts. Remove the upper bell housing bolts and upper starter bolt.

11. Install a suitable engine bridge fixture, then support the engine.

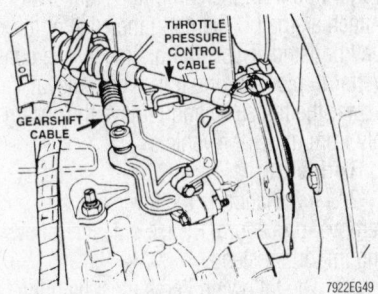

Disconnect the gear shift cable end from the transaxle shift lever—automatic transaxle

12. Raise and safely support the vehicle, then remove the front wheel and tire assemblies.

✳✳ WARNING

When installing the halfshafts, new retaining clips must be used. Do NOT reuse the old clips. Failure to use new clips could cause the inner CV-joint to disengage.

13. Remove both front halfshafts.

✳✳ WARNING

On 1998–99 vehicles, then exhaust flex joint must be disconnected from the exhaust manifold anytime the engine is lowered. If the engine is lowered while the flex pipe is attached, damage will occur.

14. On 1998–99 vehicles, remove the bolts securing the exhaust flex joint to the exhaust manifold. Disconnect the exhaust pipe from the manifold.

15. Unfasten the transaxle-to-rear lateral bending strut from the engine and transaxle.

16. Remove the front engine bracket through-bolt. Remove the front engine bracket bolts.

17. Remove the lower starter bolt.

18. Remove the lower dust shield screw.

19. Rotate the engine clockwise to get access to the converter bolts. Remove the torque converter bolts, then matchmark the converter to the flexplate for alignment during installation.

20. Support the transaxle with a transaxle jack.

21. Remove the left mount through-bolt. Remove the left mount bolts from the transaxle, then remove the left mount.

22. Remove the rear engine bolt from the transaxle.

23. Carefully work the transaxle and torque converter assembly rearward off the engine block dowels. Disengage the converter hub from the end of the crankshaft. Attach a small C-clamp to the edge of the bell housing. This will hold the torque converter in place during transaxle removal. Lower the transaxle and remove the assembly from under the vehicle.

To install:

24. Installation is the reverse of the removal procedure. Please note the following important steps.

25. The following items must be tightened to the specifications listed.
- Bell housing cover bolts: 9 ft. lbs. (12 Nm)

- Transaxle oil cooler line-to-radiator connection: 9 ft. lbs. (12 Nm).
- Transaxle oil cooler line connection: 21 ft. lbs. (28 Nm).
- Flexplate-to-crankshaft bolts: 70 ft. lbs. (95 Nm).
- Flexplate-to-torque converter bolts: 50 ft. lbs. (68 Nm).
- Left motor mount bolts: 40 ft. lbs. (54 Nm).
- Transaxle-to-cylinder block bolt: 70 ft. lbs. (95 Nm).

26. If the torque converter was removed from the transaxle, be sure to align the pump inner gear pilot flats with the torque converter impeller hub flats.

27. Adjust the gearshift and throttle cables.

28. Refill the transaxle.

29. Be sure the car's back-up lights and speedometer are working properly.

Clutch

ADJUSTMENT

The manual transaxle clutch release system has a unique self-adjusting mechanism to compensate for clutch disc wear. This adjuster mechanism is located with the clutch cable assembly. The preload spring maintains tension on the cable. This tension keeps the clutch release bearing continuously loaded against the fingers of the clutch cover assembly. No manual adjustment is obtainable.

When servicing this vehicle or if removing and installing the clutch cable, do not pull on the clutch cable housing to remove it from the dash panel. Damage to the cable self-adjuster may occur.

To check the function of the adjuster mechanism, use the following procedure:

1. With slight pressure, pull the clutch release lever end of the cable to draw the cable taut.

2. Push the clutch cable housing toward the dash panel. With less than 25 pounds of effort the cable housing should move 1.2–2.0 in. (30–50mm). This indicates proper adjuster mechanism function.

3. If the cable does not adjust, determine if the mechanism is properly seated on the bracket.

REMOVAL & INSTALLATION

➡**Vehicles made at the Toluca, Mexico assembly plant have conventional clutch and flywheel assembles. Vehicles made at the Belvidere assembly plant have modular clutch assemblies.**

If the 11th digit of the VIN code is "D", the vehicle was produced at the Belvidere assembly plant; if the VIN code is "T", the vehicle was produced at the Toluca assembly plant.

Toluca Built Vehicles

1. Remove the transaxle from the vehicle.

2. Matchmark the position of the clutch cover and flywheel for proper alignment during installation.

3. Install a suitable clutch alignment tool through the clutch disc hub to prevent the clutch disc from falling and damaging the facings.

4. Loosen the clutch cover attaching bolts, one or two turns at a time, in a criss-cross pattern. This release the spring pressure gradually, avoiding cover damage.

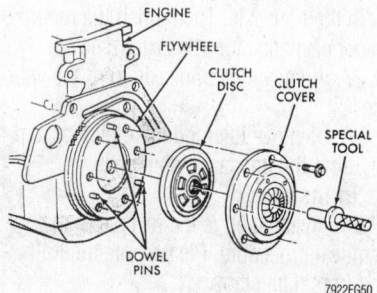

Exploded view of the conventional clutch components—vehicles built in Toluca

➡**Do NOT touch the clutch disc facing with oily or dirty hands. Oil or dirt transferred from your hands onto the clutch disc may cause clutch chatter.**

5. Remove the clutch pressure plate, cover assembly and disc from the flywheel. Handle the components carefully to avoid contaminating the friction surfaces.

6. Inspect for oil leakage through the engine rear main bearing oil seal and transaxle input shaft seal. If there is leakage, it should be fixed at this time.

7. The friction faces of the flywheel and pressure place should not have excessive discoloration, burned areas, cracks, deep grooves or ridges. Replace parts as required.

8. Clean the flywheel face with medium sandpaper, then wipe the surface with mineral spirits. If the surface is severely scored, heat checked, cracked or warped, replace the flywheel.

9. The heavy side of the flywheel is indicated by a white paint mark, near the outside diameter. To minimize the effects of flywheel unbalance, perform the following installation procedure:

a. Loosely assembly the flywheel to the crankshaft. If available, use new flywheel attaching bolts which have sealant on the threads. If new bolts are not available, apply Loctite® sealant to the threads of the original bolts. This sealant is required to prevent engine oil leakage.

b. Rotate the flywheel and crankshaft until the white paint (heavy side) is at the 12 o'clock position.

c. Tighten the flywheel attaching bolts, in a crisscross pattern, to 70 ft. lbs. (95 Nm).

10. The clutch disc should be handled without touching the facings. Replace the disc if the facings show grease or oil soakage, or wear to within less than 0.008 in. (0.20mm) of the rivet heads. The splines on the disc hub and transaxle input shaft should be a snug fit without signs of excessive wear. Metallic portions of the disc assembly should be dry, clean and not discolored from excessive heat. Each of the arched springs between the facings should be tight.

11. Wipe the friction surface of the pressure plate with mineral spirits.

12. Using a straight edge, check the pressure plate for flatness. The pressure plate friction area should be flat to slightly concave, with the inner diameter 0.000–0.0039 in. (0.0–0.1mm) below the outer diameter. It should also be free from discoloration, burned areas, cracks, grooves or ridges.

13. Using a surface plate, test the cover for flatness. All sections around the attaching bolt holes should be in contact with the surface plate within 0.015 in. (0.381mm).

14. The cover should be a snug fit on the flywheel dowels. If the clutch assembly does not meet these requirements, it should be replaced.

To install:

15. Mount the clutch assembly on the flywheel with the disc centered on the alignment tool, being careful to properly align the dowels and the alignment marks made before removal. The flywheel side of the clutch disc is marked for proper installation. If the new clutch or flywheel is installed, align the orange cover balance spot as close as possible to the orange flywheel balance spot. Apply pressure to the alignment tool. Center the tip of the tool into the crankshaft and the sliding cone into the clutch fingers. Tighten the clutch attaching bolts sufficiently to hold the disc in position.

16. To avoid distorting the clutch cover, tighten the bolts gradually, a few turns at a time. Use a crisscross pattern until all bolts

are seated. Tighten the bolts to a final torque of 21 ft. lbs. (28 Nm).

17. Remove the clutch alignment tool.

18. Install the transaxle.

Belvidere Built Vehicles

1. Disconnect the negative battery cable.

2. Remove the starter wiring, then remove the starter motor assembly.

3. Remove the rear and front transaxle brackets.

4. Unfasten the modular clutch assembly works.

5. Remove the transaxle from the vehicle. The transaxle and modular clutch come out as an assembly.

6. Remove the modular clutch assembly from the transaxle input shaft. Handle the components carefully to avoid contaminating the friction surface.

7. Inspect for oil leakage through the engine rear main bearing oil seal and transaxle input shaft seal. If any leakage is noted, it should be fixed at this time.

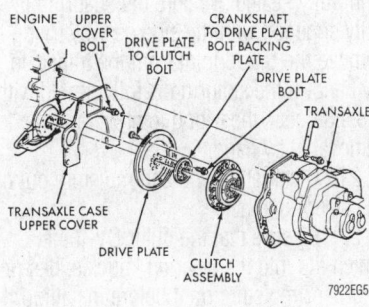

The transaxle and modular clutch are removed as an assembly

To install:

➡️**Always use new bolts when mounting the modular clutch assembly to the drive plate.**

8. Mount the modular clutch assembly onto the input shaft. Install the transaxle.

9. To avoid distorting the drive plate, tighten the bolts gradually a few turns at a time. Use a crisscross pattern, until all bolts are seated. Tighten the bolts to a final torque of 55 ft. lbs. (75 Nm).

10. Install the clutch inspection cover.

11. Install the transaxle lower support brackets.

12. Install the starter assembly, then attach the wiring.

13. Connect the negative battery cable.

Halfshafts

REMOVAL & INSTALLATION

1. Remove the cotter pin, locknut and spring washer from the end of the outer CV-joint stub axle. Discard the cotter pin.

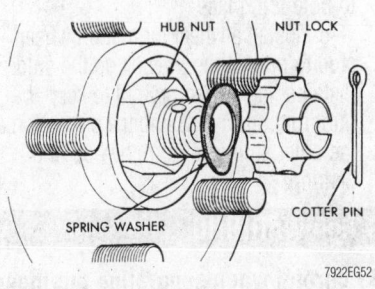

Exploded view of the cotter pin, locknut and washer—halfshaft service

2. With the vehicle on the ground and brakes applied, loosen, but do not remove the stub axle-to-hub and bearing retaining nut. The front hub and driveshaft are splined together and retained by the hub nut.

3. Raise and safely support the vehicle with jack stands.

4. Remove the front wheel and tire assembly.

5. Unfasten the front caliper-to-steering knuckle bolts.

6. Remove the caliper from the steering knuckle by lifting the bottom of the caliper away from the steering knuckle, then remove the top of the caliper out from under the steering knuckle.

7. Support the caliper out of the way by suspending it with a piece from the strut. Do NOT allow the caliper to hang by the brake hose.

8. Remove the rotor from the hub.

9. Remove the nut attaching the outer tie rod end to the steering knuckle, as follows:

a. Hold the tie rod end stud with a $1\frac{1}{32}$ in. socket while loosening and removing the nut.

10. Separate the tie rod end stud from the steering knuckle using a side puller.

11. Remove the nut and bolt holding the ball joint stud into the steering knuckle.

✳✳ WARNING

Be careful when separating the ball joint stud from the steering knuckle, so the ball joint seal does not get damaged.

12. Separate the ball joint stud from the steering knuckle by prying down on the lower control arm.

13. For 1998–99 vehicles, perform the following steps:

   a. Remove the hub and bearing-to-stub axle retaining nut.

   b. Install a suitable puller on the hub and bearing assembly, using the lug nuts to hold it in place.

   c. Install a wheel lug nut on wheel stud to protect the threads on the stud. Install a flat-bladed prytool to keep the hub from turning. Using the puller, force the outer stub axle from the hub and bearing.

✳✳ WARNING

Be careful when separating the inner CV-joint during this operation. Do not let the driveshaft hang by the inner CV-joint, the driveshaft must be supported.

14. Pull the steering knuckle assembly out and away from the outer CV-joint of the driveshaft assembly. Support the outer end of the driveshaft assembly.

15. Insert a prybar between the inner tripod joint and the transaxle case. Pry against the inner tripod joint until the joint retaining snapring is disengaged from the transaxle side gear.

➡️ **Inner tripod joint removal is easier if you apply outward pressure on the joint as you hit the punch with a hammer.**

16. For 1996–99 vehicles, remove the inner tripod joints from the side gears of the transaxle using a punch to dislodge the inner tripod joint retaining ring from the transaxle side gear. If removing the right-side inner tripod joint, position the punch against the inner tripod joint. Hit the punch sharply with a hammer to dislodge the right inner joint from the side gear. If removing the left-side inner tripod joint, position the punch in the groove of the inner tripod joint. Hit the punch sharply with a hammer to dislodge the left inner tripod joint from the side gear.

17. Hold the inner tripod joint and interconnecting shaft of the driveshaft assembly. Remove the inner tripod joint from the transaxle by pulling it straight out of the transaxle side gear and transaxle oil seal. When removing the tripod joint, do not let the spline or snapring drag across the sealing lip of the transaxle-to-tripod joint oil seal.

✳✳ WARNING

The driveshaft, when installed, acts as a bolt which secures the from hub and bearing assembly. If the vehicle is to be supported or moved on its wheels with a driveshaft removed, install a proper sized bolt and nut through the front hub. Tighten the bolt and nut to 135 ft. lbs. (183 Nm). This will ensure that the hub bearing cannot loosen.

To install:

18. Thoroughly clean the spline and oil seal sealing surface on the tripod joint. Lightly lubricate the oil seal sealing surface on the tripod joint with fresh clean transmission fluid.

19. Holding the driveshaft assembly by the tripod joint and interconnecting shaft, install the tripod joint into the transaxle side gear as far as possible by hand.

20. Carefully align the tripod joint with the transaxle side gears. Then, grasp the driveshaft interconnecting shaft and push the tripod joint into the transaxle side gear until fully seated. Be sure the snapring is fully engaged with the side gear by trying to remove the tripod joint from the transaxle by hand. If the snapring is fully seated with the side gear, the tripod joint will not be removable by hand.

21. Clean all debris and moisture out of the steering knuckle.

22. Be sure that the outer CV-joint, which fits into the steering knuckle, has no debris or moisture on it before installing into the steering knuckle.

23. Slide the driveshaft back into the front hub. Install the steering knuckle into the ball joint stud.

24. Install a NEW steering knuckle-to-ball joint stud bolt and nut. Tighten the nut and bolt to 70 ft. lbs. (95 Nm).

25. Insert the tie rod end into the steering knuckle. Start the tie rod end-to-steering knuckle nut onto the stud of the tie rod end. While holding the stud of the tie rod end stationary, tighten the nut. Then, using a crow foot and 11/32 in socket, tighten the tie rod end nut to 45 ft. lbs. (61 Nm).

26. Install the rotor back onto the hub and bearing assembly.

27. Position the caliper on the steering knuckle. Slide the top of the caliper under the top abutment on the steering knuckle, then install the bottom of the caliper against the bottom abutment of the steering knuckle.

28. Install the caliper-to-knuckle bolts and tighten to 23 ft. lbs. (31 Nm).

29. Clean all foreign matter from the threads of the outer CV-joint stub axle. Install hub nut and washer onto the threads of the stub axle and tighten the nut.

30. With the vehicle's brakes applied to prevent the axle shaft from turning, tighten the hub nut to 135 ft. lbs. (183 Nm).

31. Install the spring washer, locknut and new cotter pin into the outer CV-joint stub axle.

32. Install the front wheel and tire assembly. Install the lug nuts and tighten to 100 ft. lbs. (135 Nm).

33. Check the transaxle fluid level, lowering the vehicle as necessary.

34. If not already done, carefully lower the vehicle.

STEERING AND SUSPENSION

Air Bag

✳✳ CAUTION

Some vehicles are equipped with an air bag system, also known as the Supplemental Restraint System (SRS). The system MUST BE disabled before performing service on or around system components, steering column, instrument panel components, wiring and sensors. Failure to follow safety and disabling procedures could result in accidental air bag deployment, possible personal injury and unnecessary system repairs.

PRECAUTIONS

Several precautions must be observed when handling the inflator module to avoid accidental deployment and possible personal injury.

• Never carry the inflator module by the wires or connector on the underside of the module.

• When carrying a live inflator module, hold securely with both hands, and ensure that the bag and trim cover are pointed away.

• Place the inflator module on a bench or other surface with the bag and trim cover facing up.

• With the inflator module on the bench, never place anything on or close to the module which may be thrown in the event of an accidental deployment.

DISARMING

Proper SRS disarming can be obtained by disconnecting and isolating the negative battery cable. Allow the air bag system capacitor at least two minutes to discharge before removing air bag system components.

Rack and Pinion Steering Gear

The replacement procedure for both the manual and power steering gears is the same. The only additional steps for power steering gear removal and installation is the disconnection and connection of the power steering fluid lines from and to the steering gear

These vehicles are designed and assembled using NET BUILD front suspension alignment settings. This means that the alignment settings are determined as the vehicle is designed by the location of the front suspension components in relation to the body. This is carried out when building the vehicle, by precisely locating the front crossmember to meter gauge holes located in the underbody of the vehicle. With this method of designing and building a vehicle, it is no longer possible to adjust a vehicle's front suspension alignment settings to the required specifications. As a result, whenever the crossmember is removed from a vehicle, it MUST be replaced in the same location on the body of the vehicle it was removed from. The front suspension toe settings can still be adjusted by the outer tie rod ends.

REMOVAL & INSTALLATION

1. From inside the vehicle, disconnect the steering gear coupler, from the steering column shaft coupler.
2. Raise and safely support the vehicle with jackstands.
3. Remove both front wheel and tire assemblies.
4. If equipped, remove the engine/transaxle bobble damper, from the front suspension crossmember. The bobble strut does not have to be removed from the transaxle.
5. Unfasten the nut attaching the outer tie rod end to the steering knuckle. The nut is removed by holding the tie rod end stud with an $^{11}/_{32}$ in. socket while loosening and removing the nut with a wrench.
6. Separate the tie rod ends from the steering knuckles using a side puller, tool MB-991113 or equivalent.
7. If equipped with power steering, perform the following:

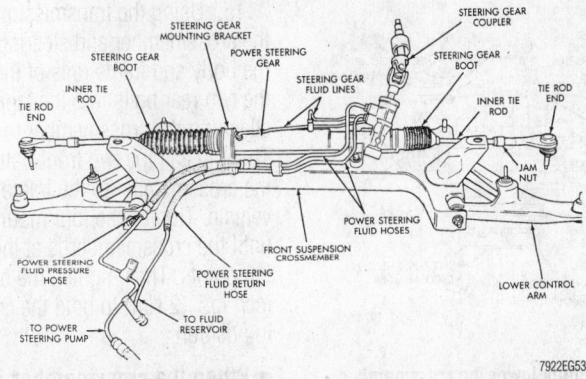

Identification of the rack and pinion steering gear components

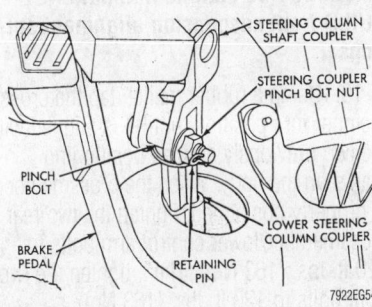

You must disconnect the steering gear coupler from inside the vehicle

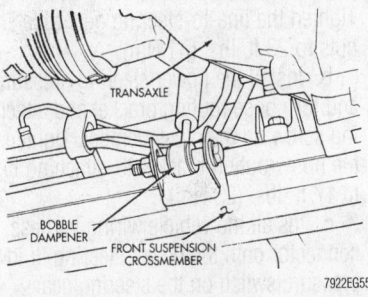

Some vehicles are equipped with a bobble damper which must be removed from the crossmember during steering gear service

a. Remove the vehicle wiring harness connector from the power steering fluid pressure switch.
b. Remove the power steering pressure and return hose routing bracket from the front crossmember. The bracket does not have to be removed from the power steering pressure and turn hoses.
c. Remove the power steering fluid, pressure and return hoses from the power steering gear assembly.

※※ WARNING

Before removing the crossmember from the vehicle, the location of the

crossmember MUST be scribed or marked on the vehicle, as shown in the accompanying figure. This has to be done so the crossmember can be installed in the exact location from which it was removed. If this is not done, the proper NET BUILD alignment specifications will not be obtained and may lead to handling and/or tire wear problems.

8. Using a paint marker or an awl or equivalent, make a line marking the location where the crossmember is mounted against the body of the vehicle.

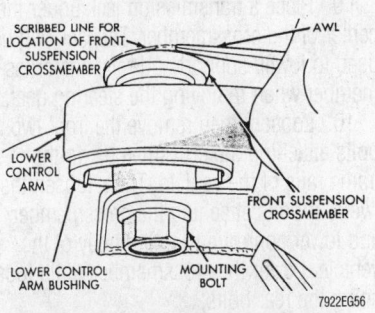

Because of the method used for aligning the vehicle, you MUST matchmark the crossmember installed position prior to steering gear removal

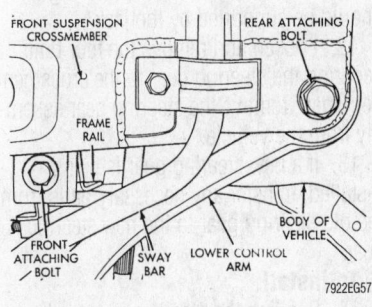

Location of the front crossmember mounting bolts

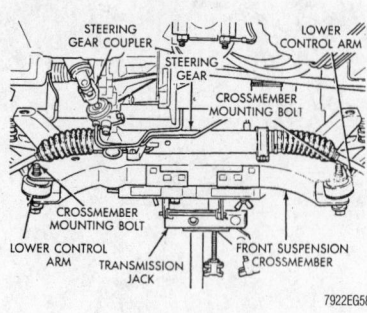

You must carefully lower the crossmember for access to the steering gear

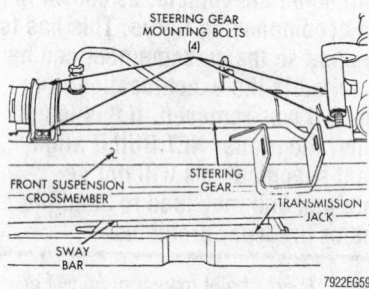

Unfasten the mounting bolts, then remove the steering gear from the vehicle

9. Place a transmission jack under the center of the crossmember. The jack will be used to lower, support and raise the crossmember when removing the steering gear.

10. Loosen, then remove the front two bolts attaching the crossmember to the frame rails of the vehicle. Then, loosen the two rear bolts attaching the crossmember and lower control arm to the body of the vehicle. Lower the crossmember while loosening the rear bolts.

11. Using the transmission jack, carefully lower the crossmember enough to allow the steering gear to be removed from the crossmember. When lowering the crossmember, do not let the crossmember hang from the lower control arms, the weight should be supported by the jack.

12. Loosen and remove the four bolts securing the steering gear to the crossmember. Then, remove the steering gear assembly from the vehicle.

13. If a new steering gear is being installed, transfer any necessary parts from the old steering gear to the new steering gear.

To install:

14. Position the steering gear on the crossmember. Install the four mounting bolts and tighten to 50 ft. lbs. (68 Nm).

15. Using the transmission jack, raise the crossmember and steering gear against the body and frame rails of the vehicle. Start the two rear bolts into the tapping plates, attaching the crossmember to the body. Then, install the two front bolts, attaching the crossmember to the frame rails of the vehicle. Tighten the four mounting bolts until the crossmember is at the four mounting points. Then, tighten the bolts to 20 inch lbs. (2 Nm) to hold the crossmember in position.

➡ **When the crossmember is reinstalled in the vehicle, it MUST be aligned with the matchmarks made during removal. This MUST be done to maintain NET BUILD front suspension alignment settings.**

16. Using a rubber mallet, tap the crossmember into position, until it is aligned with the two previously scribed positioning marks on the body. When the crossmember is properly positioned, tighten the two rear crossmember/lower control arm bolts to 120 ft. lbs. (163 Nm). Then, tighten the two front bolts to 120 ft. lbs. (163 Nm).

17. If equipped with power steering, perform the following:

a. Install the power steering fluid pressure and return hoses into the correct fluid ports on the steering gear. Tighten the line-to-steering gear tube nuts to 23 ft. lbs. (31 Nm).

b. Install the power steering pressure and turn hose routing bracket and attaching screw on the crossmember. Tighten the hose routing bracket-to-attaching bolt to 17 ft. lbs. (23 Nm).

c. Install the vehicle wiring harness connector onto the power steering fluid pressure switch on the steering gear assembly. Be sure the locking tab on the wiring harness connector is securely latched to the pressure switch.

18. Install the tie rod end into the steering knuckle. Start the tie rod end-to-steering knuckle attaching nut onto the stud of the tie rod end. While holding the stud of the tie rod still, tighten the tie rod end-to-steering knuckle attaching nut. Then, using a crowfoot wrench and 11/32 in. socket, tighten the tie rod end attaching nut to 40 ft. lbs. (55 Nm).

19. If equipped, install the engine/transaxle bobble strut back onto the crossmember bracket. Install and securely tighten the dampener-to-crossmember attaching bolt.

20. Install the wheel and tire assembly and tighten the lug nuts in a crisscross pattern to 100 ft. lbs. (135 Nm).

21. Carefully lower the vehicle.

22. From inside the vehicle, reconnect the steering gear coupler with the steering column shaft coupler. Install the steering gear coupler retaining pinch bolt and tighten to 21 ft. lbs. (28 Nm). Be sure to install the upper-to-lower steering coupler retaining bolt retention pin.

✺✺ WARNING

When filling and bleeding the power steering system, always use the proper type of fluid. NEVER substitute automatic transmission fluid for the specified fluid.

23. If equipped with power steering, fill the power steering pump fluid reservoir to the FULL-COLD level with the proper type and amount of fluid. Bleed the system.

Strut

REMOVAL & INSTALLATION

Front

1. With the vehicle on the ground, loosen the wheel lug nuts.

2. Raise and safely support the vehicle.

3. Remove the wheel and tire assembly.

4. If both strut assemblies are being removed, mark each one right or left, as applicable.

5. Remove the hydraulic brake hose routing bracket and attaching screw from the strut damper bracket. If equipped with ABS, the hydraulic hose routing bracket is combined with the speed sensor cable routing bracket.

➡ **The steering knuckle-to-strut assembly attaching bolts are serrated and must not be turned during removal. Remove the nuts while holding the bolts stationary in the steering knuckle.**

6. Hold the bolts in place, then remove the two nuts securing the strut to the steering knuckle.

7. Remove the three nuts attaching the upper mount of the strut to the strut tower of the vehicle. If necessary, partially lower the vehicle for access to the upper mounting nuts.

8. Carefully remove the strut assembly from the vehicle.

To install:

9. Install the strut assembly into the strut tower, aligning the three studs on the upper strut mount into the holes in the shock tower. Install the three upper strut

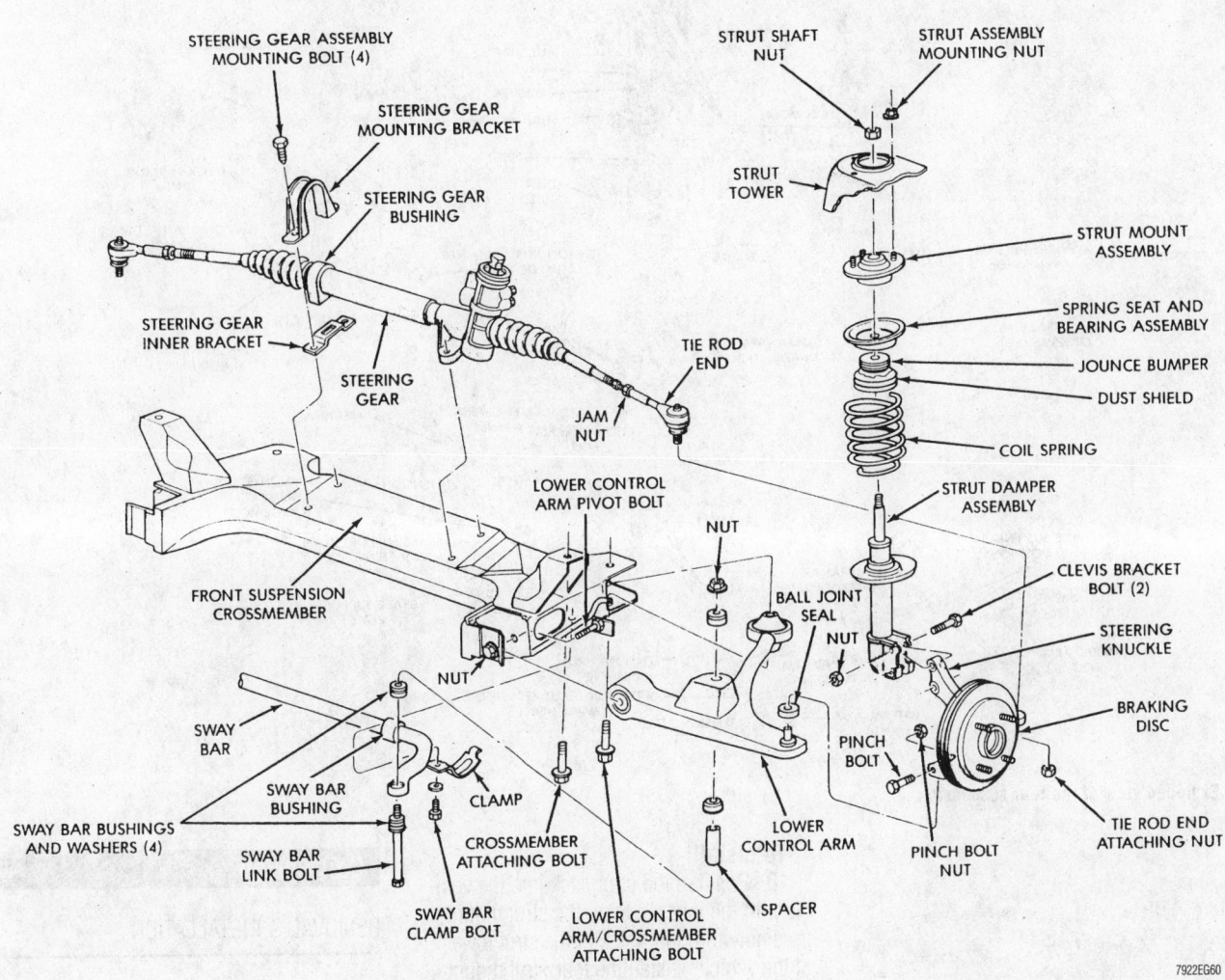

STEERING GEAR ASSEMBLY
MOUNTING BOLT (4)

STEERING GEAR
MOUNTING BRACKET

STEERING GEAR
BUSHING

STEERING GEAR
INNER BRACKET

STEERING GEAR

TIE ROD
END

JAM
NUT

LOWER CONTROL
ARM PIVOT BOLT

NUT

FRONT SUSPENSION
CROSSMEMBER

NUT

SWAY
BAR

SWAY BAR
BUSHING

CLAMP

SWAY BAR BUSHINGS
AND WASHERS (4)

SWAY BAR
LINK BOLT

SWAY BAR
CLAMP BOLT

CROSSMEMBER
ATTACHING BOLT

LOWER CONTROL
ARM/CROSSMEMBER
ATTACHING BOLT

SPACER

BALL JOINT
SEAL

NUT

LOWER
CONTROL ARM

PINCH
BOLT

PINCH BOLT
NUT

STRUT SHAFT
NUT

STRUT ASSEMBLY
MOUNTING NUT

STRUT
TOWER

STRUT MOUNT
ASSEMBLY

SPRING SEAT AND
BEARING ASSEMBLY

JOUNCE BUMPER

DUST SHIELD

COIL SPRING

STRUT DAMPER
ASSEMBLY

CLEVIS BRACKET
BOLT (2)

STEERING
KNUCKLE

BRAKING
DISC

TIE ROD END
ATTACHING NUT

7922EG60

Exploded view of the front suspension

mount retaining nut and washer assemblies. Tighten the three nuts to 23 ft. lbs. (31 Nm).

➡ **The steering knuckle-to-strut assembly attaching bolts are serrated and must not be turned during installation. Install the nuts while holding the bolts stationary in the steering knuckle.**

10. Align the strut assembly with the steering knuckle. Position the arm of the steering knuckle into the strut assembly, aligning the strut assembly to the steering knuckle mounting holes. Install the two strut-to-steering knuckle bolts. The bolts should be installed with the nuts facing the front of the vehicle. Tighten both attaching bolts to 40 ft. lbs. (53 Nm), plus an additional ¼ turn after the specified torque is met.

11. Install the hydraulic brake hose routing bracket and attaching screw onto the

strut damper bracket. If the vehicle has ABS, the hydraulic hose routing bracket is combined with the speed sensor cable routing bracket. Tighten the bracket attaching bolts to 10 ft. lbs. (13 Nm).

12. Install the wheel and tire assembly, then tighten the lug nuts in sequence hand-tight.

13. Carefully lower the vehicle, then tighten the lug nuts, in sequence, to 100 ft. lbs. (135 Nm).

Rear

1. Raise and safely support the vehicle.
2. Remove the wheel and tire assembly.
3. Unfasten the retainer(s), then remove the hydraulic flex hose bracket from the strut bracket. If equipped with ABS, the wheel speed sensor cable routing clip is also attached to the strut assembly bracket.

4. Support the rear knuckle, suspension and brake components before removing the clevis bracket-to-knuckle attaching bolts. Do NOT allow the weight of the knuckle and related components hang without support when the strut is removed.

✳✳ **WARNING**

The knuckle-to-strut attaching bolts are serrated and must not be turned during removal. Remove the nuts while holding the bolts stationary in the knuckle.

5. Hold the bolt with a wrench, then unfasten the two clevis bracket nuts attaching the strut to the knuckle.

6. Carefully lower the vehicle, then open the trunk. Access to the rear upper strut mount-to-strut tower attaching bolts is through the trunk of the vehicle.

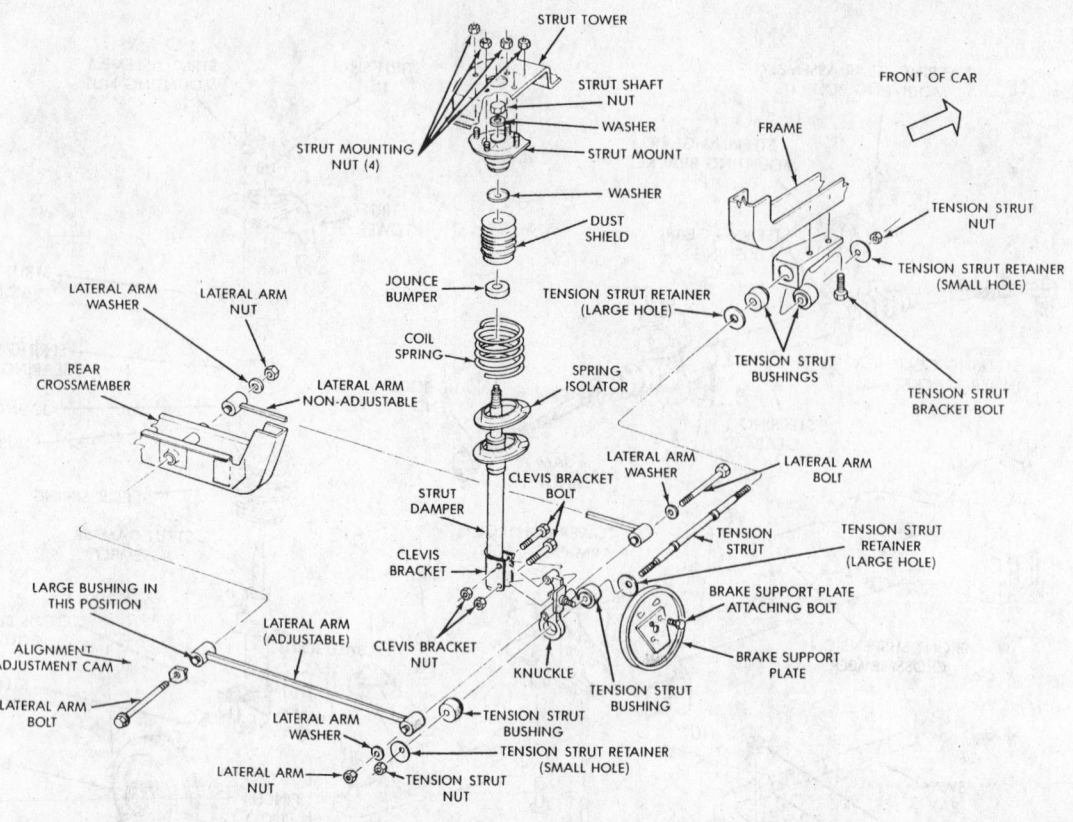

Exploded view of the rear suspension

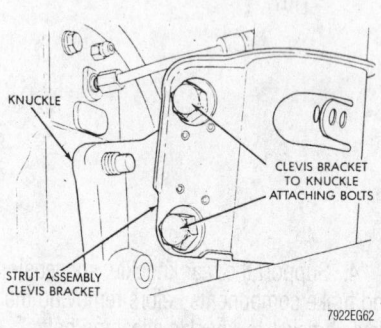

Location of the knuckle-to-clevis bracket bolts

7. If necessary, remove the carpet from the top of the strut tower. Then, remove the rubber dust shield from the top of the strut tower for easier access to the upper strut nuts.

8. Loosen, but do not remove the four upper strut mounting nuts. Then, while supporting the strut assembly, fully remove the four strut mount attaching nuts.

9. Remove the strut assembly from the knuckle by sliding the knuckle out of the clevis bracket on the strut, then remove it from the vehicle.

To install:

10. Position the strut back into the vehicle with the four studs on the strut mount assembly through holes in the strut tower of the vehicle. Install the four strut mount-to-body attaching nuts onto the mount studs. Tighten the nuts to 25 ft. lbs. (34 Nm).

11. Install the dust shield into the opening on top of the strut tower. Install the carpeting back on top of the rear strut tower.

12. Raise and safely support the vehicle.

13. Install the knuckle into the clevis bracket on the strut assembly. Install the two clevis bracket-to-knuckle attaching bolts and nuts. Hold the bolts with a wrench while tightening the nuts to 70 ft. lbs. (95 Nm).

14. Install the brake hose bracket to the strut bracket and secure with the retaining bolts. If equipped with ABS, the wheel speed sensor cable routing clip is also attached to the strut bracket.

15. Install the wheel and tire assembly on the vehicle. Tighten the lug nuts evenly, in sequence, to 100 ft. lbs. (135 Nm).

16. Carefully lower the vehicle.

17. Have the toe checked and adjusted as necessary.

Coil Spring

REMOVAL & INSTALLATION

Front

1. Remove the strut assembly from the vehicle.

2. Clamp the strut in a suitable vise, in a vertical position. When clamping the strut in the vise, do not clamp the strut using the body of the strut, only by the strut clevis bracket, as shown in the accompanying figure.

3. Mark the coil spring and strut assembly right or left, according to which side of the vehicle to strut was removed from, and which strut the coil spring was removed from.

✳✳ CAUTION

Do NOT remove the strut rod nut before the strut assembly coil spring is compressed, removing spring tension from the upper spring seat and bearing assembly. When compressing the coil spring for removal from the strut, the first full top and bottom

coil of the spring must be held by the jaws of the spring compressor.

4. Carefully compress the strut assembly coil spring, using a suitable spring compressor tool.

5. Install a suitable strut nut socket to or an open ended wrench on the strut shaft retaining nut. Then, install a 10mm socket on the hex end of the strut damper shaft. While holding the strut shaft from turning, remove the strut shaft retaining nut.

6. Remove the strut assembly mount/isolator from the strut.

7. Remove the upper spring seat, pivot bearing and dust shield as an assembly from the strut.

8. Remove the jounce bumper from the shaft of the strut assembly.

9. Remove the coil spring from the strut assembly. Mark left and right on the coil springs for their installation back on the correct side of the vehicle.

❊❊ WARNING

If a replacement coil spring is being installed on the strut, then first full top and bottom coil of the spring must be captured by the jaws of the coil spring compressor.

10. Inspect the strut for any binding of the strut shaft over the full stroke of the shaft.

11. Inspect the strut mount and upper spring seat assembly for any of the following conditions:

• Check the mount for cracks and distortion and the retaining studs for any sign of damage.

• Inspect for severe deterioration of rubber isolator, binding of the strut pivot bearing. If the pivot bearing is replaced, it is to be installed with the while side of the bearing facing up.

• Inspect the dust shield for rips and/or deterioration.

• Check the jounce bumper for cracks and/or signs of deterioration.

12. Replace any components of the strut assembly found to be worn or defective during the inspection, before assembling the strut.

To install:

13. Clamp the strut, in a vertical position, in a suitable vise. Only clamp the strut by the clevis bracket, NEVER by the body of the strut.

14. Install the compressed coil spring onto the strut. The spring should be installed with the smaller coil down, so the spring properly seats on the strut.

15. Install the jounce bumper on the strut shaft.

16. Install the dust shield, pivot bearing and upper spring seat as an assembly on the strut.

17. Position the upper spring seat alignment notch with the clevis bracket on the strut assembly.

18. Install the strut mount on the strut assembly and the strut mount retaining nut on the shaft of the strut assembly.

❊❊ CAUTION

The following two steps must be completely done before the spring compressor tool can be released from the coil spring.

19. Install strut nut socket or an open end wrench on the strut shaft retaining nut. Then, install a 10mm socket through the center of the socket an on the hex of the strut shaft. While holding the strut shaft from turning, tighten the strut shaft retaining nut to 55 ft. lbs. (75 Nm).

20. Equally loosen both spring compressor tools, until the top coil of the spring is fully seated against the upper spring seat and strut mount. Then, relieve all tensioner from the spring compressors and remove the compressors from the strut spring.

21. Install the strut into the vehicle, as outlined earlier.

Rear

1. Remove the strut requiring overhaul from the vehicle.

2. Position the strut in a vise. Using paint or a marker, matchmark the strut unit, lower spring isolator, spring and upper strut mount for installation purposes.

3. Place suitable spring compressor tools (Special tool C-4838) on the strut spring. Compress the coil spring until all load is removed from the upper strut mount.

4. Install a suitable strut nut socket to or an open ended wrench on the strut shaft retaining nut. Then, install a 10mm socket on the hex end of the strut damper shaft. While holding the strut shaft from turning, remove the strut shaft retaining nut.

5. Remove the washer between the strut shaft nut and the upper strut mount and isolator.

6. Remove the upper strut mount assembly from the strut shaft and spring.

7. Remove the washer from the strut shaft that is between the strut upper mount assembly and dust shield.

8. Remove the coil spring and spring compressor as an assembly from the strut.

9. Remove the dust shield from the strut assembly.

10. Remove the jounce bumper from the strut shaft.

11. Remove the coil spring lower isolator from the strut assembly spring seat.

12. Inspect all of the disassembled components for damage, abnormal wear or failure. Check the strut unit for excessive oil leakage and/or loss of oil charge, as follows:

a. Push the strut shaft into the body of the strut and release, the strut shaft should return to its original position.

b. If the shaft does not return to its original position, replace the strut unit.

To install:

13. Install the isolator on the lower spring seat of the strut.

14. Install the jounce bumper on the strut shaft.

15. Install the dust shield on the strut assembly.

16. Lower the coil spring onto the strut unit. Position the end of the coil spring against the edge of the spring isolator on the lower spring seat of the strut assembly.

17. Install the washer on the strut shaft with the raised edge of the washer facing upward.

18. Place the strut upper mount onto the strut shaft.

19. Install the washer on the strut upper mount, The washer must be installed with the raised edge of the washer facing down.

20. Install the upper strut mount-to-strut shaft retaining nut.

21. Use a strut rod socket or an open ended wrench and a 10mm socket, to prevent the strut shaft from rotating, tighten the strut shaft nut to 45 ft. lbs. (60 Nm).

22. Equally loosen the spring compressors until the spring is seated on the upper strut mount and all tension is relieved from the spring compressors.

23. Install the strut in the vehicle.

24. Have the toe checked and adjusted as necessary.

Lower Ball Joint

The front suspension ball joints operate with no free-play. The ball joints are replaceable ONLY as an assembly. Do not attempt any type of repair on the ball joint assembly. The ball joint is a press fit into the lower control arm with the joint stud retained in the steering knuckle by the clamp bolt. To check the ball joint, with the weight of the vehicle resting on the road wheels, grasp the grease fitting and without using any tools, attempt to move the grease fitting. If the ball

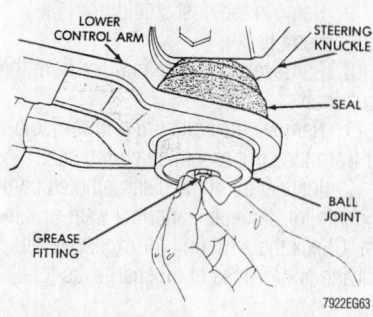

LOWER CONTROL ARM
STEERING KNUCKLE
SEAL
BALL JOINT
GREASE FITTING
7922EG63

Wiggle the grease fitting with your fingers—if it moves, the ball joint should be replaced

joint is worn the grease fitting will move easily. If movement is noted, replacement of the ball joint is recommended.

REMOVAL & INSTALLATION

1. Raise and safely support the vehicle. Remove the wheel and tire assembly.
2. Remove the steering knuckle-to-ball joint ball stud, clamping nut and bolt.
3. Remove the two attaching links connecting the stabilizer bar to the lower control arms.
4. Loosen, but do not remove the bolts holding the stabilizer bar retainers to the crossmember. Then, rotate the stabilizer bar and attaching links away from the lower control arms.

✳✳ WARNING

Pulling the steering knuckle out from the vehicle after releasing the ball joint can separate the inner CV-joint.

5. Use a prybar to separate the steering knuckle from the ball joint stud. Be careful when separating the ball joint stud from the knuckle, so the seal does not become damaged.
6. Remove the front lower control arm bushing-to-crossmember attaching nut and bolt. Remove the rear lower control arm-to-crossmember and frame rail attaching bolt. Then, remove the lower control arm from the crossmember.
7. Carefully pry the seal boot off the ball joint, using a suitable prytool.
8. Using a suitable press remove the ball joint from the lower control arm.

To install:

9. Reinstall the ball joint into the lower control arm with the notch in the ball joint stud facing the front lower control arm bushing.
10. Using a suitable press, press the ball joint into the lower control arm.

11. Reinstall the ball joint boot seal using a suitable driver such as a large socket or suitable sized piece of pipe. Do not use a shop press as was used to install the ball joint as a press exerts too much force.
12. Position the lower control arm into the front crossmember. Install the rear lower control arm-to-crossmember and frame rail attaching bolt. Do NOT tighten the rear bolt at this time. Then, install the front lower control arm-to-crossmember nut and bolt.
13. Tighten the front lower, then the rear control arm nut and bolt to 120 ft. lbs. (163 Nm).
14. Place the ball joint stud into the steering knuckle. Install the steering knuckle-to-ball joint stud clamping bolt and nut. Tighten the bolt to 70 ft. lbs. (95 Nm).
15. Assemble the stabilizer bar-to-lower control arm link assemblies and bushings.
16. Rotate the stabilizer bar into position, installing the stabilizer bar links into the lower control arms. Install the top stabilizer bar link bushings and nuts. Do NOT tighten the link yet.
17. Install the wheel and tire assembly.
18. Carefully lower the vehicle so the suspension is supporting the total weight of the vehicle.
19. Tighten the stabilizer bar-to-lower control arm links to 21 ft. lbs. (28 Nm).
20. Tighten the stabilizer bar bushing retainer-to-crossmember attaching bolts to 21 ft. lbs. (28 Nm).
21. The toe should be checked and adjusted as necessary.

Wheel Bearings

ADJUSTMENT

Neons are equipped with sealed hub and bearing assemblies. The hub and bearing assembly is non-serviceable. If the assembly is damaged, the complete unit must be replaced.

REMOVAL & INSTALLATION

Front

1995 VEHICLES

This vehicle uses a sealed for life front hub and bearing assembly attached to the front steering knuckle. The outer CV-joint assembly is splined to the front hub and bearing assembly. The front wheel bearing is called a cartridge bearing. The wheel bearing can be serviced separately from the front steering knuckle and hub assembly.

Installation and retention of the front wheel bearing in the steering knuckle is by an interference fit and retained by a snapring. If the front wheel bearing requires replacement, the hub must be removed from the original wheel bearing and transferred to the replacement bearing.

Note that the Neon does not use a rubber lip seal as on past front wheel drive cars to prevent contamination of the front wheel bearing. On this vehicle, the face of the outer CV-joint fits deeply into the steering knuckle using a close fit. This design deters direct water splash on the bearing seal while allowing any water that gets in, to run out the bottom. It is important to thoroughly clean the outer CV-joint and the wheel bearing area in the steering knuckle before it is assembled after servicing.

The steering knuckle MUST be removed to replace both the hub and the front wheel bearing.

1. Remove the front hub cotter pin, nut lock and spring washer. Discard the cotter pin.

✳✳ WARNING

Wheel bearing damage will result if, after loosening the hub nut, the vehicle is rolled on the ground or the weight of the vehicle is allowed to be supported by the tires.

2. Loosen the hub nut while the vehicle is on the floor with the brakes applied. The front hub and halfshaft are splined together through the knuckle (bearing) and retained by the hub nut. The front wheel bearing supports the front hub and weight of the vehicle.
3. Raise and safely support the vehicle. Remove the front wheels.
4. Remove the front disc brake caliper from the steering knuckle. The caliper is removed by first lifting the bottom of the caliper away from the steering knuckle, then removing the top of the caliper out from under the steering knuckle. Support the caliper using wire. Do not allow the caliper to hand by the brake hose.
5. Remove the brake rotor from the front hub/bearing assembly.
6. Remove the nut attaching the outer tie rod end to the steering knuckle.
   a. Hold the tie rod end stud with an 11/32 inch socket while loosening and removing the nut with the wrench.
   b. Remove the tie rod end from the steering knuckle using a puller. Do not hammer wedge-type tools or the steering tie rod end joint will be damaged.

7. Locate and remove the lower control arm ball joint clamping nut and bolt and separate the ball joint stud from the steering knuckle by prying down on the lower control arm. Use care not to damage the steering tie rod end or seal. In addition, use care not to allow the halfshaft to become overextended as the steering knuckle is removed. Do not allow the halfshaft to hang by the inner CV-joint boot. The halfshaft must be supported.

➡ **The steering knuckle to strut assembly attaching bolts are serrated and must NOT be turned during removal. Remove and reinstall the nuts while holding the bolts stationary in the steering knuckle.**

8. With the steering knuckle removed from the vehicle, use a suitable press to remove the wheel bearing from the steering knuckle. Use care to jig the knuckle level in the press bed and press the wheel hub and bearing slowly from the knuckle. One bearing race may come out with the hub when the hub is removed. Remove the knuckle from the press and remove the snapring retaining the hub bearing in the steering knuckle. Reposition the knuckle in the press and press the hub bearing from its bore.

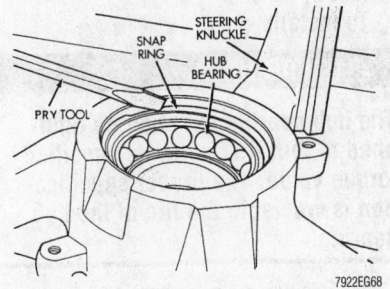

Use a prytool to carefully remove the snapring from the hub and bearing

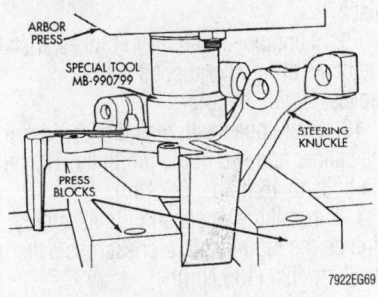

Remove the hub and bearing from the steering knuckle

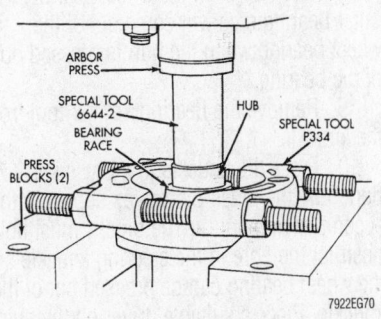

You must use a suitable press to remove the bearing race from the hub

To install:

9. Clean all parts well. Again, use care to jig the steering knuckle level in the press bed. Place the new hub bearing into the bore of the steering knuckle so it is square with the bore. Place a bearing driver on the outer race of the hub bearing and press the hub bearing into the steering knuckle until it is fully seated in the bottom of its bore. Install the hub bearing retaining snapring into its groove in the knuckle bore. Be sure it is fully seated in its groove. Use care not to damage the just-installed bearing seal when installing the snapring.

10. Again, place the knuckle assembly with the hub bearing installed on the press bed using care to align and level the assembly. Use suitable drivers and arbors to support the hub bearing on its inner race. Place the wheel hub in the bearing using care to align it square with the bearing. Using suitable drivers, press the hub into the bearing until it bottoms in the hub bearing.

11. Reinstall the steering knuckle/hub/wheel bearing assembly back into the front strut and install the through-bolts. The steering knuckle-to-strut bolts are serrated (toothed) and must not be turned in the steering knuckle during installation. Tighten the nuts (do not turn the bolt heads) to 40 ft. lbs. (54 Nm) plus an additional ¼ turn after the specified torque is met.

12. Slide the halfshaft back into the front hub and bearing assembly. Then, install the steering knuckle onto the ball joint stud.

13. Install a NEW steering knuckle-to-ball joint stud, clamp bolt and nut. Tighten the clamp bolt to 75 ft. lbs. (100 Nm).

14. Reinstall the tie rod end into the steering knuckle. Start the tie rod end-to-steering knuckle attaching nut onto the stud of the tie rod end. While holding the stud of the tie rod end stationary, tighten the tie rod end nut. Using a crow's foot wrench and 11/32 inch socket, tighten the tie rod end nut to 40 ft. lbs. (55 Nm).

15. Reinstall the brake rotor and the caliper onto the steering knuckle.

   a. The caliper is installed by first sliding the top of the caliper under the top abutment on the steering knuckle, then installing the bottom of the caliper against the bottom abutment of the steering knuckle.

   b. Install the caliper attaching bolts and tighten to 23 ft. lbs. (31 Nm).

16. Clean all foreign matter from the threads of the outer CV-joint stub axle. Install the hub nut onto the threads of the stub axle and tighten the nut.

17. With the vehicle's brakes applied to keep the brake rotor from turning, tighten the hub nut to 135 ft. lbs. (183 Nm).

18. Reinstall the wheels and lower the vehicle.

19. Reinstall the spring washer, hub nut lock and a new cotter pin. Wrap the cotter pin prongs tightly around the hub nut lock.

20. Take the vehicle to a reputable repair shop to have the front end alignment checked and the toe adjusted as required.

1996–99 VEHICLES

1. Remove the steering knuckle and hub and bearing assembly from the vehicle.

2. Use a suitable C-clamp and adapter, tool 4150A or equivalent to press one wheel lug stud out of the hub flange.

3. Rotate the hub to align the removed lug stud with the notch in the bearing retainer plate. Remove the lug stud from the hub.

4. Rotate the hub so the hole in the hub that the stud was removed from is facing away from the brake caliper lower rail on the steering knuckle. Install one half of a bearing splitter tool, Special tool 1130 or equivalent between the hub and the bearing retainer plate. The threaded hold in this half of the bearing splitter is to be aligned with the caliper rail on the steering knuckle.

5. Install the remaining pieces of the bearing splitter on the steering knuckle.

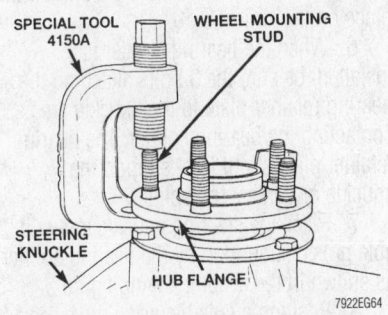

Use a proper C-clamp and adapter tool to press out one of the lug studs—1996–99 models

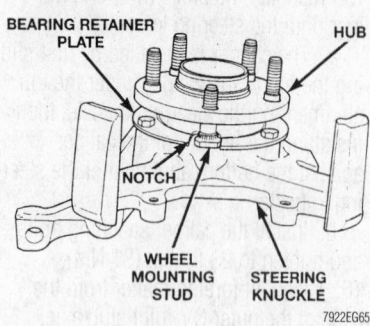

Rotate the hub in order to remove the lug stud—1996–99 models

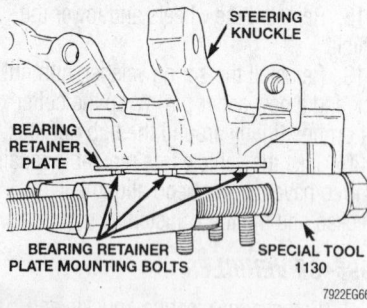

Proper installation of the bearing splitter—1996–99 models

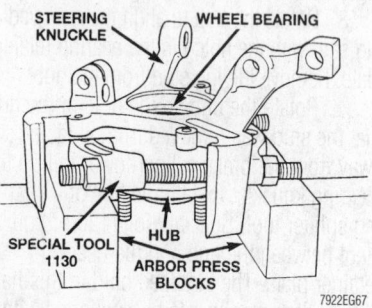

The steering knuckle must be properly supported for hub and bearing removal—1996–99 models

Hand-tighten the nuts to hold the splitter in place on the knuckle.

6. When the bearing splitter is installed, be sure the 3 bolts attaching the bearing retainer plate to the knuckle are contacting the bearing splitter. The bearing retainer plate should not support the knuckle or contact the splitter.

7. Place the steering knuckle in a suitable press, supported by the bearing splitter as shown in the accompanying figure.

8. Position a suitable sized driver on the small end of the hub. Using the press,

remove the hub from the wheel bearing. The outer bearing race will come out of the wheel bearing when the hub is pressed out of the bearing.

9. Remove the bearing splitter tool from the knuckle.

10. Place the knuckle in a press supported by the press blocks, as shown in the accompanying figure. The blocks must not obstruct the bore in the steering knuckle so the wheel bearing can be pressed out of the knuckle. Place a suitable driver on the outer race of the wheel bearing, then press the bearing out of the knuckle.

11. Install the bearing splitter on the hub. The splitter is to be installed on the hub so it is between the flange of the hub and the bearing race on the hub. Place the hub, bearing race and splitter in a press. Use a driver to press the hub out of the bearing race.

To install:

12. Use clean, dry cloth to wipe and grease or dirt from the bore of the steering knuckle.

13. Clean the rust preventative from the replacement wheel bearing using a clean, dry towel.

14. Place the new wheel bearing into the bore of the steering knuckle. Be sure the bearing is placed squarely into the bore. Place the knuckle in a press with a receiver tool, C-4698–2 supporting the steering knuckle. Place a suitable driver tool on the outer race of the wheel bearing. Press the wheel bearing into the steering knuckle until it is fully bottomed in the bore of the steering knuckle.

➡️Only the original or original equipment replacement bolts should be used to mounting the bearing retainer to the knuckle. If a bolt requires replacement when installing the bearing retainer plate, be sure to get the proper type of replacement.

15. Install the bearing retainer plate on the steering knuckle. Install the 3 bearing retainer mounting bolts. Tighten the bolts to 21 ft. lbs. (28 Nm).

16. Install the removed wheel lug stud into the hub flange.

17. Place the hub, with the lug stud installed, in a suitable press supported by adapter tool C-4698–1 or equivalent. Press the wheel lug stud into the hub flange until it is fully seated against the back side on the hub flange.

18. Place the steering knuckle, with the wheel bearing installed, in a press with spe-

cial receiver tool MB-990799 supporting the inner race of the wheel bearing. Place the hub in the wheel bearing, making sure it is square with the bearing. Press the hub into the wheel bearing until it is fully bottomed in the wheel bearing.

19. Install the steering knuckle in the vehicle.

20. Install the wheel and tire assembly, then carefully lower the vehicle.

21. Take the vehicle to a reputable repair shop to have the front end alignment checked and the toe adjusted as required.

Rear

1. Raise and safely support the vehicle.

2. Remove the wheel and tire assembly.

3. If equipped with rear drum brakes, remove the brake drum. If equipped with rear disc brakes, remove the caliper and suspend aside with a piece of wire, then remove the rotor. Do NOT allow the caliper to hang by the brake hose.

4. Remove the dust cap from the rear hub/bearing.

5. Remove the retaining nut securing the hub/bearing assembly to the knuckle/spindle. Discard the hub nut and replace with a new one during installation.

6. Remove the hub/bearing from the spindle buy pulling it off the end of the spindle by hand.

To install:

✳✳ WARNING

The hub/bearing nut must be tightened to, but NOT over, its specified torque value. The proper specification is crucial to the life of the hub bearing.

7. Position the hub/bearing assembly on the rear spindle/knuckle. Install a NEW hub nut and tighten to 160 ft. lbs. (217 Nm).

8. Install the dust cap and seat it using a soft face hammer to carefully tap it into place.

9. If equipped with drum brakes, install the brake drum. If equipped with disc brakes, install the rotor.

10. If equipped with disc brakes, install the caliper and two guide pin bolts. Tighten the bolts to 16 ft. lbs. (22 Nm).

11. Install the wheel and tire assembly. Tighten the lug nuts, in a crisscross pattern, to 100 ft. lbs. (135 Nm).

12. Carefully lower the vehicle.

CHRYSLER CORPORATION

Chrysler-Cirrus • **Sebring** Convertible • **Dodge**-Stratus • **Plymouth**-Breeze

DRIVE TRAIN **16-29**
ENGINE REPAIR **16-2**
FUEL SYSTEM **16-26**
STEERING AND
  SUSPENSION **16-32**

A

Air Bag.........................16-32
  Disarming....................16-32
  Precautions 16-32

C

Camshaft 16-18
  Removal & Installation 16-18
Clutch..........................16-30
  Adjustment...................16-30
  Removal & Installation 16-31
Coil Spring 16-34
  Removal & Installation 16-34
Cylinder Head 16-6
  Removal & Installation 16-6

D

Distributor.....................16-2
  Installation...................16-2
  Removal 16-2

E

Engine Assembly 16-2
  Removal & Installation 16-2
Exhaust Manifold................16-16
  Removal & Installation 16-16

F

Front Crankshaft Seal 16-17
  Removal & Installation 16-17

Fuel Filter 16-27
  Removal & Installation 16-27
Fuel Pump 16-28
  Removal & Installation 16-28
Fuel System Pressure 16-27
  Relieving 16-27
Fuel System Service
  Precautions....................16-26

H

Halfshafts......................16-31
  Removal & Installation 16-31

I

Ignition Timing 16-2
  Adjustment...................16-2
Intake Manifold.................16-13
  Removal & Installation 16-13

L

Lower Ball Joint.................16-35
  Removal & Installation 16-35
Lower Control Arm 16-35
  Removal & Installation 16-35

O

Oil Pan.........................16-21
  Removal & Installation 16-21
Oil Pump 16-24
  Removal & Installation 16-24

P

Power Rack and Pinion Steering
  Gear............................16-32
  Removal & Installation 16-32

R

Rear Main Seal 16-26
  Removal & Installation 16-26
Rocker Arm/Shaft................16-11
  Removal & Installation 16-11

S

Shock Absorber 16-34
  Removal & Installation 16-34

T

Transmission Assembly 16-29
  Removal & Installation 16-29

U

Upper Ball Joint.................16-35
  Removal & Installation 16-35
Upper Control Arm 16-35
  Removal & Installation 16-35

V

Valve Lash 16-21
  Adjustment...................16-21

W

Water Pump 16-5
  Removal & Installation 16-5
Wheel Bearings.................16-36
  Adjustment...................16-36
  Removal & Installation 16-36

ENGINE REPAIR

→**Disconnecting the negative battery cable on some vehicles may interfere with the functions of the on board computer systems and may require the computer to undergo a relearning process, once the negative battery cable is reconnected.**

Distributor

REMOVAL

2.5L Engine

The 2.5L engine is equipped with a camshaft driven mechanical distributor. This engine uses a fixed ignition timing system. The basic ignition timing is not adjustable. The Powertrain Control Module (PCM) determines spark advance. The crankshaft position sensor and camshaft position sensor are Hall Effect devices. The crankshaft sensor is mounted remotely from the distributor while the camshaft position sensor is mounted inside the distributor housing. Both sensors generate pulses that are inputs to the PCM. The PCM determines crankshaft position from these sensors. The PCM calculates injector sequence and ignition timing. There is a resistor built into the distributor cap. An ohmmeter connected between the center button and ignition coil terminal should read 5000 ohms.

1. Disconnect the negative battery cable from the left shock tower. The ground cable is equipped with a insulator grommet which should be placed on the stud to prevent the negative battery cable from accidentally grounding.
2. Remove the bolt attaching the air inlet resonator to the intake manifold.
3. Loosen the clamps holding the air cleaner cover to the air cleaner housing.
4. Remove the PCV make-up air hose from the air inlet tube.
5. Remove the EGR tube.
6. Mark for identification, if necessary and remove the spark plug wires from the distributor cap.
7. Remove the distributor cap.
8. Mark the rotor position. A scribe mark indicates where to position the rotor when reinstalling the distributor. Remove the rotor.

9. Detach the two electrical harness connections from the distributor.
10. Remove the two distributor hold-down nuts and washers.
11. Remove the spark plug cable mounting bracket.
12. Remove the transaxle dipstick tube.
13. Carefully remove the distributor from the engine.

INSTALLATION

2.5L Engine

TIMING NOT DISTURBED

1. Inspect the rotor for cracks or burned electrode. Replace if defective. Install the rotor onto the distributor.
2. Inspect the O-ring seal. If nicked or cracked, replace with a new one. Be sure the O-ring is properly seated on the distributor.
3. Carefully engage the distributor drive with the slotted end of the camshaft. When the distributor is installed properly, the rotor will be in line with the previously made mark.
4. Verify proper rotor alignment with mark made at disassembly.
5. Reinstall the distributor hold-down nuts and washers. Tighten the nuts to 9 ft. lbs. (13 Nm).
6. Reinstall the spark plug cable bracket.
7. Reattach the two distributor wiring connectors.
8. Reinstall the distributor cap.
9. Reinstall the spark plug cables following the identification marks made at disassembly.
10. Reinstall the transaxle dipstick tube.
11. Reinstall the EGR tube and tighten the bolts to 95 inch lbs. (11 Nm).
12. Reinstall the PCV hose.
13. Reinstall the air cleaner and tighten the hose clamps.
14. Reinstall the air inlet resonator.
15. Reconnect the negative battery cable.

TIMING DISTURBED

1. Rotate the crankshaft until No. 1 piston is at Top Dead Center of the compression stroke.
2. Rotate the rotor to the No. 1 terminal position on the distributor cap.
3. Lower the distributor into place, engaging the distributor drive with the drive on the camshaft. With the distributor fully seated on the engine, the rotor should be under the No. 1 terminal.

4. Verify proper rotor alignment with mark made at disassembly.
5. Reinstall the distributor hold-down nuts and washers. Tighten the nuts to 9 ft. lbs. (13 Nm).
6. Reinstall the spark plug cable bracket.
7. Reattach the two distributor wiring connectors.
8. Reinstall the distributor cap.
9. Reinstall the spark plug cables following the identification marks made at disassembly.
10. Reinstall the transaxle dipstick tube.
11. Reinstall the EGR tube and tighten the bolts to 95 inch lbs. (11 Nm).
12. Reinstall the PCV hose.
13. Reinstall the air cleaner and tighten the hose clamps.
14. Reinstall the air inlet resonator.
15. Reconnect the negative battery cable.

Ignition Timing

ADJUSTMENT

These engines use a fixed ignition system. The Powertrain Control Module (PCM) regulates the ignition timing. Basic ignition timing is not adjustable.

Engine Assembly

✳✳ CAUTION

Some models covered by this manual may be equipped with a Supplemental Restraint System (SRS), which uses an air bag. Whenever working near any of the SRS components, such as the impact sensors, the air bag module, steering column and instrument panel, disable the SRS.

REMOVAL & INSTALLATION

✳✳ CAUTION

The fuel injection system remains under pressure, even after the engine has been turned OFF. The fuel system pressure MUST BE relieved before disconnecting any fuel lines. Failure to do so may result in fire and/or personal injury.

1. Relieve the fuel system pressure using the recommended procedure. Disconnect the fuel line quick-connect fitting from the fuel rail by squeezing the retainer tabs together and pulling the fuel tube/quick-connect fitting assembly off the fuel tube nipple.

❊❊ CAUTION

Observe all applicable safety precautions when working around fuel. Whenever servicing the fuel system, always work in a well ventilated area. Do not allow fuel spray or vapors to come in contact with a spark or open flame. Keep a dry chemical fire extinguisher near the work area. Always keep fuel in a container specifically designed for fuel storage; also, always properly seal fuel containers to avoid the possibility of fire or explosion.

2. Remove the battery and battery tray according to the following:

 a. Be sure the ignition switch is in the **OFF** unlocked position and all vehicle accessories are turned **OFF**.

 b. Disconnect the negative battery cable from the left shock tower. The ground cable is equipped with a insulator grommet which should be placed on the stud to prevent the negative battery cable from accidentally grounding.

 c. Turn the steering wheel to the extreme left position.

 d. Release the shield by twisting the four plastic screws ¼ turn. Remove the shield.

 e. Disconnect the battery blanket heater, if equipped.

 f. Remove the negative battery cable, then remove the positive battery cable.

 g. Remove the bolt securing the battery strap to the battery hold-down bracket. Remove the hold-down bracket bolt.

 h. Slide the battery to the rear of the

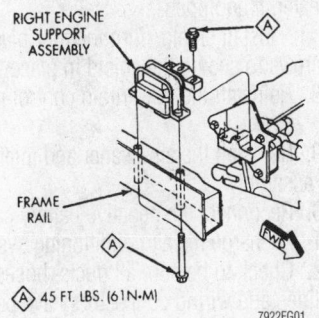

RIGHT ENGINE SUPPORT ASSEMBLY

FRAME RAIL

Ⓐ 45 FT. LBS. (61 N•M)

7922FG01

Exploded view of the right side engine mount—2.0L, 2.4L and 2.5L engines

battery tray and lift over the lip. Be careful not to tip the battery or acid will spill out.

 i. Remove the battery from the vehicle. Remove the battery blanket heater, if equipped.

 j. Remove the battery tray mounting bolts.

 k. Remove the battery tray and battery strap.

3. Remove the complete air cleaner and inlet duct assembly.

4. Unbolt the Powertrain Control Module and move it aside.

❊❊ CAUTION

Never open, service or drain the radiator or cooling system when hot; serious burns can occur from the steam and hot coolant.

5. Drain and properly contain the coolant from the engine.

6. Remove the upper and lower radiator hose, radiator and cooling fan.

7. Disconnect and plug the automatic transaxle cooler lines, if equipped.

8. Disconnect the clutch cable and transaxle shift linkage, if equipped.

9. Disconnect the throttle body linkage and the engine wiring harness.

10. Disconnect the heater hoses.

11. Recover and properly contain the refrigerant of the A/C system with an R-134a recovery unit.

12. Raise and safely support the vehicle. Remove the front wheels.

❊❊ CAUTION

The EPA warns that prolonged contact with used engine oil may cause a number of skin disorders, including cancer! You should make every effort to minimize your exposure to used engine oil. Protective gloves should be worn when changing the oil. Wash your hands and any other exposed skin areas as soon as possible after exposure to used engine oil. Soap and water, or waterless hand cleaner should be used.

13. Drain the engine oil, if necessary.

14. Remove the right side inner splash shield.

15. Remove the accessory drive belts.

16. Remove the right and left halfshaft assemblies.

17. Disconnect the exhaust pipe from the exhaust manifold.

18. Remove the front and rear engine mount brackets from the body.

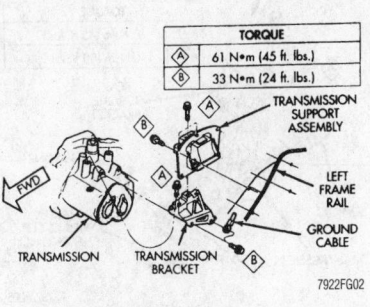

TORQUE	
Ⓐ	61 N•m (45 ft. lbs.)
Ⓑ	33 N•m (24 ft. lbs.)

TRANSMISSION SUPPORT ASSEMBLY

LEFT FRAME RAIL

GROUND CABLE

FWD

TRANSMISSION

TRANSMISSION BRACKET

7922FG02

Exploded view of the left side engine mount—Type 1

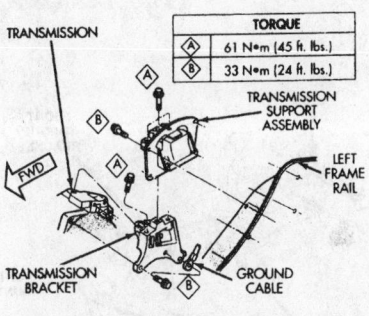

TRANSMISSION

TORQUE	
Ⓐ	61 N•m (45 ft. lbs.)
Ⓑ	33 N•m (24 ft. lbs.)

TRANSMISSION SUPPORT ASSEMBLY

LEFT FRAME RAIL

FWD

TRANSMISSION BRACKET

GROUND CABLE

7924EG03

Exploded view of the left side engine mount—Type 2

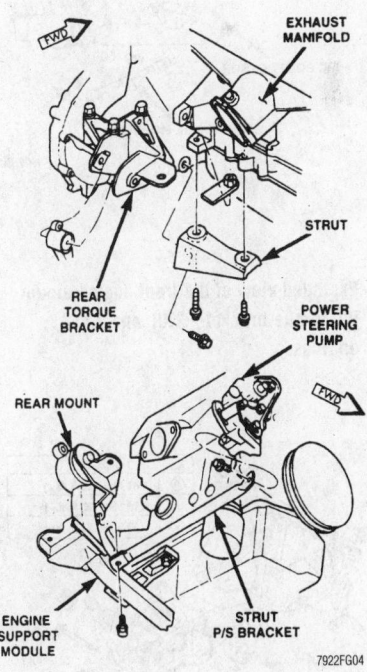

FWD

EXHAUST MANIFOLD

STRUT

REAR TORQUE BRACKET

POWER STEERING PUMP

REAR MOUNT

FWD

ENGINE SUPPORT MODULE

STRUT P/S BRACKET

7922FG04

Exploded view of the rear engine mounting torque bracket—2.0L engine

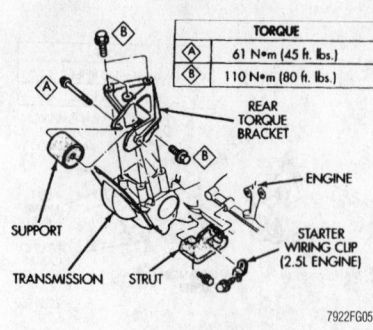

TORQUE	
Ⓐ	61 N•m (45 ft. lbs.)
Ⓑ	110 N•m (80 ft. lbs.)

REAR TORQUE BRACKET

ENGINE

SUPPORT

STARTER WIRING CLIP (2.5L ENGINE)

TRANSMISSION STRUT

7922FG05

Exploded view of the rear engine mounting torque bracket—2.4L and 2.5L engines

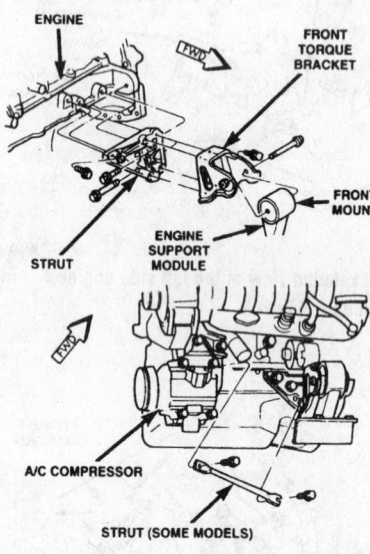

ENGINE

FRONT TORQUE BRACKET

FWD

FRONT MOUNT

STRUT

ENGINE SUPPORT MODULE

A/C COMPRESSOR

STRUT (SOME MODELS)

7922FG06

Exploded view of the front engine mounting torque bracket—2.0L and 2.4L engines

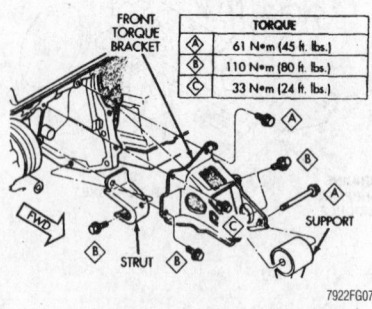

FRONT TORQUE BRACKET

TORQUE	
◇	61 N•m (45 ft. lbs.)
◇	110 N•m (80 ft. lbs.)
◇	33 N•m (24 ft. lbs.)

SUPPORT

FWD

STRUT

7922FG07

Exploded view of the front engine mounting torque bracket—2.5L engine

19. Lower the vehicle.

20. Remove the power steering pump and reservoir, set them aside.

21. Remove the A/C compressor as follows:

   a. Disconnect the compressor clutch wire lead.

   b. Disconnect and plug the refrigerant lines from the compressor.

   c. Remove the compressor mounting bolts.

   d. Remove the compressor unit from the vehicle. Be sure to plug all openings in the A/C system to prevent moisture contamination.

22. Disconnect the ground straps from the engine.

23. Raise the vehicle and install an engine dolly under the vehicle and support engine.

24. Remove the transaxle and engine mount through-bolts.

25. Raise the vehicle slowly allowing the engine and transaxle assembly to remain on the dolly.

To install:

26. Position the engine and the transaxle under the vehicle and lower the vehicle onto the engine assembly.

27. Align the engine mounts and install the right and left mount bolts.

28. Reinstall the transaxle mount.

29. Reinstall the right and left halfshaft assemblies.

30. Reinstall the transaxle and engine braces.

31. Reinstall the splash shields.

32. Reconnect the exhaust pipe to the exhaust manifold.

33. Reinstall the power steering pump and reservoir.

34. Reinstall the A/C compressor on the engine as follows:

   a. Position the compressor correctly against the engine.

   b. Reinstall the compressor mounting bolts. Tighten the compressor mounting bolts to 30 ft. lbs. (41 Nm).

   c. Reconnect the A/C refrigerant hoses with new seals.

   d. Reconnect the compressor clutch wire.

35. Reinstall the accessory drive belts and adjust.

36. Reinstall the front engine mount.

37. Reinstall the inner splash shield. Install the front wheels and lug nuts. Tighten the lug nuts, in a star pattern sequence, to 95–100 ft. lbs. (129–135 Nm).

38. If equipped with manual transaxle, reconnect the clutch cable and linkages.

39. If equipped with automatic transaxle,

reconnect the shifter and kickdown linkages.

40. Reconnect the fuel lines and heater hoses.

41. Reconnect the ground straps.

42. Reattach the engine end throttle body electrical harnesses and connections.

43. Reconnect the throttle body linkage.

44. Reinstall the radiator, cooling fan/shroud assembly and hoses. Reconnect the automatic transaxle cooler lines to the radiator, if equipped.

45. Refill the cooling system with a 50/50 mixture of clean water and ethylene glycol antifreeze.

✹✹ WARNING

Operating the engine without the proper amount and type of engine oil will result in severe engine damage.

46. If the engine oil was drained, install fresh oil and a new oil filter.

47. Reinstall the battery tray and the battery as follows:

   a. Reinstall the battery strap and battery tray into the vehicle through the left front fender well.

   b. Reinstall and tighten the battery tray mounting bolts.

   c. Reinstall the battery blanket heater onto the battery, if equipped.

   d. Reinstall the battery onto the battery tray in the proper position.

   e. Reinstall the hold-down bracket bolt and the bolt securing the battery strap to the battery hold-down bracket. Tighten the battery hold-down bracket bolt to 124 inch lbs. (14 Nm).

   f. Reconnect the positive battery cable to the battery, then attach the negative battery cable to the battery. Tighten the battery cables to 150 inch lbs. (17 Nm). Do NOT reattach the negative battery cable remote connection to the left shock tower at this time.

   g. Reconnect the battery blanket heater, if equipped.

   h. Install shield. Turn the four plastic screws to secure the shield in place.

48. Reinstall the powertrain control module.

49. Reinstall the air cleaner and inlet duct assembly.

50. Reconnect the negative battery cable.

51. Recharge the air conditioning system.

52. Check to be sure all ducts, hoses, fuel lines and wiring connectors have been properly reattached.

53. Start the engine and run until operating temperature.

54. Check for leaks and proper operation.

Water Pump

REMOVAL & INSTALLATION

2.0L and 2.4L Engines

This engine uses a die-cast aluminum body water pump with a stamped steel impeller. The water pump bolts directly to the block. The cylinder block to water pump sealing is provided by a large rubber O-ring. The water pump is driven by the timing belt which must be removed to service the water pump.

1. Disconnect the negative battery cable from the left shock tower. The ground cable is equipped with a insulator grommet which should be placed on the stud to prevent the negative battery cable from accidentally grounding.

➡This procedure requires removing the engine timing belt and the auto tensioner. The factory specifies that the timing marks should always be aligned before removing the timing belt. Set the engine at TDC on No. 1 compression stroke. This should align all timing marks on the crankshaft sprocket and both camshaft sprockets.

2. Raise and safely support the vehicle.
3. Remove the right inner splash shield.
4. Remove the accessory drive belts.

❋❋ CAUTION

Never open, service or drain the radiator or cooling system when hot; serious burns can occur from the steam and hot coolant.

5. Place a drain pan under the radiator drain plug. Drain and properly contain the cooling system.

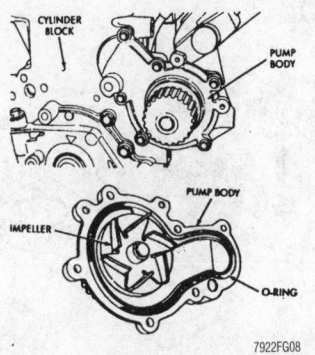

Properly install the O-ring to insure a tight seal—2.0L and 2.4L engines

6. Support the engine and remove the right motor mount.
7. Remove the power steering pump mounting bracket bolts and place the pump/bracket assembly off to one side. Do not disconnect the power steering fluid lines.
8. Remove the right engine mount bracket.
9. Remove the timing belt front covers.
10. Loosen the timing belt tensioner screws and remove the belt tensioner and timing belt.

❋❋ WARNING

With the timing belt removed, DO NOT rotate the camshaft or crankshaft or damage to the engine could occur.

11. Remove the camshaft sprockets. With the timing belt removed, remove both camshaft sprocket bolts. Do not allow the camshafts to turn when the camshaft sprockets are being removed.
12. Remove the rear timing belt cover to access the water pump.
13. Remove the water pump attaching bolts.
14. Remove the water pump.
To install:
15. Thoroughly clean all sealing surfaces. Replace the water pump if there are any cracks, signs of coolant leakage from the shaft seal, loose or rough tuning bearing, damaged impeller or sprocket or sprocket flange loose or damaged.
16. Install a new rubber O-ring into the water pump.

❋❋ WARNING

Be sure the O-ring is properly seated in the water pump groove before tightening the screws. An improperly located O-ring may cause damage to the O-ring and cause a coolant leak.

17. Install the water pump and tighten the bolts to 105 inch lbs. (12 Nm).
18. Pressurize the cooling system to 15 psi (103.4 kPa) and check for leaks. If okay, release the pressure and continue the engine assembly process.
19. Reinstall the rear timing belt cover.
20. Reinstall the camshaft sprockets and tighten the attaching bolts to 75 ft. lbs. (101 Nm). DO NOT allow the camshafts to turn while the sprockets bolts are being tighten to maintain timing mark alignment.

❋❋ WARNING

Do not attempt to compress the tensioner plunger with the tensioner assembly installed in the engine. This will cause damage to the tensioner and other related components. The tensioner MUST be compressed in a vise.

21. Reinstall the timing belt tensioner and timing belt. Be sure to properly tension the timing belt.
22. Reinstall the front upper and lower timing belt covers.
23. Reinstall the right engine mount bracket and engine mount.
24. Reinstall the crankshaft damper and tighten the center bolt to 105 ft. lbs. (142 Nm).
25. Reinstall the right inner splash shield.
26. Lower the vehicle.
27. Reinstall the power steering pump bracket and power steering pump. Tighten the bracket mounting bolts to 40 ft. lbs. (54 Nm).
28. Reinstall the drive belts. Properly tension the drive belts.
29. Refill the cooling system using a mixture of 50/50 water and ethylene glycol antifreeze. Bleed the cooling system.
30. Start the engine and check for proper operation.
31. Check and top off cooling system, if necessary.

2.5L Engine

The water pump bolts directly to the engine block using a gasket for pump-to-block sealing. The pump is serviced as a unit. The 2.5L engine uses metal piping beyond the lower radiator hose to route coolant to the suction side of the water pump, located in the "V" of the cylinder banks. These pipes also have connections for thermostat bypass and heater return coolant hoses. The pipes use O-rings for sealing.

The water pump is driven by the timing belt which must be removed to service the water pump. Timing belt covers must be removed to access the timing belt.

1. Disconnect the negative battery cable from the left shock tower. The ground cable is equipped with a insulator grommet which should be placed on the stud to prevent the negative battery cable from accidentally grounding.

✳✳ CAUTION

Never open, service or drain the radiator or cooling system when hot; serious burns can occur from the steam and hot coolant.

2. Place a large drain pan under the radiator drain plug. Drain and properly contain the engine coolant.

➡**This procedure requires removing the engine timing belt and the auto tensioner. To help assure proper alignment at assembly, it may be helpful to set the engine at TDC on No. 1 compression stroke. This should align all timing marks on the crankshaft sprocket and both camshaft sprockets.**

3. Remove the accessory drive belts and crankshaft damper.

4. Remove the right engine mount. This requires safely supporting the engine so the mount can be removed.

5. Remove the timing belt covers in this order: upper left cover, upper right cover, the lower cover.

6. Remove the timing belt and tensioner.

7. Remove the water pump mounting bolts.

8. Separate the water pump from the water inlet pipe and remove the pump.
   To install:

9. Thoroughly clean all sealing surfaces. Inspect the pump for damage or cracks, signs of coolant leakage at the vent and excessive looseness or rough turning bearing. Any problems require a new pump.

10. Install a new O-ring on the water inlet pipe. Wet the O-ring with water to make installation easier. DO NOT use oil or grease on the O-ring.

11. Install a new gasket on the water pump and fit the pump inlet opening over the water pipe. Press the assembly together to force the pipe into the water pump.

12. Reinstall the water pump to engine bolts and tighten to 20 ft. lbs. (27 Nm).

13. Reinstall the timing belt and timing belt tensioner. Set the timing belt tension.

14. Reinstall the timing belt covers. Reinstall the right engine mount.

15. Reinstall the crankshaft damper.

16. Install the accessory drive belts and set to proper tension.

17. Reconnect the negative battery cable.

18. Refill and bleed the engine cooling system.

19. Start the engine and verify proper operation.

Cylinder Head

REMOVAL & INSTALLATION

2.0L Engine

This engine uses a Single Over Head Camshaft (SOHC) 4-valves per cylinder cross flow aluminum cylinder head.

Care must be taken to be sure all valve timing marks align after cylinder head service.

✳✳ CAUTION

The fuel injection system remains under pressure, even after the engine has been turned OFF. The fuel system pressure MUST BE relieved before disconnecting any fuel lines. Failure to do so may result in fire and/or personal injury.

1. Disconnect the negative battery cable from the left shock tower. The ground cable is equipped with a insulator grommet which should be placed on the stud to prevent the negative battery cable from accidentally grounding.

✳✳ CAUTION

Observe all applicable safety precautions when working around fuel. Whenever servicing the fuel system, always work in a well ventilated area. Do not allow fuel spray or vapors to come in contact with a spark or open flame. Keep a dry chemical fire extinguisher near the work area. Always keep fuel in a container specifically designed for fuel storage; also, always properly seal fuel containers to avoid the possibility of fire or explosion.

2. Relieve the fuel system pressure using the recommended procedure.

3. Remove the air cleaner assembly.

4. Drain and properly contain the engine coolant.

5. Detach and tag all vacuum hoses and all electrical connections from the throttle body. Detach the fuel line quick-connect fitting to the fuel injectors by squeezing the retainer tabs together and pulling the fuel tube/quick-connect fitting assembly off the fuel tube nipple.

6. Disconnect the throttle linkage.

7. Remove the accessory drive belt(s).

8. Disconnect the power brake booster vacuum hose from the intake manifold.

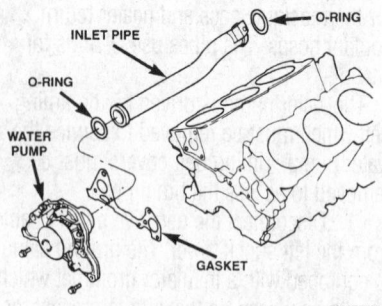

7922FG09

Exploded view of the water pump mounting—2.5L engine

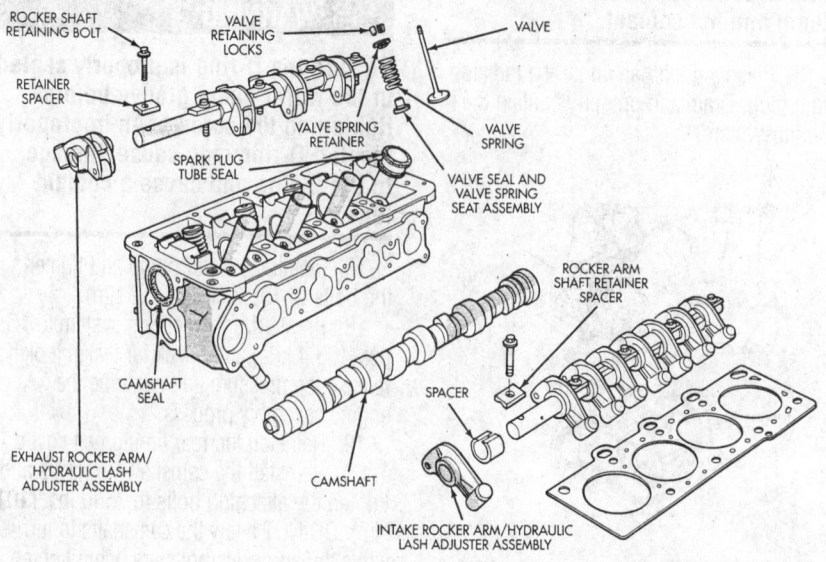

7922FG10

Exploded view of the cylinder head and valvetrain components—2.0L engine

9. Raise and safely support the vehicle.

10. Disconnect the exhaust pipe from the exhaust manifold. Lower the vehicle.

11. Remove the power steering pump and move aside.

12. Detach the coil pack wiring connector. Disconnect the spark plug wires from the spark plugs. Remove the ignition coil pack unit from the engine.

13. Remove the cylinder head cover.

14. Disconnect the cam sensor and fuel injector wiring.

15. Remove the intake and exhaust manifolds, if necessary.

16. Remove the timing belt cover, timing belt, camshaft sprocket and rear timing belt cover using the recommended procedure.

17. Remove the rocker arm/rocker arm shaft assemblies.

18. Remove the cylinder head bolts and remove the cylinder head.

To install:

➡The cylinder head bolts should be checked for stretching before reuse. If the thread area of the bolt is necked down the bolts must be replaced with new. New head bolts are recommended.

19. Thoroughly clean all parts. Clean all sealing surfaces. Use care not to scratch the aluminum cylinder head sealing surface. Check the cylinder head for flatness using a feeler gauge and a straight-edge. The cylinder head must be flat within 0.004 in. (0.1mm).

20. Check the cylinder head for cracks or other damage.

21. Install a new gasket and the cylinder head to the engine block.

22. Be sure to oil the cylinder head bolt threads with clean engine oil. Install the cylinder head bolts, the 44.330 in. (110mm) short bolts are to be installed in positions 7, 8, 9 and 10. Tighten the bolts in proper sequence.

23. Tighten the bolts in four steps as follows:

a. First: all bolts to 25 ft. lbs. (34 Nm).

b. Second: all bolts to 50 ft. lbs. (68 Nm).

c. Third: all bolts again to 50 ft. lbs. (68 Nm).

d. Fourth: all bolts an additional ¼ turn.

➡Do not use a torque wrench for the fourth step.

24. Set the crankshaft to three notches BTDC before installing the rocker arm shafts. Install the rocker arm/rocker arm shaft assemblies.

25. Reinstall the cylinder head cover with a new cylinder head cover gasket. Be sure the cover gasket mating surfaces are clean of any dirt, oil or old gasket material. Tighten the cylinder head cover mounting bolts to 105 inch lbs. (12 Nm).

26. Reinstall the timing belt rear cover and camshaft sprocket. Reinstall the timing belt using the recommended procedure to be sure the timing marks are properly aligned. Failure to do so will cause engine damage.

27. Reinstall the timing belt cover.

28. Reinstall the intake and exhaust manifolds, if removed.

29. Reconnect the cam sensor and the fuel injector wiring.

30. Reinstall the ignition coil pack onto the engine. reattach the coil pack wiring connector and the spark plug wires to the correct spark plugs.

31. Reinstall the power steering pump.

32. Raise the vehicle and reconnect the exhaust pipe to the exhaust manifold. Lower the vehicle.

33. Reconnect the brake booster vacuum line.

34. Reinstall and adjust the accessory drive belts.

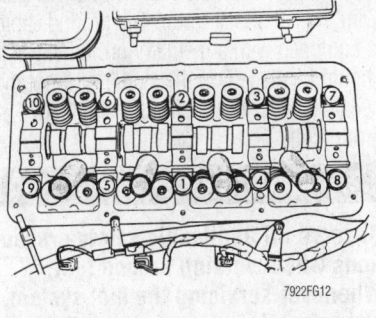

Cylinder head bolt tightening sequence— 2.0L engine

35. Reconnect the throttle linkage.

36. Reattach all vacuum hoses and wiring connectors to the throttle body.

37. Reconnect the fuel line to the fuel injectors.

38. Reinstall the air cleaner assembly.

39. Refill the cooling system with a 50/50 mixture of clean antifreeze and water. A complete oil and filter change is recommended.

40. Reconnect the negative battery cable.

41. Check to be sure all ducts, hoses, fuel lines and wiring connectors have been properly reattached.

42. Start the engine and check for leaks. Run the engine with the radiator cap off so as the engine warms and the thermostat opens, coolant can be added to the radiator. When satisfied that the cooling system is full, shut the engine **OFF** , install the radiator cap and allow the engine to cool.

43. With the engine cool, check all fluid levels. Add coolant and oil as required. Restart the engine and test drive vehicle to check for proper operation.

2.4L Engine

This engine uses a Dual Over Head Camshaft (DOHC) 4-valves per cylinder cross flow aluminum cylinder head. The valves are actuated by roller cam followers which pivot on stationary hydraulic valve adjusters. Care must be taken to be sure all valve timing marks align after cylinder head and valvetrain service.

✳✳ CAUTION

The fuel injection system remains under pressure, even after the engine has been turned OFF. The fuel system pressure MUST BE relieved before disconnecting any fuel lines. Failure to do so may result in fire and/or personal injury.

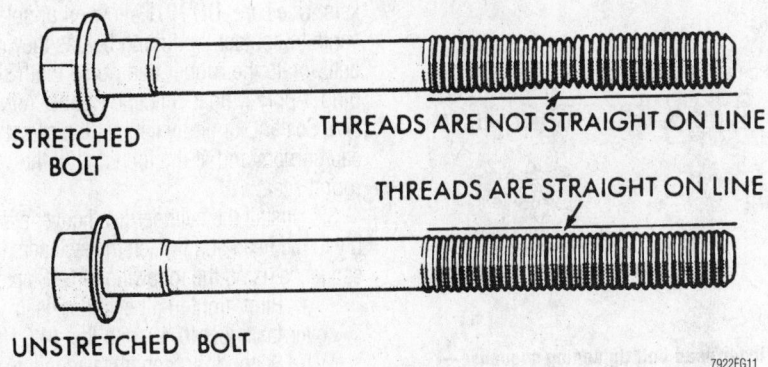

STRETCHED BOLT

THREADS ARE NOT STRAIGHT ON LINE

THREADS ARE STRAIGHT ON LINE

UNSTRETCHED BOLT

Check the cylinder head bolts for necking (stretching)—2.0L and 2.4L engines

1. Disconnect the negative battery cable from the left shock tower. The ground cable is equipped with a insulator grommet which should be placed on the stud to prevent the negative battery cable from accidentally grounding.

✳✳ CAUTION

Observe all applicable safety precautions when working around fuel. Whenever servicing the fuel system, always work in a well ventilated area. Do not allow fuel spray or vapors to come in contact with a spark or open flame. Keep a dry chemical fire extinguisher near the work area. Always keep fuel in a container specifically designed for fuel storage; also, always properly seal fuel containers to avoid the possibility of fire or explosion.

2. Relieve the fuel system pressure using the recommended procedure.

3. Place a large drain pan under the radiator drain plug. Open up the drain plug and drain the cooling system.

4. Remove the air cleaner assembly and disconnect all vacuum lines, electrical wiring and fuel lines from the throttle body.

5. Disconnect the throttle linkage.

6. Remove the accessory drive belts.

7. Disconnect the power brake vacuum hose from the intake manifold.

8. Raise and safely support the vehicle. Disconnect the exhaust pipe from the exhaust manifold.

9. Lower the vehicle as required to remove the power steering pump. Do not disconnect the fluid lines. Set the pump aside.

10. Label the spark plug wires for correct installation. detach the coil pack wiring connector and remove the coil pack and spark plug wires from the engine.

11. Detach the cam sensor and fuel injectors' wiring connectors.

12. Remove the timing belt covers, timing belt and camshaft sprockets using the recommended procedure.

13. Remove the timing belt idler pulley and rear timing belt cover.

14. Remove the cylinder head cover mounting fasteners and cylinder head cover. Remove ground strap.

15. Identify the camshafts, if they are to be reused, for later installation. The camshafts are not interchangeable. Remove the camshaft bearing caps and remove the camshafts in the prescribed sequence.

16. Remove the camshaft followers. Any components that are to be reused must be installed in their original locations. Use care to identify and mark the positions of any removed valvetrain components so they may be reinstalled correctly.

17. Remove the intake and exhaust manifolds.

18. Remove the cylinder head bolts.

19. Remove the cylinder head from the vehicle, using care not to damage the aluminum gasket surfaces.

20. Remove all gasket material from the cylinder head and engine block. Be careful not to gouge or scratch the sealing surface of the aluminum head. The cylinder head should be checked for flatness using a good straight-edge and feeler gauges. The cylinder head must be flat within 0.004 in. (0.1mm).

21. Inspect the camshaft bearing oil feed holes in the cylinder head for clogging. Inspect the camshaft bearing journals for wear or scoring. Check the cam surface for abnormal wear and damage. A visible worn groove in the roller path or on the cam lobes is cause for replacement. Valve service may be performed at this time.

To install:

22. Thoroughly clean all parts. Note that the cylinder head bolts are torqued using a new procedure. The cylinder head bolts should be checked carefully BEFORE reuse. If the threads are necked down the bolts should be replaced with new bolts. Necking can be checked by holding a steel scale or straight-edge against the threads. If all the threads do not contact the scale, the bolt should be replaced. New cylinder head bolts are recommended for any engine rebuild, especially if it is known that the engine has been disassembled before.

23. Thoroughly clean all sealing surfaces. Install a new gasket making sure all holes align with the openings in the engine block. Carefully set the cylinder head in place. A helper may be required.

24. Before installing the bolts the

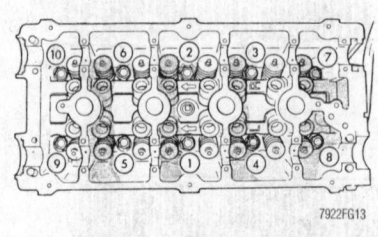

Cylinder head bolt tightening sequence—2.4L engine

threads should be oiled with clean engine oil. Install the bolts and tighten in sequence in four Steps as follows:

   a. First: tighten all bolts to 25 ft. lbs. (34 Nm).

   b. Second: tighten all bolts to 50 ft. lbs. (68 Nm).

   c. Third: tighten all bolts again to 50 ft. lbs. (68 Nm).

   d. Fourth: tighten all bolts and additional ¼ turn.

➡ **Do not use a torque wrench for the fourth step.**

25. Check the camshaft end-play using the recommended procedure, then install the camshaft.

26. Apply Mopar Gasket Maker or equivalent sealer to the No. 1 and No. 6 bearing caps. Install the bearing caps and tighten the M8 fasteners to 250 inch lbs. (28 Nm). The end caps must be installed before the seals may be installed.

27. Apply a light coating of clean engine oil to the lip of the new camshaft seal. Install the camshaft seal until it fits flush with the cylinder head.

28. Reinstall the camshaft sprockets, if removed. Reinstall the rear timing belt cover and timing belt using care to be sure all timing marks are properly aligned, using the recommended procedure. Install timing belt cover.

✳✳ WARNING

Verify that all timing marks are correct. If the timing belt or sprockets are incorrectly installed, engine damage will occur. Take time to be sure all timing marks are correctly aligned.

29. Reinstall the intake and exhaust manifolds.

30. Thoroughly clean the cylinder head cover sealing surfaces.

31. Install new cylinder head cover gaskets. Use care. DO NOT allow oil or solvents to contact the timing belt as they can deteriorate the rubber and cause tooth skipping. Apply Mopar Silicone Rubber Adhesive Sealant, or equivalent, at the camshaft cap corners and at the top edge of the ½ round seal.

32. Install the cylinder head cover assembly to the head and tighten the fasteners in sequence using the following three Steps:

   a. First: tighten all cylinder head cover fasteners to 40 inch lbs. (4.5 Nm).

   b. Second: tighten all fasteners to 80 inch lbs. (9 Nm).

c. Third: tighten all fasteners to 105 inch lbs. (12 Nm).

33. Reinstall the ground strap.

34. Reinstall the ignition coil pack and reconnect the spark plug wiring.

35. Reconnect the cam sensor and fuel injector wiring.

36. Reinstall the power steering pump assembly.

37. Reconnect the exhaust pipe to the exhaust manifold.

38. Reattach all vacuum lines and remaining wiring. Reconnect the throttle linkage and fuel lines.

39. Reinstall and adjust the accessory drive belts.

40. Refill the cooling system. An oil and filter change is recommended since coolant can enter the oil system when a head is removed.

41. Connect the remaining air ducting. Connect the negative battery cable and test run vehicle. Check for leaks and for proper operation.

2.5L Engine

This engine uses aluminum alloy cylinder heads with 4-valves per cylinder and pressed-in cast iron valve guides. The cylinders are common to either cylinder bank. Two overhead camshafts are supported by four bearing journals which are part of the head with the distributor driven off the right (firewall side) cylinder head. Right and left camshaft drive sprockets are interchangeable. The sprockets and engine water pump are driven by the timing belt. Care must be taken to be sure all valve timing marks align after cylinder head and valvetrain service.

➡Please note that for camshaft service, the cylinder head must be removed.

✳✳ CAUTION

The fuel injection system remains under pressure, even after the engine has been turned OFF. The fuel system pressure MUST BE relieved before disconnecting any fuel lines. Failure to do so may result in fire and/or personal injury.

1. Disconnect the negative battery cable from the left shock tower. The ground cable is equipped with a insulator grommet which should be placed on the stud to prevent the negative battery cable from accidentally grounding.

✳✳ CAUTION

Observe all applicable safety precautions when working around fuel. Whenever servicing the fuel system, always work in a well ventilated area. Do not allow fuel spray or vapors to come in contact with a spark or open flame. Keep a dry chemical fire extinguisher near the work area. Always keep fuel in a container specifically designed for fuel storage; also, always properly seal fuel containers to avoid the possibility of fire or explosion.

2. Relieve the fuel system pressure using the recommended procedure.

✳✳ CAUTION

Never open, service or drain the radiator or cooling system when hot; serious burns can occur from the steam and hot coolant.

3. Place a large drain pan under the radiator drain plug. Drain the cooling system.

4. Remove the accessory drive belts.

5. Remove the front timing belt covers

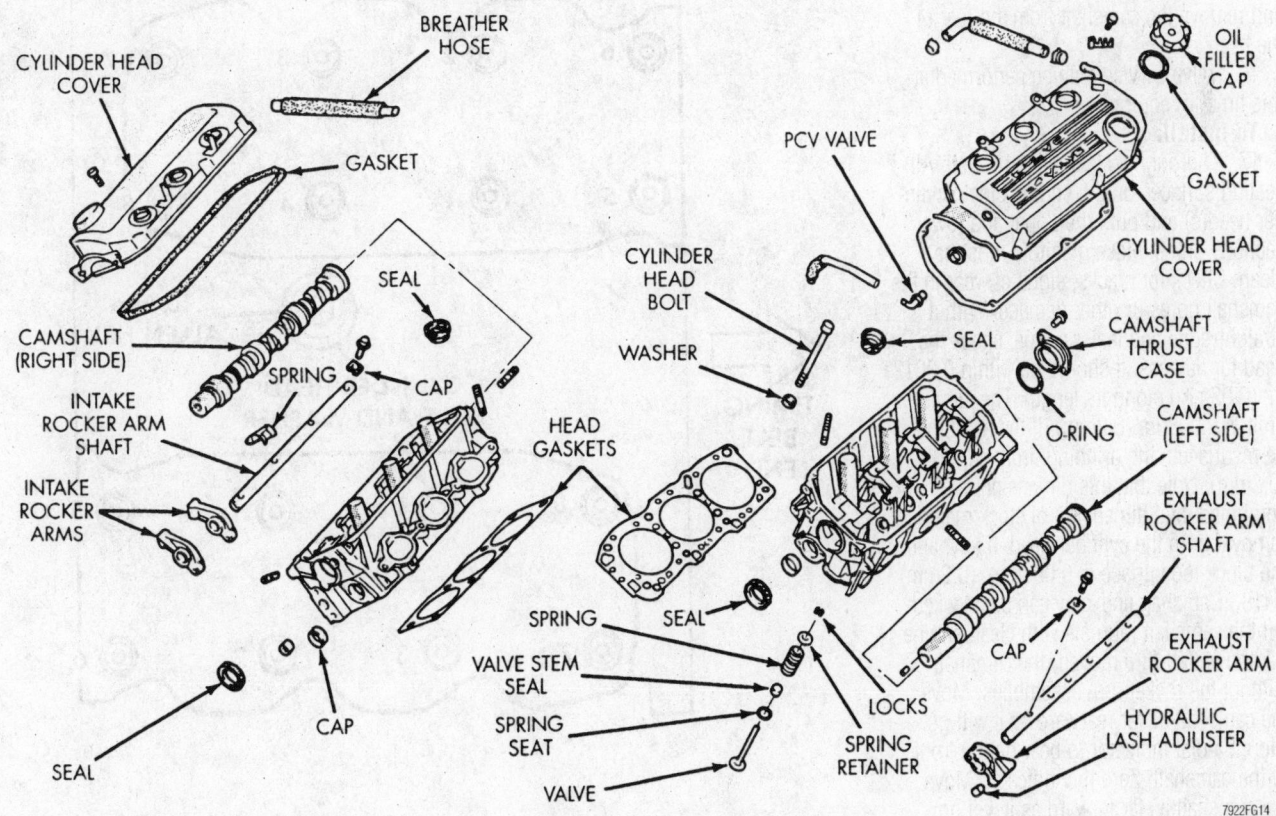

Exploded view of the cylinder head, camshafts and rocker assemblies—2.5L engine

and timing belt using the recommended procedure. Remove the camshaft sprockets.

6. The intake manifold is a 2-piece unit. The upper part is a large air intake plenum of aluminum alloy. Use care working with light alloy parts. Remove the air intake plenum first, then remove the lower intake manifold.

7. Label, disconnect and set aside the spark plug wires.

8. Remove the cylinder head cover screws and remove the cover.

9. Identify the rocker arm shaft assemblies before removal.

10. Install the auto lash adjuster retainers Special Tool MD 998443 or equivalent to keep the auto lash adjusters from falling out of the rocker arms when the rocker arm assembly is removed.

11. Loosen the attaching fasteners and remove the rocker arm shaft assemblies from the cylinder head.

12. Remove the distributor assembly.

13. Remove the exhaust manifold and crossover.

14. Remove the cylinder head bolts and remove the cylinder head from the vehicle.

15. If the camshaft(s) are to be serviced, remove the thrust case from the left head assembly and remove the camshaft from the rear of the head. If not already done, remove the distributor from the right cylinder head and remove the camshaft from the rear of the head.

16. Valve service may be performed at this time, if required.

To install:

17. Thoroughly clean all parts well. All sealing surfaces on the engine block, cylinder head(s) and both the upper and lower sections of the intake manifold must be clean. Check for cracks, signs of wear in the camshaft bores or other damage. With a straight-edge and feeler gauge, check the head for flatness. It should be within 0.0012 in. (0.03mm) along its length. The service limit is 0.008 in. (0.2mm). If the head must be resurfaced, the grinding limit is 0.008 in. (0.2mm). Note that this dimension is a combined total dimension of stock material removal from the cylinder head, if any, and the block top surface is 0.0079 in. (0.2mm).

18. Camshaft end-play can be checked. Oil the camshaft journals with clean engine oil and install (if removed) the camshaft without the rocker arm assemblies. Move the camshaft as far rearward as it will go. Mount a dial indicator to bear on the front of the camshaft. Zero the indicator. Move the camshaft as far forward as it will go. End-play should be 0.004–0.008 in.

(0.1–0.2mm). Maximum allowed end-play is 0.016 in. (0.4mm).

19. If the camshafts were removed, lubricate the camshaft journals and carefully reinstall the camshaft into the cylinder head. Install the thrust case and tighten the thrust case mounting bolts to 9 ft. lbs. (13 Nm).

20. Apply a light coating of engine oil to the lip of the camshaft seal(s). Install the camshaft seal(s). The camshaft must be installed before installing the seal. Reinstall the camshaft sprocket(s) and tighten to 65 ft. lbs. (88 Nm).

21. Reinstall a new head gasket over the locating dowels in the cylinder head. Install the head, using care to locate on top of the dowels. Install the 10mm Allen hex head bolts with washers. New head bolts are recommended. Tighten the head bolts in proper sequence. Tighten gradually, working in two or three steps and finally tighten to 80 ft. lbs. (108 Nm).

22. Reinstall the lower intake manifold using new gaskets.

23. Clean and install the rocker arm assemblies.

24. Reinstall the exhaust manifold and crossover exhaust pipe.

25. Reinstall the timing belt using the

recommended procedure. It is very important that all valve timing marks align properly or engine damage will result.

26. Reinstall the timing belt covers.

27. Inspect the spark plug tube seals located on the ends of each tube. These seals slide onto each tube to seal the cylinder head cover to the spark plug tube. If these seals show signs of hardness and/or cracks, they should be replaced.

28. Thoroughly clean all sealing surfaces. Install a new gasket and install the cover. Tighten bolts to 88 inch lbs. (10 Nm).

29. Reinstall the distributor assembly. Reconnect the spark plug wires.

30. Position a new upper intake manifold (plenum) gasket in place and install the upper intake manifold plenum. Tighten the bolts to 13 ft. lbs. (18 Nm). Install the bolts for the plenum support brackets and connect the EGR tube.

31. Reinstall the speed control cable (if equipped) and the throttle cable.

32. Reattach the TPS and the idle air control motor electrical connectors.

33. Reinstall the air cleaner cover, inlet hoses and air inlet resonator. Tighten the air tube connections.

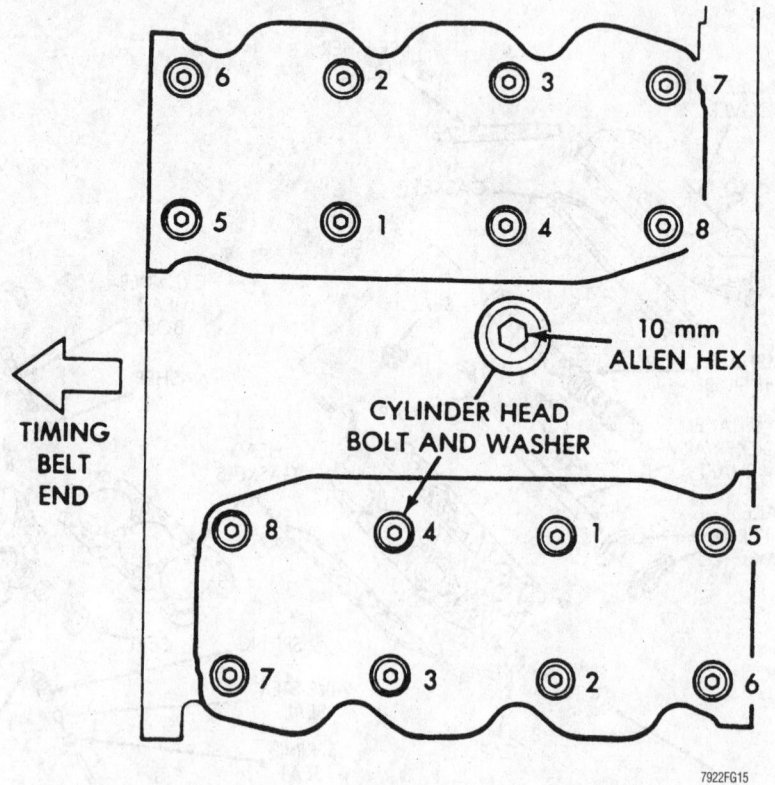

Cylinder head bolt tightening sequence—2.5L engine

34. Reattach the MAP and intake air temperature sensor wiring connectors.

35. Reinstall and adjust the accessory drive belts.

36. Refill the cooling system with a 50/50 mixture of clean antifreeze and water. An oil and filter change is recommended whenever a cylinder head has been removed since coolant can get into the oil system.

37. Reconnect the negative battery cable. Check to be sure all ducts, hoses, fuel lines and wiring connections have all been properly reattached.

38. Start the engine and check for leaks, abnormal noises and vibrations.

39. Bleed the cooling system.

Rocker Arm/Shaft

REMOVAL & INSTALLATION

The 2.0L and 2.5L engines are equipped with rocker arms/shafts. On the 2.4L engine the camshaft acts directly on the valve, therefore, no rocker arms/shafts are used.

2.0L Engine

This engine uses a Single Over Head Camshaft (SOHC) running in an aluminum cylinder head. Rocker arm shafts mount directly to the cylinder head. Care must be taken to be sure all valve timing marks align after cylinder head and valvetrain service. The hydraulic lash adjusters are located in the valve actuating end of the rocker arm and are serviced as an assembly.

✳✳ CAUTION

The fuel injection system remains under pressure, even after the engine has been turned OFF. The fuel system pressure MUST BE relieved before disconnecting any fuel lines. Failure to do so may result in fire and/or personal injury.

1. Disconnect the negative battery cable from the left shock tower. The ground cable is equipped with a insulator grommet which should be placed on the stud to prevent the negative battery cable from accidentally grounding.

✳✳ CAUTION

Observe all applicable safety precautions when working around fuel. Whenever servicing the fuel system, always work in a well ventilated area. Do not allow fuel spray or vapors to come in contact with a spark or open flame. Keep a dry chemical fire extinguisher near the work area. Always keep fuel in a container specifically designed for fuel storage; also, always properly seal fuel containers to avoid the possibility of fire or explosion.

2. Relieve the fuel system pressure using the recommended procedure.

3. Label the spark plug wires to the correct spark plugs. Disconnect the spark plug wires from each spark plug.

4. Remove the inlet duct for the air cleaner.

5. Remove the ignition coil pack.

6. Remove the cylinder head cover retaining bolts and remove the cylinder head cover. Remove and discard any gasket material. Thoroughly clean the cylinder head cover and cylinder head cover gasket mating surfaces. Be sure the gasket mating surfaces are flat.

7. Mark the rocker arm shaft assemblies to identify them for later installation.

8. Remove the rocker arm shaft bolts and remove the rocker arm assemblies from the cylinder head.

9. Mark the rocker arm spacers and retainers to identify them for correct installation. Disassemble the rocker arm/shaft assemblies by removing the attaching bolts from the rocker arm shaft.

10. Slide the rocker arm/hydraulic lash adjuster assembly and rocker arm spacers off the rocker arm shaft. Be sure the rocker arms and spacers are reassembled in the same positions they are removed from.

To install:

➡**Inspect the rocker arms and shaft for scoring and/or wear on the rollers or damage to the rocker arm. If scoring, wear or damage is present, replace the rocker arm assemblies. The rocker arm shaft is hollow and therefore, used as an oil lubrication duct. Inspect the oil**

holes for clogging, using a small wire and clean, if necessary. Inspect the location where the rocker arms mount to the shaft and replace if damaged or worn.

11. If the camshaft lobes show signs of wear, check the corresponding rocker arm roller for wear or damage. Replace rocker arms/hydraulic lash adjuster if worn or damaged. If the camshaft lobes show signs of pitting on the nose, flank or base circle, replace the camshaft.

12. Inspect the rocker arms for scoring, wear on the roller or damage to the shaft. Replace parts as necessary. Check that the rocker arm shaft is clean inside and out. Check all oil holes for clogging with a small wire and clean and required.

13. Thoroughly lubricate all rocker arm components and spacers and reinstall on the rocker arm shaft in the original locations.

14. If the vehicle exhibited a tappet-like noise, the valve lash adjusters built into the rocker arms should be cleaned and checked. Lash adjusters removed from a rocker arm should be returned to their original locations. Replacement of worn or defective lash adjusters would require the replacement of the rocker arm/hydraulic lash adjusters as an assembly. To install a lash adjuster, use the following procedure.

   a. Lubricate the lash adjuster thoroughly with clean engine oil.

   b. Reinstall the adjuster into the rocker arm making sure the adjuster is at least partially filled with oil.

   c. Place the rocker arm in clean engine oil and pump the plunger until the lash adjuster travel is taken up. If travel is not reduced, replace the adjuster with the rocker arm as an assembly.

   d. Reinstall the rocker arm back on the rocker arm shaft.

15. Before installing the rocker arm and shaft assemblies, set the crankshaft to three notches before TDC on the crankshaft sprocket.

➡**When installing the intake rocker arm/shaft assembly, be sure the plastic rocker arm spacers do not interfere with the spark plug tubes. If there is interference, rotate the plastic spacers until they are at the proper angle. Do not rotate the spacers by forcing down on the shaft assembly or damage to the spark plug tubes will occur.**

16. Reinstall the rocker arm and shaft assemblies with the small notches in the rocker shafts pointing up and toward the

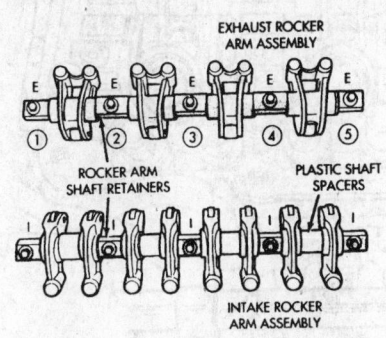

Rocker arm shaft identification—2.0L engine

timing belt side of the engine. Install the retainers in their original positions on the exhaust and intake shafts. Tighten the bolts in proper sequence to 200 inch lbs. (23 Nm).

17. Install new cylinder head cover gasket and cylinder head cover. Tighten the cylinder head cover retaining bolts to 105 inch lbs. (12 Nm).

18. Install the ignition coil pack. Tighten the ignition coil pack mounting fasteners to 200 inch lbs. (23 Nm).

19. Reconnect the spark plug wires to each spark plug.

20. Install the air cleaner inlet duct.

21. Check to be sure all electrical, vacuum and fluid connections are reattached as required.

22. An oil and filter change are recommended.

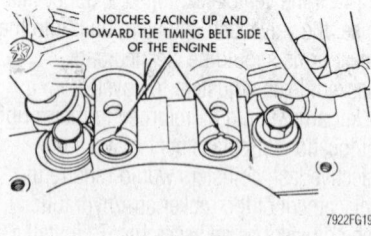

NOTCHES FACING UP AND TOWARD THE TIMING BELT SIDE OF THE ENGINE

7922FG19

Rocker arm shaft notch locations—2.0L engine

23. Reconnect the negative battery cable.

24. Start the engine and check for leaks. Test drive vehicle to check for proper operation.

2.5L Engine

1. Disconnect the negative battery cable from the left shock tower. The ground cable is equipped with a insulator grommet which should be placed on the stud to prevent the negative battery cable from accidentally grounding.

✳✳ CAUTION

Observe all applicable safety precautions when working around fuel. Whenever servicing the fuel system, always work in a well ventilated area. Do not allow fuel spray or vapors to come in contact with a spark or open flame. Keep a dry chemical fire extinguisher near the work area. Always keep fuel in a container specifically designed for fuel storage; also, always properly seal fuel containers to avoid the possibility of fire or explosion.

2. Relieve the fuel system pressure using the recommended procedure.

3. If removing the right (firewall) side rocker arm/shaft assembly, Remove the upper intake manifold (air intake plenum), which is a 2-piece unit of aluminum alloy. Use care working with light alloy parts.

Remove the air intake plenum using the following procedure.

a. Verify that the fuel system pressure relief procedure has been performed. Disconnect the fuel supply tube from the fuel rail by squeezing the retainer tabs together and pulling the fuel tube/quick-connect fitting assembly off the fuel tube nipple. Use care handling the quick-connect fittings.

b. Unplug the electrical connectors from the MAP and air intake temperature sensors.

c. Remove the plenum support bracket bolt located rearward of the MAP sensor.

d. Remove the bolt holding the air inlet resonator to the intake manifold.

e. Loosen the throttle body air inlet hose clamp. Release the snaps holding the air cleaner housing cover to the housing. Remove the air cleaner cover and inlet hoses from the engine.

f. Unplug the TPS and idle air control motor electrical connections.

g. Squeeze the retainer tab on the throttle cable and slide the cable out of the bracket. Slide the speed control cable out of its bracket, if equipped.

h. Remove the EGR tube from the intake manifold.

i. Remove the plenum support bracket bolt located rearward of the EGR tube.

j. Remove the seven bolts holding the upper intake plenum and remove the plenum.

4. Disconnect, label and set aside the spark plug wires.

5. Remove the cylinder head cover screws and remove the cover.

6. Identify the rocker arm shaft assemblies before removal.

7. Install the auto lash adjuster retainers Special Tool MD 998443 or equivalent to keep the auto lash adjusters from falling out of the rocker arms when the rocker arm assembly is removed.

8. Loosen the attaching fasteners and remove the rocker arm shaft assemblies from the cylinder head.

➡**The hydraulic automatic lash adjusters are precision units installed in the machined openings in the rocker arm units. Do not disassemble the auto lash adjusters from the rocker arms.**

To install:

9. The rocker arm shafts are hollow and used as a lubrication oil duct. Be sure all valvetrain parts are clean. Check the rocker arm mounting portion of the shafts for wear

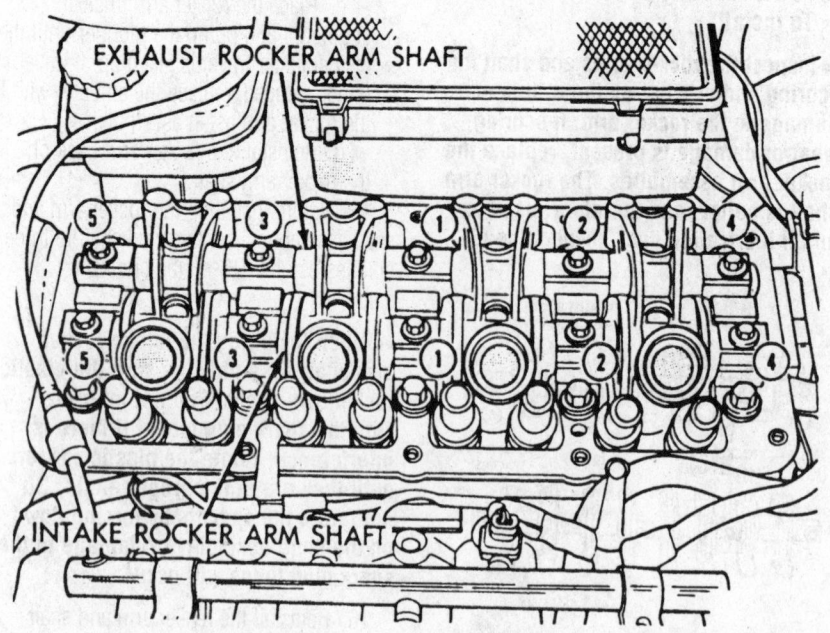

EXHAUST ROCKER ARM SHAFT

INTAKE ROCKER ARM SHAFT

7922FG20

Rocker arm shaft tightening sequence—2.0L engine

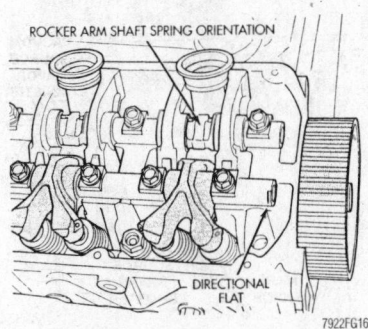

The flats on the rocker arm shaft aids proper orientation—2.5L engine

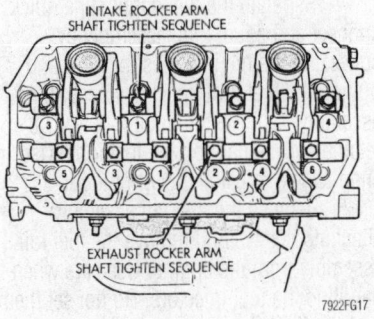

Rocker arm shaft tightening sequence—2.5L engine

or damage. Replace if necessary. Check all oil holes for clogging with a small wire and clean as required. If any rockers were removed, lubricate and install on the shafts in their original positions.

10. Install the rocker arm and shaft assemblies with the FLAT in the rocker arm shafts facing toward the timing belt side of the engine for the right cylinder head. For the left cylinder head install the rocker arm and shaft assembly with the FLAT in the rocker arm shaft facing toward the transaxle side of the engine. Install the retainers and spring clips in their original positions on the exhaust and intake shafts. Tighten the retainer bolts to 276 inch lbs. (31 Nm) working from the center, outward. Remove the valve lash retainer tools that should have been installed at disassembly.

11. Inspect the spark plug tube seals located on the ends of each tube. These seals slide onto each tube to seal the cylinder head cover to the spark plug tube. If these seals show signs of hardness and/or cracks, they should be replaced.

12. Clean the cylinder head and cover mating surfaces. Install a new gasket and install the cover. Tighten bolts to 88 inch lbs. (10 Nm). Reconnect the spark plug wires.

13. Position a new upper intake manifold (plenum) gasket in place and install the upper manifold. Tighten the bolts to 13 ft. lbs. (18 Nm). Install the bolts for the plenum support brackets and connect the EGR tube.

14. Reconnect the throttle and speed control cables.

15. Reattach the TPS and idle air control motor electrical connectors.

16. Reconnect the MAP and intake air temperature sensors.

17. Install the air inlet resonator, air inlet hose and the air cleaner housing cover.

18. Check to be sure all remaining electrical connectors have been reattached. Tighten the air tube connections.

✳✳ CAUTION

The EPA warns that prolonged contact with used engine oil may cause a number of skin disorders, including cancer! You should make every effort to minimize your exposure to used engine oil. Protective gloves should be worn when changing the oil. Wash your hands and any other exposed skin areas as soon as possible after exposure to used engine oil. Soap and water, or waterless hand cleaner should be used.

19. An oil and filter change is recommended.

✳✳ WARNING

Operating the engine without the proper amount and type of engine oil will result in severe engine damage.

20. Connect the negative battery cable. Start the engine and check for leaks, abnormal noises and vibrations.

Intake Manifold

REMOVAL & INSTALLATION

2.0L Engine

The intake manifold is a long branch design made of a molded plastic composition. It is attached to the cylinder head with 10 fasteners. Please note that all seals are to be replaced with new seals and all fasteners are to be replaced with new fasteners. Procure the necessary parts before beginning work.

1. Disconnect the negative battery cable from the left shock tower. The ground cable

is equipped with a insulator grommet which should be placed on the stud to prevent the negative battery cable from accidentally grounding.

✳✳ CAUTION

Observe all applicable safety precautions when working around fuel. Whenever servicing the fuel system, always work in a well ventilated area. Do not allow fuel spray or vapors to come in contact with a spark or open flame. Keep a dry chemical fire extinguisher near the work area. Always keep fuel in a container specifically designed for fuel storage; also, always properly seal fuel containers to avoid the possibility of fire or explosion.

2. Relieve the fuel system pressure using the recommended procedure.

3. Remove the air inlet resonator as follows:

a. Loosen the screw securing the air inlet resonator to the throttle body.

b. Loosen the clamp holding the air inlet resonator to the air inlet tube. Remove the resonator.

4. Separate the fuel supply line quick connect fitting from the fuel rail by squeezing the retainer tabs together and pulling the fuel tube/quick connect fitting from the fuel tube nipple. The retainer will remain on the fuel tube. Wrap shop towels around the fuel line openings to catch any spilling fuel.

5. Remove the fuel rail attaching screws and remove the fuel rail. Use care when handling the fuel injectors. Do not set them on their tips. Cover the fuel injector openings after fuel rail removal.

6. Remove the accelerator, kickdown and speed control cables from the throttle lever and bracket.

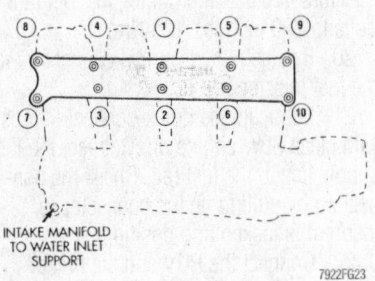

Intake manifold torque sequence—2.0L engine

7. Detach the Throttle Position Sensor (TPS) and the Idle Air Control (IAC) motor electrical connections.

8. Disconnect the vacuum hoses from the throttle body.

9. Detach the connectors from the Manifold Absolute Pressure (MAP) sensor and the intake air temperature sensors.

10. Disconnect the vapor and brake booster hoses.

11. Detach the knock sensor electrical connector, starter relay connector and the wiring harness from the tab located on the intake manifold.

12. Remove the transaxle to throttle body support bracket fasteners at the throttle body and loosen the fastener at the transaxle end.

13. Remove the throttle body assembly.

14. Remove the EGR tube bolts at the valve and at the intake manifold. Remove the tube from the engine.

15. Remove the intake manifold to inlet water tube support fastener.

16. Remove the nine intake manifold screws and washer assemblies and the one nut and washer assembly. Discard the fasteners. At assembly, they should be replaced with new fasteners. Remove the intake manifold from the vehicle.

To install:

17. Clean all sealing surfaces. Check upper and lower manifold gasket surfaces for flatness with a straight-edge. Surface must be flat within 0.006 in. (0.15mm) per foot (30 cm).

➡**All seals are to be replaced with new seals and all fasteners are to be replaced with new fasteners.**

18. Install the intake manifold with new O-ring seals. Tighten the fasteners in proper sequence to 105 inch lbs. (12 Nm).

19. Apply a light coating of engine oil to the fuel injector O-rings. Remove the covers from the fuel injector openings and install the fuel injectors into the engine. Seat the injectors in place and tighten the fuel rail bolts to 200 inch lbs. (23 Nm).

20. Reattach the electrical connectors to the fuel injectors.

21. Lubricate the quick-connect fittings with clean 30W engine oil. Connect the fuel supply line to the fuel rail. Check the connection by pulling on the connector to insure it is locked into position.

22. Connect the PCV and the brake booster hoses.

23. Reinstall the throttle body and tighten to 200 inch lbs. (23 Nm). Reinstall the transaxle to throttle body support bracket and tighten to 105 inch lbs. (11.9

Nm) at the throttle body first. Next tighten the bracket at the transaxle.

24. Reattach the MAP sensor and the air temperature sensor wiring connectors.

25. Reattach the knock sensor electrical and starter relay connectors. reattach the wiring harness to the intake manifold tab.

26. Reattach the IAC and TPS wiring connectors.

27. Reconnect the throttle body vacuum hoses.

28. Reinstall the accelerator, kickdown and speed control cables to their bracket and connect, then to the throttle lever.

29. Loosely assemble the EGR tube onto the valve and intake manifold finger-tight. Tighten the tube fasteners at the EGR valve first to 95 inch lbs. (11 Nm), then tighten the intake manifold side fasteners to 95 inch lbs. (11 Nm).

30. Reinstall the fresh air duct to the air filter housing.

31. Reinstall the air inlet resonator to the throttle body. Reconnect the air inlet tube to the resonator and tighten the clamps to 20–30 inch lbs. (2–3 Nm).

32. Reconnect the negative battery cable.

2.4L Engine

The intake manifold is a long branch design made of cast aluminum. It is attached to the cylinder head with eight fasteners.

1. Disconnect the negative battery cable from the left shock tower. The ground cable is equipped with a insulator grommet which should be placed on the stud to prevent the negative battery cable from accidentally grounding.

> ❋❋ **CAUTION**

The fuel injection system remains under pressure even after the engine has been turned OFF. The fuel system pressure MUST BE relieved before disconnecting any fuel lines. Failure to do so may result in fire and/or personal injury.

2. Relieve the fuel system pressure using the recommended procedure.

3. Remove the air inlet resonator as follows:

  a. Remove the two mounting bolts that secure the air inlet resonator to the intake manifold.

  b. Loosen the screw securing the resonator to the throttle body.

  c. Loosen the clamp securing the air inlet resonator to the air inlet tube. Remove the resonator.

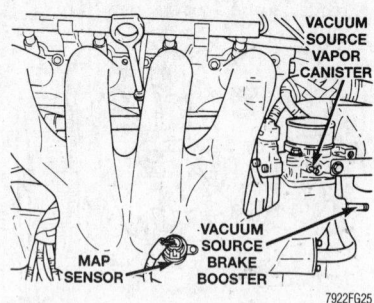

Manifold Absolute Pressure (MAP) and Intake Air Temperature (IAT) sensor locations—2.4L engine

4. Separate the fuel supply line quick connect at the fuel tube assembly by squeezing the retainer tabs together and pulling the fuel tube/quick connect fitting assembly from the fuel tube nipple. The retainer will remain on the fuel tube. Use shop towels to catch any dripping fuel.

5. Remove the fuel rail assembly attaching screws and remove the fuel rail assembly from the engine. Use care when handling the fuel injectors. Do not set them on their tips. Cover the fuel injector openings after fuel rail removal.

6. Remove the accelerator, kickdown and speed control cables from the throttle lever and bracket.

7. Detach the Idle Air Control (IAC) motor and Throttle Position Sensor (TPS) wiring connectors.

8. Disconnect the vacuum hoses from the throttle body.

9. Detach the Manifold Absolute Pressure (MAP) and Intake Air Temperature (IAT) electrical connectors. Disconnect the vapor and brake booster hoses.

10. Detach the knock sensor electrical connector and detach the wiring harness from the tab located on the intake manifold.

11. Remove the transaxle to throttle body support bracket fasteners at the throttle body and loosen the fastener at the transaxle end. Remove the throttle body from the intake manifold.

12. Remove the EGR tube bolts at the valve end and at the intake manifold. Remove the tube from the engine.

13. Remove the intake manifold support bracket. Remove the eight intake manifold fasteners and washers. Remove the intake manifold from the engine.

To install:

14. Thoroughly clean all parts. Clean all sealing surfaces

15. Install a new intake manifold gasket and position the manifold on the cylinder head. Tighten the fasteners to 200 inch lbs.

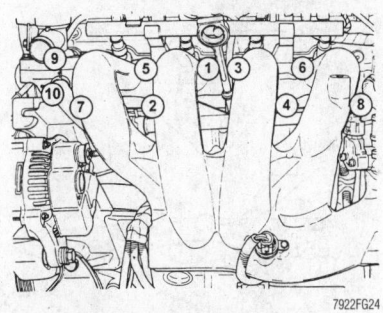

Intake manifold bolt torque sequence—2.4L engine

(23 Nm) in correct sequence starting from the center and working out.

16. Remove the covering from the fuel injector openings and be sure the openings are clean. Install the fuel rail assembly to the intake manifold. Tighten the retainer screws to 200 inch lbs. (23 Nm).

17. Connect the PCV and brake booster hoses.

18. Inspect the quick connect fittings for damage and repair as required. Lube the fuel tube with clean 30W engine oil. Reconnect the fuel supply tube hose to the fuel rail assembly. Check the connection by pulling on the connector to insure it is locked in position.

19. Reinstall the throttle body. Tighten the fasteners to 200 inch lbs. (23 Nm). Reinstall the transaxle to throttle body support bracket and tighten to 105 inch lbs. (11.9 Nm) at the throttle body first, then tighten the bracket at the transaxle.

20. Reattach the MAP and IAT wiring connectors.

21. Reconnect the knock sensor and attach the wiring harness to the tab located on the intake manifold.

22. Reattach the IAC and TPS wiring connectors.

23. Reconnect the remaining vacuum hoses to the throttle body.

24. Reconnect the accelerator, kickdown and speed control cables to the throttle lever and bracket.

25. Loosely assemble the EGR tube onto the valve and intake manifold finger-tight. Tighten the tube fasteners at the EGR valve first to 95 inch lbs. (11 Nm), then, tighten the intake manifold side fasteners to 95 inch lbs. (11 Nm).

26. Reinstall the air inlet resonator to the throttle body, then install the air inlet tube to the resonator. Tighten the clamps to 20–30 inch lbs. (2.5–3.5 Nm). Tighten the two air inlet resonator-to-intake manifold mounting bolts.

27. Reconnect the negative battery cable.

2.5L Engine

The intake manifold assembly is composed of an upper plenum and lower manifold. This aluminum alloy manifold has long runners to improve airflow inertia. The plenum chamber absorbs air pulsations created during the suction phase of each cylinder. The lower intake manifold is machined for six injectors and the fuel rail mounts.

1. Disconnect the negative battery cable from the left shock tower. The ground cable is equipped with a insulator grommet which should be placed on the stud to prevent the negative battery cable from accidentally grounding.

2. Relieve the fuel system pressure using the recommended procedure.

✳✳ CAUTION

The fuel injection system remains under pressure, even after the engine has been turned OFF. The fuel system pressure MUST BE relieved before disconnecting any fuel lines. Failure to do so may result in fire and/or personal injury.

3. Disconnect the fuel supply line from the fuel rail. This is a quick connect fitting. Squeeze the fitting retainer tabs together and separate the connection.

✳✳ CAUTION

Wrap shop towels around the connection to catch any gasoline spillage.

➡ It may be helpful to identify and tag each sensor connector as it is being removed. This may save time at assembly.

4. Detach the connectors from the Manifold Absolute Pressure (MAP) sensor and the intake air temperature sensors.

5. Remove the plenum support bracket located to the rear of the MAP sensor.

6. Remove the air inlet resonator attaching bolt.

7. Loosen the throttle body air inlet hose clamp.

8. Release the snaps holding the air cleaner housing cover to the housing. Remove the air cleaner cover and inlet hoses from the engine.

9. Detach the Throttle Position Sensor (TPS) and the Idle Air Control (IAC) motor electrical connections.

10. Squeeze the retainer tab on the throttle cable and slide the cable out of the bracket.

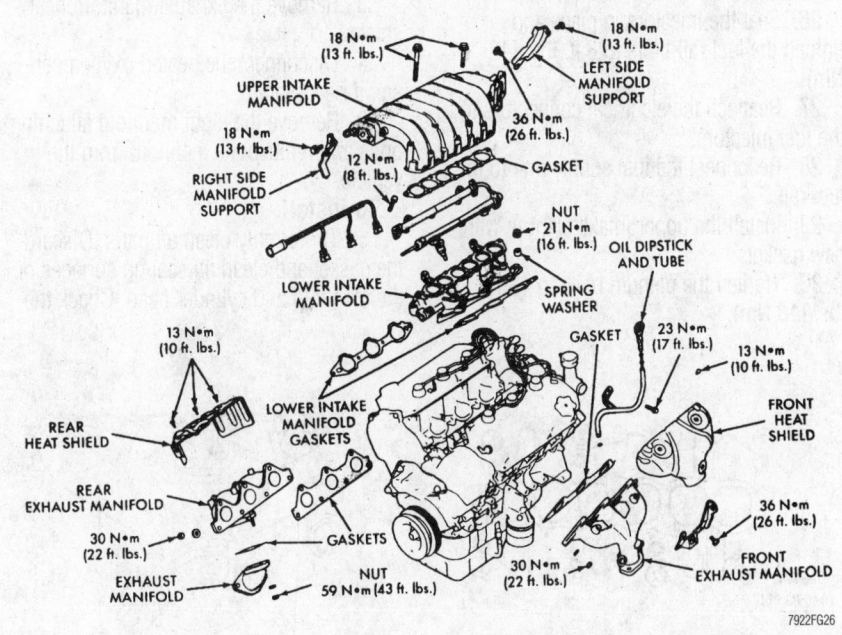

Exploded view of the intake/exhaust manifolds and related components—2.5L engine

11. If equipped with speed control, slide the speed control cable out of the bracket.

12. Remove the EGR tube from the engine intake manifold.

13. Remove the plenum support bracket located to the rear of the EGR tube.

14. Remove the seven bolts attaching the upper intake plenum to the intake manifold and remove plenum.

15. Detach the fuel injector electrical connectors.

16. Remove the four bolts attaching the fuel rail to the intake manifold. Use care. There are spacers under each fuel rail bolt. Remove the fuel rail.

17. Remove the lower intake manifold attaching bolts.

18. Remove the intake manifold and discard the old gaskets.

To install:

19. Clean all sealing surfaces.

20. Check upper and lower manifold gasket surfaces for flatness with a straight-edge.

21. Surface must be flat within 0.006 in. (0.15mm) per 12 in. (30 cm) of manifold length.

22. Install the lower intake manifold with new gaskets.

23. Tighten the bolts in correct sequence to 185 inch lbs. (21 Nm).

24. Apply a light coating of engine oil to the fuel injector O-rings.

25. Reinstall the fuel injectors into the engine.

26. Seat the injectors in place and tighten the fuel rail bolts to 8 ft. lbs. (12 Nm).

27. Reattach the electrical connectors to the fuel injectors.

28. Reconnect the fuel supply line to the fuel rail.

29. Install the upper intake plenum with new gaskets.

30. Tighten the plenum bolts to 13 ft. lbs. (18 Nm).

31. Reinstall the plenum support brackets and tighten to 13 ft. lbs. (18 Nm).

32. Reinstall the EGR tube and tighten the screws to 95 inch lbs. (11 Nm).

33. Reinstall the throttle cables.

34. Reattach the TPS and IAC electrical connections.

35. Reattach the MAP sensor and the intake air temperature sensor connectors.

36. Reinstall the air cleaner assembly and tighten the hose clamps to 25 inch lbs. (3 Nm).

37. Reinstall the air inlet resonator attaching bolt.

38. Reconnect the negative battery cable.

39. Start the engine and check for leaks.

Exhaust Manifold

REMOVAL & INSTALLATION

2.0L and 2.4L Engines

1. Disconnect the negative battery cable from the left shock tower. The ground cable is equipped with a insulator grommet which should be placed on the stud to prevent the negative battery cable from accidentally grounding.

2. Disconnect the exhaust pipe from the exhaust manifold. Apply penetrating oil on the exhaust manifold-to-exhaust pipe flange bolts to aid in removal. It may be necessary to remove the entire exhaust system.

3. Remove the exhaust manifold heat shield.

4. Disconnect the heated oxygen sensor, if necessary.

5. Remove the eight manifold attaching bolts and remove the manifold from the vehicle.

To install:

6. Thoroughly clean all parts. Discard the gasket and clean all sealing surfaces of the manifold and cylinder head. Check the

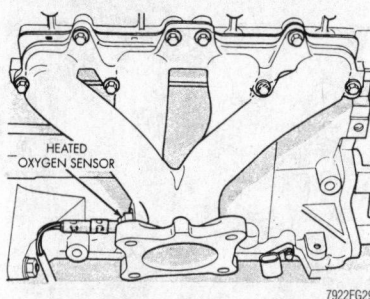

Be careful not to damage the oxygen sensor when servicing the manifold—2.4L engine

manifold gasket surface for flatness with a straight-edge and feeler gauge. The surface must be flat within 0.006 in. per foot (0.15mm per 300mm) of manifold length. Inspect the manifold for cracks or distortion. Replace if necessary.

7. Install the manifold into the vehicle with a new gasket. DO NOT APPLY SEALER.

8. Reinstall the eight manifold bolts and tighten starting at the center and working outward in both directions. Tighten to 200 inch lbs. (23 Nm).

9. Reconnect the heated oxygen sensor.

10. Reinstall the heat shield.

11. Reinstall the exhaust pipe and tighten fasteners to 250 inch lbs. (28 Nm).

12. Reconnect the negative battery cable. Start the engine and allow to idle while inspecting the manifold for exhaust leaks.

2.5L Engine

1. Disconnect the negative battery cable from the left shock tower. The ground cable is equipped with a insulator grommet which should be placed on the stud to prevent the negative battery cable from accidentally grounding.

2. Raise and safely support the vehicle.

3. Disconnect the exhaust pipe connection to the rear (cowl side) exhaust manifold at the flex joint.

➡️**It may be necessary to remove the whole exhaust system.**

4. Remove the bolts attaching the cross-over pipe to the manifolds and remove the cross-over pipe assembly.

5. Disconnect the oxygen sensor lead wire at the rear manifold. Remove the oxygen sensor at the rear exhaust manifold.

6. Remove the power steering bracket.

7. Remove the rear exhaust manifold heat shield.

8. Remove the rear manifold attaching nuts and remove the rear manifold.

9. Lower the vehicle and detach the

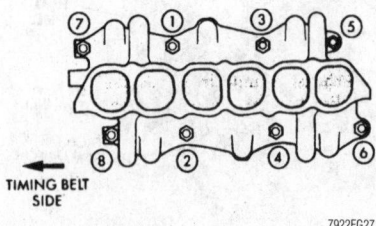

Intake manifold bolt torque sequence— 2.5L engine

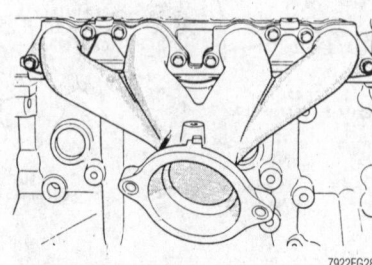

Be sure the gasket mating surfaces are clean and flat before installing the exhaust manifold—2.0L engine

front heated oxygen sensor wiring connector. Remove the front heated oxygen sensor.

10. Remove the front manifold heat shield.

11. Remove the front manifold securing nuts and remove the front manifold.

To install:

12. Thoroughly clean all parts. Inspect the exhaust manifolds for damage or cracks and check for distortion of the cylinder head sealing surface and exhaust crossover sealing surface with a straight-edge and thickness gauge.

13. Install a new front manifold gasket.

14. Install the front manifold and tighten the nuts to 22 ft. lbs. (30 Nm).

15. Reinstall the front exhaust manifold heat shield and tighten the heat shield mounting screws to 115 inch lbs. (13 Nm).

16. Reinstall the front heated oxygen sensor. reattach the oxygen sensor wiring connector.

17. Raise and safely support the vehicle.

18. Install a new rear exhaust manifold gasket. Install the rear exhaust manifold.

19. Tighten the manifold nuts to 22 ft. lbs. (30 Nm).

20. Reinstall the power steering bracket.

21. Reinstall the crossover pipe and tighten the nuts to 22 ft. lbs. (30 Nm).

22. Install the rear heated oxygen sensor. Reconnect the rear heated oxygen sensor lead.

23. Reconnect the exhaust pipe to the rear manifold. Tighten the exhaust pipe-to-rear exhaust manifold flange mounting bolts to 21 ft. lbs. (28 Nm).

24. Lower the vehicle. Reconnect the negative battery cable. Start the engine and allow the engine to idle while inspecting the vehicle for exhaust leaks at the manifold.

Front Crankshaft Seal

REMOVAL & INSTALLATION

2.0L and 2.4L Engines

The timing belt must be removed for this procedure. Use care that all timing marks are aligned after installation or, then engine will be damaged.

1. Disconnect the negative battery cable from the left shock tower. The ground cable is equipped with a insulator grommet which should be placed on the stud to prevent the negative battery cable from accidentally grounding.

2. Remove the accessory drive belts.

✳✳ CAUTION

The EPA warns that prolonged contact with used engine oil may cause a number of skin disorders, including cancer! You should make every effort to minimize your exposure to used engine oil. Protective gloves should be worn when changing the oil. Wash your hands and any other exposed skin areas as soon as possible after exposure to used engine oil. Soap and water, or waterless hand cleaner should be used.

3. Raise and safely support the vehicle. Drain the engine oil.

4. Remove the crankshaft damper/pulley using a jaw puller tool.

5. Remove the timing belt.

6. Remove the crankshaft timing belt sprocket using Sprocket Removal tool No. 6793.

✳✳ CAUTION

Be careful as not to nick the seal surface of the crankshaft or the seal bore.

7. Remove the front crankshaft seal using Seal Removal tool No. 6771 or equivalent seal puller. Be careful not to damage the seal contact area of the crankshaft.

To install:

8. Apply a light coating of clean engine oil to the lip of the new oil seal. Install the new front crankshaft oil seal by using oil seal installer tool No. 6780–1 or equivalent seal tool.

9. Place new oil seal into the opening with the seal spring facing the inside of the engine. Be sure the oil seal is installed flush with the front cover.

10. Install the crankshaft timing belt sprocket using tool No. 6792.

➡ **Be sure the word "FRONT" on the timing belt sprocket is facing outward.**

11. Reinstall the timing belt and timing belt cover using the recommended procedure.

12. Reinstall the crankshaft damper/pulley onto the crankshaft. Use thrust bearing/washer and 12M-1.75 x 150mm bolt from special tool No. 6792. Install the crankshaft damper/pulley retaining bolt and tighten to 105 ft. lbs. (142 Nm).

13. Lower the vehicle.

14. Reinstall the accessory drive belts. Adjust the belts to the proper tension.

✳✳ WARNING

Operating the engine without the proper amount and type of engine oil will result in severe engine damage.

15. Refill the engine with the correct amount of clean engine oil.

16. Reconnect the negative battery cable. Start the engine and check for leaks.

2.5L Engine

The timing belt must be removed for this procedure. Use care to be sure all timing marks are aligned after this service or the engine will be damaged.

1. Disconnect the negative battery cable from the left shock tower. The ground cable is equipped with a insulator grommet which should be placed on the stud to prevent the negative battery cable from accidentally grounding.

2. Remove the accessory drive belts.

✳✳ CAUTION

The EPA warns that prolonged contact with used engine oil may cause a number of skin disorders, including cancer! You should make every effort to minimize your exposure to used engine oil. Protective gloves should be worn when changing the oil. Wash your hands and any other exposed skin areas as soon as possible after exposure to used engine oil. Soap and water, or waterless hand cleaner should be used.

3. Raise and safely support the vehicle. Drain the engine oil.

4. Remove the right inner splash shield.

5. Remove the crankshaft damper/pulley.

6. Remove the timing belt covers and timing belt using the recommended procedure.

7. Remove the crankshaft timing belt sprocket and key.

8. Remove the front crankshaft seal by prying it out with a flat tipped prytool. Be sure to cover the end of the prytool tip with a shop towel.

✳✳ CAUTION

Be careful as not to nick the seal surface of the crankshaft or the seal bore.

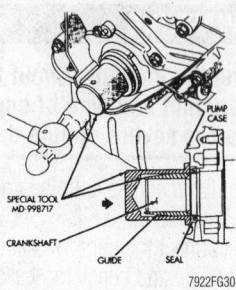

7922FG30

To prevent damage to the end of the crankshaft, use oil seal installer tool No. MD998717 or equivalent, as shown—2.5L engine

To install:

9. Apply a light coating of clean engine oil to the lip of the new oil seal. Install the new front crankshaft oil seal into the oil pump housing by using oil seal installer tool No. MD998717 or equivalent seal installer. Be sure the oil seal is installed flush with the oil pump cover.

10. Reinstall the crankshaft timing belt sprocket and key.

11. Reinstall the timing belt and timing belt covers using the recommended procedure. Verify that all timing marks are correctly aligned or the engine will be damaged.

12. Reinstall the crankshaft damper/pulley onto the crankshaft. Install the crankshaft damper/pulley retaining bolt and tighten to 134 ft. lbs. (182 Nm).

13. Reinstall the right inner splash shield.

14. Lower the vehicle.

15. Reinstall the accessory drive belts. Adjust the belts to the proper tension.

❈❈ WARNING

Operating the engine without the proper amount and type of engine oil will result in severe engine damage.

16. Refill the engine with the correct amount of clean engine oil.

17. Reconnect the negative battery cable.

18. Start the engine and check for leaks.

Camshaft

REMOVAL & INSTALLATION

2.0L Engine

1. Remove the rocker arm assemblies.

2. Remove the timing belt and camshaft sprocket using the recommended procedure.

3. Remove the cylinder head using the recommended procedure.

4. Remove the camshaft sensor and remove the camshaft from the rear of the cylinder head.

To install:

➡**The cylinder head bolts should be checked for stretching before reuse. If the thread area of the bolt is necked down the bolts must be replaced with new. New head bolts are recommended.**

5. Thoroughly clean all parts. Inspect the camshaft journals for scoring. Check the oil feed holes in the head for blockage. Check the camshaft bearing journals for scoring. If light scratches are present, they may be removed with 400 grit abrasive paper. If deep scratches are present, replace the camshaft and check the cylinder head for damage. Replace the cylinder head if worn or damaged.

6. If the camshaft lobes show signs of wear, check the corresponding rocker arm roller for wear or damage. Replace rocker arms/hydraulic lash adjuster if worn or damaged. If the camshaft lobes show signs of pitting on the nose, flank or base circle, replace the camshaft.

7. If the rocker arms and shaft are to be serviced, mark the rocker arms so any that are to be returned to service will be installed in their original locations.

8. Reinstall the rocker arm back on the rocker arm shaft.

9. To install the camshaft, lubricate the bearing journals thoroughly. Install the camshaft into the cylinder head carefully. Be sure it turns freely. If the camshaft installation is satisfactory, install the cam sensor and tighten the screws to 85 inch lbs. (9.6 Nm).

10. Camshaft end-play can be checked. Use the following procedure.

   a. Oil the camshaft journals and install the camshaft without the rocker arm assemblies. Install the cam sensor and tighten the screws to 85 inch lbs. (9.6 Nm).

   b. Setup a dial indicator to touch on the nose of the camshaft.

   c. Using a suitable prying tool, move the camshaft as far rearward as it will go. Be sure the dial indicator probe is in contact with the camshaft.

   d. Zero the dial indicator.

   e. Move the camshaft as far forward as it will go.

   f. Read the end-play on the dial indicator. Specification is 0.005–0.013 in. (0.13–0.33mm).

11. Install the camshaft seal. The camshaft must be installed before the camshaft seal is installed. The seal should be flush with the cylinder head after installation.

12. Reinstall the camshaft sprocket and tighten to the bolt to 85 ft. lbs. (115 Nm).

13. Reinstall the cylinder head using the recommended procedure. Be sure to use new cylinder head mounting bolts.

14. Before installing the rocker arm and shaft assemblies, set the crankshaft to three notches before TDC on the crankshaft sprocket.

15. Reinstall the rocker arm and shaft assemblies.

16. Reinstall the camshaft sprocket and timing belt using the recommended procedure taking care to align all valve timing marks.

2.4L Engine

This engine uses a Dual Over Head Camshaft (DOHC) 4-valves per cylinder cross flow aluminum cylinder head. The valves are actuated by roller cam followers which pivot on stationary hydraulic valve adjusters. Care must be taken to be sure all valve timing marks align after cylinder head and valvetrain service.

1. Disconnect the negative battery cable from the left shock tower. The ground cable is equipped with a insulator grommet which should be placed on the stud to prevent the negative battery cable from accidentally grounding.

❈❈ CAUTION

The fuel injection system remains under pressure, even after the engine has been turned OFF. The fuel system pressure MUST BE relieved before disconnecting any fuel lines. Failure to do so may result in fire and/or personal injury.

2. Relieve the fuel system pressure using the recommended procedure.

❈❈ CAUTION

Observe all applicable safety precautions when working around fuel. Whenever servicing the fuel system, always work in a well ventilated area. Do not allow fuel spray or vapors to come in contact with a spark or open flame. Keep a dry chemical fire extinguisher near the work area. Always keep fuel in a container specifically designed for fuel storage; also, always properly seal fuel containers to avoid the possibility of fire or explosion.

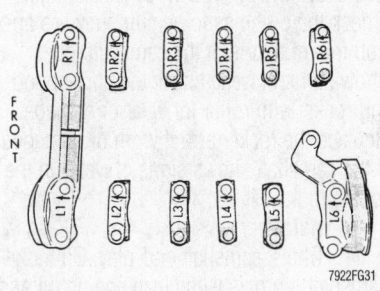

Camshaft bearing cap identification—2.4L engine

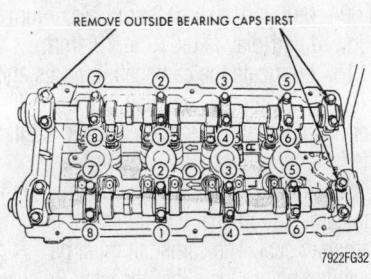

Camshaft bearing cap bolt removal sequence—2.4L engine

3. Label and disconnect the spark plug wires from the spark plugs.

4. Remove the ignition coil pack and spark plug wires.

5. Remove the cylinder head cover retaining fasteners and remove the cylinder head cover from the cylinder head. Discard the old cylinder head cover gasket.

6. Remove the ground strap.

7. Remove the timing belt covers, timing belt and camshaft sprockets using the recommended procedure.

8. Take note that the camshaft bearing caps are numbered for correct location during installation. Remove the outer bearing caps first.

9. Loosen, but do not remove, the camshaft bearing cap retaining fasteners in the correct sequence, inside working outward. Perform this step on one camshaft at a time.

10. Identify the camshafts, if they are to be reused, for later installation. The camshafts are not interchangeable. Remove the camshaft bearing caps and remove the camshafts.

11. Remove the camshaft followers. Any components that are to be reused must be installed in their original locations. Use care to identify and mark the positions of any

removed valvetrain components so they may be reinstalled correctly.

12. Inspect the camshaft bearing oil feed holes in the cylinder head for clogging. Inspect the camshaft bearing journals for wear or scoring. Check the cam surface for abnormal wear and damage. A visible worn groove in the roller path or on the cam lobes is cause for replacement.

To install:

13. Thoroughly clean all camshaft and related parts.

14. The camshaft end-play should be checked using the following procedure:

a. Oil the camshaft journals and install the camshaft **WITHOUT** the cam follower assemblies. Install the rear cam caps and tighten to 250 inch lbs. (28 Nm).

b. Carefully push the camshaft as far rearward as it will go.

c. Set up a dial indicator to bear against the front of the camshaft (the sprocket end). Zero the indicator.

d. Move the camshaft forward as far as it will go. Read the dial indicator. End-play specification is 0.002–0.010 in. (0.05–0.15mm).

e. If excessive end-play is present, inspect the cylinder head and camshaft for wear; replace if necessary.

15. If the fit and condition of the camshafts are acceptable, remove the camshafts for installation of the cam followers.

16. The hydraulic valve lash adjusters are inside the roller cam followers. Be sure they are clean, well-lubricated with clean engine oil and properly positioned. Install the cam followers in their original positions on the hydraulic adjuster and valve stem.

✳✳ WARNING

Be sure NONE of the pistons are at Top Dead Center when installing the camshafts.

17. Lubricate the camshaft bearing journals and cam followers with clean engine oil and install the camshafts. Install right and left camshaft bearing caps No. 2 through No. 5 and right side No. 6. Tighten the M6 fasteners to 105 inch lbs. (12 Nm) in correct sequence.

18. Apply Mopar® Gasket Maker or equivalent sealer to the No. 1 and left side No. 6 bearing caps. Install the bearing caps and tighten the M8 fasteners to 250 inch lbs. (28 Nm). The end caps must be installed before the seals may be installed.

19. Install the camshaft end seals.

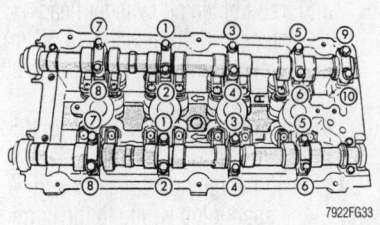

Camshaft bearing cap tightening sequence—2.4L engine

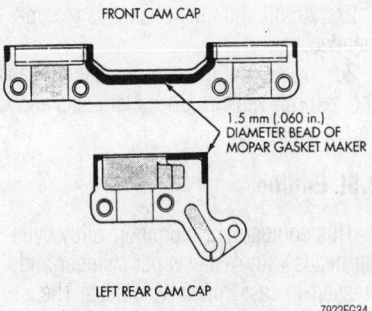

Apply sealer as shown to prevent oil leakage from the camshaft bearing end caps—2.4L engine

20. Reinstall the camshaft sprockets, if removed. Install the timing belt using care to be sure all timing marks are properly aligned, using the recommended procedure. Install the timing belt covers.

✳✳ WARNING

Verify that all timing marks are correct. If the timing belt or sprockets are incorrectly installed, engine damage will occur. Take time to be sure all timing marks are correctly aligned.

21. Clean all sealing surfaces. Make certain the rails are flat.

22. Install new cylinder head cover gaskets. Use care. DO NOT allow oil or solvents to contact the timing belt as they can deteriorate the rubber and cause tooth skipping. Apply Mopar Silicone Rubber Adhesive Sealant, or equivalent, at the camshaft cap corners and at the top edge of the ½ round seal.

➡**Inspect the spark plug well seals for cracking and/or swelling and replace if necessary.**

23. Install the cylinder head cover assembly to the head and tighten the fasten-

ers in sequence using the following three Steps:

a. First: tighten all cylinder head cover fasteners to 40 inch lbs. (4.5 Nm).

b. Second: tighten all fasteners to 80 inch lbs. (9 Nm).

c. Third: tighten all fasteners to 105 inch lbs. (12 Nm).

24. Install the ignition coil pack and connect the spark plug wiring to the correct spark plugs. Tighten the coil pack retaining fasteners to 105 inch lbs. (12 Nm).

25. Reconnect the ground strap.

26. Check to be sure all vacuum lines and remaining wiring have been reconnected.

27. An oil and filter change is recommended.

28. Reconnect the negative battery cable and test run vehicle. Check for leaks and for proper operation.

2.5L Engine

This engine uses aluminum alloy cylinder heads with 4-valves per cylinder and pressed-in cast iron valve guides. The cylinders are common to either cylinder bank. Two overhead camshafts are supported by four bearing journals which are part of the head with the distributor driven off the right (firewall side) cylinder head. Right and left camshaft drive sprockets are interchangeable. The sprockets and engine water pump are driven by the timing belt. Care must be taken to be sure all valve timing marks align after cylinder head and valvetrain service. Please note that for camshaft service, the cylinder head must be removed.

✲✲ CAUTION

The fuel injection system remains under pressure, even after the engine has been turned OFF. The fuel system pressure MUST BE relieved before disconnecting any fuel lines. Failure to do so may result in fire and/or personal injury.

1. Disconnect the negative battery cable from the left shock tower. The ground cable is equipped with a insulator grommet which should be placed on the stud to prevent the negative battery cable from accidentally grounding.

✲✲ CAUTION

Observe all applicable safety precautions when working around fuel. Whenever servicing the fuel system, always work in a well ventilated area. Do not allow fuel spray or vapors to come in contact with a spark or open flame. Keep a dry chemical fire extinguisher near the work area. Always keep fuel in a container specifically designed for fuel storage; also, always properly seal fuel containers to avoid the possibility of fire or explosion.

2. Relieve the fuel system pressure using the recommended procedure.

✲✲ CAUTION

Never open, service or drain the radiator or cooling system when hot; serious burns can occur from the steam and hot coolant.

3. Place a large drain pan under the radiator drain plug. Drain the cooling system.

4. Remove the timing belt covers, timing belt and camshaft sprockets using the recommended procedure.

5. The intake manifold is a two-piece unit. The upper part is a large air intake plenum of aluminum alloy. Use care working with light alloy parts. Remove the air intake plenum first, then remove the lower intake manifold.

6. Disconnect, label and set aside the spark plug wires.

7. Remove the cylinder head cover screws and remove the cover.

8. Identify the rocker arm shaft assemblies before removal.

9. Install the auto lash adjuster retainers Special Tool MD 998443 or equivalent to keep the auto lash adjusters from falling out of the rocker arms when the rocker arm assembly is removed.

10. Loosen the attaching fasteners and remove the rocker arm shaft assemblies from the cylinder head.

11. Remove the distributor assembly.

12. Remove the exhaust manifold and crossover.

13. Remove the cylinder head bolts and remove the cylinder head from the vehicle.

14. Remove the thrust case from the left head assembly and remove the camshaft from the rear of the head. If not already done, remove the distributor from the right cylinder head and remove the camshaft from the rear of the head.

15. Remove and discard the camshaft oil seals.

16. The camshafts should be carefully inspected for scratches or worn areas. If light scratches are seen, they may be removed with 400 grit sandpaper. If there are deep scratches, replace the camshaft and check the cylinder head for damage. Check the oil holes to be sure they are open and free of debris. If the camshaft lobes show signs of wear, check the corresponding rocker arm roller for wear or damage. Replace the rocker arm if worn or damaged. If the camshaft shows signs of wear on the lobes, replace it.

To install:

17. Check camshaft end-play. Oil the camshaft journals with clean engine oil and install the camshaft **WITHOUT** the rocker arm assemblies. Move the camshaft as far rearward as it will go. Mount a dial indicator to bear on the front of the camshaft. Zero the indicator. Move the camshaft as far forward as it will go. End-play should be 0.004–0.008 in. (0.1–0.2mm). Maximum allowed end-play is 0.016 in. (0.4mm).

18. Lubricate the camshaft journals and carefully install the camshaft into the cylinder head. Install the thrust case and tighten the fasteners to 9 ft. lbs. (13 Nm).

19. Apply a light coating of engine oil to the camshaft oil seal lip and install the camshaft seal. The camshaft must be installed before installing the seal. Be sure the seal is installed flush with the cylinder head surface. Install the camshaft sprocket and tighten to 65 ft. lbs. (88 Nm).

20. Install the cylinder head(s).

21. Reinstall the lower intake manifold using new gaskets.

22. The rocker arm shafts are hollow and used as a lubrication oil duct. Be sure all valvetrain parts are clean. Check the rocker arm mounting portion of the shafts for wear or damage. Replace if necessary. Check all oil holes for clogging with a small wire and clean as required. If any rockers were removed, lubricate and install on the shafts in their original positions.

23. Reinstall the rocker arm and shaft assemblies.

24. Reinstall the timing belt using the recommended procedure. It is very important that all valve timing marks align properly or engine damage will result.

25. Inspect the spark plug tube seals located on the ends of each tube. These seals slide onto each tube to seal the cylinder head cover to the spark plug tube. If these seals show signs of hardness and/or cracks, they should be replaced.

26. Clean all sealing surfaces. Install a new gasket and install the cover. Tighten bolts to 88 inch lbs. (10 Nm). Reconnect the spark plug wires.

27. Position a new upper intake manifold (plenum) gasket in place and install the upper manifold. Tighten the bolts to 13 ft. lbs. (18 Nm). Reinstall the bolts for the

plenum support brackets and connect the EGR tube.

28. Reconnect the throttle and speed control cables.

29. Reattach the TPS and idle air control motor electrical connectors.

30. Reattach the MAP and intake air temperature sensors.

31. Reinstall the air inlet resonator, air inlet hose and the air cleaner housing cover.

32. Check to be sure all remaining electrical connectors have been reattached. Tighten the air tube connections.

✷✷ CAUTION

The EPA warns that prolonged contact with used engine oil may cause a number of skin disorders, including cancer! You should make every effort to minimize your exposure to used engine oil. Protective gloves should be worn when changing the oil. Wash your hands and any other exposed skin areas as soon as possible after exposure to used engine oil. Soap and water, or waterless hand cleaner should be used.

33. Refill the cooling system. An oil and filter change is recommended whenever a cylinder head has been removed since coolant can get into the oil system.

✷✷ WARNING

Operating the engine without the proper amount and type of engine oil will result in severe engine damage.

34. Reconnect the negative battery cable. Start the engine and check for leaks, abnormal noises and vibrations. Bleed the cooling system.

Valve Lash

ADJUSTMENT

2.0L and 2.5L Engines

The engines are equipped with hydraulic lash adjusters which are precision units installed in machined openings in the valve actuating ends of the rocker arms. Valve clearance adjustments are not performed.

2.4L Engine

The valves are actuated by roller cam followers which pivot on stationary hydraulic lash adjusters. The hydraulic lash adjusters are precision units installed in machined openings of the cam follower. Valve clearance adjustments are not performed.

Oil Pan

REMOVAL & INSTALLATION

2.0L Engine

1. Disconnect the negative battery cable from the left shock tower. The ground cable is equipped with a insulator grommet which should be placed on the stud to prevent the negative battery cable from accidentally grounding.

2. Raise and safely support the vehicle.

✷✷ CAUTION

The EPA warns that prolonged contact with used engine oil may cause a number of skin disorders, including cancer! You should make every effort to minimize your exposure to used engine oil. Protective gloves should be worn when changing the oil. Wash your hands and any other exposed skin areas as soon as possible after exposure to used engine oil. Soap and water, or waterless hand cleaner should be used.

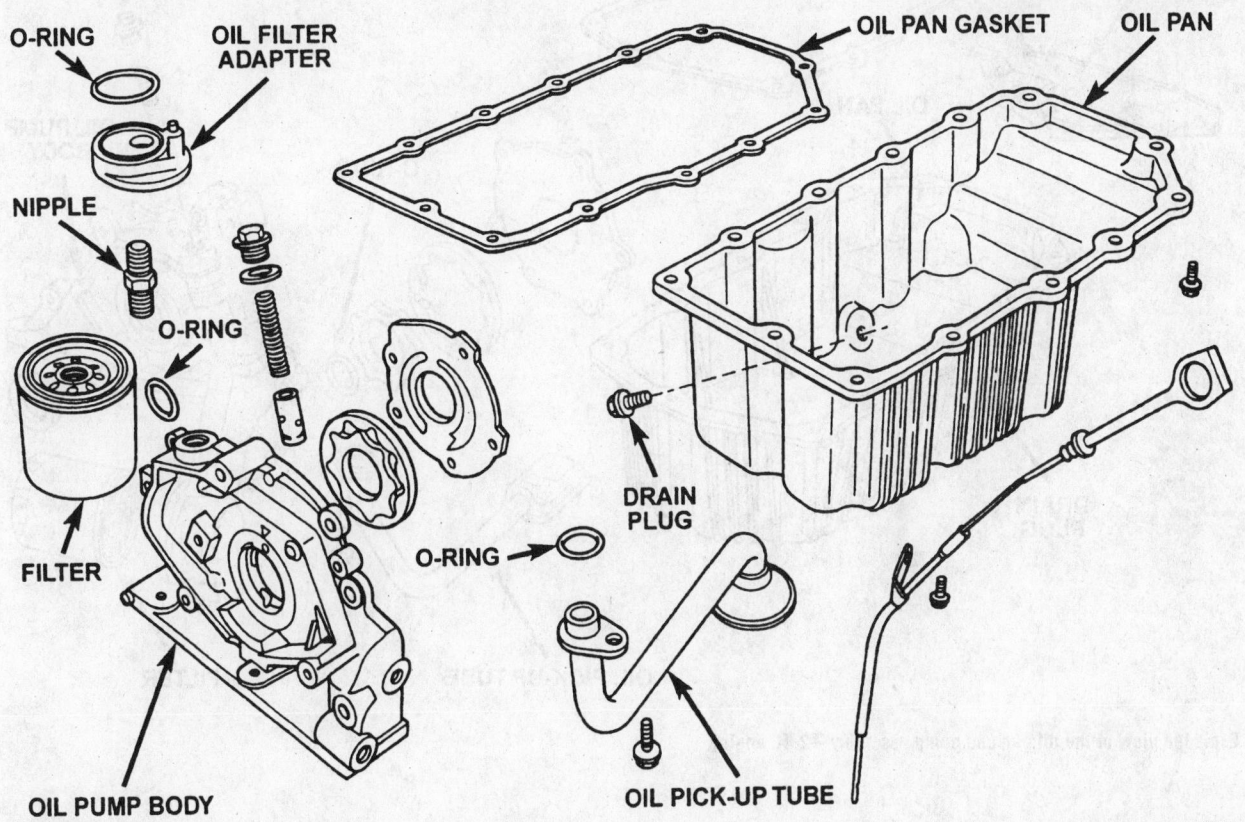

Exploded view of the oil pan and pump assembly—2.0L engine

7922FG35

3. Place a large oil pan under the oil pan drain plug. Drain the oil from the engine.

4. Remove the transaxle bending bracket.

5. Remove the front engine mount and bracket.

6. Remove the transaxle inspection cover.

7. Remove the oil filter and oil filter adapter.

8. Remove the oil pan attaching bolts.

9. Remove the oil pan.

10. Clean the oil pan as well as the oil pan gasket sealing surfaces.

To install:

11. Using a suitable rubber adhesive gasket sealant, apply a ⅛ in. (3mm) bead at the oil pump-to-engine block parting line.

12. Install the new oil pan gasket by positioning it properly onto the oil pan.

➡**If a gasket is not available, use a ⅛ in. (3mm) bead of silicone gasket maker.**

13. Install the oil pan onto the engine.

14. Tighten the oil pan attaching bolts to 105 inch lbs. (12 Nm).

15. Reinstall the oil filter adapter as follows:

a. Be sure the O-ring seal is seated in the groove on the adapter.

b. Align the locating roll pin into the engine block.

c. Tighten the retaining fastener to 60 ft. lbs. (80 Nm).

16. Reinstall a new oil filter as follows:

a. Lubricate the contact surface of the rubber oil filter gasket with a light bead of clean engine oil.

b. Be sure the gasket contact surface on the oil filter adapter is smooth, flat and clean of any debris or old pieces of rubber.

c. Rotate the oil filter clockwise until the gasket contacts the adapter base. Tighten the filter to 15 ft. lbs. (21 Nm).

17. Reinstall the transaxle inspection cover.

18. Reinstall the front engine mount and engine mount bracket.

19. Reinstall the transaxle bending bracket.

20. Reinstall the oil pan drain plug and gasket. Tighten the drain plug to 25 ft. lbs. (34 Nm).

21. Lower the vehicle.

✳✳ WARNING

Operating the engine without the proper amount and type of engine oil will result in severe engine damage.

22. Refill the engine with fresh oil to the proper level:

a. With an oil filter change—4.5 qts. (4.25L)

b. Without an oil filter change—4.0 qts. (3.8L)

23. Reconnect the negative battery cable. Start the engine and check for leaks.

2.4L Engine

1. Disconnect the negative battery cable from the left shock tower. The ground cable is equipped with a insulator grommet which should be placed on the stud to prevent the negative battery cable from accidentally grounding.

2. Raise and safely support the vehicle.

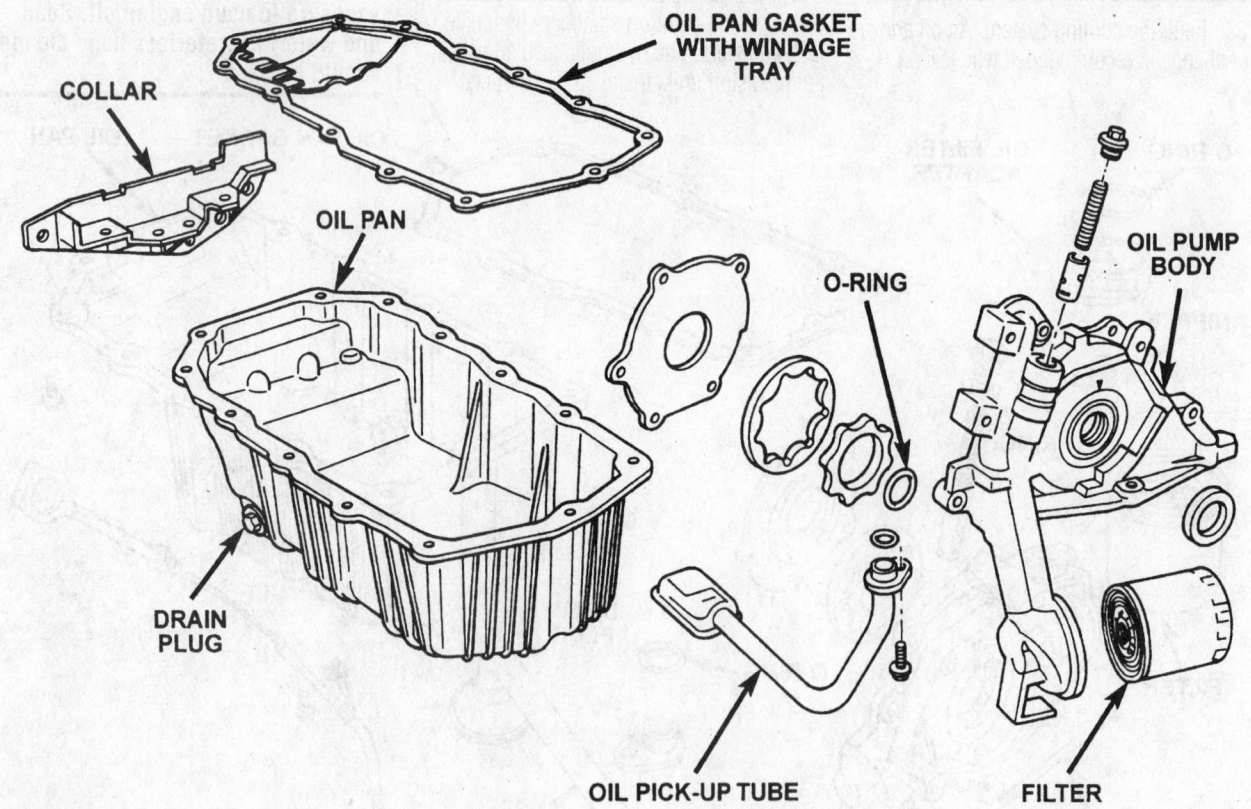

COLLAR

OIL PAN GASKET WITH WINDAGE TRAY

OIL PAN

O-RING

OIL PUMP BODY

DRAIN PLUG

OIL PICK-UP TUBE

FILTER

7922FG36

Exploded view of the oil pan and pump assembly—2.4L engine

✳✳ CAUTION

The EPA warns that prolonged contact with used engine oil may cause a number of skin disorders, including cancer! You should make every effort to minimize your exposure to used engine oil. Protective gloves should be worn when changing the oil. Wash your hands and any other exposed skin areas as soon as possible after exposure to used engine oil. Soap and water, or waterless hand cleaner should be used.

3. Place a large drain pan under the oil pan drain plug and drain the oil from the engine.

4. If necessary, remove the transaxle bending bracket.

5. Remove the front engine mount and bracket.

6. Remove the oil pan attaching bolts.

7. Remove the oil pan.

To install:

8. Using a suitable gasket sealant apply a ⅛ in. (3mm) bead at the oil pump to engine block parting line.

9. Install the new gasket.

➡**If a gasket is not available, use a ⅛ in. (3mm) bead of silicone gasket maker.**

10. Install the oil pan to the engine.

11. Tighten the oil pan attaching bolts to 105 inch lbs. (12 Nm).

12. Reinstall the front engine mount and engine mount bracket.

13. Reinstall the transaxle bending bracket, if necessary.

14. Install the oil pan drain plug and gasket. Tighten the drain plug to 25 ft. lbs. (34 Nm).

15. Lower the vehicle.

✳✳ WARNING

Operating the engine without the proper amount and type of engine oil will result in severe engine damage.

16. Refill the engine with fresh oil to the proper level. A filter change is recommended.

    a. With oil filter—5.0 qts. (4.7L)

    b. Without oil filter—4.5 qts. (4.3L)

17. Reconnect the negative battery cable. Start the engine and check for leaks.

2.5L Engine

1. Disconnect the negative battery cable from the left shock tower. The ground cable is equipped with a insulator grommet which should be placed on the stud to prevent the negative battery cable from accidentally grounding.

2. Raise and safely support the vehicle.

✳✳ CAUTION

The EPA warns that prolonged contact with used engine oil may cause a number of skin disorders, including cancer! You should make every effort to minimize your exposure to used engine oil. Protective gloves should be worn when changing the oil. Wash your hands and any other exposed skin areas as soon as possible after exposure to used engine oil. Soap and water, or waterless hand cleaner should be used.

3. Place a large drain pan under the oil pan drain plug and drain the oil from the engine.

4. Remove the engine support module as follows:

    a. Place a suitable support jack underneath the engine/transaxle assembly at the transaxle to prevent it from rotating.

    b. Remove the through-bolt at the rear mount and remove the bolts securing the support module to the crossmember.

    c. Remove the upper mounting bolt from the rear support strut bracket.

    d. Remove the front mounting bolts from the support module to the lower radiator support member.

    e. Support the radiator/cooling fan assembly. Remove the lower radiator support member.

    f. Remove the through-bolt at the front engine mount and remove the engine support module.

5. Remove the engine oil dipstick tube and dipstick.

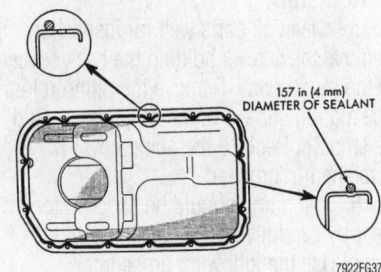

157 in (4 mm)
DIAMETER OF SEALANT

7922FG37

To ensure a tight seal, apply sealer as shown—2.5L engine

6. Remove the starter motor.

7. Remove the engine-to-transaxle struts.

8. Remove the transaxle inspection cover.

9. Remove the oil pan attaching bolts.

10. Remove the oil pan.

To install:

➡**Oil pan-to-engine block sealing is provided by using MOPAR® Silicone Rubber Adhesive Sealant or equivalent gasket material. The gasket material should be applied in a continuous bead approximately ⁵⁄₃₂ in. (4mm) in diameter with all mounting holes circled. Sealant that is not cured can be removed with a shop towel. Components should be torqued in place while the gasket sealer is still wet to the touch (within a 10 minute period). It is recommended that locating dowels be used during assembly to prevent smearing the gasket material from its location.**

11. Apply a continuous ⁵⁄₃₂ in. (4mm) bead of MOPAR® Silicone Adhesive Sealant or equivalent to the oil pan gasket surface. Be sure to circle all mounting bolt holes as well. Install the oil pan within a 10 minute period of applying the gasket material to ensure proper sealing.

12. Install the oil pan to the engine.

13. Tighten the oil pan attaching bolts to 53 inch lbs. (6 Nm).

14. Reinstall transaxle inspection cover.

15. Reinstall the engine-to-transaxle struts.

16. Reinstall the starter motor.

17. Reinstall the engine oil dipstick tube and dipstick.

18. Reinstall the engine support module as follows:

    a. Place the engine support module in proper position under the engine/transaxle assembly.

    b. Reinstall the through-bolt at the front mount. Do not tighten the through-bolt at this time.

    c. Reinstall the lower radiator support.

    d. Reinstall the mounting bolts from the engine support module to the lower radiator support.

    e. Reinstall the upper mounting bolt to the rear support strut bracket.

    f. Reinstall the through-bolt to the rear mount and tighten to 45 ft. lbs. (61 Nm).

    g. Tighten the through-bolt at the front mount to 45 ft. lbs. (61 Nm).

    h. Remove the support jack holding the engine/transaxle assembly in place.

19. Reinstall the oil pan drain plug and gasket. Tighten the drain plug to 29 ft. lbs. (40 Nm).

20. Lower the vehicle.

✳✳ WARNING

Operating the engine without the proper amount and type of engine oil will result in severe engine damage.

21. Refill the engine with fresh oil to the proper level. An oil filter change is recommended.
   - With oil filter—4.5 qts. (4.3L)
   - Without oil filter—4.0 qts. (3.8L)

22. Reconnect the negative battery cable. Start the engine and check for leaks.

➡Whenever the vehicle sub-frame is removed or lowered, the wheel alignment should be checked.

Oil Pump

REMOVAL & INSTALLATION

2.0L and 2.4L Engines

The oil drawn up through the pick-up tube is pressurized by the pump and routed through the full flow filter to the main oil gallery running the length of the cylinder block. The oil pick-up, pump and check valve provide oil flow to the main oil gallery. A vertical hole at the number 5 bulkhead routes pressurized oil through a restrictor up past a cylinder head bolt to an oil gallery running the length of the cylinder head. The camshaft journals are slotted to allow pressurized oil to pass into the bearing cap cavities. Small holes in the bearing caps direct oil to the camshaft lobes.

It is necessary to remove the oil pan, oil pick-up and oil pump housing to service the oil pump rotors. The oil pump pressure relief valve can be serviced without removing the oil pan and oil pick-up tube.

1. Disconnect the negative battery cable from the left shock tower. The ground cable is equipped with a insulator grommet which should be placed on the stud to prevent the negative battery cable from accidentally grounding.

✳✳ CAUTION

Never open, service or drain the radiator or cooling system when hot; serious burns can occur from the steam and hot coolant.

2. Raise and safely support the vehicle.

✳✳ CAUTION

The EPA warns that prolonged contact with used engine oil may cause a number of skin disorders, including cancer! You should make every effort to minimize your exposure to used engine oil. Protective gloves should be worn when changing the oil. Wash your hands and any other exposed skin areas as soon as possible after exposure to used engine oil. Soap and water, or waterless hand cleaner should be used.

3. Drain the engine oil as well as the engine coolant into suitable containers.

4. Remove the oil pan assembly. Remove the oil pump pick-up tube and O-ring.

5. Remove the right inner fender splash shield.

6. Lower the vehicle.

7. Remove the drive belts as required.

8. Using a puller, remove the crankshaft damper from the front of the crankshaft.

9. Take up the weight of the engine with a suitable lift and remove the right engine mount and bracket. Be sure the engine is safely supported.

10. Remove the timing belt cover.

11. Loosen the timing belt tensioner bolts and remove the tensioner and the timing belt.

12. Using a suitable puller, draw the crankshaft sprocket from the front of the crankshaft.

13. Loosen the oil pump bolts and remove. Take note of the location of each bolt for reassembly. Remove the oil pump from the face of the engine block. If necessary, tap lightly with a soft face mallet. Use care working with light alloy parts. Remove and discard the front oil seal.

14. Remove the relief valve from the pump body by removing the threaded plug and gasket and pulling out the spring and relief valve. Note the order of parts removal.

To install:

15. Clean all parts well for inspection. Remove the screws holding the back cover to the pump body. Remove the pump rotors. The mating surface of the oil pump should be smooth. Replace the pump cover if scratched or grooved.

16. The pump should be checked for wear by carefully measuring the components. Use the following procedure.

   a. Lay a straight-edge across the pump cover surface. If a 0.003 in. (0.076mm) feeler gauge can be inserted between the cover and straight-edge, the cover should be replaced.

   b. Measure the thickness and diameter of the outer rotor. If the outer rotor thickness measures 0.301 in. (7.6mm) or less, or if the diameter is 3.148 in. (80mm) or less, replace the outer rotor.

   c. If the inner rotor measures 0.301 in. (7.6mm) or less, replace the inner rotor.

   d. Slide the outer rotor into the pump housing, press to one side with fingers and measure the clearance between the rotor and housing. If the measurement is 0.015 in. (0.38mm) or more, replace the oil pump housing only if the outer rotor is in specification.

   e. Install the inner rotor into the pump housing. If the clearance between the inner and outer rotors is 0.008 in. (0.20mm) or more, replace both rotors.

   f. Place a straight-edge across the face of the pump housing between the bolt holes. If a feeler gauge of 0.004 in. (0.10mm) or more can be inserted between the rotors and straight-edge, replace the pump assembly only if the rotors are within specification.

   g. Inspect the oil pressure relief valve plunger for scoring and free operation in its bore. Small marks may be removed with 400 grit wet or dry sandpaper.

   h. The relief valve spring has a free length of approximately 2.39 in. (60.7mm). It should test between 18–19 lbs. (80–85mm) when compressed to 1.60 in. (40.6mm). Replace the spring if weak, damaged or fails to meet specifications.

➡If oil pressure is low and the pump is within specifications, inspect for worn engine bearings, clogged oil filter, pressure relief valve stuck open, damaged or missing oil pick-up tube O-ring, clogged oil pick-up tube screen or other reasons for oil pressure loss.

17. Clean all oil pump parts in suitable solvent before assembly. Assemble the pump with new parts as required. Install the inner rotor with the chamfer facing the cast iron oil pump cover (back of the pump). Tighten the cover screws to 105 inch lbs. (12 Nm).

18. Reinstall the relief valve first, then the spring, gasket and cover cap into the pump body. Note that installing the spring first will seriously damage the engine. The relief valve goes in first. Tighten the cover cap to 30 ft. lbs. (41 Nm) for 2.0L engine or 40 ft. lbs. (55 Nm) for 2.4L engine.

19. Prime the oil pump before installation by filling the rotor cavity with clean engine oil.

20. Insert a new oil ring seal to the oil pump counterbore on the pump body discharge passage. Apply Mopar Gasket Maker or equivalent anaerobic type gasket sealer to the oil pump body flange. This material cures in the absence of air when squeezed between two flat machined metal surfaces. For this reason, the mating surfaces of both the pump body and the engine block must be spotlessly clean so all air will be expelled when the parts are bolted together and tightened. Install the pump slowly onto the crankshaft aligning the oil pump rotor flats with the flats on the crankshaft until seated to the engine block. Tighten the fasteners to 250 inch lbs. (28 Nm).

21. Install a new front oil seal. Install the seal with the spring side towards the inside of the engine. Tap the seal into place until flush with the cover.

22. Install the crankshaft sprocket. A special tool is used to draw the sprocket onto the end of the crankshaft. Use care if using substitutes.

23. Install the timing belt and covers using the recommended procedures. Use care to be sure all valve timing marks are aligned. This is most important. Failure to properly align the timing marks will result in severe engine damage.

24. Reinstall the crankshaft damper. A special tool making use of a 12mm x 1.75 x 150mm bolt is used to draw the crankshaft damper onto the end of the crankshaft. Use care if using substitutes. Tighten the center bolt to 105 ft. lbs. (142 Nm).

25. Raise and safely support the vehicle.

26. Reinstall the oil pump pick-up tube and O-ring. Tighten the oil pump pick-up tube mounting screw to 21 ft. lbs. (28 Nm).

27. Clean the oil pan well and be sure the gasket rails are in good condition. Use Mopar Silicone Rubber Adhesive Sealant or equivalent sealer at the oil pump to engine block parting line. Use a new oil pan gasket, install the oil pan and tighten the 13 oil pan bolts to 105 inch lbs. (12 Nm).

28. Install a new oil filter.

29. Reinstall the right inner fender splash shield.

30. Lower the vehicle.

31. Reinstall the drive belts and accessories as required. Adjust the accessory drive belts.

32. Reinstall the engine mount and bracket as required.

✳✳ WARNING

Operating the engine without the proper amount and type of engine oil will result in severe engine damage.

33. Refill the engine with new, clean engine oil and coolant.

34. Test run vehicle to check for leaks. An oil pressure gauge should be installed to verify proper engine oil pressure.

2.5L Engine

The oil pump assembly is mounted on the timing belt end of the cylinder block with the inner pump

rotor indexed and installed on the crankshaft nose. The oil pump case also retains the crankshaft front oil seal and provides oil pan front end closure.

Oil pressure can be checked with a mechanical oil pressure gauge installed at the oil switch location. Oil pressure should be 6 psi (41.4 kPa) at idle and 35–75 psi (241.3–517.1 kPa) at 3000 rpm, engine at operating temperature. If an oil pressure problem is suspected or if oil pressure is zero at idle, Do not run the engine up to 3000 rpm in an attempt to raise oil pressure or the engine will be severely damaged.

Because the oil pump is driven off the nose of the crankshaft, the timing belt must be removed to access the pump. This is a lengthy process. Use care to properly align all valve timing marks.

1. Disconnect the negative battery cable

from the left shock tower. The ground cable is equipped with a insulator grommet which should be placed on the stud to prevent the negative battery cable from accidentally grounding.

2. Remove the drive belts and accessories.

✳✳ CAUTION

Never open, service or drain the radiator or cooling system when hot; serious burns can occur from the steam and hot coolant.

3. Drain the coolant and remove the radiator and cooling fan.

✳✳ CAUTION

The EPA warns that prolonged contact with used engine oil may cause a number of skin disorders, including cancer! You should make every effort to minimize your exposure to used engine oil. Protective gloves should be worn when changing the oil. Wash your hands and any other exposed skin areas as soon as possible after exposure to used engine oil. Soap and water, or waterless hand cleaner should be used.

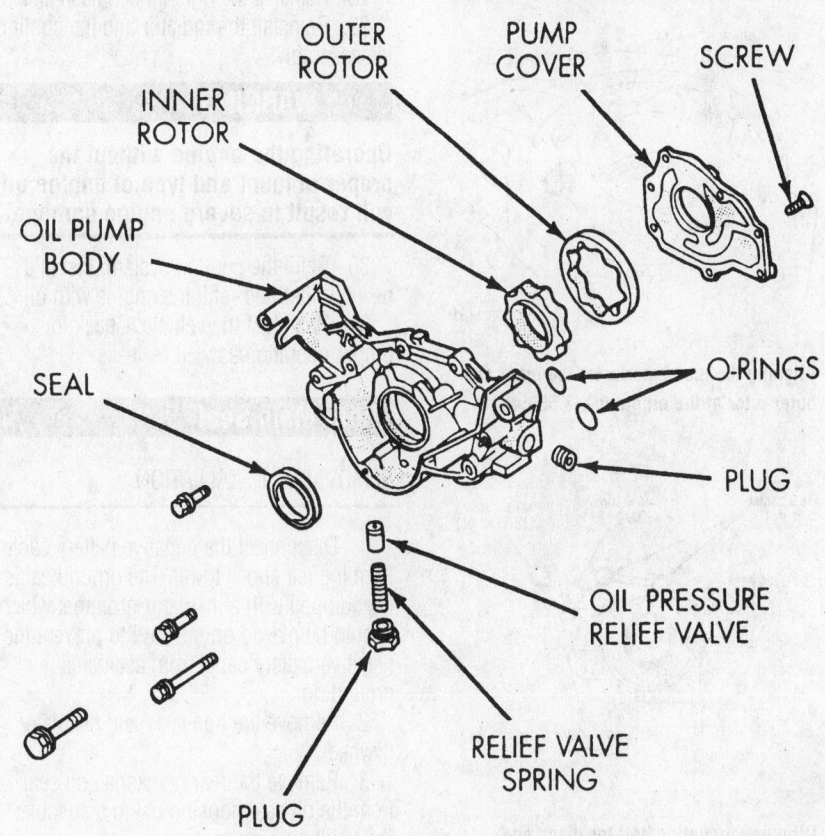

Exploded view of the oil pump assembly. Note the different bolt lengths—2.5L engine

7922FG38

4. Raise and safely support the vehicle. Drain the engine oil.

5. Remove the right inner splash shield.

6. Remove the crankshaft damper.

7. Support the engine and remove the right engine mount and the engine mount bracket

8. Remove the timing belt upper and lower covers.

9. Loosen the timing belt tensioner bolts and remove the timing belt and the tensioner.

10. Using a suitable puller, draw the crankshaft sprocket from the crankshaft.

11. Remove the five bolts that attach the oil pump to the block and remove the oil pump.

12. Inspect the oil pump case for damage and remove the rear cover.

13. Remove the pump rotors and inspect the inside of the case for excessive wear.

14. Check that the oil relief plunger slides smoothly and check for a broken spring.

To install:

15. Clean all parts well. Be sure the block and pump surfaces are clean and free of old sealer.

16. Assemble the pump using new parts

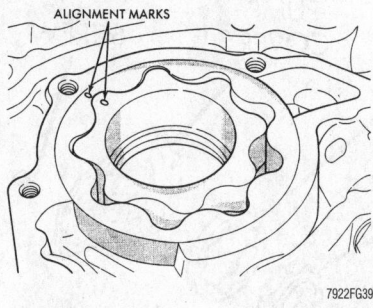

ALIGNMENT MARKS

7922FG39

Aligning the matchmarks for the inner and outer rotor of the oil pump—2.5L engine

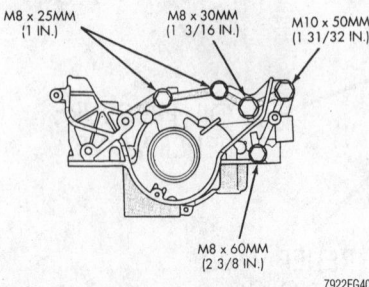

M8 x 25MM (1 IN.)   M8 x 30MM (1 3/16 IN.)   M10 x 50MM (1 31/32 IN.)

M8 x 60MM (2 3/8 IN.)

7922FG40

Oil pump mounting bolt locations and dimensions—2.5L engine

as required with clean oil. Align the marks on the inner and outer rotors when assembling.

17. Reinstall the pump back cover and tighten the screws to 88 inch lbs. (10 Nm).

18. Reinstall the pump relief valve, spring, gasket and valve cap. Tighten the valve cap to 30–33 ft. lbs. (41–44 Nm).

19. Prime the pump before installation by filling the rotor cavity with clean engine oil.

20. Apply Mopar Gasket Maker or equivalent sealer to the pump. Install the O-ring into the counterbore on the pump body discharge passage. Position the pump onto the crankshaft until seated on the block. Tighten the size M8 fasteners to 10 ft. lbs. (14 Nm) and size M10 fasteners to 30 ft. lbs. (41 Nm).

21. Reinstall the timing belt and crankshaft sprocket using the recommended procedure and following all cautions. Valve timing MUST be correct or engine damage will result.

22. Reinstall the timing belt cover.

23. Reinstall the engine mount bracket.

24. Reinstall the right engine mount.

25. Remove the engine support.

26. Reinstall the crankshaft damper and tighten to 134 ft. lbs. (182 Nm).

27. Reinstall the drive belts and accessories.

28. Reinstall the right inner splash shield.

29. Reinstall the radiator and the cooling fan assembly.

✳✳ WARNING

Operating the engine without the proper amount and type of engine oil will result in severe engine damage.

30. Refill the cooling system. Install a new oil filter and refill the engine with oil.

31. Road test the vehicle. Check for proper operation as well as leaks.

Rear Main Seal

REMOVAL & INSTALLATION

1. Disconnect the negative battery cable from the left shock tower. The ground cable is equipped with a insulator grommet which should be placed on the stud to prevent the negative battery cable from accidentally grounding.

2. Remove the transaxle and flexplate/flywheel.

3. Remove the rear crankshaft oil seal from the oil seal housing using a suitable flat-bladed prying tool.

To install:

→When installing seal there is no need to lubricate sealing surface.

4. Install seal into housing using a suitable driver or seal installer.

5. Install the flexplate/flywheel bolts and tighten to 70 ft. lbs. (95 Nm).

6. Reinstall the transaxle.

7. Reconnect the negative battery cable. Start the engine and check for leaks.

FUEL SYSTEM

Fuel System Service Precautions

Safety is the most important factor when performing not only fuel system maintenance but any type of maintenance. Failure to conduct maintenance and repairs in a safe manner may result in serious personal injury or death. Maintenance and testing of the vehicle's fuel system components can be accomplished safely and effectively by adhering to the following rules and guidelines.

• To avoid the possibility of fire and personal injury, always disconnect the negative battery cable unless the repair or test procedure requires that battery voltage be applied.

• Always relieve the fuel system pressure prior to disconnecting any fuel system component (injector, fuel rail, pressure regulator, etc.), fitting or fuel line connection. Exercise extreme caution whenever relieving fuel system pressure to avoid exposing skin, face and eyes to fuel spray. Please be advised that fuel under pressure may penetrate the skin or any part of the body that it contacts.

• Always place a shop towel or cloth around the fitting or connection prior to loosening to absorb any excess fuel due to spillage. Ensure that all fuel spillage (should it occur) is quickly removed from engine surfaces. Ensure that all fuel soaked cloths or towels are deposited into a suitable waste container.

• Always keep a dry chemical (Class B) fire extinguisher near the work area.

• Do not allow fuel spray or fuel vapors to come into contact with a spark or open flame.

• Always use a back-up wrench when loosening and tightening fuel line connection fittings. This will prevent unnecessary stress and torsion to fuel line piping.

Always follow the proper torque specifications.

• Always replace worn fuel fitting O-rings with new. Do not substitute fuel hose or equivalent, where fuel pipe is installed.

Fuel System Pressure

RELIEVING

2.0L and 2.4L Engines

✳✳ CAUTION

The fuel injection system remains under pressure, even after the engine has been turned OFF. The fuel system pressure MUST BE relieved before disconnecting any fuel lines. Failure to do so may result in fire and/or personal injury.

1. Disconnect the negative battery cable from the left shock tower. The ground cable is equipped with a insulator grommet which should be placed on the stud to prevent the negative battery cable from accidentally grounding.

✳✳ CAUTION

Observe all applicable safety precautions when working around fuel. Whenever servicing the fuel system, always work in a well ventilated area. Do not allow fuel spray or vapors to come in contact with a spark or open flame. Keep a dry chemical fire extinguisher near the work area. Always keep fuel in a container specifically designed for fuel storage; also, always properly seal fuel containers to avoid the possibility of fire or explosion.

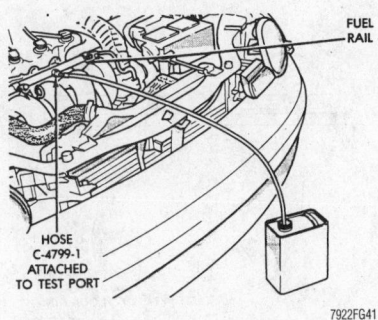

Fuel pressure test port location—2.0L and 2.4L engines

2. Remove the fuel filler cap.
3. Remove the cap on the fuel pressure test port on the fuel rail.
4. Place the open end of fuel pressure release hose special tool number C-4799–1 or equivalent into an approved gasoline container. Connect the other end of hose C-4799–1 or equivalent to the fuel pressure test port. Fuel pressure will bleed off through the hose onto the gasoline container.

2.5L Engine

✳✳ CAUTION

The fuel injection system remains under pressure, even after the engine has been turned OFF. The fuel system pressure MUST BE relieved before disconnecting any fuel lines. Failure to do so may result in fire and/or personal injury.

1. Disconnect the fuel rail electrical harness from the engine harness. This is connector C165, a black plastic connector located at the right rear of the intake manifold.
2. Circuit A142 supplies voltage for the fuel injectors while the Powertrain Control Module (PCM) controls the ground for each injector. Connect a jumper wire to the terminal for Circuit A142 (18 ga. wire, Dark Green with Orange tracer, from ASD relay).
3. Connect the other end of the jumper wire to a 12 volt power source.
4. Connect one end of a second jumper wire to a ground source.
5. Momentarily ground each of the injectors by connecting the other end of the jumper wire to the injector terminal in the harness connector. Repeat this procedure for two or three injectors.

✳✳ WARNING

Do not attempt to start the engine for several minutes to avoid hydrostatic lock.

Fuel Filter

REMOVAL & INSTALLATION

✳✳ CAUTION

Observe all applicable safety precautions when working around fuel. Whenever servicing the fuel system, always work in a well ventilated

area. **Do not allow fuel spray or vapors to come in contact with a spark or open flame. Keep a dry chemical fire extinguisher near the work area. Always keep fuel in a container specifically designed for fuel storage; also, always properly seal fuel containers to avoid the possibility of fire or explosion.**

The fuel delivery system contains a replaceable inline filter. The fuel filter mounts to the frame above the rear of the fuel tank. The fuel tank assembly must be loosened and lowered slightly to access the filter. The inlet and outlet tubes are permanently attached to the filter. Please note that the fuel system pressure must be relieved before servicing fuel system components. In addition, Chrysler uses Quick Connect fittings on fuel line connections. Specific instructions are given at the end of this procedure.

✳✳ CAUTION

The fuel injection system remains under pressure, even after the engine has been turned OFF. The fuel system pressure MUST BE relieved before disconnecting any fuel lines. Failure to do so may result in fire and/or personal injury.

1. Disconnect the negative battery cable from the left shock tower. The ground cable is equipped with a insulator grommet which should be placed on the stud to prevent the negative battery cable from accidentally grounding.
2. Relieve the fuel system pressure using the recommended procedure.
3. From inside the trunk, disconnect the fuel pump module wiring jumper from the main body harness. The 4-pin connector is located under the trunk mat on the left side of the trunk near the base of the shock tower. Locate the body grommet for the jumper near the base of the rear seat. Push the grommet out and feed the jumper completely through the hole in the body.
4. Remove the fuel cap slowly to release tank pressure.
5. Raise and safely support the vehicle.
6. Locate the drain plug on the bottom left of the fuel tank. Place an approved fuel container with a capacity of at least 16 gallons, under the drain plug. Remove the plug and drain the fuel tank. When finished draining, install the plug since there will be one to two gallons of fuel remaining. Tighten the drain plug to 32 inch lbs. (3.6 Nm).

❄❄ CAUTION

Observe all applicable safety precautions when working around fuel. Do not allow fuel spray or fuel vapors to come in contact with a spark or open flame. Keep a dry chemical (Class B) fire extinguisher near the work area. Never drain or store fuel in an open container due to the possibility of fire or explosion.

7. Remove the driver's side fuel tank strap. Loosen, but do not remove the passenger's side fuel tank strap allowing the fuel tank neck to touch the rear suspension crossmember.

❄❄ CAUTION

Wrap shop towels around the fuel hoses to catch any gasoline spillage.

8. Detach the fuel lines from the fuel pump module. These are quick connect fittings. Depress the releasing tabs on the fuel filter hose connections and detach the fuel lines.
9. Remove the fuel filter.
To install:
10. The fuel supply (to filter) tube and

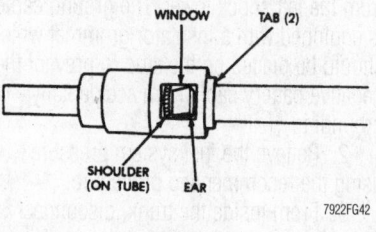

Releasing the quick connect fittings

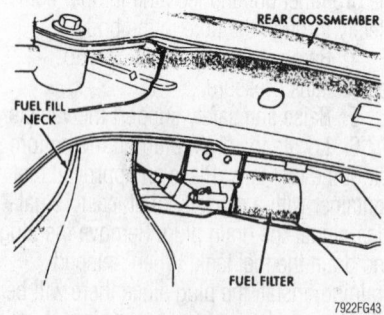

Fuel filter mounting location

the return tube (to fuel pump module) are permanently attached to the fuel filter. The ends of the fuel supply and return tubes have different size quick connect fittings. The large quick connect fitting attaches to the large nipple (supply side) on the fuel pump module. The smaller quick connect fitting attaches the small nipple (return side) on the fuel pump module. Specific Quick Connect fitting service procedures are given below.

11. Apply a light coat of clean 30W engine oil to the fuel filter nipples. Install the fuel tubes.
12. Reinstall the fuel tank, filter and fuel tank straps. Tighten the strap bolts to 250 inch lbs. (23 Nm).

➡️**Be sure the fuel pump module electrical harness grommet is installed in the body as the tank is raised into position.**

13. Lower the vehicle and reattach the pump module connector.
14. Refill the tank with fuel.
15. Reconnect the negative battery cable.

Fuel Pump

REMOVAL & INSTALLATION

The in-tank fuel pump module contains the fuel pump and pressure regulator which adjusts fuel system pressure to approximately 49 psi (337.8 kPa). Voltage to the fuel pump is supplied through the fuel pump relay.

The fuel pump is serviced as part of the fuel pump module. The fuel pump module is installed in the top of the fuel tank and contains the electric fuel pump, fuel pump reservoir, inlet strainer fuel gauge sending unit, fuel supply and return line connections and the pressure regulator. The inlet strainer, fuel pressure regulator and level sensor are the only serviceable items. If the fuel pump requires service, replace the fuel pump module.

❄❄ CAUTION

The fuel injection system remains under pressure, even after the engine has been turned OFF. The fuel system pressure MUST BE relieved before disconnecting any fuel lines. Failure to do so may result in fire and/or personal injury.

1. Disconnect the negative battery cable from the left shock tower. The ground cable is equipped with a insulator grommet which

should be placed on the stud to prevent the negative battery cable from accidentally grounding.
2. Remove the fuel filler cap and relieve the fuel system pressure using the recommended procedure.
3. Drain and remove the fuel tank following the recommended procedure.

❄❄ CAUTION

Observe all applicable safety precautions when working around fuel. Do not allow fuel spray or fuel vapors to come in contact with a spark or open flame. Keep a dry chemical (Class B) fire extinguisher near the work area. Never drain or store fuel in an open container due to the possibility of fire or explosion.

4. Clean the top of the tank to remove any loose dirt.
5. Disconnect fuel lines from the fuel pump module.
6. Using special tool 6856 or equivalent, remove the fuel pump locknut.

❄❄ CAUTION

The fuel reservoir of the fuel pump module does not empty out when the tank is drained. The fuel in the reservoir may spill out when the module is removed.

7. Remove the fuel pump and O-ring from the tank. Discard the O-ring.
To install:
8. Thoroughly clean all parts. Wipe the seal area of the tank clean. Place a new O-ring on the ledge between the tank threads and the pump module opening.
9. Position the fuel pump module in the tank. Be sure the alignment tab on the underside of the pump module flange sits in the corresponding notch in the fuel tank.

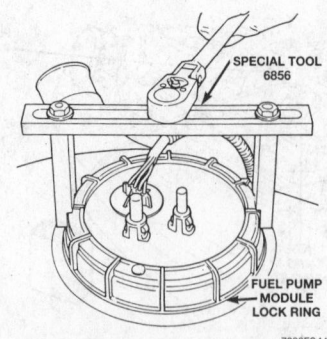

Removing the fuel pump module lock ring using special tool 6856

10. While holding the fuel pump module in place install the locking ring and tighten to 40–45 ft. lbs. (54–61 Nm) using a spanner-type tool.

11. Reinstall the fuel tank assembly.

12. Reconnect the negative battery cable.

13. Refill the fuel tank with clean fuel. Turn the ignition switch to the **ON** position to pressurize the system. Check the fuel system for leaks.

DRIVE TRAIN

Transmission Assembly

REMOVAL & INSTALLATION

Manual

1. Disconnect both battery cables, negative side first. Remove the battery, tray and stay brace.

2. Remove the air cleaner and intake hoses.

3. Drain the transaxle into a suitable waste container.

4. Remove the select and shift cables from the transaxle.

5. Disconnect the back-up light switch harness and position it aside.

6. Disengage the speedometer electrical harness, from the transaxle assembly.

7. Remove the starter motor.

8. Using special tool 7137 or C-4852 or equivalent, support the engine assembly.

9. Remove the rear roll stopper mounting bracket.

10. Remove the transaxle mount bracket.

11. Remove the upper transaxle mounting bolts.

12. Raise and safely support the vehicle.

13. Remove the front wheel assemblies.

14. Remove the under cover.

15. Remove the halfshaft assemblies. Plug the halfshaft openings in the transaxle assembly to prevent foreign material from entering.

16. Remove the cover from the transaxle bell housing.

17. Remove the engine front roll stopper through-bolt.

18. Remove the centermember.

19. Support the transaxle, using a transmission jack.

20. Rotate the engine clockwise to gain access to the driveplate clutch bolts. Remove the drive plate clutch bolts.

21. Remove the lower engine-to-transaxle mounting bolts.

22. Slide the transaxle rearward and carefully lower it from the vehicle.

To install:

23. Installation is the reverse of the removal procedure. Please note the following important steps.

24. The following items must be tightened to the specifications listed.

• Tighten the transaxle-to-engine mounting bolts to 70 ft. lbs. (95 Nm)

• Tighten the transaxle bell housing cover bolts to 7 ft. lbs. (9 Nm)

• Tighten the centermember front mounting bolts to 65 ft. lbs. (88 Nm) and the rear bolt to 54 ft. lbs. (73 Nm). Install the front engine roll stopper through-bolt and lightly tighten. Once the full weight of the engine is on the mounts, tighten the bolt to 42 ft. lbs. (57 Nm)

• Tighten the damper fork-to-lower control arm through-bolt to 65 ft. lbs. (88 Nm)

• Tighten the stabilizer link-to-damper fork nut to 29 ft. lbs. (39 Nm)

• Install the transaxle mount bracket, to the transaxle, and tighten the mounting nuts to 32 ft. lbs. (43 Nm)

• Tighten the transaxle mount through-bolt to 51 ft. lbs. (69 Nm)

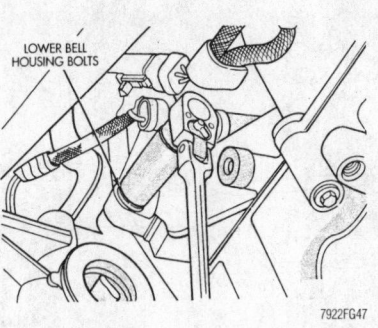

Removing the lower engine-to-transaxle mounting bolts

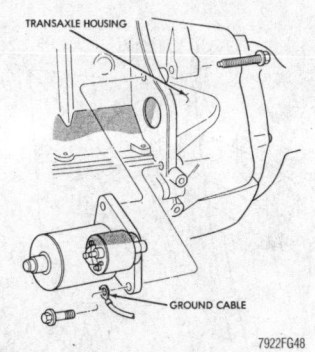

Be sure to connect the ground cable at the starter

25. Check to be sure that all fasteners are tightened and connections made.

26. Be sure the vehicle is level, and refill the transaxle.

27. Check the transaxle for proper operation. Be sure the reverse lights come on when in reverse.

Automatic

❊❊ WARNING

If the vehicle is going to be rolled on its wheels while the transaxle is out of the vehicle, obtain two outer CV-joints to install to the hubs. If the vehicle is rolled without the proper torque applied to the front wheel bearings, the bearings will no longer be usable.

1. Disconnect both battery cables, negative side first.

❊❊ CAUTION

Never open, service or drain the radiator or cooling system when hot; serious burns can occur from the steam and hot coolant.

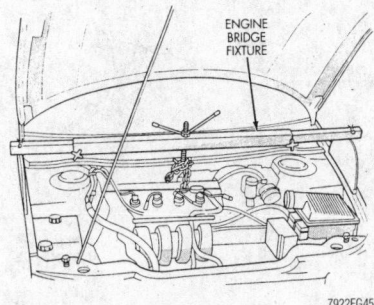

Secure the engine assembly with the appropriate support fixture

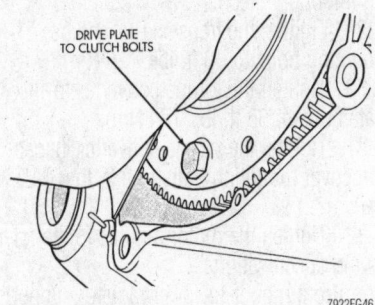

Remove the driveplate clutch bolts—rotate the engine clockwise to advance to the next bolt

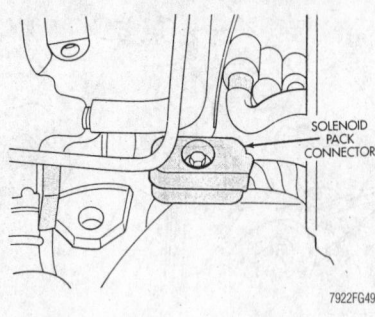

Location of the transaxle solenoid assembly 8-way connector and retaining bolt

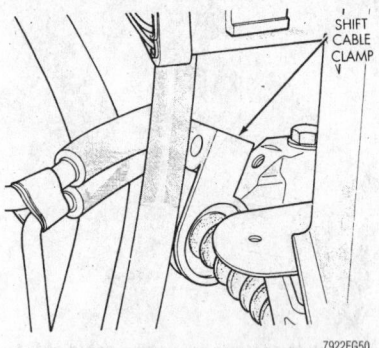

Removing the shift cable and clamp

2. If necessary, drain the coolant and remove the coolant return extension.

3. Remove the air cleaner/inlet duct assembly. Remove the upper bell housing bolts and water tube, where applicable.

4. Label and disengage all electrical connectors, cable linkages, hoses and mounting brackets required for removal of the transaxle assembly.

5. Remove the bolt securing the fluid dipstick tube, to the transaxle. Remove the dipstick and tube from the transaxle.

6. Using special tool 7137 or C-4852 or equivalent, support the engine assembly.

7. Remove the starter motor

8. Drain the transaxle fluid into a suitable waste container.

9. Raise the vehicle and support safely.

10. Remove the tire and wheel assemblies.

11. Remove splash shields.

12. Disconnect the exhaust pipe from the exhaust manifold.

13. Remove the halfshaft assemblies. Position a drain pan under the transaxle where the axles enter the differential or extension housing.

14. Unbolt the center bearing and remove the intermediate axle from the transaxle, if equipped.

15. Disconnect and tag the oil cooler lines, at the transaxle.

16. If equipped with Direct Ignition System (DIS), detach the harness connector and remove the crankshaft position sensor from the transaxle bell housing.

17. Remove the front and rear motor/transaxle mounts.

18. Remove the centermember.

19. Remove the torque converter inspection cover, matchmark the torque converter to the flexplate.

20. Remove the bolts holding the flexplate to the torque converter with a box wrench. Rotate the crankshaft to bring the bolts into a position for removal, one at a time.

21. Support the transaxle using a transmission jack (at the side of the case, NOT at the pan).

22. Remove the lower bell housing bolts.

23. Remove the transaxle mount bolts.

➡**The torque converter can become disengaged from the transaxle. Keep the front of transaxle slightly raised during removal.**

24. Carefully pry the transaxle from the engine.

25. Slide the transaxle rearward until dowels disengage from the mating holes in the transaxle case.

26. Pull the transaxle completely away from the engine and remove it from the vehicle.

27. To prepare the vehicle for rolling, support the engine with a suitable support or reinstall the front motor mount to the engine. Then, reinstall the ball joints to the steering knuckle and install the retaining bolt. Install the obtained outer CV-joints to the hubs, install the washers and tighten the axle nuts to 180 ft. lbs. (244 Nm). The vehicle may now be safely rolled.

To install:

28. Installation is the reverse of the removal procedure. Please note the following important steps.

29. Tighten the transaxle-to-engine mounting bolts to 70 ft. lbs. (95 Nm).

30. Tighten the torque converter-to-flexplate bolts to 55 ft. lbs. (74 Nm).

31. Tighten the torque converter inspection cover mounting bolts to 9 ft. lbs. (12 Nm).

32. Tighten the motor mounts to the following specifications:

• Front mount-to-lower radiator support bolts—45 ft. lbs. (61 Nm)

• Rear mount-to-front suspension crossmember—45 ft. lbs. (61 Nm)

• Rear mount through-bolt—45 ft. lbs. (61 Nm)

• Left motor mount-to-frame rail—24 ft. lbs. (33 Nm)

33. Tighten the shifter lever retaining nut to 14 ft. lbs. (19 Nm).

34. Check to be sure that all wiring harness plugs, cable linkages, and hoses have been properly connected during installation.

35. Adjust the gearshift and throttle cables.

36. Reconnect the negative battery cable.

37. Refill the transaxle and the cooling system.

38. Check the transaxle for proper operation. Be sure the car's back-up lights and speedometer are working properly.

Clutch

ADJUSTMENT

Free-Play

The manual transaxle clutch release system has a unique self-adjusting mechanism to compensate for clutch disc wear. This adjuster mechanism is located with the clutch cable assembly. The preload spring maintains tension on the cable. This tension keeps the clutch release bearing continuously loaded against the fingers of the clutch cover assembly. No manual adjustment is necessary.

When servicing this vehicle or if removing and installing the clutch cable, do not pull on the clutch cable housing to remove it from the dash panel. Damage to the cable self-adjuster may occur.

To check the function of the adjuster mechanism, use the following procedure:

1. With slight pressure, pull the clutch release lever end of the cable to draw the cable taut.

2. Push the clutch cable housing toward

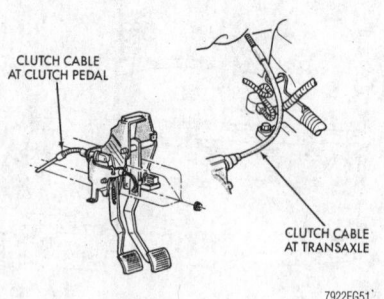

Clutch cable routing

the dash panel. With less than 25 lbs. (11 kg) of effort, the cable housing should move 1.2–2.0 in. (30–50mm). This indicates proper adjuster mechanism function.

3. If the cable does not adjust, determine if the mechanism is properly seated on the bracket.

REMOVAL & INSTALLATION

➡ **The transaxle assembly must be removed to service the clutch assembly.**

1. Disconnect the negative battery cable from the left shock tower. The ground cable is equipped with a insulator grommet which should be placed on the stud to prevent the negative battery cable from accidentally grounding.

2. Raise and safely support the vehicle.

3. Disconnect the starter wiring and remove the starter assembly.

4. Remove the rear and front transaxle support brackets.

5. Remove the clutch inspection cover.

6. Remove the bolts attaching the modular clutch to the flywheel.

7. Remove the transaxle assembly with the clutch as an assembly.

8. Remove the clutch assembly from input shaft of the transaxle.

To install:

9. Clean all parts well. Inspect for oil leakage through the engine rear crankshaft oil seal and transaxle input shaft seal. If

leakage is noted, it should be corrected at this time.

10. Examine the throwout or clutch release bearing. It is prelubricated and sealed and should not be washed in solvent. The bearing should turn smoothly when held in the hand with a light thrust load. A light drag caused by the lubricant fill is normal. If the bearing is noisy, rough or dry, replace the complete bearing assembly. In most cases where a clutch is being serviced, the complete clutch assembly and release bearing are usually replaced together.

11. Check the condition of the stud pivot spring clips on the back side of the clutch fork. If the clips are broken or distorted, replace the clutch fork. The pivot ball pocket in the fork is Teflon® coated and should be installed WITHOUT any lubricant such as grease which will break down the Teflon® coating. Be sure the ball stud and fork pocket are clean of contamination and dirt. When assembling the fork to the bearing, the small pegs on the bearing must go over the fork arms.

12. Check the flywheel for cracks, glazing or grooves. If any of these conditions exist, machine (reface) or replace the flywheel to prevent clutch chatter and premature clutch wear.

➡ **The manual transaxle is equipped with a reverse brake. It functions as a synchronizer, but only if the vehicle is not moving. When the clutch pedal is depressed to the floor and held for 3**

seconds, and the transaxle shifts to reverse, no gear clash should be present. If there is, the input shaft should be checked. When the transaxle is removed for clutch service, check the input clutch shaft, clutch disc splines and release bearing for dry rust. If present, clean rust off and apply a light coat of high temperature bearing grease to the input shaft splines. Apply grease on the input shaft splines only where the clutch disc slides. Verify that the clutch disc slides freely along the input shaft splines.

13. Install the modular clutch assembly onto the input shaft of the transaxle.

14. Install the transaxle assembly using the recommended procedure.

15. Install new clutch-to-driveplate (flywheel) bolts. Tighten the bolts in a crisscross pattern a few turns at a time to prevent distortion of the flywheel. Tighten the bolts to 55 ft. lbs. (75 Nm).

16. Install the clutch inspection cover.

17. Install the transaxle lower support brackets.

18. Install the starter assembly.

19. Lower the vehicle. Connect the negative battery cable.

20. Road test vehicle to check for proper clutch operation.

Halfshafts

REMOVAL & INSTALLATION

➡ **If the vehicle is going to be rolled while the halfshafts are out of the vehicle, obtain two outer CV-joints or proper equivalent tools and install to the hubs. If the vehicle is rolled without the proper torque applied to the front wheel bearings, the bearings will no longer be usable.**

1. Disconnect the negative battery cable.

2. Remove the cotter pin, nut lock and spring washer.

3. Loosen, but do not remove the halfshaft nut while the vehicle is on the floor with the brakes applied.

4. Raise and safely support the vehicle.

5. Remove the wheel.

6. Remove the brake caliper assembly and support it from the shock coil using a strong piece of wire.

7. Remove the brake rotor.

8. Remove the halfshaft nut and washer.

9. Using Joint Separation tool MB991113 or equivalent, disconnect the tie rod end from the steering knuckle.

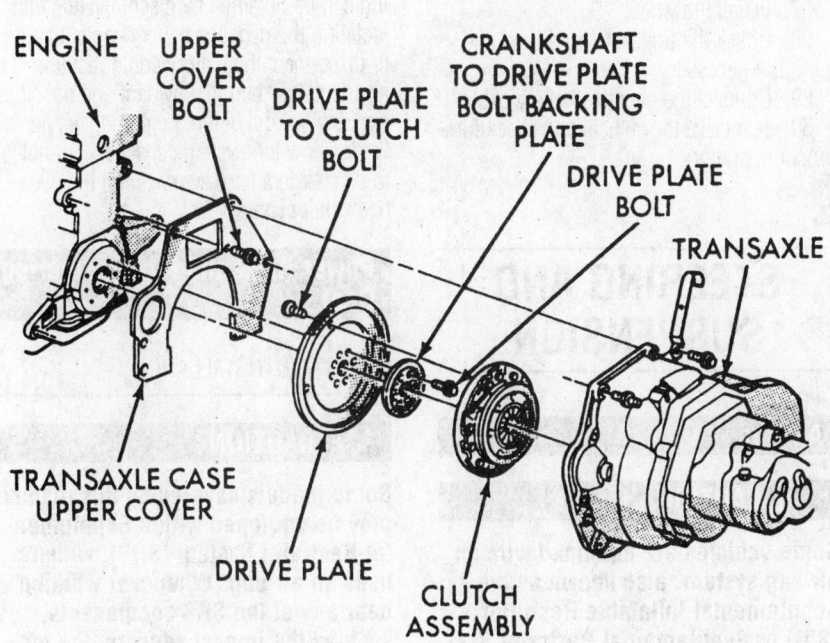

ENGINE UPPER COVER BOLT DRIVE PLATE TO CLUTCH BOLT CRANKSHAFT TO DRIVE PLATE BOLT BACKING PLATE DRIVE PLATE BOLT TRANSAXLE TRANSAXLE CASE UPPER COVER DRIVE PLATE CLUTCH ASSEMBLY

7922FG52

Exploded view of the clutch assembly

❋❋ CAUTION

Use of improper methods of joint separation can result in damage to the joint, leading to possible failure.

10. If equipped with ABS, remove the speed sensor cable routing bracket.

11. If necessary, disconnect the sway bar link from the damper fork.

12. Remove the damper fork lower through-bolts and upper pinch bolt. Remove the damper fork assembly.

13. Using a joint separation tool, disconnect the steering knuckle from the lower control arm.

14. Remove the halfshaft from the hub/knuckle by setting up a puller on the outside wheel hub, if necessary, and pushing the halfshaft from the front hub. After pressing the outer shaft, insert a prybar between the transaxle case and the halfshaft and pry the shaft from the transaxle.

➡**Do not pull on the shaft. Doing so damages the inboard joint. Do not insert the prybar too far or the oil seal in the case may be damaged.**

To install:

15. Inspect the halfshaft boot for damage

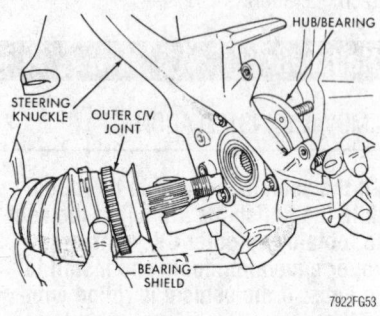

Carefully remove the outer CV joint from the steering knuckle—be careful NOT to damage the threads or splines on the joint

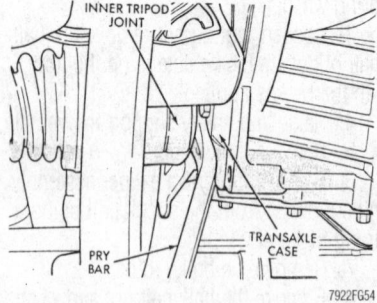

Inserting the prybar too far may damage the transaxle seal

or deterioration. Check the ball joints and splines for wear.

16. Replace the circlips on the ends of the halfshaft(s).

17. Insert the halfshaft into the transaxle. Be sure it is fully seated.

18. Pull the knuckle assembly outward and install the other end of the halfshaft into the hub.

19. Install the washer so the chamfered edge faces outward. Install the halfshaft nut and tighten temporarily.

20. Connect the control arm to the steering knuckle. Tighten the self-locking nuts to 43–52 ft. lbs. (59–71 Nm).

21. Install the damper fork. Tighten the lower through-bolt/nut to 65 ft. lbs. (88 Nm) and the upper pinch bolt to 76 ft. lbs. (103 Nm).

22. Connect the tie rod end to the steering knuckle. Tighten the retaining nut to 17–25 ft. lbs. (24–33 Nm) and install a new cotter pin.

23. Connect the sway bar link to the damper fork and tighten the link nut to 29 ft. lbs. (39 Nm).

24. Install the lockwasher and axle nut. Tighten the axle nut to 145–188 ft. lbs. (200–260 Nm).

➡**Before securely tightening the axle nut be sure there is no load on the wheel bearings.**

25. Install the brake rotor and caliper assembly.

26. Install a new cotter pin and bend to secure.

27. Install the wheel.

28. Check the transaxle fluid level and top off, if necessary.

29. Connect the negative battery cable.

30. Test drive the vehicle and check for proper operation.

STEERING AND SUSPENSION

Air Bag

❋❋ CAUTION

Some vehicles are equipped with an air bag system, also known as the Supplemental Inflatable Restraint (SIR) or Supplemental Restraint System (SRS). The system must be disabled before performing service on

or around system components, steering column, instrument panel components, wiring and sensors. Failure to follow safety and disabling procedures could result in accidental air bag deployment, possible personal injury and unnecessary system repairs.**

PRECAUTIONS

Several precautions must be observed when handling the inflator module to avoid accidental deployment and possible personal injury.

• Never carry the inflator module by the wires or connector on the underside of the module.

• When carrying a live inflator module, hold securely with both hands, and ensure that the bag and trim cover are pointed away.

• Place the inflator module on a bench or other surface with the bag and trim cover facing up.

• With the inflator module on the bench, never place anything on or close to the module which may be thrown in the event of an accidental deployment.

DISARMING

This air bag system, is a sensitive, complex, electromechanical unit. Proper SRS (also called Supplemental Inflatable Restraint, or SIR, or air bag system) disarming can be obtained by disconnecting and isolating the negative battery cable. Failure to disconnect the battery could result in accidental air bag deployment and possible personal injury. Before beginning service work, allow the system capacitor 2 minutes to discharge after disconnecting the negative battery cable.

Power Rack and Pinion Steering Gear

REMOVAL & INSTALLATION

❋❋ CAUTION

Some models covered by this manual may be equipped with a Supplemental Restraint System (SRS), which uses an air bag. Whenever working near any of the SRS components, such as the impact sensors, the air bag module, steering column and instrument panel, disable the SRS.

1. Disconnect the negative battery cable from the left shock tower. The ground cable is equipped with a insulator grommet which should be placed on the stud to prevent the negative battery cable from accidentally grounding.

➡**The negative battery is equipped with a insulated grommet that is to be placed on the shock tower connection to prevent accidental grounding.**

2. Remove the retaining pin from the intermediate shaft coupler pin bolt and remove the pin bolt.

3. Raise and safely support the vehicle.

4. Remove the front wheels.

5. Disconnect the tie rod ends by holding the tie rod end stud with a $11/32$ in. socket and loosen the retaining nut with a wrench.

6. Separate the tie rod end using a suitable tool.

✳ CAUTION

Before removing the front suspension crossmember from the vehicle you must scribe the front suspension crossmember and the vehicle body. This must be done to retain the proper alignment. The caster and camber are not adjustable.

7. Scribe a line on the body and on the crossmember on all four sides.

8. Disconnect the stabilizer from the body.

9. If equipped with anti-lock brakes remove the three bolts attaching it to the crossmember and tie the anti-lock brake controller to the vehicle body.

10. Disconnect the shock absorber clevis from the lower control arm.

11. Remove the two bolts attaching the engine support bracket to the crossmember.

12. Remove the bolt attaching the engine support bracket to the transaxle mounting bracket.

13. Place a suitable lifting device under the front suspension crossmember.

14. Remove the eight bolts attaching the crossmember to the body of the vehicle.

15. Lower the lifting device enough to gain access to the steering rack.

16. Disconnect the power steering lines and drain the fluid if equipped.

17. Disconnect the power steering pressure switch wiring.

18. If equipped with speed proportional steering, disconnect the solenoid control module wiring.

19. Remove the two steering rack isolator attaching bolts.

20. Remove the two steering rack saddle bracket attaching bolts.

21. Remove the steering from the vehicle.

To install:

22. Install the steering rack into the crossmember.

23. Reinstall the isolator and saddle bracket bolts. Tighten the bolts to 50 ft. lbs. (68 Nm).

24. If equipped with power steering reinstall the pressure and return lines. Tighten the lines to 275 inch lbs. (31 Nm).

25. Raise the crossmember against the frame rails and install the two rear bolts.

26. Reinstall the two front bolts.

27. Tighten all four bolts until the crossmember contacts the body.

28. Tighten the bolts to 20 inch lbs. (2 Nm).

29. Using a soft faced hammer tap the crossmember into position.

➡**Be sure to align the scribed marks on the crossmember.**

30. Starting with the rear bolts, tighten the crossmember bolts to 120 ft. lbs. (163 Nm).

31. Reinstall the engine support bracket.

32. Reinstall the engine support bracket bolts to crossmember.

33. Reinstall the engine support bracket bolt to the transaxle bracket.

34. Tighten the three bolts to 55 ft. lbs. (75 Nm).

35. Reconnect the power steering pressure switch.

36. Reinstall the anti-lock brake control unit.

37. Tighten the anti-lock brake bolts to 21 ft. lbs. (28 Nm).

38. Reinstall the heat shield on the tie rod ends.

39. Reconnect the shock clevis to the lower control arm.

40. Reinstall the tie rod ends and tighten to 45 ft. lbs. (61 Nm).

41. Reinstall and tighten the two stabilizer clamps.

42. Tighten the shock absorber clevis bolt to 68 ft. lbs. (92 Nm).

43. Reinstall the wheels and tighten to 95 ft. lbs. (129 Nm).

44. Lower the vehicle.

45. Reinstall the intermediate shaft pin bolt and retaining pin.

46. Tighten the pin bolt to 240 inch lbs. (27 Nm).

47. Reconnect the negative battery cable.

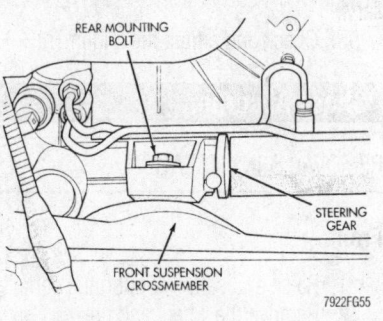

Steering gear rear mounting bolt location

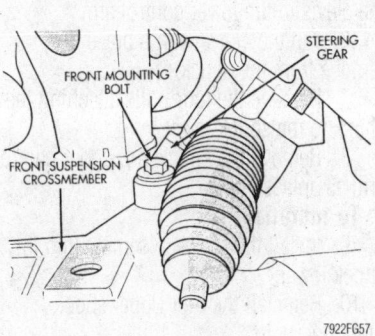

Steering gear front mounting bolt location

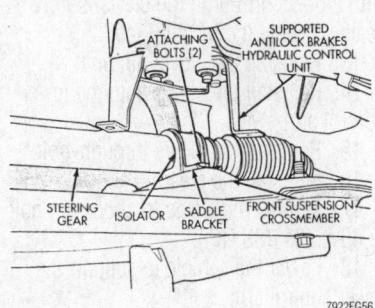

Check the condition of the isolator—if worn or oil soaked, replace

48. Refill the power steering system to the cold level with approved fluid.

49. Start the engine and allow it to run for a few minutes.

50. Shut **OFF** the engine and check the power steering fluid.

51. Add power steering fluid if necessary.

52. Raise the front wheels off the ground.

53. Start the engine and turn the wheel from stop-to-stop to bleed any air from the system.

54. Check fluid level and add if necessary.
55. Check and adjust the alignment.

Shock Absorber

REMOVAL & INSTALLATION

Front

1. Raise and safely support the vehicle.
2. Remove the wheel and tire.
3. Remove the steering knuckle.
4. Remove the pin bolt attaching the shock absorber to the shock clevis.
5. Remove the through-bolt attaching the clevis to the lower control arm.
6. Tap the clevis with a brass drift to remove from the shock.
7. Remove the four bolts attaching the shock to the shock tower.
8. Remove the shock and upper control arm as an assembly.

To install:

9. Install the shock assembly into the shock tower.
10. Reinstall the four upper shock mounting bolts.
11. Tighten the bolts to 23 ft. lbs. (31 Nm).
12. Reinstall the clevis onto the shock with a brass drift until the clevis is fully seated against the locating tab.
13. Reinstall the clevis pin bolt.
14. Reinstall the clevis onto the lower control arm.
15. Reinstall the clevis through-bolt.
16. Reinstall the steering knuckle.
17. Tighten the clevis to shock pin bolt to 65 ft. lbs. (88 Nm).
18. Lower the vehicle to support the lower control arm.
19. Tighten the clevis to lower control arm mounting bolt 40 ft. lbs. (54 Nm).
20. Reinstall the wheel and tire.
21. Lower the vehicle.

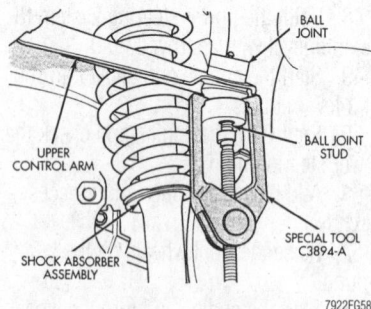

Separating the upper ball joint from the steering knuckle

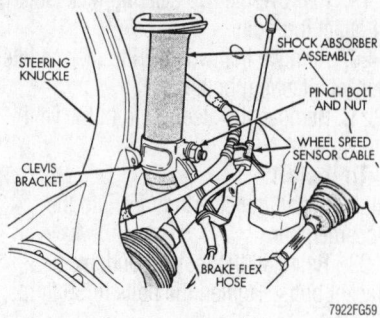

Component view of the front shock mounting

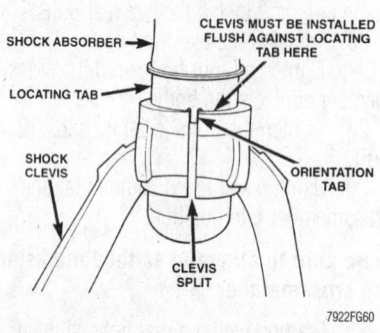

Be sure the orientation tab is situated into the clevis split

Rear

1. Pull back the carpeting from the rear shock tower.
2. Remove the plastic cover from the top of the shock tower.
3. Remove the two nuts attaching the shock assembly to the body.
4. Raise and safely support the vehicle.
5. Remove the wheel and tire.
6. Remove the bolt attaching the shock to the rear knuckle.
7. Push downward on the rear suspension and tilt the top of the shock outward.
8. Remove the shock from the vehicle.

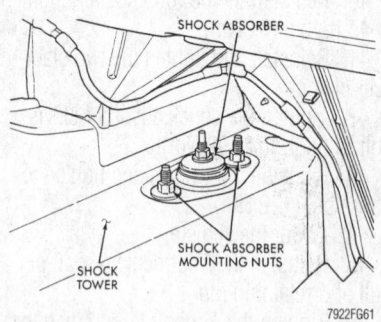

Access the rear shock upper mounting from inside the trunk

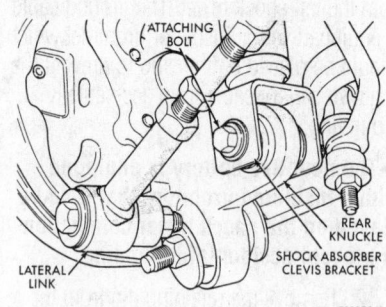

View of the rear shock lower mounting

To install:

9. Install the shock into the vehicle at the rear knuckle.
10. Push downward on the rear suspension and insert the top of the shock into the vehicle.
11. Reinstall the shock to rear knuckle attaching bolt.
12. Tighten the bolt to 70 ft. lbs. (95 Nm).
13. Lower the vehicle enough to gain access to the trunk.
14. Reinstall the shock upper mounting nuts and tighten to 25 ft. lbs. (34 Nm).
15. Reinstall the shock top cover.
16. Reinstall the rear wheel and tire and tighten to 95 ft. lbs. (125 Nm).
17. Lower vehicle and check for proper operation.

Coil Spring

REMOVAL & INSTALLATION

Front

1. Place the shock absorber assembly in a MB991237 and MB991238 spring compressor assembly or equivalent. Tighten the compressor and compress the spring slowly. Make certain the compressor is properly engaged before tightening.
2. After tension has been removed from the shock absorber assembly and shock absorber plate, remove the piston rod nut and washer from the top of the shock absorber assembly.
3. If the spring is to be replaced, slowly release the tension on the spring compressor. Allow the spring to expand fully. If only the shock absorber is being replaced the spring may remain in the compressor assembly.
4. By hand, remove the upper shock absorber bearing, washer, mount and shock absorber shield. Remove the upper insulator

ring, the spring, the bumper and lower insulator from the spring. Take notice to each components location for proper reassembly.

To install:

5. Install the lower insulator, the shock absorber bumper and the uncompressed spring. Install the upper spring insulator, shock absorber shield, mount and washer.

6. Install or align the spring compressor. Make certain the spring is correctly positioned relative to the upper and lower insulator rings. Smoothly compress the spring.

7. Install the washer and piston rod nut. Tighten the nut to 18 ft. lbs. (25 Nm). Install the dust cap.

8. Carefully release the spring compressor, watching the spring position as it seats. When the spring is properly seated, release/remove the compressor tools.

9. Reinstall the shock absorber assembly.

Rear

1. With the shock absorber removed from the vehicle, place the shock absorber assembly in a MB991237 and MB991239 spring compressor assembly or equivalent shock absorber service tool. Tighten the compressor and compress the spring slowly. Make certain the compressor is properly engaged before tightening.

2. After tension has been removed from the shock absorber assembly and shock absorber plate, remove the piston rod nut and washer from the top of the shock absorber assembly.

3. If the spring is to be replaced, slowly release the tension on the spring compressor. Allow the spring to expand fully. If only the shock absorber is being replaced the spring may remain in the compressor assembly.

4. By hand, remove the upper shock absorber bearing, washer, mount and shock absorber shield. Remove the upper insulator ring, the spring, the bumper and lower insulator from the spring. Take notice to each components location for proper reassembly.

To install:

5. Install the lower insulator, the shock absorber bumper and the uncompressed spring. Install the upper spring insulator, shock absorber shield, mount and washer.

6. Install or align the spring compressor. Make certain the spring is correctly positioned relative to the upper and lower insulator rings. Smoothly compress the spring.

7. Install the washer and piston rod nut. Tighten the nut to 16 ft. lbs. (22 Nm). Install the dust cap.

8. Carefully release the spring compres-

sor, watching the spring position as it seats. When the spring is properly seated, release/remove the compressor tools.

9. Reinstall the shock absorber assembly.

Upper Ball Joint

REMOVAL & INSTALLATION

The upper ball joint is an integrated part of the upper control arm assembly, and can not be serviced separately.

A worn or damaged ball joint requires replacement of upper control arm assembly.

Upper Control Arm

REMOVAL & INSTALLATION

1. Raise and safely support the vehicle.
2. Remove the appropriate wheel.
3. Using the Joint Separation tool, MB991113 or equivalent, disconnect the ball joint stud from the steering knuckle.
4. The ball joint can be checked using the following procedure.

   a. An adapter (MB 990326) is available that fits onto the ball joint stud and adapts to an inch-pound torque wrench. If this tool is not available, a shop-made substitute can be fabricated.

   b. Turn the ball joint stud with the torque wrench. The factory standard for breakaway torque is 3 to 13 inch lbs.

   c. If the ball joint stud is out of specification (turns too easily or is too stiff), continue with the ball joint/control arm assembly replacement. Note that the boot over the ball joint can be removed, and the joint greased. A new replacement ball joint boot is recommended.

5. Inside the engine compartment, at the shock absorber tower, locate the upper control arm mounting nuts. Remove the

nuts and using the joint separation tool, separate the upper arm shafts from the shock absorber tower.

6. Remove the control arm assembly.

To install:

7. Align the upper control arm shafts to the shock absorber tower and secure it with the mounting nuts. Tighten the mounting nuts to 62 ft. lbs. (86 Nm).

8. Connect the ball joint to the knuckle and tighten the locking nut to 20 ft. lbs. (28 Nm).

9. Install the wheel and lower the vehicle.

10. Check the wheel alignment and adjust if necessary.

Lower Ball Joint

REMOVAL & INSTALLATION

On all vehicles, the ball joint cannot be serviced separately. If the ball joint is defective it will require replacement of the lower control arm.

Lower Control Arm

REMOVAL & INSTALLATION

1. Raise and safely support the vehicle.
2. Remove the front wheels and tires.
3. If equipped with 15 in. wheels, the heat shield will need to be removed before the lower control arm can be separated from the steering knuckle.
4. Remove the ball joint clamping nut.
5. Remove the sway bar attaching bolts from both control arms.
6. Disconnect the shock clevis from the lower control arm.
7. Loosen the sway bar to crossmember attaching bolts and rotate the sway bar away from the control arm.
8. Using a prybar separate the lower control arm from the steering knuckle.

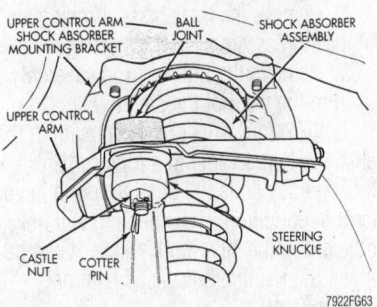

Upper control arm component identification

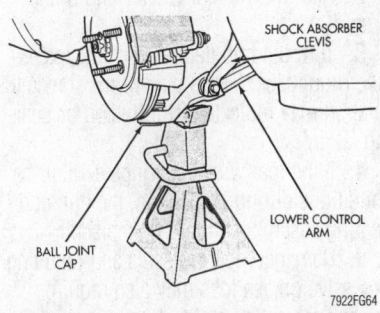

Be sure to pre-load the suspension before final tightening the control arm to the steering knuckle

9. Remove the two control arm attaching bolts.

10. Remove the control arm.

To install:

11. Install the control arm into the vehicle.

12. Reinstall the two control arm attaching bolts.

13. Tighten the control arm attaching bolts to 120 ft. lbs. (163 Nm).

14. Reinstall the control arm to the steering knuckle and tighten to 70 ft. lbs. (95 Nm).

15. Rotate the sway bar up to the control arms.

16. Reinstall the shock absorber clevis.

17. Reinstall the sway bar attaching bolts.

18. Tighten the sway bar bolts to 21 ft. lbs. (28 Nm).

19. Tighten the sway bar to crossmember attaching bolts 21 ft. lbs. (28 Nm).

20. Reinstall the wheels and tires.

Wheel Bearings

ADJUSTMENT

Front

The front hub wheel bearing is designed for the life of the vehicle and requires no type of adjustment or periodic maintenance. The bearing is a sealed unit with the wheel hub and can only be removed and/or replaced as one unit.

Rear

The rear hub and wheel bearing assembly is designed for the life of the vehicle and requires no type of adjustment or periodic maintenance. The bearing is a sealed unit with the wheel hub and can only be removed and/or replaced as one unit.

The following procedure may be used for evaluation of bearing condition.

1. Raise and safely support the vehicle.

2. Remove the rear wheels and brake drums.

3. Turn the hub flange carefully. Excessive roughness, lateral play or resistance to rotation may indicate dirt intrusion or bearing failure.

4. If the rear wheel bearings exhibit the conditions during inspection, the hub and bearing assembly should be replaced.

5. Damaged nearing seals and resulting excessive grease loss may also require bearing replacement. Moderate grease loss from the bearing is considered normal and should not require replacement of the hub and bearing assembly.

REMOVAL & INSTALLATION

Front

The front wheel bearing used on this vehicle is a bolt-in type wheel bearing.

The wheel bearing is serviced separately from the front steering knuckle and front hub assembly. Retention of the front wheel bearing into the steering knuckle is by means of three bolts installed from the rear of the steering knuckle. The three bolts attach the hub/bearing to the front surface of the steering knuckle. Removal and installation of the hub/bearing assembly from the steering knuckle must be done with the steering knuckle removed from the vehicle.

This vehicle does not use a rubber lip seal as on past front wheel drive vehicles to prevent contamination of the front wheel bearing. On this vehicle the face of the outer CV-joint has a metal bearing shield pressed on it. This design deters direct water splash on the bearing seal while allowing any water that gets in to run out the bottom of the steering knuckle. It is important to thoroughly clean the outer CV-joint and the wheel bearing area in the steering knuckle before it is assembled after servicing the front wheel bearing or driveshaft.

At no time when servicing this vehicle, can a sheetmetal screw, bolt or other metal fastener be installed in the shock tower to take the place of an original plastic clip. Also, NO holes can be drilled into the front shock tower for the installation of any metal fasteners into the shock tower. Because of the minimum clearance in this area installation of metal fasteners could damage the coil spring's protective coating and lead to corrosion failure of the spring. If a plastic clip is missing, lost or broken during servicing a vehicle, replace only with the equivalent part listed in the Mopar parts catalog.

1. Raise and safely support the vehicle.

2. Remove the front tire and wheel.

3. Remove the steering knuckle assembly using the recommended procedure.

4. Remove the three bolts attaching the hub/bearing assembly to the steering knuckle.

5. Remove the hub/bearing assembly out from the front of the steering knuckle. The bolt-in front wheel bearing used on the vehicle is transferable to a replacement steering knuckle is the bearing found in serviceable condition. If the bearing will not come out of the steering knuckle, it can be tapped out using a soft-faced hammer.

To install:

6. Clean all parts well. Thoroughly clean all the hub/bearing assembly mounting surfaces on the steering knuckle.

7. Reinstall the replacement hub/bearing assembly in the steering knuckle aligning the bolt holes in the bearing flange with the holes in the steering knuckle.

8. Reinstall the three attaching bolts and tighten evenly to be sure the bearing is square to the face of the steering knuckle. Tighten the attaching bolts to 80 ft. lbs. (110 Nm).

9. Reinstall the steering knuckle. Install the tire and wheel.

10. Lower the vehicle and check for proper operation.

Rear

All vehicles are equipped with permanently lubricated and sealed for life rear wheel bearings. There is no periodic lubrication or maintenance recommended for these units.

To evaluate the condition of the rear wheel bearings, remove the wheel and brake drum and rotate the flanged outer ring of the hub. Excessive roughness or resistance to rotation may indicate dirt intrusion or wheel bearing failure. If the rear wheel bearings exhibit these conditions during inspection, the hub and bearing assembly should be replaced. Damaged bearing seals and resulting excessive grease loss may also require bearing replacement. Moderate grease loss from the bearing is considered normal and should not require replacement of the hub and bearing assembly. If service requires removal for inspection or replacement of the rear wheel bearing and hub assembly, use the following procedure.

1. Raise and safely support the vehicle.

2. Remove the wheel and tire assembly.

3. Remove the brake drum.

4. Remove the rear hub dust cap.

5. Remove the rear hub retaining nut and discard.

6. Remove the rear hub and bearing assembly by pulling straight off the spindle.

To install:

7. Install the new bearing on the rear spindle.

8. Reinstall the hub and a new retaining nut.

9. Tighten the retaining nut to 185 ft. lbs. (250 Nm). Install the dust cap by tapping on with a soft-faced mallet.

10. Reinstall the brake drum.

11. Reinstall the wheel and tire. Tighten the lug nuts by tightening in a crisscross pattern in two steps with a final torque of 95 ft. lbs. (129 Nm).

12. Lower the vehicle.

FORD MOTOR CO.

Ford-Aspire

17

DRIVE TRAIN 17-11
ENGINE REPAIR 17-2
FUEL SYSTEM 17-9
**STEERING AND
SUSPENSION** 17-15

A
Air Bag 17-15
  Disarming 17-15
  Precautions 17-15

C
Camshaft and Valve Lifters 17-7
  Removal & Installation 17-7
Clutch 17-13
  Adjustment 17-13
  Removal & Installation 17-14
Cylinder Head 17-4
  Removal & Installation 17-4

D
Distributor 17-2
  Installation 17-2
  Removal 17-2

E
Engine Assembly 17-2
  Removal & Installation 17-2
Exhaust Manifold 17-6
  Removal & Installation 17-6

F
Front Crankshaft Seal 17-7
  Removal & Installation 17-7
Fuel Filter 17-9
  Removal & Installation 17-9
Fuel Pump 17-11
  Removal & Installation 17-11
Fuel System Pressure 17-9
  Relieving 17-9
Fuel System Service
  Precautions 17-9

H
Halfshaft 17-14
  Removal & Installation 17-14

I
Ignition Timing 17-2
  Adjustment 17-2
Intake Manifold 17-6
  Removal & Installation 17-6

L
Lower Ball Joints 17-19
  Removal & Installation 17-19
Lower Control Arm 17-19
  Removal & Installation 17-19

O
Oil Pan 17-8
  Removal & Installation 17-8

Oil Pump 17-8
  Removal & Installation 17-8

R
Rack and Pinion Steering Gear 17-15
  Removal & Installation 17-15
Rear Main Seal 17-9
  Removal & Installation 17-9
Rocker Arms/Shafts 17-5
  Removal & Installation 17-5

S
Strut and Spring 17-17
  Removal & Installation 17-17

T
Transaxle Assembly 17-11
  Removal & Installation 17-11

V
Valve Lash 17-8
  Adjustment 17-8

W
Water Pump 17-4
  Removal & Installation 17-4
Wheel Bearings 17-19
  Adjustment 17-19
  Removal & Installation 17-20

ENGINE REPAIR

➡Disconnecting the negative battery cable on some vehicles may interfere with the functioning of the on-board computer system, and may require the computer to undergo a relearning process once the negative battery cable is reconnected.

Distributor

REMOVAL

1. Disconnect the negative battery cable.
2. Position the No. 1 piston at Top Dead Center (TDC) of the compression stroke.
3. Loosen the distributor cap screws and position the cap to one side.
4. Make a matchmark with a scribe or marker on the distributor base flange and the cylinder head.
5. Mark the position of the rotor on the distributor housing.
6. Disengage the distributor electrical connections.
7. Loosen the distributor hold-down bolts.
8. Remove the distributor from the engine.
9. Inspect the distributor O-ring, if damaged or worn, replace it with a new one.

INSTALLATION

Timing Not Disturbed

➡When installing the distributor, be sure the offset drive tangs engage the camshaft slots.

1. Ensure the No. 1 piston is still at TDC on the compression stroke.

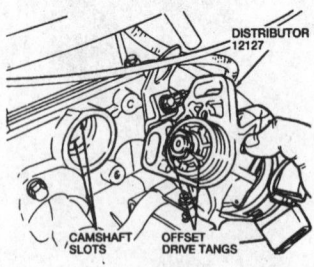

Distributor

7922GG01

View of the cylinder head and the drive side of the distributor showing the offset drive tangs and the camshaft slots

2. Install the distributor and align the matchmarks on the distributor base flange and cylinder head. Also be sure that the rotor points toward the mark on the distributor housing made previously. Make certain the rotor is pointing to the No. 1 mark on the distributor base.
3. When all the marks are aligned, install the distributor hold-down bolts and tighten them to 14–19 ft. lbs. (19–25 Nm).
4. Install the cap and tighten the screws.
5. Connect the negative battery cable.
6. Engage the electrical connections.
7. Install the No. 1 spark plug, if removed.
8. Recheck the initial timing, and adjust if necessary.

Timing Disturbed

➡When installing the distributor, be sure the offset drive tangs engage the camshaft slots.

1. Ensure the No. 1 piston is still at TDC on the compression stroke.

➡If the engine was disturbed while the distributor was removed, it will be necessary to remove the No. 1 spark plug and rotate the engine clockwise until the No. 1 piston is on the compression stroke. Align the timing marks.

2. If installing the old distributor:
   a. Install the distributor and align the matchmarks on the distributor base flange and cylinder head. Also be sure that the rotor points toward the mark on the distributor housing made previously. Make certain the rotor is pointing to the No. 1 mark on the distributor base.
3. If installing a new distributor:
   a. Install the rotor and cap (with wires still attached) on the distributor, then follow the No. 1 spark plug wire from the plug to the cap; this will be the No. 1 tower on the cap. Mark the location of the tower on the distributor housing, then remove the cap.
   b. Install the distributor and be sure the rotor aligns with the No.1 tower mark made on the distributor.
4. When all the marks are aligned, install the distributor hold-down bolts and tighten them to 14–19 ft. lbs. (19–25 Nm).
5. Install the cap and tighten the screws.
6. Connect the negative battery cable.
7. Engage the electrical connections.
8. Install the No. 1 spark plug, if removed.
9. Recheck the initial timing, and adjust if necessary.

Ignition Timing

ADJUSTMENT

1. Start the vehicle and let it reach operating temperature.
2. Turn all accessories OFF.
3. Connect a timing light according to the tool manufacturer's instructions.

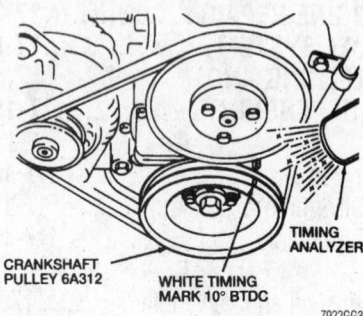

CRANKSHAFT PULLEY 6A312 WHITE TIMING MARK 10° BTDC TIMING ANALYZER

7922GG34

Clean the timing marks with a shop rag to help view them with the timing light

4. Check the timing by aiming the light at the timing pointer on the timing belt cover. The yellow timing mark on the crankshaft pulley should line up with the mark on the pointer and should read 10° Before Top Dead Center (BTDC).
5. If the timing is incorrect, loosen the distributor hold-down bolts.
6. Rotate the distributor until the desired timing is achieved.
7. Tighten the distributor hold-down bolts to 14–19 ft. lbs. (19–25 Nm).
8. Recheck the timing and if it is still correct, remove the light.
9. If the timing is still incorrect, repeat Steps 5 through 8.

Engine Assembly

REMOVAL & INSTALLATION

✳✳ CAUTION

Some models covered by this manual may be equipped with a Supplemental Restraint System (SRS), which uses an air bag. Whenever working near any of the SRS components, such as the impact sensors, the air bag module, steering column and instrument panel, properly disable the SRS.

1. Disconnect the battery cables (negative cable first).

Observe all applicable safety precautions when working around fuel. Whenever servicing the fuel system, always work in a well ventilated area. Do not allow fuel spray or vapors to come in contact with a spark or open flame. Keep a dry chemical fire extinguisher near the work area. Always keep fuel in a container specifically designed for fuel storage; also, always properly seal fuel containers to avoid the possibility of fire or explosion.

2. Relieve the fuel system pressure.
3. Remove the battery and tray.
4. Remove the hood.
5. Remove the air cleaner assembly and hoses.
6. Remove the electric cooling fan and the radiator.
7. Disconnect the accelerator cable from the throttle body.
8. Loosen the accelerator shaft bracket bolts and remove the bracket.
9. Disconnect the speedometer cable from the transaxle.
10. Disconnect the fuel tube hose and tube from the injection supply manifold.
11. Disconnect the hoses from the heater core.
12. Disconnect the power brake booster hose, vacuum modulator hose and governor hose.
13. Tag and disengage all the engine harness electrical connections and grounds.
14. On automatic transaxles, tag and disengage the Park/Neutral safety switch and the kickdown solenoid electrical connections and the transaxle ground.
15. On manual transaxles, tag and disengage the Park/Neutral safety switch electrical connection, the transaxle ground and disconnect the clutch cable.
16. On automatic transaxles, remove the shift-to-manual shaft boot, then disconnect the shift cable and bracket from the transaxle.
17. On manual transaxles, disconnect the transaxle gearshift rod and clevis. Disconnect the gearshift lever stabilizer bar and support from the transaxle.
18. Remove all the drive belts, then tag and disconnect the power lines from, then pump.
19. Raise the vehicle and safely support it.

The EPA warns that prolonged contact with used engine oil may cause a number of skin disorders, including cancer! You should make every effort to minimize your exposure to used engine oil.

20. Drain the engine oil and transaxle fluid.
21. Remove the front wheels and splash shields.
22. Loosen the front suspension lower arm clamp bolts and nuts. Pull the arms downward, separating the arms from the wheel knuckles.

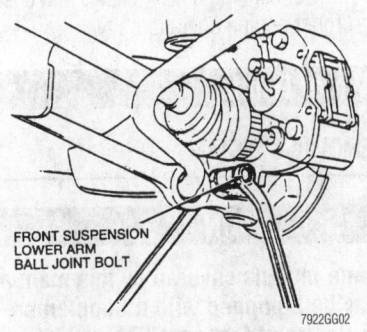

FRONT SUSPENSION LOWER ARM BALL JOINT BOLT

7922GG02

Use two wrenches (one as a back-up) to loosen the front suspension lower arm clamp bolts and nuts

23. Remove the halfshafts from each side and install plugs in the transaxle to avoid leakage or contamination.
24. Loosen the power steering pump bolts and separate the pump from the engine.

➡ Only a MVAC-trained, EPA-certified, automotive technician should service the A/C system or its components.

25. Remove the A/C compressor.
26. Disconnect the exhaust inlet pipe.
27. Loosen the front and rear transaxle support insulator nuts.

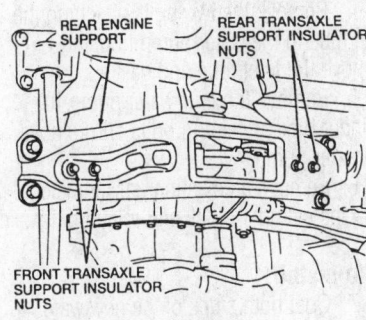

REAR ENGINE SUPPORT REAR TRANSAXLE SUPPORT INSULATOR NUTS

FRONT TRANSAXLE SUPPORT INSULATOR NUTS

7922GG03

Location of the front and rear transaxle support insulator nuts

28. Loosen the muffler pipe bracket bolts and remove the bracket.
29. Loosen the transaxle case-to-block front bracket and case rear bracket bolts. Remove the brackets.
30. Loosen the engine rear plate bolt and remove the plate.
31. On automatic transaxles, loosen the flywheel-to-torque converter nuts.
32. Lower the vehicle.
33. Tag and disconnect all vacuum lines.
34. Using an engine hoist and chain, attach lifting hooks to the lifting eyes and remove all slack in the chain.

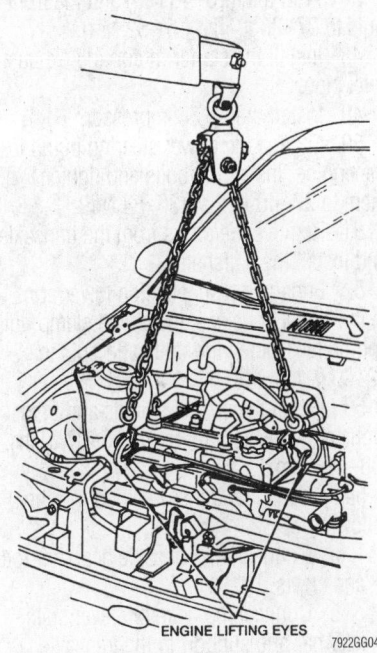

ENGINE LIFTING EYES

7922GG04

While slowly lifting the engine out of the vehicle, be sure nothing is still attached to it

35. Loosen the engine mount through-bolt and nut.
36. Remove the engine and transaxle from the engine compartment as an assembly.
37. If necessary, loosen the transaxle-to-engine bolts and separate the two components.

To install:

38. Connect the transaxle to the engine and lower them as an assembly into the engine compartment.
39. Install the engine mount through-bolt and nut. Tighten the bolt to 39–47 ft. lbs. (53–64 Nm).
40. Remove the lifting cables.
41. Connect all vacuum lines.
42. Raise the vehicle and safely support it.

43. On automatic transaxles, install the flywheel-to-torque converter nuts. Tighten the nuts to 25–36 ft. lbs. (34–49 Nm).

44. Install the engine rear plate and bolt. Tighten the bolt to 61–87 inch lbs. (7–10 Nm).

45. Install the transaxle case-to-engine block front bracket and case rear bracket bolts. Tighten the bolts to 27–38 ft. lbs. (37–52 Nm).

46. Install the muffler pipe bracket and bolts. Tighten the bolts to 28–41 ft. lbs. (38–56 Nm).

47. Install the front and rear transaxle support insulator nuts. Tighten the rear nuts to 21–34 ft. lbs. (28–46 Nm) and the front nuts to 27–38 ft. lbs. (37–52 Nm).

48. Install the starter motor and exhaust inlet pipe.

49. Install the A/C compressor.

50. Engage the power steering pump to the engine, install the bolts and tighten them to 27–40 ft. lbs. (36–54 Nm).

51. Remove the plugs from the transaxle and install the halfshafts.

52. Engage the suspension lower arms to the wheel knuckles. Install the clamp nuts and bolts. Tighten the bolts and nuts to 32–40 ft. lbs. (43–54 Nm).

53. Install the splash shields and tires. Tighten the wheel hub bolt to 65–87 ft. lbs. (88–118 Nm).

54. On manual transaxles perform the following:

  a. Connect the transaxle gearshift rod and clevis.

  b. Connect the gearshift lever stabilizer bar and support to the transaxle.

  c. Engage the clutch cable and the starter motor electrical connections.

  d. Engage the Park/Neutral safety switch, back-up lamp switch and the transaxle ground.

55. Lower the vehicle.

56. Connect the power steering lines and install the drive belts.

57. On automatic transaxles, connect the shift cable and bracket. Engage the Park/Neutral safety switch and the kickdown solenoid electrical connections and the transaxle ground.

58. Reattach all of the engine harness electrical connections and grounds.

59. Connect the power brake booster hose, vacuum modulator hose and governor hose.

60. Connect the hoses to the heater core.

61. Connect the fuel tube and hose to the injection supply manifold.

62. Connect the speedometer cable to the transaxle.

63. Install the accelerator shaft bracket and tighten the bolts.

64. Connect the accelerator cable to the throttle body.

65. Install the radiator and electric cooling fan.

66. Install the air cleaner assembly and connect the hoses.

67. Install the hood.

68. Install the battery and tray.

69. Fill and bleed the cooling system.

70. Fill the transaxle with oil.

71. Fill the crankcase with engine oil.

✱✱ WARNING

Operating the engine without the proper amount and type of engine oil will result in severe engine damage.

72. Connect the battery cables and check for proper engine operation.

Water Pump

REMOVAL & INSTALLATION

✱✱ CAUTION

Some models covered by this manual may be equipped with a Supplemental Restraint System (SRS), which uses an air bag. Whenever working near any of the SRS components, such as the impact sensors, the air bag module, steering column and instrument panel, properly disable the SRS.

1. Disconnect the negative battery cable.
2. Remove the timing belt.

✱✱ CAUTION

Never open, service or drain the radiator or cooling system when hot; serious burns can occur from the steam and hot coolant.

3. Drain the cooling system into a suitable container.

4. Remove the two bolts attaching the inlet tube to the water pump housing. Remove the inlet tube and gasket.

5. Remove the four water pump-to-cylinder block retaining bolts and remove the water pump.

6. Remove all existing gasket material from the cylinder block and inlet tube gasket surfaces.

To install:

7. Coat both sides of the new water pump and inlet tube gaskets with a suitable water resistant sealer. Apply the gaskets to the engine and inlet tube surfaces. Make

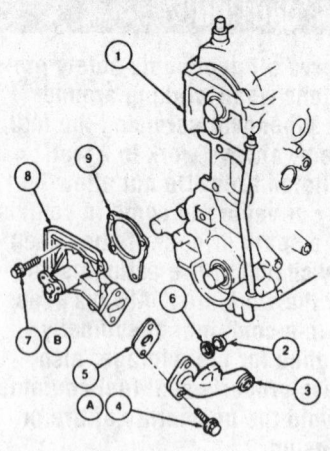

1	Cylinder Block
2	Heater Water Hose
3	Water Inlet Connection
4	Hot Water Heater Elbow Connector Bolt (2 Req'd)
5	Water Pump Inlet Gasket
6	O-Ring
7	Water Pump Bolt (4 Req'd)
8	Water Pump
9	Water Pump Housing Gasket
A	Tighten to 19-30 N-m (14-22 Lb-Ft)
B	Tighten to 19-26 N-m (14-19 Lb-Ft)

7922GG05

Exploded view of the water pump and its related components

certain the gasket holes are aligned with the bolt holes.

8. Position the water pump against the gasket. Be sure the holes in the water pump are aligned with the gasket holes and that the pump does not shift the position of the gasket.

9. Install the four water pump retaining bolts and tighten to 14–19 ft. lbs. (19–26 Nm).

10. Position the inlet tube and gasket against the water pump housing and install the attaching bolts. Tighten the bolts to 14–22 ft. lbs. (19–30 Nm).

11. Install the timing belt.

12. Fill the cooling system.

13. Connect the negative battery cable.

14. Start the engine and allow it to reach normal operating temperature. Check for coolant leaks and proper operation.

Cylinder Head

REMOVAL & INSTALLATION

✱✱ CAUTION

Some models covered by this manual may be equipped with a Supplemental Restraint System (SRS), which uses an air bag. Whenever working

near any of the SRS components, such as the impact sensors, the air bag module, steering column and instrument panel, properly disable the SRS.

1. Disconnect the negative battery cable.

❄❄ CAUTION

Never open, service or drain the radiator or cooling system when hot; serious burns can occur from the steam and hot coolant. Always drain coolant into a sealable container. Coolant should be reused unless it is contaminated or is several years old.

2. Drain the cooling system.
3. Tag and disconnect the spark plug wires from the plugs.
4. Remove the distributor.
5. Remove the timing belt and the valve cover.
6. Remove the exhaust and intake manifolds.
7. Remove the front and rear engine lift hangers. Disengage the engine block ground wire.
8. Tag and disengage all wiring harness connectors.
9. Use a pair of pliers to compress the upper radiator hose clamp and slide it away from the radiator. Disconnect the upper radiator hose from the radiator.
10. Remove the water bypass tube (or hose) and bracket.
11. Remove the ten cylinder head bolts.
12. Remove the cylinder head and gasket from the engine.

To install:
13. Use a scraper to clean any old gasket material residue from the cylinder head gasket mating surfaces.

➡ **The cylinder head gasket has marks on one of its edges that match the shape of the cylinder head, when the**

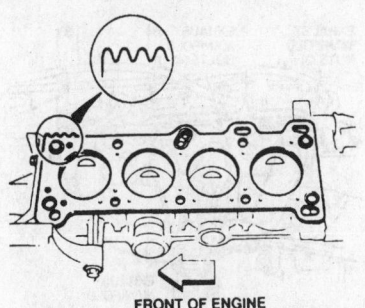

FRONT OF ENGINE   7922GG06

Be sure to position the gasket properly on the engine block before installing the cylinder head to ensure proper sealing

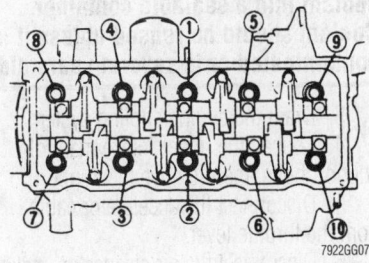

7922GG07

Tighten the cylinder head bolts in the order shown to the proper specification

gasket is installed these marks must align properly before cylinder head installation.

14. Install a new gasket and place the cylinder head in position.
15. Install NEW cylinder head bolts and tighten them with a torque wrench in the sequence illustrated and also in the following order:
• First pass: 35–40 ft. lbs. (50–60 Nm)
• Second pass: 56–60 ft. lbs. (75–81 Nm).
16. Install the water bypass tube (or hose) and bracket.
17. Connect the upper radiator hose and engage the clamp.
18. Engage all wiring harness connectors.
19. Engage the engine ground wire, then install the front and rear engine lift hangers.
20. Install the distributor, the intake and exhaust manifolds, the valve cover, and the timing belt.
21. Connect the spark plug wires to the plugs.
22. Fill the cooling system and install the radiator cap.
23. Connect the negative battery cable.
24. When the engine is warmed up, check for leaks.

Rocker Arms/Shafts

REMOVAL & INSTALLATION

1. Disconnect the negative battery cable.
2. Disconnect the accelerator cable from the throttle lever and routing bracket.
3. Remove the PCV valve and the oil separator hose.
4. Remove the air cleaner-to-intake manifold tube.
5. Remove the spark plug wires from the routing clips.
6. Remove the upper timing belt cover.
7. Remove the six valve cover retaining bolts. Remove the valve cover and discard the gasket.

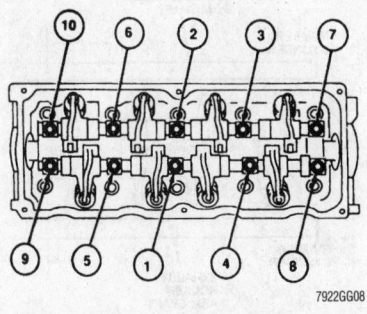

7922GG08

Rocker arm and shaft retaining bolt loosening and tightening sequence

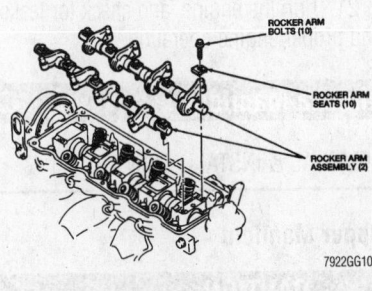

7922GG10

Exploded view of the rocker arm/shaft assemblies' mounting

8. Loosen the ten rocker arm shaft retaining bolts in the proper sequence, and remove the bolts and seats from the shafts.
9. Remove the rocker arm/shaft assemblies from the engine. If the shafts are to be disassembled, keep all parts in order so they can be assembled in their correct positions.

To install:
10. Clean all gasket mating surfaces.
11. If disassembled, coat the rocker arms and shafts with clean engine oil and reassemble.
12. The intake rocker arm shaft can be identified from the exhaust rocker arm shaft by measuring the distance between the oiling holes **A** and **B**.
13. Install the rocker arm/shaft assemblies with the shaft retaining bolts and seats. Tighten the bolts, in sequence, to 16–21 ft. lbs. (22–28 Nm).
14. Install the valve cover with a new gasket. Tighten the six valve cover retaining bolts to 44–79 inch lbs. (5–9 Nm).
15. Install the upper timing belt cover.
16. Install the spark plug wires onto the routing clips.
17. Install the PCV valve and the oil separator hose onto the valve cover.
18. Install the accelerator cable.
19. Install the air cleaner-to-intake manifold tube.

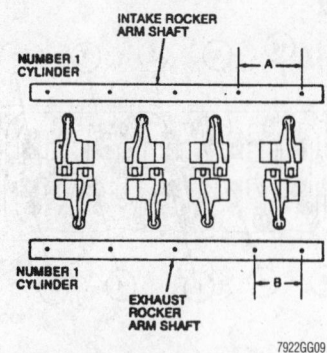

Method of identifying the intake and exhaust rocker arm shafts

20. Connect the negative battery cable.
21. Run the engine, and check for leaks and proper engine operation.

Intake Manifold

REMOVAL & INSTALLATION

Upper Manifold

> ✸✸ **CAUTION**
>
> Some models covered by this manual may be equipped with a Supplemental Restraint System (SRS), which uses an air bag. Whenever working near any of the SRS components, such as the impact sensors, the air bag module, steering column and instrument panel, properly disable the SRS.

1. Disconnect the negative battery cable.

> ✸✸ **CAUTION**
>
> The EPA warns that prolonged contact with used engine oil may cause a number of skin disorders, including cancer! You should make every effort to minimize your exposure to used engine oil. Protective gloves should be worn when changing the oil. Wash your hands and any other exposed skin areas as soon as possible after exposure to used engine oil. Soap and water, or waterless hand cleaner should be used.

2. Relieve the fuel system pressure.

> ✸✸ **CAUTION**
>
> Never open, service or drain the radiator or cooling system when hot; serious burns can occur from the steam and hot coolant. Always drain coolant into a sealable container. Coolant should be reused unless it is contaminated or is several years old.

3. Drain the engine cooling system.
4. Loosen the upper intake manifold support bolts, then remove the support.
5. Disconnect the accelerator cable from the throttle lever.
6. Disconnect the air cleaner-to-intake manifold tube.
7. Tag and disconnect the coolant hoses from the upper manifold.
8. Tag and disconnect all wiring and hoses that will interfere with manifold removal.
9. Loosen the upper intake manifold-to-lower intake manifold bolts.
10. Remove the upper intake manifold and gasket.
11. Insert clean shop rags into the lower manifold passages to prevent dirt or other debris from entering the engine.

To install:

12. Use a scraper to clean any old gasket material residue from the upper and lower manifold mating surfaces.
13. Remove the rags from the lower intake manifold.
14. Install a new gasket and place the upper intake manifold in position.
15. Install the upper intake manifold-to-lower intake manifold bolts. Tighten the bolts to 14–20 ft. lbs. (19–26 Nm).
16. Connect the coolant hoses to the upper manifold.
17. Engage all wiring and hoses that were tagged and disconnected during removal.
18. Connect the air cleaner-to-intake manifold tube.
19. Connect the accelerator cable to the throttle lever.
20. Install the upper intake manifold support, then tighten the bolts to 22–34 ft. lbs. (31–46 Nm).
21. Fill the cooling system and connect the negative battery cable.
22. Start the vehicle and check for proper engine operation.

Lower Manifold

1. Remove the upper intake manifold
2. Tag and disconnect all wiring and hoses that will interfere with lower manifold removal.
3. Loosen the lower intake manifold-to-cylinder head bolts.

➥ You may have to raise the vehicle to unfasten the lower nut on the passenger's side of the manifold.

4. Remove the intake manifold and gasket.
5. Insert clean shop rags into the cylinder head openings to prevent dirt or other debris from entering the engine.

To install:

6. Use a scraper to clean any old gasket material residue from the cylinder head and intake manifold mating surfaces.
7. Remove the rags from the cylinder head openings.
8. Install a new gasket and place the intake manifold in position.
9. Install the intake manifold-to-cylinder head bolts. Tighten the bolts to 14–20 ft. lbs. (19–26 Nm).
10. Install the intake manifold support, then tighten the bolts to 22–34 ft. lbs. (31–46 Nm).
11. Install the upper intake manifold.

Exhaust Manifold

REMOVAL & INSTALLATION

1. Disconnect the negative battery cable.
2. Raise and safely support the vehicle.
3. Remove the exhaust inlet pipe-to-exhaust manifold nuts and washers.
4. Remove the muffler pipe bracket bolts.
5. Lower the vehicle.
6. Remove the air cleaner-to-intake manifold tube.
7. Remove the exhaust manifold heat shield bolts and the shield.
8. Separate the oxygen sensor wiring connector from the routing bracket and disconnect the electrical connector.
9. Remove the oxygen sensor. Inspect the sensor gasket for damage and replace if necessary.
10. Remove the four exhaust manifold retaining bolts and three nuts.
11. Remove the exhaust manifold from the vehicle.

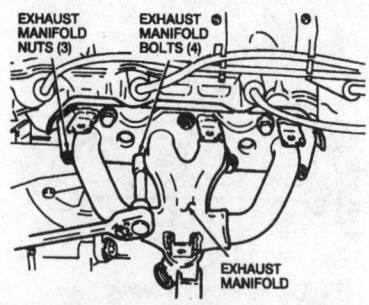

During installation, be sure the ignition wires are not pinched between the manifold and cylinder head

12. Remove the inlet pipe and exhaust manifold gaskets and discard.

To install:

13. Remove all existing gasket material from the exhaust manifold and cylinder head inlet pipe. Clean all threaded surfaces.

14. Apply a new gasket onto the cylinder head studs and position the exhaust manifold onto the gasket. Install the attaching nuts and bolts and tighten to 12–17 ft. lbs. (16–23 Nm).

15. Position a new gasket on the oxygen sensor, if needed, and install it in the exhaust manifold.

16. Position the heat shield and tighten the bolts to 12–17 ft. lbs. (16–23 Nm).

17. Engage the oxygen sensor electrical connector and secure the it in the routing bracket.

18. Install the air cleaner-to-intake manifold tube.

19. Raise and safely support the vehicle.

20. Position a new muffler inlet pipe gasket over the exhaust manifold studs.

21. Raise the exhaust inlet pipe into position on the exhaust manifold and support by hand. Install the attaching nuts and washers and tighten to 23–34 ft. lbs. (31–46 Nm).

22. Position the muffler pipe bracket and install the bolts. Tighten to 28–47 ft. lbs. (38–56 Nm).

23. Lower the vehicle.

24. Connect the negative battery cable.

25. Start the engine and inspect for exhaust leaks.

Front Crankshaft Seal

REMOVAL & INSTALLATION

➡ **The timing belt must be removed for this procedure. The crankshaft seal is located behind the crankshaft sprocket in the oil pump front housing.**

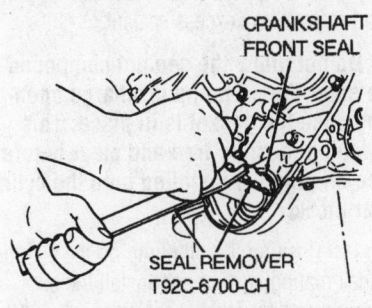

When removing the front crankshaft seal, be careful not to damage the crankshaft threads or the seal bore

1. Disconnect the negative battery cable.

2. Remove the accessory drive belts.

3. Remove the timing belt cover and the timing belt.

4. Remove the crankshaft sprocket and crankshaft key.

5. Using seal remover T92C-6700-CH or equivalent, remove the crankshaft front seal from the oil pump housing.

To install:

6. Lubricate the lip of the new seal and the crankshaft seal surface with clean engine oil.

7. Using front seal installer T87C-6019-A or equivalent, draw the front crankshaft seal into the oil pump housing.

8. Install the crankshaft sprocket and key.

9. Install the timing belt and timing belt cover.

10. Install the accessory drive belts.

11. Connect the negative battery cable.

12. Run the engine and check for oil leaks and proper engine operation.

Camshaft and Valve Lifters

REMOVAL & INSTALLATION

✳✳ CAUTION

Some models covered by this manual may be equipped with a Supplemental Restraint System (SRS), which uses an air bag. Whenever working near any of the SRS components, such as the impact sensors, the air bag module, steering column and instrument panel, properly disable the SRS.

1. Remove the battery.

2. Remove the timing belt and valve cover.

3. Remove the camshaft sprocket and the distributor.

4. Remove the rocker arm/shaft assemblies.

5. Pull the lash adjuster from the rocker arm.

6. Loosen the camshaft thrust plate bolt and remove the thrust plate.

7. Gently pull the camshaft from the left-hand side of the cylinder head.

To install:

8. Clean the camshaft and cylinder head surface.

➡ **Inspect the camshaft oil seal and replace it if necessary.**

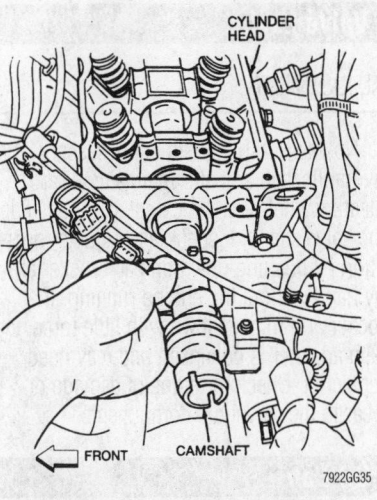

Lubricate the machined surfaces of the camshaft during installation to prevent premature wear

9. Coat the surfaces of the camshaft with clean engine oil and gently install the camshaft into the left-hand side of the cylinder head.

10. Install the camshaft thrust plate and bolt. Tighten the bolt to 71–88 inch lbs. (8–10 Nm).

11. Pour clean engine oil into the oil reservoir in the rocker arm and apply engine oil to the lash adjuster.

✳✳ CAUTION

Be careful not to damage the O-ring when installing the adjuster.

12. Insert the adjuster into the rocker arm.

13. Install the rocker arm/shaft assemblies and the distributor.

14. Install the camshaft sprocket.

15. Install the valve cover and the timing belt.

16. Adjust the camshaft timing.

17. Install the battery.

18. Start the vehicle and check for proper engine operation.

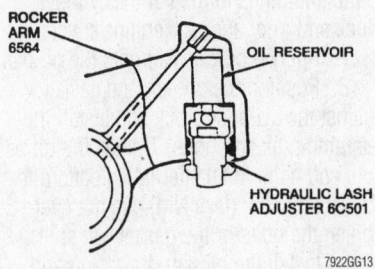

The lash adjuster is located in a bore in the underside of the rocker arm

Valve Lash

ADJUSTMENT

No valve lash adjustment necessary. The hydraulic lash adjuster maintains a zero clearance between the camshaft lobes and the valve stems. Inspect lash adjuster operation by pushing down on each rocker arm by hand without the engine running. If a rocker arm moves down with little force, the lash adjuster is collapsed and may need replacing. Check for signs of damage or wear to the valvetrain components.

Oil Pan

REMOVAL & INSTALLATION

1. Raise and safely support the vehicle.

✻✻ CAUTION

The EPA warns that prolonged contact with used engine oil may cause a number of skin disorders, including cancer! You should make every effort to minimize your exposure to used engine oil. Protective gloves should be worn when changing the oil. Wash your hands and any other exposed skin areas as soon as possible after exposure to used engine oil. Soap and water, or waterless hand cleaner should be used.

2. Drain the engine oil into a suitable container.
3. Remove the exhaust header pipe.
4. Remove the oil pan-to-cylinder block retaining nuts and bolts. Lower the oil pan and discard the old gasket.

To install:
5. Clean the oil pan and cylinder block sealing surfaces to remove all traces of existing gasket material. Thoroughly clean the oil pan.
6. Apply a suitable oil resistant sealant to the joint lines formed at the cylinder block and front and rear engine covers.
7. Apply a new gasket onto the oil pan.
8. Position the oil pan and gasket against the cylinder block and install the retaining nuts and bolts. Tighten the oil pan nuts and bolts in an alternating pattern to 69–78 inch lbs. (8–9 Nm). Do not over tighten the bolts or the gasket will split.
9. Install the oil pan drain plug and tighten to 22–30 ft. lbs. (29–41 Nm).
10. Lower the vehicle.
11. Install the exhaust header pipe.

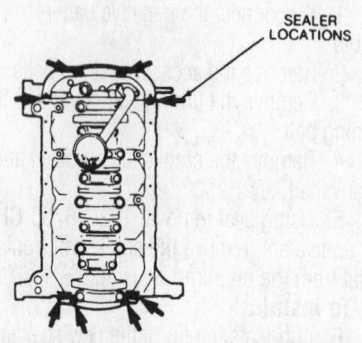

Apply sealer at these locations to ensure a tight seal

✻✻ WARNING

Operating the engine without the proper amount and type of engine oil will result in severe engine damage.

12. Fill the crankcase with the proper amount of clean engine oil.
13. Run the engine and check for oil leaks.

Oil Pump

REMOVAL & INSTALLATION

1. Raise and safely support the vehicle.
2. Remove the timing belt covers and the timing belt.
3. Remove the crankshaft sprocket.

✻✻ CAUTION

The EPA warns that prolonged contact with used engine oil may cause a number of skin disorders, including cancer! You should make every effort to minimize your exposure to used engine oil. Protective gloves should be worn when changing the oil. Wash your hands and any other exposed skin areas as soon as possible after exposure to used engine oil. Soap and water, or waterless hand cleaner should be used.

4. Drain the engine oil and remove the oil pan.
5. If equipped, remove the Crankshaft Position Sensor (CKP) bracket bolt and move the sensor and bracket aside.
6. Remove the six oil pump assembly retaining bolts and remove the oil pump assembly and gasket from the engine. Discard the gasket.
7. Remove the oil pump pick-up tube and screen, if required.

8. Disassembly the oil pump and inspect all components as necessary. Replace the oil pump if needed.

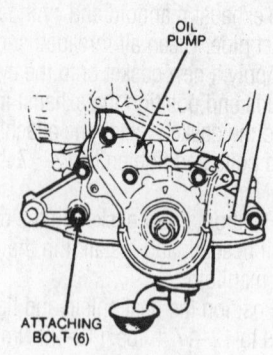

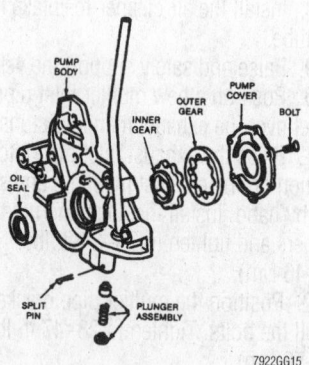

Exploded view of the oil pump assembly and mounting bolt locations

To install:
9. Clean all gasket mating surfaces. If disassembled, clean the oil pump housing and components with a suitable solvent and allow to dry. Reassemble the oil pump, lubricating all components with clean engine oil. Install a new front crankshaft seal using a suitable seal installer.
10. Carefully coat both sides of the new oil pump gasket with a suitable sealant compound. Apply the gasket to the oil pump and remove any excess sealant.

➡**Do not allow the sealant compound to enter the oil pump discharge opening once the gasket is in place. This opening must be free and clear before the oil pump is installed onto the cylinder block.**

11. Position the oil pump on the cylinder block mating surface and install the six retaining bolts. Tighten the bolts to 14–19 ft. lbs. (19–25 Nm).
12. If removed, install a new gasket on the oil pump pick-up tube and screen. Position the oil pump pick-up tube and screen

on the oil pump and install the two retaining bolts. Tighten the bolts to 69–95 inch lbs. (8–11 Nm).

13. If removed, install the CKP and tighten the bracket bolt to 25 inch lbs. (3 Nm).

14. Install the oil pan.

15. Install the crankshaft.

16. Install the timing belt and the timing belt covers.

17. Lower the vehicle.

❊❊ WARNING

Operating the engine without the proper amount and type of engine oil will result in severe engine damage.

18. Fill the crankcase to the proper level with engine oil.

19. Run the engine and check for leaks and proper engine operation.

Rear Main Seal

REMOVAL & INSTALLATION

1. Remove the transaxle from the vehicle.

2. Remove the six flywheel retaining bolts and the flywheel from the rear of the engine.

3. Remove the flywheel reinforcing plate bolt and the flywheel reinforcing plate.

4. Remove the rear oil seal using Seal Remover tool T92C-6700-CH, or equivalent.

To install:

5. Clean the sealing surface of the oil seal retainer.

6. Apply clean engine oil to the inside and outside of the new seal. Install the seal with the lip of the seal facing the engine using Seal Replacer T87C-6701-A, or equivalent.

➡**Two flywheel retaining bolts may be used along with the seal replacer tool to install the rear crankshaft seal.**

7. Install the flywheel reinforcing plate and bolt. Tighten the bolt to 71–97 inch lbs. (8–11 Nm).

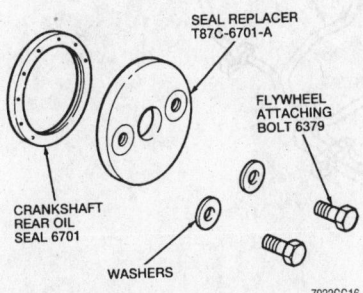

SEAL REPLACER
T87C-6701-A

FLYWHEEL
ATTACHING
BOLT 6379

CRANKSHAFT
REAR OIL
SEAL 6701

WASHERS

7922GG16

When installing a new rear main seal, be sure to use the proper tool, such as Seal Replacer T87C-6701-A

8. Install the flywheel and tighten the six bolts to 71–76 ft. lbs. (96–103 Nm).

9. Install the transaxle.

FUEL SYSTEM

Fuel System Service Precautions

Safety is the most important factor when performing not only fuel system maintenance, but any type of maintenance. Failure to conduct maintenance and repairs in a safe manner may result in serious personal injury or death. Work on a vehicle's fuel system components can be accomplished safely and effectively by adhering to the following rules and guidelines.

• To avoid the possibility of fire and personal injury, always disconnect the negative battery cable unless the repair or test procedure requires that battery voltage by applied.

• Always relieve the fuel system pressure prior to disconnecting any fuel system component (injector, fuel rail, pressure regulator, etc.) fitting or fuel line connection. Exercise extreme caution whenever relieving fuel system pressure to avoid exposing skin, face and eyes to fuel spray. Please be advised that fuel under pressure may penetrate the skin or any part of the body that it contacts.

• Always place a shop towel or cloth around the fitting or connection prior to loosening to absorb any excess fuel due to spillage. Ensure that all fuel spillage is quickly remove from engine surfaces. Ensure that all fuel-soaked cloths or towels are deposited into a flame-proof waste container with a lid.

• Always keep a dry chemical (Class B) fire extinguisher near the work area.

• Do not allow fuel spray or fuel vapors to come into contact with a spark or open flame.

• Always use a second wrench when loosening or tightening fuel line connections fittings. This will prevent unnecessary stress and torsion to fuel piping. Always follow the proper torque specifications.

• Always replace worn fuel fitting O-rings with new ones. Do not substitute fuel hose where rigid pipe is installed.

Fuel System Pressure

RELIEVING

❊❊ WARNING

Fuel lines can remain pressurized for a long period of time after the engine

has been turned off. Always relieve the fuel system pressure before performing any service on the fuel system.

1. Remove the fuel tank filler cap.

2. Unplug the fuel pump relay electrical connection, which is located behind the left-hand side of the instrument panel.

3. Turn the engine **ON** and let it idle normally.

4. The engine will stall, when it does, turn the ignition key **OFF** and engage the fuel pump relay electrical connection.

5. The fuel system pressure has now been relieved and will remain so until the engine is turned **ON**.

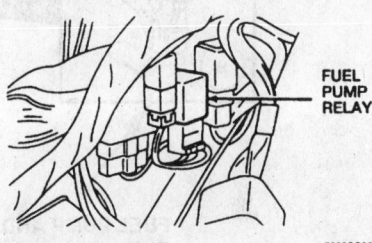

FUEL
PUMP
RELAY

7922GG36

Be sure to leave the fuel pump relay unplugged while servicing the fuel system

Fuel Filter

REMOVAL & INSTALLATION

❊❊ CAUTION

Fuel injection systems remain under pressure, even after the engine has been turned OFF. The fuel system pressure must be relieved before disconnecting any fuel lines. Failure to do so may result in fire and/or personal injury.

1. Properly relieve the fuel system pressure.

❊❊ CAUTION

Observe all applicable safety precautions when working around fuel. Whenever servicing the fuel system, always work in a well ventilated area. Do not allow fuel spray or vapors to come in contact with a spark or open flame. Keep a dry chemical fire extinguisher near the work area. Always keep fuel in a container specifically designed for fuel storage; also, always properly seal fuel containers to avoid the possibility of fire or explosion.

2. Disconnect the negative battery cable.

3. Loosen the clamp and disconnect the fuel supply line at the inlet of the fuel filter. Plug the end to prevent spillage.

4. Loosen the clamp and disconnect the fuel return line at the fuel filter outlet fitting.

5. Remove the filter bracket attaching bolt and nuts from the fuel filter.

6. Remove the fuel filter from the bracket and properly dispose.

To install:

7. Install the new fuel filter into the bracket. Be sure the filter is positioned in the proper direction of fuel flow.

8. Connect the fuel return line to the fuel filter outlet fitting. Install the fuel filter

bracket bolt and nuts and tighten to 71–97 inch lbs. (8–11 Nm).

9. Remove the plug from the fuel supply line and connect it to the fuel filter inlet. Secure it with the clamp.

10. Connect the negative battery cable.

11. Run the engine and check for fuel leaks.

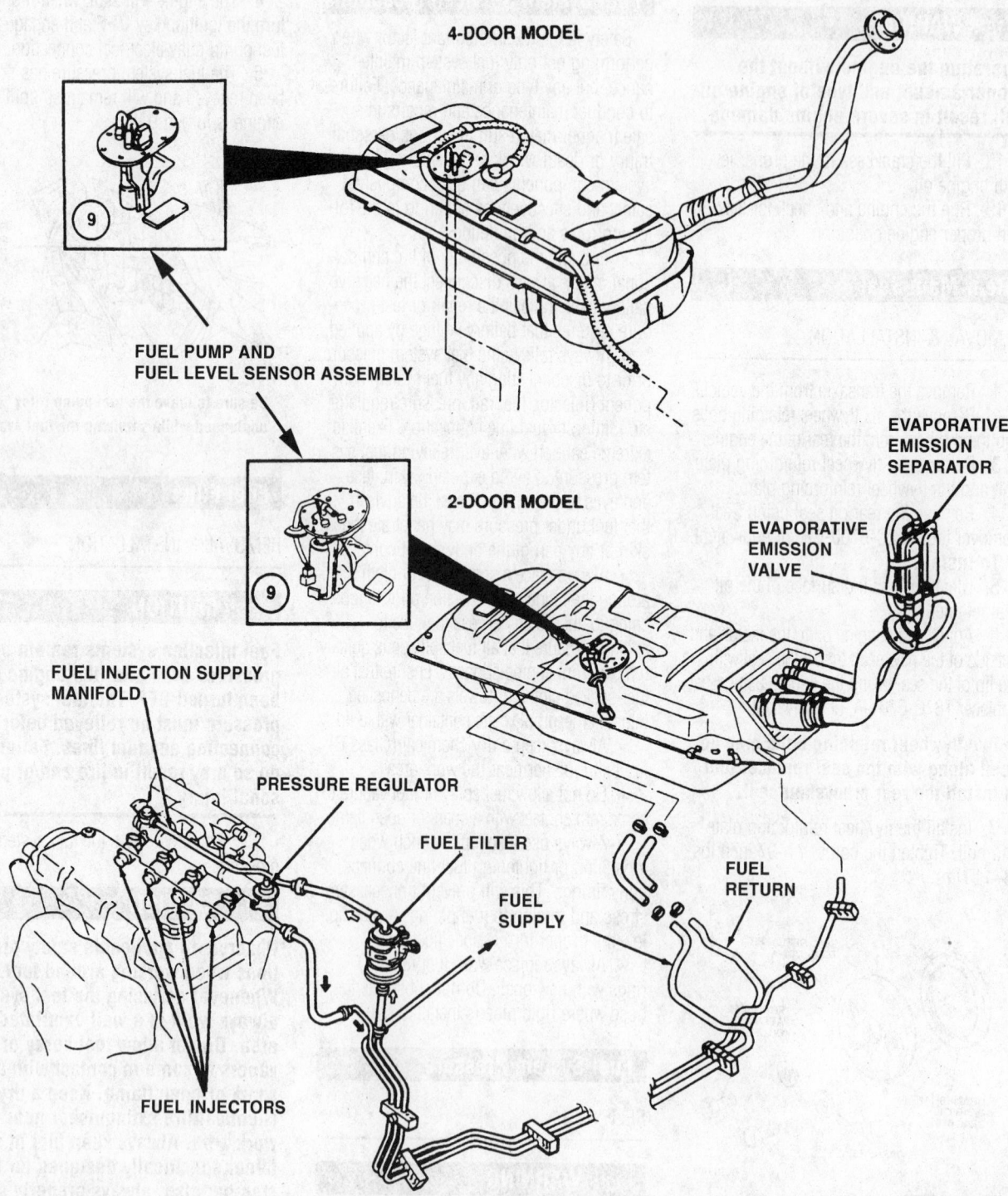

4-DOOR MODEL

FUEL PUMP AND FUEL LEVEL SENSOR ASSEMBLY

2-DOOR MODEL

EVAPORATIVE EMISSION SEPARATOR

EVAPORATIVE EMISSION VALVE

FUEL INJECTION SUPPLY MANIFOLD

PRESSURE REGULATOR

FUEL FILTER

FUEL SUPPLY

FUEL RETURN

FUEL INJECTORS

Fuel system component identification

7922GG37

Fuel Pump

REMOVAL & INSTALLATION

✳✳ CAUTION

Fuel injection systems remain under pressure, even after the engine has been turned OFF. The fuel system pressure must be relieved before disconnecting any fuel lines. Failure to do so may result in fire and/or personal injury.

1. Properly relieve the fuel system pressure.

✳✳ CAUTION

Observe all applicable safety precautions when working around fuel. Whenever servicing the fuel system, always work in a well ventilated area. Do not allow fuel spray or vapors to come in contact with a spark or open flame. Keep a dry chemical fire extinguisher near the work area. Always keep fuel in a container specifically designed for fuel storage; also, always properly seal fuel containers to avoid the possibility of fire or explosion.

2. Disconnect the negative battery cable.
3. Remove the rear seat cushion and cover.
4. Remove the three luggage compartment floor cover hold-down pins and fold the luggage compartment floor cover forward until the sending unit access plate is visible.
5. On 2-door vehicles, remove the six inner rear floor filler cover bolts and the two inner rear floor filler cover nuts.
6. Remove the inner rear floor filler cover.
7. Remove the four fuel pump assembly access plate screws, the ground lead, and the sending unit access plate from the chassis.
8. Detach the fuel pump and sending unit electrical connector.
9. On 4-door vehicles, remove the 2-way check valve bolt and the 2-way check valve from the bracket on the fuel pump and sending unit housing.
10. Loosen the two hose clamps and disconnect the two hoses from the fittings on the fuel pump and sending unit housing.
11. Remove the four screws (2-door) or eight screws (4-door) and lift the fuel pump and sending unit housing from the fuel tank.

When removing the fuel pump housing from the fuel tank, slowly pull it out to prevent fuel from splashing

12. Detach the fuel tank sending unit electrical connector.
13. Remove the two fuel tank sending unit washers and nuts and the fuel tank sending unit.
14. Remove the fuel pump screw, fuel pump grommet and the fuel pump bracket from the bottom of the fuel pump.

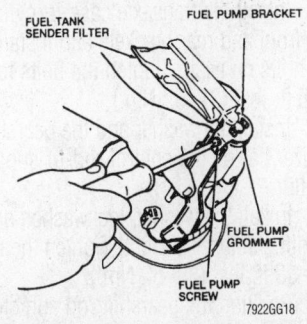

Remove the mounting screws to separate the fuel pump from the housing during pump replacement

15. Remove the fuel filter bracket, fuel filter and the retainer from the fuel pump.
16. Detach the fuel pump electrical connector.
17. Remove the clamp and the fuel pump from the housing.
To install:
18. Position the fuel pump on the housing and secure with the clamp.
19. Install the fuel tank sender filter, bracket and the retainer on the fuel pump.
20. Clip the fuel tank sender bracket onto the fuel pump.
21. Install the grommet, screw and the bracket onto the bottom of the fuel pump.
22. Position the fuel tank sending unit on the housing and install the sending unit washers and nuts.
23. Connect the fuel tank sending unit electrical connector.

24. Place the fuel pump and sending unit housing into the fuel tank opening and install the four screws (2-door) or eight screws (4-door).
25. Install the two fuel hoses on the two sending unit hose fittings and tighten the two hose clamps.
26. On 4-door vehicles, position the 2-way check valve on the fuel pump and sending unit housing and secure with the bolt.
27. Connect the fuel pump and sending unit electrical connector.
28. Connect the negative battery cable.
29. Start the engine and check for leaks at the fuel line connections and proper fuel pump operation. Turn the engine **OFF**.
30. Install the four fuel pump and sending unit access plate screws, the ground lead and the fuel pump and sending unit access plate.
31. Fold the luggage compartment floor cover over the fuel pump and sending unit access plate and install the three hold-down pins.
32. Install the rear seat cushion and cover.

DRIVE TRAIN

➡For CV-joint replacement procedures, refer to the applicable Unit Repair Section in the beginning of this manual.

Transaxle Assembly

REMOVAL & INSTALLATION

Manual

1. Disconnect the negative battery cable.
2. Unplug the two back-up light switch wiring connectors.
3. Remove the clutch cable adjusting nut and disengage the cable from the release lever. Pull the cable through the clutch cable bracket.
4. Remove the engine compartment wiring harness ground strap from the transaxle.
5. Remove the starter motor.
6. Loosen the speedometer cable retainer and disconnect the speedometer cable.
7. Remove the 2 bolts from the top of the clutch housing.
8. Install 3 Bar Engine Support tool D88L-6000-A, or equivalent. Properly secure the engine to the engine support tool.

9. Raise and safely support the vehicle.

10. Disengage the halfshafts from the differential side gears.

11. Install Differential Side Gear Plug tool T87C-7025-C, or equivalent, to prevent the side gears from moving.

12. Remove the nut and bolt attaching the shift rod to the input shift rail.

13. Remove the gearshift stabilizer bar nut, lockwasher, and flat washer, and remove the bar from the control rod-to-support bar stud.

14. Remove the three transaxle-to-engine retaining bolts from the transaxle case rear bracket and remove the bracket.

15. Remove the three transaxle-to-engine retaining bolts from the transaxle case front bracket and remove the bracket.

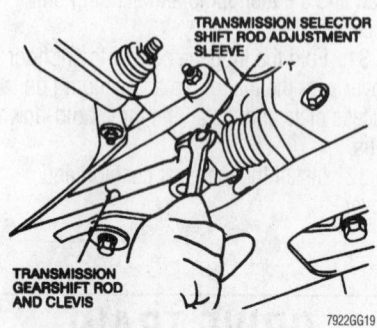

When removing the gearshift rod and clevis, note the position of the washers

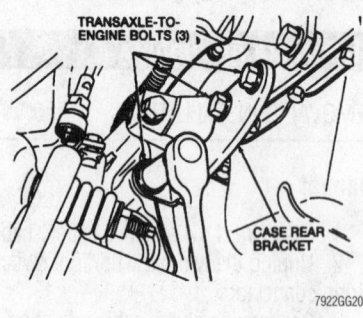

Support the transaxle before removing the transaxle case rear bracket

16. Remove the two rear and two front transaxle support insulator nuts from the rear engine support.

17. Remove the four rear engine support rebound insulator bolts and remove the rear engine support.

18. Position a suitable transmission jack under the transaxle and secure it with a safety chain or strap.

19. Remove the four flywheel reinforcing plate bolts.

20. Remove the two remaining transaxle-to-engine block retaining bolts.

21. Carefully separate the transaxle from the engine and lower the transaxle from the vehicle.

To install:

22. Raise the transaxle into position and seat it against the rear of the engine.

23. Install the four flywheel reinforcing plate bolts and tighten to 62–86 inch lbs. (7–10 Nm).

24. Install the two lower transaxle retaining bolts and tighten to 47–66 ft. lbs. (64–89 Nm).

25. Remove the transmission jack.

26. Position the rear engine support. Install the two rear transaxle support insulator nuts and tighten to 21–34 ft. lbs. (28–46 Nm.).

27. Install the two front transaxle support insulator nuts and tighten to 32–38 ft. lbs. (43–52 Nm.).

28. Install the four rear engine support rebound insulator bolts and tighten to 47–66 ft. lbs. (64–89 Nm.).

29. Install the transaxle case-to-cylinder block front and rear brackets and install the three bolts on each. Tighten the bolts to 27–38 ft. lbs. (37–52 Nm.).

30. Install the washer and the gearshift stabilizer bar on the control rod-to-support bar stud.

31. Install the washer, lockwasher, and gearshift stabilizer bar nut. Tighten the nut to 28–38 ft. lbs. (38–52 Nm.).

32. Position the gearshift rod and clevis on the main shift control shaft, and install the selector shift rod adjustment sleeve. Tighten the nut to 12–17 ft. lbs. (16–23 Nm.).

33. Route the Park/Neutral switch wiring over the rear engine support.

34. Install the halfshaft and joint assemblies.

35. Check and fill the transaxle, if needed.

36. Lower the vehicle and remove the engine support bar.

37. Install two retaining bolts at the top of the clutch housing. The top bolt is installed through the heater pipe bracket. Tighten the bolts to 47–66 ft. lbs. (64–89 Nm).

38. Connect the ground strap to the transaxle case.

39. Connect the speedometer cable onto the sleeve and hand-tighten.

40. Install the starter motor.

41. Attach the Park/Neutral and back-up light switch wiring connectors.

42. Connect the clutch cable to the release lever and adjust the clutch pedal free-play.

43. Connect the negative battery cable.

44. Road test the vehicle and check for proper transaxle operation.

Automatic

1. Disconnect the negative battery cable.

2. From the engine compartment, remove the manual control lever nut and arm.

3. Remove the shift cable and bracket from the transaxle.

4. Disconnect the speedometer cable from the transaxle.

5. Detach the transaxle electrical connectors, located next to the governor.

6. Disconnect the transaxle ground wire. Disconnect the transaxle vacuum hose and vent hose located below the distributor cap.

7. Remove the starter motor.

8. Remove the coolant pipe retaining bracket, located below the distributor cap.

9. Remove the two upper bell housing bolts.

10. Support the engine using Engine Support Bar tool D87L-6000-a or equivalent.

11. Raise and safely support the vehicle.

12. Remove the front wheel and tire assemblies.

13. Drain the transaxle fluid.

14. Remove the front fender splash shield.

15. Remove the front stabilizer bar.

16. Remove the lower arm clamp bolts and nuts. Pull the lower arms downward, separating the lower arms from the knuckles.

➥**Use care not to damage the ball joint dust boots.**

17. Remove the tie rod cotter pin and nut. Disconnect the tie rod end from the knuckle and discard the cotter pin.

18. Remove the halfshafts. Install the Differential Plug tool T87C-7025-C, or equivalent, between the differential side gears to prevent side gear movement.

19. Remove the front and rear transaxle support insulator nuts.

20. Remove the four rear engine support rebound insulator bolts and the transmission support crossmember.

21. Remove the front transaxle support insulator through-bolt and nut. Remove the front transaxle support insulator.

22. Remove the four front transaxle support bracket bolts and the front transaxle support bracket.

23. Remove the two rear transaxle support bracket bolts and the rear transaxle support bracket and insulator.

24. Remove the intake manifold support.

25. Remove the three transaxle-to-engine bolts and remove the transaxle case rear bracket.

26. Remove the three transaxle-to-engine bolts from the transaxle case-to-cylinder block front bracket and remove the bracket.

27. Remove the flywheel cover bolts and cover.

28. Using a wrench, rotate the crankshaft pulley bolt clockwise to gain access to the four torque converter-to-flywheel nuts, and remove all nuts.

29. Make alignment marks between the oil cooler tubes and hoses. Disconnect and plug the oil cooler lines.

30. Position a transmission jack under the transaxle and secure it with a chain or strap.

31. Remove the remaining engine-to-transaxle retaining bolts.

32. Carefully separate and lower the transaxle from the vehicle.

To install:

33. Raise the transaxle into position and install the two engine-to-transaxle bolts. Be sure that the torque converter is in alignment with the flexplate. Tighten to 47–66 ft. lbs. (64–89 Nm).

34. Remove the transaxle jack.

35. Install the starter motor.

36. Install the torque converter bolts and tighten to 26–36 ft. lbs. (34–49 Nm).

37. Install the flywheel cover and tighten the bolts to 71–97 inch lbs. (8–11 Nm).

38. Position the front transaxle-to-engine support and install the three retaining bolts. Tighten the bolts to 27–38 ft. lbs. (37–52 Nm.).

39. Install the intake manifold support and tighten the bolts to 27–38 ft. lbs. (37–52 Nm.).

40. Install the transaxle case rear bracket and three retaining bolts. Tighten the bolts to 27–38 ft. lbs. (37–52 Nm.).

41. Install the front transaxle support insulator bracket and four retaining bolts. Tighten the bolts to 28–37 ft. lbs. (38–51 Nm.).

42. Install the front transaxle support bracket and the through-bolt and nut. Do not tighten the through-bolt and nut until the rear engine support is installed.

43. Position the transmission support crossmember and install the rear engine support rebound insulator bolts. Tighten the rear engine support rebound insulator bolts to 47–66 ft. lbs. (64–89 Nm.).

44. Tighten the front transaxle support bracket through-bolt and nut to 69–83 ft. lbs. (93–113 Nm.).

45. Install the two front transaxle support insulator nuts and tighten them to 32–38 ft. lbs. (43–52 Nm).

46. Install the two rear transaxle support insulator nuts and tighten them to 21–34 ft. lbs. (28–46 Nm).

47. Remove the differential plugs and install the halfshafts.

48. Align the marks made on the oil cooler lines, and install the lines. Install the hose clamps.

49. Connect the tie rod ends to the steering knuckles and tighten the attaching nuts to 26–30 ft. lbs. (35–40 Nm). Install new cotter pins.

50. Attach the lower ball joints to the knuckles. Tighten the lower arm clamp bolt to 40–50 ft. lbs. (54–68 Nm).

51. Install the front stabilizer bar. Tighten the retaining nuts to 43–52 ft. lbs. (43–52 Nm).

52. Install the front fender splash shield and tighten the bolts to 65–95 inch lbs. (8–10 Nm.).

53. Install the front wheel and tire assemblies. Tighten the lug bolts to 65–87 ft. lbs. (88–118 Nm).

54. Lower the vehicle.

55. Remove the engine support tool.

56. Attach the shift cable and bracket to the transaxle. Install the manual control lever arm on the manual control lever. Install the retaining nut and tighten to 34–47 ft. lbs. (44–64 Nm).

➡**Do not use any type of power wrench to tighten the nut. Damage to the transaxle may result.**

57. Install the coolant pipe retaining bracket located below the distributor.

58. Connect the vacuum hose and vent hose located below the distributor cap.

59. Attach the transaxle electrical connectors located next to the governor.

60. Connect the ground wire to the transaxle and tighten the bolt to 65–95 inch lbs. (8–10 Nm.).

61. Connect the speedometer cable.

62. Connect the negative battery cable.

63. Fill the transaxle with Motorcraft MERCON® automatic transmission fluid to the proper level.

➡**Be sure that the gearshift lever position aligns with the manual control lever position exactly before starting the engine.**

64. Start the engine. Check for leaks and proper fluid level.

65. Road test the vehicle and check for proper operation.

Clutch

ADJUSTMENT

Clutch Cable Free-Play

1. Carefully move the clutch pedal back and forth and measure the amount of travel before the pedal activates the clutch. If the clutch pedal free-play is 0.35–0.59 in. (9–15mm), no adjustment is necessary. If the free-play is not within specification, proceed to Step 2.

2. Pull back the transaxle release lever and measure the clearance between the lever and the cable connecting link. Thread the adjuster in or out until the clearance between the connecting link and the release lever is 0.06–0.10 in. (1.5–2.5mm).

3. Check the free-play at the clutch. If it is not within specification, inspect the clutch release components for a problem.

4. After adjusting the clutch, be sure the clutch disengagement height is 2.92 inches (74mm) minimum.

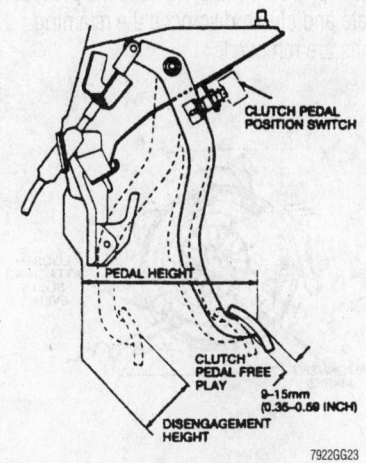

Clutch pedal free-play—specifications

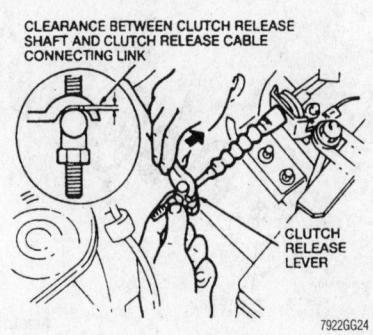

Check the clearance between the release lever and connecting link

REMOVAL & INSTALLATION

1. Disconnect the negative battery cable.
2. Raise and safely support the vehicle.
3. Remove the transaxle assembly.

➡ **During the removal procedure, do not allow oil or grease to come in contact with the clutch disc facing if the disc is to be reused. Handle the disc with clean rags wrapped around the edges and do not touch the disc facing. Even a small amount of dirt or grease may cause the clutch to grab or slip.**

4. If the pressure plate is to be reused, paint or scribe alignment marks on the pressure plate and flywheel for assembly reference.
5. Install an appropriate locking tool to prevent the flywheel from turning.
6. Install a clutch aligning tool to prevent the clutch plate from dropping when the retaining bolts are removed.
7. Loosen the six pressure plate retaining bolts in a crisscross pattern, one turn at a time. This will relieve the pressure plate spring tension evenly and prevent distortion of the pressure plate. Remove the pressure plate and clutch disc once the retaining bolts are removed.

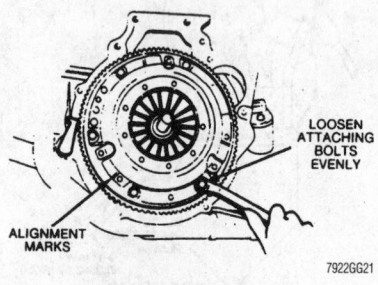

Be sure to line up the matchmarks, if re-using the pressure plate

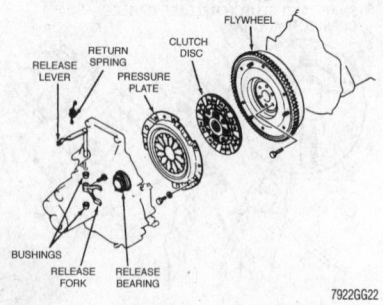

Exploded view of the clutch assembly

8. Inspect all clutch components including the clutch release fork and release bearing, and replace as required.
9. Inspect the flywheel for scoring, cracks and heat checks. Resurface or replace the flywheel, as necessary.
10. Inspect the pilot bearing for damage. Be sure the bearing turns easily. If replacement is necessary, remove the flywheel and remove the pilot bearing.

To install:

11. If necessary, install a new pilot bearing using a suitable installation tool. Use only a driver tool that contacts the bearing outer race. A driver tool that contacts the inner race or the bearing area will damage the bearing.
12. If the flywheel was removed, clean the sealant from the flywheel retaining bolts. Coat the bolt threads with a suitable sealer compound.
13. Be sure the crankshaft flange and the back of the flywheel are clean. Position the flywheel on the crankshaft and install the six retaining bolts. Tighten the bolts to 71–76 ft. lbs. (96–103 Nm).
14. Position the clutch disc on the flywheel and install a clutch alignment tool to hold the disc in place.

➡ **When installing the clutch disc, be sure the disc dampener springs are facing away from the flywheel. A new disc will be stamped FLYWHEEL to indicate the correct installation position.**

15. Align the reference marks, if present, and position the pressure plate on the flywheel and install the retaining bolts. Tighten the bolts evenly, in a crisscross pattern, to 13–20 ft. lbs. (18–26 Nm). The bolts must be tightened in this manner to prevent distortion of the pressure plate.
16. Remove the clutch alignment tool.
17. Clean the clutch disc splines on the input shaft with a dry rag and coat the spline surfaces with a light film of clutch grease.
18. Install the transaxle.
19. Connect the negative battery cable.
20. Adjust the clutch pedal free-play.
21. Road test the vehicle for proper clutch operation.

Halfshaft

REMOVAL & INSTALLATION

1. Disconnect the negative battery cable.
2. With the vehicle sitting on all four wheels, use a small cape chisel to raise the staked portion of the front wheel hub locknut.

Have an assistant apply the brakes, then loosen, but do not remove, the locknut.

3. Raise the vehicle and support it safely.
4. Drain the transaxle fluid.
5. Remove the front wheel and tire assembly.
6. Remove the ball joint clamp bolt and nut from the steering knuckle. Carefully pry the lower control arm down to disconnect the ball joint from the steering knuckle. Be careful not to tear or puncture the dust boot when disconnecting the ball joint.
7. Using a small prybar, separate the halfshaft from the transaxle.

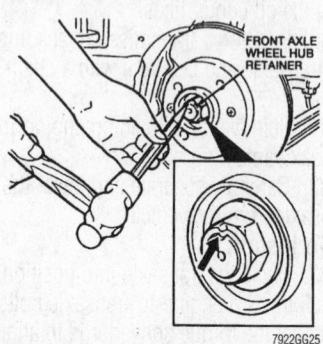

Use a chisel to raise the staked edge of the wheel hub retaining nut, then loosen the nut

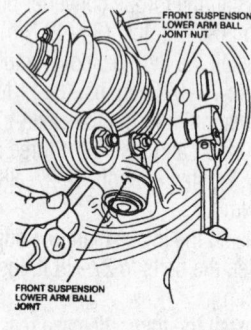

To separate the ball joint from the steering knuckle, remove the clamp bolt and nut

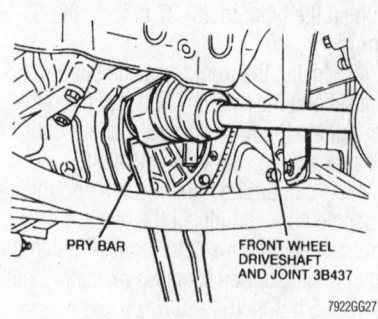

Use a small prybar to disengage the inner CV-joint from the transaxle, as shown

➡️**The halfshaft must be separated from the transaxle gradually. If the halfshaft is yanked out suddenly, the oil seal may be damaged.**

8. Install Differential Plug tool T87C-7025-C or equivalent, to prevent the differential side gear from moving.

9. Remove and discard the wheel hub locknut.

10. Withdraw the halfshaft from the hub. Be careful not to damage the oil seal. If necessary, use a suitable wheel puller to push the halfshaft out of the hub.

To install:

11. Inspect the differential and wheel hub oil seals for damage and replace as required.

12. Remove the circlip from the inboard halfshaft spline end and replace with a new clip. Lubricate the inboard and outboard halfshaft spline ends with grease.

13. Remove the differential gear holding plug.

14. Position and install the inboard end of the halfshaft into the differential side gear. Be sure the circlip snaps into place and take care not to damage the differential oil seal.

15. Position and install the outboard end of the halfshaft into the wheel hub. Take care not to damage the wheel hub oil seal.

16. Install the wheel hub locknut onto the halfshaft and tighten it by hand.

17. Raise the lower control arm and connect it to the ball joint. Take care not to damage the ball joint dust boot. Install the clamp nut and bolt and tighten the nut to 32–40 ft. lbs. (43–54 Nm).

18. Install the wheel and tire assembly. Tighten the lug bolts to 65–87 ft. lbs. (88–118 Nm).

19. Install and tighten the transaxle drain plug.

20. Lower the vehicle.

21. With the vehicle sitting on all four wheels, have an assistant apply the brakes. Tighten the wheel hub locknut to 116–174 ft. lbs. (157–235 Nm). Stake the nut using a suitable tool.

➡️**Do not stake the locking tab with a pointed tool. Be sure the locking tab is depressed at least 0.16 in. (4mm) into the locknut slot to ensure proper locking capability.**

22. Raise and safely support the vehicle.

23. Grasp the wheel hub and pull on it to ensure that the halfshaft is installed properly. Rotate the wheel hub by hand to check that the hub and halfshaft assembly turns smoothly.

24. Fill the transaxle with the proper grade and type fluid to specification.

25. Lower the vehicle.

26. Connect the negative battery cable.

27. Road test the vehicle and check for transaxle leaks and proper operation.

STEERING AND SUSPENSION

Air Bag

✳️✳️ CAUTION

Some vehicles are equipped with the Supplemental Inflatable Restraint (SIR) or air bag system. The SIR system must be disabled before performing service on or around SIR system components, steering column, instrument panel components, wiring and sensors. Failure to follow safety and disabling procedures could result in accidental air bag deployment, possible personal injury and unnecessary SIR system repairs.

PRECAUTIONS

Several precautions must be observed when handling the inflator module to avoid accidental deployment and possible personal injury.

• Never carry the inflator module by the wires or connector on the underside of the module.

• When carrying a live inflator module, hold securely with both hands, and ensure that the bag and trim cover are pointed away.

• Place the inflator module on a bench or other surface with the bag and trim cover facing up.

• With the inflator module on the bench, never place anything on or close to the module which may be thrown in the event of an accidental deployment.

DISARMING

✳️✳️ CAUTION

The Supplemental Inflatable Restraint (SIR) system must be disarmed before performing service around SIR system components or SIR system wiring. Failure to do so may cause accidental deployment of the air bag, resulting in unnecessary SIR system repairs and/or personal injury.

1. Disconnect the negative battery cable.

2. Wait one minute before proceeding with the service procedure. This is the time required for the back-up power supply in the air bag diagnostic monitor to deplete its stored energy.

3. After service is completed, reconnect the negative battery cable.

4. Turn the ignition switch to the **RUN** position. The air bag indicator should light continuously for approximately 6 seconds, then turn OFF. If the indicator fails to light, flashes or remains lit continuously, there is a fault in the air bag system.

Rack and Pinion Steering Gear

REMOVAL & INSTALLATION

Manual

1. Disconnect the negative battery cable.

2. Matchmark the steering column lower universal joint and steering rack pinion for assembly reference. Remove the steering column and intermediate shaft assembly from the vehicle.

3. Remove the floor set plate bolts and the floor set plate.

4. Cut the plastic tie wrap securing the steering column boot to the steering rack.

5. Raise and safely support the vehicle.

6. Remove the front wheel and tire assemblies.

7. Using the proper tool, separate both tie rod ends from the steering knuckles.

8. Remove the catalytic converter.

9. Remove the plastic splash shield from the right inner fender.

✳️✳️ WARNING

While maneuvering the tie rod boots in and out of the inner fender openings, guide the steering rack assembly carefully to avoid cutting or nicking the boots.

10. Remove the two steering rack mounting bolts and lower the steering rack until it is free of the steering column boot. Slide the rack to the right, through the inner fender opening until the left tie rod is clear of the left inner fender, then lower the left end until the steering rack assembly can be withdrawn from the left side of the vehicle.

11. Remove the steering column intermediate shaft coupling bolt and the coupling from the steering rack.

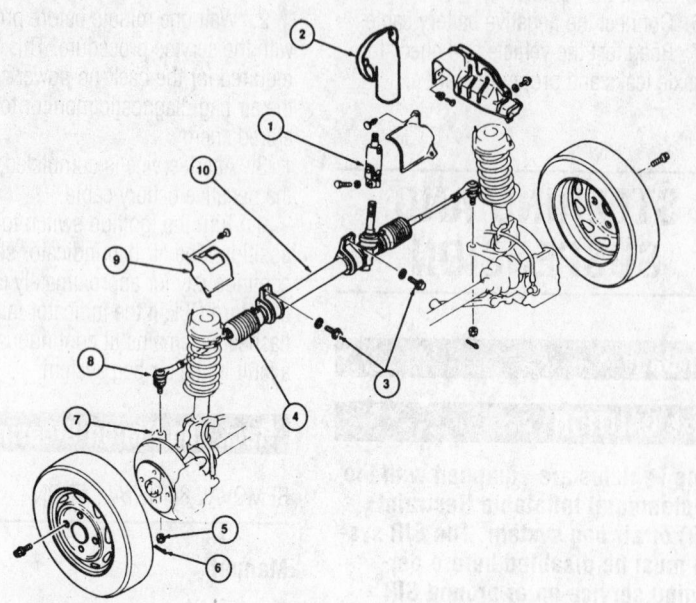

1 Steering Column Gear Input Shaft Coupling	6 Front Wheel Spindle Connecting Rod End Cotter Pin
2 Steering Column Tube Boot	7 Front Wheel Knuckle
3 Steering Gear Bolts	8 Front Wheel Spindle Connecting Rod Or End
4 Front Wheel Spindle Connecting Rod Bellow	9 Front Splash Shield
5 Front Wheel Spindle Connecting Rod End Nut	10 Steering Gear and Linkage

7922GG28

Exploded view of the manual steering gear mounting, showing related components

To install:

12. Install the steering column intermediate shaft coupling and retaining bolts to the steering rack and tighten to 13–20 ft. lbs. (18–26 Nm).

13. Position the steering rack by starting the right side tie rod end through the right inner fender opening far enough to insert the left end of the steering gear assembly into the left front fender opening. Adjust the positioning of the steering rack to the left being careful not to snag the bellows.

14. Align the intermediate shaft coupling with the steering column boot. Raise the steering rack fully into position.

15. Install the left steering rack mounting bolt followed by the right steering rack mounting bolt. Tighten the bolts to 27–38 ft. lbs. (37–52 Nm).

16. Connect the tie rod ends to the steering knuckles. Install and tighten the tie rod end nuts to 31–42 ft. lbs. (42–57 Nm). Install new cotter pins.

17. Attach the right side splash shield on the right inner fender panel.

18. Install the catalytic converter.

19. Install the tire and wheel assemblies. Tighten the lug bolts to 65–87 ft. lbs. (88–118 Nm).

20. Lower the vehicle.

21. Secure the steering column boot to the steering rack housing with a new tie wrap.

22. Install the floor set plate bolts and the floor set plate.

23. Align the matchmarks made on the steering column lower universal joint and the steering rack pinion shaft. Install the steering column when the proper alignment is achieved.

24. Connect the negative battery cable.

25. Check the front end alignment.

26. Road test the vehicle and check for proper steering rack operation.

Power

1. Disconnect the negative battery cable.

2. Remove the steering column tube boot retainer and pry up the boot.

3. Remove the intermediate shaft coupling bolt.

4. Raise and safely support the vehicle.

5. Remove the front tire and wheel assemblies.

6. Remove the power steering hose bracket. Disconnect and plug the high pressure and return lines.

7. Remove the tie rod end cotter pins and nuts. Using the proper tool, separate both tie rod ends from the steering knuckles. Discard the cotter pins.

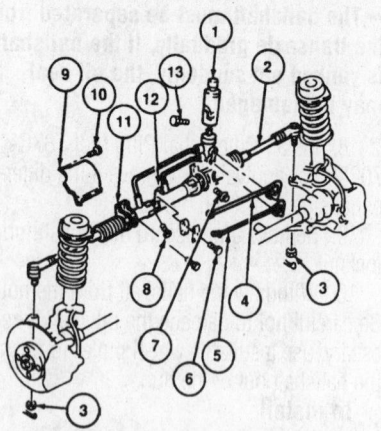

Item	Description
1	Steering Column Intermediate Shaft Coupling
2	Front Wheel Spindle Tie Rod
3	Tie Rod End Nut (2 Req'd)
4	Power Steering Pressure Hose
5	Power Steering Return Hose
6	Power Steering Hose Clamp Bolt
7	Steering Gear Bracket Bolts (4 Req'd)
8	Power Steering Hose Bracket Bolt
9	Tie Rod End Splash Shield
10	Tie Rod End Splash Shield Bolt (3 Req'd)
11	Steering Gear Mounting Bracket
12	Power Steering Hose Bracket
13	Steering Column Gear Input Shaft Coupling Bolt
A	Tighten to 42-57 N·m (31-42 Lb-Ft)

7922GG38

Exploded view of the power steering gear mounting, showing related components

8. Remove the front fender splash shield.

➡**Lowering the exhaust system will ease access to the steering gear.**

9. Remove the three exhaust inlet pipe nuts and two bracket bolts.

10. Remove the muffler inlet pipe hanger posts from the exhaust hanger insulators.

11. Place alignment marks on the right tie rod end to ease installation. Loosen the jam nut and remove the right tie rod end.

12. Remove the steering rack mounting bolts and lower the steering rack until it is free of the steering column boot. Slide the rack to the left and pull the right tie rod through the fender opening. Remove the steering gear by sliding it to the right.

To install:

13. Position the steering rack in its mounting location.

14. With an assistant lifting the steering gear, align the intermediate shaft with the universal joint and install the coupling bolt, but do not tighten it at this time.

15. Install the four steering rack bracket bolts and tighten to 27–38 ft. lbs. (37–52 Nm).

16. Tighten the intermediate shaft coupling bolt to 13–20 ft. lbs. (18–26 Nm).

17. Unplug and connect the high pressure and return lines and install the power steering hose bracket.

18. Install the muffler inlet pipe hanger posts onto the exhaust hanger insulators.

19. Raise the exhaust system and install the three exhaust inlet pipe nuts. Tighten the nuts to 28–38 ft. lbs. (38–53 Nm.).

20. Install the two exhaust inlet pipe bracket bolts.

21. Install the right tie rod end and attach the tie rod ends to the steering knuckles. Install the tie rod end nuts and tighten to 31–42 ft. lbs. (42–57 Nm). Install new cotter pins.

22. Install the small front fender splash shield.

23. Install the front wheel and tire assemblies and tighten the lug bolts to 65–87 ft. lbs. (88–118 Nm).

24. Lower the vehicle.

25. Connect the negative battery cable.

26. Add power steering fluid and allow any air to bleed from the power steering system.

27. Check for leaks.

28. Adjust the toe setting by performing a front end alignment.

29. Road test the vehicle and check for proper operation.

Strut and Spring

REMOVAL & INSTALLATION

Front

1. Raise and safely support the vehicle.

2. Remove the wheel and tire assembly.

3. Remove the brake line clip from the strut lower mounting bracket and disengage the brake line.

4. Remove the two nuts and bolts securing the strut lower bracket to the steering knuckle.

5. Working in the engine compartment, remove the two nuts securing the strut mounting block in the strut tower.

6. Disengage the strut lower bracket from the steering knuckle and lower the strut clear of the wheel well.

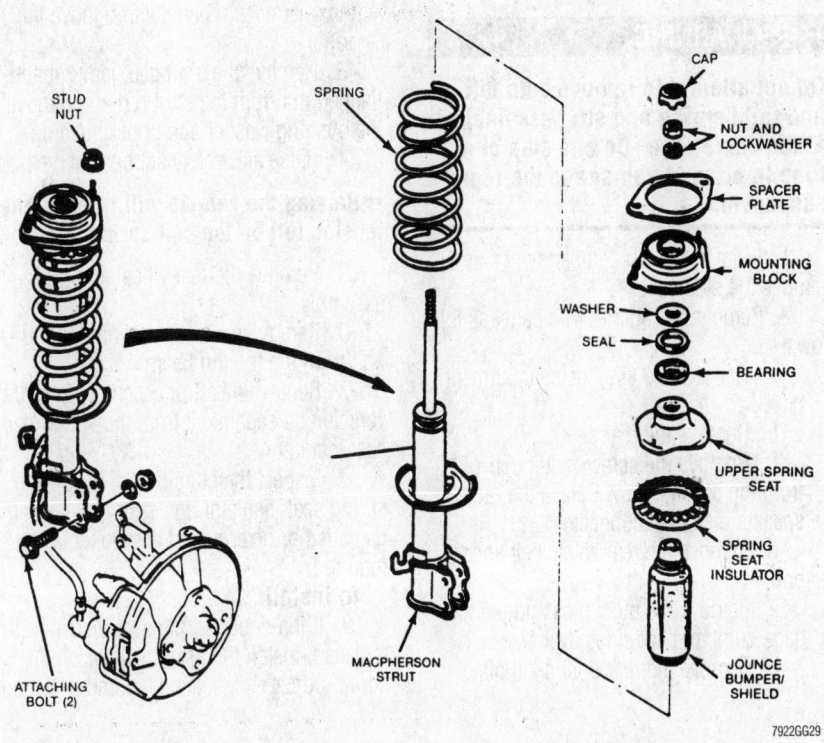

Exploded view of the front strut assembly

7. To separate the coil spring from the strut, attach Spring Compressor tool T81P-5310-A or equivalent, and compress the coil spring.

8. Pry out the mounting block cap and remove the strut upper nut and lockwasher.

9. Remove the strut mounting block and spacer plate. Remove the washer, bearing seal and bearing from the strut rod.

10. Remove the upper spring seat, seat insulator and spring. Slide the jounce bumper/shield off the strut.

➡**If replacing the spring, release the spring compressor progressively to prevent spring arching. Open the compressor jaws wide enough to grip the new spring in the same position and tighten the compressor screws progressively, compressing the spring until the strut can be assembled without interference.**

To install:

11. Check the condition of the jounce bumper and spring seat insulator and replace, as necessary. Be sure the bearing operates smoothly. Check the spring for uniform coil spacing, for nicks or burrs and compare the spring length with a new spring to check for excessive spring set; replace as necessary.

12. Slide the jounce bumper/shield onto the strut rod and over the body. Install the compressed spring, upper spring seat insulator and upper seat, positioning the spring ends against the steps in the seats.

13. Install the bearing, seal and plain washer on the strut rod. Install the strut mounting block with the white alignment spot on the same side of the strut as the steering knuckle mounting bracket.

14. Install the spacer plate. Install the lockwasher and nut and tighten to 40–50 ft. lbs. (54–67 Nm). Release and remove the spring compressor.

15. Place the strut assembly with spacer plate in the strut tower with the white alignment mark facing outward.

16. Install the two upper mounting block stud nuts and tighten to 34–46 ft. lbs. (46–63 Nm).

17. Engage the steering knuckle in the strut tower lower bracket and install the two retaining bolts and nuts. Tighten to 69–86 ft. lbs. (93–117 Nm).

18. Position the brake line into the strut lower mounting bracket cutout and install the retaining clip.

19. Install the wheel and tire assembly and tighten the lug bolts to 65–87 ft. lbs. (88–118 Nm).

20. Lower the vehicle.

21. Check the front wheel alignment.

22. Road test the vehicle and check for proper operation.

Rear

✳✳ WARNING

Do not attempt to remove both left and right spring and strut assemblies at the same time. Do one side at a time to prevent damage to the rear suspension.

1. From the cargo compartment, remove the side cover.
2. Remove the quarter trim panel as follows:
   a. Remove the luggage compartment cover.
   b. Remove the rear seat.
   c. Remove the screws and push-pins from the package tray. Detach the radio speaker electrical connector.
   d. Remove the rear safety belt anchor bolt.
   e. Remove the push pins and the luggage compartment side cover.
   f. Remove the rear door scuff plate.
   g. Pull the seaming welt away from the quarter trim panel and remove the panel.
3. Remove the strut cap, jam nut and flanged nut from the strut rod and remove the bushing washer and upper bushing.
4. Raise and safely support the vehicle.

➡ **Raising the vehicle will release any tension left on the coil spring.**

5. Remove the rear wheel and tire assembly.
6. Remove the lower strut mounting bolt from the torsion beam.
7. Remove the strut assembly from the vehicle and separate it from the spring and seat insulator.
8. Inspect the condition of the spring, spring seat insulator and strut. Replace any damaged or deteriorated components, as required.

To install:

9. If the upper spring seat insulator is replaced, install the new insulator on the spring upper end, seating the end of the coil against the step in the insulator. Position the spring on the strut, making sure the end of the coil seats against the step in the strut spring seat.
10. Guide the strut into the upper strut mounting hole through the wheel well.
11. Align the strut lower end with the mounting hole in the torsion beam. Start the mounting bolt in by hand to hold the strut in position.
12. Install the wheel and tire assembly. Tighten the lug bolts to 65–87 ft. lbs. (88–118 Nm).
13. Lower the vehicle.

➡ **Be sure that the coil spring is positioned properly on the spring seat insulator.**

14. From the cargo compartment, install the rod upper end bushing, bushing washer and flanged nut. Tighten the flanged nut to 12–18 ft. lbs. (16–24 Nm). Hold the flanged nut stationary and tighten the locknut.
15. Install the jam nut and the strut cap.

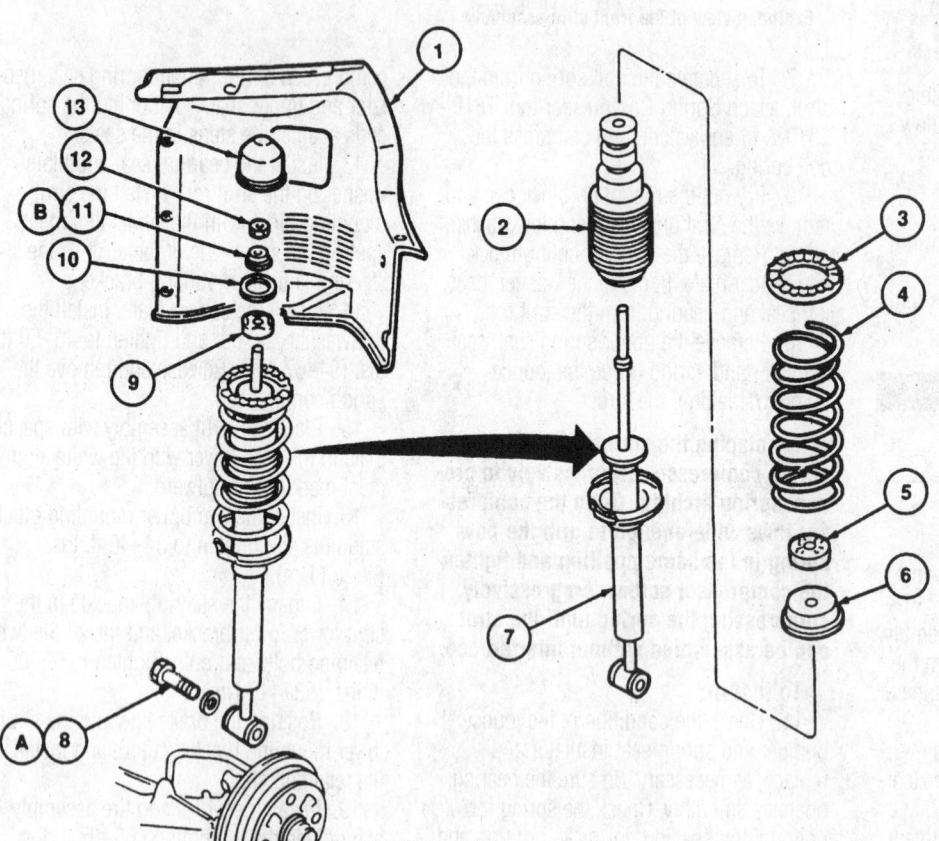

1. Luggage Compartment Side Cover
2. Rear Strut Dust Boot Cover
3. Rear Spring Anti-Squeak Insert
4. Rear Spring
5. Lower Bushing
6. Rear Strut Dust Boot Seat
7. Rear Strut
8. Strut Mounting Bolt
9. Upper Bushing
10. Bushing Washer
11. Rear Strut Mounting Bushing
12. Jam Nut
13. Strut Cap

Exploded view of the rear strut assembly

7922GG39

16. Install the side cover in the cargo compartment.

17. Raise and safely support the vehicle.

18. Tighten the lower strut mounting bolt to 50–60 ft. lbs. (68–81 Nm).

19. Lower the vehicle.

20. Check the rear wheel alignment.

Lower Ball Joints

REMOVAL & INSTALLATION

The lower ball joint is an integral component of the lower control arm. If the lower ball joint is defective, the entire lower control arm must be replaced.

Lower Control Arm

REMOVAL & INSTALLATION

➡ **The ball joint is an integral component of the lower control arm and cannot be serviced separately.**

1. Raise and safely support the vehicle.

2. Remove the front wheel and tire assembly.

3. Loosen the lower control arm-to-chassis bolt and washer at the frame bracket.

4. Remove the ball joint clamp bolt and nut from the steering knuckle assembly.

5. Remove the cotter pin and stabilizer bar bushing nut from the rear of the control arm.

6. Remove the rear bushing washer and bushing. Discard the cotter pin.

7. Lower the control arm, prying the ball joint stud out of the steering knuckle if necessary.

8. Disengage and remove the control arm from the stabilizer bar end and remove it from the vehicle.

9. If the lower control arm is to be reused, inspect the control arm bushings for damage or excessive wear. Verify that the ball joint swivels freely, but is not loose.

10. Replace the lower control arm as required.

To install:

11. Position the front bushing washer and bushing onto the stabilizer bar end. Engage the lower control arm with the stabilizer bar.

12. Raise the control arm inner end into the pivot bracket on the frame and start the

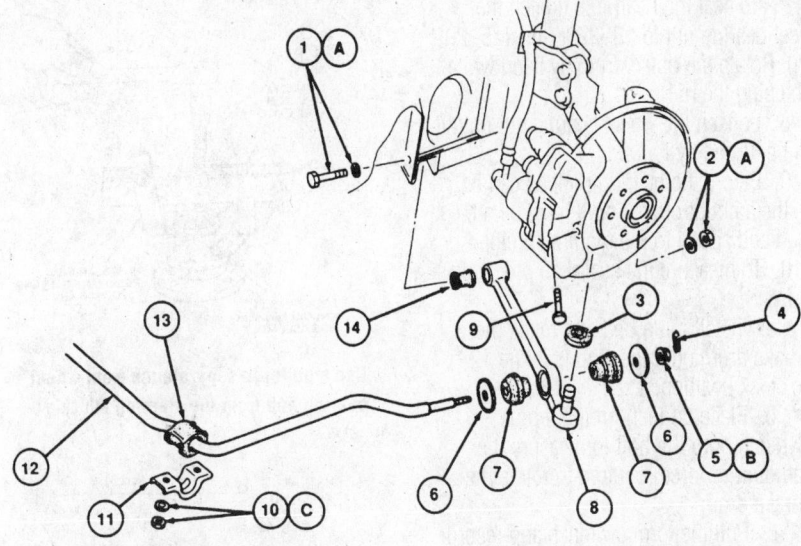

Item	Description
1	Front Suspension Lower Arm-to-Chassis Bolt and Washer
2	Front Suspension Lower Arm Ball Joint Washer and Nut
3	Front Suspension Lower Arm Ball Joint Dust Boot (Part of 3078)
4	Cotter Pin (Part of 5A486)
5	Front Stabilizer Bar Nut (Part of 5A486)
6	Front Stabilizer Bar Washers (2 Req'd) (Part of 5A486)
7	Front Stabilizer Bar Bushings (2 Req'd) (Part of 5A486)
8	Front Suspension Lower Arm

Item	Description
9	Front Suspension Lower Arm Ball Joint Bolt
10	Stabilizer Bar Bracket Washers and Nuts (2 Req'd)
11	Stabilizer Bar Bracket
12	Front Stabilizer Bar
13	Lower Suspension Arm Stabilizer Bar Insulator
14	Front Suspension Lower Arm Mounting Bolt Bushing
A	Tighten to 43-54 N·m (32-40 Lb-Ft)
B	Tighten to 64-77 N·m (47-57 Lb-Ft)
C	Tighten to 54-68 N·m (40-50 Lb-Ft)

7922GG33

Exploded view of the lower control arm assembly

pivot bolt to hold the control arm in place. Do not completely tighten the bolt at this time.

13. Engage the control arm ball joint stud with the clamp bore in the steering knuckle, and install the clamp bolt and nut. Do not tighten it yet.

14. Install the stabilizer bar rear bushing and washer onto the stabilizer bar end with the retaining nut. Tighten the retaining nut to 47–57 ft. lbs. (64–77 Nm). Install a new cotter pin.

15. Tighten the lower arm-to-chassis bolt at the frame bracket to 32–40 ft. lbs. (43–54 Nm).

16. Hold the clamp bolt stationary and tighten the clamp nut to 32–40 ft. lbs. (43–54 Nm).

17. Install the wheel and tire assembly.

18. Check the front wheel alignment.

Wheel Bearings

ADJUSTMENT

Front

The front wheel bearings on this vehicle do not require adjustment, nor is adjustment possible.

Rear

1. Be sure the parking brake is fully released.

2. Raise and safely support the vehicle.

3. Remove the wheel and tire assembly.

4. Remove the grease cap.

5. Rotate the brake drum to be sure there is no brake drag.

6. Remove the cotter pin, wheel bearing nut cover. Discard the cotter pin.

7. To seat the bearings, tighten the wheel bearing nut to 18–22 ft. lbs. (25–29 Nm). Rotate the brake drum by hand while tightening the nut.

8. Loosen the wheel bearing nut until it can be turned by hand.

9. Before the bearing preload can be set, the amount of seal drag must be measured and added to the required preload.

10. To measure the seal drag proceed as follows:

a. Install a lug bolt and rotate the brake drum until the stud is in the 12 o'clock position.

b. Place an inch pound torque wrench onto the bolt to measure the amount of force required to rotate the brake drum.

c. Pull the torque wrench and record the torque reading when rotation begins.

11. Add the oil seal drag value obtained in Step 10 to the specified value of 0.6–1.9 inch lbs. (2.6–8.5 Nm). This is the standard bearing preload.

12. Loosely tighten the bearing nut and rotate the brake drum until the nut and wheel are at the 12 o' clock position.

13. Position an inch pound torque wrench onto the nut and measure the amount of pull required to rotate the drum.

14. Tighten the wheel bearing nut until the torque shown is within the range previously calculated.

15. Turn the wheel bearing nut slowly to adjust it to the standard bearing preload.

16. Install the nut retaining cap and a new cotter pin.

17. Install the grease cap and the wheel and tire assembly.

18. Lower the vehicle.

19. Road test the vehicle and check for proper operation.

REMOVAL & INSTALLATION

Front

1. Remove the front wheel and hub assembly.

➡ **Shield replacement is not a requirement for normal bearing service.**

2. Using Puller tool T87C-1104-A and Hub/Bearing Remover Adapter T92C-1104-AH, or equivalents, separate the hub from the knuckle.

3. Remove the outer bearing retainer washer.

➡ **The outer bearing retainer washer is pre-selected to yield the correct bearing preload. Save the washer for use during assembly.**

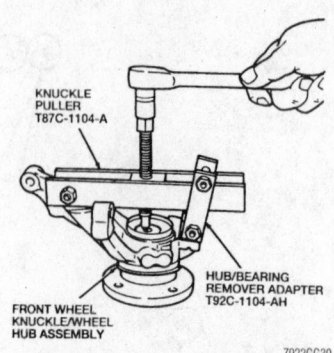

Use a puller to separate the front wheel bearing hub from the steering knuckle

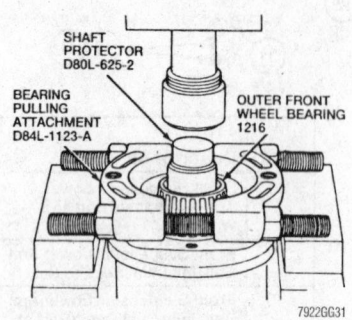

Use a bearing pulling attachment, driver and a press to remove the front outer wheel bearing, as shown

4. Remove the outer bearing from the wheel hub using a bearing pulling attachment, driver and a press.

5. Remove the grease seals from the hub and steering knuckle bore and discard. Remove the inner wheel bearing.

6. Remove the bearing races from the steering knuckle using a suitable puller and slide hammer.

7. Thoroughly clean the hub and knuckle. Inspect the hub and knuckle for wear and/or damage. Replace as necessary.

To install:

8. If the brake rotor shield was removed, install a new one using installation tools T80T-4000-W and T94C-1175-B or equivalents.

9. If the wheel bearings or steering knuckle are being replaced, bearing preload must be checked before assembly as follows:

a. Install the outer bearing races in the steering knuckle using suitable bearing cup installation tools.

b. Lubricate the bearing races and bearing with a thin film of clean grease. Install the bearings in the steering knuckle.

c. Install Spacer Selection tool T87C-1104-B or equivalent, and clamp the bolt head in a vise.

d. Tighten the center bolt in increments, to 36, 72, 108 and 145 ft. lbs. (49, 98, 147 and 196 Nm). After tightening the center bolt to each specified increment, seat the bearings by rotating the steering knuckle.

e. Remove the tool/steering knuckle from the vise. Remount the assembly in the vise, clamping it where the strut mounts.

f. Measure the amount of torque required to rotate the spacer selector tool, using an inch pound torque wrench. The torque wrench reading must be taken just as the tool starts to rotate.

g. If the torque wrench indicates 2.2–10.4 inch lbs. (0.25–1.80 Nm), the outer bearing retainer washer is the correct thickness. If the torque wrench indicates less than 2.2 inch lbs. (0.25 Nm), a thinner outer bearing retainer washer must be installed. If the torque wrench indicates more than 10.4 inch lbs. (1.8 Nm), a thicker outer bearing retainer washer must be installed.

h. Each outer bearing retainer washer has a numerical code that identifies its thickness, which is stamped onto the outer diameter of the washer. The numbers range from 1 to 21, with 1 being the thinnest washer. If the number stamped on the washer is not legible, measure the washer with a micrometer and compare it to the thickness chart to determine the number. Changing the outer bearing retainer washer thickness by 1 number, either higher or lower, will change the bearing preload by 1.7–3.5 inch lbs. (0.2–0.4 Nm).

Stamped mark	Thickness
1	6.285 mm (0.2474 in)
2	6.325 mm (0.2490 in)
3	6.365 mm (0.2506 in)
4	6.405 mm (0.2522 in)
5	6.445 mm (0.2538 in)
6	6.485 mm (0.2554 in)
7	6.525 mm (0.2570 in)
8	6.565 mm (0.2586 in)
9	6.605 mm (0.2602 in)
10	6.645 mm (0.2618 in)
11	6.685 mm (0.2634 in)
12	6.725 mm (0.2650 in)
13	6.765 mm (0.2666 in)
14	6.805 mm (0.2682 in)
15	6.845 mm (0.2698 in)
16	6.885 mm (0.2714 in)
17	6.925 mm (0.2730 in)
18	6.965 mm (0.2746 in)
19	7.005 mm (0.2762 in)
20	7.045 mm (0.2778 in)
21	7.085 mm (0.2794 in)

Front outer wheel bearing retainer washer thickness chart

10. Pack the bearings and the hub area with a suitable high temperature wheel bearing grease. Place the inner wheel bearing into the steering knuckle bore.

11. Lubricate the lip of the new inner grease seal with the bearing grease. Form the lubricant into a strip, concentrated along the edges of the seal lip. Install the bore, using a suitable installation tool.

12. Place the original outer bearing retainer washer, or the outer bearing retainer washer selected from the front wheel bearing adjustment procedure, in the steering knuckle bore. Position the outer wheel bearing in the steering knuckle bore.

13. Lubricate the lip of the new outer grease seal with the bearing grease. Form the lubricant into a strip, concentrated along the edges of the seal lip. Install the outer seal into the bore, using a suitable installation tool.

14. Position the hub in the steering knuckle bore and press it into position using a suitable driver.

15. Install the steering knuckle/hub assembly.

Rear

1. Be sure the parking brake is fully released.

2. Raise and safely support the vehicle.

3. Remove the wheel and tire assembly.

4. Remove the hub grease cap.

5. Remove the cotter pin, nut cover and the nut. Discard the cotter pin.

6. Pull the brake drum bearings and hub assembly away from the spindle shaft. Take care not to damage the spindle shaft threads.

7. Remove the outer wheel bearing assembly and washer.

8. With a small roll head prybar or equivalent, remove the bearing grease seal from the bearing hub. Discard the seal regardless of condition.

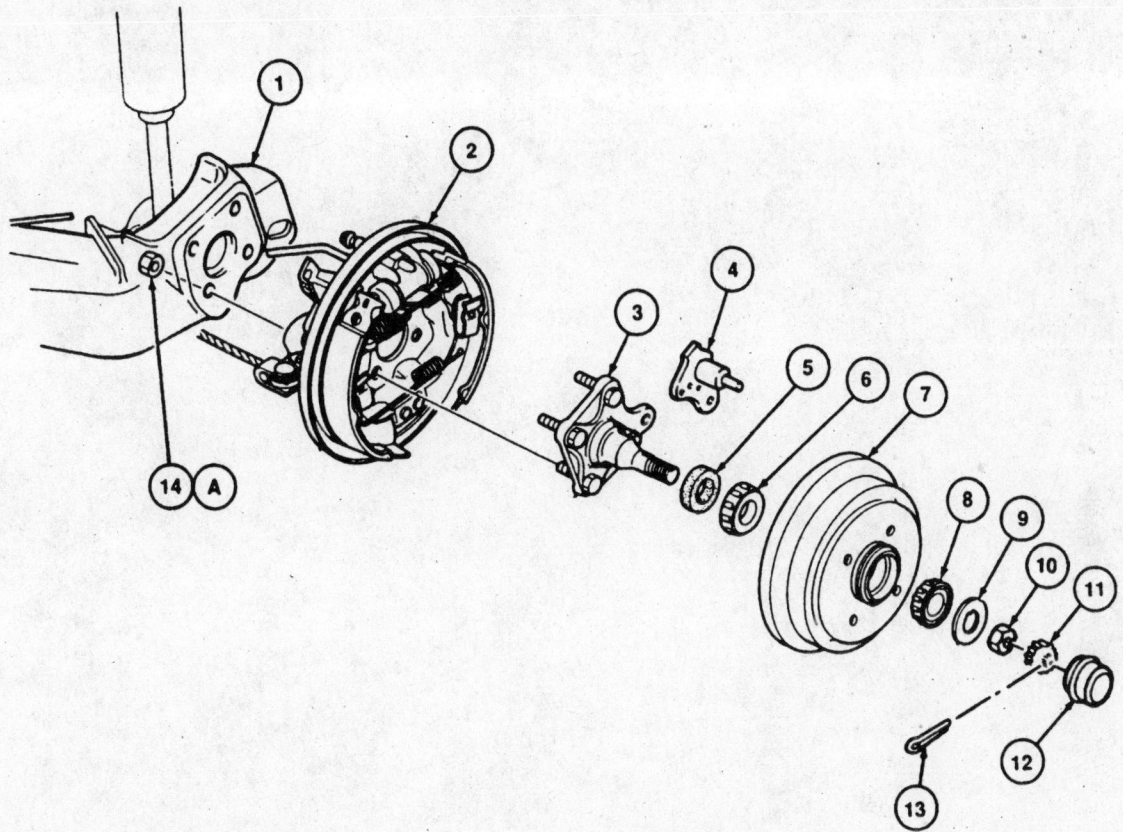

Item	Description
1	Axle Torsion Beam
2	Brake Assembly
3	Rear Wheel Spindle
4	Rear Brake Anti-Lock Sensor (If Equipped)
5	Inner Wheel Bearing Oil Seal
6	Inner Rear Wheel Bearing
7	Brake Drum
8	Outer Rear Wheel Bearing

Item	Description
9	Washer
10	Wheel Bearing Nut
11	Wheel Bearing Nut Cover
12	Hub Grease Cap
13	Cotter Pin
14	Rear Wheel Spindle Nut (4 Req'd)
A	Tighten to 43-61 N·m (31-45 Lb-Ft)

Exploded view of the rear wheel bearing and related components

7922GG40

9. Remove the inner wheel bearing assembly from the bearing hub. If the bearings are to be reused, identify and tag each bearing for installation reference.

10. Thoroughly clean the wheel bearings and hub using suitable solvent and allow to dry. Inspect the bearings and bearing races for scoring, pitting, wear or other damage and replace as necessary.

11. If replacing the bearings, the bearing races must also be replaced.

12. Remove the bearing races from the hub using a brass drift.

To install:

13. If replacing the bearing races, install new races in the hub using suitable installation tools.

14. Pack the bearings and the drum hub area with high temperature wheel bearing grease. Do not fill the entire hub with grease.

15. Position the inner bearing in the hub. Install and seat a new grease seal with a suitable driving tool. Lubricate the lip of the seal with wheel bearing grease.

16. Position the brake drum and hub assembly on the spindle. Keep the hub centered during positioning to prevent damage to the new grease seal and the spindle threads.

17. Install the outer wheel bearing, washer and nut.

18. Adjust the bearing preload.

19. Install the wheel bearing nut cover and a new cotter pin.

20. Install the hub grease cap.

21. Install the wheel and tire assembly.

22. Check and adjust the brakes as required.

23. Check for proper brake operation.

FORD MOTOR CO.

Ford-Probe

18

DRIVE TRAIN 18-22
ENGINE REPAIR 18-2
FUEL SYSTEM 18-20
**STEERING AND
 SUSPENSION** 18-29

A
Air Bag............................18-29
  Disarming.......................18-30
  Precautions.....................18-29

C
Camshaft and Valve Lifters 18-16
  Removal & Installation 18-16
Clutch..................................18-27
  Adjustments 18-27
  Removal & Installation 18-27
Cylinder Head 18-11
  Removal & Installation 18-11

D
Distributor.............................18-2
  Installation.........................18-2
  Removal 18-2

E
Engine Assembly 18-3
  Removal & Installation 18-3
Exhaust Manifold 18-14
  Removal & Installation 18-14

F
Front Crankshaft Seal 18-15
  Removal & Installation 18-15
Fuel Filter 18-21
  Removal & Installation 18-21
Fuel Pump 18-21
  Removal & Installation 18-21
Fuel System Pressure 18-20
  Relieving 18-20
Fuel System Service
  Precautions........................18-20

H
Halfshaft..............................18-29
  Removal & Installation 18-29
Hydraulic Clutch System 18-28
  Bleeding...........................18-28

I
Ignition Timing 18-3
  Adjustment 18-3
Intake Manifold 18-13
  Removal & Installation 18-13

L
Lower Ball Joint........................18-33
  Removal & Installation 18-33
Lower Control Arm 18-33
  Removal & Installation 18-33

O
Oil Pan.................................18-18
  Removal & Installation 18-18
Oil Pump 18-18
  Removal & Installation 18-18

P
Power Rack and Pinion Steering
  Gear.............................18-30
  Removal & Installation 18-30

S
Strut and Spring 18-32
  Removal & Installation 18-32

T
Transaxle Assembly 18-22
  Removal & Installation 18-22

V
Valve Lash 18-18
  Adjustment 18-18

W
Water Pump 18-10
  Removal & Installation 18-10
Wheel Bearings.........................18-33
  Adjustment 18-33
  Removal & Installation 18-33

ENGINE REPAIR

➡ **Disconnecting the negative battery cable on some vehicles may interfere with the functioning of the on-board computer system, and may require the computer to undergo a relearning process once the negative battery cable is reconnected.**

Distributor

REMOVAL

2.0L Engine

1. Disconnect the negative battery cable.
2. Remove the distributor cap and position aside, leaving the spark plug wires connected. Before removing the distributor, mark the position of the distributor cap No. 1 spark plug wire tower on the distributor.
3. Disconnect the distributor electrical harness.
4. Using a wrench on the crankshaft pulley, rotate the crankshaft to position the No. 1 piston at TDC on the compression stroke. The crankshaft pulley notch should align with the timing plate indicator and the distributor rotor should be pointing to the No. 1 spark plug tower position on the distributor cap.
5. Using chalk or paint, mark the position of the distributor housing on the cylinder head.
6. Remove the distributor hold-down bolt and remove the distributor.
7. Inspect the O-ring on the distributor housing and replace it if damaged or worn.

2.5L Engine

1. Disconnect the negative battery cable.
2. Remove the air cleaner intake tube nuts. Loosen the spring clamp at the front of the engine air cleaner assembly and slide it forward. Remove the engine air cleaner intake tube.
3. Loosen the clamp on the front of the Volume Air Flow (VAF) meter and disconnect the air duct. Disconnect the VAF meter electrical connector at the left side of the air cleaner.
4. Disconnect the evaporative emission canister vacuum hose from the routing clip on the front of the air cleaner.
5. Remove the fuel pressure regulator control solenoid from the air cleaner and position aside.

6. Remove the 2 engine air cleaner nuts and bolt and remove the air cleaner assembly.
7. Tag and disconnect the spark plug wires from the distributor cap. Disconnect the 2 electrical connectors from the top of the distributor.
8. Using chalk or paint, mark the position of the distributor housing on the cylinder head. Remove the 2 distributor hold-down bolts and remove the distributor.

INSTALLATION

2.0L Engine

TIMING NOT DISTURBED

1. Using clean engine oil, lubricate the distributor O-ring.
2. Align marks on the distributor shaft and the distributor housing.
3. Install the distributor. Be sure the distributor rotor aligns with the No. 1 spark plug tower position on the distributor cap and the distributor housing mark aligns with the cylinder head or cylinder block mark.

➡ **There are existing marks on the distributor shaft and housing, which when aligned, indicate the No. 1 position.**

4. Install the distributor hold-down bolt.
5. Connect the electrical connectors to their original locations. Install the distributor cap.
6. Connect the negative battery cable.
7. Run the engine and check the ignition timing. Adjust the ignition timing if required.

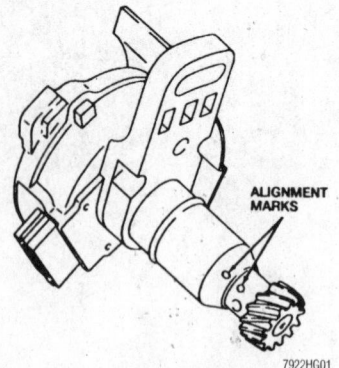

7922HG01

Distributor assembly—2.0L engine

TIMING DISTURBED

1. Using clean engine oil, lubricate the distributor O-ring.
2. Disconnect the spark plug wire from

the No. 1 cylinder spark plug. Remove the spark plug from the No. 1 cylinder and press a thumb over the spark plug hole.
3. Using a wrench on the crankshaft pulley, rotate the crankshaft until pressure is felt at the spark plug hole, indicating the piston is approaching TDC on the compression stroke. Continue rotating the crankshaft until the crankshaft pulley mark aligns with the timing cover indicator.
4. Position the distributor rotor so it aligns with the No. 1 spark plug wire tower on the distributor cap.
5. Install the distributor. Align the mark that was made on the distributor housing with the mark that was made on the cylinder block. Loosely tighten the distributor hold-down bolts.
6. Connect the electrical connectors to their original locations. Install the distributor cap.
7. Install the spark plug in the No. 1 cylinder and connect the spark plug wire.
8. Connect the negative battery cable.
9. Run the engine and check the ignition timing. Adjust the ignition timing if required.

2.5L Engine

1. Align the distributor shaft with the camshaft end and install the distributor.

➡ **The tangs on the distributor shaft are different sizes, allowing the distributor to be installed in only one position.**

2. Install the distributor hold-down bolts. Align the mark that was made on the distributor housing with the mark that was made on the cylinder head and loosely tighten the bolts.
3. Connect the electrical connectors to the distributor and the spark plug wires to the distributor cap.
4. Install the engine air cleaner assembly and tighten the 2 nuts and bolt to 14-18 ft. lbs. (19–25 Nm).
5. Install the fuel pressure regulator solenoid and connect the evaporative emission canister vacuum hose to the routing clip.
6. Connect the VAF meter electrical connector. Connect the air duct and tighten the clamp.
7. Align the engine air cleaner intake tube and install it to the engine air cleaner assembly. Loosen the spring clamp and slide it into position. Install the engine air cleaner intake tube nuts and tighten to 71–88 inch lbs. (8–10 Nm).
8. Connect the negative battery cable.

9. Run the engine and check the ignition timing. Adjust the ignition timing if required.

Ignition Timing

ADJUSTMENT

2.0L Engine

1. Apply the parking brake.
2. Be sure that the transmission is in **NEUTRAL** if equipped with manual transaxle or in **PARK** if equipped with an automatic transaxle.
3. Locate the timing marks on the crankshaft pulley and the timing indicator scale on the engine front cover. If the marks are hard to see, clean them with degreaser and a stiff brush.
4. Start the engine and bring to normal operating temperature.
5. Be sure all accessories are OFF.
6. Connect a tachometer and timing light to the engine according to the manufacturer's instructions.
7. Remove the shorting bar from the SPOUT connector.
8. Verify that the idle speed is 700 plus or minus 50 rpm; adjust if necessary.
9. Aim the timing light at the timing marks. The mark on the crankshaft pulley should line up with the 10 degrees BTDC plus or minus 1 degree timing mark on the timing indicator on the front engine cover.
10. If the timing marks are not aligned (within 1 degree), loosen the distributor hold-down bolts (2) and turn the distributor housing to adjust. When the marks align, tighten the hold-down bolt to 14–19 ft. lbs. (19–25 Nm). Recheck the timing after the bolt has been tightened.
11. Install the shorting bar to the SPOUT connector.
12. Remove all test equipment.

2.5L Engine

1. Apply the parking brake.
2. If equipped with a manual transaxle, place the shift lever in **NEUTRAL**. If equipped with an automatic transaxle, place the shift lever in **PARK**.
3. Locate the timing marks on the crankshaft pulley and timing belt cover. If the marks are hard to see, clean them off with some degreasing cleaner and a wire brush.
4. Start the engine and allow it to come to normal operating temperature.
5. Be sure all accessories are **OFF**.
6. Connect a tachometer and a timing light according to the manufacturer's instructions.
7. Connect terminals **STI (TEN)** and **GND** on the Data Link Connector (DLC) with a jumper wire.
8. Check the idle speed and verify that it is 650 plus or minus 50 rpm. Adjust the idle speed if necessary.
9. Aim the timing light at the timing marks. The mark on the crankshaft pulley should line up with the 10 degrees BTDC plus or minus 1 degree timing mark on the timing indicator on the front engine cover.

➡ Do not pinch the ignition coil-to-distributor high tension wiring when turning the distributor.

10. If the marks are not aligned, loosen the distributor hold-down bolts (2) just enough to turn the distributor housing. While aiming the timing light at the timing marks, turn the distributor until the marks are aligned. Tighten the distributor hold-down bolts to 14–19 ft. lbs. (19–25 Nm) and recheck the timing.
11. Stop the engine.
12. Remove the jumper wire and test equipment.

Engine Assembly

REMOVAL & INSTALLATION

2.0L Engine

WITH AUTOMATIC TRANSAXLE

➡The engine is lifted from the engine compartment, leaving the transaxle in the vehicle.

✴✴ CAUTION

The fuel injection system remains under pressure even after the engine has been turned OFF. The fuel system pressure must be relieved before disconnecting any fuel lines. Failure to do so may result in fire and/or personal injury.

1. Relieve the fuel system pressure using the recommended procedure.
2. Disconnect the battery cables, negative cable first. Remove the battery and battery tray.
3. Mark the position of the hood on its hinges and carefully remove the hood.
4. Drain the cooling system and the engine oil into suitable containers. Reinstall the engine oil pan plug.
5. Remove the air intake system.
6. If equipped, remove the A/C compressor and position aside, leaving the refrigerant lines attached. Support the compressor with suitable wire.
7. Label, disconnect and plug the fuel lines at the fuel rail.
8. Label and disconnect the electrical connectors from the distributor, engine coolant temperature sensor, cooling fan temperature sensor (if equipped), water temperature indicator sender unit, throttle position sensor, idle air control valve, idle switch (if equipped), fuel injector harness, EGR solenoid vacuum valve and alternator.
9. Remove the sensor harness tie wrap and retainer clip from the cylinder head cover bracket.
10. Remove the power steering pump and position aside, leaving the hoses connected.
11. Loosen the alternator adjusting bolt and remove the upper lockbolt.
12. Remove the upper and lower radiator hoses. If equipped, disconnect the cruise control vacuum hose from the back right-hand side of the intake manifold.
13. Disconnect the vacuum hose connecting the Evaporative Emission (EVAP) canister to the metal vacuum tube. If

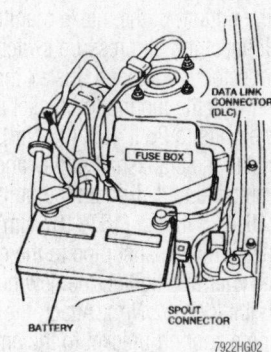

Location of the SPOUT connector-2.0L engine

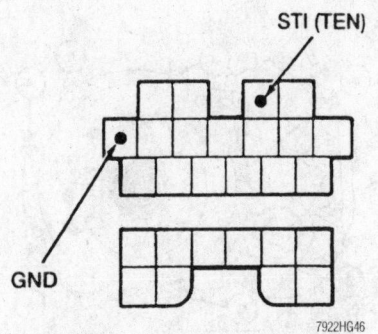

Data Link Connector terminal locations-2.5L engine

equipped, disconnect the EGR temperature sensor connector.

14. Disconnect the accelerator cable.

15. Disconnect the brake power booster vacuum line from the back left-hand side of the intake manifold.

16. Disconnect the heater hoses at the bulkhead.

17. Raise and safely support the vehicle and remove the splash shields..

18. Remove the intake manifold support bracket.

19. Remove the starter motor. Remove the torque converter access plug.

20. Remove the halfshaft support bearing bracket bolts.

21. Remove the 4 torque converter-to-flexplate nuts.

22. Remove the 3 engine-to-transaxle bolts and the 2 transaxle-to-engine mounting bolts.

23. Remove and discard the exhaust inlet pipe-to-catalytic converter nuts.

24. Remove the exhaust pipe bracket bolts. Remove and discard the exhaust inlet pipe-to-exhaust manifold nuts and remove the exhaust pipe. Support the remaining exhaust system with mechanics wire.

25. Use crankshaft pulley holder tool T92C-6316-AH or equivalent, to hold the

crankshaft pulley and remove the pulley bolt. Remove the crankshaft pulley.

26. Lower the vehicle.

27. Attach an engine sling to the engine lifting eyes and a suitable hoist. Raise the engine slightly and remove the right-hand engine support insulator.

28. Remove the remaining transaxle-to-engine mounting bolts and remove the engine from the vehicle.

29. Remove the flexplate from the crankshaft and mount the engine on a workstand, if required.

To install:

30. Remove the engine from the workstand. Remove the old sealant from the flexplate mounting bolts and bolt holes.

31. If reusing the flexplate bolts, apply silicone sealant to the bolt threads. Install the flexplate and loosely install the bolts.

➡ **New flexplate mounting bolts come with sealant already on them.**

32. Tighten the 6 flexplate bolts in 2–3 steps to 70–75 ft. lbs. (96–103 Nm) in a crisscross pattern.

33. Carefully lower the engine into the vehicle and align it to the transaxle. Be sure that the torque converter studs are aligned with the holes in the flexplate.

34. Install the upper 4 transaxle-to-engine bolts and tighten mounting bolts **A** to 50–73 ft. lbs. (68–99 Nm) for vehicles equipped with the 4EAT transaxle and 66–86 ft. lbs. (90–116 Nm) for vehicles equipped with the CD4E transaxle.

35. Raise the engine slightly and install the right-hand engine support insulator. Tighten the support insulator through-bolt to 63–86 ft. lbs. (86–116 Nm) and support insulator nuts to 54–75 ft. lbs. (74–103 Nm).

36. Remove the engine lifting equipment.

37. Raise and safely support the vehicle.

38. Install the 4 torque converter-to-flexplate nuts and tighten to 32–45 ft. lbs. (44–60 Nm). Rotate the flexplate, as necessary, to gain access to all of the nuts. Install the torque converter access plug.

39. Install the remaining transaxle-to-engine mounting bolts. Tighten mounting bolts **B** to 50–73 ft. lbs. (68–99 Nm) for vehicles equipped with the 4EAT transaxle and 28–38 ft. lbs. (38–51 Nm) for vehicles equipped with the CD4E transaxle. Tighten mounting bolt **C** to 28–38 ft. lbs. (38–51 Nm) for vehicles equipped with the 4EAT transaxle and 14–18 ft. lbs. (19–25 Nm) for vehicles equipped with the CD4E transaxle. On vehicles equipped with the 4EAT transaxle, install and tighten mounting bolt **D** to 14–18 ft. lbs. (19–25 Nm), **E** to 28–38 ft. lbs. (38–51 Nm) and **F** to 50–73 ft. lbs. (68–99 Nm).

40. Install the alternator and loosely install the through-bolt. Connect the alternator wiring and install the harness bracket to the back of the alternator.

41. Install the starter motor and tighten the bolts to 23–34 ft. lbs. (31–46 Nm). Install the intake manifold support bracket and tighten the bolts to 27–38 ft. lbs. (37–52 Nm).

42. Install the halfshaft bracket bearing bolts and tighten, in sequence, to 32–45 ft. lbs. (43–61 Nm).

43. Connect the fuel injector wiring harness to the bottom of the intake manifold.

44. Connect the oil pressure switch electrical connector. If equipped, install the A/C compressor on the mounting bracket and tighten the bolts to 26 ft. lbs. (35 Nm).

45. Install the crankshaft pulley and hold it with a suitable tool. Tighten the pulley bolt to 116–123 ft. lbs. (157–167 Nm).

46. Install the exhaust pipe to the catalytic converter and tighten the new nuts to 27–38 ft. lbs. (37–52 Nm). Attach the exhaust pipe support bracket to the engine and tighten the bolts to 27–38 ft. lbs. (37–52 Nm).

47. Install the new exhaust inlet pipe-to-

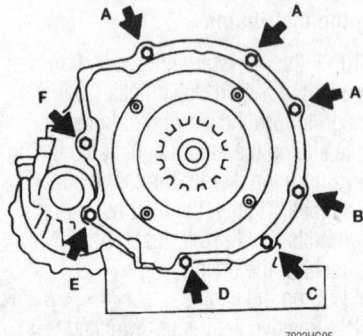

Hold the crankshaft using tool T92C-6316-AH or equivalent, to remove the pulley bolt—2.0L engine

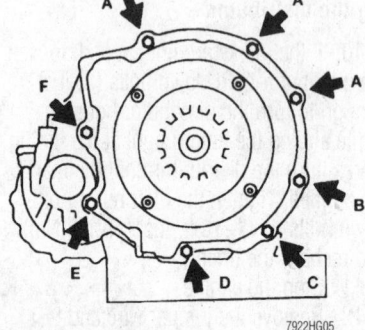

Transaxle and engine mounting bolt identification, 4EAT transaxle—2.0L engine

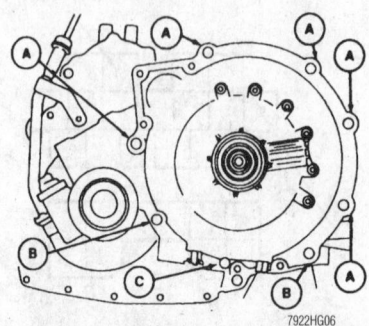

Flexplate bolt torque sequence—2.0L engine

Transaxle and engine mounting bolt identification, CD4E transaxle—2.0L engine

exhaust manifold nuts and tighten to 27–38 ft. lbs. (37–52 Nm). Install the exhaust hanger insulators and remove the mechanics wire.

48. Connect the HO2S electrical connector.

49. Lower the vehicle.

50. Loosely attach the alternator to the alternator adjuster arm. Install the alternator belt and adjust the tension. Tighten the alternator upper mounting bolt to 14–18 ft. lbs. (19–25 Nm).

51. Raise and safely support the vehicle.

52. Tighten the alternator through-bolt to 27–38 ft. lbs. (37–52 Nm).

53. Install the splash shields.

54. Lower the vehicle.

55. Install the power steering pump support and loosely install the power steering pump through-bolt and lockbolt. Connect the PSP switch electrical connector and install the power steering pump drive belt.

56. Adjust the power steering drive belt tension, then tighten the through-bolt to 27–38 ft. lbs. (37–52 Nm) and the lockbolt to 14–18 ft. lbs. (19–25 Nm).

57. Install the power steering pump belt shield and tighten the bolts to 71–88 inch lbs. (8–10 Nm). Install the power steering hose brackets to the cylinder head cover and tighten the bolts to 71–88 inch lbs. (8–10 Nm).

58. Connect the heater hoses. If equipped, connect the cruise control vacuum line to the back right-hand side of the intake manifold.

59. Connect the vacuum line connecting the EVAP canister to the metal vacuum tube.

60. Connect the power brake booster vacuum line to the back left-hand side of the intake manifold.

61. Unplug and connect the fuel lines to the fuel rail and all remaining electrical connectors.

62. Connect the sensor harness retainer clip to the cylinder head cover and the tie wrap to the water bypass hose.

63. Install the accelerator cable and the upper and lower radiator hoses.

64. Install the air intake system.

65. Carefully install the hood, aligning the marks that were made during removal.

66. Install the battery tray and battery.

67. Connect the battery cables, negative cable first.

68. Fill the engine with the proper type and quantity of oil.

69. Fill and bleed the cooling system.

70. Run the engine and bring to normal operating temperature. Check for leaks and proper engine operation.

WITH MANUAL TRANSAXLE

➡The engine and transaxle are lifted from the engine compartment as an assembly.

✳✳ CAUTION

The fuel injection system remains under pressure even after the engine has been turned OFF. The fuel system pressure must be relieved before disconnecting any fuel lines. Failure to do so may result in fire and/or personal injury.

1. Relieve the fuel system pressure using the recommended procedure.

2. Disconnect the battery cables, negative cable first. Remove the battery and battery tray.

3. Mark the position of the hood on its hinges and carefully remove the hood.

4. Drain the cooling system and the engine oil into suitable containers. Reinstall the engine oil pan plug.

5. Remove the air intake system.

6. Remove the upper and lower radiator hoses and remove the radiator.

7. If equipped, remove the A/C compressor and position aside, leaving the refrigerant lines attached. Support the compressor with suitable wire.

8. Label, disconnect and plug the fuel lines at the fuel rail and set the fuel lines aside.

9. Label and disconnect the electrical connectors from the distributor, coil, engine coolant temperature sensor, coolant temperature gauge sensor, throttle position sensor, idle air control valve, fuel injectors, EGR solenoid vacuum valve, EGR temperature sensor and the alternator.

10. Remove the power steering pump and position aside, leaving the hoses connected.

11. Loosen the alternator adjusting bolt, remove the upper lockbolt.

12. If equipped, disconnect the cruise control vacuum hose from the back right-hand side of the intake manifold.

13. Disconnect the vacuum line connecting the Evaporative Emission (EVAP) canister to the metal vacuum tube. If equipped, disconnect the EGR temperature sensor connector.

14. Disconnect the accelerator cable. Disconnect the power booster vacuum hose from the back left-hand side of the intake manifold.

15. Disconnect the heater hoses at the bulkhead and remove the upper starter

motor mounting bolts. If equipped, disconnect the speed control electrical connector, remove the 2 speed control servo mounting nuts and position the speed control servo aside.

16. Remove the ignition coil.

17. Remove the 2 fuel filter bracket bolts and position the filter and bracket aside.

18. Remove the ignition control module.

19. Remove the ground wire bracket from between the transaxle and rear transaxle support insulator (mount).

20. Remove the rear transaxle support insulator through-bolt and remove the transaxle ground from the top rear of the transaxle.

21. Label and disconnect the Brake On/Off (BOO) switch and Vehicle Speed Sensor (VSS) electrical connectors from the rear of the transaxle.

22. Disconnect and plug the lower clutch slave cylinder tube fitting at the slave cylinder. Pull the spring clips from the slave cylinder line mounting brackets, then remove the hydraulic clutch hose from the lower clutch slave cylinder tube.

23. Label and disconnect the park/neutral position switch from the front of the transaxle.

24. Raise and safely support the vehicle.

25. Remove the splash shields and the front wheels.

26. Remove the 6 crossmember bolts and the crossmember.

27. Remove the 2 lower transaxle support insulator bolts and the lower transaxle support insulator.

28. Remove the 6 rear engine support nuts and 2 bolts and remove the rear engine support.

29. Remove the halfshafts.

30. Install transaxle plug tools T88C-7025-AH or equivalent, into the differential side gears.

➡If the plugs are not installed, the differential side gears may become malpositioned. If the gears are malpositioned, the differential may have to be removed to reposition them.

31. Remove the intake manifold support bolts and the support. Remove the 3 rear transaxle support bracket bolts and remove the rear transaxle support bracket.

32. Remove the starter.

33. Label and disconnect the oil pressure switch and Heated Oxygen Sensor (HO2S) electrical connectors.

34. Remove and discard the exhaust inlet pipe-to-catalytic converter nuts. Remove the exhaust support bolts. Remove

and discard the exhaust inlet pipe-to-exhaust manifold nuts and remove the exhaust pipe. Support the exhaust system with mechanics wire.

35. Remove the control rod-to-support bar stud nut, then disengage the bar from the transaxle. Remove the transaxle shift rod adjustment sleeve through-bolt and nut, then disengage the linkage from the transaxle.

36. Remove the alternator drive belt if not already done. Remove the wiring harness bracket from the rear of the alternator and remove the alternator through-bolt. Label and disconnect the remaining alternator wiring and remove the alternator.

37. Hold the crankshaft pulley with crankshaft pulley holder T92C-6316-AH or equivalent tool and remove the pulley bolt. Remove the crankshaft pulley.

38. If equipped with A/C, remove the A/C compressor bolts and position the compressor aside by supporting the compressor with mechanics wire. Do not remove the A/C lines.

39. Lower the vehicle.

40. Attach an engine sling to the engine lifting eyes and a suitable hoist to the sling. Raise the engine slightly and remove the right-hand engine support insulator.

41. Remove the left-hand engine support insulator nuts and bolt and through-bolt and remove the left-hand engine support insulator.

42. Carefully raise and remove the engine/transaxle assembly from the vehicle.

43. Remove the transaxle-to-engine bolts and the engine-to-transaxle bolts. Separate the transaxle from the engine.

44. Remove the clutch assembly, flywheel and crankshaft rear cover plate. Mount the engine on a workstand.

To install:

45. Remove the engine from the workstand using an engine sling and hoist. Install the crankshaft rear cover plate and

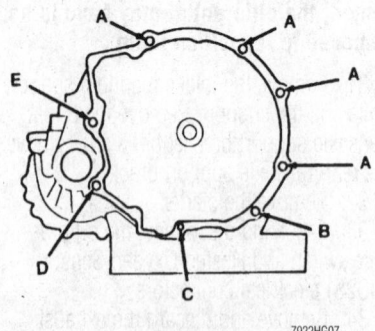

7922HG07

Transaxle and engine mounting bolt identification, MTX transaxle—2.0L engine

tighten the bolt to 71–88 inch lbs. (8–10 Nm).

46. Install the flywheel and clutch assembly.

47. Install the transaxle on the engine. Install the transaxle-to-engine bolts. Tighten bolts **A** to 66–86 ft. lbs. (90–116 Nm), **B** to 28–38 ft. lbs. (38–51 Nm) and **C** to 14–18 ft. lbs. (19–25 Nm). Install the engine-to-transaxle bolts. Tighten bolt **D** to 28–38 ft. lbs. (38–51 Nm) and bolt **E** to 66–86 ft. lbs. (90–116 Nm).

48. Carefully lower the engine/transaxle assembly into the engine compartment.

49. Install the left-hand transaxle support insulator and tighten the 2 nuts and 1 bolt to 50–68 ft. lbs. (67–93 Nm). Tighten the left-hand transaxle support insulator through-bolt to 63–86 ft. lbs. (86–116 Nm).

50. Raise the engine slightly and install the right-hand engine support insulator. Tighten the right-hand engine support insulator through-bolt to 63–86 ft. lbs. (86–116 Nm) and the right-hand engine support insulator nuts to 54–75 ft. lbs. (74–103 Nm). Remove the engine lifting equipment.

51. Raise and safely support the vehicle.

52. Install the alternator and loosely install the alternator through-bolt. Install the alternator drive belt. Connect the alternator wiring and install the wiring harness bracket to the rear of the alternator.

53. Connect the control rod-to-support bar. Tighten the nut to 28–38 ft. lbs. (38–51 Nm).

54. Connect the shift rod adjustment sleeve to the transaxle with the through-bolt and nut. Tighten the through-bolt to 14–18 ft. lbs. (19–25 Nm).

55. Install the exhaust pipe to the catalytic converter and tighten the new nuts to 27–38 ft. lbs. (37–52 Nm). Attach the exhaust pipe support bracket to the engine and tighten the bolts to 27–38 ft. lbs. (37–52 Nm).

56. Install the new exhaust pipe-to-exhaust manifold nuts and tighten to 27–38 ft. lbs. (37–52 Nm).

57. Connect the HO2S and oil pressure sensor electrical connectors.

58. Install the starter motor and tighten the bolts to 23–34 ft. lbs. (31–46 Nm).

59. Install the rear transaxle support insulator and tighten the 3 bolts to 50–68 ft. lbs. (67–93 Nm).

60. Install the intake manifold support bracket and tighten the bolts to 27–38 ft. lbs. (38–52 Nm).

61. Remove the plugs from the differential side gears and install the halfshafts.

62. Install the lower transaxle support

insulator and tighten the bolts to 41–59 ft. lbs. (55–80 Nm).

63. Install the rear engine support. Tighten bolts and nuts **B** to 50–68 ft. lbs. (67–93 Nm), nuts **A** to 55–77 ft. lbs. (75–104 Nm) and nuts **C** to 32–44 ft. lbs. (44–60 Nm).

64. Install the crossmember and tighten the 6 bolts to 68–96 ft. lbs. (94–131 Nm).

65. Install the crankshaft pulley. Hold the crankshaft pulley with a suitable tool. Install the pulley bolt and tighten to 116–123 ft. lbs. (157–167 Nm).

66. If equipped, install the A/C compressor and tighten the bolts to 18–26 ft. lbs. (24–35 Nm).

67. Lower the vehicle.

68. Loosely attach the alternator to the alternator adjuster arm. Adjust the alternator drive belt tension. Tighten the alternator upper mounting bolt to 14–18 ft. lbs. (19–25 Nm).

69. Raise and safely support the vehicle.

70. Tighten the alternator through-bolt to 27–38 ft. lbs. (38–52 Nm).

71. Install the splash shields and the wheels. Tighten the lug nuts to 65–86 ft. lbs. (88–118 Nm).

72. Lower the vehicle.

73. Connect the park/neutral position switch electrical connector.

74. Remove the plug and install the hydraulic clutch hose to the clutch slave cylinder metal tube. Install the clips to the clutch cylinder hose brackets. Install the hydraulic line fitting on the slave cylinder.

75. Connect the VSS and BOO switch electrical connectors at the rear of the transaxle.

76. Install the ground wire bracket between the transaxle and the rear transaxle support insulator. Install the rear transaxle support insulator through-bolt and tighten to 50–68 ft. lbs. (67–93 Nm). Install the transaxle ground at the top rear of the transaxle.

77. Install the ignition control module.

78. Install the fuel filter and bracket and tighten the bolts to 71–97 inch lbs. (8–11 Nm).

79. Install the ignition coil.

80. If equipped, install the speed control servo and tighten the nuts. Connect the speed control servo electrical connector.

81. Install the upper starter motor mounting bolts and tighten to 23–34 ft. lbs. (31–46 Nm).

82. Connect the heater hoses and connect the power brake booster vacuum line to the back left-hand side of the intake manifold.

83. Connect the accelerator cable. Con-

nect the EGR temperature sensor, if equipped.

84. Connect the vacuum line between the EVAP canister and the metal vacuum tube. If equipped, connect the cruise control vacuum line to the back right-hand side of the intake manifold.

85. Loosely install the power steering pump through-bolt and lockbolt. Connect the PSP switch electrical connector and install the power steering drive belt.

86. Adjust the power steering drive belt tension, then tighten the through-bolt to 32–45 ft. lbs. (43–61 Nm) and the lockbolt to 23–34 ft. lbs. (31–46 Nm).

87. Install the power steering pump drive belt pulley shield and tighten the bolts to 61–86 inch lbs. (7–9 Nm). Install the power steering hose brackets to the cylinder head cover and tighten the bolts to 71–88 inch lbs. (8–10 Nm).

88. Connect all remaining electrical connectors.

89. Unplug and connect the fuel lines.

90. Install the radiator and the upper and lower radiator hoses.

91. Install the air intake system.

92. Install the battery tray and battery.

93. Install the hood, aligning the marks that were made during removal.

94. Connect the battery cables, negative cable last.

95. Fill the engine with the proper type and quantity of oil.

96. Fill and bleed the cooling system.

97. Bleed the clutch hydraulic system.

98. Run the engine and bring to normal operating temperature. Check for leaks and proper engine operation.

2.5L Engine

WITH AUTOMATIC TRANSAXLE

➡The engine and transaxle are lifted from the engine compartment as an assembly.

❊❊ CAUTION

The fuel injection system remains under pressure even after the engine has been turned OFF. The fuel system pressure must be relieved before disconnecting any fuel lines. Failure to do so may result in fire and/or personal injury.

1. Relieve the fuel system pressure using the recommended procedure.

2. Disconnect the battery cables, negative cable first. Remove the battery and battery tray.

3. Mark the position of the hood on its hinges and carefully remove the hood.

4. Drain the cooling system and the engine oil into suitable containers. Reinstall the engine oil pan plug.

5. Remove the engine air intake system.

6. Loosen the A/C and alternator belt tensioner locknut and adjuster bolt and remove the drive belt.

7. Raise and safely support the vehicle.

8. Remove the front wheels and the splash shields.

9. Remove the 6 crossmember bolts and remove the crossmember.

10. Disconnect the front and rear Heated Oxygen Sensor (HO2S) electrical connectors. Remove the exhaust inlet pipe-to-exhaust manifold nuts.

11. Disconnect the oil pressure switch electrical connector, located near the oil filter.

12. Remove the halfshafts.

13. Loosen the power steering and water pump drive belt tensioner locknut and adjuster bolt and remove the drive belt.

14. Remove the 3 power steering pump mounting bolts through the holes in the pump pulley. Remove the power steering hose bracket-to-power steering pump bolt and the pump rear bracket bolt. Secure the pump aside with mechanics wire, leaving the hoses connected.

15. If equipped, remove the 4 A/C compressor mounting bolts and secure the compressor aside with mechanics wire, leaving the refrigerant lines attached. Do not let the compressor hang by the refrigerant lines.

16. Lower the vehicle.

17. Remove the upper and lower radiator hoses and overflow hose.

18. Disconnect the cooling fan electrical connectors. Disconnect and plug the transaxle cooler lines.

19. Remove the 2 radiator hold-down bolts and remove the radiator and cooling fan assembly.

➡Use care when lifting the radiator/cooling fan assembly not to damage the radiator and condenser cooling fins.

20. Label and disconnect the wiring from the alternator and distributor. Remove the 2 A/C and alternator wiring harness retaining bolts, then disconnect the harness from the engine block.

21. Label and disconnect the electrical connectors from the fuel injector harness, vehicle speed sensor, starter motor, throttle position sensor, knock sensor, EGR, idle air

control valve, EGR valve position sensor, neutral safety switch, engine coolant temperature sensor, cooling fan engine coolant temperature sensor, temperature gauge sending unit and crank position sensor, as equipped.

22. Label and disconnect the vacuum hoses from the speed control servo, EGR, throttle body, power brake booster, climate control assembly and fuel pressure regulator.

23. Remove the 2 heater hoses from the thermostat housing. Remove the wiring harness grounds.

24. If equipped, disconnect the speed control servo electrical connector. Remove the 2 nuts from the servo bracket and position the speed control servo and bracket aside.

25. Disconnect the fuel supply and return lines and discard the copper crush washer. Remove the 2 fuel line retaining bolts from the fuel line bracket.

26. Disconnect the accelerator cable from the throttle body. Remove the 2 nuts from the fuel filter bracket and position the filter aside, without disconnecting the fuel lines.

27. Remove the spring clip from the shift cable bracket and pull the cable from the switch. Remove the 2 bolts from the cooling fan relay bracket and position the bracket aside.

28. Raise and safely support the vehicle.

29. Remove the front and rear transaxle support insulator through-bolts.

30. Lower the vehicle.

31. Attach suitable lifting equipment to the engine lifting eyes and remove any slack using an engine hoist attached to the lifting cables.

32. Remove the left-hand transaxle support insulator through-bolt and the right-hand transaxle support insulator through-bolt and 2 nuts. Remove the right-hand engine support insulator from the vehicle.

33. Carefully lift the engine/transaxle assembly from the vehicle.

34. Separate the transaxle from the engine.

35. Remove the flexplate and mount the engine on a suitable engine stand.

To install:

36. Attach suitable lifting cables to the engine lifting eyes and remove any slack on the cables.

37. Remove the engine from the engine stand and install the flexplate. Tighten the flexplate bolts, in 2-3 steps, in sequence to 45-49 ft. lbs. (61-67 Nm).

38. Position the engine to the transaxle and install the retaining bolts. Tighten the engine-to-transaxle bolts and the transaxle-to-engine bolts to 50-73 ft. lbs. (68-99 Nm).

39. Carefully lower the engine/transaxle assembly into position in the engine compartment. Install the right-hand engine support insulator. Tighten the through-bolt to 50-68 ft. lbs. (67-93 Nm) and the 3 nuts to 54-76 ft. lbs. (74-103 Nm).

40. Install the left-hand transaxle support insulator through-bolt and tighten to 63-86 ft. lbs. (86-116 Nm). Remove the engine lifting equipment.

41. Raise and safely support the vehicle.

42. Install the front and rear transaxle support insulator through-bolts and tighten to 63-86 ft. lbs. (86-116 Nm).

43. Lower the vehicle.

44. Align the cooling fan relay bracket and install the 2 bolts. Tighten to 88 inch lbs. (10 Nm).

45. Install the shift cable and retain with the spring clip. Align the fuel filter and install the 2 nuts. Tighten to 71-88 inch lbs. (8-10 Nm).

46. If equipped, connect the vacuum line to the climate control assembly. Connect the vacuum line to the power brake booster.

Engine-To-Transaxle Bolts

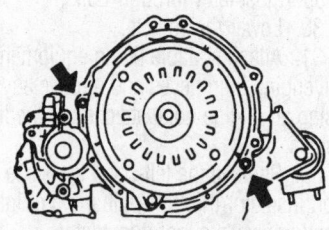

Transaxle-To-Engine Bolts

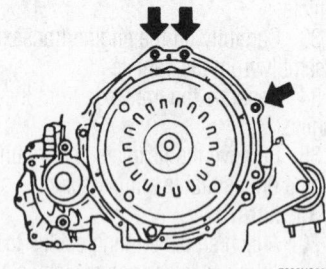

7922HG08

Engine and transaxle retaining bolt locations—2.5L engine

47. Connect the accelerator cable and vacuum lines to the throttle body. Connect the vacuum line to the fuel pressure regulator.

48. Align the fuel line bracket and install the 2 bolts. Tighten to 71-88 inch lbs. (8-10 Nm). Connect the fuel supply and return lines, using new copper crush washers. Tighten the supply line bolt to 18-25 ft. lbs. (25-34 Nm).

49. If equipped, align the speed control servo and install the nuts. Connect the speed control servo electrical connector.

50. Install the wiring harness grounds and connect the heater hoses to the thermostat housing.

51. Connect the electrical connectors for the crank position sensor, engine coolant temperature sensor, cooling fan engine coolant temperature sensor, temperature gauge sending unit, neutral safety switch, EGR valve position sensor, idle air control valve, EGR, knock sensor, throttle position sensor, starter, vehicle speed sensor and fuel injector harness, as equipped.

52. Connect the vacuum hose to the EGR and if equipped, the speed control servo.

53. Align the A/C and alternator wiring harness and install the 2 bolts. Connect the 2 electrical connectors to the top of the distributor.

54. Carefully install the radiator and cooling fan assembly and connect the cooling fan electrical connectors. Install the 2 radiator hold-down bolts and tighten to 71-88 inch lbs. (8-10 Nm).

55. Unplug and connect the transaxle oil cooler lines and install the upper and lower radiator hoses.

56. Raise and safely support the vehicle.

57. If equipped, install the A/C compressor and tighten the bolts to 28-38 ft. lbs. (38-51 Nm).

58. Position the power steering pump and install the rear bracket bolt. Tighten to 24-34 ft. lbs. (32-46 Nm). Install the power steering hose bracket bolt and tighten to 24-34 ft. lbs. (31-46 Nm).

59. Install the 3 power steering pump bolts through the pulley and tighten to 23-34 ft. lbs. (31-46 Nm). Install the power steering pump drive belt and adjust the tension.

60. Install the halfshafts.

61. Connect the oil pressure switch electrical connector.

62. Install the exhaust inlet pipe to the manifolds and tighten the nuts to 30-41 ft. lbs. (40-55 Nm). Connect the front and rear HO2S electrical connectors.

63. Position the crossmember and install the 6 bolts. Tighten to 69-93 ft. lbs. (94-126 Nm).

64. Install the splash shields and the wheels. Tighten the lug nuts to 65-87 ft. lbs. (88-118 Nm).

65. Lower the vehicle.

66. Install the alternator drive belt and adjust the tension. Be sure all electrical connectors and vacuum hose are connected.

67. Install the air intake system.

68. Carefully install the hood, aligning the marks that were made during removal.

69. Install the battery tray and battery. Connect the battery cables, negative cable last.

70. Fill the engine with the proper type and quantity of oil.

71. Add transmission fluid to the transaxle if needed.

72. Fill and bleed the cooling system.

73. Run the engine and bring to normal operating temperature. Check for leaks and proper engine operation.

74. Top off all fluids.

WITH MANUAL TRANSAXLE

➡ The engine and transaxle are lifted from the engine compartment as an assembly.

❊❊ CAUTION

The fuel injection system remains under pressure even after the engine has been turned OFF. The fuel system pressure must be relieved before disconnecting any fuel lines. Failure to do so may result in fire and/or personal injury.

1. Relieve the fuel system pressure using the recommended procedure.

2. Disconnect the battery cables, negative cable first. Remove the battery and battery tray.

3. Mark the position of the hood on its hinges and carefully remove the hood.

4. Drain the cooling system and the engine oil into suitable containers. Reinstall the engine oil pan plug.

5. Remove the air intake system.

6. Raise and safely support the vehicle.

7. Remove the front wheels and the splash shields.

8. Remove the 6 crossmember bolts and remove the crossmember.

9. Remove the 2 bolts and 6 nuts from the rear engine support and remove the rear engine support.

10. Disconnect the front and rear Heated Oxygen Sensor (HO2S) electrical connectors. Remove the exhaust inlet pipe-to-exhaust manifold nuts.

11. Remove the control rod-to-support bar stud nut, then disengage the bar from the transaxle. Remove the transaxle shift rod adjustment sleeve through-bolt and nut, then disengage the linkage from the transaxle.

12. Disconnect the A/C and oil pressure switch electrical connectors. Disconnect and plug the hydraulic line at the slave cylinder, then remove the 2 spring clips from the lower clutch slave cylinder tube.

13. Remove the halfshafts.

14. Remove the 3 bolts and 1 through-bolt from the rear transaxle support bracket and remove the rear transaxle support bracket.

15. Loosen the locknut and adjuster bolt on the power steering pump drive belt tensioner and remove the drive belt. Remove the 3 power steering pump mounting bolts working through the pulley holes.

16. Remove the rear bracket bolt from the power steering pump and secure the pump aside with mechanics wire.

17. If equipped, remove the 4 A/C compressor mounting bolts and secure the compressor aside with mechanics wire, leaving the refrigerant lines attached. Do not let the compressor hang by the refrigerant lines.

18. Remove the power steering hose bracket from the pump. Loosen the alternator drive belt tensioner locknut and adjuster bolt and remove the drive belt.

19. Remove the upper and lower radiator hoses and overflow hose. Disconnect the cooling fan electrical connectors.

20. Remove the 2 radiator hold-down bolts and remove the radiator and the cooling fan as an assembly.

➡**Use care when lifting the radiator/cooling fan assembly not to damage the radiator and condenser cooling fins.**

21. Label and disconnect the electrical connectors at the alternator. Remove the 2 A/C and alternator wiring harness bolts, then disconnect the harness from the engine block.

22. Label and disconnect the electrical connectors from the distributor, fuel rail, vehicle speed sensor, starter motor, throttle position sensor, engine coolant temperature sensor, cooling fan engine coolant temperature sensor, temperature gauge sending unit, knock sensor, crank position sensor, EGR valve, park/neutral position switch, idle air control valve, EGR valve position sensor, if equipped, speed control servo.

23. Label and disconnect the vacuum hoses from the speed control servo, if

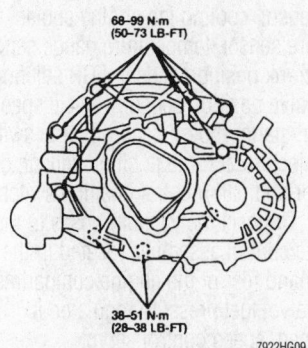

68–99 N·m
(50–73 LB-FT)

38–51 N·m
(28–38 LB-FT)

7922HG09

Engine and transaxle mounting bolt locations—2.5L engine

equipped, EGR valve and fuel pressure regulator.

24. Remove the ground-to-engine bracket bolt located near the starter. If equipped, remove the 2 nuts from the speed control servo and position aside.

25. Remove the transaxle ground and back-up lamp switch electrical connector from the rear of the transaxle. Remove the starter-to-chassis ground.

26. Disconnect the heater hoses from the engine.

27. Disconnect and plug the fuel supply and return lines. Remove the 2 fuel line retaining bolts and bracket.

28. Label and disconnect the vacuum lines and the accelerator cable from the throttle body. Label and disconnect the vacuum line from the intake manifold to the climate control assembly and the power brake booster vacuum hose.

29. Remove the 2 fuel filter mounting nuts and position the filter aside, leaving the fuel lines connected.

30. Attach suitable engine lifting equipment to the engine lifting eyes and take up any slack.

31. Remove the 2 left-hand transaxle support insulator nuts and through-bolt.

32. Remove the 3 right-hand engine support insulator nuts and the through-bolt and remove the right-hand engine support insulator.

33. Carefully lift the engine/transaxle assembly from the vehicle.

34. Remove the transaxle-to-engine bolts and the engine-to-transaxle bolts. Separate the transaxle from the engine.

35. Remove the clutch assembly, flywheel and crankshaft rear cover plate. Mount the engine on a workstand.

To install:

36. Remove the engine from the workstand using an engine sling and hoist.

37. Install the flywheel and clutch assembly.

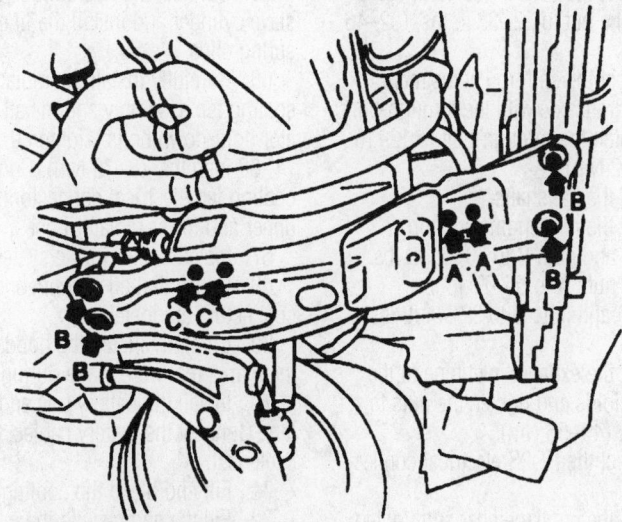

TIGHTENING TORQUE
A: 75 – 104 N·m (55 – 77 LB-FT)
B: 67 – 93 N·m (50 – 68 LB-FT)
C: 44 – 60 N·m (32 – 44 LB-FT)

7922HG10

Rear engine support mounting nut and bolt identification—2.5L engine

38. Align the engine with the transaxle and install the retaining bolts. Tighten the 6 transaxle-to-engine bolts to 50–73 ft. lbs. (68–99 Nm) and the 3 engine-to-transaxle bolts to 28–38 ft. lbs. (38–51 Nm).

39. Carefully lower the engine/transaxle assembly into position in the engine compartment.

40. Install the rear transaxle support insulator. Tighten the 3 nuts to 50–68 ft. lbs. (67–93 Nm) and the through-bolt to 63–86 ft. lbs. (86–116 Nm).

41. Install the left-hand transaxle support insulator. Tighten the 2 nuts to 55–77 ft. lbs. (75–104 Nm) and the through-bolt to 63–86 ft. lbs. (86–116 Nm).

42. Install the right-hand transaxle support insulator. Tighten the 2 nuts to 54–77 ft. lbs. (74–104 Nm) and the through-bolt to 50–68 ft. lbs. (67–93 Nm). Remove the engine lifting equipment.

43. Raise and safely support the vehicle.

44. Install the power steering pump and tighten the 3 bolts to 23–34 ft. lbs. (31–46 Nm).

45. Tighten the power steering pump rear bracket bolt to 24-34 ft. lbs. (32–46 Nm). Install the power steering pump drive belt and adjust the tension.

46. If equipped, install the A/C compressor and tighten the 4 bolts to 28–38 ft. lbs. (38–51 Nm). Install the alternator and A/C drive belt and adjust the tension.

47. Connect the control rod-to-support bar. Tighten the nut to 23–33 ft. lbs. (32–46 Nm).

48. Connect the shift rod adjustment sleeve to the transaxle with the through-bolt and nut. Tighten the through-bolt to 12–16 ft. lbs. (16–22 Nm).

49. Install the halfshafts.

50. Install the rear engine support. Tighten bolts and nuts **B** to 50–68 ft. lbs. (67–93 Nm), nuts **A** to 55–77 ft. lbs. (75–104 Nm) and nuts **C** to 32–44 ft. lbs. (44–60 Nm).

51. Install the exhaust inlet pipe to the exhaust manifolds and tighten the nuts to 30–41 ft. lbs. (40–55 Nm).

52. Connect the HO2S electrical connectors.

53. Install the crossmember and tighten the 6 bolts to 69–93 ft. lbs. (94–126 Nm).

54. Install the splash shields and the wheels. Tighten the lug nuts to 65–86 ft. lbs. (88–118 Nm).

55. Lower the vehicle.

56. Install the power steering hose bracket bolt to the pump.

57. Connect the electrical connectors to the knock sensor, engine coolant temperature sensor, cooling fan engine coolant temperature sensor, temperature gauge sending unit, crank position sensor, EGR solenoids, EGR valve position sensor, vehicle speed sensor, starter motor, back-up lamp switch, fuel injectors, throttle position sensor, distributor and park/neutral position switch.

58. Connect the vacuum hoses to the climate control assembly, located in the right-hand rear of the engine compartment, EGR valve, fuel pressure regulator, if equipped, speed control servo.

59. Install the starter-to-chassis grounds, the transaxle ground and the ground-to-engine bracket bolt.

60. Connect the heater hoses.

61. Unplug and connect the fuel supply and return lines. Tighten the 2 fuel line bracket bolts to 71–88 inch lbs. (8–10 Nm) and the fuel supply line bolt to 18–25 ft. lbs. (25–34 Nm). Be sure to use new copper crush washers.

62. Install the fuel filter to the bracket and install the 2 nuts. Install the speed control servo with the nuts and connect the electrical connector.

63. Connect the vacuum lines to the throttle body and connect the accelerator cable. Connect the power brake booster vacuum hose.

64. Connect the idle air control valve, oil pressure switch, A/C compressor and alternator electrical connectors. Connect the A/C and alternator harness bracket to the engine.

65. Connect the hydraulic line to the slave cylinder and install the line bracket spring clips.

66. Carefully install the radiator and cooling fan assembly and install the 2 radiator hold-down bolts. Tighten the 2 bolts to 71–88 inch lbs. (8–10 Nm). Connect the cooling fan electrical connectors. Install the upper and lower radiator hoses.

67. Install the air intake system.

68. Check that all vacuum and electrical connectors are installed.

69. Carefully install the hood, aligning the marks that were made during removal.

70. Install the battery tray and the battery. Connect the battery cables, negative cable last.

71. Fill and bleed the cooling system.

72. Fill the engine with the proper type and quantity of oil.

73. Fill the transaxle with the proper type and quantity of oil, if needed.

74. Bleed the clutch hydraulic system.

75. Run the engine and bring to normal operating temperature. Check for leaks and proper operation.

76. Stop the engine and check all fluid levels.

Water Pump

REMOVAL & INSTALLATION

2.0L Engine

1. Disconnect the negative battery cable.

2. Drain the cooling system into a suitable container.

3. Remove the accessory drive belts.

4. Disconnect the Power Steering Pressure (PSP) switch electrical connector.

5. Remove the power steering pump drive belt idler pulley shield bolts and remove the shield.

6. Remove the power steering pump through-bolt and lockbolt, and position it out of the way.

7. Remove the cylinder head cover.

8. Raise and safely support the vehicle.

9. Remove the water pump pulley using pulley tool T92C-6312-AH or equivalent, to hold the pulley while the bolts are removed.

10. Remove the splash shields.

11. Remove the timing belt.

12. Remove the power steering pump lower bracket from the water pump.

13. Remove the 5 water pump mounting bolts and remove the water pump.

To install:

14. Clean all gasket mating surfaces.

15. Install a new gasket on the water pump and install the water pump on the engine. Install the mounting bolts and tighten to 14–19 ft. lbs. (19–25 Nm).

16. Install the power steering pump lower bracket from the water pump.

17. Install the water pump pulley and bolts. Hold the pulley with the tool and tighten the bolts to 71–88 inch lbs. (8–10 Nm).

18. Install the timing belt.

19. Install the splash shields and tighten the bolts to 71–88 inch lbs. (8–10 Nm).

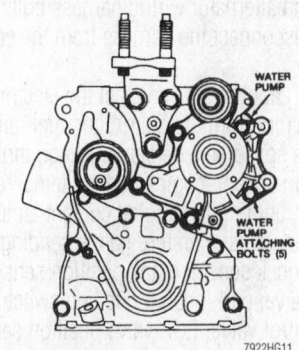

Water pump attaching bolt locations— 2.0L engine

20. Lower the vehicle and install the cylinder head cover. Tighten the bolts in 2–3 steps to 52–69 inch lbs. (6–7 Nm) in the proper sequence.

21. Place the power steering pump in position. Install the through-bolt and tighten to 32–45 ft. lbs. (43–61 Nm). Install the lockbolt and tighten to 23–34 ft. lbs. (31–46 Nm).

22. Connect the PSP switch electrical connector.

23. Install the accessory drive belts and adjust the tension.

24. Install the steering idler pulley shield and tighten the bolts to 61–86 ft. lbs. (7–9 Nm).

25. Connect the negative battery cable.

26. Fill and bleed the cooling system.

27. Run the engine and bring to normal operating temperature. Check for leaks.

2.5L Engine

1. Disconnect the negative battery cable.

2. Drain the cooling system into a suitable container.

3. Remove the timing belt covers and the timing belt.

4. Use pulley removal tool T92C-6312-AH or equivalent, to hold the water pump pulley and remove the bolts. Remove the water pump pulley.

5. Position a drain pan under the water pump.

6. Remove the 3 front engine support insulator mounting bracket bolts.

7. Remove the 5 water pump mounting bolts and remove the water pump.

To install:

8. Clean the mating surfaces of the water pump and the engine block.

9. Install a new O-ring onto the water pump.

10. Install the water pump and tighten the bolts 14–18 ft. lbs. (19–25 Nm).

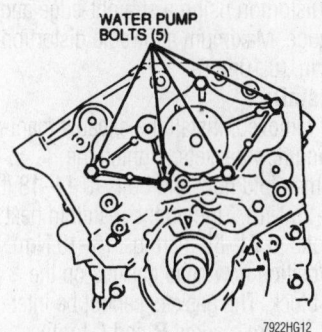

WATER PUMP BOLTS (5)

7922HG12

Water pump attaching bolt locations—2.5L engine

11. Install the 3 engine support mounting bracket bolts.

12. Install the water pump pulley with the bolts. Hold the pulley with the tool and tighten the bolts to 71–88 inch lbs. (8–10 Nm).

13. Install the timing belt and timing belt covers.

14. Connect the negative battery cable.

15. Fill and bleed the cooling system.

16. Run the engine and bring to normal operating temperature. Check for leaks.

Cylinder Head

REMOVAL & INSTALLATION

2.0L Engine

➡Before beginning this procedure, be sure new cylinder head bolts are available, if needed.

✷✷ CAUTION

The fuel injection system remains under pressure even after the engine has been turned OFF. The fuel system pressure must be relieved before disconnecting any fuel lines. Failure to do so may result in fire and/or personal injury.

1. Relieve the fuel system pressure using the recommended procedure.

2. Disconnect the negative battery cable.

3. Drain the engine cooling system into a suitable container.

4. Remove the air intake system.

5. Remove the power steering hose bracket bolts from the cylinder head cover.

6. Disconnect the Power Steering Pressure (PSP) switch.

7. Remove the 2 power steering pump idler pulley shield bolts and the shield.

8. Remove the accessory drive belts.

9. Remove the power steering pump through-bolt and lockbolt and secure the pump aside with mechanics wire, leaving the hoses attached.

10. Remove the exhaust manifold.

11. Label and disconnect the spark plug wires from the spark plugs.

12. Disconnect the hoses from the cylinder head cover and loosen the cover bolts in sequence, in 2–3 steps. Remove the cylinder head cover.

13. Remove the timing belt covers and remove the timing belt.

14. Remove the 2 intake manifold support bolts and the intake manifold support.

15. Label and disconnect the wiring from

the distributor/coil connectors, engine coolant temperature sensor and temperature gauge sensor.

16. Remove the 4 retaining bolts and the coolant temperature sensor housing from the back of the cylinder head.

17. Disconnect the fuel supply and return lines and tag for reassembly. Plug and position the fuel lines aside.

18. Label, disconnect and move aside the following electrical connectors:
- Idle switch
- Throttle position sensor
- Idle air control valve
- Fuel injector harness wiring
- EGR solenoid vacuum valve
- Alternator

19. Disconnect the vacuum hose from the vacuum fitting on the right-hand side of the intake manifold, emissions canister and the brake vacuum booster.

20. Disconnect the accelerator cable.

21. Remove the distributor.

22. Remove the camshafts.

23. Loosen the cylinder head bolts in 3 steps in the proper sequence.

24. Remove the cylinder head bolts, the cylinder head and the cylinder head gasket.

25. If necessary, remove the intake manifold from the cylinder head.

26. Clean all gasket mating surfaces. Inspect the cylinder head for damage, cracks, and fluid leakage. Check the head gasket surface for distortion (warpage) using a straight-edge and feeler gauge. Maximum allowable distortion is 0.004 in. (0.10mm).

To install:

27. If removed, install the intake manifold using a new gasket.

28. Position a new cylinder head gasket on the cylinder block and carefully install the cylinder head.

29. Measure the cylinder head bolts to determine if they are reusable or if they are stretched beyond use. If the cylinder head

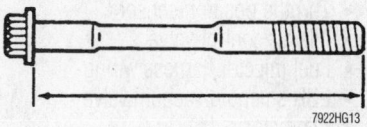

7922HG13

Measuring the cylinder head bolt length—2.0L engine

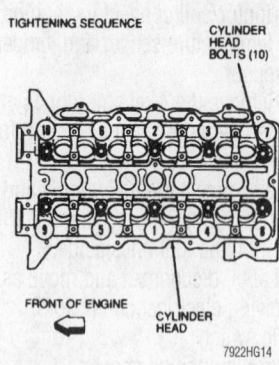

TIGHTENING SEQUENCE — CYLINDER HEAD BOLTS (10)

FRONT OF ENGINE — CYLINDER HEAD

7922HG14

Cylinder head bolt tightening sequence—2.0L engine

bolt is longer than 4 inches (105.5mm), it is stretched and must be replaced. Measurements are taken from under the shoulder to the end of the threads.

30. Tighten the cylinder head bolts in the proper sequence as follows:

a. Tighten each bolt to 10 ft. lbs. (13 Nm).

b. Tighten each bolts again to 16 ft. lbs. (22 Nm).

c. Paint a mark on the socket or the edge of each cylinder head bolt to use as a reference.

d. Using the same torque sequence, tighten each bolt 90 degrees plus or minus 5 degrees.

e. Use the same sequence and tighten each bolt an additional 90 degrees plus or minus 5 degrees.

31. Install the valve lifters, if removed and the camshafts.

➡**Be sure that none of the camshaft lobes are located directly on the hydraulic valve lifters when tightening the camshaft cap bolts.**

32. Install the distributor.

33. Connect the brake booster vacuum hose, emissions canister vacuum hose and the vacuum hose to the right-hand side of the intake manifold.

34. Connect the accelerator cable.

35. Connect the following engine wiring connectors:

• Idle switch
• Throttle position sensor
• Idle air control valve
• Fuel injector harness wiring
• EGR solenoid vacuum valve
• Alternator

36. Unplug and connect the fuel supply and return lines.

37. Connect the distributor/coil electrical connectors.

38. Install the timing belt and the timing belt covers.

39. Install the intake manifold support and bolts. Tighten the support bolts to 28–38 ft. lbs. (38–51 Nm).

40. Install a new cylinder head cover gasket on the cover. Apply sealant to the cylinder head surface in the area adjacent to the front camshaft caps, then install the cover. Tighten the bolts in 2 steps, in sequence, to 52–69 inch lbs. (6–7 Nm).

41. Connect the hoses to the cylinder head cover.

42. Connect the spark plug wires.

43. Install the exhaust manifold.

44. Install the alternator belt and adjust the tension.

45. Loosely install the power steering pump through-bolt and lockbolt.

46. Connect the PSP switch electrical connector.

47. Install the power steering pump drive belt and adjust the tension. Tighten the pump through-bolt to 32–45 ft. lbs. (43–61 Nm) and the lockbolt to 23–34 ft. lbs. (31–46 Nm).

48. Install the power steering pump idler pulley and retaining bolts. Tighten the shield retaining bolts to 61–86 inch lbs. (7–9 Nm).

49. Install the power steering hose brackets to the cylinder head cover. Tighten to 71–88 inch lbs. (8–10 Nm).

50. Install the coolant temperature sensor housing with a new gasket. Tighten the 4 bolts to 14–18 ft. lbs. (19–25 Nm).

51. Connect the engine coolant temperature sensor and temperature gauge sensor electrical connectors.

52. Install the air intake system.

53. Connect the negative battery cable.

54. Fill and bleed the cooling system.

55. Run the engine and check for leaks and proper operation.

56. Adjust the ignition timing, if necessary.

2.5L Engine

➡**The cylinder head bolts may be replaced with new, or measured and reused if they are not stretched beyond allowable limits. Before beginning this procedure, be sure new cylinder head bolts are available, if needed.**

✳✳ CAUTION

The fuel injection system remains under pressure even after the engine has been turned OFF. The fuel system pressure must be relieved before disconnecting any fuel lines. Failure to do so may result in fire and/or personal injury.

1. Relieve the fuel system pressure using the recommended procedure.

2. Disconnect the negative battery cable.

3. Drain the engine cooling system into a suitable container.

4. Remove the timing belt covers and the timing belt.

5. Remove the intake manifold.

6. Disconnect the ventilation pipe from the left cylinder head cover, remove the bolts and remove both cylinder head covers.

7. Remove the camshafts.

8. Remove the timing belt tensioner and the lower timing belt idler.

9. Remove the 2 alternator-to-alternator adjusting arm bolts.

10. Remove the 3 seal plate bolts and the seal plate from the front of the engine.

11. Remove the 4 coolant elbow bolts and the coolant elbow.

12. Raise and safely support the vehicle.

13. Disconnect the 2 Heated Oxygen Sensor (HO2S) electrical connectors. Remove the exhaust pipe-to-exhaust manifold nuts and lower the exhaust pipes.

14. If removing the rear cylinder head, remove the Exhaust Gas Recirculation (EGR) tube and bracket from the EGR valve and the cylinder head.

15. Lower the vehicle.

16. Remove the hydraulic lifters. Identify each lifter as it is removed so it can be reinstalled in the same position. If the lifters are to be reused, store them upside down in an oil filled container.

17. Loosen the cylinder head bolts, in 2–3 steps in the correct removal sequence.

18. Remove the cylinder head bolts, if equipped, head bolt washers and remove the cylinder heads.

19. If required, remove the exhaust manifolds, manifold shields and gaskets.

20. Clean all gasket mating surfaces. Inspect the cylinder head(s) for damage, cracks, and water and oil leakage.

21. Check the cylinder head gasket surface for distortion using a straight-edge and feeler gauge. Maximum allowable distortion is 0.004 in. (0.10mm).

To install:

22. If removed, install the exhaust manifolds using new gaskets. Tighten the exhaust manifold nuts and bolts to 14–18 ft. lbs. (19–25 Nm). Tighten the manifold heat shield bolts to 71–88 inch lbs. (8–10 Nm).

23. Position new head gaskets on the cylinder block. The gaskets cannot be interchanged and are marked **R** and **L** for the right and left bank.

24. Carefully position the cylinder heads on the head gaskets.

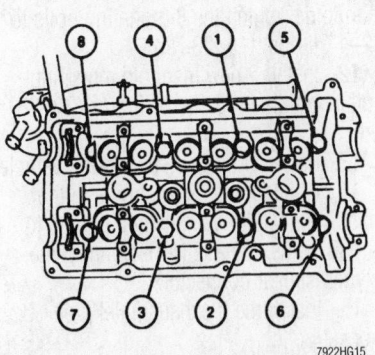

Cylinder head bolt tightening sequence (same for both heads)—2.5L engine

25. Measure the cylinder head bolts from the shoulder to the end of the threads. If the length of the bolt measures more than 5.315 inches (135.0mm) on 1995 vehicles or 5.217 in. (132.5mm) on 1996–97 vehicles, it has stretched beyond its service limit and must be replaced. Be sure that the washer is removed when measuring.

26. Apply clean engine oil to the threads of the cylinder head bolts, if equipped, install the washers.

27. Install the cylinder head bolts.

28. Tighten the new or good used cylinder head bolts using the following procedure:

   a. Tighten the cylinder head bolts in sequence to 10 ft. lbs. (13 Nm).

   b. Tighten the bolts again in sequence to 19 ft. lbs. (26 Nm).

   c. Paint a mark on the edge of the socket or each cylinder head bolt to use as a reference.

   d. Turn each bolt in sequence 90 degrees plus or minus 5 degrees.

   e. Turn each bolt in sequence an additional 90 degrees plus or minus 5 degrees.

29. Apply clean engine oil to the hydraulic lifters and install them in their original positions. Be sure they move freely in the bores.

30. Install the seal plate and the seal plate bolts. Tighten the seal plate bolts to 71–88 inch lbs. (8–10 Nm).

31. Install the coolant elbow. Tighten the 4 bolts to 14–18 ft. lbs. (19–25 Nm).

32. Install the 2 alternator-to-alternator adjusting arm bolts. Tighten the bolts to 14–18 ft. lbs. (19–25 Nm).

33. Install the lower timing belt idler and tighten the idler bolt to 28-38 ft. lbs. (38-51 Nm).

34. Install the timing belt tensioner and tighten the Allen head bolt to 27–33 ft. lbs. (37–44 Nm).

35. Install the camshafts.

36. Raise and safely support the vehicle.

37. If installing the rear cylinder head, install the EGR tube and the EGR tube bracket.

38. Connect the exhaust inlet pipes to the manifolds and tighten the nuts to 41 ft. lbs. (55 Nm).

39. Connect the 2 heated oxygen sensor electrical connectors.

40. Lower the vehicle.

41. Apply sealant to the cylinder head surface in the area of the front and rear camshaft caps. Install new gaskets and install the cylinder head covers. Tighten the bolts in 2 steps, in sequence, to 43–78 inch lbs. (5–8 Nm).

42. Install the intake manifold.

43. Install the timing belt and timing belt covers.

44. Connect the negative battery cable.

45. Fill and bleed the cooling system.

46. Run the engine and check for leaks and proper engine operation.

Intake Manifold

REMOVAL & INSTALLATION

2.0L Engine

❊❊ CAUTION

The fuel injection system remains under pressure, even after the engine has been turned OFF. The fuel system pressure must be relieved before disconnecting any fuel lines. Failure to do so may result in fire and/or personal injury.

1. Relieve the fuel system pressure using the recommended procedure.

2. Disconnect the negative battery cable.

3. Remove the air ducts and air cleaner assembly.

4. Remove and plug the fuel supply and return lines from the intake manifold.

5. Disconnect the accelerator cable.

6. Disconnect and plug the coolant lines at the Idle Air Control Bypass Air (IAC BPA) valve and the throttle body.

7. Label and disconnect the vacuum lines to and from the throttle body, the vacuum lines for the brake booster at the left-hand side of the intake manifold and disconnect the vacuum line fitting on the right-hand side of the intake manifold.

8. Label and disconnect the electrical connectors for the Throttle Position (TP)

sensor, EGR temperature sensor, if equipped, and the EGR solenoid vacuum valve.

9. Disconnect the PCV valve from the cylinder head cover.

10. Raise and safely support the vehicle.

11. Remove the intake manifold support bracket and remove the EGR pipe from the intake manifold.

12. Lower the vehicle.

13. Remove the 5 bolts and 2 nuts securing the intake manifold and remove the manifold and gasket.

   To install:

14. Clean all gasket mating surfaces.

15. Install the intake manifold, using a new gasket. Tighten the nuts and bolts to 14-19 ft. lbs. (19-25 Nm) in the proper sequence.

16. Raise and safely support the vehicle.

17. Attach the EGR pipe to the manifold and install the intake manifold support bracket. Tighten the support bracket bolts to 28–38 ft. lbs. (38–51 Nm).

18. Lower the vehicle.

19. Connect the PCV valve to the cylinder head cover. Connect the electrical connectors to the EGR solenoid vacuum valve, EGR temperature sensor, if equipped and the TP sensor.

20. Connect the vacuum lines running to and from the throttle body, the brake booster and the vacuum line fitting on the right-hand side of the intake manifold.

21. Connect the coolant lines to the IAC BPA valve and the throttle body.

22. Connect the accelerator cable and the fuel lines. Install the fuel line mounting bracket and tighten the bolt to 97 inch lbs. (11 Nm).

23. Install the air cleaner assembly and ducts.

24. Connect the negative battery cable.

25. Fill and bleed the cooling system.

26. Run the engine and check for leaks.

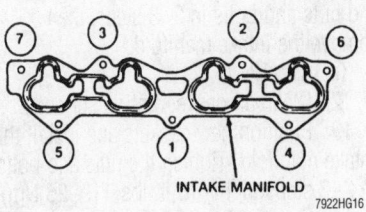

Intake manifold fastener torque sequence—2.0L engine

2.5L Engine

※ CAUTION

The fuel injection system remains under pressure, even after the engine has been turned OFF. The fuel system pressure must be relieved before disconnecting any fuel lines. Failure to do so may result in fire and/or personal injury.

1. Relieve the fuel system pressure using the recommended procedure.
2. Disconnect the negative battery cable.
3. Drain the cooling system into a suitable container.
4. Disconnect the vacuum hoses and electrical connectors from the air cleaner housing. Remove the air cleaner assembly.
5. Disconnect the Knock Sensor (KS) connector and remove the knock sensor bracket from the intake manifold. Remove the Crankshaft Position Sensor (CPS) bracket from the right side of the intake manifold.
6. Remove the right bank (rear) spark plug wires from the spark plugs and the routing clips. Remove the Variable Resonance Induction System (VRIS) solenoid connector bracket from the rear of the intake manifold.
7. Label and disconnect the necessary vacuum hoses from the rear of the intake manifold and EGR valve. Disconnect the PCV valve hose from the intake manifold, near the throttle body.
8. Label and disconnect the Throttle Position (TP) sensor and fuel rail electrical connectors. Disconnect the accelerator cable from the throttle body and the vacuum hose from the evaporative canister.
9. Disconnect and plug the fuel supply line at the fuel rails and discard the copper crush washers. Disconnect the fuel and vacuum lines from the fuel pressure regulator.
10. Disconnect the EGR breather tube.
11. Remove the intake manifold mounting nuts and bolts in 2–3 steps, then remove the intake manifold.

To install:

12. Clean all gasket mating surfaces.
13. Position new gaskets and install the intake manifold. Tighten the nuts and bolts in 2–3 steps to 14–18 ft. lbs. (19–25 Nm).
14. Connect the EGR breather tube and connect the fuel and vacuum lines to the fuel pressure regulator.
15. Connect the fuel supply line to the fuel rail, using new copper crush washers. Tighten the fuel line fittings to 18–25 ft. lbs.

(25–34 Nm).

16. Connect the vacuum hoses to the evaporative canister, intake manifold, throttle body and EGR valve.
17. Connect the TP sensor and fuel rail electrical connectors. Install the VRIS solenoid connector bracket.
18. Connect the spark plug wires to the spark plugs and routing clips. Install the CPS bracket.
19. Install the KS bracket and connect the KS electrical connector.
20. Install the air cleaner assembly and connect the vacuum hoses and electrical connectors to the air cleaner housing.
21. Connect the negative battery cable.
22. Fill and bleed the cooling system.
23. Run the engine and check for leaks.

Exhaust Manifold

REMOVAL & INSTALLATION

2.0L Engine

1. Disconnect the negative battery cable.
2. Remove the 7 exhaust manifold heat shield bolts and the heat shield.
3. Disconnect the oxygen sensor electrical connector.
4. Raise and safely support the vehicle.
5. Remove and discard the exhaust inlet pipe-to-exhaust manifold nuts. Suspend the exhaust system with wire.
6. Remove the 2 exhaust inlet pipe bracket bolts.
7. Disconnect the EGR pipe from the exhaust manifold.
8. Lower the vehicle.
9. Remove the 2 nuts and 8 bolts and remove the exhaust manifold. Discard the nuts.

To install:

10. Clean all gasket mating surfaces.
11. Position a new exhaust manifold gasket over the studs and install the exhaust

manifold. Tighten the 8 mounting bolts to 12–17 ft. lbs. (16–23 Nm).

12. Install 2 new manifold mounting nuts and tighten to 14–21 ft. lbs. (20–28 Nm).
13. Raise and safely support the vehicle.
14. Connect the exhaust pipe to the manifold. Install new nuts and tighten to 27–38 ft. lbs. (37–52 Nm). Connect the oxygen sensor connector.
15. Install the 2 exhaust inlet pipe bracket bolts.
16. Connect the EGR pipe to the back of the exhaust manifold and tighten to 24–34 ft. lbs. (32–47 Nm).
17. Lower the vehicle.
18. Install the exhaust manifold heat shield and tighten the bolts to 71–88 inch lbs. (8–10 Nm).
19. Connect the negative battery cable. Run the engine and check for exhaust leaks.

2.5L Engine

1. Disconnect the negative battery cable.
2. Raise and safely support the vehicle.
3. Disconnect the oxygen sensor connectors.
4. Remove the nuts from the front and rear exhaust pipes and lower the exhaust system. Both pipes must be disconnected, even if only one manifold is to be removed.
5. If removing the rear (right side) manifold, disconnect the EGR pipe.
6. Remove the 3 exhaust manifold shields bolts and remove the shield.
7. Remove the 2 nuts and 5 bolts and remove the exhaust manifold(s).

To install:

8. Clean all gasket mating surfaces.
9. Install the exhaust manifold, using a new gasket, and tighten the nuts and bolts to 14–18 ft. lbs. (19–25 Nm).
10. Install the exhaust manifold shield and tighten the bolts to 71–88 inch lbs. (8–10 Nm).

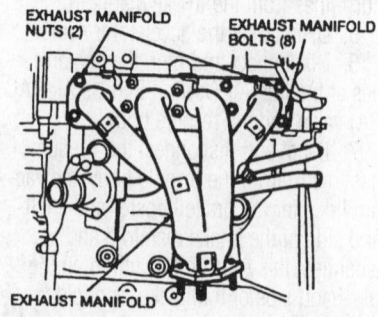

Exhaust manifold bolt and nut locations—2.0L engine

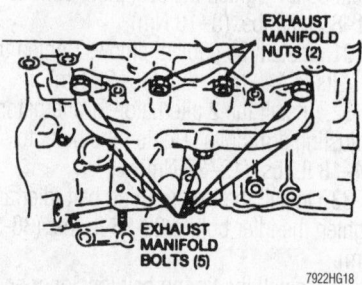

Right side exhaust manifold bolt locations—2.5L engine

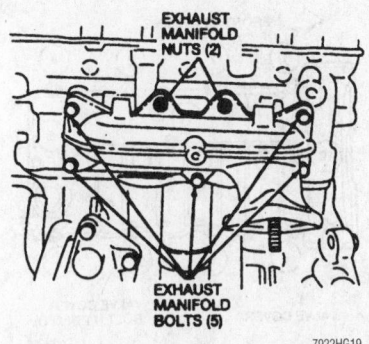

Left side exhaust manifold bolt locations —2.5L engine

11. If installing the rear (right side) manifold, connect the EGR pipe.

12. Connect the exhaust pipes to the manifolds, using new gaskets and nuts, and tighten the nuts to 30–41 ft. lbs. (40–55 Nm).

13. Connect the oxygen sensor connectors and the negative battery cable. Run the engine and check for exhaust leaks.

Front Crankshaft Seal

REMOVAL & INSTALLATION

2.0L Engine

1. Disconnect the negative battery cable.
2. Remove the accessory drive belts.
3. Raise and safely support the vehicle.
4. Remove the front splash shields.
5. Prevent the crankshaft pulley from rotating by installing crankshaft pulley holder T92C-6316-AH or equivalent, and remove the crankshaft pulley bolt.
6. Remove the timing belt.

➡ **Be careful not to turn either the crankshaft or camshaft sprockets after the belt has been removed. This is** especially important when removing the crankshaft damper retaining bolt and when using a puller on the crankshaft sprocket.

7. If necessary, remove the Crankshaft Position (CKP) sensor bolt and position the CKP sensor aside.
8. Remove the crankshaft sprocket, sprocket key and the timing chain guide.
9. Using seal remover T92C-6700-CH or equivalent, remove the front crankshaft seal.

➡ **Seal extractor tool T92C-6700-CH is recommended. It hooks the seal from inside and pulls the seal outward. If the special seal extractor recommended by the vehicle manufacturer is not available, remove the oil pump to remove the seal. Be careful not to score the crankshaft or the seal seat.**

To install:

10. Lubricate the seal lip with clean engine oil. Using oil seal installation tool T74P-6150-A or equivalent, press or drive the new seal into the oil pump cavity. Install the seal so it flush with the oil pump body.
11. Install the timing chain guide and crankshaft sprocket.
12. If necessary, position the crankshaft position (CKP) sensor and install the (CKP) bolt. Tighten the sensor bolt to 71–88 inch lbs. (8–10 Nm).
13. Install the timing belt.
14. Install the crankshaft pulley and tighten the crankshaft pulley bolt to 116–123 ft. lbs. (157–166 Nm).
15. Install the front splash shields.
16. Lower the vehicle.
17. Install the accessory drive belts.
18. Connect the negative battery cable.
19. Run the engine and check for oil leaks and proper engine operation.

2.5L Engine

1. Disconnect the negative battery cable.
2. Remove the oil pump.

➡ **Protect the oil pump housing with a rag while removing the crankshaft front seal.**

3. Using a suitable drift or punch tool protected with a rag, knock out the front crankshaft seal from the oil pump housing.

To install:

4. Clean the oil, dirt and old sealant from all contact surfaces.
5. Lubricate the new front crankshaft seal lip with clean engine oil. Using front seal replacer T74P-6150-A or equivalent, press or drive the new seal into the oil pump housing. Install the seal so it is flush with the edge of the oil pump housing or protrudes no more than 0.021 inch (0.7mm) from the edge of the pump body.
6. Install the oil pump.
7. Connect the negative battery cable.
8. Run the engine and check for leaks and proper engine operation.

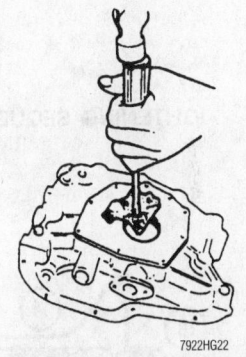

Removing the front crankshaft seal—2.5L engine

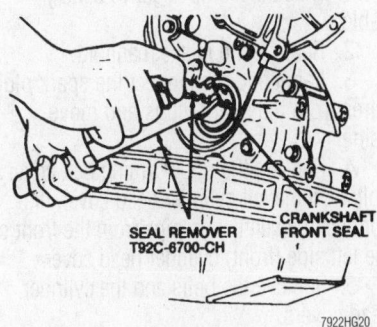

Removing the front crankshaft oil seal— 2.0L engine

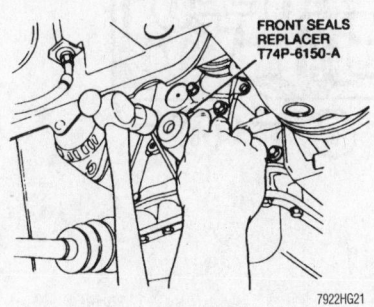

Installing the front crankshaft oil seal— 2.0L engine

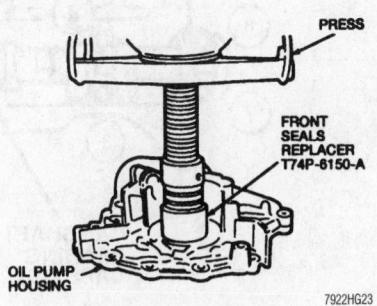

Installing the front crankshaft seal—2.5L engine

Camshaft and Valve Lifters

REMOVAL & INSTALLATION

2.0L Engine

1. Disconnect the negative battery cable.

2. Label and disconnect the spark plug wires and clips from the cylinder head cover. Remove the ignition distributor with wiring and set aside.

3. Remove the power steering hose brackets from the cylinder head cover.

4. Disconnect the breather tube and PCV valve from the cylinder head cover. Remove the cylinder head cover.

5. Remove the accessory drive belts, timing belt covers and timing belt.

6. Remove the camshaft sprockets.

7. Note the location of the numbers on top of the camshaft caps, so the caps can be reinstalled in their original positions.

8. Loosen the camshaft cap bolts in 2 steps, in the proper sequence. Remove the camshaft caps and the oil seals.

9. Remove the camshafts.

10. If they are to be removed, number each hydraulic lifter (hydraulic lash adjuster) with a paint marker or equivalent during removal.

11. Inspect the camshafts and lifters for wear and/or damage and replace, as necessary.

To install:

12. If removed, apply clean engine oil to the lifters and install. If the original lifters are to be reused, be sure to install them in their original positions.

13. Lubricate the camshaft lobes and journals with clean engine oil and install the camshafts on the cylinder head. Be sure none of the lobes are depressing any of the hydraulic lifters.

14. Apply silicone sealant to the cylinder head on the front camshaft caps mating surface. Do not get sealant on the camshaft journals.

15. Install the camshaft bearing caps in their original locations. Install the bolts and tighten, in sequence, in 3 steps:
- Step 1: 35 inch lbs. (4 Nm)
- Step 2: 71 inch lbs. (8 Nm)
- Step 3: 100–126 inch lbs. (12–14 Nm)

16. Apply clean engine oil to the lips of 2 new camshaft front seals. Install the seals

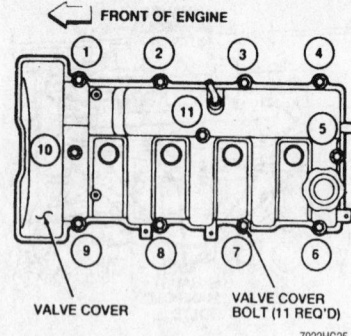

Cylinder head cover bolt torque sequence—2.0L engine

using seal replacer T90P-6256-BH or equivalent.

17. Install the camshaft sprockets, timing belt and timing belt covers.

18. Install the accessory drive belts and adjust the tension.

19. Apply silicone sealant to a new cylinder head cover gasket and install the gasket on the cylinder head cover.

20. Apply silicone sealant to the cylinder head in the area adjacent to the front camshaft caps.

21. Install the cylinder head cover. Tighten the bolts in 2 steps, in sequence, to 52–69 inch lbs. (6–7 Nm).

22. Install the power steering hose brackets and tighten the bolts to 71–88 inch lbs. (8–10 Nm).

23. Connect the breather hose and PCV valve.

24. Install the ignition distributor and connect the spark plug wires and clips.

25. Connect the negative battery cable.

26. Run the engine and check for leaks and proper engine operation.

27. Check the ignition timing and adjust if required.

2.5L Engine

1. Disconnect the negative battery cable.

2. Remove the intake manifold.

3. Label and disconnect the spark plug wires from the spark plugs and move aside.

4. Remove the upper timing belt cover bolts. On the left cylinder head cover, disconnect the ventilation pipe from the front of the left side (front) cylinder head cover.

5. Remove the bolts and the cylinder head covers.

6. Remove the timing belt.

7. Hold the camshafts on the hexagon casting and loosen the sprocket retaining bolts. Remove the camshaft sprockets.

TIGHTENING SEQUENCE

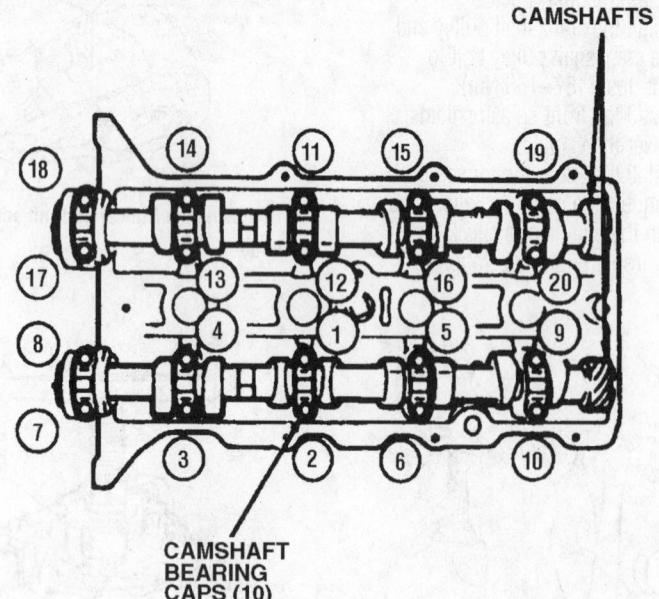

Camshaft bearing cap bolt torque sequence—2.0L engine

❊❊ WARNING

Do not remove any of the camshaft bearing caps when the camshaft lobes are depressing the valve lifters or damage to the thrust journal support or camshaft may result.

8. Turn the camshafts so the knock pins are aligned with the marks on the camshaft bearing end caps. This will reduce the pressure on the hydraulic lifters.

9. Note the markings on the camshaft bearing caps prior to removal, so they can be reinstalled in the same positions. The right bank (rear) caps are marked with numbers and the left bank (front) caps are marked with letters.

10. Loosen the camshaft bearing end cap bolts in sequence, in 5–6 steps. Remove the camshaft bearing end caps.

11. Remove the blind caps.

12. Remove the remaining camshaft bearing cap bolts, in 5–6 steps, in the proper sequence. Remove the caps, being sure to remove the thrust caps last. Do not damage the cylinder head thrust bearing support.

13. Remove the camshafts and oil seals. Tag the camshafts for identification.

14. If they are to be removed, number each hydraulic lifter (hydraulic lash adjuster) with a paint marker or equivalent during removal.

15. Inspect the camshafts and lifters for wear and/or damage; replace as necessary.

To install:

16. If removed, apply clean engine oil to the hydraulic lifters and install. If the old lifters are to be reused, be sure to install them in their original positions. Be sure that the hydraulic lifters move freely in their bore.

17. Apply clean engine oil to the camshaft lobes, journals and supports.

18. Install the camshafts so the timing marks align on the camshaft gears.

19. Apply silicone sealant to the cylinder head surface in the area forward of the camshaft gear cavity on both cylinder heads and to the left (front) cylinder head on the rear exhaust camshaft end cap mating surface.

20. Install the thrust caps. Tighten the thrust cap bolts until the caps are fully seated on the cylinder head.

❊❊ WARNING

Do not install any of the camshaft bearing caps when the camshaft lobes are depressing the valve lifters or damage to the thrust journal support or camshaft may result.

21. Install the remaining camshaft bearing caps and camshaft bearing end caps in their original positions. Tighten the caps, in sequence, in 5 equal steps, with the final step being 98–123 inch lbs. (11–14 Nm).

➡ **The right bank (rear) camshaft bearing caps are marked with numbers and the left bank (front) camshaft bearing caps are marked with letters.**

22. Apply a light coat of clean engine oil to 4 new camshaft oil seals and install the seals using a suitable socket and hammer. The camshaft oil seals should be flush with the front of the cylinder head with a maximum protrusion of 0.020 inch (0.5mm).

23. Apply sealant to 4 new blind caps and install the blind caps using a plastic hammer or equivalent.

24. Install the camshaft sprockets and retaining bolts. Hold the hexagon casting on the camshafts with a suitable wrench and tighten the retaining bolts to 90–103 ft. lbs. (123–140 Nm).

25. Install the timing belt.

26. Remove any sealant and gasket

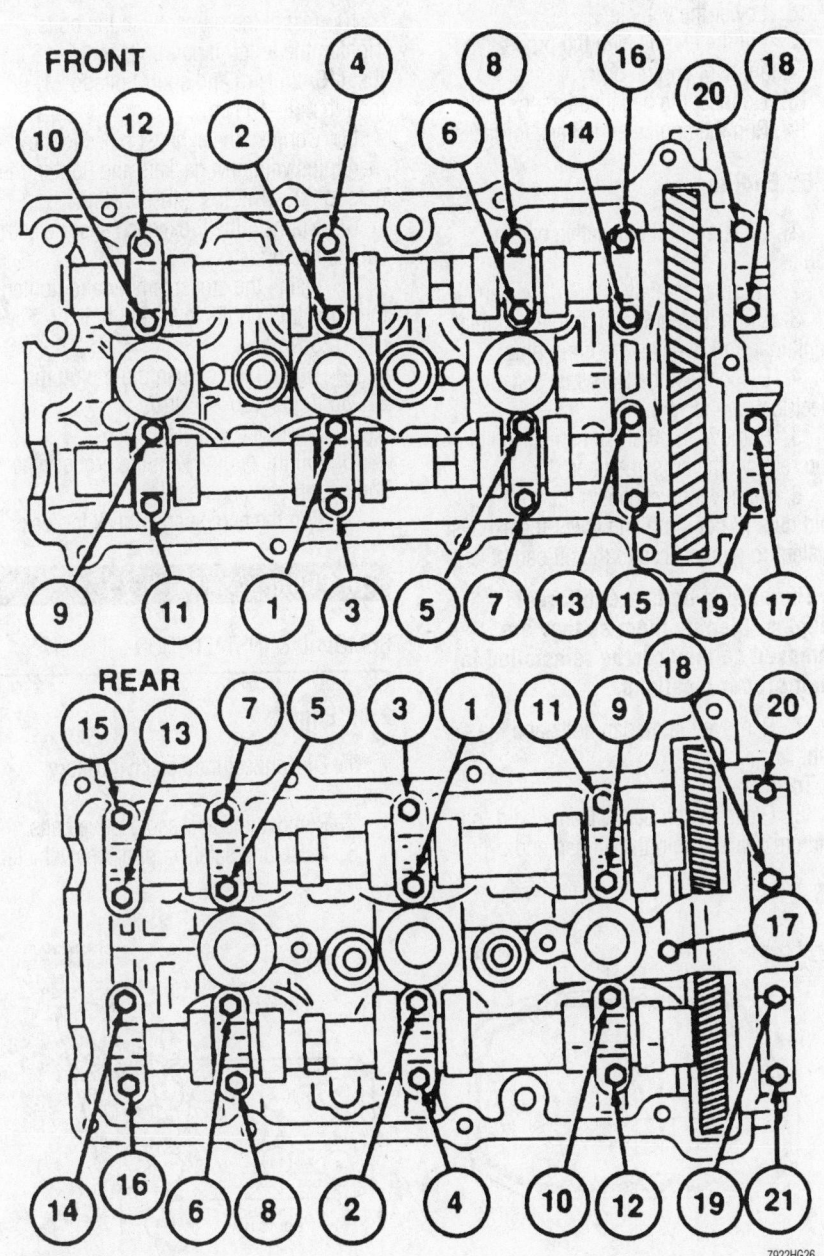

Tightening sequence for the front and rear cylinder heads—2.5L engine

7922HG26

material from the cylinder head cover contact surfaces.

27. Apply silicone sealant to the cylinder head in the area adjacent to the front and rear camshaft caps. Install a new gasket on the cylinder head.

28. Install the cylinder head cover. Tighten the bolts in 2 steps to 43–78 inch lbs. (5–8 Nm) following the proper torque sequence.

29. Tighten the upper timing cover bolts to 71–88 inch lbs. (8–10 Nm). Connect the ventilation pipe to the left side cylinder head cover.

30. Install the intake manifold.

31. Connect the spark plug wires to the spark plugs.

32. Connect the negative battery cable.

33. Run the engine and check for leaks and proper engine operation.

Valve Lash

ADJUSTMENT

The hydraulic lash adjusters cannot be adjusted. When the lash adjusters are removed from the engine, check the friction surfaces for wear or damage. Hold the lash adjuster and try to press the plunger by hand. If the lash adjuster is worn or damaged, or the plunger can be moved by hand, replace the lash adjuster.

Oil Pan

REMOVAL & INSTALLATION

2.0L Engine

1. Disconnect the negative battery cable.
2. Raise and safely support the vehicle.
3. Remove the right-hand splash shield.
4. Drain the engine oil into a suitable container. Temporarily reinstall the drain plug.
5. Disconnect the oxygen sensor electrical connector. Remove and discard the exhaust pipe-to-manifold nuts. Move the exhaust pipe aside and support it with a jack.
6. Remove the exhaust clamp from the hold-down bracket.
7. Remove the oil pan bolts and carefully pry the oil pan from the engine stiffener.

To install:

8. Clean the oil pan. Clean all dirt, oil and old sealant from the oil pan and cylinder block and stiffener contact surfaces.

9. Apply a continuous bead of silicone sealant around the oil pan, going on the inside of the bolt holes.

10. Install the oil pan and tighten the bolts to 14–19 ft. lbs. (19–25 Nm).

11. Install a new exhaust clamp and tighten the clamp nuts to 26–34 ft. lbs. (34–47 Nm).

12. Connect the exhaust pipe to the manifold with new nuts. Tighten the nuts to 27–38 ft. lbs. (37–52 Nm).

13. Connect the oxygen sensor electrical connector.

14. Install the right-hand splash shield.

15. Tighten the oil pan drain plug to 22–30 ft. lbs. (30–41 Nm).

16. Lower the vehicle.

17. Fill the engine with the proper type and quantity of engine oil.

18. Connect the negative battery cable.

19. Run the engine and check for leaks.

2.5L Engine

1. Disconnect the negative battery cable.
2. Raise and safely support the vehicle.
3. Drain the engine oil into a suitable container and reinstall the drain plug.
4. Disconnect the 2 oxygen sensor electrical connectors.
5. Remove the 6 crossmember bolts and remove the crossmember.
6. Remove the exhaust pipe-to-manifold nuts (3 per side) and lower the exhaust system to gain access to the oil pan bolts.

➡ **The oil pan bolts are different lengths. Identify them as they are removed so they can be reinstalled in their proper locations.**

7. Remove the oil pan bolts and the oil pan.

To install:

8. Clean the oil pan. Clean all dirt, oil and old sealant from the oil pan and cylinder block contact surfaces. Remove the old sealant from the threads of the oil pan bolts and the bolt holes in the block.

✳✳ WARNING

Failure to remove the old sealant from the bolts and bolt holes may cause the block to crack.

9. Apply a continuous bead of silicone sealant along the inside of the bolt holes, overlapping the ends.

➡ **Once the new silicone sealant is applied, the oil pan must be installed within 5 minutes.**

10. Install the oil pan with the bolts. Tighten the long oil pan bolts to 14–18 ft. lbs. (19–25 Nm) and short bolts to 71–88 inch lbs. (8–10 Nm).

11. Connect the exhaust pipes to the manifolds with new gaskets and tighten the nuts to 30–41 ft. lbs. (40–55 Nm).

12. Connect the 2 oxygen sensor electrical connectors.

13. Install the crossmember and tighten the 6 retaining bolts to 69–93 ft. lbs. (94–126 Nm).

14. Tighten the oil pan drain plug to 22–30 ft. lbs. (30–41 Nm).

15. Lower the vehicle.

16. Fill the engine with the proper type and quantity of engine oil.

17. Run the engine and check for leaks.

Oil Pump

REMOVAL & INSTALLATION

2.0L Engine

1. Disconnect the negative battery cable.
2. Remove the accessory drive belts.
3. Raise and safely support the vehicle.

![Oil pan silicone sealant diagram]

7922HG47

Be sure to apply the bead of sealant on the inside of the bolt holes, as shown— 2.5L engine

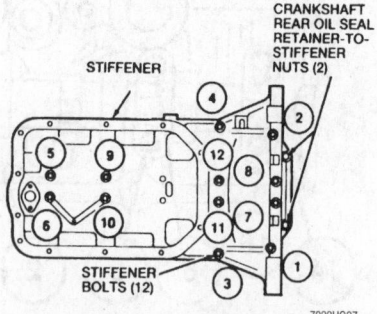

7922HG27

Stiffener bolt and nut loosening sequence—2.0L engine

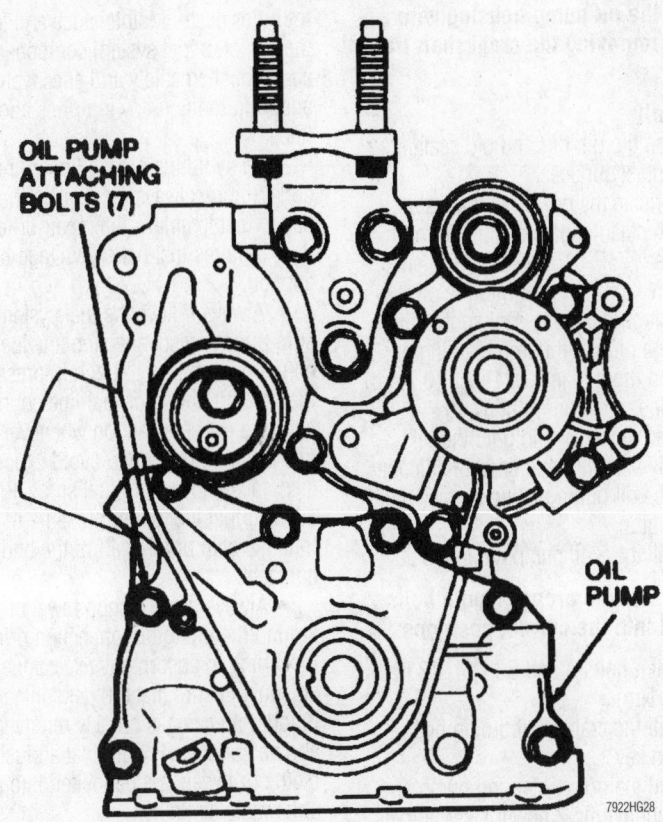

Note the oil pump attaching bolt locations—2.0L engine

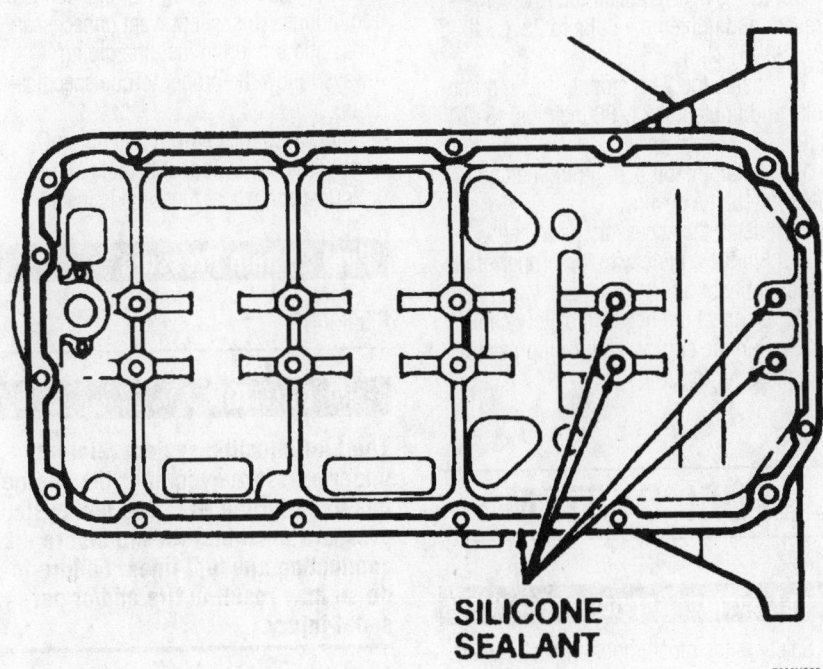

Applying sealant to the stiffener—2.0L engine

STIFFENER

SILICONE SEALANT

OIL PUMP ATTACHING BOLTS (7)

OIL PUMP

7922HG28

7922HG29

TIGHTENING SEQUENCE STIFFENER

STIFFENER MOUNTING BOLTS (12)

7922HG30

Stiffener bolt tightening sequence—2.0L engine

4. Remove the engine oil pan and the oil pump pick-up tube and screen.

5. Remove the 2 rear main seal housing-to-stiffener nuts and the 12 stiffener bolts in 2 steps, in the proper sequence, and remove the stiffener.

6. Remove the engine front covers, timing belt and crankshaft sprocket.

7. If equipped, remove the A/C compressor and secure it aside, leaving the refrigerant lines attached. Remove the compressor mounting bracket.

8. Remove the 7 oil pump bolts and remove the oil pump.

To install:

9. Clean the oil, dirt and old sealant from all contact surfaces.

10. Apply a bead of silicone to the oil pump-to-cylinder block contact surface, along the inside of the bolt holes. Be sure the sealer does not fall into the engine and form plugs that could block oil passages.

11. Install the oil pump and tighten the bolts to 14–19 ft. lbs. (19–25 Nm).

12. Clean the engine block contact area and stiffener. Apply a bead of silicone sealant to the perimeter of the stiffener, going on the inside of the bolt holes. Apply sealant to the 4 rear bolt holes. Be sure the sealer does not fall into the engine and form plugs that could block oil passages.

13. Install the stiffener and the mounting bolts. Tighten the bolts in 2 steps, in sequence, to 14–19 ft. lbs. (19–25 Nm). Tighten the rear main seal housing-to-stiffener nuts to 88 inch lbs. (10 Nm).

14. Install a new gasket and the oil pump pick-up tube and screen. Tighten the mounting bolts to 71–88 inch lbs. (8–10 Nm).

15. Install the oil pan.

16. If equipped, install the A/C compressor bracket and tighten the bolts to 38 ft. lbs. (52 Nm). Install the A/C compressor and tighten the bolts to 26 ft. lbs. (35 Nm).

17. Install the crankshaft sprocket, timing belt and the engine front covers.

18. Install the accessory drive belts.

19. Fill the engine with the proper grade and quantity of oil.

20. Connect the negative battery cable.

21. Run the engine; check oil pressure and check for leaks.

2.5L Engine

The oil pump is located behind the crankshaft timing belt sprocket. Oil pump replacement is an in-vehicle service, however it requires that the timing belt covers, timing belt, and oil pan be removed. It is recommended that the front crankshaft seal be replaced whenever the oil pump is removed from the vehicle.

1. Disconnect the negative battery cable.

2. Remove the accessory drive belts.

3. Remove the engine front covers and the timing belt.

4. Remove the engine oil pan.

5. If equipped, remove the 4 A/C compressor mounting bolts and support the compressor aside, without disconnecting the refrigerant lines. Remove the 5 A/C compressor mounting bracket bolts and remove the bracket.

6. Remove the power steering pump and tensioner bolts from the engine block. Remove the pump and tensioner and position aside.

7. Remove the crankshaft sprocket using a suitable puller. Remove the crankshaft sprocket key.

8. Remove the 9 oil pump mounting bolts and the 2 oil strainer-to-oil pump bolts. Remove the oil pump.

➡**The oil pump mounting bolts are different lengths. Identify them as they are removed so they can be reinstalled in their proper locations.**

9. Remove the oil pump O-ring. If necessary, use a small prybar or punch protected with a rag to knock out the front crankshaft seal from the oil pump housing.

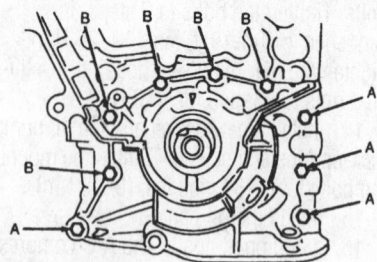

BOLT A : 40mm (1.57 IN.)
BOLT B : 25mm (0.98 IN.)

7922HG31

Notice the different length oil pump attaching bolts—2.5L engine

➡**Protect the oil pump housing with a rag while removing the crankshaft front seal.**

To install:

10. Clean the oil, dirt and old sealant from all contact surfaces.

11. Lubricate the new front crankshaft seal lip with clean engine oil. Using front seal replacer T74P-6150-A or equivalent, press or drive the new seal into the oil pump housing. Install the seal so it is flush with the edge of the oil pump housing or protrudes no more than 0.021 inch (0.7mm) from the edge of the pump body.

12. Install a new O-ring onto the oil pump. Apply a continuous bead of silicone sealant to the oil pump mating surface and install the pump.

13. Install the 9 oil pump mounting bolts.

➡**Be sure that the proper length bolts are placed into the correct positions.**

14. Tighten bolts A and B to 14–18 ft. lbs. (19–25 Nm).

15. Install the crankshaft timing belt sprocket and key.

16. Install the power steering pump and tensioner. Tighten the 2 power steering belt tensioner upper bolts and the power steering pump rear bracket bolt to 24–33 ft. lbs. (32–46 Nm). Tighten the tensioner lower bolt to 14–18 ft. lbs. (19–25 Nm).

17. If equipped, install the A/C compressor bracket and tighten the bolts to 28–38 ft. lbs. (38–51 Nm). Install the A/C compressor and tighten the bolts to 28–38 ft. lbs. (38–51 Nm).

18. Install the 2 oil strainer-to-oil pump bolts and tighten to 71–88 inch lbs. (8–10 Nm).

19. Install the oil pan, timing belt and the front engine covers.

20. Install the accessory drive belts.

21. Fill the engine with the proper grade and quantity of oil.

22. Connect the negative battery cable.

23. Run the engine; check oil pressure and check for leaks.

FUEL SYSTEM

Fuel System Service Precautions

Safety is the most important factor when performing not only fuel system maintenance but any type of maintenance. Failure to conduct maintenance and repairs in a safe manner may result in serious personal injury or death. Maintenance and testing of the vehicle's fuel system components can be accomplished safely and effectively by adhering to the following rules and guidelines.

• To avoid the possibility of fire and personal injury, always disconnect the negative battery cable unless the repair or test procedure requires that battery voltage be applied.

• Always relieve the fuel system pressure prior to disconnecting any fuel system component (injector, fuel rail, pressure regulator, etc.), fitting or fuel line connection. Exercise extreme caution whenever relieving fuel system pressure to avoid exposing skin, face and eyes to fuel spray. Please be advised that fuel under pressure may penetrate the skin or any part of the body that it contacts.

• Always place a shop towel or cloth around the fitting or connection prior to loosening to absorb any excess fuel due to spillage. Ensure that all fuel spillage (should it occur) is quickly removed from engine surfaces. Ensure that all fuel soaked cloths or towels are deposited into a suitable waste container.

• Always keep a dry chemical (Class B) fire extinguisher near the work area.

• Do not allow fuel spray or fuel vapors to come into contact with a spark or open flame.

• Always use a back-up wrench when loosening and tightening fuel line connection fittings. This will prevent unnecessary stress and torsion to fuel line piping. Always follow the proper torque specifications.

• Always replace worn fuel fitting O-rings with new. Do not substitute fuel hose or equivalent, where fuel pipe is installed.

Fuel System Pressure

RELIEVING

✳✳ CAUTION

The fuel injection system remains under pressure even after the engine has been turned OFF. The fuel system pressure must be relieved before disconnecting any fuel lines. Failure to do so may result in fire and/or personal injury.

1. Relieve the fuel system pressure as follows:

a. Start the engine and let it idle.

b. Locate and remove the fuel pump

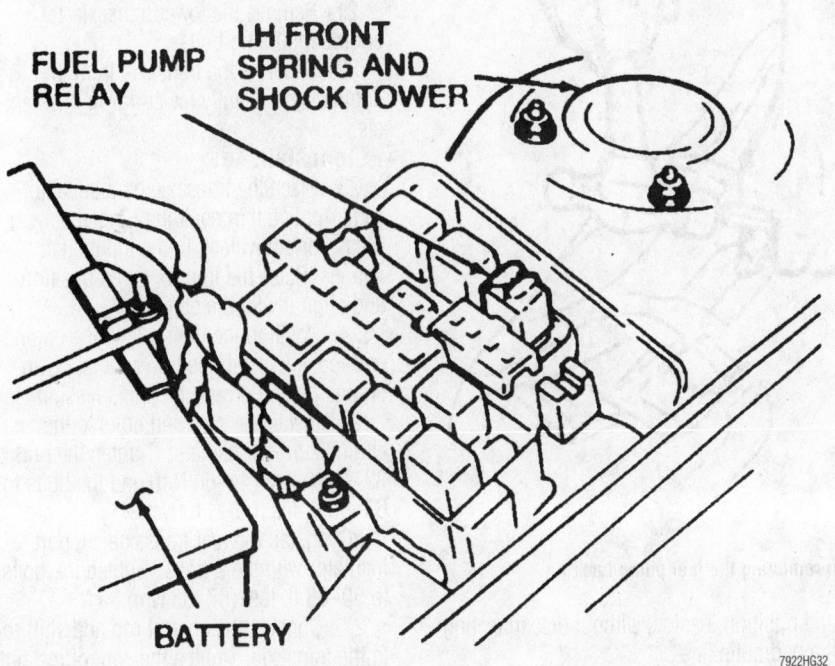

Fuel pump relay location in the fuse box

relay from the main fuse junction panel located next to the left-hand front strut tower.

   c. After the engine stalls, turn **OFF** the ignition switch.

   d. Install the fuel pump relay.

  2. Disconnect the negative battery cable.

Fuel Filter

REMOVAL & INSTALLATION

✳✳ CAUTION

The fuel injection system remains under pressure even after the engine has been turned OFF. The fuel system pressure must be relieved before disconnecting any fuel lines. Failure to do so may result in fire and/or personal injury.

  1. Properly relieve the fuel system pressure as follows:

   a. Start the engine and let it idle.

   b. Locate and remove the fuel pump relay from the main fuse junction panel located next to the left-hand front strut tower.

   c. After the engine stalls, turn **OFF** the ignition switch.

   d. Install the fuel pump relay.

  2. Disconnect the negative battery cable.

  3. If equipped with cruise control, remove the 3 speed control servo nuts and position the servo aside.

  4. Remove the 2 fuel filter bracket nuts and remove the fuel tube clamps.

  5. Disconnect the fuel lines from both ends of the fuel filter. Plug the lines to prevent leakage.

  6. Remove the filter from the mounting bracket.

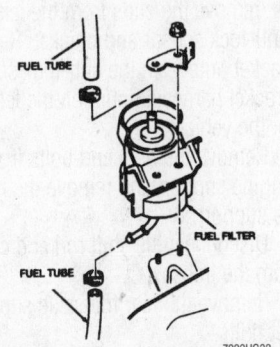

Exploded view of the fuel filter assembly mounting

To install:

  7. Position the fuel filter in the mounting bracket.

  8. Unplug the fuel lines and place new fuel line clamps onto each fuel line.

  9. Install the fuel lines to the fuel filter and position the clamps.

  10. Install the 2 bracket nuts and tighten to 71–97 inch lbs. (8–11 Nm).

  11. If equipped with cruise control, position the speed control servo and install the 3 mounting nuts.

  12. Connect the negative battery cable.

  13. Run the engine and check for fuel leaks.

Fuel Pump

REMOVAL & INSTALLATION

✳✳ CAUTION

The fuel injection system remains under pressure even after the engine has been turned OFF. The fuel system pressure must be relieved before disconnecting any fuel lines. Failure to do so may result in fire and/or personal injury.

  1. Relieve the fuel system pressure using the recommended procedure.

  2. Disconnect the negative battery cable.

  3. Remove the fuel tank and place it on a bench.

  4. Remove any dirt that has accumulated around the fuel pump retaining flange so it will not enter the tank during pump removal and installation.

  5. Turn the fuel pump locking ring counterclockwise and remove the locking ring.

✳✳ CAUTION

If the locking ring is tapped around, be sure to use a brass drift and mallet to prevent sparks.

  6. Remove the fuel pump and fuel level sensor assembly. Remove and discard the seal ring.

  To install:

  7. Clean the fuel pump mounting flange, fuel tank mounting surface and seal ring groove.

  8. Apply a light coating of grease on a new seal ring to hold it in place during assembly and install in the seal ring groove.

  9. Install the fuel pump and fuel level sensor assembly carefully to ensure the fil-

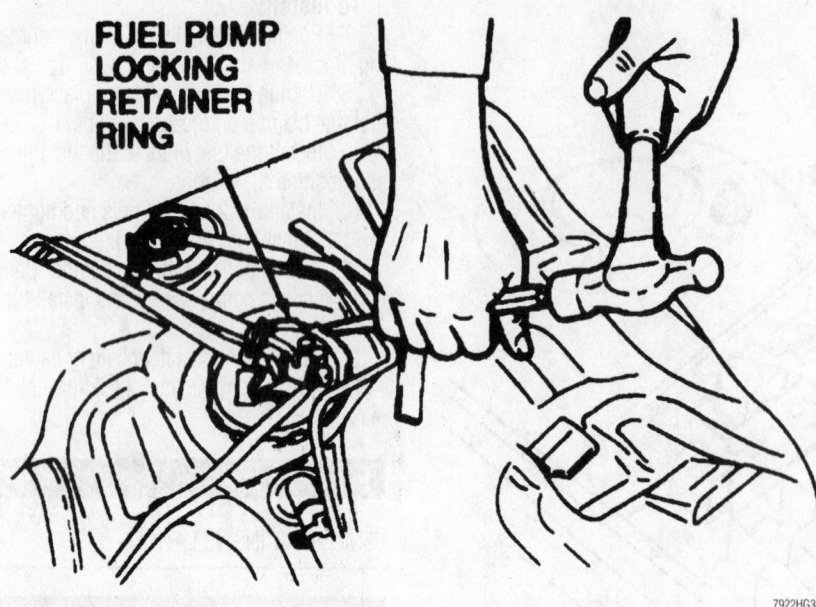

FUEL PUMP LOCKING RETAINER RING

7922HG34

Use a brass drift and mallet to prevent sparks when removing the fuel pump lockring

ter and sensor arm are not damaged. Be sure the locating keys are in the keyways and the seal ring remains in the groove.

10. Hold the pump assembly in place and install the locking ring finger-tight. Be sure all the locking tabs are under the tank lock ring tabs.

11. Rotate the locking ring clockwise until the ring is against the stops.

12. Install the fuel tank in the vehicle. Add a minimum of 10 gallons of fuel to the tank and check for leaks.

13. Connect the negative battery cable, start the engine and check for proper system operation and for fuel leaks.

DRIVE TRAIN

Transaxle Assembly

REMOVAL & INSTALLATION

Manual

1. Disconnect the battery cables, negative cable first, and remove the battery and battery tray.

2. Remove the air cleaner intake tube and the engine air cleaner.

3. Label and disengage all connectors and ground straps related to transaxle removal.

4. Disconnect the clutch slave cylinder

and position it aside, without disconnecting the hydraulic line.

5. Support the engine with engine support tool 014-00750 or equivalent.

6. Remove the upper transaxle-to-engine mounting bolts.

7. Remove the starter. Remove the fuel filter mounting nuts and position the filter aside, without disconnecting the fuel lines.

8. Remove the 2 nuts and the through-bolt from the left side engine support insulator.

9. Raise and safely support the vehicle. Remove the front wheels.

10. If equipped with 2.0L engine, remove the intake manifold support bolts and bracket.

11. Remove the drain plug and drain the transaxle fluid into a suitable container.

12. Remove the halfshafts.

13. Remove the lower splash shields.

14. Remove the crossmember bolts and the crossmember.

15. On vehicles equipped with anti-lock brakes, remove the clips from the left-hand side anti-lock sensor and bracket. Remove the bracket nuts from the anti-lock sensor and bracket harness mount on the left-hand side of the vehicle.

16. Remove the nuts and bolts from the rear engine support and remove the rear engine support.

17. Disconnect the shift rod and control rod from the transaxle.

18. Remove the rear transaxle support bracket bolts.

19. Support the transaxle with a suitable jack.

20. Remove the rear transaxle support insulator bolts and the rear transaxle support insulator.

21. Remove the lower transaxle-to-engine mounting bolts.

22. Separate the transaxle from the engine and carefully lower it from the vehicle.

To install:

23. Place the transaxle on a suitable jack. Apply a thin coating of molybdenum grease or equivalent, to the input shaft splines. Raise the transaxle into position and align it with the engine.

24. Position the transaxle to the engine and loosely install the lower transaxle-to-engine bolts. Remove the transmission jack.

25. Install the nuts and bolts to the engine support insulator. Tighten the nuts to 32–44 ft. lbs. (44–60 Nm) and the bolts to 63–86 ft. lbs. (86–116 Nm).

26. Install the rear transaxle support insulator with the 3 bolts. Tighten the bolts to 50–68 ft. lbs. (67–93 Nm).

27. Connect the control rod and shift rod to the transaxle. Tighten the control rod nut to 28–38 ft. lbs. (38–51 Nm) and the shift rod bolt and nut to 14–18 ft. lbs. (19–25 Nm).

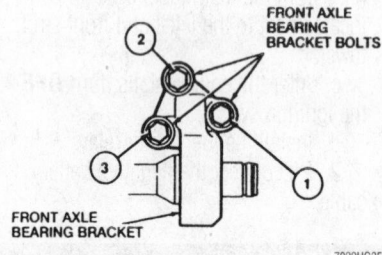

FRONT AXLE BEARING BRACKET BOLTS

FRONT AXLE BEARING BRACKET

7922HG35

Front axle bearing bracket bolt locations and tightening sequence—2.5L engine

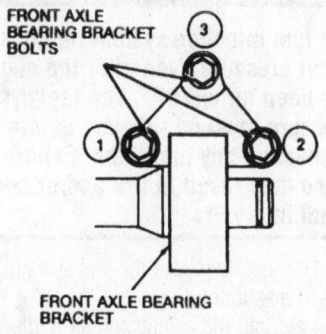

FRONT AXLE BEARING BRACKET BOLTS

FRONT AXLE BEARING BRACKET

7922HG36

Front axle bearing bracket bolt locations and tightening sequence—2.0L engine

28. Install the rear engine support and tighten the bolts and nuts as follows:

• Tighten **A** to 55–77 ft. lbs. (75–104 Nm)

• Tighten **B** to 50–68 ft. lbs. (67–93 Nm)

• Tighten **C** to 32–44 ft. lbs. (44–60 Nm)

29. Install the halfshafts.

30. Pry the lower control arm down and insert the lower ball joint stud into the knuckle. Install the pinch bolt and nut and tighten to 26–41 ft. lbs. (35–56 Nm).

31. If equipped with anti-lock brakes, install the anti-lock sensor harness mounting nuts and tighten to 71–88 inch lbs. (8–10 Nm). Install the sensor harness clips.

32. Install the 3 halfshaft support bearing bracket bolts. Tighten the bolts, in sequence, to 32–45 ft. lbs. (43–61 Nm).

33. If equipped with 2.5L engine, connect the exhaust pipes to the manifolds and tighten the new nuts to 30–41 ft. lbs. (40–55 Nm). Connect the 2 oxygen sensor connectors.

34. Position the crossmember and tighten the 6 retaining bolts to 69–96 ft. lbs. (94–131 Nm).

35. Install new halfshaft retaining nuts. Have an assistant apply the brakes to lock the hubs, then tighten the nuts to 174–235 ft. lbs. (235–319 Nm). Stake the nuts using a dull chisel or similar tool.

36. Install the wheels and tighten the lug nuts to 65–87 ft. lbs. (88–118 Nm).

37. Install a new washer on the transaxle drain plug. Install and tighten the drain plug to 29–43 ft. lbs. (40–58 Nm).

38. Position the starter motor and tighten the lower retaining bolt to 28–38 ft. lbs. (35–52 Nm). Connect the starter wiring.

39. If equipped with 2.0L engine, install the intake manifold support bracket and bolts. Tighten the bolts to 27–38 ft. lbs. (37–52 Nm).

40. Install the lower splash shields.

41. If equipped with the 2.0L engine, install the 2 lower transaxle-to-engine bolts. Tighten bolt **B** to 28–38 ft. lbs. (38–51 Nm) and tighten bolt **C** to 14–18 ft. lbs. (19–25 Nm). Install the 2 engine-to-transaxle bolts. Tighten bolt **D** to 28–38 ft. lbs. (38–51 Nm) and bolt **E** to 66–86 ft. lbs. (90–116 Nm).

42. If equipped with the 2.5L engine, install and tighten the 3 lower transaxle-to-engine bolts to 28–38 ft. lbs. (38–51 Nm).

43. Install the 3 rear transaxle support bracket bolts and tighten to 50–68 ft. lbs. (67–93 Nm).

44. Fill the transaxle with the proper fluid to a level even with the lower edge of the oil level plug port, with the vehicle level. Install the plug, using a new washer, and

tighten to 29–43 ft. lbs. (40–58 Nm) if not already done.

45. Lower the vehicle.

46. Position the fuel filter and install the 2 retaining nuts. Tighten the nuts to 71–88 inch lbs. (8–10 Nm).

47. If equipped with the 2.0L engine, install the 4 remaining transaxle-to-engine bolts. Tighten the **A** bolts to 66–86 ft. lbs. (90–116 Nm). If equipped with the 2.5L engine, install the 7 upper transaxle-to-engine bolts and tighten to 50–73 ft. lbs. (68–99 Nm).

48. Install the 2 upper starter motor bolts and tighten to 28–38 ft. lbs. (35–51 Nm).

49. Remove the engine support tool.

50. Install the clutch slave cylinder and the 2 retaining bolts. Tighten the bolts to 12–16 ft. lbs. (16–22 Nm).

51. Connect the back-up lamp switch and the vehicle speed sensor electrical connectors and the transaxle ground straps.

52. Install the battery and battery tray.

53. Install the engine air cleaner and the air cleaner intake tube.

54. Connect the battery cables, negative cable last.

55. Run the engine and check for leaks.

56. Check transaxle operation and bleed the clutch hydraulic system if required.

Automatic

4EAT TRANSAXLE

1. Disconnect the battery cables, negative cable first, and remove the battery and battery tray.

2. Remove the engine air cleaner assembly.

3. Pry the shift cable from the manual control shift outer lever. Remove the cable bracket lock tab retainer, press in on the lock tabs to release the shift cable and pull the cable through the bracket.

4. Disconnect the Manual Lever Position (MLP) sensor electrical connector. Disconnect the 2 heated oxygen sensor electrical connectors and the transaxle electrical connector.

5. Remove the wiring harness bracket from the shift cable bracket.

6. Remove the starter motor wiring and remove the starter motor.

7. Disconnect the Vehicle Speed Sensor (VSS) electrical connector. Remove the ground wire bracket and the ground wire.

8. Remove the harness support bracket to the alternator, located at the rear engine support insulator bracket.

9. Disconnect and plug the inlet and outlet oil cooler tubes.

Engine-To-Transaxle Bolts

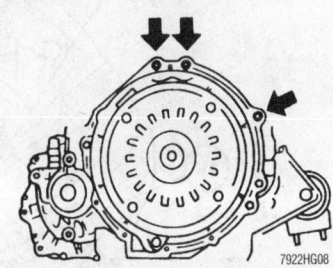

Transaxle-To-Engine Bolts

Engine and transaxle retaining bolt locations—2.5L engine

10. Remove the 3 transaxle-to-engine mounting bolts.

11. Support the engine from above with engine support tool 014-00750 or equivalent.

12. Remove the 2 left-hand engine support insulator nuts and bolt and the mount through-bolt.

13. Remove the 2 fuel filter bracket nuts from the left transaxle mount. Position the filter and bracket aside without disconnecting the fuel lines.

14. Remove the left-hand engine support insulator bracket.

15. Disconnect the Pulse Signal Generator (PSG) electrical connector.

16. Raise and safely support the vehicle.

17. Remove the front wheels and the splash shields.

18. Remove the 6 crossmember bolts and the crossmember.

19. Remove the 6 rear engine support nuts and 2 bolts and remove the rear engine support.

20. Remove the 2 lower transaxle support insulator bolts and remove the lower transaxle support insulator.

21. Remove the halfshafts and install transaxle plugs to prevent the differential gears from becoming misaligned.

22. Disconnect the transaxle vent hose and the dipstick tube.

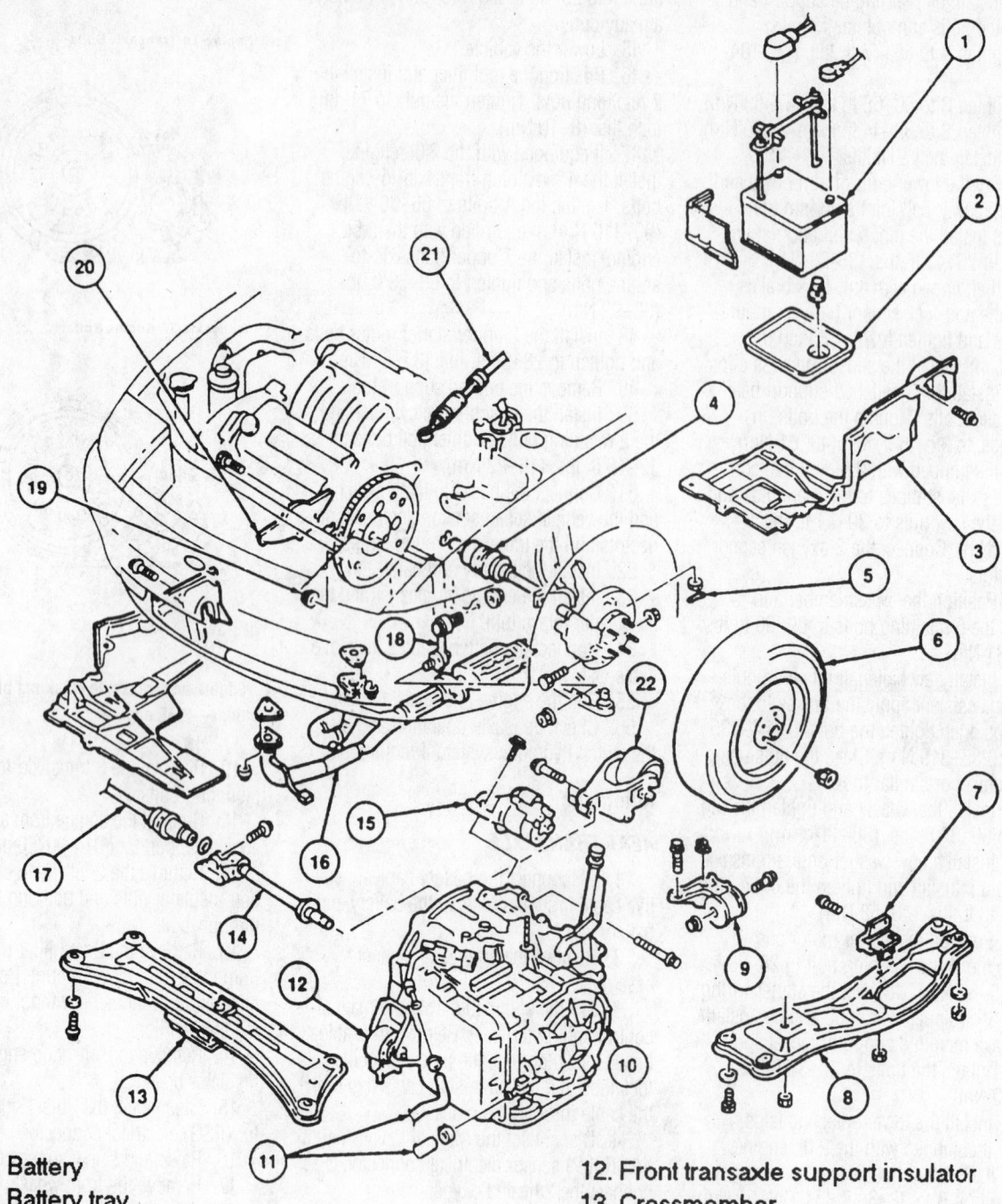

1. Battery
2. Battery tray
3. Front fender splash shield
4. LH front wheel driveshaft & joint
5. Cotter pin
6. Wheel & tire
7. Lower transaxle support insulator
8. Rear engine support
9. LH transaxle support insulator
10. Transaxle
11. Oil cooler tubes

12. Front transaxle support insulator
13. Crossmember
14. Halfshaft
15. Starter motor
16. Exhaust inlet pipe
17. RH front wheel driveshaft & joint
18. Stabilizer bar link
19. Flywheel to converter retaining nut
20. Cover plate
21. Transmission shift cable & bracket
22. Rear transaxle support insulator

7922HG49

Exploded view of all transaxle removal/installation related components—4EAT transaxle

23. Remove the 3 inspection cover bolts. Use a small prybar to hold the flexplate and remove the 4 torque converter nuts.

24. Support the transaxle with a suitable transmission jack. Secure the transaxle to the transmission jack to keep it from falling.

25. Remove the 2 engine-to-transaxle bolts.

26. Remove the 3 rear transaxle support bracket bolts.

27. Use a small prybar to separate the transaxle from the engine. Slightly tilt the transaxle and engine to ease removal.

28. Carefully remove the transaxle from the engine and lower the transaxle from the vehicle.

To install:

➡If the transaxle was rebuilt or replaced, be sure to flush the transmission oil cooler tubes to remove contaminants before installing the oil cooler tubes to the transaxle.

29. Position the transaxle onto a suitable transmission jack and secure the transaxle to the jack.

30. Raise the transaxle into position. Align the torque converter studs with the flexplate.

31. Install the 2 engine-to-transaxle bolts and tighten to 50–73 ft. lbs. (68–99 Nm).

32. Install the 3 rear transaxle support bracket bolts and tighten to 50–68 ft. lbs. (67–93 Nm).

33. Install the 4 torque converter-to-flexplate nuts and tighten to 32–45 ft. lbs. (44–60 Nm).

34. Install the inspection cover.

35. Connect the transaxle vent hose and install the dipstick tube. Tighten the dipstick tube mounting bolts to 71–88 inch lbs. (8–10 Nm).

36. Remove the transaxle plugs and install the halfshafts.

37. Install the lower transaxle support insulator and tighten the bolts to 41–59 ft. lbs. (55–80 Nm).

38. Remove the transmission jack.

39. Install the rear engine support. Tighten the rear engine support-to-body bolts and nuts to 50–68 ft. lbs. (67–93 Nm). Tighten the rear engine support-to-front mount nuts to 55–77 ft. lbs. (75–104 Nm) and the rear engine support-to-rear mount nuts to 32–44 ft. lbs. (44–60 Nm).

40. Install the crossmember and tighten the 6 bolts to 68–96 ft. lbs. (94–131 Nm).

41. Install the splash shields and the front wheels. Tighten the lug nuts to 66–86 ft. lbs. (88–118 Nm).

42. Lower the vehicle.

43. Install the 3 transaxle-to-engine support bolts and tighten to 50–73 ft. lbs. (68–99 Nm).

44. Connect the VSS and the PSG electrical connectors. Install the ground wire bracket and the ground wire.

45. Install the harness support bracket to the engine block located at the left-hand engine support insulator bracket.

46. Install the left-hand engine support insulator and 2 nuts and 1 bolt. Tighten the 2 nuts and bolt to 50–68 ft. lbs. (67–93 Nm). Install the through-bolt and tighten to 63–86 ft. lbs. (86–116 Nm).

47. Remove the engine support tool.

48. Install the fuel filter bracket and tighten the 2 nuts to 71–88 inch lbs. (8–10 Nm).

49. Unplug and connect the inlet and outlet oil cooler tubes.

50. Install the starter motor and wiring.

51. Connect the transaxle electrical connector and the 2 heated oxygen sensor electrical connectors.

52. Insert the shift cable through the cable bracket and pull the cable until the lock tabs engage. Install the lock tab retainer. Connect the shift cable to the manual control shift outer lever.

53. Connect the MLP sensor electrical connector. Snap the wiring harness bracket on the cable bracket.

54. Install the air cleaner assembly.

55. Install the battery tray and battery. Connect the battery cables, negative cable last.

56. Fill the transaxle with the proper type and quantity of fluid.

57. Run the engine and check for leaks. Road test and check for proper transaxle operation.

CD4E TRANSAXLE

1. Disconnect the battery cables, negative cable first, and remove the battery and battery tray.

2. Remove the air cleaner assembly.

3. Disconnect the Manual Lever Position (MLP) sensor electrical connector. Remove the 2 MLP sensor bolts and the sensor.

4. Remove the ground wire bracket and the ground wire.

5. Remove the shift cable from the cable bracket. Remove the 2 cable bracket bolts and remove the bracket.

6. Disconnect the transaxle electrical connector by pushing on the retaining ring and gently pulling up on the connector.

7. Disconnect and plug the oil cooler inlet and outlet hoses.

8. Remove the two upper starter motor bolts.

9. Support the engine from above with engine support tool 014-00750 or equivalent.

10. Remove the top 3 transaxle-to-engine mounting bolts.

11. Remove the 2 fuel filter bracket nuts from the left-hand transaxle mount. Position the filter and bracket aside without disconnecting the fuel lines.

12. Remove the 2 ignition coil nuts and position the ignition coil out of the way.

13. Remove the 3 speed control servo nuts and position the speed control servo out of the way.

14. Remove the 3 ignition coil mounting strap bolts.

15. Disconnect the wiring harness clips from the ignition coil mounting straps. Remove the ignition coil mounting straps.

16. Remove the 2 nut and 2 bolts from the left-hand engine support insulator. Remove the left-hand support insulator through-bolt. Remove the left-hand support insulator.

17. Raise and safely support the vehicle.

18. Remove the front wheels and the splash shields.

19. Remove the 6 crossmember bolts and the crossmember.

20. Remove the 6 rear engine support nuts and 2 bolts and remove the rear engine support.

21. Remove the halfshafts. Install transaxle plugs into the differential side gears.

22. Remove the intake manifold support bolts and the support.

23. Disconnect the starter motor wiring. Remove the lower starter motor bolt and remove the starter motor.

24. Disconnect the Transmission Speed Sensor (TSS) connector.

25. Remove the 4 torque converter-to-flywheel retaining nuts through the starter opening.

26. Remove the front transaxle support insulator through-bolt and remove the front support insulator.

27. Remove the 4 front transaxle support bracket bolts and remove the support bracket.

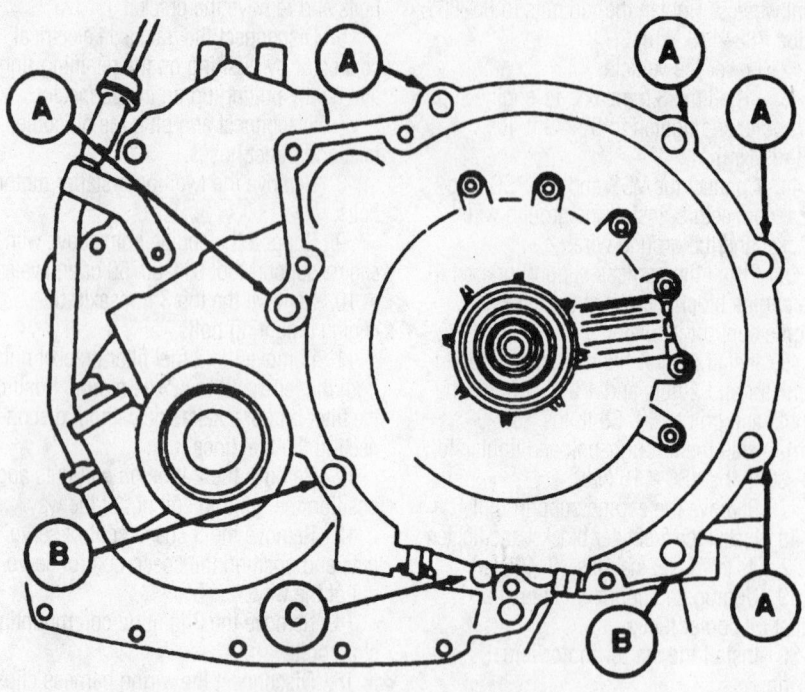

Transaxle-to-engine mounting bolts—CD4E transaxle

28. Remove the rear transaxle support insulator through-bolt.

29. Lower the vehicle.

30. Lower the transaxle by loosening the bolt on the engine support.

31. Raise and safely support the vehicle.

32. Secure the transaxle to a suitable jack using the appropriate adapters.

33. Remove the 3 engine-to-transaxle bolts.

34. Remove the 2 remaining transaxle-to-engine bolts.

35. Use a small prybar or similar tool to separate the transaxle from the engine.

36. Partially lower the transaxle from the engine.

37. Disconnect the Vehicle Speed Sensor (VSS) connector.

38. Finish lowering the transaxle and remove from the vehicle.

To install:

➡ **If the transaxle was rebuilt or is being replaced, be sure to flush the transmission oil cooler hoses to remove contaminants before installing the oil cooler hoses to the transaxle.**

39. Place the transaxle on a suitable jack using the appropriate adapters.

40. Raise the transaxle into position. Align the torque converter studs with the flexplate.

41. Install the transaxle-to-engine bolts.

Tighten bolts **B** to 28–38 ft. lbs. (38–51 Nm) and bolts **C** to 14–18 ft. lbs. (19–25 Nm).

42. Remove the jack.

43. Install the rear transaxle support insulator through-bolt. Do not tighten fully at this time.

44. Install the front transaxle support bracket. Tighten the 4 front transaxle support bracket bolts to 28–38 ft. lbs. (38–51 Nm). Install the front transaxle support insulator and through-bolt. Do not tighten the through-bolt fully at this time.

45. Connect the TSS and VSS connectors.

46. Install the 4 torque converter-to-flexplate nuts and tighten to 24–30 ft. lbs. (33–40 Nm).

47. Install the torque converter access plug.

48. Lower the vehicle.

49. Install the left-hand engine support insulator and tighten the through-bolt to 63–86 ft. lbs. (86–116 Nm). Install the 2 left-hand engine support insulator nuts and 2 bolts. Do not tighten them fully at this time.

50. Remove the engine support.

51. Raise and safely support the vehicle.

52. Install the rear engine support. Tighten the rear engine support bolts and nuts as follows:

• Tighten **A** to 55–77 ft. lbs. (75–104 Nm)

• Tighten **B** to 50–68 ft. lbs. (67–93 Nm)

• Tighten **C** to 32–44 ft. lbs. (44–60 Nm)

53. Tighten the front and rear transaxle support insulator through-bolts to 63–86 ft. lbs. (86–116 Nm).

54. Install the starter motor and the lower starter motor bolt. Tighten the lower starter motor bolt to 23–34 ft. lbs. (31–46 Nm). Connect the starter motor wiring.

55. Install the intake manifold support and bolts. Tighten the bolts to 27–38 ft. lbs. (37–52 Nm).

56. Remove the transaxle plugs and install the halfshafts.

57. Install the crossmember and 6 bolts. Tighten the crossmember bolts to 68–96 ft. lbs. (94–131 Nm).

58. Install the splash shields and the wheels. Tighten the lug nuts to 67–86 ft. lbs. (88–118 Nm).

59. Lower the vehicle.

60. Tighten the 2 left-hand support insulator nuts to 12–17 ft. lbs. (16–23 Nm). Tighten the 2 left-hand engine support insulator bolts to 28–38 ft. lbs. (38–51 Nm).

61. Position the ignition coil mounting straps and install the wiring harness clips and the 3 ignition coil strap bolts.

62. Position the speed control servo and install and tighten the 3 nuts.

63. Position the ignition coil and install the 2 nuts. Tighten the ignition coil nuts to 71–88 inch lbs. (8–10 Nm).

64. Install the 2 fuel filter bracket nuts and tighten to 71–97 inch lbs. (8–11 Nm).

65. Install the 2 upper starter motor bolts and tighten to 23–34 ft. lbs. (31–46 Nm).

66. Install the 3 remaining transaxle-to-engine bolts. Tighten the **A** bolts to 66–86 ft. lbs. (89–117 Nm).

67. Install the shift cable bracket and 2 bolts.

68. Connect the transaxle electrical connector.

69. Install the MLP sensor and the two MLP sensor bolts. Adjust the MLP sensor and tighten the bolts to 96–117 inch lbs. (11–13 Nm). Connect the MLP sensor electrical connector

70. Connect the oil cooler inlet and outlet hoses.

71. Install the air cleaner assembly.

72. Install the battery tray and battery.

73. Install the ground wire bracket and the ground wire.

74. Connect the battery cables, negative cable last.

75. Fill the transaxle with the proper type and quantity of fluid.

76. Run the engine and check for leaks.

77. Road test and check for proper transaxle operation.

Clutch

ADJUSTMENTS

Pedal Height

1. To determine if the clutch pedal height requires an adjustment, measure the distance from the bulkhead to the upper center of the pedal pad. The distance should be 7.32–8.31 inch (186–211mm).

2. If adjustment is required, loosen lock-nut **A** and turn adjusting bolt **B** until the desired pedal height is reached.

3. Tighten the locknut to 122–156 inch lbs. (14–17 Nm).

Pedal Free-Play

1. Measure the pedal height.

2. Depress the clutch pedal by hand and measure the height of the pedal when resistance is felt.

3. The free-play should be 0.04–0.12 inch (1–3mm).

4. If adjustment is necessary, proceed as follows:

   a. Loosen locknut **C** and turn clutch master cylinder pushrod **D** until the pedal play is within specifications.

   b. Measure the distance from the floor to the center of the pedal pad when the pedal is fully depressed. The distance should be 2.64 inches (67mm).

   c. Tighten the locknut to 105–147 inch lbs. (12–16 Nm).

REMOVAL & INSTALLATION

1. Disconnect the negative battery cable.

2. Remove the transaxle assembly.

3. Position a suitable clutch alignment tool through the pressure plate, clutch disc and into the pilot bearing; this will keep the assembly from dropping when the bolts are removed.

4. Install flywheel holding tool T74P-6375-A or equivalent, to keep the flywheel from turning.

5. Remove the pressure plate-to-fly-wheel bolts evenly to relieve the spring pressure.

6. Remove the pressure plate, clutch disc and alignment tool.

7. Inspect the pressure plate and clutch disc for wear and/or damage and replace, as necessary.

8. Inspect the pilot bearing for excessive wear or scoring. Remove it using a suitable puller, only if replacement is necessary.

9. Inspect the flywheel for scoring, cracks, worn or broken teeth, or other damage. Remove the flywheel if machining or replacement is necessary. Use care when removing the last bolt to prevent dropping the flywheel.

10. Remove the release bearing and fork. Inspect them for wear or damage and replace as necessary.

To install:

11. Apply molybdenum grease to the release bearing where it contacts the release fork. Apply molybdenum grease to the release fork at the pivot point and to the area where it contacts the release bearing.

12. Install the release fork and bearing.

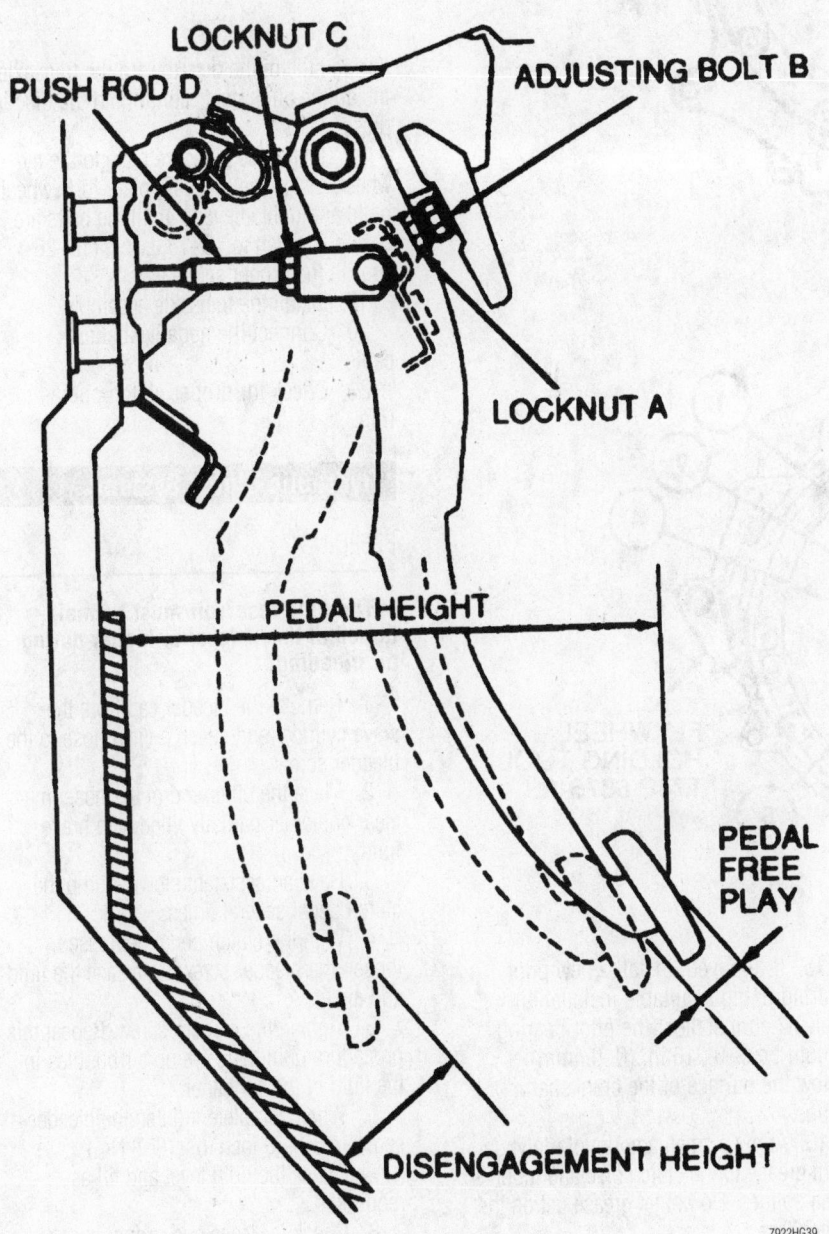

LOCKNUT C
PUSH ROD D
ADJUSTING BOLT B
LOCKNUT A
PEDAL HEIGHT
PEDAL FREE PLAY
DISENGAGEMENT HEIGHT

7922HG39

Clutch pedal adjustments

2.0L

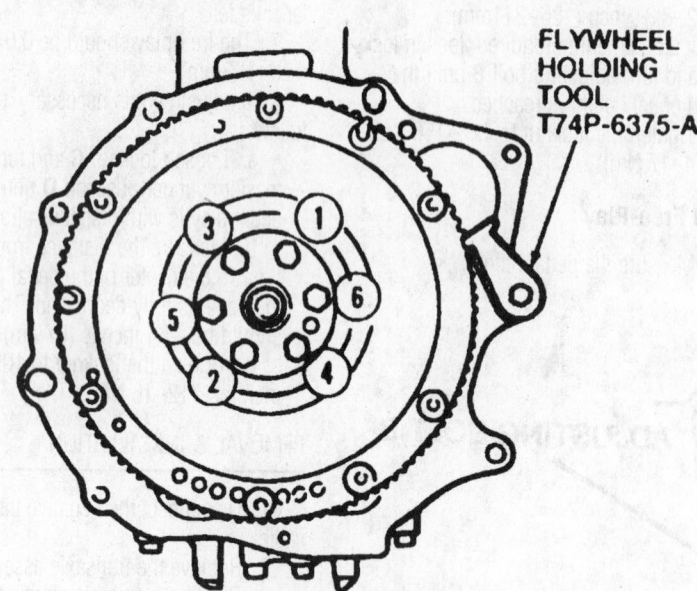

FLYWHEEL HOLDING TOOL T74P-6375-A

2.5L

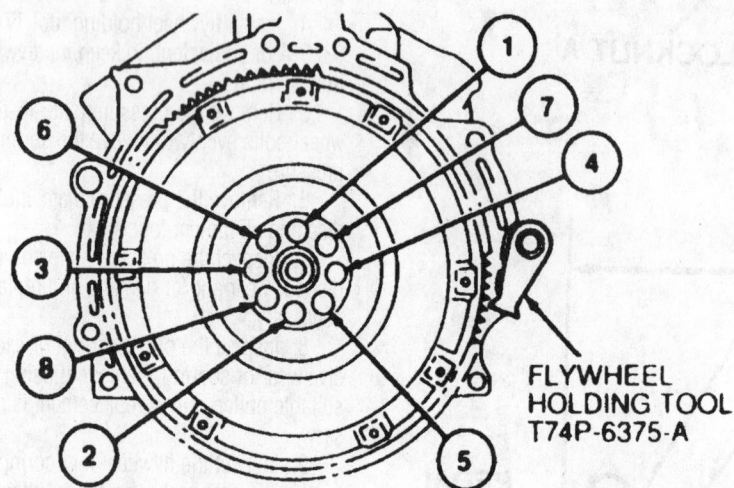

FLYWHEEL HOLDING TOOL T74P-6375-A

7922HG37

Flywheel bolt tightening sequences

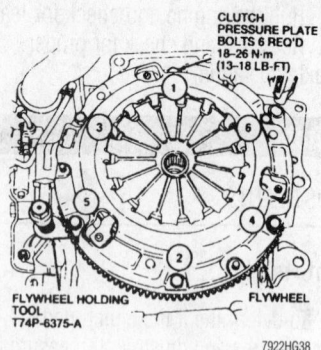

CLUTCH PRESSURE PLATE BOLTS 6 REQ'D 18-26 N·m (13-18 LB-FT)

FLYWHEEL HOLDING TOOL T74P-6375-A — FLYWHEEL

7922HG38

Clutch pressure plate bolt tightening sequence

13. If removed, install the flywheel. Be sure the crankshaft flange and flywheel mating surfaces are clean. Remove the old sealant from the flywheel bolts and apply stud and bearing mount sealant to them. If the old sealant cannot be removed, replace the bolts.

14. Install the flywheel holding tool. Tighten the flywheel bolts, in sequence, to 71–75 ft. lbs. (97–102 Nm) on 2.0L engine or 45–49 ft. lbs. (61–67 Nm) on 2.5L engine.

15. If removed, install a new pilot bearing using a suitable installation tool. When installed, the pilot bearing should be 0–0.016 in. (0–0.4mm) below the surface of the crankshaft flange.

16. Apply a small amount of molybdenum grease to the clutch disc and input shaft splines. Do not let grease get on the clutch face.

17. Install the clutch disc with the spring side of the disc toward the transaxle. Install the alignment tool to hold the disc in place.

18. Install the pressure plate to the flywheel. Install the pressure plate-to-flywheel bolts and tighten evenly until the bolts are seated. Tighten to 13–18 ft. lbs. (18–26 Nm) in the proper sequence.

19. Install the transaxle assembly.

20. Connect the negative battery cable.

21. Check for proper clutch operation.

Hydraulic Clutch System

BLEEDING

➡ **The fluid reservoir must be maintained at the ¾ level or higher during air bleeding.**

1. Remove the bleeder cap from the slave cylinder and attach a vinyl hose to the bleeder screw.

2. Place the other end of the hose in a clear container partially filled with brake fluid.

3. Have an assistant slowly pump the clutch pedal several times.

4. With the clutch pedal depressed, loosen the bleeder screw to release the fluid and air.

5. Tighten the bleeder screw. Repeat this procedure until there are no air bubbles in the fluid in the container.

6. When complete, tighten the bleeder screw to 53–78 inch lbs. (6–8 Nm).

7. Check the fluid level and fill as required.

8. Check for leaks and proper clutch operation.

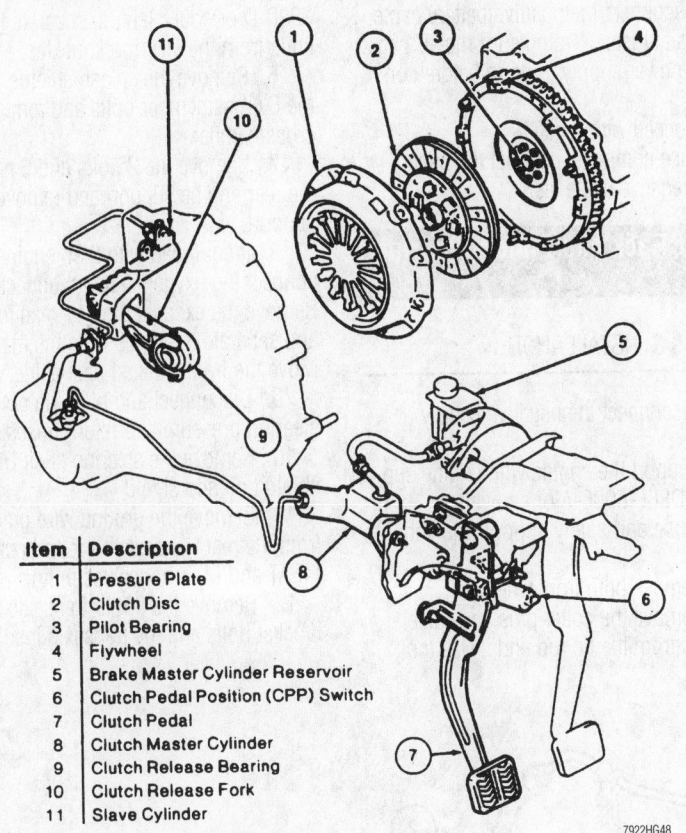

Item	Description
1	Pressure Plate
2	Clutch Disc
3	Pilot Bearing
4	Flywheel
5	Brake Master Cylinder Reservoir
6	Clutch Pedal Position (CPP) Switch
7	Clutch Pedal
8	Clutch Master Cylinder
9	Clutch Release Bearing
10	Clutch Release Fork
11	Slave Cylinder

7922HG48

Component identification for the hydraulic clutch system

Halfshaft

REMOVAL & INSTALLATION

1. Disconnect the negative battery cable.
2. With the vehicle sitting on all 4 wheels, use a chisel to raise the staked portion of the hub nut. Have an assistant apply the brakes to lock the wheels, then loosen but do not remove the hub nut.
3. Raise and safely support the vehicle.
4. Remove the wheel and the necessary inner fender splash guards.
5. Remove the stabilizer link assembly from the lower control arm.
6. Remove the ball joint clamp bolt from the lower control arm. Carefully pry the lower control arm downward to separate the ball joint from the steering knuckle.

➡**If removing the right halfshaft, remove the support bearing bracket from the cylinder block.**

7. Separate the halfshaft from the transaxle by positioning a prybar between the halfshaft and transaxle case. Pry out the halfshaft while pulling out on the steering knuckle. Be careful not to damage the transaxle case, transaxle oil seal, CV-joint or CV-joint boot.

8. Remove and discard the hub nut. Pull the halfshaft out of the wheel hub. If necessary, use a plastic hammer to tap it out or a wheel puller to press it out. Do not use a metal hammer.
9. Support the halfshaft and slide it out of the transaxle.
10. Install transaxle plug T88C-7025-AH or equivalent, into the halfshaft opening of the transaxle case and into the differential side gear; this will keep the differential side gear from falling out of place.

To install:

➡**If replacing a halfshaft and joint on an ABS equipped vehicle, install a new front brake anti-lock sensor indicator.**

11. On the end of the halfshaft, install a new circlip. Start one end of the clip in the groove and work the clip over the stub shaft end and into the groove. This will prevent over-expanding the clip. Be sure the end gap is positioned at the top of the splines.
12. Remove the transaxle plug and inspect the transaxle oil seal. Replace if necessary.
13. Lubricate the halfshaft splines with a suitable grease, align the splines with the differential side gear and push the halfshaft into the differential. Be sure the retaining clip clicks into the differential side gear groove.

14. Position the halfshaft through the wheel hub and install a new attaching nut. Do not tighten the nut at this time.
15. If installing the right halfshaft, install the halfshaft support bearing and tighten the mounting bolts to 31–46 ft. lbs. (42–62 Nm).

➡**The support bearing bolts must be tightened in the proper sequence.**

16. Position the ball joint in the steering knuckle and install the clamp bolt/nut. Tighten the nut to 25–42 ft. lbs. (34–57 Nm).
17. Install the stabilizer link assembly and tighten the nuts to 27–40 ft. lbs. (36–54 Nm).
18. Install the splash shields and wheel and lower the vehicle.
19. With the vehicle sitting on all 4 wheels, have an assistant apply the brakes and tighten the halfshaft attaching nut to 174–235 ft. lbs. (235–319 Nm). Stake the nut using a suitable chisel with a rounded cutting edge.

➡**If the nut splits or cracks after staking, it must be replaced with a new nut.**

20. Connect the negative battery cable.
21. Check the transaxle fluid level. Road test the vehicle and check for leaks.

STEERING AND SUSPENSION

Air Bag

❊❊ CAUTION

The Supplemental Restraint System (SRS) must be disarmed before removing the air bag module. Failure to do so may cause accidental deployment of the air bag, resulting in unnecessary SRS repairs and/or personal injury.

PRECAUTIONS

Several precautions must be observed when handling the inflator module to avoid accidental deployment and possible personal injury.

- Never carry the inflator module by the wires or connector on the underside of the module.
- When carrying a live inflator module, hold securely with both hands, and ensure that the bag and trim cover are pointed away.

- Place the inflator module on a bench or other surface with the bag and trim cover facing up.
- With the inflator module on the bench, never place anything on or close to the module which may be thrown in the event of an accidental deployment.

DISARMING

❊❊ CAUTION

The air bag must be disarmed before performing service around air bag components or air bag wiring. Failure to do so may cause accidental deployment of the air bag, resulting in unnecessary repairs and/or personal injury.

---

1. Position the vehicle with the front wheels in a straight ahead position.
2. Disconnect the negative battery cable.

3. Disconnect the positive battery cable.
4. Wait at least 1 minute for the air bag back-up power supply to drain before continuing.
5. Proceed with repair.
6. Once complete, connect the battery cables, negative cable last.

Power Rack and Pinion Steering Gear

REMOVAL & INSTALLATION

1. Disconnect the negative battery cable.
2. Support the engine with engine support tool D88L-6000-A or equivalent.
3. Raise and safely support the vehicle.
4. Remove both front wheels.
5. Remove the cotter pins and castellated nuts from the tie rod ends. Use tool

3290-D or equivalent, to separate the tie rod ends from the steering knuckles.
6. Remove the splash shields. Remove the 6 crossmember bolts and remove the crossmember.
7. Remove the 2 bolts and 6 nuts from the rear engine support and remove the support.
8. If equipped with 2.5L engine, disconnect the oxygen sensor connectors. Remove the exhaust pipe-to-manifold nuts and separate the pipes from the manifolds. Move the front exhaust pipe aside.
9. Disconnect and plug the power steering pressure and return hoses.
10. Remove the steering shaft U-joint shield bolt and shield.
11. Remove the ground wire bracket from the rear transaxle support bracket (4EAT and MTX transaxles only).
12. Remove the 3 rear transaxle support bracket bolts and the transaxle insulator

2.0L Engine

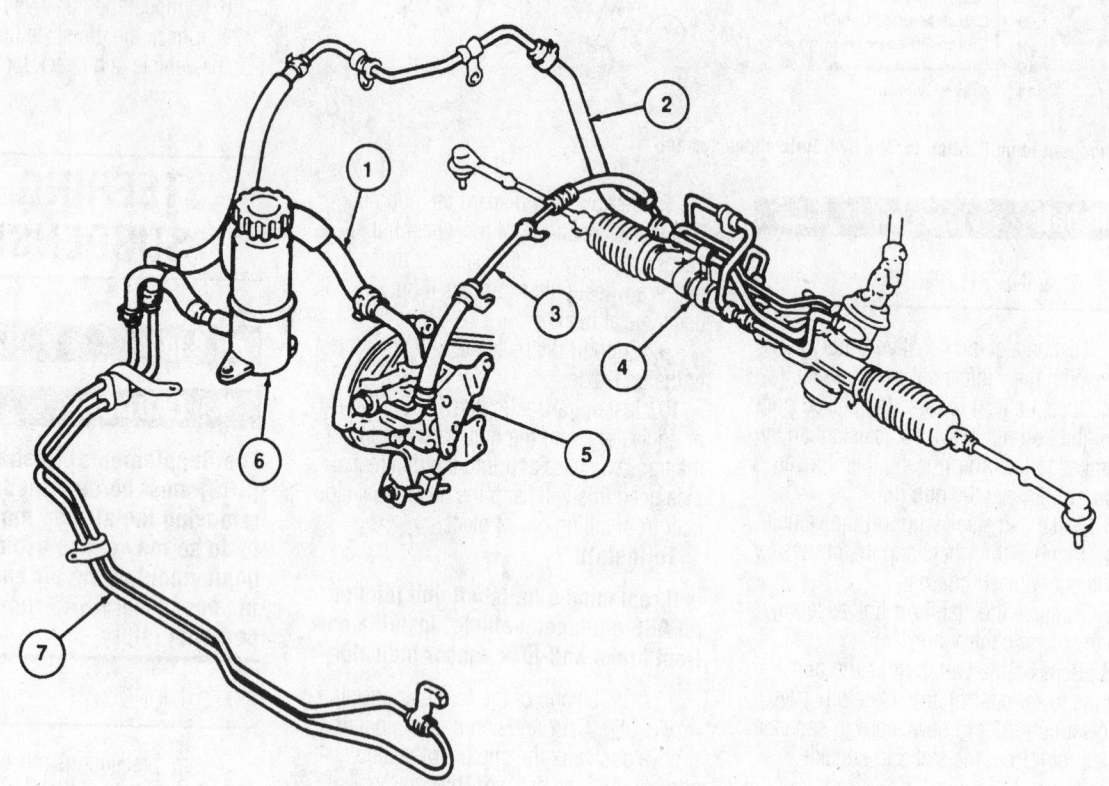

1	Power Steering Reservoir Pump Hose	4	Steering Gear
2	Power Steering Return Hose	5	Power Steering Pump
3	Power Steering Pressure Hose	6	Power Steering Pump Reservoir
		7	Power Steering Fluid Cooler

Power rack and pinion steering gear component identification—2.0L engine

2.5L Engine

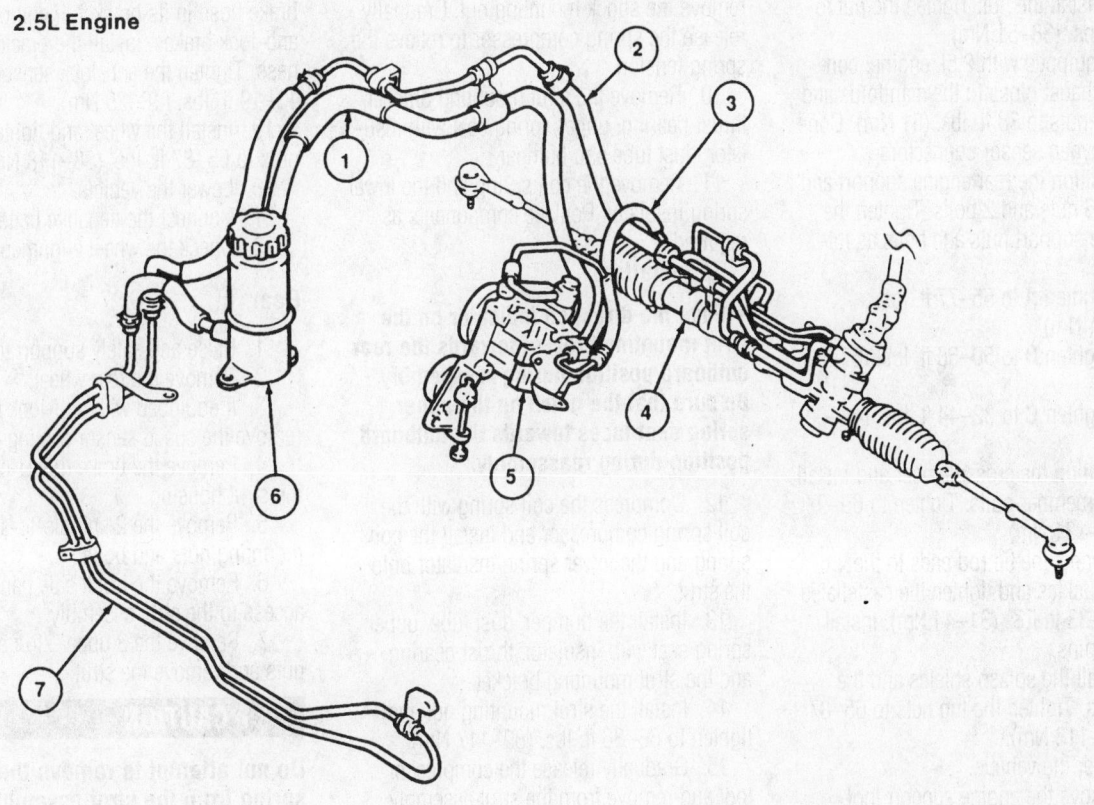

1	Power Steering Reservoir Pump Hose	4	Steering Gear
2	Power Steering Return Hose	5	Power Steering Pump
3	Power Steering Pressure Hose	6	Power Steering Pump Reservoir
		7	Power Steering Fluid Cooler

7922HG50

Power rack and pinion steering gear component identification—2.5L engine

through-bolt. Remove the transaxle support bracket (4EAT and MTX transaxles only).

13. Remove the 4 bolts from the 2 rack and pinion assembly mounting brackets and remove the brackets.

14. Remove the steering column lower yoke-to-power steering gear input shaft and bolt.

15. If equipped with a manual transaxle, remove the control rod to support bar stud nut and position the control rod aside.

16. Position a jack under the front sub-frame. Remove the 6 sub-frame bolts and 2 nuts.

17. Remove the vent tube attached to the drivers side of the sub-frame.

18. Remove the upper stabilizer bar link nuts.

19. Lower the sub-frame to allow removal of the rack and pinion assembly. Remove the rack and pinion assembly from the drivers side of the vehicle.

To install:

➡️If a new rack and pinion assembly is being installed or if the old assembly was turned, place the assembly into a soft jaw vise and rotate the steering gear input shaft counting the number of turns lock to lock. Back the steering gear input shaft up half of the number of the turns counted to center the rack and pinion. Do not damage the input shaft splines.

20. Position the rack and pinion assembly in the vehicle.

21. With the aide of an assistant, install the steering column lower yoke-to-power steering gear input shaft and bolt. Tighten the bolt to 13–20 ft. lbs. (18–26 Nm).

22. Raise the front sub-frame into position. Install the sub-frame mounting bolts and nuts and tighten to 69–97 ft. lbs. (93–131 Nm).

23. Remove the jack.

24. Install the upper stabilizer bar link nuts and tighten to 27–40 ft. lbs. (36–54 Nm). Install the vent tube.

25. Position the 2 rack and pinion assembly mounting brackets and install the 4 mounting bolts. Tighten to 28–38 ft. lbs. (38–51 Nm).

26. Position the rear transaxle support insulator and install the through-bolt, if equipped. Tighten the through-bolt to 63–86 ft. lbs. (85–117 Nm).

27. Install the 3 rear transaxle support bracket bolts if equipped, and tighten to 50–68 ft. lbs. (67–93 Nm). Install the ground wire bracket to the rear engine mount.

28. Install the steering shaft U-joint shield and bolt and tighten securely.

29. Remove the plugs and connect the power steering pressure and return lines.

30. If equipped with a manual transaxle,

install the control rod to the support bar stud and install the nut. Tighten the nut to 28–38 ft. lbs. (38–51 Nm).

31. If equipped with 2.5L engine, connect the exhaust pipes to the manifolds and tighten the nuts to 38 ft. lbs. (51 Nm). Connect the oxygen sensor connectors.

32. Position the rear engine support and install the 6 nuts and 2 bolts. Tighten the rear engine support nuts and bolts as follows:

    a. Tighten **A** to 55–77 ft. lbs. (75–104 Nm)

    b. Tighten **B** to 50–68 ft. lbs. (67–93 Nm)

    c. Tighten **C** to 32–44 ft. lbs. (44–60 Nm)

33. Position the crossmember and install the 4 crossmember bolts. Tighten to 69–97 ft. lbs. (93–131 Nm).

34. Connect the tie rod ends to the steering knuckles and tighten the castellated nuts to 23–33 ft. lbs. (31–44 Nm). Install new cotter pins.

35. Install the splash shields and the front wheels. Tighten the lug nuts to 65–87 ft. lbs. (88–118 Nm).

36. Lower the vehicle.

37. Remove the engine support tool.

38. Fill the power steering system with the proper fluid.

39. Connect the negative battery cable.

40. Run the engine and check for leaks. Bleed the air from the system.

41. Top off the power steering reservoir when complete.

42. Check the front wheel alignment.

Strut and Spring

REMOVAL & INSTALLATION

Front

1. Disconnect the negative battery cable.
2. Raise and support the vehicle safely.
3. Remove the wheel.
4. If equipped with anti-lock brakes, disconnect the electrical harness and remove the bracket.
5. Remove the U-clip from the brake line hose and slide it out of the strut bracket.
6. Remove the 2 strut-to-steering knuckle bolts and nuts.
7. Lower the vehicle enough to remove the 4 upper strut mounting bolts.
8. Remove the strut from the vehicle.
9. Place the strut assembly in a suitable holding fixture. Loosen, but do not remove the shock mounting nut. Compress the

spring with a suitable compressor tool, then remove the shock mounting nut. Gradually release the spring compressor to relieve the spring tension.

10. Remove the strut mounting bracket, thrust bearing, upper spring seat with insulator, dust tube and bumper.

11. Remove the coil spring and the lower spring insulator. Replace components as required.

To install:

➡ **Face the direction indicator on the strut mounting bracket towards the rear outboard position during reassembly. Be sure that the notch on the upper spring seat faces towards the outboard position during reassembly.**

12. Compress the coil spring with the coil spring compressor and install the coil spring and the lower spring insulator onto the strut.

13. Install the bumper, dust tube, upper spring seat with insulator, thrust bearing and the strut mounting bracket.

14. Install the strut mounting nut and tighten to 66–86 ft. lbs. (89–117 Nm).

15. Gradually release the compressor tool and remove from the strut assembly.

16. Install the strut in the shock tower with the direction indicator facing the rear outboard position. Tighten the 4 upper strut mounting nuts to 34–46 ft. lbs. (46–63 Nm).

17. Position the strut to the steering knuckle and tighten the nuts and bolts to 68–86 ft. lbs. (93–117 Nm).

18. Install the brake caliper and the brake hose in its bracket. If equipped with anti-lock brakes, install the bracket and harness. Tighten the anti-lock sensor bracket to 13–19 ft. lbs. (18–25 Nm).

19. Install the wheel and tighten the lug nuts to 65–87 ft. lbs. (88–118 Nm).

20. Lower the vehicle.

21. Connect the negative battery cable.

22. Check the wheel alignment.

Rear

1. Raise and safely support the vehicle.
2. Remove the rear wheel.
3. If equipped with anti-lock brakes, remove the speed sensor routing bracket.
4. Remove the brake line U-clip from the strut housing.
5. Remove the 2 spindle-to-strut mounting nuts and bolts.
6. Remove the trunk side panel to gain access to the strut assembly.
7. Remove the 3 upper strut attaching nuts and remove the strut.

✷✷ CAUTION

Do not attempt to remove the coil spring from the strut assembly without compressing the coil spring first.

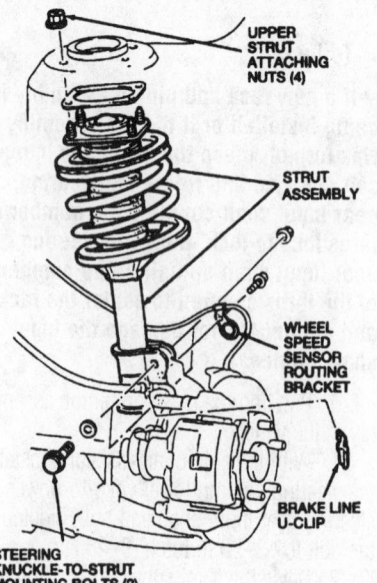

Front strut attaching nut and bolt locations

UPPER STRUT ATTACHING NUTS (4)

STRUT ASSEMBLY

WHEEL SPEED SENSOR ROUTING BRACKET

BRAKE LINE U-CLIP

STEERING KNUCKLE-TO-STRUT MOUNTING BOLTS (2)

7922HG43

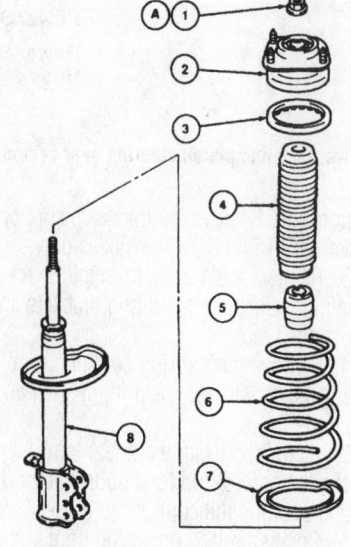

1. Shock absorber strut mounting nut
2. Rear shock absorber bracket
3. Rear spring anti-squeak insert, upper
4. Shock absorber dust tube
5. Rear shock absorber dust boot
6. Rear spring
7. Rear spring anti-squeak insert, lower
8. Rear shock absorber
A. 66–87 ft. lb.(89–117 Nm)

7922HG45

Exploded view of the rear strut assembly

8. Use spring compressor tool D85P-7178-A or equivalent, to compress the coil spring.

9. Remove the mounting nut from the strut and disassemble the strut components.

To install:

10. Assemble the strut, the compressed coil spring and related components.

11. Install the mounting nut and tighten to 66–87 ft. lbs. (89–117 Nm). Release the compressor and remove the strut.

➡**Be sure that the lower coil spring is seated properly.**

12. Position the strut in the vehicle and install the 3 upper strut nuts. Tighten to 34–46 ft. lbs. (46–63 Nm).

13. Install the trunk side panel.

14. Install the 2 spindle-to-strut bolts and nuts and tighten to 69–87 ft. lbs. (93–117 Nm).

15. Install the brake line U-clip.

16. Install the wheel speed sensor bracket, if equipped.

17. Install the wheel and tighten the lug nuts to 65–87 ft. lbs. (88–118 Nm).

18. Lower the vehicle.

19. Check the wheel alignment.

Lower Ball Joint

REMOVAL & INSTALLATION

The lower ball joint is an integral part of the lower control arm and cannot be serviced separately. If the lower ball joint is defective, the entire lower control arm must be replaced.

Lower Control Arm

REMOVAL & INSTALLATION

➡**The lower ball joint is an integral part of the lower control arm and cannot be serviced separately. If the lower ball joint is defective, the entire lower control arm must be replaced.**

1. Disconnect the negative battery cable.

2. Raise and safely support the vehicle.

3. Remove the wheel.

4. Remove the ball-joint clamp bolt and nut from the steering knuckle.

5. Remove the stabilizer bar link nut.

6. Using a prybar, pry downward to separate the ball joint from the steering knuckle.

7. Remove the 2 lower control arm rear mounting bolts.

8. Remove the lower control arm front mounting bolt and remove the lower control arm from the vehicle.

To install:

9. Position the lower control arm, install the 2 lower control arm rear mounting bolts and tighten to 69–96 ft. lbs. (93–131 Nm).

10. Install the lower control arm front mounting bolt and tighten to 58–78 ft. lbs. (78–106 Nm).

11. Install the ball joint stud into the steering knuckle and tighten the clamp bolt to 32–40 ft. lbs. (43–54 Nm).

12. Install the stabilizer bar link nut and tighten to 27–40 ft. lbs. (36–54 Nm).

13. Install the wheel and tighten the lug nuts to 65–87 ft. lbs. (88–118 Nm).

14. Lower the vehicle.

15. Check the front suspension and steering for proper operation. Check wheel alignment.

Wheel Bearings

ADJUSTMENT

➡**Wheel bearings are sealed units and are not adjustable or able to be greased. The wheel bearing unit must be replaced for noisy or rough operation or excessive end-play.**

1. Raise and safely support the vehicle.

2. Be sure the parking brake is fully released.

3. Remove the wheel.

4. On disc brakes, remove the disc brake caliper.

5. On rear disc brakes install the lug nuts to hold the rotor in place.

6. Rotate the drum or rotor to be sure there is no brake drag.

7. Position a suitable dial indicator.

8. Check the wheel bearing end-play. End-play should not exceed 0.002 in. (0.05mm).

9. If the end-play exceeds specification, replace the wheel bearing or hub/bearing assembly, as required.

10. When complete, install the calipers if removed.

11. Install the wheel and tighten the lug nuts to 65–87 ft. lbs. (88–118 Nm).

REMOVAL & INSTALLATION

Front

1. Disconnect the negative battery cable.

2. Raise and safely support the vehicle.

3. Remove the front wheel.

4. Using a small cape chisel or similar tool and a hammer, raise the staked portion of the hub retaining nut. With an assistant applying the brakes to prevent the hub from turning, loosen and remove the wheel hub retaining nut. Discard the nut after removal; it must not be reused.

5. If equipped with anti-lock brakes, remove the 2 anti-lock sensor retaining bolts and the sensor and bracket.

6. At the tie rod end, remove the cotter pin and castellated nut. Using a tie rod end

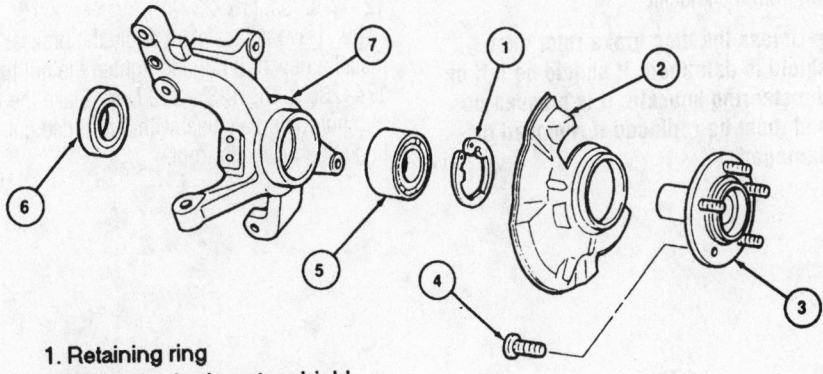

1. Retaining ring
2. Front disc brake rotor shield
3. Wheel hub
4. Wheel hub bolt
5. Front wheel bearing
6. Inner wheel bearing oil seal
7. Front wheel knuckle

7922HG44

Exploded view of the front wheel bearing assembly

separator tool or equivalent, separate the tie rod end from the steering knuckle.

7. Remove the lower control arm ball joint clamp nut/bolt. Using a prybar, pry the lower control arm downward and separate the ball joint from the steering knuckle.

8. Remove the steering knuckle-to-strut attaching bolts.

9. Remove the front disc brake caliper anchor bracket and suspend the caliper assembly from the coil spring with mechanics wire.

10. Remove the disc brake rotor.

11. Pull the steering knuckle/hub assembly off of the halfshaft. If necessary, lightly tap the end of the halfshaft CV-joint with a plastic faced hammer or use a wheel puller to press the shaft out of the hub. Support the halfshaft with wire or a stand; do not let the halfshaft hang on the inner CV-joint or the joint could be pulled apart.

12. Remove the steering knuckle, hub and wheel bearing assembly from the vehicle.

13. Using a prybar, pry the grease seal from the knuckle.

14. Position the steering knuckle in a suitable fixture and remove the hub from the knuckle. If the inner race remains on the hub, grind a section of the inner race to approximately 0.020 inch (0.5mm) and use a chisel to remove it. Wear appropriate eye protection.

15. Remove the snapring from the steering knuckle.

➡**Wheel bearings are contained within a sealed unit and are not serviceable.**

16. Position the steering knuckle in a suitable fixture and remove the wheel bearing from the knuckle.

➡**Unless the disc brake rotor dust shield is damaged, it should be left on the steering knuckle; it is pressed on and must be replaced if removed or damaged.**

To install:

17. Inspect the steering knuckle and hub for cracks, wear and scoring. Replace parts as necessary.

18. Position the steering knuckle in a suitable fixture and press in the wheel bearing. Be sure the press tool contacts only the outer bearing race or the bearing will be damaged.

19. Install the snapring.

20. Position the steering knuckle in a suitable fixture and press the hub into the steering knuckle. Be sure the inner bearing race is supported or the bearing will be damaged.

21. Apply grease to the lip of a new seal and press the seal into the knuckle, using a suitable seal installer.

22. Grease the halfshaft splines. Slide the hub/steering knuckle onto the halfshaft and position it into the strut bracket.

23. Install the strut-to-steering knuckle bolts and nuts and tighten to 68–86 ft. lbs. (93–117 Nm).

24. Push the lower control arm ball joint into the steering knuckle. Install the clamp bolt and nut and tighten to 25–42 ft. lbs. (34–57 Nm).

25. Install the disc brake rotor.

26. Position the caliper anchor bracket. Install the caliper anchor bracket-to-steering knuckle bolts and tighten to 58–72 ft. lbs. (78–98 Nm).

27. Connect the tie rod end to the steering knuckle and tighten the castellated nut to 22–33 ft. lbs. (29–44 Nm). Install a new cotter pin.

28. If equipped with anti-lock brakes, install the anti-lock brake sensor and bracket and tighten the 2 retaining bolts to 12–17 ft. lbs. (16–23 Nm).

29. Have an assistant apply the brakes. Install a new hub nut and tighten the nut to 174–235 ft. lbs. (235–319 Nm). Stake the hub nut, using a chisel with a rounded cutting edge or similar tool.

30. Install the wheel and tighten the lug nuts to 65–87 ft. lbs. (88–118 Nm).

31. Lower the vehicle.

32. Connect the negative battery cable.

33. Check the front brakes, suspension and steering for proper operation. Check the front wheel alignment.

Rear

➡**The wheel bearing cannot be disassembled from the hub and must be replaced as an assembly.**

1. Disconnect the negative battery cable.

2. Raise and safely support the vehicle.

3. Remove the rear wheel.

4. Carefully unstake the wheel hub retainer. Have an assistant apply the brakes to lock the hub, then remove the retainer. Discard the nut.

5. On vehicles equipped with rear disc brakes, remove the disc brake caliper and rotor.

6. On vehicles equipped with rear drum brakes, remove the brake drum.

7. Remove the wheel bearing and hub assembly from the wheel spindle.

To install:

8. Install the wheel bearing and hub assembly onto the wheel spindle.

9. On vehicles equipped with rear disc brakes, install the rotor and rear disc brake caliper.

10. On vehicles equipped with rear drum brakes, install the brake drum.

11. Install a new wheel hub retainer. Have an assistant apply the brakes to lock the hub, then tighten the wheel hub retainer to 130–174 ft. lbs. (177–235 Nm).

12. Stake the nut in place using a dull bladed chisel or similar tool.

13. Install the wheel and tighten the lug nuts to 65–87 ft. lbs. (88–118 Nm).

14. Lower the vehicle.

15. Connect the negative battery cable.

16. Pump the brake pedal to position the brakes, prior to moving the vehicle.

FORD MOTOR CO.

Ford-Contour • **Mercury**-Mystique • Cougar (1999)

DRIVE TRAIN 19-29
ENGINE REPAIR **19-2**
FUEL SYSTEM 19-27
**STEERING AND
SUSPENSION** **19-36**

A
Air Bag..................................19-36
  Disarming..............................19-37
  Precautions 19-36
C
Camshaft and Valve Lifters 19-16
  Removal & Installation 19-16
Clutch...................................19-34
  Adjustments..........................19-34
  Removal & Installation 19-34
Cylinder Head..........................19-8
  Removal & Installation 19-8
E
Engine Assembly 19-2
  Removal & Installation 19-2
Exhaust Manifold 19-14
  Removal & Installation 19-14
F
Front Crankshaft Seal 19-16
  Removal & Installation 19-16
Fuel Filter 19-28
  Removal & Installation 19-28

Fuel Pump 19-28
  Removal & Installation 19-28
Fuel System Pressure 19-28
  Relieving 19-28
Fuel System Service
  Precautions...........................19-27
H
Halfshaft................................19-35
  Removal & Installation 19-35
Hydraulic Clutch System 19-34
  Bleeding 19-34
I
Ignition Timing 19-2
  Adjustment 19-2
Intake Manifold 19-12
  Removal & Installation 19-12
L
Lower Ball Joint.......................19-41
  Removal & Installation 19-41
O
Oil Pan..................................19-19
  Removal & Installation 19-19
Oil Pump 19-21
  Removal & Installation 19-21
P
Power Rack and Pinion Steering
  Gear 19-37

Removal & Installation 19-37
R
Rear Main Seal 19-22
  Removal & Installation 19-22
Rocker Arms 19-11
  Removal & Installation 19-11
S
Strut and Spring 19-39
  Removal & Installation 19-39
T
Timing Chain, Sprockets, Front
  Cover and Seal 19-23
  Removal & Installation 19-23
Transaxle Assembly 19-29
  Removal & Installation 19-29
V
Valve Lash 19-19
  Adjustment 19-19
W
Water Pump............................19-7
  Removal & Installation 19-7
Wheel Bearings........................19-41
  Adjustment 19-41
  Removal & Installation 19-41

ENGINE REPAIR

➡Disconnecting the negative battery cable on some vehicles may interfere with the functions of the on board computer systems and may require the computer to undergo a relearning process, once the negative battery cable is reconnected.

Ignition Timing

ADJUSTMENT

The ignition timing is set at 10 degrees Before Top Dead Center (BTDC) and is not adjustable.

Engine Assembly

REMOVAL & INSTALLATION

2.0L Engine

✻✻ CAUTION

The fuel injection system remains under pressure, even after the engine has been turned OFF. The fuel system pressure MUST BE relieved before disconnecting any fuel lines. Failure to do so may result in fire and/or personal injury.

1. Disconnect the battery cables, negative cable first.

✻✻ CAUTION

Some models covered by this manual may be equipped with a Supplemental Restraint System (SRS), which uses an air bag. Whenever working near any of the SRS components, such as the impact sensors, the air bag module, steering column and instrument panel, properly disable the SRS.

2. Relieve the fuel system pressure using the recommended procedure.

✻✻ CAUTION

Observe all applicable safety precautions when working around fuel. Whenever servicing the fuel system, always work in a well ventilated area. Do not allow fuel spray or vapors to come in contact with a spark or open flame. Keep a dry chemical fire extinguisher near the work area. Always keep fuel in a container specifically designed for fuel storage; also, always properly seal fuel containers to avoid the possibility of fire or explosion.

3. Remove the pinch bolt and disconnect the steering shaft and joint at the cowl, inside the vehicle.
4. Remove the engine air cleaner and the engine air intake resonators.
5. Properly recover the refrigerant from the A/C system.
6. Raise and safely support the vehicle.
7. Remove the front splash shield from between the front sub-frame and the body.
8. Remove the catalytic converter.

✻✻ CAUTION

The EPA warns that prolonged contact with used engine oil may cause a number of skin disorders, including cancer! You should make every effort to minimize your exposure to used engine oil. Protective gloves should be worn when changing the oil. Wash your hands and any other exposed skin areas as soon as possible after exposure to used engine oil. Soap and water, or waterless hand cleaner should be used.

9. Drain the engine cooling system and engine oil.

✻✻ CAUTION

Never open, service or drain the radiator or cooling system when hot; serious burns can occur from the steam and hot coolant.

10. Remove the front wheel and tire assemblies.
11. Separate the left and right stabilizer bar links from the front stabilizer bar.
12. Separate the left and right outer tie rod ends from the front wheel knuckles. Discard the cotter pins.
13. Remove the pinch bolts and separate the front suspension lower control arms from the front wheel knuckles at the ball joints.
14. Remove the left and right wheel hub retainer nuts from the halfshaft ends and remove the halfshafts from the front wheel knuckles.
15. Remove the A/C accumulator retaining screws from the front sub-frame.
16. Disengage the vehicle speed sensor wiring harness at the connector.

17. Disconnect the speedometer drive cable from the transaxle.
18. If the engine is to be separated from the transaxle after removal from the vehicle, and the transaxle is an automatic, remove the right splash shield from the front fender apron.
19. Remove the access plug from the engine rear plate and remove the four torque converter retaining nuts.
20. Push the torque converter into the transaxle front pump support and gear.
21. Disconnect the wiring to the knock sensor and the oil pressure sensor located on the right side of the cylinder block.
22. Lower the vehicle.
23. Secure the radiator and fan shroud assembly to the radiator support, using safety wire or equivalent.
24. Disconnect the accelerator cable and speed control actuator from the throttle body.
25. Remove the accelerator cable bracket.
26. Disengage the wiring to the fuel injectors at the connector located near the fuel pressure regulator.
27. Remove the retaining screws from the engine control wiring at the intake manifold.
28. Remove the power steering pump auxiliary reservoir from the bracket and lay on top of the engine assembly using shop towels to absorb the fluid.
29. Disconnect the return hose from the power steering reservoir and plug the hose.

✻✻ WARNING

Do not allow the power steering fluid to come into contact with the accessory drive belts.

30. Disconnect the power steering return hose from the power steering pump.
31. Disconnect the wiring from the power steering pressure switch located on the power steering pressure hose.
32. Disconnect the wiring from the alternator and the grounding strap from the alternator mounting bracket.
33. Disconnect the vacuum supply hose from the fitting on the rear of the intake manifold.
34. Disconnect the coolant hoses from the radiator coolant recovery reservoir.
35. Disconnect the hoses from the A/C compressor and plug.
36. Disconnect the fuel return and supply lines from the fuel rail and plug.
37. Disconnect the vacuum supply hose to the EGR valve, EGR pressure sensor and the EGR valve to exhaust manifold tube.

38. If equipped with automatic transaxle, pry the end of the shift cable from the stud, remove the two retaining bolts and remove shift cable and bracket from the transaxle. Remove the wiring to the transmission range sensor and remove the wire retainers.

39. Disconnect the grounding strap from the transaxle.

40. Disconnect the wiring from the ignition coil and the radio ignition interference capacitor. Move the wiring out of the way.

41. Remove the vacuum supply line from the power brake booster.

42. Disconnect the upper radiator hose from the radiator.

43. Remove the evaporative emission hose from the connector located near the radio ignition interference capacitor.

44. Disconnect the heater hose from the connection located near the EGR valve.

45. Disconnect the positive and negative battery cable retainer from the battery tray.

46. If equipped with manual transaxle, remove the retainer and disconnect the clutch hydraulic line from the clutch actuator pipe at the transaxle case.

47. If equipped, disconnect the block heater power supply wiring from the left side of the radiator support.

48. If equipped with automatic transaxle, remove the transmission oil cooler lines from the transaxle. Remove the oil cooler return line from the bracket on the left-hand side of the transaxle.

49. If equipped with manual transaxle, remove the bolt from the shift rod and the nut from the stabilizer bar and remove from the transaxle.

50. Disconnect the lower radiator hose from the radiator.

51. Remove the four bolts retaining the lower radiator supports to the front sub-frame. Rotate the radiator supports forward.

52. Disconnect the wiring harness from the A/C compressor.

53. Disconnect the engine wiring from the heated oxygen sensor, engine coolant temperature sensor and the crankshaft position sensor.

54. Remove the two screws retaining the bumper cover braces to the left and right sides of the front sub-frame and rotate the cover braces forward.

55. Disconnect the power steering oil cooler hoses at the right front sub-frame. Drain the fluid from the hoses.

56. Partially lower the vehicle.

57. Position and secure all lines, hoses and components that will be removed with the engine.

58. Install Powertrain and Sub-frame Support Bracket 134–00250 or equivalent, with Powertrain Lift (hydraulic lift) 134–00251 or equivalent, to support the powertrain assembly for removal from the vehicle.

➡**Be sure that the powertrain and sub-frame support bracket and lift are correctly positioned for safe removal of the powertrain assembly.**

59. Remove the four sub-frame to body retaining bolts.

60. Remove the upper front engine support bracket and the engine and transmission support insulator retaining nuts.

61. With an assistant, carefully lower the powertrain assembly while checking for body interference.

62. With the powertrain assembly lowered from the vehicle, carefully roll the powertrain lift or equivalent away from the vehicle.

63. Using Floor Crane 014–00071 or equivalent, and Floor Crane Positioning Sling 014–00036 or equivalent, support the engine using the engine lifting eyes.

64. Remove the left and the right halfshafts from the transaxle.

65. Remove the left and the right front engine support insulators from the sub-frame and transaxle.

66. Using the floor crane and sling or equivalent, raise the engine and transaxle assembly and remove from the sub-frame.

67. Position the transaxle part of the assembly onto Transmission Jack 066–00016 or equivalent.

68. Remove the two starter retaining bolts and remove the starter motor.

69. Remove the battery ground cable from the engine to transaxle retaining stud bolt.

70. Remove the transaxle to engine retaining bolts and separate the engine from the transaxle.

71. If equipped with manual transaxle, remove the six clutch pressure plate retaining bolts and remove the pressure plate and the clutch disc.

72. Remove the eight flywheel retainer bolts and the flywheel from the crankshaft.

73. Remove the engine rear plate.

74. Install the engine to an engine stand for further service.

To install:

75. Remove the engine from the engine stand using the floor crane and sling or equivalent, attached to the engine lifting eyes.

76. Reinstall the engine rear plate and flywheel.

77. Tighten the flywheel retaining bolts in an alternating sequence to 79–86 ft. lbs. (107–117 Nm) for automatic transaxles and 81–89 ft. lbs. (110–120 Nm) for manual transaxles.

78. If equipped with manual transaxle, install the clutch disc and the clutch pressure plate.

79. Tighten the retaining bolts in an alternating sequence to 22 ft. lbs. (30 Nm).

80. Reinstall the engine to the transaxle. If equipped with automatic transaxle, align the torque converter to the flywheel while positioning the engine to the transaxle.

81. Reinstall the engine to transaxle retaining bolts and tighten to 25–34 ft. lbs. (34–46 Nm).

82. If removed, place the sub-frame onto the powertrain and sub-frame support bracket and hydraulic lift.

83. Using the floor crane and sling, position the engine and transaxle assembly onto the sub-frame keeping the crane attached for support.

84. Install Powertrain Alignment Gauge T94P-6000-aH or equivalent, to the left-hand front engine support bracket and sub-frame. Reinstall the through-bolt. Tighten the retaining bolts and the through-bolt to 20 ft. lbs. (27 Nm).

85. Reinstall the right engine support insulator retaining bolts and through-bolt to the sub-frame. Leave the bolts finger-tight.

86. Reinstall the battery ground cable to the engine at the transaxle stud bolt. Tighten the retaining nut to 15–22 ft. lbs. (20–30 Nm).

87. Reinstall the starter motor.

88. Reinstall the left and right halfshafts into the transaxle.

89. Install Sub-frame Alignment Pin Set 94P-2100-aH or equivalent into the sub-frame.

90. Reinstall the four sub-frame retaining bolts and tighten to 92–100 ft. lbs. (125–135 Nm).

91. Remove the sub-frame alignment pins.

➡**Be sure that the engine and transaxle are firmly seated against the front and rear insulator brackets.**

92. Using new nuts, install the upper front engine support bracket and the transmission support insulator. Tighten the nuts to 74 inch. lbs. (10 Nm).

93. Remove the sub-frame support bracket and the hydraulic lift.

94. Raise and safely support the vehicle.

95. Tighten the right front engine support insulator to the sub-frame bolts to 30–41 ft. lbs. (41–55 Nm).

➡**Check the position of the right front engine support insulator. It must be centered in its bracket and in perfect alignment front to rear.**

96. Lower the vehicle.

97. Tighten the front engine support bracket nuts to 52–70 ft. lbs. (70–95 Nm).

98. Tighten the engine and transmission support insulator nuts to 30–41 ft. lbs. (41–55 Nm) for automatic transaxles and 52–70 ft. lbs. (70–95 Nm) for manual transaxles.

99. Tighten the right front engine support insulator through-bolt to 75–102 ft. lbs. (103–137 Nm).

100. Remove the powertrain alignment gauge.

101. Reinstall the left front engine support insulator to the sub-frame. Toque the retaining bolts to 84 inch. lbs. (10 Nm).

➡**Check the position of the left front engine support insulator to ensure perfect front to rear alignment.**

102. Retighten the two retaining bolts to 30–41 ft. lbs. (41–55 Nm).

103. Reinstall the left front engine support insulator through-bolt. Tighten the through-bolt to 75–102 ft. lbs. (103–137 Nm).

104. Reconnect the power steering oil cooler hoses at the right side of the sub-frame. Tighten the hose clamps securely.

105. Reconnect the engine wiring to the heated oxygen sensor, engine coolant temperature sensor and the crankshaft position sensor.

106. Reconnect the wiring harness to the A/C compressor.

107. Reinstall the radiator supports to the sub-frame bolts. Tighten to 71–97 inch lbs. (8–11 Nm).

108. Reconnect the lower radiator hose to the radiator.

109. If equipped with manual transaxle, install the shift rod stabilizer to the transaxle.

110. Tighten the shift rod bolt to 17 ft. lbs. (23 Nm) and the stabilizer nut to 41 ft. lbs. (55 Nm).

111. If equipped with automatic transaxle, install the transmission oil cooler lines and tighten the nuts to 18–22 ft. lbs. (24–31 Nm).

112. Reconnect the heater hose to the heater water tube located under the engine.

113. Lower the vehicle.

114. If equipped, connect the block heater power supply wiring.

115. If equipped with manual transaxle, connect the clutch hydraulic line to the clutch actuator pipe at the transaxle case and install the retainer.

116. Reconnect the positive and negative battery cable retainer to the battery tray.

117. Reinstall the heater hose to the connector tube located near the EGR valve and clamp securely.

118. Reconnect the evaporative emission hose located near the radio ignition interference capacitor.

119. Reconnect the upper radiator hose to the radiator.

120. Reconnect the vacuum supply line to the power brake booster.

121. Reconnect the wiring to the ignition coil and the radio ignition interference capacitor.

122. Reconnect the ground strap to the transaxle case.

123. Reconnect the wiring harness to the transmission range sensor.

124. If equipped with automatic transaxle, install the shift cable and bracket to the transaxle. Tighten the retaining bolts to 15–19 ft. lbs. (20–25 Nm).

125. Reconnect the vacuum hoses from the EGR pressure sensor to the EGR valve to exhaust manifold tube and the vacuum supply hose to the EGR valve.

126. Unplug and connect the fuel supply and return lines to the fuel rail.

127. Unplug and connect the refrigerant hoses to the A/C compressor.

128. Reconnect the coolant hoses to the radiator coolant recovery reservoir.

129. Reconnect the vacuum supply hose for the body to the fitting at the rear of the intake manifold.

130. Reinstall the ground strap on the alternator mounting bracket.

131. Reconnect the wiring harness to the alternator.

132. Reconnect the wiring to the power steering pressure switch located on the power steering pressure hose.

133. Reconnect the power steering return hose to the power steering pump. Clamp the hose securely.

134. Reinstall the power steering pump auxiliary reservoir to its bracket.

135. Reinstall the retainer screws for the engine wiring to the intake manifold.

136. Reattach the wiring to the connector located near the fuel pressure regulator for the fuel injectors.

137. Reconnect the accelerator cable and the speed control actuator to the accelerator cable bracket and the throttle body.

138. Remove the safety wire holding the radiator and fan shroud to the radiator support.

139. Raise and safely support the vehicle.

140. Reconnect the wiring harness to the knock sensor and the oil pressure sensor located on the right side of the cylinder block.

141. If equipped with automatic transaxle, install the torque converter to flywheel retaining nuts. Tighten the retaining nuts in an alternating sequence to 54–64 ft. lbs. (73–87 Nm). Reinstall the access plug into the engine rear plate.

142. Reinstall the right splash shield onto the front fender apron.

143. Reconnect the vehicle speed sensor.

144. Reinstall the speedometer drive cable to the transaxle.

145. Reinstall the A/C accumulator retaining screws to the sub-frame.

146. Reinstall the left and right half-shafts.

147. Reinstall the left and right front wheel knuckles into the front suspension lower control arms at the ball joints. Tighten the bolts to 37–43 ft. lbs. (50–58 Nm).

148. Reinstall the left and right tie rod ends to the front wheel knuckles. Install new castellated nuts onto the tie rod end studs. Tighten the castellated nuts to 21 ft. lbs. (28 Nm). Install new cotter pins.

149. Reconnect the left and right front stabilizer bar links to the front stabilizer bar. Tighten the nuts to 35 ft. lbs. (48 Nm).

150. Reinstall the front wheel and tire assemblies. Tighten the lug nuts to 95 ft. lbs. (129 Nm).

151. Reinstall the catalytic converter.

152. Reinstall the splash shield to the front of the sub-frame and body.

153. Rotate the front bumper cover braces rearward. Reinstall the two screws to the sub-frame and tighten securely.

154. Lower the vehicle.

155. Reinstall the engine air cleaner and the engine air intake resonators.

156. Reconnect the steering shaft and joint inside the vehicle. Tighten the bolt to 18 ft. lbs. (24 Nm).

157. If equipped with automatic transaxle, fill the transmission with the proper amount and type of fluid.

158. Fill the power steering reservoir with the proper type of fluid.

159. Fill the engine cooling system.

160. Fill the engine with fresh oil.

✳✳ WARNING

Operating the engine without the proper amount and type of engine oil will result in severe engine damage.

161. Reconnect the battery cables, negative cable last.

162. Check all fluid levels.

163. Evacuate and recharge the A/C system.

➡ **Whenever the vehicle sub-frame is removed or lowered, the wheel alignment should be checked.**

164. Start the engine and check for leaks and proper operation.

2.5L Engine

✳✳ CAUTION

The fuel injection system remains under pressure, even after the engine has been turned OFF. The fuel system pressure MUST BE relieved before disconnecting any fuel lines. Failure to do so may result in fire and/or personal injury.

1. Disconnect the battery cables, negative cable first.

✳✳ CAUTION

Observe all applicable safety precautions when working around fuel. Whenever servicing the fuel system, always work in a well ventilated area. Do not allow fuel spray or vapors to come in contact with a spark or open flame. Keep a dry chemical fire extinguisher near the work area. Always keep fuel in a container specifically designed for fuel storage; also, always properly seal fuel containers to avoid the possibility of fire or explosion.

2. Relieve the fuel system pressure using the recommended procedure.

✳✳ CAUTION

Some models covered by this manual may be equipped with a Supplemental Restraint System (SRS), which uses an air bag. Whenever working near any of the SRS components, such as the impact sensors, the air bag module, steering column and instrument panel, properly disable the SRS.

3. Remove the water pump pulley shield.

4. Remove the pinch bolt and disconnect the steering shaft and joint at the cowl inside the vehicle.

5. Remove the engine air cleaner.

6. Properly recover the refrigerant from the A/C system.

7. Raise and safely support the vehicle.

8. Remove the exhaust crossover and the catalytic converter.

✳✳ CAUTION

Never open, service or drain the radiator or cooling system when hot; serious burns can occur from the steam and hot coolant.

9. Drain the engine cooling system and engine oil.

✳✳ CAUTION

The EPA warns that prolonged contact with used engine oil may cause a number of skin disorders, including cancer! You should make every effort to minimize your exposure to used engine oil. Protective gloves should be worn when changing the oil. Wash your hands and any other exposed skin areas as soon as possible after exposure to used engine oil. Soap and water, or waterless hand cleaner should be used.

10. Remove the front wheel and tire assemblies.

11. Separate the left and right stabilizer bar links from the front stabilizer bar.

12. Separate the left and right outer tie rod ends from the front wheel knuckles. Discard the cotter pins.

13. Remove the pinch bolts and separate the front suspension lower control arms from the front wheel knuckles at the ball joints.

14. Remove the left and right wheel hub retainer nuts from the halfshaft ends and remove the halfshafts from the front wheel knuckles.

15. Remove the A/C accumulator retaining screws from the front sub-frame.

16. Disconnect the vehicle speed sensor wiring.

17. Disconnect the speedometer drive cable from the vehicle speed sensor.

18. If the engine is to be separated from the transaxle after removal from the vehicle, and the transaxle is an automatic, remove the access plug from the engine rear plate and remove the four torque converter retaining nuts.

19. Push the torque converter into the transaxle front pump support and gear.

20. Lower the vehicle.

21. Secure the radiator and fan shroud

assembly to the radiator support, using safety wire.

22. Disconnect the accelerator cable and speed control actuator from the throttle body.

23. Remove the accelerator cable bracket from the throttle body.

24. Disengage the three connectors for the engine control wiring from the bracket located on the left-front fender apron and unplug the connectors.

25. Remove the retainer for the engine control wiring from the air cleaner bracket.

26. Remove the ignition control module from the bulkhead, if equipped.

27. Remove the power steering pump auxiliary reservoir from the bracket and lay on top of the engine assembly using shop towels to absorb the fluid.

28. Disconnect the return hose from the power steering reservoir and plug the hose.

✳✳ WARNING

Do not allow the power steering fluid to come into contact with the accessory drive belts.

29. Disconnect the power steering pressure hose from the power steering pump.

30. Disconnect the power steering pressure hose bracket from the upper front engine support bracket. Lay the hose on top of the engine.

31. Disconnect the wiring from the powertrain control module and retainer located on the right side of the dash panel.

32. Remove the ground strap for the engine control wiring at the right fender apron.

33. Disconnect the coolant hoses from the radiator coolant recovery reservoir.

34. Disconnect the fuel return and supply lines from the fuel rail.

35. If equipped with automatic transaxle, pry the end of the shift cable from the stud, remove the two retaining bolts and remove shift cable and bracket from the transaxle. Remove the wiring to the transmission range sensor and remove the wire retainers.

36. Disconnect the grounding strap from the transaxle.

37. Remove the vacuum supply line from the power brake booster.

38. Disconnect the upper radiator hose from the radiator.

39. Disconnect the positive and negative battery cable retainer from the battery tray.

40. If equipped with manual transaxle, remove the retainer and disconnect the clutch hydraulic line from the clutch actuator pipe at the transaxle case.

41. Disconnect the block heater power supply wiring from the right side of the radiator support if equipped.

42. Raise and safely support the vehicle.

43. Disconnect the heater hoses from the heater core.

44. Disconnect the A/C suction hose from the A/C condenser core and plug the hose.

45. Disconnect the A/C discharge hose from the A/C accumulator and plug the hose.

46. If equipped with automatic transaxle, remove the transmission oil cooler lines from the transaxle. Remove the oil cooler return line from the bracket on the left side of the transaxle.

47. If equipped with manual transaxle, remove the bolt from the shift rod and the nut from the stabilizer bar and remove from the transaxle.

48. Disconnect the lower radiator hose from the radiator.

49. Remove the four bolts retaining the lower radiator supports to the front sub-frame. Rotate the radiator supports forward.

50. Disconnect the wiring harness from the A/C compressor.

51. Remove the two screws retaining the bumper cover braces to the left and right sides of the front sub-frame and rotate the cover braces forward.

52. Partially lower the vehicle.

53. Position and secure all lines, hoses and components that will be removed with the engine.

54. Install Powertrain and Sub-frame Support Bracket 134–00250 or equivalent with Powertrain Lift (hydraulic lift) 134–00251 or equivalent to support the powertrain assembly for removal from the vehicle.

➡Be sure that the powertrain and sub-frame support bracket and lift are correctly positioned for safe removal of the powertrain assembly.

55. Remove the four sub-frame to body retaining bolts.

56. Remove the upper front engine support bracket and the engine and transmission support insulator retaining nuts.

57. With an assistant, carefully lower the powertrain assembly while checking for body interference.

58. With the powertrain assembly lowered from the vehicle, carefully roll the powertrain lift or equivalent away from the vehicle.

59. Using Floor Crane 014–00071 or equivalent, and Floor Crane Positioning Sling 014–00036 or equivalent, support the engine using the engine lifting eyes.

60. Remove the left and the right half-shafts from the transaxle.

61. Remove the left and the right front engine support insulators from the sub-frame and transaxle.

62. Using the floor crane and sling or equivalent, raise the engine and transaxle assembly and remove from the sub-frame.

63. Position the transaxle part of the assembly onto Transmission Jack 066–00016 or equivalent.

64. Remove the two starter retaining bolts and remove the starter motor.

65. Unplug the engine control wiring connectors and retaining clips at the transaxle and move aside.

66. Remove the battery ground cable from the engine to transaxle retaining stud bolt.

67. Remove the transaxle to engine retaining bolts and separate the engine from the transaxle.

68. If equipped with manual transaxle, remove the six clutch pressure plate retaining bolts and remove the pressure plate and the clutch disc.

69. Remove the eight flywheel retainer bolts and the flywheel from the crankshaft.

70. Remove the engine rear plate.

71. Install the engine to an engine stand for further service.

To install:

72. Remove the engine from the engine stand using the floor crane and sling or equivalent, attached to the engine lifting eyes.

73. Reinstall the engine rear plate and flywheel.

74. Tighten the flywheel retaining bolts in an alternating sequence to 79–86 ft. lbs. (107–117 Nm) for automatic transaxles, or 81–89 ft. lbs. (110–120 Nm) for manual transaxles.

75. If equipped with manual transaxle, install the clutch disc and the clutch pressure plate. Tighten the retaining bolts in an alternating sequence to 22 ft. lbs. (30 Nm).

76. Reinstall the engine to the transaxle. If equipped with automatic transaxle, align the torque converter to the flywheel while positioning the engine to the transaxle.

77. Reinstall the engine to transaxle retaining bolts and tighten to 25–34 ft. lbs. (34–46 Nm).

78. If removed, place the sub-frame onto the powertrain and sub-frame support bracket and hydraulic lift.

79. Using the floor crane and sling, position the engine and transaxle assembly onto the sub-frame keeping the crane attached for support.

80. Install Powertrain Alignment Gauge T94P-6000-AH or equivalent, to the left front engine support bracket and sub-frame. Reinstall the through-bolt. Tighten the retaining bolts and the through-bolt to 20 ft. lbs. (27 Nm).

81. Reinstall the right engine support insulator retaining bolts and through-bolt to the sub-frame. Leave the bolts finger-tight.

82. Reinstall the battery ground cable to the engine at the transaxle stud bolt. Tighten the retaining nut to 15–22 ft. lbs. (20–30 Nm).

83. Position the engine control wiring across the transaxle and install the retaining clips.

84. Reinstall the engine control wiring to the transaxle connectors.

85. Reinstall the starter motor.

86. Reinstall the left and right halfshafts into the transaxle.

87. Install Sub-frame Alignment Pin Set 94P-2100-aH or equivalent into the sub-frame.

88. Reinstall the four sub-frame retaining bolts and tighten to 92–100 ft. lbs. (125–135 Nm).

89. Remove the sub-frame alignment pins.

➡Be sure that the engine and transaxle are firmly seated against the front and rear insulator brackets.

90. Using new nuts, install the upper front engine support bracket and the transmission support insulator. Tighten the nuts to 84 inch. lbs. (10 Nm).

91. Remove the sub-frame support bracket and the hydraulic lift.

92. Raise and safely support the vehicle.

93. Tighten the right front engine support insulator to the sub-frame bolts to 30–41 ft. lbs. (41–55 Nm).

➡Check the position of the right front engine support insulator. It must be centered in its bracket and in perfect alignment front to rear.

94. Lower the vehicle.

95. Tighten the front engine support bracket nuts to 52–70 ft. lbs. (70–95 Nm).

96. Tighten the engine and transmission support insulator nuts to 30–41 ft. lbs. (41–55 Nm) for automatic transaxles or 52–70 ft. lbs. (70–95 Nm) for manual transaxles.

97. Tighten the right front engine support insulator through-bolt to 75–102 ft. lbs. (103–137 Nm).

98. Remove the powertrain alignment gauge.

99. Reinstall the left front engine support insulator to the sub-frame. Toque the retaining bolts to 84 inch. lbs. (10 Nm).

➡**Check the position of the left front engine support insulator to ensure perfect front to rear alignment.**

100. Retighten the two retaining bolts to 30–41 ft. lbs. (41–55 Nm).

101. Reinstall the left front engine support insulator through-bolt. Tighten the through-bolt to 75–102 ft. lbs. (103–137 Nm).

102. Reconnect the wiring harness to the A/C compressor.

103. Reinstall the front bumper cover braces to the left and right side of the sub-frame.

104. Reinstall the radiator supports to the sub-frame. Tighten the retaining bolts to 71–97 inch lbs. (8–11 Nm).

105. Reconnect the lower radiator hose to the radiator.

106. If equipped with manual transaxle, install the shift rod stabilizer to the transaxle. Tighten the shift rod bolt to 17 ft. lbs. (23 Nm) and the stabilizer nut to 41 ft. lbs. (55 Nm).

107. If equipped with an automatic transaxle, install the transmission oil cooler lines and tighten the nuts to 18–22 ft. lbs. (24–31 Nm).

108. Unplug and connect the A/C suction hose to the A/C condenser core.

109. Unplug and connect the A/C discharge hose to the A/C accumulator.

110. Lower the vehicle.

111. If equipped, connect the block heater power supply wiring.

112. If equipped with manual transaxle, connect the clutch hydraulic line to the clutch actuator pipe at the transaxle case and install the retainer.

113. Reconnect the positive and negative battery cable retainer to the battery tray.

114. Reconnect the upper radiator hose to the radiator.

115. Reconnect the vacuum supply line to the power brake booster.

116. Reconnect the ground strap to the transaxle case.

117. If equipped with automatic transaxle, install the shift cable and bracket to the transaxle. Tighten the retaining bolts to 15–19 ft. lbs. (20–25 Nm).

118. Reconnect the fuel supply and return lines to the fuel rail.

119. Reinstall the ground strap for the engine control wiring to the right cylinder head.

120. Reconnect the wiring to the powertrain control module and position the wiring into the retainer bracket.

121. Reconnect the power steering pressure hose to the power steering pump.

122. Reinstall the power steering pressure hose bracket to the upper front engine support bracket.

123. Reinstall the power steering pump auxiliary reservoir to its bracket.

124. Reinstall the ignition control module and bracket to the bulkhead, if equipped.

125. Reinstall the retainer for the engine control wiring to the air cleaner bracket.

126. Reattach the three engine control connectors and install to the bracket on the left front fender apron.

127. Reinstall the accelerator cable bracket to the throttle body and tighten to the retaining bolts to 71–106 inch lbs. (8–12 Nm).

128. Reconnect the accelerator cable and the speed control actuator to the accelerator cable bracket.

129. Remove the safety wire holding the radiator and fan shroud to the radiator support.

130. Raise and safely support the vehicle.

131. If equipped with an automatic transaxle, push the torque converter into the flywheel pilot and install four torque converter to flywheel retaining nuts. Tighten the retaining nuts in an alternating sequence to 54–64 ft. lbs. (73–87 Nm). Reinstall the access plug into the engine rear plate.

132. Reconnect the vehicle speed sensor wiring harness.

133. Reinstall the speedometer drive cable to the vehicle speed sensor.

134. Reinstall the A/C accumulator retaining screws to the sub-frame.

135. Reinstall the left and right halfshafts into the front wheel knuckles and install the front axle wheel retaining nuts.

136. Reinstall the left and right front wheel knuckles into the front suspension lower control arms at the ball joints. Tighten the bolts to 37–43 ft. lbs. (50–58 Nm).

137. Reinstall the left and right tie rod ends to the front wheel knuckles. Install new castellated nuts onto the tie rod end studs. Tighten the castellated nuts to 21 ft. lbs. (28 Nm). Install new cotter pins.

138. Reconnect the left and right front stabilizer bar links to the front stabilizer bar. Tighten the nuts to 35 ft. lbs. (48 Nm).

139. Reinstall the front wheel and tire assemblies. Tighten the lug nuts to 95 ft. lbs. (129 Nm).

140. Reinstall the exhaust crossover and the three way catalytic converter.

141. Reinstall the splash shield to the front of the sub-frame and body.

142. Rotate the front bumper cover braces rearward. Reinstall the two screws to the sub-frame and tighten securely.

143. Lower the vehicle.

144. Reinstall the engine air cleaner.

145. Reconnect the steering shaft and joint inside the vehicle. Tighten the bolt to 18 ft. lbs. (24 Nm).

146. Reinstall the water pump pulley shield.

147. If equipped with automatic transaxle, fill the transmission with the proper amount and type of fluid.

148. Fill the power steering reservoir with the proper type of fluid.

149. Fill the engine cooling system.

❈❈ WARNING

Operating the engine without the proper amount and type of engine oil will result in severe engine damage.

150. Fill the engine crankcase with clean, fresh oil.

151. Check all fluid levels.

152. Reconnect the battery cables, negative cable last.

153. Evacuate and recharge the A/C system.

➡**Whenever the vehicle sub-frame is removed or lowered, the wheel alignment should be checked.**

154. Start the engine and check for leaks and proper operation.

Water Pump

REMOVAL & INSTALLATION

❈❈ CAUTION

Some models covered by this manual may be equipped with a Supplemental Restraint System (SRS), which uses an air bag. Whenever working near any of the SRS components, such as the impact sensors, the air bag module, steering column and instrument panel, properly disable the SRS.

2.0L Engine

1. Disconnect the negative battery cable.

> ✳✳ **CAUTION**
>
> **Never open, service or drain the radiator or cooling system when hot; serious burns can occur from the steam and hot coolant.**

2. Drain the engine cooling system.
3. Raise and safely support the vehicle.
4. Remove the lower radiator hose from the water pump.
5. Lower the vehicle.
6. Remove the accessory drive belt.
7. Remove the timing belt covers and the timing belt using the recommended procedure.
8. Remove the four water pump retaining bolts.
9. Remove the water pump.

To install:

10. Thoroughly clean all sealing surfaces.
11. Install a new water pump gasket and the water pump onto the cylinder block.
12. Tighten the retaining bolts to 12–15 ft. lbs. (16–20 Nm).

13. Reinstall the timing belt and the timing belt covers using the recommended procedure.
14. Reinstall the accessory drive belt.
15. Raise and safely support the vehicle.
16. Reinstall the lower radiator hose.
17. Lower the vehicle.
18. Fill the engine cooling system.
19. Reconnect the negative battery cable.
20. Start the engine and top off the coolant as necessary. Check for leaks.

2.5L Engine

➡ **Before continuing with this procedure, be sure three new water pump retaining bolts (W701544) are available. Due to their torque-to-yield design, the bolts stretch and cannot be reused.**

1. Disconnect the negative battery cable.

> ✳✳ **CAUTION**
>
> **Never open, service or drain the radiator or cooling system when hot; serious burns can occur from the steam and hot coolant.**

2. Drain the engine cooling system.

3. Remove the water pump pulley shield.
4. Remove the water pump drive belt.
5. Remove the water pump inlet and outlet hoses from the water pump.
6. Remove the three water pump to left cylinder head retaining bolts.
7. Remove the water pump and water pump housing from the vehicle.
8. Remove the water pump to water pump housing retaining bolts and separate the water pump from the water pump housing.

To install:

9. Thoroughly clean all sealing surfaces.
10. Install the water pump to the water pump housing using a new gasket and install the retaining bolts. Tighten the retaining bolts to 16–18 ft. lbs. (22–25 Nm).
11. Position the water pump and water pump housing and install three new torque-to-yield retaining bolts into the left cylinder head.
12. Tighten the new retaining bolts to 11–13 ft. lbs. (15–18 Nm), then rotate the retaining bolts 85–95 degrees.
13. Reinstall the water pump inlet and outlet hoses to the water pump.
14. Reinstall the water pump drive belt.
15. Reinstall the water pump shield.
16. Fill the engine cooling system.
17. Reconnect the negative battery cable.
18. Start the engine and top off the coolant as necessary. Check for leaks.

Cylinder Head

REMOVAL & INSTALLATION

> ✳✳ **CAUTION**
>
> **Some models covered by this manual may be equipped with a Supplemental Restraint System (SRS), which uses an air bag. Whenever working near any of the SRS components, such as the impact sensors, the air bag module, steering column and instrument panel, properly disable the SRS.**

2.0L Engine

➡ **The cylinder head bolts are a torque-to-yield design and cannot be reused. Be sure new cylinder head bolts are available before beginning this procedure. If the cylinder head bolts are reused, engine damage may occur.**

FRONT OF ENGINE

1. Water pump
2. Water pump housing gaskets
3. Cylinder block
4. Oil pump
5. Bolt(4)
A. 12-15 ft. lb.(16–20 Nm)

7922JG01

Exploded view of the water pump mounting—2.0L engine

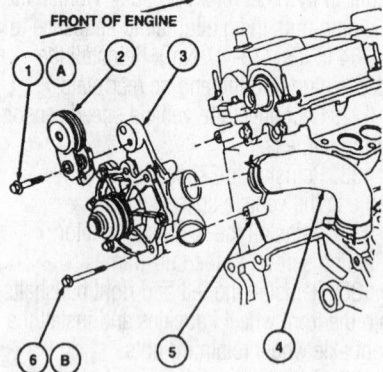

FRONT OF ENGINE

1. Bolt
2. Water pump drive belt tensioner
3. Water pump
4. Water pump outlet hose
5. LH cylinder head
6. Bolt(3)
A. 71-106 in. lb.(8–12 Nm)
B. 11-13 ft. lb.(15–18 Nm)
   then rotate 85–95°

7922JG02

Exploded view of the water pump mounting—2.5L engine

1. Disconnect the negative battery cable.

2. Drain the engine coolant from the radiator and the cylinder block drain plugs.
3. Remove the intake manifold.
4. Remove the exhaust manifold.
5. Remove the camshafts and valve tappets.
6. Support the engine with a wood block between the crankshaft pulley and the front sub-frame.
7. Remove the Three Bar Engine Support D88L-6000-A or equivalent, previously installed for the timing belt cover removal.
8. Remove the right-hand engine lifting eye retaining bolt and the lifting eye.
9. Remove the support bracket from the power steering pump mounting bracket and cylinder head.
10. Remove the camshaft timing belt tensioner pulley and the engine front cover from the front of the cylinder head.

11. Remove the thermostat housing from the rear of the cylinder head.
12. Remove the ignition coil and bracket from the cylinder head.
13. Remove the spark plugs if not already removed.
14. Remove the cylinder head retaining bolts in the reverse of the installation sequence.
15. Remove the cylinder head and gasket from the engine.
16. If the cylinder head is to be serviced, remove the left-hand engine lifting eye.

To install:

17. Clean the cylinder head and cylinder block gasket surfaces and check for flatness.
18. Install a new cylinder head gasket onto the cylinder block. Be sure the head gasket is properly positioned on the dowels.

✳✳ WARNING

Use care when positioning the cylinder head to prevent damage to the head gasket or dowels.

19. Place a light coating of engine oil onto the threads of the new cylinder head bolts and install.
20. Tighten the cylinder head bolts in sequence and in the following steps:

• Tighten all bolts to 15–22 ft. lbs. (20–30 Nm)
• Tighten all bolts to 30–37 ft. lbs. (40–50 Nm)
• Rotate all bolts 90–120 degrees.

21. Reinstall the ignition coil bracket and the ignition coil.
22. Reinstall the water thermostat housing.
23. Reinstall the engine front cover. Tighten the retaining bolts to 71–97 inch lbs. (8–11 Nm).
24. Reinstall the camshaft timing belt tensioner pulley and retaining bolt onto the front of the cylinder head.
25. Reinstall the support bracket to the power steering pump mounting bracket and the cylinder head.
26. Tighten the support bracket to 29–41 ft. lbs. (39–55 Nm).
27. Reinstall the right engine lifting eye to the cylinder head and the alternator mounting bracket. Tighten the retaining bolts to 30–41 ft. lbs. (41–55, Nm).
28. If removed, install the left-hand engine lifting eye to the cylinder head and tighten to 10–13 ft. lbs. (14–18 Nm).
29. Install the Three Bar Engine Support D88L-6000-A or equivalent to the engine lifting eyes and support the engine.
30. Remove the wood block from between the sub-frame and the crankshaft pulley.
31. Reinstall the valve tappets and camshaft into their original locations.
32. Reinstall the exhaust manifold.
33. Reinstall the intake manifold.
34. Reinstall the spark plugs.

35. Drain the engine oil and remove the engine oil filter.

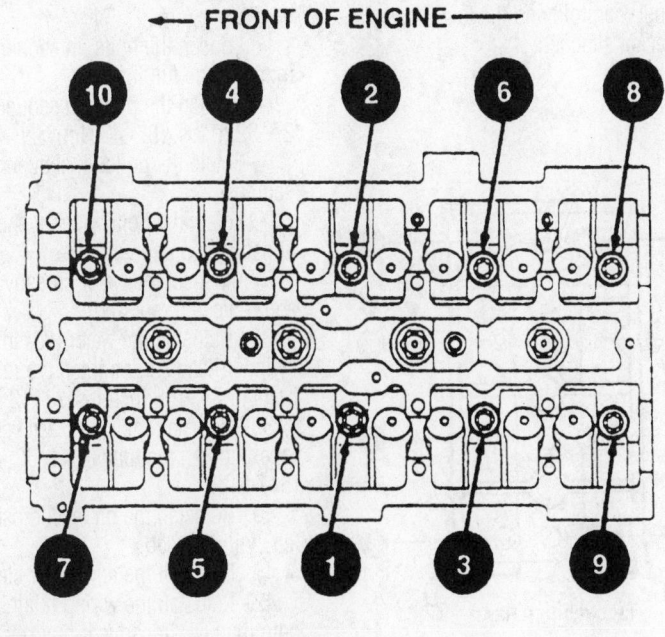

← FRONT OF ENGINE →

TIGHTEN BOLTS IN SEQUENCE SHOWN

7922JG03

Cylinder head mounting bolt tightening sequence—2.0L engine

36. Reinstall the drain plug and tighten to 15–21 ft. lbs. (21–28 Nm).

37. Reinstall a new engine oil filter and fill the crankcase with the proper amount and grade of oil.

38. Fill the engine cooling system.

39. Reconnect the negative battery cable.

40. Run the engine and check for oil and coolant leaks. Check for proper engine operation.

2.5L Engine

➡️**The cylinder head bolts are a torque-to-yield design and cannot be reused. Be sure new cylinder head bolts are available before beginning this procedure.**

1. Disconnect the negative battery cable.

✳✳ CAUTION

Never open, service or drain the radiator or cooling system when hot; serious burns can occur from the steam and hot coolant.

2. Drain the engine coolant from the radiator and cylinder block drain plugs.

3. Close the radiator draincock and install the drain plugs into the cylinder block.

4. Remove the upper and lower intake manifolds.

✳✳ CAUTION

The EPA warns that prolonged contact with used engine oil may cause a number of skin disorders, including cancer! You should make every effort to minimize your exposure to used engine oil. Protective gloves should be worn when changing the oil. Wash your hands and any other exposed skin areas as soon as possible after exposure to used engine oil. Soap and water, or waterless hand cleaner should be used.

5. Drain the engine oil and remove the oil pan.

6. Remove the alternator and the alternator mounting bracket.

7. Remove the heated oxygen sensor from the right cylinder head exhaust manifold.

8. Remove the left cylinder head exhaust manifold.

9. Remove the water pump.

10. Remove the engine front cover.

11. Install the upper front engine mount (support insulator) and the upper front engine support bracket to the front of the engine and the right front fender apron.

12. Remove Three Bar Engine Support D88L-60000-A or equivalent.

13. Remove the camshafts and valve tappets from both cylinder heads.

14. Disconnect the hoses from the EGR pressure sensor at the EGR valve, to the exhaust manifold tube.

15. Disengage the EGR pressure sensor.

16. Disconnect the interior vacuum source hose from the main emission vacuum harness.

17. Disconnect the fuel vapor hose from the PCV valve.

18. Disconnect the EGR transducer.

19. Remove the EGR valve to exhaust manifold tube from the right exhaust manifold and remove from the vehicle.

20. Remove the wiring retaining bracket from the EGR transducer bracket.

21. Remove the engine air cleaner.

22. Remove the crankcase ventilation tube from the water crossover and oil separator.

23. Remove the water crossover retaining bolt and stud bolt from the right cylinder head. Set the water crossover aside.

24. Remove the oil level dipstick from the left cylinder head.

25. Remove the cylinder head retaining bolts from the cylinder heads in the reverse of the sequence illustration.

26. Remove the right cylinder head with the exhaust manifold and the EGR transducer bracket attached.

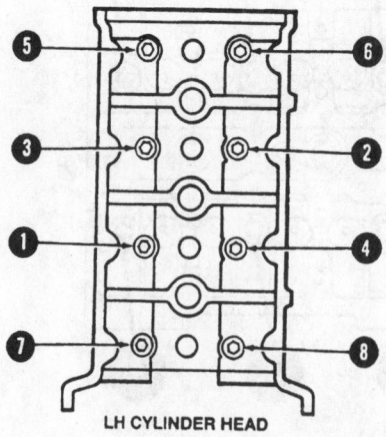

Cylinder head mounting bolt tightening sequence—2.5L engine

7922JG04

LH CYLINDER HEAD

27. If required, remove the exhaust manifold and the EGR transducer bracket from the right cylinder head.

28. Remove the left cylinder head.

29. Inspect the cylinder heads and cylinder block.

To install:

30. Clean the cylinder heads, intake manifolds, valve covers and the cylinder head gasket sealing surfaces on the cylinder block.

➡️**If the cylinder heads were removed for cylinder head gasket replacement, check the flatness of the cylinder heads and the cylinder block gasket sealing surfaces.**

31. If removed, install the right exhaust manifold and EGR transducer bracket to the right cylinder head.

32. Install new cylinder head gaskets onto the dowels of the cylinder block.

33. Position the cylinder heads into their original positions using care not to damage the heads, block or gaskets.

34. Be sure that the cylinder heads are correctly positioned on the dowels.

35. Lightly oil the threads of the new cylinder head retaining bolts and install into the cylinder heads.

36. Tighten the new cylinder head retaining bolts as follows:

• Tighten the bolts, in sequence, to 27–32 ft. lbs. (37–43 Nm)

• Rotate the bolts, in sequence, 85–95 degrees

• Loosen the bolts, in sequence, a minimum of one full turn

• Tighten the bolts, in sequence, to 27–32 ft. lbs. (37–43 Nm)

• Rotate the bolts, in sequence, 85–95 degrees

• Rotate the bolts, in sequence, an additional 85–95 degrees

37. Inspect the water crossover O-rings and replace if required.

38. Reinstall the water crossover to the cylinder heads and tighten the retaining bolts to 71–106 inch lbs. (8–12 Nm).

39. Reinstall the crankcase ventilation tube. Tighten the tube to 44–62 inch lbs. (5–7 Nm).

40. Reinstall the oil level dipstick to the left cylinder head.

41. Reinstall the engine air cleaner.

42. Reinstall the wiring retaining bracket onto the EGR transducer bracket.

43. Reinstall the EGR to exhaust manifold tube onto the right exhaust manifold.

44. Reconnect the EGR transducer and the EGR pressure sensor.

45. Reconnect the fuel vapor hose to the PCV valve.

46. Reconnect the interior vacuum source hoses to the main emission vacuum harness.

47. Reconnect the hoses from the EGR pressure sensor and EGR valve to the exhaust manifold tube.

48. Reinstall the camshafts and valve tappets.

49. Install the Three Bar Engine Support D88L-6000-A or equivalent, to the engine lifting eyes and support the engine.

50. Reinstall the engine front cover.

51. Reinstall the left exhaust manifold.

52. Reinstall the water pump.

53. Reinstall the heated oxygen sensor to the right exhaust manifold.

54. Reinstall the alternator mounting bracket and the alternator.

55. Reinstall the engine oil pan.

56. Reinstall the lower and upper intake manifolds.

57. Replace the engine oil filter.

58. Fill the engine with the proper amount and grade of engine oil.

✳✳ WARNING

Operating the engine without the proper amount and type of engine oil will result in severe engine damage.

59. Fill the engine cooling system.

60. Reconnect the negative battery cable.

61. Run the engine and check for oil and coolant leaks. Check for proper engine operation.

Rocker Arms

REMOVAL & INSTALLATION

2.5L Engine

1. Disconnect both battery cables, negative cable first.

2. Remove the valve covers as follows:

   a. Remove the ignition wires and spark plugs.

   b. Remove the ignition coil from the right valve cover.

   c. Remove the crankcase ventilation tubes from both valve covers.

   d. Remove the wiring harness and bracket to the fuel injectors and move aside.

   e. Remove the retaining nuts and engine wiring from both valve covers and move aside.

   f. Remove the valve cover retaining bolts and studs.

   g. Remove both valve covers from the engine.

3. Remove the crankshaft pulley retaining bolt.

4. Rotate the crankshaft so that the keyway is at the 11 o'clock position to locate the crankshaft at TDC for No. 1 cylinder.

5. Verify that the alignment arrows on the camshafts are aligned. If not, rotate the crankshaft one complete revolution and recheck.

6. Rotate the crankshaft so that the keyway is at the 3 o'clock position. This positions the right cylinder head camshafts to the neutral position.

7. Remove the accessory drive belts.

8. Remove the battery.

9. Remove the water pump drive pulley from the left intake camshaft using Camshaft Dampener Remover/Replacer T94P-6312-AH or equivalent, along with the shaft protector and screw or equivalent.

10. Remove the camshaft rear oil seal retainer bolts and the camshaft rear oil seal retainer and gasket from the left cylinder head.

➡**The camshaft journal caps and cylinder heads are numbered to ensure that they are assembled in their original positions. If removed, keep the camshaft journal caps together with the cylinder head that they were removed from.**

11. Remove the right cylinder head camshaft journal thrust cap retaining bolts and thrust caps.

12. Loosen the remaining camshaft journal cap bolts in sequence, releasing the bolts several revolutions at a time by making several passes to allow the camshaft to be raised from the cylinder head evenly. Do not remove the retaining bolts completely.

➡**If the valve lifters (tappets) and roller rocker arms are to be reused, mark the positions of the valve lifters and rocker arms so that they are reassembled into their original positions.**

13. With the camshafts loose, remove the rocker arms, keeping them in the order that they were removed.

14. If required, remove the valve lifters from the cylinder head.

15. Rotate the crankshaft two revolutions and locate the crankshaft keyway at the 11 o'clock position. This will position the left cylinder head camshafts to their neutral position.

16. Verify that the alignment arrows on the camshafts are aligned.

➡**The camshaft journal caps and cylinder heads are numbered to ensure that they are assembled in their original positions. If removed, keep the camshaft journal caps together with the cylinder head that they were removed from.**

17. Remove the camshaft journal thrust cap retaining bolts and thrust caps from the left cylinder head.

18. Loosen the remaining camshaft journal cap bolts in sequence, releasing the bolts several revolutions at a time by making several passes to allow the camshaft to be raised from the cylinder head evenly. Do not remove the retaining bolts completely.

➡**If the valve lifters (tappets) and roller rocker arms are to be reused, mark the positions of the valve lifters and rocker arms so that they are reassembled into their original positions.**

19. With the camshafts loose, remove the rocker arms, keeping them in the order that they were removed.

20. If required, remove the valve lifters from the cylinder head.

21. Inspect the rocker arms and valve lifters for wear and/or damage and replace as necessary.

To install:

22. Be sure that the crankshaft keyway is at the 11 o'clock position.

23. Lubricate the left cylinder head valve lifters with engine assembly lubricant and install into their correct positions in the cylinder head.

24. If the valve lifters are being replaced with new units, soak the lifters in a container of clean engine oil and/or manually pump up the lifters before installing into the cylinder head.

25. Lubricate the left cylinder head rocker arms with engine assembly lubricant

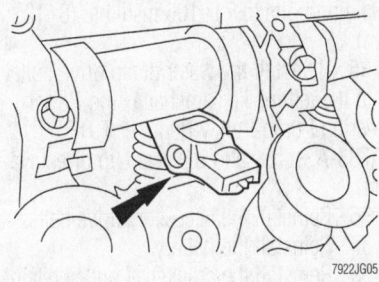

7922JG05

After the camshaft is loose, the rocker arm/cam follower (arrow) can be removed—2.5L engine

and install the left cylinder head rocker arms into their original locations.

➡ **Do not install the camshaft journal thrust caps until the camshaft journal caps are secured into position.**

26. Tighten the left cylinder head camshaft journal cap bolts in sequence (see camshaft procedure for sequence) making several passes to pull the camshafts down evenly. Tighten the bolts to 71–106 inch lbs. (8–12 Nm).

27. Reinstall the left-hand cylinder head thrust caps and bolts. Tighten to 71–106 inch lbs. (8–12 Nm).

28. Rotate the crankshaft two revolutions and position the crankshaft keyway to the 3 o'clock location. This will position the right cylinder head camshafts to the neutral position.

29. Lubricate the right cylinder head valve lifters with engine assembly lubricant and install into their original positions in the cylinder head.

30. If the valve lifters are being replaced with new units, soak the lifters in a container of clean engine oil and/or manually pump the lifters up before installing into the cylinder head.

31. Lubricate the right cylinder head rocker arms with engine assembly lubricant and install the right cylinder head rocker arms into their original locations.

➡ **Do not install the camshaft journal thrust caps until the camshaft journal caps are secured into position.**

32. Tighten the right cylinder head camshaft journal cap bolts in sequence making several passes to pull the camshafts down evenly. Tighten the bolts to 71–106 inch lbs. (8–12 Nm).

33. Reinstall the right-hand cylinder head thrust caps and bolts. Tighten to 71–106 inch lbs. (8–12 Nm).

34. Reinstall the left cylinder head camshaft rear oil seal and retainer with gasket onto the cylinder head. Tighten the retaining bolts to 71–106 inch lbs. (8–12 Nm).

35. Reinstall the water pump drive pulley onto the left intake camshaft using Power Steering Pump Pulley Replacer T91P-3A733-A, screw and replacer cup or equivalent.

36. Reinstall the accessory drive belts.

37. Reinstall the battery.

38. Reinstall the crankshaft pulley retaining bolt.

39. Tighten the crankshaft pulley retaining bolt as follows:

　　a. Tighten to 89 ft. lbs. (120 Nm).

　　b. Loosen the bolt at least one full turn.

　　c. Tighten the bolt to 35–39 ft. lbs. (47–53 Nm).

　　d. Rotate the bolt 85–95 degrees.

40. Reinstall both valve covers as follows:

　　a. Clean the valve cover gasket sealing surfaces.

　　b. Install new valve cover gaskets onto the valve covers.

　　c. For each valve cover, place a bead of silicone sealant at two places on the valve cover sealing surfaces where the engine front cover and the cylinder heads make contact and at two places on the rear of the cylinder head where the camshaft seal retainer contacts the cylinder head.

　　d. Reinstall the valve cover retaining bolts and studs and tighten in sequence to 71–106 inch lbs. (8–12 Nm).

➡ **The valve covers must be installed and properly tightened within six minutes of applying the silicone sealant.**

41. Reconnect both battery cables, negative cable last.

42. Run the engine and check for leaks and proper operation.

Intake Manifold

REMOVAL & INSTALLATION

2.0L Engine

✳✳ CAUTION

The fuel injection system remains under pressure, even after the engine has been turned OFF. The fuel system pressure must be relieved before disconnecting any fuel lines. Failure to do so may result in fire and/or personal injury.

1. Disconnect the negative battery cable.

2. Remove the engine air intake resonators.

3. Relieve the fuel system pressure.

4. Remove the accelerator cable and the speed control actuator from the throttle body.

5. Remove the accelerator cable bracket.

6. Disconnect the wiring harness to the fuel injectors and move aside.

7. Disconnect the vacuum line at the fuel pressure regulator.

8. Using the proper spring lock coupling disconnect tools (⅜ inch and ½ inch), remove the fuel supply and return hoses from the fuel injection supply manifold (fuel rail).

9. Remove the fuel supply and return hoses from the retaining bracket on the intake manifold and move aside.

10. Carefully disengage the fuel injection supply manifold with the fuel injectors attached and remove from the engine.

11. Disconnect the wiring for the engine coolant temperature sensor and the engine control sensor.

12. Remove the crankcase ventilation tube from the intake manifold fitting.

13. Remove the wiring harness connector at the camshaft position sensor located on the cylinder head.

14. Remove the camshaft position sensor retaining screw and remove the sensor.

15. Disconnect the vacuum hose at the EGR valve.

16. Use a 22mm crowfoot wrench to completely loosen the EGR valve to exhaust manifold tube nut.

17. Remove the two EGR valve retaining bolts and remove the EGR valve and gasket.

18. Remove the vacuum supply hoses for the body and brake booster from the bottom of the intake manifold.

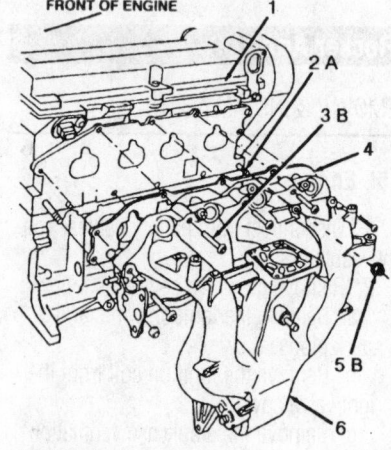

FRONT OF ENGINE

1. Cylinder head
2. Stud (2 req'd)
3. Bolt (8 req'd)
4. Intake manifold gasket
5. Nut (2 req'd)
6. Intake manifold
A. Tighten to 0-10 Nm (0-89 lb. in.)
B. Tighten to 16-20 Nm (12-15 lb. ft.)

7922JG06

Exploded view of the intake manifold mounting—2.0L engine

19. Remove the retaining bracket screws for the engine control sensor wiring.

➡ **The intake manifold can be removed without the removal of the alternator and alternator mounting bracket however their removal will make the job easier.**

20. Remove the retaining bolts and nuts from the intake manifold.

21. Remove the intake manifold and gasket from the cylinder head.

To install:

22. If the intake manifold is to be replaced, remove the throttle body and idle air control valve from the old manifold and install on the new one.

23. Clean the gasket sealing surfaces on the intake manifold and the cylinder head.

24. Install the intake manifold using a new gasket to the cylinder head.

25. Reinstall the intake manifold retaining nuts and bolts.

26. Tighten the retaining nuts and bolts in several passes to 12–15 ft. lbs. (16–20 Nm), starting at the center and working towards the ends of the cylinder head.

27. Reinstall the alternator mounting bracket and alternator, if removed.

28. Reinstall the retaining screws for the engine control wiring sensor retaining bracket.

29. Reinstall the vacuum supply hoses for the body and brake booster to the bottom of the intake manifold.

30. Reinstall the EGR valve using a new gasket and tighten the two retaining bolts.

31. Reinstall the EGR valve to exhaust manifold tube nut and tighten to 26–33 ft. lbs. (35–45 Nm).

32. Reinstall the EGR valve vacuum hose.

33. Reinstall the camshaft position sensor and its retaining screw and tighten to 13–17 ft. lbs. (18–23 Nm).

34. Reconnect the camshaft position sensor.

35. Reinstall the crankcase ventilation tube to the intake manifold fitting.

36. Reconnect the engine coolant temperature sensor and the engine control sensor wiring.

37. Carefully position the fuel injection supply manifold with injectors into the intake manifold.

38. Reinstall the fuel supply and return hoses to the fuel injection supply manifold.

39. Reinstall the vacuum line to the fuel pressure regulator.

40. Position the wiring harness for the fuel injectors and install.

41. Reinstall the accelerator cable bracket.

42. Reinstall the speed control actuator and the accelerator cable to the throttle body.

43. Reinstall the air cleaner assembly and the air intake resonators.

44. Reconnect the negative battery cable.

45. Run the engine and check for leaks and proper operation.

2.5L Engine

❊❊ CAUTION

The fuel injection system remains under pressure, even after the engine has been turned OFF. The fuel system pressure must be relieved before disconnecting any fuel lines. Failure to do so may result in fire and/or personal injury.

UPPER AND LOWER INTAKE MANIFOLDS

1. Disconnect the negative battery cable.

❊❊ CAUTION

Observe all applicable safety precautions when working around fuel. Whenever servicing the fuel system, always work in a well ventilated area. Do not allow fuel spray or vapors to come in contact with a spark or open flame. Keep a dry chemical fire extinguisher near the work area. Always keep fuel in a container specifically designed for fuel storage; also, always properly seal fuel containers to avoid the possibility of fire or explosion.

2. Relieve the fuel system pressure.

3. Remove the water pump pulley shield.

4. Depress the black retainer with a screwdriver on the upper intake manifold and disconnect the main emission vacuum control and the brake booster vacuum connector from the upper intake manifold.

5. Remove the accelerator cable and speed control actuator from the throttle body.

6. Remove the accelerator cable bracket from the intake manifold and move aside.

7. Remove the idle air control valve fresh air supply hose from the fitting on the upper intake manifold.

8. Disconnect the wiring harnesses from the throttle position sensor, idle air control valve and the EGR vacuum regulator control.

9. Remove the vacuum supply hose from the upper intake manifold to the PCV valve at the upper intake manifold.

10. Disconnect the vacuum supply hoses to the EGR vacuum regulator control and the EGR valve.

11. Loosen and remove the EGR valve to exhaust manifold tube and move aside.

12. Remove the Intake Manifold Runner Control (IMRC) vacuum solenoid linkage rod by carefully prying with a screwdriver.

13. Remove the upper intake manifold retaining bolts in the reverse of the installation sequence illustration.

➡ **When removing engine components such as manifolds and cylinder heads, always remove the retaining bolts in a reverse order of their tightening sequence to prevent warpage to the component.**

14. Remove the upper intake manifold and gaskets from the engine.

➡ **If only removing the upper intake manifold, stop at this point. The remainder of the procedure is for the lower manifold.**

15. Disconnect the fuel injector wiring harness and move aside.

16. Disconnect the vacuum line to the fuel pressure regulator and the IMRC valve and set aside.

17. Remove the spring lock coupling retainer clips from the fuel supply and return fittings.

18. Use spring lock coupling disconnect tools (⅜ inch and ½ inch) to disconnect the fuel supply and return hoses from the fuel injection supply manifold.

19. Remove the eight lower intake manifold to cylinder head retaining bolts in reverse of the tighten sequence illustration.

20. Remove the lower intake manifold and gaskets from the vehicle.

21. If the lower intake manifold is to be replaced or machined, remove the fuel injectors and the IMRC vacuum solenoid.

To install:

22. Install the IMRC vacuum solenoid and fuel injectors onto the lower intake manifold if removed. Use a hand vacuum pump to verify operation of the IMRC vacuum solenoid and plate operation at this time.

23. Thoroughly clean the gasket sealing areas and place two new intake to cylinder head gaskets into position.

24. Carefully install the lower intake manifold and install the intake manifold to cylinder head retaining bolts.

Installation Sequence

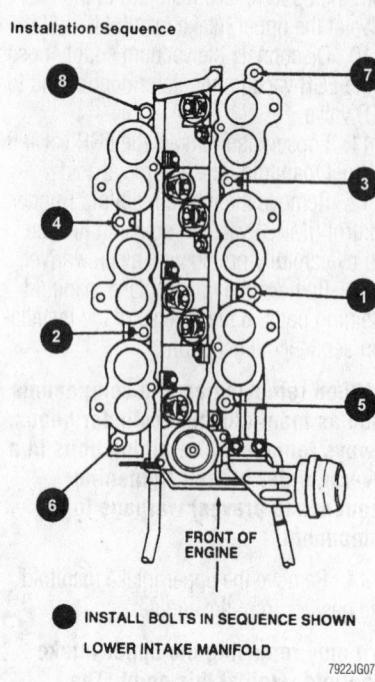

INSTALL BOLTS IN SEQUENCE SHOWN

LOWER INTAKE MANIFOLD

7922JG07

Installation Sequence

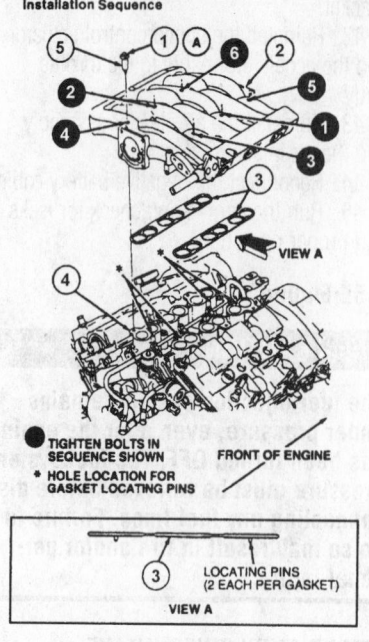

● TIGHTEN BOLTS IN SEQUENCE SHOWN

* HOLE LOCATION FOR GASKET LOCATING PINS

FRONT OF ENGINE

LOCATING PINS (2 EACH PER GASKET)

VIEW A

1. Bolt (6)
2. Intake manifold, upper
3. Intake manifold upper gasket
4. Intake manifold, lower
5. Isolator (6)
6. 71–106 in. lb.(8–12 Nm)

7922JG08

Lower intake manifold tightening sequence—2.5L engine

25. Tighten the retaining bolts in sequence to 71–106 inch lbs. (8–12 Nm).

26. Install the fuel supply and return hoses to the fuel supply manifold and ensure that the spring lock couplings are correctly installed.

27. Install the retaining clips onto the spring lock couplings.

28. Reconnect the vacuum line to the fuel pressure regulator and the IMRC vacuum solenoid.

29. Temporarily connect the negative battery cable.

30. Reconnect the fuel pressure gauge to the fuel pressure relief valve located on the fuel injection supply manifold.

31. Cycle the ignition key several times to the **RUN** position to pressurize the fuel system.

32. Watch the fuel pressure gauge for signs of leakage. If the gauge holds pressure, remove the gauge and continue with the installation of the upper intake manifold. If the pressure gauge loses pressure, remove the fuel injection supply manifold and replace the leaking O-ring(s) before continuing.

33. Disconnect the negative battery cable.

34. Reposition and install the fuel injector wiring harness.

➡**If only installing the upper intake manifold, start at this point.**

35. Install the upper intake manifold using two new gaskets onto the lower intake manifold.

36. Install the upper manifold retaining bolts and tighten following the proper sequence to 71–106 inch lbs. (8–12 Nm).

37. Install new bushings for the IMRC linkage rod and install the rod.

38. Reinstall the EGR valve to exhaust manifold tube and tighten the nut to 26–33 ft. lbs. (35–45 Nm).

39. Reconnect the vacuum supply hoses to the EGR vacuum regulator control and the EGR valve.

40. Reconnect the throttle position sensor, idle air control valve and the EGR vacuum regulator control.

41. Reinstall the idle air control valve fresh air supply hose to the fitting on the upper intake manifold.

42. Reinstall the accelerator cable bracket to the intake manifold.

43. Reinstall the speed control actuator and the accelerator cable to the throttle body.

44. Reinstall the main emission vacuum control connector and the brake booster vacuum connector to the upper intake manifold.

45. Reinstall the water pump pulley shield.

46. Reconnect the negative battery cable.

47. Run the engine and check for leaks.

Exhaust Manifold

REMOVAL & INSTALLATION

2.0L Engine

1. Disconnect the negative battery cable.

2. Remove the engine air intake resonators.

3. Disconnect the heated oxygen sensor.

4. Remove the oil level indicator tube.

5. Remove the exhaust manifold heat shield retainers and the exhaust manifold heat shield.

6. Remove the heated oxygen sensor from the exhaust manifold.

7. Remove the four catalytic converter retaining nuts.

8. Raise and safely support the vehicle.

9. Remove the EGR valve to exhaust manifold tube, retaining bracket and clamp.

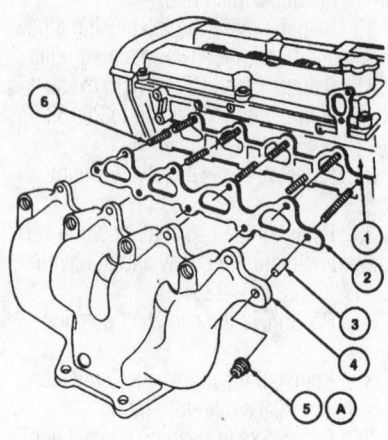

← FRONT OF ENGINE →

1	Cylinder Head
2	Exhaust Manifold Gasket
3	Spacer
4	Exhaust Manifold
5	Nut (9 Req'd)
6	Stud (9 Req'd)
A	Tighten to 14-17 N·m (13-16 Lb-Ft)

7922JG09

Exhaust manifold and related components—2.0L engine

10. Remove the catalytic converter.

11. Lower the vehicle.

12. Remove the nine exhaust manifold retaining nuts from the cylinder head studs.

13. Remove the exhaust manifold and gasket.

14. Remove the exhaust manifold from the vehicle.

15. Clean all gasket mating surfaces.

To install:

16. Position a new exhaust manifold gasket and the exhaust manifold onto the cylinder head studs.

17. Reinstall the exhaust manifold retaining nuts and tighten to 13–16 ft. lbs. (14–17 Nm).

18. Raise and safely support the vehicle.

19. Reinstall the catalytic converter using a new exhaust converter inlet gasket.

20. Reinstall the EGR valve to exhaust manifold tube, retaining bracket and clamp.

21. Tighten the EGR valve to exhaust manifold tube nut to 44 ft. lbs. (60 Nm).

22. Lower the vehicle.

23. Reinstall the catalytic converter to exhaust manifold retaining nuts.

24. Reinstall the heated oxygen sensor to the exhaust manifold and tighten to 44 ft. lbs. (60 Nm).

25. Reinstall the exhaust manifold heat shield and heat shield retainers.

26. Tighten the heat shield retainers to 71–106 inch lbs. (8–11 Nm).

27. Reinstall the oil level indicator tube.

28. Reconnect the heated oxygen sensor.

29. Reinstall the air intake resonators.

30. Reconnect the negative battery cable.

31. Run the engine and check for leaks and proper operation.

2.5L Engine

RIGHT SIDE

1. Disconnect the negative battery cable.

2. Disconnect the wiring harness from the oxygen sensor.

3. Raise and safely support the vehicle.

4. Remove the alternator and the alternator mounting bracket.

5. Remove the retaining nuts from the outlet flange on the catalytic converter.

6. Remove the muffler and the exhaust converter outlet gasket from the catalytic converter.

7. Remove the catalytic converter retaining nuts and the exhaust pipe flange hold-down springs.

8. Remove the catalytic converter.

9. Remove the halfshaft support bearing retainer bracket from the support bearing and cylinder block.

10. Remove the oxygen sensor from the exhaust manifold.

11. Loosen the EGR valve to exhaust manifold tube nuts and remove the tube.

12. Remove the exhaust manifold retaining nuts from the cylinder head studs.

13. Remove the exhaust manifold and gasket from the engine.

14. Clean all gasket mating surfaces.

To install:

15. Install the exhaust manifold with a new exhaust manifold gasket.

16. Reinstall the exhaust manifold retaining studs and tighten to 13–16 ft. lbs. (18–22 Nm) in the proper sequence.

17. Reinstall the EGR valve to exhaust manifold tube and tighten the nuts to 26–33 ft. lbs. (35–45 Nm).

18. Reinstall the oxygen sensor into the exhaust manifold and tighten to 26–34 ft. lbs. (35–46 Nm).

19. Reinstall the halfshaft support bearing retainer bracket to the support bearing and the cylinder block.

20. Reinstall the catalytic converter using a new exhaust converter inlet gasket.

21. Position the muffler and a new exhaust converter outlet gasket onto the cat-

alytic converter and loosely install the retaining nuts.

22. Align the exhaust system and tighten all nuts and bolts.

23. Tighten the catalytic converter retaining nuts with the exhaust pipe flange hold-down springs to 22–30 ft. lbs. (27.9–40.3 Nm).

24. Tighten the muffler inlet flange nuts to 26–33 ft. lbs. (34–46 Nm).

25. Reinstall the alternator mounting bracket and the alternator.

26. Lower the vehicle.

27. Reconnect the wiring harness to the oxygen sensor.

28. Reconnect the negative battery cable.

29. Run the engine and check for exhaust leaks and proper operation.

LEFT SIDE

1. Disconnect the negative battery cable.

2. Disconnect the oxygen sensor.

3. Raise and safely support the vehicle.

4. Remove the front and rear exhaust crossover tube flange fasteners from the exhaust manifolds.

5. Remove the stud and nut retainer from the engine oil pan.

6. Remove the two remaining nuts and bolts from the exhaust crossover tubes outlet connection.

7. Remove the exhaust crossover tube.

8. Remove the lower radiator hose tube retaining bracket nuts.

9. Remove the six exhaust manifold retaining nuts from the cylinder head studs.

10. Move the lower radiator hose tube to gain access for removal of the exhaust manifold.

11. Remove the exhaust manifold. If the manifold is being replaced, remove the oxygen sensor.

12. Clean all gasket mating surfaces.

To install:

13. If removed, install the oxygen sensor in the new manifold.

14. Place a new exhaust manifold to cylinder block gasket onto the cylinder head studs.

15. Move the lower radiator hose to allow installing the exhaust manifold.

16. Place the exhaust manifold onto the cylinder head studs and install the six retaining nuts.

17. Tighten the retaining nuts to 13–16 ft. lbs. (18–22 Nm), in sequence.

18. Reinstall the lower radiator hose tube retaining bracket and tighten to 71–106 inch lbs. (8–12 Nm).

19. Reinstall the exhaust crossover tube.

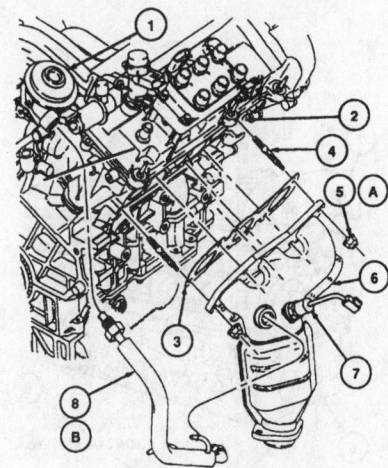

1	EGR Valve
2	RH Cylinder Head
3	Exhaust Manifold Gasket
4	Stud Bolt (6 Req'd)
5	Nut (6 Req'd)
6	RH Exhaust Manifold
7	Heated Oxygen Sensor
8	EGR Valve to Exhaust Manifold Tube
A	Tighten to 18-22 N·m (13-16 Lb-Ft)
B	Tighten to 35-45 N·m (26-33 Lb-Ft)

7922JG10

Exploded view of the right-side exhaust manifold mounting—2.5L engine

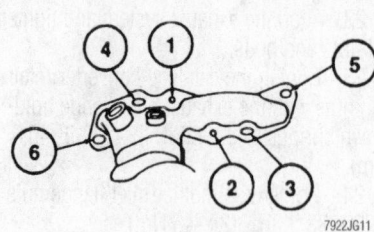

7922JG11

Exhaust manifold mounting bolt tightening sequence—2.5L engine

20. Reinstall two nuts and bolts to the exhaust crossover tube outlet connection.

21. Reinstall the stud and nut retainer at the engine oil pan.

22. Reinstall the front and rear exhaust crossover tube flange fasteners at the exhaust manifolds.

23. Lower the vehicle.

24. Reconnect the oxygen sensor.

25. Reconnect the negative battery cable.

26. Run the engine and check for exhaust leaks and proper operation.

Front Crankshaft Seal

➡The front crankshaft seal procedures apply on to engines equipped with timing belts. For the front seals on engines equipped with timing chains or gears, refer to the applicable procedure later in this section.

REMOVAL & INSTALLATION

2.0L Engine

1. Disconnect the negative battery cable.

2. Remove the camshaft timing belt and the crankshaft sprocket.

3. Using seal remover T92C-6700-CH or equivalent, remove the front crankshaft seal from the oil pump housing.

To install:

4. Clean and inspect the oil pump housing crankshaft front oil seal bore.

5. Lubricate the oil pump crankshaft oil seal bore and crankshaft front seal with engine assembly lubricant.

6. Use oil pump seal replacer T81P-6700-A or equivalent tool, and the crankshaft pulley retaining bolt, install the new crankshaft seal into the oil pump.

7. Reinstall the crankshaft sprocket and the camshaft timing belt.

8. Reconnect the negative battery cable.

9. Run the engine and check for leaks and proper operation.

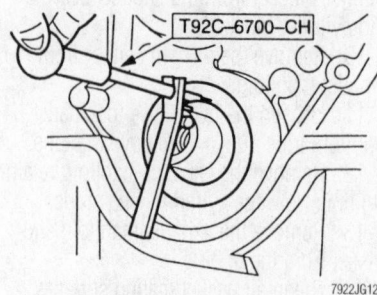

7922JG12

Be sure not to damage the crankshaft sealing surface while removing the seal— 2.0L engine

Camshaft and Valve Lifters

REMOVAL & INSTALLATION

2.0L Engine

1. Disconnect the negative battery cable.

2. Remove the valve cover as follows:

a. Remove the engine air intake resonators.

b. Disconnect the crankcase ventilation tube from the valve cover.

c. Remove the ignition wires from the valve cover and move aside.

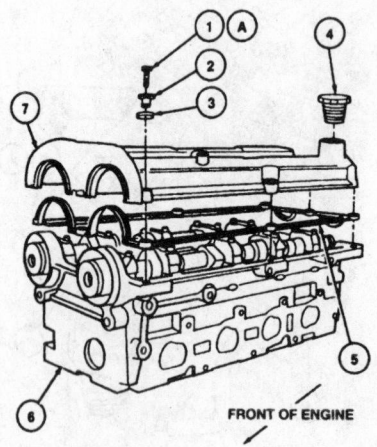

1. Bolt (10)
2. Spacer (10)
3. O-ring (10)
4. Oil filler cap
5. Valve cover gasket
6. Cylinder head
7. Valve cover
A. 53-71 in. lb.(6-8 Nm)

7922JG13

Valve cover and related components— 2.0L engine

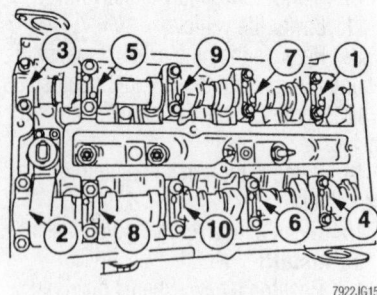

7922JG15

Remove the bearing cap bolts in pairs using the sequence shown—2.0L engine

d. Remove the retaining bolt and nut for the power steering pressure hose retaining bracket and move the hose aside.

e. Remove the bolts for the upper camshaft timing belt cover and remove the cover.

f. Remove the valve cover retaining bolts in a standard removal sequence starting from the outside of the valve cover and working toward the inside of the valve cover.

g. Remove the valve cover and gasket from the engine.

3. Remove the camshaft timing belt and sprockets.

➡Mark the camshaft journal caps to the cylinder head for installation. Do not mix the caps between the two camshafts or from another cylinder head.

4. Loosen all of the camshaft journal cap bolts in pairs and in sequence one turn at a time starting at the rear cap. This will allow the camshaft to raise up from the cylinder head evenly.

➡Remove the camshaft journal thrust caps last.

5. Remove all of the camshaft journal caps making sure that they are marked so that they can be reassembled to their original positions.

6. Remove the intake and exhaust camshafts and the camshaft front seals from the cylinder head.

7. Inspect the camshafts and cylinder head for wear.

➡If the valve lifters (tappets) are to be reused, mark their locations to ensure that they will be installed into their correct positions.

8. If required, remove the valve lifters from the cylinder head.

9. Inspect the valve lifters for wear.

To install:

➡ **Before installing the camshafts, the crankshaft must be positioned so that No. 1 cylinder is at TDC on its compression stroke.**

10. If removed, lubricate the valve lifters with engine assembly lubricant and install into the lifter bores that they were removed from.

11. If the valve lifters are new, soak the lifters in a container of clean engine oil or manually pump up the lifters before installation.

12. Lubricate the camshafts with engine assembly lubricant and place the camshafts into the cylinder head.

➡ **The intake and exhaust camshafts are marked for identification, also the intake camshaft has an extra cam lobe for the Camshaft Position (CMP) sensor.**

13. Loosely install the camshaft journal caps and retaining bolts into their original positions.

14. Install the camshaft journal thrust caps and retaining bolts last. Apply a bead of silicone sealant to the sealing surfaces of the camshaft journal thrust caps.

15. Tighten all of the camshaft journal caps in several steps pulling the camshaft down evenly following the proper sequence.

16. Once the camshaft journal caps are fully seated, tighten the retaining bolts to 13–15 ft. lbs. (17–21 Nm).

17. Install new camshaft front seals using the Camshaft Seal Replacer T92C-6700-cH or equivalent.

18. Reinstall the camshaft sprockets and the camshaft timing belt.

19. Reinstall the valve cover as follows:
   a. Clean the gasket sealing surfaces.

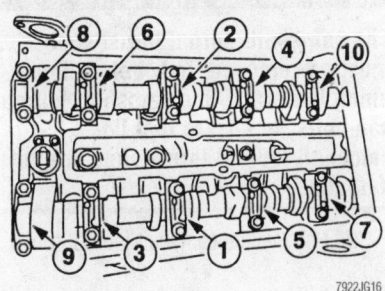

Camshaft journal cap retaining bolt tightening sequence—2.0L engine

   b. Inspect the valve cover gasket and O-rings, replace as required.
   c. Reinstall the valve cover retaining bolts and tighten in a standard sequence starting from the center and working towards the outside of the valve cover to 53–71 inch lbs. (6–8 Nm).
   d. Reinstall the upper camshaft timing belt cover and tighten the retaining bolts to 27–44 inch lbs. (3–5 Nm).
   e. Reinstall the power steering hose retaining bracket and the power steering hose.
   f. Reinstall the ignition wires.
   g. Reinstall the crankcase ventilation tube to the valve cover.
   h. Reinstall the engine air intake resonators.

20. Reconnect the negative battery cable.

21. Run the engine and check for leaks and proper operation.

2.5L Engine

1. Disconnect the negative battery cable.

2. Remove both valve covers as follows:
   a. Remove the ignition wires and spark plugs.

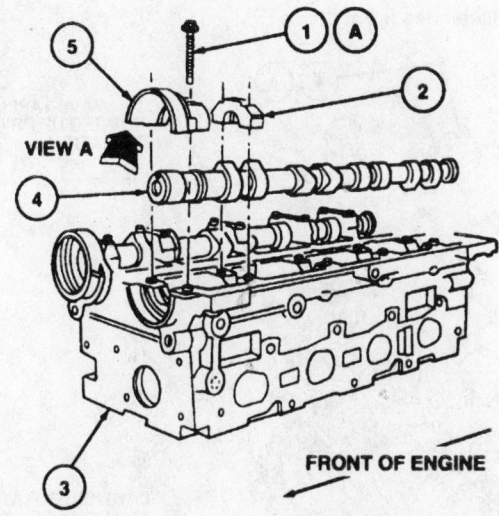

1. Bolt(20)
2. Camshaft journal cap(8)
3. Cylinder head
4. Camshaft
5. Camshaft journal thrust cap(2)
A. 13-15 ft. lb.(17-21 Nm)

VIEW A

FRONT OF ENGINE

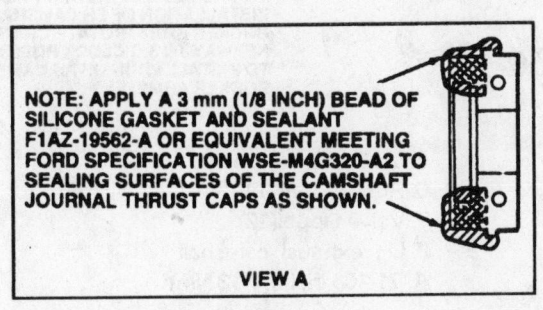

NOTE: APPLY A 3 mm (1/8 INCH) BEAD OF SILICONE GASKET AND SEALANT F1AZ-19562-A OR EQUIVALENT MEETING FORD SPECIFICATION WSE-M4G320-A2 TO SEALING SURFACES OF THE CAMSHAFT JOURNAL THRUST CAPS AS SHOWN.

VIEW A

During assembly, apply sealant to the camshaft journal thrust cap as shown—2.0L engine

Left (front) cylinder head camshaft journal cap retaining bolt loosening sequence—2.5L engine

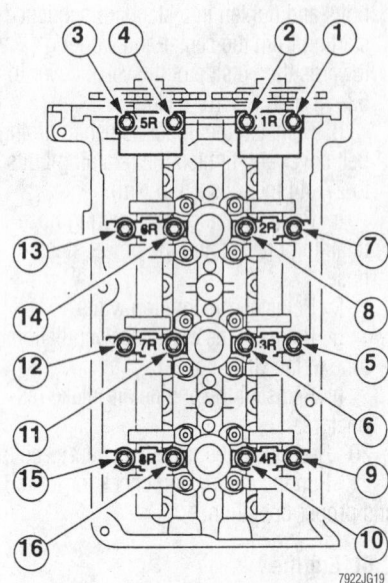

Right (rear) cylinder head camshaft journal cap retaining bolt loosening sequence—2.5L engine

7922JG19

b. Remove the ignition coil from the right valve cover.

c. Remove the crankcase ventilation tubes from both valve covers.

d. Remove the wiring harness and bracket to the fuel injectors and move aside.

e. Remove the retaining nuts and engine wiring from both valve covers and move aside.

f. Loosen the valve cover retaining bolts and studs, in sequence.

g. Remove both valve covers from the engine.

3. Remove the engine front cover.

4. Remove the timing chains.

❋❋ WARNING

The camshaft journal thrust caps must be removed first, before loosening the remaining camshaft journal cap bolts, to ensure that the camshaft journal thrust caps are not damaged.

5. Remove the camshaft journal thrust caps.

6. Loosen the camshaft journal cap bolts in sequence, in several passes, to allow the camshaft to raise off of the cylinder head evenly.

❋❋ WARNING

The camshaft journal caps and cylinder heads are numbered to ensure that they are assembled in their original positions. Keep the camshaft journal caps from each cylinder head together; do not mix them with caps from another cylinder head. Failure to do so may result in engine damage.

7. Remove the camshaft journal caps with the retaining bolts installed.

8. Remove the camshafts from the cylinder head. If necessary, remove the rocker arms, marking their position so they can be reinstalled in their original locations.

9. Repeat the procedure for both cylinder heads.

10. Inspect the camshafts and bearing journals for wear or damage.

To install:

❋❋ WARNING

The crankshaft keyway must be at the 11 o'clock position before reassembly. Failure to do so may lead to engine damage.

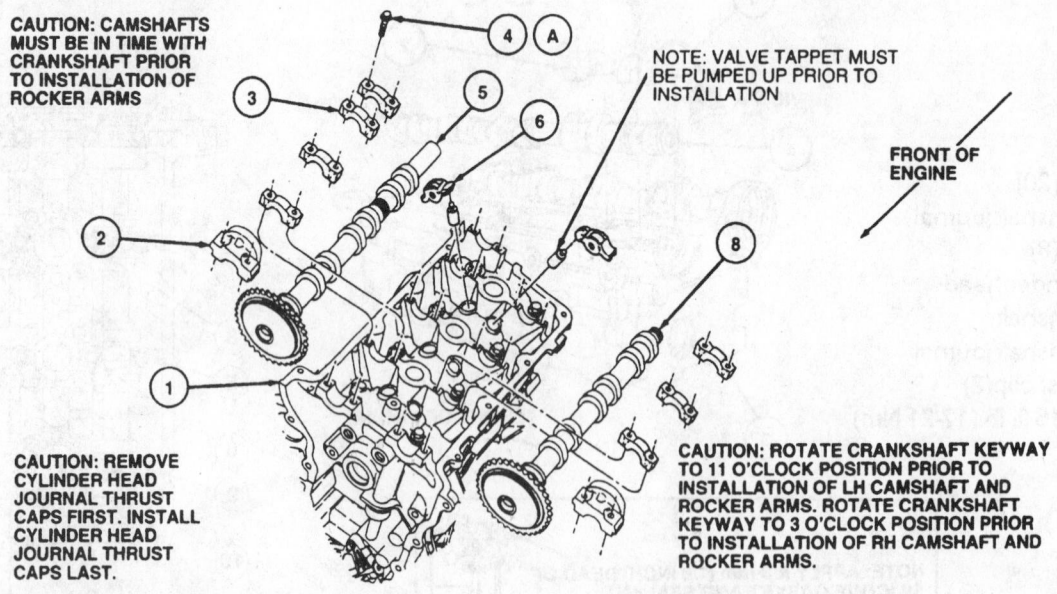

Camshafts Shown Removed From Cylinder Heads For Clarity

CAUTION: CAMSHAFTS MUST BE IN TIME WITH CRANKSHAFT PRIOR TO INSTALLATION OF ROCKER ARMS

NOTE: VALVE TAPPET MUST BE PUMPED UP PRIOR TO INSTALLATION

FRONT OF ENGINE

CAUTION: REMOVE CYLINDER HEAD JOURNAL THRUST CAPS FIRST. INSTALL CYLINDER HEAD JOURNAL THRUST CAPS LAST.

CAUTION: ROTATE CRANKSHAFT KEYWAY TO 11 O'CLOCK POSITION PRIOR TO INSTALLATION OF LH CAMSHAFT AND ROCKER ARMS. ROTATE CRANKSHAFT KEYWAY TO 3 O'CLOCK POSITION PRIOR TO INSTALLATION OF RH CAMSHAFT AND ROCKER ARMS.

1. Cylinder head
2. Camshaft journal thrust cap(2)
3. Camshaft journal cap(7)
4. Bolt(18)
5. LH intake camshaft
6. Rocker arm(12)
7. Valve tappet(12)
8. LH exhaust camshaft
A. 71-106 in. lb.(8-12 Nm)

Exploded view of camshaft mounting—2.5L engine

7922JG17

11. Rotate the crankshaft so that the keyway is at the 11 o'clock position for installation of the camshafts.

12. Reinstall the rocker arms, if removed.

13. Lubricate the camshafts with engine assembly lubricant.

14. Reinstall the camshafts into their correct positions into each cylinder head with the timing marks on the camshaft sprockets aligned.

15. Loosely install the camshaft journal caps and retaining bolts into their correct positions.

➡ **Do not install the camshaft journal thrust caps until the rocker arms and timing chains have been installed and the camshaft journal caps are secured into position.**

16. Reinstall the rocker arms.

17. Reinstall the timing chains.

18. Tighten the camshaft journal cap bolts, in the opposite order of the removal sequence, to 71–106 inch lbs. (8–12 Nm). Reinstall the thrust caps and tighten the retaining bolts to 71–106 inch lbs. (8–12 Nm).

19. Reinstall the engine front cover.

20. Reinstall both valve covers as follows:

  a. Clean the valve cover gasket sealing surfaces.

  b. Reinstall new valve cover gaskets onto the valve covers.

  c. For each valve cover, place a bead of silicone sealant at the two places on the valve cover sealing surfaces where the engine front cover and the cylinder heads make contact and at two places on the rear of the cylinder head where the camshaft seal retainer contacts the cylinder head.

  d. Reinstall the valve cover retaining bolts and studs and tighten in sequence to 71–106 inch lbs. (8–12 Nm).

➡ **The valve covers must be installed and properly tightened within six minutes of applying the silicone sealant.**

21. Reconnect the negative battery cable.

22. Run the engine and check for leaks and proper engine operation.

Valve Lash

ADJUSTMENT

The lash adjusters (valve tappets), are hydraulic and are not adjustable. It is important that all valve components are in good condition and installed and tightened properly.

Oil Pan

REMOVAL & INSTALLATION

2.0L Engine

1. Disconnect the negative battery cable.

2. Install Three Bar Engine Support D88L-6000-A or similar engine support, to the engine lifting eyes and support the engine.

3. Raise and safely support the vehicle.

4. Remove the catalytic converter system.

5. Disconnect the wiring to the low oil level sensor, if equipped.

6. Remove the heater water bolt from the bottom of the oil pan and position the tube out of the way.

✷✷ CAUTION

The EPA warns that prolonged contact with used engine oil may cause a number of skin disorders, including cancer! You should make every effort to minimize your exposure to used engine oil. Protective gloves should be worn when changing the oil. Wash your hands and any other exposed skin areas as soon as possible after exposure to used engine oil. Soap and water, or waterless hand cleaner should be used.

7. Drain the engine oil.

8. Reinstall the oil pan drain plug and tighten to 15–21 ft. lbs. (21–28 Nm).

9. Remove the oil pan retaining bolts from the transaxle housing.

10. Remove the lower engine rear plate.

11. Remove the through-bolt for the left and right support insulators.

12. Lower the vehicle to work on the top side of the engine but keep on the hoist.

➡ **Mark the location of the upper front engine support bracket before removing from the front engine support bracket.**

13. Remove the upper front engine support bracket retaining nuts from the upper front engine support insulator.

14. Raise the engine to allow room for removal of the engine oil pan, using the Three Bar Engine Support or equivalent engine brace.

15. Raise and safely support the vehicle.

16. Remove the oil pan retaining bolts from the cylinder block, working from the ends of the block toward the center.

17. Loosen and remove the oil pan and gasket.

18. Inspect the oil pump pick-up tube and screen and clean or replace as necessary.

To install:

19. If removed, install the oil pump pick-up tube and screen. Tighten the retaining bolts to 71–97 inch lbs. (8–11 Nm). Install a new self-locking oil pump pick-up tube support nut to the crankshaft main bearing cap retaining stud bolt. Tighten to 13–15 ft. lbs. (17–21 Nm).

20. Clean the gasket sealing surfaces for the oil pan at the cylinder block.

21. Clean the oil pan thoroughly leaving no traces of gasket material, grease or solvents.

22. Apply a bead of silicone gasket sealer to the oil pump parting lines and at the crankshaft rear main seal retainer on the cylinder block.

23. Install the oil pan and a new gasket into position and hold with several oil pan retaining screws.

24. Reinstall the rest of the retaining bolts and push the oil pan flush against the transaxle case before tightening the retaining bolts.

25. Tighten the 10 oil pan retaining bolts, in several passes, to 15–18 ft. lbs. (20–24 Nm), working from the center of the block towards the ends.

26. Reinstall the lower engine rear plate.

27. Reinstall the oil pan to transaxle housing retaining bolts.

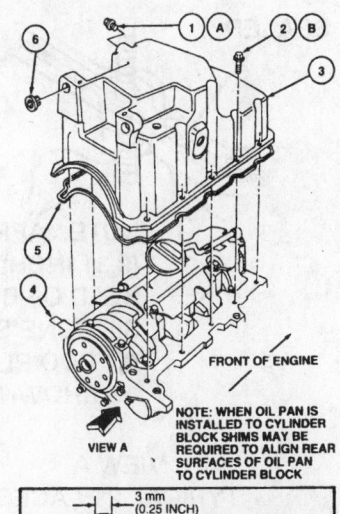

NOTE: WHEN OIL PAN IS INSTALLED TO CYLINDER BLOCK SHIMS MAY BE REQUIRED TO ALIGN REAR SURFACES OF OIL PAN TO CYLINDER BLOCK

3 mm (0.25 INCH)
SEALER
NOTE: APPLY A 3 mm (0.25 INCH) BEAD OF SILICONE GASKET AND SEALANT F1AZ-19562-A OR EQUIVALENT MEETING FORD SPECIFICATION WSE-M4G320-A2

VIEW A
TYPICAL FOUR PLACES

1 Oil pan drain plug
2 Bolt (10 req'd)
3 Oil pan
4 Cylinder block
5 Oil pan gasket
6 Oil pan spacer (as req'd)
A Tighten to 21-28 Nm (15-21 lb-ft)
B Tighten to 20-24 Nm (15-18 lb-ft)

7922JG20

Exploded view of the oil pan mounting—2.0L engine

28. Tighten the retaining bolts to 25–34 ft. lbs. (34–46 Nm).

29. Lower the vehicle but keep on the hoist.

30. Lower the engine into position by adjusting the Three Bar Engine Support or equivalent.

31. Reinstall the upper front engine support bracket retaining nuts onto the upper front engine support insulator.

32. Raise and safely support the vehicle.

33. Reinstall the through-bolts for the left and the right insulators.

34. Reconnect the low oil level sensor, if equipped.

35. Position the heater water tube and retaining bolt to the bottom of the oil pan.

36. Reinstall the catalytic converter system.

37. Lower the vehicle.

38. Remove the Three Bar Engine Support or equivalent.

✳✳ WARNING

Operating the engine without the proper amount and type of engine oil will result in severe engine damage.

39. Fill the crankcase with the proper amount of engine oil.

40. Reconnect the negative battery cable.

41. Run the engine and check for leaks and proper operation.

2.5L Engine

1. Disconnect the negative battery cable.

2. Remove the water pump pulley shield.

3. Install Three Bar Engine Support D88L-6000-A or equivalent, to the engine lifting eyes and support the engine.

4. Raise and safely support the vehicle.

5. Remove the exhaust crossover and the exhaust retaining bracket located on the right side of the oil pan.

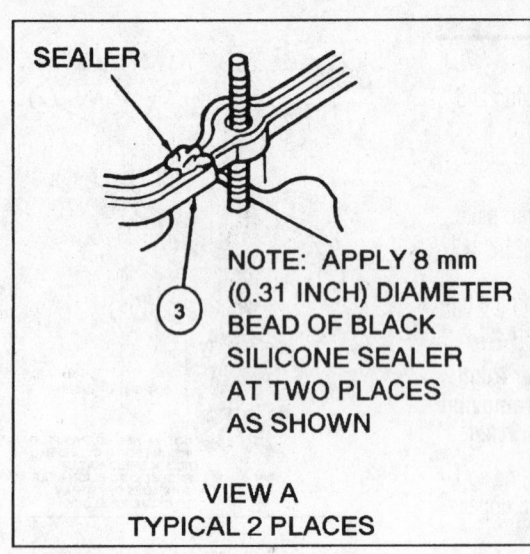

SEALER

NOTE: APPLY 8 mm (0.31 INCH) DIAMETER BEAD OF BLACK SILICONE SEALER AT TWO PLACES AS SHOWN

VIEW A
TYPICAL 2 PLACES

FRONT OF ENGINE

* **LOCATION OF STUDS**
● **TIGHTEN BOLTS/STUDS IN SEQUENCE SHOWN**

1 Upper cylinder block
2 Lower cylinder block
3 Oil pan
4 Stud bolt (5 req'd)
5 Bolt (10 req'd)
6 Oil pan gasket
7 Engine front cover
A Tighten to 20-30 Nm (15-22 lb-ft)

Exploded view of the oil pan mounting, showing the mounting bolt and stud tightening sequence—2.5L engine

7922JG21

6. Remove the exhaust heat shield retaining nuts and the exhaust heat shields from the left side of the oil pan.

✳✳ CAUTION

The EPA warns that prolonged contact with used engine oil may cause a number of skin disorders, including cancer! You should make every effort to minimize your exposure to used engine oil. Protective gloves should be worn when changing the oil. Wash your hands and any other exposed skin areas as soon as possible after exposure to used engine oil. Soap and water, or waterless hand cleaner should be used.

7. Drain the engine oil.
8. Reinstall the oil pan drain plug using a new gasket and tighten to 16–22 ft. lbs. (22–30 Nm).
9. Remove the oil pan retaining bolts from the transaxle housing.
10. If equipped with automatic transaxle, remove the access plug from the engine rear plate.
11. Remove the through-bolt for the left and right front engine support insulators.
12. Partially lower the vehicle on the hoist.

➡ **Mark the location of the upper front engine support bracket before it is removed.**

13. Remove the upper front engine support bracket retaining nuts and remove the upper front engine support bracket.
14. Using the three bar engine support or equivalent, raise the engine to allow room for removal of the engine oil pan.
15. Raise and safely support the vehicle.
16. Remove the oil pan retaining bolts and studs from the lower cylinder block following the bolt removal sequence.
17. Remove the oil pan and the oil pan gasket from the vehicle.

To install:

18. Clean the oil pan to lower cylinder block gasket sealing surfaces.
19. Thoroughly clean the oil pan.
20. Install a new oil pan gasket into the groove of the oil pan.
21. Apply a bead of silicone sealer to the gasket area where the pan meets the parting lines of the lower cylinder block and the front engine cover.
22. Carefully install the oil pan with gasket to the lower cylinder block.
23. Reinstall the bolts and studs but do not tighten.

24. Push the oil pan against the transaxle case and tighten the oil pan bolts and studs.
25. Reinstall the oil pan to transaxle case bolts and tighten to 25–34 ft. lbs. (34–46 Nm).
26. Tighten the oil pan bolts and studs in the proper sequence to 15–22 ft. lbs. (20–30 Nm).
27. If equipped with automatic transaxle, install the access plug into the engine rear plate.
28. Lower the vehicle but keep it on the hoist.
29. Using the three bar engine support or equivalent, lower the engine into its proper position.
30. Reinstall the upper front engine support bracket and its retaining nuts onto the upper front engine support insulator.
31. Raise and safely support the vehicle.
32. Reinstall the through-bolts for the left and right front engine support insulators.
33. Reinstall the exhaust crossover, exhaust retaining bracket and heat shield.
34. Replace the engine oil filter.
35. Lower the vehicle.
36. Remove the three bar engine support or equivalent.

✳✳ WARNING

Operating the engine without the proper amount and type of engine oil will result in severe engine damage.

37. Fill the crankcase with the correct amount and grade of engine oil.
38. Reconnect the negative battery cable.
39. Run the engine and check for leaks and proper operation.

Oil Pump

REMOVAL & INSTALLATION

2.0L Engine

1. Disconnect the negative battery cable.
2. Remove the accessory drive belt.
3. Remove the camshaft timing belt covers, camshaft timing belt and the crankshaft sprocket.
4. Install Three Bar Engine Support D88L-6000-A or similar engine support, to the engine lifting eyes and support the engine.
5. Raise and safely support the vehicle.
6. Remove the catalytic converter system.

✳✳ CAUTION

The EPA warns that prolonged contact with used engine oil may cause a number of skin disorders, including cancer! You should make every effort to minimize your exposure to used engine oil. Protective gloves should be worn when changing the oil. Wash your hands and any other exposed skin areas as soon as possible after exposure to used engine oil. Soap and water, or waterless hand cleaner should be used.

7. Remove the oil pan.
8. Remove the oil pump screen cover and tube retaining nut from the crankshaft main bearing cap stud bolt.
9. Remove the oil pump screen cover and tube retaining bolts from the oil pump and remove from the engine.
10. Remove the engine oil filter.
11. Remove the oil pump retaining bolts.
12. Remove the oil pump and gasket from the cylinder block.

To install:

13. Clean the oil pump gasket sealing surface on the cylinder block and oil pump.
14. Rotate the inner rotor of the oil pump to align with the flats on the crankshaft.
15. Reinstall the oil pump using a new gasket onto the cylinder block.
16. Loosely install the oil pump retaining bolts.

➡ **Clearance between the cylinder block oil pan sealing surface to the oil pump oil pan sealing surface should not exceed 0.012–0.031 in. (0.3–0.8mm).**

17. Use a straight-edge to align the oil pump oil pan sealing surface with the cylinder block oil pan sealing surface.
18. Tighten the oil pump retaining bolts to 71–102 inch lbs. (8–11.5 Nm).
19. Install a new engine oil filter.

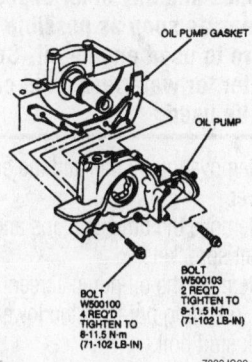

Exploded view of the oil pump mounting— 2.0L engine

20. Reinstall the oil pump screen cover and tube using a new gasket to the oil pump. Tighten the retaining bolts to 71–97 inch lbs. (8–11 Nm).

21. Reinstall a new self-locking oil pump screen cover and tube support nut to the crankshaft main bearing cap stud bolt. Tighten the retaining nut to 13–15 ft. lbs. (17–21 Nm).

22. Reinstall the engine oil pan.

23. Reinstall the catalytic converter system.

24. Lower the vehicle.

25. Remove the three bar engine support or equivalent.

26. If required, replace the front crankshaft seal at this time.

27. Reinstall the crankshaft sprocket, camshaft timing belt and the camshaft timing belt covers.

28. Reinstall the accessory drive belt.

29. Fill the crankcase with the proper amount and grade of engine oil.

✸✸ WARNING

Operating the engine without the proper amount and type of engine oil will result in severe engine damage.

30. Reconnect the negative battery cable.

31. Run the engine and check for leaks and proper operation.

2.5L Engine

1. Disconnect the negative battery cable.

✸✸ CAUTION

The EPA warns that prolonged contact with used engine oil may cause a number of skin disorders, including cancer! You should make every effort to minimize your exposure to used engine oil. Protective gloves should be worn when changing the oil. Wash your hands and any other exposed skin areas as soon as possible after exposure to used engine oil. Soap and water, or waterless hand cleaner should be used.

2. Remove the oil pan and the engine front cover.

3. Remove the timing chains and the crankshaft sprockets.

4. Remove the oil pump screen cover and tube retaining nut from the lower cylinder block stud bolt.

5. Remove the oil pump screen cover and tube retaining bolts from the oil pump and remove the tube from the engine.

6. Remove the four oil pump retaining bolts in reverse of the removal sequence.

7. Remove the oil pump from the vehicle.

To install:

8. Rotate the inner rotor of the oil pump to align with the flats on the crankshaft.

9. Reinstall the oil pump flush to the cylinder block.

10. Reinstall the oil pump retaining bolts and tighten in sequence to 71–106 inch lbs. (8–12 Nm).

11. Inspect the oil pump screen and tube O-ring and replace if needed.

12. Position the oil pump screen and tube with the O-ring to the oil pump. Tighten the retaining bolts to 71–106 inch lbs. (8–12 Nm).

13. Install a new self-locking tube support nut to the lower cylinder block stud. Tighten the nut to 15–22 ft. lbs. (20–30 Nm).

14. Reinstall the crankshaft sprockets and the timing chains.

15. Reinstall the oil pan and the engine front cover.

16. Fill the crankcase with the correct amount of engine oil.

✸✸ WARNING

Operating the engine without the proper amount and type of engine oil will result in severe engine damage.

17. Reconnect the negative battery cable.

18. Run the engine and check for leaks and proper operation.

Rear Main Seal

REMOVAL & INSTALLATION

2.0L Engine

➡The following procedure requires a specific Ford design seal remover, or equivalent aftermarket version.

1. Disconnect the negative battery cable.

2. Remove transaxle and flexplate/flywheel using the recommended procedure.

3. Using a sharp awl, punch a hole in the rear oil seal metal surface between the oil seal lip and the oil seal retainer.

4. Screw the threaded end of a seal remover into the oil seal, then install a slide hammer and remove the seal from the crankshaft oil seal retainer.

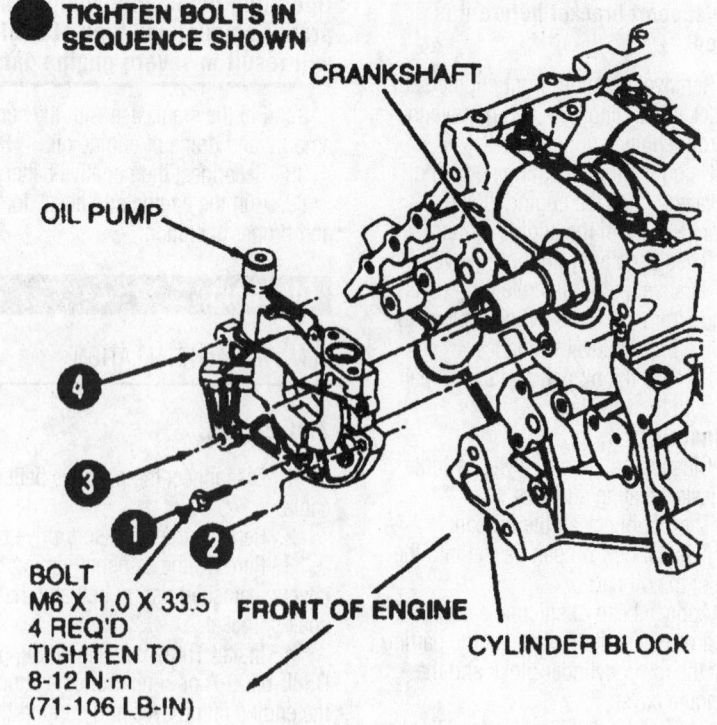

● **TIGHTEN BOLTS IN SEQUENCE SHOWN**

CRANKSHAFT

OIL PUMP

BOLT
M6 X 1.0 X 33.5
4 REQ'D
TIGHTEN TO
8-12 N·m
(71-106 LB-IN)

FRONT OF ENGINE

CYLINDER BLOCK

7922JG23

Exploded view of the oil pump mounting, showing the retaining bolt tightening sequence—2.5L engine

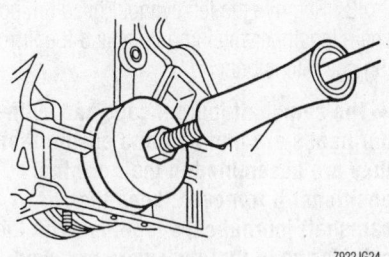

7922JG24

Thread the oil seal remover into the seal, then withdraw the seal using a slide hammer—2.0L and 2.5L engines

To install:

5. Clean all sealing surfaces.

6. Lubricate the crankshaft, oil seal bore and the lip of the seal with Engine Assembly Lubricant D9AZ-19579-D or equivalent meeting Ford specification ESR-M99C80-A.

7. Install the oil seal using an appropriate seal installer. Seat the seal flush to the rear of the crankshaft oil seal retainer.

8. Reinstall the transaxle and flexplate/flywheel using the recommended procedure.

9. Reconnect the negative battery cable.

10. Start the engine and check for leaks.

Timing Chain, Sprockets, Front Cover and Seal

REMOVAL & INSTALLATION

2.5L Engine

1. Disconnect the negative battery cable.

2. Remove the upper intake manifold.

3. Remove the valve covers as follows:

   a. Remove the ignition wires and spark plugs.

   b. Remove the ignition coil from the right valve cover.

   c. Remove the crankcase ventilation tubes from both valve covers.

   d. Remove the wiring harness and bracket to the fuel injectors and move aside.

   e. Remove the retaining nuts and engine wiring from both valve covers and move aside.

   f. Loosen the valve cover retaining bolts and studs following the proper removal sequence.

   g. Remove both valve covers from the engine.

4. Install Three Bar Engine Support D88L-6000-A or equivalent to the engine lifting eyes and support the engine.

5. Remove the power steering pressure hose bracket retainer bolt from the upper front engine support bracket.

6. Disconnect the power steering pressure hose from the power steering pump and position out of the way.

➡ **Mark the position of the upper front engine support bracket before removing.**

7. Remove the upper front engine support bracket retainer nuts and remove the bracket.

8. Disconnect the low coolant level sensor from the wiring harness.

9. Remove the radiator coolant recovery reservoir retainers and move the recovery reservoir aside.

10. Remove the upper front engine insulator.

11. Set the radiator coolant recovery reservoir back into its position but do not secure.

12. Disengage the three wiring harness connectors located at the in-line connector bracket at the front of the right cylinder head and set aside.

13. Loosen the power steering pump pulley retaining bolts but do not remove the bolts completely.

14. Remove the accessory drive belt.

15. Finish removing the power steering pump pulley retaining bolts and remove the pulley.

16. Remove the power steering pump and pump support retaining nuts and bolts and remove the power steering pump and pump support.

17. Raise and safely support the vehicle.

18. Remove the right-front wheel and tire assembly.

19. Remove the alternator from its mounting bracket and move aside.

20. Remove the alternator mounting bracket.

21. Remove the crankshaft pulley.

22. Disconnect the wiring from the Crankshaft Position (CKP) sensor and the Camshaft Position (CMP) sensor.

23. Remove the engine oil pan.

24. Loosen the A/C compressor retaining bolts and move the A/C compressor to gain access to the front cover retaining bolt.

25. Partially lower the vehicle.

26. Remove the bracket retainers and the bracket for the engine wiring and the A/C hose at the engine front cover.

➡ **It may be necessary to raise and lower the vehicle several times in order to follow the engine front cover bolt removal sequence.**

27. Remove the engine front cover bolts in reverse of the installation sequence illustration.

28. Remove the engine front cover and gasket from the vehicle.

29. Rotate the crankshaft so that the keyway is at the 11 o'clock position to locate the crankshaft at TDC for No. 1 cylinder.

30. Verify that the alignment arrows on the camshafts are aligned. If not, rotate the

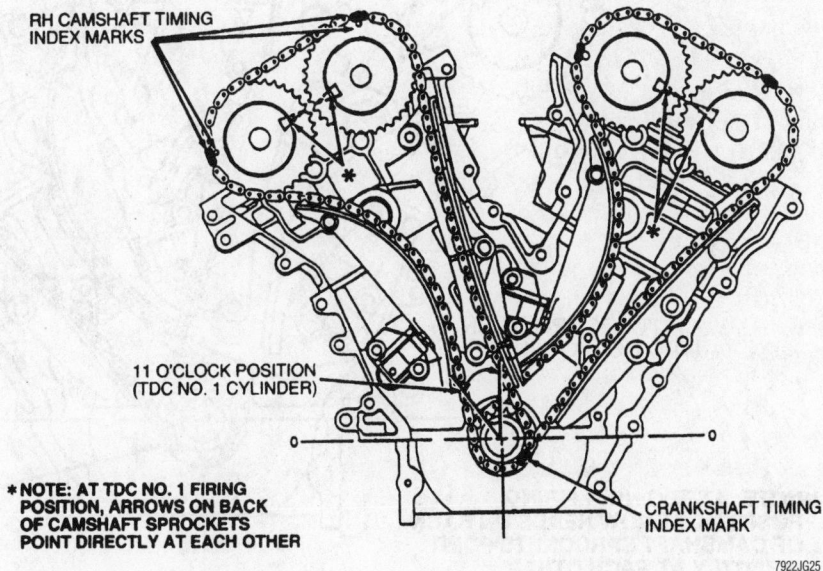

RH CAMSHAFT TIMING INDEX MARKS

11 O'CLOCK POSITION (TDC NO. 1 CYLINDER)

CRANKSHAFT TIMING INDEX MARK

* NOTE: AT TDC NO. 1 FIRING POSITION, ARROWS ON BACK OF CAMSHAFT SPROCKETS POINT DIRECTLY AT EACH OTHER

7922JG25

View of the timing chains and gears, showing the right timing chain alignment marks properly positioned—2.5L engine

crankshaft one complete revolution and recheck.

31. Rotate the crankshaft so that the keyway is at the 3 o'clock position. This positions the right cylinder head camshafts to the neutral position.

32. Remove the right cylinder head timing chain tensioner retaining bolts and the timing chain tensioner.

➡**The camshaft journal caps and cylinder heads are numbered to ensure that they are assembled in their original positions. If removed, keep the camshaft journal caps together with the cylinder head that they were removed from.**

33. Remove the right cylinder head camshaft journal thrust cap retaining bolts and thrust caps.

34. Loosen the remaining camshaft journal cap bolts in sequence, releasing the bolts several revolutions at a time by making several passes to allow the camshaft to be raised from the cylinder head evenly. Do not remove the retaining bolts completely.

➡**If the valve tappets and roller rocker arms are to be reused, mark the positions of the valve tappets and rocker arms so that they are reassembled into their original positions.**

35. With the camshafts loose, remove the rocker arms, keeping them in the order that they were removed.

➡**If the right cylinder head timing chain tensioner arm and timing chain guide are to be reused, mark the position of the timing chain tensioner arm and the timing chain guide so that they are reassembled into their original positions.**

36. Remove the right cylinder head timing chain tensioner arm and the timing chain.

37. Remove the right cylinder head timing chain guide retaining bolts and the timing chain guide.

38. If worn, replace the timing chain guide.

39. Remove the right crankshaft timing chain sprocket.

40. Remove the right cylinder head camshaft timing chain sprockets if they are to be replaced.

41. Rotate the crankshaft two revolutions and locate the crankshaft keyway at the 11 o'clock position. This will position the left cylinder head camshafts to their neutral position.

42. Verify that the alignment arrows on

the camshafts are aligned.

43. Remove the left cylinder head timing chain tensioner retaining bolts and the timing chain tensioner.

➡**The camshaft journal caps and cylinder heads are numbered to ensure that they are assembled in their original positions. If removed, keep the camshaft journal caps together with the cylinder head that they were removed from.**

44. Remove the camshaft journal thrust cap retaining bolts and thrust caps from the left cylinder head.

45. Loosen the remaining camshaft journal cap bolts in sequence, releasing the bolts several revolutions at a time by making several passes to allow the camshaft to be raised from the cylinder head evenly. Do not remove the retaining bolts completely.

➡**If the valve tappets and roller rocker arms are to be reused, mark the positions of the valve tappets and rocker arms so that they are reassembled into their original positions.**

46. With the camshafts loose, remove the rocker arms, keeping them in the order that they were removed.

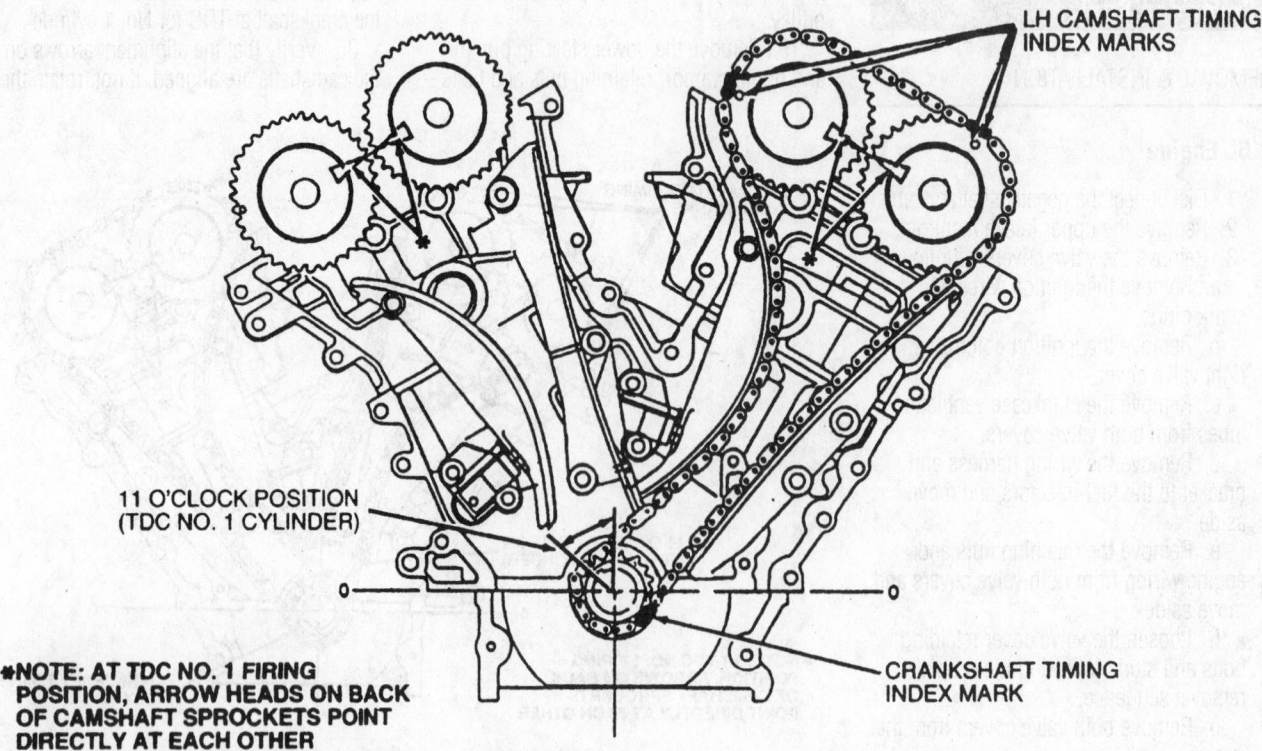

LH CAMSHAFT TIMING INDEX MARKS

11 O'CLOCK POSITION (TDC NO. 1 CYLINDER)

CRANKSHAFT TIMING INDEX MARK

✳NOTE: AT TDC NO. 1 FIRING POSITION, ARROW HEADS ON BACK OF CAMSHAFT SPROCKETS POINT DIRECTLY AT EACH OTHER

7922JG26

Left timing chain alignment mark positioning for servicing the chain—2.5L engine

➡️**If the left cylinder head timing chain tensioner arm and timing chain guide are to be reused, mark the position of the timing chain tensioner arm and the timing chain guide so that they are reassembled into their original positions.**

47. Remove the left cylinder head timing chain tensioner arm and the timing chain.

48. Remove the left cylinder head timing chain guide retaining bolts and the timing chain guide.

49. If worn, replace the timing chain guide.

50. Remove the left crankshaft timing chain sprocket.

51. If required, the left cylinder head camshaft sprockets may be removed at this time.

To install:

➡️**Inspect the timing chains, tensioners, tensioner arms, guides and sprockets for wear or damage. If any components are to be replaced for premature wear or damage, the camshaft damper should also be replaced.**

52. Reinstall or replace the camshaft sprockets, if removed.

53. Be sure that the crankshaft keyway is still at the 11 o'clock position.

54. Reinstall the left timing chain crankshaft sprocket onto the crankshaft.

55. Reinstall the left timing chain guide and retaining bolts to the engine. Tighten the retaining bolts to 15–22 ft. lbs. (20–30 Nm).

56. Verify that the alignment arrows on the left cylinder head camshafts are aligned before proceeding.

57. Reinstall the left timing chain over the left crankshaft sprocket and the left camshaft sprockets.

58. Align the timing index marks on the left cylinder head timing chain with the timing index marks on the crankshaft sprocket and the camshaft sprockets.

59. Reinstall the left timing chain tensioner arm over the alignment dowel on the left cylinder head.

➡️**Before installing the timing chain tensioner, it must be properly compressed and locked.**

60. Using a small screwdriver, release the timing chain tensioner ratchet/pawl mechanism through the access hole in the timing chain tensioner as follows:

   a. Insert a small piece of wire into the top of the piston and gently unseat the oil check ball.

   b. Compress the timing chain tensioner by hand.

   c. With the tensioner compressed, install a 0.060 inch (1.5mm) drill bit or wire into the small hole above the ratchet, engaging the lock groove in the rack of the timing chain tensioner.

61. Reinstall the compressed and locked left cylinder head timing chain tensioner and retaining bolts onto the cylinder block. Tighten the retaining bolts to 15–22 ft. lbs. (20–30 Nm).

62. Verify that the timing index marks on the left timing chain are in alignment with the timing index marks on the crankshaft sprocket and the camshaft sprockets.

63. Reinstall or replace the camshaft sprockets if removed.

64. Reinstall the right timing chain crankshaft sprocket onto the crankshaft.

65. Reinstall the right timing chain guide and retaining bolts to the engine. Tighten the retaining bolts to 15–22 ft. lbs. (20–30 Nm).

66. Verify that the alignment arrows on the right cylinder head camshafts are aligned before proceeding.

67. Reinstall the right timing chain over the right crankshaft sprocket and the right camshaft sprockets.

68. Align the timing index marks on the right cylinder head timing chain with the timing index marks on the crankshaft sprocket and the camshaft sprockets.

69. Reinstall the right timing chain tensioner arm over the alignment dowel on the right cylinder head.

➡️**Before installing the timing chain tensioner, it must be properly compressed and locked.**

70. Using a small screwdriver, release the timing chain tensioner ratchet/pawl mechanism through the access hole in the timing chain tensioner as follows:

   a. Insert a small wire into the top of the piston and gently unseat the oil check ball.

   b. Compress the timing chain tensioner by hand.

   c. With the tensioner compressed, install a 0.060 inch (1.5mm) drill bit or wire into the small hole above the ratchet, engaging the lock groove in the rack of the timing chain tensioner.

71. Reinstall the compressed and locked right cylinder head timing chain tensioner and retaining bolts onto the cylinder block. Tighten the retaining bolts to 15–22 ft. lbs. (20–30 Nm).

72. Verify that the timing index marks on the right timing chain are in alignment with the timing index marks on the crankshaft sprocket and the camshaft sprockets.

73. Be sure that the crankshaft keyway is at the 11 o'clock position.

74. Lubricate the left-side rocker arms with engine assembly lubricant and install the left cylinder head rocker arms into their original locations.

➡️**Do not install the camshaft journal thrust caps until the camshaft journal caps are secured into position.**

75. Tighten the left cylinder head camshaft journal cap bolts in sequence making several passes to pull the camshafts down evenly. Tighten the bolts to 71–106 inch lbs. (8–12 Nm).

76. Reinstall the left cylinder head thrust caps and bolts. Tighten to 71–106 inch lbs. (8–12 Nm).

77. Rotate the crankshaft two revolutions and position the crankshaft keyway to the 3 o'clock location. This will position the right cylinder head camshafts to the neutral position.

78. Lubricate the right-side rocker arms with engine assembly lubricant and install the right cylinder head rocker arms into their original locations.

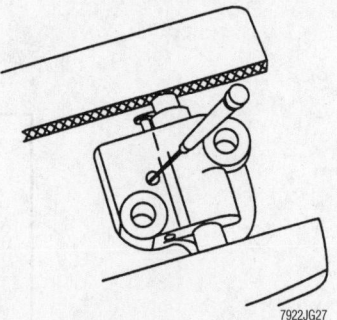

Using a thin prytool, release the timing chain tensioner ratchet/pawl mechanism—2.5L engine

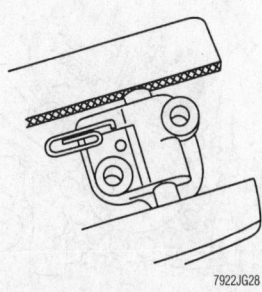

Insert a small piece of wire (such as a paper clip) into the top of the piston and gently unseat the oil check ball—2.5L engine

➡ **Do not install the camshaft journal thrust caps until the camshaft journal caps are secured.**

79. Tighten the right cylinder head camshaft journal cap bolts in sequence making several passes to pull the camshafts down evenly. Tighten the bolts to 71–106 inch lbs. (8–12 Nm).

80. Reinstall the right cylinder head thrust caps and bolts. Tighten to 71–106 inch lbs. (8–12 Nm).

81. Remove the lock pins from the timing chain tensioners.

82. Verify that the timing index marks on the timing chains are in alignment with the timing index marks on the crankshaft sprocket and the camshaft sprockets.

83. Install the crankshaft position sensor pulse ring, be sure to use the keyway for the 2.5L engine as shown.

84. Replace the crankshaft seal in the front cover with a new one. Apply clean engine oil to the seal lip.

85. Clean the engine front and the front cover-to-cylinder block gasket sealing surfaces.

➡ **The front cover must be installed and properly tightened within six minutes of the application of the sealer.**

86. Apply silicone sealer to the six critical areas shown in View **A** , to the cylinder block to prevent oil seepage.

87. Place new front cover gaskets onto the dowel pins on the cylinder block and heads.

88. Place the front cover into position by placing the front cover onto the dowel pins at the cylinder block.

89. Reinstall the six front cover retaining bolts and stud bolts where the silicone sealer was applied.

90. Tighten the bolts and stud bolts until the front cover contacts the cylinder block and heads an, then turn the bolts and stud bolts an additional ¼ turn.

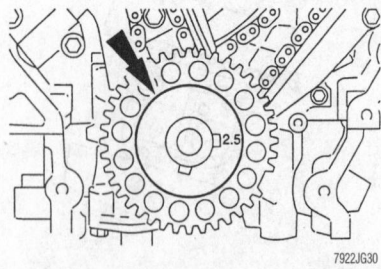

When installing the crankshaft position sensor pulse ring, be sure to use the correct keyway—2.5L engine

91. Reinstall the remaining front cover retaining bolts and stud bolts.

92. Tighten all of the front cover retaining bolts and stud bolts in proper sequence to 15–22 ft. lbs. (20–30 Nm).

93. Reinstall the bracket and bracket retainers to the front cover. Reinstall the wiring and the A/C hose to the bracket.

94. Raise and safely support the vehicle.

95. Reinstall the A/C compressor and its retaining bolts. Tighten the retaining bolts to 15–22 ft. lbs. (20–30 Nm).

96. If required, replace the crankshaft front seal at this time.

97. Reinstall the engine oil pan.

98. Reconnect the wiring harness to the CKP sensor and the CMP sensor.

99. Reinstall the crankshaft pulley.

100. Reinstall the alternator mounting bracket and the alternator.

101. Reinstall the right-front wheel and tire assembly. Tighten the lug nuts to 62 ft. lbs. (85 Nm).

102. Lower the vehicle on the hoist.

103. Lower the engine to its correct position.

104. Reinstall the power steering pump support and the power steering pump to the front of the engine.

105. Reinstall the power steering pump pulley. Loosely install the retaining bolts.

106. Reinstall the accessory drive belt.

107. Tighten the power steering retaining bolts to 15–22 ft. lbs. (20–30 Nm).

108. Position and attach the wiring to the three connectors located on the in-line connector bracket at the front of the right cylinder head.

109. Reinstall the upper front support insulator and the upper front engine support bracket.

110. Loosen and remove the Three Bar Engine Support or equivalent.

111. Reconnect the power steering pressure hose to the power steering pump.

112. Reconnect the power steering pressure hose bracket and retainer bolt to the upper front engine support bracket.

113. Reinstall both valve covers as follows:

　a. Clean the valve cover gasket sealing surfaces.

　b. Install new valve cover gaskets onto the valve covers.

　c. For each valve cover, place a bead of silicone sealant at two places on the valve cover sealing surfaces where the engine front cover and the cylinder heads make contact and at two places on the rear

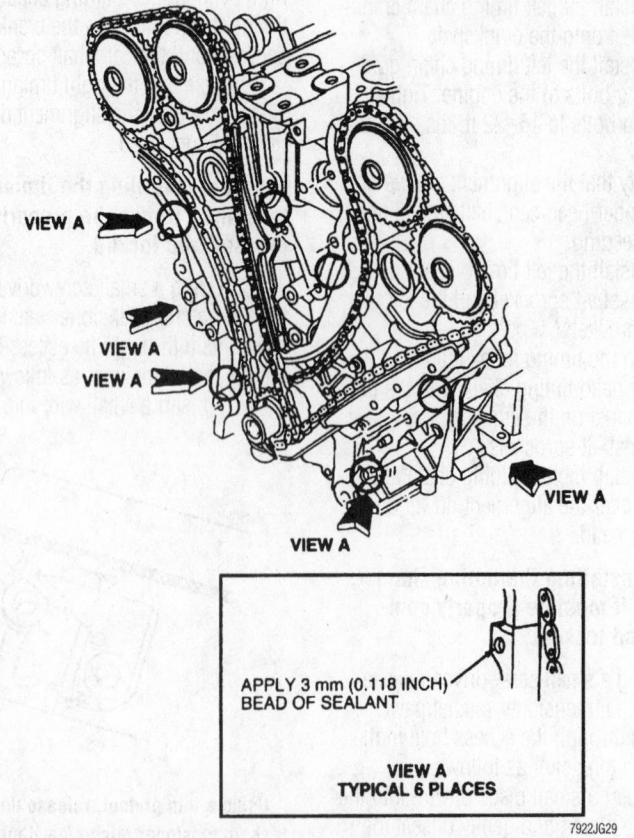

APPLY 3 mm (0.118 INCH) BEAD OF SEALANT

VIEW A TYPICAL 6 PLACES

To prevent oil leakage, apply sealant to the places indicated—2.5L engine

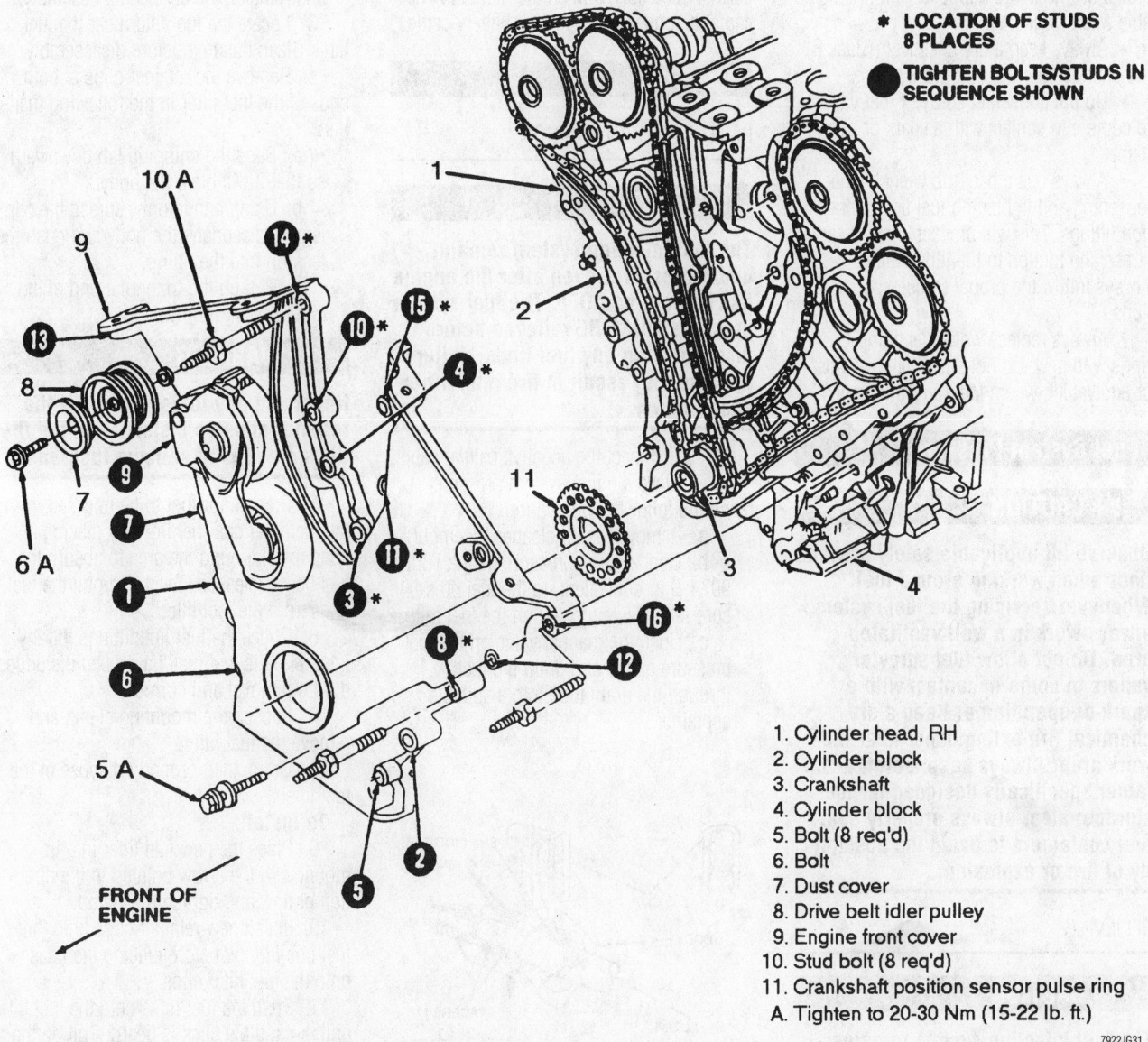

* **LOCATION OF STUDS 8 PLACES**

● **TIGHTEN BOLTS/STUDS IN SEQUENCE SHOWN**

1. Cylinder head, RH
2. Cylinder block
3. Crankshaft
4. Cylinder block
5. Bolt (8 req'd)
6. Bolt
7. Dust cover
8. Drive belt idler pulley
9. Engine front cover
10. Stud bolt (8 req'd)
11. Crankshaft position sensor pulse ring
A. Tighten to 20-30 Nm (15-22 lb. ft.)

7922JG31

FRONT OF ENGINE

Exploded view of the front cover mounting, showing the retaining bolt and nut tightening sequence—2.5L engine

of the cylinder head where the camshaft seal retainer contacts the cylinder head.

　d. Reinstall the valve cover retaining bolts and studs and tighten in sequence to 71–106 inch lbs. (8–12 Nm).

➡**The valve covers must be installed and properly tightened within six minutes of applying the silicone sealant.**

　114. Reinstall the upper intake manifold.
　115. Fill the engine with the proper amount and grade of oil.
　116. Replace any lost fluid to the power steering reservoir.
　117. Reconnect the negative battery cable.
　118. Run the engine and check for leaks and proper operation.
　119. Recheck the fluid levels.

FUEL SYSTEM

Fuel System Service Precautions

　Safety is the most important factor when performing not only fuel system maintenance but any type of maintenance. Failure to conduct maintenance and repairs in a safe manner may result in serious personal injury or death. Maintenance and testing of the vehicle's fuel system components can be accomplished safely and effectively by adhering to the following rules and guidelines.

　• To avoid the possibility of fire and personal injury, always disconnect the negative battery cable unless the repair or test procedure requires that battery voltage be applied.

　• Always relieve the fuel system pressure prior to disconnecting any fuel system component (injector, fuel rail, pressure regulator, etc.), fitting or fuel line connection. Exercise extreme caution whenever relieving fuel system pressure to avoid exposing skin, face and eyes to fuel spray. Please be advised that fuel under pressure may penetrate the skin or any part of the body that it contacts.

　• Always place a shop towel or cloth around the fitting or connection prior to loosening to absorb any excess fuel due to spillage. Ensure that all fuel spillage (should it occur) is quickly removed from engine surfaces. Ensure that all fuel soaked cloths or towels are deposited into a suitable waste container.

• Always keep a dry chemical (Class B) fire extinguisher near the work area.

• Do not allow fuel spray or fuel vapors to come into contact with a spark or open flame.

• Always use a back-up wrench when loosening and tightening fuel line connection fittings. This will prevent unnecessary stress and torsion to fuel line piping. Always follow the proper torque specifications.

• Always replace worn fuel fitting O-rings with new. Do not substitute fuel hose or equivalent, where fuel pipe is installed.

Fuel System Pressure

✳✳ CAUTION

Observe all applicable safety precautions when working around fuel. Whenever servicing the fuel system, always work in a well ventilated area. Do not allow fuel spray or vapors to come in contact with a spark or open flame. Keep a dry chemical fire extinguisher near the work area. Always keep fuel in a container specifically designed for fuel storage; also, always properly seal fuel containers to avoid the possibility of fire or explosion.

RELIEVING

✳✳ CAUTION

The fuel injection system remains under pressure, even after the engine has been turned OFF. The fuel system pressure MUST BE relieved before disconnecting any fuel lines. Failure to do so may result in fire and/or personal injury.

1. Disconnect the negative battery cable.
2. Remove the engine air cleaner assembly.
3. Loosen the fuel tank filler cap to relieve pressure in the fuel tank.
4. Connect fuel pressure gauge T80L-9974-B or equivalent, to the fuel pressure relief valve located on the fuel rail.
5. Open the manual valve on the fuel pressure gauge and drain the fuel through the drain tube into a suitable container.
6. Remove the fuel pressure gauge.
7. When service on the vehicle is com-

plete, be sure to install the engine air cleaner assembly, tighten the fuel tank filler cap and connect the negative battery cable.

Fuel Filter

REMOVAL & INSTALLATION

✳✳ CAUTION

The fuel injection system remains under pressure, even after the engine has been turned OFF. The fuel system pressure MUST BE relieved before disconnecting any fuel lines. Failure to do so may result in fire and/or personal injury.

1. Disconnect the negative battery cable.
2. Relieve the fuel system pressure using the following procedure:
   a. Remove the air cleaner assembly.
   b. Connect fuel pressure gauge T80L-9974-B or equivalent, to the fuel pressure relief valve located on the fuel rail.
   c. Open the manual valve on the fuel pressure gauge and drain the fuel through the drain tube into a suitable container.

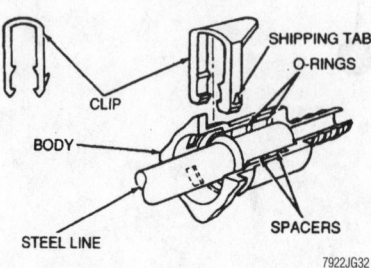

Once the retainer clip has been removed from the connector, the fuel line can be removed from the filter

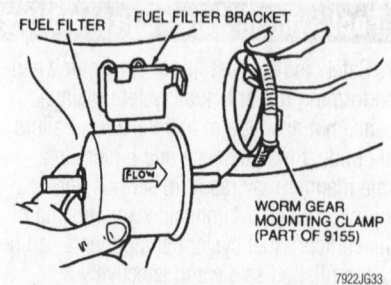

Be sure to install the fuel filter with the arrow pointing in the direction of the fuel flow

d. Remove the fuel pressure gauge and reinstall the air cleaner assembly.
3. Locate the fuel filter near the fuel tank. Clean the area before disassembly.
4. Remove the retainer clips at both ends of the fuel filter in the following manner:
   a. Bend the shipping tab downward so that it will clear the body.
   b. Using hands only, spread the clip legs to disengage the body and push the legs up into the fitting.
   c. Pull on the triangular end of the clip to finish removal.

✳✳ WARNING

Never use any tools to remove the retainer clips as distortion to the fittings may result causing fuel leaks.

5. Prepare for fuel to be released from the fuel filter and fuel lines by placing a shop towel around the area to absorb the fuel being released. Twist and pull the fuel lines from the fuel filter.
6. Check the fuel line fittings for any internal parts that may have been dislodged during removal and correct.
7. Loosen the mounting clamp and remove the fuel filter.
8. Drain, then properly dispose of the fuel filter.

To install:

9. Place the new fuel filter into its mount with the arrow pointed in the direction of flow and tighten the clamp.
10. Install new retainer clips onto the fuel line fittings before placing the lines onto the fuel filter ends.
11. Push the fuel lines onto the fuel filter until an audible click is heard. Pull on the fitting to verify a good connection.
12. Start the engine and check for fuel leaks and proper operation.

Fuel Pump

REMOVAL & INSTALLATION

✳✳ CAUTION

The fuel injection system remains under pressure, even after the engine has been turned OFF. The fuel system pressure MUST BE relieved before disconnecting any fuel lines. Failure to do so may result in fire and/or personal injury.

1. Disconnect the negative battery cable.

2. Relieve the fuel pressure using the following procedure:

a. Open the fuel tank filler cap to vent off pressure in the tank.

b. Remove the air cleaner assembly.

c. Connect fuel pressure gauge T80L-9974-B or equivalent, to the fuel pressure relief valve located on the fuel rail.

d. Open the manual valve on the fuel pressure gauge and drain the fuel through the drain tube into a suitable container.

e. Remove the fuel pressure gauge.

f. Secure the fuel fill cap and install the air cleaner assembly.

3. Remove the rear seat cushion.

4. Remove the plastic grommet from the floor pan.

5. Disconnect the fuel pump electrical harness.

6. Disconnect the fuel lines from the fuel pump by compressing the tabs on both sides of each nylon push connect fitting and easing the fuel line off of the fuel pump.

7. Using Fuel Tank Sender Wrench D84P-9275-A or equivalent, turn the fuel pump locking ring counterclockwise to loosen the ring.

8. Remove the fuel pump locking ring.

9. Remove the fuel pump being careful not to damage the fuel gauge sending unit.

10. Place a shop towel over the opening in the fuel tank to prevent dirt from contaminating the fuel.

To install:

11. Remove the shop towel over the opening in the fuel tank and clean the groove for the fuel pump seal. Be careful not to allow dirt to enter the fuel tank.

12. Apply a light coat of grease onto a new O-ring seal and install into the groove of the fuel tank.

13. Carefully install the fuel pump into the tank to prevent damage to the fuel gauge sender or the fuel pick-up filter.

➤It is recommended that the in-line fuel filter be replaced whenever a fuel pump is being replaced.

14. Ensure that the flange of the fuel pump mounting plate is located properly in its keyway and that the O-ring has not shifted out of position.

15. Keep a light downward pressure on the fuel pump while installing the fuel pump locking retainer ring.

16. Install the ring ensuring that all of the locking tabs are under the fuel tank lock ring tabs. Turn the ring clockwise finger-tight.

17. Install the fuel tank sender wrench or equivalent, over the retainer ring and finish

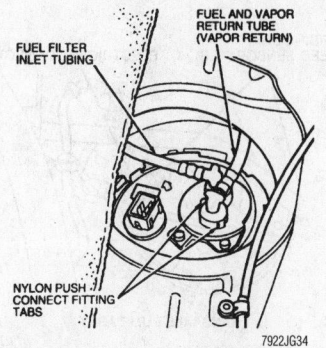

View of fuel pump fittings through the floorpan

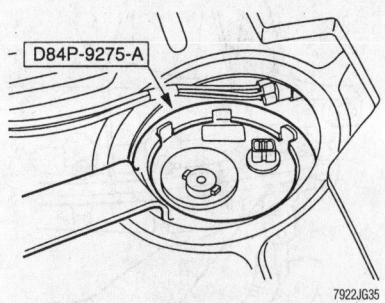

Remove the locking ring from the fuel pump sender with a special wrench such as Fuel Tank Sender Wrench D84P-9275-A

tightening until the retainer ring is resting against its stops.

18. Reconnect the fuel lines to the fuel pump.

19. Reinstall the fuel pump electrical harness connector.

20. Reinstall the plastic grommet into the floorpan.

21. Reinstall the rear seat cushion.

22. Reconnect the negative battery cable.

23. Start the engine and check for leaks and proper operation.

DRIVE TRAIN

Transaxle Assembly

REMOVAL & INSTALLATION

Automatic

1. Disconnect the negative battery cable.

2. Disconnect the positive battery cable and remove the battery.

3. On vehicles equipped with the 2.0L

engine, remove the oil level dipstick and the exhaust manifold heat shield.

4. On vehicles equipped with the 2.5L engine, remove the water pump pulley shield.

5. Secure the radiator and fan shroud with safety wire to the radiator support.

6. Remove the air cleaner assembly and mounting bracket.

7. Loosen the left and right upper strut mounting nuts five turns to allow room for removal of the halfshafts. Do not remove the nuts completely.

8. Disconnect the shift cable and remove the two retaining bolts securing the shift cable bracket to the transaxle case.

9. Disconnect the Transmission Range (TR) sensor.

10. Remove the TR sensor.

11. Disconnect the 10-pin harness from the transaxle.

12. Support the engine with Three Bar Engine Support D88L-6000-A or equivalent.

13. Remove the upper transaxle support insulator bracket mounting nuts.

14. Remove the three bolts retaining the upper transaxle support insulator to the inner fenderwell and remove the insulator.

15. Remove the upper bell housing to engine retaining bolts.

16. Raise and safely support the vehicle.

17. Remove the wheel and tire assemblies.

18. Disconnect the steering column from the rack and pinion by removing the pinch bolt.

19. Disconnect the tie rod ends from the steering knuckles. Discard the cotter pins.

20. Remove the ball joint to lower control arm pinch bolts and separate the ball joints from the lower control arms.

21. Remove the sway bar (stabilizer bar) link nuts and separate the sway bar links from the stabilizer bar.

22. Remove the splash shield at the front of the sub-frame.

23. Remove the through-bolts retaining the left and right front engine support insulators to the sub-frame.

24. Disconnect the power steering oil cooler hoses at the right front sub-frame and drain the power steering system.

25. Remove the two screws retaining the bumper cover braces to the sub-frame. Rotate the bumper cover braces forward.

26. Remove the radiator air deflector.

27. Remove the four bolts retaining the left and right lower radiator support brackets to the front of the sub-frame. Rotate the radiator supports forward.

28. Disconnect and remove the exhaust system components necessary for transaxle removal.

29. Remove the two bolts retaining the A/C accumulator to the sub-frame, located at the drivers side front corner of the vehicle.

30. Remove the bolt retaining the transaxle cooler line bracket to the front of the sub-frame.

31. Position Powertrain Lift 014–00765 or equivalent with wood blocks approximately 40 inches (1,016mm) in length secured to the lift under the sub-frame.

32. Remove the four sub-frame to body retaining bolts. Lower the sub-frame slightly and disconnect the power steering pressure and return hoses from the rack and pinion.

33. Finish lowering the sub-frame and set aside.

34. Disconnect the Turbine Shaft Speed (TSS) sensor at the transaxle oil pump.

35. Drain the transaxle oil into a suitable container for recycling.

36. Disconnect the transaxle cooler inlet line at the transaxle case.

37. Remove the transaxle cooler inlet line from the bracket at the transaxle oil pump.

38. Disconnect the transaxle cooler outlet line at the transaxle case.

39. Disconnect the transaxle cooler inlet and outlet lines at the radiator and remove from the vehicle.

40. Remove the left halfshaft and the right halfshaft and intermediate shaft from the vehicle.

41. Remove the speedometer cable and disconnect the Vehicle Speed Sensor (VSS).

42. Remove the inspection cover from the transaxle to engine spacer plate located at the right-rear corner of the engine.

43. Rotate the torque converter to align each of the four torque converter to flywheel retaining nuts and remove the nuts.

44. Lower the vehicle.

45. Remove the upper right engine support insulator (engine mount) bracket nuts.

46. Lower the engine and transaxle assembly using the three bar engine support until the transaxle assembly is level with the left frame member.

47. On vehicles equipped with A/C, lower the engine until the A/C compressor is below the right frame member.

48. Raise and safely support the vehicle.

49. Support the transaxle on a transmission jack and secure the transaxle to the jack.

50. Remove the starter motor retaining nuts and remove the starter.

51. Remove the lower transaxle to engine retaining bolts.

52. Separate the transaxle from the engine.

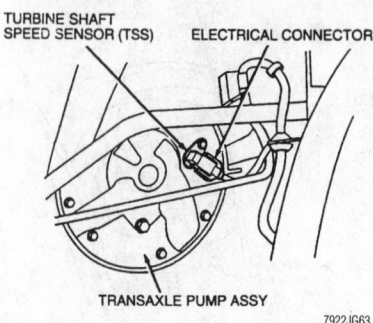

Turbine shaft speed sensor location

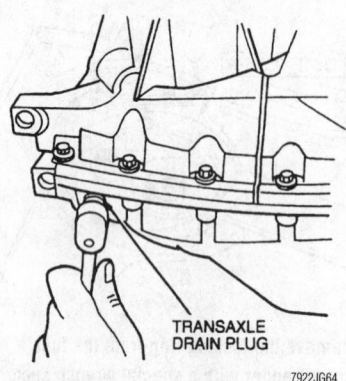

Transaxle drain plug location

➡ **Use care when removing the transaxle to prevent the torque converter from falling out.**

53. Carefully lower the transaxle from the vehicle.

To install:

54. Be sure that the torque converter is fully engaged in the transaxle.

55. There should be approximately a 7/16 inch (10mm) air gap between a straightedge across the bell housing flange and the torque converter.

56. If removed, place the transaxle on the transmission jack and secure.

➡ **Use care not to allow the torque converter to fall out of the transaxle when tilted.**

57. Raise the transaxle into position and align with the engine.

58. Align the torque converter studs with the mating holes in the flywheel.

59. Once the transaxle is fitted to the engine, install the lower transaxle to engine retaining bolts. Tighten the retaining bolts to 41–50 ft. lbs. (55–68 Nm).

60. Rotate the torque converter and install the four converter to flywheel retaining nuts. Tighten the retaining nuts to 23–39 ft. lbs. (31–53 Nm).

61. Remove the transmission jack.

62. Reinstall the transaxle to engine separator plate.

63. Reinstall the electrical connector to the TSS.

64. If removed, place the sub-frame on the powertrain lift, or equivalent with wood blocks approximately 40 inches (1,016mm) in length secured to the lift under the sub-frame.

65. Raise the sub-frame and connect the power steering pressure and return hoses to the rack and pinion.

66. Align the sub-frame. Route the power steering hoses into position at the rear of the engine.

67. Reinstall the four sub-frame retaining bolts loosely.

68. Install Sub-Frame Alignment Pin Set T95P-2100-AH or equivalent into the sub-frame and body alignment holes. After aligning the holes, slightly tighten the four sub-frame retaining bolts.

69. After the sub-frame alignment is complete, tighten the four sub-frame retaining bolts to 81–110 ft. lbs. (110–150 Nm). Remove the alignment tools.

70. Reinstall the A/C accumulator bracket to the sub-frame and tighten the screws to 48–72 inch. lbs. (6–8 Nm).

71. Reconnect the power steering oil cooler hoses to the right front side of the sub-frame.

72. Install Powertrain Alignment Gauge T94P-6000-AH or equivalent to the left-front engine support bracket and the sub-frame. Tighten the two retaining bolts to 20 ft. lbs. (27 Nm) and snug the through-bolt.

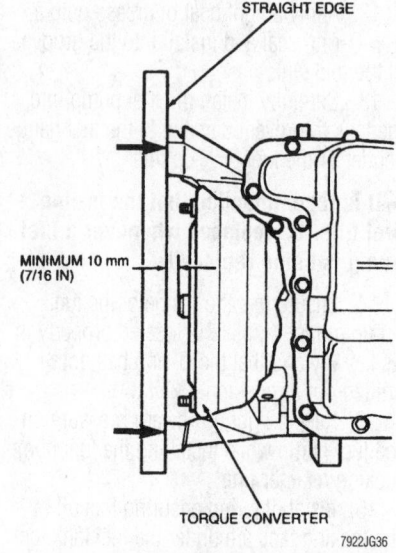

Be sure that the torque converter is properly seated in the transaxle

73. Reinstall the right engine support insulator with retaining bolts to the sub-frame and through-bolt. Tighten the two sub-frame retaining bolts to 30–41 ft. lbs. (41–55 Nm) and the through-bolt to 75–102 ft. lbs. (103–137 Nm).

74. Observe the position of the right engine support insulator. It must be centered in the bracket and in perfect alignment front to rear. Remove the powertrain alignment gauge.

75. Reinstall the left engine support insulator to the sub-frame with two retaining bolts. Tighten the retaining bolts to 84 inch. lbs. (10 Nm).

76. Observe the position of the left engine support insulator to ensure perfect alignment front to rear. Retighten the bolts to 30–41 ft. lbs. (41–55 Nm). Reinstall the left engine support insulator through-bolt and tighten to 75–102 ft. lbs. (103–137 Nm).

77. Reinstall the transaxle cooler inlet and outlet lines.

78. Reinstall the lower radiator support brackets to the sub-frame. Tighten the bolts to 20 ft. lbs. (27 Nm).

79. Reconnect the speedometer cable and the VSS.

80. Reinstall the exhaust system.

81. Reinstall the left halfshaft using a new circlip.

82. Reinstall the intermediate halfshaft and tighten the two retaining nuts to 20 ft. lbs. (27 Nm).

83. Reinstall the right-side halfshaft.

84. Reinstall the left and right lower control arms to the steering knuckles.

85. Install new pinch bolts and nuts. Tighten the pinch bolts to 61 ft. lbs. (83 Nm).

86. Reinstall the sway bar (stabilizer bar) links. Reinstall the retaining nuts and tighten to 35–48 ft. lbs. (47–65 Nm).

87. If equipped, install the ABS wiring loom retainer to the sway bar link stud. Tighten the retaining nuts to 35 ft. lbs. (47 Nm).

88. Reinstall the left and right tie rod ends to the steering knuckles. Tighten the castellated nuts to 23–35 ft. lbs. (31–47 Nm). Install new cotter pins.

89. Reinstall the front bumper cover braces and tighten the bolts securely. Reinstall the splash shield to the sub-frame.

90. Reinstall the wheel and tire assemblies. Tighten the lug nuts to 62 ft. lbs. (85 Nm).

91. Lower the vehicle.

92. Reconnect the steering yoke to the steering gear shaft. Tighten the steering yoke retaining bolt to 15–20 ft. lbs. (20–27 Nm).

93. Reinstall the upper transaxle to engine retaining bolts. Tighten the retaining bolts to 23–39 ft. lbs. (55–68 Nm).

94. Remove the safety wire securing the radiator and fan shroud to the radiator support.

95. Reinstall all wiring retaining clips that were disturbed during transaxle removal.

96. Raise the engine and transaxle assembly into position using the three bar engine support or similar tool.

97. Reinstall the engine and transmission support insulator to the left front fender apron. Tighten the bolts to 40–55 ft. lbs. (54–75 Nm).

98. Reinstall new locknuts retaining the engine and transmission support insulator to the transaxle. Tighten the locknuts to 40–55 ft. lbs. (54–75 Nm).

99. Reinstall the front engine support insulator.

100. If equipped with 2.5L engine, install the power steering line retaining bracket to the front engine support insulator. Tighten the new locknuts to 56–76 ft. lbs. (77–103 Nm). Reinstall the water pump pulley shield.

101. If equipped with the 2.0L engine, install the exhaust manifold heat shield and oil level dipstick. Tighten the retaining bolts to 71–106 inch lbs. (8–12 Nm).

102. Remove the three bar engine support, or equivalent.

103. Reinstall the TR sensor and adjust.

104. Reattach the 10-pin harness connector to the transaxle.

105. Reinstall the TR sensor electrical connector.

106. Reinstall the starter motor and retaining bolts. Tighten the starter motor retaining bolts to 43–58 ft. lbs. (59–79 Nm) for the 2.5L engine or 15–20 ft. lbs. (20–27 Nm) for the 2.0L engine.

107. Reinstall the shift cable mounting bracket. Tighten the retaining bolts to 15–19 ft. lbs. (20–25 Nm).

108. Reinstall the shift cable to the manual lever by pressing the cable end onto the stud until a click is heard.

109. Reinstall the battery tray, battery and the battery hold-down.

110. Reinstall new upper strut mounting nuts. Tighten the strut mounting nuts to 34 ft. lbs. (46 Nm).

111. Fill the power steering remote oil reservoir.

112. Reconnect the battery cables, negative cable last.

113. Fill the transaxle with the proper type and amount of transmission fluid.

114. Run the engine and check the transaxle for leaks.

115. Recheck the transmission fluid level.

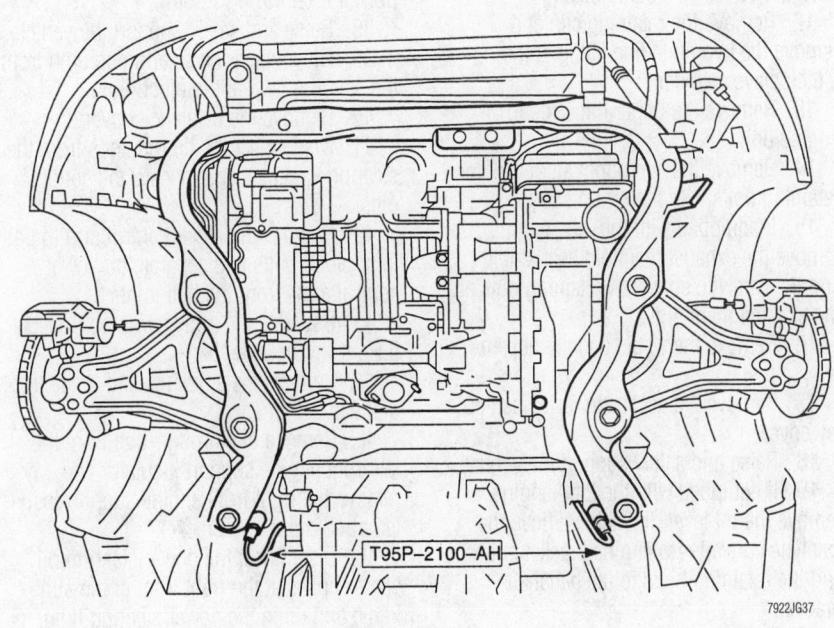

To properly align the sub-frame, install sub-frame alignment pins such as Sub-Frame Alignment Pin Set T95P-2100-AH—manual and automatic transaxles

➡**Whenever the vehicle sub-frame is removed or lowered, the wheel alignment should be checked.**

116. Check the alignment and adjust if necessary.

117. Road test the vehicle to check for proper transmission operation.

Manual

1. Disconnect the battery cables, negative cable first.

2. Remove the battery.

3. Secure the radiator and fan shroud to the radiator support using safety wire.

4. Loosen the front strut upper retaining nuts a total of five turns to allow room for the removal of the halfshafts. Do not remove the nuts.

5. Remove the Mass Air Flow sensor (MAF) and the air cleaner assembly. Remove the air cleaner lower bracket.

6. Install Three Bar Engine Support D88L-6000-A or equivalent, and support the engine.

7. Disconnect the back-up lamp switch.

8. Remove the bolts securing the ground strap to the transaxle housing.

9. Remove the engine and transmission support insulator (mount).

10. Disconnect the hydraulic line and rubber grommet from the support insulator bracket.

11. Remove the rubber inspection cover from the transaxle clutch housing.

12. Remove the retaining clip and remove the hydraulic line fitting at the clutch slave cylinder.

13. Remove the upper transaxle to engine bolt.

14. Remove the two upper starter motor retaining bolts with the ground strap.

15. If equipped with the 2.0L engine, remove the exhaust manifold heat shield and the catalytic converter retaining nuts at the exhaust manifold.

16. Remove the wheel and tire assemblies.

17. Remove the accessory drive belt pulley cover.

18. Raise and safely support the vehicle.

19. If equipped with the 2.0L engine, remove the oil level dipstick. Remove the catalytic converter to engine bracket strap and the retaining bolts to the halfshaft bracket.

20. Remove the catalytic converter.

21. If equipped with the 2.5L engine, remove the water pump pulley shield. Remove the front Y-pipe nuts and the rear Y-pipe to catalytic converter nuts and remove the Y-pipe.

22. Disconnect the Vehicle Speed Sensor (VSS).

23. Remove the speedometer cable.

24. Remove the nine screws securing the lower radiator air deflector and remove the deflector.

25. Push the shift rod forward and remove the shift rod pinch bolt.

26. Pull the shift rod back and remove it from the transaxle.

27. Remove the shift control stabilizer bar nut at the stud and remove the stabilizer bar.

28. Remove the shift control stabilizer bar and bracket from the right engine support insulator bracket.

29. Remove the underbody heat shield from under the shift control.

30. Reposition the shift rod and stabilizer bar to allow transaxle removal.

31. Remove the two screws securing the A/C accumulator to the sub-frame.

32. Remove the halfshafts and the intermediate shaft.

33. Remove the bolts and the right engine support insulator mounting nuts and remove the bracket from the transaxle.

34. Remove the left engine support insulator through-bolt.

35. Lower the vehicle.

36. Adjust the three bar engine support or equivalent, to relieve tension on the right front engine support bracket.

37. Remove the right front engine support bracket through-bolt.

38. Raise and safely support the vehicle.

39. Disconnect the steering column from the steering gear at the pinch bolt.

40. Remove the catalytic converter.

41. Disconnect the tie rod ends from the steering knuckles and discard the cotter pins.

42. Remove the lower control arm to ball joint pinch bolts and separate the lower control arms from the ball joints.

43. Separate the sway bar (stabilizer bar) links from the sway bar.

44. Remove the splash shield at the front of the sub-frame.

45. Remove the through-bolt from the left front engine support insulator and remove the right front engine support insulator and mounting bracket.

46. Disconnect the power steering oil cooler hoses at the right front of the sub-frame and drain the power steering fluid.

47. Remove the A/C accumulator retaining bracket screws from the sub-frame.

48. Remove the four bolts retaining the lower radiator supports to the sub-frame. Rotate the radiator supports forward.

49. Remove the two screws retaining the bumper cover braces to the left and right sides of the sub-frame. Rotate the bumper cover braces forward.

50. Position Powertrain Lift 014–00765 or equivalent, with wood blocks approximately 40 inches (1,016mm) in length secured to the lift under the sub-frame.

51. Remove the four sub-frame to body retaining bolts.

52. Allow the sub-frame to lower slightly and disconnect the power steering pressure and return hoses from the rack and pinion (steering gear).

53. Finish lowering the sub-frame and move aside.

54. Lower the vehicle.

55. Loosen the front mount retaining nuts five turns.

56. Place a floor jack and a block of wood under the transaxle and raise the transaxle enough to release the tension on the three bar engine support, or equivalent.

57. Back off on the three bar engine support adjustment to allow downward travel of the transaxle.

58. Slowly lower the transaxle until it reaches the limits of the front engine mount movement.

59. Adjust the three bar engine support or equivalent, to hold the transaxle in this position.

60. Remove the floor jack and wood block.

61. Raise the vehicle and safely support.

62. Position Transmission Jack 014–00210 or equivalent, to the transaxle and secure.

63. Remove the last starter motor retaining bolt and hang the starter off to the side using safety wire.

64. Remove the two bolts retaining the engine oil pan to the transaxle.

65. Remove the remaining bolts.

66. Separate the transaxle from the engine and carefully remove from the vehicle.

To install:

67. If removed, place the transaxle on the transmission jack and secure.

68. Apply a film of grease to the input shaft splines.

69. If removed, install the right engine support insulator bracket to the transaxle case. Tighten the bolts to 62 ft. lbs. (84 Nm).

70. If removed, install the shift stabilizer bar mounting bracket to the right transaxle support insulator bracket stud and transaxle case. Tighten the nut and bolt to 28–38 ft. lbs. (38–51 Nm).

71. Carefully raise the transaxle into position with the engine.

72. If required, use an 18mm socket to rotate the engine to line up the splines on the clutch disc with the input shaft.

73. Reinstall two lower side and two lower transaxle retaining bolts. Tighten the retaining bolts to 30 ft. lbs. (40 Nm).

74. Reinstall the starter motor and the lower bolt. Tighten the starter motor lower bolt to 35 ft. lbs. (48 Nm).

75. Lower the vehicle.

76. Using the floor jack and a block of wood positioned at the engine and transaxle mating area, raise the assembly into position.

77. Adjust the three bar engine support to maintain the engine in the correct position.

78. Remove the floor jack and the block of wood. Reinstall the upper transaxle retaining bolts.

79. Tighten the retaining bolts to 28–38 ft. lbs. (38–51 Nm).

80. Reinstall the nuts retaining the front engine support bracket. Tighten the retaining nuts to 61 ft. lbs. (83 Nm).

81. Reinstall the upper starter motor retaining bolts. Tighten the retaining bolts to 35 ft. lbs. (48 Nm).

82. Reinstall the ground strap to the transaxle retaining bolt.

83. Reconnect the back-up lamp switch.

84. Reinstall the engine and transaxle support insulator.

85. Raise and safely support the vehicle.

86. Reinstall the left and right halfshafts and the intermediate shaft.

87. If removed, place the sub-frame on the powertrain lift, or equivalent with wood blocks approximately 40 inches (1,016mm) in length secured to the lift under the sub-frame.

88. Raise the sub-frame and connect the power steering pressure and return hoses to the rack and pinion.

89. Align the sub-frame to the body. Route the power steering hoses into position at the rear of the engine.

90. Reinstall the four sub-frame retaining bolts loosely.

91. Reinstall Sub-Frame Alignment Pin Set T95P-2100-AH or equivalent into the sub-frame and body alignment holes. After aligning the holes, slightly tighten the four sub-frame retaining bolts.

92. After the sub-frame alignment is complete, tighten the four sub-frame retaining bolts to 81–110 ft. lbs. (110–150 Nm). Remove the alignment tools.

93. Reinstall the A/C accumulator bracket to the sub-frame and tighten the screws to 48–72 inch. lbs. (6–8 Nm).

94. Reconnect the power steering oil cooler hoses to the right front side of the sub-frame.

95. Install Powertrain Alignment Gauge T94P-6000-AH or equivalent to the left-front engine support bracket and the sub-frame. Tighten the two retaining bolts to 20 ft. lbs. (27 Nm) and snug the through-bolt.

96. Reinstall the right engine support insulator with retaining bolts to the sub-frame and through-bolt. Tighten the two retaining bolts to sub-frame to 30–41 ft. lbs. (41–55 Nm) and the through-bolt to 75–102 ft. lbs. (103–137 Nm).

97. Observe the position of the right engine support insulator. It must be centered in the bracket and in perfect alignment front to rear. Remove the powertrain alignment gauge.

98. Reinstall the left engine support insulator to the sub-frame with two retaining bolts. Tighten the retaining bolts to 84 inch. lbs. (10 Nm).

99. Observe the position of the left engine support insulator to ensure perfect alignment front to rear. Retighten the bolts to 30–41 ft. lbs. (41–55 Nm). Reinstall the left engine support insulator through-bolt and tighten to 75–102 ft. lbs. (103–137 Nm).

100. Reinstall the left and right lower control arms to the steering knuckles. Install new pinch bolts and nuts. Tighten the pinch bolts to 61 ft. lbs. (83 Nm).

101. Reconnect the steering yoke to the steering gear shaft. Tighten the steering yoke retaining bolt to 15–20 ft. lbs. (20–27 Nm).

102. Reinstall the sway bar (stabilizer bar) links. Reinstall the retaining nuts and tighten to 35–48 ft. lbs. (47–65 Nm).

103. Reinstall the left and right tie rod ends to the steering knuckles. Tighten the castellated nuts to 23–35 ft. lbs. (31–47 Nm). Install new cotter pins.

104. Reinstall the front bumper cover braces and tighten the bolts securely. Reinstall the splash shield to the sub-frame.

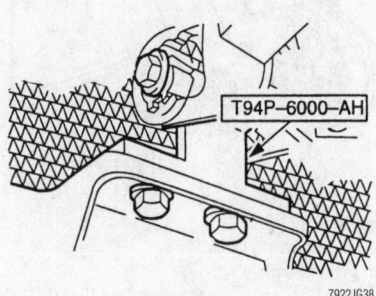

7922JG38

The Powertrain Allignment Gauge (T94P-6000-AH) tool must be installed in the correct position to ensure proper engine/transaxle orientation

105. Reinstall the two retaining screws and the A/C accumulator to the sub-frame.

106. Reinstall the radiator supports to the sub-frame. Tighten the bolts to 71–97 inch lbs. (8–11 Nm).

107. Reinstall the splash shield at the front of the sub-frame.

108. Reinstall the front wheel and tire assemblies. Tighten the lug nuts to 62 ft. lbs. (85 Nm).

109. Reinstall the shift rod to transaxle gearshift shaft. Reinstall the retaining bolt and tighten to 14–18 ft. lbs. (19–25 Nm).

110. Reinstall the shift control stabilizer bar to its mounting stud. Reinstall the mounting nut and tighten to 28–38 ft. lbs. (38–51 Nm).

111. Reinstall the underbody heat shield under the shift control.

112. Reconnect the VSS.

113. Reinstall the speedometer cable.

114. If equipped with 2.5L engine, install the Y-pipe to the exhaust manifolds and the catalytic converter using new gaskets. Reinstall the water pump pulley shield.

115. If equipped with 2.0L engine, install the catalytic converter between the exhaust manifold and the exhaust pipe using new gaskets. Reinstall the exhaust manifold heat shield and the oil level dipstick. Tighten the retaining bolts to 71–106 inch lbs. (8–11 Nm).

116. Reinstall the support bracket strap to the catalytic converter and engine.

117. Reinstall the bracket to the catalytic converter and the intermediate halfshaft bearing bracket.

118. Reinstall the lower radiator air deflector.

119. Reinstall the belt pulley cover by sliding it up and under the front fender splash shield.

120. Check the transmission fluid level and add fluid as required.

121. Lower the vehicle.

122. Remove the three bar engine support, or equivalent.

123. Reinstall the wire loom retainers removed during the transaxle removal.

124. Reinstall the hydraulic line to the clutch slave cylinder and install the clip.

125. Reinstall the rubber inspection cover to the clutch housing and the hydraulic line to the retaining grommet.

126. Remove the upper strut mounting nuts and replace with new locknuts. Tighten the strut mounting nuts to 34 ft. lbs. (46 Nm).

127. Reinstall the air cleaner lower bracket.

128. Reinstall the MAF sensor and the air cleaner assembly.

129. Remove the wire retaining the radiator and fan shroud to the radiator support.

130. Reinstall the battery and cables, negative cable last.

131. Adjust the shift linkage and bleed the hydraulic clutch system as required.

132. Road test the vehicle and check for proper operation.

Clutch

ADJUSTMENTS

Because the clutch system is hydraulic, the clutch pedal free-play is self-adjusting and requires no additional maintenance.

REMOVAL & INSTALLATION

1. Disconnect the negative battery cable.
2. Raise and safely support the vehicle.
3. Remove the starter motor.
4. Disconnect the hydraulic coupling for the slave cylinder at the transaxle by sliding the sleeve on the tube towards the slave cylinder and applying a slight pulling force to the tube.
5. Remove the transaxle.
6. Mark the assembled position of the clutch and pressure plate to the flywheel if it is to be reinstalled.

➡**The clutch pressure plate is only held in place by the retaining bolts. No dowel pins are used, therefore the pressure plate must be supported when removing the retaining bolts.**

7. Loosen the pressure plate bolts evenly until the pressure plate spring pressure is released, then finish removing the bolts while supporting the clutch and pressure plate assembly.

8. Remove the clutch and pressure plate from the vehicle.

9. Inspect the flywheel, slave cylinder and other components for wear or damage.

To install:

10. Clean the pressure plate and flywheel surfaces.

11. Install the clutch disc using Clutch Aligner T74P-7137-K or equivalent.

➡**If the clutch disc and pressure plate are being reused, align the marks made during disassembly.**

12. Install Flywheel Holding Tool T74P-6375-A or equivalent, to hold the flywheel.

13. Install the pressure plate and start the retaining bolts.

14. Tighten the retaining bolts evenly and in sequence to 13–18 ft. lbs. (18–26 Nm).

15. Remove the clutch aligner tool.

16. Reinstall the transaxle.

17. Reconnect the slave cylinder tube coupling by pushing the male coupling into the slave cylinder female coupling.

18. Lower the vehicle.

19. Reconnect the negative battery cable.

20. Bleed the hydraulic clutch system, if required.

21. Check the clutch system for proper operation.

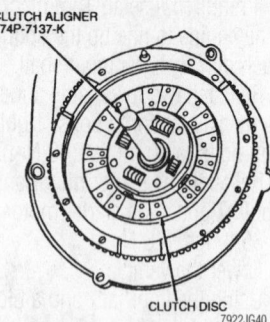

Insert an alignment tool through the clutch disc to ensure that it is centered after the pressure plate is installed

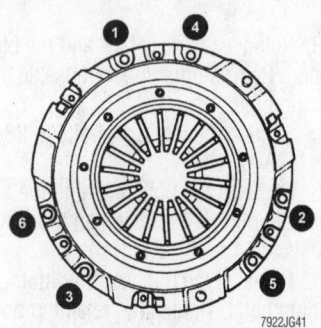

Tighten the pressure plate bolts gradually and in the sequence shown to ensure correct clutch operation

Hydraulic Clutch System

BLEEDING

1. Disconnect the negative battery cable.
2. Remove the air cleaner outlet tube and the Mass Air Flow (MAF) sensor.
3. Clean the top of the brake master cylinder fluid reservoir before opening it.

➡**The brake master cylinder fluid reservoir is also the reservoir for the hydraulic clutch master cylinder.**

4. Be sure that there is adequate fluid in the master cylinder fluid reservoir before attempting to bleed the system. Check the fluid level throughout the bleeding procedure.

5. Remove the rubber inspection cover from the bell housing.

6. Connect a hose to the bleeder valve fitting on the clutch slave cylinder. Submerge the other end of the hose into a container of clean brake fluid.

7. Push the clutch pedal down while opening the bleeder on the clutch slave cylinder. Watch for air bubbles escaping from the hydraulic system.

8. Close the bleeder before releasing the clutch pedal.

Clutch System

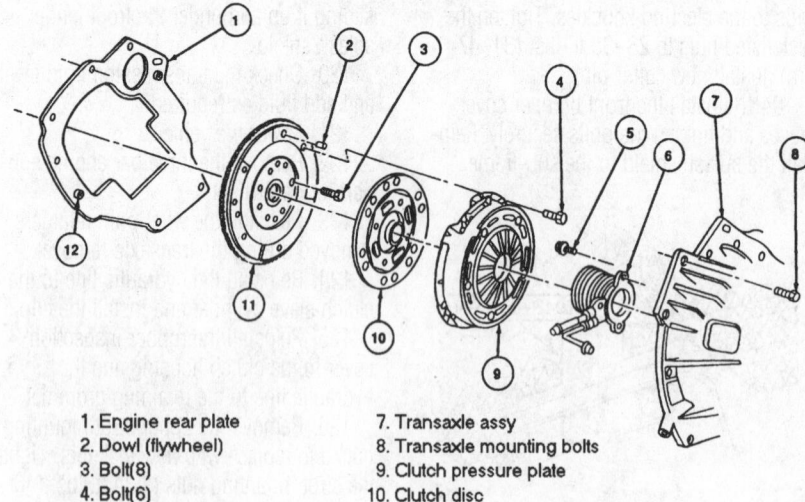

1. Engine rear plate
2. Dowl (flywheel)
3. Bolt(8)
4. Bolt(6)
5. Bolt(3)
6. Clutch slave cylinder
7. Transaxle assy
8. Transaxle mounting bolts
9. Clutch pressure plate
10. Clutch disc
11. Flywheel
12. Dowl bushing (engine plate)

Exploded view of the clutch disc, pressure plate and related component mounting

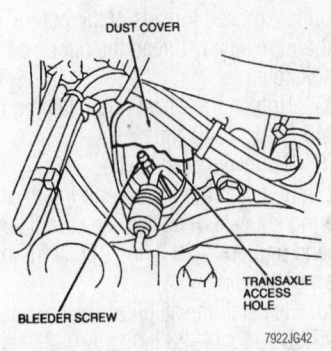

7922JG42

Remove the dust cover to gain access to the clutch slave cylinder bleeder valve

9. Repeat the procedure until no more air bubbles are seen.

10. Reinstall the rubber inspection cover to the bell housing.

11. Top off the brake master cylinder fluid reservoir and install the diaphragm and cap securely.

12. Reinstall the MAF sensor and air cleaner outlet tube.

13. Reconnect the negative battery cable.

14. Check the clutch for proper operation.

Halfshaft

REMOVAL & INSTALLATION

➡Do not begin this procedure without a new wheel hub retaining nut(s), a new lower control arm to steering knuckle pinch bolt and new retainer circlips for the CV-joints. Once removed, these parts lose their torque holding or retention capabilities and must not be reused.

Left Side

1. Raise and safely support the vehicle.

2. Remove the left front wheel and tire assembly.

3. Snug two of the lug nuts back onto the rotor.

4. Insert the tapered end of a prybar or steel rod into one of the cooling slots of the disc brake rotor and place the bar against the disc brake anchor plate, to keep the rotor from turning.

5. Loosen and remove the wheel hub retaining nut. Discard the nut.

6. Remove the nut of the stabilizer bar link and separate the stabilizer bar from the strut using a tie rod end removal tool.

7. Remove the cotter pin and castellated nut that secures the tie rod end to the steering knuckle. Discard the cotter pin.

8. Using a tie rod end removal tool,

separate the tie rod end from the steering knuckle.

9. Remove the lower control arm to steering knuckle pinch bolt and nut.

10. Using a prybar or similar tool, separate the lower control arm ball joint from the steering knuckle.

❉❉ WARNING

Never use a hammer to separate the halfshaft from the front wheel hub as damage to the threads or internal components may result.

11. Separate the outer CV-joint and halfshaft from the wheel hub using Front Hub Remover/Replacer T81P-1104-C and its associated components, or equivalent.

12. Install CV-joint puller T86P-3514-A1 or equivalent, between the inner CV-joint and the transaxle case.

➡If the right halfshaft has already been removed, install Differential Rotator T81P-4026-A or equivalent into the right side of the differential before removing the left shaft to maintain alignment within the differential.

13. Attach the corresponding extension and slide hammer to the CV-joint puller and remove the halfshaft with both CV-joints as an assembly from the transaxle case.

14. Remove the assembly from the vehicle.

To install:

15. Replace the driveshaft bearing retainer circlip. Start one end of the circlip into the groove and work the circlip over the housing end and into the groove. This will avoid over expanding the circlip.

16. Carefully align the splines of the inner CV-joint (install the halfshaft and both CV-joints as an assembly) with the splines in the transaxle case and push it into the differential side gear until the circlip is felt to seat.

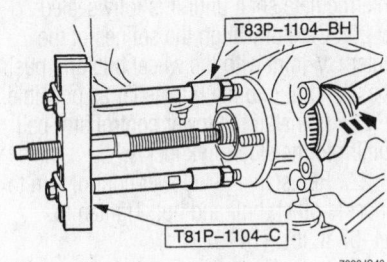

7922JG43

Use a puller such as Front Hub Remover/Replacer T81P-1104-C to press the halfshaft out of the hub assembly

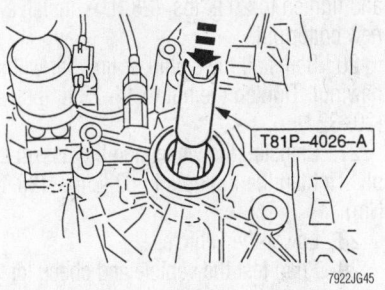

7922JG45

If the right halfshaft has been removed, install Differential Rotator T81P-4026-A or equivalent into the differential before removing the left halfshaft

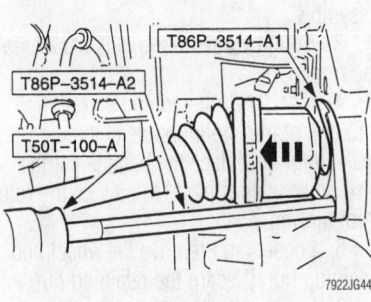

7922JG44

Use a slide hammer with the special adapter to pull the inner CV-joint from the transaxle

➡A non-metallic mallet may be used to aid in seating the inner CV-joint into the differential side gear of the transaxle case. Only tap on the outboard CV-joint stub shaft.

17. Position the outer CV-joint with halfshaft and carefully align the splines of the outer CV-joint with the splines of the wheel hub.

18. Push the CV-joint shaft into the wheel hub as far as possible.

19. Reinstall the front suspension lower control arm ball joint into the steering knuckle.

20. Reinstall a new lower control arm to knuckle pinch bolt and nut. Tighten the nut to 54–67 ft. lbs. (74–92 Nm).

21. Insert the tapered end of a prybar into one of the cooling slots in the disc brake rotor and jamb to prevent the rotor from turning.

22. Install a new wheel hub retaining nut onto the exposed threads of the outer CV-joint and manually thread the nut on as far as possible.

23. Finish running the nut up and tighten to 246 ft. lbs. (340 Nm).

24. Remove the steel rod or prybar.

25. Reinstall the tie rod end on the steering knuckle. Reinstall the castellated nut

and tighten to 20 ft. lbs. (28 Nm). Install a new cotter pin.

26. Reinstall the stabilizer link bar with a new nut. Tighten the nut to 14–23 ft. lbs. (20–32 Nm).

27. Reinstall the wheel and tire assembly. Tighten the lug nuts to 62 ft. lbs. (85 Nm).

28. Lower the vehicle.

29. Road test the vehicle and check for proper operation.

Right Side and Intermediate

1. Raise and safely support the vehicle.

2. Remove the right front wheel and tire assembly.

3. Snug two of the lug nuts back onto the rotor.

4. Insert the tapered end of a prybar into one of the cooling slots of the disc brake rotor and place the bar against the disc brake anchor plate, to prevent the rotor from turning.

5. Loosen and remove the wheel hub retaining nut. Discard the retaining nut.

6. Remove the nut of the stabilizer bar link and separate the stabilizer bar from the strut using a tie rod end removal tool.

7. Remove the cotter pin and castellated nut that secures the tie rod end to the steering knuckle. Discard the cotter pin.

8. Using a tie rod end removal tool, separate the tie rod end from the steering knuckle.

9. Remove the lower control arm to wheel spindle pinch bolt and nut.

10. Using a prybar or similar tool, separate the lower control arm ball joint from the steering knuckle.

✳✳ WARNING

Never use a hammer to separate the halfshaft from the front wheel hub as damage to the threads or internal components may result.

11. Separate the outer CV-joint and halfshaft from the wheel hub using Front Hub Remover/Replacer T81P-1104-C and its associated components, or an equivalent tool.

12. Install CV-joint puller T86P-3514-A1 or equivalent, between the inner CV-joint and the intermediate halfshaft.

13. Using the extension and slide hammer on the CV-joint puller, separate the right halfshaft with CV-joints from the intermediate halfshaft.

14. Remove the right side halfshaft from the vehicle.

15. If intermediate shaft removal is required, proceed as follows:

  a. Remove the two nuts securing the support bracket.

  b. Remove the intermediate shaft and bearing shield.

  c. On the 2.0L engine, the exhaust clamp and two bolts will need to be removed to allow removal of the intermediate shaft.

  d. Remove the intermediate shaft from the vehicle.

  e. If the intermediate shaft support bracket needs to be removed, locate the three bolts securing the support bracket and remove the bolts.

  f. Remove the support bracket.

To install:

16. If the intermediate shaft was removed, reinstall as follows:

  a. If the support bracket was removed, reinstall with the three bolts and tighten to 15–23 ft. lbs. (21–32 Nm).

  b. Carefully align the splines of the intermediate shaft with the splines of the differential side gears in the transaxle case.

  c. Push the shaft into the case until it is fully seated. The inner CV-joint is to be installed with the halfshaft and outer CV-joint attached as an assembly.

  d. Reinstall the two nuts onto the intermediate shaft support bracket and bearing shield. Tighten the nuts to 17–22 ft. lbs. (24–30 Nm).

  e. On the 2.0L engine, install the two bolts and the exhaust clamp.

17. Replace the bearing retainer circlip on the intermediate shaft. Start one end of the circlip into the groove and work the circlip over the housing end into the groove. This will prevent over-expanding the circlip.

18. Carefully align the splines of the inner CV-joint with the splines of the intermediate shaft. The inner CV-joint is to be installed as an assembly with the right halfshaft and outer CV-joint.

19. Push the inner CV-joint onto the intermediate shaft until it is fully seated.

20. Carefully align the splines of the outer CV-joint with the wheel hub and push the CV-joint into the hub as far as possible.

21. Reinstall the lower control arm ball joint into the steering knuckle.

22. Reinstall a new lower control arm to knuckle pinch bolt and nut. Tighten to 54–67 ft. lbs. (74–92 Nm).

23. Insert the tapered end of a prybar into one of the cooling slots in the front disc brake rotor and jamb the prybar to prevent the rotor from turning.

24. Install a new wheel hub retaining nut

onto the exposed threads of the outer CV-joint and manually thread the nut on as far as possible.

25. Tighten the wheel hub retaining nut to 246 ft. lbs. (340 Nm).

26. Remove the prybar.

27. Reinstall the tie rod end into the steering knuckle. Reinstall the castellated nut and tighten to 20 ft. lbs. (28 Nm). Install a new cotter pin.

28. Reinstall the stabilizer link bar and nut. Tighten to 14–23 ft. lbs. (20–32 Nm).

29. Reinstall the right front wheel and tire assembly. Tighten the lug nuts to 62 ft. lbs. (85 Nm).

30. Lower the vehicle.

31. Road test the vehicle and check for proper operation.

STEERING AND SUSPENSION

Air Bag

✳✳ CAUTION

Some vehicles are equipped with an air bag system. The system must be disabled before performing service on or around air bag system components, steering column, instrument panel components, wiring and sensors. Failure to follow safety and disabling procedures could result in accidental air bag deployment, possible personal injury and unnecessary SIR system repairs.

PRECAUTIONS

Several precautions must be observed when handling the inflator module to avoid accidental deployment and possible personal injury.

• Never carry the inflator module by the wires or connector on the underside of the module.

• When carrying a live inflator module, hold securely with both hands, and ensure that the bag and trim cover are pointed away.

• Place the inflator module on a bench or other surface with the bag and trim cover facing up.

• With the inflator module on the bench, never place anything on or close to the module which may be thrown in the event of an accidental deployment.

DISARMING

1. Position the vehicle with the front wheels in a straight ahead position.
2. Disconnect the negative battery cable.
3. Disconnect the positive battery cable.
4. Wait at least one minute for the air bag back-up power supply to drain before continuing.
5. Proceed with the repair.
6. Once complete, connect the battery cables, negative cable last.
7. Check the functioning of the air bag system by turning the ignition key to the **RUN** position and visually monitoring the air bag indicator lamp in the instrument cluster. The indicator lamp should illuminate for approximately six seconds, then turn **OFF**. If the indicator lamp does not illuminate, stays on, or flashes at any time, a fault has been detected by the air bag diagnostic monitor.

Power Rack and Pinion Steering Gear

REMOVAL & INSTALLATION

1. Disconnect the negative battery cable.
2. Working inside the vehicle, remove the clamp plate bolt retaining the steering column shaft to the flexible coupling.
3. Rotate the clamp plate to separate it from the shaft of the flexible coupling.

4. Remove the floor seal being careful not to damage the sealing lip.
5. Remove the pinch bolt securing the flexible coupling to the rack and pinion (steering gear) pinion shaft and remove the flexible coupling.
6. Remove as much of the power steering fluid as possible from the power steering auxiliary reservoir using a suction gun or similar method.
7. Disconnect the power steering return hose from the power steering pump auxiliary reservoir.

➡The front sub-frame must be removed in order to allow removal of the rack and pinion (steering gear).

8. If equipped with the 2.0L engine, remove the oil level dipstick and the exhaust manifold shield.
9. If equipped with the 2.5L engine, remove the water pump pulley shield.
10. Secure the radiator and fan shroud assembly to the radiator support using safety wire.
11. Install Three Bar Engine Support D88L-6000-a or equivalent, to the engine lifting eyes and support the engine/transaxle assembly.
12. Raise and safely support the vehicle.
13. Remove the catalytic converter.
14. Remove the front wheel and tire assemblies.
15. Separate the left and right stabilizer bar links from the front stabilizer bar.
16. Separate the left and right outer tie rod ends from the steering knuckles. Discard the cotter pins.
17. Remove the pinch bolts and separate the front suspension lower control arms from the steering knuckles at the ball joints.
18. Remove the splash shield at the front of the sub-frame.
19. If equipped with an automatic transaxle, remove the retaining through-bolts from the left and right front engine

support insulators (engine mounts) to the sub-frame.
20. If equipped with a manual transaxle, remove the through-bolt from the left front engine support insulator and remove the right front engine support insulator and mounting bracket.
21. Disconnect the power steering oil cooler hoses at the right front of the sub-frame and drain the power steering system.
22. Remove the A/C accumulator retaining screws from the front sub-frame.
23. Remove the four bolts retaining the lower radiator supports to the front sub-frame. Rotate the radiator supports forward.
24. Remove the two screws retaining the bumper cover braces to the left and right sides of the front sub-frame and rotate the cover braces forward.
25. Position Powertrain Lift (hydraulic lift) 014–00765 or equivalent, and two wood blocks approximately 40 inches in length attached to the sub-frame to support the sub-frame for removal from the vehicle.

➡Be sure that the powertrain lift and wood blocks are correctly positioned for safe removal of the sub-frame.

26. Remove the four sub-frame to body retaining bolts.
27. Lower the sub-frame slightly and disconnect the power steering pressure and return hoses from the rack and pinion.
28. Finish lowering the sub-frame.
29. Remove the six bolts and the steering gear cover plate from the sub-frame.
30. Disconnect the power steering pressure and return hose unions from the steering gear.
31. Remove the two bolts retaining the steering gear to the sub-frame and remove the rack and pinion.

To install:
32. If the rack and pinion (steering gear) is being replaced, remove the inner tie rods and boots from the old unit and install on the new one, if they are in good condition.
33. Install new plastic seals on the power steering pressure and return line fittings as required.
34. Reinstall the rack and pinion to the sub-frame and install the retaining bolts.
35. Tighten the two rack and pinion retaining bolts to 101 ft. lbs. (137 Nm).
36. Reconnect the power steering and return hose unions to the rack and pinion.
37. Tighten the unions to 23 ft. lbs. (31 Nm).
38. Reinstall the rack and pinion cover plate and install the six retaining bolts.
39. Tighten the retaining bolts to 37 ft. lbs. (50 Nm).

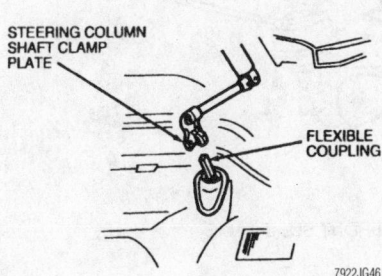

STEERING COLUMN SHAFT CLAMP PLATE

FLEXIBLE COUPLING

7922JG46

Disconnect the steering column shaft from the flexible coupling

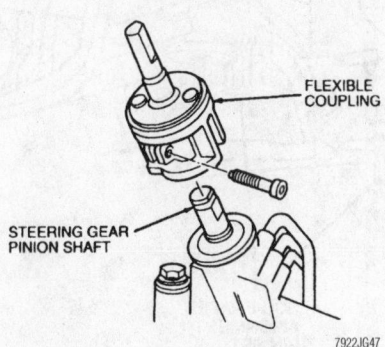

FLEXIBLE COUPLING

STEERING GEAR PINION SHAFT

7922JG47

Disengage the flexible coupling from the steering gear pinion shaft

40. If lowered, raise and safely support the vehicle.

41. If removed, position the front sub-frame onto the powertrain lift and raise.

42. Reinstall the power steering pressure and return hoses to the rack and pinion.

43. Position the front sub-frame to the body.

44. Route the power steering hoses to their correct positions.

45. Loosely install the four sub-frame to body bolts.

46. Install Sub-Frame Alignment Pin Set T94P-2100-AH or equivalent, into the front sub-frame to body alignment holes.

47. Slightly tighten the four sub-frame to body retaining bolts.

48. Move the sub-frame to complete the alignment.

49. Tighten the four sub-frame to body retaining bolts to 81–110 ft. lbs. (110–150 Nm).

50. Remove the alignment tools.

51. Reinstall the A/C bracket retaining screw to the sub-frame and secure.

52. Reconnect the power steering oil cooler hoses to the front of the sub-frame.

53. Install the Powertrain Alignment Gauge T94P-6000-AH or equivalent, to the left front engine support bracket and the sub-frame.

54. Tighten the two retaining bolts to 20 ft. lbs. (27 Nm) and snug the through-bolt.

55. Lower the vehicle.

56. Remove the engine support from the top of the engine compartment.

57. Reconnect the power steering return hose to the power steering pump auxiliary reservoir.

58. Working inside the vehicle, install the flexible coupling to the steering gear pinion shaft and install the pinch bolt.

59. Tighten the pinch bolt to 21 ft. lbs. (28 Nm).

60. Reinstall the floor seal.

61. Align the steering column shaft clamp plate with the flexible coupling and install the clamp plate bolt. Tighten the clamp plate bolt to 18 ft. lbs. (24 Nm).

62. Partially raise and safely support the vehicle.

63. Install the right front engine support insulator (engine mount). Tighten the two retaining bolts to 30–41 ft. lbs. (41–55 Nm) and the through-bolt to 75–102 ft. lbs. (103–137 Nm).

64. Check the position of the right front engine mount (support insulator). It must be centered in the transaxle bracket and in perfect front to rear alignment.

65. Remove the two retaining bolts and the through-bolt securing the powertrain

alignment gauge and remove the powertrain alignment gauge.

66. Reinstall the left front engine mount to the front sub-frame using the two retaining bolts. Tighten the retaining bolts to 84 inch lbs. (10 Nm).

67. Check the position of the left front engine mount to ensure perfect front to rear alignment.

68. Retighten the two retaining bolts to 30–40 ft. lbs. (41–55 Nm).

69. Reinstall the left front engine mount through-bolt. Tighten the through-bolt to 75–102 ft. lbs. (103–137 Nm).

70. Reinstall the stabilizer bar link to the front stabilizer bar. Tighten the retaining nuts to 35–48 ft. lbs. (47–65 Nm).

71. Reconnect the left and right lower control arms to the ball joints and install the pinch bolts. Tighten the pinch bolts to 37–43 ft. lbs. (50–58 Nm).

72. Reinstall the catalytic converter.

73. Reinstall both tie rod ends to the steering knuckles. Install new cotter pins.

74. Reinstall the wheel and tire assemblies and tighten the lug nuts to 63 ft. lbs. (85 Nm).

75. Reinstall the front bumper cover braces to both sides of the sub-frame.

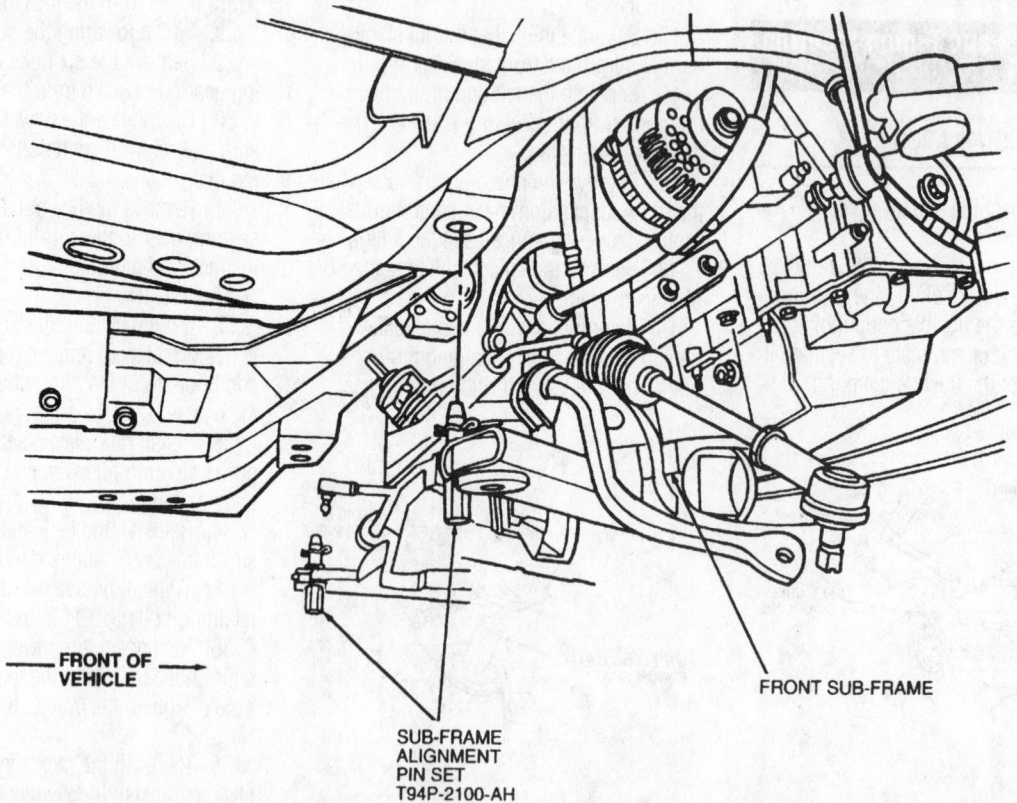

— **FRONT OF → VEHICLE**

SUB-FRAME ALIGNMENT PIN SET T94P-2100-AH

FRONT SUB-FRAME

7922JG48

Install alignment pins such as Sub-Frame Alignment Pin Set T94P-2100-AH into the sub-frame to ensure correct positioning

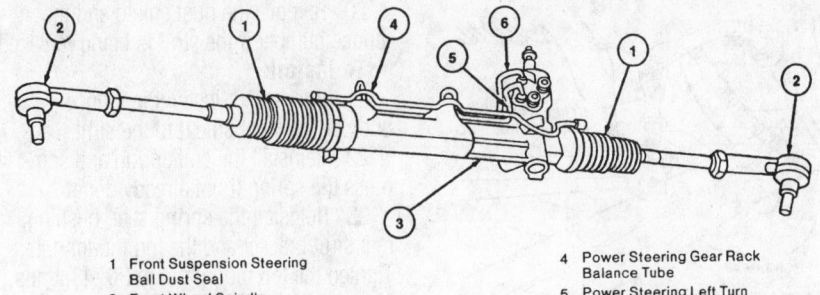

1 Front Suspension Steering Ball Dust Seal	4 Power Steering Gear Rack Balance Tube
2 Front Wheel Spindle Connecting End	5 Power Steering Left Turn Pressure Tube
3 Steering Gear	6 Power Steering Right Turn Pressure Tube

7922JG49

Power rack and pinion steering gear assembly component identification

76. Reinstall the radiator supports and the splash shield to the sub-frame.

77. Lower the vehicle.

78. Remove the safety wire supporting the radiator and fan shroud assembly.

79. If equipped with the 2.0L engine, install the exhaust manifold shield and the oil level dipstick.

80. If equipped with the 2.5L engine, install the water pump pulley shield and secure.

81. Fill the power steering system with the proper fluid.

82. Reconnect the negative battery cable.

83. Run the engine and check for leaks and proper operation.

➡Whenever the vehicle sub-frame is removed or lowered, the wheel alignment should be checked.

84. Bleed the power steering system, if needed.

Strut and Spring

REMOVAL & INSTALLATION

Front

1. Disconnect the negative battery cable.

2. Raise and safely support the vehicle.

3. Remove the wheel and tire assembly.

4. Lower the vehicle enough to gain access to the strut retaining nut.

5. From inside the engine compartment, hold the strut piston with an 8mm Allen head wrench while removing the top retaining nut.

6. Raise and safely support the vehicle.

7. Disconnect the stabilizer bar link from the strut.

8. Remove the brake hose and anti-lock wiring from the strut bracket.

9. Remove the steering knuckle to strut pinch bolt.

10. Work the strut out of the steering knuckle and lower the strut out of the strut tower.

11. Remove the strut/coil spring assembly from the vehicle.

❊❊ CAUTION

Do not attempt to remove the coil spring from the strut without first compressing the coil spring with the appropriate tool.

12. Install Spring Compressor 086–00029 or equivalent, to the coil spring and compress the spring until the spring tension is relieved from the spring seat.

13. Remove the thrust bearing retainer nut.

14. Remove the thrust bearing, spring seat and dust shield.

15. Remove the coil spring from the strut.

16. Remove the jounce bumper from the strut.

17. Replace the coil spring or the strut as needed.

To install:

18. Reinstall the jounce bumper to the strut.

19. Position the coil spring to the strut.

20. Compress the coil spring if removed from the spring compressor.

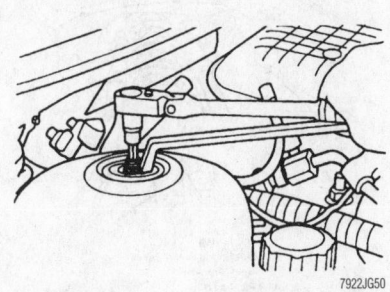

7922JG50

Hold the strut piston with an Allen wrench while removing the retaining nut

21. Reinstall the dust shield, spring seat and thrust bearing.

22. The coil spring must seat in the notch of the spring seat.

23. Reinstall the thrust bearing retainer nut. Tighten the nut to 44 ft. lbs. (59 Nm).

24. Position the strut/coil spring assembly into the strut tower and fit the lower portion of the strut into the steering knuckle.

25. Reinstall the knuckle to strut pinch bolt. Do not tighten the bolt at this time.

26. Partially lower and safely support the vehicle.

27. Reinstall the top strut mounting nut.

28. Use an 8mm Allen head wrench to prevent the strut piston rod from turning while tightening the mounting nut to 34 ft. lbs. (46 Nm).

29. Tighten the knuckle to strut pinch bolt to 40 ft. lbs. (54 Nm).

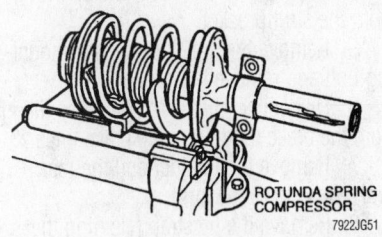

ROTUNDA SPRING COMPRESSOR

7922JG51

Properly compress the coil spring assembly before removing the retaining nut

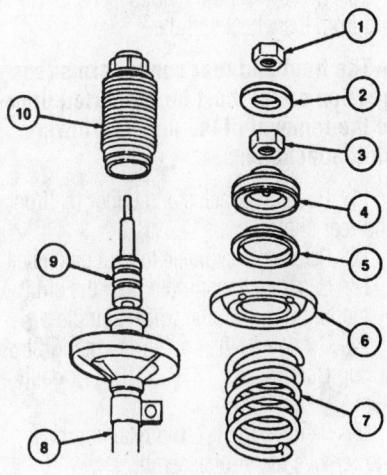

1 Nut	
2 Retainer	
3 Upper Mount Retainer Nut	
4 Upper Mount	
5 Bearing	
6 Spring Seat	
7 Front Coil Spring	
8 Front Shock Absorber	
9 Jounce Bumper	
10 Dust Shield	

7922JG52

Exploded view of the strut assembly

30. Raise and safely support the vehicle.

31. Reinstall the stabilizer bar link. Be careful not to damage the ball joint seal. Replace the stabilizer bar link if the seal is damaged.

32. Tighten the stabilizer bar link retaining nut to 37 ft. lbs. (50 Nm).

33. Position the brake hose and the anti-lock wiring to the strut bracket.

34. Reinstall the wheel and tire assembly. Tighten the lug nuts to 63 ft. lbs. (85 Nm).

35. Lower the vehicle.

36. Connect the negative battery cable.

37. Check the front wheel alignment.

38. Road test the vehicle and check for proper operation.

Rear

1. Disconnect the negative battery cable.

2. Raise and safely support the vehicle.

3. Remove the wheel and tire assembly.

4. Remove the anti-lock sensor wiring from the strut bracket.

5. Remove the anti-lock sensor mounting bolt and remove the sensor.

6. Disconnect the rear brake hose fitting from the brake tube. Plug the brake lines.

7. Remove the retainer and the rear brake hose from the strut.

8. Remove the tie strap retaining the parking brake rear cable and conduit to the rear suspension tie rod.

9. Disconnect the rear sway bar link and bushings from the rear control arm (suspension arm).

10. Disconnect the rear suspension tie rod from the wheel spindle.

➡ **The front and rear control arms (suspension arms) must be supported prior to the removal of the upper or lower strut attachments.**

11. Position a jack stand under the front and rear control arms.

12. Remove the spindle to strut pinch bolt.

13. Separate the spindle from the strut by tapping down on the wheel spindle.

14. Compress the coil spring using Strut Spring Compressor T81P-5310-A or equivalent.

15. Remove the two top retaining bolts and remove the strut assembly.

16. Place the strut assembly on a suitable workbench.

17. Compress the coil spring enough to relieve the tension on the spring seat.

18. Remove the top mount nut, rear strut bracket, bushing and spring seat.

19. Remove the coil spring and slowly relieve the tension on the spring if it is not to be reinstalled.

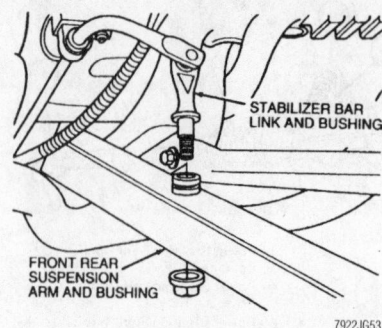

For rear strut removal, disconnect the sway bar link from the control arm . . .

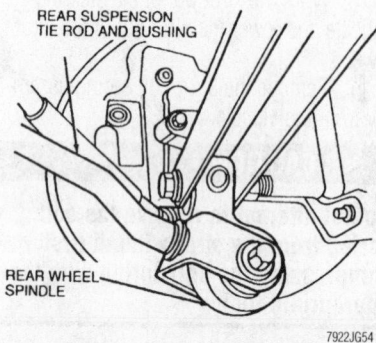

. . . and separate the tie rod from the rear wheel spindle

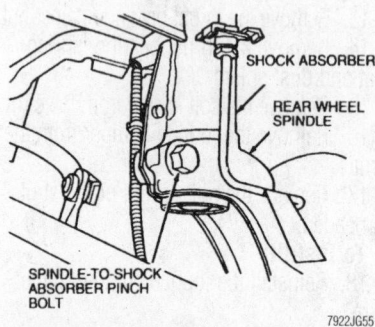

Remove the rear strut pinch bolt to release the strut assembly from the spindle

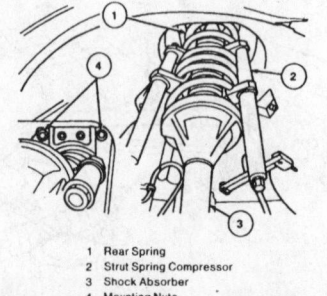

1 Rear Spring
2 Strut Spring Compressor
3 Shock Absorber
4 Mounting Nuts

Compress the coil spring, then unthread the two top retaining bolts and remove the strut from the vehicle

20. Remove the dust shield and the jounce bumper if the strut is being replaced.

To install:

21. If removed, install the jounce bumper and dust shield to the strut.

22. Reinstall the coil spring and compress the spring if not already done.

23. Reinstall the spring seat, bushing, rear strut bracket and the top mount nut. Tighten the top mount nut to 30–43 ft. lbs. (41–58 Nm).

24. With the coil spring compressed, install the strut assembly into position.

25. Reinstall the two rear strut bracket mounting bolts. Tighten the rear strut bracket mounting bolts to 17–22 ft. lbs. (23–30 Nm).

26. Position the strut to the wheel spindle.

27. Reinstall the spindle-to-strut pinch bolt. Tighten the pinch bolt to 52–72 ft. lbs. (70–98 Nm).

28. Remove the strut spring compressor.

29. Reinstall the rear suspension tie rod and bushing. Tighten the bolt to 75–102 ft. lbs. (102–138 Nm).

30. Remove the jackstand.

31. Attach the rear brake hose to the strut.

32. Reinstall the rear brake anti-lock sensor and retaining bolt. Tighten the retaining bolt to 84 inch. lbs. (9 Nm).

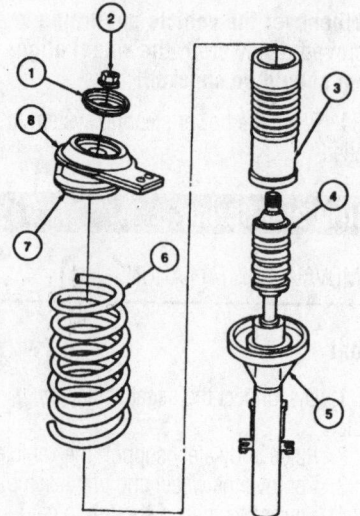

1 Rear Shock Absorber Bracket
2 Shock Absorber Mounting Nut
3 Rear Shock Absorber Dust Boot
4 Rear Suspension Jounce Bumper
5 Shock Absorber
6 Rear Spring
7 Spring Seat
8 Shock Absorber Bushing

Exploded view of the rear strut assembly

33. Reinstall the anti-lock wiring to the strut.

34. Secure the parking brake cable and conduit to the rear suspension tie rod with a tie strap.

35. Unplug and connect the rear brake hose fitting to the brake tube. Tighten the fitting securely.

36. Bleed the brake system.

37. Reinstall the wheel and tire assembly. Tighten the lug nuts to 62 ft. lbs. (85 Nm).

38. Lower the vehicle.

39. Connect the negative battery cable.

40. Check the rear wheel alignment.

41. Road test the vehicle and check for proper operation.

Lower Ball Joint

REMOVAL & INSTALLATION

If the lower ball joint requires replacement, the lower control arm and ball joint assembly must be replaced together, as the lower ball joint is not separately serviceable.

Wheel Bearings

ADJUSTMENT

Front

The front wheel bearings consist of a cartridge design and are permanently lubricated and sealed requiring no further maintenance. The bearings are preset and cannot be adjusted. If any part of a wheel bearing assembly is defective, the unit must be replaced. It is critical that the wheel hub retainer is properly tightened to 210 ft. lbs. (290 Nm) and that a new wheel hub retainer is always used.

Rear

The rear wheel bearings are not adjustable. If the bearings make noise or become loose, they must be replaced.

REMOVAL & INSTALLATION

Front

➡Before proceeding, be sure to have available new pinch bolts for the steering knuckle to lower ball joint and steering knuckle to strut as well as a new wheel hub retaining nut.

1. Disconnect the negative battery cable.
2. Raise and safely support the vehicle.
3. Remove the wheel and tire assembly.
4. Remove the disc brake caliper and rotor.

5. Support the disc brake caliper with safety wire. Do not let the caliper hang by the brake hose.

6. Remove the anti-lock brake sensor retaining bolt and remove the sensor from the steering knuckle.

7. Remove the outer tie rod end cotter pin and remove the castellated nut. Discard the cotter pin.

8. Separate the outer tie rod end from the steering knuckle using Tie Rod End Remover 3290-D or equivalent.

9. Remove the wheel hub retaining nut. Discard the nut.

10. Separate the halfshaft from the wheel hub using Front Hub Remover/Replacer T81P-1104-C or equivalent, and the related adapters.

11. Once removed, support the end of the halfshaft.

12. Remove the steering knuckle to lower ball joint pinch bolt.

13. Remove the steering knuckle to strut pinch bolt.

14. Work the steering knuckle off of the lower ball joint and out of the lower strut tube.

15. Remove the steering knuckle from the vehicle.

16. Place the steering knuckle assembly onto a suitable workbench.

17. Install Front Hub Remover/Replacer T81P-1104-C or equivalent with the appropriate adapters and separate the hub from the steering knuckle.

18. Remove the inner and outer snap rings securing the wheel bearing.

19. Remove the wheel bearing from the steering knuckle. Drive or press the old wheel bearing out as required.

To install:

20. Install the outer snap ring into the steering knuckle.

21. Install the wheel bearing using a hydraulic press with Pinion Bearing Cup Replacer T80T-4000-E or equivalent.

22. Reinstall the inner snap ring in the steering knuckle.

23. Reinstall the hub to the steering knuckle using Threaded Drawbar T75T-1176-A or equivalent.

24. Carefully align the splines of the outer CV-joint with the splines in the hub.

25. Position the steering knuckle to the lower ball joint stud.

26. Position the steering knuckle to the lower strut tube.

27. Install a new steering knuckle to lower ball joint pinch bolt. Tighten the bolt to 55–58 ft. lbs. (75–79 Nm).

28. Install a new steering knuckle to strut pinch bolt. Tighten the bolt to 40 ft. lbs. (54 Nm).

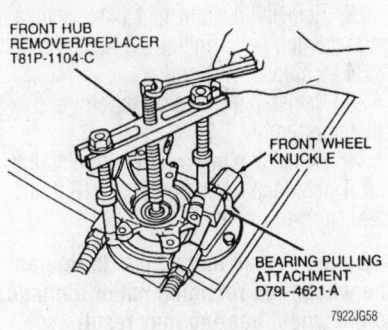

Use gear or bearing pulling tools to remove the hub from the steering knuckle

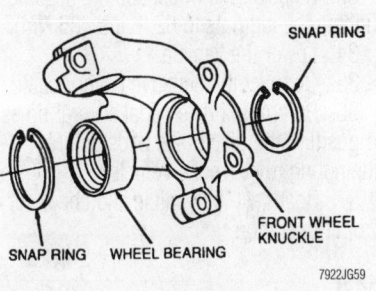

The wheel bearing is retained in the knuckle by two snaprings, as shown

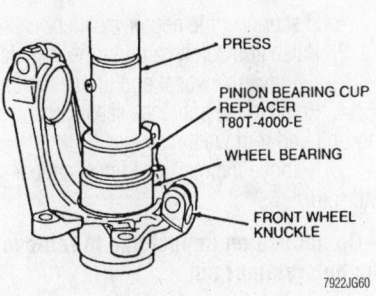

Using a press, install the new wheel bearing in the knuckle

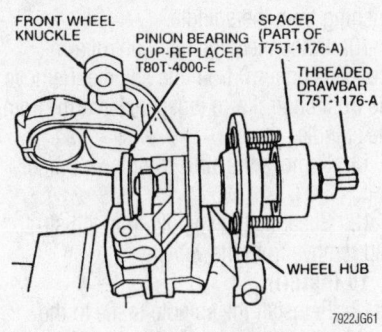

Use the special tools shown or a press to install the hub in the knuckle assembly

29. Reinstall the anti-lock brake sensor and retaining bolt. Tighten the retaining bolt to 84 inch. lbs. (10 Nm).

30. Reinstall the disc brake rotor and caliper assembly.

31. Reinstall a new wheel hub retaining nut. Tighten the retaining nut to 210 ft. lbs. (290 Nm).

➡ **Do not use an impact gun to tighten the wheel hub retaining nut or damage to the wheel bearing may result.**

32. Attach the tie rod end to the steering knuckle. Install the castellated nut and tighten the nut to 18–22 ft. lbs. (24–30 Nm). Install a new cotter pin.

33. Reinstall the wheel and tire assembly. Tighten the lug nuts to 62 ft. lbs. (85 Nm).

34. Lower the vehicle.

35. Connect the negative battery cable.

36. Pump the brake pedal several times to position the disc brake pads before attempting to move the vehicle.

37. Road test the vehicle and check for proper operation.

Rear

➡ **The wheel bearings are contained within the wheel hub and must be replaced as an assembly.**

WITH REAR DISC BRAKES

1. Disconnect the negative battery cable.
2. Raise and safely support the vehicle.
3. Remove the wheel and tire assembly.
4. Remove the anti-lock sensor retaining bolt and remove the sensor.
5. Remove the rear disc brake caliper and rotor.

➡ **Do not use an impact gun to remove the hub retainer nut.**

6. Remove the hub retainer nut.
7. Slide the hub and wheel bearing assembly off of the spindle and remove.
8. Remove the disc brake dust shield.
9. Disconnect the rear tie rod and bushing from the spindle.
10. Disconnect the rear control arm (suspension arm) from the spindle. Remove the front control arm (suspension arm) from the spindle.
11. Remove the strut-to-spindle pinch bolt.
12. Separate the spindle from the strut and remove from the vehicle.

To install:

13. Reinstall the spindle to the to the strut. Tighten the pinch nut to 52–72 ft. lbs. (70–98 Nm).
14. Connect the rear control arm (sus-

pension arm) to the spindle. Connect the front control arm (suspension arm) to the spindle. Tighten the bolt, but do not tighten at this time.

15. Connect the rear tie rod and bushing to the spindle. Tighten the bolt, but do not tighten at this time.

16. Reinstall the disc brake dust shield.

17. Reinstall the hub and bearing assembly to the spindle.

18. Reinstall the hub retainer nut. Tighten the hub retainer nut to 170–192 ft. lbs. (230–260 Nm).

➡ **Do not use an impact gun to tighten the hub retainer nut.**

19. Reinstall the rear disc brake rotor and caliper.

20. Reinstall the anti-lock sensor. Tighten the retaining bolt for the sensor to 84–96 inch. lbs. (9–11 Nm).

21. Reinstall the wheel and tire assembly. Tighten the wheel nuts to 62 ft. lbs. (85 Nm).

22. Lower the vehicle until the wheels are supporting the vehicles weight. This will properly load the suspension.

23. Tighten the spindle-to-control arm bolt to 52–79 ft. lbs. (70–98 Nm).

24. Tighten the tie rod-to-spindle bolt to 75–102 ft. lbs. (98 Nm).

25. Finish lowering the vehicle.

26. Road test the vehicle and check for proper operation.

WITH REAR DRUM BRAKES

➡ **The wheel bearings are contained within the wheel hub and must be replaced as an assembly.**

1. Disconnect the negative battery cable.
2. Raise and safely support the vehicle.
3. Remove the wheel and tire assembly.
4. Remove the anti-lock sensor retaining bolt and remove the sensor.
5. Remove the brake drum retainer and the brake drum.

➡ **Do not use an impact gun to remove the hub retainer nut.**

6. Remove the wheel hub retainer nut.
7. Slide the hub and bearing assembly off of the spindle.
8. Remove the four backing plate bolts and move the backing plate out of the way. Be careful not to damage the brake line to the wheel cylinder.
9. Disconnect the rear tie rod and bushing from the spindle.
10. Disconnect the rear control arm (suspension arm) from the spindle. Remove the front control arm (suspension arm) from the spindle.

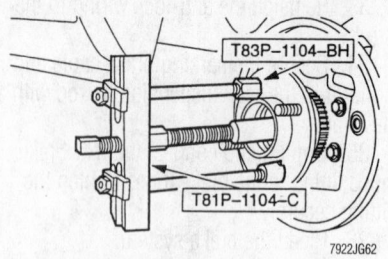

7922JG62

After removing the spindle nut, use the special tools or a puller to remove the rear hub/bearing assembly—Models equipped with rear drum brakes

11. Remove the strut-to-spindle pinch bolt.
12. Separate the spindle from the strut and remove.

To install:

13. Reinstall the spindle to the strut. Tighten the pinch bolt to 52–72 ft. lbs. (70–98 Nm).

14. Connect the rear control arm (suspension arm) to the spindle. Connect the front control arm (suspension arm) to the spindle.

15. Connect the rear tie rod and bushing to the spindle. Do not tighten the bolt at this time.

16. Place the backing plate into position. Reinstall the four backing plate bolts and tighten to 33–40 ft. lbs. (45–54 Nm).

17. Reinstall the hub and bearing assembly to the spindle.

18. Reinstall the hub retainer nut. Tighten the retainer nut to 170–192 ft. lbs. (230–260 Nm).

✳✳ WARNING

Do not use an impact gun to tighten the hub retainer nut, otherwise the bearings may be damaged.

19. Reinstall the brake drum.

20. Check the brake shoes for proper adjustment.

21. Reinstall the anti-lock sensor and the retaining bolt. Tighten the retaining bolt to 7–8 ft. lbs. (9–11 Nm).

22. Reinstall the wheel and tire assembly. Tighten the wheel nuts to 62 ft. lbs. (85 Nm).

23. Lower the vehicle until the wheels are supporting the vehicles weight. This will properly load the suspension.

24. Tighten the spindle-to-control arm bolt to 52–79 ft. lbs. (70–98 Nm).

25. Tighten the tie rod-to-spindle bolt to 75–102 ft. lbs. (98 Nm).

26. Finish lowering the vehicle.

27. Road test the vehicle and check for proper operation.

FORD MOTOR CO.

Ford-Taurus • Mercury-Sable

20

DRIVE TRAIN 20-43
ENGINE REPAIR 20-2
FUEL SYSTEM 20-41
STEERING AND
  SUSPENSION 20-47

A
Air Bag.................20-47
  Disarming.................20-47
  Precautions 20-47

C
Camshaft and Valve Lifters 20-23
  Removal & Installation 20-23
Clutch 20-45
  Adjustment 20-45
  Removal & Installation 20-45
Coil Spring 20-51
  Removal & Installation 20-51
Cylinder Head 20-7
  Removal & Installation 20-7

D
Distributor.................20-2
  Removal & Installation 20-2

E
Engine Assembly 20-2
  Removal & Installation 20-2
Exhaust Manifold.................20-20
  Removal & Installation 20-20

F
Front Crankshaft Seal 20-22
  Removal & Installation 20-22

Fuel Filter 20-41
  Removal & Installation 20-41
Fuel Pump 20-42
  Removal & Installation 20-42
Fuel System Pressure 20-41
  Relieving 20-41
Fuel System Service
  Precautions.................20-41

H
Halfshaft.................20-46
  Removal & Installation 20-46

I
Ignition Timing 20-2
  Adjustment 20-2
Intake Manifold 20-13
  Removal & Installation 20-13

L
Lower Ball Joints 20-52
  Removal & Installation 20-52

O
Oil Pan.................20-30
  Removal & Installation 20-30
Oil Pump 20-32
  Removal & Installation 20-32

P
Power Rack and Pinion
  Steering Gear.................20-47
  Removal & Installation 20-47

R
Rear Main Seal 20-33
  Removal & Installation 20-33
Rocker Arm/Shafts 20-11
  Removal & Installation 20-11

S
Shock Absorber 20-51
  Removal & Installation 20-51
Strut.................20-48
  Removal & Installation 20-48

T
Timing Chain, Sprockets, Front
  Cover and Seal 20-34
  Removal & Installation 20-34
Transaxle.................20-43
  Removal & Installation 20-43

V
Valve Lash 20-28
  Adjustment 20-28

W
Water Pump 20-4
  Removal & Installation 20-4
Wheel Bearings.................20-52
  Adjustment 20-52
  Removal & Installation 20-52

ENGINE REPAIR

➡**Disconnecting the negative battery cable on some vehicles may interfere with the functions of the on board computer systems and may require the computer to undergo a relearning process, once the negative battery cable is reconnected.**

Distributor

REMOVAL & INSTALLATION

1995 3.0L (OHV) and 3.8L Engines

1. Disconnect the negative battery cable.
2. Disconnect the wiring connector from the distributor.
3. Mark the position of the distributor cap No. 1 cylinder wire tower on the distributor base.
4. Remove distributor cap and position it and the attached wires aside.
5. Mark the position of the rotor in relation to the distributor housing and mark the position of the distributor housing on the engine.
6. Remove the distributor hold-down bolt and clamp and remove the distributor.

➡**Before installation, inspect the distributor O-ring and drive gear for wear and/or damage. Rotate the distributor shaft to be sure it moves freely, without binding.**

To install:

TIMING NOT DISTURBED

1. Install the distributor, aligning the distributor housing and rotor with the marks that were made during the removal procedure.
2. Install the distributor hold-down bolt and clamp. Only snug the bolt at this time.
3. Connect the distributor to the wiring harness.
4. Install the distributor cap. Be sure the ignition wires are securely connected to the distributor cap and spark plugs. Tighten the distributor cap screws to 18–23 inch lbs. (2.0–2.6 Nm).
5. Connect a suitable timing light and set the initial timing.
6. Tighten the distributor hold-down bolt to 14–21 ft. lbs. (19–28 Nm) on the 3.0L engine or 20–29 ft. lbs. (27–40 Nm) on the 3.8L engine.
7. Recheck the initial timing and adjust if necessary.

TIMING DISTURBED

1. Disconnect the spark plug wire from the No. 1 cylinder spark plug and remove the spark plug.
2. Place a finger over the spark plug hole. Rotate the engine clockwise until compression is felt at the spark plug hole.
3. Align the timing pointer with the TDC mark on the crankshaft damper.
4. Rotate the distributor shaft so the rotor tip is pointing to the distributor cap No. 1 spark plug tower position.
5. While installing the distributor, continue rotating the rotor slightly so the leading edge of the vane is centered in the vane switch stator assembly.
6. Rotate the distributor in the block to align the leading edge of the vane and vane switch stator assembly. Be sure the rotor is pointing to the distributor cap No. 1 spark plug tower position.

➡**If the vane and vane switch stator cannot be aligned by rotating the distributor in the block, remove the distributor just enough to disengage the distributor gear from the camshaft gear. Rotate the rotor enough to engage the distributor gear on another tooth of the camshaft gear. Repeat Steps 1 and 2, if necessary.**

7. Install the distributor hold-down bolt and clamp. Only snug the bolt at this time.
8. Connect the distributor to the wiring harness and install the distributor cap. Tighten the distributor cap hold-down screws to 18–23 inch lbs. (2.0–2.6 Nm).
9. Install the No. 1 cylinder spark plug and connect the spark plug wire.
10. Connect a suitable timing light and set the initial timing.
11. Tighten the distributor hold-down bolt to 14–21 ft. lbs. (19–28 Nm) on the 3.0L engine or 20–29 ft. lbs. (27–40 Nm) on the 3.8L engine.
12. Recheck the initial timing and adjust if necessary.

Ignition Timing

ADJUSTMENT

1995 3.0L (OHV) and 3.8L Engines

1. Apply the parking brake. If equipped with manual transaxle, place the shift lever in neutral. If equipped with automatic transaxle, place the shift lever in **P**.
2. Locate the timing marks on the crankshaft pulley and timing belt cover. If

the marks are hard to see, clean them off with some degreasing cleaner and a wire brush.
3. Connect a suitable inductive timing light according to the manufacturer's instructions.
4. Disconnect the single wire inline SPOUT connector, located near the distributor, by pulling the plug from the connector housing.
5. Start the engine and allow it to warm to operating temperature. Be sure the idle speed is correct.

➡**To set timing correctly, a remote starter should not be used. Use only the ignition key to start the vehicle. Disconnecting the start wire at the starter relay will cause the TFI module to revert to start mode timing after the vehicle is started. Reconnecting the start wire after the vehicle is running will not correct the timing.**

6. Aim the timing light at the timing marks. The timing should be at 10 degrees BTDC but always check the underhood vehicle emission label.
7. If the timing is incorrect, loosen the distributor hold-down bolt just enough to turn the distributor housing. While aiming the timing light at the timing marks, turn the distributor until the marks are aligned. Tighten the distributor hold-down bolt and recheck the timing.
8. Reconnect the single wire inline SPOUT connector. Check the timing advance to verify the distributor is advancing beyond the initial setting.
9. Remove the inductive timing light.

All Except 1995 3.0L (OHV) and 3.8L Engines

The base ignition timing is set at 10 degrees Before Top Dead Center (BTDC) and is not adjustable.

Engine Assembly

REMOVAL & INSTALLATION

✳✳ CAUTION

The fuel injection system remains under pressure, even after the engine has been turned OFF. The fuel system pressure must be relieved before disconnecting any fuel lines. Failure to do so may result in fire or personal injury.

1995 MODELS

1. Disconnect the battery cables, negative cable first.
2. Mark the position of the hood on the hinges and remove the hood.

✳✳ CAUTION

Never open, service or drain the radiator or cooling system when hot; serious burns can occur from the steam and hot coolant.

3. Drain the cooling system.
4. Properly relieve the fuel system pressure.

✳✳ CAUTION

Observe all applicable safety precautions when working around fuel. Whenever servicing the fuel system, always work in a well ventilated area. Do not allow fuel spray or vapors to come in contact with a spark or open flame. Keep a dry chemical fire extinguisher near the work area. Always keep fuel in a container specifically designed for fuel storage; also, always properly seal fuel containers to avoid the possibility of fire or explosion.

5. Properly recover the refrigerant from the air conditioning system.
6. Remove the air cleaner assembly. Remove the battery and the battery tray.
7. Remove the integrated relay controller, cooling fan and radiator with fan shroud.
8. Remove the engine bounce damper bracket on the strut tower.
9. Remove the evaporative emission line, upper radiator hose, starter brace and lower radiator hose.
10. Remove the exhaust pipes from both exhaust manifolds. Remove and plug the power steering pump lines. Remove the refrigerant lines from the air conditioner compressor and cap the openings to prevent contamination from dirt and moisture.
11. Remove the fuel lines and remove and tag all necessary vacuum lines.
12. Disconnect the ground strap, heater lines, accelerator cable linkage, throttle valve linkage and cruise control cable.
13. Label and detach the following wiring connectors: alternator, air conditioner compressor clutch, oxygen sensor, ignition coil, radio frequency suppressor, cooling fan voltage resistor, engine coolant temperature sensor, coolant temperature sending switch, ignition module, injector wiring harness, idle speed control motor wire, throttle

position sensor, oil pressure sending switch, ground wire, block heater, if equipped, knock sensor, EGR sensor and oil level sensor.
14. Raise the vehicle and support it safely.

✳✳ CAUTION

The EPA warns that prolonged contact with used engine oil may cause a number of skin disorders, including cancer! You should make every effort to minimize your exposure to used engine oil. Protective gloves should be worn when changing the oil. Wash your hands and any other exposed skin areas as soon as possible after exposure to used engine oil. Soap and water, or waterless hand cleaner should be used.

15. Drain the engine oil.
16. Remove the engine mount bolts and engine mounts.
17. Remove the flywheel-to-torque converter bolts. Remove the transaxle to engine mounting bolts and transaxle brace assembly.
18. Lower the vehicle. Install a suitable engine lifting plate onto the engine and use a suitable engine hoist to remove the engine from the vehicle. Remove the main wiring harness from the engine.

To install:

19. Install the main wiring harness on the engine. Position the engine in the vehicle and remove the engine lifting plate.
20. Raise the vehicle and support it safely.
21. Install the engine mounts and bolts and tighten to 40–55 ft. lbs. (54–75 Nm). Install the transaxle brace assembly and tighten the bolts to 40–55 ft. lbs. (54–75 Nm). Install the flywheel-to-torque converter bolts.
22. Attach all wiring connectors according to their labels.
23. Install the remaining components in the reverse order of removal.

✳✳ WARNING

Operating the engine without the proper amount and type of engine oil will result in severe engine damage.

24. Fill the cooling system with the proper type and quantity of coolant. Fill the crankcase with the correct type of motor oil to the required level.
25. Install the hood, aligning the marked made during removal.

26. Connect the negative battery cable. Start the engine and check for leaks and proper engine operation.
27. Evacuate and charge the air conditioning system.

1996–99 MODELS

1. Disconnect the battery cables, negative cable first.

✳✳ CAUTION

Never open, service or drain the radiator or cooling system when hot; serious burns can occur from the steam and hot coolant.

2. Drain the cooling system.
3. Mark the position of the hood on the hinges and remove the hood.
4. Disconnect the steering coupling at the pinch bolt joint inside the passenger compartment.
5. Disconnect the wiring from the Mass Air Flow (MAF) sensor, and the Intake Air Temperature (IAT) sensor.

➡Label all electrical connectors and vacuum hoses prior to removal so they can be reinstalled in their proper locations.

6. Remove the air cleaner outlet tube.
7. Remove the air cleaner retaining bolts at the air cleaner body.
8. Disconnect the engine intake air resonator by pushing in the top and bottom tube surfaces at the engine air cleaner and pulling the air cleaner outward. Lift the air cleaner up and out of the engine compartment.

✳✳ CAUTION

Observe all applicable safety precautions when working around fuel. Whenever servicing the fuel system, always work in a well ventilated area. Do not allow fuel spray or vapors to come in contact with a spark or open flame. Keep a dry chemical fire extinguisher near the work area. Always keep fuel in a container specifically designed for fuel storage; also, always properly seal fuel containers to avoid the possibility of fire or explosion.

9. Properly relieve the fuel system pressure.
10. Recover the refrigerant from the A/C system using the proper equipment.
11. Disconnect the chassis vacuum supply hose at the connection on the intake manifold. Position the hose aside.

12. Remove the ground straps from the dash panel.

13. Disconnect the control sensor wiring from the Powertrain Control Module (PCM) and position aside.

14. Remove the connectors for the engine control sensor wiring from the retaining bracket on the power brake booster. Disconnect the engine control sensor wiring at the two connectors.

15. Disconnect engine control sensor wiring from the evaporative emission canister purge valve.

16. Disconnect the evaporative emission hose at the crankcase vent connector and hose. Position the hose aside.

17. Remove the shield and disconnect the accelerator cable and the speed control actuator from the throttle body and from the accelerator cable bracket. Position the cables aside.

18. Remove the retaining nut and disconnect the manual control lever from the manual control lever shaft at the Transmission Range (TR) sensor.

19. Remove the connectors for the engine control sensor wiring from the retaining bracket on top of the transaxle. Disconnect the engine control sensor wiring at the two connectors.

20. Disconnect the wiring connector from the secondary air injection pump relay located on the retaining bracket on top of the transaxle.

21. Disconnect the main emission vacuum control connector at the connection near the fan shroud.

22. Disconnect the oil cooler inlet tube from the transaxle.

23. Disconnect the heater water hose from the water pump and water hose connection.

24. Disconnect the upper radiator hose and degas tube from the water hose connection.

25. Remove the power steering return hose from the power steering oil reservoir and drain.

26. Disconnect the alternator wiring harness from the alternator at the BAT terminal and stator connector plug. Remove the wiring harness retaining clip from the alternator mounting bracket.

27. Disconnect the retaining clips and the A/C compressor lines from the compressor. Cap all openings to prevent the entrance of dirt or moisture.

28. Raise and safely support the vehicle. Remove the front wheel and tire assemblies.

29. Remove both front stabilizer bar links from the front stabilizer bar.

30. Remove both halfshafts using the procedure in this manual.

31. Remove the splash shield from the radiator support and front bumper.

✳✳ CAUTION

The EPA warns that prolonged contact with used engine oil may cause a number of skin disorders, including cancer! You should make every effort to minimize your exposure to used engine oil. Protective gloves should be worn when changing the oil. Wash your hands and any other exposed skin areas as soon as possible after exposure to used engine oil. Soap and water, or waterless hand cleaner should be used.

32. Drain the engine oil.

33. Remove the converter Y-pipe.

34. Disconnect the power steering pressure hose from the power steering/transaxle oil cooler connection. Position the hose aside.

35. Disconnect the lower radiator hose at the radiator and at the radiator overflow hose.

36. Disconnect the wiring at the starter motor, and remove the starter motor.

37. Disconnect the lower cooler line from the transaxle.

38. Support the front subframe and engine/transaxle assembly using Powertrain Lift 014–00765 and Universal Powertrain Removal Bracket 014–00766 or equivalents.

39. Remove the four subframe retaining bolts.

40. Lower the engine/transaxle and front subframe from the vehicle.

41. Disconnect the power steering pressure hose from the power steering pump.

42. Install suitable engine lifting brackets on the engine and transaxle assembly.

43. Remove the front engine support insulator, rear engine support insulator and engine and transaxle support.

44. Lift the engine and transaxle from the subframe.

45. Lower the engine and transaxle. Support the transaxle on a level, stationary surface for transaxle storage.

46. Remove the transaxle-to-cylinder block mounting bolts and separate the engine from the transaxle/torque converter assembly.

47. Place the engine on a safe suitable workstand.

To install:

48. Install the transaxle/torque converter assembly to the engine. Tighten the mounting bolts to 30–44 ft. lbs. (40–60 Nm). Tighten the torque converter nuts to 20–34 ft. lbs. (27–46 Nm).

49. Remove the engine and transaxle assembly from the workstand and position it on the subframe.

50. Install the front engine support insulator, rear engine support insulator and engine and transaxle support.

51. Connect the power steering pressure hose from the power steering pump.

52. Raise the engine, transaxle and subframe into position using the powertrain lifting tool.

53. Align the front subframe to the body and install the subframe-to-body bolts. Tighten the bolts to 57–76 ft. lbs. (77–103 Nm).

54. Remove the lifting equipment and move aside.

55. Install the remaining components in the reverse and tighten the following:

- Hub retainer nuts to 170–202 ft. lbs. (230–275 Nm)
- Tie rod ends-to-steering knuckle nuts to 35–46 ft. lbs. (47–63 Nm)
- Front suspension lower arms-to-front wheel knuckle nuts to 50–68 ft. lbs. (68–92 Nm)
- Front stabilizer bar links-to-front stabilizer bar nuts to 30–40 ft. lbs. (40–55 Nm)
- Manual control lever-to-manual control lever nut to 12–16 ft. lbs. (16–22 Nm)

✳✳ WARNING

Operating the engine without the proper amount and type of engine oil will result in severe engine damage.

56. Fill the cooling system. Fill the crankcase with the proper type of motor oil to the required level.

57. Connect the battery cables, negative cable last. Run the engine and check for leaks.

58. Evacuate and recharge the air conditioning system.

59. Install and align the hood.

➡ **Whenever the vehicles subframe is removed or lowered, the wheel alignment should be checked.**

60. Check the front wheel alignment. Road test the vehicle and check the engine and transaxle for proper operation.

Water Pump

REMOVAL & INSTALLATION

✳✳ CAUTION

Do not remove the radiator cap or open the cooling system until the engine has cooled. Removing the

radiator cap or opening the cooling system prior to the engine cooling could cause severe burns from scalding engine coolant.

1995 3.0L (DOHC) and 3.2L SHO Engines

1. Disconnect the battery cables, negative cable first and remove the battery and the battery tray.

2. Drain the cooling system and remove the accessory drive belts.

3. Remove the left-hand drive belt idler pulley.

4. Disconnect the electrical connector from the ignition module and ground strap.

5. Loosen the clamps on the upper intake connector tube, then remove the retaining bolts and the connector tube.

6. Remove the upper outer timing belt cover.

7. Raise and safely support the vehicle. Remove the right wheel and tire assembly. Remove the splash shield.

8. Remove the crankshaft pulley using a suitable puller.

9. Remove the lower outer timing belt cover. Disconnect the crankshaft position sensor wiring harness and move it aside.

10. Remove the center timing belt cover. Remove the right-hand drive belt tensioner idler pulley.

11. Remove the water pump attaching bolts and remove the water pump.

To install:

12. Lightly oil all bolt threads before installation. Clean gasket surfaces on pump and engine block.

13. Position a new gasket on the water pump and use a gasket sealer to hold the gasket in place. Install water pump and retaining bolts. Tighten to 12–17 ft. lbs. (16–23 Nm).

14. To complete the installation, reverse the removal procedures.

15. Tighten the crankshaft pulley retaining bolt to 112–127 ft. lbs. (152–172 Nm). Tighten the upper intake connector tube retaining bolts to 11–17 ft. lbs. (15–23 Nm).

16. Reconnect the battery cables.

17. Be sure draincock is closed and refill cooling system. Run the engine and check for leaks.

3.0L (OHV) Engine

1. Disconnect the negative battery cable. Allow the engine to cool. Remove the radiator cap and drain the cooling system.

2. Loosen four retaining bolts securing the water pump pulley to the water pump hub.

3. Remove the accessory drive belts. Remove the idler pulley or automatic tensioner, as required.

4. Disconnect and remove the heater hose from the water pump.

5. Remove the engine control sensor wiring from the locating stud bolt, if equipped.

6. Remove the water pump-to-engine retaining bolts and lift the water pump and pulley up and out of the vehicle.

To install:

7. Clean the gasket surfaces on the water pump and engine front cover. Install a new gasket on the water pump using gasket adhesive.

8. Place the water pump in position on the engine with the pulley and four retaining bolts loosely installed on the hub.

9. Apply pipe sealant and install the bolts in the water pump housing. Tighten the bolts designated by reference No. 1 to 15–22 ft. lbs. (20–30 Nm) and the bolts designated by reference No. 2 to 72–96 inch lbs. (8–12 Nm).

➡ **The bolts are of different lengths and must be installed in the correct locations.**

10. Install the remaining components in the reverse order of removal. Tighten the water pump pulley bolts to 15–22 ft. lbs. (20–30 Nm).

11. Fill the cooling system. Connect the negative battery cable.

12. Start the engine and allow it to reach normal operating temperature. Check for leaks and proper operation.

1996–99 3.0L (DOHC) Engine

1. Disconnect the negative battery cable. Drain the engine cooling system.

2. Remove the water pump drive belt. Remove the radiator and heater hoses from the water pump.

3. Remove the four nuts securing the water pump to the engine and remove the water pump.

To install:

4. Clean the water pump to engine gasket sealing surfaces.

5. Install the water pump using a new gasket and install the four retaining nuts. Tighten the retaining nuts to 15–22 ft. lbs. (20–30 Nm).

6. Install the remaining components in the reverse order of removal.

7. Fill the engine cooling system, then connect the negative battery cable.

8. Start the engine and allow it to reach normal operating temperature, then check for coolant leaks and proper engine operation.

3.4L Engine

1. Disconnect the negative battery cable. Drain the engine cooling system.

2. Remove the two flange bolts and four cap nuts and remove the engine appearance cover.

3. Remove the battery and battery tray.

4. Remove the water pump drive belt.

5. Disconnect the throttle body return and supply hoses, heater core return hose and the oil cooler return hose from the water pump.

6. Remove the two water hose connection (thermostat housing) retaining bolts and remove the thermostat.

7. Remove the bolts and collar retaining the water pump housing to the left side cylinder head.

8. Disconnect the water outlet and inlet hoses from the water pump and remove the water pump from the engine.

9. Remove the belt idler pulley retaining nut and remove the idler pulley from the water pump.

To install:

10. Clean the water pump to engine and housing gasket sealing surfaces.

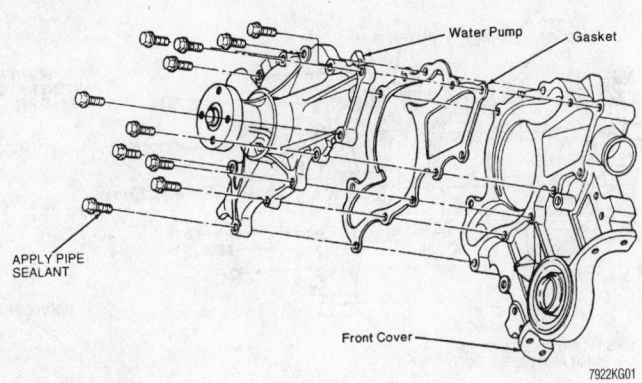

APPLY PIPE SEALANT

Water Pump   Gasket

Front Cover

7922KG01

Water pump bolt location and torque specification—3.0L (OHV) engine

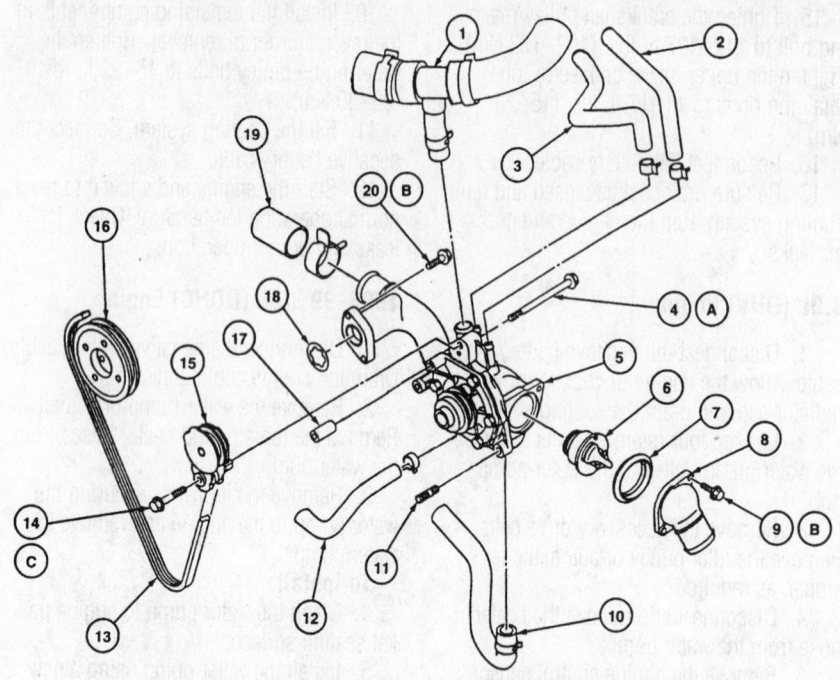

1 Water Outlet Hose	12 Oil Cooler to Water Pump Return Hose
2 Throttle Body to Water Pump Return Hose	13 Drive Belt
3 Water Pump to Throttle Body Supply Hose	14 Bolt
4 Bolt (2 Req'd)	15 Belt Idler Pulley
5 Water Pump	16 Water Pump Drive Pulley
6 Water Thermostat	17 Collar
7 Oil Cooler Return Tube Gasket	18 O-Ring
8 Water Hose Connection	19 Water Inlet Hose
9 Bolt (2 Req'd)	20 Bolt
10 Heater Core to Water Pump Return Hose	A Tighten to 18-28 N·m (14-20 Lb-Ft)
11 Stud	B Tighten to 10-16 N·m (89-141 Lb-In)
	C Tighten to 8-12 N·m (71-106 Lb-In)

7922KG03

Exploded view of the water pump and related components—3.4L engine

11. Install the belt idler pulley and retaining nut and tighten the retaining bolt to 89–141 inch lbs. (10–16 Nm).

12. Apply a silicone lubricant to the water pump inlet and outlet hoses, install them on the water pump housing and clamp securely.

13. Apply a silicone lubricant to the O-ring and install it between the water pump housing and left side cylinder head.

14. Install the bolts retaining the water pump housing to the left side cylinder head. Tighten the two retaining bolts at the water inlet to 71–106 inch lbs. (8–12 Nm). Tighten the two remaining retaining bolts to 14–20 ft. lbs. (18–28 Nm).

15. Install the water thermostat, oil cooler return tube gasket and water hose connection to the water pump housing and tighten the two retaining bolts to 71–106 inch lbs. (8–12 Nm).

16. Connect the throttle body return and supply hoses, heater core return hose and the oil cooler return hose to the water pump.

17. Install the water pump drive belt.

18. Install the engine appearance cover and tighten the two flange bolts and four cap nuts to 71–106 inch lbs. (8–12 Nm).

19. Fill the engine cooling system, then connect the negative battery cable.

20. Start the engine and allow it to reach normal operating temperature, then check for coolant leaks and proper engine operation.

3.8L (OHV) engines

1. Disconnect the negative battery cable. Allow the engine to cool before proceeding.

2. Remove the radiator cap and drain the cooling system by opening the radiator draincock.

3. Support the engine using engine support bar D88L-6000-A or equivalent. Remove the lower nut on both right engine mounts. Raise the engine.

4. Loosen the accessory drive belt idler. Remove the drive belt and water pump pulley. Remove the air suspension pump, if equipped.

5. Remove the power steering pump mounting bracket attaching bolts. Leaving hoses connected, place pump/bracket assembly aside in a position to prevent fluid from leaking out.

6. If equipped with air conditioning, remove the compressor front support bracket. Leave the compressor in place.

7. Disconnect coolant bypass and heater hoses at the water pump.

8. Remove the water pump-to-engine block attaching bolts and remove the pump from the vehicle. Discard the gasket and replace with new.

To install:

9. Lightly oil all bolt and stud threads before installation, except those that require sealant. Thoroughly clean the water pump and front cover gasket contact surfaces.

10. Apply a coating of contact adhesive to both surfaces of the new gasket. Position

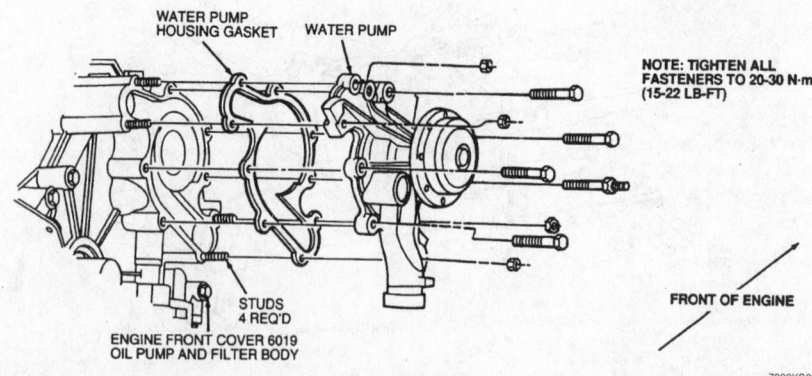

Exploded view of the water pump mounting—3.8L engine

7922KG02

a new gasket on water pump sealing surface.

11. Position water pump on the front cover and install attaching bolts. Tighten to 15–22 ft. lbs. (20–30 Nm).

12. Install the remaining components in the reverse order of removal.

13. Remove the engine support bar. Fill cooling system to the proper level. Start engine and check for coolant leaks.

Cylinder Head

REMOVAL & INSTALLATION

✳✳ CAUTION

The fuel injection system remains under pressure, even after the engine has been turned OFF. The fuel system pressure must be relieved before disconnecting any fuel lines. Failure to do so may result in fire and/or personal injury.

1995 3.0L (DOHC) and 3.2L SHO Engines

1. Disconnect the negative battery cable. Properly relieve the fuel system pressure.

2. Drain the cooling system. Remove the air cleaner outlet tube. Remove the intake manifold.

3. Loosen the accessory drive belt idlers and remove the drive belts. Remove the upper timing belt cover.

4. Remove the left idler pulley(s) and bracket assembly. Raise the vehicle and support it safely.

5. Remove the right wheel and inner fender splash shield. Remove the crankshaft damper pulley. Remove the lower timing belt cover.

6. Align both camshaft pulley timing marks with the index marks on the upper steel belt cover.

7. Release the tension on the belt by loosening the tensioner nut and rotating the tensioner with a hex head wrench. When tension is released, tighten the nut to hold the tensioner in place.

8. Disconnect the crankshaft sensor wiring assembly. Remove the center cover assembly.

9. Remove the timing belt. Note the location of the letters **KOA** on the belt. The belt must be installed to rotate in the same direction.

10. Remove the cylinder head covers. Remove the camshaft timing pulleys.

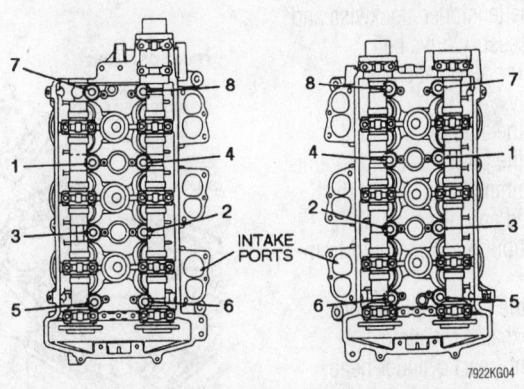

Cylinder head bolt tightening sequence—1995 3.0L (DOHC) and 3.2L SHO engines

Remove the upper rear and the center rear timing belt covers.

11. If the left cylinder head is being removed, remove the DIS coil bracket and the oil dipstick tube. If the right cylinder head is being removed, remove the coolant outlet hose.

12. Remove the exhaust manifold on the left cylinder head. On the right cylinder head the exhaust manifold must be removed with the head.

13. Remove the cylinder head bolts and remove the cylinder head.

To install:

14. Clean the bolt holes in the block with a tap. Lightly oil all bolt and stud bolt threads except those entering a coolant jacket Those bolts and studs must be sealed with a silicone sealer.

15. Clean the cylinder head and engine block mating surfaces of all gasket material.

16. Position the cylinder head and gasket on the engine block and align with the dowel pins.

17. Install the cylinder head bolts and tighten in sequence to 37–50 ft. lbs. (49–69 Nm). Repeat the sequence and tighten to 62–68 ft. lbs. (83–93 Nm).

18. When installing the left cylinder head, install the exhaust manifold, DIS coil bracket and oil dipstick tube.

19. When installing the right cylinder head, install the coolant outlet hose and connect the exhaust catalyst.

20. Install the remaining components in the reverse order of removal.

21. Connect the negative battery cable.

22. Fill the engine cooling system with the proper type and quantity of coolant. Start the engine and check for coolant, fuel or oil leaks.

3.0L (OHV) Engines

1. Rotate the crankshaft until the piston in No. 1 cylinder is at TDC on the compression stroke.

2. Disconnect the negative battery cable. Properly relieve the fuel system pressure. Drain the cooling system into a suitable container.

3. Remove the air cleaner outlet hose to throttle body. Label and disconnect the vacuum lines from the throttle body.

4. Disconnect the hoses from the EGR valve. Loosen the lower EGR tube nut and rotate the tube away from the valve.

5. Label and disconnect the wiring from the Intake Air Temperature (IAT) sensor, Throttle Position Sensor (TPS), Idle Air Control (IAC) valve and Pressure Feedback EGR (PFE) or Differential Pressure Feedback EGR (DPFE) sensors.

6. Remove the fuel line safety clips and disconnect the fuel lines from the fuel supply manifold.

7. On 1995 3.0L (OHV) engine, remove the throttle body and discard the gasket.

8. On 1996–99 3.0L (OHV) engines, remove the upper intake manifold and discard the gasket.

9. Label and disconnect the fuel injector harness from the valve cover studs and fuel injectors.

10. Remove the ignition coil and bracket from the left (front) cylinder and set aside.

11. Label and disconnect the ignition wires from the spark plugs and the valve cover studs. Disconnect the upper radiator and heater hoses.

12. On 1995 3.0L engine with non-flexible fuel, remove the distributor cap. Disconnect the wiring and remove the distributor.

13. On 1995 3.0L engine with flexible fuel, remove the Camshaft Position (CMP) sensor.

14. Disconnect the Engine Coolant Temperature (ECT) sensor and temperature sending unit electrical connectors.

15. For the left (front) cylinder head, perform the following:

   a. Disconnect the alternator electrical connectors.

b. Rotate the tensioner clockwise and remove the accessory drive belt.

c. Remove the automatic belt tensioner assembly.

d. Remove the alternator.

e. Remove the power steering mounting bracket retaining bolts. Leave the hoses connected and place the pump aside in a position to prevent fluid from leaking out.

f. Remove the engine oil dipstick tube from the exhaust manifold.

16. For the right (rear) cylinder head, perform the following:

a. Remove the alternator belt tensioner bracket.

b. Remove the heater supply tube retaining brackets from the exhaust manifold.

c. Remove the Vehicle Speed Sensor (VSS) cable retaining bolt.

d. Remove the throttle emission control solenoid and bracket.

17. Remove the rocker arm covers. Loosen the rocker arm fulcrum bolts and remove the rocker arms, fulcrums and bolts. Keep the assemblies in order so they can be reinstalled in their original locations.

➡**Regardless of the cylinder head being removed, the No. 3 cylinder intake valve pushrod must be removed to allow removal of the intake manifold.**

18. Remove the pushrods and label their positions. The pushrods must be installed in their original position during reassembly.

19. Remove the intake manifold. Remove the spark plugs. Remove the exhaust manifolds.

20. Remove and discard the cylinder head bolts and remove the cylinder heads from the engine. Remove and discard the old cylinder head gaskets.

To install:

21. Clean the cylinder head bolts holes in the block with a tap. Clean the cylinder head, intake manifold, rocker arm cover and cylinder head gasket contact surfaces.

22. Position new head gaskets on the cylinder block using the dowels in the block for alignment. If the dowels are damaged, they must be replaced.

23. Position the cylinder head on the block. Install new cylinder head bolts and tighten, in sequence, to 59 ft. lbs. (80 Nm), then back off the bolts one turn.

24. Tighten the cylinder head bolts, in sequence, to 37 ft. lbs. (50 Nm). Repeat the sequence and tighten the bolts to 68 ft. lbs. (92 Nm).

25. Install the intake manifold. Connect the ECT and coolant temperature sending unit connectors.

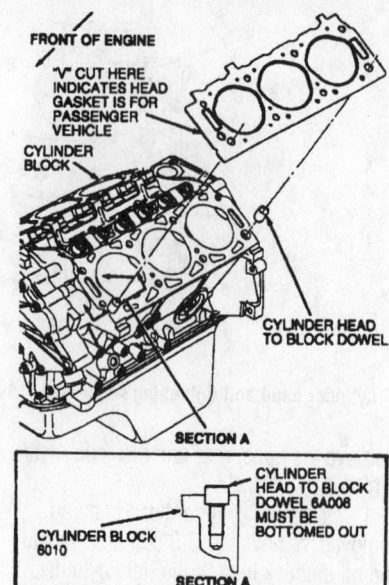

Be sure to install the cylinder head gasket properly—3.0L (OHV) engines

26. On 1995 3.0L (OHV) engine with non-flexible fuel, install the distributor.

27. Dip each pushrod end in oil conditioner or heavy engine oil. Install the pushrods in their original position.

28. Before installation, coat the valve tips, rocker arm and fulcrum contact areas with Lubriplate® or equivalent.

29. Rotate the crankshaft until the lifter is on the base circle of the cam (valve closed).

30. Install the rocker arm assemblies and tighten the rocker arm fulcrum bolts to 24 ft. lbs. (32 Nm). Be sure the lifter is on the base circle of the cam for each rocker arm as it is installed.

➡**The fulcrums must be fully seated in the cylinder head and the pushrods must be seated in the rocker arm sockets prior to the final tightening.**

31. Install the exhaust manifolds. Install the spark plugs.

32. Position the valve covers on the cylinder head and install the retaining bolts. Note the location of the ignition wire retainer stud bolts.

33. Install the fuel charging wiring to the fuel injectors and inboard valve cover stud bolts.

34. On 1995 3.0L engine, install the throttle body and intake manifold upper gasket.

35. On 1996–99 3.0L engines, install the upper intake manifold upper gasket and the upper intake manifold.

36. Install the ignition coil and bracket. Connect all sensor electrical connectors.

37. For the left (front) cylinder head, perform the following:

a. Connect the oil dipstick tube to the exhaust manifold stud and tighten the nut to 13 ft. lbs. (18 Nm).

b. Install the power steering support bracket and pump. Tighten the retaining bolt to 35 ft. lbs. (48 Nm).

c. Install the automatic belt tensioner and tighten the retaining nuts/bolt to 35 ft. lbs. (48 Nm).

38. For the right (rear) cylinder head, perform the following:

a. Install the alternator belt tensioner bracket.

b. Install the throttle emission control solenoid and bracket. Tighten the retaining bolt to 26 ft. lbs. (35 Nm).

c. Install the heater supply tube retaining brackets to the exhaust manifold and tighten the nuts to 26 ft. lbs. (35 Nm).

d. Install the VSS cable retaining bracket.

39. On flexible fuel vehicles, install the CMP sensor.

40. Install the alternator. Install the drive belt. Connect the fuel lines.

41. Connect the upper radiator hose and heater hoses. Connect all vacuum lines to pre-marked locations. Change the engine oil and filter.

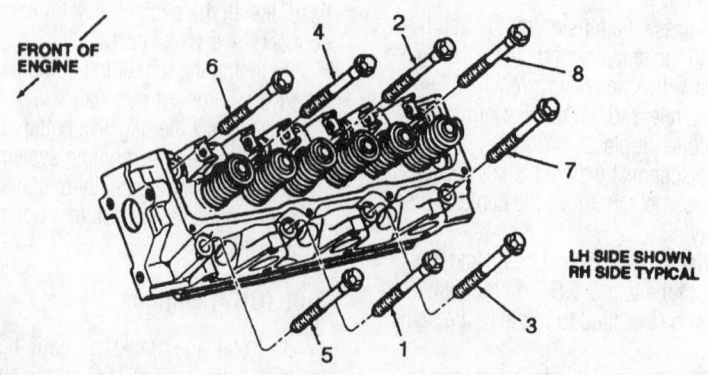

Cylinder head bolt tightening sequence—3.0L (OHV) engines

42. Install the air cleaner outlet tube to throttle body and engine air cleaner. Install crankcase ventilation tube to valve cover.

43. Fill and bleed the cooling system. Connect the negative battery cable. Start the engine and check for coolant, fuel, oil, vacuum and exhaust leaks.

44. On 1995 3.0L engine, with non-flexible fuel, check the ignition timing.

45. Check and if necessary, adjust the cruise control cable and the throttle valve cable.

1996–99 3.0L (DOHC) Engines

➡**The cylinder head bolts are a torque-to-yield design and cannot be reused. Be sure new cylinder head bolts are available before beginning this procedure.**

1. Disconnect the negative battery cable. Drain the engine coolant from the radiator and cylinder block drain plugs.

2. Remove the engine from the vehicle and position on a suitable workstand.

3. Remove the upper and lower intake manifolds. Remove the exhaust manifolds.

4. Drain the engine oil and remove the oil pan. Remove the engine front cover.

5. Remove the timing chains, camshafts and lash adjusters from both cylinder heads.

6. Remove the cylinder head retaining bolts from the cylinder heads in the proper removal sequence. Inspect the cylinder heads and cylinder block.

To install:

7. Clean the cylinder heads, intake manifolds, valve covers and the cylinder head gasket sealing surfaces on the cylinder block.

➡**If the cylinder heads were removed for cylinder head gasket replacement, check the flatness of the cylinder heads and the cylinder block gasket sealing surfaces.**

8. Install new cylinder head gaskets onto the dowels of the cylinder block.

➡**Left and right cylinder head gaskets are not interchangeable.**

9. Place the cylinder heads to their original positions using care not to damage the heads, block or gaskets.

10. Be sure the cylinder heads are correctly positioned on the dowels. Lightly oil the threads of the new cylinder head retaining bolts and install into the cylinder heads.

11. Tighten the new cylinder head retaining bolts as follows:
- Tighten the bolts, in sequence, to 27–32 ft. lbs. (37–43 Nm)
- Rotate the bolts, in sequence, 85–95 degrees
- Loosen the bolts, in sequence, a minimum of one full turn
- Tighten the bolts, in sequence, to 27–32 ft. lbs. (37–43 Nm)
- Rotate the bolts, in sequence, 85–95 degrees
- Rotate the bolts, in sequence, an additional 85–95 degrees

12. Install the EGR backpressure transducer and bracket to the rear of the right cylinder head. Tighten the retaining bolts to 71–106 inch lbs. (8–12 Nm).

13. Install the lash adjusters, camshafts and timing chains. Install the engine front cover. Install the engine oil pan.

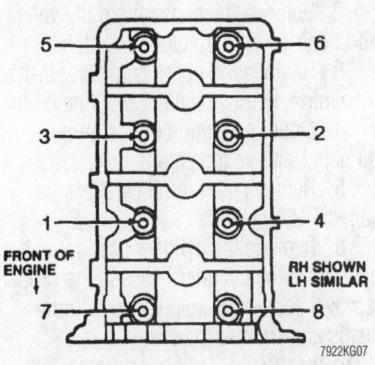

Cylinder head bolt tightening sequence— 1996–99 3.0L (DOHC) engine

14. Install the exhaust manifolds. Install the lower and upper intake manifolds. Replace the engine oil filter.

15. Install the engine assembly into the vehicle. Fill the engine with the proper amount and grade of engine oil. Fill the engine cooling system.

16. Connect the negative battery cable. Run the engine and check for leaks. Road test the vehicle and check for proper engine operation.

3.4L (DOHC) Engines

➡**The cylinder head bolts are a tighten-to-yield design and cannot be reused. Be sure new cylinder head bolts are available before beginning this procedure.**

1. Disconnect the negative battery cable. Drain the engine coolant from the radiator and cylinder block drain plugs.

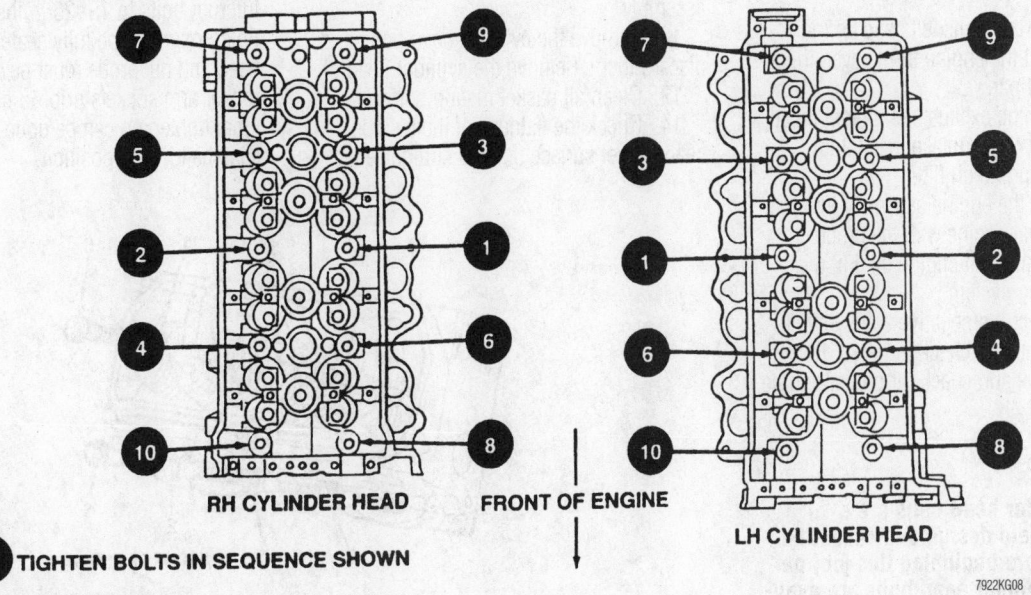

Cylinder head bolt tightening sequence—3.4L (DOHC) engine

2. Remove the engine from the vehicle and position on a suitable workstand.

3. Remove the upper and lower intake manifolds. Remove the exhaust manifolds.

4. Drain the engine oil and remove the oil pan. Remove the engine front cover.

5. Remove the timing chains and camshafts.

6. Remove the cylinder head retaining bolts from the cylinder heads in the proper removal sequence. Inspect the cylinder heads and cylinder block.

To install:

7. Clean the cylinder heads, intake manifolds, valve covers and the cylinder head gasket sealing surfaces on the cylinder block.

➡**If the cylinder heads were removed for cylinder head gasket replacement, check the flatness of the cylinder heads and the cylinder block gasket sealing surfaces.**

8. Install new cylinder head gaskets onto the dowels of the cylinder block.

➡**Left and right cylinder head gaskets are not interchangeable.**

9. Place the cylinder heads to their original positions using care not to damage the heads, block or gaskets.

10. Be sure the cylinder heads are correctly positioned on the dowels. Lightly oil the threads of the new cylinder head retaining bolts and install into the cylinder heads.

11. Tighten the new cylinder head retaining bolts as follows:

• Tighten the bolts, in sequence, to 20–23 ft. lbs. (27–32 Nm)

• Rotate the bolts, in sequence, 85–95 degrees

12. Install the camshafts and timing chains. Install the engine front cover. Install the engine oil pan.

13. Install the exhaust manifolds. Install the lower and upper intake manifolds. Replace the engine oil filter.

14. Install the engine assembly into the vehicle. Fill the engine with the proper amount and grade of engine oil. Fill the engine cooling system.

15. Connect the negative battery cable. Run the engine and check for leaks. Road test the vehicle and check for proper engine operation.

3.8L Engine

➡**The cylinder head bolts are a tighten-to-yield design and cannot be reused. Before beginning this job, be sure new cylinder head bolts are available.**

1. Disconnect the negative battery cable.

2. Relieve the fuel system pressure. Drain the cooling system.

3. Remove air cleaner assembly including the air intake duct and heat tube.

4. Loosen accessory drive belt idler. Remove drive belt.

5. For the left cylinder head, perform the following:

a. Remove oil fill cap.

b. If equipped, remove the A/C mounting bracket retaining bolts. Leaving the hoses connected position the A/C compressor aside.

c. Remove the power steering pump front mounting bracket attaching bolts.

d. Remove the alternator assembly and accessory drive belt main idler.

e. Remove the power steering/pump alternator bracket retaining bolts.

f. Leaving the hoses connected, place the power steering pump/alternator bracket assembly aside in a position to prevent the fluid from leaking out.

6. For the right cylinder head, perform the following:

a. Remove the PCV valve.

7. Remove the upper intake manifold.

8. Remove valve rocker arm cover attaching screws. Remove the fuel rail and the lower intake manifold.

9. Remove the exhaust manifold(s).

10. Loosen the rocker arm fulcrum attaching bolts enough to allow the rocker arm to be lifted off the pushrod and rotated aside.

11. Remove the pushrods. Keep the pushrods in order because they must be installed in their original position during assembly.

12. Remove the cylinder head bolts and discard them. Remove the cylinder head(s).

13. Clean all gasket mating surfaces.

14. Check the flatness of the cylinder head gasket surface using a straight-edge

and a feeler gauge. The allowable warpage is 0.003 in. for every 6.0 inches. Do not machine more than 0.010 in.

To install:

15. Clean the cylinder head bolt hole threads with a tap. Lightly oil all bolt and stud bolt threads before installation except those entering coolant jackets.

16. Position new head gasket(s) on the cylinder block using the dowels for alignment.

17. Position the cylinder head(s) on the block and install the new bolts hand tight.

18. Tighten the head bolts in the proper sequence as follows:

a. 37 ft. lbs. (50 Nm)

b. 45 ft. lbs. (60 Nm)

c. 52 ft. lbs. (70 Nm)

d. 59 ft. lbs. (80 Nm)

e. Loosen each bolt one at a time in sequence 2–3 turns, then tighten to 11–18 ft. lbs. (15–25 Nm).

f. Rotate each bolt in sequence an additional 90 degrees.

g. On 1995 3.8L engine, tighten each short bolt to 7–15 ft. lbs. (10–20 Nm), then rotate an additional 90 degrees.

➡**On 1995 3.8L engine, do not loosen more than one bolt at a time.**

19. Lubricate each pushrod with heavy engine oil and install, in their original positions.

20. For each valve, rotate the crankshaft until the lifter rests on the base circle of the camshaft lobe (pushrod all the way down). Be sure the fulcrum is seated properly, then tighten the bolt to 43 inch lbs. (5 Nm).

21. Lubricate the rocker arm assemblies with heavy engine oil and final tighten the fulcrum bolts to 19–25 ft. lbs. (25–35 Nm). Fulcrums must be fully seated in cylinder head and pushrods must be seated in rocker arm sockets prior to final tightening. Final tightening can be done with the camshaft in any position.

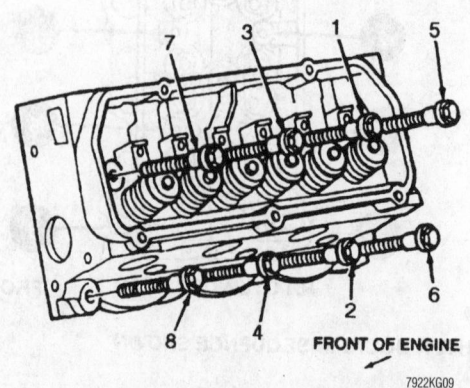

FRONT OF ENGINE

7922KG09

Install and tighten the cylinder head bolts in four steps in the sequence shown—3.8L engines

➡**If the original valvetrain components are being installed, a valve clearance check is not required. If a component has been replaced, perform a valve clearance check.**

22. Install the exhaust manifold(s).

23. Install the lower intake manifold and the fuel rail.

24. Position cover and new gasket on the cylinder head and install attaching bolts. Note the location of spark plug wire routing clip stud bolts. Tighten attaching bolts to 80–106 inch lbs. (9–12 Nm).

25. Install the upper intake manifold.

26. Install the spark plugs, if removed.

27. Connect the spark plug wires to the spark plugs.

28. For the left cylinder head, perform the following:

   a. Install the oil filler cap.

   b. Install the alternator/power steering pump mounting bracket.

   c. Install the alternator assembly.

   d. Install the main accessory drive belt tensioner assembly.

   e. Install the power steering pump assembly.

   f. Install the power steering pump support bracket.

29. For the right cylinder head, perform the following:

   a. Install PCV valve.

30. Install the accessory drive belt. If equipped, attach the thermactor tube(s) support bracket to the rear of the cylinder head. Tighten attaching bolts to 30–40 ft. lbs. (40–55 Nm).

31. Connect the negative battery cable.

32. Fill and bleed the cooling system.

33. Start engine and check for coolant, fuel and oil leaks.

34. Check and adjust the curb idle speed.

35. Install the air cleaner assembly including the air intake duct and heat tube.

Rocker Arms/Shafts

REMOVAL & INSTALLATION

3.0L (OHV) Engine

1. Disconnect the negative battery cable. Disconnect and tag the spark plug wires.

2. Remove the ignition wire/separator assembly from the rocker arm attaching bolt studs.

3. If the left rocker arm cover is being removed, remove the oil fill cap, disconnect the air cleaner closure system hose and

remove the fuel injector harness from the inboard rocker arm cover studs.

4. On 1995 3.0L engines, if the right rocker arm cover is being removed, remove the throttle body, the PCV valve, loosen the lower EGR tube, if equipped, retaining nut and rotate the tube aside and move the fuel injection harness aside.

5. On 1996–99 3.0L engines, if removing the right rocker arm cover, remove the upper intake manifold, tag and disconnect the vacuum hoses at the vacuum tee, loosen the EGR tube nuts at the EGR valve and exhaust manifold fitting. Remove, or rotate the tube aside, remove the PCV valve, the engine sensor wiring harness and move the wiring aside and the alternator brace.

6. Remove the rocker arm cover attaching screws and the covers and gaskets from the vehicle.

7. Remove the rocker arm bolts, fulcrums, rocker arms and fulcrum washers. Keep all parts in order so they can be reinstalled to their original positions.

8. Remove the pushrods, if necessary. Keep them in order so they can be reinstalled in their original positions.

9. Inspect the rocker arms, fulcrums and pushrods for wear and/or damage. Replace as necessary.

To install:

10. Install the pushrods, if removed, making sure they seat in the lifters.

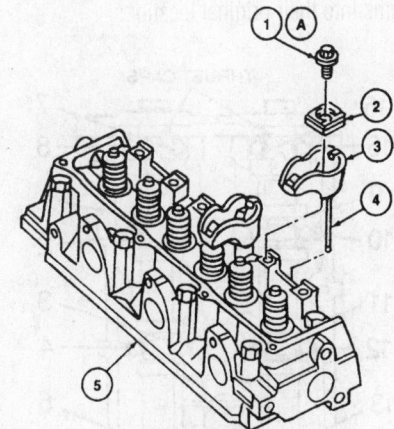

1	Bolt (12 Req'd)
2	Rocker Arm Seat (12 Req'd)
3	Rocker Arm (12 Req'd)
4	Push Rod (12 Req'd)
5	Cylinder Head (2 Req'd)
6	2.15-4.69 mm (0.085-0.185 inch)
A	Tighten in Two Steps: 7-15 N·m (5-11 Lb-Ft) 26-38 N·m (19-28 Lb-Ft)

7922KG22

Exploded view of the rocker arms and related components—3.0L (OHV) engines

11. Coat the valve and pushrod tips, rocker arm and fulcrum contact areas with Lubriplate® or equivalent. Lightly oil all the bolt and stud threads before installation.

12. Rotate the engine until the lifter is on the base circle of the cam (valve closed).

13. Install the rocker arm and components and tighten the rocker arm fulcrum bolts in two steps: the first to 8 ft. lbs. (11 Nm) and the final to 24 ft. lbs. (32 Nm). Be sure the lifter is on the base circle of the cam for each rocker arm as it is installed.

14. Clean the cylinder head and rocker arm cover sealing surfaces of all dirt and old sealer. If not equipped with integral gaskets, be sure all old gasket material is removed.

15. Apply a bead of silicone sealant at the cylinder head to intake manifold rail step. If not equipped with integral gaskets, install a new rocker arm cover gasket.

16. Install the rocker arm cover and the bolts and studs. Tighten to 9 ft. lbs. (12 Nm) in the proper sequence.

17. Install the remaining components in the reverse order of their removal.

1995 3.0L (DOHC) Engine

The 1995 3.0L (DOHC) engine does not have rocker arms. The camshaft lobes work directly on the valves.

1996–99 3.0L (DOHC) Engine

1. Disconnect the negative battery cable.

2. Remove both cylinder head covers.

3. Remove the crankshaft pulley retaining bolt from the front of the crankshaft allowing the keyway to be referenced.

4. Rotate the crankshaft so the keyway is at the 11 o'clock position to locate the crankshaft at TDC for No. 1 cylinder.

5. Verify that the alignment arrows on the camshafts are aligned. If not, rotate the crankshaft one complete revolution and recheck.

6. Rotate the crankshaft so the keyway is at the 3 o'clock position. This positions the right cylinder head camshafts to the neutral position.

7. Remove the right cylinder head camshaft journal thrust cap retaining bolts and thrust caps.

8. Loosen the remaining camshaft journal cap bolts in sequence, releasing the bolts several revolutions at a time by making several passes to allow the camshaft to be raised from the cylinder head evenly. Do not remove the retaining bolts completely.

9. With the camshafts loose, remove the rocker arms, keeping them in the order that they were removed.

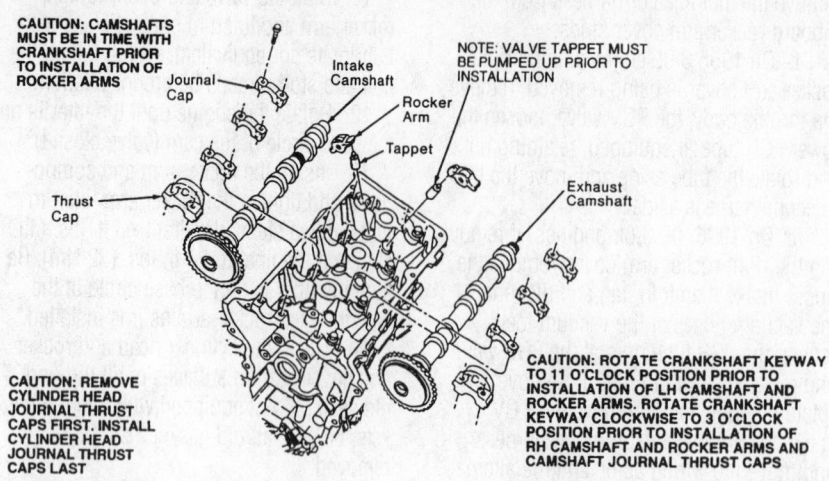

CAUTION: CAMSHAFTS MUST BE IN TIME WITH CRANKSHAFT PRIOR TO INSTALLATION OF ROCKER ARMS

Journal Cap

Intake Camshaft

Rocker Arm

Tappet

NOTE: VALVE TAPPET MUST BE PUMPED UP PRIOR TO INSTALLATION

Exhaust Camshaft

Thrust Cap

CAUTION: REMOVE CYLINDER HEAD JOURNAL THRUST CAPS FIRST. INSTALL CYLINDER HEAD JOURNAL THRUST CAPS LAST

CAUTION: ROTATE CRANKSHAFT KEYWAY TO 11 O'CLOCK POSITION PRIOR TO INSTALLATION OF LH CAMSHAFT AND ROCKER ARMS. ROTATE CRANKSHAFT KEYWAY CLOCKWISE TO 3 O'CLOCK POSITION PRIOR TO INSTALLATION OF RH CAMSHAFT AND ROCKER ARMS AND CAMSHAFT JOURNAL THRUST CAPS

7922KG10

Exploded view of camshaft mounting—1996–99 3.0L (DOHC) engine

10. If required, remove the lash adjusters from the cylinder head.

11. Rotate the crankshaft two revolutions and locate the crankshaft keyway at the 11 o'clock position. This will position the left cylinder head camshafts to their neutral position.

12. Verify that the alignment arrows on the camshafts are aligned.

➡The camshaft journal caps and cylinder heads are numbered to ensure that they are assembled in their original positions. If removed, keep the camshaft journal caps together with the cylinder head that they were removed from.

13. Remove the camshaft journal thrust cap retaining bolts and thrust caps from the left cylinder head.

14. Loosen the remaining camshaft journal cap bolts in sequence, releasing the bolts several revolutions at a time by making several passes to allow the camshaft to be raised from the cylinder head evenly. Do not remove the retaining bolts completely.

➡If the lash adjusters (tappets) and roller rocker arms are to be reused, mark the positions of the lash adjusters and rocker arms so they are reassembled into their original positions.

15. With the camshafts loose, remove the rocker arms, keeping them in the order that they were removed.

16. If required, remove the lash adjusters from the cylinder head.

17. Inspect the rocker arms and lash adjusters for wear and/or damage and replace as necessary.

To install:

18. Be sure the crankshaft keyway is at the 11 o'clock position.

19. If removed, lubricate the left cylinder head lash adjusters with engine assembly lubricant and install into their correct positions in the cylinder head.

20. If the lash adjusters are being replaced with new units, soak the adjusters in a container of clean engine oil, then manually pump up the adjusters before installing into the cylinder head.

21. Lubricate the left cylinder head rocker arms with engine assembly lubricant and install the left cylinder head rocker arms into their original locations.

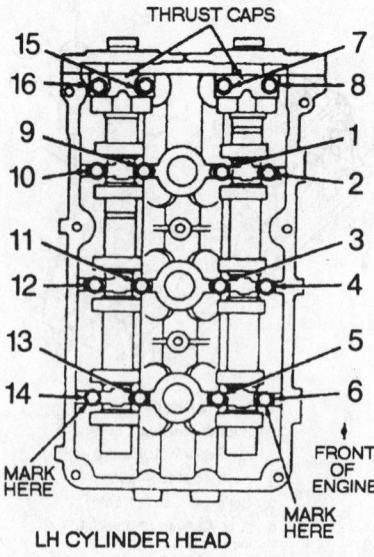

THRUST CAPS

15 — 7
16 — 8
9 — 1
10 — 2
11 — 3
12 — 4
13 — 5
14 — 6

MARK HERE

FRONT OF ENGINE

MARK HERE

LH CYLINDER HEAD

7922KG11

Tighten the journal caps gradually in the sequence shown to prevent possible camshaft warpage—1996–99 3.0L (DOHC) left cylinder head

➡Do not install the camshaft journal thrust caps until the camshaft journal caps are tightened into position.

22. Tighten the left cylinder head camshaft journal cap bolts in sequence making several passes to pull the camshafts down evenly. Tighten the bolts to 71–106 inch lbs. (8–12 Nm).

23. Install the left-hand cylinder head thrust caps and bolts. Tighten to 71–106 inch lbs. (8–12 Nm).

24. Rotate the crankshaft two revolutions and position the crankshaft keyway to the 3 o'clock location. This will position the right cylinder head camshafts to the neutral position.

25. Lubricate the right cylinder head lash adjusters with engine assembly lubricant and install into their original positions in the cylinder head.

26. If the lash adjusters are being replaced with new units, soak the adjusters in a container of clean engine oil and manually pump the adjusters up before installing them into the cylinder head.

27. Lubricate the right cylinder head rocker arms with engine assembly lubricant and install the right cylinder head rocker arms into their original locations.

➡Do not install the camshaft journal thrust caps until the camshaft journal caps are tightened into position.

28. Tighten the right cylinder head camshaft journal cap bolts in sequence making several passes to pull the camshafts down evenly. Tighten the bolts to 71–106 inch lbs. (8–12 Nm).

29. Install the right-hand cylinder head thrust caps and bolts. Tighten to 71–106 inch lbs. (8–12 Nm).

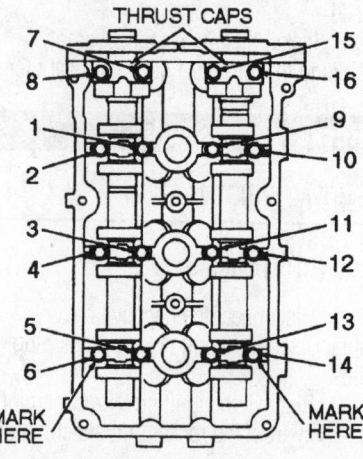

THRUST CAPS

7 — 15
8 — 16
1 — 9
2 — 10
3 — 11
4 — 12
5 — 13
6 — 14

MARK HERE

MARK HERE

7922KG12

Camshaft journal cap tightening sequence—1996–99 3.0L (DOHC) right cylinder head

30. Install the crankshaft pulley retaining bolt.

31. Tighten the crankshaft pulley retaining bolt as follows:

    a. Tighten to 89 ft. lbs. (120 Nm).

    b. Loosen the bolt at least one full turn.

    c. Tighten the bolt to 35–39 ft. lbs. (47–53 Nm).

    d. Rotate the bolt 85–95 degrees.

32. Install both valve covers as follows:

    a. Clean the valve cover gasket sealing surfaces.

    b. Install new valve cover gaskets onto the valve covers.

    c. For each valve cover, place a bead of silicone sealant at two places on the valve cover sealing surfaces where the engine front cover and the cylinder heads make contact and at two places on the rear of the cylinder head where the camshaft seal retainer contacts the cylinder head.

    d. Install the valve cover retaining bolts and studs and tighten in sequence to 71–106 inch lbs. (8–12 Nm).

➡ **The valve covers must be installed and properly tightened within six minutes of applying the silicone sealant.**

33. Connect the negative battery cable.

34. Run the engine and check for leaks and proper operation.

3.4L (DOHC) Engine

The 3.4L (DOHC) engine does not have rocker arms. The camshaft lobes work directly on the valves.

3.8L Engine

1. Disconnect the negative battery cable.

2. Tag to identify, then disconnect the spark plug wires from the spark plugs. Remove the spark plug wire routing clips from the rocker arm cover attaching bolt studs.

3. To remove the left rocker arm cover, remove the oil fill cap and the crankcase vent tube.

4. To remove the right rocker arm cover, remove the PCV valve and position the air cleaner assembly aside.

5. Remove the rocker arm cover attaching screws and remove the rocker arm covers.

6. Remove the rocker arm, fulcrum and bolt assemblies. Keep each assembly together and identify the assemblies so they may be reinstalled in their original positions.

To install:

7. Clean all gasket mating surfaces on the rocker arm covers and cylinder heads. Clean the rocker arms and fulcrums and inspect for wear or damage. Replace as necessary.

8. Apply grease to the pushrod tips and valve stem tips. Lubricate the fulcrums and rocker arms with heavy engine oil and install them over the pushrods and valve stems.

9. For each valve, rotate the crankshaft until the lifter is on the base circle of the camshaft. Install the fulcrum bolt and tighten to 5–11 ft. lbs. (7–15 Nm). Be sure the pushrod and fulcrum are fully seated prior to tightening.

10. Lubricate all rocker arm assemblies with engine oil. Final tighten the fulcrum bolts to 19–25 ft. lbs. (25–35 Nm). When final tightening, the camshaft may be in any position. Be sure the pushrod and fulcrum are fully seated prior to tightening.

11. Position new gaskets on the cylinder heads and install the rocker arm covers. Tighten the attaching bolts to 80–106 inch lbs. (9–12 Nm). Note the location of the spark plug wire routing clip stud bolts prior to installation.

12. After installing the left rocker arm cover, install the oil fill cap and the crankcase vent tube.

13. After installing the right valve cover, install the PCV valve and the air cleaner assembly.

14. Install the spark plug wire routing clips and connect the wires to the spark plugs.

15. Connect the negative battery cable, start the engine and check for leaks.

Intake Manifold

REMOVAL & INSTALLATION

✳✳ CAUTION

The fuel injection system remains under pressure, even after the engine has been turned OFF. The fuel system pressure must be relieved before disconnecting any fuel lines. Failure to do so may result in fire and/or personal injury.

1995 3.0L (DOHC) and 3.2L SHO Engines

1. Disconnect the negative battery cable. Relieve the fuel system pressure.

2. Partially drain the engine cooling system. Tag and disconnect all electrical connectors and vacuum lines from the intake assembly.

3. Remove the air cleaner tube. Disconnect the coolant lines and cables from the throttle body.

4. Remove the upper intake manifold bracket bolts. Loosen the lower bolts and remove the brackets.

5. Remove the intake manifold-to-cylinder head bolts. Remove the intake manifold and gaskets.

To install:

6. Lightly oil the attaching bolts and stud threads before installation.

➡ **The intake gasket is reusable.**

7. Position the intake manifold gasket on the cylinder heads and install the intake manifold. Tighten the bolts to 11–17 ft. lbs. (15–23 Nm). Install intake manifold support brackets. Tighten the bolts to 11–17 ft. lbs. (15–23 Nm).

8. Install the remaining components in the reverse order of removal.

9. Refill the cooling system and reconnect the negative battery cable. Run the engine and check for leaks.

1995 3.0L (OHV) Engine

1. Disconnect the negative battery cable. Drain the engine cooling system. Properly relieve the fuel system pressure.

2. Loosen the hose clamp attaching the flex hose to the throttle body. Remove the air cleaner flex hose.

3. Label and disconnect the vacuum hoses from the throttle body assembly.

➡ **The throttle body and upper intake manifold are manufactured as one assembly and will be referred to as the throttle body assembly.**

4. Loosen the lower EGR tube nut and rotate the tube away from the valve. Disconnect the accelerator and TV cables from the throttle linkage.

5. Disconnect the Throttle Position (TP) sensor, Air Charge Temperature (ACT) sensor, Engine Coolant Temperature (ECT) sensor and Idle Air Control (IAC) valve electrical connectors.

6. On flex-fuel vehicles, disconnect the Camshaft Position (CMP) sensor electrical connector.

7. Disconnect the PCV hose and disconnect the alternator support brace.

8. Remove six throttle body retaining bolts and remove the throttle body assembly.

9. Disconnect the fuel supply and return lines at the fuel supply manifold (fuel rail).

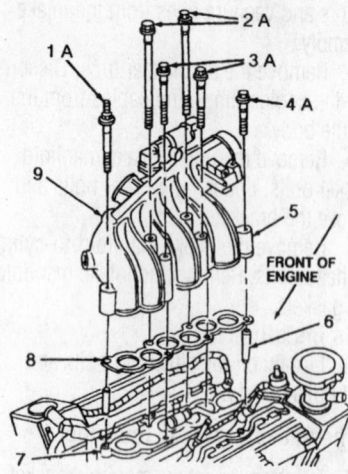

1. Stud bolt
2. Bolt (2 req'd)
3. Bolt (2 req'd)
4. Bolt
5. Throttle body
6. Guide pin (2 req'd)
7. Intake manifold
8. Intake manifold upper gasket
9. Intake manifold vacuum outlet fitting and cap
A. Tighten to 20-30 Nm (15-22 lb. ft.)

Exploded view of the upper intake manifold mounting—1995 3.0L (OHV) engine

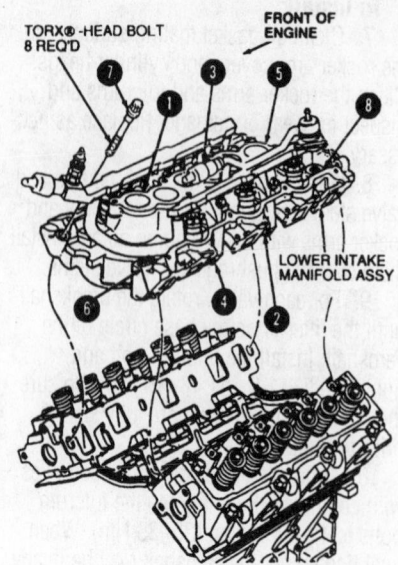

Lower intake manifold bolt tightening sequence—1995 3.0L (OHV) engines

10. Label and disconnect the fuel injection wiring harness from the engine. The manifold assembly can be removed with the fuel supply manifold and injectors in place.

11. Label and remove the ignition wires at the spark plugs.

12. Remove the rocker arm covers.

13. On non-flexible fuel vehicles, remove the distributor.

14. On flex-fuel vehicles, rotate the crankshaft until the piston for No. 1 cylinder is at TDC on its compression stroke. Note the position of the CMP electrical connector, then remove the CMP sensor housing along with the oil pump intermediate shaft.

15. Remove the ignition coil from the rear of left cylinder head.

16. Disconnect the upper radiator hose and heater hoses.

17. Loosen the intake valve rocker arm retaining bolt from No. 3 cylinder and rotate the rocker arm away from the valve stem and pushrod. Remove the pushrod.

18. Remove the intake manifold retaining bolts. Use a suitable prybar to loosen the intake manifold. Pry upward using the area between the thermostat and transaxle as a leverage point. Remove the manifold, gaskets and seals.

To install:

19. Clean the gasket mating surfaces of the intake manifold and cylinder heads. Lay a shop rag in the lifter valley to catch any gasket material. After scraping, carefully lift the cloth from the lifter valley, being careful not to let any particles enter the oil drain holes or cylinder heads. If necessary, use a suitable solvent to remove the old rubber sealant.

20. Clean and lightly oil all retaining bolts and stud threads before installation.

21. Apply a suitable silicone rubber sealer to the intersection of the cylinder block end rails and cylinder heads. Be careful not to let sealer that may block oil passages fall into the engine.

➡ **When using a silicone sealer, assembly must occur within 15 minutes after the sealer has been applied. After this time, the sealer may start to set-up and its sealing quality may be reduced. In high temperature/humidity conditions, the sealant will start to set up in approximately five minutes.**

22. Install the front and rear intake manifold end seals in place and secure. Install the intake manifold gaskets, aligning the locking tabs to the provisions on the cylinder head gaskets.

23. Carefully lower the intake manifold into position on the cylinder block and cylinder heads to prevent smearing the silicone sealer and causing gasket voids.

24. Install the intake manifold retaining bolts and tighten the bolts starting at the center and working towards the ends. Tighten the bolts in two steps, tighten to 15–22 lbs. (20–30 Nm). Tighten again to 19–24 ft. lbs. (26–32 Nm).

25. If removed, install the fuel supply manifold and injectors. Apply lubricant to the injector holes in the intake manifold and fuel supply manifold prior to injector installation. Install the fuel supply manifold retaining bolts and tighten to 7 ft. lbs. (10 Nm).

26. If removed, install the thermostat housing and a new gasket. Tighten the retaining bolts to 9 ft. lbs. (12 Nm).

27. On non-flexible fuel vehicles, install the distributor assembly, the distributor cap and ignition wires.

➡ **On non-flexible fuel vehicles, coat the distributor drive gear and on flex-fuel vehicles, coat the CMP sensor drive gear with an appropriate engine assembly lubricant.**

✱✱ WARNING

For flex-fuel vehicles, Synchro Positioning tool T93P-12200-A or equivalent, must be obtained prior to installing the CMP sensor housing. Failure to follow this procedure will result in improper CMP sensor alignment. This will result in the ignition and fuel systems being out of time with the engine, possibly causing engine damage.

28. On flex-fuel vehicles, install the CMP sensor housing as follows:

a. Engage the CMP sensor housing vane into the radial slot of Synchro Positioning tool T93P-12200-A, or equivalent. Rotate the tool on the CMP sensor housing until the tool boss engages the notch in the CMP sensor.

b. Install the CMP sensor housing along with the oil pump intermediate shaft. Install the CMP sensor housing so drive gear engagement occurs when the arrow on the locator tool is pointed approximately 30 degrees counterclockwise from the rear face of the cylinder block. This step will locate the CMP sensor electrical connector in the pre-removal position.

c. Install the hold-down clamp and tighten the bolt to 15–22 ft. lbs. (20–30 Nm). Remove the synchro positioning tool.

✱✱ WARNING

If the CMP sensor electrical connector is not positioned properly, do not reposition the connector by rotating the CMP sensor housing. This will result in the ignition and fuel sys-

tems being out of time with the engine, possibly causing engine damage. Remove the housing and repeat the installation procedure.

29. Install the No. 3 cylinder intake valve pushrod. Apply engine assembly lubricant or equivalent, to the pushrod and valve stem prior to installation. Turn the crankshaft as necessary to position the lifter on the base circle of the camshaft (pushrod all the way down). Tighten the rocker arm bolt in two steps, first to 8 ft. lbs. (11 Nm), then to 24 ft. lbs. (32 Nm).

30. Install the rocker arm covers.

31. Install the fuel injector harness and attach to the fuel injectors.

32. Install the ignition coil to the rear of left cylinder head. Tighten the retaining bolts to 30–40 ft. lbs. (40–55 Nm).

33. If removed, install the EGR valve on the intake manifold. Tighten the retaining bolts to 15–22 ft. lbs. (20–30 Nm).

34. Install the throttle body assembly and retaining bolts with a new gasket. Tighten the bolts in a cross-tightening sequence to 15–22 ft. lbs. (20–30 Nm).

35. Connect the rear crankcase ventilation hoses at the PCV valve and upper intake manifold.

36. If equipped with air conditioning, install the A/C compressor support bracket. Tighten the retaining nut and bolt to 15–22 ft. lbs. (20–30 Nm).

37. Install the accessory drive belt.

38. Connect the Throttle Position (TP) sensor, Air Charge Temperature (ACT) sensor, Engine Coolant Temperature (ECT) sensor and Idle Air Control (IAC) valve electrical connectors.

39. On flex-fuel vehicles, connect the Camshaft Position (CMP) sensor electrical connector.

40. Connect the necessary vacuum hoses.

41. Connect the heater water hose to the hot water heater elbow connection.

42. Position the heater tube support bracket and tighten the retaining nut to 15–22 ft. lbs. (20–30 Nm). Tighten the hose clamp at the hot water heater elbow connection securely.

43. Connect the heater water hose to the rear of the water bypass tube and tighten the hose clamp.

44. Connect the water bypass hose. Tighten the hose clamp securely.

45. Connect the upper radiator hose. Tighten the hose clamp securely.

46. Connect the fuel supply and return lines to the fuel injection supply manifold.

47. Connect the fuel line safety clips.

48. Position the accelerator cable bracket. Install and tighten the retaining bolts to 15–22 ft. lbs. (20–30 Nm).

49. Connect the speed control actuator to the throttle body assembly, if equipped.

50. Connect the accelerator cable to throttle body assembly.

51. Install the engine air cleaner and air cleaner outlet tube.

52. Install the fuel tank fill cap.

53. Connect the negative battery cable.

54. Turn the ignition switch to the **RUN** position several times without starting the engine to pressurize the fuel system and to check for fuel leaks.

55. Start the engine and check for fuel leaks and coolant leaks.

56. On non-flexible fuel vehicles, verify and if necessary, correct engine timing to 10 degrees BTDC. Tighten the distributor retaining bolt to 18 ft. lbs. (24 Nm).

57. Install the IAC valve shield.

58. Road test the vehicle and check for proper operation.

1996–99 3.0L (OHV) Engines

UPPER INTAKE MANIFOLD

1. Disconnect the negative battery cable.

2. Remove the air cleaner outlet tube.

3. Remove the accelerator cable shield from the from the cable bracket.

4. Remove the accelerator cable spring. Disconnect the accelerator and speed control cables from the throttle body.

5. Remove the two accelerator cable bracket retaining bolts from the side of the throttle body and position the cable bracket aside.

6. Label and disconnect the vacuum hose from the fuel pressure regulator.

7. Loosen the EGR tube nut at the EGR valve and disconnect the EGR backpressure transducer hoses from the EGR valve to exhaust manifold tube.

8. Disconnect the PCV hose, aspirator vacuum supply hose and evaporative emission return tube from the fitting underneath the upper intake manifold.

9. Disconnect the electrical connectors to the Throttle Position Sensor (TPS), Idle Air Control (IAC) valve, EGR backpressure transducer and EGR vacuum regulator solenoid.

10. Disconnect the degas tube from the radiator coolant recovery tank and lower intake manifold fitting.

11. Remove the retaining nut and bolts for the upper alternator brace and remove the brace.

12. Remove the sensor wiring bracket from the throttle body retaining stud bolt and position the wiring aside.

13. Remove the intake manifold support from the throttle body and right cylinder head.

14. Remove the upper intake manifold retaining bolts and stud bolts and note their location for installation. Remove the upper intake manifold.

15. Remove the manifold gaskets and discard.

To install:

16. Clean the gasket sealing surfaces and install the new intake manifold gaskets using locating pins as necessary to aid in gasket alignment.

17. Lightly oil all attaching bolt and stud threads before installation.

18. Position the upper intake gasket and manifold on top of the lower intake manifold. Use locating pins to secure the position of gasket between manifolds.

19. Install the retaining bolts and studs in their original locations. Tighten the stud bolts and bolts to 15–22 ft. lbs. (20–30 Nm).

20. Install the alternator brace to the upper intake manifold mounting stud and alternator mounting bracket. Tighten the nut and bolts to 9–15 ft. lbs. (12–20 Nm).

21. Install the intake manifold support to the throttle body and the right cylinder head. Tighten the top retaining bolt to 71–106 inch lbs. (8–12 Nm). Tighten the bottom bolt to 30–40 ft. lbs. (40–55 Nm).

22. Install the engine sensor wiring bracket onto the throttle body stud bolts.

23. Connect the PCV hose, aspirator vacuum supply hose and evaporative emission return tube to the fitting underneath the upper intake manifold.

24. Install the EGR tube nut to the EGR valve and tighten to 26–48 ft. lbs. (35–65 Nm).

25. Connect the vacuum hose to the fuel pressure regulator.

26. Connect the electrical connectors to the TPS, IAC, EGR backpressure transducer and EGR vacuum regulator solenoid.

27. Install the accelerator cable bracket to the side of the throttle body and install the two retaining bolts. Tighten the bolts to 13 ft. lbs. (17 Nm).

28. Connect the accelerator cable and speed control cable to the throttle body. Install the throttle retracting spring.

29. Install the accelerator cable shield and tighten the bolts to 13 inch lbs. (1.4 Nm).

30. Install the air cleaner outlet tube. Connect the negative battery cable.

31. Fill the cooling system.

32. Start the engine and check for leaks and proper operation.

LOWER INTAKE MANIFOLD

1. Disconnect the negative battery cable. Relieve the fuel system pressure.

2. Disconnect the wiring from the Mass Air Flow (MAF) sensor and the Intake Air Temperature (IAT) sensor.

3. Remove the air cleaner outlet tube. Remove the air cleaner bolts at the air cleaner body.

4. Disconnect the engine intake air resonator by pushing in the top and bottom tube surfaces at the engine air cleaner and pulling the air cleaner outward. Lift the air cleaner up and out of the engine compartment.

5. Remove the fuel line safely clips. Disconnect the fuel supply and return lines from the fuel supply manifold using the proper disconnect tools.

6. Disconnect the remaining engine wire connectors from the Camshaft Position (CMP) sensor, Throttle Position (TP) sensor, Idle Air Control (IAC) valve, Engine Coolant Temperature (ECT) sensor, ignition coil, water temperature sensor, EGR backpressure transducer and EGR vacuum regulator solenoid connector.

➥**Note the position of the CMP sensor electrical connector. The installation requires that the connector be located in the same location.**

7. With suitable pliers, slide back the upper radiator hose clamp and with a twisting motion loosen the hose from the hose connection.

8. Remove the upper intake manifold.

9. Loosen the EGR tube nut and remove the EGR valve to exhaust manifold tube from the EGR valve tube to manifold connector.

10. Disconnect the sensor wiring from the valve cover stud bolts. Carefully disconnect the electrical connectors to each fuel injector and position the sensor wiring harness aside.

11. Disconnect the heater water hoses.

12. Label and disconnect the ignition wires.

➥**Before removing the CMP sensor, position No. 1 cylinder to TDC of its compression stroke.**

13. Remove the retaining screws from the CMP sensor and remove the sensor from the sensor housing.

14. Remove the hold-down clamp and remove the CMP housing from the cylinder block.

15. Remove the cylinder head covers.

16. Remove the ignition coil from the rear of the left cylinder head.

17. Loosen the intake valve rocker arm retaining bolt from cylinder No. 3 and rotate the rocker arm away from the valve stem and pushrod. Remove the pushrod.

➥**The lower intake manifold may be removed with the fuel injection supply manifold and fuel injectors in place.**

18. Remove the intake manifold attaching bolts using a Torx® head socket. Use a suitable prybar to loosen the intake manifold. Pry upward using the area between the thermostat and transaxle as a leverage point. Remove the manifold and old gaskets and seals.

To install:

19. Clean the gasket mating surfaces of the intake manifold and the cylinder heads. Lay a shop rag in the lifter valley to catch any gasket material. After scraping, carefully lift the shop rag from the lifter valley, being careful not to let any particles enter the oil drain holes or cylinder head. If necessary, use a suitable solvent to remove old rubber sealant.

20. Lightly oil all the attaching bolts and stud threads before installation. When using a silicone rubber sealer, assembly must occur within 15 minutes after the sealer has been applied. After this time, the sealer may start to set-up and its sealing quality may be reduced. In high temperature and/or humidity conditions, the sealant will start to set up in approximately five minutes.

21. Apply a suitable silicone rubber sealer to the intersection of the cylinder block end rails and cylinder heads. Be careful not to let sealer that may block oil passages fall into the engine.

22. Install the front and rear intake manifold end seals in place and secure. Install the intake manifold gaskets, aligning the locking tabs to the provisions on the cylinder head gaskets.

23. Carefully lower the intake manifold into position on the cylinder block and cylinder heads to prevent smearing the silicone sealer and causing gasket voids.

24. Install the bolts and tighten in sequence, starting at the center and working towards the ends. Tighten the bolts in two steps, tighten to 15–22 lbs. (20–30 Nm). Tighten again in sequence to 19–24 ft. lbs. (26–32 Nm).

25. Install the fuel supply manifold and injectors, if removed. Apply a small amount of clean engine oil to the injector holes in the intake manifold and fuel supply manifold prior to injector installation. Install the fuel supply manifold retaining bolts and tighten to 7 ft. lbs. (10 Nm).

✳ WARNING

A special Synchro Positioning tool T95T-12200-A or equivalent must be used before installing the CMP sensor. If the special tool is not used the fuel system will be out of time possibly causing engine damage.

26. Attach the Synchro Position tool T95T-12200-A as follows:

    a. Engage the CMP sensor housing vane into the radial slot of the tool.

    b. Rotate the tool on the CMP sensor housing until the tool boss engages the notch in the CMP sensor housing.

27. Lube the drive gear with clean engine oil and install the CMP sensor housing so the drive gear engagement occurs when the arrow on the locator tool is pointed about 75 degrees counterclockwise from the rear face of the cylinder block. This step will locate the CMP sensor electrical connector in the same position as was noted on removal.

28. Install the hold-down clamp and tighten the bolt to 14–22 ft. lbs. (19–30 Nm).

29. Install the CMP sensor to the housing and tighten the retaining screws to 13–35 inch lbs. (1.5–4.0 Nm).

30. Install the No. 3 cylinder intake valve pushrod. Apply Lubriplate® or equivalent, to

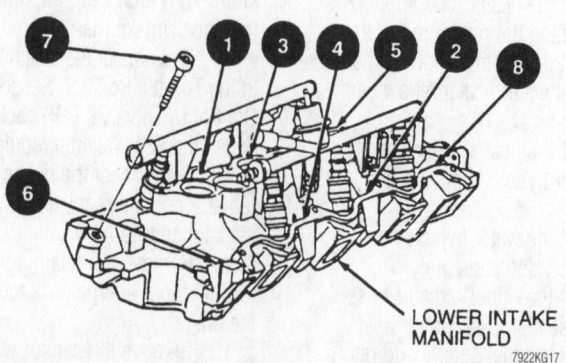

Tighten the lower intake manifold bolts, in two steps, according to the sequence shown—1996–99 3.0L (OHV) engine

7922KG17

the pushrod and valve stem prior to installation. Turn the crankshaft as necessary to position the lifter on the base circle of the camshaft (pushrod all the way down). Tighten the rocker arm bolt in two steps, first to 8 ft. lbs. (11 Nm), then to 24 ft. lbs. (32 Nm).

31. Install the cylinder head covers.

32. Install the fuel injector harness wiring and attach to the injectors.

33. Install the ignition coil to the rear of left cylinder head. Tighten the retaining bolts to 30–40 ft. lbs. (40–55 Nm).

34. Install the wiring harness to the valve cover stud bolts and connect the ignition wires to the spark plugs and ignition coil.

35. Install the upper intake manifold.

36. Install the exhaust manifold tube to the EGR valve on the intake manifold. Tighten the upper EGR tube nut to 26–48 ft. lbs. (35–65 Nm). Tighten the lower (exhaust manifold) tube nut to 26–48 ft. lbs. (35–65 Nm).

37. Install the fuel lines and safety clips.

38. Connect the water bypass hose and the upper radiator hose and properly install the squeeze clamps.

39. Connect the vacuum hoses to their pre-marked positions.

40. Connect the engine sensor wiring to the CMP sensor, IAC valve, TP sensor, ECT sensor, EGR backpressure transducer, EGR vacuum regulator solenoid, ignition coil and water temperature sender.

➡**Be sure the CMP connector is installed in the same position as it was removed for correct operation.**

41. Install engine air cleaner and air cleaner outlet tube.

42. Connect the negative battery cable.

43. Cycle the ignition switch to the **RUN** position several times without starting the engine to pressurize the fuel system and check for fuel leaks.

44. Fill the cooling system.

45. Start the engine and check for leaks and proper operation.

1996–99 3.0L (DOHC) Engine

UPPER INTAKE MANIFOLD

1. Disconnect the negative battery cable. Relieve the fuel system pressure.

2. Remove the windshield wiper motor, then remove the cowl top inner panels.

3. Remove the accelerator cable and speed control actuator from the throttle body.

4. Remove the accelerator cable bracket from the intake manifold and move aside.

5. Remove the idle air control valve fresh air supply hose from the fitting on the upper intake manifold.

6. Disconnect the wiring harnesses from the throttle position sensor and the idle air control valve.

7. Remove the vacuum supply hose from the upper intake manifold to the PCV valve at the upper intake manifold.

8. Disconnect the main emission vacuum control connector from the upper intake manifold and the secondary air injection diverter valve.

9. Remove the EGR valve.

10. Remove the secondary air injection diverter valve bracket retaining bolt and stud bolt from the upper intake manifold. Position the bracket aside.

11. Remove the upper intake manifold retaining bolts in the proper sequence.

➡**When removing engine components such as manifolds and cylinder heads, always remove the retaining bolts in a reverse order of their tightening sequence to prevent warpage to the component.**

12. Remove the upper intake manifold and gaskets from the engine.

To install:

13. Install the upper intake manifold using two new gaskets onto the lower intake manifold. Install the upper manifold

retaining bolts and tighten following the proper sequence to 71–106 inch lbs. (8–12 Nm).

14. Install the remaining components in the reverse order of removal.

15. Connect the negative battery cable. Run the engine and check for leaks and proper engine operation.

LOWER INTAKE MANIFOLD

1. Disconnect the fuel injector wiring harness and move aside.

2. Remove the spring lock coupling retainer clips from the fuel supply and return fittings.

3. Use spring lock coupling disconnect tools (⅜ inch and ½ inch) to disconnect the fuel supply and return hoses from the fuel injection supply manifold.

4. Disconnect the vacuum line from the fuel pressure regulator.

5. Disconnect the intake manifold runner actuator control cable from the intake manifold. Be careful not to loosen or bend the cable bracket, alignment is critical.

6. Disconnect the ignition wires from the left cylinder head and position the wires aside.

7. Remove the eight lower intake manifold to cylinder head retaining bolts in sequence.

8. Remove the lower intake manifold and gaskets from the vehicle.

9. If the lower intake manifold is to be replaced or machined, remove the fuel injector supply manifold (fuel rail) and the fuel injectors.

TIGHTEN BOLTS IN SEQUENCE SHOWN

LOCATING PIN TWO PLACES PER GASKET   VIEW A

1 Bolt (3 Req'd)
2 Bolt (3 Req'd)
3 Isolator (3 Req'd)
4 Isolator (3 Req'd)
5 Upper Intake Manifold
6 Gasket (2 Req'd)

7922KG15

Upper intake manifold bolt tightening sequence—1996–99 3.0L (DOHC) engine

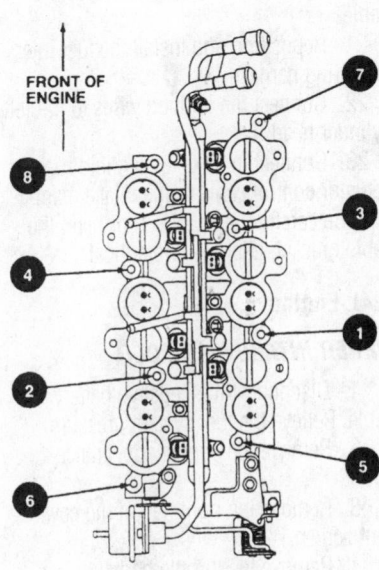

FRONT OF ENGINE

LOWER INTAKE MANIFOLD

7922KG16

Lower intake manifold bolt tightening sequence—1996–99 3.0L (DOHC) engine

To install:

10. Install the fuel injectors and the fuel supply manifold onto the lower intake manifold if removed. Verify the operation of the manifold runner control plate.

11. Thoroughly clean the gasket sealing areas and place two new intake to cylinder head gaskets into position.

12. Carefully install the lower intake manifold and install the intake manifold to cylinder head retaining bolts. Tighten the retaining bolts in sequence to 71–106 inch lbs. (8–12 Nm).

13. Install the fuel supply and return hoses to the fuel supply manifold and ensure that the spring lock couplings are correctly installed.

14. Install the retaining clips onto the spring lock couplings.

15. Connect the vacuum line to the fuel pressure regulator.

16. Temporarily connect the negative battery cable.

17. Connect the fuel pressure gauge to the fuel pressure relief valve located on the fuel injection supply manifold.

18. Cycle the ignition key several times to the **RUN** position to pressurize the fuel system.

19. Watch the fuel pressure gauge for signs of leakage. If the gauge holds pressure, remove the gauge and continue with the installation of the upper intake manifold. If the pressure gauge loses pressure, remove the fuel injection supply manifold and replace the leaking O-ring(s) before continuing.

20. Disconnect the negative battery cable.

21. Reposition and install the fuel injector wiring harness.

22. Connect the ignition wires to the left cylinder head.

23. Connect the intake manifold runner actuator control cable to the intake manifold. Be careful not to loosen or bend the cable bracket, alignment is critical.

3.4L Engine

UPPER INTAKE MANIFOLD

1. Disconnect the negative battery cable. Relieve the fuel system pressure.

2. Remove the engine appearance cover.

3. Remove the right half of the cowl vent screen.

4. Remove the throttle body.

5. Disconnect the main emission vacuum supply hose from the surge tank vacuum fitting and EGR valve and move out of the way.

6. Disconnect the vacuum tube from the intake manifold vacuum union.

7. Remove the two transducer mounting bracket-to-surge tank retaining bolts and position the bracket out of the way.

8. Remove the two intake manifold supports from the surge tank and cylinder head.

9. Remove the EGR valve.

10. Remove the retaining bolts and surge tank stay from the front of the surge tank.

11. Remove the bolt, nuts and surge tank stay from the rear of the surge tank.

12. Disconnect the crankcase ventilation tube from the valve cover.

13. Disconnect the crankcase ventilation tube from the surge tank.

14. Disconnect the vacuum tube to the pressure regulator.

15. Disconnect the right side crankcase ventilation hose and move aside.

16. Loosen the intake air connector hose clamps and remove the surge tank from the engine.

17. Remove the radio ignition interference capacitor bracket retaining bolt and position the radio ignition interference capacitor out of the way.

18. Remove the upper intake manifold retaining bolts in the proper sequence.

❄ WARNING

When removing engine components such as manifolds and cylinder heads, always remove the retaining bolts in a reverse order of their tightening sequence to prevent warpage to the component.

19. Remove the upper intake manifold and gaskets from the lower intake manifold.

To install:

20. Install the upper intake manifold using new gaskets onto the lower intake manifold. Install the upper manifold retaining bolts and tighten following the standard tightening sequence to 14–20 ft. lbs. (18–28 Nm).

21. Install the radio ignition interference capacitor bracket and retaining bolt and tighten to 71–106 inch lbs. (8–12 Nm).

22. Install the surge tank and align and tighten the intake air connector hose clamps.

23. Connect the right side crankcase ventilation hose.

24. Connect the vacuum tube to the pressure regulator.

25. Connect the crankcase ventilation tube to the surge tank.

26. Connect the crankcase ventilation tube to the valve cover.

27. Install the rear surge tank stay and tighten the retaining bolts to and nuts to 14–20 ft. lbs. (18–28 Nm).

28. Install the front surge tank stay and tighten the retaining bolts to and nuts to 14–20 ft. lbs. (18–28 Nm).

29. Install the EGR valve.

30. Install the two intake manifold supports to the surge tank and cylinder head

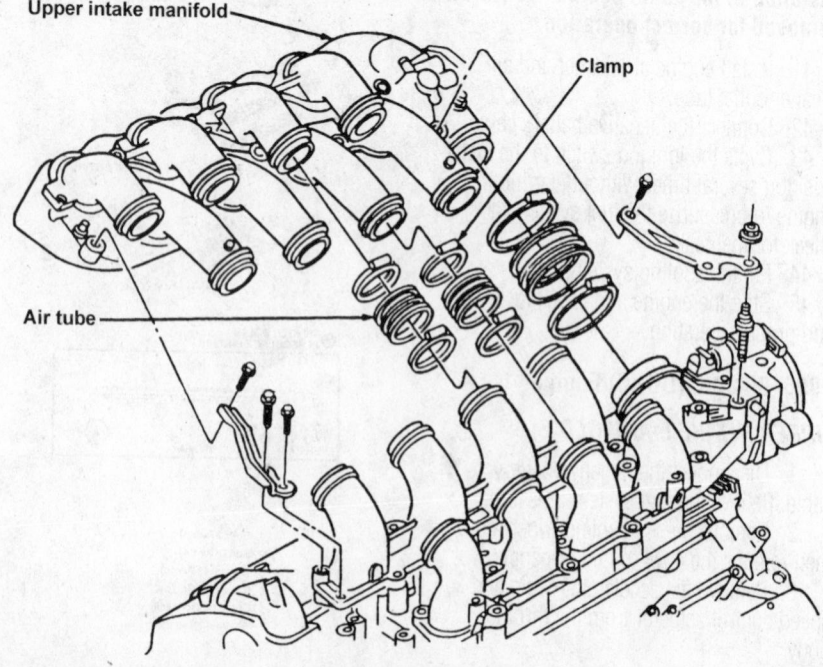

Exploded view of the upper intake manifold mounting—3.4L Engine

Upper intake manifold

Clamp

Air tube

7922KG18

and tighten the retaining bolts to 14–20 ft. lbs. (18–28 Nm).

31. Position the transducer mounting bracket to surge tank and tighten the retaining bolts to 14–20 ft. lbs. (18–28 Nm).

32. Connect the vacuum tube to the intake manifold vacuum union.

33. Connect the main emission vacuum supply hose to the surge tank vacuum fitting and EGR valve.

34. Install the throttle body.

35. Install the right half of the cowl vent screen.

36. Install the engine appearance cover.

37. Connect the negative battery cable. Run the engine and check for leaks and proper engine operation.

LOWER INTAKE MANIFOLD

1. Disconnect the negative battery cable. Relieve the fuel system pressure.

2. Disconnect the fuel injector wiring harness and move aside.

3. Remove the upper intake manifold.

4. Remove the fuel supply and return lines from the fuel injection supply manifold (fuel rail).

5. Disconnect the intake manifold runner control deactivation cable from the lower intake manifold (IMRC). Remove the cable from the bracket and position aside.

6. Disconnect the engine control sensor wiring from the fuel injectors and wiring retainers and position the sensor wiring out of the way.

7. Remove the lower intake manifold to cylinder head retaining bolts and remove the intake manifold and gaskets. Discard the gaskets.

To install:

8. Thoroughly clean the gasket sealing areas and place new intake to cylinder head gaskets into position.

9. Carefully install the lower intake manifold and install the intake manifold to cylinder head retaining bolts. Tighten the

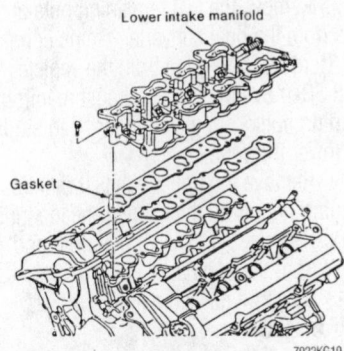

Exploded view of the lower intake manifold mounting—3.4L Engine

---

retaining bolts to 14–20 ft. lbs. (18–28 Nm).

10. Connect the engine control sensor wiring to the fuel injectors.

11. Connect the intake manifold runner control deactivation cable to the lower intake manifold (IMRC).

12. Install the fuel supply and return lines to the fuel injection supply manifold (fuel rail).

13. Install the upper intake manifold.

14. Connect the fuel injector wiring harness.

15. Connect the negative battery cable.

3.8L Engine

1. Disconnect the negative battery cable.

2. Drain the cooling system and relieve the fuel system pressure.

3. Remove the air cleaner assembly or air inlet tube.

4. Disconnect the accelerator cable at the throttle body. Disconnect the cruise control cable, if equipped.

5. If equipped with an automatic transaxle, disconnect the transaxle linkage at the upper intake manifold. Remove the retaining bolts from the accelerator cable mounting bracket and position the cables aside.

6. If equipped, disconnect the thermactor air supply hose at the check valve. The valve is located in the Y-pipe assembly.

7. Disconnect and plug the flexible fuel lines from the from steel lines over rocker arm cover.

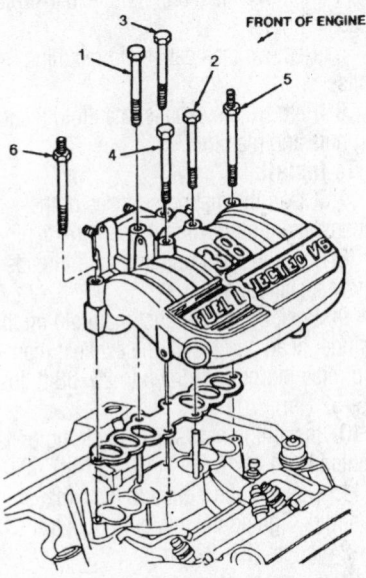

Be sure to tighten the upper intake manifold bolts in the correct sequence to prevent leaks—3.8L engine

---

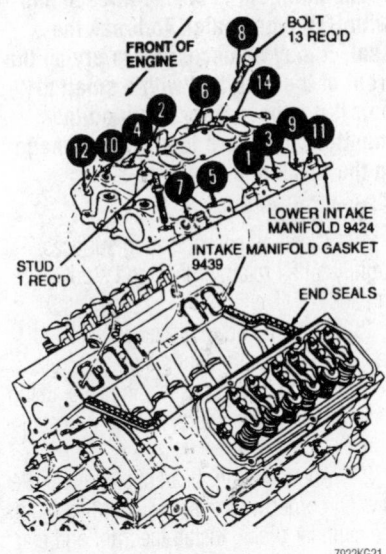

Lower intake manifold bolt tightening sequence—3.8L engine

8. Detach and plug the fuel lines at injector fuel rail assembly.

9. Disconnect the radiator hose at the thermostat housing and the coolant bypass hose at the manifold.

10. Disconnect the heater tube at the intake manifold and remove the tube support bracket retaining nut. Remove the heater hose at the rear of the heater tube. Loosen the hose clamp at the heater elbow and remove the heater tube with the hose attached. Remove the heater tube with the lines attached and set the assembly aside.

11. Tag and disconnect the vacuum lines at the fuel rail assembly and intake manifold. Tag and disconnect the necessary electrical connectors.

12. If equipped with air conditioning, remove the compressor support bracket.

13. Disconnect the PCV line at the upper intake manifold and at the valve. Remove the second PCV line from the left rocker arm cover.

14. Remove the throttle body assembly. Remove the EGR valve assembly from the upper manifold.

15. Remove the retaining nut and remove the wiring retainer bracket located at the left front of the intake manifold and set aside with the spark plug wires.

16. Unfasten the upper intake manifold retaining bolts/studs and remove the upper intake manifold.

17. Remove the injectors and fuel rail assembly. Remove the heater water outlet hose.

18. Remove the lower intake manifold retaining bolts/studs and remove the lower intake manifold.

➡ **The manifold is sealed at each end with RTV-type sealer. To break the seal, it may be necessary to pry on the front of the manifold with a small pry-bar. If it is necessary to pry on the manifold, use care to prevent damage to the machined surfaces.**

To install:

19. Clean all gasket mating surfaces. Lightly oil all retaining bolt and stud threads.

20. Apply a dab of gasket adhesive to each cylinder head mating surface. Press new intake manifold gaskets in place, using location pins as necessary to aid in installation.

21. Apply a 1/8 in. bead of silicone sealer at each corner where the cylinder head joins the cylinder block. Install the front and rear intake manifold end seals.

22. Carefully lower the intake manifold into place on the cylinder heads and cylinder block. Use locating pins as necessary to guide the manifold.

23. Install the bolts and stud bolts in their original locations and make them finger-tight.

➡ **Bolt torque depends on the type of gasket used.**

24. Graphite gaskets are usually standard. As the engine warms up, graphite gaskets allow the manifold and cylinder heads to expand at different rates without damaging the gasket and loosing the seal.

- If using the older style gasket, tighten the bolts in sequence in three steps: first to 8 ft. lbs. (10 Nm), then to 15 ft. lbs. (20 Nm), and finally to 24 ft. lbs. (32 Nm).

- Tighten the bolts in sequence in two steps, first to 13 ft. lbs. (18 Nm), then to 16 ft. lbs. (22 Nm).

25. Connect the rear PCV line to the upper intake tube. Install the front PCV tube so the mounting bracket sits over the lower intake manifold stud. Tighten the nut on the stud to 15–22 ft. lbs. (20–30 Nm).

26. Install the injectors and the fuel rail. Install the upper intake manifold assembly. Install the bolts and stud bolts in their original locations. Tighten the four center bolts, then the end bolts in three steps, first to 8 ft. lbs. (10 Nm), then to 15 ft. lbs. (20 Nm), and finally to 24 ft. lbs. (32 Nm).

27. Install the EGR valve. Install the throttle body and cross-tighten the retaining nuts to 15–22 ft. lbs. (20–30 Nm).

28. Connect the rear PCV line at the PCV valve on the upper intake manifold. If equipped with air conditioning, install the compressor support bracket.

29. Connect the necessary electrical connectors and vacuum hoses. Connect the heater tube hose to the heater elbow and position the heater tube support bracket. Tighten the retaining nut to 15–22 ft. lbs. (20–30 Nm).

30. Connect the heater hose to the heater tube and connect the coolant bypass hose and radiator upper hose.

31. Connect the fuel lines. Position the accelerator cable mounting bracket and tighten the mounting bolts to 15–22 ft. lbs. (20–30 Nm).

32. Connect the transaxle linkage at the upper intake manifold. If equipped, connect the cruise control cable.

33. Fill and bleed the cooling system. Connect the negative battery cable, start the engine and check for leaks.

34. Check and if necessary, adjust the engine idle speed, transaxle throttle linkage and cruise control.

Exhaust Manifold

REMOVAL & INSTALLATION

1995 3.0L (DOHC) and 3.2L SHO Engines

LEFT SIDE

1. Disconnect the negative battery cable.

2. Remove the oil level indicator tube support bracket.

3. Remove the power steering pump pressure and return hoses.

4. Remove the manifold to exhaust pipe attaching nuts.

5. Remove the heat shield retaining bolts.

6. Remove the exhaust manifold retaining nuts and manifold.

To install:

7. Clean the mating surfaces of the exhaust manifold, cylinder head and Y-pipe.

8. Lightly oil all bolt and stud threads before installation.

9. Position the exhaust manifold on cylinder head and install the exhaust manifold retaining nuts. Tighten to 26–38 ft. lbs. (35–52 Nm).

10. Install the heat shield retaining bolts. Tighten to 11–17 ft. lbs. (15–23 Nm).

11. Connect the Y-pipe to the exhaust manifold. Tighten the retaining nuts to 15–24 ft. lbs. (21–32 Nm).

12. Connect the power steering pressure hose and power steering return hoses.

13. Install the oil level indicator tube support bracket.

14. Connect the negative battery cable.

15. Start the engine and check for exhaust and coolant leaks.

RIGHT SIDE

1. Disconnect the negative battery cable.

2. Remove the right cylinder head.

3. Remove the heat shield retaining bolts.

4. Remove the exhaust manifold retaining nuts and manifold.

To install:

5. Clean the mating surfaces of the exhaust manifold, cylinder head and Y-pipe.

6. Lightly oil all bolt and stud threads before installation.

7. Position the exhaust manifold on the cylinder head and install the exhaust manifold retaining nuts. Tighten the nuts to 26–38 ft. lbs. (35–52 Nm).

8. Install the heat shield retaining bolts. Tighten the bolts to 11–17 ft. lbs. (15–23 Nm).

9. Install the right side cylinder head.

10. Start the engine and check for exhaust and coolant leaks.

1996–98 3.0L (DOHC) Engine

1. Disconnect the negative battery cable.

2. For the right side, perform the following procedures:

   a. Remove the upper intake manifold assembly.

   b. Remove the ignition coil assembly.

   c. Loosen the EGR valve to exhaust manifold tube nuts and remove the tube.

   d. Disconnect the wiring harness from the oxygen sensor and remove the sensor.

3. Remove the secondary air injection manifold tube from the exhaust manifold.

4. Raise and safely support the vehicle.

5. Remove the dual converter Y-pipe retaining nuts from the exhaust manifolds.

6. Remove both bolt and nut retainers from the transaxle.

7. Remove the two remaining nuts and bolts from the dual converter Y-pipe connection. Remove the Y-pipe from the vehicle.

8. Remove the lower exhaust manifold retaining nuts from the cylinder head studs and lower the vehicle.

9. Remove the upper exhaust manifold retaining nuts from the cylinder head studs.

10. Remove the exhaust manifold and gasket from the engine.

11. Clean all gasket mating surfaces.

To install:

12. Install the exhaust manifold with a new exhaust manifold gasket.

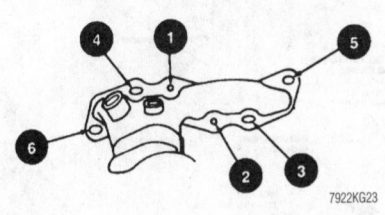

Exhaust manifold mounting bolt tightening sequence—1996–99 3.0L (DOHC) engine

13. Install the exhaust manifold retaining studs and tighten to 13–16 ft. lbs. (18–22 Nm) in sequence.

14. Raise and safely support the vehicle.

15. Position the Y-pipe assembly using a new flange gasket and install all the retaining nuts and bolts loosely.

16. Starting at the front of the system tighten the Y-pipe to exhaust manifold nuts to 26–34 ft. lbs. (34–46 Nm).

17. Tighten the converter to transaxle nut and bolt to 30 ft. lbs. (40.3 Nm). Tighten the converter outlet bolts to 26–34 ft. lbs. (34–46 Nm).

18. Lower the vehicle.

19. Install the oxygen sensor and tighten to 26–34 ft. lbs. (35–46 Nm). Connect the electrical connector.

20. Install the secondary air injection manifold tube to the exhaust manifold and tighten the nut to 28–31 ft. lbs. (38–42 Nm).

21. For the right side, perform the following procedures:

   a. Install the EGR valve to exhaust manifold tube and tighten the nuts to 26–33 ft. lbs. (35–45 Nm).

   b. Install the ignition coil assembly.

   c. Install the upper intake manifold assembly.

22. Connect the negative battery cable.

23. Run the engine and check for exhaust leaks and proper operation.

3.0L (OHV) Engine

LEFT SIDE

1. Disconnect the negative battery cable.

2. Remove oil level indicator tube support bracket retaining nut. Remove the oil level indicator tube and move engine control sensor wiring aside.

3. Raise and safely support the vehicle.

4. Remove the exhaust manifold-to-exhaust pipe retaining nuts.

5. Lower the vehicle.

6. Remove the four exhaust manifold retaining bolts and two exhaust manifold stud bolts. Remove the exhaust manifold from the vehicle.

To install:

7. Clean all mating surfaces and lightly oil all bolt and stud threads prior to installation.

8. Place the exhaust manifold into position on the cylinder head using a new gasket. Tighten the four exhaust manifold retaining bolts and two stud bolts to 15–18 ft. lbs. (20–25 Nm).

9. Install the exhaust pipe to the exhaust manifold and tighten the exhaust pipe attaching nuts to 25–34 ft. lbs. (34–47 Nm).

10. Install the oil level indicator tube. Tighten the bracket nut to 12–15 ft. lbs. (16–20 Nm). Reposition the engine control sensor wiring.

11. Connect negative battery cable.

12. Run the engine and check for exhaust leaks and proper operation.

RIGHT SIDE

1. Disconnect the negative battery cable.

2. Disconnect the EGR valve hoses. Remove the EGR tube from the exhaust manifold. Use a back-up wrench on the lower adapter.

3. Remove the three retaining bolts from the exhaust manifold heat shield and remove the shield.

4. Raise and safely support the vehicle.

5. Remove the exhaust manifold-to-exhaust pipe retaining nuts and separate the exhaust pipe from the exhaust manifold.

6. Lower the vehicle.

7. Remove the six exhaust manifold retaining bolts and remove the exhaust manifold from the vehicle.

To install:

8. Clean all mating surfaces and lightly oil all bolt threads prior to installation.

9. Place the exhaust manifold into position on the cylinder head. Tighten the six exhaust manifold retaining bolts to 15–18 ft. lbs. (20–25 Nm).

10. Raise and safely support the vehicle.

11. Position the exhaust pipe and install the retaining nuts. Tighten the retaining nuts to 25–34 ft. lbs. (34–47 Nm).

12. Install the exhaust manifold heat shield and install the three retaining bolts. Tighten the retaining bolts to 71–106 inch lbs. (8–12 Nm).

13. Lower the vehicle.

14. Install the EGR tube to the exhaust manifold. Tighten the tube nut to 26–48 ft. lbs. (35–65 Nm). Connect the EGR valve hoses.

15. Connect the negative battery cable.

16. Run the engine and check for exhaust leaks and proper operation.

3.4L Engine

LEFT SIDE

1. Disconnect the negative battery cable.

2. Raise and safely support the vehicle.

3. Remove the dual converter Y-pipe.

4. Remove the two lower exhaust manifold retaining nuts.

5. Lower the vehicle.

6. Remove the secondary air injection manifold tube from the exhaust manifold.

7. Remove the three bolts and exhaust manifold shield from the exhaust manifold.

8. Remove the oil level indicator tube retaining bolt.

9. Remove the four upper exhaust manifold retaining nuts.

10. Remove the left side exhaust manifold and gasket.

To install:

11. Position the exhaust manifold to the engine, using a new gasket.

12. Install the four upper exhaust manifold retaining nuts and tighten to 30–44 ft. lbs. (40–60 Nm).

13. Install the oil level indicator tube retaining bolt.

14. Install the three bolts and exhaust manifold shield to the exhaust manifold and tighten to 12–16 ft. lbs. (16–23 Nm).

15. Install the secondary air injection manifold tube to the exhaust manifold and tighten the nut to 29–33 ft. lbs. (40–45 Nm).

16. Raise and safely support the vehicle.

17. Install the two lower exhaust manifold retaining nuts and tighten to 30–44 ft. lbs. (40–60 Nm).

18. Install the dual converter Y-pipe.

19. Lower the vehicle.

20. Connect the negative battery cable.

RIGHT SIDE

1. Disconnect the negative battery cable.

2. Remove the secondary air injection manifold tube from the exhaust manifold.

3. Remove the three bolts and exhaust manifold shield from the exhaust manifold.

4. Raise and safely support the vehicle.

5. Remove the EGR valve to exhaust manifold tube from the exhaust manifold.

6. Disconnect the EGR valve to exhaust manifold tube from the EGR transducer.

7. Remove the dual converter Y-pipe.

8. Remove the two lower exhaust manifold retaining nuts.

9. Lower the vehicle.

10. Remove the four upper exhaust manifold retaining nuts.

11. Raise and safely support the vehicle.

12. Remove the right side exhaust manifold and gasket.

To install:

13. Position the exhaust manifold to the engine, using a new gasket.

14. Install the four upper exhaust manifold retaining nuts and tighten to 30–44 ft. lbs. (40–60 Nm).

15. Raise and safely support the vehicle.

16. Install the two lower exhaust manifold retaining nuts and tighten to 30–44 ft. lbs. (40–60 Nm).

17. Install the dual converter Y-pipe.

18. Connect the EGR valve to exhaust manifold tube to the EGR transducer.

19. Install the EGR valve to exhaust manifold tube to the exhaust manifold.

20. Lower the vehicle.

21. Install the three bolts and exhaust manifold shield to the exhaust manifold and tighten to 12–16 ft. lbs. (16–23 Nm).

22. Install the secondary air injection manifold tube to the exhaust manifold.

23. Connect the negative battery cable.

3.8L Engine

1. Disconnect the negative battery cable.

2. For the left side, perform the following procedures:

   a. Remove the oil level dipstick tube support bracket.

   b. Tag and disconnect the spark plug wires.

3. For the right side, perform the following procedures:

   a. Remove the air cleaner outlet tube assembly. If equipped, disconnect the thermactor hose from the downstream air tube check valve.

   b. Tag and disconnect the coil secondary wire from coil and the wires from spark plugs. Remove the spark plugs.

   c. Disconnect the EGR tube.

4. Raise the vehicle and support safely.

5. For the right side, remove the transaxle dipstick tube.

6. Remove the manifold-to-exhaust pipe attaching nuts.

7. Lower the vehicle.

8. Remove the exhaust manifold retaining bolts and remove the manifold from vehicle.

To install:

9. Lightly oil all bolt and stud threads before installation. Clean the mating surfaces on the exhaust manifold, cylinder head and exhaust pipe.

10. Position the exhaust manifold on the cylinder head.

11. For the left side, install the lower front bolt on No. 5 cylinder as a pilot bolt.

12. For the right side, start two attaching bolts to align the manifold with the cylinder head.

13. Install the remaining manifold retaining bolts. Tighten the bolts to 15–22 ft. lbs. (20–30 Nm).

➡ **On the left side, a slight warpage in the exhaust manifold may cause a misalignment between the bolt holes in the head and the manifold. Elongate the holes in the exhaust manifold as necessary to correct the misalignment, if apparent. Do not elongate the pilot hole, the lower front bolt on No. 5 cylinder.**

14. Raise the vehicle and support safely.

15. Connect the exhaust pipe to the manifold. Tighten the attaching nuts to 16–24 ft. lbs. (21–32 Nm).

16. For the right side, install the transaxle dipstick tube.

17. Lower the vehicle.

18. For the left side, perform the following procedures:

   a. Connect the spark plug wires.

   b. Install dipstick tube support bracket attaching nut and tighten to 15–22 ft. lbs. (20–30 Nm).

19. For the right side, perform the following procedures:

   a. Install the outer heat shroud and tighten the retaining screws to 50–70 inch lbs. (5–8 Nm).

   b. Install the spark plugs. Connect the wires to their respective spark plugs and connect coil secondary wire to coil.

   c. Connect the EGR tube. If equipped with a thermactor hose, connect the thermactor hose to the downstream air tube and secure with clamp. Install the air cleaner outlet tube assembly.

20. Start the engine and check for exhaust leaks.

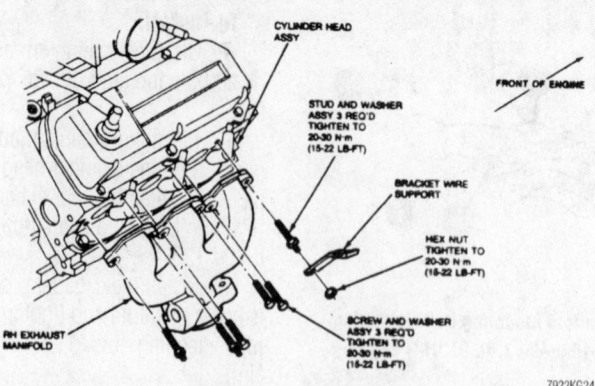

Exploded view of the right-hand side exhaust manifold mounting—3.8L engine

7922KG24

Front Crankshaft Seal

➡ **The front crankshaft oil seal procedures presented here are only for timing belt-equipped engines. For engines equipped with timing chains or gears, refer to the procedures later in this section.**

REMOVAL & INSTALLATION

1995 3.0L (DOHC) and 3.2L SHO Engines

1. Disconnect the negative battery cable, then remove the accessory drive belts.

2. Raise the vehicle and support it safely.

3. Remove the right front wheel.

4. Remove the crankshaft damper attaching bolt from the crankshaft damper. Using a suitable puller, remove the crankshaft damper from the crankshaft.

5. Remove the timing belt.

6. Remove the crankshaft timing belt sprocket using a suitable puller.

➡ **Be careful not to damage the crankshaft sensor or shutter.**

7. Remove the crankshaft front oil seal using a suitable puller.

To install:

8. Inspect the oil pump and seal surface of the crankshaft for damage, nicks, burrs or other roughness which may cause the new seal to fail. Repair or replace as necessary.

9. Using suitable tools, install a new crankshaft front oil seal.

10. Install the crankshaft sprocket.

11. Install the timing belt.

12. Install the crankshaft damper. Tighten the damper attaching bolt to 113–126 ft. lbs. (152–172 Nm).

13. Install the accessory drive belts.

14. Lower the vehicle, then start the engine and check for oil leaks.

3.4L Engine

1. Disconnect the negative battery cable.

2. Release the belt tensioner and remove the accessory drive belt.

3. Remove the crankshaft pulley attaching bolt from the crankshaft pulley. Using a suitable puller, remove the crankshaft pulley from the crankshaft.

4. Using Locknut Pin Remover T78P-3504-N or equivalent, remove the crankshaft front seal from the engine front cover.

To install:

5. Lubricate the seal bore in the front cover and the seal lip area with clean engine oil.

6. Install the new seal using Crankshaft Seal Replacer//Cover Aligner T88T-6701-A, or equivalent. Be sure the seal is installed evenly and straight.

7. Install the crankshaft pulley. Be sure to lubricate the sealing surface of the pulley with clean engine oil prior to installation. Apply a suitable silicone sealer to the crankshaft keyway and tighten the crankshaft damper retaining bolt as follows:

• Tighten the crankshaft pulley bolt to 78–99 ft. lbs. (105–135 Nm).

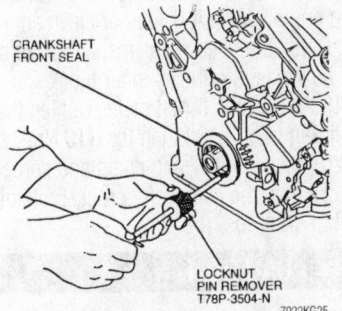

Use the special tool or a slide hammer to remove the crankshaft front oil seal—3.4L Engine

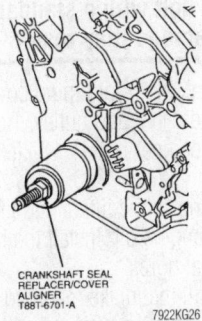

Install the new crankshaft front oil seal using a suitable seal driver or the special tool—3.4L Engine

• Loosen the crankshaft pulley bolt a minimum of one full turn.

• Tighten the crankshaft pulley bolt to 35–39 ft. lbs. (47–53 Nm).

• Rotate the crankshaft pulley bolt to an additional 85—95 degrees.

8. Install the accessory drive belt.

Camshaft and Valve Lifters

REMOVAL & INSTALLATION

✳✳ CAUTION

The fuel injection system remains under pressure, even after the engine has been turned OFF. The fuel system pressure must be relieved before disconnecting any fuel lines. Failure to do so may result in fire and/or personal injury.

1995 3.0L (DOHC) and 3.2L SHO Engines

1. Disconnect the negative battery cable. Properly relieve the fuel system pressure.

2. Rotate the crankshaft until the piston in No. 1 cylinder is at TDC.

3. Remove the intake manifold assembly. Remove the timing belt front cover and timing belt.

4. If the left cylinder head cover is being removed, remove the oil fill cap and ignition coil plastic cover. If the right cylinder head cover is being removed, disconnect the fuel lines. Remove the cylinder head cover(s).

5. Remove the camshaft pulleys, noting the location of the dowel pins.

6. Remove the upper rear timing belt cover. Uniformly loosen the camshaft bearing caps.

✳✳ WARNING

If the camshaft bearing caps are not uniformly loosened, camshaft damage may result.

7. Remove the camshaft bearing caps and note their positions for installation.

8. Remove the camshaft timing chain tensioner mounting bolts.

9. Remove the camshafts together with the timing chain and tensioner.

10. Pull the valve lifters out of its bores. Mark them so they can be installed in their original locations.

11. Remove and discard the camshaft oil seal.

12. Remove the timing chain sprocket from the camshaft.

13. Inspect the camshafts for wear and/or damage and replace, as necessary.

To install:

14. Apply a thin coat of clean engine oil to the camshaft bearing surfaces on the cylinder head and camshaft journal caps.

15. Install the valve lifters in their original positions.

16. Align the timing marks on the camshaft sprockets with the camshaft and install the sprockets. Tighten the bolts to 10–13 ft. lbs. (14–18 Nm).

17. Install the timing chain over the camshaft sprockets. Align the white painted link with the timing mark on the sprocket.

✳✳ WARNING

Left and right timing chain tensioners are not interchangeable.

18. Rotate the camshafts 60 degrees (⅙ turn) counterclockwise. Set the timing chain tensioner between the sprockets and install the camshafts on the cylinder head. The left and right tensioners are not interchangeable.

19. Apply a thin coat of clean engine oil to the camshaft journals and install bearing caps 2 through 5. Loosely install the cap retaining bolts.

➡ **The arrows on the bearing caps point to the front of the engine when installed.**

20. Apply silicone sealer to the outer diameter of the new camshaft seal and the seal seating area on the cylinder head. Install the camshaft seal using cam seal expander T89P-6256-B and cam seal replacer T89P-6256-A or equivalents.

21. Apply a 0.10 inch (2.5mm) bead of silicone sealer to the No. 1 bearing cap. Install the bearing cap while holding the camshaft seal in place with cam seal replacer T89P-6256-A or equivalent.

22. Tighten the bearing caps in sequence using a 2 step method. In the first step tighten to 71–106 inch lbs. (8–12 Nm). In the second step tighten to 12–16 ft. lbs. (16–22 Nm). For left camshaft installation, apply pressure to the timing chain tensioner to avoid damage to the bearing caps.

✳✳ WARNING

The No. 5 camshaft bearing caps function as thrust bearings for the camshaft. Always tighten No. 5 camshaft bearing caps first.

23. Position the timing chain guide and the timing chain tensioner and install the retaining bolts. Tighten the bolts to 11–14 ft. lbs. (15–19 Nm).

24. Rotate the camshafts 60 degrees clockwise and check for proper alignment of

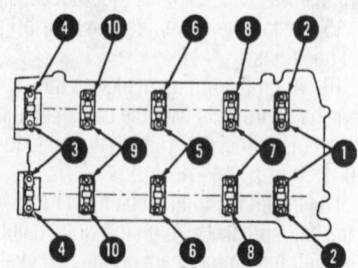

CAMSHAFT JOURNAL CAP TIGHTENING
SEQUENCE RH CYLINDER HEAD 6049

FRONT OF ENGINE

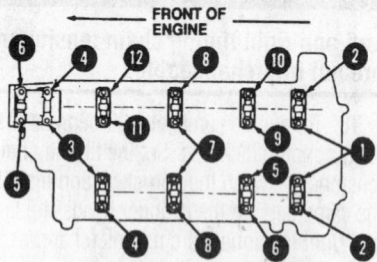

CAMSHAFT JOURNAL CAP TIGHTENING
SEQUENCE LH CYLINDER HEAD 6049

7922KG27

To prevent camshaft warpage, tighten the journal caps in two steps, according to the sequence shown—1995 3.0L (DOHC) and 3.2L SHO engines

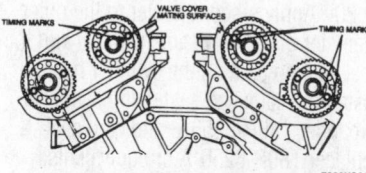

7922KG28

Align the camshaft sprocket timing marks with the valve cover mating surface—1995 3.0L (DOHC) and 3.2L SHO engines

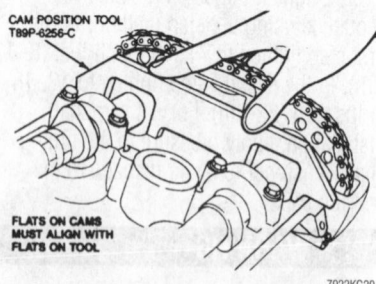

7922KG29

Be sure the camshafts are in the correct position by installing the Cam Position tool T89P-6256-C or equivalent—SHO engines

the timing marks. Marks on the camshaft sprockets should align with the valve cover mating surface.

25. Install camshaft positioning tool T89P-6256-C or equivalent, on the camshafts to check for correct positioning. The flats on the tool should align with the flats on the camshaft. If the tool does not fit and/or timing marks will not line up, repeat the installation procedure from the beginning.

26. Install the timing belt rear cover and tighten the bolts to 78 inch lbs. (8.8 Nm).

27. Install the camshaft timing belt sprockets and tighten the bolts to 15–18 ft. lbs. (21–25 Nm).

28. Install the timing belt and timing belt front cover.

29. Install the cylinder head covers and tighten the bolts to 8–11 ft. lbs. (10–16 Nm). Connect the fuel lines and install the ignition coil cover and oil fill cap.

30. Install the intake manifold assembly.

31. Connect the negative battery cable. Run the engine and check for leaks and proper engine operation.

1995 3.0L (OHV) Engine

1. Disconnect the negative battery cable. Drain the cooling system and crankcase.

2. Properly relieve the fuel system pressure.

3. Remove the engine from the vehicle and position in a suitable holding fixture.

4. Remove the accessory drive components from the front of the engine.

5. Remove the throttle body and the fuel injector harness.

6. Label and disconnect the spark plug wires from the spark plugs.

7. Remove the distributor assembly.

8. Remove the rocker arm covers.

9. Loosen the rocker arm fulcrum nuts and position the rocker arms to the side for easy access to the pushrods. Remove the pushrods and label so they may be installed in their original positions.

10. Remove the intake manifold leaving the fuel supply manifold and injectors in place.

11. Using a suitable magnet or lifter removal tool, remove the hydraulic lifters and keep them in order so they can be installed in their original positions. If the lifters are stuck in the bores by excessive varnish, use a hydraulic lifter puller to remove the lifters.

12. Remove the crankshaft pulley and damper using a suitable removal tool.

13. Remove the oil pan. Remove the front cover assembly.

14. Align the timing marks on the camshaft and crankshaft sprockets. Check the camshaft end-play as follows:

a. Push the camshaft toward the rear of the engine and install a dial indicator tool, so the indicator point is on the camshaft sprocket attaching screw.

b. Zero the dial indicator. Position a small prybar or equivalent, between the camshaft sprocket and block.

c. Pull the camshaft forward and release it. Compare the dial indicator reading with the camshaft end-play service limit specification of 0.005 in. (0.13mm).

d. If the camshaft end-play is over the amount specified, replace the thrust plate.

15. Remove the timing chain and sprockets.

16. Remove the camshaft thrust plate. Carefully remove the camshaft by pulling it toward the front of the engine. Remove it slowly to avoid damaging the bearings, journals and lobes.

17. Inspect the camshaft journals and lobes for wear and/or damage. Replace as necessary.

➡ **If the camshaft is replaced, new lifters should also be installed.**

To install:

18. Clean all gasket mating surfaces. Lubricate the camshaft lobes and journals with engine assembly lube or clean engine oil. Carefully insert the camshaft through the bearings into the cylinder block.

19. Install the thrust plate. Tighten the retaining bolts to 84 inch lbs. (10 Nm).

20. Install the timing chain and sprockets. Tighten the camshaft sprocket retaining bolt to 46 ft. lbs. (63 Nm).

❋❋ CAUTION

The camshaft bolt has a drilled oil passage in it for timing chain lubrication. Be sure the passage is clean prior to bolt installation. If the bolt is damaged, do not replace the camshaft bolt with a standard bolt or engine damage may result.

21. Install the front timing cover and crankshaft damper and pulley. Tighten the crankshaft damper bolt to 107 ft. lbs. (145 Nm).

22. Lubricate the lifters and lifter bores with clean engine oil. Install the lifters into their original bores.

23. Install the intake manifold assembly.

24. Lubricate the pushrods and rocker arms with clean engine oil. Install the pushrods and rocker arms into their original positions. Rotate the crankshaft to set each

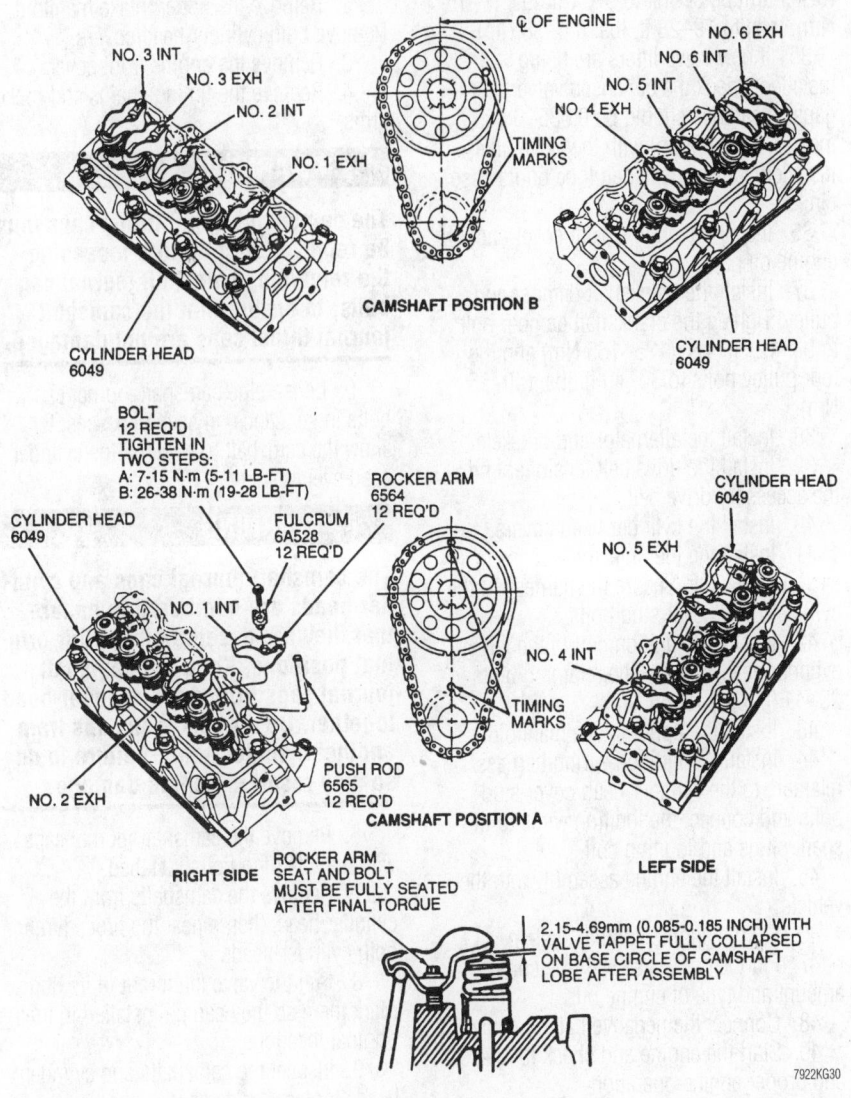

Tighten the rocker arm nuts according to the position of the camshaft—1995 3.0L (OHV) engines

lifter on its base circle, then tighten the rocker arm bolt. Tighten the rocker arm bolts to 24 ft. lbs. (32 Nm).

25. Install the oil pan and the rocker covers.

26. Install the fuel injector harness and the throttle body.

27. Install the distributor and connect the spark plug wires to the spark plugs.

28. Install the accessory drive components.

29. Install the engine assembly.

30. Connect the negative battery cable. Restore all fluid levels.

31. Start the engine and check for leaks Check and adjust the ignition timing.

1996–99 3.0L (OHV) Engine

1. Disconnect the negative battery cable. Remove the engine from the vehicle and position on a suitable holding fixture.

2. Rotate the crankshaft to TDC for No. 1 cylinder on its compression stroke.

3. Remove the upper intake manifold.

4. Disconnect the engine wiring harness connectors from the cylinder head cover stud bolts. Carefully disconnect and remove the fuel injector harness connectors from each fuel injector and position aside.

5. Label and disconnect the ignition wires from the spark plugs. Remove the ignition wire separators from the cylinder head cover stud bolts.

6. Remove the Camshaft Position (CMP) sensor housing retainer bolt and washer and remove the CMP sensor housing.

7. Remove the ignition coil from the rear of the left cylinder head.

8. Remove the cylinder head covers.

9. Loosen the tntake rocker arm fulcrum nut for No. 3 cylinder and rotate the

rocker arm off the pushrod. Remove the pushrod.

10. Remove the accessory drive belt. Remove the drive belt tensioner, alternator and alternator brackets.

11. Remove the lower intake manifold leaving the fuel supply manifold (fuel rail) and fuel injectors in place.

12. Loosen the remaining rocker arm fulcrum nuts enough to allow the rocker arms to be lifted off the pushrods. Remove the remaining pushrods, identifying each pushrod for installation.

13. Remove the valve lifter guide plate retainer bolts and the valve lifter guide plate.

14. Using a suitable magnet or lifter removal tool, remove the hydraulic valve lifters and keep them in order so they can be installed in their original positions. If the valve lifters are stuck in the bores by excessive varnish, use a hydraulic lifter puller to remove the lifters.

15. Remove the crankshaft pulley retaining bolts and the crankshaft pulley.

16. Remove the crankshaft damper retaining bolt and washer. Remove the crankshaft damper using remover T58P-6316-D or equivalent puller and adapter T82L-6316-B or equivalent.

17. Remove the engine oil pan.

18. Remove the engine front cover retaining bolts leaving the water pump attached. Remove the engine front cover.

19. Align the timing marks on the camshaft and crankshaft sprockets. Check the camshaft end-play as follows:

   a. Push the camshaft toward the rear of the engine and install a dial indicator tool, so the indicator point is on the camshaft sprocket attaching screw.

   b. Zero the dial indicator. Position a small prybar or equivalent, between the camshaft sprocket and block.

   c. Pull the camshaft forward and release it. Compare the dial indicator reading with the camshaft end-play service limit specification of 0.005 inch (0.13mm).

   d. If the camshaft end-play is over the amount specified, replace the thrust plate.

20. Remove the camshaft sprocket retaining bolt and washer.

21. Inspect the timing chain for excessive deflection.

22. Grasp the camshaft sprocket and the crankshaft sprocket and slide the timing chain and sprocket assembly off the engine.

23. Remove the two camshaft thrust plate retaining bolts and the camshaft thrust plate. Discard the thrust plate if it was found to be worn beyond specifications made during the camshaft end-play check.

24. Carefully remove the camshaft by pulling it toward the front of the engine. Remove it slowly to avoid damaging the bearings, journals and lobes.

25. Inspect the camshaft journals, lobes and bearings for wear and/or damage. Replace as necessary.

➡ **If replacing the camshaft, the valve lifters must also be replaced.**

To install:

26. Clean all gasket mating surfaces. Lubricate the camshaft lobes, journals, drive gear and bearing surfaces with engine assembly lubricant or equivalent. Carefully insert the camshaft through the bearings into the cylinder block.

27. Lubricate the camshaft thrust plate on both sides. Install the thrust plate and the two retaining bolts. Tighten the retaining bolts to 7 ft. lbs. (10 Nm).

➡ **If installing a new camshaft, check the camshaft end-play.**

28. Lubricate the timing chain and sprockets with engine assembly lubricant or equivalent and align the timing marks on the sprockets before installation.

29. Install the timing chain and sprocket assembly. Install the camshaft sprocket bolt. Tighten the camshaft sprocket bolt to 37–51 ft. lbs. (50–70 Nm).

✱✱ CAUTION

The camshaft bolt has a drilled oil passage in it for timing chain lubrication. Be sure the passage is clean prior to bolt installation. If the bolt is damaged, do not replace the camshaft bolt with a standard bolt or engine damage will result.

30. Lubricate the valve lifters and lifter bores with engine assembly lubricant or equivalent. Install the valve lifters into their original bores.

31. Align the flat surfaces on the valve lifters and install the valve lifter guide plate. The plate must be installed with the word **UP** and/or a dimple on the plate visible. Install the two retaining bolts and tighten to 8–10 ft. lbs. (10–14 Nm).

32. Install the lower intake manifold. Install the CMP sensor.

33. Lubricate the pushrods and rocker arms with engine assembly lubricant or equivalent. Install the pushrods into their original positions. Position each rocker arm onto its related pushrod.

34. Rotate the crankshaft to set each lifter on its base circle, then tighten the rocker arm bolts in two steps. Tighten the rocker arm bolts first to 5–11 ft. lbs. (7–15 Nm), then to 19–28 ft. lbs. (26–38 Nm).

35. If new valve lifters are being installed, check the collapsed valve lifter gaps. Clearance should be 0.085–0.185 inch (2.15–4.69mm) with the valve lifter installed and the camshaft lobe on its base circle.

36. Install the engine front cover and the engine oil pan.

37. Install the crankshaft damper and pulley. Tighten the crankshaft damper bolt to 93–121 ft. lbs. (125–165 Nm) and the four pulley bolts to 30–40 ft. lbs. (40–55 Nm).

38. Install the alternator and brackets.

39. Install the drive belt tensioner and the accessory drive belt.

40. Install the cylinder head covers.

41. Install the fuel injector harness to each fuel injector. Secure the harness to the cylinder head cover stud bolts.

42. Install the ignition coil to the left cylinder head. Tighten the retaining bolts to 29–41 ft. lbs. (40–55 Nm).

43. Install the upper intake manifold.

44. Install the ignition wiring harness retainers to the cylinder head cover stud bolts and connect the ignition wires to the spark plugs and ignition coil.

45. Install the engine assembly into the vehicle.

46. Fill the engine cooling system.

47. Fill the crankcase with the correct amount and type of engine oil.

48. Connect the negative battery cable.

49. Start the engine and check for leaks and proper engine operation.

50. Check and adjust the ignition timing, if needed.

1996–99 3.0L (DOHC) Engine

1. Disconnect the negative battery cable. Remove the engine assembly from the vehicle and place on an appropriate engine stand.

2. Remove the upper intake manifold. Remove both cylinder head covers.

3. Remove the engine front cover.

4. Remove the timing chains and rocker arms.

✱✱ WARNING

The camshaft journal thrust caps must be removed first, before loosening the remaining camshaft journal cap bolts, to ensure that the camshaft journal thrust caps are not damaged.

5. Loosen the camshaft journal cap bolts in sequence, in several passes, to allow the camshaft to raise off the cylinder head evenly.

✱✱ WARNING

The camshaft journal caps and cylinder heads are numbered to ensure that they are assembled in their original positions. Keep the camshaft journal caps from each cylinder head together. Do not mix with caps from another cylinder head. Failure to do so may result in engine damage.

6. Remove the camshaft journal caps with the retaining bolts installed.

7. Remove the camshafts from the cylinder head, then repeat the procedure for both cylinder heads.

8. Pull the valve lifters out of its bores. Mark them so they can be installed in their original locations.

9. Inspect the camshafts and cylinder heads for wear or damage.

To install:

✱✱ WARNING

The crankshaft keyway must be at the 11 o'clock position before reassembly. Failure to do so may lead to engine damage.

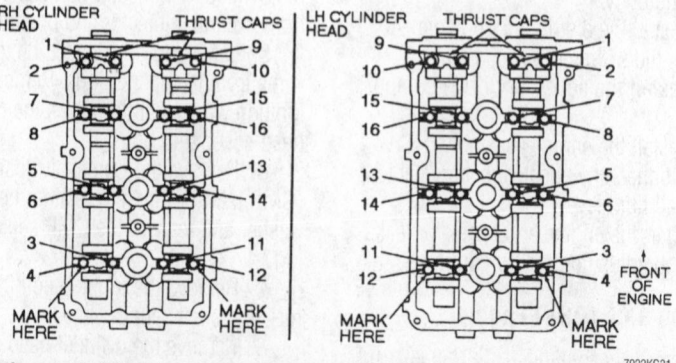

To allow the camshaft to raise off the cylinder head evenly, remove the retaining bolts in the sequence shown—1996–99 3.0L (DOHC) engine

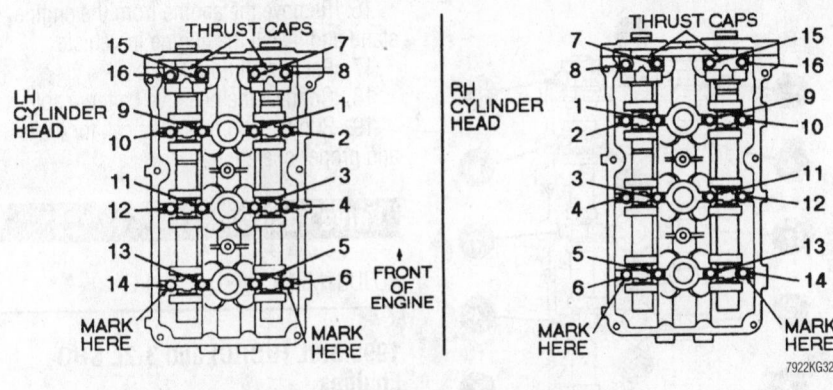

Camshaft journal bolt tightening sequence—1996–99 3.0L (DOHC) engine

10. Rotate the crankshaft so the keyway is at the 11 o'clock position for installation of the camshafts.

11. Lubricate the camshaft lobes and journals with engine assembly lubricant.

12. Install the valve lifters in their original positions.

13. Install the camshafts into their correct positions into each cylinder head with the timing marks on the camshaft sprockets aligned.

14. Loosely install the camshaft journal caps and retaining bolts into their correct positions.

➡ **Do not install the camshaft journal thrust caps until the rocker arms and timing chains have been installed and the camshaft journal caps are tightened into position.**

15. Install the timing chains.

16. Tighten the camshaft journal cap bolts, in sequence, to 71–106 inch lbs. (8–12 Nm). Install the thrust caps and tighten the retaining bolts to 71–106 inch lbs. (8–12 Nm).

17. Install the engine front cover.

18. Install both cylinder head covers, then install the upper intake manifold.

➡ **The cylinder head covers must be installed and properly tightened within six minutes of applying the silicone sealant.**

19. Install the engine assembly into the vehicle. Connect the negative battery cable.

20. Run the engine and check for leaks and proper engine operation.

3.4L Engine

❋❋ CAUTION

The fuel injection system remains under pressure, even after the engine has been turned OFF. The fuel system pressure must be relieved before dis- connecting any fuel lines. Failure to do so may result in fire and/or personal injury.

1. Disconnect the negative battery cable. Properly relieve the fuel system pressure.

2. Remove the timing chain, sprockets and front cover, as outlined earlier in this section.

3. Remove the timing chain sprocket tensioners from the cylinder heads.

4. Remove the camshaft timing chain sprocket tensioners from the cylinder heads.

5. Note the positions for reassembly and remove the cylinder head camshaft journal thrust cap retaining bolts in the proper sequence. Remove thrust caps from the cylinder heads.

➡ **The cylinder head camshaft journal caps must be installed in their original positions.**

6. Remove the timing chain and camshafts from the cylinder head.

7. Inspect the camshafts for wear and/or damage and replace, as necessary.

To install:

8. Apply a thin coat of clean engine oil to the camshaft bearing surfaces on the cylinder head and camshaft journal caps.

9. Rotate the crankshaft to position the No. 1 piston at top-dead-center TDC by aligning the crankshaft keyway groove with the oil pump mark.

10. Install the timing chain on the intake and exhaust camshafts. Match the sprocket and chain timing marks and install the camshafts into the cylinder heads with the timing marks pointing up.

11. Apply a 0.08–0.11 inch (2–3mm) bead of silicone gasket and sealant to the left side cylinder head camshaft journal cap.

➡ **Be sure the camshaft journal caps are installed in their original positions.**

12. Loosely install the cylinder head camshaft journal caps and retaining bolts.

13. Tighten the cylinder head camshaft journal caps retaining bolts in sequence and in two steps:

    a. First step: Tighten in sequence to 62–106 inch lbs. (7–12 Nm).

    b. Second step: Tighten in sequence to 12–15 ft. lbs. (16–21 Nm).

14. Install a new camshaft seal with a Cam Seal expander T89P-6256-B, or equivalent and Cam Seal Replacer T89-6256-A or equivalent.

15. Install a Cam Position Tool T96P-6256-AH, or equivalent on the left side

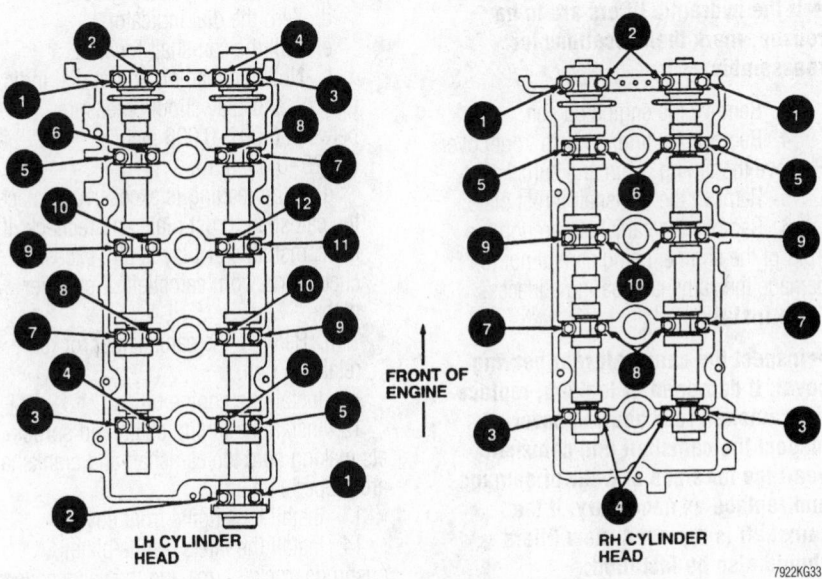

Camshaft journal thrust cap retaining bolt removal sequence—3.4L engine

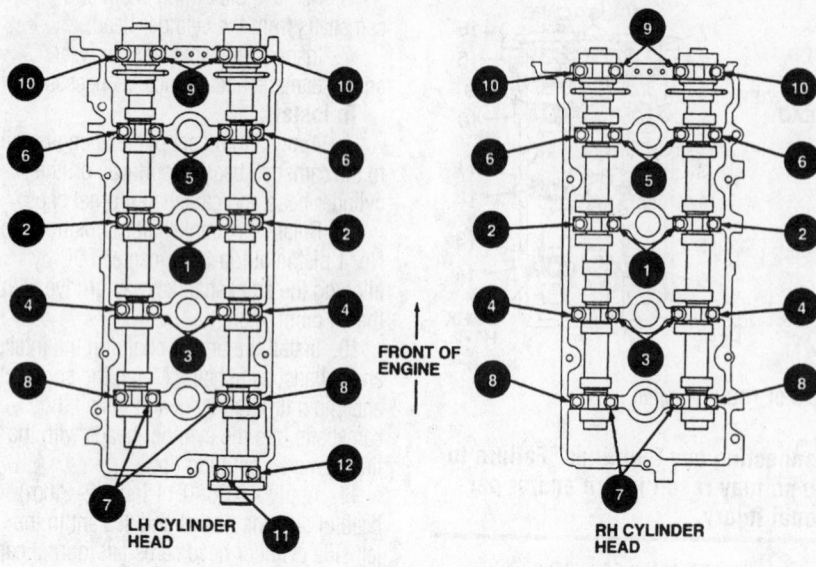

Tighten the camshaft journal caps in the sequence shown to avoid warping the camshafts—3.4L engine

intake cam and the right side exhaust cam using the existing cam chain tensioner bolts.

16. Install the timing chain, sprockets and front cover, as outlined earlier in this section.

17. Connect the negative battery cable.

3.8L Engine

1. Disconnect the negative battery cable. Remove the engine from the vehicle and place on a suitable engine stand.

2. Remove the upper plenum and the intake manifold. Remove the valve covers, rocker arms, pushrods, guide plates and lifters.

➡If the hydraulic lifters are to be reused, mark their locations for reassembly.

3. Remove the engine oil pan.

4. Remove the timing chain front cover. Remove the timing chain and sprockets.

5. Remove the camshaft thrust plate.

6. Remove the camshaft through the front of the engine, being careful not to damage the camshaft bearing surfaces.

To install:

➡Inspect the camshaft rear bearing cover. If damaged or leaking, replace the camshaft rear bearing cover. Inspect the camshaft and camshaft bearings for signs of wear or damage and replace as necessary. If the camshaft is replaced, new lifters should also be installed.

7. Lubricate the cam lobes and journals with engine assembly lubricant.

8. Install the camshaft, being careful not to damage the lobes or bearing surfaces while sliding it into position.

9. Install the camshaft thrust plate. Install and tighten the two retaining bolts to 72–120 inch lbs. (8–14 Nm).

10. Check the camshaft end-play as follows:

   a. Temporarily install the camshaft sprocket retaining bolt.

   b. Push the camshaft toward the rear of the engine.

   c. Install a dial indicator to the front of the cylinder block and position the dial indicator to rest on the face of the camshaft sprocket retaining bolt.

   d. Zero the dial indicator.

   e. Pull the camshaft forward.

   f. Note the reading on the dial indicator. The end-play should measure between 0.001–0.006 inch (0.025–0.150mm).

   g. If the reading is excessive, replace the camshaft thrust plate and recheck. If the camshaft end-play is still excessive, check for a worn camshaft or cylinder block.

   h. Remove the camshaft sprocket retaining bolt.

11. Install the engine oil pan.

12. Install the timing chain and sprockets making sure the camshaft and crankshaft are properly timed.

13. Install the engine front cover.

14. Install the lifters, guide plates, pushrods, rocker arms and the valve covers.

15. Install the intake manifold and the upper plenum.

16. Remove the engine from the engine stand and install the engine in vehicle.

17. Restore all fluid levels.

18. Connect the negative battery cable.

19. Run the engine and check for leaks and proper operation.

Valve Lash

ADJUSTMENT

1995 3.0L (DOHC) and 3.2L SHO Engines

1. Disconnect the negative battery cable. Remove the intake manifold assembly. Remove the cylinder head covers.

➡The cam lobes must be directed 90 degrees or more away from the lash adjusters (valve tappets). The engine must be COLD to check the valve clearance.

2. Insert a feeler gauge under the camshaft lobe at a 90 degree angle to the camshaft. Clearance for the intake valves should be 0.006–0.010 inch (0.15–0.25mm). Clearance for the exhaust valves should be 0.010–0.014 inch (0.25–0.35mm).

3. If no adjustments are required, install the cylinder head covers and intake manifold.

4. If adjustment is required, install Tappet Compressor T89P-6500-A or equivalent, under the camshaft next to the lobe and rotate the tool downward to depress the lash adjuster.

5. Install Tappet Holder T89P-6500-B or equivalent, and remove the compressor tool.

6. Using O-ring tool T71P-19703-C or equivalent, lift the adjusting spacer and remove the adjusting spacer with a magnet.

7. Determine the size of the adjusting spacer by the numbers on the bottom face of the spacer or by measuring with a micrometer.

8. Install the replacement adjusting spacer with the numbers down. Be sure the spacer is properly seated.

9. Release the tappet holder by installing the tappet compressor, then remove the tappet compressor.

10. Repeat the procedure for each valve requiring adjustment by rotating the crankshaft as necessary.

11. Check all adjusting spacers to ensure that they are properly seated.

12. Install the cylinder head covers. Install the intake manifold.

13. Connect the negative battery cable. Run the engine and check for proper operation.

1996–99 3.0L (DOHC) Engines

The lash adjusters (valve tappets), are hydraulic and are not adjustable. It is important that all valve components are in good condition and installed and tightened properly.

3.0L (OHV) Engines

Hydraulic valve lifters are used and no valve clearance adjustment is available. A clearance check of the rocker arm-to-valve stem gap is required when machining has been done to the cylinder heads, valves, valve seats or cylinder block head gasket surfaces or when new valvetrain components have been installed. The clearance check is also useful in determining loose, worn or damaged parts when there is a concern with the valvetrain. Clearance must be checked when the lifter is completely collapsed.

1. Disconnect the negative battery cable. Remove both rocker arm covers.

2. To check valve clearance, use Lifter Bleed Down Wrench T71P-6513-B or equivalent, to slowly push down on the pushrod end of the rocker arm and bleed the oil from the valve lifter.

3. Once the valve lifter is totally collapsed, insert the appropriate thickness feeler gauge between the rocker arm and valve stem to check the clearance. There should be 0.085–0.185 inch (2.15–4.69mm) clearance between the valve stem and rocker arm.

4. Rotate the crankshaft until No. 1 cylinder is at TDC on its compression stroke. With the engine at this position, the following valve clearances can be checked:
   a. Intake—1, 3 and 6
   b. Exhaust—1, 2 and 4

5. Rotate the crankshaft 360 degrees and check the following valves in the same manner:
   a. Intake—2, 4 and 5
   b. Exhaust—3, 5 and 6

6. If the clearance is not correct, check for loose, worn or damaged components. If no problems are found, the clearance can be adjusted by the use of shorter or longer pushrods.

3.4L Engine

1. Disconnect the negative battery cable. Remove the valve covers and intake manifold.

➡The cam lobes must be directed 90 degrees or more away from the lash adjusters (valve tappets). The engine must be COLD to check the valve clearance.

2. Insert a feeler gauge under the camshaft lobe at a 90 degree angle to the camshaft. Clearance for the intake valves should be 0.006–0.010 inch (0.15–0.25mm). Clearance for the exhaust valves should be 0.010–0.014 inch (0.25–0.35mm).

3. If no adjustments are required, install the cylinder head covers.

4. If adjustment is required, install Tappet Compressor T89P-6500-A or equivalent, under the camshaft next to the lobe and rotate the tool downward to depress the valve tappet.

5. Install Tappet Holder T96P-6500-AH or equivalent, and remove the compressor tool.

❋❋ CAUTION

Use eye protection when using compressed air to avoid possible personal injury.

6. Direct a jet of compressed air toward the hole in the face of the valve adjusting spacer to lift the valve adjusting spacer off of the valve tappet.

7. Determine the size of the adjusting spacer by the numbers on the bottom face of the spacer or by measuring with a micrometer.

8. Install the replacement adjusting spacer with the numbers down. Be sure the spacer is properly seated.

9. Release the tappet holder by installing the tappet compressor, then remove the tappet compressor.

10. Repeat the procedure for each valve requiring adjustment by rotating the crankshaft as necessary.

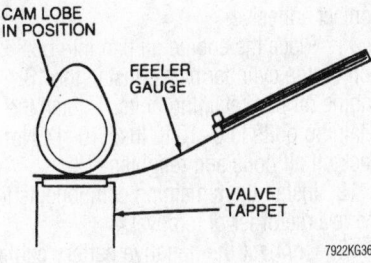

To measure the valve clearance, slide the feeler gauge between the base circle of the cam and the tappet—3.4L engine

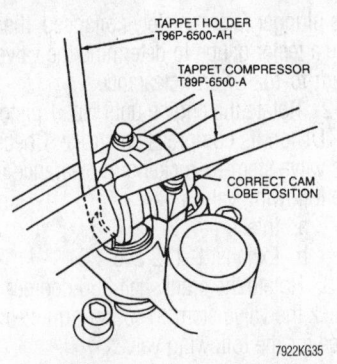

Correct positioning of the tappet compressor and holder tools—3.4L engine

11. Check all adjusting spacers to ensure that they are properly seated.

12. Install the intake manifold and valve covers.

13. Connect the negative battery cable. Run the engine and check for proper operation.

3.8L Engines

The valve stem-to-rocker arm clearance should be within specification with the valve lifter completely collapsed. If the clearance is not within specifications, check for loose, worn or damaged components and repair as necessary.

1. With the crankshaft in the designated position, install Lifter Bleeder Wrench T71P-6513-B or equivalent, on the rocker arm. Slowly apply pressure to the lifter until

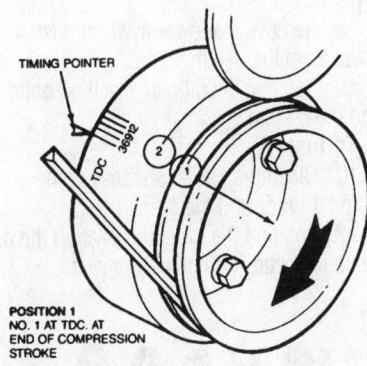

Measure the specified valve clearances with the crankshaft in the positions indicated—3.8L (OHV) engine

CYL. NO.	CRANKSHAFT POSITION	
	1	2
	SET GAP OF VALVES NOTED	
1	INT — EXH	NONE
2	EXH	INT
3	INT	EXH
4	EXH	INT
5	NONE	INT — EXH
6	INT	EXH

POSITION 1
NO. 1 AT TDC. AT
END OF COMPRESSION
STROKE

POSITION 2
ROTATE CRANKSHAFT
ONE REVOLUTION — 360
DEGREES

the plunger is completely collapsed, then use a feeler gauge to determine the valve stem-to-rocker arm clearance.

2. Rotate the engine until No. 1 piston is at TDC on its compression stroke. Check the valve stem-to-rocker arm clearance for the following valves.

    a. Intake—1, 3 and 6

    b. Exhaust—1, 2 and 4

3. Rotate the crankshaft 360 degrees and check the valve stem-to-rocker arm clearance for the following valves.

    a. Intake—2, 4 and 5

    b. Exhaust—3, 5 and 6

4. The valve stem-to-rocker arm clearance should be 0.09–0.19 inch (2.25–4.79mm) for all valves.

Oil Pan

REMOVAL & INSTALLATION

1995 3.0L (DOHC) and 3.2L SHO Engines

1. Disconnect the negative battery cable.

2. Remove the oil level dipstick.

3. Raise the vehicle and support it safely. Drain the engine oil.

4. If equipped with a low oil level sensor, remove the retainer clip and the electrical connector from the sensor.

5. Remove the starter motor.

6. Disconnect the oxygen sensors.

7. Remove the catalyst and pipe assembly.

8. Remove the lower flywheel dust cover from the bell housing.

9. Remove the oil pan attaching bolts and the oil pan.

To install:

10. Clean the gasket surfaces of the cylinder block and the oil pan.

11. Position the oil pan gasket on the oil pan and secure with silicone sealer.

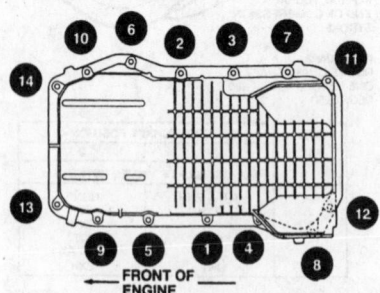

To avoid leaks, tighten the oil pan bolts according to the sequence shown—1995 3.0L (DOHC) and 3.2L SHO engines

12. Position the oil pan and tighten the retaining bolts in sequence to 11–16 ft. lbs. (15–23 Nm).

13. Install the remaining components in the reverse order of removal.

14. Connect the negative battery cable. Fill the crankcase with the proper type and quantity of oil, then start the vehicle and check for leaks.

3.0L (OHV) Engines

1. Disconnect the negative battery cable. Remove the engine oil level dipstick.

2. Raise and safely support the vehicle. Drain the engine oil.

3. If equipped with a low oil level sensor, remove the retainer clip at the sensor. Disconnect the electrical connector from the sensor.

4. Remove the starter motor and brace.

5. Detach the connector from the O_2 sensor(s).

6. Remove the catalytic converter and exhaust Y-pipe assembly.

7. Remove the engine rear plate from the torque converter housing.

8. Remove the 16 oil pan retaining bolts and carefully remove the engine oil pan from the cylinder block. Remove the oil pan gasket.

To install:

9. Clean the gasket sealing surfaces on the cylinder block and the engine oil pan. Apply a ¼ inch (6mm) bead of silicone sealer to the junction of the front cover assembly and the cylinder block and to the junction of the rear main bearing cap and cylinder block.

➡**When using a silicone sealer, the assembly process should occur within five minutes after the sealer has been applied. Be sure the sealer does not fall into the engine and form plugs that could block oil passages.**

10. Place the oil pan gasket on the engine oil pan and secure with a suitable contact adhesive.

11. Place the engine oil pan into position on the cylinder block. Install the 16 engine oil pan retaining bolts. Tighten the retaining bolts to 8–10 ft. lbs. (10–14 Nm). Back off all bolts and retighten.

12. Install the remaining components in the reverse order of removal.

13. Connect the negative battery cable. Fill the crankcase with the proper type and quantity of engine oil, then start the engine and check for leaks 'and proper operation.

1996–99 3.0L (DOHC) and 3.4L Engines

1. Disconnect the negative battery cable.

2. Raise and safely support the vehicle. Drain the engine oil.

3. Remove the dual catalytic converter Y-pipe retaining nuts from the exhaust manifolds.

4. Remove the bolt and nut retainers from the transaxle.

5. Remove the two remaining nuts and bolts from the dual converter Y-pipe connection. Remove the Y-pipe from the vehicle.

6. Reinstall the oil pan drain plug using a new gasket and tighten to 16–22 ft. lbs. (22–30 Nm).

7. Remove the oil pan retaining bolts from the transaxle housing.

8. Remove the access plug from the engine rear plate.

9. Remove the support bracket from the oil pan and transaxle.

10. Remove the oil pan retaining bolts and studs from the lower cylinder block following the correct bolt removal sequence.

11. Remove the oil pan and the oil pan gasket from the vehicle.

12. If required, remove the oil pump screen and tube assembly from the oil pump.

To install:

13. Clean the oil pan to lower cylinder block gasket sealing surfaces.

14. Thoroughly clean the oil pan and mating surfaces with soap and water and dry completely with compressed air.

15. Clean the mating surfaces with Metal Surface Cleaner F4AZ9A536-RA or equivalent to remove all the residues that may cause oil leakage.

16. If removed, install the oil pump screen and tube assembly to the oil pump using a new O-ring. Tighten the retaining bolts to 71–106 inch lbs. (8–12 Nm). Install a new self-locking nut and tighten to 71–106 inch lbs. (8–12 Nm).

17. On the 3.0L engine, install a new oil pan gasket into the groove of the oil pan. Apply silicone sealer on the gasket at the places where the engine front cover meet the engine block.

18. On the 3.4L engine, apply a 0.16 inch (4mm) bead of silicone sealer on the entire oil pan sealing surface.

19. Carefully install the oil pan with gasket to the lower cylinder block. Install the bolts and studs but do not tighten.

20. Push the oil pan against the transaxle case and tighten the oil pan bolts and studs, finger-tight.

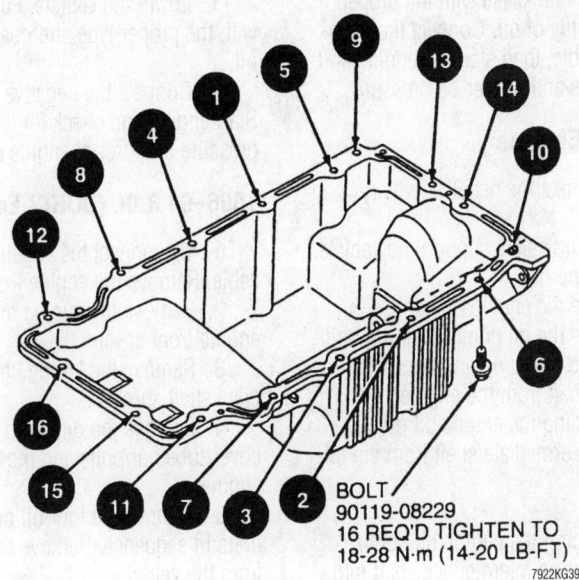

Oil pan bolt tightening sequence—1996–99 3.0L (DOHC) and 3.4L SHO engines

BOLT
90119-08229
16 REQ'D TIGHTEN TO
18-28 N·m (14-20 LB-FT)

7922KG39

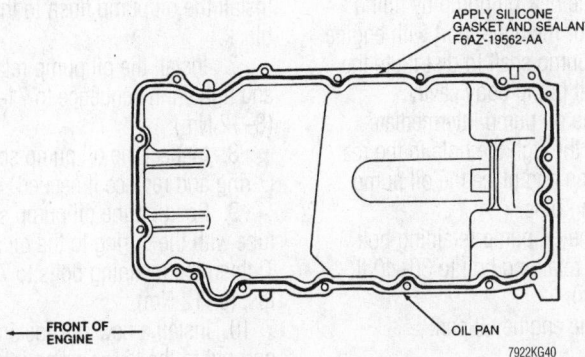

APPLY SILICONE
GASKET AND SEALANT
F6AZ-19562-AA

FRONT OF
ENGINE

OIL PAN

7922KG40

Apply silicone sealant to the entire sealing surface of the oil pan—1996–99 3.0L (DOHC) and 3.4L SHO engines

21. Tighten the oil pan retaining bolts and studs in the proper sequence to 15–22 ft. lbs. (20–30 Nm).

22. Tighten the oil pan to transaxle case bolts to 25–34 ft. lbs. (34–46 Nm).

23. Install the transaxle support bracket to the oil pan retaining stud bolts and transaxle. Tighten the retaining nuts to 71–106 inch lbs. (8–12 Nm). Tighten the retaining bolts to 15–22 ft. lbs. (20–30 Nm).

24. Install the access plug into the engine rear plate.

25. Position the Y-pipe assembly using a new flange gasket and install all the retaining nuts and bolts loosely.

26. Starting at the front of the system tighten the Y-pipe to exhaust manifold nuts to 26–34 ft. lbs. (34–46 Nm). Tighten the converter to transaxle nut and bolt to 30 ft. lbs. (40.3 Nm). Tighten the converter outlet bolts to 26–34 ft. lbs. (34–46 Nm).

27. Replace the engine oil filter.

28. Lower the vehicle.

29. Fill the crankcase with the correct amount and grade of engine oil.

✳✳ WARNING

Operating the engine without the proper amount and type of engine oil will result in severe engine damage.

30. Connect the negative battery cable. Run the engine and check for leaks and proper operation.

3.8L Engine

1. Disconnect the negative battery cable. Raise the vehicle and support safely.

2. Drain the engine oil into a suitable container and remove the oil filter element.

Install the drain plug and tighten it to 15–25 ft. lbs. (20–34 Nm) and move the drain pan aside.

3. Remove the catalytic converter Y-pipe assembly.

4. Remove the starter motor and the rear engine plate.

5. Remove the oil pan retaining bolts and remove the oil pan.

6. To remove the oil pump screen and tube, remove the two retaining bolts and the support bracket nut and remove the oil pump screen and tube and gasket.

To install:

7. If removed, clean the oil pump screen and tube mounting gasket surfaces and install a new oil pump screen and tube gasket.

8. Position the oil pump screen and tube to the mounting gasket and install the two bolts and the support bracket nut. Tighten the two retaining bolts to 15–22 ft. lbs. (20–30 Nm) and the nut to 30–40 ft. lbs. (40–55 Nm).

9. Clean the gasket surfaces on the cylinder block and the oil pan.

10. Trial fit the oil pan to cylinder block. Ensure that enough clearance has been provided to allow the oil pan to be installed without sealant being accidentally scraped off when the oil pan is positioned under the engine.

11. Be sure there is no engine oil on the gasket mating surfaces. Apply a bead of silicone sealer to the oil pan flange. Also apply a bead of sealer to the front cover/cylinder block joint and fill the grooves on both sides of the rear main seal cap.

➡ **When using silicone rubber sealer, assembly must occur within 15 minutes after sealer application. After this time, the sealer may start to harden and its sealing effectiveness may be reduced.**

12. Install the oil pan and secure to the block with the retaining bolts. Tighten the 18 retaining bolts to 84–108 inch lbs. (9–12 Nm).

13. Install a new oil filter element.

14. Install the rear engine cover and the starter motor.

15. Install the catalytic converter Y-pipe assembly.

16. Lower the vehicle.

17. Fill the crankcase with the correct grade and amount of engine oil.

18. Connect the negative battery cable. Run the engine and check for oil leaks.

Oil Pump

REMOVAL & INSTALLATION

1995 3.0L (DOHC) and 3.2L SHO Engines

1. Disconnect the negative battery cable. Raise and safely support the vehicle.
2. Drain the crankcase into a suitable container and remove the oil pan. Remove accessory drive belt.
3. Remove the timing belt from the engine.
4. Remove the crankshaft sprocket.
5. Remove the oil pump pick-up tube and screen retaining bolts. Remove the cover and tube.
6. Remove the oil pump to block bolts and remove the pump.

To install:
7. Align the oil pump on the crankshaft and install the oil pump retaining bolts. Tighten the bolts to 11–17 ft. lbs. (15–23 Nm).
8. Install the oil pump pick-up tube and screen and tighten the retaining bolts to 72–96 inch lbs. (7–11 Nm).
9. Install the retaining components in the reverse order of removal.

10. Fill the crankcase with the proper type and quantity of oil. Connect the negative battery cable, then start the engine and check for leaks and proper oil pressure.

3.0L (OHV) Engines

1. Disconnect the negative battery cable.
2. Raise and safely support the vehicle. Drain the engine oil.
3. Remove the engine oil pan.
4. Remove the oil pump retaining bolt and remove the oil pump and the oil pump intermediate shaft from the engine.
5. If replacing the engine oil pump, separate the intermediate shaft from the oil pump.

To install:
6. If replacing the engine oil pump, insert the oil pump intermediate shaft into the new oil pump assembly until the intermediate shaft retaining ring clicks into place.
7. Prime the new oil pump by filling either the inlet or the outlet port with engine oil. Rotate the pump shaft to distribute the oil within the oil pump body cavity.
8. Insert the oil pump intermediate shaft assembly through the hole in the rear main bearing cap and place the oil pump onto the locating pins.
9. Install the oil pump retaining bolt and tighten the retaining bolt to 30–40 ft. lbs. (40–55 Nm).
10. Install the engine oil pan.

11. Lower the vehicle. Fill the crankcase with the proper type and quantity of engine oil.
12. Connect the negative battery cable. Start engine and check for leaks, proper oil pressure and proper engine operation.

1996–99 3.0L (DOHC) Engine

1. Disconnect the negative battery cable. Remove the engine from the vehicle.
2. Remove the engine oil pan and the engine front cover.
3. Remove the timing chains and the crankshaft sprockets.
4. Remove the oil pump screen cover/tube nut/bolts and the tube from the engine.
5. Remove the four oil pump retaining bolts in sequence. Remove the oil pump from the vehicle.

To install:
6. Rotate the inner rotor of the oil pump to align with the flats on the crankshaft. Install the oil pump flush to the cylinder block.
7. Install the oil pump retaining bolts and tighten in sequence to 71–106 inch lbs. (8–12 Nm).
8. Inspect the oil pump screen and tube O-ring and replace if needed.
9. Position the oil pump screen and tube with the O-ring to the oil pump. Tighten the retaining bolts to 71–106 inch lbs. (8–12 Nm).
10. Install a new self-locking tube support nut to the lower cylinder block stud. Tighten the nut to 15–22 ft. lbs. (20–30 Nm).
11. Install the crankshaft sprockets and the timing chains.
12. Install the oil pan and the engine front cover.
13. Install the engine in the vehicle. Fill the crankcase with the correct amount of engine oil.

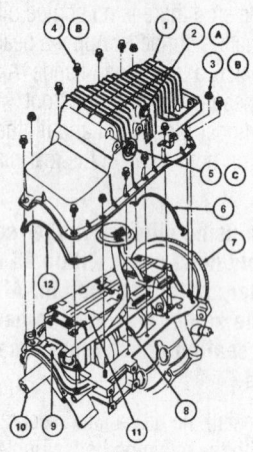

Item	Part Number	Description
1	6675	Oil Pan
2A	6730	Oil Pan Drain Plug
3B	9S702-08500	Nut (4 Req'd)
4B	97522-08525	Bolt (10 Req'd)
5C	6C824	Low Oil Level Sensor
6	6723	Oil Pan Rear Seal
7	6375	Flywheel
8	6622	Oil Pump Screen Cover and Tube
9	6600	Oil Pump
10	6303	Crankshaft
11	6687	Oil Pan Baffle
12	6722	Oil Pan Seal—Front
A		Tighten to 20-33 N·m (15-24 Lb-Ft)
B		Tighten to 15-23 N·m (11-17 Lb-Ft)
C		Tighten to 21-33 N·m (15-24 Lb-Ft)

7922KG41

Exploded view of the oil pan, oil pump pick-up tube and screen mounting—1995 3.0L (DOHC) and 3.4L SHO engines

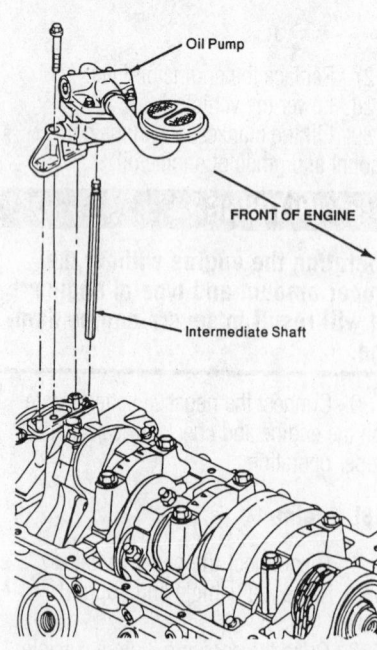

7922KG42

Exploded view of the oil pump mounting—3.0L (OHV) engines

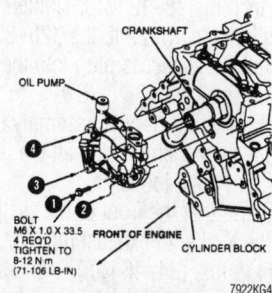

7922KG43

Tighten the oil pump retaining bolts in the proper sequence to ensure sealing to the engine block—1996–99 3.0L (DOHC) engine

14. Connect the negative battery cable. Run the engine and check for leaks and proper operation.

3.4L Engine

1. Disconnect the negative battery cable. Remove the engine from the vehicle.
2. Remove the engine oil pan and the engine front cover.
3. Remove the timing chains and the crankshaft sprockets.
4. Remove the oil pump screen cover/tube nut/bolts and the tube from the engine.
5. Remove the four oil pump retaining bolts in sequence. Remove the oil pump from the vehicle.
   To install:
6. Rotate the inner rotor of the oil pump to align with the flats on the crankshaft. Install the oil pump flush to the cylinder block.
7. Install the oil pump retaining bolts and tighten in sequence to 80–115 inch lbs. (9–13 Nm).
8. Inspect the oil pump screen and tube O-ring and replace if needed.
9. Position the oil pump screen and tube with the O-ring to the oil pump. Tighten the retaining bolts to 71–123 inch lbs. (8–14 Nm).
10. Install the crankshaft sprockets and the timing chains.
11. Install the oil pan and the engine front cover.
12. Install the engine in the vehicle. Fill the crankcase with the correct amount of engine oil.
13. Connect the negative battery cable. Run the engine and check for leaks and proper operation.

3.8L Engine

1. Disconnect the negative battery cable.
2. Raise and safely support the vehicle. Drain the oil, then remove the filter.
3. Remove the oil pump cover-to-timing chain front cover bolts and remove the oil pump cover.
4. Remove the oil pump gears.
5. Inspect the gears, oil pump cover and timing chain front cover for wear and/or damage.
   To install:
6. If reusing the oil pump cover, clean the gasket contact surface. Place a straight-edge across the oil pump cover mounting surface and check for wear or warpage using a feeler gauge. If the surface is out of flat by more than 0.0016 in. (0.04mm), replace the cover.

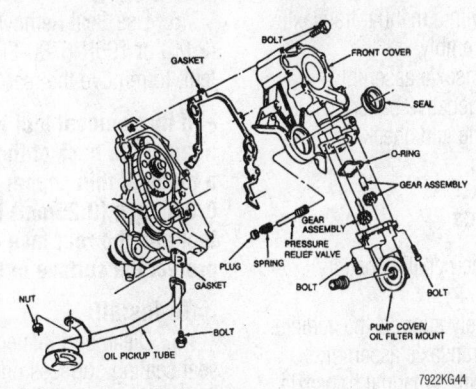

Exploded view of the oil pump and timing chain cover mounting—3.8L engine

7. Lightly pack the gear pocket with petroleum jelly or coat all pump gear surfaces with oil conditioner.
8. Install the gears in the pocket. Make certain the petroleum jelly fills the gap between the gears and the pocket.

✳✳ WARNING

Failure to properly coat the oil pump gears may result in failure of the pump to prime when the engine is started.

9. Position the oil pump cover gasket and install the oil pump cover. Tighten the oil pump cover retaining bolts to 18–22 ft. lbs. (25–30 Nm).
10. Connect the negative battery cable. Fill the crankcase with the proper type and quantity of engine oil.
11. Start the engine and check for leaks and proper oil pressure.
12. Check the ignition timing and curb idle speed, adjust as required.
13. Install the air cleaner assembly and air intake duct.

Rear Main Seal

REMOVAL & INSTALLATION

1995 3.0L (DOHC) and 3.2L SHO Engines

1. Disconnect the negative battery cable.
2. Raise and safely support the vehicle.
3. Remove the transaxle assembly.
4. If equipped with a manual transaxle, remove the clutch assembly.
5. Remove the flywheel.

➡**Use caution to avoid damaging the crankshaft seal surface.**

6. Using a sharp awl or equivalent, punch a hole in the rear crankshaft seal

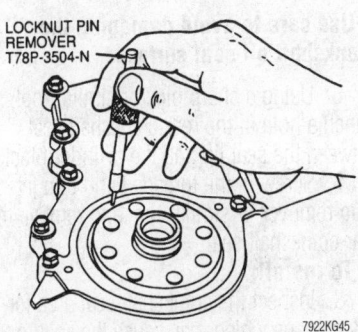

Screw the tool into the awl hole and remove the seal—1995 3.0L (DOHC) and 3.2L SHO engines

metal surface between the rear crankshaft seal lip and the rear crankshaft seal retainer.

7. Screw in the threaded end of Locknut Pin Remover T78P-3504-N or equivalent, into the rear crankshaft seal and remove the seal from the rear crankshaft seal retainer.
8. If needed, remove the retaining bolts from the rear crankshaft seal cover and remove the cover and gasket.
   To install:
9. Thoroughly clean the seal cover and cylinder block sealing surfaces.
10. Inspect the crankshaft seal area on the crankshaft for any damage which may cause the seal to leak. If damage is evident, service or replace parts as necessary.
11. If removed, install the rear crankshaft seal cover with a new gasket to the cylinder block. Install the retaining bolts and tighten to 56–83 inch lbs. (6.3–9.4 Nm).
12. Coat the rear crankshaft seal area and the seal lip with engine assembly lubricant.
13. Install the rear crankshaft seal using Seal Replacers T81P-6701-A and T88C-6701-BH with Screw Set T89P-6701-C, or equivalents.
14. Install the flywheel. Tighten the bolts in sequence to 29–43 ft. lbs. (39–50 Nm), then in sequence again to 51–58 ft. lbs. (69–78 Nm).

15. If equipped with a manual transaxle, install the clutch assembly.

16. Install the transaxle assembly.

17. Connect the negative battery cable.

18. Run the engine and check for leaks and proper operation.

3.0L (OHV) Engines

1. Disconnect the negative battery cable.

2. Raise and safely support the vehicle.

3. Remove the transaxle assembly.

4. If equipped with a manual transaxle, remove the clutch assembly.

5. Remove the flywheel.

➡**Use care to avoid damaging the crankshaft oil seal surface.**

6. Using a sharp pick or similar tool, punch a hole in the rear crankshaft seal between the seal lip and the cylinder block.

7. Screw in the threaded end of a jet plug remover or similar tool and remove the rear crankshaft seal.

To install:

8. Inspect the crankshaft seal area for any damage which may cause the seal to leak. If damage is evident, service or replace the crankshaft as necessary.

9. Coat the crankshaft seal area and the seal lip with clean engine oil.

10. Place the rear crankshaft seal on Seal Replacer T88L-6701-A or equivalent, and position the tool and seal to the rear of the cylinder block with three bolts. Alternately tighten the bolts to properly seat the seal. The seal must be flush or within 0.020 inch (0.50mm) of the cylinder block surface. Do not bottom the seal.

11. Install the flywheel. Tighten the retaining bolts to 54–64 ft. lbs. (73–87 Nm).

12. If equipped with a manual transaxle, install the clutch assembly.

13. Install the transaxle assembly.

14. Lower the vehicle.

15. Connect the negative battery cable.

16. Run the engine and check for leaks.

17. Road test the vehicle and check for proper engine and transaxle operation.

1996–99 3.0L (DOHC) and 3.4L SHO Engines

1. Disconnect the negative battery cable.
2. Raise and safely support the vehicle.
3. Remove the transaxle assembly.
4. Remove the flywheel.

➡**Use care not to scratch the crankshaft or the crankshaft retainer oil sealing surfaces when removing the rear crankshaft seal.**

5. Use Seal Remover T92C-6700-CH (3.0L) or T95P-6700-EH (3.4L) or equivalent, to remove the rear crankshaft seal.

➡**If the removal tool is going to rest against the back of the crankshaft, place a piece of thin copper or other soft metal 0.010 inch (0.25mm) thick, between the tool and the rear face of the crankshaft to protect the surface of the crankshaft.**

To install:

6. Clean and inspect the rear crankshaft seal sealing surfaces on the crankshaft and the cylinder block.

7. Lubricate the crankshaft flange and the rear crankshaft seal bore with engine assembly lubricant.

8. Install the rear main oil seal using Rear Seal Replacer T82L-6701-A and adapter T91P-6701-A or equivalents.

9. Alternate tightening the bolts to pull the rear crankshaft seal in evenly. Seat the rear crankshaft seal flush to the rear of the cylinder block.

10. Install the flywheel. Install the flywheel retaining bolts and tighten in a standard sequence to 54–64 ft. lbs. (73–87 Nm).

11. Install the transaxle.

12. Connect the negative battery cable.

13. Run the engine and check for leaks and proper operation.

3.8L Engines

1. Disconnect the negative battery cable.
2. Remove the transmission assembly.
3. If equipped with manual transmission, remove the clutch assembly.
4. Remove the flywheel.
5. Using a sharp awl or similar tool, carefully punch a hole in the rear crankshaft seal between the seal lip and the cylinder block. Install a sheet metal screw and pry the seal out or use a jet plug remover to remove the rear crankshaft seal from the cylinder block.

➡**Use extreme caution not to scratch the rear crankshaft seal surface on the crankshaft journal.**

To install:

6. Clean the rear crankshaft seal recess in the cylinder block and main bearing cap.

7. Coat the new seal and all of the seal mounting surfaces with oil. Place the seal on Rear Main Seal Installer T82L-6701-A or equivalent, and position the tool and seal to the rear of the cylinder block.

8. Alternately tighten the bolts to seat the seal properly. The rear face of the seal must be within 0.005 inch (0.127mm) of the rear face of the cylinder block.

9. Install the flywheel.

10. Install the clutch assembly, if equipped with a manual transmission.

11. Install the transmission.

12. Connect the negative battery cable.

13. Run the engine and check for leaks and proper operation.

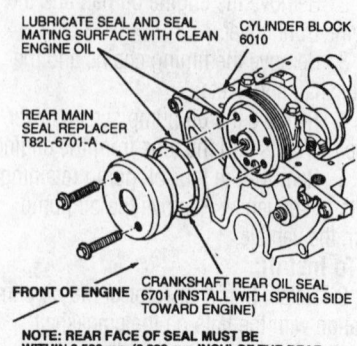

Rear main seal installation—3.8L engine

Timing Chain, Sprockets and Front Cover and Seal

REMOVAL & INSTALLATION

3.0L (OHV) Engines

1. Disconnect the negative battery cable, then drain the engine cooling system.

2. Loosen four water pump pulley bolts while the accessory drive belt is in place.

3. Remove the accessory drive belts. Remove the idler pulley or automatic tensioner, as necessary.

4. Remove the lower radiator hose and the heater hose from the water pump and front cover.

5. Remove the crankshaft pulley and damper.

6. On flexible fuel vehicles, remove the Crankshaft Position (CKP) sensor.

7. Drain the engine oil and remove the oil pan.

8. If necessary, unfasten the water pump pulley retaining bolts, then remove the pulley.

9. Remove the retaining bolts from the timing cover to the cylinder block and remove the timing cover.

10. Tap the seal out of the cover with a seal driver.

11. Remove the crankshaft damper and timing chain front cover.

12. Rotate the crankshaft until the No. 1 piston is at TDC on its compression stroke and the timing marks are aligned.

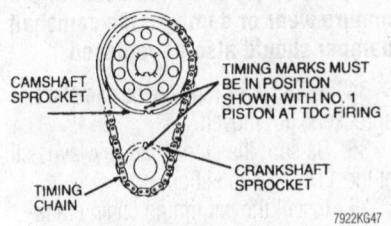

CAMSHAFT SPROCKET

TIMING MARKS MUST BE IN POSITION SHOWN WITH NO. 1 PISTON AT TDC FIRING

TIMING CHAIN

CRANKSHAFT SPROCKET

7922KG47

Be sure the timing marks are facing each other after the chain has been installed—3.0L (OHV) engines

13. Remove the camshaft sprocket attaching bolt and washer. Slide both sprockets and timing chain forward and remove as an assembly.

14. Check the timing chain and sprockets for excessive wear. Replace if necessary.

To install:

15. Before installation, clean and inspect all parts. Clean the gasket material and dirt from the oil pan, cylinder block and front cover.

16. Slide both sprockets and timing chain onto the camshaft and crankshaft with the timing marks aligned. Install the camshaft bolt and washer and tighten to 46 ft. lbs. (63 Nm). Apply clean engine oil to the timing chain and sprockets after installation.

➡**The camshaft bolt has a drilled oil passage in it for timing chain lubrication. Prior to installation, clean the passage and be sure it is clear. Never replace the camshaft bolt with a standard bolt.**

17. Lightly oil all bolt and stud threads except bolts 1, 2 and 3 that require a suitable pipe sealant.

18. Install a new seal in the timing cover.

19. Install a new timing cover gasket over the cylinder block dowels.

20. Install the timing cover/water pump assembly onto the cylinder block with the water pump pulley loosely attached to the water pump hub.

21. Apply a non-hardening sealant to bolt numbers 1, 2 and 3 and hand start them along with the rest of the cover retaining bolts. Tighten bolts 1–10 to 19 ft. lbs. (25 Nm) and bolts 11–15 to 84 inch lbs. (10 Nm).

22. Install the engine oil pan. Tighten the retaining bolts to 108 inch lbs. (12 Nm).

23. Hand-tighten the water pump pulley retaining bolts.

24. Install the crankshaft damper and pulley. Tighten the damper retaining bolt to 107 ft. lbs. (145 Nm).

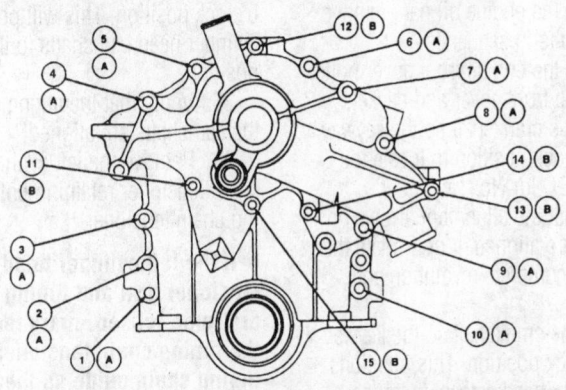

Fastener And Hole No.	Fasteners		Torque Specifications	
	Size	Fastener Application	N·m	LB-FT
1A	M8 x 1.25 x 43.5	F/C TO BLOCK	20-30	15-22
2A	M8 x 1.25 x 43.5	F/C TO BLOCK	20-30	15-22
3A	M8 x 1.25 x 73	W/P & F/C TO BLOCK	20-30	15-22
4A	M8 x 1.25 x 104 3	W/P & F/C TO BLOCK	20-30	15-22
5A	M8 x 1.25 x 73	F/C TO BLOCK	20-30	15-22
6A	M8 x 1.25 x 73	W/P & F/C TO BLOCK	20-30	15-22
7A	M8 x 1.25 x 73	W/P & F/C TO BLOCK	20-30	15-22
8A	M8 x 1.25 x 104 3	W/P & F/C TO BLOCK	20-30	15-22
9A	M8 x 1.25 x 104.3	W/P & F/C TO BLOCK	20-30	15-22
10A	M8 x 1.25 x 52	F/C TO BLOCK	20-30	15-22
11B	M6 x 1 x 28.5	W/P TO F/C	8-12	71-106 (lb-in)
12B	M6 x 1 x 28.5	W/P TO F/C	8-12	71-106 (lb-in)
13B	M6 x 1 x 28 5	W/P TO F/C	8-12	71-106 (lb-in)
14B	M6 x 1 x 28 5	W/P TO F/C	8-12	71-106 (lb-in)
15B	M6 x 1 x 28 5	W/P TO F/C	8-12	71-106 (lb-in)

W/P—Water Pump
F/C—Engine Front Cover

7922KG48

Timing chain front cover bolt location and identification—3.0L (OHV) engines

25. On flexible fuel vehicles, install the CKP sensor. Tighten the retaining bolt to 44–61 inch lbs. (5–7 Nm).

26. Install the automatic belt tensioner or idler pulley, as necessary.

27. Install the water pump and accessory drive belts. Tighten the water pump pulley retaining bolts to 16 ft. lbs. (21 Nm).

28. Install the lower radiator hose and the heater hose and tighten the clamps.

29. Fill the crankcase with the correct amount and type of engine oil. Fill and bleed the engine cooling system.

30. Connect the negative battery cable. Start the engine and check for coolant and oil leaks. Road test the vehicle and check for proper operation.

1996–99 3.0L (DOHC) Engine

1. Disconnect the negative battery cable.

2. Remove the engine from vehicle and position on a suitable workstand.

3. Remove the upper and lower intake manifolds.

4. Remove the cylinder head covers, then remove the drive belt.

5. Remove the wire connector from the water temperature sender.

6. Remove the heater water hose from the bypass tube.

7. Remove the bypass tube bolts and the tube.

8. Remove the power steering pump bolts and the pump.

9. Remove the A/C compressor bracket to water pump brace, the mounting bolts, the compressor and bracket.

➡**The A/C mounting bracket bolts are torque-to-yield bolts and must be replaced when removed.**

10. Remove the alternator and the water pump.

11. Install Flywheel Holding tool T74P-6375-A or equivalent to keep the crankshaft from rotating.

12. Remove the crankshaft accessory pulley and bracket.

➡**Rotate the pulley shaft clockwise to remove. Use a 24mm open end wrench on the inside nut to remove the pulley.**

13. Remove the crankshaft pulley bolt and washer.

14. Install a suitable damper puller and remove the crankshaft damper from the crankshaft.

15. Remove the flywheel holding tool.

16. Detach the wiring from the Crankshaft Position (CKP) sensor and the Camshaft Position (CMP) sensor connectors.

17. Remove the engine oil pan, remove the retaining bolts in sequence.

18. Remove the engine front cover bolts, in sequence, the front cover and gasket.

19. Rotate the crankshaft so the keyway is at the 11 o'clock position to locate the crankshaft at TDC for No. 1 cylinder.

20. Verify that the alignment arrows on the camshafts are aligned. If not, rotate the crankshaft one complete revolution and recheck.

21. Rotate the crankshaft so the keyway is at the 3 o'clock position. This positions the right cylinder head camshafts to the neutral position.

22. Remove the right cylinder head timing chain tensioner bolts and tensioner.

23. Remove the right cylinder head timing chain tensioner arm and the timing chain.

24. Remove the right cylinder head timing chain guide bolts and guide; if worn, replace the timing chain guide.

25. Remove the right crankshaft timing chain sprocket.

26. Remove the right cylinder head camshaft timing chain sprockets, if being replaced.

27. Rotate the crankshaft two revolutions and locate the crankshaft keyway at the 11

o'clock position. This will position the left cylinder head camshafts to their neutral position.

28. Verify that the alignment arrows on the camshafts are aligned.

29. Remove the left cylinder head timing chain tensioner retaining bolts and the timing chain tensioner.

➡ **If the left cylinder head timing chain tensioner arm and timing chain guide are to be reused, mark the position of the timing chain tensioner arm and the timing chain guide so they are reassembled into their original positions.**

30. Remove the left cylinder head timing chain tensioner arm and the timing chain.

31. Remove the left cylinder head timing chain guide bolts and guide; if worn, replace the timing chain guide.

32. Remove the left crankshaft timing chain sprocket.

33. If required, the left cylinder head camshaft sprockets may be removed at this time.

To install:

➡ **Inspect the timing chains, tensioners, tensioner arms, guides and sprockets for wear or damage. If any**

components are to be replaced for premature wear or damage, the camshaft damper should also be replaced.

34. Reinstall or replace the camshaft sprockets, if removed.

35. Be sure the crankshaft keyway is still at the 11 o'clock position.

36. Install the left timing chain crankshaft sprocket onto the crankshaft.

37. Install the left timing chain guide and retaining bolts to the engine. Tighten the retaining bolts to 15–22 ft. lbs. (20–30 Nm).

38. Verify that the arrows on the left cylinder head camshafts are aligned.

39. Install the left timing chain over the left crankshaft sprocket and the left camshaft sprockets.

40. Align the timing marks on the left cylinder head timing chain with the timing marks on the crankshaft sprocket and the camshaft sprockets.

41. Install the left timing chain tensioner arm over the alignment dowel on the left cylinder head.

➡ **Before installing the timing chain tensioner, it must be properly compressed and locked.**

42. Using a small prybar, release the timing chain tensioner ratchet/pawl mechanism through the access hole in the timing chain tensioner as follows:

a. Insert a small piece of wire into the top of the piston and gently unseat the oil check ball.

b. Compress the timing chain tensioner by hand.

c. With the tensioner compressed, install a 0.060 inch (1.5mm) drill bit or wire into the small hole above the ratchet, engaging the lock groove in the rack of the timing chain tensioner.

43. Install the compressed and locked left cylinder head timing chain tensioner and retaining bolts onto the cylinder block.

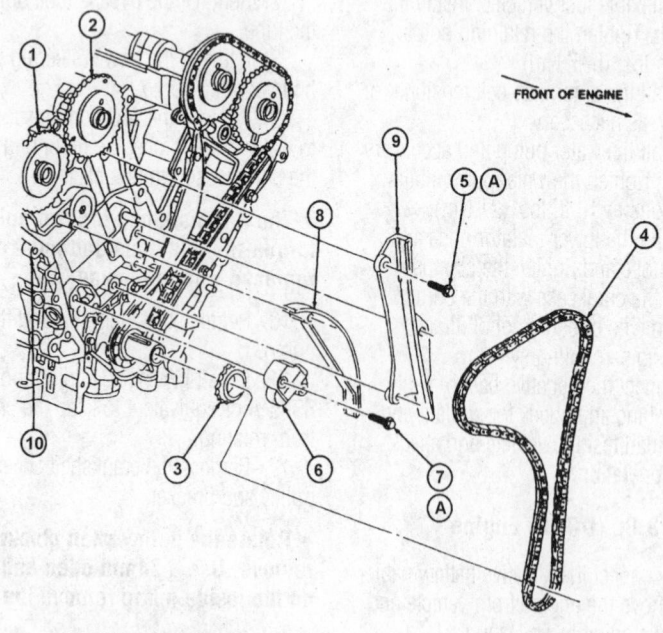

1	RH Exhaust Camshaft
2	RH Intake Camshaft
3	RH Timing Chain Crankshaft Sprocket
4	RH Timing Chain
5	Bolt (2 Req'd)
6	Timing Chain Tensioner
7	Bolt (2 Req'd)
8	Timing Chain Tensioner Arm
9	Timing Chain Guide
10	RH Cylinder Head
A	Tighten to 20-30 N-m (15-22 Lb-Ft)

7922KG49

Exploded view of the right cylinder head timing chain and related components—1996–99 3.0L (DOHC) engine—left side similar

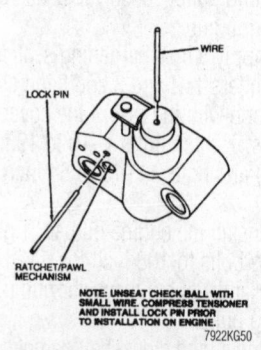

NOTE: UNSEAT CHECK BALL WITH SMALL WIRE, COMPRESS TENSIONER AND INSTALL LOCK PIN PRIOR TO INSTALLATION ON ENGINE.

7922KG50

Be sure to compress the tensioner before installing it on the engine—1996–99 3.0L (DOHC) engine

Tighten the retaining bolts to 15–22 ft. lbs. (20–30 Nm).

44. Verify that the timing marks on the left timing chain are in alignment with the timing marks on the crankshaft sprocket and the camshaft sprockets.

45. Reinstall or replace the camshaft sprockets, if removed.

46. Install the right timing chain crankshaft sprocket onto the crankshaft.

47. Install the right timing chain guide and retaining bolts to the engine. Tighten the retaining bolts to 15–22 ft. lbs. (20–30 Nm).

48. Verify that the arrows on the right cylinder head camshafts are aligned.

49. Install the right timing chain over the right crankshaft sprocket and the right camshaft sprockets.

50. Align the timing marks on the right cylinder head timing chain with the timing marks on the crankshaft sprocket and the camshaft sprockets.

51. Install the right timing chain tensioner arm over the alignment dowel on the right cylinder head.

➡ **Before installing the timing chain tensioner, it must be properly compressed and locked.**

52. Using a small prybar, release the timing chain tensioner ratchet/pawl mechanism through the access hole in the timing chain tensioner as follows:

   a. Insert a small wire into the top of the piston and gently unseat the oil check ball.

   b. Compress the timing chain tensioner by hand.

   c. With the tensioner compressed, install a 0.060 inch (1.5mm) drill bit or wire into the small hole above the ratchet, engaging the lock groove in the rack of the timing chain tensioner.

53. Install the compressed and locked right cylinder head timing chain tensioner and retaining bolts onto the cylinder block. Tighten the retaining bolts to 15–22 ft. lbs. (20–30 Nm).

54. Verify that the timing marks on the right timing chain are aligned with the timing marks on the crankshaft sprocket and the camshaft sprockets.

55. Be sure the crankshaft keyway is at the 11 o'clock position.

56. Rotate the crankshaft two revolutions and position the crankshaft keyway to the 3 o'clock location. This will position the right cylinder head camshafts to the neutral position.

57. Remove the lock pins from the timing chain tensioners.

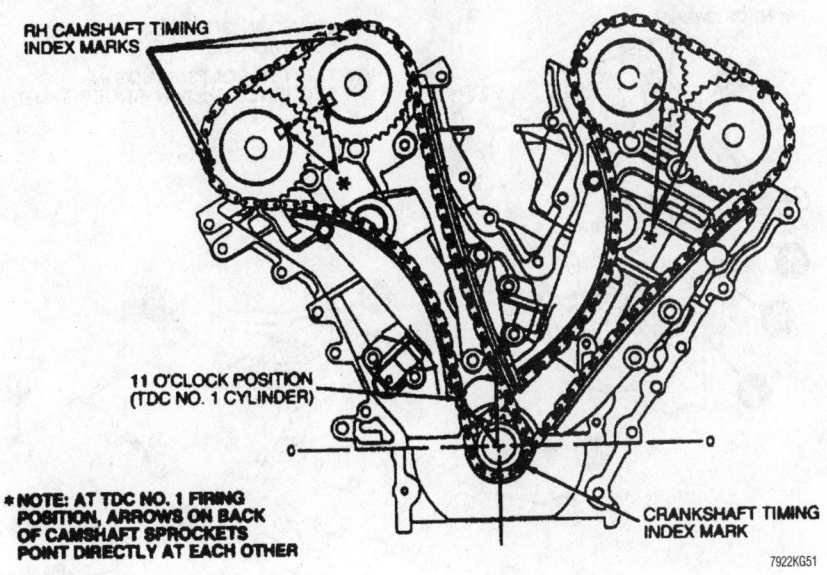

Be sure the timing marks are as shown after the chain has been installed—1996–99 3.0L (DOHC) engine

58. Verify that the timing marks on the timing chains are aligned with the timing marks on the crankshaft sprocket and the camshaft sprockets.

59. Clean the engine front and the front cover to cylinder block gasket sealing surfaces.

➡ **The front cover must be installed and properly tightened within six minutes of the application of the sealer.**

60. Apply silicone sealer to the six critical areas of the cylinder block to prevent oil seepage.

61. Place new front cover gaskets onto the dowel pins on the cylinder block and heads.

62. Place the front cover into position by placing the front cover onto the dowel pins at the cylinder block.

63. Install the six front cover retaining bolts and stud bolts where the silicone sealer was applied.

64. Tighten the bolts and stud bolts until the front cover contacts the cylinder block and heads;, then, turn the bolts and stud bolts an additional ¼ turn.

65. Install the remaining front cover retaining bolts and stud bolts. Tighten all of the front cover retaining bolts and stud bolts, in sequence, to 15–22 ft. lbs. (20–30 Nm).

66. If removed, install the belt tensioner onto the right side of the engine front cover. Tighten the bolt to 15–22 ft. lbs. (20–30 Nm).

67. Install a new oil pan gasket and apply a thin bead of silicone sealer where the timing cover meets the cylinder block.

68. Install the oil pan and install the retaining bolts loosely.

69. With the oil pan aligned to the rear of the cylinder block, tighten the oil pan bolts and studs, in sequence, to 15–22 ft. lbs. (20–30 Nm) no more than six minutes after applying the silicone sealer.

70. Connect the wiring to the CKP and CMP sensors.

71. Install Flywheel Holding tool T74P-6375-A or equivalent to keep the crankshaft from rotating.

72. Install the crankshaft damper using Crankshaft Damper Replacer T74P-6316-B, or equivalent.

73. Apply a thin coating of silicone sealer to the front of the crankshaft on the inside diameter of the damper at the keyway before installing the washer and retaining bolt.

74. Install the crankshaft damper pulley washer and bolt and tighten the bolt as follows:

   a. Tighten the bolt to 89 ft. lbs. (120 Nm) and loosen the bolt.

   b. Retighten the bolt to 39 ft. lbs. (53 Nm).

   c. Turn the retaining bolt an additional 90 degrees.

75. Install the accessory drive crankshaft pulley and bracket. Using a 34mm socket tighten the pulley counterclockwise to 70–77 ft. lbs. (95–105 Nm), tighten the bracket nuts to 15–22 ft. lbs. (20–30 Nm). Remove the flywheel holding tool.

76. Install the water pump and retaining nuts and tighten to 15–22 ft. lbs. (20–30 Nm).

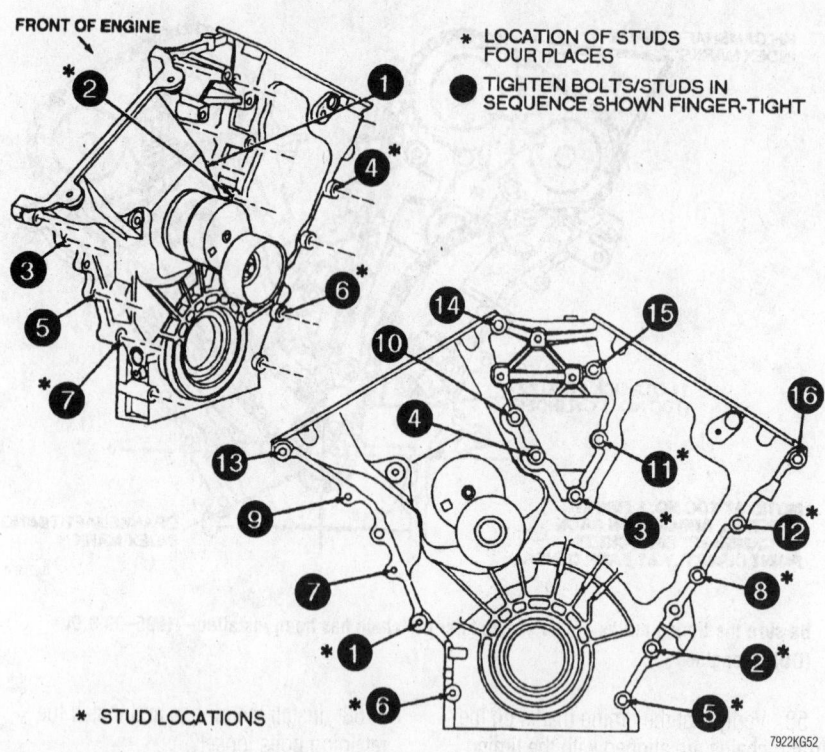

FRONT OF ENGINE

* LOCATION OF STUDS FOUR PLACES

● TIGHTEN BOLTS/STUDS IN SEQUENCE SHOWN FINGER-TIGHT

* STUD LOCATIONS

7922KG52

To prevent leaks, tighten the front cover bolts in the sequence shown—1996–99 3.0L (DOHC) engine

77. Install the alternator and the retaining bolts. Tighten the bolts to 15–22 ft. lbs. (20–30 Nm). Connect the wiring connector.

✳✳ WARNING

The A/C compressor mounting bracket must engage in the front cover dowels or an engine vibration may occur. New torque-to-yield retaining bolts must be used.

78. Install the A/C compressor mounting bracket. Tighten the bolts to 18 ft. lbs. (25 Nm), then, tighten an additional 90 degrees.

79. Install the A/C compressor and retaining bolts. Tighten the bolts to 15–22 ft. lbs. (20–30 Nm).

80. Install the A/C compressor mounting bracket to the water pump brace and tighten the nuts to 15–22 ft. lbs. (20–30 Nm).

81. Replace the water bypass tube O-ring and lubricate with clean coolant. Install the bypass tube onto the right cylinder head and install the stud and bolt. Tighten the stud and bolt to 71–106 inch lbs. (8–12 Nm.

82. Install the heater water hose and position the hose clamp. Install the wire connector to the water temperature sender.

83. Install the power steering pump and tighten the bolts to 71–106 inch lbs. (8–12 Nm).

84. Install both cylinder head covers.

➡ **The cylinder head covers must be installed and properly tightened within six minutes of applying the silicone sealant.**

85. Install the lower and upper intake manifolds.

86. Release the accessory drive belt tension by rotating the tensioner clockwise and installing the drive belt.

87. Install the engine assembly into the vehicle. Restore all fluid levels. Connect the negative battery cable.

88. Run the engine and check for leaks. Road test the vehicle and check for proper operation.

3.4L Engine

1. Disconnect the negative battery cable.

2. Remove the engine from vehicle and position on a suitable workstand.

3. Release the accessory drive belt tension by rotating the tensioner counterclockwise.

4. Remove the A/C compressor mounting bolts from the cylinder blocks and remove the A/C compressor.

5. Release the water pump drive belt tension by rotating the tensioner clockwise and remove the belt.

6. Remove the three water pump drive pulley retaining bolts from the left side camshaft.

7. Disconnect the main emission vacuum supply hose from the surge tank vacuum fitting. secondary air injection diverter valve and EGR valve.

8. Disconnect the engine control sensor wiring from the EGR backpressure transducer, EGR vacuum regulator solenoid and secondary air injection control solenoid vacuum valve.

9. Disconnect the EGR pressure sensor valve hoses from the EGR valve to exhaust manifold tube and remove the transducer mounting bracket from the engine.

10. Remove the EGR valve to exhaust manifold tube.

11. Remove the two intake manifold supports from the surge tank and cylinder head.

12. Remove the retaining bolts and surge tank stay from the front of the surge tank.

13. Remove the bolt, nuts and surge tank stay from the rear of the surge tank.

14. Loosen the intake air connector hose clamps and remove the surge tank from the engine.

15. Disconnect the engine control sensor wiring connections from the Idle Air Control (IAC) valve and the throttle position sensor.

16. Remove the right and left intake manifolds from the lower intake manifold.

17. Disconnect and tag the engine control sensor wiring connections to components, as necessary.

18. Remove the exhaust air supply tube from the exhaust manifolds and secondary air injection diverter valve.

19. Remove the secondary air injection diverter valve, as necessary.

20. Remove the water pump and hoses.

21. Remove the left and right valve covers.

22. Remove the power steering pump and pump support.

23. Remove the crankshaft pulley.

24. Remove the camshaft position sensor from the left cylinder head and the crankshaft position sensor from the engine front cover, as necessary.

25. Drain the engine oil and remove the oil pan.

26. Remove the drive belt idler pulleys from the engine front cover.

27. Remove the engine front cover.

28. Removal the crankshaft position sensor pulse ring from the crankshaft.

29. Rotate the crankshaft to position the No. 1 piston at top-dead-center TDC by aligning the crankshaft keyway groove with the oil pump mark.

30. Verify that the alignment on the camshafts sprockets are on top. If not, rotate the crankshaft one complete revolution and recheck.

31. Remove the timing chain guides, tensioner arm and tensioner retaining bolts and remove the timing chain from the engine.

32. Using a small screwdriver, release the timing chain tensioner ratchet/pawl mechanism through the access hole in the timing chain tensioner. Compress the timing chain tensioner rack and piston into the tensioner housing by inserting a small wire into the top of the piston and gently unseat the oil check ball. Compress the timing chain tensioner by hand.

✳✳ WARNING

Failure to compress and lock the timing chain tensioner prior to installation may cause damage to the engine.

33. With the tensioner compressed, install a 0.060 inch (1.5mm) drill bit or wire into the small hole above the ratchet, engaging the lock groove in the rack of the timing chain tensioner.

34. Remove the camshaft timing chain sprockets, if being replaced.

To install:

➡**Inspect the timing chains, tensioners, tensioner arms, guides and sprockets for wear or damage and replace as necessary.**

35. Rotate the crankshaft to position the No. 1 piston at top-dead-center TDC by aligning the crankshaft keyway groove with the oil pump mark.

✳✳ WARNING

Do not rotate the crankshaft more than 45 degrees counterclockwise or more than 90 degrees clockwise from the previously set No. 1 cylinder top-dead-center position or valve to piston contact could occur.

36. Match the sprocket and chain timing marks and install the timing chain with the crankshaft, camshaft and balance shaft correctly aligned as illustrated.

37. Reinstall or replace the camshaft sprockets, if removed and tighten the retaining bolt to 48–70 ft. lbs. (64–95 Nm).

38. Install the compressed timing chain guides, tensioner arm and tensioner. Tighten the timing chain tensioner pivot bolt to 25–39 ft. lbs. (34–53 Nm). Tighten the timing chain tensioner retaining bolts to 14–20 ft. lbs. (18–27 Nm).

39. Verify the timing index marks on the timing chain are in alignment with the index marks on the crankshaft sprocket, camshaft sprocket and balance shaft driven gear.

40. Install the power steering pump and pump support and tighten the three retaining bolts to 14–20 ft. lbs. (18–28 Nm).

41. Install the crankshaft position sensor pulse ring onto the crankshaft aligning the crankshaft key with the keyway on the sensor ring.

42. Clean the engine front and the front cover to cylinder block gasket sealing surfaces.

➡**The front cover must be installed and properly tightened within six minutes of the application of the sealer.**

43. Apply silicone sealer to the 13 critical areas of the cylinder block to prevent oil seepage.

44. Place new front cover gaskets onto the dowel pins on the cylinder block and heads.

45. Place the front cover into position by placing the front cover onto the dowel pins at the cylinder block.

46. Install the front cover retaining bolts and stud bolts and tighten to 14–20 ft. lbs. (18–28 Nm).

47. Install a new oil pan gasket and apply a thin bead of silicone sealer where the timing cover meets the cylinder block.

48. Install the oil pan and install the retaining bolts loosely.

49. With the oil pan aligned to the rear of the cylinder block, tighten the oil pan bolts and studs, in sequence, to 14–20 ft.

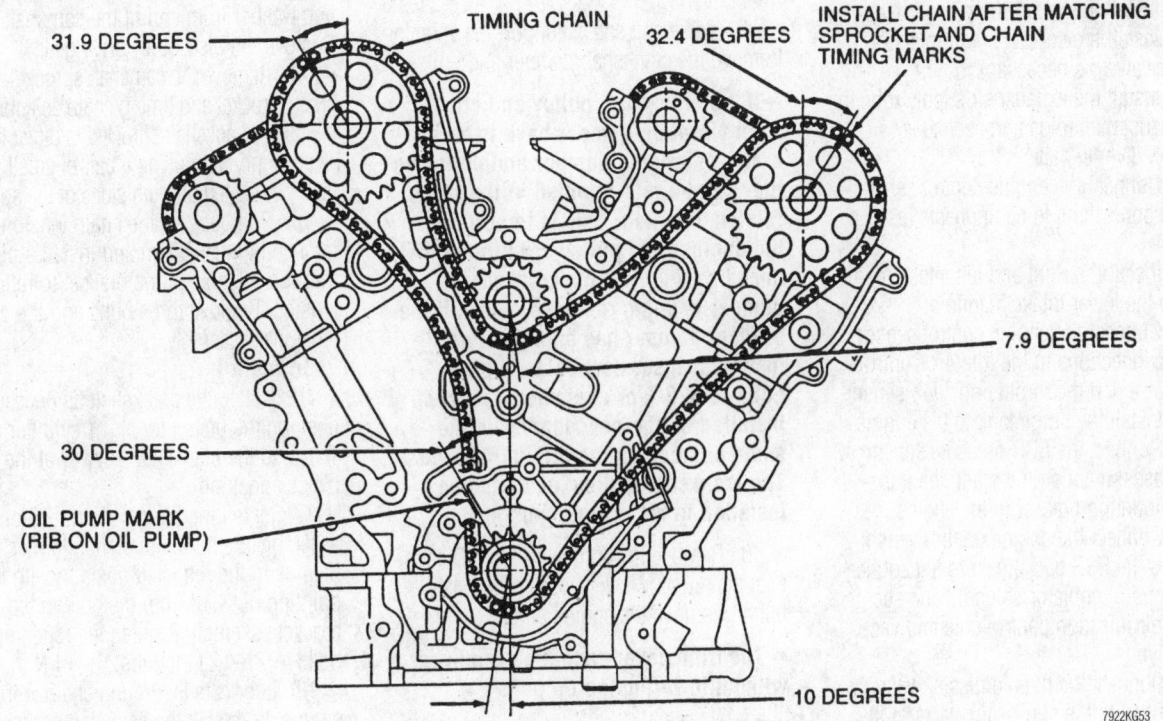

Be sure the timing marks are as shown after the chains have been installed—3.4L engine

7922KG53

lbs. (18–28 Nm) no more than six minutes after applying the silicone sealer.

50. Install the camshaft position sensor to the left cylinder head and the crankshaft position sensor to the engine front cover, as necessary and tighten to 71–106 inch lbs. (8–12 Nm).

51. Apply a thin coating of silicone sealer to the front of the crankshaft on the inside diameter of the damper at the keyway before installing the washer and retaining bolt.

52. Install the crankshaft damper pulley washer and bolt and tighten the bolt as follows:

a. Tighten the bolt to 78–99 ft. lbs. (105–135 Nm), then loosen the bolt one full turn.

b. Retighten the bolt to 35–39 ft. lbs. (47–53 Nm).

c. Turn the retaining bolt an additional 85–90 degrees.

53. Install the power steering pump pulley, plain washer, split washer and retaining nut and tighten to 41–50 ft. lbs. (55–69 Nm).

54. Install the drive belt idler pulleys to the engine front cover and tighten to 27–38 ft. lbs. (36–52 Nm).

55. Install both cylinder head covers and tighten to 71–106 inch lbs. (8–12 Nm)..

➡**The cylinder head covers must be installed and properly tightened within six minutes of applying the silicone sealant.**

56. Install the water pump and hoses.

57. Install the secondary air injection diverter valve, as necessary.

58. Install the exhaust air supply tube to the exhaust manifolds and secondary air injection diverter valve.

59. Connect the engine control sensor wiring connections to components, as necessary.

60. Install the right and left intake manifolds to the lower intake manifold.

61. Connect the engine control sensor wiring connections to the Idle Air Control (IAC) valve and the throttle position sensor.

62. Install the surge tank to the engine.

63. Connect the EGR pressure sensor valve hoses, manifold tube and the transducer mounting bracket to the engine.

64. Connect the engine control sensor wiring to the EGR backpressure transducer, EGR vacuum regulator solenoid and secondary air injection control solenoid vacuum valve.

65. Connect the main emission vacuum supply hose to the surge tank vacuum fitting. secondary air injection diverter valve and EGR valve.

66. Install the A/C compressor and mounting bolts to the cylinder blocks and tighten to 17–27 ft. lbs. (23–37 Nm).

67. Rotate the drive belt tensioner clockwise and install the belt.

68. Install the engine and connect the negative battery cable.

3.8L Engine

1. Disconnect the negative battery cable. Drain the engine cooling system. Drain the engine oil.

2. Remove the engine air cleaner assembly and air intake duct.

3. Loosen the accessory drive belt idler. Remove the drive belt and water pump pulley.

4. Remove the power steering pump mounting bracket attaching bolts. Leaving the hoses connected, place the pump/bracket assembly in a position that will prevent the loss of power steering fluid.

5. If equipped with air conditioning, remove the compressor front support bracket. Leave the compressor in place.

6. Disconnect the coolant bypass and heater hoses at the water pump. Disconnect radiator upper hose at thermostat housing.

7. Disconnect the coil wire from distributor cap and move the cap with the ignition wires aside. Remove the distributor retaining clamp and lift distributor out of the front cover.

8. Raise and safely support the vehicle. Remove the crankshaft damper and pulley.

➡**If the crankshaft pulley and crankshaft vibration damper have to be separated, mark the damper and pulley so they can be reassembled in the same relative position. This is important as the damper and pulley are initially balanced as a unit. If the crankshaft damper is being replaced, check if the original damper has balance pins installed. If so, new balance pins EOSZ-6A328-A or equivalent, must be installed on the new damper in the same position as the original damper. The crankshaft pulley must also be installed in the original position.**

9. Remove the engine oil filter. Disconnect the radiator lower hose at the water pump. Remove the engine oil pan.

➡**The front cover cannot be removed without lowering the oil pan.**

10. Lower the vehicle. Remove the front cover retaining bolts.

※※ **WARNING**

Do not overlook the front cover retaining bolt located behind the oil filter adapter. The front cover will break if all retaining bolts are not removed.

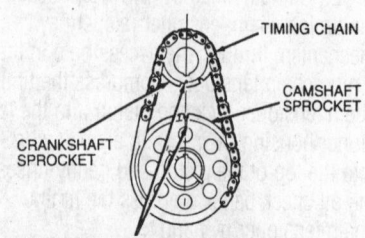

POSITIONING OF TIMING MARKS AND KEYWAYS IN CAMSHAFT AND CRANKSHAFT SPROCKETS MUST BE IN LINE AS SHOWN WITH NO. 1 PISTON AT TOP DEAD CENTER FIRING

7922KG54

Install the timing chain so the marks on the sprockets are facing each other—3.8L engine

11. Remove the ignition timing indicator.

12. Remove the front cover and water pump as an assembly. Remove the cover gasket and discard.

➡**The front cover houses the oil pump. If a new front cover is installed, remove the water pump and oil pump from the old cover and install in the new cover.**

13. Remove the camshaft retaining bolt and washer from end of the camshaft. Remove the distributor drive gear.

14. Remove the camshaft sprocket, crankshaft sprocket and timing chain as an assembly. If the crankshaft sprocket is difficult to remove, pry it off using a pair of small prybars positioned on both sides of the sprocket.

15. Pull back on the chain tensioner ratcheting mechanism and install a pin through the hole in the bracket to relieve tension. Remove three bolts and the chain tensioner assembly.

To install:

16. Rotate the crankshaft as necessary to position the piston for No. 1 cylinder at TDC and the crankshaft keyway at the 12 o'clock position.

17. Install the timing chain tensioner assembly. Be sure the ratcheting mechanism is in the retracted position with the pin pointing outward from the hole in the bracket assembly. Tighten the retaining bolts to 71–124 inch lbs. (8–14 Nm).

18. Lubricate the timing chain with clean engine oil. Install the camshaft sprocket, crankshaft sprocket and timing chain as an assembly.

19. Remove the pin from the tensioner assembly to load the tensioner arm against the chain. Make certain the timing marks are properly positioned across from each other.

20. Install the distributor drive gear. Install the washer and retaining bolt to the camshaft and tighten to 30–37 ft. lbs. (40–50 Nm).

21. Lightly oil all bolt and stud threads before installation. Clean all gasket surfaces on the front cover, cylinder block and fuel pump, if equipped. If reusing the front cover, replace the front cover oil seal.

22. If a new front cover is to be installed, complete the following:

   a. Pack the oil pump gear pocket with petroleum jelly and install the oil pump gears. Be sure the petroleum jelly fills the gap between the gears and the pocket. Install the oil pump cover using a new gasket and tighten the retaining bolts to 18–22 ft. lbs. (25–30 Nm).

   b. Clean the water pump gasket surface. Position a new water pump gasket on the front cover and install the water pump. Install the pump retaining bolts and tighten to 15–22 ft. lbs.

23. Install the distributor drive gear.

24. Lubricate the crankshaft front oil seal with clean engine oil.

25. Position a new front cover gasket on the cylinder block and install the front cover/water pump assembly using dowels for proper alignment. A suitable contact adhesive is recommended to hold the gasket in position while the front cover is installed.

26. Position the ignition timing indicator.

27. Install the front cover attaching bolts. Apply Loctite® or equivalent, to the threads of the bolt installed below the oil filter housing prior to installation. This bolt is to be installed and tightened last. Tighten all bolts to 15–22 ft. lbs. (20–30 Nm).

28. Raise and safely support the vehicle. Install the engine oil pan.

29. Connect the radiator lower hose. Install a new engine oil filter.

30. Coat the crankshaft damper sealing surface with clean engine oil. Apply a small amount of silicone sealant to the crankshaft keyway.

31. Position the crankshaft key in the crankshaft keyway.

32. Install the damper, washer and retaining bolt. Tighten the bolt to 104–132 ft. lbs. (140–180 Nm).

33. Install the crankshaft pulley and retaining bolts. Tighten the retaining bolts 19–28 ft. lbs. (26–28 Nm).

34. Lower the vehicle. Connect the coolant bypass hose.

35. Rotate the crankshaft, as necessary, to position the piston for No. 1 cylinder to TDC on its compression stroke. Install the distributor. Install the distributor cap and coil wire.

36. Connect the radiator upper hose at thermostat housing. Connect the heater hose.

37. If equipped with air conditioning, install the front compressor support bracket.

38. Install the power steering pump and mounting brackets. Position the accessory drive belt over the pulleys.

39. Install the water pump pulley. Position the accessory drive belt over water pump pulley and tighten the belt.

40. Fill the crankcase and cooling system to the proper levels.

41. Connect the negative battery cable. Start the engine and check for leaks.

42. Check the ignition timing and curb idle speed; adjust as required.

43. Install the engine air cleaner assembly and air intake duct. Road test the vehicle and check for proper operation.

FUEL SYSTEM

Fuel System Service Precautions

Safety is the most important factor when performing not only fuel system maintenance but any type of maintenance. Failure to conduct maintenance and repairs in a safe manner may result in serious personal injury or death. Maintenance and testing of the vehicle's fuel system components can be accomplished safely and effectively by adhering to the following rules and guidelines.

• To avoid the possibility of fire and personal injury, always disconnect the negative battery cable unless the repair or test procedure requires that battery voltage be applied.

• Always relieve the fuel system pressure prior to disconnecting any fuel system component (injector, fuel rail, pressure regulator, etc.), fitting or fuel line connection. Exercise extreme caution whenever relieving fuel system pressure to avoid exposing skin, face and eyes to fuel spray. Please be advised that fuel under pressure may penetrate the skin or any part of the body that it contacts.

• Always place a shop towel or cloth around the fitting or connection prior to loosening to absorb any excess fuel due to spillage. Ensure that all fuel spillage (should it occur) is quickly removed from engine surfaces. Ensure that all fuel soaked cloths or towels are deposited into a suitable waste container.

• Always keep a dry chemical (Class B) fire extinguisher near the work area.

• Do not allow fuel spray or fuel vapors to come into contact with a spark or open flame.

• Always use a back-up wrench when loosening and tightening fuel line connection fittings. This will prevent unnecessary stress and torsion to fuel line piping. Always follow the proper torque specifications.

• Always replace worn fuel fitting O-rings with new. Do not substitute fuel hose or equivalent, where fuel pipe is installed.

Fuel System Pressure

RELIEVING

✳✳ CAUTION

The fuel injection system remains under pressure, even after the engine has been turned OFF. The fuel system pressure must be relieved before disconnecting any fuel lines. Failure to do so may result in fire and/or personal injury.

1. Disconnect the negative battery cable.
2. Remove the fuel tank fill cap to relieve the pressure in the fuel tank.
3. Remove the cap from the Schrader valve located on the fuel supply manifold (fuel rail).
4. On gasoline engines, attach Fuel Pressure Gauge T80L-9974-A or equivalent, to the valve and drain the fuel through the drain tube into a suitable container.
5. On flex-fuel engines, connect fuel pressure gauge T80L-9974-A or equivalent and fuel pressure test kit 134-R0035 or equivalent, to the Schrader valve. Drain the fuel through the drain tube into a suitable container.
6. After the fuel system pressure is relieved, remove the fuel pressure gauge and install the cap on the Schrader valve.
7. Install the fuel tank fill cap.
8. Connect the negative battery cable only after system repairs are completed.

Fuel Filter

REMOVAL & INSTALLATION

✳✳ CAUTION

The fuel injection system remains under pressure, even after the engine has been turned OFF. The fuel system pressure must be relieved before disconnecting any fuel lines. Failure to do so may result in fire and/or personal injury.

1. Disconnect the negative battery cable.

2. Properly relieve the fuel system pressure.

3. Raise and safely support the vehicle.

4. If equipped with the push connect type connection, disengage the fuel lines as follows:

   a. Push the disconnect tool into the fitting to release the internal locking fingers.

   b. Gently separate the fuel line from the connector.

5. If equipped with the hairpin clip type push connect fitting, remove the connection as follows:

   a. Inspect the visible internal portion of the fitting for dirt accumulation. If more than a light coating of dust is present, clean the fitting before disassembly.

   b. Some adhesion between the seals in the fitting and the filter will occur with time. To separate, twist the fitting on the filter, then push and pull the fitting until it moves freely on the filter.

   c. Remove the hairpin clip from the fitting by first bending and breaking the shipping tab. Next, spread the two clip legs by hand about ⅛ inch each, to disengage the body and push the legs into the fitting. Lightly pull the triangular end of the clip and work it clear of the filter and fitting.

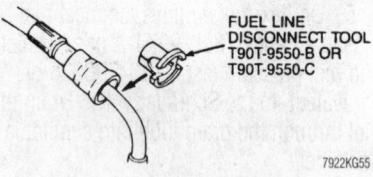

Use Fuel Line Disconnect tool (T90T-9550-B) or (T90T-9550-C) or their equivalents to disconnect the push connect type fitting

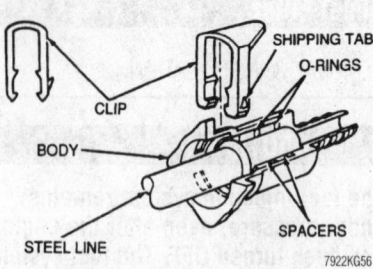

Exploded view of the hairpin clip type fuel line connection

➡**Do not use hand tools to complete this operation.**

   d. Grasp the fitting and pull in an axial direction to remove the fitting from the filter. Be careful on 90 degree elbow connectors, as excessive side loading could break the connector body.

   e. After disassembly, inspect the inside of the fitting for any internal parts such as O-rings and spacers that may have been dislodged from the fitting. Replace any damaged connector.

6. Loosen the filter retaining clamp and remove the fuel filter. Note the direction of the flow arrow on the filter, so the replacement filter can be reinstalled in the same position.

To install:

7. Install the fuel filter with the flow arrow facing the proper direction and tighten the filter retaining clamp.

8. If equipped, install the rubber insulator rings on the new filter (replace the insulator rings if the filter moves freely after the retainer is installed). Install the filter into the retainer with the flow arrow pointing out the open end of the retainer. Install the retainer on the bracket and tighten the mounting bolts to 27–44 inch lbs. (3–5 Nm) (1995) and 15–24 inch lbs. (1.7–2–8 Nm) (1996–99).

9. If equipped with the push connect fitting, insert the fuel line in the fitting until an audible click is heard, then gently try to pull the connection apart to confirm positive engagement.

10. If equipped with the hairpin clip type push connect fittings, connect the fuel lines as follows:

   a. Install a new connector if damage was found. Insert a new clip into any two adjacent openings with the triangular portion pointing away from the fitting opening. Install the clip until the legs of the clip are locked on the outside of the body. Piloting with an index finger is necessary.

   b. Before installing the fitting on the filter, wipe the filter end with a clean cloth. Inspect the inside of the fitting to be sure it is free of dirt and/or obstructions.

   c. Apply a light coating of engine oil to the filter end. Align the fitting and filter axially and push the fitting onto the filter end. When the fitting is engaged, a definite click will be heard. Pull on the fitting to be sure it is fully engaged.

11. Lower the vehicle.

12. Connect the negative battery cable.

13. Start the engine and check for fuel leaks and proper operation.

Fuel Pump

REMOVAL & INSTALLATION

✳✳ CAUTION

The fuel injection system remains under pressure, even after the engine has been turned OFF. The fuel system pressure must be relieved before disconnecting any fuel lines. Failure to do so may result in fire and/or personal injury.

1995 MODELS

1. Disconnect the negative battery cable. Properly relieve the fuel system pressure.

2. Remove the fuel tank from the vehicle and place it on a bench.

3. Remove any dirt that has accumulated around the fuel pump retaining flange so it will not enter the tank during pump removal and installation.

4. Turn the fuel pump locking ring counterclockwise and remove the locking ring.

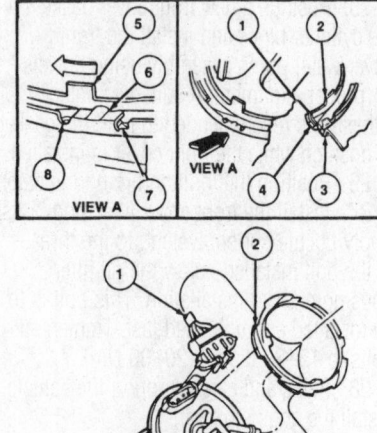

1	Fuel Pump
2	Fuel Pump Locking Retainer Ring
3	Retainer Ring
4	O-Ring
5	Locating Tabs
6	Tab
7	Stop
8	Detent

Exploded view of the fuel pump module— 1995 models

5. Remove the fuel pump/sending unit assembly. Remove and discard the seal ring.

To install:

6. Clean the fuel pump mounting flange, fuel tank mounting surface and seal ring groove.

7. Apply a light coating of grease on a new seal ring to hold it in place during assembly and install in the seal ring groove.

8. Install the fuel pump/sending unit assembly carefully to ensure the filter is not damaged. Be sure the locating keys are in the keyways and the seal ring remains in the groove.

9. Hold the pump assembly in place and install the locking ring finger-tight. Be sure all the locking tabs are under the tank lock ring tabs.

10. Rotate the locking ring clockwise until the ring is against the stops. Install the fuel tank in the vehicle. Add a minimum of 10 gallons of fuel to the tank and check for leaks.

11. Connect the negative battery cable. Turn the ignition switch to the **RUN** position several times to pressurize the fuel system. Check for fuel leaks and correct as necessary.

12. Start the engine and check for leaks. Road test the vehicle and check for proper operation.

1996–99 Models

1. Disconnect the negative battery cable. Properly relieve the fuel system pressure.

2. Raise and safely support the vehicle. Remove the fuel tank and place on a suitable work bench.

3. Remove any dirt that has accumulated around the fuel pump retaining flange so it will not enter the tank during pump removal and installation.

4. Turn the fuel pump locking ring counterclockwise and remove the locking ring using Fuel Tank Sender Wrench T74P-9275-A, or equivalent.

5. Pull the fuel pump module up and out of the fuel tank until the locking tabs for the fuel pump module are accessible. Squeeze both locking tabs together and remove the fuel pump module from the fuel tank. Remove and discard the O-ring seal.

To install:

6. Clean the fuel pump mounting flange, fuel tank mounting surface and O-ring seal groove.

7. Apply a light coating of grease on a new O-ring seal to hold it in place during assembly and install the O-ring seal.

8. Install the fuel pump module carefully to ensure the filter and hoses and float rod are not damaged.

9. Align the fuel pump module and the fuel tank retainer axially and push the fuel pump module into the fuel tank retainer. When the fuel pump module is properly engaged, a definite click will be heard engaging two locking tabs on the outside of the fuel pump.

10. Pull on the fuel pump module to ensure that both locking tabs are properly engaged.

11. Be sure the locating keys are in the keyways and the seal ring remains in the groove.

12. Hold the fuel pump module in place and install the locking ring finger-tight. Be sure all the locking tabs are under the tank lock ring tabs.

13. Using the sender wrench or equivalent, rotate the locking ring clockwise until the ring is against the stops.

14. Install the fuel tank in the vehicle. Lower the vehicle.

15. Add a minimum of 10 gallons (38 liters) of clean fuel to the tank. Connect the negative battery cable.

16. Install a suitable fuel pressure gauge to the Schrader valve on the fuel supply manifold.

17. Cycle the ignition switch from **OFF** to **ON** for three seconds. Repeat this procedure 5–10 times until the pressure gauge reads at least 30 psi (207 kPa). Check for fuel leaks.

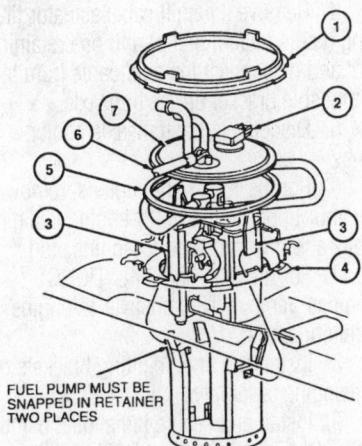

FUEL PUMP MUST BE
SNAPPED IN RETAINER
TWO PLACES

1	Fuel Pump Locking Retainer Ring
2	Fuel Tank Pressure / Vacuum Transducer
3	Locking Tab
4	Fuel Tank
5	O-Ring Seal
6	Connector
7	Fuel Pump Module

7922KG58

Exploded view of the fuel pump module mounting—1996–99 models

18. Remove the fuel pressure gauge. Start the engine and again, check for fuel leaks. Road test the vehicle and check for proper operation.

DRIVE TRAIN

Transaxle

REMOVAL & INSTALLATION

Manual

1. Disconnect the negative battery cable. Wedge a 7 in. (178mm) block of wood under the clutch pedal to hold the pedal up beyond its normal position.

2. Remove the air cleaner hose.

3. Grasp the clutch cable and pull it forward, disconnecting it from the clutch release shaft assembly.

4. Disconnect the clutch cable casing from the rib on top of the transaxle case.

5. Install engine lifting eyes.

6. Tie up the wiring harness and power steering cooler hoses.

7. Disconnect the speedometer cable and speed sensor wire.

8. Support the engine using a suitable engine support fixture.

9. Raise the vehicle and support it safely. Remove the wheel and tire assemblies.

10. Remove the nut and bolt retaining the lower control arm ball joint to the steering knuckle assembly. Discard the removed nut and bolt. Repeat on the opposite side.

11. Using a suitable prybar, pry the lower control arm away from the knuckle.

➡**Be careful not to damage or cut the ball joint boot.**

12. Remove the upper nut from the stabilizer bar and separate the stabilizer bar from the knuckle.

13. Remove the tie rod nut and separate the tie rod end from the knuckle.

14. Disconnect the oxygen sensor.

15. Remove the exhaust catalyst assembly. Disconnect the power steering cooler from the subframe and place it aside.

16. Disconnect the battery cable bracket from the subframe.

17. Using a suitable prybar, pry the left inboard CV-joint assembly from the transaxle. Install a plug into the seal to prevent fluid leakage. Remove the CV-joint from the transaxle by grasping the left steer-

ing knuckle and swinging the knuckle and halfshaft outward from the transaxle. Repeat the procedure on the right side.

➡**If the CV-joint assembly cannot be pried from the transaxle, insert a suitable tool through the left side and tap the joint out. The tool can be used from either side of the transaxle.**

18. Support the halfshaft assembly with wire in a near level position to prevent damage to the assembly during the remaining operations. Repeat the procedure on the opposite side.

19. Remove the retaining bolts from the center support bearing and remove the right halfshaft from the transaxle.

20. Remove the two steering gear retaining nuts from the subframe. Support the steering gear by wiring up the tie rod ends to the coil springs.

21. Remove the transaxle to engine retaining bolts. Disconnect the two shift rods from the transaxle.

22. Remove the engine mount bolts.

23. Position jacks under the body mount positions and remove the four bolts, lower the subframe and position it aside.

24. Remove the starter motor assembly. Remove the left engine vibration dampener lower bracket.

25. Remove the back-up light switch connector from the transaxle back-up light switch, located on top of the transaxle and remove the back-up light switch.

26. Position a suitable support jack under the transaxle. Lower the transaxle, remove it from the engine and lower it from the vehicle.

To install:

27. Raise the transaxle into position. Engage the input shaft spline into the clutch disc and work the transaxle onto the dowel sleeves. Be sure the transaxle assembly is flush with the rear face of the engine before installation of the retaining bolts.

28. Install the engine to transaxle retaining bolts. Tighten to 28–31 ft. lbs. (38–42 Nm).

29. Install the remaining components in the reverse order of removal. Tighten the following:
- Starter bolts to 30–40 ft. lbs. (41–54 Nm)
- Subframe bolts to 65–85 ft. lbs. (90–115 Nm)
- Engine mount bolts to 40–55 ft. lbs. (54–75 Nm)
- Stabilizer bolt to 35–46 ft. lbs. (47–63 Nm)
- Shift rod clamp bolt and nut to 80–106 inch lbs. (9–12 Nm)

- Engine-to-transaxle bolts to 28–31 ft. lbs. (38–42 Nm)
- Steering gear nuts to 85–100 ft. lbs. (115–135 Nm)
- Center support bearing bolts to 85–100 ft. lbs. (115–135 Nm)
- Exhaust catalyst retaining bolts to 25–34 ft. lbs. (34–47 Nm)
- Tie rod retaining nut to 35–47 ft. lbs. (47–64 Nm)
- Lower control arm ball joint-to-steering knuckle nut and bolt to 37–44 ft. lbs. (50–60 Nm)

30. Check the transaxle fluid level.

31. Lower the vehicle.

32. Remove the engine support tool.

33. Install the speedometer cable. Connect the speedometer cable and speed sensor wire.

34. Remove the engine lifting eyes.

35. Connect the clutch cable to the transaxle. Install the air cleaner hose and remove the wood block from the clutch pedal.

36. Connect the negative battery cable. Road test and check transaxle operation. Check the transaxle for fluid leaks.

Automatic

1. Disconnect both battery cables, negative cable first.

2. Remove the battery and battery tray. Remove the engine air cleaner assembly.

3. Detach the transaxle harness and the Transmission Range (TR) sensor connectors.

4. Remove the shift cable actuator fitting (cable retaining clip) and one retaining nut and disconnect the shift cable from the shift cable bracket on the transaxle.

5. Disconnect the transaxle cooler lines.

6. For the 3.0L OHV engines, remove the four upper transaxle-to-engine retaining bolts and one transaxle-to-engine stud.

7. For the 3.0L and 3.4L (DOHC) engines, remove five transaxle-to-engine retaining bolts.

8. Install two engine lifting brackets on the engine assembly.

9. Install and secure the engine using Three Bar Engine Support D88L-6000-A, or equivalent.

10. Raise and safely support the vehicle.

11. Loosen the transaxle oil pan retaining bolts and drain the transaxle fluid into a suitable container.

12. Remove both front wheel and tire assemblies.

13. Remove both halfshafts.

14. Detach the four Heated Oxygen Sensor (HO$_2$S) electrical connectors.

15. Remove three bolts and seven nuts securing the converter Y-pipe assembly and remove from the vehicle.

16. Detach the two starter motor connectors, then remove the starter.

17. Remove one bolt and one stud securing the starter motor and remove the starter motor from the transaxle.

18. For the OHV engine, remove one bolt and the transaxle housing cover.

19. Support the rack and pinion assembly using wire attached to the strut and spring assembly to hold it in position. Remove two rack and pinion assembly retaining nuts from the subframe.

20. Remove two lower control arm-to-ball joint retaining nuts and separate the lower control arms from the steering knuckles and ball joints.

21. Remove the retaining nuts from the front engine support insulators (mounts) at the subframe.

22. Remove the sway bar (stabilizer bar) link retaining nuts at each end of the sway bar and separate the links from the sway bar.

23. Remove the engine and transaxle support insulator through-bolts from the subframe.

24. Place High Lift Transmission Jack 014–00210 or equivalent, using a suitable subframe adapter under the subframe and support the subframe.

25. Remove the four subframe-to-body retaining bolts. Carefully lower the subframe and set aside.

26. Place High Lift Transmission Jack 014–00210 or equivalent, using Adapter 014–00461 or equivalent, under the transaxle and support the transaxle assembly. Secure the transaxle to the transaxle adapter using a strap or chain.

27. For the OHV engine, remove one lower engine-to-transaxle bolt.

28. For the DOHC engine, remove the four lower engine-to-transaxle bolts.

29. Remove the four flywheel-to-torque converter nuts.

30. Remove three bolts and two nuts securing the rear engine support to the transaxle and remove the rear engine support.

31. Remove one bolt from the right engine mount brace, then slowly lower the transaxle from the vehicle.

To install:

➡**Flush the transaxle cooler lines thoroughly before installing the transaxle assembly.**

32. If removed, place the transaxle assembly on High Lift Transmission Jack 014–00210 or equivalent, using Adapter 014–00461 or equivalent. Secure the

transaxle to the transaxle adapter using a strap or chain.

33. Slowly raise the transaxle assembly into place. Align the torque converter studs with the appropriate holes in the flywheel and engage the transaxle housing to the engine dowel pins.

34. Install one bolt in the right engine mount brace and tighten to 39–53 ft. lbs. (53–72 Nm).

35. Install the rear engine support to the transaxle and install three bolts and two nuts. Tighten the bolts and nuts to 39–53 ft. lbs. (53–72 Nm).

36. Install the four flywheel-to-torque converter nuts and tighten to 20–34 ft. lbs. (27–46 Nm).

37. For the OHV engine, install one lower transaxle-to-engine bolt and tighten to 39–53 ft. lbs. (53–72 Nm).

38. For the DOHC engine, install the four lower transaxle-to-engine bolts and tighten to 39–53 ft. lbs. (53–72 Nm).

39. Place the subframe on High Lift Transmission Jack 014–00210 or equivalent, using a suitable subframe adapter and raise the subframe into position.

40. Install the subframe insulators, if removed and loosely install four subframe-to-body bolts.

41. Install a ¾ inch (19mm) outside diameter pipe or similar tool into the front left subframe and body alignment holes and align the holes. Slightly tighten the front left subframe-to-body bolt.

42. Repeat the subframe alignment procedure on the front right subframe and body alignment holes. Slightly tighten the right subframe-to-body bolt.

43. Check the left alignment holes again and adjust if necessary.

44. After the subframe alignment is complete, tighten all four subframe-to-body bolts to 57–76 ft. lbs. (77–103 Nm).

45. Install the remaining components in the reverse order of removal. Tighten the following:

• Engine and transaxle support insulator-to-subframe bolts to 65–87 ft. lbs. (88–118 Nm)

• Sway bar link-to-sway bushings and nuts to 35–46 ft. lbs. (47–63 Nm)

• Rack and pinion-to-subframe nuts to 84–113 ft. lbs. (113–133 Nm)

• Front engine support insulator-to-subframe nuts to 57–76 ft. lbs. (77–103 Nm)

• Ball joint-to-lower control arm nuts to 51–67 ft. lbs. (68–92 Nm), using new nuts

• For the OHV engine: transaxle housing cover bolt to 80–106 inch lbs. (9–12 Nm)

• Starter motor bolt and stud to 15–21 ft. lbs. (21–29 Nm)

• Converter Y-pipe three bolts and seven nuts to 26–34 ft. lbs. (34–46 Nm)

• For the OHV engine: four upper transaxle-to-engine bolts and one upper transaxle-to-engine stud to 39–53 ft. lbs. (53–72 Nm)

• For the DOHC engine: five transaxle-to-engine bolts to 39–53 ft. lbs. (53–72 Nm)

• Shift actuator cable fitting nut to 14–19 ft. lbs. (19–26 Nm)

46. Connect both battery cables, negative cable last.

47. If the transaxle is empty of transaxle fluid, add several quarts of MERCON or equivalent transaxle fluid to the transaxle.

48. Start the engine and continue to fill the transaxle until the correct level is reached. Check for leaks and proper operation.

➡ **Whenever the vehicles subframe is removed or lowered, the wheel alignment should be checked.**

49. Check the front end alignment.

50. Road test the vehicle and check the transaxle for proper operation.

Clutch

ADJUSTMENT

The clutch pedal mechanism is self-adjusting. No adjustment is necessary.

REMOVAL & INSTALLATION

1. Disconnect the negative battery cable. Raise the vehicle and support it safely.

2. Remove the transaxle assembly.

3. If the pressure plate is to be reused, mark its location on the flywheel so it can be reinstalled in the same position.

4. Loosen the pressure plate bolts one turn at a time in a crisscross pattern, until spring tension is relieved, to prevent pressure plate distortion. Support the pressure plate and remove the bolts. Remove the pressure plate and clutch disc from the flywheel.

5. Inspect the flywheel, clutch disc, pressure plate, throwout bearing and the

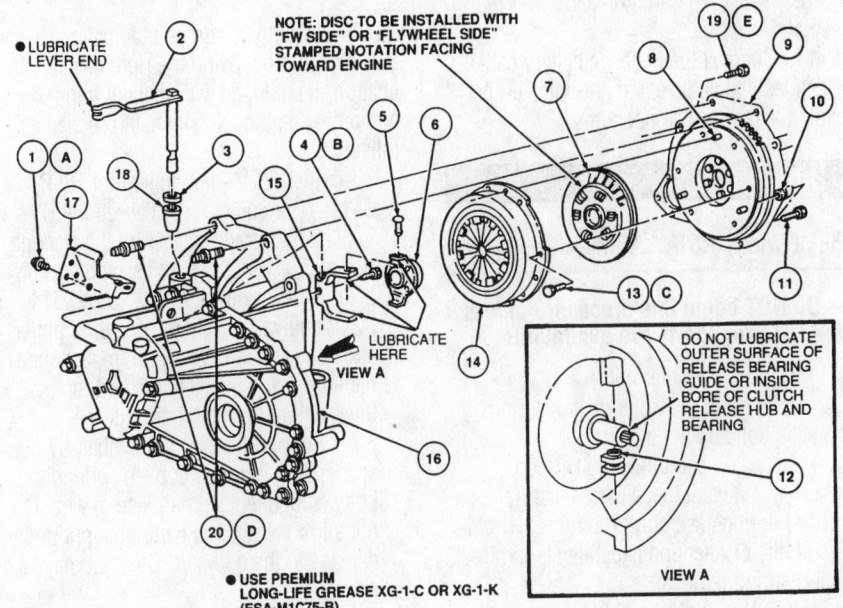

NOTE: DISC TO BE INSTALLED WITH "FW SIDE" OR "FLYWHEEL SIDE" STAMPED NOTATION FACING TOWARD ENGINE

• LUBRICATE LEVER END

LUBRICATE HERE
VIEW A

DO NOT LUBRICATE OUTER SURFACE OF RELEASE BEARING GUIDE OR INSIDE BORE OF CLUTCH RELEASE HUB AND BEARING

VIEW A

• USE PREMIUM LONG-LIFE GREASE XG-1-C OR XG-1-K (ESA-M1C75-B)

1	Bolt (2 Req'd)	14	Clutch Pressure Plate
2	Clutch Assist Lever	15	Clutch Release Lever
3	Felt Washer	16	Transaxle Assy
4	Bolt	17	Transmission Energy Bracket
5	Clutch Release Lever Stud	18	Clutch Assist Shaft Bushing
6	Clutch Release Hub and Bearing	19	Bolt
7	Clutch Disc	20	Stud (2 Req'd)
8	Flywheel-to-Clutch Pressure Plate Dowel (3 Req'd)	A	Tighten to 35-50 N·m (26-37 Lb-Ft)
9	Rear of Engine	B	Tighten to 40-55 N·m (30-40 Lb-Ft)
10	Flywheel Housing to Block Dowel (2 Req'd)	C	Tighten to 33 N·m (24 Lb-Ft)
11	Bolt (3 Req'd)	D	Tighten to 46-63 N·m (34-46 Lb-Ft)
12	Clutch Release Lever Bushing	E	Tighten to 54-92 N·m (40-68 Lb-Ft)
13	Pressure Plate Bolt (6 Req'd)		

Exploded view of the clutch assembly and related components

7922KG59

clutch fork for wear. Replace parts as required. If the flywheel shows any signs of overheating (blue discoloration) or if it is badly grooved or scored, it should be refaced or replaced.

To install:

6. Install the flywheel, if removed. Tighten the attaching bolts to 54–64 ft. lbs. (73–87 Nm) on all except the 3.0L SHO and 3.2L SHO engine. On the 3.0L SHO and 3.2L SHO engine, tighten the bolts to 51–58 ft. lbs. (69–78 Nm).

7. Clean the pressure plate and flywheel surfaces thoroughly. Place the clutch disc and pressure plate into the installed position. Align the marks made during the removal procedure if components are being reused. Support the clutch disc and pressure plate with a suitable dummy shaft or clutch aligning tool.

8. Install the pressure plate-to-flywheel bolts. Tighten them gradually in a criss-cross pattern to 12–24 ft. lbs. (17–32 Nm). Remove the alignment tool.

9. Lubricate the release bearing and install it in the fork.

10. Install the transaxle and lower the vehicle.

11. Connect the negative battery cable. Road test the vehicle and check for proper clutch and transaxle operation.

Halfshaft

REMOVAL & INSTALLATION

➡**Do NOT begin this procedure unless the following parts are available:**

- New hub retainer nut
- New lower control arm-to-steering knuckle bolt and nut
- New stub shaft/link shaft circlip

Once removed, these parts must not be reused during assembly. Their torque holding ability or retention capability is diminished during removal.

✳✳ WARNING

When removing both halfshafts on vehicles equipped with a manual transaxle, install transaxle plug tools T81P-1177-B or equivalent, to prevent dislocation of the differential side gears. Should the gears become misaligned, the differential will have to be removed from the transaxle to re-align the side gears.

1. Remove the wheel cover/hub cover from the wheel and tire assembly. Loosen the hub retainer nut and the lug nuts.

2. Raise the vehicle and support safely.

3. Remove the wheel and tire assembly. Remove the hub retainer nut and washer and discard the hub nut.

4. Remove the nut from the ball joint-to-steering knuckle attaching bolts. Drive the bolt out of the steering knuckle using a punch and hammer. Discard the bolt and nut.

5. If equipped with ABS, remove the anti-lock brake sensor and position aside. If equipped with air suspension, remove the height sensor bracket retaining bolt and wire sensor bracket to inner fender. Position the sensor link aside.

6. Separate the ball joint from the steering knuckle using a suitable prybar. Position the end of the prybar outside of the bushing pocket to avoid damage to the bushing. Use care to prevent damage to the ball joint boot. Remove the stabilizer bar link at the stabilizer bar.

➡**The remaining removal procedures differ according to transaxle application: manual transaxle or automatic transaxle.**

7. If equipped with an automatic transaxle and removing the right or left halfshaft, or if equipped with manual transaxle and removing the left halfshaft, proceed as follows:

a. Install CV-joint puller tool T86P-3514-A1 or equivalent, between the CV-joint and transaxle case. Turn the steering hub and/or wire the strut assembly aside.

b. Screw extension tool T86P-3514-A2 or equivalent, into the CV-joint puller and hand-tighten. Screw an impact slide hammer onto the extension and remove the CV-joint from the transaxle.

c. Support the end of the shaft by suspending it from a convenient underbody component with a piece of wire. Do not allow the shaft to hang unsupported; damage to the outboard CV-joint may occur.

✳✳ WARNING

Never use a hammer to separate the outer CV-joint stub shaft from the hub. Damage to the CV-joint threads and internal components may result.

d. Separate the outboard CV-joint from the hub using front hub remover tool T81P-1104-C or equivalent, metric adapter tools T83-P-1104-BH, T86P-1104-AI and front hub installer T81P-1104-A or equivalent.

e. Remove the halfshaft assembly from the vehicle.

8. If equipped with a manual transaxle and removing the right halfshaft, proceed as follows:

a. Remove the bolts attaching the bearing support to the bracket. Slide the link shaft out of the transaxle. Support the end of the shaft by suspending it from a convenient underbody component with a piece of wire. Do not allow the shaft to hang unsupported, damage to the outboard CV-joint may occur.

b. Separate the outboard CV-joint from the hub using front hub remover tool T81P-1104-C or equivalent, metric adapter tools T83-P-1104-BH, T86P-1104-AI and front hub installer T81P-1104-A or equivalent.

✳✳ WARNING

Never use a hammer to separate the outboard CV-joint stub shaft from the hub. Damage to the CV-joint threads and internal components may result. The right side link shaft and halfshaft assembly is removed as a complete unit.

To install:

9. Install a new circlip on the inboard CV-joint stub shaft and/or link shaft. The outboard CV-joint does not have a circlip. When installing the circlip, start one end in the groove and work the circlip over the stub shaft end into the groove. This will avoid over expanding the circlip.

➡**The circlip must not be re-used. A new circlip must be installed each time the inboard CV-joint is installed into the transaxle differential.**

10. Carefully align the splines of the inboard CV-joint stub shaft with the splines in the differential. Exerting some force, push the CV-joint into the differential until the circlip is felt to seat in the differential side gear. Use care to prevent damage to the differential oil seal. If equipped, tighten the link shaft bearing retaining bolts to 16–23 ft. lbs. (21–32 Nm).

➡**A non-metallic mallet may be used to aid in seating the circlip into the differential side gear groove. If a mallet is necessary, tap only on the outboard CV-joint stub shaft.**

11. Carefully align the splines of the outboard CV-joint stub shaft with the splines in the hub and push the shaft into the hub as far as possible.

12. Temporarily fasten the rotor to the hub with washers and two wheel lug nuts. Insert a steel rod into the rotor and rotate

clockwise to contact the knuckle to prevent the rotor from turning during the CV-joint installation.

13. Install the hub nut washer and a new hub nut. Manually thread the retainer onto the CV-joint as far as possible.

14. Install the remaining components in the reverse order of removal. Tighten the following:

- Control arm-to-steering knuckle nut and bolt: 40–55 ft. lbs. (54–74 Nm) for 1995 vehicles or to 50–68 ft. lbs. (68–92 Nm) for 1996–99 vehicles, using a new nut and bolt
- Stabilizer link-to-stabilizer bar/front strut to 35–48 ft. lbs. (47–65 Nm) for 1995 vehicles or to 57–75 ft. lbs. (77–103 Nm) for 1996–99 vehicles
- Hub retainer nut to 180–200 ft. lbs. (245–270 Nm)
- Lug nuts to 85–105 ft. lbs. (115–142 Nm)

15. Fill the transaxle to the proper level with the specified fluid.

STEERING AND SUSPENSION

Air Bag

✷✷ CAUTION

Some vehicles are equipped with an air bag system, also known as the Supplemental Inflatable Restraint (SIR) system. The system must be disabled before performing service on or around system components, steering column, instrument panel components, wiring and sensors. Failure to follow safety and disabling procedures could result in accidental air bag deployment, possible personal injury and unnecessary system repairs.

PRECAUTIONS

Several precautions must be observed when handling the inflator module to avoid accidental deployment and possible personal injury.

- Never carry the inflator module by the wires or connector on the underside of the module.
- When carrying a live inflator module, hold securely with both hands, and ensure that the bag and trim cover are pointed away.

- Place the inflator module on a bench or other surface with the bag and trim cover facing up.
- With the inflator module on the bench, never place anything on or close to the module which may be thrown in the event of an accidental deployment.

DISARMING

✷✷ CAUTION

The Supplemental Inflatable Restraint (SIR) system must be disarmed before performing service around SIR system components or SIR system wiring. Failure to do so may cause accidental deployment of the air bag, resulting in unnecessary SIR system repairs and/or personal injury.

1. Disconnect both battery cables from the battery, negative cable first.

2. Wait one minute before proceeding with the service procedure. This is the time required for the back-up power supply in the air bag diagnostic monitor to deplete its stored energy.

3. After service is completed, reconnect the battery cables, negative cable last.

4. Turn the ignition switch to the **RUN** position. The air bag indicator should light continuously for approximately six seconds, then turn OFF. If the indicator fails to light, flashes or remains lit continuously, there is a fault in the air bag system.

Power Rack and Pinion Steering Gear

REMOVAL & INSTALLATION

1. Disconnect the negative battery cable.
2. From inside the vehicle, remove the boot covering the shaft at the cowl panel.
3. Remove the bolts retaining the intermediate shaft to the steering column shaft. Set the weather boot aside.
4. Remove the pinch bolt at the steering gear input shaft and remove the intermediate shaft. Raise the vehicle and support safely.
5. Safely raise and support the vehicle.
6. Remove the front wheels.
7. Remove the exhaust system flex tube, then remove the dual converter Y-pipe from the vehicle.
8. Support the rear end of the front subframe with tall jackstands.
9. Disconnect the tie rod ends from the steering knuckle tie rods. Mark the position of the jam nuts for correct installation.

10. Remove the nuts securing the steering gear to the subframe.

11. Remove the bolts securing the rear of the subframe to the body.

12. Carefully raise the lift until the subframe separates from the body approximately 4 inches (102mm).

13. Remove the push-pin retainers from the power steering hose bracket, then remove the shield.

14. Remove the left turn pressure hose from the bracket, then remove the bracket.

15. Disconnect the left stabilizer link from the strut assembly.

16. Disconnect the power steering auxiliary actuator.

17. Rotate the steering gear to clear the mounting bolts and move the gear to the left to gain access to the hose fittings.

18. Position a drain pan under the steering gear and disconnect the pressure and return hoses.

19. Remove the steering gear assembly through the left wheel opening.

To install:

20. Install new Teflon® seals on the hydraulic line fittings on the gear assembly. Install the mounting bolts in the steering gear housing, then insert the steering gear through the left fender apron.

21. Install the remaining components in the reverse order of removal.

22. Align the steering gear bolts, install the nuts and tighten to 85–100 ft. lbs. (115–135 Nm). Lower the vehicle.

23. Tighten the hydraulic pressure line and return line to 15–25 ft. lbs. (20–35 Nm).

24. Tighten the tie rod end castle nuts to 35 ft. lbs. (48 Nm). If necessary, tighten the nuts a little bit more to align the slot in the nut for the cotter pin. Install the cotter pin.

25. Fill and bleed the power steering system. Check the system for leaks and proper operation. Adjust the toe setting as necessary.

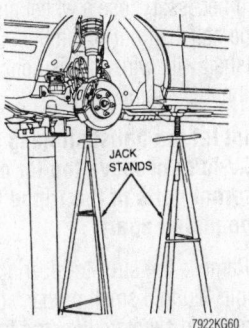

7922KG60

Support the rear of the subframe with two tall jackstands while removing the steering gear

Strut

REMOVAL & INSTALLATION

Front

1995 MODELS

➡Be sure to have available for each strut a new hub retainer nut, tie rod end castellated nut and cotter pin, lower control arm-to-steering knuckle bolt and nut and a strut-to-steering knuckle pinch bolt.

1. Place the ignition switch in the **OFF** position and the steering column in the UNLOCKED position.

2. With all wheels on the ground, remove the hub retainer nut. Discard the nut.

3. Loosen three top mount-to-strut tower retaining nuts. Do not remove the nuts at this time.

4. Raise and safely support the vehicle. Remove the wheel and tire assembly.

➡When raising the vehicle, do not lift by using the lower control arms.

5. Remove the brake caliper and hang it out of the work area with wire. Do not disconnect the brake hose from the brake caliper.

6. Remove the brake rotor.

7. Remove the cotter pin and castellated nut from the tie rod end. Discard cotter pin and nut.

8. Using Tie Rod End Remover 3290-D and Adapter T81P-3504-W or equivalents, separate the tie rod from the steering knuckle.

9. Remove the stabilizer bar link nut and the link from the strut.

10. Remove the lower control arm-to-steering knuckle pinch bolt and nut. It may be necessary to use a drift punch to remove the bolt. Discard the pinch bolt and nut.

11. Spread the pinch joint if needed and carefully pry the lower control arm down and away from the steering knuckle.

12. If necessary, use a wheel puller to press the halfshaft out of the hub. Support the halfshaft with wire so it is not hanging on the inner CV-joint.

➡Do not let the halfshaft hang by the inner CV-joint or move too far outward. The internal parts of the tripod CV-joint could be pulled apart.

13. Remove the strut-to-steering knuckle pinch bolt. Using a small prybar, spread the pinch joint and separate the strut from the steering knuckle. Remove the steering knuckle/hub assembly from the strut. Discard the pinch bolt.

14. Support the strut and remove three top mount-to-strut tower nuts. Lower the strut assembly from the vehicle.

To install:

15. Raise and position the strut assembly in the strut tower. Hand start three top mount-to-strut tower retaining nuts.

16. Install the steering knuckle and hub assembly to the strut. Install a new pinch bolt and nut and tighten to 73–97 ft. lbs. (98–132 Nm).

17. Carefully align the splines and install the halfshaft into the hub. Loosely install a new hub retainer nut.

18. Install the lower control arm to the steering knuckle, making sure the ball joint stud groove is properly positioned. Install a new bolt and nut. Tighten the nut to 40–53 ft. lbs. (53–72 Nm).

19. Install the stabilizer link to the strut and install a new nut. Tighten to 55–75 ft. lbs. (75–101 Nm).

20. Install the tie rod end to the steering knuckle using a new castellated nut. Tighten the nut 35–46 ft. lbs. (47–63 Nm) on 1995 vehicles and install a new cotter pin.

21. Install the disc brake rotor and caliper.

22. Install the wheel and tire assembly. Lower the vehicle.

23. Tighten three top mount-to-strut tower nuts to 23–30 ft. lbs. (30–40 Nm). Tighten the hub retainer nut to 180–200 ft. lbs. (244–271 Nm).

24. Apply the brake pedal several times before moving the vehicle to position the brake pads.

25. Check the front wheel alignment. Road test the vehicle and check for proper operation.

1996–99 MODELS

➡Be sure new wheel hub retainer nuts, tie rod end castellated nuts, stabilizer bar link nuts and knuckle-to-strut pinch bolt/nuts are available. These parts lose their torque holding/retention capabilities during removal and must not be reused.

1. Turn the ignition switch to the **OFF** position. Leave the steering column in the UNLOCKED position.

2. With all four wheels on the ground, remove the wheel hub retainer nut. Discard the nut.

3. On the SHO vehicles:

a. Disconnect the height sensor wiring connector.

b. Remove the wiring harness from the routing clip on the front shock absorber.

c. Remove air suspension height sensor from the height sensor ball studs.

4. Loosen the three top mount-to-shock tower nuts, but do not remove the nuts at this time.

5. Raise and safely support the vehicle. Remove the wheel and tire assembly.

➡When raising the vehicle, do not lift by using the lower control arms.

6. Remove the disc brake caliper. Hang the caliper out of the work area with wire to prevent damage to the brake hose.

7. Remove the disc brake rotor.

8. Remove the anti-lock brake sensor wiring harness clip and the mounting screw from the brake hose bracket on the strut assembly. Move the anti-lock brake sensor aside.

9. Remove the cotter pin and the castellated nut from the tie rod end. Discard the cotter pin and nut.

10. Using Removal tool 3290-D, or equivalent and Adapter tool T81P-3504-W or equivalent, separate the tie rod end from the steering knuckle.

11. On SHO vehicles, remove the vinyl cover from the upper link stud.

12. Remove the sway bar (stabilizer bar) link nut and link from the strut. Discard the link nut.

13. Remove and discard the lower ball joint retaining nut. Using Ball Joint Remover T96P-3010-A or equivalent, separate the ball joint from the lower control arm.

14. Using Spring Compressor 164-R-3571 or equivalent, compress the coil spring until the ball joint clears the lower arm.

15. Remove the pinch bolt and nut from the bottom of the steering knuckle. It may be necessary to use a drift punch to remove the bolt. Discard the pinch bolt and nut.

16. Separate the halfshaft from the wheel hub using Front Hub Remover/Replacer T81P-1104-C or equivalent and the required adapters.

17. Support the halfshaft with wire in a level position to prevent it from hanging by the inner CV-joint.

➡Do not let the halfshaft hang by the inner CV-joint or move too far outward. The internal parts of the tripod CV-joint could be pulled apart.

18. Remove the three top mount-to-strut tower nuts while supporting the strut. Lower the strut assembly from the vehicle.

To install:

19. Install the strut assembly with the spring compressor installed to the vehicle and install the three top mount-to-strut tower nuts loosely.

➡️**Further compress the coil spring if added clearance is required for installation.**

20. Install the steering knuckle and hub assembly to the strut. Install a new pinch bolt and nut. Tighten to 73–97 ft. lbs. (98–132 Nm).

21. Install the halfshaft into the hub using care to align the splines.

22. Install the sway bar link to the strut and install a new sway bar link nut. Tighten to 55–75 ft. lbs. (75–101 Nm).

23. On SHO vehicles, install the vinyl cover to the upper link stud.

24. Install the tie rod end onto the steering knuckle using a new castellated nut. Tighten the nut to 35–46 ft. lbs. (47–63 Nm). Continue to tighten the nut until a slot lines up with the opening in the tie rod end stud and install a new cotter pin.

25. Install the anti-lock brake sensor wiring routing clip and the brake hose bracket mounting screw and tighten to 11 ft. lbs. (15 Nm).

26. On the SHO vehicles:

a. Install air suspension height sensor to the height sensor ball studs.

b. Install the wiring harness to the routing clip on the front shock absorber.

c. Connect the height sensor wiring connector.

27. Install the disc brake rotor and caliper. Tighten the caliper anchor bracket bolts to 65–87 ft. lbs. (88–118 Nm).

28. Install the wheel and tire assembly. Tighten the lug nuts to 85–105 ft. lbs. (115–142 Nm).

29. Tighten the three top mount-to-shock tower nuts to 22–29 ft. lbs. (30–40 Nm).

30. Lower the vehicle.

31. Install a new wheel hub retainer nut. Tighten the nut to 170–202 ft. lbs. (230–275 Nm).

32. Pump the brake pedal several times prior to moving the vehicle, to position the brake pads.

33. Road test the vehicle and check for proper operation.

Rear

1. Remove the package tray trim panel and loosen the three nuts attaching the strut to the body.

2. Raise the vehicle and remove the rear wheel.

3. Remove the brake load sensor from the rear suspension arm.

4. Remove the clip holding the brake hose to the strut and position the hose out of the way.

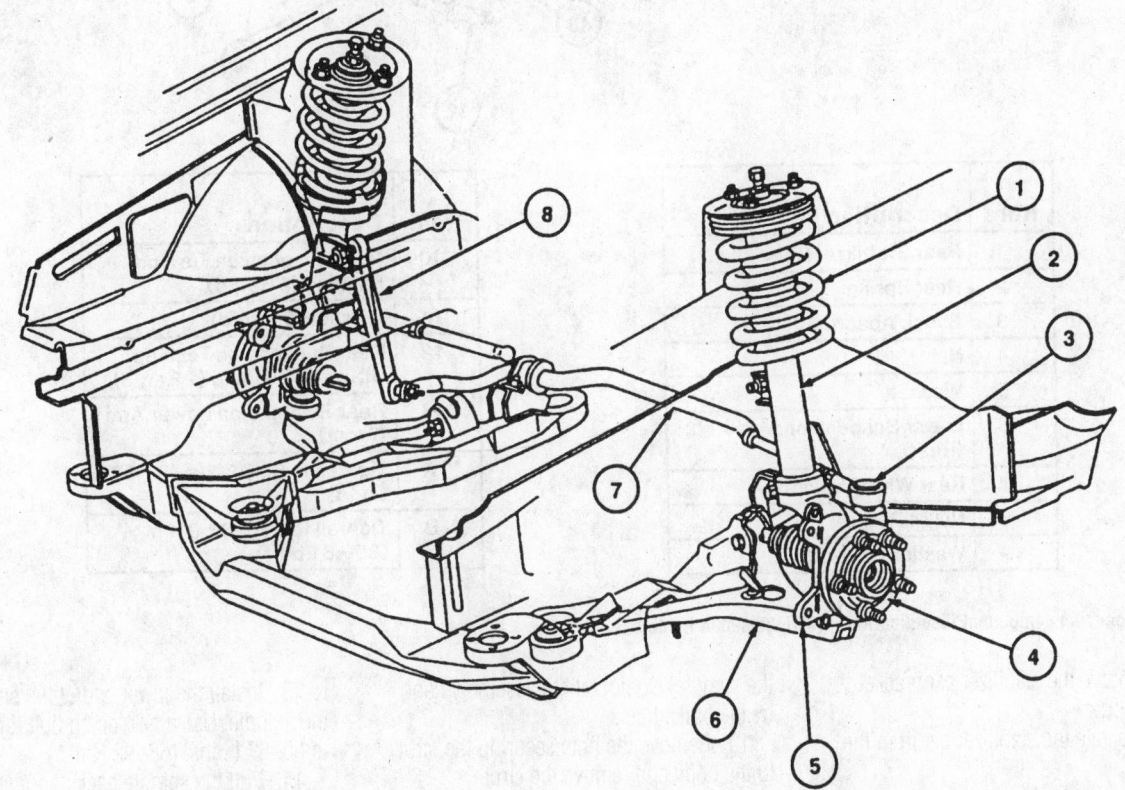

Item	Description
1	Front Coil Spring
2	Front Shock Absorber
3	Tie Rod End
4	Wheel Hub

Item	Description
5	Front Wheel Knuckle
6	Front Suspension Lower Arm
7	Front Stabilizer Bar
8	Stabilizer Bar Link

7922KG61

Front suspension component identification—1996–99 models

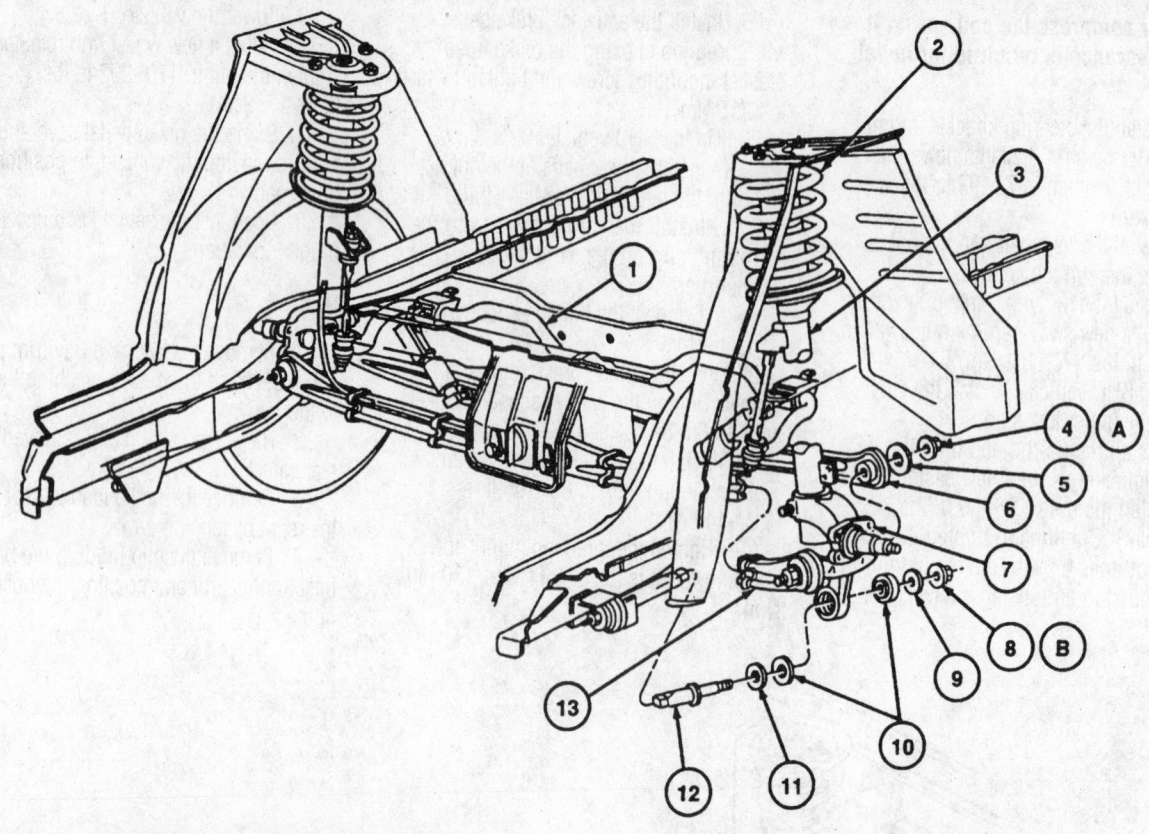

Item	Description
1	Rear Stabilizer Bar
2	Rear Spring
3	Shock Absorber
4	Nut
5	Washer
6	Lower Suspension Arm (Rear)
7	Rear Wheel Spindle
8	Nut (4 Req'd)
9	Washer (2 Req'd)

Item	Description
10	Rear Suspension Tie Rod Bushing (4 Req'd)
11	Washer (2 Req'd)
12	Rear Suspension Tension Strut and Bushing (2 Req'd)
13	Rear Suspension Lower Arm (Front)
A	Tighten to 68-92 N·m (50-67 Lb-Ft)
B	Tighten to 46.7-63.3 N·m (35-46 Lb-Ft)

7922KG62

Rear suspension component identification—1996–99 sedan models

5. Remove the stabilizer bar bracket from the body.

6. Separate the stabilizer bar from the link.

7. Remove the tension strut from the front of the rear wheel spindle. Pull the spindle back to remove the tension strut.

8. If required, remove the stabilizer link from the strut.

9. Support the spindle with a jack and remove the pinchbolt securing the strut to

the spindle. Do not let the assembly hang by the brake line.

10. Remove the nuts securing the strut to the body and remove the strut.

To install:

11. If removed, install the stabilizer link on the strut. Tighten the nut to 60–81 inch lbs. (7–9 Nm).

12. Position the strut in the vehicle and install the three upper mounting nuts. Do not tighten the nuts at this time.

13. Install the lower end of the strut in the spindle. Use a new pinch bolt tightened to 50–67 ft. lbs. (68–92 Nm).

14. Pull the spindle back and install the tension strut. Tighten the nut to 35–46 ft. lbs. (68–92 Nm).

15. Connect the link to the stabilizer bar. Tighten the nut to 62–79 inch lbs. (7–9 Nm).

16. Install the stabilizer bar bracket on the body. Tighten the bolts to 25–33 ft. lbs. (34–46 Nm).

17. Install the brake hose on the strut.

18. Connect the load sensor to the suspension arm.

19. Tighten the three upper mounting nuts to 19–25 ft. lbs. (25–34 Nm).

20. Install the rear wheel and lower the vehicle.

Shock Absorber

REMOVAL & INSTALLATION

Wagons Only

➡ **Before continuing, be sure new mounting bolts and nuts and shock absorber insulator bushings are available.**

1. Raise the vehicle enough to allow wheel and tire removal. Safely support the vehicle.

2. Remove the wheel and tire assembly.

❋❋ WARNING

The lower control arm must be supported before removal of the upper or lower shock absorber attachments to prevent injury or damage to attached components.

3. On 1994–95 models, remove the two nuts retaining the shock absorber to the lower suspension arm.

4. From inside the vehicle, remove the rear compartment access panel.

5. Remove the top shock absorber retaining nut using a crow foot wrench and ratchet while holding the shock absorber shaft stationary with an open-end wrench. Do not grip the shaft of the shock absorber if it is to be reused. Discard the retaining nut.

6. Remove the upper washer and insulator from the shock absorber.

➡ **The shock absorbers are gas filled. It will require an effort to collapse the shock to remove it from the lower control arm.**

7. On 1996–98 models, remove the lower shock absorber mounting nut and bolt.

8. Remove the shock absorber from the vehicle. Discard the nut and bolt.

To install:

9. Install a new washer and insulator on the upper shock absorber rod.

10. Maneuver the upper part of the shock absorber into the shock tower opening in the body. Push slowly on the lower part of the shock absorber until the lower bracket is aligned with the mounting holes in the lower control arm.

11. On 1996–98 models, install a new retaining bolt and nut;, then, tighten to 50–68 ft. lbs. (68–92 Nm).

12. From inside the vehicle, install a new insulator, washer and nut on top of the shock absorber shaft. Tighten the nut to 19–25 ft. lbs. (26–34 Nm.).

13. Install the rear compartment access panel.

14. On 1994–95 models, tighten the two lower attaching nuts to 15–19 ft. lbs. (19–26 Nm).

15. Install the wheel and tire assembly. Remove the jackstand.

16. Lower the vehicle. Road test the vehicle and check for proper operation.

Coil Spring

REMOVAL & INSTALLATION

Front

❋❋ CAUTION

Do not remove the strut rod nut until the spring has been compressed until it comes away from the seat.

1. Remove the strut from the vehicle.

2. Compress the coil spring using a commercially available spring compressor until the spring comes away from the seat.

3. Remove the large center nut and slowly release the spring compressor.

To install:

4. Compress the spring and install it on the strut.

5. Install the lower washer and mounting bracket.

6. Install the upper washer and a new nut. Tighten the nut to 39–53 ft. lbs. (53–72 Nm) while holding the rod with a T-50 size Torx® socket.

7. Install the strut assembly in the vehicle.

Taurus and Sable Wagons

1. Raise and safely support the vehicle. Remove the wheel and tire assembly

2. Position a floor jack under the lower suspension control arm.

❋❋ WARNING

The lower control arm must be supported before removal of the upper or lower shock absorber mounts to prevent injury or damage to the related components due to tension applied by the coil spring.

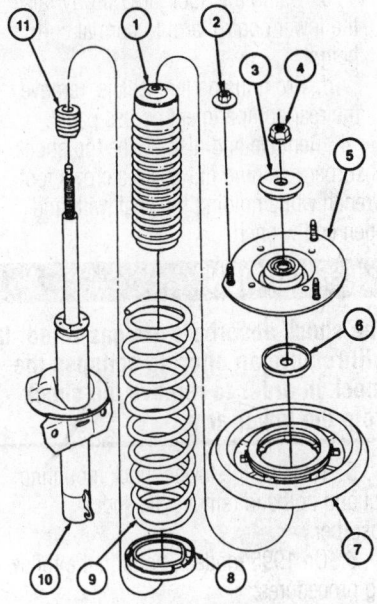

Item	Description
1	Dust Boot (Part of 18124)
2	Nut (3 Req'd)
3	Washer
4	Nut
5	Front Shock Absorber Mounting Bracket
6	Washer
7	Front Suspension Bearing and Seal
8	Front Spring Insulator (Part of 18124)
9	Front Coil Spring
10	Front Shock Absorber
11	Jounce Bumper (Part of 18124)

7922KG63

Exploded view of the front strut and coil spring assembly—1996–99 models

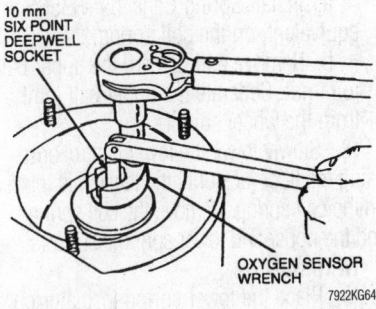

7922KG64

Hold the strut rod while loosening or tightening the nut—1996–99 models

3. On 1996–99 models, perform the following procedures:

a. Remove the bolt retaining the rear brake hose bracket to the body.

b. Remove the stabilizer bar and bracket from the lower control arm.

c. Using the floor jack, slowly raise the lower control arm to normal curb height.

d. From inside the vehicle, remove the rear compartment access panel.

4. Remove and discard the top shock absorber retaining nut using a crows foot wrench while holding the shaft with and open end wrench.

✳✳ CAUTION

The shock absorbers are gas-filled. It will require an effort to collapse the shock in order to remove the shock from the lower arm.

5. Remove the lower shock mounting nut and bolt and remove the shock absorber.

6. On 1995 models, perform the following procedures:

a. Disconnect and remove the parking brake cable and clip from the lower suspension arm.

b. If equipped with rear disc brakes, remove the ABS cable from the clips on the lower suspension arm.

c. Remove and discard the bolt and nut attaching the tension strut to the lower suspension arm.

d. Suspend the spindle and upper suspension arms from the body with a piece of wire to prevent them from dropping.

e. Remove the nut, bolt, washer and adjusting cam that retain the lower suspension arm to the spindle. Discard the nut, bolt and washer and replace with new. Set the cam aside.

7. On 1996–99 models, perform the following procedures:

a. Install Spring Cage 164-R3555 or equivalent, on the coil spring.

b. Remove and discard the upper ball joint nut. Separate the upper ball joint from the wheel spindle.

8. Slowly lower the lower control arm using the floor jack until the tension is relaxed on the coil spring. Remove the coil spring and the upper and lower spring insulators.

To install:

9. Place the lower spring insulator on the lower control arm. Press the insulator downward into place, making certain that the insulator is properly seated.

10. Position the upper insulator on top of the coil spring. Install the coil spring on the lower control arm. Make certain the spring is properly seated.

11. Using the floor jack, slowly raise lower control arm. Guide the upper spring insulator onto the upper spring seat on the underbody.

12. Position the upper ball joint into the upper control arm. Install a new nut and tighten to 50–68 ft. lbs. (68–92 Nm).

13. On 1995 models, perform the following procedures:

a. Position the spindle in the lower suspension arm with a new bolt, nut, washer, and the existing cam. Install the bolt with the head of the bolt toward the front of the vehicle. Do not tighten the bolt at this time.

b. Remove the wire supporting the spindle and suspension arms.

c. Install the tension strut in the lower suspension arm using a new nut and bolt; do not tighten at this time.

d. Attach the parking brake cable and clip to the lower suspension arm.

e. If equipped with rear disc brakes, install the ABS cable into the clips on the lower suspension arm.

14. Position the shock absorber into the tower opening with a new washer and insulator installed. Push on the lower end of the shock until the lower bracket is lined up with the mounting holes in the lower control arm. Install a new lower retaining bolt and nut. Tighten to 50–68 ft. lbs. (68–92 Nm).

15. From inside the vehicle, install a new upper shock absorber insulator and washer. Tighten the nut to 19–25 ft. lbs. (25–34 Nm).

16. On 1995 models, perform the following procedures:

a. Attach the sway bar U-bracket to the lower suspension arm using a new bolt. Tighten the bolt to 23–30 ft. lbs. (30–40 Nm).

b. Attach the flexible brake hose to the body and tighten the bolt to 8–12 ft. lbs. (11–16 Nm).

c. With the floor jack, raise the rear suspension arm and bushing to normal position when at curb height. Tighten the rear suspension arm and bushing to rear wheel spindle nut to 40–52 ft. lbs. (54–71 Nm). Tighten the rear suspension tension strut and bushing to body bracket bolt to 40–52 ft. lbs. (54–71 Nm).

17. On 1996–99 models, perform the following procedures:

a. Install the rear compartment access panel.

b. Install the stabilizer bar and bracket to the lower control arm. Tighten to 15–19 ft. lbs. (19–26 Nm).

18. Install the brake hose support bracket to the body and install the retaining bolt. Tighten the bolt to 10 ft. lbs. (12 Nm).

19. Install the wheel and tire assembly. Tighten the lug nuts to 85–105 ft. lbs. (115–142 Nm).

20. Remove the floor jack.

21. Lower the vehicle.

22. Check the rear wheel alignment and adjust if necessary.

23. Road test the vehicle and check for proper operation.

Lower Ball Joints

REMOVAL & INSTALLATION

1995 Models

The lower ball joint is an integral component of the lower control arm. If the ball joint is defective, the entire lower control arm must be replaced.

1996–99 Models

The lower ball joint is an integral part of the steering knuckle. If the lower ball joint is found to be defective, the entire steering knuckle must be replaced.

Wheel Bearings

ADJUSTMENT

There is no adjustment for the front or rear wheel bearings due to the nature of their design. These bearings are permanently lubricated and require no periodic maintenance.

REMOVAL & INSTALLATION

Front

1995 MODELS

➡Before beginning this procedure, be sure to have available a new hub retainer nut, tie rod end castellated nut and a lower control arm-to-steering knuckle pinch bolt and nut.

1. Turn the ignition switch to the **OFF** position. Position the steering wheel in the UNLOCKED position.

2. With all wheels on the ground, loosen and remove the hub retaining nut. Discard the nut.

3. Raise and safely support the vehicle. Remove the wheel and tire assembly.

4. Remove the cotter pin from the tie rod end stud and remove the castellated nut. Discard the cotter pin and nut.

5. Using a tie rod end removal tool, separate the tie rod end from the steering knuckle.

6. Remove the stabilizer bar link assembly from the strut.

7. Remove the brake caliper and wire it aside in order to gain working clearance.

8. Loosen but do not remove the three upper strut retaining nuts from the top of the strut tower.

9. Remove and discard the lower control arm-to-steering knuckle pinch bolt and nut. Using a prybar or similar tool, spread the pinch joint apart and carefully pry the lower control arm from the steering knuckle.

➡Be sure the steering column is in the UNLOCKED position. Do not use a hammer to perform this operation. Use extreme care not to damage the boot seal.

10. Remove the strut-to-steering knuckle pinch bolt.

❊❊ WARNING

Do not allow the halfshaft to move outboard or to hang by the inner CV-joint. Over extension of the CV-joint could result in separation of internal parts, causing failure of the joint.

11. Press the halfshaft from the hub with a wheel puller. Wire the halfshaft to the body to maintain a level position. If equipped, remove the rotor splash shield.

12. Remove the steering knuckle and hub assembly from the strut.

13. Install Front Hub Puller D80L-1002-L, or equivalent and Shaft Protector D80L-625–1 or equivalent, with the jaws of the puller on the knuckle bosses. Be sure the shaft protector is centered, clears the bearing inside diameter and rests on the end face of the hub journal. Remove the hub.

14. Remove the snapring that retains the bearing in the knuckle assembly and discard.

15. Using a suitable hydraulic press, place Front Bearing Spacer T86P-1104-A2 or equivalent, on the press plate with the step side facing up and position the knuckle with the outboard side up on the spacer. Install Front Bearing Remover T83P-1104-AH2 or equivalent, centered on the bearing inner race and press the bearing out of the knuckle and discard.

To install:

16. Remove all foreign material from the knuckle bearing bore and hub bearing journal to ensure correct seating of the new bearing.

➡If the hub bearing journal is scored or damaged it must be replaced. The front wheel bearings are pregreased and sealed and require no scheduled maintenance. The bearings are preset

and cannot be adjusted. If a bearing is disassembled for any reason, it must be replaced as a unit, as individual service seals, rollers and races are not available.

17. Place Front Bearing Spacer T86P-1104-A2 or equivalent, with the step side down on the hydraulic press plate and position the knuckle with the outboard side down on the spacer. Position a new bearing in the inboard side of the knuckle. Install Bearing Installer T86P-1104-A3 or equivalent, with the undercut side facing the bearing, on the bearing outer race and press the bearing into the knuckle. Be sure the bearing seats completely against the shoulder of the knuckle bore.

➡Bearing Installer T86P-1104-A3 or equivalent, must be positioned as indicated above to prevent bearing damage during installation.

18. Install a new snapring (part of the bearing kit) in the knuckle groove.

19. Place Front Bearing Spacer T86P-1104-A2 or equivalent, on the press plate and position the hub on the tool with the lugs facing downward. Position the knuckle assembly with the outboard side down on the hub barrel. Place Bearing Remover T83P-1104-AH2 or equivalent, flat side down, centered on the inner race of the bearing and press down on the tool until the bearing is fully seated onto the hub. Be sure the hub rotates freely in the knuckle after installation.

20. Prior to hub/bearing/knuckle installation, replace the bearing dust seal on the outboard CV-joint with a new seal from the bearing kit. Be sure the seal flange faces outboard toward the bearing. Use Drive Tube T83T-3132-A1 and front bearing Dust Seal Installer T86P-1104-A4, or equivalent.

21. Install the rotor splash shield using new rivets, if equipped.

22. Install the steering knuckle onto the strut. Loosely install a new pinch bolt in the knuckle to retain the strut.

23. Install the steering knuckle and hub onto the halfshaft. Loosely install a new hub retaining nut.

24. Install the lower control arm to the knuckle. Be sure the ball stud groove is properly positioned. Install a new nut and bolt. Tighten to 40–53 ft. lbs. (53–72 Nm).

25. Tighten the strut-to-knuckle pinch bolt to 70–95 ft. lbs. (98–132 Nm).

26. Install the disc brake rotor and brake caliper.

27. Position the tie rod to the steering knuckle, install a new castellated nut and

tighten to 35 ft. lbs. (47 Nm). If necessary advance the nut to align the slot and install a new cotter pin.

28. Install the stabilizer bar link to the strut. Tighten the nut to 57–75 ft. lbs. (77–103 Nm).

29. Install the wheel and tire assembly. Tighten the lug nuts to 85–105 ft. lbs. (115–142 Nm).

30. Lower the vehicle.

31. Tighten three upper strut retaining nuts to 23–29 ft. lbs. (30–40 Nm).

32. With all wheels on the ground, tighten the hub retaining nut to 180–200 ft. lbs. (245–275 Nm).

33. Pump the brake pedal several times prior to moving the vehicle, in order to position the brake pads.

34. Road test the vehicle and check for proper operation.

1996–99 MODELS

➡Be sure new wheel hub retainer nuts, tie rod end castellated nuts, hub-to-knuckle retaining bolts, knuckle-to-strut pinch bolt/nut and inboard halfshaft circlips are available. These parts lose their torque holding/retention capabilities during removal and must not be reused.

1. Turn the ignition switch to the **OFF** position. Place the steering column in the UNLOCKED position.

2. Remove the wheel hub retainer nut before raising the vehicle off the ground. Discard the wheel hub retainer nut.

3. Raise and safely support the vehicle. Remove the wheel and tire assembly.

➡When raising the vehicle, do not lift by using the lower control arms.

4. Remove the wheel and tire assembly.

5. Remove the cotter pin and the castellated nut from the tie rod end. Discard the cotter pin and nut.

6. Separate the tie rod end from the steering knuckle using Remover tool 3290-D and adapter T81P-3504-W, or equivalents.

7. On SHO vehicles, remove the vinyl cover from the upper link stud.

8. Remove the stabilizer link from the strut. Remove the disc brake caliper and hang it aside.

9. Remove the disc brake rotor.

10. Remove the anti-lock sensor and move it aside.

11. Remove and discard the lower ball joint retaining nut. Using Ball Joint Remover T96P-3010-A or equivalent, separate the ball joint from the lower control arm.

12. Using Rotunda Spring Compressor 164-R-3571 or equivalent, compress the coil spring until the ball joint clears the lower control arm.

13. Remove and discard the steering knuckle-to-strut pinch bolt and nut.

14. Separate the halfshaft from the wheel hub using Front Hub Remover/Replacer T81P-1104-C or equivalent and adapters.

15. Support the halfshaft with wire in a level position to prevent it from hanging by the inner CV-joint.

➡ **Do not let the halfshaft hang by the inner CV-joint or move too far outward. The internal parts of the tripod CV-joint could be pulled apart.**

16. Separate the steering knuckle from the strut assembly and place on a suitable workbench.

17. Remove the three hub and bearing retainer bolts from the back of the steering knuckle while using a prybar to steady the assembly. Discard the three hub and bearing retainer bolts.

✳✳ WARNING

The wheel hub is not pressed into the front wheel knuckle. DO NOT USE a slide hammer to remove a stuck wheel hub. Do not strike the back of the inner bearing race.

18. Remove the wheel hub from the steering knuckle using a suitable prybar.

19. Inspect all components and replace as necessary. The wheel bearings are not serviceable and must be replaced with a new wheel hub assembly.

To install:

20. Install the disc brake rotor shield using new rivets, if removed.

➡ **If the hub bearing journal is scored or damaged, replace the steering knuckle. If the wheel hub is damaged** or any end-play is detectable, replace the wheel hub.

21. Remove all foreign material from the knuckle bearing bore and hub bearing journal to ensure correct seating of the new hub.

➡ **The knuckle must be clean enough to allow the wheel hub to be completely seated by hand. Do not press or draw the wheel hub into place.**

22. Place the wheel hub to the steering knuckle using light oil. Push the wheel hub assembly into the steering knuckle. Install bolts and tighten to 61–78 ft. lbs. (83–107 Nm).

23. Position the steering knuckle assembly to the vehicle.

24. Install the steering knuckle to the strut and loosely install a new pinch bolt.

25. Install the steering knuckle and hub assembly onto the halfshaft. Be sure the splines are properly aligned.

26. Slowly release Rotunda Spring Compressor 164-R-3571 or equivalent, while guiding the lower ball joint into the lower control arm.

27. Remove the spring compressor.

28. Install a new nut on the lower ball joint stud and tighten to 50–67 ft. lbs. (68–92 Nm).

29. Install a new nut on the steering knuckle-to-strut pinch bolt. Tighten the pinch bolt nut to 72–97 ft. lbs. (98–132 Nm)

30. Position the tie rod end to the steering knuckle. Install a new castellated nut and tighten to 35–46 ft. lbs. (47–63 Nm). Install a new cotter pin.

31. Install the sway bar link and tighten the nut to 57–75 ft. lbs. (77–103 Nm).

32. On SHO vehicles, install the vinyl cover to the upper link stud.

➡ **Use care not to damage the sway bar link boot seals. Do not use power tools** to tighten the nuts or seal damage will result.

33. Install the disc brake rotor and disc brake caliper. Tighten the caliper anchor bracket bolts to 65–87 ft. lbs. (88–118 Nm).

34. Install the wheel and tire assembly.

35. Lower the vehicle.

36. Install a new wheel hub retainer nut. Tighten the nut to 170–202 ft. lbs. (230–275 Nm).

37. Pump the brake pedal several times prior to moving the vehicle, to position the brake pads.

38. Road test the vehicle and check for proper operation.

Rear

1. Raise and safely support the vehicle. Remove the wheel and tire assembly.

2. Remove the bolt retaining the brake hose to the strut bracket. Remove the rear disc/drum brake assembly.

3. Remove the grease cap from the bearing and hub assembly and discard the grease cap.

4. Remove the bearing and hub assembly retaining nut and discard. Remove the bearing and hub assembly from the spindle.

To install:

5. Position the wheel hub and bearing on the rear spindle.

6. Install a new wheel hub retainer nut and tighten to 188–254 ft. lbs. (255–345 Nm).

7. Install a new hub cap grease seal using Shaft Protector T89P-19623-FH, or equivalent. Tap on the tool until the hub cap grease seal is fully seated.

8. Install the rear disc/drum brake assembly.

9. Install the wheel and tire assembly. Lower the vehicle.

10. Check the front end alignment. Road test the vehicle and check for proper operation.

FORD MOTOR CO.

Lincoln-Continental

DRIVE TRAIN **21-14**
ENGINE REPAIR **21-2**
FUEL SYSTEM **21-12**
STEERING AND
  SUSPENSION **21-16**

A
Air Bag 21-16
  Disarming 21-16
  Precautions 21-16
Air Spring 21-19
  Removal & Installation 21-19

C
Camshaft and Valve Lifters 21-7
  Removal & Installation 21-7
Cylinder Head 21-3
  Removal & Installation 21-3

E
Engine Assembly 21-2
  Removal & Installation 21-2
Exhaust Manifold 21-6
  Removal & Installation 21-6

F
Fuel Filter 21-13
  Removal & Installation 21-13
Fuel Pump 21-13

Removal & Installation 21-13
Fuel System Pressure 21-13
  Relieving 21-13
Fuel System Service
  Precautions 21-12

H
Halfshaft 21-15
  Removal & Installation 21-15

I
Ignition Timing 21-2
  Adjustment 21-2
Intake Manifold 21-4
  Removal & Installation 21-4

L
Lower Ball Joint 21-19
  Removal & Installation 21-19

O
Oil Pan 21-8
  Removal & Installation 21-8
Oil Pump 21-9
  Removal & Installation 21-9

P
Power Rack and Pinion Steering
  Gear 21-16
  Removal & Installation 21-16

R
Rear Main Seal 21-9
  Removal & Installation 21-9
Rocker Arms 21-4
  Removal & Installation 21-4

S
Shock Absorber 21-18
  Removal & Installation 21-18
Strut 21-17
  Removal & Installation 21-17

T
Timing Chain, Sprockets, Front
  Cover and Seal 21-10
  Removal & Installation 21-10
Transaxle Assembly 21-14
  Removal & Installation 21-14

V
Valve Lash 21-8
  Adjustment 21-8

W
Water Pump 21-3
  Removal & Installation 21-3
Wheel Bearings 21-19
  Adjustment 21-19
  Removal & Installation 21-19

ENGINE REPAIR

➡**Disconnecting the negative battery cable on some vehicles may interfere with the functions of the on board computer systems and may require the computer to undergo a relearning process, once the negative battery cable is reconnected.**

Ignition Timing

ADJUSTMENT

The base ignition timing is set by the Powertrain Control Module (PCM) and is not adjustable.

Engine Assembly

✳✳ CAUTION

Some models covered by this manual may be equipped with a Supplemental Restraint System (SRS), which uses an air bag. Whenever working near any of the SRS components, such as the impact sensors, the air bag module, steering column and instrument panel, disable the SRS using the recommended procedure.

REMOVAL & INSTALLATION

✳✳ CAUTION

The air suspension switch, located in the left-hand side of the luggage compartment on 1995–97 models or on the right side kick panel for 1998–99 models, must be turned OFF before raising the vehicle. Failure to do so may result in unexpected inflation or deflation of the air springs which may result in shifting of the vehicle during service.

➡**Be sure to have available a New Generation Star (NGS) tester or equivalent scan tool with the proper software to properly deflate the air springs before servicing the suspension system. Be sure to have available a new wheel hub retainer nut, lower ball-joint to lower control arm nut and a halfshaft circlip, per side.**

1. Place the vehicle over a frame contact hoist. Properly deflate the air springs.

2. Disconnect both battery cables, negative cable first. Drain the engine cooling system.

3. Mark the position of the hood on the hinges and remove the hood.

4. Disconnect the steering coupling at the pinch bolt joint inside the passenger compartment.

5. Remove the engine appearance cover from the engine. Properly relieve the fuel system pressure.

6. Recover the refrigerant from the A/C system using the proper equipment.

➡**Label all electrical connectors and vacuum hoses prior to removal so they can be reinstalled in their proper locations.**

7. Disconnect the engine control sensor wiring from the Intake Air Temperature (IAT) sensor and disconnect the crankcase ventilation tube from air cleaner outlet tube.

8. Loosen the clamps on the air cleaner outlet tube to engine air cleaner and throttle body. Remove the air cleaner outlet tube.

9. Disconnect the chassis vacuum supply hose at the connection on the intake manifold. Position the hose aside.

10. Remove the ground straps from the dash panel.

11. Disconnect the engine control sensor wiring from the Powertrain Control Module (PCM) and position the engine control sensor wiring aside.

12. Remove the connectors for the engine control sensor wiring from the retaining bracket on the power brake booster. Disconnect the engine control sensor wiring at the two connectors.

13. Disconnect engine control sensor wiring from the Mass Air Flow (MAF) sensor.

14. Disconnect engine control sensor wiring from the evaporative emission canister purge valve.

15. Disconnect evaporative emission hose at crankcase vent connector and hose. Position the hose aside.

16. Remove the throttle cable shield and disconnect the accelerator cable and the speed control actuator from the throttle body and from the accelerator cable bracket. Position the cables aside.

17. Remove the retaining nut and disconnect the manual control lever from the manual control lever shaft at the Transmission Range (TR) sensor.

18. Remove the connectors for the engine control sensor wiring from the retaining bracket on top of the transaxle. Disconnect the engine control sensor wiring at the two connectors.

19. Disconnect the wiring connector from the secondary air injection pump relay located on the retaining bracket on top of the transaxle.

20. Disconnect the main emission vacuum control connector at the connection near the fan shroud.

21. Disconnect the oil cooler inlet tube from the transaxle. Remove the oil level dipstick from the indicator tube.

22. Disconnect the heater water hose from the water bypass tube. Disconnect the heater water hose at the rear of the right cylinder head. Disconnect the upper radiator hose at the water bypass tube.

23. Remove the power steering return hose from the power steering oil reservoir and drain.

24. Disconnect the alternator wiring harness from the alternator at the BAT terminal and stator connector plug. Remove the wiring harness retaining clip from the alternator mounting bracket.

25. Partially raise and safely support the vehicle. Remove the front wheel and tire assemblies.

26. Disconnect both ride height sensor links from the lower control arms. Remove both stabilizer bar links from the front stabilizer bar.

27. Separate both lower control arms from the steering knuckles at the ball joints. Separate both tie rod ends from the steering knuckles. Discard the cotter pins.

28. Remove both wheel hub retainer nuts from the halfshaft ends. Remove both halfshafts from the steering knuckles.

29. Raise and safely support the vehicle. Remove the splash shield from the radiator support and subframe.

30. Drain the engine oil. Remove the dual converter Y-pipe.

31. Disconnect the power steering pressure hose from the power steering/transaxle oil cooler connection. Position the hose aside.

32. Disconnect the lower radiator hose at the radiator and thermostat housing and remove the radiator.

33. Disconnect the wiring at the starter, and remove the starter. Disconnect the lower cooler line from the transaxle.

34. Disconnect the retaining clips and the A/C compressor lines from the compressor. Cap all openings to prevent the entrance of dirt or moisture.

35. Support the subframe, engine and transaxle assembly using Powertrain Lift 014–00765 and Universal Powertrain Removal Bracket 014–00766, or equivalents.

36. Remove the four subframe-to-body retaining bolts. Slowly lower the engine, transaxle and subframe from the vehicle as one assembly.

37. Disconnect the power steering pressure hose from the power steering pump.

38. Install suitable engine lifting brackets on the engine and transaxle assembly.

39. Remove the front engine support insulator, rear engine support insulator and engine and transaxle support. Lift the engine and transaxle from the subframe.

40. Lower the engine and transaxle. Support the transaxle on a level, stationary surface for transaxle storage.

41. Remove the transaxle-to-cylinder block mounting bolts and separate the engine from the transaxle/torque converter assembly. Place the engine on a suitable workstand.

To install:

42. Install the transaxle/torque converter assembly to the engine. Tighten the mounting bolts to 30–44 ft. lbs. (40–60 Nm). Tighten the torque converter nuts to 20–34 ft. lbs. (27–46 Nm).

43. Place the engine and transaxle assembly on the subframe. Install the front engine support insulator, rear engine support insulator and engine and transaxle support.

44. Connect the power steering pressure hose from the power steering pump.

45. Raise the engine, transaxle and subframe assembly into position using the powertrain lifting tool. Loosely install the four subframe to body bolts.

46. Align the subframe using a 0.75 inch (19mm) outside diameter pipe or similar tool installed through both alignment holes. Tighten the bolts to 55–75 ft. lbs. (75–102 Nm).

47. Remove the lifting equipment and move aside. Connect the retaining clips and the A/C compressor lines to the compressor.

48. Install the remaining components in the reverse order of removal, then tighten the following:
- Hub retainer nuts to 170–202 ft. lbs. (230–275 Nm)
- Tie rod end-to-steering knuckle castellated nuts to 35–46 ft. lbs. (47–63 Nm)
- Lower control arm-to-steering knuckle nuts to 50–68 ft. lbs. (68–92 Nm)
- Stabilizer bar links-to-stabilizer bar nuts to 30–40 ft. lbs. (40–55 Nm)
- Manual control lever-to-manual control lever shaft nut to 12–16 ft. lbs. (16–22 Nm)

- Air cleaner outlet tube clamps to 12–22 inch lbs. (1.4–2.5 Nm)

49. Install the hood and align as needed. Install the fuel tank fill cap.

50. Fill the cooling system with the proper type and quantity of coolant and fill the crankcase with the proper type of motor oil to the required level. Check the power steering and transaxle fluid levels and fill as needed.

✳✳ WARNING

Operating the engine without the proper amount and type of engine oil will result in severe engine damage.

51. Connect both battery cables, negative cable last. Fill the air springs. Run the engine and check for leaks.

52. Evacuate, pressure test and recharge the air conditioning system.

53. Install the engine appearance cover. Tighten the retaining nuts to 61–79 inch lbs. (7–9 Nm).

➡ **Whenever the vehicles subframe is removed or lowered, the wheel alignment should be checked.**

54. Check the front wheel alignment. Road test the vehicle and check the engine for proper operation.

Water Pump

REMOVAL & INSTALLATION

✳✳ CAUTION

Never open, service or drain the radiator or cooling system when hot; serious burns can occur from the steam and hot coolant.

1. Disconnect the negative battery cable. Drain the engine cooling system.

2. Remove the coolant recovery reservoir assembly.

3. Loosen the four bolts retaining the water pump pulley to the water pump.

4. Release the drive belt tensioner and remove the drive belt.

5. Remove the four bolts retaining the water pump pulley to the water pump and remove the pulley.

6. Remove the four bolts retaining the water pump to the cylinder block and remove the water pump.

To install:

7. Replace the water pump O-ring and

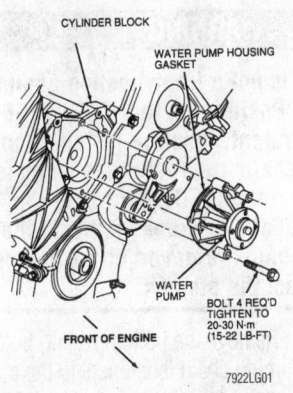

Exploded view of the water pump mounting

clean the sealing surface of the cylinder block and the water pump.

8. Lubricate the water pump O-ring with fresh coolant and install the water pump into position. Be sure the water pump is fully seated. Install the four water pump retaining bolts and tighten to 15–22 ft. lbs. (20–30 Nm).

9. Install the water pump pulley on the water pump with the four retaining bolts. Tighten to 15–22 ft. lbs. (20–30 Nm).

10. Install the remaining components in the reverse order of removal.

11. Fill the cooling system to the proper level. Connect the negative battery cable, then start the engine and check for coolant leaks.

Cylinder Head

REMOVAL & INSTALLATION

➡ **The cylinder head bolts are a torque-to-yield design and cannot be reused. Be sure new cylinder head bolts are available before performing this procedure.**

1. Disconnect the negative battery cable. Remove the engine from the vehicle and position on a suitable workstand.

2. Remove both cylinder head covers. Remove the engine front cover.

3. Remove the intake manifold. Remove the Crankshaft Position (CKP) sensor pulse wheel.

4. Remove the rocker arms. Remove the exhaust manifolds. Drain the coolant from the cylinder block.

5. Rotate the engine to position the piston for No. 1 cylinder to TDC on its compression stroke.

6. Install Camshaft Positioning tool T93P-6256-A or equivalent, in the rear D-slots of the camshafts.

✳✳ WARNING

This is not a freewheeling engine. Cam Positioning tool T93P-6256-A or equivalent, must be installed on the camshafts to prevent the camshafts from rotating. Do not rotate the camshafts or crankshaft with the timing chains removed or the valves will contact the pistons.

7. Remove the bolts retaining both primary timing chain tensioners to the cylinder heads and remove the timing chain tensioners.

8. Remove both primary timing chains, timing chain tensioner arms and timing chain guides. Do not loosen the camshaft timing sprocket retaining bolts.

9. Remove the outlet heater water hose retaining bolts from the rear of the right cylinder head.

10. Remove 10 bolts retaining the cylinder head to the engine block in the opposite order of the installation sequence and remove the cylinder head.

11. Discard the cylinder head bolts. Remove one or both cylinder heads as required.

12. Clean all gasket mating surfaces and bolt holes. Check the cylinder head and cylinder block for flatness. Check the cylinder head for scratches near the coolant passage and combustion chamber that could provide leak paths.

To install:

13. Rotate the crankshaft counterclockwise 45 degrees. The crankshaft keyway should be at the 9 o'clock position as viewed from the front of the engine. This ensures that all pistons are below the top of the cylinder block deck face.

14. Position new head gaskets on the cylinder block.

15. Position the cylinder heads on the engine block dowels, being careful not to

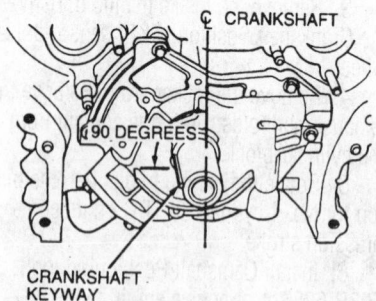

Turn the crankshaft counterclockwise to the 9 o'clock position before installing the cylinder heads

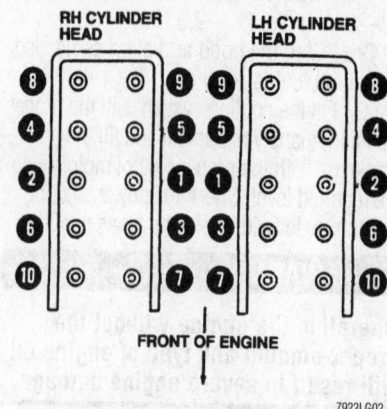

Be sure to tighten the cylinder head bolts in the correct sequence to avoid leakage

score the surface of the head face. Apply clean engine oil to the new cylinder head bolts and install them hand-tight.

16. Tighten the cylinder head bolts as follows:

a. Tighten the bolts, in sequence, to 27–32 ft. lbs. (37–43 Nm).

b. Rotate each bolt, in sequence, 85–95 degrees.

c. Rotate each bolt, in sequence, an additional 85–95 degrees.

17. Position the outlet heater water hose on the right cylinder head. Install and tighten the bolts to 15–22 ft. lbs. (20–30 Nm).

18. Install the primary timing chains and set the valve timing. Install the rocker arms.

19. Install the CKP sensor pulse wheel. Install the intake manifold.

➡ **When cleaning the sealing surfaces of the timing chain cover and oil pan-to-cylinder block joints, be careful not to damage the rubber bead of the oil pan gasket. If damaged, the oil pan gasket must be replaced.**

20. Install the engine front cover. Install the cylinder head covers. Install the exhaust manifolds.

21. Install the engine in vehicle. Connect the negative battery cable.

22. Start the engine and bring to normal operating temperature and check for leaks.

23. Check all fluid levels. Road test the vehicle and check for proper engine operation.

Rocker Arms

REMOVAL & INSTALLATION

1. Disconnect the negative battery cable.
2. Remove the camshaft covers.

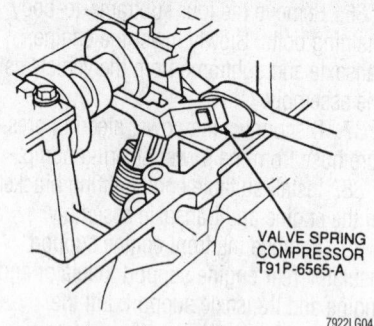

Using the proper tool, compress the valve spring and remove the rocker arm

3. Position the cylinder being serviced at the bottom of it's travel.

➡ **Two different valve spring compressor tool are used for this procedure. Valve Spring Compressor (T91P-6565-A) is used on the exhaust camshaft and Valve Spring Compressor (T93P-6565-A) is used on the intake camshaft.**

4. Compress the valve spring and remove the rocker arm.

To install:

5. Position the cylinder being serviced at the bottom of it's travel.

6. Apply clean engine oil to the rocker arm, valve stem tip and tappet bore.

➡ **Valve tappet should have no more than 1/16 inch (1.5mm) of travel before installing the rocker arm.**

7. Compress the valve spring using the correct tool and install the rocker arm.

Intake Manifold

REMOVAL & INSTALLATION

✳✳ CAUTION

Never open, service or drain the radiator or cooling system when hot; serious burns can occur from the steam and hot coolant.

1. Disconnect the negative battery cable. Drain the engine cooling system.

2. Remove engine appearance cover. Remove the air cleaner outlet tube.

➡ **Label all electrical connectors and vacuum hoses as they are removed, so they can be reinstalled in their original locations.**

Observe all applicable safety precautions when working around fuel. Whenever servicing the fuel system, always work in a well ventilated area. Do not allow fuel spray or vapors to come in contact with a spark or open flame. Keep a dry chemical fire extinguisher near the work area. Always keep fuel in a container specifically designed for fuel storage; also, always properly seal fuel containers to avoid the possibility of fire or explosion.

3. Relieve the fuel system pressure. Disconnect the fuel supply and return lines from the fuel injection supply manifold.

4. Remove 8 retaining nuts on both ignition wire covers and remove the ignition wire covers.

➡**Do not pull on the ignition wire as the wire may separate from the connector in the ignition wire boot.**

5. Tag and disconnect the ignition wires from the spark plugs with a gentle twist/pull motion on the spark plug boot.

6. Disconnect the ignition wires from the ignition coils.

7. Detach the two center ignition wire separators from the retaining studs.

8. Remove the ignition wires.

9. Disconnect the alternator wiring harness from the alternator at the battery terminal and disconnect the stator connector plug. Remove the wiring harness retaining clip from the alternator mounting bracket.

10. Remove the two bolts and two nuts attaching the alternator mounting bracket to the intake manifold. Remove the mounting bracket.

11. Disconnect the upper radiator hose, the heater water hose and the water bypass tube to water thermostat hose at the water bypass tube.

12. Disconnect the wiring from the Engine Coolant Temperature (ECT) sensor at the water bypass tube.

13. Remove the two retaining nuts and spacers from the water bypass tube hold-down braces to intake manifold studs and remove the water bypass tube.

14. Disconnect the accelerator cable and the speed control actuator from the throttle body and from the accelerator cable bracket. Position the accelerator cable and speed control actuator aside.

15. Disconnect the chassis vacuum sup-ply hose from the intake manifold connector and move it aside.

16. Disconnect the vacuum harness from the fuel pressure regulator, right and left secondary air injection diverter valves, EGR control valve, EGR vacuum regulator control, and secondary air injection vacuum control solenoid and intake manifold connection. Remove the vacuum harness.

17. Disconnect the Intake Manifold Runner Control (IMRC) actuator from the IMRC assemblies at the levers. Remove the IMRC actuator retainers and the IMRC actuator.

18. Disconnect the crankcase vent connector and hose from the throttle body. Remove the PCV valve from the left side cylinder head cover. Move the crankcase vent connector and hose and PCV valve aside.

19. Disconnect the engine wiring harness retaining clips from the intake manifold studs.

20. Disconnect 8 wiring connectors from the fuel injectors.

21. Disconnect the engine control sensor wiring from the Idle Air Control (IAC) valve, Throttle Position (TP) sensor or TP sensors (if equipped with traction control). Position the wiring harness aside.

22. Remove the two retaining bolts attaching the EGR valve to the intake manifold.

23. Loosen and remove 20 bolts and studs retaining the intake manifold to the cylinder heads in sequence.

24. Lift the engine control sensor wiring upward for intake manifold removal clearance.

25. Lift the intake manifold off the engine.

26. Remove the four fuel injection supply manifold retaining bolts from the intake manifold. Remove the fuel injection supply manifold from the fuel injectors.

27. Remove the four bolts retaining both lower intake manifolds (IMRC housing assemblies) to the intake manifold and remove both lower intake manifolds.

➡**The upper intake manifold gaskets must be replaced when the manifold retaining bolts have been loosened.**

28. Remove the intake manifold upper gaskets and all load limiting spacers from both lower intake manifolds (IMRC assemblies) and discard.

29. Remove the lower intake manifold gaskets from the cylinder heads and discard.

30. Remove the EGR valve gasket and discard. Clean the EGR gasket sealing surface on the EGR valve and the intake manifold.

To install:

31. Clean and inspect all gasket mating surfaces. The upper sealing surfaces of the Intake Manifold Runner Controls (IMRC) housing assemblies and intake manifold mating surfaces should be cleaned with a

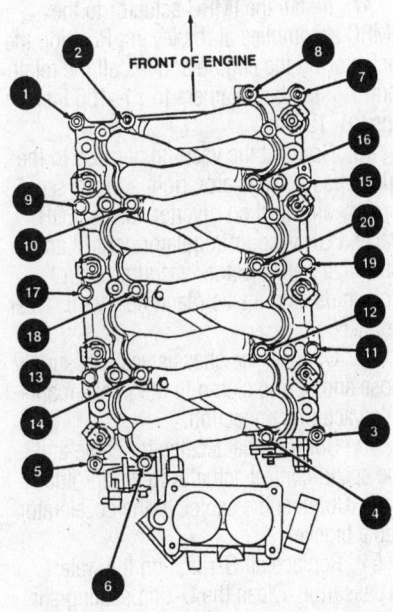

7922LG05

To prevent warping or cracking the manifold, remove the intake manifold bolts in the sequence shown

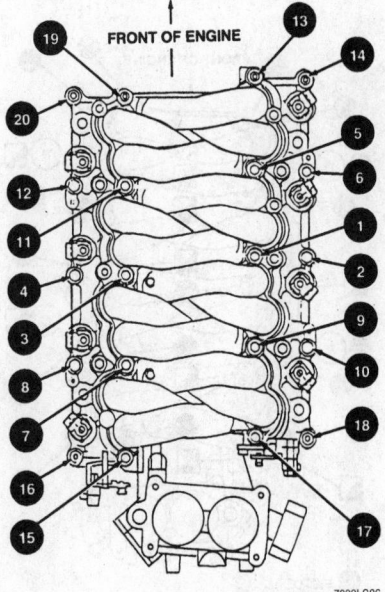

7922LG06

Tighten the intake manifold bolts in the sequence shown to ensure proper sealing

suitable solvent to remove adhesive residue. Inspect the intake manifold mating surfaces and upper IMRC assembly surfaces to be sure no load limiting spacers (washers) are stuck to the sealing surfaces. The load limiting spacers are part of the intake manifold upper gasket. New load limit spacers will be included with the replacement intake manifold upper gasket as an assembly.

32. If a new intake manifold or IMRC housing assembly is being installed, transfer all necessary components to the new intake manifold or lower intake manifold using new gaskets.

33. Install the upper intake manifold gasket on the right IMRC housing using the tapered pins at the end bolt locations for proper gasket alignment. Repeat the procedure for the left IMRC housing.

34. Install the IMRC assemblies to the intake manifold and install the four IMRC retaining bolts finger-tight.

35. Install the fuel injection supply manifold to the fuel injectors and install the four retaining bolts. Tighten to 71–106 inch lbs. (8–12 Nm).

36. Position new intake manifold gaskets onto the cylinder heads, using the gaskets integral locating pins to align the gaskets.

37. Place the intake manifold under the engine control sensor wiring and connect the fuel charging wiring to the IAC valve and TP sensor, or sensors, depending upon application.

38. Move the intake manifold into position and install a new EGR valve gasket and the EGR retaining bolts. Finger-tighten the bolts.

39. Install the longer bolts and stud bolts in the outer bolt holes of the intake manifold, but do not tighten yet.

40. Install the shorter bolts and stud bolts in the inner bolt holes of the intake manifold. Hand-tighten all 20 fasteners.

41. Following the sequence shown, tighten fasteners 1 through 20 in numerical sequence as follows:

    a. No. 5, 7, 9 and 11—9–11 ft. lbs. (12–15 Nm)

    b. All others—13–16 ft. lbs. (18–22 Nm)

    c. Then, tighten all fasteners in numerical sequence, by rotating an additional 85–95 degrees.

42. Tighten the four intake manifold to IMRC housing retaining bolts A-D to 71–89 inch lbs. (8–10 Nm). Tighten fasteners A through D, in sequence, by rotating an additional 85–95 degrees.

43. Tighten the two EGR valve retaining bolts to 15–22 ft. lbs. (20–30 Nm).

44. Connect the wiring to the 8 fuel injectors.

45. Connect the fuel charging wiring harness retaining clips to the intake manifold studs.

46. Connect the crankcase vent connector and hose to the throttle body connection and install the PCV valve into the grommet on the right cylinder head cover.

47. Install the IMRC actuator to the IMRC assemblies at the levers. Position the actuator on the engine and install the retainers. Tighten the retainers to 71–106 inch lbs. (8–12 Nm).

48. Connect the vacuum harness to the fuel pressure regulator, right and left secondary air injection diverter valves, EGR valve, EGR vacuum regulator control and secondary air injectors vacuum control solenoids, and intake manifold vacuum rear connection.

49. Connect the chassis vacuum supply hose and spring clamp to the intake manifold vacuum connection.

50. Connect the accelerator cable and the speed control actuator to the throttle body. Connect the cables to the accelerator cable bracket.

51. Replace all O-rings on the water bypass tube. Clean the O-ring sealing surface on the cylinder heads.

52. Coat all O-rings with clean coolant.

53. Install the water bypass tube support braces over the intake manifold inner studs and install the water bypass tube into position.

54. Install the two water bypass tube spacers and retaining nuts. Tighten the nuts to 71–106 inch lbs. (8–12 Nm).

55. Connect the wiring connector to the ECT sensor.

56. Connect the upper radiator hose, heater water hose and water bypass tube to thermostat housing hose to the water bypass tube.

57. Position the alternator mounting bracket and install the two front retaining bolts and the two rear nuts. Tighten to 71–106 inch lbs. (8–12 Nm).

58. Connect the alternator wiring harness connector plug and battery terminal wire to the alternator. Install the retaining clip to the alternator mounting bracket. Tighten the retaining screw securely.

59. Position the ignition wires and connect the two center ignition wire separators to the retaining studs on the intake manifold.

60. Connect the ignition wires to the ignition coils.

61. Connect the ignition wires to the proper spark plugs.

62. Install the ignition wires into the cylinder head cover slots and install the ignition wire covers. Install 8 retaining nuts and tighten to 18–35 inch lbs. (2–4 Nm).

63. Connect the fuel supply and return lines to the fuel injection supply manifold.

64. Install the engine air cleaner outlet tube. Tighten the air cleaner outlet tubes clamps to 12–22 inch lbs. (1.4–2.5 Nm).

65. Install the engine appearance cover. Tighten the retaining nuts to 61–79 inch lbs. (7–9 Nm).

66. Install the fuel tank fill cap.

67. Connect the negative battery cable.

68. Fill and bleed the cooling system.

69. Start the engine and check for leaks.

70. Road test the vehicle and check for proper operation.

Exhaust Manifold

REMOVAL & INSTALLATION

✳✳ CAUTION

The air suspension switch, located in the left-hand side of the luggage compartment on 1995–97 models or on the right side kick panel for 1998–99 models, must be turned OFF before raising the vehicle. Failure to do so may result in unexpected inflation or deflation of the air springs which may result in shifting of the vehicle during service.

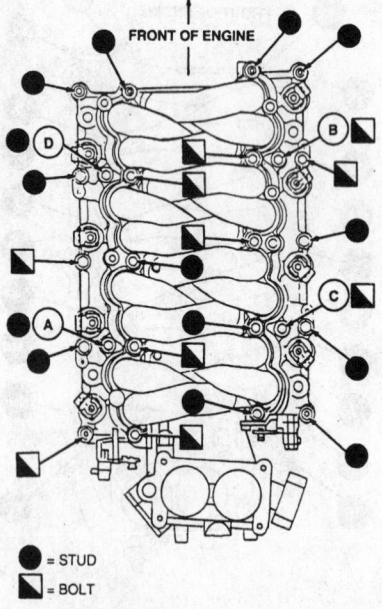

FRONT OF ENGINE

● = STUD
◼ = BOLT

7922LG07

Tighten the IMRC bolts in alphabetical order for proper sealing

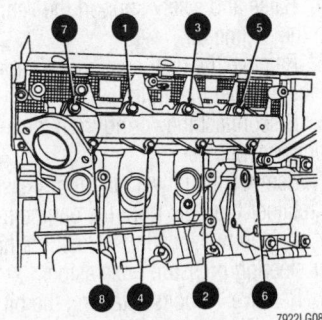

Be sure to tighten the exhaust manifold bolts according to the sequence shown—right side shown, left side is the same

1. Turn the air suspension switch, located in the left side of the luggage compartment, to the OFF position.
2. Disconnect the negative battery cable.
3. Raise and safely support the vehicle.
4. For the right side, remove the wiring connectors from the Heated Oxygen Sensors (HO2S).
5. For the left side, remove the splash shield from the lower radiator support and subframe.
6. Remove the dual converter Y-pipe from the exhaust manifolds.
7. For the right side, perform the following procedures:
   a. Remove the four exhaust connector retaining bolts and the exhaust connector and muffler gasket from the exhaust manifold.
   b. Remove the EGR valve to exhaust manifold tube from the EGR valve tube to manifold connector using a 22mm crowsfoot wrench.
8. Loosen the retaining nut and remove the secondary air injection manifold tube from the exhaust manifold.
9. Remove 8 exhaust manifold-to-cylinder head retaining nuts.
10. Remove the exhaust manifold and gasket.

➡For the right side, access may be gained through the right wheel opening area. It may be necessary to remove the EGR valve tube to manifold connector from the right exhaust manifold.

To install:
11. Clean the mating surfaces of the manifold and the cylinder head.
12. For the right side, if removed, install the EGR valve tube to manifold connector to the right exhaust manifold. Tighten to 33–48 ft. lbs. (45–65 Nm).
13. Position a new exhaust manifold gasket and the exhaust manifold to the

cylinder head. Install 8 retaining nuts. Tighten the nuts in sequence to 13–16 ft. lbs. (18–22 Nm) following the same pattern for both manifolds.
14. Connect the secondary air injection manifold tube to the exhaust manifold and tighten the nut to 25–34 ft. lbs. (34–46 Nm).
15. For the right side, perform the following procedures:
   a. Connect the EGR valve to exhaust manifold tube to the EGR valve tube to manifold connector. Tighten the nut to 30–33 ft. lbs. (40–45 Nm).
   b. Install the exhaust connector with a new inlet pipe gasket.
   c. Install the inlet pipe retaining bolts and tighten to:
   • 1st—13–17 ft. lbs. (17–23 Nm)
   • 2nd—30–40 ft. lbs. (41–54 Nm)
16. Install the dual converter Y-pipe to the exhaust manifolds.
17. For the right side, connect the HO2S wiring connectors.
18. For the left side, install the splash shield to the subframe and lower radiator support.
19. Lower the vehicle.
20. Connect the negative battery cable.
21. Turn the air suspension switch to the ON position.
22. Start the engine and check for exhaust leaks and proper engine operation.

Camshaft and Valve Lifters

REMOVAL & INSTALLATION

1. Disconnect the negative battery cable. Remove the engine from the vehicle and position on a suitable workstand.
2. Remove the cylinder head covers and the engine front cover.
3. Remove the timing chains.

➡If the rocker arms are to be reused, label them as they are removed so they can be reinstalled in their original locations.

4. Remove the rocker arms from the cylinder head and the camshaft that is being serviced. If both bank camshafts are to be serviced, all the rocker arms must be removed.
5. Remove 13 bolts retaining the exhaust camshaft cap cluster assemblies to the cylinder head to remove the exhaust camshaft.
6. Remove 12 bolts retaining the intake camshaft cap cluster assemblies to the cylinder head to remove the intake camshaft.

7. Tap upward lightly on the camshaft cap clusters and gradually lift the camshaft cap clusters from the cylinder head.
8. Lift the camshaft straight upward to avoid bearing surface damage.

➡The previous Steps will remove only one bank of camshafts. If both banks are being serviced, repeat the removal procedure for the remaining camshafts.

9. Inspect the camshaft lobes and journals for wear and/or damage. Replace as necessary.

To install:

➡When cleaning the sealing surfaces of the engine front cover and the oil pan-to-cylinder block joints, use extreme caution not to damage the rubber bead of the oil pan gasket. If damaged, the oil pan gasket must be replaced.

10. Clean and inspect the cylinder head cover, engine front cover, and cylinder head sealing surfaces.
11. If removed, install the lash adjusters (valve tappets).
12. Apply clean engine oil to the journals and lobes of the camshaft. Install the camshaft to the cylinder head.
13. Install and seat the camshaft cap cluster assemblies. Hand-start 12 (52mm long) bolts into the intake camshaft caps, seven (52mm long) bolts into the inboard side of the exhaust camshaft and the six (42mm long) bolts into the outboard side of the exhaust camshaft caps.

➡Each camshaft cap cluster is tightened individually.

14. Tighten the camshaft cap cluster retaining bolts, in sequence, to 71–106 inch lbs. (8–12 Nm).
15. Loosen the camshaft cap cluster retaining bolts approximately two turns or until the head of the bolt is free.
16. Tighten all bolts again, in sequence, to 71–106 inch lbs. (8–12 Nm).

➡Once installed, the camshaft should turn freely but with a slight drag.

17. Check the camshaft end-play using a dial indicator.
18. Install the timing chains.
19. Install Valve Spring Compressor T91P-6565-A or equivalent, under the exhaust camshaft and on top of the exhaust valve spring retainer. Compress the valve spring far enough to install the rocker arm.
20. Install Valve Spring Compressor T93P-6565-A or equivalent, under the

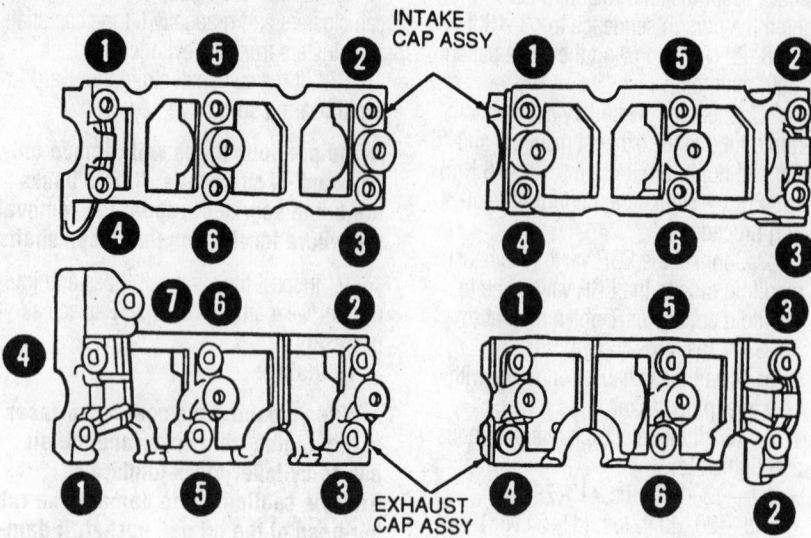

Tighten the camshaft cap clusters in the order shown to prevent warpage

intake camshaft and on top of the primary intake valve spring retainer. Compress the valve spring far enough to install the rocker arm.

21. Install Valve Spring Compressor T93P-6565-A or equivalent, under the intake camshaft and on top of the secondary intake valve spring retainer. Compress the valve spring far enough to install the rocker arm.

➡ **The previous installation Steps must be repeated if both banks are being serviced to install the remaining camshafts.**

22. Inspect and replace the crankshaft front seal and timing chain front cover gasket.

23. Install the engine front cover and the cylinder head covers.

24. Remove the engine from the workstand and install in the vehicle.

25. Start the engine and check for leaks.

26. Road test the vehicle and check for proper engine operation.

Valve Lash

ADJUSTMENT

The 4.6L DOHC engine uses hydraulic valve lash adjusters that do no require any maintenance. If the valvetrain becomes noisy, inspect the camshaft, rocker arm and valve for excessive wear or damage.

Oil Pan

REMOVAL & INSTALLATION

✳✳ CAUTION

The air suspension switch, located in the left-hand side of the luggage compartment on 1995–97 models or on the right side kick panel for 1998–99 models, must be turned OFF before raising the vehicle. Failure to do so may result in unexpected inflation or deflation of the air springs which may result in shifting of the vehicle during service.

1. Turn the air suspension switch, located in the left side of the luggage compartment, to the OFF position.

2. Disconnect the negative battery cable.

3. Remove the oil level dipstick.

✳✳ CAUTION

The EPA warns that prolonged contact with used engine oil may cause a number of skin disorders, including cancer! You should make every effort to minimize your exposure to used engine oil. Protective gloves should be worn when changing the oil. Wash your hands and any other exposed skin areas as soon as possible after exposure to used engine oil. Soap and water, or waterless hand cleaner should be used.

4. Raise and safely support the vehicle. Drain the engine oil.

5. Remove the dual converter Y-pipe from the exhaust manifolds.

6. Disconnect the wiring from the low oil sensor connector.

7. Remove the power steering pressure hose retainer brackets from the engine front cover studs in two locations. Position the power steering pressure hose aside.

8. Remove 16 bolts retaining the oil pan to the cylinder block and remove the oil pan and gasket.

To install:

9. Clean the gasket surfaces on the cylinder block, oil pan and front cover with a clean cloth. If scraping is required, only use plastic-tipped scrapers to prevent damage to aluminum surfaces. Thoroughly clean the inside of the oil pan.

10. Apply a bead of silicone sealer to the oil pan flange. Also apply a bead of sealer to the front cover/cylinder block joint and fill the grooves on both sides of the rear main seal cap.

➡ **When using silicone rubber sealer, assembly must occur within 15 minutes after sealer application. Be sure the sealer does not fall into the engine where it may plug oil passages.**

11. Install a new gasket on the oil pan.

12. Carefully install the oil pan using 16 retaining bolts. Tighten the bolts, in sequence, first to 14 ft. lbs. (20 Nm), then turn the bolts an additional 60 degrees in the same sequence.

13. Install the power steering pressure hose retainer bracket onto the engine front cover studs. Install and tighten the retaining nuts to 71–106 inch lbs. (8–12 Nm).

14. Install the low oil sensor wiring on the sensor connector.

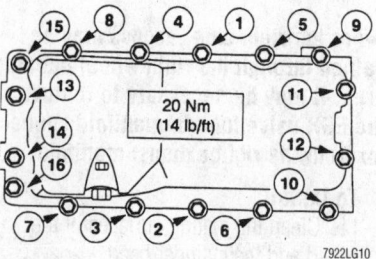

Tighten the oil pan bolts in sequence to ensure a good seal with the engine block

15. Install the dual converter Y-pipe to the exhaust manifolds.

16. Lower the vehicle. Install the oil level dipstick.

❊❊ WARNING

Operating the engine without the proper amount and type of engine oil will result in severe engine damage.

17. Fill the crankcase with the correct type and quantity of engine oil.

18. Connect the negative battery cable. Start the engine and check for leaks.

19. Turn the air suspension switch to the ON position.

Oil Pump

REMOVAL & INSTALLATION

1. Disconnect the negative battery cable. Remove the engine from the vehicle and position on a suitable workstand.

2. Remove the engine front cover. Remove the engine oil pan.

3. Remove the oil pump screen cover/tube bolts and the tube; discard the O-ring.

4. Remove the timing chains. Remove the crankshaft sprockets.

5. Remove the oil pump-to-cylinder block bolts and the oil pump.

To install:

6. Thoroughly clean the oil pump and cylinder block mounting surfaces and the oil pump screen cover and tube.

7. Turn the inner rotor of the oil pump to align with the flats on the crankshaft and install the oil pump flush with the cylinder block. Install the four retaining bolts and tighten to 71–106 inch lbs. (8–12 Nm).

8. Replace the engine oil filter. Install the crankshaft sprockets. Install both timing chains.

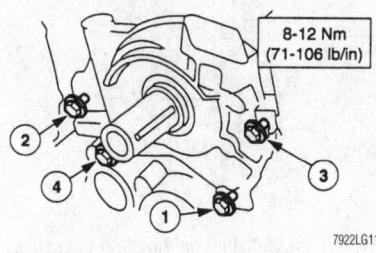

Tighten the oil pump mounting bolts in the sequence shown

9. Position the oil pump screen cover and tube on the oil pump with a new O-ring and hand-start the two retaining bolts.

10. Install the bolt retaining the oil pump screen cover and tube to the main bearing stud spacer finger-tight.

11. Tighten the oil pump screen cover and tube-to-oil pump bolts to 71–106 inch lbs. (8–12 Nm). Tighten the oil pump screen cover and tube-to-main bearing stud spacer bolt to 15–22 ft. lbs. (20–30 Nm).

❊❊ WARNING

Operating the engine without the proper amount and type of engine oil will result in severe engine damage.

12. Install the engine oil pan. Install the engine front cover.

13. Remove the engine from the workstand and install in the vehicle.

14. Restore all fluid levels. Connect the negative battery cable.

15. Run the engine and check for leaks and proper oil pressure. Road test the vehicle and check for proper operation.

Rear Main Seal

REMOVAL & INSTALLATION

❊❊ CAUTION

The air suspension switch, located in the left-hand side of the luggage compartment on 1995–97 models or on the right side kick panel for 1998–99 models, must be turned OFF before raising the vehicle. Failure to do so may result in unexpected inflation or deflation of the air springs which may result in shifting of the vehicle during service.

1. Turn the air suspension switch, located in the left-hand side of the luggage compartment, to the OFF position.

2. Disconnect the negative battery cable.

3. Raise and safely support the vehicle.

4. Properly support the engine.

5. Remove the sub-frame and transaxle from the vehicle.

6. Remove the flywheel from the crankshaft.

7. Remove the rear crankshaft seal retainer bolts and remove the retainer from the cylinder block.

8. Support the seal retainer securely and remove the seal, using a punch or similar tool.

To install:

9. Clean and inspect the retainer and cylinder block mating surfaces.

10. Apply a 0.060 inch (1.5mm) continuous bead of a silicone gasket sealer to the cylinder block.

11. Install the seal retainer and retaining bolts. Tighten the bolts, in sequence, to 71–106 inch lbs. (8–12 Nm).

12. Install a new rear crankshaft seal using Seal Installer T82L-6701-A and adapter T91P-6701-A or equivalents.

13. Install the flywheel to the engine. Install the six retaining bolts and tighten in an alternating sequence to 54–64 ft. lbs. (73–87 Nm).

14. Install the transaxle and sub-frame.

15. Lower the vehicle.

16. Remove the engine support.

17. Connect the negative battery cable.

18. Turn the air suspension switch to the ON position.

19. Start the engine and check for leaks.

➥**Whenever the vehicles sub-frame is removed or lowered, the wheel alignment should be checked.**

20. Align the front wheels.

21. Road test the vehicle and check for proper operation.

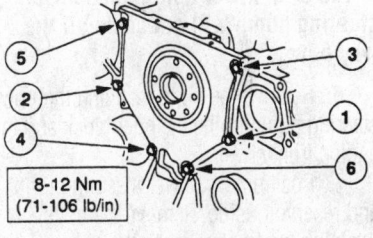

Be sure to tighten the rear main oil seal retainer bolts in the correct sequence

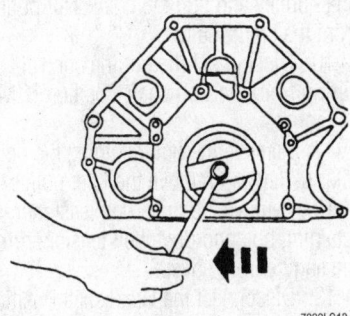

Use the Seal Installer T82L-6701-A and adapter T91P-6701-A or equivalents to install the seal evenly in the retainer

Timing Chain, Sprockets and Front Cover and Seal

REMOVAL & INSTALLATION

✳✳ WARNING

This is an interference engine. When the timing chains are removed and the cylinder heads are installed, the crankshaft and/or camshafts must not be rotated unless as directed in this procedure. Failure to follow these instructions will result in valve and/or piston damage.

1. Disconnect the negative battery cable. Remove the engine from the vehicle and position on a suitable workstand.

2. Remove the support brackets for the engine sensor wiring from the engine front cover.

3. Loosen the water pump pulley bolts. Remove the accessory drive belt.

4. Remove the water pump pulley and lower water pump-to-cylinder block retaining bolt for engine front cover removal clearance.

5. Using Pulley Remover T69L-10300-B or equivalent, remove the power steering pump pulley.

➡**The front lower bolt on the power steering pump will not come all the way out.**

6. Remove the bolts retaining the power steering pump to the cylinder block and the engine front cover.

7. Position the power steering pump and reservoir aside. Remove the bolts retaining the front cover to the oil pan.

8. Remove the crankshaft pulley retaining bolt and washer from the crankshaft.

9. Install Crankshaft Damper Remover T58P-6316-D or equivalent, on the crankshaft pulley and pull the crankshaft pulley from the crankshaft.

10. Disconnect the engine control wiring from the Camshaft Position (CMP) sensor.

11. Remove the bolt retaining the belt idler pulley and remove the idler pulley.

12. Remove the drive belt tensioner retaining bolt and drive belt tensioner from the engine front cover.

13. Disconnect the Crankshaft Position (CKP) sensor and move the wiring aside.

14. Remove the engine front cover.

Remove both cylinder head covers. Remove the CKP sensor tooth wheel.

15. Rotate the crankshaft to place the piston for No. 1 cylinder at Top Dead Center (TDC) on its compression stroke.

✳✳ WARNING

Camshaft Positioning tool T93P-6256-A or equivalent, and Camshaft Holding tool T93P-6256-AH or equivalent, must be installed to prevent accidental rotation of the camshafts and possible engine damage.

16. Install Camshaft Positioning tool T93P-6256-A or equivalent, in the rear D-slots of the camshaft.

✳✳ WARNING

Failure to use Camshaft Holding tool T93P-6256-AH or equivalent, while assembling or disassembling the timing chains will result in damage to the D-slots or Camshaft Positioning tool T93P-6256-A.

17. Install Camshaft Holding tool T93P-6256-AH or equivalent, onto the camshafts to keep the camshafts from rotating and to prevent damaging the camshaft positioning tool.

18. Remove the right primary timing chain tensioner bolts and the tensioner.

19. Remove the right tensioner arm bolt and the arm.

➡**The two bolts retaining the timing chain guide to the cylinder head are longer than the one to the block.**

20. Remove the right timing chain guide bolts and guide. Remove the right primary timing chain.

21. Remove the right crankshaft sprocket. Remove the camshaft sprocket retaining bolts.

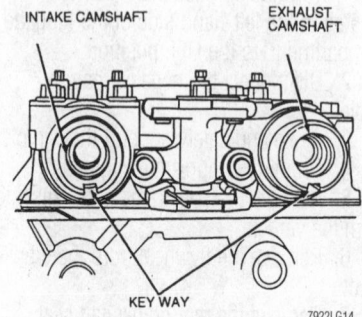

Install the camshaft positioning tool on the rear of the camshafts to position both keyways down toward the crankshaft

✳✳ WARNING

The secondary timing chain tensioner plunger is spring loaded. Care must be taken to prevent the plunger from dropping out of the tensioner during disassembly.

22. Unlock and compress the secondary timing chain tensioners and lock the timing chain tensioner in the compressed position using a paper clip or stiff wire. Remove the secondary timing chains and camshaft sprockets.

23. Remove the left primary timing chain tensioner bolts tensioner.

24. Remove the bolt retaining the left timing chain tensioner arm and remove the left timing chain tensioner arm.

25. Remove the left timing chain guide bolts and the guide. Remove the left primary timing chain. Remove the left crankshaft sprocket.

✳✳ WARNING

Failure to use Camshaft Holding tool T93P-6256-AH or equivalent, while removing or tightening the camshaft bolts may result in damage to the camshaft D-slots or the camshaft positioning tool.

26. Remove the camshaft sprocket retaining bolts.

✳✳ WARNING

The secondary timing chain tensioner plunger is spring loaded. Care must be taken to prevent the plunger from dropping out of the tensioner during disassembly.

27. Unlock and compress the secondary chain tensioner and lock the chain tensioner

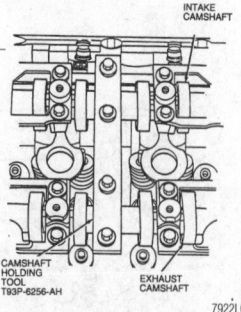

Install the Camshaft holding tool to secure the camshafts and prevent damage to the Camshaft Positioning Tool

in the compressed position using a paper clip or stiff wire. Remove the secondary timing chain and camshaft sprockets.

※※ WARNING

Do not rotate the crankshaft and/or camshafts or engine damage may occur.

28. Inspect the friction material on the timing chain tensioner arms and timing chain guides. If worn or damaged, remove and clean the oil pan and replace the oil pump screen cover and tube.

To install:

29. If the engine has jumped time, be sure all service to the necessary engine components and/or valvetrain has been made.

30. Be sure the primary and secondary timing chain tensioners are in the collapsed position and retained with paper clips or equivalent.

31. Install Camshaft Positioning tool T93P-6256-A or equivalent, into the D-slots on the rear of the camshafts to position both intake and exhaust camshafts with the keyways pointing down towards the crankshaft. Install Camshaft Holding tool T93P-6256-AH or equivalent, on the center of the camshafts and do not remove until all parts are installed and tightened.

32. Install the left secondary timing chain tensioner and tighten the bolts to 70–106 inch lbs. (8–12 Nm).

33. Install the left secondary camshaft sprockets and secondary timing chain as an assembly. Be sure the hubs of the sprocket are facing the proper direction.

34. Install the left primary camshaft sprocket and camshaft sprocket spacer on the left camshaft.

35. Install the washers and camshaft sprocket retaining bolts, finger-tight only.

➡ **The secondary camshaft sprockets must be free to turn.**

36. Install Timing Chain Tensioning tool T93P-6256-BH or equivalent, on the left secondary timing chain tensioner.

37. Tighten the camshaft sprocket bolts to 81–95 ft. lbs. (110–130 Nm).

38. Install the right secondary timing chain tensioner and tighten the retaining bolts to 70–106 inch lbs. (8–12 Nm).

39. Install the right secondary camshaft sprockets and secondary timing chain as an assembly. Be sure the hubs of the sprocket are facing the proper direction.

40. Install the right primary camshaft sprocket and camshaft sprocket spacer on the right camshaft.

41. Install the washers and camshaft sprocket retaining bolts, finger-tight only.

➡ **The secondary camshaft sprockets must be free to turn.**

42. Install Timing Chain Tensioning tool T93P-6256-BH or equivalent, on the right secondary timing chain tensioner.

43. Tighten the camshaft sprocket bolts to 81–95 ft. lbs. (110–130 Nm).

44. Install the left crankshaft sprocket, making sure the tapered part of the sprocket faces away from the cylinder block.

45. Install the primary timing chain on the left primary camshaft sprocket. Be sure the copper link of the timing chain aligns with the timing mark on the camshaft sprocket.

➡ **If the copper links of the timing chain are not visible, pull the chain taught until the opposite sides of the chain contact one another and lay it on a flat surface. Mark the links at each end of the chain and use them in place of the copper links.**

46. Install the left primary timing chain on the left crankshaft sprocket. Be sure the copper link of the timing chain aligns with the timing mark on the crankshaft sprocket.

47. Install the right crankshaft sprocket, making sure the tapered part of the sprocket faces toward the cylinder block.

48. Install the primary timing chain on the right primary camshaft sprocket. Be sure the copper link of the timing chain aligns with the timing mark on the camshaft sprocket.

49. Install the right primary timing chain on the right crankshaft sprocket. Be sure the copper link of the timing chain aligns with the timing mark on the crankshaft sprocket.

➡ **The two timing chain guide bolts to the cylinder head are longer than the bolt to the cylinder block.**

50. Install both timing chain guides and tighten the bolts to 71–106 inch lbs. (8–12 Nm).

51. Lubricate the timing chain tensioner arm contact surfaces with clean engine oil. Install both tensioner arms and tighten the bolts to 84–132 inch lbs. (10–15 Nm).

52. Install right and left primary timing chain tensioners and tighten the bolts to 15–22 ft. lbs. (20–30 Nm).

53. Remove the locking pins from the timing chain tensioners and be sure all timing marks are aligned. Remove the camshaft positioning tool and the camshaft holding tool.

➡ **When cleaning the sealing surfaces of the engine front cover and oil pan-to-cylinder block joints, be extremely careful not to damage the rubber bead of the oil pan gasket. If the oil pan gasket is damaged, it must be replaced.**

54. Clean the sealing surfaces of the cylinder block, cylinder heads and engine front cover; remove all traces of oil, dirt and old sealant. Sealing surfaces must be clean and dry before applying sealant.

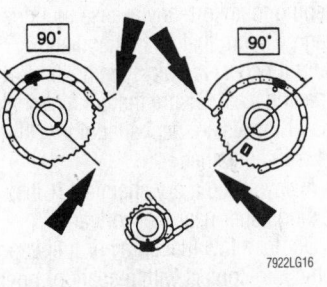

Be sure the copper links colored black on the timing chain align with the marks on the gears and the keyways are positioned 90 degrees from the camshaft cover surface

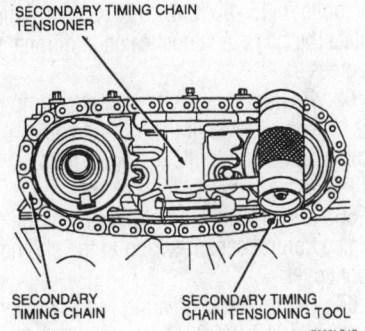

Secondary timing chain tensioning tool installed

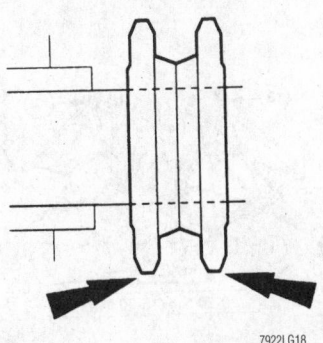

Be sure to position the crankshaft sprockets as shown to prevent engine damage

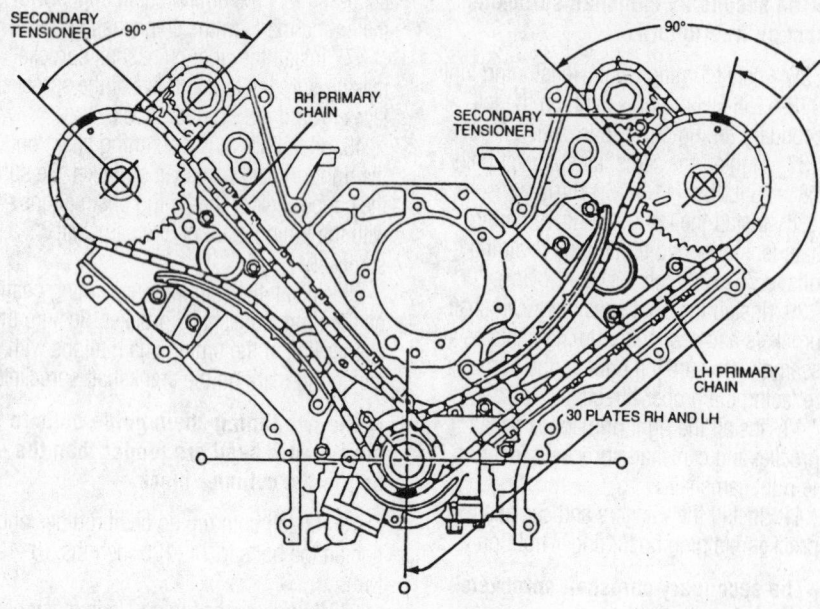

Correct alignment of the primary timing chains and sprockets

55. Replace the engine front cover gaskets and the crankshaft front seal. Apply silicone sealant to the proper locations on the cylinder block and heads.

➡**The engine front cover must be rolled into place. DO NOT slide on the oil pan gasket.**

56. Carefully install the front cover to the engine.

57. Install the five studs and 10 bolts retaining the engine front cover to the engine. Tighten, in sequence, to 15–22 ft. lbs. (20–30 Nm). Tighten the stud bolts and bolts numbered 6, 7, 8, 9, 10, and 11 within four minutes after applying the sealer.

58. Install the drive belt tensioner and bolt. Tighten the bolt to 15–22 ft. lbs. (20–30 Nm).

59. Install the belt idler pulley and bolt. Tighten the bolt to 15–22 ft. lbs. (20–30 Nm).

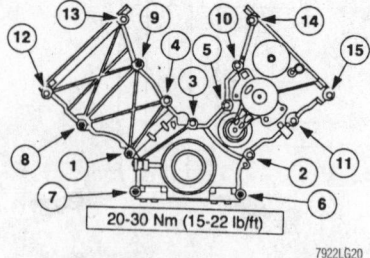

To prevent leakage, tighten the front cover mounting bolts in the sequence shown

60. Connect the engine control sensor wiring to the CKP and CMP sensors. Install the left and the right cylinder head covers.

61. Install the crankshaft pulley on the crankshaft, using Crankshaft Damper Replacer T74P-6316-B or equivalent.

62. Apply silicone sealant in the keyway of the crankshaft pulley.

63. Install the pulley bolt and washer. On 1995 vehicles, tighten to 114–121 ft. lbs. (155–165 Nm). On 1996–99 vehicles, tighten as follows:

   a. Tighten the bolt to 89 ft. lbs. (120 Nm), then loosen the bolt.

   b. Retighten the bolt to 39 ft. lbs. (53 Nm).

   c. Turn the retaining bolt an additional 90 degrees.

64. Install the four bolts retaining the engine front cover to the oil pan. On 1995 vehicles, tighten the bolts to 15–22 ft. lbs. (20–30 Nm). On 1996–99 vehicles, tighten the bolts to 15–22 ft. lbs. (20–30 Nm), then rotate the bolts in sequence an additional 60 degrees.

65. Position the power steering pump on the engine and install the four retaining bolts. Tighten the bolts to 15–22 ft. lbs. (20–30 Nm).

66. Install the retainer brackets for the engine control sensor wiring to the engine front cover.

67. Using Power Steering Pump Pulley Replacer T91P-3A733-A or equivalent, install the power steering pump pulley.

68. Install the water pump lower bolt removed for engine front cover clearance.

Tighten the bolt to 15–22 ft. lbs. (20–30 Nm).

69. Install the water pump pulley with the four bolts. Tighten the bolts to 15–22 ft. lbs. (20–30 Nm).

70. Install the accessory drive belt. Install the engine assembly in the vehicle. Restore all fluid levels.

71. Connect the negative battery cable. Start the engine and check for leaks and proper engine operation.

FUEL SYSTEM

Fuel System Service Precautions

Safety is the most important factor when performing not only fuel system maintenance but any type of maintenance. Failure to conduct maintenance and repairs in a safe manner may result in serious personal injury or death. Maintenance and testing of the vehicle's fuel system components can be accomplished safely and effectively by adhering to the following rules and guidelines.

• To avoid the possibility of fire and personal injury, always disconnect the negative battery cable unless the repair or test procedure requires that battery voltage be applied.

• Always relieve the fuel system pressure prior to disconnecting any fuel system component (injector, fuel rail, pressure regulator, etc.), fitting or fuel line connection. Exercise extreme caution whenever relieving fuel system pressure to avoid exposing skin, face and eyes to fuel spray. Please be advised that fuel under pressure may penetrate the skin or any part of the body that it contacts.

• Always place a shop towel or cloth around the fitting or connection prior to loosening to absorb any excess fuel due to spillage. Ensure that all fuel spillage (should it occur) is quickly removed from engine surfaces. Ensure that all fuel soaked cloths or towels are deposited into a suitable waste container.

• Always keep a dry chemical (Class B) fire extinguisher near the work area.

• Do not allow fuel spray or fuel vapors to come into contact with a spark or open flame.

• Always use a back-up wrench when loosening and tightening fuel line connection fittings. This will prevent unnecessary stress and torsion to fuel line piping.

Always follow the proper torque specifications.

• Always replace worn fuel fitting O-rings with new. Do not substitute fuel hose or equivalent, where fuel pipe is installed.

Fuel System Pressure

RELIEVING

✲✲ CAUTION

The fuel injection system remains under pressure, even after the engine has been turned OFF. The fuel system pressure must be relieved before disconnecting any fuel lines. Failure to do so may result in fire and/or personal injury.

1. Disconnect the negative battery cable.
2. Remove the fuel tank fill cap to relieve the pressure in the fuel tank.
3. Remove the cap from the Schrader valve located on the fuel supply manifold (fuel rail).
4. On gasoline engines, attach Fuel Pressure Gauge T80L-9974-A or equivalent, to the valve and drain the fuel through the drain tube into a suitable container.
5. On flex-fuel engines, connect fuel pressure gauge T80L-9974-A or equivalent and fuel pressure test kit 134-R0035 or equivalent, to the Schrader valve. Drain the fuel through the drain tube into a suitable container.
6. After the fuel system pressure is relieved, remove the fuel pressure gauge and install the cap on the Schrader valve.
7. Install the fuel tank fill cap.
8. Connect the negative battery cable only after system repairs are completed.

Fuel Filter

REMOVAL & INSTALLATION

✲✲ CAUTION

The air suspension switch, located in the left-hand side of the luggage compartment on 1995–97 models or on the right side kick panel for 1998–99 models, must be turned OFF before raising the vehicle. Failure to do so may result in unexpected inflation or deflation of the air springs which may result in shifting of the vehicle during service.

1. Turn the air suspension service switch OFF.
2. Disconnect the negative battery cable.

✲✲ CAUTION

Observe all applicable safety precautions when working around fuel. Whenever servicing the fuel system, always work in a well ventilated area. Do not allow fuel spray or vapors to come in contact with a spark or open flame. Keep a dry chemical fire extinguisher near the work area. Always keep fuel in a container specifically designed for fuel storage; also, always properly seal fuel containers to avoid the possibility of fire or explosion.

3. Properly relieve the fuel system pressure.
4. Raise and safely support the vehicle.
5. Using Fuel Line Disconnect tool T90T-9550-B or T90T-9550-C or equivalent, disconnect the push connect fittings at the fuel filter.

➡**Grasp the fitting and pull in an axial direction to remove the fitting from the fuel filter. Use care, as excessive side loading could break the connector body.**

6. Loosen the fuel filter retaining clamp and remove the fuel filter. Note the direction of the flow arrow on the filter, so the replacement filter can be reinstalled in the same position.

To install:

7. Install the fuel filter with the flow arrow facing the proper direction. Tighten the filter retaining clamp to 15–25 inch lbs. (1.7–2.8 Nm).
8. Apply a light coating of clean engine oil to the fuel filter ends. Align the fittings and filter axially and push the fittings onto

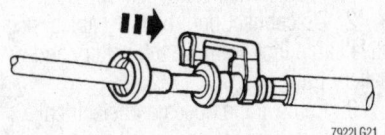

7922LG21

Slide the special tool into the fitting to disengage it—push connect type fitting shown

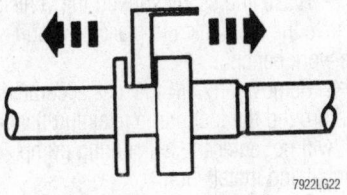

7922LG22

Remove the hairpin clip and pull the connection apart—hairpin clip type fitting shown

the filter ends. When the fittings are properly engaged, a definite click will be heard. Be sure the fittings are fully engaged.

9. Lower the vehicle.
10. Connect the negative battery cable.
11. Start the engine and check for fuel leaks.
12. Turn the air suspension switch to the ON position.

Fuel Pump

REMOVAL & INSTALLATION

✲✲ CAUTION

The air suspension switch, located in the left-hand side of the luggage compartment on 1995–97 models or on the right side kick panel for 1998–99 models, must be turned OFF before raising the vehicle. Failure to do so may result in unexpected inflation or deflation of the air springs which may result in shifting of the vehicle during service.

1. Turn the air suspension switch, located in the left side of the luggage compartment, to the OFF position.

✲✲ CAUTION

Observe all applicable safety precautions when working around fuel. Whenever servicing the fuel system, always work in a well ventilated area. Do not allow fuel spray or vapors to come in contact with a spark or open flame. Keep a dry chemical fire extinguisher near the work area. Always keep fuel in a container specifically designed for fuel storage; also, always properly seal fuel containers to avoid the possibility of fire or explosion.

2. Disconnect the negative battery cable. Properly relieve the fuel system pressure.

3. Raise and safely support the vehicle. Remove the fuel tank and place on a suitable work bench.

4. Remove any dirt that has accumulated around the fuel pump retaining flange so it will not enter the tank during pump removal and installation.

5. Turn the fuel pump locking ring counterclockwise and remove the locking ring using Fuel Tank Sender Wrench T74P-9275-A, or equivalent.

6. Pull the fuel pump module up and out of the fuel tank until the locking tabs for the fuel pump module are accessible. Squeeze both locking tabs together and remove the fuel pump module from the fuel tank. Remove and discard the O-ring seal.

To install:

7. Clean the fuel pump mounting flange, fuel tank mounting surface and O-ring seal groove.

8. Apply a light coating of grease on a new O-ring seal to hold it in place during assembly and install the O-ring seal.

9. Install the fuel pump module carefully to ensure the filter and hoses and float rod are not damaged.

10. Align the fuel pump module and the fuel tank retainer axially and push the fuel pump module into the fuel tank retainer. When the fuel pump module is properly engaged, a definite click will be heard engaging the two locking tabs on the outside of the fuel pump.

11. Pull on the fuel pump module to ensure that both locking tabs are properly engaged.

12. Be sure the locating keys are in the keyways and the seal ring remains in the groove.

13. Hold the fuel pump module in place and install the locking ring finger-tight. Be sure all the locking tabs are under the tank lock ring tabs.

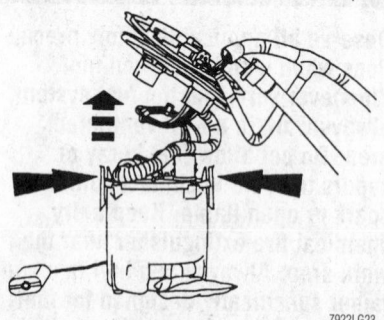

7922LG23

To remove the fuel pump, reach into the tank and press the locking tabs inward

14. Using the sender wrench or equivalent, rotate the locking ring clockwise until the ring is against the stops.

15. Install the fuel tank in the vehicle. Lower the vehicle.

16. Add a minimum of 10 gallons (38 liters) of clean fuel to the tank. Connect the negative battery cable.

17. Turn the air suspension switch to the ON position.

18. Install a suitable fuel pressure gauge to the Schrader valve on the fuel supply manifold.

19. Cycle the ignition switch from **OFF** to **ON** for three seconds. Repeat this procedure 5–10 times until the pressure gauge reads at least 30 psi (207 kPa). Check for fuel leaks.

20. Remove the fuel pressure gauge. Start the engine and again, check for fuel leaks. Road test the vehicle and check for proper operation.

DRIVE TRAIN

Transaxle Assembly

REMOVAL & INSTALLATION

❉❉ CAUTION

The air suspension switch, located in the left-hand side of the luggage compartment on 1995–97 models or on the right side kick panel for 1998–99 models, must be turned OFF before raising the vehicle. Failure to do so may result in unexpected inflation or deflation of the air springs which may result in shifting of the vehicle during service.

➡Be sure to have available new halfshaft circlips and wheel hub retainer nuts.

1. Properly deflate the air springs.

2. Disconnect both battery cables, negative cable first. Remove the battery and battery tray.

3. Remove the hoses and electrical connectors from the engine air cleaner, and remove the air cleaner assembly.

4. Remove the two screws and the accelerator control splash shield from the throttle body.

5. Disconnect the transaxle harness electrical connector and the Transmission

Range (TR) sensor electrical connectors.

6. Remove the nut from shift lever, remove the clip from the cable bracket and lift the cable from the bracket.

7. Disconnect the transaxle oil cooler lines from the transaxle.

8. Remove the four transaxle housing-to-engine bolts from the top of the transaxle.

9. Install the two engine lifting eyes to the front and rear locations of the engine. Install the three Bar Engine Support Bracket Set D88L-6000-A or equivalent, and suitably support the engine.

10. Raise and safely support the vehicle. Remove the front wheel and tire assemblies.

11. Loosen the transaxle oil pan and drain the fluid into a suitable container. When most of the fluid has drained, install the pan bolts.

12. Disconnect the height sensors.

13. Remove the cotter pins and castellated nuts from both tie rod ends. Separate the tie rod ends from the steering knuckles. Discard the cotter pins.

14. Disconnect the sway bar links from the sway bar.

15. Separate the lower control arms from the steering knuckles.

16. Remove the wheel hub retainer nuts and remove both halfshafts from the vehicle. Discard the nuts.

17. Disconnect the Heated Oxygen Sensor (HO2S) electrical connectors.

18. Remove the dual converter Y-pipe assembly from the vehicle.

19. Remove the starter motor and the dust cover.

20. Support the subframe using Powertrain Lift 014–00765 and Removal Bracket 014–00766, or equivalents.

21. Remove the four subframe-to-body bolts and lower the subframe.

22. Remove the transaxle housing cover from the transaxle.

23. Remove the four torque converter-to-flywheel retaining nuts.

24. Remove the five bolts retaining the rear engine support and remove the support from the transaxle.

25. Support the transaxle with High Lift Transmission Jack 014–00210 and Adapter 014–00461, or equivalents.

26. Remove the two lower engine-to-transaxle bolts.

27. Remove the two bolts and one nut from the right side of engine mount to transaxle case.

28. Carefully separate the transaxle from the engine and slowly lower the transaxle out of the vehicle.

To install:

➡**Verify the transaxle cooler lines are thoroughly cleaned before installing the transaxle assembly.**

29. If removed, place the transaxle on High Lift Transmission Jack 014–00210 and Adapter 014–00461, or equivalents.

30. Slowly raise the transaxle assembly into position until the transaxle housing engages the engine dowel pins. Align the torque converter studs to the flywheel.

31. Install the two lower engine-to-transaxle bolts. Alternately tighten the bolts while making sure the transaxle pulls in flush with the engine. Tighten the bolts to 80–106 inch lbs. (9–12 Nm).

➡**When installing the transaxle to the engine, verify that the converter-to-transaxle engagement is maintained. Prevent the converter from moving forward and disengaging during installation.**

32. Install the four flywheel-to-torque converter nuts. Tighten to 20–34 ft. lbs. (27–46 Nm).

33. Mount the rear engine/transaxle support to the transaxle. Tighten the five retaining bolts to 44–60 ft. lbs. (60–80 Nm).

34. Install the transaxle housing cover and securely tighten the bolts.

35. Install the two bolts and one nut to the right side of the engine mount to transaxle case.

36. Install the starter motor and tighten the bolt and stud to 15–21 ft. lbs. (21–29 Nm).

37. Connect the electrical connectors to the starter motor.

38. If removed, place the subframe on Powertrain Lift 014–00765 and Removal Bracket 014–00766, or equivalents.

39. Raise the subframe into position.

40. Loosely install the four subframe to body bolts.

41. Align the subframe using a 0.75 inch (19mm) outside diameter pipe or similar tool installed through both alignment holes. Tighten the bolts to 55–75 ft. lbs. (75–102 Nm).

42. Remove the powertrain lift.

43. Install the remaining components in the reverse order of removal. Tighten the following:
- Ball joint stud retaining nuts to 50–68 ft. lbs. (68–92 Nm)
- Sway bar link-to-sway bar nuts to 30–40 ft. lbs. (40–55 Nm)
- Tie rod end-to-steering knuckle nuts to 35–46 ft. lbs. (47–63 Nm)
- Manual control lever nut to 12–16 ft. lbs. (16–22 Nm)

44. Install the battery tray and the battery. Connect both battery cables, negative cable last.

45. Refill the air springs. Install the engine appearance cover. Tighten the retaining nuts to 61–79 inch lbs. (7–9 Nm).

46. Fill the transaxle with the proper type and quantity of fluid. Check the front end alignment.

47. Check the transaxle for leaks and proper shift selector functioning. Road test the vehicle and check for proper operation.

Halfshaft

REMOVAL & INSTALLATION

➡**Do NOT begin this procedure unless the following parts are available:**

- New hub retainer nut
- New lower control arm-to-steering knuckle bolt and nut
- New stub shaft/link shaft circlip

Once removed, these parts must not be reused during assembly. Their torque holding ability or retention capability is diminished during removal.

➡**Be sure to have available a New Generation Star (NGS) tester or equivalent scan tool with the proper software to properly deflate the air springs before servicing the suspension system.**

✳✳ CAUTION

Do not remove an air spring under any circumstances when pressurized. Do not remove any components supporting an air spring without exhausting the air. Failure to follow these instructions may result in unexpected inflation or deflation of the air springs which may result in shifting of the vehicle during service.

1. Place the vehicle over a frame contact hoist.

2. Properly deflate the air springs.

3. Turn the ignition switch to the **OFF** position. Leave the steering column in the UNLOCKED position.

4. Disconnect the negative battery cable.

5. Loosen and remove the wheel hub retainer nut. Discard the nut.

6. Raise and safely support the vehicle. Remove the wheel and tire assembly.

7. Remove the brake anti-lock sensor mounting bolt and the anti-lock sensor from the steering knuckle. Move the sensor aside.

8. Remove the air suspension height sensor at the lower control arm ball stud attachment. Move the sensor aside.

9. Remove the nuts from the sway bar link at the strut. Remove the sway bar link from the strut and move aside.

10. Using Hub Remover/Replacer T81P-1104-C, Stud Adapter T86P-1104-A1, Hub Remover Adapters T83P-1104-BH and Hub Replacer T81P-1104-A or equivalents, push the CV-joint only to the point that the splines are free in the hub.

11. Remove the lower ball joint-to-lower control arm nut. Using Puller T64P-3590-F or equivalent, separate the lower ball joint from the lower control arm. Discard the nut.

12. Pull down on the lower control arm until the lower ball joint clears the lower control arm.

➡**Do not allow the halfshaft to hang unsupported. Damage to the CV-joint or boots may result.**

13. Push the outboard CV-joint stub shaft in while placing a wooden dowel between the frame pocket and steering knuckle. Remove the outer CV-joint from the hub.

14. Assemble CV-Joint Puller Adapter T89P-3415-B, Puller Extension T86P-3514-A2 or equivalents, and a slide hammer. Insert the puller adapter behind the inner CV-joint housing at the 4 o'clock position on the left side or at the 5 o'clock position on the right side.

15. Use the slide hammer to release the halfshaft bearing retainer circlip.

16. Carefully remove halfshaft and tools from the vehicle. Discard the circlip.

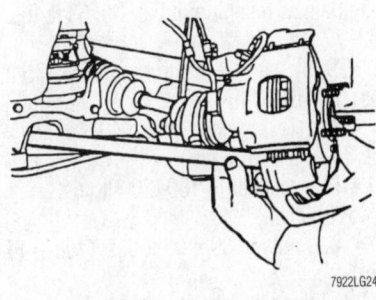

7922LG24

Place the wooden dowel between the frame and the steering knuckle to keep it out of your way

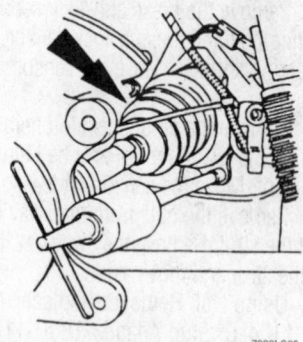

7922LG25

Use a slide hammer and adapter to remove the halfshaft from the transaxle

To install:

17. Install a new circlip on the inner CV-joint stub shaft. When installing the circlip, start one end in the groove and work the circlip over the stub shaft end into the groove. This will avoid over expanding the circlip.

18. Carefully align the splines of the inner CV-joint stub shaft with the splines in the transaxle. Exerting some force, push the CV-joint into the transaxle until the circlip is felt to seat in the differential side gear. Use care to prevent damage to the oil seal.

➡**A non-metallic (plastic or rawhide) mallet may be used to aid in seating the halfshaft bearing retainer circlip into the differential side gear groove. If a mallet is necessary, tap only on the outer CV-joint.**

19. Remove the wooden dowel and carefully align the splines of the outer CV-joint with the splines in the wheel hub.

20. Carefully push the shaft into the wheel hub as far as possible.

21. Install a new wheel hub retainer nut. Manually thread the hub retainer nut onto the outer CV-joint as far as possible.

22. Install the remaining components in the reverse order of removal:
- Lower ball joint-to-lower control arm, using a new retaining nut, to 39–53 ft. lbs. (53–72 Nm)
- Sway bar link-to-strut nut to 57–75 ft. lbs. (77–107 Nm)
- Anti-lock sensor bolt to 40–60 inch lbs. (4.5–6.8 Nm)
- Hub retainer to 180–200 ft. lbs. (245–270 Nm)
- Lug nuts to 80–105 ft. lbs. (108–144 Nm)

23. Lower the vehicle. Connect the negative battery cable. Refill the air spring.

24. Check the transaxle fluid level and add, as necessary. Road test the vehicle and check for proper operation.

STEERING AND SUSPENSION

Air Bag

✳✳ CAUTION

Some vehicles are equipped with an air bag system, also known as the Supplemental Inflatable Restraint (SIR) system. The system must be disabled before performing service on or around system components, steering column, instrument panel components, wiring and sensors. Failure to follow safety and disabling procedures could result in accidental air bag deployment, possible personal injury and unnecessary system repairs may result.

PRECAUTIONS

Several precautions must be observed when handling the inflator module to avoid accidental deployment and possible personal injury.

- Never carry the inflator module by the wires or connector on the underside of the module.
- When carrying a live inflator module, hold securely with both hands, and ensure that the bag and trim cover are pointed away.
- Place the inflator module on a bench or other surface with the bag and trim cover facing up.
- With the inflator module on the bench, never place anything on or close to the module which may be thrown in the event of an accidental deployment.

DISARMING

✳✳ CAUTION

The Supplemental Inflatable Restraint (SIR) system must be disarmed before performing service around SIR system components or SIR system wiring. Failure to do so may cause accidental deployment of the air bag, resulting in unnecessary SIR system repairs and/or personal injury.

1. Disconnect both battery cables from the battery, negative cable first.
2. Wait one minute before proceeding with the service procedure. This is the time required for the back-up power supply in the air bag diagnostic monitor to deplete its stored energy.

3. After service is completed, reconnect the battery cables, negative cable last.

4. Turn the ignition switch to the **RUN** position. The air bag indicator should light continuously for approximately six seconds, then turn OFF. If the indicator fails to light, flashes or remains lit continuously, there is a fault in the air bag system.

Power Rack and Pinion Steering Gear

REMOVAL & INSTALLATION

✳✳ CAUTION

The air suspension switch, located in the left side of the luggage compartment, must be turned OFF before raising the vehicle. Failure to do so may result in unexpected inflation or deflation of the air springs which may result in shifting of the vehicle during service.

1. Turn the air suspension switch, located in the left side of the luggage compartment on 1995–97 models or on the right kick panel for 1998–99 models, to the OFF position.

2. Disconnect the negative battery cable.

3. Disconnect the engine control sensor wiring from the Powertrain Control Module (PCM) and position the wiring aside.

4. Directly above the engine control sensor wiring, remove the ground straps from the dash panel.

5. Working from inside the vehicle, remove the nuts retaining the steering shaft weather boot to the dash panel.

6. Remove the two bolts retaining the intermediate shaft to the steering column shaft. Set the weather boot aside.

7. Remove the pinch bolt at the steering gear input shaft and remove the intermediate shaft. Raise the vehicle and support safely.

8. Support the rear of the subframe with Powertrain Lift 014–00765 and Bracket 014–00766 or equivalents.

9. Remove the front wheels.

10. Remove the tie rod cotter pins and nuts. Remove the tie rod ends from the spindle. Remove the tie rod ends from the shaft. Mark the position of the jam nut to maintain the alignment.

11. Remove both air suspension and height sensor attachments.

12. Remove the nuts from the gear-to-subframe attaching bolts. Remove both height sensor attachments.

13. Remove the dual converter Y-pipe.

14. Remove the rear subframe-to-body attaching bolts and loosen the front subframe bolts.

15. Lower the subframe until it separates from the body; approximately 4 in. Remove the heat shield band and fold down the shield.

16. Disconnect the VAPS electrical connector from the actuator assembly.

17. Disconnect the electrical connectors from the auxiliary actuator and the Power Steering Pressure (PSP) switch.

18. Rotate the gear to clear the bolts from the subframe and pull to the left to facilitate line fitting removal.

19. Position a drain pan under the vehicle and remove the line fittings. Remove the O-rings from the fitting connections and replace with new.

20. Remove the steering gear assembly through the left wheel well.

To install:

21. Install new O-rings into the line fittings on the steering gear.

22. Place the gear attachment bolts in the gear housing.

23. Install the steering gear assembly through the left wheel well.

24. Connect and tighten the line fittings to the steering gear assembly.

  a. Connect the power steering pressure and return hoses to the fittings on the rack and pinion assembly and tighten to 24–30 ft. lbs. (33–41 Nm).

  b. Install the power steering left turn pressure hose and screw. Tighten to 80–106 inch lbs. (9–12 Nm).

25. Connect the PSP switch and auxiliary actuator electrical connectors.

26. Position the steering gear into the subframe. Install the tie rod ends onto the shaft. Install the heat shield band.

27. Attach the tie rod ends onto the spindle. Install the nuts and secure with new cotter pins. Attach the sway bar link.

28. Raise the vehicle until the subframe contacts the body. Install the subframe attaching bolts.

29. Install the gear-to-subframe nuts. Tighten the nuts to 84–112 ft. lbs. (113–153 Nm).

30. Install the dual converter Y-pipe.

31. Connect the air suspension height sensors.

32. Attach the exhaust pipe to the catalytic converter.

33. Tighten the following items to:

  a. Power steering line fittings-to-steering gear: 24–30 ft. lbs. (33–41 Nm)

  b. Tie rod ends-to-steering knuckles: 35–46 ft. lbs. (47–63 Nm)

  c. Subframe bolts: 100–144 ft. lbs. (135–195 Nm)

  d. Intermediate shaft-to-steering gear input shaft bolt: 30–38 ft. lbs. (41–51 Nm)

  e. Intermediate shaft-to-steering column shaft nuts: 15–24 ft. lbs. (21–33 Nm)

34. Install the wheel and tire assemblies and lower the vehicle. Fill the power steering system.

35. Working from inside the vehicle, pull the weather boot end out of the vehicle and install it over the valve housing. Install the intermediate shaft to the steering gear input shaft. Install the inner weather boot to the floor pan.

36. Connect engine control sensor wiring to the PCM.

37. Install the screw retaining the ground straps to the dash panel directly above the engine control sensor wiring.

38. Install the intermediate shaft to the steering column shaft. Fill the power steering system.

39. Turn the air suspension switch to the ON position.

40. Check the system for leaks and proper operation. Adjust the toe setting as necessary.

Strut

REMOVAL & INSTALLATION

Front

✳✳ CAUTION

Do not remove an air spring under any circumstances when pressurized. Do not remove any components supporting an air spring without exhausting the air. Failure to follow these instructions may result in unexpected inflation or deflation of the air springs which may result in shifting of the vehicle during service.

1. Turn the air suspension switch to the OFF position. The switch is located in the

1 Tie Rod End
2 Nut
3 Front Wheel Spindle Tie Rod (2 Req'd)
4 Front Suspension Steering Ball Stud Dust Seal
5 Power Steering Hose Bracket
6 Bolt (2 Req'd)
7 Screw (2 Req'd)
8 Steering Gear
9 Screw (2 Req'd)
10 Nut (2 Req'd)
11 Front Wheel Knuckle (LH)
12 Crossmember
13 Nut (2 Req'd)
14 Steering Shaft U-Joint Shield
15 Front Wheel Knuckle (RH)
16 Power Steering Gear Input Shaft and Control

7922LG26

Exploded view of the power rack and pinion steering gear mounting

trunk on 1995–97 models or on the right kick panel for the 1998–99 models.

2. Turn the ignition switch to the **OFF** position.

➡ **Be sure to leave the steering column in the UNLOCKED position.**

3. Loosen, but do not remove, the three top mount-to-strut tower retaining nuts.

4. Raise and safely support the vehicle. Remove the wheel and tire assembly.

5. Remove the brake line bracket from the strut assembly.

6. Remove the vinyl cover from the upper link stud, if equipped.

7. Remove the stabilizer bar link nut and the link from the strut.

8. If equipped, disconnect the air line from the solenoid valve.

9. Disconnect the electrical connector at the air spring solenoid valve. Disconnect the front spring and strut actuator electrical connector at the body harness connector.

10. Disconnect the air suspension height sensor from the lower height sensor ball stud and position aside.

✳✳ CAUTION

Do not fully release the solenoid until the air is completely bled from the air spring.

11. If equipped, remove the solenoid valve as follows:

a. Remove the air spring solenoid retainer.

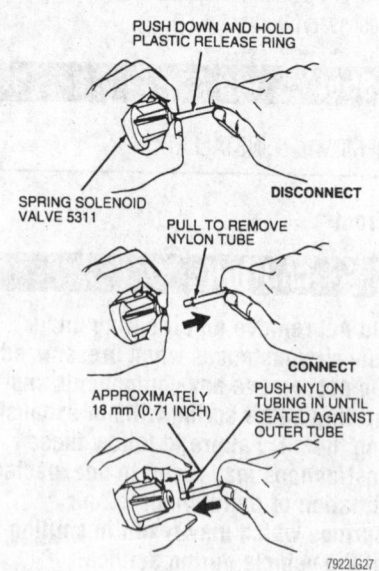

PUSH DOWN AND HOLD
PLASTIC RELEASE RING

SPRING SOLENOID VALVE 5311

DISCONNECT

PULL TO REMOVE NYLON TUBE

CONNECT

APPROXIMATELY 18 mm (0.71 INCH)

PUSH NYLON TUBING IN UNTIL SEATED AGAINST OUTER TUBE

7922LG27

Air line disconnect/connect procedure

b. Rotate the solenoid valve counter-clockwise to the first stop.

c. Pull the solenoid valve straight out slowly to the second stop to bleed the air from the system.

d. After the air is fully bled from the system, rotate the solenoid valve counterclockwise to the third stop and remove the solenoid valve from the housing.

12. Remove and discard the front suspension arm-to-ball joint nut. Using Tie Rod End Remover tool 3290-D or equivalent, separate the ball joint from the front suspension lower arm.

✳✳ CAUTION

Do not allow the halfshaft to move outward. This could result in separation of the internal parts of the tripod CV-joint, causing failure of the joint.

13. Remove the strut-to-steering knuckle pinch bolt and nut. Using a large prybar, slightly spread the strut-to steering knuckle pinch joint, if required, for removal.

✳✳ WARNING

Do not remove the front spring and strut large center nut while the front spring and strut is in the vehicle as this may result in a permanent air leak through the top of the air spring.

14. Remove the three top mount-to-strut tower retaining nuts and remove the strut/air spring assembly from the vehicle.

To install:

15. Install the strut with the three top mount-to-shock tower nuts, leaving the nuts loose.

16. Install the steering knuckle and hub assembly to the strut. Install a new strut-to-steering knuckle pinch bolt and nut. Tighten the bolt to 73–97 ft. lbs. (98–132 Nm). Install a new control arm-to-ball joint nut. Tighten to 50–68 ft. lbs. (68–92 Nm).

17. Install the stabilizer bar link to the strut and install a new stabilizer bar link nut. Tighten to 57–75 ft. lbs. (77–101 Nm).

18. Install the height sensor link on the ball stud pin on the control arm.

19. If equipped, install the solenoid valve as follows:

a. Check the solenoid O-ring for abrasions or cuts and replace, as necessary.

b. Lightly grease the solenoid O-ring prior to installation using a suitable silicone dielectric compound.

c. Insert the solenoid O-ring into the air spring end cap and rotate clockwise to

the third stop. Push in to the second stop, then rotate clockwise to the first stop.

d. Install the air solenoid valve retainer.

20. Connect the electrical connector to the solenoid valve.

21. Connect the air line to the solenoid valve.

22. Connect the front spring and shock actuator electrical connector.

23. Install the screw and brake hose bracket to the front spring and shock. Tighten the screw to 11 ft. lbs. (15 Nm).

24. Install the front wheels.

25. Lower the vehicle enough to gain access to the three top mount-to-strut tower nuts.

26. Tighten the three top mount-to-strut tower nuts to 23–29 ft. lbs. (30–40 Nm).

27. Lower the vehicle. Turn the air suspension switch to the **ON**position. Road test the vehicle and check for proper operation.

Shock Absorber

REMOVAL & INSTALLATION

Rear

1. Turn the air suspension switch to the OFF position. The switch located in the luggage compartment on 1995–97 models or on the right kick panel for 1998–99 models.

2. Turn the ignition switch to the **OFF** position.

3. Remove the side trim panel in the luggage compartment and remove the mass damper retaining clip.

4. Raise and safely support the vehicle.

5. Remove the rear wheels.

6. If equipped, unplug the wiring connector and retaining clips.

7. Remove the two upper shock absorber mounting bolts.

8. Remove the lower mounting nut and

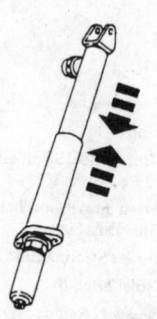

7922LG28

To bleed the air from the shock, turn it upside down, then compress and expand it several times

bolt and remove the shock absorber, mounting bracket and mass damper.

To install:

9. Install the mounting bracket and mass damper on the shock absorber.

10. Turn the shock absorber upside down and compress and expand the piston to remove the air from the cylinder.

➡**If equipped, the actuator should face the rear after the shock has been installed.**

11. Insert the top of the shock through the upper mounting hole, then compress the shock and install the lower mounting bolt and nut.

12. Install the two upper mounting bolts. Tighten the bolts to 58–68 ft. lbs. (68–92 Nm).

13. If equipped, connect the wiring to the actuator and install the retaining clips on the lower arm.

14. Install the wheels and lower the vehicle.

15. Turn the air suspension system ON and replace the trim panels.

Air Spring

REMOVAL & INSTALLATION

❋❋ CAUTION

Do not remove an air spring under any circumstances when pressurized. Do not remove any components supporting an air spring without exhausting the air. Failure to follow these instructions may result in unexpected inflation or deflation of the air springs which may result in shifting of the vehicle during service.

1. Properly deflate the air springs using the New Generation Star tester or equivalent scan tool.

2. Turn the air suspension switch to the OFF position. The switch located in the luggage compartment on 1995–97 models or on the right kick panel for 1998–99 models.

3. Turn the ignition switch to the OFF position.

4. Raise and safely support the vehicle. Remove the wheel and tire assembly.

5. Disconnect the air spring solenoid electrical connector and air line.

6. On the 1995–97 models, remove the three bolts securing the air spring to the suspension. On 1998–99 models, depress the plastic tabs while lifting up on

the air spring to disengage it from the suspension.

To install:

7. Inspect the O-rings for cuts or abrasions and replace as required.

8. Collapse the air spring until the bottom of the air spring sleeve (black rubber) is within ½ inch (12mm) of the bottom of the tapered piston.

9. Position the air spring in the vehicle.

10. On 1995–97 models, install the three air spring-to-lower suspension arm bolts. Tighten them to 50–59 ft. lbs. (68–82 Nm). On 1998–99 models, press the plastic tab through the suspension until it clicks into place.

11. Engage the air spring cap to the body bracket.

12. Connect the solenoid air line and the electrical connector.

13. Fill the air spring.

14. Install the wheel and tire assembly. Install and tighten the lug nuts to 80–105 ft. lbs. (108–144 Nm).

15. Lower the vehicle.

16. Turn the air suspension switch to the ON position.

17. Road test the vehicle and check for proper operation.

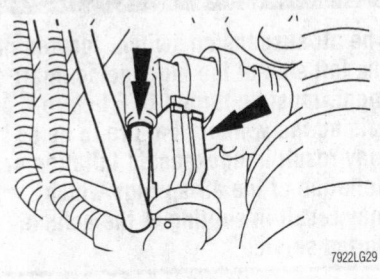

Be sure to unplug the connections to the air spring before removing it

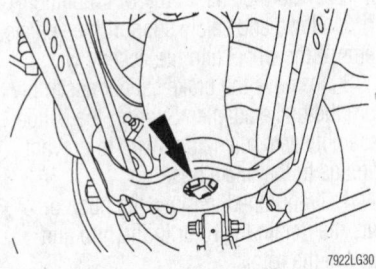

On 1998–99 models, depress the plastic locking tabs to disengage the air spring from the suspension

Lower Ball Joint

REMOVAL & INSTALLATION

The lower ball joint is an integral part of the steering knuckle. If the lower ball joint is found to be defective, the entire steering knuckle must be replaced.

Wheel Bearings

ADJUSTMENT

There is no adjustment for the front or rear wheel bearings due to the nature of their design which contains both inner and outer bearings.

REMOVAL & INSTALLATION

Front

❋❋ CAUTION

Do not remove an air spring under any circumstances when pressurized. Do not remove any components supporting an air spring without exhausting the air. Failure to follow these instructions may result in unexpected inflation or deflation of the air springs which may result in shifting of the vehicle during service.

➡**Be sure to have available a new wheel hub retainer nut, tie rod end castellated nut and cotter pin, lower control arm-to-ball joint nut, steering knuckle-to-strut pinch bolt and nut, stabilizer bar link nut and the three hub and bearing retaining bolts, per side.**

The air suspension switch, located on the left side of the luggage compartment for 1995–97 models or on the right kick panel for 1998–99 models, must be turned OFF before raising the vehicle. Failure to do so may result in unexpected inflation or deflation of the air springs which may result in shifting of the vehicle during service.

1. Properly deflate the air springs using the New Generation Star tester or equivalent scan tool.

2. Remove and discard the wheel hub retainer nut.

3. Raise and safely support the vehicle. Remove the wheel and tire assembly.

4. Remove the brake caliper and rotor. Wire the caliper aside.

➡**Do not allow the brake caliper to hang by the brake hose or damage to the hose may result.**

5. Remove the brake rotor splash shield, if damaged.

6. Using Front Hub Remover/Replacer T81P-1104-C or equivalent and adapters, press the outer CV-joint from the hub.

7. Support the halfshaft with wire to maintain a level position.

8. Remove and discard the three hub and bearing retainer bolts from the steering knuckle.

9. Using a small prybar, slightly spread the strut-to-knuckle pinch joint and remove the knuckle and hub from the strut.

❈❈ WARNING

The hub and bearing assembly is not pressed into the steering knuckle. DO NOT use a slide hammer to remove a stuck wheel hub. Do not strike the back of inner bearing race.

10. Remove the wheel hub from the steering knuckle.

To install:

➡ **If the hub bearing journal is scored or damaged, replace the knuckle. If the wheel hub is damaged or any end-play is detectable, replace the wheel hub and bearing assembly.**

11. Remove all foreign material from the knuckle bearing bore and hub bearing journal to ensure correct seating of the new hub.

➡ **The steering knuckle must be clean enough to allow the wheel hub to be completely seated by hand. Do not press or draw the wheel hub into place.**

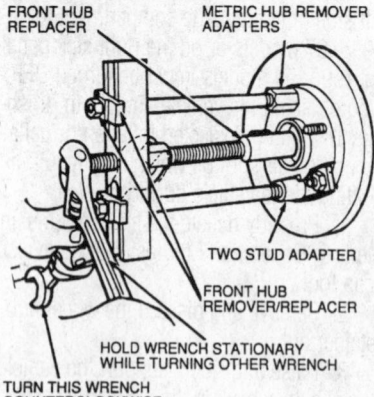

MAKE SURE THE HUB REMOVER ADAPTER IS FULLY THREADED ONTO THE HUB STUD AND IS POSITIONED OPPOSITE THE TWO STUD ADAPTER

FRONT HUB REPLACER

METRIC HUB REMOVER ADAPTERS

TWO STUD ADAPTER

FRONT HUB REMOVER/REPLACER

HOLD WRENCH STATIONARY WHILE TURNING OTHER WRENCH

TURN THIS WRENCH COUNTERCLOCKWISE

7922LG32

Use a puller such as the one shown to press the halfshaft through the hub assembly

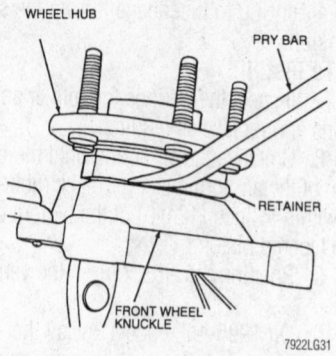

WHEEL HUB

PRY BAR

RETAINER

FRONT WHEEL KNUCKLE

7922LG31

Pry the hub assembly from the knuckle if it is stuck

12. Install the wheel hub and tighten bolts to 61–79 ft. lbs. (83–107 Nm).

13. Install a new brake rotor splash shield using new rivets, if necessary.

14. Install a new wheel hub retainer nut and tighten to 170–202 ft. lbs. (230–275 Nm).

15. Install the brake rotor and brake caliper.

16. Fill the air springs. Lower the vehicle.

17. Pump the brake pedal several times prior to moving the vehicle, to position the brake pads to the rotor. Road test the vehicle and check for proper operation.

Rear

❈❈ CAUTION

The air suspension switch, located in the left side of the luggage compartment, must be turned OFF before raising the vehicle. Failure to do so may result in unexpected inflation or deflation of the air springs which may result in shifting of the vehicle during service.

➡ **Be sure to have available a new wheel hub retainer nut.**

1. Turn the air suspension switch, located in the luggage compartment on 1995–97 models or on the right kick panel for 1998–99 models, to the OFF position.

2. Raise and safely support the vehicle. Remove the wheel and tire assembly.

3. Remove the brake caliper assembly from the brake adapter. Support the caliper assembly with a length of wire to prevent damage to the brake hose.

4. If equipped, remove the push on nuts that retain the rotor to the hub and remove the rotor.

5. Remove the grease cap from the bearing and hub assembly.

6. Remove the wheel hub retainer nut, then pull the hub assembly from the spin-

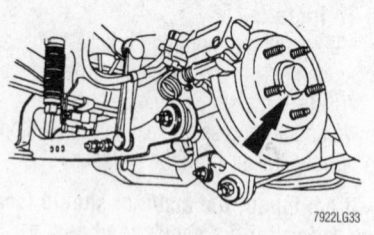

7922LG33

Remove the grease cap to expose the wheel hub retaining nut

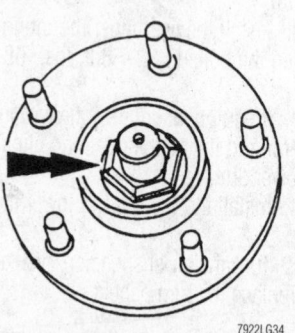

7922LG34

Remove the retaining nut, then pull the hub assembly from the rear wheel spindle

dle. Be sure to use a new retainer nut when installing the hub.

➡ **The wheel bearings are permanently sealed and lubricated. If wheel bearing replacement is required, the bearing and hub must be replaced as an assembly.**

To install:

7. Position the wheel hub and bearing on the rear spindle.

❈❈ WARNING

Damage to the bearing assembly may result if power tools are used to install it. Do not back the nut off after reaching the required torque..

8. Install a new wheel hub retainer nut and tighten to 188–254 ft. lbs. (255–345 Nm).

9. Install the rear disc/drum brake assembly.

10. Install the wheel and tire assembly. Lower the vehicle.

11. Turn the air suspension switch to the ON position.

12. Pump the brake pedal several times to position the brake pads to the rotor.

13. Road test the vehicle and check for proper operation.

FORD MOTOR CO.

Ford-Escort • Escort ZX2 • Mercury-Tracer

22

DRIVE TRAIN 22-36
ENGINE REPAIR 22-2
FUEL SYSTEM 22-34
STEERING AND
 SUSPENSION 22-48

A

Air Bag.................22-48
 Disarming.................22-48
 Precautions.................22-48

C

Camshaft(s) and Valve Lifters 22-24
 Removal & Installation 22-25
Clutch.................22-41
 Adjustment.................22-41
 Removal & Installation 22-43
Coil Spring.................22-51
 Removal & Installation 22-51
Cylinder Head.................22-9
 Piston Squish Height 22-15
 Removal & Installation 22-9

D

Distributor.................22-2
 Installation.................22-2
 Removal 22-2

E

Engine Assembly 22-3
 Removal & Installation 22-3
Exhaust Manifold.................22-20
 Removal & Installation 22-20

F

Front Crankshaft Seal 22-24
 Removal & Installation 22-24
Fuel Filter 22-34
 Removal & Installation 22-34
Fuel Pump 22-35
 Removal & Installation 22-35
Fuel System Pressure 22-34
 Relieving 22-34
Fuel System Service
 Precautions.................22-34

H

Halfshafts.................22-44
 Removal & Installation 22-44
Hydraulic Clutch System 22-44
 Bleeding 22-44

I

Ignition Timing 22-3
 Adjustment.................22-3
Intake Manifold 22-16
 Removal & Installation 22-16

L

Lower Ball Joint.................22-53
 Removal & Installation 22-53

O

Oil Pan.................22-28
 Removal & Installation 22-28
Oil Pump 22-30
 Removal & Installation 22-30

R

Rack and Pinion Steering
 Gear.................22-48
 Removal & Installation 22-48
Rear Main Seal 22-33
 Removal & Installation 22-33
Rocker Arms 22-16
 Removal & Installation 22-16

S

Strut.................22-50
 Removal & Installation 22-50

T

Transmission Assembly 22-36
 Removal & Installation 22-36

V

Valve Lash 22-27
 Adjustment.................22-27

W

Water Pump.................22-7
 Removal & Installation 22-7
Wheel Bearings.................22-53
 Adjustment.................22-53
 Removal & Installation 22-53

ENGINE REPAIR

➡Disconnecting the negative battery cable on some vehicles may interfere with the functions of the on board computer systems and may require the computer to undergo a relearning process, once the negative battery cable is reconnected.

Distributor

REMOVAL

1. Disconnect the negative battery cable.
2. Disengage the distributor electrical connector and coil wire at the cap.
3. Remove the distributor cap and rotor.
4. Position it to the side. If the distributor unit is not being replaced, scribe a reference mark across the distributor base flange and the cylinder head.
5. Loosen the distributor mounting bolts and remove the distributor from the engine.
6. If necessary, remove the distributor base gasket from the distributor.
7. If necessary, remove the distributor O-ring.

➡Do not rotate the engine while the distributor assembly is removed. Do not attempt to disassemble and service the distributor unit. The distributor unit should be replaced as an assembly.

8. If removed, install a new distributor base gasket.
9. If removed, install a new O-ring on the distributor unit.

INSTALLATION

Engine Not Disturbed

➡The distributor can only be installed in one direction. Do not force the distributor into position and be sure the drive tangs engage the camshaft slots.

1. Install the distributor unit into the engine. Make certain the drive tangs engage with the camshaft slots.
2. If installing the old distributor, be sure the reference marks made during removal are aligned.
3. Tighten the mounting bolts finger-tight.
4. Install the rotor and distributor cap.
5. Engage the distributor electrical connector and coil wire to the cap.
6. Connect the negative battery cable and check ignition timing.

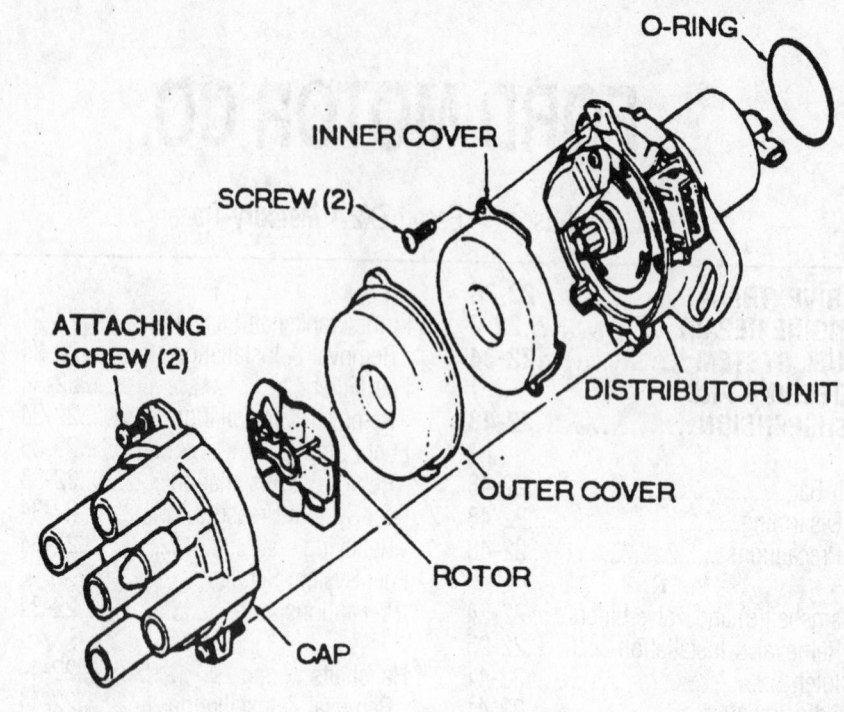

Exploded view of the distributor assembly—1.8L engine

7. If necessary, adjust the ignition timing.
8. Tighten the distributor mounting bolt to 14–19 ft. lbs. (19–25 Nm).

Engine Disturbed

➡When installing the distributor be sure the offset drive tangs engage the camshaft slots.

1. Be sure that the engine is still with the No. 1 piston up on TDC of its compression stroke.

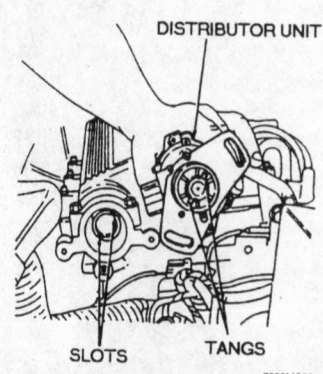

When installing the distributor be sure the drive tangs engage the camshaft slots—1.8L engine

➡If the engine was disturbed while the distributor was removed, it will be necessary to remove the No. 1 spark plug and rotate the engine clockwise until the No. 1 piston is on the compression stroke. Align the timing marks.

2. If installing the old distributor:
   a. Install the distributor and align the marks on the distributor base flange-to-cylinder head and the rotor points toward the mark on the distributor housing made previously. Make certain the rotor is pointing to the No. 1 mark on the distributor base.
3. If installing a new distributor:
   a. Install the rotor and cap (with wires still attached) on the distributor, follow the No. 1 spark plug wire from the plug to the cap, this will be the No.1 tower on the cap. Mark the location of the tower on the distributor housing, then remove the cap.
   b. Install the distributor and be sure the rotor aligns with the No.1 tower mark made on the distributor.
4. When all the marks are aligned, install the distributor hold-down bolts and tighten them to 14–19 ft. lbs. (19–25 Nm).
5. Install the cap and tighten the screws.
6. Connect the negative battery cable.

7. Engage the electrical connections.

8. Install the No. 1 spark plug, if removed.

9. Recheck the initial timing and adjust if necessary.

Ignition Timing

ADJUSTMENT

1.8L Engine

1. Clean any dirt from around the timing scale on the timing cover and the yellow timing mark on the crankshaft pulley. This will make it easier to see when you are inspecting the timing.

2. Place the gear selector lever in the **PARK** or **NEUTRAL** position. Apply the parking brake.

3. Turn all accessories and loads **OFF**.

4. Start the engine and allow it to reach operating temperature.

5. Connect an inductive timing light to the engine, following the tool manufacturers instructions.

6. Using a jumper wire, connect the GROUND terminal to the TEN terminal of the Diagnostic Link Connector (DLC).

7. Using Rotunda tachometer 059–00010 or its equivalent, connect the tachometer positive lead to the IG terminal on the DLC and the tachometer negative lead to the battery negative post.

8. Point the timing light at the timing scale.

9. Check the ignition timing. Timing should be 10° BTDC at 700–800 rpm. The mark on the pulley should be aligned with the corresponding mark on the timing scale.

10. If not, loosen the distributor mounting bolt and turn the distributor until the marks are aligned.

11. Tighten the distributor mounting bolt to 14–19 ft. lbs. (19–25 Nm).

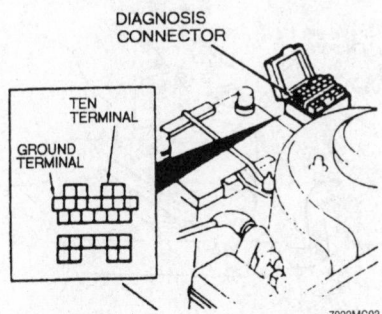

Connect a jumper wire between the ground and ten terminals on the DLC—1.8L engine

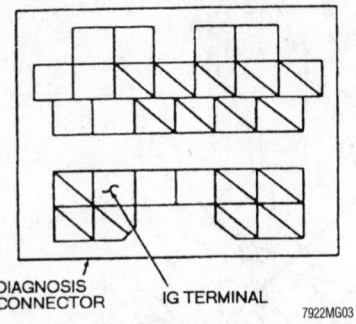

Connect the tachometer positive lead to the IG terminal on the DLC—1.8L engine

12. Check the timing again to be sure it did not change when tightening the distributor mounting bolts

13. Remove the jumper wire, timing light and tachometer.

1.9L Engine

The 1.9L engine is equipped with an Electronic Distributorless Ignition System (EDIS). Ignition timing adjustment is not possible on the EDIS system.

2.0L Engines

The Powertrain Control Module controls ignition timing on these models. No adjustment is necessary or possible.

Engine Assembly

REMOVAL & INSTALLATION

1. On all engines, perform the following basic steps for engine removal.

   a. Remove the hood from the vehicle.

   b. Properly relieve the fuel system pressure.

   c. Disconnect the negative battery cable.

➡ If your vehicle is equipped with air conditioning, refer to Section 1 for information regarding the implications of servicing your A/C system yourself. Only a MVAC-trained, EPA-certified, automotive technician should service the A/C system or its components.

   d. If equipped with A/C, have the system discharged by a qualified technician.

   e. Tag and disconnect all the wires and hoses that would interfere with engine removal.

   f. Tag and disconnect all cables such as the accelerator and kickdown cables (etc.) that would interfere with engine removal.

   g. Remove the drive belt(s).

   h. Drain the cooling system.

※※ CAUTION

Never open, service or drain the radiator or cooling system when hot; serious burns can occur from the steam and hot coolant. Also, when draining engine coolant, keep in mind that cats and dogs are attracted to ethylene glycol antifreeze and could drink any that is left in an uncovered container or in puddles on the ground. This will prove fatal in sufficient quantities. Always drain coolant into a sealable container. Coolant should be reused unless it is contaminated or is several years old.

   i. Drain the engine oil.

※※ CAUTION

The EPA warns that prolonged contact with used engine oil may cause a number of skin disorders, including cancer! You should make every effort to minimize your exposure to used engine oil. Protective gloves should be worn when changing the oil. Wash your hands and any other exposed skin areas as soon as possible after exposure to used engine oil. Soap and water, or waterless hand cleaner should be used.

2. On 1.8L engines with a automatic transaxle, complete the generic steps for all engines at the start of this procedure, then perform the following steps for removal.

   a. Remove the starter motor.

   b. Remove the splash shields and the starter motor.

   c. Disconnect the transaxle cooling lines from the radiator and plug the lines.

   d. Remove the axle bearing bracket.

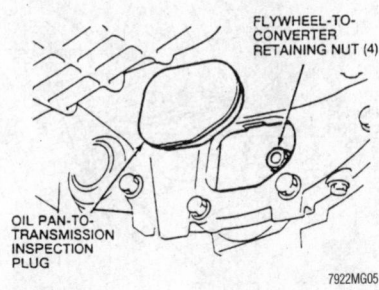

Remove flywheel inspection plate from the oil pan—1.8L engine with a automatic transaxle

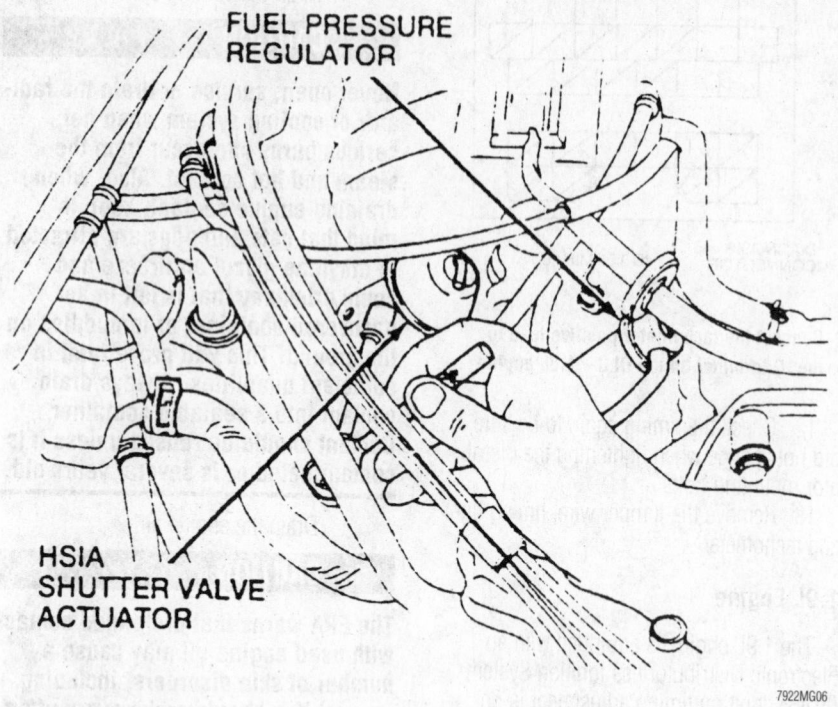

FUEL PRESSURE REGULATOR

HSIA SHUTTER VALVE ACTUATOR

7922MG06

Remove the fuel pressure regulator and High Speed Inlet Air (HSIA) shutter valve actuator—1.8L engine with a automatic transaxle

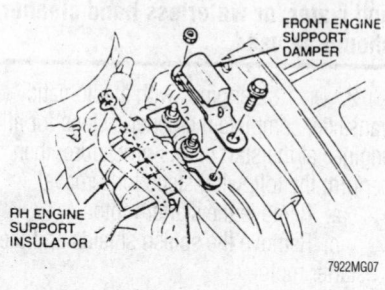

FRONT ENGINE SUPPORT DAMPER

RH ENGINE SUPPORT INSULATOR

7922MG07

Remove the right-hand support insulator and the front support damper—1.8L engine with a automatic transaxle

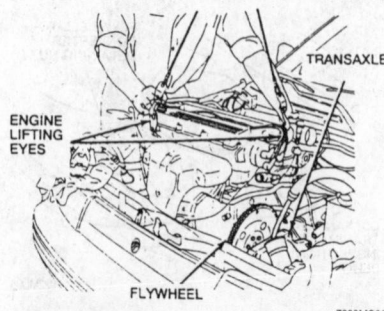

TRANSAXLE

ENGINE LIFTING EYES

FLYWHEEL

7922MG08

Attach a sling to the lifting eyes on the engine and connect the sling to a hoist to remove the engine

e. Remove flywheel inspection plate from the oil pan.

f. Place a wrench on the crankshaft pulley and rotate the crankshaft to gain access to the converter nuts, then unfasten the nuts.

g. If equipped, remove the A/C compressor and set it aside with the hoses still attached.

h. Remove the power steering pump, transaxle-to-engine upper bolts, radiator, vacuum reservoir, fuel pressure regulator and the High Speed Inlet Air (HSIA) shutter valve actuator.

i. Remove the alternator.

j. Connect an engine sling to the lifting eyes on the engine.

k. Using an engine hoist, take up the slack on the sling until the engine is supported.

l. Unfasten the oil pan-to-transaxle bolts and the transaxle-to-engine bolts from the block.

m. Remove the right-hand support insulator and the front support damper.

n. Separate the engine from the transaxle and remove the engine from the vehicle.

3. On 1.8L engines with a manual transaxle, complete the generic steps for all

engines at the start of this procedure, then perform the following steps for removal.

a. Remove the battery and tray.

b. Remove the radiator upper support brackets.

c. Remove the splash shields.

d. Remove the clutch cylinder hose bracket with the hose still attached and position the assembly aside.

e. Disconnect the transaxle gearshift rod, clevis and lever stabilizer bar from the transaxle.

f. Unbolt the power steering pump and bracket assembly and set it aside with the hoses still connected.

g. Unbolt the A/C compressor and set it aside, if equipped.

h. Remove the A/C accumulator hoses clip from the crossmember and move the hose away from the work area.

i. Remove the A/C hose routing support clips from the radiator and position the hose aside.

j. Disconnect the speedometer cable from the transaxle.

k. Disconnect the exhaust pipe from the manifold.

l. Disconnect the starter motor wiring.

m. Remove the stabilizer bar.

n. Separate the tie-rod ends from the steering knuckles.

o. Remove the halfshafts.

p. Unbolt the transaxle front and rear mount nuts and remove the mount from the support crossmember.

q. Remove the radiator and fan.

r. Connect an engine sling to the lifting eyes on the engine.

s. Using an engine hoist, take up the slack on the sling until the engine is supported.

t. Remove a crankshaft pulley.

u. Remove the front engine support damper nut and nut.

v. Remove the engine insulator mounting bolt and nut.

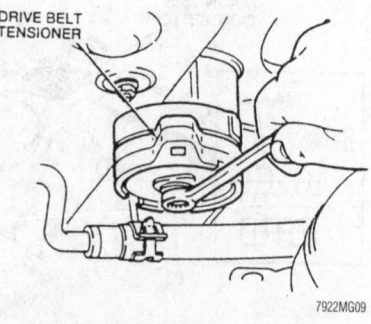

DRIVE BELT TENSIONER

7922MG09

Loosen the drive belt tensioner mounting bolt and remove the pulley—1.9L engine

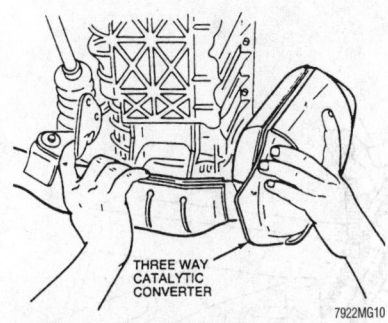

Remove the catalytic converter to access engine-to-transaxle mounting bolts—1.9L engine with an automatic transaxle

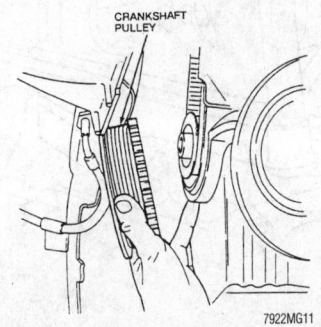

Remove the crankshaft pulley because it will interfere with the inner fender liner—1.9L engine with an automatic transaxle

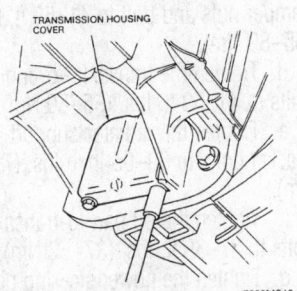

Remove the transaxle housing cover to access the torque converter-to-flexplate mounting bolts—1.9L engine with an automatic transaxle

w. Remove the right-hand engine support insulator-to-engine nuts.

x. Remove the left-hand transaxle insulator and bracket.

y. Remove the engine and transaxle from the vehicle.

z. Separate the transaxle from the engine and place the engine on a suitable stand.

4. On 1.9L engines, complete the generic steps for all engines at the start of this procedure, then perform the following steps for removal.

a. Remove the Idle Air Control valve (IAC)

b. Remove the splash shields.

c. Disconnect the transaxle cooling lines from the radiator and plug the lines.

d. Remove the radiator from the engine compartment.

e. Disconnect the A/C suction manifold and tube from the accumulator/drier.

f. If equipped, disconnect the power steering return hose from the reservoir and the pressure line from the pump.

g. Unbolt the power steering and A/C bracket bolts from the compressor bracket and set the hoses aside.

h. Remove the drive belt tensioner and pulley.

i. Remove the A/C compressor and set it aside with the lines still attached.

j. On models wit an automatic transaxle, remove the catalytic converter.

k. On models with an automatic transaxle, remove the transaxle housing cover, four flywheel-to-converter nuts, crankshaft pulley and the starter motor.

l. On models with an automatic transaxle, unbolt the lower transaxle-to-engine mounting bolts.

m. On models with a manual transaxle, disconnect the transaxle gearshift rod, clevis and lever stabilizer bar from the transaxle.

n. On models with an automatic transaxle, unbolt the upper transaxle-to-engine mounting bolts.

o. On models with a manual transaxle, remove the halfshafts. Install transaxle plugs into the differential side gears to prevent them from becoming misaligned.

p. On models with a manual transaxle, remove the halfshafts.

q. On models with a manual transaxle, unbolt the clutch slave cylinder and with the line still attached, set it aside.

r. On models with a manual transaxle, unbolt the transaxle front, rear and left-hand support insulator bolts.

s. On models with a manual transaxle, remove the engine cooling fan motor.

t. Connect an engine sling to the lifting eyes on the engine.

u. Using an engine hoist, take up the slack on the sling until the engine is supported.

v. On models with a manual transaxle, remove the front engine support damper nut and nut.

w. On models with a manual transaxle, remove the engine insulator mounting bolt and nut.

x. On models with a manual

transaxle, remove the right-hand engine support bracket and support insulator bolt.

y. On models with a manual transaxle, remove the right-hand insulator.

z. On models with a manual transaxle, remove the engine and transaxle from the vehicle.

aa. On models with a manual transaxle, separate the transaxle from the engine and place the engine on a suitable stand.

bb. On models with an automatic transaxle, remove the front engine support damper and right-hand engine support insulator, then remove the engine from the vehicle.

5. On 2.0L SOHC engines, complete the generic steps for all engines at the start of this procedure, then perform the following steps for removal.

a. Remove the battery and tray.

b. Remove the air cleaner assembly, then tag and disengage all wiring, hoses, cables and sensors, etc., that would interfere with engine removal.

c. Loosen the power steering pressure hose bracket bolts and set the hose aside.

d. Remove the exhaust manifold heat shield, exhaust manifold-to-catalytic converter nuts and one upper starter motor bolt. Position the starter motor bracket aside.

e. If equipped with an automatic transaxle, disconnect the transaxle cooler lines from the transaxle.

f. Disconnect the A/C manifold tube spring lock coupling at the accumulator/drier and the condenser.

g. Remove the drive belt auto tensioner and if equipped, the accessory drive belt tensioner.

h. Drain the power steering fluid and disconnect the lines from the reservoir.

i. Remove the splash shield, crossmember and the catalytic converter.

j. Remove the drive shafts and the A/C compressor with the lines still attached. Set the compressor aside.

k. Remove the radiator, fan and shroud.

l. Attach a lifting bracket on the left rear side of the cylinder head.

m. Install a three-bar engine support as illustrated.

n. Remove the transaxle crossmember.

o. Unfasten the transaxle mount nuts and remove the mount.

p. Remove the right-hand engine support insulator through-bolt.

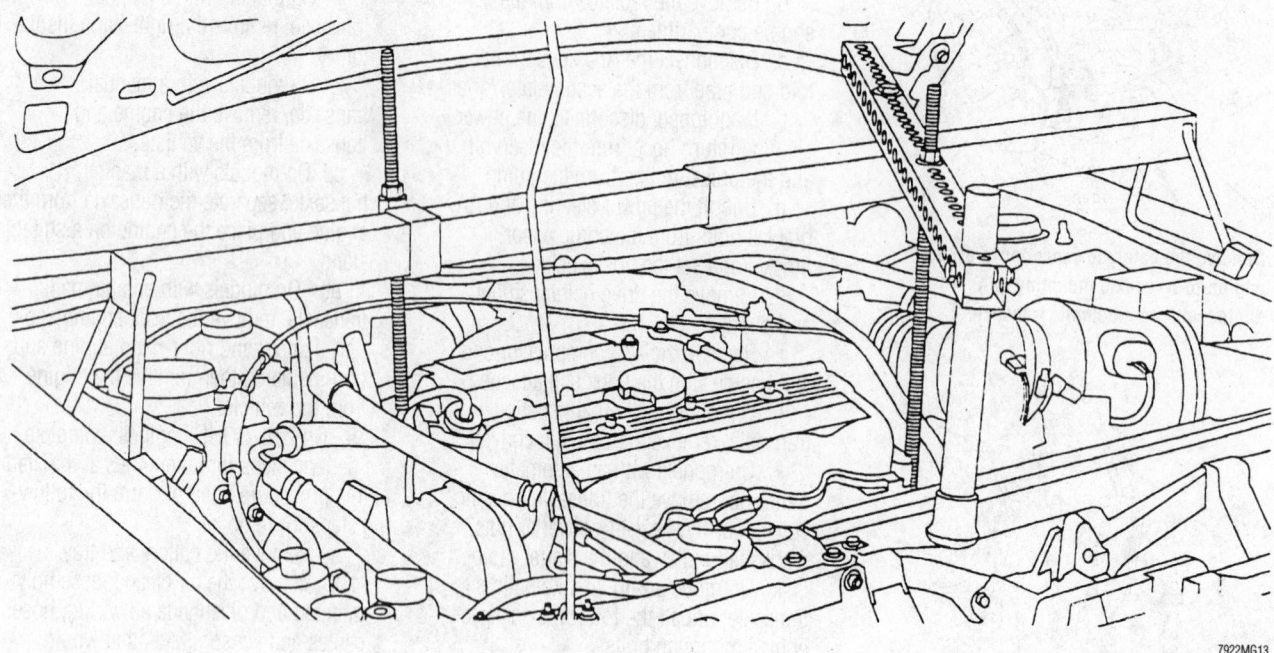

7922MG13

Properly support the engine as shown—2.0L SOHC engines

q. Attach a suitable lifting device to the engine.

r. Remove the engine and the transaxle as an assembly.

6. On 2.0L DOHC engines, complete the generic steps for all engines at the start of this procedure, then perform the following steps for removal.

a. Remove the battery and tray.

b. Remove the air cleaner assembly, then tag and disengage all wiring, hoses, cables and sensors that would interfere with engine removal.

c. If equipped with an automatic

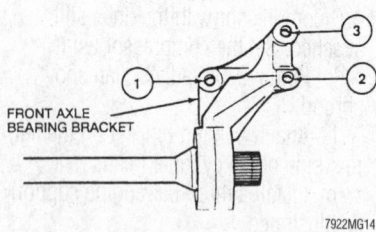

7922MG14

Tighten the front axle bearing bracket bolts in the sequence illustrated—1.8L engine with an automatic transaxle

transaxle, remove the torque converter cover and loosen the converter nuts

d. Remove the radiator, fan and shroud.

e. Disconnect the power steering return hose and the A/C compressor-to-condenser discharge line.

f. Remove the catalytic converter and the front roll restrictor

g. Remove the driveshafts and disconnect the hydraulic clutch line.

h. Support the engine with a floor jack.

i. Unfasten all bolts that attach the engine to its compartment.

j. Remove the engine and transaxle as an assembly.

To install:

7. Installation is the reverse of removal, but please note the following important steps.

8. On 1.8L engines with automatic transaxles pay special attention to the following torque specifications:

a. Tighten the transaxle-to-engine upper right-hand bolt to 41–59 ft. lbs. (55–80 Nm).

b. Tighten the right-hand insulator bolt to 49–69 ft. lbs. (67–93 Nm).

c. Tighten the front engine support

damper nuts and bolt to 41–59 ft. lbs. (55–80 Nm).

d. Tighten the transaxle-to-engine bolts to 41–59 ft. lbs. (55–80 Nm).

e. Tighten the radiator support upper bracket bolts to 69–95 inch lbs. (7.8–11 Nm).

f. Tighten the oil pan-to-transaxle bolts to 27–38 ft. lbs. (37–52 Nm).

g. Tighten the power steering pump bracket bolts to 27–38 ft. lbs. (37–52 Nm).

h. Tighten the A/C accumulator tube support clip nuts to 56–82 inch lbs. (6–9 Nm).

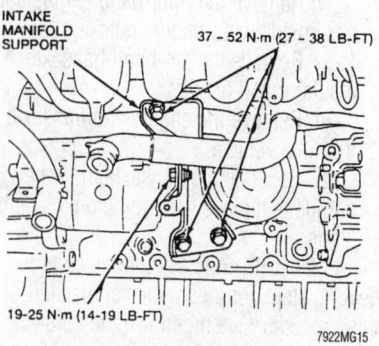

7922MG15

Tighten the intake manifold support bolts to the specifications illustrated

i. Tighten the flywheel-to-converter nuts to 25–36 ft. lbs. (34–49 Nm).

j. Tighten the front axle bearing bracket bolts in the sequence illustrated to 31–46 ft. lbs. (42–62 Nm).

k. Tighten the exhaust inlet pipe-to-exhaust manifold nuts to 23–34 ft. lbs. (31–46 Nm).

l. Tighten the splash shield bolts to 69–95 inch lbs. (7–11Nm).

m. Tighten the throttle cable bracket bolts to 69–95 inch lbs. (7–11Nm).

9. On 1.8L engines with manual transaxles pay special attention to the following torque specifications:

a. Tighten the transaxle-to-engine bolts to 47–66 ft. lbs. (64–89 Nm).

b. Tighten the oil pan-to-transaxle bolts to 27–38 ft. lbs. (37–52 Nm).

c. Tighten the front transaxle support insulator bolts to 27–38 ft. lbs. (37–52 Nm).

d. Tighten the intake manifold bolts to the specification shown. Refer to the accompanying illustration.

e. Tighten the engine mount through-bolt and nut to 49–69 ft. lbs. (67–93 Nm).

f. Tighten the engine mount-to-engine attaching nuts to 54–76 ft. lbs. (74–103 Nm).

g. Tighten the vibration damper bolt and nut to 41–59 ft. lbs. (55–80 Nm).

h. Tighten the transaxle support bracket bolts to 41–59 ft. lbs. (55–80 Nm).

i. Tighten the transaxle upper mount bolts to 32–45 ft. lbs. (43–61 Nm).

j. Tighten the transaxle upper mount nuts to 49–69 ft. lbs. (67–93 Nm).

k. Tighten the radiator mounting bracket nuts to 69–95 inch lbs. (7.8–11 Nm).

l. Tighten the exhaust mounting flange-to-manifold nuts to 23–34 ft. lbs. (31–46 Nm).

m. Tighten the exhaust pipe support bracket bolts to 27–38 ft. lbs. (37–52 Nm).

n. If equipped, tighten the A/C compressor mounting bolts to 15–22 ft. lbs. (20–30 Nm). And the A/C hose-to-crossmember routing bracket bolt to 56–82 inch lbs. (6.4–9.3 Nm).

o. Tighten the power steering mounting bolts to 27–38 ft. lbs. (37–52 Nm).

p. Tighten the shift control rod attaching nut to 12–17 ft. lbs. (16–23 Nm).

q. Tighten the clutch slave cylinder mounting bolts and the routing bracket bolt to 12–17 ft. lbs. (16–23 Nm).

r. Tighten the splash shield bolts to 69–95 inch lbs. (7.8–11 Nm).

10. On 1.9L engines, pay special attention to the following torque specifications:

a. On models with an automatic transaxle, tighten the crankshaft pulley bolt to 81–96 ft. lbs. (110–130 Nm).

b. If equipped, tighten the A/C compressor-to-bracket bolts to 15–20 ft. lbs. (20–30 Nm).

c. On models with an automatic transaxle, tighten the two upper transaxle-to-engine bolts to 40–59 ft. lbs. (55–80 Nm).

d. On models with an automatic transaxle, tighten the five lower transaxle-to-engine bolts to 27–38 ft. lbs. (37–52 Nm).

e. Tighten the engine insulator mounting bolt to 49–69 ft. lbs. (67–93 Nm).

11. On all models, perform the following steps.

a. Engage all wires and hoses.

b. Have the A/C system evacuated and charged by a qualified technician using approved equipment.

※※ WARNING

Operating the engine without the proper amount and type of engine oil will result in severe engine damage.

c. Fill the engine with the proper type and quantity of oil.

d. Fill the cooling system with the proper amount and mix of anti-freeze.

e. Install the drive belt(s).

f. Fill the transaxle with the proper type and amount of fluid.

12. On 2.0L SOHC engines pay special attention to the following torque specifications:

a. Tighten the right-hand insulator through-bolt to 50–69 ft. lbs. (67–93 Nm).

b. Tighten the motor mount bolts to 50–69 ft. lbs. (67–93 Nm).

c. Tighten the transaxle support crossmember inner retainers to 28–38 ft. lbs. (38–51 Nm) and the outer retainers to 48–65 ft. lbs. (64–89 Nm).

d. Tighten the A/C compressor mounting bolts to 15–22 ft. lbs.)

e. Tighten the muffler-to-catalytic converter to 30–40 ft. lbs. (40–55 Nm).

f. Install the crossmember and tighten the bolts to 69–97 ft. lbs. (94–131 Nm).

g. Tighten the power steering pressure hose connection to 15–18 ft. lbs. (20–25 Nm).

h. Tighten the A/C line bracket bolt to 15–18 ft. lbs. (20–25 Nm).

i. Tighten the exhaust manifold-to-catalytic converter nuts to 26–34 ft. lbs. (34–47 Nm).

13. On 2.0L DOHC engines pay special attention to the following torque specifications:

a. Tighten the front isolator nuts and bolts to 50–68 ft. lbs. (67–93 Nm).

b. Tighten the front engine support isolator through-bolt to 50–68 ft. lbs. (67–93 Nm).

c. Tighten the front roll restrictor bolts to 48–65 ft. lbs. (64–89 Nm) and the nuts to 50–69 ft. lbs. (67N93 Nm).

d. Tighten the rear roll restrictor nuts to 28–38 ft. lbs. (38–51 Nm).

e. If equipped with an automatic transaxle, tighten the torque converter nuts to 27 ft. lbs. (37 Nm).

Water Pump

REMOVAL & INSTALLATION

※※ CAUTION

Never open, service or drain the radiator or cooling system when hot; serious burns can occur from the steam and hot coolant. Also, when draining engine coolant, keep in mind that cats and dogs are attracted to ethylene glycol antifreeze and could drink any that is left in an uncovered container or in puddles on the ground. This will prove fatal in sufficient quantities. Always drain coolant into a sealable container. Coolant should be reused unless it is contaminated or is several years old.

1.8L Engine

1. Disconnect the negative battery cable.
2. Drain the cooling system.
3. Remove the timing belt.
4. Raise and safely support the vehicle.
5. Remove the engine oil dipstick tube bracket bolt(s) from the water pump.
6. Remove the two bolts and the gasket from the water inlet pipe.
7. Remove all but the uppermost water pump mounting bolt.
8. Lower the vehicle.
9. Remove the remaining bolts and the water pump assembly.
10. If the water pump is being reused, remove all gasket material from the water pump.
11. Remove all gasket material from the engine block.

To install:

12. Install a new gasket onto the water pump.

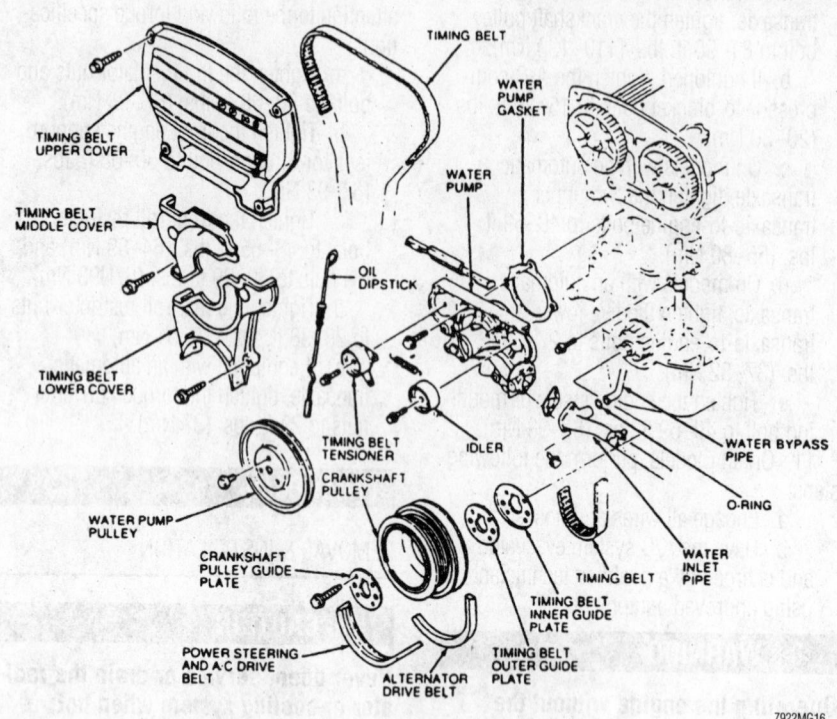

Exploded view of water pump and related components—1.8L engine

13. Place the water pump into its mounting position, then install the uppermost bolt.

14. Raise and safely support the vehicle.

15. Install the remaining water pump mounting bolts and tighten all bolts to 14–19 ft. lbs. (19–25 Nm).

16. Install a new gasket onto the water inlet pipe.

17. Install the two bolts from the water inlet pipe to the water pump and tighten to 14–19 ft. lbs. (19–25 Nm).

18. Install the bolt to the engine oil dipstick tube bracket.

19. Lower the vehicle.

20. Install the timing belt.

21. Fill the cooling system.

22. Connect the negative battery cable.

23. Start the engine and allow it to reach operating temperature. Check for coolant leaks.

24. Check the coolant level and add coolant, as necessary.

1.9L Engine

1. Disconnect the negative battery cable.

2. Drain the cooling system.

3. Remove the timing belt cover and the timing belt.

4. Raise and safely support the vehicle.

5. Remove the lower radiator hose.

6. Remove the heater hose from the water pump.

7. Lower the vehicle.

8. Support the engine with a suitable floor jack.

9. Remove the right engine mount attaching bolts and roll the engine mount aside.

10. Remove the water pump attaching bolts.

11. Using the floor jack, raise the engine enough to provide clearance for removing the water pump.

12. Remove the water pump and the gasket from the engine through the top of the engine compartment.

To install:

13. Be sure the mating surfaces of the cylinder block and water pump are clean and free of gasket material.

14. If the water pump is to be replaced, transfer the timing belt tensioner components to the new water pump.

15. With the engine supported and raised with a suitable floor jack, place the water pump and the gasket on the cylinder block and install the four attaching bolts. Tighten the bolts to 15–22 ft. lbs. (20–30 Nm).

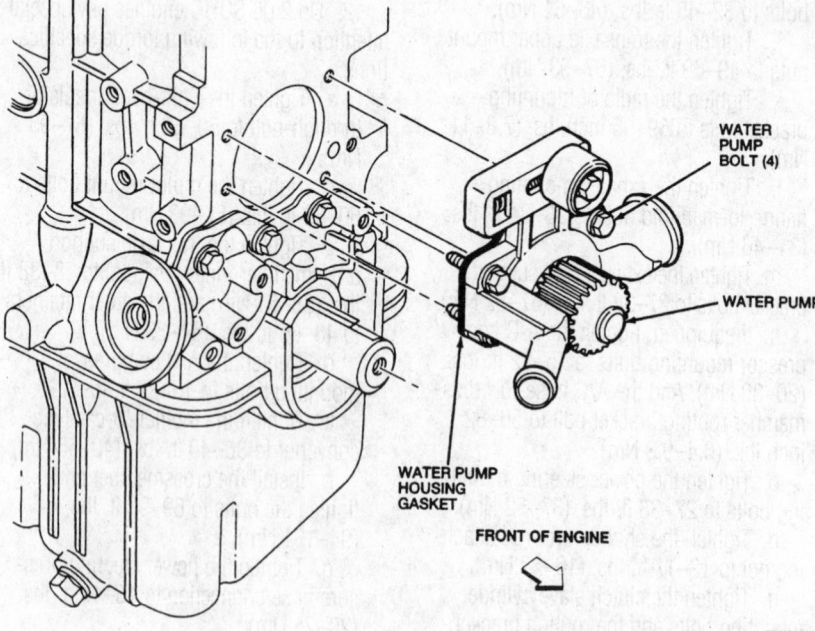

Exploded view of water pump mounting—1.9L engine

16. Install the timing belt and cover.

17. Roll the engine mount into position and install the mount bolts. Remove the floor jack.

18. Raise and safely support the vehicle.

19. Install the lower radiator hose and connect the heater hose to the pump.

20. Lower the vehicle.

21. Connect the negative battery cable.

22. Refill the cooling system.

23. Start the engine, allow it to reach normal operating temperature and check for coolant leaks.

24. Check the coolant level and add as necessary.

2.0L Engine

SOHC SPLIT PORT INJECTION (SPI)

1. Disconnect the negative battery cable.

2. Raise the vehicle and support it with safety stands.

❋❋ CAUTION

Never open, service or drain the radiator or cooling system when hot; serious burns can occur from the steam and hot coolant. Also, when draining engine coolant, keep in mind that cats and dogs are attracted to ethylene glycol antifreeze and could drink any that is left in an uncovered container or in puddles on the ground. This will prove fatal in sufficient quantities. Always drain coolant into a sealable container. Coolant should be reused unless it is contaminated or is several years old.

3. Drain and recycle the engine coolant.

4. Remove the timing belt.

5. Unfasten the timing belt tensioner bolt and remove the tensioner.

6. Disconnect the lower radiator hose from the water pump.

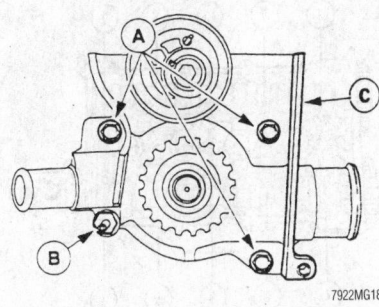

Location of the water pump (C) mounting bolts (A) and stud (B)—2.0L SOHC engine

7. Lower the vehicle and disconnect the heater hose from the water pump.

8. Unfasten the three bolts and one stud from the water pump.

9. Remove the water pump.

To install:

❋❋ CAUTION

Do not use any abrasive grinding discs to remove gasket material. Use a plastic manual gasket scraper to remove the gasket residue. Be careful not to scratch or gouge the aluminum sealing surfaces when cleaning them.

10. Clean the gasket surfaces thoroughly until all traces of the old gasket residue are removed. Inspect the gasket mating surfaces, both must be clean and flat.

11. Install a new gasket and the water pump.

12. Install the water pump bolts and stud to 15–22 ft. lbs. (20–30 Nm).

13. Connect the heater hose to the pump.

14. Raise the vehicle and support it with safety stands.

15. Connect the lower radiator hose to the pump.

16. Install the timing belt tensioner and tighten the bolt to 15–22 ft. lbs. (20–30 Nm).

17. Install the timing belt and lower the vehicle.

18. Fill the cooling system with the proper amount and mixture of coolant.

19. Start the engine and check for leaks.

DOHC ZETEC

1. Disconnect the negative battery cable.

❋❋ CAUTION

Never open, service or drain the radiator or cooling system when hot; serious burns can occur from the steam and hot coolant. Also, when draining engine coolant, keep in mind that cats and dogs are attracted to ethylene glycol antifreeze and could drink any that is left in an uncovered container or in puddles on the ground. This will prove fatal in sufficient quantities. Always drain coolant into a sealable container. Coolant should be reused unless it is contaminated or is several years old.

2. Drain and recycle the engine coolant.

3. Raise the vehicle and support it with safety stands.

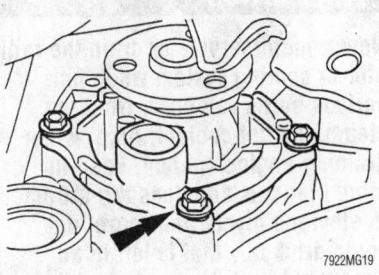

Remove the water pump mounting bolts— 2.0L Zetec engine

4. Unfasten the splash shield bolts and remove the shield.

5. Loosen the water pump pulley retaining bolts, then remove the drive belt.

6. Remove the A/C compressor bolts and move the compressor aside.

7. Unfasten the water pump bolts and remove the pump from the middle timing belt cover.

To install:

8. Install the water pump and tighten the bolts to 17 ft. lbs. (24 Nm).

9. Place the A/C compressor in position and tighten the mounting bolts.

10. Install the drive belt and tighten the water pump pulley bolts.

11. Install the splash shield and tighten the bolts.

12. Fill the cooling system with the proper amount and mixture of coolant.

13. Connect the negative battery cable.

14. Start the engine and check for leaks.

Cylinder Head

REMOVAL & INSTALLATION

❋❋ WARNING

To reduce the possibility of cylinder head warpage and/or distortion, do not remove the cylinder head while the engine is warm. Always, allow the engine to cool entirely before disassembly.

1.8L Engine

1. Properly relieve the fuel system pressure.

2. Disconnect the negative battery cable.

3. Drain the cooling system.

✳✳ CAUTION

Never open, service or drain the radiator or cooling system when hot; serious burns can occur from the steam and hot coolant. Also, when draining engine coolant, keep in mind that cats and dogs are attracted to ethylene glycol antifreeze and could drink any that is left in an uncovered container or in puddles on the ground. This will prove fatal in sufficient quantities. Always drain coolant into a sealable container. Coolant should be reused unless it is contaminated or is several years old.

4. Unfasten the bolts from the timing belt upper and middle covers, then remove the covers and gaskets.

5. Manually rotate the crankshaft in the direction of normal engine rotation (clockwise) and align the timing marks located on the camshaft pulleys and seal plate.

6. Loosen the timing belt tensioner lockbolt and temporarily secure the tensioner spring in the fully extended position.

7. Remove the timing belt from the camshaft pulleys and secure it aside to pre-vent damage during the removal and installation of the cylinder head.

➡ **Do not allow the timing belt to become contaminated by oil or grease.**

8. Tag and remove the vacuum hoses from the valve head cover.

9. Tag and disconnect the spark plug wires from the spark plugs and position the wires aside.

10. Remove the rocker arm (valve) cover and gasket.

11. Loosen the air cleaner outlet tube clamps and disconnect it from the resonance chamber and throttle body.

12. Disconnect the accelerator cable, if equipped with automatic transaxle, the kickdown cable from the throttle cam. Remove the cable bracket from the intake manifold.

13. Tag and remove all vacuum lines from the intake manifold.

14. Tag and unplug all necessary electrical connectors from the cylinder head, exhaust manifold, intake manifold, and throttle body.

15. Disconnect the ground straps.

16. Loosen the upper radiator hose clamps and remove the hose.

17. Unfasten the transaxle-to-engine block upper-right bolt.

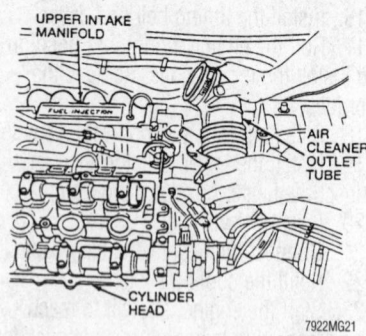

Disconnect the air cleaner outlet tube from the resonance chamber and throttle body

18. If equipped, remove the safety clips from the fuel rail.

19. Disconnect the fuel pressure and return lines, then plug the lines.

20. Disconnect the ignition coil high-tension lead from the distributor.

21. Tag and disengage the necessary hoses connected to the cylinder head and intake manifold.

22. Remove the two bolts from the transaxle vent tube routing brackets.

23. Raise and safely support the vehicle.

24. Remove the bolt from the water pump-to-cylinder head hose bracket.

25. Remove the front exhaust mounting flange and exhaust pipe support bracket from the exhaust manifold.

26. Remove the intake manifold support bracket.

27. Lower the vehicle.

28. Remove the cylinder head bolts in the sequence illustrated.

29. Remove the cylinder head assembly, with the intake manifold, exhaust manifold and distributor still attached, from the vehicle.

30. Remove the intake manifold and exhaust manifold.

To install:

31. Remove all dirt, oil and old gasket material from all gasket contact surfaces.

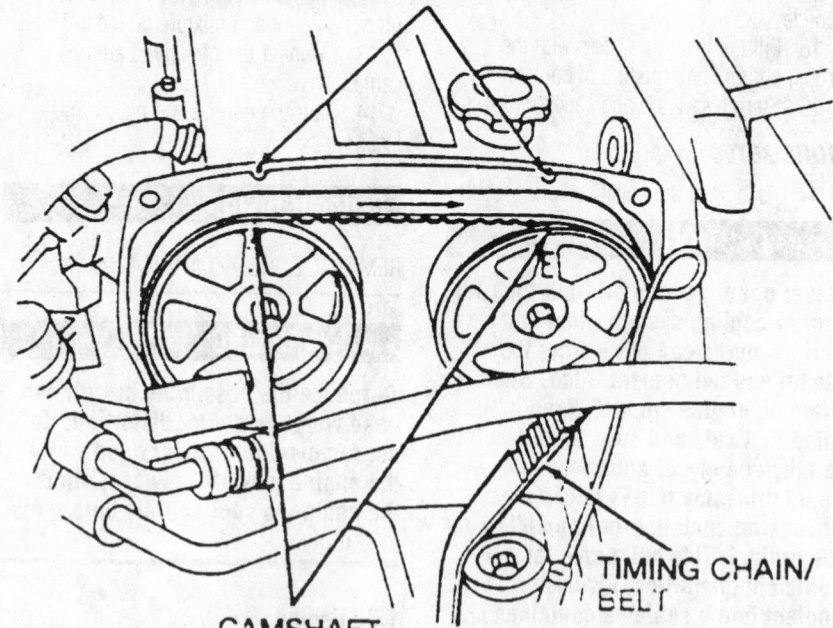

Be sure the timing marks on the camshaft pulleys and seal plate are aligned—1.8L engine

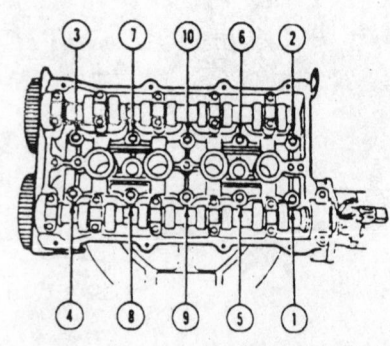

Removal sequence for cylinder head bolts—1.8L engines

32. Install the intake manifold and exhaust manifold.

33. Install a new head gasket onto the top of the engine block, using the dowel pins for reference.

34. Place the cylinder head into its mounting position on top of the engine block.

35. Lubricate the cylinder head bolts with engine oil and install them finger-tight. Tighten the bolts in the proper sequence (refer to the accompanying illustration) in two or three steps to 56–60 ft. lbs. (76–81 Nm).

36. Install the two bolts to the transaxle vent tube routing brackets.

37. Connect the heater hoses to the cylinder head and tighten the clamps.

38. Connect the ignition coil high-tension lead to the distributor.

39. Connect the fuel pressure and return lines to the fuel supply manifold and install the safety clips.

40. Install the transaxle-to-engine block upper-right bolt. If equipped with a manual transaxle, tighten the bolt to 47–66 ft. lbs. (64–89 Nm). If equipped with an automatic transaxle, tighten the bolt to 41–59 ft. lbs. (55–80 Nm).

41. Install the upper radiator hose and tighten the clamps.

42. Connect the ground straps.

43. Attach the electrical connectors that were unplugged at the cylinder head, exhaust manifold, intake manifold, and throttle body.

44. Connect the vacuum lines to the intake manifold.

45. Install the accelerator and kickdown cable bracket onto the intake manifold and tighten the bolts to 69–95 inch lbs. (7.8–11.0 Nm). Connect the cable(s) to the throttle cam.

46. Install the valve cover and gasket, then connect the hose running from the manifold to the valve cover. Tighten the

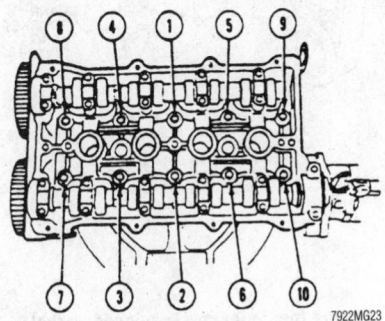

Tighten the cylinder head bolts in this sequence to the proper specification— 1.8L engine

7922MG23

cover bolts to 43–78 inch lbs. (4.9–8.8 Nm).

47. Install and connect the spark plug wires.

48. Connect the air cleaner outlet tube to the resonator and throttle body and tighten the clamps.

49. Connect the hose going from the air duct to the cylinder head cover.

50. Raise and safely support the vehicle.

51. Install the intake manifold support bracket. Tighten the bolts to 27–38 ft. lbs. (37–52 Nm) and the nut to 14–19 ft. lbs. (19–25 Nm).

52. Install the bolt to the water pump-to-cylinder head hose bracket.

53. Install the exhaust front mounting flange with a new gasket to the exhaust manifold. Tighten the flange-to-manifold attaching nuts to 23–34 ft. lbs. (31–46 Nm).

54. Install the exhaust pipe support bracket. Tighten the bracket attaching bolts to 27–38 ft. lbs. (37–52 Nm).

55. Be sure the yellow ignition timing mark on the crankshaft pulley is aligned with the TDC mark on the timing belt cover.

56. Lower the vehicle.

57. Be sure the timing marks on the camshaft pulleys and seal plate are aligned. Install the timing belt so there is no looseness at the idler pulley side or between the two camshaft pulleys.

➡ **Do not turn the crankshaft counterclockwise.**

58. Turn the crankshaft two turns clockwise by hand and verify that the yellow ignition timing mark on the crankshaft pulley is aligned with the timing mark on the timing belt cover. Verify that the timing marks on the camshaft pulley and seal plate are aligned.

➡ **If the timing marks are not aligned, remove the timing belt and repeat the procedure.**

59. Turn the crankshaft 1⅚ turns clockwise by hand and align the 4th tooth to the right of the **I** and **E** timing marks on the camshaft pulleys with the seal plate alignment marks.

60. Loosen the timing belt tensioner lockbolt and apply tension to the timing belt. Tighten the tensioner lockbolt to 27–38 ft. lbs. (37–52 Nm).

61. Turn the crankshaft 2⅙ (780 degrees) turns clockwise and verify that the timing marks on the camshaft pulleys and the seal plate are aligned.

62. Install new gaskets onto the timing belt upper and middle covers and install the covers. Tighten the mounting bolts to 69–95 inch lbs. (8–11 Nm).

63. Fill the cooling system.

64. Connect the negative battery cable.

65. Start the engine and check for leaks.

1.9L Engine

✳✳ CAUTION

Observe all applicable safety precautions when working around fuel. Whenever servicing the fuel system, always work in a well ventilated area. Do not allow fuel spray or vapors to come in contact with a spark or open flame. Keep a dry chemical fire extinguisher near the work area. Always keep fuel in a container specifically designed for fuel storage; also, always properly seal fuel containers to avoid the possibility of fire or explosion.

1. Properly relieve the fuel system pressure.

2. Disconnect the negative battery cable.

3. Drain the cooling system.

✳✳ CAUTION

Never open, service or drain the radiator or cooling system when hot; serious burns can occur from the steam and hot coolant. Also, when draining engine coolant, keep in mind that cats and dogs are attracted to ethylene glycol antifreeze and could drink any that is left in an uncovered container or in puddles on the ground. This will prove fatal in sufficient quantities. Always drain coolant into a sealable container. Coolant should be reused unless it is contaminated or is several years old.

4. Remove the air intake duct.

5. Remove the crankcase breather hose from the rocker arm cover and the vacuum hose from the bottom of the throttle body.

6. Remove the power brake supply hose.

7. Tag and unplug the following the electrical connectors:

   a. Fuel charging harness (right-hand strut tower).

   b. Alternator harness (back of the alternator).

   c. Crank angle sensor.

   d. Oxygen sensor.

   e. Ignition coil.

   f. Radio suppresser (mounted on the coil bracket).

   g. Engine Coolant Temperature (ECT)

sensor, cooling fan sensor and temperature sending unit.

8. Remove the ground strap from the stud on the left side of the cylinder head.

9. Disconnect the accelerator and the transaxle kickdown cables from the throttle lever and remove the cable bracket from the intake manifold.

10. Disconnect the heater hose containing the coolant temperature switches at the bulkhead.

11. Remove the upper radiator hose.

12. Disconnect and plug the fuel supply and return lines.

13. Remove the oil level indicator tube mounting nut from the cylinder head stud.

14. Remove the power steering hose and the air conditioner line retainer bracket bolts from the alternator bracket.

15. Remove the accessory drive belt, alternator, and the drive belt automatic tensioner.

16. Raise and safely support the vehicle.

17. Remove the catalytic converter inlet pipe.

18. Remove the starter wiring harness from the retaining clip below the intake manifold.

19. Set the engine No. 1 cylinder on TDC.

20. Lower the vehicle.

21. Remove the timing belt.

22. Remove the heater hose support bracket retaining bolt and the alternator bracket-to-cylinder head mounting bolt.

23. Remove the rocker arm cover.

24. Remove the cylinder head bolts. The old bolts can be used to check the piston squish height before discarding them.

25. Remove the cylinder head with the exhaust and intake manifolds attached.

26. Remove the cylinder head gasket. Keep the gasket to check the piston squish height, then after checking the squish height, discard the gasket.

➡**Do not lay the cylinder head flat. Damage to the spark plugs, valves or gasket surfaces may result.**

To install:

27. Clean all gasket material from the mating surfaces on the cylinder head and block and clean out the head bolt holes in the block.

28. Check the piston squish height. Refer to the procedure in this section.

➡**No cylinder block deck machining or use of replacement crankshaft, piston or connecting rod causing the assembled squish height to be over or under tolerance specification, is permitted. If**

no parts other than the head gasket are replaced, the squish height should be within specification. If parts other than the head gasket are replaced, check the squish height. If the squish height is out of specification, replace the parts again and recheck the squish height.

29. Install the dowels in the cylinder block, if removed. Check the dowel height, it should be 0.407–0.40 in. (10.40–11.75mm) above the surface of the block. A dowel that is too long will not allow the cylinder head to sit properly.

30. Position the cylinder head gasket on the cylinder block.

➡**The cylinder head attaching bolts cannot be tightened to the specified torque more than once and must therefore be replaced when installing a cylinder head.**

31. Install the cylinder head and install new bolts and washers in the following order:

a. Apply a light coat of engine oil to the threads of the new cylinder head bolts and install the new bolts into the head.

b. Tighten the cylinder head bolts in sequence (refer to the illustration) to 44 ft. lbs. (60 Nm).

c. Loosen the cylinder head bolts approximately two turns, then tighten them again in the same sequence to 44 ft. lbs. (60 Nm).

d. After bolts have all been tightened, turn the head bolts 90° in sequence and to complete the head bolt installation, turn the head bolts an additional 90° in the same torque sequence.

32. Install the rocker arm cover and the alternator bracket-to-cylinder head bolt.

33. Be sure the timing marks aligned so that the engine is at Top Dead Center (TDC), install the timing belt and the timing belt cover.

34. Raise and safely support the vehicle.

35. Install the starter wiring harness on the retaining clip below the intake manifold.

36. Install the catalytic converter inlet pipe.

37. Lower the vehicle.

38. Install the alternator and the accessory drive belt automatic tensioner. Install the accessory drive belt.

39. Install both the power steering hose and air conditioner line retainer bracket bolts. Install the oil level indicator tube retainer bolt.

40. Connect the fuel supply and return lines.

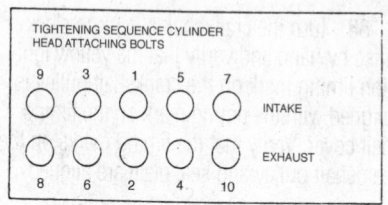

Cylinder head bolt tightening sequence—1.9L engine

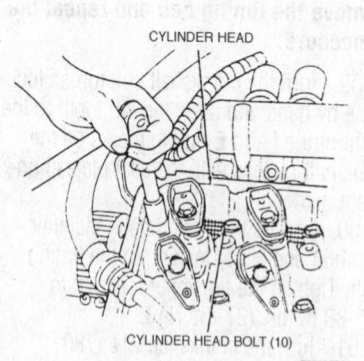

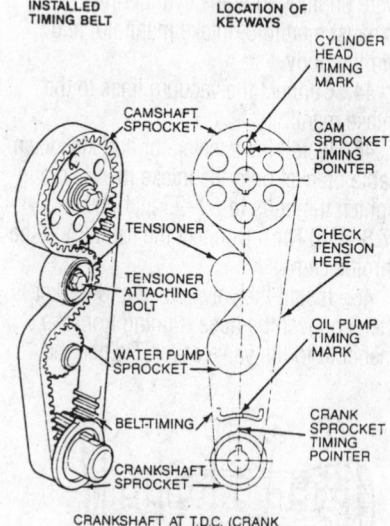

Be sure the timing marks aligned so that the engine is at Top Dead Center (TDC) before installing the timing belt—1.9L engine

41. Install the upper radiator hose and connect the heater hose at the engine compartment bulkhead.

42. Install the heater hose support bracket retaining bolt.

43. Install the accelerator cable bracket on the intake manifold and connect the accelerator and kickdown cables to the throttle lever.

44. Install the ground strap at the left side of the cylinder head.

45. Attach all the remaining electrical connectors according to their positions marked during the removal procedure.

46. Connect the power brake supply hose, crankcase breather hose and the vacuum line at the bottom of the throttle body.

47. Install the air intake duct.

48. Connect the negative battery cable.

49. Fill and bleed the cooling system.

50. Start the engine and check for leaks. Stop the engine and check the coolant level. Add fluid as necessary.

2.0L Engine

SOHC SPLIT PORT INJECTION (SPI)

➡Never work on aluminum engine parts when the engine is warm.

1. Disconnect the negative battery cable.

2. Properly relieve the fuel system pressure.

3. Remove the timing belt and engine air cleaner intake tube.

4. Tag and disconnect the vacuum hoses from the following components:
- EGR valve
- PCV valve
- Throttle body
- Fuel pressure regulator
- Intake manifold

5. Disconnect the accelerator cable, throttle control lever and if equipped, the speed control cable.

6. If equipped, remove the speed control cable bracket bolt.

7. Unplug the two fuel charging wiring electrical connections.

8. Unplug the Crankshaft Position (CKP) sensor and Heated Oxygen sensor (HO2S) electrical connections.

9. Using tool D87L-9280-A or its equivalent, disconnect the fuel supply line from the fuel rail.

10. Unfasten the power steering pressure hose bracket bolts, then position the bracket and hose aside.

11. Loosen the alternator lower bolt, unfasten the alternator upper bolt and pivot the alternator forward.

12. Unfasten the engine oil dipstick tube bracket bolt that attaches the tube to the intake manifold.

13. Disconnect the EGR manifold tube located below the EGR valve.

14. Unfasten the exhaust manifold heat shield nuts and remove the heat shield.

15. Unfasten the catalytic converter-to-exhaust manifold nuts and separate the components.

16. Raise the vehicle and support it with safety stands.

✳✳ CAUTION

Never open, service or drain the radiator or cooling system when hot; serious burns can occur from the steam and hot coolant. Also, when draining engine coolant, keep in mind that cats and dogs are attracted to ethylene glycol antifreeze and could drink any that is left in an uncovered container or in puddles on the ground. This will prove fatal in sufficient quantities. Always drain coolant into a sealable container. Coolant should be reused unless it is contaminated or is several years old.

17. Drain the cooling system into a suitable container.

18. Unfasten the right-hand splash shield bolts and remove the shield.

19. Remove the engine oil dipstick tube from the cylinder block.

20. Unplug the A/C compressor electrical connection.

21. Unbolt the A/C compressor retainers and set the compressor aside with the lines still attached.

22. Loosen but do not remove, the four bolts and nut that attach the front engine accessory drive bracket about four turns $FR3/8 in.

23. Lower the vehicle and unfasten the uppermost front engine accessory drive bracket-to-cylinder head bolt.

24. Unfasten the A/C line bracket-to-front engine accessory drive bracket bolt.

25. Remove the valve cover, then disconnect the upper radiator hose from the thermostat housing water hose connection.

26. Disconnect the heater hose from the thermostat housing.

✳✳ CAUTION

The removal of the cylinder head requires two people, please do not remove the head without the aid of a helper.

➡Maintain a balanced pressure when removing the cylinder head bolts.

27. Unfasten the cylinder head bolts, remove the head and gasket.

28. Discard the old bolts and the gasket.

To install:

➡Always use new bolts when installing the cylinder head.

✳✳ CAUTION

Do not use any abrasive grinding discs to remove gasket material. Use a plastic manual gasket scraper to remove the gasket residue. Be careful not to scratch or gouge the aluminum sealing surfaces when cleaning them.

29. Clean the gasket surfaces thoroughly until all traces of the old gasket residue are removed. Inspect the gasket mating surfaces, both must be clean and flat.

30. Install a new head gasket on the cylinder block.

✳✳ CAUTION

Do not attempt to install the cylinder head without the aid of an assistant, it takes two people.

31. Lubricate the new cylinder bolts when engine oil.

32. Install the cylinder head and tighten the bolts in the sequence illustrated in the following order:

a. Tighten all the bolts in sequence to 30–44 ft. lbs. (40–60 Nm).

b. Back all the bolts off ½ a turn.

c. Retighten all the bolts to 30–44 ft. lbs. (40–60 Nm).

33. Rotate all the bolts in sequence an additional 180° in two steps of 90°.

34. Install and tighten the uppermost accessory drive bracket bolt to 30–40 ft. lbs. (40–55 Nm).

35. Install the A/C line bracket and tighten the bolt to 15–18 ft. lbs. (20–25 Nm).

36. Install the valve cover, then raise and support the vehicle on safety stands.

37. Install and tighten the four front engine accessory drive bracket bolts and the nut to 30–40 ft. lbs. (40–55 Nm).

38. Install the A/C compressor and tighten the retaining bolts.

39. Install the right-hand splash shield and tighten its retainers.

40. Install the engine oil dipstick tube and lower the vehicle.

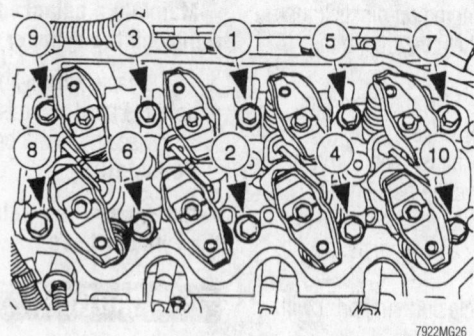

7922MG26

Tighten the cylinder head bolts in the proper sequence in the order specified—2.0L SOHC engines

41. Connect the heater and upper radiator hoses.

42. Install the timing belt and alternator.

43. Install the power steering pressure hose.

44. Attach the CKP sensor and two main fuel charging electrical connections

45. Install and tighten the engine oil dipstick tube bolt.

46. Install and tighten the EGR valve manifold tube connection.

47. Connect the EGR valve manifold tube to the exhaust manifold.

48. Connect the fuel supply line to the fuel rail.

49. Connect the catalytic converter to the exhaust manifold using a new gasket and tighten the nuts.

50. Install the heat shield and tighten the nuts.

51. If equipped, install the speed control cable bracket.

52. Connect the throttle control lever, accelerator cable and if equipped, the speed control cable.

53. Connect the vacuum hoses tagged and removed to the following components:
- EGR valve
- PCV valve
- Throttle body
- Fuel pressure regulator
- Intake manifold

54. Install the air cleaner outlet tube.

55. Fill the cooling system.

56. Connect the negative battery cable.

57. Start the engine and check for leaks.

DOHC ZETEC

1. Disconnect the negative battery cable.

2. Properly relieve the fuel system pressure.

3. Remove the air cleaner assembly outlet tube.

4. Remove the timing belt.

5. Tag and disconnect the vacuum hoses from the following components:
- PCV valve
- Throttle body
- Fuel pressure sensor
- Intake manifold

6. Disconnect the speed control and accelerator cables from the control lever.

7. Unplug the fuel charging electrical connectors at the main engine connector.

8. Unplug the Crankshaft Position (CKP) sensor and Heated Oxygen sensor (HO2S) electrical connections.

9. Disconnect the fuel line.

10. Unbolt the power steering pump and bracket, then set them aside with the lines still attached.

11. Remove the alternator and engine oil dipstick tube.

12. Raise the vehicle and support it with safety stands.

> ❋❋ **CAUTION**
>
> **Never open, service or drain the radiator or cooling system when hot; serious burns can occur from the steam and hot coolant. Also, when draining engine coolant, keep in mind that cats and dogs are attracted to ethylene glycol antifreeze and could drink any that is left in an uncovered container or in puddles on the ground. This will prove fatal in sufficient quantities. Always drain coolant into a sealable container. Coolant should be reused unless it is contaminated or is several years old.**

13. Drain the cooling system into a suitable container.

14. Unfasten the splash shield bolts and remove the shield.

15. Unplug the A/C compressor electrical connection.

16. Unbolt the A/C compressor retainers and set the compressor aside with the lines still attached.

17. Lower the car, tag and disconnect the spark plug wires and remove the plugs.

18. Remove the valve cover, then disconnect the upper radiator hose from the thermostat housing water hose connection.

19. Disconnect the heater hose from the thermostat housing.

20. Remove the camshafts and the ignition coil.

21. Remove the thermostat housing.

22. Unfasten the cylinder head bolts, then remove the head and the gasket.

To install:

➡ **Always use new bolts when installing the cylinder head.**

> ❋❋ **CAUTION**
>
> **Do not use any abrasive grinding discs to remove gasket material. Use a plastic manual gasket scraper to remove the gasket residue. Be careful not to scratch or gouge the aluminum sealing surfaces when cleaning them.**

23. Clean the gasket surfaces thoroughly until all traces of the old gasket residue are removed. Inspect the gasket mating surfaces, both must be clean and flat.

24. Install a new head gasket on the cylinder block.

> ❋❋ **CAUTION**
>
> **Do not attempt to install the cylinder head without the aid of an assistant, it takes two people.**

25. Lubricate the new cylinder bolts when engine oil.

26. Install the cylinder head and tighten the bolts in the sequence illustrated in the following steps:
- Step 1: Tighten all the bolts in sequence to 12–18 ft. lbs. (15–25 Nm)
- Step 2: Tighten all the bolts to 26–33 ft. lbs. (35–45 Nm)
- Step 3: On Escort/Tracer models, tighten all the bolts an additional 105°. On Escort coupe models, tighten the bolts an additional 90°

27. Install the thermostat housing.

28. Install the ignition coil and the camshafts.

29. Connect the heater and upper radiator hoses.

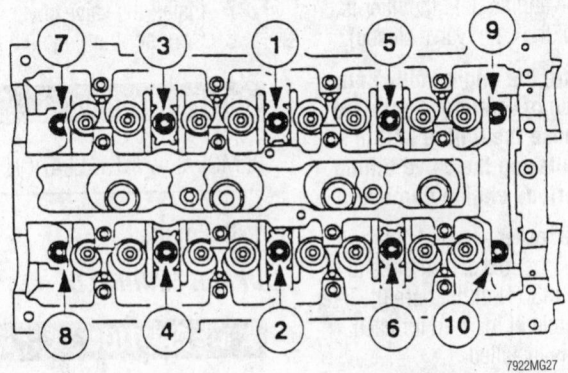

Tighten the cylinder head bolts in the order specified—2.0L Zetec engines

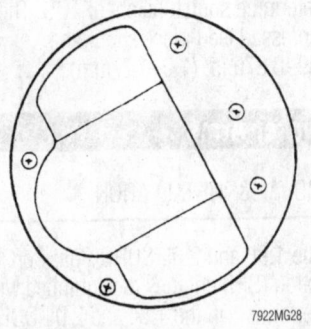

Place the lead shot or solder on the positions illustrated to measure the piston squish height

30. Install the valve cover, spark plugs and connect the spark plug wires.

31. Install the A/C compressor and tighten the retaining bolts.

32. Attach the A/C compressor electrical connection.

33. Install the splash shield and tighten its retainers.

34. Install the engine oil dipstick tube.

35. Install the alternator and the power steering reservoir and bracket.

36. Connect the fuel line.

37. Attach the CKP sensor, HO2S sensor, and two main fuel charging electrical connections.

38. Connect accelerator cable and if equipped, the speed control cable to the control lever.

39. Connect the vacuum hoses tagged and removed to the following components:
- PCV valve
- Throttle body
- Fuel pressure sensor
- Intake manifold

40. Install the timing belt.

41. Install the air cleaner outlet tube.

42. Fill the cooling system.

43. Connect the negative battery cable.

44. Start the engine and check for leaks.

PISTON SQUISH HEIGHT

1.9L Engines

Before final installation of the cylinder head to the engine, piston ``squish height'' must be checked. Squish height is the clearance of the piston dome to the cylinder head dome at piston TDC. No rework of the head gasket surfaces (slabbing) or use of replacement parts (crankshaft, piston and connecting rod) that causes the assembled squish height to be over or under the tolerance specification is permitted.

➡ **If no parts other than the head gasket are replaced, the piston squish height should be within specification. If parts other than the head gasket are replaced, check the squish height. If out of specification, replace the parts again and recheck the squish height.**

1. Clean all gasket material from the mating surfaces on the cylinder head and engine block.

2. Place a small amount of soft lead solder or lead shot of an appropriate thickness on the piston spherical areas.

3. Rotate the crankshaft to lower the piston in the bore and install the head gasket and cylinder head.

➡ **A compressed (used) head gasket is preferred for checking squish height.**

4. Install used head bolts and tighten the head bolts to 30–44 ft. lbs. (40–60 Nm) following proper sequence.

5. Rotate the crankshaft to move the piston through its TDC position.

6. Remove the cylinder head and measure the thickness of the compressed solder

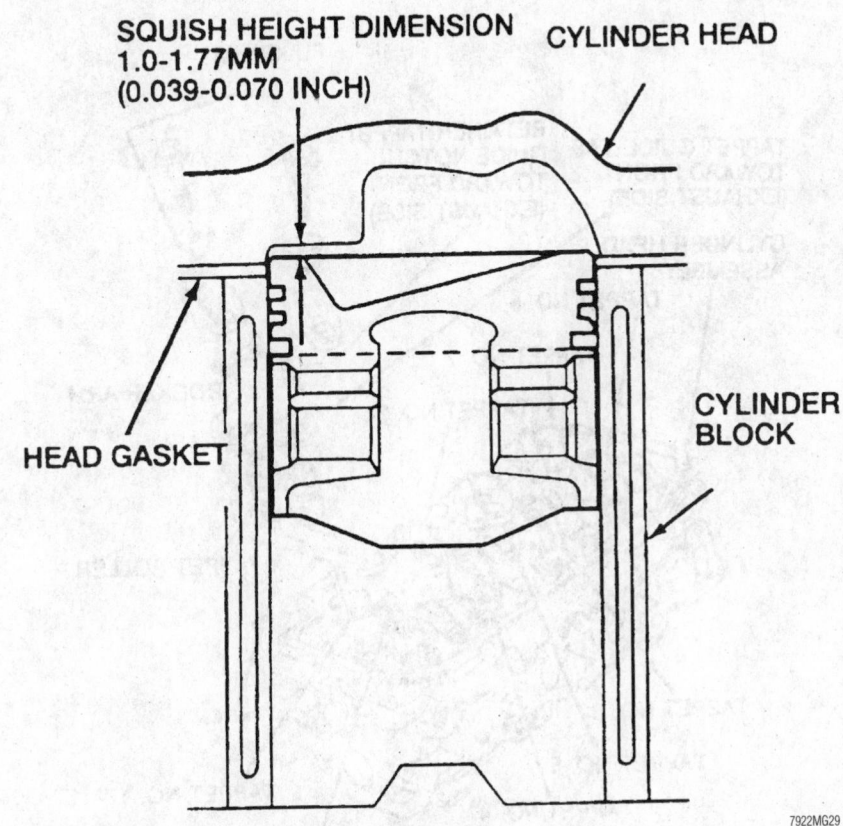

Cross-sectional view of squish height measurements—1.9L engines

to determine squish height at TDC. The compressed lead piece should be 0.039–0.070 in. (1.0–1.77mm).

Rocker Arms

REMOVAL & INSTALLATION

The 1.9L and 2.0L SOHC Split Port Injection (SPI) engines are equipped with rocker arms. On the 1.8L and 2.0L DOHC ZETEC the engines the camshafts acts directly on the valve, therefore, no rocker arms/shafts are used.

1.9L Engine

1. Disconnect the negative battery cable.
2. Place fender covers on the aprons.
3. Remove the rocker arm (valve) cover.

➡**Keep all parts in order so they can be reinstalled in their original position.**

4. Unfasten and remove the rocker arm bolts.
5. Remove the rocker arms and fulcrums.
To install:
6. Before installation, coat the valve tips, rocker arm and fulcrum contact areas with Lubriplate® or equivalent.

7. Rotate the engine until the lifter is on the base circle of the cam (valve closed).

➡**Be sure to turn the engine only in the normal direction of rotation. Backward rotation will cause the timing belt to slip or lose teeth, altering the valve timing and causing serious engine damage.**

8. Install the rocker arm and components and tighten the rocker arm bolts to 17–22 ft. lbs. (23–30 Nm). Be sure the lifter is on the base circle of the cam for each rocker arm as it is installed.
9. Install the rocker arm cover.
10. Connect the negative battery cable.

2.0L Engine

SOHC SPLIT PORT INJECTION (SPI)

1. Disconnect the negative battery cable.
2. Remove the valve cover.
3. Mark the location of the rocker arms, then unfasten the rocker arm bolts.
4. Remove the rocker arms and seats.
To install:
5. Install the seats and rocker arms in their original positions.
6. Install and tighten the rocker arm bolts to 17–22 ft. lbs. (23–30 Nm).

7. Install the valve cover.
8. Connect the negative battery cable.

Intake Manifold

REMOVAL & INSTALLATION

1.8L Engine

UPPER MANIFOLD

✳✳ CAUTION

Observe all applicable safety precautions when working around fuel. Whenever servicing the fuel system, always work in a well ventilated area. Do not allow fuel spray or vapors to come in contact with a spark or open flame. Keep a dry chemical fire extinguisher near the work area. Always keep fuel in a container specifically designed for fuel storage; also, always properly seal fuel containers to avoid the possibility of fire or explosion.

1. Properly relieve the fuel system pressure.
2. Disconnect the negative battery cable.

✳✳ CAUTION

Never open, service or drain the radiator or cooling system when hot; serious burns can occur from the steam and hot coolant.

3. Partially drain the cooling system.
4. Remove the air cleaner outlet tube from the throttle body and intake air resonator.
5. Disconnect the throttle cable and throttle valve control actuating cable from the throttle body control lever.
6. Unfasten the accelerator cable throttle valve control actuating cable bracket bolt and remove the bracket from the manifold.
7. Unplug the throttle body electrical connections.
8. Tag and disconnect all necessary vacuum hoses from the upper manifold.
9. Disconnect the air bypass and idle air control hoses from the manifold.
10. Remove the idle air control hose.
11. Unfasten the upper intake manifold upper bolts.
12. Raise and safely support the vehicle.
13. Unfasten the upper intake lower bolts and lower the vehicle.
14. Remove the manifold with the throttle body attached.

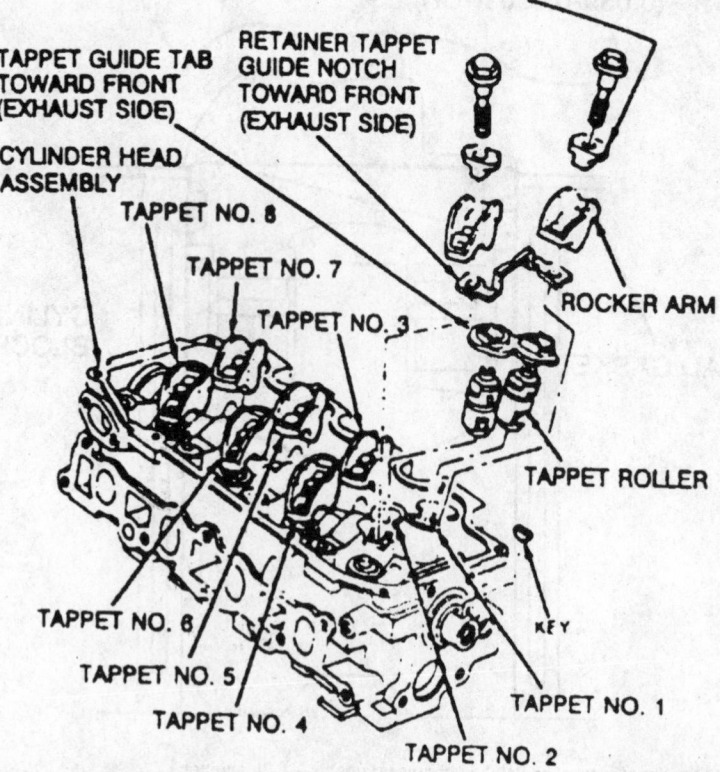

FULCRUM

TAPPET GUIDE TAB TOWARD FRONT (EXHAUST SIDE)

RETAINER TAPPET GUIDE NOTCH TOWARD FRONT (EXHAUST SIDE)

CYLINDER HEAD ASSEMBLY

TAPPET NO. 8

TAPPET NO. 7

TAPPET NO. 3

ROCKER ARM

TAPPET ROLLER

TAPPET NO. 6

TAPPET NO. 5

TAPPET NO. 4

TAPPET NO. 2

KEY

TAPPET NO. 1

7922MG30

Exploded view of the upper valvetrain—1.9L engine

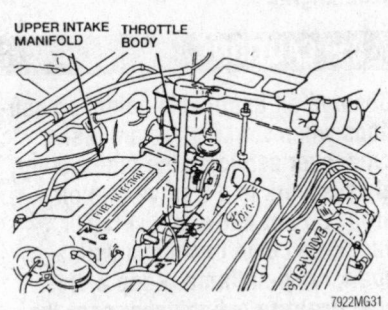

Loosen the upper intake manifold
bolts . . .

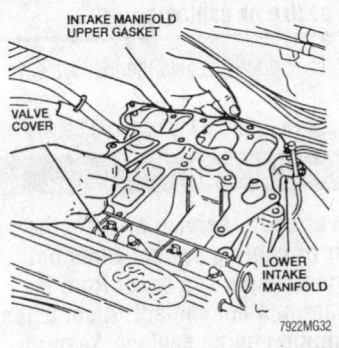

. . . then remove the upper intake manifold and gasket—1.8L engine

15. Remove the upper intake manifold
gasket.

To install:

16. Use a gasket scraper to clean the
upper intake mounting surfaces.

17. Install a new upper intake manifold
gasket.

18. Place the upper intake manifold in
position.

19. Install the manifold upper bolts and
nuts, then tighten them to 14–19 ft. lbs.
(19–25 Nm).

20. Raise the vehicle and support it with
safety stands.

21. Install the manifold lower bolts
and nuts, then tighten them to 14–
19 ft. lbs. (19–25 Nm) and lower the
vehicle.

22. Install the idle air control hose.

23. Connect the air bypass and idle air
control hoses to the manifold.

24. Connect all the vacuum hoses to the
upper manifold.

25. Engage the throttle body electrical
connections.

26. Install the accelerator cable throttle
valve control actuating cable bracket and
tighten the bolt to 69–95 inch lbs. (7.8–11
Nm).

27. Connect the throttle cable and throttle valve control actuating cable to the throttle body control lever.

28. Install the air cleaner outlet tube.

29. Be sure all wires and hoses are connected.

30. Fill the cooling system

31. Connect the negative battery cable.

32. Start the vehicle and check for
proper operation.

LOWER MANIFOLD

❋❋ CAUTION

Observe all applicable safety precautions when working around fuel.
Whenever servicing the fuel system,
always work in a well ventilated
area. Do not allow fuel spray or
vapors to come in contact with a
spark or open flame. Keep a dry
chemical fire extinguisher near the
work area. Always keep fuel in a container specifically designed for fuel
storage; also, always properly seal
fuel containers to avoid the possibility of fire or explosion.

1. Properly relieve the fuel system
pressure.

2. Disconnect the negative battery cable.

❋❋ CAUTION

Never open, service or drain the radiator or cooling system when hot;
serious burns can occur from the
steam and hot coolant. Also, when
draining engine coolant, keep in
mind that cats and dogs are attracted
to ethylene glycol antifreeze and
could drink any that is left in an
uncovered container or in puddles on
the ground. This will prove fatal in
sufficient quantities. Always drain
coolant into a sealable container.
Coolant should be reused unless it is
contaminated or is several years old.

3. Partially drain the cooling system.

4. Tag and disconnect all necessary
vacuum hoses from the upper manifold and
throttle body.

5. Remove the vacuum reservoir from
the upper intake manifold.

6. Tag and disconnect the coolant
hoses from the valves.

7. Disconnect the throttle cable and
throttle valve control actuating cable from
the throttle body control lever.

8. Unfasten the accelerator cable throttle valve control actuating cable bracket

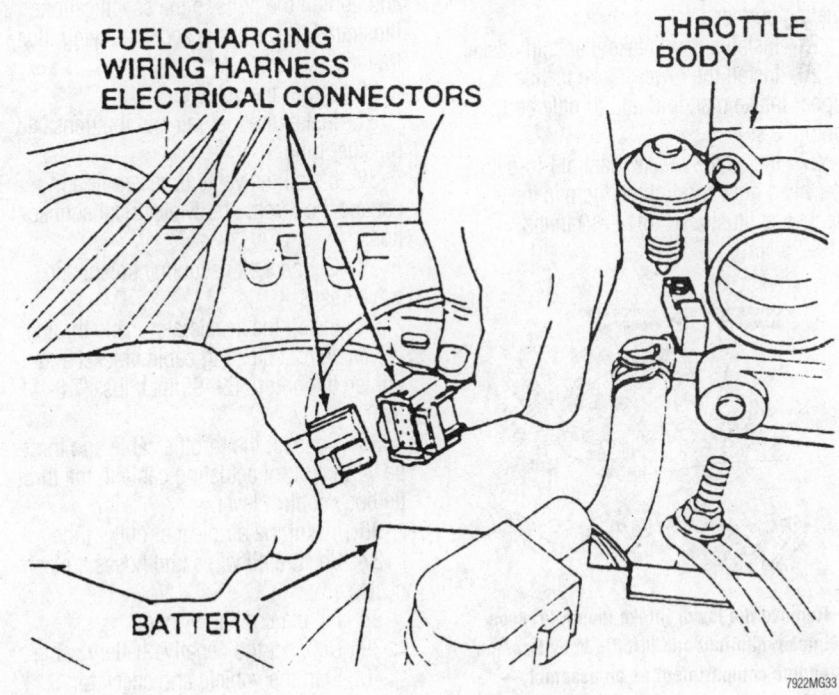

Unplug the fuel charging harness electrical connectors—1.8L engine

bolt and remove the bracket from the manifold.

9. Unplug the throttle body electrical connections.

10. Disconnect the fuel and vapor line spring lock couplings.

11. Disconnect the PCV hose from the upper intake manifold and valve cover.

12. Disconnect the vacuum hose from the fuel pressure regulator.

13. Unplug the fuel charging harness electrical connectors.

14. Unfasten the fuel rail retaining bolts and remove the fuel rail and injectors.

15. Unfasten the transaxle breather tube and separate the tube from the manifold.

16. Unfasten the five uppermost intake manifold-to-cylinder head nuts.

17. Raise the vehicle and support it with safety stands.

18. Unfasten the bolts attaching the intake manifold support and remove the support.

19. Unfasten the four lowermost intake manifold-to-cylinder block nuts.

20. Lower the vehicle.

21. Remove the lower intake manifold, upper intake manifold and throttle body from the engine compartment as an assembly.

22. Remove the intake manifold gasket.

23. If necessary, separate the upper manifold from the lower manifold.

To install:

24. Use a scraper to clean the manifold mating surfaces.

25. Install a new intake manifold gasket.

26. Install the lower intake manifold, upper intake manifold and throttle body onto the studs.

27. Install the intake manifold-to-cylinder head nuts and tighten them in the sequence illustrated to 14–19 ft. lbs. (19–25 Nm).

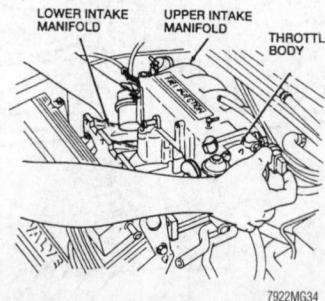

Remove the lower intake manifold, upper intake manifold and throttle body from the engine compartment as an assembly—1.8L engine

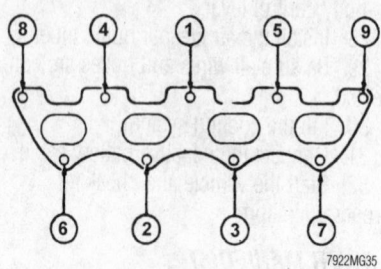

Tighten the intake manifold-to-cylinder head nuts as illustrated—1.8L engine

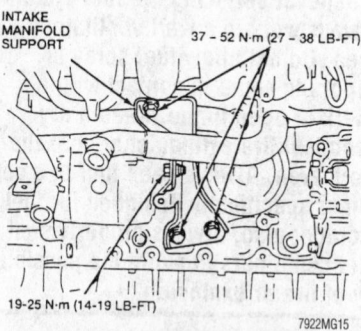

Tighten the intake manifold support bolts to the specifications illustrated—1.8L engine

28. Raise the vehicle and support it with safety stands.

29. Install the intake manifold support and tighten the bolts to the specifications illustrated. Refer to the accompanying illustration.

30. Lower the vehicle.

31. Install the fuel rail and the transaxle breather tube.

32. Install the vacuum reservoir and engage the throttle body electrical connections.

33. Connect all vacuum, coolant and valve hoses

34. Install the accelerator cable throttle valve control actuating cable bracket and tighten the bolt to 69–95 inch lbs. (7.8–11 Nm).

35. Connect the throttle cable and throttle valve control actuating cable to the throttle body control lever.

36. Install the air cleaner outlet tube.

37. Be sure all wires and hoses are connected.

38. Fill the cooling system

39. Connect the negative battery cable.

40. Start the vehicle and check for proper operation.

1.9L Engine

1. Properly relieve the fuel system pressure.

2. Disconnect the negative battery cable.

3. Partially drain the cooling system.

4. Remove the air intake tube.

5. Unplug the fuel injector harness from the engine control harness at the right shock tower.

6. Unplug the crankshaft position sensor electrical connection.

7. Disconnect and plug the fuel supply and return lines.

8. Unplug the camshaft position sensor electrical connection.

9. Remove the accelerator cable, if equipped with automatic transaxle, the kickdown cable from the throttle lever.

10. Remove the cable bracket from the intake manifold and position the cables aside.

11. Remove the power brake supply hose, PCV line and the vacuum line from the bottom of the throttle body.

12. Remove the seven attaching nuts from the intake manifold studs.

13. Slide the manifold assembly off of the studs and remove it from the cylinder head.

14. Remove and discard the intake manifold gasket.

To install:

15. Use a scraper to clean and inspect the mounting faces of the intake manifold and cylinder head. Both surfaces must be clean and flat.

16. Clean and oil the manifold studs and position a new gasket over them.

17. Install the intake manifold and the attaching nuts. Tighten the nuts to 12–15 ft. lbs. (16–20 Nm).

18. Install the vacuum line on the bottom of the throttle body, the power brake supply hose and the PCV line.

19. Install the accelerator cable bracket and connect the accelerator cable, if equipped, kickdown cable on the throttle lever.

20. Connect the fuel supply and return lines. Install the fuel line retaining clips.

21. Attach the two fuel injector harness connectors to the engine control harness at the right shock tower.

22. Attach the crankshaft position sensor and camshaft position sensor electrical connections.

23. Install the air intake tube.

24. Refill the cooling system.

25. Connect the negative battery cable.

26. Start the engine and bring to normal operating temperature, then check for fuel and coolant leaks.

27. Stop the engine and check the coolant level.

2.0L Engine

SOHC SPLIT PORT INJECTION (SPI)

1. Disconnect the negative battery cable.

> ### ✷✷ CAUTION
>
> **Never open, service or drain the radiator or cooling system when hot; serious burns can occur from the steam and hot coolant. Also, when draining engine coolant, keep in mind that cats and dogs are attracted to ethylene glycol antifreeze and could drink any that is left in an uncovered container or in puddles on the ground. This will prove fatal in sufficient quantities. Always drain coolant into a sealable container. Coolant should be reused unless it is contaminated or is several years old.**

2. Partially drain the cooling system.

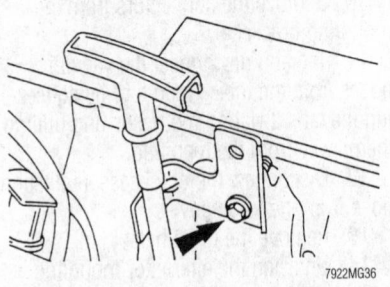

Remove the engine oil dipstick tube bracket bolt that attaches the tube to the intake manifold—2.0L SOHC engine

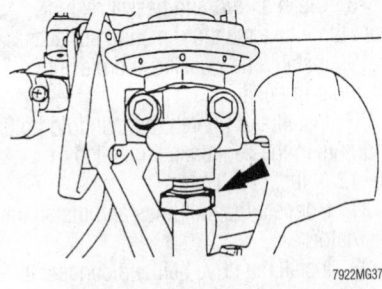

Disconnect the EGR manifold tube located below the EGR valve—2.0L SOHC engine

3. Remove the air cleaner outlet tube.

4. Tag and disconnect the vacuum hoses from the following components:
- EGR valve
- PCV valve
- Throttle body
- Fuel pressure regulator.
- Intake manifold

5. Disconnect the accelerator cable, throttle control lever and if equipped, the speed control cable.

6. If equipped, remove the speed control cable bracket bolt.

7. Unplug the Idle Air Control (IAC) valve and Throttle Position (TP) sensor electrical connections.

8. Unfasten the engine oil dipstick tube bracket bolt that attaches the tube to the intake manifold.

9. Disconnect the EGR manifold tube located below the EGR valve.

10. Raise the vehicle and support it with safety stands.

11. Remove the engine oil dipstick tube from the block.

12. Unfasten the intake manifold lower nuts, then lower the vehicle.

13. Unfasten the intake manifold upper nuts and remove the nuts. Discard the manifold gasket.

To install:

> ### ✷✷ CAUTION
>
> **Do not use any abrasive grinding discs to remove gasket material. Use a plastic manual gasket scraper to remove the gasket residue. Be careful not to scratch or gouge the aluminum sealing surfaces when cleaning them.**

14. Clean the gasket surfaces thoroughly until all traces of the old gasket residue are removed. Inspect the gasket mating surfaces, both must be clean and flat.

15. Clean and oil the intake manifold mounting studs.

16. Install the intake manifold and hand-tighten the upper nuts.

17. Raise the vehicle and support it with safety stands.

18. Install the lower manifold nuts and tighten them to 15–22 ft. lbs. (20–30 Nm).

19. Connect the engine oil dipstick tube to the block.

20. Lower the vehicle and tighten the manifold upper nuts to 15–22 ft. lbs. (20–30 Nm).

21. Connect the EGR manifold tube and tighten it to 15–20 ft. lbs. (20–28 Nm).

22. Install and tighten the engine oil dipstick tube bolt to 71–97 inch lbs. (8–11 Nm).

23. Attach the Idle Air Control (IAC) valve and Throttle Position (TP) sensor electrical connections.

24. If equipped, install and tighten the speed control cable bracket bolt to 71–88 inch lbs. (8–10 Nm).

25. Connect the throttle control lever, accelerator cable and if equipped, the speed control cable.

26. Connect the vacuum hoses tagged and removed to the following components:
- EGR valve
- PCV valve
- Throttle body
- Fuel pressure regulator.
- Intake manifold

27. Install the air cleaner outlet tube.

28. Fill the cooling system.

29. Connect the negative battery cable.

30. Start the engine and check for coolant leaks.

DOHC ZETEC

1. Properly relieve the fuel system pressure.

2. Disconnect the negative battery cable.

3. Remove the air cleaner outlet tube.

4. Unplug the Throttle Position (TP) sensor electrical connection.

5. Raise the vehicle and support it with safety stands.

✳✳ CAUTION

Never open, service or drain the radiator or cooling system when hot; serious burns can occur from the steam and hot coolant. Also, when draining engine coolant, keep in mind that cats and dogs are attracted to ethylene glycol antifreeze and could drink any that is left in an uncovered container or in puddles on the ground. This will prove fatal in sufficient quantities. Always drain coolant into a sealable container. Coolant should be reused unless it is contaminated or is several years old.

6. Drain the cooling system into a suitable container.

7. Remove the bolt that secures the pipe located by the crankshaft pulley, then lower the vehicle.

8. Disconnect the heater hoses from the core.

9. Unplug the main engine control sensor wiring.

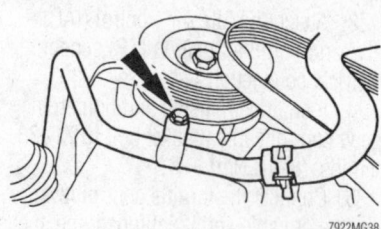

7922MG38

Unbolt the pipe bracket located by the crankshaft pulley—2.0L Zetec engine

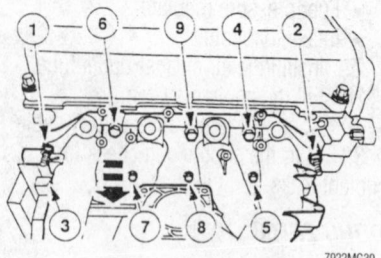

7922MG39

Remove the intake manifold bolts and nuts in this sequence—2.0L Zetec engine

10. Remove the connectors from the mounting bracket.

11. Tag and disconnect the vacuum hoses from the intake manifold by squeezing the tabs, twisting the hoses and pulling them away from the manifold.

12. Disconnect the crankcase ventilation hose from the valve cover.

13. Remove the drive belt.

14. Unfasten the alternator mounting bolts and move the alternator aside.

15. Disconnect the fuel lines.

16. Unfasten the intake manifold nuts and bolts in the sequence illustrated.

17. Remove the intake manifold and gasket.

To install:

18. Clean all dirt and gasket residue from the intake manifold mating surfaces.

19. Install a new gasket and place the manifold in position.

20. Install and tighten the manifold bolts and nuts in the sequence illustrated to 10–12 ft. lbs. (14–17 Nm).

21. Connect the fuel lines and install the alternator.

22. Install the drive belt and connect the crankcase ventilation hose to the valve cover.

23. Connect the vacuum hoses to the manifold making sure the tabs are firmly engaged.

24. Install the connectors in the mounting bracket and attach the main engine control sensor wiring.

25. Connect the heater hoses to the heater core.

26. Install the bolt that secures the pipe located by the crankshaft pulley. Tighten the bolt to 71–97 inch lbs. (8–11 Nm).

27. Attach the Throttle Position (TP) sensor electrical connection.

28. Install the air cleaner outlet tube.

29. Fill the cooling system.

30. Connect the negative battery cable.

Exhaust Manifold

REMOVAL & INSTALLATION

1.8L Engine

1. Disconnect the negative battery cable.

2. Partially drain the cooling system and disconnect the upper radiator hose.

3. Remove the engine intake air resonator (duct).

4. Disconnect the upper radiator hose.

5. Remove the cooling fan motor.

6. Raise and safely support the vehicle.

7. If equipped, remove the engine and transaxle splash shield.

8. Remove the exhaust inlet pipe from the exhaust manifold and remove the gasket.

9. Remove the two bolts from the exhaust pipe support bracket.

10. Unfasten the cooling fan lower mounting bolts

11. Lower the vehicle.

12. Unplug the oxygen sensor electrical connector.

13. Unfasten the exhaust manifold heat shield mounting bolts and remove the shield.

14. Unfasten the exhaust manifold mounting nuts and remove the assembly.

15. Using a scraper, remove all gasket material from the cylinder head and exhaust manifold.

To install:

16. Install a new gasket onto the exhaust manifold mounting studs.

17. Place the exhaust manifold onto the mounting studs and install the manifold mounting nuts. Tighten the nuts to 28–34 ft. lbs. (38–46 Nm).

18. Place the heat shield into its mounting position and install the shield mounting bolts. Tighten the bolts to 69–95 inch lbs. (7.8–11.0 Nm).

19. Attach the oxygen sensor electrical connector.

20. Install the cooling fan.

21. Connect the upper radiator hose.

22. Install the intake air duct.

23. Raise and safely support the vehicle.

24. Install the exhaust pipe support bracket.

25. Install a new gasket and attach the exhaust pipe to the exhaust manifold. Tighten the attaching nuts to 23–34 ft. lbs. (31–46 Nm).

26. Install the splash shields and tighten the bolts to 69–95 inch lbs. (7.8–11.0 Nm).

27. Lower the vehicle.

28. Refill the cooling system.

29. Connect the negative battery cable.

1.9L Engine

1. Disconnect the negative battery cable.

2. Remove the accessory drive belt.

3. Remove the alternator.

4. Remove the radiator cooling fan motor and the shroud assembly.

5. On 1995–96 models perform the following steps to remove the EGR valve-to-exhaust manifold tube:

　a. Hold the EGR valve tube-to-manifold connector with a wrench and loosen the valve-to-exhaust manifold tube nut.

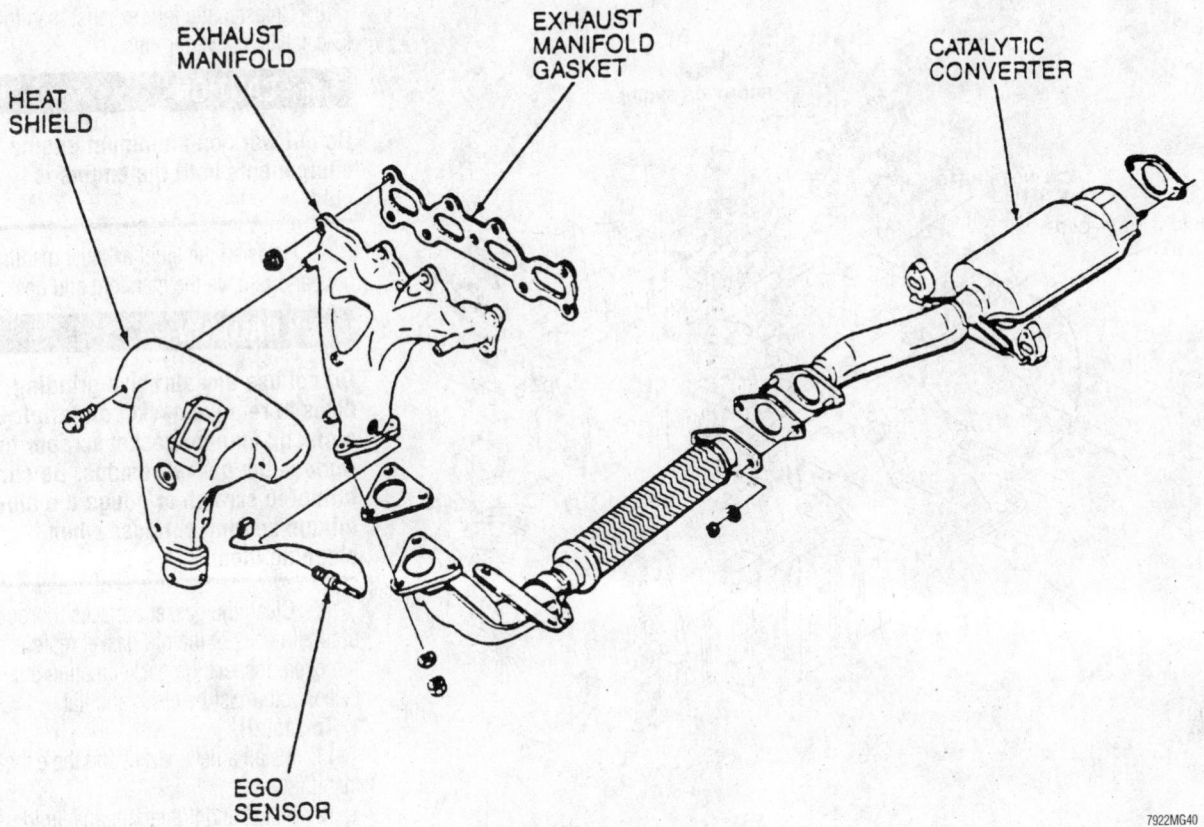

Exploded view of the exhaust manifold assembly—1.8L and 2.0L Zetec engines

b. Unfasten the two EGR valve-to-exhaust manifold tube bolts from the ignition coil bracket.

c. Remove the EGR valve tube.

6. Remove the exhaust manifold heat shield.

7. Raise and safely support the vehicle.

8. Remove the two catalytic converter inlet pipe-to-exhaust manifold attaching nuts.

9. Lower the vehicle.

10. Remove the eight exhaust manifold attaching nuts and remove the exhaust manifold and gasket.

To install:

11. Use a scraper to clean the cylinder head and exhaust manifold gasket surfaces.

12. Position the new gasket onto the manifold mounting studs.

13. Position the exhaust manifold on the cylinder head and install the attaching nuts. Tighten the nuts to 16–19 ft. lbs. (21–26 Nm).

14. Raise and safely support the vehicle.

15. Install the catalytic converter inlet pipe-to-exhaust manifold attaching nuts. Tighten the nuts to 25–33 ft. lbs. (34–47 Nm).

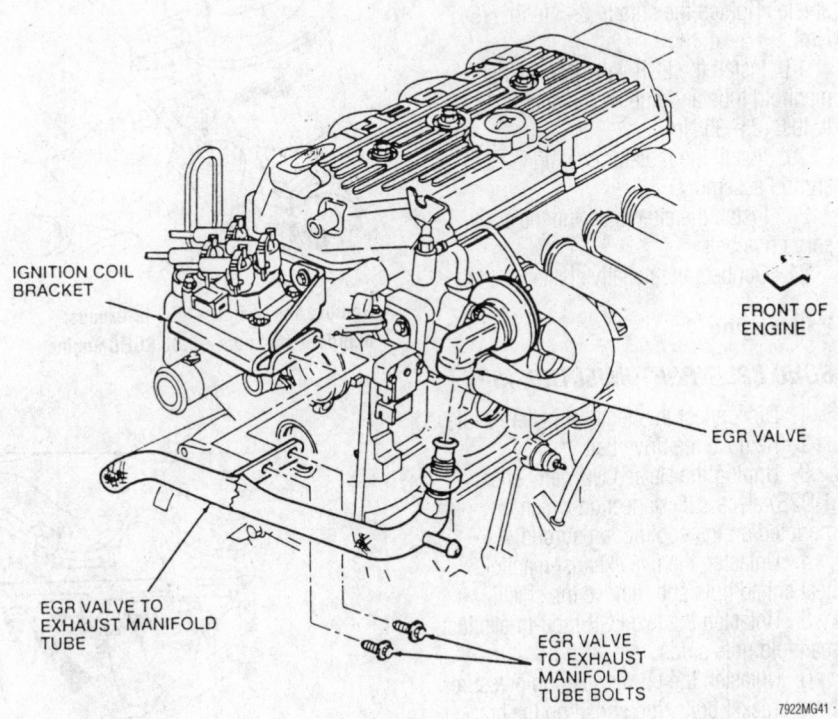

EGR valve tube-to-exhaust manifold routing—1995–96 1.9L engine

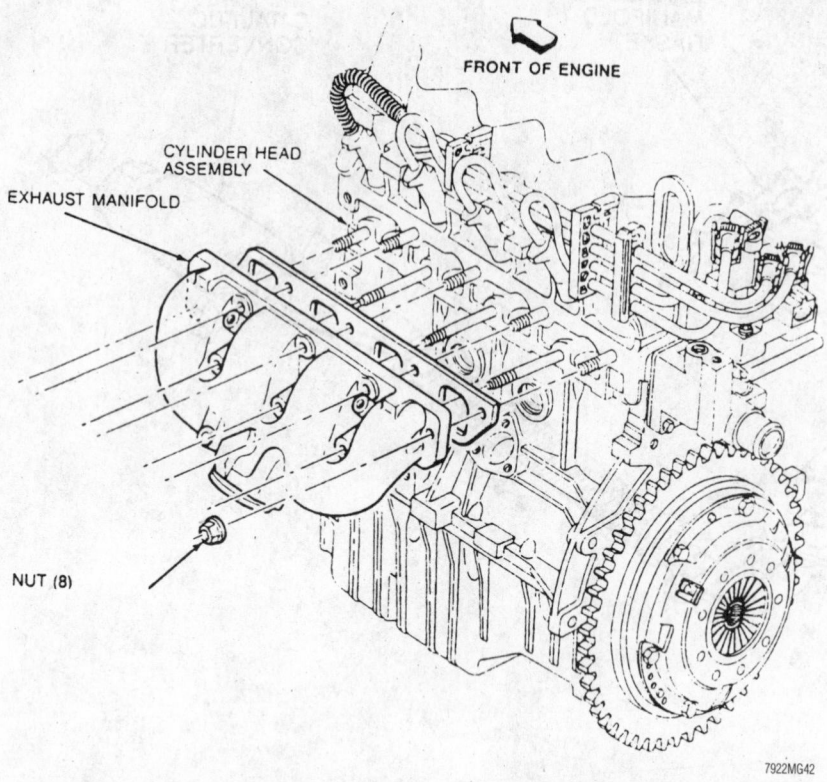

Exploded view of the exhaust manifold mount-

16. Lower the vehicle.

17. Install the exhaust manifold heat shield. Tighten the nuts to 3–5 ft. lbs. (5–7 Nm).

18. Install the EGR valve-to-exhaust manifold tube and tighten the nuts to 18–25 ft. lbs. (25–35 Nm).

19. Install the radiator cooling fan and shroud assembly.

20. Install the alternator and the accessory drive belt.

21. Connect the negative battery cable.

2.0L Engine

SOHC SPLIT PORT INJECTION (SPI)

1. Disconnect the negative battery cable.

2. Remove the drive belt.

3. Unplug the Heated Oxygen Sensor (HO2S) electrical connection which is mounted on the cooling fan shroud.

4. Unfasten the five exhaust manifold heat shield nuts and remove the shield.

5. Unfasten the two EGR tube-to-exhaust manifold nuts studs.

6. Unfasten the power steering pressure hose bracket bolts, then position the bracket and hose aside.

7. Loosen the alternator lower bolt, unfasten the alternator upper bolt and pivot the alternator forward.

8. Unfasten the four exhaust manifold-to-catalytic converter nuts.

> **✷✷ CAUTION**
>
> **Do not work on aluminum engine components until the engine is cold.**

9. Unfasten the eight exhaust manifold nuts and remove the manifold and gasket.

> **✷✷ CAUTION**
>
> **Do not use any abrasive grinding discs to remove gasket material. Use a plastic manual gasket scraper to remove the gasket residue. Be careful not to scratch or gouge the aluminum sealing surfaces when cleaning them.**

10. Clean the gasket surfaces thoroughly until all traces of the old gasket residue are removed. Inspect the gasket mating surfaces, both must be clean and flat.

To install:

11. Install a new gasket and the exhaust manifold.

12. Install and tighten the manifold retaining nuts to 14.7–17.7 ft. lbs.(20–24 Nm).

13. Install and tighten the exhaust manifold-to-catalytic converter nuts to 26–34 ft. lbs. (34–47 Nm).

14. Place the alternator in position and tighten the upper and lower mounting bolts.

15. Place the power steering pressure hose and bracket in position, then tighten the retaining bolts.

16. Tighten the two EGR tube-to-exhaust manifold nuts studs to 44.6–62 inch lbs. (5–7 Nm).

17. Install the exhaust manifold heat shield and tighten the nuts to 45–61 inch lbs. (5–7 Nm).

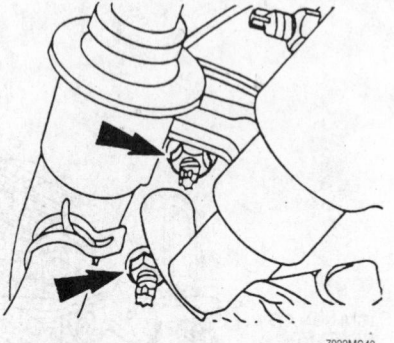

Remove the two EGR tube-to-exhaust manifold stud nuts—2.0L SOHC engine

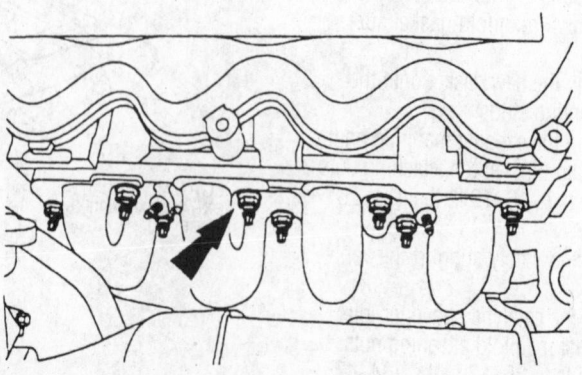

Remove the exhaust manifold retaining nuts—2.0L SOHC engine

18. Attach the Heated Oxygen Sensor (HO2S) electrical connection.

19. Install the drive belt.

20. Connect the negative battery cable.

DOHC ZETEC

1. Disconnect the negative battery cable.

2. Unplug the fan motor electrical connection.

3. Remove the fan shroud.

4. Raise the vehicle and support it with safety stands..

5. Unplug the Heated Oxygen Sensor (HO2S) electrical connection.

6. Unfasten the catalytic converter-to-exhaust manifold bolt and nut, the disconnect the converter from the manifold.

7. Lower the vehicle.

❋❋ CAUTION

Do not break the threadlock seal on the engine oil dipstick tube. If the seal is broken, relock it to prevent any oil leaks from the block.

8. Unfasten the engine oil dipstick tube bracket bolt.

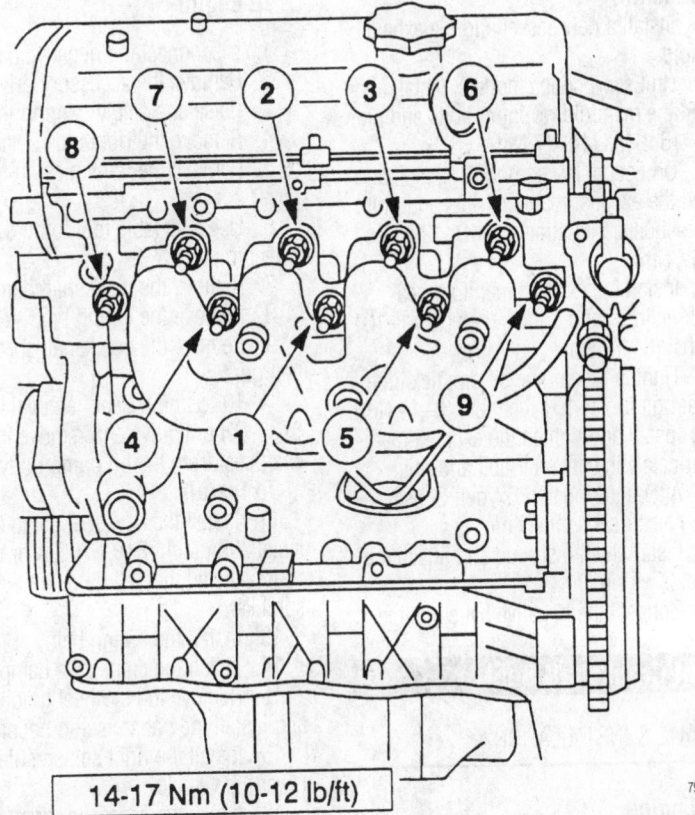

14-17 Nm (10-12 lb/ft)

7922MG46

Tighten the exhaust manifold nuts and bolts in the sequence illustrated to the proper specification—2.0L Zetec engine

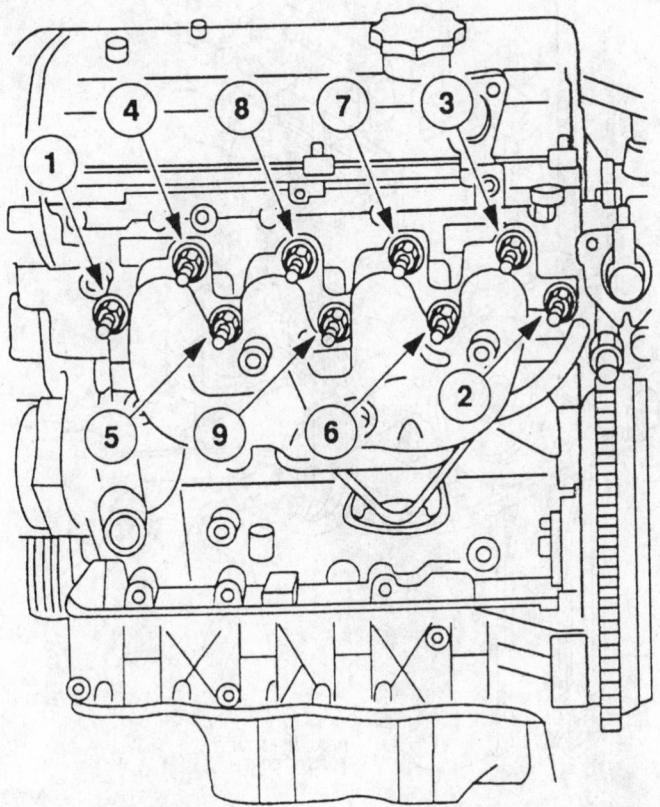

7922MG45

Remove the exhaust manifold nuts and bolts in the sequence illustrated—2.0L Zetec engine

9. Unfasten the exhaust manifold heat shield bolts and nuts, then remove the shield.

❋❋ CAUTION

Do not work on aluminum engine components until the engine is cold.

10. On Escort coupe models, unfasten the seven exhaust manifold bolts and two studs, then remove the manifold and gasket.

11. On Escort/Tracer models, unfasten the exhaust manifold nuts and bolts in the sequence illustrated.

❋❋ CAUTION

Do not use any abrasive grinding discs to remove gasket material. Use a plastic manual gasket scraper to remove the gasket residue. Be careful not to scratch or gouge the aluminum sealing surfaces when cleaning them.

12. Clean the gasket surfaces thoroughly until all traces of the old gasket residue are removed. Inspect the gasket mating surfaces, both must be clean and flat.

To install:

13. Install a new gasket and the exhaust manifold.

14. On Escort coupe models, install and tighten the manifold retaining bolts and nuts to 13–16 ft. lbs.(14–17 Nm).

15. On Escort/Tracer models, install and tighten the exhaust manifold nuts and bolts in the sequence illustrated to 10–12 ft. lbs. (14–17 Nm).

16. Install the exhaust manifold heat shield retainers and tighten them to 71–101 inch lbs. (8–11 Nm).

17. Tighten the engine oil dipstick tube bracket bolt to 71–101 inch lbs. (8–11 Nm).

18. Install and tighten the exhaust manifold-to-catalytic converter bolt and nut.

19. Attach the Heated Oxygen Sensor (HO2S) electrical connection.

20. Install the fan shroud and attach the fan motor wiring.

21. Connect the negative battery cable.

Front Crankshaft Seal

REMOVAL & INSTALLATION

1.8L Engine

1. Disconnect the negative battery cable.
2. Remove the crankshaft sprocket.
3. If necessary, cut the lip of the front oil seal with a razor to ease removal.
4. Use Seal Removal tool T92C-6700-CH or its equivalent to remove the seal.

To install:

5. Lubricate the lip of the new oil seal with clean engine oil.

6. Using Crankshaft Front Seal Replacer T88C-6701-AH or equivalent tool, install the seal evenly until it is flush with the edge of the oil pump body.

7. Install the crankshaft sprocket.
8. Connect the negative battery cable.

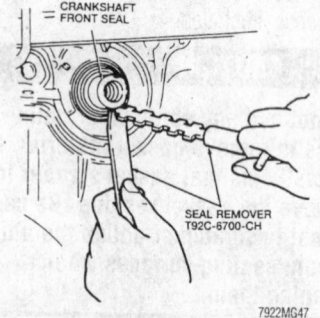

Use an appropriate seal removal tool to remove the crankshaft front oil seal—1.8L engine

1.9L Engine

1. Disconnect the negative battery cable.
2. Remove the accessory drive belt.
3. Raise and safely support the vehicle.
4. Remove the right side splash shield.
5. Remove the transmission housing cover.
6. Use a suitable tool to hold the flywheel in place.
7. Remove the crankshaft dampener.
8. Remove the timing belt.
9. Remove the crankshaft sprocket and belt guide.
10. Using a Seal Removal tool T92C-6700-CH or equivalent, remove the crankshaft seal from the oil pump body.

To install:

11. Install the new seal using Seal Installation tool T81P-6700-A or equivalent.

12. Install the belt guide and crankshaft sprocket.

13. Install the timing belt.
14. Install the crankshaft dampener.
15. Remove the flywheel holding tool and install the transmission housing cover.
16. Install the right splash shield and lower the vehicle.
17. Install the accessory drive belt.

18. Connect the negative battery cable.
19. Start the engine and check for leaks.

2.0L Engines

1. Remove the timing belt and crankshaft sprocket.

⁂ CAUTION

Be careful not to damage the crankshaft surface when removing the seal.

2. Using Seal Remover tool T92C-6700-CH or its equivalent, remove the crankshaft front oil seal.

To install:

3. Use Seal Replacer tool T81P-6700-A or its equivalent, install the new seal.

4. Install the crankshaft sprocket and the timing belt.

Camshaft(s) and Valve Lifters

All engines covered in this manual except the 2.0L Zetec engine, employ hydraulic (oil pressurized) valve lifters. These lifters operate more quietly and do not require periodic adjustments.

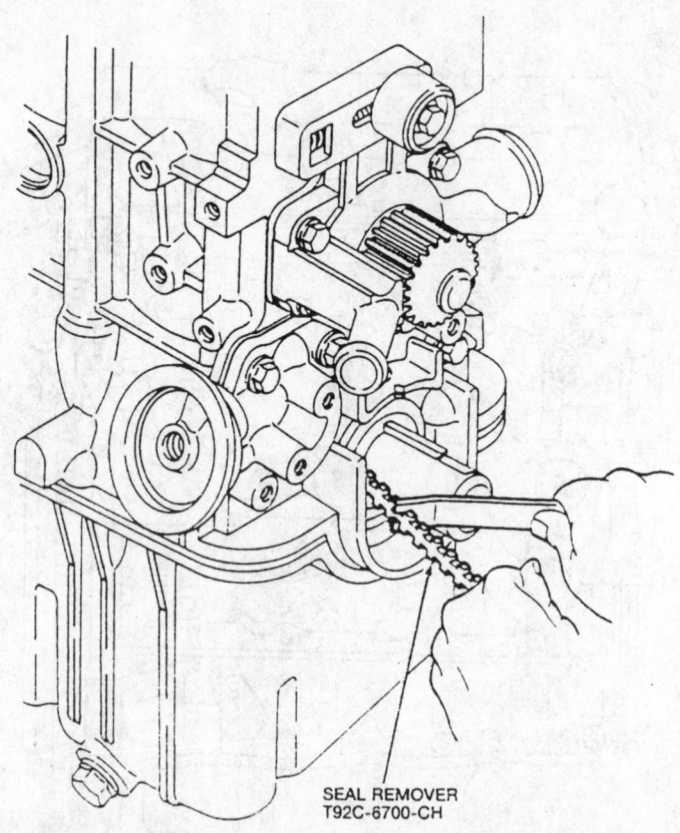

Removing the front crankshaft oil seal using tool T92C-6700-CH or equivalent—1.9L engine

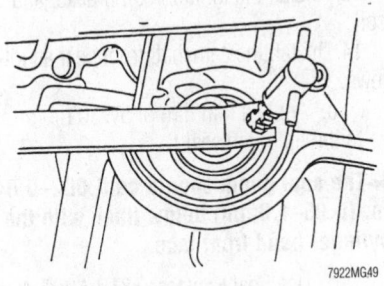

Use an appropriate tool to remove the crankshaft front oil seal—2.0L engines

REMOVAL & INSTALLATION

1.8L Engine

1. Disconnect the negative battery cable.
2. Remove the distributor assembly.
3. Remove the camshaft sprockets.
4. Unbolt the seal plate mounting bolts and remove the seal plate.
5. Loosen the camshaft cap bolts in the correct sequence (see accompanying illustration).
6. Remove the camshaft caps and note their mounting locations for installation reference.

➡The camshaft caps are numbered and have arrow marks for installation and direction reference.

7. Remove the camshaft and camshaft oil seal.

To install:

8. Apply clean engine oil to the camshaft journals and bearings.
9. Place the camshaft into its mounting position.

➡The exhaust camshaft has a groove which must be aligned with the distributor drive gear.

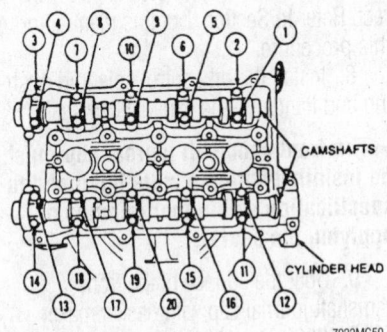

Camshaft cap bolt loosening sequence—1.8L engines

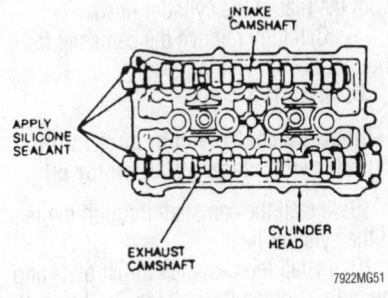

Apply silicone sealer as shown—1.8L engines

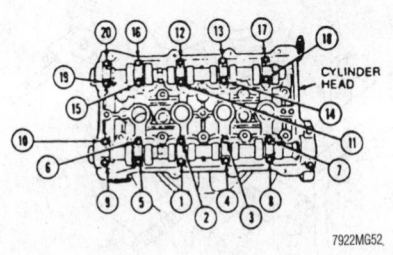

Tighten the cap bolts in the sequence shown—1.8L engines

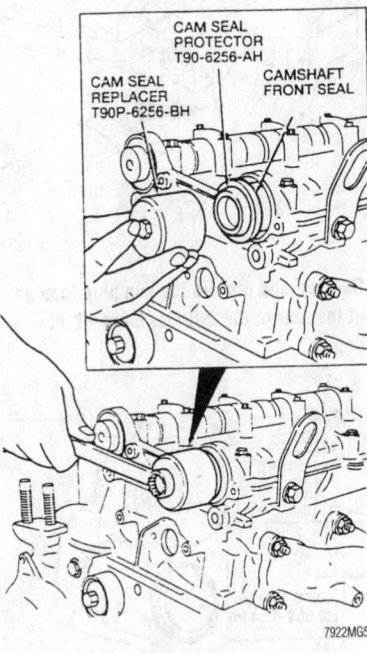

Use a suitable installation tool to install the new camshaft seal—1.8L engine

10. Apply silicone sealant to the required areas (see accompanying illustration).
11. Install the camshaft caps according to the cap numbers and arrow marks.
12. Install the camshaft cap bolts and tighten them in the proper sequence to 100–126 inch lbs. (11.3–14.2 Nm).
13. Apply a small amount of clean engine oil to the lip of a new camshaft oil seal. Using a suitable installation tool, install the new seal. Refer to the accompanying illustration.
14. Place the seal plate into its mounting position and install the mounting bolts. Tighten the bolts to 69–95 inch lbs. (7.8–11.0 Nm).
15. Install the camshaft sprockets and the distributor assembly.
16. Connect the negative battery cable.

1.9L Engine

1. Disconnect the negative battery cable.
2. Remove the air cleaner intake duct.
3. Remove the accessory drive belts and the crankshaft pulley.
4. Remove the timing belt cover and the rocker arm cover.
5. Set the engine No. 1 cylinder at TDC prior to removing the timing belt.

➡Be sure the crankshaft is positioned at TDC. Do not turn the crankshaft until the timing belt is installed.

6. Remove the rocker arms and lifters.
7. Remove the ignition coil assembly.
8. Remove the timing belt.
9. Remove the camshaft sprocket bolt, sprocket and key.
10. Remove the camshaft thrust plate.
11. Remove the cup plug from the back of the cylinder head.
12. Remove the camshaft through the back of the head toward the transaxle.
13. Remove and discard the camshaft seal.

To install:

14. Thoroughly coat the camshaft bearing journals, cam lobe surfaces and thrust plate groove with a suitable lubricant.

➡Before installing the camshaft, apply a thin film of lubricant to the lip of the camshaft seal.

15. Install a new camshaft seal and install the camshaft through the rear of the cylinder head. Rotate the camshaft during installation.
16. Install the camshaft thrust plate. Tighten attaching bolts to 6–9 ft. lbs. (8–13 Nm).

17. Align and install the cam sprocket over the cam key. Install the attaching washer and bolt. While holding the camshaft stationary, tighten the bolt to 70–85 ft. lbs. (95–115 Nm).

18. Install the cup plug using a sealer EOAZ-19554-BA. Use the sealer sparingly, as excess sealer may clog the oil holes in the camshaft.

19. Install the timing belt.
20. Install the timing belt cover.
21. Install the ignition coil assembly.
22. Install a new rocker arm cover gasket, if required.

➡ **Be sure the surfaces on the cylinder head and rocker arm cover are clean and free of sealant material.**

23. Install the valve cover.
24. Install the air cleaner intake tube.
25. Connect the negative battery cable.

2.0L Engine

SOHC SPLIT PORT INJECTION (SPI)

1. Disconnect the negative battery cable.
2. Remove the air cleaner and valve cover.
3. Remove the camshaft front seal as follows:

   a. Align the timing marks and remove the timing belt.

   b. Use cam sprocket holding/removing tool T74P-6256-B or its equivalent, and remove the camshaft sprocket.

   c. Unfasten the timing belt tensioner bolt and remove the tensioner.

   d. Remove the inner engine front cover.

   e. Use Seal Removal tool T92C-6700-CH or its equivalent to remove the camshaft front seal.

4. Remove the ignition coil and bracket.
5. Remove the rocker arms and valve tappets.
6. Unfasten the camshaft thrust plate bolts and remove the plate.

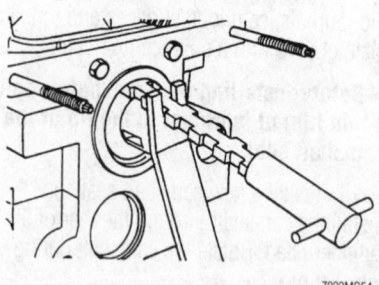

Removing the camshaft front seal—2.0L SOHC engine

7. Remove and discard the cup plug from the rear of the cylinder head.
8. Carefully remove the camshaft from the rear of the cylinder head.

To install:

➡ **Liberally coat the cam bore in the cylinder with clean 5W30 motor oil.**

9. Install the camshaft through the rear of the cylinder head.
10. Install the camshaft thrust plate and the bolts. Tighten the bolts to 71–115 inch lbs. (8–13 Nm).
11. Install a new cup plug and the tappets.
12. Install the rocker arms.

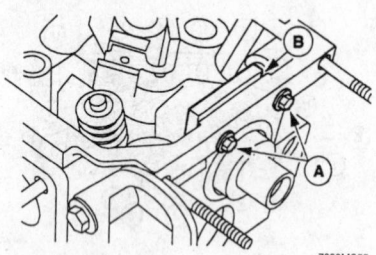

Remove the camshaft thrust plate retaining bolts (A) and the plate (B)—2.0L SOHC engine

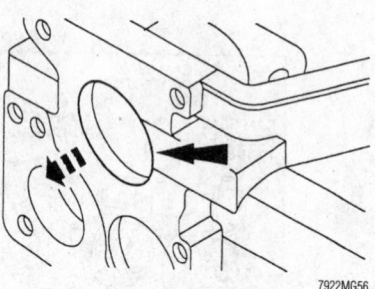

Remove and discard the cup plug located at the rear of the cylinder head—2.0L engine

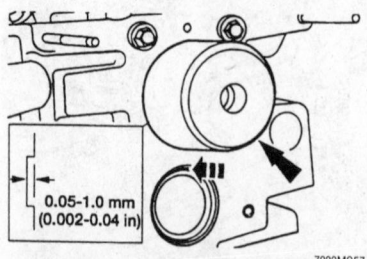

Use Seal Replacer T81P-6292-A or its equivalent, to install the camshaft front seal 0.002–0.04 in. (0.05–1.0mm) below flush with the cylinder head front face—2.0L SOHC engine

13. Install the ignition coil bracket and coil.
14. Install the camshaft front seal as follows:

   a. Apply a thin film of 5W30 motor oil to the lip of the seal.

➡ **The seal depth should be 0.002–0.04 in. (0.05–1.0mm) below flush with the cylinder head front face.**

   b. Use Seal Replacer T81P-6292-A or its equivalent, to install the camshaft front seal.

   c. Install the inner engine front cover.

   d. Install the timing belt tensioner. Tighten the tensioner bolt to15–22 ft. lbs. (20–30 Nm).

   e. Install the camshaft sprocket and the timing belt

15. Install the valve cover and the air cleaner assembly.
16. Connect the negative battery cable.

DOHC ZETEC

1. Remove the timing belt, valve cover and camshaft sprockets.

➡ **It may be necessary to rotate the oil control solenoid flange 90° prior to removal.**

2. Unfasten the oil control solenoid flange bolts and remove the flange.

➡ **Mark the camshaft journal caps with a number to identify their location as they must be replaced in their original position.**

3. Loosen the camshaft journal caps in several passes in the sequence illustrated, then remove the bolts and caps.
4. Remove the camshafts from the cylinder head.
5. Remove the oil control sensor and bushing.
6. Inspect the camshafts for damage and wear.

To install:

7. Be sure the valve clearance is correct. Refer to Section 1 of this manual for this procedure.
8. Install the oil control solenoid bushing and flange on the exhaust camshaft.

➡ **The front camshaft journal cap must be installed with the bolts tightened to specification within four minutes of applying the sealer.**

9. Coat the surface of the front camshaft journal cap with gasket maker E2AZ-19562-B or its equivalent.
10. Place the camshafts in position and lubricate the bearing surfaces with engine

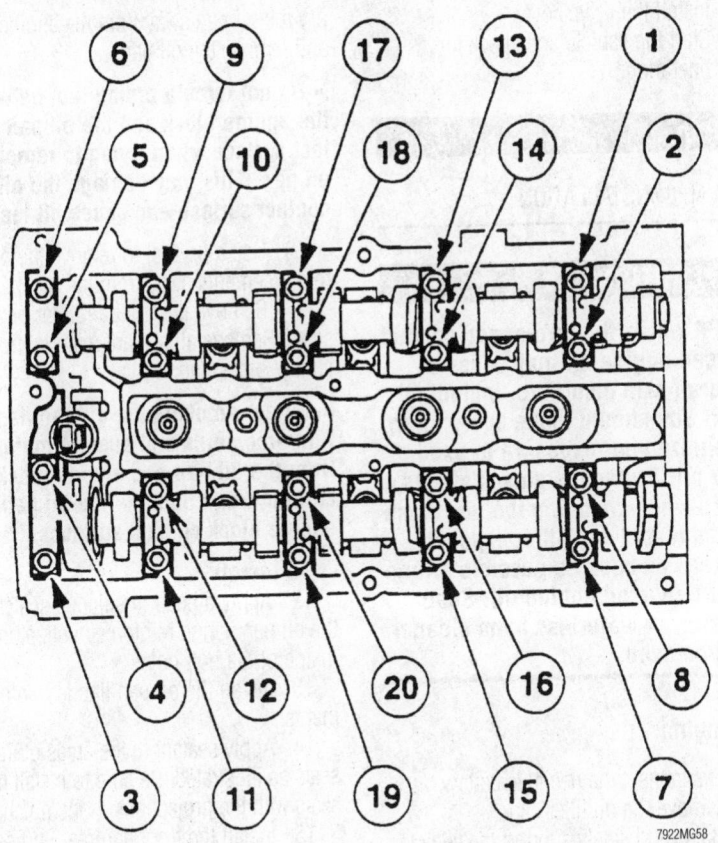

Loosen the camshaft journal caps in several passes in this sequence—2.0L Zetec engine

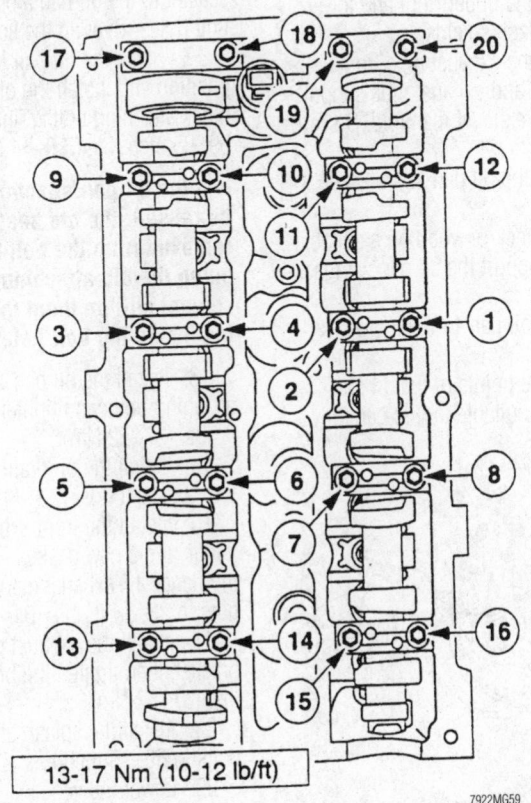

13-17 Nm (10-12 lb/ft)

Tighten the camshaft journal caps in the sequence illustrated to the proper specification—2.0L Zetec engine

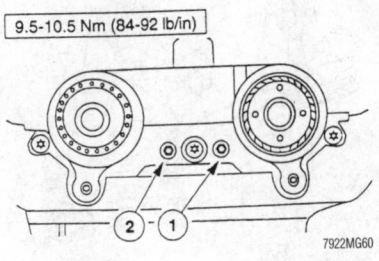

9.5-10.5 Nm (84-92 lb/in)

Tighten the oil control solenoid flange bolts in this sequence—2.0L Zetec engine

assembly lubricant D9AZ-19579-D or its equivalent.

11. Install new camshaft front oil seals.

12. Apply a thin coat of silicone gasket and sealant F6AZ-19562-AA or its equivalent to the sealing surface of the of the front camshaft journal bearing cap.

13. Install the caps and tighten the bolts in several 2-turn passes in the sequence illustrated to 10–12 ft. lbs. (13–17 Nm).

� CAUTION

The oil control O-ring seals must not fall out of position in the oil control solenoid flange or poor engine performance may occur.

14. Inspect the oil control solenoid flange O-rings for damage or wear and replace as necessary.

15. Install the oil control solenoid flange and tighten the bolts to 84–92 inch lbs. (9.5–10.5 Nm) in the sequence illustrated.

16. Rotate the camshafts a full turn and check for binding.

17. Install the camshaft sprockets and timing belt.

18. Install the valve cover.

Valve Lash

ADJUSTMENT

2.0L Zetec DOHC Engine

1. Remove the valve cover.
2. Remove the timing belt.

➡**Measure each valve's clearance at the base circle before removing the camshafts. The shims are not serviceable with the camshafts in place. Failure to measure all valve clearances prior to removing the camshafts, will result in repeated camshaft removal and installation and a lot of wasted time.**

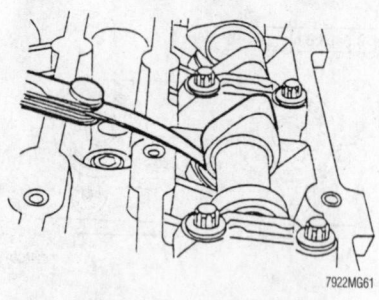

Measure each valve's clearance at the base circle before removing the camshafts—2.0L Zetec DOHC engine

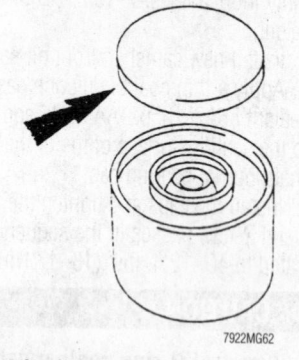

Example of a valve lifter shim (arrow)—2.0L Zetec DOHC engine

3. Measure the valve clearances, then remove the camshafts and shims.

➡**The shims are marked for thickness. For example: 2.2mm = 222 on shim.**

The correct shims allow the following valve clearances:
- Intake valve clearance: 0.0043–0.0071 in. (0.11–0.18mm)
- Exhaust valve clearance: 0.0106–0.0134 in. (0.27–0.34mm)

➡**A midrange clearance is the most desirable.**

The midrange clearances should be as follows:
- Intake valve clearance: 0.006 in. (0.15mm)
- Exhaust valve clearance: 0.012 in. (0.3mm)

4. Select the shims using the following formula: shim thickness = measured clearance plus the base shim thickness minus the most desirable thickness.

5. Select the correct shims and mark their installation location (where they are going).

6. Replace the shims and install the camshafts.

7. Check the new valve clearances.

8. Install the timing belt and covers.

9. Install the valve cover.

10. Start the vehicle and check for proper operation.

Oil Pan

REMOVAL & INSTALLATION

✳✳ CAUTION

The EPA warns that prolonged contact with used engine oil may cause a number of skin disorders, including cancer! You should make every effort to minimize your exposure to used engine oil. Protective gloves should be worn when changing the oil. Wash your hands and any other exposed skin areas as soon as possible after exposure to used engine oil. Soap and water, or waterless hand cleaner should be used.

1.8L Engine

1. Disconnect the negative battery cable. Remove the oil filler cap.

2. Raise and safely support the vehicle.

3. Remove the drain plug and drain the engine oil into a suitable container.

4. Remove the upper right and lower right and left splash shields.

5. Remove the exhaust pipe front mounting flange and exhaust pipe support bracket from the exhaust manifold. Discard the pipe bracket.

6. Unfasten the oil pan-to-transaxle attaching bolts.

7. Place a block of wood on a suitable jackstand and support the oil pan with jackstand.

8. Unfasten oil pan-to-engine block attaching bolts.

9. Only at the points of the oil pan illustrated, use a suitable tool to carefully

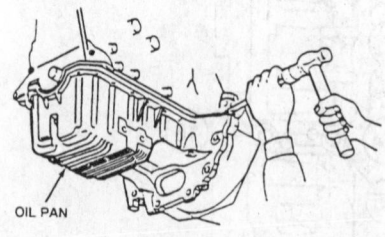

Use a suitable tool to carefully pry the oil pan away from the engine block and remove the oil pan—1.8L engine

pry the oil pan away from the engine block and remove the oil pan.

➡**Do not force a prying tool between the engine block and the oil pan contact surface when trying to remove the oil pan. This may damage the oil pan contact surface and cause oil leakage.**

10. Use a prytool to remove the oil pan reinforcements away from the block.

11. Remove the front and rear oil pan seals. Remove all sealant material from the engine block and oil pan.

➡**When removing the oil pan flange reinforcements and sealant material from the oil pan and engine block, be careful not to damage the oil pan and engine block contact surfaces.**

To install:

12. Apply a bead of silicone sealant to the oil pan flange reinforcements along the inside of the bolt holes.

13. Install the oil pan flange reinforcements.

14. Apply sealant to the areas of the end seals as illustrated. Be sure to install the end seals with the projections in the notches.

15. Install the front and rear end seals onto the oil pan.

16. Apply a continuous bead of silicone sealant to the oil pan along the inside of the bolt holes. Overlap the sealant ends.

17. Place the oil pan into its mounting position and install the oil pan-to-engine block attaching bolts. Tighten the bolts to 69–95 inch lbs. (7.8–11.0 Nm).

➡**If the oil pan attaching bolts are to be reused, the old sealant must be removed from the bolt threads. Tightening the old attaching bolts with old sealant still on them may cause cracking inside the bolt holes.**

18. Install the oil pan-to-transaxle attaching bolts and tighten them to 27–38 ft. lbs. (37–52 Nm).

19. Install the oil drain plug and tighten it to 22–30 ft. lbs. (29–41 Nm).

20. Install the front exhaust mounting flange and a new gasket. Tighten the mounting flange-to-exhaust manifold attaching nuts to 23–34 ft. lbs. (31–46 Nm).

21. Install the exhaust pipe support bracket, then tighten the bolts to 27–38 ft. lbs. (37–52 Nm).

22. Install the splash shields. Tighten the bolts to 69–95 inch lbs. (7.8–11.0 Nm).

23. Lower the vehicle.

24. Fill the crankcase with the proper type and quantity of engine oil. Install the filler cap.

25. Connect the negative battery cable.

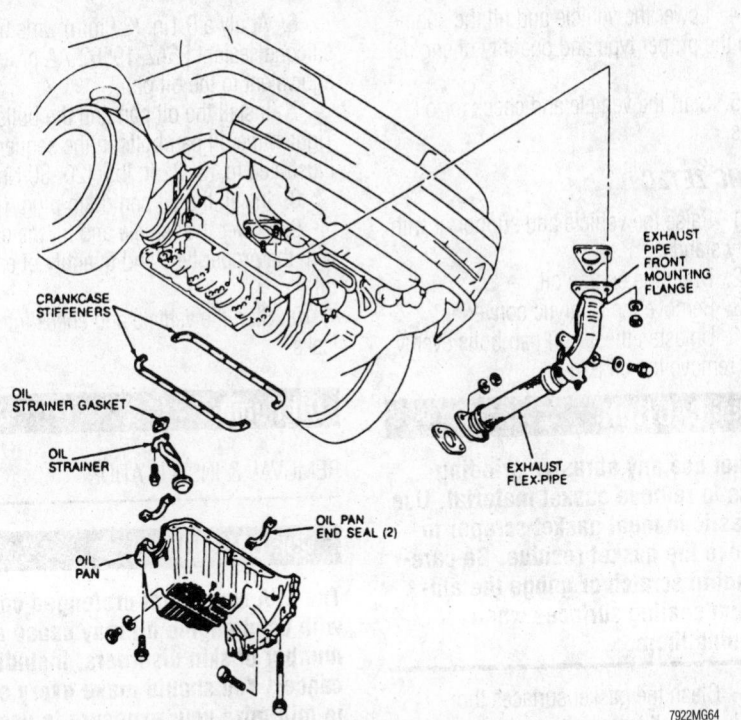

Exploded view of oil pan and related components—1.8L engine

1.9L Engines

1. Disconnect the negative cable at the battery.
2. Raise the vehicle and support safely.
3. Remove the drain plug and drain the engine oil into a suitable container.
4. Remove the catalytic converter from the exhaust manifold and from the inlet pipe and resonator.
5. Unfasten the two oil pan-to-transaxle bolts.
6. Unfasten the ten oil pan-to-engine bolts.
7. Gently pry the oil pan from the block and remove the oil pan
8. Remove the oil pan gasket and discard.

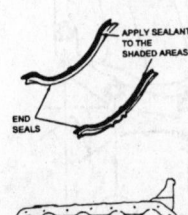

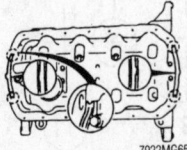

After thoroughly cleaning the gasket mating surfaces, apply sealant to the areas shown—1.8L engine

To install:

9. Use a scraper to clean the oil pan gasket surface and the mating surface on the cylinder block. Wipe the oil pan rail with a solvent-soaked, lint-free cloth to remove all traces of oil.
10. Remove the two oil pump screen cover screws to clean the oil pump pick up tube and screen assembly. Install tube and screen assembly using a new tube gasket.

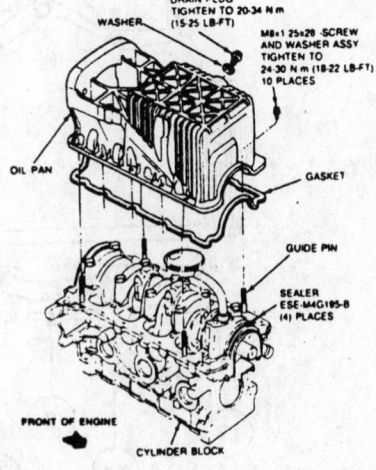

Exploded view of oil pan mounting—1.9L engine

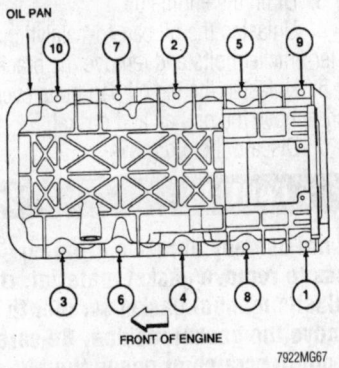

Oil pan tightening sequence—1.9L engine

Tighten the screen screws and tube bolts to 7–9 ft. lbs. (10–13 Nm).

➡**Install the pan within 10 minutes of applying the sealant.**

11. Apply a bead of suitable silicone rubber sealer approximately 0.125 in. (3.0mm) wide at the corners of the block and at the oil pump-to-block and the crankshaft rear oil seal retainer-to-block joints.
12. Install the oil pan gasket in the pan ensuring the press fit tabs are fully engaged in the gasket channel.
13. Install the oil pan and attaching bolts. Tighten the bolts lightly until the two oil pan-to-transaxle bolts can be installed.

➡**If the oil pan is installed on the engine outside of the vehicle, a transaxle case or equivalent, fixture must be bolted to the block to line-up the oil pan flush with the rear face of the block.**

14. Tighten the two pan-to-transaxle bolts to 30–40 ft. lbs. (40–54 Nm), then loosen them one half turn.
15. Tighten the oil pan flange-to-cylinder block bolts to 15–22 ft. lbs. (20–30 Nm) in the sequence illustrated.
16. Retighten the two pan-to-transaxle bolts to 30–40 ft. lbs. (40–54 Nm).
17. Install the catalytic converter.
18. Install the oil pan drain plug and tighten it to 15–22 ft. lbs. (20–30 Nm).
19. Lower the vehicle and fill the crankcase.
20. Connect negative battery cable.
21. Start the engine and check for oil leaks.

2.0L Engine

SOHC SPLIT PORT INJECTION (SPI)

1. Raise the vehicle and support it with safety stands.
2. Remove the catalytic converter.

3. Drain the engine oil.

4. Unfasten the oil pan-to-catalytic converter bracket bolts and remove the bracket.

5. Unfasten the ten oil pan bolts evenly, then remove the oil pan and gasket.

6. Discard the old gasket.

✳✳ CAUTION

Do not use any abrasive grinding discs to remove gasket material. Use a plastic manual gasket scraper to remove the gasket residue. Be careful not to scratch or gouge the aluminum sealing surfaces when cleaning them.

7. Clean the gasket surfaces thoroughly until all traces of the old gasket residue are removed. Inspect the gasket mating surfaces, both must be clean, flat and dry.

To install:

➡️**Install the oil pan within ten minutes of applying the silicone sealer.**

8. Apply a 0.125 in. (3.0mm) wide bead of silicone sealant F6AZ-19562-AA or its equivalent at the oil pump-to-cylinder block joints and the crankshaft rear oil seal retainer-to-cylinder block joints.

➡️**Be sure the press fit tabs are fully engaged in the oil pan gasket channel when installing the pan and gasket.**

9. Install the gasket.

10. When installing the oil pan, be sure the rear of the pan is aligned with the cylinder block by using a straight edge as a guide.

11. Install and tighten the oil pan bolts in the sequence illustrated to 15–22 ft. lbs. (20–30 Nm).

12. Install the oil pan-to-catalytic converter bracket and tighten the bolts to 30–40 ft. lbs. (40–55 Nm).

13. Install the catalytic converter and oil pan drain plug. Tighten the drain plug to 15–22 ft. lbs. (20–30 Nm).

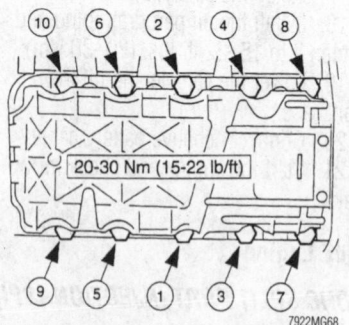

Tighten the oil pan bolts in this sequence to the proper specification—2.0L SOHC engine

14. Lower the vehicle and fill the engine with the proper type and quantity of engine oil.

15. Start the vehicle and check for oil leaks.

DOHC ZETEC

1. Raise the vehicle and support it with safety stands.

2. Drain the engine oil.

3. Remove the catalytic converter.

4. Unfasten the 17 oil pan bolts evenly, then remove the oil pan.

✳✳ CAUTION

Do not use any abrasive grinding discs to remove gasket material. Use a plastic manual gasket scraper to remove the gasket residue. Be careful not to scratch or gouge the aluminum sealing surfaces when cleaning them.

5. Clean the gasket surfaces thoroughly until all traces of the old gasket residue are removed. Inspect the gasket mating surfaces, both must be clean, flat and dry.

To install:

➡️**Install the oil pan within four minutes of applying the silicone sealer.**

6. Apply a 0.1in. (3.0mm) wide bead of silicone sealant F6AZ-19562-AA or its equivalent to the oil pan.

7. Install the oil pan and the bolts. Tighten the oil pan bolts in the sequence illustrated to 15–22 ft. lbs. (20–30 Nm).

8. Install the oil pan drain plug.

9. Lower the vehicle and fill the engine with the proper type and quantity of engine oil.

10. Start the vehicle and check for oil leaks.

Oil Pump

REMOVAL & INSTALLATION

✳✳ CAUTION

The EPA warns that prolonged contact with used engine oil may cause a number of skin disorders, including cancer! You should make every effort to minimize your exposure to used engine oil. Protective gloves should be worn when changing the oil. Wash your hands and any other exposed skin areas as soon as possible after exposure to used engine oil. Soap and water, or waterless hand cleaner should be used.

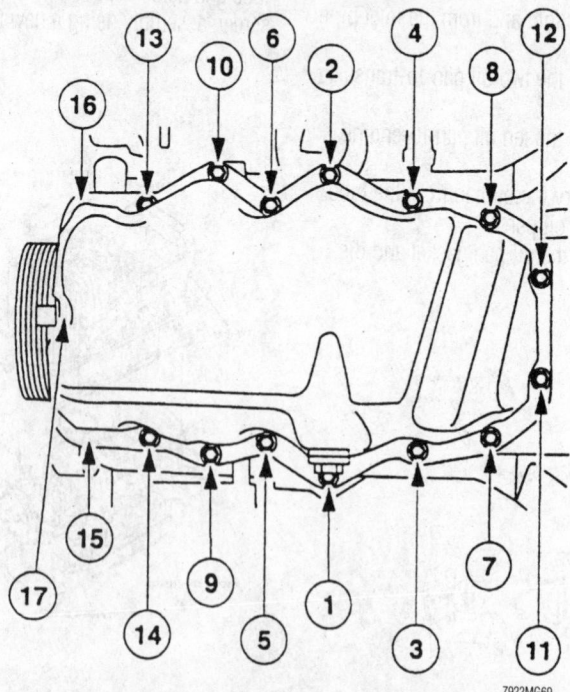

Tighten the oil pan bolts in the proper sequence to the correct torque specification—2.0L Zetec engine

1.8L Engine

1. Disconnect the negative battery cable.
2. Remove the timing belt and timing belt pulley.
3. Remove the oil pan.
4. Remove the oil pump screen cover and tube.
5. If equipped, remove the A/C compressor mounting bolts and position the compressor so it is free from the work area.
6. Remove the A/C compressor mounting bracket.
7. Unbolt the mounting bolt from the engine oil dipstick tube bracket and the alternator lower mounting bolt.
8. Unfasten all oil pump mounting bolts and remove the oil pump.
9. Use a scraper to remove all gasket material from the oil pump mounting surfaces.

To install:

10. Install a new gasket onto the oil pump.
11. Place the oil pump into its mounting position and install the pump mounting bolts. Tighten the bolts to 14–19 ft. lbs. (19–25 Nm).
12. Place the dipstick tube bracket bolt into its mounting position and install the mounting bolt. Tighten the bolt to 68–95 inch lbs. (7.8–11 Nm).
13. Install the alternator lower mounting bolt and tighten it to 27–38 ft. lbs. (37–52 Nm).
14. Install a new gasket onto the oil strainer, place the strainer into its mounting position and install the mounting bolts. Tighten the bolts to 69–95 inch lbs. (7.8–11.0 Nm).
15. Install the oil pan.
16. If equipped, place the A/C compressor bracket into its mounting position and install the mounting bolts. Tighten the bolts to 30–40 ft. lbs. (40–55 Nm).
17. If equipped, install A/C compressor side bolt. Tighten the bolts to 14–19 ft. lbs. (19–25 Nm).
18. Install the timing belt pulley and timing belt.
19. Connect the negative battery cable.
20. Start the engine and check for oil leaks. Be sure the oil pressure indicator lamp has gone out after a second or two. If the lamp remains on after several seconds, immediately shut off the engine. Determine the cause and correct the condition.

1.9L Engine

1. Be sure the No. 1 cylinder is at TDC.
2. Disconnect the negative battery cable.
3. Remove the accessory drive belt and the automatic tensioner.

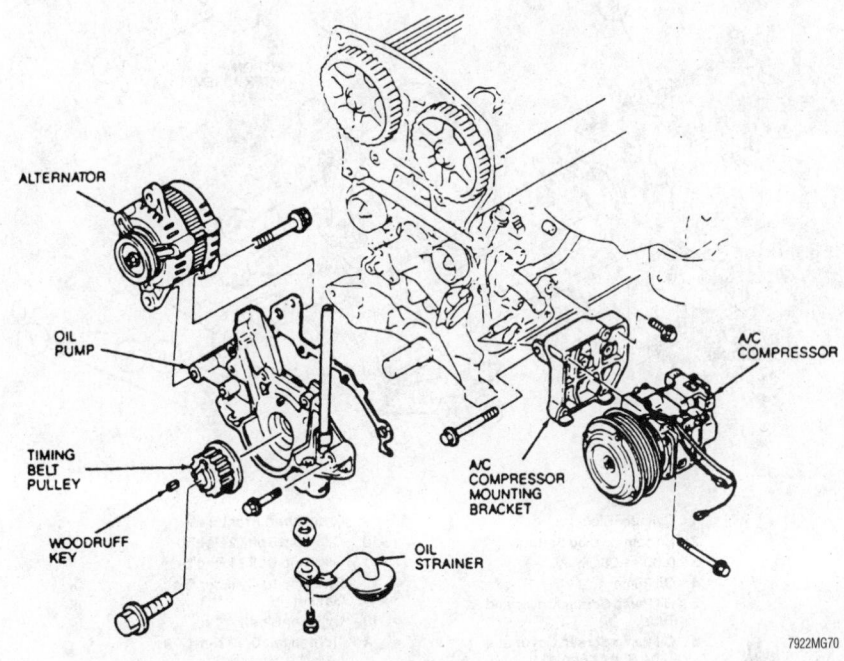

Exploded view of oil pump mounting—1.8L engine

4. Support the engine with a suitable floor jack.
5. Remove the right engine mount dampener and remove the right engine mount bolts from the mount bracket.
6. Loosen the mount through-bolt and roll the mount aside.
7. Remove the timing belt cover.
8. Roll the engine mount back into place and install the two mount bolts. Remove the floor jack.
9. Loosen the belt tensioner attaching bolt and pry the tensioner to the rear of the engine. Tighten the attaching bolt.
10. Raise and safely support the vehicle.
11. Remove the right side splash shield.
12. Remove the catalytic converter inlet pipe.
13. Drain and remove the oil pan. Remove the oil filter.
14. Remove the crankshaft dampener and the timing belt.
15. Remove the crankshaft sprocket and the timing belt guide from the crankshaft.
16. Disconnect the crankshaft position sensor.
17. Unfasten the 6 oil pump-to-engine bolts and remove the oil pump assembly from the engine. Remove and discard the gasket.

18. Remove the crankshaft seal from the pump and discard.

To install:

19. Be sure the pump mating surfaces on the cylinder block and oil pump are clean and free of gasket material.
20. Remove the oil pick-up tube and screen assembly from the pump for cleaning.
21. Lubricate the outside diameter of the crankshaft seal with engine oil and install the seal with a suitable installation tool. Lubricate the seal lip with clean engine oil.
22. Position the oil pump gasket on the cylinder block.
23. Using a suitable tool, position the pump drive gear to allow the pump to pilot over the crankshaft and seat firmly on the cylinder block.

➡**The pump drive gear can be accessed through the oil pick-up hole in the body of the pump. Do not install the oil pump pick-up tube and screen until the pump has been correctly installed on the cylinder block.**

24. Install the six oil pump bolts and tighten to 8–12 ft. lbs. (11–16 Nm).

➡**When the oil pump bolts are tightened, the gasket must not be below the cylinder block sealing surface.**

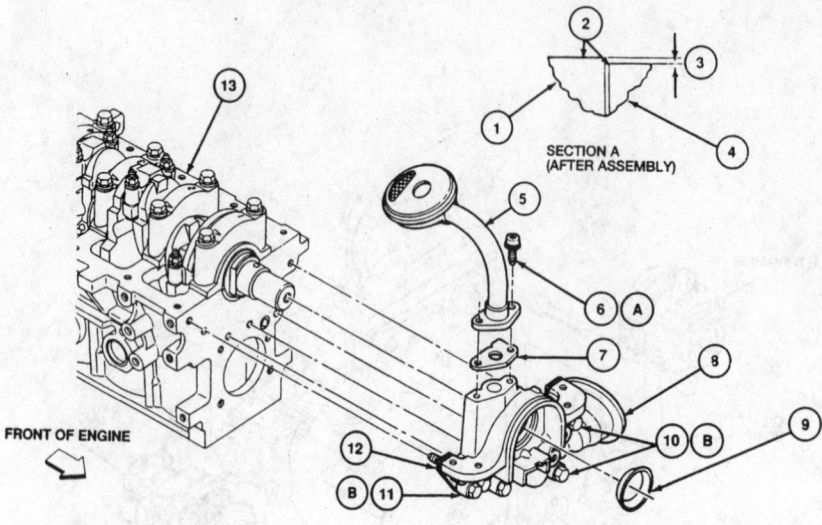

1 Cylinder Block	9 Crankshaft Front Seal
2 Bottom Sealing Surface	10 Oil Pump Bolt (2 Req'd)
3 0.00 - 1.08mm	11 Oil Pump Bolt (4 Req'd)
4 Oil Pump	12 Oil Pump to Cylinder Block
5 Oil Pump Screen Cover and	Gasket
Tube	13 Cylinder Block
6 Oil Pump Screen Cover and	A Tighten to 10-13 N·m (7-9
Tube Bolt (2 Req'd)	Lb-Ft)
7 Oil Pump Inlet Tube Gasket	B Tighten to 11-16 N·m (8-12
8 Oil Filter Mounting Position	Lb-Ft)

7922MG71

Exploded view of oil pump mounting, including related components—1.9L engine

25. Install the pick-up tube and screen assembly on the oil pump using a new gasket. Tighten the attaching screws to 7–9 ft. lbs. (10–13 Nm).

26. Install the timing belt guide over the end of the crankshaft and install the crankshaft sprocket.

27. Be sure the No. 1 cylinder is at TDC.

28. Position the timing belt over the sprockets.

29. Connect the crank angle sensor.

30. Install the oil pan and the crankshaft dampener.

31. Install the catalytic converter inlet pipe.

32. Install the splash shield and lower the vehicle.

33. Install the timing belt. Tighten the tensioner attaching bolt to 17–22 ft. lbs. (23–30 Nm).

34. Support the engine with a suitable floor jack.

35. Remove the right engine mount bolts and roll the mount back.

36. Install the timing belt cover.

37. Roll the engine mount back into place and install the attaching bolts. Tighten the mount through-bolt and install the mount dampener.

38. Remove the floor jack.

39. Install the accessory drive belt automatic tensioner and the accessory drive belt.

40. Fill the crankcase with the proper type and amount of engine oil.

41. Connect the negative battery cable, start the engine and check for leaks.

2.0L Engine

SOHC SPLIT PORT INJECTION (SPI)

1. Remove the timing belt.

2. Raise the vehicle and support it with safety stands.

3. Remove the oil pan.

4. Unplug the Crankshaft Position (CKP) sensor electrical connection.

5. Unfasten the oil pump screen cover and tube bolts and remove the cover and tube.

6. Unfasten the six oil pump retaining bolts.

7. Remove the old pump and gasket. Discard the gasket.

To install:

✳✳ CAUTION

Do not use any abrasive grinding discs to remove gasket material. Use a plastic manual gasket scraper to remove the gasket residue. Be careful not to scratch or gouge the aluminum sealing surfaces when cleaning them.

8. Clean the gasket surfaces thoroughly until all traces of the old gasket residue are removed. Inspect the gasket mating surfaces, both must be clean, flat and dry.

9. Lubricate the crankshaft front oil seal lip with clean engine oil.

➡️**When the oil pan bolts are tightened, the oil pump-to-cylinder block gasket must be below the cylinder block sealing surface.**

10. Install the pump gasket and the pump.

11. Install and tighten the oil pump bolts to 8–12 ft. lbs. (11–16 Nm).

12. Install the oil pump screen cover and tube. Tighten the retaining bolts to 71–97 inch lbs. (8–11 Nm).

13. Attach the Crankshaft Position (CKP) sensor electrical connection.

14. Install the oil pan.

15. Lower the vehicle and install the timing belt.

16. Fill the engine with the proper type and quantity of engine oil.

17. Start the vehicle and check for oil leaks.

DOHC ZETEC

1. Remove the timing covers and belt.

2. Raise the vehicle and support it with safety stands.

3. Remove the crankshaft pulley, sprocket and the timing chain guide.

4. Remove the oil pan.

5. Unfasten the oil pump cover and screen bolts and remove the cover and screen.

6. If necessary, remove the lower cylinder block shims.

7. If necessary, remove the lower cylin-

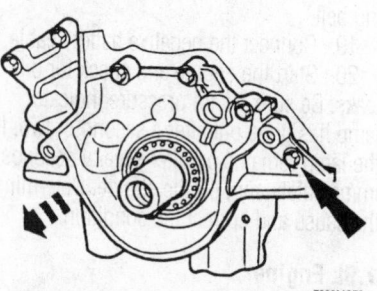

7922MG72

Remove the oil pump mounting bolts— 2.0L Zetec engine

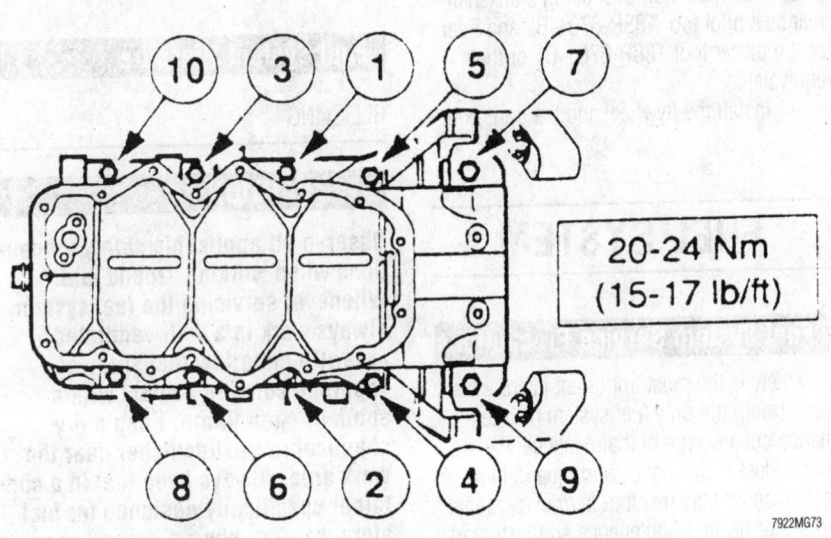

20-24 Nm
(15-17 lb/ft)

7922MG73

Tighten the lower cylinder block bolts in the sequence illustrated to the proper torque specification—2.0L Zetec engine

der block, the gasket and the crankshaft front oil seal.

8. Unfasten the oil pump retaining bolts and remove the pump and gasket.

To install:

✱✱ CAUTION

Do not use any abrasive grinding discs to remove gasket material. Use a plastic manual gasket scraper to remove the gasket residue. Be careful not to scratch or gouge the aluminum sealing surfaces when cleaning them.

9. Clean the gasket surfaces thoroughly until all traces of the old gasket residue are removed. Inspect the gasket mating surfaces, both must be clean, flat and dry.

➡**The clearance between the lower cylinder block sealing surfaces on the oil pump and the cylinder block cannot exceed 0.012–0.031 in. (0.3–0.8mm).**

10. Install a new oil pump gasket and the pump.

11. Install and tighten the oil pump bolts to 88–97 inch lbs. (10–11 Nm).

12. If removed, install the crankshaft front oil seal.

13. If removed, install the lower cylinder block and gasket. Tighten the lower cylinder block bolts to 15–17 ft. lbs. (20–24 Nm) in the sequence illustrated.

14. Install the oil pump screen cover and tube. Tighten the retaining bolts to 71–97 inch lbs. (8–11 Nm).

15. Install the oil pan, timing chain guide, crankshaft sprocket and the pulley.

16. Lower the car, then install the timing belt and covers.

17. Fill the engine with the proper type and quantity of engine oil.

18. Start the vehicle and check for oil leaks.

Rear Main Seal

REMOVAL & INSTALLATION

1.8L engines

1. Remove the flywheel (MT) or flexplate (AT).

2. If necessary, unfasten the rear cover attaching bolts and remove the cover.

3. Using a prytool with a rag over the end (refer to the accompanying illustration) or Seal Removal tool such as T92C-6700-

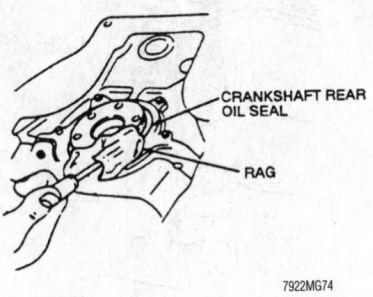

CRANKSHAFT REAR OIL SEAL

RAG

7922MG74

Use a suitable prytool and a rag (as shown) or . . .

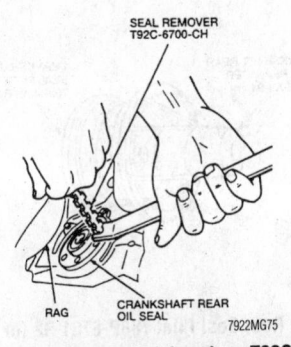

SEAL REMOVER T92C-6700-CH

RAG

CRANKSHAFT REAR OIL SEAL

7922MG75

. . . a seal removal tool such as T92C-6700-CH to remove the rear crankshaft oil seal—1.8L engine

CH or its equivalent, pry out the rear oil seal.

To install:

4. If removed install the rear cover and tighten its retaining bolts to 69–95 inch lbs. (7.8–11 Nm).

5. Coat the lip of the new seal with clean engine oil.

6. Use seal Replacer T87C-6701-A or its equivalent, and tighten the bolts to install the seal. Keep tightening the bolts until the seal is flush with the rear cover.

7. Install the flywheel or flexplate.

1.9L Engine

1. Remove the flywheel.

2. Remove the engine or transmission housing cover..

➡**Use care to avoid damaging the crankshaft surface.**

3. Using a seal removal tool such T92C-6700-CH or its equivalent, pry out the rear oil seal.

To install:

4. Inspect the oil seal area for damage. If damage is found, service or replace the crankshaft

5. Coat the lip of the new seal wit clean engine oil.

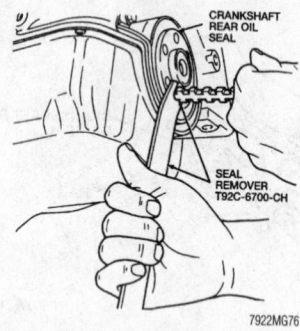

Use a seal removal tool such as T92C-6700-CH or its equivalent to pry out the rear oil seal—1.9L engine

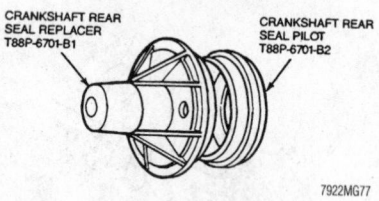

Use Rear Seal Pilot T88P-6701-B2 (or equivalent) and Crankshaft Rear Seal Replacer T88P-6701-B1 (or equivalent) when installing the seal—1.9L engine

➡ **Be sure the rear oil seal is properly aligned and that the edges are not rolled up.**

6. Place Crankshaft Rear Seal Pilot T88P-6701-B2 (or equivalent) into the Crankshaft Rear Seal Replacer T88P-6701-B1 (or equivalent) and lubricate the pilot replacer with clean engine oil.

7. Slide the rear oil seal over the pilot onto the replacer.

8. Remove the Rear Seal Pilot T88P-6701-B2 (or equivalent) from Seal Replacer T88P-6701-B1 (or equivalent).

9. Place the rear oil seal and replacer over the crankshaft and install the seal.

10. Remove the seal installer tool and install the flywheel.

2.0L Engine

1. Remove the transaxle and flywheel.

➡ **Be careful not to damage the crankshaft surface when removing the seal.**

2. Use seal replacer tool T92C-6700-Ch or its equivalent to remove the old seal.

To install:

Coat the lip of the new seal with clean 5W30 engine oil.

➡**Be sure the crankshaft rear oil seal is on correctly and that the edges are not rolled over.**

3. Install the new seal using crankshaft rear seal pilot tool T88P-6701-B2 and Rear seal replacer tool T88P-6701-B1 or their equivalents.

4. Install the flywheel and the transaxle.

FUEL SYSTEM

Fuel System Service Precautions

Safety is the most important factor when performing not only fuel system maintenance but any type of maintenance. Failure to conduct maintenance and repairs in a safe manner may result in serious personal injury or death. Maintenance and testing of the vehicle's fuel system components can be accomplished safely and effectively by adhering to the following rules and guidelines.

• To avoid the possibility of fire and personal injury, always disconnect the negative battery cable unless the repair or test procedure requires that battery voltage be applied.

• Always relieve the fuel system pressure prior to disconnecting any fuel system component (injector, fuel rail, pressure regulator, etc.), fitting or fuel line connection. Exercise extreme caution whenever relieving fuel system pressure to avoid exposing skin, face and eyes to fuel spray. Please be advised that fuel under pressure may penetrate the skin or any part of the body that it contacts.

• Always place a shop towel or cloth around the fitting or connection prior to loosening to absorb any excess fuel due to spillage. Ensure that all fuel spillage (should it occur) is quickly removed from engine surfaces. Ensure that all fuel soaked cloths or towels are deposited into a suitable waste container.

• Always keep a dry chemical (Class B) fire extinguisher near the work area.

• Do not allow fuel spray or fuel vapors to come into contact with a spark or open flame.

• Always use a back-up wrench when loosening and tightening fuel line connection fittings. This will prevent unnecessary stress and torsion to fuel line piping. Always follow the proper torque specifications.

• Always replace worn fuel fitting O-rings with new. Do not substitute fuel hose or equivalent, where fuel pipe is installed.

Fuel System Pressure

RELIEVING

❊❊ CAUTION

Observe all applicable safety precautions when working around fuel. Whenever servicing the fuel system, always work in a well ventilated area. Do not allow fuel spray or vapors to come in contact with a spark or open flame. Keep a dry chemical fire extinguisher near the work area. Always keep fuel in a container specifically designed for fuel storage; also, always properly seal fuel containers to avoid the possibility of fire or explosion.

1995–96 Models

1. Start the engine.
2. Remove the rear seat cushion.
3. Disconnect the fuel pump electrical connectors.
4. Wait for the engine to stall, then turn the ignition switch **OFF**.
5. Connect the fuel pump electrical connectors.
6. Install the rear seat cushion.
7. Disconnect the negative battery cable.

1997–99 Models

1. Remove the Schrader valve cap at the end of the fuel rail and a attach a fuel pressure gauge.
2. Open the manual relief valve on the fuel pressure gauge slowly to relive the fuel pressure

Fuel Filter

REMOVAL & INSTALLATION

1.8L and 1.9L Engines

The inline fuel filter is located in the engine compartment between the fuel tank and the fuel rail.

1. Properly relieve the fuel system pressure.
2. Disconnect the negative battery cable.

3. Position a suitable container below the fuel filter to collect any excess fuel that may leak from the filter and lines.

❊❊ CAUTION

Observe all applicable safety precautions when working around fuel. Whenever servicing the fuel system, always work in a well ventilated area. Do not allow fuel spray or vapors to come in contact with a spark or open flame. Keep a dry chemical fire extinguisher near the work area. Always keep fuel in a container specifically designed for fuel storage; also, always properly seal fuel containers to avoid the possibility of fire or explosion.

4. Remove the retaining clip from the fuel filter upper hose.

5. Disconnect the upper hose from the fuel filter and drain any excess fuel into the container. Plug the hose.

6. Loosen the fuel filter mounting clamp.

7. Raise and safely support the vehicle.

8. Remove the retaining clip from the fuel filter lower hose.

9. Disconnect the lower hose from the fuel filter and drain any excess fuel into the container. Plug the hose.

10. Lower the vehicle.

11. Remove the fuel filter.

To install:

12. Position the fuel filter and tighten the filter mounting clamp.

13. Remove the plug, then connect the upper hose to the filter and install the hose retaining clip.

14. Raise and safely support the vehicle.

15. Remove the plug, connect the lower hose to the filter and install the hose retaining clip.

16. Lower the vehicle.

17. Connect the negative battery cable.

18. Start the engine and check for leaks.

2.0L Engines

The inline fuel filter is located in the engine compartment between the fuel tank and the fuel rail.

1. Properly relieve the fuel system pressure.

2. Disconnect the negative battery cable.

3. Position a suitable container below the fuel filter to collect any excess fuel that may leak from the filter and lines.

❊❊ CAUTION

Observe all applicable safety precautions when working around fuel. Whenever servicing the fuel system, always work in a well ventilated area. Do not allow fuel spray or vapors to come in contact with a spark or open flame. Keep a dry chemical fire extinguisher near the work area. Always keep fuel in a container specifically designed for fuel storage; also, always properly seal fuel containers to avoid the possibility of fire or explosion.

4. Loosen the fuel filter mounting clamp.

5. Remove the filter from the mounting bracket.

6. Remove the retaining clips from the upper fuel filter hose.

7. Disconnect the upper hose from the fuel filter and drain any excess fuel into the container. Plug the hose.

8. Remove the retaining clip from the fuel filter lower hose.

9. Disconnect the lower hose from the fuel filter and drain any excess fuel into the container. Plug the hose.

10. Remove the fuel filter.

To install:

11. Remove the plug from the lower hose and connect the hose to the filter, then install the hose retaining clip.

12. Remove the plug from the upper hose and connect the hose to the filter, then install the hose retaining clip.

13. Position the fuel filter and tighten the filter mounting clamp.

14. Connect the negative battery cable.

15. Start the engine and check for leaks.

➡ On newer model vehicles when the battery is disconnected it may cause some abnormal drive symptoms until the Powertrain Control Module (PCM) relearns its adaptive strategy. The vehicle may need to be driven 10 miles (16 km) or more for the PCM to relearn its strategy.

Fuel Pump

The fuel pump is located inside the fuel tank.

REMOVAL & INSTALLATION

1995–96 Engines

❊❊ WARNING

Extreme caution should be taken when removing the fuel tank from the vehicle! Ensure that all removal procedures are conducted in a well ventilated area! Have a sufficient amount of absorbent material in the vicinity of the work area to quickly contain any fuel spillage. Never store waste fuel in an open container, as it presents a serious fire hazard!

This procedure will require a new fuel pump gasket for pump installation, so be sure to have one before starting.

1. Relieve the fuel system pressure.

2. Disconnect the negative battery cable.

3. Remove the rear seat cushion and unplug the electrical connector from the fuel pump.

4. Unfasten and remove the four pump cover screws and the ground strap.

5. Unfasten the fuel tube clips and disconnect the tubes from the pump.

6. Use a fuel pump locking ring removal tool or a brass drift and a hammer to unfasten the pump locking retaining ring.

7. Remove the fuel pump and the gasket from the tank.

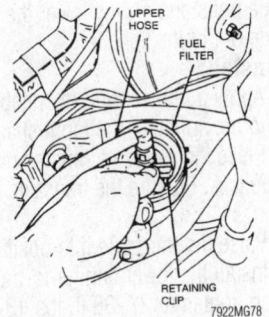

To detach the upper fuel hose, remove the fuel filter hose retaining clip—1995–96 models

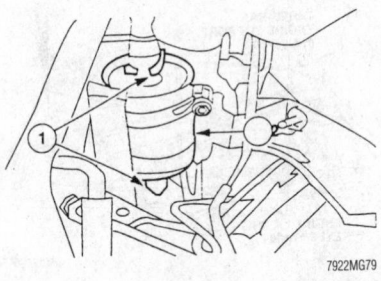

Detach the upper and lower hose clips before disconnecting hoses (1) from the fuel filter

To install:

8. Install a new gasket and place the pump into position in the tank.

9. Install the pump locking retaining ring.

10. Connect the fuel tubes to the pump and install the tube retaining clips.

11. Install the ground strap and fuel pump cover, then tighten the retainers.

12. Attach the electrical connector to the pump.

13. Install the rear seat cushion.

14. Connect the negative battery cable.

15. Start the vehicle and check for proper operation.

1997–99 Engines

✳✳ WARNING

Extreme caution should be taken when removing the fuel tank from the vehicle! Ensure that all removal procedures are conducted in a well ventilated area! Have a sufficient amount of absorbent material in the vicinity of the work area to quickly contain any fuel spillage. Never store waste fuel in an open container, as it presents a serious fire hazard!

This procedure will require a new fuel pump gasket for pump installation, so be sure to have one before starting.

1. Relieve the fuel system pressure.

2. Disconnect the negative battery cable.

3. Remove the rear seat cushion and unplug the electrical connectors from the fuel pump.

4. Unfasten the four pump access cover screws and remove the cover.

5. If equipped, unplug the pump electrical connector located under the cover.

6. Unfasten the fuel line clip(s) and disconnect the line(s) from the pump.

7. Use a fuel pump locking ring removal tool or a brass drift and a hammer to unfasten the pump locking retaining ring.

8. Remove the fuel pump and the gasket from the tank.

To install:

9. Install a new gasket and place the pump into position in the tank.

10. Install the pump locking retaining ring.

11. Connect the fuel line(s) to the pump and install the line retaining clip(s).

12. If equipped, attach the fuel pump electrical connection(s).

13. Install access cover, then tighten the retainers.

14. Attach the pump electrical connector(s).

15. Install the rear seat cushion.

16. Connect the negative battery cable.

17. Start the vehicle and check for proper operation.

DRIVE TRAIN

Transmission Assembly

REMOVAL & INSTALLATION

Manual

1.8L AND 1.9L ENGINES

1. Disconnect both battery cables, negative cable first.

2. Remove the battery and battery tray.

3. Remove the engine air cleaner tube and the air intake resonator.

4. Disconnect the speedometer cable at the transaxle.

5. Remove the slave cylinder line-to-slave cylinder hose retaining clip, then disconnect the slave cylinder line from the slave cylinder hose and plug the hose.

6. Disconnect the ground strap from the transaxle.

7. Remove the tie wrap and unplug the three electrical connectors located above the transaxle.

8. Remove the electrical connector support bracket.

9. Install Engine Support Bar 014–00750 or equivalent, and attach it to the engine lifting eyes with suitable chains or cables.

10. If equipped, remove the three nuts from the left-hand engine support bracket.

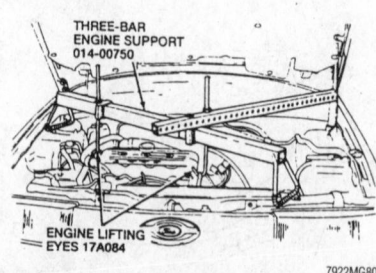

Support the weight of the engine as shown, because the engine will be unstable when the transaxle is removed—1.8L and 1.9L engines

11. Loosen the mount pivot nut and rotate the mount out of the way.

12. Remove the three bolts and the left-hand engine support mount.

13. Remove the two upper transaxle-to-engine bolts.

14. Raise and safely support the vehicle.

15. Remove the front wheel and tire assemblies.

16. Remove the inner fender splash shields.

17. Drain the transaxle fluid and install the drain plug.

18. Remove the halfshafts.

19. Install Transaxle Plugs T88C-7025-AH or equivalent, between the differential side gears.

✳✳ WARNING

Failure to install the transaxle plugs may cause the differential side gears to become improperly positioned. If the gears become misaligned, the differential will have to be removed from the transaxle to align them.

20. If equipped with the 1.8L engine, remove the intake manifold support bolts and the support.

21. Remove the starter motor.

22. Unfasten the gearshift stabilizer bar nut and washers. Remove the stabilizer bar and bracket from the transaxle.

23. Unfasten the shift control rod-to-transaxle bolt and nut and remove the shift control rod from the transaxle.

24. Remove both lower splash shields.

25. Unfasten the transaxle mount-to-crossmember bolts and nuts and remove the lower crossmember (rear engine support).

26. Position and secure a suitable transaxle jack under the transaxle.

27. Remove the front transaxle mount and bracket.

28. Remove the lower engine-to-transaxle bolts and slowly lower the transaxle out of the vehicle.

To install:

29. Apply a thin coating of suitable grease to the spline of the input shaft.

30. Place the transaxle onto a suitable transaxle jack. Be sure the transaxle is secure.

31. Raise the transaxle into position.

32. Install the lower engine-to-transaxle bolts and tighten to 27–38 ft. lbs. (37–52 Nm).

33. Install the front transaxle mount and bracket. Tighten the bolts to 12–17 ft. lbs. (16–23 Nm).

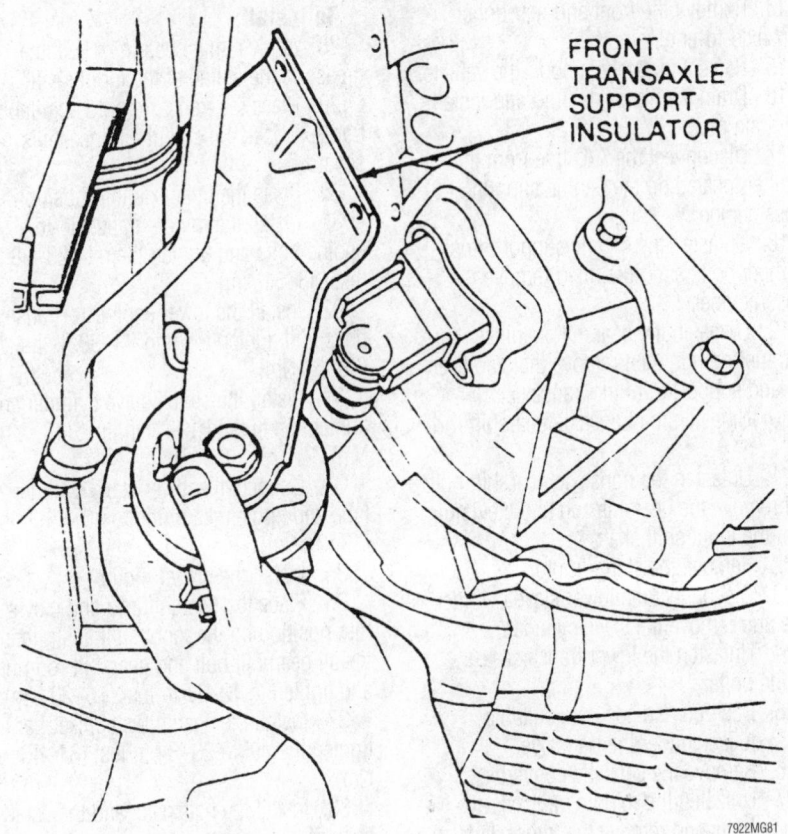

FRONT
TRANSAXLE
SUPPORT
INSULATOR

7922MG81

Removing the front transaxle mount and bracket—1.8L and 1.9L engines

34. Remove the transaxle jack.

35. Install the lower crossmember. Tighten the support bolts to 47–66 ft. lbs. (64–89 Nm) and the transaxle support insulator-to-rear engine support nuts to 27–38 ft. lbs. (37–52 Nm).

36. Install both lower splash shields.

37. Install the shift control rod bolt and nut and tighten to 12–17 ft. lbs. (16–23 Nm).

38. Install the stabilizer bracket and stabilizer bar with the nut and washers and tighten to 23–34 ft. lbs. (31–46 Nm).

39. Install the starter motor.

40. If equipped with the 1.8L engine, install the intake manifold support and tighten the retaining bolts to 27–38 ft. lbs. (37–52 Nm).

41. Remove the transaxle plugs and install the halfshafts.

42. Install the inner fender splash shields.

43. Install the wheel and tire assemblies.

44. Lower the vehicle.

45. Install the upper engine-to-transaxle bolts and tighten to 47–66 ft. lbs. (64–89 Nm).

46. Install the left-hand engine support mount and tighten 3 bolts to 32–45 ft. lbs. (43–61 Nm) for 1.8L engine or 41–59 ft. lbs. (55–80 Nm) for 1.9L engine.

47. Rotate the left-hand engine support bracket into position and tighten the pivot nut.

48. Install and tighten the left-hand engine support bracket nuts and tighten to 47–66 ft. lbs. (64–89 Nm).

49. Remove the engine support bar.

50. Install the electrical connector support bracket. Attach the electrical connectors and secure with a new tie wrap.

51. Connect the ground strap to the transaxle.

52. Connect the slave cylinder line to the slave cylinder hose and install the retaining clip.

53. Add the proper type and amount of fluid to the transaxle.

54. Connect the speedometer cable.

55. Install the engine air cleaner tube and the air intake resonator.

56. Install the battery tray and the battery.

57. Connect the battery cables, negative cable last.

58. Check for fluid leaks and proper clutch operation.

59. Road test the vehicle and check for proper transaxle operation.

2.0L ENGINES

1. Disconnect both battery cables, negative cable first.

2. Remove the battery and battery tray.

3. Remove the engine air cleaner outlet tube.

4. If necessary, unplug the Constant Control Relay Module (CCRM) electrical connection, loosen its retainers and remove the module and bracket as an assembly.

5. Disconnect the slave cylinder line from the slave cylinder hose and plug the hose.

6. Unfasten the slave cylinder line retaining clip and remove the line from the bracket.

7. Unplug the heated oxygen sensor electrical connection.

8. Unplug the back-up light electrical connection.

9. Unfasten the electrical connector bracket bolt and remove the bracket.

10. Unplug the Vehicle Speed Sensor (VSS) electrical connection.

11. Remove the halfshafts.

12. Install Engine Support Bar D88–6000A or equivalent, and attach it to the engine lifting eyes with suitable chains or cables.

13. Remove the insulator nuts from the left-hand engine support bracket.

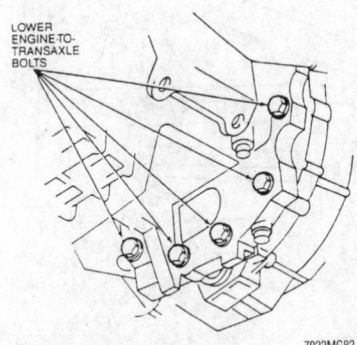

LOWER
ENGINE-TO-
TRANSAXLE
BOLTS

7922MG82

Location of the lower engine-to-transaxle bolts—1.8L and 1.9L engines

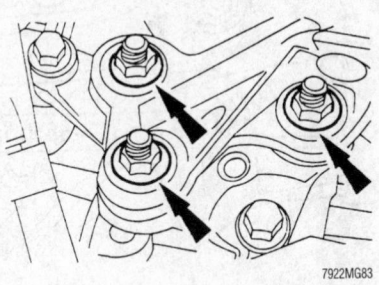

7922MG83

Location of the left-hand engine support bracket nuts—2.0L engines

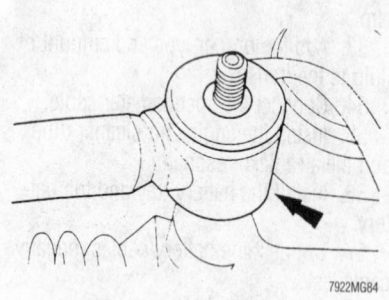

Remove the gearshift stabilizer bar and support from the transaxle—2.0L engines

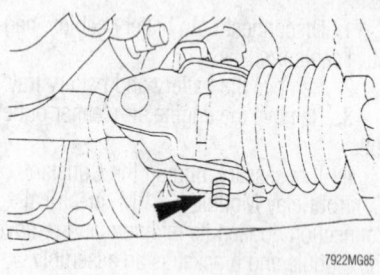

Loosen the transaxle gearshift bolt and remove the gearshift rod and clevis pin from the input shift shaft—2.0L engines

14. Remove the front and rear upper transaxle-to-engine bolts.

15. Raise and safely support the vehicle.

16. Drain the transaxle fluid and install the drain plug.

17. Disconnect the A/C line from the retainer located on the engine support crossmember.

18. Unfasten the engine support crossmember bolts and nuts, then remove the crossmember.

19. Unfasten the gearshift stabilizer bar-to-transaxle nut, then remove the stabilizer bar and support from the transaxle.

20. Unfasten the transaxle gearshift rod nut.

21. Unfasten the transaxle gearshift bolt and remove the gearshift rod and clevis pin from the input shift shaft.

22. Remove the starter motor.

23. Disconnect the lower slave cylinder tube and remove the slave cylinder.

24. Unfasten the lower transaxle-to-engine bolts.

25. Position and secure a suitable transaxle jack under the transaxle.

26. Remove the catalytic converter.

27. Unfasten the middle transaxle-to-engine bolts and remove the transaxle from the vehicle.

To install:

28. Apply a thin coating of suitable grease to the spline of the input shaft.

29. Place the transaxle onto a suitable transaxle jack. Be sure the transaxle is secure.

30. Raise the transaxle into position.

31. Install the middle transaxle-to-engine bolts and tighten them to 23–38 ft. lbs. (38–51 Nm).

32. Install the lower engine-to-transaxle bolts and tighten them to 23–38 ft. lbs. (38–51 Nm).

33. Install the clutch slave cylinder and tighten the nuts to 12–17 ft. lbs. (16–23 Nm).

34. Connect the lower slave cylinder tube and tighten the fitting to 10–16 ft. lbs. (13–21 Nm).

35. Install the starter motor.

36. Place the gearshift rod and clevis into position on the input shift shaft, then install gearshift bolt and gearshift rod nut and tighten to 12–17 ft. lbs. (16–23 Nm).

37. Install the gearshift stabilizer bar and tighten the nut to 23–34 ft. lbs. (31–46 Nm).

38. Install the halfshafts and the catalytic converter.

39. Install the engine support cross-

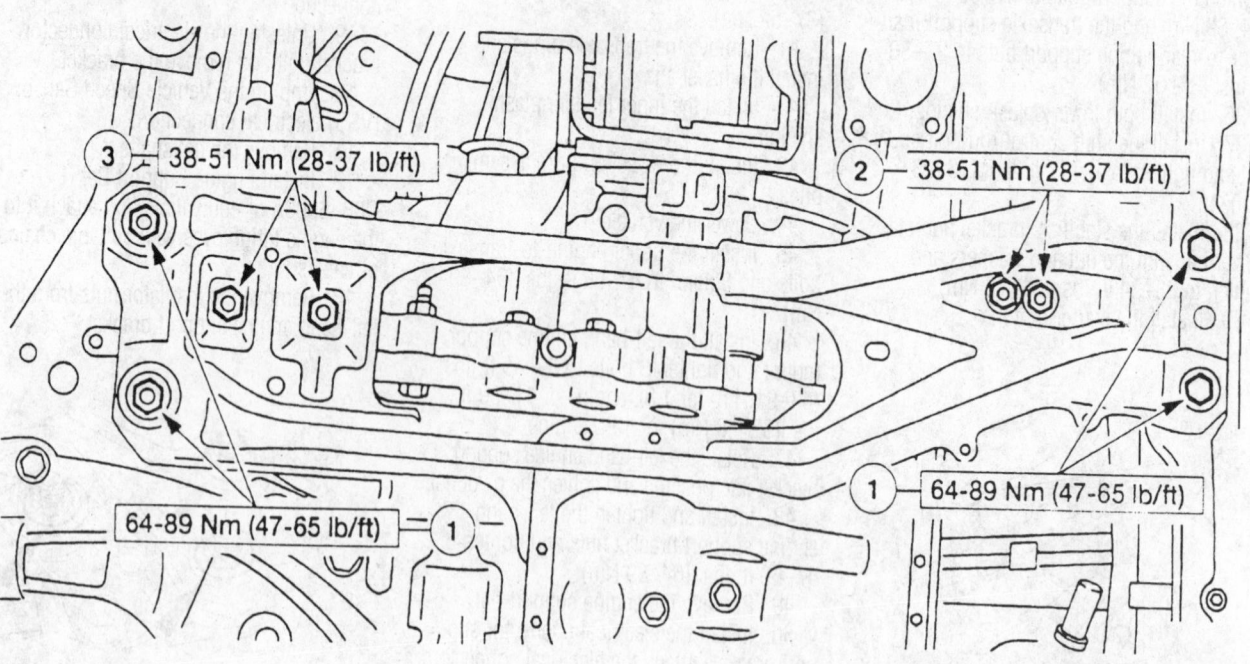

Install the engine support crossmember and tighten the retainers to specification—2.0L engines

member. Tighten the crossmember bolts and nuts to 47–65 ft. Lbs. (64–89 Nm) and the insulator nuts to 28–37 ft. Lbs. (38–51 Nm).

40. Install the front and rear upper transaxle-to-engine bolt and tighten it to 28–38 ft. Lbs. (38–51 Nm).

41. Install the left-hand engine support insulator nuts and tighten them to 50–68 ft. lbs. (67–93 Nm).

42. Remove the engine support bar.

43. Install the electrical connector support bracket.

44. Attach all the electrical connectors.

45. Connect the slave cylinder line to the slave cylinder hose and install the retaining clip. Tighten the fitting to 10–16 ft. lbs. (13–21 Nm).

46. Bleed the clutch hydraulic system.

47. Add the proper type and amount of fluid to the transaxle.

48. If removed, install the CCRM and bracket assembly. Tighten the retainers and attach the electrical connection.

49. Install the engine air cleaner tube.

50. Install the battery tray and the battery.

51. Connect the battery cables, negative cable last.

52. Check for fluid leaks and proper clutch operation.

53. Road test the vehicle and check for proper transaxle operation.

Automatic

1.8L AND 1.9L ENGINES

1. Disconnect both battery cables, negative cable first.

2. Remove the battery and battery tray.

3. Disconnect the wiring harness retaining clip from the battery tray.

4. Remove the engine air cleaner assembly.

5. Disconnect the shift control cable from the manual shift lever on the transaxle.

6. Disconnect the speedometer cable from the transaxle by unsnapping the cable at the Vehicle Speed Sensor (VSS).

7. Unplug the transaxle electronic control electrical connectors and separate the harness from the transaxle clips.

8. Remove the Manual Lever Position (MLP) switch wiring brackets and disconnect the ground cables from the top of the transaxle.

9. Remove the starter motor.

10. Unplug the MLP switch wiring connectors.

11. Install Engine Support D88L-6000-A or equivalent, to the engine. The engine must be properly supported for transaxle removal.

12. Disconnect the kickdown cable (also known as the throttle valve cable) at the throttle cam on the throttle body.

13. Place a suitable drain pan under the transaxle and disconnect the transaxle cooler lines at the transaxle.

➡In some later models, the upper transaxle cooler line has a spring and a check ball under the fitting.

14. Remove the left-hand engine mount bolts and the mount.

15. Remove the upper transaxle housing bolts.

16. Unplug the Oxygen Sensor (O2S) and/or Heated Oxygen Sensor (HO2S) electrical connector, the transaxle vent hose and the electrical connector at the VSS.

17. Raise and safely support the vehicle.

18. Separate the halfshafts from both steering knuckles.

19. Remove the three engine/transaxle

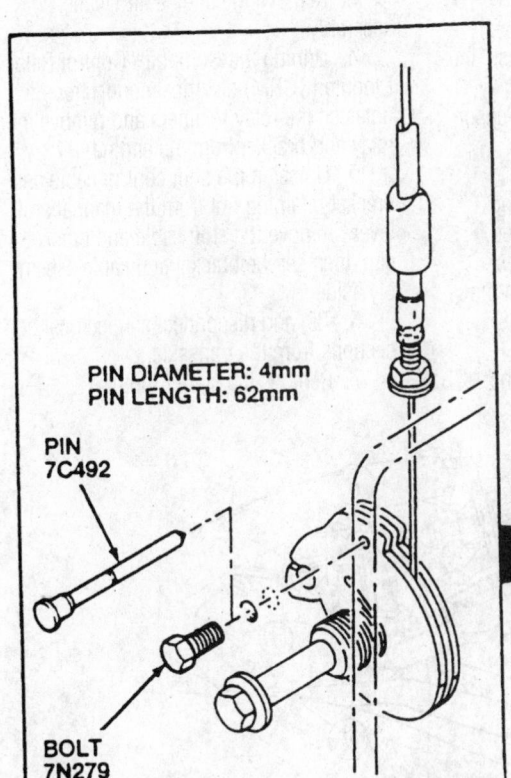

PIN DIAMETER: 4mm
PIN LENGTH: 62mm

PIN
7C492

BOLT
7N279

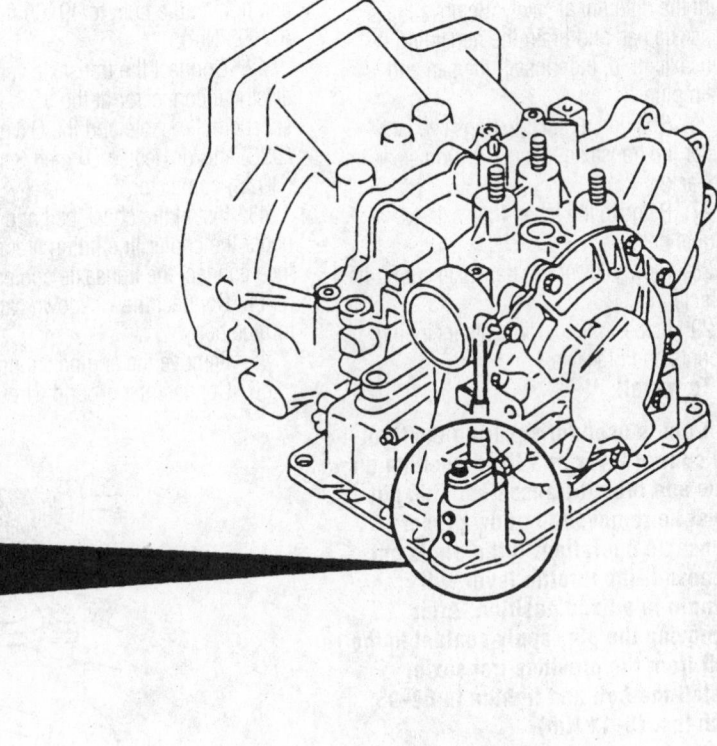

The pin used for securing the throttle control lever in a fixed position on new and rebuilt transaxles must be removed—1995–96 models

7922MG87

lower splash shield bolts and the torque converter inspection plate.

20. Remove the four nuts securing the torque converter to the flexplate.

21. Remove the bolts securing the lower transaxle to the engine oil pan.

22. Unfasten the rear engine support-to-vehicle chassis bolt and remove the transaxle support crossmember from the vehicle chassis.

23. Unfasten the rear engine support-to-transaxle mount nuts and remove the transaxle support crossmember from the transaxle mounts.

24. Remove both halfshafts from the transaxle. Install two transaxle plugs T88C-7025-AH or equivalent, into the differential side gears.

❊❊ WARNING

Failure to install the transaxle plugs may cause the differential side gears to become improperly positioned. If the gears become misaligned, the differential will have to be removed from the transaxle to align them.

25. Position a drain pan and remove the drain plug from the transaxle. Drain the fluid from the differential cavity. Remove the transaxle pan and drain the remaining transaxle fluid, then install the pan and drain plug.

26. Position a suitable transaxle jack under the transaxle. Secure the transaxle to the jack.

27. Remove the lower transaxle-to-engine bolts.

28. Slowly lower the transaxle out of the vehicle.

29. Inspect all components including mounts and brackets.

To install:

➡**A pin is used for securing the throttle control lever in a fixed position on new and rebuilt transaxles. This pin must be removed to allow proper transaxle operation. If the pin is not removed, the throttle lever will remain in a fixed position. After removing the pin, apply sealant to the bolt from the previous transaxle. Install the bolt and tighten to 69–95 inch lbs. (8–11 Nm).**

30. If rebuilding or replacing the transaxle, be sure to completely flush the transaxle cooler and cooler lines before installing the transaxle.

31. Remove the pin securing the throttle control lever, if installed.

32. Secure the transaxle on the transaxle jack.

33. Carefully raise the transaxle into position and install the lower transaxle-to-engine bolts. Tighten the bolts to 41–59 ft. lbs. (55–80 Nm).

34. Position the torque converter to the flexplate and install the retaining nuts. Tighten the nuts to 25–36 ft. lbs. (34–49 Nm). Install the torque converter inspection plate.

35. Remove the transaxle plugs and install the halfshafts.

36. Connect the crossmember to the transaxle mounts and the chassis. Tighten the crossmember-to-transaxle mount nuts to 27–38 ft. lbs. (37–52 Nm). Tighten the crossmember-to-chassis nuts and bolts to 47–66 ft. lbs. (64–89 Nm).

37. Install the lower transaxle-to-engine oil pan bolts and tighten to 27–38 ft. lbs. (37–52 Nm). Install the engine/transaxle splash shields.

38. Install the starter motor.

39. Lower the vehicle.

40. Install the upper transaxle-to-engine bolts and tighten to 41–59 ft. lbs. (55–80 Nm).

41. Install the left-hand engine mount and tighten the nuts to 49–69 ft. lbs. (67–93 Nm).

42. Connect the transaxle vent hose, the electrical connector at the VSS, the speedometer cable and the Oxygen Sensor (O2S) and/or Heated Oxygen Sensor (HO2S) connector.

43. Install the check ball and spring under the cooler line fitting if equipped, then connect the transaxle cooler lines.

44. Connect the kickdown cable at the throttle body.

45. Remove the engine support.

46. Connect the ground wires to the transaxle and connect the MLP sensor bracket and wiring connectors.

47. Connect the shift control cable to the manual shift lever on the transaxle. Tighten the selector lever attaching locknut to 33–47 ft. lbs. (44–64 Nm).

➡**Do not use any type of power wrench to tighten the locknut. Damage to the transaxle may result.**

48. Install the battery tray and battery. Connect the wiring harness retaining clip to the battery tray.

49. Install the engine air cleaner assembly.

50. Connect both battery cables, negative cable last.

51. Add the proper type and quantity of transaxle fluid.

52. Check the transaxle for leaks and for proper operation.

53. Check the manual lever position sensor for proper adjustment.

54. Road test the vehicle and check for proper operation.

2.0L ENGINES

1. Disconnect both battery cables, negative cable first.

2. Remove the battery and battery tray.

3. Remove the engine air cleaner assembly.

4. Unplug the Computer Control Relay Module (CCRM) electrical connections, unfasten the relay retainers and remove the relay and bracket from the engine.

5. Unfasten the shift control cable and bracket retaining nut from the manual shift lever , remove the shift cable and bracket clip, then set the bracket and cable assembly aside.

6. Tag and disconnect all electrical connections from the transaxle.

7. Remove the starter motor.

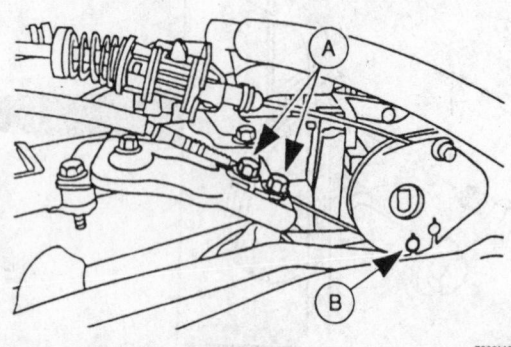

Connect the throttle valve cable at the throttle cam (B) and tighten the cable retaining bolts (A)—1997–99 models

7922MG88

8. If equipped, unfasten the throttle valve actuating cable bolts and disconnect the cable from the throttle cam.

9. Install Engine Support D88L-6000-A or equivalent, to the engine. The engine must be properly supported for transaxle removal.

10. Place a suitable drain pan under the transaxle and disconnect the transaxle cooler lines at the transaxle.

11. Remove the left-hand engine mount bolts and the mount.

12. Remove the upper transaxle housing bolts.

13. Raise and safely support the vehicle.

14. If equipped, remove the left-hand splash shields.

15. Remove the transaxle plug and drain the fluid.

16. Remove the halfshafts. Install two transaxle plugs T88C-7025-AH or equivalent, into the differential side gears.

✳✳ WARNING

Failure to install the transaxle plugs may cause the differential side gears to become improperly positioned. If the gears become misaligned, the differential will have to be removed from the transaxle to align them.

17. Remove the engine support crossmember and the catalytic converter.

18. Disconnect the A/C line from the retainer located on the engine support crossmember.

19. Unfasten the transaxle support crossmember bolts and nuts, the left-hand engine isolator nuts,, then remove the crossmember.

20. Unfasten the transaxle housing cover bolts and remove the cover.

21. Remove the four flywheel-to-torque converter nuts.

22. Position a suitable transaxle jack under the transaxle. Secure the transaxle to the jack.

23. Remove the remaining transaxle-to-engine bolts.

➡ **The upper front transaxle-to-engine bolt on 1997 models is equipped with a retainer and will remain in the engine block.**

24. On 1997 models, unfasten the upper front transaxle-to-engine bolt.

25. Slowly lower the transaxle out of the vehicle.

26. Inspect all components including mounts and brackets.

To install:

➡ **Prior to installing the transaxle, lubricate the torque converter pilot hub with multi-purpose grease.**

27. Align the torque converter studs to the flywheel.

28. Secure the transaxle on the transaxle jack.

29. Carefully raise the transaxle into position and install the middle and lower transaxle-to-engine bolts. Tighten the bolts to 40–58 ft. lbs. (55–80 Nm).

30. On 1997 models, install the upper front transaxle-to-engine bolt. Tighten the bolt to 40–58 ft. lbs. (55–80 Nm).

31. Install the flywheel-to-converter nuts and tighten them to 26–36 ft. Lbs. (35–49 Nm).

32. Install the transaxle housing cover and tighten the bolts to40–58 ft. Lbs. (55–80 Nm).

33. Install the engine support crossmember. Tighten the crossmember bolts and nuts to 47–65 ft. Lbs. (64–89 Nm) and the insulator nuts to 28–37 ft. Lbs. (38–51 Nm).

34. Install the front and rear upper transaxle-to-engine bolt and tighten it to 28–38 ft. Lbs. (38–51 Nm).

35. Remove the transaxle plugs and install the halfshafts.

36. Connect the crossmember to the transaxle mounts and the chassis. Tighten the crossmember-to-transaxle mount nuts to 27–38 ft. lbs. (37–52 Nm). Tighten the crossmember-to-chassis nuts and bolts to 47–66 ft. lbs. (64–89 Nm).

37. Install the lower transaxle-to-engine oil pan bolts and tighten to 27–38 ft. lbs. (37–52 Nm).

38. Install the engine/transaxle splash shields.

39. Install the starter motor.

40. Attach all the electrical connections.

41. Lower the vehicle.

42. Install the upper transaxle-to-engine bolts and tighten to 40–58 ft. lbs. (55–80 Nm).

43. Install the left-hand engine mount and tighten the nuts to 50–68 ft. lbs. (67–93 Nm).

44. If equipped, connect the throttle valve cable at the throttle cam. Adjust the cable as outlined in this section.

45. Remove the engine support.

46. Attach the shift cable and bracket to the manual shift lever, install the shift cable and bracket clip. Tighten the nut to 12–16 ft. lbs. (16–22 Nm).

47. Install the engine air cleaner assembly. Adjust the cable as outlined in this section.

48. Install the battery tray and battery.

49. Connect both battery cables, negative cable last.

50. Add the proper type and quantity of transaxle fluid.

51. Check the transaxle for leaks and for proper operation.

52. Check the manual lever position sensor for proper adjustment.

53. Road test the vehicle and check for proper operation.

Clutch

ADJUSTMENT

Pedal Free-Play

1995–96 MODELS

To determine if the clutch pedal free-play needs adjusting, depress the clutch pedal by hand until resistance is felt. Using a ruler, measure the distance between the upper clutch pedal height and where the resistance is felt. Free-play should be 0.20–0.51 in. (5–13mm). If not, proceed as follows:

1. Loosen the clutch pedal to clutch master cylinder rod locknut.

2. Turn the clutch pedal to clutch master cylinder rod until the free-play is within specification.

3. Check that the disengagement height is within specification. Minimum disengagement height is 1.6 in. (41mm).

4. Tighten the clutch pedal to clutch master cylinder rod locknut to 9–12 ft. lbs. (12–17 Nm).

5. Check clutch pedal free-play to verify proper adjustment.

1997–99 MODELS

1. Depress the clutch pedal until resistance can be felt, and measure the distance between the upper clutch height and where the resistance is felt. The free-play should be 0.0–0.40 in. (5.0–13.9mm).

2. If an adjustment is necessary, turn the locknut and equalizer bar-to-clutch release lever rod.

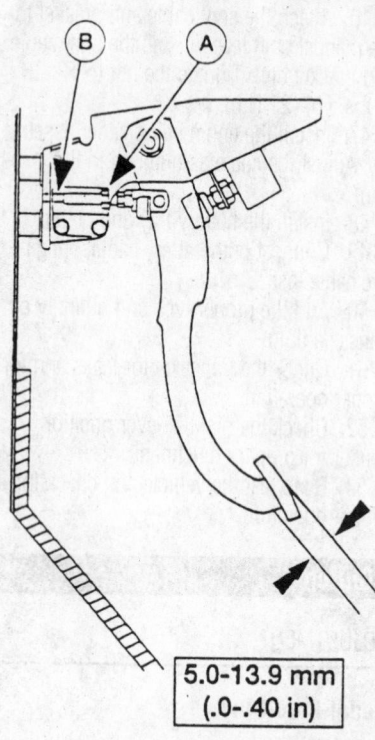

5.0-13.9 mm
(.0-.40 in)

7922MG89

Adjust the pedal free-play by turning the locknut (A) and equalizer bar-to-clutch release lever rod (B)—1997-99 models

3. After adjustment, measure the disengagement height from the upper surface of the clutch pedal pad to the carpet. The distance should be 2.3 in. (59mm).

Pedal Height

1995-96 MODELS

To determine if the pedal height requires adjustment, measure the distance from the bulkhead to the upper center of the pedal pad. The distance should be 7.72-8.03 in. (196-204mm). If adjustment is necessary, proceed as follows:

1. Disconnect the negative battery cable.
2. Unplug the clutch switch electrical connector.
3. Loosen the clutch switch locknut.
4. Turn the clutch switch until the correct height is achieved.
5. Tighten the locknut to 10-13 ft. lbs. (14-18 Nm).
6. Measure the pedal free-play.
7. Attach the electrical connector.
8. Connect the negative battery cable.

1997-99 MODELS

1. Measure the distance from the upper surface of the pedal pad to the carpet. The measurement should be 8.35-8.54 in. (212-217mm).

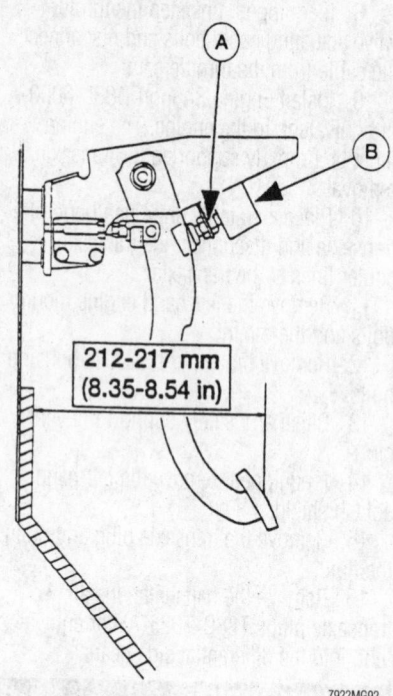

212-217 mm
(8.35-8.54 in)

7922MG92

To adjust pedal height, turn the locknut (A) and the Clutch Pedal Position (CPP) switch (B)—1997-99 models

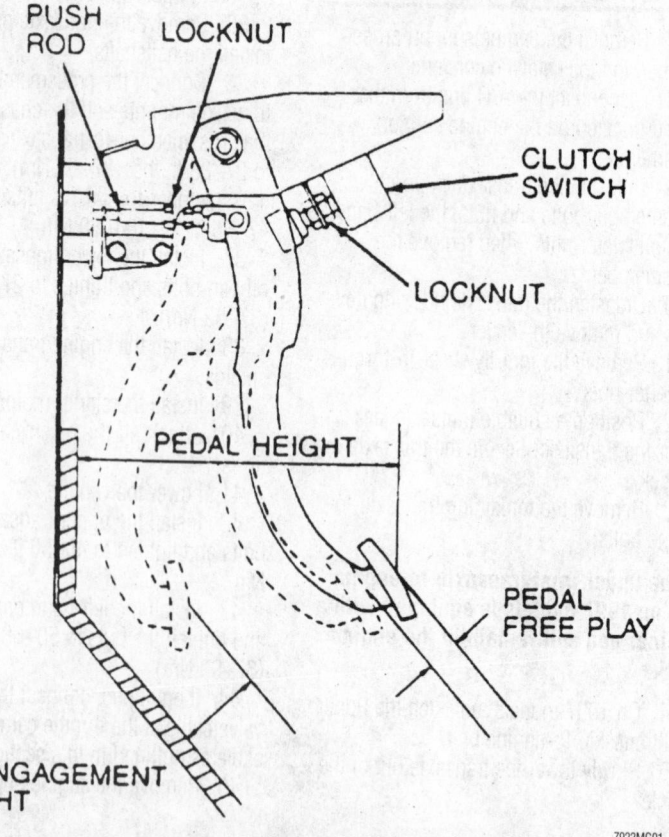

59 mm (2.3 in)

7922MG90

The disengagement height measurement from the upper surface of the clutch pedal pad to the carpet should be 2.3 in. (59mm)—1997-99 models

PUSH ROD
LOCKNUT
CLUTCH SWITCH
LOCKNUT
PEDAL HEIGHT
PEDAL FREE PLAY
DISENGAGEMENT HEIGHT

7922MG91

Clutch pedal height measurement and adjustment points—1995—96 models

2. If an adjustment is necessary, turn the locknut and the Clutch Pedal Position (CPP) switch until the pedal height is correct.

REMOVAL & INSTALLATION

1995–96 Models

1. Disconnect the negative battery cable.

2. Raise and safely support the vehicle.

3. Remove the transaxle assembly.

4. If the clutch assembly is to be reused, matchmark the pressure plate and the flywheel so they can be assembled in the same position.

5. Install flywheel holding tool T84P-6375-a or equivalent, in a transaxle mounting hole on the engine and engage the tooth of the holding tool into the flywheel ring gear.

6. Loosen the pressure plate-to-flywheel retaining bolts one turn at a time, in a crisscross pattern, until the spring tension is relieved, to prevent pressure plate cover distortion.

7. Support the pressure plate and unfasten the retaining bolts. Remove the pressure plate and clutch disc from the flywheel.

➡ If the flywheel shows any signs of overheating (blue discoloration) or if it is badly grooved or scored, it should be refaced or replaced.

8. Inspect the flywheel, clutch disc, pressure plate, release bearing, pilot bearing and the clutch fork for wear. Replace parts as needed.

To install:

9. If removed, install a new pilot bearing using a suitable installation tool.

10. If removed, install the flywheel. Be sure the flywheel and crankshaft flange mating surfaces are clean. Tighten the flywheel retaining bolts to 71–76 ft. lbs. (96–103 Nm) on the 1.8L engine or 54–67 ft. lbs. (73–91 Nm) on the 1.9L engine.

11. Clean the pressure plate and flywheel surfaces thoroughly. Position the clutch disc and pressure plate into the installed position and support them with a dummy shaft or clutch aligning tool. If the clutch assembly is being reused, align the matchmarks that were made during the removal procedure.

12. Install the pressure plate-to-flywheel retaining bolts. Tighten the bolts in the correct sequence to 13–20 ft. lbs. (18–26 Nm). Remove the alignment tool.

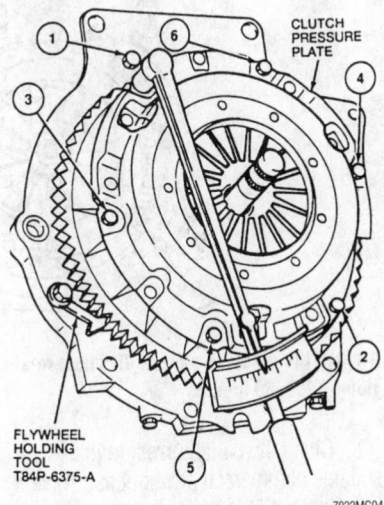

Tighten the pressure plate-to-flywheel retaining bolts in the sequence illustrated to specification—1995–96 models

13. If the release bearing was removed, lubricate the release fork where it contacts the bearing and install the bearing in the fork.

14. Install the transaxle assembly.

15. Lower the vehicle.

16. Bleed the hydraulic clutch system, if needed.

17. Connect the negative battery cable.

18. Road test the vehicle and check the clutch for proper operation.

1997–99 Models

1. Disconnect the negative battery cable.

2. Raise and safely support the vehicle.

3. Remove the transaxle assembly.

4. If the clutch assembly is to be reused, matchmark the pressure plate and the flywheel so they can be assembled in the same position.

5. Install flywheel holding tool T74P-7337-K or equivalent, in a transaxle mounting hole on the engine and engage the tooth of the holding tool into the flywheel ring gear.

6. Install clutch alignment tool T74P-7137-k or equivalent.

7. Loosen the pressure plate-to-flywheel retaining bolts one turn at a time, in a crisscross pattern, until the spring tension is relieved, to prevent pressure plate damage.

✽✽ CAUTION

Do not use any cleaners with a petroleum base and do not immerse the clutch pressure plate in solvent.

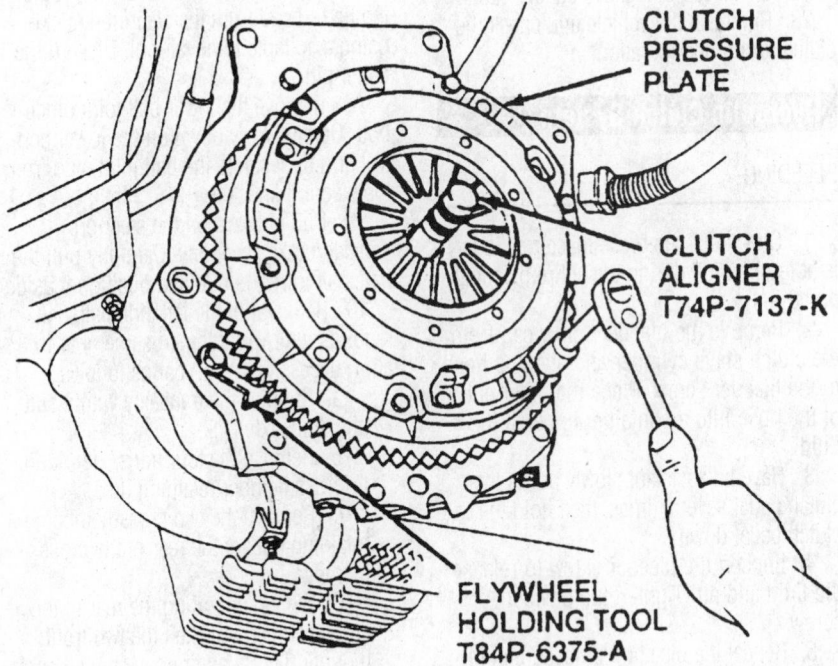

CLUTCH PRESSURE PLATE

CLUTCH ALIGNER T74P-7137-K

FLYWHEEL HOLDING TOOL T84P-6375-A

7922MG93

To hold the crankshaft from turning, install a flywheel holding tool in a transaxle mounting hole on the engine and engage the tooth of the tool into the flywheel ring gear—1995–96 models

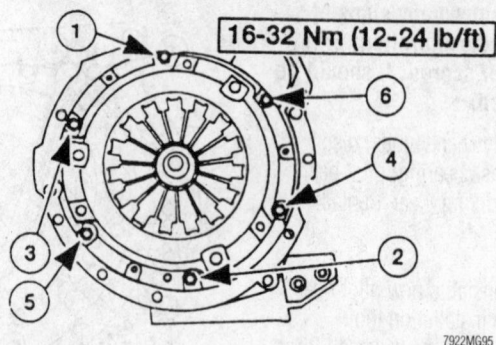

16-32 Nm (12-24 lb/ft)

Tighten the pressure plate-to-flywheel retaining bolts in the sequence illustrated to specification—1997-99 models

8. Clean the clutch pressure plate with a suitable commercial alcohol base solvent.

9. Inspect the pressure plate surface for burns, scores, flatness or ridges. Reface or replace the pressure plate.

10. Inspect the pressure plate diaphragm fingers for wear. Replace the pressure plate if necessary.

11. Using a slide caliper, measure the depth of the rivet heads. If the rivet head is within 0.012 in. (0.3mm) from the clutch surface, replace the clutch.

✳✳ CAUTION

If the clutch disc is saturated with oil, inspect the rear engine crankshaft seal for leakage. If the seal is leaking, it must be replaced before installing the clutch.

➡**Use emery cloth to remove minor imperfections from the clutch disc lining surface.**

12. Inspect the clutch disc for the following:
- Oil or grease saturation
- Worn or loose facings
- Warpage or loose rivets at the hub
- Loose or broken torsion dampening springs
- Wear or rust on the splines

13. If the clutch disc shows any of these conditions, it should be replaced.

14. Use a dial indicator mounted on a metal base to measure the clutch disc run-out. If the run-out exceeds 0.0276 in. (0.700mm), replace the disc.

15. Inspect the clutch release for distortion, cracks, excessive release bearing surface wear or damaged tines and replace as necessary.

To install:

16. Clean the pressure plate and flywheel surfaces thoroughly. Position the clutch disc and pressure plate into the installed position and support them with clutch aligning tool T74P-7137-k or equivalent. If the clutch assembly is being reused, align the matchmarks that were made during the removal procedure.

17. Install the pressure plate-to-flywheel retaining bolts. Tighten the bolts in the correct sequence to 12–24 ft. lbs. (16–32 Nm). Remove the alignment tool.

18. Install the transaxle assembly.

19. Lower the vehicle.

20. Bleed the hydraulic clutch system.

21. Adjust the clutch pedal free-play.

22. Connect the negative battery cable.

23. Road test the vehicle and check the clutch for proper operation.

Hydraulic Clutch System

BLEEDING

1. Check that the brake master cylinder is at least ¾ full during the entire bleeding process.

2. Remove the bleeder screw cap from the clutch slave cylinder and attach a hose to the bleeder screw. Place the other end of the hose into a container to catch the fluid.

3. Have an assistant slowly pump the clutch pedal several times, then hold the clutch pedal down.

4. Loosen the bleeder screw to release the fluid and air. Tighten the bleeder screw.

5. Repeat the bleeding procedure until no more air bubbles are seen in the fluid.

6. Tighten the bleeder screw to 52–78 inch lbs. (6–9 Nm).

7. Top off the brake master cylinder to the full line.

8. Check for proper clutch system operation.

Halfshafts

REMOVAL & INSTALLATION

1995–96 Models

1.8L ENGINE—LEFT SIDE

➡**Before continuing with any halfshaft procedure, be sure to have available, new halfshaft retaining nuts and circlips. Once removed, these parts loose their torque holding ability or retention capability and must not be reused.**

1. Disconnect the negative battery cable.

2. With the vehicle sitting on the ground, carefully raise the staked portion of the halfshaft retaining nut using a suitable small chisel. Loosen the nut.

3. Raise and safely support the vehicle.

4. Remove the wheel and tire assembly.

5. Remove the splash shield.

6. Remove and discard the halfshaft retaining nut.

7. Remove the cotter pin and castellated nut from the tie rod end and separate the tie rod end from the steering knuckle using a suitable removal tool. Discard the cotter pin.

8. Remove the lower ball joint pinch bolt. Carefully pry down on the lower control arm to separate the ball joint stud from the steering knuckle.

9. Pull outward on the steering knuckle/brake assembly. Carefully pull the halfshaft from the hub and position it aside.

10. Removal of the left side halfshaft requires removal of the crossmember to allow access with a prybar as follows:

   a. Support the transaxle with a suitable transaxle jack.

   b. Remove the four transaxle mount-to-crossmember retaining nuts.

   c. Remove the two crossmember retaining nuts at the rear of the crossmember.

   d. While supporting the rear of the crossmember, unfasten the two front mounting bolts and remove the crossmember.

11. Position a drain pan under the transaxle.

12. Insert a prybar between the halfshaft and the transaxle case. Gently pry outward to release the halfshaft from the differential side gear. Be careful not to damage the transaxle case, oil seal, CV-joint or CV-joint boot.

13. Remove the halfshaft.

➡Install suitable plugs after removing the halfshafts to prevent the differential side gears from moving out of place. Should the gears become misaligned,

the differential will have to be removed from the transaxle to align the gears.

To install:

14. Position a new circlip on the inner CV-joint spline so the circlip gap is at the top. Lubricate the splines lightly with a suitable grease.

15. Remove the plugs that were installed in the differential side gears.

16. Position the halfshaft so the CV-joint splines are aligned with the differential side gear splines. Push the halfshaft into the differential. When seated properly, the circlip

can be felt snapping into the differential side gear groove.

17. Pull outward on the steering knuckle/brake assembly and insert the halfshaft into the hub.

18. Pry downward on the lower control arm and position the lower ball joint stud in the steering knuckle.

19. Install the crossmember and the crossmember-to-frame nuts and bolts. Tighten the nuts to 27–38 ft. lbs. (37–52 Nm) and the bolts to 47–66 ft. lbs. (64–89 Nm).

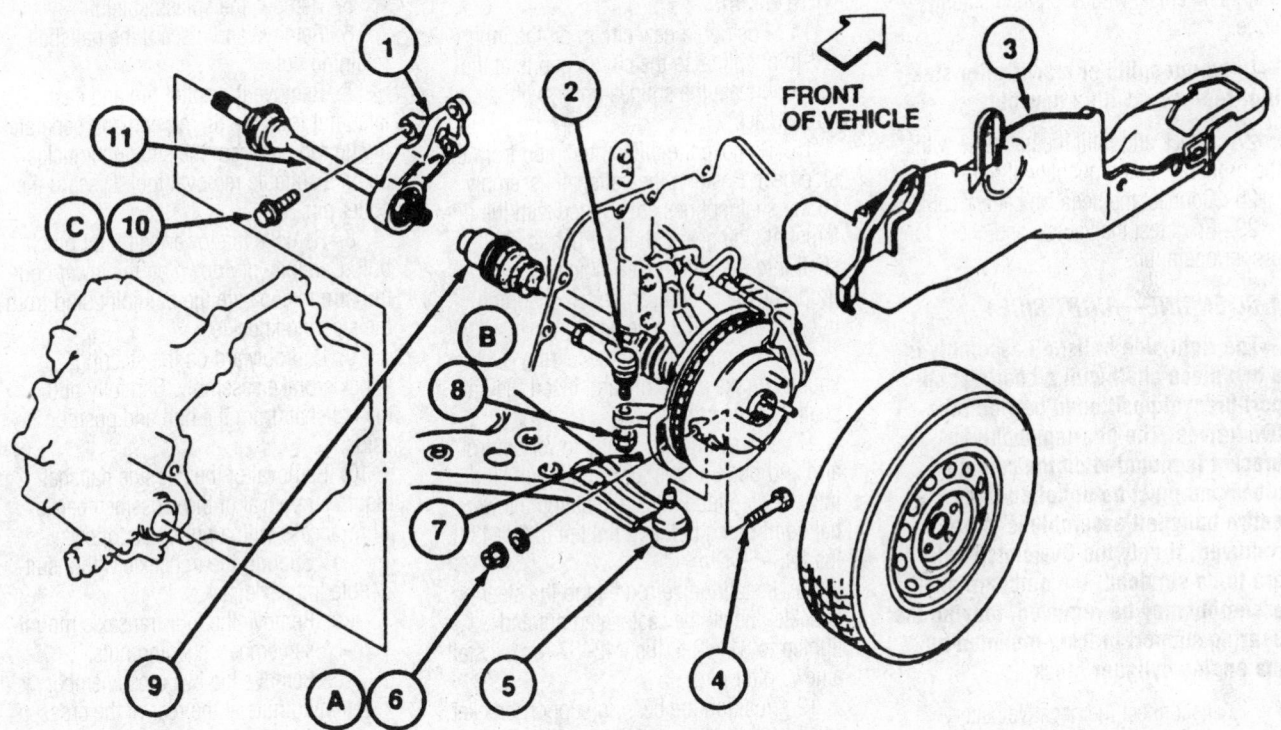

1	Front Axle Bearing Bracket (Part of 3A329)
2	Tie Rod End
3	Front Fender Splash Shield
4	Ball Joint Bolt
5	Front Suspension Lower Arm Ball Joint
6	Ball Joint Bolt Nut
7	Cotter Pin
8	Tie Rod End Nut
9	Front Wheel Driveshaft Joint (1.8L)
10	Front Axle Bearing Bracket Bolts (3 Req'd)
11	Halfshaft (1.8L)
A	Tighten to 43-59 N·m (32-43 Lb-Ft)
B	Tighten to 34-46 N·m (25-33 Lb-Ft)
C	Tighten to 42-62 N·m (31-46 Lb-Ft)

7922MG96

Exploded view of typical halfshaft positioning—1.8 and 1.9L engines

20. Install the four transaxle mount-to-crossmember nuts and tighten to 27–38 ft. lbs. (37–52 Nm).

21. Remove the transaxle jack.

22. Install the lower ball joint pinch bolt and tighten to 32–43 ft. lbs. (43–59 Nm).

23. Attach the tie rod end to the steering knuckle. Install the castellated nut and tighten to 31–42 ft. lbs. (42–57 Nm). Install a new cotter pin.

24. Install the splash shield.

25. Install the wheel and tire assembly.

26. Install a new halfshaft retaining nut and tighten to 174–235 ft. lbs. (235–319 Nm). Stake the halfshaft retaining nut using a suitable chisel with a rounded cutting edge.

➡️If the nut splits or cracks after staking, replace it with a new nut.

27. Check and refill the transaxle with the proper type and quantity of fluid.

28. Connect the negative battery cable.

29. Road test the vehicle and check for proper operation.

1.8L ENGINE—RIGHT SIDE

➡️**The right side halfshaft assembly is a two piece shaft with a bearing support bracket positioned between the two halves. The bearing support bracket is mounted on the cylinder block and must be unbolted if the entire halfshaft assembly is to be removed. If only the CV-joints/boots are to be serviced, the outboard shaft assembly may be removed, leaving the bearing support bracket mounted on the engine cylinder block.**

1. Disconnect the negative battery cable.

2. With the vehicle sitting on the ground, carefully raise the staked portion of the halfshaft retaining nut using a suitable small chisel. Loosen the nut.

3. Raise and safely support the vehicle.

4. Remove the right front wheel and tire assembly.

5. Remove the splash shield.

6. Remove and discard the halfshaft retaining nut.

7. Remove the cotter pin and castellated nut from the tie rod end and separate the tie rod end from the steering knuckle using a suitable removal tool. Discard the cotter pin.

8. Remove the lower ball joint pinch bolt. Carefully pry down on the lower control arm to separate the ball joint stud from the steering knuckle.

9. Pull outward on the steering knuckle/brake assembly. Carefully pull the halfshaft from the hub and position it aside.

10. Position a drain pan under the transaxle.

11. Remove the three bearing support bracket mounting bolts.

12. Insert a prybar between the bearing support bracket and the starter bracket. Gently pry outward on the damper until the halfshaft disengages from the differential side gear.

13. Remove the halfshaft assembly. Install an appropriate differential plug in the differential side gear.

To install:

14. Position a new circlip on the inner CV-joint spline so the circlip gap is at the top. Lubricate the splines lightly with a suitable grease.

15. Remove the differential plug from the side gear. Position the halfshaft assembly so the shaft splines are aligned with the differential side gear splines. Push the halfshaft into the differential. When seated properly, the circlip can be felt snapping into the differential side gear groove.

16. Pull outward on the steering knuckle/brake assembly and insert the halfshaft into the hub.

17. Pry downward on the lower control arm and position the lower ball joint stud into the steering knuckle. Install the lower ball joint pinch bolt and tighten to 32–43 ft. lbs. (43–59 Nm).

18. Install the tie rod end to the steering knuckle. Install the castellated nut and tighten to 31–42 ft. lbs. (42–57 Nm). Install a new cotter pin.

19. Position the bearing support bracket and install the three retaining bolts. Tighten the outer bolt first, then the top inner, then the bottom inner. Tighten the bolts to 31–46 ft. lbs. (42–62 Nm).

20. Install the splash shield.

21. Install the wheel and tire assembly.

22. Lower the vehicle.

23. Install a new halfshaft retaining nut and tighten to 174–235 ft. lbs. (235–319 Nm). Stake the retaining nut using a suitable chisel with the cutting edge rounded off.

➡️If the nut splits or cracks after staking, it must be replaced with a new nut.

24. Check and refill the transaxle with the proper type and quantity of fluid.

25. Connect the negative battery cable.

26. Road test the vehicle and check for proper operation.

1.9L ENGINE

➡️**Before continuing with any halfshaft procedure, be sure to have available, new halfshaft retaining nuts and circlips. Once removed, these parts loose their torque holding ability or retention capability and must not be reused.**

1. Disconnect the negative battery cable.

2. With the vehicle sitting on the ground, carefully raise the staked portion of the halfshaft retaining nut using a suitable small chisel. Loosen the nut.

3. Raise and safely support the vehicle.

4. Remove the wheel and tire assembly.

5. Remove the splash shield.

6. Remove and discard the halfshaft retaining nut.

7. Remove the cotter pin and castellated nut from the tie rod end and separate the tie rod end from the steering knuckle using a suitable removal tool. Discard the cotter pin.

8. Remove the lower ball joint pinch bolt. Carefully pry down on the lower control arm to separate the ball joint stud from the steering knuckle.

9. Pull outward on the steering knuckle/brake assembly. Carefully pull the halfshaft from the hub and position it aside.

10. Removal of the left side halfshaft requires removal of the crossmember to allow access with a prybar as follows:

a. Support the transaxle with a suitable transaxle jack.

b. Remove the four transaxle mount-to-crossmember retaining nuts.

c. Remove the two crossmember retaining nuts at the rear of the crossmember.

d. While supporting the rear of the crossmember, unfasten the two front mounting bolts and remove the crossmember.

11. Position a drain pan under the transaxle.

12. Insert a prybar between the halfshaft and the transaxle case. Gently pry outward to release the halfshaft from the differential side gear. Be careful not to damage the transaxle case, oil seal, CV-joint or CV-joint boot.

13. Remove the halfshaft.

➡️**Install suitable plugs after removing the halfshafts to prevent the differential side gears from moving out of place. Should the gears become misaligned, the differential will have to be removed from the transaxle to align the gears.**

To install:

14. Position a new circlip on the inner CV-joint spline so the circlip gap is at the top. Lubricate the splines lightly with a suitable grease.

15. Remove the plugs that were installed in the differential side gears.

16. Position the halfshaft so the CV-joint splines are aligned with the differential side gear splines. Push the halfshaft into the differential. When seated properly, the circlip can be felt snapping into the differential side gear groove.

17. Pull outward on the steering knuckle/brake assembly and insert the halfshaft into the hub.

18. Pry downward on the lower control arm and position the lower ball joint stud in the steering knuckle.

19. Install the crossmember and the crossmember-to-frame nuts and bolts. Tighten the nuts to 27–38 ft. lbs. (37–52 Nm) and the bolts to 47–66 ft. lbs. (64–89 Nm).

20. Install the four transaxle mount-to-crossmember nuts and tighten to 27–38 ft. lbs. (37–52 Nm).

21. Remove the transaxle jack.

22. Install the lower ball joint pinch bolt and tighten to 32–43 ft. lbs. (43–59 Nm).

23. Attach the tie rod end to the steering knuckle. Install the castellated nut and tighten to 31–42 ft. lbs. (42–57 Nm). Install a new cotter pin.

24. Install the splash shield.

25. Install the wheel and tire assembly.

26. Install a new halfshaft retaining nut and tighten to 174–235 ft. lbs. (235–319 Nm). Stake the halfshaft retaining nut using a suitable chisel with a rounded cutting edge.

➡️**If the nut splits or cracks after staking, replace it with a new nut.**

27. Check and refill the transaxle with the proper type and quantity of fluid.

28. Connect the negative battery cable.

29. Road test the vehicle and check for proper operation.

1997–99 Models

➡️**Before continuing with any halfshaft procedure, be sure to have available, new halfshaft retaining nuts and circlips. Once removed, these parts loose their torque holding ability or retention capability and must not be reused.**

1. Disconnect the negative battery cable.

2. With the vehicle sitting on the ground, carefully raise the staked portion of the halfshaft retaining nut using a suitable small chisel. Loosen the nut.

3. Raise and safely support the vehicle.

4. Remove the wheel and tire assembly.

5. Remove and discard the halfshaft retaining nut.

6. Remove the cotter pin and nut from the tie rod end, then separate the tie rod end from the steering knuckle using a suitable removal tool. Discard the cotter pin.

7. On 1998–99 models, remove the stabilizer bar link as follows:

a. Unfasten the stabilizer bar end nut (1) and bolt (2).

b. Remove the stabilizer bar end retainer (3).

c. Remove the end bushing (4) above the stabilizer bar.

d. Remove the end bushing (5) below the stabilizer bar.

e. Remove the stabilizer bar spacer (6).

f. Remove the stabilizer bar end bushings from above (7) and below (8) the sub-frame

g. Remove the lower stabilizer bar end retainer (9).

h. Remove the stabilizer bar.

8. Remove the ball joint bolt and nut. Carefully pry down on the lower control arm to separate the ball joint stud from the steering knuckle.

9. Pull outward on the steering knuckle/brake assembly. Carefully pull the halfshaft from the hub and position it aside.

➡️**Removal of the left side halfshaft requires removal of the crossmember to allow access with a prybar.**

10. Support the transaxle with a suitable transaxle jack.

11. Unfasten the four transaxle crossmember retainers and remove the crossmember.

12. If removing the right side halfshaft,

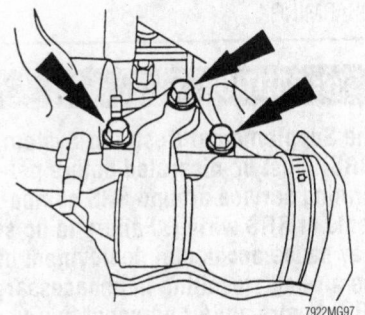

On 1998–99 2.0L models with a manual transaxle, unfasten the center support bearing bolts

remove the right-hand shield and splash shield.

13. Position a drain pan under the transaxle.

14. Remove the left-hand halfshaft on 1997–99 models and right-hand halfshaft on 1997 models as follows:

a. Insert a prybar between the halfshaft and the transaxle case. Gently pry outward to release the halfshaft from the differential side gear. Be careful not to damage the transaxle case, oil seal, CV-joint or CV-joint boot.

b. Remove the halfshaft.

15. On 1998–99 models, remove the right-hand halfshaft as follows:

a. On models with a manual transaxle, unfasten the center support bearing bolts.

b. Lower the halfshaft and remove it from the transaxle.

c. Separate the halfshaft from the center support bearing and remove the halfshaft from the vehicle.

d. Inspect the center support bearing for damage and replace as necessary.

➡️**Install suitable plugs after removing the halfshafts to prevent the differential side gears from moving out of place. Should the gears become misaligned, the differential will have to be removed from the transaxle to align the gears.**

To install:

16. Position a new circlip on the inner CV-joint spline so the circlip gap is at the top. Lubricate the splines lightly with a suitable grease.

17. Remove the plugs that were installed in the differential side gears.

18. On all models except 1998–99 models wit a manual transaxle, position the halfshaft so the CV-joint splines are aligned

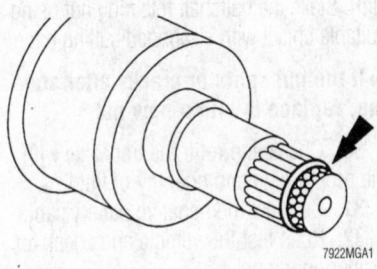

During assembly, be sure to install a new circlip on the inner CV-joint spline—2.0L engine

with the differential side gear splines. Push the halfshaft into the differential.

➡**When seated properly, the circlip can be felt snapping into the differential side gear groove.**

19. Install the right-hand halfshaft on 1998–99 models with a manual transaxle as follows:

a. Position the halfshaft and joint so that the splines line up with the splines in the halfshaft, and push the halfshaft, joint and halfshaft together with the center support bearing.

b. Install the halfshaft in the transaxle.

20. Place the center support bearing into position and tighten it retaining bolts to 32–46 ft. Lbs. (46–62 Nm).

a. Install the right-hand shield and splash shield.

21. Pull outward on the steering knuckle/brake assembly and insert the halfshaft into the hub.

22. Pry downward on the lower control arm and position the lower ball joint stud in the steering knuckle.

23. Install the crossmember and the crossmember-to-frame bolts. Tighten the bolts to 69–93 ft. lbs. (94–126 Nm).

24. Remove the transaxle jack.

25. Install the steering knuckle and ball joint nut and bolt. Tighten the nut and bolt to 32–43 ft. Lbs. (43–59 Nm).

26. On 1998–99 models, install the stabilizer bar link in reverse order of removal and tighten the bar end nut until the protruding bar end bolt length is 0.67–75 in. (17–19mm).

27. Install the tie rod end to the steering knuckle. Install the nut and tighten to 25–33 ft. Lbs. (34–46 Nm) on 1997 models and 32–41 ft. Lbs. (43–56 Nm) on 1998–99 models. Install a new cotter pin.

28. Install the wheel and tire assembly.

29. Install a new halfshaft retaining nut and tighten to 174–235 ft. lbs. (235–319 Nm). Stake the halfshaft retaining nut using a suitable chisel with a rounded cutting edge.

➡**If the nut splits or cracks after staking, replace it with a new nut.**

30. Check and refill the transaxle with the proper type and quantity of fluid.

31. Connect the negative battery cable.

32. Road test the vehicle and check for proper operation.

STEERING AND SUSPENSION

Air Bag

✳✳ CAUTION

Some vehicles are equipped with an air bag system, also known as the Supplemental Inflatable Restraint (SIR) or Supplemental Restraint System (SRS). The system must be disabled before performing service on or around system components, steering column, instrument panel components, wiring and sensors. Failure to follow safety and disabling procedures could result in accidental air bag deployment, possible personal injury and unnecessary system repairs.

PRECAUTIONS

Several precautions must be observed when handling the inflator module to avoid accidental deployment and possible personal injury.

• Never carry the inflator module by the wires or connector on the underside of the module.

• When carrying a live inflator module, hold securely with both hands, and ensure that the bag and trim cover are pointed away.

• Place the inflator module on a bench or other surface with the bag and trim cover facing up.

• With the inflator module on the bench, never place anything on or close to the module which may be thrown in the event of an accidental deployment.

DISARMING

✳✳ CAUTION

The Supplemental Restraint System (SRS) must be disarmed before performing service around SRS components or SRS wiring. Failure to do so may cause accidental deployment of the air bag, resulting in unnecessary SRS repairs and/or personal injury.

1. Disconnect both battery cables, negative cable first.

2. Wait at least 1 minute. This allows time for the back-up power supply to deplete its stored energy.

3. Remove the drivers side air bag module, then the passenger side if required.

4. Use caution when carrying live air bags. Always place the air bag with the cover up.

✳✳ CAUTION

When carrying a live air bag, be sure the bag and trim cover are pointed away from the body. In the unlikely event of an accidental deployment, the bag will, then deploy with minimal chance of injury. When placing a live air bag on a bench or other surface, always face the bag and trim cover up, away from the surface. This will reduce the motion of the module if it is accidentally deployed.

5. If the battery needs to be reconnected while one or both of the air bags are removed from the system, install Air Bag Simulator 105–00010 or equivalent, to the drivers side and/or passenger side air bag harness connectors as required. Before removing either air bag simulator, disconnect both battery cables and wait at least 1 minute before continuing.

6. Once service is completed and the air bag modules are back in place, connect the negative battery cable and prove out the air bag system by turning the ignition key to the **RUN** position and visually monitoring the air bag indicator lamp in the instrument cluster. The indicator lamp should illuminate for approximately 6 seconds, then turn OFF. if the indicator lamp does not illuminate, stays ON, or flashes at any time, a fault has been detected by the air bag diagnostic monitor requiring immediate attention.

Rack and Pinion Steering Gear

REMOVAL & INSTALLATION

Manual

1995–96 MODELS

1. Disconnect the negative battery cable.

2. Working from inside the vehicle, unfasten the five nuts securing the steering column tube boot and remove the boot.

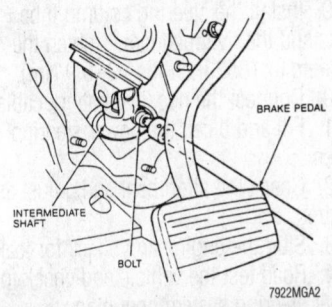

Unfasten the intermediate shaft-to-pinion shaft coupling bolt from inside the vehicle—1995–96 models with manual rack and pinion steering gears

3. Remove the intermediate shaft-to-pinion shaft coupling bolt from inside the vehicle.

4. Raise and safely support the vehicle.

5. Remove the front wheel and tire assemblies.

6. Remove the cotter pins and castellated nuts securing the tie rod ends to the steering knuckles. Separate the tie rod ends from the steering knuckles. Discard the cotter pins.

7. If equipped with a manual transaxle, unfasten the gearshift lever stabilizer bar nut and disconnect the bar and support from the transaxle.

8. Unfasten the nuts securing the rack and pinion (steering gear) mounting brackets to the bulkhead. Remove the mounting brackets.

9. Remove the rack and pinion assembly from the vehicle.

To install:

10. Position the rack and pinion assembly to its mounting position and install the mounting brackets and retaining nuts. Tighten the nuts to 27–38 ft. lbs. (37–52 Nm).

11. If equipped with a manual transaxle, connect the gearshift lever stabilizer bar. Tighten the nut to 23–34 ft. lbs. (31–46 Nm).

12. Install the tie rod ends to the steering knuckles. Install the castellated nuts and tighten to specifications. Install new cotter pins.

13. Install the front wheel and tire assemblies.

14. Lower the vehicle.

15. Install the intermediate shaft-to-pinion shaft bolt and tighten to 33 ft. lbs. (45 Nm).

16. Install the steering column tube boot and 5 retaining nuts. Tighten to 35 inch lbs. (4 Nm).

17. Connect the negative battery cable.

18. Check the alignment and adjust the toe as required.

19. Road test the vehicle and check the steering system for proper operation.

Power

1995–96 MODELS

1. Disconnect the negative battery cable.

2. From inside the passenger compartment, remove the steering column tube boot nuts at the base of the column and remove the tube boots.

3. Remove the intermediate shaft-to-pinion shaft bolt.

4. Raise and safely support the vehicle.

5. Remove the front wheel and tire assemblies.

6. Remove the left-hand splash shield.

7. Remove the cotter pins and castellated nuts from the tie rod ends. Using a suitable tool, separate the tie rod ends from the steering knuckles. Discard the cotter pins.

8. Remove the strap that holds the power steering lines to the rack and pinion (steering gear) housing and discard the strap.

9. Disconnect the pressure and return lines from the rack and pinion assembly and plug the lines.

10. If equipped with a manual transaxle, disconnect the gearshift stabilizer bar and shift control rod from the transaxle.

11. Remove the retaining nuts from the rack and pinion mounting brackets.

12. Remove the rack and pinion assembly from the left-hand side of the vehicle.

To install:

13. Place the rack and pinion assembly in its mounting location.

14. Install the rack and pinion mounting brackets. Install the retaining nuts and tighten to 28–38 ft. lbs. (37–57 Nm).

15. If equipped with a manual transaxle, connect the gearshift stabilizer bar and shift control rod. Tighten the gearshift stabilizer bar nut to 23–34 ft. lbs. (31–46 Nm) and the shift control rod nut to 12–17 ft. lbs. (16–23 Nm).

16. Remove the plugs and connect the pressure and return lines to the rack and pinion assembly. Tighten the flare nuts to 22–28 ft. lbs. (29–39 Nm).

17. Install a new strap to hold the power steering lines to the rack and pinion housing.

18. Install the tie rod ends to the steering knuckles and install the castellated nuts. Tighten to specification. Install new cotter pins.

19. Install the left-hand splash shield.

20. Install the wheel and tire assemblies.

21. Lower the vehicle.

22. From inside the vehicle, install the intermediate shaft-to-pinion shaft bolt. Tighten the bolt to 30–36 ft. lbs. (40–50 Nm).

23. Install the steering column tube boot

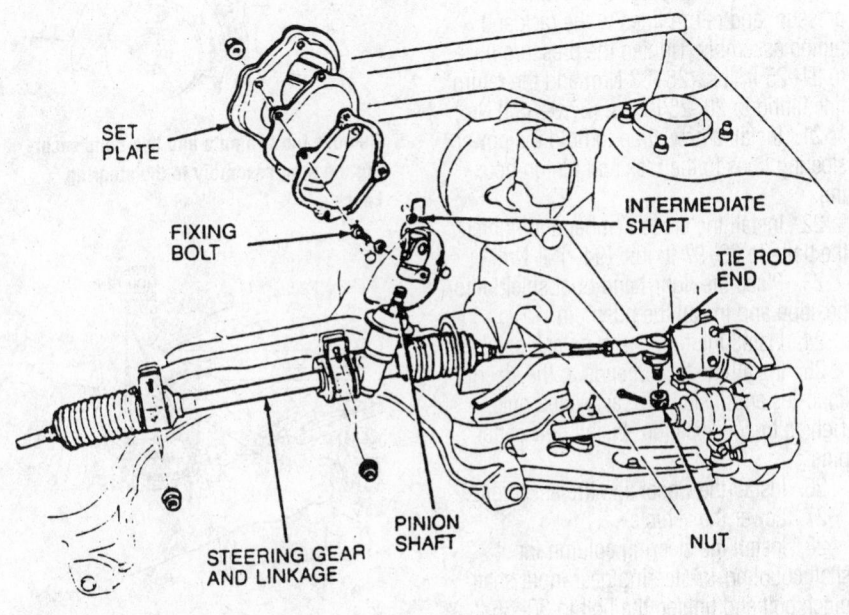

Exploded view of the steering rack mounting—1995–96 models

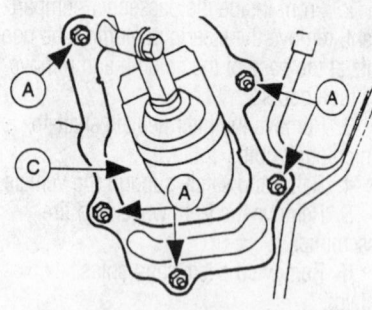

Unfasten the steering column tube boot nuts (A) at the base of the column (C) and remove the tube boots—1997–99 models

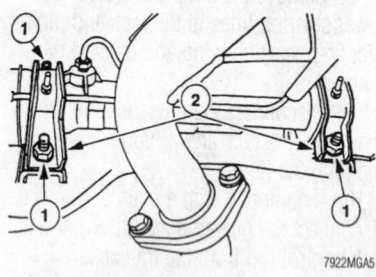

Unfasten the retaining nuts (1) from the rack and pinion mounting brackets (2) and remove the brackets—1997–99 models

and the five retaining nuts. Tighten the retaining nuts to 35 inch lbs. (4 Nm).

24. Connect the negative battery cable.

25. Fill and bleed the power steering system.

26. Check the alignment and adjust as required.

27. Start the engine and check for leaks.

28. Road test the vehicle and check for proper steering system operation.

1997–99 MODELS

1. Turn the key to the **ACC** position.

2. From inside the passenger compartment, remove the steering column tube boot nuts at the base of the column and remove the tube boots.

3. Remove the steering column input shaft coupling-to-steering gear input shaft pinch bolt.

4. Raise and safely support the vehicle.

5. Remove the front wheel and tire assemblies.

6. Using a suitable tool, separate the tie rod ends from the steering knuckles. Discard the cotter pins.

7. Remove the right-hand lower splash shield.

8. Unfasten the bolts and remove the crossmember.

9. Disconnect the pressure and return lines from the rack and pinion assembly and plug the lines.

10. Remove and discard the strap holding the hoses to the steering gear.

11. If equipped with a manual transaxle, disconnect the gearshift rod and clevis from the transaxle.

12. If equipped with a manual transaxle, unfasten the extension bar nut and disconnect the gearshift lever stabilizer bar and support from the transaxle.

13. Unfasten the retaining nuts from the rack and pinion mounting brackets and remove the brackets.

14. Remove the pushpin and position the right-hand boot aside.

15. Remove the rack and pinion assembly from the right-hand side of the vehicle.

To install:

16. Place the rack and pinion assembly in its mounting location.

17. Align the steering column input shaft coupling and the steering gear input shaft.

18. Install the rack and pinion mounting brackets. Install the retaining nuts and tighten to 28–38 ft. lbs. (37–57 Nm).

19. If equipped with a manual transaxle, connect the gearshift stabilizer bar, support and the gearshift rod and clevis.

➡**Install new Teflon seals on the power steering pressure hose fitting and be sure the threads are clean before connecting the hose.**

20. Remove the plugs and connect the pressure and return lines to the rack and pinion assembly. Tighten the pressure hose to 21–25 ft. lbs. (28–33 Nm) and the return line fitting to 20–25 ft. lbs. (27.3–33.9 Nm).

21. Install a new strap to hold the power steering lines to the rack and pinion housing.

22. Install the crossmember and tighten the bolts to 69–97 ft. lbs. (94–131 Nm).

23. Place the right-hand boot shield into position and install the push pin.

24. Install the right-hand splash shield.

25. Install the tie rod ends to the steering knuckles and install the castellated nuts. Tighten to specification. Install new cotter pins.

26. Install the wheel and tire assemblies.

27. Lower the vehicle.

28. Install the steering column input shaft coupling-to-steering gear input shaft pinch bolt and tighten the bolt to 30–36 ft. lbs. (40–50 Nm).

29. Install the steering column tube boots and the five retainers. Tighten the retainers to 1852 inch lbs. (2–5.9 Nm).

30. Connect the negative battery cable.

31. Fill and bleed the power steering system.

32. Check the alignment and adjust as required.

33. Start the engine and check for leaks.

34. Road test the vehicle and check for proper steering system operation.

Strut

REMOVAL & INSTALLATION

Front

1. Disconnect the negative battery cable.

2. Raise and safely support the vehicle.

3. Remove the front wheel and tire assembly.

4. Remove the clip securing the brake hose to the strut (spring and shock) assembly. If equipped with anti-lock brakes, remove the anti-lock brake harness cable and clip.

5. Unfasten the two nuts and two bolts

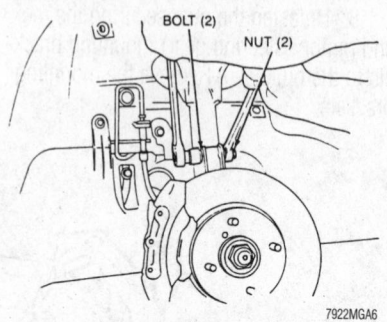

Remove the two nuts and two bolts securing the strut assembly to the steering knuckle

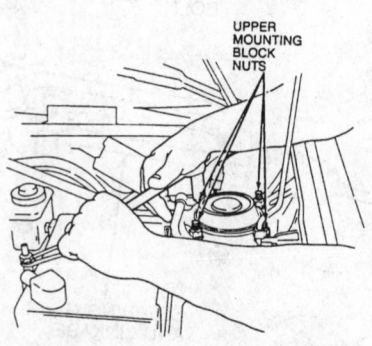

Unfasten the front strut assembly upper mounting bracket retaining nuts

securing the strut assembly to the steering knuckle.

6. Unfasten the four upper strut retaining nuts and remove the strut assembly from the vehicle.

✳✳ WARNING

Never remove the strut piston rod nut unless the coil spring is compressed. Always wear safety glasses when using a spring compressor.

7. Inspect all components and replace as needed.

To install:

8. Position the strut assembly into the wheel housing. Be sure the direction indicator on the upper mounting bracket faces inboard.

9. Secure the upper mounting bracket to the strut tower with the retaining nuts. Tighten the nuts to 22–30 ft. lbs. (29–40 Nm).

10. Attach the strut assembly to the steering knuckle and install the retaining bolts and nuts. Tighten to 69–93 ft. lbs. (93–127 Nm).

11. Position the brake hose on the strut assembly and secure it with the brake hose clip. If equipped with anti-lock brakes, install the anti-lock harness cable and clip.

12. Install the front wheel and tire assembly.

13. Lower the vehicle.

14. Connect the negative battery cable.

15. Check the front wheel alignment.

16. Road test the vehicle and check for proper operation.

Rear

1995–96 MODELS

1. Disconnect the negative battery cable.

2. Raise and safely support the vehicle.

3. Remove the wheel and tire assembly.

4. Remove the clip securing the brake hose to the rear strut assembly.

5. If equipped with anti-lock brakes, remove the ABS harness from the clip on the strut assembly.

6. Remove the nuts and bolts securing the rear strut assembly to the wheel spindle.

7. Partially lower the vehicle.

8. On hatchback and wagon models, remove the quarter trim panel.

9. Unfasten the two upper strut insulator retaining nuts and remove the rear strut assembly from the vehicle.

To install:

10. Position the strut assembly into the vehicle wheel housing.

11. Install the two upper strut insulator retaining nuts and tighten to 22–27 ft. lbs. (29–40 Nm).

12. On hatchback and wagon models, install the quarter trim panel.

13. Install the nuts and bolts securing the strut assembly to the rear wheel spindle assembly. Tighten the lower strut bolts and nuts to 69–93 ft. lbs. (93–127 Nm).

14. Install the clip securing the flexible brake hose to the rear strut assembly.

15. If equipped with anti-lock brakes, install the ABS harness to the clip.

16. Install the wheel and tire assembly.

17. Lower the vehicle.

18. Connect the negative battery cable.

19. Check the rear wheel alignment.

20. Road test the vehicle and check for proper operation.

1997–99 MODELS

1. Disconnect the negative battery cable.

2. On sedan and coupe models, remove the package tray trim panel.

3. On wagon models, remove the quarter trim panel

4. Unfasten the two upper strut retaining nuts.

5. Raise and safely support the vehicle.

6. Remove the wheel and tire assembly.

7. Remove the clip securing the brake hose to the rear strut assembly.

8. If equipped with anti-lock brakes, unfasten the ABS sensor bolt.

9. Remove the nuts and bolts securing the rear strut assembly to the wheel spindle.

10. Remove the strut assembly from the vehicle.

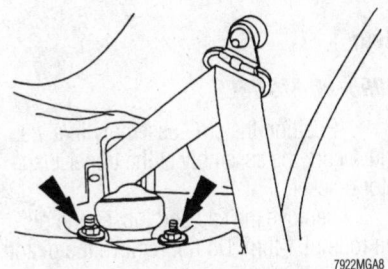

Location of the two rear strut upper retaining nuts (arrows) inside the vehicle—1997–99 models

To install:

11. Position the strut assembly into the vehicle wheel housing.

12. Install the nuts and bolts securing the strut assembly to the rear wheel spindle assembly. Tighten the lower strut bolts and nuts to 76–100 ft. lbs. (103–136 Nm).

13. If equipped with anti-lock brakes, tighten the ABS sensor bolt.

14. Install the clip securing the flexible brake hose to the rear strut assembly.

15. Install the wheel and tire assembly.

16. Lower the vehicle.

17. Install the two upper strut retaining nuts and tighten to 34–46 ft. lbs. (47–62 Nm).

18. Install the trim panel.

19. Connect the negative battery cable.

20. Check the rear wheel alignment.

21. Road test the vehicle and check for proper operation.

Coil Spring

REMOVAL & INSTALLATION

Front

1995–96 MODELS

1. Remove the strut assembly.

2. Remove the cap from the top of the strut.

3. Place the strut assembly in a vise, hold the piston rod in position and loosen the piston rod nut.

✳✳ WARNING

Always wear goggles when using a spring compressor

4. Install a suitable spring compressor tool onto the coil spring.

5. Compress the spring using the tool.

6. Remove the strut piston rod nut, front mounting block, thrust bearing, upper spring seat, rubber spring seat, coil spring and bound stopper.

To install:

7. Place the bound stopper onto the strut.

8. With coil spring compressed, position the spring onto the strut assembly.

9. Install the rubber spring seat, upper spring seat, thrust bearing, mounting block and piston rod nut. Tighten the piston rod nut to 58–81 ft. lbs. (79–110 Nm).

10. After the piston rod nut has been tightened to specification, carefully remove

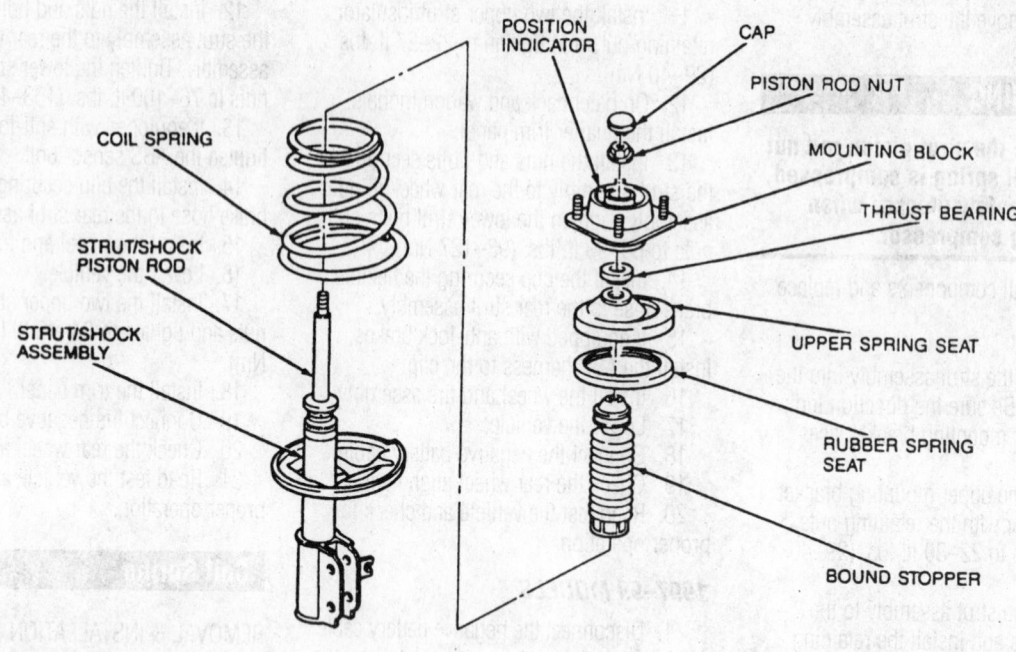

Exploded view of the strut assembly—1995–96 models

the compressor tool from the spring while making sure the spring is properly seated in the upper and lower spring seats.

11. Install the cap on top of the strut.

➡**The position indicator on the mounting block faces inboard during installation.**

12. Install the strut assembly in the vehicle.

13. Have the vehicle alignment checked and if necessary, adjusted.

1997–99 MODELS

1. Remove the strut assembly.
2. Place the strut assembly in a vise.

❊❊ WARNING

Always wear goggles when using a spring compressor

3. Install a suitable spring compressor tool onto the coil spring and compress the spring.

4. Unfasten the piston rod nut and remove the upper mounting bracket.

5. Remove the bound stopper, dust boot, rubber spring seat, spring, upper spring seat and thrust bearing.

To install:

6. Install the thrust bearing, upper spring seat, spring, rubber spring seat, dust boot and the bound stopper.

7. Install the piston rod nut and tighten it to 58–81 ft. lbs. (79–110 Nm).

8. After the piston rod nut has been tightened to specification, carefully remove the compressor tool from the spring while making sure the spring is properly seated in the upper and lower spring seats.

9. Install the strut assembly in the vehicle.

10. Have the vehicle alignment checked and if necessary, adjusted.

Rear

1995–96 MODELS

1. Position the strut assembly in a vise and secure the assembly at the upper insulator bracket.

2. Remove the cap and loosen the piston rod nut 1 turn. Do not remove the piston rod nut at this time.

❊❊ CAUTION

Attempting to remove the spring from the strut without first compressing

the spring with a tool designed for that purpose could cause bodily injury.

3. Install an appropriate coil spring compressor and compress the coil spring.

4. Remove the piston rod nut, washer, retainer, anti-rattle plate and insulator.

5. Remove the coil spring.

6. Remove the stopper seat and dust boot from the strut piston.

To install:

7. Position the strut assembly in a vise and secure.

8. Install the dust boot and the stopper seat onto the strut piston rod.

9. Compress the coil spring and install on the strut assembly.

10. Install the upper strut insulator, then align the upper strut insulator mounting studs and the lower bracket of the strut assembly.

11. Install the retainer, washer and piston rod nut. Tighten the nut to 41–50 ft. lbs. (55–68 Nm).

12. Be sure the spring is properly aligned and carefully release the spring into the seats of the strut.

13. Remove the spring compressor from the coil spring and install the cap.

1997–99 MODELS

1. Remove the strut assembly.
2. Place the strut assembly in a vise.

✳✳ WARNING

Always wear goggles when using a spring compressor

3. Install a suitable spring compressor tool onto the coil spring and compress the spring.
4. Remove the rear strut assembly top mounting cover
5. Unfasten the piston rod nut and remove the retainer.
6. Remove the strut insulator.
7. Carefully remove the spring compressor.
8. Remove the rear strut dust boot and stopper seat.
9. Remove the spring and the rear strut insulator.

To install:

10. Place the strut assembly in a vise.
11. Install the strut insulator and spring.
12. Install the stopper seat and dust boot.
13. Use the spring compressor tool to compress the strut spring.
14. Remove the strut insulator and the retainer.
15. Install the piston rod nut and tighten it to 41–49 ft. lbs. (55–67 Nm).
16. Install the top mounting cover.
17. Be sure the spring is properly aligned and carefully release the spring into the seats of the strut.
18. Remove the spring compressor from the coil spring.
19. Install the strut assembly.

Lower Ball Joint

REMOVAL & INSTALLATION

1. Raise the front of the vehicle and support it with safety stands.
2. Remove the wheel assembly.
3. Unfasten the nut and bolt attaching the lower ball joint to the steering knuckle.
4. Separate the ball joint from the steering knuckle.
5. Unfasten the lower control arm ball joint nuts and bolt. Remove the ball joint.

To install:

6. Place the ball joint into position on

the lower control arm. Install the ball joint-to-lower control arm retainers. Tighten the nuts and bolt to 69–86 ft. lbs. (93–117 Nm).
7. Apply Loctite® 290 thread locking compound or equivalent to the ball joint nut and threads.
8. Attach the ball joint to the steering knuckle and install the nut and bolt. Tighten the nut and bolt to 32–43 ft. lbs. (43–59 Nm).
9. Install the wheel assembly and lower the vehicle.
10. Have the vehicle alignment checked and if necessary, adjusted.

Wheel Bearings

ADJUSTMENT

The bearings on the front and rear wheels are a lone piece cartridge design and cannot be adjusted. Wheel bearing play can be checked with a dial indicator. If wheel bearing play exceeds 0.002 in. (0.05mm) check the wheel hub retainer nut for proper torque. If the torque is correct, replacement of the wheel bearing is required.

REMOVAL & INSTALLATION

Front

1. With the vehicle sitting on the ground, carefully raise the staked portion of the halfshaft retaining nut using a suitable small chisel. Loosen the nut.
2. Raise the front of the vehicle and support it with safety stands.
3. Remove the wheel and tire assembly.

➡**Do not disconnect the brake hose from the caliper**

4. Remove the brake caliper and secure it out of the way with a piece of mechanics wire. Do not let the caliper hang on the hose.
5. Remove the brake rotor.
6. Remove the halfshaft retaining nut and discard it.
7. Remove the cotter pin and castellated nut from the tie rod end and separate the tie rod end from the steering knuckle using a suitable removal tool. Discard the cotter pin.
8. Use the appropriate tool to separate the tie rod end from the wheel knuckle.
9. If equipped with an Anti-lock Brake

System (ABS), unfasten the ABS sensor bolt and remove the sensor.
10. Remove the strut mounting nuts and the studs which attach the strut assembly to the steering knuckle.
11. Separate the strut from the steering knuckle.
12. Remove the lower ball joint pinch bolt. Carefully pry down on the lower control arm to separate the ball joint stud from the steering knuckle.
13. Remove the wheel hub, knuckle and bearing assembly from the vehicle.

To install:

14. Apply Loctite® 290 thread locking compound or equivalent to the ball joint nut and threads.
15. Install the wheel hub, knuckle and bearing assembly onto the ball joint and tighten the pinch bolt and nut to 32–43 ft. lbs. (43–59 Nm).
16. Install the tie rod end and tighten the nut to 25–33 ft. lbs. (34–46 Nm). Install a new cotter pin.
17. If equipped with an Anti-lock Brake System (ABS), install the ABS sensor and tighten the bolt.
18. Attach the knuckle to the strut assembly and tighten the strut mounting nuts to 69–93 ft. lbs. (93–127 Nm).
19. Install a new halfshaft retaining nut and tighten to 174–235 ft. lbs. (235–319 Nm). Stake the retaining nut using a suitable chisel with the cutting edge rounded off.

➡**If the nut splits or cracks after staking, it must be replaced with a new nut.**

20. Install the brake rotor and caliper.
21. Install the wheel and tire assembly.
22. Lower the vehicle.
23. Check the front wheel alignment.
24. Road test the vehicle and check for proper operation.

Rear

➡**The wheel bearings are a cartridge design and are not serviceable. If bearing replacement is required, the bearings and hub must be replaced as an assembly. Do not continue with this procedure without having available a new wheel hub retainer nut. Once removed, the nut looses its torque holding ability or retention capability and must not be reused.**

1. Raise and safely support the vehicle.
2. Remove the wheel and tire assembly.
3. Remove the hub grease cap.

4. Remove the brake drum or disc brake caliper and rotor, as necessary.

5. Unstake and remove the wheel hub retainer nut securing the hub to the spindle and remove the hub and bearing assembly. Discard the hub retainer nut.

6. If equipped with disc brakes, remove the disc brake shield retaining bolts and the shield.

7. If equipped with drum brakes, remove the backing plate.

8. Remove the strut-to-spindle retaining bolts and nuts.

9. Remove the trailing arm bolt securing the trailing arm to the spindle.

10. Remove the stabilizer bar link nuts, retainers, bushings, sleeves and bolts.

11. Remove the control arm nut and bolt securing both control arms to the wheel spindle and remove the spindle from the vehicle.

12. Inspect all components. If the wheel spindle is damaged, replace it. Wheel bearings are sealed and must be replaced if damaged with the wheel hub.

To install:

13. Place the wheel spindle in position to the strut and install the retaining bolts and nuts. Tighten to 69–93 ft. lbs. (93–127 Nm).

14. Install the trailing arm to the spindle. Install the retaining bolt and tighten to 69–93 ft. lbs. (93–127 Nm).

15. Place both control arms in position to the spindle and install the retaining bolt and nut. Tighten the nut to 63–86 ft. lbs. (85–117 Nm).

16. Install the stabilizer bar link nuts, retainers, bushings, sleeves and bolts.

17. If equipped with drum brakes, install the brake backing plate.

18. If equipped with disc brakes, install the brake shield and the disc brake shield retaining bolts. Tighten securely.

19. Install the wheel hub and bearing assembly to the wheel spindle.

20. Install a new wheel hub retainer nut and tighten to 130–174 ft. lbs. (177–235 Nm).

21. Stake the wheel hub retainer nut using a cape or round end chisel. Do not use a sharp chisel to stake the hub nut.

22. Install the brake drum or the disc brake rotor and caliper, as equipped.

23. Install the hub grease cap.

24. Install the wheel and tire assembly.

25. Lower the vehicle.

26. Pump the brake pedal several times to position the brake lining before attempting to move the vehicle.

27. Road test the vehicle and check for proper operation.

FORD MOTOR CO.

Ford-Mustang

DRIVE TRAIN 23-40
ENGINE REPAIR 23-2
FUEL SYSTEM 23-39
STEERING AND
SUSPENSION 23-42

A
Air Bag.................................23-42
  Disarming the System 23-43
  Precautions 23-42

C
Camshaft and Valve Lifters 23-22
  Removal & Installation 23-22
Clutch.....................................23-40
  Adjustments 23-40
  Removal & Installation 23-41
Coil Spring 23-45
  Removal & Installation 23-45
Cylinder Head 23-7
  Removal & Installation 23-7

D
Distributor.................................23-2
  Removal & Installation 23-2

E
Engine Assembly 23-3
  Removal & Installation 23-3
Exhaust Manifold 23-19
  Removal & Installation 23-19

F
Fuel Filter 23-39
  Removal & Installation 23-39
Fuel Pump 23-40
  Removal & Installation 23-40
Fuel System Pressure 23-39
  Relieving 23-39
Fuel System Service
  Precautions............................23-39

I
Ignition Timing 23-2
  Adjustment 23-2
Intake Manifold 23-13
  Removal & Installation 23-13

L
Lower Ball Joint.........................23-46
  Removal & Installation 23-46
Lower Control Arm 23-46
  Removal & Installation 23-46

O
Oil Pan....................................23-26
  Removal & Installation 23-26
Oil Pump 23-30
  Removal & Installation 23-30

P
Power Rack and Pinion Steering
  Gear......................................23-43
  Removal & Installation 23-43

R
Rear Main Seal 23-31
  Removal & Installation 23-31
Rocker Arms 23-11
  Removal & Installation 23-11

S
Shock Absorber 23-44
  Removal & Installation 23-44
Strut.......................................23-43
  Removal & Installation 23-43

T
Timing Chain, Sprockets, Front
  Cover and Seal 23-32
  Removal & Installation 23-32
Transmission Assembly 23-40
  Removal & Installation 23-40

W
Water Pump..............................23-4
  Removal & Installation 23-4
Wheel Bearings.........................23-47
  Adjustment 23-47
  Removal & Installation 23-47

ENGINE REPAIR

➡Disconnecting the negative battery cable on some vehicles may interfere with the functioning of the on-board computer system, and may require the computer to undergo a relearning process once the negative battery cable is reconnected.

Distributor

➡Only the 5.0L engine is equipped with an adjustable distributor. The 3.8L and 4.6L engines use distributorless ignitions.

REMOVAL & INSTALLATION

1. Rotate the engine until the No. 1 piston is at Top Dead Center (TDC) of its compression stroke.
2. Disconnect the negative battery cable. Disconnect the vehicle wiring harness connector from the distributor. Before removing the distributor cap, mark the position of the No. 1 wire tower on the cap for reference.
3. Loosen the distributor cap hold-down screws and remove the cap. Match-mark the position of the rotor to the distributor housing. Position the cap and wires out of the way.
4. Scribe a mark in the distributor body and the engine block to indicate the position of the distributor in the engine.
5. Remove the distributor hold-down bolt and clamp.

➡Some engines may be equipped with a security-type distributor hold-down bolt. If this is the case, use distributor wrench T82L-12270-A or equivalent, to remove the retaining bolt and clamp.

6. Remove the distributor assembly from the engine. Be sure not to rotate the engine while the distributor is removed.

To install:
7. Be sure that the engine still has the No. 1 piston at TDC of its compression stroke.

➡If the engine was disturbed while the distributor was removed, it will be necessary to remove the No. 1 spark plug and rotate the crankshaft clockwise until the No. 1 piston is on its compression stroke. Align the timing pointer with TDC on the crankshaft damper or flywheel, as required.

8. Check that the O-ring is installed and in good condition on the distributor body.
9. On all vehicles:
   a. Rotate the distributor shaft so the rotor points toward the mark on the distributor housing made previously.
   b. Turn the rotor slightly so the leading edge of the vane is centered in the vane switch state assembly.
   c. Rotate the distributor in the block to align the leading edge of the vane with the vane switch stator assembly. Make certain the rotor is pointing to the No. 1 mark on the distributor base.

➡If the vane and vane switch stator cannot be aligned by rotating the distributor in the cylinder block, remove the distributor enough to just disengage the distributor gear from the camshaft gear. Rotate the rotor enough to engage the distributor gear on another tooth of the camshaft gear. Repeat Step 9 if necessary.

10. Install the distributor hold-down clamp and bolt(s); tighten them slightly.
11. Attach the vehicle wiring harness connector to the distributor.
12. Install the cap and wires. Install the No. 1 spark plug, if removed.
13. Recheck the initial timing.
14. Tighten the hold-down clamp and recheck the timing. Adjust if necessary.

Ignition Timing

➡No periodic checking or adjustment of the ignition timing is necessary for any of the vehicles covered by this manual. However, the distributor ignition system used by the 5.0L engine does allow for both, should the distributor be removed and installed or otherwise disturbed.

ADJUSTMENT

3.8L and 4.6L Engines

The 3.8L and 4.6L engines utilize the Distributorless Ignition System (DIS). On this system, ignition coil packs fire the spark plugs directly through the spark plug wires. All spark timing and advance is determined by the ignition control module and engine control computer. No ignition timing adjustments are necessary or possible.

5.0L Engine

➡Specific instructions and specifications for setting initial timing can be found in the Vehicle Emission Control Information (VECI) label in the engine compartment. Because this label contains information regarding any specific calibration requirements for YOUR vehicle, those instructions and specifications should be followed if they differ from the following.

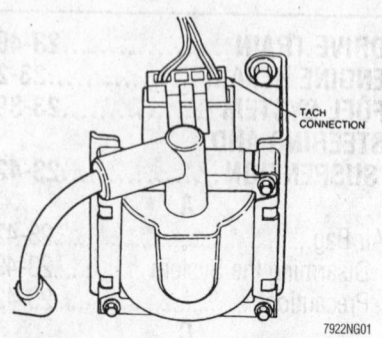

Using an alligator clip, a tachometer may be hooked up to the ignition coil connector

This procedure should not be used as a periodic maintenance adjustment. Timing should only be set after the distributor has been disturbed (removed and reinstalled) in some way.

➡Do not change the ignition timing by the use of a different octane rod without having the proper authority to do so. Federal emission requirements will be affected.

1. Start the engine and allow it to run until it reaches normal operating temperature.

✱✱ CAUTION

NEVER run an engine in a garage or building without proper ventilation. Carbon monoxide will quickly enter the body, excluding oxygen from the blood stream. This condition will cause dizziness, sleepiness and eventually death.

2. Once normal operating temperature has been reached, shut the engine **OFF**.
3. Firmly apply the parking brake and block the drive wheels. Place the transmission in **P** (A/T) or **NEUTRAL** (M/T), as applicable.
4. Be sure the heater and A/C, along with all other accessories, are in the OFF position.
5. Connect an inductive timing light, such as the Rotunda 059–00006 or equivalent, to the No. 1 spark plug wire, according the tool manufacturer's instructions.
6. Connect a tachometer to the ignition coil connection using an alligator clip. This

can be done by inserting the alligator clip into the back of the connector, onto the dark green/yellow dotted wire.

➡**DO NOT allow the alligator clip to accidentally ground to a metal surface while attached to the coil connector, as that could permanently damage the ignition coil.**

7. Disconnect the single wire inline SPOUT connector which connects the control computer (usually terminal 36) to the ignition control module. This will prevent the electronic ignition from advancing the timing during the setting procedure.

8. Using a suitable socket or wrench, loosen the distributor hold-down bolt slightly at this time, BUT DO NOT ALLOW THE DISTRIBUTOR TO MOVE, or timing will have to be set regardless of the current conditions.

➡**A remote starter must NOT be used to start the vehicle when setting the initial ignition timing. Disconnecting the start wire at the starter relay will cause the ignition control module to revert to Start Mode timing after the vehicle is started. Reconnecting the start wire after the vehicle is running WILL NOT correct the timing.**

9. Start the engine (using the ignition key and NOT a remote starter to assure timing will be set correctly) and allow the engine to return to normal operating temperature.

10. With the engine running at the specified rpm, check the initial timing. If adjustments must be made, rotate the distributor while watching the timing marks. Once proper adjustment has been reached, tighten the hold-down without moving the distributor.

11. Reconnect the single wire inline SPOUT connector and check the timing to verify that the distributor is now advancing beyond the initial setting.

12. Shut the engine **OFF** and tighten the distributor bolt while CAREFULLY holding the distributor from turning. If the distributor moves, you will have to start the engine and reset the timing.

13. Restart the engine and repeat the procedure to check the timing and verify that it did not change.

14. Shut the engine **OFF** , then disconnect the tachometer and timing light.

Engine Assembly

REMOVAL & INSTALLATION

1. Properly discharge the A/C system into an EPA approved recovery and recycling system.

2. Disconnect the battery cables.

✳✳ CAUTION

Never open, service or drain the radiator or cooling system when hot; serious burns can occur from the steam and hot coolant.

3. Remove the cooling fan, radiator and all cooling system hoses.

4. If equipped, remove the Mass Air Flow (MAF) sensor and intake air duct.

5. Label and detach all engine wiring and vacuum hoses which will interfere with engine removal.

6. Remove the hood.

7. If equipped, remove the engine compartment brace from the front fender apron and dash panel.

8. Detach the accelerator and transmission control cables from the throttle body, and the control cables' mounting bracket from the engine.

✳✳ CAUTION

Observe all applicable safety precautions when working around fuel. Whenever servicing the fuel system, always work in a well ventilated area. Do not allow fuel spray or vapors to come in contact with a spark or open flame. Keep a dry chemical fire extinguisher near the work area. Always keep fuel in a container specifically designed for fuel storage; also, always properly seal fuel containers to avoid the possibility of fire or explosion.

9. Relieve the fuel system pressure, then disconnect the fuel supply and return lines from the engine.

10. Remove the A/C compressor mounting bracket and drive belt tensioner.

11. Raise and safely support the vehicle.

12. Remove the power steering pump and hoses.

✳✳ CAUTION

The EPA warns that prolonged contact with used engine oil may cause a number of skin disorders, including cancer! You should make every effort to minimize your exposure to used engine oil. Protective gloves should be worn when changing the oil. Wash your hands and any other exposed skin areas as soon as possible after exposure to used engine oil. Soap and water, or waterless hand cleaner should be used.

13. Drain the engine oil and remove the oil filter.

14. Detach the exhaust system from the exhaust manifolds.

15. Label and detach any undervehicle engine wiring which will interfere with engine removal.

16. Remove the transmission.

17. If equipped, remove the transmission oil cooler line retainers from the right-hand, front engine support insulator.

18. Remove the front engine support insulator-to-crossmember fasteners.

19. Remove the starter motor and starter motor wiring from the engine.

20. Partially lower the vehicle and support it with jackstands in this new position.

21. On the 3.8L engine, install Engine Lifting Brackets 014–00791 and Separator Bar 014–00793 or their equivalents.

22. On the 4.6L SOHC engine, install the Engine Lifting Brackets D91P-6001-A, or their equivalents, onto the cylinder heads with M-12 x 1.75 x 20mm bolts.

23. On the 4.6L DOHC engine, install the lifting eyes from the Rotunda Engine Lift Bracket Set 014–00340, or their equivalents, onto the cylinder heads. Tighten the attaching bolts to 30–36 ft. lbs. (40–50 Nm).

24. On the 5.0L engine, attach Engine Lifting Eyes 014–00791 or equivalent to the exhaust manifolds.

25. Using an engine crane or hoist, lift the engine out of the vehicle. Be sure to lift the engine slowly and check often that nothing (such as wires, hoses, etc.) will cause the engine to hang up on the vehicle.

26. At this point, the engine can be installed on an engine stand.

To install:

➡**Lightly oil all bolts and stud threads, except those specifying special sealant, prior to installation.**

27. Using the hoist or engine crane, slowly and carefully position the engine in the vehicle; lower the engine onto the front engine support insulators.

28. On the 3.8L engine, tighten the front engine support insulator nuts to 72–98 ft. lbs. (97–133Nm).

29. On the 4.6L engine, install the support through-bolts and tighten them to 15–22 ft. lbs. (20–30 Nm).

30. On the 5.0L engine, tighten the front engine support insulator nuts to 44–60 ft. lbs. (59–81Nm).

31. Detach the engine crane or hoist from the engine, and remove the lifting brackets from the cylinder heads.

32. Install the transmission.

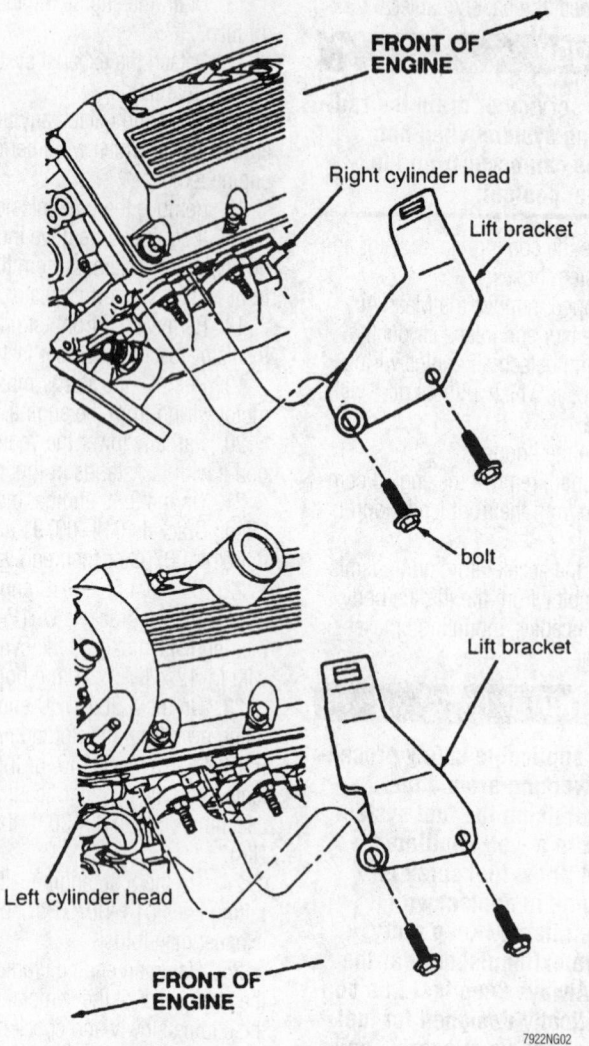

FRONT OF ENGINE

Right cylinder head

Lift bracket

bolt

Lift bracket

Left cylinder head

FRONT OF ENGINE

7922NG02

Install the engine lifting brackets on both sides of the engine—4.6L SOHC engine shown, other engines similar

33. The remainder of installation is the reverse of the removal procedure.

✳✳ WARNING

Do NOT start the engine without first filling it with the proper type and amount of clean engine oil, and installing a new oil filter. Otherwise, severe engine damage will result.

34. Fill the crankcase with the proper type and quantity of engine oil. Adjust the transmission throttle linkage.

35. Install the air intake duct assembly.

36. Connect the negative battery cable, then fill and bleed the cooling system.

37. Bring the engine to normal operating temperature, then check for leaks.

38. Stop the engine and check all fluid levels.

39. Install the hood, aligning the marks that were made during removal.

40. If equipped, have the A/C system properly leak-tested, evacuated and charged by an MVAC-trained, EPA-certified automotive technician.

Water Pump

REMOVAL & INSTALLATION

3.8L Engine

✳✳ CAUTION

Never open, service or drain the radiator or cooling system when hot; serious burns can occur from the steam and hot coolant.

1. Drain the engine cooling system.
2. Remove the electric cooling fan from the radiator.

3. Slightly loosen the water pump pulley bolts.

4. Remove the accessory drive belt.

5. Remove the water pump pulley retaining bolts, then separate the pulley from the water pump flange.

6. Remove the ignition coil bracket mounting nuts and bolts, then position the ignition coil (with all wires still attached) aside.

7. Remove the power steering pump pulley, then remove the water pump-to-power steering pump brace.

8. Remove the heater water outlet tube retaining bolts, then separate the outlet tube from the water pump.

9. Detach the lower radiator hose from the water pump.

✳✳ WARNING

If it is necessary to utilize a prytool to separate the water pump from the engine, use caution not to damage the water pump and engine mating surfaces.

10. Loosen the water pump bolts, then pull the water pump off of the engine. Remove and discard the old water pump gasket.

To install:

➡ **Lightly oil all bolt and stud threads prior to assembly, except those specifying sealant, with clean engine oil.**

11. Clean the gasket mating surfaces of the water pump and engine of all old gasket material and dirt.

12. Apply adhesive, such as Ford Gasket and Trim Adhesive F3AZ-19B508-B or equivalent, to the water pump gasket, then position the water pump and the new gasket on the engine.

➡ **The threads of the No. 1 water pump retaining bolt must be coated with a Teflon® sealant, such as Ford Pipe Sealant with Teflon® D8AZ-19554-A or equivalent, prior to installation.**

13. Install the water pump retaining bolts, studs and nuts, then tighten, in the sequence shown in the accompanying illustration, the retaining bolts and studs to 15–22 ft. lbs. (20–30 Nm) and the nuts to 71–106 inch lbs. (8–12 Nm).

14. Inspect the O-ring seal on the heater water outlet tube for damage, and replace it if necessary. Then, install the outlet tube to the water pump. Tighten the retaining bolts to 71–106 inch lbs. (8–12 Nm).

15. Attach the lower radiator hose to the water pump. Ensure that the radiator hose clamps are tightened properly.

16. Install the water pump-to-power steering pump brace, then install the power steering pump pulley.

17. Install the water pump pulley and retaining bolts. Tighten the bolts to 15–21 ft. lbs. (21–29 Nm).

18. Install the accessory drive belt.
19. Install the electric cooling fan.
20. Fill the cooling system with the proper amount and type of coolant.
21. Start the engine and check for coolant leaks.

4.6L Engines

❉❉ CAUTION

Never open, service or drain the radiator or cooling system when hot; serious burns can occur from the steam and hot coolant.

1. Drain the engine cooling system.
2. Remove the electric cooling fan from the radiator.
3. Remove the accessory drive belt.
4. Remove the water pump pulley by loosening the retaining bolts and pulling it off of the water pump flange.

❉❉ WARNING

If it is necessary to utilize a prytool to separate the water pump from the engine, use caution not to damage the water pump and engine mating surfaces.

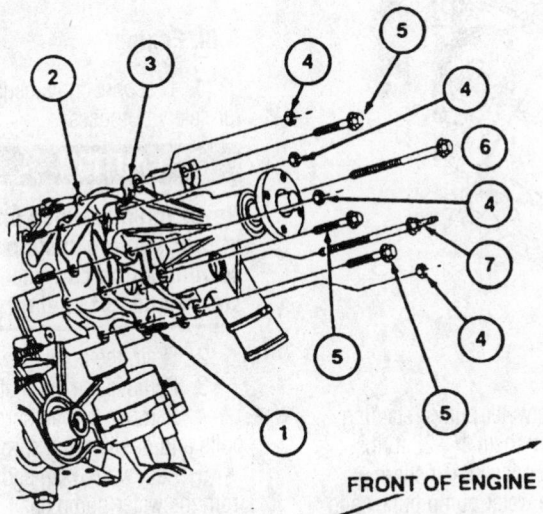

FRONT OF ENGINE

1. Mounting stud
2. Water pump housing gasket
3. Water pump
4. Mounting nuts
5. Short mounting bolts
6. Long mounting bolt
7. Mounting stud bolt

7922NG03

Exploded view of the water pump mounting on the 3.8L engine—during removal, note the original positions of the different length bolts for reassembly

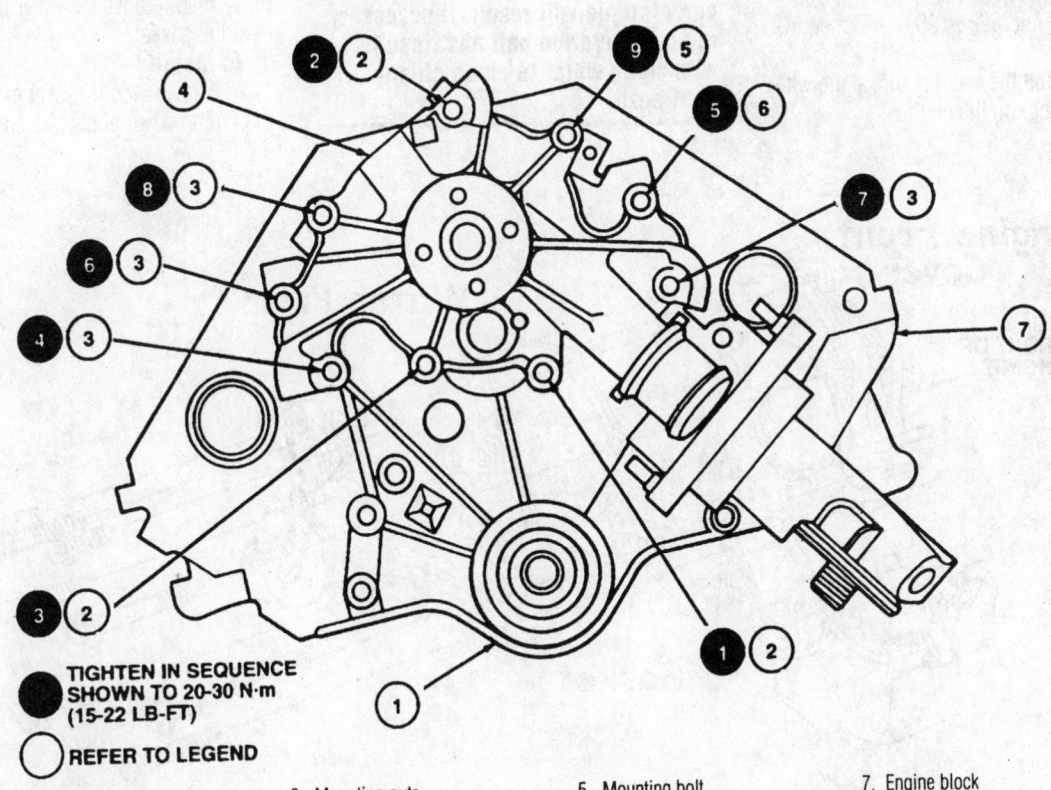

● TIGHTEN IN SEQUENCE SHOWN TO 20-30 N·m (15-22 LB-FT)

○ REFER TO LEGEND

1. Engine front cover
2. Mounting bolts
3. Mounting nuts
4. Water pump
5. Mounting bolt
6. Mounting stud bolt
7. Engine block

7922NG04

When tightening the mounting fasteners, be sure to adhere to the sequence shown for proper sealing—3.8L engine

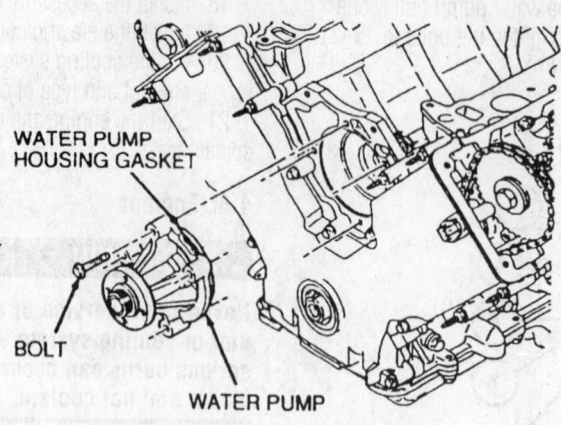

WATER PUMP
HOUSING GASKET

BOLT

WATER PUMP

7922NG05

Exploded view of the water pump mounting on all 4.6L engines

5. Loosen the water pump bolts, then separate the water pump from the engine. Remove and discard the old water pump O-ring.

To install:

➡ **Lightly oil all bolt and stud threads prior to assembly, except those specifying sealant, with clean engine oil.**

6. Clean the mating surfaces of the water pump and engine of all corrosion and dirt.

7. Install a new O-ring in the groove on the water pump, then lubricate the O-ring with coolant, such as Ford Premium Cooling System Fluid E2Fz-19549-AA or equivalent.

8. Position the water pump, along with the new O-ring, on the engine.

9. Install the water pump retaining bolts, then tighten them 15–22 ft. lbs. (20–30 Nm) in a crisscross pattern.

10. Install the water pump pulley and retaining bolts. Tighten the bolts to 15–21 ft. lbs. (21–29 Nm).

11. Install the accessory drive belt.

12. Install the electric cooling fan.

✳✳ WARNING

Use care to prevent engine coolant from spilling on the accessory drive belt, otherwise drive belt squeal and early fatigue will result. If necessary, remove the drive belt and rinse it with clean water to clean off any antifreeze.

13. Fill the cooling system with the proper amount and type of coolant.

14. Start the engine and check for coolant leaks.

5.0L Engine

1. Disconnect the negative battery cable for safety purposes.

✳✳ CAUTION

Never open, service or drain the radiator or cooling system when hot; serious burns can occur from the steam and hot coolant.

2. Drain the cooling system.

3. Remove the air inlet tube.

4. Remove the fan shroud attaching bolts and position the shroud over the fan.

5. Remove the fan and clutch assembly from the water pump shaft, then remove the clutch, fan and shroud from the vehicle.

6. Remove the accessory drive belt, then remove the water pump pulley.

7. Remove all accessory brackets that attach to the water pump.

8. Disconnect the lower radiator hose, heater hose and water pump bypass hose from the water pump.

9. Remove the water pump attaching bolts, then remove the water pump and discard the gasket.

To install:

10. Clean all old gasket material from the timing cover and water pump.

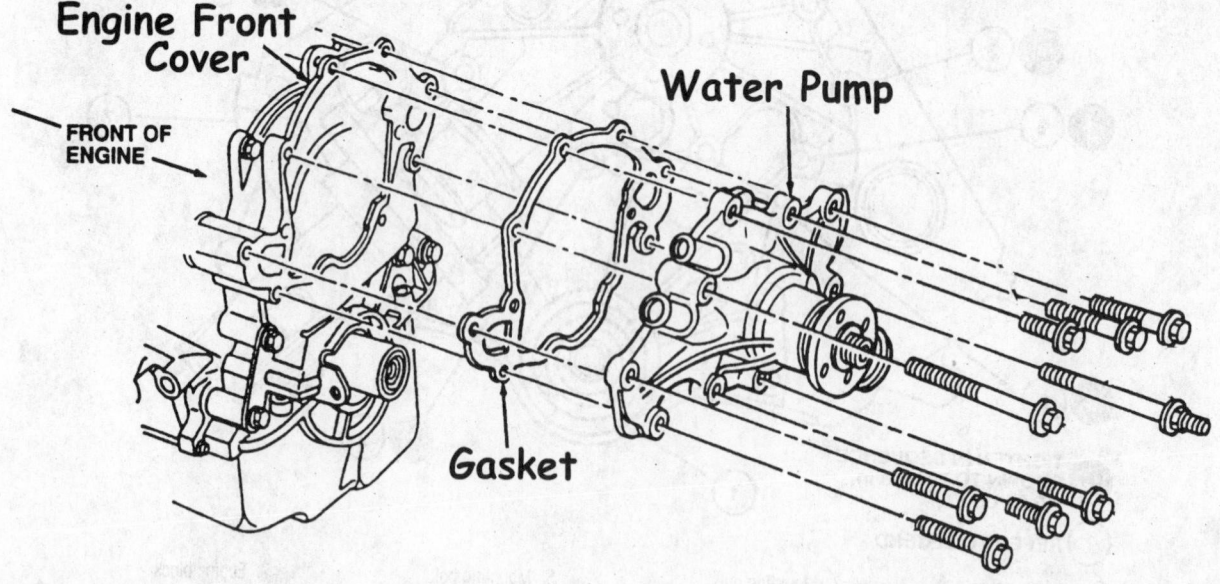

Engine Front Cover

FRONT OF ENGINE

Water Pump

Gasket

7922NG06

Exploded view of the water pump mounting. Note the positions of the bolts when removing them so they can be installed in their original positions—5.0L engines

11. Apply a suitable waterproof sealing compound to both sides of a new gasket, then position the gasket on the timing cover.

12. Position the water pump carefully over the gasket (making sure not to dislodge it). Install and tighten the pump mounting bolts to 12–18 ft. lbs. (16–24 Nm).

13. Connect the hoses and accessory brackets to the water pump.

14. Install the pulley on the water pump shaft.

15. Install the shroud along with the clutch and fan assembly.

16. Route and install the accessory drive belt.

17. Connect the negative battery cable and refill the cooling system.

18. Run the engine and check for leaks.

Cylinder Head

REMOVAL & INSTALLATION

3.8L Engine

1. Disconnect the negative battery cable.
2. Remove the upper and lower intake manifolds.
3. Remove the valve cover(s) and rocker arms.

➡Be sure to label or separate all of the cylinder head components (such as pushrods, rocker arms, etc.) so that they can be installed in their original positions on the cylinder heads.

4. Remove the exhaust manifolds.
5. If not already done, perform the following for left-hand cylinder head removal:

  a. Remove the Power Steering (PS) pump front mounting brace retaining nuts from the water pump.

  b. Remove the alternator.

  c. Remove the idler pulley.

  d. Remove the PS pump/alternator bracket mounting fasteners, then (leaving all PS hoses attached) position the PS pump and bracket aside in a position so that PS fluid does not leak out.

6. If not already done, perform the following for right-hand cylinder head removal:

  a. If equipped with A/C, remove the compressor mounting bracket bolts, then (leaving all A/C lines attached) position the compressor aside.

  b. If not equipped with A/C, remove the idler pulley.

7. Remove and discard the cylinder head mounting bolts.

8. Lift the cylinder head(s) up and off of the engine block.

9. Remove and discard the old cylinder head gasket(s).

To install:

➡Lightly oil all bolt and stud threads prior to assembly, except those specifying sealant, with clean engine oil.

10. Clean the cylinder head, intake manifold and valve cover gasket mating surfaces thoroughly.

11. If a cylinder head was removed for head gasket replacement, inspect the cylinder head for warpage.

➡If a cylinder head is warped, have it ground by a competent machine shop before installing.

12. Position the new cylinder head gasket(s) on the engine block, then install the cylinder head(s).

✳✳ WARNING

Always use new cylinder head bolts to ensure a leak-tight seal. Torque values with used cylinder head bolts can vary, which may cause coolant or compression leakage.

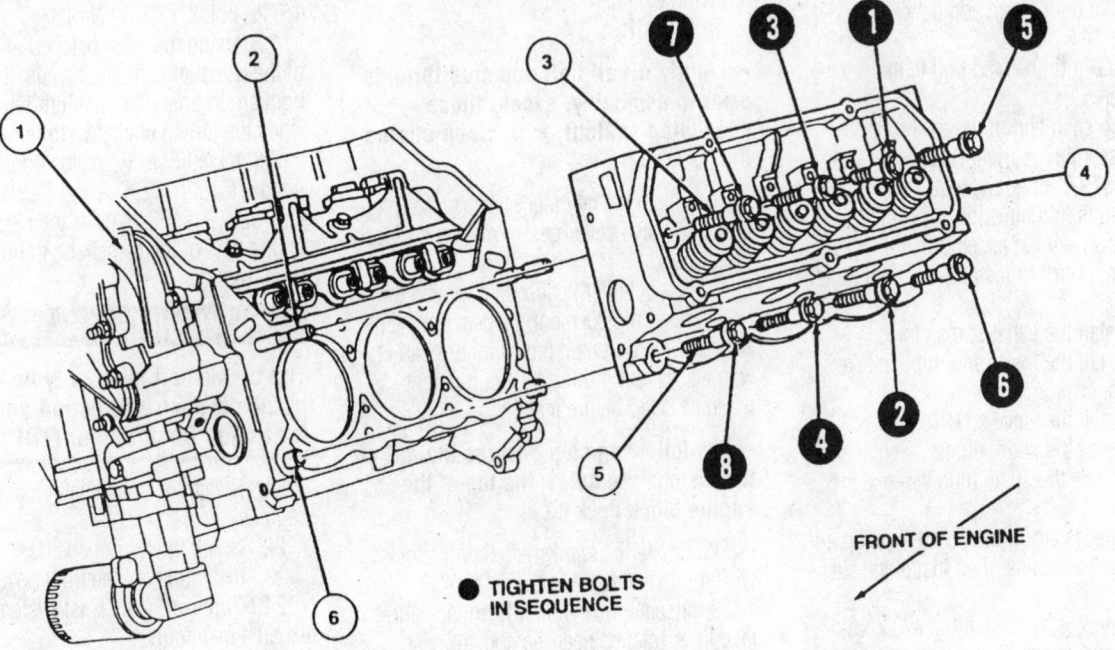

1. Engine block
2. Locating pin
3. Long cylinder head mounting bolts
4. Cylinder head
5. Short cylinder head mounting bolts
6. Alignment dowel

● TIGHTEN BOLTS IN SEQUENCE

FRONT OF ENGINE

7922NG07

Be sure to tighten the cylinder head mounting bolts in the sequence shown (black numbers) to ensure proper cooling system and combustion chamber sealing—3.8L engine

13. Tighten the cylinder head bolts in the sequence shown, and in the following steps to the proper values:

 a. Tighten the bolts to 15 ft. lbs. (20 Nm).

 b. Next, tighten the bolts to 29 ft. lbs. (40 Nm).

 c. Then, tighten the bolts to 37 ft. lbs. (50 Nm).

✳✳ WARNING

Do not loosen all of the bolts at the same time in the following step; only work on one bolt at a time, then move on the next bolt in the sequence.

 d. Loosen the bolts, one at a time, two or three revolutions, then retighten them as follows:

1995 Models

• Tighten the bolt to 133–221 inch lbs. (15–25 Nm) for a long bolt, or to 88–177 inch lbs. (10–20 Nm).

• Rotate the bolt an additional 85–95 degrees.

• Proceed to the next bolt in the sequence.

1996–99 Models

• Tighten the bolt to 29–37 ft. lbs. (40–50 Nm) for a long bolt, or to 133–221 inch lbs. (15–25 Nm) for a short bolt.

• Rotate the bolt an additional 180 degrees.

• Proceed to the next bolt in the sequence.

14. Dip each end of the pushrods in engine assembly lubricant, such as Ford D9AZ-19579-D or equivalent, then install the pushrods in their original positions.

15. Apply engine assembly lube to the rocker arms prior to installation, then install the rocker arms.

16. Install the exhaust manifolds.

17. Install the lower and upper intake manifolds.

18. Install the valve cover(s).

19. Install the spark plugs, if removed. Then, connect the spark plug wires to the plugs.

20. If necessary, install the PS pump and bracket on the engine, then install the alternator.

21. If necessary, install the A/C compressor and bracket on the engine.

22. If not already performed, install the accessory drive belt pulleys and the drive belt.

23. Connect the negative battery cable.

24. Fill the engine with the proper type and amount of coolant.

25. Start the engine and check for coolant leaks.

4.6L SOHC Engine

1. Disconnect the negative battery cable, drain the cooling system and relieve fuel system pressure.

2. Remove the valve cover(s).

3. Remove the throttle body.

4. Remove the intake manifold.

5. Disconnect the exhaust system from the exhaust manifolds.

6. Remove the engine front cover.

7. Remove the timing chains. Be sure to note the positions of the flats on the ends of the two camshafts before removing the cylinder heads.

8. Remove or detach any remaining wires, hoses or cables which will interfere with cylinder head removal.

✳✳ WARNING

Do not set the cylinder head down on its engine block mating surface; if any valves are open, they may be damaged.

9. Remove the cylinder head mounting bolts and lift the cylinder heads off of the engine block. If necessary, at this time the heater water hose can be removed from the left-hand cylinder head.

To install:

➡ **Lightly oil all bolt and stud threads prior to assembly, except those specifying sealant, with clean engine oil.**

10. Clean the cylinder head, intake manifold and valve cover gasket mating surfaces thoroughly.

11. If a cylinder head was removed for head gasket replacement, inspect the cylinder head for warpage. If a cylinder head is warped, have it ground by a competent machine shop before installing.

➡ **The following step ensures that all of the pistons are below the top of the engine block deck face.**

12. Rotate the crankshaft counterclockwise 45 degrees.

13. Inspect the cylinder head face surface for scratches near the coolant passages and combustion chambers.

14. Rotate the camshaft to a stable position where the valves do not extend below the cylinder head face.

15. Position the new cylinder head gasket(s) on the engine block, then install the cylinder head(s).

✳✳ WARNING

Cylinder head mounting bolts must be replaced with new bolts; do NOT reuse original cylinder head mounting bolts. They are torque-to-yield bolts and cannot be reused; otherwise, engine damage may occur.

16. Apply clean engine oil to the cylinder head mounting bolts' spot faces. Then, install the mounting bolts hand-tight.

17. Tighten the cylinder head bolts (starting with the left-hand cylinder head first) in the sequence shown in the accompanying illustration and following these steps:

 a. Tighten the bolts to 27–31 ft. lbs. (37–43 Nm).

 b. Tighten the bolts an additional 85–95 degrees.

 c. Loosen the bolts a minimum of one full revolution (360 degrees).

 d. Retighten the bolts to 27–31 ft. lbs. (37–43 Nm).

 e. Tighten the bolts an additional 85–95 degrees.

 f. Finally, tighten the bolts an additional 85–95 degrees.

18. Install the heater water hose on the left-hand cylinder head, then tighten the lower mounting bolt to 71–106 inch lbs. (8–12 Nm) and the upper mounting bolt to 15–22 ft. lbs. (20–30 Nm).

19. Using the flats matched at the center of the camshafts, rotate the camshafts until both are in time (the position they were in when the timing chains were removed).

20. Install Ford Cam Positioning Tool Adapters T92P-6256-A and Cam Positioning Tool T91P-6256-A (or their equivalents) on the flats of the camshafts to prevent them from rotating.

✳✳ WARNING

The crankshaft must only be rotated in the clockwise direction and only as far as Top Dead Center (TDC).

21. Rotate the crankshaft clockwise 45 degrees.

22. Install the timing chains.

23. Install the engine front cover.

24. Connect the exhaust system to the exhaust manifolds.

25. Install the intake manifold.

26. Install the throttle body.

27. Install the valve cover(s).

28. Install or attach any remaining wires, hoses or cables which were removed earlier.

29. Connect the negative battery cable and fill the cooling system.

30. Start the engine and check for leaks.

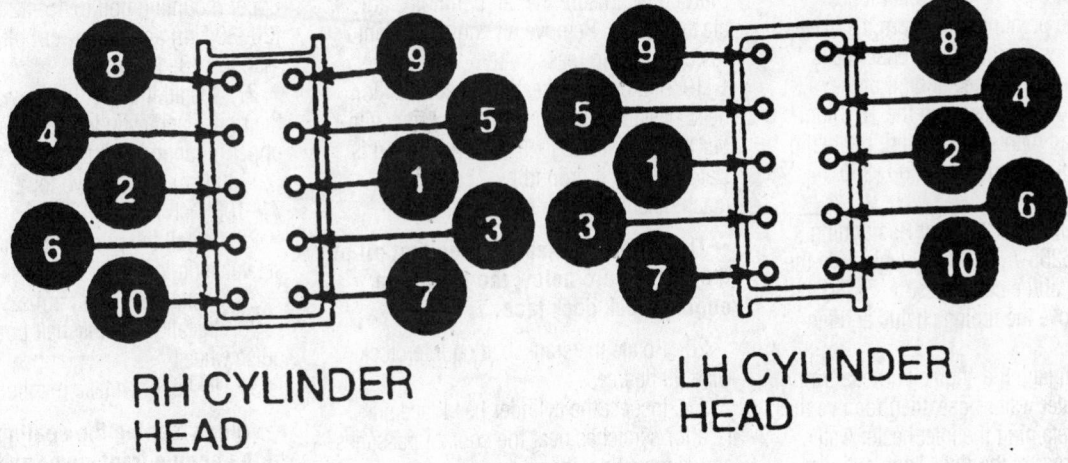

RH CYLINDER HEAD   LH CYLINDER HEAD

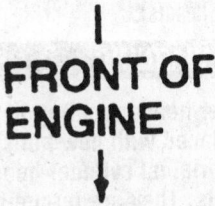

FRONT OF ENGINE

7922NG08

When tightening the cylinder head mounting bolts, be sure to follow the order shown—4.6L SOHC engine

4.6L DOHC Engine

➥**According to the manufacturer, the cylinder heads are not removable with the engine in the vehicle; therefore, this procedure is written with the assumption that the engine is removed from the vehicle.**

1. Remove the engine from the vehicle, as described earlier in this section.

2. Remove the valve covers.

3. Remove the engine front cover.

4. Remove the intake manifolds.

5. Remove the crankshaft position sensor pulse wheel.

6. Remove the rocker arms.

7. Remove the exhaust manifolds.

8. If not already done, drain the coolant from the engine block.

9. Rotate the engine to Top Dead Center (TDC) as follows:

   a. Remove the spark plugs from the cylinder heads.

   b. Install the crankshaft damper center bolt in the crankshaft.

   c. Position your thumb over the No. 1 cylinder's spark plug hole, so that it seals the hole completely.

➥**Only rotate the crankshaft in the clockwise direction (when looking at the front of the engine).**

   d. Have an assistant rotate the crankshaft using a breaker bar and socket on the crankshaft center bolt until you start to feel air escaping past your thumb (you will not have the strength to keep the cylinder sealed—air will escape).

➥**It may take a couple of tries to get the feeling for the air down.** Have

patience and take your time, it is vitally important to get the engine at TDC.

   e. From that point, rotate the crankshaft until the TDC marks are aligned. The timing chain and crankshaft gears' timing marks should be aligned at this point. Also, all four valves should be

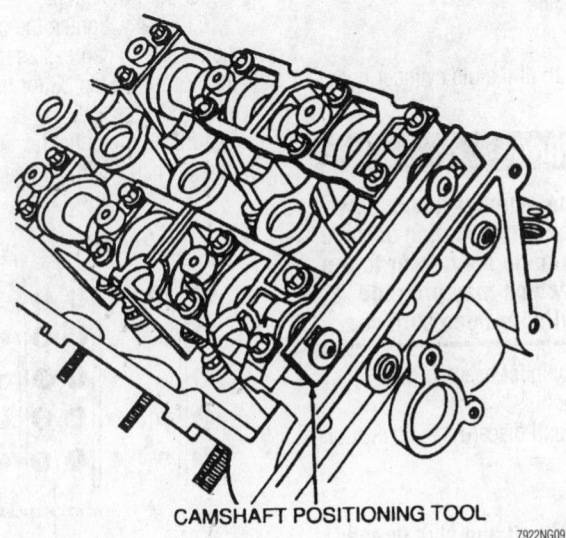

CAMSHAFT POSITIONING TOOL

7922NG09

Use a camshaft positioning tool to hold the camshafts from rotating once the timing chains are removed—4.6L DOHC engine

closed (that is, their camshaft lobes should be pointing away from the valves and springs). If the timing chain gear marks are not aligned and all of the valves are not closed, but the TDC ignition timing marks are aligned, rotate the crankshaft one full revolution (360 degrees).

10. Install Ford Camshaft Positioning Tool T93P-6256-A, or the equivalent, in the rear D-slots of the camshafts.

11. Remove the timing chains and tensioners.

12. Disengage the engine harness wiring from the heater water hose, then remove the three bolts retaining the inlet heater water hose to the rear of the right-hand cylinder head.

13. Remove the outlet heater water hose mounting bolts from the rear of the left-hand cylinder head.

14. Using the removal sequence, shown in the accompanying illustration, remove the cylinder head mounting bolts.

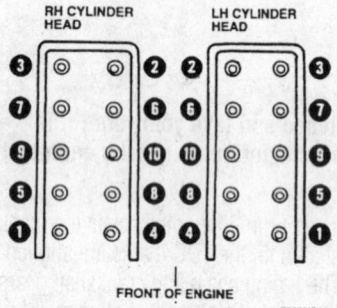

7922NG10

Be sure to loosen the cylinder head mounting bolts in the sequence shown—4.6L DOHC engine

15. Discard all of the old cylinder head mounting bolts.

✷✷ WARNING

Do NOT set the cylinder heads down on their engine block mating surface; otherwise, damage may occur to the valve heads, which may protrude beyond the cylinder head surface.

16. Lift the cylinder heads up and off of the engine block.

17. Remove and discard the old cylinder head gaskets.

To install:

➡**Lightly oil all bolt and stud threads prior to assembly, except those specifying sealant, with clean engine oil.**

18. Clean the cylinder head, intake manifold, valve cover gasket mating surfaces

thoroughly. Ensure that all bolt holes are clean and dry. Remove all engine coolant residue prior to reassembly.

19. If a cylinder head was removed for head gasket replacement, inspect the cylinder head for warpage. If a cylinder head is warped, have it ground by a competent machine shop before installing.

➡**The following step ensures that all of the pistons are below the top of the engine block deck face.**

20. Rotate the crankshaft counterclockwise 90 degrees.

21. Inspect the cylinder head face surface for scratches near the coolant passages and combustion chambers.

22. Position the new cylinder head gasket(s) on the engine block, then install the cylinder head(s).

✷✷ WARNING

Cylinder head mounting bolts must be replaced with new bolts; do NOT reuse original cylinder head mounting bolts. They are torque-to-yield bolts and cannot be reused; otherwise, engine damage may occur.

23. Apply clean engine oil to both sides of the cylinder head mounting bolts' washer faces.

24. Tighten the left cylinder head bolts in the sequence presented in the accompanying illustration, as follows:

   a. Tighten the bolts to 27–32 ft. lbs. (37–43 Nm).

   b. Then, tighten the bolts an additional 85–95 degrees.

   c. Finally, tighten the bolts an additional 85–95 degrees again.

25. Perform Step 23 for the right-hand cylinder head as well.

26. Install the outlet heater water hose on the left-hand cylinder head. Tighten the

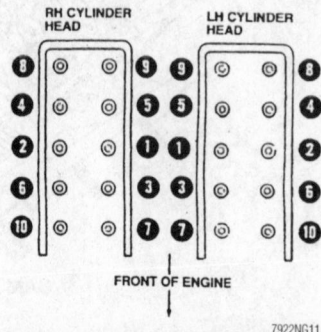

7922NG11

Tighten the new cylinder head mounting bolts ONLY in the order shown. This will ensure proper cooling system and combustion chamber sealing—4.6L DOHC engine

upper mounting bolt to 15–22 ft. lbs. (20–30 Nm) and the lower bolt to 71–106 inch lbs. (8–12 Nm).

27. Install the inlet heater water hose on the right-hand cylinder head. Tighten the upper mounting bolt to 15–22 ft. lbs. (20–30 Nm) and the two lower bolts to 71–106 inch lbs. (8–12 Nm).

28. Install the primary timing chains and set valve timing.

29. Install the rocker arms.

30. Install the crankshaft position sensor pulse wheel.

31. Install the intake manifolds.

➡**When cleaning the sealing surfaces of the engine front cover and the oil pan-to-engine block joints, be careful to avoid damaging the oil pan gasket rubber bead. If a leak develops after reassembly, the oil pan must be removed and the oil pan gasket replaced.**

32. Install the engine front cover.

33. Install the exhaust manifolds.

34. Install the engine in the vehicle.

✷✷ WARNING

It is very important to fill the engine with oil prior to starting it. Excessive engine damage will result if the engine is run without oil.

35. If necessary, fill the engine with the proper type and amount of clean engine oil.

36. Fill the engine cooling system.

37. Connect the positive, then the negative battery cables.

38. Start the engine and check for exhaust, coolant, fuel and oil leaks.

39. Shut the engine **OFF** and check the engine oil level; add oil, if necessary.

5.0L Engine

➡**Only an MVAC-trained, EPA-certified, automotive technician should service the A/C system or its components.**

1. For left-hand cylinder head removal, have the A/C compressor removed.

✷✷ WARNING

Ensure that the technician immediately caps or plugs all openings to the air conditioning system in order to prevent system contamination and damage.

2. Disconnect the negative battery cable for safety purposes.

3. Drain the cooling system and properly relieve the fuel system pressure.

4. Remove the upper and lower intake manifold assemblies, along with the throttle body.

5. If you are removing the left cylinder head, perform the following:

　a. If not done earlier, remove the drive belt from the power steering pump pulley, then disconnect the power steering pump bracket from the cylinder head. Position the pump out of the way in a position that will prevent the oil from draining out.

　b. Disconnect the oil level indicator tube bracket from the exhaust manifold stud, if necessary.

6. Remove the Thermactor/secondary air injection crossover tube from the rear of the cylinder heads.

7. If you are removing the right cylinder head, perform the following steps:

　a. Disconnect the alternator or alternator and air pump mounting bracket from the cylinder head, as equipped.

　b. Remove the fuel line from the clip at the front of the cylinder head.

8. Apply the parking brake and block the rear wheels, then raise and safely support the front of the vehicle.

9. Disconnect the exhaust manifolds from the muffler inlet pipes, then lower the vehicle.

10. Loosen the rocker arm fulcrum bolts so the rocker arms can be rotated to the side. Remove the pushrods and tag or arrange them in sequence so they may be installed in their original positions.

➡**A piece of wood or cardboard may be drilled and labeled to help sort the pushrods and be sure they are only installed to the rocker arms and lifters from which they were removed.**

11. Remove the cylinder head attaching bolts and the cylinder head(s). Remove and discard the head gasket(s).

➡**It may be necessary to remove the exhaust manifolds for access to the lower head bolts. If so, be sure to thoroughly clean the gasket mating surfaces and to use new exhaust manifold gaskets upon installation.**

12. Clean all gasket mating surfaces. Check the flatness of the cylinder head using a straightedge and a feeler gauge. The cylinder head must not be warped any more than 0.003 in. (0.076mm) in any 6 in. (152mm) span; 0.006 in. (0.152mm) overall. Machine if necessary.

To install:

13. Position the new cylinder head gasket(s) over the dowels on the block. Posi-

tion the cylinder head(s) on the block and install the head bolts.

✳✳ WARNING

Cylinder head mounting bolts must be replaced with new bolts; do NOT reuse original cylinder head mounting bolts. They are torque-to-yield bolts and cannot be reused; otherwise, engine damage may occur.

14. The cylinder head bolts should be tightened in three steps of the proper sequence, as follows:

　a. Tighten the bolts to 25–35 ft. lbs. (34–47 Nm).

　b. Next, tighten the bolts to 45–55 ft. lbs. (61–75 Nm).

　c. Finally, tighten all bolts an additional ¼ turn (85–95° if a torque angle meter is available).

15. If removed, install the exhaust manifolds using new gaskets.

16. Clean the pushrods, making sure the oil passages are clean. If you have access to compressed air, blow out the passages to be sure they are clear. Check the ends of the pushrods for wear. Visually check the pushrods for straightness or check for run-out using a dial indicator. Replace pushrods, as necessary.

17. Apply a suitable multi-purpose grease to the ends of the pushrods and install them in their original positions. Position the rocker arms over the pushrods and the valves.

➡**If all the original valvetrain parts are reinstalled and no cylinder head milling was performed, a valve clearance check is not necessary. If any valvetrain components are replaced, a valve clearance check must be performed.**

18. Raise and safely support the vehicle. Connect the exhaust manifolds to the muffler inlet pipes, then lower the vehicle.

19. If the right cylinder head was removed, reposition and install the alternator, or alternator and air pump bracket. Install the alternator assembly.

20. Install the valve cover(s) using new gasket(s).

21. If the left cylinder head was removed, reposition and install the A/C compressor and/or the power steering pump, as equipped.

22. Install the drive belt, making sure that the belt tensioner is maintaining the proper tension and that the belt is properly aligned on all of the pulleys.

23. Install the Thermactor/secondary air injection crossover tube at the rear of the cylinder heads.

24. Install the upper and lower intake manifold assemblies.

25. Connect the negative battery cable, then fill and bleed the cooling system.

26. Bring the vehicle to normal operating temperature. Check for leaks and check all fluid levels.

27. If the A/C system was discharged, take the vehicle to a reputable service facility and have the refrigerant system leak-tested, evacuated and charged according to the proper procedures.

Rocker Arms

REMOVAL & INSTALLATION

3.8L Engine

➡**During this procedure, always keep the parts in order so that they may be reinstalled in their original positions.**

1. Remove the valve cover(s) from the engine.

2. Loosen the rocker arm fulcrum (seat) mounting bolts from the cylinder head, then lift the rocker arms, fulcrums and bolts off of the head.

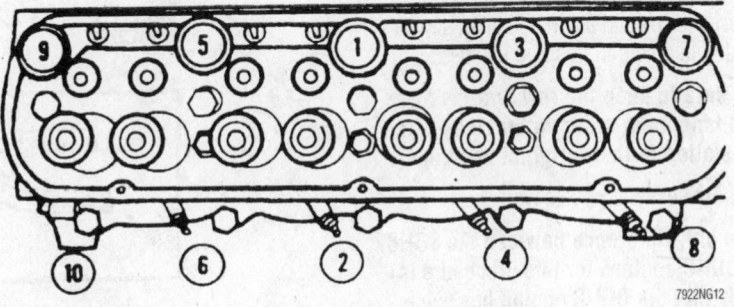

Cylinder head torque sequence—5.0L engine

7922NG12

3. If necessary, the pushrods may be removed from the engine at this time. If removing more than one pushrod at a time, be sure to label them so that they can be installed in their original positions.

To install:

4. If removed, install the pushrods in their original positions, after lubricating their tips with engine assembly lube (such as Ford D9AZ-19579-D).

5. Lubricate the rocker arms and fulcrums with engine assembly lube prior to installation.

6. Install each rocker arm, one at a time, as follows:

    a. Rotate the crankshaft so that the valve lifter (tappet), which corresponds to the rocker arm being installed, rests on the base circle of the camshaft (the camshaft lobe points away from the valve lifter). This position can be found by depressing the pushrod while having an assistant rotate the crankshaft with a socket and ratchet wrench on the crankshaft pulley/damper center bolt; when the pushrod is at its lowest point, the lobe is pointing away from the lifter.

    b. Position the rocker arm over the pushrods and mounting bolt holes.

    c. Install the rocker arm fulcrum and mounting bolt. Tighten the mounting bolt to a maximum of 44 inch lbs. (5 Nm).

7. Perform Step 4 for each of the rocker arms requiring installation.

➡**For the final-tightening of the rocker arm mounting bolts, the camshaft can be in any position.**

8. Final-tighten the rocker arm fulcrum bolts to 19–25 ft. lbs. (25–35 Nm) for 1994–95 models, and to 23–29 ft. lbs. (30–40 Nm) for 1996–98 models.

9. Install the valve cover(s).

4.6L Engines

On this engine, the cam followers (rocker arms) are part of the hydraulic valve lash adjustment assembly. The hydraulic lash adjusters are placed at the fulcrum point of the rocker arms and act in a manner similar to hydraulic lifters in pushrod engines.

➡**Be sure to keep the rocker arms and valve tappets in order so that they can be installed in their original positions.**

1. Remove the valve covers.

➡**The only difference between the SOHC and DOHC engines for this procedure is simply that the DOHC engine has twice as many valves, rocker arms and valve tappets. The removal and installation**

technique is the same, regardless of which valve is being serviced, or which engine is being serviced.

2. Rotate the crankshaft so that the position of the piston in the cylinder being serviced is at the bottom of its stroke, and the base circle of the camshaft is resting against the rocker arm.

✳✳ WARNING

A valve spring spacer, such as Ford Tool T91P-6565-AH or equivalent, must be used in order to prevent the valve spring retainer from contacting the valve stem seal, which can result in damage to the seal.

3. Install the valve spring spacer between the valve spring coils.

4. Install a valve spring compressor, such as Ford Tool T91P-6565-A or equivalent, under the camshaft and on top of the valve spring retainer.

5. Slowly push down on the compressor tool until the spring is depressed far enough to allow the rocker arm to be removed, then pull the rocker arm out from underneath the camshaft.

6. Remove the valve spring compressor and spacer tools.

7. Repeat Steps 2 through 6 for all of the rocker arms needing removal.

8. If necessary, pull the hydraulic valve tappets out of the cylinder head(s).

9. Thoroughly clean the valve tappets and rocker arms with clean solvent, then wipe them with a lint-free cloth. Do not immerse the valve tappet in solvent.

10. Inspect the valve tappet, and discard it if any part shows signs of pitting, scoring, or excessive wear.

To install:

11. Apply clean engine oil to the valve stems and tips, the rocker arm roller surfaces, the valve tappets, and the valve tappet bores in the cylinder head(s).

➡**The valve tappets must not have more than 0.059 in. (1.5mm) of plunger travel prior to installation.**

12. If removed earlier, insert the tappets in their original bores in the cylinder head(s).

13. Rotate the crankshaft so that the position of the piston in the cylinder being serviced is at the bottom of its stroke, and the base circle of the camshaft is facing where the rocker arm will be installed.

14. Install the valve spring spacer and compressor tools as before.

15. Carefully depress the valve spring with the compressor tool until the rocker arm can be positioned between the valve stem and valve tappet, and the camshaft.

16. Slowly allow the valve spring to decompress, and ensure that the rocker arm is properly retained.

17. Remove the valve spring compressor and spacer tools.

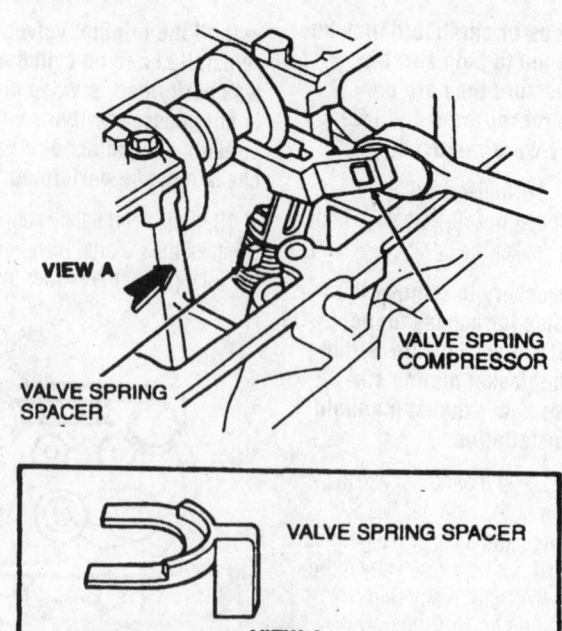

To properly depress the valve and spring, use both the valve spring spacer and compressor tools

18. Repeat Steps 13 through 17 for all of the rocker arms.

19. Install the valve covers.

20. Start the engine and check for oil leaks from the valve cover gaskets.

5.0L Engine

1. Disconnect the negative battery cable for safety purposes.

2. Remove the upper intake manifold assembly.

3. Remove the valve covers.

4. For non-Cobra models, loosen the rocker arm fulcrum bolt, then remove the bolt, fulcrum (seat), rocker arm and guide. KEEP ALL PARTS IN ORDER FOR INSTALLATION!

➡**Label and/or arrange all rocker arm and fulcrum components to assure installation in their original locations.**

5. On the Cobra, loosen the fulcrum bolt, then remove the bolt, roller rocker arm assembly and pedestal assembly.

To install:

6. Inspect the fulcrum bolts for damage. Replace any bolt on which damage is found.

7. Inspect the rocker arm and fulcrum seat or roller contact surfaces for wear

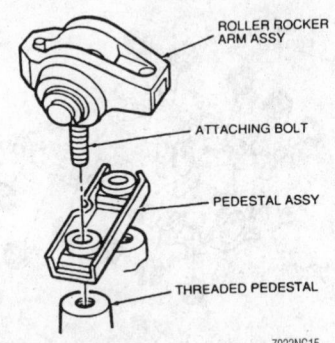

Exploded view of the rocker arm assembly—5.0L engine (Cobra)

and/or damage. Also check the rocker arm for wear on the valve stem tip contact surface and the pushrod socket. Replace complete rocker arm assemblies, as necessary.

8. Inspect the pushrod end and the valve stem tip. Replace pushrods, as necessary. If the valve stem tip is worn, the cylinder head must be removed to replace or machine the valve.

9. Lubricate the rocker arms and fulcrum seats with heavy engine oil or pre-assembly lube before installation.

10. Apply multi-purpose grease to the valve stem tips, the underside of the fulcrum seats and the sockets.

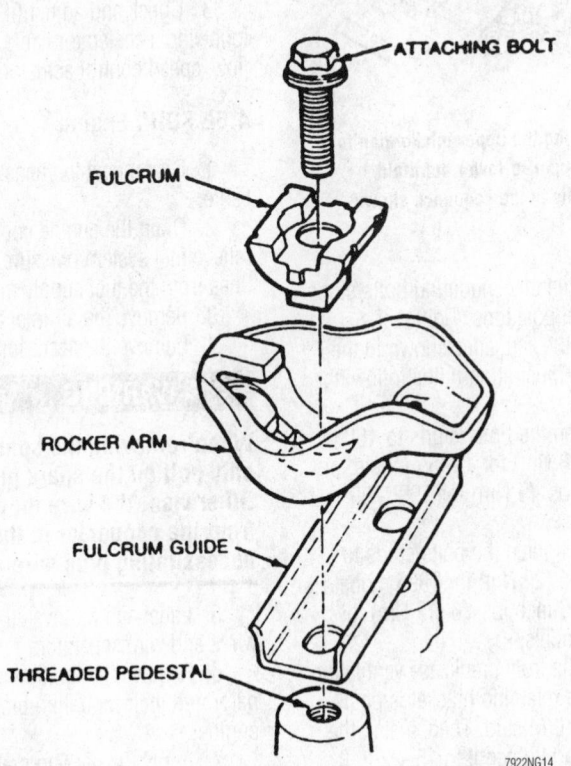

Exploded view of the rocker arm assembly—5.0L engine (except Cobra)

11. Rotate the crankshaft until the tappet is on the base circle of the camshaft lobe (all the way down), then install the rocker arm assembly. Tighten the fulcrum bolt to 18–25 ft. lbs. (24–34 Nm).

12. Install the valve covers.

13. Install the upper intake manifold assembly.

14. Connect the negative battery cable.

Intake Manifold

The 3.8L, 4.6L DOHC and 5.0L engines covered by this manual utilize an upper and lower intake manifold assembly. If necessary, only the upper intake manifold may be removed by following the intake manifold procedure up to that point. Obviously, installation would also begin at the upper intake steps.

REMOVAL & INSTALLATION

➡**Any time the upper or lower intake manifold has been removed, cover all openings with a rag or a sheet of plastic to prevent dirt, debris or, more importantly, a loose nut from falling into the engine.**

➡**Only an MVAC-trained, EPA-certified automotive technician should service the A/C system or its components.**

3.8L Engine

1. Remove the air inlet tube from the throttle body.

2. Detach the accelerator cable from the throttle body, then remove the cable bracket bolts and position the cable and bracket aside.

3. If equipped, remove the speed control actuator.

4. Detach and label any necessary engine harness wiring connectors and vacuum lines.

5. If necessary, remove the throttle body at this time.

6. Remove the EGR valve from the Upper Intake Manifold (UIM).

7. Remove the engine support and wiring retainer bracket (located on the front, left-hand side of the engine).

➡**Note the original locations of the UIM mounting bolts/studs for reinstallation.**

8. Remove the UIM mounting bolts/studs.

9. Lift the UIM off of the Lower Intake Manifold (LIM). Remove the old UIM-to-LIM gasket.

➡ If only the UIM is to be serviced, stop at this point (and resume installation at Step 23).

10. Remove the fuel injectors and fuel supply manifold.

11. Remove the heater water outlet hose from the LIM.

➡ Note the original locations of the LIM mounting bolts/studs for reinstallation.

12. Remove the LIM mounting bolts/studs.

✳✳ WARNING

The LIM may be sealed at the front and rear with silicone sealant. It may be necessary to use a prytool to loosen the intake manifold. If it is necessary to use a prytool, use care to avoid damaging the manifold, cylinder head and engine block machined surfaces.

13. Lift the LIM up and off of the engine. Remove and discard the old gaskets and end seals.

14. If a new manifold is to be installed, transfer the following components to the new manifold:
- Water hose connection and thermostat
- Engine Coolant Temperature (ECT) sensor
- Coolant outlet elbow
- All vacuum and electrical fittings

To install:

➡ **Lightly oil all mounting bolt/stud threads before installation.**

15. Clean the UIM and LIM, cylinder head and engine block gasket surfaces of all old gasket material or dirt. Use Ford Metal Surface Cleaner F4AZ-19A536-RGH or equivalent to clean the gasket surfaces of all residue which may interfere with the new sealant's adhesion.

16. Apply a dab of Ford Gasket and Trim Adhesive D7AZ-19B508-B or equivalent to each cylinder head mating surface, then press the new LIM gasket into place.

➡ **When using silicone sealant, assembly must occur within 15 minutes after the sealant is applied, otherwise the sealing ability may be jeopardized.**

17. Apply a 1/8 in. (3–4mm) bead of Ford Silicone Rubber D6AZ-19562-BA or equivalent to each corner where the cylinder head meets the engine block.

18. Install the new front and rear LIM end seals.

19. Carefully lower the LIM onto the cylinder heads and engine block.

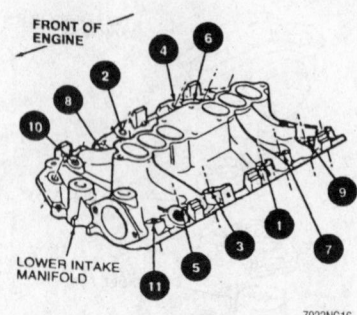

Be sure to tighten the lower intake manifold mounting bolts in the sequence shown—3.8L engine

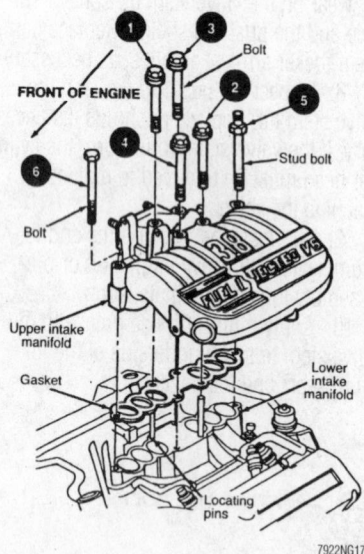

When installing the upper intake manifold, tighten the upper-to-lower manifold attaching bolts in the sequence shown—3.8L engine

20. Install the LIM mounting bolts/studs in their original positions. Tighten the bolts/studs in the sequence shown in the accompanying illustration in the following two steps:

a. Tighten the bolts/studs to 160 inch lbs. (18 Nm) for 1995 models, or to 45 inch lbs. (5 Nm) for 1996–99 models.

b. Then, tighten the bolts/studs to 195 inch lbs. (22 Nm) for 1995 models, or to 71–106 inch lbs. (8–12 Nm) for 1996–99 models.

21. Install the front crankcase ventilation tube so that the retaining bracket is positioned over the LIM stud. Then, install the retaining nut and tighten it to 15–22 ft. lbs. (20–30 Nm).

22. Install the fuel injectors and fuel supply manifold.

23. Position the new UIM-to-LIM gasket on the LIM, then set the UIM on the gasket.

24. Apply a light coating of Ford Pipe Sealant with Teflon® D8AZ-19554-A, or equivalent, to the threads of the UIM mounting bolts/studs.

25. Install the bolts/studs in their original positions. Tighten the bolts/studs in the sequence shown in the accompanying illustration in the following three steps:

a. Tighten the bolts/studs to 88 inch lbs. (10 Nm).

b. Then, tighten the bolts/studs to 177 inch lbs. (20 Nm).

c. Finally, tighten them to 24 ft. lbs. (32 Nm).

26. Install the engine support and wiring retainer bracket. Tighten the mounting bolt to 15–22 ft. lbs. (20–30 Nm) and the nut to 71–97 inch lbs. (8–11 Nm).

27. Install the EGR valve and the throttle body (if removed).

28. Attach all necessary wiring and vacuum lines.

29. Install the accelerator cable bracket and attach the cable to the throttle body. Tighten the bracket mounting bolts to 124–177 inch lbs. (14–20 Nm).

30. If applicable, attach the speed control actuator.

31. Install the air inlet tube.

32. Start the engine and check for leaks.

33. Check and adjust (if necessary) the following: accelerator cable, engine idle airflow, speed control actuator.

4.6L SOHC Engine

1. Disconnect the negative battery cable.

2. Drain the engine cooling system and relieve fuel system pressure. Detach the fuel lines from the fuel supply manifold.

3. Remove the air inlet tube.

4. Remove the accessory drive belt.

✳✳ WARNING

When removing the spark plug wires, only pull on the spark plug wire boot. Otherwise, the wire may separate from the connector in the boot, necessitating plug wire replacement.

5. Label and remove all spark plug wires and wire separators.

6. Remove the ignition coils, the alternator and their mounting brackets from the engine.

7. Apply the parking brake and block the rear wheels, then raise and safely support the front of the vehicle.

8. Disconnect all necessary wiring.

9. Disconnect the EGR valve-to-exhaust manifold tube from the right-hand exhaust manifold.

10. Lower the vehicle.

11. Label and disconnect all applicable wiring, vacuum lines, cooling system hoses and throttle body cables from the intake manifold or engine which will interfere with manifold removal.

➡**Two water outlet connection retaining bolts also retain the intake manifold.**

12. Remove the two bolts retaining the water hose connection to the manifold, then position the upper radiator hose and connection out of the way.

13. Remove the remaining bolts, then lift the manifold up and off of the engine.

14. Remove and discard the old gaskets.

➡**If a replacement intake manifold is being installed, transfer all necessary components to it from the old manifold.**

To install:

15. Thoroughly clean the cylinder head, engine block and intake manifold gasket surfaces.

16. Position the new intake manifold gaskets on the cylinder heads, then place the intake manifold on the heads.

➡**After positioning the intake manifold, be sure that the alignment tabs on the**

intake manifold gasket are aligned with the holes in the cylinder heads.

17. Install and tighten intake manifold mounting bolts Nos. 1 through 9 one at a time to 15–22 ft. lbs. (20–30 Nm), in the sequence shown.

18. Install a new O-ring on the water hose connection, then position the connection on the intake manifold and install the two mounting bolts to 15–22 ft. lbs. (20–30 Nm).

19. Attach all applicable wiring, vacuum lines, cooling system hoses and throttle body cables which were removed earlier.

20. Apply the parking brake and block the rear wheels, then raise and safely support the front of the vehicle.

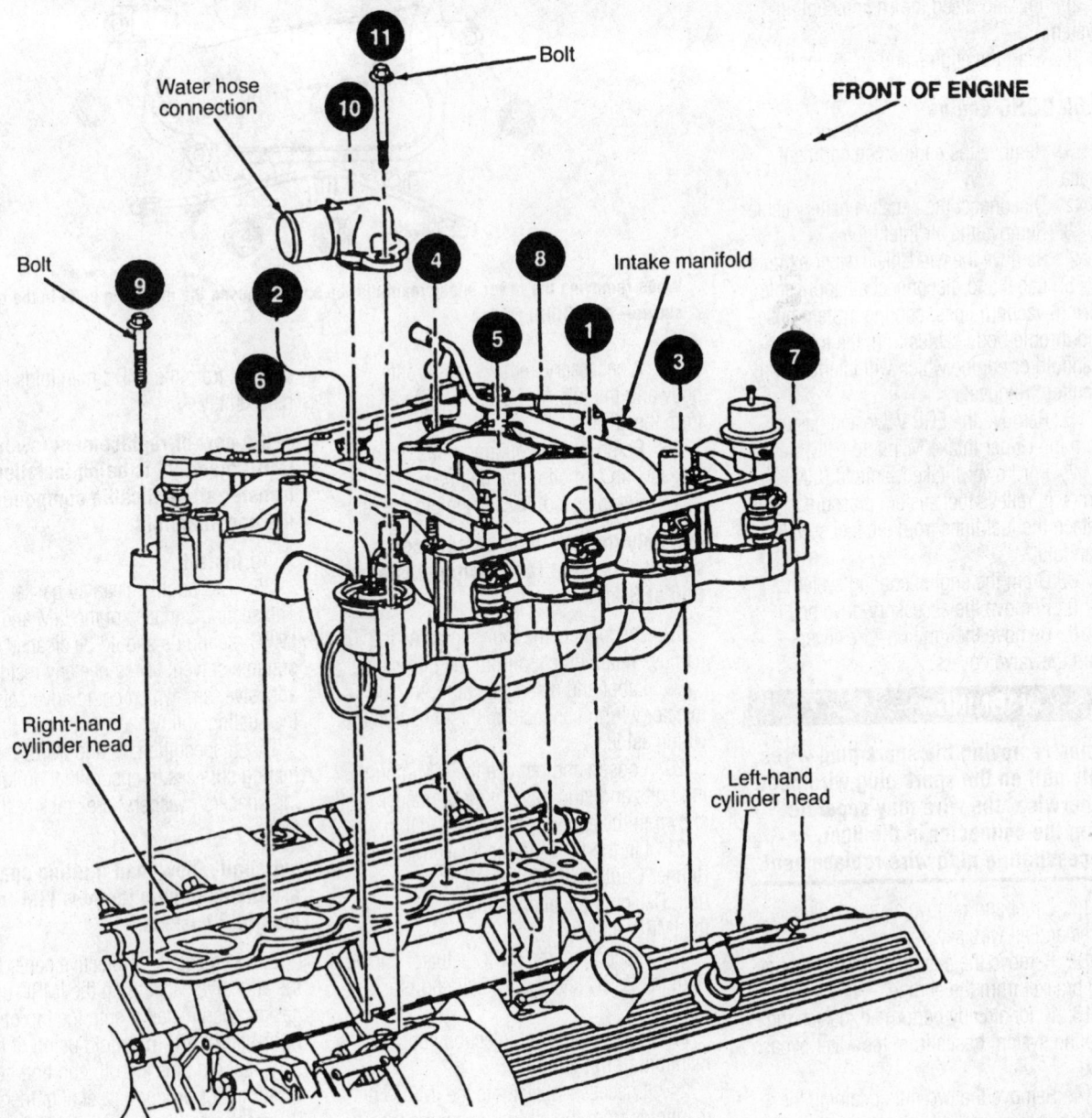

Be sure to tighten the intake manifold mounting bolts in the sequence shown—4.6 SOHC engine

7922NG18

21. Connect the EGR valve-to-exhaust manifold tube to the right-hand exhaust manifold; tighten the nut to 26–33 ft. lbs. (35–45 Nm).

22. Reconnect all applicable wiring, which was disconnected earlier.

23. Lower the vehicle and remove the blocks from the wheels.

24. Install the alternator, the ignition coils and their mounting brackets.

25. Install the spark plug wires, being sure to connect them to their applicable plugs and coil terminal towers.

26. Install the accessory drive belt and the air inlet tube.

27. Attach the fuel supply and return lines to the fuel supply manifold.

28. Connect the negative battery cable.

29. Fill and bleed the engine cooling system.

30. Start the engine and check for leaks.

4.6L DOHC Engine

1. Remove the engine compartment brace.

2. Disconnect the negative battery cable.

3. Remove the air inlet tube.

4. Remove the windshield wiper module.

5. Label and disconnect all applicable wiring, vacuum lines, cooling system hoses and throttle body cables from the intake manifold or engine which will interfere with manifold removal.

6. Remove the EGR valve and gasket from the Upper Intake Manifold (UIM).

7. For Lower Intake Manifold (LIM) removal, relieve fuel system pressure. Detach the fuel lines from the fuel supply manifold.

8. Drain the engine cooling system.

9. Remove the accessory drive belt.

10. Remove the ignition wire covers from the valve covers.

❊❊ WARNING

When removing the spark plug wires, only pull on the spark plug wire boot. Otherwise, the wire may separate from the connector in the boot, necessitating plug wire replacement.

11. Label and remove all spark plug wires and all wire separators.

12. Remove the alternator and its mounting bracket from the engine.

13. If not already performed, detach the cooling system hoses from the water bypass tube.

14. Remove the two nuts retaining the water bypass tube to the manifold, then lift the bypass tube off of the UIM.

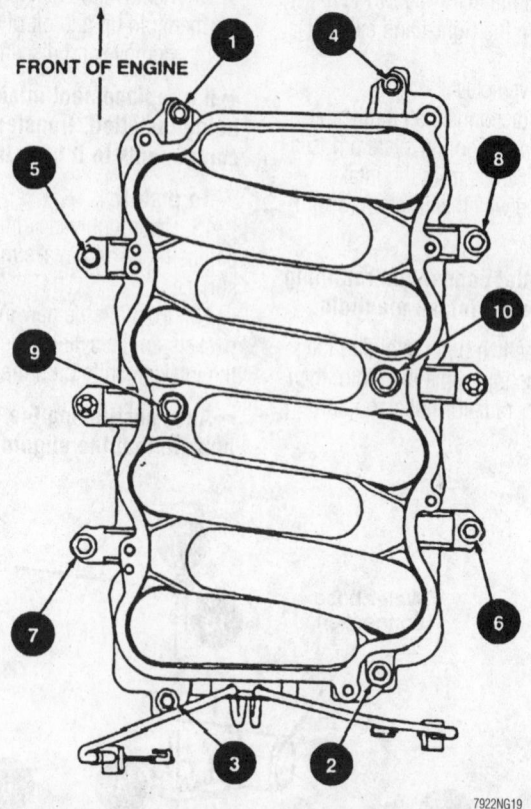

7922NG19

When removing the lower intake manifold, be sure to loosen the mounting bolts in the order shown—4.6L DOHC engine

15. If necessary, remove the throttle body and the Idle Air Control (IAC) valve from the UIM.

16. Remove the mounting bolts, then lift the UIM up and off of the engine.

17. Remove and discard the old gaskets.

➡️**If only the UIM is to be serviced, stop at this point (and resume installation at Step 37).**

18. Disconnect the four engine wiring harness retaining clips from the LIM studs.

19. Label and disengage all of the fuel injector wiring connectors, then position the wiring aside.

20. Loosen and remove the LIM mounting bolts and stud bolts in the sequence shown in the accompanying illustration.

21. Lift the LIM and the Intake Manifold Runner Control (IMRC) manifolds as one unit. Detach the fuel charging wiring from the IMRC actuator.

22. Disconnect the IMRC actuator cables from the IMRC levers and retaining brackets.

23. Remove the fuel injection supply manifold from the LIM.

24. If necessary, separate the IMRC manifolds from the LIM by removing the attaching bolts. Retain the load limiting

spacers from the IMRC manifolds for reassembly.

➡️**If a new or replacement lower or IMRC manifold is being installed, transfer all applicable components to the new manifold.**

To install:

25. Thoroughly clean all gasket surfaces. The sealing surfaces of the LIM and the IMRC manifolds should be cleaned with a suitable solvent to remove any residual adhesive film, which could adversely affect the sealing ability.

26. Inspect the LIM and IMRC manifold mating surfaces to ensure that no load limiting spacers (washers) are stuck to the surfaces.

➡️**Usually, new load limiting spacers are included with the new LIM-to-IMRC manifold gaskets.**

27. Remove the protective paper from the adhesive backing on the IMRC upper gasket, then install it onto the top of the right-hand IMRC manifold using, if necessary, tapered pins at both end bolt hole locations for proper gasket alignment. Repeat this for the left-hand IMRC manifold as well.

28. Assemble the IMRC manifolds to the LIM, and install the four attaching bolts finger-tight.

29. Install the fuel injection supply manifold and fuel injectors. Install the four retaining bolts to 71–106 inch lbs. (8–12 Nm).

30. Position new LIM gaskets on the cylinder heads using the integral locating pins to properly align the gaskets.

31. Attach the fuel charging wiring to the IMRC actuator.

32. Set the LIM/IMRC manifold assembly on the engine. Install the bolts and stud bolts in locations Nos. 7, 8, 9 and 10 (refer to the accompanying torque sequence illustration. Do NOT tighten them yet. Then, install the bolts and stud bolts in positions Nos. 1, 2, 3, 4, 5 and 6. Now hand-tighten all bolts and stud bolts.

33. Following the sequence shown in the accompanying illustration, tighten fasteners 1 through 10 to 15–22 ft. lbs. (20–30 Nm).

34. Tighten the four IMRC-to-LIM attaching bolts to 71–106 inch lbs. (8–12 Nm).

35. Connect the IMRC actuator cables to the retaining brackets and IMRC levers.

36. Reattach all fuel injection wiring to the injectors.

37. Clean the LIM-to-UIM gasket surfaces of all old gasket material and dirt, then install a new UIM-to-LIM gasket on the LIM.

38. Install and tighten the UIM mounting bolts in the sequence shown in the accompanying illustration to 71–106 inch lbs. (8–12 Nm).

39. Install the EGR valve.

40. If necessary, position the accelerator cable bracket on the UIM and install the attaching bolts to 71–106 inch lbs. (8–12 Nm).

41. If applicable, install the throttle body.

42. Install new O-rings on the water bypass tube, then install the bypass tube on the intake manifold as follows:

✳✳ WARNING

While installing the water bypass tube, be careful not to accidentally damage the new O-ring. Otherwise, coolant leakage may result.

a. Clean the O-ring sealing surface on the cylinder heads and coat the water bypass tube O-rings with a rubber lubricant (such as Ford ESE-M99B 176-A).

b. Position the bypass tube support braces over the UIM inner studs and press the bypass tube into position.

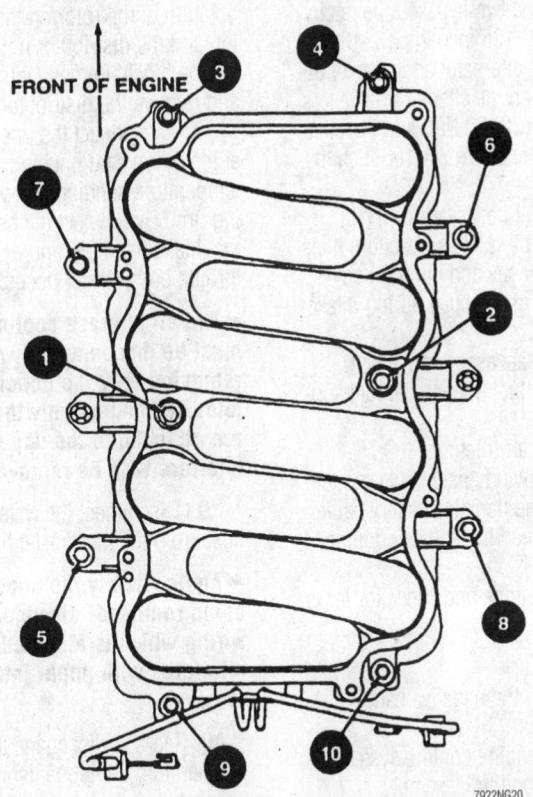

When installing the lower intake manifold, tighten the mounting bolts only in the order shown—4.6L DOHC engine

7922NG20

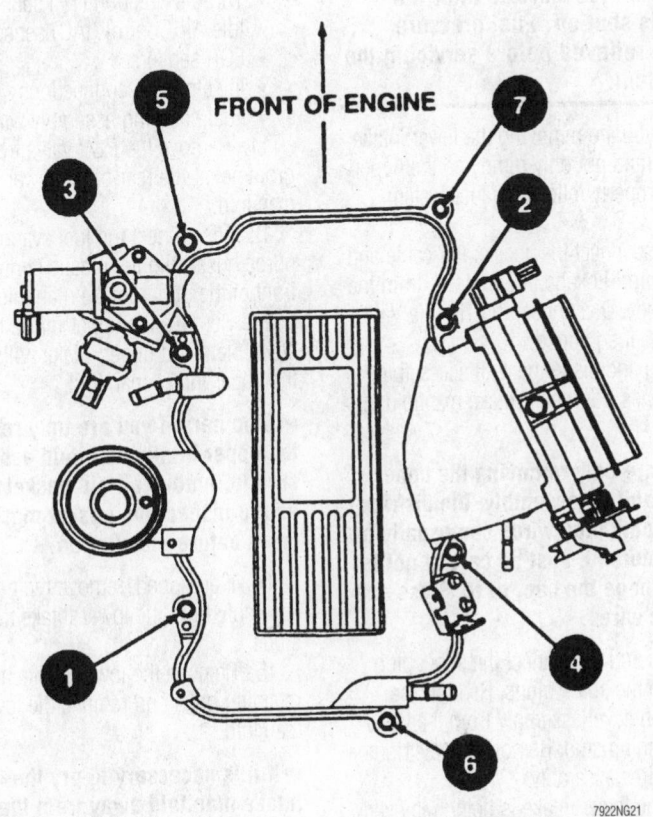

Tighten the upper-to-lower intake manifold mounting bolts in the sequence shown—4.6L DOHC engine

7922NG21

c. Install the two bypass tube retaining nuts to 71–106 inch lbs. (8–12 Nm).

43. Install the alternator and bracket on the engine. Tighten the alternator fasteners to 15–22 ft. lbs. (20–30 Nm) and the alternator bracket fasteners to 71–106 inch lbs. (8–12 Nm).

44. Reattach all necessary wiring (including spark plug wires), vacuum lines, cooling system hoses and throttle body cables, which were detached during intake manifold removal.

45. Install the accessory drive belt.

46. Attach the fuel lines to the fuel supply manifold.

47. Install the air inlet tube.

48. Install the windshield wiper module.

49. Connect the negative battery cable.

50. Fill and bleed the engine cooling system.

51. Start the engine and check for leaks.

5.0L Engine

1. Disconnect the negative battery cable for safety purposes.

2. Drain the engine cooling system to a level below the manifold assembly.

✳✳ CAUTION

Fuel lines on fuel injected vehicles will remain pressurized after the engine is shut off. Fuel pressure must be relieved before servicing the fuel system.

3. If you are removing the lower intake manifold (and not only removing the upper intake), properly relieve the fuel system pressure.

4. Disconnect the accelerator cable and cruise control linkage, if equipped, from the throttle body. Disconnect the Throttle Valve (TV) cable, if equipped.

5. Tag and disconnect all accessible vacuum lines from their intake manifold fittings.

➡**If you are only removing the upper intake manifold assembly, the distributor and spark plug wires can usually be left undisturbed. Just be careful not to hit or damage the cap, or to stress and break the wires.**

6. Tag and disconnect the spark plug wires from the spark plugs. Remove the wires and bracket assembly from the valve cover attaching stud. Remove the distributor cap and wires assembly.

7. If the lower intake is being removed, disconnect the fuel lines, then disengage the distributor wiring connector. Mark the

position of the rotor on the distributor housing and the distributor housing on the engine block. Remove the hold-down bolt and remove the distributor.

8. Disconnect the upper radiator hose at the thermostat housing and the water temperature sending unit wire at the sending unit. Disconnect the heater hose from the intake manifold and disconnect the two throttle body cooler hoses.

➡**Not all of these cooling system hoses must be disconnected when you are removing only the upper intake manifold. Disconnect only the hoses which are attached to the upper manifold or interfere with its removal.**

9. Disconnect the water pump bypass hose from the thermostat housing.

➡**Again, if only the upper intake is being removed, disengage only the wiring which is attached to or which interferes with upper intake manifold removal.**

10. Tag and disengage the wiring from the manifold components including the connectors from the:
- Engine Coolant Temperature (ECT) sensor
- Intake Air Temperature (IAT) sensor
- Throttle Position (TP) sensor
- Idle Air Control (IAC) sensor
- EGR sensors
- Injector wire connections
- Fuel charging assembly wiring

11. Remove the PCV valve from the grommet at the rear of the lower intake manifold.

12. Disconnect the fuel evaporative purge hose from the plastic connector at the front of the upper intake manifold.

13. Remove the upper intake manifold cover plate and upper intake bolts. Remove the upper intake manifold.

➡**Stop here if you are only removing the upper intake manifold assembly. Be sure to remove all old gasket material and to inspect the gasket mating surfaces before installation.**

14. If equipped, remove the heater tube assembly from the lower intake manifold studs.

15. Remove the lower intake manifold retaining bolts and remove the lower intake manifold.

➡**If it is necessary to pry the lower intake manifold away from the cylinder heads, be careful to avoid damaging the gasket sealing surfaces.**

To install:

16. Clean all gasket mating surfaces. Apply a 1/8 in. (3mm) bead of silicone sealer to the points where the cylinder block rails meet the cylinder heads.

17. Position new seals on the cylinder block and new gaskets on the cylinder heads with the gaskets interlocked with the seal tabs. Be sure the holes in the gaskets are aligned with the holes in the cylinder heads.

18. Apply a 3/16 in. (5mm) bead of sealer to the outer end of each intake manifold seal for the full width of the seal. Be sure the silicone sealer will not fall into the engine and possibly block oil passages.

19. Using guide pins to ease installation, carefully lower the intake manifold into position on the cylinder block and cylinder heads.

➡**After the intake manifold is in place, run a finger around the seal area to be sure the seals are in place. If the seals are not in place, remove the intake manifold and position the seals.**

20. Be sure the holes in the manifold gaskets and the manifold are in alignment. Remove the guide pins.

21. Install the intake manifold attaching bolts and tighten them, in the indicated sequence, using 3 passes:
   a. First, tighten the bolts to 8 ft. lbs. (11 Nm).
   b. Next, tighten the bolts to 16 ft. lbs. (22 Nm).
   c. Finally, tighten the bolts to 23–25 ft. lbs. (31–34 Nm).

22. Install the heater tube assembly to the lower intake manifold studs.

23. Install the water pump bypass hose on the thermostat housing. Install the hoses to the heater tubes. Connect the upper radiator hose.

24. Connect the fuel lines, then temporarily connect the negative battery cable. Cycle the ignition to pressurize the fuel system and check for leaks. Turn the ignition key back and forth (from **ON** to **OFF**) at least 6 times, leaving the key **ON** for 5 seconds each time. If no leaks are found, disconnect the negative battery cable and continue the installation.

25. If removed, install the distributor, aligning the housing and rotor with the marks that were made during removal. Install the distributor cap. Position the spark plug wires in the harness brackets on the valve cover attaching stud, and connect the wires to the spark plugs.

➡**If the engine was moved with the distributor out, rotate the crankshaft until the No. 1 piston is at TDC of the com-**

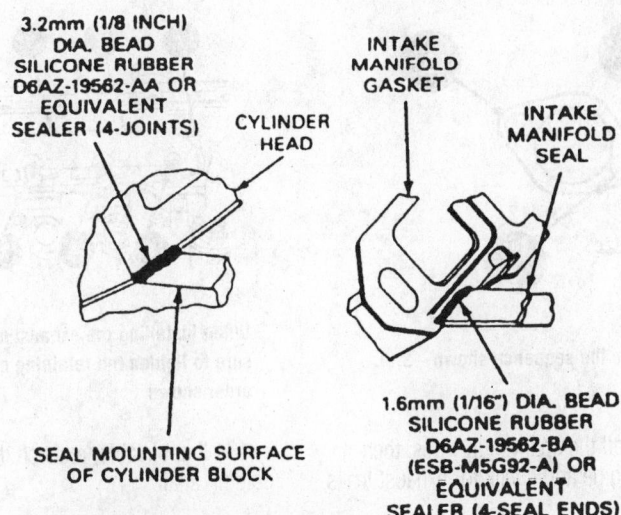

Apply sealer at the intersection between the engine block and cylinder head—5.0L engine

NOTE: THIS SEALER SETS UP WITHIN 15 MINUTES AFTER APPLICATION. TO ASSURE EFFECTIVE SEALING, ASSEMBLY SHOULD PROCEED PROMPTLY.

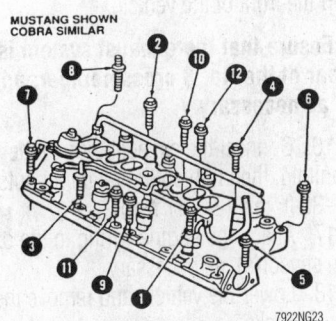

Follow the illustrated tightening sequence when installing the lower intake manifold—5.0L engine

pression stroke. Align the correct initial timing mark with the pointer, then position the distributor in the block with the rotor at the No. 1 firing position and install the hold-down clamp. For more information on distributor removal and installation, and for helpful hints on figuring out when the engine is at TDC, please refer to the distributor and timing procedures in Section 1 of this manual.

26. Install a new gasket and the upper intake manifold. Tighten the bolts to 12–18 ft. lbs. (16–24 Nm). Install the cover plate and connect the crankcase vent tube.

27. Connect the accelerator cable, TV cable and cruise control cable, as equipped, to the throttle body.

28. Engage the wiring connectors and vacuum lines as tagged during removal.

29. Connect the coolant hoses to the EGR spacer.

30. Install the air intake duct assembly and the crankcase vent hose.

31. Connect the negative battery cable, then fill and bleed the cooling system.

32. Run the engine to normal operating temperature and check for leaks.

33. Operate the engine at fast idle. When engine temperatures have stabilized, check the intake manifold bolt torque values.

34. If the lower intake was removed and/or the distributor was disturbed, check the ignition timing. Adjust as necessary.

Exhaust Manifold

REMOVAL & INSTALLATION

➡Exhaust fasteners often rust in position and are easily rounded, broken or otherwise stripped. If possible, allow the engine to thoroughly cool, then apply a penetrating lubricant to the retainers and allow it to soak in before removal. Always use 6-point sockets on rusty fasteners to prevent rounding off the corners.

✳✳ CAUTION

Use extra caution when working around rusted exhaust manifold fasteners, since they can break loose or just plain break, suddenly causing your hand and tools to jerk. Always

position yourself properly to prevent a fall and be sure to pull on the wrench and not push on it. Wearing heavy gloves can help protect your hands from injury.

3.8L Engine

1. For the left-hand manifold, remove the oil dipstick tube support bracket.

2. Label and detach the spark plug wires from the spark plugs and the ignition coil.

3. For the left-hand manifold, detach the oxygen sensor wiring from the engine wiring harness.

4. For the right-hand manifold, disconnect the EGR valve-to-exhaust manifold tube from the exhaust manifold. Loosen the retaining nut, then remove the heater inlet tube and heater water hose.

5. Apply the parking brake and block the rear wheels, then raise and safely support the front of the vehicle.

6. Separate the exhaust system pipe from the exhaust manifold by removing the retaining nuts.

7. For the right-hand manifold, remove the heated oxygen sensor from the exhaust manifold.

8. Lower the vehicle.

9. Remove the exhaust manifold mounting bolts, then lift the manifold up and off of the engine.

10. Remove and discard the old exhaust manifold gasket.

To install:

➡**Lightly oil all bolt and stud threads, except any requiring special sealant, with clean oil prior to installation.**

11. If a new exhaust manifold is being installed, or if a new manifold-to-exhaust pipe stud is needed, install the stud until it is securely seated in the manifold.

12. Clean the gasket surfaces of the exhaust manifold, cylinder head and exhaust pipe.

✳✳ WARNING

Do not allow anti-seize compound to enter the oxygen sensor flutes, otherwise possible damage to the sensor may occur.

13. For the left-hand manifold, apply high-temperature anti-seize compound to the oxygen sensor threads, then install the oxygen sensor in the manifold. Tighten the sensor to 28–33 ft. lbs. (37–45 Nm).

14. Install the exhaust manifold, along with a new gasket, on the cylinder head.

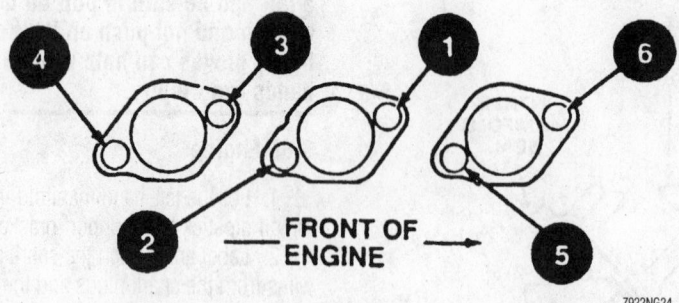

To prevent possible exhaust leaks, tighten the manifold bolts in the sequence shown—3.8L engine

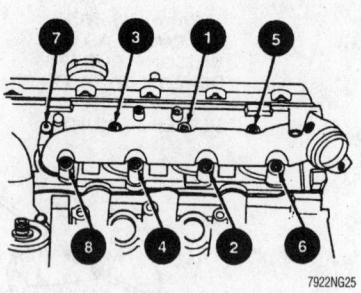

When installing the exhaust manifolds, be sure to tighten the retaining nuts in the order shown

Install two new retaining bolts loosely to hold them in place.

➡️**According to the manufacturer, if slight warpage of the exhaust manifold causes a misalignment between the bolt holes in the cylinder head and exhaust manifold, the holes in the exhaust manifold should be elongated to provide proper alignment. However, do NOT elongate the pilot hole (lower rear bolt hole on the No. 2 cylinder).**

15. Start the remaining mounting bolts, then (for the right-hand manifold only) loosely connect the EGR valve-to-exhaust manifold tube.

16. Tighten the exhaust manifold bolts in the sequence shown in the accompanying illustration to 22–26 ft. lbs. (30–36 Nm).

17. Apply the parking brake and block the rear wheels, then raise and safely support the front of the vehicle.

18. Connect the exhaust pipe to the manifold, then tighten the retaining nuts to 16–23 ft. lbs. (21–32 Nm).

✳✳ WARNING

Do not allow anti-seize compound to enter the oxygen sensor flutes, otherwise possible damage to the sensor may occur.

19. For the right-hand exhaust manifold, apply high-temperature anti-seize compound to the oxygen sensor threads, then install the oxygen sensor in the manifold. Tighten the sensor to 28–33 ft. lbs. (37–45 Nm).

20. Lower the vehicle and remove the wheel blocks.

21. For the right-hand exhaust manifold, install the heater inlet tube and the heater water hose. Tighten the retaining nut to 15–22 ft. lbs. (20–30 Nm). Then, tighten the EGR valve-to-exhaust manifold tube nuts to 25–34 ft. lbs. (34–47 Nm).

22. Install the spark plug wires, then start the engine and check for exhaust leaks.

4.6L SOHC Engine

1. Disconnect the negative battery cable.

2. For the left-hand manifold, remove the oil dipstick tube support bracket retaining bolt.

3. Apply the parking brake and block the rear wheels, then raise and safely support the front of the vehicle.

4. Separate the exhaust system pipe from the exhaust manifold by removing the retaining nuts. Support the exhaust system pipe from a crossmember with a piece of strong string or wire.

5. Disconnect the engine wiring from the heated oxygen sensor, installed in the exhaust manifold.

6. For the left-hand exhaust manifold, detach the steering column shaft and position it out of the way.

7. Lower the vehicle.

8. Remove the exhaust manifold retaining nuts, then lift the manifold up and off of the engine.

9. Remove and discard the old exhaust manifold gaskets.

To install:

➡️**Lightly oil all bolt and stud threads, except any requiring special sealant, with clean oil prior to installation.**

10. If a new exhaust manifold is being installed, transfer the EGR valve tube to the new manifold connector. Tighten the tube fastener to 33–47 ft. lbs. (45–65 Nm).

11. Clean the gasket surfaces of the exhaust manifold, cylinder head and exhaust pipe.

12. Install the exhaust manifold, along with a new gasket, on the cylinder head. Install the retaining nuts and tighten them, in sequence, to 159–195 inch lbs. (18–22 Nm).

13. If applicable, reattach the steering column shaft.

➡️**Loosen the line nut on the EGR valve before installing it onto the manifold.**

14. Attach the EGR valve-to-exhaust manifold tube to the left-hand exhaust manifold; tighten the tube nut to 26–33 ft. lbs. (35–45 Nm).

15. Apply the parking brake and block the rear wheels, then raise and safely support the front of the vehicle.

➡️**Ensure that the exhaust system is clear of the No. 3 crossmember; adjust it, as necessary.**

16. Connect the exhaust pipe to the manifold, then tighten the retaining nuts to 20–30 ft. lbs. (27–41 Nm).

17. Attach the engine wiring to the oxygen sensor(s), as necessary.

18. Lower the vehicle and remove the wheel blocks.

19. Connect the negative battery cable, then start the engine and check for exhaust leaks.

4.6L DOHC Engine

RIGHT-HAND MANIFOLD

1. Disconnect the negative battery cable.

2. Remove the air inlet tube from the air cleaner housing and the throttle body.

3. Detach the secondary air injection manifold tube from the exhaust manifold.

4. Remove the four retaining nuts from the upper side of the exhaust manifold.

5. Apply the parking brake and block the rear wheels, then raise and safely support the front of the vehicle.

6. Separate the exhaust system pipe from the exhaust manifold by removing the retaining nuts. Support the exhaust system pipe from a crossmember with a piece of strong string or wire.

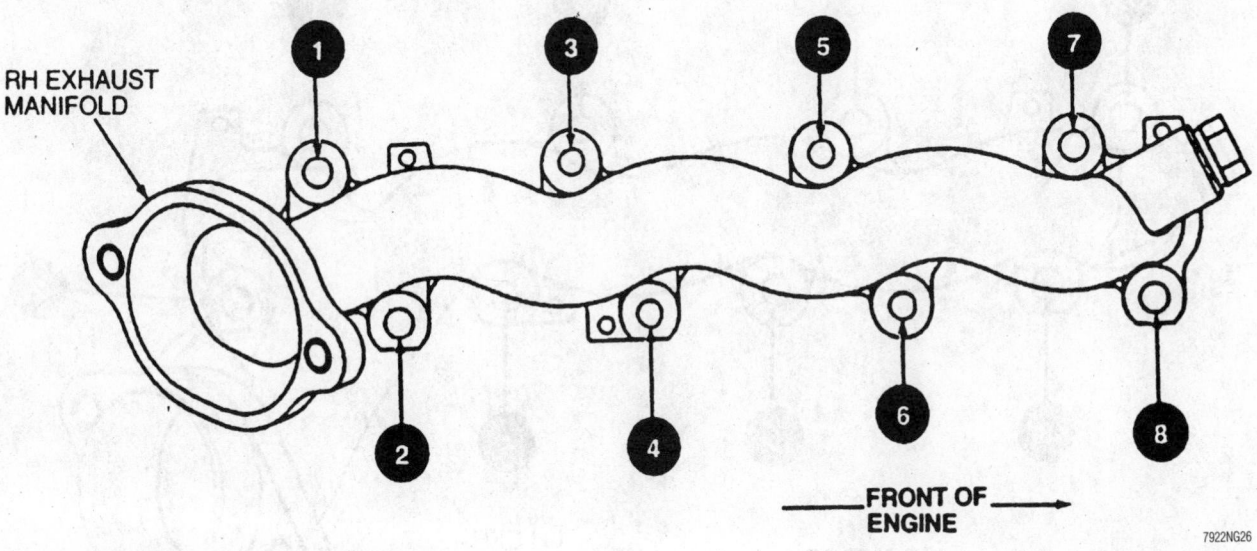

Tighten the right-hand exhaust manifold retaining nuts in the order shown—4.6L engine

7. Remove the four retaining nuts from the lower side of the manifold, then remove the manifold from the engine.

8. Remove and discard the old exhaust manifold gasket.

To install:

→Lightly oil all bolt and stud threads, except any requiring special sealant, with clean oil prior to installation.

9. Clean the gasket surfaces of the exhaust manifold, cylinder head and exhaust pipe. ·

10. Install the exhaust manifold, along with a new gasket, on the cylinder head. Install the eight retaining nuts and tighten them, in the sequence shown in the accompanying illustration, to 159–195 inch lbs. (18–22 Nm).

11. Connect the exhaust pipe to the manifold, then tighten the retaining nuts to 20–30 ft. lbs. (27–41 Nm).

12. Lower the vehicle and remove the wheel blocks.

13. Reattach the secondary air injection manifold tube to the manifold. Tighten the retaining nut to 28–31 ft. lbs. (38–42 Nm).

14. Install the air inlet tube.

15. Connect the negative battery cable, then start the engine and check for exhaust leaks.

LEFT-HAND MANIFOLD

→A Ford Three-Bar Engine Support D88L-6000-A and Engine Lifting Bracket Set 014–00340, or equivalent tools, are necessary for this procedure.

1. Open the trunk and turn the rear suspension leveler compressor switch OFF.

2. Disconnect the negative battery cable.

3. Install the engine lifting brackets on the engine, then position the three-bar engine support on the engine and attach it to the lifting brackets.

4. Loosen the front wheel lug nuts slightly, then apply the parking brake and block the rear wheels. Raise and safely support the front of the vehicle.

5. Remove the front wheels.

6. Detach the secondary air injection manifold tube and the EGR valve-to-exhaust manifold tube from the exhaust manifold.

7. Remove the front exhaust system pipe.

8. Detach both front suspension lower control arms and outer tie rod ends from the front wheel hub/spindle assemblies.

9. Detach the steering column shaft at the pinch bolt joint and position it out of the way.

10. Support the front sub-frame assembly with a floor jack.

11. Remove the front engine support insulator through-bolts.

12. Remove the front sub-frame bolts.

13. Carefully lower the front sub-frame.

14. Remove the retaining nuts from the exhaust manifold, then remove and discard the old exhaust manifold gasket.

To install:

→Lightly oil all bolt and stud threads, except any requiring special sealant, with clean oil prior to installation.

15. Clean the gasket surfaces of the exhaust manifold, cylinder head and exhaust pipe.

16. Install the exhaust manifold, along with new gaskets, on the cylinder head. Install the eight retaining nuts and tighten them, in the sequence shown in the accompanying illustration, to 159–195 inch lbs. (18–22 Nm).

17. Attach the EGR valve-to-exhaust manifold tube to the exhaust manifold. Tighten the nut to 30–33 ft. lbs. (40–45 Nm).

18. Connect the secondary air injection manifold tube to the exhaust manifold. Tighten the nut to 28–31 ft. lbs. (38–42 Nm).

19. Raise the front sub-frame assembly into position, then install and tighten the sub-frame bolts to 70–96 ft. lbs. (95–131 Nm).

20. Attach the steering column shaft to the pinch bolt joint.

21. Reattach the front suspension control arms and tie rod ends to the spindle/hub assemblies.

22. Install the front exhaust system pipe.

23. Install the wheels, then lower the vehicle and remove the wheel blocks.

24. Remove the three-bar engine support and lifting brackets from the engine.

25. Connect the negative battery cable, then start the engine and check for exhaust leaks.

26. Turn the rear suspension leveler compressor switch ON, then shut the trunk lid.

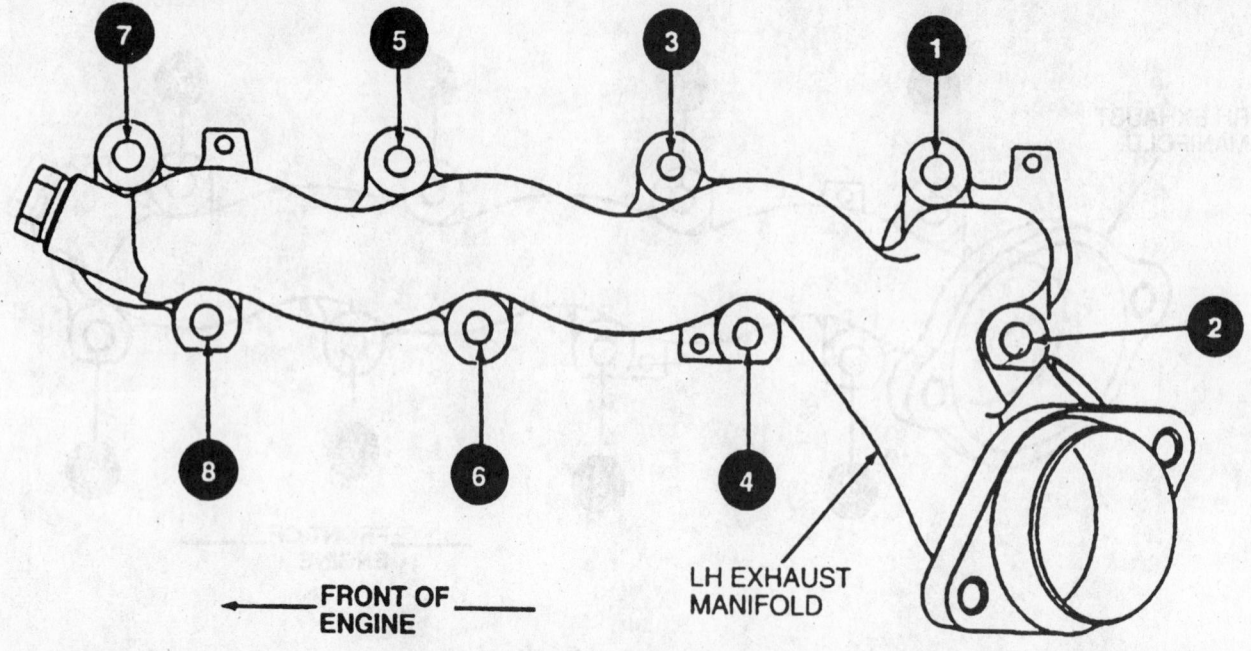

FRONT OF ENGINE ←

LH EXHAUST MANIFOLD

7922NG27

When installing the left-hand exhaust manifold, be sure to tighten the mounting nuts in the sequence shown—4.6L DOHC engine

5.0L Engine

1. Disconnect the negative battery cable for safety purposes.
2. Remove the Thermactor/secondary air injection hardware from the right exhaust manifold.
3. If necessary, remove the air cleaner and inlet duct.
4. Tag and disconnect the spark plug wires, then remove the spark plugs from the cylinder heads.
5. If necessary, disconnect the engine oil dipstick tube from the exhaust manifold stud.

➡If the vehicle is supported at a level that allows access to both the top and bottom of the engine, you will not have to keep raising and lowering the vehicle for this procedure.

6. Raise and support the vehicle safely using jackstands.
7. Disconnect the exhaust pipe(s) from the exhaust manifold(s).
8. If necessary, remove the engine oil dipstick tube by carefully tapping upward on the tube.
9. If necessary, disengage the oxygen sensor connector. If the manifold or sensor is being replaced, remove the sensor.
10. Unless upper engine access is possible in the current position, remove the jackstands and carefully lower the vehicle.

11. Remove the attaching bolts and washers, then remove the exhaust manifold(s).

To install:

12. Clean and inspect all gasket mating surfaces.
13. Position the exhaust manifold to the cylinder head using a new gasket. Retain the manifold to the engine by threading and finger-tightening the mounting bolts.
14. Working from the center to the ends, tighten the exhaust manifold attaching bolts to 26–32 ft. lbs. (35–44 Nm).
15. If lowered for access, raise and support the vehicle safely using jackstands.
16. As applicable, install the engine oil dipstick tube, then install the oxygen sensor and engage the connector.
17. Position the exhaust pipe(s) to the manifold(s). Alternately tighten the exhaust pipe flange nuts to 20–30 ft. lbs. (27–41 Nm).
18. Remove the jackstands and carefully lower the vehicle.
19. Install the spark plugs and connect the spark plug wires.
20. If removed, install the Thermactor/air injection hardware to the right exhaust manifold.
21. If removed, install the air cleaner and inlet duct.
22. Connect the negative battery cable, then start the engine and check for exhaust leaks.

Camshaft and Valve Lifters

REMOVAL & INSTALLATION

3.8L Engine

VALVE LIFTERS

➡Before replacing a valve lifter because of noisy operation, ensure that the noise is not caused by improper valve-to-rocker arm clearance, or by worn rocker arms or pushrods.

1. Remove the valve covers.
2. Remove the upper and lower intake manifolds.

➡Be sure to label or arrange the rocker arms, pushrods and valve tappets so that they can be reinstalled in their original locations.

3. Remove the rocker arms and pushrods.
4. Remove the bolts holding the two tappet retainers in place (the bolts are held captive in retainers). Remove the six valve tappet guide plates from adjacent tappets.
5. Using a magnet, remove the tappets from their bores. If the valve tappets are stuck in their bores due to excessive varnish or gum build-up, use a claw-type tappet removal tool or equivalent. When using such a tool, rotate the tappets back and forth to loosen the build-up.

To install:

➡**Lightly oil all bolt and stud threads prior to assembly, except those specifying sealant, with clean engine oil.**

6. Clean the cylinder head and valve cover sealing surfaces of all old gasket material, dirt and oil residue.

7. Lubricate each valve tappet and bore with engine assembly lubricant.

8. Insert each valve tappet into its original bore. If new tappets are being installed, check for a free fit in the bore into which they are to be installed.

9. Align the flats on the sides of the valve tappets, then install the six guide plates between adjacent tappets, making sure that the word **UP** is showing.

10. Install the valve tappet retainers and tighten the mounting bolts to 88–124 inch lbs. (10–14 Nm).

11. Install the pushrods and rocker arms.

12. Install the lower and upper intake manifolds.

13. Install the valve covers.

14. Start the engine and check for leaks.

CAMSHAFT

➡**Only a MVAC-trained, EPA-certified, automotive technician should service the A/C system or its components.**

1. If your vehicle is equipped with A/C, have the condenser core removed.

2. Remove the radiator and radiator grille.

3. Remove the Camshaft Position (CMP) sensor and housing.

4. Remove the valve tappets.

5. Remove the engine front cover, timing chain and camshaft sprocket spacer.

6. Remove the oil pan.

7. Remove the mounting bolts, then remove the camshaft thrust plate.

8. Carefully slide the camshaft out of the front of the engine block, making sure not to damage the camshaft bearing surfaces and lobes.

To install:

➡**Lightly oil all bolt and stud threads prior to assembly, except those specifying sealant, with clean engine oil.**

9. Lubricate the camshaft lobes and bearing journals with engine assembly lubricant.

10. Carefully slide the camshaft into the engine block, making sure not to damage the bearing or lobe surfaces.

11. Install the camshaft thrust plate, and tighten the mounting bolts to 71–124 inch lbs. (8–14 Nm).

12. Install the camshaft gear spacer, timing chain and engine front cover.

13. Install the oil pan.

14. Install the valve tappets and related components.

15. Install the upper and lower intake manifolds.

16. Install the valve covers.

17. Install the CMP sensor and housing.

18. Install the radiator grille and radiator.

✳✳ WARNING

It is extremely important to fill the engine with oil prior to starting it, otherwise severe engine damage will occur.

19. Fill the engine with the proper amount and type of clean engine oil.

20. Fill the cooling system.

21. Connect the negative battery cable, then start the engine and check for oil, coolant and fuel leaks.

22. After 3–5 minutes, stop the engine.

23. Wait 10 minutes, then check the oil level with the dipstick. Add more oil, if necessary.

24. If necessary, have the A/C condenser core installed and the system recharged.

4.6L Engines

➡**The 4.6L engines do not possess removable camshaft bearings. If the bearing surfaces are damaged, the cylinder head must either be serviced by a qualified machine shop, or replaced.**

1. Remove the valve covers.

2. Remove the engine front cover.

3. Remove the timing chains.

✳✳ WARNING

The crankshaft must be positioned so that the keyway is 45 degrees (SOHC engines) or 90 degrees (DOHC engines) counterclockwise from the 12 o'clock position in order to avoid damaging the valves when the camshafts are rotated.

4. Rotate the crankshaft 45 degrees (SOHC engines) or 90 degrees (DOHC engines) counterclockwise.

5. Rotate the camshaft so that the base circle of the camshaft is positioned against the rocker arm being serviced.

✳✳ WARNING

A valve spring spacer, such as Ford Tool T91P-6565-AH, must be used in order to prevent the valve spring

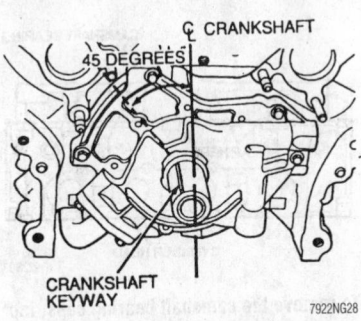

On the SOHC engine, rotate the crankshaft 45 degrees counterclockwise before depressing the valve and springs—4.6L SOHC engine

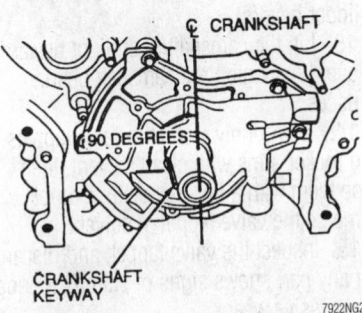

On the DOHC engine, the crankshaft must be rotated 90 degrees counterclockwise to ensure that the valves will not contact the pistons when depressed—4.6L DOHC engine

retainer from contacting the valve stem seal, which can result in damage to the seal.

6. Install the valve spring spacer between the valve spring coils.

7. Install a valve spring compressor, such as Ford Tool T91P-6565-A, under the camshaft and on top of the valve spring retainer.

8. Slowly push down on the compressor tool until the spring is depressed far enough to allow the rocker arm to be removed, then pull the rocker arm out from underneath the camshaft.

9. Remove the valve spring compressor and spacer tools.

10. Repeat Steps 5 through 9 until all of the rocker arms are removed from the cylinder head(s).

11. If necessary, pull the hydraulic valve tappets out of the cylinder head(s).

➡**On the DOHC engine, note the positions of the bearing cap bolts because they are various lengths.**

12. Remove the camshaft bearing cap bolts from the cylinder head(s), then tap

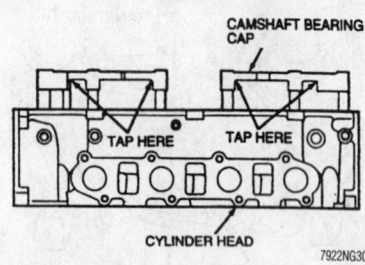

To remove the camshaft bearing caps, tap the caps with a rubber or leather mallet where shown—4.6L engines

upward on the bearing caps at the positions shown in the accompanying illustration. Gradually lift the bearing caps off of the cylinder head(s).

13. Lift the camshaft(s) straight upward to avoid damaging the camshaft bearing surfaces.

14. Thoroughly clean the valve tappets and rocker arms with clean solvent, then wipe them with a lint-free cloth. Do not immerse the valve tappet in solvent.

15. Inspect the valve tappet, and discard it if any part shows signs of pitting, scoring, or excessive wear.

To install:

16. Clean and inspect the valve cover, engine front cover and cylinder head gasket mating surfaces.

17. Apply clean engine oil to the camshaft journals on the cylinder head(s), then gently lay the camshaft(s) on the cylinder head(s).

18. Apply clean engine oil to the camshaft lobes and journals.

➡ **The bearing cap bolts on DOHC engines should be installed as follows:**

• 2 in. (52mm) long bolts—intake camshaft bearing caps and inboard side of the exhaust camshaft bearing caps

• 1.6 in. (42mm) long bolts—outboard side of the exhaust camshaft bearing caps

19. Install and seat the camshaft bearing caps. Install and hand-tighten the bearing cap retaining bolts.

➡ **Each camshaft bearing cap is tightened individually.**

20. Tighten the bearing cap bolts, in the sequence shown in the accompanying illustration, to 71–106 inch lbs. (8–12 Nm).

21. Loosen the bearing cap bolts approximately 2 revolutions, or until the head of the bolt turns freely.

➡ **The camshaft should spin freely with a slight drag.**

22. Retighten the bearing cap bolts, in sequence, to 71–106 inch lbs. (8–12 Nm).

23. Check camshaft end-play.

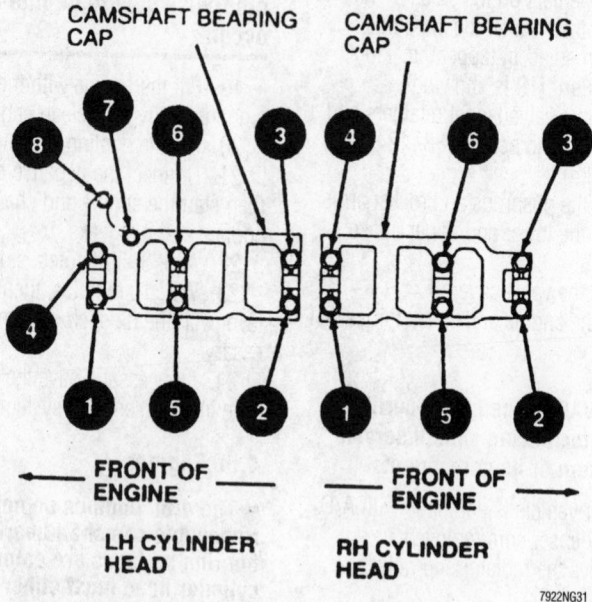

Tighten the camshaft bearing cap mounting bolts in the sequence indicated—4.6L SOHC engine

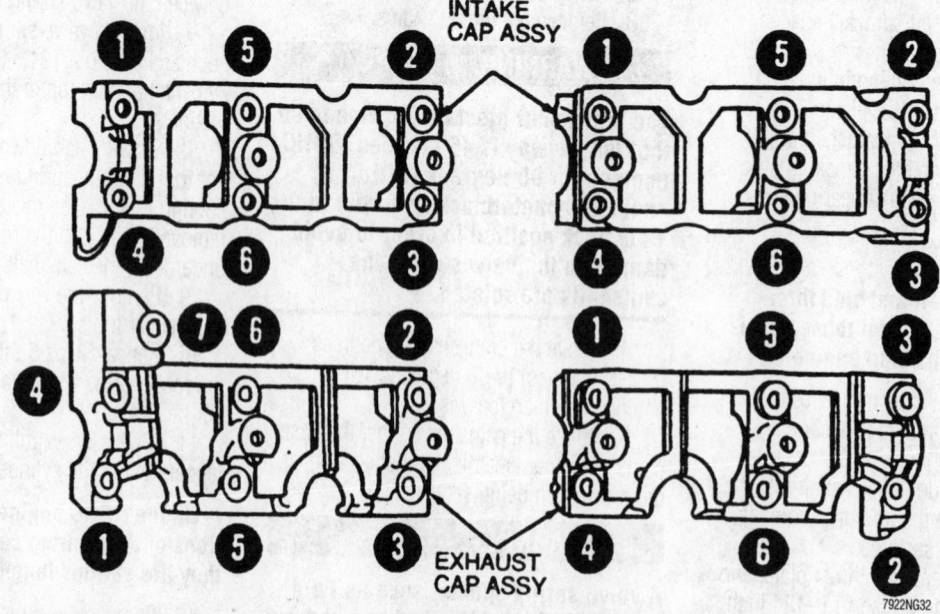

DOHC engine camshaft bearing cap mounting bolt torque sequence—4.6L DOHC engine

24. Apply clean engine oil to the valve stems and tips, the rocker arm roller surfaces, the valve tappets, and the valve tappet bores in the cylinder head(s).

➡**The valve tappets must not have more than 0.059 in. (1.5mm) of plunger travel prior to installation.**

25. If removed earlier, insert the tappets in their original bores in the cylinder head(s).

26. Rotate the crankshaft clockwise 45 degrees (for SOHC engines) or 90 degrees (for DOHC engines)—until the keyway is at the 12 o'clock position.

27. Install the timing chains and the engine front cover.

28. Install the rocker arms, as follows:

   a. Rotate the crankshaft so that the position the piston of the cylinder being serviced is at the bottom of its stroke and the base circle of the camshaft is facing where the rocker arm will be installed.

   b. Install the valve spring spacer and compressor tools as before.

   c. Carefully depress the valve spring with the compressor tool until the rocker arm can be positioned between the valve stem and valve tappet, and the camshaft.

   d. Slowly allow the valve spring to decompress, and ensure that the rocker arm is properly retained.

   e. Remove the valve spring compressor and spacer tools.

   f. Repeat Steps **a** through **e** for all of the rocker arms.

29. Install the valve covers.

30. Start the engine and check for leaks.

5.0L Engine

VALVE LIFTERS

1. Disconnect the negative battery cable for safety.

2. Remove the valve covers.

3. Loosen the rocker arm fulcrum bolts and rotate the rocker arms to the side.

4. Remove the valve pushrods and identify them so that they can be installed in their original position.

5. Remove the lifter guide retainer bolts. Remove the retainer and lifter guide plates. Identify the guide plates so they may be reinstalled in their original positions.

➡**On roller lifters it is very important to note not only the original location, but the position as well, to assure that the roller rotates in the same direction.**

6. Using a magnet, remove the lifters and place them in a rack so that they can be installed in their original bores.

➡**If the lifters are stuck in the bores due to excessive varnish or gum deposits, it may be necessary to use a claw-type tool to aid removal. When using a remover tool, rotate the lifter back and forth to loosen it from gum or varnish that may have formed on the lifter.**

To install:

7. Lubricate the lifters and install them in their original bores. If new lifters are being installed, check them for free fit in their respective bores.

8. Install the lifter guide plates in their original positions, then install the guide plate retainer.

9. Install the pushrods in their original positions. Apply grease to the ends prior to installation.

10. Lubricate the rocker arms and fulcrum seats with heavy engine oil, then install the rocker arms and check for proper valve clearance.

11. Install the valve covers.

12. Connect the negative battery cable, then start the engine and check for leaks.

CAMSHAFT

➡**If your vehicle is equipped with A/C, you will have to take the vehicle to a service station to have the refrigerant discharged and recovered (using a proper recycling/recovery station) before beginning this procedure.**

1. Properly discharge and recover the A/C refrigerant using a recovery station.

2. Disconnect the negative battery cable for safety.

3. Drain the cooling system and relieve the fuel system pressure.

4. Remove the radiator. If equipped with air conditioning, remove the condenser.

5. Remove the grille.

6. Remove the upper and lower intake manifold assembly.

7. Remove the valve covers.

8. Remove the pushrods and lifters.

➡**Keep track of all valvetrain components. Any components which are to be reused, such as pushrods or lifters, must be reinstalled in their original locations. If necessary sort or label them during removal to assure installation in their correct positions.**

9. Remove the timing chain front cover.

10. Remove the timing chain and gears.

➡**Two or three long bolts screwed into the camshaft will provide a convenient handle during removal and installation.**

11. Remove the thrust plate. Remove the camshaft, being careful not to damage the bearing surfaces.

To install:

12. Lubricate the camshaft lobes and journals with heavy engine oil or a suitable engine assembly lube. An engine assembly lube is preferable if some time may pass before the job is completed and the engine is started. Install the camshaft, being careful not to damage the bearing surfaces while sliding it into position.

13. Install the camshaft thrust plate with the groove toward the cylinder block. Tighten the bolts to 9–12 ft. lbs. (12–16 Nm).

14. Check the camshaft play to determine whether or not the thrust plate must be replaced.

15. Install the lifters and pushrods.

➡**When reinstalling a camshaft, be sure the lifters and pushrods are installed in their original locations. If a new camshaft is being installed, new lifters and pushrods should be used.**

16. Install the timing chain and gears.

17. Install the engine front cover.

18. Install the valve covers.

19. Install the upper and lower intake manifolds.

20. Install the grille. If equipped with air conditioning, install the condenser.

21. Install the radiator.

22. Connect the negative battery cable, then fill and bleed the engine cooling system.

23. Run the engine and check for leaks.

➡**This procedure is applicable only to pushrod engines (3.8L and 5.0L engines); the OHC engines (4.6L engines) utilize automatic hydraulic lash adjusters.**

No periodic valve lash adjustments are necessary or possible on these engines. The 3.8L and 5.0L engines utilize hydraulic valvetrains to automatically maintain proper valve lash, if one of these engines is determined to have a valve tap, the following inspection procedures can help determine whether the valve tappet is to blame.

The valve lash is not adjustable. If the collapsed lifter clearance is incorrect, different length replacement pushrods are available to compensate. However, the reason for the incorrect clearance should be investigated first. The clearance could be wrong for a number of reasons, such as: worn camshaft lobe, bent pushrod, bent rocker arm, worn valve stem tip, etc.

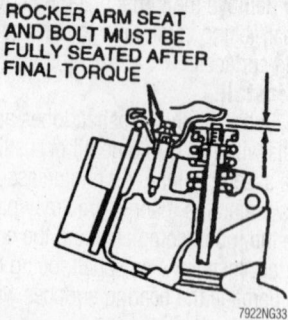

ROCKER ARM SEAT AND BOLT MUST BE FULLY SEATED AFTER FINAL TORQUE

7922NG33

Measure the clearance between the tip of the rocker arm and valve stem tip as shown

24. Remove the valve cover(s) for access to the valves whose lash is being checked.

➡**An easy way to tell if a cylinder is at TDC of the compression stroke is to watch that cylinder's valves as the engine is cranked or rotated. If the valves remain closed (the rocker arms don't move) as the piston approaches the top of its travel, that piston is on its compression stroke (the closed valves were allowing compression to build). If, instead, a valve opens as the piston travels upward (releasing compression in that cylinder), then that piston is on its exhaust stroke.**

25. Either have an assistant help by cranking the engine or install a remote starter switch. Crank the engine with the ignition switch **OFF** until No.1 piston is at TDC on the compression stroke.

➡**Follow the tool manufacturer's instructions when installing the remote starter switch. In most cases, the BROWN lead (terminal I) and the RED/BLUE lead (terminal S) at the starter relay should be disconnected. Then, install the remove starter switch between the battery and terminal S of the relay.**

26. With the crankshaft in the positions designated in the steps below, position a lifter bleed-down wrench, such as Tool T71P-6513-B or equivalent, on the rocker arm. Slowly apply pressure to bleed down the lifter until the plunger is completely bottomed. Hold the lifter in this position and check the available clearance between the rocker arm and the valve stem tip with a feeler gauge.

27. The clearance should be 0.09–0.19 in. (2.25–4.79mm) for 3.8L engines, or 0.123–0.146 in. (3.1–3.7mm) for 5.0L engines. If the clearance is less or more than specified, investigate the reason. If everything checks out OK, install a shorter

(too little clearance) or longer (too much clearance) pushrod to compensate.

28. The following valves can be checked with the crankshaft at position No. 1, with the No. 1 piston at TDC on the compression stroke:

3.8L engine
- No. 1 intake and exhaust
- No. 2 exhaust
- No. 3 intake
- No. 4 exhaust
- No. 6 intake

5.0L engine
- No. 1 intake and exhaust
- No. 3 exhaust
- No. 4 intake
- No. 7 exhaust
- No. 8 intake

29. Rotate the engine 360° (1 full revolution) from the 1st position (No. 1 is now on its exhaust stroke) and check the following valves:

3.8L engine
- No. 2 intake
- No. 3 exhaust
- No. 4 intake
- No. 5 intake and exhaust
- No. 6 exhaust

5.0L engine
- No. 2 exhaust
- No. 3 intake
- No. 6 exhaust
- No. 7 intake

30. For the 5.0L engine, rotate the engine 90° (¼ revolution) from the 2nd position and check the following valves:
- No. 2 intake and No. 4 exhaust
- No. 5 intake and No. 5 exhaust
- No. 6 intake and No. 8 exhaust

Oil Pan

REMOVAL & INSTALLATION

3.8L Engine

1. Disconnect the negative battery cable.
2. Remove the air inlet tube from the throttle body and air cleaner housing.
3. Remove the upper radiator plastic cover and the hood weather seal.
4. Remove the windshield wiper arms.
5. Remove the left-hand cowl vent screen and the windshield wiper module.
6. Install an engine support tool, such as the Ford Three Bar Engine Support D88L-6000-A. Do not support the weight of the engine yet.
7. Apply the parking brake and block the rear wheels, then raise and safely support the front of the vehicle.

8. Remove the front engine mount through-bolts.
9. Using the engine support tool, raise the engine slightly.
10. Remove the starter motor.
11. If equipped, remove the automatic transmission cooler lines.
12. Drain the engine oil and remove the engine oil filter.
13. Remove the oil pan-to-bell housing bolts and the bolts at the crankshaft position sensor lower shield.
14. Remove the remaining oil pan mounting bolts.
15. Separate the steering column shaft at the pinch bolt joint.
16. Position a transmission jack under the front sub-frame, then support the front sub-frame with the jack.
17. Matchmark the sub-frame position in relation to the vehicle chassis for reinstallation purposes.
18. Remove the six rear and the two front mounting bolts from the front sub-frame.
19. Remove the lower shock absorber-to-lower control arm bolts and nuts on both sides of the vehicle.
20. Cautiously lower the front sub-frame.
21. Remove the oil pan. Pour any residual oil out of the oil pan.
22. If necessary, remove the oil pump screen and tube retainer bolts, and support bracket nut. Remove the oil pump screen cover and tube. Discard the old oil pump inlet tube gasket.

To install:
23. Install the oil pump screen cover and tube, along with a new gasket. Tighten the retaining bolts to 15–22 ft. lbs. (20–30 Nm) and the bracket nut to 30–40 ft. lbs. (40–55 Nm).
24. Trial fit the oil pan on the engine block to ensure that there is adequate clearance to prevent the new sealer from being scraped off of the engine block by the oil pan.
25. Clean the mating surfaces of the oil pan and engine block of all old sealant and oil residue with Ford Metal Surface Cleaner F4AZ-19A536-RA or equivalent.

➡**It is imperative to use the proper solvent when cleaning the engine block-to-oil pan surfaces, otherwise the new sealer may not adhere properly. Also, when using silicone sealer, assembly must be performed within 15 minutes of applying the sealer, otherwise the sealer will start to set-up and lose its sealing effectiveness.**

26. Apply a bead of silicone sealer, such as Ford Silicone Gasket and Sealant F6AZ-19652-AA, to the oil pan.

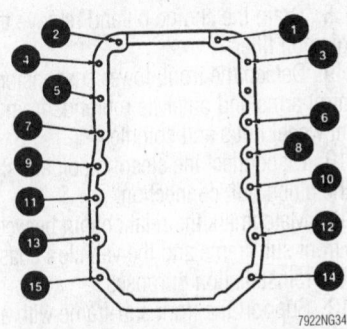

Be sure to tighten the oil pan mounting bolts in the sequence shown to ensure proper sealing—3.8L engine

27. Install the oil pan, then install all of the oil pan mounting bolts. Tighten the bolts, in the sequence shown in the accompanying illustration, to 80–106 inch lbs. (9–12 Nm).

28. Raise the front sub-frame into position, then install the lower shock absorber mounting bolts and nuts to 103–143 ft. lbs. (140–195 Nm).

29. Install the two front and six rear sub-frame bolts loosely.

30. Adjust the position of the front sub-frame assembly until the matchmarks are correctly aligned. Then, tighten the large sub-frame bolts to 83–113 ft. lbs. (113–153 Nm) and the small bolts to 72–97 ft. lbs. (98–132 Nm).

➡It is important that the front sub-frame be realigned as when removed to ensure proper chassis geometry.

31. Connect the steering column shaft at the pinch bolt joint. Tighten the pinch bolt to 30–42 ft. lbs. (41–57 Nm).

32. Install the starter motor and the automatic transmission cooler lines, when applicable.

33. Install the engine oil drain plug and a new oil filter.

➡When seating the engine on the mounts, set it on the left-hand side mount first.

34. Slowly lower the engine until it is resting on the front engine mounts, then install the mount through-bolts. Tighten them to 35–50 ft. lbs. (47–68 Nm).

35. Lower the vehicle and remove the rear wheel blocks.

36. Remove the engine support tool.

37. Install the wiper module and left-hand cowl vent screen.

38. Install the windshield wiper arms and the hood weather seal.

39. Install the radiator upper plastic cover and the air inlet tube.

✳✳ WARNING

It is vitally important to fill the engine with oil prior to starting it, otherwise severe engine damage will occur.

40. Fill the engine with the proper type and amount of clean engine oil.

41. Connect the negative battery cable, then start the engine and check for oil leaks.

42. After 3–5 minutes, stop the engine and check the oil level; add more oil if necessary.

4.6L SOHC Engine

➡Only an MVAC-trained, EPA-certified, automotive technician should service the A/C system or its components.

1. Have your A/C system discharged and evacuated and the A/C compressor outlet hose disconnected and sealed by a qualified automotive technician.

➡It is a good idea to place the vehicle's ignition keys in the box with the new oil containers so that you do not accidentally start the vehicle without first filling the engine with oil.

2. Disconnect the negative battery cable.

3. Remove the air inlet tube from the throttle body and air cleaner housing.

✳✳ CAUTION

Never open, service or drain the radiator or cooling system when hot; serious burns can occur from the steam and hot coolant. Always drain coolant into a sealable container. Coolant should be reused unless it is contaminated or is several years old.

4. Drain the engine cooling system.

5. Remove the cooling fan assembly.

6. Relieve fuel system pressure, then detach the fuel lines from the engine.

7. Remove the upper radiator hose.

8. Remove the windshield wiper governor and support bracket.

9. Remove the A/C hose-to-right-hand ignition coil bracket retaining bolt.

10. Label and disengage the engine wiring harness 42-pin and 8-pin connectors.

11. Remove the EGR backpressure transducer, and disconnect the heater water hose.

12. Detach the heater water hose from the left-hand cylinder head, then position it aside.

13. Remove the heater blower switch resistor.

---

14. Remove the right-hand front engine mount-to-front sub-frame bolt.

15. Disconnect the vacuum hose from the EGR valve.

16. Detach the EGR backpressure transducer hoses from the EGR valve-to-exhaust manifold tube.

17. Remove the EGR valve from the throttle body.

18. Apply the parking brake and block the rear wheels, then raise and safely support the front of the vehicle.

19. Drain the engine oil.

20. Remove both front engine mount through-bolts.

21. Remove the EGR valve-to-manifold tube nut from the left-hand exhaust manifold, then remove the tube.

22. Detach the exhaust system pipes from the exhaust manifolds. Suspend the exhaust pipes with strong cord or wire from the crossmember.

23. Position a floor jack and a block of wood beneath the oil pan, just rearward of the oil drain plug, then raise the engine approximately 4 in. (10cm). Insert two wood blocks (approximately 2.5–2.75 in./ 60–70mm thick) under each front engine mount. Carefully lower the engine onto the wood blocks, once the engine is securely settled, remove the floor jack from beneath the oil pan.

➡It may be necessary to loosen, not remove, the two transmission mount nuts.

24. Using a floor jack (a bottle jack may be necessary because of the lack of clearance), raise the transmission extension housing slightly.

25. Remove the oil pan mounting bolts, then lower the oil pan out of the vehicle.

26. If necessary, the oil pump screen cover and tube may be removed. Remove the old cover and tube gasket.

To install:

27. Clean the oil pan and inspect it for damage; replace it if necessary.

28. Clean the mating surfaces of the oil pan and engine block of all old sealant and oil residue with Ford Metal Surface Cleaner F4AZ-19A536-RA or equivalent.

➡It is imperative to use the proper solvent when cleaning the engine block-to-oil pan surfaces, otherwise the new sealer may not adhere properly.

29. Inspect the oil pump screen cover, tube and O-ring; replace any of them if necessary.

30. If necessary, position the oil pump screen cover and tube on the oil pump.

Hand start the mounting bolts. Install the oil pump screen cover and tube-to-main bearing stud spacer retaining bolt hand-tight.

31. Tighten the oil pump screen cover and tube-to-oil pump bolts to 71–106 inch lbs. (8–12 Nm) and the oil pump screen cover and tube-to-main bearing stud spacer bolt to 15–22 ft. lbs. (20–30 Nm).

32. Install a new oil pan gasket on the oil pan.

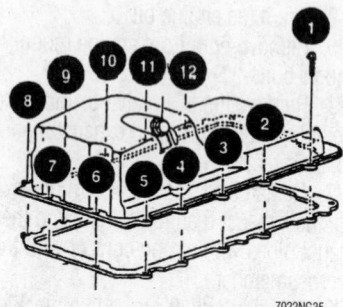

Be sure to tighten the oil pan mounting bolts in the order shown for proper sealing—4.6L SOHC engine

➡ When using silicone sealer, assembly must be performed within four minutes of applying the sealer, otherwise the sealer will start to set-up and lose its sealing effectiveness.

33. Apply a bead of silicone sealer, such as Ford Silicone Gasket and Sealant F6AZ-19652-AA, where the engine front cover, and the crankshaft rear oil seal retainer, meet the engine block.

34. Within four minutes of applying the silicone sealer, install the oil pan and retaining bolts. Tighten the bolts, in the sequence shown in the accompanying illustration, to 177 inch lbs. (20 Nm), then tighten all of the bolts an additional 60 degrees.

35. Position the floor jack under the oil pan and raise the engine to remove the wood blocks.

36. Lower the engine onto the front engine mounts, then remove the floor jack.

37. Install the front engine mounts' through-bolts and tighten them to 15–22 ft. lbs. (20–30 Nm).

➡ Prior to installation, loosen the nut at the EGR valve. This will allow adequate movement to align the EGR valve bolts.

38. Install the EGR valve-to-exhaust manifold tube. Tighten the line nut to 26–33 ft. lbs. (35–45 Nm).

39. Attach the exhaust system pipes to the exhaust manifolds.

40. Install a new engine oil filter, and install the oil pan drain plug.

41. Lower the vehicle and remove the rear wheel blocks.

42. Install the right-hand front engine mount-to-front sub-frame retaining bolt and tighten it to 15–22 ft. lbs. (20–30 Nm).

43. Install the EGR valve along with a new gasket.

44. Install the heater blower motor switch resistor.

45. Install the EGR backpressure transducer bracket and heater water hose. Tighten the upper bolts t o 15–22 ft. lbs. (20–30 Nm) and the lower bolts to 71–106 inch lbs. (8–12 Nm).

46. Connect the heater water hose.

47. Engage the 42-pin and 8-pin engine wiring harness connectors.

48. Install the upper radiator hose.

49. Connect the fuel lines.

50. Install the windshield wiper governor and retaining bracket.

51. Install the engine cooling fan.

52. Install the air inlet tube.

❈❈ WARNING

It is vitally important to fill the engine with oil prior to starting it, otherwise severe engine damage will occur.

53. Fill the engine with the proper type and amount of clean engine oil.

54. Connect the negative battery cable, then start the engine and check for oil leaks.

55. After 3–5 minutes, stop the engine and check the oil level; add more oil if necessary.

56. Have the A/C compressor line reattached and the system evacuated and recharged by a qualified MVAC technician.

4.6L DOHC Engine

1. Open the trunk lid and turn the suspension leveler compressor switch OFF.

2. Disconnect the negative battery cable.

3. Remove the engine oil dipstick.

4. Install an engine support device, such as the Ford Three Bar Engine Support D88L-6000-A, in the engine compartment, and attach it to engine lifting eyes, such as those found in the Rotunda Engine Lift Bracket Set 014–00340.

5. Slightly loosen the front wheel lug nuts.

6. Apply the parking brake and block the rear wheels, then raise and safely support the front of the vehicle.

7. Remove the front wheels.

8. Drain the engine oil and remove the engine oil filter.

9. Detach the front, lower suspension control arms and outer tie rod ends from the front wheel hubs and spindles.

10. Disconnect the steering column shaft at the pinch bolt connection.

11. Matchmark the relationship between the front sub-frame and the vehicle's chassis for reinstallation purposes.

12. Support the front sub-frame with a floor jack and jack stands, then remove the front engine mounts' through-bolts.

13. Detach the front suspension struts from the lower control arms.

14. Remove the sub-frame mounting bolts.

15. Remove the jack stands, then carefully lower the front sub-frame with the floor jack.

16. If equipped, disengage the wiring from the low oil level sensor connector.

17. Detach the wiring from the oil pan rail.

18. Remove the oil pan mounting bolts, then separate the pan from the engine block and front engine cover.

19. If necessary, remove the oil pump screen cover and tube.

To install:

20. Clean the oil pan and inspect it for damage; replace it if necessary.

21. Clean the mating surfaces of the oil pan and engine block of all old sealant and oil residue with Ford Metal Surface Cleaner F4AZ-19A536-RA or equivalent. If scraping is required, only use a plastic tool.

➡ It is imperative to use the proper solvent when cleaning the engine block-to-oil pan surfaces, otherwise the new sealer may not adhere properly.

22. Inspect the oil pump screen cover and tube; replace either of them if necessary.

23. If necessary, position the oil pump screen cover and tube on the oil pump, along with a new gasket. Hand start the mounting bolts. Install the oil pump screen cover and tube-to-main bearing stud spacer retaining bolt hand-tight.

24. Tighten the oil pump screen cover and tube-to-oil pump bolts to 71–106 inch lbs. (8–12 Nm) and the oil pump screen cover and tube-to-main bearing stud spacer bolt to 15–22 ft. lbs. (20–30 Nm).

25. Install a new oil pan gasket on the oil pan.

➡ When using silicone sealer, assembly must be performed within four minutes of applying the sealer, otherwise the sealer will start to set-up and lose its sealing effectiveness.

26. Apply a bead of silicone sealer, such as Ford Silicone Gasket and Sealant F6AZ-19652-AA, where the engine front cover, and the crankshaft rear oil seal retainer, meet the engine block.

27. Within four minutes of applying the silicone sealer, install the oil pan and retaining bolts. Tighten the bolts, in the sequence shown in the accompanying illustration, to 15–22 ft. lbs. (20–30 Nm).

28. If equipped, attach the wiring to the low oil level sensor connector, and attach the wiring harness to the oil pan rail.

29. Raise the sub-frame into position, install the front engine mount through-bolts, then install the front sub-frame mounting bolts.

30. Adjust the position of the front sub-frame assembly until the matchmarks are correctly aligned. Then, tighten the large sub-frame bolts to 83–113 ft. lbs. (113–153 Nm) and the small bolts to 72–97 ft. lbs. (98–132 Nm).

➡️It is important that the front sub-frame be realigned as when removed to ensure proper chassis geometry.

31. Connect the steering column shaft at the pinch bolt joint. Tighten the pinch bolt to 30–42 ft. lbs. (41–57 Nm).

32. Reconnect the lower control arms and outer tie rod ends to the front wheel hubs and spindles.

33. Install the wheels and tighten the lug nuts until snug.

34. Install the oil pan drain plug to 97–142 inch lbs. (11–16 Nm) and a new engine oil filter hand-tight (follow the oil filter manufacturer's instructions).

35. Lower the vehicle and remove the rear wheel blocks. Fully tighten the lug nuts.

36. Remove the engine support device from the engine and engine compartment. Remove the lifting eye brackets from the engine.

✱✱ WARNING

It is extremely important to fill the engine with oil prior to starting it, otherwise severe engine damage will occur.

37. Fill the engine with the proper amount and type of clean engine oil.

38. Connect the negative battery cable, then start the engine and check for oil leaks.

39. After 3–5 minutes, stop the engine.

40. Wait 10 minutes, then check the oil level with the dipstick. Add more oil, if necessary.

41. Turn the rear suspension leveler compressor switch ON, then shut the trunk lid.

5.0L Engine

1. Disconnect the negative battery cable for safety.

2. Remove the air cleaner tube.

3. Remove the oil level indicator from the left side of the cylinder block.

4. Remove the fan shroud retaining bolts, then position the shroud over the fan.

5. Raise and support the vehicle safely using jackstands.

6. Drain the crankcase and remove the oil level sensor wiring from the oil pan.

7. Remove the starter motor assembly.

8. Remove the catalytic converter and muffler inlet pipes.

9. Remove the engine mount-to-No. 2 crossmember attaching bolts or nuts. Support the transmission and remove the No. 3 crossmember and rear insulator support assemblies.

10. Remove the steering gear attaching bolts and position the steering gear forward, out of the way.

11. Raise and support the engine to a position which allows clearance for oil pan removal.

➡️The best way to raise the engine is using an engine support fixture or an engine hoist. But, if necessary, a floor jack can be used under the oil pan if a large block of wood is used to distribute the load and protect the pan.

12. With the engine raised, install wood blocks between the engine mounts and frame. Then, lower the engine onto the wood blocks. If you are using an engine hoist or support fixture, leave it in place for additional safety.

13. Remove the oil pan attaching bolts and lower the pan to the No. 2 crossmember.

14. Loosen the retaining bolts, then lower the oil pump and pick-up tube assembly into the pan.

15. Remove the pan, along with the oil pump assembly, from the vehicle.

To install:

16. Clean the oil pan and the gasket mating surfaces.

17. Refer to the instructions included with the new oil pan gasket to determine if sealer is necessary for that type of gasket. Position the new gasket on the pan or block, as applicable.

18. With the oil pump and pick-up tube assembly positioned in the oil pan, raise the pan onto the crossmember. Install the oil pump assembly.

19. With the pump properly secured, raise the oil pan into position and install the retaining bolts. Tighten the oil pan bolts to 110–144 inch lbs. (12–16 Nm).

20. Carefully raise the engine and remove the wood blocks, then lower the engine and remove the lifting device.

21. Install the engine mount-to-No. 2 crossmember attaching nuts. Tighten the retainers to 72–98 ft. lbs. (97–133 Nm).

22. Position the steering gear and install the retaining bolts.

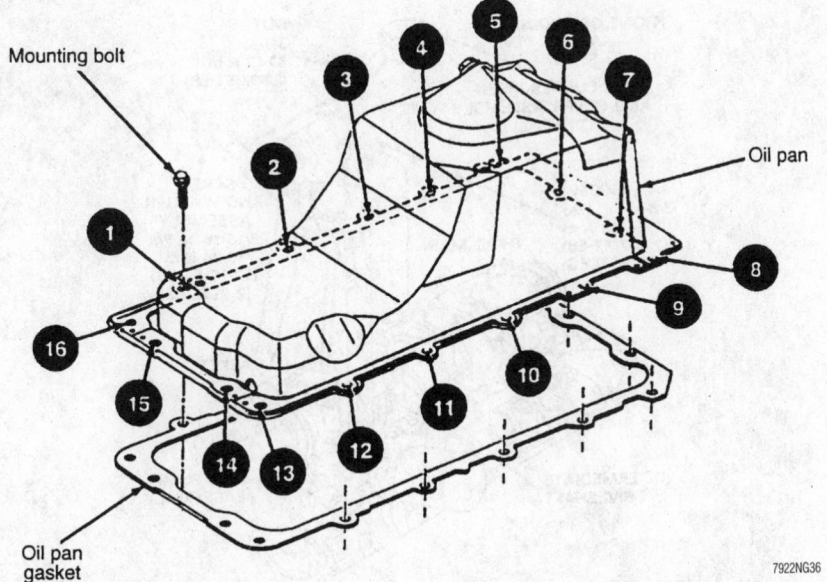

Mounting bolt

Oil pan

Oil pan gasket

7922NG36

When installing the oil pan, be sure to tighten the mounting bolts in the order shown

23. Install the starter motor assembly.

24. Connect the oil level sensor wire to the oil pan.

25. Install the rear insulator and the No. 3 crossmember. Tighten the retainers to 80–106 ft. lbs. (108–144 Nm).

26. Install the catalytic converter and muffler inlet pipes.

27. Remove the jackstands and carefully lower the vehicle.

28. IMMEDIATELY refill the engine crankcase to prevent an accidental attempt to start the engine without oil in it. Install the oil level indicator.

29. Install the fan shroud.

30. Install the air cleaner tube.

31. Connect the negative battery cable, then start the engine and allow it to run for a few minutes. If there is no indication of oil pressure after the engine runs for a few seconds, shut it **OFF** and investigate the problem.

32. Check for leaks.

Oil Pump

REMOVAL & INSTALLATION

3.8L Engine

1. If necessary, remove the engine oil filter.

2. Remove the oil pump and filter body-to-engine front cover retaining bolts, then separate the oil pump from the front cover.

3. Inspect the oil pump sealing O-ring for damage or distortion.

To install:

4. Position the oil pump on the engine front cover, then install the mounting bolts.

5. Tighten the four large bolts to 17–23 ft. lbs. (23–32 Nm) and the two small bolts to 71–97 inch lbs. (8–11 Nm).

6. If necessary, install a new engine oil filter. Be sure to coat the new oil filter's gasket with a film of clean engine oil. Tighten the oil filter according to the manufacturer's instructions.

4.6L Engines

1. For the SOHC engine, remove the valve covers.

2. Remove the engine front cover.

3. Remove the oil pan, and oil pump screen cover and pick-up tube.

4. Remove the timing chain(s).

5. For the DOHC engine, remove the crankshaft sprockets.

6. Remove the four oil pump mounting bolts, then separate the oil pump from the engine block.

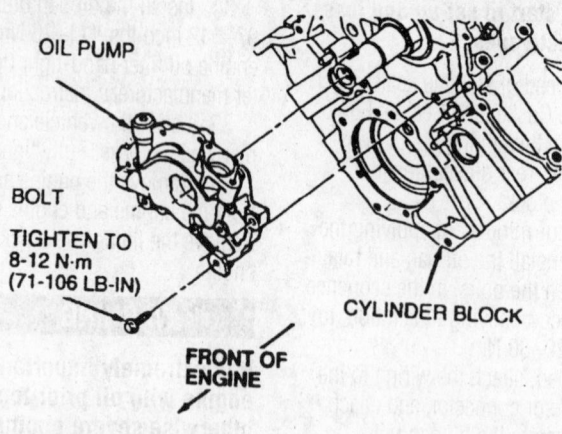

OIL PUMP

BOLT

TIGHTEN TO 8-12 N·m (71-106 LB-IN)

FRONT OF ENGINE

CYLINDER BLOCK

7922NG37

Exploded view of the oil pump mounting on the 4.6L SOHC and DOHC engines

To install:

7. Rotate the inner rotor of the oil pump to align it with the flats on the crankshaft, then install the oil pump on the engine block.

8. Install the four mounting bolts and tighten them to 71–106 inch lbs. (8–12 Nm).

9. Install a new engine oil filter. Be sure to coat the new oil filter's gasket with a film of clean engine oil. Tighten the oil filter according to the manufacturer's instructions.

10. If applicable, install the crankshaft sprockets.

11. Install the timing chain.

12. Install the oil pump screen cover and pick-up tube.

13. Install the oil pan.

14. Install the engine front cover.

15. If applicable, install the valve covers.

✱✱ WARNING

It is extremely important to fill the engine with oil prior to starting it, otherwise severe engine damage will occur.

16. Fill the engine with the proper amount and type of clean engine oil.

17. Start the engine and check for oil leaks.

18. After 3–5 minutes, stop the engine.

19. Wait 10 minutes, then check the oil level with the dipstick. Add more oil, if necessary.

5.0L Engine

➡The oil pump and pick-up tube assembly must be unbolted and positioned in the oil pan, in order to

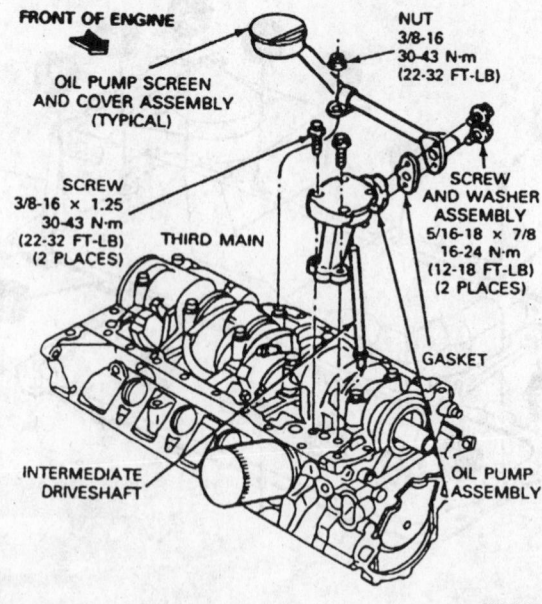

FRONT OF ENGINE

OIL PUMP SCREEN AND COVER ASSEMBLY (TYPICAL)

SCREW 3/8-16 × 1.25 30-43 N·m (22-32 FT-LB) (2 PLACES)

THIRD MAIN

INTERMEDIATE DRIVESHAFT

NUT 3/8-16 30-43 N·m (22-32 FT-LB)

SCREW AND WASHER ASSEMBLY 5/16-18 × 7/8 16-24 N·m (12-18 FT-LB) (2 PLACES)

GASKET

OIL PUMP ASSEMBLY

7922NG38

Exploded view of the oil pump mounting on the 5.0L engine

remove the pan while the engine is installed in the vehicle.

1. Disconnect the negative battery cable for safety.

2. Remove the oil pan, along with the oil pump and pick-up tube assembly.

3. If not done for pan removal, loosen the retainers, then remove the oil pump pick-up (inlet) tube and screen assembly, along with the oil pump assembly.

To install:

4. Prime the oil pump by filling either the inlet or outlet ports with engine oil and rotating the pump shaft to distribute the oil within the pump body.

5. Position the intermediate driveshaft into the distributor socket. With the shaft firmly seated in the distributor socket, the stop on the shaft should touch the roof of the crankcase. Remove the shaft and position the stop, as necessary.

➡**Remember, that on most vehicles covered by this manual, you will have to position the oil pan over the crossmember (with the pump in the pan), then raise the pump into position and install.**

6. Position a new gasket on the pump body, insert the intermediate shaft into the oil pump and install the pump and shaft as an assembly.

➡**Do not attempt to force the pump or shaft into position if it will not seat readily. The driveshaft hex may be misaligned with the distributor shaft. To align, rotate the intermediate shaft into a new position.**

7. Tighten the oil pump attaching screws to 14–21 ft. lbs. (19–29 Nm) on the 2.3L engine and 22–32 ft. lbs. (30–43 Nm) on the 5.0L engine.

8. If separated, clean and install the oil pump inlet tube and screen assembly.

9. Install the oil pan.

10. Connect the negative battery cable.

Rear Main Seal

REMOVAL & INSTALLATION

3.8L and 5.0L Engines

1. Disconnect the negative battery cable for safety.

2. Remove the flywheel or flexplate for access to the rear main seal.

✳✳ WARNING

Use extreme caution not to scratch the crankshaft oil seal surface.

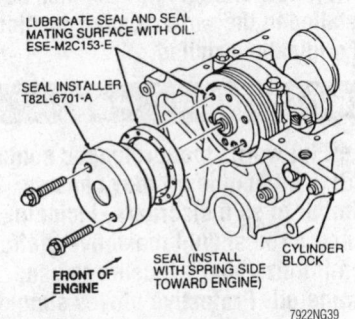

The one piece rear main seal on these engines is installed with a special tool—3.8L and 5.0L engines

3. Carefully punch 2 holes in the crankshaft rear oil seal on opposite sides of the crankshaft, just above the bearing cap-to-cylinder block split line. Install a sheet metal screw in each of the holes or use a small slide hammer to pry the crankshaft rear main oil seal from the block.

To install:

4. Clean the oil seal recess in the cylinder block and main bearing cap.

5. Coat the seal and the seal mounting surfaces with oil.

6. Position the seal on rear main seal installer T82L-6701-A, or equivalent, then position the tool and seal to the rear of the engine.

7. Alternate tightening the seal installer tool bolts until the seal is properly seated. The rear face of the seal must be within 0.005 in. (0.127mm) of the rear face of the block for 5.0L engines, or within 0.020 in. (0.508mm) for 3.8L engines.

8. Install the flywheel or flexplate, as applicable.

9. Connect the negative battery cable.

4.6L Engines

1. Disconnect the negative battery cable for safety.

2. Remove the flywheel or flexplate for access to the rear main seal.

3. Remove the crankshaft rear main oil seal retainer.

✳✳ WARNING

Use care to avoid damaging the retainer when removing the oil seal;

---

a damaged retainer may allow oil to leak.

4. Securely support the oil seal retainer, then remove the oil seal from the retainer using a punch and a hammer.

To install:

5. Clean the crankshaft rear oil seal retainer and retainer-to-engine block mating surfaces of all dirt, grime and oil residue. The oil residue should be cleaned from the surfaces with an appropriate metal cleaner, such as Ford Metal Surface Cleaner F4AZ-19A536-RA. Also inspect for any damage.

➡**Failure to clean the surfaces with the appropriate cleaning solvent can retard the new sealant's ability to adhere to the surfaces.**

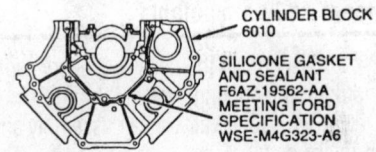

Apply silicone gasket sealant to the engine block (as shown) for to prevent oil leakage from the oil seal retainer

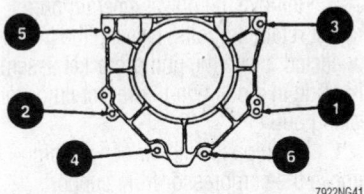

Position the oil seal retainer against the engine block, then install and tighten the mounting bolts in the sequence shown

6. Apply a continuous 0.06 in. (1.5mm) diameter bead of silicone gasket sealant, such as Ford Sealant F6AZ-19562-A, to the engine block as shown in the accompanying illustration.

7. Install the rear oil seal retainer on the engine block, then tighten the mounting fasteners to 71–106 inch lbs. (8–12 Nm) in the sequence shown.

8. Coat the seal and the seal mounting surfaces with oil.

9. Position the seal on rear main seal installer tool (Ford T82L-6701-A or equivalent) and the adapters (Ford T91P-6701-A

or equivalent), then position the tool and seal on the rear of the engine.

10. Tighten the bolts alternately until the seal is properly seated.

11. Install the flywheel or flexplate, as applicable.

12. Connect the negative battery cable.

Timing Chain, Sprockets and Front Cover

REMOVAL & INSTALLATION

3.8L Engine

> ✳ CAUTION
>
> **Never open, service or drain the radiator or cooling system when hot; serious burns can occur from the steam and hot coolant.**

1. Disconnect the negative battery cable and drain the cooling system.

2. Remove the air cleaner assembly and air intake duct.

3. On non-supercharged engines, remove the fan/clutch assembly and shroud. On supercharged engines, remove the electric cooling fan assembly.

4. Remove the accessory drive belt idlers, drive belts and the water pump pulley.

5. Remove the power steering pump bracket retaining bolts. Leaving the hoses connected, place the pump/bracket assembly aside in a position to prevent fluid from leaking out.

6. If equipped with air conditioning, remove the compressor front support bracket but leave the compressor in place.

7. Disconnect the coolant bypass hose and heater hose at the water pump. Disconnect the upper radiator hose at the thermostat housing.

8. Disconnect the coil wire from the distributor cap and remove the cap with the secondary wires attached.

9. Remove the distributor hold-down clamp and lift the distributor out of the front cover.

10. Raise and safely support the vehicle. Remove the crankshaft damper and pulley using a puller.

➡ **If the crankshaft pulley and vibration damper have to be separated, mark the damper and pulley so they may be reassembled in the same relative position. This is important as the damper and pulley are initially balanced as a unit. If the crankshaft damper is being**

replaced, check if the original damper has balance pins installed. If so, new balance pins must be installed on the new damper in the same position as the original damper. The crankshaft pulley, new or original, must also be installed in the same relative position as originally installed.

> ✳ CAUTION
>
> **The EPA warns that prolonged contact with used engine oil may cause a number of skin disorders, including cancer! You should make every effort to minimize your exposure to used engine oil. Protective gloves should be worn when changing the oil. Wash your hands and any other exposed skin areas as soon as possible after exposure to used engine oil. Soap and water, or waterless hand cleaner should be used.**

11. Remove the oil filter. On supercharged engines, remove the oil cooler.

12. Disconnect the lower radiator hose at the water pump. Remove the oil pan.

➡ **The front cover cannot be removed without lowering the oil pan.**

13. Lower the vehicle. Remove the front cover retaining bolts. It is not necessary to remove the water pump.

➡ **Do not overlook the cover retaining bolt located behind the oil filter adapter. The front cover will break if pried on and all retaining bolts are not removed.**

14. Remove the front cover and water pump as an assembly. Remove and discard the cover gasket.

➡ **The front cover contains the oil pump and water pump. If a new front cover is to be installed, remove the water pump and oil pump from the old front cover.**

15. Remove the camshaft bolt and washer from the end of the camshaft.

16. Remove the distributor drive gear, camshaft sprocket, crankshaft sprocket and timing chain.

➡ **If the crankshaft sprocket is difficult to remove, pry the sprocket off the shaft using a pair of large prybars positioned on both sides of the sprocket.**

To install:

17. Clean all gasket mating surfaces. If reusing the front cover, replace the front cover oil seal.

18. Rotate the crankshaft to position the No. 1 piston at TDC and the crankshaft keyway at the 12 o' clock position.

19. Lubricate the timing chain with engine oil.

20. Install the camshaft sprocket, crankshaft sprocket and timing chain. Be sure the timing marks align.

21. Install the distributor drive gear. Install the bolt and washer assembly on the end of the camshaft and tighten to 30–37 ft. lbs. (40–50 Nm).

22. Position a new gasket on the cylinder block and install the front cover using dowels for proper alignment. Install the front cover retaining bolts and tighten to 15–22 ft. lbs. (20–30 Nm).

23. Raise and safely support the vehicle. Install the oil pan. Connect the lower radiator hose and install the oil filter.

24. Coat the crankshaft damper sealing surface with clean engine oil. Apply a small amount of silicone sealer to the crankshaft keyway.

25. Position the crankshaft pulley key in the crankshaft keyway and install the damper, using a suitable installation tool.

26. Install the damper washer and retaining bolt and tighten to 103–132 ft. lbs. (140–180 Nm). Install the crankshaft pulley and tighten the retaining bolts to 20–28 ft. lbs. (26–38 Nm).

27. Lower the vehicle. Connect the coolant bypass hose.

28. Install the distributor with the rotor pointing at the No. 1 distributor cap tower. Install the distributor cap and coil wire. On supercharged engines, install the camshaft synchronizer.

29. Connect the upper radiator hose at the thermostat housing. Connect the heater hose.

30. If equipped with air conditioning, install the compressor and mounting brackets. Tighten retaining bolts to 30–45 ft. lbs. (41–61 Nm).

31. Install the power steering pump and mounting bracket. Tighten the retaining bolts to 30–45 ft. lbs. (41–61 Nm).

32. Install the water pump pulley. Position the accessory drive belts over the pulleys.

33. Install the fan/clutch assembly and fan shroud. Cross-tighten the fan/clutch assembly retaining bolts to 12–18 ft. lbs. (16–24 Nm).

> ✳ WARNING
>
> **Operating the engine without the proper amount and type of engine oil will result in severe engine damage.**

34. Fill the crankcase with the proper type and quantity of engine oil. Fill and bleed the cooling system.

35. Connect the negative battery cable.

36. Start the engine and check for leaks. Check the ignition timing and curb idle speed and adjust, as necessary.

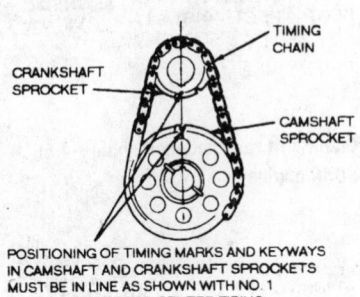

CRANKSHAFT SPROCKET

TIMING CHAIN

CAMSHAFT SPROCKET

POSITIONING OF TIMING MARKS AND KEYWAYS IN CAMSHAFT AND CRANKSHAFT SPROCKETS MUST BE IN LINE AS SHOWN WITH NO. 1 PISTON AT TOP DEAD CENTER FIRING.

7922NG53

Be sure that the timing marks are facing each other after the chain is installed— 3.8L engine

4.6L SOHC Engine

➡**This is not a free wheeling engine. If it has "jumped time" there will be damage to the valves and/or pistons and will require the removal of the cylinder heads.**

1. Disconnect the negative battery cable.

2. Remove the cooling fan and shroud. Loosen the water pump pulley bolts, remove the accessory drive belt and remove the water pump pulley.

3. Raise and safely support the vehicle.

4. Remove the bolts retaining the power steering pump to the engine block and cylinder front cover. The lower front bolt on the power steering pump will not come all the way out. Wire the power steering pump out of the way.

5. Remove the 4 bolts retaining the oil pan to the front cover. Remove the crankshaft damper retaining bolt and washer. Remove the damper using a puller.

6. Lower the vehicle. Remove the bolt retaining the air conditioner high pressure line to the right coil bracket.

7. Remove the front bolts and loosen the remaining bolts on the camshaft covers. Using plastic wedges or similar tools, prop up both camshaft covers. Disconnect both ignition coils and CID sensor.

8. Remove the 3 nuts retaining the right coil bracket to the front cover. Position the power steering hose out of the way.

9. Remove the 4 nuts retaining the left coil bracket to the front cover. Slide both coil brackets and ignition wires off the mounting studs and lay the assembly on top of the engine.

10. Disconnect the High Data Rate (HDR) sensor. Remove the 7 stud bolts and 4 bolts retaining the front cover to the engine and remove the front cover.

11. Remove the High Data Rate (HDR) wheel.

12. Rotate the engine to set the No. 1 piston at TDC on the compression stroke.

13. Install cam positioning tools T91P-6256-A or equivalent, on the flats of the camshaft. This will prevent accidental rotation of the camshafts.

14. Remove the 2 bolts retaining the right tensioner to the cylinder head and remove the tensioner. Remove the right tensioner arm.

15. Remove the 2 bolts retaining the right chain guide to the cylinder head and remove the chain guide. Remove the right chain and right crankshaft sprocket. Remove the right camshaft sprocket retaining bolt, washer, sprocket and spacer.

➡**Cam positioning tools T91P-6256-A or equivalent, must be installed on the camshaft to prevent the camshaft from rotating.**

16. Remove the 2 bolts retaining the left tensioner to the cylinder head and remove the tensioner. Remove the left tensioner arm.

17. Remove the 2 bolts retaining the left chain guide to the cylinder head and remove the chain guide. Remove the left chain and left crankshaft sprocket. Remove the left camshaft sprocket retaining bolt, washer, sprocket and spacer.

➡**Cam positioning tools T91P-6256-A or equivalent, must be installed on the camshaft to prevent the camshaft from rotating.**

18. Inspect the friction material on the tensioner arms and chain guides. If worn or damaged, remove and clean the oil pan and replace the oil pick-up tube.

➡**At no time, when the timing chains are removed and the cylinder heads are installed, may the crankshaft and/or camshafts be rotated. Failure to follow these directions will result in valve and/or piston damage.**

To install:

19. Be sure cam positioning tools T91P-6256-A or equivalent, are installed on the camshafts to prevent them from rotating.

20. Position the camshaft spacers and sprockets on the camshafts and install the washers and retaining bolts. Tighten the retaining bolts to 81–96 ft. lbs. (110–130 Nm).

21. Install the left crankshaft sprocket with the tapered part of the sprocket facing away from the engine block.

➡**The crankshaft sprockets are identical. They may only be installed one way, with the tapered part of the sprocket facing each other.**

22. Install the left timing chain on the camshaft and crankshaft sprockets. Be sure the copper links of the chain line up with the timing marks of the sprockets.

➡**If the copper links of the timing chain are not visible, pull the chain taught until the opposite sides of the chain contact one another and lay it on a flat surface. Mark the links at each end of the chain and use them in place of the copper links.**

23. Install the right crankshaft sprocket with the tapered part of the sprocket facing the left crankshaft sprocket.

24. Install the right timing chain on the camshaft and crankshaft sprockets. Be sure the copper links of the chain line up with the timing marks of the sprockets.

25. It is necessary to bleed the timing chain tensioners before installation. Proceed as follows:

    a. Position the timing chain tensioner in a soft-jawed vice.

    b. Using a small pick or similar tool, hold the ratchet lock mechanism away from the ratchet stem and slowly compress the tensioner plunger by rotating the vise handle.

➡**The tensioner must be compressed slowly or damage to the internal seals will result.**

    c. Once the tensioner plunger bottoms in the tensioner bore, continue to hold the ratchet lock mechanism and push down on the ratchet stem until flush with the tensioner face.

    d. While holding the ratchet stem flush to the tensioner face, release the ratchet lock mechanism and install a paper clip or similar tool in the tensioner body to lock the tensioner in the collapsed position.

    e. The paper clip must not be removed until the timing chain, tensioner, tensioner arm and timing chain guide are completely installed on the engine.

26. Lubricate the tensioner arm contact surfaces with engine oil and install the right and left tensioner arms on their dowels.

27. Install the right and left timing chain tensioners and secure with 2 bolts on each. Tighten the bolts to 15–22 ft. lbs. (20–30 Nm).

28. Install the right and left timing chain guides and secure with 2 bolts on each. Tighten the bolts to 6.0–8.8 ft. lbs. (8–12 Nm).

29. Remove the paper clips from the timing chain tensioners and be sure all timing marks are aligned.

30. Remove the camshaft positioning tools.

31. Inspect and replace the front cover seal as necessary and clean the sealing surfaces of the cylinder block. Apply silicone sealer to the oil pan where it meets the cylinder block and to the points where the cylinder head meets the cylinder block.

32. Install the front cover and the attaching studs and bolts. Tighten to 15–22 ft. lbs. (20–30 Nm). Connect the HDR sensor.

33. Position the coil brackets and ignition wires as an assembly onto the mounting studs. Position the power steering hose and install the 7 nuts retaining the coil brackets to the front cover. Tighten the nuts to 15–22 ft. lbs. (20–30 Nm). Connect both ignition coils and CID sensor.

34. Remove the plastic wedges holding up the camshaft covers. Apply silicone sealer where the front cover meets the cylin-

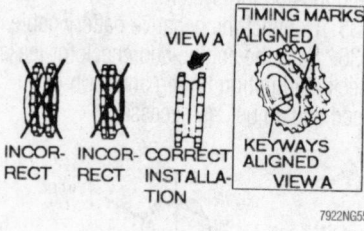

Crankshaft sprocket positioning—4.6L SOHC engine

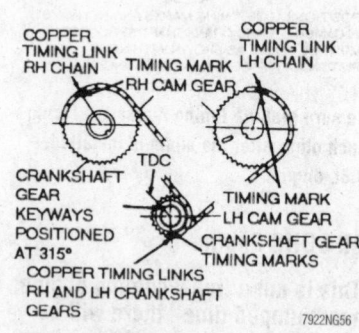

Timing chain and sprocket alignment—4.6L SOHC engine

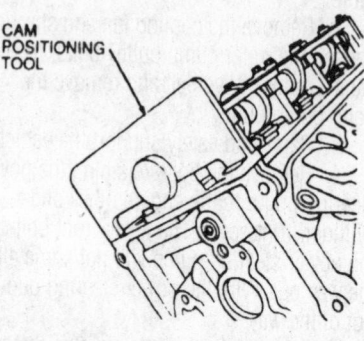

Camshaft positioning tool—4.6L SOHC engine

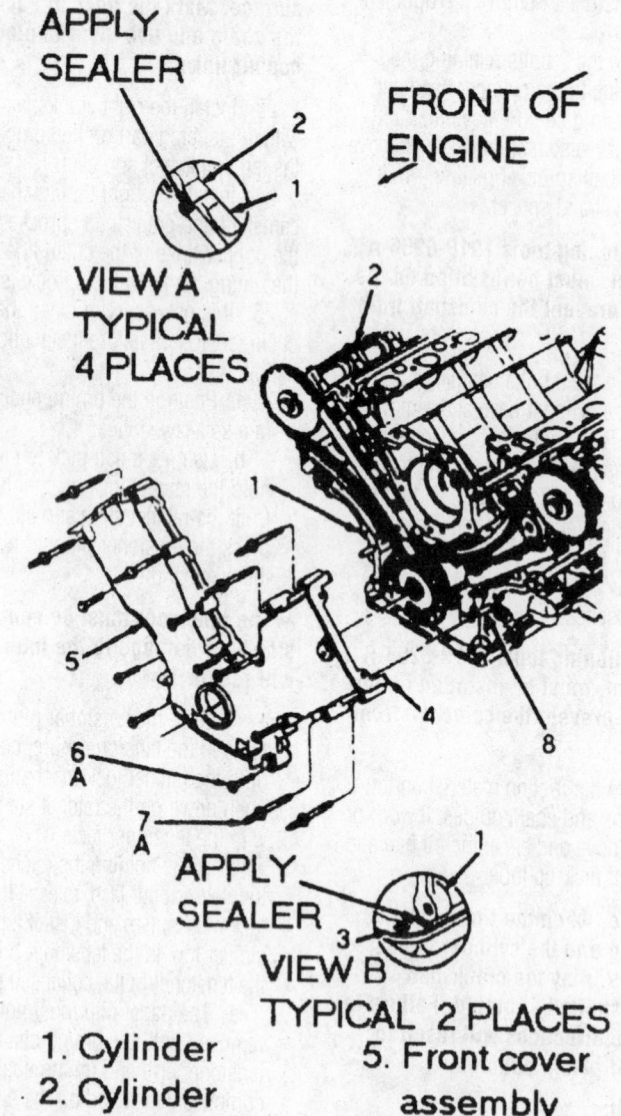

1. Cylinder
2. Cylinder
3. Oil pan gasket
4. Gasket
5. Front cover assembly
6a. Bolts
7a. Studs
8. Dowel

Timing chain front cover installation—4.6L SOHC engine

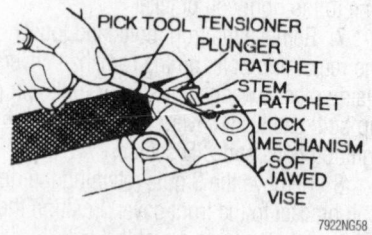

Timing chain tensioner bleeding procedure—4.6L SOHC engine

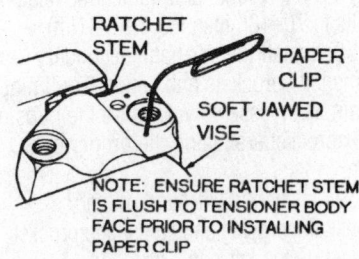

RATCHET STEM — PAPER CLIP — SOFT JAWED VISE

NOTE: ENSURE RATCHET STEM IS FLUSH TO TENSIONER BODY FACE PRIOR TO INSTALLING PAPER CLIP

7922NG59

Timing chain tensioner locking procedure—4.6L SOHC engine

der head and be sure the camshaft cover gaskets are properly positioned. Install the front retaining bolts into the camshaft cover and tighten the bolts to 6.0–8.8 ft. lbs. (8–12 Nm).

35. Position the air conditioner high pressure line on the right coil bracket and install the bolt. Raise and safely support the vehicle.

36. Apply a small amount of silicone sealer in the rear of the keyway in the damper. Position the damper on the crankshaft and install, using a suitable installation tool. Install the damper bolt and washer and tighten to 114–121 ft. lbs. (155–165 Nm).

37. Install the 4 bolts retaining the oil pan to the front cover. Tighten to 15–22 ft. lbs. (20–30 Nm).

38. Position the power steering pump on the engine and install the 4 retaining bolts. Tighten to 15–22 ft. lbs. (20–30 Nm). Lower the vehicle.

39. Install the water pump pulley with the 4 bolts. Tighten to 15–22 ft. lbs. (20–30 Nm). Install the accessory drive belt and the cooling fan and shroud.

40. Connect the negative battery cable, start the engine and check for leaks.

4.6L DOHC Engine

✳✳ WARNING

This is not a free-wheeling engine. When the timing chains are removed and the cylinder heads are installed, the crankshaft and/or camshafts must not be rotated unless as directed in this procedure. Failure to follow these instructions will result in valve and/or piston damage.

1. Disconnect the negative battery cable.
2. Remove the windshield wiper module assembly.
3. Remove the engine appearance cover, if equipped.

4. Remove the engine air cleaner outlet tube.
5. Drain the engine cooling system.
6. Remove the water bypass tube and the upper radiator hose.
7. Remove the engine cooling fan motor and shroud assembly.
8. Remove the engine control wiring support bracket and straps from the left-hand center of the engine front cover.
9. Loosen the water pump pulley bolts.
10. If equipped with air injection, disconnect the vacuum and electrical harness connectors. Remove the necessary tube brackets, hoses and gaskets. Remove the air injection diverter valve.
11. Remove the accessory drive belt.
12. Remove the water pump pulley and lower water pump-to-cylinder block retaining bolt for engine front cover removal clearance.
13. Remove the bolts and stud bolt retaining the power steering pump reservoir to the left-hand ignition coil bracket.
14. Raise and safely support the vehicle.
15. Using Pulley Remover T69L-10300-B or equivalent, remove the power steering pump pulley.
16. Remove the bolt retaining the power steering pressure hose to the power steering pump bracket.
17. Remove the bolts retaining the power steering pump to the cylinder block and the engine front cover.

➡ **The front lower bolt on the power steering pump will not come all the way out.**

18. Position the power steering pump and reservoir out of the way.
19. Remove 4 bolts retaining the oil pan to the engine front cover.
20. Remove the crankshaft pulley retaining bolt and washer from the crankshaft.
21. Install Crankshaft Damper Remover T58P-6316-D or equivalent, on the crankshaft pulley and pull the crankshaft pulley from the crankshaft.
22. Lower the vehicle.
23. Remove both cylinder head covers.
24. Disconnect the engine control wiring from both ignition coils and the Camshaft Position (CMP) sensor.
25. Remove 3 bolts retaining the right-hand ignition coil bracket to the engine front cover.
26. Remove 3 nuts and 1 bolt retaining the left-hand ignition coil bracket to the engine front cover.
27. Slide both ignition coil brackets with the ignition wires attached off of the front cover studs and move aside.

28. Remove the bolt retaining the belt idler pulley and remove the idler pulley.
29. Remove the drive belt tensioner retaining bolt and drive belt tensioner from the engine front cover.
30. Disconnect the Crankshaft Position (CKP) sensor and move the wiring out of the way.
31. Remove the stud bolts and bolts retaining the engine front cover to the engine.
32. Remove the engine front cover.
33. Remove the left-hand and right-hand cylinder head covers.
34. Remove the CKP sensor tooth wheel.
35. Rotate the crankshaft to place the piston for No. 1 cylinder at Top Dead Center (TDC) on its compression stroke.

✳✳ WARNING

Camshaft Positioning Tool T93P-6256-A or equivalent, and Camshaft Holding Tool T93P-6256-AH or equivalent, must be installed to prevent accidental rotation of the camshafts and possible engine damage.

36. Install Camshaft Positioning Tool T93P-6256-A or equivalent, in the rear D-slots of the camshaft.

✳✳ WARNING

Failure to use Camshaft Holding Tool T93P-6256-AH or equivalent, while assembling or disassembling the timing chains will result in damage to the D-slots or Camshaft Positioning Tool T93P-6256-A.

37. Install Camshaft Holding Tool T93P-6256-AH or equivalent, onto the camshafts to keep the camshafts from rotating and to prevent damaging the camshaft positioning tool.
38. Remove 2 bolts retaining the right-hand primary timing chain tensioner to the cylinder head and remove the primary timing chain tensioner.
39. Remove the bolt retaining the right-hand tensioner arm and remove the right-hand tensioner arm.

➡ **The 2 bolts retaining the timing chain guide to the cylinder head are longer than the one to the block.**

40. Remove 3 bolts retaining the right-hand timing chain guide and remove the timing chain guide.
41. Remove the right-hand primary timing chain from the camshaft and crankshaft sprockets.
42. Remove the right-hand crankshaft sprocket.

43. Remove the camshaft sprocket retaining bolts.

✳ WARNING

The secondary timing chain tensioner plunger is spring loaded. Care must be taken to prevent the plunger from dropping out of the tensioner during disassembly.

44. Unlock and compress the secondary timing chain tensioners and lock the timing chain tensioner in the compressed position using a paper clip or stiff wire. Remove the secondary timing chains and camshaft sprockets.

45. Remove 2 bolts retaining the left-hand primary timing chain tensioner to the cylinder head and remove the primary timing chain tensioners.

46. Remove the bolt retaining the left-hand timing chain tensioner arm and remove the left-hand timing chain tensioner arm.

47. Remove 3 bolts retaining the left-hand timing chain guide and remove the timing chain guide.

48. Remove the left-hand primary timing chain from the camshaft sprocket and crankshaft sprocket.

49. Remove the left-hand crankshaft sprocket.

✳ WARNING

Failure to use Camshaft Holding Tool T93P-6256-AH or equivalent, while removing or tightening the camshaft bolts may result in damage to the camshaft D-slots or the camshaft positioning tool.

50. Remove the camshaft sprocket retaining bolts.

✳ WARNING

The secondary timing chain tensioner plunger is spring loaded. Care must be taken to prevent the plunger from dropping out of the tensioner during disassembly.

51. Unlock and compress the secondary chain tensioner and lock the chain tensioner in the compressed position using a paper clip or stiff wire. Remove the secondary timing chain and camshaft sprockets.

✳ WARNING

Do not rotate the crankshaft and/or camshafts or engine damage may occur.

52. Inspect the friction material on the timing chain tensioner arms and timing chain guides. If worn or damaged, remove and clean the oil pan and replace the oil pump screen cover and tube.

To install:

53. If the engine has jumped time, be sure that all service to the necessary engine components and/or valvetrain have been made.

54. Be sure the primary and secondary timing chain tensioners are in the collapsed position and retained with paper clips or equivalent.

55. If the cylinder heads were removed from the cylinder block, position the crankshaft keyway to 270 degrees (9 o'clock position), before reinstalling the cylinder heads.

➡ **The rocker arms must not be installed in the cylinder heads at this time.**

56. Install Camshaft Positioning Tool T93P-6256-A or equivalent, into the D-slots on the rear of the camshafts to position both intake and exhaust camshafts with the keyways pointing down towards the crankshaft. Install Camshaft Holding Tool T93P-6256-AH or equivalent, on the center of the camshafts and do not remove until all parts are installed and tightened.

57. Install the left-hand secondary timing chain tensioner and tighten the retaining bolts to 70–106 inch lbs. (8–12 Nm).

58. Install the left-hand secondary camshaft sprockets and secondary timing chain as an assembly. Be sure the hubs of the sprocket are facing the proper direction.

59. Install the left-hand primary camshaft sprocket and camshaft sprocket spacer on the left-hand camshaft.

60. Install the washers and camshaft sprocket retaining bolts, finger-tight only.

➡ **The secondary camshaft sprockets must be free to turn.**

61. Install Timing Chain Tensioning Tool T93P-6256-BH or equivalent, on the left-hand secondary timing chain tensioner.

62. Tighten the camshaft sprocket bolts to 81–95 ft. lbs. (110–130 Nm).

63. Install the right-hand secondary timing chain tensioner and tighten the retaining bolts to 70–106 inch lbs. (8–12 Nm).

64. Install the right-hand secondary camshaft sprockets and secondary timing chain as an assembly. Be sure the hubs of the sprocket are facing the proper direction.

65. Install the right-hand primary camshaft sprocket and camshaft sprocket spacer on the right-hand camshaft.

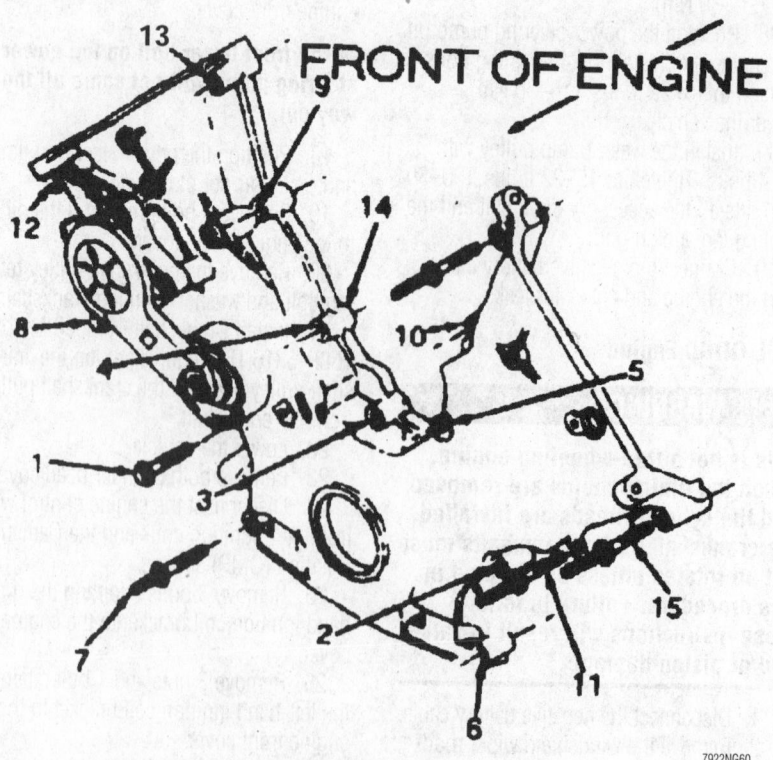

Front cover bolt torque sequence—4.6L DOHC engine

7922NG60

66. Install the washers and camshaft sprocket retaining bolts, finger-tight only.

➡️**The secondary camshaft sprockets must be free to turn.**

67. Install Timing Chain Tensioning Tool T93P-6256-BH or equivalent, on the right-hand secondary timing chain tensioner.

68. Tighten the camshaft sprocket bolts to 81–95 ft. lbs. (110–130 Nm).

69. Install the left-hand crankshaft sprocket, making sure the tapered part of the sprocket faces away from the cylinder block.

70. Install the primary timing chain on the left-hand primary camshaft sprocket. Be sure the copper link of the timing chain aligns with the timing mark on the camshaft sprocket.

➡️**If the copper links of the timing chain are not visible, pull the chain taught until the opposite sides of the chain contact one another and lay it on a flat surface. Mark the links at each end of the chain and use them in place of the copper links.**

71. Install the left-hand primary timing chain on the left-hand crankshaft sprocket. Be sure the copper link of the timing chain aligns with the timing mark on the crank-shaft sprocket.

72. Install the right-hand crankshaft sprocket, making sure the tapered part of the sprocket faces toward the cylinder block.

73. Install the primary timing chain on the right-hand primary camshaft sprocket. Be sure the copper link of the timing chain aligns with the timing mark on the camshaft sprocket.

74. Install the right-hand primary timing chain on the right-hand crankshaft sprocket. Be sure the copper link of the timing chain aligns with the timing mark on the crank-shaft sprocket.

➡️**The 2 timing chain guide bolts to the cylinder head are longer than the bolt to the cylinder block.**

75. Install the right-hand and left-hand timing chain guides and tighten the bolts to 71–106 inch lbs. (8–12 Nm).

76. Lubricate the timing chain tensioner arm contact surfaces with clean engine oil. Install the right-hand and left-hand tensioner arms and tighten the bolts to 84–132 inch lbs. (10–15 Nm).

77. Install right-hand and left-hand primary timing chain tensioners and tighten the bolts to 15–22 ft. lbs. (20–30 Nm).

78. Remove the locking pins from the timing chain tensioners and be sure all timing marks are aligned. Remove the camshaft

positioning tool and the camshaft holding tool.

➡️**When cleaning the sealing surfaces of the engine front cover and oil pan-to-cylinder block joints, be extremely careful not to damage the rubber bead of the oil pan gasket. If the oil pan gasket is damaged, it must be replaced.**

79. Clean the sealing surfaces of the cylinder block, cylinder heads and engine front cover; remove all traces of oil, dirt and old sealant. Sealing surfaces must be clean and dry before applying sealant.

80. Replace the engine front cover gaskets and the crankshaft front seal.

81. Apply silicone sealant to the proper locations on the cylinder block and heads.

➡️**The engine front cover must be rolled into place. DO NOT slide on the oil pan gasket.**

82. Carefully install the front cover to the engine.

83. Install the bolts and stud bolts retaining the front cover to the engine. Tighten in sequence to 15–22 ft. lbs. (20–30 Nm). Tighten the stud bolts and bolts numbered 6, 7, 8, 9, 10, and 11 within 4 minutes after applying the sealer.

84. Install the drive belt tensioner and bolt. Tighten the bolt to 15–22 ft. lbs. (20–30 Nm).

85. Install the belt idler pulley and bolt. Tighten the bolt to 15–22 ft. lbs. (20–30 Nm).

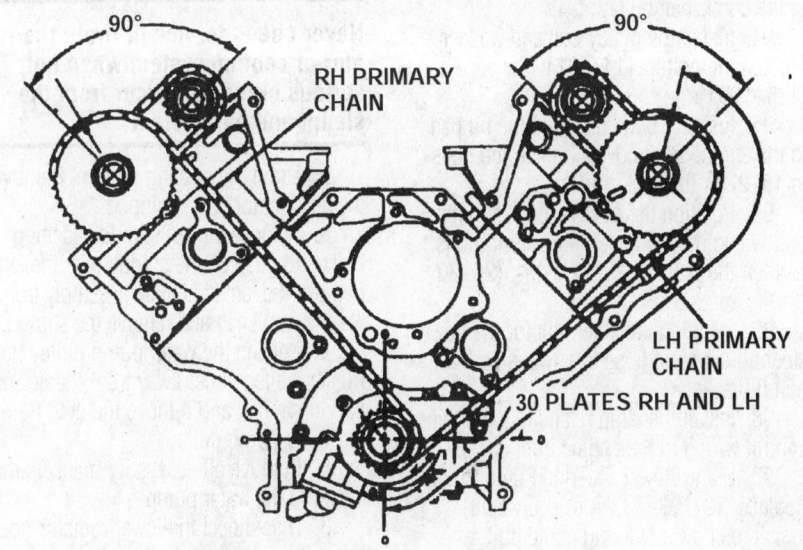

Timing chain alignment to TDC—4.6L DOHC engine

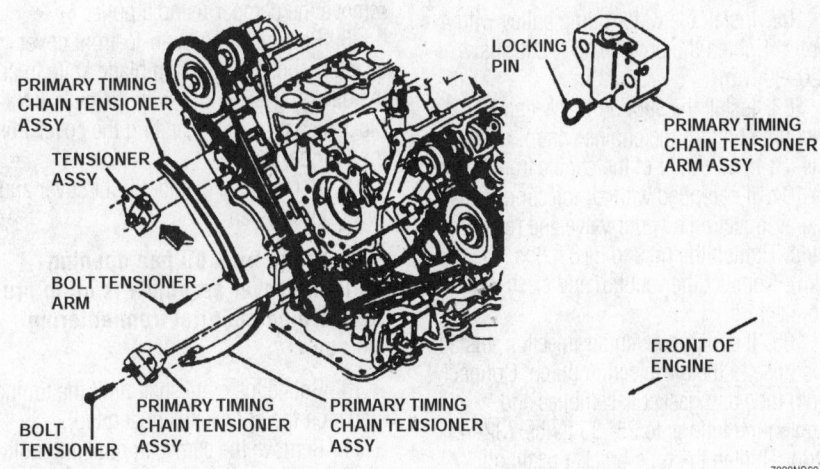

Timing chain tensioner installation—4.6L DOHC engine

86. Connect the engine control sensor wiring to the CKP sensor.

87. Place the left-hand and right-hand ignition coil brackets in position on the engine front cover. Install all retaining nuts and bolts and tighten to 15–22 ft. lbs. (20–30 Nm).

88. Install the power steering pump reservoir to the left-hand ignition coil bracket. Tighten the retaining bolts to 71–106 inch lbs. (8–12 Nm).

89. Connect the engine control wiring to both ignition coils and the CMP sensor.

90. Install both cylinder head covers.

91. Raise the vehicle.

92. Install the crankshaft pulley on the crankshaft, using Crankshaft Damper Replacer T74P-6316-B or equivalent.

93. Apply silicone sealant in the keyway of the crankshaft pulley.

94. Install the pulley bolt and washer. Tighten the bolt to 114–121 ft. lbs. (155–165 Nm).

95. Install 4 bolts retaining the oil pan to the engine front cover. Tighten the bolts to 15–22 ft. lbs. (20–30 Nm).

96. Position the power steering pump on the engine and install 4 retaining bolts. Tighten the bolts to 15–22 ft. lbs. (20–30 Nm).

97. Install the retainer bolt for the power steering pressure hose to power steering pump bracket.

98. Install the strap retaining the engine control wiring to the engine front cover.

99. Using Power Steering Pump Pulley Replacer T91P-3A733-A or equivalent, install the power steering pump pulley.

100. Lower the vehicle.

101. Install the lower water pump retaining bolt removed for engine front cover clearance. Tighten the bolt to 15–22 ft. lbs. (20–30 Nm).

102. Install the water pump pulley with 4 bolts. Tighten the bolts to 15–22 ft. lbs. (20–30 Nm).

103. Install the support bracket and retaining bolt for the engine control wiring to the left-hand center of the engine front cover.

104. If equipped with air injection, install the air injection diverter valve and retaining nuts. Tighten the nuts to 6–8 ft. lbs. (8–12 Nm). Connect the vacuum and electrical connectors.

105. If equipped with air injection, install the hose to the air injection pump. Connect both tube and gasket assemblies and tighten the fittings to 25–35 ft. lbs. (32–48 Nm). Tighten the tube bracket retaining bolts to 15–22 ft. lbs. (20–30 Nm).

106. Install the engine cooling fan motor and shroud assembly.

107. Install the water bypass tube and the upper radiator hose.

108. Install the accessory drive belt.

109. Restore all fluid levels.

110. Install the windshield wiper module assembly.

111. Connect the engine air cleaner outlet tube.

112. Connect the negative battery cable.

113. Start the engine and check for leaks.

114. Install the engine appearance cover, if equipped.

115. Road test the vehicle and check for proper engine operation.

5.0L Engine

1. Disconnect the negative battery cable.

✱✱ CAUTION

Never open, service or drain the radiator or cooling system when hot; serious burns can occur from the steam and hot coolant.

2. Drain the cooling system. Remove the air inlet tube, if equipped.

3. Remove the fan shroud attaching bolts and position the shroud over the fan. Remove the fan and clutch assembly from the water pump shaft and remove the shroud.

4. Loosen the water pump pulley bolts. Rotate the tensioner away from the accessory drive belt and remove the belt. Remove the water pump pulley.

5. Remove all accessory brackets that attach to the water pump.

6. Disconnect the lower radiator hose, heater hose and water pump bypass hose at the water pump.

7. Remove the crankshaft pulley from the crankshaft vibration damper. Remove the damper attaching bolt and washer and remove the damper using a puller.

8. Remove the oil pan-to-front cover attaching bolts. Use a thin blade knife to cut the oil pan gasket flush with the cylinder block face prior to separating the cover from the cylinder block.

9. Remove the cylinder front cover and water pump as an assembly.

➡**Cover the front oil pan opening while the cover assembly is off to prevent foreign material from entering the pan.**

10. Rotate the crankshaft until the timing marks on the sprockets are aligned.

11. Remove the camshaft retaining bolt, washer and eccentric, if equipped. Slide both sprockets and the timing chain forward and remove them as an assembly.

To install:

12. Position the sprockets and timing chain on the camshaft and crankshaft simultaneously. Be sure the timing marks on the sprockets are aligned.

13. Install the washer, eccentric if equipped, and camshaft sprocket retaining bolt. Tighten the bolt to 40–45 ft. lbs. (54–61 Nm).

14. If a new front cover is to be installed, remove the water pump from the old front cover and install it on the new front cover.

15. Clean all gasket mating surfaces. Pry the old oil seal from the front cover and install a new 1, using a seal installer.

16. Coat the gasket surface of the oil pan with sealer, cut and position the required sections of a new gasket on the oil pan and apply silicone sealer at the corners. Apply sealer to a new front cover gasket and install on the block.

17. Position the front cover on the cylinder block. Use care to avoid seal damage or gasket mislocation. It may be necessary to force the cover downward to slightly compress the pan gasket. Use front cover aligner tool T61P-6019-B or equivalent to assist the operation.

18. Coat the threads of the front cover attaching screws with pipe sealant and

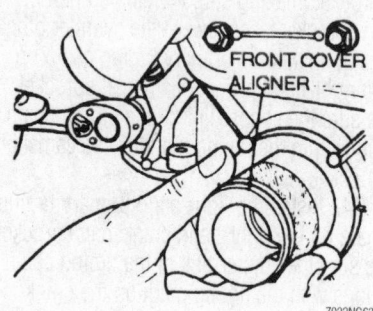

Timing chain front cover alignment—5.0L engine

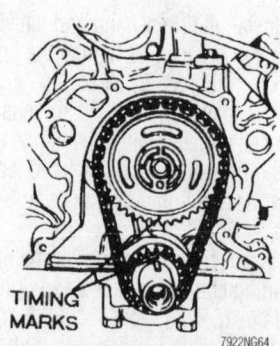

Timing chain sprocket alignment—5.0L engine

install. While pushing in on the alignment tool, tighten the oil pan to cover attaching screws to 9–12 ft. lbs. (12–16 Nm).

19. Tighten the front cover to cylinder block attaching bolts to 12–18 ft. lbs. (16–24 Nm). Remove the alignment tool.

20. Apply multi-purpose grease to the sealing surface of the vibration damper. Apply silicone sealer to the keyway of the vibration damper.

21. Line up the vibration damper keyway with the crankshaft key and install the damper using a suitable installation tool. Tighten the retaining bolt to 70–90 ft. lbs. (95–122 Nm). Install the crankshaft pulley.

22. Install the remaining components in the reverse order of their removal.

✳✳ WARNING

Operating the engine without the proper amount and type of engine oil will result in severe engine damage.

23. Fill the crankcase with the proper type and quantity of engine oil. Fill and bleed the cooling system.

24. Connect the negative battery cable, start the engine and check for leaks.

FUEL SYSTEM

Fuel System Service Precautions

Safety is the most important factor when performing not only fuel system maintenance, but any type of maintenance. Failure to conduct maintenance and repairs in a safe manner may result in serious personal injury or death. Work on a vehicle's fuel system components can be accomplished safely and effectively by adhering to the following rules and guidelines.

• To avoid the possibility of fire and personal injury, always disconnect the negative battery cable unless the repair or test procedure requires that battery voltage be applied.

• Always relieve the fuel system pressure prior to disconnecting any fuel system component (injector, fuel rail, pressure regulator, etc.) fitting or fuel line connection. Exercise extreme caution whenever relieving fuel system pressure to avoid exposing skin, face and eyes to fuel spray. Please be advised that fuel under pressure may penetrate the skin or any part of the body that it contacts.

• Always place a shop towel or cloth around the fitting or connection prior to loosening to absorb any excess fuel due to spillage. Ensure that all fuel spillage is quickly removed from engine surfaces. Ensure that all fuel-soaked cloths or towels are deposited into a flame-proof waste container with a lid.

• Always keep a dry chemical (Class B) fire extinguisher near the work area.

• Do not allow fuel spray or fuel vapors to come into contact with a spark or open flame.

• Always use a second wrench when loosening or tightening fuel line connection fittings. This will prevent unnecessary stress and torsion on fuel piping. Always follow the proper torque specifications.

• Always replace worn fuel fitting O-rings with new ones. Do not substitute fuel hose where rigid pipe is installed.

Fuel System Pressure

RELIEVING

All SFI engines are equipped with a pressure relief valve located on the fuel supply manifold. Remove the fuel tank cap and attach fuel pressure gauge T80L-9974-B, or equivalent, to the valve on the fuel rail. Place the tube from the tool into a small container and open the valve on the tool to release the fuel pressure. Be sure to drain the fuel into a suitable container and to avoid gasoline spillage. If a pressure gauge is not available, disconnect the vacuum hose from the fuel pressure regulator and attach a hand-held vacuum pump. Apply about 25 in. Hg (84 kPa) of vacuum to the regulator to vent the fuel system pressure into the fuel tank through the fuel return hose.

➡ **This procedure will remove the fuel pressure from the lines, but not the fuel. Take precautions to avoid the risk of fire and use clean rags to soak up any spilled fuel when the lines are disconnected.**

Fuel Filter

REMOVAL & INSTALLATION

✳✳ CAUTION

Because of its toxicity, refrain from breathing gasoline vapors and avoid exposing unprotected skin to liquid gasoline. Since gasoline fumes are extremely flammable and volatile, NEVER smoke when working around or near gasoline. Be sure that there is no possible ignition source (such

as sparks or open flames) near your work area. Always work on the fuel system in a well-ventilated area.

This procedure applies only to the inline fuel filter.

➡ **To prevent the siphoning of fuel from the tank when the filter is removed, raise the front of the vehicle slightly above the level of the tank, but be sure to properly support the vehicle using jackstands. Fuel line clamps may be used to prevent leakage, but ONLY WITH CAUTION. Old, brittle fuel lines may be damaged by the clamps and could leak afterwards. If a clamp is used, BE SURE to thoroughly inspect the lines after installation of the fuel filter. Any damaged line should be replaced immediately.**

1. Disconnect the negative battery cable.
2. Properly relieve the fuel system pressure using a test gauge at the fuel pressure relief valve.
3. Raise and safely support the rear of the vehicle.
4. Detach the fuel lines from both ends of the fuel filter by disengaging both push-connect fittings. Install new retainer clips in each push-connect fitting prior to reassembly.

➡ **Position a rag below the filter to catch any fuel which may spill as it is disconnected and removed.**

5. Remove the fuel filter from the bracket by loosening the worm gear clamp. Note the

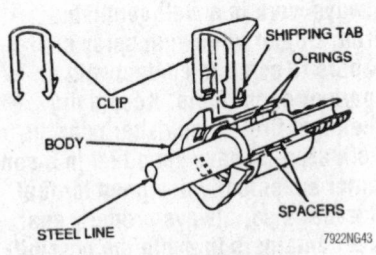

Always use new hairpin clips when assembling connections using them

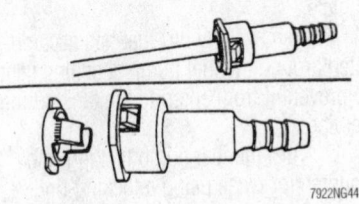

To disengage the duckbill clip fitting, depress the two tabs until one tube can be pulled from the other

orientation of the direction-of-flow arrow on the fuel filter, as installed in the bracket, to ensure proper fuel flow through the replacement filter.

To install:

6. Install the fuel filter into the bracket, ensuring the proper direction of flow. Tighten the worm gear clamp to 15–25 inch lbs. (2–3 Nm).

7. Install the push-connect fittings onto the filter ends. Connect the negative battery cable, then start the engine and check for fuel leaks.

✳✳ CAUTION

Use extreme caution when starting and running an engine which is supported by jackstands. ENSURE that no drive wheels are on the ground. Also, ensure that the wheels which are on the ground are properly blocked so the vehicle cannot move.

8. Lower the vehicle.

Fuel Pump

REMOVAL & INSTALLATION

➡ **To gain access to the fuel pump, it is necessary to remove the fuel tank.**

✳✳ CAUTION

Observe all applicable safety precautions when working around fuel. Whenever servicing the fuel system, always work in a well ventilated area. Do not allow fuel spray or vapors to come in contact with a spark or open flame. Keep a dry chemical fire extinguisher near the work area. Always keep fuel in a container specifically designed for fuel storage; also, always properly seal fuel containers to avoid the possibility of fire or explosion.

1. Depressurize the fuel system, then drain and remove the fuel tank from the vehicle.

2. Remove any dirt that has accumulated around the fuel pump attaching flange, to prevent it from entering the tank during service.

3. Turn the fuel pump locking ring counterclockwise using a locking ring removal tool, then remove the ring.

4. Remove the fuel pump and bracket assembly.

5. Remove the seal gasket and discard it.

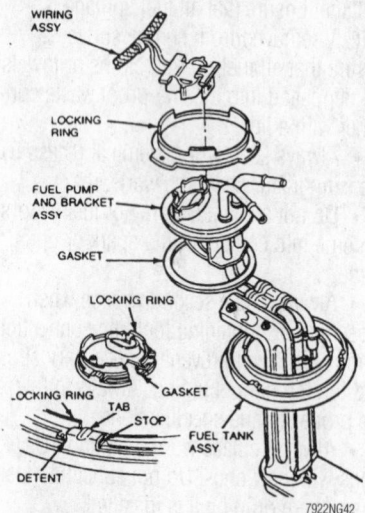

Exploded view of the electric fuel pump assembly, which is mounted in the fuel tank

To install:

6. Put a light coating of heavy grease on a new seal ring to hold it in place during assembly. Install it in the fuel tank ring groove.

7. Insert the fuel pump assembly into the fuel tank, then secure it in place with the locking ring. Tighten the ring until secure.

8. Install the fuel tank in the vehicle.

9. Add a minimum of 10 gallons of fuel and check for leaks.

10. Install a pressure gauge on the throttle body valve and turn the ignition **ON** for 3 seconds. Turn the key **OFF**, then repeat the key cycle five to ten times until the pressure gauge shows at least 30 psi (207 kPa). Reinspect for leaks at the fittings.

11. Remove the pressure gauge. Start the engine and check for fuel leaks.

DRIVE TRAIN

➡ **For information on driveshaft and U-joint service, refer to the unit repair section.**

Transmission Assembly

REMOVAL & INSTALLATION

1. Disconnect the negative battery cable.

2. On manual transmissions, remove the shifter boot and lever from inside the vehicle.

3. Raise and safely support the vehicle.

4. Disconnect all cables, connectors and fluid lines that may interfere with trans-

mission removal. Tag them if helpful for installation.

5. On automatic transmissions, unbolt the torque converter from the flexplate and disconnect the shift linkage.

6. Place a drain pan under the transmission and drain the fluid.

7. Position a transmission jack under the transmission and safety-chain the case to the jack.

8. Matchmark and remove the driveshaft.

9. Disconnect the exhaust pipe(s) from the manifold(s) and use wire to position the pipe(s) out of the way.

10. Remove the transmission rear mount.

11. Slightly raise the transmission and remove the crossmember.

12. On automatic transmissions, remove the transmission-to-engine block bolts. For manual transmissions, remove the bolts securing the transmission to the bell housing.

✳✳ CAUTION

The torque converter will fall out of the transmission if it is tilted forward. Keep a hand on it while lowering the transmission out of the vehicle.

13. Roll the transmission rearward until the input shaft clears, lower the jack and remove the transmission.

To install:

14. Carefully raise the transmission to the engine or bell housing.

15. Roll the transmission forward and into position.

16. On automatic transmissions tighten the bolts to 28–38 ft. lbs. (38–51 Nm. On manual transmissions, tighten the bolts to 45–65 ft. lbs. (61–68 Nm).

17. Install the crossmember, tighten the bolts to 36–50 ft. lbs. (47–68 Nm).

18. Lower the transmission onto the crossmember and remove the jack.

19. Install the transmission rear insulator and lower retainer. Tighten the nuts to 26–35 ft. lbs. (34–48 Nm).

20. Connect the exhaust pipe(s) to the manifold(s).

21. The rest of the installation is the reverse of removal.

22. Refill the transmission with the correct type and amount of fluid.

Clutch

ADJUSTMENTS

The clutch free-play is adjusted automatically by a built-in clutch control mecha-

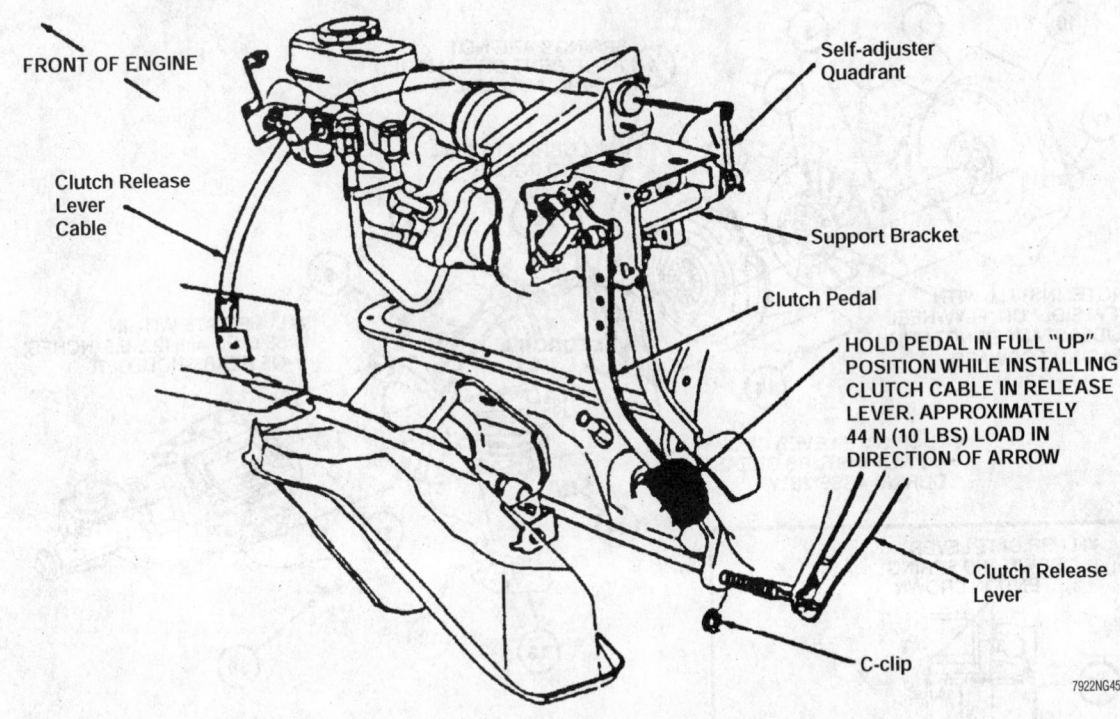

FRONT OF ENGINE

Self-adjuster Quadrant

Clutch Release Lever Cable

Support Bracket

Clutch Pedal

HOLD PEDAL IN FULL "UP" POSITION WHILE INSTALLING CLUTCH CABLE IN RELEASE LEVER. APPROXIMATELY 44 N (10 LBS) LOAD IN DIRECTION OF ARROW

Clutch Release Lever

C-clip

7922NG45

Clutch self-adjusting system component identification—4.6L engine shown, 3.8L and 5.0L engine are similar

nism. This device allows the clutch controls to self-adjust during normal operation.

The system consists of a spring-loaded gear quadrant, a spring-loaded pawl and a clutch cable which is spring-loaded to pre-load the release lever bearing. This compensates for movement of the release lever, as the clutch disc wears. The pawl, located at the top of the clutch pedal, engages the gear quadrant when the clutch pedal is depressed and pulls the cable through its continuously adjusted stroke. Clutch cable adjustments are not required because of this feature.

REMOVAL & INSTALLATION

1. Disconnect the negative battery cable for safety purposes.

⁂ WARNING

The clutch release lever cable should never be removed from the clutch and brake pedal pivot shaft with a prying instrument such as a prybar.

2. Lift the clutch pedal to its uppermost position to disengage the pawl and quadrant. Push the quadrant forward, unhook the cable from the quadrant and allow it to slowly swing rearward.

3. Raise and support the vehicle safely using jackstands. Remove the clutch release lever dust shield.

4. Disconnect the cable from the release lever. Remove the retaining clip and remove the clutch cable from the flywheel housing.

5. Remove the starter motor and the bolts holding the engine rear plate to the lower part of the flywheel housing.

6. Remove the transmission from the vehicle. Remove the bolts attaching the flywheel housing to the engine block, then remove the flywheel housing from the vehicle.

7. Remove the clutch release lever from the housing by pulling it through the window in the housing until the retainer spring is disengaged from the pivot. Remove the release bearing from the release lever.

8. Loosen the pressure plate cover attaching bolts evenly, to release spring tension gradually and avoid distorting the cover. If the same pressure plate and cover are to be reinstalled, matchmark the cover and flywheel so that the pressure plate can be installed in its original position. Remove the pressure plate and clutch disc from the engine.

9. Inspect the flywheel for scoring, cracks or other damage. Machine or replace as necessary. Inspect the pilot bearing for damage and free movement. Replace as necessary.

To install:

10. If removed, install the flywheel. Be sure the mating surfaces of the flywheel and the crankshaft flange are clean prior to installation. Tighten the flywheel bolts to

54–64 ft. lbs. (73–87 Nm) for 3.8L and 4.6L engines, or to 75–85 ft. lbs. (102–115 Nm) for the 5.0L engine.

11. Position the clutch disc and pressure plate assembly on the flywheel. The three dowel pins on the flywheel must be properly aligned with the pressure plate. Bent, damaged or missing dowels must be replaced. Start the pressure plate bolts, but do not tighten them.

12. Align the clutch disc using a disc alignment tool inserted into the pilot bearing. Alternately tighten the bolts a few turns at a time, until they are all tight. Then, tighten all the bolts to 12–24 ft. lbs. (17–32 Nm) for 1994–95 models, to 20–28 ft. lbs. (27–39 Nm) for 1996–98 3.8L engines, or to 19–24 ft. lbs. (25–33 Nm) for 4.6L engines, then remove the alignment tool.

13. Apply a light coat of multi-purpose grease to the release lever pivot pocket, the release lever fork and the flywheel housing pivot ball. Fill the grease groove of the release bearing hub with the same grease. Clean all excess grease from the inside bore of the bearing hub.

14. Install the release bearing on the release lever and install the lever in the flywheel housing.

15. Install the flywheel housing. Tighten the bolts to 38–55 ft. lbs. (52–74 Nm) on 1995 models, or to 28–38 ft. lbs. (38–52 Nm) on 1996–99 models.

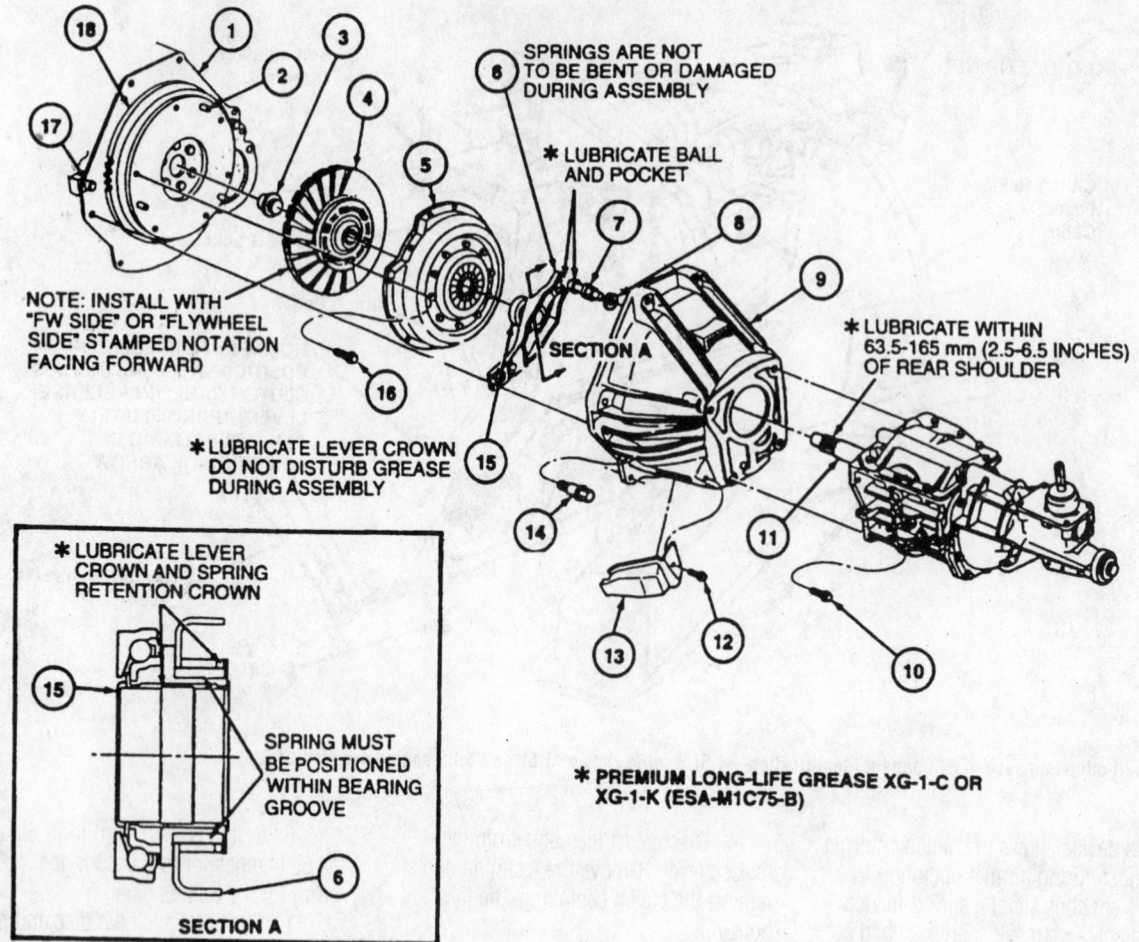

SPRINGS ARE NOT
TO BE BENT OR DAMAGED
DURING ASSEMBLY

✱ LUBRICATE BALL
AND POCKET

SECTION A

✱ LUBRICATE WITHIN
63.5-165 mm (2.5-6.5 INCHES)
OF REAR SHOULDER

NOTE: INSTALL WITH
"FW SIDE" OR "FLYWHEEL
SIDE" STAMPED NOTATION
FACING FORWARD

✱ LUBRICATE LEVER CROWN
DO NOT DISTURB GREASE
DURING ASSEMBLY

✱ LUBRICATE LEVER
CROWN AND SPRING
RETENTION CROWN

SPRING MUST
BE POSITIONED
WITHIN BEARING
GROOVE

✱ PREMIUM LONG-LIFE GREASE XG-1-C OR
XG-1-K (ESA-M1C75-B)

SECTION A

1. Rear face of engine block
2. Flywheel-to-clutch cover alignment dowel
3. Pilot bearing
4. Clutch disc
5. Clutch pleasure plate
6. Clutch release lever
7. Clutch release lever stud
8. Lockwasher
9. Flywheel housing
10. Bolt
11. Input shaft
12. Bolt
13. Clutch release lever dust shield
14. Bolt
15. Clutch release bearing
16. Bolt
17. Flywheel housing-to-engine block dowel

7922NG46

Exploded view of the clutch system components used on all 1996–99 Mustang models

16. Install the remaining components.

17. Remove the jackstands and carefully lower the vehicle.

✱✱ WARNING

The clutch pedal must be lifted to disengage the adjusting mechanism during clutch release lever cable installation. If this is not done, damage to the self-adjuster mechanism will occur.

18. Install the clutch release lever cable by lifting the clutch pedal to disengage the clutch and brake pedal pivot shaft, then push the pivot shaft forward and hook the end of the clutch release cable over the rear of the pivot shaft.

19. Depress and lift the clutch several times to allow the self-adjusting mechanism to properly set the free-play.

20. Connect the negative battery cable, then check for proper clutch operation.

STEERING AND SUSPENSION

Air Bag

✱✱ CAUTION

Some models covered by this manual may be equipped with a Supplemental Restraint System (SRS), or Supplemental Inflatable Restraint (SIR) which uses an air bag. Whenever working near any of the SRS/SIR components, such as the impact sensors, the air bag module, steering column and instrument panel, disable the SRS/SIR.

PRECAUTIONS

• Always wear safety glasses when servicing an air bag vehicle, and when handling an air bag.

• Never attempt to service the steering wheel or steering column on an air bag-equipped vehicle without first properly disarming the air bag system. The air bag system should be properly disarmed whenever ANY service procedure in this manual indicates that you should do so.

• When carrying a live air bag module, always be sure the bag and trim cover are pointed away from your body. In the unlikely event of an accidental deployment, the bag will, then deploy with minimal chance of injury.

• When placing a live air bag on a bench or other surface, always face the bag and trim cover up, away from the surface. This will reduce the motion of the air bag if it is accidentally deployed.

• If you should come in contact with a deployed air bag, be advised that the air bag surface may contain deposits of sodium hydroxide, which is a product of the gas combustion and is irritating to the skin. Always wear gloves and safety glasses when handling a deployed air bag, and wash your hands with mild soap and water afterwards.

DISARMING THE SYSTEM

1. Disconnect the negative battery cable from the battery.
2. Disconnect the positive battery cable from the battery.
3. Wait one minute. This time is required for the back-up power supply in the air bag diagnostic monitor to completely drain. The system is now disarmed.

If you are disarming the system with the intent of testing the system, do not! The SRS is a sensitive, complex system and should only be tested or serviced by a qualified automotive technician. Also, specific tools are needed for SRS testing.

After the necessary repairs are completed, re-enable the system as follows:

4. Connect the positive battery cable.
5. Connect the negative battery cable.
6. Stand outside the vehicle and carefully turn the ignition to the **RUN** position. Be sure that no part of your body is in front of the air bag module on the steering wheel, to prevent injury in case of an accidental air bag deployment.
7. Ensure the air bag indicator light turns off after approximately 6 seconds. If the light does not illuminate at all, does not turn off, or starts to flash, have the system tested by a qualified automotive technician. If the light does turn off after 6 seconds and does not flash, the SRS is working properly.

Power Rack and Pinion Steering Gear

REMOVAL & INSTALLATION

1. Disconnect the negative battery cable. Turn the ignition switch to the **RUN** position so that the front wheels can be turned while working on the power rack and pinion.
2. Loosen all of the wheel lug nuts ½ turn to break them free.
3. Raise and support the vehicle safely using jackstands. Position a drain pan to catch the fluid from the power steering lines.
4. Remove the one bolt retaining the steering column shaft flexible coupling to the input shaft.
5. Remove the front wheel and tire assemblies. Remove the cotter pins and nuts from the tie rod ends, then separate the tie rod studs from the spindles.

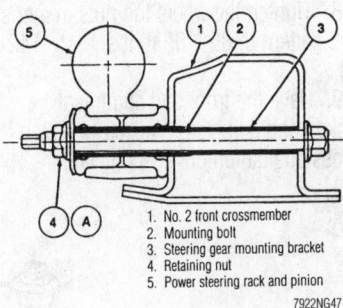

1. No. 2 front crossmember
2. Mounting bolt
3. Steering gear mounting bracket
4. Retaining nut
5. Power steering rack and pinion

7922NG47

Power rack and pinion steering gear mounting on the No. 2 crossmember

6. Remove the two nuts, insulator washers and bolts retaining the steering rack to the crossmember. Remove the front rubber insulators.
7. Position the rack to allow access to the hydraulic lines and disconnect the lines.
8. Remove the steering rack from the vehicle.

To install:

9. Install new plastic seals on the hydraulic line fittings.
10. Install the rack on the mounting studs, and connect the hydraulic lines to the rack and pinion. Tighten the fittings to 20–25 ft. lbs. (27–34 Nm). Tighten the nut on the studs to 30–40 ft. lbs. (41–54 Nm).

➡The hoses are designed to swivel when properly tightened. Do not attempt to eliminate looseness by over-tightening the fittings.

11. Install the front rubber insulators. Be sure all rubber insulators are pushed completely inside the gear housing before installing the mounting bolts.
12. Insert the input shaft into the steering column shaft flexible coupling. Install the mounting bolts, insulator washers and nuts. Tighten the nuts to 20–30 ft. lbs. (28–40 Nm).
13. Connect the tie rod ends to the spindle arms and install the retaining nuts. Tighten the nut to 35–47 ft. lbs. (47–64 Nm). Continue tightening the nut to align the next castellation of the nut with the cotter pin hole in the stud. Install a new cotter pin.
14. Remove the jackstands and carefully lower the vehicle.
15. Turn the ignition switch **OFF** and connect the negative battery cable.
16. Fill the power steering system with the proper type and quantity of fluid.
17. If the tie rod ends were loosened, have the front end alignment checked by a qualified automotive suspension technician.

Strut

REMOVAL & INSTALLATION

Front

✳✳ CAUTION

During this procedure when the strut is detached from the spindle, if the floor jack under the lower control arm is lowered, the coil spring, under a great amount of tension, can be abruptly released and cause severe physical injury. As a precaution, chain the bottom end of the coil spring to the lower control arm so that it cannot accidentally spring out of its seat.

1. Disconnect the negative battery cable.
2. Place the ignition switch in the unlocked position to permit free movement of the front wheels.
3. Raise the vehicle by the lower control arms until the wheels are just off the ground.

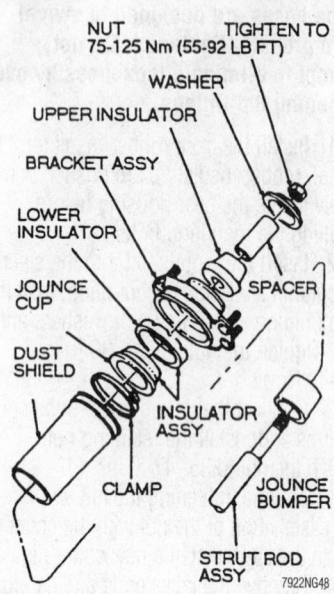

NUT TIGHTEN TO
75-125 Nm (55-92 LB FT)
WASHER
UPPER INSULATOR
BRACKET ASSY
LOWER INSULATOR
JOUNCE CUP
SPACER
DUST SHIELD
INSULATOR ASSY
CLAMP
JOUNCE BUMPER
STRUT ROD ASSY
7922NG48

Exploded view of the front strut upper mounting

4. From the engine compartment, remove the strut's central upper mount nut. Do not remove the three small retaining nuts around the larger central nut, and NEVER remove the pop rivet holding the camber plate in position. Slide the large washer off of the strut shaft.

5. Continue to raise the front of the vehicle by the lower control arm and position a jackstand under the frame jacking pad, rearward of the front wheel. Do NOT remove the jack from beneath the lower control arm!

6. Remove the wheel and tire assembly, then remove the brake caliper. Support the caliper with a length of wire; do not let the caliper hang by the brake hose.

7. If equipped, remove the brake anti-lock sensor and bracket.

8. Chain the lower end of the coil spring to the lower control arm for safety.

9. Remove the two lower nuts that attach the strut to the spindle, leaving the bolts in place. Carefully remove both spindle-to-strut bolts, then push the bracket free of the spindle and remove the strut.

10. Compress the strut to clear the upper mount of the body mounting pad. If necessary, remove the upper mount and jounce bumper.

➡ **The upper insulator mounts, if damaged and in need of replacement, can be removed after the strut is removed from the vehicle. Pull the upper and lower insulators out of the bracket assembly.**

To install:

11. If removed, install the upper insulators in the bracket assembly.

12. Position the upper strut shaft through the insulator mounts and camber plate, then start the central mounting nut.

13. Compress the strut and position the lower end in the spindle. Install two new lower retaining bolts and hand-start the nuts. Remove the suspension load from the control arm by slowly lowering the floor jack until the arm hangs without any support from the jack. Tighten the lower retaining nuts 141–191 ft. lbs. (190–259 Nm).

14. Raise the suspension control arm (to position the suspension at normal ride height) and tighten the upper mount retaining nut to 55–92 ft. lbs. (75–125 Nm).

15. Remove the safety chain from the lower control arm and coil spring.

16. Install the brake anti-lock sensor bracket, along with the anti-lock sensor, caliper and front wheel.

17. Remove the jackstands and carefully lower the vehicle to the ground.

18. Tighten the wheel lug nuts in a criss-cross pattern to 85–105 ft. lbs. (115–142 Nm).

19. Have the front end alignment checked and adjusted, if necessary, by a professional automotive technician.

Shock Absorber

REMOVAL & INSTALLATION

Rear

➡ **Some Ford Mustangs use Torx® head bolts to retain the shocks at the lower mounts. Check to be sure you have the proper drivers before beginning this procedure.**

1. Open the luggage compartment (trunk) lid.

2. Remove the trim panels, as necessary, to gain access to the upper shock absorber mount.

3. Remove the shock absorber retaining nut washer and insulator.

4. Raise the vehicle and support it safely using jackstands under the rear axle housing.

5. From under the vehicle, remove the shock absorber bolt, washer and nut at the lower arm and remove the shock absorber.

➡ **These vehicles are equipped with gas pressurized shock absorbers which will extend unassisted.**

To install:

6. Prime the new shock absorber as follows:

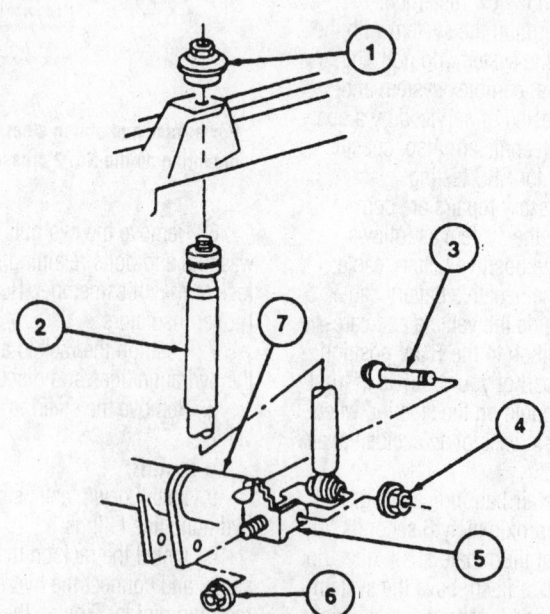

1. Insulator nut
2. Shock absorber
3. Mounting bolt
4. Mounting nut
5. Shock absorber lower mount bracket
6. Mounting nut
7. Rear axle housing

7922NG49

Exploded view of the rear shock absorber mounting

a. With the shock absorber right side up, extend it fully.

b. Turn the shock absorber upside down and fully compress it.

c. Repeat the previous two steps at least three times to be sure any trapped air has been expelled.

7. From under the vehicle, place the inner washer and insulator on the upper retaining stud and position the stud through the shock tower mounting hole.

8. Attach the lower end of the shock absorber with the retaining bolt and nut. Tighten the bolt to 57–75 ft. lbs. (76–103 Nm).

9. Remove the jackstands and carefully lower the vehicle.

10. Install the upper insulator, washer and retaining nut, then tighten the nut to 25–34 ft. lbs. (34–46 Nm).

Coil Spring

✷✷ CAUTION

Always use extreme caution when working with coil springs. Always use the proper spring compression tools, since the springs are VERY strong, if pressure is released suddenly (and without control), serious personal injury could result. Also, ALWAYS be sure the vehicle is very well supported when working around springs.

REMOVAL & INSTALLATION

Front

➡This procedure REQUIRES the use of a coil spring compression tool. This tool can usually be rented for a one-time use; otherwise, it can be purchased from your local parts store.

1. Loosen the front wheel lug nuts only slightly.

2. Raise and support the vehicle safely using jackstands, but allow the control arms to hang free.

3. Remove the wheel and tire assembly.

4. Remove the brake caliper. Suspend the caliper with a length of wire; DO NOT let the caliper hang by the brake hose.

5. Disconnect the tie rod end from the steering spindle and disconnect the stabilizer link from the lower arm.

6. If necessary, remove the steering gear (rack and pinion) bolts and reposition the gear so the suspension arm-to-No. 2 crossmember mounting bolts can be removed.

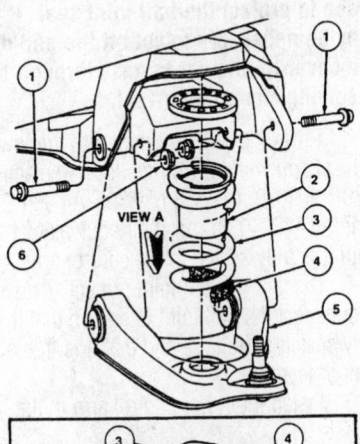

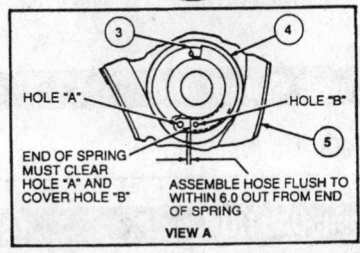

1. Mounting bolt
2. Damper
3. Front coil spring
4. Insulator
5. Front suspension lower arm
6. Nuts

7922NG50

Exploded view of the front coil spring mounting

7. On models equipped with the 3.8L engine, use a coil spring compressor, such as Ford Coil Spring Compressor T82P-5310-A, and install the upper plate in the spring pocket cavity on the crossmember. The hooks on the plate should be facing toward the center of the vehicle.

8. On Mustangs with the 4.6L or 5.0L engines, use spring compressor tool D78P-5310-A or equivalent, to install a plate between the coils near the toe of the spring. Mark the location of the upper plate on the coils for installation.

9. Install the compression rod into the lower arm spring pocket hole, through the coil spring and into the upper plate.

10. Install the lower plate, lower ball nut, thrust washer and bearing, and forcing nut onto the compression rod. Tighten the forcing nut until a drag on the nut is felt.

11. Remove the suspension arm-to-crossmember nuts and bolts. The compressor tool forcing nut may have to be tightened or loosened for easy bolt removal.

12. Pull the lower control arm down and remove the spring from the suspension.

13. Loosen the compression rod forcing nut until spring tension is relieved and remove the forcing nut. Remove the compression rod and coil spring.

To install:

14. Place the insulator on top of the spring. Position the spring into the lower arm pocket. Be sure the spring pigtail is positioned between the two holes in the lower arm spring pocket.

15. Position the spring into the upper spring seat in the crossmember.

16. On Mustangs with the 3.8L engine, insert the compression rod through the control arm and spring, then hook it to the upper plate. The upper plate is installed with the hooks facing the center of the vehicle.

17. On Mustangs with either the 4.6L or 5.0L engines, install the upper plate between the coils in the location marked during removal.

18. Install the lower plate, ball nut, thrust washer and bearing, and forcing nut onto the compression rod.

19. Tighten the forcing nut, position the lower arm into the crossmember and install new lower arm-to-crossmember bolts and nuts. Do not tighten at this time.

20. Remove the spring compressor tool from the vehicle. Raise the suspension arm to a normal attitude position with a jack. With the suspension in the normal ride-height position, tighten the lower arm-to-crossmember attaching nuts to 141–191 ft. lbs. (191–259 Nm). Carefully lower the suspension again and remove the jack.

21. If repositioning, install the steering gear (rack and pinion)-to-crossmember bolts and nuts. Hold the bolts and tighten the nuts to 30–40 ft. lbs. (41–54 Nm).

22. Connect the stabilizer bar link to the lower suspension arm. Tighten the attaching nut to 11–16 ft. lbs. (16–22 Nm).

23. Position the tie rod into the steering spindle and install the retaining nut. Tighten the nut to the specified torque value and continue tightening the nut to align the next castellation with the hole in the stud. Install a new cotter pin.

24. Install the brake caliper.

25. Install the wheel and tire assembly, then remove the jackstands and carefully lower the vehicle.

26. Tighten the wheel lug nuts to the specified torque value.

Rear

➡It is easier to replace one spring at a time; otherwise, the axle housing will be disconnected from both lower control arms.

1. Raise and support the vehicle safely under the frame. Support the body at the rear body crossmember.

2. If equipped, remove the stabilizer bar.

3. Support the axle with a suitable jack.

4. Place another jack under the lower arm axle pivot bolt. Remove and discard the bolt and nut. Lower the jack slowly until the coil spring load is relieved.

5. Remove the coil spring and insulator from the vehicle.

To install:

6. Place the upper spring insulator on top of the spring. Place the lower spring insulator on the lower arm.

7. Position the coil spring on the lower arm spring seat, so the pigtail on the lower arm is at the rear of the vehicle and pointing toward the left side of the vehicle.

8. Slowly raise the jack until the arm is in position. Insert a new rear pivot bolt and nut, with the nut facing outward. Do not tighten them at this time.

9. Raise the axle to curb height. Tighten the lower arm-to-axle pivot bolt to 71–97 ft. lbs. (97–132 Nm).

10. If equipped, install the stabilizer bar.

11. Remove the crossmember supports and carefully lower the vehicle.

Lower Ball Joint

REMOVAL & INSTALLATION

1. Remove the front strut and spindle from the vehicle.

2. Position the lower control arm in a vise so that the ball joint is easy to service.

3. Using a large C-clamp and adapters, such as Ford Tools T74P-4635-C, D89P-3010-A and D84P-3395-A4, carefully press the old ball joint out of the lower control arm.

To install:

➡**When installing a new front ball joint, leave the protective cover in**

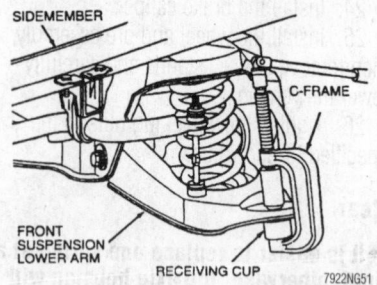

Use the C-clamp, cup and adapters to press the old ball joint out of the lower control arm

---

place to protect the ball joint seal. It may be necessary to cut off the end of the cover to allow it to pass through the receiving cup adapter tool.

4. Press the new ball joint into the lower control arm with Ford's Ball Joint Replacer D89P-3010-B, Cup D84P-3395-A4 and C-Frame T74P-4635C, or their equivalents, until it is fully seated in the control arm.

5. Discard the ball joint protective cover and inspect the ball joint to ensure that it is fully seated in its bore and that it is free of cuts or tears.

6. Install the lower control arm in the vehicle.

Lower Control Arm

REMOVAL & INSTALLATION

✳✳ CAUTION

Always use extreme caution when working with coil springs. Always use the proper spring compression tools, since the springs are VERY strong and, if pressure is released suddenly (and without control), serious personal injury could result. Also, ALWAYS be sure the vehicle is very well supported when working around springs.

1. Raise and safely support the vehicle.

2. Remove the wheel and tire assembly.

3. If necessary, remove the brake caliper and suspend it with a length of wire; do NOT let the caliper hang by the brake hose. Remove the brake rotor and dust shield.

4. Disconnect the tie rod end from the steering spindle. Disconnect the stabilizer bar link from the lower control arm.

5. If necessary for suspension arm mounting bolt removal, remove the steering gear (rack and pinion) bolts and lower the gear out of the way to provide access.

6. Remove the cotter pin and loosen the lower ball joint stud nut 1–2 turns. **Do not completely remove the nut at this time.** Tap the spindle boss sharply with a brass mallet to relieve the stud pressure; the lower control arm and ball joint should "pop" loose from the spindle (leaving the nut on the ball joint prevents the arm from completely disengaging from the spindle, which would allow the coil spring to jump out of the lower arm cup, possibly causing severe personal injury).

7. Install a suitable spring compressor and compress the coil spring until it can be moved freely in the lower control arm seat.

➡**For more information on coil spring compression or removal, refer to the coil spring procedure earlier in this section.**

8. Remove and discard the ball joint nut, then raise the entire strut and spindle assembly up and off of the ball joint stud. Wire the strut/spindle out of the way to obtain working room.

9. Remove and discard the suspension arm-to-crossmember nuts and bolts. Remove the lower control arm and coil spring from the vehicle.

To install:

10. Position the compressed coil spring into the lower arm pocket. Make sure the spring pigtail is positioned between the two holes in the pocket.

11. Position the lower arm ends in the crossmember brackets and install new arm-to-crossmember bolts and nuts. Do not tighten at this time.

12. Remove the retaining wire from the strut/spindle assembly, then position the spindle over the ball joint stud.

13. Raise the control arm with a jack to the normal vehicle resting position, ensuring the ball joint stud is inserted into the spindle hole. Install the ball joint nut loosely.

14. With the jack in place, tighten the lower arm-to-crossmember attaching nuts to 141–191 ft. lbs. (191–259 Nm).

➡**When tightening the ball joint nut, do not loosen the nut to install the cotter pin. Tighten the nut until the next set of grooves are aligned with the stud hole, then install the new cotter pin.**

15. Tighten the ball joint stud nut to 109–149 ft. lbs. (148–202 Nm) and install a new cotter pin; bend the cotter pin ends over. Remove the floor jack.

16. If removed, install the dust shield, rotor and brake caliper.

17. If removed, install the steering gear (rack and pinion)-to-crossmember bolts and nuts. Hold the bolts and tighten the nuts to 30–40 ft. lbs. (41–54 Nm).

➡**When tightening the tie rod end nut, do not loosen the nut to install the cotter pin. Tighten the nut until the next set of grooves are aligned with the stud hole, then install the new cotter pin.**

18. Position the tie rod into the steering spindle and install the retaining nut. Tighten the nut to 35–47 ft. lbs. (47–64 Nm) and continue tightening the nut to align the next castellation with the hole in the stud. Install a new cotter pin and bend its ends over.

19. Connect the stabilizer bar link to the lower control arm. Tighten the retaining nut to 11–16 ft. lbs. (16–22 Nm).
20. Install the wheel and tire assembly, then lower the vehicle.
21. Tighten the wheel lug nuts to 85–105 ft. lbs. (115–142 Nm).
22. Check the front end alignment.

Wheel Bearings

ADJUSTMENT

➡The wheel bearings are an integral part of the hub assembly. They require no periodic maintenance or adjustment. If the bearings are found to be defective, they must be replaced along with the hub assembly.

REMOVAL & INSTALLATION

Front

➡The front wheel bearings, hub and anti-lock sensor ring (where applicable) are one unit and cannot be separately serviced or replaced. Therefore, when the wheel bearings are faulty, the entire assembly must be replaced.

1. Raise and safely support the vehicle.
2. Using a prytool, remove the plastic hub grease cap and discard it.
3. Remove the brake caliper and suspend it with strong cord or wire from the chassis.

⁂ WARNING

Do not allow the brake caliper to hang from its brake hose.

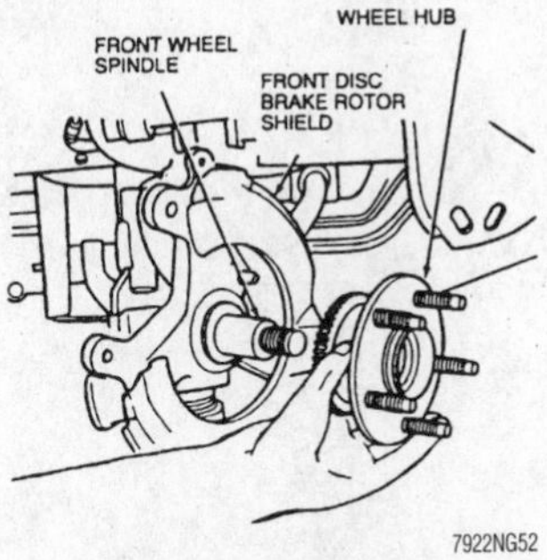

Note that the hub assembly shown here is equipped with an anti-lock brake sensor ring (the serrated disc affixed to the inboard side of the hub assembly)

4. Remove the front disc brake rotor.
5. Remove the front axle wheel hub retainer and discard it.
6. Pull the wheel hub and bearing assembly off of the spindle. If the assembly cannot be removed by hand, a two or three jaw puller may be used. Take care not to damage the spindle.

To install:

7. Install the hub/bearing assembly on the wheel spindle.
8. Install a new front axle wheel hub retainer and tighten it to 221–295 ft. lbs. (300–400 Nm).
9. Install the front disc brake rotor and new push-on nuts.
10. Install a new front hub grease cap, being cautious to avoid damaging or distorting the grease cap.
11. Install the brake caliper.
12. Install the wheel and tire assembly, and snug the wheel lug nuts.
13. Lower the vehicle and remove the rear wheel blocks.
14. Tighten the wheel lug nuts in a crisscross pattern to 85–105 ft. lbs. (115–143 Nm).

Rear

1. Block the front wheels, then loosen the lug nuts on the rear wheel that is being removed.
2. Raise and support the vehicle.
3. Remove the wheel, then remove the rear brake caliper and rotor.
4. If equipped, remove the anti-lock brake speed sensor.
5. Clean all dirt from the area of the axle housing cover. Drain the axle lubricant by removing the housing cover. For details, please refer to the Fluid and Lubricant information in Section 1 of this manual.
6. Remove the differential pinion shaft lockbolt and pinion shaft.
7. Push the flanged end of the axle shaft toward the center of the vehicle (to create the necessary play and free the C-lock), then remove the C-lock from the button end of the axle shaft.
8. Slowly withdraw the axle shaft from the housing, being careful not to damage the oil seal (unless you are replacing it anyway).
9. If the seal is being replaced (or if you damaged it on the way out) insert a wheel bearing and seal replacement tool, such as T85L-1225-AH or equivalent, in the bore and position it behind the bearing so the tangs on the tool engage the bearing outer race. Remove the bearing and seal as a unit, using an impact slide hammer.

➡If only the seal is being replaced, use a seal removal tool to pry ONLY THE SEAL from the axle housing.

To install:

10. If removed, lubricate the new bearing with rear axle lubricant. Install the bearing into the housing bore with a bearing installer.
11. If removed, install a new axle seal using a seal installer. Essentially, the installation tool is a driver of the right diameter; a smooth socket or piece of pipe can also be used as a driver, just be careful not to damage the seal.

➡Check for the presence of an axle shaft O-ring on the spline end of the shaft; install one if none is found.

12. Carefully slide the axle shaft STRAIGHT into the axle housing, without damaging the bearing or seal assembly. Start the splines into the side gear and push firmly until the shaft splines engage. It may be necessary to rotate the axle slightly to align the splines.
13. Install the C-lock, then pull outward slightly on the axle shaft and be sure the C-lock seats in the counterbore of the differential side gear.
14. Insert the differential pinion shaft through the case and pinion gears, aligning the hole in the shaft with the lockbolt hole. Apply a suitable locking compound to the lockbolt and insert it through the case and pinion shaft. Tighten the lockbolt to 15–30 ft. lbs. (20–41 Nm).

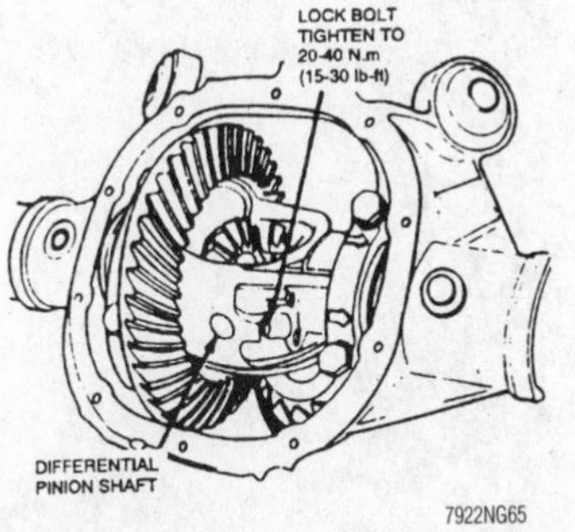

Be sure to tighten the pinion shaft lockbolt as indicated to prevent it from accidentally loosening

15. Cover the inside of the differential case with a shop rag and clean the machined surface of the carrier and cover. Remove the shop rag.

16. Carefully clean the gasket mating surfaces of the cover and axle housing of any remaining gasket or sealer. A putty knife is a good tool to use for this. You may want to cover the differential gears using a rag or piece of plastic to prevent contaminating them with dirt or pieces of the old gasket.

17. Install the rear cover using a new gasket and sealant. Tighten the retaining bolts using a crosswise pattern.

➡ **Be sure the vehicle is level before attempting to add fluid to the rear axle or an incorrect fluid level will result. You may have to lift all four corners of the vehicle and support it using 4 jackstands in order to do this.**

18. Refill the rear axle housing using the proper grade and quantity of lubricant as detailed in Section 1. Install the filler plug.

19. If removed, install the anti-lock speed sensor and tighten the retaining bolt to 40–60 inch lbs. (5–7 Nm).

20. Install the brake caliper and rotor.

21. Install the wheel, then remove the jackstands and carefully lower the vehicle.

FORD MOTOR CO.

Lincoln-Mark VIII

24

DRIVE TRAIN 24-21
ENGINE REPAIR 24-2
FUEL SYSTEM 24-18
**STEERING AND
  SUSPENSION** 24-23

A

Air Bag.........................24-23
  Disarming...................24-23
  Enabling....................24-23
  Precautions.................24-23
Air Spring.....................24-26
  Removal & Installation.......24-26

C

Camshaft and Valve Lifters.....24-12
  Removal & Installation.......24-12
Cylinder Head..................24-3
  Removal & Installation.......24-3

E

Engine Assembly................24-2
  Removal & Installation.......24-2
Exhaust Manifold...............24-8
  Removal & Installation.......24-8

F

Fuel Filter....................24-19
  Removal & Installation.......24-19
Fuel Pump......................24-20
  Removal & Installation.......24-20
Fuel System Pressure...........24-19

Relieving......................24-19
Fuel System Service
  Precautions..................24-18

H

Halfshaft......................24-22
  Removal & Installation.......24-22

I

Ignition Timing................24-2
  Adjustment...................24-2
Intake Manifold................24-6
  Removal & Installation.......24-6

L

Lower Ball Joint...............24-28
  Removal & Installation.......24-28
Lower Control Arm..............24-28
  Removal & Installation.......24-28

O

Oil Pan........................24-13
  Removal & Installation.......24-13
Oil Pump.......................24-14
  Removal & Installation.......24-14

P

Power Rack and Pinion Steering
  Gear.........................24-24
  Removal & Installation.......24-24

R

Rear Main Seal.................24-14
  Removal & Installation.......24-14

Rocker Arms....................24-5
  Removal & Installation.......24-5

S

Shock Absorber.................24-25
  Removal & Installation.......24-25

T

Timing Chain, Sprockets, Front
  Cover and Seal...............24-15
  Removal & Installation.......24-15
Transmission Assembly..........24-21
  Removal & Installation.......24-21

U

Upper Ball Joint...............24-27
  Removal & Installation.......24-27
Upper Arm Control..............24-27
  Removal & Installation.......24-27

V

Valve Lash.....................24-13
  Adjustment...................24-13

W

Water Pump.....................24-3
  Removal & Installation.......24-3
Wheel Bearings.................24-29
  Adjustment...................24-29
  Removal & Installation.......24-29

ENGINE REPAIR

➤Disconnecting the negative battery cable on some vehicles may interfere with the functions of the on board computer systems and may require the computer to undergo a relearning process, once the negative battery cable is reconnected.

Ignition Timing

ADJUSTMENT

The ignition timing is controlled by the engine control module. No provision is made for timing adjustment.

Engine Assembly

✳✳ CAUTION

All models covered in this section are equipped with a Supplemental Restraint System (SRS), which uses an air bag. Whenever working near any of the SRS components, such as the impact sensors, the air bag module, steering column and instrument panel, disable the SRS using the recommended procedure.

REMOVAL & INSTALLATION

✳✳ CAUTION

The EPA warns that prolonged contact with used engine oil may cause a number of skin disorders, including cancer! You should make every effort to minimize your exposure to used engine oil. Protective gloves should be worn when changing the oil. Wash your hands and any other exposed skin areas as soon as possible after exposure to used engine oil. Soap and water, or waterless hand cleaner should be used.

1. Disconnect the negative battery cable. Drain the crankcase and the cooling system.

✳✳ CAUTION

Never open, service or drain the radiator or cooling system when hot; serious burns can occur from the steam and hot coolant.

2. Remove the engine appearance cover.

✳✳ CAUTION

Observe all applicable safety precautions when working around fuel. Whenever servicing the fuel system, always work in a well ventilated area. Do not allow fuel spray or vapors to come in contact with a spark or open flame. Keep a dry chemical fire extinguisher near the work area. Always keep fuel in a container specifically designed for fuel storage; also, always properly seal fuel containers to avoid the possibility of fire or explosion.

3. Properly relieve the fuel system pressure and discharge the A/C system. Air conditioning system should only be serviced by an EPA-certified, MVAC-trained technician.

4. Disconnect the IAT sensor and the crankcase vent tube from air cleaner outlet tube. Loosen the air cleaner outlet to throttle body bolt and disconnect the tube.

5. Remove the remaining bolt from support bracket clamp on right valve cover and remove air cleaner outlet tube assembly and resonator assembly.

6. Remove the hush panel to expose the windshield wiper module, remove the screw, lower module and bracket assembly and unplug electrical connector. Remove the wiper module.

7. Remove the cooling fan and shroud.

8. Disconnect the fuel lines. Remove the 42 pin connector from the retaining bracket on the left fender well. Disconnect connector and position aside.

9. Remove the power distribution box and remove the alternator **B+** connector from inside of box.

10. Disconnect the engine harness connector from the canister purge solenoid assembly. Disconnect the accelerator and speed control cables from the throttle body and accelerator cable bracket.

11. Disconnect the canister purge vacuum line at the throttle body and chassis vacuum supply hose at connection on the cowl support.

12. Disconnect the heater supply and return hose at the rear of the right cylinder head and upper radiator hose at the coolant crossover tube.

13. Remove the power steering return and supply hoses from reservoir and drain the fluid into an appropriate container. Remove the reservoir retaining bolt and stud from the left coil bracket and remove the reservoir.

14. Remove the upper transmission cooler line from the radiator.

15. Install suitable engine lifting eyes. Disconnect and plug the air conditioning compressor lines and disconnect retaining clips from pump.

16. Raise and safely support the vehicle. Remove the wheel and tire assemblies.

17. Disconnect right and left ride height sensor electrical connectors.

18. Remove right and left caliper bolts and remove calipers from rotors. Support calipers with mechanics wire.

19. Disconnect right and left upper control arms from spindles.

20. Remove Y-pipe from both exhaust manifolds and resonator. Disconnect ground strap from right fender apron.

21. Disconnect power steering pressure line from steering rack at power steering pump connection.

22. Remove lower radiator hose. Remove right and left lower strut to control arm bolts and nuts.

23. Disconnect transmission wiring harness and shift linkage. Disconnect lower transmission cooler line from radiator.

24. Index driveshaft to rear axle companion flange.

25. Remove the four bolts connecting driveshaft to rear axle companion flange. Support rear axle housing with jackstand.

26. Remove the two nut and bolt assemblies retaining the front of the differential to the undercarriage. Loosen the rear differential bracket-to-body bolts.

27. Slide driveshaft rearward until it is free of transmission extension housing.

28. Remove the two nuts from bottom front cover studs and remove starter cable.

29. Disconnect the low oil level sensor connector and wiring harness from oil pan. Remove starter.

30. Support the subframe with Rotunda Powertrain Lift 014–00765 and adapter 014–00341, or equivalent. Remove the rear transmission mount bolts.

31. Disconnect the steering shaft coupler at the rag joint. Remove the eight subframe bolts.

32. Lower the engine and transmission assembly from the vehicle.

33. Remove the motor mount through-bolts.

34. Install suitable engine lifting equipment on the engine lifting eyes and lift the engine and transmission from the subframe.

35. Lower the engine and transmission. Support the transmission on a level, stationary surface for transmission storage.

36. Remove the transmission to cylinder

block mounting bolts and separate the engine from transmission/torque converter assembly. Place the engine on suitable workstand. Remove the engine lifting equipment.

37. Remove the right and left motor mounts from the block. Drain the coolant and remove the right and left water jacket pipe plugs from the cylinder block. Reinstall the plugs.

38. Disconnect the EGR to exhaust manifold tube from the left exhaust manifold connector, differential pressure feedback EGR hose connections and loosen the EGR tube connector at the EGR control valve. Remove EGR valve to exhaust manifold tube.

39. Remove the exhaust manifolds and discard the gaskets.

To install:

40. Install the exhaust manifolds with new gaskets. Tighten the nuts in sequence to 15–22 ft. lbs. (20–30 Nm).

41. Install the EGR to exhaust manifold tube, to EGR control valve, left exhaust manifold and differential pressure feedback EGR hose.

42. Install the left and right motor mounts to the block. Install the engine lifting bracket.

43. Remove the engine from the stand using suitable lifting equipment.

44. Connect the transmission to cylinder block. Tighten the bolts to 30–44 ft. lbs. (40–60 Nm).

45. Position the engine and transmission on the subframe assembly. Install mount through-bolts and tighten them to 15–22 ft. lbs. (20–30 Nm).

46. Raise the engine, transmission and subframe assembly into position.

47. Align the subframe to body and install the bolts. Tighten them to 73–100 ft. lbs. (95–130 Nm).

48. Connect the steering shaft coupler at rag joint. Install the rear transmission mount bolt and tighten to 15–22 ft. lbs. (20–30 Nm). Remove the lifting equipment.

49. Install the starter, low oil level sensor and lower engine wiring harness.

50. Install the starter cable and two retaining nuts to lower front cover studs.

51. Install the driveshaft to the transmission. Lift the rear differential into position. Install the two nut and bolt assemblies retaining the front of the differential to the rear subframe and tighten to 72–89 ft. lbs. (98–120 Nm).

52. Tighten the two differential mounts to 75–89 ft. lbs. (102–127 Nm). Align the driveshaft to the companion flange and install the four bolts. Tighten them to
p 3 270–96 ft. lbs. (95–130 Nm).

53. Connect the lower transmission

cooler line to the radiator. Connect the transmission wiring harness and shift linkage.

54. Position the lower control arms to the strut and install the bolt. Tighten the bolt to 118–162 ft. lbs. (160–220 Nm).

55. Install the lower radiator hose and power steering pressure line. Connect the ground strap to the right fender apron.

56. Install the exhaust Y-pipe.

57. Connect the upper control arms to the spindles. Tighten the bolt to 50–68 ft. lbs. (68–92 Nm). Connect the right and left ride height sensors.

58. Install the brake calipers. Install the front wheels. Lower the vehicle.

59. Unplug and connect the air conditioning lines. Remove the engine lifting brackets and eyelets.

60. Connect the upper transmission cooler line.

61. Install the power steering reservoir, power steering return line and power steering supply hose.

62. Install the coolant supply hose to the reservoir. Connect the upper radiator hose at the crossover pipe and connect the heater supply and return hoses.

63. Connect the chassis vacuum supply and canister purge vacuum hoses.

64. Connect the accelerator and speed control cables. Connect the engine wiring harness to canister purge solenoid assembly.

65. Connect the alternator **B+** connector and install the power distribution box. Connect the 42 pin connector and attach to retaining bracket.

66. Connect the fuel lines. Install the cooling fan and shroud.

67. Install the windshield wiper module. Install the engine air cleaner outlet tube assembly and resonator assembly.

68. Connect the IAT sensor and crankcase vent tube.

✳✳ WARNING

Operating the engine without the proper amount and type of engine oil will result in severe engine damage.

69. Fill the engine with the proper type and quantity of engine oil. Fill and bleed the cooling system.

70. Fill the power steering system with the proper type and quantity of fluid. Connect the negative battery cable.

71. Start the engine and bring to normal operating temperature. Check for leaks. Check all fluid levels.

72. Leak test, evacuate and charge the air conditioning system. Observe all safety precautions.

73. Install the engine appearance cover.

Water Pump

REMOVAL & INSTALLATION

✳✳ CAUTION

Never open, service or drain the radiator or cooling system when hot; serious burns can occur from the steam and hot coolant.

1. Disconnect the negative battery cable and drain the cooling system.

2. Remove the four bolts retaining the pulley to the water pump shaft.

3. Release belt tensioner and remove the drive belts. Remove the four bolts retaining water pump pulley and remove the pulley.

4. Remove the four water pump retaining bolts and remove the water pump.

To install:

5. Install a new water pump O-ring seal.

6. Installation is the reverse of the removal procedure. Clean all mating surfaces prior to installation. Tighten the water pump bolts and pulley retaining bolts to 15–22 ft. lbs. (20–30 Nm).

7. Fill and bleed the cooling system. Operate the engine until normal operating temperatures have been reached and check for leaks.

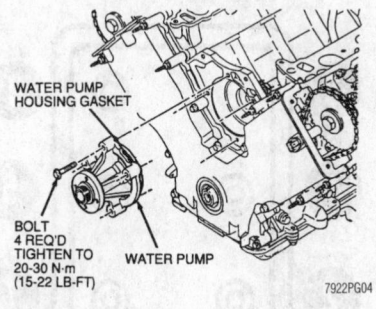

Exploded view of the water pump mounting

Cylinder Head

REMOVAL & INSTALLATION

1. Remove the engine from the vehicle.

2. Remove the valve covers and the engine front cover.

3. Remove the intake manifold.

4. Remove the crankshaft position sensor and pulse wheel.

5. Remove the rocker arms using the recommended procedure.

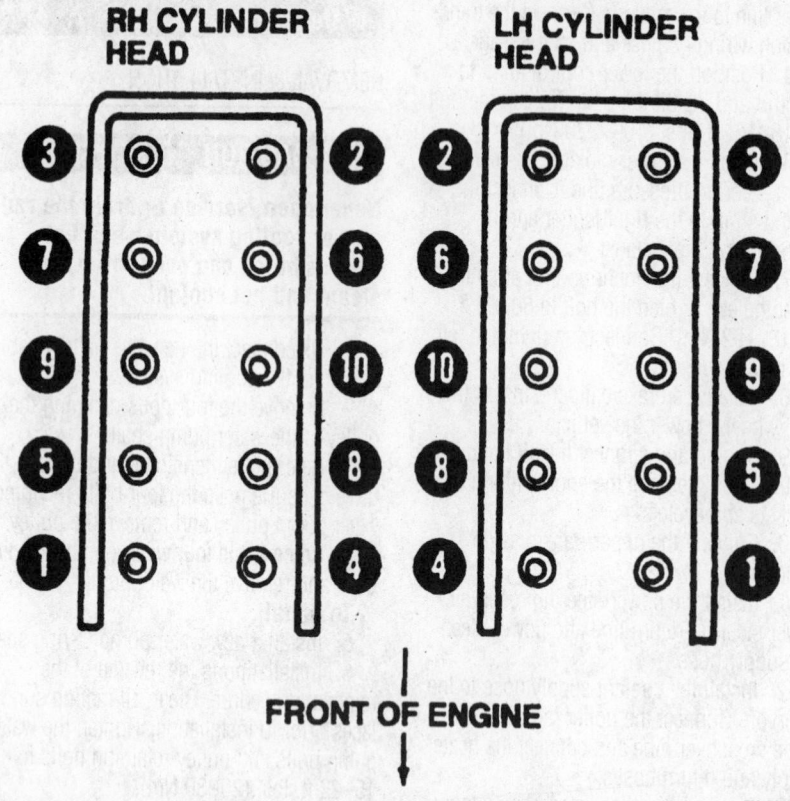

RH CYLINDER HEAD **LH CYLINDER HEAD**

FRONT OF ENGINE

7922PG01

Be sure to remove the cylinder head bolts in the correct sequence to prevent damage to the head

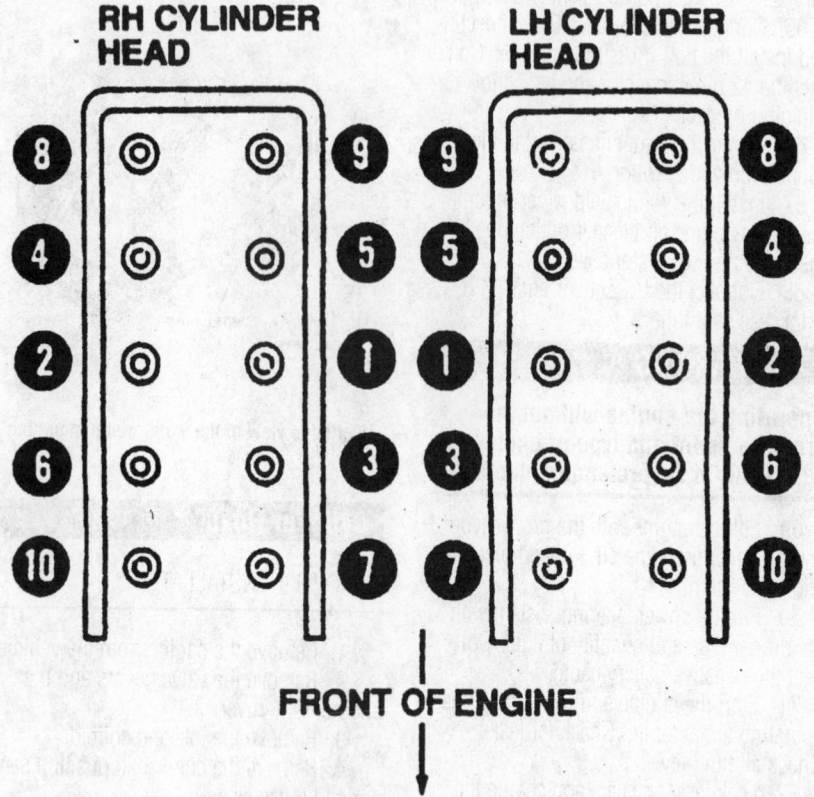

RH CYLINDER HEAD **LH CYLINDER HEAD**

FRONT OF ENGINE

7922PG02

For proper cylinder sealing, be sure to tighten the cylinder head bolts in the correct sequence

✳✳ CAUTION

Never open, service or drain the radiator or cooling system when hot; serious burns can occur from the steam and hot coolant.

6. Remove the exhaust manifolds and drain the coolant.

7. Position the No. 1 cylinder at TDC on the compression stroke. Be sure the timing marks on the camshaft sprockets are on top.

8. Install the Camshaft Positioning tool (T93P-6256-A) into the D-shaped slots at the rear of the camshafts.

9. Remove both left and right primary timing chain tensioners.

10. Remove both primary timing chains, tensioner arms and guides but do not loosen the camshaft sprocket mounting bolts.

11. Remove the one nut and two bolts attaching the inlet heater coolant tube to the rear of the right cylinder head and remove the tube.

12. Remove the outlet heater coolant tube from the rear of the right cylinder head.

13. Loosen the 10 cylinder head bolts in the correct sequence, then remove the head. Discard all cylinder head bolts. New bolts must be used for installation.

✳✳ WARNING

Do not place the bottom of the cylinder head on the workbench or table. Damage to the valves may occur on the valve that are open.

To install:

14. Clean and dry all cylinder head bolt holes.

15. Turn the crankshaft counterclockwise 90 degrees to the 9 O'clock position. This will position all of the pistons below the deck surface of the engine block.

16. Place new head gaskets on the engine block.

17. Carefully place the cylinder head on the engine block, be sure to align the dowels.

➡ New cylinder head bolts must be used. Cylinder head bolts are of a torque-to-yield design and cannot be reused.

18. Apply clean engine oil to the new bolt threads and both sides of the bolt washer. Install the cylinder head bolts finger-tight.

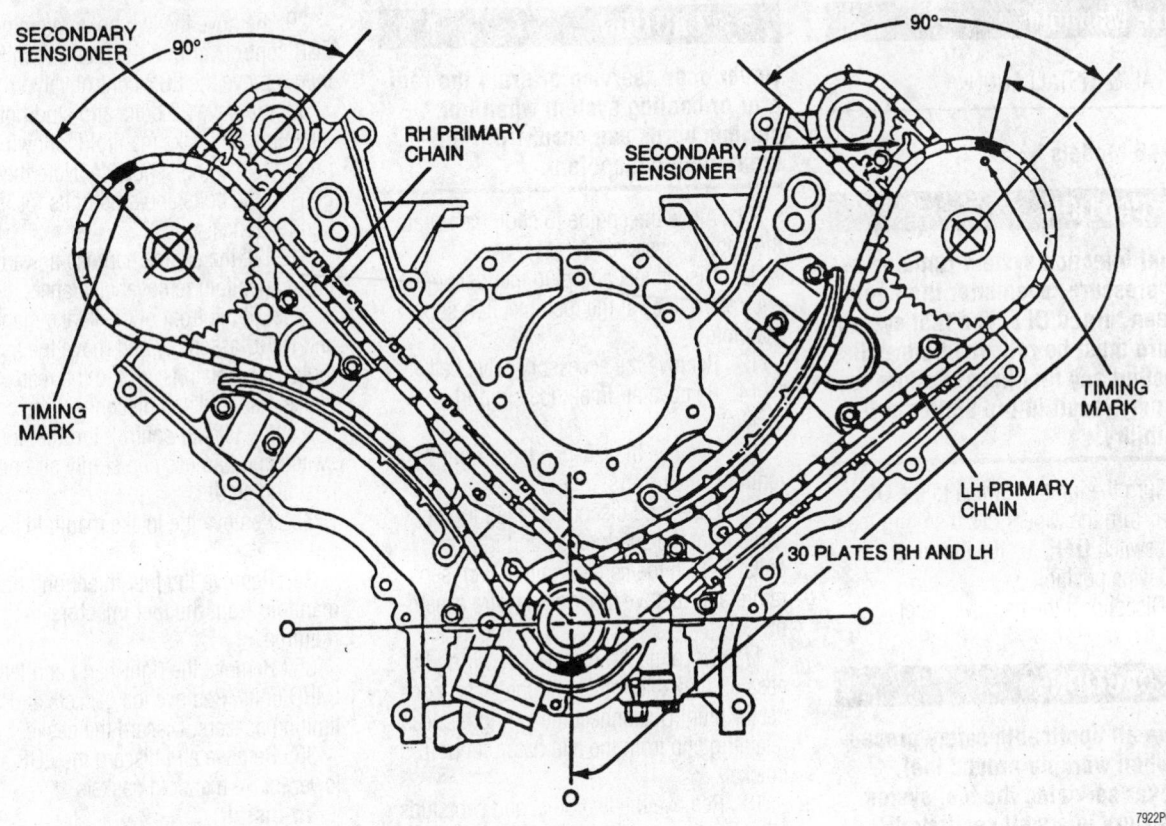

The timing marks should be positioned as shown after the primary chains have been installed

➡ **The cylinder head bolts will be tightened in three stages using the proper sequence.**

19. Tighten cylinder head bolts in three steps:
- Tighten bolts in sequence to 27–32 ft. lbs. (37–43 Nm).
- Rotate bolts, in sequence 85–95 degrees.
- Rotate bolts, in sequence an additional 85–95 degrees.

20. Install the outlet heater coolant tube on the rear of the right cylinder head. Tighten the stud bolts to 15–22 ft. lbs. (20–30 Nm).

21. Install the inlet heater coolant tube on the right cylinder head. Tighten the two bolts and one nut to 71–106 inch lbs. (8–12 Nm).

22. Install the primary timing chains. Be sure the valve timing is correct.

23. Install the rocker arms.

24. Install the crankshaft position sensor and pulse wheel.

25. Install the intake manifold.

26. Install the engine front cover.

27. Install the valve covers.

28. Install the exhaust manifolds.

29. Install the engine assembly in the vehicle.

✳✳ WARNING

Operating the engine without the proper amount and type of engine oil will result in severe engine damage.

30. Refill the engine and cooling system with the correct type and amount of fluid.

31. Start the engine and check for leaks.

Rocker Arms

REMOVAL & INSTALLATION

1. Disconnect the negative battery cable.
2. Remove the camshaft covers.
3. Position the cylinder being serviced at the bottom of it's travel.

➡ **Two different valve spring compressor tool are used for this procedure. Valve Spring Compressor (T91P-6565-A) or equivalent is used on the exhaust camshaft and Valve Spring Compressor (T93P-6565-A) or equivalent is used on the intake camshaft.**

4. Compress the valve spring and remove the rocker arm.

To install:

5. Position the cylinder being serviced at the bottom of it's travel.

6. Apply clean engine oil to the rocker arm, valve stem tip and tappet bore.

➡ **Valve tappet should have no more than 1/16 in. (1.5mm) of travel before installing the rocker arm.**

7. Compress the valve spring using the correct tool and install the rocker arm.

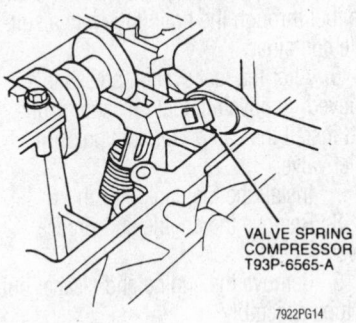

Using the proper tool to compress the valve springs and remove the rocker arms

Intake Manifold

REMOVAL & INSTALLATION

1995–96 Models

✳✳ CAUTION

The fuel injection system remains under pressure, even after the engine has been turned OFF. The fuel system pressure must be relieved before disconnecting any fuel lines. Failure to do so may result in fire and/or personal injury.

1. Turn the ignition switch to the **ON** position. Turn the wipers ON, then turn the ignition switch **OFF**, to stop the wipers in the mid-wipe position.
2. Disconnect the negative battery cable.

✳✳ CAUTION

Observe all applicable safety precautions when working around fuel. Whenever servicing the fuel system, always work in a well ventilated area. Do not allow fuel spray or vapors to come in contact with a spark or open flame. Keep a dry chemical fire extinguisher near the work area. Always keep fuel in a container specifically designed for fuel storage; also, always properly seal fuel containers to avoid the possibility of fire or explosion.

3. Remove the fuel tank fill cap to relieve the pressure in the fuel tank.
4. Remove the cap from the fuel pressure relief valve located on the fuel injection supply manifold.
5. Attach Fuel Pressure Gauge T80L-9974-B or equivalent, to the valve and drain the fuel through the drain tube into a suitable container.
6. After the fuel system pressure is relieved, remove the fuel pressure gauge and install the cap on the fuel pressure relief valve.
7. Install the fuel tank fill cap.
8. Remove the engine appearance cover.
9. Remove the engine and cleaner outlet tube assembly.
10. Remove the hush panel to expose the windshield wiper module, remove screw, lower module and bracket assembly and disconnect electrical connector. Remove the wiper module.

✳✳ CAUTION

Never open, service or drain the radiator or cooling system when hot; serious burns can occur from the steam and hot coolant.

11. Allow the engine to cool, then drain the cooling system.
12. Disconnect and plug the fuel supply and return lines at the fuel injection supply manifold.
13. Remove the accessory drive belt.
14. Remove air inlet tube support bracket.
15. Remove the left-hand and right-hand ignition wire covers.
16. Label and disconnect the ignition wires from the spark plugs and ignition coils. Detach the center ignition wire separators and remove the ignition wire assembly.
17. Disconnect the alternator wiring harness at the **B+** terminal and the stator connector plug. Disconnect the wiring harness retaining clip from the alternator support bracket.
18. Remove the two bolts and two studs retaining the support bracket to the alternator and intake manifold.
19. Disconnect the upper radiator hose and coolant bypass hose at the coolant crossover tube.
20. Remove the coolant crossover tube.
21. Remove the alternator assembly.
22. Disconnect the accelerator and speed control cables from the throttle body lever and from the cable mounting bracket.
23. Disconnect the chassis vacuum supply hose from the intake manifold and position aside.
24. Tag and disconnect the vacuum lines from the fuel pressure regulator, right and left IMRC vacuum control, EGR control valve, EGR vacuum regulator control, IMRC vacuum control solenoids and the intake manifold.
25. Disconnect the PCV hose connector from the throttle body. Remove the PCV valve from the left-hand cylinder head cover and move the PCV tube assembly aside.
26. Disconnect the four engine harness retaining clips from the intake manifold studs and one clip from the fuel regulator bracket.
27. Disconnect the fuel injector electrical harness connectors.
28. Remove the upper bolt retaining the accelerator cable bracket to the left-hand cylinder head. Loosen the lower bolt and move the bracket against the left-hand cylinder head cover.

29. Remove the two bolts retaining the EGR control valve to the intake manifold, then remove the EGR control valve.
30. Remove 20 bolts and stud bolts retaining the intake manifold following the proper loosening sequence. Note the locations of the bolts and stud bolts for installation reference.
31. Lift the engine harness upward for intake manifold removal clearance.
32. Lift the front of the intake manifold and IMRC assembly and move the assembly forward to obtain access to the rear of the intake manifold and disconnect the connectors at the idle air control, throttle position switch and harness clip at idle air control retaining stud.
33. Remove the intake manifold assembly.
34. Remove the fuel injection supply manifold from the fuel injectors, if required.
35. Remove the right-hand and left-hand IMRC units. Remove the gaskets and load limiting spacers. Discard the gaskets.
36. Remove and discard the EGR and lower intake manifold gaskets.

To install:

37. Clean and inspect all gasket sealing surfaces.
38. Remove the paper from the adhesive backing on a new upper intake manifold gasket. Install the gasket on the top of the right-hand IMRC unit. Use tapered pins at the end bolt hole locations for proper alignment. Repeat the gasket installation for the left-hand IMRC unit.
39. Assemble the IMRC units onto the intake manifold and hand-tighten the four retaining bolts.
40. Install the fuel injection supply manifold onto the fuel injectors and tighten the four retaining screws to 26–45 inch lbs. (3–5 Nm). Install the two fuel pressure regulator retaining bracket screws and tighten to 72–108 inch lbs. (8–12 Nm).
41. Place new IMRC gaskets onto the cylinder heads. Use the gaskets integral location pins to align the gaskets.
42. Place the intake manifold under the engine harness and connect the wiring harness plugs to the idle air control valve, throttle position sensor and connect wiring clip to idle air control valve retaining clip.
43. Move the intake manifold into position and install a new EGR valve gasket. Hand-tighten the two retaining bolts.
44. Install the 10 long bolts and stud bolts in the outer holes of the intake manifold. Do not tighten.
45. Install the 10 short bolts and stud

bolts in the inner holes of the intake manifold. Hand-tighten all 20 fasteners.

46. Tighten the intake manifold bolts and stud bolts in sequence as follows:

    a. Bolt numbers 5, 7, 9, 11: tighten to 8.8–11 ft. lbs. (12–15 Nm)

    b. All remaining intake manifold bolts: tighten to 13–16 ft. lbs. (18–22 Nm)

    c. Rotate all intake manifold bolts in sequence, 85–95 degrees.

47. Tighten the four IMRC retaining bolts to 6–8 ft. lbs. (8–10 Nm). Rotate, in sequence, 85–95 degrees.

48. Tighten the two EGR retaining bolts to 15–22 ft. lbs. (20–30 Nm).

49. Install the upper retaining bolt into the accelerator cable bracket and tighten both upper and lower bolts to 15–22 ft. lbs. (20–30 Nm).

50. Connect the fuel injector electrical harness connectors.

51. Connect the engine harness clips to the intake manifold studs and connect the harness clip to the fuel pressure regulator bracket. Connect the PCV system.

52. Connect the vacuum harness connectors to the fuel pressure regulator, right and left IMRC vacuum controls, EGR vacuum regulator control, IMRC vacuum control solenoids and intake manifold vacuum at the rear connection.

53. Connect the chassis vacuum supply hose and spring clamp to the intake manifold vacuum at the forward connection.

54. Connect the accelerator and speed control cables to the throttle body lever and to the cable mounting bracket.

55. Replace the O-rings on the coolant crossover tubes, if necessary. Insert the crossover tube support tube support braces over the intake manifold inner studs and press the crossover tube into place. Install the retaining nuts and tighten to 6–9 ft. lbs. (8–12 Nm).

56. Connect the ECT and temperature gauge sensor electrical harness connectors. Connect the upper radiator hose and coolant bypass hose at the coolant crossover tube.

57. Install the alternator and tighten two bolts to 15–22 ft. lbs. (20–30 Nm). Install the alternator support bracket and tighten the two bolts and two stud bolts to 6–9 ft. lbs. (8–12 Nm).

58. Connect the alternator wiring harness connector plug and the **B+** terminal wire to the alternator. Connect the wiring harness retaining clip to the support bracket.

59. Install the center ignition wire separators and the ignition wire assembly. Connect the ignition wires to the spark plugs and ignition coils.

60. Install the accessory drive belt.

61. Install the engine air inlet tube support bracket. Tighten the retaining nuts to 6–9 ft. lbs. (8–12 Nm).

62. Unplug and connect the fuel supply and return lines at the fuel injection supply manifold.

63. Install the engine air cleaner outlet tube and resonator assembly.

64. Install windshield wiper module.

65. Fill and bleed the engine cooling system.

66. Connect the negative battery cable.

67. Start the engine and check for fuel, coolant and oil leaks.

68. Install the engine appearance cover.

69. Road test the vehicle and check for proper engine operation.

1997–99 Models

※※ CAUTION

The fuel injection system remains under pressure, even after the engine has been turned OFF. The fuel system pressure must be relieved before disconnecting any fuel lines. Failure to do so may result in fire and/or personal injury.

1. Disconnect the negative battery cable.

※※ CAUTION

Observe all applicable safety precautions when working around fuel. Whenever servicing the fuel system, always work in a well ventilated area. Do not allow fuel spray or vapors to come in contact with a spark or open flame. Keep a dry chemical fire extinguisher near the work area. Always keep fuel in a container specifically designed for fuel storage; also, always properly seal fuel containers to avoid the possibility of fire or explosion.

2. Relieve fuel system pressure as follows.

3. Remove the water bypass tube.

4. Disconnect the accelerator and cruise control cables.

5. Disconnect the evaporator emission hose from the upper intake manifold.

6. Disconnect the throttle position sensor electrical connector.

7. Disconnect the crankcase ventilation hose from the intake manifold.

8. Disconnect the idle air control valve electrical connector.

9. Loosen the EGR tube nut from the EGR valve.

10. Disconnect the EGR valve vacuum hose.

11. Remove the bolts and remove the upper intake manifold.

12. Remove the fuel injection supply manifold.

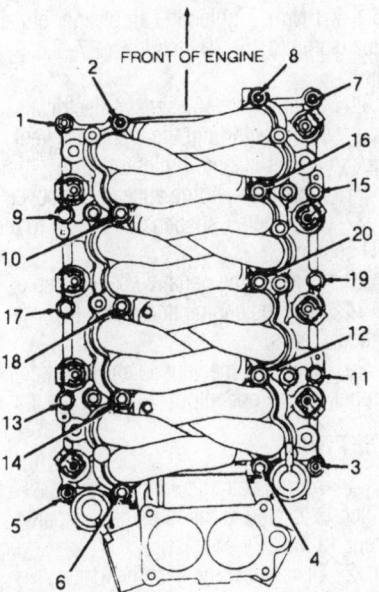

FRONT OF ENGINE

7922PG05

Be sure to tighten the intake manifold fasteners in the sequence shown to ensure sealing—1995-96 models

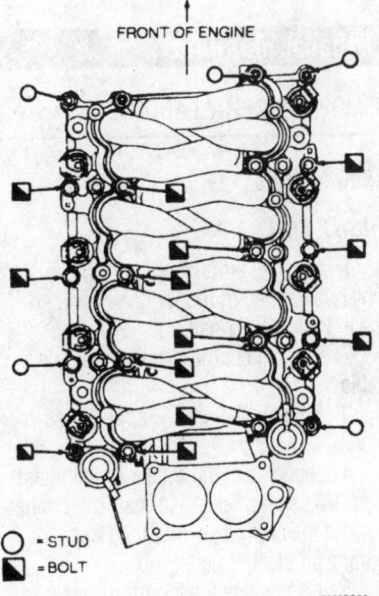

FRONT OF ENGINE

○ = STUD
⬛ = BOLT

7922PG06

Install the bolts and studs in the correct locations—1995-96 models

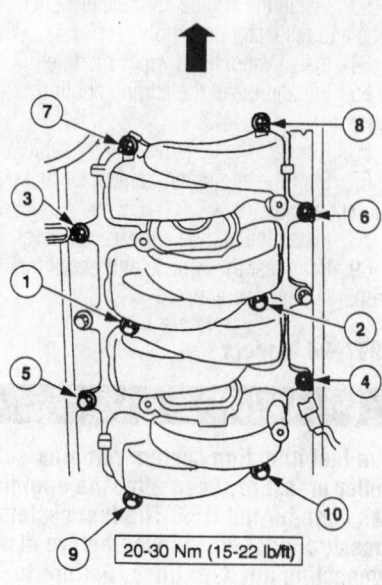

Tighten the intake manifold bolts according to the sequence shown to prevent leakage—1997–99 models

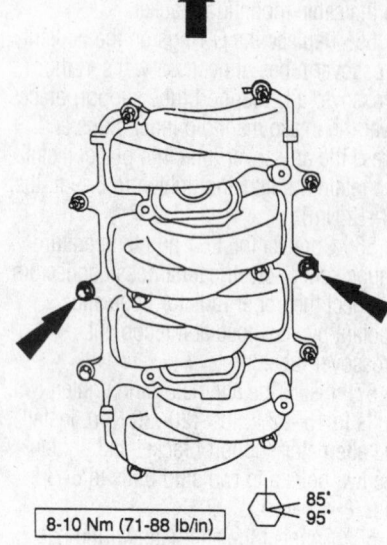

After tightening the 10 lower intake manifold bolts in sequence, tighten these two remaining bolts to specifications—1997–99 models

13. Remove the bolts and remove the lower intake manifold.

14. Disconnect the right and left intake manifold runner control (IMRC).

15. Remove the intake manifold runner control and the upper and lower gaskets. The lower runner control gaskets are housed within the grooves on the bottom surface of the IMRC. Discard the lower gaskets and use new ones.

To install:

16. Install the right and left intake manifold runner control (IMRC) and gaskets and tighten the retaining bolts to 71–89 inch lbs. (8–10 Nm) plus an additional 85–95 degree rotation.

17. Install the lower intake manifold and following the sequence shown, tighten the 10 retaining fasteners to 15–22 ft. lbs. (20–30 Nm).

18. Tighten the two remaining lower intake manifold retaining bolts shown, to 71–88 inch lbs. (8–10 Nm) plus an additional 85–95 degree rotation.

19. Install the fuel injection supply manifold and tighten the retaining bolts to 71–106 inch lbs. (8–12 Nm).

20. Install the upper intake manifold and tighten the retaining bolts to 15–22 ft. lbs. (20–30 Nm).

21. Tighten the EGR valve tube nut and connect the EGR valve vacuum hose.

22. Connect the idle air control valve electrical connector.

23. Connect the crankcase ventilation hose to the intake manifold.

24. Connect the throttle position sensor electrical connector.

25. Connect the evaporator emission hose to the upper intake manifold.

26. Connect the accelerator and cruise control cables.

27. Install the water bypass tube.

28. Disconnect the negative battery cable.

Exhaust Manifold

REMOVAL & INSTALLATION

1995 Models

RIGHT SIDE

1. Turn the air suspension switch, which is located in the luggage compartment, to the OFF position.

2. Disconnect the negative battery cable.

3. Remove the engine appearance cover.

4. Disconnect the Intake Air Temperature (IAT) sensor electrical harness connector and the crankcase vent tube from the engine air cleaner outlet tube.

5. Completely loosen the air cleaner outlet-to-throttle body bolt.

6. Remove the remaining bolt from the support bracket clamp on the right-hand cylinder head cover-to-air cleaner outlet tube assembly and remove the assembly and resonator.

7. Disconnect the heated oxygen sensor electrical harness connector.

8. Raise and safely support the vehicle.

9. Remove the four exhaust manifold-to-cylinder head retaining nuts from the upper side of the exhaust manifold.

10. Disconnect the dual converter Y-pipe from both exhaust manifolds and remove the dual converter Y-pipe assembly.

11. Remove the four lower exhaust manifold-to-cylinder head retaining nuts and remove the manifold. Remove the exhaust manifold gasket and discard it.

To install:

12. Clean the exhaust manifold gasket sealing surfaces.

13. Place a new exhaust manifold gasket and the exhaust manifold in position on the right-hand cylinder head.

14. Install the eight upper and lower exhaust manifold retaining nuts and tighten in sequence, to 15–22 ft. lbs. (20–30 Nm).

15. Install the dual converter Y-pipe assembly to both exhaust manifolds.

16. Lower the vehicle.

17. Connect the heated oxygen sensor electrical harness connector.

18. Install the engine air cleaner outlet tube and resonator assembly. Install the outlet tube to throttle body bolt and tighten to 16–23 inch lbs. (1.8–2.6 Nm).

19. Install the engine air cleaner outlet tube bracket and tighten to 45–63 inch lbs. (5.1–7.1 Nm). Tighten the air cleaner outlet tube clamp to 15–29 inch lbs. (1.7–3.3 Nm).

20. Connect the IAT sensor electrical harness connector and the crankcase vent tube to the air cleaner outlet tube.

21. Install the engine appearance cover.

22. Turn the air suspension switch to the ON position.

23. Connect the negative battery cable.

24. Start the engine and check for exhaust leaks.

25. Road test the vehicle and check for proper engine operation.

LEFT SIDE

1. Turn the air suspension switch, which is located in the luggage compartment, to the OFF position.

2. Disconnect the negative battery cable.

3. Disconnect the heated oxygen sensor electrical harness connector.

4. Install 3-Bar Engine Support D88L-6000-A or equivalent to the engine compartment and secure to suitable engine lifting brackets on the engine.

5. Raise and safely support the vehicle.

6. Remove the front wheel and tire assemblies.

7. Disconnect the EGR valve-to-exhaust manifold tube at the dual converter Y-pipe.

8. Disconnect the left-hand and right-hand exhaust manifold connections at the dual converter Y-pipe.

9. Disconnect the left-hand and right-hand ride height sensor wiring connectors.

10. Remove the right-hand and left-hand disc brake caliper bolts. Remove the brake calipers and support them with mechanic's wire.

11. Disconnect the left-hand and right-hand upper control arms from the wheel spindles.

12. Disconnect the steering shaft coupling at the pinch bolt joint.

13. Remove the power steering line clamp from the sub-frame.

14. Support the sub-frame with a suitable jack.

15. Remove the two front engine support insulator (mount) through-bolts. Remove the left-hand and right-hand lower strut-to-control arm retaining nuts and bolts.

16. Remove the eight sub-frame retaining bolts, then carefully lower the sub-frame.

17. Remove the eight exhaust manifold retaining nuts and remove the left-hand exhaust manifold. Remove and discard the gaskets.

To install:

18. If the exhaust manifold is being replaced, transfer the heated oxygen sensor and tighten to 27–33 ft. lbs. (37–45 Nm).

19. Clean the left-hand exhaust manifold gasket sealing surfaces.

20. Place new front and rear exhaust manifold gaskets on the cylinder head studs.

21. Place the left-hand exhaust manifold in position and install the eight retaining nuts. Tighten the nuts in sequence, to 15–22 ft. lbs. (20–30 Nm).

22. Connect the EGR valve-to-exhaust manifold tube. Tighten the nuts to 26–33 ft. lbs. (35–45 Nm).

23. Raise the sub-frame into position and install the eight sub-frame retaining bolts Only hand-tighten the bolts at this time.

24. Install a ¾ in. (19mm) outside diameter length of suitable pipe into both the left-hand and right-hand front sub-frame and body alignment holes to align the sub-frame to the body. Tighten one bolt on each corner securely.

25. Remove the alignment tool, then tighten all sub-frame retaining bolts to 57–75 ft. lbs. (77–103 Nm).

26. Install the two front engine support insulator through-bolts.

27. Connect the power steering line clamp to the sub-frame.

28. Connect the left-hand and right-hand upper control arms to the wheel spindles. Tighten the pinch bolt retaining nut to 50–68 ft. lbs. (68–92 Nm).

29. Connect the steering shaft coupling at the pinch bolt joint. Tighten the pinch bolt to 30–42 ft. lbs. (41–57 Nm).

30. Install the disc brake calipers and bolts. Tighten the brake pin retainers to 16–23 ft. lbs. (22–32 Nm).

31. Connect the left-hand and right-hand ride height sensor wiring connectors.

32. Connect the dual converter Y-pipe to the exhaust manifolds.

33. Install the front wheel and tire assemblies. Tighten the lug nuts to 85–105 ft. lbs. (115–142 Nm).

34. Remove the jack.

35. Lower the vehicle.

36. Connect the heated oxygen sensor electrical harness connector.

37. Remove the 3-Bar Engine Support or equivalent, from the engine compartment.

38. Connect the negative battery cable.

39. Turn the air suspension switch to the ON position.

40. Start the vehicle and check for exhaust leaks.

➡Whenever the vehicles sub-frame is removed or lowered, the wheel alignment should be checked.

41. Check the alignment and adjust if not within specification.

42. Road test the vehicle and check for proper engine operation.

1996 Models

RIGHT SIDE

1. Turn the air suspension switch to the OFF position which is located in the luggage compartment.

2. Disconnect the negative battery cable.

3. Remove the engine appearance cover. Disconnect the Intake Air Temperature (IAT) sensor electrical harness connector and the crankcase vent tube from the engine air cleaner outlet tube.

4. Completely loosen the air cleaner outlet-to-throttle body bolt.

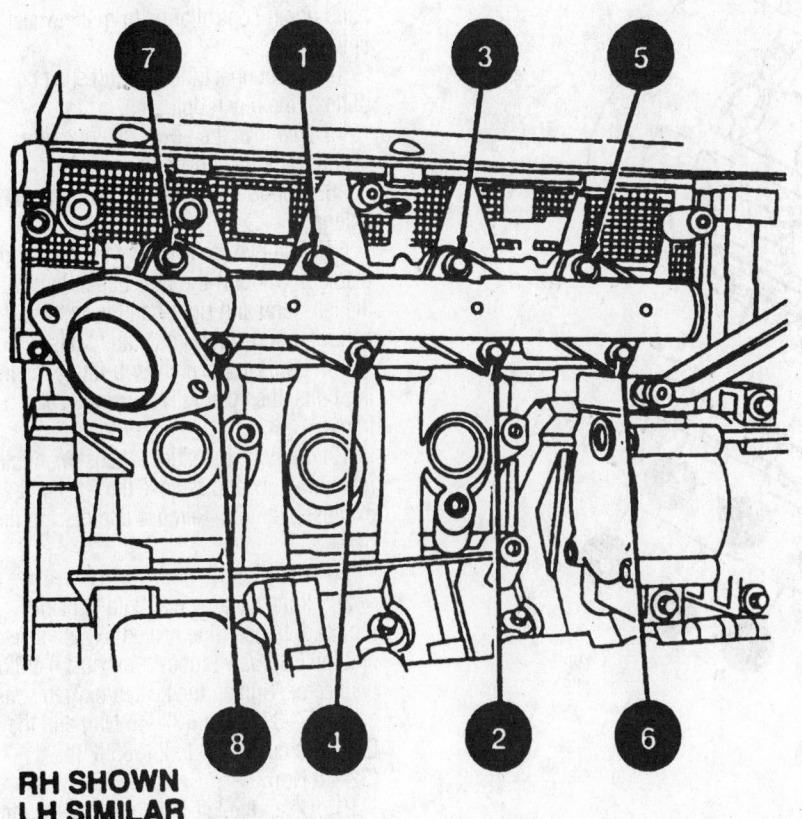

RH SHOWN LH SIMILAR

Exhaust manifold bolt torque sequence—1995 models, right side shown

7922PG09

5. Remove the remaining bolt from the support bracket clamp on the right-hand cylinder head cover-to-air cleaner outlet tube assembly and remove the assembly and resonator.

6. Disconnect the heated oxygen sensor electrical harness connector and the secondary air injection manifold tube.

7. Raise and safely support the vehicle.

8. Remove the four exhaust manifold-to-cylinder head retaining nuts from the upper side of the exhaust manifold.

9. Disconnect the dual converter Y-pipe from both exhaust manifolds and remove the dual converter Y-pipe assembly.

10. Remove the four lower exhaust manifold-to-cylinder head retaining nuts and remove the manifold. Remove the exhaust manifold gasket and discard it.

To install:

11. If the exhaust manifold is being replaced, transfer the EGR valve tube-to-exhaust manifold connector. Tighten the connector to 24–35 ft. lbs. (32–48 Nm).

12. Clean the exhaust manifold gasket sealing surfaces.

13. Place a new exhaust manifold gasket and the exhaust manifold in position on the right-hand cylinder head.

14. Install the eight upper and lower exhaust manifold retaining nuts and tighten in sequence, to 14–16 ft. lbs. (18–22 Nm).

15. Install the dual converter Y-pipe assembly to both exhaust manifolds.

16. Lower the vehicle.

17. Connect the heated oxygen sensor electrical harness connector and the secondary air injection manifold tube.

18. Install the engine air cleaner outlet tube and resonator assembly. Install the outlet tube to throttle body bolt and tighten to 16–23 inch lbs. (1.8–2.6 Nm).

19. Install the engine air cleaner outlet tube bracket and tighten to 45–63 inch lbs. (5.1–7.1 Nm). Tighten the air cleaner outlet tube clamp to 15–29 inch lbs. (1.7–3.3 Nm).

20. Connect the IAT sensor electrical harness connector and the crankcase vent tube to the air cleaner outlet tube.

21. Install the engine appearance cover.

22. Turn the air suspension switch to the ON position.

23. Connect the negative battery cable.

24. Start the engine and check for exhaust leaks.

25. Road test the vehicle and check for proper engine operation.

LEFT SIDE

1. Turn the air suspension switch to the OFF position which is located in the luggage compartment.

2. Disconnect the negative battery cable.

3. Disconnect the heated oxygen sensor electrical harness connector.

4. Disconnect the left-hand secondary air injection manifold tube.

5. Remove the left-hand secondary air injection tube bracket bolt from the exhaust manifold.

6. Install 3-Bar Engine Support D88L-6000-A or equivalent to the engine compartment and secure to suitable engine lifting brackets on the engine.

7. Raise and safely support the vehicle.

8. Remove the front wheel and tire assemblies.

9. Disconnect the EGR valve-to-exhaust manifold tube at the dual converter Y-pipe.

10. Disconnect the left-hand and right-hand exhaust manifold connections at the dual converter Y-pipe.

11. Disconnect the left-hand and right-hand ride height sensor wiring connectors.

12. Remove the right-hand and left-hand disc brake caliper bolts. Remove the brake calipers and support them with mechanic's wire.

13. Disconnect the left-hand and right-hand upper control arms from the wheel spindles.

14. Disconnect the steering shaft coupling at the pinch bolt joint.

15. Remove the power steering line clamp from the sub-frame.

16. Support the sub-frame with a suitable jack.

17. Remove the two front engine support insulator (mount) through-bolts. Remove the left-hand and right-hand lower strut-to-control arm retaining nuts and bolts.

18. Remove the eight sub-frame retaining bolts, then carefully lower the sub-frame.

19. Remove the eight exhaust manifold retaining nuts and remove the left-hand exhaust manifold. Remove and discard the gaskets.

To install:

20. If the exhaust manifold is being replaced, transfer the heated oxygen sensor and the EGR valve tube-to-exhaust manifold connector. Tighten the heated oxygen sensor to 27–33 ft. lbs. (37–45 Nm) and the EGR tube connector to 24–35 ft. lbs. (32–48 Nm).

21. Clean the left-hand exhaust manifold gasket sealing surfaces.

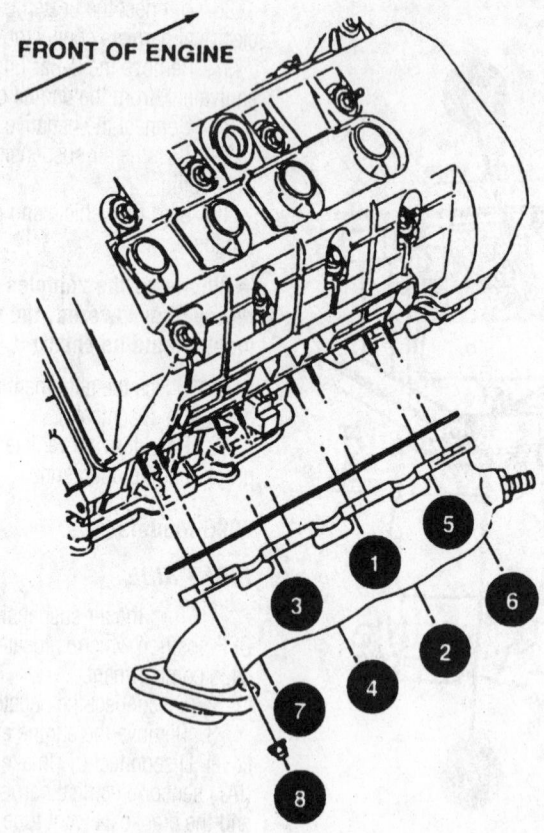

FRONT OF ENGINE

7922PG10

Tighten the exhaust manifold bolts in the sequence shown—1996 models

22. Place new front and rear exhaust manifold gaskets on the cylinder head studs.

23. Place the left-hand exhaust manifold in position and install the eight retaining nuts. Tighten the nuts in the correct sequence, to 14–16 ft. lbs. (18–22 Nm).

24. Connect the EGR valve-to-exhaust manifold tube. Tighten the nuts to 30–33 ft. lbs. (40–45 Nm).

25. Raise the sub-frame into position and install the eight sub-frame retaining bolts Only hand-tighten the bolts at this time.

26. Install a ¾ in. (19mm) outside diameter length of suitable pipe into both the left-hand and right-hand front sub-frame and body alignment holes to align the sub-frame to the body. Tighten one bolt on each corner securely.

27. Remove the alignment tool, then tighten all sub-frame retaining bolts to 57–75 ft. lbs. (77–103 Nm).

28. Install the two front engine support insulator through-bolts.

29. Connect the power steering line clamp to the sub-frame.

30. Connect the left-hand and right-hand upper control arms to the wheel spindles. Tighten the pinch bolt retaining nut to 50–68 ft. lbs. (68–92 Nm).

31. Connect the steering shaft coupling at the pinch bolt joint. Tighten the pinch bolt to 30–42 ft. lbs. (41–57 Nm).

32. Install the disc brake calipers and bolts. Tighten the brake pin retainers to 16–23 ft. lbs. (22–32 Nm).

33. Connect the left-hand and right-hand ride height sensor wiring connectors.

34. Connect the dual converter Y-pipe to the exhaust manifolds.

35. Install the front wheel and tire assemblies. Tighten the lug nuts to 85–105 ft. lbs. (115–142 Nm).

36. Remove the jack.

37. Lower the vehicle.

38. Connect the heated oxygen sensor electrical harness connector.

39. Install the left-hand secondary air injection manifold tube to the exhaust manifold and install the secondary air tube bracket bolt. Tighten the bracket bolt to 15–22 ft. lbs. (20–30 Nm) and the tube nuts to 24–35 ft. lbs. (32–48 Nm).

40. Remove the 3-Bar Engine Support or equivalent, from the engine compartment.

41. Connect the negative battery cable.

42. Turn the air suspension switch to the ON position.

43. Start the vehicle and check for exhaust leaks.

➡Whenever the vehicles sub-frame is removed or lowered, the wheel alignment should be checked.

44. Check the alignment and adjust if not within specification.

45. Road test the vehicle and check for proper engine operation.

1997–99 Models

RIGHT SIDE

1. Turn the air suspension switch to the OFF position which is located in the luggage compartment.

2. Disconnect the negative battery cable.

3. Disconnect the heated oxygen sensor electrical harness connector and the secondary air injection manifold tube.

4. Raise and safely support the vehicle.

5. Disconnect the dual converter Y-pipe.

6. Disconnect the exhaust air tube from the exhaust manifold.

7. Remove the four exhaust manifold-to-cylinder head retaining nuts from the upper side of the exhaust manifold.

8. Remove the four lower exhaust manifold-to-cylinder head retaining nuts and remove the manifold. Remove the exhaust manifold gasket and discard it.

To install:

9. Clean the exhaust manifold gasket sealing surfaces.

10. Place a new exhaust manifold gasket and the exhaust manifold in position on the right-hand cylinder head.

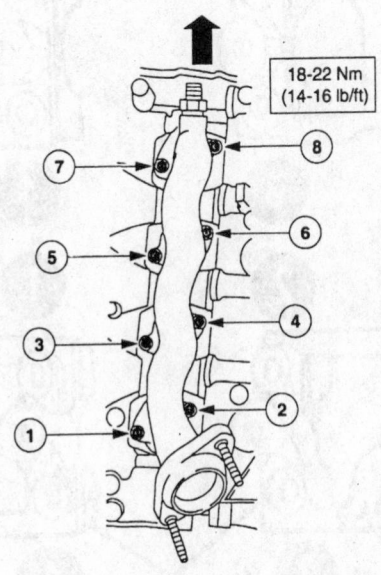

18-22 Nm (14-16 lb/ft)

7922PG12

Be sure to tighten the bolts according to the sequence shown to prevent leaks— 1997–99 models, right side shown

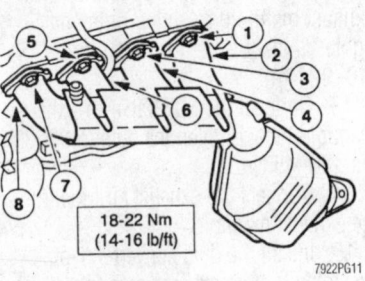

18-22 Nm (14-16 lb/ft)

7922PG11

Be sure to tighten the bolts according to the sequence shown to prevent leaks— 1996–99 models, left side shown

11. Install the eight upper and lower exhaust manifold retaining nuts and tighten in sequence, to 14–16 ft. lbs. (18–22 Nm).

12. Connect the exhaust air tube to the exhaust manifold.

13. Install the dual converter Y-pipe assembly to both exhaust manifolds.

14. Lower the vehicle.

15. Connect the heated oxygen sensor electrical harness connector and the secondary air injection manifold tube.

16. Turn the air suspension switch to the ON position.

17. Connect the negative battery cable.

18. Start the engine and check for exhaust leaks.

19. Road test the vehicle and check for proper engine operation.

LEFT SIDE

1. Turn the air suspension switch to the OFF position which is located in the luggage compartment.

2. Disconnect the negative battery cable.

3. Disconnect the left side heated oxygen sensor electrical harness connector.

4. Raise and safely support the vehicle.

5. Disconnect the dual converter Y-pipe.

6. Disconnect the exhaust air tube from the exhaust manifold.

7. Disconnect the EGR valve-to-exhaust manifold tube.

8. Remove the four exhaust manifold-to-cylinder head retaining nuts from the upper side of the exhaust manifold.

9. Remove the four lower exhaust manifold-to-cylinder head retaining nuts and remove the manifold. Remove the exhaust manifold gasket and discard it.

To install:

10. Clean the exhaust manifold gasket sealing surfaces.

11. Place a new exhaust manifold gasket and the exhaust manifold in position on the right-hand cylinder head.

12. Install the eight upper and lower exhaust manifold retaining nuts and tighten in sequence, to 14–16 ft. lbs. (18–22 Nm).

13. Connect the EGR valve-to-exhaust manifold tube. Tighten the nuts to 30–33 ft. lbs. (40–45 Nm).

14. Connect the exhaust air tube from the exhaust manifold.

15. Install the dual converter Y-pipe assembly to both exhaust manifolds.

16. Lower the vehicle.

17. Connect the heated oxygen sensor electrical harness connector.

18. Turn the air suspension switch to the ON position.

19. Connect the negative battery cable.

20. Start the engine and check for exhaust leaks.

21. Road test the vehicle and check for proper engine operation.

Camshaft and Valve Lifters

REMOVAL & INSTALLATION

1. Disconnect the negative battery cable. Remove the engine from the vehicle and position on a suitable workstand.

2. Remove the cylinder head covers and the engine front cover.

3. Remove the timing chains.

➡ **If the rocker arms are to be reused, label them as they are removed so they can be reinstalled in their original locations.**

4. Remove the rocker arms from the cylinder head and the camshaft that is being serviced. If both bank camshafts are to be serviced, all the rocker arms must be removed.

5. Remove 13 bolts retaining the exhaust camshaft cap cluster assemblies to the cylinder head to remove the exhaust camshaft.

6. Remove 12 bolts retaining the intake camshaft cap cluster assemblies to the cylinder head to remove the intake camshaft.

7. Tap upward lightly on the camshaft cap clusters and gradually lift the camshaft cap clusters from the cylinder head.

8. Lift the camshaft straight upward to avoid bearing surface damage.

9. If required, lift off the rocker arm and pull the lash adjuster (lifter) out of its bore. Be sure to return the lash adjuster and rocker to the same location that it was remove from.

➡ **The previous Steps will remove only one bank of camshafts. If both banks are being serviced, repeat the removal procedure for the remaining camshafts.**

10. Inspect the camshaft lobes and journals for wear and/or damage. Replace as necessary.

To install:

➡ **When cleaning the sealing surfaces of the engine front cover and the oil pan-to-cylinder block joints, use extreme caution not to damage the rubber bead of the oil pan gasket. If damaged, the oil pan gasket must be replaced.**

11. Clean and inspect the cylinder head cover, engine front cover, and cylinder head sealing surfaces.

12. If removed, install the lash adjusters (lifters) and rocker arms.

13. Apply clean engine oil to the journals and lobes of the camshaft. Install the camshaft to the cylinder head.

14. Install and seat the camshaft cap cluster assemblies. Hand-start twelve 2 in. (52mm) long bolts into the intake camshaft caps, seven 2 in. (52mm) long bolts into the inboard side of the exhaust camshaft and the

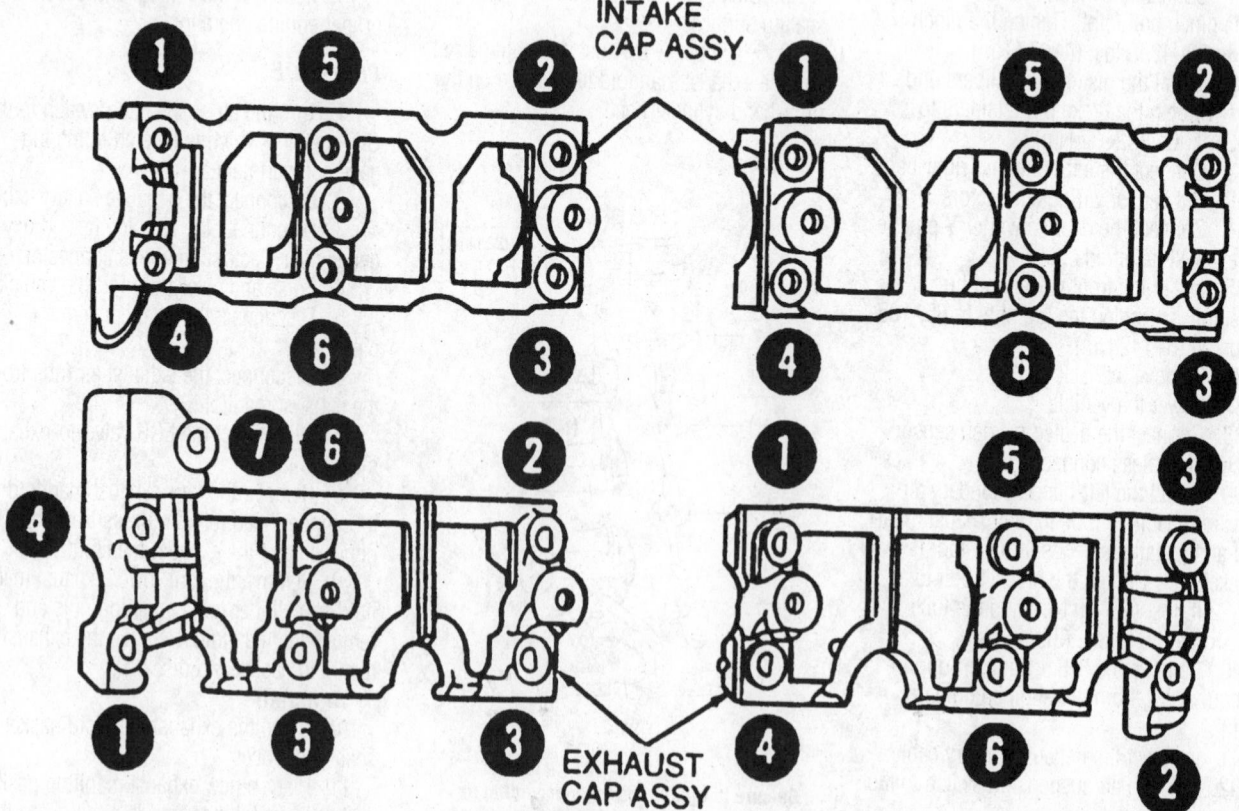

Tighten the camshaft cap clusters in the order shown to prevent camshaft damage

7922PG13

six 1 11/16 (42mm) long bolts into the out-board side of the exhaust camshaft caps.

➡**Each camshaft cap cluster is tightened individually.**

15. Tighten the camshaft cap cluster retaining bolts, in sequence, to 71–106 inch lbs. (8–12 Nm).

16. Loosen the camshaft cap cluster retaining bolts approximately two turns or until the head of the bolt is free.

17. Tighten all bolts again, in sequence, to 71–106 inch lbs. (8–12 Nm).

➡**Once installed, the camshaft should turn freely but with a slight drag.**

18. Check the camshaft end-play using a dial indicator.

19. Install the timing chains.

20. Install Valve Spring Compressor T91P-6565-A or equivalent, under the exhaust camshaft and on top of the exhaust valve spring retainer. Compress the valve spring far enough to install the rocker arm.

21. Install Valve Spring Compressor T93P-6565-A or equivalent, under the intake camshaft and on top of the primary intake valve spring retainer. Compress the valve spring far enough to install the rocker arm.

22. Install Valve Spring Compressor T93P-6565-A or equivalent, under the intake camshaft and on top of the secondary intake valve spring retainer. Compress the valve spring far enough to install the rocker arm.

➡**The previous installation steps must be repeated if both banks are being serviced to install the remaining camshafts.**

23. Inspect and replace the crankshaft front seal and timing chain front cover gasket.

24. Install the engine front cover and the cylinder head covers.

25. Remove the engine from the workstand and install in the vehicle.

26. Start the engine and check for leaks.

27. Road test the vehicle and check for proper engine operation.

Valve Lash

ADJUSTMENT

The 4.6L DOHC engine uses hydraulic valve lash adjusters that do no require any maintenance. If the valvetrain becomes noisy, inspect the camshaft, rocker arm and valve for excessive wear or damage.

Oil Pan

REMOVAL & INSTALLATION

1. Turn the air suspension switch, located in the luggage compartment, to the OFF position.

2. Disconnect the negative battery cable.

3. Remove the engine appearance cover.

4. Remove the engine oil level dipstick.

5. Install Three Bar Engine Support D88L-6000-A or equivalent, into the engine compartment and attach to the engine using suitable lifting brackets.

6. Raise and safely support the vehicle.

7. Remove the front wheel and tire assemblies.

✸✸ CAUTION

The EPA warns that prolonged contact with used engine oil may cause a number of skin disorders, including cancer! You should make every effort to minimize your exposure to used engine oil. Protective gloves should be worn when changing the oil. Wash your hands and any other exposed skin areas as soon as possible after exposure to used engine oil. Soap and water, or waterless hand cleaner should be used.

8. Drain the engine oil.

9. Disconnect the left-hand and right-hand ride height sensor electrical harness connectors.

10. Remove the right-hand and left-hand disc brake caliper retaining bolts. Remove the brake calipers from the rotors and support with mechanics wire. Do not allow the brake calipers to hang by the brake hoses.

11. Remove the pinch bolts and nuts and separate the left-hand and right-hand upper control arms from the wheel spindles.

12. Remove the pinch bolt at the steering column coupling joint and separate the coupling.

13. Remove the power steering hose clamp from the sub-frame.

14. Support the sub-frame with adjustable jack stands, or equivalent.

15. Remove the two front engine support insulator (mount) through-bolts.

16. Remove the left-hand and right-hand strut-to-lower control arm retaining nuts. Separate the struts from the lower control arms.

17. Remove the eight sub-frame retaining bolts.

18. Lower the sub-frame enough to access the engine oil pan.

19. Disconnect the low oil level sensor electrical harness connector and sensor wiring clips from the engine oil pan rail in two places.

20. Remove 16 engine oil pan bolts and remove the oil pan.

21. If removing the oil pump pick-up tube, remove the two bolts retaining the tube to the oil pump and one nut securing the tube to the main bearing stud. Remove and discard the gasket at the oil pump.

To install:

22. Clean the engine oil pan gasket sealing surfaces and inspect for damage.

23. Clean the oil pump pick-up tube screen, if required.

24. If the oil pump pick-up tube was removed, install the tube to the oil pump housing using a new gasket and install the two retaining bolts. Tighten the bolts to 71–106 inch lbs. (8–12 Nm). Install the nut retaining the tube to the main bearing stud and tighten to 15–22 ft. lbs. (20–30 Nm).

25. Place a new gasket on the engine oil pan.

26. Apply silicone sealer where the engine front cover meets the cylinder block and where the rear crankshaft seal retainer meets the cylinder block.

27. Carefully place the engine oil pan in position to the cylinder block and install 16 oil pan bolts. Tighten the bolts, in sequence, to 15–22 ft. lbs. (20–30 Nm) on 1995 vehicles. For 1996–99 vehicles, tighten the bolts in sequence, to 15 ft. lbs. (20 Nm), then rotate the bolts an additional 60 degrees.

28. Connect the low oil level sensor electrical harness connector and the sensor wiring clips to the engine oil pan rail.

29. Raise the sub-frame into position and install the eight sub-frame retaining bolts Only hand-tighten the bolts at this time.

30. Install a ¾ in. (19mm) outside diameter length of suitable pipe into both the left-hand and right-hand front sub-frame and body alignment holes to align the sub-frame to the body. Tighten one bolt on each corner securely.

31. Remove the alignment tools, then tighten all sub-frame retaining bolts to 57–75 ft. lbs. (77–103 Nm).

32. Place the struts into the lower control arms and install the retaining nuts. Tighten the nuts to 83–113 ft. lbs. (113–153 Nm).

33. Install the two front engine support insulator through-bolts.

34. Connect the power steering hose clamp to the sub-frame.

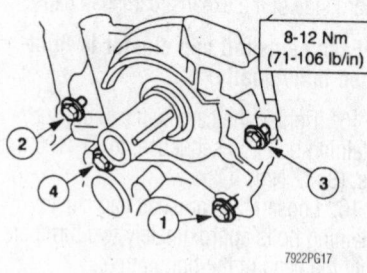

Tighten the oil pump mounting bolts in the sequence shown

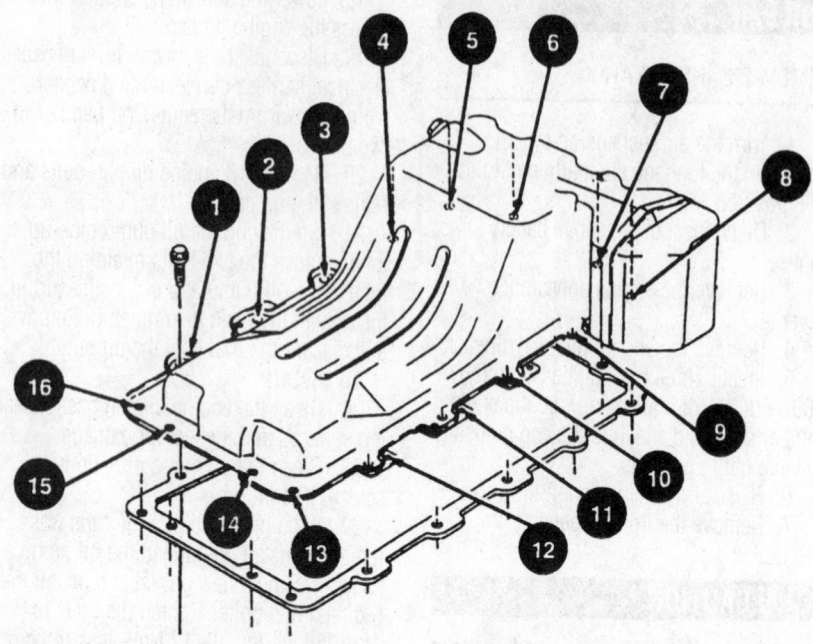

Oil pan bolt tightening sequence—1995–96 models

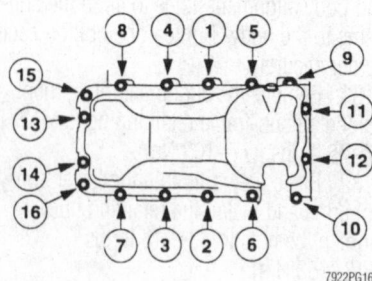

Oil pan bolt tightening sequence—1997–99 models

35. Connect the left-hand and right-hand upper control arms to the wheel spindles. Tighten the pinch bolt retaining nut to 50–68 ft. lbs. (68–92 Nm).

36. Connect the steering shaft coupling at the pinch bolt joint. Tighten the pinch bolt to 30–42 ft. lbs. (41–57 Nm).

37. Install the disc brake calipers and bolts. Tighten the brake pin retainers to 16–23 ft. lbs. (22–32 Nm).

38. Connect the left-hand and right-hand ride height sensor electrical harness connectors.

39. Install the front wheel and tire assemblies. Tighten the lug nuts to 85–105 ft. lbs. (115–142 Nm).

40. Remove the jack stands, or equivalent.

41. Lower the vehicle.

42. Remove the 3-Bar Engine Support or equivalent, from the engine compartment.

43. Install the engine oil level dipstick.

44. Fill the crankcase with the proper type and quantity of engine oil.

45. Connect the negative battery cable.

46. Turn the air suspension switch to the ON position.

47. Start the engine and check for oil leaks.

➥Whenever the vehicles sub-frame is removed or lowered, the wheel alignment should be checked.

48. Check the alignment and adjust if not within specification.

49. Road test the vehicle and check for proper engine operation.

Oil Pump

REMOVAL & INSTALLATION

1. Disconnect the negative battery cable. Remove the engine from the vehicle and position on a suitable workstand.

2. Remove the engine front cover. Remove the engine oil pan.

3. Remove the oil pump screen cover/tube bolts and the tube; discard the O-ring.

4. Remove the timing chains. Remove the crankshaft sprockets.

5. Remove the oil pump-to-cylinder block bolts and the oil pump.

To install:

6. Thoroughly clean the oil pump and cylinder block mounting surfaces and the oil pump screen cover and tube.

7. Turn the inner rotor of the oil pump to align with the flats on the crankshaft and install the oil pump flush with the cylinder block. Install the four retaining bolts and tighten to 71–106 inch lbs. (8–12 Nm).

8. Replace the engine oil filter. Install the crankshaft sprockets. Install both timing chains.

9. Position the oil pump screen cover and tube on the oil pump with a new O-ring and hand-start the two retaining bolts.

10. Install the bolt retaining the oil pump screen cover and tube to the main bearing stud spacer finger-tight.

11. Tighten the oil pump screen cover and tube-to-oil pump bolts to 71–106 inch lbs. (8–12 Nm). Tighten the oil pump screen cover and tube-to-main bearing stud spacer bolt to 15–22 ft. lbs. (20–30 Nm).

12. Install the engine oil pan. Install the engine front cover.

13. Remove the engine from the workstand and install in the vehicle.

✳✳ WARNING

Operating the engine without the proper amount and type of engine oil will result in severe engine damage.

14. Restore all fluid levels. Connect the negative battery cable.

15. Run the engine and check for leaks and proper oil pressure. Road test the vehicle and check for proper operation.

Rear Main Seal

REMOVAL & INSTALLATION

1. Remove the transmission assembly.

2. Remove the flexplate.

3. With a sharp awl, carefully punch a small hole in the metal portion of the seal.

4. Remove the seal using a slide hammer with a sheet metal screw attached.

To install:

5. Lubricate the seal and the crankshaft with clean engine oil.

6. Install the seal with the spring side toward the engine using Rear Crankshaft Seal Replacer (T95P-6701-CH) or equivalent.

7. Remove the installation tool and install the flexplate.

8. Install the transmission assembly.

Timing Chain, Sprockets and Front Cover and Seal

REMOVAL & INSTALLATION

✳✳ WARNING

This is not a free-wheeling engine. When the timing chains are removed and the cylinder heads are installed, the crankshaft and/or camshafts must not be rotated unless as directed in this procedure. Failure to follow these instructions will result in valve and/or piston damage.

1. Disconnect the negative battery cable.

2. Remove the windshield wiper module assembly.

3. Remove the engine appearance cover, if equipped.

4. Remove the engine air cleaner outlet tube.

5. Drain the engine cooling system.

6. Remove the water bypass tube and the upper radiator hose.

7. Remove the engine cooling fan motor and shroud assembly.

8. Remove the engine control wiring support bracket and straps from the left-hand center of the engine front cover.

9. Loosen the water pump pulley bolts.

10. If equipped with air injection, disconnect the vacuum and electrical harness connectors. Remove the necessary tube brackets, hoses and gaskets. Remove the air injection diverter valve.

11. Remove the accessory drive belt.

12. Remove the water pump pulley and lower water pump-to-cylinder block retaining bolt for engine front cover removal clearance.

13. Remove the bolts and stud bolt retaining the power steering pump reservoir to the left-hand ignition coil bracket.

14. Raise and safely support the vehicle.

15. Using Pulley Remover T69L-10300-B or equivalent, remove the power steering pump pulley.

16. Remove the bolt retaining the power steering pressure hose to the power steering pump bracket.

17. Remove the bolts retaining the power steering pump to the cylinder block and the engine front cover.

➡ **The front lower bolt on the power steering pump will not come all the way out.**

18. Position the power steering pump and reservoir out of the way.

19. Remove the four bolts retaining the oil pan to the engine front cover.

20. Remove the crankshaft pulley retaining bolt and washer from the crankshaft.

21. Install Crankshaft Damper Remover T58P-6316-D or equivalent, on the crankshaft pulley and pull the crankshaft pulley from the crankshaft.

22. Lower the vehicle.

23. Remove both cylinder head covers.

24. Disconnect the engine control wiring from both ignition coils and the Camshaft Position (CMP) sensor.

25. Remove the three bolts retaining the right-hand ignition coil bracket to the engine front cover.

26. Remove the three nuts and one bolt retaining the left-hand ignition coil bracket to the engine front cover.

27. Slide both ignition coil brackets with the ignition wires attached off of the front cover studs and move aside.

28. Remove the bolt retaining the belt idler pulley and remove the idler pulley.

29. Remove the drive belt tensioner retaining bolt and drive belt tensioner from the engine front cover.

30. Disconnect the Crankshaft Position (CKP) sensor and move the wiring out of the way.

31. Remove the stud bolts and bolts retaining the engine front cover to the engine.

32. Remove the engine front cover.

Install the camshaft holding tool to maintain correct alignment of the camshafts while the chain is removed

33. Remove the left-hand and right-hand cylinder head covers.

34. Remove the CKP sensor tooth wheel.

35. Rotate the crankshaft to place the piston for No. 1 cylinder at Top Dead Center (TDC) on its compression stroke.

✳✳ WARNING

Camshaft Positioning Tool T93P-6256-A or equivalent, and Camshaft Holding Tool T93P-6256-AH or equivalent, must be installed to prevent accidental rotation of the camshafts and possible engine damage.

36. Install Camshaft Positioning Tool T93P-6256-A or equivalent, in the rear D-slots of the camshaft.

✳✳ WARNING

Failure to use Camshaft Holding Tool T93P-6256-AH or equivalent, while assembling or disassembling the timing chains will result in damage to the D-slots or Camshaft Positioning Tool T93P-6256-A.

37. Install Camshaft Holding Tool T93P-6256-AH or equivalent, onto the camshafts to keep the camshafts from rotating and to prevent damaging the camshaft positioning tool.

38. Remove the two bolts retaining the right-hand primary timing chain tensioner to the cylinder head and remove the primary timing chain tensioner.

39. Remove the bolt retaining the right-hand tensioner arm and remove the right-hand tensioner arm.

➡ **The two bolts retaining the timing chain guide to the cylinder head are longer than the one to the block.**

40. Remove the three bolts retaining the right-hand timing chain guide and remove the timing chain guide.

41. Remove the right-hand primary timing chain from the camshaft and crankshaft sprockets.

42. Remove the right-hand crankshaft sprocket.

43. Remove the camshaft sprocket retaining bolts.

✳✳ WARNING

The secondary timing chain tensioner plunger is spring loaded. Care must be taken to prevent the plunger from dropping out of the tensioner during disassembly.

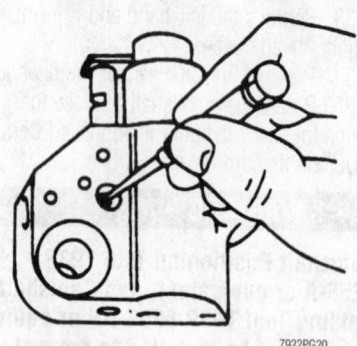

Release the ratchet lock to compress the chain tensioner

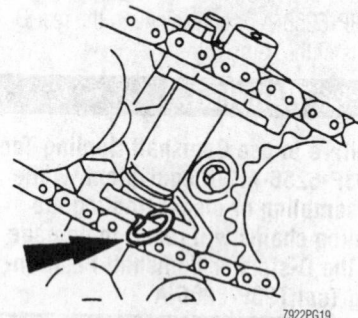

After the secondary tensioner has been compressed, install a pin through the tensioner to retain the piston

44. Unlock and compress the secondary timing chain tensioners and lock the timing chain tensioner in the compressed position using a paper clip or stiff wire. Remove the secondary timing chains and camshaft sprockets.

45. Remove the two bolts retaining the left-hand primary timing chain tensioner to the cylinder head and remove the primary timing chain tensioners.

46. Remove the bolt retaining the left-hand timing chain tensioner arm and remove the left-hand timing chain tensioner arm.

47. Remove the three bolts retaining the left-hand timing chain guide and remove the timing chain guide.

48. Remove the left-hand primary timing chain from the camshaft sprocket and crankshaft sprocket.

49. Remove the left-hand crankshaft sprocket.

✳✳ WARNING

Failure to use Camshaft Holding Tool T93P-6256-AH or equivalent, while removing or tightening the camshaft bolts may result in damage to the camshaft D-slots or the camshaft positioning tool.

50. Remove the camshaft sprocket retaining bolts.

✳✳ WARNING

The secondary timing chain tensioner plunger is spring loaded. Care must be taken to prevent the plunger from dropping out of the tensioner during disassembly.

51. Unlock and compress the secondary chain tensioner and lock the chain tensioner in the compressed position using a paper clip or stiff wire. Remove the secondary timing chain and camshaft sprockets.

✳✳ WARNING

Do not rotate the crankshaft and/or camshafts or engine damage may occur.

52. Inspect the friction material on the timing chain tensioner arms and timing chain guides. If worn or damaged, remove and clean the oil pan and replace the oil pump screen cover and tube.

To install:

53. If the engine has jumped time, be sure that all service to the necessary engine components and/or valvetrain have been made.

54. Be sure the primary and secondary timing chain tensioners are in the collapsed position and retained with paper clips or equivalent.

55. If the cylinder heads were removed from the cylinder block, position the crankshaft keyway to 270 degrees (9 o'clock position), before reinstalling the cylinder heads.

➡**The rocker arms must not be installed in the cylinder heads at this time.**

56. Install Camshaft Positioning Tool T93P-6256-A or equivalent, into the D-slots

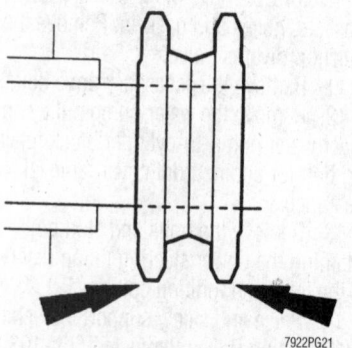

Be sure the crankshaft sprockets are installed correctly as shown

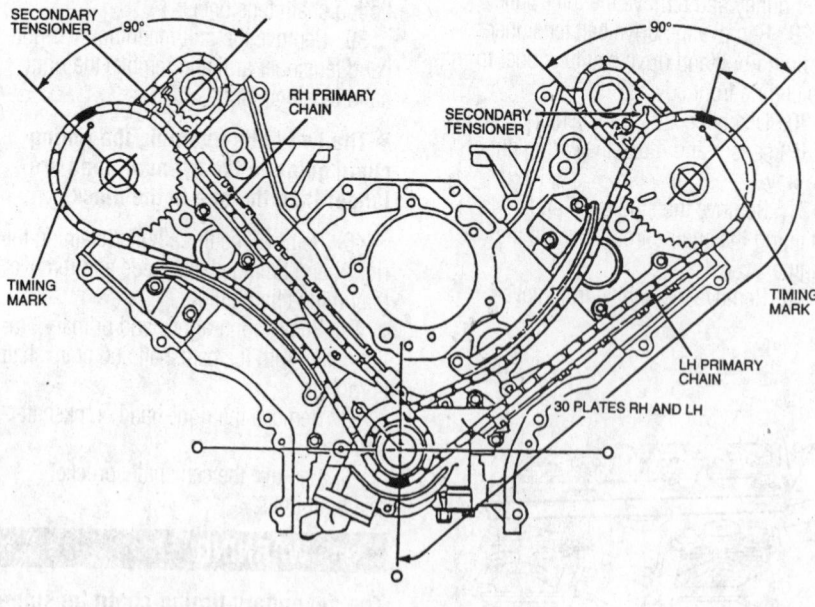

Be sure that the copper links on the timing chains align with the marks on the sprockets

on the rear of the camshafts to position both intake and exhaust camshafts with the keyways pointing down towards the crankshaft. Install Camshaft Holding Tool T93P-6256-AH or equivalent, on the center of the camshafts and do not remove until all parts are installed and tightened.

57. Install the left-hand secondary timing chain tensioner and tighten the retaining bolts to 70–106 inch lbs. (8–12 Nm).

58. Install the left-hand secondary camshaft sprockets and secondary timing chain as an assembly. Be sure the hubs of the sprocket are facing the proper direction.

59. Install the left-hand primary camshaft sprocket and camshaft sprocket spacer on the left-hand camshaft.

60. Install the washers and camshaft sprocket retaining bolts, finger-tight only.

➡**The secondary camshaft sprockets must be free to turn.**

61. Install Timing Chain Tensioning Tool T93P-6256-BH or equivalent, on the left-hand secondary timing chain tensioner.

62. Tighten the camshaft sprocket bolts to 81–95 ft. lbs. (110–130 Nm).

63. Install the right-hand secondary timing chain tensioner and tighten the retaining bolts to 70–106 inch lbs. (8–12 Nm).

64. Install the right-hand secondary camshaft sprockets and secondary timing chain as an assembly. Be sure the hubs of the sprocket are facing the proper direction.

65. Install the right-hand primary camshaft sprocket and camshaft sprocket spacer on the right-hand camshaft.

66. Install the washers and camshaft sprocket retaining bolts, finger-tight only.

➡**The secondary camshaft sprockets must be free to turn.**

67. Install Timing Chain Tensioning Tool T93P-6256-BH or equivalent, on the right-hand secondary timing chain tensioner.

68. Tighten the camshaft sprocket bolts to 81–95 ft. lbs. (110–130 Nm).

69. Install the left-hand crankshaft sprocket, making sure the tapered part of the sprocket faces away from the cylinder block.

70. Install the primary timing chain on the left-hand primary camshaft sprocket. Be sure the copper link of the timing chain aligns with the timing mark on the camshaft sprocket.

➡**If the copper links of the timing chain are not visible, pull the chain taught until the opposite sides of the chain contact one another and lay it on a flat surface. Mark the links at each end of the chain and use them in place of the copper links.**

71. Install the left-hand primary timing chain on the left-hand crankshaft sprocket. Be sure the copper link of the timing chain aligns with the timing mark on the crankshaft sprocket.

72. Install the right-hand crankshaft sprocket, making sure the tapered part of the sprocket faces toward the cylinder block.

73. Install the primary timing chain on the right-hand primary camshaft sprocket. Be sure the copper link of the timing chain aligns with the timing mark on the camshaft sprocket.

74. Install the right-hand primary timing chain on the right-hand crankshaft sprocket. Be sure the copper link of the timing chain aligns with the timing mark on the crankshaft sprocket.

➡**The two timing chain guide bolts to the cylinder head are longer than the bolt to the cylinder block.**

75. Install the right-hand and left-hand timing chain guides and tighten the bolts to 71–106 inch lbs. (8–12 Nm).

76. Lubricate the timing chain tensioner arm contact surfaces with clean engine oil. Install the right-hand and left-hand tensioner arms and tighten the bolts to 84–132 inch lbs. (10–15 Nm).

77. Install right-hand and left-hand primary timing chain tensioners and tighten the bolts to 15–22 ft. lbs. (20–30 Nm).

78. Remove the locking pins from the timing chain tensioners and be sure all timing marks are aligned. Remove the camshaft positioning tool and the camshaft holding tool.

➡**When cleaning the sealing surfaces of the engine front cover and oil pan-to-cylinder block joints, be extremely careful not to damage the rubber bead of the oil pan gasket. If the oil pan gasket is damaged, it must be replaced.**

79. Clean the sealing surfaces of the cylinder block, cylinder heads and engine front cover; remove all traces of oil, dirt and old sealant. Sealing surfaces must be clean and dry before applying sealant.

80. Replace the engine front cover gaskets and the crankshaft front seal.

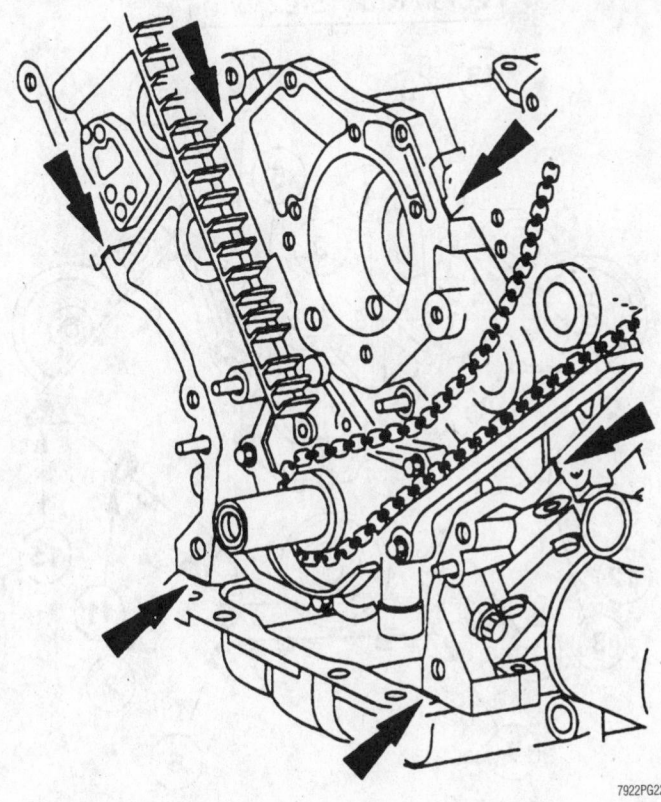

7922PG22

Be sure to apply silicone sealant to the areas on the engine where two components come together and the sealing surface is irregular as shown

81. Apply silicone sealant to the proper locations on the cylinder block and heads.

➡ **The engine front cover must be rolled into place. DO NOT slide on the oil pan gasket.**

82. Carefully install the front cover to the engine.

83. Install the bolts and stud bolts retaining the front cover to the engine. Tighten in sequence to 15–22 ft. lbs. (20–30 Nm). Tighten the stud bolts and bolts numbered 6, 7, 8, 9, 10, and 11 within four minutes after applying the sealer.

84. Install the drive belt tensioner and bolt. Tighten the bolt to 15–22 ft. lbs. (20–30 Nm).

85. Install the belt idler pulley and bolt. Tighten the bolt to 15–22 ft. lbs. (20–30 Nm).

86. Connect the engine control sensor wiring to the CKP sensor.

87. Place the left-hand and right-hand ignition coil brackets in position on the engine front cover. Install all retaining nuts and bolts and tighten to 15–22 ft. lbs. (20–30 Nm).

88. Install the power steering pump reservoir to the left-hand ignition coil bracket. Tighten the retaining bolts to 71–106 inch lbs. (8–12 Nm).

89. Connect the engine control wiring to both ignition coils and the CMP sensor.

90. Install both cylinder head covers.

91. Raise the vehicle.

92. Install the crankshaft pulley on the crankshaft, using Crankshaft Damper Replacer T74P-6316-B or equivalent.

93. Apply silicone sealant in the keyway of the crankshaft pulley.

94. Install the pulley bolt and washer. Tighten the bolt to 114–121 ft. lbs. (155–165 Nm).

95. Install the four bolts retaining the oil pan to the engine front cover. Tighten the bolts to 15–22 ft. lbs. (20–30 Nm).

96. Position the power steering pump on the engine and install the four retaining bolts. Tighten the bolts to 15–22 ft. lbs. (20–30 Nm).

97. Install the retainer bolt for the power steering pressure hose to power steering pump bracket.

98. Install the strap retaining the engine control wiring to the engine front cover.

99. Using Power Steering Pump Pulley Replacer T91P-3A733-A or equivalent, install the power steering pump pulley.

100. Lower the vehicle.

101. Install the lower water pump retaining bolt removed for engine front cover clearance. Tighten the bolt to 15–22 ft. lbs. (20–30 Nm).

102. Install the water pump pulley with the four bolts. Tighten the bolts to 15–22 ft. lbs. (20–30 Nm).

103. Install the support bracket and retaining bolt for the engine control wiring to the left-hand center of the engine front cover.

104. If equipped with air injection, install the air injection diverter valve and retaining nuts. Tighten the nuts to 6–8 ft. lbs. (8–12 Nm). Connect the vacuum and electrical connectors.

105. If equipped with air injection, install the hose to the air injection pump. Connect both tube and gasket assemblies and tighten the fittings to 25–35 ft. lbs. (32–48 Nm). Tighten the tube bracket retaining bolts to 15–22 ft. lbs. (20–30 Nm).

106. Install the engine cooling fan motor and shroud assembly.

107. Install the water bypass tube and the upper radiator hose.

108. Install the accessory drive belt.

109. Restore all fluid levels.

110. Install the windshield wiper module assembly.

111. Connect the engine air cleaner outlet tube.

112. Connect the negative battery cable.

113. Start the engine and check for leaks.

114. Install the engine appearance cover, if equipped.

115. Road test the vehicle and check for proper engine operation.

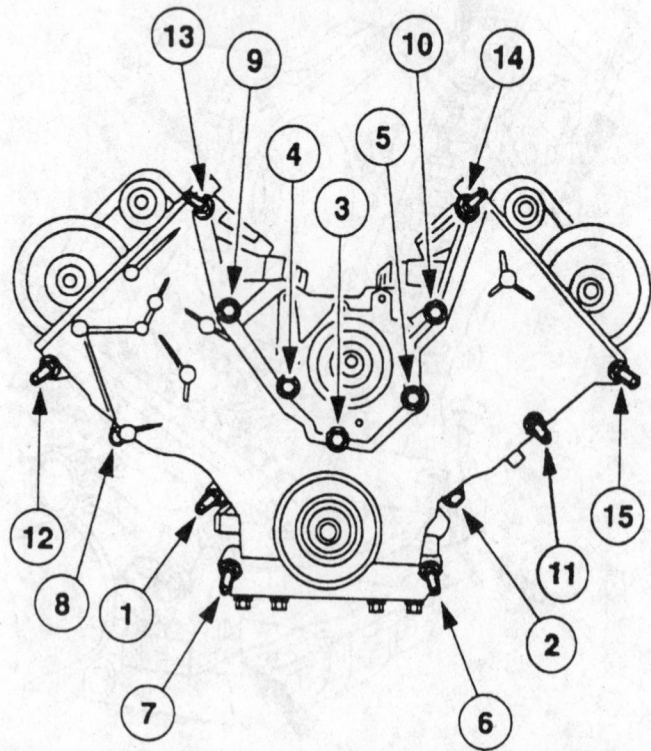

20-30 Nm (15-22 lb/ft)

7922PG23

Tighten the front cover mounting bolts in the proper sequence to ensure a good seal

FUEL SYSTEM

Fuel System Service Precautions

Safety is the most important factor when performing not only fuel system maintenance but any type of maintenance. Failure to conduct maintenance and repairs in a safe manner may result in serious personal injury or death. Maintenance and testing of the vehicle's fuel system components can be accomplished safely and effectively by adhering to the following rules and guidelines.

• To avoid the possibility of fire and personal injury, always disconnect the negative battery cable unless the repair or test procedure requires that battery voltage be applied.

• Always relieve the fuel system pressure prior to disconnecting any fuel system component (injector, fuel rail, pressure regulator, etc.), fitting or fuel line connection. Exercise extreme caution whenever relieving fuel system pressure to avoid exposing skin, face and eyes to fuel spray. Please be advised that fuel under pressure may penetrate the skin or any part of the body that it contacts.

• Always place a shop towel or cloth around the fitting or connection prior to loosening to absorb any excess fuel due to spillage. Ensure that all fuel spillage (should it occur) is quickly removed from engine surfaces. Ensure that all fuel soaked cloths or towels are deposited into a suitable waste container.

• Always keep a dry chemical (Class B) fire extinguisher near the work area.

• Do not allow fuel spray or fuel vapors to come into contact with a spark or open flame.

• Always use a back-up wrench when loosening and tightening fuel line connection fittings. This will prevent unnecessary stress and torsion to fuel line piping. Always follow the proper torque specifications.

• Always replace worn fuel fitting O-rings with new. Do not substitute fuel hose or equivalent, where fuel pipe is installed.

Fuel System Pressure

RELIEVING

Fuel supply lines on all fuel injected engines will remain pressurized for some period of time after the engine is shut **OFF**. This pressure must be relieved before servicing the fuel system. Pressure is relieved through the fuel pressure relief valve. To relieve the fuel system pressure, first remove the fuel tank cap to relieve pressure in the tank, then remove the cap on the fuel pressure relief valve, located on the fuel rail. Attach fuel pressure gauge T80L-9974-A or equivalent and open the manual valve on the pressure gauge to release the pressure and drain the system through the drain tube into a suitable container. Remove the fuel pressure gauge and replace the cap on the relief valve.

Fuel Filter

REMOVAL & INSTALLATION

✳✳ CAUTION

Observe all applicable safety precautions when working around fuel. Whenever servicing the fuel system, always work in a well ventilated area. Do not allow fuel spray or vapors to come in contact with a spark or open flame. Keep a dry chemical fire extinguisher near the work area. Always keep fuel in a container specifically designed for fuel storage; also, always properly seal fuel containers to avoid the possibility of fire or explosion.

1. Disconnect the negative battery cable and relieve the fuel system pressure.
2. Raise and safely support the vehicle.

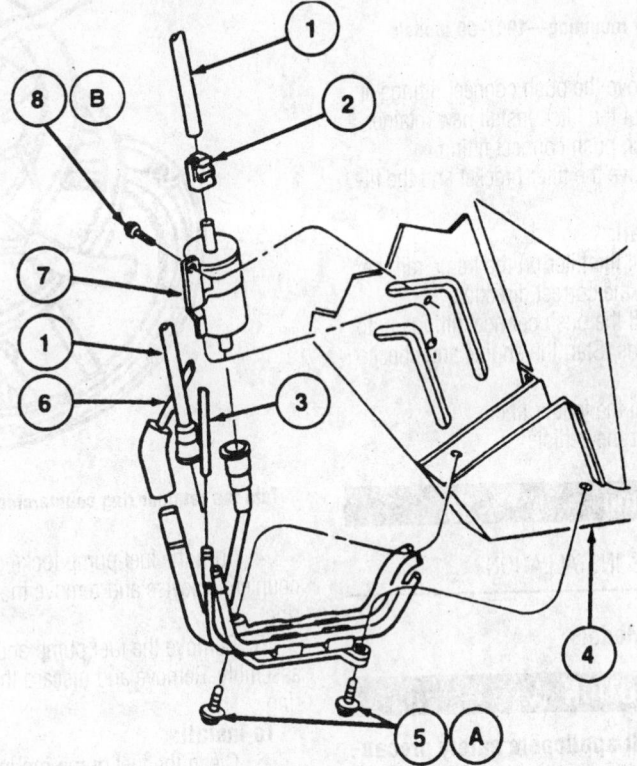

1. Front Fuel Supply Return and Vapor Tube
2. Fuel Tube Clip
3. Fuel Tank Vent Tube
4. Body
5. Screw (2 Req'd)
6. Atmospheric Vapor Line
7. Fuel Filter and Base
8. Screw
A. Tighten to 5.2-7.2 N·m (46-64 Lb-In)
B. Tighten to 7.6-10.4 N·m (67-92 Lb-In)

7922PG24

Fuel filter mounting—1995–96 models

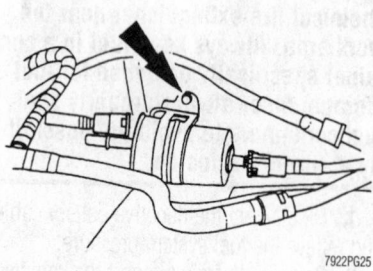

Fuel filter mounting—1997–99 models

3. Remove the push connect fittings at both ends of the filter. Install new retainer clips in each push connect fitting.

4. Remove the filter bracket and the filter.

To install:

5. Install the filter on the frame rail, be sure it is in the correct direction.

6. Install the push connect fittings onto the filter ends. Start the engine and check for leaks.

7. Install the fender liner.

8. Lower the vehicle.

Fuel Pump

REMOVAL & INSTALLATION

1995–96 Models

> ✵✵ **CAUTION**
>
> Observe all applicable safety precautions when working around fuel. Whenever servicing the fuel system, always work in a well ventilated area. Do not allow fuel spray or vapors to come in contact with a spark or open flame. Keep a dry chemical fire extinguisher near the work area. Always keep fuel in a container specifically designed for fuel storage; also, always properly seal fuel containers to avoid the possibility of fire or explosion.

1. Disconnect the negative battery cable and properly relieve the fuel system pressure.

2. Remove the fuel tank and place it on a bench.

3. Remove any dirt that has accumulated around the fuel pump retaining flange so it will not enter the tank during pump removal and installation.

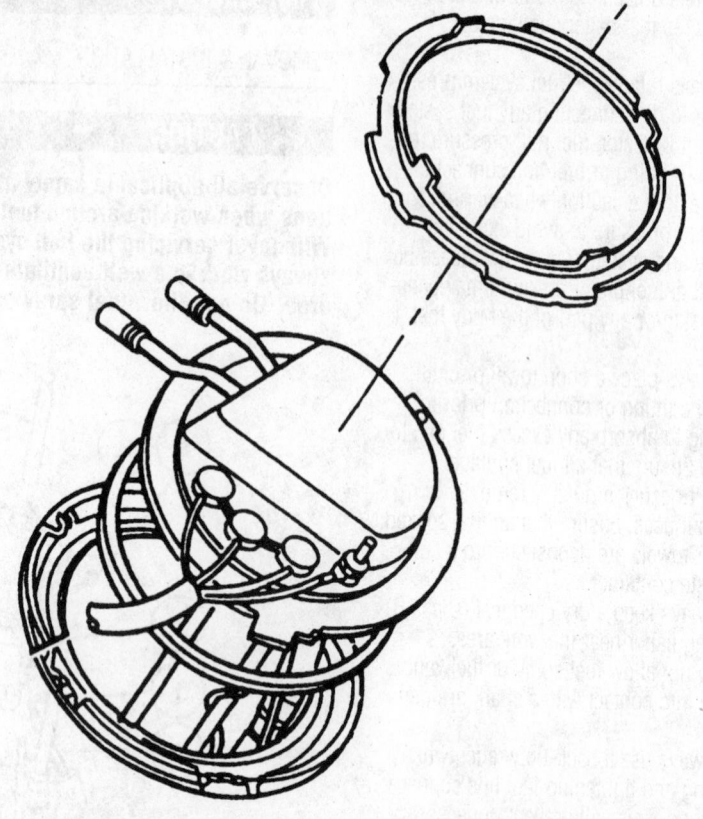

Turn the retainer ring counterclockwise to unlock it and remove the pump from the tank

4. Turn the fuel pump locking ring counterclockwise and remove the locking ring.

5. Remove the fuel pump and bracket assembly. Remove and discard the seal ring.

To install:

6. Clean the fuel pump mounting flange, fuel tank mounting surface and seal ring groove.

7. Apply a light coating of grease on a new seal ring to hold it in place during assembly and install in the seal ring groove.

8. Install the fuel pump and bracket assembly carefully to ensure the filter is not damaged. Be sure the locating keys are in the keyways and the seal ring remains in the groove.

9. Hold the pump assembly in place and install the locking ring finger-tight. Be sure all the locking tabs are under the tank lock ring tabs.

10. Rotate the locking ring clockwise until the ring is against the stops.

11. Install the fuel tank in the vehicle. Add a minimum of 10 gallons of fuel to the tank and check for leaks.

12. Install a suitable fuel pressure gauge to the valve on the fuel rail.

13. Turn the ignition switch from **OFF** to

ON for three seconds. Repeat this procedure 5–10 times until the pressure gauge shows at least 35 psi. (241 kPa). Check for fuel leaks.

14. Remove the pressure gauge, start the engine and check for leaks.

1997–98 Models

> ✵✵ **CAUTION**
>
> Observe all applicable safety precautions when working around fuel. Whenever servicing the fuel system, always work in a well ventilated area. Do not allow fuel spray or vapors to come in contact with a spark or open flame. Keep a dry chemical fire extinguisher near the work area. Always keep fuel in a container specifically designed for fuel storage; also, always properly seal fuel containers to avoid the possibility of fire or explosion.

1. Disconnect the negative battery cable and properly relieve the fuel system pressure.

2. Remove the fuel tank and place it on a bench.

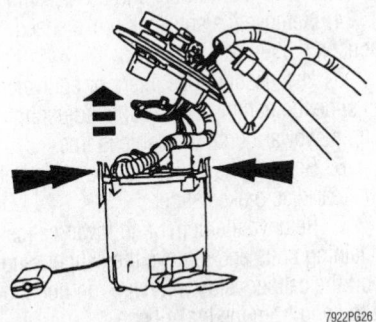

Squeeze the locking tabs together to release the pump from the tank.

3. Remove any dirt that has accumulated around the fuel pump retaining flange so it will not enter the tank during pump removal and installation.

4. Remove the fuel pump retaining bolts and remove the top flange plate.

5. Reach into the tank and squeeze the locking tabs together, then remove the pump.

To install:

6. Clean the fuel pump mounting flange, fuel tank mounting surface.

7. Install the fuel pump in the tank and tighten the mounting screws to 71–88 inch lbs. (8–10 Nm).

8. Install the fuel tank in the vehicle. Add a minimum of 10 gallons of fuel to the tank and check for leaks.

DRIVE TRAIN

➡ **Refer to the Unit Repair Section for information on halfshaft and CV-joint boot service.**

Transmission Assembly

REMOVAL & INSTALLATION

1. Turn the air suspension switch to the OFF position.

2. Disconnect the negative battery cable.

3. Raise and safely support the vehicle.

4. Drain the fluid from the transmission by removing all of the oil pan bolts except the two at the front of the pan. Loosen the remaining bolts at the front of the pan and slowly lower the oil pan at the rear to allow the fluid to drain into a container. When drained, install a few of the bolts at the rear of the pan to hold it in place.

5. Remove the exhaust system as required for transmission removal.

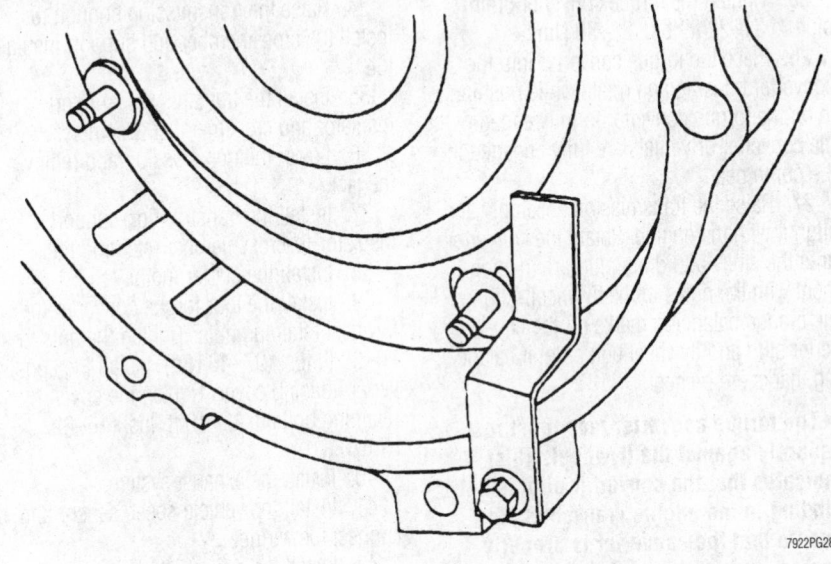

Install a torque converter holding tool to keep the converter from falling out of the transmission

6. Remove the converter inspection cover and remove the torque converter drain plug to allow the converter to drain. After the torque converter has drained, reinstall the drain plug and tighten.

7. Crank the engine over with a wrench on the crankshaft pulley retaining bolt to access and remove the four torque converter-to-flywheel retaining nuts.

➡ **Never rotate the crankshaft pulley in a counterclockwise direction, as viewed from the front of the engine, to prevent possible engine damage.**

8. Matchmark and remove the driveshaft as follows:

a. Loosen the differential housing assembly rear mounting nuts approximately ¼ inch (6.4mm).

b. Place a jack stand under the front of the differential housing and remove the forward mounting nuts and bushings. Pull the vent tube from the hole in the sub-frame.

c. Lower the front of the differential housing with the jack stand and slide the driveshaft out of the transmission above the axle housing. Let the drive shaft rest on the front driveshaft support and axle assembly.

9. Disconnect and remove the vehicle speed sensor from the transmission extension housing.

10. Disconnect the shift cable from the manual control lever.

11. Remove the starter cable and remove the starter motor.

12. Remove the transmission electrical harness connectors.

13. Place a suitable transmission jack under the transmission and raise it slightly.

14. Remove the rear crossmember and transmission support insulator.

15. Lower the transmission allowing it to hang.

16. Place a suitable jack under the front of the engine and raise the engine enough to gain access to the two upper converter housing-to-cylinder block retaining bolts. Do not remove the bolts at this time.

17. Disconnect and plug the transmission cooler lines at the transmission.

18. Remove the lower converter housing-to-cylinder block retaining bolts.

19. Remove the transmission fluid dipstick tube.

20. Secure the transmission to the transmission to the jack with a safety strap or chain.

21. Remove the two upper converter housing-to-cylinder block retaining bolts.

22. Move the transmission rearward and down from under the vehicle. Hold the converter in place to avoid having it drop from the transmission.

23. Remove the torque converter from the transmission.

To install:

➡ **If overhauling the transmission, be sure to flush the transmission oil cooler and the cooler lines of any contaminates before transmission installation.**

24. Place the transmission on a suitable transmission jack and secure with a strap or chain.

25. Tighten the torque converter drain plug to 21–22 ft. lbs. (28–30 Nm).

26. Place the torque converter into the converter housing and rotate while pushing in on the torque converter to fully engage the converter drive flats are fully engaged in the pump gear.

27. Raise the transmission assembly and align it with the engine. Rotate the converter until the studs and drain plug are in alignment with the holes in the flywheel. Align the orange balancing marks on the converter stud and flywheel bolt hole, if balancing marks are present.

➡**The torque converter face must rest squarely against the flywheel. This indicates that the converter pilot is not binding in the engine crankshaft. To ensure that the converter is properly seated, grasp a converter stud. It should move freely back and forth in the flywheel hole. If the converter will not move, the transmission must be removed and the converter repositioned so the impeller hub is properly engaged in the pump gear.**

28. Move the transmission forward and install to the cylinder block dowels. Be sure that the torque converter studs are properly aligned with the flywheel.

29. Install the transmission-to-cylinder block retaining bolts. Tighten the bolts to 30–40 ft. lbs. (40–55 Nm).

30. Remove the safety chain or strap from around the transmission.

31. Install the transmission fluid dipstick tube and secure it to the cylinder block with one retaining bolt. Tighten the bolt to 28–38 ft. lbs. (38–51 Nm).

32. Unplug and connect the oil cooler lines to the right-hand side of the transmission case. Tighten the cooler line fittings to 15–19 ft. lbs. (20–26 Nm).

33. Remove the jack supporting the front of the engine.

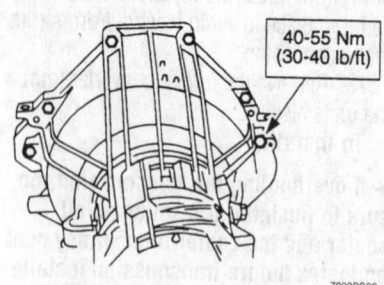

40-55 Nm (30-40 lb/ft)

7922PG29

Transmission mounting bolt location and torque specification

34. Raise the transmission enough to install the crossmember and support insulator.

35. Install the transmission support insulator and crossmember assembly.

36. Lower the transmission and remove the jack.

37. Install the transmission support insulator-to-crossmember retaining nut.

38. Install the starter motor.

39. Install the four torque converter-to-flywheel retaining nuts. Tighten the nuts to 20–33 ft. lbs. (27–46 Nm). Install the converter housing cover. Tighten the cover retaining bolts to 12–16 ft. lbs. (16–22 Nm).

40. Install the exhaust system.

41. Install the vehicle speed sensor and connect the wiring.

42. Install the driveshaft, making sure the matchmarks are aligned as follows:

a. Raise the differential housing with the jack stand and install the bushings and retaining nuts. Tighten the nuts to 68–100 ft. lbs. (92–136 Nm). Remove the jack stand.

b. Install the vent tube in the hole of the sub-frame.

c. Align the driveshaft yoke and companion flange and install the retaining bolts. Tighten the retaining nuts to 70–95 ft. lbs. (95–129 Nm).

43. Safely lower the vehicle.

44. Connect the negative battery cable.

45. Turn the air suspension switch to the ON position.

46. Fill the transmission with the proper type and quantity of fluid.

47. Start the engine and add transmission fluid as needed until the proper fluid level is reached while checking for leaks.

48. Road test the vehicle and check for proper operation.

49. Recheck the transmission fluid level.

Halfshaft

REMOVAL & INSTALLATION

➡**Before continuing with this procedure, be sure to have available one new inboard CV-joint stub shaft pilot bearing housing seal, one shaft bearing retaining clip and one wheel hub retainer nut, per side. Once removed, these parts lose their torque holding ability or retention capability and must not be reused.**

1. With the vehicle sitting on the ground, remove the wheel hub retainer nut.

2. Raise and safely support the vehicle.

3. Remove the wheel and tire assembly.

4. Remove the anti-lock brake speed sensor, if equipped.

5. Use needle-nose pliers or equivalent, to slide the parking brake cable adjusting clip downward until the cable is free.

6. Remove the parking brake cable from the rear disc brake caliper.

7. Remove the upper and lower caliper retaining bolts and remove the caliper. Support the caliper aside with wire, do not allow it to hang from the brake hose.

8. Remove the upper control arm retaining nut and bolt. Wire the upper control arm to the top of the shock absorber.

9. Using a paint marker, matchmark the position of the lower control arm in relation to the knuckle with the lower bushings in the relaxed position.

➡**Failure to matchmark this relationship will result in bushing wind-up on assembly and incorrect ride height, causing misalignment and premature tire wear.**

10. Use a suitable puller to free the halfshaft from the hub.

11. Remove the lower control arm to knuckle retaining nuts. Push the halfshaft through the hub while positioning the CV-joint and knuckle to allow the front lower bolt to clear the CV-joint and remove the bolt. Remove and save the washers.

12. Remove the rear bolt, washer, and knuckle assembly. Be careful not to damage the differential oil seal, differential housing and/or CV-joint boot.

13. Remove the halfshaft from the vehicle.

14. Insert a plug into the differential housing to prevent fluid loss.

To install:

15. Remove the differential plug and install a new differential oil seal.

16. Install a new circlip on the halfshaft. Start the ends in the groove and push the circlip into the groove, to prevent over expanding the circlip.

17. Lightly lubricate the stub shaft splines and carefully align the splines on the shaft with the splines in the differential.

18. Push the halfshaft inward to seat the circlip in the differential side gear groove. Use care not to damage the seal.

19. Engage the hub splines with the outboard CV-joint splines.

20. Install the lower control arm retaining bolts and nuts. Align the paint marks and tighten the bolts to 94–131 ft. lbs. (128–178 Nm).

21. Install a new wheel hub retaining nut

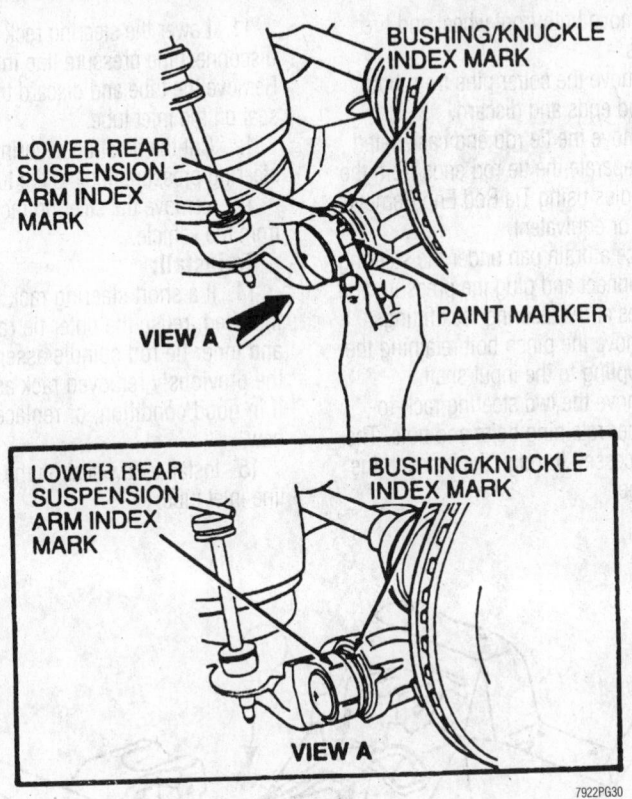

LOWER REAR SUSPENSION ARM INDEX MARK

BUSHING/KNUCKLE INDEX MARK

PAINT MARKER

VIEW A

LOWER REAR SUSPENSION ARM INDEX MARK

BUSHING/KNUCKLE INDEX MARK

VIEW A

7922PG30

Be sure to matchmark the lower control arm and the knuckle assembly to maintain correct alignment

and use it to draw the CV-joint into the hub as far as possible by hand.

22. Install the upper arm retaining bolt and nut and tighten to 117–142 ft. lbs. (158–193 Nm).

23. Install the brake caliper assembly to the rotor with the outer brake shoe against the rotor's braking surface. This prevents pinching the piston boot between the inner brake shoe and the piston.

24. Install the upper and lower caliper retaining bolts and tighten to 64–88 ft. lbs. (87–119 Nm).

25. Install the parking brake cable to the brake caliper. Install the cable adjustment clip.

26. Install the anti-lock brake speed sensor, if equipped. Tighten the retaining bolts to 15–19 ft. lbs. (19–27 Nm).

27. Check the inboard CV-joint circlip engagement by attempting to pull the inboard CV-joint from the axle. If the CV-joint circlip is not seated, push the CV-joint in until the circlip is fully engaged in the side gear.

28. Check the axle lubricant level and fill, as necessary.

29. Install the wheel and tire assembly. Tighten the lug nuts to 85–105 ft. lbs. (115–142 Nm).

30. Lower the vehicle.

31. Pump the brake pedal several times to position the brake pads before moving the vehicle.

32. Tighten the wheel hub retainer nut to 250 ft. lbs. (340 Nm).

33. Check the rear alignment and adjust if not within specification.

34. Road test the vehicle and check for proper operation.

STEERING AND SUSPENSION

Air Bag

✳✳ CAUTION

Some vehicles are equipped with an air bag system, also known as the Supplemental Inflatable Restraint (SIR) or Supplemental Restraint System (SRS). The system must be disabled before performing service on or around system components, steering column, instrument panel compo-

nents, wiring and sensors. Failure to follow safety and disabling procedures could result in accidental air bag deployment, possible personal injury and unnecessary system repairs.

PRECAUTIONS

Several precautions must be observed when handling the inflator module to avoid accidental deployment and possible personal injury.

- Never carry the inflator module by the wires or connector on the underside of the module.
- When carrying a live inflator module, hold securely with both hands, and ensure that the bag and trim cover are pointed away.
- Place the inflator module on a bench or other surface with the bag and trim cover facing up.
- With the inflator module on the bench, never place anything on or close to the module which may be thrown in the event of an accidental deployment.

DISARMING

1. Disconnect the positive battery cable. Wait one minute for the back-up power supply in the diagnostic monitor to deplete its stored energy.

2. Remove the four nut and washer assemblies (2 screws on Mark VIII) retaining the driver air bag module to the steering wheel.

✳✳ CAUTION

When carrying a live air bag, be sure the bag and trim cover are pointed away from the body. In the unlikely event of an accidental deployment, the bag will, then deploy with minimal chance of injury. When placing a live air bag on a bench or other surface, always face the bag and trim cover up, away from the surface. This will reduce the motion of the module if it is accidentally deployed.

3. Disconnect the driver air bag connector. Connect air bag simulator tool 105–00008 or equivalent, to the vehicle harness at the top of the steering wheel.

ENABLING

1. Install the air bag module.
2. Connect the negative battery cable.

3. Turn the ignition **ON** and be sure the air bag warning light does not remain ON.

Power Rack and Pinion Steering Gear

REMOVAL & INSTALLATION

1. Turn the air suspension switch to the **OFF** position.
2. Disconnect the negative battery cable.
3. Leave the ignition key in the **ON** position to prevent the steering column from locking.
4. Raise and safely support the vehicle.

5. Remove both front wheel and tire assemblies.
6. Remove the cotter pins from both outer tie rod ends and discard.
7. Remove the tie rod end castellated nuts and separate the tie rod ends from the wheel spindles using Tie Rod End Remover T-3290-D, or equivalent.
8. Place a drain pan under the steering rack. Disconnect and plug the pressure and return hoses at the steering rack fittings.
9. Remove the pinch bolt retaining the flexible coupling to the input shaft.
10. Remove the two steering rack-to-crossmember retaining bolts and nuts. The nuts are accessed through the holes in the crossmember.

11. Lower the steering rack enough to disconnect the pressure line inlet tube. Remove the tube and discard the plastic seal on the inlet tube.
12. Cut the tie strap securing the power steering pressure hose to the tube.
13. Remove the steering rack assembly from the vehicle.

To install:

14. If a short steering rack is being installed, reuse the outer tie rod ends and inner tie rod spindle assemblies from the previously removed rack assembly if in good condition, or replace with new.
15. Install a new seal on the pressure line inlet tube.

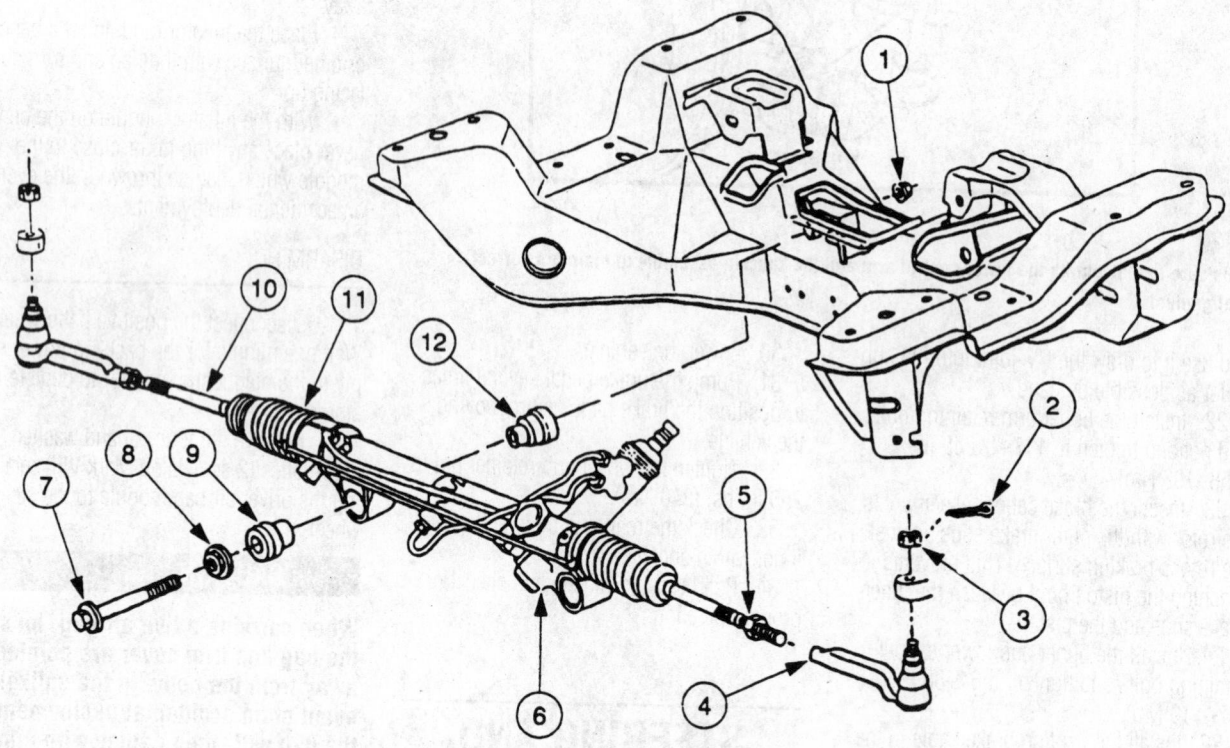

Item	Part Number	Description	Item	Part Number	Description
1	N800237-S1050	Nut (2 Req'd)	8	385935-S36	Washer (2 Req'd)
2	72044-S36	Cotter Key	9	3F638	Insulator (2 Req'd)
3	N803972-S150	Nut (2 Req'd)	10	3280	Front Wheel Spindle Tie Rod (2 Req'd)
4	3A130	Tie Rod End	11	3332	Front Suspension Steering Ball Stud Dust Seal
5	N803637-S100	Jam Nut (2 Req'd)			
6	3504	Steering Gear	12	3K576	Insulator (2 Req'd)
7	N807147-S190	Bolt (2 Req'd)			

Exploded view of the steering gear mounting and related components

7922PG31

16. Install the two steering rack mounting insulators, if not equipped.

➡**The rubber insulators must be pushed completely inside the gear housing before the installation of the gear housing on the No. 2 crossmember.**

17. Place the steering rack in position on the crossmember so that the pressure line inlet tube can be installed and install the inlet tube.

18. Align the steering shaft to allow the steering rack to completely seat on the crossmember.

19. Install the two steering rack-to-crossmember retaining bolts and nuts. Tighten the bolts to 100–143 ft. lbs. (135–195 Nm).

20. Install the steering shaft pinch bolt retaining the flexible coupling to the input shaft. Tighten the pinch bolt to 30–42 ft. lbs. (41–57 Nm).

21. Secure the power steering pressure hose to the tube with a new tie strap.

22. Connect the power steering pressure hose to the steering rack. Tighten the fitting to 42–53 ft. lbs. (57–73 Nm). The design allows the hoses to swivel when tightened properly. do not attempt to eliminate looseness by overtightening, since this can cause damage to the fittings.

23. Connect the power steering return hose to the steering rack. Tighten the clamp to 12–17 inch lbs. (1.4–2.0 Nm).

24. Place the outer tie rod end studs to the wheel spindles. Install the castellated nuts and tighten to 40–53 ft. lbs. (53–72 Nm). If required, tighten the nuts to the next cotter pin castellation. Install new cotter pins.

25. Install the wheel and tire assemblies. Tighten the lug nuts to 85–105 ft. lbs. (115–142 Nm).

26. Lower the vehicle.

27. Turn the ignition key to the **OFF** position.

28. Connect the negative battery cable.

29. Turn the air suspension switch to the ON position.

30. Fill the power steering pump reservoir.

31. Start the engine and turn the steering wheel slowly from lock-to-lock several times to purge any air from the power steering system while checking for leaks.

32. Check the fluid level and add as required.

33. Check the alignment and adjust if not within specification.

34. Road test the vehicle and check for proper steering system operation.

Shock Absorber

REMOVAL & INSTALLATION

Front

➡**The front air spring and shock absorber are incorporated into one assembly and are not serviceable separately.**

1. Turn the air suspension switch, located in the luggage compartment, to the OFF position.

2. Raise and safely support the vehicle.

➡**The front suspension system must be allowed to hang at full rebound.**

3. Remove the wheel and tire assembly.

4. Remove the air spring height sensor as follows:

a. Carefully remove the forward section of the front fender splash shield enough to access the height sensor electrical harness connector and the push-in clips.

b. Pull out the height sensor clips.

c. Disconnect the height sensor electrical harness connector from the wiring harness.

d. Remove the air spring height sensor from its retaining studs by gently pulling back on the height sensor spring clips and removing the height sensor.

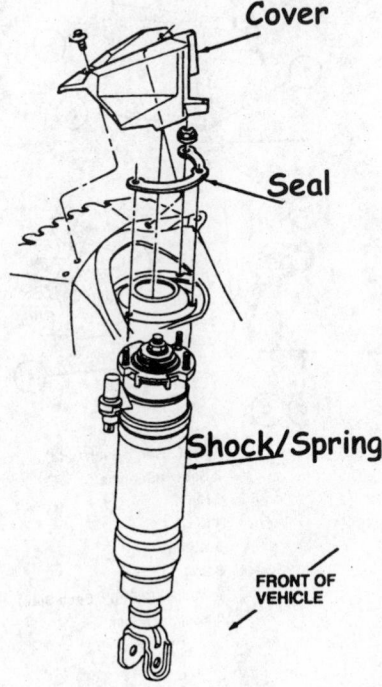

Exploded view of the front shock absorber/ air spring assembly

5. Disconnect the air spring solenoid electrical harness connector.

6. Remove the air line from the air spring solenoid valve by pushing down and holding the plastic retaining ring on the valve while pulling on the air line.

7. Remove the air spring solenoid valve as follows:

a. Remove the solenoid retainer clip.

b. Rotate the solenoid valve counterclockwise to the 1st stop.

c. Pull the solenoid valve straight out slowly to the 2nd stop to bleed air from the system.

➡**Do not fully release the solenoid until the air is completely bled from the air spring.**

d. After the air is fully bled from the system, rotate the solenoid valve counterclockwise to the third stop and remove the solenoid valve from the solenoid housing. Remove the large O-ring from the solenoid housing.

8. Remove the clip retaining the spring to the lower arm. Push down on the spring clip on the collar of the air spring and rotate the collar counterclockwise to release the spring from the body spring seat.

9. Remove the shock tower cover from the engine compartment.

10. Remove three retaining nuts and the collar securing the air spring to the shock tower.

➡**Do not remove the large center nut which may result in a permanent air leak.**

11. Remove the lower nut and bolt retaining the air spring and shock assembly to the lower control arm.

12. Remove the air spring and shock assembly from the vehicle.

To install:

13. Install the air spring solenoid valve as follows:

a. Check the solenoid valve O-rings for cuts or abrasion. Replace the O-rings as required. Lightly grease the O-ring area of the solenoid valve and the larger solenoid housing O-ring with silicone dielectric compound.

b. Insert the solenoid valve into the air spring end cap and rotate clockwise to the third stop, push in to the second stop, then rotate clockwise to the first stop.

c. Install the air spring solenoid retainer clip. Inspect the electrical harness connector and ensure the rubber gasket is in place at the bottom of the connector cavity.

14. Place the upper air spring and shock

mounting studs into position and loosely install the three upper air spring stud retaining nuts and collar at the shock tower.

➡ **The solenoid valve socket for either front air spring should be oriented forward of the air spring.**

15. Place the lower end of the air spring and shock assembly into position over the lower control arm support and loosely install the retaining bolt and nut. Tighten the three upper retaining nuts over the shock tower to 17–24 ft. lbs. (23–32 Nm).

16. Install the appearance cover over the top of the shock tower.

17. Connect the air line and the electrical harness connector to the air spring solenoid valve.

18. Install the air spring height sensor as follows:

   a. Snap the height sensor into the upper and lower ball studs with the height sensor wiring oriented downward.

   b. Install the air spring height sensor electrical harness connector.

   c. Install the two harness wiring retaining clips into their original locations.

   d. Install the front fender splash shield.

19. Verify that the air suspension switch if OFF , then connect the negative battery cable.

20. Fill the air spring as follows:

   a. Connect Super Star II Tester 007–0041B or equivalent scan tool to the Data Link Connector (DLC) located on the front side of the right-hand shock tower.

   b. Install a batter charger to reduce battery drain.

   c. Enter the appropriate function test to fill the required air spring:
   • 212 left-front air spring
   • 214 right-front air spring

   d. Remove the tester and battery charger.

➡ **Any further inflation should be done with the vehicle on the ground and the air suspension switch in the ON position.**

21. Inspect the air spring folds and creases.

22. Install the wheel and tire assembly. Tighten the lug nuts to 85–105 ft. lbs. (115–142 Nm).

23. Lower the vehicle.

24. Turn the air suspension switch to the ON position.

25. Place the vehicle on an alignment rack or equivalent, to raise the vehicle off of the ground but keep the suspension loaded.

26. Neutralize the front suspension bushings by pushing the front of the vehicle down several times and allowing the suspension to settle. Tighten the lower air spring and shock assembly retaining bolt and nut to 125–170 ft. lbs. (170–230 Nm).

27. Check the alignment and adjust if not within specification.

28. Road test the vehicle and check for proper operation.

Rear

✱✱ CAUTION

The rear shocks act as the rebound stops for the rear suspension. When the rear shocks are removed, the suspension arms must be supported. The best method for rear shock removal is to use a drive-on hoist or an alignment pit. Portable vehicle ramps can be substituted, but be sure the ramps are secure, the parking brake is set, and the front wheels are chocked properly before proceeding.

1. Place the vehicle on a drive-on hoist or over an alignment pit or equivalent.

2. Turn the air suspension switch,

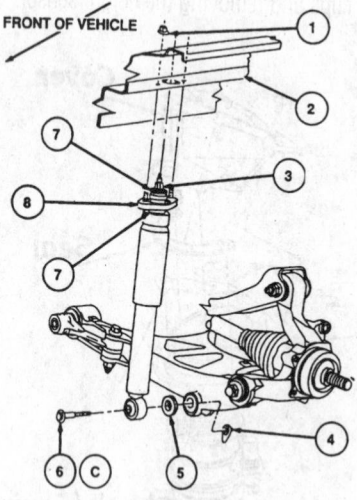

FRONT OF VEHICLE

1 Nut (2 Req'd Each Side)
2 Body Reference
3 Nut
4 Nut
5 Washer
6 Bolt
7 Washer (2 Req'd Each Side)
8 Insulator Mount

7922PG33

Exploded view of the rear shock absorber mounting

located in the luggage compartment, to the OFF position.

3. From inside the luggage compartment, remove the two upper shock absorber retaining nuts.

4. Remove the lower shock absorber retaining bolt and nut at the lower control arm.

5. Remove the shock absorber and insulator mount from the vehicle.

6. Grasp the flats on the insulator mount upper washer and loosen the insulator retaining nut.

7. Remove the upper washer, insulator mount, retaining nut and lower washer for use on the new shock absorber assembly.

   To install:

8. Install the upper insulator lower washer and insulator mount on the shock absorber. Align the upper washer with the flat on the shock absorber rod and install the washer and retaining nut.

9. Grasp the upper washer flats firmly and rotate the insulator mount studs 90 degrees from the upper washer flats.

10. Holding the insulator mount studs in this position, tighten the insulator retaining nut to 53–65 ft. lbs. (72–88 Nm).

11. Properly prime the new shock absorber to remove air from the working cylinder by inverting and compressing, then releasing the shock to its fully extended position at least three times.

12. Install the shock absorber rod and insulator mount studs through the shock tower mounting holes.

13. Compress the shock absorber and align the lower mounting hole with the lower control arm. Install the retaining bolt, washer and nut to the lower control arm. While holding the nut, tighten the bolt to 83–113 ft. lbs. (113–153 Nm).

14. Install the retaining nuts to the upper insulator studs and tighten to 17–23 ft. lbs. (23–31 Nm).

15. Install the luggage compartment panels, if removed.

16. Lower vehicle.

17. Turn the air suspension switch to the ON position.

18. Road test the vehicle and check for proper operation.

Air Spring

REMOVAL & INSTALLATION

Rear

1. Turn the air suspension switch to the OFF position.

2. Disconnect the negative battery cable.

3. Raise and safely support the vehicle.

➡**The rear suspension system must be allowed to hang at full rebound.**

4. Remove the wheel and tire assembly.

5. Disconnect the air spring solenoid electrical harness connector.

6. Remove the air line from the air spring solenoid valve by pushing down and holding the plastic retaining ring on the valve while pulling on the air line.

7. Remove the air spring solenoid valve as follows:

   a. Remove the solenoid retainer clip.

   b. Rotate the solenoid valve counterclockwise to the first stop.

   c. Pull the solenoid valve straight out slowly to the second stop to bleed air from the system.

➡**Do not fully release the solenoid until the air is completely bled from the air spring.**

   d. After the air is fully bled from the system, rotate the solenoid valve counterclockwise to the 3rd stop and remove the solenoid valve from the solenoid housing. Remove the large O-ring from the solenoid housing.

8. Depress the four plastic locking fingers in the bottom of the air springs piston to disconnect the air spring from the lower control arm.

9. Depress the metal locking tab at the top of the air spring and rotate to disengage the air spring cap from the body bracket.

10. Carefully remove the air spring from the vehicle.

To install:

11. Install the air spring solenoid as follows:

   a. Check the solenoid O-rings for cuts or abrasion. Replace the O-rings as required. Lightly grease the O-ring area of the solenoid and the larger solenoid

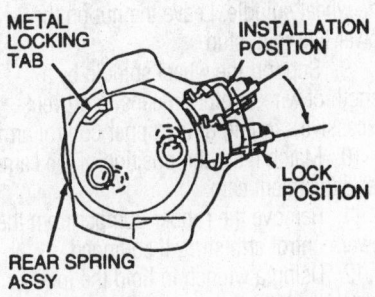

METAL LOCKING TAB — INSTALLATION POSITION — LOCK POSITION — REAR SPRING ASSY

7922PG34

Depress the metal locking tab and rotate the spring to remove it from the body

housing O-ring with silicone dielectric compound.

   b. Insert the solenoid valve into the air spring end cap and rotate clockwise to the 3rd stop, push in to the 2nd stop, then rotate clockwise to the first stop.

   c. Install the air spring solenoid retainer clip. Inspect the wiring harness connector and ensure the rubber gasket is in place at the bottom of the connector cavity.

12. Insert the air spring upper cap into the two stud holes in the body bracket and twist to lock in place.

➡**Rear air springs are the same for left-hand or right-hand installation.**

13. Insert the four plastic locking fingers into the lower control arm and lock them into position.

✳✳ WARNING

Do not inflate the air spring until it is properly installed.

14. Connect the air line and electrical connector to the solenoid.

15. Verify that the air suspension switch is OFF , then connect the negative battery cable.

16. Fill the air spring as follows:

   a. Connect the Super Star II Tester 007–0041B or equivalent scan tool to the Data Link Connector (DLC) located on the front side of the right-hand shock tower.

   b. Install a batter charger to reduce battery drain.

➡**The following function test numbers are for the Star Tester II only. The Star Tester II is the scan tool used in the Ford dealerships. Follow the instructions included with your scan tool to fill the air spring.**

   c. Enter the appropriate function test to fill the required air spring:
   - 216 left-rear air spring
   - 218 right-rear air spring

   d. Remove the tester and battery charger.

➡**Any further inflation should be done with the vehicle on the ground and the air suspension switch in the ON position.**

17. Inspect the air spring folds and creases.

18. Install the wheel and tire assembly. Tighten the lug nuts to 85–105 ft. lbs. (115–142 Nm).

19. Lower the vehicle.

20. Turn the air suspension switch to the ON position.

21. Check the alignment and adjust if not within specification.

22. Road test the vehicle and check for proper operation.

Upper Ball Joint

REMOVAL & INSTALLATION

The ball joint is an integral part of the upper control arm. If the ball joint is defective, the entire upper control arm must be replaced.

Upper Control Arm

REMOVAL & INSTALLATION

➡**The following procedure is the same for both the right-hand and left-hand upper control arm assemblies.**

1. Turn the air suspension switch to the OFF position which is located in the luggage compartment.

2. Raise and safely support the vehicle.

3. Remove the wheel and tire assembly.

4. Remove and discard the upper wheel spindle-to-ball joint pinch bolt and nut. Slightly spread the wheel spindle at the slot opening to remove the upper ball joint stud from the wheel spindle.

5. Lower the vehicle enough to gain access to the upper control arm pivot bolts. Support the vehicle in this position.

➡**A 6-point socket must be used on the bolt due to the fact that the corners of the heads have been shaved off. Discard the nuts and bolts after removing them. Replacement bolts do not require metal flags.**

6. Break off the small metal flags from the upper control arm pivot bolt heads. Remove the bolts and nuts.

7. Remove the upper control arm from the vehicle. Inspect the upper control arm ball joint. If worn the entire upper control arm assembly must be replaced.

To install:

8. Place the upper control arm in position and install new pivot bolts and nuts. Holding the control arm in the horizontal position, tighten the nuts to 65–88 ft. lbs. (88–119 Nm).

➡**If nuts cannot be torqued due to limited access on some models, Tighten the pivot bolts to 82–88 ft. lbs. (110–120 Nm).**

9. Raise and safely support the vehicle.

10. Install the upper control arm ball

joint stud to the wheel spindle and install a new pinch bolt and nut, making sure to install the bolt from the front of the vehicle. Tighten the nut to 51–67 ft. lbs. (68–92 Nm).

11. Install the wheel and tire assembly. Using a torque wrench, tighten the lug nuts in a star pattern to 85–105 ft. lbs. (115–142 Nm).

12. Lower the vehicle.

13. If equipped, turn the air suspension switch to the ON position.

14. If required, check the alignment and adjust if not within specification.

15. Road test the vehicle and check for proper operation.

Lower Ball Joint

REMOVAL AND INSTALLATION

1. If equipped, turn the air suspension switch to the OFF.

2. Raise and safely support the vehicle.

3. Remove the wheel and tire assembly.

4. Remove the lower control arm assembly.

5. Remove and discard the lower ball joint boot seal.

6. Clamp the lower control arm in a suitable vise.

7. Press out the lower control arm ball joint using Ball Joint Remover D89P-3010-A and Receiving Cup D84P-3395-A4 or equivalents and a suitable U-joint tool or hydraulic press.

To install:

➡**When installing a new lower ball joint, it is advisable that the protective cover be left in place during installation to protect the ball joint seal. It may be necessary to cut off the end of the cover to allow it to pass through the receiving cup.**

8. Install the lower ball joint into the lower control arm using Ball Joint Replacer D89P-3010-B and Receiving Cup D84P-3395-A4 or equivalents and a suitable U-joint tool or hydraulic press.

9. Remove and discard the protective cover. Check the lower ball joint to be sure that it is fully seated in the lower control arm.

10. Inspect the ball joint seal and ensure that it is free of cuts or tears.

11. Install the lower control arm assembly into the vehicle.

12. Install the wheel and tire assembly.

Tighten the lug nuts to 85–105 ft. lbs. (115–142 Nm).

13. Lower the vehicle.

14. If equipped, turn the air suspension switch to the ON position.

15. Check the alignment and adjust if not within specification.

16. Road test the vehicle and check for proper operation.

Lower Control Arm

REMOVAL & INSTALLATION

1. Disconnect the negative battery cable.

2. Turn the air suspension switch to the OFF position which is located in the luggage compartment.

3. Raise and safely support the vehicle.

4. Remove the wheel and tire assembly.

5. Remove the air line from the air spring solenoid valve by pushing down and holding the plastic retaining ring on the valve while pulling on the air line.

6. Remove the air spring solenoid valve as follows:

   a. Remove the solenoid retainer clip.

   b. Rotate the solenoid valve counter-clockwise to the 1st stop.

   c. Pull the solenoid valve straight out slowly to the 2nd stop to bleed air from the system.

➡**Do not fully release the solenoid until the air is completely bled from the air spring.**

   d. After the air is fully bled from the system, rotate the solenoid valve counterclockwise to the 3rd stop and remove the solenoid valve from the solenoid housing. Remove the large O-ring from the solenoid housing.

7. Loosen the lower ball joint retaining nut 3 to 4 turns.

8. Rap on the wheel spindle with a hammer to separate the lower ball joint from the wheel spindle. Leave the nut on the lower ball joint stud.

9. Support the wheel spindle by a length of wire or other means to prevent excessive sagging of the upper control arm.

10. Match mark the position of the camber adjustment cam.

11. Remove the rubber bumper from the lower control arm strut, if equipped.

12. Using a wrench to hold the lower control arm strut, remove the nut retaining the lower control arm strut to the lower control arm and discard the nut.

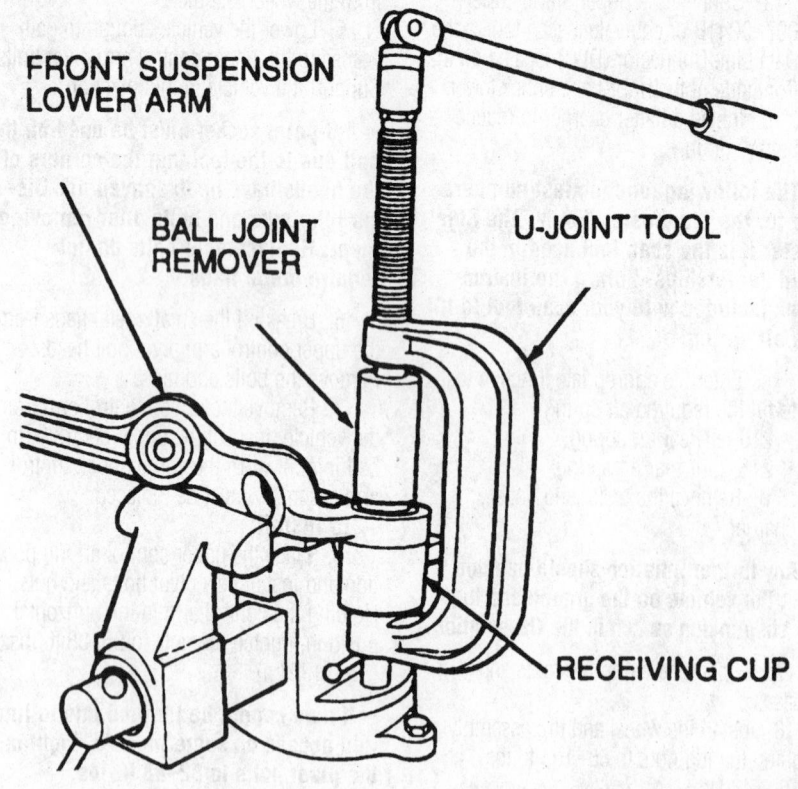

FRONT SUSPENSION LOWER ARM

BALL JOINT REMOVER

U-JOINT TOOL

RECEIVING CUP

7922PG35

The lower ball joint must be pressed out of the control arm using the appropriate tools as shown

✳✳ CAUTION

Do not hold the lower control arm strut with a wrench on any curved surface of the strut. Use only the flat spots, as damage to the tension strut may result.

13. Verify that all air is exhausted from the air spring, then remove the lower air spring/shock assembly retaining bolt and nut.

14. Remove the pivot (camber) bolt and nut.

15. Remove the nut from the lower ball joint stud and remove the lower control arm.

To install:

16. Install the air spring solenoid valve as follows:

    a. Check the solenoid valve O-rings for cuts or abrasion. Replace the O-rings as required. Lightly grease the O-ring area of the solenoid valve and the larger solenoid housing O-ring with silicone dielectric compound.

    b. Insert the solenoid valve into the air spring end cap and rotate clockwise to the 3rd stop, push in to the 2nd stop, then rotate clockwise to the 1st stop.

    c. Install the air spring solenoid retainer clip. Inspect the electrical harness connector and ensure the rubber gasket is in place at the bottom of the connector cavity.

17. Connect the air line and the electrical harness connector to the air spring solenoid valve.

18. Place the lower control arm into the vehicle and loosely install the pivot (camber) bolt and nut.

19. Place the lower control arm strut to the lower control arm and install the washer, bushings and retaining nut. Do not tighten the nut at this time.

20. Install the lower control arm ball joint nut loosely.

21. Install the lower air spring/shock assembly retaining bolt and nut to the lower control arm. Do not tighten at this time.

22. Tighten the lower ball joint retaining nut to 83–113 ft. lbs. (113–153 Nm).

23. Tighten the lower control arm strut retaining nut to 89–118 ft. lbs. (120–160 Nm).

24. Install the rubber bumper to the lower control arm strut, if equipped.

25. Remove the wire or other type of support holding the wheel spindle.

26. Verify that the air suspension switch is OFF, then connect the negative battery cable.

27. Fill the air spring as follows:

    a. Connect Super Star II Tester 007-0041B or equivalent, to the Data Link

Connector (DLC) located on the front side of the right-hand shock tower.

    b. Install a batter charger to reduce battery drain.

    c. Enter the appropriate function test to fill the required air spring:
    • 212 left-front air spring
    • 214 right-front air spring
    • 216 left-rear air spring
    • 218 right-rear air spring

    d. Remove the tester and battery charger.

➡ **Any further inflation should be done with the vehicle on the ground and the air suspension switch in the ON position.**

28. Inspect the air spring folds and creases.

29. Install the wheel and tire assembly. Torque the lug nuts to 85–105 ft. lbs. (115–142 Nm).

30. Lower the vehicle.

31. Turn the air suspension switch to the ON position.

32. Place the vehicle on an alignment rack or equivalent, to raise the vehicle off of the ground but to keep the suspension loaded.

33. Neutralize the front suspension bushings by pushing the front of the vehicle down several times and allowing the suspension to settle. Torque the lower air spring and shock assembly retaining bolt and nut to 125–170 ft. lbs. (170–230 Nm).

34. Align the camber match marks at the pivot bolt. Using a torque wrench, tighten the retaining nut to 82–114 ft. lbs. (110–155 Nm).

35. Check the alignment and adjust if not within specification.

36. Road test the vehicle and check for proper operation.

Wheel Bearings

ADJUSTMENT

The wheel bearings are not adjustable. If the bearings feel rough or make noise, they must be replaced.

REMOVAL & INSTALLATION

Front

➡ **Before continuing with this procedure, be sure to have available, a new wheel hub retainer nut, two wheel spindle-to-strut nuts and bolts, one lower ball joint stud nut and one wheel hub grease cap, per side. Once these**

parts are removed, they lose their torque holding ability or retention capability and must not be reused.

✳✳ CAUTION

The air spring must be completely drained of air before disassembling the suspension components.

1. Raise and safely support the vehicle.

2. Place supports under both frame rails just behind the lower control arms.

3. Remove the wheel and tire assembly.

4. Disconnect the air spring solenoid electrical harness connector.

5. Remove the air line from the air spring solenoid valve by pushing down and holding the plastic retaining ring on the valve while pulling on the air line.

6. Remove the air spring solenoid valve as follows:

    a. Remove the solenoid retainer clip.

    b. Rotate the solenoid valve counterclockwise to the first stop.

    c. Pull the solenoid valve straight out slowly to the second stop to bleed air from the system.

➡ **Do not fully release the solenoid until the air is completely bled from the air spring.**

    d. After the air is fully bled from the system, rotate the solenoid valve counterclockwise to the third stop and remove the solenoid valve from the solenoid housing. Remove the large O-ring from the solenoid housing.

7. Remove the anti-lock speed sensor from the wheel spindle and move aside.

8. Remove and discard the hub cap grease seal.

9. Remove the two bolts from the disc brake caliper and caliper anchor plate and remove as an assembly. Support the assembly with wire. Do not allow the caliper assembly to hang by the brake hose.

10. Matchmark the brake rotor to one of the lug bolts and remove the disc brake rotor.

11. Remove the disc brake rotor dust shield.

12. Remove and discard the wheel hub retainer nut.

➡ **The front wheel bearings are part of the wheel hub and cannot be replaced separately.**

13. Remove the wheel hub and bearing assembly. If necessary, use Front Hub Remover T81P-1104-C or equivalent, to remove the wheel hub from the spindle.

14. Remove the sway bar link retaining

nut at the wheel spindle. Separate the sway bar link from the wheel spindle using Ball Joint Remover D88L-3006-A, or equivalent.

15. Remove the tie rod end cotter pin and castellated nut. Separate the tie rod end from the wheel spindle using Tie Rod End Remover T-3290-D, or equivalent.

16. Loosen the lower ball joint retaining nut one or two turns. Do not remove the nut at this time.

17. Tap the wheel spindle boss sharply to separate the lower ball joint stud from the wheel spindle. Remove and discard the lower ball joint retaining nut.

18. Remove and discard the upper control arm-to-wheel spindle pinch bolt and nut.

19. Spread the slot in the wheel spindle slightly and remove the wheel spindle from the upper ball joint stud.

20. Remove the wheel spindle from the vehicle.

To install:

21. Install the air spring solenoid valve as follows:

  a. Check the solenoid valve O-rings for cuts or abrasion. Replace the O-rings as required. Lightly grease the O-ring area of the solenoid valve and the larger solenoid housing O-ring with silicone dielectric compound.

  b. Insert the solenoid valve into the air spring end cap and rotate clockwise to the 3rd stop, push in to the 2nd stop, then rotate clockwise to the 1st stop.

  c. Install the air spring solenoid retainer clip. Inspect the electrical harness connector and ensure the rubber gasket is in place at the bottom of the connector cavity.

22. Connect the air line and the electrical harness connector to the air spring solenoid valve.

23. Place the wheel spindle in position to the lower and upper control arm ball joint studs. Install a new upper ball joint pinch bolt and nut. Tighten the nut to 50–68 ft. lbs. (68–92 Nm). Install a new retaining nut on the lower control arm ball joint stud and tighten to 83–113 ft. lbs. (113–153 Nm).

24. Install the disc brake rotor dust shield.

25. Clean the wheel spindle stem, then install the wheel hub and bearing assembly onto the spindle.

➡️**Any time a wheel hub retainer nut is loosened or removed, it must be replaced with a new nut. Do not use an impact tool to install the new nut.**

26. Install a new wheel hub retainer nut and tighten to 188–254 ft. lbs. (255–345 Nm).

27. Install the disc brake rotor. Check the brake rotor runout to verify proper installation.

28. Install a new hub cap grease seal.

29. Install the disc brake caliper and anchor plate assembly to the wheel spindle. Install the two anchor plate retaining bolts and tighten to 88 ft. lbs. (119 Nm).

30. Install the anti-lock speed sensor and mounting bolt. Tighten the bolt to 40–60 inch lbs. (4.5–6.8 Nm).

31. Connect the sway bar to the wheel spindle and install the retaining nut. Tighten the nut to 39–53 ft. lbs. (53–72 Nm).

32. Attach the tie rod end stud to the wheel spindle. Install the castellated nut and tighten to 39 ft. lbs. (53 Nm). Continue tightening the nut to align the next castellation with the hole in the stud. Install a new cotter pin.

33. Verify that the air suspension switch if OFF, then connect the negative battery cable.

34. Fill the air spring as follows:

  a. Connect Super Star II Tester 007-0041B or equivalent to the data link connector (DLC) located on the front side of the right-hand shock tower.

  b. Install a batter charger to reduce battery drain.

  c. Enter the appropriate function test to fill the required air spring:
  - 212 left-front air spring
  - 214 right-front air spring
  - 216 left-rear air spring
  - 218 right-rear air spring

  d. Remove the tester and battery charger.

➡️**Any further inflation should be done with the vehicle on the ground and the air suspension switch in the ON position.**

35. Inspect the air spring folds and creases.

36. Install the wheel and tire assembly. Using a torque wrench, tighten the lug nuts in a star pattern, to 85–105 ft. lbs. (115–142 Nm).

37. Lower the vehicle.

38. Turn the air suspension switch to the ON position.

39. Pump the brake pedal several times to position the brake pads.

40. Check the alignment and adjust if not within specification.

41. Road test the vehicle and check for proper operation.

HUB AND BEARING ASSY

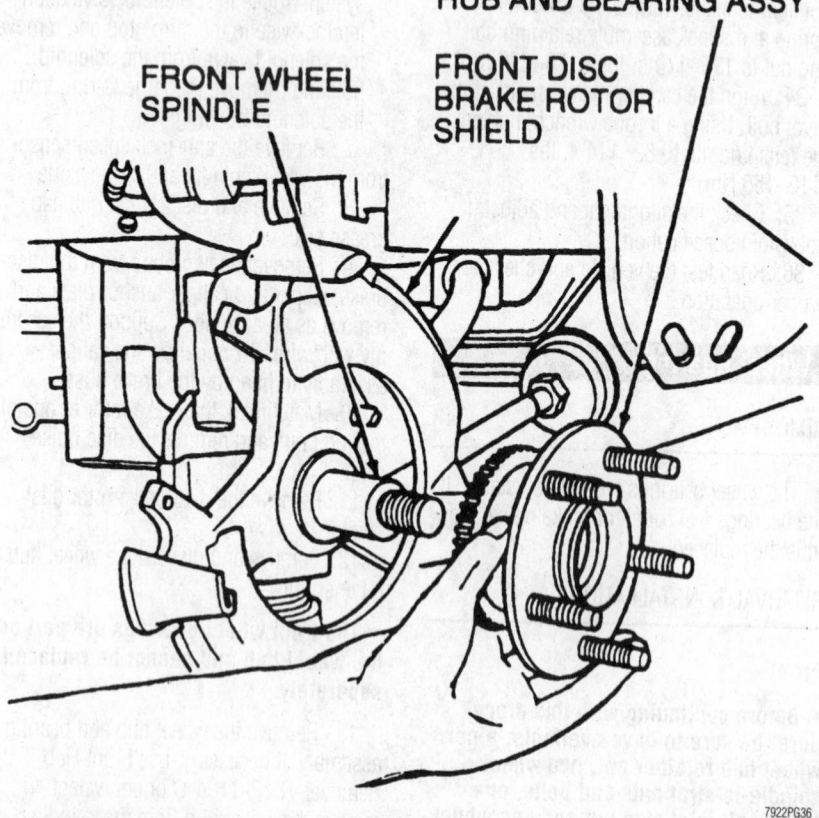

FRONT WHEEL SPINDLE

FRONT DISC BRAKE ROTOR SHIELD

7922PG36

Front hub/bearing assembly

Rear

➡ **Before continuing with this procedure, be sure to have available, one new wheel hub retainer nut, one inboard halfshaft circlip and one inboard CV-joint stub shaft seal, per side. Once removed, these parts lose their torque holding ability or retention capability and must not be reused.**

1. Turn the air suspension switch, located in the luggage compartment, to the OFF position.

2. Remove the hub cap grease seal.

3. With all wheels on the ground, remove and discard the wheel hub retainer nut.

4. Raise and safely support the vehicle.

5. Remove the wheel and tire assembly.

6. Use needle-nose pliers to slide the parking brake cable adjusting clip downward until the cable is free.

7. Remove the parking brake cable from the rear disc brake caliper.

8. Remove the disc brake caliper from the brake rotor, leaving the brake hose connected. Wire the brake caliper aside. Do not let it hang by the brake hose.

9. Remove the brake rotor.

10. Remove the upper control arm nut and bolt at the wheel knuckle. Wire the upper control arm to the body to prevent damage to the CV-joint boots when the knuckle and hub assembly is removed.

11. Mark the position of the control arm in relation to the knuckle with the bushings in the relaxed position. When the upper control arm bolt is removed from the knuckle, the lower arm bushings will return to the relaxed position.

➡ **Failure to mark the position will cause bushing wind up at assembly resulting in improper ride height. This can cause incorrect alignment and tire wear.**

12. Install Hub Remover/Replacer T81P-1104-C or equivalent to the hub studs and turn the tool shaft until the halfshaft is free in the hub.

13. Remove the lower control arm-to-knuckle retaining bolts and nuts and remove the knuckle assembly from the halfshaft.

14. Remove and save the washers used between the lower control arm bushings and the wheel knuckle for installation.

15. Place the knuckle and hub assembly in a suitable vise.

16. Wire the halfshaft to a suitable location to prevent damage to the CV-joint boot.

17. Place a suitable two or three jaw puller and Step Plate Adapter D80L-630-5 or equivalent, on the rear wheel knuckle and press out the hub from the wheel knuckle and bearing assembly.

18. Remove the disc brake backing plate from the wheel knuckle.

19. Remove the snapring retaining the wheel bearing assembly.

20. Place the wheel knuckle and bearing assembly on a press bed using Hub Support T89P-1104-A and Front Hub Bearing Remover/Replacer T83P-1104-AH2 or equivalents and press the wheel bearing assembly from the wheel knuckle.

21. Inspect all components for wear and/or damage and replace as needed.

To install:

22. Place the wheel knuckle on the press bed and position the wheel bearing assembly in the wheel knuckle bore. Press the wheel bearing assembly into the knuckle bore using Bearing Replacer T86P-1104-A3 or equivalent. Be sure the bearing assembly is fully seated.

23. Install the wheel bearing retainer snapring.

24. Place the disc brake backing plate in position on the wheel knuckle and install the three retaining bolts. Tighten the bolts to 22-37 ft. lbs. (30-50 Nm).

✳✳ WARNING

Before proceeding with the following step, the wheel bearing must be supported or damage will result.

25. Place the wheel knuckle on Hub Support T89P-1104-A or equivalent and place on a suitable press bed plate. Place the wheel hub in position with the wheel bearing and knuckle assembly and press into position using Front Hub Bearing Remover/Replacer T83P-1104-AH2 or equivalent. Be sure the inner bearing race is properly supported or the bearing will be damaged.

26. Using a hammer and chisel, drive the bearing dust seal from the outer CV-joint. Using Axle Seal Replacer T83T-3132-A1 and Bearing Dust Seal Replacer T89P-1249-A or equivalents, install a new dust seal, making sure the seal flange faces out toward the bearing.

27. Remove the wire holding the halfshaft.

28. Engage the wheel knuckle and hub assembly on the halfshaft splines and install the rear lower bolt, washer and nut, finger-tight.

29. Install the washers between the front and rear lower control arm/wheel knuckle

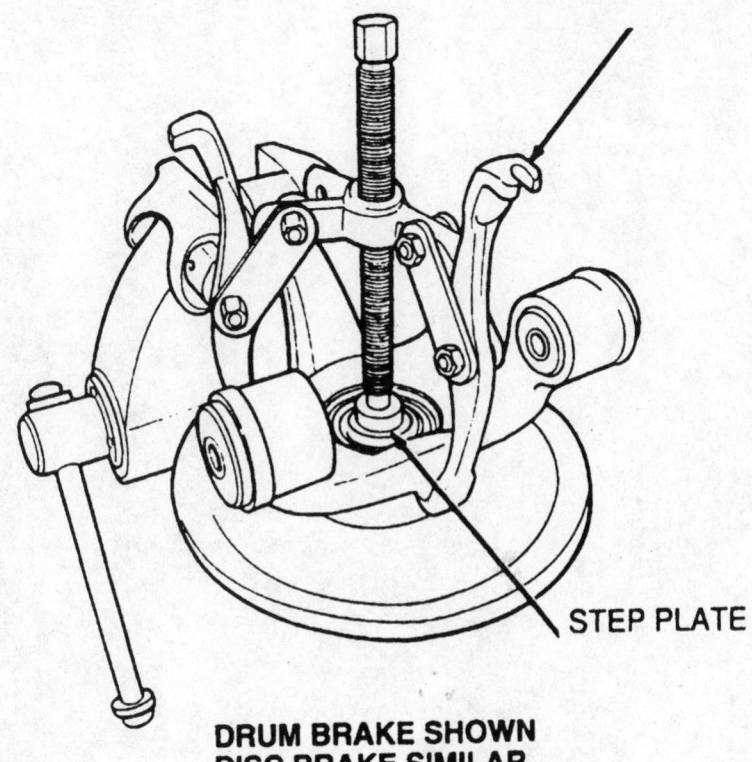

2/3-JAW PULLER

STEP PLATE

**DRUM BRAKE SHOWN
DISC BRAKE SIMILAR**

7922PG37

Use a two or three jaw puller to bearing out of the hub assembly as shown

locations and install the bolts with the bolt heads facing the rear of the vehicle.

30. Push the halfshaft inward through the hub assembly while placing the wheel knuckle in position to install the front lower bolt, washer and nut.

31. Align the lower control arm to wheel knuckle matchmarks made during removal. If a new wheel knuckle is being installed, set the knuckle at the approximate angle noted during removal. Tighten the lower control arm-to-wheel knuckle retaining bolts to 94–131 ft. lbs. (128–178 Nm) using Holding Tool T93P-5494-A, or equivalent.

32. Install the brake rotor.

33. Install a new wheel hub retainer nut and hand-tighten only.

34. Install the upper control arm retaining bolt and nut. Tighten the nut to 117–142 ft. lbs. (158–193 Nm).

35. Install the disc brake caliper over the brake rotor with the outer brake pad against the brake rotor, to prevent pinching the caliper piston boot between the inner pad and piston. Install the two disc brake caliper retaining bolts and tighten to 64–88 ft. lbs. (87–119 Nm).

36. Connect the parking brake rear cable to the disc brake caliper and install the cable adjustment clip.

37. Install the wheel and tire assembly. Using a torque wrench, tighten the lug nuts to 85–105 ft. lbs. (115–142 Nm).

38. Lower the vehicle.

39. Turn the air suspension switch to the ON position.

40. Apply the parking brake several times and adjust if necessary.

41. With all wheels on the ground, tighten the wheel hub retainer nut to 188–254 ft. lbs. (255–345 Nm).

42. Install the hub cap grease seal.

43. Pump the brake pedal several times to position the brake pads before moving the vehicle.

44. Check the alignment and adjust if not within specification.

45. Road test the vehicle and check for proper operation.

FORD MOTOR CO.

25

Ford-Thunderbird • Mercury-Cougar

DRIVE TRAIN 25-28
ENGINE REPAIR 25-2
FUEL SYSTEM 25-27
STEERING AND
SUSPENSION 25-31

A
Air Bag 25-31
 Disarming 25-32
 Precautions 25-31

C
Camshaft and Valve Lifters 25-18
 Removal & Installation 25-18
Clutch 25-30
 Removal & Installation 25-30
Coil Spring 25-34
 Removal & Installation 25-34
Cylinder Head 25-5
 Removal & Installation 25-5

D
Distributor 25-2
 Installation 25-2
 Removal 25-2

E
Engine Assembly 25-3
 Removal & Installation 25-3
Exhaust Manifold 25-16
 Removal & Installation 25-16

F
Fuel Filter 25-28
 Removal & Installation 25-28

Fuel Pump 25-28
 Removal & Installation 25-28
Fuel System Pressure 25-28
 Relieving 25-28
Fuel System Service
 Precautions 25-27

H
Halfshaft 25-31
 Removal & Installation 25-31
Hydraulic Clutch System 25-31
 Bleeding 25-31

I
Ignition Timing 25-2
 Adjustment 25-2
Intake Manifold 25-12
 Removal & Installation 25-12

L
Lower Ball Joint 25-36
 Removal & Installation 25-36

O
Oil Pan 25-20
 Removal & Installation 25-20
Oil Pump 25-22
 Removal & Installation 25-22

P
Power Rack and Pinion
 Steering Gear 25-32
 Removal & Installation 25-32

R
Rear Main Seal 25-23
 Removal & Installation 25-23
Rocker Arms 25-10
 Removal & Installation 25-10

S
Shock Absorber 25-33
 Removal & Installation 25-33
Supercharger 25-11
 Removal & Installation 25-11

T
Timing Chain, Sprockets,
 Front Cover and Seal 25-23
 Removal & Installation 25-23
Transmission Assembly 25-28
 Removal & Installation 25-28

U
Upper Ball Joint 25-36
 Removal & Installation 25-36

V
Valve Lash 25-19
 Adjustment 25-19

W
Water Pump 25-5
 Removal & Installation 25-5
Wheel Hub and Bearings 25-37
 Adjustment 25-37
 Removal & Installation 25-37

ENGINE REPAIR

→Disconnecting the negative battery cable on some vehicles may interfere with the functions of the on board computer systems and may require the computer to undergo a relearning process, once the negative battery cable is reconnected.

Distributor

REMOVAL

1. Disconnect the negative battery cable.
2. Mark the position of the No. 1 cylinder wire tower on the distributor base.

→This reference is necessary in case the engine is disturbed while the distributor is removed.

3. Remove the distributor cap and position the cap and ignition wires to the side. Disconnect the wiring harness plug from the distributor connector.
4. Scribe a mark on the distributor body to indicate the position of the rotor tip. Scribe a mark on the distributor housing and engine block or timing cover to indicate the position of the distributor in the engine.
5. Remove the hold-down bolt and clamp located at the base of the distributor. Remove the distributor from the engine. Note the direction the rotor tip points if it moves from the No. 1 position when the drive gear disengages. For reinstallation purposes, the rotor should be at this point to insure proper gear mesh and timing.
6. Cover the distributor opening in the engine to prevent the entry of dirt or foreign material.
7. Avoid turning the crankshaft, if possible, while the distributor is removed. If the engine is disturbed, the No. 1 cylinder piston will have to be brought to TDC on the compression stroke before the distributor is installed.

INSTALLATION

→Before installing, visually inspect the distributor. The drive gear should be free of nicks, cracks and excessive wear. The distributor driveshaft should move freely, without binding. The O-ring should fit tightly and be free of cuts.

Timing Not Disturbed

1. Position the distributor in the engine, aligning the rotor and distributor housing with the marks that were made during removal. If the distributor does not fully seat in the engine block or timing cover, it may be because the distributor is not engaging properly with the oil pump intermediate shaft. Remove the distributor, using a suitable tool, turn the intermediate shaft until the distributor will seat properly.
2. Install the hold-down clamp and bolt. Snug the mounting bolt so the distributor can be turned for ignition timing purposes.
3. Install the distributor cap and connect the distributor to the wiring harness.
4. Connect the negative battery cable. Check, if necessary, adjust the ignition timing.
5. After the timing has been set, tighten the distributor hold-down clamp bolt to15–22 ft. lbs. (20–30 Nm).
6. Recheck the ignition timing after tightening the bolt.

Timing Disturbed

1. Disconnect the No. 1 cylinder spark plug wire and remove the No. 1 cylinder spark plug.
2. Place a finger over the spark plug hole and crank the engine slowly until compression is felt.
3. Align the TDC mark on the crankshaft pulley with the pointer on the timing cover. This places the piston in No. 1 cylinder at TDC on the compression stroke.
4. Turn the distributor shaft until the rotor points to the distributor cap No. 1 spark plug tower, as marked during the removal procedure.
5. Install the distributor in the engine, aligning the rotor and distributor housing with the marks that were made during removal. If the distributor does not fully seat in the engine block or timing cover, it may be because the distributor is not engaging properly with the oil pump intermediate shaft. Remove the distributor, using a suitable tool, turn the intermediate shaft until the distributor will seat properly.
6. Install the hold-down clamp and bolt. Snug the mounting bolt so the distributor can be turned for ignition timing purposes.
7. Install the No. 1 cylinder spark plug and connect the spark plug wire. Install the distributor cap and connect the distributor to the wiring harness.

8. Connect the negative battery cable and set the ignition timing.
9. After the timing has been set, tighten the distributor hold-down clamp bolt to15–22 ft. lbs. (20–30 Nm).
10. Recheck the ignition timing after tightening the bolt.

Ignition Timing

ADJUSTMENT

→Always refer to the Vehicle Emission Control Information (VECI) label to verify the timing adjustment procedure.

Distributorless Ignition System

Base timing for distributorless engines is set from the factory at 10 degrees BTDC and is not adjustable.

Distributor Ignition System

1. Locate the timing marks and pointer on the crankshaft pulley and the timing cover. Clean the marks so they will be visible with a timing light. Apply chalk or bright-colored paint, if necessary.
2. Place the transaxle in **P** or **N**. The air conditioning and heater controls should be in the **OFF** position.
3. Connect a suitable tachometer and inductive timing light according to the manufacturer's instructions.

→The tachometer can be connected to the ignition coil without removing the coil connector. Insert an alligator clip into the back of the connector, onto the dark green/yellow dotted wire. Do not let the clip accidentally ground to a metal surface as it may permanently damage the coil.

4. Disconnect the single wire inline SPOUT connector or remove the shorting bar from the double wire SPOUT connector.
5. Start the engine and allow it to warm up to operating temperature.

→To set timing correctly, a remote starter should not be used. Use the ignition key only to start the vehicle. Disconnecting the start wire at the starter relay will cause the TFI module to revert to start mode timing after the vehicle is started. Reconnecting the start wire after the vehicle is running will not correct the timing.

6. With the engine at the timing rpm specified, check the initial timing by aiming the timing light at the timing marks and pointer. Refer to the underhood Vehicle Emission Information Label for specifications.

7. If the marks align, shut **OFF** the engine and proceed to Step 8. If the marks do not align, shut **OFF** the engine and loosen the distributor hold-down clamp bolt. Start the engine, aim the timing light and turn the distributor until the timing marks align. Shut **OFF** the engine and tighten the distributor hold-down clamp bolt. Recheck the timing after the bolt has been tightened.

8. Reconnect the single wire inline SPOUT connector or reinstall the shorting bar on the double wire SPOUT connector. Check the timing advance to verify the distributor is advancing beyond the initial setting.

9. Remove the inductive timing light and tachometer.

Engine Assembly

REMOVAL & INSTALLATION

3.8L Engines

❊❊ CAUTION

Never open, service or drain the radiator or cooling system when hot; serious burns can occur from the steam and hot coolant. Also, when draining engine coolant, keep in mind that cats and dogs are attracted to ethylene glycol antifreeze and could drink any that is left in an uncovered container or in puddles on the ground. This will prove fatal in sufficient quantities. Always drain coolant into a sealable container. Coolant should be reused unless it is contaminated or is several years old.

1. Disconnect the negative battery cable.
2. Drain the crankcase and the cooling system.

❊❊ CAUTION

Observe all applicable safety precautions when working around fuel. Whenever servicing the fuel system, always work in a well ventilated area. Do not allow fuel spray or vapors to come in contact with a spark or open flame. Keep a dry chemical fire extinguisher near the work area. Always keep fuel in a container specifically designed for fuel storage; also, always properly seal fuel containers to avoid the possibility of fire or explosion.

3. Relieve the fuel system pressure and have the A/C system discharged by an EPA certified technician.
4. Disconnect the electrical connector to the underhood lamp. Mark the position of the hood on the hinges and remove the hood.
5. Remove the left cowl vent screen and wiper module. On non-supercharged engines, disconnect the alternator to voltage regulator wiring assembly.
6. On supercharged engines, remove the upper charge air cooler tube at the supercharger and cooler assemblies. Remove the bolt retaining the cooler tube to the alternator bracket and remove the tube.
7. Remove the radiator upper sight shield. Release the belt tension and remove the drive and accessory/supercharger belts. Remove the air cleaner-to-throttle body tube.
8. On supercharged engines, disconnect the cooling fan and remove the cooling fan/shroud assembly. On non-supercharged engines, remove the fan and shroud.
9. Remove the upper radiator hose and disconnect the heater hoses. If equipped with an automatic transmission, disconnect the oil cooler lines from the radiator.
10. Disconnect the lower radiator hose at the water pump. Remove the radiator. On supercharged engines it will also be necessary to remove the two push pins retaining the charge air cooler to the radiator assembly.
11. Disconnect the power steering pressure hose assembly. On non-supercharged engines, remove the power steering pump and bracket assembly and position aside.
12. Disconnect the air conditioner compressor clutch wire. Disconnect and plug the refrigerant lines. Remove the compressor.
13. Remove the coolant recovery reservoir and remove the wiring shield. Remove the accelerator cable mounting bracket and position aside.
14. Disconnect the fuel lines from the fuel rail. Tag and disconnect the engine control module (PCM) wiring, engine feed harnesses and vacuum hoses.
15. On non-supercharged engines, disconnect the ground and coil wires. On supercharged engines, disconnect the DIS module wiring, remove the coil pack retaining bolts and position the coil pack aside.

16. On supercharged engines, remove the nuts retaining the lower charge air cooler tube to the supercharger elbow and lower charge air cooler tube bracket and remove the charge air cooler tube retaining bolt and nut at the alternator bracket.
17. On supercharged engines, remove the alternator bracket bolts, disconnect the alternator wiring and remove the alternator. Remove the power steering pump bracket assembly and position aside.
18. Disconnect the canister purge line and disconnect one end of the throttle control valve cable.
19. Raise and safely support the vehicle. Remove the oil filter element.
20. On supercharged engines, remove the two nuts retaining the lower charge air cooler tube to the charge air cooler and remove the charge air cooler and charge air cooler tube.
21. Remove the exhaust pipe-to-manifold nuts and remove the left exhaust shield. Disconnect the oxygen sensors.
22. If equipped with an automatic transmission, remove the inspection plug and remove the torque converter bolts.
23. Remove the engine-to-transmission bolts and remove the engine mount through-bolts. On supercharged engines, remove the left mount retaining strap bolt.
24. Remove the crankshaft pulley assembly.

➡ If the crankshaft pulley and vibration damper have to be separated, mark the damper and pulley so they may be reassembled in the same relative position. This is important as the damper and pulley are initially balanced as a unit. If the crankshaft damper is being replaced, check if the original damper has balance pins installed. If so, new balance pins must be installed on the new damper in the same position as the original damper.

25. Remove the starter. Remove the ground cable and remove the left and right starter harness retainers.
26. Disconnect the oil level indicator sensor and partially lower the vehicle. Disconnect the oil pressure sending unit gauge assembly.
27. Position a floor jack under the transmission and position suitable engine lifting equipment.
28. Remove the engine from the vehicle and position on a workstand.

To install:
29. Remove the engine assembly from the workstand and install engine lifting equipment.

30. Position the engine in the vehicle and install the two engine-to-transmission bolts. Lower the engine onto the mounting seats, left side first, and remove the lifting equipment. Remove the jacks.

31. Tighten the two engine-to-transmission bolts to 40–50 ft. lbs. (55–68 Nm) and connect the oil pressure sending unit gauge assembly. Raise and safely support the vehicle.

32. Install the remaining engine-to-transmission bolts and tighten to 40–50 ft. lbs. (55–68 Nm).

33. Install the torque converter bolts and tighten to 20–34 ft. lbs. (27–46 Nm). Install the inspection plug.

34. Install and tighten the engine mount through-bolts to 35–50 ft. lbs. (47–68 Nm). On supercharged engines, install the left mount retaining strap bolt and tighten to 33–45 ft. lbs. (45–61 Nm).

35. Install the starter. Install the starter harness retainer, ground cable and transmission oil cooler line bracket. Install the exhaust pipe-to-manifold nuts.

36. Install the crankshaft pulley assembly and tighten the bolts to 20–28 ft. lbs. (26–30 Nm).

37. Connect the oxygen sensors and the oil level indicator sensor. Install a new oil filter and lower the vehicle.

38. Connect the throttle control valve cable and the canister purge line.

39. On supercharged engines, perform the following:

    a. Install the lower charge air cooler tube, charge air cooler and power steering pump bracket assembly.

    b. Install the alternator, connect the wiring and install the alternator bracket bolts.

    c. Install the charge air cooler tube bolts at the power steering bracket and install the nuts retaining the lower charge air cooler tube to the lower charge air cooler tube bracket and supercharger elbow.

    d. Install the coil pack and retaining bolts.

40. Install the coolant recovery reservoir.

41. Connect the alternator-to-voltage regulator wiring, the engine control module (PCM) wiring assembly and engine feed harnesses. Connect the vacuum hoses.

42. On non-supercharged engines, connect the wiring assembly ground and coil wire. On supercharged engines, connect the DIS module wiring.

43. Connect the fuel lines to the fuel rail. Install the accelerator cable mounting bracket and the wiring shield.

44. Install the air conditioning compressor and retaining bolts. Tighten the bolts to 30–45 ft. lbs. (41–61 Nm).

45. Remove the plugs from the air conditioner compressor lines and connect the lines to the compressor. Connect the compressor clutch wire.

46. On non-supercharged engines, install the power steering pump bracket assembly. Connect the power steering hoses.

47. Install the radiator. On supercharged engines, install the charge air cooler to the radiator and install the retaining push pins.

48. Connect the lower radiator hose to the water pump and install the heater hoses. If equipped with an automatic transmission, install the oil cooler lines to the radiator.

49. Install the upper radiator hose and the fan and fan shroud. On supercharged engines, connect the cooling fan electrical connector.

50. Position the drive belts and the accessory/supercharger belts. Install the radiator sight shield.

51. On supercharged engines, install the charge air cooler tube and bolts retaining the tube to the power steering bracket. Install the upper charge air cooler tube to the supercharger and cooler assemblies.

52. Install the cowl vent screen and wiper module. Install the hood, aligning the marks that were made during removal. Connect the underhood lamp wiring.

✳✳ WARNING

Operating the engine without the proper amount and type of engine oil will result in severe engine damage.

53. Fill the crankcase with the proper type and quantity of engine oil. Fill and bleed the cooling system.

54. Connect the negative battery cable, start the engine and bring to normal operating temperature. Check for leaks. Check all fluid levels.

55. Leak test, evacuate and charge the air conditioning system. Observe all safety precautions.

4.6L Engine

✳✳ CAUTION

Never open, service or drain the radiator or cooling system when hot; serious burns can occur from the steam and hot coolant. Also, when draining engine coolant, keep in mind that cats and dogs are attracted to ethylene glycol antifreeze and could drink any that is left in an uncovered container or in puddles on the ground. This will prove fatal in sufficient quantities. Always drain coolant into a sealable container. Coolant should be reused unless it is contaminated or is several years old.

1. Disconnect the negative battery cable, drain the cooling system and have the A/C system discharged by an EPA approved technician.

2. Remove the air cleaner outlet tube and air cleaner assembly.

3. Remove the fan blade and shroud.

✳✳ CAUTION

Observe all applicable safety precautions when working around fuel. Whenever servicing the fuel system, always work in a well ventilated area. Do not allow fuel spray or vapors to come in contact with a spark or open flame. Keep a dry chemical fire extinguisher near the work area. Always keep fuel in a container specifically designed for fuel storage; also, always properly seal fuel containers to avoid the possibility of fire or explosion.

4. Relieve the fuel system pressure.

5. Disconnect the 42-pin and 8-pin connectors and position out of the way.

6. Label all components for reassembly. Disconnect the accelerator cable, speed control actuator, throttle valve control cable, canister purge electrical connector and vacuum lines.

7. Disconnect the power supply from the power distribution box and starter relay.

8. Disconnect the transmission oil cooler tubes from the transmission, upper radiator hose and heater hoses.

9. Disconnect the engine-to-frame ground straps. Partially raise the vehicle and support safely. Remove the front wheels.

10. Disconnect the right and left front anti-lock sensor and brackets.

11. Remove the right and left disc brake caliper bolts. Remove the calipers and support with wire to a frame member.

12. Disconnect the right and left front suspension upper arms from the spindles.

13. Disconnect the front springs and shocks from the lower arms.

14. Raise the vehicle and support safely.

✳✳ CAUTION

The EPA warns that prolonged contact with used engine oil may cause a number of skin disorders, including cancer! You should make every effort to minimize your exposure to used engine oil. Protective gloves should be worn when changing the oil. Wash your hands and any other exposed skin areas as soon as possible after exposure to used engine oil. Soap and water, or waterless hand cleaner should be used.

15. Drain the engine oil.

16. Disconnect the dual converter Y-pipe from the manifolds. Disconnect the transmission shift cable and bracket.

17. Index the driveshaft centering socket yoke to the rear axle universal joint flange.

18. Remove the four bolts connecting the driveshaft centering socket yoke to the rear axle universal joint flange. Support the rear axle assembly with a jackstand.

19. Remove the rear axle assembly to rear sub-frame bolts. Loosen the rear differential bracket-to-body bolts and lower.

20. Slide the driveshaft rearward until it is free of the extension housing.

21. Remove the lower radiator hose, power steering lines and steering oil cooler.

22. Disconnect the wiring connector at the bulkhead. Support the front sub-frame using Rotunda Powertrain Lift 014–00765 and Adapter 014–00341 or equivalent.

23. Remove the rear engine support insulator bolts and disconnect the steering coupling at the pinch bolt joint.

24. Remove the eight sub-frame bolts and lower the engine/transmission assembly.

25. Label and remove all needed components. Separate the engine from the transmission and place the engine on a suitable workstand.

To install:

26. Install engine lifting brackets 014–00334. Install the transmission to the engine. Be sure the torque converter studs align with the holes in the flywheel. Tighten the bell housing retaining bolts to 30–44 ft. lbs. (40–60 Nm). Tighten the torque converter bolts to 25 ft. lbs. (35 Nm).

27. Install the transmission housing cover, starter motor and transmission-to-engine block brackets.

28. Position the transmission oil cooler tube bracket to the transmission case.

29. Raise the engine/transmission assembly and carefully lower onto the front sub-frame.

30. Install the right and left front engine support insulator through-bolts. Tighten the bolts to 22 ft. lbs. (30 Nm).

31. Install the power steering lines and remove the engine lift brackets from the cylinder heads.

32. Raise the engine/transmission/sub-frame assembly using the Powertrain Lift 014–00765 and Adapter 014–00341 into the vehicle.

33. Align the sub-frame-to-body and install the bolts. Tighten bolts to 100 ft. lbs. (130 Nm).

34. Connect the steering pinch joint and driveshaft. Raise the rear axle assembly into the vehicle and tighten the rear sub-frame bolts to 89 ft. lbs. (120 Nm). Tighten the two rear axle differential insulator nuts to 94 ft. lbs. (127 Nm).

35. Align the driveshaft centering socket yoke to the rear axle universal joint flange and install the four retaining bolts and tighten to 162 ft. lbs. (220 Nm).

36. Connect the transmission shift linkage, front suspension arms and lower radiator hose.

37. Connect the power steering hoses to steering cooler, ground straps and dual converter Y-pipe.

38. Connect the front suspension upper arms to the spindles and tighten to 98 ft. lbs. (92 Nm).

39. Connect the anti-lock sensor and install the calipers. Install the front wheels.

40. Lower the vehicle and connect the A/C lines, transmission cooler lines and radiator hoses.

41. Connect all remaining hoses, lines, electrical connectors and cables.

42. Install the fan/ shroud and air cleaner.

43. Refill the cooling system. Evacuate, recharge and leak test the A/C system.

✳✳ WARNING

Operating the engine without the proper amount and type of engine oil will result in severe engine damage.

44. Connect the negative battery cable, fill the engine with oil, start the engine and check for leaks.

Water Pump

REMOVAL & INSTALLATION

3.8L Engines

✳✳ CAUTION

Never open, service or drain the radiator or cooling system when hot; serious burns can occur from the steam and hot coolant. Also, when draining engine coolant, keep in mind that cats and dogs are attracted to ethylene glycol antifreeze and could drink any that is left in an uncovered container or in puddles on the ground. This will prove fatal in sufficient quantities. Always drain coolant into a sealable container. Coolant should be reused unless it is contaminated or is several years old.

1. Disconnect the negative battery cable and drain the cooling system.

2. On all except supercharged engine, remove the fan/clutch assembly and shroud.

3. Rotate the main accessory drive belt tensioner. Remove the main drive belt and water pump pulley.

4. Remove the power steering pump pulley and remove the water pump to power steering pump brace.

➡**On supercharged engines, it may be necessary to remove the charge air cooler to gain access to the power steering pump pulley.**

5. On all except supercharged engine, disconnect the coolant bypass hose(s) and the heater hose at the water pump. On supercharged engine, disconnect the oil cooler tube and bypass hose and remove the upper crankshaft sensor cover.

6. Disconnect the lower radiator hose. Remove the water pump retaining bolts and the pump. If a prybar is used to assist removal, be careful not to damage the mating surfaces.

7. Installation is the reverse of the removal procedure. Clean all gasket mating surfaces prior to installation. Tighten the water pump retaining bolts to 15–22 ft. lbs. (20–30 Nm). Fill and bleed the cooling system. Operate the engine until normal operating temperatures have been reached and check for leaks.

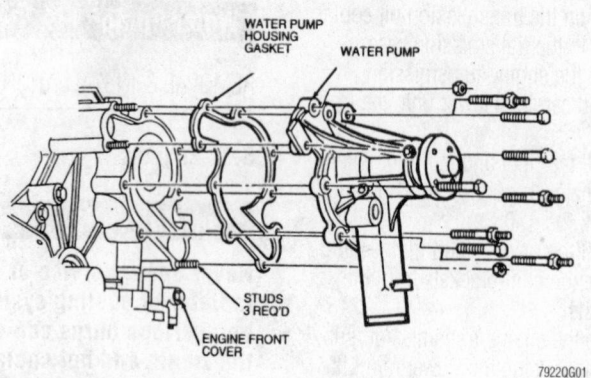

Exploded view of the water pump mounting—3.8L engines

➡The threads of the No. 1 water pump retaining bolt must be coated with pipe sealant before installing.

4.6L Engine

✳✳ CAUTION

Never open, service or drain the radiator or cooling system when hot; serious burns can occur from the steam and hot coolant. Also, when draining engine coolant, keep in mind that cats and dogs are attracted to ethylene glycol antifreeze and could drink any that is left in an uncovered container or in puddles on the ground. This will prove fatal in sufficient quantities. Always drain coolant into a sealable container. Coolant should be reused unless it is contaminated or is several years old.

1. Disconnect the negative battery cable and drain the cooling system.
2. Release the drive belt tensioner and remove the belt.
3. Remove the water pump pulley bolts and pulley.
4. Remove the pump-to-block bolts and pump.
5. Installation is the reverse of the removal procedure. Clean all gasket mating surfaces prior to installation. Tighten the water pump bolts and pulley retaining bolts to 15–22 ft. lbs. (20–30 Nm).

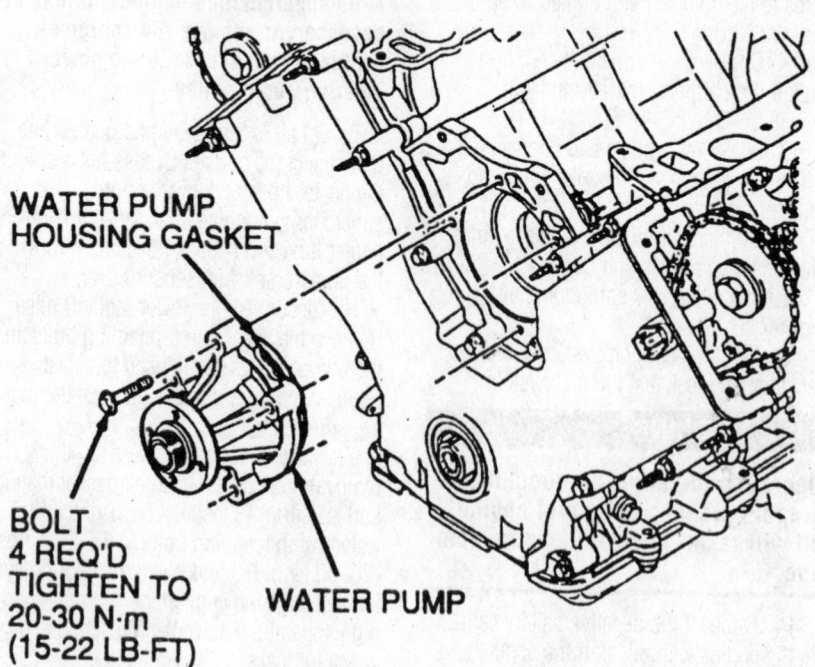

WATER PUMP HOUSING GASKET

BOLT 4 REQ'D TIGHTEN TO 20-30 N·m (15-22 LB-FT)

WATER PUMP

Exploded view of the water pump mounting—4.6L engine

6. Fill and bleed the cooling system. Operate the engine until normal operating temperatures have been reached and check for leaks.

Cylinder Head

REMOVAL & INSTALLATION

3.8L Engines

✳✳ CAUTION

The fuel injection system remains under pressure, even after the engine has been turned OFF. The fuel system pressure must be relieved before disconnecting any fuel lines. Failure to do so may result in fire and/or personal injury.

➡Before continuing with this procedure, be sure to have available eight new cylinder head retaining bolts, per cylinder head. Once removed, these parts lose their torque holding ability or retention capability and must not be reused.

1. Disconnect the negative battery cable.
2. Remove the fuel tank fill cap to relieve the pressure in the fuel tank.
3. Remove the cap from the fuel pressure relief valve located on the fuel injection supply manifold.
4. Attach Fuel Pressure Gauge T80L-9974-B or equivalent, to the relief valve and drain the fuel through the drain tube into a suitable container.
5. After the fuel system pressure is relieved, remove the fuel pressure gauge and install the cap on the relief valve.
6. Install the fuel tank fill cap.

✳✳ CAUTION

Never open, service or drain the radiator or cooling system when hot; serious burns can occur from the steam and hot coolant.

7. Allow the engine to cool, then drain the cooling system.
8. Remove the engine air cleaner outlet tube from the throttle body.
9. Remove the accessory drive belts.
10. If removing the left-hand cylinder head, perform the following:
   a. Remove engine oil fill cap.
   b. Remove the power steering pump front mounting bracket retaining bolts.

c. Remove the alternator and the drive belt idler pulley.

d. Remove the alternator/power steering pump bracket retaining bolts. Leaving the hoses connected, place the power steering pump/alternator bracket assembly aside in a position to prevent the fluid from leaking out.

11. If removing the right-hand cylinder head, perform the following:

a. Remove the accessory drive belt.

b. If equipped with air conditioning, remove the A/C compressor mounting bracket retaining bolts. Leave the hoses connected and move the A/C compressor aside.

c. If not equipped with air conditioning, remove the idler pulley.

d. Remove the PCV valve.

12. Remove the upper intake manifold.

13. Remove both cylinder head covers.

14. Disconnect and plug the fuel supply and return lines at the fuel injection supply manifold.

15. Remove the fuel rail and fuel injectors.

16. Remove the lower intake manifold.

17. Remove the exhaust manifolds.

18. Loosen the rocker arm seat retaining bolts enough to allow the rocker arms to be lifted off the pushrods and rotated aside.

19. Remove the pushrods. Identify the position of each rod for installation reference.

20. Remove eight cylinder head retaining bolts, per head and discard.

21. Remove the cylinder heads and gaskets.

To install:

22. Clean all gasket sealing surfaces.

23. Check the flatness of the cylinder head gasket surface using a straight-edge and a feeler gauge. The allowable warpage is 0.003 inch (0.08mm) in 6.0 inches (152.0mm). Do not machine more than 0.010 inch (0.254mm)

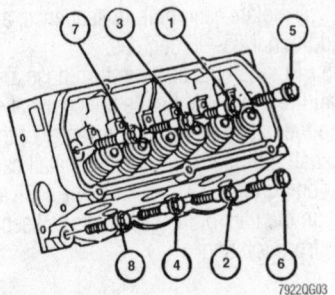

Tighten the cylinder head bolts in the sequence shown to ensure good sealing—3.8L engines

24. Lightly oil the new cylinder head bolt threads except those entering coolant jackets which require pipe sealant applied to the threads.

25. Place new cylinder head gaskets on the cylinder block using the dowels for alignment.

26. Carefully place the cylinder heads on the cylinder block.

27. Install 16 new cylinder head bolts and tighten in sequence, as follows:

a. 15 ft. lbs. (20 Nm)

b. 29 ft. lbs. (40 Nm)

c. 37 ft. lbs. (50 Nm)

28. Loosen each cylinder head bolt, one at a time, 2–3 turns, then retighten as follows:

a. Tighten each long bolt to 11–19 ft. lbs. (15–25 Nm), then rotate an additional 85–95 degrees.

b. Tighten each short bolt to 7–15 ft. lbs. (10–20 Nm), then rotate an additional 85–95 degrees.

c. Continue to the next bolt in sequence.

29. Lubricate each pushrod tip with a suitable engine assembly lubricant.

30. Install the pushrods to their original positions and rotate the rocker arms into position.

31. For each valve, rotate the crankshaft until the valve lifter rests on the heel (base circle) of the camshaft lobe (pushrod all the way down), then tighten the rocker arm fulcrum bolt to 44 inch lbs. (5 Nm). The fulcrum must be fully seated and the pushrod must be seated in rocker arm socket before final tightening.

32. Lubricate the rocker arm assemblies with a suitable engine assembly lubricant. Tighten the rocker arm fulcrum bolts to 19–25 ft. lbs. (25–35 Nm). Final tightening can be done with the camshaft in any position.

➡ If the original valvetrain components are being installed, a valve clearance check is not required. If a component has been replaced, perform a valve clearance check.

33. Install the exhaust manifolds.

34. Install the lower intake manifold using new gaskets.

35. Install the fuel injection supply manifold and connect the fuel supply and return lines.

36. Install the cylinder head covers using new gaskets. Note the location of ignition wire routing clip stud bolts. Tighten the retaining bolts to 80–106 inch lbs. (9–12 Nm).

37. Install the upper intake manifold.

38. Install the spark plugs, if removed.

39. Connect the ignition wires to the spark plugs.

40. If the left-hand cylinder head is being installed, perform the following:

a. Install the engine oil filler cap.

b. Install the alternator/power steering pump mounting bracket.

c. Install the alternator assembly.

d. Install the accessory drive belt tensioner.

e. Install the power steering pump.

f. Install the power steering pump support bracket. Tighten the retaining bolts to 30–45 ft. lbs. (40–62 Nm).

41. If the right-hand cylinder head is being installed, perform the following:

a. Install the PCV valve.

b. If equipped with air conditioning, install the A/C compressor mounting and support brackets. Install the A/C compressor.

c. If not equipped with air conditioning, install the idler pulley.

42. Install the accessory drive belt.

43. Install the engine air cleaner outlet tube.

44. Connect the negative battery cable.

45. Fill and bleed the engine cooling system.

✳✳ WARNING

Operating the engine without the proper amount and type of engine oil will result in severe engine damage.

46. Start the engine and check for coolant, fuel and oil leaks.

47. Road test the vehicle and check for proper operation.

4.6L Engine

1. Disconnect the negative battery cable.

2. Drain the cooling system and remove the cooling fan and shroud.

✳✳ CAUTION

Observe all applicable safety precautions when working around fuel. Whenever servicing the fuel system, always work in a well ventilated area. Do not allow fuel spray or vapors to come in contact with a spark or open flame. Keep a dry chemical fire extinguisher near the work area. Always keep fuel in a container specifically designed for fuel storage; also, always properly seal fuel containers to avoid the possibility of fire or explosion.

3. Relieve the fuel system pressure and disconnect the fuel lines.

4. Remove the air inlet tube and the wiper module. Release the belt tensioner and remove the accessory drive belt.

5. Tag and disconnect the ignition wires from the spark plugs. Disconnect the ignition wire brackets from the camshaft cover studs and remove the two bolts retaining the ignition wire tray to the coil brackets.

6. Remove the bolt retaining the air conditioner high pressure line to the right coil bracket. Disconnect both ignition coils and CID sensor.

7. Remove the nuts retaining the coil brackets to the front cover. Slide the ignition coil brackets and ignition wire assembly off the mounting studs and remove from the vehicle.

8. Remove the water pump pulley. Disconnect the generator wiring harness from the junction block, fender apron and generator. Disconnect the bolts retaining the generator to the intake manifold and engine block and remove the generator.

9. Disconnect the positive battery cable at the power distribution box. Remove the retaining bolt from the positive battery cable bracket located on the side of the right cylinder head.

10. Disconnect the vent hose from the canister purge solenoid and position the positive battery cable out of the way. Disconnect the canister purge solenoid vent hose from the PCV valve and remove the PCV valve from the camshaft cover.

11. Remove the 42-pin engine harness connector from the retaining bracket on the brake vacuum booster, disconnect and position out of the way.

12. Disconnect the crankshaft position sensor, air conditioning compressor clutch and canister purge solenoid connectors.

13. Raise and safely support the vehicle.

14. Remove the bolts retaining the power steering pump to the engine block and front cover. The front lower bolt on the power steering pump will not come all the way out. Wire the power steering pump out of the way.

15. Remove the four bolts retaining the oil pan to the front cover. Remove the crankshaft damper retaining bolt and remove the damper, using a suitable puller.

16. Disconnect the oil sending unit. Position the oil pressure sending unit harness out of the way.

17. Disconnect the EGR tube from the right exhaust manifold. Disconnect the exhaust pipes from the exhaust manifolds. Lower the exhaust pipes and hang with wire from the crossmember.

18. Remove the bolt retaining the starter wiring harness to the rear of the right cylinder head. Lower the vehicle.

19. Remove the bolts and stud bolts retaining the camshaft covers to the cylinder heads and remove the covers.

20. Disconnect the accelerator, cruise control and throttle valve cables. Remove the accelerator cable bracket from the intake manifold and position out of the way.

21. Disconnect the vacuum hose from the throttle body elbow vacuum port, both oxygen sensors and the heater supply hose.

22. Remove the two bolts retaining the thermostat housing to the intake manifold and position the upper hose and thermostat housing out of the way.

➡ **Two thermostat housing bolts also retain the intake manifold.**

23. Remove the nine bolts retaining the intake manifold to the cylinder heads and remove the intake manifold and gaskets.

24. Remove the seven stud bolts and four bolts retaining the front cover to the engine and remove the front cover.

25. Remove the timing chains.

26. Remove the 10 bolts retaining the left cylinder head to the engine block and remove the head.

➡ **The lower rear bolt cannot be removed due to interference with the brake vacuum booster. Use a rubber band to hold the bolt away from the engine block.**

27. Remove the ground strap, one stud and one bolt retaining the heater return line to the right cylinder head.

28. Remove the 10 bolts retaining the right cylinder head to the engine block and remove the head.

➡ **The lower rear bolt cannot be removed due to interference with the evaporator housing. Use a rubber band to hold the bolt away from the engine block.**

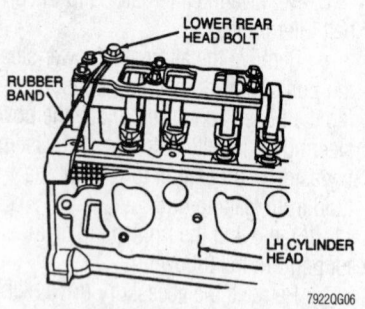

79220G06

A rubber band should be used to secure the lower rear bolt in the up position— 4.6L engine

29. Clean all gasket mating surfaces. Check the cylinder head and engine block for flatness. Check the cylinder head for scratches near the coolant passage and combustion chamber that could provide leak paths. Machine as necessary.

To install:

30. Rotate the crankshaft counterclockwise 45 degrees. The crankshaft keyway should be at the 9 o'clock position viewed from the front of the engine. This ensures that all pistons are below the top of the engine block deck face.

31. Rotate the camshaft to a stable position where the valves do not extend below the head face.

32. Position new head gaskets on the engine block. Install the lower rear bolts on both cylinder heads and retain with rubber bands as explained during the removal procedure.

33. Position the cylinder heads on the engine block dowels, being careful not to score the surface of the head face. Apply clean oil to the head bolts, remove the rubber band from the lower rear bolt and install all bolts hand-tight.

34. Tighten the head bolts as follows:

    a. Tighten the bolts, in sequence, to 15–22 ft. lbs. (20–30 Nm).

    b. Rotate each bolt, in sequence, 85–96 degrees.

    c. Rotate each bolt, in sequence, an additional 85–96 degrees.

35. Position the heater return hose and install the two bolts. Rotate the camshafts using the flats matched at the center of the camshaft until both are in time. Install cam positioning tools T91P-6256-A or equivalent, on the flats of the camshafts to keep them from rotating.

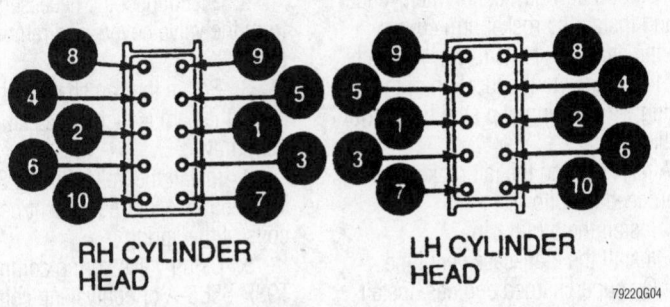

RH CYLINDER HEAD

LH CYLINDER HEAD

7922QG04

Be sure to tighten the cylinder head bolts in the sequence shown to ensure good sealing—4.6L engine

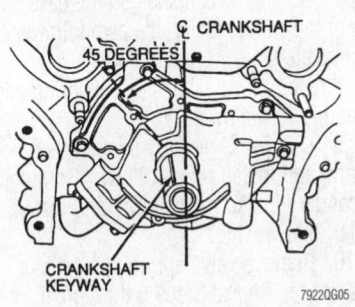

7922QG05

Rotate the crankshaft clockwise 45 degrees, then install the timing chains—4.6L engine

36. Rotate the crankshaft clockwise 45 degrees to position the crankshaft at TDC on No. 1 cylinder.

➡The crankshaft must only be rotated in the clockwise direction and only as far as TDC.

37. Install the timing chains according to the proper procedure.

38. Install a new front cover seal and gasket. Apply silicone sealer to the lower corners of the cover where it meets the junction of the oil pan and cylinder block and to the points where the cover contacts the junction of the cylinder block and cylinder head.

39. Install the front cover and the stud bolts and bolts. Tighten to 15–22 ft. lbs. (20–30 Nm).

40. Position new intake manifold gaskets on the cylinder heads. Be sure the alignment tabs on the gaskets are aligned with the holes in the cylinder heads.

➡Before installing the intake manifold, inspect it for nicks and cuts that could provide leak paths.

41. Position the intake manifold on the cylinder heads and install the retaining bolts. Tighten the bolts, in sequence, to 15–22 ft. lbs. (20–30 Nm).

42. Install the thermostat and O-ring, then position the thermostat housing and install the two bolts. Tighten to 15–22 ft. lbs. (20–30 Nm). Connect the hose to the thermostat housing.

43. Connect the heater supply hose and both oxygen sensors. Connect the vacuum hose to the throttle body adapter vacuum port.

44. Connect, if necessary, adjust the throttle valve cable. Install the accelerator cable bracket on the intake manifold and connect the accelerator and cruise control cables to the throttle body.

45. Apply silicone sealer to both places where the front cover meets the cylinder head. Install new gaskets on the camshaft covers.

46. Install the camshaft covers on the cylinder heads. Install the bolts and stud bolts and tighten to 6.0–8.8 ft. lbs. (8–12 Nm).

47. Raise and safely support the vehicle. Position the starter wiring harness to the right cylinder head and install the retaining bolt.

48. Cut the wire and position the exhaust pipes to the exhaust manifolds. Tighten the four nuts to 20–30 ft. lbs. (27–41 Nm).

➡Be sure the exhaust system clears the No. 3 crossmember. Adjust as necessary.

49. Connect the EGR tube to the right exhaust manifold and tighten the line nut to 26–33 ft. lbs. (35–45 Nm). Connect the oil sending unit.

50. Apply a small amount of silicone sealer in the rear of the keyway on the damper. Position the damper on the crankshaft, making sure the crankshaft key and keyway are aligned.

51. Using damper installer T74P-6316-B or equivalent, install the crankshaft damper. Install the damper bolt and washer and tighten to 114–121 ft. lbs. (155–165 Nm).

52. Install the four bolts retaining the oil pan to the front cover and tighten to 15–22 ft. lbs. (20–30 Nm).

53. Position the power steering pump on the engine and install the four retaining bolts. Tighten to 15–22 ft. lbs. (20–30 Nm). Lower the vehicle.

54. Connect the air conditioning compressor, crankshaft position sensor and canister purge solenoid.

55. Connect the 42-pin engine harness connector and transmission harness connector. Install the 42-pin connector on the retaining bracket on the vacuum brake booster.

56. Install the PCV valve in the right camshaft cover and connect the canister purge solenoid vent hose.

57. Position the positive battery cable harness on the right cylinder head and install the bolt retaining the cable bracket to the cylinder head. Connect the positive battery cable at the power distribution box and battery.

58. Position the generator and install the two retaining bolts. Tighten to 15–22 ft. lbs. (20–30 Nm). Install the two bolts retaining the generator brace to the intake manifold and tighten to 53–71 inch lbs. (6–12 Nm).

59. Install the water pump pulley and tighten the bolts to 15–22 ft. lbs. (20–30 Nm).

60. Position the ignition coil brackets and ignition wire assembly onto the mounting studs. Install the seven nuts retaining the coil brackets to the front cover and tighten to 15–22 ft. lbs. (20–30 Nm).

61. Install the two bolts retaining the ignition wire tray to the coil bracket and tighten to 6.0–8.8 ft. lbs. (8–12 Nm). Connect both ignition coils and CID sensor.

62. Position the air conditioner high pressure line on the right coil bracket and install the bolt. Connect the ignition wires to the spark plugs and install the bracket onto the camshaft cover studs.

63. Install the accessory drive belt and the wiper module. Connect the fuel lines and install the cooling fan and shroud. Fill and bleed the cooling system.

64. Install the air inlet tube and connect the negative battery cable. Start the engine and bring to normal operating temperature. Check for leaks. Check all fluid levels.

Rocker Arms

REMOVAL & INSTALLATION

3.8L Engines

1. Disconnect the negative battery cable.
2. Disconnect the spark plug wires from the spark plugs. Remove the spark plug wire routing clips from the rocker arm cover attaching bolt studs.
3. To remove the left rocker arm cover, proceed as follows:
   a. Remove the oil fill cap.
   b. Remove the crankcase vent tube.
   c. On supercharged engines, remove the charge air cooler tubes and the oil cooler inlet tube.
4. To remove the right rocker arm cover, proceed as follows:
   a. Remove the PCV valve.
   b. Position the air cleaner assembly aside, if necessary.
   c. On supercharged engines, remove the air inlet tube and remove the throttle body assembly.
5. Remove the rocker arm cover attaching screws and remove the rocker arm covers.
6. Remove the rocker arm, fulcrum and bolt assemblies. Keep each assembly together and identify the assemblies so they may be reinstalled in their original positions.

To install:

7. Clean all gasket mating surfaces on the rocker arm covers and cylinder heads. Clean the rocker arms and fulcrums and inspect for wear or damage. Replace as necessary.
8. Apply grease to the pushrod tips and valve stem tips. Lubricate the fulcrums and rocker arms with heavy engine oil and install them over the pushrods and valve stems.
9. For each valve, rotate the crankshaft until the lifter is on the base circle of the camshaft. Install the fulcrum bolt and tighten to 5–11 ft. lbs. (7–15 Nm). Be sure the pushrod and fulcrum are fully seated prior to tightening.
10. Lubricate all rocker arm assemblies with engine oil. Final tighten the fulcrum bolts to 19–25 ft. lbs. (25–35 Nm). When final tightening, the camshaft may be in any position. Be sure the pushrod and fulcrum are fully seated prior to tightening.

11. Position new gaskets on the cylinder heads and install the rocker arm covers. Tighten the attaching bolts to 80–106 inch lbs. (9–12 Nm). Note the location of the spark plug wire routing clip stud bolts prior to installation.
12. After installing the left rocker arm cover, proceed as follows:
   a. Install the oil fill cap.
   b. Install the crankcase vent tube.
   c. On supercharged engines, install the charge air cooler tubes and the oil cooler inlet tube.
13. After installing the right valve cover, proceed as follows:
   a. Install the PCV valve.
   b. Install the air cleaner assembly, if necessary.
   c. On supercharged engines, install the air inlet tube and the throttle body assembly.
14. Install the spark plug wire routing clips and connect the wires to the spark plugs.
15. Connect the negative battery cable, start the engine and check for leaks.

4.6L Engine

1. Disconnect the negative battery cable.

2. Disconnect the necessary hoses from the valve covers and remove the covers.
3. Rotate the camshaft so the base circle of the cam is facing the cam follower to be removed.
4. Install the Valve Spring Spacer T91P-6565-AH or equivalent between the coils of the spring.
5. Using valve spring compressor tool T93P-6565-A or equivalent, compress the lash adjuster as required and remove the roller follower.
6. Repeat Steps 3 and 4 for followers being removed.

To install:

7. Rotate the camshaft so the base circle of the cam is facing the cam follower to be installed.
8. Compress exhaust valve spring and install roller follower.
9. Compress primary intake valve spring and install roller follower. Compress secondary intake valve spring and install roller follower.
10. Remove valve spring compressor and spacer. Repeat Steps 6 through 10 until all followers are installed.
11. Install the rocker arm covers and necessary hoses. Connect the negative battery cable, start the engine and check for leaks.

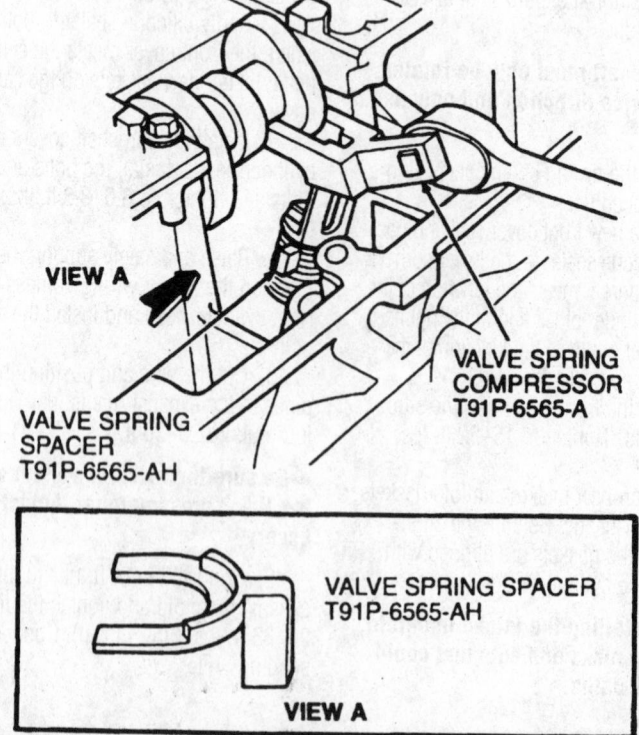

VALVE SPRING COMPRESSOR
T91P-6565-A

VALVE SPRING SPACER
T91P-6565-AH

VIEW A

VALVE SPRING SPACER
T91P-6565-AH

VIEW A

79220G07

With the valve spring spacer installed, compress the valve spring and remove the rocker arm (cam follower)—4.6L engine

Supercharger

REMOVAL & INSTALLATION

➡️**Before beginning any supercharger service, clean the area around the supercharger assembly. Cover the engine and supercharger openings while the supercharger is removed, to prevent damage by foreign material.**

1. Disconnect the negative battery cable and partially drain the cooling system.

2. Remove the throttle body air inlet tube and the cowl vent screens.

3. Tag and disconnect the right side spark plug wires at the coil and position aside. Tag and disconnect the electrical connections at the idle air bypass valve, throttle position sensor and air charge temperature sensors.

4. Tag and disconnect the vacuum lines from the inlet/plenum assembly. If equipped, remove the EGR transducer from the bracket and disconnect the vacuum line. Disconnect the PCV tube.

5. Disconnect the throttle linkage at the throttle housing. Remove the linkage bracket retaining bolts and position the bracket aside. Disconnect the cruise control, if equipped.

6. Remove the two EGR valve attaching bolts and move the EGR valve away from the intake assembly, if equipped. Disconnect the coolant hoses from the throttle body, if equipped.

7. Remove the supercharger drive belt. Remove the charge air cooler inlet and outlet tubes as follows:

   a. Disconnect the inlet tube from the supercharger outlet adapter using spanner nut wrench tool T89P-6634-A or equivalent. Remove the four nuts retaining the inlet and outlet tubes to the charge air cooler.

   b. Remove the nut and push-on nut retaining the inlet tube to the alternator-power steering pump bracket. Remove the stud from the alternator-power steering pump bracket. Remove the inlet tube.

➡️**Use extreme care during removal and installation of the charge air cooler tubes so as not to scratch, nick or contaminate the sealing surfaces.**

   c. Remove the two nuts retaining the outlet tube to the intake elbow assembly. Raise and safely support the vehicle.

   d. Remove the bolt retaining the outlet tube to the cylinder block front upper support bracket. Loosen, but do not remove the support bracket.

➡️**The bracket must be close to the front face of the cylinder block to allow the bracket to pivot during outlet tube reinstallation.**

   e. Remove the nut and push-on nut retaining the outlet tube to the alternator-power steering pump bracket. Remove the power steering pump drive belt.

   f. Tag and disconnect the spark plug wires from the coil. Remove the power steering pump bracket brace to water pump retaining stud nuts. Remove the two power steering pump bracket to cylinder head retaining bolts and one stud nut.

   g. Install a 10 x 1.5mm x 170mm bolt, 6½ in. long into the top hole in the power steering pump bracket. Thread the bolt into the cylinder head approximately five turns. This will aid in holding the power steering pump bracket in position.

   h. Remove the power steering pump filler cap. Slide the power steering pump bracket assembly forward on the stud and bolt that was installed in the previous step.

   i. Remove the outlet tube by pulling underneath the power steering pump bracket assembly and up through the engine compartment. It may be necessary to pivot the outlet tube clamping connector to gain clearance during removal.

8. Remove the three intake elbow retaining bolts and the three supercharger retaining bolts. Lift the supercharger and intake elbow assembly from the vehicle as a unit.

To install:

9. Clean and inspect all gasket surfaces. Position a new gasket on the intake manifold using guide pins, if available.

10. Install the supercharger, throttle body and intake elbow as an assembly. Tighten the two 8mm bolts to 15–22 ft. lbs. (20–30 Nm). Tighten the 12mm bolt to 52–70 ft. lbs. (70–96 Nm).

11. Install the three intake elbow retaining bolts and tighten to 20–28 ft. lbs. (26–38 Nm). Install the charge air cooler tubes as follows:

   a. Clean and inspect the sealing surfaces of the supercharger outlet adapter intake elbow, charge air cooler and tubes.

➡️**Be sure there are no foreign particles on the sealing surfaces of the tubes. It is important that the charge air cooler tubes seal completely. Any air leak will cause poor operation and performance.**

   b. Install gasket sealant tape ESE-M4G168-B or equivalent, to the spherical seat surfaces of the charge air cooler tubes. Install the tape approximately ⅛ in. (3mm) from the inner diameter of the tubes. Overlap the tape ends approximately ¼ in. (6mm). Do not stretch the tape during installation or the seal may leak. During proper installation, a slight wrinkling will occur on the tape edge at the inner diameter.

➡️**The system must be tightened in sequence and to the specification for that step. This is required for proper alignment of the system to ensure sealing of the charge air cooler tubes.**

   c. Guide the outlet tube down through the engine compartment and underneath the power steering pump bracket assembly. It may be necessary to rotate the lower outlet tube clamping connector to gain clearance.

➡️**Use extreme care during installation of the charge air cooler tubes so as not to scratch, nick or contaminate the sealing surfaces.**

   d. Slide the power steering pump bracket assembly into position. Install the power steering pump bracket retaining stud nut and tighten to 30–40 ft. lbs. (40–55 Nm).

   e. Remove the bolt installed in Step 7g of the removal procedure. Install the power steering pump bracket to cylinder head bolts and tighten to 30–40 ft. lbs. (40–55 Nm).

   f. Install the power steering pump bracket brace to water pump retaining stud nuts and tighten to 15–22 ft. lbs. (20–30 Nm). Install the outlet tube over the lower stud on the alternator-power steering pump bracket.

   g. Install the push-on nut onto the stud, tight enough to retain the tube against the alternator-power steering pump bracket surface but free enough to allow tube movement to ensure seating of the spherical seat on the outlet tube to intake elbow assembly.

   h. Install the outlet tube clamping connector over the studs on the intake elbow assembly and secure with the two nuts. Tighten both nuts to 15–22 ft. lbs. (20–30 Nm). The clamping connector should be installed so it is visually parallel to the stud mounting face of the intake elbow assembly.

   i. Install the nut to the stud on the alternator-power steering pump bracket and tighten to 30–40 ft. lbs. (40–55 Nm).

Install the bolt to secure the outlet tube to the cylinder block support bracket and tighten to 30–40 ft. lbs. (40–55 Nm). Tighten the support bracket to front of cylinder block retaining nut to 15–22 ft. lbs. (20–30 Nm) and bolt to 52–70 ft. lbs. (70–96 Nm).

j. Apply anti-seize compound to the inner backside spherical seat surface and threads of the supercharger outlet adapter collar. Position the inlet tube, then install the upper stud into the alternator-power steering pump bracket.

k. Install the push-on nut onto the stud, tight enough to retain the tube against the alternator-power steering pump bracket surface but free enough to allow tube movement to ensure seating of the spherical seat on the inlet tube to supercharger outlet adapter.

l. Fully hand-tighten the supercharger outlet adapter collar onto the threaded tube end of the inlet tube assembly. Install the charge air cooler assembly to the inlet and outlet tubes. Install the nuts to the studs tight enough to retain the charge air cooler and tubes together but free enough to allow movement on the spherical seats. Do not tighten at this time.

m. Tighten the supercharger outlet adapter collar to inlet tube to 148 ft. lbs. (200 Nm).

n. Wait 10 minutes minimum and retighten the supercharger outlet collar to 148 ft. lbs. (200 Nm).

➡**When 1st compressed, the sealant tape flows and forms to the sealing surface. If the collar is not retightened, it will loosen and cause a leak at the joint.**

o. Tighten the inlet and outlet tube to charge air cooler nuts to 15–22 ft. lbs. (20–30 Nm). The clamping connectors should be installed so they are visually parallel to the stud mounting face of the charge air cooler assembly.

p. Install the nut retaining the inlet tube to the alternator-power steering pump support bracket and tighten to 30–40 ft. lbs. (40–55 Nm).

12. Install the supercharger drive belt. Connect the coolant hoses to the throttle body, if equipped.

13. Connect the EGR valve with a new gasket to the intake manifold, if equipped. Tighten the retaining bolts to 14–22 ft. lbs. (20–30 Nm).

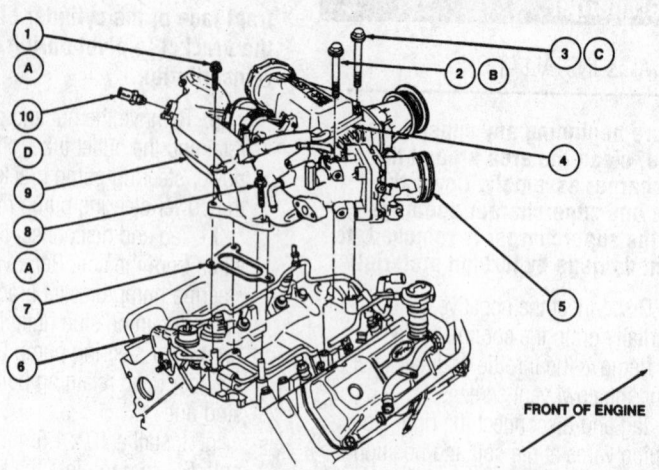

Item	Part Number	Description
1	N605909-S2	Bolt (2 Req'd)
2	N805964-S36B	Bolt (2 Req'd)
3	N606590-S40	Bolt
4	6F066	Supercharger
5	9E926	Throttle Body
6	9424	Lower Intake Manifold
7	9E936	Gasket
8	N806514-S2	Stud Bolt

Item	Part Number	Description
9	6C643	Charge Air Cooler to Intake Manifold Adapter
10	12A697	Intake Air Temperature Sensor
A	—	Tighten to 26-38 N·m (20-28 Lb-Ft)
B	—	Tighten to 20-30 N·m (15-22 Lb-Ft)

79220G08

Exploded view of the supercharger and related component mounting—3.8L (VIN R) engine

14. Install the throttle linkage bracket and connect the throttle linkage. Tighten to 10–15 ft. lbs. (14–20 Nm).

15. Connect the vacuum lines to the inlet assembly and connect the PCV tube. If equipped, connect the vacuum line to the EGR transducer and install the transducer in the bracket.

16. Install the right side spark plug wires. Connect the electrical connectors at the idle air bypass valve, throttle position sensor and air charge temperature sensor.

17. Install the cowl covers and the throttle body air inlet tube.

18. Fill and bleed the cooling system. Connect the negative battery cable. Start the engine and check for leaks and proper operation.

Intake Manifold

REMOVAL & INSTALLATION

3.8L Engines

1. Disconnect the negative battery cable.
2. Drain the cooling system and relieve the fuel system pressure.
3. Remove the air cleaner assembly or air inlet tube.

4. Disconnect the accelerator cable at the throttle body. Disconnect the cruise control cable, if equipped.

5. If equipped with an automatic transmission, disconnect the transmission linkage at the upper intake manifold. Remove the retaining bolts from the accelerator cable mounting bracket and position the cables aside.

6. If equipped, disconnect the thermactor air supply hose at the check valve. The valve is located in the Y-pipe assembly.

7. Disconnect the fuel lines. If equipped, remove the supercharger.

8. Disconnect the radiator hose at the thermostat housing and the coolant bypass hose at the manifold.

9. Disconnect the heater tube at the intake manifold and remove the tube support bracket retaining nut. Remove the heater hose at the rear of the heater tube. Loosen the hose clamp at the heater elbow and remove the heater tube with the hose attached. Remove the heater tube with the lines attached and set the assembly aside.

10. Tag and disconnect the vacuum lines at the fuel rail assembly and intake manifold. Tag and disconnect the necessary electrical connectors.

11. If equipped with air conditioning, remove the compressor support bracket. Disconnect the PCV line at the upper intake manifold and at the valve. Remove the 2nd PCV line from the left rocker arm cover.

12. Remove the throttle body assembly. Remove the EGR valve assembly from the upper manifold.

13. Remove the retaining nut and remove the wiring retainer bracket located at the left front of the intake manifold and set aside with the spark plug wires.

14. Remove the upper intake manifold retaining bolts/studs and remove the upper intake manifold.

15. Remove the injectors and fuel rail assembly. Remove the heater water outlet hose.

16. Remove the lower intake manifold retaining bolts/studs and remove the lower intake manifold.

➡The manifold is sealed at each end with RTV-type sealer. To break the seal, it may be necessary to pry on the front of the manifold with a small prybar. If it is necessary to pry on the manifold, use care to prevent damage to the machined surfaces.

To install:

17. Clean all gasket mating surfaces. Lightly oil all retaining bolt and stud threads.

18. Apply a dab of gasket adhesive to each cylinder head mating surface. Press new intake manifold gaskets in place, using location pins as necessary to aid in installation.

19. Apply a ⅛ in. bead of silicone sealer at each corner where the cylinder head joins the cylinder block. Install the front and rear intake manifold end seals.

20. Carefully lower the intake manifold into place on the cylinder heads and cylinder block. Use locating pins as necessary to guide the manifold.

21. Install the bolts and stud bolts in their original locations.

22. On all supercharged engines tighten the bolts, in sequence, in two steps, 1st to 8 ft. lbs. (11 Nm), then to 11 ft. lbs. (15 Nm).

23. Connect the rear PCV line to the upper intake tube. Install the front PCV tube so the mounting bracket sits over the lower intake manifold stud. Tighten the nut on the stud to 15–22 ft. lbs. (20–30 Nm).

24. Install the injectors and the fuel rail. On non-supercharged engines, install the upper intake manifold assembly. Install the bolts and stud bolts in their original locations. Tighten the four center bolts, then the end bolts in three steps, 1st to 8 ft. lbs. (10 Nm), then to 15 ft. lbs. (20 Nm), and finally to 24 ft. lbs. (32 Nm).

25. On supercharged engines, install the supercharger.

26. Install the EGR valve. Install the throttle body and cross-tighten the retaining nuts to 15–22 ft. lbs. (20–30 Nm).

27. Connect the rear PCV line at the PCV valve on the upper intake manifold. If equipped with air conditioning, install the compressor support bracket.

28. Connect the necessary electrical connectors and vacuum hoses. Connect the heater tube hose to the heater elbow and position the heater tube support bracket. Tighten the retaining nut to 15–22 ft. lbs. (20–30 Nm).

29. Connect the heater hose to the heater tube and connect the coolant bypass hose and radiator upper hose.

30. Connect the fuel lines. Position the accelerator cable mounting bracket and tighten the mounting bolts to 15–22 ft. lbs. (20–30 Nm).

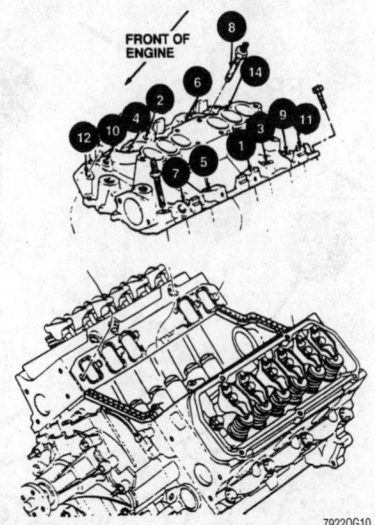

Lower intake manifold bolt tightening sequence—1995 3.8L (non-supercharged) engine

31. Connect the transmission linkage at the upper intake manifold. If equipped, connect the cruise control cable.

32. Fill and bleed the cooling system. Connect the negative battery cable, start the engine and check for leaks.

33. Check and if necessary, adjust the engine idle speed, transmission throttle linkage and cruise control.

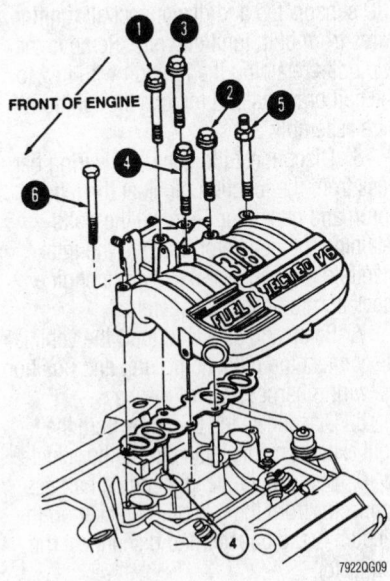

To avoid leaks, tighten the upper intake manifold bolts in the sequence shown—3.8L engines

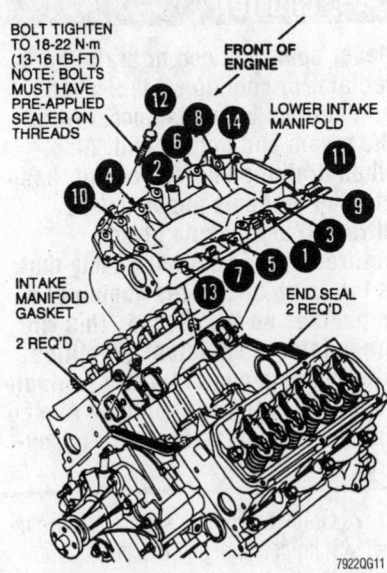

Lower intake manifold bolt tightening sequence—1995 3.8L (supercharged) engine

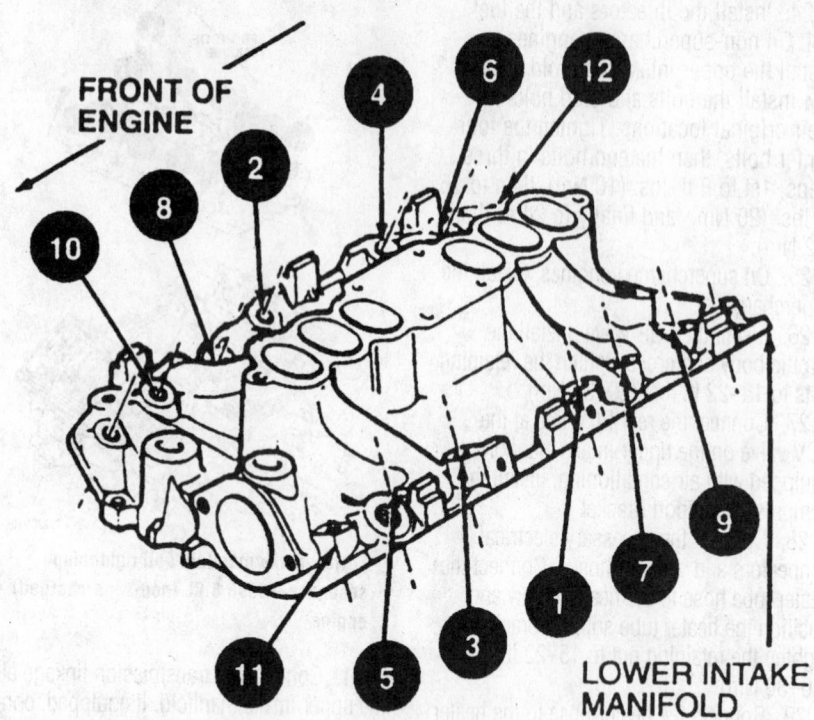

FRONT OF ENGINE

LOWER INTAKE MANIFOLD

79220G12

Be sure to tighten the bolts according to the sequence shown—1996–99 3.8L engines

4.6L Engine

1995 MODELS

1. Disconnect the negative battery cable.

> ❊❊ **CAUTION**
>
> **Never open, service or drain the radiator or cooling system when hot; serious burns can occur from the steam and hot coolant. Also, when draining engine coolant, keep in mind that cats and dogs are attracted to ethylene glycol antifreeze and could drink any that is left in an uncovered container or in puddles on the ground. This will prove fatal in sufficient quantities. Always drain coolant into a sealable container. Coolant should be reused unless it is contaminated or is several years old.**

2. Drain the cooling system. Relieve the fuel system pressure and disconnect the fuel lines.

3. Remove the wiper module and the air inlet tube. Release the belt tensioner and remove the accessory drive belt.

4. Tag and disconnect the ignition wires from the spark plugs. Disconnect the ignition wire brackets from the camshaft cover studs.

5. Disconnect both ignition coils and CID sensor. Tag and disconnect all ignition wires from both ignition coils. Remove the two bolts retaining the ignition wire tray to the coil brackets and remove the ignition wire assembly.

6. Disconnect the generator wiring harness from the junction block at the fender apron and generator. Remove the bolts retaining the generator brace to the intake manifold and the generator to the engine block and remove the generator.

7. Raise and safely support the vehicle. Disconnect the oil sending unit and position the wiring harness out of the way.

8. Disconnect the EGR tube from the right exhaust manifold and lower the vehicle.

9. Remove the 42-pin engine harness connector from the retaining bracket on the vacuum brake booster and disconnect the connector.

10. Disconnect the air conditioning compressor, crankshaft position sensor and canister purge solenoid.

11. Remove the PCV valve from the camshaft cover and disconnect the canister purge vent hose from the PCV valve.

12. Disconnect the accelerator and cruise control cables from the throttle body. Remove the accelerator cable bracket from the intake manifold and position out of the way.

13. Disconnect the throttle valve cable from the throttle body and the vacuum hose from the throttle body adapter port.

14. Disconnect both oxygen sensors and the heater supply hose.

15. Remove the two bolts retaining the thermostat housing to the intake manifold and position the upper hose and thermostat housing out of the way.

➡**The two thermostat housing bolts also retain the intake manifold.**

16. Remove the bolts retaining the intake manifold to the cylinder heads and remove the intake manifold. Remove and discard the gaskets.

To install:

17. Clean all gasket mating surfaces. Position new intake manifold gaskets on the cylinder heads. Be sure the alignment tabs on the gaskets are aligned with the holes in the cylinder heads.

18. Install the intake manifold and the retaining bolts. Tighten the bolts, in sequence, to 15–22 ft. lbs. (20–30 Nm).

19. Inspect and if necessary, replace the O-ring seal on the thermostat housing. Position the housing and upper hose and install the two bolts. Tighten to 15–22 ft. lbs. (20–30 Nm).

20. Connect the heater supply hose and connect both oxygen sensors.

21. Connect the vacuum hose to the throttle body adapter vacuum port. Connect, if necessary, adjust the throttle valve cable.

22. Install the accelerator cable bracket on the intake manifold and connect the accelerator and cruise control cables to the throttle body.

23. Install the PCV valve in the camshaft cover and connect the canister purge solenoid vent hose. Connect the air conditioning compressor, crankshaft position sensor and canister purge solenoid.

24. Connect the 42-pin engine harness connector. Install the connector on the retaining bracket on the vacuum brake booster.

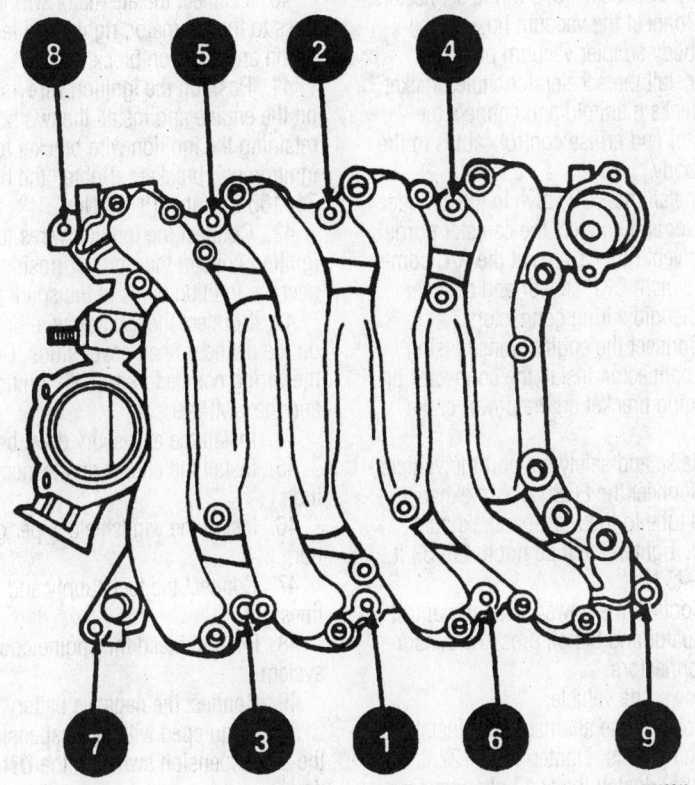

Tighten the intake manifold bolts in sequence to prevent leakage—1995 4.6L engine

79220G13

25. Raise and safely support the vehicle. Connect the EGR tube to the right exhaust manifold and tighten the line nut to 26–33 ft. lbs. (35–45 Nm).

26. Connect the oil sending unit. Lower the vehicle.

27. Position the generator and install the retaining bolts. Tighten to 15–22 ft. lbs. (20–30 Nm). Install the two bolts retaining the generator brace to the intake manifold and tighten to 6.0–8.8 ft. lbs. (8–12 Nm).

28. Connect the generator wiring harness to the generator, right-hand fender apron and junction block.

29. Position the ignition wire assembly on the engine and install the two bolts retaining the ignition wire tray to the coil brackets. Tighten the bolts to 6.0–8.8 ft. lbs. (8–12 Nm).

30. Connect the ignition wires to the ignition coils in their proper positions. Connect the ignition wires to the spark plugs.

31. Connect the ignition wire brackets on the camshaft cover studs. Connect both ignition coils and CID sensor.

32. Install the accessory drive belt and the air inlet tube. Install the wiper module and connect the fuel lines.

33. Fill and bleed the cooling system. Connect the negative battery cable, start the engine and check for leaks.

1996–99 MODELS

❋❋ CAUTION

The fuel injection system remains under pressure, even after the engine has been turned OFF. The fuel system pressure must be relieved before disconnecting any fuel lines. Failure to do so may result in fire and/or personal injury.

1. If equipped with air suspension, the air suspension switch, located on the right-hand side of the luggage compartment, must be turned to the **OFF** position before raising the vehicle.

2. Disconnect the negative battery cable.

❋❋ CAUTION

Never open, service or drain the radiator or cooling system when hot; serious burns can occur from the steam and hot coolant. Also, when draining engine coolant, keep in mind that cats and dogs are attracted to ethylene glycol antifreeze and could drink any that is left in an uncovered container or in puddles on the ground. This will prove fatal in

sufficient quantities. Always drain coolant into a sealable container. Coolant should be reused unless it is contaminated or is several years old.

3. Drain the engine cooling system.

4. Relieve the fuel system pressure.

5. Disconnect the fuel supply and return lines.

6. Remove the windshield wiper governor (module).

7. Remove the engine air cleaner outlet tube.

8. Release the drive belt tensioner and remove the accessory drive belt.

9. Tag and disconnect the ignition wires from the spark plugs. Disconnect the ignition wire brackets from the cylinder head cover studs.

10. Disconnect the wiring from both ignition coils and the Camshaft Position (CMP) sensor. Tag and disconnect all ignition wires from both ignition coils. Remove the two bolts retaining the ignition wire bracket to the ignition coil brackets and remove the ignition wire assembly.

11. Disconnect the alternator wiring harness from the junction block at the fender apron and alternator. Remove the bolts retaining the alternator brace to the intake manifold and the alternator to the cylinder block and remove the alternator.

12. Raise and safely support the vehicle.

13. Disconnect the oil pressure sensor and power steering control valve actuator wiring and position the wiring harness out of the way.

14. Disconnect the EGR valve to exhaust manifold tube from the right-hand exhaust manifold.

15. Lower the vehicle.

16. Remove and disconnect the engine/transmission harness connector from the retaining bracket on the power brake booster.

17. Disconnect the A/C compressor clutch, Crankshaft Position (CKP) sensor and the canister purge solenoid wiring connectors.

18. Remove the PCV valve from the cylinder head cover and disconnect the canister purge vent hose from the PCV valve.

19. Disconnect the accelerator and cruise control cables from the throttle body. Remove the accelerator cable bracket from the intake manifold and position out of the way.

20. Disconnect the vacuum hose from the throttle body adapter port.

21. Disconnect both Heated Oxygen Sensors (HO_2S) and the heater water hose.

22. Remove the two bolts retaining the thermostat housing to the intake manifold and position the upper hose and thermostat housing out of the way.

➡ **The two thermostat housing bolts are also used to retain the intake manifold.**

23. Remove nine bolts retaining the intake manifold to the cylinder heads and remove the intake manifold. Remove and discard the gaskets.

24. If replacing the intake manifold, swap over the necessary parts.

To install:

25. Clean all gasket mating surfaces.

26. Position new intake manifold gaskets on the cylinder heads. Be sure the alignment tabs on the gaskets are aligned with the holes in the cylinder heads.

27. Install the intake manifold and nine retaining bolts. Hand-tighten the right-rear bolt (viewed from the front of the engine) before final tightening, then tighten the bolts, in sequence, to 15–22 ft. lbs. (20–30 Nm).

28. Inspect and if necessary, replace the O-ring seal on the thermostat housing. Position the housing and upper hose and install the two retaining bolts. Tighten to 15–22 ft. lbs. (20–30 Nm).

29. Connect the heater water hose.

30. Connect both HO2S wiring connectors.

31. Connect the vacuum hose to the throttle body adapter vacuum port.

32. Install the accelerator cable bracket on the intake manifold and connect the accelerator and cruise control cables to the throttle body.

33. Install the PCV valve in the cylinder head cover and connect the canister purge solenoid vent hose. Connect the A/C compressor clutch, CKP sensor and canister purge solenoid wiring connectors.

34. Connect the engine/transmission harness connector. Install the connector on the retaining bracket on the power brake booster.

35. Raise and safely support the vehicle.

36. Connect the EGR valve to exhaust manifold tube to the right-hand exhaust manifold. Tighten the tube nut to 26–33 ft. lbs. (35–45 Nm).

37. Connect the power steering control valve actuator and the oil pressure sensor wiring connectors.

38. Lower the vehicle.

39. Position the alternator and install the two retaining bolts. Tighten to 15–22 ft. lbs. (20–30 Nm). Install the two bolts retaining the alternator brace to the intake manifold and tighten to 71–106 inch lbs. (8–12 Nm).

40. Connect the alternator wiring harness to the alternator, right-hand fender apron and junction block.

41. Position the ignition wire assembly on the engine and install the two bolts retaining the ignition wire bracket to the ignition coil brackets. Tighten the bolts to 71–106 inch lbs. (8–12 Nm).

42. Connect the ignition wires to the ignition coils in their proper positions. Connect the ignition wires to the spark plugs.

43. Connect the ignition wire brackets on the cylinder head cover studs. Connect the wiring connectors to both ignition coils and the CMP sensor.

44. Install the accessory drive belt.

45. Install the engine air cleaner outlet tube.

46. Install the windshield wiper governor.

47. Connect the fuel supply and return lines.

48. Fill and bleed the engine cooling system.

49. Connect the negative battery cable.

50. If equipped with air suspension, turn the air suspension switch to the **ON** position.

51. Start the engine and check for leaks.

52. Road test the vehicle and check for proper operation.

Exhaust Manifold

REMOVAL & INSTALLATION

3.8L Engines

LEFT SIDE

1. Disconnect the negative battery cable. Remove oil level dipstick tube support bracket.

2. Disconnect the oxygen sensor at the wiring connector.

3. Tag and disconnect the wires from the spark plugs.

4. Raise and safely support the vehicle.

5. Remove the manifold to exhaust pipe attaching nuts.

6. Lower the vehicle.

7. On supercharged engines, remove the charge air cooler tubes and remove the oil cooler tube and dipstick tube support brackets from the studs.

8. Remove exhaust manifold attaching bolts and manifold.

9. Installation is the reverse of the removal procedure. Tighten the manifold retaining bolts to 15–22 ft. lbs. (20–30 Nm).

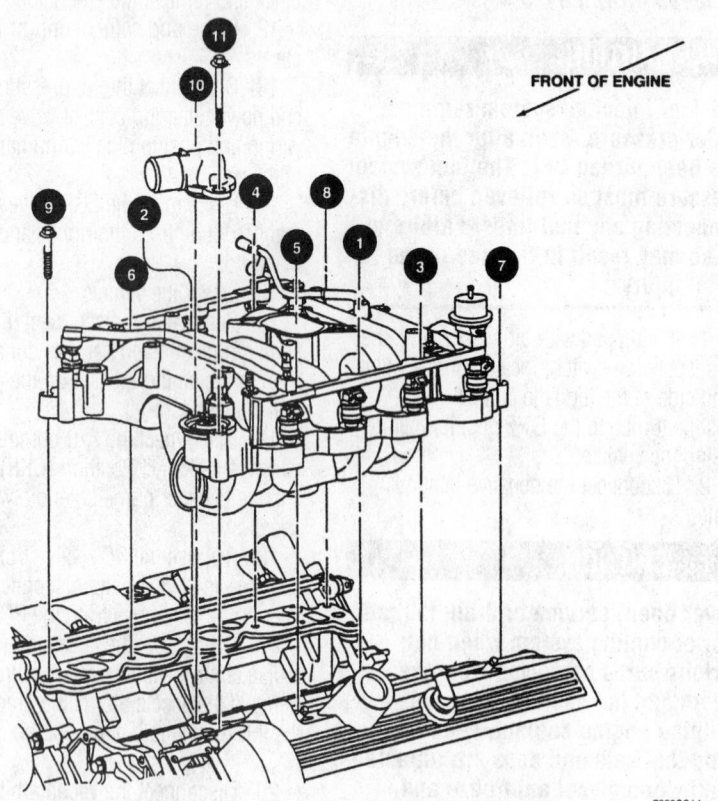

FRONT OF ENGINE

7922QG14

To avoid leakage, tighten the bolts according to the sequence shown—1996–99 with 4.6L engine

RIGHT SIDE

1. Disconnect the negative battery cable. On supercharged engines, remove the air cleaner inlet tube.

2. On non-supercharged engines, disconnect the coil secondary wire from the coil. Tag and disconnect the wires from the spark plugs.

3. On non-supercharged engines, remove the spark plugs and the outer heat shield.

4. Raise and safely support the vehicle. Disconnect the EGR tube.

5. If equipped with automatic transmission, remove the dipstick tube.

6. Remove the manifold-to-exhaust pipe retaining nuts and lower the vehicle.

7. Remove the exhaust manifold retaining bolts and remove the manifold.

8. Installation is the reverse of the removal procedure. Tighten the exhaust manifold retaining bolts to 15–22 ft. lbs. (20–30 Nm).

4.6L Engine

1. Disconnect the battery cables. Remove the air inlet tube.

✳✳ CAUTION

Never open, service or drain the radiator or cooling system when hot; serious burns can occur from the steam and hot coolant. Also, when draining engine coolant, keep in mind that cats and dogs are attracted to ethylene glycol antifreeze and could drink any that is left in an uncovered container or in puddles on the ground. This will prove fatal in sufficient quantities. Always drain coolant into a sealable container. Coolant should be reused unless it is contaminated or is several years old.

2. Drain the cooling system and remove the cooling fan and shroud. Relieve the fuel system pressure and disconnect the fuel lines.

3. Remove the upper radiator hose. Remove the wiper module and support bracket.

4. Have the A/C system discharged be an EPA certified technician. Disconnect and plug the compressor outlet hose at the compressor and remove the bolt retaining the

hose assembly to the right coil bracket. Cap the compressor opening.

5. Remove the 42-pin engine harness connector from the retaining bracket on the brake vacuum booster. Disconnect the connector.

6. Disconnect the throttle valve cable from the throttle body. Disconnect the heater outlet hose.

7. Remove the nut retaining the ground strap to the right cylinder head. Remove the upper stud and lower bolt retaining the heater outlet hose to the right cylinder head and position out of the way.

8. Remove the blower motor resistor and remove the bolt retaining the right engine insulator to the lower engine bracket. Disconnect both oxygen sensors.

9. Raise and safely support the vehicle. Remove the engine mount through-bolts.

10. Remove the EGR tube line nut from the right exhaust manifold.

11. Disconnect the exhaust pipes from the manifolds. Lower the exhaust system and hang it from the crossmember with wire.

12. To remove the left exhaust manifold, remove the engine mount from the engine block and remove the eight bolts retaining the exhaust manifold.

13. Position a jack and a block of wood under the oil pan, rearward of the oil drain hole. Raise the engine approximately 4 in. (100mm).

14. Remove the eight bolts retaining the right exhaust manifold and remove the manifold.

To install:

15. If the exhaust manifolds are being replaced, transfer the oxygen sensors and

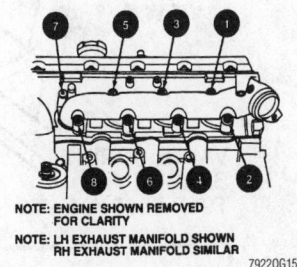

NOTE: ENGINE SHOWN REMOVED FOR CLARITY
NOTE: LH EXHAUST MANIFOLD SHOWN RH EXHAUST MANIFOLD SIMILAR
79220G15

Tighten the exhaust manifold bolts in the order shown to prevent leaks—4.6L engine, left side shown, right side is similar

tighten to 27–33 ft. lbs. (37–45 Nm). On the right manifold, transfer the EGR tube connector and tighten to 33–48 ft. lbs. (45–65 Nm).

16. Clean the mating surfaces of the exhaust manifolds and cylinder heads.

17. Position the exhaust manifolds to the cylinder heads and install the retaining bolts. Tighten, in sequence, to 15–22 ft. lbs. (20–30 Nm).

18. Position and connect the EGR valve and tube assembly to the exhaust manifold. Tighten the line nut to 26–33 ft. lbs. (35–45 Nm).

19. Install the left engine mount and tighten the bolts to 15–22 ft. lbs. (20–30 Nm). Lower the engine onto the mounts and remove the jack. Install the engine mount through-bolts and tighten to 15–22 ft. lbs. (20–30 Nm).

20. Cut the wire and position the exhaust system. Tighten the nuts to 20–30 ft. lbs. (27–41 Nm).

➡**Be sure the exhaust system clears the No. 3 crossmember. Adjust as necessary.**

21. Lower the vehicle. Connect both oxygen sensors and install the bolt retaining the right engine mount to the frame. Tighten to 15–22 ft. lbs. (20–30 Nm).

22. Install the blower motor resistor. Position the heater outlet hoses. Install the upper stud and lower bolt and tighten to 15–22 ft. lbs. (20–30 Nm). Install the ground strap onto the stud and tighten the nut to 15–22 ft. lbs. (20–30 Nm).

23. Connect the heater outlet hose. Connect and if necessary, adjust the throttle valve cable.

24. Connect the 42-pin connector and transmission harness connector. Install the connector to the retaining bracket on the brake vacuum booster.

25. Connect the air conditioning compressor outlet hose to the compressor and install the bolt retaining the hose assembly to the right coil bracket.

26. Install the upper radiator hose and connect the fuel lines. Install the wiper module and retaining bracket.

27. Install the cooling fan and shroud. Install the air inlet tube. Connect the battery cables, start the engine and check for leaks.

28. Leak test, evacuate and charge the air conditioning system according to the proper procedure. Observe all safety precautions.

Camshaft and Valve Lifters

REMOVAL & INSTALLATION

3.8L Engines

1. Disconnect the negative battery cable and drain the cooling system.

❋❋ CAUTION

Observe all applicable safety precautions when working around fuel. Whenever servicing the fuel system, always work in a well ventilated area. Do not allow fuel spray or vapors to come in contact with a spark or open flame. Keep a dry chemical fire extinguisher near the work area. Always keep fuel in a container specifically designed for fuel storage; also, always properly seal fuel containers to avoid the possibility of fire or explosion.

2. Relieve the fuel system pressure and have the A/C system discharged by an EPA certified technician.

3. Remove the radiator. If equipped with air conditioning, remove the condenser.

4. Remove the grille.

5. Remove the camshaft position sensor.

6. Remove the upper intake manifold.

7. Remove the valve covers.

8. Remove the lower intake manifold.

9. Loosen each rocker arm enough to remove the pushrods. Remove the pushrods and keep them in order so they can be returned to their original positions.

10. Remove the lifter guide plates.

11. Remove the valve lifters using a magnet if necessary. Keep them in order so they can be returned to their original positions.

12. Remove the engine front cover, timing chain and camshaft sprocket spacer.

13. Remove the oil pan.

14. Remove the thrust plate. Remove the camshaft, being careful not to damage the bearing surfaces.

To install:

15. Lubricate the cam lobes and journals with heavy engine oil. Install the camshaft, being careful not to damage the bearing surfaces while sliding into position.

16. Install the thrust plate. Tighten the bolts to 71–123 inch lbs. (8–14 Nm).

17. Install the timing chain and sprockets. Install the engine front cover.

18. Install the oil pan.

19. Install the lifters, guide plate and retainers.

20. Install the lower intake manifold.

21. Install the pushrods and tighten the rocker arms.

22. Install the valve cover and the upper intake manifold.

23. Install the camshaft position sensor.

24. Install the grille. If equipped with air conditioning, install the condenser.

25. Install the radiator. Fill and bleed the cooling system.

26. Connect the negative battery cable. Start the engine and check for leaks.

4.6L Engine

1. Disconnect the negative battery cable and drain the cooling system.

❋❋ CAUTION

Observe all applicable safety precautions when working around fuel. Whenever servicing the fuel system, always work in a well ventilated area. Do not allow fuel spray or vapors to come in contact with a spark or open flame. Keep a dry chemical fire extinguisher near the work area. Always keep fuel in a container specifically designed for fuel storage; also, always properly seal fuel containers to avoid the possibility of fire or explosion.

2. Relieve the fuel system pressure.

3. Remove the right and left camshaft covers.

4. Remove the timing chain front cover. Remove the timing chains.

5. Rotate the crankshaft counterclockwise 45 degrees from TDC to be sure all pistons are below the top of the engine block deck face.

➡The crankshaft must be in this position prior to rotating the camshafts or damage to the pistons and/or valvetrain will result.

6. Install valve spring compressor tool T91P-6565-A or equivalent, under the camshaft and on top of the valve spring retainer.

➡Valve spring spacer tool T91P-6565-AH or equivalent, must be installed between the spring coils and the camshaft must be at the base circle before compressing the valve spring.

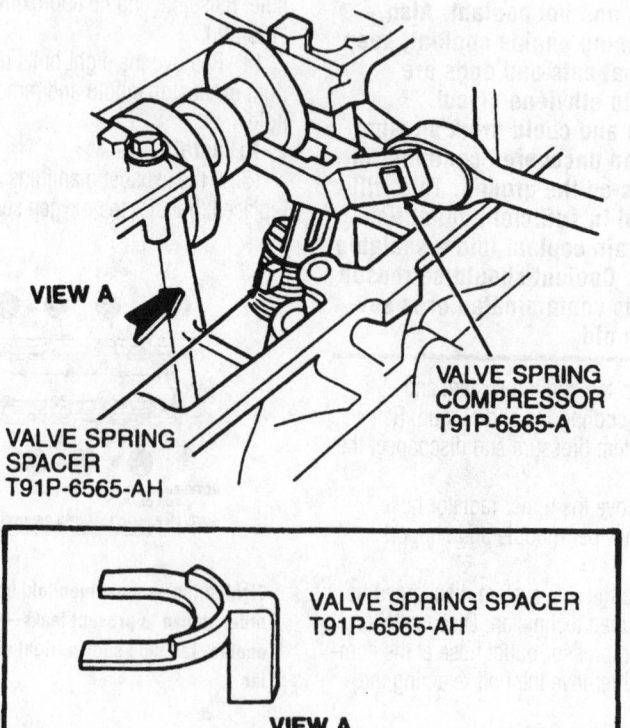

VIEW A

VALVE SPRING SPACER T91P-6565-AH

VALVE SPRING COMPRESSOR T91P-6565-A

VALVE SPRING SPACER T91P-6565-AH

VIEW A

7922QG07

To remove the rocker arm, install the spacer and compress the valve spring using the special tools—4.6L engine

7. Compress the valve spring far enough to remove the roller follower. Repeat Steps 5 and 6 until all roller followers are removed.

8. Remove the bolts retaining the camshaft cap cluster assemblies to the cylinder heads. Tap upward on the camshaft caps at points near the upper bearing halves and gradually lift the camshaft clusters from the cylinder heads.

9. Lift the camshafts straight upward to avoid bearing damage.

To install:

10. Apply heavy engine oil to the camshaft journals and lobes. Position the camshafts on the cylinder heads.

11. Install and seat the camshaft cap cluster assemblies. Hand start the bolts.

12. Tighten the camshaft cluster retaining bolts in sequence to 6.0–8.8 ft. lbs. (8–12 Nm).

➡**Each camshaft cap cluster assembly is tightened individually.**

13. Loosen the camshaft cap cluster retaining bolts approximately two turns or until the heads of the bolts are free.

Retighten all bolts, in sequence, to 71–106 inch lbs. (8–12 Nm).

➡**The camshafts should turn freely with a slight drag.**

14. Install cam positioning tools T91P-6256-A or equivalent, on the flats of the camshafts and install the spacers and camshaft sprockets. Install the bolts and washers and tighten to 81–96 ft. lbs. (110–130 Nm).

15. Install valve spring compressor T91P-6565-A or equivalent, under the camshaft and on top of the valve spring retainer.

➡**Valve spring spacer tool T91P-6565-AH or equivalent, must be installed between the spring coils and the camshaft must be at the base circle before compressing the valve spring.**

16. Compress the valve spring far enough to install the roller followers.

17. Repeat Steps 14 and 15 until all roller followers are installed.

18. Rotate the crankshaft clockwise 45 degrees to position the crankshaft at TDC.

➡**The crankshaft must only be rotated in the clockwise direction and only as far as TDC.**

19. Install the timing chains and install the timing chain front cover. Install the camshaft covers.

20. Install the remaining components in the reverse order of removal.

21. Connect the negative battery cable. Start the engine and check for leaks.

Valve Lash

ADJUSTMENT

3.8L Engines

The valve lash is not adjustable. If the collapsed lifter clearance is found to be incorrect, there are replacement pushrods available to compensate for excessive or insufficient clearance.

1. Disconnect the negative battery cable.

2. Remove the valve cover assembly on the side to be checked.

3. Turn the engine until the No. 1 piston is at TDC on the compression stroke.

4. The following valves can be checked with the engine in this position:
   a. No. 1 intake—No. 1 exhaust
   b. No. 3 intake—No. 2 exhaust
   c. No. 6 intake—No. 4 exhaust

5. Rotate the engine 360 degrees and check the following valves:
   a. No. 2 intake—No. 3 exhaust
   b. No. 4 intake—No. 5 exhaust
   c. No. 5 intake—No. 6 exhaust

6. Check each of the lifters by placing hydraulic lifter compressor tool T71P-6513-B or equivalent, on the rocker arm and slowly applying pressure to the lifter, until the lifter is collapsed.

7. Hold the lifter in this position and check the clearance between the rocker arm and the valve stem tip. The clearance should be 0.09–0.19 in. (2.25–4.79mm).

8. Repeat this operation for each valve to be checked.

9. If the clearance is greater than specification, replace the pushrod with a longer one. If the clearance is less than specified, replace the pushrod with a shorter one.

CAMSHAFT BEARING CAP   CAMSHAFT BEARING CAP

FRONT OF ENGINE   FRONT OF ENGINE

LH CYLINDER HEAD   **RH CYLINDER HEAD**

79220G16

To avoid damage to the camshaft and related components, tighten the cap cluster bolts in the order shown—4.6L engine

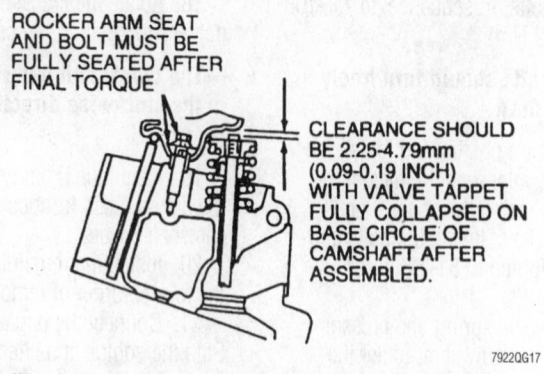

ROCKER ARM SEAT AND BOLT MUST BE FULLY SEATED AFTER FINAL TORQUE

CLEARANCE SHOULD BE 2.25-4.79mm (0.09-0.19 INCH) WITH VALVE TAPPET FULLY COLLAPSED ON BASE CIRCLE OF CAMSHAFT AFTER ASSEMBLED.

79220G17

Collapse the valve lifter before measuring the valve clearance—3.8L engines

4.6L Engine

The 4.6L SOHC engine requires no periodic valve adjustment. Hydraulic lash adjusters are used in the valve train to compensate for excessive valve lash.

Oil Pan

REMOVAL & INSTALLATION

3.8L Engines

1. Disconnect the negative battery cable and remove the air inlet tube.
2. Remove the two bolts retaining the sight shield and position aside. Remove the hood weather seal.
3. Remove the wipers. Remove the left cowl vent screen and the wiper module. On supercharged engines, remove the charge air cooler tubes.
4. Install engine support fixture tool D88L-6000-A or equivalent. Raise and safely support the vehicle.
5. Remove the engine mount through-bolts. On supercharged engine, remove the left mount retaining strap bolt.
6. Partially lower the vehicle and raise the engine with the support fixture.
7. Raise and safely support the vehicle. Remove the starter.

8. Drain the crankcase and remove the oil filter.
9. Remove the wire loom, ground strap and automatic transmission oil cooler lines, if equipped.
10. Remove the oil pan-to-bell housing bolts and the bolts at the crankshaft position sensor shield, if equipped. Remove the remaining oil pan retaining bolts.
11. Remove the steering shaft pinch bolts and separate the steering shaft. Position a jack under the front of the sub-frame.
12. Remove the six rearward bolts on the front of the sub-frame. Loosen the two front sub-frame bolts.
13. Remove the lower strut-to-control arm bolts and nuts and lower the sub-frame. Remove the oil pan.

To install:
14. Clean the gasket mating surfaces and the oil pan. Apply silicone sealer to the oil pan.
15. Fit the oil pan to the cylinder block. Be sure enough clearance has been provided to allow the oil pan to be installed without sealer being scraped off under the cylinder block.
16. Install the oil pan retaining bolts at the cylinder block and bell housing and install the lower crankshaft sensor shield, if equipped. Tighten the bolts to 80–106 inch lbs. (9–12 Nm).
17. Raise the sub-frame into position and install the lower strut mount-to-control arm bolts. Tighten to 103–144 ft. lbs. (140–195 Nm).
18. Install the two front sub-frame bolts and the six bolts at the rear of the front sub-frame member. Install a ¾ in. outside diameter pipe or equivalent, into both front left and right sub-frame and body alignment holes. Tighten one bolt at each corner. Remove the alignment tools and tighten the bolts to 70–96 ft. lbs. (95–130 Nm).
19. Connect the steering shaft and install the pinch bolt. Tighten to 30–42 ft. lbs. (41–57 Nm).

20. Install the transmission cooler lines, wire loom and ground strap. Install a new oil filter.
21. Install the starter and partially lower the vehicle.
22. Lower the engine with the support fixture. Seat the left side locating pin before the right. Partially raise the vehicle and support safely.
23. Install the engine mount through-bolts and tighten to 35–50 ft. lbs. (47–68 Nm). On supercharged engine, install the left mount retaining strap bolt and tighten to 33–45 ft. lbs. (45–61 Nm). Lower the vehicle.
24. Remove the engine support fixture. On supercharged engine, install the charge air cooler tubes.
25. Install the wiper module and the left cowl vent screen. Install the wipers and the hood weather seal.

❋❋ WARNING

Operating the engine without the proper amount and type of engine oil will result in severe engine damage.

26. Install the sight shield and the two retaining bolts. Install the air duct assembly. Fill the crankcase with the proper type and quantity of engine oil.
27. Connect the negative battery cable, start the engine and check for leaks.

4.6L Engine

1. Disconnect the battery cables and remove the air inlet tube.
2. Drain the cooling system and remove the cooling fan and shroud. Relieve the fuel system pressure and disconnect the fuel lines.
3. Remove the upper radiator hose. Remove the wiper module and support bracket.
4. Have the A/C system discharged by an EPA certified technician. Disconnect and plug the compressor outlet hose at the compressor and remove the bolt retaining the hose assembly to the right coil bracket. Cap the compressor outlet.
5. Remove the 42-pin engine harness connector from the retaining bracket on the brake vacuum booster and disconnect the connector and transmission harness connector.
6. Disconnect the throttle valve cable from the throttle body and disconnect the heater outlet hose.
7. Remove the nut retaining the ground strap to the right cylinder head. Remove the upper stud and loosen the

lower bolt retaining the heater outlet hose to the right cylinder head and position out of the way.

8. Remove the blower motor resistor. Remove the bolt retaining the right engine mount to the lower engine bracket.

9. Disconnect the vacuum hoses from the EGR valve and tube. Remove the two bolts retaining the EGR valve to the intake manifold.

10. Raise and safely support the vehicle. Drain the crankcase and remove the engine mount through-bolts.

11. Remove the EGR tube line nut from the right exhaust manifold and remove the EGR valve and tube assembly.

12. Disconnect the exhaust from the exhaust manifolds. Lower the exhaust system and support it with wire from the crossmember.

13. Position a jack and a block of wood under the oil pan, rearward of the oil drain hole. Raise the engine approximately 4 in. and insert two wood blocks approximately 2 ½ in. thick under each engine mount. Lower

17. Position the jack and wood block under the oil pan, rearward of the oil drain hole and raise the engine enough to remove the wood blocks. Lower the engine and remove the jack.

18. Install the engine mount through-bolts and tighten to 15–22 ft. lbs. (20–30 Nm).

19. Position the EGR valve and tube assembly in the vehicle and connect to the exhaust manifold. Tighten the line nut to 26–33 ft. lbs. (35–45 Nm).

➡**Loosen the line nut at the EGR valve prior to installing the assembly into the vehicle. This will allow enough movement to align the EGR valve retaining bolts.**

20. Cut the wire and position the exhaust system to the manifolds. Install the four nuts and tighten to 20–30 ft. lbs. (27–41 Nm). Be sure the exhaust system clears the crossmember. Adjust as necessary.

21. Install a new oil filter and lower the vehicle.

22. Install the bolt retaining the right engine mount to the lower engine bracket. Tighten to 15–22 ft. lbs. (20–30 Nm).

23. Install a new gasket on the EGR valve and position on the intake manifold. Install the two bolts retaining the EGR valve to the intake manifold and tighten to 15–22 ft. lbs. (20–30 Nm). Tighten the EGR tube line nut at the EGR valve to 26–33 ft. lbs. (35–45 Nm). Connect the vacuum hoses to the EGR valve and tube.

24. Install the blower motor resistor. Position the heater outlet hose, install the upper stud and tighten the upper and lower bolts to 15–22 ft. lbs. (20–30 Nm). Install the ground strap on the stud and tighten to 15–22 ft. lbs. (20–30 Nm).

25. Connect the heater outlet hose and the throttle valve cable. If necessary, adjust the throttle valve cable.

26. Connect the 42-pin connector and transmission harness connector. Install the harness connector on the brake vacuum booster.

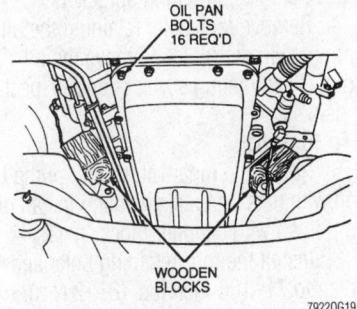

Place wooden blocks between the engine mounts and the frame to allow access to the oil pan—4.6L engine

the engine onto the wood blocks and remove the jack.

14. Remove the 16 bolts retaining the oil pan to the engine block and remove the oil pan.

➡**It may be necessary to loosen, but not remove, the two nuts on the rear transmission mount and with a jack, raise the transmission extension housing slightly to remove the pan.**

To install:

15. Clean the oil pan and the gasket mating surfaces.

16. Position a new gasket on the oil pan. Apply silicone sealer to where the front cover meets the cylinder block and rear seal retainer meets the cylinder block. Position the oil pan on the engine and install the bolts. Tighten the bolts, in sequence, to 15–22 ft. lbs. (20–30 Nm).

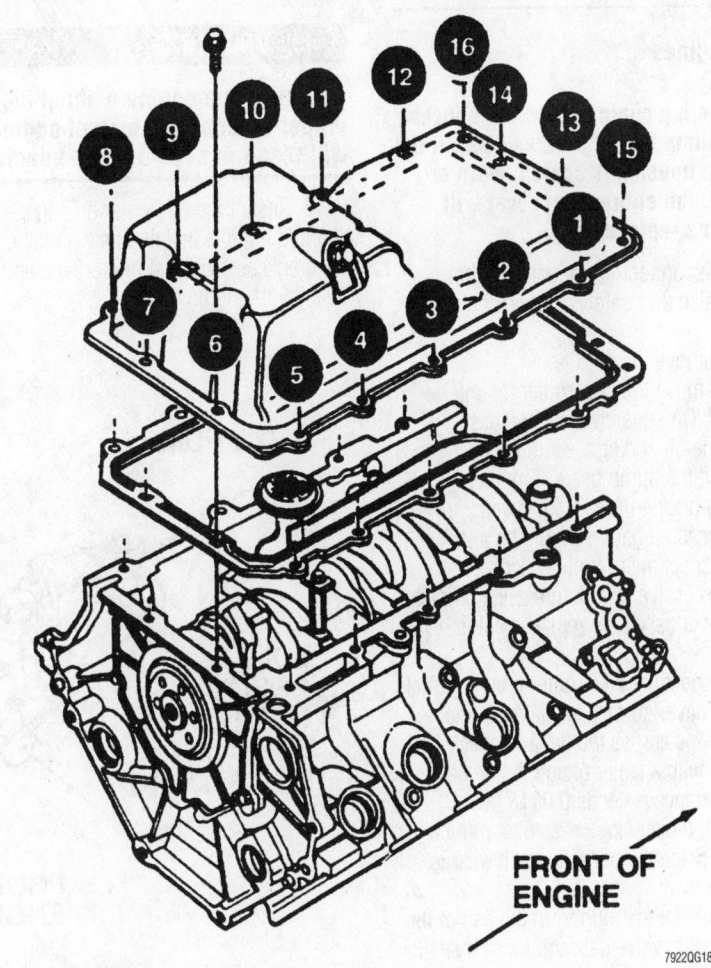

FRONT OF ENGINE

To prevent oil leaks, tighten the bolts in the order shown—4.6L engine

27. Connect the air conditioning compressor outlet hose to the compressor and install the bolt retaining the hose to the right coil bracket.

28. Install the upper radiator hose and connect the fuel lines. Install the wiper module and retaining bracket.

✸✸ WARNING

Operating the engine without the proper amount and type of engine oil will result in severe engine damage.

29. Install the cooling fan and shroud and fill the cooling system. Fill the crankcase with the proper type and quantity of engine oil.

30. Connect the negative battery cable and install the air inlet tube. Start the engine and check for leaks.

31. Evacuate and recharge the air conditioning system.

Oil Pump

REMOVAL & INSTALLATION

3.8L Engines

➡ **The timing chain front cover houses the oil pump on the 3.8L engines. If the oil pump housing is scored, worn or grooved, the entire front cover will have to be replaced.**

1. Disconnect the negative battery cable. Raise and safely support the vehicle.

2. Remove the oil filter.

3. Remove the cover/filter mount assembly. On supercharged engines, remove the oil cooler assembly.

4. Lift the pump gears from their mounting pocket in the front cover.

5. Clean all gasket mounting surfaces.

6. Inspect the mounting pocket for wear. If excessive wear is present, complete timing cover assembly replacement is necessary.

7. Inspect the cover/filter mount gasket to timing cover surface for flatness. Place a straight-edge across the flat and check clearance with a feeler gauge. If the measured clearance exceeds 0.0016 in. (0.04mm), replace the cover/filter mount.

8. Replace the pump gears if wear is excessive.

9. Remove the plug from the end of the pressure relief valve passage using a small drill and slide hammer. Use caution when drilling.

10. Remove the spring and valve from the bore. Clean all dirt and metal chips from the bore and valve. Inspect all parts for wear. Replace as necessary.

To install:

11. Install the valve and spring after lubricating them with engine oil. The end with the smaller diameter goes in first.

12. Install a new plug. The plug can be tapped into the bore using a plastic tipped hammer. Be sure the plug is 0–0.010 in. (0–0.25mm) below the machined surface.

13. Lightly pack the gear pocket with petroleum jelly. Install the gears in the cover pocket, making sure petroleum jelly fills all the voids between the gears and pockets.

➡ **Failure to properly coat the oil pump gears may result in failure of the pump to prime when the engine is started.**

14. Position the pump body O-ring seal and install the pump body to the front cover using alignment dowels on the front cover.

15. Tighten the pump body retaining bolts to 18–22 ft. lbs. (25–30 Nm) for M8 bolts and 30–40 ft. lbs. (40–55 Nm) for M10 bolts.

✸✸ WARNING

Operating the engine without the proper amount and type of engine oil will result in severe engine damage.

16. Install the oil cooler on supercharged engine. Install a new oil filter. Fill the crankcase with the proper type and quantity of engine oil.

17. Connect the negative battery cable, start the engine and check for leaks and proper oil pressure.

4.6L Engine

1. Disconnect the negative battery cable. Remove the front cover.

✸✸ CAUTION

The EPA warns that prolonged contact with used engine oil may cause a number of skin disorders, including cancer! You should make every effort to minimize your exposure to used engine oil. Protective gloves should be worn when changing the oil. Wash your hands and any other exposed skin areas as soon as possible after exposure to used engine oil. Soap and water, or waterless hand cleaner should be used.

2. Remove the oil pan and pick-up tube.

3. Remove primary timing chain only. Remove crankshaft timing sprockets.

4. Remove four bolts retaining the oil pump to cylinder block. Remove the oil pump. Clean mating surfaces and inspect for damage.

To install:

5. Rotate the inner rotor of oil pump to align with flats on crankshaft and install oil pump flush with cylinder block.

6. Install the four retaining bolts and tighten to 71–106 inch lbs. (8–12 Nm).

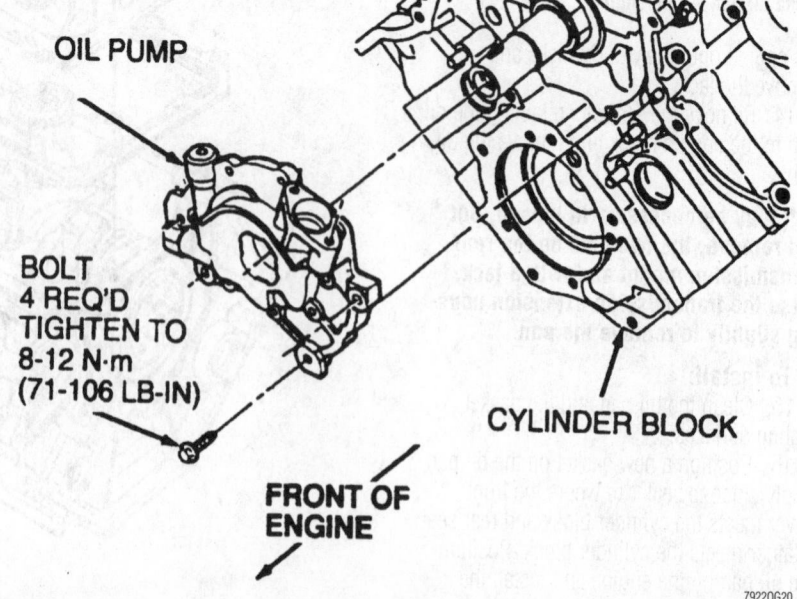

OIL PUMP

BOLT
4 REQ'D
TIGHTEN TO
8-12 N·m
(71-106 LB-IN)

CYLINDER BLOCK

FRONT OF ENGINE

Exploded view of the oil pump mounting—4.6L engine

7922QG20

7. Replace oil filter. Install timing chains.

8. Install pick-up tube and oil pan.

※※ WARNING

Operating the engine without the proper amount and type of engine oil will result in severe engine damage.

9. Install front cover. Fill the crankcase with the proper type and quantity of engine oil.

10. Connect the negative battery cable, start the engine and check for leaks and proper oil pressure.

Rear Main Seal

REMOVAL & INSTALLATION

➡**Special tools are available for installing rear main oil seals. In most cases, the seals can be installed using a common seal and bearing driver set.**

1. Remove the transmission assembly.

2. Remove the flexplate or flywheel.

3. With a sharp awl, carefully punch a small hole in the metal portion of the seal.

4. Remove the seal using a slide hammer with a sheet metal screw attached.

To install:

5. Lubricate the seal and the crankshaft with clean engine oil.

6. Install the seal with the spring side toward the engine.

7. Remove the installation tool and install the flexplate or flywheel.

8. Install the transmission assembly.

Timing Chain, Sprockets, Front Cover and Seal

REMOVAL & INSTALLATION

3.8L Engines

1. Disconnect the negative battery cable and drain the cooling system.

2. Remove the air cleaner assembly and air intake duct.

3. On non-supercharged engines, remove the fan/clutch assembly and shroud. On supercharged engines, remove the electric cooling fan assembly.

4. Remove the accessory drive belt idlers, drive belts and the water pump pulley.

5. Remove the power steering pump bracket retaining bolts. Leaving the hoses connected, place the pump/bracket assembly aside in a position to prevent fluid from leaking out.

6. If equipped with air conditioning, remove the compressor front support bracket but leave the compressor in place.

7. Disconnect the coolant bypass hose and heater hose at the water pump. Disconnect the upper radiator hose at the thermostat housing.

8. On non-supercharged engines, disconnect the coil wire from the distributor cap and remove the cap with the secondary wires attached.

9. On non-supercharged engines, remove the distributor hold-down clamp and lift the distributor out of the front cover.

10. On supercharged engines, remove the hold-down clamp and lift the camshaft synchronizer from the front cover.

11. Raise and safely support the vehicle. Remove the crankshaft damper and pulley using a puller.

➡**If the crankshaft pulley and vibration damper have to be separated, mark the damper and pulley so they may be reassembled in the same relative position. This is important as the damper and pulley are initially balanced as a unit. If the crankshaft damper is being replaced, check if the original damper has balance pins installed. If so, new balance pins must be installed on the new damper in the same position as the original damper. The crankshaft pulley, new or original, must also be installed in the same relative position as originally installed.**

※※ CAUTION

The EPA warns that prolonged contact with used engine oil may cause a number of skin disorders, including cancer! You should make every effort to minimize your exposure to used engine oil. Protective gloves should be worn when changing the oil. Wash your hands and any other exposed skin areas as soon as possible after exposure to used engine oil. Soap and water, or waterless hand cleaner should be used.

12. Remove the oil filter. On supercharged engines, remove the oil cooler.

13. Disconnect the lower radiator hose at the water pump. Remove the oil pan.

➡**The front cover cannot be removed without lowering the oil pan.**

14. Lower the vehicle. Remove the front cover retaining bolts. It is not necessary to remove the water pump.

➡**Do not overlook the cover retaining bolt located behind the oil filter adapter. The front cover will break if pried on and all retaining bolts are not removed.**

15. Remove the front cover and water pump as an assembly. Drive the crankshaft seal out of the front cover with a suitable seal driver. Remove and discard the cover gasket.

➡**The front cover contains the oil pump, water pump and crankshaft seal. If a new front cover is to be installed, remove the water pump and oil pump from the old front cover and install them on the new cover along with a new crankshaft seal.**

16. Remove the camshaft bolt and washer from the end of the camshaft.

17. Remove the distributor drive gear, camshaft sprocket, crankshaft sprocket and timing chain.

➡**If the crankshaft sprocket is difficult to remove, pry the sprocket off the shaft using a pair of large prybars positioned on both sides of the sprocket.**

To install:

18. Clean all gasket mating surfaces. If reusing the front cover, replace the front cover oil seal.

19. If removed, install the timing chain vibration damper. Tighten the mounting bolts to 71–123 inch lbs. (8–14 Nm).

20. Rotate the crankshaft to position the No. 1 piston at TDC and the crankshaft keyway at the 12 o' clock position.

21. Lubricate the timing chain with engine oil.

22. Install the camshaft sprocket, crankshaft sprocket and timing chain. Be sure the timing marks align.

23. Install the distributor drive gear. Install the bolt and washer assembly on the end of the camshaft and tighten to 30–37 ft. lbs. (40–50 Nm).

24. Install a new crankshaft seal in the front cover and lubricate the seal lip with engine oil.

25. Position a new gasket on the cylinder block and install the front cover using dowels for proper alignment. Install the front cover retaining bolts and tighten to 15–22 ft. lbs. (20–30 Nm).

26. Raise and safely support the vehicle. Install the oil pan. Connect the lower radiator hose and install the oil filter.

27. Coat the crankshaft damper sealing surface with clean engine oil. Apply a small amount of silicone sealer to the crankshaft keyway.

28. Position the crankshaft pulley key in the crankshaft keyway and install the damper, using a suitable installation tool.

29. Install the damper washer and retaining bolt and tighten to 103–132 ft. lbs. (140–180 Nm). Install the crankshaft pulley and tighten the retaining bolts to 20–28 ft. lbs. (26–38 Nm).

30. Lower the vehicle. Connect the coolant bypass hose.

31. On non-supercharged engines, install the distributor with the rotor pointing at the No. 1 distributor cap tower. Install the distributor cap and coil wire. On supercharged engines, install the camshaft synchronizer.

32. Connect the upper radiator hose at the thermostat housing. Connect the heater hose.

33. If equipped with air conditioning, install the compressor and mounting brackets. Tighten retaining bolts to 30–45 ft. lbs. (41–61 Nm).

34. Install the power steering pump and mounting bracket. Tighten the retaining bolts to 30–45 ft. lbs. (41–61 Nm).

35. Install the water pump pulley. Position the accessory drive belts over the pulleys.

36. On non-supercharged engines, install the fan/clutch assembly and fan shroud. Cross-tighten the fan/clutch assembly retaining bolts to 12–18 ft. lbs. (16–24 Nm).

37. On supercharged engines, install the electric cooling fan assembly and connect the harness connector to the fan motor.

❊❊ WARNING

Operating the engine without the proper amount and type of engine oil will result in severe engine damage.

38. Fill the crankcase with the proper type and quantity of engine oil. Fill and bleed the cooling system. Connect the negative battery cable.

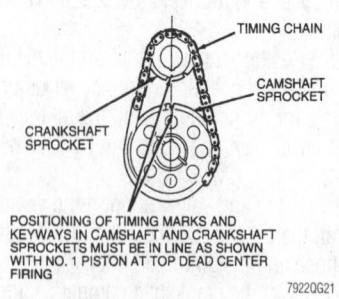

POSITIONING OF TIMING MARKS AND KEYWAYS IN CAMSHAFT AND CRANKSHAFT SPROCKETS MUST BE IN LINE AS SHOWN WITH NO. 1 PISTON AT TOP DEAD CENTER FIRING

79220G21

The timing marks should be facing each other, when the timing chain is installed correctly—3.8L engines

39. Start the engine and check for leaks. Check the ignition timing and curb idle speed and adjust, as necessary.

4.6L Engine

➡ **This is not a free wheeling engine. If it has "jumped time" there will be damage to the valves and/or pistons and will require the removal of the cylinder heads.**

1. Disconnect the negative battery cable.
2. Remove the cooling fan and shroud. Loosen the water pump pulley bolts, remove the accessory drive belt and remove the water pump pulley.
3. Raise and safely support the vehicle.
4. Remove the bolts retaining the power steering pump to the engine block and cylinder front cover. The lower front bolt on the power steering pump will not come all the way out. Wire the power steering pump out of the way.
5. Remove the four bolts retaining the oil pan to the front cover. Remove the crankshaft damper retaining bolt and washer. Remove the damper using a puller.
6. Lower the vehicle. Remove the bolt retaining the air conditioner high pressure line to the right coil bracket.
7. Remove the front bolts and loosen the remaining bolts on the camshaft covers. Using plastic wedges or similar tools, prop up both camshaft covers. Disconnect both ignition coils and CID sensor.

8. Remove the three nuts retaining the right coil bracket to the front cover. Position the power steering hose out of the way.

9. Remove the four nuts retaining the left coil bracket to the front cover. Slide both coil brackets and ignition wires off the mounting studs and lay the assembly on top of the engine.

10. Disconnect the crankshaft position sensor. Remove the seven stud bolts and four bolts retaining the front cover to the engine and remove the front cover. Remove the crankshaft seal with a suitable seal driver.

11. Remove the crankshaft position sensor wheel.

12. Rotate the engine to set the No. 1 piston at TDC on the compression stroke.

13. Install cam positioning tools T91P-6256-A or equivalent, on the flats of the camshaft. This will prevent accidental rotation of the camshafts.

14. Remove the two bolts retaining the right tensioner to the cylinder head and remove the tensioner. Remove the right tensioner arm.

15. Remove the two bolts retaining the right chain guide to the cylinder head and remove the chain guide. Remove the right chain and right crankshaft sprocket. Remove the right camshaft sprocket retaining bolt, washer, sprocket and spacer.

➡ **Cam positioning tools T91P-6256-A or equivalent, must be installed on the camshaft to prevent the camshaft from rotating.**

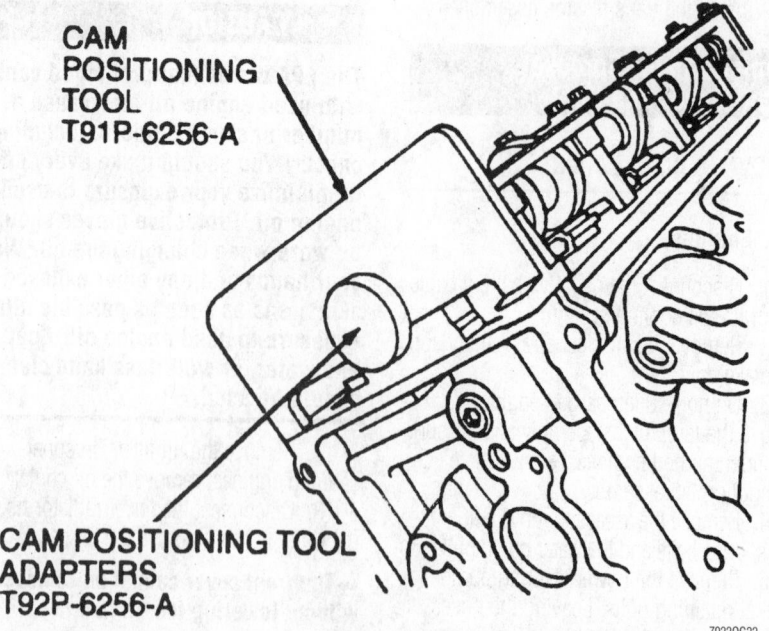

CAM POSITIONING TOOL T91P-6256-A

CAM POSITIONING TOOL ADAPTERS T92P-6256-A

79220G22

Install the Camshaft Positioning tool in order to keep the camshaft in the correct position—4.6L engine

16. Remove the two bolts retaining the left tensioner to the cylinder head and remove the tensioner. Remove the left tensioner arm.

17. Remove the two bolts retaining the left chain guide to the cylinder head and remove the chain guide. Remove the left chain and left crankshaft sprocket. Remove the left camshaft sprocket retaining bolt, washer, sprocket and spacer.

➡ **Cam positioning tools T91P-6256-A or equivalent, must be installed on the camshaft to prevent the camshaft from rotating.**

18. Inspect the friction material on the tensioner arms and chain guides. If worn or damaged, remove and clean the oil pan and replace the oil pick-up tube.

➡ **At no time, when the timing chains are removed and the cylinder heads are installed, may the crankshaft and/or camshafts be rotated. Failure to follow these directions will result in valve and/or piston damage.**

To install:

19. Be sure cam positioning tools T91P-6256-A or equivalent, are installed on the camshafts to prevent them from rotating.

20. Position the camshaft spacers and sprockets on the camshafts and install the washers and retaining bolts. Tighten the retaining bolts to 81–96 ft. lbs. (110–130 Nm).

21. Install the left crankshaft sprocket with the tapered part of the sprocket facing away from the engine block.

➡ **The crankshaft sprockets are identical. They may only be installed one way, with the tapered part of the sprocket facing each other.**

22. Install the left timing chain on the camshaft and crankshaft sprockets. Be sure the copper links of the chain line up with the timing marks of the sprockets.

➡ **If the copper links of the timing chain are not visible, pull the chain taught until the opposite sides of the chain contact one another and lay it on a flat surface. Mark the links at each end of the chain and use them in place of the copper links.**

23. Install the right crankshaft sprocket with the tapered part of the sprocket facing the left crankshaft sprocket.

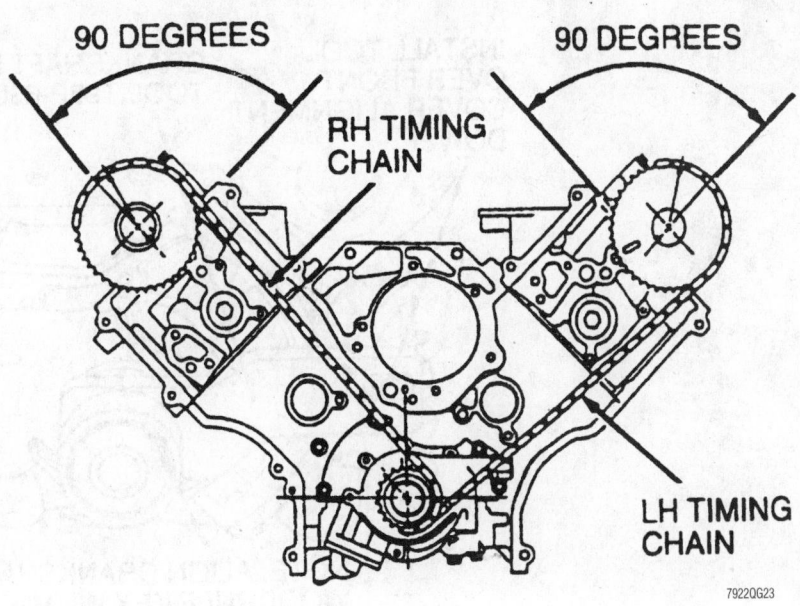

Be sure the marks on the sprockets align with the copper colored links on the chain—4.6L engine

24. Install the right timing chain on the camshaft and crankshaft sprockets. Be sure the copper links of the chain line up with the timing marks of the sprockets.

25. It is necessary to bleed the timing chain tensioners before installation. Proceed as follows:

a. Position the timing chain tensioner in a soft-jawed vice.

b. Using a small pick or similar tool, hold the ratchet lock mechanism away from the ratchet stem and slowly compress the tensioner plunger by rotating the vise handle.

➡ **The tensioner must be compressed slowly or damage to the internal seals will result.**

c. Once the tensioner plunger bottoms in the tensioner bore, continue to hold the ratchet lock mechanism and push down on the ratchet stem until flush with the tensioner face.

d. While holding the ratchet stem flush to the tensioner face, release the ratchet lock mechanism and install a paper clip or similar tool in the tensioner body to lock the tensioner in the collapsed position.

e. The paper clip must not be removed until the timing chain, tensioner, tensioner arm and timing chain guide are completely installed on the engine.

26. Lubricate the tensioner arm contact surfaces with engine oil and install the right and left tensioner arms on their dowels.

27. Install the right and left timing chain tensioners and secure with two bolts on each. Tighten the bolts to 15–22 ft. lbs. (20–30 Nm).

28. Install the right and left timing chain guides and secure with two bolts on each. Tighten the bolts to 6.0–8.8 ft. lbs. (8–12 Nm).

29. Install the Crankshaft Holding tool T93P-6303-A or equivalent to the crankshaft and the alignment dowel on the front cover mounting surface.

30. Remove the slack from the timing chain using a suitable C-clamp positioned on the timing chain guide and tensioner arm.

31. Remove the paper clips from the timing chain tensioners and be sure all timing marks are aligned.

32. Remove the camshaft positioning tools.

33. Replace the front cover seal with a new one and clean the sealing surfaces of the cylinder block. Apply silicone sealer to the oil pan where it meets the cylinder block and to the points where the cylinder head meets the cylinder block.

34. Install the front cover and the attaching studs and bolts. Tighten to 15–22 ft. lbs. (20–30 Nm). Connect the crankshaft position sensor.

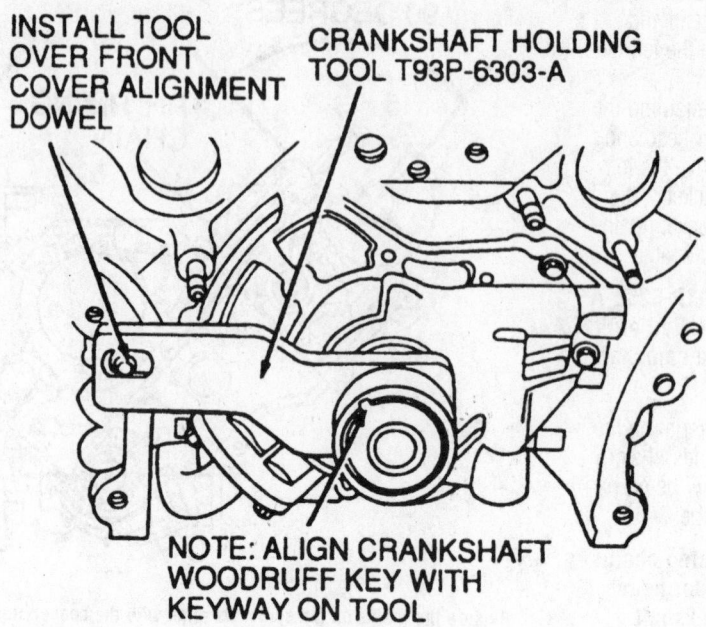

INSTALL TOOL OVER FRONT COVER ALIGNMENT DOWEL

CRANKSHAFT HOLDING TOOL T93P-6303-A

NOTE: ALIGN CRANKSHAFT WOODRUFF KEY WITH KEYWAY ON TOOL

7922QG25

Install the Crankshaft Holding tool T93P-6303-A or equivalent to maintain crankshaft position while removing the slack from the timing chain—4.6L engine

NOTE: USE C-CLAMP TO REMOVE SLACK FROM CAMSHAFT TIMING CHAIN

7922QG26

Remove the slack from the timing chain with a C-clamp before removing the pin from the tensioner—4.6L engine

35. Position the coil brackets and ignition wires as an assembly onto the mounting studs. Position the power steering hose and install the seven nuts retaining the coil brackets to the front cover. Tighten the nuts to 15–22 ft. lbs. (20–30 Nm). Connect both ignition coils and CID sensor.

36. Remove the plastic wedges holding up the camshaft covers. Apply silicone sealer where the front cover meets the cylinder head and be sure the camshaft cover gaskets are properly positioned. Install the front retaining bolts into the camshaft cover and tighten the bolts to 6.0–8.8 ft. lbs. (8–12 Nm).

37. Position the air conditioner high pressure line on the right coil bracket and install the bolt. Raise and safely support the vehicle.

38. Apply a small amount of silicone sealer in the rear of the keyway in the damper. Position the damper on the crankshaft and install, using a suitable installation tool. Install the damper bolt and washer and tighten to 114–121 ft. lbs. (155–165 Nm).

39. Install the four bolts retaining the oil pan to the front cover. Tighten to 15–22 ft. lbs. (20–30 Nm).

40. Position the power steering pump on the engine and install the four retaining bolts. Tighten to 15–22 ft. lbs. (20–30 Nm). Lower the vehicle.

41. Install the water pump pulley with the four bolts. Tighten to 15–22 ft. lbs. (20–30 Nm). Install the accessory drive belt and the cooling fan and shroud.

42. Connect the negative battery cable, start the engine and check for leaks.

FUEL SYSTEM

Fuel System Service Precautions

Safety is the most important factor when performing not only fuel system maintenance but any type of maintenance. Failure to conduct maintenance and repairs in a safe manner may result in serious personal injury or death. Maintenance and testing of the vehicle's fuel system components can be accomplished safely and effectively by adhering to the following rules and guidelines.

• To avoid the possibility of fire and personal injury, always disconnect the negative battery cable unless the repair or test procedure requires that battery voltage be applied.

• Always relieve the fuel system pressure prior to disconnecting any fuel system component (injector, fuel rail, pressure regulator, etc.), fitting or fuel line connection. Exercise extreme caution whenever relieving fuel system pressure to avoid exposing skin, face and eyes to fuel spray. Please be advised that fuel under pressure may penetrate the skin or any part of the body that it contacts.

• Always place a shop towel or cloth around the fitting or connection prior to loosening to absorb any excess fuel due to spillage. Ensure that all fuel spillage (should it occur) is quickly removed from engine surfaces. Ensure that all fuel soaked cloths or towels are deposited into a suitable waste container.

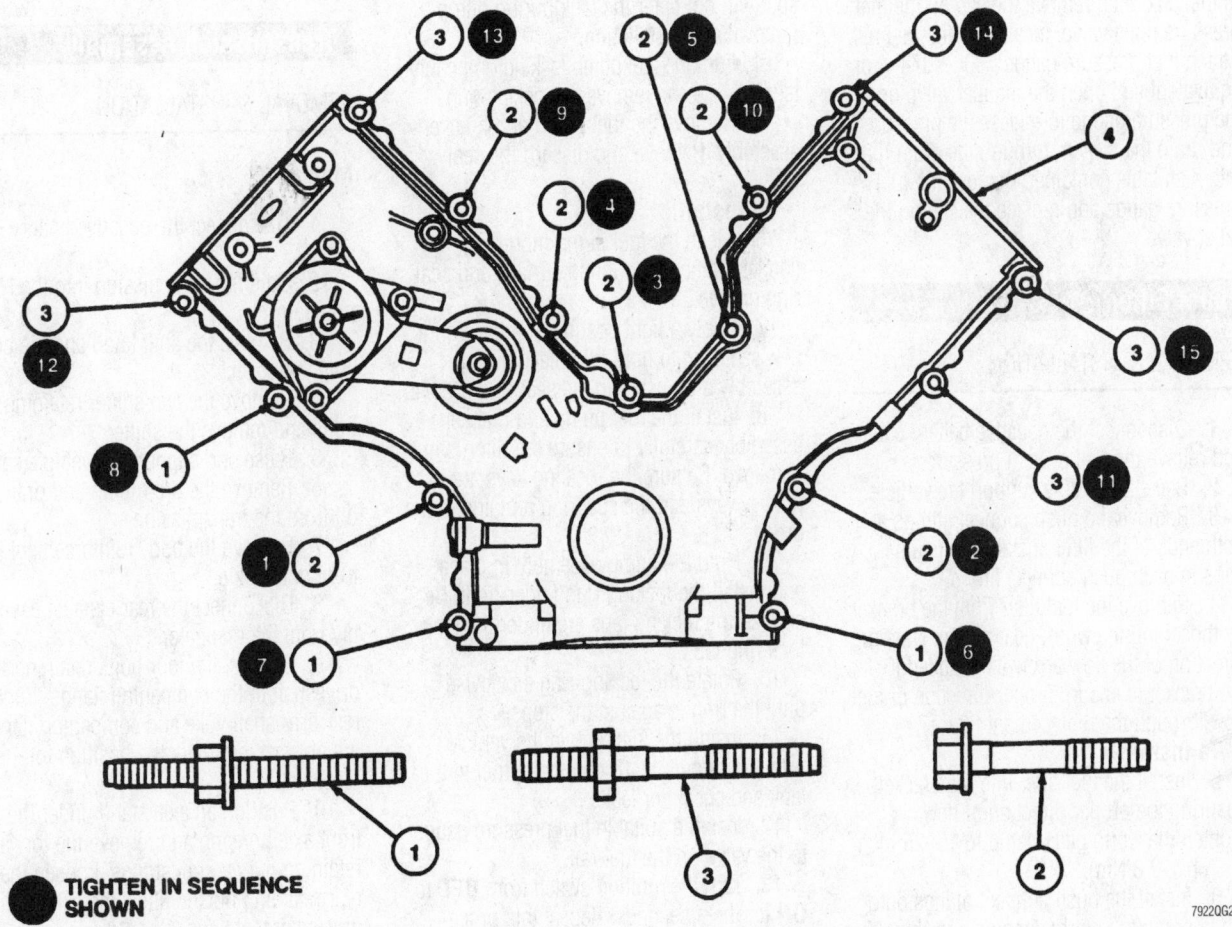

TIGHTEN IN SEQUENCE SHOWN

7922QG24

Be sure to install the fasteners in the correct positions and tighten them in the sequence shown—4.6L engine

• Always keep a dry chemical (Class B) fire extinguisher near the work area.

• Do not allow fuel spray or fuel vapors to come into contact with a spark or open flame.

• Always use a back-up wrench when loosening and tightening fuel line connection fittings. This will prevent unnecessary stress and torsion to fuel line piping. Always follow the proper torque specifications.

• Always replace worn fuel fitting O-rings with new. Do not substitute fuel hose or equivalent, where fuel pipe is installed.

Fuel System Pressure

RELIEVING

Fuel supply lines on all fuel injected engines will remain pressurized for some period of time after the engine is shut OFF. This pressure must be relieved before servicing the fuel system. Pressure is relieved through the fuel pressure relief valve. To relieve the fuel system pressure, first remove the fuel tank cap to relieve pressure in the tank, then remove the cap on the fuel pressure relief valve, located on the fuel rail. Attach fuel pressure gauge T80L-9974-A or equivalent and open the manual valve on the pressure gauge to release the pressure and drain the system through the drain tube into a suitable container. Remove the fuel pressure gauge and replace the cap on the relief valve.

Fuel Filter

REMOVAL & INSTALLATION

1. Disconnect the negative battery cable and relieve the fuel system pressure.
2. Raise and safely support the vehicle.
3. Remove the push connect fittings at both ends of the filter. Install new retainer clips in each push connect fitting.
4. Remove the fuel filter from the bracket by loosening the worm gear clamp. Note the direction of the flow arrow as installed in the bracket to ensure proper direction of fuel flow through the replacement filter.

   To install:
5. Install the fuel filter into the bracket, ensuring the proper direction of flow. Tighten the worm gear clamp to 15–25 inch lbs. (1.7–2.8 Nm).
6. Install the push connect fittings onto the filter ends. Start the engine and check for leaks.
7. Lower the vehicle.

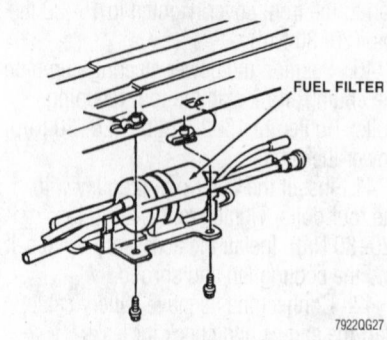

Exploded view of the fuel filter mounting

Fuel Pump

REMOVAL & INSTALLATION

1. Disconnect the negative battery cable and properly relieve the fuel system pressure.
2. Remove the fuel tank and place it on a bench.
3. Remove any dirt that has accumulated around the fuel pump retaining flange so it will not enter the tank during pump removal and installation.
4. Turn the fuel pump locking ring counterclockwise and remove the locking ring.
5. Remove the fuel pump and bracket assembly. Remove and discard the seal ring.

   To install:
6. Clean the fuel pump mounting flange, fuel tank mounting surface and seal ring groove.
7. Apply a light coating of grease on a new seal ring to hold it in place during assembly and install in the seal ring groove.
8. Install the fuel pump and bracket assembly carefully to ensure the filter is not damaged. Be sure the locating keys are in the keyways and the seal ring remains in the groove.
9. Hold the pump assembly in place and install the locking ring finger-tight. Be sure all the locking tabs are under the tank lock ring tabs.
10. Rotate the locking ring clockwise until the ring is against the stops.
11. Install the fuel tank in the vehicle. Add a minimum of 10 gallons of fuel to the tank and check for leaks.
12. Install a suitable fuel pressure gauge to the valve on the fuel rail.
13. Turn the ignition switch from **OFF** to **ON** for three seconds. Repeat this procedure 5–10 times until the pressure gauge shows at least 35 psi. Check for fuel leaks.

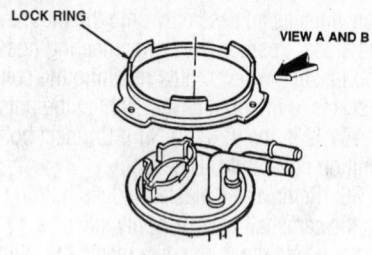
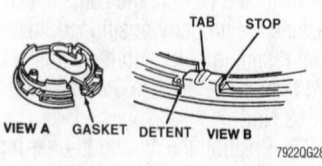

To remove the fuel pump, turn the locking ring counterclockwise

14. Remove the pressure gauge, start the engine and check for leaks.

DRIVE TRAIN

Transmission Assembly

REMOVAL & INSTALLATION

Manual

1. Disconnect the negative battery cable.
2. Shift the transmission into the **N** position.
3. Remove the shift knob and the console top cover.
4. Remove the two shifter retaining bolts and remove the shifter.
5. Raise and support the vehicle safely.
6. Remove the drain plug and drain the oil from the transmission.
7. Remove the body reinforcement in front of the axle.
8. Disconnect the rear exhaust assembly from the resonator.
9. Remove the four bolts retaining the driveshaft to the companion flange. If the rear driveshaft yoke and companion flange are not marked, mark the position for reassembly.
10. Position an axle stand under the front axle housing and remove the forward retaining nuts and bushings. Loosen the rear retaining nuts to allow the axle to tilt for driveshaft removal.
11. Pull the vent tube from the hole in the sub-frame.

12. Lower the front of the axle housing with the axle stand and slide the driveshaft out of the transmission above the axle housing. Let the driveshaft rest on the front driveshaft support and axle assembly.

13. Remove the catalytic converter.

14. Disconnect the hydraulic clutch line.

15. Disconnect the electrical connectors and remove the starter.

16. Position a transmission jack under the transmission. Remove the crossmember and the bell housing to engine bolts.

17. Move the transmission to the rear until the input shaft clears the flywheel and lower the transmission from the vehicle.

To install:

18. Install guide studs in the engine block and raise the transmission until the input shaft splines are aligned with the clutch disc splines.

19. Slide the transmission forward on the guide studs until it is against the bell housing. Install the bell housing-to-engine retaining bolts and tighten to 28–38 ft. lbs. (38–51 Nm).

20. Install the crossmember and tighten the bolts to 35–50 ft. lbs. (47–68 Nm). Remove the transmission jack.

21. Install the starter and connect the electrical connectors. Connect the hydraulic clutch line.

22. Install the catalytic converter assembly.

23. Lubricate the splines with grease and slide the driveshaft into the transmission.

24. Raise the axle housing with the axle stand and install the bushings and retaining nuts. Tighten the retaining nuts to 68–100 ft. lbs. (92–136 Nm) and remove the axle stand.

25. Position the vent tube in the hole of the sub-frame.

26. Align the driveshaft yoke and companion flange and install the retaining bolts. Tighten to 71–96 ft. lbs. (95–129 Nm).

27. Connect the exhaust pipe muffler assembly to the resonator. Lower the vehicle.

28. Position the shifter and install the retaining bolts. Tighten to 18–24 ft. lbs. (24–33 Nm). Install the console top cover and the shifter knob.

29. Connect the negative battery cable. Check transmission operation.

Automatic

1. Disconnect the negative battery cable.

2. Raise and safely support the vehicle.

3. Drain the fluid from the transmission by removing all of the oil pan bolts except the two at the front of the pan. Loosen the bolts and slowly lower the rear of the pan to allow the fluid to drain into a suitable container. When drained, reinstall a few of the bolts to hold the transmission oil pan in place.

4. Remove the inspection cover and remove the torque converter drain plug if equipped, allowing the torque converter to drain. After the converter has drained, reinstall the drain plug and tighten.

5. Crank the engine over with a wrench on the crankshaft pulley retaining bolt to access and remove four torque converter-to-flywheel retaining nuts.

➡️**To prevent possible engine damage, never rotate the crankshaft pulley in a counterclockwise direction as viewed from the front of the engine.**

6. Remove the exhaust system as required for transmission removal.

7. Match mark and disconnect the driveshaft assembly, proceed as follows:

 a. Loosen the differential housing assembly rear mounting nuts approximately ¼ in.

 b. Position an axle stand under the front of the differential housing and remove the forward mounting nuts and bushings. Pull the vent tube from the hole in the sub-frame.

 c. Lower the front of the differential housing with the axle stand and slide the driveshaft out of the transmission above the axle housing. Let the drive shaft rest on the front driveshaft support and axle assembly.

8. Remove the speedometer cable or Vehicle Speed Sensor (VSS) as equipped, from the extension housing.

9. Disconnect the shift cable from the transmission manual control lever.

10. Remove the starter cable and remove the starter motor.

11. Remove the electrical wires and vacuum lines as required from the transmission assembly.

12. Place a suitable transmission jack under the transmission and raise it slightly.

13. Remove the transmission support insulator (mount) to crossmember bolts.

14. Remove the crossmember-to-frame side rail retaining bolts. Remove the crossmember and the transmission support insulator.

15. Lower the transmission jack allowing the transmission to hang.

16. Place a suitable jack under the front of the engine and raise the engine slightly to gain access to two upper transmission bell housing-to-cylinder block retaining bolts. Do not remove the bolts at this time.

17. Disconnect and plug the oil cooler lines at the transmission.

18. Remove the lower transmission bell housing-to-cylinder block retaining bolts.

19. Remove the transmission fluid fill tube.

20. Secure the transmission to the transmission jack with a safely chain or strap.

21. Remove the two upper transmission bell housing-to-cylinder block retaining bolts.

22. Carefully move the transmission to the rear of the vehicle disengaging the transmission from the cylinder block dowels and the torque converter from the flywheel.

23. Lower the transmission from the vehicle.

24. Remove the torque converter from the transmission.

To install:

➡️**Thoroughly flush the transmission oil cooler and cooler lines of any contaminates before installing the transmission.**

25. Place the transmission on a suitable transmission jack and secure with a safety chain or strap.

26. Tighten the torque converter drain plug to 8–28 ft. lbs. (11–38 Nm).

27. Place the torque converter to the transmission and rotate into position to be sure the drive flats are fully engaged in the pump gear.

28. Raise the converter and transmission assembly into position. Rotate the converter until the studs and drain plug are in alignment with the holes in the flywheel. Align the orange balancing marks on the converter stud and flywheel bolt hole if balancing marks are present.

➡️**The torque converter face must rest squarely against the flywheel. This indicates that the converter pilot is not binding in the engine crankshaft. To ensure the converter is properly seated, grasp a converter stud. It should move freely back and forth in the flywheel hole. If the converter will not move, the transmission must be removed and the converter repositioned so the impeller hub is properly engaged in the pump gear.**

29. Install the transmission bell housing-to-cylinder block retaining bolts. Tighten the bolts to 40–50 ft. lbs. (55–68 Nm)

30. Remove the safety chain or strap from around the transmission.

31. Install a new O-ring on the lower end of the transmission fill tube and install the tube to the transmission case.

32. Unplug and connect the oil cooler lines to the transmission case.

33. Raise the transmission slightly and place the crossmember on the side supports. Position the transmission support insulator on the crossmember and install the retaining bolts and/or nuts to the frame side rails and the transmission extension housing.

34. Lower the transmission jack and remove it from under the vehicle.

35. Install the exhaust system components removed for transmission removal.

36. Connect the speedometer or VSS cable to the transmission extension housing.

37. Install the starter motor and wiring.

38. Connect the shift cable to the manual control lever.

39. Connect all transmission electrical connectors.

40. Install the four torque converter-to-flywheel retaining nuts. Tighten the nuts to 20–34 ft. lbs. (27–46 Nm).

41. Install the transmission inspection cover.

42. Install the driveshaft, making sure the index marks are aligned. Proceed as follows:

    a. Raise the differential housing with the axle stand and install the bushings and retaining nuts. Tighten to 68–100 ft. lbs. (92–136 Nm). Remove the axle stand.

    b. Install the vent tube in the hole of the sub-frame.

    c. Align the driveshaft yoke and companion flange and install the retaining bolts. Tighten to 70–95 ft. lbs. (95–129 Nm).

43. Remove all supports, if applicable, and lower the vehicle.

44. Connect the negative battery cable.

45. Fill the transmission with the proper type and quantity of fluid.

46. Start the engine and check the transmission for leakage.

47. Check and adjust the shift linkage as required.

48. Road test the vehicle and check the transmission for proper operation.

Clutch

REMOVAL & INSTALLATION

1. Disconnect the negative battery cable.

2. Disconnect the clutch hydraulic system master cylinder from the clutch pedal.

3. Raise and support the vehicle safely.

4. Remove the starter.

5. Disconnect the hydraulic coupling at the transmission with tool T88T-70522-A or equivalent, by sliding the white plastic sleeve toward the slave cylinder and applying a slight tug on the tube.

6. Remove the transmission.

7. Matchmark the assembled position of the pressure plate to the flywheel.

8. Loosen the pressure plate attaching bolts evenly until the pressure plate springs are expanded, and remove the bolts. Be sure to support the pressure plate before removing the last bolt.

9. Remove the pressure plate and clutch disc from the flywheel.

10. Inspect the flywheel for scoring, cracks or other damage and machine or replace, as necessary. Inspect the pilot bearing for damage and free movement. Replace, as necessary.

To install:

11. If removed, install the flywheel. Be sure the mating surfaces of the flywheel and the crankshaft flange are clean prior to installation. Tighten the flywheel bolts to 54–64 ft. lbs. (73–87 Nm).

12. Position the clutch disc on the flywheel so a suitable alignment tool can enter the clutch pilot bearing and align the disc.

13. If reinstalling the original pressure plate, align the matchmarks. Position the pressure plate on the flywheel and install the retaining bolts hand-tight. Tighten the bolts, in sequence, to 20–28 ft. lbs. (27–39 Nm). Remove the alignment tool.

14. Install the remaining components in the reverse order of removal. Tighten the flywheel housing-to-engine bolts to 40–49 ft. lbs. (54–67 Nm).

➡**Reuse the aluminum washers under the attaching bolts to prevent galvanic corrosion.**

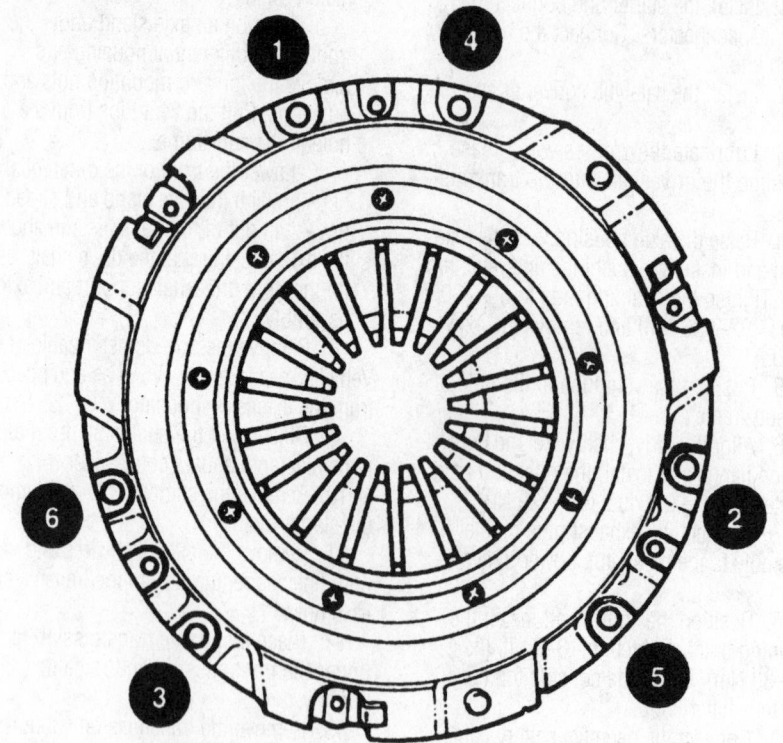

Tighten the pressure plate bolts gradually in the correct sequence

79220G29

Hydraulic Clutch System

BLEEDING

➡**Be sure to pump the clutch at least 30 times to be sure air is in the system. If the slave cylinder is pushed off the clutch plate, a similar pedal feel may occur. Pumping the clutch pushes fluid from the clutch reservoir into the slave cylinder, pushing it out to meet the clutch plate.**

1. Clean all dirt and grease from the cap to be sure no foreign substances enter the system.
2. Remove the cap and diaphragm and fill the reservoir to the top with the proper fluid.
3. Raise and support the vehicle safely.
4. Attach a hose to the bleeder valve at the slave cylinder.

➡**Keep the clutch fluid reservoir full at all times to prevent air from being pulled into the system.**

5. While the clutch pedal is being depressed, slightly open the bleeder valve and observe air bubbles in the clutch fluid at the end of the hose.
6. Close the bleeder valve before releasing the clutch pedal.
7. Repeat Steps 5 and 6, as necessary, until no air bubbles are observed.
8. Lower the vehicle and fill the reservoir. Road test the vehicle.

Halfshaft

REMOVAL & INSTALLATION

➡**Before continuing with this procedure, be sure to have available one new inboard CV-joint stub shaft pilot bearing housing seal, one shaft bearing retaining clip and one wheel hub retainer nut, per side. Once removed, these parts lose their torque holding ability or retention capability and must not be reused.**

1. With the vehicle sitting on the ground, remove the wheel hub retainer nut.
2. Raise and safely support the vehicle.
3. Remove the wheel and tire assembly.
4. Remove the anti-lock brake speed sensor, if equipped.
5. Use needle-nose pliers or equivalent, to slide the parking brake cable adjusting clip downward until the cable is free.
6. Remove the parking brake cable from the rear disc brake caliper.
7. Remove the upper and lower caliper retaining bolts and remove the caliper. Support the caliper aside with wire, do not allow it to hang from the brake hose.
8. Remove the upper control arm retaining nut and bolt. Wire the upper control arm to the top of the shock absorber.
9. Using a paint marker, match mark the position of the lower control arm in relation to the knuckle with the lower bushings in the relaxed position.

➡**Failure to match mark this relationship will result in bushing wind-up on assembly and incorrect ride height, causing misalignment and premature tire wear.**

10. Use a suitable puller to free the halfshaft from the hub.
11. Remove the lower control arm to knuckle retaining nuts. Push the halfshaft through the hub while positioning the CV-joint and knuckle to allow the front lower bolt to clear the CV-joint and remove the bolt. Remove and save the washers.
12. Remove the rear bolt, washer, and knuckle assembly. Be careful not to damage the differential oil seal, differential housing and/or CV-joint boot.
13. Remove the halfshaft from the vehicle.
14. Insert a plug into the differential housing to prevent fluid loss.

To install:

15. Remove the differential plug and install a new differential oil seal.
16. Install a new circlip on the halfshaft. Start the ends in the groove and push the circlip into the groove, to prevent over expanding the circlip.
17. Lightly lubricate the stub shaft splines and carefully align the splines on the shaft with the splines in the differential.
18. Push the halfshaft inward to seat the circlip in the differential side gear groove. Use care not to damage the seal.
19. Engage the hub splines with the outboard CV-joint splines.
20. Install the lower control arm retaining bolts and nuts. Align the paint marks and tighten the bolts to 94–131 ft. lbs. (128–178 Nm).
21. Install a new wheel hub retaining nut and use it to draw the CV-joint into the hub as far as possible by hand.
22. Install the upper arm retaining bolt and nut and tighten to 117–142 ft. lbs. (158–193 Nm).
23. Install the brake caliper assembly to the rotor with the outer brake shoe against the rotor's braking surface. This prevents pinching the piston boot between the inner brake shoe and the piston.
24. Install the upper and lower caliper retaining bolts and tighten to 64–88 ft. lbs. (87–119 Nm).

25. Install the parking brake cable to the brake caliper. Install the cable adjustment clip.
26. Install the anti-lock brake speed sensor, if equipped. Tighten the retaining bolts to 15–19 ft. lbs. (19–27 Nm).
27. Check the inboard CV-joint circlip engagement by attempting to pull the inboard CV-joint from the axle. If the CV-joint circlip is not seated, push the CV-joint in until the circlip is fully engaged in the side gear.
28. Check the axle lubricant level and fill, as necessary.
29. Install the wheel and tire assembly. Tighten the lug nuts to 85–105 ft. lbs. (115–142 Nm).
30. Lower the vehicle.
31. Pump the brake pedal several times to position the brake pads before moving the vehicle.
32. Tighten the wheel hub retainer nut to 250 ft. lbs. (340 Nm).
33. Check the rear alignment and adjust if not within specification.
34. Road test the vehicle and check for proper operation.

STEERING AND SUSPENSION

Air Bag

✳✳ CAUTION

Some vehicles are equipped with an air bag system, also known as the Supplemental Inflatable Restraint (SIR) or Supplemental Restraint System (SRS). The system must be disabled before performing service on or around system components, steering column, instrument panel components, wiring and sensors. Failure to follow safety and disabling procedures could result in accidental air bag deployment, possible personal injury and unnecessary system repairs.

PRECAUTIONS

Several precautions must be observed when handling the inflator module to avoid accidental deployment and possible personal injury.

• Never carry the inflator module by the wires or connector on the underside of the module.

• When carrying a live inflator module, hold securely with both hands, and ensure that the bag and trim cover are pointed away.

• Place the inflator module on a bench or other surface with the bag and trim cover facing up.

• With the inflator module on the bench, never place anything on or close to the module which may be thrown in the event of an accidental deployment.

DISARMING

1. Disconnect the positive battery cable. Wait one minute for the back-up power supply in the diagnostic monitor to deplete its stored energy.

2. Remove the four nut and washer assemblies retaining the driver air bag module to the steering wheel.

✽✽ CAUTION

When carrying a live air bag, be sure the bag and trim cover are pointed away from the body. In the unlikely event of an accidental deployment, the bag will, then deploy with minimal chance of injury. When placing a live air bag on a bench or other surface, always face the bag and trim cover up, away from the surface. This will reduce the motion of the module if it is accidentally deployed.

3. Disconnect the driver air bag connector. Connect air bag simulator tool 105–00008 or equivalent, to the vehicle harness at the top of the steering wheel.

After the applicable service is accomplished, re-enable the air bag system as follows:

4. Remove the air bag simulator from the air bag connector.

5. Attach the connector and install the air bag.

6. Connect the positive battery cable.

7. Turn the ignition key **ON** and be sure the air bag light does not stay on.

Power Rack and Pinion Steering Gear

REMOVAL & INSTALLATION

1. Disconnect the negative battery cable.

2. Turn the ignition key to the **ON** position.

3. Raise and safely support the vehicle.

4. Place a drain pan under the rack and pinion assembly.

5. Remove the bolt retaining the flexible coupling to the rack and pinion input shaft.

6. Remove the cotter pins and castellated nuts from both tie rod ends. Discard the cotter pins.

7. Separate the tie rod end studs from the wheel spindles using Tie Rod End Remover T-3290-D, or equivalent.

8. Remove the two nuts and insulator washers retaining the rack and pinion assembly to the front crossmember.

Steering Gear

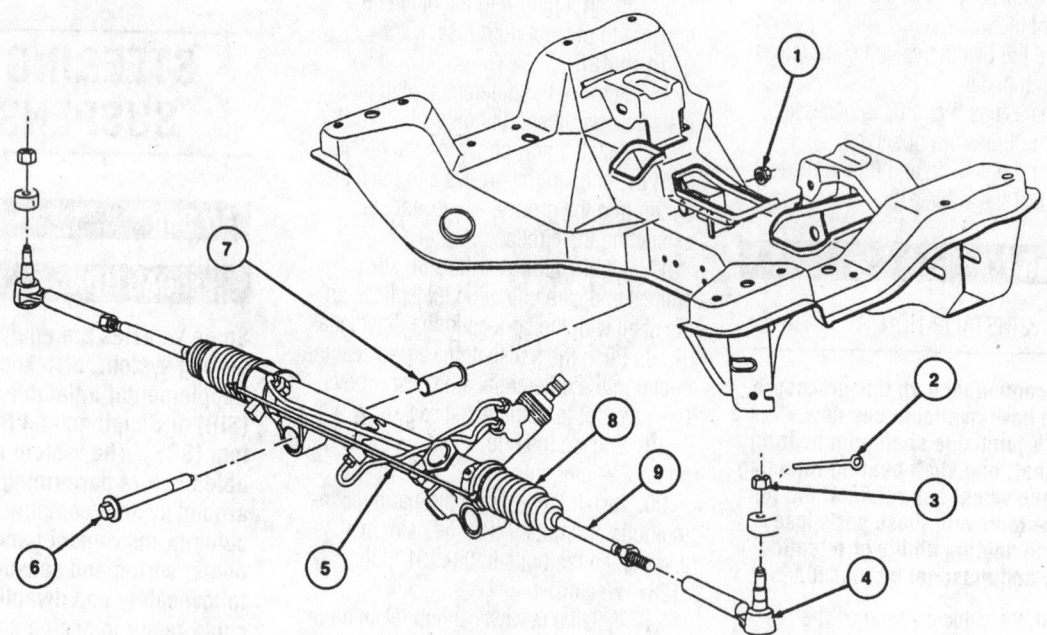

Item	Description
1	Nut (2 Req'd)
2	Pin (2 Req'd)
3	Nut (2 Req'd)
4	Tie Rod End (2 Req'd)
5	Steering Gear
6	Bolt and Washer Assy (2 Req'd)

Item	Description
7	Steering Gear Insulator (2 Req'd)
8	Front Suspension Steering Ball Stud Dust Seal
9	Front Wheel Spindle Tie Rod (2 Req'd)

7922QG30

Exploded view of the power rack and pinion steering gear mounting

9. Remove the rubber insulators, if needed.

10. Move the rack and pinion assembly to gain access to the pressure and return lines at the rack.

11. Disconnect the pressure and return lines allowing the power steering fluid to drain into the drain pan.

12. Disconnect the electrical harness connector at the power steering pressure switch on the rack and pinion assembly, if equipped.

13. Remove the rack and pinion assembly from the vehicle.

14. Remove the inner tie rod spindles and install on the new rack and pinion assembly if they are in good condition.

To install:

15. Install the rack and pinion assembly into the vehicle allowing access for connecting the power steering pressure and return lines to the rack.

16. Install new plastic seals on the power steering line fittings.

17. Be sure that the rubber insulators are fully seated in the rack housing.

18. Place the rack and pinion assembly in position on the mounting bolts.

19. Install the power steering pressure and return lines. Tighten the line fittings to 20–25 ft. lbs. (27–34 Nm).

20. Connect the rack and pinion input shaft with the flexible coupling of the steering column. The design allows the hoses to swivel when tightened properly. Do not attempt to eliminate looseness by overtightening, since this can cause damage to the fittings and leaks.

21. Install the two insulator washers and nuts securing the rack and pinion assembly to the crossmember. Tighten the nuts to 30–40 ft. lbs. (41–54 Nm).

22. Install the flexible coupling bolt and tighten to 20–30 ft. lbs. (28–40 Nm).

23. Install the tie rod end studs to the wheel spindles and install the castellated nuts. Tighten the nuts to 35–47 ft. lbs. (48–63 Nm), then continue to tighten until a new cotter pin can be installed. Install new cotter pins.

24. Lower the vehicle.

25. Turn the ignition key to the **OFF** position.

26. Connect the negative battery cable.

27. Fill the power steering pump reservoir.

28. Bleed the power steering system and check for leaks.

29. Check and adjust the wheel alignment.

30. Road test the vehicle and check for proper steering system operation.

Shock Absorber

REMOVAL & INSTALLATION

Front

1. Disconnect the negative battery cable.

2. Remove the plastic cover from the top of the shock assembly in the engine compartment.

3. If equipped with automatic ride control, remove the actuator and actuator mounting washer at top shock mount in the engine compartment.

4. Remove three upper retaining nuts and the housing seal from the mounting studs.

5. Raise and safely support the vehicle.

6. Remove the front wheels.

7. Remove the lower mounting bolt and nut.

8. Remove the nut at the sway (stabilizer) bar link upper mounting stud.

9. Separate the sway bar link from the wheel spindle using Ball Joint Remover D88L-3006-A, or equivalent.

10. Support the lower control arm with a transmission jack or equivalent.

11. Raise the lower control arm and wheel spindle with the jack until the sway bar link can be completely separated from the wheel spindle, then move the sway bar link out of the way.

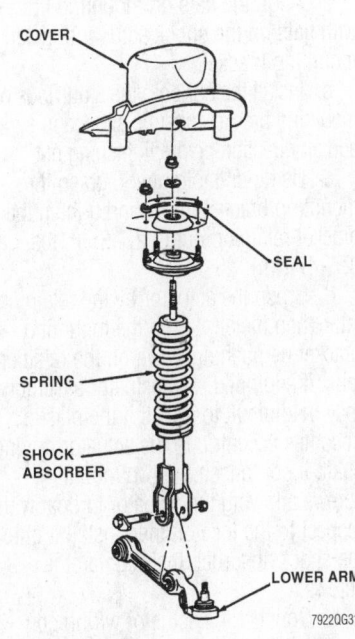

COVER

SEAL

SPRING

SHOCK ABSORBER

LOWER ARM

79220G31

Exploded view of the front shock absorber mounting

12. Remove the wheel spindle-to-upper ball joint pinch bolt and nut. Discard the bolt and nut.

13. Lower the jack to separate the wheel spindle from the upper ball joint stud. Do not allow the wheel spindle to hang free, support it with wire or other means.

14. Remove the jack from under the lower control arm.

15. Remove the shock absorber assembly from the vehicle.

To install:

16. Place the shock assembly over the lower control arm and insert the lower bolt through the shock into the lower control arm.

17. Using a transmission jack or equivalent, raise the lower control arm and shock into position aligning the upper shock mounting studs with the mounting holes.

18. Remove the wire or other support holding the wheel spindle and place the wheel spindle to the upper ball joint stud. Raise the lower control arm with the jack and install a new wheel spindle-to-upper ball joint stud pinch bolt and nut. Tighten the nut to 51–67 ft. lbs. (68–92 Nm).

19. Place the sway bar link and the wheel spindle assembly into the proper position to install the nut onto the link. Tighten the nut to 30–40 ft. lbs. (40–55 Nm).

20. Remove the jack from under the lower control arm.

21. Install the lower shock nut. Do not tighten at this time.

22. Install the wheel and tire assembly. Using a torque wrench, tighten the lug nuts in a start pattern, to 85–105 ft. lbs. (115–142 Nm).

23. Lower the vehicle.

24. Be sure the upper shock mounting studs are aligned with the mounting holes.

25. Install the housing seal and three upper shock retaining nuts onto the mounting studs. Tighten the nuts to 17–22 ft. lbs. (22–31 Nm).

26. Install the automatic ride actuator, if equipped.

27. Install the plastic shock cover, if equipped.

28. Place the vehicle on a drive-on hoist or alignment rack. Neutralize the lower control arm bushings by pushing down on the front of the vehicle, then releasing it.

29. Connect the negative battery cable.

30. Install the lower shock retaining nut onto the bolt and tighten to 126–169 ft. lbs. (170–230 Nm).

31. Road test the vehicle and check for proper operation.

Rear

> ✳✳ **CAUTION**
>
> **The rear shock absorbers act as the rebound stops for the rear suspension. When the rear shocks are removed, the suspension arms must be supported. The best method for rear shock removal is to use a drive on hoist or an alignment pit. Portable vehicle ramps can be substituted, but be sure the ramps are secure, the parking brake is set, and the front wheels are chocked properly before proceeding.**

1. Place and safely support the vehicle on a drive-on hoist, alignment pit, or suitable ramps.

2. Remove the side trim panels from inside the luggage compartment.

3. If equipped with Automatic Ride Control:

   a. Disconnect the actuator wiring connector from the harness connector.

   b. Squeeze the two actuator retaining tabs firmly inward with one hand and lift the actuator off with the other hand.

   c. Using suitable pliers, grasp the mounting bracket and hold firmly. Loosen the upper shock absorber retaining nut and remove mounting bracket.

4. With standard shock absorbers, remove the upper retaining nut, washer and insulator.

5. Remove the lower shock absorber-to-lower control arm retaining bolt, washer and nut.

6. Remove the shock absorber from the vehicle.

To install:

7. Properly prime the new shock absorber by inverting and compressing, then releasing the shock absorber to its fully extended position. Repeat this procedure several times.

➡ **When replacing a shock absorber on an Automatic Ride Control equipped vehicle, an exact replacement adjustable shock absorber must be used. Failure to do so will result in a vehicle with dangerous handling qualities.**

8. Install the lower washer and insulator on the shock absorber. Place the shock absorber into position in the upper mounting hole.

9. Compress the shock absorber enough to align the lower mounting hole with the lower control arm, then install the

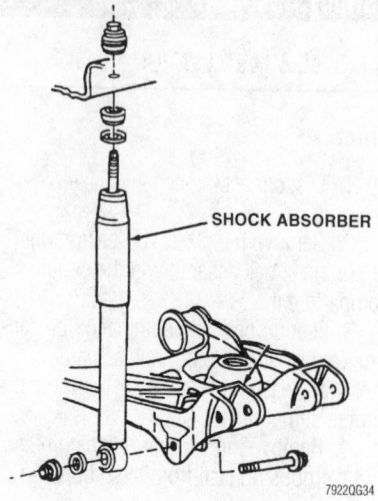

SHOCK ABSORBER

79220G34

Exploded view of the rear shock absorber mounting

retaining bolt, washer and nut. While holding the nut, tighten the lower shock absorber retaining bolt to 83–113 ft. lbs. (113–153 Nm) for 1994–95 models and 95–126 ft. lbs. (128–172 Nm) for 1996–98 models.

10. With the upper shock absorber stud protruding through the shock tower hole into the luggage compartment, install the insulator, washer or actuator mounting bracket and the upper shock absorber retaining nut. Tighten the nut to 27–40 ft. lbs. (37–54 Nm).

11. If equipped with Automatic Ride Control:

   a. Align the flats on mounting bracket with flats on the shock stud, and install mounting bracket.

   b. Install the nut onto the stud. Tabs on mounting bracket must be aligned in a fore and aft direction before tightening nut.

   c. Using suitable pliers, grasp the mounting bracket firmly and tighten the bracket retaining nut to 27–35 ft. lbs. (27–47 Nm).

   d. Grasp the actuator by the retaining tabs, then install it onto the mounting bracket by pushing down on the retainer tabs, if equipped. The actuator assembly may be difficult to install if the plastic becomes eccentric to the actuator driving shaft. If so, loosen the two mounting screws allowing the actuator to float with respect to the locator, then install it onto the shock absorber, then tighten the screws.

   e. Connect the actuator wiring connector, and harness mating connector. Slide the connector onto the rail and push it into the secured position.

12. Install the luggage compartment panels.

13. Lower vehicle.

14. Road test the vehicle and check for proper operation.

Coil Spring

REMOVAL & INSTALLATION

Front

1. Remove the shock absorber assembly from the vehicle.

2. Support the bottom of the shock assembly in a suitable vise. Mark the position of the coil spring to the shock upper mount with a marker or chalk.

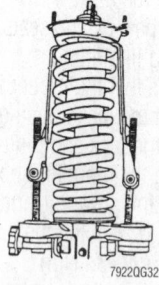

79220G32

Safely compress the coil spring in a spring compressor such as Ford tool No. 086–0029B

➡ **If the coil spring is not positioned properly, the assembly will not mount into position in the vehicle.**

3. Using Spring Compressor 086–0029B or equivalent, compress the coil spring, then remove the retaining nut, washer and the upper shock mounting bracket.

4. Slowly release the spring compressor and remove the coil spring from the shock.

5. Inspect all components for damage or wear and replace parts as necessary.

To install:

➡ **When installing a new coil spring or upper mount, transfer the reference (chalk) marks from the old part to the new part. Always work on only one shock and coil spring assembly at a time to avoid possible mixing of parts. Also if the reference marks are missing, the opposite assembly may be used as reference for aligning the coil spring to the upper mount. The right-hand and left-hand upper mount orientations are identical.**

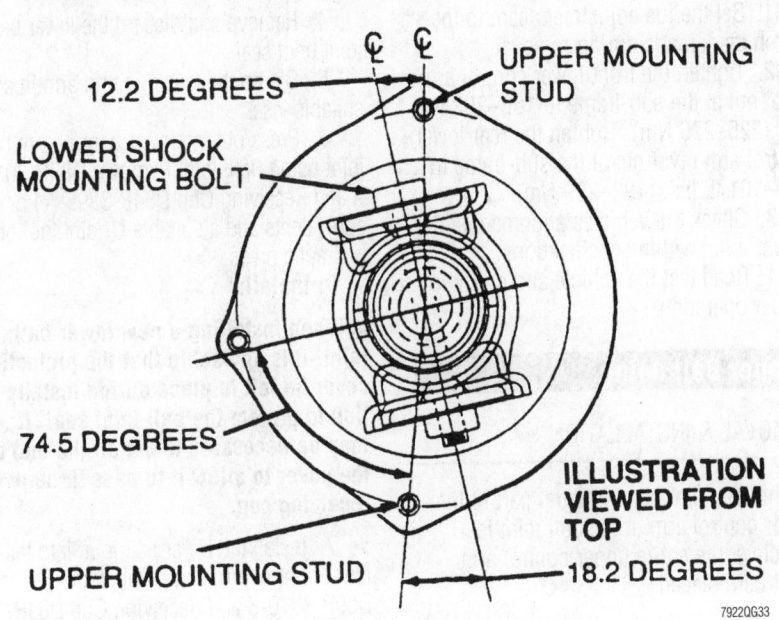

12.2 DEGREES

UPPER MOUNTING STUD

LOWER SHOCK MOUNTING BOLT

74.5 DEGREES

UPPER MOUNTING STUD

ILLUSTRATION VIEWED FROM TOP

18.2 DEGREES

79220G33

Be sure to install the mounting bracket in the correct orientation

6. Place the coil spring on the shock assembly. Using Spring Compressor 086-0029B or equivalent, compress the coil spring and install the upper mount, aligning the previously made marks. Install the washer and retaining nut and tighten to 37–52 ft. lbs. (50–71 Nm).

7. Release the spring compressor and remove.

8. Verify that the coil spring is seated properly at the top and bottom of the shock assembly.

9. Install the shock absorber assembly into the vehicle.

Rear

1. Raise and safely support the vehicle.

2. Remove the rear wheel and tire assembly.

3. Remove the sway (stabilizer) bar link nuts at both ends of the sway bar. Rotate the sway bar up and out of the way.

4. Disconnect the parking brake cable at the disc brake caliper, if equipped with rear disc brakes.

5. Install three Spring Cages 086-00031 or equivalent, as follows:

   a. Install one spring cage without an adjuster link, to the inboard side (the innermost bend of the spring) of the coil spring.

   b. Install the two more spring cages with adjusters, at 120 degree angles to the previously installed cage.

6. Place a transmission jack or equivalent, under the lower control arm as far outboard as possible.

7. Use a wire or other suitable means to secure the upper control arm to the frame or body. The upper control arm must remain in place when the lower control arm is disconnected from the wheel knuckle.

8. Remove the lower shock absorber mounting bolt, washer and nut.

→The lower control arm must not be lowered until the pivot bolts are loose. Do not attempt to remove the plastic cap on the front pivot nut.

9. Match mark the toe adjustment cam positioning to the sub-frame with a suitable marker. Loosen both inboard pivot bolts on the lower control arm.

10. Remove the two bolts and nuts retaining the lower control arm to the wheel knuckle.

11. Lower the control arm by carefully lowering the transmission jack, or equivalent. Be sure the spring cages are securely holding the coil spring as the control arm is lowered.

12. Remove the jack once the coil spring pressure is relieved, then pull the lower control arm down fully by hand and remove the coil spring with the cages in place.

13. Remove the coil spring mounting insulators if necessary.

14. Before removing the cages from the spring:

   a. Measure the length of the coil spring with the cages installed. This dimension is needed for installation.

   b. Install Coil Spring Compressor D78P-5310-a or equivalent, to the coil spring and compress the spring.

   c. Remove the cages, then the spring compressor.

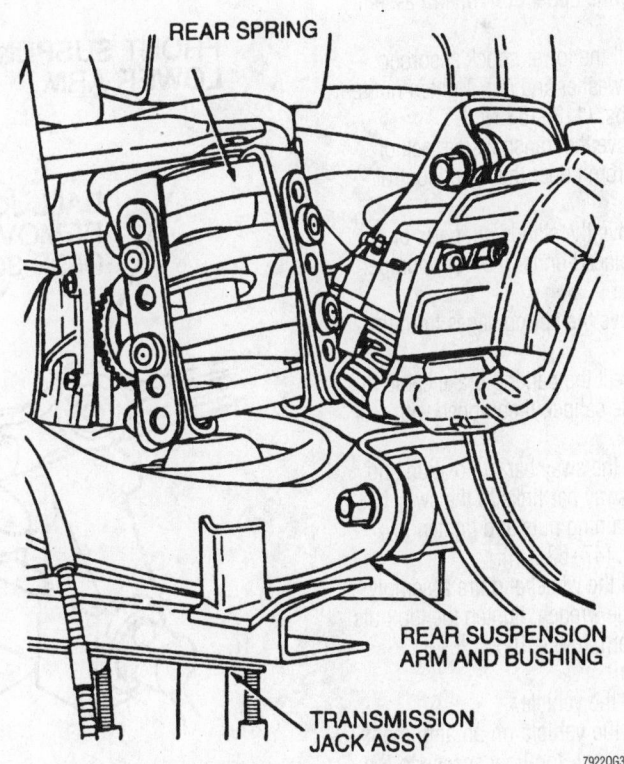

REAR SPRING

REAR SUSPENSION ARM AND BUSHING

TRANSMISSION JACK ASSY

79220G35

Be sure the spring compressor is installed correctly before lowering the control arm

To install:

15. If the cages were removed from the coil spring, use the spring compressor to compress the spring to the dimension measured during removal. If the original dimension is not available, compress the spring to a length of 10.5 inches (267mm) not including the mounting insulators. Install the spring cages and remove the spring compressor.

16. Install the spring mounting insulators, if removed.

17. Place the caged coil spring onto the upper and lower control arm seats. The cage with no adjuster must face inward and cages must be closer to the bottom of spring. The spring pigtails can be in any position.

18. Place two supports/stands under the front bumper reinforcement of the vehicle. This will prevent the rear of the vehicle from lifting off of the hoist, or equivalent.

19. Place a transmission jack or equivalent, under the lower control arm and carefully raise the arm enough to line up the mounting holes to the wheel knuckle, ensuring that the coil spring is properly seated.

20. Install the two bolts and nuts retaining the lower control arm to the wheel knuckle. Tighten the bolts to 110–148 ft. lbs. (149–201 Nm).

21. Remove the wire or equivalent support holding the upper control arm assembly.

22. Install the lower shock absorber mount bolt, washer and nut. Tighten nut to 83–113 ft. lbs. (113–153 Nm).

23. Remove the transmission jack or equivalent, from under the lower control arm.

24. Remove the stands/supports or equivalent, placed under the front bumper reinforcement.

25. Remove the spring cages from the coil spring.

26. Connect the parking brake cable to the disc brake caliper, if equipped with rear disc brakes.

27. Place the sway bar in position and connect the sway bar links to the sway bar. Install the retaining nuts and tighten to 35–47 ft. lbs. (47–63 Nm).

28. Install the wheel and tire assembly. Using a torque wrench, tighten the lug nuts in a star pattern, to 85–105 ft. lbs. (115–142 Nm).

29. Lower the vehicle.

30. Place the vehicle on an alignment rack or equivalent, to allow access to the suspension while the suspension is loaded.

31. Set the toe adjustment cam to the match mark made during removal.

32. Tighten the front lower control arm pivot nut at the sub-frame to 166–202 ft. lbs. (225–275 Nm). Tighten the rear lower control arm pivot nut at the sub-frame to 142–191 ft. lbs. (192–259 Nm).

33. Check the vehicles alignment and adjust if not within specifications.

34. Road test the vehicle and check for proper operation.

Upper Ball Joint

REMOVAL & INSTALLATION

The ball joint is an integral part of the upper control arm. If the ball joint is defective, the entire upper control arm must be replaced.

Lower Ball Joint

REMOVAL & INSTALLATION

1. Raise and safely support the vehicle.
2. Remove the wheel and tire assembly.
3. Remove the lower control arm assembly.

4. Remove and discard the lower ball joint boot seal.

5. Clamp the lower control arm in a suitable vise.

6. Press out the lower control arm ball joint using Ball Joint Remover D89P-3010-A and Receiving Cup D84P-3395-A4 or equivalents and a suitable U-joint tool or hydraulic press.

To install:

➡ When installing a new lower ball joint, it is advisable that the protective cover be left in place during installation to protect the ball joint seal. It may be necessary to cut off the end of the cover to allow it to pass through the receiving cup.

7. Install the lower ball joint into the lower control arm using Ball Joint Replacer D89P-3010-B and Receiving Cup D84P-3395-A4 or equivalents and a suitable U-joint tool or hydraulic press.

8. Remove and discard the protective cover. Check the lower ball joint to be sure that it is fully seated in the lower control arm.

9. Inspect the ball joint seal and ensure that it is free of cuts or tears.

10. Install the lower control arm assembly into the vehicle.

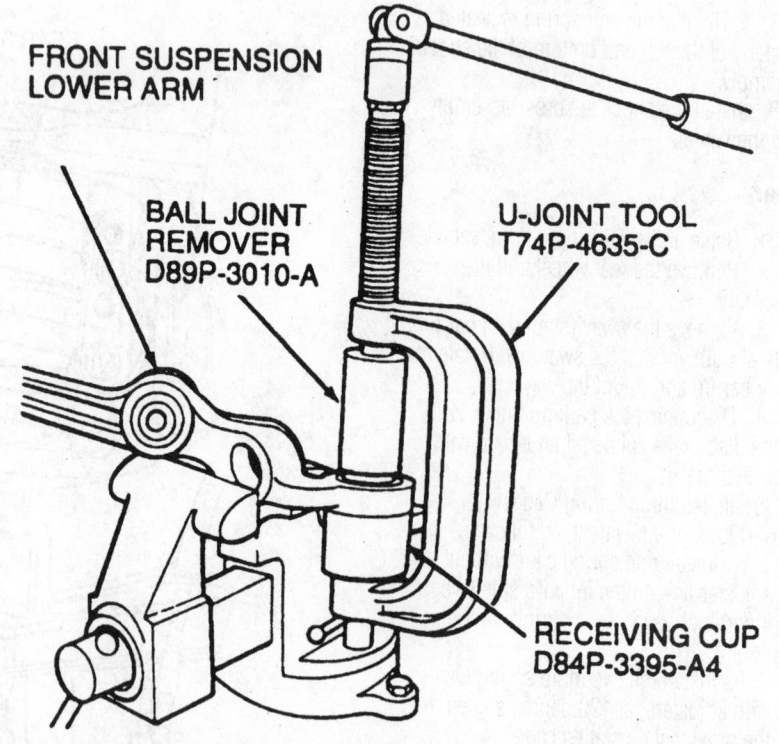

FRONT SUSPENSION LOWER ARM

BALL JOINT REMOVER D89P-3010-A

U-JOINT TOOL T74P-4635-C

RECEIVING CUP D84P-3395-A4

79220G36

Use a ball joint press and adapters such as Ball Joint Remover D89P-3010-A and Receiving Cup D84P-3395-A4 to remove the ball joint from the control arm

11. Install the wheel and tire assembly. Tighten the lug nuts to 85–105 ft. lbs. (115–142 Nm).

12. Lower the vehicle.

13. Check the alignment and adjust if not within specification.

14. Road test the vehicle and check for proper operation.

Wheel Hub and Bearings

ADJUSTMENT

The front and rear wheel bearings are pregreased, sealed and require no periodic maintenance. The bearings are preset and cannot be adjusted.

REMOVAL & INSTALLATION

Front

➡ **Before continuing with this procedure, be sure to have available, one new wheel hub retainer nut, one upper ball joint pinch bolt and nut, one lower ball joint stud nut and one wheel hub grease cap, per side. Once these parts are removed, they lose their torque holding ability or retention capability and must not be reused.**

1. Raise and safely support the vehicle.

2. Place safety stand under both frame rails just behind the lower control arms.

3. Remove the wheel and tire assembly.

4. Remove and discard the hub cap grease seal.

5. Remove the two retaining bolts and the disc brake caliper. Support the brake caliper with wire or equivalent. Do not allow the caliper assembly to hang by the brake hose.

6. Match mark the disc brake rotor to a lug bolt for proper indexing during reassembly, then remove the brake rotor.

7. Remove the brake rotor dust shield.

8. Remove and discard the wheel hub retainer nut.

➡ **The front wheel bearings are part of the wheel hub and cannot be replaced separately.**

9. Remove the wheel hub and bearing assembly. If necessary, use Hub Puller T81P-1104-C or equivalent, to remove the wheel hub from the spindle.

10. If equipped with ABS, remove the speed sensor from the wheel spindle.

11. Remove the outer tie rod end cotter pin and castellated nut. Separate the tie rod end from the wheel spindle using Tie Rod End Remover T-3290-D, or equivalent.

12. Separate the sway bar link from the wheel spindle using Ball Joint Remover D88L-3006-A, or equivalent.

➡ **Use care when separating the sway bar link. If the ball joint seal is damaged, a new link must be used.**

13. Loosen the lower ball joint nut one or two turns. Do not remove the nut at this time.

14. Tap the wheel spindle boss sharply to separate the lower ball joint stud from the wheel spindle. Remove the lower ball joint nut and discard.

15. Remove and discard the upper ball joint pinch bolt and nut.

16. Spread the slot slightly and separate the wheel spindle from the upper ball joint stud.

17. Separate the wheel spindle from the lower ball joint stud and remove from the vehicle.

To install:

18. Place the wheel spindle on the lower ball joint stud. Install a new lower ball joint stud nut and tighten to 83–113 ft. lbs. (113–153 Nm).

19. Place the wheel spindle to the upper ball joint stud and install a new pinch bolt and nut. Tighten the nut to 51–67 ft. lbs. (68–92 Nm).

20. Attach the outer tie rod end to the wheel spindle. Install the castellated nut and tighten to 39–53 ft. lbs. (53–73 Nm). continue to tighten the nut until a slot in the nut lines up with the hole through the stud. Install a new cotter pin.

21. Place the sway bar link to the wheel spindle. Install the retaining nut and tighten to 30–40 ft. lbs. (40–55 Nm).

22. Install the wheel hub and bearing assembly onto the wheel spindle.

23. Install a new wheel hub retainer nut. Tighten the nut to 188–254 ft. lbs. (255–345 Nm).

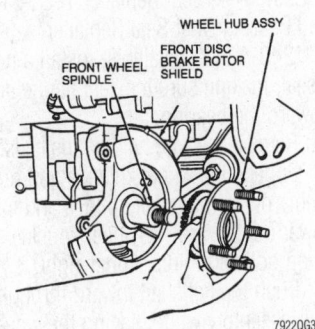

The hub/bearing assembly can be pulled from the spindle after removing the retaining nut

➡ **Any time a wheel hub retainer nut is loosened or removed, it must be replaced with a new nut. Do not use an impact tool to install the new nut.**

24. Install the brake anti-lock speed sensor, if equipped.

25. Install the disc brake rotor splash shield, if removed.

26. Install the brake rotor making sure to align the match marks made during removal. Install new push on nuts, if available.

27. Install a new wheel hub grease cap.

28. Place the disc brake caliper in position on the brake rotor and install the two retaining bolts. Tighten the bolts to 97 ft. lbs. (132 Nm).

29. Install the wheel and tire assembly. Using a torque wrench, tighten the lug nuts in a star pattern, to 85–105 ft. lbs. (115–142 Nm).

30. Remove the safely stands.

31. Lower the vehicle.

32. Pump the brake pedal several times to position the brake pads.

33. Road test the vehicle and check for proper operation.

Rear

➡ **Before continuing with this procedure, be sure to have available, one new wheel hub retainer nut, one inboard halfshaft circlip and one inboard CV-joint stub shaft seal, per side. Once removed, these parts lose their torque holding ability or retention capability and must not be reused.**

1. Remove the hub cap grease seal.

2. With all wheels on the ground, remove and discard the wheel hub retainer nut.

3. Raise and safely support the vehicle.

4. Remove the wheel and tire assembly.

5. Use needle-nose pliers to slide the parking brake cable adjusting clip downward until the cable is free.

6. If equipped with rear disc brakes, remove the parking brake cable from the brake caliper. Remove the disc brake caliper from the brake rotor, leaving the brake hose connected. Wire the brake caliper aside. Do not let it hang by the brake hose.

7. Remove the brake rotor or brake drum, as equipped.

8. If equipped with disc brakes, remove the brake dust shield.

9. If equipped with drum brakes, disconnect the parking brake cable and remove and plug the brake line at the wheel cylinder.

10. Remove the upper control arm nut and bolt at the wheel knuckle. Wire the upper control arm to the body to prevent damage to the CV-joint boots when the knuckle and hub assembly is removed.

11. Mark the position of the control arm in relation to the knuckle with the bushings in the relaxed position. When the upper control arm bolt is removed from the knuckle, the lower arm bushings will return to the relaxed position.

➡**Failure to mark the position will cause bushing wind up at assembly resulting in improper ride height. This can cause incorrect alignment and tire wear.**

12. Install Hub Remover/Replacer T81P-1104-C or equivalent to the hub studs and turn the tool shaft until the halfshaft is free in the hub.

13. Remove the lower control arm-to-knuckle retaining bolts and nuts and remove the knuckle assembly from the halfshaft.

14. Place the wheel knuckle and hub assembly in a suitable vise.

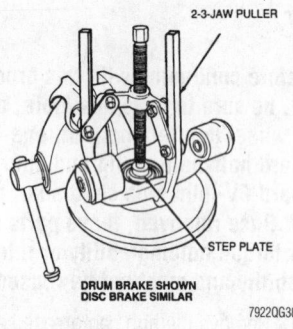

Use a two or three jaw puller to press the hub out of the bearing

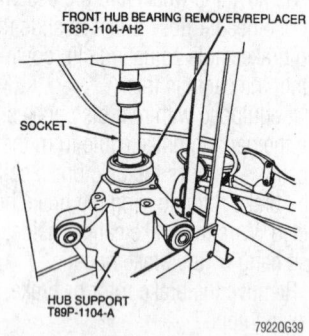

Support the knuckle while removing the bearing assembly with a hydraulic press

15. Wire the halfshaft to a suitable location to prevent damage to the CV-joint boot.

16. Place a suitable two or three jaw puller and Step Plate Adapter D80L-630–5 or equivalent, on the rear wheel knuckle and press out the hub from the wheel knuckle and bearing assembly.

17. Remove the disc brake backing plate from the wheel knuckle.

18. Remove the snapring retaining the wheel bearing assembly.

19. Place the wheel knuckle and bearing assembly on a press bed using Hub Support T89P-1104-A and Front Hub Bearing Remover/Replacer T83P-1104-AH2 or equivalents and press the wheel bearing assembly from the wheel knuckle.

20. Inspect all components for wear and/or damage and replace as needed.

To install:

21. Place the wheel knuckle on the press bed and position the wheel bearing assembly in the wheel knuckle bore. Press the wheel bearing assembly into the knuckle bore using Bearing Replacer T86P-1104-A3 or equivalent. Be sure the bearing assembly is fully seated.

22. Install the wheel bearing retainer snapring.

❋ WARNING

Before proceeding with the following step, the wheel bearing must be supported or damage will result.

23. Place the wheel knuckle on Hub Support T89P-1104-A or equivalent and place on a suitable press bed plate. Place the wheel hub in position with the wheel bearing and knuckle assembly and press into position using Front Hub Bearing Remover/Replacer T83P-1104-AH2 or equivalent. Be sure the inner bearing race is properly supported or the bearing will be damaged.

24. Using a hammer and chisel, drive the bearing dust seal from the outer CV-joint. Using Axle Seal Replacer T83T-3132-A1 and Bearing Dust Seal Replacer T89P-1249-A or equivalents, install a new dust seal, making sure the seal flange faces out toward the bearing.

25. Remove the wire holding the halfshaft.

26. Engage the wheel knuckle and hub assembly on the halfshaft splines and install the lower control arm-to-wheel knuckle retaining bolts and nuts, finger-tight.

27. Push the halfshaft inward through the hub assembly while placing the wheel

knuckle in position to install the front lower bolt, washer and nut.

28. Align the lower control arm to wheel knuckle match marks made during removal. If a new wheel knuckle is being installed, set the knuckle at the approximate angle noted during removal. Tighten the lower control arm-to-wheel knuckle retaining bolts to 94–131 ft. lbs. (128–178 Nm) using Holding Tool T93P-5494-A, or equivalent.

29. Install a new wheel hub retainer nut and hand-tighten only.

30. If equipped with rear disc brakes, place the disc brake backing plate in position on the wheel knuckle and install three retaining bolts. Tighten the bolts to 22–37 ft. lbs. (30–50 Nm).

31. If equipped with drum brakes, install the brake backing plate to the wheel knuckle and install four retaining bolts. Tighten the bolts to 44–59 ft. lbs. (59–80 Nm).

32. Connect the rear parking brake cable.

33. If equipped with drum brakes, connect the brake line to the wheel cylinder.

34. Install the brake rotor or brake drum, as equipped.

35. Install the upper control arm retaining bolt and nut. Tighten the nut to 117–142 ft. lbs. (158–193 Nm).

36. If equipped, install the disc brake caliper over the brake rotor with the outer brake pad against the brake rotor, to prevent pinching the caliper piston boot between the inner pad and piston. Install the two disc brake caliper retaining bolts and tighten to 64–88 ft. lbs. (87–119 Nm).

37. Connect the parking brake rear cable to the disc brake caliper and install the cable adjustment clip.

38. Install the wheel and tire assembly. Using a torque wrench, tighten the lug nuts in a star pattern, to 85–105 ft. lbs. (115–142 Nm).

39. Lower the vehicle.

40. Apply the parking brake several times and adjust if necessary.

41. With all wheels on the ground, tighten the wheel hub retainer nut to 188–254 ft. lbs. (255–345 Nm).

42. Install the hub cap grease seal.

43. Pump the brake pedal several times to position the brake pads before moving the vehicle.

44. Check the alignment and adjust if not within specification.

45. Road test the vehicle and check for proper operation.

FORD MOTOR CO.

26

Ford-Crown Victoria • **Lincoln-**Town Car • **Mercury-**Grand Marquis

DRIVE TRAIN **26-20**
ENGINE REPAIR **26-2**
FUEL SYSTEM **26-18**
STEERING AND
SUPPLEMENT **26-21**

A

Air Bag 26-21
  Disarming 26-22
  Precautions 26-21

C

Camshaft and Valve Lifters 26-11
  Removal & Installation 26-11
Coil Spring 26-23
  Removal & Installation 26-23
Cylinder Head 26-4
  Removal & Installation 26-4

E

Engine Assembly 26-2
  Removal & Installation 26-2
Exhaust Manifold 26-10
  Removal & Installation 26-10

F

Fuel Filter 26-19
  Removal & Installation 26-19

Fuel Pump 26-20
  Removal & Installation 26-20
Fuel System Pressure 26-19
  Relieving 26-19
Fuel System Service
  Precautions 26-18

I

Ignition Timing 26-2
  Adjustment 26-2
Intake Manifold 26-7
  Removal & Installation 26-7

L

Lower Ball Joint 26-26
  Removal & Installation 26-26

O

Oil Pan 26-13
  Removal & Installation 26-13
Oil Pump 26-15
  Removal & Installation 26-15

R

Rear Main Seal 26-15
  Removal & Installation 26-15
Recirculating Ball Power Steering
  Gear 26-22

Removal & Installation 26-22
Rocker Arms 26-6
  Removal & Installation 26-6

S

Shock Absorber 26-23
  Removal & Installation 26-23

T

Timing Chain, Sprockets, Front
  Cover and Seal 26-16
  Removal & Installation 26-16
Transmission Assembly 26-20
  Removal & Installation 26-20

U

Upper Ball Joint 26-26
  Removal & Installation 26-26

V

Valve Lash 26-13
  Adjustment 26-13

W

Water Pump 26-3
  Removal & Installation 26-3
Wheel Bearings 26-27
  Adjustment 26-27
  Removal & Installation 26-27

ENGINE REPAIR

→Disconnecting the negative battery cable on some vehicles may interfere with the functions of the on board computer systems and may require the computer to undergo a relearning process, once the negative battery cable is reconnected.

Ignition Timing

ADJUSTMENT

The 4.6L engine used in the Ford Crown Victoria, Lincoln Town vehicle and Mercury Grand Marquis utilizes a distributorless ignition system (DIS). The DIS consists of the following components: crankshaft sensor, ignition module, ignition coil packs, the spark angle portion of the ECU and the related wiring.

The DIS eliminates the need for a distributor by using multiple ignition coils. Each coil fires two spark plugs at the same time. The plugs are paired so as one fires during the compression cycle, the other fires during the exhaust stroke. The next time the coil is fired, the plug that was on exhaust will be on compression and the one that was on compression will be on exhaust. The spark in the exhaust cylinder is wasted but little of the coil energy is lost. The ignition coils are mounted together in coil packs. There are two coil packs used, each containing two ignition coils.

The crankshaft sensor is a variable reluctance-type sensor triggered by a 36-minus-1 tooth trigger wheel located inside the front cover. The signal generated by this sensor is called a Variable Reluctance Sensor (VRS) signal. The VRS signal provides engine position and rpm information to the ignition module.

The ignition module is a microprocessor that receives input from the crankshaft sensor in regards to engine position, base timing and engine speed and input from the ECU pertaining to spark advance. The ignition module uses this information to direct which coil to fire and to calculate the turn ON and turn OFF times of the coils required to achieve the correct dwell and spark advance.

Base ignition timing is referenced to the position of the crankshaft sensor, and is set at 8–12° BTDC and is not adjustable.

Engine Assembly

✶✶ CAUTION

Some models covered by this manual may be equipped with a Supplemental Restraint System (SRS), which uses an air bag. Whenever working near any of the SRS components, such as the impact sensors, the air bag module, steering column and instrument panel, properly disable the SRS.

REMOVAL & INSTALLATION

✶✶ CAUTION

The fuel injection system remains under pressure, even after the engine has been turned OFF. The fuel system pressure must be relieved before disconnecting any fuel lines. Failure to do so may result in fire and/or personal injury.

1. Disconnect both battery cables, negative cable first.
2. Mark the position of the hood on the hinges and remove the hood.

✶✶ CAUTION

Never open, service or drain the radiator or cooling system when hot; serious burns can occur from the steam and hot coolant.

3. Drain the engine cooling system into a suitable container.
4. Recover the refrigerant from the A/C system using approved recovery equipment.

✶✶ CAUTION

Observe all applicable safety precautions when working around fuel. Whenever servicing the fuel system, always work in a well ventilated area. Do not allow fuel spray or vapors to come in contact with a spark or open flame. Keep a dry chemical fire extinguisher near the work area. Always keep fuel in a container specifically designed for fuel storage; also, always properly seal fuel containers to avoid the possibility of fire or explosion.

5. Relieve the fuel system pressure as follows:

a. Remove the fuel tank fill cap to relieve the pressure in the fuel tank.
b. Remove the cap from the Schrader valve located on the fuel injection supply manifold.
c. Attach Fuel Pressure Gauge T80L-9974-A or equivalent, to the Schrader valve and drain the fuel through the drain tube into a suitable container.
d. After the fuel system pressure is relieved, remove the fuel pressure gauge and install the cap on the Schrader valve. Secure the fuel tank fill cap.

6. Remove the engine cooling fan, shroud and radiator.
7. Remove the windshield wiper governor (module) and support bracket.
8. Remove the engine air cleaner outlet tube.
9. Remove and disconnect the engine/transmission harness connector from the retaining bracket on the power brake booster and move aside.
10. Disconnect the accelerator and cruise control cables at the throttle body.
11. Disconnect the electrical connector and vacuum hose from the evaporative emission canister purge valve.
12. Disconnect the positive battery cable from the power distribution box and harness.
13. Disconnect the vacuum supply hose from the throttle body adapter vacuum port.
14. Disconnect both heater hoses.
15. Disconnect the alternator harness from the front fender apron and the power distribution box.
16. Disconnect the A/C hoses from the A/C compressor using the appropriate spring-lock disconnect tools.
17. Disconnect the power steering control valve harness connector. Disconnect the body ground strap from the dash panel.
18. Raise and safely support the vehicle.
19. Disconnect the exhaust system from the exhaust manifolds and support with wire hung from the crossmember.
20. Remove the retaining nut from the transmission line bracket and remove the three bolts and one stud retaining the engine to the transmission knee braces.
21. Remove the starter motor.
22. Remove the four bolts retaining the power steering pump to the cylinder block and position aside.
23. Remove the transmission housing cover from the cylinder block to access the torque converter retaining nuts. Rotate the crankshaft until each of the four nuts is accessible and remove the nuts.

24. Remove the six transmission-to-engine retaining bolts.

25. Remove the engine support insulator (mount) through-bolts.

26. Lower the vehicle.

27. Support the transmission with a suitable floor jack and a block of wood.

28. Remove the bolt retaining the right-hand front engine support insulator to the front engine mount insulator support bracket.

29. Install an engine lifting bracket to the front of the left-hand cylinder head and to the rear of the right-hand cylinder head. Connect suitable engine lifting equipment to the lifting brackets.

30. Raise the engine slightly using a suitable floor crane and carefully separate the engine from the transmission. Do not let the torque converter fall out of the transmission.

31. Carefully lift the engine out of the engine compartment and position on a workstand. Remove the engine lifting equipment.

To install:

32. Install the engine lifting brackets. Support the engine using a suitable floor crane installed to the lifting equipment and remove the engine from the workstand.

33. Carefully lower the engine into the engine compartment. Start the converter pilot into the flywheel and align the paint marks on the flywheel and torque converter. Be sure the studs on the torque converter align with the holes in the flywheel.

34. Fully engage the engine to the transmission and lower onto front engine support insulators.

35. Remove the engine lifting equipment and brackets.

36. Install the bolt retaining the right-hand front engine support insulator to the front engine mount insulator support bracket.

37. Raise and safely support the vehicle.

38. Install the six engine-to-transmission retaining bolts and tighten to 30–44 ft. lbs. (40–60 Nm).

39. Install the front engine support insulator through-bolts and tighten to 15–22 ft. lbs. (20–30 Nm).

40. Install the four torque converter retaining nuts and tighten to 22–25 ft. lbs. (20–30 Nm).

41. Install the transmission housing cover to the cylinder block.

42. Position the power steering pump on the cylinder block and install the four retaining nuts. Tighten to 15–22 ft. lbs. (20–30 Nm).

43. Install the starter motor.

44. Position the engine to transmission brace and install the three bolts and one stud. Tighten the bolts and stud to 18–31 ft. lbs. (25–43 Nm).

45. Position the transmission line bracket to the brace stud and install one retaining nut. Tighten to 15–22 ft. lbs. (20–30 Nm).

46. Position the exhaust system to the exhaust manifolds. Install the four nuts and tighten to 20–30 ft. lbs. (27–41 Nm). Be sure the exhaust system clears the No. 3 crossmember. Adjust as necessary.

47. Lower the vehicle.

48. Reconnect the power steering valve harness connector. Reconnect the ground strap to the dash panel.

49. Reconnect the A/C lines to the A/C compressor.

50. Reconnect the alternator harness at the front fender apron and the power distribution box.

51. Reconnect both heater hoses.

52. Reconnect the vacuum supply hose to the throttle body adapter vacuum port.

53. Reconnect the positive battery cable to the power distribution box and harness. Reconnect the electrical connector and vacuum hose to the evaporative emission canister purge valve.

54. Reconnect the accelerator and cruise control cables at the throttle body.

55. Reconnect and install the engine/transmission harness connector to the retaining bracket on the power brake booster.

56. Install the windshield wiper governor and support bracket.

57. Reconnect the fuel supply and return lines.

58. Install the radiator, cooling fan and shroud.

59. Install the engine air cleaner outlet tube.

60. If needed, fill the crankcase with the proper type and quantity of engine oil.

61. Fill and bleed the engine cooling system.

62. Reconnect both battery cables, negative cable last.

63. Start the engine and allow to reach normal operating temperature.

64. Check for leaks and proper fluid levels.

65. Properly evacuate and recharge the A/C system.

66. Install the hood, aligning the marks that were made during removal.

67. Road test the vehicle and check the engine and transmission for proper operation.

Water Pump

REMOVAL & INSTALLATION

1. Disconnect the negative battery cable.

❋❋ CAUTION

Never open, service or drain the radiator or cooling system when hot; serious burns can occur from the steam and hot coolant.

2. Drain the cooling system, remove the cooling fan and the shroud.

3. Release the belt tensioner and remove the accessory drive belt.

4. Remove the four bolts retaining the water pump pulley to the water pump and remove the pulley.

5. Remove the four bolts retaining the water pump to the engine assembly and remove the water pump.

To install:

6. Installation is the reverse of the removal procedure. Be sure to clean the sealing surfaces of the water pump and block and use a new O-ring. Lubricate the O-ring with clean anti-freeze prior to installation.

7. Tighten the water pump-to-engine bolts and the pulley-to-water pump bolts to 15–22 ft. lbs. (20–30 Nm). Fill and bleed the cooling system. Operate the engine until normal operating temperatures have been reached and check for leaks.

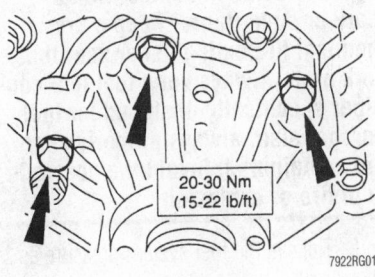

20-30 Nm
(15-22 lb/ft)

7922RG01

Be sure to tighten the water pump mounting bolts to the specification

Cylinder Head

REMOVAL & INSTALLATION

> ✳✳ **CAUTION**
>
> The fuel injection system remains under pressure, even after the engine has been turned OFF. The fuel system pressure must be relieved before disconnecting any fuel lines. Failure to do so may result in fire and/or personal injury.

➡ The cylinder head bolts are a torque-to-yield design and cannot be reused. Before beginning this procedure, be sure new cylinder head bolts are available.

1. If equipped with air suspension, the air suspension switch, located on the right-hand side of the luggage compartment, must be turned to the OFF position before raising the vehicle.
2. Disconnect the negative battery cable.

> ✳✳ **CAUTION**
>
> Never open, service or drain the radiator or cooling system when hot; serious burns can occur from the steam and hot coolant.

3. Drain the engine cooling system.
4. Remove the cooling fan and shroud assembly.

> ✳✳ **CAUTION**
>
> Observe all applicable safety precautions when working around fuel. Whenever servicing the fuel system, always work in a well ventilated area. Do not allow fuel spray or vapors to come in contact with a spark or open flame. Keep a dry chemical fire extinguisher near the work area. Always keep fuel in a container specifically designed for fuel storage; also, always properly seal fuel containers to avoid the possibility of fire or explosion.

5. Relieve the fuel system pressure as follows:

    a. Remove the fuel tank fill cap to relieve the pressure in the fuel tank.

    b. Remove the cap from the Schrader valve located on the fuel injection supply manifold.

    c. Attach Fuel Pressure Gauge T80L-9974-B or equivalent, to the Schrader valve and drain the fuel through the drain tube into a suitable container.

    d. After the fuel system pressure is relieved, remove the fuel pressure gauge and install the cap on the Schrader valve. Secure the fuel tank fill cap.

6. Remove the engine air cleaner outlet tube.
7. Remove the windshield wiper governor (module).
8. Release the drive belt tensioner and remove the accessory drive belt.
9. Tag and disconnect the ignition wires from the spark plugs. Disconnect the ignition wire brackets from the cylinder head cover studs and remove the two bolts retaining the ignition wire tray to the ignition coil brackets.
10. Remove the bolt retaining the A/C pressure line to the right-hand ignition coil bracket.
11. Disconnect the wiring to both ignition coils and the Camshaft Position (CMP) sensor.
12. Remove the nuts retaining the ignition coil brackets to the engine front cover. Slide the ignition coil brackets and ignition wire assemblies off the mounting studs and remove from the vehicle.
13. Remove the water pump pulley.
14. Disconnect the alternator wiring harness from the junction block, fender apron and alternator. Disconnect the bolts retaining the alternator to the intake manifold and cylinder block and remove the alternator.
15. Disconnect the positive battery cable at the power distribution box. Remove the retaining bolt from the positive battery cable bracket located on the side of the right-hand cylinder head.
16. Disconnect the vent hose from the canister purge solenoid and position the positive battery cable out of the way. Disconnect the canister purge solenoid vent hose from the PCV valve and remove the PCV valve from the cylinder head cover.

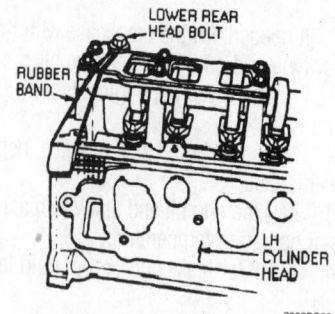

Use a rubber band to support the rear cylinder head bolt to ease removal of the head

17. Remove and disconnect the engine/transmission harness connector from the retaining bracket on the power brake booster.
18. Disconnect the Crankshaft Position (CKP) sensor, A/C compressor clutch and canister purge solenoid electrical connectors.
19. Raise and safely support the vehicle.
20. Remove the bolts retaining the power steering pump to the cylinder block and engine front cover. The front lower bolt on the power steering pump will not come all the way out. Wire the power steering pump out of the way.

> ✳✳ **CAUTION**
>
> The EPA warns that prolonged contact with used engine oil may cause a number of skin disorders, including cancer! You should make every effort to minimize your exposure to used engine oil. Protective gloves should be worn when changing the oil. Wash your hands and any other exposed skin areas as soon as possible after exposure to used engine oil. Soap and water, or waterless hand cleaner should be used.

21. Remove the engine oil pan and oil pan gasket.
22. Remove the crankshaft pulley retaining bolt and remove the pulley, using Damper Remover T58P-6316-D, or equivalent.
23. Disconnect the power steering control valve actuator and oil pressure sensor wiring connectors and position out of the way.
24. Disconnect the EGR tube from the right-hand exhaust manifold.
25. Disconnect the exhaust pipes from the exhaust manifolds. Lower the exhaust pipes and hang with wire from the crossmember.
26. Remove the bolt retaining the starter wiring harness to the rear of the right-hand cylinder head.
27. Lower the vehicle.
28. Remove the bolts and stud bolts retaining the cylinder head covers to the cylinder heads and remove the covers.
29. Disconnect the accelerator and cruise control cables. Remove the accelerator cable bracket from the intake manifold and position out of the way.
30. Disconnect the vacuum hose from the throttle body elbow vacuum port, both Heated Oxygen Sensors (HO2S) and the heater water hose.
31. Remove the two bolts retaining the thermostat housing to the intake manifold and position the upper hose and thermostat housing out of the way.

➡ **The two thermostat housing bolts also retain the intake manifold.**

32. Remove the nine bolts retaining the intake manifold to the cylinder heads and remove the intake manifold and gaskets.

33. Remove seven stud bolts and the four bolts retaining the engine front cover to the engine and remove the front cover.

34. Remove both timing chains.

✱✱ WARNING

This is an interference engine. (For a definition of an interference engine, please refer to the Timing Belt unit repair section in the front of this manual.) Cam Positioning Tools T92P-6256-A or equivalent, must be installed on the camshafts to prevent the camshafts from rotating. Do not rotate the camshafts or crankshaft with the timing chains removed or the valves will contact the pistons.

35. Remove 10 bolts retaining the left-hand cylinder head to the cylinder block and remove the head. The lower rear cylinder head bolt must stay in the cylinder head until the cylinder head is removed due to lack of clearance for removal in the vehicle. Use a rubber band to secure the cylinder head bolt in the cylinder head during removal and installation of the cylinder head and to prevent the bolt from damaging the cylinder block or head gasket.

➡ **The lower rear cylinder head bolt cannot be removed due to interference with the power brake booster. Use a rubber band to hold the bolt away from the cylinder block.**

36. Remove the ground strap, one stud and one bolt retaining the heater return line to the right-hand cylinder head.

37. Remove 10 bolts retaining the right-hand cylinder head to the cylinder block and

remove the head. The lower rear cylinder head bolt must stay in the cylinder head until the cylinder head is removed due to lack of clearance for removal in the vehicle. Use a rubber band to secure the cylinder head bolt in the cylinder head during removal and installation of the cylinder head and to prevent the bolt from damaging the cylinder block or head gasket.

➡ **The lower rear cylinder head bolt cannot be removed due to interference with the evaporator housing. Use a rubber band to hold the bolt away from the cylinder block.**

38. Clean all gasket mating surfaces. Check the cylinder heads and cylinder block for flatness. Check the cylinder heads for scratches near the coolant passages and combustion chambers that could provide leak paths.

To install:

39. Rotate the crankshaft counterclockwise 45 degrees. The crankshaft keyway should be at the 9 o'clock position viewed from the front of the engine. This ensures that all pistons are below the top of the engine block deck face.

40. Rotate the camshaft to a stable position where the valves do not extend below the head face.

41. Position new head gaskets on the cylinder block. Install new bolts in the lower rear bolt holes on both cylinder heads and retain with rubber bands as explained during the removal procedure.

➡ **New cylinder head bolts must be used whenever the cylinder head is removed and reinstalled. The cylinder head bolts are a torque-to-yield design and cannot be reused.**

42. Position the cylinder heads on the cylinder block dowels, being careful not to score the surface of the head face. Apply clean oil to the new cylinder head bolts, remove the rubber bands from the lower rear bolts and install all bolts hand-tight.

43. On 1995 models, tighten the new cylinder head bolts as follows:

   a. Tighten the bolts in sequence, to 22–30 ft. lbs. (30–40 Nm).

   b. Rotate each bolt in sequence 85–95 degrees.

   c. Rotate each bolt in sequence an additional 85–95 degrees.

44. On 1996–99 models, tighten the new cylinder head bolts as follows:

   a. Tighten the bolts in sequence, to 28–31 ft. lbs. (37–43 Nm).

   b. Rotate each bolt in sequence 85–95 degrees.

   c. Loosen all bolts at least one full turn.

   d. Tighten the bolts in sequence, to 27–32 ft. lbs. (37–43 Nm).

   e. Rotate each bolt in sequence 85–95 degrees.

   f. Rotate each bolt in sequence an additional 85–95 degrees.

45. Position the heater return hose and install the two retaining bolts.

46. Rotate the camshafts using the flats matched at the center of the camshaft until both are in time. Install Cam Positioning Tools T91P-6256-A or equivalent, on the flats of the camshafts to keep them from rotating.

47. Rotate the crankshaft clockwise 45 degrees to position the crankshaft at TDC for No. 1 cylinder.

➡ **The crankshaft must only be rotated in the clockwise direction and only as far as TDC.**

48. Install both timing chains.

49. Install a new engine front cover seal and gasket. Apply silicone sealer to the lower corners of the cover where it meets the junction of the engine oil pan and cylinder block and to the points where the cover contacts the junction of the cylinder block and the cylinder heads.

50. Install the engine front cover and the stud bolts and bolts. Tighten to 15–22 ft. lbs. (20–30 Nm).

51. Position new intake manifold gaskets on the cylinder heads. Be sure the alignment tabs on the gaskets are aligned with the holes in the cylinder heads.

➡ **Before installing the intake manifold, inspect it for nicks and cuts that could provide leak paths.**

52. Position the intake manifold on the cylinder heads and install the retaining bolts. Tighten the bolts in sequence, to 15–22 ft. lbs. (20–30 Nm).

53. Install the thermostat and O-ring, then position the thermostat housing and upper hose and install the two retaining bolts. Tighten to 15–22 ft. lbs. (20–30 Nm).

54. Reconnect the heater water hose and both HO2S sensors.

55. Reconnect the vacuum hose to the throttle body adapter vacuum port.

56. Install the accelerator cable bracket on the intake manifold and connect the accelerator and cruise control cables to the throttle body.

57. Apply silicone sealer to both places where the engine front cover meets the cylinder heads. Install new gaskets on the cylinder head covers.

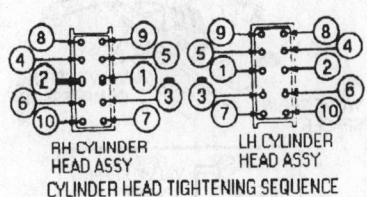

RH CYLINDER
HEAD ASSY

LH CYLINDER
HEAD ASSY

CYLINDER HEAD TIGHTENING SEQUENCE

7922RG03

Tighten the cylinder head bolts in the proper sequence to prevent damage to the head and possible leaks

58. Install the cylinder head covers on the cylinder heads. Install the bolts and stud bolts and tighten to 71–106 inch lbs. (8–12 Nm).

59. Raise and safely support the vehicle.

60. Place the starter motor wiring harness to the right-hand cylinder head and install the retaining bolt.

61. Fit the exhaust pipes to the exhaust manifolds. Install and tighten the four nuts to 20–30 ft. lbs. (27–41 Nm).

➡**Be sure the exhaust system clears the No. 3 crossmember. Adjust as necessary.**

62. Reconnect the EGR tube to the right-hand exhaust manifold and tighten the line nut to 26–33 ft. lbs. (35–45 Nm).

63. Reconnect the power steering control valve actuator and oil pressure sensor electrical connectors.

64. Apply a small amount of silicone sealer in the rear of the keyway on the crankshaft pulley. Position the pulley on the crankshaft, making sure the crankshaft key and keyway are aligned.

65. Using Damper Installer T74P-6316-B or equivalent, install the crankshaft pulley. Install the pulley bolt and washer and tighten to 114–121 ft. lbs. (155–165 Nm).

66. Install the engine oil pan and a new gasket.

67. Place the power steering pump in position on the cylinder block and install the four retaining bolts. Tighten the bolts to 15–22 ft. lbs. (20–30 Nm).

68. Lower the vehicle.

69. Reconnect the A/C compressor, CKP sensor and canister purge solenoid electrical connectors.

70. Reconnect the engine/transmission harness connector and install on the retaining bracket on the power brake booster.

71. Install the PCV valve in the right-hand cylinder head cover and connect the canister purge solenoid vent hose.

72. Position the positive battery cable harness on the right-hand cylinder head and install the bolt retaining the cable bracket to the cylinder head. Reconnect the positive battery cable at the power distribution box and battery.

73. Position the alternator and install the two retaining bolts. Tighten the bolts to 15–22 ft. lbs. (20–30 Nm). Install the two bolts retaining the alternator brace to the intake manifold and tighten to 6–8 ft. lbs. (8–12 Nm).

74. Install the water pump pulley and tighten the bolts to 15–22 ft. lbs. (20–30 Nm).

75. Position the ignition coil brackets and ignition wire assemblies onto the mounting studs. Install seven nuts retaining the ignition coil brackets to the engine front cover and tighten to 15–22 ft. lbs. (20–30 Nm).

76. Install the two bolts retaining the ignition wire tray to the ignition coil bracket and tighten to 71–106 inch lbs. (8–12 Nm). Reconnect both ignition coil and CMP sensor harness connectors.

77. Place the A/C pressure line on the right-hand ignition coil bracket and install the retaining bolt. Reconnect the ignition wires to the spark plugs and install the bracket onto the cylinder head cover studs.

78. Install the accessory drive belt and the windshield wiper governor.

79. Reconnect the fuel supply and return lines.

80. Install the cooling fan and shroud.

81. Fill the engine cooling system.

82. Install the engine air cleaner outlet tube.

83. Reconnect the negative battery cable.

84. If equipped with air suspension, turn the air suspension switch to the ON position.

※※ WARNING

Operating the engine without the proper amount and type of engine oil will result in severe engine damage.

85. Refill the engine with the correct amount of oil and replace the filter.

86. Start the engine and bring to normal operating temperature while checking for leaks.

87. Road test the vehicle and check for proper engine operation.

Rocker Arms

REMOVAL & INSTALLATION

1. Disconnect the negative battery cable.

2. Remove the right camshaft cover as follows:

   a. Disconnect the positive battery cable at the battery and at the power distribution box. Remove the retaining bolt from the positive battery cable bracket located on the side of the right cylinder head.

   b. Disconnect the crankshaft position sensor, air conditioning compressor clutch and canister purge solenoid connectors. Position the harness out of the way.

   c. Disconnect the vent hose from the purge solenoid and position the positive battery cable out of the way.

   d. Disconnect the ignition wires from the spark plugs. Remove the ignition wire brackets from the camshaft cover studs and position the wires out of the way.

   e. Remove the PCV valve from the camshaft cover grommet and position out of the way.

   f. Remove the bolts and stud bolts and remove the camshaft cover.

3. Remove the left camshaft cover as follows:

   a. Remove the air inlet tube. Relieve the fuel system pressure and disconnect the fuel lines.

   b. Raise and safely support the vehicle.

   c. Disconnect the EVO sensor and oil pressure sending unit and position the harness out of the way. Lower the vehicle.

   d. Remove the 42-pin engine harness connector from the retaining bracket on the brake vacuum booster. Disconnect and position out of the way.

   e. Remove the windshield wiper module.

   f. Disconnect the ignition wires from the spark plugs. Remove the ignition wire brackets from the studs and position the wires out of the way.

   g. Remove the bolts and stud bolts and remove the camshaft cover.

4. Position the piston of the cylinder being serviced at the bottom of its stroke and position the camshaft lobe on the base circle.

5. Install valve spring spacer tool T91P-6565-AH or equivalent, between the spring coils to prevent valve seal damage.

➡**If the valve spring spacer tool is not used, the retainer will hit the valve stem seal and damage the seal.**

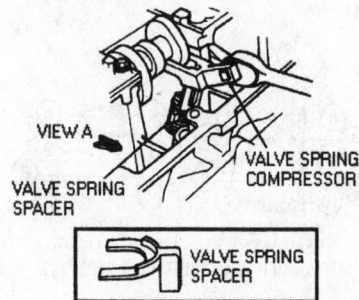

After installing the spring spacer, compress the valve spring and remove the rocker arm

6. Install valve spring compressor tool T91P-6565-A or equivalent, under the camshaft and on top of the valve spring retainer.

7. Compress the valve spring and remove the roller follower. Remove the valve spring compressor and spacer.

To install:

8. Apply engine oil to the valve stem and tip and roller follower contact surfaces.

9. Install valve spring spacer tool T91P-6565-AH or equivalent, between the spring coils. Compress the valve spring using valve spring compressor tool T91P-6565-A or equivalent, and install the roller follower.

➡ **The piston must be at the bottom of its stroke and the camshaft at the base circle.**

10. Remove the valve spring compressor and spacer.

11. Clean the sealing surfaces of the camshaft covers and cylinder heads. Apply silicone sealer to the places where the front cover meets the cylinder head.

12. Position new gaskets onto the camshaft covers and install the covers. Install the bolts and stud bolts and tighten to 6.0–8.8 ft. lbs. (8–12 Nm).

13. When installing the right camshaft cover, proceed as follows:

    a. Install the PCV into the camshaft cover grommet.

    b. Install the ignition wire brackets on the studs and connect the wires to the spark plugs.

    c. Position the harness and connect the canister purge solenoid, air conditioning compressor clutch and crankshaft position sensor.

    d. Position the positive battery cable harness on the right cylinder head. Install the bolt retaining the cable bracket to the cylinder head.

    e. Reconnect the positive battery cable at the power distribution box and the battery.

14. When installing the left camshaft cover, proceed as follows:

    a. Install the ignition wire brackets on the studs and connect the wires to the spark plugs.

    b. Install the windshield wiper module.

    c. Reconnect the 42-pin connector and transmission harness connector. Install the connector on the retaining bracket.

    d. Raise and safely support the vehicle. Position and connect the EVO sensor and oil pressure sending unit harness.

e. Lower the vehicle. Reconnect the fuel lines.

15. Reconnect the negative battery cable. Start the engine and check for leaks.

Intake Manifold

REMOVAL & INSTALLATION

1995 Models

> ✳✳ **CAUTION**
>
> **The fuel injection system remains under pressure, even after the engine has been turned OFF. The fuel system pressure must be relieved before disconnecting any fuel lines. Failure to do so may result in fire and/or personal injury.**

1. If equipped with air suspension, the air suspension switch, located on the right-hand side of the luggage compartment, must be turned to the OFF position before raising the vehicle.

2. Disconnect the negative battery cable.

> ✳✳ **CAUTION**
>
> **Never open, service or drain the radiator or cooling system when hot; serious burns can occur from the steam and hot coolant.**

3. Drain the engine cooling system.

> ✳✳ **CAUTION**
>
> **Observe all applicable safety precautions when working around fuel. Whenever servicing the fuel system, always work in a well ventilated area. Do not allow fuel spray or vapors to come in contact with a spark or open flame. Keep a dry chemical fire extinguisher near the work area. Always keep fuel in a container specifically designed for fuel storage; also, always properly seal fuel containers to avoid the possibility of fire or explosion.**

4. Relieve the fuel system pressure as follows:

    a. Remove the fuel tank fill cap to relieve the pressure in the fuel tank.

    b. Remove the cap from the Schrader valve located on the fuel injection supply manifold.

    c. Attach Fuel Pressure Gauge T80L-9974-B or equivalent, to the Schrader

valve and drain the fuel through the drain tube into a suitable container.

    d. After the fuel system pressure is relieved, remove the fuel pressure gauge and install the cap on the Schrader valve. Secure the fuel tank fill cap.

5. Disconnect the fuel supply and return lines.

6. Remove the windshield wiper governor (module).

7. Remove the engine air cleaner outlet tube.

8. Release the drive belt tensioner and remove the accessory drive belt.

9. Tag and disconnect the ignition wires from the spark plugs. Disconnect the ignition wire brackets from the cylinder head cover studs.

10. Disconnect the wiring from both ignition coils and the Camshaft Position (CMP) sensor. Tag and disconnect all ignition wires from both ignition coils. Remove the two bolts retaining the ignition wire bracket to the ignition coil brackets and remove the ignition wire assembly.

11. Disconnect the alternator wiring harness from the junction block at the fender apron and alternator. Remove the bolts retaining the alternator brace to the intake manifold and the alternator to the cylinder block and remove the alternator.

12. Raise and safely support the vehicle.

13. Disconnect the oil pressure sensor and power steering control valve actuator wiring and position the wiring harness out of the way.

14. Disconnect the EGR valve to exhaust manifold tube from the right-hand exhaust manifold.

15. Lower the vehicle.

16. Remove and disconnect the engine/transmission harness connector from the retaining bracket on the power brake booster.

17. Disconnect the A/C compressor clutch, Crankshaft Position (CKP) sensor

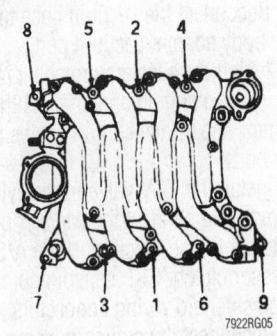

To prevent leaks, be sure to tighten the intake manifold bolts in the sequence shown—1995 models

and the canister purge solenoid wiring connectors.

18. Remove the PCV valve from the cylinder head cover and disconnect the canister purge vent hose from the PCV valve.

19. Disconnect the accelerator and cruise control cables from the throttle body. Remove the accelerator cable bracket from the intake manifold and position out of the way.

20. Disconnect the vacuum hose from the throttle body adapter port.

21. Disconnect both Heated Oxygen Sensors (HO2S) and the heater water hose.

22. Remove the two bolts retaining the thermostat housing to the intake manifold and position the upper hose and thermostat housing out of the way.

➡ **The two thermostat housing bolts are also used to retain the intake manifold.**

23. Remove the nine bolts retaining the intake manifold to the cylinder heads and remove the intake manifold. Remove and discard the gaskets.

24. If replacing the intake manifold, swap over the necessary parts.

To install:

25. Clean all gasket mating surfaces.

26. Position new intake manifold gaskets on the cylinder heads. Be sure the alignment tabs on the gaskets are aligned with the holes in the cylinder heads.

27. Install the intake manifold and the nine retaining bolts. Hand-tighten the right-rear bolt (viewed from the front of the engine) before final tightening, then tighten the bolts, in sequence, to 15–22 ft. lbs. (20–30 Nm).

28. Inspect and if necessary, replace the O-ring seal on the thermostat housing. Position the housing and upper hose and install the two retaining bolts. Tighten to 15–22 ft. lbs. (20–30 Nm).

29. Reconnect the heater water hose.

30. Reconnect both HO2S wiring connectors.

31. Reconnect the vacuum hose to the throttle body adapter vacuum port.

32. Install the accelerator cable bracket on the intake manifold and connect the accelerator and cruise control cables to the throttle body.

33. Install the PCV valve in the cylinder head cover and connect the canister purge solenoid vent hose. Reconnect the A/C compressor clutch, CKP sensor and canister purge solenoid wiring connectors.

34. Reconnect the engine/transmission harness connector. Install the connector on the retaining bracket on the power brake booster.

35. Raise and safely support the vehicle.

36. Reconnect the EGR valve to exhaust manifold tube to the right-hand exhaust manifold. Tighten the tube nut to 26–33 ft. lbs. (35–45 Nm).

37. Reconnect the power steering control valve actuator and the oil pressure sensor wiring connectors.

38. Lower the vehicle.

39. Position the alternator and install the two retaining bolts. Tighten to 15–22 ft. lbs. (20–30 Nm). Install the two bolts retaining the alternator brace to the intake manifold and tighten to 71–106 inch lbs. (8–12 Nm).

40. Reconnect the alternator wiring harness to the alternator, right-hand fender apron and junction block.

41. Position the ignition wire assembly on the engine and install the two bolts retaining the ignition wire bracket to the ignition coil brackets. Tighten the bolts to 71–106 inch lbs. (8–12 Nm).

42. Reconnect the ignition wires to the ignition coils in their proper positions. Reconnect the ignition wires to the spark plugs.

43. Reconnect the ignition wire brackets on the cylinder head cover studs. Reconnect the wiring connectors to both ignition coils and the CMP sensor.

44. Install the accessory drive belt.

45. Install the engine air cleaner outlet tube.

46. Install the windshield wiper governor.

47. Reconnect the fuel supply and return lines.

48. Fill and bleed the engine cooling system.

49. Reconnect the negative battery cable.

50. If equipped with air suspension, turn the air suspension switch to the ON position.

51. Start the engine and check for leaks.

52. Road test the vehicle and check for proper operation.

1996–99 Models

✳✳ CAUTION

The fuel injection system remains under pressure, even after the engine has been turned OFF. The fuel system pressure must be relieved before disconnecting any fuel lines. Failure to do so may result in fire and/or personal injury.

1. If equipped with air suspension, the air suspension switch, located on the right-hand side of the luggage compartment, must be turned to the OFF position before raising the vehicle.

2. Disconnect the negative battery cable.

✳✳ CAUTION

Never open, service or drain the radiator or cooling system when hot; serious burns can occur from the steam and hot coolant.

3. Drain the engine cooling system.

✳✳ CAUTION

Observe all applicable safety precautions when working around fuel. Whenever servicing the fuel system, always work in a well ventilated area. Do not allow fuel spray or vapors to come in contact with a spark or open flame. Keep a dry chemical fire extinguisher near the work area. Always keep fuel in a container specifically designed for fuel storage; also, always properly seal fuel containers to avoid the possibility of fire or explosion.

4. Relieve the fuel system pressure as follows:

   a. Remove the fuel tank fill cap to relieve the pressure in the fuel tank.

   b. Remove the cap from the Schrader valve located on the fuel injection supply manifold.

   c. Attach Fuel Pressure Gauge T80L-9974-B or equivalent, to the Schrader valve and drain the fuel through the drain tube into a suitable container.

   d. After the fuel system pressure is relieved, remove the fuel pressure gauge and install the cap on the Schrader valve. Secure the fuel tank fill cap.

5. Disconnect the fuel supply and return lines.

6. Remove the windshield wiper governor (module).

7. Remove the engine air cleaner outlet tube.

8. Release the drive belt tensioner and remove the accessory drive belt.

9. Tag and disconnect the ignition wires from the spark plugs. Disconnect the ignition wire brackets from the cylinder head cover studs.

10. Disconnect the wiring from both ignition coils and the Camshaft Position (CMP) sensor. Tag and disconnect all ignition wires from both ignition coils. Remove the two bolts retaining the ignition wire bracket to the ignition coil brackets and remove the ignition wire assembly.

11. Disconnect the alternator wiring harness from the junction block at the fender apron and alternator. Remove the bolts retaining the alternator brace to the intake manifold and the alternator to the cylinder block and remove the alternator.

12. Raise and safely support the vehicle.

13. Disconnect the oil pressure sensor and power steering control valve actuator wiring and position the wiring harness out of the way.

14. Disconnect the EGR valve to exhaust manifold tube from the right-hand exhaust manifold.

15. Lower the vehicle.

16. Remove and disconnect the engine/transmission harness connector from the retaining bracket on the power brake booster.

17. Disconnect the A/C compressor clutch, Crankshaft Position (CKP) sensor and the canister purge solenoid wiring connectors.

18. Remove the PCV valve from the cylinder head cover and disconnect the canister purge vent hose from the PCV valve.

19. Disconnect the accelerator and cruise control cables from the throttle body. Remove the accelerator cable bracket from the intake manifold and position out of the way.

20. Disconnect the vacuum hose from the throttle body adapter port.

21. Disconnect both Heated Oxygen Sensors (HO2S) and the heater water hose.

22. Remove the two bolts retaining the thermostat housing to the intake manifold and position the upper hose and thermostat housing out of the way.

➡ **The two thermostat housing bolts are also used to retain the intake manifold.**

23. Remove the nine bolts retaining the intake manifold to the cylinder heads and remove the intake manifold. Remove and discard the gaskets.

24. If replacing the intake manifold, swap over the necessary parts.

To install:

25. Clean all gasket mating surfaces.

26. Position new intake manifold gaskets on the cylinder heads. Be sure the alignment tabs on the gaskets are aligned with the holes in the cylinder heads.

27. Install the intake manifold and the nine retaining bolts. Hand-tighten the right-rear bolt (viewed from the front of the engine) before final tightening, then tighten the bolts, in sequence, to 15–22 ft. lbs. (20–30 Nm).

28. Inspect and if necessary, replace the

O-ring seal on the thermostat housing. Position the housing and upper hose and install the two retaining bolts. Tighten to 15–22 ft. lbs. (20–30 Nm).

29. Reconnect the heater water hose.

30. Reconnect both HO2S wiring connectors.

31. Reconnect the vacuum hose to the throttle body adapter vacuum port.

32. Install the accelerator cable bracket on the intake manifold and connect the accelerator and cruise control cables to the throttle body.

33. Install the PCV valve in the cylinder head cover and connect the canister purge solenoid vent hose. Reconnect the A/C compressor clutch, CKP sensor and canister purge solenoid wiring connectors.

34. Reconnect the engine/transmission harness connector. Install the connector on the retaining bracket on the power brake booster.

35. Raise and safely support the vehicle.

36. Reconnect the EGR valve to exhaust manifold tube to the right-hand exhaust manifold. Tighten the tube nut to 26–33 ft. lbs. (35–45 Nm).

37. Reconnect the power steering control valve actuator and the oil pressure sensor wiring connectors.

38. Lower the vehicle.

39. Position the alternator and install the two retaining bolts. Tighten to 15–22 ft. lbs. (20–30 Nm). Install the two bolts retaining the alternator brace to the intake manifold and tighten to 71–106 inch lbs. (8–12 Nm).

40. Reconnect the alternator wiring harness to the alternator, right-hand fender apron and junction block.

41. Position the ignition wire assembly on the engine and install the two bolts retaining the ignition wire bracket to the ignition coil brackets. Tighten the bolts to 71–106 inch lbs. (8–12 Nm).

42. Reconnect the ignition wires to the ignition coils in their proper positions. Reconnect the ignition wires to the spark plugs.

43. Reconnect the ignition wire brackets on the cylinder head cover studs. Reconnect the wiring connectors to both ignition coils and the CMP sensor.

44. Install the accessory drive belt.

45. Install the engine air cleaner outlet tube.

46. Install the windshield wiper governor.

47. Reconnect the fuel supply and return lines.

48. Fill and bleed the engine cooling system.

49. Reconnect the negative battery cable.

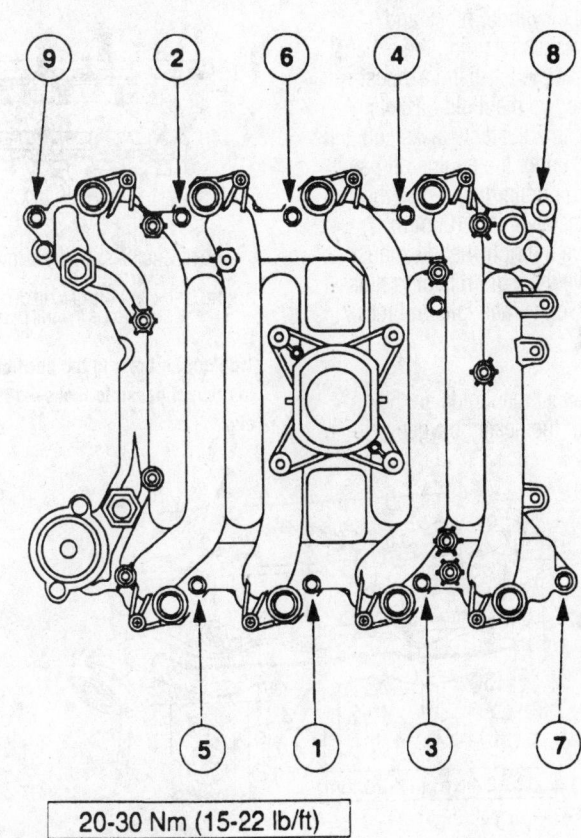

20-30 Nm (15-22 lb/ft)

7922RG06

Tighten the intake manifold in the sequence shown—1996–99 models

50. If equipped with air suspension, turn the air suspension switch to the ON position.

51. Start the engine and check for leaks.

52. Road test the vehicle and check for proper operation.

Exhaust Manifold

REMOVAL & INSTALLATION

✳✳ CAUTION

The fuel injection system remains under pressure, even after the engine has been turned OFF. The fuel system pressure must be relieved before disconnecting any fuel lines. Failure to do so may result in fire and/or personal injury.

1. Disconnect both battery cables, negative cable first.

2. Remove the engine air inlet tube.

✳✳ CAUTION

Never open, service or drain the radiator or cooling system when hot; serious burns can occur from the steam and hot coolant.

3. Drain the engine cooling system.

4. Remove the cooling fan and shroud assembly.

✳✳ CAUTION

Observe all applicable safety precautions when working around fuel. Whenever servicing the fuel system, always work in a well ventilated area. Do not allow fuel spray or vapors to come in contact with a spark or open flame. Keep a dry chemical fire extinguisher near the work area. Always keep fuel in a container specifically designed for fuel storage; also, always properly seal fuel containers to avoid the possibility of fire or explosion.

5. Relieve the fuel system pressure using the recommended procedure.

6. Disconnect the fuel supply and return lines.

7. Remove the upper radiator hose.

8. Remove the windshield wiper governor and support bracket.

9. Discharge the air conditioning system using proper recovery equipment. Disconnect the compressor outlet hose at the compressor and remove the bolt retaining

the hose assembly to the right-hand ignition coil bracket. Plug both openings.

10. Remove and disconnect the engine/transmission harness connector from the retaining bracket on the power brake booster.

11. Disconnect the heater hose.

12. Remove the nut retaining the ground strap to the right-hand cylinder head.

13. Remove the upper stud and lower bolt retaining the heater hose to the right cylinder head and position out of the way.

14. Remove the heater blower motor switch resistor.

15. Remove the bolt retaining the right-hand front engine support insulator to the sub-frame.

16. Disconnect both Heated Oxygen Sensor (HO2S) electrical connectors.

17. Raise and safely support the vehicle.

18. Remove the front engine support insulator through-bolts.

19. Remove the EGR valve-to-exhaust manifold tube nut from the right-hand exhaust manifold.

20. Disconnect the catalytic converter pipes from both exhaust manifolds. Lower the exhaust system and hang it from the crossmember with wire.

21. To remove the left-hand exhaust manifold, remove the front engine support insulator from the cylinder block and remove the eight nuts retaining the exhaust manifold. Remove the left-hand exhaust manifold and the two manifold gaskets.

22. Position an adjustable jackstand and a block of wood under the engine oil pan, rearward of the oil drain hole. Raise the engine approximately 4 in. (100mm).

23. To remove the right-hand exhaust manifold, remove the eight retaining nuts retaining. Remove the manifold and the manifold gasket.

To install:

24. If the exhaust manifolds are being replaced, transfer the heated oxygen sen-

sors and tighten to 27–33 ft. lbs. (37–45 Nm). On the right-hand exhaust manifold, transfer the EGR tube connector and tighten to 33–48 ft. lbs. (45–65 Nm).

25. Clean the mating surfaces of the exhaust manifolds and cylinder heads.

26. Position the exhaust manifolds to the cylinder heads, using new gaskets and install the retaining bolts. Tighten the bolts in sequence to 15–22 ft. lbs. (20–30 Nm).

27. Reconnect the EGR valve and tube assembly to the exhaust manifold. Tighten the line nut to 26–33 ft. lbs. (35–45 Nm).

28. Install the left-hand front engine support insulator to the cylinder block and tighten the bolts to 15–22 ft. lbs. (20–30 Nm). Lower the engine onto the front engine support insulator and remove the jack.

29. Install the left-hand and right-hand engine support insulator through-bolts and tighten to 15–22 ft. lbs. (20–30 Nm).

30. Reconnect the catalytic converter pipes to both exhaust manifolds. Install the four retaining nuts and tighten to 20–30 ft. lbs. (27–41 Nm).

➡Be sure the exhaust system clears the No. 3 crossmember. Adjust as necessary.

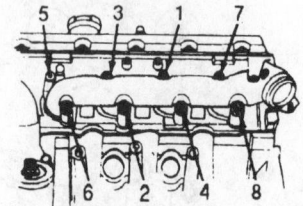

NOTE: ENGINE SHOWN REMOVED FOR CLARITY
NOTE: LH EXHAUST MANIFOLD SHOWN RH EXHAUST MANIFOLD TYPICAL

7922RG07

Tighten the bolts in the correct sequence to prevent possible leaks—1995–96 models

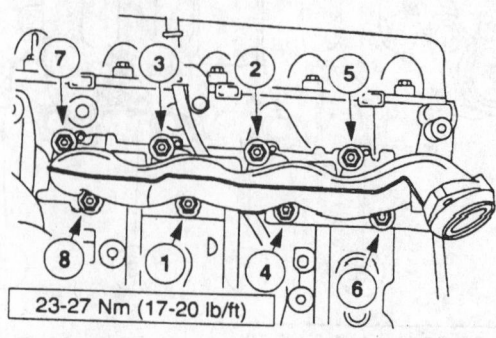

23-27 Nm (17-20 lb/ft)

7922RG08

To prevent warpage, be sure to tighten the exhaust manifold bolts according to the sequence shown—1997–99 models (left side)

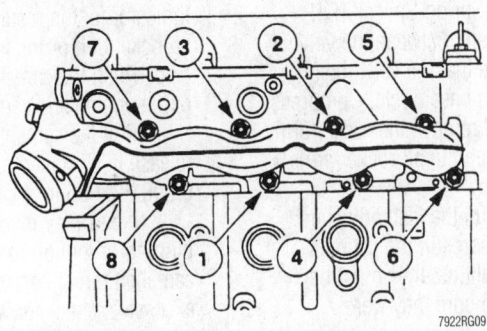

7922RG09

To prevent warpage, be sure to tighten the exhaust manifold bolts according to the sequence shown—1997–99 models (right side)

31. Lower the vehicle.

32. Reconnect both HO2S electrical connectors and install the bolt retaining the right-hand front engine support insulator to the sub-frame. Tighten to 15–22 ft. lbs. (20–30 Nm).

33. Install the heater blower motor switch resistor with the two retaining screws.

34. Place the heater hose in position. Install the upper stud and lower bolt and tighten to 15–22 ft. lbs. (20–30 Nm). Install the ground strap onto the stud and tighten the nut to 15–22 ft. lbs. (20–30 Nm).

35. Reconnect the heater hose.

36. Reconnect the engine/transmission harness connector. Install the connector to the retaining bracket on the power brake booster.

37. Unplug and connect the air conditioning compressor outlet hose to the compressor and install the bolt retaining the hose assembly to the right-hand ignition coil bracket.

38. Install the upper radiator hose.

39. Reconnect the fuel supply and return lines.

40. Install the windshield wiper governor and retaining bracket.

41. Install the engine cooling fan blade and fan shroud.

42. Fill and bleed the engine cooling system.

43. Install the engine air inlet tube.

44. Reconnect both battery cables, negative cable last.

45. Start the engine and check for leaks.

46. Properly evacuate and charge the air conditioning system.

47. Road test the vehicle and check for proper operation.

Camshaft and Valve Lifters

REMOVAL & INSTALLATION

✳✳ CAUTION

The fuel injection system remains under pressure, even after the engine has been turned OFF. The fuel system pressure must be relieved before disconnecting any fuel lines. Failure to do so may result in fire and/or personal injury.

1. If equipped with air suspension, the air suspension switch, must be turned to the OFF position before raising the vehicle.

2. Disconnect the negative battery cable.

✳✳ CAUTION

Observe all applicable safety precautions when working around fuel. Whenever servicing the fuel system, always work in a well ventilated area. Do not allow fuel spray or vapors to come in contact with a spark or open flame. Keep a dry chemical fire extinguisher near the work area. Always keep fuel in a container specifically designed for fuel storage; also, always properly seal fuel containers to avoid the possibility of fire or explosion.

3. Relieve the fuel system pressure by performing the following:

a. Remove the fuel tank cap to relieve the pressure in the fuel tank.

b. Remove the cap from the Schrader valve located on the fuel injection supply manifold.

c. Attach Fuel Pressure Gauge T80L-9974-B or equivalent, to the Schrader valve and drain the fuel through the drain tube into a suitable container.

d. After the fuel system pressure is relieved, remove the fuel pressure gauge and install the cap on the Schrader valve. Tighten the fuel tank fill cap.

4. Remove the fan blade and fan shroud assembly.

5. Disconnect the fuel supply and return lines from the fuel injection supply manifold.

6. Remove the windshield wiper governor (module) assembly from the vehicle.

7. Remove the engine air cleaner outlet tube.

8. Release the drive belt tensioner and remove the accessory drive belt.

9. Tag and disconnect all ignition wires from the spark plugs. Disconnect the ignition wire brackets from the cylinder head cover studs.

10. Remove the two bolts retaining the ignition wire separator to the ignition coil brackets and the bolt retaining the A/C pressure line to the right-hand ignition coil bracket.

11. Disconnect the electrical connectors from both ignition coils and the Camshaft Position (CMP) sensor.

12. Remove both ignition coils with brackets attached.

13. Disconnect the electrical connector from the alternator and at the power distribution box.

14. Remove the water pump pulley.

15. Disconnect the positive battery cable at the power distribution box.

16. Remove the bolt from the positive battery cable bracket located on the right-hand cylinder head.

17. Disconnect the fuel vapor hose from the evaporative emission canister purge valve (EVAP canister purge valve) and position the positive battery cable out of the way.

18. Disconnect the PCV valve from the cylinder head cover and position out of the way.

19. Remove and disconnect the engine/transmission harness connector from the bracket on the power brake booster.

20. Disconnect the Crankshaft Position (CKP) sensor and A/C clutch harness connectors.

21. Raise and safely support the vehicle.

22. Remove the bolts retaining the power steering pump to the cylinder block and wire the pump out of the way.

➡The front lower bolt on the power steering pump will not come all the way out.

✲✲ CAUTION

The EPA warns that prolonged contact with used engine oil may cause a number of skin disorders, including cancer! You should make every effort to minimize your exposure to used engine oil. Protective gloves should be worn when changing the oil. Wash your hands and any other exposed skin areas as soon as possible after exposure to used engine oil. Soap and water, or waterless hand cleaner should be used.

23. Drain the engine oil and remove the oil pan.

24. Remove the crankshaft pulley bolt and washer. Remove the crankshaft pulley using Pulley Remover T58P-6316-D, or equivalent.

25. Remove the engine oil filter.

26. Disconnect the power steering control valve actuator and oil pressure sensor.

27. Remove the oil filter adapter.

28. Lower the vehicle.

29. Remove both cylinder head covers.

30. Remove the engine front cover.

31. Remove the timing chains.

32. Rotate the crankshaft counterclockwise no more than 45 degrees from TDC to ensure that all pistons are below the top of the engine block deck face.

✲✲ WARNING

The crankshaft must be in this position prior to rotating the camshafts or damage to the pistons and/or valve-train will result.

33. Install Valve Spring Compressor T91P-6565-A or equivalent, under the camshaft and on top of one of the valve spring retainers.

✲✲ WARNING

Valve Spring Spacer Tool T91P-6565-AH or equivalent, must be installed between the spring coils and the camshaft lobe must be at the base circle before compressing the valve spring for each valve to prevent damage.

34. Install Valve Spring Spacer T91P-6565-AH or equivalent, between the valve spring coils. Be sure that the valve being compressed is on its base circle. Compress the valve spring and remove the rocker arm. Repeat the procedure until all rocker arms are removed.

35. If required, pull the lash adjusters (lifters) out of their bores in the cylinder head. Note their locations, they must be installed in the same bore they were removed from.

➡Do not mix the camshaft bearing caps. Note the camshaft bearing cap locations for installation.

36. To remove each camshaft, remove the 14 bolts retaining the camshaft bearing caps (cluster assemblies) to the cylinder head. Tap upward on the camshaft bearing caps at points near the upper bearing halves and gradually lift the camshaft bearing cap clusters from the cylinder head.

37. Repeat the removal procedure for the opposite cylinder head.

38. Remove the camshaft straight upward to avoid bearing damage.

39. Clean and inspect the camshafts and related components for unusual wear or damage.

To install:

40. Clean and inspect the cylinder head covers, engine front cover and cylinder head sealing surfaces.

41. Apply clean engine oil to the camshaft journals and lobes. Position the camshafts on the cylinder heads.

42. Install and seat the camshaft bearing cap cluster assemblies. Install and hand start the retaining bolts. Tighten the camshaft cluster retaining bolts in sequence to 71–106 inch lbs. (8–12 Nm). Be sure to tighten each camshaft bearing cap cluster individually.

➡Each camshaft bearing cap cluster assembly is tightened individually.

43. Loosen the camshaft bearing cap cluster retaining bolts approximately two turns or until the heads of the bolts are free. Tighten all bolts, again in sequence, to 71–106 inch lbs. (8–12 Nm).

44. Repeat the camshaft bearing cap installation for the opposite cylinder head.

➡The camshafts should turn freely but with a slight drag.

45. Check camshaft end-play as follows:

   a. Install a suitable dial indicator on the front of the engine. Position it so the indicator foot is resting on the camshaft sprocket bolt or the front of the camshaft.

   b. Push the camshaft toward the rear of the engine and zero the dial indicator.

   c. Pull the camshaft forward and release it. Specified end-play is 0.0901–0.006 in. (0.025–0.190mm).

   d. If end-play is too tight, check for binding or foreign material in the camshaft thrust bearing. If end-play is excessive, check for worn camshaft thrust plate and replace the cylinder head, as required.

   e. Remove the dial indicator.

46. If removed, install the lash adjusters in their original positions.

47. If necessary, install Cam Positioning Tools T92P-6256-A or equivalent, on the flats of the camshafts and install the spacers and camshaft sprockets. Install the bolts and washers and tighten to 81–95 ft. lbs. (110–130 Nm).

48. Install Valve Spring Compressor T91P-6565-A or equivalent, under the camshaft and on top of the valve spring retainer.

✲✲ WARNING

Valve Spring Spacer Tool T91P-6565-AH or equivalent, must be installed between the spring coils and the camshaft lobe must be at the base circle before compressing the valve spring for each valve to prevent damage.

49. Install Valve Spring Spacer T91P-6565-AH or equivalent, between the valve spring coils. Be sure that the valve being compressed is on its base circle. Compress the valve spring and install the rocker arm. Repeat the procedure until all rocker arms are installed.

50. Rotate the crankshaft clockwise 45 degrees to position the crankshaft at TDC.

➡The crankshaft must only be rotated in the clockwise direction and only as far as TDC.

51. Install the timing chains.

52. Install engine front cover.

53. Install both cylinder head covers. Tighten the cylinder head cover bolts to 71–106 inch lbs. (8–12 Nm).

54. Raise and safely support the vehicle.

55. Reconnect the power steering control valve actuator and oil pressure sensor harness connectors.

56. Apply silicone sealer to the crankshaft keyway, and install the crankshaft pulley, using Damper Replacer T74P-6316-B

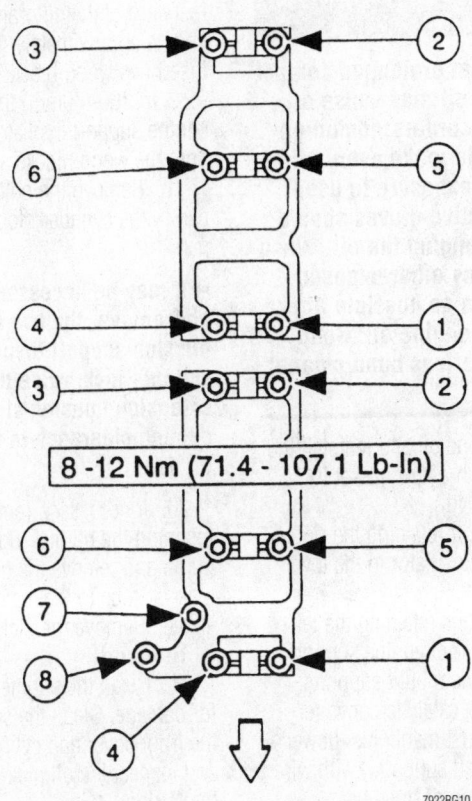

8 -12 Nm (71.4 - 107.1 Lb-In)

7922RG10

Be sure to tighten the camshaft bearing cap cluster bolts in the sequence shown to avoid damage to the camshaft or bearings

or equivalent. Tighten the crankshaft pulley bolt to 114–121 ft. lbs. (155–165 Nm).

57. Install the engine oil pan.

58. Position the power steering pump on the engine and install the four retaining bolts. Tighten the bolts to 15–22 ft. lbs. (20–30 Nm).

59. Safely lower the vehicle.

60. Reconnect the A/C clutch, CKP sensor and evaporative emission canister purge valve harness connectors.

61. Reconnect the engine/transmission harness connectors and install the connectors on the power brake booster retaining bracket.

62. Install the PCV valve to the right-hand cylinder head cover.

63. Install the positive battery cable harness on the right-hand cylinder head.

64. Install the bolt retaining the battery cable bracket to the cylinder head.

65. Reconnect the evaporative emission hose to the canister purge valve.

66. Reconnect the positive battery cable at the power distribution box.

67. Install the water pump pulley and tighten the bolts to 15–22 ft. lbs. (20–30 Nm).

68. Install the ignition coil brackets and ignition wires to the engine front cover.

Tighten the retaining nuts to 15–22 ft. lbs. (20–30 Nm).

69. Reconnect the harness connectors to the ignition coils and the CMP sensor.

70. Position the A/C pressure line on the right-hand ignition coil bracket and install the retaining bolt.

71. Reconnect the ignition wires to the spark plugs and install the brackets onto the cylinder head cover studs.

72. Release the drive belt tensioner and install the accessory drive belt.

73. Install the windshield wiper governor.

74. Properly connect the fuel supply and return lines.

75. Install the fan and shroud assembly.

76. Fill the engine cooling system to the proper level.

✳✳ WARNING

Operating the engine without the proper amount and type of engine oil will result in severe engine damage.

77. Fill the crankcase to the proper level and grade of engine oil.

78. Reconnect the negative battery cable.

79. If equipped with air suspension, turn the air suspension switch to the ON position.

80. Start the engine and check for leaks.

81. Road test the vehicle and check for proper engine operation.

Valve Lash

ADJUSTMENT

The valve lash is not adjustable. If the collapsed lash adjuster clearance is incorrect, check the camshaft, roller follower and valve for wear or damage.

1. Disconnect the negative battery cable.

2. Remove the camshaft covers.

3. Rotate the crankshaft until the camshaft base circle is contacting the roller follower.

4. Use a suitable tool to bleed down the lash adjuster. Slowly compress the lash adjuster until the plunger is bottomed.

5. Use a feeler gauge to check the clearance between the camshaft and the roller follower. The clearance should be 0.018–0.033 in. (0.45–0.85mm).

Oil Pan

REMOVAL & INSTALLATION

✳✳ CAUTION

The fuel injection system remains under pressure, even after the engine has been turned OFF. The fuel system pressure must be relieved before disconnecting any fuel lines. Failure to do so may result in fire and/or personal injury.

1. Disconnect the negative battery cable.

2. Remove the engine air cleaner outlet tube.

✳✳ CAUTION

Observe all applicable safety precautions when working around fuel. Whenever servicing the fuel system, always work in a well ventilated area. Do not allow fuel spray or vapors to come in contact with a spark or open flame. Keep a dry chemical fire extinguisher near the work area. Always keep fuel in a container specifically designed for fuel storage; also, always properly seal fuel containers to avoid the possibility of fire or explosion.

3. Relieve the fuel system pressure.

4. Disconnect the fuel supply and return lines at the fuel injection supply manifold.

5. Drain the engine cooling system and remove the cooling fan and fan shroud.

6. Remove the upper radiator hose.

7. Remove the wiper governor and support bracket.

8. Properly discharge the air conditioning system using suitable recovery equipment. Disconnect and plug the compressor outlet hose at the compressor and remove the bolt retaining the hose assembly to the right-hand ignition coil bracket. Cap the compressor outlet.

9. Remove the engine/transmission electrical harness connector from the retaining bracket on the power brake booster and disconnect.

10. Disconnect the heater water hose.

11. Remove the nut retaining the ground strap to the right-hand cylinder head. Remove the upper stud and loosen the lower bolt retaining the heater outlet hose to the right-hand cylinder head and position out of the way.

12. Remove the heater blower motor switch resistor.

13. Raise and safely support the vehicle.

✳✳ CAUTION

The EPA warns that prolonged contact with used engine oil may cause a number of skin disorders, including cancer! You should make every effort to minimize your exposure to used engine oil. Protective gloves should be worn when changing the oil. Wash your hands and any other exposed skin areas as soon as possible after exposure to used engine oil. Soap and water, or waterless hand cleaner should be used.

14. Drain the engine oil and reinstall the oil pan drain plug with a new gasket to 8–12 ft. lbs. (11–16 Nm).

15. Remove the bolt retaining the right-hand engine support insulator to the lower front sub-frame.

16. Remove the bolts retaining the left-hand and right-hand front engine support insulators to the engine mount supports.

17. Disconnect the catalytic converter pipes from both exhaust manifolds. Lower the exhaust system and support it with wire from the transmission crossmember.

18. Position a screw-type jack, or equivalent and a block of wood under the oil pan, rearward of the oil drain hole. Raise the engine approximately 4 in. (100mm) and insert two wood blocks approximately 2½ –2¾ in. (60–70mm) thick under each front engine support insulator. Lower the engine onto the wood blocks and remove the jack.

19. Remove the bolts retaining the oil pan to the cylinder block and remove the oil pan.

➡️It may be necessary to loosen, but not remove, the two nuts on the transmission support insulator and with a suitable jack, raise the transmission extension housing slightly to allow enough clearance to remove the engine oil pan.

20. If necessary, remove the two bolts retaining the oil pick-up tube to the oil pump and remove the bolt retaining the pick-up tube to the main bearing stud spacer. Remove the pick-up tube.

To install:

21. Clean the engine oil pan and inspect for damage. Clean the sealing surfaces of the front cover and cylinder block. Clean and inspect the oil pick-up tube and replace the O-ring.

22. If removed, position the oil pick-up tube on the oil pump and hand start the two retaining bolts. Install the bolt retaining the pick-up tube to the main bearing stud spacer hand tight.

23. Tighten the pick-up tube-to-oil pump bolts to 72–108 inch lbs. (8–12 Nm), then tighten the pick-up tube-to-main bearing stud spacer bolt to 15–22 ft. lbs. (20–30 Nm).

24. Position a new gasket on the oil pan. Apply silicone sealer to where the front cover meets the cylinder block and the crankshaft rear oil seal and retainer meets the cylinder block. Position the oil pan to the engine and install the retaining bolts. Tighten the bolts in sequence, to 18–26 ft. lbs. (25–35 Nm) for 1995 vehicles and for 1996–99 vehicles tighten the bolts in sequence to 14 ft. lbs. (20 Nm), then rotate the oil pan retaining bolts, in sequence an additional 60 degrees within four minutes of applying the silicone sealer.

25. Position the jack and wood block under the engine oil pan, rearward of the oil drain hole, and raise the engine enough to remove the wood blocks. Lower the engine and remove the jack.

26. Install the left-hand and right-hand engine support insulator through-bolts and tighten to 15–22 ft. lbs. (20–30 Nm).

27. Install the bolt retaining the right-hand engine support insulator to the lower front sub-frame. Tighten the bolt to 15–22 ft. lbs. (20–30 Nm).

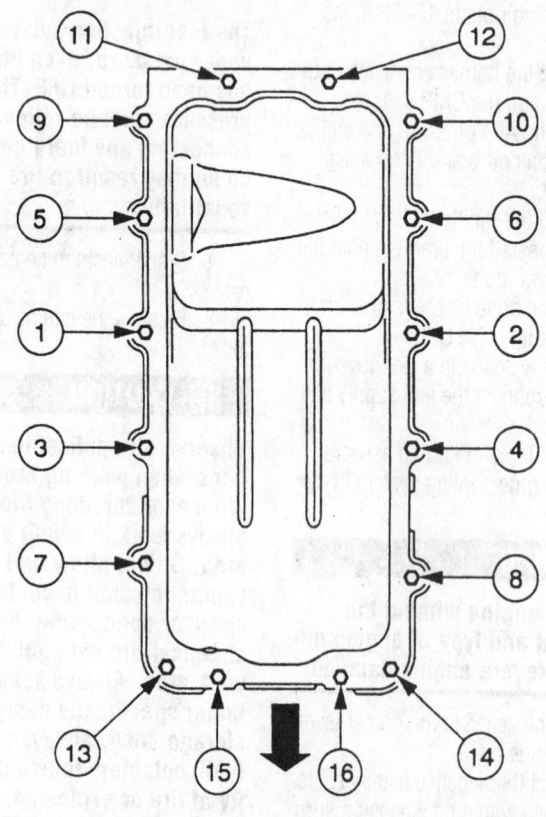

7922RG11

To prevent oil leaks, tighten the oil pan bolts in the sequence shown

28. Position the exhaust system to the exhaust manifolds. Install the four retaining nuts and tighten to 20–30 ft. lbs. (27–41 Nm). Be sure the exhaust system clears the crossmember. Adjust as necessary.

29. Install a new engine oil filter.

30. Lower the vehicle.

31. Install the heater blower motor switch resistor using the two retaining screws.

32. Position the heater water hose, install the upper stud and tighten the upper and lower bolts to 15–22 ft. lbs. (20–30 Nm). Install the ground strap on the stud and tighten to 15–22 ft. lbs. (20–30 Nm).

33. Reconnect the heater water hose.

34. Reconnect the throttle valve cable, if equipped. If necessary, adjust the throttle valve cable.

35. Reconnect the engine/transmission electrical harness connector. Install the harness connector on the power brake booster bracket.

36. Unplug and connect the air conditioning compressor outlet hose to the compressor and install the bolt retaining the hose to the right-hand ignition coil bracket.

37. Install the upper radiator hose.

38. Reconnect the fuel supply and return lines.

39. Install the wiper governor and retaining bracket.

40. Install the engine cooling fan and fan shroud.

41. Fill the engine cooling system.

✳✳ WARNING

Operating the engine without the proper amount and type of engine oil will result in severe engine damage.

42. Fill the engine crankcase with the proper type and quantity of engine oil.

43. Install the engine air cleaner outlet tube.

44. Reconnect the negative battery cable.

45. Start the engine and check for leaks.

46. Properly evacuate and recharge the air conditioning system.

47. Road test the vehicle and check for proper engine operation.

Oil Pump

REMOVAL & INSTALLATION

1. Disconnect the negative battery cable.

2. Remove both cylinder head covers.

3. Remove the engine front cover.

✳✳ CAUTION

The EPA warns that prolonged contact with used engine oil may cause a number of skin disorders, including cancer! You should make every effort to minimize your exposure to used engine oil. Protective gloves should be worn when changing the oil. Wash your hands and any other exposed skin areas as soon as possible after exposure to used engine oil. Soap and water, or waterless hand cleaner should be used.

4. Remove the engine oil pan.

5. Remove both timing chains.

6. Remove the two bolts retaining the oil pick-up tube to the oil pump and remove the bolt retaining the oil pick-up tube to the main bearing stud spacer. Remove the pick-up tube.

7. Remove the four bolts retaining the oil pump to the cylinder block and remove the oil pump.

To install:

8. Rotate the inner rotor of the oil pump to align with the flats on the crankshaft and install the oil pump flush with the cylinder block. Install the four retaining bolts and tighten to 72–106 inch lbs. (8–12 Nm).

9. Clean the oil pick-up tube and replace the O-ring.

10. Place the pick-up tube on the oil pump and hand start the two retaining bolts.

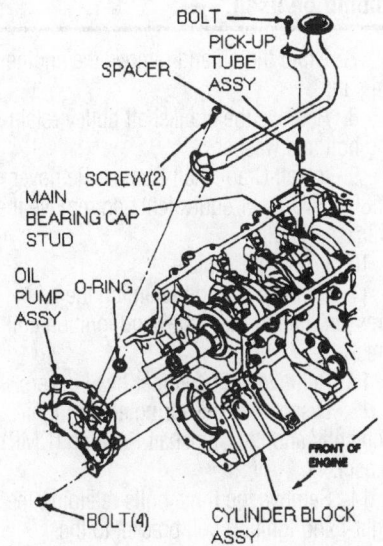

SCREW(2)

MAIN BEARING CAP STUD

OIL PUMP ASSY   O-RING

SPACER

BOLT

PICK-UP TUBE ASSY

FRONT OF ENGINE

BOLT(4)

CYLINDER BLOCK ASSY

7922RG12

The oil pump is mounted on the crankshaft at the front of the engine

Install the bolt retaining the pick-up tube to the main bearing stud spacer hand tight. Tighten the pick-up tube-to-oil pump bolts to 72–106 inch lbs. (8–12 Nm). Tighten the pick-up tube to main bearing stud spacer bolt to 15–22 ft. lbs. (20–30 Nm).

11. Install a new engine oil filter.

12. Install both timing chains.

13. Install the engine oil pan.

14. Install the engine front cover.

15. Install both cylinder head covers.

✳✳ WARNING

Operating the engine without the proper amount and type of engine oil will result in severe engine damage.

16. Fill the crankcase with the proper type and quantity of engine oil.

17. Reconnect the negative battery cable.

18. Start the engine and check for leaks and proper engine oil pressure.

19. Road test the vehicle and check for proper engine operation.

Rear Main Seal

REMOVAL & INSTALLATION

➡**Special tools are available for installing rear main oil seals. In most cases, the seals can be installed using a common seal and bearing driver set.**

1. Remove the transmission assembly.

2. Remove the flexplate or flywheel.

3. With a sharp awl, carefully punch a small hole in the metal portion of the seal.

4. Remove the seal using a slide hammer with a sheet metal screw attached.

➡**If the oil leak is coming from around the seal retainer, the retainer must also be removed and resealed.**

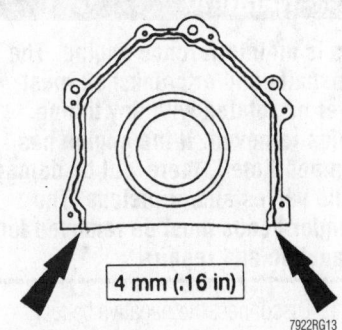

4 mm (.16 in)

7922RG13

Apply a continuos bead of silicone sealant to the back of the seal retainer before installing it on the engine

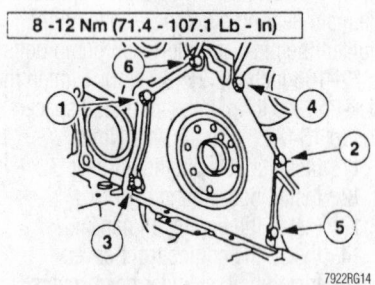

8 -12 Nm (71.4 - 107.1 Lb - In)

7922RG14

To avoid leakage, be sure to tighten the crankshaft rear oil seal retainer bolts in the correct sequence

To install:

5. If the seal retainer was removed, carefully clean the sealant from the retainer and engine block using a plastic scraper. Remove any oil or grease residue from the sealing surfaces with a solvent.

6. Apply silicone sealant to the back of the retainer and immediately install it on the engine block. Tighten the bolts in the proper sequence to 71–107 inch lbs. (8–12 Nm).

7. Lubricate the seal and the crankshaft with clean engine oil.

8. Install the seal with the spring side toward the engine.

9. Remove the installation tool and install the flexplate or flywheel. Tighten the bolts, in a crisscross pattern, to 54–64 ft. lbs. (73–87 Nm).

10. Install the transmission assembly.

11. Lower the vehicle.

12. Check the engine oil level.

13. Reconnect the negative battery cable.

14. Start the engine and check for leaks.

Timing Chain, Sprockets, Front Cover and Seal

REMOVAL & INSTALLATION

✳✳ WARNING

This is an interference engine. The camshafts and/or crankshaft must never be rotated with any timing chains removed. If the engine has "jumped time", there will be damage to the valves and/or pistons. The cylinder heads must be removed for inspection and repairs.

1. Disconnect the negative battery cable.

2. Remove the cooling fan and shroud.

3. Loosen the water pump pulley bolts, then remove the accessory drive belt.

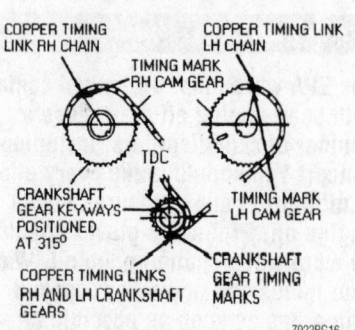

Be sure that the timing marks are aligned when the No. 1 piston is at TDC on compression

4. Remove the water pump pulley.

5. Raise and safely support the vehicle.

6. Remove the bolts retaining the power steering pump to the cylinder block and engine front cover. The lower front bolt on the power steering pump will not come all the way out. Wire the power steering pump out of the way.

✳✳ CAUTION

The EPA warns that prolonged contact with used engine oil may cause a number of skin disorders, including cancer! You should make every effort to minimize your exposure to used engine oil. Protective gloves should be worn when changing the oil. Wash your hands and any other exposed skin areas as soon as possible after exposure to used engine oil. Soap and water, or waterless hand cleaner should be used.

7. Drain the oil and remove the engine oil pan.

8. Remove the crankshaft pulley retaining bolt and washer.

9. Install Crankshaft Damper Remover T58P-6316-D, or equivalent and remove the crankshaft pulley.

10. Lower the vehicle.

11. Remove the bolt retaining the A/C pressure line to the right-hand ignition coil bracket.

12. Remove both cylinder head covers.

13. Disconnect the wiring at both ignition coils and the Camshaft Position (CMP) sensor.

14. Remove the three bolts retaining the right-hand ignition coil bracket to the engine front cover. Position the power steering hose out of the way.

15. Remove the three nuts retaining the left-hand ignition coil bracket to the engine front cover. Slide both ignition coil brackets

and ignition wires off the mounting studs and lay the assembly on top of the engine.

16. Remove the bolts retaining the drive belt idler pulley and remove the pulley.

17. Disconnect the wiring to the Crankshaft Position (CKP) sensor and remove the sensor.

18. If equipped, remove the retainers for the oil cooler from the engine front cover retaining stud bolts and position the oil cooler out of the way.

19. Remove the nine stud bolts and the six standard bolts retaining the engine front cover to the engine and remove the cover.

20. Remove the crankshaft oil seal from the cover with a suitable seal driver.

21. Remove the crankshaft position sensor pulse wheel.

22. Rotate the engine to set the piston for No. 1 to TDC on its compression stroke.

➤**Cam Positioning Adapters T92P-6256-A or equivalent, must be installed on the camshafts to prevent the camshafts from rotating.**

23. Install Cam Positioning Adapters T92P-6256-A or equivalent, on the flats of both camshafts. This will prevent accidental rotation of the camshafts.

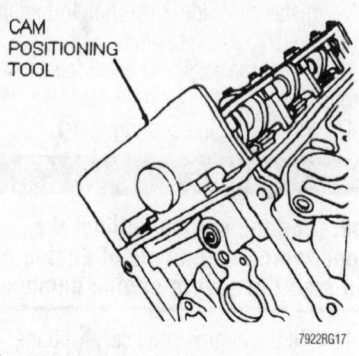

Use the special tool to maintain camshaft position while installing the timing chains

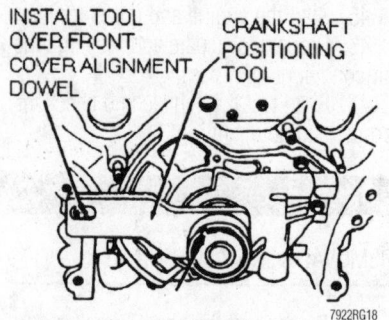

Install the crankshaft positioning tool to be sure the crankshaft does not turn while installing the timing chains

24. Remove the two bolts retaining the tensioner to the right-hand cylinder head and remove the tensioner. Remove the right-hand timing chain tensioner arm.

25. Remove the two bolts retaining the right-hand timing chain guide to the cylinder head and remove the timing chain guide.

26. Remove the right-hand timing chain from the camshaft and crankshaft sprockets.

27. If necessary, remove the right-hand camshaft sprocket retaining bolt, washer, sprocket and spacer.

28. Remove the two bolts retaining the timing chain tensioner to the left-hand cylinder head and remove the timing chain tensioner. Remove the left-hand timing chain tensioner arm.

29. Remove the two bolts retaining the timing chain guide to the left-hand cylinder head and remove the timing chain guide.

30. Remove the left-hand timing chain from the camshaft and crankshaft sprockets.

31. If necessary, remove the left-hand camshaft sprocket retaining bolt, washer, sprocket and spacer.

32. If necessary, note the position of the crankshaft sprockets and remove the crankshaft sprockets by sliding them off the front of the crankshaft.

33. Inspect the plastic running face on the tensioner arms and chain guides. If worn or damaged, inspect the engine oil pan for contamination and thoroughly clean the oil pan. Replace the oil pick-up tube.

To install:

34. Examine the timing chains, looking for the copper links. If the copper links are not visible, lay the chain on a flat surface and pull the chain taught until the opposite sides of the chain contact one another. Mark the links at each end of the chain and use these marks in place of the copper links.

➡ **If the engine jumped time, damage has been done to valves and possibly pistons and/or connecting rods. Any damage must be corrected before installing the timing chains.**

35. Be sure Cam Positioning Adapters T92P-6256-A or equivalent, are installed on the flats of the camshafts to prevent them from rotating.

36. Install the left-hand and right-hand timing chain guides and retaining bolts. Tighten the retaining bolts to 71–106 inch lbs. (8–12 Nm).

37. If removed, position the left-hand and right-hand camshaft spacers and sprockets on the camshafts. Install the washers and retaining bolts but do not tighten at this time.

38. If removed, install the left-hand crankshaft sprocket with the tapered part of the sprocket facing away from the engine block.

➡ **The crankshaft sprockets are identical. They may only be installed one way, with the tapered part of the sprockets facing each other. Ensure that the keyway and timing marks on the crankshaft sprockets are aligned.**

39. Install the left-hand timing chain on the camshaft and crankshaft sprockets. Be sure the copper links of the timing chain line up with the timing marks on both sprockets.

40. If removed, install the right-hand crankshaft sprocket with the tapered part of the sprocket facing the left-hand crankshaft sprocket.

41. Install the right-hand timing chain on the camshaft and crankshaft sprockets. Be sure the copper links of the timing chain line up with the timing marks on both sprockets.

42. It is necessary to bleed the timing chain tensioners before installation. Proceed as follows:

    a. Position the timing chain tensioner in a soft-jawed vise.

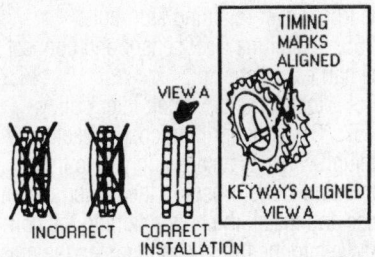

Be sure to install the crankshaft sprockets with the tapered sides facing each other as shown

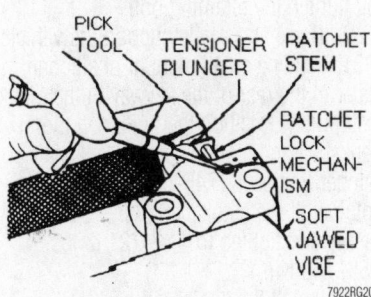

Slowly compress the timing chain tensioner while holding the ratchet lock away from the stem with a suitable tool

b. Using a small pick or similar tool, hold the ratchet lock mechanism away from the ratchet stem and slowly compress the tensioner plunger by rotating the vise handle.

❋❋ WARNING

The tensioner must be compressed slowly or damage to the internal seals will result.

    c. Once the tensioner plunger bottoms in the tensioner bore, continue to hold the ratchet lock mechanism and push down on the ratchet stem until flush with the tensioner face.

    d. While holding the ratchet stem flush to the tensioner face, release the ratchet lock mechanism and install a paper clip or similar tool in the tensioner body to lock the tensioner in the collapsed position.

    e. The paper clip must not be removed until the timing chain, tensioner, tensioner arm and timing chain guide are completely installed on the engine.

43. Install the right-hand and left-hand timing chain tensioners and secure with two bolts on each. Tighten the bolts to 15–22 ft. lbs. (20–30 Nm).

44. Install Crankshaft Positioning Tool T93P-6265-A or equivalent, over the crankshaft and the engine front cover alignment dowel to position the crankshaft.

45. Lubricate the timing chain tensioner arm contact surfaces with clean engine oil and install the right-hand and left-hand tensioner arms on their dowel pins.

46. Position a suitable C-clamp around the timing chain tensioner arm and timing chain guide to remove all slack from the timing chain. Use care not to bend the timing chain guide.

47. Remove the locking pins or paper clips from the timing chain tensioners and be sure that all timing marks are aligned.

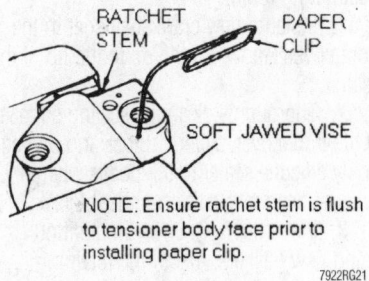

NOTE: Ensure ratchet stem is flush to tensioner body face prior to installing paper clip.

Install a paper clip or wire into the tensioner to hold the plunger in during assembly

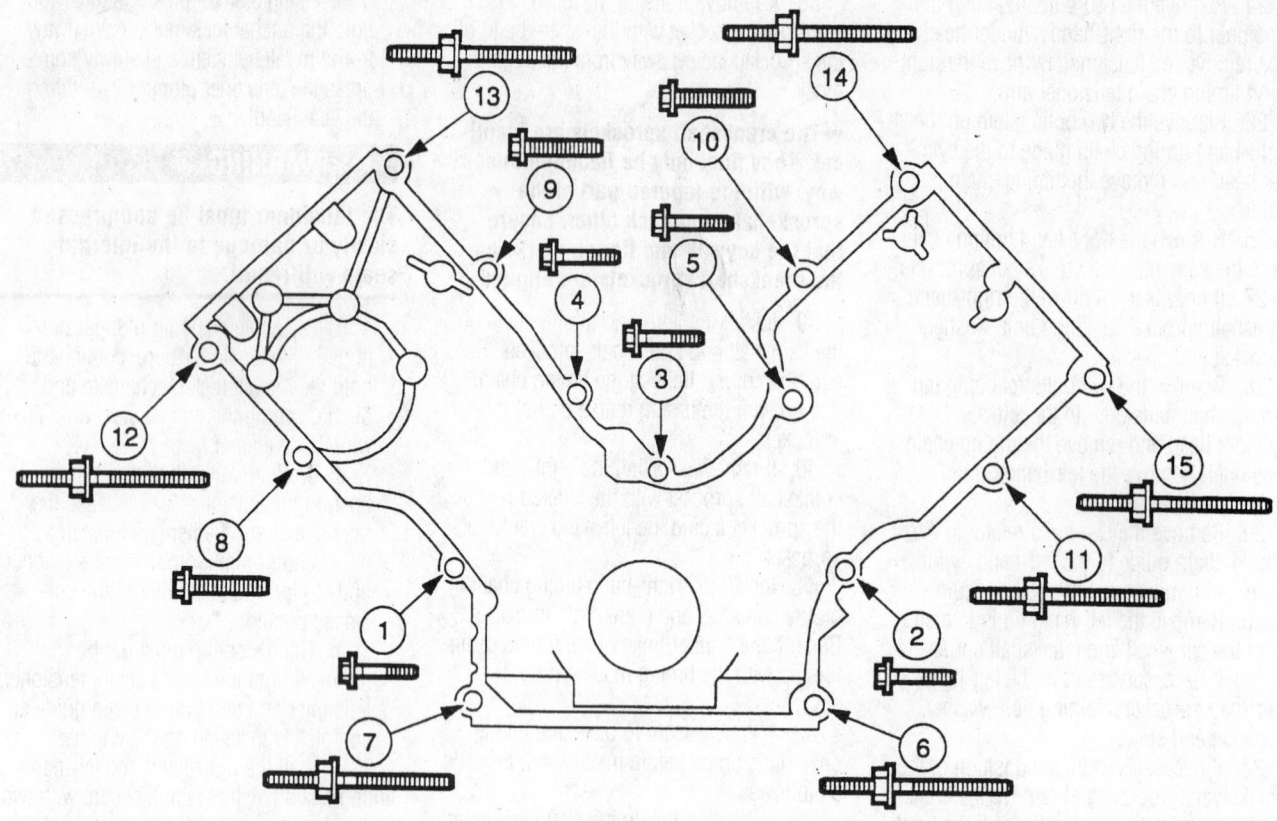

Timing chain front cover bolt tightening sequence

7922RG15

48. Using Camshaft Positioning Adapters T92P-6265-A or equivalent to align and hold the camshafts, tighten the camshaft sprocket retaining bolts to 81–95 ft. lbs. (110–130 Nm).

49. Position a suitable dial indicator in the No. 1 cylinder spark plug hole to measure intake valve lift. The intake valve should be at maximum lift when the crankshaft is at 114 degrees after TDC. If the intake valve lift is not at maximum lift, loosen the camshaft sprocket bolt and repeat the steps detailing the installation of the timing chain tensioners to the tightening of the camshaft sprockets.

50. Remove the camshaft and crankshaft positioning tools.

51. Install a new crankshaft seal in the front cover. Apply engine oil to the lip of the seal.

52. Thoroughly clean the sealing surfaces of the front cover, cylinder block and oil pan. Apply silicone sealer to the points where the cylinder head meets the cylinder block.

53. Place the front cover in position using new gaskets. Install the retaining bolts and studs in their proper locations. Tighten in sequence to 15–22 ft. lbs. (20–30 Nm) within four minutes of applying the silicone sealer.

54. If equipped, install the oil cooler to the front cover retaining stud bolts.

55. Install the CKP sensor and connect the harness connector.

56. Install the drive belt idler pulley.

57. Place the ignition coil brackets and ignition wires as an assembly onto the mounting studs. Position the power steering hose and install the nuts retaining the coil brackets to the front cover. Tighten the nuts to 15–22 ft. lbs. (20–30 Nm). Reconnect the wiring to both ignition coils and the CMP sensor.

58. Install both cylinder head covers.

59. Position the A/C pressure line on the right-hand ignition coil bracket and install and tighten the retaining bolt.

60. Raise and safely support the vehicle.

61. Apply a small amount of silicone sealer in the rear of the keyway in the crankshaft pulley. Position the pulley on the crankshaft and install, using Crankshaft Damper Replacer T74P-6316-B or equivalent. Install the crankshaft pulley bolt and washer and tighten to 114–121 ft. lbs. (155–165 Nm).

62. Install the engine oil pan.

63. Place the power steering pump on the engine and install the four retaining bolts. Tighten to 15–22 ft. lbs. (20–30 Nm).

64. Lower the vehicle.

65. Install the water pump pulley with the four retaining bolts. Tighten to 15–22 ft. lbs. (20–30 Nm).

66. Install the accessory drive belt.

67. Install the engine cooling fan and shroud.

✳✳ WARNING

Operating the engine without the proper amount and type of engine oil will result in severe engine damage.

68. Fill the engine with the proper type and quantity of engine oil.

69. Reconnect the negative battery cable.

70. Start the engine and check for leaks.

71. Road test the vehicle and check for proper engine operation.

FUEL SYSTEM

Fuel System Service Precautions

Safety is the most important factor when performing not only fuel system maintenance but any type of maintenance. Failure to conduct maintenance and repairs in a safe manner may result in serious personal injury or death. Maintenance and testing of the vehicle's fuel system components can be accomplished safely and effectively by adhering to the following rules and guidelines.

- To avoid the possibility of fire and personal injury, always disconnect the negative battery cable unless the repair or test procedure requires that battery voltage be applied.

- Always relieve the fuel system pressure prior to disconnecting any fuel system component (injector, fuel rail, pressure regulator, etc.), fitting or fuel line connection. Exercise extreme caution whenever relieving fuel system pressure to avoid exposing skin, face and eyes to fuel spray. Please be advised that fuel under pressure may penetrate the skin or any part of the body that it contacts.

- Always place a shop towel or cloth around the fitting or connection prior to loosening to absorb any excess fuel due to spillage. Ensure that all fuel spillage (should it occur) is quickly removed from engine surfaces. Ensure that all fuel soaked cloths or towels are deposited into a suitable waste container.

- Always keep a dry chemical (Class B) fire extinguisher near the work area.

- Do not allow fuel spray or fuel vapors to come into contact with a spark or open flame.

- Always use a back-up wrench when loosening and tightening fuel line connection fittings. This will prevent unnecessary stress and torsion to fuel line piping. Always follow the proper torque specifications.

- Always replace worn fuel fitting O-rings with new. Do not substitute fuel hose or equivalent, where fuel pipe is installed.

Fuel System Pressure

RELIEVING

Fuel supply lines on all fuel injected engines will remain pressurized for some period of time after the engine is shut **OFF**. This pressure must be relieved before servicing the fuel system. Pressure is relieved through the fuel pressure relief valve, located on the fuel rail.

To relieve the fuel system pressure, first remove the fuel tank cap to relieve pressure in the tank, then remove the cap on the fuel pressure relief valve. Attach fuel pressure gauge T80L-9974-a or equivalent, and drain

the system through the drain tube into a suitable container. Remove the fuel pressure gauge and replace the cap on the relief valve.

Fuel Filter

REMOVAL & INSTALLATION

✳✳ CAUTION

The fuel injection system remains under pressure, even after the engine has been turned OFF. The fuel system pressure must be relieved before disconnecting any fuel lines. Failure to do so may result in fire and/or personal injury.

1. Disconnect the negative battery cable.

✳✳ CAUTION

Observe all applicable safety precautions when working around fuel. Whenever servicing the fuel system, always work in a well ventilated area. Do not allow fuel spray or vapors to come in contact with a spark or open flame. Keep a dry chemical fire extinguisher near the work area. Always keep fuel in a container specifically designed for fuel storage; also, always properly seal fuel containers to avoid the possibility of fire or explosion.

2. Relieve the fuel system pressure.
3. If equipped with air suspension, turn the air suspension switch to the OFF position.
4. Raise and safely support the vehicle.
5. Remove the hairpin clip push connect fittings from both ends of the fuel filter as follows:

   a. Inspect the visible internal portion of the fitting for dirt accumulation. If more than a light coating of dust is present, clean the fitting before disassembly.

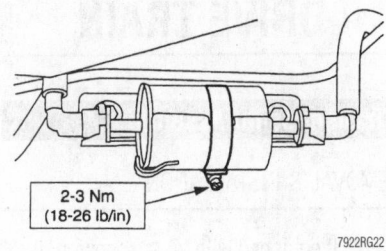

2-3 Nm
(18-26 lb/in)

7922RG22

The fuel filter is located near the center of the vehicle on the frame rail

b. Some adhesion between the seals in the fitting and the filter will occur with time. To separate, twist the fitting on the filter, then push and pull the fitting until it moves freely on the filter.

c. Remove the hairpin clip from the fitting by first bending and breaking the shipping tab. Next, spread the two clip legs by hand about ⅛ in. each, to disengage the body and push the legs into the fitting. Lightly pull the triangular end of the clip and work it clear of the filter and fitting.

➡**Do not use hand tools to complete this operation.**

d. Grasp the fitting and pull in an axial direction to remove the fitting from the filter. Be careful on 90 degree elbow connectors, as excessive side loading could break the connector body.

e. After disassembly, inspect the inside of the fitting for any internal parts such as O-rings and spacers that may have been dislodged from the fitting. Replace any damaged connector.

6. Loosen the filter retaining clamp and remove the fuel filter. Note the direction of the flow arrow on the filter, so the replacement filter can be reinstalled in the same position.

To install:

7. Install the fuel filter with the flow arrow facing the proper direction and tighten the filter retaining clamp.

8. Install the rubber insulator rings on the new filter (replace the insulator rings if the filter moves freely after the retainer is installed). Install the filter into the retainer with the flow arrow pointing out the open end of the retainer. Install the retainer on the bracket and tighten the mounting bolts to 27–44 inch lbs. (3–5 Nm).

9. Install the hairpin clip push connect fittings at both ends of the fuel filter as follows:

   a. Install a new connector if damage was found. Insert a new clip into any two adjacent openings with the triangular portion pointing away from the fitting opening. Install the clip until the legs of the clip are locked on the outside of the body. Piloting with an index finger is necessary.

   b. Before installing the fitting on the filter, wipe the filter end with a clean cloth. Inspect the inside of the fitting to be sure it is free of dirt and/or obstructions.

   c. Apply a light coating of engine oil to the filter end. Align the fitting and filter axially and push the fitting onto the filter end. When the fitting is engaged, a definite click will be heard. Pull on the fitting to be sure it is fully engaged.

10. Lower the vehicle.

11. If equipped with air suspension, turn the air suspension switch to the ON position.

12. Reconnect the negative battery cable.

13. Start the engine and check for fuel leaks and proper operation.

Fuel Pump

REMOVAL & INSTALLATION

> ※ **CAUTION**
>
> **The fuel injection system remains under pressure, even after the engine has been turned OFF. The fuel system pressure must be relieved before disconnecting any fuel lines. Failure to do so may result in fire and/or personal injury.**

1. Disconnect the negative battery cable.

> ※ **CAUTION**
>
> **Observe all applicable safety precautions when working around fuel. Whenever servicing the fuel system, always work in a well ventilated area. Do not allow fuel spray or vapors to come in contact with a spark or open flame. Keep a dry chemical fire extinguisher near the work area. Always keep fuel in a container specifically designed for fuel storage; also, always properly seal fuel containers to avoid the possibility of fire or explosion.**

2. Relieve the fuel system pressure.

3. Raise and safely support the vehicle.

4. Install a hose into the fuel filler pipe and drain or siphon the fuel into a suitable storage tank designed for fuel storage.

5. Remove any dirt that has accumulated around the fuel pump and fuel lines to prevent the entry of contaminants into the tank during fuel pump removal and installation.

6. Disconnect the fuel supply and return line fittings at the fuel pump using Fuel Line Disconnect Tools T90T-9550-B and T90T-9559-C, or equivalents.

7. Disconnect the fuel pump module electrical connector.

8. Remove the six retaining bolts around the perimeter of the fuel pump module.

9. Remove the fuel pump module and seal from the fuel tank.

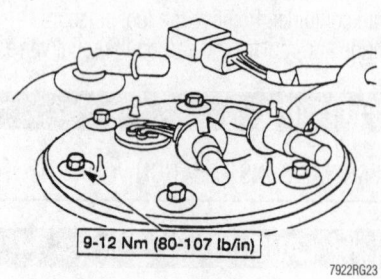

9-12 Nm (80-107 lb/in)

7922RG23

Tighten the fuel pump mounting bolts to 80–107 inch lbs. (9–12 Nm)

To install:

10. Clean the fuel pump module mounting flange and fuel tank mounting surface.

11. Install a new seal and the fuel pump module using care not to damage the inlet filter and fuel sending unit float arm.

12. Install and tighten the six retaining bolts to 80–107 inch lbs. (9–12 Nm).

13. Reconnect the fuel pump module electrical connector.

14. Install the fuel supply and return lines to the fuel pump module. Pull on the fuel line fittings to verify engagement.

15. Lower the vehicle.

16. Add a minimum of 10 gallons (38L) of clean fuel to the fuel tank and check for leaks.

17. Install Fuel Pressure Gauge T80L-9974-B or equivalent, to the Schrader valve on the fuel injection supply manifold.

18. Reconnect the negative battery cable.

19. Cycle the ignition switch from the **OFF** to **ON** position 5–10 times for three second intervals or until the fuel pressure gauge shows at least 35 psi (241 kPa).

20. Check for fuel leaks.

21. Remove the fuel pressure gauge.

22. Start the engine and recheck for fuel leaks.

23. Road test the vehicle and check for proper operation.

DRIVE TRAIN

Transmission Assembly

REMOVAL & INSTALLATION

1. If equipped with air suspension, the air suspension switch, located on the right-hand side of the luggage compartment, must be turned to the OFF position before raising the vehicle.

2. Disconnect the negative battery cable.

3. Raise and safely support the vehicle.

4. Remove the exhaust system as necessary for transmission removal.

5. Drain the fluid from the transmission by removing all oil pan bolts except the two bolts at the front of the pan. Loosen the two remaining bolts and if necessary, gently pry on the opposite end of the pan to lower the pan and allow the fluid to drain into a suitable container. When drained, reinstall a few of the bolts to hold the pan in place.

6. Remove the converter bottom access cover and adapter plate bolts.

7. If equipped, remove the torque converter drain plug, to allow the converter to drain into a suitable container. After the converter has drained, reinstall the drain plug and tighten.

8. Remove the four torque converter-to-flywheel retaining nuts.

➡ **Crank the engine over with a wrench on the crankshaft pulley retaining bolt to gain access to each torque converter-to-flywheel retaining bolt. Never turn the crankshaft in a counterclockwise direction, as viewed from the front of the vehicle.**

9. Mark the position of the driveshaft on the rear axle flange and remove the driveshaft. Install a suitable plug in the transmission extension housing to prevent fluid leakage.

10. Disconnect and remove the Vehicle Speed Sensor (VSS) or if equipped, the speedometer cable from the transmission extension housing.

11. Carefully disconnect the shift cable from the transmission manual control lever. If equipped, disconnect the throttle valve cable from the transmission throttle valve lever.

12. Disconnect all transmission wiring harness connectors.

13. Remove the starter motor retaining bolts and place the starter motor aside.

14. Position a suitable transmission jack under the transmission and raise it enough to allow crossmember removal.

15. Remove the engine rear support-to-crossmember bolts and the crossmember-to-frame side support retaining bolts. Remove the crossmember and transmission support insulator.

16. Lower the transmission jack and allow the transmission to hang.

17. Place a suitable jack to the front of the engine and raise the engine enough to gain access to the two upper transmission-to-cylinder block retaining bolts. Do not remove the bolts at this time.

18. Disconnect the transmission cooler lines at the transmission. Plug all openings to keep dirt out.

19. Remove the lower transmission-to-cylinder block retaining bolts.

20. Remove the transmission fluid fill tube and plug the opening in the transmission.

21. Secure the transmission to the transmission jack with a safety strap or chain.

22. Remove the two upper transmission-to-cylinder block retaining bolts.

23. Carefully move the transmission rearward to disengage from the dowel pins and the torque converter studs from the flywheel.

24. Lower the transmission from the vehicle.

25. Remove the torque converter to prevent the converter from dropping out of the transmission causing possible damage or personal injury.

26. If the transmission is to be removed for more than a speedy repair, support the rear of the engine with a safety stand and a block of wood.

To install:

➡ **Verify the transmission cooler lines are thoroughly cleaned before installing the transmission assembly.**

27. Remove the safety stand and block of wood supporting the rear of the engine, if installed.

28. If equipped, tighten the torque converter drain plug to 21–23 ft. lbs. (28–30 Nm).

29. If removed, position the torque converter on the transmission and rotate into position to be sure the drive flats are fully engaged in the pump gear. When fully seated, the center of the torque converter should be about 7/16–9/16 in. (10.2–14.4mm) below the transmission mounting surface

30. If removed, install the transmission on a suitable transmission jack and secure

DIMENSION A TO BE 10.23–14.43 mm (7/16-9/16 INCH) APPROXIMATELY

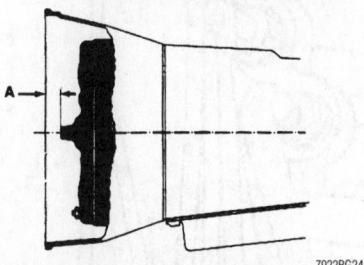

7922RG24

To prevent transmission damage, be sure that the torque converter is fully seated in the front pump of the transmission

with a safely strap or chain. Raise the transmission and align with the cylinder block dowel pins.

➡ **Do not allow the transmission to get in a nose-down position causing possible torque converter disengagement from the pump gear.**

31. Rotate the converter until the studs and drain plug are in alignment with the holes in the flywheel. Align the orange balancing marks on the converter stud and flywheel bolt hole, if balancing marks are present.

32. Move the transmission assembly forward into position, being careful not to damage the flywheel and converter pilot.

➡ **The converter face must rest squarely against the flywheel. This indicates that the converter pilot is not binding in the engine crankshaft. To ensure the converter is properly seated, grasp a converter stud. It should move freely back and forth in the flywheel hole. If the converter will not move, the transmission must be removed and the converter repositioned so the impeller hub is properly engaged in the pump gear.**

33. Install the two transmission housing-to-cylinder block retaining bolts at the engine dowel pin locations. Tighten the bolts to 41–50 ft. lbs. (55–68 Nm).

34. Install the remaining transmission housing-to-cylinder block retaining bolts. Tighten the bolts to 41–50 ft. lbs. (55–68 Nm).

35. Remove the safety strap or chain from around the transmission.

36. Install the transmission fluid fill tube and secure with the retaining bolt. Tighten the bolt to 28–38 ft. lbs. (38–51 Nm).

37. Reconnect the oil cooler lines to the transmission case. Tighten the cooler line fittings to 15–19 ft. lbs. (20–26 Nm).

38. Remove the jack supporting the front of the engine.

39. Using the transmission jack, raise the transmission enough to install the crossmember.

40. Place the crossmember and transmission support insulators in position. Install the engine rear support-to-crossmember retaining bolts and the crossmember-to-frame side support retaining bolts.

41. Remove the transmission jack.

42. Reconnect all transmission wiring harness connectors.

43. Install the starter motor and wiring.

44. Install the four torque converter-to-flywheel retaining nuts. Tighten to 20–33 ft. lbs. (27–46 Nm).

45. Install the torque converter access cover and cover plate bolts. Tighten the bolts to 12–16 ft. lbs. (16–22 Nm).

46. Install the exhaust system.

47. Install the Vehicle Speed Sensor (VSS) and connect the wiring or if equipped, the speedometer cable to the transmission extension housing.

48. Install the driveshaft aligning the marks that were made during removal.

49. Reconnect the shift cable to the transmission manual control lever. If equipped, connect the throttle valve cable to the transmission throttle valve lever.

50. Lower the vehicle.

51. If equipped with air suspension, turn the air suspension switch to the ON position.

52. Fill the transmission with the proper type and quantity of fluid.

53. Start the engine and check the transmission for leakage.

54. Adjust the shift linkage and manual position sensor, as required.

55. Road test the vehicle and check for proper transmission operation.

STEERING AND SUSPENSION

Air Bag

✳✳ CAUTION

Some vehicles are equipped with an air bag system, also known as the Supplemental Inflatable Restraint (SIR) or Supplemental Restraint System (SRS). The system must be disabled before performing service on or around system components, steering column, instrument panel components, wiring and sensors. Failure to follow safety and disabling procedures could result in accidental air bag deployment, possible personal injury and unnecessary system repairs.

PRECAUTIONS

Several precautions must be observed when handling the inflator module to avoid accidental deployment and possible personal injury.

• Never carry the inflator module by the wires or connector on the underside of the module.

• When carrying a live inflator module, hold securely with both hands, and ensure that the bag and trim cover are pointed away.

• Place the inflator module on a bench or other surface with the bag and trim cover facing up.

• With the inflator module on the bench, never place anything on or close to the module which may be thrown in the event of an accidental deployment.

DISARMING

✳✳ CAUTION

The Supplemental Inflatable Restraint (SIR) system must be disarmed before performing service around SIR system components or SIR system wiring. Failure to do so may cause accidental deployment of the air bag, resulting in unnecessary SIR system repairs and/or personal injury.

1. Position the vehicle with the front wheels in a straight ahead position.
2. Disconnect both battery cables, negative cable first.
3. Wait at least one minute for the air bag back-up power supply to deplete its stored energy before continuing.

✳✳ CAUTION

When carrying a live air bag, be sure the air bag and trim cover are pointed away from the body. In the unlikely event of an accidental deployment, the bag will, then deploy with minimal chance if injury. When placing a live air bag on a bench or other surface, always face the bag and trim cover up, away from the surface. This will reduce the motion of the module if it is accidentally deployed.

4. Proceed with the repair.

✳✳ WARNING

If the vehicles battery must be connected while the air bag module is removed, install Air Bag Simulator 105–00010 or equivalent, to the air bag module harness connector.

5. Disconnect the battery cables, if connected.
6. Remove the air bag simulator, if installed.

7. Once the repair is complete, install the air bag module.
8. Reconnect both battery cables, negative cable last.
9. Prove out the air bag system by turning the ignition key to the **RUN** position and visually monitoring the air bag indicator lamp in the instrument cluster. The indicator lamp should illuminate for approximately six seconds, then turn OFF. If the indicator lamp does not illuminate, stays on, or flashes at any time, a fault has been detected by the air bag diagnostic monitor.

Recirculating Ball Power Steering Gear

REMOVAL & INSTALLATION

1. If equipped with air suspension, the air suspension switch must be turned to the OFF position before raising the vehicle.
2. Center the steering wheel and turn the key to the locked position.
3. Disconnect the negative battery cable.
4. On the Town Car, separate the two halves and remove the steering gear cover.
5. Remove the bolt and separate the intermediate shaft from the steering gear.
6. Tag the power steering pressure and return lines so they may be reassembled in their original positions.
7. Place a drain pan under the steering gear and disconnect the pressure and return lines. Plug the lines and ports in the gear to prevent the entry of dirt.
8. Raise and safely support the vehicle.
9. Disconnect the Pitman arm from the center link using a suitable puller. It is not necessary to remove the Pitman arm from the steering gear.
10. Support the steering gear and remove the steering gear-to-frame rail retaining bolts.
11. Remove the steering gear from the vehicle.

To install:
12. Position the steering gear on the frame rail. Install the retaining bolts and tighten them to 50–67 ft. lbs. (66–90 Nm).
13. Install the Pitman arm to the center link. Tighten the retaining nut to 52–60 ft. lbs. (70–81 Nm).
14. Lower the vehicle.
15. Unplug and connect the power steering pressure and return lines to the steering gear and tighten the lines to 12–18 ft. lbs. (16–24 Nm).
16. Connect the intermediate shaft to the steering gear. Tighten the bolt to 31–41 ft. lbs. (41–55 Nm).
17. Reconnect the negative battery cable.
18. If equipped with air suspension, turn the air suspension switch to the ON position.
19. Fill the reservoir with the correct power steering fluid and turn the steering wheel from stop-to-stop to distribute the

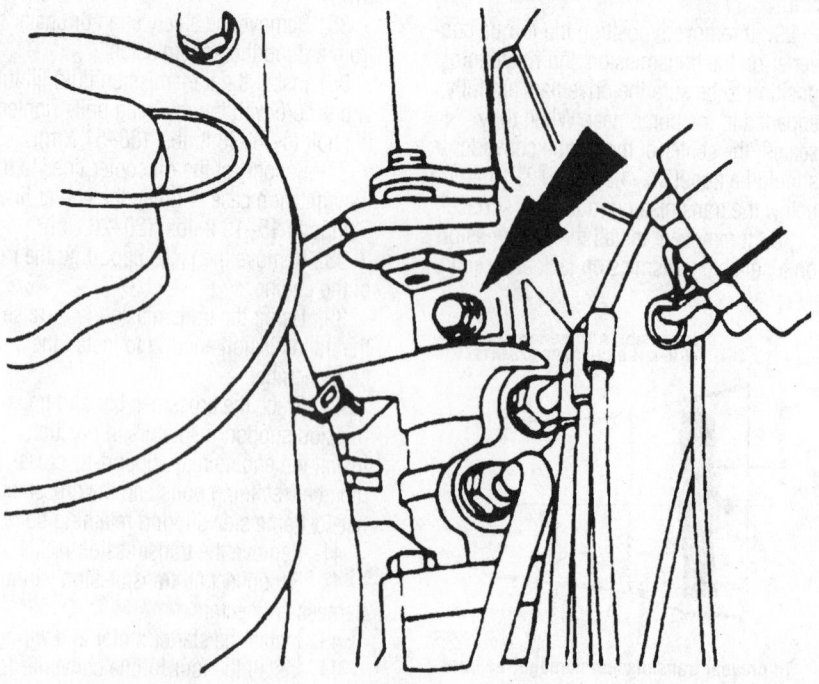

Remove the bolt and separate the intermediate shaft from the steering gear input shaft

7922RG25

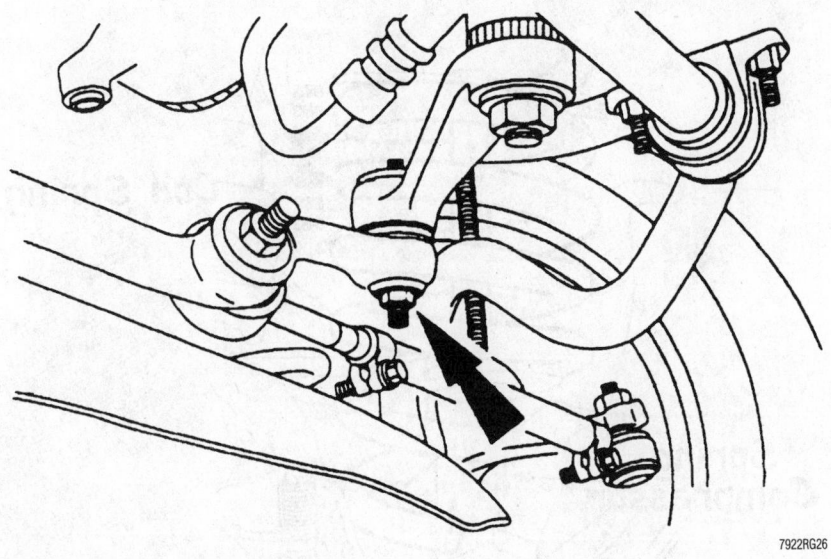

Remove the locknut and separate the Pitman arm from the center link using the appropriate puller

fluid. Check the fluid level and add fluid, if necessary.

20. Start the engine and turn the steering wheel from left to right. Check for leaks.

21. On the Town Car, install the steering gear cover.

22. Road test the vehicle and check the steering system for proper operation.

Shock Absorber

REMOVAL & INSTALLATION

Front

1. If equipped with air suspension, the air suspension switch, located on the right-hand side of the luggage compartment, must be turned to the OFF position before raising the vehicle.

2. Remove the nut, washer and bushing from the upper end of the shock absorber.

3. Raise and safely support the vehicle.

4. Remove the two bolts retaining the shock absorber to the lower control arm and remove the shock absorber.

To install:

5. Prior to installation, prime the new shock absorber. Fully extend the shock absorber while in the right side up (installed) position. Turn the shock absorber upside down and fully compress it. Repeat the procedure at least three times to purge any air trapped in the shock absorber.

6. Install a new bushing and washer on the stud on the top of the new shock absorber and position the unit inside the front coil spring.

7. Install the two lower retaining bolts and tighten them to 10–12 ft lbs. (13–17 Nm).

8. Lower the vehicle.

9. If equipped with air suspension, turn the air suspension switch to the ON position.

10. Place a new bushing and washer on the shock absorber top stud and install a new retaining nut. Tighten the retaining nut to 25–34 ft. lbs. (34–46 Nm).

11. Road test the vehicle and check for proper operation.

Rear

> ⁂ **WARNING**
>
> **When removing and installing rear shock absorbers on vehicles with air springs, it is very important that this procedure be followed exactly. Failure to do so may result in damaged shock absorbers.**

1. If equipped with air suspension, turn the air suspension service switch OFF.

2. Be sure the ignition switch is in the **OFF** position.

3. Raise and safely support the vehicle.

4. Support the rear axle assembly with a jack.

➡️**To assist in removing the upper retainer on shock absorbers using a plastic dust tube, place an open end wrench on the hex stamped into the dust tube's metal cap. For shock absorbers with a steel dust tube, simply grasp the tube to prevent stud rotation when loosening the retaining nut.**

5. Remove the top retaining nut, washer and bushing.

6. Remove the bottom retaining nut and washer. Remove the shock absorber.

To install:

7. Position the shock absorber so the upper stud enters the hole in the frame. Install the top bushing, washer and retaining nut. Tighten to 26–34 ft. lbs. (34–46 Nm).

8. Extend the shock absorber and place the lower stud through the hole in the bracket. Install the bottom retaining washer and nut. Tighten to 57–75 ft. lbs. (76–103 Nm).

9. Remove the jack from the axle assembly.

10. Safely lower the vehicle to the floor.

11. Turn the air suspension service switch to the ON position.

12. Road test the vehicle and check for proper operation.

Coil Spring

REMOVAL & INSTALLATION

Front

1. If equipped with air suspension, turn the air suspension service switch to the OFF position before raising the vehicle.

2. Raise and safely support the vehicle.

3. Remove the wheel and tire assembly.

4. Remove the shock absorber.

5. Separate the center link from the Pitman arm using Tie Rod End Remover 3290-D, or equivalent.

6. Using Spring Compressor D78P-5310-A or equivalent, install one plate with the pivot ball seat facing downward into the coils of the spring. Rotate the plate so it is flush with the upper surface of the lower arm.

7. Install the other plate with the pivot ball seat facing upward into the coils of the spring. Insert the upper ball nut through the coils of the spring, so the nut rests in the upper plate.

8. Insert the compression rod into the opening in the lower arm, through the upper and lower plate and upper ball nut. Insert the securing pin through the upper ball nut and compression rod.

➡️**This pin can only be inserted one way into the upper ball nut because of a stepped hole design.**

9. With the upper ball nut secured, turn the upper plate so it walks up the coil until it contacts the upper spring seat. Then, back off ½ turn.

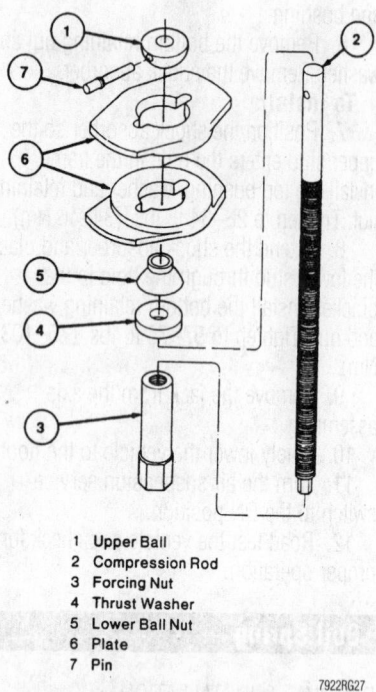

1 Upper Ball
2 Compression Rod
3 Forcing Nut
4 Thrust Washer
5 Lower Ball Nut
6 Plate
7 Pin

7922RG27

Exploded view of Spring Compressor D78P-5310-A—similar compressors are commercially available that will do the same job

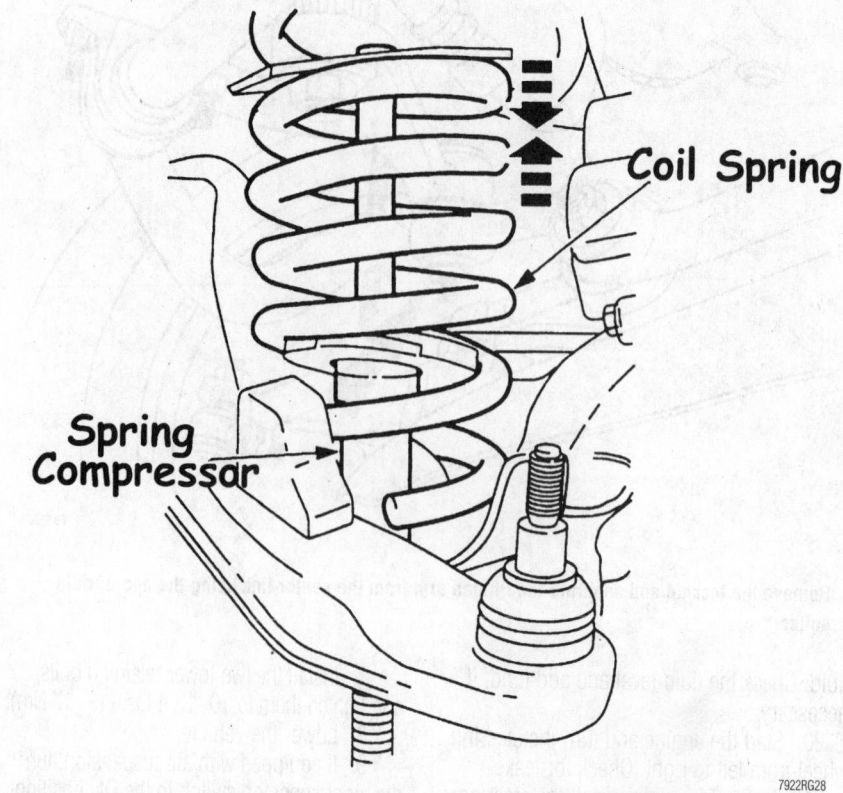

7922RG28

Compress the coil spring until it moves away from it's seat

10. Install the lower ball nut and thrust washer on the compression rod and screw on the forcing nut. Tighten the forcing nut until the spring is compressed enough so it is free in its seat.

11. Remove the two lower control arm pivot bolts, disengage the lower arm from the frame crossmember and remove the coil spring.

12. If a new coil spring is to be installed, mark the position of the upper and lower plates on the spring with chalk. With an assistant, compress a new spring for installation and measure the compressed length and the amount of curvature of the old spring.

13. Loosen the forcing nut to relieve the spring tension and remove the tools from the spring.

To install:

14. Assemble the spring compressor and locate in the same position as marked during disassembly.

15. Before compressing the coil spring, be sure the upper ball nut securing the pin is inserted properly.

16. Compress the coil spring until the spring height reaches the dimension measured during disassembly.

17. Position the coil spring assembly into the lower arm and position the lower arm into the frame crossmember.

18. Install both the front and rear lower control arm pivot bolts through the frame and lower arm bushings. Tighten the bolts and nuts to 109–148 ft. lbs. (148–201 Nm).

19. Remove the spring compressor from the coil spring.

20. Install the drag link to the Pitman arm. Tighten the retaining nut to 60 ft. lbs. (80 Nm).

21. Position the shock absorber inside the coil spring and install the retaining bolts.

22. Install the wheel and tire assembly. Tighten the lug nuts in a star pattern to 85–105 ft. lbs. (115–142 Nm).

23. Lower the vehicle.

24. If equipped with air suspension, turn the air suspension switch to the ON position.

25. Place a washer and retaining nut on the shock absorber top stud. Tighten the nut to 25–34 ft. lbs. (34–46 Nm).

26. Check the front end alignment.

27. Road test the vehicle and check for proper operation.

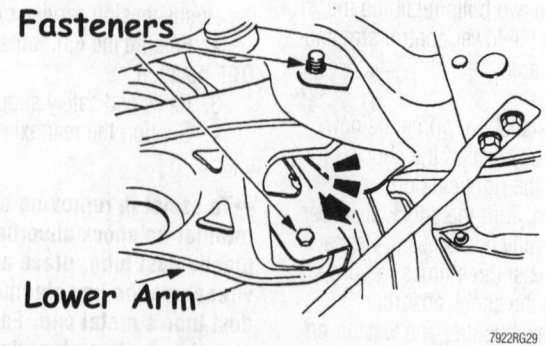

7922RG29

Remove the fasteners attaching the lower arm to the frame, then lower the arm and remove the spring with the compressor

Rear

AIR SPRING

> ※ **CAUTION**
>
> **Before servicing any air suspension component, disconnect power to the system by turning the air suspension service switch OFF or by disconnecting the negative battery cable. Do not remove an air spring under any circumstances when there is pressure in the air spring. Do not remove any components supporting an air spring without either exhausting the air or providing support for the air spring.**

1. Turn the air suspension switch to the OFF position.
2. Raise and safely support the vehicle. The suspension must be fully down with no load.
3. Remove the heat shield, as required. Remove the spring retainer clip.
4. Disconnect the air spring solenoid valve electrical connector, then disconnect the air line.
5. Remove the air spring solenoid retainer.
6. Rotate the solenoid valve counterclockwise to the first stop.
7. Pull the solenoid valve straight out slowly to the second stop to bleed air from the system.

> ※ **CAUTION**
>
> **Do not fully release the solenoid until the air is completely bled from the air spring or personal injury may result.**

8. After the air is fully bled from the system, rotate the solenoid valve counterclockwise to the third stop and remove the solenoid valve from the solenoid housing.

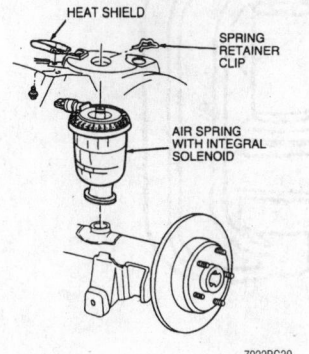

HEAT SHIELD

SPRING RETAINER CLIP

AIR SPRING WITH INTEGRAL SOLENOID

7922RG30

Exploded view of the air spring mounting

Remove the large O-ring from the solenoid housing.

9. On 1995–97 models, insert Air Spring Removal Tool T90P-5310-A or equivalent, between the axle tube and the spring seat on the forward side of the axle. Position the tool so its flat end rests on the piston knob. Push downward, forcing the piston and retainer clip off the axle spring seat.
10. On 1998–99 models, lift the air spring off of the rear axle.
11. Remove the air spring.

To install:

12. Check the solenoid valve O-rings for cuts or abrasion. Replace the O-rings as required. Lightly grease the O-ring area of the solenoid valve and the larger solenoid housing O-ring with silicone dielectric compound.
13. Insert the solenoid into the air spring end cap and rotate clockwise to the third stop, push in to the second stop, then rotate clockwise to the first stop.
14. Install the air spring solenoid retainer. Inspect the wiring harness connector and ensure the rubber gasket is in place at the bottom of the connector cavity.
15. Install the air spring into the frame (upper) spring seat, taking care to keep the solenoid air and electrical connections clean and free of damage.
16. Reconnect the push-on ring spring retainer clip to the knob of the spring cap from the top side of the frame spring seat.
17. Reconnect the air line and electrical connector to the solenoid. Install the heat shield to the frame spring seat, if removed.
18. Align the air spring piston-to-axle (lower) seat. Squeeze to increase pressure and push downward on the piston, snapping the piston to the axle seat at rebound and supported by the shock absorber.

> ※ **WARNING**
>
> **The air springs may be damaged if the suspension is allowed to compress before the spring is inflated.**

19. Reconnect the negative battery cable.
20. Refill the air spring as follows:
    a. Turn the air suspension switch to the ON position. The ignition switch must be **ON** and the engine running or a battery charger must be connected to the battery to reduce battery drain.
    b. Fold back or remove the right-hand luggage compartment trim panel and connect SUPER STAR II Tester 007–0041-A or equivalent, to the air suspension data link connector, which

is located near the air suspension switch.
    c. Set the tester to EEC-IV/MCU mode. Also set the tester to FAST mode. Release the tester button to the HOLD (up) position and turn the tester ON.
    d. Depress the tester button to TEST (down) position. A Code 10 will be displayed. Within two minutes a Code 13 will be displayed. After Code 13 is displayed, release the tester button to the HOLD (up) position, wait five seconds and depress the tester button to TEST (down) position. Ignore any codes displayed.
    e. Release the tester button to the HOLD (up) position. Wait at least 20 seconds, then depress the tester button to TEST (down) position. Within 10 seconds, the codes will be displayed in the order shown.
    f. Within four seconds after Code 26 is displayed, release the tester button to the HOLD (up) position. Waiting longer than four seconds may result in Functional Test 31 being entered. The compressor will fill the air springs with air as long as the tester button is in the HOLD (up) position. To stop filling the air springs, depress the tester button to TEST (down) position.

➡ **It is possible to overheat the compressor during this operation. If the compressor overheats, the self-resetting circuit breaker in the compressor will open and remain open for about 15 minutes. This allows the compressor to cool down.**

    g. To exit Functional Test 26, disconnect the tester and turn the ignition switch to the OFF position.
21. Install the luggage trim panel, if removed.
22. Lower the vehicle.
23. Road test the vehicle and check for proper operation.

COIL SPRING

1. Place a hoist under the rear axle housing and raise and safely support the vehicle.
2. Support the frame side rails with two jackstands.

➡ **If the vehicle is raised by the frame rails, place a jack under the rear axle housing.**

3. Remove the rear sway (stabilizer) bar.
4. Disconnect the lower studs of both rear shock absorbers from the mounting brackets on the axle tube.

5. Unsnap the parking brake cable from the upper arm retainer before lowering the axle housing.

6. Lower the axle housing until the coil springs are released. If the axle housing is supported by the hoist, lower the hoist allowing the rear of the vehicle to rest on the jackstands. If the vehicles axle housing is supported by the jackstands, leave the hoist stationary and lower the jackstands or raise the hoist to release the tension on the coil springs.

7. Remove the coil springs and insulators.

To install:

8. Position the coil spring in the upper and lower seats with an insulator between the upper end of the spring and frame seat.

9. Raise the axle housing and connect the lower studs of the shock absorbers to the mounting brackets.

10. Snap the parking cable into the upper arm retainer.

11. Install the sway bar.

12. Remove the jackstands.

13. Lower the vehicle.

14. Road test the vehicle and check for proper operation.

Upper Ball Joint

REMOVAL & INSTALLATION

1. If equipped with air suspension, the air suspension switch to the OFF position before raising the vehicle.

2. Raise and safely support the vehicle.

3. Place supports under both sides of the frame just behind the lower control arms.

4. Remove the wheel and tire assembly.

5. Place a floor jack under the lower control arm at the lower ball joint area. The floor jack will support the spring load on the lower control arm.

6. Remove the retaining nut and pinch bolt from the upper ball joint stud.

7. Mark the position of the alignment cams. When replacing the upper ball joint this will approximate the current alignment.

8. Remove the two nuts retaining the upper ball joint to the upper control arm.

9. Separate the upper ball joint from the upper control arm and spread the slot in the wheel spindle with a suitable prybar to remove the ball joint stud from the wheel spindle.

To install:

10. Place the upper ball joint to the upper control arm and insert the ball stud into the wheel spindle.

11. Install the upper ball joint pinch bolt and retaining nut. Tighten to 56–77 ft. lbs. (76–104 Nm).

12. Install the alignment cams to the approximate position at removal. If not marked, install in the neutral positions.

13. Install the two nuts retaining the upper ball joint to the upper control arm. Hold the cams and tighten the nuts to 107–129 ft. lbs. (145–175 Nm).

14. Remove the floor jack from under the lower control arm.

15. Install the wheel and tire assembly. Tighten the lug nuts in a star pattern to 85–105 ft. lbs. (115–142 Nm).

16. Remove the supports.

17. Lower the vehicle.

18. If equipped with air suspension, turn the air suspension switch to the ON position.

19. Check and adjust the front wheel alignment.

20. Road test the vehicle and check for proper operation.

Lower Ball Joint

REMOVAL & INSTALLATION

1. If equipped with air suspension, the air suspension service switch to the OFF position before raising the vehicle.

2. Raise and safely support the vehicle.

3. Place supports under both sides of the frame behind the lower control arms.

4. Remove the wheel and tire assembly.

5. Remove the wheel spindle.

6. Remove and discard the ball joint boot seal.

7. Press out the ball joint using U-Joint Tool T74P-4635-C, Ball Joint Remover D89P-3395-A4 and Receiving Cup D84P-3395-A4, or equivalents.

To install:

8. When installing a new ball joint, the protective cover should be left on to protect the ball joint seal during installation. It may be necessary to trim the cover so it can pass through the installation tool.

9. Install the ball joint with Ball Joint Replacer D89P-3010-B, Receiving Cup D84P-3395-A4 and U-Joint Tool T74P-4635-C, or equivalents.

10. Discard the protective cover and be sure the new ball joint is fully seated in the lower control arm. ensure that the ball joint seal is not damaged.

11. Install the wheel spindle.

12. Install the wheel and tire assembly. Tighten the lug nuts in a star pattern to 85–105 ft. lbs. (115–142 Nm).

13. Lower the vehicle.

14. If equipped with air suspension, turn the air suspension service switch to the ON position.

15. Check the front end alignment.

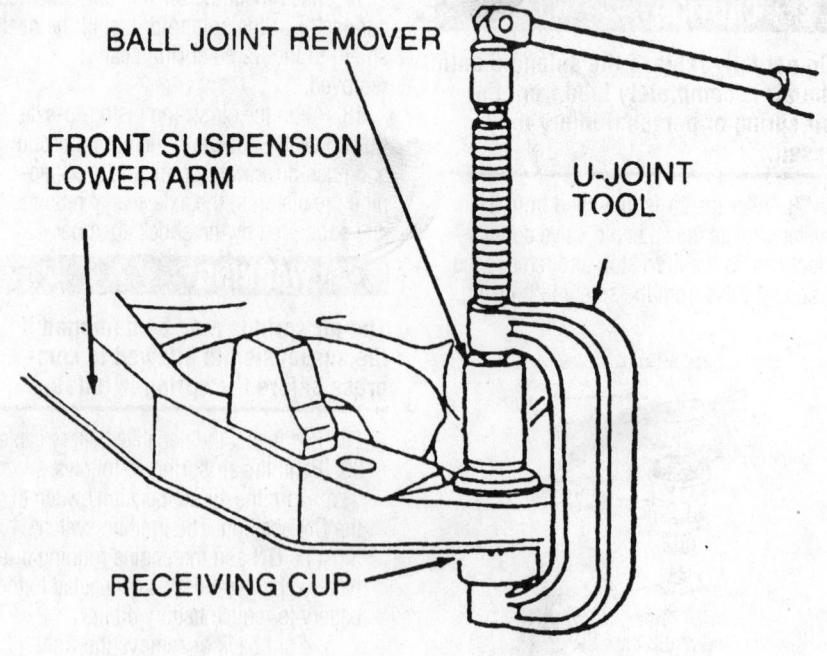

Use a ball joint press such as the one shown to remove the ball joint from the lower control arm

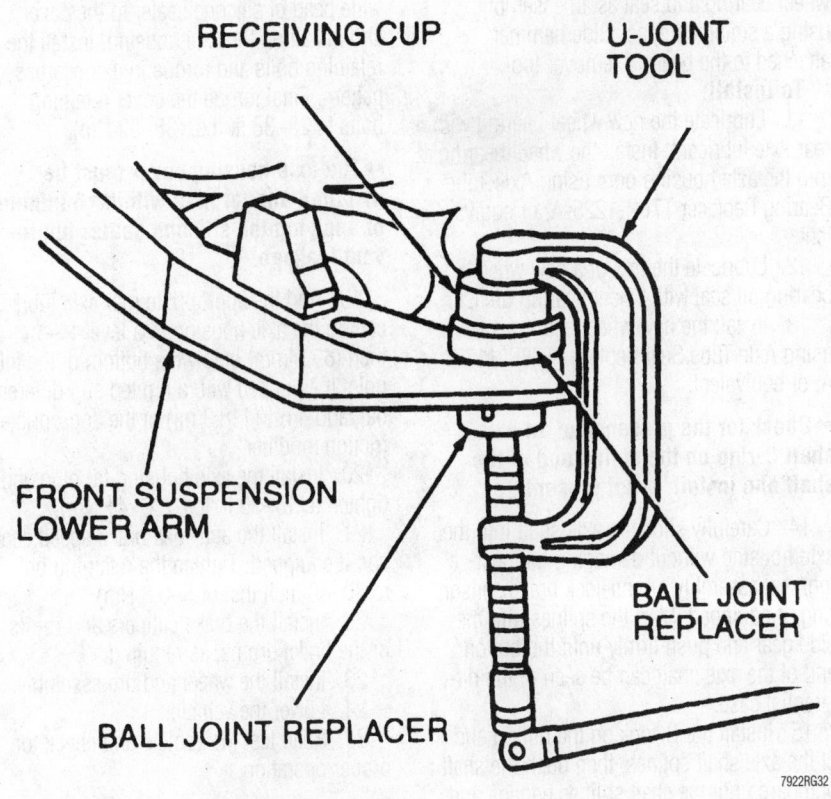

Use a ball joint press such as the one shown to install the new ball joint into the lower control arm

16. Road test the vehicle and check for proper operation.

Wheel Bearings

ADJUSTMENT

The front wheel bearings are of a hub unit design and are pregreased, sealed and require no maintenance. The bearings are preset and cannot be adjusted. No adjustment is possible for the rear axle bearing. If either the front or rear bearings make noise or become loose, replacement is necessary.

REMOVAL & INSTALLATION

Front

➡ Before continuing with this procedure, be sure that the two new caliper mounting bolts, one sway bar link nut, one lower ball joint stud nut and one hub grease cap are available, per side. Once removed, these parts lose their torque holding ability or retention capability and must not be reused.

1. If equipped, turn the air suspension service switch to the OFF position before raising the vehicle.

2. Raise and safely support the vehicle.
3. Remove the front wheel.
4. Remove grease cap from the hub.
5. Remove the disc brake caliper. Suspend the caliper with a length of wire. Do not let it hang from the brake hose. Discard the disc brake caliper mounting bolts.
6. Remove the disc brake rotor. If the factory installed push on nuts are installed, remove them first.
7. Remove the wheel hub retainer nut and discard.
8. Remove the hub and bearing assembly.

➡ The wheel bearings are permanently greased and sealed. If replacement is necessary, the wheel hub and bearings must be replaced as an assembly.

To install:

9. Install the hub and bearing assembly. Install a new wheel hub retainer nut and tighten to 189–254 ft. lbs. (255–345 Nm).
10. Install the disc brake rotor and push on nuts, if equipped.
11. Install a new grease cap seal.
12. Install the disc brake caliper using the two new disc brake caliper mounting bolts. Tighten the bolts to 125–170 ft. lbs. (170–230 Nm).

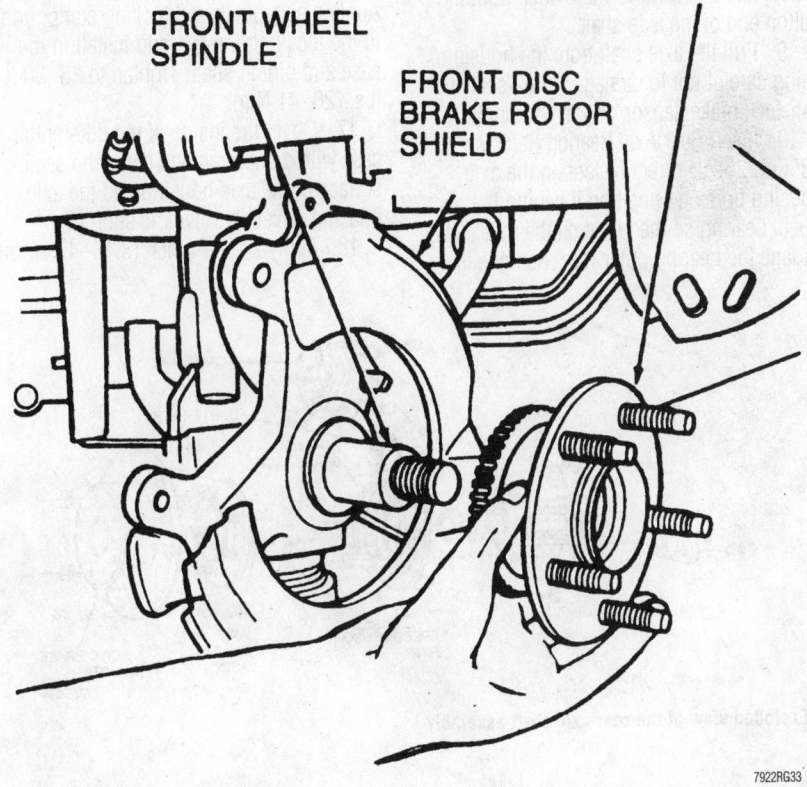

Front hub and bearing assembly

13. Install the wheel and tire assembly. Tighten the lug nuts in a star pattern to 85–104 ft. lbs. (115–142 Nm).

14. Remove the safely stands.

15. Lower the vehicle.

16. If equipped with air suspension, turn the air suspension switch to the ON position.

17. Pump the brake pedal several times to position the brake pads prior to moving the vehicle.

18. Check the front end alignment.

19. Road test the vehicle and check for proper operation.

Rear

1. Raise and safely support the vehicle.

2. Remove the wheel and tire assembly.

3. Remove the brake drum or brake rotor.

4. If equipped, remove the anti-lock brake speed sensor.

5. Clean all dirt from the area of the axle housing cover.

6. Place a suitable drain pan under the axle housing. Remove the axle housing cover retaining bolts and the cover, draining the axle lubricant from the housing.

7. Remove the differential pinion shaft lock bolt and the differential pinion shaft.

8. Push the flanged end of the axle shaft being removed toward the center of the vehicle and remove the C-lock from the button end of the axle shaft.

9. Pull the axle shaft from the housing, being careful not to damage the oil seal and anti-lock brake sensor ring, if equipped.

10. Insert Rear Axle Bearing Remover T85L-1225-AH or equivalent, in the axle housing bore and position it behind the wheel bearing so the tangs on the tool engage the bearing outer race. Remove the

wheel bearing and seal as an assembly using a suitable impact slide hammer attached to the bearing remover tool.

To install:

11. Lubricate the new wheel bearing with rear axle lubricant. Install the wheel bearing into the axle housing bore using Axle Tube Bearing Replacer T78P-1225-A, or equivalent.

12. Lubricate the lips of a new wheel bearing oil seal with wheel bearing grease.

13. Install the new wheel bearing seal using Axle Tube Seal Replacer T78P-1177-A, or equivalent.

➡ **Check for the presence of an axle shaft O-ring on the spline end of the shaft and install, if not present.**

14. Carefully slide the axle shaft into the axle housing without damaging the bearing/seal assembly or anti-lock brake sensor ring, if equipped. Start the splines into the side gear and push firmly until the button end of the axle shaft can be seen in the differential case.

15. Install the C-lock on the button end of the axle shaft splines, then push the shaft outboard until the shaft splines engage and the C-lock seats in the counterbore of the differential side gear.

16. Insert the differential pinion shaft through the case and pinion gears, aligning the hole in the shaft with the lock bolt hole. Apply a suitable thread locking compound to the lock bolt threads and install in the case and pinion shaft. Tighten to 15–30 ft. lbs. (20–41 Nm).

17. Cover the inside of the differential case with a shop rag and clean the sealing surface of the axle housing and the axle housing cover. Remove the shop rag.

18. Apply a 1/8–3/16 inch (3.18–4.76mm)

wide bead of silicone sealer to the cover and install on the axle housing. Install the retaining bolts and torque in a crisscross pattern. Final torque the cover retaining bolts to 28–38 ft. lbs. (38–52 Nm).

➡ **The axle housing cover must be installed and torqued within 15 minutes of applying the silicone sealer to prevent leakage.**

19. Add the appropriate rear axle lubricant to the axle housing to a level 1/4–9/16 inch (6–14mm) below the bottom of the fill hole. If equipped with a limited slip differential, add 4 oz. (118.3 ml) of the appropriate friction modifier.

20. Install the axle housing fill plug and tighten to 15–30 ft. lbs. (20–41 Nm).

21. Install the anti-lock brake speed sensor, if equipped. Tighten the retaining bolt to 40–60 inch lbs. (4.5–6.8 Nm).

22. Install the brake calipers and rotors or the brake drums, as required.

23. Install the wheel and tire assembly.

24. Lower the vehicle.

25. Road test the vehicle and check for proper operation.

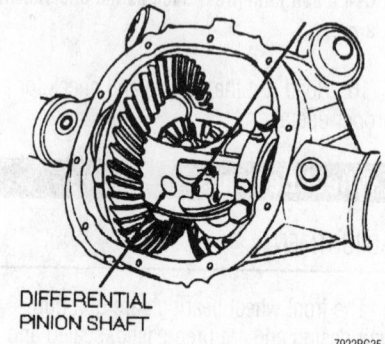

DIFFERENTIAL PINION SHAFT

Removal of differential pinion shaft

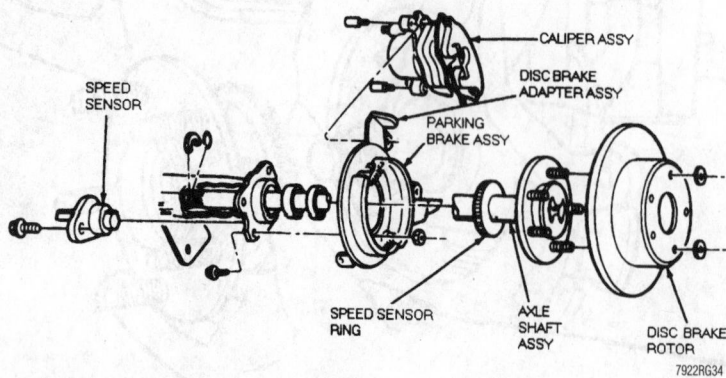

Exploded view of the rear axle shaft assembly

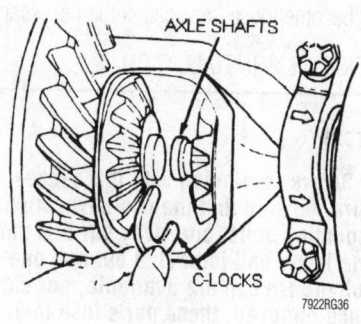

AXLE SHAFTS

C-LOCKS

Removing axle shaft C-lock clips

GENERAL MOTORS A BODY

Buick-Century • **Oldsmobile**-Cutlass Ciera • Cutlass Cruiser

DRIVE TRAIN 27-24
ENGINE REPAIR 27-2
FUEL SYSTEM 27-22
STEERING AND
SUSPENSION 27-26

C
Camshaft and Valve Lifters 27-13
Removal & Installation 27-13
Coil Spring 27-30
Removal & Installation 27-30
Cylinder Head.............................. 27-4
Removal & Installation 27-5

E
Engine Assembly 27-2
Removal & Installation 27-2
Exhaust Manifold........................27-12
Removal & Installation 27-12

F
Fuel Filter 27-22
Removal & Installation 27-22
Fuel Pump 27-0..........................27-23
Removal & Installation 27-23
Fuel System Pressure.................27-22
Relieving 27-22

Fuel System Service
Precautions................................27-22
H
Halfshaft......................................27-25
Removal & Installation 27-25
I
Ignition Timing 27-2
Adjustment............................... 27-2
Intake Manifold...........................27-9
Removal & Installation 27-9
L
Lower Ball Joint..........................27-30
Removal & Installation 27-30
O
Oil Pan..27-16
Removal & Installation 27-16
Oil Pump 27-18
Removal & Installation 27-18
P
Power Rack and Pinion Steering
Gear...27-26
Removal & Installation 27-26
R
Rear Main Seal 27-19

Removal & Installation 27-19
Rocker Arms 27-7
Removal & Installation 27-7
S
Shock Absorber 27-29
Removal & Installation 27-29
Strut and Spring 27-27
Removal & Installation 27-27
T
Timing Chain, Sprockets, Front
Cover and Seal 27-20
Removal & Installation 27-20
Transaxle Assembly 27-24
Removal & Installation 27-24
V
Valve Lash 27-16
Adjustment.............................. 27-16
W
Water Pump................................. 27-4
Removal & Installation 27-4
Wheel Bearings...........................27-31
Adjustment............................... 27-31
Removal & Installation 27-31

ENGINE REPAIR

➡️**Disconnecting the negative battery cable on some vehicles may interfere with the functioning of the on-board computer system, and may require the computer to undergo a relearning process once the negative battery cable is reconnected.**

Ignition Timing

ADJUSTMENT

The ignition timing on these vehicles is controlled by the engine control module, no adjustment is possible or necessary.

Engine Assembly

REMOVAL & INSTALLATION

✳✳ CAUTION

Some models covered by this manual may be equipped with a Supplemental Restraint System (SRS), which uses an air bag. Whenever working near any of the SRS components, such as the impact sensors, the air bag module, steering column and instrument panel, properly disable the SRS.

2.2L Engine

✳✳ CAUTION

The fuel injection system remains under pressure, even after the engine has been turned OFF. The fuel system pressure must be relieved prior to disconnecting any fuel lines. Failure to do so may result in fire and/or personal injury.

1. Relieve the fuel system pressure.

✳✳ CAUTION

Observe all applicable safety precautions when working around fuel. Whenever servicing the fuel system, always work in a well ventilated area. Do not allow fuel spray or vapors to come in contact with a spark or open flame. Keep a dry chemical fire extinguisher near the work area. Always keep fuel in a con-
tainer specifically designed for fuel storage; also, always properly seal fuel containers to avoid the possibility of fire or explosion.

2. Disconnect the negative battery cable.

✳✳ CAUTION

Never open, service or drain the radiator or cooling system when hot; serious burns can occur from the steam and hot coolant.

3. Drain the cooling system into a suitable container.
4. While supporting the hood, disconnect the hydraulic shock from the cowl pan and secure the hood in the fully open position.
5. Remove the air cleaner and air duct assembly.
6. Disconnect the control cables from the throttle body and remove the control cable bracket from the intake manifold and rocker arm cover. Set the assembly out of the way.
7. Disconnect and cap the fuel lines from the throttle body and manifold mounting bracket.
8. Tag and disconnect any vacuum hoses that will interfere with engine removal.
9. Remove the upper and lower radiator hoses.
10. Disconnect the heater hoses from the intake manifold and water pump and secure out of the way.
11. Remove the torque strut.
12. Disconnect the engine harness connector.
13. Rotate the engine forward.
14. Remove the power steering pump from the engine and support out of the way with the power steering lines attached.
15. Disconnect the electrical connectors from the rear of the engine.
16. Remove the transaxle oil fill tube.
17. Remove the transaxle-to-engine bolts leaving in only the upper two bolts.
18. Rotate the engine to its normal position.
19. Raise and safely support the vehicle.
20. Remove the right front tire and wheel assembly.
21. Remove the right inner fender well splash shield.
22. Disconnect the exhaust pipe from the exhaust manifold.
23. Remove the flywheel cover.
24. Disconnect and remove the starter.

25. Remove the engine mount-to-frame nuts.
26. Remove the torque converter-to-flywheel bolts.
27. Remove the A/C compressor from the mounting bracket and secure out of the way to the side without disconnecting the refrigerant lines.
28. Remove the front pipe support bracket from the transaxle.
29. Remove the transaxle support bracket from the transaxle and frame.
30. Lower the vehicle.
31. Remove the two remaining engine-to-transaxle mounting bolts.
32. Attach a suitable lifting device and slowly remove the engine. While lifting out the engine be sure that no hoses or electrical connectors are still attached.
To install:
33. Lower the engine into the vehicle and connect the engine to the transaxle.
34. Install the two upper engine-to-transaxle bolts. Tighten the bolts only until snug.
35. Raise and safely support the vehicle.
36. Install the transaxle support bracket and tighten the mounting bolts to 38 ft. lbs. (52 Nm).
37. Install the front exhaust pipe-to-transaxle mounting bolt.
38. Install the A/C compressor in the mounting bracket.
39. Install the torque converter-to-flywheel bolts and tighten to 46 ft. lbs. (62 Nm).
40. Install the engine mount-to-frame nuts and tighten to 33 ft. lbs. (45 Nm).
41. Connect and install the starter and tighten the mounting bolts to 33 ft. lbs. (45 Nm).
42. Install the flywheel cover.
43. Connect the exhaust pipe to the exhaust manifold and tighten the mounting bolts to 22 ft. lbs. (30 Nm).
44. Install the right inner fender well splash shield.
45. Install the tire and wheel assembly and tighten to specification.
46. Lower the vehicle.
47. Install the remainder of the engine-to-transaxle bolts and tighten all the bolts to 37 ft. lbs. (50 Nm).
48. Install the transaxle fill tube.
49. Connect the electrical connectors to the rear of the engine.
50. Install the power steering pump on the mounting bracket.
51. Connect the engine harness.
52. Install the torque strut and tighten the mounting bolts to 41 ft. lbs. (56 Nm).

53. Connect the heater hoses to the intake manifold and water pump.

54. Install the upper and lower radiator hoses.

55. Connect any vacuum hoses disconnected for engine removal.

56. Connect the fuel lines to the throttle body and fuel line brackets.

57. Install the throttle cable bracket on the intake manifold and rocker arm cover and connect the control cables to the throttle body.

58. Install the air cleaner assembly and air inlet duct work.

59. Connect the hood hydraulic shock to the cowl panel.

60. Refill the cooling system. An oil and filter change is recommended.

61. Connect the negative battery cable.

62. Start the vehicle and verify no leaks.

3.1L Engine

✳✳ CAUTION

The fuel injection system remains under pressure, even after the engine has been turned OFF. The fuel system pressure must be relieved before disconnecting any fuel lines. Failure to do so may result in fire and/or personal injury.

1. Relieve the fuel system pressure.

✳✳ CAUTION

Observe all applicable safety precautions when working around fuel. Whenever servicing the fuel system, always work in a well ventilated area. Do not allow fuel spray or vapors to come in contact with a spark or open flame. Keep a dry chemical fire extinguisher near the work area. Always keep fuel in a container specifically designed for fuel storage; also, always properly seal fuel containers to avoid the possibility of fire or explosion.

2. Disconnect the negative battery cable.

3. Scribe reference marks at the hood supports and remove the hood. Install covers on both fenders.

4. Remove the air cleaner assembly.

✳✳ CAUTION

Never open, service or drain the radiator or cooling system when hot; serious burns can occur from the steam and hot coolant.

5. Drain the cooling system.

6. Remove the radiator hoses from the engine.

7. Remove the engine torque strut.

8. Remove the serpentine belt.

9. Remove the heater hoses.

10. Remove the throttle body bracket and cable.

11. Disconnect the fuel lines.

12. Tag and disconnect electrical connections, from the engine wiring harness and position the harness aside.

13. Tag and disconnect the vacuum lines, from the engine, to non-engine mounted components.

14. Disconnect the power steering lines.

15. Remove the power steering brace.

16. Remove the coolant reservoir.

17. Remove the two A/C compressor upper bolts.

18. Remove the electrical connections from transaxle.

19. Remove the electrical grounds from transaxle mounting bolts.

20. Remove the four transaxle top bolts.

21. Safely raise and support the vehicle.

22. Disconnect the front exhaust pipe from manifold.

23. Remove the oil drip shield bolts and shield.

24. Remove the flywheel cover bolts and cover.

25. Disconnect and remove the starter.

26. Disconnect the electrical connection to the lower engine.

27. Remove the flywheel to converter bolts.

28. Remove the A/C compressor, position aside.

➡ Do not disconnect the A/C compressor lines when removing.

29. Remove transaxle support bracket to transaxle bolts.

30. Remove engine mounts.

31. Remove two remaining transaxle bolts.

32. Lower the vehicle.

33. Attach a lifting device to the engine.

34. Carefully remove the engine assembly.

To install:

35. Install engine assembly into vehicle with lifting device.

36. Start two transaxle to engine bolts.

37. Remove lifting device and transaxle support.

38. Safely raise and support the vehicle.

39. Install remaining transaxle bolts and tighten to 37 ft. lbs. (50 Nm).

40. Install the engine mounts.

41. Install transaxle support bracket to transaxle bolts.

42. Properly position A/C compressor.

43. Install flywheel to converter bolts and tighten to 47 ft. lbs. (63 Nm).

44. Install flywheel cover and bolts. Tighten bolts to 89 inch lbs. (10 Nm).

45. Install the electrical grounds to the transaxle mounting bolts.

46. Connect the lower engine wiring to the engine.

47. Install the oil drip shield and bolts.

48. Connect the front exhaust pipe to manifold.

49. Safely lower the vehicle.

50. Install the four transaxle top bolts and tighten to 37 ft. lbs. (50 Nm).

51. Connect the electrical grounds to transaxle mounting bolts.

52. Connect the upper engine electrical connections, to the engine.

53. Install the A/C compressor upper bolts.

54. Install the coolant reservoir.

55. Install the power steering brace.

56. Connect the power steering lines.

57. Connect all previously removed vacuum lines to the engine.

58. Connect the fuel lines.

59. Install the throttle body bracket and cable.

60. Install the heater hoses and secure with clamps.

61. Install the serpentine belt.

62. Connect the radiator hoses and secure with clamps.

63. Align the hood-to-hinge mating marks and install the hood.

64. Connect the negative battery cable.

65. Install air cleaner assembly.

66. Fill the coolant and engine oil.

67. Start the engine and allow to reach normal operating temperature. Check for leaks and refill the cooling system.

Water Pump

REMOVAL & INSTALLATION

2.2L Engine

1. Disconnect the negative battery cable.

❋❋ CAUTION

Never open, service or drain the radiator or cooling system when hot; serious burns can occur from the steam and hot coolant.

2. Drain the cooling system into a suitable container.
3. Loosen, but do not remove, the water pump pulley bolts.
4. Remove the serpentine belt.
5. Remove the alternator mounting bolts and set the alternator aside.
6. Remove the water pump pulley bolts and remove the water pump pulley.
7. Remove the four water pump mounting bolts and remove the water pump.

To install:
8. Clean all the gasket surfaces completely.
9. Apply a thin bead of sealer around the outer edge of the water pump gasket seating area and place he gasket on the pump.
10. Install the water pump on the engine and tighten the four mounting bolts to 18 ft. lbs. (25 Nm).
11. Install the water pump pulley and tighten the mounting bolts finger-tight.
12. Install the alternator in the mounting bracket.
13. Install the serpentine belt.
14. Tighten the water pump pulley mounting bolts to 22 ft. lbs. (30 Nm).
15. Connect the negative battery cable.
16. Refill and bleed the cooling system.

3.1L Engine

1. Disconnect the negative battery cable.

❋❋ CAUTION

Never open, service or drain the radiator or cooling system when hot; serious burns can occur from the steam and hot coolant.

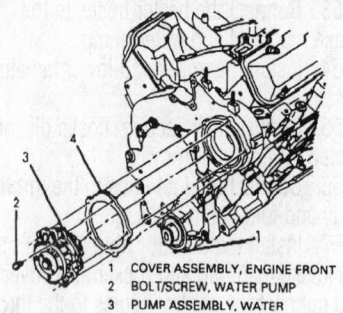

1. COVER ASSEMBLY, ENGINE FRONT
2. BOLT/SCREW, WATER PUMP
3. PUMP ASSEMBLY, WATER
4. GASKET, WATER PUMP

7922SG02

Exploded view of the water pump mounting—3.1L engine

2. Drain the cooling system into a suitable container.
3. Loosen, but do not remove, the water pump pulley bolts.
4. Remove the serpentine belt.
5. Remove the water pump pulley bolts and remove the water pump pulley.
6. Remove the five water pump mounting bolts and remove the water pump.

To install:
7. Clean all the gasket surfaces completely.
8. Apply a thin bead of sealer around the outside edge of the water pump along the gasket sealing area and install the gasket onto the water pump.
9. Install the water pump on the engine and tighten the water pump mounting bolts to 89 inch lbs. (10 Nm).
10. Install the water pump pulley and tighten the pulley bolts finger-tight.
11. Install the serpentine belt.
12. Tighten the water pump pulley bolts to 18 ft. lbs. (25 Nm).
13. Connect the negative battery cable.
14. Refill and bleed the cooling system.

Cylinder Head

❋❋ CAUTION

Some models covered by this manual may be equipped with a Supplemental Restraint System (SRS), which uses an air bag. Whenever working near any of the SRS components, such as the impact sensors, the air bag module, steering column and instrument panel, properly disable the SRS.

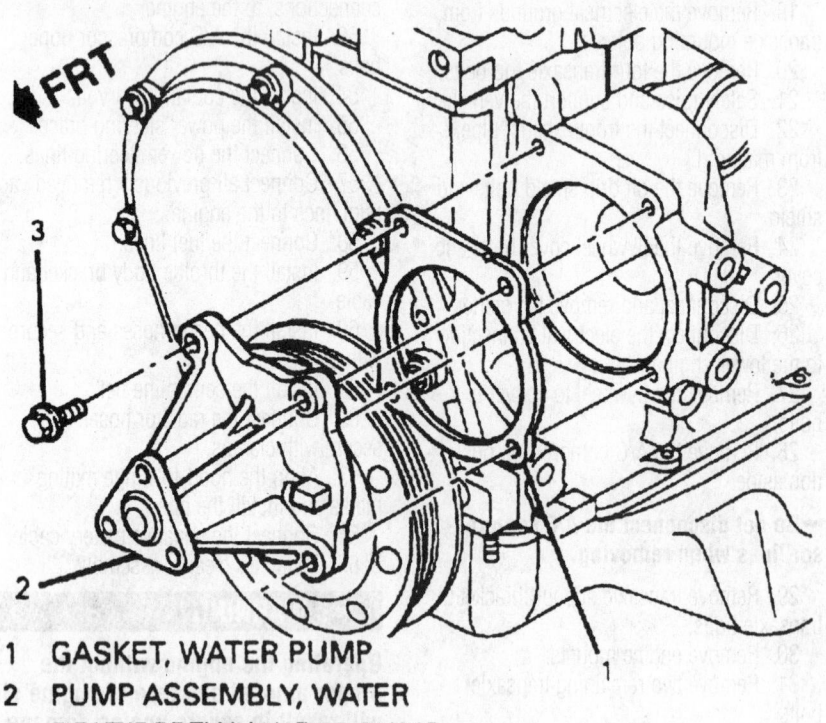

FRT

1 GASKET, WATER PUMP
2 PUMP ASSEMBLY, WATER
3 BOLT/SCREW, WATER PUMP

7922SG01

Exploded view of the water pump mounting—2.2L engine

REMOVAL & INSTALLATION

2.2L Engine

> **※ CAUTION**
>
> **The fuel injection system remains under pressure, even after the engine has been turned OFF. The fuel system pressure must be relieved before disconnecting any fuel lines. Failure to do so may result in fire and/or personal injury.**

1. Disconnect the negative battery cable.

> **※ CAUTION**
>
> **Observe all applicable safety precautions when working around fuel. Whenever servicing the fuel system, always work in a well ventilated area. Do not allow fuel spray or vapors to come in contact with a spark or open flame. Keep a dry chemical fire extinguisher near the work area. Always keep fuel in a container specifically designed for fuel storage; also, always properly seal fuel containers to avoid the possibility of fire or explosion.**

2. Relieve the fuel system pressure.

> **※ CAUTION**
>
> **Never open, service or drain the radiator or cooling system when hot; serious burns can occur from the steam and hot coolant.**

3. Drain the cooling system into a suitable container.
4. Remove the air cleaner and air duct assembly.
5. Remove the air inlet resonator upper tie bar.
6. Remove the lower air inlet.
7. Remove the serpentine belt.
8. Remove the alternator.
9. Remove the power steering pump and position it aside without disconnect the power steering lines.
10. Disconnect the spark plug wires and lay them out of the way.
11. Disconnect the control cables from the throttle body and remove the cable bracket at the throttle body and rocker arm cover.

➡Use care when removing valvetrain components. Parts to be reused must be returned to their original locations.

12. Remove the rocker arm cover, rocker arm nuts, rocker arms and pushrods.
13. Disconnect the electrical connectors from the intake manifold, throttle body and cylinder head.
14. Disconnect the oxygen (O$_2$) sensor connector.
15. Remove the power steering pump bracket from the intake manifold brace, located under the intake manifold.
16. Remove the torque strut and engine-side torque strut mounting bracket.
17. Remove the alternator rear bracket.
18. Tag and disconnect the vacuum lines at the intake manifold and cylinder head.
19. Disconnect the upper radiator hose from the engine.
20. Raise and safely support the vehicle.
21. Disconnect the exhaust pipe from the exhaust manifold.
22. Lower the vehicle.
23. Disconnect and cap the fuel lines at the quick disconnects.
24. Remove the transaxle fill tube.
25. Remove the cylinder head bolts.
26. Remove the cylinder head with both manifolds. Remove the intake and exhaust manifolds from the cylinder head.

  To install:
27. Clean all the gasket surfaces completely. Clean the threads on the cylinder head bolts and the block threads.
28. Install the intake and exhaust manifolds on the cylinder head.

29. Place a new cylinder head gasket in position over the dowel pins on the block. Carefully guide the cylinder head into position.
30. New head bolts are recommended. Install all the cylinder head bolts finger-tight. The long bolts go in bolt positions 1, 4, 5, 8 and 9. The short bolt are in positions 2, 3, 6 and 7. The stud is in position 10.
31. Tighten the bolts in sequence. The long bolts to 46 ft. lbs. (63 Nm) and the short bolts and stud to 43 ft. lbs. (58 Nm). Make second pass tightening the long bolts to 46 ft. lbs. (63 Nm) and the short bolts to 43 ft. lbs. (58 Nm). Make a final pass over all bolts tightening each an additional 90 degrees (¼ turn).
32. Install the transaxle fill tube.
33. Connect the fuel lines to the throttle body.
34. Raise and safely support the vehicle.
35. Connect the exhaust pipe to the exhaust manifold. Tighten the mounting bolts to 22 ft. lbs. (30 Nm).
36. Lower the vehicle.
37. Connect the upper radiator hose.
38. Connect the vacuum lines to the intake manifold.
39. Install the engine-side torque strut bracket and torque strut.
40. Install the alternator rear bracket.
41. Install the power steering pump bracket to the intake manifold brace, located under the intake manifold.

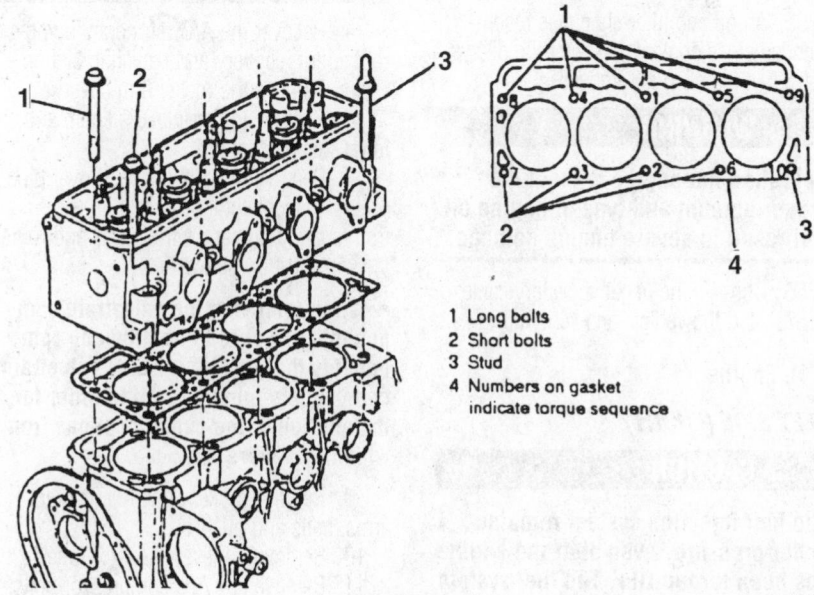

1 Long bolts
2 Short bolts
3 Stud
4 Numbers on gasket indicate torque sequence

7922SG03

Tighten the cylinder head bolts in the sequence shown for proper cylinder sealing and gasket crush—2.2L engine

42. Connect the electrical connections at the intake manifold, throttle body and cylinder head.

43. Connect the oxygen (O_2) sensor connector.

44. Install the pushrods, rocker arms and rocker arm nuts and tighten the nuts to 22 ft. lbs. (30 Nm).

45. Install the rocker arm cover.

46. Connect the control cables to the throttle body and install the cable brackets at the throttle body and rocker arm cover.

47. Connect the spark plug wires.

48. Install the power steering pump in the mounting bracket.

49. Install the alternator.

50. Install the serpentine belt.

51. Install the lower air inlet duct.

52. Install the air inlet resonator tie bar.

53. Install the air cleaner and duct assembly.

54. Refill the cooling system.

✷✷ CAUTION

The EPA warns that prolonged contact with used engine oil may cause a number of skin disorders, including cancer! You should make every effort to minimize your exposure to used engine oil. Protective gloves should be worn when changing the oil. Wash your hands and any other exposed skin areas as soon as possible after exposure to used engine oil. Soap and water, or waterless hand cleaner should be used.

55. An oil and filter change is recommended since coolant can enter the oil system when the head is removed.

✷✷ WARNING

Operating the engine without the proper amount and type of engine oil will result in severe engine damage.

56. Connect the negative battery cable.

57. Start the vehicle and verify no leaks.

3.1L Engine

LEFT SIDE (FRONT)

✷✷ CAUTION

The fuel injection system remains under pressure, even after the engine has been turned OFF. The fuel system pressure must be relieved before disconnecting any fuel lines. Failure to do so may result in fire and/or personal injury.

1. Disconnect the negative battery cable.

✷✷ CAUTION

Observe all applicable safety precautions when working around fuel. Whenever servicing the fuel system, always work in a well ventilated area. Do not allow fuel spray or vapors to come in contact with a spark or open flame. Keep a dry chemical fire extinguisher near the work area. Always keep fuel in a container specifically designed for fuel storage; also, always properly seal fuel containers to avoid the possibility of fire or explosion.

2. Relieve the fuel system pressure.

✷✷ CAUTION

Never open, service or drain the radiator or cooling system when hot; serious burns can occur from the steam and hot coolant.

3. Drain the cooling system into a suitable container.

✷✷ CAUTION

When servicing the A/C system the correct tools and procedures must be used. Protective eye covering must be worn. DO NOT smoke or expose the refrigerant to any open flames or sparks.

4. Recover the A/C refrigerant using a refrigerant recovery and recycling station.

5. Remove the rocker arm covers.

6. Remove upper intake plenum and lower intake manifold.

7. Remove the exhaust crossover pipe.

8. Disconnect the spark plug wires from spark plugs and wire looms and route the wires out of the way.

➡When removing the valvetrain components use care to identify any components that will be reused. Valvetrain components must be kept in order for installation in the same locations from which they were removed.

9. Remove rocker arms nut, rocker arms, balls and pushrods.

10. Remove oil level indicator tube.

11. Remove any A/C compressor bolts accessible from the top.

12. Raise and safely support the vehicle.

13. Remove the lower A/C compressor mounting bolts.

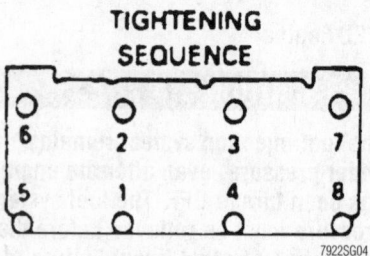

TIGHTENING SEQUENCE

7922SG04

Tighten the cylinder head bolts in the proper sequence to ensure good cylinder sealing—3.1L engine

14. Remove the refrigerant lines from the rear of the compressor.

15. Disconnect the A/C compressor electrical connections and remove the A/C compressor.

16. Remove the A/C compressor lower bracket bolts.

17. Lower the vehicle.

18. Remove the A/C compressor upper bracket bolts.

19. Remove the compressor brackets.

20. Remove the cylinder head bolts evenly.

21. Remove the cylinder head.

To install:

22. Clean all the gasket surfaces completely. Clean the threads on the cylinder head bolts and block threads.

23. Place the cylinder head gasket in position over the dowel pins on the cylinder block so the words THIS SIDE UP or other gasket identification are showing.

24. Coat the bolt threads with sealer and install finger-tight.

25. Tighten the cylinder head bolts in sequence to 33 ft. lbs. (45 Nm). With all the bolts tightened make a second pass tightening all the bolts an additional 90 degrees (¼ turn).

26. Install compressor bracket and tighten the upper bracket bolts to 35 ft. lbs. (47 Nm).

27. Raise vehicle and safely support.

28. Install the compressor lower bracket bolts and tighten to 35 ft. lbs. (47 Nm).

29. Connect the electrical connection to the rear of compressor.

30. Install the A/C compressor in the mounting bracket.

31. Connect the A/C lines to the rear of the compressor with NEW seals.

32. Install the compressor lower mounting bolts and tighten to 18 ft. lbs. (25 Nm).

33. Lower the vehicle.

34. Tighten the compressor upper mounting bolts to 18 ft. lbs. (25 Nm).

35. Install the oil level indicator tube.

36. Install the pushrods, rocker arms, balls and rocker arm nuts. Tighten the rocker arm nuts to 20 ft. lbs. (27 Nm).

37. Connect the spark plug wires to spark plugs and wire looms.

38. Install the exhaust crossover pipe.

39. Install the lower intake manifold and upper intake plenum.

40. Install the rocker arm covers.

41. Refill the cooling system.

42. Evacuate and charge A/C system.

43. Connect negative battery cable.

✳✳ CAUTION

The EPA warns that prolonged contact with used engine oil may cause a number of skin disorders, including cancer! You should make every effort to minimize your exposure to used engine oil. Protective gloves should be worn when changing the oil. Wash your hands and any other exposed skin areas as soon as possible after exposure to used engine oil. Soap and water, or waterless hand cleaner should be used.

44. An oil and filter change are recommended since coolant can enter the oil system when the head is being removed.

✳✳ WARNING

Operating the engine without the proper amount and type of engine oil will result in severe engine damage.

45. Start vehicle and verify no leaks.

RIGHT SIDE (REAR)

✳✳ CAUTION

The fuel injection system remains under pressure, even after the engine has been turned OFF. The fuel system pressure must be relieved before disconnecting any fuel lines. Failure to do so may result in fire and/or personal injury.

1. Disconnect the negative battery cable.

✳✳ CAUTION

Observe all applicable safety precautions when working around fuel. Whenever servicing the fuel system, always work in a well ventilated area. Do not allow fuel spray or vapors to come in contact with a spark or open flame. Keep a dry chemical fire extinguisher near the work area. Always keep fuel in a container specifically designed for fuel storage; also, always properly seal fuel containers to avoid the possibility of fire or explosion.

2. Relieve the fuel system pressure.

✳✳ CAUTION

Never open, service or drain the radiator or cooling system when hot; serious burns can occur from the steam and hot coolant.

3. Drain the cooling system into a suitable container.

4. Remove the rocker arm covers.

5. Remove upper intake plenum and lower intake manifold.

6. Disconnect the electrical connector from the ignition assembly.

7. Remove the alternator.

8. Remove the exhaust crossover pipe.

9. Disconnect the oxygen (O2) sensor connector.

10. Raise and safely support the vehicle.

11. Disconnect the exhaust pipe from the exhaust manifold.

12. Lower the vehicle.

13. Remove the exhaust manifold.

14. Disconnect the spark plug wires from spark plugs and wire looms and route the wires out of the way.

➡ **When removing the valvetrain components use care to identify any components that will be reused. Valvetrain components must be kept in order for installation in the same locations from which they were removed.**

15. Remove rocker arms nut, rocker arms, balls and pushrods.

16. Remove the cylinder head bolts evenly.

17. Remove the cylinder head.

To install:

18. Clean all the gasket surfaces completely. Clean the threads on the cylinder head bolts and block threads.

19. Place the cylinder head gasket in position over the dowel pins on the cylinder block so the words **THIS SIDE UP** or other gasket identification is showing.

20. Coat the bolt threads with sealer and install finger-tight.

21. Tighten the cylinder head bolts in sequence to 33 ft. lbs. (45 Nm). With all the bolts tightened make a second pass tightening all the bolts an additional 90 degrees (¼ turn).

22. Install the pushrods, rocker arms, balls and rocker arm nuts. Tighten the rocker arm nuts to 20 ft. lbs. (27 Nm).

23. Install the exhaust manifold.

24. Raise vehicle and safely support.

25. Connect the exhaust pipe to the exhaust manifold.

26. Lower the vehicle.

27. Connect the oxygen (O2) sensor connector.

28. Connect the spark plug wires to spark plugs and wire looms.

29. Install the exhaust crossover pipe.

30. Install the alternator.

31. Connect the electrical connector to the ignition assembly.

32. Install the lower intake manifold and upper intake plenum.

33. Install the rocker arm covers.

34. Refill the cooling system.

✳✳ CAUTION

The EPA warns that prolonged contact with used engine oil may cause a number of skin disorders, including cancer! You should make every effort to minimize your exposure to used engine oil. Protective gloves should be worn when changing the oil. Wash your hands and any other exposed skin areas as soon as possible after exposure to used engine oil. Soap and water, or waterless hand cleaner should be used.

35. An oil and filter change are recommended since coolant can enter the oil system when the head is being removed.

✳✳ WARNING

Operating the engine without the proper amount and type of engine oil will result in severe engine damage.

36. Connect negative battery cable.

37. Start vehicle and verify no leaks.

Rocker Arms

REMOVAL & INSTALLATION

2.2L Engine

1. Disconnect the negative battery cable.

2. Remove the air cleaner and air duct assembly.

3. Tag and disconnect the spark plug wires from the spark plugs and disconnect the spark plug wire clips from the rocker arm cover and position aside.

4. Disconnect the throttle cables from

the throttle body and remove the throttle cable bracket from the intake plenum and move aside.

5. Remove the rocker arm cover bolts.

6. Remove the rocker arm cover.

➡**If any valvetrain components are to be reused, keep in order. They should be installed in the same locations from which they were removed.**

7. Remove the rocker arm nut(s), ball(s) and rocker arms.

To install:

8. Clean the gasket surfaces completely.

9. Coat the bearing surfaces of the rocker arm(s) and the rocker arm ball(s) with Molykote® or its equivalent.

10. Seat the pushrods in the lifters.

11. Install the rocker arm(s), ball(s) and nut(s) in the same positions they were removed from and tighten the rocker arm nut(s) to 22 ft. lbs. (30 Nm).

12. Install a new gasket in the cut out in the rocker arm cover.

13. Install the rocker arm cover on the cylinder head and tighten the rocker arm cover bolts to 89 inch lbs. (10 Nm).

14. Install the throttle cable bracket on the intake plenum and connect the throttle control cables to the throttle body.

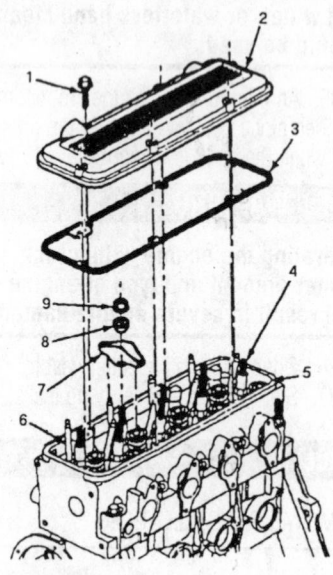

1 Bolt
2 Rocker cover
3 Gasket
4 Rocker stud
5 Flange, must be free of oil upon rocker cover gasket installation.
6 Pushrod
7 Rocker arm
8 Ball
9 Nut

7922SG05

The rocker arms are attached to the studs by means of a ball and nut—2.2L engine

15. Connect the spark plug wire clips to the rocker arm cover and connect the spark plug wires to the spark plugs.

16. Install the air cleaner and air duct assembly.

17. Connect the negative battery cable.

18. Start the engine and verify no oil leaks.

3.1L Engine
LEFT SIDE (FRONT)

1. Disconnect the negative battery cable.

> **✳✳ CAUTION**
>
> **Never open, service or drain the radiator or cooling system when hot; serious burns can occur from the steam and hot coolant.**

2. Drain the cooling system to a level below the coolant pipe on the front of the engine.

3. Remove the coolant bypass hose clamp at the coolant tube.

4. Remove the two bolts and nut securing the coolant tube to the cylinder head and position the tube out of the way.

➡**If any valvetrain components are to be reused, keep in order. They should be installed in the same locations from which they were removed.**

5. Remove the four rocker arm cover bolts and remove the rocker arm cover.

6. Remove the rocker arm nut(s), ball(s) and rocker arm(s).

To install:

7. Clean the gasket surfaces completely.

8. Coat all the valvetrain components with engine oil prior to installation.

9. Install the rocker arm(s) on the stud(s). Install the rocker arm ball(s) and mounting nuts. Be sure the pushrods are properly seated in the lifter and rocker arm.

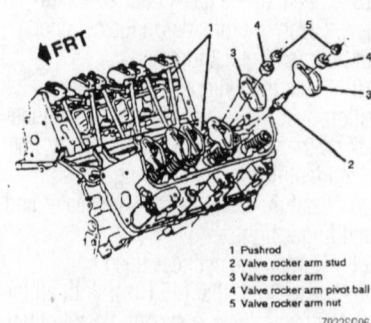

1 Pushrod
2 Valve rocker arm stud
3 Valve rocker arm
4 Valve rocker arm pivot ball
5 Valve rocker arm nut

7922SG06

Rocker arms and related components— 3.1L engine

10. Tighten the rocker arm nuts to 89 inch lbs. (10 Nm), then an additional 30 degrees.

11. Install the rocker arm cover using a new gasket and tighten the rocker cover bolts to 90 inch lbs. (10 Nm).

12. Position the coolant tube and connect the thermostat bypass hose.

13. Install the coolant tube mounting nut and bolts. Tighten the screw at the water pump to 106 inch lbs. (12 Nm), the bolt at the corner of the cylinder head to 18 ft. lbs. (25 Nm) and the nut to 18 ft. lbs. (25 Nm).

14. Refill the cooling system.

15. Connect the negative battery cable.

16. Start the vehicle and verify no oil or coolant leaks.

RIGHT SIDE (REAR)

1. Disconnect the negative battery cable.

> **✳✳ CAUTION**
>
> **Never open, service or drain the radiator or cooling system when hot; serious burns can occur from the steam and hot coolant.**

2. Drain the cooling system into a suitable container.

3. Remove the serpentine belt.

4. Remove the alternator.

5. Remove the alternator mounting bracket.

6. Remove the four rocker arm cover bolts and remove the rocker arm cover.

➡**If any valvetrain components are to be reused, keep in order. They should be installed in the same locations from which they were removed.**

7. Remove the rocker arm nut(s), ball(s) and rocker arm(s).

To install:

8. Clean all the gasket surfaces completely.

9. Coat all the valvetrain components with engine oil prior to installation.

10. Install the rocker arm(s) on the stud(s). Install the rocker arm ball(s) and mounting nuts. Be sure the pushrods are properly seated in the lifter and rocker arm.

11. Tighten the rocker arm nuts to 89 inch lbs., then an additional 30 degrees.

12. Install the rocker arm cover using a new gasket and tighten the rocker cover bolts to 90 inch lbs. (10 Nm).

13. Install the alternator mounting bracket.

14. Install the alternator.

15. Install the serpentine belt.

16. Refill the cooling system.

17. Connect the negative battery cable.
18. Start the vehicle and verify no oil or coolant leaks.

Intake Manifold

REMOVAL & INSTALLATION

2.2L Engine

These vehicles use a two-piece intake manifold. The upper half, sometimes called a plenum, contains the throttle body and control cable connections. The lower half has individual port runners to each intake port on the cylinder head. The lower half of the manifold bolts to the cylinder head and houses the fuel injectors. Note that these pieces are cast aluminum. Care should be exercised when working with any light alloy component.

> ※※ **CAUTION**
>
> **The fuel injection system remains under pressure, even after the engine has been turned OFF. The fuel system pressure must be relieved before disconnecting any fuel lines. Failure to do so may result in fire and/or personal injury.**

1. Relieve the fuel system pressure.

> ※※ **CAUTION**
>
> **Observe all applicable safety precautions when working around fuel. Whenever servicing the fuel system, always work in a well ventilated area. Do not allow fuel spray or vapors to come in contact with a spark or open flame. Keep a dry chemical fire extinguisher near the work area. Always keep fuel in a container specifically designed for fuel storage; also, always properly seal fuel containers to avoid the possibility of fire or explosion.**

2. Disconnect the negative battery cable.
3. Remove the throttle body air intake duct.

> ※※ **CAUTION**
>
> **Never open, service or drain the radiator or cooling system when hot; serious burns can occur from the steam and hot coolant.**

4. Drain the cooling system.
5. Identify, tag and disconnect all necessary vacuum lines.

6. Disconnect the control cables from the throttle body lever and remove the control cable bracket from the intake manifold.
7. Remove the serpentine belt.
8. Remove the power steering pump and lay it aside, without disconnecting the fluid lines.
9. Remove the transaxle fluid fill tube.
10. Identify, tag and disconnect the electrical connectors from the MAP sensor, EGR solenoid valve, Idle Air Control (IAC) valve, Throttle Position Sensor (TPS) and fuel injectors.
11. Remove the MAP sensor.

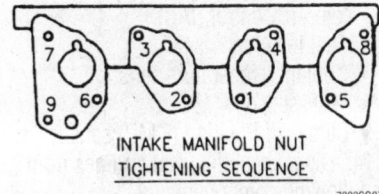

INTAKE MANIFOLD NUT
TIGHTENING SEQUENCE

7922SG07

Lower intake manifold bolt tightening sequence—2.2L engine

12. Remove the upper intake manifold mounting bolts and the upper intake manifold.
13. Disconnect the fuel lines from the fuel rail.
14. Remove the EGR valve injector.
15. Remove the EGR valve.
16. Remove the fuel injector retainer bracket, regulator and injectors.
17. Remove the control cable bracket.
18. Remove the six intake manifold nuts.
19. Remove the intake manifold.

To install:

20. Clean the gasket mounting surfaces.
21. Install a new gasket and position the lower intake manifold. Tighten the lower intake manifold nuts in the proper sequence to 24 ft. lbs. (33 Nm).
22. Connect the control cables and cable bracket.
23. Install the EGR valve.
24. Connect the fuel lines to the fuel rail.
25. Install the fuel injectors, regulator and injector retainer bracket and tighten the retaining bolts to 22 inch lbs. (3.5 Nm).
26. Install the EGR valve injector so that the port is facing directly towards the throttle body.
27. Install the upper intake manifold assembly. Tighten the upper intake manifold

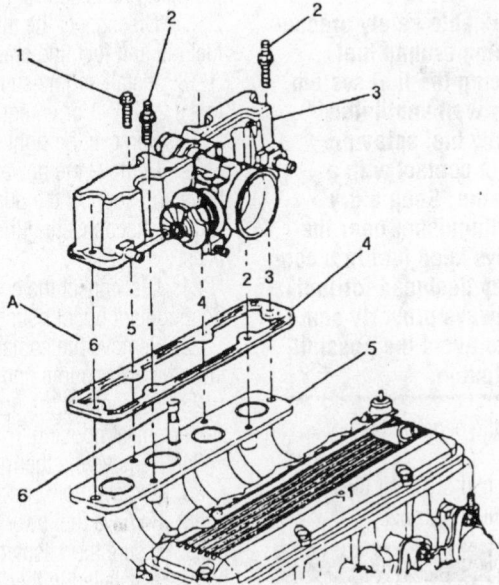

1 Bolt
2 Stud
3 Upper intake manifold assembly
4 Gasket
5 Lower intake manifold
6 EGR valve injector
A Upper intake manifold assembly tightening sequence

7922SG08

Upper intake manifold bolt tightening sequence—2.2L engine

nuts in the proper sequence to 22 ft. lbs. (30 Nm).

28. Install the MAP sensor.

29. Connect the electrical connectors to the MAP sensor, EGR solenoid valve, Idle Air Control (IAC) valve, Throttle Position Sensor (TPS), and the fuel injectors.

30. Install the transaxle fill tube.

31. Install the power steering pump.

32. Install the serpentine belt.

33. Connect the vacuum lines.

34. Install the air intake duct.

35. Connect the negative battery cable.

36. Refill the coolant system.

37. Start the vehicle and verify no coolant or vacuum leaks.

3.1L Engine

1995 MODELS

> ✳✳ **CAUTION**
>
> **The fuel system is under pressure and must be properly relieved before disconnecting the fuel lines. Failure to properly relieve the fuel system pressure can lead to personal injury and component damage.**

1. Relieve the fuel system pressure.

> ✳✳ **CAUTION**
>
> **Observe all applicable safety precautions when working around fuel. Whenever servicing the fuel system, always work in a well ventilated area. Do not allow fuel spray or vapors to come in contact with a spark or open flame. Keep a dry chemical fire extinguisher near the work area. Always keep fuel in a container specifically designed for fuel storage; also, always properly seal fuel containers to avoid the possibility of fire or explosion.**

2. Disconnect the negative battery cable.

3. Remove top half of the air cleaner assembly and throttle body duct.

> ✳✳ **CAUTION**
>
> **Never open, service or drain the radiator or cooling system when hot; serious burns can occur from the steam and hot coolant.**

4. Drain and recover the cooling system.

5. Remove the EGR pipe from exhaust manifold.

6. Remove the serpentine belt.

7. Remove the brake vacuum pipe at the intake plenum.

8. Disconnect the control cables from the throttle body and intake plenum mounting bracket.

9. Remove the power steering lines at the alternator bracket.

10. Remove the alternator.

11. Disconnect the spark plug wires from the spark plugs and wire retainers on the intake plenum.

12. Remove the Ignition assembly and the EVAP canister purge solenoid together.

13. Disconnect the upper engine wiring harness connectors at the following components:

- Throttle Position Sensor (TPS)
- Idle Air Control (IAC)
- Fuel Injectors
- Coolant temperature sensor
- MAP sensor
- Camshaft Position (CMP) sensor

14. Disconnect the vacuum lines from the following components:

- Vacuum modulator
- Fuel pressure regulator
- PCV valve

15. Remove the MAP sensor from upper intake manifold.

16. Remove the upper intake plenum mounting bolts and remove the plenum.

17. Disconnect the fuel lines from the fuel rail and fuel line bracket.

18. Install engine support fixture special tool J 28467-A or an equivalent.

19. Remove the right side engine mount.

20. Remove the power steering mounting bolts and support the pump out of the way without disconnecting the power steering lines.

21. Disconnect the coolant inlet pipe from coolant outlet housing.

22. Remove the coolant bypass hose from the water pump and the cylinder head.

23. Disconnect the upper radiator hose at thermostat housing.

24. Remove the thermostat housing.

25. Remove both rocker arm covers.

26. Remove the lower intake manifold bolts. Be sure the washers on the four center bolts are installed in their original locations.

➡ **When removing the valvetrain components they should be kept in order for installation the original locations.**

27. Remove the rocker arm retaining nuts and remove the rocker arms and pushrods.

28. Remove the intake manifold from the engine.

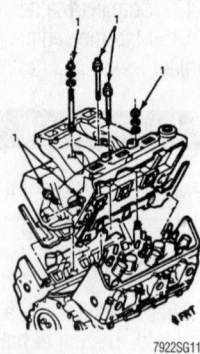

| 1 | 22 N·m (16 LB. FT.)
THEN 32 N·m (23 LB. FT.)
RETORQUE 32 N·m
(23 LB. FT.) IN SEQUENCE |

⑦ ④ ③ ⑥
⑧ ① ② ⑤

7922SG11

Intake manifold bolt tightening sequence—1995 3.1L engine

To install:

29. Clean gasket material from all mating surfaces. Remove all excess RTV sealant from front and rear ridges of cylinder block.

30. Place a 3mm bead of RTV, on each ridge, where the front and rear of the intake manifold contact the block.

31. Using a new gasket, install the intake manifold on the engine. Tighten the bolts evenly and gradually in the proper sequence first to 16 ft. lbs. (22 Nm), then to 23 ft. lbs. (32 Nm).

32. Install the pushrods, rocker arms and mounting nuts. Be sure the pushrods are properly seated in the valve lifters and rocker arms.

33. Install rocker arm nuts and tighten the rocker arm nuts to 18 ft. lbs. (24 Nm).

34. Install lower the intake manifold attaching bolts. Apply sealant PN 12345739 or equivalent to the threads of bolts, and torque bolts to 115 inch lbs. (13 Nm).

35. Install the front rocker arm cover.

36. Install the thermostat housing.

37. Connect the upper radiator hose to the thermostat housing.

38. Install the coolant inlet pipe to thermostat housing.

39. Install coolant bypass pipe at the water pump and cylinder head.

40. Install the power steering pump in the mounting bracket.

41. Connect the right side engine mount.

42. Remove the special engine support tool.

43. Connect the fuel lines to fuel rail and bracket.

44. Install the upper intake manifold and torque the mounting bolts to 18 ft. lbs. (25 Nm).

45. Install the MAP sensor.

46. Connect the upper engine wiring harness connectors to the following components:

- Throttle Position Sensor (TPS)

- Idle Air Control (IAC)
- Fuel Injectors
- Coolant temperature sensor
- MAP sensor
- Camshaft Position (CMP) sensor

47. Connect the vacuum lines to the following components:
- Vacuum modulator
- Fuel pressure regulator
- PCV valve

48. Install the EVAP canister purge solenoid and ignition assembly.

49. Install the alternator assembly.

50. Connect the power steering line to the alternator bracket.

51. Install the serpentine belt.

52. Connect the spark plug wires to the spark plugs and intake plenum wire retainer.

53. Install the EGR pipe to the exhaust manifold.

54. Connect the control cables to the throttle body lever and upper intake plenum mounting bracket.

55. Install air intake assembly and top half of the air cleaner assembly.

56. Install the brake vacuum pipe.

57. Fill the cooling system.

58. Connect the negative battery cable.

59. Start the vehicle and verify no leaks.

1996 MODELS

These vehicles use a two-piece intake manifold. These pieces are aluminum and care should be exercised when working with these components.

> ※※ **CAUTION**
>
> **The fuel injection system remains under pressure, even after the engine has been turned OFF. The fuel system pressure must be relieved before disconnecting any fuel lines. Failure to do so may result in fire and/or personal injury.**

1. Disconnect the negative battery cable.

> ※※ **CAUTION**
>
> **Observe all applicable safety precautions when working around fuel. Whenever servicing the fuel system, always work in a well ventilated area. Do not allow fuel spray or vapors to come in contact with a spark or open flame. Keep a dry chemical fire extinguisher near the work area. Always keep fuel in a container specifically designed for fuel storage; also, always properly seal fuel containers to avoid the possibility of fire or explosion.**

2. Relieve the fuel system pressure.

3. Remove the air cleaner assembly.

4. Remove the throttle cables from the throttle body and bracket.

5. Remove the fuel line retaining clamp from control cable bracket.

6. Remove the control cable bracket from the manifold.

7. Tag and disconnect the vacuum lines from the upper intake manifold.

8. Remove the EGR valve.

9. Position the heater inlet pipe hose clamps out of the way.

10. Remove the rear ignition coil nuts.

11. Remove the front ignition coil bolts.

12. Remove the power steering line clip from the alternator brace.

13. Remove the alternator and brackets.

14. Disconnect the electrical connections from the throttle body.

15. Remove the upper intake manifold bolts and remove upper intake manifold with throttle body.

16. Remove the heater hoses from the upper intake manifold pipes.

17. Remove the fuel rail.

18. Disconnect the coolant hose.

19. Remove the power steering pump.

20. Remove the heater hose from the thermostat housing.

21. Remove the ignition coil.

22. Remove the engine mount strut.

23. Remove the valve covers.

24. Remove the lower intake manifold bolts.

25. Remove the lower intake manifold.

To install:

26. Clean all gasket material from mating surfaces. Then, clean all sealing surfaces with a degreaser product and blow dry with compressed air.

27. Apply a small bead (0.08–0.11 inch) of RTV sealer on each ridge where the intake manifold contacts the engine block.

28. Install the gasket and position the lower intake manifold into place.

29. Apply sealant, GM P/N 12345382 or equivalent to threads of the manifold bolts.

➡ **When installing the manifold bolts, you must tighten the vertical bolts before the diagonal bolts. Failure to follow this step may result in an oil leak.**

30. Hand-tighten the vertical bolts.

31. Hand-tighten the diagonal bolts.

32. Tighten the vertical bolts to 115 inch lbs. (13 Nm).

33. Tighten the diagonal bolts to 115 inch lbs. (13 Nm).

34. Install the valve covers.

35. Install the engine mount strut.

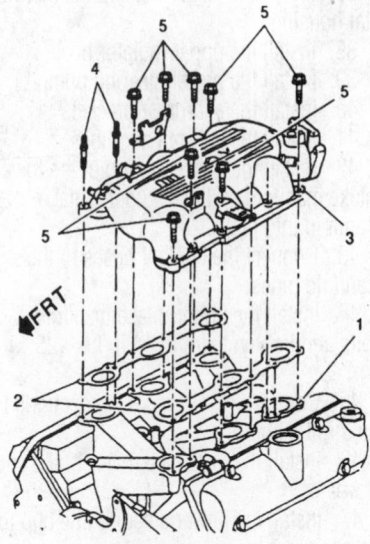

1. Manifold, lower intake
2. Gasket, upper intake manifold
3. Manifold, upper intake
4. Stud, upper intake manifold
5. Bolt, upper intake manifold

7922SG09

Exploded view of the upper intake manifold mounting—1996 3.1L engine

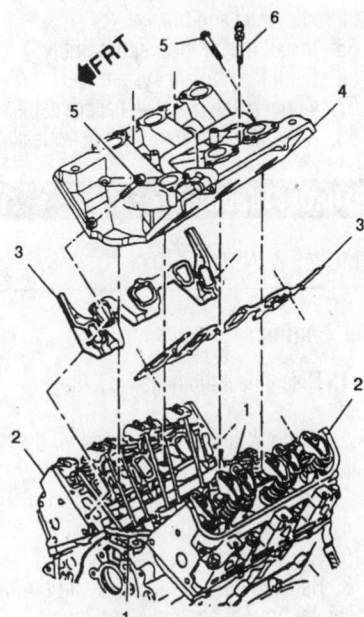

1. Apply sealant
2. Head, cylinder
3. Gasket, lower intake manifold
4. Manifold, lower intake
5. Bolt, lower intake manifold
6. Bolt, lower intake manifold

7922SG10

Exploded view of the lower intake manifold mounting—1996 3.1L engine

36. Install the ignition coil.
37. Install the heater hose to the thermostat housing.
38. Install the upper radiator hose.
39. Install the power steering pump.
40. Install the alternator brackets.
41. Install the fuel rail and lines.
42. Install the gasket on top of the lower intake manifold and place upper intake manifold into place.
43. Connect the coolant hoses to the manifold pipes.
44. Install the upper intake manifold bolts and tighten them to 18 ft. lbs. (25 Nm).
45. Connect the electrical connections to the throttle body.
46. Install the upper alternator brace and alternator.
47. Install the power steering line clip to the alternator brace.
48. Install the mounting nuts for the coil.
49. Properly position the heater inlet pipe hose clamps.
50. Install the EGR valve.
51. Connect the vacuum lines to the upper intake manifold.
52. Install the control cable bracket and return spring.
53. Install the fuel line retaining clamp to the control cable bracket.
54. Install the throttle control cables to the throttle body and bracket.
55. Install the air cleaner assembly.
56. Refill the coolant system.
57. Connect the negative battery cable.
58. Start the vehicle and verify no leaks.

Exhaust Manifold

REMOVAL & INSTALLATION

2.2L Engine

1. Disconnect the negative battery cable.
2. Remove the air cleaner and air duct assembly.
3. Remove the lower air inlet duct.
4. Remove the engine drive belt.
5. Remove the alternator.
6. Remove the engine torque strut from the engine bracket and radiator support bracket.
7. Remove the torque strut bracket from the cylinder head.
8. Remove the alternator rear support bracket.
9. Raise and safely support the vehicle.
10. Remove the two bolts securing the exhaust pipe to the exhaust manifold.

11. Lower the vehicle.
12. Remove the dipstick tube mounting bolt and dipstick tube and dipstick.
13. Disconnect the electrical connector from the oxygen sensor (O_2).
14. Remove the exhaust manifold mounting bolts.
15. Remove the exhaust manifold.

To install:

16. Clean all gasket surfaces completely.
17. Install the exhaust manifold and tighten the mounting bolts to 116 inch lbs. (13 Nm) starting on the center two ports and working outwards.
18. Connect the O_2 sensor electrical connector.
19. Install the dipstick tube and mounting bolt.
20. Raise and safely support the vehicle.
21. Connect the front exhaust pipe to the exhaust manifold and tighten the mounting bolts to 18 ft. lbs. (25 Nm).
22. Lower the vehicle.
23. Install the alternator rear support bracket and tighten the three mounting bolts to 74 ft. lbs. (100 Nm).
24. Install the engine torque strut bracket to the engine and tighten the nuts to 41 ft. lbs. (56 Nm) and the bolt to 40 ft. lbs. (55 Nm).
25. Install the torque strut and tighten the through-bolts to 40 ft. lbs. (55 Nm).
26. Install the alternator.
27. Install the drive belt.
28. Install the lower air inlet duct.
29. Install the air inlet resonator to the upper tie bar.
30. Install the air cleaner and duct work.
31. Connect the negative battery cable.
32. Start the vehicle and verify no exhaust leaks.

3.1L Engine

✳✳ CAUTION

The fuel injection system remains under pressure, even after the engine has been turned OFF. The fuel system pressure must be relieved before disconnecting any fuel lines. Failure to do so may result in fire and/or personal injury.

LEFT SIDE (FRONT)

✳✳ CAUTION

Observe all applicable safety precautions when working around fuel. Whenever servicing the fuel system,

always work in a well ventilated area. Do not allow fuel spray or vapors to come in contact with a spark or open flame. Keep a dry chemical fire extinguisher near the work area. Always keep fuel in a container specifically designed for fuel storage; also, always properly seal fuel containers to avoid the possibility of fire or explosion.

1. Relieve the fuel system pressure.
2. Disconnect the negative battery cable.

✳✳ CAUTION

Never open, service or drain the radiator or cooling system when hot; serious burns can occur from the steam and hot coolant.

3. Drain the engine coolant.
4. Disconnect the electrical connectors and the heater hoses from the throttle body.
5. Remove the throttle body from the intake.
6. Disconnect the fuel lines and position aside.
7. Disconnect the coolant hose from the thermostat housing.
8. Remove the crossover pipe heat shield, then remove the crossover pipe.
9. Remove the engine torque strut.
10. If equipped with air conditioning, perform the following steps:
   a. Recover the air conditioning refrigerant, using an approved recovery system.
   b. Remove the serpentine belt cover and belt.
   c. Remove the compressor front mounting bolts.
   d. Remove the two bolts at the top of the compressor mounting bracket.
   e. Raise and properly support the vehicle.
   f. Disconnect the A/C lines from the rear of the compressor.
   g. Disconnect the electrical connector from the compressor.
   h. Remove the compressor rear mounting bolts.
   i. Remove the compressor assembly from the vehicle.
   j. Remove the mounting bracket lower bolts.
   k. Remove the compressor mounting bracket.
   l. Lower the vehicle.
11. Remove the exhaust manifold retaining nuts and manifold.

To install:

12. Thoroughly clean all gasket material from the manifold and cylinder head mating surfaces.

13. Install the exhaust manifold and retaining nuts. Tighten the retaining nuts to 18 ft. lbs. (25 Nm).

14. If equipped with air conditioning, perform the following steps:

a. Raise and properly support the vehicle.

b. Install the mounting bracket lower bolts and tighten to 35 ft. lbs. (47 Nm).

c. Position the compressor assembly in the mounting bracket.

d. Install the rear compressor mounting bolts and tighten to 18 ft. lbs. (25 Nm).

e. Connect the electrical connector to the compressor.

f. Using new O-rings, connect the A/C lines to the rear of the compressor. Tighten the mounting bolt to 24 ft. lbs. (35 Nm).

g. Lower the vehicle.

h. Install the front compressor mounting bolts and tighten to 37 ft. lbs. (50 Nm).

i. Install the two bolts at the top of the compressor mounting bracket and tighten to 35 ft. lbs. (47 Nm).

j. Install the serpentine belt and belt cover.

15. Install the torque strut and tighten to 39 ft. lbs. (53 Nm).

16. Connect the exhaust crossover pipe to the manifold and tighten the retaining nuts to 18 ft. lbs. (25 Nm).

17. Install the heat shield and tighten the bolts to 89 inch lbs. (10 Nm).

18. Connect the coolant hose to the thermostat housing and secure with retaining clamp.

19. Connect the fuel lines.

20. Install the throttle body to the intake manifold.

21. Connect the coolant hoses and electrical wiring to the throttle body.

22. Refill the engine coolant and bleed the cooling system as required.

23. Following proper procedures, evacuate and recharge the A/C system.

24. Connect the negative battery cable.

25. Start the engine and check for coolant, fuel, refrigerant, and exhaust leaks.

RIGHT SIDE (REAR)

✳✳ CAUTION

Observe all applicable safety precautions when working around fuel. Whenever servicing the fuel system, always work in a well ventilated area. Do not allow fuel spray or vapors to come in contact with a spark or open flame. Keep a dry chemical fire extinguisher near the work area. Always keep fuel in a container specifically designed for fuel storage; also, always properly seal fuel containers to avoid the possibility of fire or explosion.

1. Relieve the fuel system pressure.
2. Disconnect the negative battery cable.

✳✳ CAUTION

Never open, service or drain the radiator or cooling system when hot; serious burns can occur from the steam and hot coolant.

3. Drain the engine coolant.
4. Disconnect the electrical connectors and the heater hoses from the throttle body.
5. Remove the throttle body from the intake.
6. Disconnect the fuel lines and position aside.
7. Disconnect the upper radiator hose from the thermostat housing.
8. Disconnect the EGR tube from the manifold.
9. Remove the crossover pipe heat shield, then remove the crossover pipe.
10. Disconnect the oxygen sensor electrical connector and remove the sensor.
11. Remove the exhaust manifold upper heat shield.
12. Raise and properly support the vehicle.
13. Disconnect the front (converter) exhaust pipe at the manifold.
14. Remove the automatic transaxle dipstick and fill tube from the transaxle.
15. Remove the exhaust manifold lower heat shield.
16. Remove the exhaust manifold retaining nuts and manifold.

To install:

17. Thoroughly clean all gasket material from the manifold and cylinder head mating surfaces.

18. Install the exhaust manifold and retaining nuts. Tighten the retaining nuts to 18 ft. lbs. (25 Nm).

19. Install the manifold lower heat shield and tighten the bolts to 89 inch lbs. (10 Nm).

20. Install the transaxle fill tube and dipstick.

21. Connect the front exhaust pipe to the manifold assembly and tighten the bolts to 22 ft. lbs. (30 Nm).

22. Lower the vehicle.

23. Install the upper manifold heat shield and tighten the bolts to 89 inch lbs. (10 Nm).

24. Install the oxygen sensor and connect the wiring.

25. Connect the exhaust crossover pipe to the manifold and tighten the retaining nuts to 18 ft. lbs. (25 Nm).

26. Install the crossover pipe heat shield and tighten the bolts to 89 inch lbs. (10 Nm).

27. Connect the upper radiator hose to the thermostat housing and secure with retaining clamp.

28. Connect the fuel lines.

29. Connect the EGR tube to the manifold.

30. Install the throttle body to the intake manifold.

31. Connect the coolant hoses and electrical wiring to the throttle body.

32. Refill the engine coolant and bleed the cooling system as required.

33. Connect the negative battery cable.

34. Start the engine and check for coolant, fuel and exhaust leaks.

Camshaft and Valve Lifters

REMOVAL & INSTALLATION

✳✳ CAUTION

Some models covered by this manual may be equipped with a Supplemental Restraint System (SRS), which uses an air bag. Whenever working near any of the SRS components, such as the impact sensors, the air bag module, steering column and instrument panel, properly disable the SRS.

2.2L Engine

Please note that the engine must be removed from the vehicle for this procedure. Use care when disassembling valvetrain components. Any part that is to be reused must be returned to its original location. In addition, if the camshaft is being replaced, all new lifters must also be installed. Installing used lifters on a new camshaft will fail the new camshaft.

❋❋ CAUTION

The fuel injection system remains under pressure, even after the engine has been turned OFF. The fuel system pressure must be relieved before disconnecting any fuel lines. Failure to do so may result in fire and/or personal injury.

1. Disconnect the negative battery cable.

❋❋ CAUTION

Observe all applicable safety precautions when working around fuel. Whenever servicing the fuel system, always work in a well ventilated area. Do not allow fuel spray or vapors to come in contact with a spark or open flame. Keep a dry chemical fire extinguisher near the work area. Always keep fuel in a container specifically designed for fuel storage; also, always properly seal fuel containers to avoid the possibility of fire or explosion.

2. Relieve the fuel system pressure.
3. Remove the engine assembly from the vehicle using the recommended procedure. Mount engine assembly on a suitable engine stand.
4. Remove the serpentine belt.
5. Remove the serpentine belt tensioner assembly with the alternator attached.
6. Remove the strut bracket and the rear alternator bracket.
7. Remove the front engine mount bracket.
8. Remove the oil level indicator tube.

❋❋ CAUTION

The EPA warns that prolonged contact with used engine oil may cause a number of skin disorders, including cancer! You should make every effort to minimize your exposure to used engine oil. Protective gloves should be worn when changing the oil. Wash your hands and any other exposed skin areas as soon as possible after exposure to used engine oil. Soap and water, or waterless hand cleaner should be used.

9. Drain the engine oil into a suitable container.
10. Remove the oil pan.
11. Remove the crankshaft balancer and front cover.
12. Remove the timing chain and camshaft sprocket.
13. Disconnect the spark plug wires.
14. Remove the camshaft position sensor.
15. Remove the rocker arm cover.

➡**When removing the valvetrain components they must be kept in order for installation in the same locations from which they were removed.**

16. Remove the rocker arm nuts, rocker arms, balls and pushrods.
17. Remove the power steering pump pencil brace.
18. Remove the cylinder head with the intake and exhaust manifolds attached.
19. Remove the anti-rotation bracket and the valve lifters. A magnet may be helpful when removing the lifters from the cylinder head. Be sure to keep the lifters in order. They must be returned to their original locations.
20. Remove the camshaft thrust plate mounting bolts and remove the thrust plate.
21. Remove the oil pump drive assembly.
22. Remove the camshaft carefully from the engine.

To install:

23. Coat the camshaft lobes with and bearings with GM Engine Oil Supplement (E.O.S.) part number 1051396 or equivalent camshaft break-in lube and insert the camshaft carefully into the engine.
24. Install the oil pump drive assembly.
25. Install the thrust plate and tighten the mounting bolts to 106 inch lbs. (12 Nm).
26. Install the valve lifters in their original locations.
27. Install the anti-rotation brackets on the lifters. Tighten the bolt to 97 in. lbs. (11 Nm).
28. Install the cylinder head and manifold assemblies.
29. Install the power steering pump pencil brace.
30. Install the pushrods, rocker arms, balls and rocker arm nuts. Tighten the nuts to 22 ft. lbs. (30 Nm).

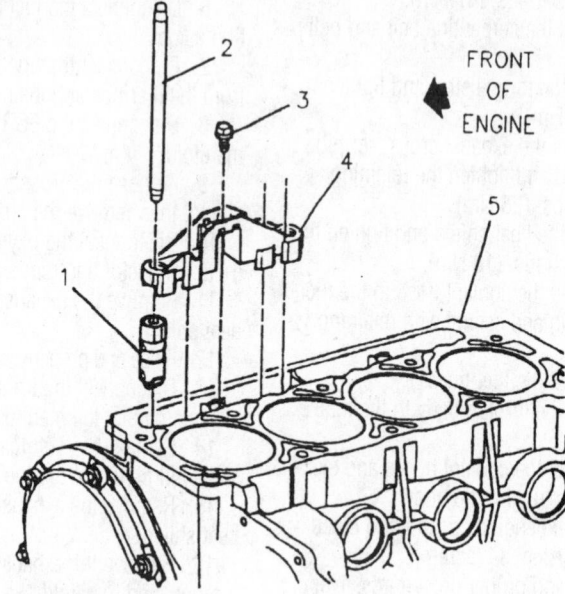

FRONT
OF
ENGINE

1   LIFTER, VALVE ROLLER
2   ROD, PUSH
3   BOLT – 11 N·m (97 LBS. IN.)
4   BRACKET, ANTI-ROTATION
5   BLOCK, CYLINDER

7922SG28

Remove the lifters after removing the anti-rotation bracket—2.2L engine

31. Install the rocker arm cover.

32. Connect the spark plug wires.

33. Install the camshaft position sensor.

34. Install the timing chain and camshaft sprocket. Verify that the camshaft and crankshaft sprocket timing marks are correctly aligned.

35. Install the timing chain front cover and crankshaft balancer.

36. Install the oil pan.

37. Install the oil level indicator tube.

38. Install the front engine bracket, rear alternator bracket and strut bracket.

39. Install the drive belt tensioner and alternator assembly and install the serpentine belt.

40. Install the engine assembly in the vehicle.

41. Install the oil filter and refill with oil.

❋ WARNING

Operating the engine without the proper amount and type of engine oil will result in severe engine damage.

42. Connect the negative battery cable.

43. Refill the crankcase with the recommended engine oil and the cooling system with 50–50 water antifreeze mix.

44. Start the vehicle and verify no leaks.

A TIMING ALIGNMENT MARKS
24 SPROCKET, CRANKSHAFT
25 CHAIN, TIMING
50 BOLT, TIMING CHAIN DAMPER
51 DAMPENER, TIMING CHAIN
52 BLOCK, ENGINE
91 BOLT, CAMSHAFT SPROCKET
92 SPROCKET, CAMSHAFT
93 BOLT, THRUST PLATE
94 PLATE, THRUST

3.1L Engine

Please note that the engine must be removed from the vehicle to perform this procedure. When removing valvetrain components, any parts that are to be reused must be returned to their original locations. Lay all parts out in an orderly fashion and mark them for identification. In addition, if the camshaft is being replaced, all of the valve lifters must also be replaced with new parts. Installing used lifters on a new camshaft will quickly wear the camshaft.

❋ CAUTION

The fuel injection system remains under pressure, even after the engine has been turned OFF. The fuel system pressure must be relieved before disconnecting any fuel lines. Failure to do so may result in fire and/or personal injury.

1. Relieve the fuel system pressure following the recommended procedure.

❋ CAUTION

Observe all applicable safety precautions when working around fuel. Whenever servicing the fuel system, **always work in a well ventilated area. Do not allow fuel spray or vapors to come in contact with a spark or open flame. Keep a dry chemical fire extinguisher near the work area. Always keep fuel in a container specifically designed for fuel storage; also, always properly seal fuel containers to avoid the possibility of fire or explosion.**

2. Disconnect the negative battery cable.

3. Remove the engine assembly.

4. Remove the intake manifold, valve cover, rocker arms, pushrods and valve lifters.

5. Remove the crankshaft balancer and front cover.

6. Remove the timing chain and sprockets.

7. Remove the oil pump driven gear mounting bolt and remove the oil pump driven gear.

8. Remove the two bolts and remove the camshaft thrust plate.

9. Carefully remove the camshaft. Avoid marring the camshaft bearing surfaces.

To install:

10. Coat the camshaft with lubricant GM part number 1052365 or equivalent

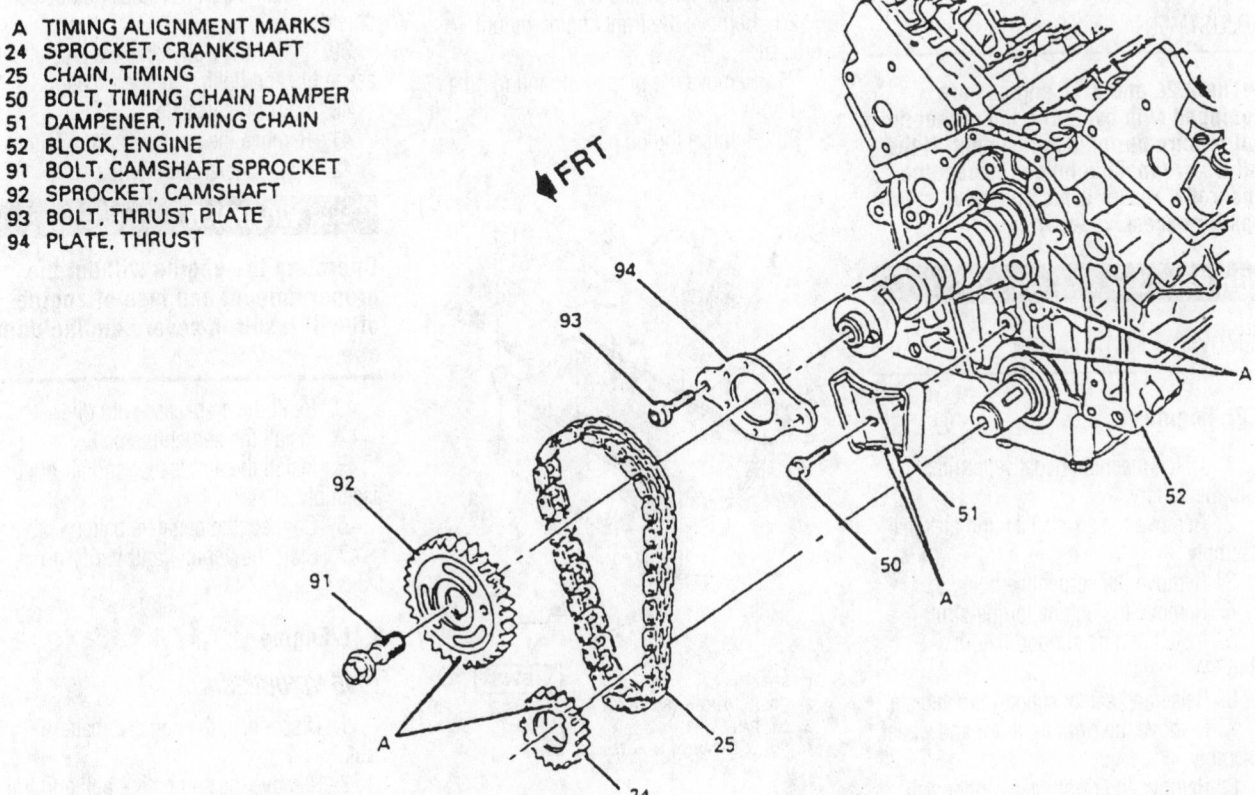

Exploded view of the camshaft and timing chain mounting—3.1L engine

7922SG29

camshaft break-in lubricant or quality engine oil supplement, and install the camshaft.

11. Install the camshaft thrust plate and tighten the mounting bolts to 89 inch lbs. (10 Nm).

12. Install the oil pump driven gear and tighten the mounting bolt to 27 ft. lbs. (36 Nm).

13. Install the timing chain and sprocket.

14. Install the camshaft thrust button and front cover.

15. Install the crankshaft balancer and tighten the bolt to 76 ft. lbs. (103 Nm).

16. Install the intake manifold, valve cover, rocker arms, pushrods and valve lifters.

17. Install the engine assembly.

18. Connect the negative battery cable.

✳✳ WARNING

Operating the engine without the proper amount and type of engine oil will result in severe engine damage.

19. Fill the crankcase with fresh oil.

20. Adjust the valves, as required.

21. Start the engine and verify no oil leaks.

Valve Lash

ADJUSTMENT

➡ **The 2.2L and 3.1L engines are equipped with hydraulic lifters that do not require periodic adjustment. If the valves are making noise, inspect the pushrods, rocker arms and valve stem tips for excessive wear.**

Oil Pan

REMOVAL & INSTALLATION

2.2L Engine

1. Disconnect the negative terminal from the battery.

2. Remove the air cleaner and air duct assembly.

3. Remove the serpentine belt.

4. Remove the engine torque strut.

5. Install engine support fixture J-28467-A.

6. Raise and safely support the vehicle.

7. Remove the right front tire and wheel assembly.

8. Remove the right inner fender well splash shield.

9. Remove the flywheel cover bolts.

10. Disconnect and remove the starter and starter bracket.

11. Remove the flywheel cover.

12. Remove the exhaust pipe and converter.

13. Remove the A/C compressor mounting bolts and set the compressor aside without disconnecting the refrigerant lines.

14. Remove the front engine mount nuts from the frame.

15. Remove the front engine mount bolts from the engine.

16. Lower the vehicle.

17. Raise the engine about 3 in. (76mm) using the support fixture.

18. Raise and safely support the vehicle.

✳✳ CAUTION

The EPA warns that prolonged contact with used engine oil may cause a number of skin disorders, including cancer! You should make every effort to minimize your exposure to used engine oil. Protective gloves should be worn when changing the oil. Wash your hands and any other exposed skin areas as soon as possible after exposure to used engine oil. Soap and water, or waterless hand cleaner should be used.

19. Drain the engine oil.

20. Remove the front engine mount bracket.

21. Remove the oil pan mounting nuts and bolts.

22. Remove the oil pan.

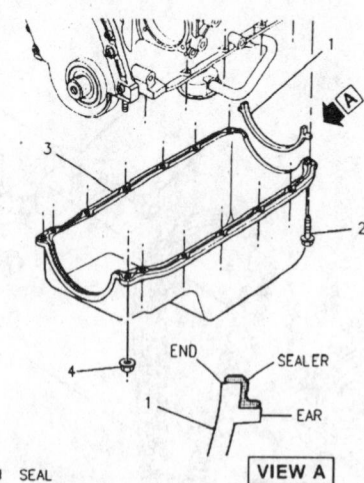

1 SEAL
2 BOLT, 10 N•m (89 LB. IN.)
3 OIL PAN
4 NUT, OIL PAN 10 N•m (89 LB. IN.)

7922SG30

Exploded view of the oil pan mounting—2.2L engine

To install:

23. Clean all the gasket surfaces completely.

24. Apply a thin bead of sealer around the outside edge of the oil pan and on the corners of the end seals.

25. Position the oil pan gasket onto the sealer.

26. Install the oil pan onto the engine and loosely install all the fasteners.

27. Tighten the nuts and bolts to 89 inch lbs. (10 Nm).

28. Install the front engine mount bracket and loosely install the mount-to-engine bolts.

29. Lower the vehicle.

30. Lower the engine into place.

31. Raise and safely support the vehicle.

32. Tighten the front engine mount bolts.

33. Install the engine mount nuts to 33 ft. lbs. (45 Nm).

34. Install the A/C compressor in the mounting bracket and tighten to 37 ft. lbs. (50 Nm).

35. Install the exhaust pipe and converter.

36. Connect the starter and install the starter and support bracket.

37. Install the flywheel cover and cover mounting bolts.

38. Install the right fender well splash shield.

39. Install the right front tire and wheel assembly and tighten to specification.

40. Lower the vehicle.

41. Remove the engine support fixture.

42. Install the engine torque strut.

✳✳ WARNING

Operating the engine without the proper amount and type of engine oil will result in severe engine damage.

43. Refill the crankcase with oil.

44. Install the serpentine belt.

45. Install the air cleaner and air duct assembly.

46. Connect the negative battery cable.

47. Start the vehicle and verify no leaks.

3.1L Engine

1995 VEHICLES

1. Disconnect the negative battery cable.

2. Remove the serpentine belt and the tensioner.

3. Support the engine with tool J-28467 or equivalent.

✳✳ CAUTION

The EPA warns that prolonged contact with used engine oil may cause a number of skin disorders, including cancer! You should make every effort to minimize your exposure to used engine oil. Protective gloves should be worn when changing the oil. Wash your hands and any other exposed skin areas as soon as possible after exposure to used engine oil. Soap and water, or waterless hand cleaner should be used.

4. Raise and safely support the vehicle. Drain the engine oil.

5. Remove the right tire and wheel assembly. Remove the right inner fender splash shield.

6. Remove the steering gear pinch bolt. Remove the transaxle mount retaining bolts. Failure to disconnect intermediate shaft from rack and pinion stub shaft can result in damage to the steering gear and/or intermediate shaft. This could cause a loss of steering control which could result in personal injury.

7. Remove the engine-to-cradle mounting nuts. Remove the front engine collar bracket from the block.

8. Remove the starter shield and the flywheel cover. Remove the starter.

9. Loosen, but do not remove the rear engine cradle bolts. Remove electrical connector at DIS sensor, if equipped.

10. Remove the front cradle bolts and lower front of frame. Remove the oil pan retaining bolts and nuts. Remove the oil pan.

To install:

11. Clean the gasket mating surfaces.

12. Install a new gasket on the oil pan. Apply silicon sealer to the portion of the pan that contacts the rear of the block.

13. Install the oil pan, nuts and retaining bolts. Tighten rear bolts to 18 ft. lbs. (18–25 Nm), and remaining nuts and bolts to 89 inch lbs. (10 Nm).

14. Install the front cradle bolts and tighten the rear cradle bolts. Install DIS connector, if equipped. Install the starter and splash shield. Install the flywheel shield.

15. Attach the collar bracket to the block, install the engine-to-cradle nuts. Install the transaxle mount nuts.

16. Install the steering pinch bolt. Install the right inner fender splash shield and tire assembly. Lower the vehicle.

17. Remove the engine support tool. Install the serpentine belt and tensioner.

✳✳ WARNING

Operating the engine without the proper amount and type of engine oil will result in severe engine damage.

18. Fill the crankcase to the correct level. Connect the negative battery cable. Run the engine to normal operating temperature and check for leaks.

1996 VEHICLES

1. Disconnect the negative battery cable.

2. Install engine support fixture J-28467-A or equivalent.

3. Remove the engine mount strut bolt.

4. Raise and safely support the vehicle.

✳✳ CAUTION

The EPA warns that prolonged contact with used engine oil may cause a number of skin disorders, including cancer! You should make every effort to minimize your exposure to used engine oil. Protective gloves should be worn when changing the oil. Wash your hands and any other exposed skin areas as soon as possible after exposure to used engine oil. Soap and water, or waterless hand cleaner should be used.

5. Drain the engine oil into a suitable container.

6. Remove the oil drip shield bolts and shield.

7. Remove the engine mount.

8. Remove the transaxle mount nuts.

9. Disconnect the exhaust pipe from the exhaust manifold.

10. Raise the engine using the special tools J-28467-A and J-36462 or equivalents.

11. Lower the vehicle.

12. Place supports under the frame at the front and rear center crossmembers.

13. Loosen the rear frame bolts, but do not remove.

14. Remove the front frame bolts and lower front of frame.

15. Remove the front engine mount bracket bolts, bracket and mount.

16. Disconnect the electrical leads from the starter and remove the starter.

17. Remove the brackets from the oil pan.

18. Remove the oil pan retaining bolts and remove the oil pan.

19. If required, remove the oil deflector.

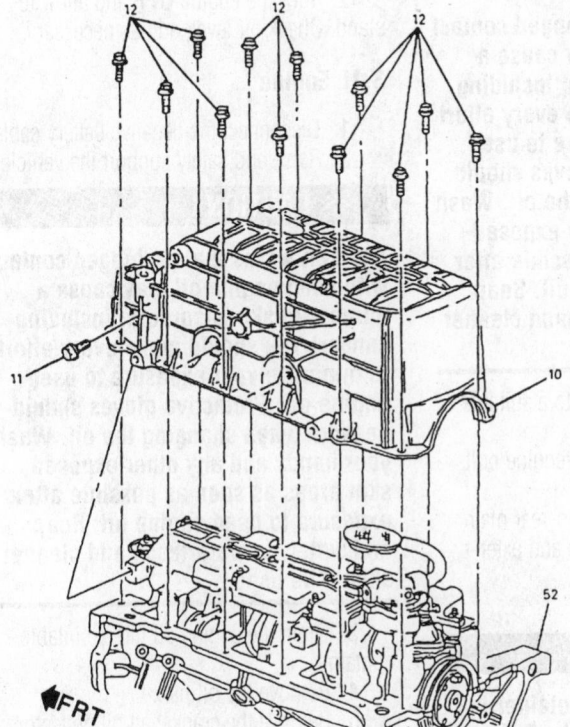

```
10   PAN, OIL
11   BOLT, OIL PAN SIDE
12   BOLT, OIL PAN RETAINING
52   BLOCK, ENGINE
```

7922SG31

Exploded view of the oil pan mounting—3.1L engine

➡Clean all gasket mating surfaces.

To install:

➡Apply a small amount of RTV sealer on either side of the rear main bearing cap, where the seal surface on the cap meets the cylinder block.

20. If removed, install the oil deflector and tighten the mounting nuts to 18 ft. lbs. (25 Nm).

21. Install the gasket and oil pan and hand-tighten retaining bolts.

22. After all bolts are hand tight, tighten to 18 ft. lbs. (25 Nm). Tighten the side bolts to 37 ft. lbs. (50 Nm).

23. Install the electrical brackets to the pan.

24. Install the starter and connect the electrical leads.

25. Install the engine mount bracket and mount.

26. Install the front engine mount bracket bolts.

27. Raise the frame to proper position. Using new bolts tighten to 76 ft. lbs. (103 Nm).

28. Remove the supports.

29. Raise and safely support the vehicle.

30. Lower the engine to the correct position.

31. Install the exhaust pipe to the manifold.

32. Install the transaxle mount nuts.

33. Install the engine mount.

34. Install the oil drip shield.

35. Lower the vehicle.

❋❋ WARNING

Operating the engine without the proper amount and type of engine oil will result in severe engine damage.

36. Fill the crankcase with new engine oil.

37. Install the engine mount strut bolt.

38. Remove the engine support fixture.

39. Connect the negative battery cable.

40. Start the vehicle and verify no leaks.

Oil Pump

REMOVAL & INSTALLATION

2.2L Engine

1. Disconnect the negative battery cable.

2. Raise and safely support the vehicle.

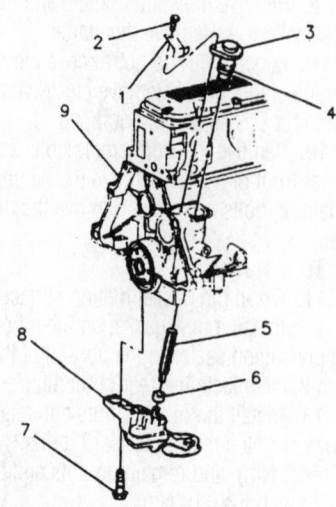

1 BRACKET
2 BOLT
3 OIL PUMP DRIVE ASSEMBLY
4 O–RING
5 SHAFT
6 RETAINER; HEAT AND WATER SOAK PRIOR TO INSTALLATION.
7 BOLT
8 OIL PUMP
9 CYLINDER BLOCK

7922SG13

Exploded view of the oil pump mounting—2.2L engine

❋❋ CAUTION

The EPA warns that prolonged contact with used engine oil may cause a number of skin disorders, including cancer! You should make every effort to minimize your exposure to used engine oil. Protective gloves should be worn when changing the oil. Wash your hands and any other exposed skin areas as soon as possible after exposure to used engine oil. Soap and water, or waterless hand cleaner should be used.

3. Drain the engine oil into a suitable container.

4. Remove the oil pan-to-engine bolts and the oil pan.

5. Remove the oil pump-to-rear main bearing cap bolt, the oil pump and extension shaft.

To install:

❋❋ WARNING

Heat the extension shaft retainer in hot water prior to assembly. Be sure the retainer does not crack upon installation.

6. Install the extension shaft, oil pump and pump-to-rear main cap bolt. Tighten the oil pump-to-bearing cap bolt to 32 ft. lbs. (43 Nm) and the upper oil pump drive bolt to 18 ft. lbs. (25 Nm).

❋❋ WARNING

To avoid engine damage, all oil pump cavities must be filled with petroleum jelly before installing the gears into the pump body. This seals the gears, acts like a prime and allows the pump to start drawing oil as soon as the engine begins to crank the first time after pump service. Also use only original equipment gaskets. Gasket thickness is critical to proper oil pump operation.

7. Install the oil pan and attaching bolts.

8. Lower the vehicle.

❋❋ WARNING

Operating the engine without the proper amount and type of engine oil will result in severe engine damage.

9. Fill the crankcase with clean engine oil.

10. Connect the negative battery cable.

11. Start the engine and check oil pressure and check for leaks.

12. Turn the engine **OFF** and allow to stand. Check oil level, add as necessary.

3.1L Engine

1. Disconnect the negative battery cable.

2. Raise and safely support the vehicle.

❋❋ CAUTION

The EPA warns that prolonged contact with used engine oil may cause a number of skin disorders, including cancer! You should make every effort to minimize your exposure to used engine oil. Protective gloves should be worn when changing the oil. Wash your hands and any other exposed skin areas as soon as possible after exposure to used engine oil. Soap and water, or waterless hand cleaner should be used.

3. Drain the engine oil into a suitable container.

4. Remove the oil pan.

5. Remove the crankshaft oil deflector bolts.

6. Remove the crankshaft oil deflector.

7. Remove the oil pump retaining bolts

14 BOLT/SCREW, OIL PUMP
15 PUMP, OIL
16 DRIVESHAFT, OIL PUMP
18 CAP, REAR CRANKSHAFT BEARING
19 PIN, OIL PUMP
52 BLOCK, ENGINE
240 BOLT/SCREW, OIL PUMP DRIVE
241 CLAMP, OIL PUMP DRIVE
242 SEAL, OIL PUMP DRIVE
243 DRIVE, OIL PUMP

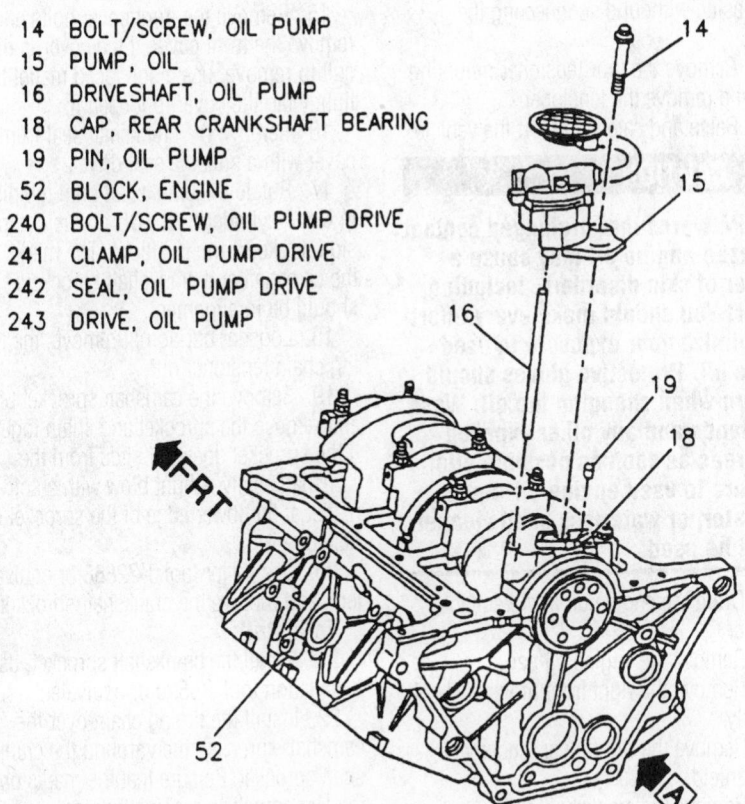
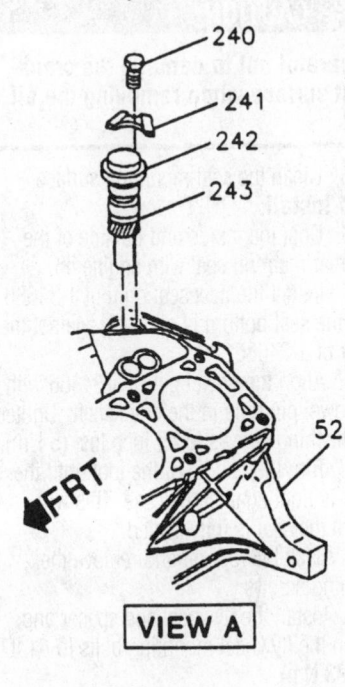

Exploded view of the oil pump and related components—3.1L engine

and remove the oil pump and pump drive-shaft.

To install:

8. Install the oil pump and pump drive-shaft. Tighten the oil pump mounting bolts to 30 ft. lbs. (41 Nm).

9. Install the crankshaft oil deflector and mounting bolts. Tighten the mounting bolts to 18 ft. lbs. (25 Nm).

10. Install the oil pan.

11. Lower the vehicle.

❊❊ WARNING

Operating the engine without the proper amount and type of engine oil will result in severe engine damage.

12. Fill the crankcase to the correct level with oil.

13. Start the engine, check the oil pressure and check for leaks.

Rear Main Seal

REMOVAL & INSTALLATION

2.2L Engine

The transaxle must be removed from the vehicle to perform this service.

1. Disconnect the negative battery cable.

2. Remove the transaxle using the recommended procedure.

3. Remove the flywheel mounting bolts and remove the flywheel and retainer.

4. Remove crankshaft seal by inserting a suitable prying tool in through the dust lip. Pry out seal moving tool around seal as required until is removed.

❊❊ CAUTION

Care must be taken not to damage crankshaft seal surface with prytool.

To install:

5. Coat the inside and outside of the new rear main oil seal with engine oil.

6. Install the new seal on tool J-34686 until the seal bottom is squarely against the collar of J-34686.

7. Align the dowel pin of J-34686 with the dowel pin hole in the crankshaft. Tighten the attaching screws to 45 inch lbs. (6 Nm).

8. Turn the handle of the tool until the collar is tight against the case. This will ensure the seal is fully seated.

9. Back the tool off and remove the attaching screws.

10. Install the flywheel and retainer and

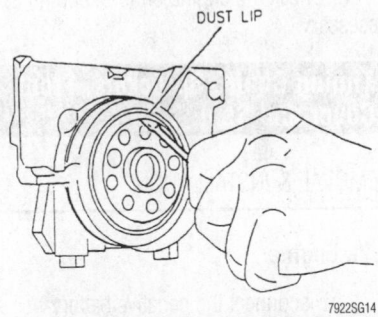

Carefully pry the oil seal from the rear of the block, do not damage the sealing surface of the crankshaft—all engines

tighten the flywheel mounting bolts to 55 ft. lbs. (75 Nm).

11. Install the transaxle assembly using the recommended procedure.

12. Connect the negative battery cable.

13. Check the engine oil level and fill as necessary.

3.1L Engine

The transaxle must be removed from the vehicle to perform this service.

1. Disconnect the negative battery cable.

2. Remove the transaxle using the recommended procedure.

3. Remove the flywheel mounting bolts and remove the flywheel and spacer.

4. Using a small prybar, pry the seal from the block.

❉❉ CAUTION

Be careful not to damage the crankshaft surface when removing the oil seal.

5. Clean the seal mounting surface.

To install:

6. Coat the inside and outside of the new rear main oil seal with engine oil.

7. Install the new seal on tool J-34686 until the seal bottom is squarely against the collar of J-34686.

8. Align the dowel pin of J-34686 with the dowel pin hole in the crankshaft. Tighten the attaching screws to 45 inch lbs. (5 Nm).

9. Turn the handle of the tool until the collar is tight against the case. This will ensure the seal is fully seated.

10. Back the tool off and remove the attaching screws.

11. Install the flywheel and spacer and tighten the flywheel mounting bolts to 61 ft. lbs. (83 Nm).

12. Install the transaxle assembly using the recommended procedure.

13. Connect the negative battery cable.

14. Check the engine oil level and fill as necessary.

Timing Chain, Sprockets, Front Cover and Seal

REMOVAL & INSTALLATION

2.2L Engine

1. Disconnect the negative battery cable.

❉❉ CAUTION

Never open, service or drain the radiator or cooling system when hot, serious burns can occur from the steam and hot coolant.

2. Drain the engine coolant into a suitable container.

3. Remove the serpentine belt.

4. Remove the coolant reservoir.

5. Remove the front alternator mounting bolts.

6. Remove the three mounting bolts from the power steering pump. These bolts can be reached by going through the holes in the drive pulley on the pump. Lay the pump aside without disconnecting the hoses.

7. Remove the four tensioner mounting bolts and remove the tensioner.

8. Raise and safely support the vehicle.

❉❉ CAUTION

The EPA warns that prolonged contact with used engine oil may cause a number of skin disorders, including cancer! You should make every effort to minimize your exposure to used engine oil. Protective gloves should be worn when changing the oil. Wash your hands and any other exposed skin areas as soon as possible after exposure to used engine oil. Soap and water, or waterless hand cleaner should be used.

9. Drain the engine oil into a suitable container.

10. Remove the engine oil pan.

11. Remove the right front tire and wheel assembly.

12. Remove the right inner fender well splash shield.

13. Remove the crankshaft pulley.

14. Remove the crankshaft balancer.

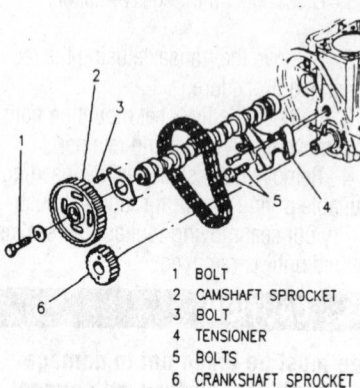

1 BOLT
2 CAMSHAFT SPROCKET
3 BOLT
4 TENSIONER
5 BOLTS
6 CRANKSHAFT SPROCKET

A ALIGN TABS ON TENSIONER WITH MARKS ON CAMSHAFT & CRANKSHAFT SPROCKETS.

7922SG15

Exploded view of the timing chain components and timing mark alignment—2.2L engine

15. Remove the front cover bolts and remove the front cover. If the cover is difficult to remove, use a soft faced mallet to lightly tap the cover to loosen it.

16. Remove the crankshaft seal from the cover with a suitable seal driver.

17. Rotate the crankshaft until the piston in No. 1 cylinder is at TDC on the compression stroke (firing position). The marks on the camshaft and crankshaft sprockets should be in alignment.

18. Loosen, but do not remove, the timing chain tensioner nut.

19. Remove the camshaft sprocket bolt and remove the sprocket and chain together. If the sprocket does not slide from the camshaft easily, a light blow with a soft mallet at the lower edge of the sprocket will dislodge it.

20. Use puller tool J-22888 or equivalent, and remove the crankshaft sprocket.

To install:

21. Install the crankshaft sprocket, using installation tool J 5590 or equivalent.

22. Install the timing chain over the camshaft sprocket, then around the crankshaft sprocket. Be sure that the marks on the two sprockets are in alignment. Lubricate the thrust surface with Molykote® or its equivalent.

23. Align the dowel in the camshaft with the dowel hole in the sprocket, then install the sprocket onto the camshaft. Use the mounting bolt to draw the sprocket onto the camshaft, then tighten to 77 ft. lbs. (105 Nm).

24. Lubricate the timing chain with clean engine oil. Tighten the bolts on the chain tensioner to 18 ft. lbs. (24 Nm).

25. Clean all gasket surfaces completely.

26. Install a new seal in the front cover using a suitable seal driver. Apply engine oil to the lip of the seal.

27. Apply a thin layer of sealer to the front cover and install a new gasket onto the cover.

28. Install the cover to the engine making sure the dowel pins line up with the holes in the front cover.

29. Install the cover mounting bolts and tighten them to 97 inch lbs. (11 Nm).

30. Install the crankshaft and pulley and tighten the pulley bolts to 37 ft. lbs. (50 Nm) and the center balancer bolt to 77 ft. lbs. (105 Nm).

31. Install the right side splash shield.

32. Install the tire and wheel assembly.

33. Install the oil pan.

34. Lower the vehicle.

35. Install the belt adjuster and tighten the bolts to 37 ft. lbs. (50 Nm).

36. Install the power steering pump and tighten the mounting bolts to 25 ft. lbs. (34 Nm).

37. Install the alternator mounting bolts and tighten the upper bolt to 22 ft. lbs. (30 Nm) and tighten the lower to 37 ft. lbs. (50 Nm).

38. Install the coolant reservoir.

39. Install the serpentine belt.

✳✳ WARNING

Operating the engine without the proper amount and type of engine oil will result in severe engine damage.

40. Fill the engine crankcase with clean oil.

41. Refill the coolant system.

42. Connect the negative battery cable.

43. Start the vehicle and verify no leaks.

44. Road test the vehicle and ensure proper operation.

3.1L Engine

1. Disconnect the negative battery cable.

✳✳ CAUTION

Never open, service or drain the radiator or cooling system when hot; serious burns can occur from the steam and hot coolant.

2. Drain the cooling system into a suitable container.

3. Remove the right engine mount bracket.

4. Remove the serpentine belt.

5. Remove the crankshaft balancer as follows:

   a. Raise and safely support the vehicle.

   b. Remove the right front tire and wheel assembly.

   c. Remove the right inner fender well splash shield.

   d. Remove the flywheel cover and install a flywheel holding tool.

   e. Remove the balancer mounting bolt and washer.

   f. Using a suitable puller, J-24420-B or the equivalent remove the balancer from the crankshaft.

6. Remove the serpentine belt tensioner mounting bolt and tensioner.

✳✳ CAUTION

The EPA warns that prolonged contact with used engine oil may cause a number of skin disorders, including cancer! You should make every effort to minimize your exposure to used engine oil. Protective gloves should be worn when changing the oil. Wash your hands and any other exposed skin areas as soon as possible after exposure to used engine oil. Soap and water, or waterless hand cleaner should be used.

7. Remove the oil pan following the recommended procedure.

8. Remove coolant bypass pipe from the water pump and the intake manifold.

9. Disconnect the lower radiator hose to from the front cover outlet.

10. Remove the front cover mounting bolts and remove the front cover.

11. Rotate the crankshaft until the timing marks on the camshaft and crankshaft sprockets are in alignment at their closest approach.

12. Remove the camshaft sprocket mounting bolt and remove the camshaft sprocket and timing chain.

13. Remove the crankshaft sprocket, using a gear puller J-5825-A or equivalent.

14. Remove the two mounting bolts from the timing chain damper and remove the damper.

To install:

15. Install the timing chain damper and tighten the mounting bolts to 15 ft. lbs. (21 Nm).

16. Install the crankshaft sprocket onto the crankshaft making sure the notch in the sprocket fits over the crankshaft key. Fully seat the sprocket on the crankshaft using J-38612, or an equivalent gear installer.

17. Be sure the timing mark on the crankshaft sprocket is pointing straight up.

18. Install the timing chain over the camshaft sprocket and hold the sprocket in such a way, that the timing mark is pointing down, and the timing chain is hanging down off the sprocket.

19. Loop the timing chain under the crankshaft sprocket and install the camshaft sprocket on the camshaft. The sprocket will only fit on the camshaft if, the dowel on the camshaft lines up with the hole in the sprocket.

20. Verify that the marks are aligned (the camshaft sprocket will be at the 6 o'clock position and the crankshaft sprocket will be in the 12 o'clock position).

21. On 1995 vehicles, tighten the camshaft sprocket mounting bolt to 74 ft. lbs. (100 Nm). On 1996 vehicles tighten the bolt to 81 ft. lbs. (110 Nm).

22. Lubricate the timing chain components with engine oil.

23. Clean all gasket surfaces completely.

24. Apply a thin bead of sealer around the gasket sealing area of the front cover. Install a new front cover seal on the front cover.

25. Install the front cover on the engine and install the mounting bolts. Tighten the small bolts to 18 ft. lbs. (24 Nm) and the

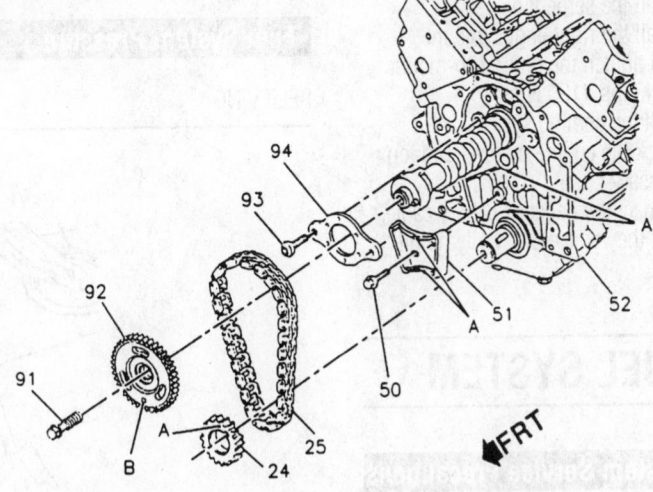

A	TIMING ALIGNMENT MARKS	52	BLOCK, ENGINE
B	LOCATOR HOLE	91	BOLT, CAMSHAFT SPROCKET
24	SPROCKET, CRANKSHAFT	92	SPROCKET, CAMSHAFT
25	CHAIN, TIMING	93	BOLT, THRUST PLATE
50	BOLT, TIMING CHAIN DAMPENER	94	PLATE, THRUST
51	DAMPENER, TIMING CHAIN		

Exploded view of the timing chain mounting—3.1L engine

7922SG16

large bolts to 41 ft. lbs. (55 Nm) for 1995 models, or tighten the small bolts to 15 ft. lbs. (21 Nm) and tighten the large bolts to 35 ft. lbs. (47 Nm) for 1996 models.

26. Connect the radiator hose to the coolant outlet.

27. Install coolant bypass pipe to the water pump and the intake manifold.

✳✳ WARNING

Operating the engine without the proper amount and type of engine oil will result in severe engine damage.

28. Install the oil pan following the recommended procedure.

29. Install crankshaft balancer as follows:

   a. Coat the seal contact surface of the crankshaft balancer with clean engine oil.

   b. Line up the notch in the balancer with the crankshaft key and slide the balancer on until the key is in the balancer.

   c. Using J-29113 or an equivalent puller, seat the balancer on the crankshaft.

   d. Install the balancer mounting bolt and washer and tighten to 76 ft. lbs. (103 Nm).

   e. Install the flywheel cover.

   f. Install the right inner fender well splash shield.

   g. Install the tire and wheel assembly and tighten to specification.

30. Install the serpentine belt tensioner and tighten the mounting bolt to 40 ft. lbs. (54 Nm).

31. Install the serpentine belt.

32. Install the right engine mount bracket and tighten the bracket-to-mount bolts to 96 ft. lbs. (130 Nm).

33. Refill the cooling system.

34. Check the engine oil level and top off as necessary.

35. Connect the negative battery cable.

36. Start the vehicle and verify no oil leaks.

FUEL SYSTEM

Fuel System Service Precautions

Safety is the most important factor when performing not only fuel system maintenance but any type of maintenance. Failure to conduct maintenance and repairs in a safe manner may result in serious personal injury or death. Maintenance and testing of the vehicle's fuel system components can be accomplished safely and effectively by adhering to the following rules and guidelines.

• To avoid the possibility of fire and personal injury, always disconnect the negative battery cable unless the repair or test procedure requires that battery voltage be applied.

• Always relieve the fuel system pressure prior to disconnecting any fuel system component (injector, fuel rail, pressure regulator, etc.), fitting or fuel line connection. Exercise extreme caution whenever relieving fuel system pressure to avoid exposing skin, face and eyes to fuel spray. Please be advised that fuel under pressure may penetrate the skin or any part of the body that it contacts.

• Always place a shop towel or cloth around the fitting or connection prior to loosening to absorb any excess fuel due to spillage. Ensure that all fuel spillage (should it occur) is quickly removed from engine surfaces. Ensure that all fuel soaked cloths or towels are deposited into a suitable waste container.

• Always keep a dry chemical (Class B) fire extinguisher near the work area.

• Do not allow fuel spray or fuel vapors to come into contact with a spark or open flame.

• Always use a back-up wrench when loosening and tightening fuel line connection fittings. This will prevent unnecessary stress and torsion to fuel line piping. Always follow the proper torque specifications.

• Always replace worn fuel fitting O-rings with new. Do not substitute fuel hose or equivalent, where fuel pipe is installed.

Fuel System Pressure

RELIEVING

✳✳ CAUTION

Observe all applicable safety precautions when working around fuel. Whenever servicing the fuel system, always work in a well ventilated area. Do not allow fuel spray or vapors to come in contact with a spark or open flame. Keep a dry chemical fire extinguisher near the work area. Always keep fuel in a container specifically designed for fuel storage; also, always properly seal fuel containers to avoid the possibility of fire or explosion.

1. Disconnect the negative battery cable to avoid possible fuel discharge if an accidental attempt is made to start the engine.

2. Loosen the fuel filler cap to relieve the tank pressure.

3. Connect a suitable fuel pressure gauge to the fuel pressure test fitting. Wrap a shop towel around the fitting while connecting gauge to avoid spillage.

4. Place the bleed hose in an approved container and open the valve on the pressure gauge to relieve system pressure.

5. Dispose of the discharged liquid fuel promptly.

Fuel Filter

REMOVAL & INSTALLATION

✳✳ CAUTION

The fuel injection system remains under pressure, even after the engine has been turned OFF. The fuel system

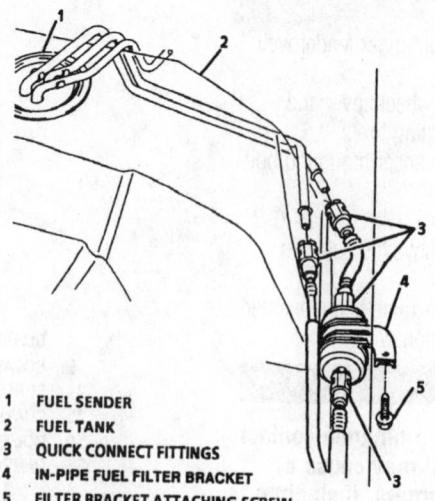

1	FUEL SENDER
2	FUEL TANK
3	QUICK CONNECT FITTINGS
4	IN-PIPE FUEL FILTER BRACKET
5	FILTER BRACKET ATTACHING SCREW

7922SG17

Typical fuel filter mounting

pressure must be relieved before disconnecting any fuel lines. Failure to do so may result in fire and/or personal injury.

1. Disconnect the negative battery cable.

✳✳ CAUTION

Observe all applicable safety precautions when working around fuel. Whenever servicing the fuel system, always work in a well ventilated area. Do not allow fuel spray or vapors to come in contact with a spark or open flame. Keep a dry chemical fire extinguisher near the work area. Always keep fuel in a container specifically designed for fuel storage; also, always properly seal fuel containers to avoid the possibility of fire or explosion.

2. Relieve the fuel system pressure.
3. Raise and safely support the vehicle.
4. Place a pan under the fuel filter.
5. Rotate each of the fuel filter quick-Connects ¼ turn to loosen any dirt trapped under the fittings.
6. Squeeze the plastic tabs on the male ends of the quick-Connects and disconnect the fuel lines from the filter.
7. Remove the filter bracket mounting bolt.
8. Remove the filter from the vehicle.

To install:
9. Install the filter assembly in position and install the filter bracket mounting bolt.
10. Apply a few drops of clean engine oil to the male ends of the fuel filter assembly and fuel line.
11. Snap the fuel lines onto the fuel filter. The lock tabs on the quick-Connects will lock in place. Give each line a light tug to be sure the fittings are tight.
12. Lower the vehicle.
13. Connect the negative battery cable.
14. Pressurize the fuel system and verify no leaks.

Fuel Pump

REMOVAL & INSTALLATION

The fuel pump is located in the fuel tank. The fuel tank must be removed from the vehicle for this procedure. Note that the fuel pump has a strainer attached to reduce sediment and debris that might enter the pump. These strainers may become clogged on high-mileage vehicles or where the vehicles has had contamination in the tank. A clogged strainer can keep the pump from drawing suf-

ficient fuel to maintain enough fuel pressure to keep the engine running smoothly. Keep this in mind when troubleshooting a driveability and/or low fuel pressure complaint. This strainer should be replaced whenever the tank is removed for any kind of service.

✳✳ CAUTION

The fuel injection system remains under pressure, even after the engine has been turned OFF. The fuel system pressure must be relieved before disconnecting any fuel lines. Failure to do so may result in fire and/or personal injury.

1. Properly relieve the fuel system pressure.
2. Disconnect the negative battery cable.

✳✳ CAUTION

Observe all applicable safety precautions when working around fuel. Do not allow fuel spray or fuel vapors to come into contact with a spark or open flame. Keep a dry chemical (Class B) fire extinguisher near the work area. Never drain or store fuel in an open container due to the possibility of fire or explosion.

3. Remove the fuel tank from the vehicle using the recommended procedure and place in a suitable work area.
4. Remove the fuel sender assembly locking cam using a spanner wrench, J-24187, or an equivalent.
5. Remove the fuel sender assembly from the fuel tank.
6. Place the sender assembly on a work bench with the fuel strainer facing up.
7. Note the direction the fuel strainer is facing and remove the strainer from the fuel pump by pulling it off the pump.
8. Disconnect the electrical connectors from the fuel pump.
9. Push the fuel pump toward the fuel outlet hose to unseat the pump from the lower mounting bracket. Once the pump is clear of the mounting bracket, tilt the bottom of the pump outward away from the mounting bracket.
10. Remove the plastic clamp on the outlet line and disconnect the fuel pump from the outlet hose.

To install:
11. Connect the new pump to the outlet hose and install the plastic clamp.
12. Install the rubber insulator on the fuel pump.
13. Push the pump toward the outlet hose until the bottom of the pump can be seated in the mounting bracket.

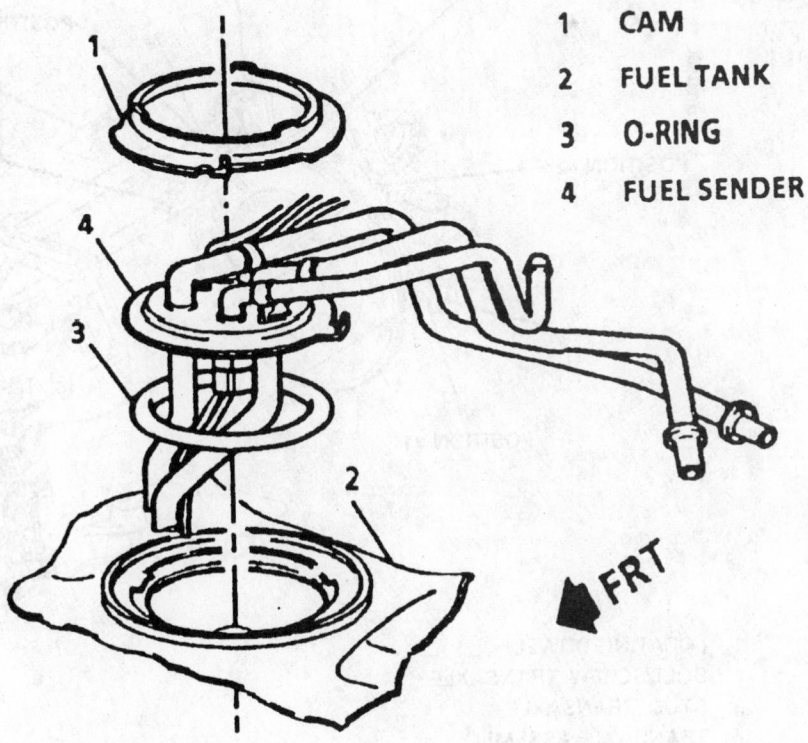

1	CAM
2	FUEL TANK
3	O-RING
4	FUEL SENDER

7922SG18

Fuel pump and sender assembly is located in the fuel tank

14. Connect the electrical connectors to the fuel pump.

15. Install a new fuel strainer on the pump making sure the direction the strainer is facing is the same as on the old pump.

16. Install the fuel sender into the fuel tank using a new O-ring on top of the tank. It may help to lubricate the O-ring with a silicone grease.

17. Install the locking cam and rotate into the locked position with a spanner wrench, J-24187, or an equivalent.

18. Install the fuel tank in the vehicle. Refill with gasoline. Never attempt to test run a fuel pump dry, even for a short period of time. It is made to be immersed in fuel for both lubrication and cooling.

19. Connect the negative battery cable.

20. Pressurize the fuel system and verify no fuel leaks. This can be done by rotating the ignition switch to the **ON** position without starting the vehicle.

DRIVE TRAIN

➡**CV-joint boot service can be found in the Unit Repair Section of this manual.**

Transaxle Assembly

REMOVAL & INSTALLATION

1. Disconnect the negative battery cable.
2. Remove the air cleaner assembly.
3. Remove the engine torque strut from the engine.
4. Remove the shift control cable bracket from the transaxle case and lever.
5. Disengage the electrical connector from the torque converter clutch solenoid.
6. Disconnect the Throttle Valve (TV) cable from the throttle bracket and transaxle.
7. Remove the upper transaxle bolts, including the grounds.
8. Install engine support fixture J-28467-A or equivalent.
9. Raise and safely support the vehicle.
10. Remove the front tire and wheel assemblies.
11. Remove the engine splash shields.
12. Remove the pinch bolts from the control arms.
13. Remove the pinch bolt from the intermediate steering shaft.
14. Remove the stabilizer shaft bolts and reinforcement plates from the frame.
15. Remove the stabilizer shaft nuts and

bracket from the control arm and separate the stabilizer shaft from the control arm.

16. Using a 7/16 in. drill bit, drill through two spot welds located between the front and rear holes of the left front stabilizer shaft mounting.
17. Remove the front and rear transaxle mounting nuts.
18. Disconnect the engine wiring harness from the frame.
19. Remove the power steering cooler line bolts.
20. Remove the right frame to left frame retaining bolt. Position a jackstand under the frame for support.
21. Loosen the two right frame mounts and discard the bolts.
22. Remove the two left frame bolts from the frame.
23. Remove the left frame with the aid of an assistant.
24. Disconnect the right lower ball joint from the steering knuckle.
25. Remove the transaxle support bracket bolts from the transaxle.
26. Remove the power steering cooler line support from the transaxle.
27. Remove the torque converter cover.
28. Remove the starter.
29. Remove the torque converter bolts.

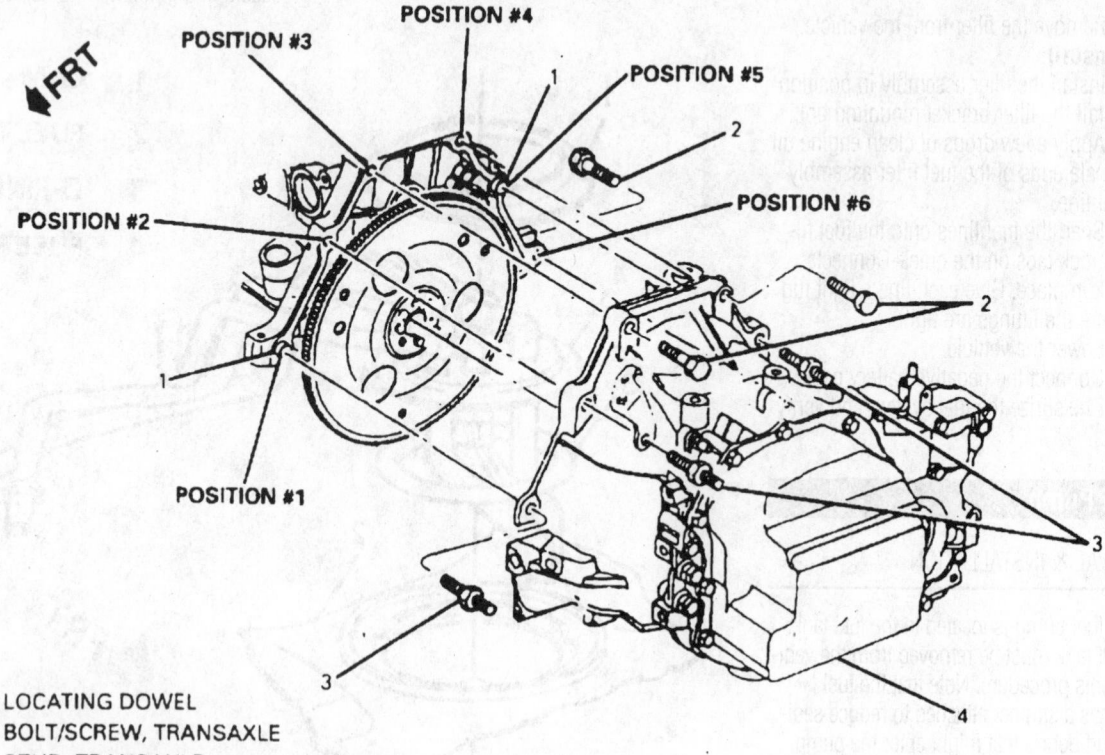

1 LOCATING DOWEL
2 BOLT/SCREW, TRANSAXLE
3 STUD, TRANSAXLE
4 TRANSAXLE ASSEMBLY

Transaxle-to-engine mounting bolt locations

7922SG19

30. Remove the transaxle mount bolts from the transaxle case and remove the mount.

31. Disconnect the transaxle range switch electrical connectors.

32. Remove both drive axles from the transaxle.

33. Disconnect the vehicle speed sensor from the transaxle.

34. Position transaxle jack under the transaxle.

35. Remove the oil cooler lines and plug openings.

36. Remove the remaining transaxle bolts.

37. Remove the transaxle.

To install:

38. Install the transaxle into the vehicle and install the transaxle bolts. Tighten to 55 ft. lbs. (75 Nm).

39. Install the oil cooler lines.

40. Remove the transaxle jack.

41. Connect the vehicle speed sensor connector to the transaxle.

42. Install the drive axles.

43. Connect the transaxle range switch electrical connector.

44. Install the transaxle mount and mounting bolts to the transaxle case.

45. Install the torque converter bolts and tighten them to 47 ft. lbs. (63 Nm).

46. Install the starter motor.

47. Install the torque converter cover.

48. Install the power steering cooler line support to the transaxle.

49. Install the transaxle brace and the bolts to the transaxle.

50. Install the pinch bolts to the control arms.

51. Install the left frame (with the aid of an assistant).

52. Tighten the frame bolts to 40 ft. lbs. (54 Nm).

53. Install the right frame bolts to the body and tighten them to 40 ft. lbs. (54 Nm).

54. Remove the jackstand.

55. Install the right frame to the left frame retaining bolts, and tighten to 40 ft. lbs. (54 Nm).

56. Install the engine wiring harness to the frame.

57. Install the power steering cooler lines.

58. Install the transaxle rear mount nuts to frame and tighten to 39 ft. lbs. (53 Nm).

59. Install the left stabilizer shaft to the control arm.

60. Install the stabilizer shaft bolts and reinforcement plates to frame using support.

61. Install the steering shaft pinch bolt and tighten to 35 ft. lbs. (48 Nm).

62. Install the engine splash shields.

63. Install the front tire and wheel assemblies.

64. Lower the vehicle.

65. Remove the engine support fixture.

66. Install the three upper transaxle bolts and the ground wires to the engine and tighten to 55 ft. lbs. (75 Nm).

67. Connect the TV cable to the transaxle and throttle bracket.

68. Connect the torque converter clutch switch electrical connector.

69. Install the transaxle shift cable, mounting bracket and lever.

70. Install the engine torque strut.

71. Install the air cleaner assembly.

72. Connect the negative battery cable.

73. Start the vehicle and check transaxle oil level.

➡ **Whenever the vehicle subframe is removed or lowered, the wheel alignment should be checked.**

74. Check the front end alignment and adjust as required.

75. Road test the vehicle and verify proper operation.

Halfshaft

REMOVAL & INSTALLATION

1. Raise and safely support the vehicle.
2. Remove the tire and wheel assembly.

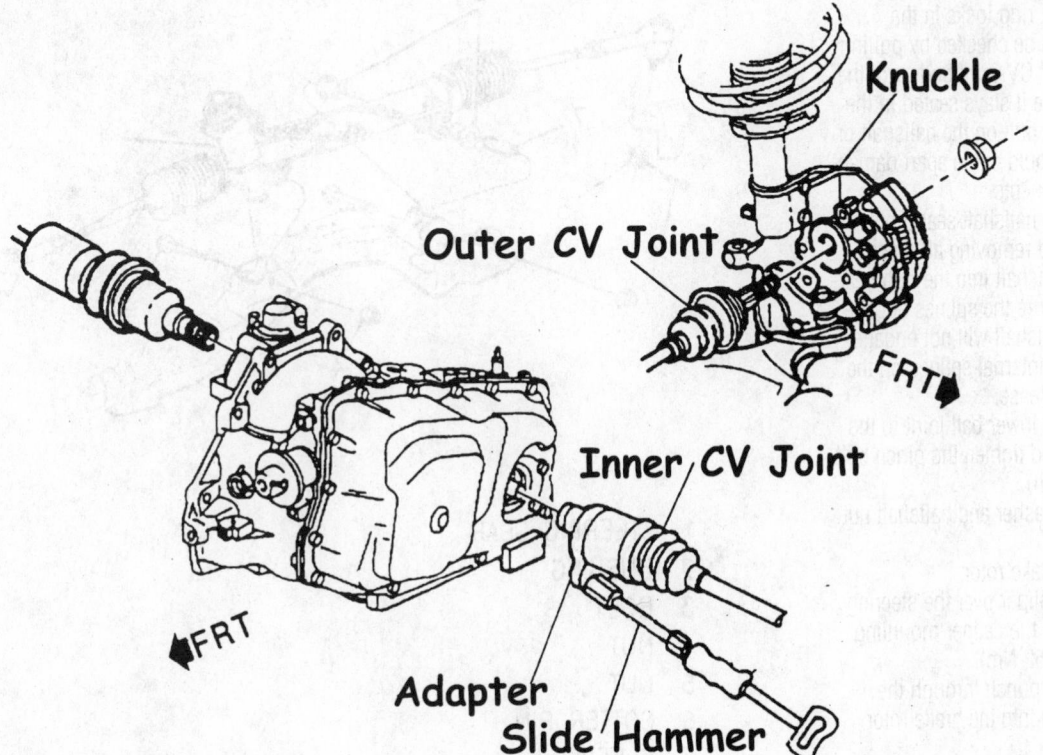

A slide hammer with a special adapter is used to remove the halfshaft from the transaxle

7922SG20

3. Because the hub nut is torqued so tightly, the halfshaft must be kept from turning while the hub nut is being loosened. Insert a drift punch through the opening in the top of the caliper and down into the brake rotor cooling fins to keep the assembly from turning.

4. Remove the halfshaft nut and washer.

5. Remove the drift punch.

6. Remove the caliper mounting bolts and remove the caliper from the steering knuckle and support out of the way. Hang the caliper on a stiff piece of wire. DO NOT allow the brake hose to support the caliper.

7. Remove the brake rotor.

8. Remove the lower ball joint pinch bolt and separate the ball joint from the steering knuckle.

9. Separate the halfshaft from the hub bearing using J-28733-B, or an equivalent puller.

10. Pull the steering knuckle outward to slip the halfshaft out of the hub bearing.

11. Disconnect the halfshaft from the transaxle using J-33008, J-29794 and J-2619–01 or their equivalents.

To install:

12. If installing the right side halfshaft install J-37292-B, or an equivalent halfshaft seal protector over the end of the CV-joint. The handle of the tool should be between 5 and 7 o'clock for ease of removal.

13. Seat the halfshaft into the transaxle until the inner lock ring locks in the transaxle. This can be checked by pulling lightly on the inner CV-joint body, not the halfshaft, to be sure it stays seated in the transaxle. DO NOT pull on the halfshaft or the inboard joint could come apart damaging internal components.

14. Remove the halfshaft seal protector by cutting it off and removing all pieces.

15. Seat the halfshaft into the hub bearing assembly. Be sure the splines engage smoothly. If the halfshaft will not engage smoothly, coat the internal splines on the hub bearing with grease.

16. Connect the lower ball joint to the steering knuckle and tighten the pinch bolt to 38 ft. lbs. (52 Nm).

17. Install the washer and halfshaft nut loosely.

18. Install the brake rotor.

19. Install the caliper over the steering knuckle and tighten the caliper mounting bolts to 38 ft. lbs. (52 Nm).

20. Insert a drift punch through the caliper opening and into the brake rotor cooling fins.

21. Tighten the halfshaft nut to 103 ft. lbs. (140 Nm) plus 20 degrees additional rotation.

22. Remove the drift.

23. Install the tire and wheel assembly and tighten the wheel nuts to 100 ft. lbs. (140 Nm).

24. Lower the vehicle.

STEERING AND SUSPENSION

Power Rack and Pinion Steering Gear

REMOVAL & INSTALLATION

With 2.2L Engine

1. Disconnect the negative battery cable.

2. Install an engine support fixture J-28467-A, or the equivalent.

3. Raise and safely support the vehicle.

4. Remove the tire and wheel assemblies.

5. Remove the intermediate shaft lower pinch bolt.

6. Disconnect the intermediate shaft from stub shaft.

✳✳ CAUTION

Failure to disconnect the intermediate shaft from the rack and pinion stub shaft can result in damage to the steering gear and/or intermediate shaft. This damage can cause loss of steering control which could result in personal injury.

7. On the 2.2L engine, remove the exhaust pipe hanger bracket near rear of frame including the brake line retainer and rubber exhaust pipe hangers.

8. Remove the engine and transaxle mount nuts at the subframe.

9. Support the rear of the subframe with jack stands.

10. Remove the rear subframe bolts and discard.

11. Lower the subframe for access to rack and pinion.

➡**Do not lower rear of cradle too far as damage to engine components nearest to the cowl may result.**

12. Remove the cotter pins and castle nuts from the outer tie rod ends and separate the tie rod ends from the steering knuckles using a suitable puller.

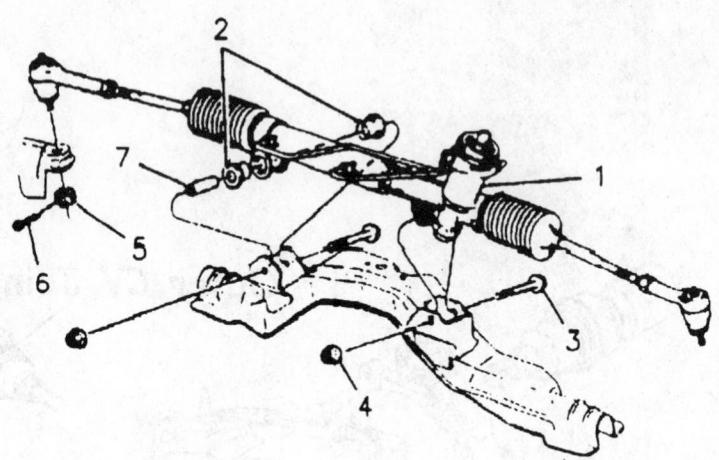

1	STEERING GEAR
2	BUSHING
3	BOLT
4	NUT
5	NUT
6	COTTER PIN
7	SLEEVE

Rack and pinion steering gear mounting

7922SG32

13. Disconnect the power steering lines from the steering gear.

14. Remove the rack and pinion mounting bolts and nuts.

15. Remove the rack and pinion through left wheel opening.

To install:

16. Install the rack and pinion through left wheel opening.

17. Install the rack and pinion mounting bolts and nuts and tighten to 66 ft. lbs. (90 Nm).

18. Connect the power steering lines to the rack using new O-rings and tighten the fittings to 13 ft. lbs. (17 Nm) on the return line and 21 ft. lbs. (28 Nm). on the pressure line.

19. Install the power steering line clip to steering gear.

20. Connect the tie rod ends to the steering knuckles. Tighten the castle nuts to 31 ft. lbs. (42 Nm). Install new cotter pins.

21. Raise the subframe into position and install new frame mounting bolts. Tighten the bolts to 103 ft. lbs. (140 Nm).

22. Remove the jack.

23. Install the engine mount and transaxle mount nuts and tighten to 39 ft. lbs. (53 Nm).

24. On the 2.2L engine, install the exhaust pipe hanger bracket near the rear of frame including brake line retainer and rubber exhaust pipe hanger.

25. Connect the intermediate shaft to the stub shaft and tighten the pinch bolt to 29 ft. lbs. (40 Nm).

26. Install the tire and wheel assemblies.

27. Lower the vehicle.

28. Remove the support fixture.

29. Connect the negative battery cable.

30. Refill the power steering reservoir and bleed the system.

→**Whenever the vehicle subframe is removed or lowered, the wheel alignment should be checked.**

31. Check the front end alignment and adjust as necessary.

With 3.1L Engine

1. Disconnect the negative battery cable.

2. Remove engine torque strut from engine.

3. Install an engine support fixture J-28467-A, or the equivalent.

4. Raise and safely support the vehicle.

5. Remove the tire and wheel assemblies.

6. Remove the cotter pins and castle nuts from the outer tie rod ends and sepa-

rate the tie rod ends from the steering knuckles using a suitable puller.

7. Remove the center engine and rear transaxle mounts from the frame.

8. Remove the intermediate shaft lower pinch bolt.

9. Disconnect the intermediate shaft from stub shaft.

✳✳ CAUTION

Failure to disconnect the intermediate shaft from the rack and pinion stub shaft can result in damage to the steering gear and/or intermediate shaft. This damage can cause loss of steering control which could result in personal injury.

10. Remove the brace bolts and brace, including brake line brace.

11. Support the rear of the subframe with jack stands.

12. Remove the rear subframe bolts and discard.

13. Lower the subframe for access to rack and pinion.

14. Remove the steering gear heat shield.

15. Remove the clip holding lines at rack assembly.

16. Disconnect the power steering lines and O-rings.

17. Remove the rack and pinion mounting bolts and nuts.

18. Remove the rack and pinion unit through the left wheel well.

To install:

19. Install the rack and pinion through left wheel opening.

20. Install the rack and pinion mounting bolts and nuts and tighten the mounting bolts to 66 ft. lbs. (90 Nm).

21. Connect the power steering lines to the rack using new O-rings and tighten the fittings to 13 ft. lbs. (17 Nm) on the return line and 21 ft. lbs. (28 Nm). on the pressure line.

22. Install the power steering line clip to steering gear.

23. Install the steering gear heat shield.

24. Raise the subframe into position and install new frame mounting bolts. Tighten the bolts to 103 ft. lbs. (140 Nm).

25. Remove the jackstands.

26. Install the engine mount and transaxle mount nuts and tighten to 39 ft. lbs. (53 Nm).

27. Install the brace and bolts including brake line brace.

28. Connect the intermediate shaft to the

stub shaft and tighten the pinch bolt to 29 ft. lbs. (40 Nm).

29. Connect the tie rod ends to the steering knuckles. Tighten the castle nuts to 31 ft. lbs. (42 Nm). Install new cotter pins.

30. Install the tire and wheel assemblies.

31. Lower the vehicle.

32. Remove the support fixture.

33. Connect the negative battery cable.

34. Refill the power steering reservoir and bleed the system.

→**Whenever the vehicle subframe is removed or lowered, the wheel alignment should be checked.**

35. Check the front end alignment and adjust as necessary.

Strut and Spring

REMOVAL & INSTALLATION

Front

1. Raise and safely support the vehicle.

2. Remove the tire and wheel assembly.

3. Remove the bolt securing the brake hose bracket to the strut and position the hose out of the way.

4. Matchmark the strut to the steering knuckle to preserve the camber setting when the strut is installed.

5. Remove the lower strut mounting nuts and through-bolts.

6. Remove the upper mounting nuts and remove the strut from the vehicle.

→**If the spring is to be removed, make a note of the orientation of the upper spring seat in relation to the strut assembly. The seat must be installed in the same position.**

7. To remove the spring from the strut, compress the spring using a good quality spring compressor until the spring is approximately ½ in. (13mm) away from the seat.

8. Hold the center shaft with a socket and remove the shaft nut with a line wrench or crows foot.

9. Remove the bearing assembly, upper spring seat, bumper, shield, upper spring insulator and spring from the strut.

To install:

10. If the spring was removed, position the spring on the lower spring seat. Be sure the end of the spring is against the notch in the lower seat.

11. Install the upper spring insulator, shield, bumper, upper spring seat and bearing assembly.

1. Front wheel drive shaft kit
2. Front suspension strut mount washer
3. Front suspension strut mount nut
4. Prevailing torque nut (M 12 x 1.75)
5. Front suspension strut bolt
6. Front suspension strut nut
7. Front suspension strut mount
8. Front spring seat
9. Front suspension strut bumper
10. Front suspension strut shield
11. Front spring upper insulator
12. Front spring

29. Front wheel drive shaft kit
30. Front lower control arm reinforcement
31. Front lower control arm reinforcement bolt
32. Washer flat (M 12.2 x 24 x 3.38)
33. Front lower control arm bolt
34. Front lower control arm
35. Front lower control arm ball stud kit

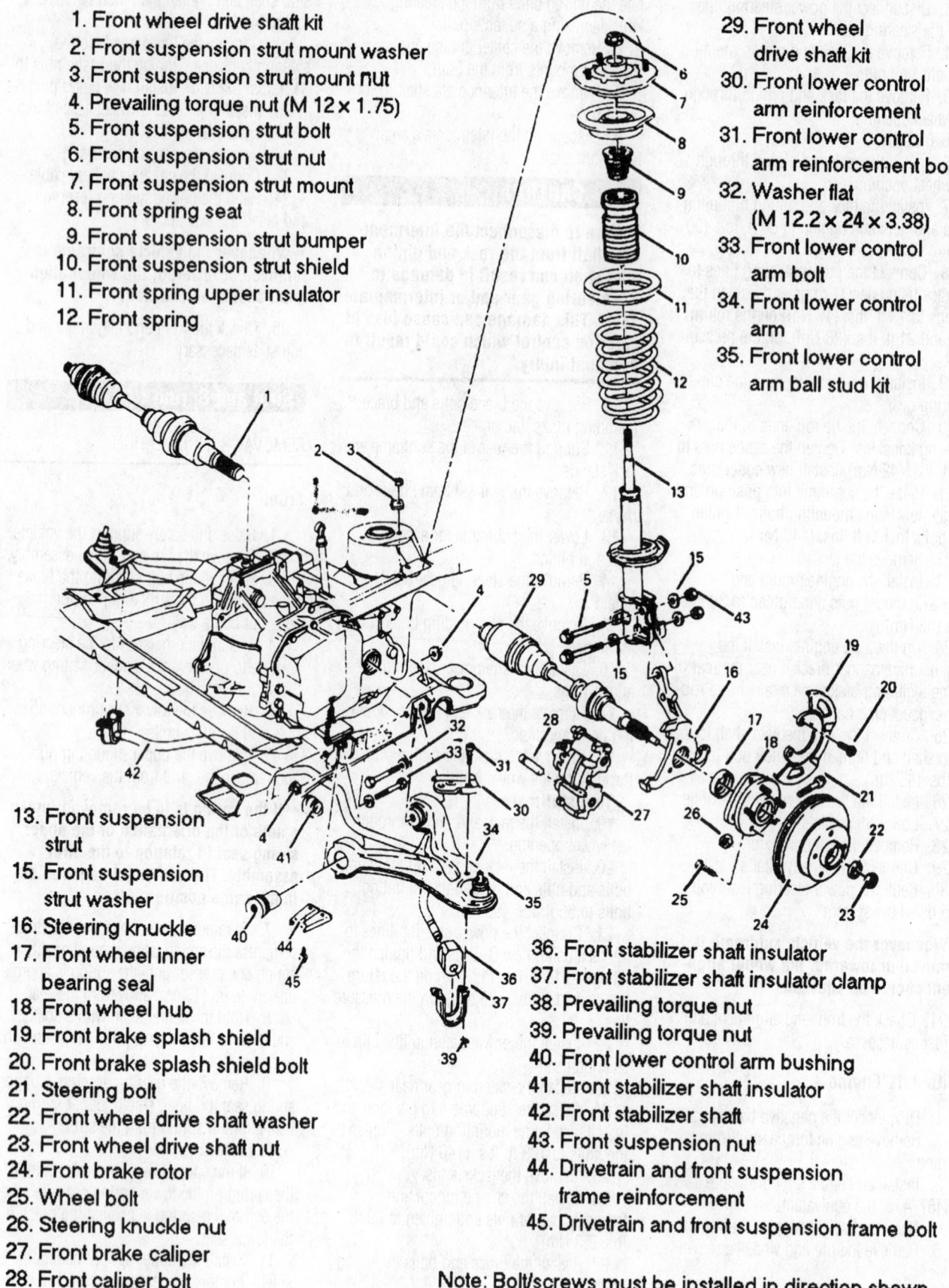

13. Front suspension strut
15. Front suspension strut washer
16. Steering knuckle
17. Front wheel inner bearing seal
18. Front wheel hub
19. Front brake splash shield
20. Front brake splash shield bolt
21. Steering bolt
22. Front wheel drive shaft washer
23. Front wheel drive shaft nut
24. Front brake rotor
25. Wheel bolt
26. Steering knuckle nut
27. Front brake caliper
28. Front caliper bolt

36. Front stabilizer shaft insulator
37. Front stabilizer shaft insulator clamp
38. Prevailing torque nut
39. Prevailing torque nut
40. Front lower control arm bushing
41. Front stabilizer shaft insulator
42. Front stabilizer shaft
43. Front suspension nut
44. Drivetrain and front suspension frame reinforcement
45. Drivetrain and front suspension frame bolt

Note: Bolt/screws must be installed in direction shown

Exploded view of the front suspension

7922SG33

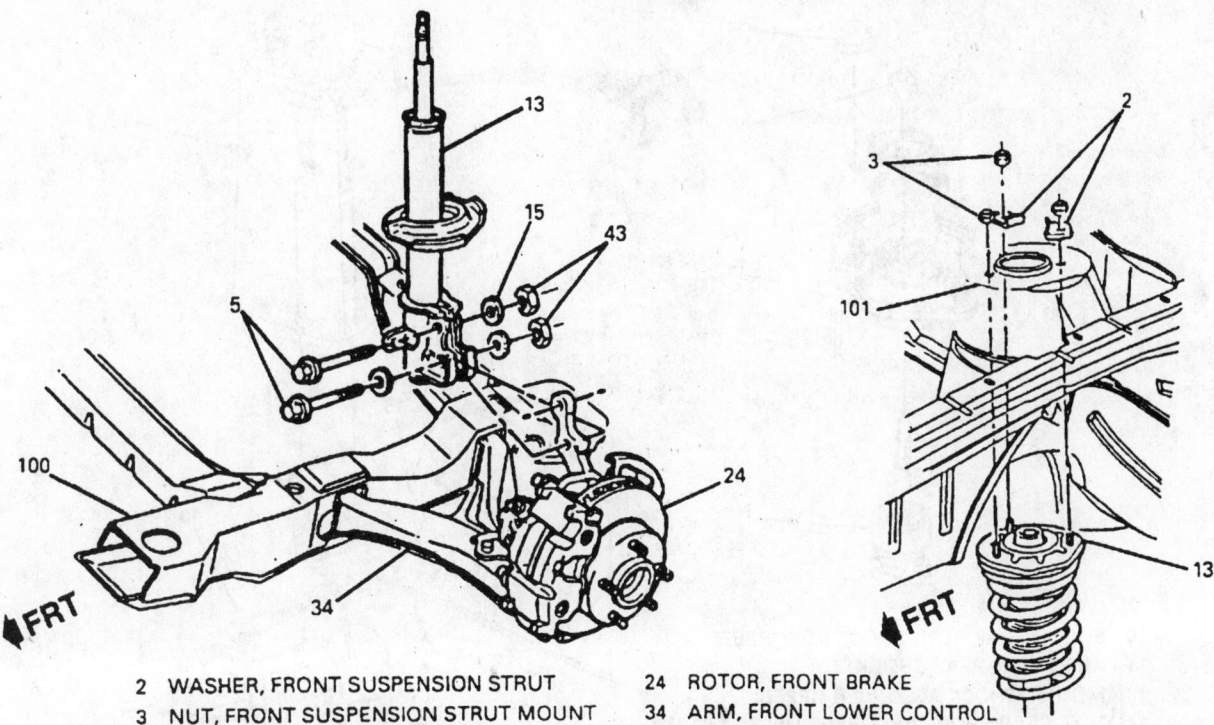

2	WASHER, FRONT SUSPENSION STRUT	24	ROTOR, FRONT BRAKE
3	NUT, FRONT SUSPENSION STRUT MOUNT	34	ARM, FRONT LOWER CONTROL
5	BOLT, FRONT SUSPENSION STRUT	100	FRAME ASSEMBLY
13	STRUT, FRONT SUSPENSION	101	SHOCK TOWER
15	WASHER, FRONT SUSPENSION STRUT	43	NUT, FRONT SUSPENSION STRUT

7922SG21

Exploded view of the upper and lower strut mounting

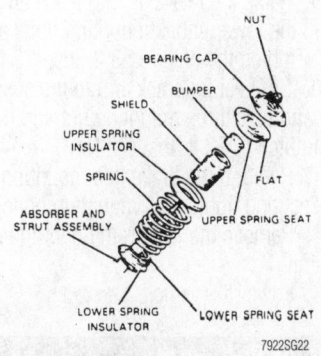

7922SG22

Exploded view of the coil spring mounting and related components

12. Install the shaft nut on the strut shaft. Tighten the nut to 80 ft. lbs. (108 Nm).

13. Install the strut into the vehicle so the upper plate studs pass through the body. Install the mounting nuts loosely.

14. Install the lower strut bracket through-bolts. Install the mounting nuts and with the matchmarks in alignment tighten the nuts to 122 ft. lbs. (165 Nm).

15. Connect the brake line bracket to the strut and tighten the mounting bolt to 15 ft. lbs. (21 Nm).

16. Install the tire and wheel assembly.

17. Lower the vehicle.

18. Tighten the upper strut plate mounting nuts to 18 ft. lbs. (25 Nm).

19. Check the front end alignment and adjust as necessary.

Shock Absorber

REMOVAL & INSTALLATION

Rear

➡**If the vehicle is equipped with Super Lift® shocks, bleed the air out of the system before disconnecting the air lines.**

1. Raise and safely support the vehicle at a height where the upper shock plate mounting nuts in the trunk can be reached.

2. Support the rear axle with a jack.

3. Pull the trunk side trim away from the rear shock tower.

4. Remove the cap from the top of the shock and remove the two shock plate mounting nuts.

5. Disconnect the air line from the shock absorber, if equipped with Super Lift® shocks.

6. Remove the lower shock through-bolt and nut and remove the shock assembly from the vehicle.

7. Remove the center shock nut and washer and remove the shock plate from the shock.

To install:

8. Install the shock plate on the new shock absorber and install the washer and mounting nut. Tighten the nut to 21 ft. lbs. (28 Nm).

9. Position the shock absorber in the vehicle so the upper plate studs line up with the holes in the shock tower. Loosely install the mounting nuts.

10. Install the lower shock through-bolt and nut and tighten to 50 ft. lbs. (72 Nm).

11. Remove the jack under the suspension and lower the vehicle.

12. Tighten the shock plate mounting nuts to 18 ft. lbs. (25 Nm).

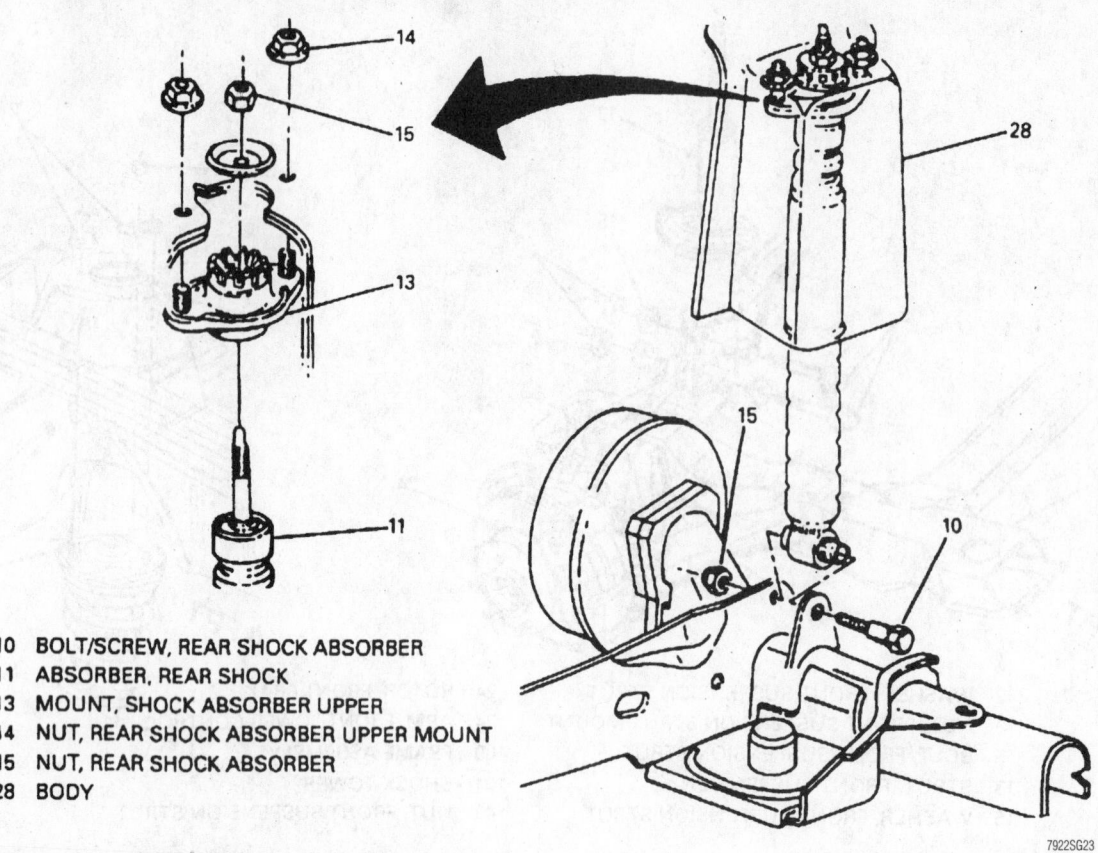

10 BOLT/SCREW, REAR SHOCK ABSORBER
11 ABSORBER, REAR SHOCK
13 MOUNT, SHOCK ABSORBER UPPER
14 NUT, REAR SHOCK ABSORBER UPPER MOUNT
15 NUT, REAR SHOCK ABSORBER
28 BODY

7922SG23

Exploded view of the rear shock absorber mounting and related components

13. Install the cap on the top of the shock and position the trunk side trim in place.

Coil Spring

REMOVAL & INSTALLATION

Front

Refer to strut removal and installation for the front coil spring procedure.

Rear

1. Raise and safely support the vehicle.
2. Support the rear axle with a suitable jack.
3. Remove the left and right side brake line bracket attaching screws and allow the brake lines to hang freely.
4. Remove the lower track arm bolt from the suspension and position the bar out of the way.
5. Remove the lower shock absorber mounting bolts.
6. Lower the rear axle.
7. Remove the coil springs and insulators.

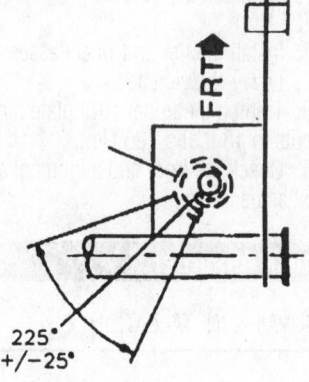

POSITION RIGHT HAND AND LEFT HAND REAR SPRINGS SUCH THAT THE LOWER PIGTAIL (END OF SPRING) IS POINTING TOWARD THE REAR OF VEHICLE.

7922SG24

Proper alignment of the rear springs

To install:

8. Install the springs and insulators in the vehicle and orient the springs so the lower spring ends point toward the rear of the vehicle.

9. Raise the rear axle into place and install the lower shock mounting bolts and nuts. Tighten the nuts to 53 ft. lbs. (72 Nm).
10. Connect the track arm to the suspension and install the mounting nut and bolt and tighten to 53 ft. lbs. (72 Nm).
11. Position the brake line mounting brackets and tighten the mounting bolts.
12. Remove the jack from under the suspension.
13. Lower the vehicle.

Lower Ball Joint

REMOVAL & INSTALLATION

1. Raise and safely support the vehicle under the frame so the control arm is unsupported.
2. Remove the tire and wheel assembly.
3. Remove the nut from the ball joint pinch bolt nut and remove the pinch bolt.
4. Loosen the stabilizer shaft at the lower control arm.
5. Pull the control arm down to disengage the ball joint stud from the steering knuckle.
6. Drill a pilot hole into each of the three ball joint rivets using a ⅛ in. drill bit.

7. Using a ½ in. drill bit, drill off the rivet heads.

8. With a hammer and punch, remove the remaining part of the rivets.

9. Remove the ball joint from the control arm.

To install:

10. Install the ball joint in the control arm and install the mounting nuts and bolts provided with the service kit. Tighten the nuts to the specifications in the service kit.

11. Connect the ball joint to the steering knuckle and install a new pinch bolt and nut. Tighten the nut to 38 ft. lbs. (52 Nm). If the replacement joint is supplied with a grease fitting, lubricate the ball joint with quality chassis grease.

12. Tighten the stabilizer shaft bracket bolts to 32 ft. lbs. (43 Nm).

13. Install the tire and wheel assembly and tighten the wheel nuts to 100 ft. lbs. (140 Nm).

14. Lower the vehicle.

Wheel Bearings

ADJUSTMENT

The front and rear wheel bearings on these vehicles are not adjustable. The wheel bearings are pressed into the hub assembly. If the wheel bearings are noisy or defective, the hub and bearing assembly must be replaced.

REMOVAL & INSTALLATION

Front

The front wheel bearings on these vehicles are not serviced separately. The wheel bearings are pressed into the hub. If the wheel bearings are noisy or defective, the hub and bearing assembly must be replaced.

1. Disconnect the negative battery cable.

2. Raise and safely support the vehicle.

3. Remove the front tire and wheel assembly.

4. Matchmark the lower strut bracket to the steering knuckle.

5. To keep the hub and rotor from turning when removing the axle nut, insert a drift punch through the caliper and into the brake rotor cooling fins to lock the assembly.

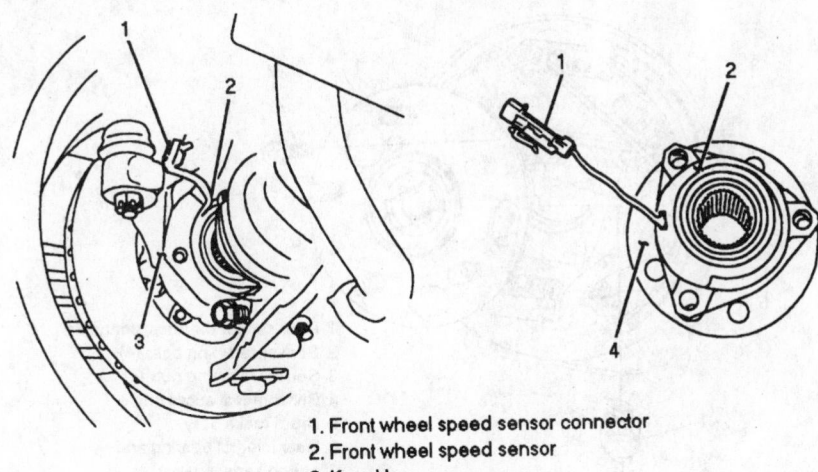

1. Front wheel speed sensor connector
2. Front wheel speed sensor
3. Knuckle
4. Hub and bearing assy

7922SG27

Don't forget to disconnect the ABS sensor harness when removing the front hub assembly

6. Remove the axle nut and washer. Remove the drift.

7. Remove the caliper mounting bolts and remove the caliper from the steering knuckle and support it out of the way. DO NOT allow the caliper to hang unsupported from the brake hose.

8. Remove the brake rotor.

9. Remove the three hub and bearing assembly mounting bolts and remove the backing plate.

10. If equipped, disengage the ABS speed sensor connector.

11. Using a puller, separate the axle from the hub bearing assembly. Remove the hub bearing assembly.

12. If the steering knuckle must also be removed, continue this procedure. Remove the cotter pin and castle nut from the outer tie rod end and using a steering linkage puller, separate the tie rod from the steering knuckle.

13. Remove the strut mounting nuts and through-bolts.

14. Remove the ball joint pinch bolt and lift the steering knuckle off of the ball joint.

To install:

15. If removed, install the steering knuckle onto the lower ball joint and install a new pinch bolt and nut and tighten to 38 ft. lbs. (52 Nm).

16. Connect the steering knuckle to the lower strut bracket and install the through-bolts and nuts. With the matchmarks in alignment, tighten the nuts to 122 ft. lbs. (165 Nm).

17. Connect the tie rod end to the steering knuckle and tighten the castle nut to 30

ft. lbs. (41 Nm). If necessary to align the cotter pin holes tighten the castle nut up to 60° additional. Install a new cotter pin.

18. Install the hub bearing onto the steering knuckle and position the backing plate. Be sure the ABS sensor is properly routed.

19. Install the hub bearing mounting bolts and tighten to 63 ft. lbs. (85 Nm).

20. Install the brake rotor.

21. Install the caliper onto the steering knuckle and tighten the mounting bolts to 38 ft. lbs. (52 Nm).

22. Insert a drift punch into the rotor cooling fins. Install the washer and axle nut onto the axle and tighten to 103 ft. lbs. (140 Nm) plus 20 degrees additional rotation.

23. Install the tire and wheel assembly.

24. Lower the vehicle.

25. Connect the negative battery cable.

26. Check the front end alignment and adjust as necessary.

Rear

The rear wheel bearings on this vehicle are not serviced separately. The wheel bearings are pressed into the rear hub. If the wheel bearings are noisy or defective, the hub and bearing assembly must be replaced.

1. Disconnect the negative battery cable.

2. Raise and safely support the vehicle.

3. Remove the tire and wheel assembly.

4. Remove the brake drum.

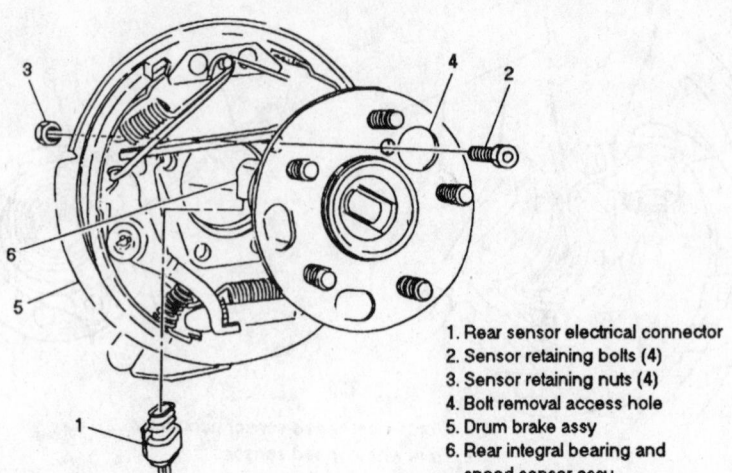

1. Rear sensor electrical connector
2. Sensor retaining bolts (4)
3. Sensor retaining nuts (4)
4. Bolt removal access hole
5. Drum brake assy
6. Rear integral bearing and speed sensor assy

7922SG26

Remove the hub mounting bolts through the large holes in the wheel mounting flange—rear hub assembly shown

➡When the hub bearing mounting bolts and hub bearing are removed from the axle the backing plate assembly must be supported.

5. Remove the four bolts securing the hub and bearing assembly to the axle.

6. Remove the hub and bearing assembly and support the backing plate.

7. Disconnect the rear ABS sensor.

To install:

8. Connect the ABS speed sensor.

9. Position the hub bearing assembly along with the backing plate onto the axle. Install and tighten the mounting bolts to 60 ft. lbs. (82 Nm).

10. Install the brake drum.

11. Install the tire and wheel assembly.

12. Lower the vehicle.

13. Connect the negative battery cable.

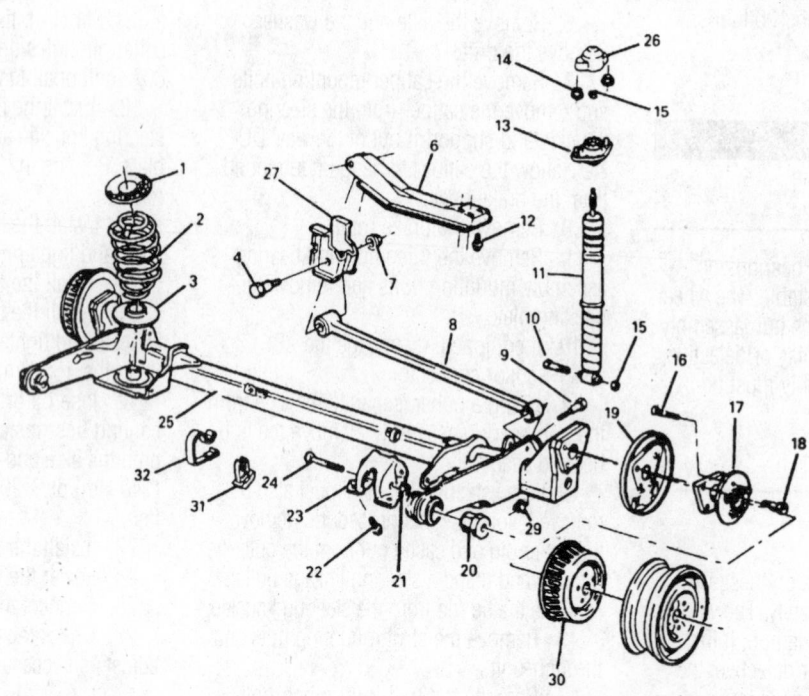

1 REAR SPRING UPPER INSULATOR	17 HUB AND BEARING ASSEMBLY, REAR WHEEL
2 REAR SPRING	18 BOLT/SCREW REAR WHEEL HUB AND BEARING
3 REAR SPRING LOWER INSULATOR	19 PLATE, REAR BRAKE BACKING
4 BOLT/SCREW, REAR AXLE TIE ROD BRACKET BRACE	20 BUSHING, REAR SUSPENSION CONTROL ARM
6 BRACE, REAR AXLE TIE ROD BRACKET (WAGON)	21 NUT, REAR SUSPENSION CONTROL ARM BRACKET
7 NUT, REAR AXLE TIE ROD BRACKET BRACE	22 BOLT/SCREW, REAR SUSPENSION CONTROL ARM BRACKET
8 ROD, REAR AXLE TIE (TRACK BAR)	23 BRACKET, REAR SUSPENSION CONTROL ARM
9 BOLT/SCREW REAR AXLE TIE ROD	24 BOLT/SCREW, REAR SUSPENSION CONTROL ARM
10 BOLT/SCREW, REAR SHOCK ABSORBER	25 AXLE, REAR
11 ABSORBER, REAR SHOCK	26 COVER, REAR SHOCK ABSORBER UPPER
12 BOLT/SCREW, REAR AXLE TIE ROD BRACKET BRACE	27 BRACKET, REAR AXLE TIE ROD
13 MOUNT, SHOCK ABSORBER UPPER	29 NUT, REAR AXLE TIE ROD
14 NUT, REAR SHOCK ABSORBER UPPER MOUNT	30 DRUM, REAR BRAKE
15 NUT, REAR SHOCK ABSORBER	31 INSULATOR, REAR STABILIZER SHAFT
16 STUD, WHEEL	32 CLAMP, REAR STABILIZER SHAFT INSULATOR

7922SG25

Exploded view of the rear suspension and related components

GENERAL MOTORS B BODY

28

Buick-Roadmaster • **Cadillac**-Fleetwood • **Chevrolet**-Caprice • Impala SS

DRIVE TRAIN 28-15
ENGINE REPAIR 28-2
FUEL SYSTEM 28-13
**STEERING AND
SUSPENSION** 28-16

A
Air Bag 28-16
  Precautions 28-16
  Disarming 28-16

C
Camshaft and Valve Lifters 28-8
  Removal & Installation 28-8
Coil Spring 28-18
  Removal & Installation 28-18
Cylinder Head 28-5
  Removal & Installation 28-5

D
Distributor 28-2
  Removal 28-2
  Installation 28-2

E
Engine Assembly 28-3
  Removal & Installation 28-3
Exhaust Manifold 28-8
  Removal & Installation 28-8

F
Fuel Filter 28-14
  Removal & Installation 28-14

Fuel Pump 28-14
  Removal & Installation 28-14
Fuel System Pressure 28-14
  Relieving 28-14
Fuel System Service
  Precautions 28-13

I
Ignition Timing 28-3
  Adjustment 28-3
Intake Manifold 28-7
  Removal & Installation 28-7

L
Lower Ball Joint 28-20
  Removal & Installation 28-20

O
Oil Pan 28-10
  Removal & Installation 28-10
Oil Pump 28-10
  Removal & Installation 28-10

R
Reciprocating Ball Power Steering
  Gear 28-16
  Removal & Installation 28-16
Rear Main Seal 28-11
  Removal & Installation 28-11
Rocker Arms 28-6
  Removal & Installation 28-6

S
Shock Absorber 28-17
  Removal & Installation 28-17

T
Timing Chain, Sprockets, Front
  Cover and Seal 28-11
  Removal & Installation 28-11
Transmission Assembly 28-15
  Removal & Installation 28-15

V
Valve Lash 28-9
  Adjustment 28-9

U
Upper Ball Joint 28-19
  Removal & Installation 28-19

W
Water Pump 28-4
  Removal & Installation 28-4
Wheel Bearings 28-20
  Adjustment 28-20
  Removal & Installation 28-21

ENGINE REPAIR

➡ **Disconnecting the negative battery cable on some vehicles may interfere with the functioning of the on-board computer system, and may require the computer to undergo a relearning process once the negative battery cable is reconnected.**

Distributor

All models in this section use a new distributor ignition system which was originally developed for use on the Corvette. The new system known as the Opti-Spark ignition system consists of a distributor assembly which is mounted on the front engine cover, under the water pump assembly. The system consists of a distributor assembly, control circuitry and an external coil. In the Opti-Spark system, all ignition timing is controlled by the PCM based on signals from the distributor's internally mounted optical camshaft position sensor. There is no way to bypass the PCM control or to adjust/set ignition timing on this system.

REMOVAL

1. Be sure the ignition is in the **OFF** or **LOCK** position, then disconnect the negative battery cable.

2. Disengage the wiring harness from the engine cooling fan assembly, then remove the cooling fan from the vehicle.

3. Remove the block drain plug and the knock sensor, then drain the engine cooling system.

4. Remove the air intake duct and the air cleaner assembly.

5. Remove the coolant and heater hoses from the water pump assembly and the throttle body.

6. Unplug the wiring harness from the Engine Coolant Temperature (ECT) sensor, then reposition the ignition coil and bracket.

7. Remove the water pump assembly.

8. Remove the serpentine drive belt from the vehicle.

9. If not done already to remove the pump or the belt, raise and safely support the vehicle for access.

10. Remove the torsional damper bolts, then remove the damper from the crankshaft hub.

11. Disconnect the spark plug wires from the distributor. Be sure to twist each boot ½ turn and pull only on the boot to remove each wire. The wire numbers should be molded into the distributor housing. If not, be sure to tag the wires before disconnection.

12. Unplug the 4-terminal PCM connector from the distributor.

13. Remove the distributor mounting bolts and pull the distributor forward until the driveshaft disengages from the engine. Mark the top of the shaft for alignment during reassembly.

INSTALLATION

1. With the mark made on the distributor shaft earlier on top, install the distributor to the engine. Tighten the distributor bolts to 8 ft. lbs. (11 Nm).

2. Install the 4-terminal PCM connector and the spark plug wires to the distributor.

3. Position the crankshaft damper to the hub, then install the damper bolts and tighten to 60 ft. lbs. (81 Nm).

4. Install the serpentine drive belt.

5. Install the water pump assembly and engage the ECT wiring harness.

6. Reposition the ignition coil and bracket, then install the coolant and heater hoses.

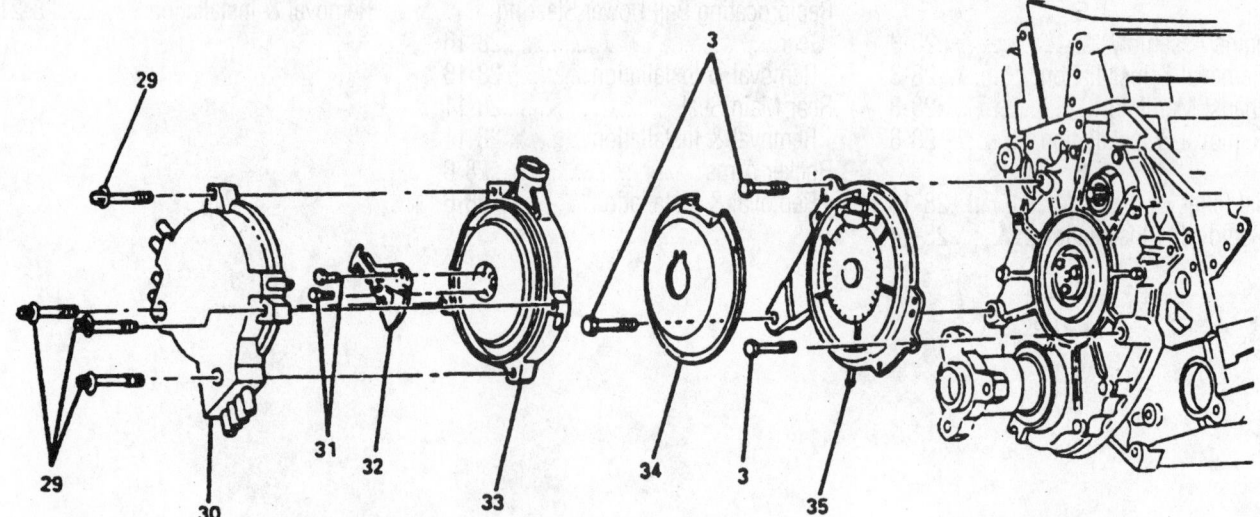

3	BOLT/SCREW, DISTRIBUTOR
29	BOLT/SCREW, DISTRIBUTOR CAP
30	CAP, DISTRIBUTOR
31	BOLT/SCREW, DISTRIBUTOR ROTOR
32	ROTOR, DISTRIBUTOR
33	COVER
34	SHIELD, DISTRIBUTOR
35	BASE, DISTRIBUTOR

Exploded view of the distributor assembly which is located under the water pump

7922TG01

7. Install the air cleaner assembly and air intake duct.

8. Install the engine block drain plug and the knock sensor.

9. Install the cooling fan assembly, then connect the wiring harness.

10. Connect the negative battery cable and fill the engine cooling system.

11. Start and run the engine, then check for leaks.

Ignition Timing

ADJUSTMENT

When checking ignition timing NEVER pierce a secondary ignition wire. Use either a timing light with an inductive type pick-up or connect an adapter between the wire and the spark plug. If the secondary wire insulation is pierced, current will eventually arc and cause engine misfiring.

➡**Some engines incorporate a magnetic timing probe hole for the use with electronic timing equipment. Be sure to consult the tool manufacture's instructions for the use of this equipment.**

On these vehicles, base timing is preset when the engine is manufactured. All timing changes are, then controlled directly by the PCM based on information from the ignition and knock sensor systems. No adjustments are necessary or possible.

Engine Assembly

REMOVAL & INSTALLATION

Except Fleetwood Brougham

Disconnect the negative cable, then the positive cable from the battery.

✷✷ CAUTION

Observe all applicable safety precautions when working around fuel. Whenever servicing the fuel system, always work in a well ventilated area. Do not allow fuel spray or vapors to come in contact with a spark or open flame. Keep a dry chemical fire extinguisher near the work area. Always keep fuel in a container specifically designed for fuel storage; also, always properly seal fuel containers to avoid the possibility of fire or explosion.

1. Relieve the fuel system, then scribe alignment marks and remove the hood from the vehicle.

2. Remove the air cleaner assembly and drain the engine cooling system into a suitable container.

3. Remove the radiator hoses and upper fan shroud.

✷✷ CAUTION

Never open, service or drain the radiator or cooling system when hot; serious burns can occur from the steam and hot coolant.

4. Remove the engine cooling fan, then remove the radiator from the vehicle.

5. Disconnect the heater hoses at the engine

6. Disconnect the power steering pump and air conditioning compressor brackets and position out of the way. Be careful not to kink or damage the fluid or refrigerant lines.

7. Disconnect the accelerator, TV, and cruise control cables.

8. Tag and disconnect all necessary vacuum hoses.

9. Disconnect the PCM wiring harness, the engine wiring harness at the engine bulkhead, engine-to-bulkhead ground strap and any remaining wires between body and engine.

10. Remove the distributor assembly from the engine.

11. Remove the windshield wiper motor assembly.

12. Remove the MAP sensor.

13. Remove the negative battery cable from the cylinder head.

14. If not done already, remove the brake pipe from the intake manifold.

15. Raise and support the vehicle safely.

✷✷ WARNING

Never raise the engine using a jack under the oil pan, crankshaft pulley or any sheet metal. Because there only is a small clearance between the oil pan and the oil pump screen, if the pan is bent even slightly, damage could occur to the pump screen and pick-up unit.

16. Disconnect the battery positive cable and wires at the starter motor. Be sure to tag the wires for installation purposes.

17. Disconnect the crossover pipe and catalytic converter as an assembly.

18. Remove the flywheel cover and torque converter-to-flywheel bolts.

19. Remove the engine mount through-bolts.

20. Disconnect the front fuel hoses from the front fuel pipes.

21. Disconnect the transmission converter clutch wiring at the transmission and the transmission oil cooler lines at the clip on the oil pan.

22. Disconnect the equalizer rod from the transmission.

23. Remove the transmission-to-engine bolts.

24. Lower the vehicle.

25. Support the transmission and connect a suitable lifting device to the engine.

26. Remove the engine.

To install:

27. With the engine safely supported, lower into position and align with the motor mounts and transmission.

28. Install motor mount through-bolts and the transmission-to-engine bolts. Tighten either the nuts to 59 ft. lbs. (80 Nm) or the through-bolts to 70 ft. lbs. (95 Nm). Tighten the transmission-to-engine bolts to 35 ft. lbs. (47 Nm).

29. Raise and support the vehicle safely.

30. If applicable, connect the AIR pipe to the exhaust manifold or the equalizer rod to the transmission.

31. Connect the transmission converter clutch wiring to the transmission and the transmission oil cooler lines to the clip on the oil pan.

32. Connect the front fuel hoses to the front fuel pipes.

33. Install the torque converter-to-flywheel bolts and the flywheel housing cover.

34. Connect the crossover pipe and catalytic converter assembly.

35. Connect the battery positive cable and wires to the starter motor.

36. Lower vehicle.

37. Connect the brake pipe to the intake manifold, then connect the negative battery cable to the cylinder head.

38. As applicable, install the EGR solenoid, the MAP sensor and the windshield wiper motor.

39. Install the distributor assembly.

40. Connect the PCM wiring harness, the engine wiring harness at the engine bulkhead, engine to bulkhead ground straps and all other wires between body and engine.

41. Connect all vacuum hoses as noted during removal.

42. Connect the accelerator, TV and cruise control cables.

43. Connect the power steering pump and air conditioning compressor brackets.

44. Connect the heater hoses to the engine.

45. Install the engine cooling fan.

46. Install the radiator, hoses and fan shroud.

47. Install the air cleaner assembly.

48. Install the hood, aligning the marks made during removal.

49. Connect the negative battery cable, then fill and bleed the engine cooling system.

50. Inspect vehicle fluid levels, specifications and verify there are no fluid leaks.

Fleetwood Brougham

✳✳ CAUTION

Observe all applicable safety precautions when working around fuel. Whenever servicing the fuel system, always work in a well ventilated area. Do not allow fuel spray or vapors to come in contact with a spark or open flame. Keep a dry chemical fire extinguisher near the work area. Always keep fuel in a container specifically designed for fuel storage; also, always properly seal fuel containers to avoid the possibility of fire or explosion.

1. Disconnect the battery cables and properly relieve fuel system pressure.

2. Mark the hood hinge outline for proper reassembly alignment and remove the hood. Remove the air cleaner assembly.

✳✳ CAUTION

Never open, service or drain the radiator or cooling system when hot; serious burns can occur from the steam and hot coolant.

3. Drain the cooling system. Disconnect the radiator hoses. Disconnect the heater hose from the radiator. Disconnect and plug the transmission and engine oil cooler lines.

4. Remove the radiator cover and tie struts. Disconnect the fan shroud from the radiator assembly and position it aside. Remove the radiator from the vehicle.

5. Remove the serpentine drive belt. Remove the cooling fan assembly and the fan shroud from the vehicle.

6. Disconnect the heater hose at the rear of the intake manifold. Disconnect and plug the power steering hoses at the power steering gear.

7. Remove the air conditioning compressor and position it aside.

8. Disconnect the accelerator, cruise control and throttle valve cables from their mountings and position out of the way. Remove the vacuum pipe and fuel lines from the throttle body.

9. Remove the generator assembly. Disconnect the fuel line clips at the thermostat housing and air pump. Position the fuel lines aside. As required, remove the air pump assembly.

10. Disconnect and plug all required electrical connectors. Remove the distributor cap. Remove the negative battery cable from the cylinder head.

11. Raise and support the vehicle safely. Disconnect the crossover pipe at both manifolds.

12. Disconnect the starter electrical connectors and the positive battery cable. If necessary, remove the starter retaining bolts and remove the starter from the vehicle.

13. Remove the flywheel cover. Remove the torque converter to flywheel retaining bolts. Remove the motor mount through-bolts.

14. Disconnect the transmission oil cooler lines at the clip on the oil pan. Disconnect the oil pressure, knock and oxygen sensor connectors. Remove the oil cooler hose shield.

15. Remove the ground wires from the rear of the cylinder head at both sides.

16. Remove the transmission to engine retaining bolts. Lower the vehicle.

17. Install the lifting equipment to the engine. Support the transmission properly.

18. Raise the engine slightly and pull it forward to disengage it from the transmission. Remove the engine from the vehicle.

To install:

19. Lower the engine assembly into the engine compartment; align the transmission bell housing dowels and motor mounts.

20. Loosely install 2 transmission-to-engine bolts.

21. Remove the engine and transmission supports.

22. Raise and safely support the vehicle, install the engine mount through-bolts and tighten to 70 ft. lbs. (95 Nm).

23. Route the wiring harness into its original location and reconnect the oil pressure, knock sensor and oxygen sensor connectors.

24. Reinstall the oil cooler line bracket and heat shield.

25. Connect the ground straps to the back of the cylinder heads.

26. Install and torque all transmission-to-engine bolts to 55 ft. lbs. (75 Nm).

27. Install the starter assembly, if removed and/or reconnect the wiring. Clip the transmission cooler lines to the oil pan bracket.

28. Install the flywheel-to-torque converter bolts and torque the bolts to 45 ft. lbs. (62 Nm). Install the flywheel cover.

29. Reconnect the exhaust and exhaust hangers. Tighten the crossover pipe bolts to 15 ft. lbs. (20 Nm).

30. Lower the vehicle.

31. Install heater hose to the right rear of intake manifold. Reconnect the throttle cable brackets and cables.

32. Install the distributor cap and coil wires.

33. Install the generator with wiring, but leave the rear brace disconnected. Connect the negative battery cable at the cylinder head.

34. Unplug and connect the power steering lines at the power steering gear.

35. Route the fuel lines and connect at the throttle body. Install the fuel line clips at the thermostat housing and the AIR pump.

36. Connect the rear generator brace. Connect all vacuum hoses to the throttle body. Connect all electrical connections to the intake manifold and the throttle body.

37. Connect the AIR hose from the diverter valve to the converter.

38. Install the fan and fan shroud assembly.

39. Install the radiator assembly. Connect the transmission and oil cooler lines.

40. Connect the heater hose to the radiator tank and the radiator hoses.

41. Install the radiator cover and secure the fan shroud. Install the radiator tie struts.

42. Install the air conditioning compressor and serpentine belt. Install the air cleaner assembly.

43. Fill the cooling system and connect the battery cables.

44. Check all fluid levels, start engine and inspect for leaks.

45. Align the marks made earlier and install the hood assembly.

Water Pump

REMOVAL & INSTALLATION

1. Disconnect the negative battery cable.

2. Disengage the wiring harness from the cooling fan assembly, then remove the assembly from the vehicle.

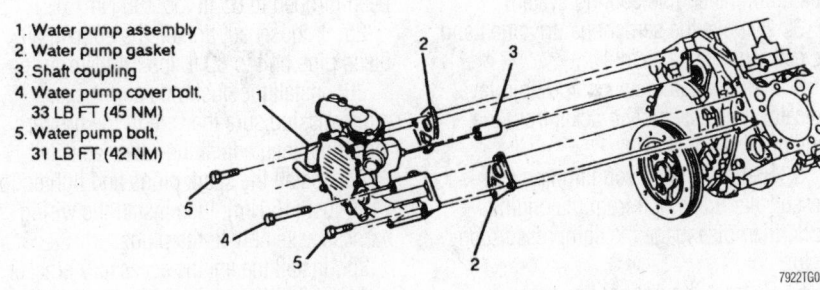

1. Water pump assembly
2. Water pump gasket
3. Shaft coupling
4. Water pump cover bolt.
   33 LB FT (45 NM)
5. Water pump bolt,
   31 LB FT (42 NM)

7922TG02

Exploded view of the water pump assembly mounting

✳✳ CAUTION

Never open, service or drain the radiator or cooling system when hot; serious burns can occur from the steam and hot coolant.

3. Drain the engine cooling system, removing the block drain plug and the knock sensor to assure proper draining. Reinstall the drain plug and knock sensor, as soon as the system is empty.

4. Disconnect the upper and lower radiator hoses from the water pump assembly.

5. Remove the heater hose assemblies from the water pump and from the throttle body.

6. Disengage the coolant sensor wiring harness, then reposition the ignition coil and bracket assembly.

7. Remove the shorter water pump retaining bolt from the center of each pump mating flange, then remove the longer pump bolts from either side of the center bolts.

8. Carefully remove the water pump assembly and gaskets along with the pump shaft coupling.

To install:

9. Thoroughly clean the gasket mating surfaces of any remaining gasket material.

10. Install the water pump shaft coupling along with the water pump and gaskets.

11. Install the longer pump bolts and tighten 33 ft. lbs. (45 Nm), then install the shorter bolts and tighten to 31 ft. lbs. (42 Nm).

12. Reposition the ignition coil and bracket assembly.

13. Engage the coolant sensor electrical connector.

14. Install the heater and radiator hoses to the throttle body and water pump, as applicable.

15. Install the air cleaner and intake duct assemblies.

16. Install the engine cooling fan assembly and engage the wiring harness connector.

17. Connect the negative battery cable and properly fill the engine cooling system.

Cylinder Head

REMOVAL & INSTALLATION

Left Side

1. Disconnect the negative battery cable, then raise and support the vehicle safely.

✳✳ CAUTION

Never open, service or drain the radiator or cooling system when hot; serious burns can occur from the steam and hot coolant.

2. Drain the engine cooling system, then disconnect the crossover pipe from the exhaust manifold.

3. Lower the vehicle.

4. Remove the intake manifold assembly.

5. Disconnect the secondary air injection hose from the check valve assembly.

6. Disconnect the coolant air bleed pipe and bolt from the left cylinder head assembly using a back-up wrench on the pipe fitting.

7. Remove the ignition coil assembly.

8. Remove the left exhaust manifold assembly.

9. Remove the spark plug wiring harness assembly from the clips, then disconnect the harness from the spark plugs and remove the plugs from the left cylinder head.

10. Disengage the coolant temperature sensor connector.

11. Remove the left rocker arm cover.

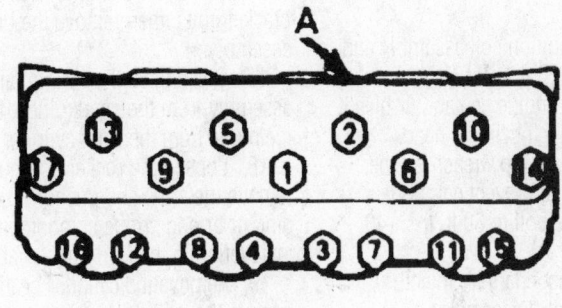

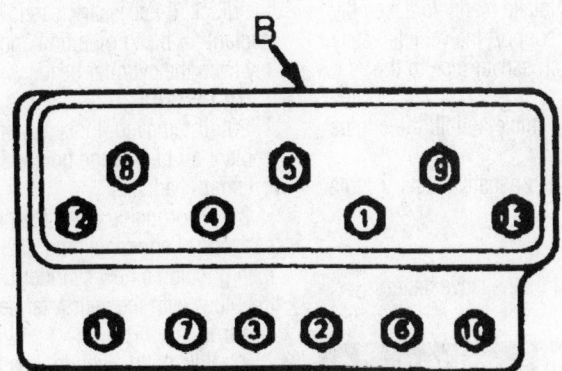

A. V8 cylinder head
B. V6 cylinder head

Cylinder head bolt torque sequences

7922TG03

12. Loosen the rocker arm nuts, then remove the arms and pushrods, either tagging or arranging the components to assure installation in their original locations.

13. Remove the cylinder head bolts, then remove the cylinder head and gasket from the block.

To install:

14. Thoroughly clean the gasket mating surfaces of any remaining gasket material, then position a new cylinder head gasket on the block with the yellow tab facing upwards.

15. Install the cylinder head over the locator pins and the new gasket.

16. Coat the cylinder head bolts with a sealing compound, then install the bolts finger-tight.

17. Torque the bolts using 3 passes of the proper sequence until all bolts have been torqued to 65 ft. lbs. (88 Nm).

18. Install the pushrods and rocker arms, making sure they are in the proper locations, then adjust the valve lash.

19. Install the left rocker arm cover and tighten the retaining bolts to 100 inch lbs. (11 Nm).

20. Install the spark plugs and tighten to 11 ft. lbs. (15 Nm), then install the wiring harness assembly and secure the assembly to the clips.

21. Install the left exhaust manifold assembly.

22. Install the ignition coil assembly and tighten the bolts to 24 ft. lbs. (33 Nm).

23. Connect the engine coolant air bleed pipe and bolt to the left cylinder head assembly using a back-up wrench on the pipe fitting in order to prevent component damage. Tighten the bolt to 30 ft. lbs. (40 Nm).

24. Install the secondary air injector hose to the check valve assembly.

25. Install the intake manifold assembly.

26. Raise and support the vehicle safely, then connect the crossover pipe to the exhaust manifold.

27. Lower the vehicle and fill the engine cooling system.

28. Connect, then negative battery cable.

Right Side

1. Disconnect the negative battery cable.

❊❊ CAUTION

Never open, service or drain the radiator or cooling system when hot; serious burns can occur from the steam and hot coolant.

2. Raise and support the vehicle safely, then drain the engine cooling system.

3. Remove the serpentine drive belt and the belt tensioner assembly.

4. Remove the transmission fluid level indicator tube assembly bracket from the transmission housing.

5. Remove the air conditioning compressor rear brace bolt from the engine block, then disengage the compressor connector.

6. Remove the front compressor mounting bolts, then position the compressor aside taking care not to kink or damage the lines.

7. Remove the right exhaust manifold assembly.

8. Lower the vehicle.

9. Remove the alternator assembly.

10. Remove the right rocker arm cover.

11. Remove the intake manifold assembly.

12. Remove the coolant air bleed pipe bolt from the left cylinder head assembly.

13. Disconnect the lower radiator hose and the heater hose from the water pump assembly, then position the hoses aside.

14. Remove the coolant air bleed pipe hose from the radiator.

15. Remove the power steering pump assembly.

16. Remove the engine accessory bracket bolts, then remove the bracket assembly.

17. Disconnect the wiring harness assembly from the spark plugs, then remove the plugs from the right cylinder head.

18. Loosen the rocker arm nuts, then remove the arms and pushrods, either tagging or arranging the components to assure installation in their original locations.

19. Remove the cylinder head bolts, then remove the cylinder head and gasket from the block. If necessary, carefully remove the coolant air bleed pipe bolt and pipe assembly from the cylinder head.

To install:

20. If removed, loosely install the coolant air bleed pipe bolt and pipe to the cylinder head.

21. Thoroughly clean the gasket mating surfaces of any remaining gasket material, then position a new cylinder head gasket on the block with the yellow tab facing upwards.

22. Install the cylinder head over the locator pins and the new gasket.

23. Coat the cylinder head bolts with a sealing compound, then install the bolts finger-tight.

24. Torque the bolts using 3 passes of

the proper sequence until all bolts have been torqued to 65 ft. lbs. (88 Nm).

25. If loosened, tighten the coolant air bleed pipe bolt to 30 ft. lbs. (40 Nm).

26. Install the pushrods and rocker arms, making sure they are in the proper locations, then adjust the valve lash.

27. Install the spark plugs and tighten to 11 ft. lbs. (15 Nm), then install the wiring harness assembly to the plugs.

28. Install the engine accessory bracket and tighten the retaining bolts to 31 ft. lbs. (42 Nm).

29. Install the right rocker arm cover and tighten the retaining bolts to 100 inch lbs. (11 Nm).

30. Install the alternator assembly.

31. Install the power steering pump assembly.

32. Connect the coolant air bleed pipe hose to the radiator, then connect the heater hose and the lower radiator hose to the water pump.

33. Connect the coolant air bleed pipe bolt to the left cylinder head and tighten to 30 ft. lbs. (41 Nm) while using a back-up wrench to prevent component damage.

34. Install the intake manifold assembly.

35. Raise and support the vehicle safely.

36. Install the right exhaust manifold assembly.

37. Position the compressor and install the front mounting bolts, then engage the electrical connector. Install the rear compressor brace bolt and tighten to 24 ft. lbs. (33 Nm).

38. Install the transmission fluid level indicator tube assembly to the transmission housing.

39. Install the serpentine drive belt, then lower the vehicle.

40. Fill the engine cooling system, then connect the negative battery cable.

Rocker Arms

REMOVAL & INSTALLATION

1. Disconnect the negative battery cable.

2. Remove the left valve cover:

   a. Remove the brake booster vacuum hose.

   b. Remove the secondary AIR injection hose from the pump to check valve assembly.

   c. Remove the valve cover retaining bolts, then remove the cover and gasket.

3. Remove the right valve cover:

   a. Raise and support the vehicle safely, then remove the serpentine drive belt.

27 PUSHROD ASSEMBLY, VALVE
28 STUD, VALVE ROCKER ARM BALL
29 ARM, VALVE ROCKER
30 BALL, VALVE ROCKER ARM
31 NUT, VALVE ROCKER ARM

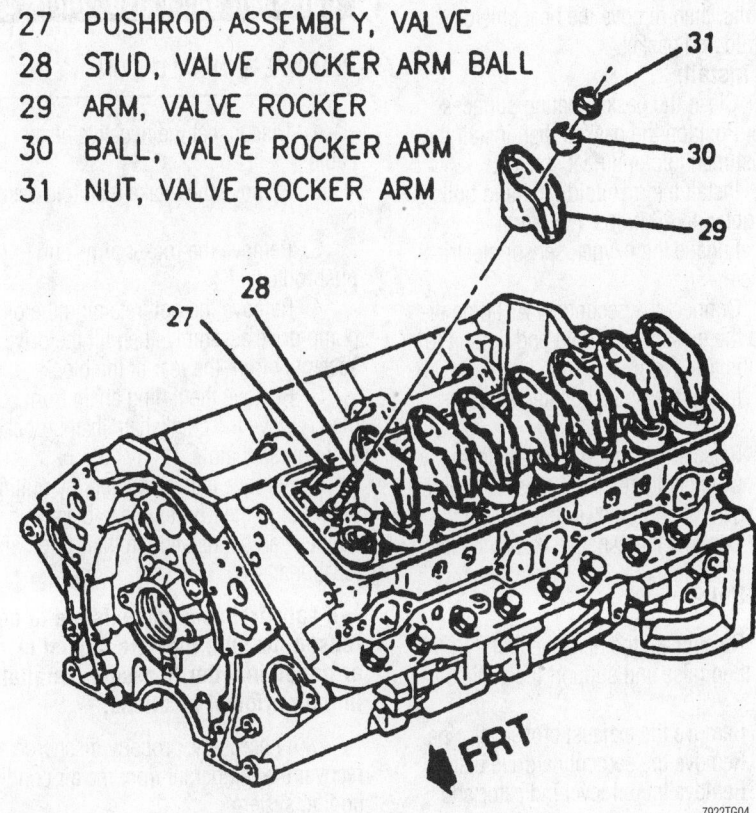

Exploded view of a typical rocker arm mounting

b. Remove the transmission fluid level indicator tube assembly from the bracket on the transmission housing.

c. Lower the vehicle, then remove the crankcase vent hose.

d. Remove the alternator and rear alternator brace.

e. Remove the valve cover retaining bolts, then remove the cover and gasket.

4. Remove the valve rocker arm nuts and balls, then remove the rocker arms and pushrods. Tag or arrange all valvetrain components to assure installation in their original locations.

To install:

5. Coat the bearing surfaces of the rocker arms, balls and pushrods with prelube.

6. Install the pushrods, making sure they are properly seated in the lifter sockets, then install the rocker arms, balls and nuts. If components are being reused, be sure they are installed in their original locations.

7. Adjust the valve lash.

8. Install the valve covers and gaskets, then tighten the retainers to 100 inch lbs. (11 Nm).

9. Install the components which were removed to access the valve covers in the reverse order of removal.

10. Connect the negative battery cable.

Intake Manifold

REMOVAL & INSTALLATION

✵✵ CAUTION

Observe all applicable safety precautions when working around fuel. Whenever servicing the fuel system, always work in a well ventilated area. Do not allow fuel spray or vapors to come in contact with a spark or open flame. Keep a dry chemical fire extinguisher near the work area. Always keep fuel in a container specifically designed for fuel storage; also, always properly seal fuel containers to avoid the possibility of fire or explosion.

1. Disconnect the negative battery cable and relieve the fuel system pressure.

✵✵ CAUTION

Never open, service or drain the radiator or cooling system when hot; serious burns can occur from the steam and hot coolant.

2. Drain the engine cooling system into a suitable container.

3. Remove the throttle body air duct.

4. Disengage the wiring harness connectors from the fuel injectors. Disengage and reposition the left and right wiring harnesses.

5. Remove the accelerator cable bracket retainers, then disconnect the cable and bracket assembly from the throttle body.

6. Disconnect the secondary AIR diverter valve hoses.

7. Disengage the fuel pipe connectors from the fuel rail assembly.

8. Remove the fuel rail bolts and disconnect the vacuum hose from the fuel pressure regulator.

9. Carefully remove the fuel rail and injector assembly from the manifold and position aside.

10. Disconnect the vacuum and crankcase vent hoses.

11. Remove the EGR solenoid assembly and the fuel EVAP canister solenoid assembly.

12. Remove the EGR valve.

13. Remove the AIR pipe from the intake and the right exhaust manifold.

14. Remove the alternator rear brace.

15. Disconnect the coolant hoses from the throttle body.

16. Remove the throttle body bolts, the throttle body and gasket from the intake.

17. Remove the intake manifold bolts and studs.

18. Remove the intake manifold and discard the old gaskets.

To install:

19. Thoroughly clean the intake manifold bolts and studs. Inspect and clean all gasket mating surfaces.

20. Apply a ³⁄₁₆ in. (5mm) bead of RTV sealer to the front and rear of the cylinder block. Extend the bead ½ in. (13mm) up each cylinder head to seal and retain the gaskets.

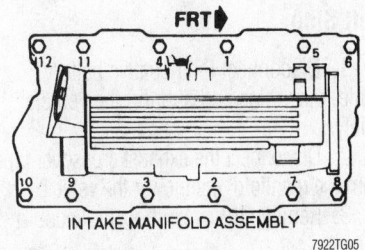

INTAKE MANIFOLD ASSEMBLY

Tighten the intake manifold bolts in sequence to help prevent vacuum leaks

21. Position the new gaskets and install the intake manifold.

22. Install the manifold bolts and studs, then tighten using 2 passes of the proper sequence. First, tighten the bolts/studs to 71 inch lbs. (8 Nm), then tighten them to 35 ft. lbs. (48 Nm).

23. Install the throttle body, gasket and retaining bolts. Tighten the throttle body bolts to 19 ft. lbs. (26 Nm).

24. Connect the coolant hoses to the throttle body, then install the alternator rear brace.

25. Install the accelerator cables and bracket, then tighten the bracket bolts to 90 inch lbs. (10 Nm).

26. Install the secondary AIR pipe. Tighten the exhaust manifold fitting to 25 ft. lbs. (34 Nm) and tighten the flange-to-intake manifold bolts to 19 ft. lbs. (26 Nm).

27. Install the EGR valve, then EGR solenoid and bracket. Tighten valve nuts and the solenoid bracket nut to 16 ft. lbs. (22 Nm).

28. Install the fuel EVAP canister purge solenoid and bracket, then tighten the bolt to 53 inch lbs. (6 Nm).

29. Connect the vacuum and crankcase vent hoses.

30. Install the fuel injector and fuel rail assembly to the intake manifold, connect the fuel pressure regulator vacuum hose and install the fuel rail bolts. Tighten the bolts to 15 ft. lbs. (20 Nm). Engage the fuel pipe connections to the fuel rail assembly.

31. Connect the secondary AIR diverter valve hoses.

32. Position the left and right wiring harnesses, then engage the fuel injector electrical connectors.

33. Install the throttle body air duct.

34. Properly fill the engine cooling system.

35. Connect the negative battery cable.

Exhaust Manifold

REMOVAL & INSTALLATION

Left Side

1. Disconnect the negative battery cable, then raise and support the vehicle safely.

2. Disconnect the exhaust crossover pipe from the manifold, then lower the vehicle.

3. Remove the brake booster vacuum hose.

4. Disconnect the secondary AIR pipe fitting from the exhaust manifold.

5. Disengage the oxygen sensor electrical connector.

6. Remove the exhaust manifold retaining bolts, then remove the heat shields, manifold and gasket.

To install:

7. Clean the gasket mating surfaces.

8. Position the gasket, then install the exhaust manifold and heat shields.

9. Install the manifold retaining bolts and tighten to 26 ft. lbs. (35 Nm).

10. Engage the oxygen sensor electrical connector.

11. Connect the secondary AIR pipe fitting to the exhaust manifold and tighten to 25 ft. lbs. (34 Nm).

12. Install the brake booster vacuum hose.

13. Raise and support the vehicle safely.

14. Connect the exhaust crossover pipe to the manifold, then lower the vehicle.

15. Connect the negative battery cable.

Right Side

1. Disconnect the negative battery cable, then raise and support the vehicle safely.

2. Remove the exhaust crossover pipe.

3. Remove the serpentine drive belt.

4. Remove the oil level indicator and tube assembly, then disengage the oxygen sensor electrical connector.

5. Remove the 3 rear exhaust manifold retaining bolts, then lower the vehicle.

6. Disconnect the secondary AIR pipe fitting from the exhaust manifold.

7. Remove the alternator rear lower brace.

8. Remove the remaining exhaust manifold retaining bolts, then remove the heat shields, manifold and gasket.

To install:

9. Clean the gasket mating surfaces.

10. Position the gasket, then install the exhaust manifold and heat shields.

11. Install the front 3 manifold retaining bolts and tighten to 26 ft. lbs. (35 Nm).

12. Install the alternator rear lower brace.

13. Raise and support the vehicle safely.

14. Install the remaining manifold retaining bolts and tighten to 26 ft. lbs. (35 Nm).

15. Engage the oxygen sensor electrical connector.

16. Install the oil level indicator and tube assembly.

17. Install the serpentine drive belt.

18. Install the exhaust crossover pipe.

19. Lower the vehicle.

20. Connect the secondary AIR pipe fitting to the exhaust manifold and tighten to 25 ft. lbs. (34 Nm).

21. Connect the negative battery cable.

Camshaft and Valve Lifters

REMOVAL & INSTALLATION

1. Disconnect the negative battery cable.

2. Remove the intake manifold assembly.

3. Remove the rocker arms and pushrods.

4. Remove the bolt retaining the oil pump drive assembly, then lift the drive assembly from the rear of the block.

5. Remove the timing chain front cover, then remove the crankshaft shaft sprocket and timing chain.

6. Remove the valve lifters by pulling them out of their bores. Keep them in order so they can be installed in their original locations.

➡**If valvetrain components are to be reused, be sure they are tagged or arranged in order to assure installation in their original locations.**

7. If necessary, properly discharge and recover the refrigerant from the air conditioning system.

8. Remove the air conditioning compressor and condenser hose from the condenser, then remove the receiver and dehydrator hose from the condenser. Plug all of the openings in order to prevent system contamination.

9. Remove the radiator and condenser assembly, then remove the air conditioning condenser support.

10. Remove the camshaft retainer bolts, then remove the camshaft retainer from the front of the block.

11. Install three ⁵⁄₁₆ -18 x 4 in. bolts or equivalent, in the camshaft bolt holes, then using the bolts to pull and rotate the camshaft, carefully pull the camshaft from the bearings. All camshaft journals are the same diameter and care must be used to avoid damage to the bearings.

To install:

12. If installing a new camshaft, be sure to coat all camshaft lobes with Molykote® or an equivalent prelube and to replace all lifters in order to assure camshaft durability.

13. Lubricate the camshaft journals with clean engine oil, then carefully insert the camshaft into the engine.

14. Install the camshaft retainer and tighten the bolts to 105 inch lbs. (12 Nm).

15. Install the condenser support, then install the radiator and condenser assembly.

16. Unplug the openings, then connect the condenser refrigerant lines.

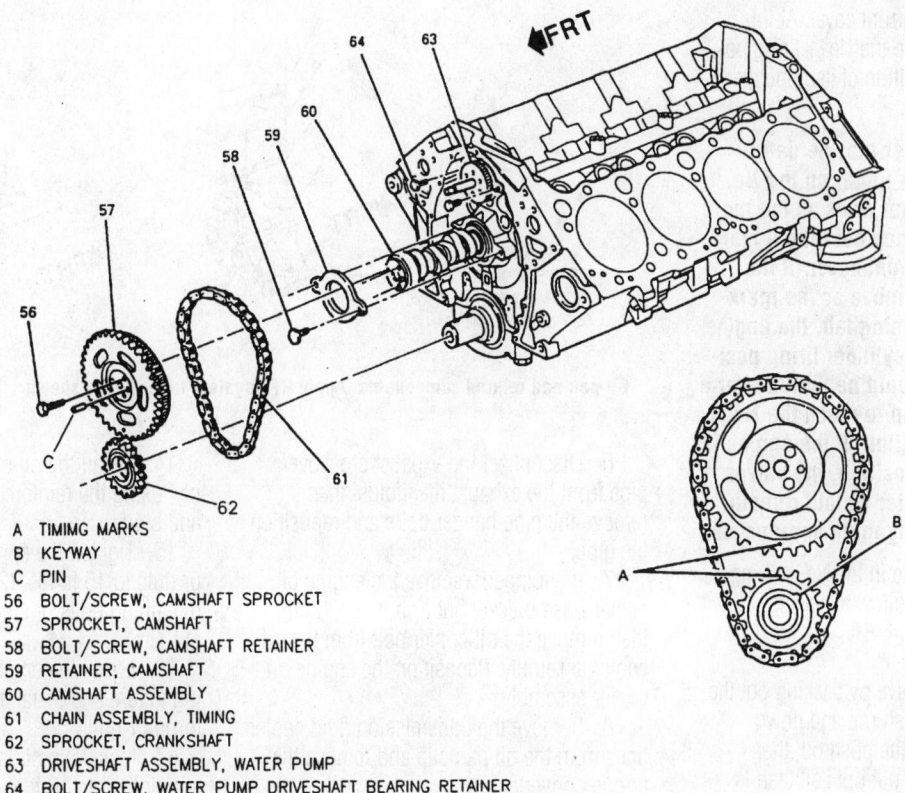

A TIMIMG MARKS
B KEYWAY
C PIN
56 BOLT/SCREW, CAMSHAFT SPROCKET
57 SPROCKET, CAMSHAFT
58 BOLT/SCREW, CAMSHAFT RETAINER
59 RETAINER, CAMSHAFT
60 CAMSHAFT ASSEMBLY
61 CHAIN ASSEMBLY, TIMING
62 SPROCKET, CRANKSHAFT
63 DRIVESHAFT ASSEMBLY, WATER PUMP
64 BOLT/SCREW, WATER PUMP DRIVESHAFT BEARING RETAINER

7922TG06

The camshaft is withdrawn through the front of the engine block, after the timing chain and sprocket is removed

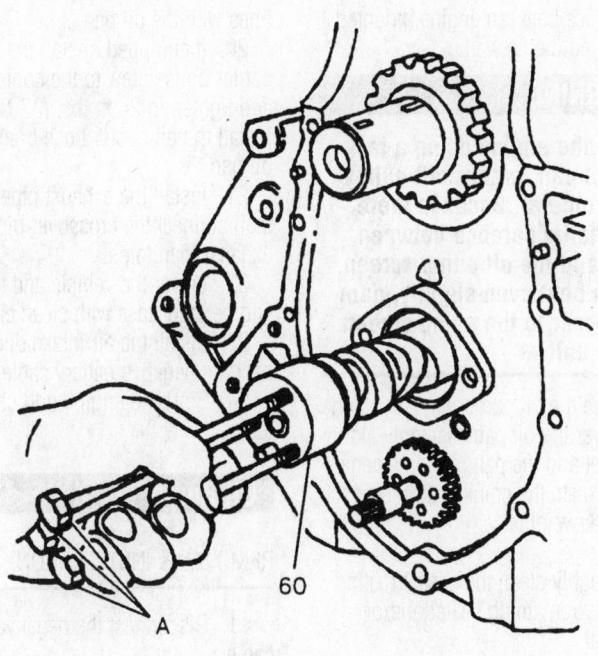

A 5/16–18 X 4" BOLTS
60 CAMSHAFT ASSEMBLY

7922TG07

Temporarily install three long bolts in the camshaft to use as a handle, then carefully remove the camshaft without damaging the bearings

17. Install the valve lifters. If reusing the camshaft and lifters, they must be installed into their original bores.

18. Install the camshaft sprocket and timing chain, then install the timing chain front cover.

19. Install the oil pump drive assembly and tighten the retaining bolt to 13 ft. lbs. (18 Nm).

20. Install the rocker arms and pushrods.

21. Install the intake manifold assembly.

22. Connect the negative battery cable, then if necessary, recharge the A/C system.

Valve Lash

ADJUSTMENT

These engines do not require any routine valve lash adjustments. However, if the rocker arms are removed, the initial valve lash must be adjusted before the engine is started. Use the following procedure:.

1. With the valve covers removed and the rocker arm assemblies loosely installed, position the engine at the No. 1 cylinder Top Dead Center (TDC) position.

2. To determine TDC, slowly turn the engine until the mark on the vibration

damper aligns with the center or **0** mark on the timing tab of the front cover. At this point the engine is on the No. 1 firing position or the firing position of its opposite cylinder No. 6.

➡**The firing cylinder may be determined by placing a finger on the No. 1 cylinder valve rocker arms as the mark on the damper comes near the 0 mark on the crankcase front cover. If the valve rocker arms move as the mark comes up to the timing tab, the engine is on the opposite cylinder firing position, No. 6 and should be turned over a complete revolution to reach the No. 1 cylinder firing position. If the engine is in the No. 1 TDC position, the valves for the No. 1 cylinder should remain closed as the timing mark approaches.**

3. With the engine in the No. 1 firing position, adjust the following valves:
  • Exhaust—1, 3, 4, 8
  • Intake—1, 2, 5, 7
4. Adjust each valve by backing out the adjusting nut until lash (up and down movement) is felt at the pushrod, then tighten the adjusting nut until all lash is removed. This can be determined by rotating pushrod while turning the adjusting nut. When play has been removed, turn adjusting nut one full additional turn clockwise. The lifter plunger will now be centered.
5. Turn the engine 1 revolution until the pointer **0** mark and the vibration damper mark are again in alignment. This is the No. 6 firing position. As the timing mark approaches the pointer, the No. 1 cylinder valves should move.
6. With the engine in this position, adjust the following valves:
  • Exhaust—2, 5, 6, 7
  • Intake—3, 4, 6, 8
7. Install the rocker arm covers or valve covers.
8. Start the engine and check/adjust the minimum the idle speed, as required.

Oil Pan

REMOVAL & INSTALLATION

1. Disconnect the negative battery cable.
2. Remove the air intake duct.
3. Raise and support the vehicle safely, then drain the engine crankcase of oil.
4. Drain the engine cooling system.
5. Disengage the wiring harness connector from the oil level sensor, then remove the sensor.

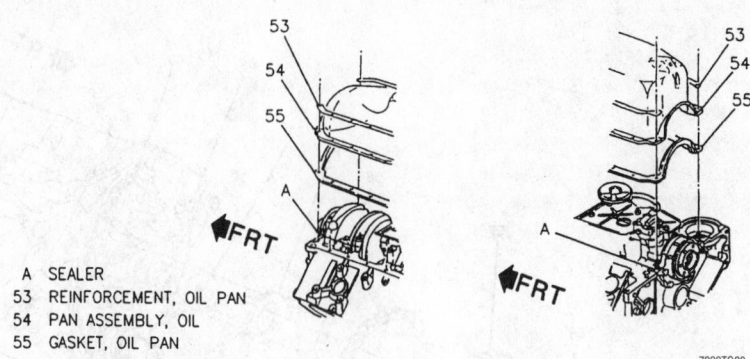

A SEALER
53 REINFORCEMENT, OIL PAN
54 PAN ASSEMBLY, OIL
55 GASKET, OIL PAN

7922TG08

Oil pan and related components. Apply RTV sealant to the areas shown

6. Disconnect the exhaust crossover pipe from the exhaust manifolds, then remove the pipe hanger bolts and reposition the pipe.
7. If equipped, remove the engine oil cooler hose bracket nut from the oil pan, then remove the oil cooler bolt from the oil cooler assembly. Reposition the engine oil cooler assembly.
8. Remove the transmission fluid cooler lines from the oil pan clip and remove the torque converter cover.
9. Remove the start motor assembly.
10. Remove the engine mount through-bolts, then install a suitable engine jacking fixture and carefully raise the engine, watching the clearance between engine mounted components and the firewall.

※※ WARNING

Never raise the engine using a jack under the oil pan, crankshaft pulley or any sheet metal. Because there only is a small clearance between the oil pan and the oil pump screen, if the pan is bent even slightly, damage could occur to the pump screen and pick-up unit.

11. Remove the oil pan bolts, studs and nuts, then lower the oil pan assembly along with the gasket and the pan reinforcements. If necessary, rotate the crankshaft to reposition the counterweights.
  To install:
12. Thoroughly clean the gasket mating surfaces of any remaining sealer and/or gasket material.
13. Apply a small amount of RTV sealer, 1052914 or equivalent, to the front cover-cylinder block junction and to the rear seal retainer-cylinder block junction. Continue the bead of sealer for 1 in. (25mm) in either direction of the radius cavity of these junctions.

14. Install the oil pan and gasket assembly, using the reinforcements, nuts, bolts and studs.
15. Tighten the corner oil pan bolts, stud or nuts to 15 ft. lbs. (20 Nm) and the remaining bolts or studs to 100 inch lbs. (11 Nm).
16. Lower the engine, remove the jacking fixture and install the engine mount through-bolts.
17. Install the oil level sensor assembly and tighten to 16 ft. lbs. (22 Nm), then engage the wiring harness connector.
18. Install the starter motor assembly.
19. Install the torque converter cover, then secure the transmission fluid cooler lines with the oil pan clip.
20. If equipped, install the engine oil cooler bolts screw to the cooler assembly and tighten to 24 ft. lbs. (33 Nm), then install the oil cooler hose bracket nut to the oil pan.
21. Install the exhaust pipe hanger bolt, then connect the crossover pipe to the exhaust manifolds.
22. Lower the vehicle and refill the engine crankcase with clean engine oil.
23. Install the air intake duct and connect the negative battery cable.
24. Start the engine and check for leaks.

Oil Pump

REMOVAL & INSTALLATION

1. Disconnect the negative battery cable.
2. Remove the oil pan.
3. Remove the bolt(s) attaching the pump to the rear main bearing cap, then remove the pump and driveshaft extension.
4. Remove the oil pump baffle retaining nuts and remove the baffle.

To install:

➡The oil pump pick-up should be submerged in oil and the pump primed prior to installation. Failure to prime the pump may result in oil pump failure or internal engine damage. Also, if the pick-up screen and pipe assembly was removed from the pump, they must be replaced to assure a proper interference fit.

5. If the pick-up screen and pipe was removed, it should be replaced as an assembly with a new part. Using a suitable tool, install a new pick-up screen and pipe to the oil pump

6. Align the slot or hexagon head on the end of the shaft extension with the drive tang or the hexagon socket on the distributor shaft and position the oil pump on the rear main bearing cap. Do not install the bolt at this time.

7. Position the oil pump baffle before installing the pump retaining bolt.

8. Install the baffle nuts and tighten to 25 ft. lbs. (34 Nm). Tighten the oil pump bolt to 66 ft. lbs. (90 Nm).

9. Install the oil pan assembly to the engine.

10. Connect the negative battery cable, then check engine oil level and add, as necessary.

11. Start the engine while watch the indicator light or oil pressure gauge to ensure immediate oil pump operation.

Rear Main Seal

REMOVAL & INSTALLATION

The transmission assembly must be removed to perform this service.

1. Disconnect the negative battery cable.
2. Raise and safely support the vehicle.
3. Remove the transmission assembly using the recommended procedure.
4. Remove the flywheel mounting bolts and remove the flywheel.
5. Remove the rear crankshaft oil seal by prying the seal out of the seal carrier using a suitable tool in the notches provided.

To install:

6. A special tool is used to remove the seal and install it. If not available, use the following procedure.

a. Unbolt the crankshaft rear oil seal housing assembly.

b. Locate the notches in the housing. Using a suitable drift pin, carefully tap the old seal out of the housing.

c. Clean the housing well. Tap in a

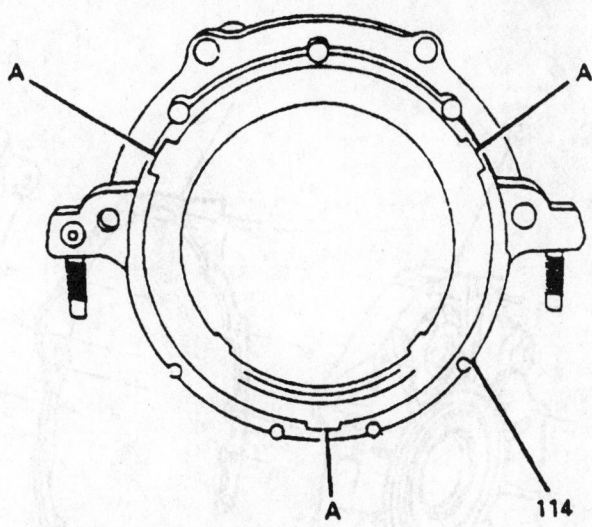

A SEAL REMOVAL NOTCHES
114 HOUSING ASSEMBLY, CRANKSHAFT REAR OIL SEAL

7922TG09

Rear main seal housing and removal notches

new seal using care to keep the seal square to the housing.

d. Install the housing and seal back onto the rear of the block and crankshaft. Oil the seal well with clean engine oil to help it slide over the end of the crankshaft.

7. If tool J-35621 or equivalent is available, use the following procedure.

a. Install the seal assembly on tool J-35621 or the equivalent seal installation tool. Tighten the tool retainer screws snugly to ensure the seal will be installed squarely over the end of the crankshaft.

b. Install the rear crankshaft oil seal by tightening the wing nut on the special tool, if used, until the tool bottoms out.

c. Remove J-35621.

8. Install the flywheel and tighten the mounting bolts to 74 ft. lbs. (100 Nm). New replacement bolts are recommended.

9. Install the transmission in the vehicle using the recommended procedure.

10. Lower the vehicle.

11. Check engine oil level.

12. Connect the negative battery cable.

Timing Chain, Sprockets, Front Cover and Seal

REMOVAL & INSTALLATION

1. Disconnect the negative battery cable.
2. Drain the engine oil and coolant into suitable containers.

3. Remove the throttle body air intake duct.
4. Remove the serpentine drive belt.
5. Remove the water pump assembly.
6. Remove the crankshaft balancer and hub.

a. If not done already, raise and support the vehicle safely.

b. Remove the crankshaft balancer retaining bolts, then remove the balancer from the hub.

c. Matchmark the crankshaft hub to the engine front cover, then remove the hub bolt and washer.

d. Remove the crankshaft hub using J-39046 or an equivalent hub removal/installation tool. To preserve the relationship between the hub and crankshaft, DO NOT crank the engine over once the hub has been removed. If the hub is not matchmarked and installed in the original position, an engine imbalance could result.

7. Remove the distributor assembly.
8. Remove the oil pan assembly.
9. Remove the engine front cover bolts.
10. Remove the engine front cover and gasket.
11. Remove the crankshaft seal from the front cover with a suitable seal driver.
12. Rotate the crankshaft until the timing marks on the timing chain sprockets are aligned nearest each other. The camshaft sprocket mark should be at the 6 o'clock position while the mark on the crankshaft sprocket should be at the 12 o'clock position.

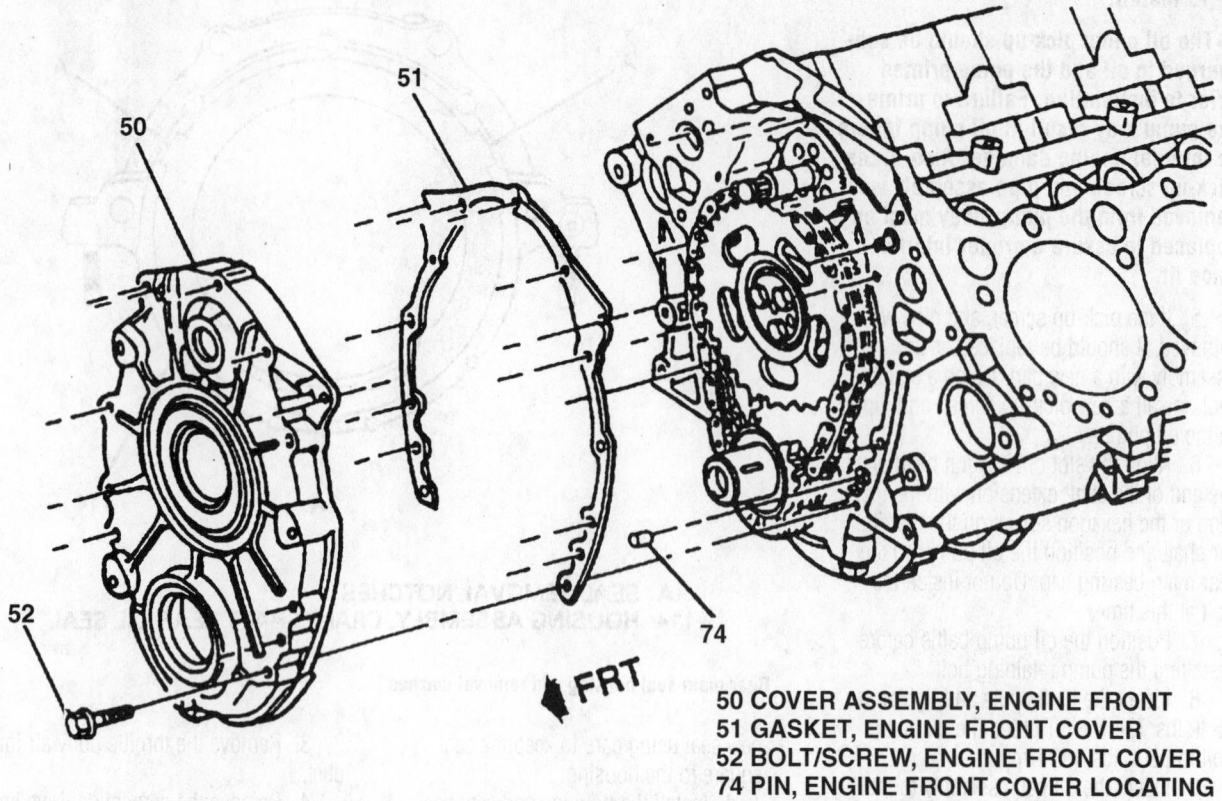

50 COVER ASSEMBLY, ENGINE FRONT
51 GASKET, ENGINE FRONT COVER
52 BOLT/SCREW, ENGINE FRONT COVER
74 PIN, ENGINE FRONT COVER LOCATING

7922TG11

Exploded view of the timing chain front cover mounting—the crankshaft seal is easier to replace while the front cover is off

13. Remove the camshaft sprocket bolts.

14. Remove the camshaft sprocket and timing chain.

➡**To prevent piston or valve damage, do not turn the crankshaft after the timing chain has been removed.**

15. Remove the water pump bearing retainer bolts, then remove the driveshaft assembly using J-39243 or equivalent driven gear assembly remover.

16. Remove the crankshaft sprocket using J-5825-A or equivalent crankshaft sprocket remover.

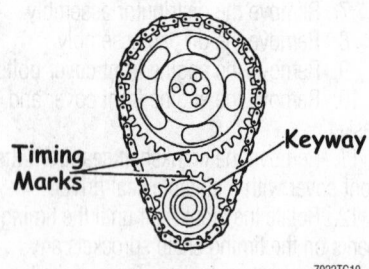

7922TG10

Align the timing marks on the gears before removing the chain

17. If necessary, remove the crankshaft key.

To install:

18. If removed, install the crankshaft key.

19. Install the crankshaft sprocket using J-5590 or an equivalent installation tool.

20. Install the water pump driveshaft assembly using J-39092 or an equivalent installer tool. Install the retainer bolts and tighten to 105 inch lbs. (12 Nm).

21. Align the timing marks and install the camshaft sprocket and timing chain. The gears of the camshaft sprocket and water pump driveshaft must mesh or damage to the thrust plate retainer could occur.

22. Install the camshaft sprocket bolts and tighten to 21 ft. lbs. (28 Nm).

23. Install a new O-ring to the water pump driven gear shaft using J-39089 or an equivalent seal installation tool.

24. Thoroughly clean the engine front cover and cylinder block gasket mating surfaces. Inspect the engine front cover for damage, replace as necessary. Install a new seal in the front cover.

25. Using J-39087 or equivalent front cover seal protector on the water pump driveshaft, install the gasket and front cover

into position over the shafts and guide pins.

26. Install the engine front cover bolts and tighten to 100 inch lbs. (11 Nm).

27. Install the oil pan and gasket.

28. Install the distributor assembly.

29. Install the crankshaft hub and torsional damper assembly.

a. Align the matchmarks made earlier and install the crankshaft hub using the hub tool. If the engine was cranked and the matchmarks were lost, set the engine to No. 1 TDC, then install the crankshaft hub with the cast arrow in the 12 o'clock position.

b. Install the hub washer and bolt, but do not torque at this time.

c. Install the crankshaft balancer to the hub, then tighten the crankshaft hub bolt to 75 ft. lbs. (102 Nm) and the balancer bolts to 60 ft. lbs. (81 Nm).

➡ **If a new balancer is installed, new balancer weights of the same size must be installed in the same hole locations as the original balancer.**

30. Install the water pump assembly.

31. Install the serpentine drive belt.

32. Install the throttle body air duct.

33. Properly fill the engine crankcase with clean engine oil.

34. Properly fill the engine cooling system.

35. Connect the negative battery cable, operate the engine and check for leaks.

FUEL SYSTEM

Fuel System Service Precautions

Safety is the most important factor when performing not only fuel system maintenance but any type of maintenance. Failure to conduct maintenance and repairs in a safe manner may result in serious personal injury or death. Maintenance and testing of the vehicle's fuel system components can be accomplished safely and effectively by adhering to the following rules and guidelines.

• To avoid the possibility of fire and personal injury, always disconnect the negative battery cable unless the repair or test procedure requires that battery voltage be applied.

• Always relieve the fuel system pres-

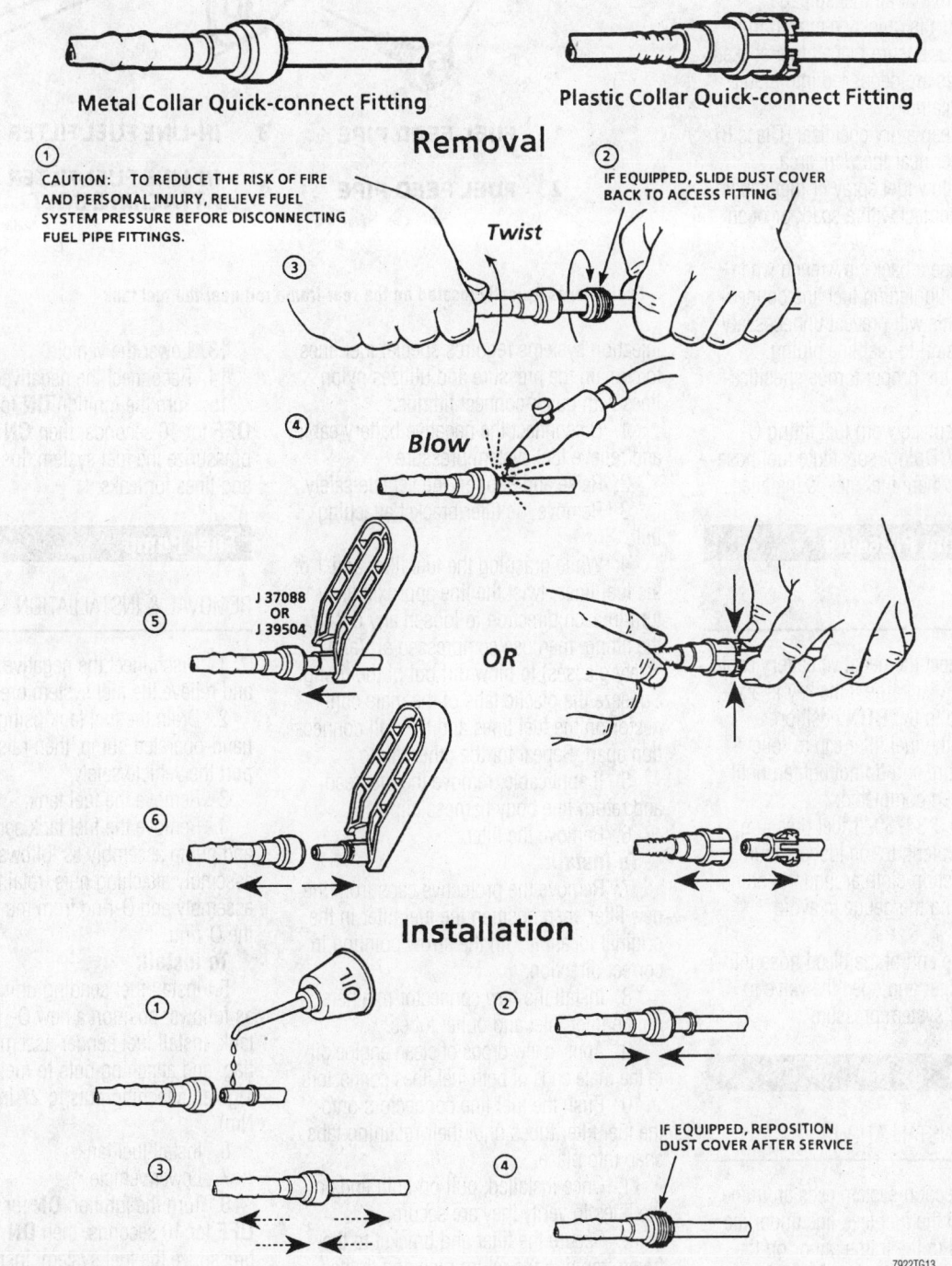

Metal Collar Quick-connect Fitting

Plastic Collar Quick-connect Fitting

Removal

① CAUTION: TO REDUCE THE RISK OF FIRE AND PERSONAL INJURY, RELIEVE FUEL SYSTEM PRESSURE BEFORE DISCONNECTING FUEL PIPE FITTINGS.

② IF EQUIPPED, SLIDE DUST COVER BACK TO ACCESS FITTING

③ *Twist*

④ *Blow*

⑤ J 37088 OR J 39504

OR

⑥

Installation

① OIL

②

③

④ IF EQUIPPED, REPOSITION DUST COVER AFTER SERVICE

7922TG13

Two different types of fuel line connections may be used, use the diagrams to determine which type you have and how to disconnect them

sure prior to disconnecting any fuel system component (injector, fuel rail, pressure regulator, etc.), fitting or fuel line connection. Exercise extreme caution whenever relieving fuel system pressure to avoid exposing skin, face and eyes to fuel spray. Please be advised that fuel under pressure may penetrate the skin or any part of the body that it contacts.

• Always place a shop towel or cloth around the fitting or connection prior to loosening to absorb any excess fuel due to spillage. Ensure that all fuel spillage (should it occur) is quickly removed from engine surfaces. Ensure that all fuel soaked cloths or towels are deposited into a suitable waste container.

• Always keep a dry chemical (Class B) fire extinguisher near the work area.

• Do not allow fuel spray or fuel vapors to come into contact with a spark or open flame.

• Always use a back-up wrench when loosening and tightening fuel line connection fittings. This will prevent unnecessary stress and torsion to fuel line piping. Always follow the proper torque specifications.

• Always replace worn fuel fitting O-rings with new. Do not substitute fuel hose or equivalent, where fuel pipe is installed.

Fuel System Pressure

RELIEVING

1. Disconnect the negative battery cable to prevent fuel discharge if the key is accidentally turned to the **RUN** position.
2. Loosen the fuel filler cap to relieve the tank pressure and do not tighten until service has been completed.
3. Connect J-34730–1 fuel pressure gauge or equivalent, to the fuel pressure valve. Wrap a shop cloth around the fitting while connecting the gauge to avoid spillage.
4. Place the end of the bleed hose into a suitable container and open the valve to relieve the fuel system pressure.

Fuel Filter

REMOVAL & INSTALLATION

The fuel injection system uses an inline filter located in the fuel feed line under the hood, attached to the frame rail or on the rear crossmember of the vehicle. The high pressure fuel system used with all fuel

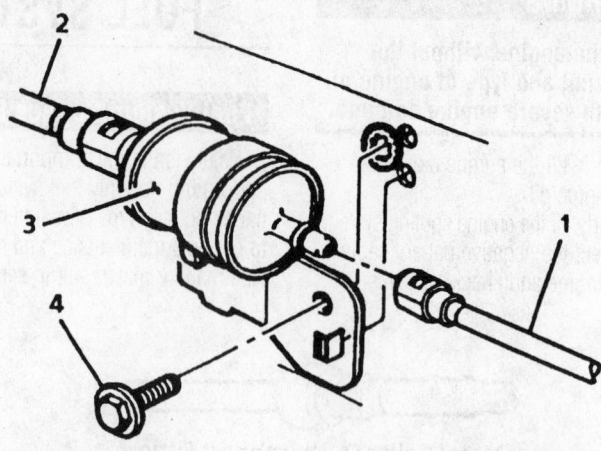

| 1 | FUEL FEED PIPE | 3 | IN-LINE FUEL FILTER |
| 2 | FUEL FEED PIPE | 4 | IN-LINE FUEL FILTER ATTACHING BOLT |

7922TG12

Fuel filter is typically located on the rear frame rail near the fuel tank

injection systems requires special fuel lines to contain the pressure and utilizes nylon lines with quick-connect fittings.

1. Disconnect the negative battery cable and relieve fuel system pressure.
2. Raise and support the vehicle safely.
3. Remove the filter bracket attaching bolt.
4. While grasping the fuel filter and 1 of the fuel lines, twist the line approximately ¼ turn in each direction to loosen any dirt in the fitting, then use compressed air (and safety glasses) to blow dirt out of the fitting. Squeeze the plastic tabs of the male connector on the fuel lines and the pull connection apart. Repeat for the other fitting.
5. If applicable, remove the fuel feed and return line body harness clips.
6. Remove the filter.

To install:
7. Remove the protective caps from the new filter, then position the fuel filter in the original location with the arrow pointing in correct direction.
8. Install the new connector retainers on the filter inlet and outlet tubes.
9. Apply a few drops of clean engine oil to the male ends of both fuel lines connectors.
10. Push the fuel line connectors onto the fuel filter tubes until their retaining tabs snap into place.
11. Once installed, pull on both ends of the lines to verify they are secure.
12. Secure the filter and bracket to the frame trapping the return pipe and tighten the attaching bolt to 89 inch lbs. (10 Nm).

13. Lower the vehicle.
14. Reconnect the negative battery cable.
15. Turn the ignition **ON** for 2 seconds, **OFF** for 10 seconds, then **ON** again to pressurize the fuel system. Inspect the tank and lines for leaks.

Fuel Pump

REMOVAL & INSTALLATION

1. Disconnect the negative battery cable and relieve the fuel system pressure.
2. Drain the fuel tank using a suitable hand-operated pump, then raise and support the vehicle safely.
3. Remove the fuel tank.
4. Remove the fuel tank sending unit and pump assembly as follows: remove the assembly attaching nuts, retaining flag, assembly and O-ring from the tank. Discard the O-ring.

To install:
5. Install fuel sending unit in fuel tank as follows: position a new O-ring on fuel tank. Install fuel sender assembly, retaining flag, and attaching nuts to fuel tank. Tighten attaching nuts to 27 inch lbs. (3 Nm).
6. Install fuel tank.
7. Lower vehicle.
8. Turn the ignition **ON** for 2 seconds, **OFF** for 10 seconds, then **ON** again to pressurize the fuel system. Inspect the fuel system for leaks.

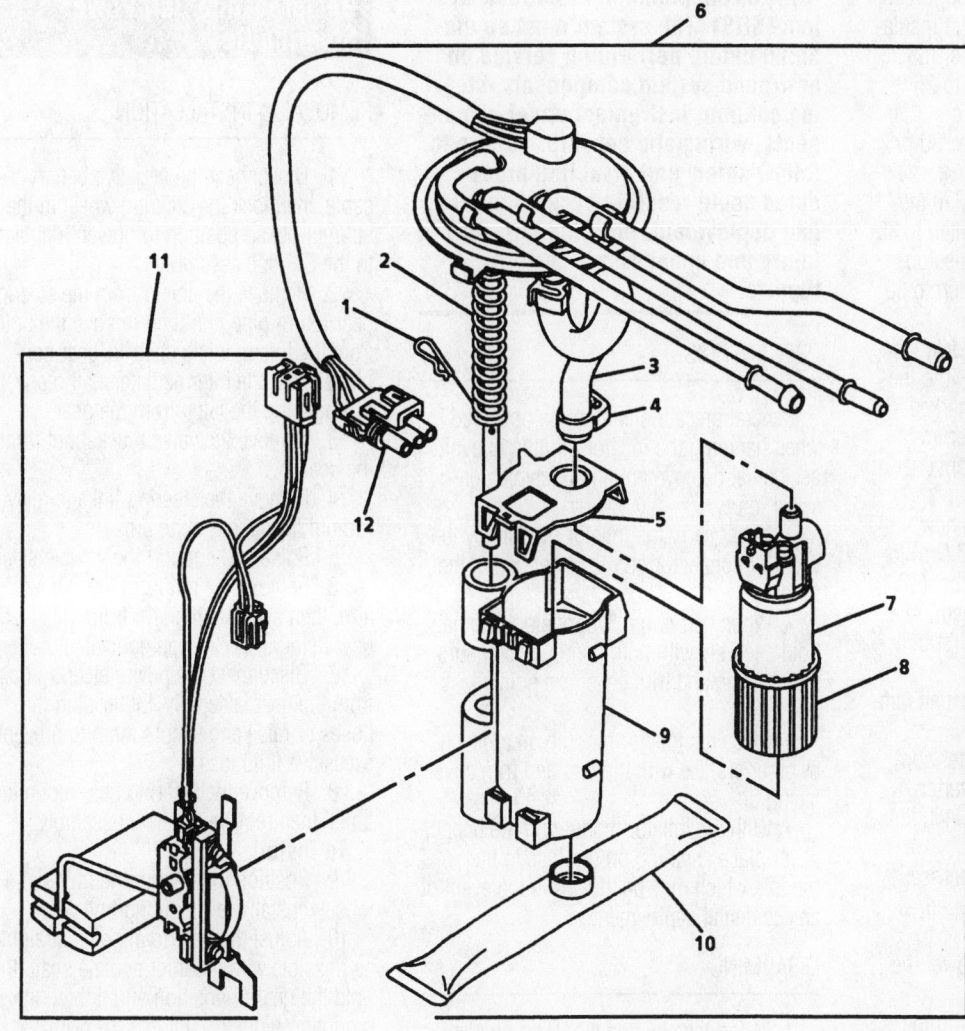

1	**RETAINING CLIP**
2	**SPRING**
3	**FUEL FEED HOSE**
4	**CLAMP**
5	**HOUSING COVER**
6	**FUEL SENDER ASSEMBLY**
7	**FUEL PUMP**
8	**FUEL PUMP ISOLATOR**
9	**FUEL PUMP HOUSING**
10	**FUEL PUMP STRAINER**
11	**FUEL LEVER SENSOR ASSEMBLY**
12	**FUEL SENDER ELECTRICAL CONNECTOR**

7922TG14

Exploded view of the fuel pump, showing all related components

DRIVE TRAIN

Transmission Assembly

REMOVAL & INSTALLATION

Automatic

1. Disconnect the battery negative cable and remove the air cleaner.
2. Disconnect the Throttle Valve (TV) cable at the throttle lever.
3. Remove the transmission dipstick, then remove the indicator tube from the transmission.
4. Raise and support the vehicle safely.
5. Remove the driveshaft.
6. Disconnect the shift linkage at the transmission.

7. Disengage all electrical leads at the transmission and any clips that retain the leads to the transmission case.
8. Remove the retaining bolts, then remove the flywheel cover.
9. Matchmark the flywheel and converter for installation purposes.
10. Remove and discard the torque converter-to-flywheel bolts.
11. Remove the catalytic converter support bracket.
12. Support and raise the transmission slightly using a suitable transmission jack.
13. Remove the transmission mount-to-support nut, washer, and bolt.
14. Remove the transmission support-to-frame bolts, nuts, if used, insulators.
15. Slide the transmission support rearward.
16. Lower the transmission to gain access to the oil cooler lines and TV cable attachments.

17. Disconnect the lines and cap all openings to prevent excessive fluid loss or system contamination, then disconnect the TV cable.
18. Support the engine with a suitable tool, then and remove the transmission-to-engine bolts.
19. Install tool J-21366 or equivalent, to the torque converter or converter clutch in order to hold it in place.
20. Remove the transmission assembly from the vehicle.
   To install:
21. Raise the transmission into place and remove tool J-21366.
22. Install the transmission-to-engine bolts and tighten to 35 ft. lbs. (47 Nm).
23. Unplug and install the oil cooler pipes, then connect the TV cable.
24. Install the fluid level tube using a new seal, then tighten tube retaining bolt to 35 ft. lbs. (47 Nm).

25. Install the transmission support-to-frame bolts, nuts, if applicable, the insulators. Tighten the bolts to 41 ft. lbs. (55 Nm) for Fleetwood Brougham or to 25 ft. lbs. (34 Nm) except for Fleetwood Brougham, if equipped the nuts to 30 ft. lbs. (41 Nm).

26. If removed or replaced, install the transmission mount bolts and tighten to 35 ft. lbs. (47 Nm). Install the transmission support nut and washer, then tighten to 30 ft. lbs. (41 Nm).

27. Remove the transmission jack, then position the converter by aligning it to the flywheel in the original position marked. Be sure the weld nuts on the converter are flush with the flywheel. Test the converter or clutch for freedom of movement.

28. Install and finger-tighten 3 new bolts, then tighten o 35 ft. lbs. (47 Nm) for Fleetwood Brougham or to 46 ft. lbs. (62 Nm) except for Fleetwood Brougham. After tightening all bolts, retorque the first bolt tightened.

29. If removed, install the floor pan reinforcement.

30. Install the catalytic converter support bracket, then install the converter cover and tighten the bolts to 89 inch lbs. (10 Nm).

31. Install the shift linkage, electrical leads, retaining clips, and if equipped, speedometer cable.

32. Install the driveshaft and lower the vehicle.

33. Install the TV cable to the throttle lever, then install the fluid level indicator.

34. Install the air cleaner, then connect the negative battery cable.

35. Adjust the shift linkage and the TV cable.

36. Flush the transmission and cooler system to prevent damage to the system components.

STEERING AND SUSPENSION

Air Bag

❊❊ CAUTION

Some vehicles are equipped with an air bag system, also known as the Supplemental Inflatable Restraint (SIR) or Supplemental Restraint System (SRS). The system must be disabled before performing service on or around system components, steering column, instrument panel components, wiring and sensors. Failure to follow safety and disabling procedures could result in accidental air bag deployment, possible personal injury and unnecessary system repairs.

PRECAUTIONS

Several precautions must be observed when handling the inflator module to avoid accidental deployment and possible personal injury.

• Never carry the inflator module by the wires or connector on the underside of the module.

• When carrying a live inflator module, hold securely with both hands, and ensure that the bag and trim cover are pointed away.

• Place the inflator module on a bench or other surface with the bag and trim cover facing up.

• With the inflator module on the bench, never place anything on or close to the module which may be thrown in the event of an accidental deployment.

DISARMING

1. Align the steering wheel so the vehicle wheels are pointing in the straight-ahead position.

2. Turn the ignition switch to the **LOCK** position.

3. Remove the SIR or AIR BAG fuse from the fuse block.

4. Remove the Connector Position Assurance (CPA) device, then disengage the yellow 2-way SIR wiring harness connector at the base of the steering column.

After the appropriate repairs have been made, re-enable the air bag system as follows:

5. Turn the ignition switch to the **LOCK** position.

6. Engage the yellow 2-way connector at the base of the steering column, then install the CPA device.

7. Reinstall the SIR or AIR BAG fuse.

8. Turn the ignition switch to the **RUN** position.

9. Verify the SIR indicator light flashes 7–9 times, if not, inspect system for malfunction.

Reciprocating Ball Power Steering Gear

REMOVAL & INSTALLATION

1. Disconnect the negative battery cable, then lock the steering wheel in the straight-ahead position to prevent damage to the SIR coil assembly.

2. Remove the shield from the steering gear return pipe nut, then remove the bolt from the intermediate shaft-to-gear coupling. Push the intermediate shaft rearward, disengaging the latch from the gear.

3. Remove the valve bracket nut from the gear.

4. Remove the steering linkage relay rod nut from the Pitman arm.

5. Raise and support the vehicle safely.

6. Remove the nut from the Pitman arm, then separate the arm from the steering gear using J-9172 or an equivalent puller.

7. Disconnect the power steering hoses from the gear assembly. Either plug the hoses or raise and secure them to prevent excessive fluid loss.

8. Remove the steering gear mounting bolts, then remove the gear assembly.

To install:

9. Position the gear to the frame and loosely install the mounting bolts.

10. Adjust the gear to align as straight as possible with the intermediate shaft, then hold the gear in position and tighten the mounting bolts to 70 ft. lbs. (95 Nm).

11. Install the power steering hose assemblies to the gear and tighten the fittings to 21 ft. lbs. (28 Nm).

12. Install the Pitman arm to the steering gear using a lockwasher and a new nut. Tighten the nut to 185 ft. lbs. (250 Nm) for Fleetwood Brougham or to 179 ft. lbs. (243 Nm) except for Fleetwood Brougham.

13. Lower the vehicle, then install the steering linkage relay rod nut and tighten to 35 ft. lbs. (47 Nm).

14. Install the valve bracket nut to the gear assembly and tighten to 18 ft. lbs. (24 Nm).

15. Install the intermediate shaft coupling bolt and nut, then tighten to 30 ft. lbs. (40 Nm) for Fleetwood Brougham or to 40 ft. lbs. (54 Nm) except for Fleetwood Brougham.

16. Position the shield, making sure the latch is seated around the gear return pipe nut.

17. Connect the negative battery cable, then properly bleed the steering system.

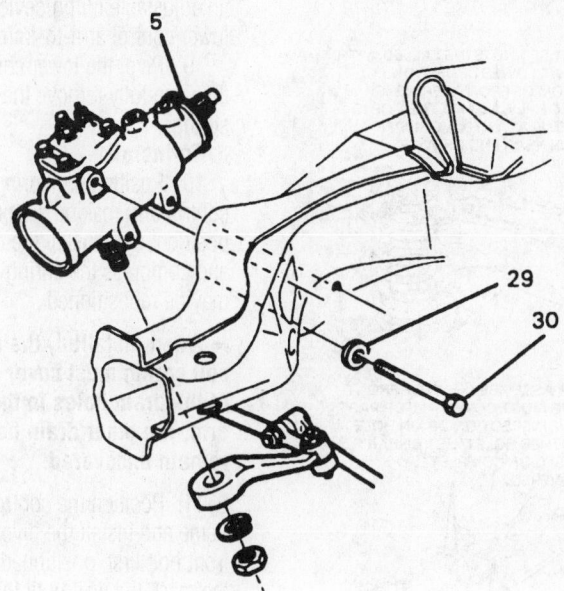

5   GEAR, POWER STEERING
29   WASHER, STEERING GEAR
30   BOLT/SCREW, STEERING GEAR

7922TG15

The steering gear is mounted on the left frame rail

Shock Absorber

REMOVAL & INSTALLATION

Front

1. Raise and support the vehicle safely.

2. Hold the shock absorber upper stem from turning and remove the upper nut, then remove the retainer and grommet.

3. Remove the 2 bolts and lockwashers securing the shock to the lower control arm, then lower the shock from the vehicle.

To install:

4. With the lower retainer and grommet in place over the upper stem, install the fully extended shock up through the lower control arm and spring. Be sure the upper stem passes through the mounting hole in the frame bracket.

5. Install the upper rubber insulator, retainer and attaching nut over the shock, then tighten the nut to 97 inch lbs. (11 Nm).

6. Install the shock lower pivot to the lower control arm and tighten the 2 attaching bolts to 20 ft. lbs. (27 Nm).

7. Lower the vehicle.

Rear

1. Raise and safely support the vehicle safely, making sure to properly support the rear axle housing.

2. If equipped with adjustable shocks, disconnect the air line by turning the spring clip 90 degrees and pulling gently on air line housing.

3. Remove the upper nuts and bolts from the shock absorber at the frame.

4. Using a wrench to hold the stud in place, remove the lower nut and washer from the shock at the rear axle housing. The stud must not be allowed to turn during this operation or damage may result in the bond between the bushing and stud.

5. Remove the shock from the vehicle.

To install:

6. Install the shock absorber and loosely connect the upper frame bolts and nuts.

7. Position the stud into the bracket on the axle housing, then install the nut and washer.

8. Either tighten the upper bolts at the frame to 20 ft. lbs. (27 Nm) or the nuts at the frame to 12 ft. lbs. (16 Nm), whichever is easier. Then, while holding the stud steady with a wrench, tighten the lower

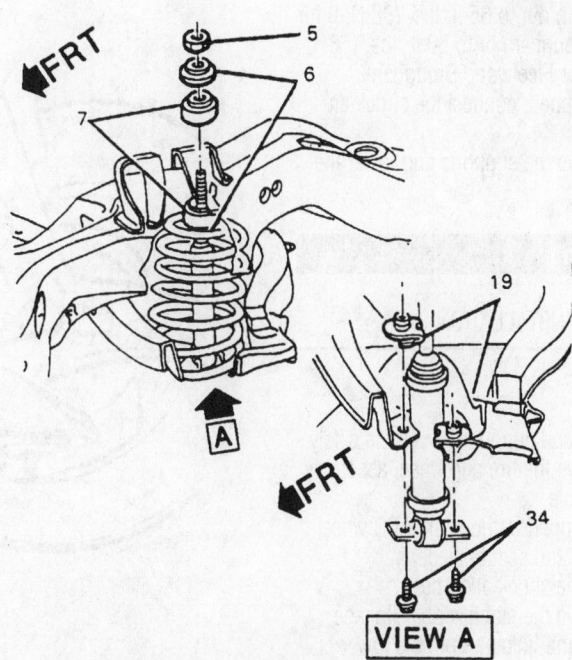

5   NUT, FRONT SHOCK ABSORBER, 11 N•m (97 LB. IN.)
6   RETAINER, FRONT SHOCK ABSORBER
7   INSULATOR, FRONT SHOCK ABSORBER
19   NUT, FRONT SHOCK ABSORBER
34   BOLT/SCREW, FRONT SHOCK ABSORBER, 27 N•m (20 LB. FT.)

7922TG16

One nut and two bolts secure the front shock absorber to the suspension and frame

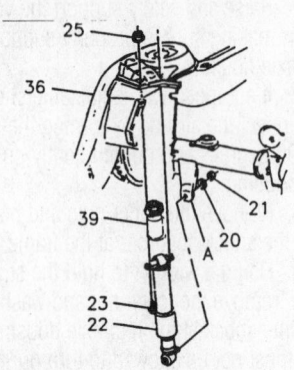

A BRACKET, REAR AXLE TUBE
20 WASHER, REAR SHOCK ABSORBER
21 NUT, REAR SHOCK ABSORBER,
     85 N•m (63 LB. FT.)
22 ABSORBER, REAR SHOCK
23 BOLT/SCREW, REAR SHOCK ABSORBER
25 NUT, REAR SHOCK ABSORBER,
     22 N•m (16 LB. FT.)
36 FRAME
39 SHIELD, REAR SHOCK ABSORBER BOOT
     HEAT (WITH ELC ONLY)

7922TG17

Two upper bolts and one through-bolt secure the rear shock absorber to the frame and axle assembly

shock retaining nut to 65 ft. lbs. (88 Nm) for Fleetwood Brougham or to 50 ft. lbs. (68 Nm) except for Fleetwood Brougham.

9. If equipped, connect the shock air line.

10. Remove the supports and lower the vehicle.

Coil Spring

REMOVAL & INSTALLATION

Front

1. Raise and support the vehicle safely.
2. Remove the tire and wheel assembly from the vehicle.
3. If equipped, remove the ABS wheel speed sensor and secure aside.
4. Remove shock absorber.
5. Remove the stabilizer bar linkage nut, retainer and linkage from the lower control arm.
6. Remove the cotter pin and castellated nut, then separate the tie rod from the steering knuckle, using a suitable puller tool.
7. Compress the coil spring using a universal spring compressor tool.
8. Support the lower control arm using

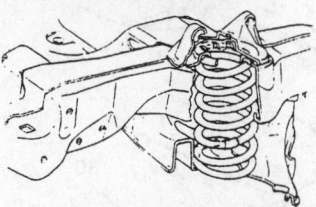

FRONT SPRING IS INSTALLED WITH TAPE AT LOWEST POSITION. BOTTOM OF FRONT SPRING IS COILED HELICAL AND THE TOP IS COILED FLAT WITH A GRIPPER NOTCH NEAR END OF WIRE.

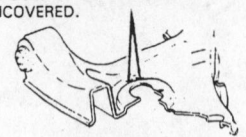

AFTER ASSEMBLY, END OF FRONT SPRING MUST COVER ALL OR PART OF ONE INSPECTION DRAIN HOLE. THE OTHER HOLE MUST BE PARTLY EXPOSED OR COMPLETELY UNCOVERED.

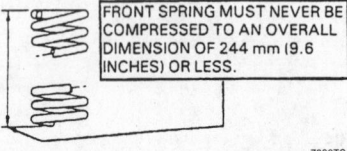

FRONT SPRING MUST NEVER BE COMPRESSED TO AN OVERALL DIMENSION OF 244 mm (9.6 INCHES) OR LESS.

7922TG18

Position the front coil spring as shown

an adjustable lifting device, then remove the lower control arm-to-frame pivot bolts.

9. Pivot the lower control arm rearward, then carefully remove the compressor and spring.

To install:

10. Position the spring onto the lower control arm making sure the insulator is in position, then install the compressor tool and compress the spring so the control arm may be repositioned.

➡**When installed, the lower end of the coil spring must cover all or part of one of the drain holes in the lower control arm, the other drain hole should remain uncovered.**

11. Position the control arm into the frame and install the pivot bolts (install the front bolt first, positioned from the front to the rear), but wait until the suspension is supporting the vehicle's weight before tightening the control arm fasteners to specification at the end of the procedure.

12. Remove the spring compressor tool, then install the tie rod end to the steering knuckle and tighten the nut to 35 ft. lbs. (47 Nm). Tighten the nut additionally, as necessary to align the hole, then install a new cotter pin.

A UNIVERSAL SPRING COMPRESSOR
10 SPRING, FRONT
39 ARM ASSEMBLY, FRONT LOWER CONTROL

7922TG19

Use a coil spring compressor installed through the lower control arm when removing the front coil spring

13. Remove the support from the lower control arm and install the stabilizer bar linkage.

14. Install the shock absorber. Tighten the lower attaching bolts to 20 ft. lbs. (27 Nm) and the upper attaching nut to 97 inch lbs. (11 Nm).

15. If equipped, install the ABS wheel speed sensor.

16. Install the wheel and lower the vehicle.

17. Tighten the lower control arm nuts to 92 ft. lbs. (125 Nm).

Rear

➡**If both springs are to be replaced, only disconnect 1 control arm at a time in order to prevent the axle from rolling or slipping sideways.**

1. Raise and support the vehicle safely, then place an adjustable support under the axle housing.

2. If equipped, disconnect the ABS rear speed sensor.

3. If equipped, disconnect the height sensor link from the upper control arm by removing the attaching nut and sliding the sensor link stud out of the hole in the upper control arm.

4. Disconnect the brake line fitting bolt at the center of the axle housing. No brake lines need to be disconnected, therefore brake bleeding should not be necessary.

5. Remove the nut and washer from the shock absorber, then disconnect the shock absorber from the axle bracket.

6. Carefully lower the axle housing sufficiently to remove the spring. Be careful not to stretch the brake hose.

7. Remove the spring, upper and lower insulator, as equipped.

To install:

8. Install the upper insulator, lower insulator, as equipped, and the rear spring to the bracket on the frame seat. Point the coil leg toward the left side of the vehicle, at a right angle from the centerline of the vehicle within 5 degrees rearward and 15 degrees forward.

➡**If the spring is being replaced, be sure to position the tape facing the top or spring noises could occur.**

9. Raise the rear axle back into position using the adjustable lifting device.

10. Install the rear shock absorber to the bracket, and tighten the nut to 50 ft. lbs. (68 Nm).

11. Connect the rear brake line fitting bolt and tighten to 20 ft. lbs. (27 Nm).

12. If equipped, connect the height sensor link to the upper control arm and tighten the nut to 27 inch lbs. (3 Nm).

13. Remove the adjustable support from under the rear axle assembly.

14. If equipped, connect the ABS rear axle speed sensor.

15. Lower vehicle, if necessary, adjust the height sensor.

Upper Ball Joint

REMOVAL & INSTALLATION

1. Raise and safely support the vehicle; place supports under the lower control arm between the spring seats and the ball joints.

➡**Leave the jack under the spring seat during removal and installation, in order to retain the coil spring and relieve spring tension from the upper control arm. The weight of the vehicle is used to relieve spring tension on the upper control arm.**

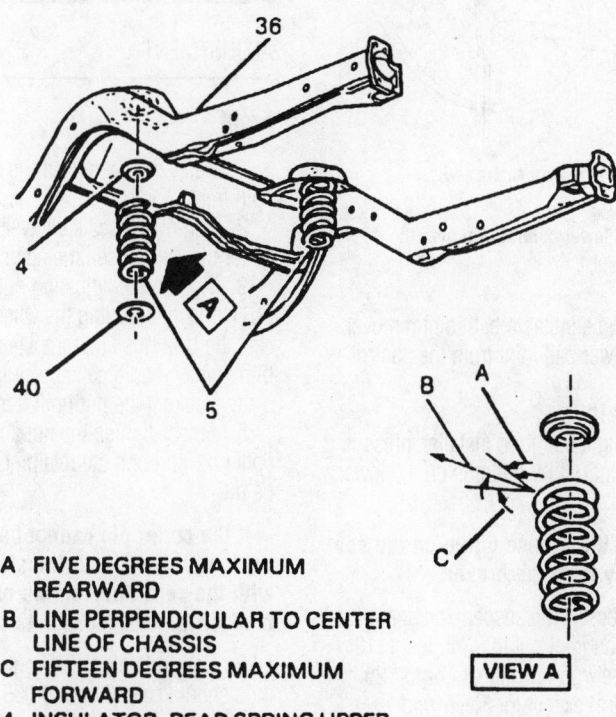

A FIVE DEGREES MAXIMUM REARWARD
B LINE PERPENDICULAR TO CENTER LINE OF CHASSIS
C FIFTEEN DEGREES MAXIMUM FORWARD
4 INSULATOR, REAR SPRING UPPER
5 SPRING, REAR
36 FRAME ASSEMBLY
40 INSULATOR, REAR SPRING LOWER (WAGON ONLY)

7922TG20

Position the coil spring leg perpendicular to the centerline of the chassis

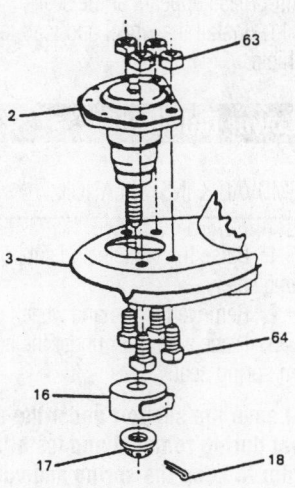

2 STUD, FRONT UPPER CONTROL ARM BALL
3 ARM, FRONT UPPER CONTROL
16 KNUCKLE, STEERING
17 NUT, STEERING KNUCKLE
    170 N•m (125 LB. FT.)
18 PIN, STEERING KNUCKLE COTTER
63 SERVICE KIT NUT
64 SERVICE KIT BOLT

7922TG21

The replacement kit comes with nuts and bolts to mount the ball joint in the control arm

2. Remove the wheel.

3. Remove the cotter pin and nut from the upper ball joint.

4. Using a ball joint separator tool, break the stud loose and pull the stud out of the knuckle. Support the steering knuckle to prevent damage to the brake line.

5. With the control arm in a raised position, use a ⅛ in. diameter bit and drill into each of the 4 rivet heads to a depth of ¼ in. (6mm).

6. Drill off the rivet heads with a ½ in. diameter bit.

7. Punch out the rivets using a suitable driver or punch, then remove the ball joint.

To install:

8. Place the new ball joint in the upper control arm and secure it with 4 bolts and nuts in place of rivets. Torque the nuts to specifications provided with the ball joint kit.

9. Connect the ball joint to steering knuckle. Torque the nut to 61 ft. lbs. (83 Nm), then insert a new cotter pin. Do not back off the specified torque in order to install the cotter pin.

➡️**When replacing the ball joints, use only high-quality replacement parts; bolts and nuts specified to be strong enough to endure the stress.**

10. Attach the grease fitting and lubricate until grease appears at the seal.

11. Install the wheel and road test the vehicle.

Lower Ball Joint

REMOVAL & INSTALLATION

1. Raise the vehicle and support the frame safely.

2. Remove the tire and wheel assembly.

3. Place a support under the control arm spring seat.

➡️**Leave the support under the spring seat during removal and installation, in order to keep the spring and control arm positioned.**

4. Remove the cotter pin and nut from the ball joint stud, then using a ball joint separator, remove the ball joint stud from the steering knuckle.

5. Guide the control arm down, past the brake shield (backing plate) with a flat bladed tool.

6. Block the steering knuckle aside using a block of wood between the frame and the upper control arm.

7. Remove the grease fittings.

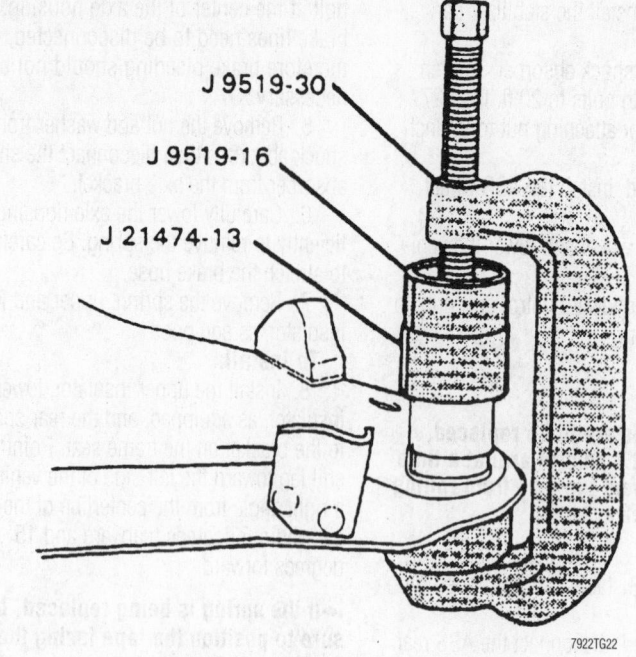

A special tool is needed to remove and install the lower ball joint in the control arm

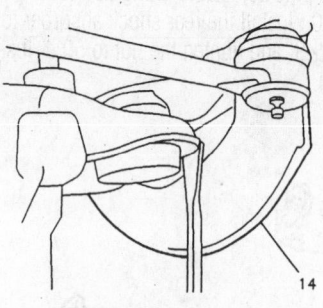

14 SHIELD, FRONT BRAKE

7922TG25

Guide the lower control arm over the brake shield

8. Using a suitable ball joint remover, drive the lower ball joint from the control arm.

To install:

9. Using a ball joint installer, press in a new ball joint until it bottoms on the lower control arm.

➡️**Be sure the grease purge on the seal faces away from the brakes.**

10. Assemble the suspension and torque the lower ball joint nut to 83 ft. lbs. (113 Nm) for Fleetwood Brougham or to 125 ft. lbs. (170 Nm) except for Fleetwood Brougham. Install the cotter pin and bend it to the side, not over the top of the nut.

11. Install the ball joint fitting and grease the new ball joint.

12. Install the tire and wheel assembly.

13. Check and adjust the wheel alignment, as necessary.

14. Lower and road test the vehicle.

Wheel Bearings

ADJUSTMENT

Front

1. Raise the vehicle so the wheel can spin freely.

2. Remove the wheel cover, dust cap, cotter pin and loosen the adjusting nut.

3. Tighten the adjusting nut to 12 ft. lbs. (16 Nm) while turning the wheel, this will seat the bearings and remove any grease or burrs which could cause play later.

4. Back off the nut until it is just loose.

5. Finger-tighten the nut and install the cotter pin through the retaining ring or castle nut.

➡️**If the cotter pin cannot be installed, back off the nut until the slot aligns with the serrations on the nut. Do not back off the nut more than ¼ of a turn.**

6. Once adjusted, the front wheel bearings should have 0.001–0.005 in. (0.03–0.13mm) end-play.

7. When adjusted properly cut off any extra length from the cotter pin to prevent interference, then install the dust cap and wheel cover.

8. Lower the vehicle.

Rear

The rear axle bearings are not adjustable. If the bearings become loose or make noise, they must be replaced.

REMOVAL & INSTALLATION

Front

1. Raise and support the vehicle safely.
2. Remove the tire and wheel assembly.
3. Remove the rotor and hub assembly.
4. Pry the inner bearing seal from the hub, then remove the inner roller bearing assembly.
5. If necessary, remove the inner and outer bearing races using tool J-29117-A or a suitable brass punch inserted behind the races.

To install:

6. Using fresh solvent, clean all old grease from hub, spindle and bearing.
7. If the inner and outer races were removed, press or drive the races into the hub using a suitable sized driver.
8. Pack the bearings with a high temperature wheel bearing grease and reassemble the hub. Do not mix greases.
9. Install a new inner bearing seal using a flat plate to assure the seal is flush with the rotor.
10. Apply a thin coat of grease to the spindle, then install the rotor and hub assembly on the steering knuckle.

11. Adjust the wheel bearings, install a new cotter pin and replace the dust cap.
12. Install the caliper assembly.
13. Install the tire and wheel assembly.
14. Lower the vehicle.

Rear

1. Raise and support the vehicle safely.
2. Remove the tire and wheel assembly, then remove the brake drum.
3. Clean all dirt from the rear carrier cover, then loosen the bolts and remove the cover while draining the gear oil. Discard the old gasket.
4. Remove the shaft lock bolt from the differential case located in the housing, then withdraw the pinion gear shaft.
5. Push the flanged end of axle shaft toward center of the vehicle, then remove C-lock from the end of the shaft located in housing.
6. Remove axle shaft from the housing. If replacement of the oil seal is not planned, be very careful not to damage the seal.
7. Remove the brake backing plate.
8. Remove seal from housing using a prybar behind the seal's steel case, being careful not to damage housing.
9. Insert an appropriately sized bearing remover into the bore and position it behind the bearing so tangs on tool engage bearing outer race. Remove the bearing, using a slide hammer.

To install:

10. Lubricate the new bearing with gear lubricant and install bearing using a suitable driver so the tool bottoms against the shoulder in the housing.
11. Lubricate seal lips with gear lubricant, then position the seal on a suitably sized driver and position seal into housing bore. Tap seal into place so it is flush with axle tube.
12. If removed, install the brake backing plate.
13. Insert the axle into the place while engaging the splines on the end of the shaft with the splines of the rear axle side gear. Be careful when inserting the axle not to damage the seal.
14. Install the C-lock on the bottom of the axle shaft and push the shaft outward so the lock seats in the counterbore of the rear axle side gear.
15. Install the rear axle pinion gear shaft through the differential case, thrust washers and pinions, align the hole in the shaft with the lock bolt hole. Install the lock bolt and tighten to 24 ft. lbs. (31 Nm) for 7½ in. ring gears or 20 ft. lbs. (27 Nm) for 8½ in. ring gear.
16. Position a new gasket, then install the carrier cover and tighten the bolt to 22 ft. lbs. (30 Nm) using a crosswise pattern.
17. Fill the rear assembly with the proper grade and type gear oil.
18. Install the brake drum, then install the tire and wheel assembly.
19. Lower the vehicle.

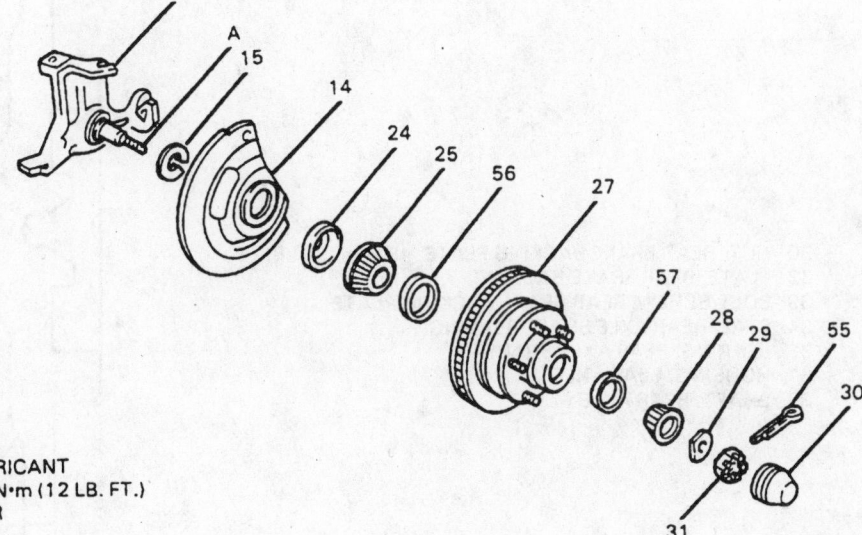

A   SPINDLE
14 SHIELD, FRONT BRAKE
15 GASKET, FRONT BRAKE SHIELD
16 KNUCKLE, STEERING
24 SEAL, FRONT WHEEL HUB
25 BEARING, FRONT WHEEL INNER
27 ROTOR, FRONT BRAKE
28 BEARING, FRONT WHEEL OUTER
29 WASHER, FRONT WHEEL BEARING
30 CAP, FRONT WHEEL BEARING LUBRICANT
31 NUT, FRONT WHEEL BEARING, 16 N•m (12 LB. FT.)
55 PIN, STEERING KNUCKLE COTTER
56 RACE, FRONT WHEEL INNER
57 RACE, FRONT WHEEL OUTER

Exploded view of the front wheel bearings

7922TG23

3   GEAR, DIFFERENTIAL RING
7   CASE, DIFFERENTIAL
28 BOLT/SCREW, DIFFERENTIAL BEARING CAP, 75 N•m
    (55 LB. FT.)
29 CAP, DIFFERENTIAL BEARING
37 HOUSING, REAR AXLE
38 LOCK, REAR AXLE SHAFT

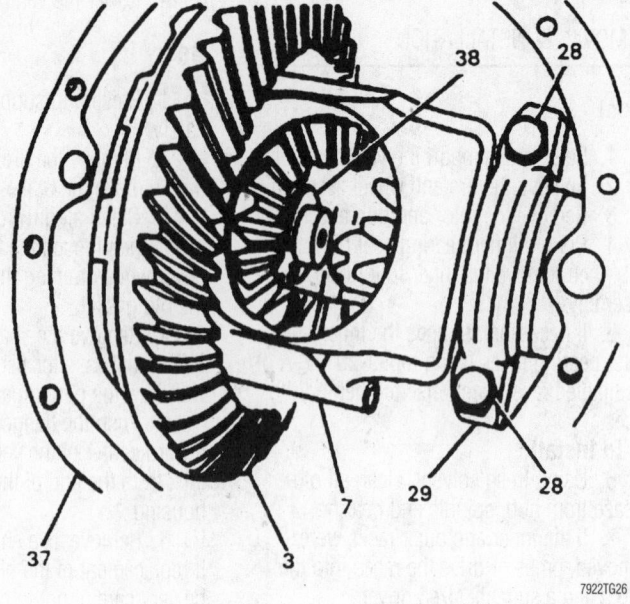

Differential component identification

30 NUT, REAR BRAKE BACKING PLATE, 48 N•m (35 LB. FT.)
32 PLATE, REAR BRAKE BACKING
33 BOLT/SCREW, REAR BRAKE BACKING PLATE
34 SEAL, REAR AXLE SHAFT BEARING
35 BEARING, REAR AXLE SHAFT
37 HOUSING, REAR AXLE
39 SHAFT, REAR AXLE

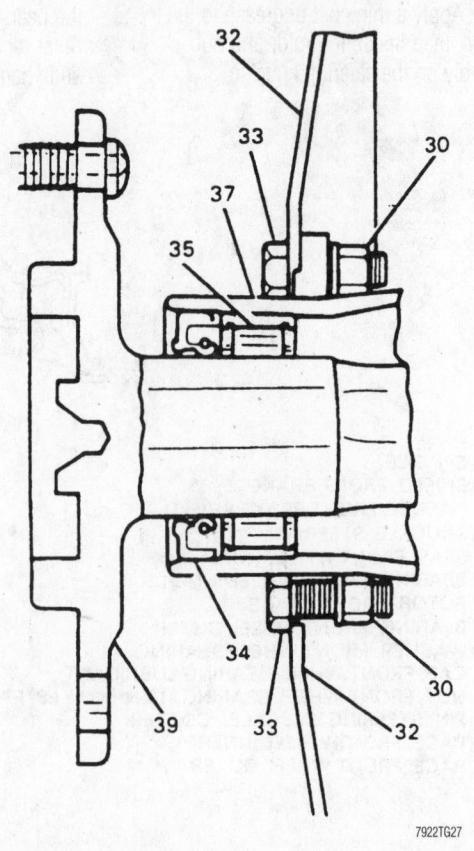

Cut-away view of the rear axle bearing and seal

GENERAL MOTORS C & H BODIES

Buick-Le Sabre • Park Ave. • **Oldsmobile-**Eighty Eight
Eighty Eight LS • LSS • Regency • **Pontiac-**Bonneville

29

DRIVE TRAIN 29-16
ENGINE REPAIR 29-2
FUEL SYSTEM 29-14
STEERING AND
  SUSPENSION 29-18

A
Air Bag.....................................29-18
  Disarming................................29-18
  Precautions 29-18
C
Camshaft and Valve Lifters 29-11
  Removal & Installation 29-11
Coil Spring 29-22
  Removal & Installation 29-22
Cylinder Head 29-5
  Removal & Installation 29-5
E
Engine Assembly 29-2
  Removal & Installation 29-2
Exhaust Manifold 29-10
  Removal & Installation 29-10
F
Fuel Filter 29-14
  Removal & Installation 29-14
Fuel Pump 29-15
  Removal & Installation 29-15

Fuel System Pressure 29-14
  Relieving 29-14
Fuel System Service
  Precautions.............................29-14
H
Halfshaft................................29-17
  Removal & Installation 29-17
I
Ignition Timing 29-2
  Adjustment.............................29-2
Intake Manifold 29-8
  Removal & Installation 29-8
L
Lower Ball Joint.........................29-24
  Removal & Installation 29-24
Lower Control Arm 29-25
  Removal & Installation 29-25
O
Oil Pan...................................29-12
  Removal & Installation 29-12
Oil Pump 29-12
  Removal & Installation 29-12
P
Power Rack and Pinion
  Steering Gear..........................29-19
  Removal & Installation 29-19

R
Rear Main Seal 29-13
  Removal & Installation 29-13
Rocker Arms 29-5
  Removal & Installation 29-5
S
Strut......................................29-20
  Removal & Installation 29-20
Supercharger 29-6
  Removal & Installation 29-6
T
Timing Chain, Sprockets,
  Front Cover and Seal.................29-13
  Removal & Installation 29-13
Transaxle Assembly 29-16
  Removal & Installation 29-16
V
Valve Lash 29-12
  Adjustment.............................29-12
W
Water Pump 29-4
  Removal & Installation 29-4
Wheel Bearings.........................29-25
  Adjustment.............................29-25
  Removal & Installation 29-25

ENGINE REPAIR

→Disconnecting the negative battery cable on some vehicles may interfere with the functions of the on board computer systems and may require the computer to undergo a relearning process, once the negative battery cable is reconnected.

Ignition Timing

The 3.8L (VIN L, K and 1) engines utilizes a Distributorless Ignition System (DIS). This system uses three twin tower coils which fire two spark plugs simultaneously. One spark plug is located in a cylinder on the compression stroke and the other in a cylinder on the exhaust stroke. This is known as the waste spark method of distribution.

ADJUSTMENT

The ignition timing is not adjustable, and is set according to engine demand electronically. The Powertrain Control Module (PCM) controls the ignition timing for all driving conditions.

Engine Assembly

REMOVAL & INSTALLATION

✳✳ CAUTION

The fuel injection system remains under pressure, even after the engine has been turned OFF. The fuel system pressure MUST BE relieved before disconnecting any fuel lines. Failure to do so may result in fire and/or personal injury.

1995 3.8L (VIN L and 1) Engines

1. Disconnect the negative battery cable.
2. Relieve the fuel system pressure using the recommended procedure.
3. Mark the position of the hood hinges for re-installation reference and remove the hood.

✳✳ CAUTION

Never open, service or drain the radiator or cooling system when hot; serious burns can occur from the steam and hot coolant.

4. Drain the cooling system.

✳✳ CAUTION

The EPA warns that prolonged contact with used engine oil may cause a number of skin disorders, including cancer! You should make every effort to minimize your exposure to used engine oil. Protective gloves should be worn when changing the oil. Wash your hands and any other exposed skin areas as soon as possible after exposure to used engine oil. Soap and water, or waterless hand cleaner should be used.

5. Drain the engine oil.
6. Remove the strut tower cross brace, if necessary.
7. Disconnect the heater hoses, the upper and lower radiator hoses from the engine.
8. Remove the radiator assembly.
9. Disconnect the starter wiring.
10. Disconnect the main wiring harness to the engine and the battery harness connectors located near the relay center.
11. Remove the serpentine belt(s).
12. Remove the power steering pump.
13. Remove the air cleaner and air inlet duct.
14. Disconnect the control cables from the throttle body and cable mounting bracket.
15. Disconnect the following sensors:
    a. Throttle Position Sensor (TPS).
    b. Idle Air Control (IAC) valve.
    c. MAT Sensor.
    d. Oxygen Sensor (O$_2$).
16. Disconnect the ignition coil assembly ground wire from the inner fender panel.
17. Disconnect and cap the fuel lines from the fuel rail.
18. Disconnect and tag all necessary vacuum lines from the engine.
19. Install engine support fixture J-28467-A or the equivalent, and raise the engine slightly to remove the weight from the engine mounts.
20. Raise and safely support the vehicle.
21. Disconnect the front exhaust pipe from the exhaust manifold.
22. Remove vacuum hoses from cruise control servo.
23. Remove the A/C compressor from the mounting bracket and support the compressor out of the way. DO NOT disconnect the refrigerant lines from the compressor.
24. Disconnect and cap the oil cooler lines, if equipped.
25. Remove the left front engine mount.
26. Remove the right front engine-to-transaxle bracket.

27. Remove the transaxle-to-engine bolts.
28. Remove the flexplate cover and the starter motor.
29. Remove the torque converter-to-flexplate bolts.

→Use a scribe to mark the flexplate and the torque converter relation. This will aid in proper reassembly.

30. Lower the vehicle and remove the torque axis engine mount.
31. Support the transaxle with a suitable jack.
32. Install a suitable lifting device and remove the support fixture.
33. Remove the engine from the vehicle.
To install:
34. Carefully install the engine in the vehicle.
35. Install the transaxle-to-engine bolts and tighten to 55 ft. lbs. (75 Nm).
36. Install the engine support fixture and remove the lifting device.
37. Raise and safely support the vehicle.
38. Install the torque axis engine mount.
39. Install the right front engine-to-transaxle bracket.
40. Install the left front engine mount.
41. Install the torque converter-to-flexplate bolts and tighten to 46 ft. lbs. (62 Nm).
42. Install the starter motor.
43. Install the flexplate cover.
44. Connect the engine oil cooler lines.
45. Install the A/C compressor.
46. Connect vacuum line to cruise control servo.
47. Connect the exhaust pipe to the exhaust manifold.
48. Lower the vehicle.
49. Connect the following sensors:
    a. Throttle Position Sensor (TPS).
    b. Idle Air Control (IAC) valve.
    c. MAT Sensor.
    d. Oxygen (O$_2$) Sensor.
50. Connect the control cables to the throttle body and cable bracket.
51. Install the air cleaner and air duct assembly.
52. Install the power steering pump.
53. Connect the main engine harness and battery harness connectors.
54. Connect the wiring to the starter.
55. Connect the fuel lines to the fuel rail.
56. Connect the engine vacuum lines.
57. Install the radiator assembly.
58. Connect the heater hoses and install the upper and lower radiator hoses.
59. Install the strut tower cross brace, if necessary.

60. Refill and bleed the cooling system following the proper procedure.

✳✳ WARNING

Operating the engine without the proper amount and type of engine oil will result in severe engine damage.

61. Refill the engine crankcase.
62. Connect the negative battery cable.
63. Install the hood, aligning the marks made during removal.
64. Start the engine and check for leaks.
65. Test all systems for proper operation.

3.8L (VIN K) and 1996–99 (VIN 1) Engines

1. Disconnect the negative battery cable.
2. Mark the position of the hood hinges, for installation reference, and remove the hood from the vehicle.
3. Relieve the fuel system pressure using the recommended procedure.

✳✳ CAUTION

The EPA warns that prolonged contact with used engine oil may cause a number of skin disorders, including cancer! You should make every effort to minimize your exposure to used engine oil. Protective gloves should be worn when changing the oil. Wash your hands and any other exposed skin areas as soon as possible after exposure to used engine oil. Soap and water, or waterless hand cleaner should be used.

4. Drain the engine coolant and engine oil.
5. Remove the radiator and heater supply hoses.
6. Remove the negative battery cable from the engine block.
7. Disconnect the engine harness at the bulkhead.
8. Remove the drive belts.
9. Remove the power steering pump and set aside.
10. Remove the air flow duct.
11. Remove the throttle cable from the throttle mounting bracket and other applicable cables.
12. Disconnect the following sensors:
    a. MAT sensor.
    b. Throttle Position Sensor (TPS).
    c. Idle Air Control valve (IAC).
    d. Oxygen sensor.
    e. Oil pressure switch.
    f. Power Steering cutoff switch.
    g. Vehicle Speed Sensor (VSS).
    h. Low oil level sensor.
13. Remove the ignition assembly ground strap from the fender inner panel attaching screws.
14. Remove the fuel feed and return lines from the fuel rail and fuel pressure regulator.
15. Remove the emission control canister hoses from the throttle body connections.
16. Remove the brake booster and heater control hoses from the engine vacuum connections.
17. Disconnect the vacuum hoses at the cruise control servo assembly.
18. Raise and safely support the vehicle.
19. Remove the exhaust pipe from the right manifold.
20. Remove the A/C compressor and tie back the compressor away from the engine. Do not disconnect the refrigerant lines.
21. Remove the right front engine-to-transaxle bracket.
22. Remove the flexplate cover.
23. Remove the starter motor.
24. Remove the torque converter-to-flexplate bolts.

➡ **Use a scribe to mark the flexplate to torque converter relationship for reassembly.**

25. Lower the vehicle.
26. Attach a suitable lifting hook and chain to the engine lifting brackets. Raise the engine slightly to take the weight off the engine mounts.
27. Remove the engine torque axis engine mount.
28. Support the transaxle and remove the engine-to-transaxle bolts.
29. Separate the engine from the transaxle, raise the engine and remove from the vehicle.
To install:
30. Lower the engine assembly into the vehicle.
31. Install and tighten the engine-to-transaxle bolts to 55 ft. lbs. (75 Nm).
32. Install and tighten the torque axis engine mount bolts to 52 ft. lbs. (87 Nm).
33. Raise and safely support the vehicle.
34. Align the torque converter and flexplate with the marks made during removal.
35. Install and tighten the converter to flexplate bolts to 46 ft. lbs. (62 Nm). .

36. Install the starter motor.
37. Install the flexplate cover.
38. Install the engine to transaxle bracket.
39. Remount the A/C compressor.
40. Connect the exhaust pipe to the manifold.
41. Lower the vehicle.
42. Connect the brake booster and heater control hoses to the engine vacuum connections.
43. Connect the vacuum hoses to the cruise control servo assembly.
44. Connect the emission control canister hoses to the throttle body connections.
45. Connect the fuel feed and return lines from the fuel rail and fuel pressure regulator.
46. Connect the ignition assembly ground strap to the fender inner panel attaching screws.
47. Connect the following sensors:
    a. MAT sensor.
    b. Throttle Position Sensor (TPS).
    c. Idle Air Control valve (IAC).
    d. Oxygen sensor.
    e. Oil pressure switch.
    f. Power Steering cutoff switch.
    g. Vehicle Speed Sensor (VSS).
    h. Low oil level sensor.
48. Connect the throttle cable to the throttle mounting bracket and connect other applicable cables.
49. Install the air flow duct.
50. Install the power steering pump.
51. Install the drive belts.
52. Connect the engine harness at the bulkhead.
53. Install the negative battery cable to the engine block.
54. Connect the radiator and heater supply hoses.

✳✳ WARNING

Operating the engine without the proper amount and type of engine oil will result in severe engine damage.

55. Refill engine oil.
56. Refill and bleed the engine cooling system following the recommended procedure.
57. Install the hood, aligning the marks made during removal.
58. Connect the negative battery cable.
59. Start the engine and check for leaks.
60. Road test the vehicle and check operation.

Water Pump

REMOVAL & INSTALLATION

3.8L (VIN L and K) Engines

1. Disconnect the negative battery cable.

> **✳✳ CAUTION**
>
> **Never open, service or drain the radiator or cooling system when hot; serious burns can occur from the steam and hot coolant.**

2. Drain the cooling system into an approved container.
3. Remove the accessory drive belt following the procedure in the belt removal section.
4. Disconnect the coolant hoses from the water pump.
5. Remove the water pump pulley bolts. (The long bolt can be removed by lining the head of the bolt up with the hole in the frame rail). Remove the pulley.
6. The following step is necessary only on 1995 engines:
   a. Support the engine using engine support fixture tool J 28467-A or equivalent.
   b. Remove the front engine mount.
7. Remove the water pump mounting bolts and remove the water pump.

To install:

8. Clean all the sealing surfaces.
9. Apply a thin bead of sealer around the outside edge of the water pump and install the gasket on the pump.
10. Install the water pump on the engine. Tighten the water pump-to-engine block bolts to 22 ft. lbs. (30 Nm) and the water pump-to-front cover bolts to 11 ft. lbs. (15 Nm) plus an additional 80 degree turn.
11. Install the front engine mount, if removed.

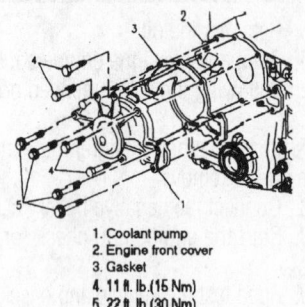

1. Coolant pump
2. Engine front cover
3. Gasket
4. 11 ft. lb. (15 Nm)
5. 22 ft. lb. (30 Nm)

7922UG01

Exploded view of the water pump mounting—3.8L (VIN L, K and 1) engines shown

12. Install the water pump pulley. Tighten the water pump pulley bolts to 115 inch lbs. (13 Nm).
13. Connect the coolant hoses to the water pump and secure clamps.
14. Install the drive belt following the procedure in the belt section.
15. Refill and bleed the cooling system following proper procedures.
16. Run the engine and check for leaks.
17. Recheck the coolant level when the engine has cooled.

3.8L (VIN 1) Engine

1995 MODELS

1. Disconnect the negative battery cable.

> **✳✳ CAUTION**
>
> **Never open, service or drain the radiator or cooling system when hot; serious burns can occur from the steam and hot coolant.**

2. Drain the cooling system into an approved container.
3. Remove the accessory drive and supercharger belts.
4. Remove the alternator and brace.
5. Disconnect the hoses and pipes from the water pump.
6. Remove the pulley bracket assembly.
7. Raise and safely support the vehicle.
8. Remove the power steering pump and lines.
9. Lower the vehicle.
10. Support the engine using engine support fixture J 28467-A or equivalent, and remove the front engine mount.
11. Remove the water pump pulley.
12. Remove the water pump mounting bolts and remove the water pump from the vehicle.

To install:

13. Clean all sealing surfaces.
14. Apply a thin bead of sealant to the gasket mating surface of the water pump and install a new gasket to the pump.
15. Install the water pump on the engine. Tighten the pump-to-block bolts to 22 ft. lbs. (30 Nm) and the pump-to-front cover bolts to 11 ft. lbs. (15 Nm) plus an additional 80 degree turn.
16. Install the water pump pulley and tighten the bolts to 115 inch lbs. (13 Nm).
17. Install the front engine mount following the proper procedure.
18. Raise and safely support the vehicle.
19. Install the power steering pump and lines following the proper procedure.

20. Lower the vehicle.
21. Install the pulley bracket assembly.
22. Connect the hoses and pipes from the water pump.
23. Install the alternator and brace.
24. Install the supercharger and accessory drive and belts.
25. Connect the negative battery cable.
26. Fill and bleed the cooling system following the proper procedure.
27. Run the engine and check for leaks.
28. Recheck the coolant level when the engine has cooled.

1996–99 MODELS

1. Disconnect the negative battery cable.
2. Drain the cooling system into an approved container.
3. Remove the A/C compressor splash shield.
4. Remove the supercharger and accessory drive belts.
5. Remove the coil pack and position out of the way.
6. Remove the supercharger belt tensioner.
7. Support the engine using engine support fixture J 28467-A or equivalent, and remove the front engine mount.
8. Remove the power steering pump.
9. Remove the engine mount bracket and the idler pulley.
10. Remove the water pump pulley.
11. Remove the water pump mounting bolts and remove the water pump.

To install:

12. Clean all sealing surfaces.
13. Apply a thin bead of sealer around the outside edge of the water pump and install a new gasket on the pump.
14. Install the water pump on the engine and tighten the short bolts to 11 ft. lbs. (15 Nm). Tighten the long bolts to 22 ft. lbs. (30 Nm).
15. Install the water pump pulley and tighten bolts to 115 inch lbs. (13 Nm).
16. Install the engine mount bracket and the idler pulley.
17. Install the power steering pump.
18. Install the front engine mount.
19. Install the supercharger belt tensioner.
20. Install the ignition coil pack assembly.
21. Install the supercharger and accessory drive belts.
22. Install the A/C compressor splash shield.
23. Connect the negative battery cable.
24. Fill and bleed the cooling system following the proper procedure.

25. Run the engine and check for leaks.

26. Recheck the coolant level when the engine has cooled.

Cylinder Head

REMOVAL & INSTALLATION

✳✳ CAUTION

The fuel injection system remains under pressure, even after the engine has been turned OFF. The fuel system pressure MUST BE relieved before disconnecting any fuel lines. Failure to do so may result in fire and/or personal injury.

3.8L Engines

1. Disconnect the negative battery cable.

2. Relieve the fuel system pressure using the recommended procedure.

✳✳ CAUTION

Never open, service or drain the radiator or cooling system when hot; serious burns can occur from the steam and hot coolant. Also, when draining engine coolant, keep in mind that cats and dogs are attracted to ethylene glycol antifreeze and could drink any that is left in an uncovered container or in puddles on the ground. This will prove fatal in sufficient quantities. Always drain coolant into a sealable container. Coolant should be reused unless it is contaminated or is several years old.

3. Drain the cooling system.

4. Remove the supercharger assembly (VIN 1 only).

5. Remove the intake manifold (VIN K only).

6. Remove the valve covers.

7. Tag and disconnect the ignition wires and remove the ignition coil/ module assembly.

8. Remove the alternator front mounting bracket.

9. Remove the one bolt securing the A/C bracket to the cylinder head.

10. Remove the power steering pump and set to the side. Complete removal of the steering pump is not needed.

11. Remove the accessory drive belt tensioner. For VIN 1 engines, remove the supercharger belt tensioner.

12. Remove the fuel pipe heat shield.

13. Remove the rocker arm assemblies and take note of their original position.

14. Remove the pushrods.

15. Remove the cylinder head bolts and discard.

16. Remove the cylinder head gaskets.

17. Clean all sealing surfaces and the cylinder head bolt holes in the block.

To install:

18. Place the cylinder head gasket on the engine block dowels with the note **THIS SIDE UP** facing the cylinder head and the arrow facing the front of the engine. Position the cylinder head on the engine block.

✳✳ WARNING

In order to prevent damage to the gasket, when installing the cylinder head, do not slide the cylinder head on the gasket. Head gaskets are not interchangeable. Failure to install with arrow pointing to the front will cause gasket failure and possible engine damage. Gaskets are identified by either an L or an R stamped on it next to the arrow.

19. Install the cylinder head bolts and tighten them as follows:

➥This engine uses special torque to yield head bolts. The procedure must be followed carefully and new bolts must be used whenever the head is removed. Total bolt torque should not exceed 60 ft. lbs. (81 Nm).

   a. Tighten the cylinder head bolts, in the proper sequence, to 37 ft. lbs. (50 Nm).

   b. Rotate each bolt 130 degrees, in sequence.

   c. Rotate the center four bolts an additional 30 degrees, in sequence.

20. Install the pushrods and guide plate.

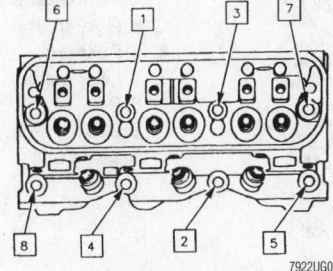

7922UG02

To avoid leakage and to prevent damage to the cylinder head, always tighten the bolts according to the sequence shown— 3.8L (VIN L, K and 1) engines

21. Install the rocker arm assemblies into their original location.(Apply approved thread lock compound to the rocker arm pedestal bolts before assembly).

22. Install the valve covers.

23. Install the fuel pipe heat shield.

24. Install the accessory drive belt tensioner. For VIN 1 engines, install the supercharger belt tensioner.

25. Install the power steering pump.

26. Install the A/C compressor bracket bolt, and tighten to 52 ft. lbs. (70 Nm).

27. Install the alternator front mounting bracket.

28. Install the ignition coil/ module assembly. Install the spark plug wires.

29. Install the supercharger assembly. (VIN 1 ONLY).

30. Install the intake manifold. (VIN K ONLY).

31. Fill and bleed the cooling system following the proper procedure.

32. Connect the negative battery cable.

33. Start the engine and check for leaks and proper operation.

Rocker Arms

REMOVAL & INSTALLATION

➥When removing valvetrain components, it is very important that they are marked for installation reference, so that they can be reinstalled in the same location from which they were removed.

3.8L Engines

1. Disconnect the negative battery cable.

2. If removing the left (front) rocker arm cover, proceed as follows:

   a. Disconnect the spark plug wires from the spark plugs and position the wires aside, out of the way.

   b. Remove the rocker arm cover bolts and pull off valve cover.

3. If removing the right (rear) rocker arm cover, proceed as follows:

   a. Disconnect the spark plug wires from the spark plugs and position the wires aside, out of the way.

4. Remove the rocker arm cover bolts and the rocker arm cover.

5. Remove the rocker arm pedestal bolts and remove the rocker arm and pedestal assembly. Keep all parts in order so they can be reinstalled in their original locations.

6. Inspect the rocker arms and pedestals for wear and/or damage; replace as necessary.

7. Remove the pushrods. Keep them in order so they can be reinstalled in their original locations.

8. Inspect the pushrod tips for wear and/or damage. Roll the pushrods on a flat surface to check for a bent condition. Replace any pushrod that is bent and/or damaged.

9. Clean all parts and all sealing surfaces. Be sure all thread adhesive is removed from the rocker arm pedestal bolts and rocker arm cover bolts.

To install:

10. Lubricate the pushrod tips with assembly lubricant and install them in their proper locations.

11. Lubricate the rocker arms and pedestals with assembly lubricant and install. Be sure the pushrod tips are properly seated in the rocker arms.

12. Apply suitable thread locking compound to the rocker arm pedestal bolt threads and install. Tighten the bolts to 11 ft. lbs. (15 Nm) plus an additional 90 degree turn.

13. Install the rocker arm cover using a new gasket. Apply suitable thread locking compound to the rocker arm cover bolts and install. Tighten to 89 inch lbs. (10 Nm).

14. Reposition the spark plug wires and connect them to the spark plugs.

15. Reinstall the power steering pump and/or alternator brace, as required.

16. Install the accessory drive belt(s).

17. Connect the negative battery cable.

18. Run the engine and check for leaks and proper engine operation.

Supercharger

REMOVAL & INSTALLATION

➡A small amount of oil seepage through the front seal, behind the pulley, of the supercharger is normal. This seepage is caused by minute traces of oil escaping around the seal due to normal pressure build up in the oil cavity within the supercharger. A build up of dust can stick to the thin oil film, which causes the oil seepage to appear worse than it really is. The supercharger should not be replaced for this seepage, however, if supercharger oil is visually dripping from the supercharger front seal, the supercharger will need to be replaced. The supercharger oil level should be checked every 30,000 miles or every 36 months.

✳✳ CAUTION

The fuel injection system remains under pressure, even after the engine has been turned OFF. The fuel system pressure must be relieved before disconnecting any fuel lines. Failure to do so may result in fire and/or personal injury.

1995 Engines

1. Disconnect the negative battery cable.

✳✳ CAUTION

Observe all applicable safety precautions when working around fuel. Whenever servicing the fuel system, always work in a well ventilated area. Do not allow fuel spray or vapors to come in contact with a spark or open flame. Keep a dry chemical fire extinguisher near the work area. Always keep fuel in a container specifically designed for fuel storage; also, always properly seal fuel containers to avoid the possibility of fire or explosion.

2. Relieve the fuel system pressure using the recommended procedure.

3. Drain the cooling system.

4. Remove the drive belt from the supercharger pulley.

5. Remove the engine dress up cover.

6. Disconnect and cap the fuel lines from the fuel rail.

7. Label and disconnect the vacuum lines from the supercharger.

8. Disconnect the electrical connectors from the fuel injector.

9. Remove the electrical harness shield from the front of the supercharger and detach the electrical connector from the supercharger.

10. Remove the fuel rail mounting bolts and remove the fuel rail.

11. Tag and detach the following connectors:

   a. IAC valve.

   b. Throttle Position Sensor (TPS).

   c. MAP sensor.

   d. MAF sensor.

   e. EGR valve.

   f. Boost control solenoid.

   g. Coolant Temperature Sensor.

12. Disconnect the air intake duct from the throttle body.

13. Disconnect the EGR pipe from the supercharger.

14. Disconnect the control cables from the throttle body.

15. Remove the boost pressure manifold and vacuum block.

16. Remove the control cable bracket.

17. Remove the tensioner bracket-to-supercharger mounting stud using the double-nut method.

➡The stud must be removed or the supercharger can not be lifted off the lower intake manifold mounting dowels.

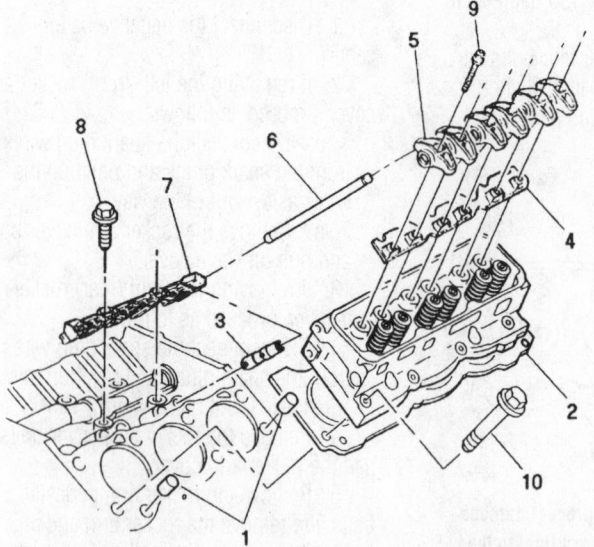

1. Dowel pin
2. Head gasket
3. Valve lifter
4. Pivot retainer
5. Rocker arm
6. Pushrod
7. Lifter guide
8. Bolt
9. Bolt
10. Head bolt

7922UG03

Exploded view of the rocker arms and related component mounting—3.8L engines

18. Remove the throttle body from the supercharger.

19. Remove the supercharger-to-intake manifold bolts.

20. Remove the supercharger, gasket and coolant passage O-rings.

To install:

21. Thoroughly clean the intake manifold and supercharger sealing surfaces. Be sure the locator pins are in their proper location on the intake manifold.

22. Install new coolant passage O-rings and new supercharger-to-intake manifold gasket. Do not use any sealant on the gasket.

23. Install the supercharger and the mounting bolts. Only tighten the bolts finger-tight at this time.

24. Install the tensioner bracket-to-supercharger stud and tighten to 88 inch lbs. (10 Nm).

25. Tighten the supercharger mounting bolts, gradually and evenly, to 19 ft. lbs. (26 Nm).

26. Install the tensioner bracket nut and tighten to 37 ft. lbs. (50 Nm).

27. Install the throttle body, using a new gasket, and tighten the mounting nuts to 11 ft. lbs. (15 Nm).

28. Install the boost pressure manifold.

29. Install the vacuum block with a new gasket and tighten the mounting bolt to 62 inch lbs. (7 Nm).

30. Install the control cable bracket.

31. Connect the control cables to the throttle body.

32. Connect the EGR pipe to the supercharger.

33. Connect the air intake duct to the throttle body.

34. Attach the following connectors:
   a. IAC valve.
   b. Throttle Position Sensor (TPS).
   c. MAP sensor.
   d. MAF sensor.
   e. EGR valve.
   f. Boost control solenoid.
   g. Coolant Temperature Sensor.

35. Install the fuel rail mounting and bolts.

36. Attach the electrical connector to the supercharger. Install the electrical harness shield onto the front of the supercharger.

37. Attach the electrical connectors to the fuel injectors.

38. Connect the vacuum lines to the supercharger.

39. Connect the fuel lines to the fuel rail.

40. Install the engine dress up cover.

41. Install the supercharger drive belt.

42. Connect the negative battery cable.

43. Fill and bleed the cooling system using the recommended procedure.

44. Run the engine and check for leaks and proper engine operation.

1996–99 Engines

1. Disconnect the negative battery cable.

2. Relieve the fuel system pressure using the recommended procedure.

✳✳ CAUTION

Never open, service or drain the radiator or cooling system when hot; serious burns can occur from the steam and hot coolant. Coolant should be reused unless it is contaminated or is several years old.

3. Drain the cooling system.

4. Remove the drive belt from the supercharger pulley.

5. Remove the engine dress up cover.

6. Remove the air duct from the throttle body.

7. Label and disconnect the right side spark plug wires from the ignition module and set aside.

8. Remove the alternator brace with purge solenoid.

9. Remove EGR wiring harness and shield.

10. Disconnect the fuel injectors.

11. Remove the MAP sensor bracket.

12. Disconnect and cap the fuel lines from the fuel rail.

13. Remove the fuel rail mounting bolts and remove the fuel rail.

14. Remove the boost control solenoid.

15. Remove the throttle body nuts and remove the throttle body.

16. Remove the bolts from the supercharger assembly.

17. Remove the Supercharger.

To install:

18. Thoroughly clean the supercharger and intake manifold sealing surfaces.

19. Install a new supercharger-to-intake manifold gasket. Do not use any sealer on this gasket.

20. Install the supercharger with the mounting bolts. Tighten the bolts, gradually and evenly, to 17 ft. lbs. (23 Nm).

21. Install the MAP sensor bracket.

22. Install the throttle body to the supercharger. Tighten the nuts to 89 inch lbs. (10 Nm).

23. Install the boost control solenoid. Tighten the nut to 72 inch lbs. (8 Nm).

24. Install the fuel rail and connect the fuel lines.

25. Attach the electrical connectors to the fuel injectors.

26. Install the alternator brace with purge solenoid.

27. Install the EGR wiring harness and shield.

1. Supercharger
2. Bolts
3. Stud
4. Bolt
5. Gasket

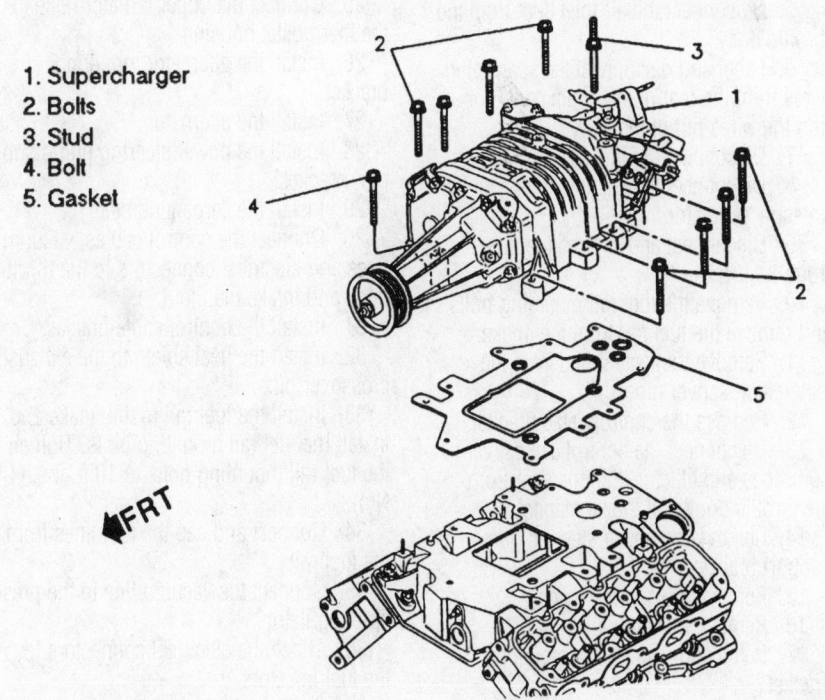

Exploded view of the supercharger installation

7922UG04

28. Connect the right side spark plug wires to the ignition module.

29. Install the supercharger belt.

30. Install the air duct to the throttle body.

31. Install the engine dress up cover.

32. Connect the negative battery cable.

33. Fill and bleed the cooling system using the recommended procedure.

34. Run the engine and check for leaks and proper engine operation.

Intake Manifold

REMOVAL & INSTALLATION

✳✳ CAUTION

The fuel injection system remains under pressure, even after the engine has been turned OFF. The fuel system pressure MUST BE relieved before disconnecting any fuel lines. Failure to do so may result in fire and/or personal injury.

3.8L (VIN L) Engine

1. Disconnect the negative battery cable.

2. Relieve the fuel system pressure using the recommended procedure.

3. Drain the cooling system.

4. Remove the plastic engine cover clipped to the fuel rail.

5. Disconnect the air inlet duct from the throttle body.

6. Label and disconnect the spark plug wires from the rear of the engine and position the wires out of the way.

7. Disconnect the fuel injectors.

8. Disconnect the vacuum line from the pressure regulator.

9. Disconnect and cap the fuel lines from the fuel rail.

10. Remove the fuel rail mounting bolts and remove the fuel rail from the intake.

11. Remove the heat shield from the exhaust crossover pipe.

12. Remove the control cable bracket.

13. Disconnect the control cables, vacuum lines and electrical connectors from the throttle body and intake manifold.

14. Remove the power steering pump support bracket.

15. Remove the serpentine belt.

16. Remove the alternator.

17. Remove the alternator mounting bracket.

18. Disconnect the upper radiator hose from the thermostat housing.

19. Disconnect the heater and bypass hoses from the intake manifold.

20. Remove the intake manifold bolts and remove the manifold and gasket.

To install:

21. Thoroughly clean all sealing surfaces.

22. Install new intake manifold gaskets and position the intake manifold on the engine.

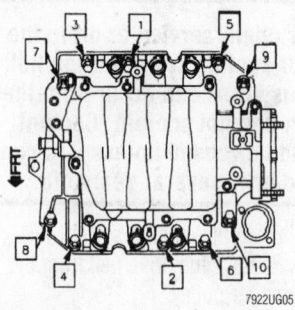

Tighten the intake manifold bolts in the sequence shown to ensure a good seal—3.8L (VIN L) engine

23. Install the intake manifold bolts and tighten, in sequence, to 89 inch lbs. (10 Nm). After all the bolts have been tightened, make a second pass making sure all bolts are still tightened to 89 inch lbs. (10 Nm).

24. Connect the heater and bypass hoses to the intake manifold.

25. Connect the upper radiator hose to the thermostat housing.

26. Install the alternator mounting bracket.

27. Install the alternator.

28. Install the power steering pump support bracket.

29. Install the serpentine belt.

30. Connect the control cables, vacuum lines and electrical connectors to the throttle body and intake manifold.

31. Install the control cable bracket.

32. Install the heat shield to the exhaust crossover pipe.

33. Install the fuel rail to the intake and install the fuel rail mounting bolts. Tighten the fuel rail mounting bolts to 10 ft. lbs. (14 Nm).

34. Connect and cap the fuel lines from the fuel rail.

35. Connect the vacuum line to the pressure regulator.

36. Attach the electrical connectors to the fuel injectors.

37. Connect the spark plug wires to the engine.

38. Connect the air inlet duct to the throttle body.

39. Install the plastic engine cover to the fuel rail.

40. Fill and bleed the cooling system using the recommended procedure.

41. Connect the negative battery cable.

42. Turn the ignition **ON**, to pressurize the fuel system, and check for leaks.

43. Run the engine and check for leaks and proper engine operation.

3.8L (VIN K) Engine

1. Disconnect the negative battery cable.

2. Relieve the fuel system pressure using the recommended procedure.

3. Remove the fuel injector sight shield and air inlet duct.

4. Disconnect the spark plug wires from the right side spark plugs and route the wires out of the way.

5. Label and disconnect the vacuum lines from the intake manifold.

6. Disconnect the fuel lines, vacuum lines and electrical connectors from the fuel rail.

7. Remove the fuel rail mounting bolts and remove the fuel rail from the intake manifold.

8. Remove the EGR heat shield.

9. Remove the throttle cable bracket from the cylinder head mounting bracket and disconnect the cables from the throttle body lever.

10. Remove the throttle body support bracket.

11. Remove the upper intake plenum mounting bolts and remove the plenum and gasket.

12. Drain the cooling system.

13. Disconnect the upper radiator hose from the thermostat housing.

14. Remove the serpentine belt and the alternator.

15. Remove the four bolts and remove the drive belt tensioner assembly.

16. Disconnect the EGR valve outlet pipe, if necessary.

17. Remove the lower intake manifold mounting bolts and remove the intake manifold and gaskets.

To install:

18. Thoroughly clean all of the sealing surfaces.

19. Install the intake manifold using new manifold gaskets and install the intake manifold bolts finger-tight. With all the bolts in place, tighten the bolts in sequence to 11 ft. lbs. (15 Nm). With all bolts tightened, make a second pass and tighten each bolt again to 11 ft. lbs. (15 Nm).

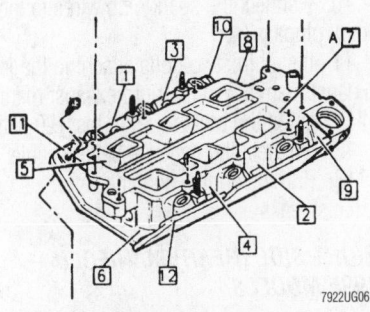

Be sure to tighten the lower intake manifold mounting bolts in the correct sequence—3.8L (VIN K) engine

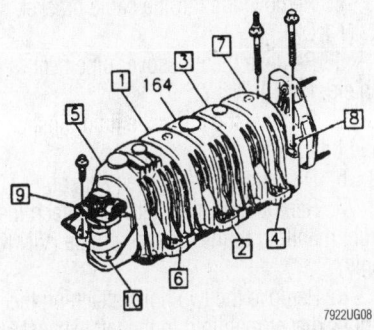

Upper intake manifold mounting bolt torque sequence—3.8L (VIN K) engine

20. Connect the EGR valve outlet pipe, if necessary.

21. Install the drive belt tensioner assembly and tighten the tensioner bolts to 37 ft. lbs. (50 Nm).

22. Install the alternator and serpentine belt.

23. Connect the upper radiator hose to the thermostat housing.

24. Install the upper intake plenum and mounting bolts. Tighten the intake plenum bolts to 11 ft. lbs. (15 Nm).

25. Install the throttle body support bracket.

26. Install the throttle cable bracket to the cylinder head mounting bracket and connect the cables to the throttle body lever.

27. Install the EGR heat shield.

28. Install the fuel rail and the mounting bolts on the intake manifold. Tighten the fuel rail bolts to 88 inch. lbs. (10 Nm).

29. Connect the fuel lines, vacuum lines and electrical connectors to the fuel rail.

30. Connect the vacuum lines to the intake manifold.

31. Connect the spark plug wires to the right side spark plugs and route the wires.

32. Install the fuel injector sight shield and air inlet duct.

33. Connect the negative battery cable.

34. Turn the ignition **ON** , to pressurize the fuel system, and check for leaks.

35. Refill and bleed the cooling system using the recommended procedure.

36. Run the engine and check for leaks and proper engine operation.

3.8L (VIN 1) Engine

1995 MODELS

1. Disconnect the negative battery cable.

2. Remove the supercharger dress up cover and air intake duct.

3. Remove fuel injector sight shield.

4. Relieve the fuel system pressure using the recommended procedure.

5. Drain the cooling system.

6. Label and disconnect the spark plug wires on the right side of the engine and set aside.

7. Remove the manifold vacuum source harness.

8. Disconnect the fuel supply and return lines from the fuel rail.

9. Disconnect the vacuum hoses. Identify and tag as they are removed for installation reference.

10. Disconnect the upper radiator hose and bypass hose.

11. Detach the following electrical connectors:

   a. EGR valve.
   b. Throttle Position Sensor (TPS).
   c. Idle Air Control (IAC) solenoid.
   d. Fuel injectors.
   e. Manifold Absolute Pressure (MAP) sensor.

12. Remove the EGR outlet pipe.

13. Disconnect the throttle and cruise control cables.

14. Remove the throttle bracket with the steering reservoir and set aside.

15. Remove the inner accessory drive belt.

16. Disconnect the heater hose pipe from the intake.

17. Remove the tensioner bracket-to-supercharger retaining stud using the double-nut method.

18. Remove the manifold bolts and remove the supercharger and intake manifold as an assembly. Discard the gasket and seals.

19. If the intake manifold is damaged or is to be replaced with another, the supercharger must be removed.

To install:

20. Clean the cylinder block, heads and intake manifold sealing surface of all old gasket material and oil.

21. Clean the intake manifold bolts and bolt holes of adhesive compound. Apply thread lock compound GM No. 12345493 or equivalent to the intake manifold bolt threads before assembly.

22. Install new gaskets and seals. Apply sealant to the ends of the manifold seals.

23. Install the intake manifold/supercharger assembly. Tighten the bolts, in sequence, to 11 ft. lbs. (15 Nm). Repeat the sequence and be sure all bolts are tightened to 11 ft. lbs. (15 Nm).

24. Install the tensioner bracket-to-supercharger retaining stud.

25. Connect the heater hose pipe to the intake.

26. Install the inner accessory drive belt.

27. Install the throttle bracket with the steering reservoir.

28. Connect the throttle and cruise control cables.

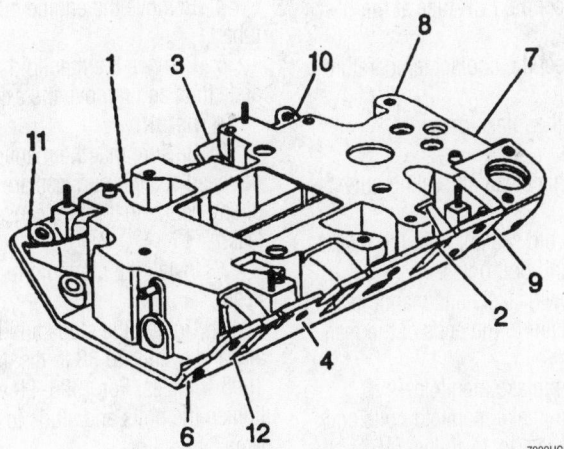

Tighten the lower intake manifold mounting bolts in the correct sequence to prevent warping the manifold—3.8L (VIN 1) engine

29. Install the EGR outlet pipe.

30. Attach the following electrical connectors:

    a. EGR valve.

    b. Throttle Position Sensor (TPS).

    c. Idle Air Control (IAC) solenoid.

    d. Fuel injectors.

    e. Manifold Absolute Pressure (MAP) sensor.

31. Connect the upper radiator hose and bypass hose.

32. Connect the vacuum hoses to the intake manifold.

33. Connect the fuel supply and return lines to the fuel rail.

34. Install the manifold vacuum source harness.

35. Connect the spark plug wires on the right side of the engine.

36. Install the fuel injector sight shield.

37. Install the supercharger dress up cover and air intake duct.

38. Connect the negative battery cable.

39. Fill and bleed the cooling system using the recommended procedure.

40. Turn the ignition **ON**, to pressurize the fuel system, and check for leaks.

41. Run the engine and check for proper engine operation.

1996–99 MODELS

1. Disconnect the negative battery cable.

2. Relieve the fuel system pressure using the recommended procedure.

3. Raise and safely support vehicle.

4. Remove the front splash shield.

5. Drain the radiator coolant into a suitable container. Close radiator drain.

6. Reinstall the front splash shield.

7. Lower the vehicle.

8. Remove the supercharger assembly.

9. Remove the thermostat housing.

10. Disconnect the EGR tube at the intake manifold.

11. Disconnect the coolant temperature sensor.

12. Remove the intake manifold.

To install:

13. Thoroughly clean all sealing surfaces.

14. Clean all old sealant from the intake manifold bolts and bolt holes.

15. Install new gaskets and manifold seals. Apply sealant to the ends of the manifold seals.

16. Install the intake manifold.

17. Install the intake manifold bolts and tighten, in sequence, to 11 ft. lbs. (15 Nm).

18. Install the electrical connector to the temperature sensor.

19. Install the EGR tube to the intake manifold.

20. Install the thermostat housing.

21. Install the supercharger assembly.

22. Connect the negative battery cable.

23. Fill and bleed the cooling system using the recommended procedure.

24. Run the engine and check for leaks and proper engine operation.

Exhaust Manifold

REMOVAL & INSTALLATION

3.8L Engines

LEFT SIDE (FRONT) MANIFOLD

1. Disconnect the negative battery cable.

2. Remove the two bolts attaching the left exhaust manifold to the right exhaust manifold.

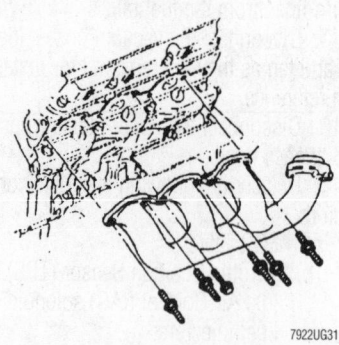

7922UG31

Left exhaust manifold mounting—3.8L engine

3. Label and disconnect the spark plug wires from the spark plugs and position out of the way.

4. Remove the engine oil dipstick and tube.

5. Remove the manifold mounting bolts and studs and remove the exhaust manifold.

To install:

6. Be sure that the manifold and cylinder head sealing surfaces are clean and free of any debris that might cause an exhaust leak.

7. Install the manifold to the cylinder head.

8. Tighten the studs and bolts, gradually and evenly, to 38 ft. lbs. (52 Nm) for 1995 vehicles. For 1996–99 vehicles, tighten the bolts and studs to 22 ft. lbs. (30 Nm).

9. Install the engine oil dipstick and tube.

10. Connect the spark plug wires to the spark plugs.

11. Install the two bolts attaching the left exhaust manifold to the right exhaust manifold. Tighten the bolts to 15 ft. lbs. (20 Nm).

12. Connect the negative battery cable.

13. Run the engine and check for exhaust leaks.

RIGHT SIDE (REAR) MANIFOLD— 1995 MODELS

1. Disconnect the negative battery cable.

2. Label and disconnect the spark plug wires from the spark plugs and position out of the way.

3. Remove the throttle cable bracket. (VIN 1 ONLY).

4. Remove the crossover pipe heat shield. (VIN 1 ONLY).

5. Remove the transaxle fluid dipstick and tube.

6. Disconnect the oxygen sensor.

7. Remove the two bolts that attach the right manifold to the crossover pipe (VIN K only).

8. Remove the two bolts attaching the right exhaust manifold to the left exhaust manifold (VIN 1 only).

9. Remove the vacuum reservoir from the cowl.

10. Raise and safely support the vehicle.

11. Remove the catalytic converter heat shield and pipe hanger.

12. Remove the exhaust pipe-to-manifold mounting nuts and disconnect the exhaust pipe from the manifold.

13. Lower the vehicle.

14. Remove the engine lift bracket.

15. Remove the exhaust manifold bolts and studs and remove the exhaust manifold.

To install:

16. Be sure that the manifold and cylinder head sealing surfaces are clean and free of any debris that might cause an exhaust leak.

17. Install the manifold to the cylinder head and to the left exhaust manifold.

18. Install the manifold mounting studs and bolts. Tighten the studs and bolts, gradually and evenly, to 38 ft. lbs. (52 Nm), beginning at the center of the manifold and working towards the ends.

19. Install the engine lift bracket.

20. Raise and safely support the vehicle.

21. Connect the exhaust pipe to the manifold and install the exhaust pipe-to-manifold mounting nuts. Tighten the exhaust pipe-to-manifold nuts to 18 ft. lbs. (25 Nm).

22. Install the catalytic converter heat shield and pipe hanger.

23. Lower the vehicle.

24. Install the vacuum reservoir to the cowl.

25. Install the two bolts attaching the right exhaust manifold to the left exhaust manifold. (VIN 1 ONLY). Tighten the bolts to 15 ft. lbs. (20 Nm).

26. Install the two bolts that attach the right manifold to the crossover pipe. (VIN K ONLY). Tighten the bolts to 15 ft. lbs. (20 Nm).

27. Attach the oxygen sensor electrical connector.

28. Install the transaxle fluid dipstick and tube.

29. Install the crossover pipe heat shield. (VIN 1 ONLY).

30. Install the throttle cable bracket. (VIN 1 ONLY).

31. Connect the spark plug wires to the spark plugs.

32. Connect the negative battery cable.

33. Run the engine and check for leaks.

RIGHT SIDE (REAR) MANIFOLD— 1996–99 MODELS

1. Disconnect the negative battery cable.

2. Label and disconnect the spark plug wires from the spark plugs.

3. Remove the transaxle fluid dipstick and tube.

4. Disconnect the oxygen sensor.

5. Remove the two bolts attaching the right exhaust manifold to the crossover pipe.

6. Raise and safely support the vehicle.

7. Remove the front exhaust pipe-to-exhaust manifold nuts and disconnect the exhaust pipe from the manifold.

8. Lower the vehicle.

9. Remove the engine lift bracket.

10. Remove the exhaust manifold mounting studs and bolts and remove the exhaust manifold.

11. Remove and discard the manifold-to-crossover pipe gasket and manifold-to-front exhaust pipe gasket.

To install:

12. Be sure that the manifold, cylinder head and crossover pipe sealing surfaces are clean and free of any debris that might cause an exhaust leak.

13. Install the manifold to the cylinder head and crossover pipe using a new manifold-to-crossover pipe gasket.

14. Install the manifold mounting studs and bolts. Tighten the studs and bolts, gradually and evenly, to 22 ft. lbs. (30 Nm), beginning at the center of the manifold and working towards the ends.

15. Install the engine lift bracket.

16. Raise and safely support the vehicle.

17. Connect the exhaust pipe from the manifold and install the front exhaust pipe-to-manifold nuts. Tighten the nuts to 18 ft. lbs. (25 Nm).

18. Lower the vehicle.

19. Install the two bolts attaching the right exhaust manifold to the crossover pipe. Tighten the manifold-to-crossover pipe bolts and tighten to 15 ft. lbs. (20 Nm).

20. Attach the oxygen sensor electrical connector.

21. Install the transaxle fluid dipstick and tube.

22. Connect the spark plug wires to the spark plugs.

23. Connect the negative battery cable.

24. Run the engine and check for exhaust leaks.

Camshaft and Valve Lifters

REMOVAL & INSTALLATION

✳✳ CAUTION

The fuel injection system remains under pressure, even after the engine has been turned OFF. The fuel system pressure must be relieved before disconnecting any fuel lines. Failure to do so may result in fire and/or personal injury.

3.8L Engines

1. Disconnect the negative battery cable.

2. Relieve the fuel system pressure using the recommended procedure.

3. Remove the engine assembly from the vehicle and mount on a suitable engine stand.

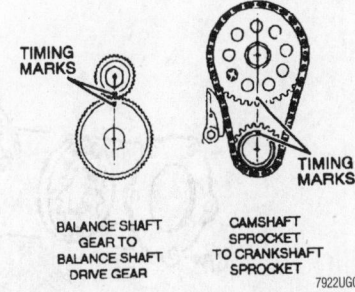

TIMING MARKS

TIMING MARKS

BALANCE SHAFT GEAR TO BALANCE SHAFT DRIVE GEAR

CAMSHAFT SPROCKET TO CRANKSHAFT SPROCKET

7922UG09

The timing marks should face each other when the chain and gears are installed properly

4. If equipped, remove the supercharger.

5. Remove the intake manifold.

6. Remove the rocker arm covers.

7. Remove the rocker arm assemblies, pushrods and lifters. A magnet may be helpful when pulling the lifters out of their bores. Identify all parts as they are removed, so they can be reinstalled in their original locations.

8. Remove the crankshaft balancer center bolt, using a suitable puller, remove the balancer from the crankshaft.

9. Remove the timing chain front cover.

10. Set the engine to Top Dead Center (TDC) No. 1 cylinder (firing position) to align the timing marks, before disassembling the timing chain and sprockets.

➡**Align the timing marks of the camshaft and crankshaft sprockets to avoid burring the camshaft journals by the crankshaft.**

11. Remove the camshaft sprocket and timing chain.

12. Remove the camshaft thrust plate bolts and remove the thrust plate.

13. Carefully remove the camshaft from the engine block.

14. Inspect the camshaft lobes and journals for wear and/or damage; replace as necessary.

➡**If the camshaft was replaced the lifters must also be replaced. The old lifters have developed a wear pattern and will cause the new camshaft to wear prematurely.**

To install:

15. Coat the camshaft lobes and bearings with lubricant GM part number 1052365 or equivalent camshaft break-in prelube prior to installation.

16. Carefully install the camshaft into the engine.

17. Install the camshaft thrust plate and tighten the mounting bolts to 10 ft. lbs. (14 Nm).

18. Install the camshaft sprocket and timing chain. Be sure the timing marks are aligned. Tighten the camshaft sprocket retaining bolt to 74 ft. lbs. (100 Nm) plus an additional 90 degree (¼) turn.

19. Install the timing chain front cover.

20. Install the crankshaft balancer and tighten the mounting bolt to 111 ft. lbs. (150 Nm) plus an additional 76 degree turn.

21. Coat the valve lifters with camshaft prelube and install the lifters in the lifter bores.

22. Install the lifter guides and lifter guide retainer. Tighten the retainer mounting bolts to 22 ft. lbs. (30 Nm).

23. Install the pushrods and rocker arms and tighten the rocker arm bolts to 11 ft. lbs. (15 Nm) plus an additional 90 degree turn.

24. Install the rocker arm covers.

25. Install the intake manifold.

26. If equipped, install the supercharger.

27. Install the engine in the vehicle.

28. Connect the negative battery cable.

29. Verify that all fluid levels are full and correct.

30. Start the engine and check for leaks. Check engine operation.

Valve Lash

ADJUSTMENT

The valve clearance cannot be adjusted on these engines. The hydraulic lifters function to maintain a zero clearance when the valves are opening and closing. Any clearance is instantaneously taken up by the hydraulic action. "Valve lifter noise" complaints may require cleaning of sludge coated or varnished valve lifters, or replacement.

Oil Pan

REMOVAL & INSTALLATION

3.8L Engines

✳✳ WARNING

The oil level sensor, located in the oil pan, must be removed prior to removal of the oil pan. If the oil pan is removed first, damage to the oil level sensor may occur.

1. Disconnect the negative battery cable.

2. Raise and safely support the vehicle.

3. Remove the oil drain plug and drain the engine oil into a suitable container.

4. For 1996–99 vehicles, remove the flexplate cover.

5. Disconnect the oil level sensor and remove the sensor.

6. For 1996–99 vehicles, remove the oil filter.

7. Remove the oil pan mounting bolts and remove the oil pan.

To install:

✳✳ WARNING

The oil level sensor, located in the oil pan, must be installed after the oil pan has been installed. If the oil level sensor is installed first, damage to the sensor may occur.

8. Clean all sealing surfaces completely. Thoroughly clean the inside of the oil pan.

9. Install the oil pan with a new gasket and tighten the mounting bolts to 125 inch lbs. (14 Nm).

10. If necessary, install a new oil filter and the flexplate cover.

11. Install the oil level sensor and attach the connector.

12. Install the oil drain plug and tighten to 30 ft. lbs. (40 Nm).

13. Lower the vehicle.

14. Refill the crankcase with the proper type and quantity of engine oil.

15. Connect the negative battery cable.

16. Run the engine and check for leaks.

Oil Pump

REMOVAL & INSTALLATION

3.8L Engines

1. Disconnect the negative battery cable.

2. Remove the engine drive belts and tensioner assembly.

3. Remove the drive belt idler pulley and bracket, if necessary.

4. Raise and safely support the vehicle.

5. Support the engine using engine support fixture J 28467 or equivalent, and remove the torque axis mount and bracket assembly.

6. Remove the engine front cover assembly.

7. Remove the four bolts securing the oil filter adapter to the front cover assembly and remove the adapter, pressure regulator valve and spring.

8. Remove the four oil pump cover, on the inside of the front cover, attaching screws and remove the cover.

9. Remove the inner and outer pump gears and inspect.

To install:

10. Lubricate the oil pump gears with petroleum jelly and install the gears into the housing.

11. Pack the gear cavity with petroleum jelly after the gears have been installed in the housing.

➡**Petroleum jelly seals the pump and acts like a "prime" so oil will begin to**

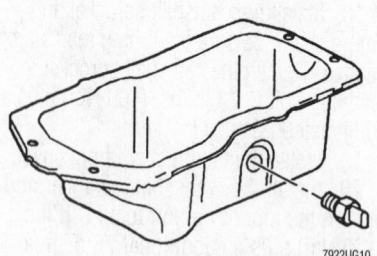

If equipped, be sure to remove the oil level sensor before removing the pan

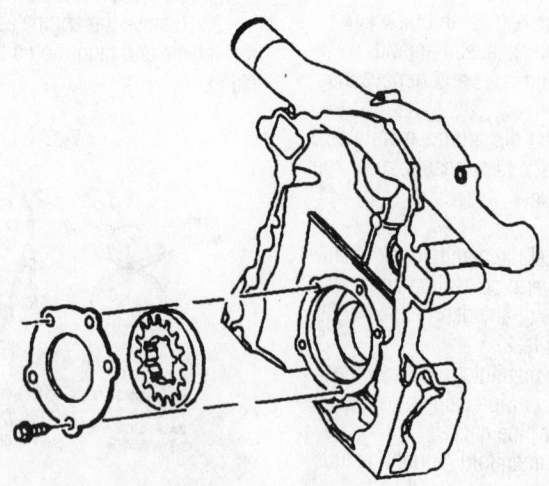

The oil pump is located inside the front engine cover—3.8L (VIN L, K and 1) engines

be drawn from the oil pan as soon as the engine begins to turn. Do not neglect this step. DO NOT use any type of grease. Petroleum jelly has a low melting point and will correctly dissipate when oil begins to flow and it is no longer needed.

12. Install the oil pump cover and screws and tighten to 97 inch lbs. (11 Nm).

13. Install the oil filter adapter with new gasket, pressure regulator valve and spring. Apply sealant to the bolt threads and tighten the mounting bolts to 24 ft. lbs. (33 Nm).

14. Install the front cover assembly.

15. Install the tensioner assembly.

16. Install the drive belt idler pulley and bracket, if removed.

17. Install the drive belts.

18. Install the torque axis mount assembly and remove the engine support fixture.

19. Verify the correct engine oil level. A new oil filter is recommended.

20. Connect the negative battery cable.

21. Start the vehicle and verify no leaks and proper oil pressure.

Rear Main Seal

REMOVAL & INSTALLATION

3.8L Engines

1. Remove the transaxle assembly.

2. Remove the flexplate from the crankshaft.

3. Carefully pry the old seal out of the engine block. Do not damage or scratch the sealing surface of the crankshaft or the seal bore.

To install:

4. Lubricate the lip and the outer edge of the new seal with clean engine oil.

5. Slide the oil seal on the mandrel until the back of the seal is seated squarely against the collar of the tool.

6. Attach Seal Installer J-38196 or equivalent to the rear of the crankshaft with the two mounting bolts, then turn the T-handle until the oil seal is fully seated into the rear of the engine.

7. Loosen the T-handle of the tool completely.

8. Remove the two attaching bolts and remove the tool

9. Install the flexplate. Tighten the bolts to 11 ft. lbs. (15 Nm), then turn the bolts an additional 50°.

10. Install the transaxle using the recommended procedure.

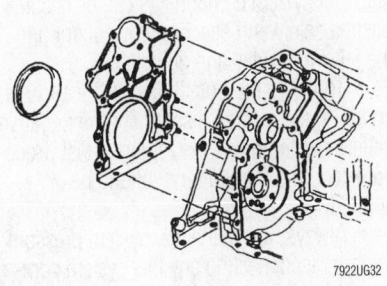

Rear main oil seal and rear cover—3.8L engine

Timing Chain, Sprockets, Front Cover and Seal

REMOVAL & INSTALLATION

3.8L Engines

1. Disconnect the negative battery cable.

✳✳ CAUTION

Never open, service or drain the radiator or cooling system when hot; serious burns can occur from the steam and hot coolant.

2. Drain the cooling system.

3. Disconnect the coolant hoses from the timing chain front cover.

4. Support the engine using support fixture J 28467-A or equivalent, and remove the engine mount.

5. Remove the drive belt(s).

6. Raise and safely support the vehicle.

7. Remove the right front wheel.

8. Remove the right inner fender access panel.

9. Remove the drive belt idler pulley and bracket. (VIN 1).

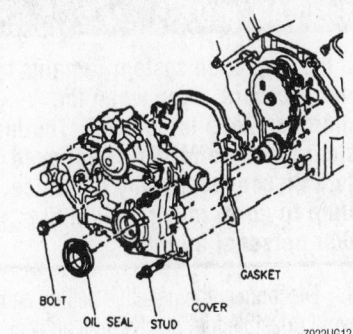

Timing chain front cover—3.8L (VIN L, K and 1) engines

10. Remove the drive belt tensioner.

11. Keep the flexplate from turning using holder tool J 37096 or equivalent. Remove the crankshaft balancer retaining bolt, then use puller tool J 38197 or equivalent, to remove the balancer from the crankshaft.

12. Remove the crankshaft position sensor shield and the crankshaft position sensor.

13. Remove the oil pan-to-front cover bolts.

14. Remove the front cover attaching bolts and remove the timing chain front cover.

15. Align the timing marks on the camshaft and crankshaft sprockets so they are as close together as possible.

16. Remove the timing chain damper.

17. Remove the camshaft sprocket retaining bolt and remove the camshaft sprocket and timing chain.

18. Remove the crankshaft sprocket.

➡**Do not rotate the camshaft or crankshaft while the timing chain and sprockets are removed.**

19. Thoroughly clean all sealing surfaces.
To install:

20. Assemble the timing chain and sprockets with the timing marks aligned.

21. Install the timing chain and sprockets to the crankshaft first,, then to the camshaft.

22. Install the camshaft sprocket bolt and tighten to 74 ft. lbs. (100 Nm) plus an additional 90 degree turn. Recheck the camshaft and crankshaft sprocket timing marks to be sure they are still aligned.

23. Install the timing chain damper and tighten the bolt to 16 ft. lbs. (22 Nm).

✳✳ WARNING

The oil pump is built into the front cover. When the cover is removed, oil drains from the pump. Since the pump "looses its prime" it may not establish oil pressure as soon as the engine starts. Therefore, it is important to remove the oil pump cover from the back of the timing chain front cover and pack the space around the oil pump gears completely full of petroleum jelly. If this is not done, the oil pump may not pump engine oil when the engine is started, resulting in severe engine damage.

24. Remove the screws and the oil pump cover from the back of the timing chain front cover. Pack the space around the oil pump gears completely full of petroleum jelly. There must be no air space left inside the pump.

25. Reinstall the pump cover and tighten the screws to 97 inch lbs. (11 Nm).

26. Using new gaskets, install the timing chain front cover to the block. Tighten the front cover-to-engine block bolts to 11 ft. lbs. (15 Nm) plus an additional 40 degrees.

27. Install the oil pan-to-front cover bolts and tighten the bolts to 125 inch lbs. (14 Nm).

28. Install the crankshaft position sensor and tighten the bolts to 14–28 ft. lbs. (20–40 Nm). Install the crankshaft position sensor shield.

29. Keep the flexplate from turning using holder tool J 37096 or equivalent. Install the crankshaft balancer and tighten the bolt to 111 ft. lbs. (150 Nm) plus an additional 76 degree turn.

30. Install the drive belt tensioner assembly.

31. Install the drive belt idler pulley. (VIN 1).

32. Install the right inner fender access panel and the right front wheel.

33. Lower the vehicle.

34. Install the drive belt(s).

35. Install the engine mount and remove the engine support fixture.

36. Connect the coolant hoses.

37. Connect the negative battery cable.

38. Fill and bleed the cooling system using the recommended procedure.

39. Start the vehicle and check for leaks and proper engine operation.

 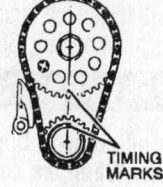

BALANCE SHAFT GEAR TO BALANCE SHAFT DRIVE GEAR

CAMSHAFT SPROCKET TO CRANKSHAFT SPROCKET

7922UG09

Timing chain sprocket and balance shaft gear alignment—3.8L (VIN L, K and 1) engines

FUEL SYSTEM

Fuel System Service Precautions

Safety is the most important factor when performing not only fuel system maintenance but any type of maintenance. Failure to conduct maintenance and repairs in a safe manner may result in serious personal injury or death. Maintenance and testing of the vehicle's fuel system components can be accomplished safely and effectively by adhering to the following rules and guidelines.

• To avoid the possibility of fire and personal injury, always disconnect the negative battery cable unless the repair or test procedure requires that battery voltage be applied.

• Always relieve the fuel system pressure prior to disconnecting any fuel system component (injector, fuel rail, pressure regulator, etc.), fitting or fuel line connection. Exercise extreme caution whenever relieving fuel system pressure to avoid exposing skin, face and eyes to fuel spray. Please be advised that fuel under pressure may penetrate the skin or any part of the body that it contacts.

• Always place a shop towel or cloth around the fitting or connection prior to loosening to absorb any excess fuel due to spillage. Ensure that all fuel spillage (should it occur) is quickly removed from engine surfaces. Ensure that all fuel soaked cloths or towels are deposited into a suitable waste container.

• Always keep a dry chemical (Class B) fire extinguisher near the work area.

• Do not allow fuel spray or fuel vapors to come into contact with a spark or open flame.

• Always use a back-up wrench when loosening and tightening the fuel line connection fittings. This will prevent unnecessary stress and torsion to fuel line piping. Always tighten the fastener to the proper specifications.

• Always replace worn fuel fitting O-rings with new. Do not substitute fuel hose or equivalent, where fuel pipe is installed.

Fuel System Pressure

RELIEVING

✳✳ CAUTION

The fuel injection system remains under pressure, even when the engine has been turned OFF. The fuel system pressure MUST BE relieved before disconnecting any fuel lines. Failure to do so may result in fire and/or personal injury.

1. Disconnect the negative battery cable to avoid possible fuel discharge if an accidental attempt is made to start the engine.

2. Remove the fuel tank cap to relieve tank pressure. Do not tighten until service procedure has been completed.

3. Connect a suitable fuel pressure gauge with bleed valve such as J-34730–1 or equivalent, to the fuel pressure test port. Wrap a shop towel around the fitting while connecting the gauge to catch any spilled fuel.

4. Install the bleed hose into an approved container and open the valve to bleed off the fuel system pressure.

5. Drain any fuel remaining in the gauge into an approved container.

✳✳ CAUTION

There may still be residual fuel in the system, and a small amount of fuel may be released when servicing fuel lines or connections. In order to reduce the chance of personal injury, cover the fuel line fittings with a shop towel before disconnecting to catch any fuel that may leak out.

Fuel Filter

REMOVAL & INSTALLATION

✳✳ CAUTION

The fuel injection system remains under pressure, even after the engine has been turned OFF. The fuel system pressure MUST BE relieved before disconnecting any fuel lines. Failure to do so may result in fire and/or personal injury.

3.8L Engines

1995–96 MODELS

1. Disconnect the negative battery cable.

2. Relieve the fuel system pressure using the recommended procedure.

3. Raise and safely support the vehicle.

4. Detach the inlet and outlet quick connect fuel lines from the filter following these steps:

   a. Twist quick connector ¼ turn in each direction to loosen ant dirt that may have accumulated in the connector.

   b. Use compressed air to remove any dirt in the connector.

   c. Squeeze plastic tabs of male connector and pull apart.

5. Remove the filter bracket mounting bolt and remove the fuel filter from the vehicle.

To install:

6. Position the fuel filter in the mounting bracket, making sure it is facing in the proper direction.

7. Attach the inlet and outlet quick connect lines to the fuel filter following these steps:

a. Be sure the connectors are clean and that new plastic retainers are used on the filter.

b. Apply a couple of drops of engine oil to the male pipe ends of filter.

c. Push the fuel lines onto the fuel filter until the plastic retainer snaps into place.

d. Check that the connectors are locked into place by trying to pull the connector from the filter.

8. Install the filter bracket mounting bolt and tighten to 10 ft. lbs. (14 Nm).

9. Lower the vehicle.

10. Connect the negative battery cable.

11. Pressurize the fuel system by turning the ignition switch to the **ON** position, then check for leaks.

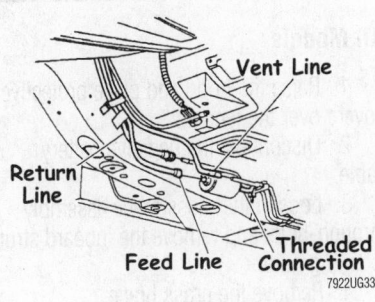

Common fuel filter mounting location

1997–99 MODELS

1. Disconnect the negative battery cable.

2. Relieve the fuel system pressure using the recommended procedure.

3. Raise and safely support the vehicle.

4. Detach the quick connect fuel line from the filter following these steps:

a. Twist quick connector ¼ turn in each direction to loosen ant dirt that may have accumulated in the connector.

b. Use compressed air to remove any dirt in the connector.

c. Squeeze plastic tabs of male connector and pull apart.

5. Remove the threaded connection from the inlet side of the filter.

6. Remove the filter.

To install:

7. Position the fuel filter, making sure it is facing in the proper direction.

8. Attach the outlet quick connect line to the fuel filter following these steps:

a. Be sure the connector is clean and that a new plastic retainer is used on the filter.

b. Apply a couple of drops of engine oil to the male pipe end of filter.

c. Push the fuel line onto the fuel filter until the plastic retainer snaps into place.

d. Check that the connector is locked into place by trying to pull the connector from the filter.

9. Install the threaded connection to the inlet side of filter.

10. Lower the vehicle.

11. Connect the negative battery cable.

12. Pressurize the fuel system by turning the ignition switch to the **ON** position, then check for leaks.

Fuel Pump

REMOVAL & INSTALLATION

❋❋ CAUTION

The fuel injection system remains under pressure, even after the engine has been turned OFF. The fuel system pressure MUST BE relieved before disconnecting any fuel lines. Failure to do so may result in fire and/or personal injury.

Except 1997–99 Park Avenue

1. Disconnect the negative battery cable.

2. Relieve the fuel system pressure using the recommended procedure.

3. Raise and safely support the vehicle.

4. Remove the fuel tank from the vehicle following these steps:

a. Drain the fuel tank with a hand operated pumping device.

b. Remove the fuel tank filler pipe and the EVAP pipe from the tank.

c. Remove the quick connect fuel lines from the fuel tank.

d. Remove the rubber exhaust hangers in the rear to allow room to lower the fuel tank.

e. Use a jacking device, such as a transmission jack, to support the fuel tank.

f. Remove the fuel tank strap retaining bolts and let the straps hang freely.

g. Lower the fuel tank enough to remove fuel sender connector.

h. Remove the fuel tank and place on a suitable workbench.

5. Clean the fuel tank in the area of the fuel sender assembly, to prevent dirt and debris from entering the tank when the fuel sender is removed.

6. Rotate the lock ring on top of the tank counterclockwise using a fuel sender spanner wrench and remove the fuel sender assembly from the tank.

7. Note the direction the strainer is pointing and remove the strainer from the pump by pulling it down and twisting.

8. Disconnect the pump electrical wires and hoses.

9. Pull the pump assembly out of the rubber connectors.

To install:

10. Transfer any insulators and grommets from the old pump to the new one.

11. Connect the pump to the fuel hose and tilt the bottom of the pump into the mounting bracket.

12. Install a new strainer on the pump so it points in the same direction as noted during removal.

13. Attach the electrical connectors and fuel lines to the pump.

14. Install a new O-ring on top of the fuel tank and install the fuel sender assembly carefully into the tank.

15. Install the lock ring and rotate clockwise until the tabs are against the stops.

16. Install the fuel tank in the vehicle following these steps:

a. Raise and safely support the fuel tank into position.

b. Attach fuel sender electrical connector to the sender.

c. Place fuel tank straps into position and install retaining bolts. Tighten retaining bolts to 25 ft. lbs. (34 Nm).

d. Remove support.

e. Install the rubber exhaust hangers to rear exhaust.

f. Install the quick connect lines onto the fuel tank.

g. Install the fuel tank EVAP pipe to the tank and tighten the hose clamp to 25 inch lbs. (2.8 Nm).

h. Install the fuel tank filler pipe to the tank and tighten the hose clamp to 25 inch lbs. (2.8 Nm).

17. Lower the vehicle.

18. Connect the negative battery cable.

19. Refill the fuel tank.

20. Turn the ignition switch to the **ON** position, to pressurize the fuel system, and check for leaks.

21. Run the engine and check for leaks and proper engine operation.

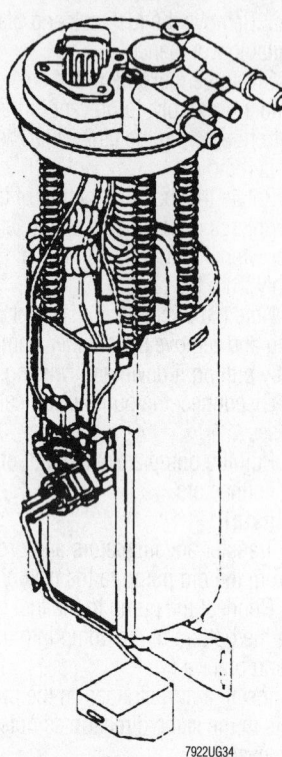

Fuel pump and level sender assembly

1997–99 Park Avenue

1. Disconnect the negative battery cable.
2. Relieve the fuel system pressure using the recommended procedure.
3. Drain the fuel tank with a hand operated pumping device to at least a ¼ of tank of fuel.

4. Open trunk and remove the spare tire cover, jack and spare tire.
5. Pull back the trunk lining to access the fuel sender access panel.
6. Remove the retaining screws in the sender access panel and remove.
7. Clean the area around the connection before removing sender.
8. Remove the quick connect fittings from the sender.
9. Remove the electrical connector from the sender and position harness and hoses aside.

➡ **When removing the fuel sender from the tank, the reservoir bucket is full of fuel. Use caution in containing the fuel with some shop towels.**

10. Remove the sender retaining ring with a fuel sender spanner wrench.
11. Remove the sender and take note of it's removal position.
12. Remove and discard fuel sender O-ring.
13. Note the direction the strainer is pointing and remove the strainer from the pump by pulling it down and twisting.
14. Disconnect the pump electrical wires and hoses.
15. Pull the pump assembly out of the rubber connectors.

To install:

16. Transfer any insulators and grommets from the old pump to the new one.
17. Connect the pump to the fuel hose and tilt the bottom of the pump into the mounting bracket.

18. Install a new strainer on the pump so it points in the same direction as noted during removal.
19. Attach the electrical connectors and fuel lines to the pump.
20. Install a new O-ring on top of the fuel tank and install the fuel sender assembly carefully into the tank.
21. Install the lock ring and rotate clockwise until the tabs are against the stops.
22. Attach the fuel line quick connectors and the sender electrical connector.
23. Install the fuel sender access cover.
24. Reposition trunk liner and install the spare tire, jack and spare tire cover.
25. Fill with fuel and check for leaks.

DRIVE TRAIN

Transaxle Assembly

REMOVAL & INSTALLATION

All Models

1. Raise the hood and place protective covers over the fenders.
2. Disconnect the negative battery cable.
3. Loosen the cross brace assembly through-bolts and remove the inboard strut retaining nuts.
4. Remove the cross brace.
5. Reinstall the inboard strut retaining nuts.
6. Remove the air intake duct.
7. Remove the cruise control cable at throttle body.
8. Remove the shift control linkage and mounting bracket at the transaxle.
9. Detach the wiring connectors at the following switches:
    a. Transaxle park/neutral position switch.
    b. Back-up light switch.
    c. Transaxle electrical connector.
    d. Vehicle speed sensor.
10. Remove the fuel pipe retainers.
11. Remove the three top transaxle-to-engine bolts.
12. Install special tool J 28467 A or equivalent engine support fixture.
13. Load the engine support fixture by tightening the wing nuts several turns to relieve tension on the frame and mounts.
14. Turn the steering wheel to the full left position.

The fuel pump service cover is located in the luggage compartment under the spare tire—1997–99 Park Avenue

15. Raise and safely support the vehicle.
16. Remove both front wheels.
17. Remove the right and left front ball joint nuts.
18. Separate the right and left control arms from the steering knuckle.
19. Remove the right halfshaft from the transaxle only. Do not remove the halfshaft from the steering knuckle assembly.
20. Remove the left halfshaft from both the transaxle and steering knuckle assembly.
21. Support the transaxle with a suitable jack.
22. Remove the left front transaxle mount.
23. Remove the torque strut bracket from the transaxle.
24. Remove the left rear transaxle mount-to-transaxle bolts.
25. Remove the transaxle brace from the engine bracket.
26. Remove the stabilizer shaft link-to-control arm bolt.
27. Remove the flexplate cover bolts and the flexplate cover.
28. Mark the flexplate to the converter with a reference line and remove the flexplate-to-converter bolts.
29. Remove the bolts attaching the rear frame member to the front dog leg.
30. Remove the front left frame-to-body attaching bolts.
31. Remove the frame assembly by swinging it aside and supporting with a suitable stand.
32. Remove the oil cooler lines from the transaxle.
33. Remove the remaining lower transaxle-to-engine block bolts.

➡ **One transaxle bolt is located between the transaxle case and the engine block and is installed in the opposite direction.**

34. Remove the transaxle assembly and lower it away from the vehicle.

To install:

35. Install the transaxle into the vehicle.
36. Support the transaxle with a suitable jack.
37. Install the lower transaxle-to-engine block bolts and tighten to 55 ft. lbs. (76 Nm).
38. Connect the cooler lines at the transaxle.
39. Raise the frame assembly into place and tighten the frame-to-body bolts to 83 ft. lbs. (112 Nm).

40. Install and tighten the front frame dog leg-to-right frame member bolts.
41. Install the left frame-to-body attaching bolts and tighten to 83 ft. lbs. (112 Nm).
42. Install the bolts attaching the rear frame member to the front frame dog leg.
43. Install the flexplate-to-converter bolts, aligning the marks made during removal, and tighten to 46 ft. lbs. (62 Nm).
44. Install the oil cooler lines to the transaxle.
45. Install the flexplate cover bolts and the flexplate cover.
46. Install the stabilizer shaft link-to-control arm bolt.
47. Install the transaxle brace to the engine bracket. Tighten the transaxle brace-to-engine bracket bolts to 70 ft. lbs.(95Nm).
48. Install the left rear transaxle mount-to-transaxle bolts.
49. Install the torque strut bracket to the transaxle.
50. Install the left front transaxle mount.
51. Install the left halfshaft to both the transaxle and steering knuckle assembly.
52. Install the right halfshaft to the transaxle.
53. Connect the right and left control arms to the steering knuckle.
54. Install the right and left front ball joint nuts.
55. Install both front wheels.
56. Lower the vehicle.
57. Turn the steering wheel to the straight ahead position.
58. Remove the support fixture from the engine.
59. Install the three top transaxle-to-engine bolts. Tighten the bolts to 55 ft. lbs. (76 Nm).
60. Install the fuel pipe retainers.
61. Attach the wiring connectors at the following switches:
   a. Transaxle park/neutral position switch.
   b. Back-up light switch.
   c. Transaxle electrical connector.
   d. Vehicle speed sensor.
62. Install the shift control linkage and mounting bracket at the transaxle.
63. Install the cruise control cable at throttle body.
64. Install the air intake duct.
65. Remove the inboard strut retaining nuts.
66. Install the cross brace.
67. Install the cross brace assembly through-bolts and remove the inboard strut retaining nuts.
68. Connect the negative battery cable.
69. Check the transaxle fluid level.
70. Start the engine and check for leaks.

71. Adjust the transaxle shifter cable, as necessary.
72. Road test the vehicle.

Halfshaft

REMOVAL & INSTALLATION

All Models

1. Raise and safely support the vehicle.
2. Remove the front wheel. Mark the position of the wheel to the wheel stud, prior to removal, for installation reference.
3. Install J-34754 boot protector, or equivalent on the outer CV-joint boot.
4. If necessary, loosen and remove the stabilizer shaft link assembly bolt.
5. Remove the ball joint cotter pin and nut. Loosen the joint using tool J-36226 or equivalent. The grease fitting may have to be removed from ball joint for tool access.
6. If removing the right halfshaft, turn the wheel to the left. If removing the left halfshaft, turn the wheel to the right.
7. Use a suitable prybar between the suspension support and the lower control arm to separate the joint.
8. Remove the hub nut. A large amount of torque is required to loosen the hub nut. Insert a drift pin or punch through the opening in the caliper into the ventilation openings in the brake rotor to keep the rotor from turning as the nut is loosened. Discard the hub nut; it must not be reused.

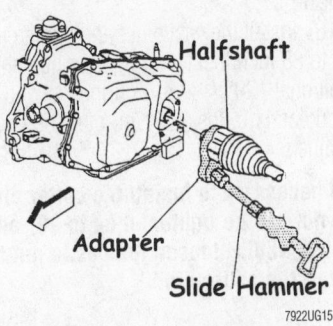

7922UG15

Use a slide hammer with the special adapter J-3308 or equivalent to remove the halfshaft from the transaxle

9. Partially install the hub nut to protect the threads, then remove the halfshaft from the hub using tool J-28733–B or equivalent.

10. Move the strut and knuckle rearward.

11. Remove the halfshaft from the transaxle using special tools J-33008 and J-2619-01 or their equivalents.

➡️**If equipped with anti-lock brakes, care must be used to prevent damage to the toothed sensor ring on the halfshaft and the wheel speed sensor on the steering knuckle.**

To install:

12. If installing the right side halfshaft, install J-37292-B seal protector tool or equivalent, so that it can be pulled out after the halfshaft is installed.

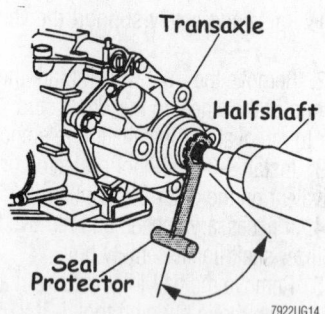

Insert the seal protector in the transaxle to prevent seal damage while the halfshaft is being installed

13. Install the halfshaft into the transaxle by placing a drift pin or punch into the groove on the joint housing and tapping lightly until seated. Verify that the halfshaft is seated by grasping the inner joint housing and pulling. DO NOT pull on the halfshaft.

14. Install the halfshaft into the hub and bearing assembly. Loosely install a new hub nut.

15. Install the ball joint into the steering knuckle.

16. Install the castle nut and tighten the nut to 88 inch lbs. (10 Nm), then tighten an additional 120° (⅓ turn) during which a torque of 41 ft. lbs. (55 Nm) must be obtained.

➡️**If necessary to install the cotter pin, the nut can be tightened up to 20° additional. NEVER loosen the castle nut to install the cotter pin.**

17. Insert a drift in the brake rotor cooling fins to keep the rotor from turning and tighten the nut to 107 ft. lbs. (145 Nm), unless equipped the J55 brake option, in which case the torque is 130 ft. lbs. (177 Nm).

18. If removed, install the stabilizer shaft link assembly. Tighten the nut to 14 ft. lbs. (17 Nm).

19. If J-37292-B seal protector was installed, remove it by pulling in line with the handle.

20. Install the wheel, aligning the marks made during removal.

21. Lower the vehicle and road test for proper operation.

STEERING AND SUSPENSION

Air Bag

❋❋ CAUTION

Some vehicles are equipped with an air bag system, also known as the Supplemental Inflatable Restraint (SIR) or Supplemental Restraint System (SRS). The system must be disabled before performing service on or around system components, steering column, instrument panel components, wiring and sensors. Failure to follow safety and disabling procedures could result in accidental air bag deployment, possible personal injury and unnecessary system repairs.

PRECAUTIONS

Several precautions must be observed when handling the inflator module to avoid accidental deployment and possible personal injury.

• Never carry the inflator module by the wires or connector on the underside of the module.

• When carrying a live inflator module, hold securely with both hands, and ensure that the bag and trim cover are pointed away.

• Place the inflator module on a bench or other surface with the bag and trim cover facing up.

• With the inflator module on the bench, never place anything on or close to the module which may be thrown in the event of an accidental deployment.

DISARMING

❋❋ CAUTION

The Supplemental Restraint System (SRS) must be disarmed before performing service around the air bag or SRS wiring. Failure to do so may cause accidental deployment of the air bag, resulting in unnecessary SRS repairs and/or personal injury.

1. Turn the steering wheel so the front wheels are in the straight ahead position.

2. Turn the ignition switch to the **LOCK** position.

3. Disconnect the negative battery cable.

4. Remove the Air Bag fuse from the fuse panel.

➡️**The position of the fuse on the panel varies according to model and year. Consult the vehicle owner's manual for fuse location.**

5. Remove the left-hand sound insulator (trim panel under the instrument panel).

6. Remove the Connector Position Assurance (CPA) clip and detach the yellow 2-way connector at the base of the steering column.

7. Remove the CPA and detach the passenger side yellow 2-way connector, if equipped. Positions for the connector vary from behind the glove box to removing the right side sound insulator and finding the yellow connector.

After the necessary repairs have been made, re-enable the air bag system as follows:

8. Turn the steering wheel so the front wheels are in the straight ahead position.

9. Turn the ignition switch to the **LOCK** position.

10. Disconnect the negative battery cable.

11. Attach the yellow 2-way connector at the base of the steering column and install the CPA.

12. Install the left-hand sound insulator.

13. Attach the yellow 2-way connector on the right side and install the CPA, if equipped. Install the sound insulator and/or glove box.

14. Install the air bag fuse.

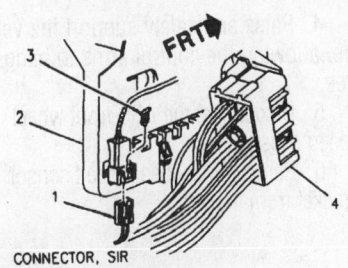

1 CONNECTOR, SIR
2 BRACKET, MULTIUSE MODULE
3 CONNECTOR POSITION ASSURANCE (CPA)
4 CONNECTOR, STEERING COLUMN
  WIRING HARNESS

7922UG35

Drivers side air bag connector

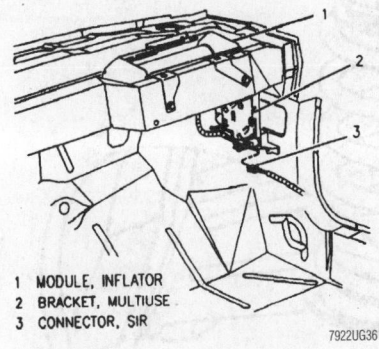

1 MODULE, INFLATOR
2 BRACKET, MULTIUSE
3 CONNECTOR, SIR

7922UG36

Passenger's side air bag connector

15. Connect the negative battery cable.
16. Turn the ignition switch to the **RUN** position. Verify that the INFLATABLE RESTRAINT indicator lamp flashes 7–9 times, then remains OFF. If the lamp does not function as specified, there is a malfunction in the air bag system.

Power Rack and Pinion Steering Gear

REMOVAL & INSTALLATION

✳✳ WARNING

The wheels of the vehicle must be straight ahead and the steering column in the LOCK position before disconnecting the steering column or intermediate shaft from the steering gear. Failure to do so will cause the Supplemental Inflatable Restraint (SIR) coil assembly in the steering column to become uncentered which will cause damage to the coil assembly.

All Models

1. Disconnect the negative battery cable.

2. Raise and safely support the vehicle.
3. Remove both front wheels.
4. Remove the pinch bolt from the intermediate shaft at the rack and pinion assembly and separate the shaft from the rack.

✳✳ CAUTION

Failure to disconnect the intermediate shaft from the rack and pinion stub shaft can result in damage to the steering gear and/or intermediate shaft. This damage can cause loss of steering control which could result in personal injury.

5. Remove the cotter pins and castle nuts from the tie rod ends and separate the tie rods from the steering knuckles using special tool J 24319–01 or equivalent.
6. Remove the power steering line retainers and retaining clips.
7. Disconnect the power steering lines from the rack and pinion unit.
8. Support the rear of the subframe with a suitable jack. Loosen the front frame bolts and remove the rear bolts, then lower the frame about 3 in. (76mm) for clearance purposes.

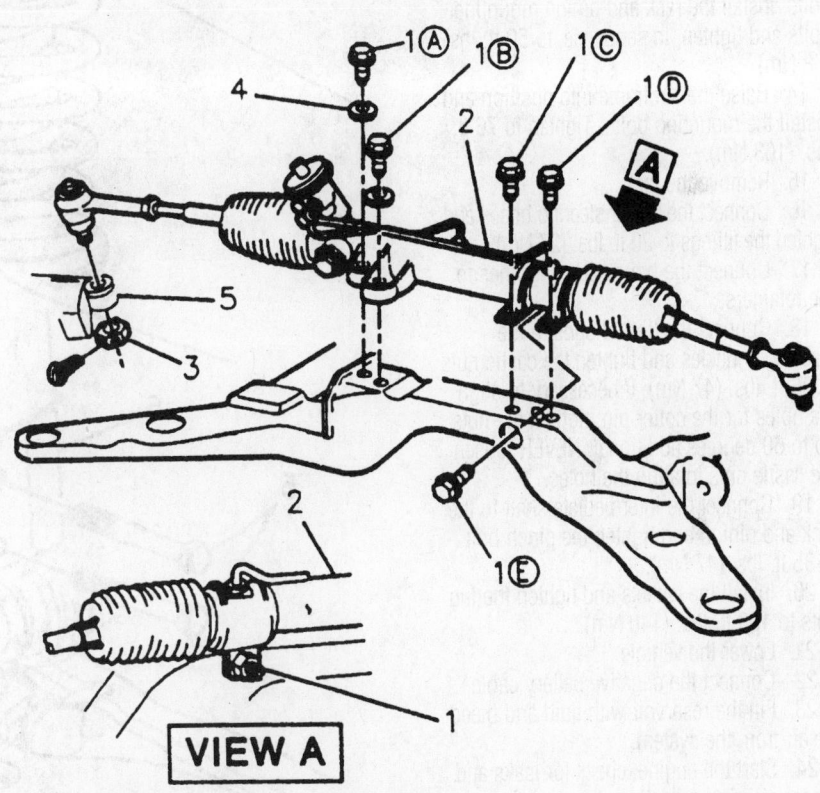

VIEW A

1 BOLT; 68 N•m (50 LB. FT.).
   TIGHTEN IN SEQUENCE A THRU E.
2 STEERING GEAR
3 NUT; 47 N•m (35 LB. FT.). MAXIMUM
   PERMISSIBLE TORQUE TO ALIGN COTTER
   PIN SLOT IS 70 N•m (52 LB. FT.).
4 WASHER
5 STEERING KNUCKLE

7922UG16

Be sure to tighten the steering gear mounting bolts in the sequence shown

9. Remove the rack and pinion mounting bolts.

10. Remove the rack and pinion through the left wheel opening.

To install:

11. Install the rack and pinion through the left wheel opening.

12. Raise the rear of the frame.

13. Apply Loctite® or equivalent, to the threads of the rack and pinion mounting bolts. Install the rack and pinion mounting bolts and tighten, in sequence, to 50 ft. lbs. (68 Nm).

14. Raise the subframe into position and install the mounting bolts. Tighten to 76 ft. lbs. (103 Nm).

15. Remove the jack.

16. Connect the power steering hoses and tighten the fittings to 20 ft. lbs. (27 Nm).

17. Connect the power steering lines to the retainers.

18. Connect the tie rod ends to the steering knuckles and tighten the castle nuts to 35 ft. lbs. (47 Nm). If necessary to align the holes for the cotter pin, tighten the nuts up to 60 degrees additional. NEVER loosen the castle nuts to align the holes.

19. Connect the intermediate shaft to the rack and pinion and tighten the pinch bolt to 35 ft. lbs. (47 Nm).

20. Install the wheels and tighten the lug nuts to 100 ft. lbs. (140 Nm).

21. Lower the vehicle.

22. Connect the negative battery cable.

23. Fill the reservoir with fluid and bleed the air from the system.

24. Start the engine, check for leaks and proper steering operation.

25. Check the wheel alignment.

Strut

REMOVAL & INSTALLATION

Front

1. Loosen the strut housing tie bar through-bolts on both ends of the tie bar.

2. If equipped with electronic ride control option, detach the electrical connection.

3. Remove the three strut-to-body nuts at the strut mount.

4. Raise and safely support the vehicle, allowing the control arms to hang free.

5. Disconnect the ABS front wheel speed sensor.

6. Remove the wheel speed sensor bracket from the strut.

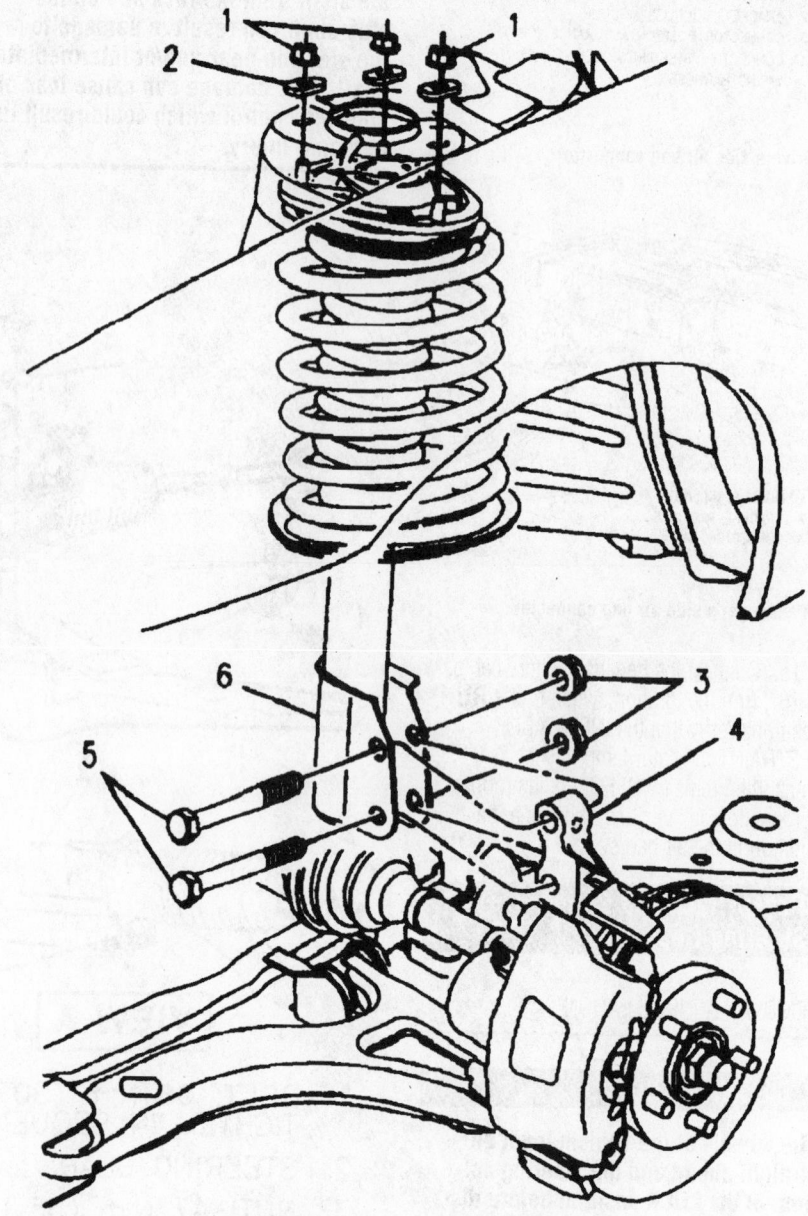

1	NUTS, 47 N·m (35 LB. FT.)
2	WASHER
3	NUTS, 185 N·m (136 LB. FT.)
4	KNUCKLE
5	BOLT
6	STRUT

The strut assembly is mounted between the steering knuckle and the body—front strut shown

7922UG17

7. Remove the brake line bracket from the strut.

8. Remove the strut-to-steering knuckle bolts and remove the strut from the vehicle.

9. Install the strut assembly into compressor tool J 34013-B or equivalent, to compress the coil spring.

10. Compress spring slightly.

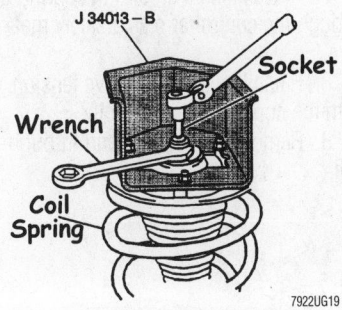

Use a Torx® socket to keep the piston rod from turning while removing the upper nut-front strut shown

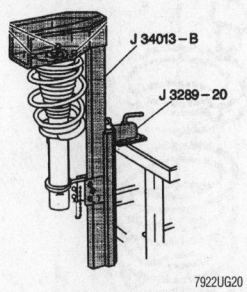

Install the strut assembly into a suitable compressor such as compressor tool J 34013-B to safely remove the strut from the coil spring-front strut shown

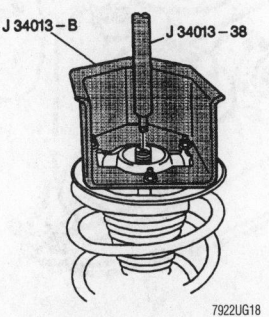

Install rod J 34013-38 or equivalent, to help guide the strut shaft out of the upper mount assembly—front strut shown

11. Hold the strut shaft from turning using a number 50 Torx® socket and remove the 24mm nut on the top end of the strut.

12. Install rod J 34013-38 or equivalent, to help guide the strut shaft out of the upper mount assembly.

13. Loosen the spring compressor tool until the coil spring and mount can be removed as an assembly. Remove the lower spring insulator, if equipped.

To install:

14. Install the strut into spring compressor tool J 34013-B or equivalent. Install the lower insulator, coil spring and upper mount.

15. Compress the coil spring while guiding the strut shaft through the upper mount, using tool J 34013-38 or equivalent.

16. Install and tighten the upper nut to 55 ft. lbs. (75 Nm) while holding the strut shaft with a socket.

17. Remove the strut assembly from the spring compressor.

18. Install the strut in the vehicle.

19. Install and tighten the three strut mount nuts to 18 ft. lbs. (24 Nm).

20. Attach the electronic ride control option electrical connector, if removed.

21. Install the strut-to-knuckle bolts and tighten to 140 ft. lbs. (190 Nm).

22. Install the brake line bracket to the strut.

23. Install the wheel speed sensor bracket from the strut.

24. Attach the ABS front wheel speed sensor connector.

25. Install the front wheel, aligning the marks made during removal, and tighten the lug nuts to 100 ft. lbs. (140 Nm).

26. Lower the vehicle.

27. Check and adjust the wheel alignment.

Rear

1. Remove the trunk side cover or the rear seat cushion and seat back, as required, to gain access to the upper strut mounting nuts.

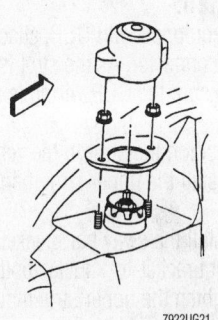

Remove the cover to gain access to the upper strut mounting components—rear strut shown

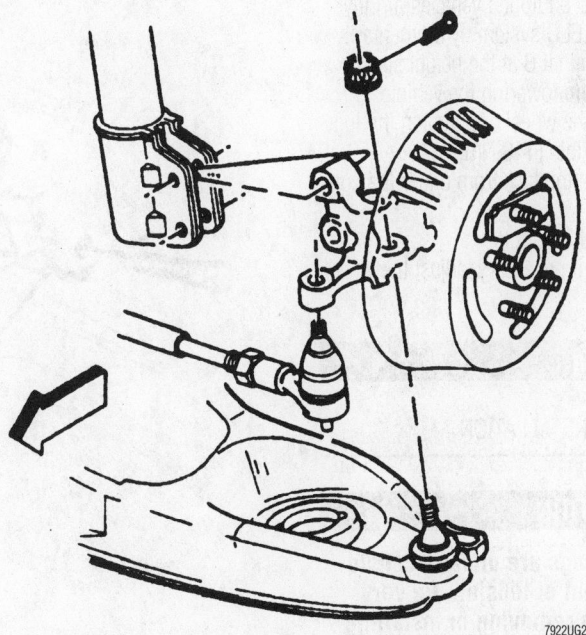

Exploded view of the lower strut mounting to the knuckle assembly—rear strut shown

2. Raise and safely support the vehicle so the rear wheels are just off the ground.

3. Remove the rear wheel.

4. Using a suitable floor jack and a block of wood, support the lower control arm.

5. Remove the two upper strut mounting nuts.

6. If equipped with Electronic Leveling Control (ELC), disconnect the air tube from the strut.

7. Remove the lower strut bracket-to-knuckle bolts.

8. If equipped with Computer Control Ride (CCR), lower the strut and separate the electrical connector.

9. Remove the strut from the vehicle.

To install:

10. If equipped with CCR, attach the electrical connector. As the strut is positioned, be sure the CCR wiring is routed properly.

11. Position the strut in the vehicle and loosely install the upper strut mounting nuts.

12. Install the sway bar bracket and the lower strut bracket-to-knuckle bolts.

13. Tighten the upper strut mounting nuts to 30 ft. lbs. (41 Nm).

14. Tighten the strut bracket-to-knuckle bolts to 140 ft. lbs. (190 Nm).

15. If equipped with ELC, connect the air tube.

16. For ELC equipped vehicles, lightly pressurize the ELC system by momentarily grounding terminal B at the height sensor connector before lowering the vehicle.

17. Install the wheel and tighten the lug nuts to 100 ft. lbs. (140 Nm).

18. Remove the jack from under the control arm.

19. Lower the vehicle.

20. Check, if necessary, adjust the wheel alignment.

Coil Spring

REMOVAL & INSTALLATION

※※ CAUTION

The coil springs are under a considerable amount of tension. Be very careful when removing or installing them; they can exert enough force to cause very serious injury.

Front

For front coil spring service, please refer to the front strut procedure.

Rear

1. Raise and safely support the vehicle.

2. Remove the rear wheel.

3. If removing the right side coil spring, disconnect the height sensor link from the right control arm, if equipped with Electronic Leveling Control (ELC).

4. If removing the left side coil spring, disconnect the parking brake cable retaining clip from the left control arm.

5. If equipped, remove the rear sway bar link from the bracket on the knuckle.

6. Mount tool J-23028-01 or an equivalent control arm support adapter, on a transmission jack and position to cradle the control arm bushings.

※※ CAUTION

Tool J-23028–01 or equivalent, must be secured to the jack or personal injury could result.

7. Place a chain around the spring and through the control arm as a safety measure.

8. Raise the jack to remove tension from the control arm pivot bolts.

9. Remove the rear nut and through-bolt.

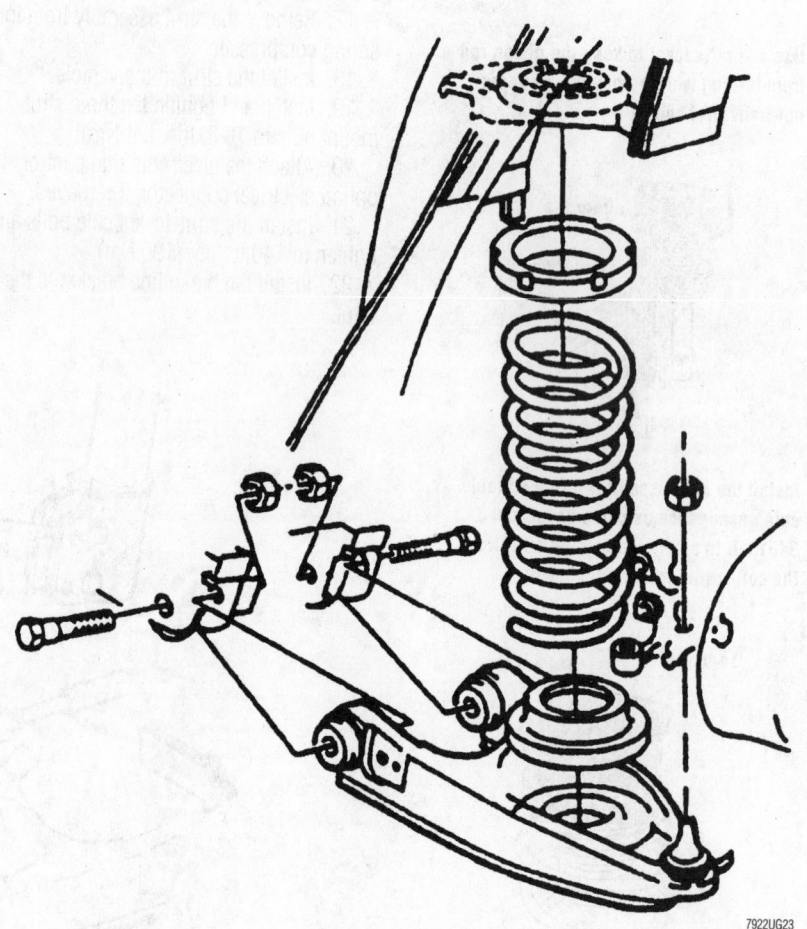

7922UG23

Exploded view of the rear coil spring mounting—H body

10. Slowly maneuver the jack to relieve tension from the front control arm bolt.

11. Remove the front nut and through-bolt.

➡**Do not apply force to the control arm and/or ball joint to remove the spring. Proper maneuvering of the spring will allow for easy removal.**

12. Lower the jack to pivot the control arm downward. When all compression is removed from the spring, remove the safety chain, spring and insulators.

13. Inspect the spring insulators and replace them if they are cut or torn. If the vehicle has been driven more than 50,000 miles, replace them regardless of condition.

To install:

14. Snap the upper insulator onto the spring.

15. Position the lower insulator and spring in the vehicle.

16. Using the jack and J-23028–01 or equivalent, raise the control arm into place.

17. Slowly maneuver the jack to permit installation of the control arm bolts and nuts.

➡**DO NOT tighten the nuts until the weight of the vehicle is on the suspension.**

18. If equipped, connect the rear sway bar to the knuckle bracket with the link assembly. Do not tighten the link bolt yet.

19. Connect the height sensor link to the right control arm or connect the parking brake cable retaining clip to the left control arm, as required.

20. Install the rear wheel and tighten the lug nuts to 100 ft. lbs. (140 Nm).

21. Lower the vehicle.

22. With the vehicle resting on its wheels, tighten the control arm through-bolt nuts to 85 ft. lbs. (115 Nm).

23. Tighten the sway bar link bolt to 13 ft. lbs. (17 Nm).

24. Check, if necessary, adjust the rear wheel alignment.

25. Road test the vehicle for proper operation.

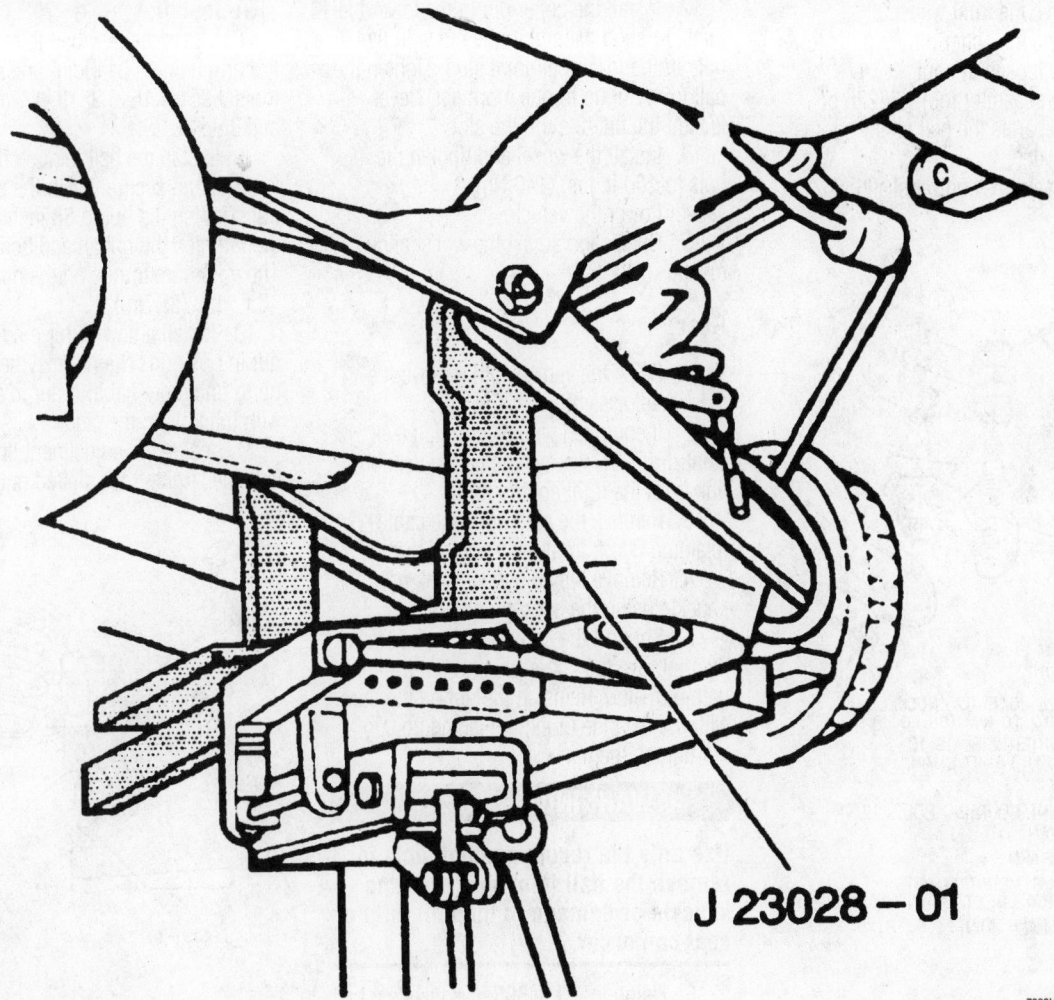

J 23028 – 01

7922UG24

Use a support bracket such as tool J-23028-01 mounted on a jack to support the rear lower control arm—H body

Lower Ball Joint

REMOVAL & INSTALLATION

Front

EXCEPT PARK AVENUE

❋❋ WARNING

If the ball joint is separated for related suspension/driveline service, the ball joint seal should be inspected for damage. A damaged seal will cause ball joint failure. The ball joint should be replaced if seal damage is found.

1. Raise and safely support the vehicle, allowing the control arms to hang free.
2. Remove the front wheel.
3. Remove the cotter pin from the lower ball joint and loosen the nut.
4. Using separator tool J 36226 or equivalent, separate the ball joint from the steering knuckle.
5. Remove the stabilizer shaft link assembly.

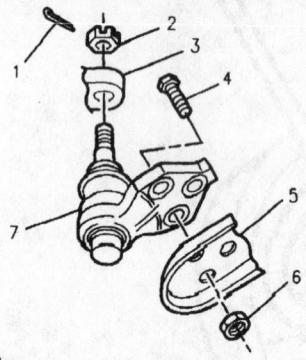

1 PIN
2 NUT, BALL JOINT TO KNUCKLE; TIGHTEN TO 10 N·m (88 LB. IN.) THEN TIGHTEN 2 FLATS TO 55 N·m (41 LB. FT.), MIN.
3 KNUCKLE
4 BALL JOINT MOUNTING BOLTS MUST FACE DOWN
5 CONTROL ARM
6 BALL JOINT MOUNTING NUTS 68 N·m (50 LB. FT.)
7 SERVICE BALL JOINT

7922UG25

The replacement ball joint should be attached to the control arm using three bolts and nuts—except Park Avenue

6. Drill out the three rivets retaining the lower ball joint to the lower control arm.
7. Remove the lower ball joint from the steering knuckle and control arm.

To install:

8. Install the lower ball joint to the steering knuckle and align the holes with the holes in the lower control arm.
9. Attach the ball joint to the lower control arm with bolts and nuts, making sure the bolts face down. Tighten the nuts to 50 ft. lbs. (68 Nm).
10. Install the stabilizer shaft link assembly. Tighten the nut to 13 ft. lbs. (17 Nm).
11. Tighten the lower ball joint nut to 88 inch lbs. (10 Nm). Then, tighten an additional 120 degree turn, during which a torque of 41 ft. lbs. (55 Nm) must be obtained.
12. Install the cotter pin to the lower ball joint. To align the slot in the nut with the hole in the lower ball joint stud, Tighten the ball joint nut up to one more flat. Never loosen the nut to align the slot.
13. Install the wheel and tighten the lug nuts to 100 ft. lbs. (140 Nm).
14. Lower the vehicle.
15. Check and adjust the wheel alignment.

Rear

1. Raise the vehicle and remove the rear wheel assembly.
2. If equipped with Electronic Level Control (ELC), remove the height sensor link from the right control arm.
3. Remove the parking brake cable retaining clip from the left control arm.
4. Remove the adjustment link from the knuckle using the appropriate puller.
5. Support the control arm with a jack, then remove the cotter pin and nut from the ball joint stud. Reinstall the nut on the stud with the flat side of the nut facing up. Do not tighten the nut.

❋❋ WARNING

Use only the recommended tools to remove the ball joint stud from the knuckle or damage to the ball joint or seal can occur.

6. Using tool J 34505 or equivalent ball joint separator, remove the stud from the knuckle. Remove the slotted hex nut from the ball joint stud.

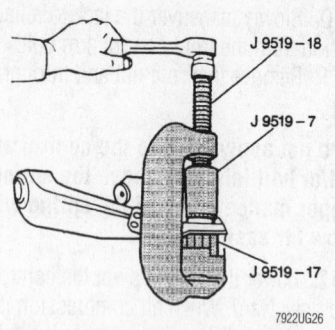

Use the ball joint press to remove the ball joint from the lower control arm—except Park Avenue

7. Press the ball joint out of the control arm using a suitable ball joint press such as tools J 9519–18, J 9519–7 and J 9519–17.

To install:

8. Press the new ball joint into the control arm using a ball joint press such as tools J 9519–16, J 9519–17, J 9519–18 and J 9519–23.
9. Install the ball joint stud into the knuckle and secure it with the slotted hex nut. Tighten the nut to 88 inch lbs. (10 Nm), then tighten the nut an additional four flats. The minimum torque on the nut should be 40 ft. lbs. (55 Nm).
10. Install a new cotter pin to secure the nut in position. The nut may be tightened up to one more flat in order to align the slot with the hole in the stud.
11. Install the adjustment link to the knuckle. Tighten the slotted nut to 33 ft. lbs. (45 Nm).

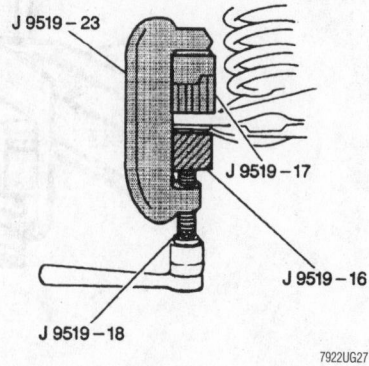

Use the ball joint press to install the ball joint into the lower control arm—except Park Avenue

12. Install a new cotter pin to secure the nut. The nut must only be tightened to align the cotter pin hole with the slot in the nut.

13. If equipped with ELC, connect the height sensor link to the right control arm.

14. Install the parking brake cable retainer to the left control arm.

15. Install the rear wheel assembly and lower the vehicle to the floor.

PARK AVENUE

The ball joint on the Park Avenue must be replaced with the lower control arm as an assembly.

Lower Control Arm

REMOVAL & INSTALLATION

❊❊ WARNING

If the ball joint is separated for related suspension/driveline service, the ball joint seal should be inspected for damage. A damaged seal will cause ball joint failure. The ball joint should should be replaced if seal damage is found.

1. Raise and safely support the vehicle, allowing the control arms to hang free.

2. Remove the front wheel.

3. Remove the stabilizer shaft link assembly from the control arm.

4. Remove the cotter pin from the lower ball joint and remove the nut.

5. Using separator tool J 36226 or equivalent, remove the ball joint from the steering knuckle.

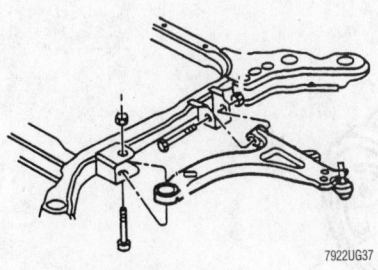

Lower control arm mounting—Park Avenue

6. Remove the lower control arm mounting bolts, then remove the lower control arm from the frame.

To install:

7. Install the lower control arm to the frame with the attaching bolts, washers and nuts.

➡**Do not tighten the lower control arm nuts at this time. The weight of the vehicle must be supported by the control arms such that the vehicle design trim heights are obtained before tightening the lower control arm mounting nuts.**

8. Connect the ball joint stud to the steering knuckle and torque the lower ball joint nut to 88 inch lbs. (10 Nm). Then tighten an additional 120 degree turn, during which a torque of 41 ft. lbs. (55 Nm) must be obtained.

9. Install the cotter pin to the lower ball joint. To align the slot in the nut with the hole in the lower ball joint stud, Tighten the ball joint nut up to one more flat. Never loosen the nut to align the slot.

10. Install the stabilizer shaft link assembly and torque the nuts to 13 ft. lbs. (17 Nm).

11. Install the wheel and torque the lug nuts to 100 ft. lbs. (140 Nm).

12. Lower the vehicle.

13. Torque the front lower control arm mounting nut to 140 ft. lbs. (190 Nm), then torque the rear lower control arm nut to 91 ft. lbs. (123 Nm).

14. Check and adjust wheel alignment.

Wheel Bearings

ADJUSTMENT

The wheel bearings are not adjustable. If a wheel bearing is out of specifications, it must be replaced. Using a dial indicator, check for looseness. If play exceeds 0.005 in. (0.127mm) the bearing wear is excessive and the hub and bearing should be replaced.

REMOVAL & INSTALLATION

Front

1. Raise and safely support the vehicle.

2. Remove the front wheel.

3. Insert a drift punch through the caliper and into the rotor cooling fins to prevent the rotor from turning.

4. Remove the axle nut and washer.

5. Remove the caliper mounting bolts. Remove the caliper from the steering knuckle and support out of the way.

➡**DO NOT allow the brake hose to support the weight of the caliper.**

6. Remove the brake rotor.

7. Disconnect the ABS speed sensor and unclip from the dust shield.

8. Remove the three hub and bearing bolts and remove the dust shield.

9. Place the transaxle selector in the **P** detent.

10. Install J-28733 or an equivalent puller, and separate the hub and bearing from the drive axle.

11. Remove the hub and bearing assembly from the steering knuckle.

To install:

➡**The hub and bearing are replaced only as an assembly.**

12. Install the hub and bearing over the half shaft splines. Be sure the splines engage smoothly.

13. Apply a light coating of grease to the steering knuckle bore.

14. Slide the hub assembly onto the axle as far as possible. If the hub will not bottom out on the axle, install the hub mounting bolts and use the axle nut to draw the hub onto the axle.

15. Once the hub is flush with the steering knuckle, remove the mounting bolts and install the dust shield.

16. Reinstall the mounting bolts and tighten to 70 ft. lbs. (95 Nm).

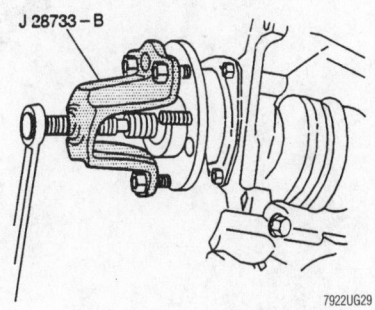

Use a puller such as J 2873-B to push the axle shaft out of the hub assembly

17. Place the transaxle in **N**.

18. Attach the ABS front wheel speed sensor connector and clip to the dust shield.

19. Install the brake rotor.

20. Install the caliper and tighten the mounting bolts to 38 ft. lbs. (51 Nm).

21. Insert a drift punch through the rotor to prevent the axle from turning.

22. Tighten the drive axle shaft nut to 107 ft. lbs. (145 Nm).

23. Remove the drift punch.

24. Install the wheels and tighten the lug nuts to 100 ft. lbs. (140 Nm).

25. Lower the vehicle.

26. Road test the vehicle.

Rear

1. Raise and safely support the vehicle.

2. Remove the rear wheel.

3. Remove the brake drum.

4. Disconnect the ABS sensor wire, if equipped.

❋❋ WARNING

The hub assembly mounting bolts also secure the backing plate assembly. Once the bolts are removed, the backing plate must be supported with wire or other means. Do not let the brake line or ABS electrical wire support the brake assembly.

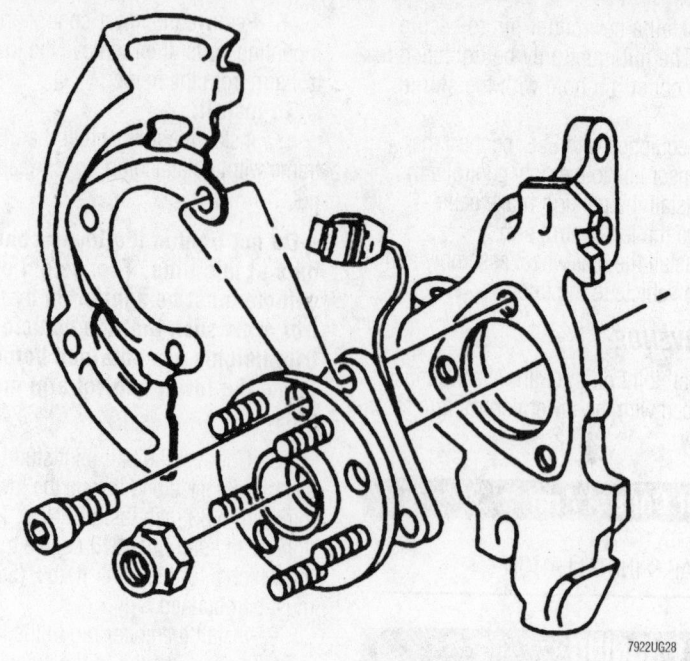

Exploded view of the front hub and wheel bearing assembly

5. Remove the four hub and bearing mounting bolts and remove the hub assembly.

To install:

6. Install hub and bearing assembly onto the rear knuckle.

7. Install the four mounting bolts and tighten to 52 ft. lbs. (70 Nm).

8. Attach the ABS sensor connector, if equipped.

9. Install the brake drum.

10. Install the wheel and tighten the nuts to 100 ft. lbs. (140 Nm).

11. Lower the vehicle and road test for proper operation.

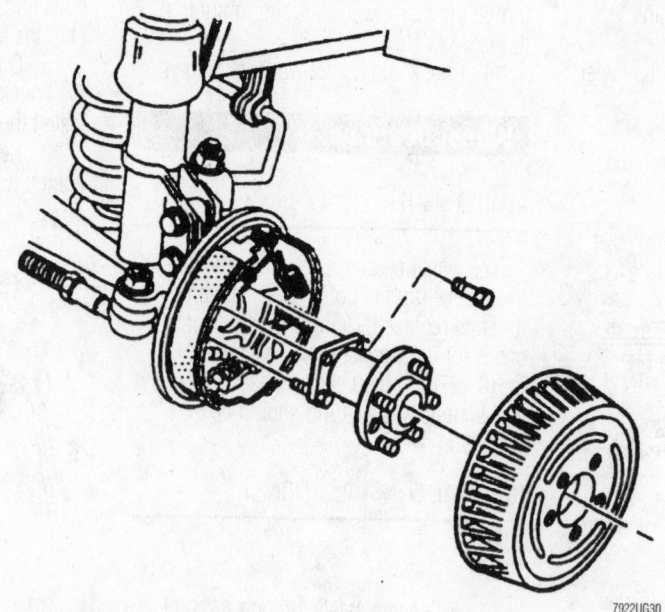

Exploded view of the rear hub/wheel bearing assembly

GENERAL MOTORS E & K BODIES 30

Cadillac-Deville • Concours • Eldorado • Seville

DRIVE TRAIN 30-22
ENGINE REPAIR 30-2
FUEL SYSTEM................30-19
STEERING AND
SUSPENSION..................30-25

A

Air Bag.........................30-25
  Disarming..................30-25
  Precautions 30-25

C

Camshaft and Valve Lifters 30-10
  Removal & Installation 30-10
Coil Spring........................30-28
  Removal & Installation 30-28
Cylinder Head.....................30-5
  Removal & Installation 30-5

D

Distributor.........................30-2
  Removal & Installation 30-2

E

Engine Assembly 30-2
  Removal & Installation 30-2
Exhaust Manifold 30-9
  Removal & Installation 30-9

F

Fuel Filter 30-20

Removal & Installation 30-20
Fuel Pump........................30-21
  Removal & Installation 30-21
Fuel System Pressure 30-19
  Relieving 30-19
Fuel System Service
  Precautions.....................30-19

H

Halfshaft..........................30-24
  Removal & Installation 30-24

I

Ignition Timing 30-2
  Adjustment 30-2
Intake Manifold 30-7
  Removal & Installation 30-7

L

Lower Ball Joint.....................30-29
  Removal & Installation 30-29

O

Oil Pan...........................30-13
  Removal & Installation 30-13
Oil Pump 30-15
  Removal & Installation 30-15

P

Power Rack and Pinion Steering
  Gear...........................30-26

Removal & Installation 30-26

R

Rear Main Seal 30-16
  Removal & Installation 30-16

S

Shock Absorber 30-27
  Removal & Installation 30-27
Strut.............................30-27
  Removal & Installation 30-27

T

Timing Chain, Sprockets, Front
  Cover and Seal 30-16
  Removal & Installation 30-16
Transaxle Assembly 30-22
  Removal & Installation 30-22

V

Valve Lash 30-13
  Adjustment 30-13

W

Water Pump.......................30-4
  Removal & Installation 30-4
Wheel Bearings....................30-29
  Adjustment 30-29
  Removal & Installation 30-30

ENGINE REPAIR

➡ **Disconnecting the negative battery cable on some vehicles may interfere with the functions of the on board computer systems and may require the computer to undergo a relearning process, once the negative battery cable is reconnected.**

Distributor

REMOVAL & INSTALLATION

4.9L Engine

1. Disconnect the negative battery cable.
2. Disconnect the feed wire and coil connection from the distributor cap.
3. Remove the distributor cap mounting bolts, then move the cap out of the way. It is not necessary to disconnect the ignition wires.
4. Disconnect the PCM harness from the distributor.
5. Mark the position of the rotor with respect to the distributor housing and scribe the position of the distributor housing with respect to the engine.
6. Remove the distributor hold-down nut and clamp.
7. Remove the distributor.

➡ **Do not crank the engine after the distributor is removed. If the engine is accidentally cranked, the engine will have to be positioned at No. 1 cylinder TDC.**

To install:

8. Install the distributor, aligning the marks made during removal.
9. If the engine was accidentally cranked after the distributor was removed, proceed as follows:
    a. Remove the No. 1 cylinder spark plug.
    b. Place a finger over the No. 1 cylinder spark plug hole and crank the engine slowly until compression is felt.
    c. Align the timing mark on the crankshaft pulley to **0** on the engine timing indicator.
    d. Position the distributor rotor to point towards the No. 1 spark plug wire terminal on the distributor cap.
    e. Install the distributor, aligning the mark on the distributor housing with the mark on the engine.
    f. Install the spark plug in the No. 1 cylinder.
10. Install the distributor hold-down clamp and nut. DO NOT tighten the nut at this time.
11. Connect the PCM connector.
12. Install the distributor cap and mounting screws.
13. Connect the coil leads and feed wire to the cap.
14. Connect the negative battery cable.

15. Set the ignition timing.
16. Tighten the distributor hold-down nut to 20 ft. lbs. (26 Nm).

Ignition Timing

ADJUSTMENT

4.6L Engines

The 4.6L Northstar engine is equipped with a Distributorless Ignition System (DIS). The system consists of two crankshaft position sensors, crankshaft reluctor ring, camshaft position sensor, ignition control module, four ignition coils, eight plug wires and spark plugs, knock sensor and the Powertrain Control Module (PCM).

The base ignition timing is determined by the relationship of the crankshaft position sensors to the crankshaft reluctor ring. This relationship is not adjustable. Base ignition timing is 10 degrees Before Top Dead Center (BTDC).

The PCM controls spark advance under all driving conditions. The PCM incorporates a permanent spark control override, which electronically lowers the base timing if spark knock (detonation) is encountered during normal operation due to the use of low octane fuel.

4.9L Engine

➡ **Check the Vehicle Emission Control Information (VECI) label, located in the engine compartment. If the procedure below differs from that described on the label, follow the instructions on the label. The information on the VECI label often reflects changes made during production.**

1. With the ignition **OFF**, connect a timing light to the No. 1 spark plug wire. Connect the power leads for the timing light according to the manufacturers directions.
2. Jumper pins **A** and **B** together at the Assembly Line Diagnostic Link (ALDL) connector while not in diagnostic display.
3. Start the engine and allow it to reach normal operating temperature. Aim the timing light at the timing indicator and crankshaft pulley. Refer to the VECI label for ignition timing specifications.
4. If adjustment is required, loosen the distributor hold-down bolt and rotate the distributor until the timing marks are in alignment.
5. Tighten the hold-down bolt and check the timing to be sure the distributor didn't move.

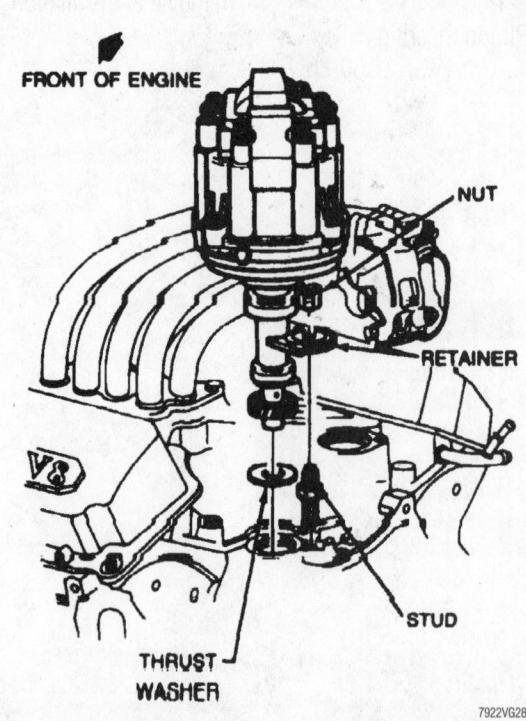

FRONT OF ENGINE

NUT

RETAINER

STUD

THRUST WASHER

7922VG28

Distributor and mounting components—4.9L engine

6. Turn **OFF** the engine and disconnect the timing light.

Engine Assembly

REMOVAL & INSTALLATION

✳✳ CAUTION

Some models covered by this manual may be equipped with a Supplemental Restraint System (SRS), which uses an air bag. Whenever working near any of the SRS components, such as the impact sensors, the air bag module, steering column and instrument panel, properly disable the SRS.

4.6L Engines

✳✳ CAUTION

The fuel injection system remains under pressure, even after the engine has been turned OFF. The fuel system pressure must be relieved before disconnecting any fuel lines. Failure to do so may result in fire and/or personal injury.

1. Disconnect the negative battery cable.

✳✳ CAUTION

Observe all applicable safety precautions when working around fuel. Whenever servicing the fuel system, always work in a well ventilated area. Do not allow fuel spray or vapors to come in contact with a spark or open flame. Keep a dry chemical fire extinguisher near the work area. Always keep fuel in a container specifically designed for fuel storage; also, always properly seal fuel containers to avoid the possibility of fire or explosion.

2. Relieve the fuel system pressure.
3. Remove the air cleaner assembly.
4. Remove the left and right torque struts. Install the left front strut bolt back into the bracket.

✳✳ CAUTION

Never open, service or drain the radiator or cooling system when hot; serious burns can occur from the steam and hot coolant.

5. Disconnect the radiator hoses at the water crossover.
6. Remove both cooling fans from the engine.
7. Properly remove the refrigerant from the A/C system.
8. Detach the cruise control servo connections.
9. Detach the ISC motor connectors.
10. Disconnect the throttle cable from the throttle body cam. Disconnect the shift cable from the park/neutral switch. Remove the cable bracket at the transaxle.
11. Remove the park/neutral switch connector and detach the power brake vacuum hose.
12. Disconnect the fuel inlet and fuel return lines using special tool J37088 or equivalent.
13. Remove the fuel line retainer at the transaxle case.
14. Disconnect the hoses from the coolant reservoir. Remove the reservoir.
15. Disconnect the heater hoses from the front of the right cylinder head. Disconnect the temperature switch.
16. Remove the starter.
17. Disconnect the power steering pump pressure and return lines at the cooler. Remove power steering line retainer from the right front of the crankcase.
18. Detach the engine harness connectors at the PCM.
19. From under the hood, remove the wiring harness retainer screws at the cowl and pull the engine harness through.
20. Disconnect the refrigerant high temperature switch.
21. To remove the engine harness on the left wheel housing, remove one screw holding the connector halves together and separate. The engine portion of the harness will be removed with the engine.
22. Remove the serpentine drive belt.
23. Raise and safely support the vehicle.
24. Remove both front wheels.

✳✳ CAUTION

The EPA warns that prolonged contact with used engine oil may cause a number of skin disorders, including cancer! You should make every effort to minimize your exposure to used engine oil. Protective gloves should be worn when changing the oil. Wash your hands and any other exposed skin areas as soon as possible after exposure to used engine oil. Soap and water, or waterless hand cleaner should be used.

25. Disconnect the oil cooler lines at the oil filter adapter.
26. Disconnect the exhaust Y-pipe.
27. Disconnect the coupling between the steering rack and the column.
28. Disconnect the speed sensitive steering solenoid and the knock sensor.
29. Disconnect the power steering switch.
30. Separate the lower ball joints and stabilizer links (the struts will stay in the vehicle).
31. Disconnect the A/C hoses from the accumulator and the condenser.
32. Move Powertrain Dolly J-36295 or equivalent, into position and lower the vehicle onto the table.
33. Remove the six engine cradle mounting bolts and remove the powertrain assembly by lifting the vehicle or lowering the table.
34. Remove the torque converter splash shield and remove the four converter-to-flywheel bolts.
35. Separate the engine and transaxle assemblies.

To install:

36. Install the engine to the transaxle and tighten the bolts to 55 ft. lbs. (75 Nm).
37. Install the exhaust manifolds and tighten to 18 ft. lbs. (25 Nm).
38. Install the exhaust system Y-pipe and the transaxle to oil pan brace.
39. With the powertrain on the dolly, move the assembly into approximate position under the vehicle.
40. Lower the vehicle onto the powertrain.
41. Align the engine cradle to the body and install the six engine cradle mounting bolts. Tighten the bolts to 75 ft. lbs. (100 Nm).
42. Raise the vehicle and remove the dolly.
43. Connect the oil cooler lines at the oil filter adapter.
44. Install the exhaust Y-pipe.
45. Connect the coupling between the steering rack and the column.
46. Connect the speed sensitive steering solenoid and the knock sensor.
47. Connect the power steering switch.
48. Install the A/C hoses to the accumulator and the condenser.
49. Connect the lower ball joint and the stabilizer shaft link.
50. Position the ABS/TCS assembly to the engine cradle.
51. Lower the vehicle.
52. Attach the engine harness connectors at the PCM.
53. Connect the refrigerant high temperature switch.
54. Position the engine harness connector on the left wheel housing and install the one screw holding the connector halves together.

55. Connect or install the remaining components.

56. Install the right and left torque struts and tighten the retainer bolts as follows:

→It is important during installation that the engine torque struts are not pre-loaded in their installed position. Adjustment is provided at the point the strut fastens to the core support bracket. Be sure this bolt is loose during assembly. Tighten to 45 ft. lbs. (60 Nm) as the final step of assembly.

    a. Tighten the strut bracket to cylinder head (M10) bolt: 35 ft. lbs. (50 Nm).

    b. Tighten the strut bracket to cylinder head (M10) stud: 35 ft. lbs. (50 Nm).

    c. Tighten the strut bracket to water manifold (M8) bolts: 20 ft. lbs. (25 Nm).

    d. Tighten the strut bracket to cylinder head (M10) bolt: 35 ft. lbs. (50 Nm).

    e. Tighten the strut to core support bracket bolt: 45 ft. lbs. (60 Nm).

57. Connect the negative battery cable.

58. Evacuate and recharge the A/C system.

59. Install the air cleaner.

❄❄ WARNING

Operating the engine without the proper amount and type of engine oil will result in severe engine damage.

60. Refill the cooling system and engine crankcase.

61. Run the engine and check for leaks and proper vehicle operation.

→A wheel alignment is recommended after the removal of the sub-frame assembly.

4.9L Engine

❄❄ CAUTION

Fuel Injection systems remain under pressure, even after the engine has been turned OFF. The fuel system pressure must be relieved before disconnecting any fuel lines. Failure to do so may result in fire and/or personal injury.

1. Disconnect the negative battery cable.

2. Relieve the fuel system pressure.

3. Mark the position of the hood hinges and remove the hood.

4. Drain the cooling system and engine oil into suitable containers.

5. Disconnect the heater and both radiator hoses from the engine.

6. Remove the starter.

7. Disconnect the harnesses to the engine and the battery near the relay center.

8. Remove the serpentine belt(s).

9. Remove the power steering pump and set aside.

10. Remove the air cleaner and air inlet duct.

11. Disconnect the control cables from the throttle body and cable mounting bracket.

12. Disconnect the Throttle Position Sensor (TPS), Idle Air Control (IAC) valve, MAT Sensor and Oxygen Sensor (O₂) connectors.

13. Disconnect the ignition coil assembly ground wire from the inner fender panel.

14. Disconnect and cap the fuel lines from the fuel rail.

15. Install engine support fixture tool J-28467-A or the equivalent. Raise the engine slightly to remove the weight from the engine mounts.

16. Raise and safely support the vehicle.

17. Disconnect the front exhaust pipe from the exhaust manifold.

18. Remove the A/C compressor from the bracket and support out of the way. Do not disconnect the refrigerant lines from the compressor.

19. Remove the left front engine mount.

20. Remove the right front engine-to-transaxle bracket.

21. Remove the lower transaxle-to-engine bolts.

22. Remove the flywheel cover.

23. Remove the torque converter-to-flywheel bolts.

24. Lower the vehicle.

25. Remove the torque strut.

26. Remove the vibration absorber.

27. Remove the transaxle-to-engine bolts.

28. Support the transaxle with a suitable jack.

29. Install a suitable engine lifting device and remove the support fixture.

30. Remove the engine from the vehicle.

To install:

31. Install the engine in the vehicle, install the transaxle-to-engine bolts and tighten to 55 ft. lbs. (75 Nm).

32. Install the engine support fixture and remove the engine lifting device.

33. Raise and safely support the vehicle.

34. Install the right front engine-to-transaxle bracket.

35. Install the left front engine mount.

36. Install the torque strut. Install the vibration absorber.

37. Install the torque converter-to-flywheel bolts and tighten to 46 ft. lbs. (62 Nm).

38. Install the remaining components.

❄❄ WARNING

Operating the engine without the proper amount and type of engine oil will result in severe engine damage.

39. Refill the cooling system and engine crankcase.

40. Connect the negative battery cable.

41. Install the hood.

42. Turn the ignition key **ON** and inspect for any fuel leaks.

43. Run the engine and check for leaks and proper vehicle performance.

Water Pump

REMOVAL & INSTALLATION

→On 4.6L engines, there was a design change to the water pump inlet housing. If it is a black plastic housing, the housing must be replaced. There are no seals available for the plastic housings. The new housings are made of aluminum. When ordering parts, be sure to ask for a water pump, water pump seal and an inlet housing seal and housing if needed.

1. Disconnect the negative battery cable.

2. Drain the coolant into a suitable container.

3. Remove the accessory drive belt. Remove the water pump pulley.

4. On 4.6L engines, remove the following:

    a. Remove the air cleaner.

    b. Remove the lower radiator hose and remove the thermostat bypass hose from the coolant inlet housing.

5. On 4.9L engines, remove the coolant overflow tank.

6. Remove the water pump mounting bolts.

7. On 4.6L engines, remove the water pump from the vehicle by rotating clockwise with tool J-38816-A or equivalent. Remove the O-ring and clean out the groove.

To install:

8. Clean all gasket mating surfaces.

9. On 4.6L engines, install the O-ring seal into the groove, then install the water pump, turning it counterclockwise until it stops, using tool J-38816-A or equivalent.

10. Install the water pump housing and tighten the bolts to 88 inch lbs. (10 Nm).

11. On 4.9L engines, install the new water pump gasket on the front cover studs with the raised sealing surface facing outward.

12. Install the water pump on the engine and tighten the mounting bolts.

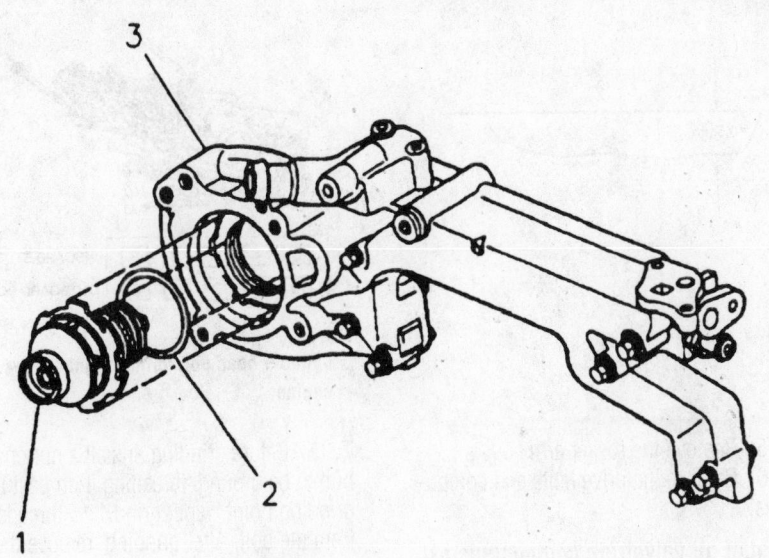

1 WATER PUMP ASSEMBLY
2 O-RING SEAL
3 WATER PUMP HOUSING ASSEMBLY

7922VG01

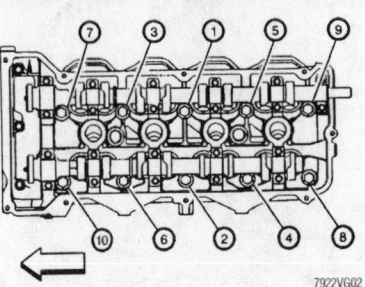

Left cylinder head bolt tightening
sequence—4.6L engine

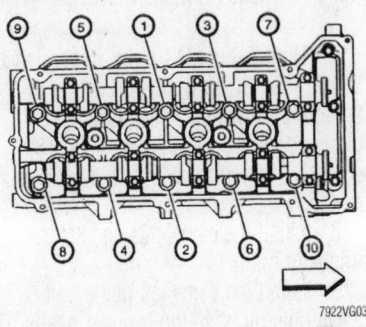

7922VG03

Right cylinder head bolt tightening
sequence—4.6L engine

To ensure proper operation, be sure to install a new O-ring

13. Install the water pump pulley. Install the pulley bolts finger-tight.

14. Install the accessory drive belt.

15. Tighten the water pump pulley bolts to 115 inch lbs. (13 Nm).

16. On 4.6L engines, install the air cleaner.

17. On 4.9L engines, install and connect the coolant reservoir.

18. Connect the negative battery cable.

19. Refill and bleed the cooling system.

20. Run the engine and check for leaks.

Cylinder Head

REMOVAL & INSTALLATION

4.6L Engines

➡**The manufacturer recommends that the entire powertrain be removed from the vehicle before removing the cylinder heads.**

✳✳ CAUTION

The fuel injection system remains under pressure, even after the engine has been turned OFF. The fuel system pressure must be relieved before disconnecting any fuel lines. Failure to do so may result in fire and/or personal injury.

1. Disconnect the negative battery cable.

2. Properly relieve the fuel system pressure.

3. Drain the cooling system into a suitable container.

4. Remove the powertrain assembly.

5. Remove the intake manifold, cam covers, harmonic balancer, timing chain front cover and oil pump.

✳✳ WARNING

Align all timing marks before performing the next step.

6. Remove the chain tensioner from the timing chain.

7. Remove the cam sprockets. The timing chain remains in the chain case.

8. Removing the timing chain guides. Access for the retaining screws is through the plugs at the front of the cylinder head.

9. Remove the water crossover.

10. Remove the exhaust manifold.

11. Remove the cylinder head bolts, a little at a time, in the reverse order of the tightening sequence.

12. Remove the cylinder head and gasket.

✳✳ WARNING

With the camshafts remaining in the cylinder head, some valves will be open at all times. Do not rest the cylinder head on a flat service with

the cylinder face down, or valve damage will result.

13. Clean all gasket mating surfaces. Clean the head bolt holes in the crankcase with compressed air.

✳✳ WARNING

Be careful when cleaning aluminum gasket surfaces to prevent damage to the sealing surfaces. Use only plastic, wood or "dull" gasket scrapers. Chemical agents to dissolve gasket materials can also be used by following the manufacturers directions.

14. Check the cylinder head for warpage using a straightedge and feeler gauge. Measure along each edge, at the center and across both ends.

15. If warpage is less than 0.002 in. (0.05mm), the cylinder head surface is usable. If warpage is 0.002-0.008 in. (0.05-0.2mm), the cylinder head must be resurfaced. After resurfacing, the dimension between the combustion chamber gauge pad and the deck surface must be at least 10.5mm.

To install:

16. Using a new cylinder head gasket, install the cylinder head and the M11 and M6 head bolts. Lube the washer and the underside of the bolt head with engine oil prior to installation.

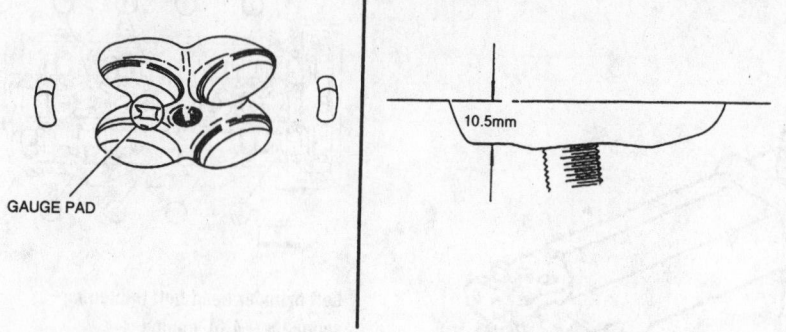

Minimum head resurface dimension—4.6L engine

7922VG04

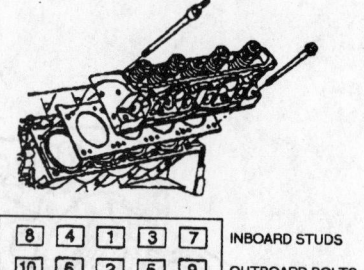

7922VG29

Cylinder head bolt torque sequence—4.9L engine

17. Tighten the M11 bolts, in sequence, to 22 ft. lbs. (30 Nm) plus 60 degrees. Repeat the sequence twice, turning each bolt an additional 60 degrees (total 180 degrees). Tighten the M6 bolts to 106 inch lbs. (12 Nm).

18. Install the camshafts and set the camshaft timing.

19. Install the camshaft guide bolt access hole plugs in the cylinder heads. The plugs should be seated and snug.

20. Install the intake cam covers, oil pump, timing chain front cover, harmonic balancer, intake manifold and water crossover.

21. Install the exhaust manifold. Tighten the nuts to 22 ft. lbs. (30 Nm) or the bolts to 18 ft. lbs. (25 Nm).

22. Install the powertrain assembly.

23. Connect the negative battery cable.

24. Fill the cooling system and check all fluid levels.

25. Properly charge the A/C system.

26. Run the engine and check for leaks and proper engine performance.

➡A wheel alignment is recommended after removal of the sub-frame assembly.

4.9L Engine

✳✳ CAUTION

The fuel injection system remains under pressure, even after the engine has been turned OFF. The fuel system pressure must be relieved before disconnecting any fuel lines. Failure to do so may result in fire and/or personal injury.

1. Disconnect the negative battery cable.
2. Relieve the fuel system pressure.
3. Drain the cooling system into a suitable container.
4. Disconnect and label the spark plug wires and harnesses.

5. Remove the rocker arm covers.
6. Remove the drive belts and components.

➡**Label all valvetrain components as they are removed, so they can be reinstalled in their original locations.**

Remove the rocker arm support retaining nuts/bolts and remove the rocker arm support.

✳✳ WARNING

The rocker arm support must be removed with the rocker arms and pivots attached. The pivot assemblies or support bar threads may be damaged if pivot bolt torque is removed against valve spring pressure. If necessary, secure the support in a vise and remove the rocker arms and pivots.

7. Remove the pushrods.
8. Remove the intake manifold.
9. Disconnect the exhaust pipe from the manifold and detach the oxygen sensor.
10. Remove the exhaust manifold.
11. Remove the cylinder head bolts, loosening them a little at a time in the reverse order of the torque sequence. Note the locations of the stud headed bolts, so they can be reinstalled in their proper locations.
12. Remove the cylinder head.
13. Clean all gasket mating surfaces. Clean the head bolt threads and the corresponding bolt holes in the cylinder block. Be sure there are no accumulations of oil or coolant in the bolt holes.
14. Measure the thread length of the inboard and outboard head bolts/studs to determine whether they are "short" or "long". Measure from the first thread (closest to the bolt head) to the end of the bolt. The "short" bolts are 39–41mm long and must be taper ground to prevent them from bottoming out in the engine block. The "long" bolts are 47–50mm long and should not be modified.

15. Before grinding, install a nut on the bolt to be ground, threading it on past the grinding point. Taper grind 1 ½ threads from the bolt. After grinding, remove the nut; the nut will clean the threads of the bolt as it is removed, easing installation.

To install:

16. Install a new cylinder head gasket on the engine block. Be sure the gasket aligns with the dowel pins on the block.

17. Install the cylinder head.

18. Lubricate the cylinder head bolts with GM lubricant 1052356 or equivalent. Do not use engine oil.

19. If equipped with the "short" head bolts, install the cylinder head bolts finger-tight, then tighten, in sequence, as follows:

a. Tighten all bolts and studs, in sequence, to 29 ft. lbs. (40 Nm).

b. Tighten all bolts and studs, in sequence, to 51 ft. lbs. (70 Nm).

c. Tighten all bolts and studs, in sequence, to 85 ft. lbs. (115 Nm).

d. Tighten the 3 center inboard studs, 1, 3 and 4 to 88 ft. lbs. (120 Nm).

20. If equipped with the "long" head bolts, install the cylinder head bolts finger-tight, then tighten, in sequence, as follows:

a. Tighten all bolts and studs, in sequence, to 29 ft. lbs. (40 Nm).

b. Tighten all bolts and studs, in sequence, to 51 ft. lbs. (70 Nm).

c. Tighten all bolts and studs, in sequence, to 81 ft. lbs. (110 Nm).

d. Tighten all inboard studs, in sequence, to 88 ft. lbs. (120 Nm).

e. Tighten the 3 center inboard studs, 1, 3 and 4 to 96 ft. lbs. (130 Nm).

21. Install the engine lift bracket.

22. Install the exhaust manifold and the oxygen sensor connector.

23. Connect the exhaust system pipe to the manifold and the cross-over pipe.

24. Install the intake manifold.

25. Install the pushrods, making sure they are seated in the lifters.

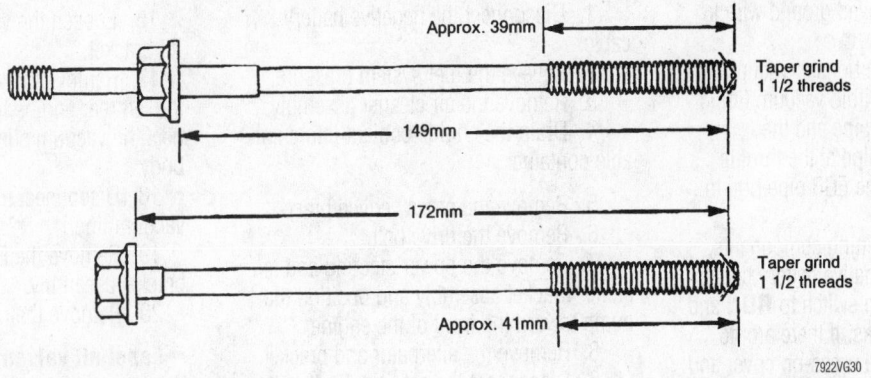

7922VG30

Taper grind the "short" cylinder head bolts—4.9L engine

26. Install the rocker arms and pivots to the rocker arm support. Loosely install the pivot bolts, then tighten the pivot bolts to 22 ft. lbs. (30 Nm).

27. Install the rocker arm support, with the rocker arms and pivots over the 5 studded head bolts. Be sure each pushrod is seated in a rocker arm.

28. Loosely install the rocker arm support retaining nuts and bolts.

29. Tighten the rocker arm support retaining nuts alternately and evenly until snug.

➡**Check the pushrods from time to time during the tightening procedure to be sure they are correctly positioned.**

30. Tighten the rocker arm support retaining nuts to 37 ft. lbs. (50 Nm).

31. Tighten the rocker arm retaining bolts to 7 ft. lbs. (10 Nm).

32. Install the wedges and coat them with RTV for seal ability.

33. Install the rocker arm covers.

34. Install the belt driven components and drive belts.

35. Refill the cooling system.

36. Connect the negative battery cable.

37. Run the engine and check for leaks and proper engine operation.

Intake Manifold

REMOVAL & INSTALLATION

4.6L Engines

✳✳ CAUTION

The fuel injection system remains under pressure, even after the engine has been turned OFF. The fuel system pressure must be relieved before disconnecting any fuel lines. Failure to do so may result in fire and/or personal injury.

1. Disconnect the negative battery cable.

2. Remove the dress-up cover by removing the four nuts.

3. Relieve the fuel system pressure.

4. Drain the cooling system into a suitable container.

5. Remove the intake duct from the throttle body.

6. Remove the transaxle vent hose and the transaxle shift cable at the bracket.

7. Remove the Throttle Position (TP) sensor and the Idle Air Control (IAC) valve connectors.

8. Disconnect the throttle cable and the cruise control cable at the throttle body.

9. Remove the throttle body coolant hoses and the surge tank pipe.

10. Remove the EGR pipe and the crankcase ventilation pipe at the throttle body spacer.

11. Remove the brake booster vacuum hose at the intake manifold.

12. Remove the fuel rail ground wire at the rear cylinder head.

13. Remove the quick-disconnect fuel rail fittings using tool J-37088-A or equivalent, insert the tool into the female connector and push inward to release the locking tabs and pull the connection apart.

14. Disconnect the fuel rail bracket at the EGR valve.

15. Disconnect the PCV hose at the intake manifold.

16. Detach the main injector harness connector.

17. Remove the six bolts and the four studs in the intake manifold.

➡**The intake manifold carrier gaskets are attached to the intake manifold through a snap-lock feature. When removing the intake manifold, the carrier gaskets will remain attached to the intake manifold. DO NOT replace the intake manifold gaskets after the intake manifold removal.**

The gaskets are reusable. The gaskets should only be replaced if the plastic housing or the rubber seals are damaged.

To install:

18. Install the intake manifold to the engine assembly.

19. Install the six bolts and four studs in their proper locations. Tighten the bolts and studs to 89 inch lbs. (10 Nm). When tightening the intake manifold bolts and studs, start at the center of the manifold and work outward in a circular pattern. DO NOT tighten the intake manifold bolts when the engine is HOT or at operating temperature. This repair should only be done on a cool engine.

20. Connect the injector main wiring harness.

21. Install the PCV hose to the intake manifold.

22. Install the fuel bracket at the EGR valve.

23. Install the fuel lines to the intake manifold. Apply a few drops of clean engine oil to the male ends of the fuel rail inlet and outlet tubes prior to assembly. When installed, pull outward to ensure a good connection.

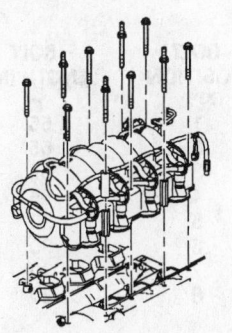

7922VG05

Exploded view of the intake manifold mounting—4.6L engine

24. Install the removed ground wire to the rear cylinder head.

25. Install the brake booster vacuum hose to the intake manifold vacuum fitting.

26. Install the EGR pipe and the crankcase ventilation pipe at the throttle body spacer. Tighten the EGR pipe bolt to 21 ft. lbs. (28 Nm).

27. Install the remaining components.

28. Connect the negative battery cable.

29. Turn the ignition switch to **RUN** and inspect for any fuel leaks. If there are no leaks, install the engine dress-up cover and tighten the cover nuts to 89 inch lbs. (10 Nm).

30. Fill and bleed the cooling system.

31. Road test the vehicle for proper operation.

4.9L Engine

✳✳ CAUTION

The fuel injection system remains under pressure, even after the engine has been turned OFF. The fuel system pressure must be relieved before disconnecting any fuel lines. Failure to do so may result in fire and/or personal injury.

1. Disconnect the negative battery cable.

2. Relieve the fuel system pressure.

3. Remove the air cleaner assembly.

4. Drain the engine coolant into a suitable container.

5. Remove the cross vehicle brace.

6. Remove the drive belt.

7. Remove the power steering and tensioner bracket assembly and position the pump toward the front of the engine.

8. Remove the alternator and bracket.

9. Disconnect the cable at the throttle body.

10. Label and disconnect the wiring at the distributor, oil pressure switch, coolant sensor, EGR solenoid, Idle Speed Control (ISC) motor, Throttle Position Sensor (TPS), fuel injectors and Manifold Air Temperature (MAT) sensor and Manifold Absolute Pressure (MAP) sensor hoses.

11. Disconnect the upper radiator and heater hoses at the thermostat housing.

12. Remove the A/C hose bracket.

13. Remove the distributor.

14. Disconnect the fuel lines at the transaxle bracket.

15. Label and disconnect the vacuum and fuel lines at the throttle body.

16. Loosen the vacuum line clip at the lift bracket.

17. Remove the vacuum line bracket at the left rear engine lift bracket, and disconnect the vacuum supply line at the throttle body.

18. Disconnect the transaxle modulator vacuum line.

19. Remove the EGR solenoid and bracket assembly.

20. Remove the rocker arm covers.

➡**Label all valvetrain components as they are removed, so they can be reinstalled in their original locations.**

21. Remove the rocker support retaining nuts/bolts and remove the rocker arm support.

✳✳ WARNING

The rocker arm support must be removed with the rocker arms and pivots attached. The pivot assemblies or support bar threads may be damaged if pivot bolt torque is removed against valve spring pressure. If necessary, secure the support in a vise and remove the rocker arms and pivots.

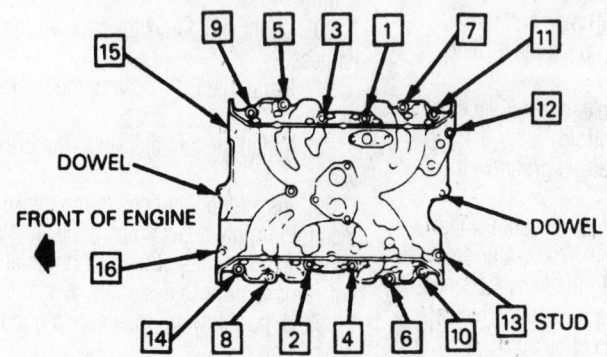

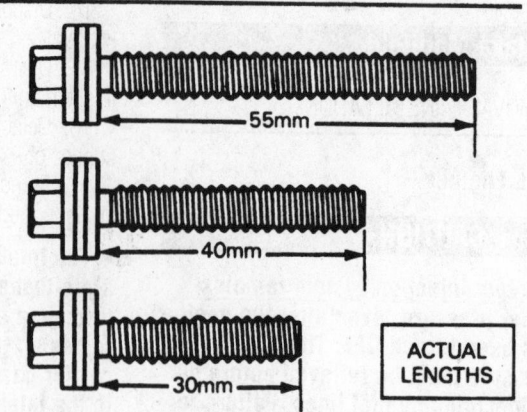

BOLT TIGHTENING SEQUENCE

1. TIGHTEN BOLTS 1, 2, 3, & 4 IN SEQUENCE TO 12.0 N·m (8 FT-LBS).

2. TIGHTEN BOLTS 5 THRU 16 IN SEQUENCE TO 12.0 N·m (8 FT-LBS).

3. RETIGHTEN ALL BOLTS IN SEQUENCE TO 16.0 N·m (12 FT-LBS).

4. REPEAT STEP 3 UNTIL TORQUE LEVEL IS MAINTAINED.

BOLT POSITION	BOLT LENGTH (MM)	BOLT POSITION	BOLT LENGTH (MM)
1	55	9	40
2	55	10	40
3	55	11	40
4	55	12	55
5	30	13	40 W/Studhead
6	30	14	40
7	30	15	55
8	30	16	40

Intake manifold bolt identification and torque sequence—4.9L engine

7922VG31

22. Remove the pushrods.

23. Remove the power steering bracket.

24. Remove the right front engine lift bracket.

25. Remove the intake manifold bolts and remove the intake manifold.

26. Remove and discard the gaskets and seals. Clean all gasket mating surfaces.

To install:

27. Install new intake manifold end seals.

28. Apply RTV sealant to the four corners where the end seals meet the side gaskets.

29. Install new intake manifold gaskets.

30. Install the intake manifold bolts and tighten, in sequence.

31. Install the pushrods, making sure they are seated in the lifters.

32. Install the rocker arms and pivots to the rocker arm support. Loosely install the pivot bolts, then tighten the pivot bolts to 22 ft. lbs. (30 Nm).

33. Install the rocker arm support, with the rocker arms and pivots installed, over the 5 stud headed head bolts. Be sure each pushrod is seated in a rocker arm.

34. Loosely install the rocker arm support retaining nuts and bolts.

35. Tighten the 5 rocker arm support retaining nuts alternately and evenly until snug.

➡**Check the pushrods from time to time during the tightening procedure to be sure they are correctly positioned.**

36. Tighten the 4 rocker arm retaining bolts until snug.

37. Tighten the 5 rocker arm support retaining nuts to 37 ft. lbs. (50 Nm).

38. Tighten the 4 rocker arm retaining bolts to 7 ft. lbs. (10 Nm).

39. Install the rocker arm covers.

40. Install the remaining components.

41. Install the air cleaner.

42. Refill the cooling system.

43. Connect the negative battery cable.

44. Start the engine and set the ignition timing. Run the engine and check for leaks and proper engine operation.

Exhaust Manifold

REMOVAL & INSTALLATION

4.6L Engines

LEFT SIDE

1. Disconnect the negative battery cable.

2. Remove the radiator cover panel.

3. Remove the air cleaner assembly.

4. Disconnect the left and right engine torque struts and position out of the way.

5. Remove the engine cooling fans.

6. Support the engine using engine support fixture J-28467-A or equivalent.

7. Raise and safely support the vehicle.

8. Remove the nuts securing the engine mount to the engine cradle.

9. Remove the bolts securing the engine mount bracket to the crankcase.

10. Remove the bolts securing the engine mount bracket to the cylinder head.

11. Remove the nuts securing the engine mount to the mount bracket.

12. Disconnect the Y-pipe from the front of the catalytic converter and position the converter out of the way.

13. Lower the vehicle.

14. Raise the engine by adjusting the engine support fixture.

15. Remove the engine mount and bracket.

16. Raise and safely support the vehicle.

17. Remove the rear alternator bracket.

18. Remove the bolts at the manifold outlet flange.

19. Disconnect the oxygen sensor.

20. Remove the exhaust manifold from the cylinder head, then remove the manifold from the vehicle.

21. Remove the gasket. Thoroughly clean the cylinder head and exhaust manifold mating surfaces.

To install:

22. Position the exhaust manifold by inserting the outlet pipe partially into the exhaust crossover pipe and moving the manifold into position.

23. Install a new gasket to the manifold. Insert two bolts to hold the gasket in place.

24. Install the remainder of the bolts and tighten all the bolts to 18 ft. lbs. (25 Nm).

25. Coat the oxygen sensor threads with Hi temperature anti-seize compound and install. Tighten the sensor nut to 30 ft. lbs. (40 Nm). Connect the oxygen sensor harness.

26. Install the rear alternator bracket. Tighten the crankcase bolts to 40 ft. lbs. (60 Nm) and the alternator bolts to 25 ft. lbs. (30 Nm).

27. Install two new bolts at the manifold outlet flange and tighten to 25 ft. lbs. (30 Nm).

28. Position the engine mount and bracket.

29. Loosely install the two nuts securing the mount to the bracket.

30. Lower the vehicle.

31. Loosely install the two bolts securing the mount bracket at the cylinder head.

32. Lower the engine into position guiding the engine mount studs in the cradle holes.

33. Raise and safely support the vehicle.

34. Loosely install the two nuts to the bottom of the engine mount.

35. Loosely install the two bolts securing the mount bracket to the crankcase.

36. Tighten the two nuts securing the mount to the bracket, the two nuts at the bottom of the engine mount, and the two bolts securing the mount bracket to the crankcase to 22 ft. lbs. (30 Nm).

37. Install the two bolts at the converter to exhaust Y-pipe and tighten to 20 ft. lbs. (25 Nm).

38. Lower the vehicle.

39. Remove the engine support fixture.

40. Tighten the two bolts securing the mount bracket to the cylinder head to 22 ft. lbs. (30 Nm).

41. Install the engine cooling fans.

42. Install the air cleaner assembly.

43. Connect the left and right engine torque struts. Be sure to tighten the bolts that attach the struts to the core support bracket last, and tighten them to 44 ft. lbs. (60 Nm).

44. Install the radiator trim panel.

45. Connect the negative battery cable.

46. Run the engine and check for exhaust leaks.

RIGHT SIDE

1. Disconnect the negative battery cable.

2. Disconnect the oxygen sensor at the rear of the right cam cover. Disconnect the harness clip.

3. Raise and safely support the vehicle.

4. Disconnect the Y-pipe from the front of the catalytic converter.

5. Disconnect the suspension position sensor at the lower control arm from both sides.

6. Disconnect the intermediate shaft from the steering gear.

7. Place a support below the rear crossmember of the engine cradle and remove the four cradle to body bolts.

8. Lower the rear of the engine cradle and disconnect the Y-pipe from the exhaust crossover and from the manifold.

9. Remove the manifold nuts and manifold.

10. Remove the gasket from the manifold.

11. Thoroughly clean all cylinder head and exhaust manifold contact surfaces.

To install:

12. Coat the oxygen sensor threads with hi-temperature anti-seize compound. Tighten the sensor to 30 ft. lbs. (40 Nm).

13. Install the gasket, manifold and nuts. Tighten the nuts to 25 ft. lbs. (30 Nm).

14. Install the exhaust Y-pipe and install four new bolts. Tighten the M10 bolts to 35 ft. lbs. (50 Nm) and the M8 bolts to 25 ft. lbs. (30 Nm).

15. Raise the engine cradle into position and tighten the bolts to 75 ft. lbs. (100 Nm).

16. Connect the intermediate shaft to the steering gear and tighten to 35 ft. lbs. (50 Nm).

17. Connect the exhaust Y-pipe to the catalytic converter and tighten the bolts to 35 ft. lbs. (50 Nm).

18. Connect the suspension position sensors to the lower control arms.

19. Lower the vehicle and connect the oxygen sensor. Install the harness retainer.

20. Connect the negative battery cable.

21. Run the engine and check for exhaust leaks.

4.9L Engine

RIGHT SIDE MANIFOLD

1. Disconnect the negative battery cable.

2. Remove the air cleaner assembly.

3. Remove the 2 heat shield screws.

4. Raise and safely support the vehicle.

5. Disconnect the exhaust Y-pipe from the right side manifold.

6. Disconnect the engine mount brace from the right side of the exhaust manifold.

7. Disconnect the oxygen sensor wire.

8. Remove the heat shield.

9. Support the engine cradle with screw jacks and remove the rear cradle bolts on both sides. Loosen the front cradle bolts to act as a pivot.

10. Slightly lower the engine cradle. Be careful not to lower the cradle more than 2 inches, as the rack and pinion intermediate shaft may become disconnected.

11. Remove the exhaust manifold bolts and manifold.

To install:

12. Thoroughly clean all mounting surfaces.

13. Install the exhaust manifold and torque the bolts to 16 ft. lbs. (20 Nm).

14. Raise the cradle back into position, install the frame bolts and torque to 75 ft. lbs. (100 Nm).

15. Connect the oxygen sensor wire.

16. Install the heat shield and screws.

17. Connect the engine mount brace to the front of the exhaust manifold.

18. Connect the exhaust Y-pipe to the right side manifold.

19. Lower the vehicle.

20. Install the air cleaner assembly.

21. Connect the negative battery cable.

22. Run the engine and check for exhaust leaks.

LEFT SIDE MANIFOLD

1. Disconnect the negative battery cable.

2. Remove the air cleaner assembly.

3. Disconnect the oxygen sensor wire.

4. Remove the power steering and tensioner bracket from the front of the manifold.

5. Remove the A/C hose bracket.

6. Remove the cooling fans.

7. Label and disconnect the spark plug wires from the left side spark plugs.

8. Raise and safely support the vehicle.

9. Disconnect the Y-pipe from the left exhaust manifold.

10. Remove the engine and A/C brace from the front of the manifold.

11. Remove the exhaust manifold bolts and remove the manifold.

To install:

12. Thoroughly clean all mounting surfaces.

13. Install the exhaust manifold and tighten the bolts to 16 ft. lbs. (20 Nm).

14. Install the engine and A/C brace to the front of the manifold.

15. Connect the Y-pipe to the left side exhaust manifold.

16. Lower the vehicle.

17. Connect the spark plug wires to the plugs.

18. Install the cooling fans.

19. Install the A/C hose bracket.

20. Install the power steering and tensioner bracket.

21. Connect the oxygen sensor wire.

22. Install the air cleaner assembly.

23. Connect the negative battery cable.

24. Run the engine and check for exhaust leaks.

Camshaft and Valve Lifters

REMOVAL & INSTALLATION

4.6L Engines

LEFT SIDE

1. Disconnect the negative battery cable.

2. Drain the coolant from the radiator.

3. Disconnect the upper radiator hose at the water crossover.

4. Label and disconnect the spark plug wires.

5. Remove the right-side fan.

6. Disconnect the battery cable at the alternator. Disconnect the cable harness at the cam cover and move out of the way.

7. Disconnect the PCV fresh air tube from the cam cover.

8. Remove the right and left torque struts.

9. Disconnect the water pump drive belt and pulley.

10. Remove the camshaft seal retainer screws and remove the seal.

11. Remove the cam cover screws and remove the cam cover by moving the cam drive end of the cover up, then pivot the entire cover around the water pump drive shaft. Continue moving the cover upward and pivoting so that the edge of the cover closely follows the left edge of the intake manifold cover.

12. Discard the cam cover seal if it is damaged. The spark plug seals may be reused if undamaged.

13. Secure the cam sprocket to the timing chain by installing tie-wraps through the cam sprocket holes. Use four tie-wraps per sprocket.

➡**The sprocket/chain relationship must be maintained throughout this procedure or camshaft timing will be lost and require further engine disassembly to retime.**

14. Working from behind the sprockets, install cam chain holder J-38222 so that it is positioned between the chain tensioner and chain guide. Apply tension to the tool by tightening the tension adjusting screw.

15. Remove both cam sprocket bolts. Note the relative location of the cam drive pins in the end of the camshafts.

16. Work the sprockets off the cams using play in the chain.

17. Alternately loosen the cam bearing cap screws a few turns at a time until all valve spring pressure has been released. Remove the bolts and caps.

18. Remove the camshaft.

19. Remove the valve lifters. Store the lifters on their camshaft face so that the residual oil is retained.

➡**Retain the valve lifters in order so that they can be installed in the same bores.**

20. Inspect the lifters for wear and/or damage; replace as necessary.

21. Inspect the camshaft for excessive lobe wear. Check the bearing journals, making sure they are not scored or burned. Replace the camshaft, as necessary.

To install:

22. Install the valve lifters in the same bore from which they were removed.

23. Lubricate the camshaft lobes with Camshaft Prelube 1052365 or equivalent. Lubricate the camshaft journals with clean engine oil.

24. Install the camshaft.

25. Position the cam bearing caps to the cylinder head.

➡**Each cap is identified for position and direction. The arrow points towards the front of the engine. An "E" indicates a cap for the exhaust cam. An "I" indicates a cap for the intake cam. Position No. 1 is towards the front of the engine.**

26. Loosely install the cam bearing cap bolts.

27. Alternately tighten the cam bearing cap bolts a few turns at a time against valve spring pressure until all the bolts are snug. Tighten the bolts to 106 inch lbs. (12 Nm).

28. Using the hex cast into the camshaft, rotate the cams until the drive pins are in position to engage the cam sprockets over the cams, and install the retaining bolts.

29. Attach the cam sprockets and install the retaining bolts. Tighten the bolts to 90 ft. lbs. (120 Nm).

30. Remove the chain holder J-38222.

31. Remove the tie-wraps from the cam sprockets.

32. Install the remaining components.

33. Install the cam cover screws and tighten to 89 inch lbs. (10 Nm).

34. Connect the battery cable retainer to the front of the cam cover.

35. Connect the battery cable at the alternator.

36. Lubricate the seal lips and install the camshaft seal to the end of the intake cam. Seal the screw threads with sealer.

37. Install the water pump pulley with tool J-38825 or equivalent, the install the drive belt.

38. Install the right and left torque struts and tighten the retaining bolts as follows:

➡**It is important during installation that the engine torque struts are not pre-loaded in their installed position. Adjustment is provided at the point the strut fastens to the core support bracket. Be sure this bolt is loose during assembly. Tighten to 45 ft. lbs. (60 Nm) as the final step of assembly.**

a. Tighten the strut bracket to cylinder head (M10) bolt: 35 ft. lbs. (50 Nm).

b. Tighten the strut bracket to cylinder head (M10) stud: 35 ft. lbs. (50 Nm).

c. Tighten the strut bracket to water manifold (M8) bolts: 20 ft. lbs. (25 Nm).

d. Tighten the strut bracket to cylinder head (M10) bolt: 35 ft. lbs. (50 Nm).

e. Tighten the strut to core support bracket bolt: 45 ft. lbs. (60 Nm) (see note above).

39. Connect the PCV fresh air tube to the cam cover.

40. Install the right-side fan.

41. Connect the spark plug wires.

42. Connect the upper radiator hose to the water crossover.

43. Fill the cooling system.

44. Connect the negative battery cable.

45. Run the engine and check for leaks and proper engine operation.

RIGHT SIDE

1. Disconnect the negative battery cable.

2. Raise and support the vehicle safely.

3. Disconnect the exhaust Y-pipe at the converter. Position the converter out of the way.

4. Lower the vehicle.

5. Remove the tower-to-tower brace.

6. Disconnect the ICM wiring harnesses and mounting bolts.

7. Remove the ICM and the spark plug wires on the right bank. Tag wires for installation reference.

8. Disconnect the PCV valve.

9. Remove the purge canister solenoid from the rear of the cover.

10. Remove the three screws retaining the wiring harness to the cover.

11. Remove the cam cover screws.

12. Support the front of the engine cradle and remove the mounting screws at the front of the cradle.

13. Remove the right and left torque struts.

14. Lower the engine cradle (or raise the vehicle) to provide clearance at the rear of the engine compartment.

15. Remove the cam cover.

16. Discard the cover seal if damaged. The spark plug seals may be reused if undamaged.

17. Secure the cam sprocket to the timing chain by installing tie-wraps through the cam sprocket holes. Use four tie-wraps per sprocket.

➡**The sprocket/chain relationship must be maintained throughout this procedure or camshaft timing will be lost and require further engine disassembly to retime.**

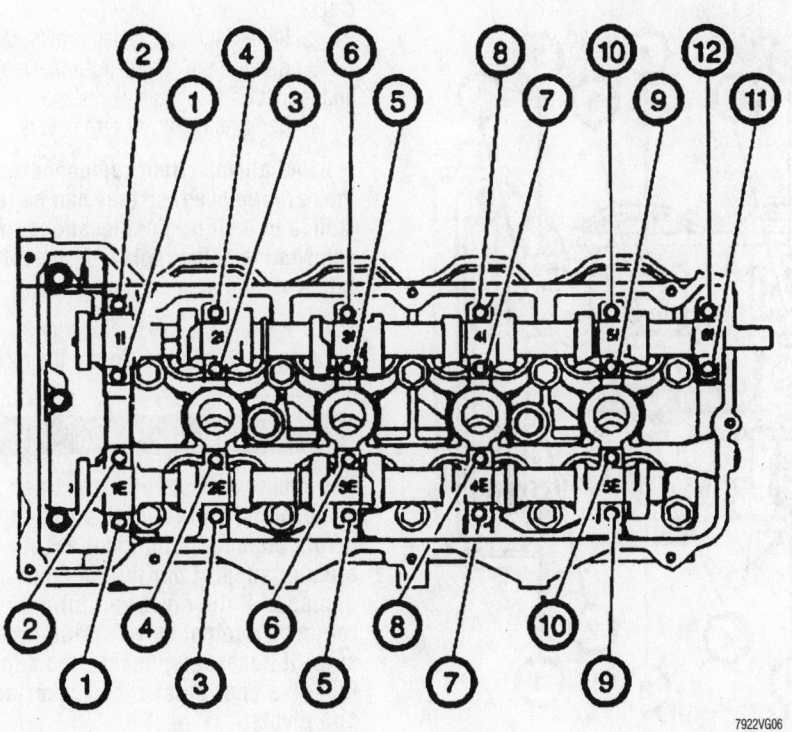

7922VG06

Left cylinder head camshaft bearing cap tightening sequence—4.6L engine

18. Working from behind the sprockets, install cam chain holder J-38222 so that it is positioned between the chain tensioner and chain guide. Apply tension to the tool by tightening the tension adjusting screw.

19. Remove both cam sprocket bolts. Note the relative location of the cam drive pins in the end of the camshafts.

20. Work the sprockets off the cams.

21. Alternately loosen the cam bearing cap screws a few turns at a time until all valve spring pressure has been released. Remove the bolts and caps.

22. Remove the camshaft.

23. Inspect the camshaft for excessive lobe wear. Check the bearing journals, making sure they are not scored or burned. Replace the camshaft, as necessary.

To install:

24. Lubricate the camshaft lobes with Camshaft Prelube 1052365 or equivalent. Lubricate the camshaft journals with clean engine oil.

25. Install the camshaft.

26. Position the cam bearing caps to the cylinder head.

➡**Each cap is identified for position and direction. The arrow points towards the front of the engine. An "E"**

indicates a cap for the exhaust cam. An "I" indicates a cap for the intake cam. Position No. 1 is towards the front of the engine.

27. Loosely install the cam bearing cap bolts.

28. Alternately tighten the cam bearing cap bolts a few turns at a time against valve spring pressure until all the bolts are snug. Tighten the bolts to 106 inch lbs. (12 Nm).

29. Using the hex cast into the camshaft, rotate the cams until the drive pins are in position to engage the cam sprockets over the cams and install the retaining bolts.

30. Attach the cam sprockets and install the retaining bolts. Tighten the bolts to 90 ft. lbs. (120 Nm).

31. Remove the chain holder J-3822-2.

32. Remove the tie-wraps from the cam sprockets.

33. Install spark plug and cam cover seals, as required.

34. Install the cam cover and tighten the screws to 7 ft. lbs. (10 Nm).

35. Raise the engine cradle into position. Install and tighten the two mounting bolts to 75 ft. lbs. (100 Nm).

36. Install the right and left torque struts and tighten the retaining bolts as follows:

➡It is important during installation that the engine torque struts are not pre-loaded in their installed position. Adjustment is provided at the point the strut fastens to the core support bracket. Be sure this bolt is loose during assembly. Tighten to 45 ft. lbs. (60 Nm) as the final step of assembly.

   a. Tighten the strut bracket to cylinder head (M10) bolt: 35 ft. lbs. (50 Nm).

   b. Tighten the strut bracket to cylinder head (M10) stud: 35 ft. lbs. (50 Nm).

   c. Tighten the strut to core support bracket bolt: 45 ft. lbs. (60 Nm) (see note above).

37. Install the remaining components.

38. Connect the negative battery cable.

39. Run the engine and check for leaks and proper engine operation.

4.9L Engine

✳✳ CAUTION

The fuel injection system remains under pressure, even after the engine has been turned OFF. The fuel system pressure must be relieved before disconnecting any fuel lines. Failure to do so may result in fire and/or personal injury.

1. Disconnect the negative battery cable.

2. Relieve the fuel system pressure.

3. Remove the engine from the vehicle and mount on a suitable workstand.

4. Remove the rocker arm covers.

➡Label all valvetrain components as they are removed, so they can be reinstalled in their original locations. If the camshaft is being replaced, the valve lifters should also be replaced.

5. Remove the rocker arm support retaining nuts/bolts and remove the rocker arm support.

✳✳ WARNING

The rocker arm support must be removed with the rocker arms and pivots attached. The pivot assemblies or support bar threads may be damaged if the pivot bolt torque is removed against valve spring pressure. If necessary, secure the support in a vise and remove the rocker arms and pivots.

6. Remove the pushrods.

7. Remove the timing chain front cover.

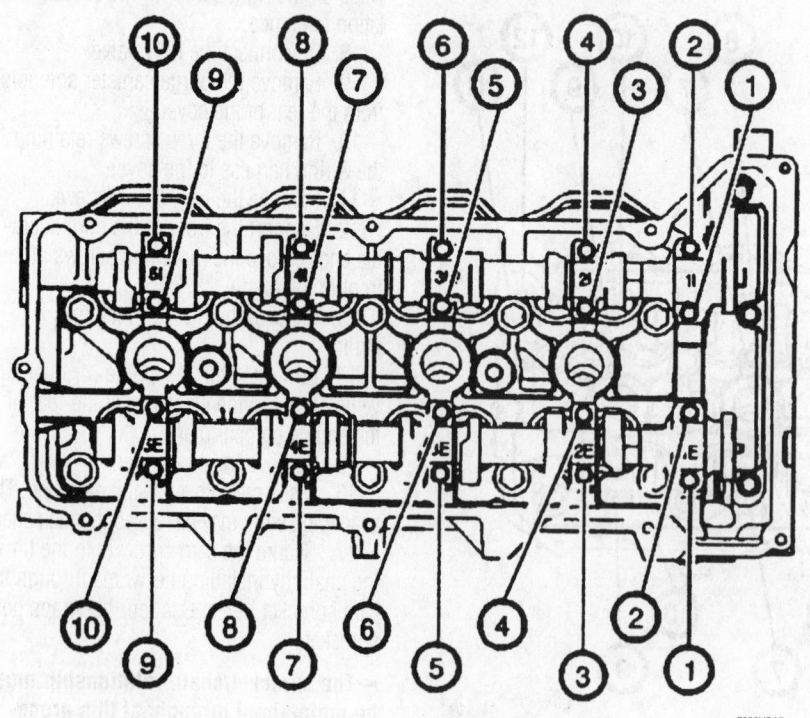

7922VG07

Right cylinder head camshaft bearing cap tightening sequence—4.6L engine

8. Rotate the crankshaft until the timing chain sprocket timing marks are aligned.

9. Remove the distributor.

10. Remove the intake manifold.

11. Remove the valve lifters.

12. Remove the timing chain and camshaft sprocket.

13. Temporarily reinstall the camshaft sprocket to the camshaft (to provide a suitable hand hold). Remove the camshaft, being careful not to damage the lobes, journals or camshaft bearings.

14. Inspect the camshaft journals and lobes for wear and/or damage. Replace as necessary.

To install:

➡**If a new camshaft is required, new lifters and a new distributor gear must also be installed.**

15. Temporarily reinstall the camshaft sprocket to the camshaft (to provide a suitable hand hold). Lubricate the camshaft journals, lobes and distributor gear with prelube 1052365 or equivalent.

16. Install the camshaft, being careful not to damage the lobes, journals or camshaft bearings. Remove the camshaft sprocket.

17. Install the timing chain and sprocket. Install the camshaft sprocket retaining bolt and tighten to 37 ft. lbs.

18. Install the timing chain front cover.

19. Install the valve lifters and intake manifold.

20. Install the distributor.

21. Install the pushrods and rocker arms. Loosely install the pivot bolts, then tighten the pivot bolts to 22 ft. lbs. (30 Nm). Install the rocker arm support. Be sure each pushrod is seated in a rocker arm.

22. Loosely install the rocker arm support retaining nuts and bolts.

23. Tighten the rocker arm support retaining nuts alternately and evenly until snug.

➡**Check the pushrods from time to time during the tightening procedure to be sure they are correctly positioned.**

24. Tighten the rocker arm retaining bolts until snug.

25. Tighten the rocker arm support retaining nuts to 37 ft. lbs. (50 Nm).

26. Tighten the rocker arm retaining bolts to 7 ft. lbs. (10 Nm).

27. Install the rocker arm covers.

28. Install the engine in the vehicle.

29. Connect the negative battery cable.

30. Restore all fluid levels and make all necessary adjustments.

31. Run the engine and check for leaks and proper engine performance. Check the ignition timing.

Valve Lash

ADJUSTMENT

The valve clearance cannot be adjusted on these engines. The hydraulic lifters function to maintain a zero clearance when the valves are opening and closing. Any clearance is instantaneously taken up by the hydraulic action. The hydraulic lifter consists of 3 main parts: the body, plunger and valve. Oil under pressure from the engine lubricating system passes through the check valve and forces the plunger upwards in the body of the lifter and takes up any excess clearance. "Valve lifter noise" complaints may require cleaning of sludged or varnished valve lifters, or replacement.

Oil Pan

REMOVAL & INSTALLATION

❈❈ CAUTION

The EPA warns that prolonged contact with used engine oil may cause a number of skin disorders, including cancer! You should make every effort to minimize your exposure to used engine oil. Protective gloves should be worn when changing the oil. Wash your hands and any other exposed skin areas as soon as possible after exposure to used engine oil. Soap and water, or waterless hand cleaner should be used.

1. Disconnect the negative battery cable.

2. Raise and safely support the vehicle.

3. Remove the oil pan drain plug and drain the engine oil into a suitable container.

4. Disconnect the oil level indicator harness, if equipped.

5. Remove the exhaust Y-pipe, if needed.

6. On 4.6L engines, remove the transaxle assembly from the vehicle.

7. On 4.9L engines, remove the flywheel cover.

❈❈ WARNING

Attempting to remove the oil pan without first dropping the oil pump will result in breaking the oil pump housing.

8. On the 4.6L engine, remove the oil pan mounting bolts and remove the oil pan.

9. On the 4.9L engine, remove the oil pan mounting bolts, then drop the oil pan down about 2 inches, then remove the oil pump housing mounting bolts and nut. Drop the oil pump down into the oil pan, then remove the oil pan from the vehicle.

❈❈ WARNING

Attempting to remove the oil pan without first dropping the oil pump will result in breaking the oil pump housing.

To install:

10. On 4.6L engines, do the following:

a. The oil pan gasket is reusable unless it is damaged. Do not remove the gasket from the oil pan groove unless gasket replacement is required.

b. Thoroughly clean the inside of the oil pan and the cylinder block contact surface. If the oil pan gasket is being reused, be careful not to damage it. Do not expose the gasket to cleaning solvents.

c. If a new gasket is being installed, start the gasket into the oil pan groove and work the gasket into the groove in both directions. Once the gasket is exposed to oil, it will expand and no longer stay in the groove without wrinkles. If this condition exists, replace the gasket.

d. Install the oil pan and secure using the mounting bolts. Tighten, in sequence, to 89 inch lbs. (10 Nm).

11. On 4.9L engines, do the following:

a. Before installing the oil pan, the excess material on the rear edge must be ground off 1/16 in. (1.5mm) on both sides for oil pump clearance.

❈❈ WARNING

Failure to grind off the excess material will result in breaking the oil pump housing upon installation.

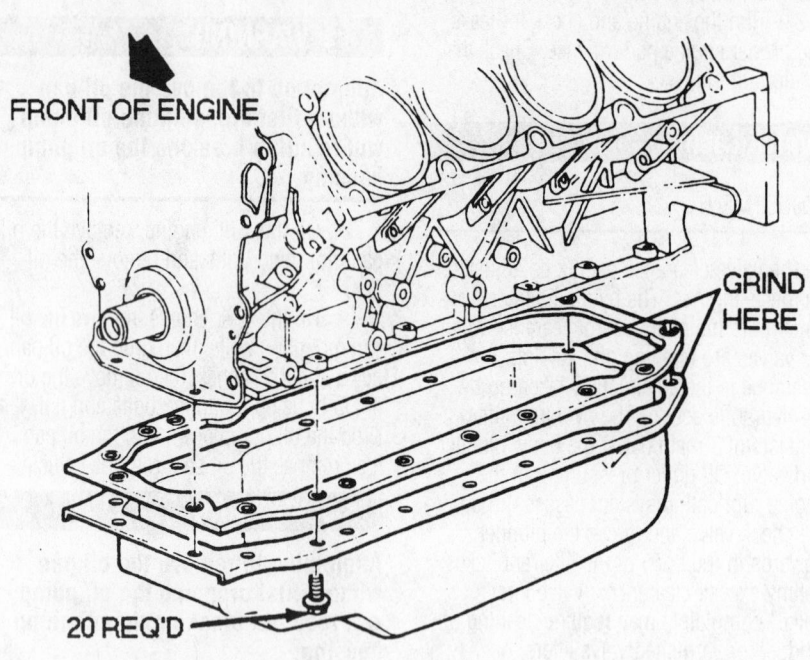

FRONT OF ENGINE

GRIND HERE

20 REQ'D

7922VG32

Oil pan grinding location—4.9L engine

b. Clean all gasket surfaces completely. Thoroughly clean the inside of the oil pan.

c. Install a new O-ring on the oil pump outlet pipe. Install the oil pump, making sure the pump driveshaft engages the distributor gear. Tighten the nut to 22 ft. lbs. (30 Nm) and the bolts to 15 ft. lbs. (20 Nm).

d. Apply a ¼ in. bead of sealant at the rear main bearing cap and at the front cover-to-block joints.

e. Install the oil pan using a new gasket.

f. Install the oil pan bolts and nuts and tighten to 14 ft. lbs. (18 Nm).

12. Connect the oil level indicator connector, if removed.

13. Install the transaxle assembly, if removed.

14. Install the Y-pipe, if removed.

15. Install the flywheel cover, if removed.

16. Install the oil pan drain plug and tighten.

17. Lower the vehicle.

18. Refill the crankcase with the proper type and quantity of engine oil.

19. Connect the negative battery cable.

20. Run the engine and check for leaks.

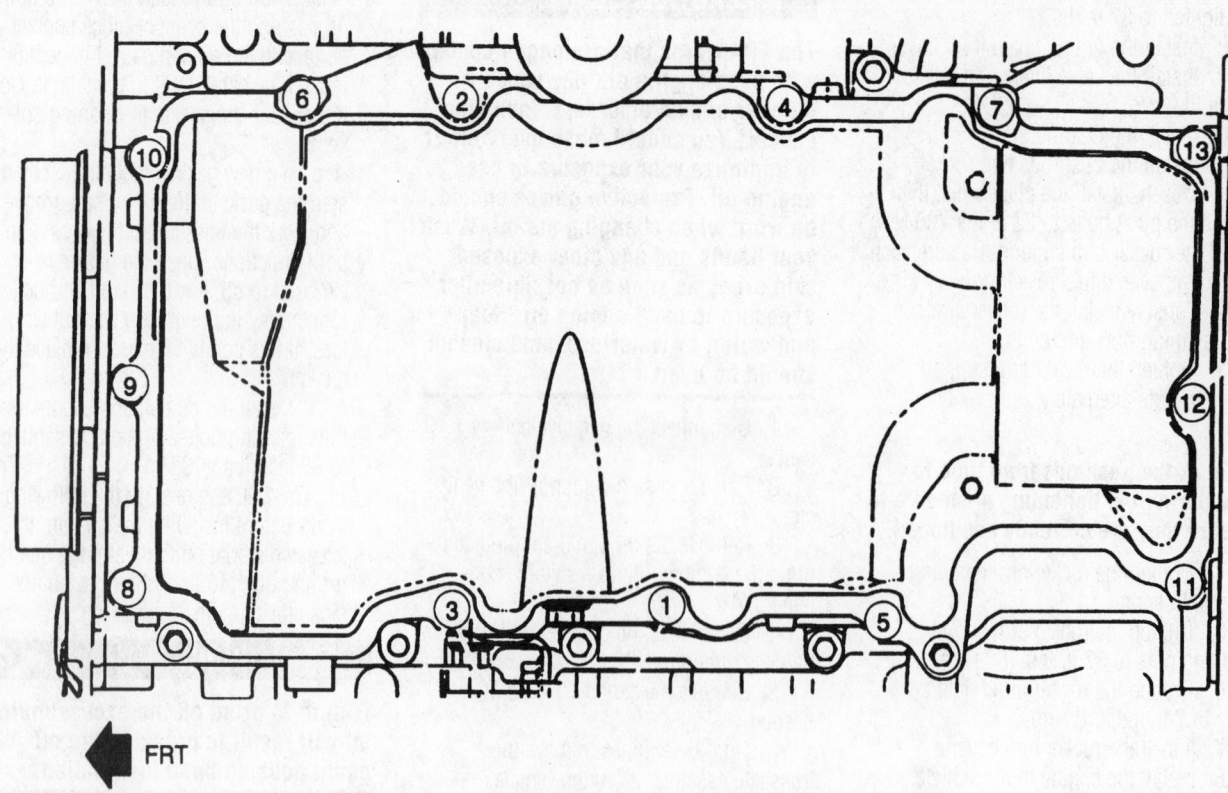

FRT

7922VG08

Oil pan bolt tightening sequence—4.6L engines

Oil Pump

REMOVAL & INSTALLATION

✳✳ CAUTION

The EPA warns that prolonged contact with used engine oil may cause a number of skin disorders, including cancer! You should make every effort to minimize your exposure to used engine oil. Protective gloves should be worn when changing the oil. Wash your hands and any other exposed skin areas as soon as possible after exposure to used engine oil. Soap and water, or waterless hand cleaner should be used.

4.6L Engines

1. Disconnect the negative battery cable.
2. Remove the drive belt.
3. Remove the bolt securing the power steering hose.
4. Raise and safely support the vehicle.
5. Remove the right front wheel.
6. Remove the two splash shields from the wheel well.

7. Remove the brace between the engine oil pan and transaxle case.
8. Install Flywheel Holder tool J-39411 or equivalent, to keep the crankshaft from turning.
9. Remove the crankshaft balancer bolt.
10. Support the engine cradle with a screw type jack and remove the three right-side engine cradle bolts.
11. Disconnect the RSS sensor from the right lower control arm.
12. Lower the cradle to gain access for the crankshaft balancer puller.
13. Using Puller tool J-38416 or equivalent, remove the crankshaft balancer.
14. Remove the accessory drive belt tensioner and idler pulley.
15. Remove the front cover bolts and remove the front cover and gasket.

➡**The front cover gasket is reusable as long as it is not damaged.**

16. Remove the three oil pump mounting bolts and remove the oil pump and drive spacer.
17. If necessary, disassemble and inspect the pump as follows:
    a. Remove the drive spacer from the pump housing.

b. Remove the two screws holding the pump housing halves together.
    c. Remove the inner (drive) and outer (driven) rotors from the housing. Indicate the mating surfaces (dimples).
    d. Remove the pressure relief valve.
    e. Inspect the pump housing for nicks, burrs, chips or debris that might cause a leak or binding condition in the rotor pocket.
    f. Inspect the drive and driven rotors for nicks or burrs.
    g. Check the pump cover and interior surface for excessive wear or score marks. Check for flatness.
    h. If any components show signs of excessive wear or damage, replace the pump assembly.
To install:
18. If the pump was disassembled, reassemble as follows:
    a. Install the inner and outer rotors to the pump cover in the same orientation as removed.
    b. Install the pressure relief valve seat, spring and pilot in the pump housing.
    c. Pack the pump housing halves with Amojell® or white petroleum grease to ensure pump priming.
    d. Assemble the housing and cover over the locating dowel.
    e. Insert a 9mm drill in the pump mounting hole on the opposite side to aid alignment of the housing and cover. Install the two screws and tighten to 108 inch lbs. (12 Nm).
19. Install the oil pump drive spacer into the oil pump from the rear so the drive flat engages the pump rotor.
20. Install the oil pump over the crankshaft and loosely install the mounting bolts.
21. Hold the pump in its furthest up position and tighten the mounting bolts (1, 2 and 3) in sequence, in 2 Steps. First tighten in sequence to 89 inch lbs. (10 Nm), then tighten in sequence to 20 ft. lbs. (26 Nm).
22. Place a small amount of RTV sealant at the split line of the upper and lower crankcases.
23. Install the front cover gasket on the dowel pins on the block.
24. Install the front cover on the dowel pins and install the front cover mounting bolts. Tighten the bolts to 89 inch lbs. (10 Nm).
25. Install the drive belt idler pulley and tighten the mounting bolt to 37 ft. lbs. (50 Nm).
26. Install the drive belt tensioner and tighten the tensioner mounting bolt to 37 ft. lbs. (50 Nm).

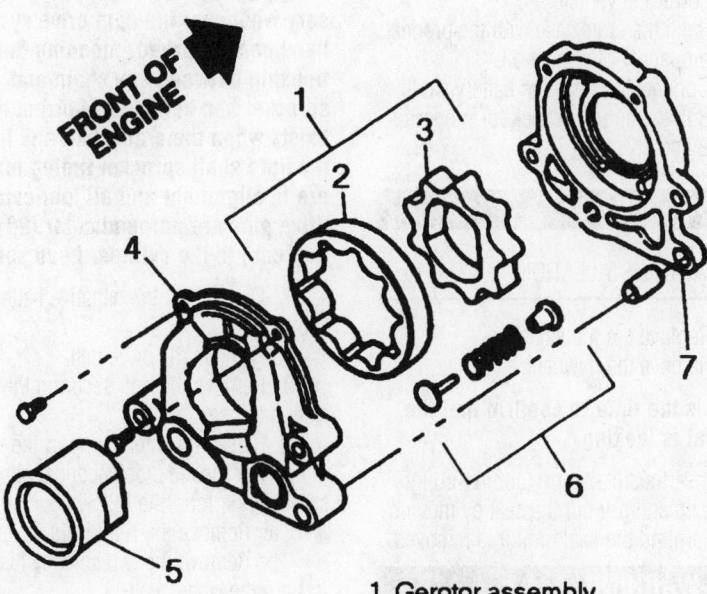

1. Gerotor assembly
2. Outer gear
3. Inner gear
4. Housing
5. Drive spacer
6. Relief valve
7. Cover

7922VG09

Exploded view of the oil pump—4.6L engine

27. Install the crankshaft balancer using tool J-39344 or equivalent.

28. Apply engine oil to the balancer bolt threads and tighten the bolt to 37 ft. lbs. (50 Nm) plus an additional 120 degrees

29. Raise the screw jack until the three cradle bolts can be installed. Tighten the bolts to 75 ft. lbs. (100 Nm).

30. Connect the RSS sensor.

31. Remove the flywheel holding tool and install the oil pan-to-transaxle brace. Tighten the bolts to 37 ft. lbs. (50 Nm).

32. Install the wheel well splash shields and the front wheel.

33. Lower the vehicle.

34. Install the power steering hose retainer.

35. Install the accessory drive belt.

36. Connect the negative battery cable.

37. Run the engine and check for proper engine oil pressure. Check for leaks.

4.9L Engine

1. Disconnect the negative battery cable.

2. Raise and safely support the vehicle.

3. Remove the oil pan drain plug and drain the engine oil into a suitable container.

4. Remove the flywheel covers.

5. Remove the oil pan bolts and nuts and loosen the oil pan from the cylinder block.

6. Drop the oil pan down about 2 inches, then remove the oil pump housing mounting bolts and nut. Drop the oil pump down into the oil pan, then remove the oil pan from the vehicle.

❋❋ WARNING

Attempting to remove the oil pan without first dropping the oil pump will result in breaking the oil pump housing.

7. Remove the oil pump from the pan and discard the O-ring.

8. If necessary, disassemble and inspect the oil pump for wear and/or damage. Replace as necessary.

To install:

9. Reassemble the oil pump, as required, lubricating all internal parts with clean engine oil during assembly. Tighten the pump cover-to-housing screws to 96 inch lbs. (10 Nm) and the pipe-to-housing screws to 120 inch lbs. (12 Nm). Be sure to use a new O-ring when installing the outlet pipe.

10. Temporarily install the driveshaft to the pump. Prime the pump by pouring clean engine oil into the pump pick-up screen and rotating the driveshaft until oil emerges from the passage in the pump.

11. Install the oil pump and driveshaft to the block, engaging the driveshaft to the distributor gear. Use a new O-ring at the pump outlet pipe.

12. Install the mounting nut and tighten to 22 ft. lbs. (30 Nm). Install the mounting bolts and tighten to 15 ft. lbs. (20 Nm).

13. Before installing the oil pan, the excess material on the rear edge must be ground off 1/16 in. (1.5mm) on both sides for oil pump clearance.

❋❋ WARNING

Failure to grind off the excess material will result in breaking the oil pump housing upon installation.

14. Clean all gasket surfaces completely. Thoroughly clean the inside of the oil pan.

15. Apply a ¼ in. bead of sealant at the rear main bearing cap and at the front cover-to-block joints.

16. Install the oil pan using a new gasket.

17. Install the oil pan bolts and nuts and tighten to 14 ft. lbs. (18 Nm).

18. Install the flywheel cover.

19. Install the oil pan drain plug and tighten to 22 ft. lbs. (30 Nm).

20. Lower the vehicle.

21. Refill the crankcase with the proper type and quantity of engine oil.

22. Connect the negative battery cable.

23. Run the engine: check for proper oil pressure and check for leaks.

Rear Main Seal

REMOVAL & INSTALLATION

1. Remove the transaxle.
2. Remove the flywheel.

➡**Now is the time to confirm that the rear seal is leaking.**

3. Insert a suitable prytool in through the dust lip and pry out the seal by moving the tool around the seal until it is removed.

❋❋ WARNING

Use care not to damage the crankshaft seal surface with a prytool.

To install:

4. Before installing, lubricate the seal bore to seal surface with clean engine oil.

5. Install the new seal using Seal Installer tool J-38817 or an equivalent rear main seal installation tool.

6. Slide the new seal over the mandrel until the dust lip bottoms squarely against the tool collar.

7. Thread the tool into the crankshaft flange and install the seal by turning the T-handle until the tool bottoms against the crankcase.

8. Loosen and remove the tool from the crankshaft.

➡**Check to see that the seal is squarely seated in the bore.**

9. Install the flywheel and tighten the bolts to 11 ft. lbs. (15 Nm) plus 50∫, then install the transaxle.

10. Start the engine and check for leaks.

Timing Chain, Sprockets, Front Cover and Seal

REMOVAL & INSTALLATION

4.6L Engines

The left and right-side secondary timing chains can be removed with the engine in the vehicle. If the primary timing chain or intermediate shaft sprocket need to be replaced, the engine must be removed from the vehicle and supported on an engine stand.

➡**Setting the camshaft timing is necessary whenever the cam drive system has been disturbed, meaning the relationship between any chain and sprocket has been lost. Correct timing exists when the crankshaft and intermediate shaft sprocket timing marks are in alignment and all four camshaft drive pins are perpendicular (90 degrees) to the cylinder head surface.**

1. Disconnect the negative battery cable.

2. Remove the drive belt.

3. Remove the bolt securing the power steering hose.

4. Raise and safely support the vehicle.

5. For the left side secondary chain and sprocket, perform the following:

    a. Remove the right front wheel.

    b. Remove the two splash shields from the wheel well.

    c. Remove the flywheel cover and install a suitable flywheel holder.

    d. Remove the crankshaft balancer bolt.

    e. Support the engine cradle with a screw type jack and remove the three right-side engine cradle bolts.

    f. Disconnect the RSS sensor from the right control arm.

g. Lower the cradle to gain access for the crankshaft damper puller.

h. Using Puller J-38416 or equivalent, remove the crankshaft damper.

i. Remove the drive belt tensioner.

j. Remove the drive belt idler pulley.

k. Remove the front cover bolts.

6. Remove the front cover and gasket.

➡ **The front cover gasket is reusable as long as it is not damaged.**

a. Partially drain the coolant from the radiator.

b. Disconnect the upper radiator hose at the water crossover.

c. Label and disconnect the spark plug wires.

d. Remove the right side fan.

e. Disconnect the battery cable at the alternator and disconnect the cable harness at the cam cover.

f. Disconnect the PCV fresh air tube from the cam cover.

g. Remove the right and left torque struts.

h. Remove the water pump pulley with tool J-38825 or equivalent.

i. Remove the camshaft seal retainer screws and remove the seal.

j. Disconnect the battery cable retainer at the front of the cam cover.

k. Remove the cam cover screws and remove the cam cover by pivoting the entire cover around the water pump drive shaft. Continue moving the cover upward and pivoting so that the edge of the cover closely follows the left edge of the intake manifold cover.

➡ **The cam cover gasket is reusable as long as it is not damaged.**

l. Remove the left side secondary chain tensioner.

m. Remove the left side chain guide. Access the upper chain guide mounting bolt through the hole in the cylinder head capped with the plastic plug.

n. Remove the left side cam sprocket bolts and sprockets.

o. Remove the secondary drive chain.

7. For the right side secondary chain and sprocket, perform the following:

a. Disconnect the exhaust Y-pipe at the converter.

b. Lower the vehicle.

c. Remove the tower-to-tower brace.

d Detach the ICM wiring connectors and mounting bolts.

e. Remove the ICM and the plug wires on the right bank.

f. Disconnect the PCV valve.

g. Remove the purge canister solenoid from the rear of the cover.

h. Remove the cam cover screws.

i. Safely support the front of the engine cradle and remove the two mounting bolts at the front of the cradle.

j. Remove the right and left torque struts.

k. Lower the engine cradle (or raise the vehicle) to provide clearance at the rear of the engine compartment.

l. Remove the cam cover.

➡ **The cam cover gasket is reusable as long as it is not damaged.**

m. Remove the right side secondary chain tensioner.

n. Remove the right side chain guide. Access the upper chain guide mounting bolt through the hole in the cylinder head capped with the plastic plug.

o. Remove the right side cam sprocket bolts and cam sprockets.

8. Remove the secondary drive chain.

9. If only servicing the secondary chains, proceed to Step 19 of the installation procedure.

10. Remove the engine.

11. Remove the intermediate shaft

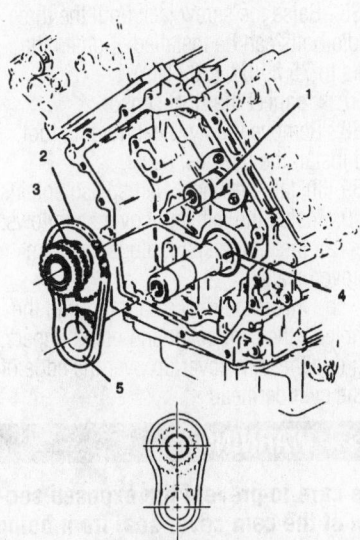

1 INTERMEDIATE SHAFT
2 PRIMARY CHAIN
3 INTERMEDIATE SHAFT SPROCKET
4 CRANKSHAFT SPROCKET KEY
5 SPROCKET

7922VG10

Primary drive chain components—4.6L engine

sprocket-to-intermediate shaft bolt, then remove the sprocket.

12. Slide the primary timing sprockets and primary chain off the engine.

To install:

➡ **The following procedure must be followed to set the camshaft timing on the vehicle.**

13. Install the primary and secondary chain guide.

14. Rotate the crankshaft until the sprocket drive key is at the 1 o'clock position. Use tool J-39946 or the equivalent, to rotate the crankshaft.

15. Install the crankshaft sprocket and intermediate shaft sprocket in the primary timing chain so the timing marks are aligned. Install the assembly in position on the engine. The crankshaft sprocket key way will have to slide over the key on the crankshaft. If it is necessary to turn the crankshaft sprocket, the intermediate shaft sprocket will also have to be turned so the timing mark still lines up with the crankshaft sprocket.

16. Install the intermediate shaft sprocket-to-intermediate shaft bolt and tighten to 45 ft. lbs. (61 Nm).

17. Install the primary timing chain tensioner. Tighten the tensioner mounting bolts to 20 ft. lbs. (27 Nm).

18. Install a suitable flywheel holder to lock the crankshaft in position.

19. Install the secondary timing chain over the inner row of teeth on the intermediate shaft sprocket. Route the chain over the chain guide and install the exhaust cam sprocket so the **RE** (Right Head Exhaust) pin engages the sprocket notch. There should be no slack in the lower section of the timing chain and the cam drive pin **must** be perpendicular to the cylinder head face.

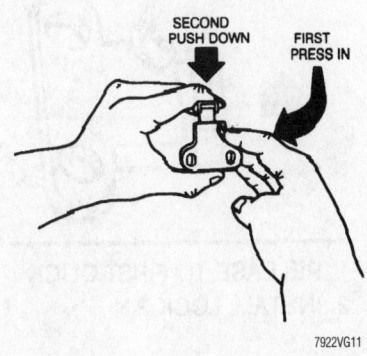

SECOND PUSH DOWN   FIRST PRESS IN

7922VG11

Rotating tensioner release lever—4.6L engine

20. Install the intake cam sprocket into the chain so the sprocket notch **RI** (Right Head Intake) engages the cam and the camshaft drive pin remains perpendicular to the cylinder head face. A hex is cast into the camshafts behind the lobes for cylinder No. 1, so an open end wrench may be used to provide minor repositioning of the cams.

21. Loosely install the exhaust cam sprocket bolt and intake sprocket bolt.

22. Install the chain tensioner and tighten the mounting bolts to 20 ft. lbs. (27 Nm).

23. Tighten the camshaft sprocket bolts to 90 ft. lbs. (120 Nm).

24. Route the secondary timing chain for the left side over the outer row of intermediate sprocket teeth.

25. Install the secondary timing chain over the inner row of teeth on the intermediate shaft sprocket. Route the chain over the chain guide and install the exhaust cam sprocket so the **LE** (Left Head Exhaust) pin engages the sprocket notch. There should be no slack in the lower section of the timing chain and the cam drive pin **must** be perpendicular to the cylinder head face.

26. Install the intake cam sprocket into the chain so the sprocket notch **LI** (Left Head Intake) engages the cam and the camshaft drive pin remains perpendicular to the cylinder head face. A hex is cast into the

camshafts behind the lobes for cylinder No. 2, so an open end wrench may be used to provide minor repositioning of the cams.

27. Loosely install the exhaust cam sprocket bolt and intake sprocket bolt.

28. Install the chain tensioner and tighten the mounting bolts to 20 ft. lbs. (27 Nm).

29. Tighten the camshaft sprocket bolts to 90 ft. lbs. (120 Nm).

➡**The RE cam sprocket must contain the cam position sensor pick-up.**

30. Install the front cover gasket on the dowel pins on the block.

31. Install the front cover on the dowel pins and install the front cover mounting bolts. Tighten the bolts to 89 inch lbs. (10 Nm). Apply a dab of RTV to the split line between the upper and lower crankcase assemblies.

32. Install the drive belt idler pulley and tighten the mounting bolt to 35 ft. lbs. (47 Nm).

33. Install the drive belt tensioner and tighten the tensioner mounting nut to 35 ft. lbs. (47 Nm).

34. Install the crankshaft balancer using tool J-39344 or the equivalent.

35. Apply engine oil to the balancer bolt threads and tighten the bolt to 44 ft. lbs. (60 Nm) plus an additional 120 degree turn.

36. Raise the screw jack until the three cradle bolts can be installed. Tighten the bolts to 75 ft. lbs. (102 Nm).

37. Connect the RSS sensor.

38. Remove the flywheel holding tool and install the flywheel cover.

39. Install the wheel well splash shields.

40. Install the left cam cover as follows:

   a. Install the spark plugs and cam cover seals.

   b. Insert the intake cam through the hole in the cam cover and using fingers, guide the cam cover up over the edge of the cylinder head.

❋❋ WARNING

Use care to prevent the exposed section of the cam cover seal from being damaged by the edge of the cylinder head casting.

   c. Work the cover into position by allowing the top edge of the cover to follow the left side edge of the intake manifold.

   d. Install the cam cover screws and tighten to 7 ft. lbs. (10 Nm).

   e. Connect the battery cable retainer to the front of the cam cover.

   f. Connect the battery cable at the alternator.

   g. Lubricate the seal lips and install the camshaft seal to the end of the intake cam. Seal the screw threads with sealer.

   h. Install the water pump pulley with tool J-38825 or equivalent.

   i. Connect the PCV fresh air tube to the cam cover.

   j. Install the right side fan.

   k. Connect the spark plug wires.

   l. Connect the upper radiator hose to the water crossover.

   m. Refill the cooling system.

41. Install the right cam cover as follows:

   a. Install spark plug and cam cover seals.

   b. Install the cam cover and tighten the screws to 7 ft. lbs. (10 Nm).

   c. Raise the engine cradle into position and install and tighten the two mounting bolts to 75 ft. lbs. (100 Nm).

   d. Install the right and left torque struts and tighten the retaining bolts as follows:

➡**It is important during installation that the engine torque struts are not preloaded in their installed position. Adjustment is provided at the point the strut fastens to the core support bracket. Be sure this bolt is loose during assembly. Tighten to 45 ft. lbs. (60 Nm) as the final step of assembly.**

   e. Tighten the strut bracket to cylinder head (M10) bolt: 35 ft. lbs. (50 Nm).

   f. Tighten the strut bracket to cylinder head (M10) stud: 35 ft. lbs. (50 Nm).

   g. Tighten the strut bracket to water manifold (M8) bolts: 20 ft. lbs. (25 Nm).

   h. Tighten the strut bracket to cylinder head (M10) bolt: 35 ft. lbs. (50 Nm).

   i. Tighten the strut to core support bracket bolt: 45 ft. lbs. (60 Nm) (see note above).

   j. Install the screws retaining wiring harness to the cover.

   k. Install the purge canister solenoid to the rear of the cover.

   l. Connect the PCV valve.

   m. Install the ICM and the spark plug wires on the right-bank.

   n. Attach the ICM wiring connectors.

   o. Install the tower-to-tower brace.

   p. Raise and support the vehicle safely. Connect the exhaust Y-pipe to the converter and tighten the bolts to 20 ft. lbs. (25 Nm).

42. Connect the negative battery cable.

43. Start the engine. Check for leaks and inspect for proper operation.

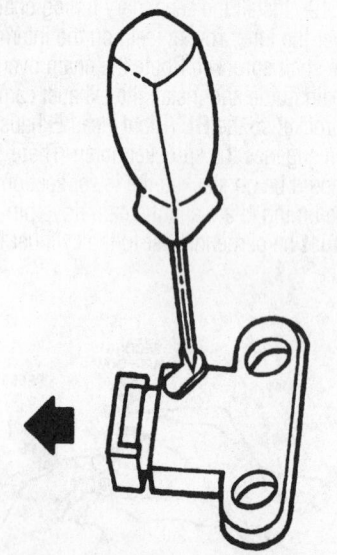

1 RELEASE TO FIRST CLICK
2 INSTALL LOCK PIN

7922VG12

Locking the tensioner in the collapsed position—4.6L engine

4.9L Engine

1. Disconnect the negative battery cable.
2. Drain the engine coolant into a suitable container.
3. Remove the cross vehicle brace.
4. Remove the coolant reservoir.
5. Remove the drive belt.
6. Remove the water pump and pulley.
7. Raise and safely support the vehicle.
8. Remove the right front wheel and the right front air deflector.
9. Support the vehicle and the right side of the engine cradle.
10. Remove the right side cradle body bolts and lower the right side of the cradle for balancer clearance.
11. Remove the bolt from the end of the crankshaft. Using J-24420-B crankshaft damper puller or equivalent, remove the crankshaft damper.
12. Remove the timing chain front cover bolts and cover.
13. Remove the oil slinger from the crankshaft.
14. Rotate the engine until the timing marks are in proper alignment.
15. Remove the thrust button and the camshaft sprocket mounting bolt. Discard the thrust button.
16. Remove the camshaft sprocket and timing chain together.
17. Remove the crankshaft sprocket.

To install:

18. Thoroughly clean all gasket mating surfaces.
19. Install the camshaft and crankshaft sprockets and timing chain as an assembly. Be sure the timing marks are in alignment and the sprockets are properly seated on the crankshaft and camshaft.
20. Install the camshaft bolt and tighten to 37 ft. lbs. (50 Nm).
21. Install a new thrust button.
22. Install the crankshaft oil slinger against the crankshaft sprocket.
23. Apply a bead of RTV sealant to the front cover lip on the oil pan sealing surface. The bead must be placed along the front cover lip behind the 2 oil pan-to-front cover bolts, in order to prevent leakage through the bolt threads.
24. Apply a ¼ inch bead of RTV sealant on the oil pan where the oil pan, block and front cover join together. The front corner has a rounder corner, instead of square; if the RTV sealant is not applied an oil leak may occur.
25. Install the front cover and bolts, using a new gasket.
26. Lubricate the crankshaft damper hub bore and the front cover seal with EP lubri-

cant. Position the damper on the crankshaft, aligning the crankshaft key with the hub keyway.

27. Install the crankshaft damper using installer tool J-29774 or equivalent, until the damper hub bottoms on the crankshaft. Remove the installer tool and attach the bolt. Torque the bolt to 70 ft. lbs. (95 Nm) on 1993-94 vehicles or 118 ft. lbs. (160 Nm) on 1995 vehicles.
28. Raise the right side of the cradle and install the cradle-to-body bolts. Tighten to 66 ft. lbs. (90 Nm).
29. Install the remaining components.
30. Connect the negative battery cable.
31. Refill and bleed the cooling system.
32. Start the engine and check for leaks and proper engine operation.

FUEL SYSTEM

Fuel System Service Precautions

Safety is the most important factor when performing not only fuel system maintenance but any type of maintenance. Failure to conduct maintenance and repairs in a safe manner may result in serious personal injury or death. Maintenance and testing of the vehicle's fuel system components can be accomplished safely and effectively by adhering to the following rules and guidelines.

• To avoid the possibility of fire and personal injury, always disconnect the negative battery cable unless the repair or test procedure requires that battery voltage be applied.

• Always relieve the fuel system pressure prior to disconnecting any fuel system component (injector, fuel rail, pressure regulator, etc.), fitting or fuel line connection. Exercise extreme caution whenever relieving fuel system pressure to avoid exposing skin, face and eyes to fuel spray. Please be advised that fuel under pressure may penetrate the skin or any part of the body that it contacts.

• Always place a shop towel or cloth around the fitting or connection prior to loosening to absorb any excess fuel due to spillage. Ensure that all fuel spillage (should it occur) is quickly removed from engine surfaces. Ensure that all fuel soaked cloths or towels are deposited into a suitable waste container.

• Always keep a dry chemical (Class B) fire extinguisher near the work area.

• Do not allow fuel spray or fuel vapors to come into contact with a spark or open flame.

• Always use a back-up wrench when loosening and tightening fuel line connection fittings. This will prevent unnecessary stress and torsion to fuel line piping. Always follow the proper torque specifications.

• Always replace worn fuel fitting O-rings with new. Do not substitute fuel hose or equivalent, where fuel pipe is installed.

Fuel System Pressure

RELIEVING

✳✳ CAUTION

The fuel injection system remains under pressure, even when the engine has been turned OFF. The fuel system pressure must be relieved before disconnecting any fuel lines. Failure to do so may result in fire and/or personal injury.

1. Loosen the fuel filler cap to relieve tank vapor pressure.

✳✳ CAUTION

Observe all applicable safety precautions when working around fuel. Whenever servicing the fuel system, always work in a well ventilated area. Do not allow fuel spray or vapors to come in contact with a spark or open flame. Keep a dry chemical fire extinguisher near the work area. Always keep fuel in a container specifically designed for fuel storage; also, always properly seal fuel containers to avoid the possibility of fire or explosion.

2. Be sure the ignition switch is in the **OFF** position.
3. Disconnect the negative battery cable.
4. Remove the engine dress-up cover.

✳✳ CAUTION

There may still be residual fuel in the system, and a small amount of fuel may be released when servicing fuel lines or connections. In order to reduce the chance of personal injury, cover the fuel line fittings with a shop towel before disconnecting to catch any fuel that may leak out.

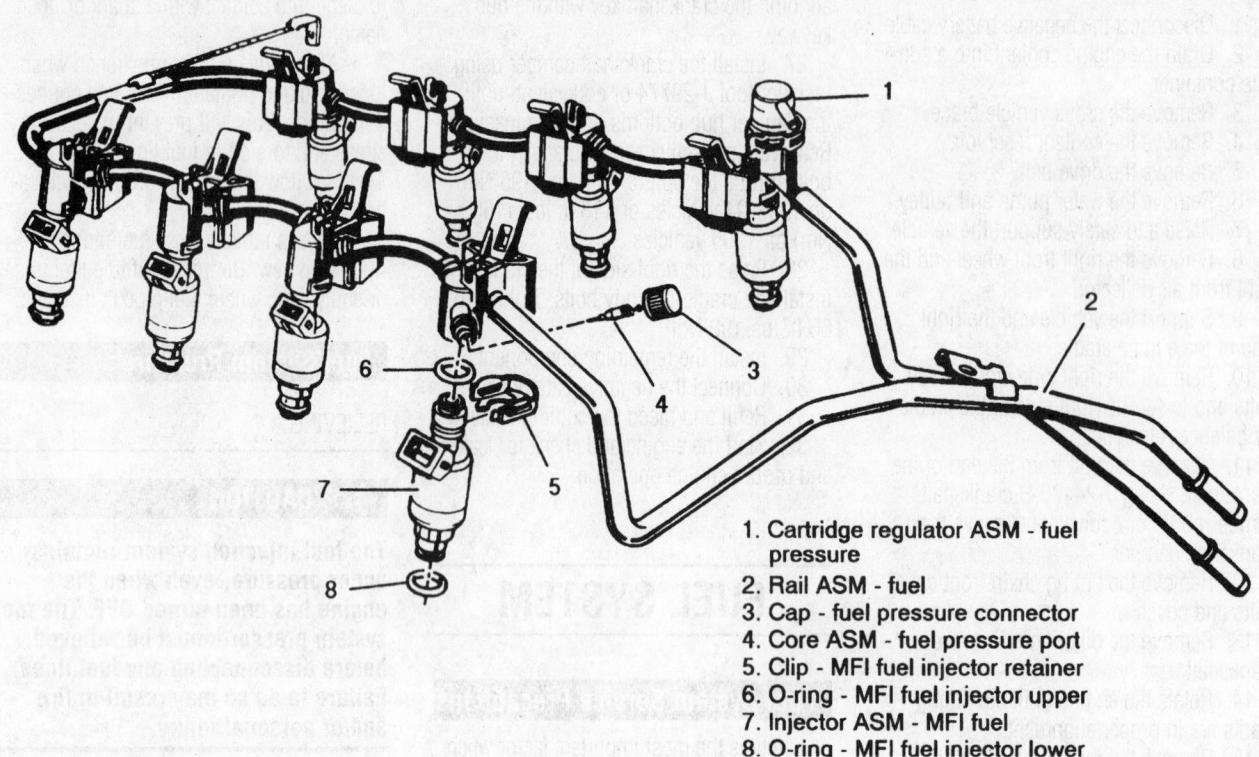

1. Cartridge regulator ASM - fuel pressure
2. Rail ASM - fuel
3. Cap - fuel pressure connector
4. Core ASM - fuel pressure port
5. Clip - MFI fuel injector retainer
6. O-ring - MFI fuel injector upper
7. Injector ASM - MFI fuel
8. O-ring - MFI fuel injector lower

7922VG13

View of the fuel rail assembly showing fuel system service port location

5. Install a fuel pressure gauge with a drain hose attached, J-34730-1 or equivalent. Wrap a shop towel around the fitting while connecting the gauge to avoid spillage.

6. Install the drain hose into an approved container and open the valve to drain the system pressure. Fuel connections are now safe for servicing.

7. Drain any remaining fuel from inside the gauge into the approved container.

Fuel Filter

REMOVAL & INSTALLATION

※ CAUTION

The fuel injection system remains under pressure, even after the engine has been turned OFF. The fuel system pressure must be relieved before disconnecting any fuel lines. Failure to do so may result in fire and/or personal injury.

1. Properly relieve the fuel system pressure.

※ CAUTION

Observe all applicable safety precautions when working around fuel. Whenever servicing the fuel system, always work in a well ventilated area. Do not allow fuel spray or vapors to come in contact with a spark or open flame. Keep a dry chemical fire extinguisher near the work area. Always keep fuel in a container specifically designed for fuel storage; also, always properly seal fuel containers to avoid the possibility of fire or explosion.

2. Disconnect the negative battery cable.

3. Release the locking tabs on the fuel filter retainer.

4. Disconnect the fuel lines from the fuel filter by releasing the locking tabs on the fuel filter quick-connects. Wrap a shop towel around the fitting when removing to catch excess fuel leaking out of the filter.

5. Remove the fuel filter.

To install:

6. Apply a few drops of engine oil to the tips of the fuel filter and install the fuel filter in the mounting bracket.

7. Connect the fuel lines to the fuel filter and snap the quick-connects into place. Be sure the tabs on the quick-connects lock into place.

8. Engage the locking tabs on the fuel filter retainer.

9. Connect the negative battery cable.

10. Turn the ignition key **ON** for 2 seconds, then **OFF** for 5 seconds. Again turn the ignition key **ON** and check for fuel leaks.

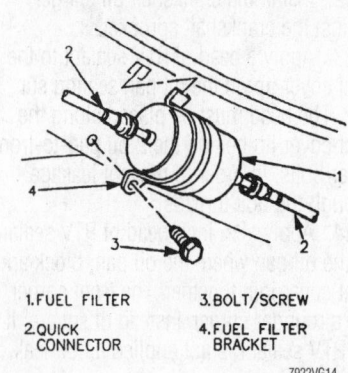

1. FUEL FILTER
2. QUICK CONNECTOR
3. BOLT/SCREW
4. FUEL FILTER BRACKET

7922VG14

Exploded view of the fuel filter mounting

Fuel Pump

REMOVAL & INSTALLATION

✳ CAUTION

The fuel injection system remains under pressure, even after the engine has been turned OFF. The fuel system pressure must be relieved before disconnecting any fuel lines. Failure to do so may result in fire and/or personal injury.

➡ The modular fuel sender assembly must be disassembled in the exact order described.

1. Disconnect the negative battery cable.

✳ CAUTION

Observe all applicable safety precautions when working around fuel. Whenever servicing the fuel system, always work in a well ventilated area. Do not allow fuel spray or vapors to come in contact with a spark or open flame. Keep a dry chemical fire extinguisher near the work area. Always keep fuel in a container specifically designed for fuel storage; also, always properly seal fuel containers to avoid the possibility of fire or explosion.

2. Relieve the fuel system pressure.
3. Remove the fuel tank from the vehicle and position on a suitable workbench.
4. Clean the fuel tank in the area of the modular fuel sender assembly.
5. Remove the locking nut by turning counterclockwise using tool J-39348 or equivalent.
6. Remove the modular fuel sender assembly from the fuel tank.

✳ CAUTION

The modular fuel sender assembly may spring up from its position. When removing the assembly, be aware that the reservoir bucket is full of fuel. Tip the assembly slightly during removal to avoid damaging the float. Have a shop towel ready to absorb any leakage.

7. Slide the fuel sender seal downward, past the reservoir and carefully over the float arm assembly. Discard the seal.

✳ CAUTION

Observe all applicable safety precautions when working around gasoline. Do not allow fuel spray or fuel vapors to come in contact with a spark or open flame. Keep a dry chemical (Class B) fire extinguisher near the work area. Never drain or store fuel in an open container due to the possibility of fire or explosion.

8. Remove the Connector Position Assurance (CPA) clip from the wiring harness under the modular fuel sender assembly cover. Depress the black connector tabs to remove the electrical connector from the cover.
9. Locate the curved side of the modular unit's reservoir. Beginning at locking tab 1, squeeze the reservoir to release the first locking tab.
10. Moving clockwise to locking tab 2, apply gentle pressure to the guide rod to release the second locking tab.
11. At locking tab 3, gently twist and squeeze to release the reservoir from the retainer.
12. Remove the cover and the retainer from the reservoir. Be careful not to damage

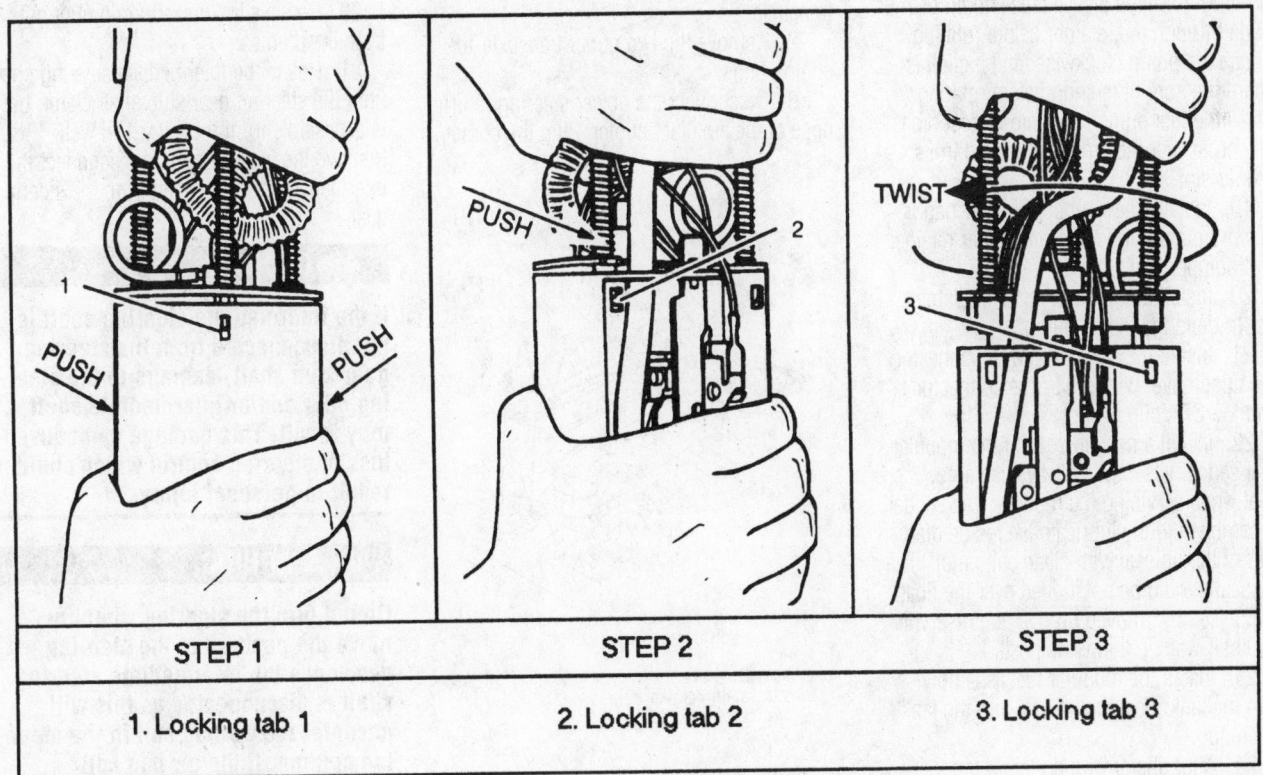

STEP 1	STEP 2	STEP 3
1. Locking tab 1	2. Locking tab 2	3. Locking tab 3

Modular fuel sender disassembly

7922VG15

the crossover tube; the unit will still be attached by the convoluted fuel pipe and the crossover tube.

13. Remove the external strainer by carefully prying the strainer ferrule off the reservoir. Excessive force may dislodge the jet pump. Note the position of the strainer for installation reference. Discard the strainer.

14. Remove the rubber bumper pad and discard. The fuel pump and the sleeve assembly are attached to the retainer when pulled from the reservoir. Depress the flex member on the pump sleeve and rotate the sleeve counterclockwise to remove the fuel pump from the retainer. Note the orientation of the pump to the retainer.

15. Slide the lower connector assembly out of the retainer to remove the fuel pulse dampener from the lower connector. Note the orientation of the seal (the modular unit is now held together by the crossover tube only). Discard the fuel pulse dampener.

To install:

16. Install the fuel pulse dampener. Always use a new dampener when installing a new fuel pump.

17. Slide the lower connector into the retainer.

18. Push the pump outlet tube into the fuel pulse dampener and rotate the flex member back to its original position. Line up the pump outlet tube into the retainer opening. All three sleeve tabs should protrude through the retainer before rotating. Rotate the pump clockwise until a click is heard, be sure fit is snug before rotating. Place the fuel pump back into its reservoir. The crossover tube must be placed in its proper slot.

19. Install a new rubber bumper pad. Insert the drain tube into the proper retainer and bumper pad slots.

20. Install a new strainer, being careful not to dislodge the jet pump.

21. Install the fuel pump wire connector, the undercover wiring harness connector and the CPA clip.

22. Install a new lip seal on the modular fuel sender assembly. Always use a new seal when servicing the modular fuel sender assembly. Lightly lubricate the inside diameter of the lip seal with clean engine oil. The lip seal should be positioned over the float arm assembly, moved up over the reservoir and half-way up the guide posts.

23. Install the modular fuel assembly into the tank. Seat the lip seal into the tank opening.

24. Align the arrows on top of the fuel tank to the arrow on the modular assembly.

25. Slowly apply pressure to the top of

the spring loaded sender until the lip seal is flush between the fuel tank and the modular cover.

26. Install the locking nut, and tighten to 37 ft. lbs. (50 Nm).

27. Install the fuel tank in the vehicle.

28. Replace the fuel that was drained before removal.

29. Connect the negative battery cable.

30. Pressurize the fuel system and verify there are no fuel leaks.

DRIVE TRAIN

Transaxle Assembly

REMOVAL & INSTALLATION

1. Disconnect the negative battery cable.

2. Remove the headlight housing upper filler panel and diagonal brace.

3. Remove the air cleaner assembly.

4. Disconnect the shift control cable and bracket at the transaxle.

5. Remove the torque struts.

6. Disconnect the oil cooler lines at the cooler and the oil sending line at the transaxle.

7. Remove the two upper transaxle-to-engine bolts.

8. Disconnect the power steering return hose at the auxiliary cooler. Plug the cooler

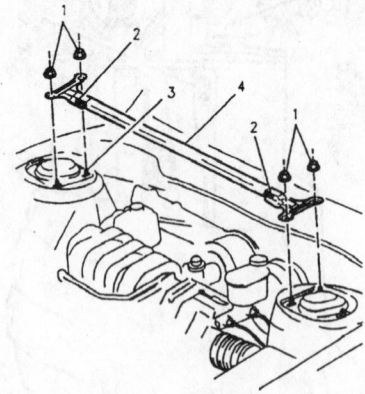

1	NUT
2	THROUGH-BOLTS
3	STUD, STRUT INBOARD
4	BAR, CROSS BRACE

7922VG16

Exploded view of the cross brace to strut towers mounting

and return hose to prevent leakage.

9. Support the engine with Engine Support Fixture J-28467 or equivalent. Tighten the wing nuts several turns to take the weight of the powertrain off of the mounts and frame.

10. Raise and safely support the vehicle. Remove the front wheels.

11. Remove the splash shields from both front wheel wells.

12. Disconnect both front suspension position sensors from the lower control arms and position aside.

13. Remove both stabilizer links from the struts.

14. Separate the tie rod ends from the steering knuckles.

15. Separate the lower ball joints from the steering knuckles.

16. Remove the halfshafts.

17. Remove the power steering filter at the cradle and the A/C splash shield from the frame.

18. Remove the ABS modulator from the bracket and support. Remove the engine oil pan-to-transaxle bracket.

19. Remove the torque converter cover, then remove the torque converter-to-flexplate bolts. Prior to bolt removal, mark the torque converter position in relation to the flexplate so they can be reassembled in the same position.

20. Remove the powertrain mount nuts from the cradle.

21. Rotate the intermediate steering shaft until the steering gear stub shaft clamp bolt is accessible from the left wheel well. Remove the clamp bolt and disconnect the intermediate steering shaft from the steering gear.

✳✳ CAUTION

If the intermediate steering shaft is not disconnected from the steering gear stub shaft, damage to the steering gear and/or intermediate shaft may result. This damage can cause loss of steering control which could result in personal injury.

✳✳ WARNING

Do not turn the steering wheel or move the position of the steering gear once the intermediate steering shaft is disconnected as this will uncenter the air bag coil in the steering column. If the air bag coil becomes uncentered, it may be damaged during vehicle operation.

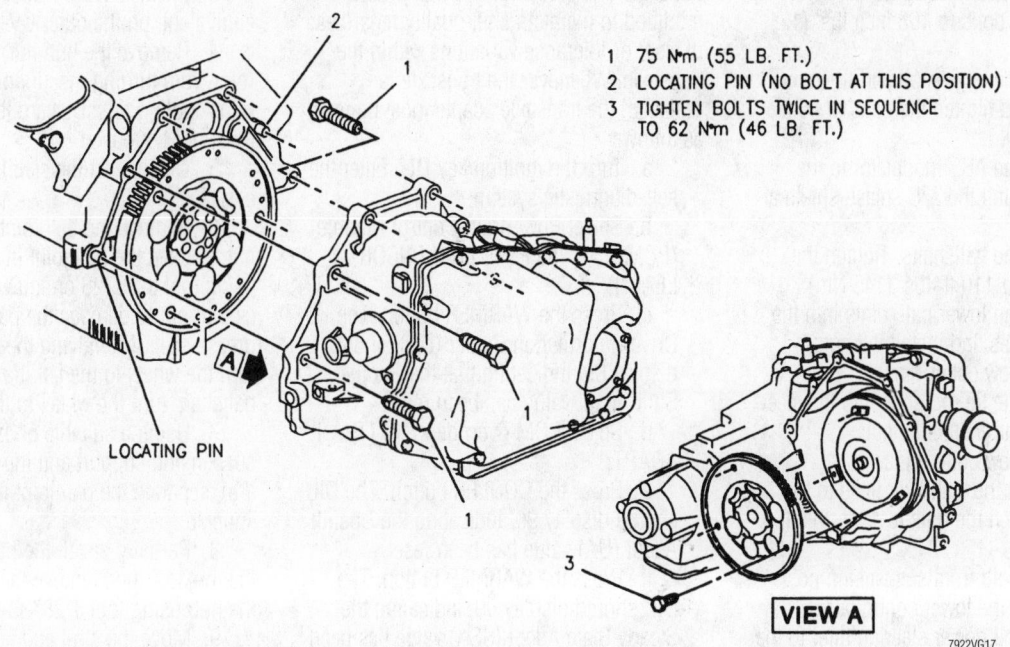

1 75 N·m (55 LB. FT.)
2 LOCATING PIN (NO BOLT AT THIS POSITION)
3 TIGHTEN BOLTS TWICE IN SEQUENCE
   TO 62 N·m (46 LB. FT.)

LOCATING PIN

VIEW A

7922VG17

Exploded view of the engine to transaxle attachments

22. Remove the electrical harness and connector from the front of the cradle.

23. Support the rear of the cradle with a suitable jack, then remove the four rear cradle bolts.

24. Lower the jack a few in. to gain access to the power steering gear heat shield and return line fitting. Remove the heat shield and disconnect the return line. Plug the line and the opening in the gear to prevent fluid leakage.

25. Detach the power steering electrical connector.

26. Raise the jack and reinstall one rear cradle bolt on each side finger-tight to support the cradle. Remove the jack.

27. Support the frame with a suitable jack and remove the six frame mount bolts. Lower the frame and/or raise the vehicle with the steering gear attached.

28. Label and detach the electrical connectors to the transaxle, vehicle speed sensor and ground. Remove the transaxle harness from the transaxle clip.

29. Remove the fuel line bundle from the transaxle.

30. Remove the left and right transaxle mount and bracket from the transaxle.

31. Support the transaxle with a suitable jack.

32. Remove the engine-to-transaxle heat shield and bracket and the remaining transaxle-to-engine bolts. Lower the transaxle.

33. Remove the manual shaft linkage and neutral safety switch. Remove the vehicle speed sensor and oil return line.

To install:

34. Install the oil return line and the vehicle speed sensor.

35. Install the neutral safety switch and tighten the bolts to 106 inch lbs. (12 Nm).

36. Install the manual shaft linkage and tighten the manual shaft nut to 15 ft. lbs. (20 Nm).

37. Raise the transaxle into position and install the two lower transaxle-to-engine bolts. Tighten to 35 ft. lbs. (47 Nm).

38. Install the engine-to-transaxle bracket and heat shield. Tighten the bolts to 35 ft. lbs. (47 Nm).

39. Remove the transaxle jack.

40. Install the right and left transaxle bracket and mount to the transaxle. Tighten the bolts and nuts to 35 ft. lbs. (47 Nm).

41. Install the fuel line bundle to the transaxle.

42. Attach the electrical connectors to the transaxle, vehicle speed sensor and ground. Install the transaxle harness to the transaxle clip.

43. Raise the frame and/or lower the vehicle while locating the engine and transaxle mount studs into the frame, harnesses at the cradle, and frame mount bolt holes to the underbody.

44. Install two front and two rear cradle bolts finger-tight to support the cradle, then remove the cradle support.

45. Support the rear of the cradle with a suitable jack and remove the two rear cradle bolts.

46. Lower the jack a few in. to gain access to the power steering gear. Connect the hose at the steering gear and tighten the fitting to 20 ft. lbs. (27 Nm). Attach the power steering gear electrical connector and install the steering gear heat shield.

47. Raise the cradle with the jack. Install the six frame mount bolts beginning with the No. 2 mount bolt into the body, followed by the No. 1 mount bolt into the body, followed by the remaining frame mount bolts. Tighten the bolts to 74 ft. lbs. (100 Nm).

48. Install the electrical harness to the front of the cradle.

49. Connect the intermediate steering shaft to the steering gear and install the clamp bolt. Tighten the bolt to 35 ft. lbs. (47 Nm).

✳✳ WARNING

Do not turn the steering wheel or move the position of the steering gear while the intermediate steering shaft is disconnected as this will uncenter the air bag coil in the steering column. If the air bag coil becomes uncentered, it may be damaged during vehicle operation.

50. Install the left and right transaxle mount nuts and right engine mount nuts at the frame. Tighten the nuts to 35 ft. lbs. (47 Nm).

51. Align the flexplate and torque converter using the marks made during the removal procedure. Install the flexplate-to-converter bolts and tighten to 35 ft. lbs. (47 Nm).

52. Install the torque converter cover and tighten the bolts to 106 inch lbs. (12 Nm).

53. Install the engine oil pan-to-transaxle bracket and tighten the bolts to 35 ft. lbs. (47 Nm).

54. Install the ABS modulator to the bracket and install the A/C splash shield at the frame.

55. Install the halfshafts. Tighten the halfshaft nuts to 110 ft. lbs. (145 Nm).

56. Install the lower ball joints into the steering knuckles and install the nuts.

57. Install new cotter pins.

58. Install the tie rod ends into the steering knuckles and install the nuts.

59. Install new cotter pins.

60. Connect the stabilizer links to the struts and tighten the nuts to 49 ft. lbs. (65 Nm).

61. Install both front suspension position sensors to the lower control arms.

62. Install the power steering filter to the cradle and install the splash shields in the wheel wells.

63. Install the wheels and lower the vehicle. Remove the engine support fixture.

64. Connect the power steering hose at the auxiliary cooler.

65. Install the remaining transaxle-to-engine bolts and tighten to 35 ft. lbs. (47 Nm).

66. Flush the transaxle oil cooler using flushing tool J-35944 and flushing solution J-35944-20. The transaxle oil cooler and lines should be flushed before the oil cooler lines are connected to the transaxle.

67. Connect the transaxle oil cooler lines to the transaxle. Start the fittings by hand and tighten them finger-tight, then tighten the fittings to 16 ft. lbs. (22 Nm).

68. Install the torque struts.

69. Adjust the neutral safety switch.

70. Install the shift control cable and bracket to the transaxle and tighten the bracket bolts to 106 inch lbs. (12 Nm). Adjust the shift control cable.

71. Install the air cleaner assembly. Install the headlight housing upper filler panel and diagonal brace.

72. Connect the negative battery cable.

73. Fill the transaxle with the proper type and quantity of transaxle fluid. Bleed the power steering system.

74. Check the front suspension alignment and adjust as necessary.

75. The Powertrain Control Module (PCM) maintains three types of transaxle adapt parameters which are used to modify transaxle line pressure. The line pressure is modified to maintain shift quality regardless of wear or tolerance variations within the transaxle. Whenever the transaxle is replaced, the transaxle adapts must be reset as follows:

a. Turn the ignition key **ON**. Enter the self-diagnostic system.

b. Select Powertrain Control Module (PCM) override PS13 (TP SENSOR LEARN).

c. Press the WARMER button. The Driver Information Center (DIC) should display 09, indicating that the Garage Shift Adapt value has been reset.

d. Select PCM override PS14 (TRAN ADAPT).

e. Press the COOLER button. The DIC should display 90, indicating the Upshift Adapt (UA) value has been reset.

f. Press the WARMER button. The DIC should display 09, indicating the Steady State Adapt (SSA) value has been reset.

76. The PCM maintains a value for transaxle oil life. This value indicates the percentage of oil life remaining and is calculated based on transaxle temperature and speed. When the vehicle is new, the transaxle oil life value is 100 As the vehicle operates, the percentage will decrease. Whenever the transaxle is replaced, the transaxle oil life indicator should be reset to 100 as follows:

a. Turn the ignition key **ON**, but leave the engine OFF.

b. Press and hold the OFF and REAR DEFOG buttons on the DIC until the message TRANSAXLE OIL LIFE RESET is displayed on the DIC.

Halfshaft

REMOVAL & INSTALLATION

❋❋ WARNING

Use care when removing the halfshaft to prevent the inner CV-joint from becoming over-extended. Over-extension of the joint could result in separation of internal components and possible joint failure.

1. Raise and safely support the vehicle.
2. Remove the front wheel. Mark the position of the wheel to the wheel studs, prior to removal, for installation reference.

3. Install J-34754 Boot Protector, or the equivalent, on the outer CV-joint boot.

4. Remove the hub nut. To prevent the rotor from turning insert a drift punch in the rotor cooling fins. Discard the hub nut; it must not be reused.

5. Disconnect the stabilizer link, if necessary.

6. Remove the ball joint cotter pin and nut. Loosen the ball joint in the knuckle using tool J-36226 or equivalent, being careful not to damage the ball joint and grease seal. If removing the right halfshaft, turn the wheel to the left. If removing the left halfshaft, turn the wheel to the right.

7. Using a suitable prybar between the suspension support and the lower control arm, separate the ball joint from the steering knuckle.

8. Partially install the hub nut to protect the threads, then remove the halfshaft from the hub using tool J-28733-B or equivalent.

9. Move the strut and knuckle rearward.

10. Remove the halfshaft from the transaxle using a suitable prying tool and a wood block fulcrum to protect the transaxle case.

➡**If equipped with anti-lock brakes, care must be used to prevent damage to the toothed sensor ring on the halfshaft and the wheel speed sensor on the steering knuckle.**

To install:

11. If installing the right-side halfshaft, install tool J-37292-B or equivalent, so it can be pulled out after the halfshaft is installed.

12. Install the halfshaft into the transaxle. To verify the halfshaft is properly seated, grasp the inner CV-joint housing and pull it outward. DO NOT pull on the halfshaft. If the CV-joint is properly seated, the halfshaft will not pull back out.

13. Install the halfshaft into the hub and bearing assembly. Loosely install a new hub nut.

14. Install the ball joint into the steering knuckle.

15. Install the castle nut and tighten to 84 inch lbs. plus 120 degrees (1/3(/frac) turn). A minimum of 37 ft. lbs. (51 Nm) of torque must be attained. If necessary to install the cotter pin, the nut can be tightened up to 20 degrees additional. NEVER loosen the castle nut to install the cotter pin.

16. With a drift in the brake rotor cooling fins to keep the rotor from turning, tighten

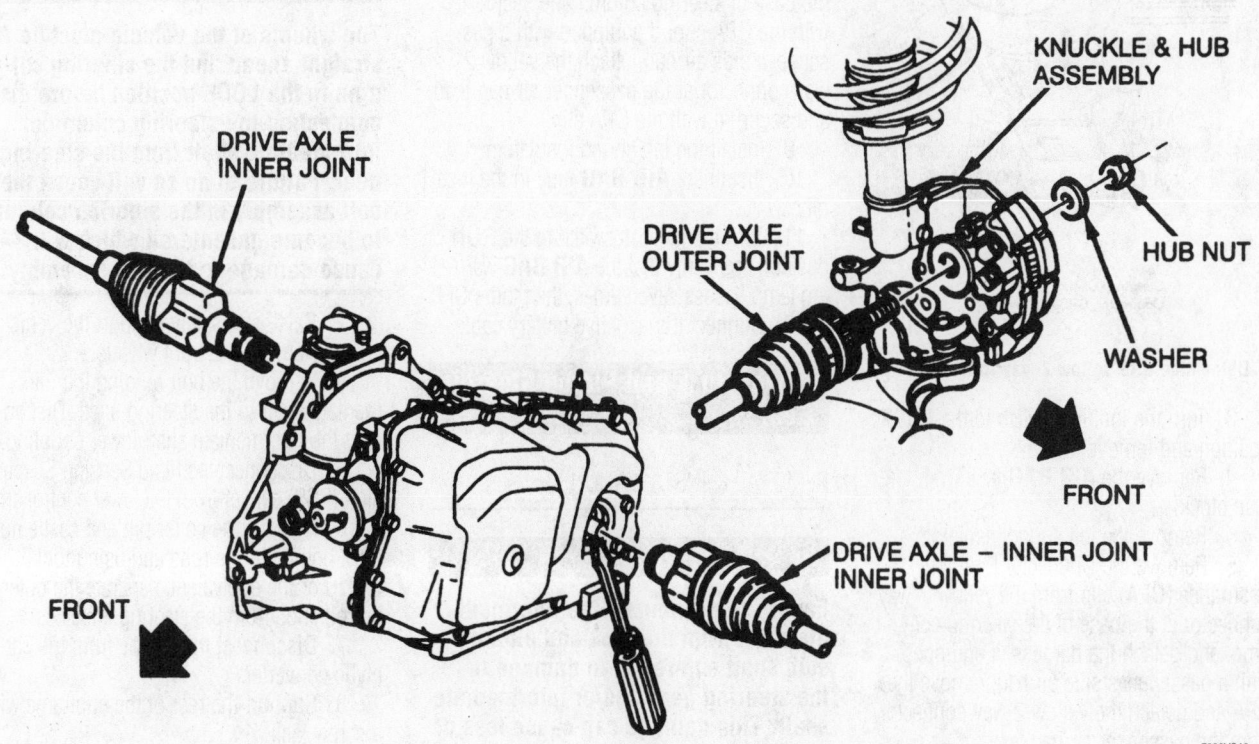

Exploded view of the removal of the halfshaft from the transaxle

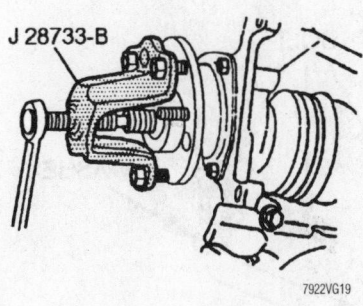

Removing the halfshaft from the hub

the nut to 110 ft. lbs. (149 Nm) on all models, except DeVille with J55 brake option, in which case the torque is 130 ft. lbs. (177 Nm).

17. Connect the stabilizer link, if removed.

18. Remove the boot protector.

19. If tool J-37292-B was installed, remove it by pulling in line with the handle.

20. Install the wheel, aligning the marks made during removal.

21. Lower the vehicle.

22. Road test and check vehicle operation.

STEERING AND SUSPENSION

Air Bag

✳✳ CAUTION

Some vehicles are equipped with an air bag system, also known as the Supplemental Inflatable Restraint (SIR) or Supplemental Restraint System (SRS). The system must be disabled before performing service on or round system components, steering column, instrument panel components, wiring and sensors. Failure to follow safety and disabling procedures could result in accidental air bag deployment, possible personal injury and unnecessary system repairs.

PRECAUTIONS

Several precautions must be observed when handling the inflator module to avoid accidental deployment and possible personal injury.

• Never carry the inflator module by the wires or connector on the underside of the module.

• When carrying a live inflator module, hold securely with both hands, and ensure that the bag and trim cover are pointed away.

• Place the inflator module on a bench or other surface with the bag and trim cover facing up.

• With the inflator module on the bench, never place anything on or close to the module which may be thrown in the event of an accidental deployment.

DISARMING

✳✳ CAUTION

The air bag system must be disarmed before performing service procedures around the air bag or wiring. Failure to do so may cause accidental deployment, resulting in unnecessary repairs and/or personal injury.

1. Disconnect the negative battery cable.
2. Turn the steering wheel so that the vehicle's wheels are pointing straight ahead.

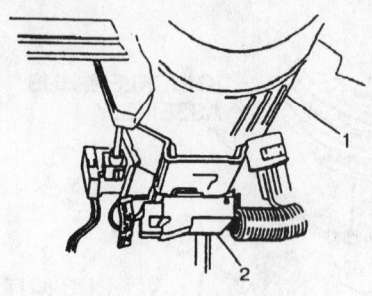

1. Steering column
2. Connector, SRS (yellow)

7922VG20

Detach the SRS yellow 2-way connector

3. Turn the ignition switch to the **LOCK** position and remove the key.

4. Remove the **AIR BAG** fuse from the fuse block.

5. Remove the left sound insulator.

6. Remove the Connector Position Assurance (CPA) clip from the yellow 2-way connector at the base of the steering column, and detach the harness. If equipped with a passenger's side air bag, remove the CPA and detach the yellow 2-way connector from the passenger air bag lead.

After the applicable service is concluded, enable the air bag system as follows:

7. Turn the ignition switch to the **LOCK** position and remove the key.

8. Attach the yellow 2-way connector at the base of steering column and secure it with the CPA clip. If equipped with a passenger's side air bag, attach the yellow 2-way connector at the passenger air bag lead and secure it with the CPA clip.

9. Install the left sound insulator.

10. Install the **AIR BAG** fuse in the fuse block.

11. Turn the ignition switch to the **RUN** position and verify that the **AIR BAG** warning lamp flashes seven times, then turns OFF.

12. Connect the negative battery cable.

Power Rack and Pinion Steering Gear

REMOVAL & INSTALLATION

※※ CAUTION

Failure to disconnect the intermediate shaft from the rack and pinion stub shaft can result in damage to the steering gear and/or intermediate shaft. This damage can cause loss of steering control which could result in personal injury.

1. Disconnect the negative battery cable.

※※ WARNING

The wheels of the vehicle must be straight ahead and the steering column in the LOCK position before disconnecting the steering column or intermediate shaft from the steering gear. Failure to do so will cause the coil assembly in the steering column to become uncentered which will cause damage to the coil assembly.

2. Raise and safely support the vehicle.

3. Remove the front wheels.

4. Remove the bolt holding the intermediate shaft to the steering shaft. Disconnect the intermediate shaft lower coupling.

5. Disconnect the Road Sensing Suspension (RSS) sensor from the lower control arm.

6. Remove the cotter pin and castle nuts from both outer tie rods and using tool J-24319 or the equivalent, separate the outer tie rod ends from the steering knuckles.

7. Disconnect the Y-pipe from the catalytic converter.

8. Support the rear of the subframe with a screw jack.

9. Remove the rear subframe bolts.

10. Slowly lower the subframe to gain access.

11. Remove the heat shield.

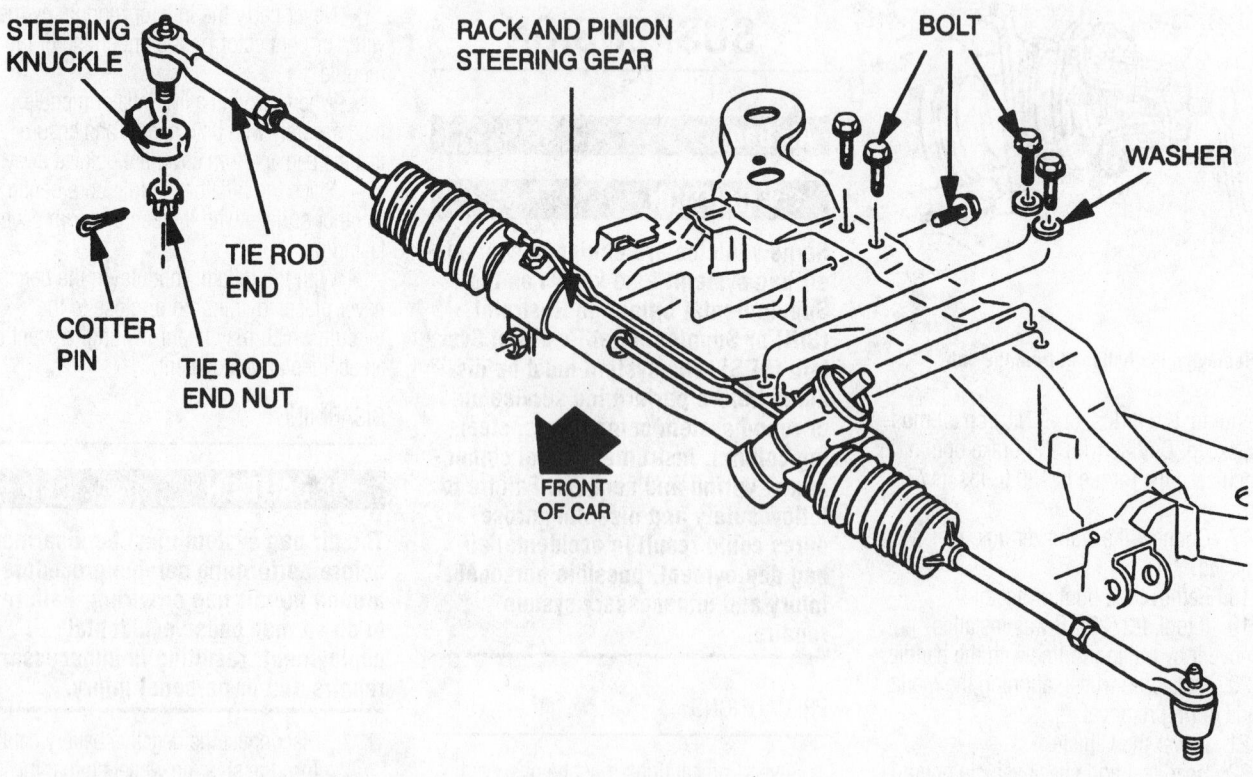

Exploded view of the power steering rack components

7922VG21

12. Remove the power steering line retainer.

13. Place a catch pan under the power steering rack and the line fittings.

14. Disconnect the power steering pressure and return lines from the rack and pinion unit.

15. Detach the Speed Sensitive Steering (SSS) solenoid valve connector.

16. Remove the five power steering rack-to-subframe bolts.

17. Slide the rack and pinion unit out the left wheel well.

To install:

18. Install the rack and pinion through the left wheel well and into position on the subframe.

19. Install the five mounting bolts and tighten to 50 ft. lbs. (68 Nm).

20. Attach the Speed Sensitive Steering (SSS) solenoid valve connector.

21. Connect the power steering pressure and return hoses and tighten the fittings to 20 ft. lbs. (27 Nm).

22. Install the power steering line retainer.

23. Install the heat shield.

24. Raise the subframe into position and install the subframe mounting bolts. Tighten to 76 ft. lbs. (103 Nm).

25. Connect the Y-pipe to the catalytic converter.

26. Connect the outer tie rod ends to the steering knuckle. Install the castle nuts and tighten. Install the cotter pins.

27. Connect the RSS sensor to the lower control arm.

28. Connect the intermediate shaft and install the intermediate shaft-to-steering shaft bolt. Tighten the bolt to 30 ft. lbs. (41 Nm).

29. Install the front wheels.

30. Lower the vehicle.

31. Connect the negative battery cable.

32. Refill the power steering reservoir and bleed the power steering system.

33. Check the wheel alignment. Road test the vehicle.

Strut

REMOVAL & INSTALLATION

Front

✳✳ WARNING

When working near the halfshafts, use care to prevent the inner Tripot CV-joint from being overextended. If the joint is overextended, the internal joint components could separate, resulting in CV-joint failure.

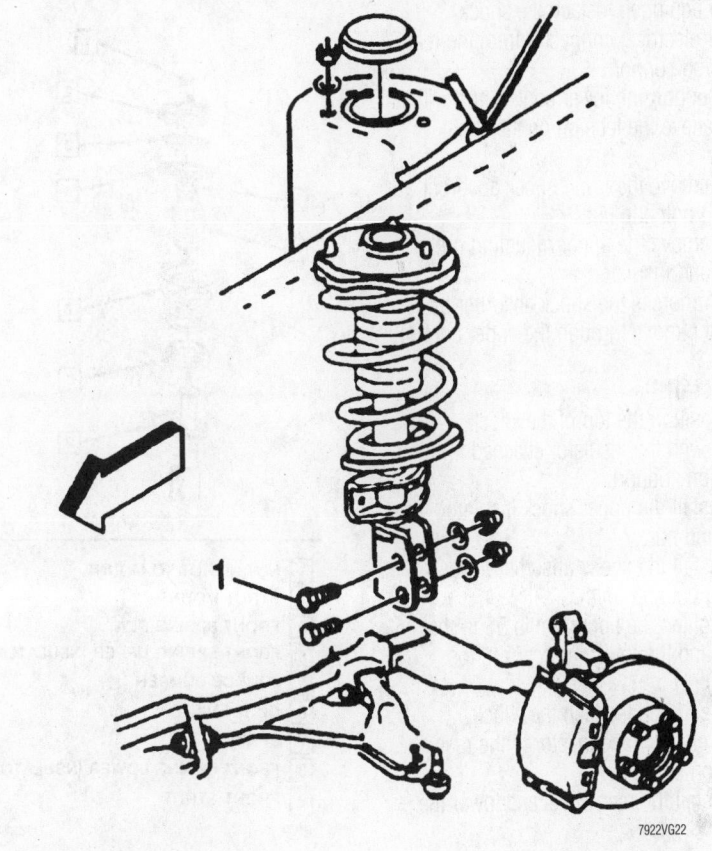

7922VG22

Exploded view of the strut mounting

1. Disconnect the negative battery cable.

2. Raise and safely support the vehicle.

3. Remove the front wheel.

4. If equipped, detach the electrical connector from the top of the strut.

5. Remove the upper strut mounting nuts.

6. Remove the Road Sensing Suspension (RSS) position sensor from the lower control arm, if equipped.

7. Disconnect the ABS wheel speed sensor, if equipped.

8. Remove the wheel speed sensor from the bracket on the strut, if equipped.

9. Remove the brake line bracket from the strut.

10. Remove the stabilizer link from the strut.

11. Scribe a mark on the strut referencing the lower strut bracket to the steering knuckle.

12. Remove the nuts and through-bolts from the lower strut bracket and remove the strut from the vehicle.

To install:

13. Install the strut assembly into the vehicle. Install the strut-to-knuckle bolts and nuts, but do not tighten yet.

14. Connect the stabilizer link to the strut assembly, but do not tighten the nuts yet.

15. Connect the brake line bracket to the strut.

16. Install the speed sensor bracket on the strut, if equipped.

17. Connect the ABS sensor, if equipped.

18. Install the RSS sensor to the lower control arm, if equipped.

19. Install the upper strut mounting nuts. If equipped, attach the electrical connector to the top of the strut.

20. Tighten the upper strut mounting nuts to 18 ft. lbs. (24 Nm). Tighten the stabilizer link nuts to 48 ft. lbs. (65 Nm). Align the marks made during removal and tighten the strut-to-knuckle bolt nuts to 140 ft. lbs. (190 Nm).

21. Install the front wheel and lower the vehicle.

22. Check the wheel alignment.

Shock Absorber

REMOVAL & INSTALLATION

Rear

1. Disconnect the negative battery cable.

2. Raise and safely support the vehicle.

3. Remove the rear wheel.

4. If equipped, detach the shock absorber electrical connector from the rear suspension support.

5. Support the lower control arm with a jack to relieve the tension on the shock absorber.

6. Remove the lower shock absorber mounting bolt and nut.

7. Remove the upper mounting nut, retainer, and insulator.

8. Compress the shock absorber by hand and remove through the upper control arm.

To install:

9. Position the top of the shock absorber with the insulator attached into the suspension support.

10. Install the upper shock insulator, retainer and nut.

11. Install the shock absorber lower mounting nut and bolt.

12. Tighten the upper nut to 55 ft. lbs. (74 Nm) and the lower nut to 75 ft. lbs. (102 Nm).

13. If equipped, attach the shock absorber electrical connector to the rear suspension support.

14. Install the rear wheel and lower the vehicle.

15. Connect the negative battery cable.

Coil Spring

REMOVAL & INSTALLATION

Front

1. Remove the strut from the vehicle.

✳✳ WARNING

Be careful not to scratch or crack the protective coating on the coil spring, as damage may cause premature spring failure.

2. Mount the strut assembly in Strut Compressor tool J-34013 or equivalent strut compressor.

3. Turn the compressor forcing screw until the spring compresses slightly.

4. Use a T-50 Torx® bit to keep the strut shaft from turning, then remove the nut on the top of the strut shaft.

5. Loosen the compressor screw while guiding the strut shaft out of the assembly. Continue loosening the compressor screw until the strut and spring can be removed.

To install:

6. Mount the strut in the compressor tool. Use clamping tool J-34013-20 or the

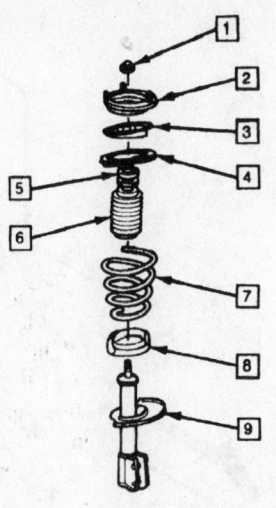

1	NUT, STRUT TO MOUNT
2	STRUT MOUNT
3	FRONT SPRING SEAT
4	FRONT SPRING UPPER INSULATOR
5	JOUNCE BUMPER
6	DUST SHIELD
7	SPRING
8	FRONT SPRING LOWER INSULATOR
9	FRONT STRUT

7922VG23

Disassembled view of strut

equivalent, to hold the strut shaft in place.

7. Install the spring over the strut. The flat on the upper spring seat must face out from the centerline of the vehicle, or when mounted in the strut compressor, the spring seat must face the same direction as the steering knuckle mounting flange.

➡**If the bearing was removed from the upper spring seat, it must be reinstalled in the spring seat in the same position before attaching to the strut mount.**

8. Turn the compressor screw to compress the spring, while guiding the strut shaft through the top of the strut assembly. If necessary, use tool J-34013-38 or equivalent, to guide the shaft.

9. When the strut shaft threads are visible through the top of the strut assembly, install the nut.

10. Remove clamping tool J-34013-20 from the strut shaft.

11. Tighten the strut shaft nut to 55 ft. lbs. (75 Nm) while holding the strut shaft with a T-50 Torx® bit.

12. Remove the strut assembly from the compressor tool, then install the strut into the vehicle.

Rear

1. Raise and safely support the vehicle.

2. Remove the rear wheel.

3. Support the inboard end of the lower control arm with a transmission jack.

4. Remove the sway bar link lower mounting bolt.

5. Remove the shock absorber lower mounting bolt and push the shock up and out of the way.

6. Remove the lower control arm inner mounting nuts and bolts.

7. Slowly lower the transmission jack until the spring tension has been released.

8. Remove the coil spring.

To install:

9. Install the coil spring and spring insulators.

10. Raise the control arm with the transmission jack until the lower control arm bolts can be installed.

11. Install the mounting nuts. Do not tighten at this time.

12. Remove the transmission jack.

13. Extend the shock absorber and install the mounting nuts. Tighten the nut to 75 ft. lbs. (102 Nm).

14. Install the link kit and tighten the nut to 44 ft. lbs. (60 Nm).

15. Install the rear wheel and lower the vehicle.

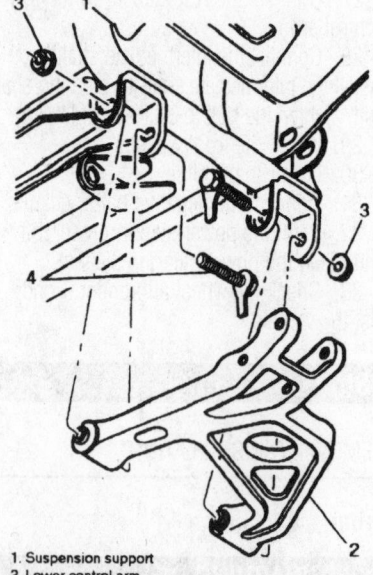

1. Suspension support
2. Lower control arm
3. Lower control arm inner nut - 102 Nm (75 lb. ft.)
4. Lower control arm inner bolt

7922VG24

Exploded view of the lower control arm mounting

16. Tighten the lower control arm inner bolts to 75 ft. lbs. (102 Nm).

17. Check the alignment and adjust, as necessary.

Lower Ball Joint

REMOVAL & INSTALLATION

1. Raise and safely support the vehicle, allowing the front suspension to hang free.

2. Remove the front wheel.

❊❊ CAUTION

Be careful when working in the area of the CV-boot. Damage to the boot could result in eventual joint failure. Install CV-boot protector tool J-34754 or take equivalent precautions to protect the soft boot.

3. If equipped, remove the Road Sensing Suspension (RSS) position sensor from the lower control arm.

4. Separate the ball joint from the steering knuckle.

5. Drill out the three rivets retaining the

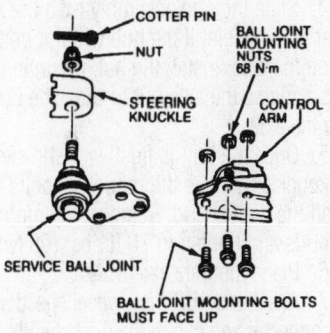

Exploded view of the lower ball joint mounting

ball joint to the lower control arm and remove the ball joint.

To install:

6. Attach the new ball joint to the lower control arm with three mounting bolts and nuts. The bolts must be installed from the bottom of the control arm. Tighten the nuts to 50 ft. lbs. (68 Nm).

7. Connect the ball joint to the steering knuckle and install the castellated nut.

8. Tighten the ball joint nut to 84 inch lbs. (10 Nm), then tighten the nut an addi-

tional 120 degrees during which a minimum torque of 37 ft. lbs. (50 Nm) must be obtained. If the minimum torque is not obtained, check for stripped threads. If the threads are okay, replace the ball joint and knuckle.

9. Install a new cotter pin. If the cotter pin cannot be installed because the hole in the stud does not align with a nut slot, tighten the nut up to an additional 60 degrees to allow for installation. NEVER loosen the nut to provide for cotter pin installation.

10. If equipped, install the RSS position sensor to the lower control arm.

11. If used, remove the CV-Joint boot protector tool.

12. Install the front wheel and lower the vehicle.

13. Check the wheel alignment.

Wheel Bearings

ADJUSTMENT

The wheel bearings are not adjustable. If a wheel bearing is out of specifications, it

WHEEL BEARING LOOSENESS DIAGNOSIS

DRUM BRAKE

Mount dial indicator. Grasp bearing flange; using a push-pull movement, note indicator readings.

If looseness exceeds 0.1270 mm (0.005 inch), replace hub and bearing assembly.

DISC BRAKE

Free shoes from the disc, or remove calipers. Reinstall two wheel nuts to secure disc. Mount dial indicator. Grasp disc; using a push-pull movement, note indicator readings.

If looseness exceeds 0.1270 mm (0.005 inch), replace hub and bearing assembly.

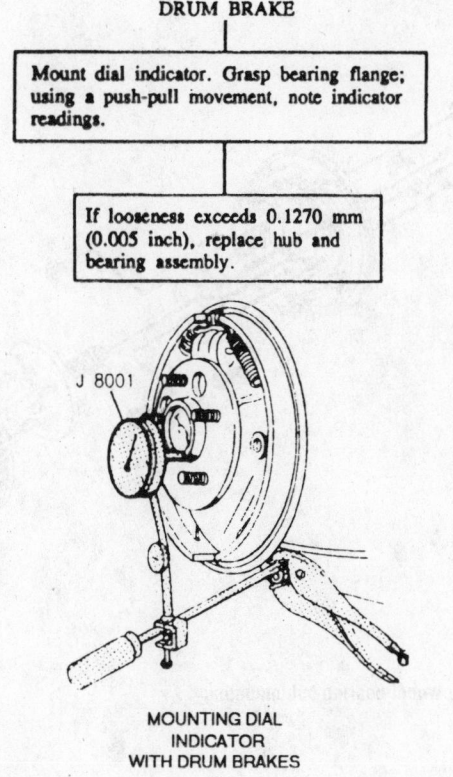

MOUNTING DIAL INDICATOR WITH DRUM BRAKES

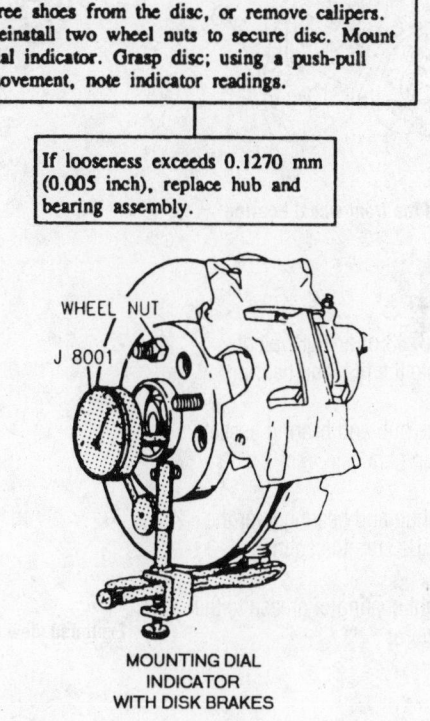

MOUNTING DIAL INDICATOR WITH DISK BRAKES

7922VG33

Wheel bearing check

must be replaced. Using a dial indicator, check for looseness. If play exceeds 0.005 in. (0.127mm) the bearing wear is excessive and the hub and bearing should be replaced.

REMOVAL & INSTALLATION

Front

1. Raise and safely support the vehicle.
2. Remove the front wheel.
3. Insert a drift punch through the caliper and into the rotor cooling fins to prevent the rotor from turning.
4. Remove the axle nut and washer.
5. Remove the caliper mounting bolts. Remove the caliper from the steering knuckle and support out of the way. DO NOT allow the brake hose to support the weight of the caliper.
6. Remove the brake rotor.
7. Detach the ABS speed sensor connector and unclip from the dust shield.
8. Remove the three hub and bearing bolts and remove the dust shield.
9. Place the transaxle selector in the **P**

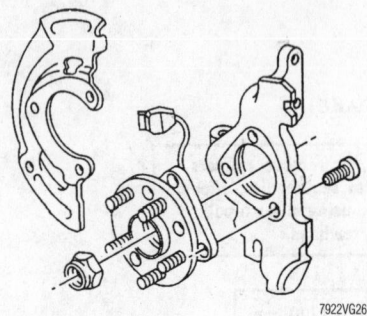

Exploded view of the front wheel bearing mounting

detent.

10. Install J-28733 or an equivalent puller, and separate the hub and bearing from the drive axle.
11. Remove the hub and bearing assembly from the steering knuckle.

To install:

12. Install the hub and bearing over the half shaft splines. Be sure the splines engage smoothly.
13. Apply a light coating of grease to the steering knuckle bore.

14. Slide the hub assembly onto the axle as far as possible. If the hub will not bottom out on the axle, install the hub mounting bolts and use the axle nut to draw the hub onto the axle.
15. Once the hub is flush with the steering knuckle, remove the mounting bolts and install the dust shield. Reinstall the mounting bolts and tighten to 70 ft. lbs. (95 Nm).
16. Place the transaxle in **N**.
17. Attach the ABS front wheel speed sensor connector and clip to the dust shield.
18. Install the brake rotor.
19. Install the caliper and tighten the mounting bolts to 38 ft. lbs. (51 Nm).
20. Insert a drift punch through the rotor to prevent the axle from turning.
21. Tighten the drive axle shaft nut to 107 ft. lbs. (145 Nm).
22. Remove the drift punch.
23. Install the wheel and lower the vehicle.
24. Road test for proper operation.

Rear

➡**The wheel bearing and hub are serviced as an assembly. The individual components are not serviceable separately.**

1. Raise and safely support the vehicle.
2. Remove the rear wheel.
3. Remove the caliper bracket mounting bolts and remove the caliper assembly from the knuckle. Support the caliper out of the way on a wire. Do not disconnect the brake hose from the caliper.
4. Remove the brake rotor. Mark the relationship between the wheel stud and the brake rotor for installation reference.
5. Remove the four hub and bearing assembly mounting bolts.
6. Remove the hub and bearing assembly.

To install:

7. Install the hub and bearing assembly.
8. Install the mounting bolts and tighten to 52 ft. lbs. (70 Nm).
9. Install the brake rotor, aligning the marks made during removal.
10. Install the caliper and new caliper bracket mounting bolts. Tighten to 83 ft. lbs. (113 Nm).
11. Install the wheel and lower the vehicle.
12. Road test the vehicle for proper operation.

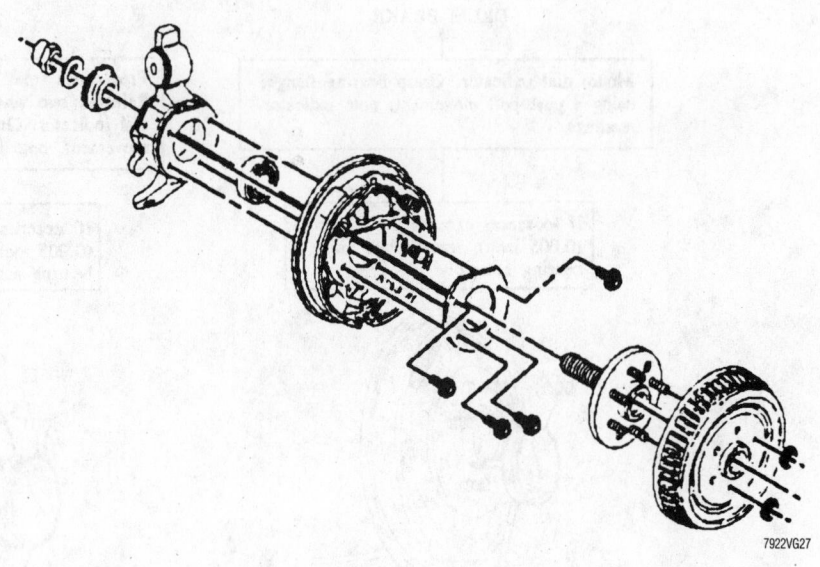

Exploded view of the rear wheel bearing hub mounting

GENERAL MOTORS F BODY

31

Chevrolet-Camaro • **Z28** • **Pontiac**-Firebird • Trans Am

DRIVE TRAIN 31-27
ENGINE REPAIR 31-2
FUEL SYSTEM 31-25
STEERING AND
  SUSPENSION 31-28

A

Air Bag 31-28
  Disarming 31-29
  Precautions 31-29
  Rearming 31-29

C

Camshaft and Valve Lifters 31-16
  Removal & Installation 31-16
Clutch 31-27
  Removal & Installation 31-27
Coil Spring 31-32
  Removal & Installation 31-32
Cylinder Head 31-6
  Removal & Installation 31-6

D

Distributor 31-2
  Removal 31-2

E

Engine Assembly 31-3
  Removal & Installation 31-3
Exhaust Manifold 31-14
  Removal & Installation 31-14

F

Fuel Filter 31-25
  Removal & Installation 31-25
Fuel Pump 31-26
  Removal & Installation 31-26
Fuel System Pressure 31-25
  Relieving 31-25
Fuel System Service
  Precautions 31-25

H

Hydraulic Clutch System 31-28
  Bleeding 31-28

I

Ignition Timing 31-2
  Adjustment 31-2
Intake Manifold 31-11
  Removal & Installation 31-11

L

Lower Ball Joint 31-33
  Removal & Installation 31-33

O

Oil Pan 31-18
  Removal & Installation 31-18
Oil Pump 31-19
  Removal & Installation 31-19

P

Power Rack and Pinion Steering
  Gear 31-29

Removal & Installation 31-29

R

Rear Main Seal 31-20
  Removal & Installation 31-20
Rocker Arms 31-10
  Removal & Installation 31-10

S

Shock Absorber 31-29
  Removal & Installation 31-29

T

Timing Chain, Sprockets, Front
  Cover and Seal 31-21
  Removal & Installation 31-21
Transmission 31-27
  Removal & Installation 31-27

U

Upper Ball Joint 31-33
  Removal & Installation 31-33

V

Valve Lash 31-17
  Adjustment 31-17

W

Water Pump 31-4
  Removal & Installation 31-4
Wheel Bearings 31-34
  Adjustment 31-34
  Removal & Installation 31-34

ENGINE REPAIR

➡Disconnecting the negative battery cable on some vehicles may interfere with the functioning of the on-board computer system, and may require the computer to undergo a relearning process once the negative battery cable is reconnected.

Distributor

All 1998–99 5.7L (VIN G) engines use a direct ignition system. This system eliminates the need for a distributor and has several advantages. Some of the advantages are, no moving parts, no load on the engine, more coil cool down time between firing and increased coil saturation time which produces a stronger spark.

The direct ignition system consists of:
• Eight ignition coils and modules (one for each cylinder)
• Eight ignition control circuits
• Camshaft position sensor
• 1X Camshaft reluctor wheel
• 24X Crankshaft reluctor wheel
• Crankshaft position sensor
• Powertrain Control Module (PCM)

The 24X crankshaft position sensor is the most critical part of the system, if it is damaged, the engine will not start. Ignition timing is controlled by the PCM. There are no timing marks on the crankshaft balancer or timing chain cover.

REMOVAL

➡This procedure is for 5.7L VIN P engines only. The 5.7 VIN G engine does not utilize a distributor.

1. Be sure the ignition is in the **OFF** or **LOCK** position. Disconnect the negative battery cable.

2. Remove the water pump and crankshaft balancer.

➡The spark plug wire boots should be twisted ½ turn in each direction while removing. Do not pull on the wires to remove them from the spark plugs. Pull on the boots or use a tool specifically designed for this purpose.

3. Remove the spark plug wiring harnesses from the distributor, as follows:
  a. Raise and safely support the vehicle.
  b. Remove the power steering pump.
  c. Unfasten the spark plug wiring harness from the wiring harness clips.

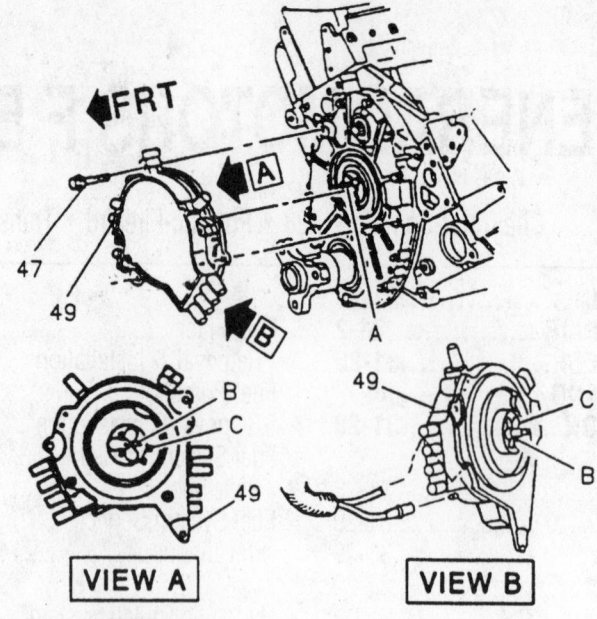

VIEW A **VIEW B**

A   CAMSHAFT PIN
B   CAMSHAFT PIN SLOT
C   DISTRIBUTOR BASE TIMING MARK
24   DISTRIBUTOR VACUUM HARNESS
47   BOLT/SCREW DISTRIBUTOR
49   DISTRIBUTOR

7922WG01

The distributor assembly is located under the water pump—5.7L (VIN P) engine

  d. Remove the spark plug wiring harness from the distributor.

4. Unplug the 4-terminal electrical connector from the distributor.

5. If equipped, disconnect the distributor vacuum harness from the distributor.

6. Loosen and remove the distributor attaching bolts.

7. Pull the distributor assembly forward until the driveshaft disengages from the end of the camshaft.

8. Mark the top surface of the driveshaft for proper alignment during installation.

To install:

➡Replace the O-rings on the coupling shaft or ignition system performance may suffer. Lubricate the O-rings and the end of the camshaft.

❊❊ WARNING

Don't try to fully seat the distributor using the distributor retainers. If the distributor will not seat by hand, it's not properly aligned with the camshaft. Rotate the crankshaft until the engine is at the number 1 cylinder TDC (camshaft sprocket pin at the 9 o'clock position). Rotate the distributor coupling until the camshaft

sprocket pin slot aligns with the distributor base timing mark. Install the distributor using hand pressure to fully seat the distributor.

9. Install the distributor assembly into position with the driveshaft in the end of the camshaft. Rotate the distributor coupling until the camshaft sprocket pin slot aligns with the camshaft sprocket pin. Slide the distributor onto the end of the camshaft until fully seated on the engine front cover.

10. Install the distributor mounting bolts. Tighten to 97–106 inch lbs. (11–12 Nm).

11. Engage all electrical connections and/or vacuum hoses on the distributor that were unplugged earlier.

12. Install the crankshaft balancer and water pump.

13. Connect the negative battery cable.

Ignition Timing

ADJUSTMENT

When checking ignition timing NEVER pierce a secondary ignition wire. Use either a timing light with an inductive type pick-up or connect an adapter between the wire and the spark plug. If the secondary wire insulation is pierced, current will eventually arc and cause engine misfiring.

➡**Some engines incorporate a magnetic timing probe hole for the use with electronic timing equipment. Be sure to consult the tool manufacture's instructions for the use of this equipment.**

On these vehicles, base timing is preset when the engine is manufactured. All timing changes are, then controlled directly by the PCM based on information from the ignition and knock sensor systems. No adjustments are necessary or possible.

Engine Assembly

REMOVAL & INSTALLATION

➡**The engine, transmission and suspension assembly is removed from the bottom of the vehicle. After the assembly is removed, separate the engine from the transmission and frame.**

1. Discharge the A/C system.

➡**Only EPA-certified, MVAC-trained technicians should service the A/C system.**

2. Disconnect the negative and positive battery cables.
3. Raise and safely support the vehicle.
4. Remove the front wheels.

✳✳ CAUTION

Never open, service or drain the radiator or cooling system when hot; serious burns can occur from the steam and hot coolant.

5. Drain the coolant and engine oil.

✳✳ CAUTION

Observe all applicable safety precautions when working around fuel. Whenever servicing the fuel system, always work in a well ventilated area.

6. Relieve the fuel system pressure.
7. Remove the exhaust pipes front exhaust pipe from the engine.
8. Remove the front fascia lower deflectors.
9. Remove the stabilizer bar.
10. Remove the drive belt.
11. If equipped, disconnect the transmission cooler lines from the radiator.

12. Remove the radiator and heater hoses.
13. Unplug all electrical connections from the engine and wheel speed sensors.
14. Disconnect the right front brake line from the caliper brake hose.
15. Unplug the wiring harness from the transmission assembly.
16. Disconnect the shift linkage from the transmission.
17. Remove the driveshaft from the vehicle after matchmarking.
18. Remove the torque arm from the transmission assembly.
19. Remove the intermediate steering shaft from the rack and pinion assembly.
20. Disconnect the ground straps from the left side frame rail.
21. Lower the vehicle.
22. Remove the air intake duct.
23. Disconnect the fuel pipes from the engine.
24. Disconnect the throttle and cruise control cables from the throttle body.
25. Unplug the radiator fan electrical connections and remove the fan.
26. Disconnect the brake booster vacuum hose.
27. If equipped, remove the "Y" brace from the right side exhaust manifold.
28. Remove the alternator and air conditioning compressor bracket, then reposition out of the way.
29. Remove the brake master cylinder and reposition it out of the way.
30. Detach both upper strut assemblies from the body.
31. Remove the right front brake line from the modulator valve assembly and clips.
32. Raise and safely support the vehicle.
33. Position a lift table under the engine and engine frame assembly.
34. Remove the engine frame and transmission support screws.
35. Raise the vehicle from the engine, transmission and engine frame assemblies.
36. Secure the strut assemblies to the engine frame.
37. Remove the transmission assembly from the engine.

To install:
38. If removed, install the engine assembly onto the frame. Tighten the through-bolt to 70 ft. lbs. (95 Nm).
39. Install the transmission assembly.
40. Install the engine, transmission, and frame assembly.
41. Install the frame and transmission support bolts.
42. Remove the engine lift table.

43. Lower the vehicle.
44. Install the bolts and nuts to the upper strut assemblies.
45. Connect the right front brake line to the modulator valve assembly and clips.
46. Install the brake master cylinder.
47. Engage the engine wiring harness connectors.
48. Install the alternator and compressor bracket.
49. If removed, install the "Y" brace to the right exhaust manifold assembly.
50. Connect the brake booster vacuum line.
51. Install the radiator fan assembly.
52. Install the lower and upper radiator hoses.
53. Connect the cruise control and accelerator cables to the throttle body.
54. Connect the engine fuel pipes.
55. Raise and suitably support the vehicle.
56. Install the intermediate steering shaft to the rack and pinion assembly.
57. Install the torque arm to the transmission assembly.
58. Install the driveshaft.
59. Connect the wiring and shift linkage to the transmission assembly.
60. Connect the right front brake line to the caliper brake hose.
61. Engage all electrical connectors that were unplugged for engine removal.
62. Connect the heater hoses to the engine.
63. If equipped, connect the transmission fluid cooler lines to the radiator.
64. Install the drive belt.
65. Install the stabilizer bar.
66. Install the front fascia lower deflectors.
67. If removed, install the catalytic converter on A/T cars.
68. Install the front exhaust pipe(s).
69. Install the wheels.
70. Lower the vehicle.

✳✳ WARNING

Operating the engine without the proper amount and type of engine oil will result in severe engine damage.

71. Refill the engine with motor oil.
72. Refill radiator with coolant.
73. Connect the negative battery cable.
74. Bleed the brake system.
75. Bleed the power steering system.
76. Install the air intake duct.
77. Wheel alignment must be performed.

Water Pump

REMOVAL & INSTALLATION

3.4L Engine

1. Disconnect the negative battery cable.

> ✳✳ **CAUTION**
>
> **Never open, service or drain the radiator or cooling system when hot; serious burns can occur from the steam and hot coolant.**

2. Drain the cooling system into a suitable container.
3. Remove the air intake duct.
4. Remove the top coil of the ignition coil pack.
5. Loosen the tensioner pulley bolt.
6. Disconnect the heater hose from the water pump.
7. Loosen the water pump pulley bolts.
8. Unfasten the power steering bracket bolts.
9. Remove the serpentine drive belt.
10. Remove the water pump pulley.
11. Remove the power steering pump bracket and swing the bracket and pump aside. Do NOT disconnect the power steering lines from the pump.
12. Unfasten the water pump and engine front cover bolts, then remove the water pump assembly from the vehicle.
13. Remove and discard the gasket and thoroughly clean the gasket mating surfaces.

To install:

14. Position a new water pump gasket, then install the pump. Secure using all of the water pump retaining bolts, except the bottom three, and tighten to 33 ft. lbs. (45 Nm).
15. Install the engine front cover bolts and tighten to 97 inch lbs. (11 Nm).
16. Install the three bottom water pump retaining screws and tighten to 89 inch lbs. (10 Nm).
17. Place the power steering pump into position and install the pump bracket.
18. Install the water pump pulley.
19. Install the serpentine drive belt.
20. Install the power steering pump bracket retaining bolts and tighten to 30 ft. lbs. (40 Nm).
21. Connect the heater hose to the water pump.
22. Install the top coil to the ignition coil pack.
23. Attach the electrical connector to the alternator.
24. Install the air inlet duct and air cleaner assembly.
25. Connect the negative battery cable.
26. Fill the cooling system. Start the engine and check for leaks.

3.8L Engine

1. Disconnect the negative battery cable.

> ✳✳ **CAUTION**
>
> **Never open, service or drain the radiator or cooling system when hot; serious burns can occur from the steam and hot coolant.**

2. Drain the cooling system into a suitable container.
3. Remove the serpentine belt.
4. Disconnect the radiator inlet hose from the water pump.
5. Unfasten the water pump pulley bolts and remove the pulley.
6. Remove the water pump pulley retaining bolts, then remove the pump from the vehicle.
7. Remove and discard the gasket and thoroughly clean the gasket mating surfaces.

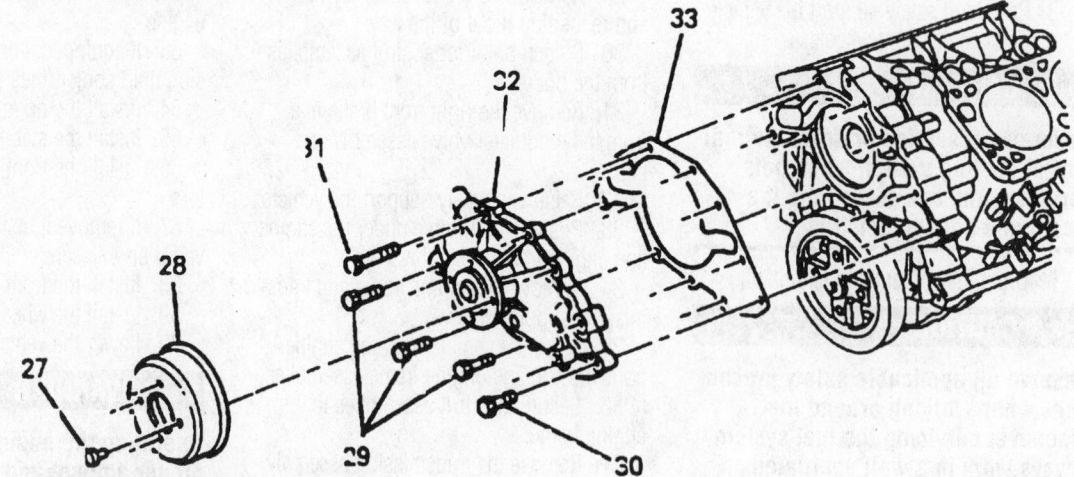

27 BOLT/SCREW, WATER PUMP PULLEY
28 PULLEY, WATER PUMP
29 BOLT/SCREW, WATER PUMP
30 BOLT/SCREW, ENGINE FRONT COVER
31 BOLT/SCREW, WATER PUMP COVER
32 PUMP ASSEMBLY, WATER
33 GASKET, WATER PUMP

7922WG02

Exploded view of the water pump mounting—3.4L engine

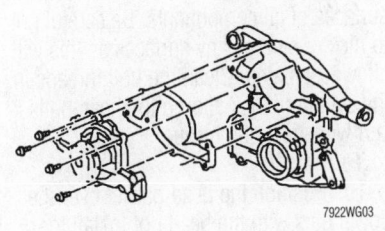

Exploded view of the water pump mounting—3.8L engine

To install:

8. Position a new water pump gasket, then install the pump. Secure using the water pump retaining bolts and tighten to 11 ft. lbs. (15 Nm), plus an additional 80 degree rotation.

9. Install the water pump pulley and retaining bolts. Tighten to 115 inch lbs. (13 Nm).

10. Install the serpentine drive belt.

11. Connect the radiator inlet hose to the water pump.

12. Connect the negative battery cable.

13. Fill the cooling system. Start the engine and check for leaks.

5.7L Engines

1. Unplug the electrical connector from the cooling fan.

2. Remove both cooling fan assemblies.

✴✴ CAUTION

Never open, service or drain the radiator or cooling system when hot; serious burns can occur from the steam and hot coolant.

3. Place a suitable drain pan under the vehicle, then remove the block coolant drain plug and the knock sensor. Drain the radiator.

4. Install the drain plug and knock sensor.

5. Remove the air intake duct and air cleaner.

6. Disconnect the upper and lower radiator hoses from the water pump.

7. Remove the heater hoses from the water pump and throttle body.

8. Unplug the electrical connector from the coolant sensor.

9. Unfasten the retainers and reposition the ignition coil and bracket.

10. If equipped, remove the air pump and bracket.

11. Unfasten the water pump attaching bolts, then remove the water pump and discard the gasket. Remove the shaft coupling and water pump driveshaft seals, if necessary.

12. Thoroughly clean the gasket mating surfaces

To install:

13. If removed, install the shaft coupling and water pump driveshaft using J 39089 or equivalent water pump driveshaft O-ring installer.

14. Install the water pump with a new gasket. Tighten the bolts to 30–32 ft. lbs. (41–43 Nm).

15. If equipped, install the air pump and bracket.

16. Reposition and secure the ignition coil and bracket.

17. Attach the coolant sensor connector.

18. Connect the heater hoses to the water pump and throttle body.

19. Connect the upper and lower radiator hoses to the water pump.

20. Install the air cleaner and air inlet duct.

21. Install the electric cooling fan(s), then attach the electrical connector.

22. Refill the cooling system.

23. Connect the negative battery cable, then start the engine and check for leaks.

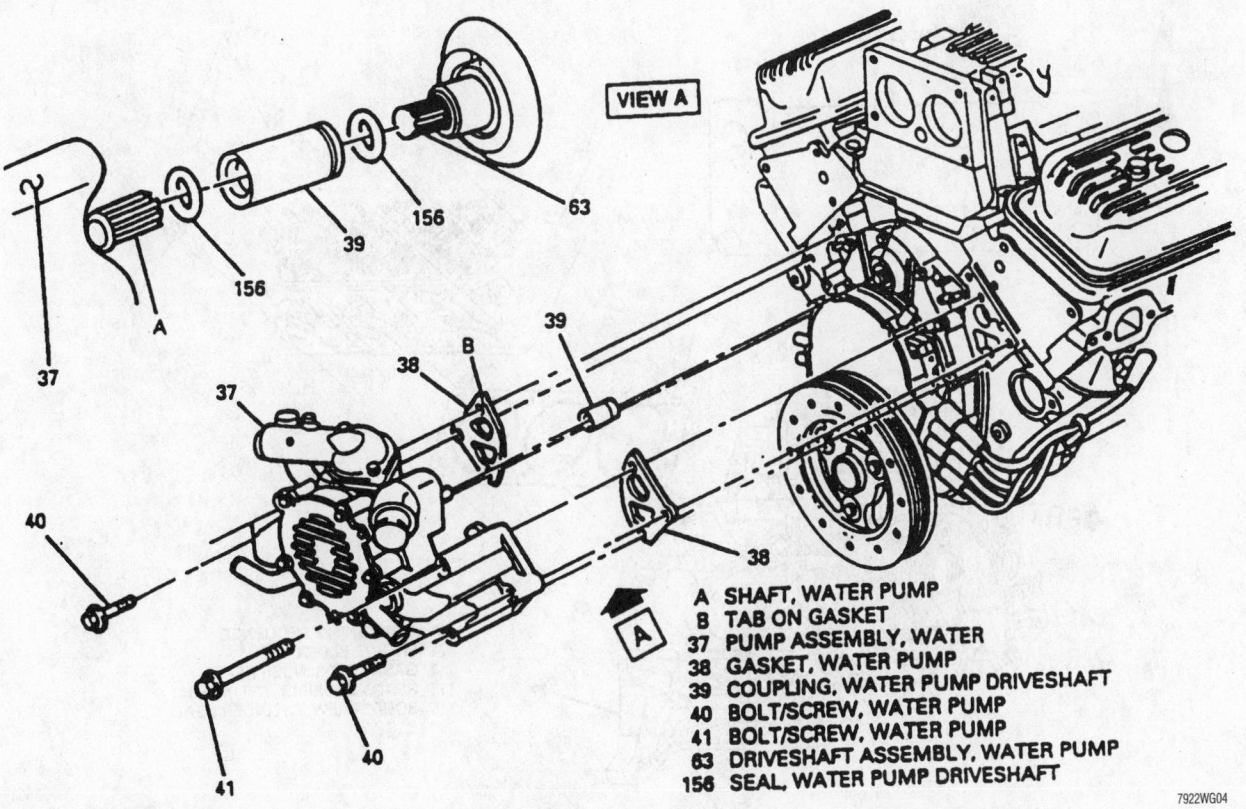

VIEW A

A SHAFT, WATER PUMP
B TAB ON GASKET
37 PUMP ASSEMBLY, WATER
38 GASKET, WATER PUMP
39 COUPLING, WATER PUMP DRIVESHAFT
40 BOLT/SCREW, WATER PUMP
41 BOLT/SCREW, WATER PUMP
63 DRIVESHAFT ASSEMBLY, WATER PUMP
156 SEAL, WATER PUMP DRIVESHAFT

Exploded view of the water pump mounting and related components—5.7L engines

Cylinder Head

REMOVAL & INSTALLATION

➡ **When servicing the engine, be absolutely sure to mark or tag vacuum hoses and wiring so that these items may be properly reconnected during installation. Also, when disconnecting fittings of metal lines (fuel, power brake vacuum), always use two flare nut (or line) wrenches. Hold the wrench on the large fitting with pressure on the wrench as if you were tightening the fitting (clockwise), THEN loosen and disconnect the smaller fitting from the larger fitting. If this is not done, damage to the line will result.**

3.4L Engine

LEFT SIDE

1. Disconnect the negative battery cable.
2. Raise and safely support the vehicle.

✳✳ CAUTION

Never open, service or drain the radiator or cooling system when hot; serious burns can occur from the steam and hot coolant.

3. Place a drain pan under the radiator, then drain the cooling system.
4. Carefully lower the vehicle.
5. Remove the upper intake manifold.
6. Remove the rocker arm (valve) cover assemblies.
7. Remove the lower intake manifold.
8. Remove the left exhaust manifold.
9. Remove the oil level dipstick and tube assembly.
10. Remove the serpentine drive belt.
11. Detach the coolant temperature sensor connector.
12. Tag and disconnect the spark plug wires from the plugs. Remove the spark plugs.
13. Remove the engine lift bracket.
14. Remove the wiring harness clip from the rear of the cylinder head.
15. Unfasten the secondary air injection pipe bracket bolt.
16. Remove the power steering pump assembly and brackets and position aside. Do NOT disconnect the power steering pump lines.
17. Loosen the rocker arms until the pushrods can be removed.
18. Unfasten the cylinder head bolts, then remove the cylinder head. Remove and discard the gasket. The cylinder head

should be properly cleaned and inspected before installation. Clean the gasket mating surfaces of all components. Be careful not to nick or scratch any surfaces as this will allow leak paths. Clean the bolt threads in the cylinder block and on the head bolts. Dirt will affect bolt torque.

To install:

19. Position the head gasket over the dowel pins with the bead up. Install the cylinder head over the dowel pins and gasket.
20. Coat the threads of the head bolts with GM 1052080 thread sealer or equivalent. Install the head bolts and tighten in sequence to 41 ft. lbs. (55 Nm). Then, tighten an additional 90 degrees (¼ turn) using tool J 36660 or equivalent torque angle meter.
21. Install the pushrods and rocker arms.
22. Install the power steering pump brackets.
23. Install the secondary air injection pipe bracket bolt.
24. Install the spark plugs, tighten to 11 ft. lbs. (15 Nm), then connect the wires as tagged during removal.
25. Attach the wiring harness clip to the rear of the cylinder head.

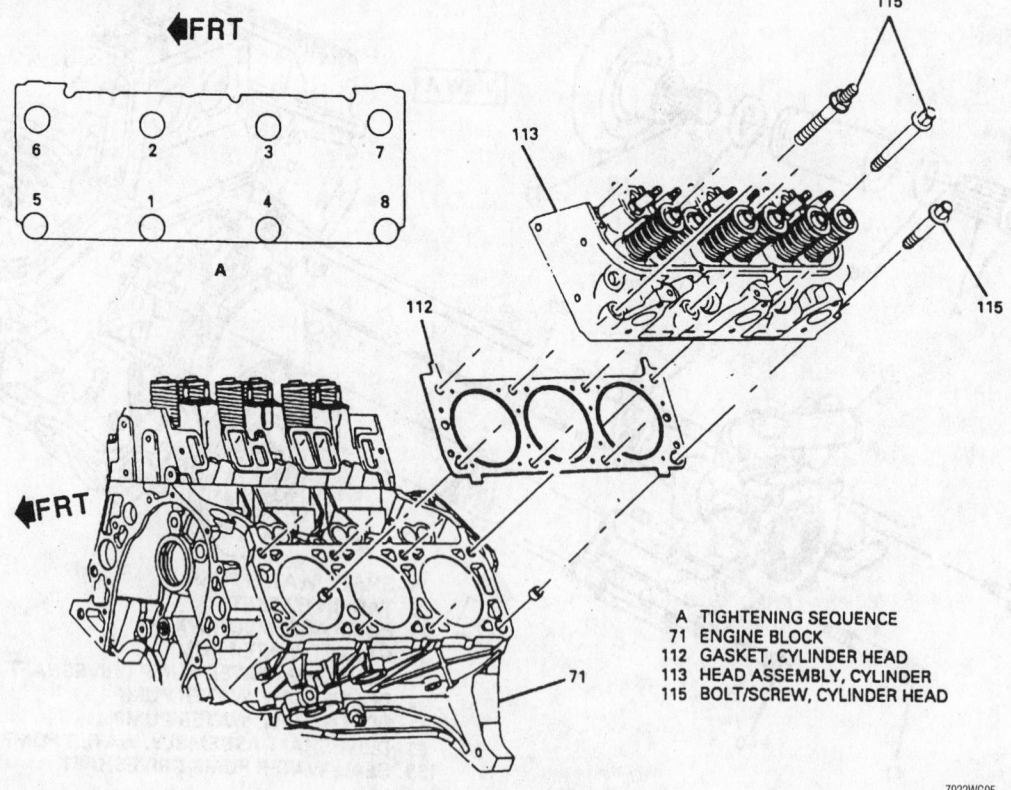

	A	TIGHTENING SEQUENCE
	71	ENGINE BLOCK
	112	GASKET, CYLINDER HEAD
	113	HEAD ASSEMBLY, CYLINDER
	115	BOLT/SCREW, CYLINDER HEAD

7922WG05

Exploded view of the cylinder head mounting showing the bolt torque sequence—3.4L engine

26. Install the engine lift bracket.

27. Attach the coolant temperature sensor connector.

28. Install the serpentine drive belt.

29. Install the oil level dipstick and tube assembly.

30. Install the left exhaust manifold.

31. Install the lower intake manifold.

32. Install the rocker arm (valve) cover.

33. Install the upper intake manifold.

34. Fill the cooling system with the proper type and amount of coolant. Connect the negative battery cable.

35. Start the engine and check for leaks.

RIGHT SIDE

1. Disconnect the negative battery cable.

2. Raise and safely support the vehicle.

> **❋❋ CAUTION**
>
> **Never open, service or drain the radiator or cooling system when hot; serious burns can occur from the steam and hot coolant.**

3. Place a drain pan under the radiator, then drain the cooling system.

4. Carefully lower the vehicle.

5. Remove the serpentine drive belt tensioner.

6. Remove the upper intake manifold.

7. Remove the rocker arm (valve) cover assemblies.

8. Remove the lower intake manifold.

9. Remove the right exhaust manifold.

10. Remove the alternator and brackets.

11. Tag and disconnect the spark plugs wires from the plugs, then remove the spark plugs.

12. Loosen the rocker arms until the pushrods can be removed.

13. Unfasten the cylinder head bolts, then remove the cylinder head. Remove and discard the gasket. The cylinder head should be properly cleaned and inspected before installation. Clean the gasket mating surfaces of all components. Be careful not to nick or scratch any surfaces as this will allow leak paths. Clean the bolt threads in the cylinder block and on the head bolts. Dirt will affect bolt torque. Refer to Engine Reconditioning for further details.

To install:

14. Position the head gasket over the dowel pins with the bead up. Install the cylinder head over the dowel pins and gasket.

15. Coat the threads of the head bolts with GM 1052080 thread sealer or equivalent. Install the head bolts hand-tight.

Tighten the bolts in sequence to 41 ft. lbs. (55 Nm). Then, tighten them an additional 90 degrees (¼ turn) using tool J 36660 or equivalent torque angle meter.

16. Install the pushrods and rocker arms.

17. Install the spark plugs, tighten to 11 ft. lbs. (15 Nm), then connect the spark plugs, as tagged during removal.

18. Install the alternator and brackets.

19. Install the right exhaust manifold assembly.

20. Install the rocker arm (valve) covers.

21. Install the upper intake manifold assembly.

22. Install the serpentine drive belt assembly.

23. Fill the cooling system with the proper type and amount of coolant. Connect the negative battery cable.

24. Start the engine and check for leaks.

3.8L Engine

LEFT SIDE

> **❋❋ CAUTION**
>
> **Observe all applicable safety precautions when working around fuel. Whenever servicing the fuel system, always work in a well ventilated area. Do not allow fuel spray or vapors to come in contact with a spark or open flame. Keep a dry chemical fire extinguisher near the work area. Always keep fuel in a container specifically designed for fuel storage; also, always properly seal fuel containers to avoid the possibility of fire or explosion.**

1. Disconnect the negative battery cable. Properly relieve the fuel system pressure.

2. Disconnect the fuel lines from the fuel rail.

> **❋❋ CAUTION**
>
> **Never open, service or drain the radiator or cooling system when hot; serious burns can occur from the steam and hot coolant.**

3. Drain the cooling system into a suitable container.

4. Remove the upper and lower intake manifolds.

5. Tag and detach the spark plug wires, then remove the spark plugs.

6. Remove the left exhaust manifold.

7. Remove the rocker arm (valve) cover and the rocker arms and pushrods.

8. Unfasten the cylinder head retaining bolts, then remove the cylinder head. Remove and discard the gasket and bolts. The cylinder head should be properly cleaned and inspected before installation. Clean the gasket mating surfaces of all components. Be careful not to nick or scratch any surfaces as this will allow leak paths. Clean the bolt threads in the cylinder block and on the head bolts. Dirt will affect bolt torque.

To install:

9. Place a new cylinder head gasket on the engine with the arrow pointing to the front of the engine.

10. Position the cylinder head on top of the gasket, and secure with NEW head bolts. Tighten the bolts as follows:

   a. Tighten, in sequence, to 37 ft. lbs. (50 Nm).

   b. Rotate the bolts 130 degrees, in sequence, using tool J 36660 or equivalent torque angle meter.

   c. Rotate the center four bolts an additional 30 degrees, in sequence, using tool J 36660.

11. Install the rocker arms and pushrods, then install the valve cover.

12. Install the left exhaust manifold.

13. Install the spark plugs, tighten to 20 ft. lbs. (27 Nm) and connect the spark plugs as tagged during removal.

14. Install the lower and upper intake manifolds.

15. Fill the cooling system with the proper type and amount of coolant. Connect the negative battery cable.

16. Start the engine and check for leaks.

RIGHT SIDE

1. Disconnect the negative battery cable.

> **❋❋ CAUTION**
>
> **Observe all applicable safety precautions when working around fuel. Whenever servicing the fuel system, always work in a well ventilated area. Do not allow fuel spray or vapors to come in contact with a spark or open flame. Keep a dry chemical fire extinguisher near the work area. Always keep fuel in a container specifically designed for fuel storage; also, always properly seal fuel containers to avoid the possibility of fire or explosion.**

2. Properly relieve the fuel system pressure.

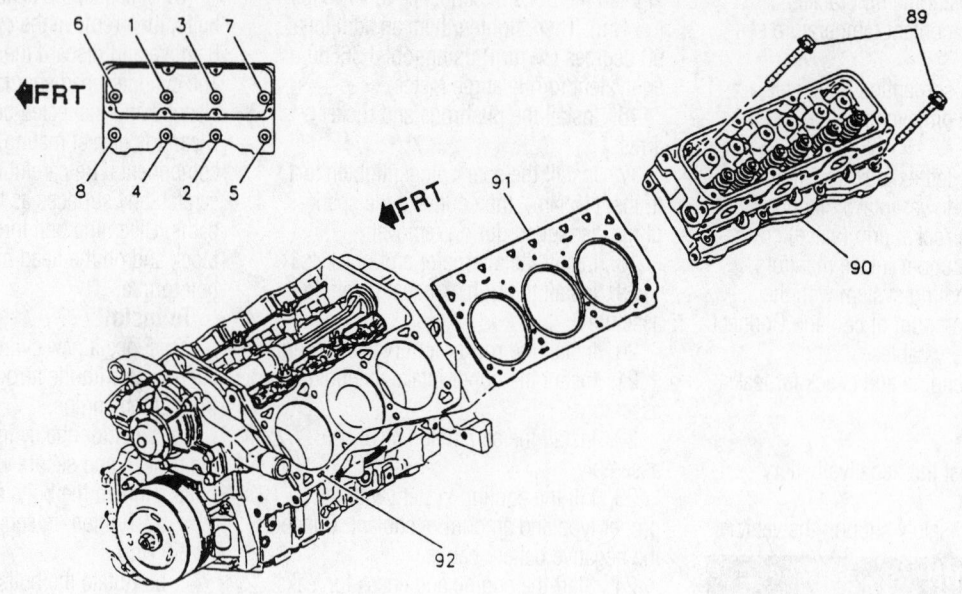

A   TIGHTENING SEQUENCE
89 BOLT/SCREW, CYLINDER HEAD
90 HEAD, CYLINDER
91 GASKET, CYLINDER HEAD
92 LOCATOR PINS

7922WG06

Exploded view of the cylinder head mounting showing the bolt torque sequence—3.8L engine

✲✲ CAUTION

Never open, service or drain the radiator or cooling system when hot; serious burns can occur from the steam and hot coolant.

3. Disconnect the fuel lines from the fuel rail.

4. Drain the cooling system into a suitable container.

5. Remove the upper and lower intake manifolds.

6. Tag and detach the spark plug wires, then remove the spark plugs.

7. Remove the right exhaust manifold.

8. Remove the rocker arm (valve) cover and the rocker arms and pushrods.

9. Remove the serpentine drive belt tensioner.

10. Remove the rear alternator brace.

11. Unfasten the cylinder head retaining bolts, then remove the cylinder head. Remove and discard the gasket and bolts. The cylinder head should be properly cleaned and inspected before installation. Clean the gasket mating surfaces of all components. Be careful not to nick or scratch any surfaces as this will allow leak paths. Clean the bolt threads in the cylinder block and on the head bolts. Dirt will affect bolt torque.

To install:

12. Place a new cylinder head gasket on the engine with the arrow pointing to the front of the engine.

13. Position the cylinder head on top of the gasket, and secure with NEW head bolts. Tighten the bolts as follows:

    a. Tighten, in sequence, to 37 ft. lbs. (50 Nm).

    b. Rotate the bolts 130 degrees, in sequence, using tool J 36660 or equivalent torque angle meter.

    c. Rotate the center four bolts an additional 30 degrees, in sequence, using tool J 36660.

14. Install the rocker arms and pushrods, then install the valve cover.

15. Install the right exhaust manifold.

16. Install the spark plugs, tighten to 20 ft. lbs. (27 Nm) and connect the spark plugs as tagged during removal.

17. Install the lower and upper intake manifolds.

18. Fill the cooling system with the proper type and amount of coolant. Connect the negative battery cable.

19. Start the engine and check for leaks.

5.7L Engines

LEFT SIDE

1. Disconnect the negative battery cable.

2. Raise and safely support the vehicle.

✲✲ CAUTION

Never open, service or drain the radiator or cooling system when hot; serious burns can occur from the steam and hot coolant.

3. Place a suitable pan under the vehicle and drain the cooling system.

4. Disconnect the exhaust crossover pipe from the exhaust manifold on 1995 vehicles. For 1996–99 vehicles, remove the catalytic converter.

5. Carefully lower the vehicle.

6. Remove the intake manifold.

7. Disconnect the secondary air injection hose from the check valve.

8. Using a back-up wrench on the pipe fitting, remove the engine coolant bleed pipe bolt from the left cylinder head.

9. Unfasten the ignition coil bolts, then remove the coil.

10. Remove the left exhaust manifold.

11. Detach the spark plug wiring harness

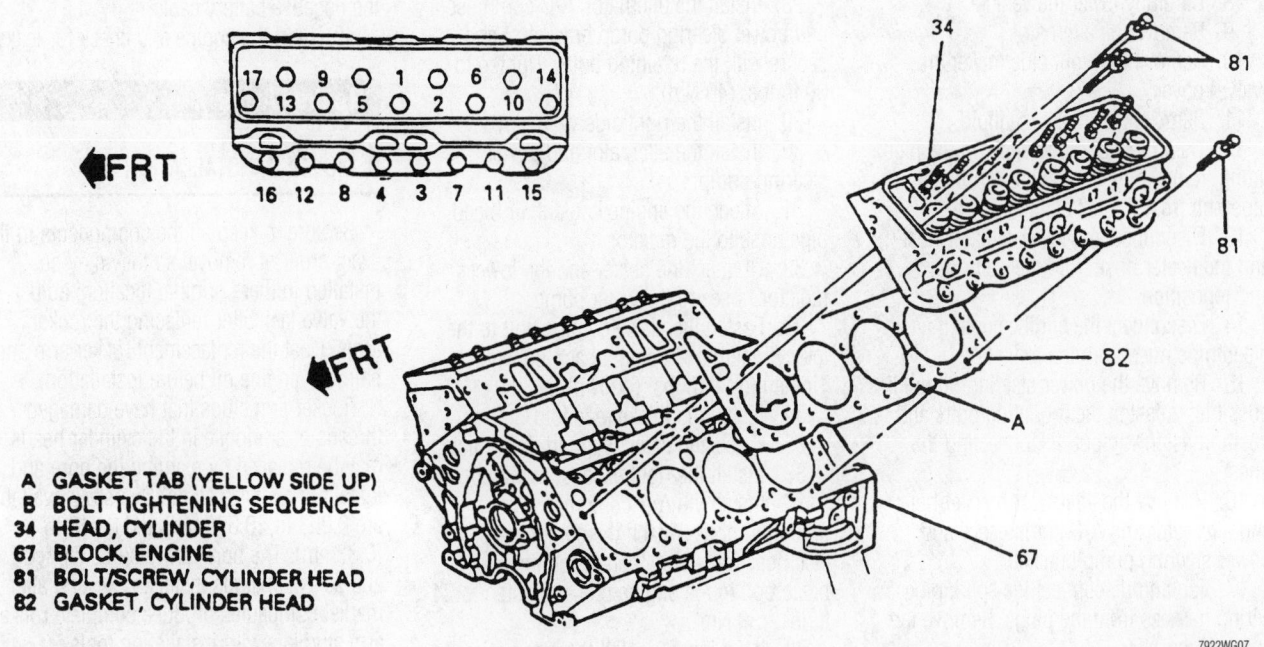

A GASKET TAB (YELLOW SIDE UP)
B BOLT TIGHTENING SEQUENCE
34 HEAD, CYLINDER
67 BLOCK, ENGINE
81 BOLT/SCREW, CYLINDER HEAD
82 GASKET, CYLINDER HEAD

Exploded view of the cylinder head mounting showing the bolt torque sequence—5.7L engine

from the clips, then tag and disconnect the spark plug wiring harness from the plugs. Remove the spark plugs.

12. Detach the coolant temperature sensor electrical connector.

13. Remove the rocker arm (valve) cover. Remove the rocker arm nuts, rocker arms and pushrods.

14. Unfasten the cylinder head bolts, then remove the cylinder head and gasket. Discard the gasket.

The cylinder head should be properly cleaned and inspected before installation. Clean the gasket mating surfaces of all components. Be careful not to nick or scratch any surfaces as this will allow leak paths. Clean the bolt threads in the cylinder block and on the head bolts. Dirt will affect bolt torque. Refer to Engine Reconditioning for further details.

To install:

15. Position a new cylinder head gasket with the yellow tab facing up. Place the cylinder head over the locator pins and head gasket.

➡**When installing the head bolts, be sure the mark the short bolts to be sure the correct tightening specifications are achieved.**

16. Coat the threads of the head bolts with GM 1052080 or equivalent, then install the bolts finger-tight.

17. Tighten the cylinder head bolts in sequence to 22 ft. lbs. (30 Nm), then tighten the long and medium length bolts an additional 80 degrees and the short bolts an additional 67 degrees.

18. Install the pushrods and rocker arms. Install the rocker arm (valve) cover. Tighten the retaining bolts to 106 inch lbs. (12 Nm).

19. Attach the coolant temperature sensor electrical connector.

20. Install the spark plugs to 15 ft. lbs. (20 Nm). Connect the spark plug wiring harness to the spark plugs. Fasten the harness to the retaining clips.

21. Install the left exhaust manifold.

22. Install the ignition coil and secure with the retaining bolts/studs. Tighten to 30 ft. lbs. (40 Nm).

23. Install the engine coolant air bleed pipe and bolt to the left cylinder head, using a back-up wrench on the pipe fitting. Tighten to 30 ft. lbs. (40 Nm).

24. Connect the secondary air injection hose to the check valve.

25. Install the intake manifold.

26. Raise and safely support the vehicle.

27. Install the catalytic converter, then carefully lower the vehicle.

28. Fill the cooling system with the proper type and amount of coolant. Connect the negative battery cable.

29. Start the engine and check for leaks.

RIGHT SIDE

1. Disconnect the negative battery cable.

2. Raise and safely support the vehicle.

❋❋ CAUTION

Never open, service or drain the radiator or cooling system when hot; serious burns can occur from the steam and hot coolant.

3. Place a suitable pan under the vehicle and drain the cooling system.

4. Remove the serpentine drive belt and belt tensioner.

5. If equipped, remove the automatic transmission fluid level indicator tube bracket from the transmission housing.

6. If equipped, remove the A/C compressor rear brace bolt from the engine block. Detach the compressor electrical connector. Remove the A/C compressor front mounting bolts, then position the

compressor aside. Do NOT disconnect the refrigerant lines.

7. Remove the right exhaust manifold.

8. Carefully lower the vehicle.

9. Remove the alternator.

10. Remove the right side rocker arm (valve) cover.

11. Remove the intake manifold.

12. Using a back-up wrench on the pipe fitting, remove the engine coolant bleed pipe bolt from the left cylinder head.

13. Disconnect the lower radiator hose and the heater hose from the water pump and reposition.

14. Disconnect the engine coolant air bleed pipe hose from the radiator.

15. Remove the power steering pump. If possible, unfasten the mounting bolts and position aside without disconnecting the lines.

16. Remove the alternator bracket bolts and alternator and A/C compressor and power steering pump bracket.

17. Tag and disconnect the spark plug wiring harness from the plugs. Remove the spark plugs.

18. Remove the rocker arms and pushrods.

19. Unfasten the cylinder head bolts, then remove the cylinder head and gasket. Discard the gasket.

20. Using a back-up wrench on the pipe fitting, remove the engine coolant air bleed pipe bolt, and the nut and fitting from the cylinder head.

To install:

21. Install the engine coolant air bleed pipe, nut and bolt, using a back-up wrench on the pipe fitting, to the cylinder head hand-tight.

22. Position a new cylinder head gasket with the yellow tab facing up. Place the cylinder head over the locator pins and head gasket.

➡**When installing the head bolts, be sure the mark the short bolts to be sure the correct tightening specifications are achieved.**

23. Coat the threads of the head bolts with GM 1052080 or equivalent, then install the bolts finger-tight.

24. Tighten the cylinder head bolts in sequence to 22 ft. lbs. (30 Nm), then tighten the long and medium length bolts an additional 80 degrees and the short bolts an additional 67 degrees.

25. Tighten the engine coolant air bleed pipe bolt to 30 ft. lbs. (40 Nm) and the nut to 13 ft. lbs. (17 Nm).

26. Install the pushrods and rocker arms.

27. Install the spark plugs, tighten to 15 ft. lbs. (20 Nm) and attach the spark plug wiring harness to the spark plugs.

28. Install the alternator, A/C compressor and power steering pump bracket(s) and secure with the retaining bolts. Tighten to 30 ft. lbs. (40 Nm).

29. Install the right side valve cover.

30. Install the alternator and power steering pump.

31. Attach the engine coolant air bleed pipe hose to the radiator.

32. Connect the heater and the lower radiator hoses to the water pump.

33. Fasten the air bleed pipe bolt to the left cylinder head using a back-up wrench. Tighten to 30 ft. lbs. (40 Nm).

34. Install the intake manifold.

35. Raise and safely support the vehicle.

36. Install the right exhaust manifold.

37. Install the A/C compressor front mounting bolts. Attach the compressor electrical connector, then fasten the rear brace bolt to the engine block. Tighten to 24 ft. lbs. (33 Nm).

38. If equipped, install the automatic transmission fluid level indicator tube bracket to the transmission housing.

39. Install the serpentine drive belt and tensioner.

40. Carefully lower the vehicle.

41. Fill the cooling system with the proper type and amount of coolant. Connect the negative battery cable.

42. Start the engine and check for leaks.

Rocker Arms

REMOVAL & INSTALLATION

Be sure to keep all the components in the exact order of removal so they may be installed in there original location; adjust the valve lash after replacing the rocker arms. Coat the replacement rocker arm and ball with engine oil before installation.

Rocker arm studs that have damaged threads or are loose in the cylinder heads may be replaced by reaming the bore and installing oversize studs. Oversizes available are 0.003 in. (0.076mm) and 0.013 in (0.33mm). The bore may also be tapped and screw-in studs installed. Several aftermarket companies produce complete rocker arm stud kits with installation tools.

➡**The 3.4L and 5.7L engines use press fit studs.**

1. Disconnect the negative battery cable.

2. Remove the rocker arm cover.

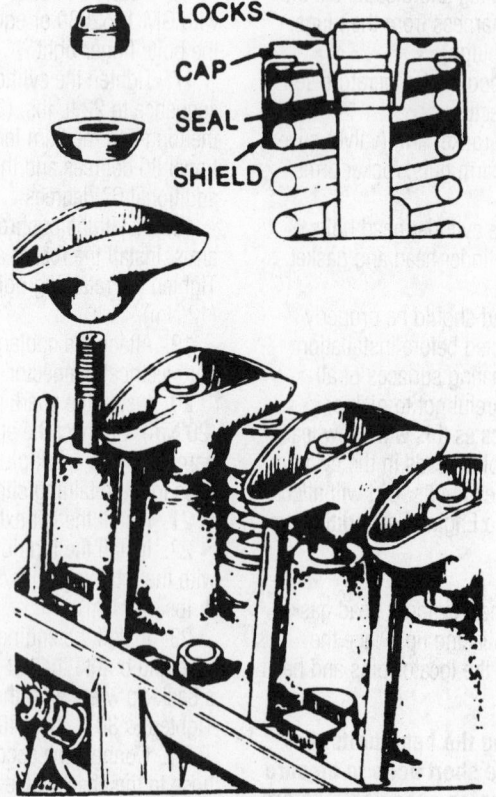

Exploded view of common rocker arm components and valve spring retainer

7922WG08

3. Remove the rocker arm nuts, balls and rocker arms. Place components in a rack so they can be reinstalled in the same location.

To install:

4. Coat the bearings surfaces with a thin coating of Molykote(r) or its equivalent.

5. Install the pushrods and be sure the rod is in the lifter seat.

6. Install the rocker arm, balls and nut. Tighten the nut until all lash is eliminated.

7. The engine must be on the No. 1 firing position before proceeding. This may be determined by placing your fingers on the No. 1 rocker arms as the mark crankshaft damper is rotated towards the "0" on the timing tab. If the arms did not move, it is in the No. 1 firing position. If they did move, turn the crankshaft one full revolution to reach the No. 1 position. Remember, the mark on the crankshaft balancer must be aligned with the "0" on the timing tab.

8. Adjust the valves as follows:

V6 Engines:

With the engine on the number 1 firing position, exhaust valves 1, 2 and 3 and intake valves 1, 5 and 6 may be adjusted. Back out the adjusting nut until lash is felt at the pushrod. Tighten the adjusting nut until all lash is removed, then tighten the nut an additional 1½ turns to center the lifter plunger. Turn the engine one revolution until the "0" timing mark is once again aligned. Exhaust valves 4, 5 and 6 and intake valves 2, 3 and 4 may be adjusted.

V8 Engines:

With the engine on the number 1 firing position, exhaust valves 1, 3, 4 and 8, intake valves 1, 2, 5 and 7 may be adjusted. Back out the adjusting nut until lash is felt at the pushrod. Tighten the adjusting nut until all lash is removed, then tighten an additional 1 turn to center the lifter plunger. Turn the engine one revolution until the 0 timing mark is once again aligned. Exhaust valves 2, 5, 6 and 7, intake valves 3, 4, 6 and 8 may be adjusted.

9. Install the rocker arm cover.

10. Connect the negative battery cable.

Intake Manifold

REMOVAL & INSTALLATION

3.4L Engine

UPPER MANIFOLD

1. Disconnect the negative battery cable.

❊❊ CAUTION

Observe all applicable safety precautions when working around fuel. Whenever servicing the fuel system, always work in a well ventilated area. Do not allow fuel spray or vapors to come in contact with a spark or open flame. Keep a dry chemical fire extinguisher near the work area. Always keep fuel in a container specifically designed for fuel storage; also, always properly seal fuel containers to avoid the possibility of fire or explosion.

2. Relieve the fuel system pressure.

❊❊ CAUTION

Never open, service or drain the radiator or cooling system when hot; serious burns can occur from the steam and hot coolant.

3. Drain the coolant into a suitable container.

4. Remove the throttle body air duct.

5. Detach the fuel rail injector wiring harness connection.

6. Remove the wiring harness retaining nut and position the harness aside.

7. Remove the accelerator cable bracket bolts, bracket and cables from the throttle body assembly.

8. Detach the fuel pipe connectors. Unfasten the fuel pipe bracket bolts. Remove the fuel pipe hold-down plate bolt, then remove the fuel pipe assembly and position out of the way.

9. Disconnect the coolant hoses from the throttle body assembly.

10. Remove the fuel rail stud.

11. Disconnect the fuel pressure regulator vacuum hose.

12. Detach the electrical connectors from the Idle Air Control (IAC), Throttle Position (TP) and Manifold Absolute Pressure (MAP) sensors and the evaporative emission canister purge solenoid valve.

13. Use compressed air to blow all dirt from the injector bores.

14. Remove the fuel rail assembly.

15. If equipped, remove the vacuum harness assembly.

16. Unfasten the EGR flexible pipe bolts and reposition the pipe and gasket.

17. Remove the upper intake manifold bolts, then remove the manifold and gasket Discard the old gaskets and clean the mating surfaces.

To install:

18. Position a new gasket, then install the manifold. Tighten the bolts to 18 ft. lbs. (25 Nm).

19. Install the EGR pipe. Use a new gasket and wrap the bolt threads with Teflon® tape. Tighten the bolts to 19 ft. lbs. (26 Nm). The gasket must be installed with the text "tube side" facing the tube.

20. Install the vacuum harness.

➡Do not press on the regulator cover or piedmont valve when installing the fuel rail.

21. Install the fuel rail. Lightly lubricate the fuel injector O-rings using clean engine oil. Install the injectors of the fuel rail into the bores of the manifold. Carefully press on the fuel rail with the palms of your hands until the injectors are fully seated.

22. Install the fuel rail stud and tighten to 18 ft. lbs. (25 Nm).

23. Engage the wiring connectors to the Idle Air Control (IAC), Throttle Position (TP) and Manifold Absolute Pressure (MAP) sensors and the evaporative emission canister purge solenoid valve.

24. Connect the fuel pressure regulator vacuum hose. Install the fuel rail wiring harness retaining nut.

25. Attach the fuel pipe assembly to the fuel rail. Slide the fuel line hold-down plate over the fuel pipe.

26. Apply Teflon® sealant to the fuel pipe hold-down plate bolt and tighten to 18 ft. lbs. (25 Nm). Install the fuel pipe bracket bolts.

27. Install the accelerator cable bracket, bolts and cables. Tighten the accelerator cable bracket bolts to 90 inch lbs. (10 Nm). Tighten the fuel pipe bracket-to-front brace bolt to 71 inch lbs. (8 Nm) and the fuel pipe bracket-to-intake manifold assembly bolt to 89 inch lbs. (10 Nm).

28. Attach the fuel pipe and fuel injector electrical connections.

29. Connect the coolant hoses to the throttle body assembly.

30. Install the throttle body air duct.

31. Fill the coolant system.

32. Connect the negative battery cable, then start the engine and check for leaks.

LOWER MANIFOLD

1. Disconnect the negative battery cable.

2. Remove the upper intake manifold.

3. Remove the valve covers.

4. Disconnect the upper radiator hose and the heater hose from the lower intake manifold.

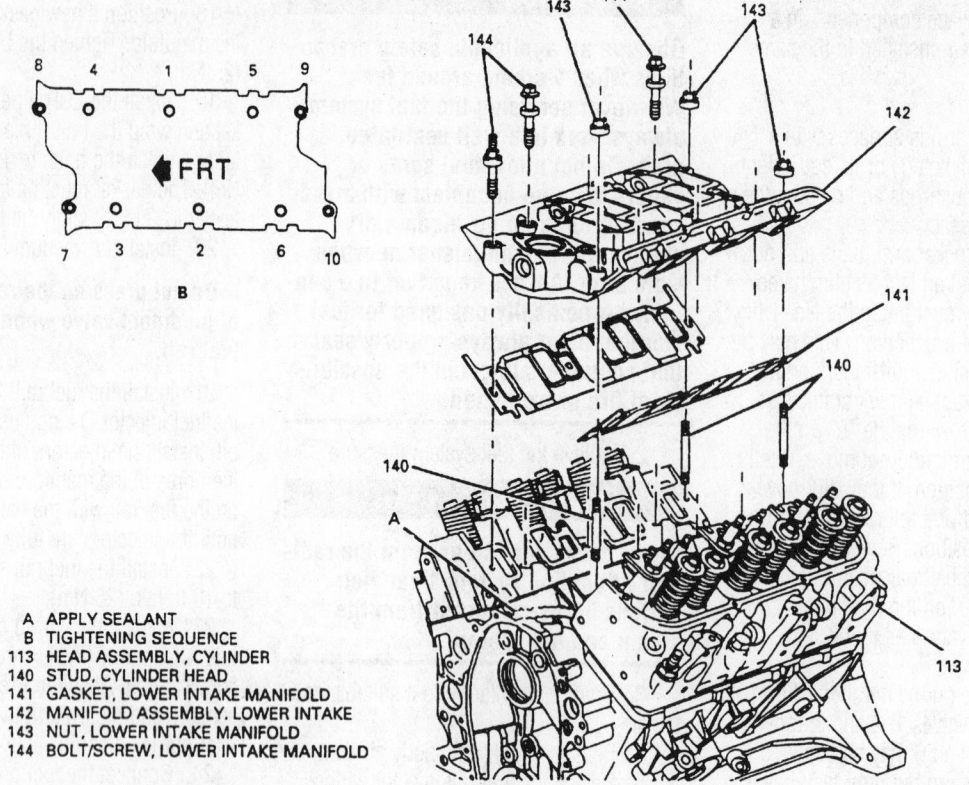

A APPLY SEALANT
B TIGHTENING SEQUENCE
113 HEAD ASSEMBLY, CYLINDER
140 STUD, CYLINDER HEAD
141 GASKET, LOWER INTAKE MANIFOLD
142 MANIFOLD ASSEMBLY, LOWER INTAKE
143 NUT, LOWER INTAKE MANIFOLD
144 BOLT/SCREW, LOWER INTAKE MANIFOLD

7922WG09

Exploded view of the lower intake manifold mounting showing the bolt torque sequence—3.4L engine

5. Detach wiring harness ground leads.

6. Unfasten the lower manifold bolts and nuts. Remove the lower manifold and gaskets. Discard the old gaskets and clean the sealing surfaces of the manifold and engine.

To install:

7. Install new lower intake manifold gaskets.

8. Apply a ³⁄₁₆ in. (5mm) bead of RTV sealant at the engine block to manifold mating surface.

9. Install the intake manifold and secure with the retaining bolts. Tighten the bolts in sequence to 22 ft. lbs. (30Nm).

10. Attach the wiring harness ground leads.

11. Connect the upper radiator hose and the heater hose.

12. Install the valve covers.

13. Install the upper intake manifold.

14. Connect the negative battery cable.

3.8L Engine

UPPER MANIFOLD

1. Disconnect the negative battery cable.

✳✳ CAUTION

Observe all applicable safety precautions when working around fuel. Whenever servicing the fuel system, always work in a well ventilated area. Do not allow fuel spray or vapors to come in contact with a spark or open flame. Keep a dry chemical fire extinguisher near the work area. Always keep fuel in a container specifically designed for fuel storage; also, always properly seal fuel containers to avoid the possibility of fire or explosion.

2. Relieve the fuel system pressure.

✳✳ CAUTION

Never open, service or drain the radiator or cooling system when hot; serious burns can occur from the steam and hot coolant.

3. Drain the coolant into a suitable container.

4. Remove the serpentine drive belt.

5. If necessary, remove the transmission fluid level indicator.

6. Remove the alternator and the serpentine drive belt tensioner.

7. Remove the Manifold Absolute Pressure (MAP) sensor and vacuum source from the upper intake manifold.

8. Remove the evaporative emission canister purge solenoid valve.

9. Remove the fuel pressure regulator and canister purge harness.

10. Remove the Idle Air Control (IAC) valve from the upper intake manifold.

11. Disconnect the fuel lines from the fuel rail.

12. Remove the accelerator control cable bracket bolts, bracket and accelerator control cables from the throttle body.

13. Remove the ignition coil pack/control module assembly.

14. Disconnect the brake booster hose from the upper manifold.

15. Remove the alternator brace from the upper manifold.

16. Detach the electrical connectors from the idle air control, throttle position, mass airflow and intake air temperature sensors.

17. Unfasten the wiring harness from the rosebud clips at the fuel rail.

18. Remove the air cleaner, outlet rear duct and resonator duct.

19. If necessary, remove the throttle body from the upper intake manifold.

20. Remove the EGR valve outlet pipe.

➡**The thermostat must be removed to get to the "hidden" bolt located under the housing. This bolt retains the upper and lower manifolds.**

21. Remove the thermostat and thermostat housing.

22. Detach the electrical connectors from the fuel injectors. Clean the fuel injector bores using compressed air.

23. Unfasten the fuel rail nuts, then remove the fuel rail.

24. Unfasten the upper intake manifold retaining bolts, then remove the upper manifold and gasket. Discard the gasket and clean the mating surfaces and bolt threads.

To install:

25. Position a new gasket, then install the upper intake manifold. Tighten the bolts as follows:

  a. Tighten the ten vertical intake manifold bolts 1–10 to 11 ft. lbs. (15 Nm) using the sequence in the accompanying figure.

  b. Tighten the two coolant outlet bolts to 20 ft. lbs. (27 Nm).

  c. Tighten the side bolts to 22 ft. lbs. (30 Nm).

25. Install the fuel rail:

  a. Lightly lubricate the fuel injector O-rings using clean engine oil.

  b. Install the injectors into the bores of the manifold.

  c. Carefully press the fuel rail with the palms your hands until the injectors are properly seated.

26. Install the fuel rail nuts and tighten to 89 inch lbs. (10 Nm). Attach the fuel injector electrical connectors.

27. Attach install the thermostat and thermostat housing.

28. Install the EGR valve outlet pipe.

29. If necessary, install the throttle body to the intake manifold.

30. Install the air cleaner, resonator duct and air cleaner outlet duct.

31. Fasten the wiring harness to the rosebud clips at the fuel rail.

32. Attach the electrical connectors to the idle air control, throttle position, mass airflow and intake air temperature sensors.

33. Install the alternator brace.

34. Connect the brake booster hose to the upper intake manifold.

35. Install the ignition control module.

36. Install the accelerator cable bracket, bolts and accelerator control cables to the throttle body. Tighten the bracket bolts to 89 inch lbs. (10 Nm).

37. Connect the fuel lines to the fuel rail.

38. Install the IAC valve to the upper intake manifold.

39. Install the fuel pressure regulator and the canister purge harness. Install the evaporative emissions canister purge solenoid valve.

40. Install the MAP sensor and vacuum source to the upper intake manifold.

41. Install the serpentine drive belt tensioner.

42. Install the alternator.

43. If necessary, install the transmission level indicator.

44. Install the serpentine drive belt.

45. Fill the coolant system.

46. Connect the negative battery cable, then start the engine and check for leaks.

LOWER MANIFOLD

1. Disconnect the negative battery cable.

➡**There are two bolts which are hidden under the upper manifold. They are located in the front and left rear corners of the lower manifold. Its necessary to remove the upper manifold to service the lower manifold.**

2. Remove the upper intake manifold.

3. Remove the rocker arm (valve) covers.

✳✳ CAUTION

Never open, service or drain the radiator or cooling system when hot; serious burns can occur from the steam and hot coolant.

4. Remove the Engine Coolant Temperature (ECT) sensor.

5. Unfasten the lower intake manifold bolts, then remove the manifold, gaskets and seals. Discard the gaskets and seals.

6. Thoroughly clean the gasket mating surfaces.

To install:

7. Install new gaskets and seals. Apply GM part no. 12345336 or equivalent to the front and rear manifold seals. Coat the bolt threads with GM part no. 12345493, or equivalent.

8. Position the lower manifold and install the retaining bolts. The two lower intake manifold bolts are located under the upper intake manifold (right front and left rear corners) and must be installed before upper manifold installation. Tighten the manifold bolts, in sequence, to 11 ft. lbs. (15 Nm).

9. Install the ECT sensor.

10. Install the valve covers and the upper intake manifold.

11. Connect the negative battery cable.

5.7L Engines

1. Disconnect the negative battery cable.

✳✳ CAUTION

Observe all applicable safety precautions when working around fuel. Whenever servicing the fuel system, always work in a well ventilated area. Do not allow fuel spray or vapors to come in contact with a spark or open flame. Keep a dry chemical fire extinguisher near the work area. Always keep fuel in a container specifically designed for fuel storage; also, always properly seal fuel containers to avoid the possibility of fire or explosion.

2. Relieve the fuel system pressure.

3. Remove the air duct.

✳✳ CAUTION

Never open, service or drain the radiator or cooling system when hot; serious burns can occur from the steam and hot coolant.

4. Drain the coolant system.

5. Unplug the connectors from the fuel injectors. Lay the wiring harnesses aside.

6. Disconnect the control cables from the throttle body.

7. Remove the secondary air injection diverter valve hoses.

8. Detach the fuel pipe connectors from the fuel rail. Remove the fuel rail bolts.

9. Detach the fuel pressure regulator vacuum tube.

10. Remove the fuel rail from the intake manifold and position aside.

11. Disconnect the vacuum and crankcase vent hoses.

12. Remove the EGR valve control valve relay nut and control valve relay.

13. Remove the emission canister purge solenoid bracket nut and solenoid.

14. Remove the EGR nuts, then remove the valve. Remove the EGR valve pipe nuts and studs.

15. Remove the secondary air injection pipe fitting, nuts and pipe from the intake and right exhaust manifold.

16. Remove the alternator rear brace.

17. Unplug the hoses and electrical connections from the throttle body. Remove the throttle body and gasket.

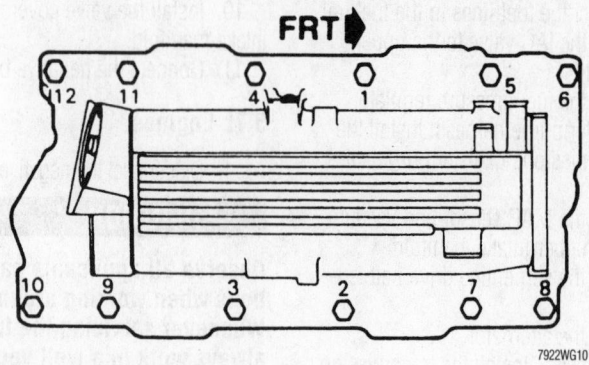

Intake manifold bolt tightening sequence for 5.7L engines

18. Remove the manifold bolt and studs. Remove the intake manifold and gaskets. Discard the gaskets and thoroughly clean the gasket mating surfaces, bolt and stud threads.

To install:

19. Be sure the sealing surfaces are clean.

20. Apply a 3/16 in. (5mm) bead of RTV sealant on the front and rear ridge of the cylinder block.

21. Install the new gaskets on the cylinder heads. Hold the gaskets in place by extending the RTV bead up onto the gasket ends.

22. Install the intake manifold along with the intake manifold bolts. Tighten bolts in sequence to 71 inch lbs. (8 Nm) on the first pass and 35 ft. lbs. (48 Nm) on the second pass.

23. Install the throttle body, tightening the throttle body bolts to 18 ft. lbs. (26 Nm). Engage all cable, hose and wiring connections on the throttle body.

24. Install the alternator rear brace.

25. If equipped, install the accelerator control/cruise control servo cable adjuster. Install the accelerator cable bracket and tightening the retaining screws to 90 inch lbs. (10 Nm). Connect the cables.

26. Install the secondary air injection pipe and nuts. Tighten the exhaust manifold fitting to 41 ft. lbs. (55 Nm).

27. Install the EGR valve pipe studs and nuts. Install the EGR valve and nuts and the control valve relay and nuts. Tighten as follows:

  a. EGR valve pipe studs to 53 inch lbs. (6 Nm).

  b. EGR valve pipe nuts, valve nuts and control valve relay nut to 18 ft. lbs. (25 Nm).

28. Install the emission canister purge solenoid and bracket bolt. Tighten to 53 inch lbs. (6 Nm).

29. Connect the vacuum and crankcase vent hoses.

30. Install the fuel rail, fuel pressure regulator vacuum tube, and fuel rail bolts. Tighten to 15 ft. lbs. (20 Nm). Attach the fuel pipe connectors to the fuel rail.

31. Connect the secondary air injection diverter valve hoses.

32. Install the left and right wiring harnesses and engage the wiring harness connectors.

33. Install the air duct. Connect the negative battery cable. Fill the cooling system.

Exhaust Manifold

REMOVAL & INSTALLATION

3.4L Engine

LEFT SIDE

1. Disconnect the negative battery cable.

2. Raise and safely support the vehicle.

3. Remove the exhaust crossover pipe.

4. Carefully lower the vehicle.

5. Detach the oxygen sensor electrical connector.

6. On vehicles equipped with manual transmission, remove the secondary air injection check valve pipe bracket nut and reposition the pipe.

7. Unfasten the exhaust manifold bolts, then remove the manifold and heat shields.

8. Remove the manifold gasket, inspect for damage and replace if necessary. Clean the exhaust manifold and cylinder head mating surfaces.

To install:

9. Position the exhaust manifold gasket, then install the exhaust manifold and heat shields and secure with the retaining bolts. Tighten the bolts to 18 ft. lbs. (25 Nm).

10. Attach the oxygen sensor electrical connector.

11. If equipped with a manual transmis-sion, install the secondary air injection check valve pipe bracket nut and tighten to 18 ft. lbs. (25 Nm).

12. Raise and safely support the vehicle.

13. Install the exhaust crossover pipe, then carefully lower the vehicle.

14. Connect the negative battery cable.

RIGHT SIDE

1. Disconnect the negative battery cable.

2. Raise and safely support the vehicle.

3. Remove the exhaust crossover pipe.

4. Remove the serpentine drive belt.

5. Detach the oxygen sensor electrical connector.

6. Remove the alternator.

7. Unfasten the A/C compressor mounting bolts and position the compressor aside. Do NOT disconnect the refrigerant lines.

8. Remove the alternator rear "Y" brace.

9. Remove the EGR valve, adapter and flexible pipe.

10. If equipped with a manual transmission, remove the secondary air injection pipe assembly from the exhaust manifold.

11. Remove the exhaust manifold front and middle bolts/screw, then remove the manifold and heat shields.

12. Remove the manifold gasket, inspect for damage and replace if necessary. Clean the exhaust manifold and cylinder head mating surfaces.

To install:

13. Position the exhaust manifold gasket, then install the exhaust manifold and heat shields and secure with the front and middle retaining bolts. Tighten the bolts to 18 ft. lbs. (25 Nm).

14. Install the EGR valve adapter and flexible pipe.

15. Install the alternator rear "Y" brace and the alternator.

16. Install the A/C compressor.

17. Attach the oxygen sensor electrical connector.

18. Install the serpentine drive belt.

19. Raise and safely support the vehicle.

20. If equipped, install the transmission filler tube.

21. Install the A/C compressor rear bracket bolts.

22. Install the exhaust crossover pipe and the exhaust manifold rear bolts. Tighten to 18 ft. lbs. (25 Nm).

23. Carefully lower the vehicle.

24. Connect the negative battery cable. Refill the transmission with the proper type and amount of fluid.

3.8L Engine

LEFT SIDE

1. Disconnect the negative battery cable.
2. Remove the EGR valve adapter from the exhaust manifold.
3. Remove the oil level dipstick and tube.
4. Unfasten the retaining nuts, then remove the exhaust manifold heat shields.
5. Raise and safely support the vehicle.
6. Remove the exhaust catalytic converter pipe.
7. Detach the oxygen sensor electrical connector.
8. Tag and disconnect the spark plug wires from the plugs.
9. Carefully lower the vehicle.
10. Unfasten the bolt and the studs, the remove the exhaust manifold and gasket from the vehicle.
11. Inspect the manifold gasket for damage and replace if necessary. Clean the exhaust manifold and cylinder head mating surfaces.

To install:

12. Position the exhaust manifold gasket, then install the exhaust manifold and heat shields and secure with the studs. Tighten the studs to 106 inch lbs. (12 Nm).
13. Install the retaining bolt and tighten to 106 inch lbs. (12 Nm).
14. Raise and safely support the vehicle.
15. Attach the oxygen sensor electrical connector.
16. Install the exhaust catalytic converter pipe.
17. Connect the wires to the spark plugs, as tagged during removal.
18. Carefully lower the vehicle.
19. Install the exhaust manifold heat shields and tighten the retaining nuts to 80 inch lbs. (9 Nm).
20. Install the oil level dipstick and tube.
21. Install the EGR valve adapter to the exhaust manifold. Tighten the retaining bolt to 18 ft. lbs. (25 Nm).
22. Connect the negative battery cable.

RIGHT SIDE

1. Disconnect the negative battery cable.
2. Unfasten the retaining nuts, then remove the exhaust manifold heat shields.
3. Raise and safely support the vehicle.
4. Remove the exhaust catalytic converter pipe.
5. Detach the oxygen sensor electrical connector.
6. Tag and disconnect the spark plug wires from the plugs.
7. Carefully lower the vehicle.
8. Unfasten the bolt and the studs, then remove the exhaust manifold and gasket from the vehicle.
9. Inspect the manifold gasket for damage and replace if necessary. Clean the exhaust manifold and cylinder head mating surfaces.

To install:

10. Position the exhaust manifold gasket, then install the exhaust manifold and heat shields and secure with the studs and bolt. Tighten to 106 inch lbs. (12 Nm).
11. Raise and safely support the vehicle.
12. Attach the oxygen sensor electrical connector.
13. Install the exhaust catalytic converter pipe.
14. Connect the wires to the spark plugs, as tagged during removal.
15. Carefully lower the vehicle.
16. Install the exhaust manifold heat shields and tighten the retaining nuts to 80 inch lbs. (9 Nm).
17. Connect the negative battery cable.

5.7L Engines

LEFT SIDE

1. Disconnect the negative battery cable.
2. Raise and safely support the vehicle.
3. Remove the catalytic converter.
4. Disconnect the brake booster vacuum hose.
5. Remove the secondary air injection pipe fitting from the exhaust manifold.
6. If equipped, remove the accelerator control/cruise control servo cable adjuster.
7. Detach the oxygen sensor electrical connector.
8. Unfasten the exhaust manifold retaining bolts, then remove the manifold from the vehicle.
9. Remove and inspect the manifold gasket for damage and replace if necessary. Clean the exhaust manifold and cylinder head mating surfaces.

To install:

10. Position the exhaust manifold gasket and manifold. Secure with the retaining bolts and tighten to 30 ft. lbs. (40 Nm).
11. Attach the oxygen sensor electrical connector.
12. If equipped, install the accelerator control/cruise control servo cable adjuster.
13. Fasten the secondary air injection pipe fitting from the exhaust manifold. Tighten the fitting to 41 ft. lbs. (55 Nm).
14. Connect the brake booster vacuum hose.
15. Raise and safely support the vehicle.
16. Install the catalytic converter, then carefully lower the vehicle.
17. Connect the negative battery cable.

RIGHT SIDE

1. Disconnect the negative battery cable.
2. Raise and safely support the vehicle.
3. Remove the catalytic converter.
4. Remove the serpentine drive belt.
5. Remove the oil level dipstick and tube assembly.
6. Detach the oxygen sensor electrical connector.
7. Remove the three rear exhaust manifold bolts.
8. Carefully lower the vehicle.
9. Remove the secondary air injection pipe fitting from the exhaust manifold.
10. Remove the EGR valve pipe.
11. Remove the alternator and rear lower brace.
12. Remove the remaining exhaust manifold mounting studs and bolt.
13. Remove the exhaust manifold and gasket from the vehicle.
14. Inspect the manifold gasket for damage and replace if necessary. Clean the exhaust manifold and cylinder head mating surfaces.

To install:

15. Position the exhaust manifold gasket, then install the exhaust manifold. Secure with the front three studs and bolt screw.
16. Install the alternator and the rear lower brace.
17. Raise and safely support the vehicle.
18. Install the remaining exhaust manifold bolts. Tighten to 30 ft. lbs. (40 Nm).
19. Attach the oxygen sensor electrical connector.
20. Install the oil level dipstick and tube assembly.
21. Install the serpentine drive belt.
22. Install the catalytic converter.
23. Carefully lower the vehicle.
24. Install the EGR valve pipe.
25. Fasten the secondary air injection pipe fitting to 41 ft. lbs. (55 Nm).
26. Connect the negative battery cable.

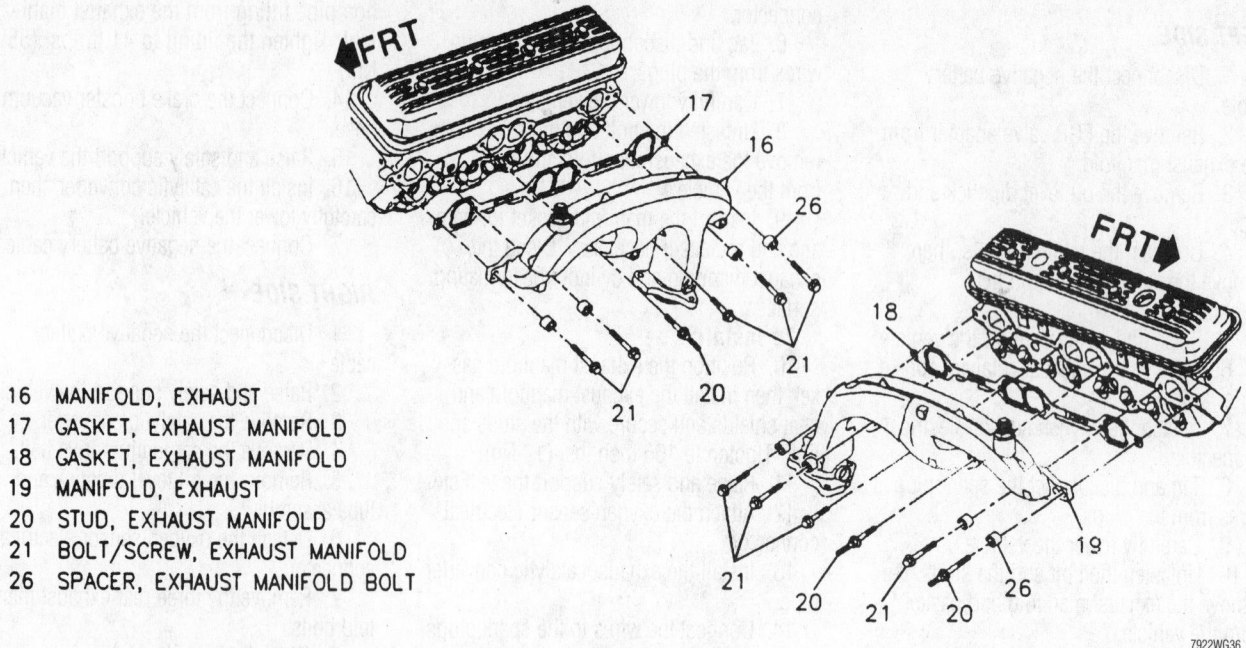

16 MANIFOLD, EXHAUST
17 GASKET, EXHAUST MANIFOLD
18 GASKET, EXHAUST MANIFOLD
19 MANIFOLD, EXHAUST
20 STUD, EXHAUST MANIFOLD
21 BOLT/SCREW, EXHAUST MANIFOLD
26 SPACER, EXHAUST MANIFOLD BOLT

7922WG36

Exploded view of the right and left exhaust manifolds—5.7L engine

Camshaft and Valve Lifters

REMOVAL & INSTALLATION

3.4L and 5.7L Engines

1. Discharge the A/C system.

➡**Only EPA-certified, MVAC-trained technicians should service the A/C system.**

2. Disconnect the negative battery cable.

3. Remove the upper intake manifold assembly.

4. Remove the rocker arms and pushrod assemblies.

5. Remove the lower intake manifold assembly.

6. Unfasten the bolt, clamp and oil pump drive assembly.

7. Remove the timing chain (front) cover assembly.

8. Remove the camshaft sprocket.

9. Remove the valve lifters.

10. Disconnect the A/C compressor and condenser hose from the condenser.

11. Disconnect the A/C receiver and dehydrator hose from the condenser.

12. Remove the radiator with the condenser. Remove the A/C condenser support assembly.

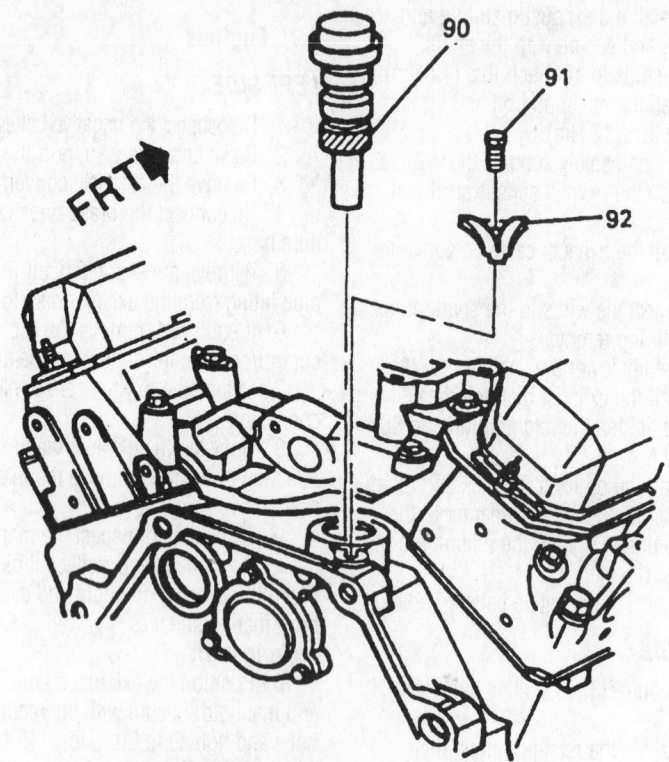

90 DRIVE ASSEMBLY, OIL PUMP
91 BOLT/SCREW, OIL PUMP DRIVE CLAMP
92 CLAMP, OIL PUMP DRIVE

7922WG12

Unfasten the retaining bolt and clamp, then remove the oil pump drive

➥**All camshaft journals are the same diameter so care must be used when removing the camshaft to avoid damaging the bearings.**

13. Remove the camshaft assembly, as follows:

   a. Install three ⁵⁄₁₆–18 x 4-inch bolts in the camshaft bolt holes.

   b. Carefully rotate and pull the camshaft assembly out of the bearings.

To install:

14. When installing a new camshaft assembly, coat the camshaft lobes with "Molykote" or equivalent. When installing a new camshaft, replace all valve lifters to ensure durability of the camshaft lobes and lifters.

15. Lubricate all camshaft journals with engine oil, then install the camshaft assembly.

16. Install the A/C condenser support assembly.

17. Install the radiator with the A/C condenser.

18. Connect the A/C receiver and dehydrator hose to the condenser. Attach the A/C compressor and condenser hose to the condenser.

19. Install the valve lifters.

20. Install the camshaft sprocket.

21. Install the timing chain (front) cover.

22. Lubricate the oil pump drive gear with prelube 1052365 and coat the assembly with engine oil, then install the oil pump drive assembly, clamp and bolt. Tighten the bolt to 25 ft. lbs. (34 Nm).

23. Install the lower intake manifold.

24. Install the pushrods and rocker arms.

25. Install the upper intake manifold.

26. Connect the negative battery cable.

27. Have the A/C system recharged by an EPA certified technician.

3.8L Engine

1. Discharge the A/C system.

➥**Only EPA-certified, MVAC-trained technicians should service the A/C system.**

2. Disconnect the negative battery cable.

3. Remove the radiator with the A/C condenser assembly.

4. Remove the valve lifters.

5. Remove the timing chain (engine) front cover.

6. Remove the camshaft sprocket and timing chain.

7. Remove the camshaft thrust plate bolts, then remove the thrust plate.

8. Remove the camshaft assembly, as follows:

   a. Install three ⁵⁄₁₆–18 x 4-inch bolts in the camshaft bolt holes.

   b. Carefully rotate and pull the camshaft assembly out of the bearings.

9. Inspect the camshaft for damage and replace if necessary.

To install:

10. Coat the camshaft with prelube 12345501 or equivalent, then install the camshaft.

11. Install the thrust plate and secure with the retaining bolts. Tighten to 11 ft. lbs. (15 Nm).

12. Install the timing chain (front) cover.

13. Install the valve lifters.

14. Install the radiator with the A/C condenser.

15. Connect the negative battery cable.

16. Evacuate and charge the A/C system.

➥**Only EPA-certified, MVAC-trained technicians should service the A/C system.**

Valve Lash

ADJUSTMENT

➥**The engines in this section utilize hydraulic valve lifter which do not require periodic adjustment, however, if the rocker arms have been removed, use the following procedure for initial adjustment.**

1. Install the pushrods and be sure the rod is in the lifter seat.

2. Install the rocker arm, balls and nut. Tighten the nut until all lash is eliminated.

3. The engine must be on the No. 1 firing position before proceeding. This may be determined by placing your fingers on the No. 1 rocker arms as the mark crankshaft damper is rotated towards the "0" on the timing tab. If the arms did not move, it is in the No. 1 firing position. If they did move, turn the crankshaft one full revolution to reach the No. 1 position. Remember, the mark on the crankshaft balancer must be aligned with the "0" on the timing tab.

4. Adjust the valves as follows:

V6 Engines:

With the engine on the number 1 firing position, exhaust valves 1, 2 and 3 and intake valves 1, 5 and 6 may be adjusted. Back out the adjusting nut until lash is felt at the pushrod. Tighten the adjusting nut until all lash is removed, then tighten the nut an additional 1½ turns to center the lifter plunger. Turn the engine one revolution

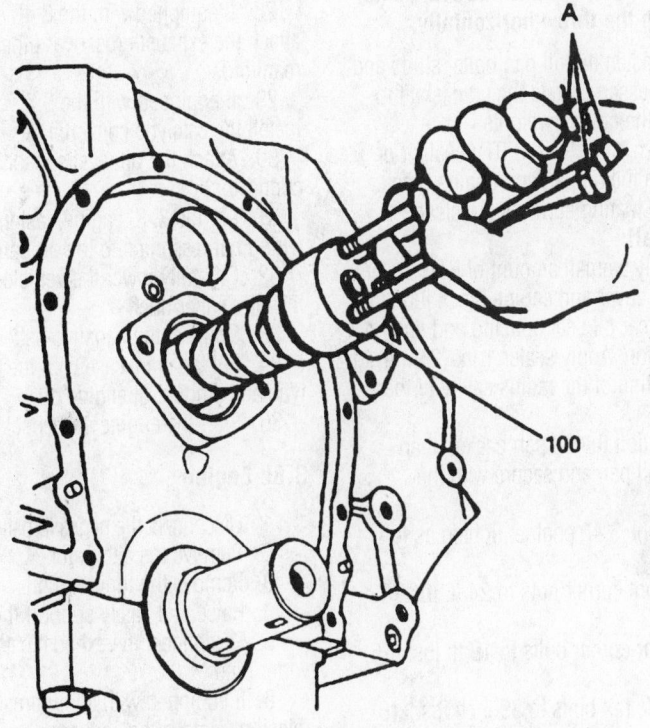

A   **BOLTS**
100   **CAMSHAFT ASSEMBLY**

7922WG11

Carefully pull the camshaft out, using the bolts threaded into the camshaft

until the "0" timing mark is once again aligned. Exhaust valves 4, 5 and 6 and intake valves 2, 3 and 4 may be adjusted.

V8 Engines:

With the engine on the number 1 firing position, exhaust valves 1, 3, 4 and 8, intake valves 1, 2, 5 and 7 may be adjusted. Back out the adjusting nut until lash is felt at the pushrod. Tighten the adjusting nut until all lash is removed, then tighten an additional 1 turn to center the lifter plunger. Turn the engine one revolution until the 0 timing mark is once again aligned. Exhaust valves 2, 5, 6 and 7, intake valves 3, 4, 6 and 8 may be adjusted.

    5. Install the rocker arm cover.

    6. Connect the negative battery cable.

Oil Pan

REMOVAL & INSTALLATION

✴✴ CAUTION

The EPA warns that prolonged contact with used engine oil may cause a number of skin disorders, including cancer! You should make every effort to minimize your exposure to used engine oil. Protective gloves should be worn when changing the oil. Wash your hands and any other exposed skin areas as soon as possible after exposure to used engine oil. Soap and water, or waterless hand cleaner should be used.

3.4L and 5.7L Engines

    1. Disconnect the negative battery cable. Remove the air intake duct.

    2. Raise the vehicle and support it safely.

    3. Drain the engine oil into a suitable container.

    4. For 3.4L engines, drain the cooling system.

    5. For 3.4L engines, remove the wiring harness clips from the left side of the oil pan.

    6. Detach the oil level sensor connector, then remove the sensor.

✴✴ CAUTION

Be sure that the exhaust system is cool before proceeding.

    7. For the 3.4L engine, disconnect the exhaust crossover pipes from the exhaust manifold.

    8. For the 5.7L engine, remove the catalytic converter.

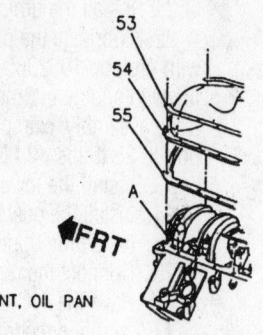
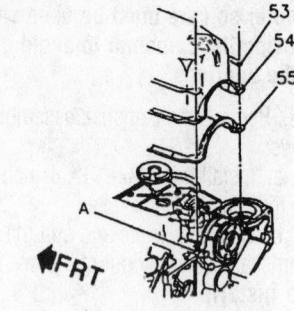

A   SEALER
53 REINFORCEMENT, OIL PAN
54 PAN, OIL
55 GASKET, OIL PAN

7922WG13

Oil pan sealer and gasket location—5.7L engines

    9. Remove the exhaust pipe hanger bolt and reposition the pipe.

    10. If equipped with an automatic transmission, detach the cooler lines from the clip at the oil pan.

    11. Remove the starter motor.

    12. If equipped with an automatic transmission, unfasten the converter cover bolts, then remove the cover.

    13. Remove the engine mount through-bolts.

    14. Use a suitable jacking fixture to raise the engine enough to provide sufficient clearance for oil pan removal.

➡ **If the front crankshaft throw prohibits removal of the pan, turn the crankshaft to position the throw horizontally.**

    15. Unfasten the oil pan bolts, studs and nuts. If necessary, rotate the crankshaft to reposition the counterweights.

    16. Clean all of the old RTV sealant or gasket from the oil pan and engine block. Be sure the mating surfaces are clean.

To install:

    17. Apply a small amount of RTV sealer to the front cover and engine block junction and to the rear oil seal housing and engine block junction. Apply sealer 1 in. (25mm) in either direction of the radius cavity of these junctions.

    18. Position the oil pan gasket, then install the oil pan and secure with the retainers.

    19. For the 3.4L engine, tighten as follows:

      a. Front corner nuts to 24 ft. lbs. (33 Nm).

      b. Rear corner bolts to 18 ft. lbs. (25 Nm).

      c. Side rail bolts to 89 inch lbs. (10 Nm).

      d. Oil pan studs to 53 inch lbs. (6 Nm).

    20. For the 5.7L engine tighten as follows:

      a. Corner oil and bolts or stud and nuts to 15 ft. lbs. (20 Nm).

      b. Remainder of oil pan bolts and studs to 106 inch lbs. (12 Nm).

    21. Carefully lower the engine.

    22. Install the engine mount through-bolts. Tighten to 70 ft. lbs. (95 Nm).

    23. Install the oil level sensor. Tighten to 16 ft. lbs. (22 Nm).

    24. If equipped, install the converter cover and secure with the retaining bolts. Tighten to 89 inch lbs. (10 Nm).

    25. Install the starter motor.

    26. If equipped, fasten the transmission fluid cooler lines to the clip at the oil pan.

    27. Install the exhaust pipe hanger bolt.

    28. If equipped with the 3.4L engine, attach the exhaust crossover pipes to the manifolds.

    29. If equipped with the 5.7L engine, install the catalytic converter.

    30. Attach the oil level sensor electrical connector.

    31. For the 3.4L engine, fasten the wiring harness clips to the oil pan.

    32. Carefully lower the vehicle. Install the air cleaner duct.

    33. Connect the negative battery cable.

    34. Fill the crankcase with the proper type and amount of engine oil.

    35. Start the engine and check for leaks.

3.8L Engine

    1. Disconnect the negative battery cable.

    2. Remove the alternator.

    3. Remove the ignition coil.

    4. Raise and safely support the vehicle.

    5. Disconnect the exhaust catalytic converter pipe.

    6. If equipped with an automatic transmission, remove the converter cover.

    7. Unfasten the engine mount through-bolts and nuts.

    8. Raise the engine using tools J 28467 and J 41044.

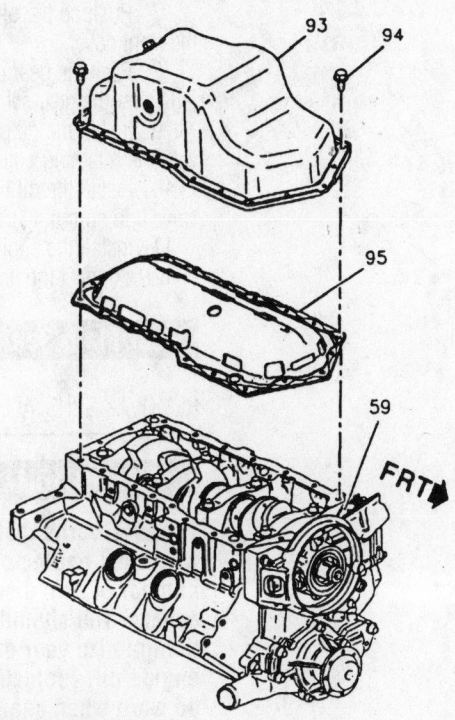

59 COVER, FRONT
93 PAN, OIL
94 BOLT/SCREW, OIL PAN
95 GASKET, OIL PAN

7922WG14

Exploded view of the oil pan and gasket—3.8L engine

➡The oil level switch, located in the oil pan, must be removed before removing the pan. If the pan is removed first, the switch may be damaged.

9. Detach the oil level indicator switch electrical connector, then remove the switch.

10. Unfasten the oil pan retaining bolts, then remove the oil pan and gasket. Lower the rear of the oil pan while rotating outward.

11. Clean all of the old RTV sealant or gasket from the oil pan and engine block. Be sure the mating surfaces are clean.

To install:

12. Position the oil pan gasket, then install the oil pan. Tighten the retainers to 125 inch lbs. (14 Nm).

13. Install the oil level switch and attach the switch electrical connector.

14. Carefully lower the engine. Install the engine mount through-bolts.

15. If equipped, install the automatic transmission converter cover.

16. Connect the exhaust catalytic converter pipe.

17. Carefully lower the vehicle.

18. Install the ignition coil.

19. Install the alternator.

20. Connect the negative battery cable.

21. Fill the crankcase with the proper type and amount of engine oil.

22. Start the engine and check for leaks.

Oil Pump

REMOVAL & INSTALLATION

3.4L and 5.7L Engines

1. Drain and remove the oil pan.

✳✳ CAUTION

The EPA warns that prolonged contact with used engine oil may cause a number of skin disorders, including cancer! You should make every effort to minimize your exposure to used engine oil. Protective gloves should be worn when changing the oil. Wash your hands and any other exposed skin areas as soon as possible after exposure to used engine oil. Soap and water, or waterless hand cleaner should be used.

2. For the 5.7L engine, unfasten the oil pan deflector/baffle nuts.

3. Remove the oil pump-to-rear main

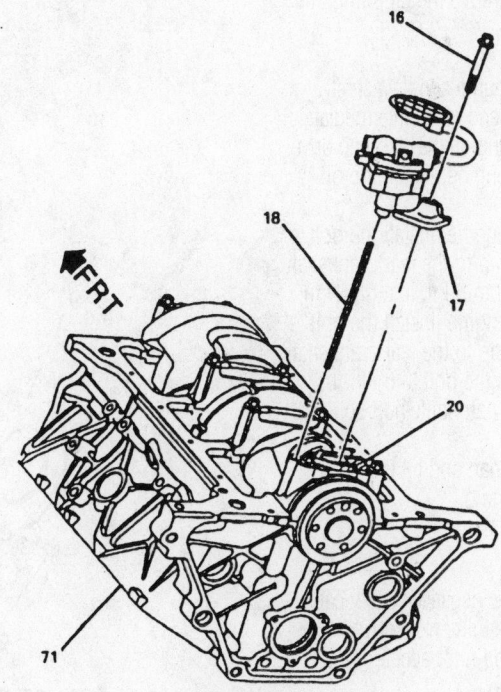

16 BOLT/SCREW, OIL PUMP
17 PUMP ASSEMBLY, OIL
18 DRIVESHAFT, OIL PUMP
20 CAP, CRANKSHAFT BEARING
71 ENGINE BLOCK

7922WG15

Exploded view of the oil pump and driveshaft mounting—3.4L engine

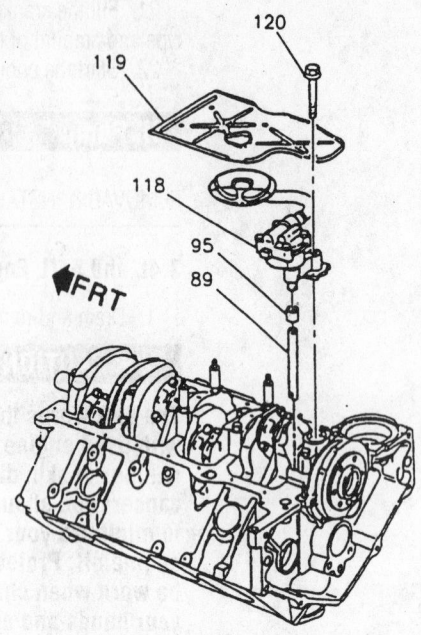

89 DRIVESHAFT, OIL PUMP
95 RETAINER, OIL PUMP DRIVESHAFT
118 PUMP, OIL
119 DEFLECTOR, CRANKSHAFT OIL
120 BOLT/SCREW, OIL PUMP

7922WG17

Exploded view of the oil pump and driveshaft mounting—5.7L engine

bearing cap bolt. Remove the oil pump and the driveshaft.

To install

4. Install the oil pump and driveshaft. Align the slot on the end of the intermediate shaft with the drive tang on the oil pump drive.

5. For the 5.7L engine, install the oil pan deflector/baffle.

6. For the 3.4L engine, install the bolt attaching the oil pump to the rear crankshaft bearing cap. Tighten to 30 ft. lbs. (41 Nm).

7. For the 5.7L engine, install the bolt attaching the oil pump to the rear crankshaft bearing cap. Tighten the bolt to 66 ft. lbs. (90 Nm) and the oil pan baffle nuts to 30 ft. lbs. (40 Nm).

8. Install the oil pan and fill the crankcase with engine oil.

3.8L Engine

1. Disconnect the negative battery cable.
2. Remove the engine front cover.
3. Remove the oil filter adapter, pressure valve and spring.
4. Remove the oil pump cover attaching bolts and cover.
5. Remove the oil pump gear set.

To install:

6. Lubricate the gear set with petroleum jelly.

7. Position the oil pump gear set into the front cover.

8. Pack the gear cavity with petroleum jelly after the gear set has been installed.

9. Install the oil pump cover and secure with the retaining bolts.

10. Install the oil filter adapter, pressure valve and spring.

11. Install the engine front cover.

12. Connect the negative battery cable.

Rear Main Seal

REMOVAL & INSTALLATION

✳✳ CAUTION

The EPA warns that prolonged contact with used engine oil may cause a number of skin disorders, including cancer! You should make every effort to minimize your exposure to used engine oil. Protective gloves should be worn when changing the oil. Wash your hands and any other exposed skin areas as soon as possible after exposure to used engine oil. Soap and water, or waterless hand cleaner should be used.

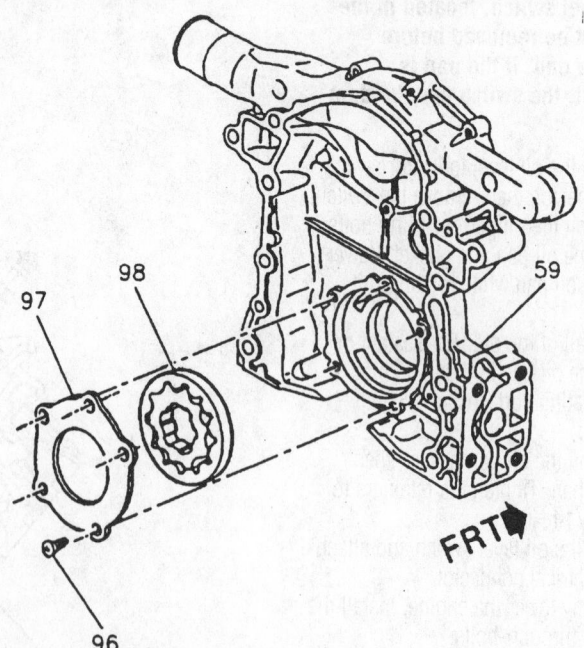

59 COVER, FRONT
96 BOLT/SCREW, OIL PUMP COVER
97 COVER, OIL PUMP
98 GEAR SET, OIL PUMP

7922WG16

Exploded view of the oil pump gear set—3.8L engine

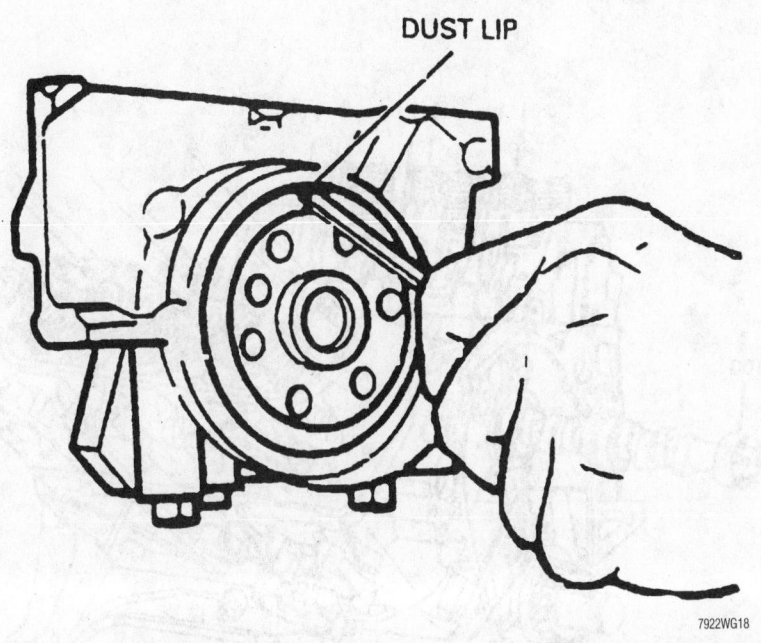

DUST LIP

7922WG18

When prying the rear main seal out, be careful not to damage the crankshaft sealing surface or the seal bore

➡ **The rear main seal is a one piece unit. It can be removed or installed without removing the oil pan or crankshaft.**

1. Raise and safely support the vehicle with jackstands.
2. Remove the transmission.
3. If equipped with a manual transmission, remove the clutch and pressure plate.
4. Remove the flywheel assembly.
5. Using a suitable prytool, carefully pry the old seal out.
6. Inspect the crankshaft for nicks or burrs, correct as required.

To install:

7. Clean the area and coat the seal with engine oil. Install the seal onto tool J-34686 for 3.4L engines, J 41349 for 3.8L engines or J 35621 for 5.7L engines.
8. Install the tool (with the seal mounted to it) onto the rear of the crankshaft. Tighten the screws snugly to be sure the seal will be installed squarely over the crankshaft.
9. Tighten the wing nut on the installation tool until it bottoms out.
10. Remove the tool from the crankshaft or rear oil seal housing, as applicable.
11. Install the flywheel.
12. If equipped with a manual transmission, install the clutch and pressure plate.
13. Install the transmission.

14. Check the fluid levels, start the engine and check for leaks.

Timing Chain, Sprockets, Front Cover and Seal

REMOVAL & INSTALLATION

3.4L Engine

1. Disconnect the negative battery cable.
2. Remove the throttle body air intake duct.
3. Remove the serpentine drive belt.

❊❊ **CAUTION**

Never open, service or drain the radiator or cooling system when hot; serious burns can occur from the steam and hot coolant.

4. Remove the water pump assembly.
5. Remove the crankshaft balancer bolt and washer, then remove the pulley bolts and the pulley. Using J-24420-B, or an equivalent damper remover, pull the balancer from the end of the crankshaft assembly.
6. Remove the power steering pump and bracket assembly.
7. Remove the oil pan assembly.
8. Disconnect the lower radiator hose from the front cover assembly.
9. Remove the crankshaft sensor.

10. Remove the front cover retaining bolts, then remove the cover and gasket from the engine.
11. Remove the crankshaft seal from the front cover using a suitable seal driver.
12. Rotate the crankshaft until the timing marks punched on the crankshaft and camshaft sprockets are aligned to the marks on the engine block or timing chain dampener. This is the No. 4 piston at Top Dead Center (TDC).
13. Remove the camshaft sprocket bolts.
14. Remove the camshaft sprocket and timing chain assembly.
15. If necessary, remove the crankshaft sprocket using tool J 23444-A, or equivalent gear/sprocket remover.
16. If necessary, unfasten the timing chain dampener retaining bolts, then remove the dampener. If necessary remove the crankshaft sprocket key.
17. Clean the chain and sprockets. Inspect all components for damage, and replace as necessary.

To install:

18. If removed, install the crankshaft sprocket key.
19. Install the crankshaft sprocket using tool J 38612 or equivalent installer.
20. Install the camshaft sprocket assembly and timing chain assembly. Be sure the timing marks on the camshaft and crankshaft sprockets are aligned with the marks on the engine block.
21. Install the camshaft sprocket bolts and tighten to 18 ft. lbs. (24 Nm).
22. Install the timing chain dampener and tighten the retaining bolts to 15 ft. lbs. (20 Nm).
23. Thoroughly clean the mating surfaces of any remaining gasket material.
24. Install a new crankshaft seal in the front cover.
25. Coat both sides of the gasket lower tabs with sealer, then install the front cover with the gasket.
26. Install the front cover retaining bolts and tighten to 15 ft. lbs. (21 Nm).
27. Install the oil pan assembly.
28. Install the crankshaft sensor.
29. Connect the lower radiator hose to the front cover.

➡ **If a new balancer is installed, new balancer weights of the same size must be installed in the same hole locations as the original balancer.**

30. Using J-29113, or an equivalent installer tool, pull the balancer onto the end of the crankshaft. Install the crankshaft pulley and the retaining bolts, then install crankshaft damper bolt and washer. Tighten

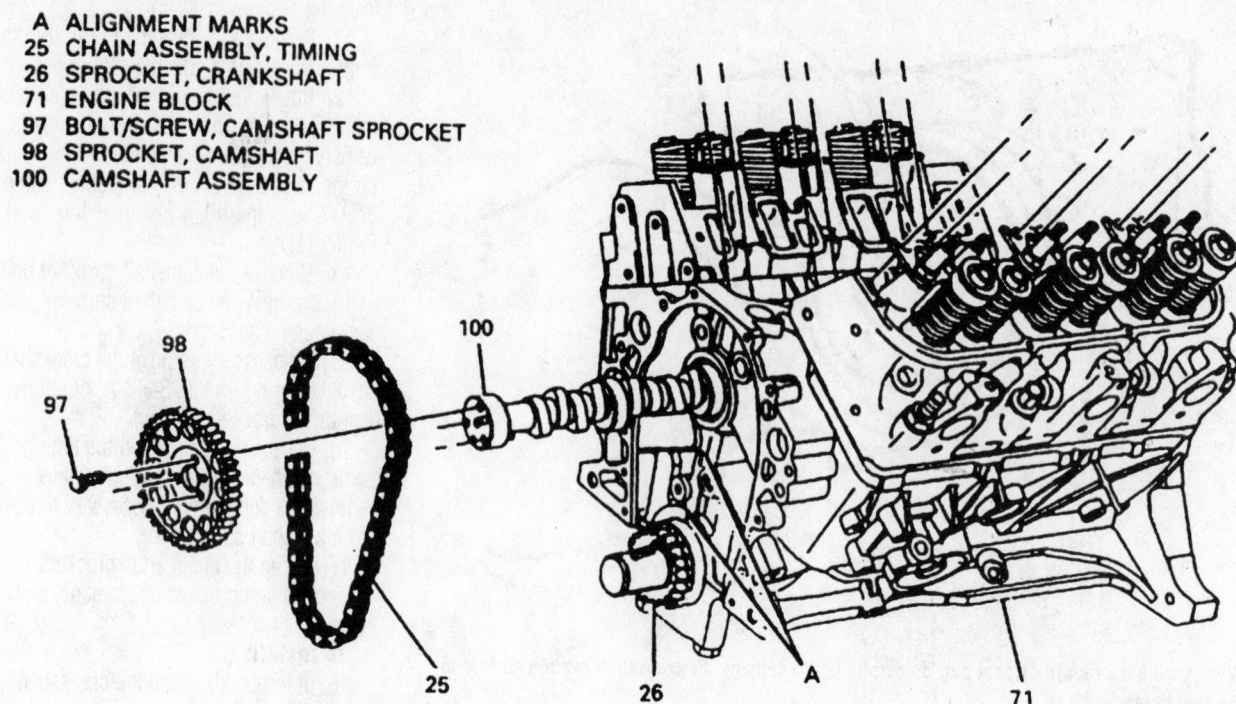

A ALIGNMENT MARKS
25 CHAIN ASSEMBLY, TIMING
26 SPROCKET, CRANKSHAFT
71 ENGINE BLOCK
97 BOLT/SCREW, CAMSHAFT SPROCKET
98 SPROCKET, CAMSHAFT
100 CAMSHAFT ASSEMBLY

7922WG19

Exploded view of the timing chain and camshaft sprocket (note the alignment marks)—3.4L engine

the damper bolt to 58 ft. lbs. (78 Nm), then tighten the pulley retaining bolts to 37 ft. lbs. (50 Nm).

31. Lower the vehicle.

32. Install the water pump assembly.

33. Install the power steering pump and bracket assembly.

34. Install the serpentine drive belt, then install the throttle body air duct.

35. Connect the negative battery cable, then refill the engine cooling system.

3.8L Engine

1. Disconnect the negative battery cable.

✳✳ CAUTION

Never open, service or drain the radiator or cooling system when hot; serious burns can occur from the steam and hot coolant.

2. Remove the air cleaner and intake air duct.

3. Remove the radiator hose from the front cover.

4. Raise and safely support the vehicle.

5. Remove the crankshaft balancer.

6. Remove the oil pan-to-front cover bolts and lower the vehicle.

7. Remove the crankshaft sensor shield.

8. Remove the water pump.

9. Remove the timing chain (front) cover.

10. Remove the crankshaft seal from the cover using a suitable seal driver.

11. Align the timing marks on the sprockets so they are as close together as possible, as shown in the accompanying figure.

12. Remove the timing chain dampener.

13. Unfasten the camshaft sprocket retaining bolt, then pull the timing chain and sprocket from the camshaft.

14. If necessary, remove the crankshaft

sprocket. If the sprocket does not come off easily, a light blow on the edge of the sprocket with a plastic mallet should dislodge it.

15. Clean the chain and sprockets. Inspect all components for damage, and replace as necessary.

To install:

16. If the pistons have been moved in the engine, perform the following:

a. Turn the crankshaft so the No. 1 piston is at Top Dead Center (TDC).

b. Turn the camshaft so that, with the sprocket temporarily installed, the timing mark is straight down.

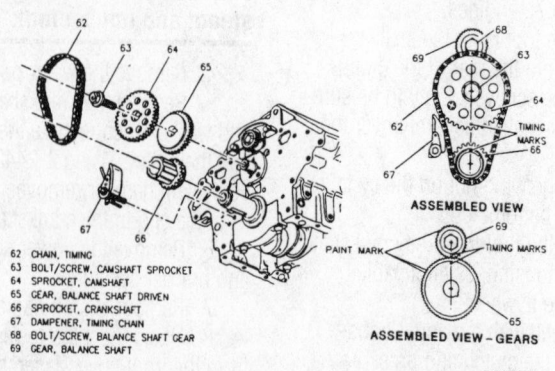

62 CHAIN, TIMING
63 BOLT/SCREW, CAMSHAFT SPROCKET
64 SPROCKET, CAMSHAFT
65 GEAR, BALANCE SHAFT DRIVEN
66 SPROCKET, CRANKSHAFT
67 DAMPENER, TIMING CHAIN
68 BOLT/SCREW, BALANCE SHAFT GEAR
69 GEAR, BALANCE SHAFT

ASSEMBLED VIEW

TIMING MARKS

PAINT MARK

ASSEMBLED VIEW – GEARS

7922WG20

Exploded view of the timing chain and sprockets and timing mark alignment—3.8L engine

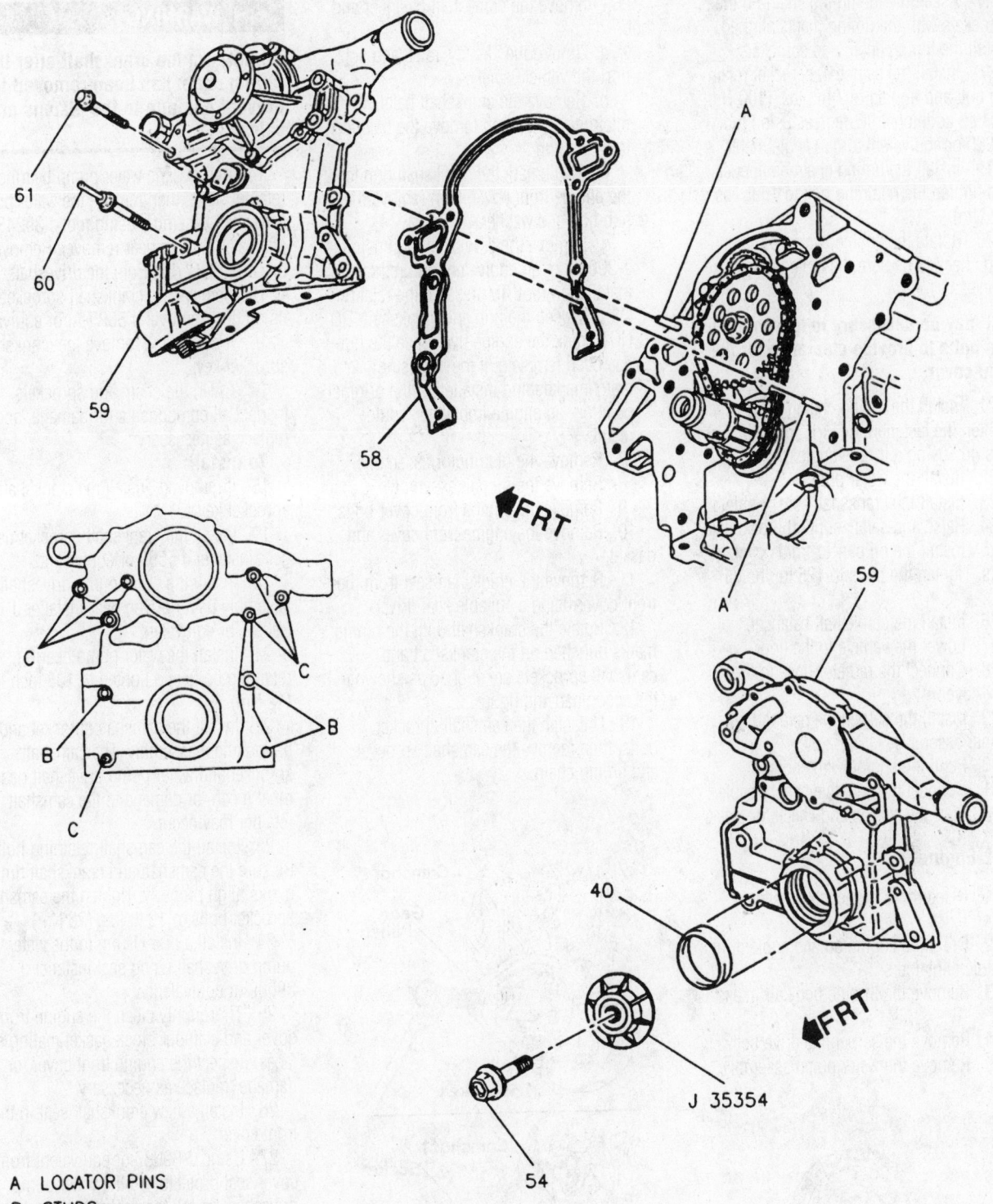

A LOCATOR PINS
B STUDS
C BOLTS/SCREWS
40 SEAL, CRANKSHAFT FRONT OIL
54 BOLT/SCREW, CRANKSHAFT BALANCER
58 GASKET, FRONT COVER
59 COVER, FRONT
60 STUD, FRONT COVER
61 BOLT/SCREW, FRONT COVER

7922WG37

Exploded view of the front cover mounting and bolt locations—3.8L engine.

17. Assemble the timing chain on the sprockets with the timing marks aligned. Install the timing chain and sprockets.

18. Install the camshaft sprocket retaining bolt and tighten to 74 ft. lbs. (100 Nm) plus an additional 90 degrees using tool J 36660 or equivalent torque angle meter.

19. Install the timing chain dampener and tighten the retaining bolt to 16 ft. lbs. (22 Nm).

20. Rotate the engine two revolutions, then check to be sure the timing marks are aligned.

➡ **It may be necessary to loosen the oil pan bolts to provide clearance for the front cover.**

21. Install the timing chain (front) cover. Tighten the fasteners to 11 ft. lbs. (15 Nm) plus 40° using a torque angle meter.

22. Install the water pump.

23. Install the crankshaft sensor shield.

24. Raise and safely support the vehicle.

25. Install the oil pan-to-front cover bolts. Tighten the bolts to 125 in. lbs. (14 Nm).

26. Install the crankshaft balancer.

27. Lower the vehicle to the floor.

28. Connect the radiator hose to the front cover.

29. Install the serpentine belt and air cleaner assembly.

30. Refill the engine with coolant.

31. Connect the negative battery cable.

32. Start the engine and check for leaks.

5.7L Engine

1. Disconnect the negative battery cable.

2. Drain the engine oil and coolant into suitable containers.

3. Remove the throttle body air intake duct.

4. Remove the serpentine drive belt.

5. Remove the water pump assembly.

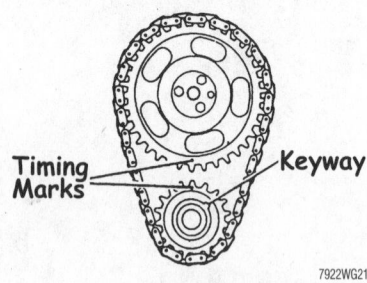

When assembled, be sure the marks on the gears are facing each other—5.7L engine

6. Remove the crankshaft balancer and hub.

a. If not done already, raise and support the vehicle safely.

b. Remove the crankshaft balancer retaining bolts, then remove the balancer from the hub.

c. Matchmark the crankshaft hub to the engine front cover, then remove the hub bolt and washer.

d. Remove the crankshaft hub using J-39046, or an equivalent hub removal/installation tool. To preserve the relationship between the hub and crankshaft, DO NOT crank the engine over once the hub has been removed. If the hub is not matchmarked and installed in the original position, an engine imbalance could result.

7. Remove the distributor assembly.

8. Remove the oil pan assembly.

9. Remove the engine front cover bolts.

10. Remove the engine front cover and gasket.

11. Remove the crankshaft seal from the front cover using a suitable seal driver.

12. Rotate the crankshaft until the timing marks punched on the crankshaft and camshaft sprockets are aligned as shown in the accompanying figure.

13. Unfasten the camshaft sprocket bolts, then remove the camshaft sprocket and timing chain.

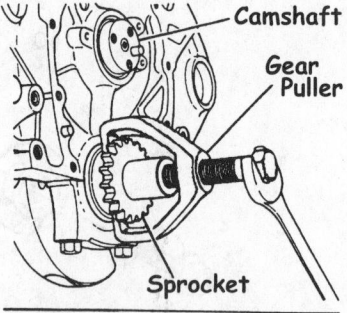

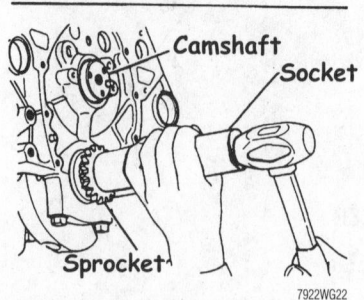

Use a gear puller to remove, and a large socket to install, the crankshaft sprocket—5.7L engine shown

Do not turn the crankshaft after the timing chain has been removed to prevent damage to the pistons or valves.

14. Remove the water pump bearing retainer bolts, then remove the water pump driveshaft assembly using tool J 39243 or equivalent driven gear remover. Remove and discard the O-ring from the driveshaft.

15. Remove the crankshaft sprocket using gear remover J 5825-A, or equivalent.

16. If necessary, remove the crankshaft sprocket key.

17. Clean the chain and sprockets. Inspect all components for damage, and replace as necessary.

To install:

18. If removed, install the crankshaft sprocket key.

19. Install the crankshaft sprocket using gear installer J 5590 or equivalent.

20. Install the water pump driveshaft assembly using driven gear installer J 39092, or equivalent.

21. Install the water pump bearing retaining bolts and tighten to 105 inch lbs. (12 Nm).

22. Install the camshaft sprocket and timing chain assembly. The camshaft sprocket and water pump driveshaft gears must mesh, or damage to the camshaft retainer may occur.

23. Install the camshaft retaining bolts. Be sure the camshaft and crankshaft timing marks align properly. Tighten the camshaft sprocket bolts to 18 ft. lbs. (25 Nm).

24. Install a new O-ring to the water pump driveshaft using seal installer J 39089 or equivalent.

25. Thoroughly clean the engine front cover and cylinder block gasket mating surfaces. Inspect the engine front cover for damage, replace as necessary.

26. Install a new crankshaft seal in the front cover.

27. Using J-39087 or equivalent front cover seal protector on the water pump driveshaft, install the gasket and front cover into position over the shafts and guide pins.

28. Install the engine front cover bolts and tighten to 100 inch lbs. (11 Nm).

29. Install the oil pan and gasket.

30. Install the distributor assembly.

31. Install the crankshaft hub and torsional damper assembly.

a. Align the matchmarks made earlier and install the crankshaft hub using the hub tool. If the engine was cranked and the matchmarks were lost, set the engine to No. 1 TDC, then install the crankshaft hub with the cast arrow in the 12 o'clock position.

b. Install the hub washer and bolt, but do not torque at this time.

c. Install the crankshaft balancer to the hub, then tighten the crankshaft hub bolt to 75 ft. lbs. (102 Nm) and the balancer bolts to 60 ft. lbs. (81 Nm).

➡ **If a new balancer is installed, new balancer weights of the same size must be installed in the same hole locations as the original balancer.**

32. Install the water pump assembly.
33. Install the serpentine drive belt.
34. Install the throttle body air duct.
35. Properly fill the engine crankcase with clean engine oil.
36. Properly fill the engine cooling system.
37. Connect the negative battery cable, operate the engine and check for leaks.

FUEL SYSTEM

Fuel System Service Precautions

Safety is the most important factor when performing not only fuel system maintenance but any type of maintenance. Failure to conduct maintenance and repairs in a safe manner may result in serious personal injury or death. Maintenance and testing of the vehicle's fuel system components can be accomplished safely and effectively by adhering to the following rules and guidelines.

• To avoid the possibility of fire and personal injury, always disconnect the negative battery cable unless the repair or test procedure requires that battery voltage be applied.

• Always relieve the fuel system pressure prior to disconnecting any fuel system component (injector, fuel rail, pressure regulator, etc.), fitting or fuel line connection. Exercise extreme caution whenever relieving fuel system pressure to avoid exposing skin, face and eyes to fuel spray. Please be advised that fuel under pressure may penetrate the skin or any part of the body that it contacts.

• Always place a shop towel or cloth around the fitting or connection prior to loosening to absorb any excess fuel due to spillage. Ensure that all fuel spillage (should it occur) is quickly removed from

engine surfaces. Ensure that all fuel soaked cloths or towels are deposited into a suitable waste container.

• Always keep a dry chemical (Class B) fire extinguisher near the work area.

• Do not allow fuel spray or fuel vapors to come into contact with a spark or open flame.

• Always use a back-up wrench when loosening and tightening fuel line connection fittings. This will prevent unnecessary stress and torsion to fuel line piping. Always follow the proper torque specifications.

• Always replace worn fuel fitting O-rings with new. Do not substitute fuel hose or equivalent, where fuel pipe is installed.

Fuel System Pressure

RELIEVING

1. Disconnect the negative battery cable to prevent fuel discharge if the key is accidentally turned to the **RUN** position.
2. Loosen the fuel filler cap to relieve the tank pressure and do not tighten until service has been completed.
3. Connect J-34730-1 fuel pressure gauge or equivalent, to the fuel pressure valve. Wrap a shop cloth around the fitting while connecting the gauge to avoid spillage.
4. Place the end of the bleed hose into a suitable container and open the valve to relieve the fuel system pressure.

Fuel Filter

REMOVAL & INSTALLATION

❋❋ CAUTION

Never smoke when working around gasoline! Avoid all sources of sparks or ignition. Gasoline vapors are EXTREMELY volatile!

The inline fuel filter is located on the fuel feed pipe, before the fuel injection system, and mounted directly in front of the rear axle. The filter housing is constructed of steel with quick-connect inlet and threaded outlet fittings. The threaded fitting is sealed with an O-ring. In order to disengage quick-connect fittings, a fuel line quick-connect separator tool set, such as J-37088-A or equivalent, is required. There is no service interval for fuel filter replacement. Only replace the fuel filter if it is restricted.

❋❋ CAUTION

To reduce the risk of fire and personal injury, it is necessary to relieve the fuel system pressure before servicing fuel system components. After relieving system pressure, a small amount of fuel may be released when servicing fuel lines or connections. In order to reduce the chance of personal injury, cover the fuel pipe fittings with a shop rag before disconnecting to catch any fuel that may leak out. Place the towel in an approved container.

1. Relieve the fuel system pressure.
2. Raise and safely support the vehicle.
3. Clean both the inlet and the outlet fittings on the fuel filter.
4. Disengage the quick-connect fittings at the fuel filter inlet as follows:

a. Slide the dust covers from the quick-connect fittings, if equipped.

b. Grasp both sides of the fitting. Twist female connector ¼ turn in each direction to loosen any dirt within the fitting. Using compressed air and safety glasses, blow dirt out of fitting.

c. If equipped with the plastic hand releasable fitting, squeeze the plastic retainer release tabs and pull the connection apart.

d. If equipped with metal fitting, choose correct size tool from tool set J-37088-A. Insert tool into female connec-

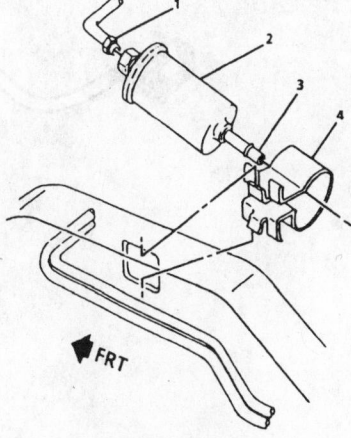

1	THREADED FITTING
2	IN-LINE FUEL FILTER
3	QUICK-CONNECT FITTING
4	IN-LINE FUEL FILTER BRACKET

7922WG23

The inline fuel filter is mounted in a bracket located under the vehicle, directly in front of the rear axle

tor, then push inward to release locking tabs. Pull the connector apart.

e. Use a clean lint free rag to clean male pipe ends. Inspect both ends of the fitting for dirt and burrs. Clean or replace components as required. If it is necessary to remove rust or burrs from the fuel pipe, use emery cloth in a radial motion with the pipe end to prevent damage to the O-ring sealing surface.

5. Remove the threaded outlet fitting from the chassis fuel pipe. Slide the fuel filter from the bracket.

6. Inspect the fuel pipe O-ring for cuts, nicks, swelling or distortion. Replace if necessary.

To install:

7. Slide the fuel filter into the bracket. Thread the outlet fitting to the chassis fuel pipe.

8. Tighten the inline fitting to 20 ft. lbs. (27 Nm).

9. Engage the quick-connect inlet fitting as follows:

a. Apply a few drops of clean engine oil to the male pipe end. This will ensure proper reconnection and prevent a possible fuel leak.

b. Push both sides of the fitting together to cause the retainer tabs to snap in place. Once installed, pull on both sides of the fitting to ensure connection is secure.

c. Reposition dust cover over the quick-connect fitting, if equipped.

10. Lower the vehicle. Tighten the fuel filler cap.

11. Confirm that the ignition is in the **OFF** position. Connect the negative battery cable.

12. Pressurize the fuel system by cycling the ignition without attempting to start the engine. Turn the ignition switch to the **ON** position for two seconds, then turn to the **OFF** position for ten seconds. Again, turn to the **ON** position and check for fuel leaks.

Fuel Pump

REMOVAL & INSTALLATION

The fuel pump is part of the fuel sender assembly located inside the fuel tank.

1. Release the fuel system pressure and disconnect the negative battery cable.

2. Drain the fuel tank, then raise and safely support the vehicle.

3. Remove the fuel tank from the vehicle.

4. Clean the area surrounding the sender assembly to prevent contamination of the fuel system.

5. To remove the fuel sender from the tank, remove the fuel sender assembly nuts and retaining ring. Carefully remove the sending unit from the fuel tank. Discard the O-rings.

6. If necessary, separate the fuel pump from the sending unit assembly.

To install:

7. If removed, install the fuel pump to the sending unit. If the strainer was removed, it must be replaced with a new one.

8. Inspect and clean the O-ring mating surfaces.

9. Install a new O-ring in the groove around the tank opening. If applicable, install a new O-ring on the fuel sender feed tube.

10. Install the fuel sender assembly as follows:

a. The fuel pump strainer must be in a horizontal position, and when installed, must not block the travel of the float arm. Gently fold the strainer over itself and slowly position the sending assembly in the tank so the strainer is not damaged or trapped by the sump walls.

11. Install the retaining ring and nuts, then tighten the nuts to 63 inch lbs. (7 Nm).

12. Install the fuel tank assembly.

13. Lower the vehicle.

14. Fill the fuel tank, tighten the fuel filler cap and connect the negative battery cable.

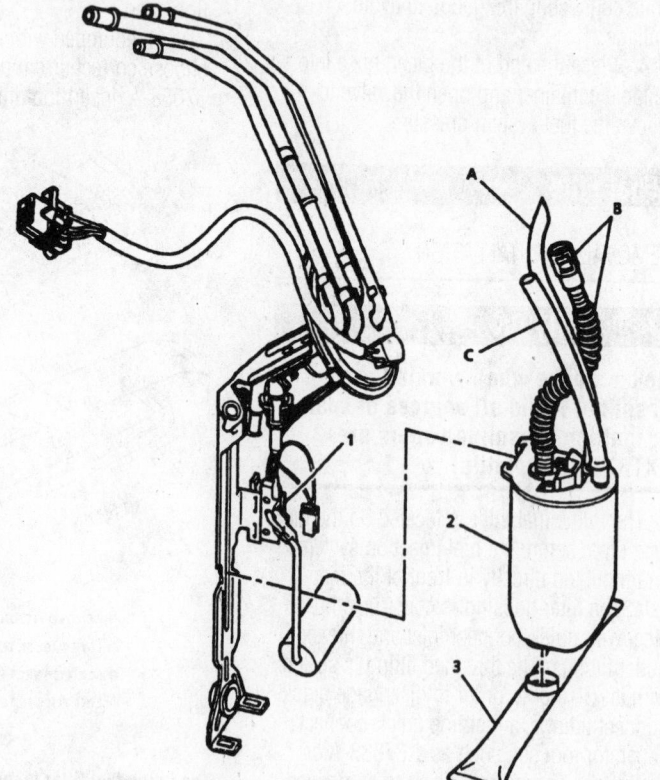

1 **SENDER – FUEL**

2 **FUEL PUMP AND RESERVOIR ASSEMBLY**
A **VAPOR VENT HOSE**
B **FUEL PUMP FLEX PIPE AND QUICK – CONNECT FITTING**
C **FUEL RETURN HOSE**

3 **STRAINER – FUEL PUMP**

Exploded view of the fuel sender/pump assembly

7922WG24

15. Turn the ignition switch to the **ON** position for 2 seconds, **OFF** for 10 seconds, then back to the **ON** position. Check for fuel leaks.

DRIVE TRAIN

➡️**Refer the unit repair section for information on driveshaft service.**

Transmission

REMOVAL & INSTALLATION

1. Disconnect the negative battery cable.
2. On manual transmissions, remove the shift lever boot and the shift lever from inside the vehicle before raising the vehicle.
3. Raise the vehicle and support it safely with jackstands.

➡️**Disconnect the clutch master cylinder pushrod before disconnecting the actuator cylinder. If not disconnected, permanent damage to the actuator cylinder will occur if the clutch pedal is depressed while the actuator is disconnected.**

4. On manual transmissions, disconnect the clutch actuator line from the actuator.
5. Position a suitable pan under the vehicle, then drain the lubricant from the transmission.
6. On automatic transmissions, disconnect the fluid cooler lines, throttle valve cable and remove the dipstick tube.
7. Support the rear axle with a suitable jackstand.
8. Matchmark and remove the driveshaft.
9. Remove the rear axle torque arm from the vehicle.
10. Remove the catalytic converter hanger.
11. Detach all electrical connectors from the transmission and position them out of the way.
12. On manual transmissions, disengage the clutch fork from the release bearing. The fork MUST be detached from the release bearing to prevent damage to the clutch system.
13. On automatic transmissions, remove the starter, then remove the bolts securing the torque converter to the flexplate. Push the torque converter away from the flexplate into the transmission.
14. Support the engine with a suitable jackstand. Support the transmission with a jackstand.

15. Remove the transmission rear crossmember.
16. On automatic transmissions, remove the bolts securing the transmission to the engine. On manual transmissions, remove the bolts securing the transmission to the bell housing. Pull the transmission rearward and carefully lower the assembly away from the vehicle. Place the transmission in a suitable holding fixture.
17. Installation is the reverse of the removal procedure. Please note the following important steps:
18. On manual transmissions, position the splined input shaft into the clutch disc hub and pilot bearing.
19. For the 5-speed manual transmission, tighten the transmission mounting bolts to 55 ft. lbs. (75 Nm) and the transmission control assembly bolts to 15 ft. lbs. (20 Nm).
20. For the 6-speed manual transmission, tighten the transmission mounting bolts to 26 ft. lbs. (35 Nm). Tighten the actuator cylinder nuts to 15 ft. lbs. (25 Nm).
21. On manual transmissions, when connecting the actuator line, be sure it is not twisted or kinked and will not rub against any other components. Also, the quick-connect fitting must be pushed on,

then pulled back to be sure of proper engagement.
22. For 3.4L and 3.8L engines, tighten the transmission mounting bolts to 70 ft. lbs. (95 Nm).
23. For 5.7L engine, tighten the transmission mounting bolts to 35 ft. lbs. (47 Nm).
24. Fill the transmission with the proper type and amount of fluid. For manual transmissions, bleed the clutch system if any hydraulic clutch components were removed.

Clutch

REMOVAL & INSTALLATION

❋❋ CAUTION

The clutch disc may contain asbestos, which has been determined to be a cancer causing agent. Never clean clutch surfaces with compressed air! Avoid inhaling any dust from any clutch surface! When cleaning clutch surfaces, use a commercially available brake cleaning fluid.

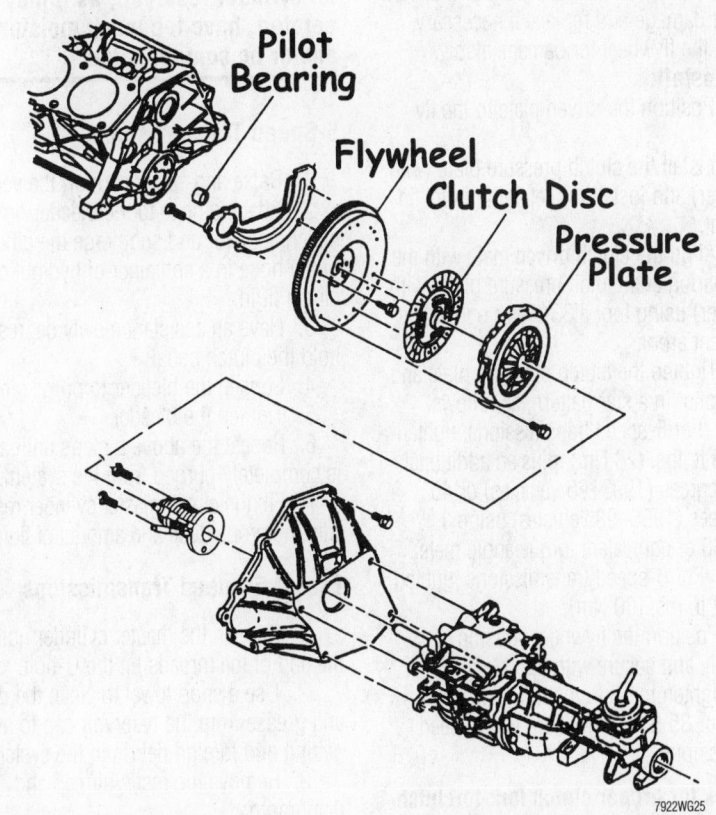

Exploded view of the clutch disc, pressure plate and related components—5-speed transmission shown

7922WG25

1. Disconnect the negative battery cable.

2. Remove the instrument panel knee bolster assembly.

3. Disconnect the clutch master cylinder pushrod from the clutch pedal.

4. Raise and safely support the vehicle.

5. Remove the transmission assembly.

6. Remove the clutch actuator cylinder nuts.

➡**Do NOT let the actuator cylinder hang by the fluid lines, as this could damage them.**

7. Remove the clutch actuator cylinder and position aside, supporting with a suitable piece of wire.

8. Unfasten the transmission brace nut and bolt, then remove the brace.

9. Remove the flywheel housing cover retaining bolts, then remove the cover.

10. Unfasten the flywheel housing bolts, then remove the housing.

11. Install a clutch disc alignment tool through the center of the disc and into the pilot bearing to prevent the disc from falling when the pressure plate is removed.

12. Remove the pressure plate retaining bolts.

13. Remove the pressure plate with clutch disc from the flywheel.

14. Inspect the pressure plate and driven disc for damage and replace if necessary. Inspect the flywheel for damage also.

To install:

15. Position the driven plate to the flywheel.

16. Install the clutch pressure plate (with the cover) and install the retaining bolts finger-tight.

17. Align the clutch driven plate with the pilot bearing and clutch pressure plate (with the cover) using tool J 33169 or equivalent alignment arbor.

18. Tighten the clutch pressure plate and cover bolts, in a star pattern, as follows:

  a. For 5-speed transmissions, tighten to 15 ft. lbs. (20 Nm), plus an additional 30 degrees (1993–95 vehicles) or 45 degrees (1996–98 vehicles) using J 36660 or equivalent torque angle meter.

  b. For 6-speed transmissions, tighten to 22 ft. lbs. (30 Nm).

19. Position the flywheel housing assembly and secure with the retaining bolts. Tighten to 55 ft. lbs. (75 Nm) for 5-speeds or 35 ft. lbs. (47 Nm) for 6-speed transmissions.

➡**Check for proper clutch fork-to-clutch release bearing engagement.**

20. Install the flywheel housing cover and retain with the bolts. Tighten to 75 inch lbs. (8.5 Nm).

21. Position the transmission brace, install the retaining nut and bolt, then tighten to 37 ft. lbs. (50 Nm).

22. Install the clutch actuator cylinder and secure with retaining nuts. Tighten the nuts to 15 ft. lbs. (20 Nm).

23. Install the transmission assembly.

24. Carefully lower the vehicle.

25. Connect the clutch master cylinder pushrod to the clutch pedal.

26. Install the instrument panel knee bolster.

27. Connect the negative battery cable.

Hydraulic Clutch System

BLEEDING

Bleeding air from the hydraulic clutch system is necessary whenever any part of the system has been disconnected or the fluid level (in the reservoir) has been allowed to fall so low that air has been drawn into the master cylinder.

❊❊ WARNING

NEVER use fluid which has been bled from a clutch system to fill the master cylinder reservoir, as it may be aerated, have too much moisture and/or be contaminated.

5-Speed Transmissions

1. Raise and safely support the vehicle.

2. Attach a hose to the bleeder on the clutch actuator and submerge the other end of the hose in a container of hydraulic clutch fluid.

3. Have an assistant slowly depress and hold the clutch pedal.

4. Loosen the bleeder to purge air.

5. Tighten the bleeder.

6. Repeat the above 2 steps until all air is completely purged from the system.

7. Fill the clutch master cylinder reservoir with the proper type and amount of fluid.

Except 5-Speed Transmissions

1. Loosen the master cylinder nuts to the end of the threads on the U-bolt.

2. Use a shop towel to clean the dirt and grease from the reservoir cap to avoid getting and foreign debris in the system.

3. Remove the reservoir cap and diaphragm.

4. Wrap a suitable piece of wire around the left-hand hood strut bracket, making sure the wire can be accessed from under the vehicle.

5. Raise and safely support the vehicle.

6. Remove the actuator cylinder nuts, then remove the actuator cylinder. Secure the actuator cylinder in the engine compartment using the wire previously installed.

7. Carefully lower the vehicle.

8. Grasp the actuator cylinder and depress the actuator cylinder pushrod about 0.0787 in. (20mm) into the cylinder bore and hold.

9. Have an assistant install the diaphragm and reservoir cap while holding the actuator pushrod in. Release the pushrod.

10. Hold the actuator cylinder lower than the master cylinder, vertically with the pushrod end facing down.

11. Press the pushrod into the actuator cylinder bore with short 0.0390 in. (10mm) strokes. Check the master cylinder reservoir for bubbles.

12. Continue until bubbles are not longer entering the reservoir.

13. Raise and safely support the vehicle.

14. Remove the actuator cylinder from the wire.

15. Install the actuator cylinder and secure with the retaining nuts. Tighten to 15 ft. lbs. (20 Nm).

16. Carefully lower the vehicle, then remove the wire from the hood strut.

17. Fill the clutch master cylinder reservoir with the proper type and amount of fluid.

STEERING AND SUSPENSION

Air Bag

❊❊ CAUTION

Some vehicles are equipped with an air bag system, also known as the Supplemental Inflatable Restraint (SIR) or Supplemental Restraint System (SRS). The system must be disabled before performing service on or around system components, steering column, instrument panel components, wiring and sensors. Failure to follow safety and disabling procedures could result in accidental air bag deployment, possible personal injury and unnecessary system repairs.

PRECAUTIONS

Several precautions must be observed when handling the inflator module to avoid accidental deployment and possible personal injury.

- Never carry the inflator module by the wires or connector on the underside of the module.
- When carrying a live inflator module, hold securely with both hands, and ensure that the bag and trim cover are pointed away.
- Place the inflator module on a bench or other surface with the bag and trim cover facing up.
- With the inflator module on the bench, never place anything on or close to the module which may be thrown in the event of an accidental deployment.

DISARMING

1. Align the steering wheel so the vehicle wheels are pointing in the straight-ahead position.
2. Turn the ignition switch to the **LOCK** position.
3. Remove the SIR or AIR BAG fuse from the fuse block.
4. Remove the Connector Position Assurance (CPA) device, then disengage the yellow 2-way SIR wiring harness connector at the base of the steering column.

REARMING

1. Turn the ignition switch to the **LOCK** position.
2. Engage the yellow 2-way connector at the base of the steering column, then install the CPA device.
3. Reinstall the SIR or AIR BAG fuse.
4. Turn the ignition switch to the **RUN** position.
5. Verify the SIR indicator light flashes 7–9 times, if not, inspect system for malfunction.

Power Rack and Pinion Steering Gear

REMOVAL & INSTALLATION

1. Raise and safely support the vehicle.
2. Remove the wheel and tire assemblies.
3. Place a drain pan under the steering gear unit.
4. Disconnect the inlet and outlet hoses from the steering gear.

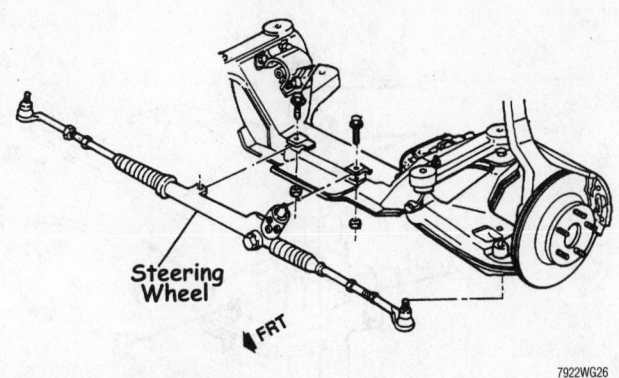

Exploded view of the power steering gear mounting

5. Separate the outer tie rod ends from the knuckles.
6. Remove the steering gear coupling shaft from the steering gear.
7. Remove the steering gear mounting nuts and bolts, then remove the steering gear from the vehicle.

To install:

8. Position the steering gear to the crossmember. Adjust the steering gear so it aligns as straight as possible with the steering gear coupling shaft.
9. Hand-start the retaining bolts and nuts. Using a back-up wrench on the nuts, tighten the bolts to 63 ft. lbs. (85 Nm).
10. Position the steering gear to the coupling shaft.
11. Fasten the outer tie rod ends to the steering knuckle.
12. Connect the inlet and outlet hoses to the steering gear.
13. Install the tire and wheel assemblies.
14. Carefully lower the vehicle, then bleed the power steering system.

Shock Absorber

While each shock absorber can be replaced individually, it is recommended that they be changed as a pair (both front or both rear) to maintain equal response on both sides of the vehicle.

REMOVAL & INSTALLATION

Front

1. If removing the driver's side spring and shock assembly, unbolt the brake master cylinder from the booster and position it aside, but do not disconnect the brake lines.
2. Remove the upper spring/shock mounting bolts and nuts.
3. Raise and safely support the vehicle with jackstands.
4. Remove the wheel and tire assembly.

5. Disconnect the stabilizer bar (shaft link) from the control arm.
6. Carefully matchmark the upper and lower shock/spring mounting positions using chalk, or preferably, paint.

✳✳ WARNING

If the position of the shock and spring assembly is not carefully marked, it cannot be reinstalled!

7. Remove the lower end shock absorber nuts bolts.
8. Separate the lower ball joint (stud) from the steering knuckle.
9. Remove the spring/shock absorber unit from the vehicle.
10. To remove the spring from the shock absorber assembly perform the following steps.

➡ **If using other than a GM spring compressor, follow the manufacturers instructions regarding the use of the specific tool you are using.**

11. Assemble spring compressor tool J-34013-B and adapter J-34013-114 on the spring unit. Use the wing nuts to secure the tool to mounting holes **C-H** (lower left corner) and **P** (upper right corner) for the driver's side shock and to mounting holes **A-X-P** (upper left) and **C-H** (lower right) for the passenger's side shock.
12. Install tools J-34013-114 and J-34013-88.

➡ **Make certain that J-34013-114 and J-34013-88 are aligned so that they can open and close together. If not properly aligned, they will not function.**

13. Attach the shock unit to the tools.

➡ **Make certain that the top of the shock is flat against J-34013-114!**

14. Close the tools and install the locking pin.

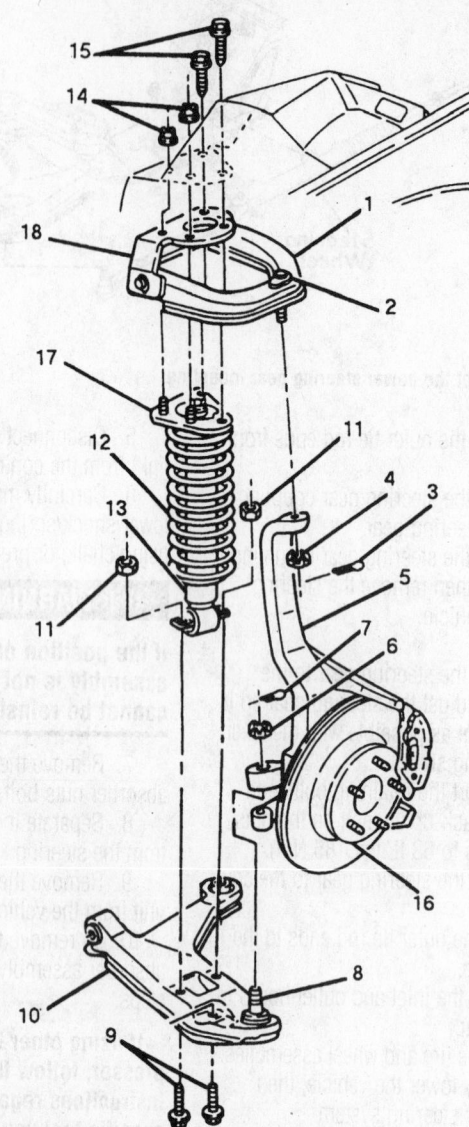

1. ARM ASSEMBLY, FRONT UPPER CONTROL
2. STUD ASSEMBLY, FRONT UPPER CONTROL ARM BALL
3. PIN, FRONT UPPER CONTROL ARM COTTER
4. NUT, FRONT UPPER CONTROL ARM, 53 N•m (39 LB. FT.)
5. KNUCKLE ASSEMBLY, STEERING
6. NUT, FRONT LOWER CONTROL ARM, 110 N•m (81 LB. FT.)
7. PIN, FRONT LOWER CONTROL ARM COTTER
8. STUD ASSEMBLY, FRONT LOWER CONTROL ARM BALL
9. BOLT/SCREW, FRONT SHOCK ABSORBER, 65 N•m (48 LB. FT.)
10. ARM ASSEMBLY, FRONT LOWER CONTROL
11. NUT, FRONT SHOCK ABSORBER LOWER BRACKET, 65 N•m (48 LB. FT.)
12. SPRING ASSEMBLY, FRONT
13. ABSORBER ASSEMBLY, FRONT SHOCK
14. NUT, FRONT SHOCK ABSORBER UPPER MOUNT, 43 N•m (32 LB. FT.)
15. BOLT/SCREW, FRONT SHOCK ABSORBER UPPER MOUNT, 50 N•m (37 LB. FT.)
16. HUB ASSEMBLY, FRONT WHEEL
17. MOUNT ASSEMBLY, FRONT UPPER SHOCK ABSORBER
18. SUPPORT, FRONT UPPER CONTROL ARM

7922WG27

Exploded view of the spring & shock absorber unit and related suspension components

➡ Make certain that the mounting ears of the shock are facing downward towards the rear of J-34013-B, or the shock will not align properly.

15. Turn the screw of J-34013-B counterclockwise to raise the shock up to J-34013-114. Be sure that the studs go through the guide holes in J-34013-114 and the top of the shock is flat against the tool.

✳✳ CAUTION

Do not over-compress the spring! Over-compression can cause tool failure, resulting in severe bodily injury!

16. Compress the spring about ½ in. (13mm), or 3–4 complete turns of the screw on J-34013-114.
17. Insert J-39642-1 on the shock nut, then insert J-39642-2 through J-39642-1 to hold the shock absorber rod in place.
18. Remove the shock absorber nut with J-39642-1, while holding the shock rod from rotating, with J-39642-2.
19. Discard the shock absorber nut.
20. Turn J-34013-B clockwise to fully relieve spring pressure and remove the spring from the shock.

To install:
21. Assemble tools J-34013-B spring compressor and J-34013-114 adapter on the spring unit. Use wing nuts to secure the tool to mounting holes **C-H** (lower left corner and **P** (upper right corner) for the driver's side shock and to mounting holes **A-X-P** (upper left) and **C-H** (lower right) for the passenger's side shock.
22. Install tools J-34013-114 and J-34013-88.
23. Attach the shock unit to the tools.

➡ Make certain that the mounting ears of the shock are facing downward towards the rear of J-34013-B!

24. Close the tools and install the locking pin.
25. Be sure that the spring seats are positioned properly.

➡ Make certain that the top of the shock is flat against J-34013-114!

26. Turn the screws of J-34013-B counterclockwise to raise the shock up to J-34013-114. Be sure that the studs go through the guide holes in J-34013-114 and the top of the shock is flat against the tool.

➡ Turn the screw only enough to secure the shock. DO NOT compress the spring.

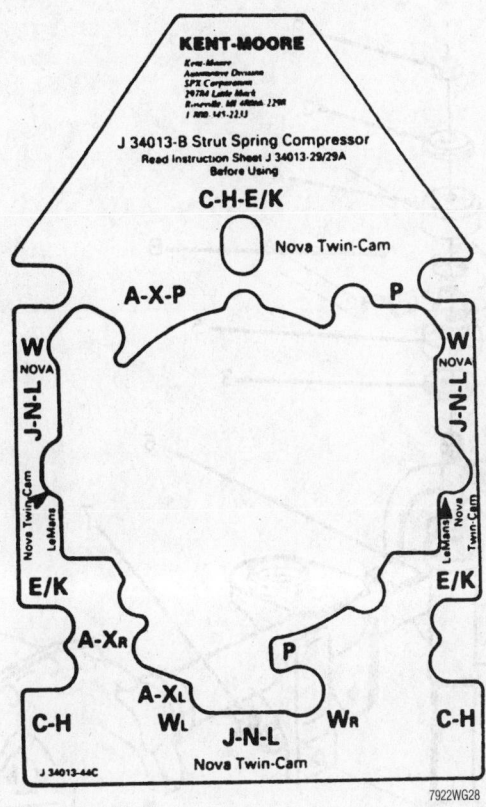

Shock absorber compressor mounting hole locations

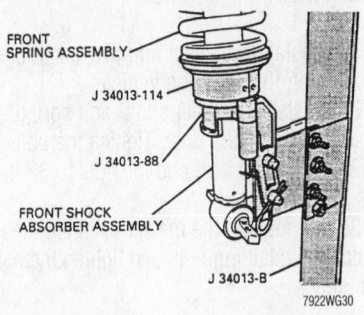

Installing the shock absorber on the compressor

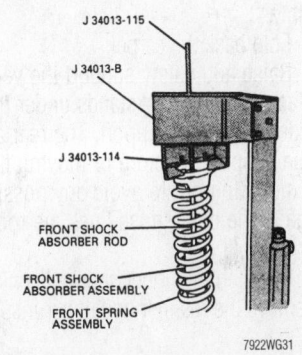

Inserting the shock assembly alignment rod

27. Place J-34013-115 down through the top of J-34013-B, through the top of the shock absorber and onto the rod.

➡ **Make certain that J-34013-115 is straight with the shock.**

✳✳ CAUTION

Do not over-compress the spring! Over-compression can cause tool failure, resulting in severe bodily injury!

28. Turn the operating screw clockwise to compress the spring until the threaded portion of the rod is through the top of the shock. Remove J-34013-115.

29. Install a NEW shock absorber nut.

30. Install J-39642-1 on the nut, insert J-39642-2 though J-39642-1 and tighten the nut while holding J-39642-2.

31. Remove the spring and shock assembly from the tool.

32. Position the unit on the lower control arm.

33. Install the lower mounting nuts and bolts, then tighten them to 48 ft. lbs. (65 Nm).

34. Attach the lower ball joint to the steering knuckle.

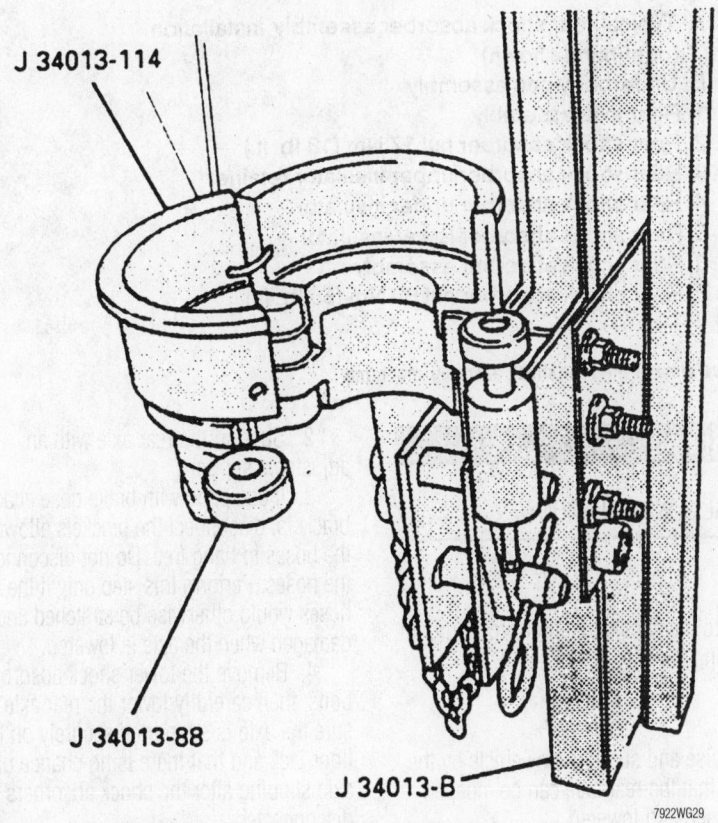

Install the strut compressor adapter on the spring compressor

35. Install the stabilizer shaft link.

36. Install the wheel and tire assembly, then carefully lower the vehicle.

37. Install the upper shock and spring assembly bolts and nuts. Tighten the bolts to 37 ft. lbs. (50 Nm) and the nuts to 32 ft. lbs. (43 Nm).

38. Reposition the master cylinder, then install the retaining nuts and tighten them securely.

Rear

1. Fold down the rear seatback.
2. Remove the quarter panel trim assembly.
3. Fold back the carpet.
4. Raise and safely support the vehicle. Place a jack or jackstands under the rear axle to provide support. The rear axle must be supported before re-moving the upper mounting nut to avoid any possible damage to the brake hose lines, tie rod and driveshaft.
5. Unfasten the upper shock attaching nut. Remove the retainer and upper insulator.
6. Remove the retainer and lower insulator.
7. Remove the lower shock-to-rear axle mounting nut.
8. Remove the shock absorber from the vehicle.

To install:

9. Place the shock in position.
10. Install the lower shock-to-axle mounting nut and tighten to 66 ft. lbs. (90 Nm).
11. Install the lower insulator and retainer to the shock absorber.
12. Position the shock absorber through the underbody pan and seat the insulator.
13. Place the upper insulator and retainer to the shock absorber. Hand-start the upper retaining nut.

➡**Turning the shock absorber while tightening the nut could damage the shock. To prevent damage, keep the shock absorber stationary when tightening the nut.**

14. Tighten the upper nut to 13 ft. lbs. (17 Nm).
15. Remove the rear axle support, then carefully lower the vehicle.
16. Reposition the carpet.
17. Install the quarter panel trim assembly.
18. Raise the rear seat seatback.

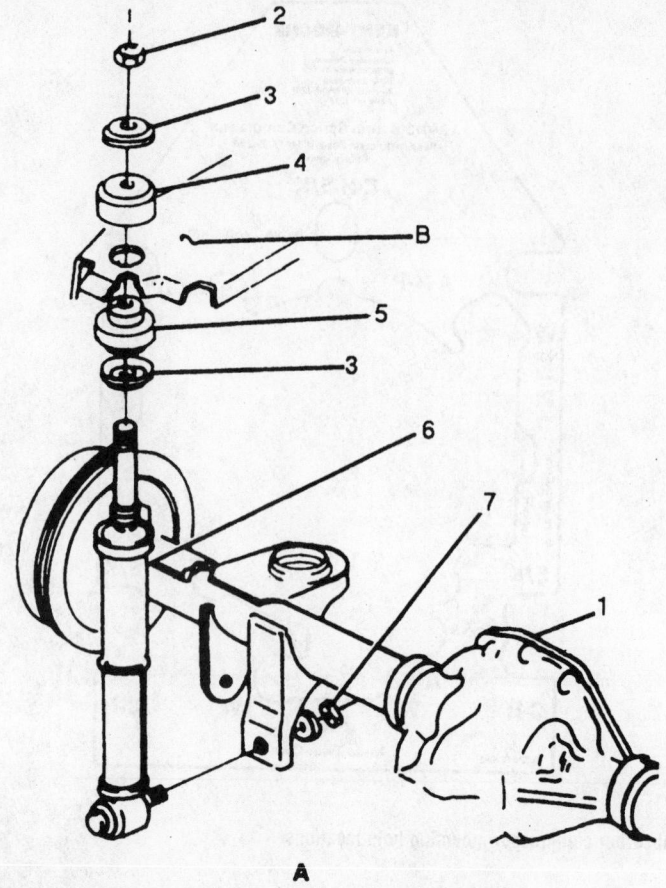

A

A Typical rear shock absorber assembly installation (right-hand shown)
B Underbody pan assembly
1 Rear axle assembly
2 Rear shock absorber nut 17 Nm (13 lb. ft.)
3 Rear shock absorber upper insulator retainer
4 Rear shock absorber upper insulator
5 Rear shock absorber lower insulator
6 Rear shock absorber assembly
7 Rear shock absorber nut 90 Nm (66 lb. ft.)

7922WG32

Exploded view of the rear shock absorber mounting

Coil Spring

REMOVAL & INSTALLATION

Front

Refer to the shock absorber procedure for front spring service.

Rear

1. Raise and support the vehicle by the frame so that the rear axle can be independently raised and lowered.

2. Support the rear axle with an adjustable support.

3. If equipped with brake hose attaching brackets, disconnect the brackets allowing the hoses to hang free. Do not disconnect the hoses. Perform this step only if the hoses would otherwise be stretched and damaged when the axle is lowered.

4. Remove the lower shock absorber bolts, then carefully lower the rear axle. Be sure the axle is supported securely on the floor jack and that there is no chance of the axle slipping after the shock absorbers are disconnected.

➡**Be sure that the brake hoses are not stretched.**

5. Lower the axle and remove the upper insulator.

6. Remove the spring assembly.

➡**The springs are painted with a protective coating. Take care to avoid damaging this coating. If the coating is chipped or damaged, paint the exposed spring to prevent rust.**

To install:

7. Position the spring on the axle with the open lower end facing forward.

8. Install the upper insulator.

9. Raise the rear axle and install the lower shock absorber bolts.

10. Connect the brake hose attaching brackets, if removed.

11. Remove the support from the rear axle.

12. Carefully lower the vehicle.

Upper Ball Joint

REMOVAL & INSTALLATION

1. Raise and safely support the vehicle, then remove the wheel and tire assembly.

2. Position a floor jack under the shock mounting location on the lower control arm for support.

✳✳ CAUTION

The jack must remain in place for the entire duration of the procedure to hold the spring and lower control arm in proper position.

3. Remove the cotter pin and loosen, then remove the ball stud nut. Discard the cotter pin.

4. Support the steering knuckle with jackstands.

5. Separate the ball joint from the upper control arm using tool J 39549 or equivalent separator tool.

6. With the upper control arm still in the raised position, use a ⅛ in. (3.175mm) bit to drill the four rivets about 0.25 in. (6mm) deep. Now use a ½ in. (12.7mm) bit to drill off the rivet heads.

7. Use a small drift to punch out the rivets, then remove upper ball joint from the lower control arm.

To install:

8. Position the new upper ball joint on the control arm, then secure using the nuts and bolts supplied with the kit. Be sure to

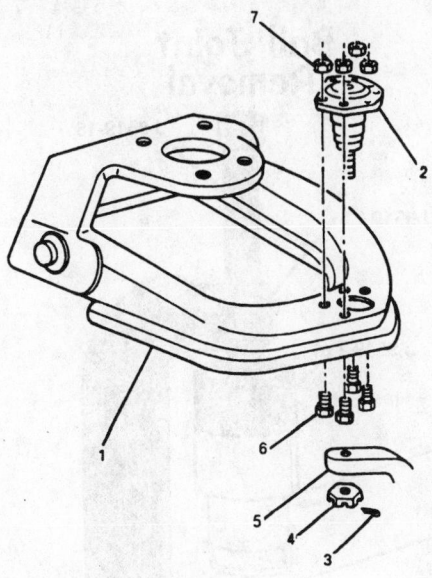

1 Front upper control arm assembly
2 Front upper control ball stud assembly
3 Front upper control arm cotter pin
4 Front upper control arm nut 53 Nm (39 lb. ft.)
5 Steering knuckle assembly
6 Service kit bolt/screw
7 Service kit nut

7922WG33

Position the new upper ball joint (stud) in the control arm and secure with the replacement nuts and bolts

follow the specifications given in the replacement kit.

9. Remove the jackstand from under the knuckle and connect the ball joint to the steering knuckle.

10. Hand-start the nut, then tighten to 39 ft. lbs. (53 Nm). Advance the nut to align the nearest cotter pin hole. NEVER back off the nut to align a hole! Install a new cotter pin.

11. Remove the floor jack from under the lower control arm shock mounting point.

12. Install the wheel and tire assembly, then carefully lower the vehicle.

Lower Ball Joint

REMOVAL & INSTALLATION

➡**To prevent component damage, an on-car ball joint press, such as Kent-Moore tool J-9519-23 should be used.**

1. Raise and safely support the vehicle with jackstands under the frame.

2. Remove the wheel and tire assembly.

3. Position a floor jack under the shock mounting location on the lower control arm for support.

✳✳ CAUTION

The jack must remain in place for the entire duration of the procedure to hold the spring and lower control arm in proper position.

4. Remove and discard the lower ball joint cotter pin, then loosen the lower ball joint nut.

5. Separate the lower ball joint from the steering knuckle using J 39549 or an equivalent separator.

6. Press the lower ball joint out of the lower control arm using tools J 9519-7, J 9519-18 and 9518-23.

To install:

7. Position a new lower ball joint into the lower control arm, and press in using tools J 9519-7, J 9519-18 and 9518-23. The ball joint must firmly press into the lower control arm. If it doesn't the lower control arm must be replaced.

8. Attach the ball joint (stud) to the steering knuckle.

9. Hand-start the lower control arm nut, then tighten to 81 ft. lbs. (110 Nm). Continue to tighten the nut just until the cotter pin holes align, then install a new

Ball Joint Removal

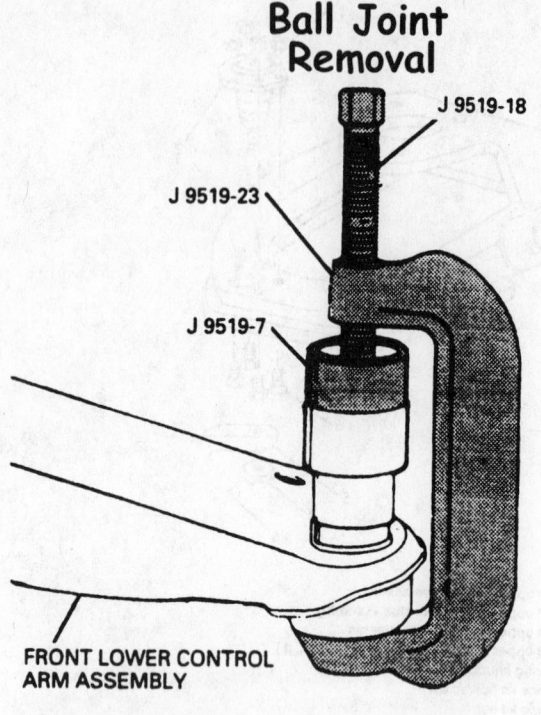

J 9519-18
J 9519-23
J 9519-7
FRONT LOWER CONTROL ARM ASSEMBLY

Ball Joint Installation

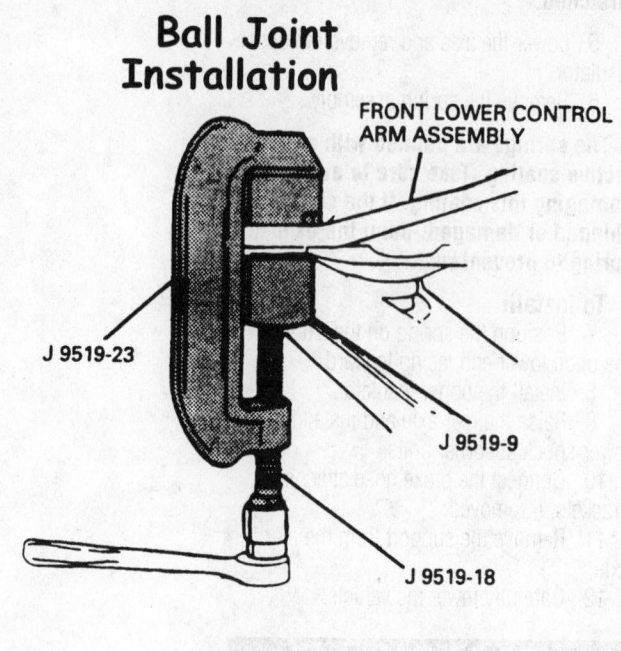

FRONT LOWER CONTROL ARM ASSEMBLY
J 9519-23
J 9519-9
J 9519-18

7922WG34

Lower ball joint replacement requires the use of special tools, such as the ones shown

pin. NEVER back off the nut to align the holes.

10. Remove the floor jack from under the control arm shock absorber mounting location.

11. Install the tire and wheel assembly, then carefully lower the vehicle.

12. Take the vehicle to a reputable repair shop to have the alignment checked.

Wheel Bearings

ADJUSTMENT

The front wheel bearing assembly used on these vehicles is a sealed, non-serviceable unit. No wheel bearing adjustments (front or rear bearings) are necessary or possible.

REMOVAL & INSTALLATION

Front

1. Raise and safely support the vehicle.
2. Remove the wheel and tire assembly.
3. Remove the brake caliper and rotor.
4. Detach the wheel speed sensor electrical connector and position it aside.
5. Unfasten the hub retaining bolts, then pull the hub and bearing assembly from the spindle.

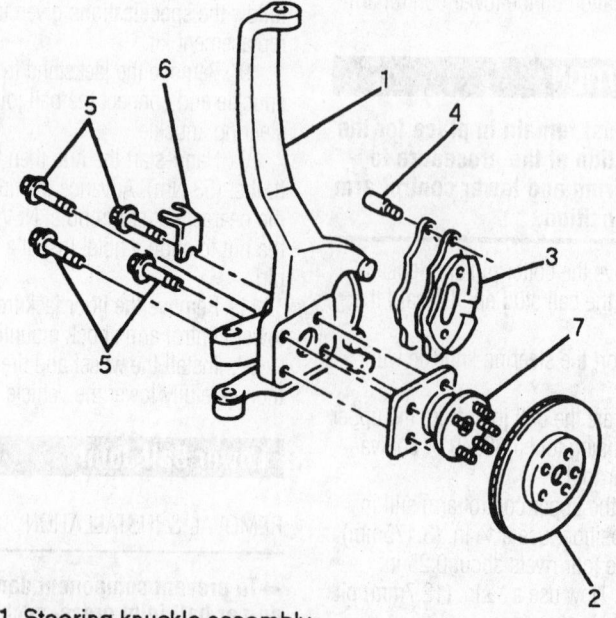

1 Steering knuckle assembly
2 Front brake rotor assembly
3 Front brake caliper assembly
4 Front brake caliper bolt/screw
5 Front wheel bolt/screw 86 Nm (63 lb. ft.)
6 Wheel speed sensor wire bracket
7 Front wheel hub assembly

7922WG35

Exploded view of the wheel hub/bearing unit mounting

To install:

6. Position the hub and bearing assembly on the spindle. Install the retaining bolts and tighten to 63 ft. lbs. (86 Nm).

7. Attach the wheel speed sensor electrical connector.

➡**Be sure the wheel speed sensor electrical connector is reattached to the sensor wire bracket and sensor or the wires could be damaged.**

8. Install the brake rotor and caliper.

9. Install the tire and wheel assembly, then carefully lower the vehicle.

Rear

1. Raise and support the vehicle safely.

2. Remove the rear tire and wheel assembly.

3. Remove the brake rotor or the brake drum and components, as equipped.

4. Clean the carrier cover and surrounding area to prevent dirt or contamination from entering the housing, then remove the carrier cover and drain the gear oil into a suitable container.

5. Install J-39446, or an equivalent ABS executor ring protector kit.

6. Remove the rear axle pinion shaft lockscrew and pinion shaft.

7. Push the flanged end of the axle shaft into the axle housing and remove the C-clip shaft lock from the differential case end of the shaft.

8. Remove the axle shaft from the axle housing.

9. If necessary to service the seal or bearing, use a small suitable prytool to remove the oil seal from the axle housing. Be careful not to score or damage the housing.

10. If necessary, install tool J-22813-01 or equivalent axle bearing remover, into the bore of the axle housing and position it behind the bearing, ensure the tangs of the tool engage the outer race. Remove the bearing using a slide hammer.

To install:

11. If removed, lubricate the new bearing and sealing lips with gear lubricant and install the bearing with a suitable driver so the tool bottoms against the shoulder in axle housing.

12. If removed, Position seal on suitable seal installer, then insert the seal into the housing bore. Position the seal flush with the axle tube.

13. Taking care not to damage the seal, slide the axle shaft into place so the splines engage with the splines of the side gear.

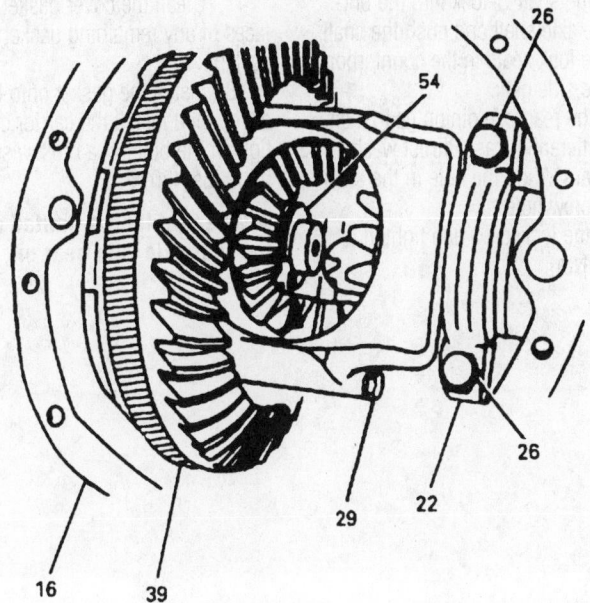

16 HOUSING, REAR AXLE
22 CAP, DIFFERENTIAL CARRIER BEARING
26 BOLT/SCREW, DIFFERENTIAL BEARING CAP
29 BOLT/SCREW, DIFFERENTIAL PINION GEAR SHAFT LOCK
39 WHEEL, REAR WHEEL SPEED SENSOR RELUCTOR
54 LOCK, REAR AXLE SHAFT

7922WG38

Differential component identification

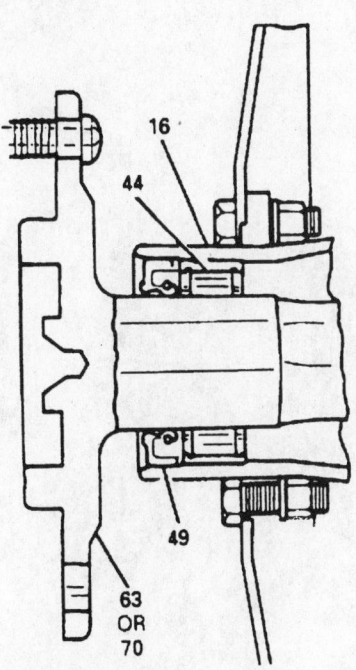

16 HOUSING, REAR AXLE
44 BEARING, REAR AXLE SHAFT
49 SEAL, REAR AXLE SHAFT BEARING
63 SHAFT, REAR AXLE (DRUM BRAKE ASSEMBLY)
70 SHAFT, REAR AXLE (DISC BRAKE ASSEMBLY)

7922WG39

Cut away view of the rear axle bearing and seal

14. Insert the shaft C-lock into the bottom end of the axle shaft and push the shaft outward so the lock seats in the counterbore of the rear axle side gear.

15. Insert the rear axle pinion gear shaft through the differential case, thrust washer and pinion gears. Align the hole in the shaft with the lockscrew hole.

16. Install the lockscrew and tighten to 27 ft. lbs. (36 Nm).

17. Clean the cover gasket mating surfaces of any remaining gasket or old sealant.

18. Install the gasket onto the carrier cover, then install the carrier cover and tighten the bolts in a crosswise pattern to 22 ft. lbs. (30 Nm).

➡ **When refilling a limited slip differential rear axle with gear oil, 4 oz (118ml) of limited slip additive should be added.**

19. Fill the differential carrier with SAE 80W-90 GL-5 gear lubricant or equivalent, then install the plug.

20. Install the rear brake assemblies and rear wheels. Lower the vehicle.

GENERAL MOTORS G BODY

32

Buick-Riviera • **Oldsmobile**-Aurora

DRIVE TRAIN 32-27
ENGINE REPAIR 32-2
FUEL SYSTEM 32-26
STEERING AND
  SUSPENSION 32-29

A
Air Bag...............................32-29
  Disarming...........................32-29
  Precautions 32-29

C
Camshaft and Valve Lifters 32-9
  Removal & Installation 32-9
Coil Spring 32-31
  Removal & Installation 32-31
Cylinder Head 32-5
  Removal & Installation 32-5

E
Engine Assembly 32-2
  Removal & Installation 32-2
Exhaust Manifold....................32-15
  Removal & Installation 32-15

F
Fuel Filter 32-26
  Removal & Installation 32-26
Fuel Pump 32-26

Removal & Installation 32-26
Fuel System Pressure 32-26
  Relieving 32-26
Fuel System Service
  Precautions........................32-26

H
Halfshaft..............................32-29
  Removal & Installation 32-29

I
Ignition Timing 32-2
  Adjustment 32-2
Intake Manifold 32-12
  Removal & Installation 32-12

L
Lower Ball Joint......................32-32
  Removal & Installation 32-32

O
Oil Pan..............................32-17
  Removal & Installation 32-17
Oil Pump 32-18
  Removal & Installation 32-18

P
Power Rack and Pinion Steering
  Gear...............................32-30
    Removal & Installation 32-30

R
Rear Main Seal 32-21
  Removal & Installation 32-21
Rocker Arms 32-7
  Removal & Installation 32-7

S
Shock Absorber 32-31
  Removal & Installation 32-31
Strut..................................32-30
  Removal & Installation 32-30
Supercharger 32-8
  Removal & Installation 32-8

T
Timing Chain, Sprockets, Front
  Cover and Seal 32-21
  Removal & Installation 32-21
Transaxle Assembly 32-27
  Removal & Installation 32-27

W
Water Pump..........................32-4
  Removal & Installation 32-4
Wheel Bearings......................32-32
  Adjustment........................32-32
  Removal & Installation 32-32

ENGINE REPAIR

→Disconnecting the negative battery cable on some vehicles may interfere with the functions of the on board computer systems and may require the computer to undergo a relearning process, once the negative battery cable is reconnected.

Ignition Timing

ADJUSTMENT

The 3.8L and 4.0L engine are equipped with a Distributorless Ignition System (DIS). The system consists of two crankshaft position sensors, crankshaft reluctor ring, camshaft position sensor, ignition control module, four ignition coils, eight plug wires and spark plugs, knock sensor and the Powertrain Control Module (PCM).

The base ignition timing is determined by the relationship of the crankshaft position sensors to the crankshaft reluctor ring. This relationship is not adjustable. Base ignition timing is 10 degrees Before Top Dead Center (BTDC).

The PCM controls spark advance under all driving conditions. The PCM incorporates a permanent spark control override, which electronically lowers the base timing if spark knock (detonation) is encountered during normal operation due to the use of low octane fuel.

Engine Assembly

REMOVAL & INSTALLATION

3.8L (VIN 1 and K) Engines

✱✱ CAUTION

The fuel injection system remains under pressure, even after the engine has been turned OFF. The fuel system pressure MUST BE relieved before disconnecting any fuel lines. Failure to do so may result in fire and/or personal injury.

1. Relieve the fuel system pressure using the recommended procedure.
2. Disconnect the negative battery cable.
3. Matchmark the hood to the hood hinges and remove the hood.

✱✱ CAUTION

Never open, service or drain the radiator or cooling system when hot; serious burns can occur from the steam and hot coolant.

4. Drain the cooling system.
5. Remove the radiator hoses and disconnect the heater hoses from the heater core.
6. Disconnect the negative battery cable from the engine block.
7. Detach the engine harness connector at the bulkhead.
8. Remove the serpentine belt(s).
9. Remove the power steering pump from the mounting bracket and position the pump aside. DO NOT disconnect the power steering lines from the pump.
10. Remove the air inlet duct.
11. Disconnect the throttle cables from the linkage mounting bracket and disconnect all the cables from the throttle body lever.
12. Detach the following wiring connectors from their related components:
- MAT sensor
- Throttle Position Sensor
- Idle Air Control valve
- Oxygen Sensor
- A/C compressor
- Oil Pressure Switch
- Power steering cutout switch
- Vehicle Speed Sensor
- Low oil level sensor
13. Remove the screws securing the ignition assembly ground strap to the inner fender panel.
14. Disconnect and cap the fuel feed and sender lines from the fuel rail and pressure regulator.
15. Disconnect the EVAP canister hoses from the throttle body.
16. Disconnect the vacuum lines from the power brake booster and heater control hoses.
17. Disconnect and remove the cruise control servo.
18. Raise and safely support the vehicle.

✱✱ CAUTION

The EPA warns that prolonged contact with used engine oil may cause a number of skin disorders, including cancer! You should make every effort to minimize your exposure to used engine oil. Protective gloves should be worn when changing the oil. Wash your hands and any other exposed skin areas as soon as possible after exposure to used engine oil. Soap and water, or waterless hand cleaner should be used.

19. Drain the engine oil.
20. Disconnect the exhaust pipe from the right side exhaust manifold.
21. Remove the A/C compressor from the mounting bracket and support the compressor out of the way. DO NOT disconnect the refrigerant lines from the compressor.
22. Remove the right front engine-to-transaxle brace.
23. Remove the flywheel cover mounting screws and remove the cover.
24. Disconnect and remove the starter.
25. Matchmark the flywheel to the converter. Remove the torque converter-to-flywheel bolts.
26. Lower the vehicle.
27. Install a suitable engine lifting device and slightly raise the engine to take the weight off the front engine mount.
28. Remove the torque axis mount.
29. Support the transaxle with a floor jack. Remove the engine-to-transaxle bolts.
30. Carefully remove the engine from the vehicle. Be sure as the engine is being raised that there are no electrical or vacuum connections still attached.

To install:

31. Lower the engine into the vehicle. Guide the engine into position against the transaxle and be sure the alignment dowels are correctly seated in the transaxle.
32. Reinstall the engine-to-transaxle bolts and tighten to 55 ft. lbs. (75 Nm).
33. Reinstall the torque axis mount. Tighten the mount-to-frame bolts to 64 ft. lbs. (87 Nm), the engine side through-bolt to 65 ft. lbs. (87 Nm) and the frame side through-bolt to 52 ft. lbs. (70 Nm).
34. Remove the engine lifting assembly and floor jack.
35. Raise and safely support the vehicle.
36. Line up the matchmarks on the torque converter and flywheel and install the mounting bolts. Tighten the mounting bolts to 46 ft. lbs. (62 Nm).
37. Reconnect and install the starter.
38. Reinstall the flywheel cover and tighten the mounting screws to 88 inch lbs. (10 Nm).
39. Reinstall the right side engine-to-transaxle bracket.
40. Position the A/C compressor in the mounting bracket and tighten the front mounting bolts to 44 ft. lbs. (60 Nm), the rear mounting bolts to 18 ft. lbs. (25 Nm) and the rear mounting nut to 18 ft. lbs. (25 Nm).

41. Reconnect the exhaust pipe to the rear exhaust manifold and tighten the mounting nuts to 18 ft. lbs. (25 Nm).
42. Lower the vehicle.
43. Reinstall the cruise control servo and connect the vacuum lines.
44. Reconnect the vacuum lines to the power brake booster and heater control hoses.
45. Reconnect the EVAP canister hoses to the throttle body.
46. Reconnect the fuel feed and return lines to the fuel rail and pressure regulator.
47. Reconnect the ignition assembly ground strap to the inner fender well mounting screw.
48. Reconnect the following wiring connectors to their related components:
- MAT sensor
- Throttle Position Sensor
- Idle Air Control valve
- Oxygen Sensor
- A/C compressor
- Oil Pressure Switch
- Power steering cutout switch
- Vehicle Speed Sensor
- Low oil level sensor
49. Reconnect the control cables to the throttle body lever and connect the throttle cable to the mounting bracket.
50. Reinstall the air inlet duct.
51. Reinstall the power steering pump in the mounting bracket.
52. Reinstall the serpentine belt(s).
53. Reconnect the main engine harness at the bulkhead connector.
54. Reconnect the negative battery cable to the engine block.
55. Reconnect the heater hoses to the heater core and install the upper and lower radiator hoses.

✳✳ WARNING

Operating the engine without the proper amount and type of engine oil will result in severe engine damage.

56. Refill the crankcase with engine oil.
57. Reconnect the negative battery cable.
58. Reinstall the hood on the vehicle.
59. Pressurize the fuel system and verify no leaks.
60. Refill and bleed the cooling system.

4.0L (VIN C) Engine

On this vehicle, some engine-related service procedures require that the entire Powertrain Assembly be removed from the vehicle. The requires specialized lifts and supports to lower the powertrain as the vehicle body is lifted up and off of the powertrain support cradle. Use care during this operation.

✳✳ CAUTION

The fuel injection system remains under pressure, even after the engine has been turned OFF. The fuel system pressure MUST BE relieved before disconnecting any fuel lines. Failure to do so may result in fire and/or personal injury.

1. Relieve the fuel system pressure using the recommended procedure.
2. Disconnect the negative battery cable.
3. Remove the air inlet duct.

✳✳ CAUTION

Never open, service or drain the radiator or cooling system when hot; serious burns can occur from the steam and hot coolant.

4. Drain the cooling system.

✳✳ CAUTION

The EPA warns that prolonged contact with used engine oil may cause a number of skin disorders, including cancer! You should make every effort to minimize your exposure to used engine oil. Protective gloves should be worn when changing the oil. Wash your hands and any other exposed skin areas as soon as possible after exposure to used engine oil. Soap and water, or waterless hand cleaner should be used.

5. Drain the engine oil.
6. Recover the refrigerant from the A/C system.
7. Disconnect and cap the fuel feed and return lines from the fuel rail.
8. Disconnect the vacuum harness from the rear of the intake manifold.
9. Disconnect the throttle cables from the throttle body lever and intake manifold bracket and position out of the way.
10. Detach the engine harness connector at the bulkhead.
11. Disconnect the coolant hoses from the radiator surge tank.
12. Disconnect the upper radiator hose from the coolant crossover.
13. Disconnect the lower radiator hose from the thermostat housing.
14. Disconnect and cap the upper transaxle cooler line from the radiator and the lower cooler line from the transaxle.

15. Disconnect the heater hoses from the engine compartment heater pipes.
16. Disconnect the vacuum supply line from the power booster.
17. Disconnect the shift lever from the transaxle manual shift lever and remove the transaxle cable bracket and position the cable assembly out of the way.
18. Remove the vacuum reservoir.
19. Detach the electrical connectors and vacuum lines from the cruise control servo.
20. Install an Engine Support Fixture, J-28467-A, or the equivalent.
21. Disconnect the positive battery cable at the junction block.
22. Remove the through-bolt from the torque axis mount.
23. Remove the front engine mount bolts from the torque axis mount bracket and set the mount aside.
24. Disconnect the negative battery cable from the engine block.
25. Raise and safely support the vehicle.
26. Remove both front wheels.
27. Remove the left and right inner fender well splash shields.
28. Detach the electrical connector from the power steering rack.
29. Remove the cotter pins and castle nuts from the lower ball joints and separate the ball joints from the steering knuckles using J-36226, or an equivalent puller.
30. Remove the cotter pins and castle nuts from the outer tie rod ends and separate the ends from the steering knuckles using J-36226, or the equivalent.
31. Remove the axle nuts and separate the halfshafts from the wheel bearing and hub assemblies.
32. Remove the pinch bolt from the power steering rack at the intermediate shaft.
33. Remove the complete exhaust system.
34. Disconnect the engine oil cooler lines at the adapter.
35. Remove the engine-to-transaxle brace located at the oil pan, right cylinder bank in the rear of the engine and the left bank on the front of the engine.
36. Remove the flywheel cover.
37. Matchmark the converter to the flywheel and remove the mounting bolts.
38. Remove the oil cooler adapter from the engine.
39. Disconnect the A/C hose and muffler from the rear of the A/C compressor.
40. Position a frame support table under the powertrain assembly.
41. Remove the left transaxle mount.
42. Remove the subframe mounting bolts.

43. Release the engine support fixture.

44. Raise the vehicle off of the engine table slowly and verify no electrical or vacuum connections are still hooked to the engine.

45. Remove the transaxle-to-engine.

46. Using a suitable engine lift, separate the engine from the transaxle and subframe assembly.

To install:

47. Position the engine assembly onto the subframe. Line up the engine dowels with the holes in the transaxle and install the mounting bolts. Tighten the bolts to 55 ft. lbs. (75 Nm).

48. Remove the engine lift and position the powertrain assembly under the vehicle.

49. Lower the vehicle and line up the subframe with the vehicle and install the mounting bolts and tighten to 142 ft. lbs. (192 Nm).

50. Reinstall the engine support fixture.

51. Raise the vehicle and remove the engine dolly.

52. Reinstall the left side transaxle mount.

53. Reconnect the A/C hose and muffler to the A/C compressor.

54. Reinstall the oil filter adapter to the engine block and tighten the mounting bolts to 12 ft. lbs. (16 Nm).

55. With the converter-to-flywheel matchmarks in alignment, tighten the mounting bolts to 35 ft. lbs. (47 Nm).

56. Reinstall the rear engine-to-transaxle brace and tighten the mounting bolts to 44 ft. lbs. (60 Nm).

57. Reinstall the front engine-to-transaxle brace and tighten the mounting bolts to 44 ft. lbs. (60 Nm).

58. Reinstall the oil pan brace and tighten the mounting bolts to 37 ft. lbs. (50 Nm).

59. Reinstall the flywheel cover.

60. Reconnect the oil cooler hoses to the adapter and tighten the fittings to 12 ft. lbs. (18 Nm).

61. Reinstall the exhaust system.

62. Reconnect the halfshafts to the hub bearing assemblies and tighten the axle nuts to 107 ft. lbs. (148 Nm).

63. Reconnect the tie rod ends to the steering knuckle and tighten the castle nuts to 41 ft. lbs. (55 Nm). If necessary tighten the castle nuts up to 60° additional to align the cotter pin holes. NEVER loosen the castle nut and DO NOT exceed 52 ft. lbs. (70 Nm) to make the holes align. Install new cotter pins.

64. Reconnect the ball joints to the steering knuckles and tighten the castle nuts to 88 inch lbs. (10 Nm) plus 120° additional. If necessary tighten the castle nuts up to 60° additional to align the cotter pin holes. NEVER loosen the castle nuts to make the holes align. Install new cotter pins.

65. Reconnect the intermediate shaft to the rack and pinion and tighten the pinch bolt to 35 ft. lbs. (47 Nm).

66. Reconnect the electrical connector to the rack and pinion assembly.

67. Reinstall the left and right inner fender splash shields.

68. Reinstall the front tire and wheel assemblies and tighten the wheel nuts to 100 ft. lbs. (140 Nm).

69. Lower the vehicle.

70. Reconnect the negative battery cable to the engine block.

71. Reinstall the front engine mount bracket-to-torque axis mount bolts.

72. Reinstall the torque axis mount through-bolt.

73. Reconnect the positive battery cable to the junction block.

74. Remove the engine support fixture.

75. Reconnect the electrical and vacuum connectors to the cruise control servo.

76. Install the vacuum reservoir.

77. Reconnect the shift cable to the transaxle manual shift lever and install the cable mounting bracket.

78. Reconnect the vacuum hose to the power booster.

79. Reconnect the heater hoses at the rear of the engine.

80. Reconnect the transaxle cooler lines.

81. Reconnect the lower radiator hose to the thermostat housing.

82. Reconnect the upper radiator hose to the coolant crossover.

83. Reconnect the hoses to the coolant surge tank.

84. Reconnect the main engine harness to the bulkhead connector.

85. Reconnect the throttle cables to the throttle body lever and install the intake manifold bracket.

86. Reconnect the vacuum hoses to the rear of the intake manifold.

87. Reconnect the fuel feed and return lines.

88. Refill the cooling system, engine crankcase and recharge the A/C system.

89. Reinstall the air inlet duct.

✳✳ WARNING

Operating the engine without the proper amount and type of engine oil will result in severe engine damage.

90. Refill the crankcase with engine oil.

91. Reconnect the negative battery cable.

92. Start the vehicle and verify no leaks.

Water Pump

REMOVAL & INSTALLATION

3.8L (VIN 1 and K) Engines

1. Disconnect the negative battery cable.

✳✳ CAUTION

Never open, service or drain the radiator or cooling system when hot; serious burns can occur from the steam and hot coolant.

2. Drain the cooling system into a suitable container.

3. Remove the serpentine belt.

4. Disconnect the heater and bypass hoses from the water pump.

5. Remove the water pump pulley bolts. The long bolt can be removed by lining the head of the bolt up with the hole in the frame rail. Remove the pulley.

6. Install an engine support fixture and remove the torque axis mount.

7. Remove the water pump mounting bolts and remove the water pump.

To install:

8. Thoroughly clean all the sealing surfaces.

9. Apply a thin bead of sealer around the outside edge of the water pump and install the gasket on the pump.

10. Install the water pump on the engine and tighten the water pump bolts on the sides of the water inlet and outlet to 29 ft. lbs. (39 Nm) and the remainder of the bolts to 97 inch lbs. (11 Nm).

11. Reinstall the torque axis mount and remove the support fixture.

12. Reinstall the water pump pulley and tighten the bolts finger-tight.

13. Reconnect the hoses to the water pump.

14. Reinstall the serpentine belt.

15. Tighten the water pump pulley bolts to 115 inch lbs. (13 Nm).

16. Refill the cooling system.

17. Reconnect the negative battery cable.

18. Start the vehicle and check for proper operation.

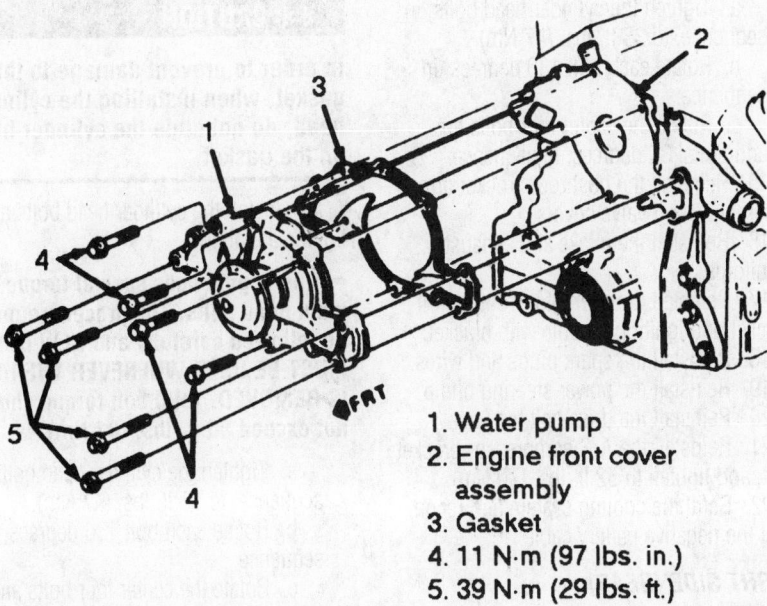

1. Water pump
2. Engine front cover assembly
3. Gasket
4. 11 N·m (97 lbs. in.)
5. 39 N·m (29 lbs. ft.)

7922XG01

Exploded view of the water pump and mounting bolt locations—3.8L engines

4.0L (VIN C) Engine

1. Disconnect the negative battery cable.

✳✳ CAUTION

Never open, service or drain the radiator or cooling system when hot; serious burns can occur from the steam and hot coolant.

2. Drain the cooling system.
3. Remove the air inlet duct.
4. Remove the water pump drive belt cover.
5. Remove the water pump drive belt and set aside.
6. Disconnect the lower radiator hose and bypass hose.
7. Remove the thermostat housing from the water pump housing.
8. Remove the water pump from the water pump housing by rotating turning the locking ring with J-38816, or an equivalent water pump remover/installer.

To install:

9. Install the water pump and seat the locking ring using J-38816.

10. Reinstall the thermostat housing to the water pump housing.
11. Reconnect the lower radiator hose and coolant bypass hose.
12. Reinstall the drive belt and drive belt cover.
13. Reinstall the air inlet duct.
14. Reconnect the negative battery cable.
15. Refill and bleed the cooling system.

Cylinder Head

REMOVAL & INSTALLATION

3.8L (VIN 1 and K) Engines

➡Head gaskets are not interchangeable. Failure to install them with the arrow pointing to the front, will cause gasket failure and possible engine failure. The left-hand gasket can be identified with the letter "L" punched in it (next to the arrow).

LEFT SIDE (FRONT)

1. Disconnect the negative battery cable.
2. Relieve the fuel system pressure using the recommended procedure.

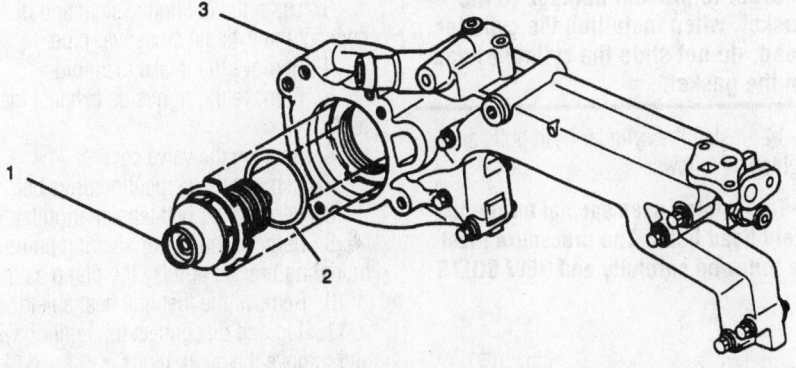

1	WATER PUMP ASM.
2	O-RING SEAL
3	WATER PUMP HOUSING ASM.

7922XG02

Exploded view of the water pump housing and water pump—4.0L (VIN C) engine

Never open, service or drain the radiator or cooling system when hot; serious burns can occur from the steam and hot coolant.

3. Drain the cooling system and remove the intake manifold.

4. Remove the left exhaust manifold.

5. Remove the valve covers, the rocker arm assemblies and pushrods, keeping everything in order for reinstallation.

6. Tag and disconnect the ignition wires and remove the spark plugs.

7. Remove the alternator front mounting bracket and the ignition module with bracket.

8. Remove the one bolt securing the A/C bracket to the cylinder head.

9. Remove the power steering pump.

10. Remove the drive belt tensioner.

11. Remove the cylinder head bolts in the reverse order of installation, and remove the cylinder head.

12. Thoroughly clean all sealing surfaces and the cylinder head bolt holes in the block.

To install:

13. Place the cylinder head gasket on the engine block dowels with the note **THIS SIDE UP** facing the cylinder head and the arrow facing the front of the engine. Position the cylinder head on the engine block.

In order to prevent damage to the gasket, when installing the cylinder head, do not slide the cylinder head on the gasket.

14. Install the cylinder head bolts and tighten as follows:

➡This engine uses special torque to yield head bolts. The procedure must be followed carefully and NEW BOLTS

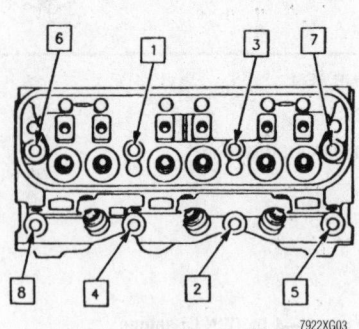

Cylinder head bolt tightening sequence—
3.8L (VIN 1 and K) engines

7922XG03

MUST BE USED WHENEVER THE HEAD IS REMOVED. Total bolt torque should not exceed 60 ft. lbs. (81 Nm).

   a. Tighten the cylinder head bolts, in sequence, to 35 ft. lbs. (47 Nm).

   b. Rotate each bolt 130 degrees, in sequence.

   c. Rotate the center four bolts an additional 30 degrees, in sequence.

15. Reinstall the pushrods, rocker arm assemblies and valve covers.

16. Reinstall the intake and exhaust manifolds.

17. Reinstall the alternator front mount bracket and ignition module with bracket.

18. Reinstall the spark plugs and wires.

19. Reinstall the power steering pump.

20. Reinstall the drive belt tensioner

21. Reinstall the A/C compressor bracket bolt, and tighten to 52 ft. lbs. (70 Nm).

22. Refill the cooling system and reconnect the negative battery cable.

RIGHT SIDE (REAR)

1. Disconnect the negative battery cable.

2. Relieve the fuel system pressure using the recommended procedure.

Never open, service or drain the radiator or cooling system when hot; serious burns can occur from the steam and hot coolant.

3. Drain the cooling system and disconnect the exhaust crossover pipe.

4. Remove the intake manifold.

5. Remove the right side exhaust manifold.

6. Remove the valve covers.

7. Remove the serpentine drive belt.

8. Remove the belt tensioner pulley.

9. Remove the power steering pump mounting bracket and lay the pump aside.

10. Remove the fuel line heat shield.

11. Tag and disconnect the ignition wires and remove the spark plugs.

12. Remove the rocker arm assemblies and pushrods, keeping everything in order for reinstallation.

13. Remove the cylinder head bolts in reverse order of installation and remove the cylinder head.

14. Thoroughly clean all sealing surfaces and the cylinder head bolt holes in the block.

To install:

15. Place the cylinder head gasket on the engine block dowels with the note **THIS SIDE UP** facing the cylinder head

and the arrow facing the front of the engine. Position the cylinder head on the engine block.

In order to prevent damage to the gasket, when installing the cylinder head, do not slide the cylinder head on the gasket.

16. Install the cylinder head bolts and tighten as follows:

➡This engine uses special torque to yield head bolts. The procedure must be followed carefully and NEW BOLTS MUST BE USED WHENEVER THE HEAD IS REMOVED. Total bolt torque should not exceed 60 ft. lbs. (81 Nm).

   a. Tighten the cylinder head bolts, in sequence, to 35 ft. lbs. (47 Nm).

   b. Rotate each bolt 130 degrees, in sequence.

   c. Rotate the center four bolts an additional 30 degrees, in sequence.

17. Reinstall the exhaust manifold and intake manifold.

18. Reinstall the pushrods and rocker arm assemblies.

19. Reinstall the valve cover(s).

20. Reinstall the spark plugs and wires.

21. Reinstall the power steering pump bracket and tighten the bolts to 35 ft. lbs. (47 Nm).

22. Reinstall the belt tensioner pulley and serpentine belt.

23. Reinstall the exhaust crossover pipe.

24. Refill the cooling system and connect the negative battery cable.

4.0L (VIN C) Engine

➡The manufacturer recommends that the entire powertrain be removed from the vehicle before removing the cylinder heads.

The fuel injection system remains under pressure, even after the engine has been turned OFF. The fuel system pressure must be relieved before disconnecting any fuel lines. Failure to do so may result in fire and/or personal injury.

1. Disconnect the negative battery cable.

2. Relieve the fuel system pressure using the recommended procedure.

✳✳ CAUTION

Never open, service or drain the radiator or cooling system when hot; serious burns can occur from the steam and hot coolant.

3. Drain the cooling system into a suitable container.

4. Remove the powertrain assembly.

5. Remove the intake manifold, cam covers, harmonic balancer, timing chain front cover and oil pump.

✳✳ WARNING

Align all timing marks before performing the next Step.

6. Remove the chain tensioner from the timing chain for the cylinder head being removed.

7. Remove the cam sprockets from the head being removed. The timing chain remains in the chain case.

8. Removing the timing chain guides. Access for the retaining screws is through the plugs at the front of the cylinder head.

9. Remove the water crossover.

10. Remove the exhaust manifold.

11. Remove the cylinder head bolts, a little at a time, in the reverse order of the tightening sequence.

12. Remove the cylinder head and gasket.

✳✳ CAUTION

With the camshafts remaining in the cylinder head, some valves will be open at all times. Do not rest the cylinder head on a flat service with the cylinder face down, or valve damage will result.

13. Clean all gasket sealing surfaces. Clean the head bolt holes in the crankcase with compressed air and clean the head bolt bosses in the cylinder head.

✳✳ CAUTION

Be careful when cleaning aluminum gasket surfaces to prevent damage to the sealing surfaces. Use only plastic, wood or "dull" gasket scrapers. Chemical agents to dissolve gasket materials can also be used by following the manufacturers directions.

14. Check the cylinder head for warpage using a straightedge and feeler gauge. Measure along each edge, at the center and across both ends.

15. If warpage is less than 0.002 in. (0.05mm), the cylinder head surface is usable. If warpage is 0.002–0.008 in. (0.05–0.2mm), the cylinder head must be resurfaced. After resurfacing, the dimension between the combustion chamber gauge pad and the deck surface must be at least 10.5mm.

To install:

16. Using a new cylinder head gasket, install the cylinder head and the ten M11 and three M6 head bolts. Lube the washer and the underside of the bolt head with engine oil prior to installation. New replacement head bolts are recommended.

17. Tighten the ten M11 bolts, in sequence, to 22 ft. lbs. (30 Nm) plus 90 degrees ¼ turn). Repeat the sequence, turning each bolt an additional 75 degrees (total 165 degrees). Tighten the three M6 bolts to 10 ft. lbs. (12 Nm).

18. Reinstall the camshafts and set the camshaft timing.

19. Reinstall the camshaft guide bolt access hole plugs in the cylinder heads. The plugs should be seated and snug.

20. Reinstall the intake cam covers, oil pump, timing chain front cover, harmonic balancer, intake manifold and water crossover.

21. Reinstall the exhaust manifold. Tighten the nuts to 22 ft. lbs. (30 Nm) or the bolts to 18 ft. lbs. (25 Nm).

22. Reinstall the powertrain assembly into the vehicle.

23. Reconnect the negative battery cable.

24. Refill the cooling system and check all fluid levels.

25. Properly charge the A/C system.

26. Run the engine and check for leaks and proper engine performance.

Rocker Arms

REMOVAL & INSTALLATION

➡**All valvetrain components should be kept in the order that they were removed, so that they can be reinstalled in the their original position.**

3.8L (VIN 1 and K) Engines

LEFT SIDE (FRONT)

1. Disconnect the negative battery cable.

2. Disconnect the spark plug wires from the plugs and route the wires out of the way.

3. Remove the rocker arm cover bolts and remove the rocker arm cover.

4. Remove the rocker arm mounting bolt and remove the pedestal and rocker arm.

5. Remove the pushrod.

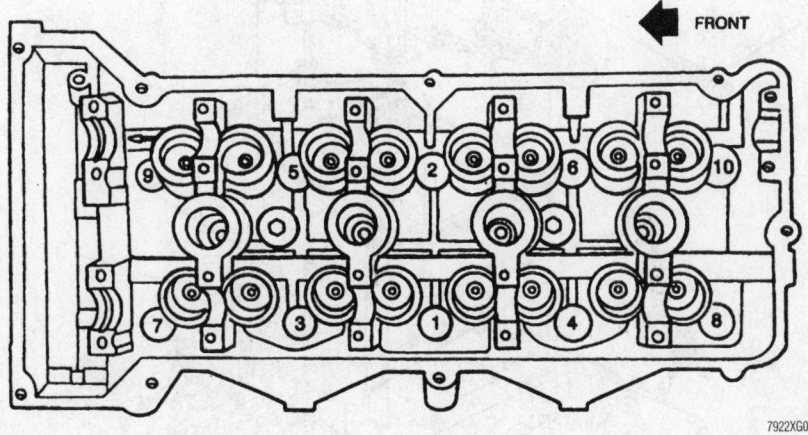

Cylinder head bolt tightening sequence—4.0L (VIN C) engine

To install:

6. Install the pushrod into the lifter and install the rocker arm and pedestal. Tighten the mounting bolt to 19 ft. lbs. (25 Nm) plus 70 degrees additional rotation.

7. Reinstall the rocker arm cover with a new gasket and tighten the mounting bolt to 88 ft. lbs. (10 Nm).

8. Reconnect the spark plug wires to the front plugs.

9. Reconnect the negative battery cable.

RIGHT SIDE (REAR)

1. Disconnect the negative battery cable.
2. Remove the serpentine belt.
3. Remove the alternator rear brace.
4. Disconnect the spark plug wires from the plugs and route the wires out of the way.
5. Remove the rocker arm cover bolts and remove the rocker arm cover.
6. Remove the rocker arm mounting bolt and remove the pedestal and rocker arm.
7. Remove the pushrod.

To install:

8. Install the pushrod into the lifter and install the rocker arm and pedestal. Tighten the mounting bolt to 19 ft. lbs. (25 Nm) plus 70 degrees additional rotation.

9. Reinstall the rocker arm cover with a new gasket and tighten the mounting bolt to 88 ft. lbs. (10 Nm).

10. Reconnect the spark plug wires to the front plugs.

11. Reinstall the alternator rear brace.

12. Reinstall the serpentine belt.

13. Reconnect the negative battery cable.

4.0L (VIN C) Engine

The 4.0L engine is not equipped with rocker arms. The camshaft directly actuates the valves.

Supercharger

REMOVAL & INSTALLATION

3.8L (VIN 1) Engine
1995–96 MODELS

The supercharger oil level should be checked every 30,000 miles or 36 months. Remove the oil only when the engine is cold. The oil level should be maintained to the bottom of the threads in the inspection hole. Do not use petroleum based oil. Use only GM 12345982 synthetic oil or equivalent 5W-30 synthetic oil. Use of the wrong oil can cause the supercharger to fail.

✳✳ CAUTION

The fuel injection system remains under pressure, even after the engine has been turned OFF. The fuel system pressure MUST BE relieved before disconnecting any fuel lines. Failure to do so may result in fire and/or personal injury.

1. Relieve the fuel system pressure.
2. Disconnect the negative battery cable.
3. Remove the drive belt from the supercharger pulley. It is not necessary to remove the drive belt from the remainder of the pulleys.
4. Remove the plastic fuel injector shield.
5. Disconnect and cap the fuel lines from the fuel rail.
6. Disconnect the vacuum lines from the supercharger.
7. Detach the electrical connectors from the fuel injector.
8. Remove the electrical harness shield from the front of the supercharger and detach the electrical connector from the supercharger.
9. Remove the fuel rail mounting bolts and remove the fuel rail.

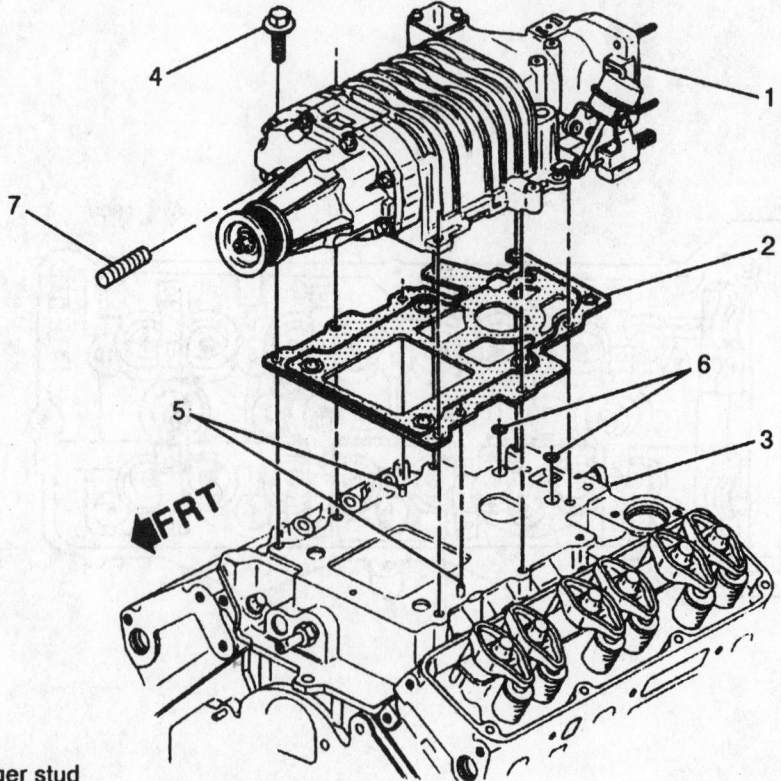

1 Supercharger
2 Supercharger gasket
3 Lower intake manifold
4 Supercharger bolts (8)
5 Locator pins
6 Coolant passage o-rings (2)
7 Tensioner bracket to supercharger stud

Exploded view of the supercharger mounting—3.8L (VIN 1) Engine

7922XG05

10. Detach the following electrical connectors and lay the harness aside:
- Idle Air Control (IAC) valve
- Throttle Position Sensor (TPS)
- MAP sensor
- MAF sensor
- EGR valve
- Boost control solenoid
- Coolant temperature sensor

11. Disconnect the air intake duct from the throttle body.

12. Disconnect the EGR pipe from the supercharger.

13. Disconnect the control cables from the throttle body.

14. Remove the boost pressure manifold and vacuum block.

15. Remove the control cable bracket.

16. Remove the tensioner bracket-to-supercharger mounting stud.

➥The stud must be removed or the supercharger can not be lifted off the lower intake manifold mounting dowels.

17. Disconnect the throttle body from the supercharger.

18. Remove the supercharger-to-intake manifold bolts.

19. Remove the supercharger, gasket and coolant passage O-rings.

To install:

20. Install new coolant passage O-rings and new supercharger-to-intake manifold gasket.

21. Install the supercharger and the mounting bolts. Only tighten the bolts finger-tight.

22. Install the tensioner bracket-to-supercharger stud and tighten to 88 inch lbs. (10 Nm).

23. Tighten the supercharger mounting bolts to 19 ft. lbs. (26 Nm).

24. Install the tensioner bracket nut and tighten to 37 ft. lbs. (50 Nm).

25. Install the throttle body and tighten the mounting nuts to 11 ft. lbs. (15 Nm).

26. Install the boost pressure manifold.

27. Install the vacuum block with a new gasket and tighten the mounting bolt to 62 inch lbs. (7 Nm).

28. Install the control cable bracket.

29. Reconnect the control cables to the throttle body.

30. Reconnect the EGR pipe to the supercharger.

31. Install the air intake duct.

32. Reconnect the electrical connectors to the following components:
- Idle Air Control (IAC) valve
- Throttle Position Sensor (TPS)

- MAP sensor
- MAF sensor
- EGR valve
- Boost control solenoid
- Coolant temperature sensor

33. Install the fuel rail and mounting bolts.

34. Reconnect the electrical connectors to the fuel injectors.

35. Reconnect the vacuum hoses to the supercharger.

36. Reconnect the harness to the front of the supercharger and install the harness shield.

37. Install the drive belt.

38. Reconnect the negative battery cable.

39. Start the vehicle and verify no leaks.

40. Install the fuel injector sight shield.

1997–99 MODELS

1. Disconnect the negative battery cable.

2. Remove the engine cover.

3. Relieve the fuel system pressure using the recommended procedure.

4. Remove the drive belt from the supercharger pulley. It is not necessary to remove the drive belt from the remainder of the pulleys.

5. Disconnect the right side spark plug wires from the ignition module and set aside.

6. Remove the alternator brace.

7. Detach the electrical connectors from the fuel injectors.

8. Remove the MAP sensor bracket.

9. Remove the fuel rail mounting bolts and the fuel rail with the injectors.

10. Remove the boost control solenoid.

11. Remove the throttle body nuts and the supercharger.

To install:

12. Clean all sealing surfaces.

13. Install the new supercharger to the intake gasket and install the supercharger. Tighten the bolts to 17 ft. lbs. (23 Nm).

14. Reinstall the MAP sensor bracket.

15. Reinstall the throttle body to the supercharger.

16. Reinstall the boost control solenoid.

17. Reinstall the fuel rail and reconnect the lines.

18. Reinstall the alternator brace.

19. Reconnect the right side spark plug wires to the ignition module.

20. Reinstall the supercharger belt.

21. Reinstall the engine cover.

22. Reconnect the negative battery cable.

23. Start the engine and check for proper operation and no leaks.

Camshaft and Valve Lifters

REMOVAL & INSTALLATION

➥All valvetrain components should be kept in the order that they were removed, so that they can be reinstalled in the their original position.

3.8L (VIN 1 and K) Engines

✳✳ CAUTION

The fuel injection system remains under pressure, even after the engine has been turned OFF. The fuel system pressure MUST BE relieved before disconnecting any fuel lines. Failure to do so may result in fire and/or personal injury.

1. Relieve the fuel system pressure using the recommended procedure.

2. Disconnect the negative battery cable.

3. Remove the engine from the vehicle and mount the engine on a suitable engine stand.

4. Remove the intake manifold.

5. Remove the rocker arm covers.

6. Remove the rocker arm assemblies, and pushrods.

7. Remove the lifter guide retainer mounting bolts and retainer.

8. Remove the lifter guides.

9. Remove the lifters from their bores.

10. Remove the crankshaft balancer center bolt and using a puller remove the balancer from the crankshaft.

11. Remove the crankshaft sensor cover and disconnect the crankshaft sensor.

12. Remove the timing chain front cover.

➥Align the timing marks of the camshaft and crankshaft sprockets to avoid burring the camshaft journals by the crankshaft.

13. Remove the camshaft sprocket and timing chain.

14. Remove the camshaft thrust plate bolts and remove the thrust plate.

15. Carefully remove the camshaft.

To install:

16. Coat the camshaft lobes and bearings with prelube prior to installation.

17. Install the camshaft into the engine.

18. Reinstall the camshaft thrust plate and tighten the mounting bolts to 10 ft. lbs. (14 Nm).

19. Reinstall the camshaft sprocket and timing chain.

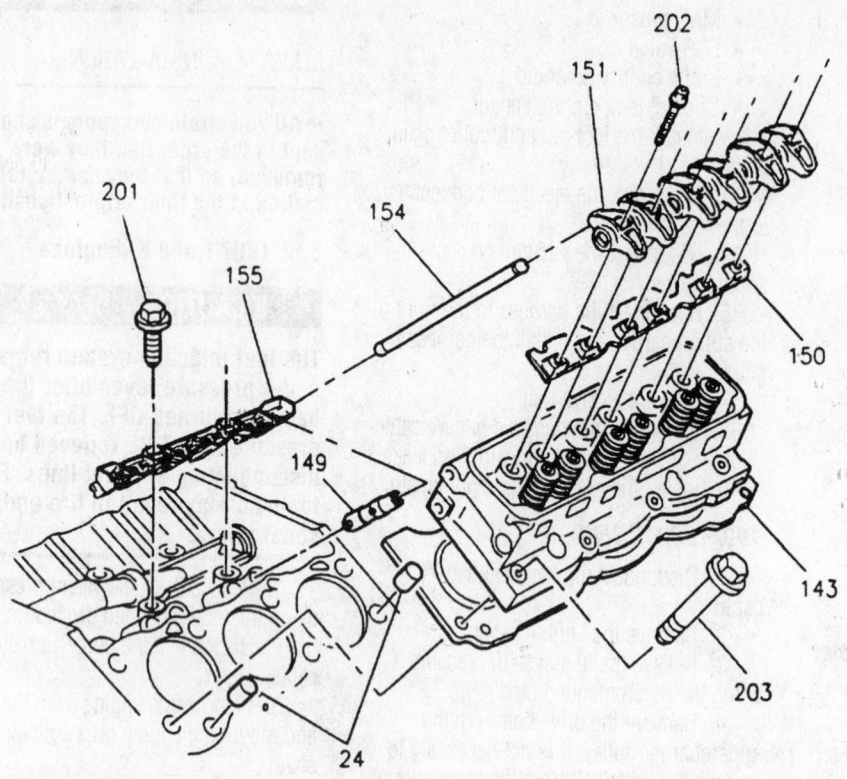

24	DOWEL PIN
143	HEAD GASKET
149	VALVE LIFTER
150	PIVOT RETAINER
151	ROCKER ARM
154	PUSHROD
155	LIFTER GUIDE
201	BOLT
202	BOLT
203	HEAD BOLT

Exploded view of the cylinder head and valvetrain components—3.8L (VIN 1 and K) engines

20 Reinstall the front timing chain cover.

21. Reconnect the crankshaft sensor and install the crankshaft sensor cover.

22. Reinstall the crankshaft balancer and tighten the mounting bolt to:

a. 1995 3.8L (VIN 1 and K) engines: 103 ft. lbs. (140 Nm) plus 56 degrees additional rotation.

b. 1996–99 3.8L (VIN 1 and K) engines: 111 ft. lbs. (150 Nm) plus 76 degrees additional rotation.

➡**If the camshaft was replaced the lifters should also be replaced. The old lifters have developed a wear pattern and will cause the new camshaft to wear prematurely. New lifters MUST be installed with a new camshaft.**

23. Coat the valve lifters with prelube and install the lifters in the lifter bores.

24. Reinstall the lifter guides and lifter guide retainer. Tighten the retainer mounting bolts to 27 ft. lbs. (37 Nm).

25. Reinstall the pushrods and rocker arms and tighten the rocker arm bolts to 28 ft. lbs. (38 Nm).

26. Reinstall the rocker arm covers.

27. Reinstall the intake manifold.

28. Reinstall the engine in the vehicle.

29. Reconnect the negative battery cable, start the engine and check for leaks.

4.0L (VIN C) Engine

LEFT SIDE (FRONT)

1. Disconnect the negative battery cable.

2. Remove the intake manifold sight shield.

3. Raise and safely support the vehicle.

✴✴ CAUTION

Never open, service or drain the radiator or cooling system when hot; serious burns can occur from the steam and hot coolant.

4. Partially drain the cooling system to a level below the water pump assembly.

5. Remove the oil level indicator tube.

6. Lower the vehicle.

7. Disconnect the upper radiator hose from the thermostat housing.

8. Disconnect the spark plug wires from the front plugs and route the wires out of the way.

9. Remove the upper radiator support assembly.

10. Disconnect the PCV fresh air tube from the left side camshaft cover.

11. Remove the air inlet duct.

12. Disconnect the EGR outlet pipe.

13. Remove the water pump drive belt shield.

14. Remove the water pump drive belt and drive belt tensioner.

15. Remove the coolant pump pulley using J-38825, or an equivalent pulley remover.

16. Remove the camshaft seal retainer screws and remove the camshaft seal.

17. Remove the camshaft cover bolts.

18. Remove the camshaft cover by pivoting up the intake manifold side of the cover 10 inches. Lift up the exhaust manifold side of the cover 2 inches. Swing the oil fill cap end of the cover up over the intake manifold and slide the cover over the camshafts.

19. Secure the cam sprocket to the timing chain by installing tie-wraps through the cam sprocket holes. Use four tie-wraps per sprocket.

✴✴ CAUTION

The sprocket/chain relationship must be maintained throughout this procedure or camshaft timing will be lost and require further engine disassembly to retime.

20. Working from behind the sprockets, install Cam Chain Holder J 38815 so that it is positioned between the chain tensioner and chain guide. Apply tension to the tool by tightening the tension adjusting screw.

21. Remove both cam sprocket bolts. Note the relative location of the cam drive pins in the end of the camshafts.

22. Work the sprockets off the cams using play in the chain.

23. Alternately loosen the cam bearing cap screws a few turns at a time until all valve spring pressure has been released. Remove the bolts and caps.

24. Remove the camshaft.

25. Inspect the camshaft for excessive lobe wear such as the evidence of grooves, scoring or flaking. Check the bearing journals, making sure they are not scored or burned. Replace the camshaft, as necessary.

To install:

26. Lubricate the camshaft lobes with camshaft prelube 1052365 or equivalent. Lubricate the camshaft journals with clean engine oil.

27. Install the camshaft.

28. Position the cam bearing caps to the cylinder head.

➡**Each cap is identified for position and direction. The arrow points towards the front of the engine. An "E" indicates a cap for the exhaust cam. An "I" indicates a cap for the intake cam. Position No. 1 is towards the front of the engine.**

29. Loosely install the cam bearing cap bolts.

30. Alternately tighten the cam bearing cap bolts a few turns at a time against valve spring pressure until all the bolts are snug. Tighten the bolts to 9 ft. lbs. (12 Nm).

31. Using the hex cast into the camshaft, rotate the cams until the drive pins are in position to engage the cam sprockets over the cams and install the retaining bolts.

32. Work the cam sprockets over cams and install the retaining bolts. Tighten the bolts to 90 ft. lbs. (120 Nm).

33. Remove the chain holder J 38815.

34. Remove the tie-wraps from the cam sprockets.

35. Install a new camshaft cover gasket and spark plug seals in the camshaft cover.

36. Reinstall the back end of the camshaft cover first while keeping the front end of the cover high. Pivot the oil fill end of the cover into place once the back edge of the cover is clear of the tensioner assembly. Hold the seal in place and position the cover so it is square with the cylinder head. Slide the cover left and down onto the cylinder head.

37. Install the camshaft cover bolts and tighten to 89 inch lbs. (10 Nm).

38. Lubricate the seal lip with petroleum jelly and install the camshaft seal retainer over the end of the intake camshaft. Install the mounting bolts and tighten to 10 inch lbs. (1.1 Nm).

39. Reinstall the water pump pulley using J-38823, or an equivalent pulley installer.

40. Reinstall the drive belt tensioner and drive belt.

41. Reinstall the water pump belt shield.

42. Reconnect the EGR outlet pipe.

43. Reconnect the upper radiator hose to the thermostat housing.

44. Reconnect the PCV hose to the camshaft cover.

45. Reinstall the air inlet duct.

46. Reconnect the spark plug wires to the front plugs.

47. Reinstall the upper radiator support.

48. Raise and safely support the vehicle.

49. Reinstall the oil level indicator tube.

50. Lower the vehicle.

51. Refill the cooling system.

52. Reinstall the intake manifold sight shield.

53. Reconnect the negative battery cable.

54. Run the engine and check for leaks and proper engine operation.

RIGHT SIDE (REAR)

1. Disconnect the negative battery cable.

2. Remove the intake manifold sight shield.

3. Remove the vacuum reservoir.

4. Remove the cruise control servo from its mount and position out of the way.

5. Detach the ignition assembly electrical connectors.

6. Disconnect the spark plug wires from the right side plugs and remove the ignition assembly from its mount and position out of the way.

7. Remove the PCV valve from the camshaft cover.

8. Remove the EVAP canister solenoid from the right camshaft cover.

9. Raise and safely support the vehicle.

10. Disconnect the knock sensor, vehicle speed sensor and power steering pressure switch.

11. Lower the vehicle.

12. Disconnect the wiring harness retainers from the cover.

13. Remove the camshaft cover bolts and remove the camshaft cover.

14. Secure the cam sprocket to the timing chain by installing tie-wraps through the cam sprocket holes. Use four tie-wraps per sprocket.

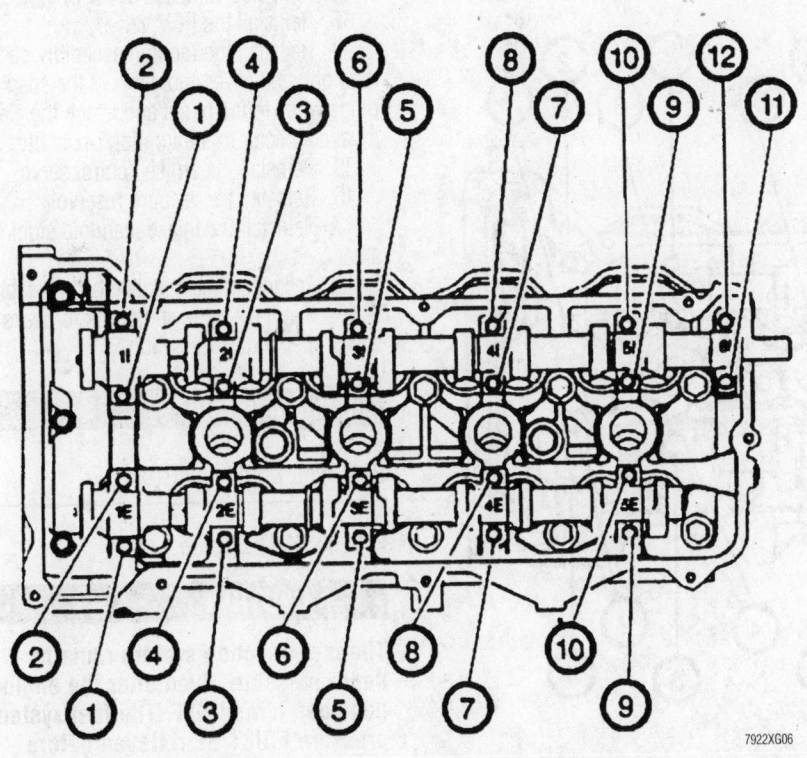

7922XG06

Left cylinder head camshaft bearing cap tightening sequence—4.0L (VIN C) engine

❋❋ CAUTION

The sprocket/chain relationship must be maintained throughout this procedure or camshaft timing will be lost and require further engine disassembly to retime.

15. Working from behind the sprockets, install Cam Chain Holder J 38815 so that it is positioned between the chain tensioner and chain guide. Apply tension to the tool by tightening the tension adjusting screw.

16. Remove both cam sprocket bolts. Note the relative location of the cam drive pins in the end of the camshafts.

17. Work the sprockets off the cams using play in the chain.

18. Alternately loosen the cam bearing cap screws a few turns at a time until all valve spring pressure has been released. Remove the bolts and caps.

19. Remove the camshaft.

20. Inspect the camshaft for excessive lobe wear such as the evidence of grooves, scoring or flaking. Check the bearing journals, making sure they are not scored or burned. Replace the camshaft, as necessary.

To install:

❋❋ CAUTION

The EPA warns that prolonged contact with used engine oil may cause a number of skin disorders, including cancer! You should make every effort to minimize your exposure to used engine oil. Protective gloves should be worn when changing the oil. Wash your hands and any other exposed skin areas as soon as possible after exposure to used engine oil. Soap and water, or waterless hand cleaner should be used.

21. Lubricate the camshaft lobes with Camshaft Prelube 1052365 or equivalent. Lubricate the camshaft journals with clean engine oil.

22. Install the camshaft.

23. Position the cam bearing caps to the cylinder head.

➡Each cap is identified for position and direction. The arrow points towards the front of the engine. An "E" indicates a cap for the exhaust cam. An "I" indicates a cap for the intake cam. Position No. 1 is towards the front of the engine.

24. Loosely install the cam bearing cap bolts.

25. Alternately tighten the cam bearing cap bolts a few turns at a time against valve spring pressure until all the bolts are snug. Tighten the bolts to 9 ft. lbs. (12 Nm).

26. Using the hex cast into the camshaft, rotate the cams until the drive pins are in position to engage the cam sprockets over the cams and install the retaining bolts.

27. Work the cam sprockets over cams and install the retaining bolts. Tighten the bolts to 90 ft. lbs. (120 Nm).

28. Remove the Chain Holder special tool J 38815.

29. Remove the tie-wraps from the cam sprockets.

30. Install a new camshaft cover gasket and spark plug seals in the camshaft cover.

31. Reinstall the camshaft cover on the cylinder head and tighten the mounting bolts to 89 inch lbs. (10 Nm).

32. Reconnect the wiring harness retainers to the camshaft cover.

33. Raise and safely support the vehicle.

34. Reconnect the knock sensor, vehicle speed sensor and power steering pressure switch.

35. Lower the vehicle.

36. Reinstall the EVAP canister solenoid on the right side camshaft cover.

37. Reinstall the PCV valve.

38. Reinstall the ignition assembly on the right side cover and connect the spark plugs wires to the plugs and attach the electrical connector to the ignition assembly.

39. Reinstall the cruise control servo.

40. Reinstall the vacuum reservoir.

41. Reinstall the intake manifold sight shield.

42. Reconnect the negative battery cable.

43. Run the engine and check for leaks and proper engine operation.

Intake Manifold

REMOVAL & INSTALLATION

3.8L (VIN 1) Engine

❋❋ CAUTION

The fuel injection system remains under pressure, even after the engine has been turned OFF. The fuel system pressure MUST BE relieved before disconnecting any fuel lines. Failure to do so may result in fire and/or personal injury.

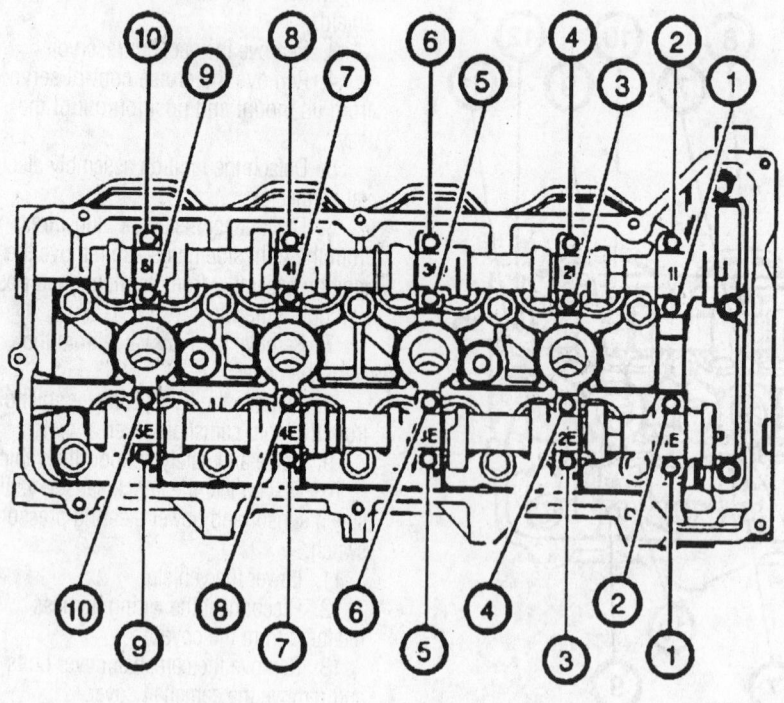

Right cylinder head camshaft bearing cap tightening sequence—4.0L (VIN C) engine

7922XG07

1. Relieve the fuel system pressure using the recommended procedure.
2. Disconnect the negative battery cable.
3. Raise and safely support the vehicle.

❊❊ CAUTION

Never open, service or drain the radiator or cooling system when hot; serious burns can occur from the steam and hot coolant.

4. Remove the front splash shield and drain the cooling system.
5. Close the drain cock, reinstall the splash shield and lower the vehicle.
6. Remove the supercharger using the recommended procedure.
7. Remove the thermostat housing.
8. Remove the EGR tube at the intake manifold.
9. Detach the electrical connection at the temperature sensor.
10. Remove the intake manifold mounting bolts and the intake manifold and gaskets.

To install:

11. Thoroughly clean all sealing surfaces.
12. Install the intake manifold using new manifold gaskets and install the intake manifold bolts finger-tight. With all the bolts in place tighten the bolts in sequence to 11 ft. lbs. (15 Nm). With all the bolts tightened, make a second pass tightening each bolt to 11 ft. lbs. (15 Nm).
13. Reinstall the electrical connector at the temperature sensor.
14. Reinstall the EGR tube to the intake manifold.
15. Reinstall the thermostat housing.
16. Reinstall the supercharger assembly using the recommended procedure.
17. Reconnect the negative battery cable, start the vehicle and check for proper operation.
18. Pressurize the fuel system, refill and bleed the cooling system.
19. Start the vehicle, check for proper operation and no leaks.

3.8L (VIN K) Engine

➡The 3.8L (VIN K) engine has two bolts which are hidden beneath the upper intake manifold. These bolts are located in the right front and left rear corners of the lower intake manifold. It is necessary to remove the upper intake manifold to service the lower intake manifold.

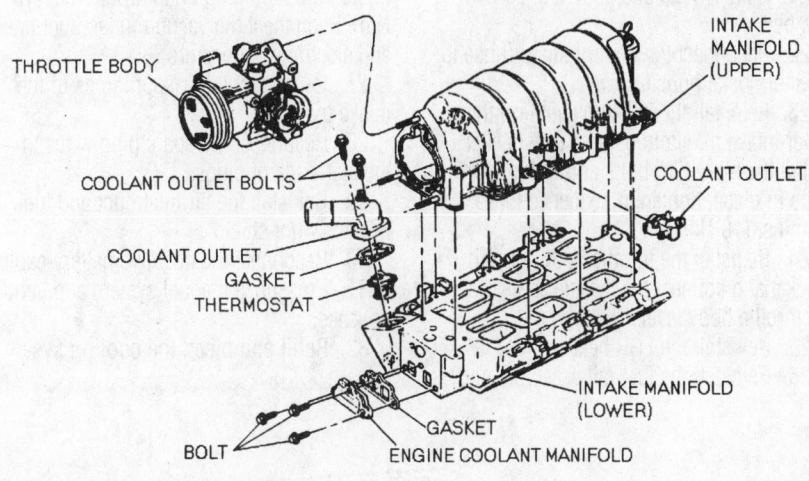

Exploded view of the upper intake plenum—3.8L (VIN K) engine

❊❊ CAUTION

The fuel injection system remains under pressure, even after the engine has been turned OFF. The fuel system pressure MUST BE relieved before disconnecting any fuel lines. Failure to do so may result in fire and/or personal injury.

1. Relieve the fuel system pressure using the recommended procedure.
2. Disconnect the negative battery cable.
3. Remove the fuel injector sight shield and air inlet duct.
4. Remove the air intake duct.
5. Disconnect the spark plug wires from the right side of the engine and set aside.
6. Remove the manifold vacuum source.
7. Remove the fuel rail and the EGR heat shield.

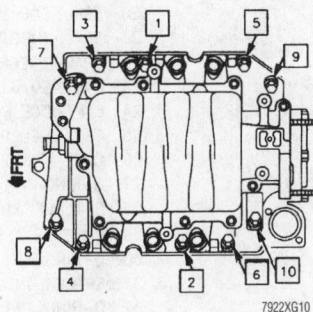

Lower intake manifold mounting bolt tightening sequence—3.8L (VIN 1 and K) engines

8. Remove the throttle cable bracket from the cylinder head mounting bracket and disconnect the cables from the throttle body lever.
9. Remove the throttle body support bracket.
10. Remove the upper intake plenum mounting bolts and remove the plenum and gasket.

❊❊ CAUTION

Never open, service or drain the radiator or cooling system when hot; serious burns can occur from the steam and hot coolant.

11. Drain the cooling system.
12. Disconnect the upper radiator hose from the thermostat housing.
13. Remove the alternator.
14. Remove the four bolts and remove the drive belt tensioner assembly.
15. Disconnect the EGR valve outlet pipe.
16. Remove the lower intake manifold mounting bolts and remove the intake manifold and gaskets.

To install:

17. Thoroughly clean all sealing surfaces.
18. Install the intake manifold using new manifold gaskets and install the intake manifold bolts finger-tight. With all the bolts in place tighten the bolts in sequence to 11 ft. lbs. (15 Nm). With all the bolts tightened, make a second pass tightening each bolt to 11 ft. lbs. (15 Nm).
19. Reconnect the EGR valve outlet pipe.
20. Reinstall the drive belt tensioner and tighten the mounting bolts to 37 ft. lbs. (50 Nm).

21. Reinstall the alternator and serpentine belt.

22. Reconnect the upper radiator hose to the thermostat housing.

23. Reinstall the intake plenum on the lower intake manifold using a new gasket. Install the mounting bolts and with all the bolts in place, tighten in sequence to 89 inch lbs. (10 Nm).

24. Reinstall the throttle body support bracket and connect the throttle cables to the throttle body lever.

25. Reinstall the EGR heat shield.

26. Reinstall the fuel rail assembly and tighten the mounting bolts to 7 ft. lbs. (10 Nm). Reconnect the vacuum lines, fuel lines and electrical connectors.

27. Reconnect the vacuum lines to the intake manifold.

28. Reconnect the spark plug wires to the rear bank of plugs.

29. Reinstall the air inlet duct and fuel injector sight shield.

30. Reconnect the negative battery cable.

31. Pressurize the fuel system and verify no leaks.

32. Refill and bleed the cooling system.

4.0L (VIN C) Engine

✳✳ CAUTION

The fuel injection system remains under pressure, even after the engine has been turned OFF. The fuel system pressure MUST BE relieved before disconnecting any fuel lines. Failure to do so may result in fire and/or personal injury.

1. Relieve the fuel system pressure using the recommended procedure.
2. Disconnect the negative battery cable.

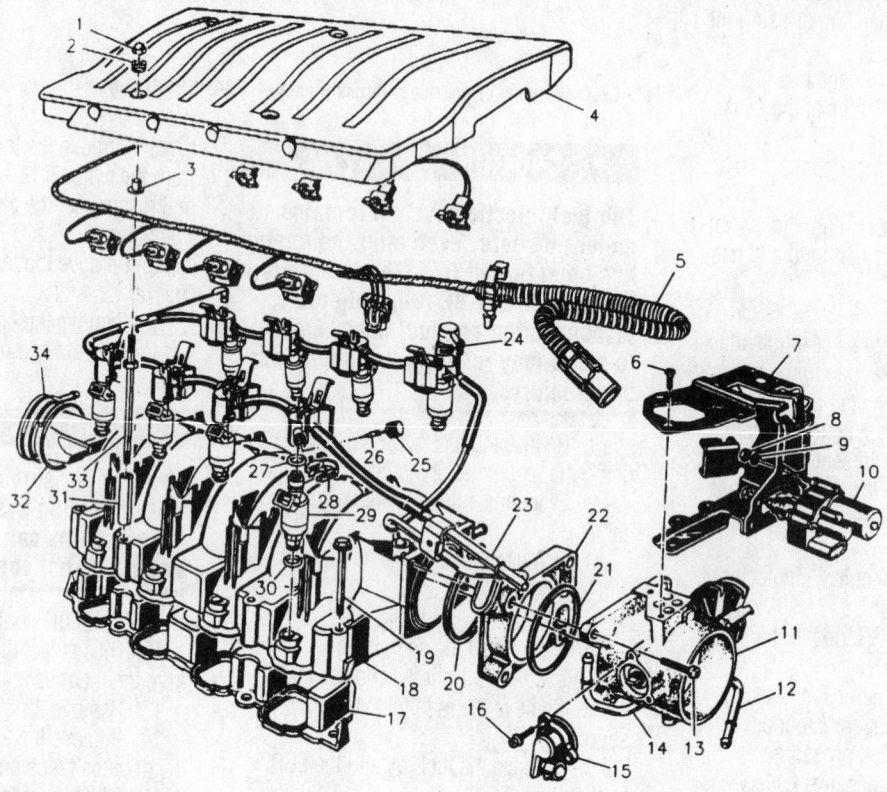

1   NUT, ATTACHING COVER
2   GROMMET, COVER
3   WASHER, COVER GROMMET
4   COVER, INTAKE MANIFOLD
5   WIRING HARNESS ASM
6   SCREW, ISC BRACKET ASM ATTACHING
7   BRACKET ASM, IDLE SPEED CONTROL (ISC)
8   NUT, HEX
9   WASHER, LOCK
10 ACTUATOR ASM, IDLE SPEED CONTROL (ISC)
11 BODY ASM, THROTTLE
12 TUBE, COOLANT OUTLET
13 BOLT, THROTTLE BODY ATTACHING
14 TUBE, COOLANT INLET
15 SENSOR, THROTTLE POSITION (TP)
16 SCREW ASM, TP SENSOR ATTACHING
17 GASKET, INTAKE MANIFOLD

18 HOUSING ASM, INTAKE MANIFOLD
19 BOLT, INTAKE MANIFOLD ATTACHING
20 SEAL, EGR TRANSFER SPACER TO INTAKE MANIFOLD
21 SEAL, THROTTLE BODY TO EGR TRANSFER SPACER
22 SPACER, EGR TRANSFER
23 RAIL ASM, FUEL
24 CARTRIDGE REGULATOR ASM, FUEL PRESSURE
25 CAP, FUEL PRESSURE CONNECTION
26 CORE ASM, FUEL PRESSURE CONNECTION VALVE
27 O-RING, MFI FUEL INJECTOR UPPER
28 CLIP, MFI FUEL INJECTOR RETAINER
29 INJECTOR ASM, MFI FUEL
30 O-RING, MFI FUEL INJECTOR LOWER
31 SPACER, INTAKE MANIFOLD
32 O-RING, PRESSURE RELIEF VALVE
33 STUD, INTAKE MANIFOLD/COVER ATTACHING
34 VALVE ASM, PRESSURE RELIEF

7922XG11

Exploded view of the intake manifold and related components —4.0L (VIN C) engine

3. Remove the four cap nut and washers and remove the intake manifold sight shield.

4. Disconnect the PCV hose from the intake manifold.

5. Disconnect the front spark plug wires and route them out of the way.

6. Disconnect the main fuel injector harness.

7. Detach the electrical connectors from the ISC motor, TPS switch and the MAP sensor.

8. Disconnect the fuel rail ground wires from the right cylinder head.

9. Disconnect the vacuum hoses from the power brake booster and throttle body.

10. Disconnect the crankcase vent hose at the air inlet duct.

11. Disconnect the EGR outlet tube.

12. Disconnect and cap the fuel lines from the fuel rail and remove the line retaining bolt near the throttle body.

13. Remove the air inlet duct.

14. Disconnect and cap the throttle body coolant hoses.

15. Disconnect the control cables from the throttle body lever.

16. Take note of the positions of the four studs and remove the six intake manifold mounting bolts and four mounting studs.

17. Remove the intake manifold.

To install:

18. Install a new intake manifold gasket and position the manifold in place and install the six mounting bolts and four studs. With all the bolts in place tighten the bolts to 89 inch lbs. (10 Nm) starting in the center and working in a circular pattern.

19. Reconnect the control cables to the throttle body lever.

20. Reconnect the throttle body coolant hoses and add coolant as necessary.

21. Reinstall the air inlet duct.

22. Reconnect the fuel lines to the fuel rail and install the line retaining bolt near the throttle body.

23. Reconnect the EGR outlet pipe.

24. Reconnect the crankcase vent pipe to the air inlet duct.

25. Reconnect the vacuum hoses to the power booster and throttle body.

26. Reconnect the fuel rail ground wires to the right cylinder head.

27. Reconnect the electrical connectors to the ISC motor, the TPS switch and MAP sensor.

28. Reconnect the main fuel injector harness.

29. Reconnect the spark plug wires to the front plugs.

30. Reconnect the PCV hose to the intake manifold.

31. Reconnect the negative battery cable.

32. Pressurize the fuel system and verify no leaks.

33. Reinstall the intake manifold sight shield.

Exhaust Manifold

REMOVAL & INSTALLATION

3.8L (VIN 1 and K) Engines

LEFT SIDE (FRONT)

1. Disconnect the negative battery cable.

2. Remove the two bolts attaching the left exhaust manifold to the crossover pipe.

3. Disconnect the spark plug wires from the plugs and position out of the way.

4. Remove the oil level indicator tube.

5. Remove the exhaust manifold bolts and the exhaust manifold.

To install:

6. Install the exhaust manifold with a new gasket and loosely install the mounting bolts.

7. Reinstall the oil level indicator tube.

8. Tighten the exhaust manifold mounting bolts to 22 ft. lbs. (30 Nm).

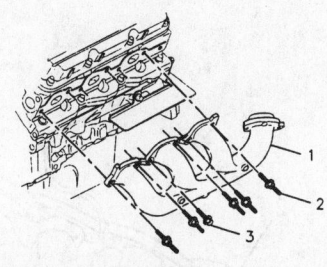

1 LEFT (FRONT) EXHAUST MANIFOLD
2 STUD 30 N•m (22 LB. FT.)
3 BOLT 30 N•m (22 LB. FT.)

7922XG30

Exploded view of the left exhaust manifold mounting—3.8L (VIN 1 and K) engines

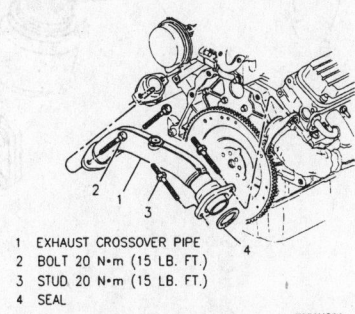

1 EXHAUST CROSSOVER PIPE
2 BOLT 20 N•m (15 LB. FT.)
3 STUD 20 N•m (15 LB. FT.)
4 SEAL

7922XG31

Exploded view of the crossover pipe mounting—3.8L (VIN 1 and K) engines

9. Reconnect the spark plug wires to the plugs.

10. Reinstall the bolts connecting the left exhaust manifold to the crossover pipe and tighten to 15 ft. lbs. (20 Nm).

11. Reconnect the negative battery cable.

12. Start the vehicle and verify no leaks.

RIGHT SIDE (REAR)

1. Disconnect the negative battery cable.

2. Disconnect the spark plug wires from the plugs and position out of the way.

3. Remove the transaxle level indicator tube.

4. Detach the oxygen sensor (O_2) electrical connector.

5. Remove the two bolts connecting the right exhaust manifold to the crossover pipe.

6. Remove the plastic vacuum tank mounted on the cowl.

7. Raise and safely support the vehicle.

8. Remove the converter heat shield and pipe hanger.

9. Disconnect the exhaust pipe from the manifold.

10. Lower the vehicle.

11. Remove the rear engine lift bracket.

12. Remove the exhaust manifold bolts.

13. Remove the exhaust manifold.

To install:

14. Install the exhaust manifold with a new gasket and tighten the exhaust manifold mounting bolts to 22 ft. lbs. (30 Nm).

15. Reinstall the rear engine lift bracket.

16. Raise and safely support the vehicle.

17. Reconnect the front pipe to the manifold and tighten the mounting bolts to 15 ft. lbs. (20 Nm).

18. Reinstall the exhaust hanger and converter heat shield.

19. Lower the vehicle.

20. Reinstall the vacuum tank on the cowl.

21. Reinstall the bolts connecting the

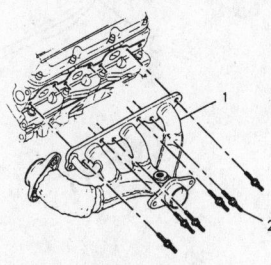

1 RIGHT (REAR) EXHAUST MANIFOLD
2 STUD 30 N•m (22 LB. FT.)

7922XG29

Exploded view of the right exhaust manifold mounting—3.8L (VIN 1 and K) engines

right exhaust manifold to the crossover pipe and tighten the bolts to 15 ft. lbs. (20 Nm).

22. Reinstall the transaxle level indicator tube.

23. Reconnect the oxygen sensor (O2) electrical connector.

24. Reconnect the spark plug wires to the plugs.

25. Reconnect the negative battery cable.

26. Start the vehicle and verify no leaks.

4.0L (VIN C) Engine

LEFT SIDE (FRONT)

1. Disconnect the negative battery cable.

2. Release the tension on the serpentine belt and remove the belt from the power steering and alternator pulleys.

3. Remove the alternator upper mounting bolt.

4. Raise and safely support the vehicle.

5. Remove the right inner fender well splash shield.

6. Remove the lower center air deflector.

7. Remove the alternator rear bracket.

8. Remove the lower alternator mounting bolt and position the alternator aside.

9. Remove the exhaust manifold-to-exhaust crossover pipe bolts.

10. Detach the electrical connector from the oxygen sensor.

11. Remove the bolts securing the power steering line retainers.

12. Remove the exhaust manifold nuts and remove the exhaust manifold.

13. If the manifold is being replaced, remove the oxygen sensor.

To install:

14. Thoroughly clean all sealing surfaces.

15. Coat the oxygen sensor threads with high temperature anti-seize and install the sensor in the manifold and tighten to 30 ft. lbs. (40 Nm).

16. Reinstall the exhaust manifold gasket over the cylinder head studs.

17. Insert the outlet pipe on the manifold partially into the crossover pipe and position the manifold over the cylinder head studs. Tighten the mounting nuts starting in the center and working outward to 18 ft. lbs. (24 Nm).

18. Reconnect the electrical connector to the oxygen sensor.

19. Reinstall the power steering line retainers and tighten the mounting bolts to 10 ft. lbs. (14 Nm).

20. Position the alternator and loosely reinstall the lower retaining bolt and the rear

mounting bracket. Tighten the bolts to 35 ft. lbs. (47 Nm) and the nuts to 28 ft. lbs. (38 Nm).

21. Reinstall the crossover pipe mounting bolts and tighten to 37 ft. lbs. (50 Nm).

22. Reinstall the lower center air deflector.

23. Reinstall the right inner fender well splash shield.

24. Lower the vehicle.

25. Reinstall the upper alternator retaining bolt and tighten to 35 ft. lbs. (47 Nm).

26. Reinstall the serpentine belt around the alternator and power steering pulleys.

27. Reconnect the negative battery cable.

28. Start the vehicle and verify no leaks.

RIGHT SIDE (REAR)

1. Disconnect the negative battery cable.

2. Raise and safely support the vehicle.

3. Remove the exhaust system.

4. Remove the bolts securing the connector pipe to the exhaust manifold and crossover pipe.

5. Remove the heat shield from the knock sensor.

6. Detach the electrical connector from the oxygen sensor.

7. Remove the exhaust manifold nuts and remove the exhaust manifold.

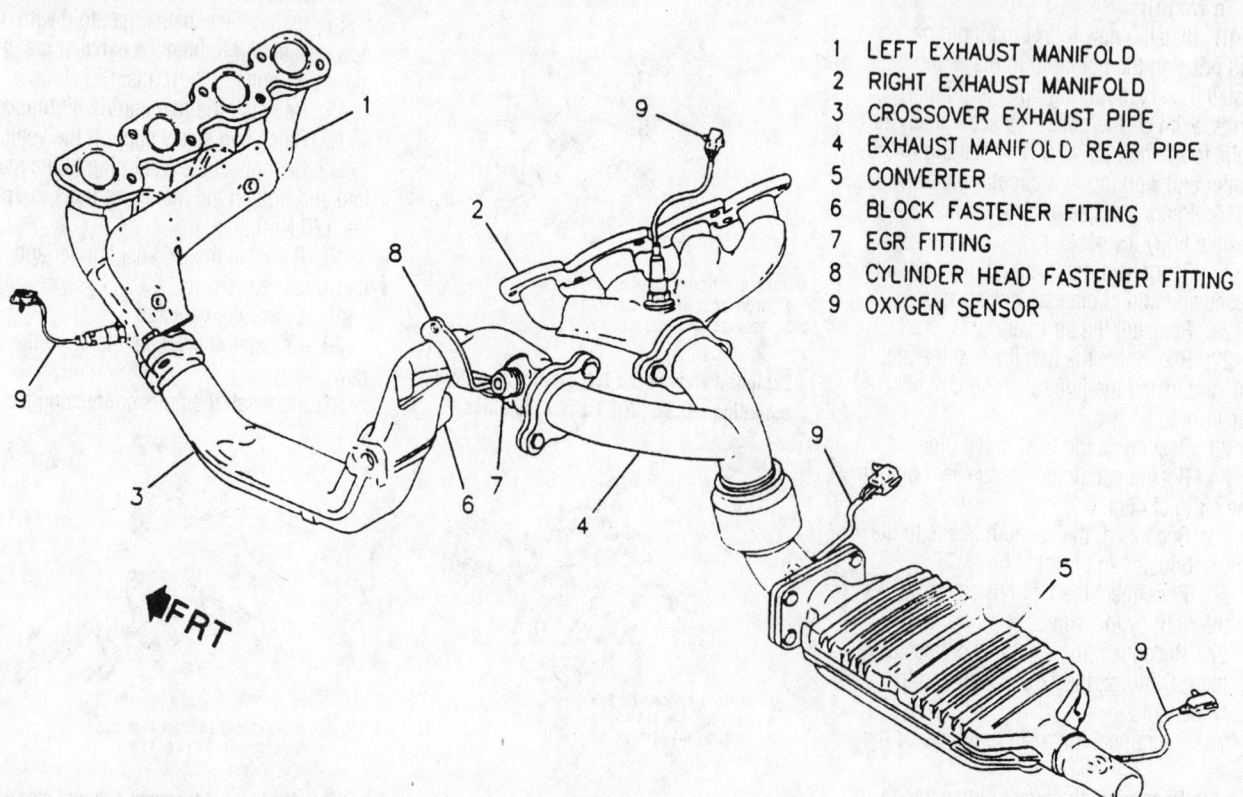

1	LEFT EXHAUST MANIFOLD
2	RIGHT EXHAUST MANIFOLD
3	CROSSOVER EXHAUST PIPE
4	EXHAUST MANIFOLD REAR PIPE
5	CONVERTER
6	BLOCK FASTENER FITTING
7	EGR FITTING
8	CYLINDER HEAD FASTENER FITTING
9	OXYGEN SENSOR

Exhaust system component identification—4.0L (VIN C) engine

8. If the manifold is being replaced, remove the oxygen sensor.

To install:

9. Thoroughly clean all sealing surfaces.

10. Coat the oxygen sensor threads with high temperature anti-seize and install the sensor in the manifold and tighten to 30 ft. lbs. (40 Nm).

11. Reinstall the exhaust manifold gasket over the cylinder head studs.

12. Position the manifold over the cylinder head studs. Tighten the mounting nuts starting in the center and working outward to 18 ft. lbs. (24 Nm).

13. Reinstall the knock sensor heat shield.

14. Reconnect the electrical connector to the oxygen sensor.

15. Reinstall the connector pipe and install the mounting bolts connecting it to the exhaust manifold and crossover pipe. Tighten the mounting bolts to 30 ft. lbs. (40 Nm).

16. Reinstall the exhaust system and tighten the spring loaded nuts to 18 ft. lbs. (25 Nm).

17. Lower the vehicle.

18. Reconnect the negative battery cable.

19. Start the vehicle and verify no leaks.

Oil Pan

REMOVAL & INSTALLATION

3.8L (VIN 1 and K) Engines

1. Disconnect the negative battery cable.

2. Raise and safely support the vehicle.

❊❊ CAUTION

The EPA warns that prolonged contact with used engine oil may cause a number of skin disorders, including cancer! You should make every effort to minimize your exposure to used engine oil. Protective gloves should be worn when changing the oil. Wash your hands and any other exposed skin areas as soon as possible after exposure to used engine oil. Soap and water, or waterless hand cleaner should be used.

3. Drain the engine oil.

4. Detach the oil level indicator connector.

5. Remove the oil pan mounting bolts and remove the oil pan.

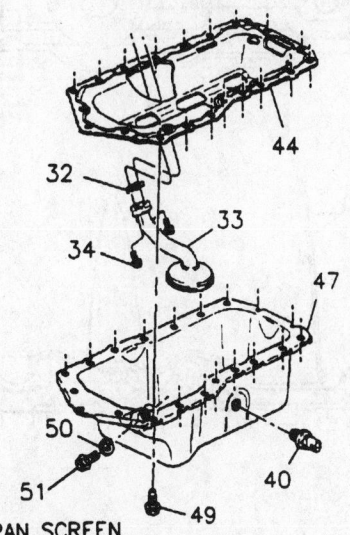

32	GASKET, OIL PAN SCREEN
33	SCREEN, OIL PAN
34	BOLT/SCREW, OIL PAN SCREEN
40	SENSOR, ENGINE OIL LEVEL
44	GASKET, OIL PAN (INCLUDES BAFFLE)
47	PAN, OIL
49	BOLT/SCREW, OIL PAN
50	GASKET, OIL PAN DRAIN PLUG
51	PLUG, OIL PAN DRAIN

7922XG33

Exploded view of the oil pan mounting and related components—3.8L (VIN 1 and K) engines

To install:

6. Clean all the gasket surfaces completely.

7. Install the oil pan with a new gasket and tighten the mounting bolts to 10 ft. lbs. (14 Nm).

8. Reconnect the oil level indicator connector.

9. Lower the vehicle.

❊❊ WARNING

Operating the engine without the proper amount and type of engine oil will result in severe engine damage.

10. Refill the crankcase with engine oil.

11. Reconnect the negative battery cable.

12. Start the vehicle and verify no oil leaks.

4.0L (VIN C) Engine

On the 4.0L (VIN C) Engine Oldsmobile Aurora, remove the transaxle assembly, not the engine, for engine oil pan removal.

1. Disconnect the negative battery cable. The battery is located under the passenger side of the rear seat cushion. Raise the rear seat cushion to access the battery.

2. Raise and safely support the vehicle.

❊❊ CAUTION

The EPA warns that prolonged contact with used engine oil may cause a number of skin disorders, including cancer! You should make every effort to minimize your exposure to used engine oil. Protective gloves should be worn when changing the oil. Wash your hands and any other exposed skin areas as soon as possible after exposure to used engine oil. Soap and water, or waterless hand cleaner should be used.

3. Remove the oil pan drain plug and drain the engine oil into a suitable container.

4. If necessary, remove the exhaust crossover pipe retaining bolts from the left exhaust manifold flange and the exhaust manifold rear pipe.

5. Remove the transaxle assembly.

6. Remove the oil pan bolts and remove the oil pan from the vehicle.

➡The oil pan gasket is reusable unless it is damaged. Do not remove the gasket from the oil pan groove unless gasket replacement is required.

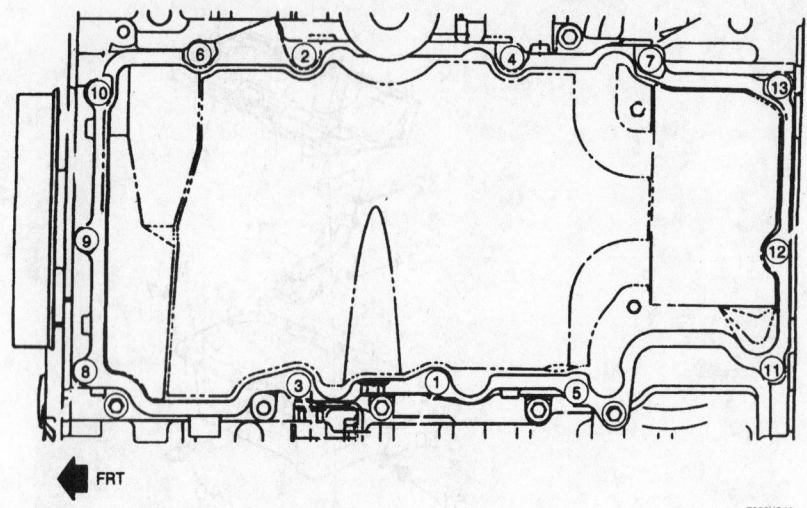

Oil pan bolt tightening sequence—4.0L (VIN C) engine

To install:

7. Install the oil pan and seal in the vehicle.

8. Reinstall the oil pan mounting bolts and tighten in sequence to 108 inch. lbs. (13 Nm).

9. Reinstall the transaxle assembly.

10. Reinstall the exhaust crossover pipe retaining bolts on the left exhaust manifold flange and the exhaust manifold rear pipe if removed.

11. Reinstall the oil pan drain plug and tighten to 15 ft. lbs. (20 Nm).

12. Lower the vehicle.

✳✳ WARNING

Operating the engine without the proper amount and type of engine oil will result in severe engine damage.

13. Refill the crankcase with the proper type and quantity of engine oil.

14. Reconnect the negative battery cable.

Oil Pump

REMOVAL & INSTALLATION

3.8L (VIN 1 and K) Engines

1. Disconnect the negative battery cable.

2. Raise and safely support the vehicle.

✳✳ CAUTION

The EPA warns that prolonged contact with used engine oil may cause a number of skin disorders, including cancer! You should make every effort to minimize your exposure to used

engine oil. Protective gloves should be worn when changing the oil. Wash your hands and any other exposed skin areas as soon as possible after exposure to used engine oil. Soap and water, or waterless hand cleaner should be used.

3. Drain the engine oil.

4. Remove the front cover assembly.

5. Remove the four bolts securing the oil filter adapter to the front cover assembly and oil filter adapter, pressure regulator valve and spring.

6. Remove the four oil pump cover attaching screws and remove the cover.

7. Remove the inner and outer pump gears.

8. Make the following measurements and replace any components not within specification:

 a. Gear pocket depth: 0.461–0.4265 in. (11.71–11.75mm)

 b. Gear pocket diameter: 3.508–3.512 in. (89.10–89.20mm)

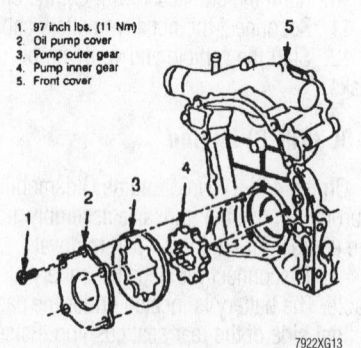

1. 97 inch lbs. (11 Nm)
2. Oil pump cover
3. Pump outer gear
4. Pump inner gear
5. Front cover

Exploded view of the oil pump assembly—3.8L (VIN 1 and K) engines

 c. Inner gear tip clearance: 0.006 in. (0.152mm)

 d. Outer gear diameter clearance: 0.008–0.015 in. (0.203–0.381mm)

 e. Gear end clearance: 0.001–0.0035 in. (0.025–0.089mm)

9. If measurement **A** or **B** is out of specification the front cover assembly must be replaced. If measurement **C** , **D** or **E** is out of specification replace the gears.

To install:

10. Lubricate the gears with petroleum jelly and install the gears into the housing.

11. Pack the gear cavity with petroleum jelly after the gears have been installed in the housing. This seals the gears and acts as a prime to allow the oil pump to draw oil as soon as it starts to turn as the engine is cranked. Do not overlook this step. DO NOT use chassis grease. Use only petroleum jelly since its low-temperature nature means it will dissolve as the engine warms up.

12. Reinstall the oil pump cover and screws and tighten to 97 inch lbs. (11 Nm).

13. Reinstall the oil filter adapter with new gasket, pressure regulator valve and spring. Tighten the mounting bolts to 24 ft. lbs. (33 Nm).

14. Reinstall the front cover assembly.

✳✳ WARNING

Operating the engine without the proper amount and type of engine oil will result in severe engine damage.

15. Refill the crankcase with clean engine oil.

16. Reconnect the negative battery cable.

17. Start the vehicle and verify no leaks and proper oil pressure.

4.0L (VIN C) Engine

1. Disconnect the negative battery cable.

2. Install an Engine Support Fixture J-28467-A or the equivalent.

3. Remove the engine mount-to-body through-bolt nut and through-bolt.

4. Remove the through-bolt nut and bolt from the engine side of the mount.

5. Raise and safely support the vehicle.

6. Remove the front tire and wheel assembly.

7. Remove the right inner fender well splash shield.

8. Remove the lower center air deflector.

9. Remove the left transaxle mount-to-frame through-bolt.

10. Remove the lower retaining bolt and nut from the engine mount bracket.

11. Remove the bolt and disconnect the power steering line retainer from the bracket.

12. Lower the vehicle.

13. Remove the upper retaining nut and bolt from the engine mount bracket.

14. Remove the serpentine belt from the power steering pump pulley.

15. Remove the fuel plastic sight shield on the intake manifold.

16. Remove the power steering pump from the mounting bracket and set aside. DO NOT disconnect the power steering lines from the pump.

17. Raise the engine using the support fixture.

18. Remove the engine mount bracket from the engine.

19. Remove the torque axis mount from the frame rail.

20. Lower the engine using the support fixture until clearance for J-38416-B, or an equivalent puller is attained.

21. Raise and safely support the vehicle.

22. Remove the crankshaft balancer retaining bolt.

23. Remove the crankshaft balancer using J-38416-B.

24. Remove the serpentine belt tensioner.

25. Remove the serpentine belt idler pulley.

26. Remove the front cover bolts and remove the cover and gasket. DO NOT discard the gasket, if it is undamaged it can be re-used.

27. Remove the three oil pump mounting bolts and remove the oil pump and drive spacer.

28. If necessary, disassemble and inspect the pump as follows:

   a. Remove the drive spacer from the pump housing.

   b. Remove the two screws holding the pump housing halves together.

   c. Remove the inner (drive) and outer (driven) rotors from the housing. Indicate the mating surfaces (dimples).

   d. Remove the pressure relief valve.

   e. Inspect the pump housing for nicks, burrs, chips or debris that might cause a leak or binding condition in the rotor pocket.

   f. Inspect the drive and driven rotors for nicks or burrs.

   g. Check the pump cover and interior surface for excessive wear or score marks. Check for flatness.

   h. If any components show signs of excessive wear or damage, replace the pump assembly.

To install:

29. If the pump was disassembled, reassemble as follows:

   a. Reinstall the inner and outer rotors to the pump cover in the same orientation as removed.

   b. Reinstall the pressure relief valve seat, spring and pilot in the pump housing.

   c. Pack the pump housing halves with Amojell® or white petroleum grease to ensure pump priming.

   d. Assemble the housing and cover over the locating dowel.

   e. Insert a 9mm drill in the pump mounting hole on the opposite side to aid alignment of the housing and cover. Install the two screws and tighten to 108 inch lbs. (12 Nm).

30. Reinstall the oil pump drive spacer into the oil pump from the rear so the drive flat engages the pump rotor.

31. Install the oil pump over the crankshaft and loosely install the mounting bolts.

32. Hold the pump in its furthest up position and tighten the mounting bolts to 89 inch lbs. (10 Nm) plus 35 degrees additional rotation.

33. Place a small amount of RTV sealant at the split line of the upper and lower crankcases.

34. Reinstall the front cover gasket over the dowel pins on the block.

35. Reinstall the front cover on the engine and tighten the mounting bolts to 89 inch lbs. (10 Nm).

36. Reinstall the serpentine belt idler pulley and tighten the mounting bolt to 37 ft. lbs. (50 Nm).

37. Reinstall the serpentine belt tensioner and tighten the mounting bolt to 37 ft. lbs. (50 Nm).

38. Coat the seal contact area on the balancer with engine oil and install the balancer onto the end of the crankshaft with the notch lined up with the crankshaft key. Using J-39344 or the equivalent push the balancer into place.

39. Install the balancer center bolt and tighten to 44 ft. lbs. (60 Nm) plus 120 degrees ⅔ turn) additional rotation.

40. Lower the vehicle.

41. Raise the engine with the support fixture.

42. Reinstall the torque axis mount on the body.

43. Reinstall the engine mount bracket on the engine and loosely install the mounting nuts and bolts. With all the fasteners in place tighten the nuts to 30 ft. lbs. (40 Nm) and the bolts to 41 ft. lbs. (55 Nm).

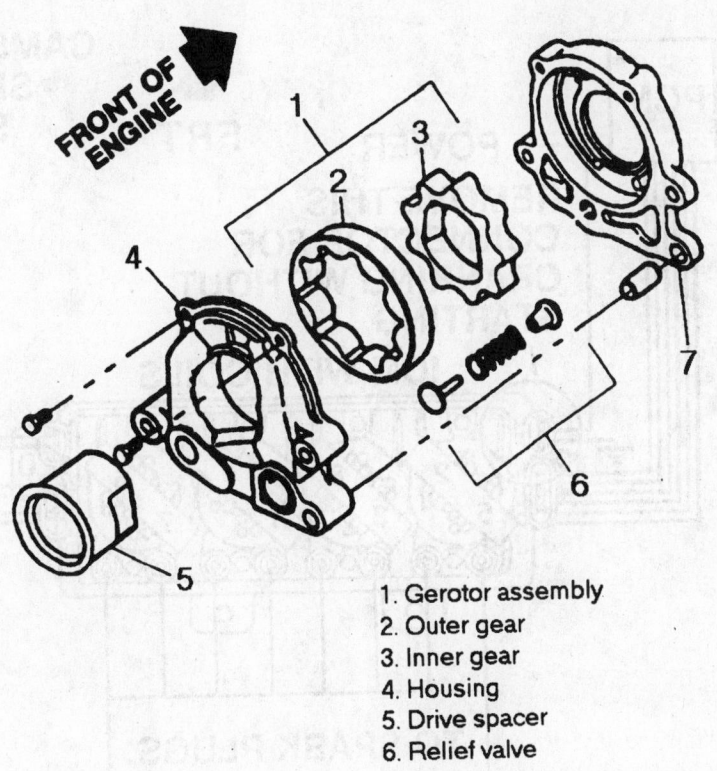

FRONT OF ENGINE

1. Gerotor assembly
2. Outer gear
3. Inner gear
4. Housing
5. Drive spacer
6. Relief valve
7. Cover

7922XG14

Exploded view of the oil pump—4.0L (VIN C) engine

44. Lower the engine support fixture until the engine is at its normal height.

45. Reinstall the power steering pump in the mounting bracket.

46. Reinstall the serpentine belt.

47. Reconnect the power steering line retainer to the engine mount bracket and install the mounting bolt.

48. Raise and safely support the vehicle.

49. Reinstall the left transaxle mount-to-frame through-bolt and tighten to 63 ft. lbs. (85 Nm).

50. Reinstall the lower center air deflector.

51. Reinstall the right inner fender well splash shield.

52. Reinstall the tire and wheel assembly and tighten the wheel nuts to 100 ft. lbs. (140 Nm).

53. Lower the vehicle.

54. Reinstall the torque axis mount-to-engine mount bracket through-bolt and nut and tighten the bolt to 70 ft. lbs. (95 Nm).

55. Reinstall the torque axis mount-to-frame through-bolt and nut and tighten the through-bolt to 37 ft. lbs. (50 Nm).

56. Remove the engine support fixture.

57. Reconnect the negative battery cable.

58. Run the engine and check for leaks and proper engine operation.

ENGINE LUBRICATION SYSTEM PRIMING PROCEDURE

After completing service to an Aurora 4.0L (VIN C) engine requiring engine oil pump removal, the following priming procedures MUST be performed before engine start-up.

The factory recommends a coat of GM Prelube No. 1052367 be applied to all bearing surfaces and crankshaft journals (cover completely) whenever servicing connecting rod and/or main crankshaft bearings. Also perform the following when servicing the internal components of an Aurora engine. These steps will aid in priming the lubrication system: Store the valve lifters (also called tappets or lash adjusters) with the camshaft contact surface down, so engine oil will not drain from the lifters; Apply liberal amounts of GM Camshaft Prelube, No. 1052365 to the camshaft lobes, bearing caps and lifter surfaces; Piston rings and pistons should be completely covered with proper specification motor oil during installation.

To perform the factory-required Engine Lubrication System Priming Procedure, use the following procedures.

1. Thoroughly pack the oil pump with white petroleum jelly during reassembly and fill the oil filter with correct specification engine oil before installation.

2. Verify engine oil is at the proper level. System capacity is 7 quarts with the oil filter full. Fill to proper level if necessary.

3. Detach the right front connector (power lead) from the Distributorless Ignition System (DIS) module and crank the engine for 30 seconds.

4. Connect the power lead to the DIS and start the engine. Check the Driver Information Center (DIC) for **Low Oil Pressure** message and listen for any audible noise such as lifters "ticking".

5. If engine noises persist, stop the engine and remove the Oil Pressure Switch from the Oil Filter Adapter and install a mechanical oil pressure gauge. **Be careful not to damage the threads in the adapter.** If is made of soft magnesium and is easily damaged. Also note that the ports are sealed with O-rings which must be in place and in good condition to seal properly. The bypass valves in the adapter are non-serviceable.

6. If oil is indicated on the gauge and no unusual sounds are heard, oil pressure is present and both the oil pump and engine lubrication system are primed. If no oil

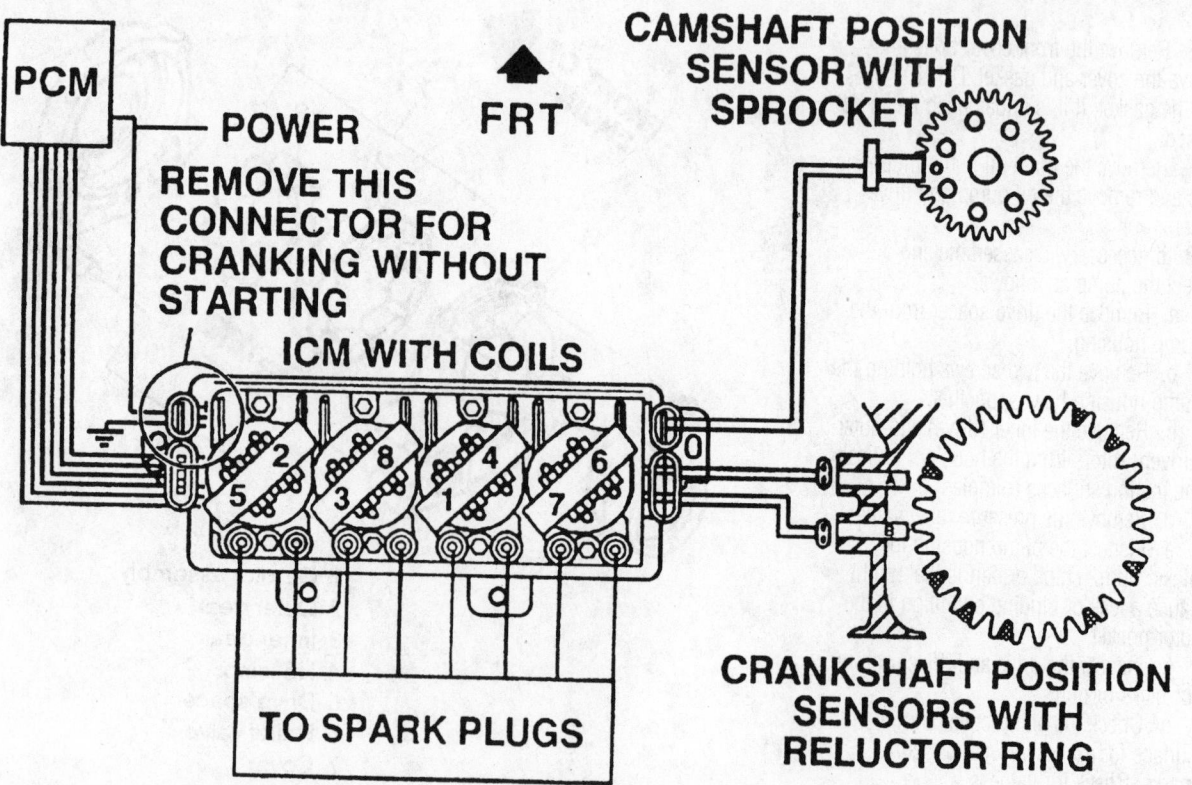

Remove the connector shown to disable the ignition system—4.0L (VIN C) engine

7922XG15

pressure is recorded, repeat Step 3 of this procedure, then proceed to Step 7.

7. If no oil pressure is recorded after repeating the process given above, remove the Oil Filter Adapter and force engine oil under pressure (using shop air) in the engine block outlet port (the port closest to the front of the engine). Reinstall the Oil Filter Adapter with the mechanical oil pressure gauge installed in the sender port and start the engine.

8. If oil pressure is obtained, stop the engine and reinstall the Oil Pressure Switch in the Oil Filter Adapter. Once connected, check the instrument display for no oil pressure or **Low Oil Pressure** message. If that message is present, check the switch connection or begin low oil pressure complaint troubleshooting.

Rear Main Seal

REMOVAL & INSTALLATION

1. Disconnect the negative battery cable.

2. Remove the transaxle and flexplate using the recommended procedure.

3. Carefully pry the oil seal from the housing using a suitable prying tool taking care not to damage the housing or the crankshaft sealing surface.

→**Extreme care must be taken in order to avoid damage to the crankshaft sealing surface.**

To install:

4. On 4.0L engines place a small amount of GM Gasket Maker® or equivalent at the top of the crankcase split line across the end of the upper\lower crankcase seal.

5. Apply clean engine oil to the inside and outside diameter of the oil seal.

6. Install the seal into the housing using tool J-38196 for 3.8L engines or tool J-38817 for 4.0L engines or equivalent.

7. Reinstall the flexplate and transaxle using the recommended procedure.

8. Reconnect the negative battery cable, start the engine and check for leaks.

Timing Chain, Sprockets, Front Cover and Seal

REMOVAL & INSTALLATION

3.8L (VIN 1 and K) Engines

1. Disconnect the negative battery cable.

⁕⁕⁕ CAUTION

Never open, service or drain the radiator or cooling system when hot; serious burns can occur from the steam and hot coolant.

2. Drain the coolant into a suitable container.

3. Remove the vacuum reservoir.

4. Install an engine support fixture, J-28467-A or equivalent.

5. Raise the engine so that the weight is removed from the torque axis mount.

6. Remove the two lower torque axis-to-frame bolts.

7. Remove the torque axis mount.

8. Remove the serpentine belt(s).

9. Remove the water pump on vehicles equipped with supercharger.

10. Raise the engine with the support fixture and remove the engine mount bracket bolts and bracket.

11. Remove the alternator and the three bolts securing the drive belt tensioner and remove the tensioner.

12. Remove the crankshaft balancer.

13. Remove the crankshaft sensor shield.

14. Detach the electrical connector from the crankshaft sensor.

---

15. Remove the front oil pan-to-front cover bolts.

16. Remove the front cover bolts and remove the front cover.

17. Using the appropriate seal driver, remove the seal from the front cover.

18. Rotate the crankshaft until the timing mark on the camshaft sprocket is lined up with the crankshaft sprocket timing mark.

19. Remove timing chain damper mounting bolt and remove the chain damper assembly.

20. Remove camshaft sprocket bolt.

21. Remove camshaft sprocket along with the timing chain.

22. Using a suitable gear puller remove the crankshaft sprocket from the crankshaft.

To install:

23. Line up the notch in the crankshaft sprocket with the crankshaft key and slide the gear on the crankshaft. It may be necessary to use a gear installer to fully seat the gear. Be sure the timing mark on the crankshaft gear is pointing straight up.

24. Insert the camshaft gear into the timing chain. Hold the sprocket with the timing mark straight down and with the chain hanging down off of the sprocket.

25. Loop the chain under the crankshaft sprocket.

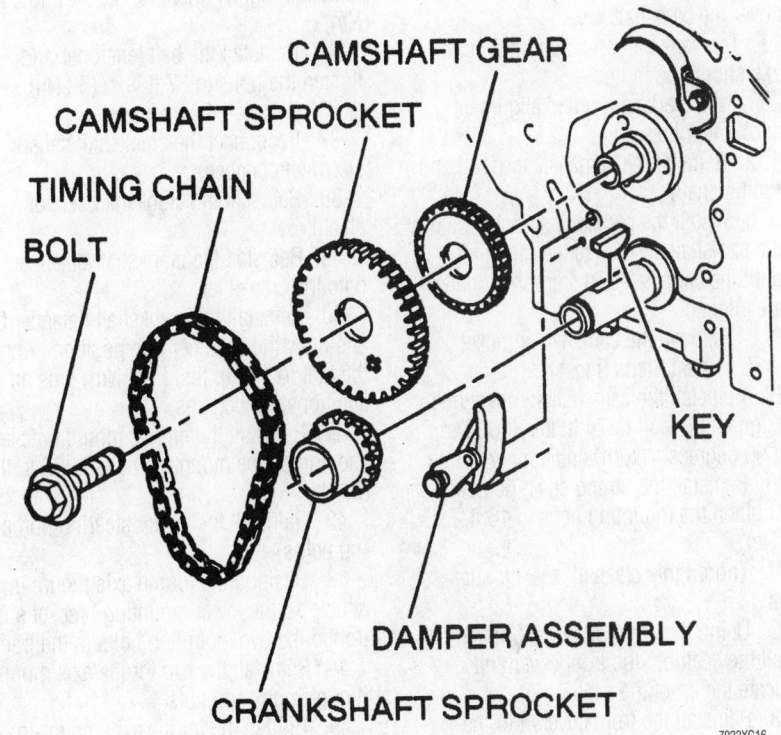

Exploded view of the timing chain and sprockets—3.8L (VIN 1 and K) engines

7922XG16

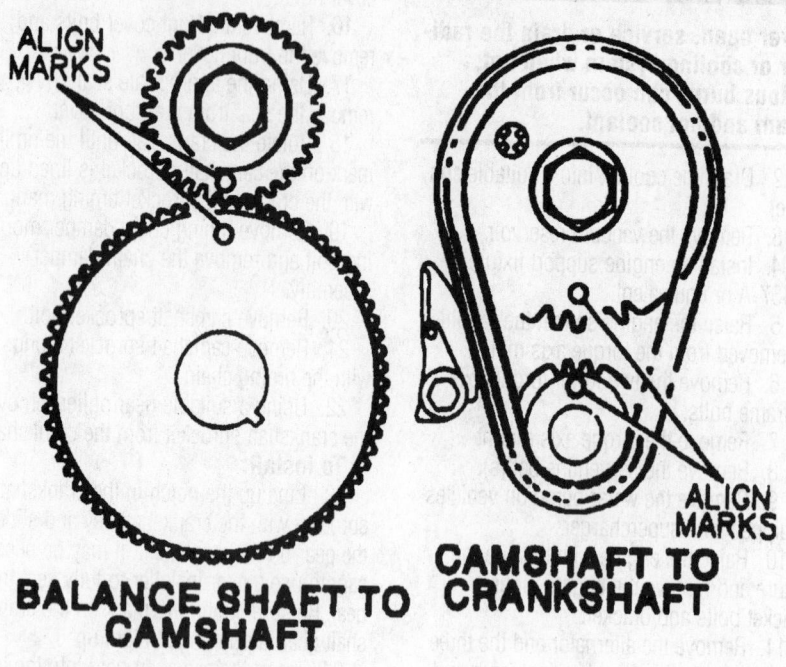

BALANCE SHAFT TO CAMSHAFT

CAMSHAFT TO CRANKSHAFT

ALIGN MARKS

ALIGN MARKS

7922XG17

Balance shaft-to-camshaft and camshaft-to-crankshaft timing mark alignment—3.8L (VIN 1 and K) engines

26. Install the camshaft sprocket onto the camshaft so the notch in the sprocket fits over the camshaft key.

27. The camshaft and crankshaft timing marks should be in line.

28. If the marks are not in alignment perform the following:

　a. Remove the camshaft sprocket and timing chain.

　b. Install the camshaft sprocket onto the camshaft and rotate the camshaft until the camshaft and crankshaft marks are aligned.

　c. Remove the camshaft sprocket.

　d. Repeat Steps 9 to 11.

29. Reinstall the camshaft sprocket bolt and tighten the bolt to 74 ft. lbs. (100 Nm) plus 90 degrees ¼ turn) additional rotation.

30. Reinstall the timing chain damper and tighten the mounting bolts to 16 ft. lbs. (22 Nm).

31. Thoroughly clean all sealing surfaces.

32. Using the appropriate seal driver, install the seal into the front cover and lubricate the lip of the seal.

33. Reinstall the front cover onto the engine using a new gasket. Reinstall the mounting bolts and tighten them to 22 ft. lbs. (30 Nm).

34. Reinstall the oil pan-to-front cover bolts and tighten them to 125 inch lbs. (14 Nm).

35. Reinstall the belt tensioner and tighten the bolts to 37 ft. lbs. (50 Nm).

36. Reinstall the alternator.

37. Reconnect the crankshaft sensor electrical connector.

38. Reinstall the crankshaft sensor shield.

39. Reinstall the crankshaft balancer onto the crankshaft.

40. Reinstall the crankshaft balancer bolt and draw the balancer into position. Tighten the bolt to 111 ft. lbs. (150 Nm) plus an additional 75 degrees.

41. Reinstall the engine mount bracket and tighten the mounting bolts to 65 ft. lbs. (87 Nm).

42. Reinstall the power steering pump and belt(s).

43. Position the torque axis mount in the vehicle so the lower mounting bracket slips around the two mounting bolts in the frame.

44. Reinstall the two torque axis mount through-bolts and nuts.

45. Tighten the torque axis mount-to-frame bolts to 52 ft. lbs. (70 Nm) and tighten the through-bolts to 65 ft. lbs. (87 Nm).

46. Remove the engine support fixture.

47. Reinstall the vacuum reservoir.

48. Reconnect the negative battery cable.

49. Refill the coolant system.

50. Start the vehicle and verify no leaks.

51. Road test the vehicle and ensure proper operation.

4.0L (VIN C) Engine

The following procedure covers removal and installation of the crankshaft sprocket, intermediate shaft sprocket, primary timing chain, left side secondary timing chain, right side secondary timing chain, left side cam sprockets, right side cam sprockets, primary timing chain tensioner, both secondary timing chain tensioners, and both timing chain guides.

The left and right side secondary timing chains can be removed with the engine in the vehicle. If the primary timing chain or intermediate shaft sprocket need to be replaced, the engine must be removed from the vehicle and supported on an engine stand.

➡Setting the camshaft timing is necessary whenever the cam drive system has been disturbed, meaning the relationship between any chain and sprocket has been lost. Correct timing exists when the crankshaft and intermediate shaft sprocket timing marks are in alignment and all four camshaft drive pins are perpendicular (90 degrees) to the cylinder head surface.

1. Disconnect the negative battery cable.

2. Remove the drive belt.

3. Remove the bolt securing the power steering hose.

4. Raise and safely support the vehicle.

5. For the left side secondary chain and sprocket, perform the following:

　a. Remove the right front wheel.

　b. Remove the two splash shields from the wheel well.

　c. Remove the flywheel cover and install a suitable flywheel holder.

　d. Remove the crankshaft balancer bolt.

　e. Support the engine cradle with a screw type jack and remove the three right-side engine cradle bolts.

　f. Disconnect the RSS sensor from the right control arm.

　g. Lower the cradle to gain access for the crankshaft damper puller.

　h. Using Puller J-38416 or equivalent, remove the crankshaft damper.

　i. Remove the drive belt tensioner.

j. Remove the drive belt idler pulley.

k. Remove the front cover bolts.

6. Remove the front cover and gasket.

➡**The front cover gasket is reusable as long as it is not damaged.**

7. Partially drain the coolant from the radiator.

8. Disconnect the upper radiator hose at the water crossover.

9. Label and disconnect the spark plug wires.

10. Remove the right side fan.

11. Disconnect the battery cable at the alternator and disconnect the cable harness at the cam cover.

12. Disconnect the PCV fresh air tube from the cam cover.

13. Remove the right and left torque struts.

14. Remove the water pump pulley with tool J-38825 or equivalent.

15. Remove the camshaft seal retainer screws and remove the seal.

16. Disconnect the battery cable retainer at the front of the cam cover.

17. Remove the cam cover screws and remove the cam cover by pivoting the entire cover around the water pump drive shaft. Continue moving the cover upward and pivoting so that the edge of the cover closely follows the left edge of the intake manifold cover.

➡**The cam cover gasket is reusable as long as it is not damaged.**

a. Remove the left side secondary chain tensioner.

b. Remove the left side chain guide. Access the upper chain guide mounting bolt through the hole in the cylinder head capped with the plastic plug.

c. Remove the left side cam sprocket bolts and sprockets.

d. Remove the secondary drive chain.

18. For the right side secondary chain and sprocket, perform the following:

a. Disconnect the exhaust Y-pipe at the converter.

b. Lower the vehicle.

c. Remove the tower-to-tower brace.

d. Detach the ICM wiring connectors and mounting bolts.

e. Remove the ICM and the plug wires on the right bank.

f. Disconnect the PCV valve.

g. Remove the purge canister sole-noid from the rear of the cover.

h. Remove the cam cover screws.

i. Safely support the front of the engine cradle and remove the two mounting bolts at the front of the cradle.

j. Remove the right and left torque struts.

k. Lower the engine cradle (or raise the vehicle) to provide clearance at the rear of the engine compartment.

l. Remove the cam cover.

➡**The cam cover gasket is reusable as long as it is not damaged.**

m. Remove the right side secondary chain tensioner.

n. Remove the right side chain guide. Access the upper chain guide mounting bolt through the hole in the cylinder head capped with the plastic plug.

o. Remove the right side cam sprocket bolts and cam sprockets.

19. Remove the secondary drive chain.

20. If only servicing the secondary chains, proceed to Step 19 of the installation procedure.

21. Remove the engine.

22. Remove the intermediate shaft sprocket-to-intermediate shaft bolt, then remove the sprocket.

23. Slide the primary timing sprockets and primary chain off the engine.

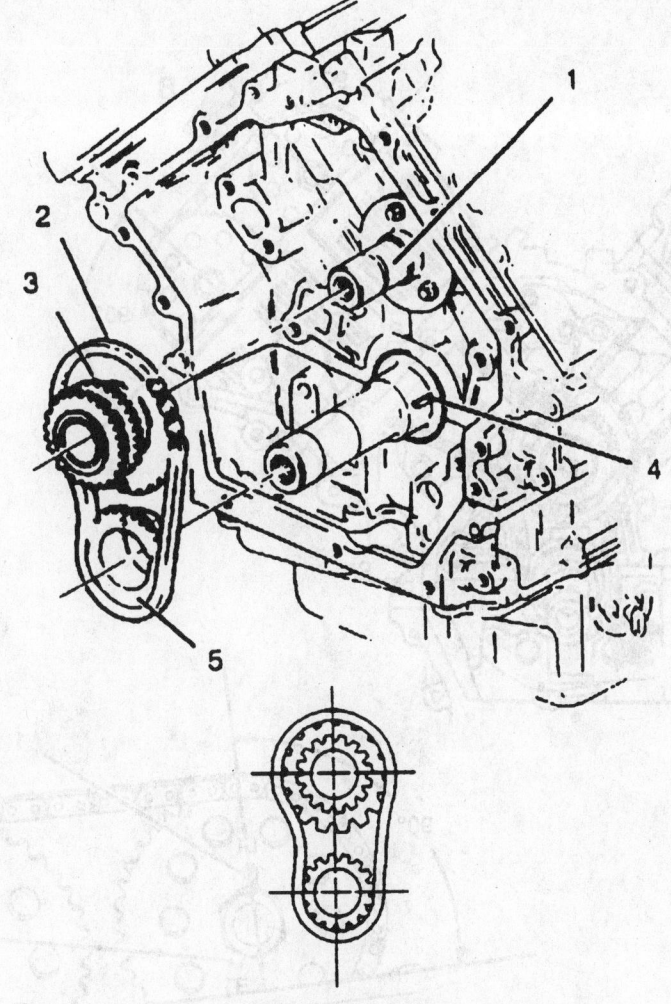

1 INTERMEDIATE SHAFT
2 PRIMARY CHAIN
3 INTERMEDIATE SHAFT SPROCKET
4 CRANKSHAFT SPROCKET KEY
5 SPROCKET

7922XG25

Primary drive chain components—4.0L (VIN C) engine

To install:

➡**The following procedure must be followed to set the camshaft timing on the vehicle.**

24. Install the primary and secondary chain guide.

25. Rotate the crankshaft until the sprocket drive key is at the 1 o'clock position. Use tool J-39946 or the equivalent, to rotate the crankshaft.

26. Install the crankshaft sprocket and intermediate shaft sprocket in the primary timing chain so the timing marks are aligned. Install the assembly in position on the engine. The crankshaft sprocket key way will have to slide over the key on the crankshaft. If it is necessary to turn the crankshaft sprocket, the intermediate shaft sprocket will also have to be turned so the timing mark still lines up with the crankshaft sprocket.

27. Install the intermediate shaft sprocket-to-intermediate shaft bolt and tighten to 45 ft. lbs. (61 Nm).

28. Install the primary timing chain tensioner. Tighten the tensioner mounting bolts to 20 ft. lbs. (27 Nm).

29. Install a suitable flywheel holder to lock the crankshaft in position.

30. Install the secondary timing chain over the inner row of teeth on the intermediate shaft sprocket. Route the chain over the chain guide and install the exhaust cam sprocket so the **RE** (Right Head Exhaust) pin engages the sprocket notch. There should be no slack in the lower section of the timing chain and the cam drive pin **must** be perpendicular to the cylinder head face.

31. Install the intake cam sprocket into the chain so the sprocket notch **RI** (Right

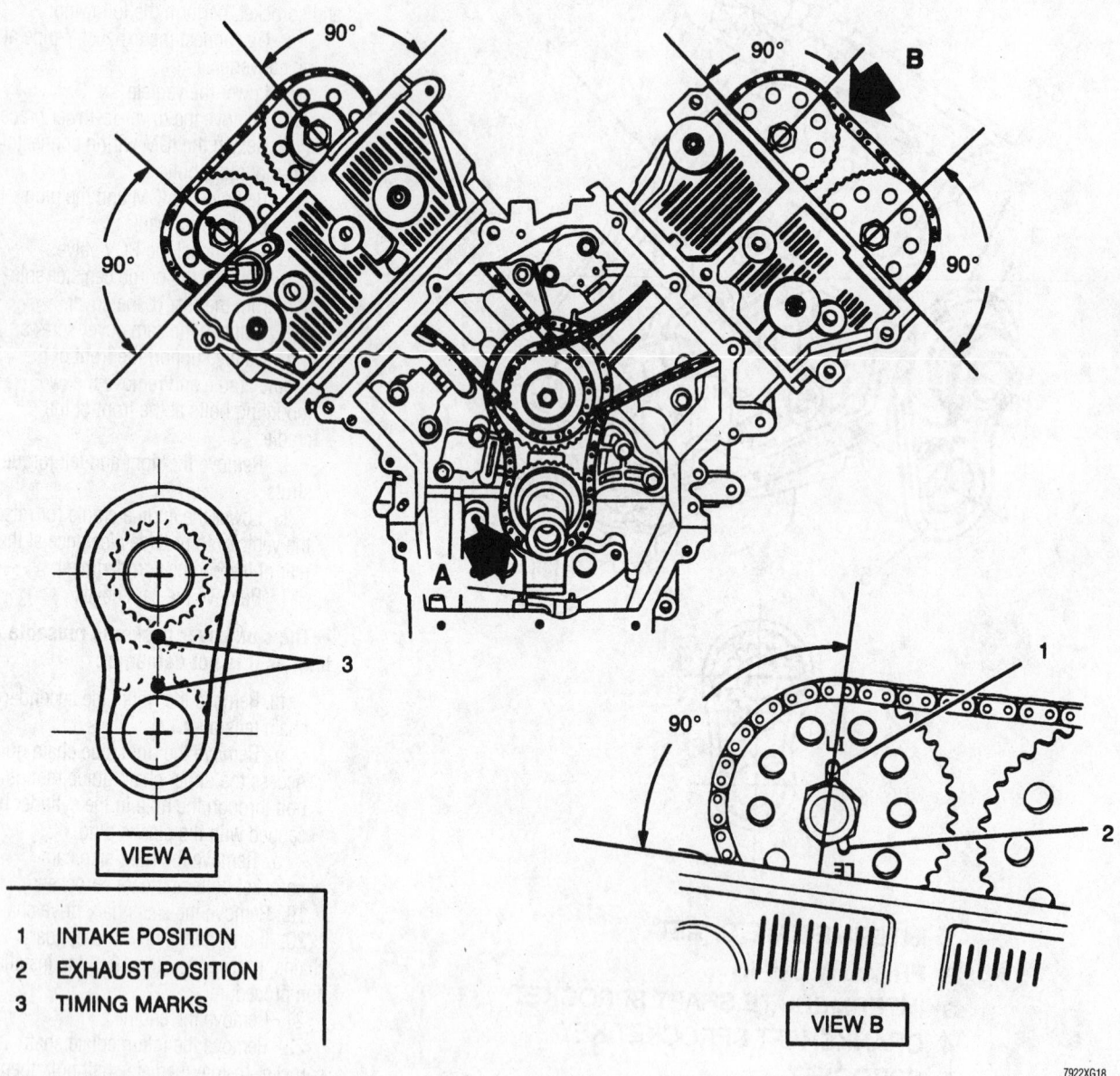

1   INTAKE POSITION
2   EXHAUST POSITION
3   TIMING MARKS

Primary and secondary timing mark alignment—4.0L (VIN C) engine

7922XG18

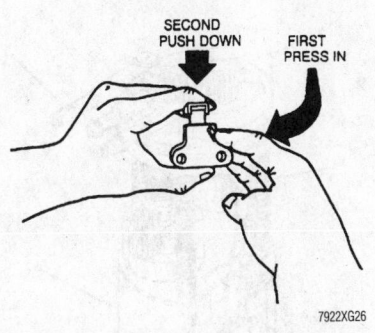

SECOND PUSH DOWN FIRST PRESS IN

7922XG26

Rotating tensioner release lever—4.0L (VIN C) engine

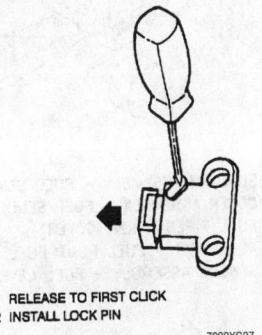

1 RELEASE TO FIRST CLICK
2 INSTALL LOCK PIN

7922XG27

Locking the tensioner into position—4.0L (VIN C) engine

Head Intake) engages the cam and the camshaft drive pin remains perpendicular to the cylinder head face. A hex is cast into the camshafts behind the lobes for cylinder No. 1, so an open end wrench may be used to provide minor repositioning of the cams.

32. Loosely install the exhaust cam sprocket bolt and intake sprocket bolt.

33. Install the chain tensioner and tighten the mounting bolts to 20 ft. lbs. (27 Nm).

34. Tighten the camshaft sprocket bolts to 90 ft. lbs. (120 Nm).

35. Route the secondary timing chain for the left side over the outer row of intermediate sprocket teeth.

36. Install the secondary timing chain over the inner row of teeth on the intermediate shaft sprocket. Route the chain over the chain guide and install the exhaust cam sprocket so the **LE** (Left Head Exhaust) pin engages the sprocket notch. There should be no slack in the lower section of the timing chain and the cam drive pin **must** be perpendicular to the cylinder head face.

37. Install the intake cam sprocket into the chain so the sprocket notch **LI** (Left Head Intake) engages the cam and the camshaft drive pin remains perpendicular to the cylinder head face. A hex is cast into the

camshafts behind the lobes for cylinder No. 2, so an open end wrench may be used to provide minor repositioning of the cams.

38. Loosely install the exhaust cam sprocket bolt and intake sprocket bolt.

39. Install the chain tensioner and tighten the mounting bolts to 20 ft. lbs. (27 Nm).

40. Tighten the camshaft sprocket bolts to 90 ft. lbs. (120 Nm).

➡ **The RE cam sprocket must contain the cam position sensor pick-up.**

41. Install the front cover gasket on the dowel pins on the block.

42. Install the front cover on the dowel pins and install the front cover mounting bolts. Tighten the bolts to 89 inch lbs. (10 Nm). Apply a dab of RTV to the split line between the upper and lower crankcase assemblies.

43. Install the drive belt idler pulley and tighten the mounting bolt to 35 ft. lbs. (47 Nm).

44. Install the drive belt tensioner and tighten the tensioner mounting nut to 35 ft. lbs. (47 Nm).

45. Install the crankshaft balancer using tool J-39344 or the equivalent.

46. Apply engine oil to the balancer bolt threads and tighten the bolt to 44 ft. lbs. (60 Nm) plus an additional 120 degree turn.

47. Raise the screw jack until the three cradle bolts can be installed. Tighten the bolts to 75 ft. lbs. (102 Nm).

48. Connect the RSS sensor.

49. Remove the flywheel holding tool and install the flywheel cover.

50. Install the wheel well splash shields.

51. Install the left cam cover as follows:

a. Install the spark plugs and cam cover seals.

b. Insert the intake cam through the hole in the cam cover and using fingers, guide the cam cover up over the edge of the cylinder head.

✳✳ WARNING

Use care to prevent the exposed section of the cam cover seal from being damaged by the edge of the cylinder head casting.

52. Work the cover into position by allowing the top edge of the cover to follow the left side edge of the intake manifold.

53. Install the cam cover screws and tighten to 7 ft. lbs. (10 Nm).

54. Connect the battery cable retainer to the front of the cam cover.

55. Connect the battery cable at the alternator.

56. Lubricate the seal lips and install the camshaft seal to the end of the intake cam. Seal the screw threads with sealer.

57. Install the water pump pulley with tool J-38825 or equivalent.

58. Connect the PCV fresh air tube to the cam cover.

59. Install the right side fan.

60. Connect the spark plug wires.

61. Connect the upper radiator hose to the water crossover.

62. Refill the cooling system.

63. Install the right cam cover as follows:

a. Install spark plug and cam cover seals.

b. Install the cam cover and tighten the screws to 7 ft. lbs. (10 Nm).

c. Raise the engine cradle into position and install and tighten the two mounting bolts to 75 ft. lbs. (100 Nm).

d. Install the right and left torque struts and tighten the retaining bolts as follows:

➡ **It is important during installation that the engine torque struts are not preloaded in their installed position. Adjustment is provided at the point the strut fastens to the core support bracket. Be sure this bolt is loose during assembly. Tighten to 45 ft. lbs. (60 Nm) as the final step of assembly.**

e. Tighten the strut bracket to cylinder head (M10) bolt: 35 ft. lbs. (50 Nm).

f. Tighten the strut bracket to cylinder head (M10) stud: 35 ft. lbs. (50 Nm).

g. Tighten the strut bracket to water manifold (M8) bolts: 20 ft. lbs. (25 Nm).

h. Tighten the strut bracket to cylinder head (M10) bolt: 35 ft. lbs. (50 Nm).

i. Tighten the strut to core support bracket bolt: 45 ft. lbs. (60 Nm) (see note above).

j. Install the screws retaining wiring harness to the cover.

k. Install the purge canister solenoid to the rear of the cover.

l. Connect the PCV valve.

m. Install the ICM and the spark plug wires on the right-bank.

n. Attach the ICM wiring connectors.

o. Install the tower-to-tower brace.

p. Raise and support the vehicle safely. Connect the exhaust Y-pipe to the converter and tighten the bolts to 20 ft. lbs. (25 Nm).

64. Connect the negative battery cable.

65. Start the engine. Check for leaks and inspect for proper operation.

FUEL SYSTEM

Fuel System Service Precautions

Safety is the most important factor when performing not only fuel system maintenance but any type of maintenance. Failure to conduct maintenance and repairs in a safe manner may result in serious personal injury or death. Maintenance and testing of the vehicle's fuel system components can be accomplished safely and effectively by adhering to the following rules and guidelines.

• To avoid the possibility of fire and personal injury, always disconnect the negative battery cable unless the repair or test procedure requires that battery voltage be applied.

• Always relieve the fuel system pressure prior to disconnecting any fuel system component (injector, fuel rail, pressure regulator, etc.), fitting or fuel line connection. Exercise extreme caution whenever relieving fuel system pressure to avoid exposing skin, face and eyes to fuel spray. Please be advised that fuel under pressure may penetrate the skin or any part of the body that it contacts.

• Always place a shop towel or cloth around the fitting or connection prior to loosening to absorb any excess fuel due to spillage. Ensure that all fuel spillage (should it occur) is quickly removed from engine surfaces. Ensure that all fuel soaked cloths or towels are deposited into a suitable waste container.

• Always keep a dry chemical (Class B) fire extinguisher near the work area.

• Do not allow fuel spray or fuel vapors to come into contact with a spark or open flame.

• Always use a back-up wrench when loosening and tightening fuel line connection fittings. This will prevent unnecessary stress and torsion to fuel line piping. Always follow the proper torque specifications.

• Always replace worn fuel fitting O-rings with new. Do not substitute fuel hose or equivalent, where fuel pipe is installed.

Fuel System Pressure

RELIEVING

1. Disconnect the negative battery cable.
2. Remove the fuel filler cap from the filler neck.
3. Connect J-34730–1, or an equivalent fuel pressure gauge to the fuel pressure test port. Wrap a shop towel around the fitting while connecting the gauge to prevent fuel spillage.

4. Install the gauge bleed hose into a suitable container and open the valve to bleed the system.
5. Drain any remaining fuel from the gauge into the container and remove the gauge from the test port.

Fuel Filter

REMOVAL & INSTALLATION

❊❊ CAUTION

The fuel injection system remains under pressure, even after the engine has been turned OFF. The fuel system pressure MUST BE relieved before disconnecting any fuel lines. Failure to do so may result in fire and/or personal injury.

1. Relieve the fuel system pressure using the recommended procedure.
2. Disconnect the negative battery cable.
3. Raise and safely support the vehicle.
4. Detach the quick connect fitting at the fuel filter inlet.
5. Disconnect the fuel filter outlet fitting from the fuel filter while holding the filter fitting with a back-up wrench.
6. Remove the fuel filter from the vehicle.

To install:
7. Install a new plastic retainer on the fuel inlet line.
8. Apply a drop of oil on the fuel filter inlet fitting and snap the fitting onto the fuel filter.
9. Reconnect the fuel outlet line to the filter and while holding the filter with a back-up wrench tighten the line fitting to 22 ft. lbs. (30 Nm).
10. Lower the vehicle.
11. Pressurize the fuel system and verify no leaks.

Fuel Pump

REMOVAL & INSTALLATION

❊❊ CAUTION

The fuel injection system remains under pressure, even after the engine has been turned OFF. The fuel system pressure MUST BE relieved before disconnecting any fuel lines. Failure to do so may result in fire and/or personal injury.

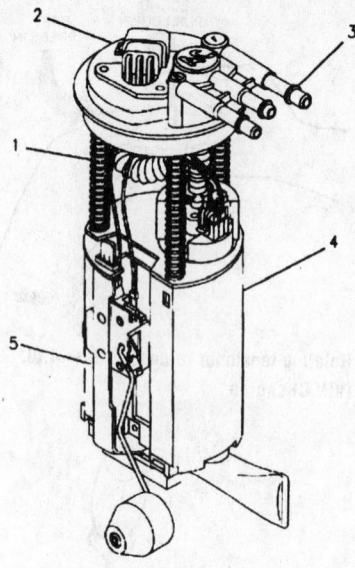

1 SUPPORT ASSEMBLY – FUEL SENDER
2 COVER ASSEMBLY – FUEL SENDER
3 FUEL PIPES (ABOVE COVER)
4 RESERVOIR – FUEL PUMP FUEL
5 SENSOR ASSEMBLY – FUEL LEVEL

7922XG19

Fuel pump and sending unit module assembly

1. Relieve the fuel system pressure using the recommended procedure.
2. Disconnect the negative battery cable.

❊❊ CAUTION

Observe all applicable safety precautions when working around fuel. Do not allow fuel spray or fuel vapors to come into contact with a spark or open flame. Keep a dry chemical (Class B) fire extinguisher near the work area. Never drain or store fuel in an open container due to the possibility of fire or explosion.

3. Drain the fuel tank with a hand held siphon until the level is less than ¼ full.
4. Working in the trunk, remove the spare tire cover, jack and spare tire.
5. Pull back the floor trunk liner.
6. Remove the cover from the fuel sender access cover.
7. Detach the quick connect fittings from the fuel sender assembly.
8. Detach the electrical connector from the fuel sender assembly.

✳✳ CAUTION

When the lock-ring is removed from the fuel sender the sender assembly will spring up. Downward pressure should be kept on the assembly and slowly released to ensure the sender assembly does not get damaged.

9. Remove lock-ring from the fuel sender using J-39765, or an equivalent spanner wrench.

10. Slowly release the spring pressure on the sender.

✳✳ CAUTION

The reservoir bucket on the fuel sender assembly will be full of fuel when it is removed from the tank. Be sure to have a catch pan nearby to drain the sender into.

11. Remove the sender assembly from the tank. It will have to be tilted slightly about half way out to be sure the fuel level float clears the side of the tank.

To install:

12. Install a new O-ring on top of the tank and carefully install the sender assembly into the tank.

13. Reinstall the retainer on top of the tank and compress the sender until the retainer can be engaged.

14. Lock the sender in place with J-39765, or the equivalent.

15. Reconnect the quick connect fittings to the fuel sender assembly.

16. Reconnect the electrical connector to the sender assembly.

17. Reconnect the negative battery cable.

18. Pressurize the fuel system and verify no leaks.

19. Reinstall the fuel sender access panel.

20. Reposition the trunk liner.

21. Reinstall the spare tire, jack and spare tire cover.

22. Refill the fuel tank.

DRIVE TRAIN

Transaxle Assembly

REMOVAL & INSTALLATION

Riviera

1. Disconnect the negative battery cable.
2. Remove the air intake duct.

3. Disconnect the cruise control cable from the throttle body lever, disconnect the vacuum hose at the control servo and remove the servo from the engine.

4. Remove the nut from the shift linkage at the manual shaft on the transaxle. Remove the two linkage bracket bolts and remove the cable and bracket from the transaxle and position out of the way.

5. Disconnect the vacuum line from the modulator.

6. Detach the electrical connectors from the park neutral switch, back-up light switch, transaxle harness and vehicle speed sensor.

7. Remove the left-to-right exhaust manifold bolts.

8. Remove the vacuum reservoir.

9. Remove the heater hose retainer bracket.

10. Detach the oxygen sensor connector.

11. Install an engine support fixture, J-28467-A, or the equivalent. Be sure the support fixture is tight and the weight of the powertrain is removed from the mounts.

12. Remove the upper transaxle-to-engine bolts.

13. Raise and safely support the vehicle.

14. Remove the front tire and wheel assemblies.

15. Remove the left side power steering rack mounting bolts.

16. Remove the left front splash shield.

17. Remove the cotter pins and castle nuts from the lower ball joints and using a suitable ball joint separator, disconnect the ball joints from the steering knuckles.

18. Remove the front lower air deflector.

19. Remove the power steering line retaining clamp from the frame.

20. Remove the remaining rack mounting bolt and raise the rack off its frame mount and support.

21. Remove the left and right transaxle mount through-bolts.

22. Remove the complete exhaust system.

23. Support the subframe assembly.

24. Remove the subframe mounting bolts and remove the subframe assembly.

25. Remove the torque converter cover mounting screws and remove the cover.

26. Matchmark the converter to the flywheel and remove the three converter mounting bolts.

27. Disconnect the starter.

28. Remove the starter.

29. Remove the halfshafts.

30. Support the transaxle with a suitable jack.

31. Remove the rear transaxle mount-to-bracket retaining nuts and remove the rear transaxle mount bracket from the vehicle.

32. Disconnect the rear spark plug wires

from the plugs and route them out of the way.

33. Remove the transaxle filler tube.

34. Remove the right side exhaust manifold.

35. Remove the lower transaxle-to-engine mounting bolts.

36. Disconnect and cap the transaxle oil cooler lines from the transaxle.

37. Lower the transaxle from the vehicle.

To install:

38. Raise the transaxle assembly into the vehicle and guide the transaxle onto the engine alignment dowels.

39. Reconnect the upper and lower transaxle oil cooler lines.

40. Reinstall the transaxle-to-engine bolts and with all the bolts in place tighten them alternately an evenly to 55 ft. lbs. (75 Nm).

41. Reinstall the right side exhaust manifold and tighten the mounting bolts to 38 ft. lbs. (52 Nm).

42. Reinstall the transaxle filler tube.

43. Reconnect the spark plug wires to the rear plugs.

44. Reinstall the rear transaxle mount to the body and tighten the mounting bolts to 37 ft. lbs. (50 Nm), the transaxle mount-to-transaxle bolts to 55 ft. lbs. (75 Nm) and the mount-to-transaxle bracket nuts to 29 ft. lbs. (40 Nm).

45. Remove the transaxle jack.

46. Reinstall the halfshafts.

47. Reinstall the starter and attach the electrical connectors.

48. Be sure the matchmarks on the converter and flywheel are in alignment. Install the converter mounting bolts and tighten to 44 ft. lbs. (60 Nm).

49. Reinstall the flywheel cover and cover mounting bolts.

50. Raise the subframe into position and tighten the mounting bolts to 142 ft. lbs. (192 Nm).

51. Reinstall the exhaust system.

52. Reinstall the left transaxle mount-to-frame bolt and tighten to 63 ft. lbs. (85 Nm).

53. Reinstall the right transaxle mount-to-transaxle bolt and tighten to 75 ft. lbs. (102 Nm).

54. Lower the steering rack into position and install the through-bolt on the right side of the rack.

55. Reinstall the power steering line retaining bracket to the subframe.

56. Reconnect the ball joints to the steering knuckles and tighten the castle nuts to 41 ft. lbs. (55 Nm). If necessary tighten the nuts up to 60 degrees ⅙ turn) additional to align the cotter pin holes. NEVER loosen the nuts to make the holes align.

57. Reinstall the left side splash shield.

58. Reinstall the left side rack and pinion mounting bolts.

59. Reinstall the front tire and wheel assemblies and tighten the wheel nuts to 100 ft. lbs. (140 Nm).

60. Lower the vehicle.

61. Remove the engine support fixture.

62. Reconnect the oxygen sensor electrical connector.

63. Reinstall the heater hose retainer bracket.

64. Reinstall the vacuum reservoir.

65. Reinstall the right-to-left exhaust manifold mounting bolts.

66. Reconnect the electrical connectors to the park neutral switch, back-up light switch, transaxle harness and vehicle speed sensor.

67. Reconnect the vacuum line to the modulator.

68. Reconnect the shift cable to the manual shaft and install the cable mounting bracket. Tighten the manual shaft nut to 15 ft. lbs. (20 Nm).

69. Reinstall the cruise control servo, connect the vacuum line and linkage.

70. Reinstall the air inlet duct.

71. Reconnect the negative battery cable.

72. Refill the transaxle with fluid. Start the vehicle and check the fluid level again and top off as necessary.

73. Road test the vehicle and verify proper transaxle operation and no fluid leaks.

Aurora

1. Disconnect the negative battery cable.

2. Remove the nut from the shift linkage at the manual shaft on the transaxle. Remove the two linkage bracket bolts and remove the cable and bracket from the transaxle and position out of the way.

3. Detach the electrical connectors from the park neutral switch.

4. Install an engine support fixture, J-28467-A, or the equivalent. Be sure the support fixture is tight and the weight of the powertrain is removed from the mounts.

5. Drain the cooling system.

6. Disconnect the vacuum line at the brake booster.

7. Disconnect the transaxle vent hose.

8. Detach the speed sensor connector and power steering gear connector.

9. Disconnect and cap the upper transaxle oil cooler line from the radiator.

10. Disconnect and cap the lower transaxle oil cooler line from the transaxle.

11. Remove the retaining bracket nut for the cooler lines at the transaxle.

12. Disconnect the coolant bypass pipe from the thermostat housing and position out of the way.

13. Remove the left and right transaxle mount bolts.

14. Raise and safely support the vehicle.

15. Remove the left front tire and wheel assembly.

16. Remove the left front splash shield.

17. Remove the cotter pins and castle nuts from the outer tie rod end and using a suitable ball joint separator, disconnect the tie rod ends from the steering knuckles.

18. Remove the cotter pins and castle nuts from the lower ball joints and using a suitable ball joint separator, disconnect the ball joints from the steering knuckles.

19. Remove the halfshafts.

20. Remove the engine oil pan-to-transaxle bracket.

21. Remove the torque converter cover mounting screws and remove the cover.

22. Matchmark the converter to the flywheel and remove the three converter mounting bolts.

23. Remove the complete exhaust system.

24. Remove the exhaust manifold rear pipe.

25. Remove the steering rack-to-right transaxle mount bolts.

26. Remove the bolt from the right transaxle mount.

27. Remove the frame-to-right transaxle mount and remove the right side mount.

28. Remove the power steering line retaining clamp from the frame.

29. Remove the remaining rack mounting bolt and raise the rack off its frame mount and support.

30. Support the subframe assembly.

31. Remove the knock sensor shield.

32. Remove the engine-to-transaxle brace.

33. Remove the rear transaxle mount-to-bracket retaining nuts and remove the rear transaxle mount bracket from the vehicle.

34. Remove the right and left lower transaxle-to-engine bolts.

35. Remove the subframe mounting bolts and remove the subframe assembly.

36. Support the transaxle with a suitable jack.

37. Remove the remainder of the transaxle-to-engine bolts.

38. Lower the transaxle from the vehicle.

To install:

39. Raise the transaxle assembly into the vehicle and guide the transaxle onto the engine alignment dowels.

40. Reinstall the transaxle-to-engine bolts and with all the bolts in place tighten them alternately an evenly to 55 ft. lbs. (75 Nm).

41. Reinstall the rear transaxle mount to the body and tighten the mounting bolts to 37 ft. lbs. (50 Nm).

42. Reinstall the rear transaxle mount bracket to the transaxle and tighten the mounting bolts to 43 ft. lbs. (58 Nm).

43. Reinstall the engine-to-transaxle brace and tighten the brace mounting bolts to 35 ft. lbs. (47 Nm).

44. Reinstall the knock sensor shield.

45. Reconnect the electrical connectors to the speed and knock sensors.

46. Be sure the matchmarks on the converter and flywheel are in alignment. Install the converter mounting bolts and tighten to 44 ft. lbs. (60 Nm).

47. Reinstall the flywheel cover and cover mounting bolts.

48. Reinstall the transaxle-to-oil pan brace.

49. Raise the subframe into position and tighten the mounting bolts to 142 ft. lbs. (192 Nm).

50. Reinstall the left transaxle mount-to-frame bolt and tighten to 63 ft. lbs. (85 Nm).

51. Lower the steering rack into position and install the through-bolt on the right side of the rack.

52. Reinstall the power steering line retaining bracket to the subframe.

53. Reinstall the right transaxle mount to the frame and tighten the mounting bolts to 54 ft. lbs. (73 Nm).

54. Reinstall the right transaxle mount-to-transaxle bolt and tighten to 81 ft. lbs. (110 Nm).

55. Reinstall the rear exhaust manifold pipe.

56. Reinstall the exhaust system.

57. Reinstall the halfshafts.

58. Reconnect the ball joints to the steering knuckles and tighten the castle nuts to 41 ft. lbs. (55 Nm). If necessary tighten the nuts up to 60 degrees ⅙ turn) additional to align the cotter pin holes. NEVER loosen the nuts to make the holes align.

59. Reconnect the tie rod ends to the steering knuckles and tighten the castle nuts to 52 ft. lbs. (70 Nm). If necessary tighten the nuts up to 60 degrees ⅙ turn) additional to align the cotter pin holes. NEVER loosen the nuts to make the holes align.

60. Reinstall the left side splash shield.

61. Reinstall the front tire and wheel assemblies and tighten the wheel nuts to 100 ft. lbs. (140 Nm).

62. Lower the vehicle.

63. Remove the engine support fixture.

64. Reconnect the coolant bypass pipe to the thermostat housing.

65. Reinstall the transaxle oil cooler line bracket and mounting bolt.

66. Reconnect the upper and lower transaxle oil cooler lines.

67. Reconnect the shift cable to the man-

ual shaft and install the cable mounting bracket. Tighten the manual shaft nut to 15 ft. lbs. (20 Nm).

68. Reconnect the vacuum line to the brake booster.

69. Reconnect the transaxle vent.

70. Refill the cooling system.

71. Reconnect the negative battery cable.

72. Refill the transaxle with fluid. Start the vehicle and check the fluid level again and top off as necessary. Service this vehicle with DEXRON® II or DEXRON® IIE automatic transmission fluid.

73. Road test the vehicle and verify proper transaxle operation and no fluid leaks.

Halfshaft

REMOVAL & INSTALLATION

1. Raise and safely support the vehicle.
2. Remove the tire and wheel assembly.
3. Remove the sway bar link kit.
4. Remove the cotter pin and castle nut from the lower ball joint. Using a suitable ball joint separator, J-36226 or the equivalent, disconnect the ball stud from the steering knuckle.
5. If removing the right halfshaft turn the wheel to the left and if removing the left halfshaft turn the wheel to the right.
6. Using a suitable prying tool, pry the lower control arm down away from the steering knuckle.
7. Insert a drift punch through the caliper and into the rotor cooling fins. Remove the hub nut from the end of the halfshaft.
8. Separate the halfshaft from the hub using J-28733-B, or an equivalent puller. Once the halfshaft is clear of the knuckle assembly, swing the strut assembly toward the rear of the vehicle.
9. Separate the halfshaft from the

transaxle using J-33008 and J-2619-01, or an equivalent slide hammer and adapter.

10. Remove the halfshaft.

To install:

11. If installing the right side halfshaft, install J-37292-B, or an equivalent tear away axle seal protector over the seal.

12. Install the halfshaft into the transaxle and seat in place by inserting a suitable prybar in the groove in the inboard joint and tapping the joint into place. Be sure the joint is properly seated by grasping the inboard joint and making sure it wont pull out of the transaxle. DO NOT pull on the shaft itself or the inboard joint can become damaged.

13. Insert the halfshaft through the hub and install the washer and mounting nut. Insert a drift through the caliper and into the rotor cooling fins and tighten the mounting nut to 107 ft. lbs. (145 Nm).

14. Reconnect the ball joint to the steering knuckle and tighten the castle nut to 41 ft. lbs. (55 Nm). If necessary to align the cotter pin holes rotate the nut up to 60 degrees (⅙ turn) additional. NEVER loosen the nut to align the holes.

15. Install a new cotter pin.

16. Install the sway bar link kit and tighten the bolt to 13 ft. lbs. (17 Nm).

17. Remove the tear away seal protector. Be sure no pieces of the tool remain in the transaxle.

18. Install the tire and wheel assembly and tighten the wheel nuts to 100 ft. lbs. (140 Nm).

19. Lower the vehicle.

STEERING AND SUSPENSION

Air Bag

✳✳ CAUTION

Some vehicles are equipped with the Supplemental Inflatable Restraint (SIR) or air bag system. The SIR system must be disabled before performing service on or around SIR system components, steering column, instrument panel components, wiring and sensors. Failure to follow safety and disabling procedures could result in accidental air bag deployment, possible personal injury and unnecessary SIR system repairs.

PRECAUTIONS

Several precautions must be observed when handling the inflator module to avoid accidental deployment and possible personal injury.

• Never carry the inflator module by the wires or connector on the underside of the module.

• When carrying a live inflator module, hold securely with both hands, and ensure that the bag and trim cover are pointed away.

• Place the inflator module on a bench or other surface with the bag and trim cover facing up.

• With the inflator module on the bench, never place anything on or close to the module which may be thrown in the event of an accidental deployment.

DISARMING

1. Turn the steering wheel so the vehicle wheels are pointing straight ahead.
2. Turn the ignition key to the **LOCK** position and remove the key.
3. Remove the AIR BAG fuse from the fuse block.
4. Remove the left side sound insulator.
5. Detach the Connector Position Assurance (CPA) and yellow 2-way SIR

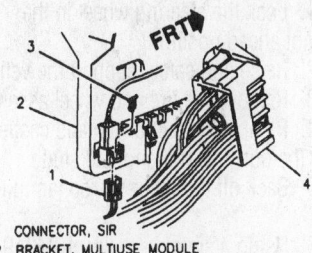

1 CONNECTOR, SIR
2 BRACKET, MULTIUSE MODULE
3 CONNECTOR POSITION ASSURANCE (CPA)
4 CONNECTOR, STEERING COLUMN WIRING HARNESS

7922XG21

SRS 2-way connector location—driver's side

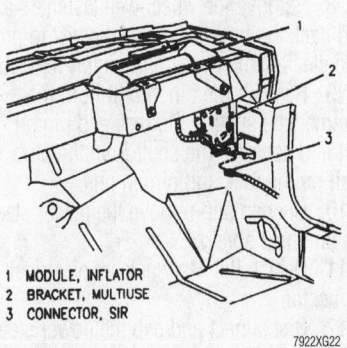

1 MODULE, INFLATOR
2 BRACKET, MULTIUSE
3 CONNECTOR, SIR

7922XG22

SRS 2-way connector location—passenger's side

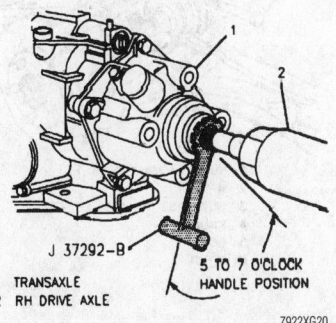

J 37292-B

5 TO 7 O'CLOCK HANDLE POSITION

1 TRANSAXLE
2 RH DRIVE AXLE

7922XG20

Installing the right side axle with the tear away seal protector

connector at the multi-use bracket near the base of the steering column. The driver's side air bag is now disabled.

6. Remove the right side sound insulator.

7. Detach the Connector Position Assurance (CPA) and yellow 2-way SIR connector at the DERM mounting bracket. The passenger's side air bag is now disabled.

After the necessary repairs are made, re-enable the air bag system as follows:

8. Be sure the ignition is locked and the key is removed.

9. Attach the yellow 2-way SIR connector and Connector Position Assurance (CPA) at the DERM mounting bracket.

10. Install the right side sound insulator.

11. Attach the yellow 2-way SIR connector and Connector Position Assurance (CPA) at the multi-use bracket at the base of the column.

12. Install the left side sound insulator.

13. Install the AIR BAG fuse.

14. Turn the ignition switch to the **RUN** position and verify the AIR BAG light flashes 7 times, then shuts off.

Power Rack and Pinion Steering Gear

REMOVAL & INSTALLATION

1. Lock the steering wheel in the straight ahead position.

2. Raise and safely support the vehicle.

3. Remove the tire and wheel assembly.

4. Remove the cotter pin and castle nut from the outer tie rod end ball stud.

5. Back off the inner tie rod jam nut ½ turn.

6. Using a suitable puller, J-24319–01 or the equivalent, separate the outer tie rod end from the steering knuckle.

7. Disconnect the exhaust pipe from the rear manifold and remove the intermediate pipe hangers. Lower the exhaust system to provide clearance.

8. Remove the wheel well fasteners and fold back the inner wheel well panel to provide clearance for rack and pinion removal.

9. Remove the pinch bolt from the intermediate shaft at the rack and pinion unit and separate the shaft from the stub shaft on the rack and pinion unit.

10. Unsnap and remove the power steering unit heat shield.

11. Detach the Magnasteer® electrical connector.

12. Disconnect and cap the power steering lines from the rack and pinion unit.

13. Remove the rack and pinion mounting bolts.

14. Support the rear of the subframe. Remove the two rear subframe mounting bolts and lower the frame assembly. Only lower the frame enough to remove the rack and pinion assembly.

15. Remove the rack and pinion assembly through the right side wheel well.

To install:

16. Install the rack and pinion unit through the right side wheel well.

17. Loosely install the three mounting bolts.

18. Raise the subframe into position and tighten the frame mounting bolts to 142 ft. lbs. (192 Nm). Remove the support.

19. Tighten the rack and pinion mounting bolts in sequence to 48 ft. lbs. (65 Nm). Start with the vertically installed bolt closest to the pinion housing and work toward the passenger side of the vehicle.

20. Reconnect the power steering hoses to the rack and pinion unit and tighten the fittings to 20 ft. lbs. (27 Nm).

21. Reconnect the Magnasteer® electrical connector.

22. Reinstall the rack and pinion heat shield.

23. Reconnect the intermediate shaft to the rack and pinion stub shaft and tighten the pinch bolt to 35 ft. lbs. (47 Nm).

24. Reconnect the tie rod end to the steering knuckle and tighten the castle nut to 35 ft. lbs. (47 Nm). If necessary to align the cotter pin holes, tighten the nut up to 60° additional rotation. NEVER loosen the castle nut to align the cotter pin holes.

25. Install a new cotter pin.

26. Reconnect the exhaust pipe to the exhaust manifold and connect the intermediate pipe hangers.

27. Reinstall the wheel well molding.

28. Reinstall the tire and wheel assembly and tighten the wheel nuts to 100 ft. lbs. (140 Nm).

29. Lower the vehicle.

➡ **Whenever the vehicle sub-frame is removed or lowered, the front wheel alignment should be checked.**

30. Refill and bleed the power steering system. Verify no leaks.

31. Check the front end alignment and adjust as necessary.

Strut

REMOVAL & INSTALLATION

Front

1. Disconnect the negative battery cable.

2. Raise and safely support the vehicle.

3. Remove the tire and wheel assembly.

4. Disconnect the ABS wheel speed sensor.

5. Remove the ABS speed sensor bracket from the strut.

6. If removing the left side strut, disconnect the brake line bracket from the strut.

7. Support the steering knuckle so when the strut is disconnected the brake line does not get stretched.

8. Scribe a mark on the strut referencing the lower strut bracket to the steering knuckle.

9. Remove the nut and through-bolts from the lower strut bracket.

10. From under the hood, remove the three upper strut plate mounting nuts and washers,

11. Remove the strut from the vehicle.

12. Place the strut in an approved fixture to disassemble the coil spring from the strut. With the spring compressed, hold the strut shaft from turning using a socket and remove the 24mm strut shaft nut. Relieve the pressure on the spring and separate the front coil spring from the strut.

To install:

13. If the spring and strut were disassembled, assemble using an approved fixture.

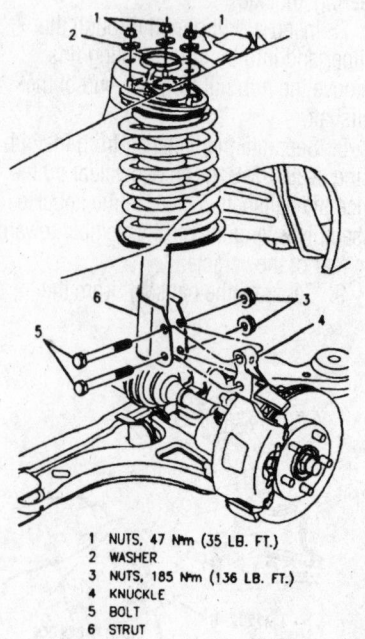

1 NUTS, 47 Nm (35 LB. FT.)
2 WASHER
3 NUTS, 185 Nm (136 LB. FT.)
4 KNUCKLE
5 BOLT
6 STRUT

7922XG23

Exploded view of the upper and lower strut mounting components

14. Position the strut in the vehicle guiding the upper strut plate studs into the body and loosely install the nuts.

15. Reinstall the lower strut through-bolts. With the matchmarks in alignment, tighten the nuts to 136 ft. lbs. (185 Nm).

16. Remove the support from the steering knuckle.

17. Reconnect the brake line to the strut, if removed.

18. Install the speed sensor bracket on the strut.

19. Reconnect the ABS sensor.

20. Reinstall the tire and wheel assembly and tighten the wheel nuts to 100 ft. lbs. (140 Nm).

21. Lower the vehicle.

22. Tighten the upper strut plate mounting nuts to 35 ft. lbs. (47 Nm).

23. Check the front end alignment and adjust as necessary.

24. Reconnect the negative battery cable.

Shock Absorber

REMOVAL & INSTALLATION

Rear

1. Raise and safely support the vehicle.

2. Remove the rear tire and wheel assembly.

3. Support the lower control arm on a safety stand at such a height that the upper shock bolts will still be accessible.

4. Disconnect the Electronic Level Control (ELC) air tube from the shock.

5. Remove the two bolts from under the control arm securing the lower shock mount.

6. From inside the trunk, remove the trunk trim panel to access the upper shock mount bolts.

7. Remove the upper shock cap.

8. Remove the two nuts from the upper shock mount and remove the reinforcement.

9. Remove the shock from the vehicle.

To install:

10. Install the shock in the vehicle.

11. Reinstall the reinforcement and upper shock mounting nuts. Tighten the mounting nuts to 15 ft. lbs. (20 Nm).

12. Reinstall the upper shock cap and reposition the inner trunk trim.

13. Extend the shock and install the lower shock mounting bolts and tighten to 18 ft. lbs. (24 Nm).

14. Reconnect the ELC air tube to the shock.

15. Raise the vehicle off the safety stand and remove the stand.

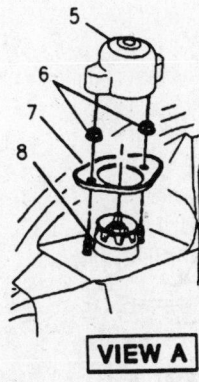

VIEW A

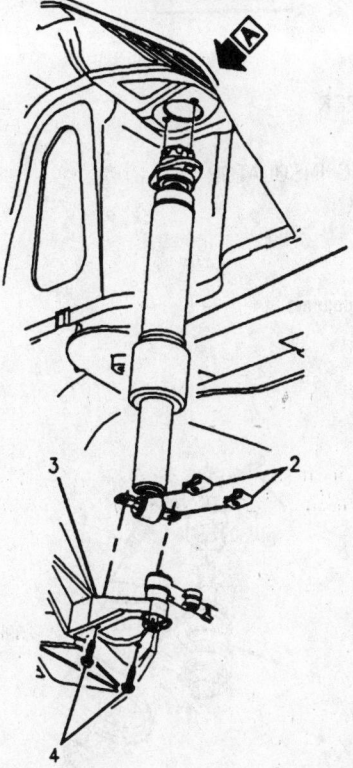

1	SHOCK
2	U–NUTS
3	CONTROL ARM
4	BOLTS 24 N·m (18 LB. FT.)
5	COVER
6	NUTS 20 N·m (15 LB. FT.)
7	REINFORCEMENT
8	MOUNT, UPPER

7922XG34

Exploded view of the rear shock mounting

16. Reinstall the tire and wheel assembly and tighten the wheel nuts to 100 ft. lbs. (140 Nm).

17. Lower the vehicle.

Coil Spring

REMOVAL & INSTALLATION

Front

For front coil spring removal and installation, please refer to the front strut procedure.

Rear

1. Raise and safely support the vehicle.

2. Remove the rear tire and wheel assembly.

3. Support the lower control arm with a suitable screw type jack.

4. Disconnect the ELC air tube from the shock absorber.

5. Remove the two lower shock absorber-to-control arm bolts.

6. Remove the cotter pin and castle nut from the adjustment link outer ball stud.

7. Using J-24319-B, or an equivalent ball joint separator, disconnect the ball stud from the knuckle.

8. Lower the control arm until the arm bottoms out on the rear suspension support.

9. Using a suitable prying tool, pry under the lower spring insulator to unseat it from the control arm and remove the insulator and coil spring.

10. Remove the upper spring insulator if needed.

To install:

11. Reinstall the upper spring insulator if removed, and engage the retainer on the back of the insulator in the upper mount hole.

12. Install the coil spring and lower insulator in the vehicle and seat the lower insulator in the control arm hole.

13. Raise the control arm until the shock bolts can be installed. Tighten the two bolts to 18 ft. lbs. (24 Nm).

14. Remove the jack.

15. Reconnect the adjustment link ball stud to the knuckle and tighten the castle nut to 88 inch lbs. (10 Nm) plus 180 degrees ½ turn) additional rotation. If necessary to align the cotter pin holes tighten the nut up to 60 degrees ⅙ turn) more. NEVER loosen the castle nut to align the cotter pin holes.

16. Install a new cotter pin.

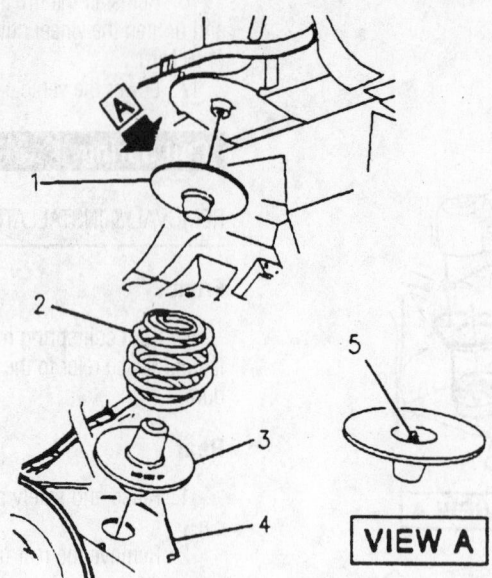

1 JOUNCE BUMPER
2 SPRING
3 LOWER SPRING INSULATOR
4 CONTROL ARM
5 RETAINER

7922XG35

Exploded view of the rear coil spring and related components

17. Reconnect the ELC air tube to the shock absorber.

18. Reinstall the tire and wheel assembly and tighten the wheel nuts to 100 ft. lbs. (140 Nm).

19. Lower the vehicle.

Lower Ball Joint

REMOVAL & INSTALLATION

Ball joints must be replaced if any looseness is detected in the joint or if the ball joint seal is cut. To inspect the ball joints, raise the front of the vehicle allowing the suspension to hang free. Grasp the tire at the top and bottom and move the top of the tire in an in-And-out motion. Check for any horizontal movement of the knuckle relative to the control arm. If movement is in the wheel bearing, the bearing and hub must be replaced. If the ball joint stud is disconnected from the knuckle and looseness can be detected or if the ball stud can be twisted in its socket using finger pressure, replace the ball joint.

Ball joint tightness in the knuckle boss should also be checked. This may be done by shaking the wheel and feeling for movement of the stud end or nut at the knuckle

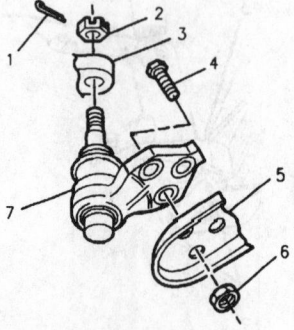

1 PIN
2 NUT, BALL JOINT TO KNUCKLE;
   TIGHTEN TO 10 N•m (88 LB. IN.)
   THEN TIGHTEN 2 FLATS TO
   55 N•m (41 LB. FT.), MIN.
3 KNUCKLE
4 BALL JOINT MOUNTING BOLTS MUST
   FACE DOWN
5 CONTROL ARM
6 BALL JOINT MOUNTING NUTS
   68 N•m (50 LB. FT.)
7 SERVICE BALL JOINT

7922XG24

Exploded view the lower ball joint and replacement mounting bolts and nuts

boss. Worn or damaged ball joints and knuckles must be replaced.

1. Raise and safely support the vehicle.

2. Remove the front tire and wheel assembly.

3. Remove the sway bar link kit. Take note of the positions of the washers and insulators for installation purposes.

4. Remove the cotter pin and castle nut from the lower ball joint and using a suitable ball joint separator, J-36226 or the equivalent separate the ball joint from the steering knuckle.

5. Using a ½ in. drill bit, drill the heads off of the three rivets securing the lower ball joint to the control arm. With the rivet heads removed, drive the bodies of the rivets out of the control arm using a suitable punch.

6. Remove the ball joint.

To install:

7. Position the lower ball joint on the control arm and install the three mounting bolts so the bolt threads point down. Install the mounting nuts and tighten to 50 ft. lbs. (68 Nm).

8. Reconnect the lower ball joint to the steering knuckle and tighten the castle nut to 41 ft. lbs. (55 Nm). If necessary tighten the castle nut up to 60 degrees (⅙ turn) additional rotation to align the cotter pin holes. NEVER loosen the nut to make the alignment. Install a new cotter pin.

9. Install the sway bar link kit and tighten the mounting nut to 13 ft. lbs. (17 Nm).

10. Install the tire and wheel assembly and tighten the wheel nuts to 100 ft. lbs. (140 Nm).

11. Lower the vehicle.

Wheel Bearings

ADJUSTMENT

The wheel bearings are not adjustable. If the wheel bearings are defective, the hub and bearing assembly must be replaced.

REMOVAL & INSTALLATION

Front

The front wheel bearings are not serviced separately. If the front wheel bearings are defective, the hub and bearing assembly must be replaced.

1. Disconnect the negative battery cable.

2. Raise and safely support the vehicle.

3. Remove the tire and wheel assembly.

4. Lubricate the threads on the drive axle with clean engine oil.

5. Install a drift punch through the caliper and into the brake rotor cooling fins. This keeps the hub from turning when removing the hub nut. Remove the hub nut from the drive axle.

6. Remove the caliper mounting bolts and remove the caliper from the steering knuckle. Support the caliper out of the way. DO NOT allow the brake hose to support the weight of the caliper.

7. Remove the brake rotor.

8. Disconnect the ABS sensor and unclip the sensor from the backing plate.

9. Remove the three hub and bearing mounting bolts and remove the backing plate.

10. Using axle puller J-28733 or equivalent hub puller, separate the hub and bearing assembly from the halfshaft.

11. Remove the hub and bearing assembly.

12. If the steering knuckle is also to be removed, remove the cotter pin and castle nut from the outer tie rod end. Using a puller J-24319-B or equivalent, separate the tie rod end from the steering knuckle.

13. Scribe matchmarks on the steering knuckle and strut for reference on installation.

14. Remove the cotter pin and castle nut from the lower ball joint and separate the ball joint from the steering knuckle using J-36226, or the equivalent.

15. Remove the two strut through-bolt nuts and remove the through-bolts.

16. Lift the steering knuckle off of the lower ball joint.

To install:

17. Install the steering knuckle onto the lower ball joint and tighten the castle nut to 41 ft. lbs. (55 Nm). If necessary to align the cotter pin holes tighten the nut up to 60° additional. NEVER loosen the nut to align the holes. Install a new cotter pin.

18. Line up the lower strut bracket with the steering knuckle and install the through-bolts and nuts. With the matchmarks in alignment tighten the strut through-bolts to 136 ft. lbs. (185 Nm).

19. Reconnect the outer tie rod end to the steering knuckle and tighten the castle to 35 ft. lbs. (48 Nm). If necessary to align the cotter pin holes tighten the nut up to 60 degrees ⅙ turn) additional rotation. NEVER loosen the nut to align the holes. Install a new cotter pin.

20. Insert the ABS sensor wire through the opening in the steering knuckle and slide the axle shaft into the splined opening of the hub and bearing assembly.

21. Install a new axle nut and draw the hub and bearing assembly in to place.

22. Reinstall the backing plate and install the three mounting bolts. Tighten the bolts alternately and evenly to 70 ft. lbs. (95 Nm).

23. Reconnect the ABS sensor to the backing plate and attach the wiring harness connector.

24. Reinstall the brake rotor.

25. Reinstall the caliper on the steering knuckle and tighten the mounting bolts to 38 ft. lbs. (51 Nm).

26. Insert a drift punch through the caliper and into the brake rotor cooling fins and tighten the axle nut to 107 ft. lbs. (145 Nm).

27. Reinstall the tire and wheel assembly and tighten the wheel nuts to 100 ft. lbs. (140 Nm).

28. Lower the vehicle.

29. Pump the brakes to obtain a firm pedal before attempting to move the vehicle.

30. Check the front end alignment and adjust as necessary.

31. Reconnect the negative battery cable.

Rear

The rear wheel bearings are not serviced separately. If the rear wheel bearings are defective, the hub and bearing assembly must be replaced.

1. Disconnect the negative battery cable.

2. Raise and safely support the vehicle.

3. Remove the tire and wheel assembly.

4. Remove the caliper mounting bracket bolts and remove the caliper and bracket assembly from the brake rotor. Support the assembly out of the way. DO NOT allow the brake hose to support the weight of the caliper and bracket.

5. Remove the brake rotor.

6. Detach the ABS sensor electrical connector.

7. Remove the four hub assembly mounting bolts and remove the hub and bearing assembly along with the backing plate from the rear control arm.

To install:

8. Install the hub and bearing assembly along with the backing plate onto the control arm and loosely install the mounting bolts.

9. Be sure the ABS sensor is properly routed and connect the ABS sensor harness.

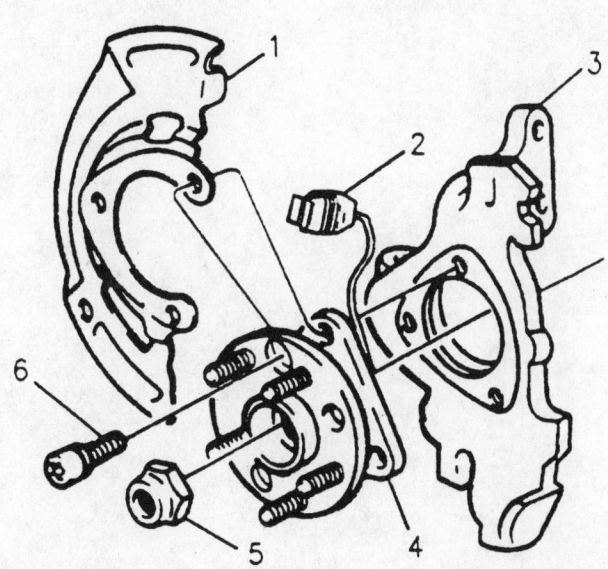

1	DUST SHIELD	4	HUB AND BEARING
2	WHEEL SPEED SENSOR CONNECTOR	5	NUT, DRIVE AXLE, 145 N•m (107 LB. FT.)
3	STEERING KNUCKLE	6	RETAINING BOLT, 95 N•m (75 LB. FT.)

7922XG36

Exploded view of the front hub mounting and related components

10. Tighten the hub and bearing mounting bolts alternately and evenly to 52 ft. lbs. (70 Nm).

11. Reinstall the brake rotor.

12. Reinstall the caliper and bracket assembly and tighten the bracket bolts to 35 ft. lbs. (48 Nm).

13. Reinstall the tire and wheel assembly and tighten the wheel nuts to 100 ft. lbs. (140 Nm).

14. Lower the vehicle.

15. Reconnect the negative battery cable.

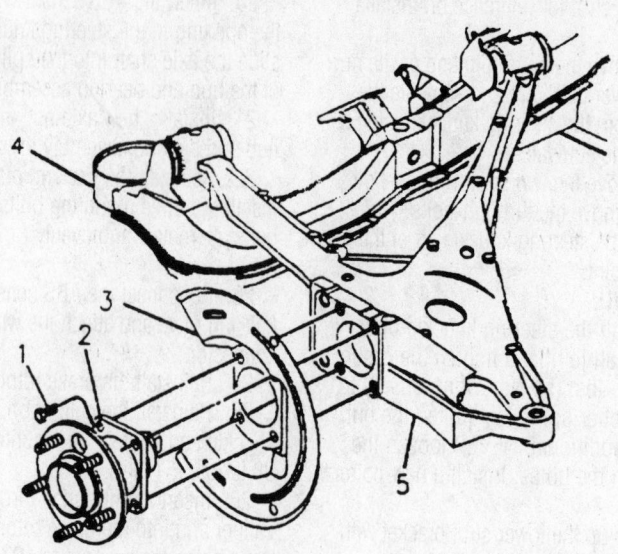

1 BOLT
2 HUB & BEARING
3 BRAKE SHIELD
4 REAR SUSPENSION SUPPORT ASSEMBLY
5 CONTROL ARM

7922XG37

Exploded view of the rear hub mounting and related components

GENERAL MOTORS J BODY

33

Chevrolet-Cavalier • **Pontiac**-Sunfire

DRIVE TRAIN 33-20
ENGINE REPAIR 33-2
FUEL SYSTEM 33-18
STEERING AND
  SUSPENSION 33-23

A
Air Bag........................33-23
  Disarming....................33-23
  Precautions..................33-23
  Rearming.....................33-23

C
Camshaft and Valve Lifters 33-12
  Removal & Installation 33-12
Clutch...........................33-21
  Removal & Installation 33-21
Coil Spring......................33-26
  Removal & Installation 33-26
Cylinder Head....................33-6
  Removal & Installation 33-6

E
Engine Assembly 33-2
  Removal & Installation 33-2
Exhaust Manifold 33-11
  Removal & Installation 33-11

F
Fuel Filter 33-18
  Removal & Installation 33-18

Fuel Pump 33-19
  Removal & Installation 33-19
Fuel System Pressure 33-18
  Relieving 33-18
Fuel System Service
  Precautions....................33-18

H
Halfshaft.......................33-22
  Removal & Installation 33-22
Hydraulic Clutch System 33-22
  Bleeding 33-22

I
Ignition Timing 33-2
  Adjustment 33-2
Intake Manifold 33-8
  Removal & Installation 33-8

L
Lower Ball Joint...................33-27
  Removal & Installation 33-27

O
Oil Pan.........................33-14
  Removal & Installation 33-14
Oil Pump 33-15
  Removal & Installation 33-15

R
Rack and Pinion Steering Gear 33-23
  Removal & Installation 33-23
Rocker Arms 33-8
  Removal & Installation 33-8

S
Strut...........................33-24
  Removal & Installation 33-24

T
Timing Chain, Sprockets,
  Front Cover and Seal.................33-16
  Removal & Installation 33-16
Transaxle Assembly 33-20
  Removal & Installation 33-20

V
Valve Lash 33-13
  Adjustment 33-13

W
Water Pump.......................33-4
  Removal & Installation 33-4
Wheel Bearings...................33-27
  Adjustment 33-27
  Removal & Installation 33-27

ENGINE REPAIR

→Disconnecting the negative battery cable on some vehicles may interfere with the functions of the on board computer systems and may require the computer to undergo a relearning process, once the negative battery cable is reconnected.

Ignition Timing

ADJUSTMENT

Ignition timing is controlled by the PCM. No adjustment is necessary nor possible.

Engine Assembly

REMOVAL & INSTALLATION

2.2L Engine

※※ CAUTION

The fuel injection system remains under pressure, even after the engine has been turned OFF. The fuel system pressure must be relieved before disconnecting any fuel lines. Failure to do so may result in fire and/or personal injury.

1. Relieve the fuel system pressure.
2. Disconnect the negative, then the positive battery cables.

※※ CAUTION

Never open, service or drain the radiator or cooling system when hot; serious burns can occur from the steam and hot coolant.

3. Drain the cooling system into a suitable container.
4. Remove the throttle body air inlet duct.
5. Remove the battery from the vehicle.
6. Remove the air cleaner assembly.
7. Remove the upper radiator.
8. Disconnect the vacuum hose from power brake booster.
9. Remove the alternator upper brace and disconnect the wiring from the alternator.
10. Tag and disconnect the electrical wiring from the following components:
- Oxygen (O$_2$) sensor
- Fuel injector harness

- Idle Air Control (IAC)
- Throttle Position Sensor (TPS)
- Coolant Temperature Sender
- Park/neutral switch
- TCC solenoid
- Transaxle shift solenoid
- Transaxle ground
- MAP sensor
- EGR
- Cooling fan

11. If equipped, properly discharge and recover the refrigerant from the A/C system.
12. Disconnect the compressor-to-condenser and accumulator lines.
13. Remove the slave cylinder from the transaxle, if equipped with a manual transaxle.
14. Disconnect and cap the fuel lines.
15. Disconnect the shift linkage from the transaxle bracket.
16. Disconnect the control cables from the throttle body lever and remove the cable bracket from the rocker arm cover and intake manifold.
17. Place a pan under the power steering pump and disconnect and cap the power steering lines.
18. Install an engine support fixture, J-28467, or the equivalent.
19. Safely raise and support the vehicle.

※※ CAUTION

The EPA warns that prolonged contact with used engine oil may cause a number of skin disorders, including cancer! You should make every effort to minimize your exposure to used engine oil. Protective gloves should be worn when changing the oil. Wash your hands and any other exposed skin areas as soon as possible after exposure to used engine oil. Soap and water, or waterless hand cleaner should be used.

20. Drain and recycle the engine oil.
21. Remove the front tire and wheel assemblies.
22. Remove the right and left inner fender splash shields.
23. Disconnect the front exhaust pipe from manifold and remove the front exhaust pipe.
24. Tag and detach the following electrical connectors from the lower engine components:
- Ignition assembly
- Starter
- Vehicle Speed Sensor (VSS)
- Transaxle ground wire
25. Remove the flywheel cover.

26. Remove the lower radiator hose.
27. Disconnect the heater hoses from the heater core pipes.
28. Disconnect the transaxle cooler lines at the radiator.
29. Remove the mounting bolt from the lower engine strut at the suspension crossmember.
30. Detach the both front ABS wheel sensor connectors and unfasten the harnesses from the suspension crossmember.
31. Remove the cotter pins and castle nuts from the lower ball joints, then separate the joints from the steering knuckles.
32. Remove the suspension crossmembers.
33. Disconnect the halfshafts from the transaxle and support out of the way.
34. Position a suitable table under the engine and transaxle assembly.
35. Lower the vehicle until the powertrain assembly is on the table.
36. Remove the transaxle-to-frame mount bolts.
37. Remove the intermediate bracket from the right engine mount support bracket.
38. Remove the engine support fixture.
39. Raise the vehicle, leaving the powertrain assembly on the table. When raising the vehicle, take it up slowly and verify that no lines are still connected to the powertrain assembly.
40. Remove the torque converter-to-flywheel bolts and transaxle-to-engine mounting bolts and remove the transaxle from the engine.

To install:
41. Install the transaxle on the engine and tighten the mounting bolts to 68 ft. lbs. (93 Nm).
42. Install flywheel-to-converter bolts and tighten to 46 ft. lbs. (62 Nm).
43. Position the engine on the table and lower the vehicle until the engine is in the engine bay.
44. Install an engine support fixture, J-28467, or the equivalent.
45. Raise and safely support the vehicle and remove the engine table.
46. Connect the transaxle cooler lines to the radiator.
47. Attach the heater hoses to the heater core outlet pipes.
48. Install the lower radiator hose.
49. Connect the engine mount support bracket to the intermediate bracket and tighten the bolts to 96 ft. lbs. (103 Nm).
50. Install the transaxle mount-to-body bolts and tighten to 40 ft. lbs. (54 Nm).
51. Connect the halfshafts to the transaxle.

52. Install the suspension crossmembers and control arms in the vehicle and tighten the mounting bolts to 89 ft. lbs. (120 Nm).

53. Connect the lower ball joints to the steering knuckles and tighten the mounting bolts to 48 ft. lbs. (65 Nm) and install 2 new cotter pins.

54. Attach the front ABS wheel speed sensors to the harness connector and the subframe clips.

55. Connect the engine strut to the suspension support and tighten the mounting bolt to 89 ft. lbs. (120 Nm).

56. Install the transaxle cover.

57. Attach the following electrical connectors to the lower engine components:
- Ignition assembly
- Starter
- Vehicle Speed Sensor (VSS)
- Transaxle ground wire

58. Install the front exhaust pipe and tighten the pipe-to-manifold bolts to 22 ft. lbs. (30 Nm).

59. Install the left and right inner fender splash shields.

60. Install the front tire and wheel assemblies and tighten to 100 ft. lbs. (140 Nm).

61. Lower the vehicle.

62. Attach the power steering lines to the pump.

63. Connect the control cables to the throttle body lever and mounting bracket.

64. Attach the shift linkage to the transaxle and mounting bracket.

65. Connect the fuel lines.

66. Install the clutch slave cylinder, if equipped.

67. Connect the A/C compressor-to-accumulator and condenser lines.

68. Recharge the A/C system.

69. Connect the electrical wiring to the following components:
- Oxygen sensor (O$_2$)
- Fuel injector harness
- Idle Air Control (IAC)
- Throttle Position Sensor (TPS)
- Coolant Temperature Sender
- Park/neutral switch
- TCC solenoid
- Transaxle shift solenoid
- Transaxle ground
- MAP sensor
- EGR
- Cooling fan

70. Install the alternator upper mounting bracket and connect the alternator wiring.

71. Connect the vacuum hose to the power brake booster.

72. Install the upper radiator hoses.

73. Install the air cleaner.

74. Install the battery cable and connect the positive cable.

75. Install the throttle body air inlet duct.

76. Connect the negative battery cable.

❊❊ WARNING

Operating the engine without the proper amount and type of engine oil will result in severe engine damage.

77. Fill the engine with the correct amounts and types of coolant and engine oil.

78. Refill and bleed the power steering system.

79. Start the vehicle and verify no leaks.

➡Whenever the vehicle sub-frame is removed or lowered, the wheel alignment should be checked.

2.3L and 2.4L Engines

1. If equipped with A/C, properly recover the refrigerant.

2. Disconnect the negative battery.

❊❊ CAUTION

Never open, service or drain the radiator or cooling system when hot; serious burns can occur from the steam and hot coolant.

3. Properly drain the cooling system into an approved container.

4. Relieve the fuel system pressure.

5. Remove the left sound insulator, then disconnect the clutch pushrod from the pedal assembly.

6. Disconnect the heater hose at the thermostat assembly, then detach the radiator inlet (upper) hose.

7. Remove the air cleaner assembly and the coolant fan.

8. If equipped with A/C, disconnect the compressor/condenser hose assembly at the compressor, then discard the O-rings.

9. Disconnect the 2 vacuum hoses from the front of the engine.

10. Tag and detach the following electrical connectors:
- Alternator
- A/C compressor (if equipped)
- Fuel injector harness
- Idle Air Control (IAC) and TP sensor at the throttle body
- Manifold Absolute Pressure (MAP) sensor
- Intake Air Temperature (IAT) sensor
- EVAP canister purge solenoid
- Starter solenoid
- Ground connections
- Negative battery cable from the transaxle

- Electronic ignition coil and module assembly
- Engine Coolant Temperature (ECT) sensor(s)
- Oil pressure sensor/switch
- Oxygen (O$_2$) sensor
- Crankshaft Position (CKP) sensor
- Back-up lamp switch, then position the harness aside

11. Disconnect the power brake vacuum hose from the throttle body. Detach the power brake vacuum tube-to-check valve hose from the tube.

12. Remove the throttle cable and bracket.

13. Unfasten the power steering pump rear bracket, then remove the bracket and vacuum tube as an assembly.

14. Unfasten the power steering pivot bolt, then remove the pump and drive belt. Position the pump aside, with the lines still attached.

15. Carefully disconnect the fuel lines.

16. Disconnect the shift cables. Detach the clutch actuator line.

17. Remove the exhaust manifold and heat shield.

18. Disconnect the radiator outlet (lower) hose from the radiator.

19. Install J-28467-A or equivalent engine support fixture.

20. Unfasten the bolt attaching the coolant recovery/surge tank, then position the tank aside with the hoses still connected.

21. Remove the engine mount assembly.

22. Raise and safely support the vehicle, then remove the front wheel and tire assemblies. Remove the right splash shield.

❊❊ CAUTION

The EPA warns that prolonged contact with used engine oil may cause a number of skin disorders, including cancer! You should make every effort to minimize your exposure to used engine oil. Protective gloves should be worn when changing the oil. Wash your hands and any other exposed skin areas as soon as possible after exposure to used engine oil. Soap and water, or waterless hand cleaner should be used.

23. Drain and recycle the engine oil.

24. Remove the radiator air deflector.

25. Tag and detach the following electrical connections:
- Vehicle Speed Sensor (VSS)
- Knock sensor
- Starter solenoid

- If equipped, both front ABS wheel speed sensors

26. Remove the engine mount strut and the transaxle mount.

27. Separate the ball joints from the steering knuckles.

28. Remove the suspension supports, crossmember, and stabilizer shaft as an assembly.

29. Disconnect the heater outlet hose from the radiator outlet pipe.

30. Remove the axle shaft from the transaxle and intermediate shaft, then position aside.

31. If equipped, disconnect the A/C lines from the oil pan.

32. Remove the flywheel housing cover.

33. Position a suitable support below the engine, then carefully lower the vehicle onto the support.

34. Matchmark the threads on the support fixture hooks so the setting can be duplicated when reinstalling the engine. Remove the engine support fixture J-hooks.

35. Raise the vehicle slowly off the engine and transaxle assembly. If may be necessary to move the engine/transaxle assembly rearward to clear the intake manifold.

36. Noting the position of the bolts, separate the engine from the transaxle.

To install:

✸✸ WARNING

Be sure the retaining bolts are in their correct locations. If not, engine damage may occur.

37. Assemble the engine to the transaxle.

38. Position the engine and transaxle assembly under the engine compartment, then slowly lower the vehicle over the assembly until the transaxle mount is indexed, then install the retaining bolt.

39. Install engine support fixture J-28467-A or equivalent, making sure to adjust it to the previous setting.

40. Install the engine mount assembly and transaxle mount.

41. Carefully raise the vehicle off the support.

42. Attach the axle shafts to the transaxle.

43. Connect the heater outlet hose to the radiator outlet pipe.

44. Install the suspension supports, crossmember and stabilizer shaft assembly.

45. Attach the ball joints to the steering knuckles, then secure with the nuts.

46. Install the engine strut mount.

47. If equipped, connect the A/C line to the oil pan.

48. Attach the following electrical connectors, as tagged during removal:
- Vehicle Speed Sensor (VSS)
- Knock sensor
- Starter solenoid
- If equipped, both front ABS wheel speed sensors

49. Install the flywheel housing cover.

50. Fasten the radiator air deflector.

51. Connect the lower radiator hose.

52. Install the right splash shield, then the front wheel and tire assemblies.

53. Carefully lower the vehicle, then remove the engine support fixture.

54. Install the coolant recovery/surge tank, then secure using the retaining bolt.

55. Attach the following electrical connections, as tagged during removal:
- Alternator
- A/C compressor (if equipped)
- Fuel injector harness
- Idle Air Control (IAC) and TP sensor at the throttle body
- Manifold Absolute Pressure (MAP) sensor
- Intake Air Temperature (IAT) sensor
- EVAP canister purge solenoid
- Starter solenoid
- Ground connections
- Negative battery cable to the transaxle
- Electronic ignition coil and module assembly
- Engine Coolant Temperature (ECT) sensor(s)
- Oil pressure sensor/switch
- Oxygen (O$_2$) sensor
- Crankshaft Position (CKP) sensor
- Back-up lamp switch

56. Attach the vacuum hoses.

57. If equipped with A/C, attach the compressor/condenser hose assembly to the compressor.

58. Fasten the clutch actuator line.

59. Install the exhaust manifold and heat shield.

60. Connect the fuel lines.

61. Connect the positive battery cable.

62. Fasten the power steering pump pivot-to-block bolt. Install the power steering pump rear bracket and tension belt.

63. Connect the vacuum hoses to the intake manifold and to the tube from the brake booster.

64. Install the throttle cable and bracket.

65. Install the coolant fan and air cleaner assembly.

66. Attach the radiator outlet (upper) hose. Fill the cooling system with the proper type and quantity of coolant.

67. Connect the clutch pushrod to the pedal assembly, then install the left sound insulator.

68. Attach the heater hose at the thermostat housing.

✸✸ WARNING

Operating the engine without the proper amount and type of engine oil will result in severe engine damage.

69. Fill the transaxle with fluid, then fill the crankcase with oil.

70. Connect the negative battery cable.

71. If equipped with A/C, evacuate, charge and leak test the system.

72. Start the engine and check the fluid levels, proper operation of the engine and/or fluid leakage.

Water Pump

REMOVAL & INSTALLATION

2.2L Engine

✸✸ WARNING

When adding coolant, it is important that you use GM Goodwrench DEX-COOL® coolant meeting GM Specification 6277M.

1. Disconnect the negative battery cable.

✸✸ CAUTION

The EPA warns that prolonged contact with used engine oil may cause a number of skin disorders, including cancer! You should make every effort to minimize your exposure to used engine oil. Protective gloves should be worn when changing the oil. Wash your hands and any other exposed skin areas as soon as possible after exposure to used engine oil. Soap and water, or waterless hand cleaner should be used.

2. Drain the cooling system into a suitable container.

3. Loosen, but do not remove, the water pump pulley bolts.

4. Remove the serpentine belt.

5. Remove the alternator mounting bolts and set the alternator aside.

6. Remove the water pump pulley bolts, then remove the pulley.

7. Remove the 4 water pump mounting bolts, then remove the water pump.

To install:

8. Clean all the gasket surfaces completely.

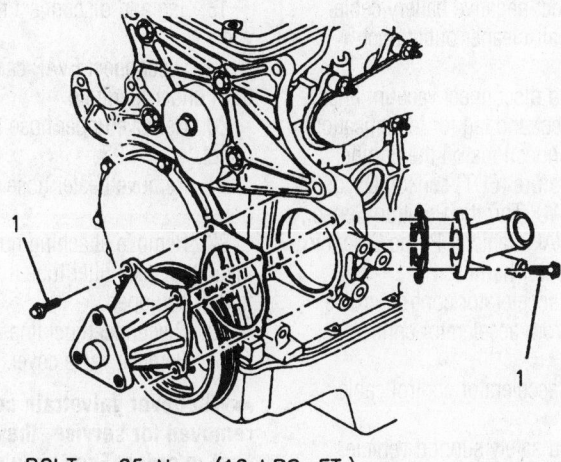

1 BOLT – 25 N·m (18 LBS. FT.)

7922YG01

Exploded view of the water pump mounting—2.2L engine

9. Apply a thin bead of sealer around the outer edge of the water pump gasket seating area and place he gasket on the pump.

10. Install the water pump on the engine and tighten the 4 mounting bolts to 18 ft. lbs. (25 Nm).

11. Install the water pump pulley and tighten the mounting bolts finger-tight.

12. Install the alternator in the mounting bracket.

13. Install the serpentine belt.

14. Tighten the water pump pulley mounting bolts to 22 ft. lbs. (30 Nm).

15. Connect the negative battery cable.

16. Refill and bleed the cooling system.

2.3L and 2.4L Engines

1. Disconnect the negative battery cable

2. Detach the oxygen sensor connector.

❈❈ CAUTION

Never open, service or drain the radiator or cooling system when hot; serious burns can occur from the steam and hot coolant.

3. Properly drain the engine coolant into a suitable container. Remove the heater hose from the thermostat housing for more complete coolant drain.

4. Remove upper exhaust manifold heat shield.

5. Remove the bolt that attaches the exhaust manifold brace to the manifold.

6. Remove the lower exhaust manifold heat shield.

7. Break loose the manifold to exhaust pipe spring loaded bolts using a 13mm box wrench.

8. Raise and safely support the vehicle.

➡**It is necessary to relieve the spring pressure from 1 bolt prior to removing the second bolt. If the spring pressure is not relieved, it will cause the exhaust pipe to twist and bind up the bolt as it is removed.**

9. Unfasten the 2 radiator outlet pipe-to-water pump cover bolts.

10. Remove the manifold to exhaust pipe bolts from the exhaust pipe flange as follows:

   a. Unscrew either bolt clockwise 4 turns.

   b. Remove the other bolt.

   c. Remove the first bolt.

❈❈ WARNING

On the 2.4L engines, DO NOT rotate the flex coupling more than 4 degrees or damage may occur.

11. Pull down and back on the exhaust pipe to disengage it from the exhaust manifold bolts.

12. Remove the radiator outlet pipe from the oil pan and transaxle. If equipped with a manual transaxle, remove the exhaust manifold brace. Leave the lower radiator hose attached and pull down on the outlet pipe to remove it from the water pump. Leave the radiator outlet pipe hang.

13. Carefully lower the vehicle.

14. Unfasten the exhaust manifold-to-cylinder head retaining nuts, then remove the exhaust manifold, seals and gaskets.

15. For the 2.4L engine, remove the front timing chain cover and the chain tensioner.

16. Unfasten the water pump-to-cylinder block bolts. Remove the water pump-to-timing chain housing nuts. Remove the water pump and cover mounting bolts and nuts. Remove the water pump and cover as an assembly, then separate the 2 pieces.

To install:

17. Thoroughly clean and dry all mounting surfaces, bolts and bolt holes. Using a new gasket, install the water pump to the cover and tighten the bolts finger-tight.

18. Lubricate the splines of the water pump with clean grease and install the assembly to the engine using new gaskets. Install the mounting bolts and nuts finger-tight.

19. Lubricate the radiator outlet pipe O-ring with antifreeze and slid the pipe onto the water pump cover. Install the bolts finger-tight.

20. With all gaps closed, tighten the bolts, in the following sequence, to the proper values:

   a. Pump assembly-to-chain housing nuts-19 ft. lbs. (26 Nm).

   b. Pump cover-to-pump assembly-106 inch lbs. (12 Nm).

   c. Cover-to-block, bottom bolt first-19 ft. lbs. (26 Nm).

   d. Radiator outlet pipe assembly-to-pump cover-125 inch lbs. (14 Nm).

21. Using new gaskets, install the exhaust manifold.

22. Raise and safely support the vehicle.

1 TIMING CHAIN HOUSING
2 GASKET, TIMING CHAIN HOUSING TO WATER PUMP COVER
3 NUT (3)
4 WATER PUMP BODY ASM.
5 GASKET, WATER PUMP BODY TO WATER PUMP COVER
6 WATER PUMP COVER
7 BOLT (M6 X 1 X 65) – 3 LOWER POSITIONS
8 BOLT (M6 X 1 X 25)
9 BOLT (M6 X 1 X 90)
10 GASKET, WATER PUMP COVER TO BLOCK
11 BOLTS, WATER PUMP COVER TO BLOCK (2)

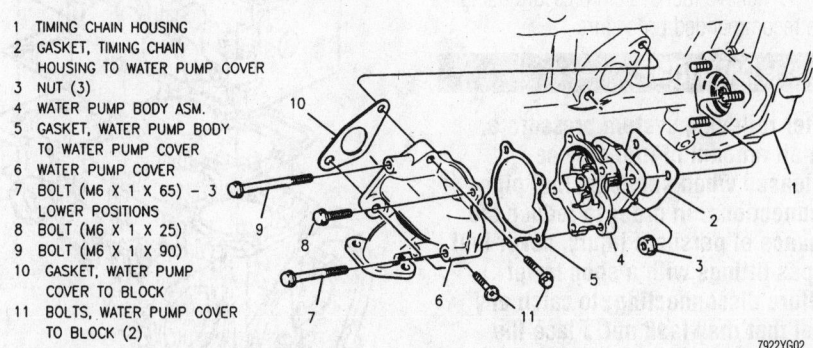

7922YG02

Exploded view of the water pump mounting and related components—2.3L and 2.4L engines

23. Index the exhaust manifold bolts into the exhaust pipe flange.

24. Connect the exhaust pipe to the manifold. Install the exhaust pipe flange bolts evenly and gradually to avoid binding. Turn the bolts in until fully seated.

25. Connect the radiator outlet pipe to the transaxle and oil pan. Install the exhaust manifold brace, if removed.

26. On the 2.4L engine, install the timing chain tensioner and front cover.

27. Install the lower heat shield.

28. Carefully lower the vehicle.

29. Fasten the bolt that attaches the exhaust manifold brace to the manifold.

30. Tighten the manifold-to-exhaust pipe nuts to 26 ft. lbs. (35 Nm).

31. Install the upper heat shield.

32. Attach the oxygen sensor connector.

33. Fill the radiator with coolant until it comes out the heater hose outlet at the thermostat housing. Then, connect the heater hose. Leave the radiator cap off.

34. Connect the negative battery cable, then start the engine. Run the vehicle until the thermostat opens, fill the radiator and recovery tank to their proper levels, then turn the engine **OFF**.

35. Once the vehicle has cooled, recheck the coolant level.

Cylinder Head

REMOVAL & INSTALLATION

2.2L Engine

❋❋ CAUTION

The fuel injection system remains under pressure, even after the engine has been turned OFF. The fuel system pressure must be relieved before disconnecting any fuel lines. Failure to do so may result in fire and/or personal injury.

1. Relieve fuel system pressure using the recommended procedure.

❋❋ CAUTION

After relieving system pressure a small amount of fuel may be released when servicing fuel pipes or connections. In order to reduce the chance of personal injury, cover fuel pipes fittings with a shop towel before disconnecting, to catch any fuel that may leak out. Place the towel in an approved container when disconnect is complete.

2. Disconnect negative battery cable.

3. Remove air cleaner outlet duct assembly.

4. Label and disconnect vacuum lines.

5. Disconnect and tag for identification the electrical connections on the Engine Coolant Temperature (ECT) sensor, Oxygen Sensor (O_2S), IAC, Throttle Position Sensor, MAP sensor, EVAP Canister Purge Solenoid and the fuel injector harness.

6. Remove accelerator control, cruise and TV cables from accelerator control bracket.

7. Remove accelerator control cable bracket.

8. Raise and safely support vehicle.

9. Remove exhaust pipe from exhaust manifold.

❋❋ CAUTION

Never open, service or drain the radiator or cooling system when hot; serious burns can occur from the steam and hot coolant.

10. Drain and recover coolant into a suitable container.

11. Lower vehicle.

12. Remove serpentine drive belt.

13. Remove the alternator.

14. Remove the power steering pump and position aside with lines attached.

15. Remove power steering pump bracket.

16. Install engine support fixture J-28467-A or equivalent.

17. Remove serpentine drive belt tensioner bracket.

18. Tag and disconnect the spark plug wires.

19. Disconnect EVAP canister purge line, from under manifold.

20. Remove upper hose from coolant outlet.

21. Remove heater hose from coolant outlet.

22. Remove attaching nut holding automatic transaxle filler tube from intake manifold, if equipped.

23. Disconnect fuel lines.

24. Remove valve cover.

➡**Whenever valvetrain components are removed for service, they should be kept in order. They should be installed in the same locations and with the same mating surfaces as when removed.**

25. Remove rocker arms and pushrods.

26. Remove cylinder head bolts. Two sizes of bolts are used. Note the location of each. These bolts are called "torque-to-yield." This means that, at assembly, after the bolts are tightened to a specific torque, they are tightened another quarter turn. This stretches the bolts slightly. Therefore, new cylinder head bolts are recommended.

27. Remove cylinder head with both manifolds attached.

28. Remove the intake and exhaust manifolds from the cylinder head.

To install:

29. Clean all the gasket surfaces completely. Clean the threads on cylinder head bolts and be sure all bolt holes are clean and free of foreign material. It is good prac-

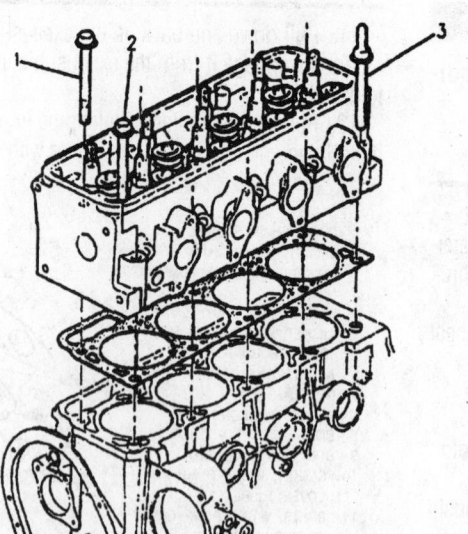

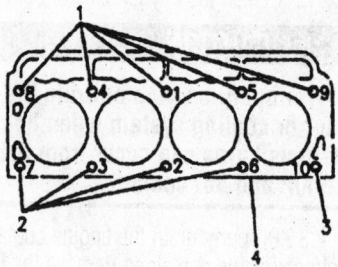

1. Long bolts
2. Short bolts
3. Stud
4. Numbers on gasket indicate torque sequence

Cylinder head tightening sequence—2.2L engine

7922YG03

tice to clean all internally threaded openings with the proper size thread cutting tap. This removes rust, dirt and old sealer build-up that can prevent getting a proper torque reading when tightening bolts.

30. Inspect cylinder head and block surface for cracks, nicks, heavy scratches and flatness.

31. Install exhaust and intake manifolds on cylinder head prior to installing cylinder head.

32. Place a new cylinder head gasket in position over the dowel pins on the engine block. Carefully guide the cylinder head into position.

33. Install cylinder head bolts finger-tight. New cylinder head bolts are recommended.

34. Tighten bolts in sequence, tighten the long bolts to 46 ft. lbs. (63 Nm) plus 90 degrees. Tighten the short bolts to 43 ft. lbs. (58 Nm) plus 90 degrees.

35. Install pushrods and rocker arms and rocker arm nuts. Tighten nuts to 22 ft. lbs. (30 Nm).

36. Install valve cover and tighten bolts to 89 inch lbs. (10 Nm).

37. Connect fuel lines.

38. Install transaxle filler tube nut and tighten to 20 ft. lbs. (27 Nm).

39. Install heater hose to coolant outlet.

40. Install upper radiator hose to coolant.

41. Connect EVAP canister purge line.

42. Attach the ignition wires to spark plugs, as tagged during removal.

43. Install serpentine drive belt tensioner bracket and bolts. Tighten bolts to 37 ft. lbs. (27 Nm).

44. Remove engine support fixture.

45. Install power steering pump bracket and power steering pump.

46. Install alternator and brace.

47. Install serpentine drive belt.

48. Raise and safely support the vehicle.

49. Connect the exhaust pipe to the exhaust manifold.

50. Lower the vehicle.

51. Install the accelerator control cable bracket and bolts. Tighten bolts to 18 inch lbs. (25 Nm).

52. Connect the accelerator control, cruise and TV cables to control bracket.

53. Attach all of the electrical connections for the sensors.

54. Connect the vacuum lines.

55. Install air cleaner outlet duct assembly.

56. Refill the coolant system.

57. Connect negative battery cable.

58. Start vehicle and inspect for leaks.

59. Bleed air from coolant system as follows:

a. Loosen the engine coolant air bleed screw, (located on the top side of the engine coolant outlet) and add coolant until all of the air is evacuated through the air bleed. Tighten the air bleed screw.

b. When filling the coolant system, use coolant meeting GM specifications.

2.3L and 2.4L Engines

✳✳ CAUTION

The fuel injection system remains under pressure, even after the engine has been turned OFF. The fuel system pressure must be relieved before disconnecting any fuel lines. Failure to do so may result in fire and/or personal injury.

1. Relieve the fuel system pressure.

✳✳ CAUTION

After relieving system pressure, a small amount of fuel may be released when servicing fuel pipes or connections. In order to reduce the chance of personal injury, cover fuel pipe fittings with a shop towel before disconnecting, to catch any fuel that may leak out. Place the towel in an approved container when disconnect is complete.

2. Disconnect the negative battery cable.

✳✳ CAUTION

Never open, service or drain the radiator or cooling system when hot; serious burns can occur from the steam and hot coolant.

3. Drain and recover the coolant into a suitable container.

4. Disconnect the heater inlet and throttle body heater hoses from water outlet.

5. Remove the exhaust manifold.

6. Remove the intake camshaft housing and lifters, then remove the exhaust camshaft housing and lifters.

7. Remove the oil fill tube.

8. Remove the throttle body-to-air cleaner duct.

9. Disconnect the power brake vacuum hose from throttle body.

10. Remove the throttle cable bracket.

11. Remove the throttle body from intake manifold, with electrical harness and throttle cable attached. Position it aside.

12. Disconnect the MAP sensor vacuum hose from intake manifold.

13. Remove the intake manifold brace.

14. Disconnect electrical connections from the following sensors: MAP sensor, intake air temperature sensor and EVAP canister purge solenoid.

15. Disconnect the upper radiator hose from water outlet.

16. Detach the coolant temperature sensors connectors.

17. Unfasten the cylinder head bolt, then remove cylinder head and gasket.

To install:

This is an aluminum cylinder head and must be treated with care. Do not use abrasive pads to clean the cylinder head or block surfaces. An abrasive pad may damage the cylinder head and block. GM says that abrasive pads should not be used for the following reasons:

a. Abrasive pads will produce a fine grit that the oil filter will not be able to remove from the oil. This grit is abrasive and has been known to cause internal engine damage.

b. Abrasive pads can easily remove enough metal to round cylinder head edges. This has been known to affect the gasket's ability to seal, especially in the narrow areas between the combustion chambers and coolant jackets. The cylinder head gasket is likely to leak if these edges are rounded.

c. Abrasive pads can also remove enough metal to affect cylinder head flatness. It takes only about 15 seconds to remove 0.008 in. (0.20mm) of metal from the cylinder head with an abrasive pad. If the cylinder head flatness is out of specification, the gasket will not be able to seal and the gasket will leak.

18. Use a razor blade gasket scraper to clean the cylinder head and cylinder block gasket surfaces. Be careful not to gouge or scratch the gasket surfaces. Do not gouge or scrape the combustion chamber surfaces. Use a new razor blade for each cylinder head. Hold the scraper so the razor blade is as parallel to the gasket surface as possible. Do not use any other method or technique to clean these gasket surfaces. In addition, GM warns not to use a tap to clean cylinder head bolt holes.

19. When working on an aluminum head, do not remove spark plugs from an aluminum cylinder head until the cylinder head has cooled. Always clean all dirt and debris from the spark plug recess area. If the spark plug opening threads are damaged and NOT restorable with a Thread Chaser, replace the cylinder head. GM **DOES NOT** approve of the installation of thread inserts into the spark plug openings

on this engine. If threads are installed into the spark plug openings, severe engine damage will occur.

20. Clean all the gasket surfaces completely. Clean the threads on cylinder head bolts and be sure all bolt holes are clean and free of debris. New bolts are recommended.

21. Inspect the cylinder head and block surface for cracks, nicks, heavy scratches and flatness.

22. Place a new cylinder head gasket on the block. Do not use any sealing material.

23. Carefully place the cylinder head on dowel pins, being careful not to disturb the gasket.

24. Apply a small amount of clean engine oil to the threads of the cylinder head bolts, and install finger-tight.

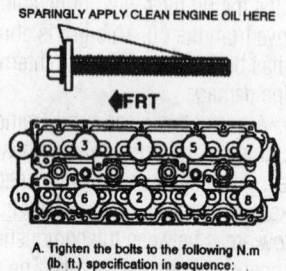

Head bolt torque sequence—2.3L and 2.4L engines

25. Tighten head bolts in sequence. Tighten bolts 1 through 8 to 40 ft. lbs. (65 Nm);, then, tighten bolts 9 and 10 to 30 ft. lbs. (40 Nm). Turn all 10 bolts an additional 90 degrees (¼ turn) in sequence.

26. Attach the coolant temperature sensor connections.

27. Connect upper radiator hose to coolant outlet.

28. Install manifold brace and tighten to 19 ft. lbs. (26 Nm).

29. Attach all sensor connections.

30. Connect the MAP sensor vacuum hose to intake manifold.

31. Install throttle body to intake manifold, using a new gasket.

32. Install accelerator control cable bracket to the throttle body, and tighten the bolts to 106 inch lbs. (12 Nm). Tighten the nut to 19 ft. lbs. (26 Nm).

33. Install the throttle body-to-air cleaner duct.

34. Install oil fill tube, tighten attaching bolt to 71 inch lbs. (8 Nm).

35. Install the lifters and camshaft housing.

36. Install the exhaust manifold, then tighten the exhaust nuts to 26 ft. lbs. (35 Nm).

37. Connect negative battery.

38. Fill coolant system and bleed off air from system. An oil and filter change is recommended.

39. Check and verify that vehicle has no coolant or vacuum leaks.

Rocker Arms

REMOVAL & INSTALLATION

➡**Place the components in a rack in order to be sure they are installed at the same location and with the same mating surface as when removed.**

1. Disconnect the negative battery cable.

2. Remove the rocker (valve) arm cover(s).

3. Unfasten the rocker arm nuts.

4. Remove the rocker arm pivot ball(s).

5. Remove the rocker arm(s).

6. Remove the pushrods. For the 3.1L engine, the intake pushrods are marked orange and are 6 in. (15.2cm) long and the exhaust pushrods are marked blue and are 6 ⅜ in. (16.2cm) long.

To install:

7. Install the pushrods. Be sure to install the pushrods in the correct positions, and be sure they seat properly in the lifters.

8. Coat the bearing surfaces of the rocker arms and pivot balls with camshaft and lifter prelube GM specification 1052365 or equivalent. Install the rocker arm(s).

9. Install the rocker arm pivot ball(s).

10. Install the rocker arm nuts, and tighten to 22 ft. lbs. (30 Nm) for 2.2L engine or 18 ft. lbs. (25 Nm) for 3.1L engine.

11. Install the rocker arm covers.

12. Connect the negative battery cable.

Intake Manifold

REMOVAL & INSTALLATION

2.2L Engine

1995–97 MODELS

These vehicles use a two-piece intake manifold. The upper half, sometimes called a plenum, contains the throttle body and the control cable connections. The lower half has individual port runners to each intake port on the cylinder head. The lower half of the manifold bolts to the cylinder head and houses the fuel injectors. Note that these pieces are cast aluminum. Care should be exercised when working with any light alloy component.

1. Properly relieve the fuel system pressure.

2. Disconnect the negative battery cable.

3. Remove the throttle body air intake duct.

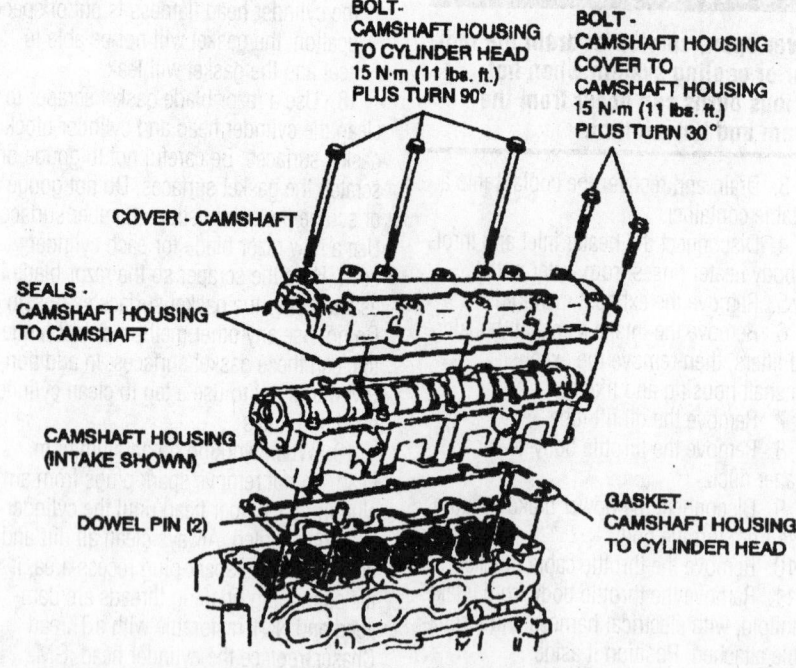

Exploded view of the camshaft housing cover mounting—2.3L and 2.4L engines

> ✳✳ **CAUTION**
>
> **Never open, service or drain the radiator or cooling system when hot; serious burns can occur from the steam and hot coolant.**

4. Drain the cooling system into an approved container.

5. Identify, tag and disconnect all necessary vacuum lines.

6. Disconnect the control cables from the throttle body lever and remove the control cable bracket from the intake manifold.

7. Remove the serpentine belt.

8. Remove the power steering pump and lay it aside with the fluid lines attached.

9. Remove the transaxle fill tube.

10. Tag and disconnect the following electrical wires:
- Idle Air Control (IAC) valve
- Throttle Position (TP) sensor
- Manifold Absolute Pressure (MAP) sensor
- EVAP Emission solenoid
- Fuel injector harness
- Exhaust Gas Recirculation (EGR) valve

11. Remove the MAP sensor.

12. Unfasten the upper intake manifold mounting bolts, then remove the upper intake manifold.

13. Disconnect the fuel lines from the fuel rail.

14. Remove the EGR valve injector, then remove the EGR valve.

15. Remove the fuel injector retainer bracket, regulator and injectors.

16. Unfasten and remove the control cable bracket.

17. If necessary for access, raise and safely support the vehicle.

18. Unfasten the 6 intake manifold nuts, then remove the manifold.

19. Clean the gasket mounting surfaces.

To install:

20. Install a new gasket, then position the lower intake manifold. Tighten the lower intake manifold nuts in the proper sequence to 24 ft. lbs. (33 Nm).

21. Connect the control cables and cable bracket.

22. Install the EGR valve.

23. Attach the fuel lines to the fuel rail.

24. Install the fuel injectors, regulator and injector retainer bracket, then tighten the retaining bolts to 22 inch lbs. (3.5 Nm).

25. Install the EGR valve injector that the port is facing directly towards the throttle body.

26. Install the upper intake manifold assembly. Tighten the upper intake manifold nuts in the proper sequence to 22 ft. lbs. (30 Nm).

27. Install the MAP sensor.

28. Attach the electrical connectors to the MAP sensor, EGR solenoid valve, Idle Air Control (IAC) valve, Throttle Position (TP) sensor, and the fuel injectors.

29. Install the transaxle fill tube.

30. Install the power steering pump, then install the serpentine belt.

31. Connect the vacuum lines, as tagged during removal.

32. Install the air intake duct.

33. Refill the coolant system.

34. Connect the negative battery cable, then start the engine and check for leaks.

1998–99 MODELS

1. Disconnect the negative battery cable.

2. Properly relieve the fuel system pressure.

3. Remove the air cleaner inlet duct.

4. Remove the air inlet resonator and bracket.

5. Remove the throttle and cruise control cable from the throttle body.

6. Label and detach the following electrical connectors:
- Manifold Absolute Pressure (MAP) sensor
- Throttle Position (TP) sensor
- Idle Air Control (IAC) valve

7. Unbolt and remove the throttle body.

8. Disconnect the fuel supply line and the fuel inlet pipe.

9. Remove the intake manifold mounting nuts/bolts, then remove the manifold from the engine.

10. Clean the gasket mating surfaces on the cylinder head, intake manifold and throttle body. Inspect the manifold for cracks, broken flanges, and gasket surface damage.

To install:

11. Install the intake manifold using a new gasket.

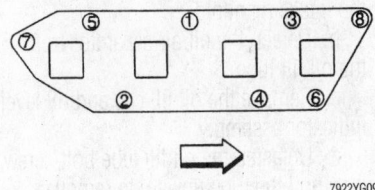

Intake manifold tightening sequence—1998–99 2.2L engine

12. Tighten the mounting bolts/nuts, in sequence, to 17 ft. lbs. (24 Nm).

13. Install the throttle body and tighten the mounting bolts to 89 inch lbs. (10 Nm).

14. Mount the fuel pipe and connect the fuel supply line.

15. Attach the MAP sensor, TP sensor, and the IAC valve electrical connectors.

16. Connect the cruise control and throttle cables to the throttle body.

17. Install the air inlet resonator bracket and resonator.

18. Install the air cleaner inlet duct.

19. Connect the negative battery cable.

2.3L and 2.4L Engines

1. Properly relieve the fuel system pressure.

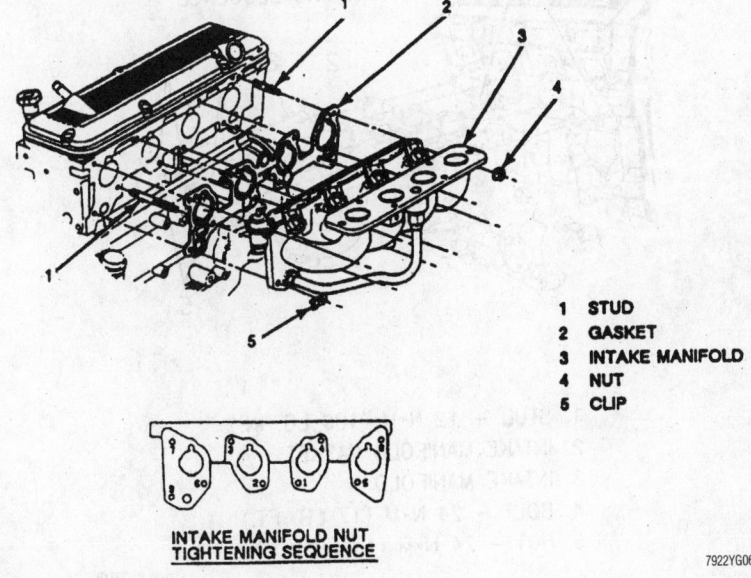

1	STUD
2	GASKET
3	INTAKE MANIFOLD
4	NUT
5	CLIP

INTAKE MANIFOLD NUT TIGHTENING SEQUENCE

Lower intake manifold tightening sequence—2.2L engine

Never open, service or drain the radiator or cooling system when hot; serious burns can occur from the steam and hot coolant.

2. Disconnect the negative battery cable, then properly drain the cooling system.

3. Tag and detach the following electrical connectors:

• Manifold Absolute Pressure (MAP) sensor
• Intake Air Temperature (IAT) sensor
• EVAP canister purge solenoid
• Fuel injector harness

4. Label and disconnect the vacuum hoses from the fuel regulator and EVAP canister purge solenoid to canister.

5. Unfasten the air cleaner duct.

6. Remove the accelerator control cable bracket.

7. For the 2.4L engine, remove the stud-ended alternator mount bolt, then detach the EGR pipe from the EGR adapter.

8. For the 2.3L engine, perform the following procedures:

a. Remove the oil air separator (crankcase ventilation system) as an assembly. Leave the hoses attached to the separator. Disconnect the hoses from the oil fill, chain cover, intake duct and the intake manifold.

b. Detach the oil/air separator from the oil fill tube.

c. Remove the oil fill cap and oil level indicator assembly.

d. Unfasten the oil fill tube bolt/screw, then pull the tube upward to remove.

9. Remove the fill tube out the top, rotating as necessary to gain clearance for the oil/air separator nipple between the intake tubes and fuel rail electrical harness.

10. For the 2.4L engine, raise and safely support the vehicle.

11. Remove the intake manifold support brace.

12. If raised, carefully lower the vehicle.

13. Unfasten the manifold retaining nuts and bolts, then remove the intake manifold from the engine.

➡️**If installing a new intake manifold, transfer all necessary parts from the old manifold to the new one.**

14. Using a suitable scraping tool, clean the old gasket material from the intake manifold mating surfaces. Do not let any debris fall into the engine!

To install:

15. Install the manifold with a new gasket.

➡️**Be sure the numbers stamped on the gasket are facing towards the manifold surface.**

16. Follow the tightening sequence in the accompanying figure, then tighten the bolts/nuts to 19 ft. lbs. (26 Nm) for 2.3L engine or 18 ft. lbs. (24 Nm) for 2.4L engine.

17. For the 2.4L engine, raise and safely support the vehicle.

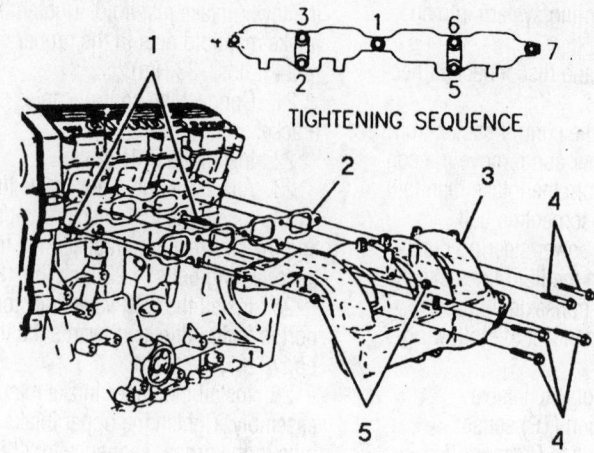

1. STUD – 11 N·m (96 LBS. IN.)
2. INTAKE MANIFOLD GASKET
3. INTAKE MANIFOLD
4. BOLT – 26 N·m (19 LBS. FT.)
5. NUT – 26 N·m (19 LBS. FT.)

7922YG07

Exploded view of the intake manifold, showing the torque sequence—2.3L engine

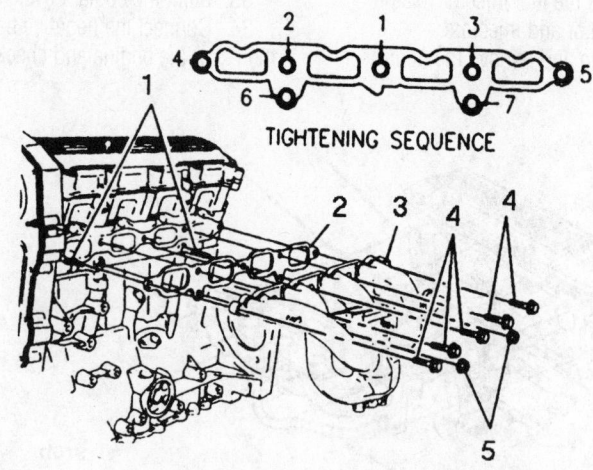

1. STUD – 12 N·M (100 LB. IN.)
2. INTAKE MANIFOLD GASKET
3. INTAKE MANIFOLD
4. BOLT – 24 N·M (17 LB. FT.)
5. NUT – 24 N·M (17 LB. FT.)

7922YG08

Exploded view of the intake manifold, showing the torque sequence—2.4L engine

18. Install the intake manifold brace and retainers.

19. If raised, carefully lower the vehicle.

20. For the 2.3L engine, install the oil/air separator assembly.

21. For the 2.3L engine, lubricate a new oil fill tube O-ring seal with clean engine oil, then install the tube down between intake manifold. Rotate as needed to gain clearance for the oil/air separator nipple on the fill tube.

22. If removed, position the oil fill tube in its cylinder block opening. Align the fill tube so it is in about its proper position. Place the palm of your hand over the oil fill opening and press straight down to seat the fill tube and O-ring into the cylinder block.

23. If necessary, connect the oil/air separator hose to the oil fill tube. You can lubricate the hose as necessary to ease installation. Install the oil fill tube bolt/screw. Fasten the cap.

24. For the 2.4L engine, attach the EGR pipe to the adapter; tighten the fasteners to 19 ft. lbs. (26 Nm). Install the stud-ended alternator bolt.

25. Install the accelerator control cable bracket.

26. Connect the vacuum hoses to the fuel regulator and EVAP canister purge solenoid.

27. Attach all electrical connectors, as tagged during removal.

28. Install the air cleaner duct.

29. Refill the coolant to it's proper level.

30. Connect the negative battery cable, then start the engine and inspect for leaks.

Exhaust Manifold

REMOVAL & INSTALLATION

2.2L Engine

1. Disconnect the negative battery cable.

2. Detach the oxygen sensor wire.

3. Remove the serpentine belt or alternator drive belt.

4. Remove the alternator-to-bracket bolts, then support the alternator (with the wires attached) out of the way.

5. Raise and support the vehicle safely.

6. Unfasten the exhaust pipe-to-exhaust manifold bolts, then carefully lower the vehicle.

7. If necessary for access to remove the manifold, remove the oil fill tube and disconnect the heater outlet hose assembly nut from the exhaust manifold.

8. Remove the exhaust manifold-to-cylinder head bolts.

9. Detach the exhaust manifold from the exhaust pipe flange.

10. Unfasten the retaining nuts, then remove the exhaust manifold from the vehicle. Remove and discard the gasket(s).

To install:

11. Using a gasket scraper, carefully clean the gasket mounting surfaces.

12. To install, use new gaskets and reverse the removal procedures. Tighten the exhaust manifold-to-cylinder head nuts to 3–12 ft. lbs. (4–16 Nm) and the bolts to 6–13 ft. lbs. (8–18 Nm).

13. Start the engine and check for exhaust leaks.

2.3L and 2.4L Engines

1. Disconnect the negative battery cable

2. Detach the Oxygen (O_2) sensor connector.

3. Raise and safely support the vehicle.

4. Unfasten the exhaust manifold brace-to-manifold bolt and the oil pan nuts, if necessary.

5. For 2.3L engine, remove the manifold-to-exhaust pipe spring loaded nuts.

➡️**Do not bend the exhaust flex coupler more than necessary to remove it. Excessive movement will damage the flex coupler.**

6. For 2.4L engine, remove the manifold-to-exhaust flex coupler fasteners.

7. Pull down and back on the exhaust pipe to disengage it from the exhaust manifold bolts.

8. Carefully lower the vehicle.

9. Unfasten the exhaust manifold-to-cylinder head retaining nuts/bolts, then remove the manifold. Remove and discard the gaskets and/or seals. Clean the mating surfaces.

To install:

10. Use gaskets, then position the exhaust manifold. Tighten the retaining nuts to 31 ft. lbs. (42 Nm) for 2.3L engine or to 110 inch lbs. (12.5 Nm) for 2.4L engine, in the sequence.

11. Raise and safely support the vehicle.

12. Install the heat shield. Tighten the bolts to 124 inch lbs. (14 Nm).

13. Fasten the exhaust manifold brace-to-manifold bolt and the oil pan nuts. Tighten the bolts to 41 ft. lbs. (56 Nm) and the nuts to 19 ft. lbs. (26 Nm).

14. For 2.3L engine, install the manifold-to-exhaust pipe nuts. Be sure to turn

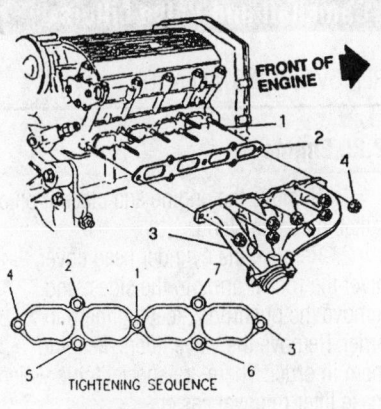

1. STUD, EXHAUST MANIFOLD
2. GASKET, EXHAUST MANIFOLD
3. MANIFOLD, EXHAUST
4. NUT, EXHAUST MANIFOLD
   42 N•m (31 LB. FT.)

7922YG10

Exploded view of the exhaust manifold, showing the torque sequence—2.3L engine

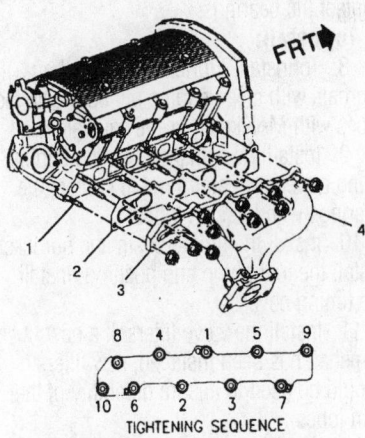

1. STUD, EXHAUST MANIFOLD
2. GASKET, EXHAUST MANIFOLD
3. MANIFOLD, EXHAUST
4. NUT, EXHAUST MANIFOLD, MUST BE TIGHTENED IN SEQUENCE SHOWN TO 12.5 N•m (110 LB. IN.)

7922YG11

Exploded view of the exhaust manifold, showing the torque sequence—2.4L engine

both nuts in evenly to avoid cocking the exhaust pipe and binding the nuts.

15. For 2.4L engine, install the manifold-to-flex coupler fasteners. Tighten the bolts to 26 ft. lbs. (35 Nm).

16. Carefully lower the vehicle.

17. Attach the O_2 connector. Coat the threads of the sensor with anti-seize compound 5613695 or equivalent.

18. Connect the negative battery cable and check for leaks.

Camshaft and Valve Lifters

REMOVAL & INSTALLATION

2.2L Engine

1. Remove the engine and place it on a suitable engine stand.

2. Remove the cylinder head cover, pivot the rocker arms to the sides, and remove the pushrods, keeping them in order. Remove the valve lifters, keeping them in order. There are special tools which make lifter removal easier.

3. Remove the front cover.

4. If necessary, remove the oil pump drive.

5. Remove the fuel pump and it's pushrod.

6. Remove the timing chain and sprocket.

7. Carefully pull the camshaft from the block, being sure the camshaft lobes do not contact the bearings.

To install:

8. To install, lubricate the camshaft journals with clean engine oil. Lubricate the lobes with Molykote® or the equivalent.

9. Install the camshaft into the engine, being extremely careful not to contact the bearings with the cam lobes.

10. Install the timing chain and sprocket. Install the fuel pump and pushrod. Install the timing cover.

11. Install the valve lifters. If a new camshaft has been installed, new lifters should be used to ensure durability of the cam lobes.

12. Install the pushrods and rocker arms and the intake manifold. Adjust the valve lash after installing the engine. Install the cylinder head cover.

2.3L and 2.4L Engines

INTAKE CAMSHAFT

➡**Any time the camshaft housing to cylinder head bolts are loosened or removed, the camshaft housing to cylinder head gasket must be replaced.**

1. Relieve the fuel system pressure. Disconnect the negative battery cable.

2. Label and detach the ignition coil and module assembly electrical connections.

3. Unfasten the ignition coil and module assembly to camshaft housing bolts, then remove the assembly by pulling straight up. Use a special spark plug boot wire remover tool to remove connector assemblies, if they have stuck to the spark plugs.

4. If equipped, remove the idle speed power steering pressure switch connector.

5. Loosen the 3 power steering pump pivot bolts and remove drive belt.

6. Disconnect the 2 rear power steering pump bracket-to-transaxle bolts.

7. Remove the front power steering pump bracket to cylinder block bolt.

8. Disconnect the power steering pump assembly, then position it aside.

9. Using the special tool, remove the power steering pump drive pulley from the intake camshaft.

10. Remove oil/air separator bolts and hoses. Leave the hoses attached to the separator, disconnect from the oil fill, chain housing and intake manifold. Remove as an assembly.

11. Remove vacuum line from fuel pressure regulator and detach the fuel injector harness connector.

12. Disconnect fuel line attaching clamp from bracket on top of intake camshaft housing.

13. Unfasten the fuel rail-to-camshaft housing attaching bolts, then remove the fuel rail from the cylinder head. Cover or plug injector openings in cylinder head and the injector nozzles. Leave the fuel lines attached, then position fuel rail aside.

14. Disconnect the timing chain and housing, but do NOT remove from the engine.

15. Remove the intake camshaft housing cover-to-camshaft housing attaching bolts.

16. Unfasten the intake camshaft housing-to-cylinder head attaching bolts. Use the reverse of the tightening sequence when loosening the bolts. Leave 2 of the bolts loosely in place to hold the camshaft housing while separating the camshaft cover from housing.

17. Push the cover off the housing by threading 4 of the housing-to-head attaching bolts into the tapped holes in the cam housing cover. Tighten the bolts evenly so the cover does not bind on the dowel pins.

18. Remove the 2 loosely installed camshaft housing to head bolts and remove the cover. Discard the gaskets.

19. Note the position of the chain sprocket dowel pin for reassembly.

20. Remove intake camshaft oil seal from camshaft and discard seal. This seal must be replaced any time the housing and cover are separated.

21. Remove the camshaft carrier from the cylinder head and remove the gasket. Discard the gasket.

To install:

22. Thoroughly clean the mating surfaces of the camshaft carrier and the cylinder head, bolts and bolt holes. Install a new gasket and place the housing on the head. Install one bolt loosely to hold it in place.

23. Install the lifters into their bores. If the camshaft is being replaced, the lifters must also be replaced. Lubricate camshaft lobes, journals and lifters with camshaft and lifter prelube. The camshaft lobes and journals must be adequately lubricated or engine damage could occur upon start up.

24. Install the camshaft in the same position as when removed. The timing chain sprocket dowel pin should be straight up and align with the centerline of the lifter bores.

25. Install new camshaft housing to camshaft housing cover seals into cover; do not use sealer. Be sure the correct color seal is placed in each groove. Install the cover to the housing.

26. Apply thread locking compound to the camshaft housing and cover attaching bolt threads.

27. Install the bolts, then tighten to 11 ft. lbs. (15 Nm). Rotate the bolts (except the 2 rear bolts that hold the fuel pipe to the camshaft housing) an additional 75 degrees, in sequence. Tighten the 2 rear bolts to 16 ft. lbs. (15 Nm), then rotate an additional 25 degrees.

28. Install the timing chain housing and the timing chain.

29. Uncover fuel injectors, then install new fuel injector O-ring seals lubricated with oil. Install the fuel rail.

30. Fasten the fuel line attaching clamp and retainer to bracket on top of the intake camshaft housing.

31. Connect the vacuum line to the fuel pressure regulator.

32. Attach the fuel injectors harness connector.

33. Install the oil/air separator assembly.

34. Lubricate the inner sealing surface of the intake camshaft seal with oil and install the seal to the housing.

35. Install the power steering pump pulley onto the intake camshaft.

36. Install the power steering pump assembly and drive belt.

37. Attach the idle speed power steering pressure switch connector.

38. Clean any loose lubricant that is present on the ignition coil and module assembly to camshaft housing bolts. Apply Loctite® 592 or equivalent, onto the ignition coil and module assembly to camshaft housing bolts. Install the bolts and tighten to 13 ft. lbs. (18 Nm).

39. Attach the electrical connectors to ignition coil and module assembly.

40. Connect the negative battery cable, then start the engine and check for leaks.

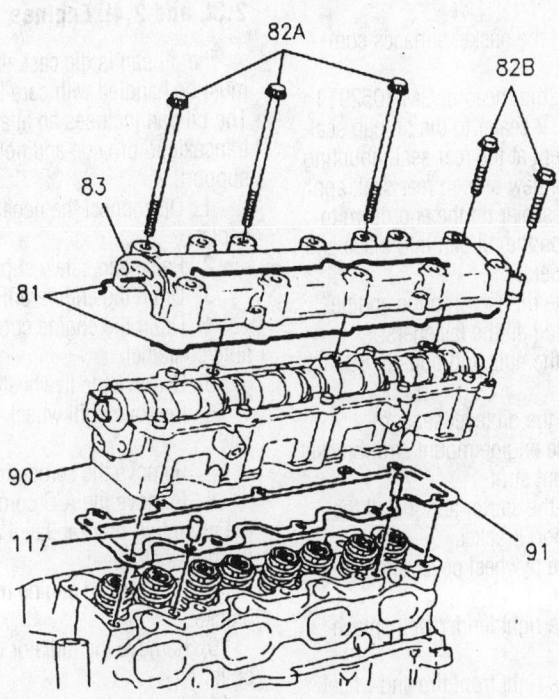

81   SEALS – CAMSHAFT HOUSING TO CAMSHAFT

82A   BOLT – CAMSHAFT HOUSING TO CYLINDER
        HEAD – 15 N·m (11 LBS. FT.) PLUS TURN 90°

82B   BOLT – CAMSHAFT HOUSING COVER TO
        CAMSHAFT HOUSING – 15 N·m (11 LBS. FT.)
        PLUS TURN 30°

83   COVER – CAMSHAFT

90   CAMSHAFT HOUSING (INTAKE SHOWN)

91   GASKET – CAMSHAFT HOUSING TO CYLINDER
       HEAD

117   DOWEL PIN (2)

7922YG31

Exploded view of the camshaft housing, cover and gaskets—2.3L and 2.4L engines

EXHAUST CAMSHAFT

➡**Any time the camshaft housing-to-cylinder head bolts are loosened or removed, the camshaft housing to cylinder head gasket must be replaced.**

1. Relieve the fuel system pressure. Disconnect the negative battery cable.

2. Label and disconnect the ignition coil and module assembly electrical connections.

3. Unfasten the ignition coil and module assembly-to-camshaft housing bolts, then remove the assembly by pulling straight up. Use a special tool to remove connector assemblies if they have stuck to the spark plugs.

4. If equipped, remove the idle speed power steering pressure switch connector.

5. Remove the transaxle fluid level indicator tube assembly from exhaust camshaft cover and position aside.

6. Remove exhaust camshaft cover and gasket.

7. Disconnect the timing chain and housing but do not remove from the engine.

8. Remove exhaust camshaft housing to cylinder head bolts. Use the reverse of the tightening procedure when loosening camshaft housing while separating camshaft cover from housing.

9. Push the cover off the housing by threading 4 of the housing to head attaching bolts into the tapped holes in the camshaft cover. When threading the bolt, tighten them evenly so the cover does not bind on the dowel pins.

10. Remove the 2 loosely installed camshaft housing to cylinder head bolts and remove cover, discard gaskets.

11. Loosely reinstall one camshaft housing to cylinder head bolt to retain the housing during camshaft and lifter removal.

12. Note the position of the chain sprocket dowel pin for reassembly. Remove camshaft being careful not to damage the camshaft or journals.

13. Remove the camshaft carrier from the cylinder head and remove the gasket. Discard the gasket.

To install:

14. Thoroughly clean the mating surfaces of the camshaft carrier and the cylinder head, bolts and bolt holes. Install a new gasket and place the housing on the head. Install 1 bolt loosely to hold in place.

15. Install the lifters into their bores. If the camshaft is being replaced, the lifters must also be replaced. Lubricate camshaft lobes, journals and lifters with camshaft and lifter prelube. The camshaft lobes and journals must be adequately lubricated or engine damage could occur upon start up.

16. Install camshaft in same position as when removed. The timing chain sprocket dowel pin should be straight up and align with the centerline of the lifter bores.

17. Install new camshaft housing-to-camshaft housing cover seals into the cover; do not use sealer. Be sure the correct color seal is placed in each groove. Install the cover to the housing.

18. Apply thread locking compound to the camshaft housing and cover attaching bolt threads.

19. Install bolts, then tighten, in sequence, to 11 ft. lbs. (15 Nm). Then, rotate the bolts an additional 75 degrees, in sequence.

20. Install timing chain housing and timing chain.

21. Install the transaxle fluid level indicator tube assembly to the exhaust camshaft cover.

22. Attach the idle speed power steering pressure switch connector.

23. Clean any loose lubricant that is present on the ignition coil and module assembly to camshaft housing bolts. Apply Loctite® 592 or equivalent, onto the ignition coil and module assembly to camshaft housing bolts. Install the bolts and tighten to 13 ft. lbs. (18 Nm).

24. Attach the electrical connectors to ignition coil and module assembly.

25. Connect the negative battery cable, then start the engine and check for leaks.

Valve Lash

ADJUSTMENT

All of the engines are equipped with hydraulic valve lifters that do not require periodic valve lash adjustment. Adjustment to zero lash is maintained automatically by hydraulic pressure in the lifters. Also, the rocker arm retaining nuts are tightened to a specific torque value (refer to the rocker arm procedure) to provide proper rocker arm placement.

Oil Pan

REMOVAL & INSTALLATION

2.2L Engine

1. Disconnect the negative terminal from the battery.
2. Raise and safely support the vehicle.
3. Drain the engine oil.
4. Remove the right front tire and wheel assembly.
5. Remove the right inner fender well splash shield.
6. Remove the starter and starter bracket.
7. Remove the flywheel cover bolts and flywheel cover.
8. Remove the engine support strut and support strut bracket.
9. Disconnect the oil level sensor.
10. Unfasten the oil pan mounting nuts and bolts, then remove the oil pan.

To install:

11. Clean all the gasket surfaces completely.
12. Place a 2mm bead of GM 1052914 or equivalent RTV sealer to the oil pan sealing surface except at the rear seal mounting surface. Using a new oil pan rear seal, apply a thin coat RTV sealer on the end down to the ears. Position the oil pan into place and install the fasteners.
13. Install the oil pan onto the engine and loosely install all the fasteners.
14. Tighten the nuts and bolts to 89 inch lbs. (10 Nm).
15. Connect the oil level sensor.
16. Install the engine mount strut bracket and engine mount strut.
17. Connect the starter and install the starter and support bracket.
18. Install the flywheel cover and cover mounting bolts.
19. Install the right fender well splash shield.
20. Install the right front tire and wheel assembly and tighten to 100 ft. lbs. (140 Nm).
21. Lower the vehicle.
22. Refill the crankcase with oil.
23. Connect the negative battery cable.
24. Start the vehicle and verify no leaks.

2.3L and 2.4L Engines

The oil pan is die cast aluminum and must be handled with care to avoid damage. The oil pan includes an attachment to the transaxle to provide additional structural support.

1. Disconnect the negative battery cable.
2. Raise and safely support the vehicle.
3. Drain the engine oil.
4. Drain the engine coolant into a suitable container.
5. Remove the flywheel/converter cover.
6. Remove right wheel and tire assembly.
7. Remove the serpentine drive belt.
8. Remove the A/C compressor from the mounting bracket, lay it aside leaving the hoses attached.
9. Remove the engine mount strut bracket.
10. Remove the radiator outlet pipe bolts.
11. Remove the air conditioning and radiator outlet pipes from the oil pan.
12. Remove the exhaust manifold brace.
13. Remove the oil pan-to-flywheel cover bolt and nut.
14. Remove the flywheel cover stud for clearance.
15. Disconnect the radiator outlet pipe from the lower radiator hose and oil pan.
16. Detach the oil level sensor connector.
17. Remove the oil pan bolts and remove the oil pan.

To install:

18. Inspect the oil pan gasket for damage. The oil pan gasket is reusable, if it is not damaged.
19. With the gasket in place, position the pan to the engine and install oil pan bolts.
20. Install the flywheel cover stud, spacer and nut. Tighten this nut to 19 ft. lbs. (26 Nm).
21. Attach the oil level sensor connector.
22. Connect the radiator outlet pipe to lower radiator hose and oil pan.
23. Install the exhaust manifold brace.
24. Install the air conditioning and radiator outlet pipes to the oil pan.
25. Install the radiator outlet pipe bolts and tighten them to 124 inch lbs. (14 Nm).
26. Install the engine mount strut bracket and tighten bolts to 55 ft. lbs. (75 Nm).
27. Install the A/C compressor into the mounting bracket.
28. Install the serpentine drive belt.
29. Install the right splash shield.
30. Install the right front wheel and tire assembly.

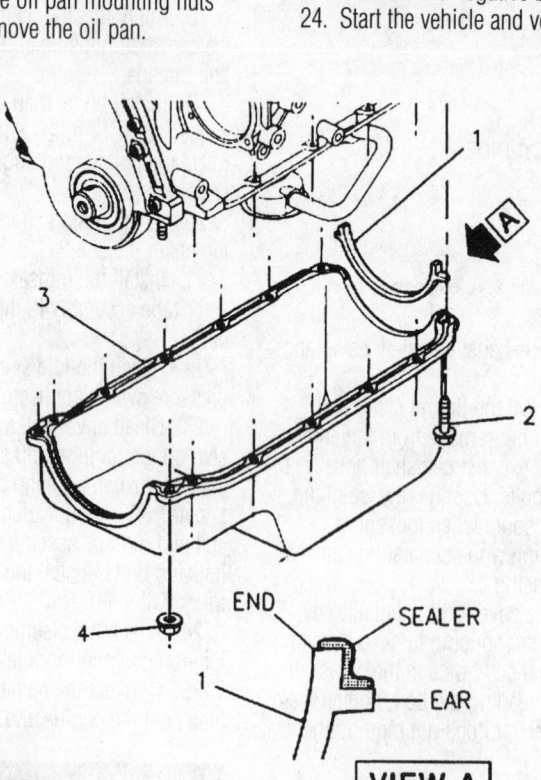

1 SEAL
2 BOLT, 10 N•m (89 LB. IN.)
3 OIL PAN
4 NUT, OIL PAN 10 N•m (89 LB. IN.)

VIEW A

END · SEALER · EAR

7922YG32

Exploded view of the oil pan mounting and related components—2.2L engine

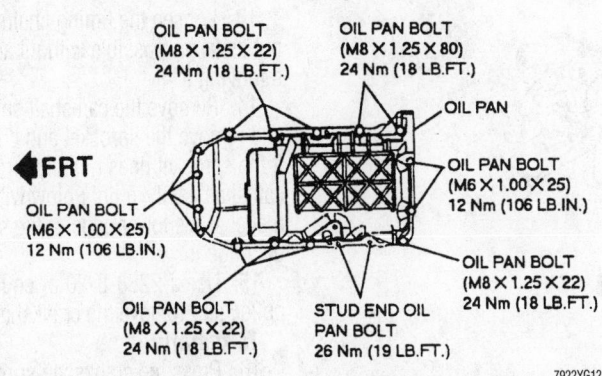

OIL PAN BOLT
(M8 × 1.25 × 22)
24 Nm (18 LB.FT.)

OIL PAN BOLT
(M8 × 1.25 × 80)
24 Nm (18 LB.FT.)

OIL PAN

◀FRT

OIL PAN BOLT
(M6 × 1.00 × 25)
12 Nm (106 LB.IN.)

OIL PAN BOLT
(M6 × 1.00 × 25)
12 Nm (106 LB.IN.)

OIL PAN BOLT
(M8 × 1.25 × 22)
24 Nm (18 LB.FT.)

OIL PAN BOLT
(M8 × 1.25 × 22)
24 Nm (18 LB.FT.)

STUD END OIL
PAN BOLT
26 Nm (19 LB.FT.)

7922YG12

Oil pan fastener torque specifications—2.3L and 2.4L engines

31. Install the flywheel/converter cover.
32. Lower the vehicle.
33. Fill the crankcase with clean oil.
34. Fill the coolant system.
35. Connect the negative battery cable.
36. Start the vehicle and verify no leaks.

Oil Pump

REMOVAL & INSTALLATION

2.2L Engine

1. Disconnect the negative battery cable.
2. Raise and safely support the vehicle.
3. Drain the engine oil into a suitable container.

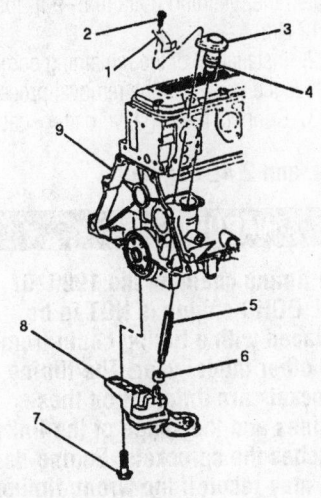

1 Bracket
2 Bolt
3 Oil pump drive assembly
4 O-ring
5 Shaft
6 Retainer; Heat and water soak prior to installation
7 Bolt
8 Oil pump
9 Cylinder block

7922YG13

Exploded view of the oil pump mounting to engine block—2.2L engine

4. Remove the oil pan-to-engine bolts and the oil pan.
5. Remove the oil pump-to-rear main bearing cap bolt, the oil pump and extension shaft.

To install:

❋❋ WARNING

A plastic sleeve called the extension shaft retainer connects the oil pump drive shaft to the oil pump. Heat the extension shaft retainer in hot water prior to assembly. Be sure the retainer does not crack upon installation.

6. Install the extension shaft, oil pump and pump-to-rear main cap bolt. Tighten the oil pump-to-bearing cap bolt to 32 ft. lbs. (43 Nm) and the upper oil pump drive bolt to 18 ft. lbs. (25 Nm).

❋❋ WARNING

To avoid engine damage, all oil pump cavities must be filled with petroleum jelly before installing the gears into the pump body. Also, use only original equipment gaskets. Gasket thickness is critical to proper oil pump operation.

7. Install the oil pan and attaching bolts.
8. Lower the vehicle.
9. Fill the crankcase with clean engine oil.
10. It is good practice to install a reliable mechanical oil pressure gauge so actual pump pressure can be read after startup. If no oil pressure is seen within a short time of startup, shut down the engine to determine why there is no oil pressure.
11. Connect the negative battery cable.
12. Start the engine and check oil pressure and check for leaks.
13. Turn the engine **OFF** and allow to stand. Check oil level, add as necessary.

2.3L and 2.4L Engines

Please note that the transaxle must be removed from the vehicle to service the oil pump.

1. Disconnect the negative battery cable.
2. Install engine support fixture J-28467-A or equivalent.
3. Drain the engine oil.
4. Remove the oil pan.
5. Remove the transaxle.
6. Remove the flywheel.
7. Remove the balance shaft chain cover and guide.
8. Remove the oil pump bolts, then remove the oil pump cover.
9. Pull the housing to disconnect pump gear from the balance shaft. Remove the housing assembly from the balance shaft assembly.
10. Disassemble the oil pump gerotor from the oil pump housing.
11. Disassemble the oil pump from the balance shaft housing.
12. Disassemble the pressure relief valve.
13. Remove the roll pin (drive it out with a small punch).

To install:

14. Clean all of the parts in suitable cleaning solvent. Remove all varnish, sludge and dirt.
15. Inspect the pump cover and housing for cracks and excessive wear, replace as necessary.
16. Lubricate the gears with clean engine oil.
17. Assemble the gerotor gear into the housing.

➡Fill oil pump cavities with petroleum jelly prior to installation. This will ensure that there is oil pressure immediately on start-up and will prevent engine damage.

18. Install the pressure relief valve, use a ⁹⁄₁₆ in. deep well socket to seat the valve.

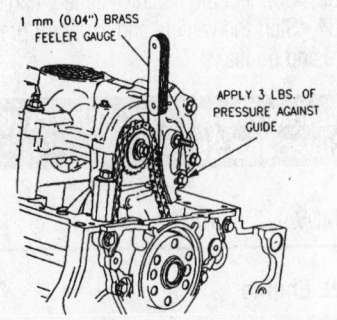

1 mm (0.04") BRASS
FEELER GAUGE

APPLY 3 LBS. OF
PRESSURE AGAINST
GUIDE

7922YG14

Using a feeler gauge to check the chain tension—2.3L and 2.4L engines

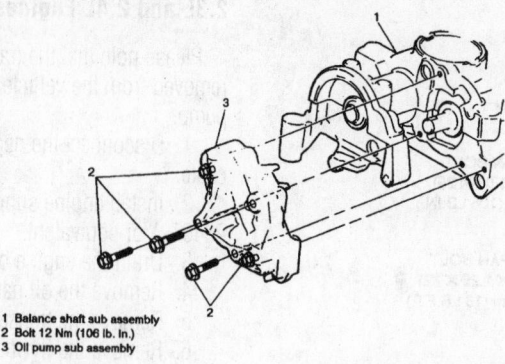

1 Balance shaft sub assembly
2 Bolt 12 Nm (106 lb. in.)
3 Oil pump sub assembly

7922YG15

Exploded view of the oil pump assembly mounting—2.3L and 2.4L engines

19. Install the roll pin.

20. Install the pump housing to the balance shaft assembly.

21. Install the pump cover to the oil pump housing.

22. Install the oil pump to block bolts and tighten 40 ft. lbs. (54 Nm).

23. Install the balance shaft chain guide and chain. Adjust the chain tension by inserting a 0.40 in. (1mm) brass feeler, between the chain guide and the chain.

➡**A brass feeler gauge must be used to ensure that correct measurements are obtained. If a steel gauge is used, it will not bend to conform to the guide and will allow for incorrect measurements.**

24. Press the guide against the chain using about 3 pounds of force.

25. Tighten the chain tensioner fastener to 115 inch lbs. (13 Nm).

26. Install the balance shaft chain cover and tighten the nut and bolt to 115 inch lbs. (13 Nm).

27. Install the flywheel. Tighten the flywheel bolts to 22 ft. lbs. (30 Nm) plus an additional 45 degrees.

28. Install the transaxle.

29. Install the oil pan.

30. Lower the vehicle.

31. Fill the crankcase with clean oil.

32. Remove the engine support fixture.

33. Connect the negative battery cable.

34. Start the vehicle and verify oil pressure and no leaks.

Timing Chain, Sprockets, Front Cover and Seal

REMOVAL & INSTALLATION

2.2L Engine

➡**The following procedure requires the use of a special centering tool (J-23042).**

1. Disconnect the negative battery cable.

2. Remove the serpentine belt and tensioner.

➡**Although not absolutely necessary, removal of the right front inner fender splash shield will facilitate access to the front cover.**

3. Install engine support fixture J-28467-A or equivalent.

4. Remove the engine mount assembly.

5. Remove the alternator rear brace, then remove the alternator.

6. Remove the power steering pump, then position it aside with the lines still attached.

7. Raise and safely support the vehicle.

8. Remove the oil pan.

9. Unscrew the center bolt from the crankshaft pulley and slide the pulley and hub from the crankshaft.

10. Unfasten the front cover-to-block bolts, then remove the front cover. If the cover is difficult to remove, use a plastic mallet to carefully loosen the cover.

11. Using a suitable seal driver, tap the oil seal out of the front cover.

12. Place the No. 1 piston at TDC of the compression stroke that the marks on the camshaft and crankshaft sprockets are in alignment.

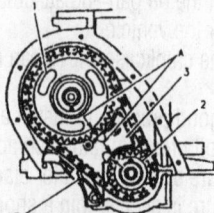

B
1. Camshaft sprocket
2. Crankshaft sprocket
3. Timing chain tensioner
A. Line up timing marks on sprockets with tabs on timing chain tensioner
B. Remove pin after timing chain is installed

7922YG16

Align the sprocket timing marks with the alignment tabs on the tensioner during timing chain installation—2.2L engine

13. Loosen the timing chain tensioner nut as far as possible without actually removing it.

14. Remove the camshaft sprocket bolts and remove the sprocket and chain together. If the sprocket does not slide from the camshaft easily, a light blow with a soft mallet at the lower edge of the sprocket will dislodge it.

15. Use J-2288-8-20 or equivalent gear puller, and remove the crankshaft sprocket.

To install:

16. Press the crankshaft sprocket back onto the crankshaft.

17. Install the timing chain over the camshaft sprocket, then around the crankshaft sprocket. Be sure the marks on the 2 sprockets are in alignment. Lubricate the thrust surface with Molykote® or equivalent.

18. Align the dowel in the camshaft with the dowel hole in the sprocket, then install the sprocket onto the camshaft. Use the mounting bolts draw the sprocket onto the camshaft, then tighten to 66–68 ft. lbs. (89–92 Nm).

19. Lubricate the timing chain with clean engine oil. Tighten the chain tensioner.

20. Using a suitable seal installer, tap the new seal into the front cover and lubricate the lip of the seal.

21. The surfaces of the block and front cover must be clean and free of oil. Install a new gasket, then position the front cover on the block using a centering tool (J-23042). Tighten the retaining bolts to 6–9 ft. lbs. (8–12 Nm).

22. Installation of the remaining components is the reverse of the removal procedure.

23. Connect the negative battery cable.

2.3L and 2.4L Engines

✳✳ WARNING

The timing chain on the 1996–97 2.4L DOHC engine is NOT to be replaced with a timing chain from any other model year. The timing sprockets are different on these engines and the shape of the links matches the sprockets. Engine damage may result if the wrong timing chain is used.

1. Disconnect the negative battery cable.

2. Remove the coolant recovery reservoir.

3. Remove the serpentine belt, using a 13mm wrench that is at least 24 in. (61cm) long.

4. For the 2.3L engine, remove the alternator, then position it aside. Install

engine support J-28467-A or equivalent. Reinstall the alternator through-bolt, then attach the engine support fixture.

5. For the 2.4L engine, install tool J-28467-400 onto the alternator stud-ended bolt, and attach the fixture.

6. Remove the upper cover fasteners.

7. Detach the cover vent hose.

8. Remove the right engine mount and the engine mount bracket or bracket adapter. Whenever the engine mounting bracket adapter is removed, the bolts must be replaced.

9. Raise and safely support the vehicle.

10. Remove the right front wheel and tire assembly and the splash shield.

11. Remove the crankshaft balancer assembly.

➡**Do not install an automatic transaxle equipped engine balancer on a manual transaxle equipped engine or vice-versa.**

12. Remove the lower cover fasteners.

13. Carefully lower the vehicle.

14. Remove the front cover and gasket. Inspect the gasket for damage and replace if necessary.

15. Using a suitable seal driver, tap the oil seal out of the front cover.

16. Rotate the crankshaft clockwise, as viewed from the front of engine/normal rotation, until the camshaft sprocket timing dowel pin holes line up with the holes in the

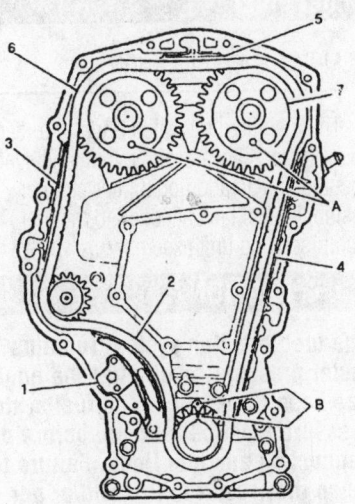

A Camshaft timing alignment pin location
B Crankshaft gear timing mark
1 Shoe assembly timing chain tensioner
2 Timing chain
3 R.H. timing chain guide
4 L.H. timing chain guide
5 Upper timing chain guide
6 Exhaust camshaft sprocket
7 Intake camshaft sprocket

7922YG17

After installation the chain must be in the "timed" position—2.3L and 2.4L engines

timing chain housing. The crankshaft sprocket keyway should point upwards and line up with the centerline of the cylinder bores; this is the "timed" position.

17. Remove the timing chain guides.

18. Raise and safely support the vehicle.

19. Be sure all of the slack in the timing chain is above the tensioner assembly, then remove the tensioner. The timing chain must be disengaged from any wear grooves in the tensioner shoe in order to remove the shoe. Slide a suitable prytool under the timing chain while pulling the shoe outward.

✳✳ WARNING

Do NOT attempt to pry the socket off the camshaft or damage to the sprocket or chain housing could occur.

20. If difficulty is encountered in removing the chain tensioner shoe, remove the intake camshaft sprocket, as follows:

a. Carefully lower the vehicle.

b. Hold the intake camshaft sprocket with tool J-39579, or equivalent, and remove the sprocket bolt and washer.

c. Remove the washer from the bolts and thread the bolt back into the camshaft by hand. The bolt provides a surface to push against.

d. Remove the camshaft sprocket using a three-jaw puller in the 3 relief holes in the sprocket.

21. Unfasten the tensioner assembly retaining bolts, then remove the tensioner.

➡**The timing chain and crankshaft sprocket MUST be marked before removal. If the chain or sprocket is installed with the wear pattern in the opposite direction, noise and increased wear may occur.**

22. Mark the crankshaft sprocket and timing chain outer surface for reassembly, then remove the chain.

23. Clean the old sealant off the bolt with a wire brush. Clean the threaded hole in the camshaft with a round nylon brush. Inspect the parts for wear and replace as necessary. Note that some scoring of the chain shoe and guides is normal.

To install:

✳✳ WARNING

Failure to follow this procedure may result in severe engine damage.

24. Position the intake camshaft sprocket onto the camshaft with the surface marked during removal showing.

25. Install the intake camshaft sprocket retaining bolt and washer, tighten to 52 ft. lbs. (70 Nm) while holding the sprocket with tool J-39579, if removed. Use sealant 12345493 or equivalent on the camshaft sprocket bolt.

26. Place tool J-36008, or equivalent camshaft aligning pins, through the holes in the camshaft sprockets into the holes in the timing chain housing. This positions the cams for correct timing.

27. If the camshafts are out of position and must be rotated more than 1/8 turn in order to install the alignment dowel pins, proceed as follows:

a. The crankshaft MUST be rotated 90 degrees clockwise off TDC in order to give the valves adequate clearance to open.

b. Once the camshafts are in position and the dowels installed, rotate the crankshaft counterclockwise back to TDC.

✳✳ WARNING

Do not rotate the crankshaft clockwise to TDC; valve or piston damage could result.

28. Place the timing chain over the exhaust camshaft sprockets, around the idler sprocket and around the camshaft sprocket.

29. Set the camshafts at the timed position and install the timing chain. Remove the alignment dowel pin from the intake camshaft. Using tool J-39579, rotate the intake camshaft sprocket counterclockwise enough to slide the timing chain over the intake camshaft sprocket. Release the camshaft sprocket wrench J-39579. The length of the chain between the 2 camshaft sprockets will tighten. If properly timed, the intake camshaft alignment dowel pin should slide in easily. If the dowel pin does not fully index, the camshafts are NOT timed correctly and the procedure must be repeated.

30. Leave the alignment dowel pins installed. Raise and safely support the vehicle.

31. With the slack removed from the chain between the intake camshaft sprocket and the crankshaft sprocket, the timing marks on the crankshaft and cylinder block should be aligned. If the marks are not aligned, move the chain one tooth forward or rearward, remove the slack and recheck the marks.

32. Reload the timing chain tensioner assembly to it "zero" position as follows:

a. Form a keeper from a piece of heavy gauge wire.

b. Apply slight force on the tensioner blade to compress the plunger.

c. Insert a small prytool into the reset access hole, and pry the ratchet pawl away from the ratchet teeth while forcing the plunger completely in the hole.

d. Install the keeper between the access hole and the blade.

33. Install the tensioner assembly to the timing chain housing. Recheck the plunger assembly installation, it is correctly installed when the long end is toward the crankshaft. Install the tensioner retaining bolts; tighten to 89 inch lbs. (10 Nm).

34. Carefully lower the vehicle enough to reach and remove the alignment dowel pins.

35. Rotate the crankshaft clockwise (normal rotation) 2 full rotations. Align the crankshaft keyway with the mark on the cylinder block and reinstall the alignment dowel pins. The pins will slide in easily if the engine of correctly timed.

36. Using a suitable seal installer, tap the new seal into the front cover and lubricate the lip of the seal.

37. Install the timing chain guides, then install the front (timing chain) cover. Tighten the timing chain cover fasteners to 106 inch lbs. (12 Nm). Tighten the balancer attaching bolt to 74 ft. lbs. (100 Nm).

38. Installation is the reverse of the removal procedure.

39. Connect the negative battery cable.

FUEL SYSTEM

Fuel System Service Precautions

Safety is the most important factor when performing not only fuel system maintenance but any type of maintenance. Failure to conduct maintenance and repairs in a safe manner may result in serious personal injury or death. Maintenance and testing of the vehicle's fuel system components can be accomplished safely and effectively by adhering to the following rules and guidelines.

• To avoid the possibility of fire and personal injury, always disconnect the negative battery cable unless the repair or test procedure requires that battery voltage be applied.

• Always relieve the fuel system pressure prior to disconnecting any fuel system component (injector, fuel rail, pressure regulator, etc.), fitting or fuel line connection. Exercise extreme caution whenever relieving fuel system pressure to avoid exposing skin, face and eyes to fuel spray. Please be advised that fuel under pressure may penetrate the skin or any part of the body that it contacts.

• Always place a shop towel or cloth around the fitting or connection prior to loosening to absorb any excess fuel due to spillage. Ensure that all fuel spillage (should it occur) is quickly removed from engine surfaces. Ensure that all fuel soaked cloths or towels are deposited into a suitable waste container.

• Always keep a dry chemical (Class B) fire extinguisher near the work area.

• Do not allow fuel spray or fuel vapors to come into contact with a spark or open flame.

• Always use a back-up wrench when loosening and tightening fuel line connection fittings. This will prevent unnecessary stress and torsion to fuel line piping. Always follow the proper torque specifications.

• Always replace worn fuel fitting O-rings with new. Do not substitute fuel hose or equivalent, where fuel pipe is installed.

Fuel System Pressure

RELIEVING

1995 Models

1. Disconnect the negative battery cable.
2. Release the fuel vapor pressure in the fuel tank by momentarily removing the tank filler cap.
3. Connect J-34730-1 or equivalent fuel pressure gauge, to the fuel pressure connection located on the end of the fuel rail assembly. Wrap a cloth around the fitting to absorb any fuel leakage.
4. Install the bleed hose into an approved container and open the valve to bleed system pressure. The fuel pipe connections are now safe for servicing.
5. Drain any fuel remaining in the fuel pressure gauge into an approved container.
6. Once the tests or repairs are completed, prime the fuel system by cycling the ignition switch **ON** for 2 seconds, **OFF** for 10 seconds, then **ON** again. Repeat, if necessary to build system pressure.

1996–99 Models

1. Loosen the fuel filler cap in order to relieve the pressure in the tank (do not tighten at this time).
2. Raise and safely support the vehicle.
3. Detach the fuel pump electrical connector.
4. Start and run the vehicle until it stalls, the engage the starter for an additional 3 seconds to ensure the relief of any remaining pressure.
5. Disconnect the negative battery cable.
6. Once the tests or repairs are completed, reattach the fuel pump electrical connector.
7. Connect the negative battery cable.
8. Lower the vehicle.
9. Tighten the fuel filler cap.
10. Prime the fuel system by cycling the ignition switch **ON** for 2 seconds, **OFF** for 10 seconds, then **ON** again. Repeat, if necessary to build system pressure.

Fuel Filter

REMOVAL & INSTALLATION

The fuel filter is located under the rear of the vehicle, rearward of the fuel tank. Note that there is an additional filter/strainer inside the fuel tank attached to the fuel pump/sending unit assembly.

1. Relieve the fuel system pressure using the recommended procedure.
2. Disconnect the negative battery cable.
3. Raise and safely support the vehicle.
4. Using a back-up wrench on the fuel filter, disconnect the fuel line from the filter.

✳✳ CAUTION

If a nylon fuel line becomes kinked and cannot be straightened, replace it. Some technicians use compressed air to blow dirt from the filter's quick-connect fittings; be sure to wear safety glasses.

5. Grasp the filter and nylon connection line fitting. Twist the quick-connect fitting ¼ turn in each direction to loosen any dirt within the fitting. Disconnect the quick-connect fitting from the fuel filter by compressing the tabs while pulling outward on the line. GM also has available special tool J-38778, that is placed between the fuel filter and quick-connect fitting release mechanism to force the 2 apart.

6. Remove the fuel filter from the mounting bracket.

To install:

7. Before installing a new filter, always apply a few drops of clean engine oil to the male tube end of the filter and to the fuel sending unit assembly connection. This will help ensure proper connection and prevent possible fuel leaks. During normal operation, the O-rings located in the female connector will swell and may prevent proper connection if not lubricated.

8. Install the fuel filter in the mounting bracket.

9. Connect the quick-connect fitting to the fuel filter using the following procedure:

a. Apply a few drops of clean engine oil to the male ends of the filter and the fuel sender assembly.

b. Push the connectors together to cause the retaining tabs/fingers to snap into place.

c. Once installed, pull on both ends of each connection to be sure they are secure.

10. Using a new O-ring, tighten the fuel line using a back-up wrench on the fuel filter. Tighten the fuel line fitting to 20 ft. lbs. (27 Nm).

11. Lower the vehicle.

12. Connect the negative battery cable.

13. Pressurize the fuel system and verify no leaks.

Fuel Pump

REMOVAL & INSTALLATION

✳✳ CAUTION

The fuel injection system remains under pressure, even after the engine has been turned OFF. The fuel system pressure must be relieved before disconnecting any fuel lines. Failure to do so may result in fire and/or personal injury.

1. Relieve the fuel system pressure using the recommended procedure.

2. Disconnect the negative battery cable.

3. Drain the fuel tank, then remove the fuel tank from the vehicle.

4. While holding the modular fuel sender assembly down, remove the snapring from the designated slots located on the retainer.

✳✳ WARNING

The modular fuel sender assembly may spring up from its position. When removing the modular fuel sender from the tank, be aware that the reservoir bucket is full of fuel. It must be tipped slightly during removal to avoid damage to the float.

5. Remove the external fuel strainer.

6. Detach the Connector Position Assurance (CPA) piece from the electrical connector and detach the fuel pump electrical connector.

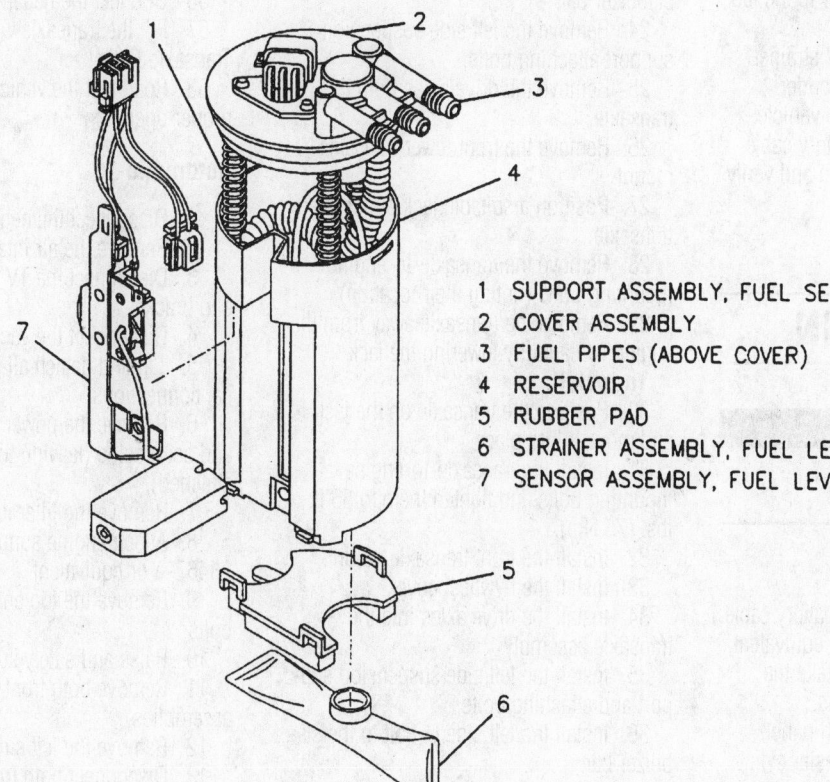

1. SUPPORT ASSEMBLY, FUEL SENDER
2. COVER ASSEMBLY
3. FUEL PIPES (ABOVE COVER)
4. RESERVOIR
5. RUBBER PAD
6. STRAINER ASSEMBLY, FUEL LEVEL FUEL
7. SENSOR ASSEMBLY, FUEL LEVEL

Modular fuel pump component identification

7922YG33

7. Gently release the tabs on the sides of the fuel sender at the cover assembly. Begin by squeezing the sides of the reservoir and releasing the tab opposite the fuel level sensor. Move clockwise to release the second and third tabs in the same manner.

8. Lift the cover assembly out far enough to detach the fuel pump electrical connection.

9. Rotate the fuel pump baffle counterclockwise and remove the baffle and pump assembly from the retainer.

10. Slide the fuel pump outlet out of slot, then remove the fuel pump outlet seal.

To install:

11. Install the fuel pump outlet seal, then slide the fuel pump outlet in the slots of reservoir cover.

12. Install the fuel pump and baffle assembly onto the reservoir retainer and rotate clockwise until seated.

13. Install the lower retainer assembly partially into the reservoir. Line up all 3 sleeve tabs. Press the retainer onto the reservoir making sure all 3 tabs are firmly seated.

➡**Gently pull on the fuel pump reservoir from retainer to assure a secure fastening. If not secure, replace the entire fuel sender.**

14. Attach the fuel pump connector.

15. Fasten the CPA connector to the fuel sender cover.

16. Install a new external fuel strainer.

17. Install the modular fuel sender.

18. Install the fuel tank in the vehicle.

19. Connect the negative battery cable.

20. Pressurize the fuel system and verify no leaks.

DRIVE TRAIN

Transaxle Assembly

REMOVAL & INSTALLATION

Manual

1. Disconnect the negative battery cable.

2. Install tool J-28467-A or equivalent, and raise the engine enough to take the pressure off the transaxle mounts.

3. Remove the left side hush panel.

4. Disconnect the clutch master cylinder pushrod from the clutch pedal.

5. Remove the air cleaner and duct assembly from the throttle body.

6. Remove the wiring harness from the mount bracket.

7. Remove the upper transaxle mount-to-transaxle bolts.

8. Remove the clutch master cylinder from the clutch actuator.

9. Disconnect the ground cables from the transaxle mounting studs.

10. Detach the back-up light switch connector.

11. Disconnect the transaxle vent tube.

12. Remove the rear transaxle-to-engine bolts.

13. Lower the engine support fixture enough to ease removal and installation of the transaxle.

14. Raise and safely support the vehicle.

15. Drain the transaxle into a suitable container.

16. Remove the tire and wheel assemblies.

17. Remove the left side splash shield.

18. Disconnect both front ABS wheel speed sensor harness and move out of the way.

19. Remove the flywheel cover.

20. Disconnect the vehicle speed sensor at the transaxle.

21. Remove the left and right ball joint nuts and separate them from the steering knuckle.

22. Remove the left stabilizer link pin.

23. Remove the left side U-bolt from the stabilizer bar.

24. Remove the left side suspension support attaching bolts.

25. Remove the drive axles from the transaxle.

26. Remove the front lower transaxle mount.

27. Position a suitable jack under the transaxle.

28. Remove the transaxle-to-engine mounting bolts (noting their location).

29. Remove the transaxle away from the engine by carefully lowering the jack.

To install:

30. Position the transaxle on the jack and move it into place.

31. Install the transaxle-to-engine mounting bolts and tighten them to 55 ft. lbs. (75 Nm).

32. Install the front transaxle mount.

33. Install the flywheel cover.

34. Install the drive axles into the transaxle assembly.

35. Install the left side suspension support and attaching bolts.

36. Install the left side U-bolt to the stabilizer bar.

37. Connect the ball joints to the steering knuckle and install the nuts, tighten to 48 ft. lbs. (65 Nm).

38. Install the left side stabilizer link pin assembly.

39. Route the left side ABS wheel speed sensor wiring harness and connect both front wheel speed sensor connectors.

40. Install the inner splash shield.

41. Connect the vehicle speed sensor to the transaxle.

42. Install both tire and wheel assemblies.

43. Lower the vehicle.

44. Install the ground cables to the transaxle mounting studs.

45. Install the vent tube to the transaxle.

46. Attach the back-up light switch connector.

47. Install the upper transaxle mounting bolts and tighten to 55 ft. lbs. (75 Nm).

48. Install the clutch master cylinder to clutch actuator cylinder.

49. Install the rear transaxle mount. Tighten the bolts to 55 ft. lbs. (75 Nm).

50. Clip the wiring harness to the mount bracket.

51. Remove the engine support fixture.

52. Connect the shift cables clamp and nut. Tighten the nut to 89 inch lbs. (10 Nm).

53. Install the air cleaner and duct assembly to the throttle body.

54. Connect the pushrod to the clutch pedal.

55. Install the left side hush panel.

56. Connect the negative battery cable.

57. Fill the transaxle with Synchromesh® Transaxle Fluid.

58. Road test the vehicle and verify proper operation.

Automatic

1. Disconnect the negative battery cable.

2. Remove the air intake duct.

3. Disconnect the TV cable, shift cable and bracket.

4. Disconnect the vacuum lines.

5. Tag and detach all necessary electrical connections.

6. Remove the power steering pump and set it aside (leaving the hoses attached).

7. Remove the filler tube.

8. Attach engine support fixture J-28467-A or equivalent.

9. Remove the top engine to transaxle bolts.

10. Raise and safely support the vehicle.

11. Remove both front tire and wheel assemblies.

12. Remove the left side splash shield.

13. Disconnect both front ABS wheel speed sensors and harness from left suspension support.

14. Disconnect both lower ball joints.
15. Disconnect the stabilizer shaft links.
16. Remove the front air deflector.
17. Remove the left suspension support.
18. Remove both drive axles (halfshafts).
19. Remove the engine to transaxle brace.
20. Remove the transaxle converter cover.
21. Remove the starter motor.
22. Remove the converter bolts.
23. Disconnect the transaxle cooler lines and remove brace.
24. Disconnect the ground wires going to transaxle.
25. Remove the exhaust brace.
26. Remove the bolts from engine and transaxle mount.
27. Support transaxle with a jack and remove transaxle mount-to-body bolts.
28. Disconnect heater core hose brace from transaxle.
29. Remove the remaining engine to transaxle bolts, then remove the transaxle assembly.

To install:

30. Be sure to properly seat the torque converter in the oil pump.
31. Install the transaxle into position with jack while installing right drive axle.
32. Install the lower engine-to-transaxle bolts and tighten to 71 ft. lbs. (96 Nm).
33. Install the transaxle mount-to-body bolts.
34. Install the engine and transaxle mount bolts.
35. Install the exhaust brace.
36. Install the cooler line brace.
37. Attach the ground wires to transaxle bolt.
38. Connect the cooler lines.
39. Install and tighten the converter bolts to 46 ft. lbs. (62 Nm).
40. Install the transaxle converter cover.
41. Install the starter.
42. Install the engine-to-transaxle brace and tighten the bolts to 32 ft. lbs. (43 Nm).
43. Install the drive axles.
44. Install the left suspension support.
45. Install the front air deflector.
46. Install the stabilizer links.
47. Connect the lower ball joints.
48. Connect both ABS wheel speed sensors.
49. Install the left splash shield.
50. Install the heater core pipe brace nut and bolt.
51. Install the front wheel and tire assemblies.
52. Lower the vehicle.
53. Install the top engine-to-transaxle bolts and tighten to 71 ft. lbs. (96 Nm).

54. Remove the engine support fixture.
55. Install the filler tube.
56. Install the power steering pump assembly.
57. Attach the electrical connections, as tagged during removal.
58. Connect the vacuum lines.
59. Attach the shift cable and bracket.
60. Attach the TV cable and adjust as necessary.
61. Install the intake air duct.
62. Connect the negative battery cable.
63. Fill transaxle and start vehicle, verify that there are no leaks.
64. Road test the vehicle.

Clutch

REMOVAL & INSTALLATION

The manual transaxle assembly must be removed from the vehicle to service the clutch assembly.

➡**Prior to any vehicle service that requires the removal of the actuator cylinder, the master cylinder pushrod must be disconnected from the clutch pedal. If not disconnected, permanent damage to the actuator cylinder will occur if the clutch pedal is depressed while the actuator cylinder is disconnected.**

1. Disconnect the negative battery cable.
2. Remove the clutch master cylinder pushrod from the clutch pedal.
3. Remove the transaxle.
4. If any of the parts are to be reused, mark the pressure plate assembly and the flywheel so they can be assembled in the same position. They were balanced as an assembly at the factory.
5. Loosen the attaching bolts one turn at a time until spring tension is relieved.
6. Support the pressure plate and remove the bolts. Remove the pressure plate and clutch disc. Do not disassemble the pressure plate assembly. Replace it if defective.
7. Inspect the flywheel, clutch disc, pressure plate, throwout bearing and the clutch fork and pivot shaft assembly for wear. Replace the parts as required. If the flywheel shows any signs of overheating or if it is badly grooved or scored, it should be resurfaced or replaced.
8. Clean the pressure plate and flywheel mating surfaces thoroughly.

To install:

9. Clean all parts well. Apply a small amount of high temperature grease to the pilot bearing inside the end of the crankshaft.

10. Position the clutch disc and pressure plate into the installed position, and support with clutch aligning tool J-29074 or equivalent. The clutch plate is assembled with the damper springs offset toward the transaxle. One side of the factory supplied clutch disc should be stamped "Flywheel Side."
11. Install the pressure plate-to-flywheel bolts. Tighten them gradually in a cross pattern as follows:
  a. Install lightly seat all bolts.
  b. Tighten bolts 1, 2, 3, then 4, 5, and 6 to 12 ft. lbs. (16 Nm).
  c. Final torque bolts 1, 2, 3, then 4, 5, 6 to 15 ft. lbs. (20 Nm).
12. Lubricate the outside groove and the inside recess of the release bearing with high temperature grease. Wipe off any excess. Install the release bearing.
  a. On the NVG-T550 transaxle, lubricate the inside diameter of the bearing with clutch bearing lubricant.
  b. On the Isuzu transaxle, pack the inside recess of the release bearing completely full of chassis grease.

➡**On the Isuzu transaxle, be sure the bearing pads are located on the fork ends and both spring ends are in the fork holes with the spring completely seated in bearing groove.**

13. Install the transaxle.
14. Install clutch master cylinder pushrod to the clutch pedal and install the retaining clip.
15. If equipped with cruise control, check switch adjustment at clutch pedal bracket.

➡**When adjusting the cruise control switch, do not exert an upward force on the clutch pedal pad of more than 20 ft. lbs. (27 Nm) or damage to the master cylinder pushrod retaining ring can result.**

16. Connect the negative battery cable.
17. Bleed clutch system as necessary and road test vehicle.

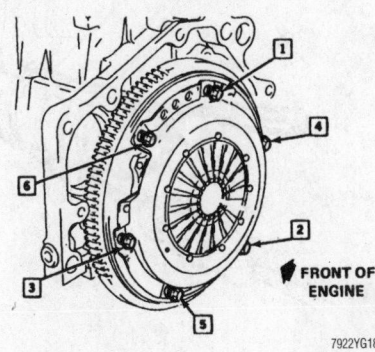

7922YG18

Clutch cover bolt tightening sequence

Hydraulic Clutch System

BLEEDING

Do not use fluid which has been bled from a system to fill the reservoir as it may be aerated, have too much moisture content or possibly be contaminated. Clean the dirt and grease from the cap to ensure that no foreign substances enter the system. It is also important to maintain the fluid level in the clutch reservoir to the top step with hydraulic clutch fluid GM part number 12345347 or equivalent.

1. Attach a hose to the bleeder screw on the clutch actuator assembly and submerge the other end of the hose in a container of hydraulic clutch fluid.
2. Depress the clutch pedal slowly and hold.
3. Loosen the bleeder screw to purge air.
4. Tighten the bleeder screw to 18 inch lbs. (2 Nm).
5. Let up on the clutch pedal.
6. Repeat Steps 2 through 5 until all air is purged from the system.
7. Fill the reservoir to the top step with hydraulic clutch fluid.
8. Repeat this bleeding procedure if there is a grinding noise during the clutch spin down procedure.

Halfshaft

REMOVAL & INSTALLATION

Some manual transaxle applications may also use an intermediate shaft.

1. Disconnect the negative battery cable.
2. With the weight of the vehicle still on the wheels, loosen, but do not remove the front hub nut. This may require an assistant holding the brakes to keep the front halfshaft from turning. It is good practice to wire-brush the exposed threads on the outer CV-joint stub shaft and apply a generous amount of penetrating oil before attempting to loosen the hub nut.
3. Raise and safely support the vehicle.
4. Remove the tire and wheel assembly.
5. Remove the hub nut and washer.
6. Install the axle boot seal protector J-33162 or the equivalent on the right-hand inner boot, if equipped.
7. Remove and support the brake caliper.
8. Remove the brake rotor.
9. Remove the lower ball joint cotter pin and nut and loosen the joint. If removing the right halfshaft, turn the wheel to the left. If removing the left halfshaft turn the wheel to the right.
10. Disconnect the ABS sensor, if equipped.

11. Disconnect the stabilizer bar link.
12. Separate the lower ball joint from the steering knuckle.
13. Disengage the halfshaft stub end from the front wheel bearing and hub assembly using a suitable press-type tool, pressing until the halfshaft splines are just loose.
14. Separate the hub and bearing assembly from the halfshaft. Move the strut and knuckle assembly rearward.
15. Separate the inner joint from the transaxle using the proper puller tools such as J-33008 and J-29764 or their equivalents.
16. Remove the halfshaft from the transaxle. Do not pull the halfshaft by the CV-joint boot or on the joint itself.

To install:

17. Prior to installation, cover all sharp edges in the area of the halfshaft with shop towels so the CV-joint boots will be protected from damage. When a halfshaft is removed for any reason, the transaxle (the halfshaft male and female shank) and knuckle sealing surfaces should be inspected for debris and corrosion. If debris or corrosion are present, clean with 320 grit crocus cloth or equivalent. Transmission fluid may be used to clean off any remaining debris. The surface should be wiped clean and dry before attempting to install the halfshaft.

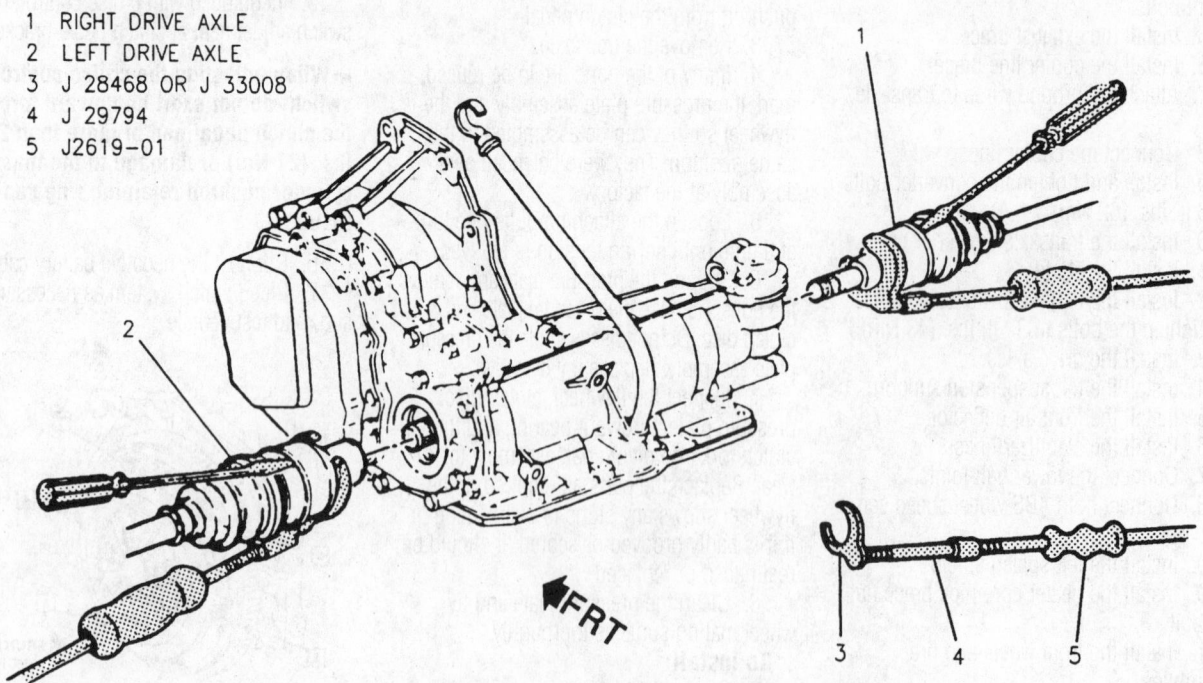

```
1   RIGHT DRIVE AXLE
2   LEFT DRIVE AXLE
3   J 28468 OR J 33008
4   J 29794
5   J2619-01
```

7922YG34

To prevent damaging the transaxle or halfshaft, use the tools as shown to remove the halfshafts

18. Install the halfshaft into the transaxle (or intermediate shaft, if equipped) by placing a brass drift pin into the groove on the joint housing and tapping until seated. Be careful not to damage the axle seal or dislodge the seal garter spring when installing the axle.

➡**Be sure the halfshaft is fully engaged in the transaxle. Verify that the half-shaft is seated by grasping the inner joint housing and pulling outward. Do not pull on the shaft or the boot, but on the inner joint housing only.**

19. Install the drive axle into the hub and bearing assembly.

20. Install the lower ball joint to the steering knuckle. Tighten the ball joint-to-steering knuckle nut to 41–48 ft. lbs. (55–65 Nm) to install the cotter pin. Do not loosen the nut at any time during installation. Install a new cotter pin.

21. Install the washer and a new hub nut. To keep the hub from turning while the hub nut is being torqued, insert a drift pin through the caliper opening into one of the ventilation openings in the brake rotor. This should lock the assembly together. Tighten the hub nut to 185 ft. lbs. (260 Nm).

22. Install the tire and wheel assembly.

23. Lower the vehicle.

24. Test drive vehicle to verify no front drive noise.

STEERING AND SUSPENSION

Air Bag

⁂ CAUTION

Some vehicles are equipped with an air bag system, also known as the Supplemental Inflatable Restraint (SIR) system. The system must be disabled before performing service on or around system components, steering column, instrument panel components, wiring and sensors. Failure to follow safety and disabling procedures could result in accidental air bag deployment, possible personal injury and unnecessary system repairs.

PRECAUTIONS

Several precautions must be observed when handling the inflator module to avoid accidental deployment and possible personal injury.

• Never carry the inflator module by the wires or connector on the underside of the module.

• When carrying a live inflator module, hold securely with both hands, and ensure that the bag and trim cover are pointed away.

• Place the inflator module on a bench or other surface with the bag and trim cover facing up.

• With the inflator module on the bench, never place anything on or close to the module which may be thrown in the event of an accidental deployment.

DISARMING

⁂ CAUTION

The Supplemental Inflatable Restraint (SIR) system must be disarmed before performing many in-vehicle service procedures. Failure to do so may cause accidental deployment of the air bag, resulting in unnecessary SIR system repairs and/or personal injury.

1. Turn the steering wheel so the vehicle's wheels are pointing straight ahead.

2. Turn the ignition switch to the **LOCK** position and remove the key.

3. Remove the AIR BAG fuse from the instrument panel fuse block.

4. Remove the left-hand sound insulator.

5. Detach the Connector Position Assurance (CPA) and yellow 2-way connector at the base of the steering column.

6. If equipped with passenger side air bags, remove the right-hand sound insulator, then disconnect the CPA and yellow 2-way connector from the passenger inflator module pigtail.

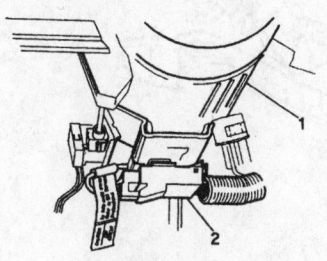

1. Steering column
2. Connector,sir(yellow)

7922YG19

Driver's side Yellow 2-way air bag CPA connector location

REARMING

1. Turn the ignition switch to the **LOCK** position and remove the key.

2. If equipped with passenger side air bags, connect the yellow 2-way connector and CPA to the passenger inflator module pigtail, then install the right-hand sound insulator.

3. Attach the yellow 2-way connector and CPA at the base of the steering column. After installing the CPA, clip the connector to flange on the steering column support.

4. Install the left-hand insulator.

5. Install the AIR BAG fuse into the instrument panel fuse block.

6. Turn the ignition switch to the **RUN** position and verify that the AIR BAG warning lamp flashes 7–9 times, then the light turns OFF.

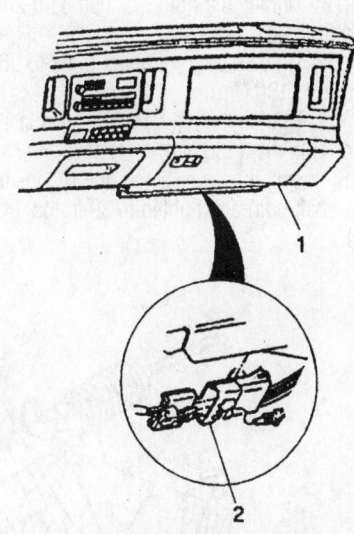

1. I/P compartment
2. Connector,sir(yellow)

7922YG20

Passenger's side Yellow 2-way air bag CPA connector location

Rack and Pinion Steering Gear

REMOVAL & INSTALLATION

1. Disconnect the negative battery cable.

2. Remove the left side sound insulator.

3. Remove the upper pinch bolt on the inter-shaft assembly.

4. Remove the line retainer, if applicable.

5. Raise and safely support the vehicle.

6. Remove the left front tire and wheel assembly.

7. Disconnect the tie rod ends from the struts using tool J-2431-01 or equivalent.

8. Remove the left and right mounting bolts.

9. Disconnect the gear inlet and outlet hose assemblies from the rack and pinion.

10. Remove the lower pinch bolt from the flange inter-shaft assembly.

11. Remove the inter-shaft assembly.

12. Loosen the crossmember bolts to gain additional clearance for removal.

13. Remove the rack and pinion through the wheel opening.

To install:

14. Install the rack and pinion through the left wheel opening.

15. Install the crossmember bolts as follows:

   a. Tighten the left rear outboard 1st to 96 ft. lbs. (130 Nm).

   b. Tighten the right rear outboard 2nd to 96 ft. lbs. (130 Nm).

   c. Tighten the front upper bolts to 96 ft. lbs. (130 Nm).

   d. Tighten the rear inboard bolts last to 96 ft. lbs. (130 Nm).

16. Install the lower pinch bolt (flange to inter-shaft bolt) and tighten to 30 ft. lbs. (41 Nm).

17. Connect the gear inlet and outlet pipes to the rack and pinion and tighten to 20 ft. lbs. (27 Nm).

18. Hand start bolts and nuts. Tighten left side bolt to 89 ft. lbs. (120 Nm), then tighten right side to 89 ft. lbs. (120 Nm).

19. Connect the tie rod ends to the struts and tighten to 44 ft. lbs. (60 Nm) and install new cotter pins.

20. Install left tire and wheel assembly.

21. Install line retainer, if applicable.

22. Lower the vehicle.

23. Install the upper pinch bolt and tighten to 30 ft. lbs. (41 Nm).

24. Install the left side sound insulator.

25. Connect the negative battery cable.

26. Fill with fluid and bleed air from system.

27. Check toe setting and adjust as required.

28. Road test vehicle and verify no leaks.

Strut

REMOVAL & INSTALLATION

Front

1. From inside the engine compartment, pry off the cover on the strut tower, if equipped, then unfasten the upper strut-to-body nuts and/or bolts.

2. Loosen the wheel lug nuts, then raise and safely support the vehicle.

3. Place jackstands under the front crossmember. Lower the vehicle slightly so the weight of the vehicle rests on the jackstands and NOT the control arms.

4. Remove the wheel and tire assembly.

5. Before removing front suspension components, their positions should be marked so they may assemble correctly.

❊❊ WARNING

Whenever working near the drive axles, use care to prevent damage from over extension of the drive shaft joints. When either end of the shaft is disconnected, over extension of the joint could result in separation of the internal components and possible joint failure.

6. Install a drive axle joint protective cover (modified), such as J-34754 or equivalent.

7. If necessary, remove the brake line bracket.

8. Remove the cotter pin and nut, then press the tie rod out of the strut bracket using J-24319-01 or equivalent two-armed puller. Discard the cotter pin.

9. Unfasten and remove the strut-to-steering knuckle bolts and carefully lift out the strut.

➥The steering knuckle MUST be supported to prevent axle joint overextension.

10. Remove the strut assembly from the vehicle. Be careful to avoid chipping or cracking the spring coating when handling the front suspension coil spring assembly.

To install:

11. Move the strut into position, then install the nuts and/or bolts connecting the strut assembly to the body.

12. Align the steering knuckle with the strut flange scribe mark made during removal, then install the bolts and nuts. Tighten to 133 ft. lbs. (180 Nm).

13. Position the tie rod end into the strut assembly, then secure with the tie rod end bolt and new cotter pin. Tighten the tie rod end bolt to 44 ft. lbs. (60 Nm).

14. Tighten the nuts and/or bolts attaching the top of the strut to the body to 18–20 ft. lbs. (25–27 Nm).

15. Install the brake line bracket.

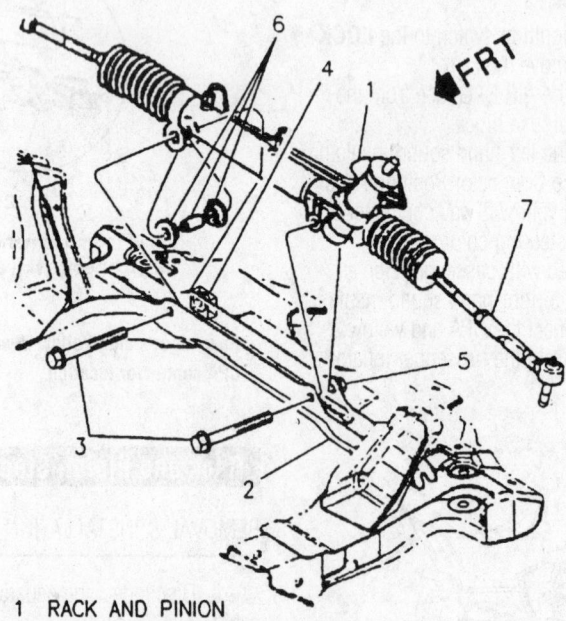

1 RACK AND PINION
2 CROSSMEMBER, SUSPENSION
3 BOLTS, STEERING GEAR
4 CAGE NUT
5 WELD NUTS
6 BUSHING AND SLEEVE
7 TIE ROD

7922YG21

Exploded view of the rack and pinion steering gear mounting

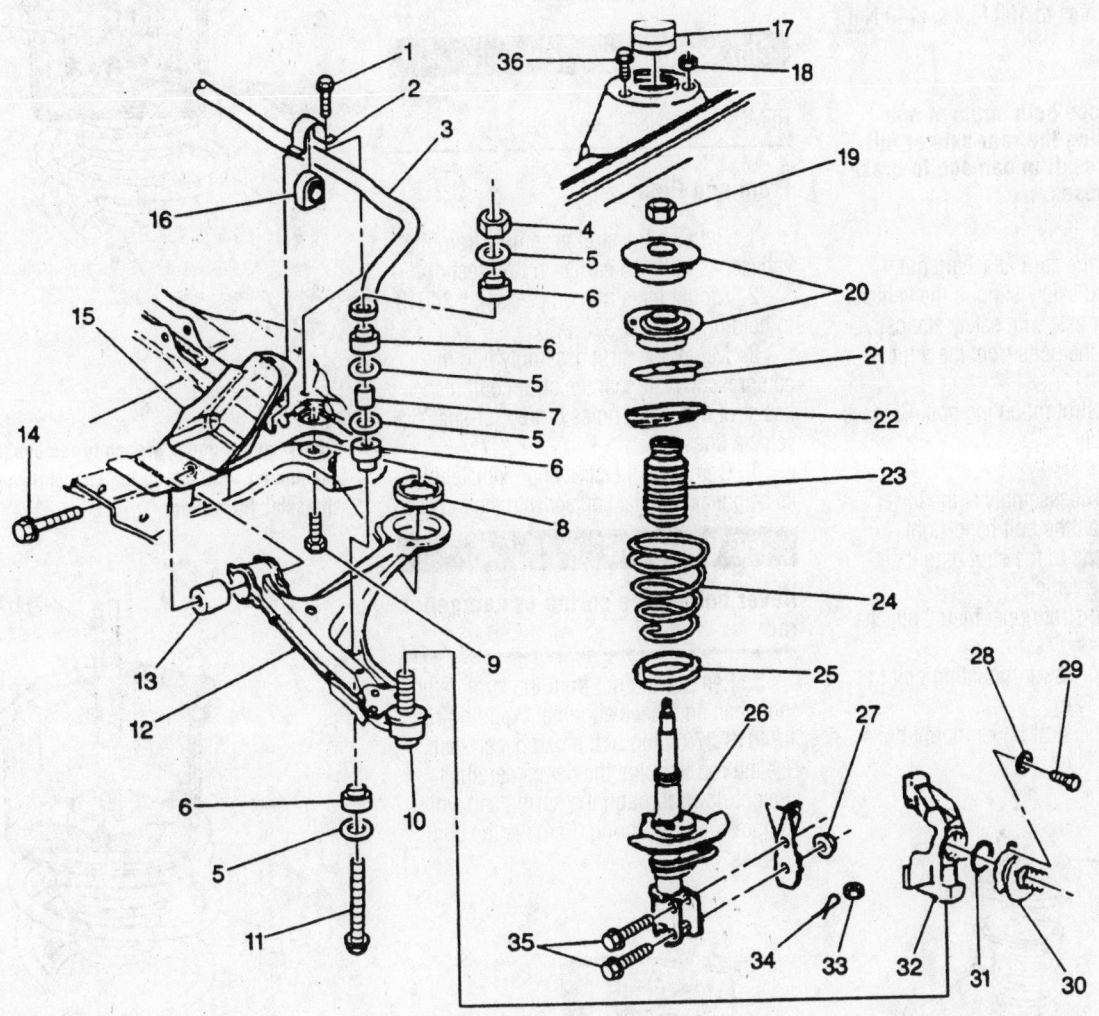

Legend

- (1) Bolt
- (2) Clamp, Stabilizer Shaft
- (3) Stabilizer Shaft
- (4) Nut, Stabilizer Link
- (5) Washer
- (6) Insulator, Stabilizer Link
- (7) Spacer Stabilizer Link
- (8) Bushing, Vertical
- (9) Bolt, Vertical Bushing
- (10) Ball Joint
- (11) Bolt, Stabilizer Link
- (12) Control Arm
- (13) Bushing, Control Arm
- (14) Lower Spring Insulator
- (15) Suspension Support
- (16) Insulator, Stabilizer Shaft
- (17) Cover, Strut Mount
- (18) Nut
- (19) Nut, Strut Dampener Shaft
- (20) Strut Mount and Rate Washer Assembly
- (21) Spring Seat
- (22) Upper Spring Insulator
- (23) Strut Bumper and Shield
- (24) Spring
- (25) Lower Spring Insulator
- (26) Strut
- (27) Nut
- (28) Washer
- (29) Bolt
- (30) Hub and Bearing Assembly
- (31) Seal (Part of 5)
- (32) Steering Knuckle
- (33) Nut, Ball Joint
- (34) Cotter Pin
- (35) Bolt
- (36) Bolt

Exploded view of the front suspension

7922YG22

16. Slightly raise the vehicle, then remove the jackstands from under the suspension supports.

17. Install the tire and wheel assembly.

18. Carefully lower the vehicle, then final tighten the lug nuts to 100 ft. lbs. (140 Nm).

Rear

➡**Do not remove both struts at one time. Suspending the rear axle at full length could result in damage to brake lines and/or hoses.**

1. Open the deck lid.

2. Remove the strut attaching nut.

3. Raise and safely support the vehicle. Support the rear axle with safety stands.

4. Remove the bolts from the strut upper mount.

5. Remove strut mounting bolt. Remove the strut assembly.

To install:

6. Install strut assembly at the lower attachment. Install the bolt hand-tight.

7. Install bolts to the strut mount.

8. Lower the vehicle.

9. Install the strut upper mount attaching nut.

10. Tighten the lower mounting bolt to 125 ft. lbs. (170 Nm).

11. Tighten the strut upper mount bolts to 21 ft. lbs. (28 Nm).

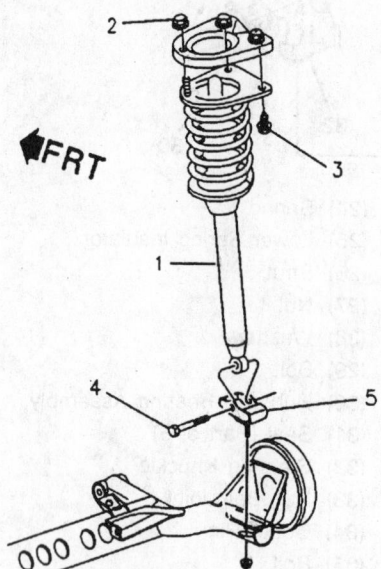

1	SHOCK, COIL—OVER
2	NUT
3	BOLT, UPPER STRUT (2)
4	BOLT, AXLE
5	ADAPTER

7922YG23

Exploded view of the rear strut mounting

12. Remove the safety stands and lower the vehicle.

13. Tighten the upper strut mounting nut to 15 ft. lbs. (20 Nm).

14. Road test the vehicle.

Coil Spring

REMOVAL & INSTALLATION

Front and Rear

1. Remove the strut assembly from the vehicle, as outlined earlier in this section.

2. Mount the strut compressor J-34013 in holding fixture J-3289-20.

3. Mount the strut assembly into the compressor. Note that the strut compressor has strut mounting holes drilled for specific vehicle lines.

4. Compress the strut approximately ½ its height after initial contact with the top cap.

✳✳ WARNING

Never bottom the spring or dampener rod!

5. Remove the nut from the strut dampener shaft and place alignment/guiding rod J-34013-27 on top of the dampener shaft. Use the rod to guide the dampener shaft straight down through the spring cap while compressing the spring. Remove the components.

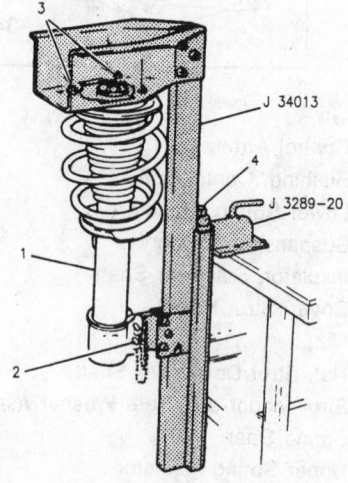

1	STRUT ASSEMBLY
2	INSTALL LOCKING PINS THROUGH STRUT ASSEMBLY
3	TIGHTEN NUTS UNTIL FLUSH WITH STRUT COMPRESSOR
4	COMPRESSOR FORCING SCREW

7922YG24

View of a typical strut assembly mounted in a compressor

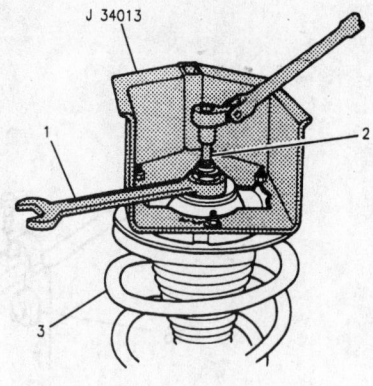

1	WRENCH
2	SOCKET
3	STRUT ASSEMBLY

7922YG25

Use a socket and a wrench to remove the dampener shaft nut spring cap while compressing the spring

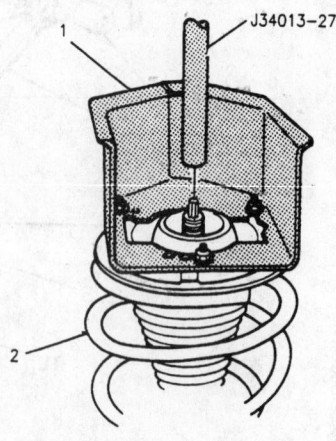

1	STRUT COMPRESSOR
2	STRUT ASSEMBLY

7922YG26

Install the rod to guide the dampener shaft straight down through the spring cap while compressing the spring

➡**Be careful to avoid chipping or cracking the spring coating when handling the front suspension coil spring assembly.**

To install:

6. Install the bearing cap into the strut compressor if previously removed.

7. Mount the strut assembly in strut compressor, using bottom locking pin only. Extend the dampener shaft and install clamp J-34013–20 on the dampener shaft.

8. Install the spring over the dampener and swing the assembly up so the upper locking pin can be installed.

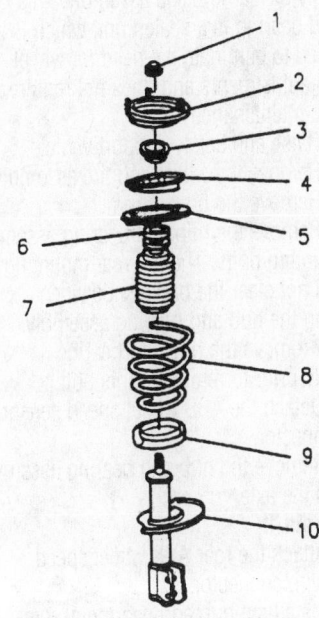

1 STRUT MOUNT NUT
2 STRUT MOUNT
3 RATE WASHER
4 SPRING SEAT
5 SPRING UPPER INSULATOR
6 JOUNCE BUMPER
7 STRUT DUST SHIELD
8 SPRING
9 SPRING LOWER INSULATOR
10 STRUT

7922YG27

Exploded view of the front strut assembly

9. Install all shields, bumpers and insulators on the spring seat. Install the spring seat on top of the spring. Be sure the flat on the upper spring seat is facing in the proper direction. The spring seat flat should be facing the same direction as the centerline of the strut assembly spindle.

10. Install the guiding rod and turn the forcing screw while the guiding rod centers the assembly. When the threads on the dampener shaft are visible, remove the guiding rod and install the nut. Tighten the nut to 52 ft. lbs. (70 Nm). Use a crow's foot line wrench while holding the dampener shaft with a socket.

11. Remove the clamp.

Lower Ball Joint

REMOVAL & INSTALLATION

1. Raise and safely support the vehicle.
2. If suspension contact hoist is used, place safety stands under the crossmember.

Lower the vehicle slightly so the weight of the vehicle rests on the crossmember.

3. Remove the tire and wheel assembly.

➡**Care must be exercised to prevent the axle shaft joints from being over-extended. When either end of the shaft is disconnected, over-extension of the joint could result in separation of internal components and possible joint failure. Failure to observe this can result in interior joint or boot damage and possible joint failure.**

4. Remove the cotter pin and nut from the ball joint.
5. Separate the ball joint from the steering knuckle using tool J-38892 or equivalent.
6. Drill out the 3 rivets retaining ball joint to the lower control arm. Use an 1/8 in. (3mm) drill bit to make a pilot hole through the rivets. Finish drilling rivets with a 1/2 in. (13mm) drill bit.
7. Remove the nut attaching link to the stabilizer shaft.
8. Remove the ball joint from the steering knuckle and control arm.

To install:

9. Install ball joint into the control arm.
10. Install 3 new bolts and nuts (supplied with new ball joint) and tighten.

11. Position ball joint stud through the steering knuckle and tighten nut to 50 ft. lbs. (65 Nm) and install cotter pin.
12. Connect stabilizer link to stabilizer shaft and tighten nut to 13 ft. lbs. (17 Nm).
13. Install tire and wheel assembly.
14. Lower the vehicle.
15. Check front wheel alignment and adjust, if necessary.

Wheel Bearings

ADJUSTMENT

These vehicles are equipped with sealed hub and bearing assemblies. The hub and bearing assemblies are non-serviceable. If the assembly is damaged, the complete unit must be replaced.

REMOVAL & INSTALLATION

Front

1. Raise and safely support the vehicle.
2. Remove the tire and wheel assembly.
3. Remove the drive axle nut.
4. Remove the brake caliper bolts and support the brake caliper to the side.
5. Remove the brake rotor.

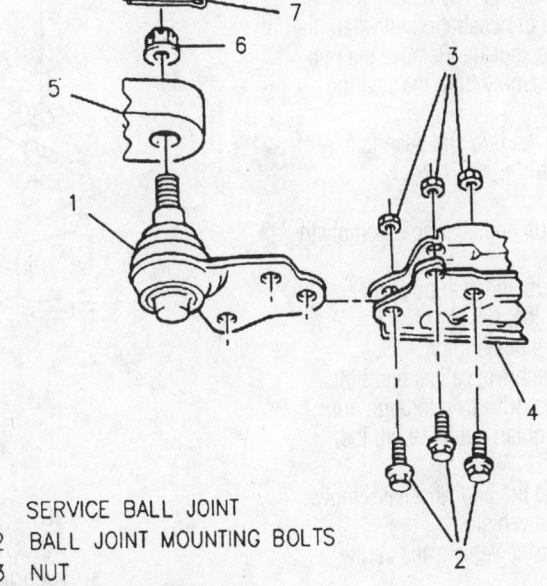

1 SERVICE BALL JOINT
2 BALL JOINT MOUNTING BOLTS
3 NUT
4 LOWER CONTROL ARM
5 STEERING KNUCKLE
6 NUT — 55 N·m (41 LBS. FT.) MINIMUM TORQUE
       65 N·m (48 LBS. FT.) MAXIMUM TORQUE
       TO INSTALL PIN
7 PIN

7922YG28

Exploded view of the ball joint mounting

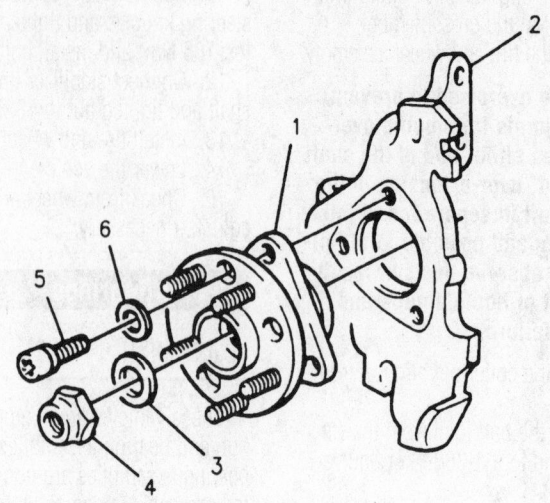

1 HUB AND BEARING ASSEMBLY
2 STEERING KNUCKLE
3 WASHER
4 DRIVE AXLE NUT – 260 N·m (192 LBS. FT.)
5 HUB AND BEARING RETAINING BOLT
6 WASHER

7922YG29

Exploded view of the front hub and bearing assembly

6. Remove the 3 bolts that go through the steering knuckle from the back of the knuckle. Rust buildup may require a generous application of penetrating oil where the hub fits into the knuckle. Remove the hub and bearing assembly from the steering knuckle.

7. Remove the hub and bearing assembly from the vehicle.

To install:

8. Install hub and bearing assembly to steering knuckle.

9. Install hub and bearing bolts and tighten to 70 ft. lbs. (95 Nm).

10. Install the brake rotor.

11. Install the brake caliper and bolts.

12. Install the drive axle through hub assembly and tighten nut to 192 ft. lbs. (260 Nm).

13. Install the tire and wheel assembly.

14. Lower the vehicle.

15. Road test the vehicle and verify proper operation.

Rear

A single-unit hub and bearing assembly is bolted to both ends of the rear axle

assembly or rear knuckle assembly. This hub and bearing is a sealed unit which is supposed to eliminate the need for wheel bearing adjustments and does not require periodic maintenance.

1. Raise and safely support vehicle.

2. Remove the wheel and tire assembly.

3. Remove the brake drum.

4. Remove the hub and bearing assembly mounting bolts. The top rear mounting bolt will not clear the brake shoe when removing the hub and bearing assembly. Partially remove the hub and bearing assembly prior to removing this bolt.

5. Detach the ABS wheel speed sensor wire connector.

6. Remove the hub and bearing assembly from the axle.

To install:

7. Attach the rear ABS wheel speed sensor wire connector.

8. Install the hub and bearing assembly. Position the top rear mounting bolt in the hub and bearing assembly prior to installation to the axle housing.

9. Tighten the mounting bolts to 44 ft. lbs. (60 Nm).

10. Install the brake drum.

11. Install the wheel and tire assembly.

12. Lower the vehicle.

13. Road test the vehicle.

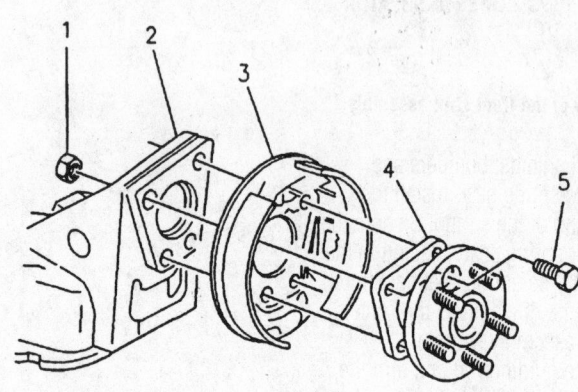

1 NUT
2 REAR AXLE ASSEMBLY
3 BACKING PLATE
4 HUB AND BEARING ASSEMBLY
5 BOLT

7922YG30

Exploded view of the rear hub and bearing assembly

GENERAL MOTORS L BODY **34**

Chevrolet-Beretta • Corsica

DRIVE TRAIN 34-19
ENGINE REPAIR 34-2
FUEL SYSTEM 34-17
STEERING AND
SUSPENSION 34-23

A

Air Bag 34-23
 Disarming 34-23
 Precautions 34-23
 Rearming 34-24

C

Camshaft and Valve Lifters 34-12
 Removal & Installation 34-12
Clutch 34-21
 Removal & Installation 34-21
Coil Spring 34-27
 Removal & Installation 34-27
Cylinder Head 34-4
 Removal & Installation 34-4

D

Distributor 34-2

E

Engine Assembly 34-2
 Removal & Installation 34-2
Exhaust Manifold 34-10
 Removal & Installation 34-10

F

Fuel Filter 34-18
 Removal & Installation 34-18

Fuel Pump 34-18
 Removal & Installation 34-18
Fuel System Pressure 34-17
 Relieving 34-17
Fuel System Service
 Precautions 34-17

H

Halfshaft 34-22
 Removal & Installation 34-22
Hydraulic Clutch System 34-22
 Bleeding 34-22

I

Ignition Timing 34-2
 Adjustment 34-2
Intake Manifold 34-8
 Removal & Installation 34-8

L

Lower Ball Joint 34-28
 Removal & Installation 34-28

O

Oil Pan 34-13
 Removal & Installation 34-13
Oil Pump 34-15
 Removal & Installation 34-15

P

Power Rack and Pinion Steering
 Gear 34-24
 Removal & Installation 34-24

R

Rear Main Seal 34-16
 Removal & Installation 34-16
Rocker Arms 34-7
 Removal & Installation 34-7

S

Shock Absorber 34-27
 Removal & Installation 34-27
Strut 34-26
 Removal & Installation 34-26

T

Transaxle Assembly 34-19
 Removal & Installation 34-19

V

Valve Lash 34-13
 Adjustment 34-13

W

Water Pump 34-4
 Removal & Installation 34-4
Wheel Bearings 34-29
 Adjustment 34-29
 Removal & Installation 34-29

ENGINE REPAIR

➡️Disconnecting the negative battery cable on some vehicles may interfere with the functioning of the on-board computer system, and may require the computer to undergo a relearning process once the negative battery cable is reconnected.

Distributor

The 2.2L and 3.1L engines utilize the Direct Ignition System (DIS). This system is called the Electronic Ignition (EI) system in later model years. This system is designed with a distributorless ignition.

Ignition Timing

ADJUSTMENT

All of the vehicles covered by this manual are equipped with distributorless ignition systems. Accordingly, ignition timing is controlled by the Engine/Powertrain Control Module (ECM/PCM) and is not adjustable.

Engine Assembly

REMOVAL & INSTALLATION

2.2L Engine

➡️The following procedure is for the engine and transaxle assembly.

1. Disconnect the battery.

✳✳ CAUTION

Never open, service or drain the radiator or cooling system when hot; serious burns can occur from the steam and hot coolant.

2. Drain the cooling system.
3. Relieve the fuel system pressure.
4. Disconnect the hood lamp wiring, if so equipped and remove the hood.
5. Disconnect the throttle body intake duct.
6. Remove the rear sight shields.
7. Disconnect the upper radiator hose.
8. Disconnect the brake booster vacuum hose.
9. Disconnect the alternator top brace and wiring.
10. Disconnect and tag the upper engine harness from the engine.

➡️If your vehicle is equipped with air conditioning, refer to Section 1 for information regarding the implications of servicing your A/C system yourself. Only a MVAC-trained, EPA-certified, automotive technician should service the A/C system or its components.

11. Discharge the A/C system.
12. Disconnect the A/C compressor-to-condenser and accumulator lines.
13. Raise and support the vehicle safely.

✳✳ CAUTION

The EPA warns that prolonged contact with used engine oil may cause a number of skin disorders, including cancer! You should make every effort to minimize your exposure to used engine oil. Protective gloves should be worn when changing the oil. Wash your hands and any other exposed skin areas as soon as possible after exposure to used engine oil. Soap and water, or waterless hand cleaner should be used.

14. Drain and recycle the engine oil.
15. Remove the left splash shield.
16. Disconnect the exhaust system.
17. Disconnect and tag the lower engine wiring.
18. Remove the flywheel inspection cover.
19. Remove the front wheels.
20. Disconnect the lower radiator hose.
21. Disconnect the heater hoses from the heater core.
22. Remove the brake calipers from the steering knuckle and wire up out of the way.
23. Disconnect the tie rods from the struts.
24. Carefully lower the vehicle.
25. Remove the clutch slave cylinder.
26. With the fuel system pressure released, place an absorbent shop towel around the connections and disconnect the fuel lines.
27. Disconnect the transaxle linkage at the transaxle.
28. Disconnect the accelerator cables from the TBI unit.
29. Disconnect the cruise control cables from the TBI unit.
30. Disconnect the throttle valve cables from the TBI, on vehicles equipped with an automatic transaxle.
31. Disconnect the automatic transaxle cooling lines.
32. Disconnect the power steering hoses from the power steering pump.

33. Remove the center suspension support bolts.
34. Align Engine/Transaxle Frame Handler tool No. J 36295 under the suspension supports, engine and transaxle; lower vehicle to dolly and add support under the engine.
35. Safely support the rear of the vehicle.
36. Disconnect the upper transaxle mount.
37. Remove the upper strut bolts and nuts.
38. Disconnect the front engine mount.
39. Disconnect the rear engine mount.
40. Remove the 4 rear suspension support bolts.
41. Remove the 4 front suspension support bolts and wire the bolt holes together to prevent axle separation.
42. Raise the vehicle and remove the engine and transaxle assembly on tool No. J 36295.

To install:

43. Carefully lower the vehicle and install the engine and transaxle assembly using tool No. J 36295.
44. Install the suspension supports bolts and tighten to 65 ft. lbs. (88 Nm) for the front and rear suspension supports and 66 ft. lbs. (89 Nm) for the center suspension support.
45. Install the transaxle mount but do not tighten.
46. Install the rear engine mount but do not tighten.
47. Install the front engine mount but do not tighten.
48. Tighten the manual transaxle mounting bolts as follows:
    a. Front transaxle strut-to-body bolts to 40 ft. lbs. (54 Nm).
    b. Rear transaxle mount-to-body bolts to 23 ft. lbs. (31 Nm).
49. Tighten the automatic transaxle mount bolts to 22 ft. lbs. (30 Nm).
50. Tighten the front and rear engine mount bolts.

➡️All engine mount bolts that have been removed must be cleaned and a new thread locking compound applied to the threads before reinstallation.

51. Install the power steering hoses.
52. Connect the accelerator, cruise control and TV cables to the TBI.
53. Connect the transaxle cooling lines to the automatic transaxle.
54. Connect the transaxle linkage.
55. Reconnect the fuel lines.
56. Reconnect the clutch slave cylinder.
57. Raise and support the vehicle safely.

58. Install the tie rods.
59. Install the calipers to the steering knuckle.
60. Attach the heater hoses to the heater core.
61. Connect the lower radiator hose.
62. Install the A/C compressor.
63. Install the flywheel inspection cover.
64. Install the engine splash shield.
65. Install the front wheel and tighten the wheel stud nuts to 100 ft. lbs. (136 Nm).
66. Lower the vehicle.
67. Install the upper engine wiring.
68. Install the compressor to condenser and accumulator lines.
69. Connect the brake booster vacuum hose.
70. Attach the upper radiator hose.
71. Raise and support the vehicle safely.
72. Install the lower engine wiring.
73. Reconnect the exhaust system.
74. Lower the vehicle.
75. Connect the TBI wiring.
76. Install the air cleaner assembly.
77. Close the radiator cock and refill the cooling system.

✳✳ WARNING

Operating the engine without the proper amount and type of engine oil will result in severe engine damage.

78. Fill the crankcase with oil.
79. Recharge the A/C system.
80. Check and adjust the wheel alignment.
81. Install the hood and connect the positive, then the negative battery cables.

3.1L Engine

1. Relieve the fuel pressure. Disconnect the negative, then the positive battery cables. Remove the battery from the vehicle.
2. Remove the air cleaner, the air inlet hose and the Mass Air Flow (MAF) sensor.

✳✳ CAUTION

Never open, service or drain the radiator or cooling system when hot; serious burns can occur from the steam and hot coolant.

3. Position a clean drain pan under the radiator, open the drain cock and drain the cooling system. Remove the exhaust manifold crossover assembly bolts and separate the assembly from the exhaust manifolds.
4. Remove the serpentine belt tensioner and the drive belt. Remove the power steering pump-to-bracket bolts and support the pump aside.
5. Disconnect the radiator hose from the engine.
6. Disconnect the TV and accelerator cables from the throttle valve bracket on the plenum.
7. Detach the electrical connectors. Remove the alternator-to-bracket bolts and the alternator. Label and disconnect the electrical wiring harness from the engine.
8. Disconnect and plug the fuel hoses. Remove the coolant overflow and bypass hoses from the engine.
9. From the charcoal canister, disconnect the purge hose. Label and disconnect all the necessary vacuum hoses.
10. Using a engine holding fixture tool, support the engine.
11. Raise and safely support the vehicle.

✳✳ CAUTION

The EPA warns that prolonged contact with used engine oil may cause a number of skin disorders, including cancer! You should make every effort to minimize your exposure to used engine oil. Protective gloves should be worn when changing the oil. Wash your hands and any other exposed skin areas as soon as possible after exposure to used engine oil. Soap and water, or waterless hand cleaner should be used.

12. Drain and recycle the engine oil.
13. Remove the right inner fender splash shield. Remove the crankshaft pulley-to-crankshaft bolt. Using a wheel puller, press the crankshaft pulley from the crankshaft.
14. Remove the flywheel cover. Label and disconnect the starter wires. Remove the starter-to-engine bolts and the starter.
15. Disconnect the wires from the oil pressure sending unit.
16. Remove the air conditioning compressor-to-bracket bolts and the bracket-to-engine bolts. Support the compressor so it will not interfere with the engine; do not disconnect the refrigerant lines.
17. Disconnect the exhaust pipe from the rear of the exhaust manifold.
18. If equipped with an automatic transaxle, remove the torque converter-to-flywheel bolts and push the converter into the transaxle.
19. Remove the front and rear engine mount bolts along with the mount brackets.
20. Remove the intermediate shaft bracket from the engine.
21. Disconnect the shifter cable from the transaxle.
22. Remove the lower engine-to-transaxle bolts and lower the vehicle.
23. Disconnect the heater hoses from the engine.
24. Using an vertical engine lift, install it to the engine and lift it slightly. Remove the engine holding fixture. Using a floor jack, support the transaxle.
25. Remove the upper engine-to-transaxle bolts. Remove the front engine mount bolts and transaxle mounting bracket.
26. Remove the engine from the vehicle.

To install:
27. Secure the engine on a engine suitable lifting device.
28. Carefully lower the engine into the vehicle, aligning it to the transaxle.
29. Install the upper engine-to-transaxle bolts. Tighten bolts to 55 ft. lbs. (75 Nm).
30. Install the transaxle mount bracket and front engine mount attaching bolts. Tighten the bolts to 65 ft. lbs. (88 Nm).
31. Using a floor jack, support the transaxle and remove the engine lifting device from the engine.
32. Install the lower engine-to-transaxle.
33. Connect the heater hoses to the engine.
34. Connect the shifter cable to the transaxle.
35. Install the intermediate shaft bracket to the engine.
36. Install the front and rear engine mount bolts along with the mount brackets.
37. Lower the jack and remove it from the transaxle.
38. Raise the vehicle and support it safely.
39. If equipped with an automatic transaxle, install the torque converter-to-flywheel bolts.
40. Install the flywheel cover and attaching bolts.
41. Connect the exhaust pipe to the exhaust manifold and install the attaching bolts.
42. Lower the vehicle.
43. Position the air conditioning compressor, with the lines attached, in place and install the compressor-to-bracket bolts.
44. Install the compressor bracket-to-engine bolts.
45. Connect the wires to the oil pressure sending unit.
46. Connect the starter wires. Position the starter in place and install the starter-to-engine bolts.

47. Install the crankshaft pulley and install the pulley-to-crankshaft bolt. Install the right inner fender splash shield.

48. Connect the purge hose to the charcoal canister. Connect all the necessary vacuum hoses.

49. Connect the coolant overflow and bypass hoses to the engine.

50. Connect the fuel delivery hoses to the engine.

51. Position the alternator in place and install the alternator-to-bracket bolts. Connect the electrical connectors to the alternator.

52. Connect all electrical wiring harnesses to the engine.

53. Connect the TV and accelerator cables to the throttle valve bracket on the plenum.

54. Connect the radiator hoses to the engine.

55. Install the serpentine belt tensioner and the drive belt.

56. Position the power steering pump in place and install the power steering pump-to-bracket bolts.

57. Connect the crossover pipe to the exhaust manifold and install the attaching bolts.

58. Install the air cleaner, air inlet hose and the mass air flow sensor.

59. Close the radiator cock and refill the cooling system.

❋❋ WARNING

Operating the engine without the proper amount and type of engine oil will result in severe engine damage.

60. Fill the crankcase with oil.

61. Install the battery and secure it in place. Connect the battery cables (the negative cable last).

62. Start the engine, allow it to reach normal operating temperatures and check for leaks.

Water Pump

REMOVAL & INSTALLATION

2.2L and 3.1L Engines

❋❋ CAUTION

Never open, service or drain the radiator or cooling system when hot; serious burns can occur from the steam and hot coolant.

1. Disconnect the negative battery cable. Drain the cooling system.

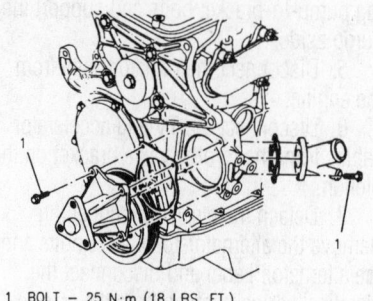

1 BOLT — 25 N·m (18 LBS. FT.)

79227120

Exploded view of the water pump mounting—2.2L engine

2. Loosen, but do not remove, the water pump pulley bolts.

3. Remove the serpentine belt.

4. Remove the alternator mounting bolts and set the alternator aside.

5. Remove the water pump pulley bolts and remove the water pump pulley.

6. Remove the 4 water pump mounting bolts and remove the water pump.

To install:

7. Clean all the gasket surfaces completely.

8. Apply a thin bead of sealer around the outer edge of the water pump gasket seating area and place he gasket on the pump.

9. Install the water pump on the engine and tighten the mounting bolts to 18 ft. lbs. (25 Nm) for the 2.2L engine or to 89 inch lbs. (10 Nm) on the 3.1L engine.

10. Install the remaining components in the reverse order of removal. Tighten the water pump pulley mounting bolts to 22 ft. lbs. (30 Nm) for the 2.2L engine or to 18 ft. lbs. (25 Nm) on the 3.1L engine.

11. Connect the negative battery cable. Refill the cooling system with the proper coolant. Bleed the cooling system and check for leaks with the engine running at idle.

Cylinder Head

REMOVAL & INSTALLATION

❋❋ CAUTION

The fuel injection system remains under pressure even after the engine has been turned OFF. The fuel system pressure must be relieved before disconnecting any fuel lines. Failure to do so may result in fire and/or personal injury.

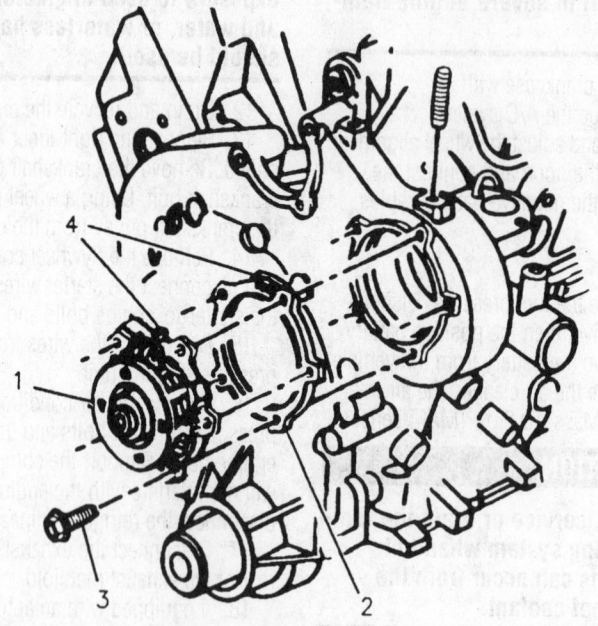

1	COOLANT PUMP
2	GASKET
3	BOLT — 10 N·m (89 LBS. IN.)
4	LOCATOR (MUST BE VERTICAL)

79227121

Exploded view of the water pump mounting—3.1L engine

➡When removing the valvetrain components, they must be kept in order for installation in the same locations from which they were removed.

2.2L Engine

✳✳ CAUTION

Never open, service or drain the radiator or cooling system when hot; serious burns can occur from the steam and hot coolant.

1. Relieve the fuel system pressure. Disconnect the negative battery cable. Drain the cooling system.
2. Remove the air inlet duct.
3. Tag and disconnect the vacuum lines at the intake manifold and cylinder head.
4. Tag and disconnect the following electrical connectors:
- Coolant Temperature Sensor
- Oxygen (O_2) Sensor
- Idle Air Control (IAC)
- Throttle Position Sensor (TPS)
- MAP sensor
- EVAP canister purge solenoid
- Fuel injector harness.

5. Disconnect the control cables from the throttle body and remove the cable bracket at the throttle body and rocker arm cover.
6. Remove the coolant reservoir, if necessary.
7. Install an engine support fixture, J-28467-a, or the equivalent.
8. Remove the right engine mount, if necessary.
9. Remove the serpentine belt.
10. Remove the alternator and alternator mounting bracket.
11. Disconnect and cap the power steering lines from the power steering pump and remove the power steering pump.
12. Remove the serpentine belt tensioner assembly.
13. Remove the right engine mount bracket.
14. Disconnect the spark plug wires and lay them aside.

15. Disconnect the EVAP canister purge line under the intake manifold.
16. Disconnect the upper radiator hose from the engine.
17. Disconnect the coolant inlet pipe brackets at the exhaust·manifold and transaxle.
18. Remove the intake manifold brace from the power steering bracket.
19. Disconnect and cap the fuel lines at the quick disconnects.
20. Remove the rocker arm cover, rocker arm nuts, rocker arms and pushrods.
21. Remove the ignition wire bracket.
22. Remove the engine lift bracket.
23. Raise and safely support the vehicle. Disconnect the exhaust pipe from the exhaust manifold.
24. Lower the vehicle.
25. Remove the transaxle fill tube.
26. Remove the cylinder head bolts.
27. Remove the cylinder head with both manifolds. Remove the intake and exhaust manifolds from the cylinder head.

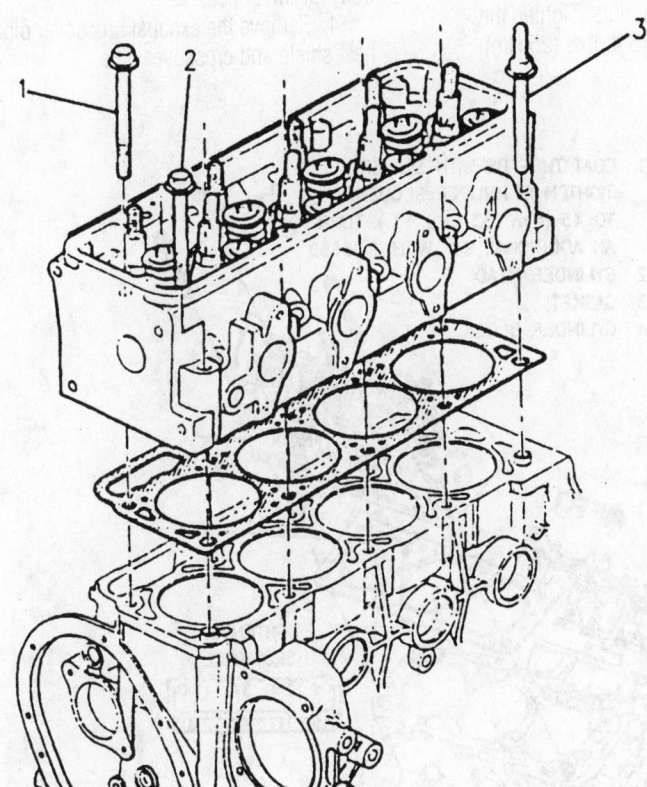

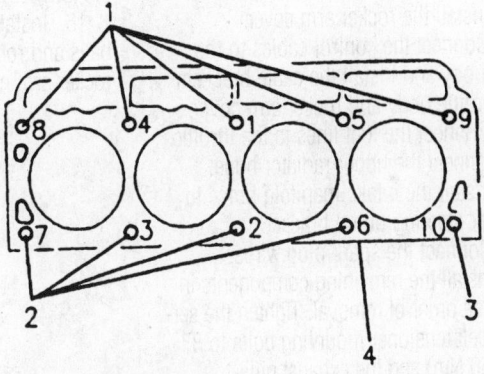

1 LONG BOLTS
2 SHORT BOLTS
3 STUD
4 NUMBERS ON GASKET INDICATE TORQUE SEQUENCE

CYLINDER HEAD BOLT TORQUE PROCEDURE

1 TIGHTEN BOLTS IN SEQUENCE (ITEM 4) TO:
LONG BOLTS: 63 N·m (46 LBS. FT.)
SHORT BOLTS: 58 N·m (43 LBS. FT.)
2 TIGHTEN ALL BOLTS AN ADDITIONAL ANGLE OF 90° IN SEQUENCE (ITEM 4) USING J 36660 OR EQUIVALENT

7922Z101

Cylinder head bolt tightening sequence—2.2L engine

To install:

28. Clean all the gasket surfaces completely. Clean the threads on the cylinder head bolts and the block threads.

29. Install the intake and exhaust manifolds on the cylinder head.

30. Place a new cylinder head gasket in position over the dowel pins on the block. Carefully guide the cylinder head into position.

31. Install all the cylinder head bolts finger-tight. The long bolts go in bolt positions 1, 4, 5, 8 and 9. The short bolt are in positions 2, 3, 6 and 7. The stud is in position 10.

32. Tighten the bolts in sequence. The long bolts to 23 ft. lbs. (32 Nm) and the short bolts and stud to 22 ft. lbs. (29 Nm). Make second pass tightening the long bolts to 46 ft. lbs. (63 Nm) and the short bolts to 43 ft. lbs. (58 Nm). Make a final pass over all bolts tightening each an additional 90 degrees.

33. Install the engine lift bracket.

34. Install the ignition wire bracket.

35. Install the transaxle fill tube.

36. Install the pushrods, rocker arms and rocker arm nuts and tighten the nuts to 22 ft. lbs. (30 Nm).

37. Install the rocker arm cover.

38. Connect the control cables to the throttle body and install the cable brackets at the throttle body and rocker arm cover.

39. Connect the fuel lines to the throttle body. Connect the upper radiator hose.

40. Install the intake manifold brace to the power steering pump bracket.

41. Connect the spark plug wires.

42. Install the remaining components in the reverse order of removal. Tighten the serpentine belt tensioner mounting bolts to 37 ft. lbs. (50 Nm) and the exhaust pipe-to-exhaust manifold bolts to 22 ft. lbs. (30 Nm).

43. Refill the cooling system.

44. Connect the negative battery cable. Start the vehicle and verify no leaks.

3.1L Engine

LEFT (FRONT) CYLINDER HEAD

1. Relieve the fuel system pressure. Disconnect the negative battery cable.

✷✷ CAUTION

Never open, service or drain the radiator or cooling system when hot; serious burns can occur from the steam and hot coolant.

2. Drain the cooling system.

3. Remove the top half of the air cleaner assembly and remove the throttle body air inlet duct.

4. Remove the exhaust crossover pipe heat shield and crossover pipe.

5. Disconnect the spark plug wires from spark plugs and wire looms and route the wires aside.

6. Remove the rocker arm covers.

7. Remove upper intake plenum and lower intake manifold.

8. Remove the left side exhaust manifold.

9. Remove oil level indicator tube.

10. Remove rocker arms nut, rocker arms, balls and pushrods.

11. Remove the cylinder head bolts evenly. Remove the cylinder head.

To install:

12. Clean all the gasket surfaces completely. Clean the threads on the cylinder head bolts and block threads.

13. Place the cylinder head gasket in position over the dowel pins on the cylinder block so the words **THIS SIDE UP** showing.

14. Coat the bolt threads with sealer and install finger-tight.

15. Tighten the cylinder head bolts in sequence to 33 ft. lbs. (45 Nm). With all the bolts tightened make a second pass tightening all the bolts an additional 90 degrees (¼ turn).

16. Install the pushrods, rocker arms, balls and rocker arm nuts. Tighten the rocker arm nuts to 18 ft. lbs. (25 Nm).

17. Install the lower intake manifold and upper intake plenum.

18. Install the rocker arm covers.

19. Install the oil level indicator tube.

20. Connect the spark plug wires to spark plugs and wire looms.

21. Install the left side exhaust manifold.

22. Install the exhaust crossover pipe and crossover pipe heat shield.

23. Refill the cooling system.

24. Install the top half of the air cleaner assembly and the throttle body air inlet duct.

25. Connect negative battery cable.

26. Start vehicle and verify no leaks.

RIGHT (REAR) CYLINDER HEAD

1. Properly relieve the fuel system pressure. Disconnect the negative battery cable.

✷✷ CAUTION

Never open, service or drain the radiator or cooling system when hot; serious burns can occur from the steam and hot coolant.

2. Drain the cooling system.

3. Remove the top half of the air cleaner assembly and remove the throttle body air inlet duct.

4. Remove the exhaust crossover pipe heat shield and crossover pipe.

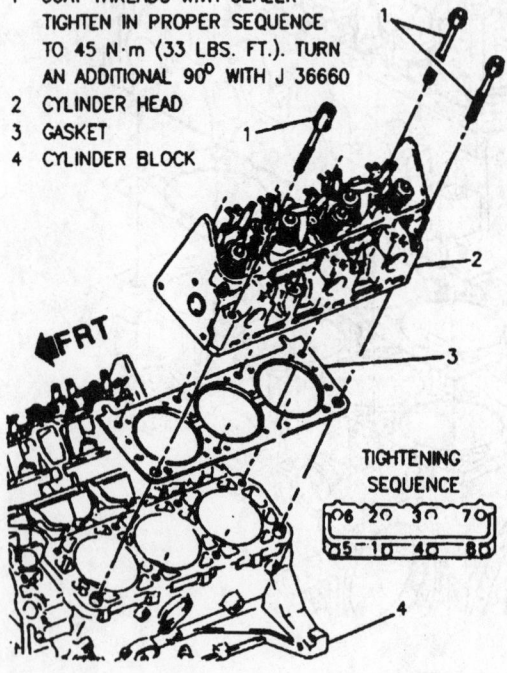

1 COAT THREADS WITH SEALER
  TIGHTEN IN PROPER SEQUENCE
  TO 45 N·m (33 LBS. FT.). TURN
  AN ADDITIONAL 90° WITH J 36660
2 CYLINDER HEAD
3 GASKET
4 CYLINDER BLOCK

TIGHTENING SEQUENCE

79222Z102

Exploded view of the cylinder head mounting, showing the bolt tightening sequence—3.1L engine

5. Raise and safely support the vehicle.

6. Disconnect the Oxygen (O₂) sensor connector.

7. Disconnect the exhaust pipe from the exhaust manifold.

8. Remove the right side exhaust manifold. Lower the vehicle.

9. Disconnect the spark plug wires from spark plugs and wire looms and route the wires aside.

10. Remove the rocker arm covers.

11. Remove upper intake plenum and lower intake manifold.

12. Remove rocker arms nut, rocker arms, balls and pushrods.

13. Remove the cylinder head bolts evenly.

14. Remove the cylinder head.

To install:

15. Clean all the gasket surfaces completely. Clean the threads on the cylinder head bolts and block threads.

16. Place the cylinder head gasket in position over the dowel pins on the cylinder block so the words **THIS SIDE UP** showing.

17. Coat the bolt threads with sealer and install finger-tight.

18. Tighten the cylinder head bolts in sequence to 33 ft. lbs. (45 Nm). With all the bolts tightened make a second pass tightening all the bolts an additional 90 degrees (¼ turn).

19. Install the pushrods, rocker arms, balls and rocker arm nuts. Tighten the rocker arm nuts to 18 ft. lbs. (25 Nm).

20. Install the lower intake manifold and upper intake plenum.

21. Install the rocker arm covers.

22. Connect the spark plug wires to spark plugs and wire looms.

23. Raise vehicle and safely support.

24. Install the exhaust manifold.

25. Connect the exhaust pipe to the exhaust manifold.

26. Lower the vehicle.

27. Connect the Oxygen (O₂) sensor connector.

28. Install the exhaust crossover pipe and heat shield.

29. Refill the cooling system.

30. Install the top half of the air cleaner assembly and the throttle body air inlet duct.

31. Connect negative battery cable.

32. Start vehicle and verify no leaks.

Rocker Arms

REMOVAL & INSTALLATION

2.2L Engine

This engine uses a simple ball pivot-type rocker arm. Motion is transmitted from the camshaft through the hydraulic roller lifters and pushrod to the rocker arm. The rocker arm pivots on its ball and transmits the camshaft motion to the valve. The rocker arm ball is located on a stud threaded onto the head and is retained by a nut. The pushrod is located by a guide plate held under the rocker arm stud.

1. Disconnect the negative battery cable. Disconnect the air intake hose from the air cleaner and throttle body.

2. Remove the control cable bracket from the rocker arm cover.

3. Disconnect the PCV hose.

4. Remove the rocker arm cover bolts, then remove the cover.

➡Place components in a rack in order at disassembly so they can be reinstalled at the same location and with the same mating surface as when removed.

5. Remove the rocker arm nut(s), ball(s) and rocker arms. Pushrods may be removed by simply pulling out of the block.

To install:

6. Clean the gasket surfaces completely.

7. Coat the bearing surfaces of the rocker arm(s) and the rocker arm ball(s) with Molykote®, engine assembly lube or equivalent.

8. Seat the pushrods in the lifters.

9. Install the rocker arm(s), ball(s) and nut(s) in the same positions they were removed from and tighten the rocker arm nut(s) to 22 ft. lbs. (30 Nm). Valve lash adjustment is not required.

10. Install a new gasket in the cut out of the rocker arm cover.

11. Install the rocker arm cover on the cylinder head and tighten the rocker arm cover bolts to 89 inch lbs. (10 Nm).

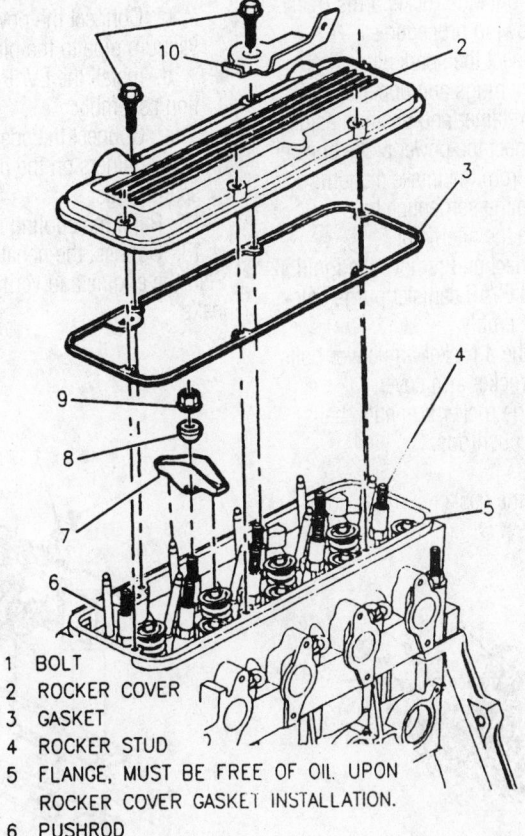

1 BOLT
2 ROCKER COVER
3 GASKET
4 ROCKER STUD
5 FLANGE, MUST BE FREE OF OIL UPON ROCKER COVER GASKET INSTALLATION.
6 PUSHROD
7 ROCKER ARM
8 BALL
9 NUT
10 SUPPORT, SPARK PLUG WIRE

79222122

Exploded view of the valve cover and rocker arm mounting—2.2L engine

12. Install the remaining components in the reverse order of removal.

13. Connect the negative battery cable. Start the engine and verify no oil leaks.

3.1L Engine

1. Disconnect the negative battery cable.

2. For the left side rocker arms (front), perform the following procedures:

✲✲ CAUTION

Never open, service or drain the radiator or cooling system when hot; serious burns can occur from the steam and hot coolant.

a. Drain the cooling system to a level below the coolant pipe on the front of the engine.

b. Remove the coolant bypass hose clamp at the coolant tube.

c. Remove the 2 bolts and nut securing the coolant tube to the cylinder head and position the tube aside.

d. Disconnect the PCV valve from the rocker arm cover.

3. For the right side rocker arms (rear), perform the following procedures:

a. Disconnect the spark plug wires from the spark plugs and upper intake plenum wire retainer and position aside.

b. Disconnect the power brake booster vacuum pipe from the intake plenum.

c. Remove the serpentine belt.

d. Remove the alternator.

e. Disconnect and remove the ignition assembly and EVAP canister purge solenoid as an assembly.

4. Remove the 4 rocker arm cover bolts and remove the rocker arm cover.

5. Remove the rocker arm nuts, balls, rocker arms and pushrods.

To install:

6. Clean all the gasket surfaces completely.

7. Coat all the valvetrain components with engine oil prior to installation.

8. Install the pushrods and install the rocker arms on the studs. Install the rocker arm balls and mounting nuts. Be sure the pushrods are properly seated in the lifter and rocker arm. Tighten the mounting nuts to 18 ft. lbs. (24 Nm).

9. Install the rocker arm cover using a new gasket and bolt grommet, tighten the rocker cover bolts to 90 inch lbs. (10 Nm).

10. For the left side rocker arms (front), perform the following procedures:

a. Connect the PCV valve to the rocker arm cover.

b. Position the coolant tube and connect the thermostat bypass hose.

c. Install the coolant tube mounting nut and bolts. Tighten the screw at the water pump to 106 inch lbs. (12 Nm), the bolt at the corner of the cylinder head to 18 ft. lbs. (25 Nm) and the nut to 18 ft. lbs. (25 Nm).

11. For the right side rocker arms (rear), perform the following procedures:

a. Install the alternator.

b. Install the serpentine belt.

c. Connect the power brake booster vacuum pipe to the plenum.

d. Install the EVAP solenoid and ignition assembly.

e. Connect the spark plug wires to the wire retainers on the plenum and the spark plugs.

12. Refill the cooling system.

13. Connect the negative battery cable. Start the engine and verify that there are no leaks.

Intake Manifold

REMOVAL & INSTALLATION

2.2L Engine

1. Relieve the fuel system pressure.

2. Disconnect the negative battery cable.

3. Remove the air intake duct.

✲✲ CAUTION

Never open, service or drain the radiator or cooling system when hot; serious burns can occur from the steam and hot coolant.

4. Position a suitable drain pan under the radiator, then drain the cooling system.

5. Tag and disconnect all necessary vacuum lines and wires.

6. Disconnect the throttle linkage.

7. Remove the serpentine drive belt.

8. Unbolt the power steering pump, then position it aside without disconnecting the fluid lines.

9. Remove the transaxle fluid level indicator and fill tube.

10. Detach the electrical connectors from the following components:

a. MAP sensor

b. EGR solenoid valve

c. Idle Air Control (IAC) valve

d. Throttle Position (TP) sensor

e. Fuel injector wiring harness

11. Remove the MAP sensor and EGR solenoid valve.

12. Remove the upper intake manifold assembly.

13. Remove the EGR valve injector.

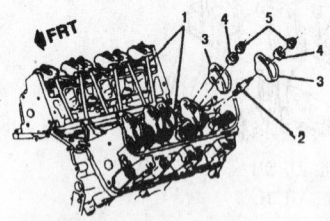

1 Pushrod
2 Valve rocker arm stud
3 Valve rocker arm
4 Valve rocker arm pivot ball
5 Valve rocker arm nut

7922Z103

Rocker arm component identification— 3.1L engine

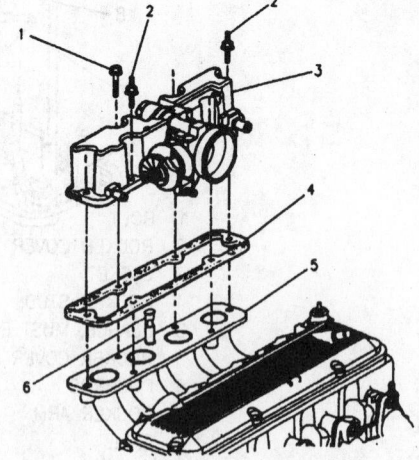

Exploded view of the upper intake manifold mounting, showing the bolt tightening sequence— 2.2L engines

1 BOLT
2 STUD
3 UPPER INTAKE MANIFOLD ASSEMBLY
4 GASKET
5 LOWER INTAKE MANIFOLD
6 EGR VALVE INJECTOR

7922Z104

14. Remove the fuel injector retainer bracket, regulator and injectors.

15. Disconnect the accelerator and TV cables, then remove the cable bracket.

16. Remove the EGR valve pipe.

17. Remove the lower intake manifold-to-cylinder head retaining nuts and studs, then remove the lower intake manifold from the vehicle.

18. Remove and discard the gasket. Use a suitable gasket to carefully and thoroughly clean the mating surfaces on the manifold and cylinder head.

To install:

19. Position a new intake manifold gasket on the cylinder, then place the lower intake manifold over the new gasket. Install the retaining nuts and studs. Tighten the studs to 89 inch lbs. (10 Nm) and the nuts, in sequence, to 24 ft. lbs. (33 Nm).

20. Install the EGR valve pipe.

21. Connect the fuel lines.

22. Install the fuel injectors, regulator and injector retainer bracket and tighten the retaining bolts to 22 inch lbs. (3.5 Nm).

23. Install the cable bracket, then connect the accelerator and TV cables.

24. Install the EGR valve injector so that the port is facing directly towards the throttle body.

25. Install the upper intake manifold assembly. Tighten the upper intake manifold nuts in the proper sequence to 22 ft. lbs. (30 Nm).

26. Install the MAP sensor and EGR solenoid valve.

27. Install the serpentine drive belt.

28. Install the accelerator cable bracket and throttle linkage. Tighten the cable bracket bolts to 18 ft. lbs. (25 Nm).

29. Connect the vacuum lines and wires as tagged during removal.

30. Attach the air intake duct.

31. Position the power steering pump and secure with retaining bolts.

32. Install the transaxle fluid level indicator and fill tube.

33. Connect the negative battery cable.

34. Fill the engine cooling system with the proper type and amount of coolant.

3.1L Engine

These vehicles use a two-piece intake manifold. Note that these pieces are cast aluminum. Use care when working with light alloy components.

1. Disconnect the negative battery cable. Relieve the fuel system pressure.

2. Remove the top half of the air cleaner assembly and throttle body duct.

❋❋ CAUTION

Never open, service or drain the radiator or cooling system when hot; serious burns can occur from the steam and hot coolant.

3. Drain the cooling system.

4. Remove the EGR pipe from exhaust manifold.

5. Remove the serpentine belt.

6. Remove the brake vacuum pipe at the intake plenum.

7. Disconnect the control cables from the throttle body and intake plenum mounting bracket.

8. Remove the power steering lines at the alternator bracket.

9. Remove the alternator.

10. Disconnect the spark plug wires from the spark plugs and wire retainers on the intake plenum.

11. Remove the ignition assembly and the EVAP canister purge solenoid together.

12. Detach the connectors from the following components:
   - Throttle Position Sensor (TPS)
   - Idle Air Control (IAC)
   - Fuel Injectors
   - Coolant temperature sensor
   - MAP sensor
   - Camshaft Position (CMP) sensor

13. Detach the vacuum lines from the vacuum modulator, fuel pressure regulator, and the PCV valve.

14. Remove the MAP sensor from upper intake manifold.

15. Remove the upper intake plenum mounting bolts and remove the plenum.

16. Disconnect the fuel lines from the fuel rail and fuel line bracket.

17. Install engine support fixture special tool J-28467-A or an equivalent.

18. Remove the right side engine mount.

19. Remove the power steering mounting bolts and support the pump aside without disconnecting the power steering lines.

20. Disconnect the coolant inlet pipe from coolant outlet housing.

21. Remove the coolant bypass hose from the water pump and the cylinder head.

22. Disconnect the upper radiator hose at thermostat housing.

23. Remove the thermostat housing.

24. Remove both rocker arm covers.

25. Remove the lower intake manifold bolts. Be sure the washers on the 4 center bolts are installed in their original locations.

➡**When removing the valvetrain components they should be kept in order for installation the original locations.**

26. Remove the rocker arm retaining nuts and remove the rocker arms and pushrods.

27. Remove the intake manifold from the engine.

To install:

28. Clean gasket material from all mat-

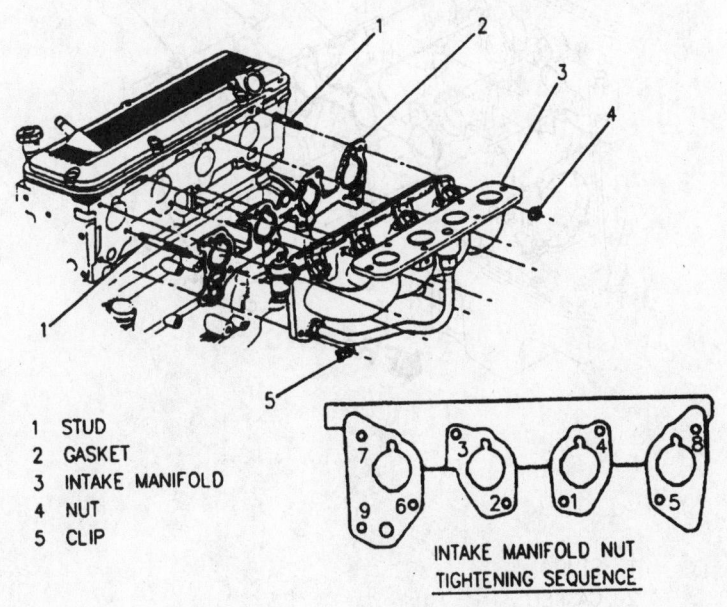

1 STUD
2 GASKET
3 INTAKE MANIFOLD
4 NUT
5 CLIP

INTAKE MANIFOLD NUT
TIGHTENING SEQUENCE

7922Z105

Exploded view of the lower intake manifold mounting, showing the bolt tightening sequence—2.2L engines

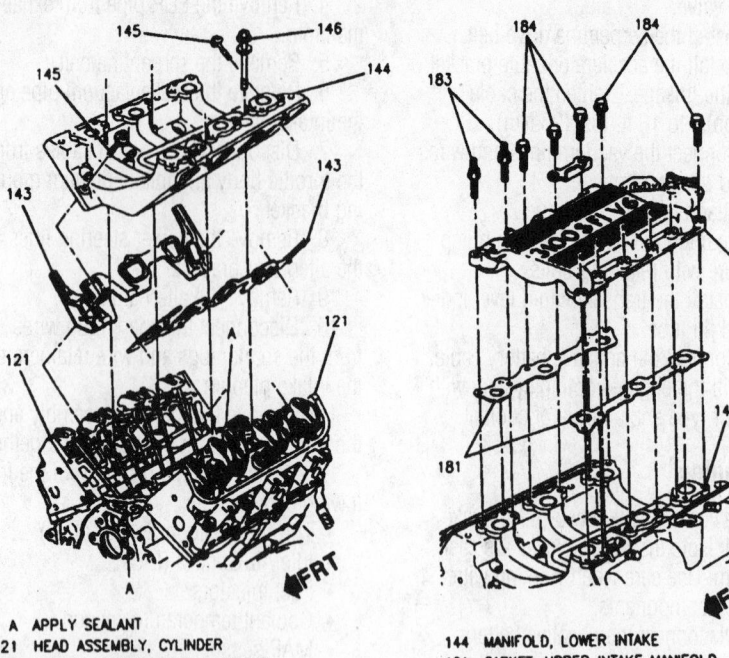

A APPLY SEALANT
121 HEAD ASSEMBLY, CYLINDER
143 GASKET, LOWER INTAKE MANIFOLD
144 BOLT, LOWER INTAKE
145 BOLT, LOWER INTAKE MANIFOLD
146 BOLT, LOWER INTAKE MANIFOLD

144 MANIFOLD, LOWER INTAKE
181 GASKET, UPPER INTAKE MANIFOLD
182 MANIFOLD, UPPER INTAKE
183 STUD, UPPER INTAKE MANIFOLD
184 BOLT, UPPER INTAKE MANIFOLD

79222Z106

Exploded view of the upper and lower intake manifolds—3.1L engine

ing surfaces. Remove all excess RTV sealant from front and rear ridges of cylinder block.

29. Place a 3mm bead of RTV, on each ridge, where the front and rear of the intake manifold contact the block.

30. Using a new gasket, install the intake manifold to the engine.

31. Install the pushrods, rocker arms and mounting nuts. Be sure the pushrods are properly seated in the valve lifters and rocker arms.

32. Install rocker arm nuts and tighten the rocker arm nuts to 18 ft. lbs. (24 Nm).

33. Install lower the intake manifold attaching bolts. Apply sealant PN 12345739 or equivalent to the threads of bolts, and torque bolts to 115 inch lbs. (13 Nm).

34. Install the front rocker arm cover.

35. Install the thermostat housing.

36. Connect the upper radiator hose to the thermostat housing.

37. Install the coolant inlet pipe to thermostat housing.

38. Install coolant bypass pipe at the water pump and cylinder head.

39. Install the power steering pump in the mounting bracket.

40. Connect the right side engine mount. Remove the special engine support tool.

41. Connect the fuel lines to fuel rail and bracket.

42. Install the upper intake manifold and torque the mounting bolts to 18 ft. lbs. (25 Nm).

43. Install the remaining components in the reverse order of removal.

44. Fill the cooling system. An engine oil and filter change is recommended.

45. Connect the negative battery cable. Start the vehicle and verify no leaks.

Exhaust Manifold

REMOVAL & INSTALLATION

2.2L Engine

1. Disconnect the negative battery cable. Detach the oxygen sensor connector.

2. Remove the serpentine drive belt. Remove the alternator.

3. Raise and support the vehicle safely. Remove the retainers and separate the exhaust pipe from the manifold.

4. Lower the vehicle.

5. Remove the bolt securing the oil fill tube to the engine block and pull the tube from the engine.

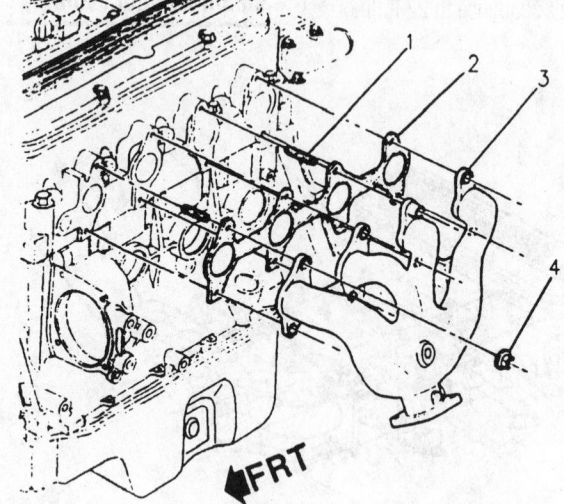

1 STUD
2 GASKET
3 EXHAUST MANIFOLD
4 NUT

79222Z123

Exploded view of the exhaust manifold mounting—2.2L engine

6. If the coolant inlet pipe interferes with removal, disconnect the mounting bracket from the transaxle stud.

7. Remove the manifold retaining nuts, then remove the manifold from the cylinder head. Remove and discard the gasket from the mating surfaces.

To install:

8. Install the manifold to the cylinder head using a new gasket. Tighten the retaining nuts to 115 inch lbs. (13 Nm).

9. Raise and support the vehicle safely, then install the exhaust pipe to the manifold.

Tighten the retaining nuts to 18 ft. lbs. (25 Nm), then lower the vehicle.

10. Install the alternator and the serpentine drive belt.

11. Connect the coolant inlet pipe, if disconnected.

12. Lubricate the oil fill tube seal and install the oil fill tube.

13. Connect the wiring harness to the oxygen sensor.

14. Connect the negative battery cable. Start the engine and check for exhaust leaks.

3.1L Engine

LEFT (FRONT) SIDE MANIFOLD

1. Disconnect the negative battery cable.
2. Remove the top half of the air cleaner assembly and throttle cable duct.

✷✷ CAUTION

Never open, service or drain the radiator or cooling system when hot; serious burns can occur from the steam and hot coolant.

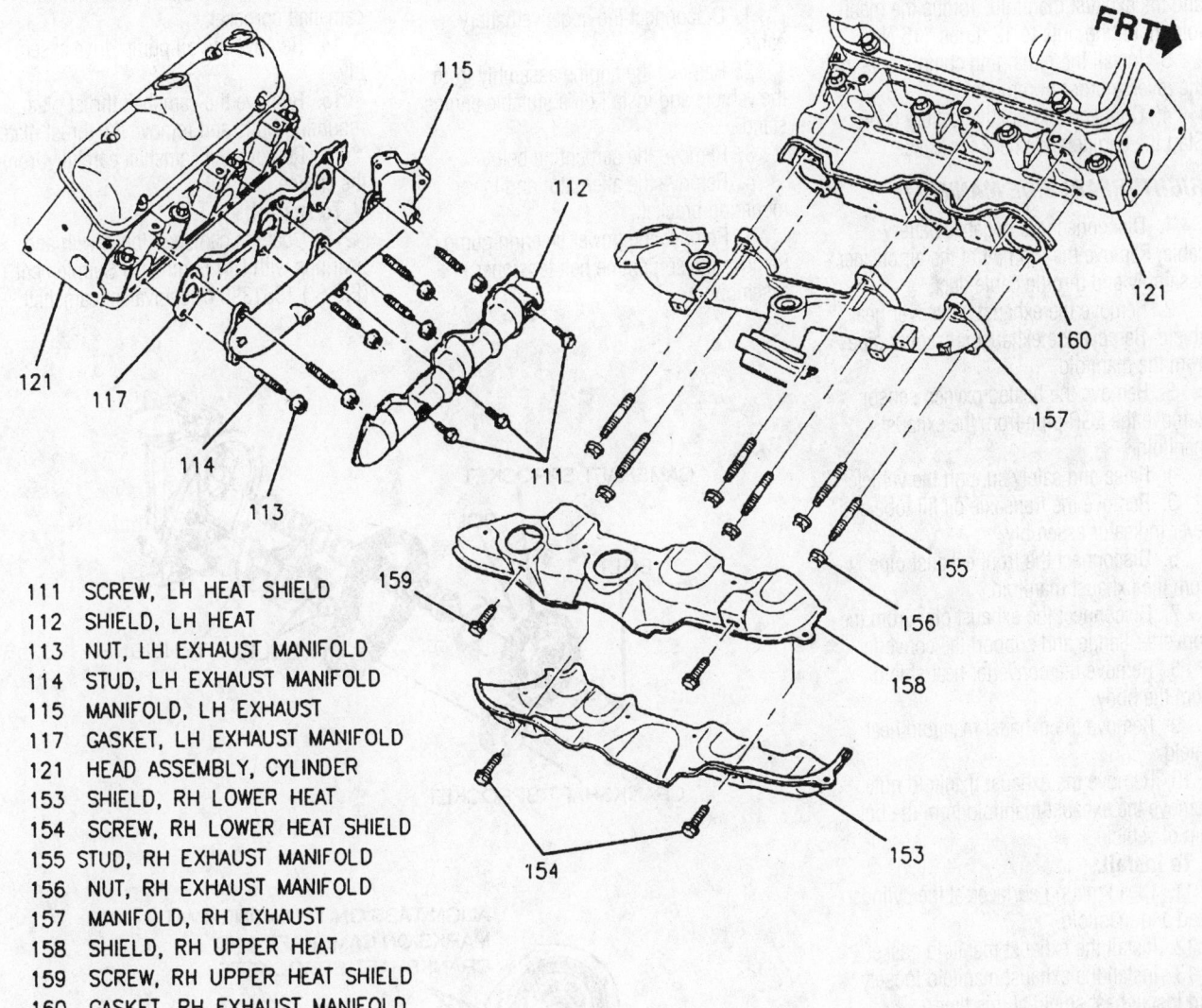

111	SCREW, LH HEAT SHIELD
112	SHIELD, LH HEAT
113	NUT, LH EXHAUST MANIFOLD
114	STUD, LH EXHAUST MANIFOLD
115	MANIFOLD, LH EXHAUST
117	GASKET, LH EXHAUST MANIFOLD
121	HEAD ASSEMBLY, CYLINDER
153	SHIELD, RH LOWER HEAT
154	SCREW, RH LOWER HEAT SHIELD
155	STUD, RH EXHAUST MANIFOLD
156	NUT, RH EXHAUST MANIFOLD
157	MANIFOLD, RH EXHAUST
158	SHIELD, RH UPPER HEAT
159	SCREW, RH UPPER HEAT SHIELD
160	GASKET, RH EXHAUST MANIFOLD

Exploded view of the exhaust manifold mounting—3.1L engine

79222124

3. Partially drain the cooling system. Remove the radiator hose from the thermostat housing.

4. Remove the coolant bypass hose at the coolant pump and from the exhaust manifold.

5. Remove the exhaust crossover heat shield. Remove the exhaust crossover pipe from the manifold.

6. Tag and disconnect the spark plug wires.

7. Remove the exhaust manifold heat shield. Remove the exhaust manifold retaining nuts.

8. Remove the exhaust manifold.

To install:

1. Clean mating surfaces at the cylinder head and manifold.

2. Install the exhaust manifold gasket and the exhaust manifold. Torque the manifold mounting nuts to 12 ft. lbs. (16 Nm).

3. Install the remaining components in the reverse order of removal.

4. Connect the negative battery cable. Start the engine and check for leaks.

RIGHT (REAR) SIDE MANIFOLD

1. Disconnect the negative battery cable. Remove the top half of the air cleaner assembly and throttle cable duct.

2. Remove the exhaust crossover heat shield. Remove the exhaust crossover pipe from the manifold.

3. Remove the heated oxygen sensor. Remove the EGR pipe from the exhaust manifold.

4. Raise and safely support the vehicle.

5. Remove the transaxle oil fill tube and lever indicator assembly.

6. Disconnect the front exhaust pipe from the exhaust manifold.

7. Disconnect the exhaust pipe from the converter flange and support the converter.

8. Remove the converter heat shield from the body.

9. Remove the exhaust manifold heat shield.

10. Remove the exhaust manifold nuts. Remove the exhaust manifold from the bottom of vehicle.

To install:

11. Clean mating surfaces at the cylinder head and manifold.

12. Install the exhaust manifold gasket.

13. Install the exhaust manifold loosely and install heat shield at this time.

14. Install the manifold nuts and torque to 12 ft. lbs. (16 Nm).

15. Install the remaining components in the reverse order of removal.

16. Connect the negative battery cable, then start the engine and check for leaks.

Camshaft and Valve Lifter

REMOVAL & INSTALLATION

✽✽ CAUTION

The fuel system is under pressure and must be properly relieved before any service procedures are performed. Failure to properly relieve the system pressure can lead to personal injury and component damage.

2.2L Engine

Please note that the engine must be removed from the vehicle for camshaft service.

1. Disconnect the negative battery cable.

2. Remove the engine assembly from the vehicle and install on a suitable engine stand.

3. Remove the serpentine belt.

4. Remove the alternator and lower mounting bracket.

5. Remove the power steering pump. Remove the serpentine belt tensioner assembly.

6. Remove the water pump pulley.

7. Remove the engine oil filter.

8. Remove the crankshaft pulley from the crankshaft hub.

9. Remove the rocker arm cover.

➡ When removing the valvetrain components they must be kept in order for installation in the same locations they were removed from. In addition, a new camshaft requires that new lifters be installed.

10. Remove the rocker arm nuts, rocker arms, balls and pushrods.

11. Remove the cylinder head with the intake and exhaust manifolds attached.

12. Remove the valve lifters.

13. Remove the timing chain front cover. Remove the timing chain, tensioner and camshaft sprocket.

14. Remove the oil pump drive assembly.

15. Remove the camshaft thrust plate mounting bolts and remove the thrust plate.

16. Remove the camshaft carefully from the engine.

To install:

17. Coat the camshaft lobes with and bearings with GM Engine Oil Supplement (E.O.S.) 1051396 or equivalent camshaft

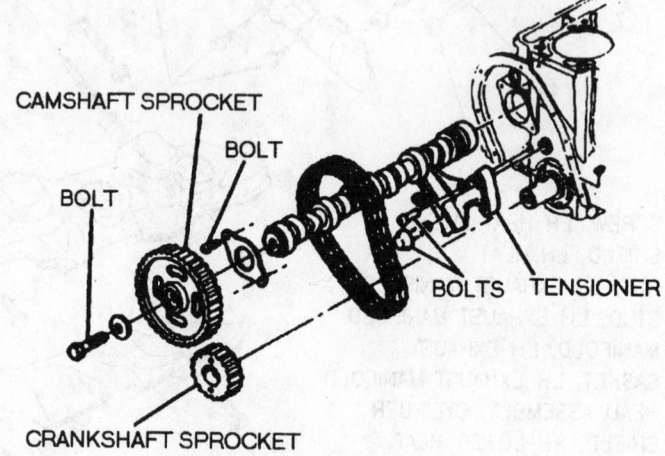

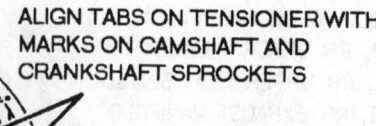

ALIGN TABS ON TENSIONER WITH MARKS ON CAMSHAFT AND CRANKSHAFT SPROCKETS

Camshaft and timing chain components—2.2L engine

79222107

lubricant and insert the camshaft carefully into the engine.

18. Install the thrust plate and tighten the mounting bolts to 106 inch lbs. (12 Nm).
19. Install the timing chain, tensioner and camshaft sprocket.
20. Install the timing chain front cover.

➡**If the camshaft was replaced with a new one, the valve lifters MUST also be replaced. The lifters have already developed a wear pattern from the old camshaft and installing them on a new camshaft can cause the new camshaft to wear prematurely.**

21. Install the valve lifters.
22. Install the cylinder head and manifold assemblies.
23. Install the oil pump drive assembly.
24. Install the pushrods, rocker arms, balls and rocker arm nuts. Tighten the nuts to 22 ft. lbs. (30 Nm).
25. Install the rocker arm cover.
26. Install the crankshaft pulley hub and tighten the mounting bolt to 77 ft. lbs. (105 Nm).
27. Install the crankshaft pulley and tighten the mounting bolts to 37 ft. lbs. (50 Nm).
28. Install the water pump pulley and tighten the mounting bolts to 15 ft. lbs. (21 Nm).
29. Install the drive belt tensioner and tighten the mounting bolts to 37 ft. lbs. (50 Nm).
30. Install the power steering pump. Install the alternator mounting bracket and alternator.
31. Install the engine assembly in the vehicle.
32. Coat the seal on the oil filter with clean engine oil and install the oil filter on the engine.

✻✻ WARNING

Operating the engine without the proper amount and type of engine oil will result in severe engine damage.

33. Refill the engine with oil.
34. Connect the negative battery cable. Refill the cooling system.
35. Start the vehicle and verify no leaks.

3.1L Engine

If the camshaft is being replaced, all of the valve lifters must also be replaced with new parts. Installing used lifters on a new camshaft will quickly wear the camshaft.

1. Relieve the fuel system pressure. Disconnect the negative battery cable.

2. Remove the engine assembly from the vehicle, and mount on a suitable engine stand.
3. Remove the intake manifold, valve cover, rocker arms, pushrods and valve lifters.
4. Remove the using a puller to draw the crankshaft balancer from the crankshaft nose.
5. Remove the timing chain front cover, timing chain and sprockets.
6. Remove the oil pump driven gear mounting bolt and hold-down. Remove the oil pump driven gear so the camshaft can be pulled out.
7. Remove the 2 bolts and remove the camshaft thrust plate.
8. Carefully remove the camshaft. Avoid marring the camshaft bearing surfaces.

To install:

9. Coat the camshaft with lubricant GM part No. 1052365 or equivalent camshaft break-in lubricant or quality engine oil supplement, and install the camshaft.
10. Install the camshaft thrust plate and tighten the mounting bolts to 89 inch lbs. (10 Nm).
11. Install the oil pump driven gear and tighten the mounting bolt to 27 ft. lbs. (36 Nm).
12. Install the timing chain and sprocket.
13. Verify that all timing marks are aligned. It is good practice to turn the crankshaft several complete rotations to be sure the timing marks still align when the crankshaft is returned to Top Dead Center No. 1 cylinder compression stroke (firing position). This is most important. If the valvetrain is not correctly timed, the engine will be damaged on start-up. When satisfied that all marks are correctly aligned, install the camshaft thrust button and front cover.
14. Install the crankshaft balancer and tighten the bolt to 75 ft. lbs. (103 Nm).
15. Install the intake manifold, valve cover, rocker arms, pushrods and valve lifters.
16. Install the engine assembly.
17. Connect the negative battery cable.

✻✻ WARNING

Operating the engine without the proper amount and type of engine oil will result in severe engine damage.

18. Verify that all fluid levels (coolant, engine oil, etc.) are full and correct. An oil and filter change is recommended.
19. Adjust the valves, as required.
20. Start the engine and verify no oil leaks.

Valve Lash

ADJUSTMENT

These vehicles are originally equipped with hydraulic valve lifters and do not require adjustment. Please note that if they are removed for service, they must be kept in order so they may be reinstalled in their original positions.

Oil Pan

REMOVAL & INSTALLATION

✻✻ CAUTION

The EPA warns that prolonged contact with used engine oil may cause a number of skin disorders, including cancer! You should make every effort to minimize your exposure to used engine oil. Protective gloves should be worn when changing the oil. Wash your hands and any other exposed skin areas as soon as possible after exposure to used engine oil. Soap and water, or waterless hand cleaner should be used.

2.2L Engine

1. Disconnect the negative battery cable.
2. Raise and safely support the vehicle.
3. Remove the right front tire and wheel assembly, then remove the right engine splash shield.
4. Position a suitable drain pan under the oil pan, then drain the engine oil. Don't forget to install the drain plug, after the oil is fully drained.
5. If equipped, remove the A/C brace.
6. Remove the starter bracket from the block, then remove the starter motor and position it out of the way.
7. Remove the transaxle converter cover or flywheel housing cover.
8. Remove the engine mount strut bracket. Remove the oil level sensor.
9. Remove the oil pan nuts and bolts, then lower the oil pan and remove it from the vehicle.

➡**Before installing the oil pan, be sure the sealing surfaces on the pan, cylinder block and front cover are clean and free of oil. If installing the old pan, be sure that all old RTV has been removed.**

To install:

10. Apply a 1/10 in. (2mm) wide bead of RTV sealant to the oil pan sealing surface, except for at the rear seal mounting surface. Apply a thin coat of RTV sealer on a new oil pan rear seal, on the ends down to the ears, then install the pan against the case and install bolts. Tighten the bolts 71–89 inch lbs. (8–10 Nm).

11. Install the oil level sensor and the engine mount strut bracket.

12. If removed, raise the right support into position, then secure using the four retaining bolts.

13. Install the starter motor and bracket.

14. Install the transaxle converter cover or flywheel housing cover.

15. If equipped, remove the A/C brace.

16. If removed, install the right front engine shield, and the right front wheel and tire assembly.

17. Carefully lower the vehicle.

⁑ WARNING

Operating the engine without the proper amount and type of engine oil will result in severe engine damage.

18. Fill the crankcase with the proper type and amount of engine oil.

19. Connect the negative battery cable, then start the engine and check for leaks. Recheck the oil level and add if necessary.

3.1L Engine

1. Disconnect the negative battery cable.

2. Remove the serpentine drive belt.

3. If equipped, loosen but do not remove the upper A/C compressor.

4. Raise and safely support the vehicle.

5. Position a suitable container under the oil pan, then drain the engine oil. Don't forget to install the drain plug, after the oil has completely drained.

6. Remove the right front wheel and tire assembly, then remove the splash shield.

7. Remove the engine mount strut from the suspension support.

8. Detach the right front ABS wheel speed sensor harness from the right suspension support.

9. Remove the right front ball joint.

10. Remove the right side stabilizer link. Remove the right side suspension support.

11. If equipped, remove the lower A/C compressor bolts, then position the compressor aside and support. Do NOT disconnect the refrigerant lines.

12. Remove the engine mount strut bracket.

13. Remove the engine-to-transaxle brace.

14. Remove the oil filter and the oil filter adapter, if equipped.

15. Remove the starter motor from the vehicle.

16. Remove the flywheel inspection cover.

17. Detach the oil level sensor electrical connector.

18. Remove and discard the oil pan gasket. Thoroughly clean the oil pan flanges, rail, front cover, rear main bearing cap and the threaded bolt holes.

To install:

19. Position a new gasket in the groove on the oil pan. If the rear main bearing cap is being installed, then GM sealant 1052080 or equivalent, must be placed on the oil pan gasket tabs that insert into the gasket groove of the outer surface on the rear main bearing cap.

20. Position the oil pan and secure with the retaining bolts. Tighten the retaining bolts to 18 ft. lbs. (25 Nm).

21. Install the oil pan side bolts, then tighten to 37 ft. lbs. (50 Nm).

22. Attach the oil level sensor electrical connector.

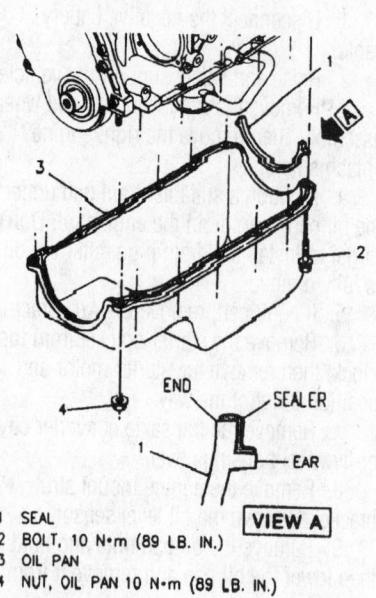

1 SEAL
2 BOLT, 10 N•m (89 LB. IN.)
3 OIL PAN
4 NUT, OIL PAN 10 N•m (89 LB. IN.)

79222109

Exploded view of the oil pan and seal mounting—2.2L engine

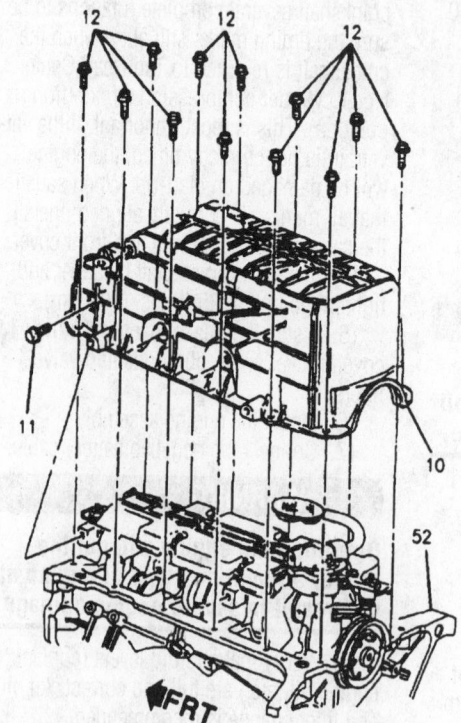

10 PAN, OIL
11 BOLT, OIL PAN SIDE
12 BOLT, OIL PAN RETAINING
52 BLOCK, ENGINE

79222110

Exploded view of the oil pan mounting—3.1L engine

23. Install the transaxle converter cover.

24. Install the starter motor.

25. Install the oil filter.

26. Install the engine-to-transaxle brace.

27. Install the engine mount strut bracket.

28. Loosely assembly the A/C compressor spacer and upper bolts, then secure in place.

29. Install the A/C compressor to the engine.

30. Install the right side suspension support. Install the right side stabilizer bar.

31. Install the right side ball joint.

32. Attach the right front ABS wheel speed sensor wiring harness.

33. Install the engine mount strut to the suspension support.

34. Install the right splash shield, then install the wheel and tire assembly.

35. Carefully lower the vehicle.

36. Install the serpentine drive belt.

❋❋ WARNING

Operating the engine without the proper amount and type of engine oil will result in severe engine damage.

37. Fill the crankcase with the proper type and amount of engine oil. Connect the negative battery cable, then start the engine and check for leaks.

38. Turn the engine off and allow to stand. Check the fluid levels and add as necessary to obtain the proper levels.

Oil Pump

REMOVAL & INSTALLATION

2.2L Engine

1. Remove the engine oil pan, as outlined earlier in this section.

2. Remove the pump-to-rear bearing cap bolt, then remove the pump and extension shaft.

3. Remove the extension shaft and retainer, being careful not to crack the retainer.

To install:

4. Heat the retainer in hot water prior to assembling the extension shaft.

5. Install the extension to the oil pump.

➡**Be sure the retainer does not crack upon installation.**

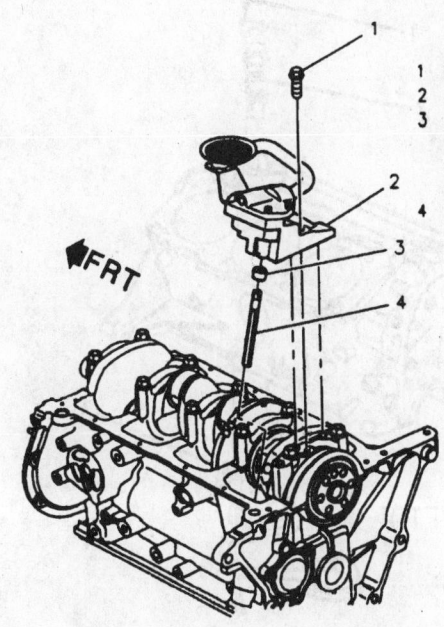

1 BOLT, 43 N•m (32 LBS. FT.)
2 PUMP ASSEMBLY, OIL
3 RETAINER, OIL PUMP SHAFT
   THE RETAINER MUST BE HEATED AND SOAKED
   IN WATER PRIOR TO INSTALLATION. THE
   RETAINER MUST NOT HAVE ANY SPLITS IN IT
   AFTER INSTALLATION.
4 SHAFT, OIL PUMP DRIVE

79222Z111

Exploded view of the oil pump assembly mounting—2.2L engine

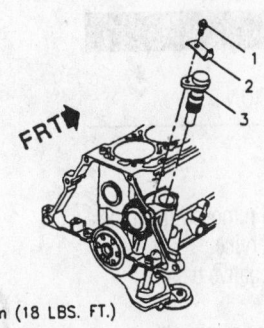

1 BOLT, 25 N•m (18 LBS. FT.)
2 SUPPORT
3 DRIVE ASSEMBLY, OIL PUMP

79222Z112

The oil pump drive is mounted to the top of the engine—2.2L engine

6. Install the pump-to-rear bearing cap bolt and tighten to 32 ft. lbs. (43 Nm).

7. Install the oil pan.

3.1L Engine

1. Raise and safely support the vehicle.

2. Remove the oil pan as described earlier in this section.

3. Unfasten and remove the oil pump and driveshaft extension.

To install:

4. Install the oil pump and driveshaft extension.

5. Engage the driveshaft extension into the drive gear.

6. Install the pump-to-rear bearing cap bolt and tighten to 30 ft. lbs. (41 Nm).

7. Install the oil pan, then carefully lower the vehicle.

8. Fill the crankcase with the proper type and amount of engine oil. Connect the negative battery cable, then start the engine and check for leaks.

9. Turn the engine off and allow to stand. Check oil level, add as necessary.

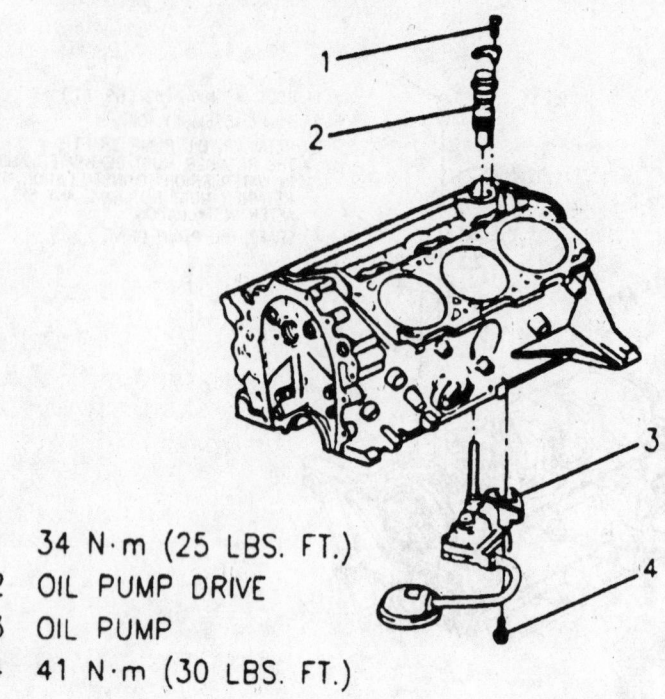

1 34 N·m (25 LBS. FT.)
2 OIL PUMP DRIVE
3 OIL PUMP
4 41 N·m (30 LBS. FT.)

79222113

Exploded view of the oil pump and drive mounting—3.1L engine

Rear Main Seal

REMOVAL & INSTALLATION

2.2L Engine

The transaxle must be removed from the vehicle to perform this service.

1. Disconnect the negative battery cable.
2. Remove the transaxle using the recommended procedure.
3. Remove the flywheel mounting bolts and remove the flywheel and retainer.
4. Remove crankshaft seal by inserting a suitable prying tool in through the dust lip. Pry out seal moving tool around seal as required until is removed.

✳✳ CAUTION

Care must be taken not to damage crankshaft seal surface with prytool.

To install:
5. Coat the inside and outside of the new rear main oil seal with engine oil.
6. Install the new seal on tool J-34686 until the seal bottom is squarely against the collar of J-34686.

7. Align the dowel pin of J-34686 with the dowel pin hole in the crankshaft. Tighten the attaching screws to 45 inch lbs. (6 Nm).
8. Turn the handle of the tool until the collar is tight against the case. This will ensure the seal is fully seated.
9. Back the tool off and remove the attaching screws.
10. Install the flywheel and retainer and tighten the flywheel mounting bolts to 55 ft. lbs. (75 Nm).
11. Install the transaxle assembly using the recommended procedure.
12. Connect the negative battery cable.
13. Check the engine oil level and fill as necessary.

3.1L Engine

The transaxle must be removed from the vehicle to perform this service.

1. Disconnect the negative battery cable.
2. Remove the transaxle using the recommended procedure.
3. Remove the flywheel mounting bolts and remove the flywheel and spacer.
4. Using a small prybar, pry the seal from the block.

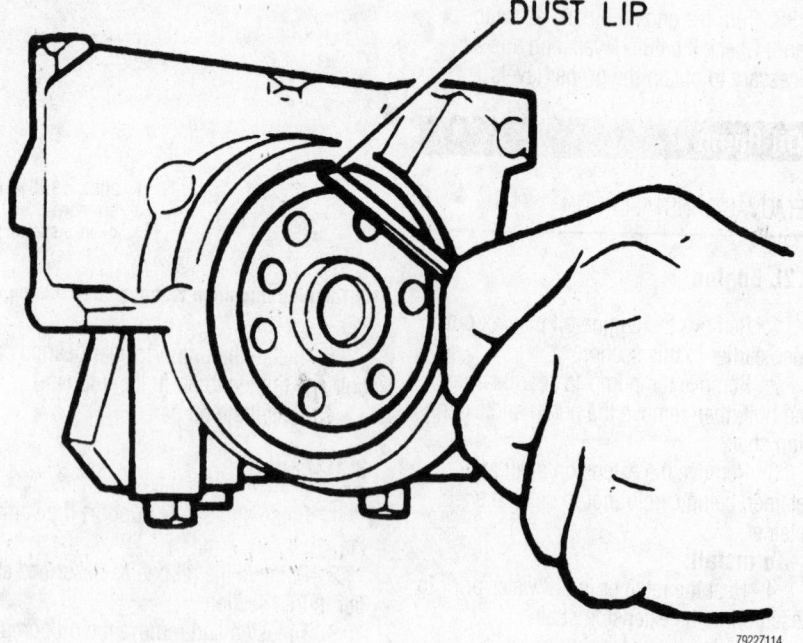

DUST LIP

79222114

When removing the rear main seal, be sure not to damage the bore with the prytool—all engines

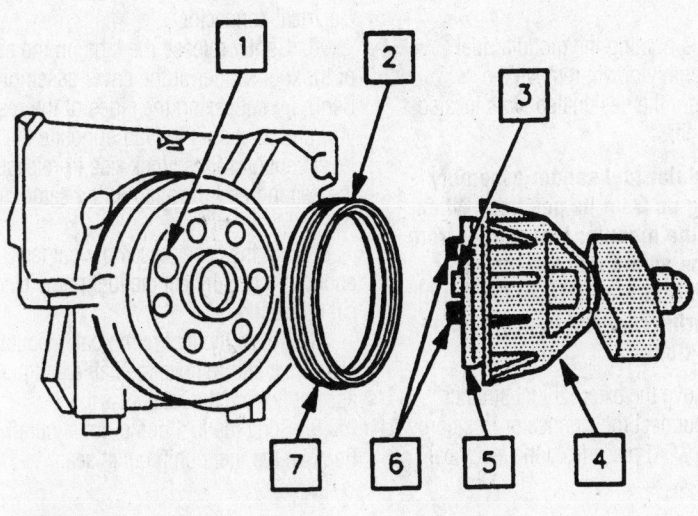

1	ALIGNMENT HOLE
2	DUST LIP
3	DOWEL PIN
4	COLLAR

5	MANDRIL
6	ATTACHING SCREWS
7	SEAL

7922Z115A

Rear main seal installation—all engines

✳✳ CAUTION

Be careful not to damage the crankshaft surface when removing the oil seal.

5. Clean the seal mounting surface.
6. **To install:**
7. Coat the inside and outside of the new rear main oil seal with engine oil.
8. Install the new seal on tool J-34686 until the seal bottom is squarely against the collar of J-34686.
9. Align the dowel pin of J-34686 with the dowel pin hole in the crankshaft. Tighten the attaching screws to 45 inch lbs. (5 Nm).
10. Turn the handle of the tool until the collar is tight against the case. This will ensure the seal is fully seated.
11. Back the tool off and remove the attaching screws.
12. Install the flywheel and spacer and tighten the flywheel mounting bolts to 61 ft. lbs. (83 Nm).
13. Install the transaxle assembly using the recommended procedure.
14. Connect the negative battery cable.
15. Check the engine oil level and fill as necessary.

FUEL SYSTEM

Fuel System Service Precautions

Safety is the most important factor when performing not only fuel system maintenance but any type of maintenance. Failure to conduct maintenance and repairs in a safe manner may result in serious personal injury or death. Maintenance and testing of the vehicle's fuel system components can be accomplished safely and effectively by adhering to the following rules and guidelines.

• To avoid the possibility of fire and personal injury, always disconnect the negative battery cable unless the repair or test procedure requires that battery voltage be applied.

• Always relieve the fuel system pressure prior to disconnecting any fuel system component (injector, fuel rail, pressure regulator, etc.), fitting or fuel line connection. Exercise extreme caution whenever relieving fuel system pressure to avoid exposing skin, face and eyes to fuel spray. Please be advised that fuel under pressure may penetrate the skin or any part of the body that it contacts.

• Always place a shop towel or cloth around the fitting or connection prior to loosening to absorb any excess fuel due to spillage. Ensure that all fuel spillage (should it occur) is quickly removed from engine surfaces. Ensure that all fuel soaked cloths or towels are deposited into a suitable waste container.

• Always keep a dry chemical (Class B) fire extinguisher near the work area.

• Do not allow fuel spray or fuel vapors to come into contact with a spark or open flame.

• Always use a back-up wrench when loosening and tightening fuel line connection fittings. This will prevent unnecessary stress and torsion to fuel line piping. Always follow the proper torque specifications.

• Always replace worn fuel fitting O-rings with new. Do not substitute fuel hose or equivalent, where fuel pipe is installed.

Fuel System Pressure

RELIEVING

✳✳ CAUTION

The fuel injection system remains under pressure, even after the engine has been turned OFF. The fuel system pressure must be relieved before disconnecting any fuel lines. Failure to do so may result in fire and/or personal injury.

1. Disconnect the negative battery cable. Remove the fuel filler cap to relieve tank vapor pressure.
2. Connect a fuel gauge to the fuel pressure test fitting.

➡**Be sure to wrap a shop cloth around the fuel line fitting when connecting the fuel gauge tool to the fuel pressure connector.**

3. Place the bleeder hose and shop cloth in an approved fuel container. Open the pressure valve to bleed the fuel pressure from the system.

✳✳ CAUTION

Observe all applicable safety precautions when working around fuel. Do not allow fuel spray or fuel vapors to come in contact with a spark or open flame. Keep a dry chemical (Class B) fire extinguisher near the work area. Never drain or store fuel in an open container due to the possibility of fire or explosion.

4. After the fuel pressure is bled, retighten the fuel pressure valve.

Fuel Filter

REMOVAL & INSTALLATION

✳✳ CAUTION

The fuel injection system remains under pressure, even after the engine has been turned OFF. The fuel system pressure must be relieved before disconnecting any fuel lines. Failure to do so may result in fire and/or personal injury.

1. Relieve the fuel system pressure. Disconnect the negative battery cable.

2. Raise and safely support the vehicle.

3. Disconnect the fuel inlet line from the fuel filter using a back-up wrench on the filter to prevent damaging the fuel line or filter.

4. Disconnect the fuel line quick-connect from the filter outlet by squeezing the tabs together and pulling the line off the filter.

5. Remove the fuel filter from the bracket.

To install:

6. Install the fuel filter in the mounting bracket.

7. Connect the quick-connect to the fuel filter. Be sure the tabs lock in place and the line can not be pulled off the filter.

8. Connect the outlet line to the filter using a back-up wrench on the filter body to prevent damaging the fuel line or filter. Tighten the line fitting to 20 ft. lbs. (27 Nm).

9. Lower the vehicle.

10. Connect the negative battery cable. Pressurize the fuel system and verify no leaks.

Fuel Pump

REMOVAL & INSTALLATION

✳✳ CAUTION

The fuel injection system remains under pressure, even after the engine has been turned OFF. The fuel system pressure must be relieved before disconnecting any fuel lines. Failure to do so may result in fire and/or personal injury.

1. Relieve the fuel system pressure. Disconnect the negative battery cable.

2. Drain, then remove the fuel tank from the vehicle.

3. While holding the modular fuel sender assembly down, remove the snapring from the designated slots located on the retainer.

➡**The modular fuel sender assembly may spring up from its position. When removing the modular fuel sender from the tank, be aware that the reservoir bucket is full of fuel. It must be tipped slightly during removal to avoid damage to the float.**

4. Remove the external fuel strainer.

5. Disconnect the Connector Position Assurance (CPA) piece from the electrical connector and disconnect the fuel pump electrical connector.

6. Gently release the tabs on the sides of the fuel sender at the cover assembly. Begin by squeezing the sides of the reservoir and releasing the tab opposite the fuel level sensor. Move clockwise to release the second and third tab in the same manner.

7. Lift the cover assembly out far enough to disconnect the fuel pump electrical connection.

8. Rotate the fuel pump baffle counterclockwise and remove the baffle and pump assembly from the retainer.

9. Slide the fuel pump outlet out of slot. Remove the fuel pump outlet seal.

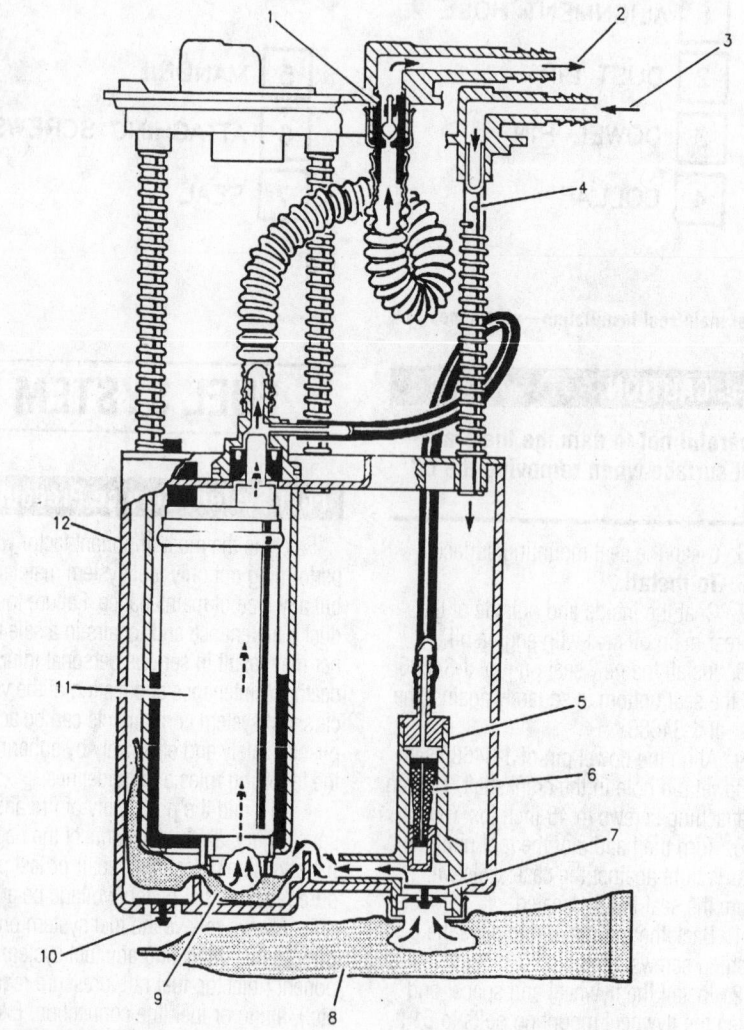

1 CHECK VALVE
2 FUEL FLOW TO ENGINE
3 RETURN FUEL FROM ENGINE
4 ANTI-SIPHON HOLE
5 PUMP ASSEMBLY (JET PUMP ASSEMBLY)–
    FUEL PUMP RESERVOIR
6 STRAINER – JET PUMP
7 UMBRELLA VALVE – PRIMARY
8 STRAINER (EXTERNAL) – FUEL SENDER
9 STRAINER – FUEL PUMP FUEL (ROLLERVANE)
10 UMBRELLA VALVE – SECONDARY
11 PUMP ASSEMBLY (ROLLERVANE) – FUEL
12 RESERVOIR

Cut-away view of the modular sender—showing fuel flow

79222125

To install:

10. Install the fuel pump outlet seal. Slide the fuel pump outlet in the slots of reservoir cover.

11. Install the fuel pump and baffle assembly onto the reservoir retainer and rotate clockwise until seated.

12. Install the lower retainer assembly partially into the reservoir. Line up all 3 sleeve tabs. Press the retainer onto the reservoir making sure all 3 tabs are firmly seated.

➡**Gently pull on the fuel pump reservoir from retainer to assure a secure fastening. If not secure replace entire fuel sender.**

13. Connect the fuel pump connector. Attach the CPA connector to the fuel sender cover.

14. Install a new external fuel strainer.

15. Install the modular fuel sender.

16. Install the fuel tank in the vehicle.

17. Connect the negative battery cable. Pressurize the fuel system and verify no leaks.

DRIVE TRAIN

Transaxle Assembly

REMOVAL & INSTALLATION

Manual

➡**Before performing any maintenance that requires the removal of the slave cylinder, transaxle or clutch housing, the clutch master cylinder pushrod must first be disconnected from the clutch pedal. Failure to disconnect the pushrod will result in permanent damage to the slave cylinder if the clutch pedal is depressed with the slave cylinder disconnected.**

1. Disconnect the negative terminal from the battery.

2. Using the Engine Support Fixture tool No. J-28467 or equivalent and Adapter tool No. J-35953 or equivalent, install them on the engine and raise the engine enough to take the engine weight off of the engine mounts.

3. Remove the left side sound insulator.

4. Disconnect the clutch master cylinder pushrod from the clutch pedal.

5. Disconnect the clutch slave cylinder-to-transaxle support bolts and position the cylinder aside.

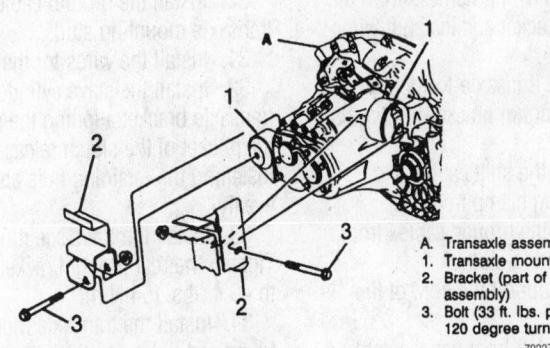

A. Transaxle assembly
1. Transaxle mount
2. Bracket (part of body assembly)
3. Bolt (33 ft. lbs. plus 120 degree turn)

7922Z115B

Transaxle mount-to-body attachment—all vehicles

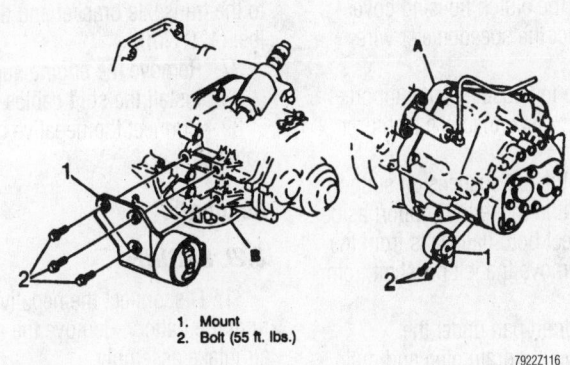

1. Mount
2. Bolt (55 ft. lbs.)

7922Z116

Front transaxle mount-to-transaxle attachment and bolt torque values

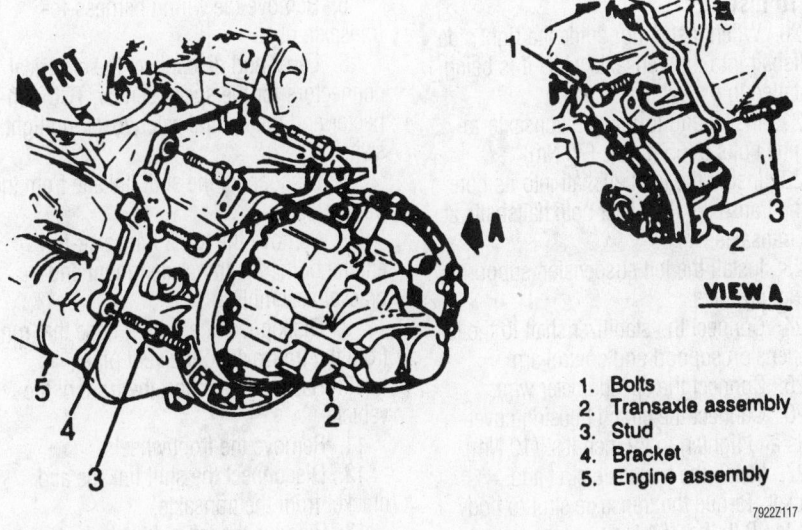

1. Bolts
2. Transaxle assembly
3. Stud
4. Bracket
5. Engine assembly

7922Z117

Transaxle-to-engine attachment—all vehicles

6. Remove the wiring harness from the transaxle mount bracket and the shift wire electrical connector.

7. Remove the transaxle-to-mount bolts and the transaxle mount bracket-to-chassis nuts/bolts.

8. Disconnect the shift cables and remove the retaining clamp from the transaxle. Remove the ground cables from the transaxle mounting studs.

9. Raise and support the front of the vehicle.

10. Remove the left front tire assembly and the left side inner splash shield.

11. Remove the transaxle front strut and bracket.

12. Remove the clutch housing cover bolts. Disconnect the speedometer wire connector.

13. From the left suspension support and control arm, disconnect the stabilizer shaft.

14. Remove the left suspension support mounting bolts and move the support aside.

15. Disconnect both halfshafts from the transaxle and remove the left halfshaft from the vehicle.

16. Place a drain pan under the transaxle, remove the drain plug and drain the fluid from the transaxle.

17. Using a transmission jack, attach it to and support the transaxle.

18. Remove the transaxle-to-engine bolts.

19. Slide the transaxle away from the engine, lower it and remove the right side halfshaft.

To install:

20. When installing, guide the right side halfshaft into the transaxle while it is being installed in the vehicle.

21. Install and torque the transaxle-to-engine bolts to 55 ft. lbs. (75 Nm)

22. Install the left halfshaft into its bore at the transaxle, then seat both halfshafts at the transaxle.

23. Install the left suspension support mounting bolts.

24. Connect the stabilizer shaft to the left suspension support and control arm.

25. Connect the speedometer wire.

26. Connect the clutch housing cover bolts and tighten to 89 inch lbs. (10 Nm).

27. Install the transaxle strut and bracket. Torque the transaxle strut to body bolt to 40 ft. lbs. (54 Nm) and the transaxle strut to transaxle to 50 ft. lbs. (68 Nm). See applicable illustration.

28. Install the left front tire assembly and the left side inner splash shield.

29. Lower the vehicle.

30. Install the ground cables at the transaxle mounting studs.

31. Install the wires for the shift light.

32. Install the slave cylinder to the transaxle bracket aligning the pushrod into the pocket of the clutch release lever and installing the retaining nuts and tighten evenly.

33. Install the transaxle mount bracket. Tighten the rear mount bracket to transaxle to 40 ft. lbs. (54 Nm).

34. Install the transaxle mount to side frame and tighten to 23 ft. lbs. (30 Nm).

35. Install the wiring harness at the mount bracket.

36. Install the bolt attaching the mount to the transaxle bracket and tighten to 88 ft. lbs. (120 Nm).

37. Remove the engine support.

38. Install the shift cables.

39. Connect the negative cable at the battery.

Automatic

2.2L ENGINE

1. Disconnect the negative terminal from the battery. Remove the air cleaner and air intake assembly.

2. Disconnect the Throttle Valve (TV) cable from the throttle lever and the transaxle.

3. Remove the fluid level indicator (dipstick) and the filler tube.

4. Install Engine Support Fixture tool No. J-28467 or equivalent and the Adapter tool No. J-35953 or equivalent, on the engine.

5. Remove the wiring harness-to-transaxle nut.

6. Label and disconnect the electrical connectors for the speed sensor, TCC connector and the neutral safety/back-up light switch.

7. Disconnect the shift linkage from the transaxle.

8. Remove the top 2 transaxle-to-engine bolts, the transaxle mount and bracket assembly.

9. Disconnect the rubber hose that runs from the transaxle to the vent pipe.

10. Raise and support the front of the vehicle.

11. Remove the front wheels.

12. Disconnect the shift linkage and bracket from the transaxle.

13. Remove the left side splash shield.

14. Using a modified Drive Axle Seal Protector tool No. J-34754 or equivalent, install one on each drive axle to protect the seal from damage and the joint from possible failure.

15. Using care not to damage the half-shaft boots, disconnect the halfshafts from the transaxle.

16. Remove the transaxle strut. Remove the left side stabilizer link pin bolt and bushing clamp nuts from the support.

17. Remove the left frame support bolts and move it aside.

18. Disconnect the speedometer wire from the transaxle.

19. Remove the transaxle converter cover and matchmark the torque converter-to-flywheel for reassembly.

20. Disconnect and plug the transaxle cooler pipes.

21. Remove the transaxle-to-engine support.

22. Using a transmission jack, position and secure the jack to the transaxle. Remove the remaining transaxle-to-engine bolts.

23. Remove the transaxle from the vehicle. Use care not to let the torque converter fall from the front of the transaxle.

➡**The transaxle cooler and lines should be flushed any time the transaxle is removed for overhaul or replacing the pump, case or converter.**

To install:

24. Put a small amount of grease on the pilot hub of the converter and be sure the converter is properly engaged with the pump.

25. Raise the transaxle to the engine while guiding the right side halfshaft into the transaxle.

26. Install the lower transaxle mounting bolts and remove the jack.

27. Align the converter with the marks made previously on the flywheel and install the bolts hand-tight.

28. Torque the converter bolts to 45 ft. lbs. (61 Nm). Retorque the first bolt after the others.

29. Install the transaxle converter cover. Attach the speedometer wire connector.

30. Position the LH halfshaft into the transaxle. Install the left frame support assembly.

31. Install the left stabilizer shaft frame bushing nuts. Install the left stabilizer bar link pin bolt.

32. Install the transaxle strut.

33. Seat the drive axles in the transaxle. Remove the drive axle seal protectors.

34. Install the remaining components in the reverse order of removal.

35. Install the negative battery cable.

36. Fill with automatic transmission fluid and check for leaks.

3.1L ENGINE

1. Disconnect the negative terminal from the battery.

2. Remove the air cleaner, bracket, Mass Air Flow (MAF) sensor and air tube as an assembly.

3. Disconnect the exhaust crossover from the right side manifold and remove the left side exhaust manifold. Raise and support the manifold/crossover assembly.

4. Disconnect the TV cable from the throttle lever and the transaxle.

5. Remove the vent hose and the shift cable from the transaxle.

6. Remove the fluid level indicator and the filler tube.

7. Using the Engine Support Fixture tool No. J-28467 or equivalent and the Adapter tool No. J-35953 or equivalent, install them on the engine. Use care to be sure the engine is safely supported and that the engine support fixture does not damage any bodywork.

8. Remove the wiring harness-to-transaxle nut.

9. Label and disconnect the wires for the speed sensor, TCC connector and the neutral safety/back-up light switch.

10. Remove the upper transaxle-to-engine bolts.

11. Remove the transaxle-to-mount through-bolt, the transaxle mount bracket and the mount.

12. Raise and safely support the vehicle. Remove the front wheels.

13. Disconnect the shift cable bracket from the transaxle.

14. Remove the left side splash shield.

15. Using a modified Drive Axle Seal Protector tool No. J-34754 or equivalent, install one on each drive axle to protect the seal from damage and the joint from possible failure.

16. Using care not to damage the halfshaft boots, disconnect the halfshafts from the transaxle.

17. Remove the torsional and lateral strut from the transaxle. Remove the left side stabilizer link pin bolt.

18. Remove the left frame support bolts and move it aside.

19. Disconnect the speedometer wire from the transaxle.

20. Remove the transaxle converter cover and matchmark the converter-to-flywheel for assembly.

21. Disconnect and plug the transaxle cooler pipes.

22. Remove the transaxle-to-engine support.

23. Using a transmission jack, position and secure it to the transaxle. Remove the remaining transaxle-to-engine bolts.

24. Remove the transaxle from the vehicle using care to be sure the torque converter does not fall out.

➡ The transaxle cooler and lines should be flushed any time the transaxle is removed for overhaul, to replace the pump, case or converter.

To install:

25. Put a small amount of grease on the pilot hub of the converter and be sure the converter is properly engaged with the pump.

26. Raise the transaxle to the engine while guiding the right side halfshaft into the transaxle.

27. Install the lower transaxle mounting bolts and remove the jack.

28. Install the cooler lines at the transmission.

29. Position the left side halfshaft into the transaxle.

30. Install the left frame support bolts.

31. Install the left stabilizer shaft bushing clamp nuts at the support.

32. Install the left stabilizer bar link pin bolt.

33. Seat the drive axles in the transaxle. Remove the drive axle seal protectors.

34. Install the shift linkage bracket to the transaxle.

35. Install the speedometer wire connector.

36. Install the transaxle brace bolts. Install the transaxle lateral strut.

37. Install the transaxle torsional strut.

38. Align the converter with the marks made previously on the flywheel and install the bolts hand-tight.

39. Torque the converter bolts to 45 ft. lbs. Retorque the first bolt after the others.

40. Install the transaxle converter cover. Install the splash shield.

41. Install both front wheels.

42. Lower the vehicle.

43. Install the transaxle mount.

44. Install the transaxle mount through-bolt.

45. Install the wiring harness and nut securing it to the transaxle.

46. Remove the engine support fixture.

47. Install the fluid level indicator and fill tube.

48. Install the neutral start and TCC connectors.

49. Install the shift cable.

50. Install the TV cable at the transaxle and throttle lever.

51. Install the rubber hoses to the transaxle vent pipe.

52. Install the left side exhaust manifold bolts and exhaust crossover bolts.

53. Install the air cleaner mounting bracket, MAF sensor and air tube as an assembly.

54. Install the negative battery cable.

55. Fill transaxle with automatic transmission fluid and check for leaks.

Clutch

❊❊ CAUTION

The clutch driven disc contains asbestos, which has been determined to be a cancer causing agent. Never clean clutch surfaces with compressed air! Avoid inhaling any dust from any clutch surface! When cleaning clutch surfaces, use a commercially available brake cleaning fluid.

REMOVAL & INSTALLATION

1. Disconnect the negative battery cable.

2. From inside the vehicle, remove the hush panel.

3. Disconnect the clutch master cylinder pushrod from the clutch pedal.

4. With the transaxle removed, matchmark the pressure plate and flywheel assembly to insure proper balance during reassembly.

5. Loosen the pressure plate-to-flywheel bolts (one turn at a time) until the spring pressure is removed.

6. Support the pressure plate and remove the bolts.

7. Remove the pressure plate and disc assembly; be sure to note the flywheel side of the clutch disc.

8. Clean and inspect the clutch assembly, flywheel, release bearing, clutch fork and pivot shaft for signs of wear. Replace any necessary parts.

To install:

9. Position the clutch disc and pressure plate in the appropriate position, align the "Heavy Side" of the flywheel assembly stamped with an **X** with the clutch cover "Light Side" marked with paint. Support the assembly with Alignment tool No. J-290742 or equivalent.

➡ **The clutch disc is installed with the damper springs offset towards the transaxle. Stamped letters on the clutch disc identify "Flywheel Side". Be sure the clutch disc is facing the same direction it was when removed. If the same pressure plate is being reused, align the marks made during the removal.**

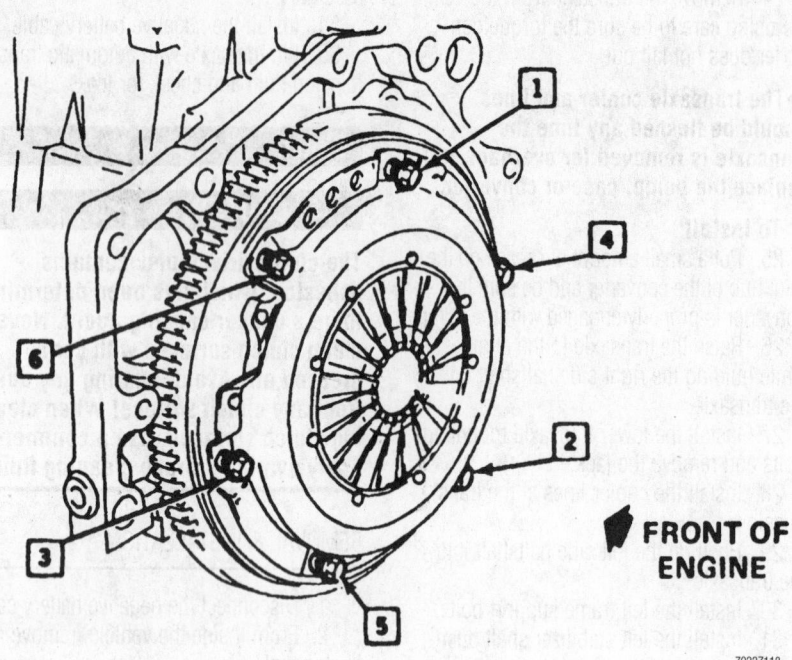

Tighten the clutch cover bolts in the proper sequence

10. Install the pressure plate-to-flywheel retaining bolts.

11. Tighten the flywheel-to-pressure plate bolts as follows:

   a. Install and lightly seat bolts 1, 2, 3, then 4, 5, 6.

   b. Tighten bolts 1, 2, 3 to 12 ft. lbs. (16 Nm).

   c. Tighten bolts 4, 5, 6 to 12 ft. lbs. (16 Nm).

   d. Tighten bolts 1, 2, 3, then 4, 5, 6 to 15 ft. lbs. (20mm) plus 30 degree rotation.

12. Remove the alignment tool.

13. Lightly lubricate the clutch fork ends. Fill the recess ends of the release bearing with grease. Lubricate the input shaft with a light coat of grease.

14. To complete the installation, reverse the removal procedures.

➡ **The clutch lever must not be moved towards the flywheel until the transaxle is bolted to the engine. Damage to the transaxle, release bearing and clutch fork could occur if this is not followed.**

15. Bleed the clutch system and check the clutch operation when finished.

Hydraulic Clutch System

BLEEDING

1. Remove any dirt or grease around the reservoir cap so dirt cannot enter the system. Fill the reservoir with an approved DOT 3 brake fluid. Maintain the fluid level while bleeding the system.

2. Attach a hose to the bleeder screw on the clutch actuator assembly and submerge the other end of the hose in a container of hydraulic clutch fluid GM P/N 12345347 or equivalent DOT 3 brake fluid.

3. Depress the clutch pedal slowly and hold. Loosen the bleeder screw to purge air.

4. Tighten the bleeder screw to 18 inch lbs. (2 Nm).

5. Repeat the bleeding process until all the air is purged from the system.

6. Fill the master cylinder to the proper level.

7. To test the system, start the engine and push the clutch pedal to the floor. Wait 10 seconds and select reverse gear. There should be no gear clash. If clash is present, air may still be present in the system. Repeat bleeding procedure.

Halfshaft

REMOVAL & INSTALLATION

1. Raise and safely support vehicle. Remove the wheel and tire assembly.

2. Install drive seal protector J-34754 or equivalent, on the outer joint.

3. Insert a drift pin through the opening in the caliper and into the ventilation openings in the brake rotor to prevent the rotor from turning.

4. Remove the drive shaft hub nut and washer.

5. Remove the lower ball joint cotter pin and nut and loosen the joint using tool J-38892 or equivalent. If removing the right axle, turn the wheel to the left. If removing the left axle, turn the wheel to the right.

6. Separate the joint, with a prybar between the suspension support.

7. Disengage the axle from the hub and bearing using J-28733-A or equivalent puller.

8. Separate the hub and bearing assembly from the drive axle and move the strut and knuckle assembly rearward.

9. Disconnect the inner joint from the transaxle using tool J-28468 or J-33008 attached to J-29794 and J-2619–01 or from the intermediate shaft (V6 and 2.3L engines), if equipped.

To install:

10. Install axle seal protector J-37292-A into the transaxle.

11. Insert the drive axle into the transaxle or intermediate shaft, if equipped, by placing a suitable tool into the groove on the joint housing and tapping until seated.

✳✳ CAUTION

Be careful not to damage the axle seal or dislodge the transaxle seal garter spring when installing the axle.

12. Verify that the drive axle is seated into the transaxle by grasping on the housing and pulling outward.

13. Install the drive axle into the hub and bearing assembly.

14. Install the lower ball joint to the knuckle. Tighten the ball joint to steering knuckle nut to 41 ft. lbs. (55 Nm) and install the cotter pin.

15. Install the washer and new drive shaft nut. Again insert a drift pin into the caliper and rotor to prevent the rotor from turning and tighten the drive shaft nut to

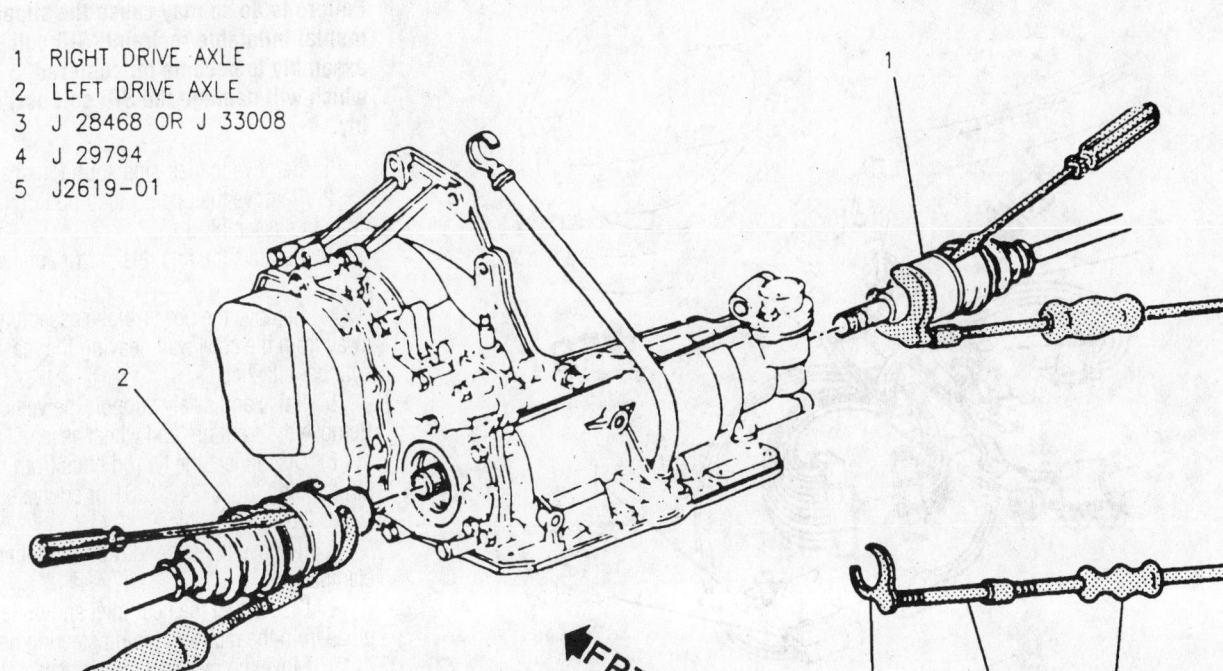

1 RIGHT DRIVE AXLE
2 LEFT DRIVE AXLE
3 J 28468 OR J 33008
4 J 29794
5 J2619-01

Use the tools as shown to prevent damaging the transaxle or halfshaft

185 ft. lbs. (260 Nm). Install the brake caliper.

16. Remove both J-37292-B and J-34754 seal protectors.

17. Install the tire and wheel assembly. Lower the vehicle.

STEERING AND SUSPENSION

Air Bag

✳✳ CAUTION

Some vehicles are equipped with an air bag system, also known as the Supplemental Inflatable Restraint (SIR). The system must be disabled before performing service on or around system components, steering column, instrument panel components, wiring and sensors. Failure to follow safety and disabling procedures could result in accidental air bag deployment, possible personal bag deployment, possible personal injury and unnecessary system repairs.

PRECAUTIONS

Several precautions must be observed when handling the inflator module to avoid accidental deployment and possible personal injury.

• Never carry the inflator module by the wires or connector on the underside of the module.

• When carrying a live inflator module, hold securely with both hands, and ensure that the bag and trim cover are pointed away.

• Place the inflator module on a bench or other surface with the bag and trim cover facing up.

• With the inflator module on the bench, never place anything on or close to the module which may be thrown in the event of an accidental deployment.

DISARMING

✳✳ CAUTION

The Supplemental Inflatable Restraint (SIR) system must be disarmed before performing many service functions. Failure to do so may cause accidental deployment of the air bag, resulting in unnecessary SIR system repairs and/or personal injury.

1. Turn the steering wheel so the vehicle's wheels are pointing straight ahead.

2. This helps protect the clockspring mechanism under the steering wheel hub which powers the air bag module.

3. Turn the ignition switch to **LOCK** and remove the key.

4. Remove the AIR BAG fuse from the instrument panel fuse block.

5. Remove the left side sound insulator.

6. Remove the air bag connector CPA and disconnect the yellow 2-way SIR harness connector located near the base of the steering column.

➡ With the AIR BAG fuse removed, and if, for some reason, the negative battery cable should be connected and the ignition switch turned to the ON (not recommended), the air bag warning lamp in the instrument cluster will be ON. This is normal operation and does not indicate an SIR system malfunction.

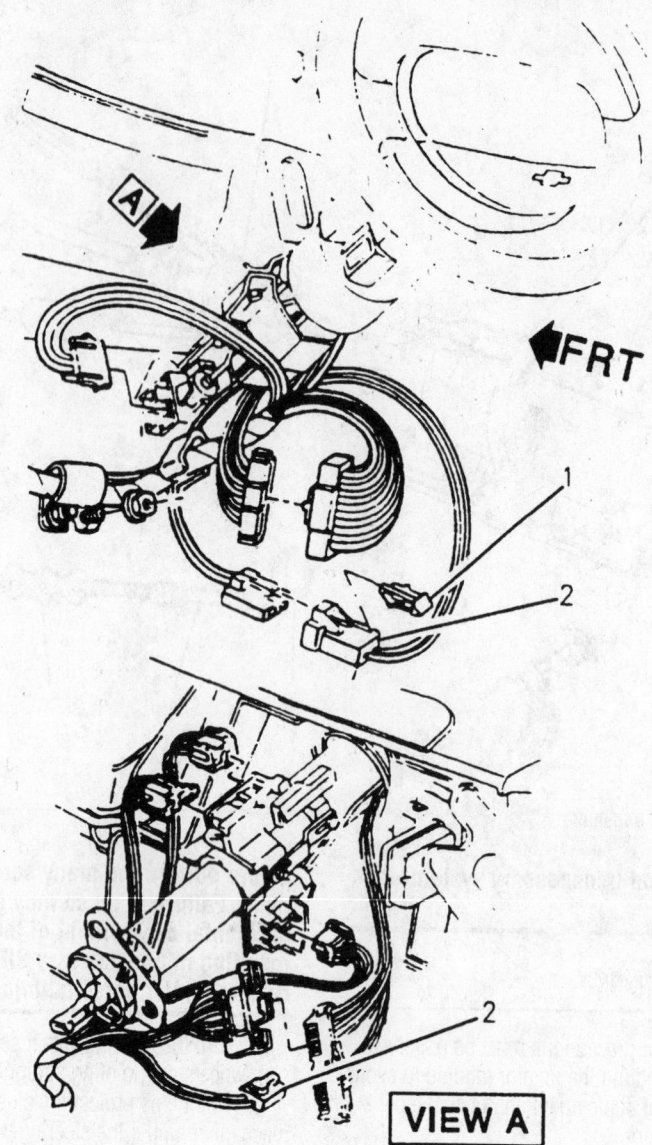

1   CONNECTOR POSITION ASSURANCE (CPA)
2   YELLOW TWO—WAY SIR HARNESS CONNECTOR

79222127

Air bag module connector location

REARMING

1. Turn the ignition switch to the **LOCK** position and remove the key.

2. At the base of the steering column, connect the yellow 2-way connector and install the CPA bar to verify that the connection is complete and tight.

3. Install the left side sound insulator.

4. Install the AIR BAG fuse into the instrument panel fuse block.

5. Turn the ignition switch to the **RUN** position and verify that AIR BAG warning lamp flashes 7 times and turns OFF. If it does not operate as described, perform SIR Diagnostic System Check.

Power Rack and Pinion Steering Gear

REMOVAL & INSTALLATION

➡**The wheels of the vehicle must be straight ahead and the ignition switch must be in the LOCK position, before disconnecting the flange and coupling assembly from the steering column.**

Failure to do so may cause the supplemental inflatable restraint (SIR coil assembly to become off centered, which will damage the SIR coil assembly.

1. Remove the left side sound insulator.

2. Remove the upper pinch bolt on the coupling assembly.

3. Remove the line retainer (if applicable).

4. Remove the power brake assembly away from the cowl wall, leaving the master cylinder attached.

5. Raise and safely support the vehicle. Remove the front tire and wheel assemblies.

6. Disconnect the tie rod ends from the struts using tool J-24319-01 or equivalent.

7. Lower the vehicle.

8. Remove the left and right mounting clamps.

9. Disconnect the gear inlet and outlet pipes from the rack and pinion steering gear.

10. Move the rack and pinion assembly forward and remove the lower pinch bolt from flange on the coupling assembly.

11. Disconnect the coupling from the rack and pinion assembly.

12. Remove the dash seal from the rack and pinion assembly.

13. Remove the rack and pinion assembly through the left wheel opening.

➡**If the studs were removed with the mounting clamps, apply some thread locking compound on them and reinstall studs into cowl panel and tighten until the studs are fully seated against the dash panel. The torque should not exceed 13 ft. lbs. (18 Nm).**

To install:

14. Install the rack and pinion assembly into the vehicle through the left wheel opening.

15. Attach the dash seal onto the rack and pinion assembly.

16. Move the rack and pinion assembly forward and install the coupling onto the rack and pinion.

17. Install the flange and coupling lower pinch bolt. Tighten this bolt to 30 ft. lbs. (41 Nm).

18. Connect the gear inlet and outlet pipes to the rack and pinion assembly and tighten them to 20 ft. lbs. (27 Nm).

19. Install the mounting clamp nuts and tighten (left side first) to 22 ft. lbs. (30 Nm).

20. Raise the vehicle.

21. Connect the tie rods to the struts and tighten to 44 ft. lbs. (60 Nm) and install cotter pins through the nuts.

22. Install the tire and wheel assemblies.

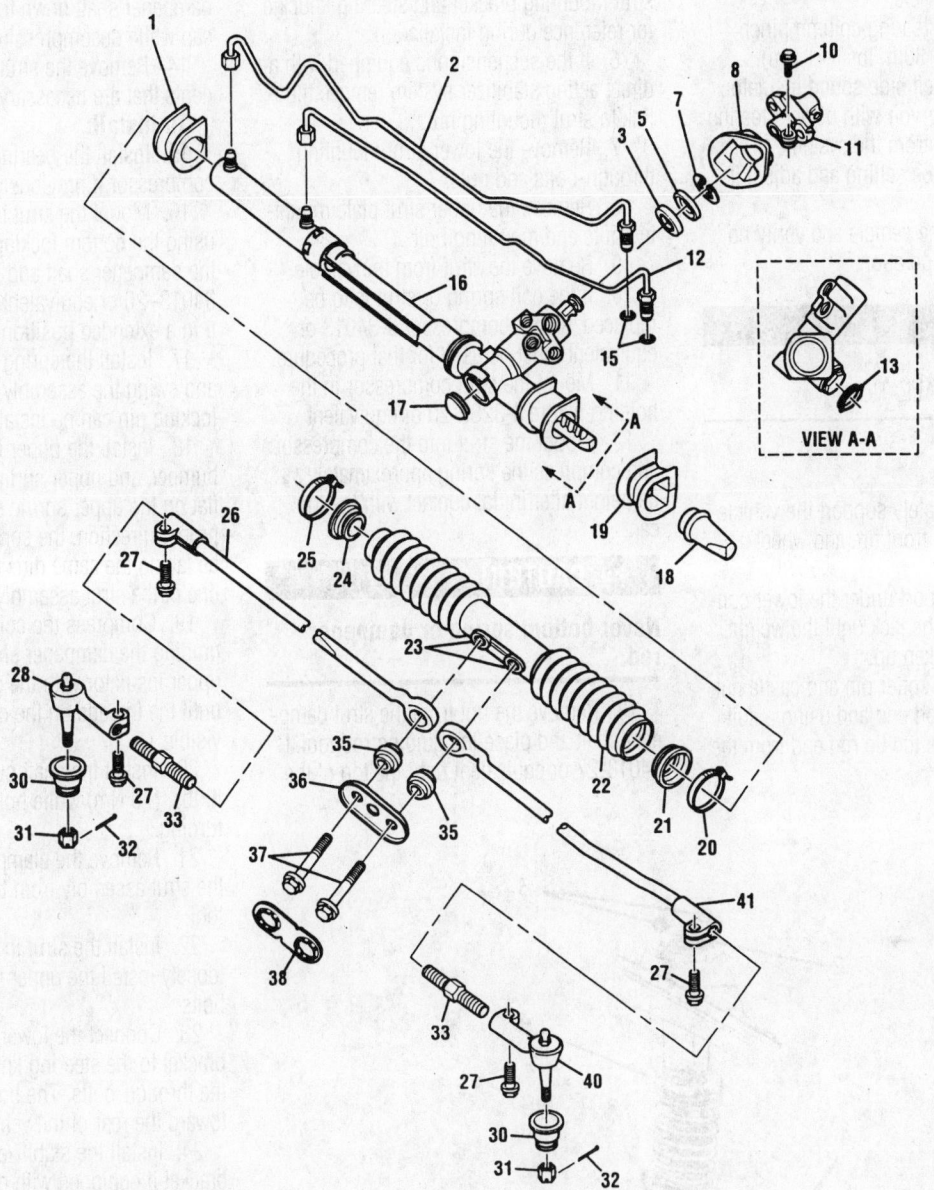

VIEW A-A

1 - GROMMET, MOUNTING
2 - LINE ASM, CYLINDER (LT)
3 - ANNULUS ASM, BEARING
5 - SEAL, STUB SHAFT
7 - RING, RETAINING
8 - SEAL, DASH
10 - BOLT, PINCH
11 - COUPLING, FLANGE & STEERING
12 - LINE ASM, CYLINDER (RT)
13 - NUT, ADJUSTER PLUG LOCK
15 - SEAL, O-RING
16 - HOUSING, RACK & PINION
17 - COVER, DUST
18 - COVER, HOUSING END
19 - GROMMET, MOUNTING
20 - CLAMP, BOOT
21 - BUSHING, BOOT RETAINING

22 - BOOT, RACK & PINION
23 - WASHER, CENTER HOUSING COVER (PART OF 22)
24 - BUSHING, BOOT RETAINING
25 - CLAMP, BOOT
26 - ASSEMBLY, INNER TIE ROD
27 - BOLT, PINCH
28 - ASSEMBLY, OUTER TIE ROD
30 - SEAL, TIE ROD
31 - NUT, HEX SLOTTED
32 - PIN, COTTER
33 - ADJUSTER, TIE ROD
35 - BUSHING, INNER PIVOT
36 - PLATE, BOLT SUPPORT
37 - BOLT, INNER TIE ROD
38 - PLATE, LOCK
40 - ASSEMBLY, OUTER TIE ROD
41 - ASSEMBLY, INNER TIE ROD

7922Z128

Exploded view of the rack and pinion steering gear

23. Install the line retainer (if applicable). Lower the vehicle.

24. Install the steering column pinch bolt and tighten to 30 ft. lbs. (41 Nm).

25. Install the left side sound insulator.

26. Fill the reservoir with power steering fluid and bleed air from the system.

27. Check the toe setting and adjust if necessary.

28. Road test the vehicle and verify no leaks and proper operation.

Strut

REMOVAL & INSTALLATION

Front

1. Raise and safely support the vehicle.
2. Remove the front tire and wheel assembly.
3. Place a support under the lower control arm and raise the jack until the weight of suspension is taken up.
4. Remove the cotter pin and castle nut from the outer tie rod end and using a suitable puller, separate the tie rod end from the strut steering arm.

5. Scribe matching marks on the lower strut mounting bracket and steering knuckle for reference during installation.

6. If the suspension is equipped with a direct acting stabilizer system remove the link to strut mounting nut.

7. Remove the lower strut mounting through-bolts and nuts.

8. Remove the upper strut plate mounting nuts and mounting bolt.

9. Remove the strut from the vehicle.

10. If the coil spring or strut is to be replaced, a strut compressor J-34013 or equivalent must be used for that procedure.

11. Mount the strut compressor in the holding fixture J-3289-20 or equivalent.

12. Mount the strut into the compressor and compress the spring approximately ½ its height after initial contact with the top cap.

✳✳ CAUTION

Never bottom spring or dampener rod.

13. Remove the nut from the strut dampener shaft and place the guiding rod tool J-34013-27 or equivalent onto the top of the

damper shaft. Use this rod tool to guide the dampener shaft down through the bearing cap while decompressing the spring.

14. Remove the strut and any components that are necessary.

To install:

15. Install the bearing cap into the strut compressor if previously installed.

16. Mount the strut into the compressor using the bottom locking pin only. Extend the dampener shaft and install clamp J-34013–20 or equivalent on the shaft to hold it in a extended position.

17. Install the spring over the dampener and swing the assembly up so the upper locking pin can be installed.

18. Install the upper insulator, shield, bumper, and upper spring seat. Be sure the flat on the upper spring seat is facing in the proper direction. the spring seat flat should be facing the same direction as the centerline of the strut assembly.

19. Compress the coil spring while guiding the dampener shaft through the upper insulator with the guiding rod tool until the threads on the dampener shaft are visible.

20. Install the shaft nut and tighten to 52 ft. lbs. (70 Nm) while holding the shaft from turning.

21. Remove the clamp and remove the strut assembly from the compressor tool.

22. Install the strut in the vehicle and loosely install the upper mounting nuts and bolts.

23. Connect the lower strut mounting bracket to the steering knuckle and insert the through-bolts. The bolts should point toward the rear of the vehicle.

24. Install the stabilizer link to the strut bracket if equipped with direct acting suspension. Install the mounting nut and tighten to 70 ft. lbs. (95 Nm).

25. Install the through-bolt nuts loosely. Position the strut so the matching marks are aligned and tighten the nuts to 133 ft. lbs. (180 Nm).

26. Connect the tie rod end to the strut steering arm and tighten the castle nut to 44 ft. lbs. (60 Nm). If necessary tighten the nut up to 60 degrees ⅙ turn) additional rotation to align the cotter pin holes. NEVER loosen the nut to make the holes align.

27. Remove the support jack from under the suspension.

28. Install the tire and wheel assembly and tighten the mounting nuts to 100 ft. lbs. (140 Nm).

29. Lower the vehicle.

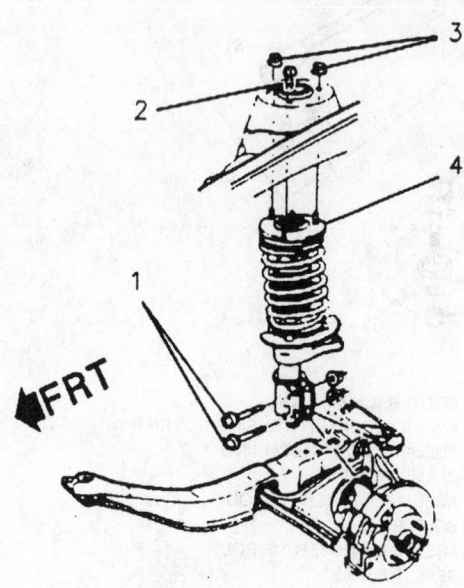

1 BOLT
2 BOLT — 25 N·m (18 LBS. FT.)
3 NUT — 25 N·m (18 LBS. FT.)
4 STRUT ASSEMBLY

79222Z129

Exploded view of the front strut mounting

30. Tighten the upper strut mounting nuts and bolt and tighten to 18 ft. lbs. (25 Nm).

31. Check the front end alignment and adjust as necessary.

Shock Absorber

REMOVAL & INSTALLATION

Rear

1. Open the deck lid.
2. Remove the shock absorber upper cover and remove the upper mounting nut.

➡**Remove one shock absorber at a time if both shocks are to be replaced. This will ensure that the axle assembly will not be suspended by the brake hoses.**

3. Raise and safely support the vehicle.
4. Support the rear axle with safety stands. When lifting vehicle with body hoist, it will be necessary to use adjustable safety stands.
5. Remove the shock absorber nut and remove the shock absorber.

   To install:
6. Install the shock absorber at the lower attachment. Install the nut hand-tight.
7. Lower the vehicle enough to guide upper stud through body opening and install upper shock mounting nut.
8. Tighten the lower shock absorber mounting nuts to 35 ft. lbs. (47 Nm).
9. Remove the axle support and lower vehicle all the way, then tighten shock

absorber upper nut to 21 ft. lbs. (29 Nm).

10. Install the shock absorber upper cover.

Coil Spring

REMOVAL & INSTALLATION

Front

1. Raise and safely support the vehicle. Remove the tire and wheel assembly.
2. Remove the strut assembly.
3. Place the strut assembly into strut compressor tool J-34013-A or equivalent.
4. With the strut firmly mounted in the strut compressor tool, compress the spring and remove the center nut for strut plate. Carefully back off the strut com-

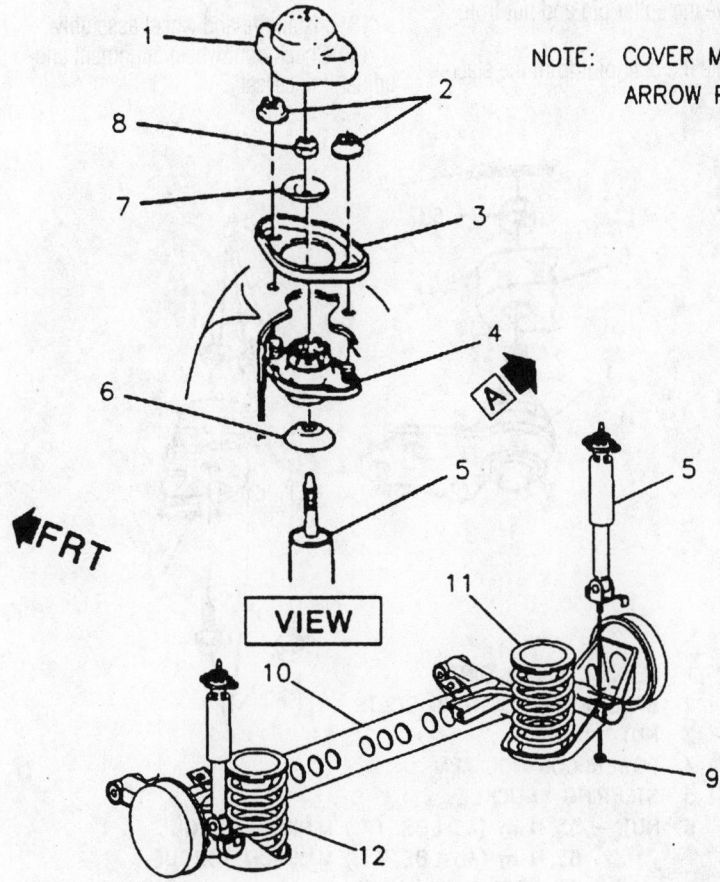

NOTE: COVER MUST BE INSTALLED SO THAT
ARROW POINTS TO LEFT SIDE OF VEHICLE

1 COVER (MUST BE INSTALLED
   WITH ARROW POINTING TO
   LEFT OF VEHICLE)
2 NUT
3 REINFORCEMENT
4 UPPER SHOCK ABSORBER
   MOUNT
5 SHOCK ABSORBER
6 WASHER
7 WASHER
8 NUT
9 NUT
10 REAR AXLE; SPRING-ON-CENTER TYPE
11 INSULATOR, COIL SPRING
12 COIL SPRING

VIEW

FRT

Exploded view of the rear shock mounting

79222130

pressor tool, after tension is fully relieved remove the coil spring from its mounting.

To install:

5. Place the coil spring into its mounting perch.

6. Place the strut mounting plate on top of the spring and attach the strut compressor.

7. Compress the coil spring enough so the mounting nut can be installed. Tighten this nut to 52 ft. lbs. (70 Nm).

8. Carefully back off the tension from strut compressor tool and remove the strut assembly.

9. Install the strut assembly into the vehicle.

10. Install the tire and wheel assembly. Lower the vehicle.

11. Check wheel alignment and road test vehicle.

Rear

✳✳ CAUTION

When removing rear springs do not use a twin-post type hoist. When certain fasteners are removed from the rear axle assembly, the axle may have a tendency to cause the hoist to slip, which may cause personal injury. Perform operation on the floor if necessary.

1. Raise and safely support the vehicle. Use safety stands to support the rear axle.

2. Remove the tire and wheel assemblies.

3. Remove the right and left brake line bracket attaching screws from the body and allow brake line to hang free.

4. Remove the lower mounting bolts from the shock absorbers.

➡**Do not suspend rear axle by the brake hoses. Damage to hoses could result.**

5. Lower the rear axle and remove the spring(s).

To install:

6. Position the coil spring in its mounting perch and raise axle. The end of the lower coil spring must be positioned in the spring seat and within 9/16 in. (15mm) of the spring stop.

7. Connect the shock absorbers to the rear axle. Tighten the nuts to 35 ft. lbs. (47 Nm).

8. Install the brake line brackets to body and tighten screws to 97 inch lbs. (11 Nm).

9. Install the tire and wheel assemblies. Remove the safety stands and lower the vehicle.

Lower Ball Joint

REMOVAL & INSTALLATION

1. Raise and safely support the vehicle.

2. If suspension contact hoist is used, place safety stands under the crossmember. Lower the vehicle slightly so the weight of the vehicle rests on the crossmember.

3. Remove the tire and wheel assembly.

➡**Care must be exercised to prevent the halfshaft joints from being over-extended. When either end of the shaft is disconnected, over-extension of the joint could result in separation of internal components and possible joint failure. Failure to observe this can result in interior joint or boot damage and possible joint failure.**

4. Remove the cotter pin and nut from the ball joint.

5. Separate the ball joint from the steering knuckle using tool J-38892 or equivalent.

6. Drill out the 3 rivets retaining ball joint to the lower control arm. Use an 1/8 in. (3mm) drill bit to make a pilot hole through the rivets. Finish drilling rivets with a 1/2 in. (13mm) drill bit.

7. Remove the nut attaching link to the stabilizer shaft.

8. Remove the ball joint from the steering knuckle and control arm.

To install:

9. Install ball joint into the control arm.

10. Install 3 new bolts and nuts (supplied with new ball joint). Check the ball joint service kit for an instruction sheet listing bolt torque. If none is supplied, tighten the bolts to 40 ft. lbs. (55 Nm).

11. Position ball joint stud through the steering knuckle and tighten nut to 50 ft. lbs. (65 Nm) and install cotter pin.

12. Connect stabilizer link to stabilizer shaft and tighten nut to 13 ft. lbs. (17 Nm).

13. Install tire and wheel assembly.

14. Check front wheel alignment and adjust if necessary.

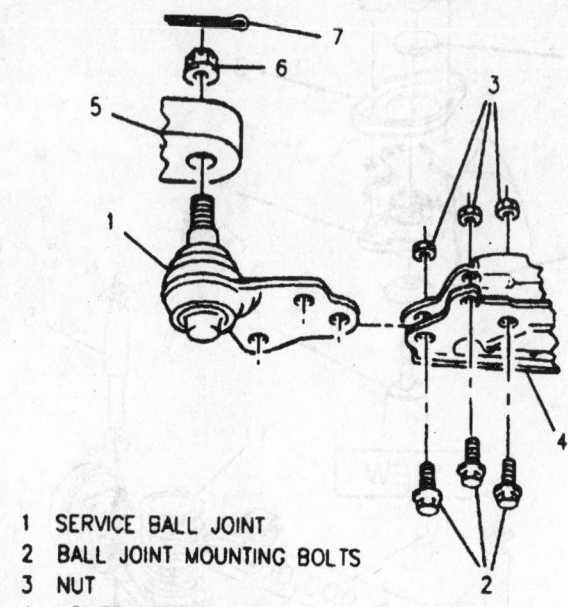

1	SERVICE BALL JOINT
2	BALL JOINT MOUNTING BOLTS
3	NUT
4	LOWER CONTROL ARM
5	STEERING KNUCKLE
6	NUT — 55 N·m (41 LBS. FT.) MINIMUM TORQUE 65 N·m (48 LBS. FT.) MAXIMUM TORQUE TO INSTALL PIN
7	PIN

7922Z119

Exploded view of the replacement ball joint assembly mounting

Wheel Bearings

ADJUSTMENT

Front

The wheel bearings are not adjustable. If a wheel bearing is out of specifications, it must be replaced. Using a dial indicator, check for looseness. If it exceeds 0.005 in. (0.1270mm) on drum or disc brakes the bearing wear is excessive and the hub and bearing should be replaced.

Rear

The wheel bearings are not adjustable. If a wheel bearing is out of specifications, it must be replaced. Using a dial indicator, check for looseness. If it exceeds 0.005 in. (0.1270mm) on drum or disc brakes the bearing wear is excessive and the hub and bearing should be replaced.

REMOVAL & INSTALLATION

Front

The front wheel bearings on these vehicles are not serviced separately. The wheel bearings are pressed into the hub. If the wheel bearings are noisy or defective, the hub and bearing assembly must be replaced.

1. Loosen the axle hub nut.
2. Raise and safely support the vehicle. Remove the tire and wheel assembly.
3. Install a special boot cover J-33162 or equivalent on the L4 vehicles with automatic transaxle only. This is to protect the soft CV-joint boot.
4. Remove the axle hub nut.
5. Remove the brake caliper and rotor.
6. Remove the 3 hub and bearing mounting bolts.
7. Remove the splash shield.

➡️**If the hub and bearing assembly is to be reused, mark the attaching bolt and corresponding hole for installation in the same position. Use of a hammer or direct heat on the halfshaft should be avoided during removal, as this could result in internal bearing damage.**

8. Install special tool J-28733 or equivalent hub puller and turn the bolt to press the hub and bearing assembly off the end of the drive axle (halfshaft).

9. Disconnect the stabilizer link bolt at the lower control arm.
10. Separate the lower ball joint using special tool J-29330 or equivalent.
11. Remove the drive axle from the steering knuckle and support it aside.
12. If removing the steering knuckle use the following procedure:
    a. Using a sharp tool, scribe the outline of the strut onto the steering knuckle.
    b. Remove the cotter pin and castle nut from the outer tie rod end and using a steering linkage puller, separate the tie rod from the steering knuckle.
    c. Remove the bolts attaching the strut to the steering knuckle and remove the knuckle.
    d. Using a file, elongate the lower strut-to-knuckle hole for camber alignment adjustment.

To install:
13. Clean and inspect the hub and bearing surfaces and the steering knuckle bore.
14. If the steering knuckle was removed use the following procedure.
    a. Install the steering knuckle onto the lower ball joint and tighten the ball joint nut to 48 ft. lbs. (65 Nm).
    b. Connect the steering knuckle to the lower strut bracket and install the through-bolts and nuts. With the matchmarks in alignment, tighten the nuts to 122 ft. lbs. (165 Nm).
    c. Connect the stabilizer link bolt at the lower control arm and tighten the nut to 22 ft. lbs. (30 Nm).
15. Install the hub and bearing assembly and torque the 3 attaching bolts to 70 ft. lbs. (90 Nm).
16. If removed, connect the tie rod end to the steering knuckle and tighten the castle nut to 30 ft. lbs. (41 Nm). If necessary to align the cotter pin holes tighten the castle nut up to 60 degrees (⅙ turn) additional. Install a new cotter pin.
17. Install the drive axle (halfshaft) to the hub and bearing assembly being careful to align the axle splines to the hub and not to damage the seal.
18. Install a new hub and bearing nut on the drive axle and apply a partial torque of 75 ft. lbs. (100 Nm).
19. Install the caliper and rotor assembly.
20. Install the wheel and tire assembly.
21. Safely lower the vehicle and apply a final torque to the axle hub nut to 190 ft. lbs. (260 Nm).
22. Perform a front-end alignment check, and adjust if required. Road test vehicle.

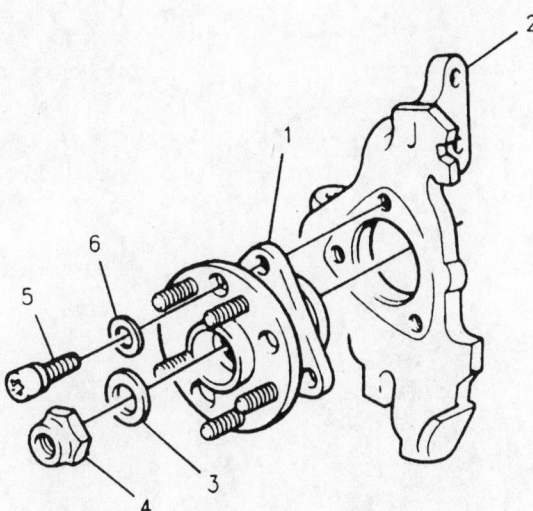

1 HUB AND BEARING ASSEMBLY
2 STEERING KNUCKLE
3 WASHER
4 DRIVE AXLE NUT – 260 N·m (192 LBS. FT.)
5 HUB AND BEARING RETAINING BOLT
6 WASHER

79222131

Exploded view of the front wheel bearing mounting

Rear

1. Raise and safely support the vehicle. Remove the tire and wheel assembly.

2. Remove the brake drum.

3. Remove the nuts from the hub bearing mounting bolts. With the nuts removed, push the mounting bolts out of the hub bearing assembly. The top rear bolt will not clear the brake shoe and has to be left in the hub flange when removing the hub assembly.

4. Remove the rear ABS wheel speed sensor wiring harness.

5. Remove the bearing assembly.

To install:

6. Connect the rear ABS wheel speed sensor wire connector.

7. Install the bearing assembly onto the backing plate with the top rear bolt already installed through the bearing bolt hole.

8. Install the remainder of the mounting bolts.

9. Install the mounting nuts and tighten to 44 ft. lbs. (60 Nm).

10. Install the brake drum.

11. Install the tire and wheel assembly and tighten the wheel nuts to 100 ft. lbs. (140 Nm).

12. Lower the vehicle.

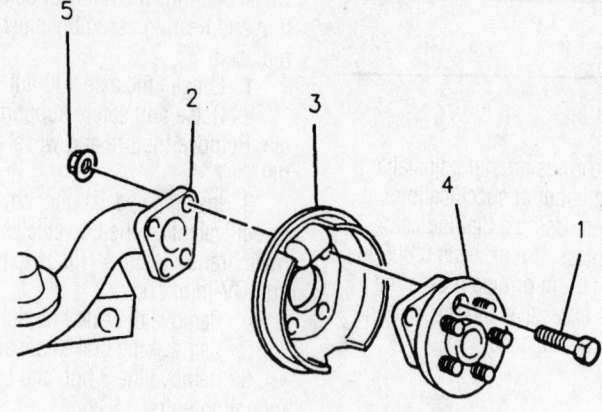

1 BOLT
2 REAR AXLE ASSEMBLY
3 BACKING PLATE
4 HUB AND BEARING ASSEMBLY
5 LOCKNUT

79222132

Exploded view of the rear wheel bearing

GENERAL MOTORS L/N BODY 35

Chevrolet-Malibu • Oldsmobile-Cutlass

DRIVE TRAIN **35-24**
ENGINE REPAIR **35-2**
FUEL SYSTEM **35-20**
STEERING AND
  SUSPENSION **35-26**

A
Air Bag 35-26
  Disarming 35-26
  Precautions 35-26
  Rearming 35-26

C
Camshaft and Valve Lifters 35-13
  Removal & Installation 35-13
Coil Spring 35-29
  Removal & Installation 35-29
Cylinder Head 35-5
  Removal & Installation 35-5

E
Engine Assembly 35-2
  Removal & Installation 35-2
Exhaust Manifold 35-11
  Removal & Installation 35-11

F
Fuel Filter 35-21
  Removal & Installation 35-21

Fuel Pump 35-22
  Removal & Installation 35-22
Fuel System Pressure 35-21
  Relieving 35-21
Fuel System Service
  Precautions 35-20

H
Halfshaft 35-25
  Removal & Installation 35-25

I
Ignition Timing 35-2
  Adjustment 35-2
Intake Manifold 35-8
  Removal & Installation 35-8

L
Lower Ball Joint 35-30
  Removal & Installation 35-30

O
Oil Pan 35-15
  Removal & Installation 35-15
Oil Pump 35-17
  Removal & Installation 35-17

P
Power Rack and Pinion Steering
  Gear 35-26

  Removal & Installation 35-26

R
Rear Main Seal 35-17
  Removal & Installation 35-17
Rocker Arms 35-7
  Removal & Installation 35-7

S
Strut 35-27
  Removal & Installation 35-27

T
Timing Chain, Sprockets, Front
  Cover and Seal 35-18
  Removal & Installation 35-18
Transaxle 35-24
  Removal & Installation 35-24

V
Valve Lash 35-15
  Adjustment 35-15

W
Water Pump 35-4
  Removal & Installation 35-4
Wheel Bearings 35-31
  Adjustment 35-31
  Removal & Installation 35-31

ENGINE REPAIR

➡**Disconnecting the negative battery cable on some vehicles may interfere with the functions of the on board computer systems and may require the computer to undergo a relearning process, once the negative battery cable is reconnected.**

Ignition Timing

ADJUSTMENT

Ignition timing is controlled by the PCM. No adjustment is necessary nor possible.

Engine Assembly

REMOVAL & INSTALLATION

2.4L Engine

1. If equipped with A/C, properly recover the refrigerant.
2. Disconnect the negative battery.

✳✳ CAUTION

Never open, service or drain the radiator or cooling system when hot; serious burns can occur from the steam and hot coolant.

3. Properly drain the cooling system into an approved container.
4. Relieve the fuel system pressure.
5. Remove the left sound insulator, then disconnect the clutch pushrod from the pedal assembly.
6. Disconnect the heater hose at the thermostat assembly, then detach the radiator inlet (upper) hose.
7. Remove the air cleaner assembly and the coolant fan.
8. If equipped with A/C, disconnect the compressor/condenser hose assembly at the compressor, then discard the O-rings.
9. Disconnect the 2 vacuum hoses from the front of the engine.
10. Tag and detach the following electrical connectors:
- Alternator
- A/C compressor (if equipped)
- Fuel injector harness
- Idle Air Control (IAC) and TP sensor at the throttle body
- Manifold Absolute Pressure (MAP) sensor

- Intake Air Temperature (IAT) sensor
- EVAP canister purge solenoid
- Starter solenoid
- Ground connections
- Negative battery cable from the transaxle
- Electronic ignition coil and module assembly
- Engine Coolant Temperature (ECT) sensor(s)
- Oil pressure sensor/switch
- Oxygen (O₂) sensor
- Crankshaft Position (CKP) sensor
- Back-up lamp switch, then position the harness aside

11. Disconnect the power brake vacuum hose from the throttle body. Detach the power brake vacuum tube-to-check valve hose from the tube.
12. Remove the throttle cable and bracket.
13. Unfasten the power steering pump rear bracket, then remove the bracket and vacuum tube as an assembly.
14. Unfasten the power steering pivot bolt, then remove the pump and drive belt. Position the pump aside, with the lines still attached.
15. Carefully disconnect the fuel lines.
16. Disconnect the shift cables. Detach the clutch actuator line.
17. Remove the exhaust manifold and heat shield.
18. Disconnect the radiator outlet (lower) hose from the radiator.
19. Install J-28467-A or equivalent engine support fixture.
20. Unfasten the bolt attaching the coolant recovery/surge tank, then position the tank aside with the hoses still connected.
21. Remove the engine mount assembly.
22. Raise and safely support the vehicle, then remove the front wheel and tire assemblies. Remove the right splash shield.

✳✳ CAUTION

The EPA warns that prolonged contact with used engine oil may cause a number of skin disorders, including cancer! You should make every effort to minimize your exposure to used engine oil. Protective gloves should be worn when changing the oil. Wash your hands and any other exposed skin areas as soon as possible after exposure to used engine oil. Soap and water, or waterless hand cleaner should be used.

23. Drain and recycle the engine oil.
24. Remove the radiator air deflector.

25. Tag and detach the following electrical connections:
- Vehicle Speed Sensor (VSS)
- Knock sensor
- Starter solenoid
- If equipped, both front ABS wheel speed sensors
26. Remove the engine mount strut and the transaxle mount.
27. Separate the ball joints from the steering knuckles.
28. Remove the suspension supports, crossmember, and stabilizer shaft as an assembly.
29. Disconnect the heater outlet hose from the radiator outlet pipe.
30. Remove the axle shaft from the transaxle and intermediate shaft, then position aside.
31. If equipped, disconnect the A/C lines from the oil pan.
32. Remove the flywheel housing cover.
33. Position a suitable support below the engine, then carefully lower the vehicle onto the support.
34. Matchmark the threads on the support fixture hooks so the setting can be duplicated when reinstalling the engine. Remove the engine support fixture J-hooks.
35. Raise the vehicle slowly off the engine and transaxle assembly. If may be necessary to move the engine/transaxle assembly rearward to clear the intake manifold.
36. Noting the position of the bolts, separate the engine from the transaxle.

To install:

✳✳ WARNING

Be sure the retaining bolts are in their correct locations. If not, engine damage may occur.

37. Assemble the engine to the transaxle.
38. Position the engine and transaxle assembly under the engine compartment, then slowly lower the vehicle over the assembly until the transaxle mount is indexed, then install the retaining bolt.
39. Install engine support fixture J-28467-A or equivalent, making sure to adjust it to the previous setting.
40. Install the engine mount assembly and transaxle mount.
41. Carefully raise the vehicle off the support.
42. Attach the axle shafts to the transaxle.
43. Connect the heater outlet hose to the radiator outlet pipe.

44. Install the suspension supports, crossmember and stabilizer shaft assembly.

45. Attach the ball joints to the steering knuckles, then secure with the nuts.

46. Install the engine strut mount.

47. If equipped, connect the A/C line to the oil pan.

48. Attach the following electrical connectors, as tagged during removal:
- Vehicle Speed Sensor (VSS)
- Knock sensor
- Starter solenoid
- If equipped, both front ABS wheel speed sensors

49. Install the flywheel housing cover.

50. Fasten the radiator air deflector.

51. Connect the lower radiator hose.

52. Install the right splash shield, then the front wheel and tire assemblies.

53. Carefully lower the vehicle, then remove the engine support fixture.

54. Install the coolant recovery/surge tank, then secure using the retaining bolt.

55. Attach the following electrical connections, as tagged during removal:
- Alternator
- A/C compressor (if equipped)
- Fuel injector harness
- Idle Air Control (IAC) and TP sensor at the throttle body
- Manifold Absolute Pressure (MAP) sensor
- Intake Air Temperature (IAT) sensor
- EVAP canister purge solenoid
- Starter solenoid
- Ground connections
- Negative battery cable to the transaxle
- Electronic ignition coil and module assembly
- Engine Coolant Temperature (ECT) sensor(s)
- Oil pressure sensor/switch
- Oxygen (O$_2$) sensor
- Crankshaft Position (CKP) sensor
- Back-up lamp switch

56. Attach the vacuum hoses.

57. If equipped with A/C, attach the compressor/condenser hose assembly to the compressor.

58. Fasten the clutch actuator line.

59. Install the exhaust manifold and heat shield.

60. Connect the fuel lines.

61. Connect the positive battery cable.

62. Fasten the power steering pump pivot-to-block bolt. Install the power steering pump rear bracket and tension belt.

63. Connect the vacuum hoses to the intake manifold and to the tube from the brake booster.

64. Install the throttle cable and bracket.

65. Install the coolant fan and air cleaner assembly.

66. Attach the radiator outlet (upper) hose. Fill the cooling system with the proper type and quantity of coolant.

67. Connect the clutch pushrod to the pedal assembly, then install the left sound insulator.

68. Attach the heater hose at the thermostat housing.

✴✴ WARNING

Operating the engine without the proper amount and type of engine oil will result in severe engine damage.

69. Fill the transaxle with fluid, then fill the crankcase with oil.

70. Connect the negative battery cable.

71. If equipped with A/C, evacuate, charge and leak test the system.

72. Start the engine and check the fluid levels, proper operation of the engine and/or fluid leakage.

3.1L Engine

Please note that the engine and transaxle are removed as an assembly from under the vehicle.

1. Relieve the fuel system pressure. Disconnect the negative battery cable.

✴✴ CAUTION

Never open, service or drain the radiator or cooling system when hot; serious burns can occur from the steam and hot coolant.

2. Remove the top half of the air cleaner assembly and the throttle body inlet duct. Drain the cooling system.

3. Remove the upper and lower radiator hoses. Disconnect the coolant inlet line from the coolant surge tank.

4. Disconnect the vacuum hoses from the EVAP canister purge valve, vacuum modulator and power brake booster.

5. Disconnect the heater outlet hose from the water pump. Remove the serpentine drive belt.

6. Disconnect the control cables from the throttle body lever and intake manifold bracket and position the cables aside.

7. Label and detach the electrical connectors from the necessary components.

8. Remove the alternator. Disconnect and cap the power steering lines from the power steering pump.

9. Disconnect and cap the fuel lines from the fuel rail. Remove the cooling fan assembly.

10. Disconnect the shift control cable from the transaxle shift lever and cable bracket. Disconnect the transaxle vent tube from the transaxle.

11. Disconnect the vacuum hose from the vacuum reservoir. Install an engine support fixture, J-28467-A, or the equivalent.

12. Loosen, but do not remove the top two A/C compressor mounting bolts. Raise and safely support the vehicle.

13. Remove the front tire and wheel assemblies. Remove the right and left inner fender splash shields.

✴✴ CAUTION

The EPA warns that prolonged contact with used engine oil may cause a number of skin disorders, including cancer! You should make every effort to minimize your exposure to used engine oil. Protective gloves should be worn when changing the oil. Wash your hands and any other exposed skin areas as soon as possible after exposure to used engine oil. Soap and water, or waterless hand cleaner should be used.

14. Drain and recycle the engine oil.

15. Remove the engine mount strut. Disconnect the ABS sensor wires from the wheel sensors and suspension member supports, if equipped.

16. Remove the cotter pins and castle nuts from the lower ball joints. Using a suitable tool, separate the lower ball joints from the steering knuckles.

17. Remove the lower suspension support assemblies with the lower control arms attached. Disconnect the halfshafts from the transaxle and support aside.

18. Remove the oil filter and oil filter adapter. Remove the flywheel cover. Remove the starter. Disconnect the heater hoses from the heater core.

19. Remove the A/C compressor lower mounting bolts and remove the compressor from the mounting bracket and position aside. DO NOT disconnect the refrigerant lines from the compressor or allow the lines top support the weight of the compressor.

20. Remove the vacuum reservoir tank. Disconnect the exhaust pipe from the exhaust manifold and position the pipe aside.

21. Remove the engine mount strut bracket from the engine. Disconnect the transaxle cooler lines from the radiator.

22. Remove the transaxle oil fill tube.

Lower the vehicle until the power train assembly is resting on a suitable engine table.

23. Remove the transaxle mount-to-body bolts, and the intermediate bracket from the right engine mount.

24. Remove the engine support fixture. Raise the vehicle leaving the powertrain assembly on the engine table.

25. Separate the engine and transaxle assemblies.

To install:

26. Connect the transaxle to the engine and tighten the mounting bolts to 55 ft. lbs. (75 Nm).

27. Position the powertrain assembly under the vehicle and lower the vehicle into position. Loosely install the serpentine belt.

28. Complete the installation by reversing the removal procedures.

29. Tighten the following items to:
• Engine strut bracket-to-engine bolts: 44 ft. lbs. (60 Nm)
• Exhaust pipe-to-exhaust manifold bolts: 18 ft. lbs. (25 Nm)
• Ball joints-to-steering knuckle nuts: 41 ft. lbs. (55 Nm)

✳✳ WARNING

Operating the engine without the proper amount and type of engine oil will result in severe engine damage.

30. Connect the negative battery cable. Check and fill all the engine fluids as necessary. Start the vehicle and bleed the power steering system.

Water Pump

REMOVAL & INSTALLATION

2.4L Engine

1. Disconnect the negative battery cable.

✳✳ CAUTION

Never open, service or drain the radiator or cooling system when hot; serious burns can occur from the steam and hot coolant.

2. Drain the cooling system.

3. Detach the oxygen sensor (O_2) electrical connector.

4. Remove the upper and lower exhaust manifold heat shields.

5. Raise and safely support the vehicle.

6. Remove the exhaust manifold brace-to-manifold bolt.

7. Loosen the exhaust pipe-to-manifold spring loaded nuts.

8. Remove the radiator outlet pipe-to-water pump cover bolts.

9. Raise and safely support the vehicle.

10. Disconnect the exhaust pipe from the manifold.

11. Disconnect the radiator outlet pipe from the oil pan and transaxle.

12. If equipped with a manual transaxle, remove the exhaust manifold brace. If equipped with an automatic transaxle, leave the lower radiator hose attached and pull down on the radiator pipe to disconnect it from the water pump.

13. Lower the vehicle.

14. Remove the exhaust manifold.

15. Remove the water pump cover-to-engine bolts.

16. Remove the three water pump-to-timing chain housing nuts.

17. Remove the water pump and cover assembly.

18. Remove the five water pump cover-to-pump bolts and remove the cover.

To install:

19. Install the water pump-to-engine components and tighten the bolts finger-tight.

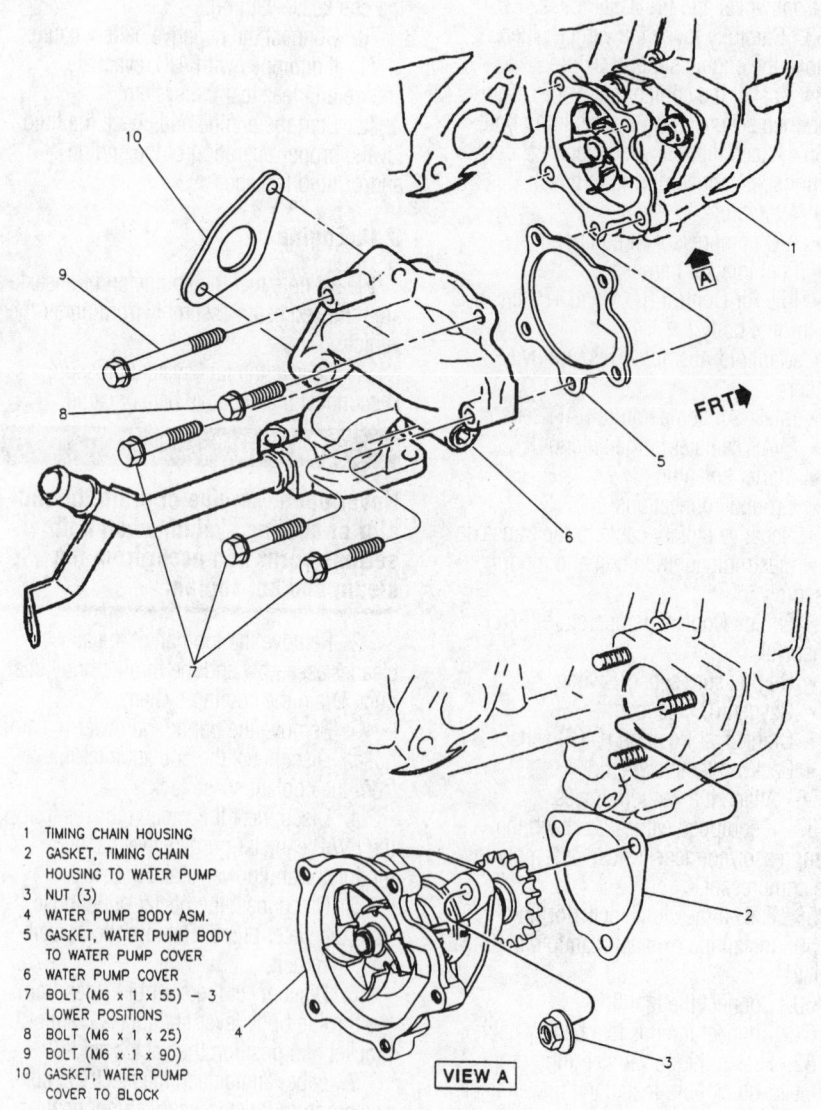

1 TIMING CHAIN HOUSING
2 GASKET, TIMING CHAIN HOUSING TO WATER PUMP
3 NUT (3)
4 WATER PUMP BODY ASM.
5 GASKET, WATER PUMP BODY TO WATER PUMP COVER
6 WATER PUMP COVER
7 BOLT (M6 x 1 x 55) – 3 LOWER POSITIONS
8 BOLT (M6 x 1 x 25)
9 BOLT (M6 x 1 x 90)
10 GASKET, WATER PUMP COVER TO BLOCK

VIEW A

7922Z201

Exploded view of the water pump mounting—2.4L engine

20. Lubricate the splines of the water pump drive with chassis grease.

21. To complete the installation, reverse the removal procedures.

22. With all the bolts in place tighten in the following sequence:
- Water pump-to-timing chain housing: 19 ft. lbs. (26 Nm)
- Water pump cover-to-water pump: 106–124 inch lbs. (12–14 Nm)
- Water pump cover-to-engine block (bottom first): 19 ft. lbs. (26 Nm)
- Radiator outlet pipe-to-pump cover: 10 ft. lbs. (14 Nm).

23. Refill the cooling system. Connect the negative battery cable. Start the vehicle and verify no leaks.

3.1L Engine

> **✳✳ CAUTION**
>
> **Never open, service or drain the radiator or cooling system when hot; serious burns can occur from the steam and hot coolant.**

1. Disconnect the negative battery cable. Drain the cooling system into a suitable container.

2. Loosen, but do not remove, the water pump pulley bolts. Remove the serpentine belt.

3. Remove the water pump pulley bolts and pulley. Remove the five water pump mounting bolts and pump.

To install:

4. Clean all the gasket surfaces completely. Apply a thin bead of sealer around the outside edge of the water pump along the gasket sealing area and install the gasket onto the water pump.

5. Install the water pump and tighten the bolts to 89 inch lbs. (10 Nm).

6. Install the water pump pulley and tighten the pulley bolts finger-tight. Install the serpentine belt.

7. Tighten the water pump pulley bolts to 18 ft. lbs. (25 Nm). Connect the negative battery cable.

8. Refill and bleed the cooling system. Test run the engine and check for leaks.

Cylinder Head

REMOVAL & INSTALLATION

> **✳✳ CAUTION**
>
> **The fuel injection system remains under pressure, even after the engine has been turned OFF. The fuel system pressure must be relieved before disconnecting any fuel lines. Failure to do so may result in fire and/or personal injury.**

2.4L Engine

> **✳✳ CAUTION**
>
> **Never open, service or drain the radiator or cooling system when hot; serious burns can occur from the steam and hot coolant.**

1. Disconnect the negative battery cable. Drain the cooling system.

2. Disconnect the heater inlet and throttle body heater hoses from water outlet. Remove the power brake vacuum hose from the throttle body.

3. Disconnect and tag for identification, if necessary, the electrical connections from the following components:
- MAP sensor
- MAT sensor
- EVAP purge solenoid
- Camshaft sensor

4. Remove the stud-ended bolt from the alternator. Remove the intake manifold brace. Remove the intake manifold.

5. Install the stud-ended alternator bolt back into the engine. Install engine support fixtures J-28467-400 and J-28467-A or equivalent.

6. Remove the exhaust manifold. Remove the ignition coil and module assembly.

7. Disconnect the camshaft position sensor. Remove the power steering pump.

8. Disconnect the vacuum line from the fuel pressure regulator, fuel injector and fuel injector wiring harness.

9. Disconnect the fuel line clamp from the bracket on top of the intake camshaft housing. Remove the fuel line rail and position it aside (leaving the fuel lines attached).

10. Disconnect the timing chain housing at the intake camshaft housing, but do not remove from the vehicle.

11. Remove the electrical connection from the oil switch.

12. If equipped with an automatic

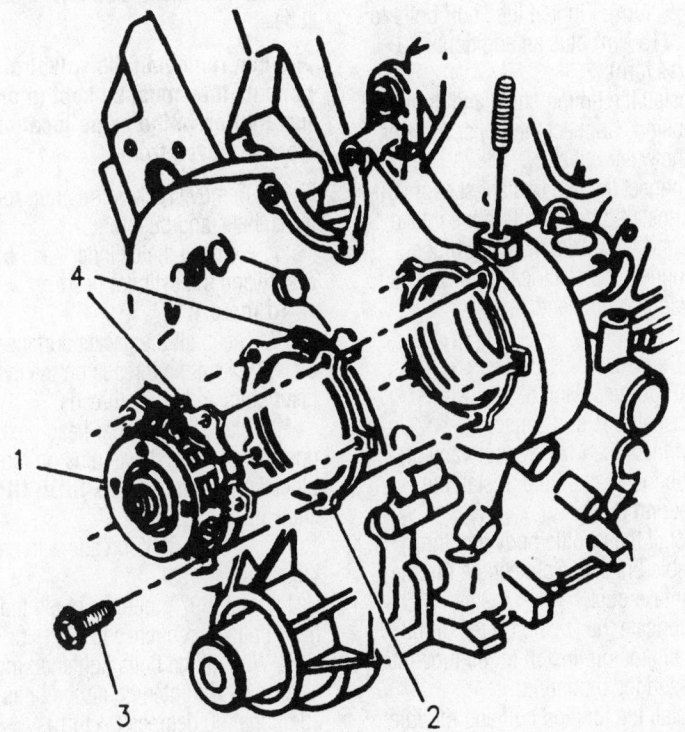

1	COOLANT PUMP
2	GASKET
3	BOLT – 10 N·m (89 LBS. IN.)
4	LOCATOR (MUST BE VERTICAL)

79222202

Exploded view of the water pump mounting—3.1L engine

79222Z203

Cylinder head bolt tightening sequence—2.4L engine

transaxle, remove the transaxle fluid level indicator tube assembly from the exhaust camshaft cover and position it aside.

➡️**Any time the camshaft housing to cylinder head bolts are loosened or removed, the camshaft housing to cylinder head gasket must be replaced.**

13. Remove the intake camshaft housing bolts, the housing and gasket.

➡️**Turn the camshaft housing upside down as soon as it is removed from the cylinder head. The lifters will fall out of the camshaft housing if it is not turned upside down. The lifters can be damaged if they fall out and hit a hard surface.**

14. Remove the exhaust camshaft housing from the cylinder head (remove the bolts in reverse of the tightening sequence).

15. Remove the radiator inlet (upper) hose from the coolant outlet. Disconnect the coolant temperature sensor electrical connections.

16. Remove the cylinder head bolts (in reverse of the tightening sequence), the cylinder head and gasket.

➡️**Do not use abrasive pads to clean the cylinder head or block surfaces. An abrasive pad may damage the cylinder head and block. Use a razor or scraper to clean the gasket surfaces.**

To install:

17. Clean all gasket surfaces completely. Be sure the threaded holes in the engine block for the cylinder head bolts are clean.

18. Install a new cylinder head gasket on the cylinder block and carefully position the cylinder head in place.

19. New replacement head bolts are recommended. Coat the cylinder head bolt

threads with clean engine oil and allow the oil to drain off before installing.

20. Install the cylinder head bolts and tighten, in sequence, as follows:

   a. No's 1 through 8: 40 ft. lbs. (65 Nm)

   b. No's 9 and 10: 30 ft. lbs. (40 Nm)

   c. All 10 bolts: an additional 90 degrees (¼ turn).

21. Install the intake and exhaust camshaft housings. Tighten the long bolts to 11 ft. lbs. (15 Nm) plus an additional 90 degrees (¼ turn). Tighten the short bolts to 11 ft. lbs. (15 Nm) plus an additional 30 degrees (⅙ turn).

22. Install the timing chain and timing chain housing. Connect the upper radiator hose to the water outlet.

23. Connect the two coolant sensor connections. Install the fuel rail on the intake manifold and the intake manifold brace.

24. Connect the electrical connections to the following components:
   • MAP sensor
   • MAT sensor
   • EVAP purge solenoid
   • Camshaft position sensor

25. Connect the MAP sensor vacuum hose to the intake manifold. Install the power steering pump.

26. Install the throttle body and the throttle cable bracket. Connect the throttle body air intake duct.

27. Lubricate the O-ring on the oil fill tube with engine oil. Install oil fill tube into the block and the mounting.

28. Install the ignition coil and module assembly. Attach the injector harness electrical connector.

29. Remove the engine support fixture. Install the exhaust manifold.

30. Fill all fluids to their proper levels. An oil and filter change is recommended.

31. Connect the negative battery cable. Start the vehicle and verify no leaks.

3.1L Engine

LEFT (FRONT) CYLINDER HEAD

✳✳ CAUTION

Never open, service or drain the radiator or cooling system when hot; serious burns can occur from the steam and hot coolant.

1. Relieve the fuel system pressure. Disconnect the negative battery cable. Drain the cooling system.

2. Remove the top half of the air cleaner assembly and the throttle body air inlet duct.

3. Remove the exhaust crossover pipe heat shield and crossover pipe.

4. Disconnect the spark plug wires from spark plugs and wire looms and route the wires aside. Remove the rocker arm covers.

5. Remove upper intake plenum and lower intake manifold. Remove the left side exhaust manifold and oil level indicator tube.

➡️**When removing the valvetrain components they must be kept in order for installation in the same locations they were removed from.**

6. Remove rocker arms nut, rocker arms, balls and pushrods.

7. Remove the cylinder head bolts evenly and the cylinder head.

To install:

8. Clean all the gasket surfaces completely. Clean the threads on the cylinder head bolts and block threads.

9. Place the cylinder head gasket in position over the dowel pins on the cylinder block so the words **THIS SIDE UP** are showing.

10. Coat the bolt threads with sealer and install finger-tight.

11. On 1997 models, tighten the cylinder head bolts in sequence to 33 ft. lbs. (45 Nm). With all the bolts tightened make a second pass tightening all the bolts an additional 90 degrees (¼ turn).

12. On 1998–99 models, tighten the cylinder head bolts in sequence to 37 ft. lbs. (50 Nm). With all the bolts tightened make a second pass tightening all the bolts an additional 90 degrees (¼ turn).

13. Install the pushrods, rocker arms,

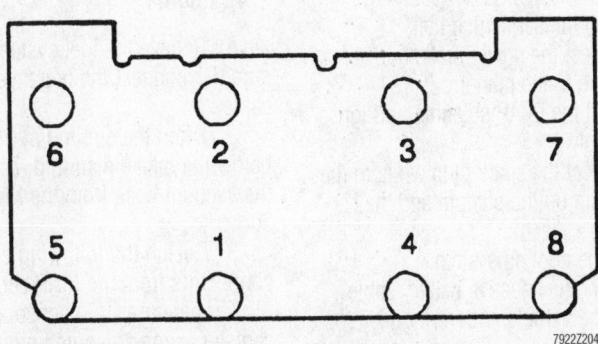

Cylinder head bolt tightening sequence—3.1L engine

balls and rocker arm nuts. Tighten the rocker arm nuts to 18 ft. lbs. (25 Nm).

14. Install the lower intake manifold, upper intake plenum and the rocker arm covers.

15. Install the oil level indicator tube. Connect the spark plug wires to spark plugs and wire looms.

16. Install the left side exhaust manifold. Install the exhaust crossover pipe and crossover pipe heat shield.

17. Refill the cooling system. Drain the engine oil and refill with the proper type of engine oil. A filter change is recommended.

18. Install the top half of the air cleaner assembly and the throttle body air inlet duct.

19. Connect negative battery cable. Start the engine and verify no leaks.

RIGHT (REAR) CYLINDER HEAD

✽✽ CAUTION

Never open, service or drain the radiator or cooling system when hot; serious burns can occur from the steam and hot coolant.

1. Relieve the fuel system pressure. Disconnect the negative battery cable. Drain the cooling system.

2. Remove the top half of the air cleaner assembly and remove the throttle body air inlet duct.

3. Remove the exhaust crossover pipe heat shield and crossover pipe. Raise and safely support the vehicle.

4. Disconnect the Oxygen (O_2) sensor. Disconnect the exhaust pipe from the exhaust manifold.

5. Remove the right side exhaust manifold. Lower the vehicle.

6. Disconnect the spark plug wires from spark plugs and wire looms and route the wires aside.

7. Remove the rocker arm covers, upper intake plenum and lower intake manifold.

➡️ **When removing the valvetrain components they must be kept in order for installation in the same locations they were removed from.**

8. Remove rocker arms nut, rocker arms, balls and pushrods.

9. Remove the cylinder head bolts evenly and the cylinder head.

To install:

10. Clean all the gasket surfaces completely. Clean the threads on the cylinder head bolts and block threads.

11. Place the cylinder head gasket in position over the dowel pins on the cylinder block so the words **THIS SIDE UP** are showing.

12. Coat the bolt threads with sealer and install finger-tight.

13. On 1997 models, tighten the cylinder head bolts in sequence to 33 ft. lbs. (45 Nm). With all the bolts tightened make a second pass tightening all the bolts an additional 90 degrees (¼ turn).

14. On 1998–99 models, tighten the cylinder head bolts in sequence to 37 ft. lbs. (50 Nm). With all the bolts tightened make a second pass tightening all the bolts an additional 90 degrees (¼ turn).

15. Install the pushrods, rocker arms, balls and rocker arm nuts. Tighten the rocker arm nuts to 18 ft. lbs. (25 Nm).

16. Install the lower intake manifold and upper intake plenum. Install the rocker arm covers.

17. Connect the spark plug wires to spark plugs and wire looms. Raise vehicle and safely support.

18. Install the exhaust manifold. Connect the exhaust pipe to the exhaust manifold.

19. Lower the vehicle. Connect the Oxygen (O_2) sensor.

20. Install the exhaust crossover pipe and heat shield. Refill the cooling system.

21. Drain the engine oil and refill with the proper type of engine oil. A filter change is recommended.

22. Install the top half of the air cleaner assembly and the throttle body air inlet duct.

23. Connect negative battery cable. Start the engine and verify no leaks.

Rocker Arms

REMOVAL & INSTALLATION

2.4L Engine

The 2.4L engine is not equipped with rocker arms. The camshafts directly actuates the valves.

3.1L Engine

1. Disconnect the negative battery cable.

2. For the left side (front), perform the following procedures:

✽✽ CAUTION

Never open, service or drain the radiator or cooling system when hot; serious burns can occur from the steam and hot coolant.

a. Drain the cooling system to a level below the coolant pipe on the front of the engine.

b. Remove the coolant bypass hose clamp at the coolant tube.

c. Remove the two bolts and nut securing the coolant tube to the cylinder head and position the tube aside.

d. Disconnect the PCV valve from the rocker arm cover.

3. For the right side (rear), perform the following procedures:

a. Disconnect the spark plug wires from the spark plugs and upper intake plenum wire retainer and position aside.

b. Disconnect the power brake booster vacuum pipe from the intake plenum.

c. Remove the serpentine belt.

d. Remove the alternator.

e. Disconnect and remove the ignition assembly and EVAP canister purge solenoid as an assembly.

4. Remove the four rocker arm cover bolts and remove the rocker arm cover.

5. Remove the rocker arm nuts, balls, rocker arms and pushrods.

To install:

6. Clean all the gasket surfaces completely.

7. Coat all the valvetrain components with engine oil prior to installation.

8. Install the pushrods and install the rocker arms on the studs. Install the rocker arm balls and mounting nuts. Be sure the pushrods are properly seated in the lifter and rocker arm. Tighten the mounting nuts to 89 inch lbs. (10 Nm) plus an additional 30 degrees.

9. Install the rocker arm cover using a new gasket and tighten the rocker cover bolts to 90 inch lbs. (10 Nm).

10. For the left side (front), perform the following procedures:

a. Connect the PCV valve to the rocker arm cover.

b. Position the coolant tube and connect the thermostat bypass hose.

c. Install the coolant tube mounting nut and bolts. Tighten the screw at the water pump to 106 inch lbs. (12 Nm), the bolt at the corner of the cylinder head to 18 ft. lbs. (25 Nm) and the nut to 18 ft. lbs. (25 Nm).

11. For the right side (rear), perform the following procedures:

a. Install the alternator.

b. Install the serpentine belt.

c. Connect the power brake booster vacuum pipe to the plenum.

d. Install the EVAP solenoid and ignition assembly.

e. Connect the spark plug wires to the wire retainers on the plenum and the spark plugs.

12. Refill the cooling system.

13. Connect the negative battery cable.

14. Start the vehicle and verify no leaks.

Intake Manifold

REMOVAL & INSTALLATION

❄ CAUTION

The fuel injection system remains under pressure, even after the engine has been turned OFF. The fuel system pressure must be relieved before disconnecting any fuel lines. Failure to do so may result in fire and/or personal injury.

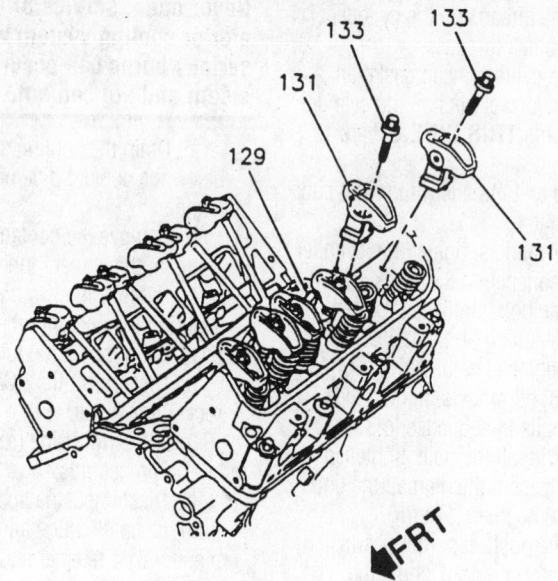

129 PUSHROD
131 ARM, ROLLER ROCKER
133 BOLT

79222Z205

Rocker arm components—3.1L engine

2.4L Engine

1. Relieve the fuel system pressure. Disconnect the negative battery cable.

2. Drain the cooling system to a level below the intake manifold. Disconnect the vacuum hose from the MAP sensor.

3. Detach the following electrical connectors from the MAP sensor, Intake Air Temperature (IAT) sensor, EVAP canister purge solenoid and fuel injectors.

4. Disconnect the vacuum hoses from the intake manifold, fuel pressure regulator and EVAP canister purge solenoid to the canister.

5. Remove the air cleaner duct. Remove the vent tube-to-air cleaner duct. Remove the control cable mounting bracket from the intake manifold.

6. Disconnect the power brake vacuum line from the intake manifold and remove the vacuum line mounting bracket at the power steering pump.

7. Disconnect the coolant lines from the throttle body. Remove the crankcase air/oil separator mounting bolts from the intake manifold.

8. Disconnect the hoses from the separator and the separator. Remove the oil fill tube.

9. Remove the intake manifold support brace, the intake manifold nuts/bolts and discard the old gasket.

To install:

10. Clean all the gasket surfaces completely. Install the intake manifold gasket and intake manifold.

11. Tighten the intake manifold fasteners in sequence, to 18 ft. lbs. (25 Nm).

12. Install the intake manifold brace and tighten the mounting bolts to 19 ft. lbs. (26 Nm). The brace-to-block bolts must be tightened first, then the brace-to-manifold bolt.

13. Complete the installation by reversing the removal procedures.

14. Lubricate a new oil fill tube O-ring seal with engine oil and install tube between No. 1 and 2 intake tubes. Rotate as necessary to gain clearance for oil/air separator nipple on fill tube.

15. Refill the cooling system. Connect the negative battery cable.

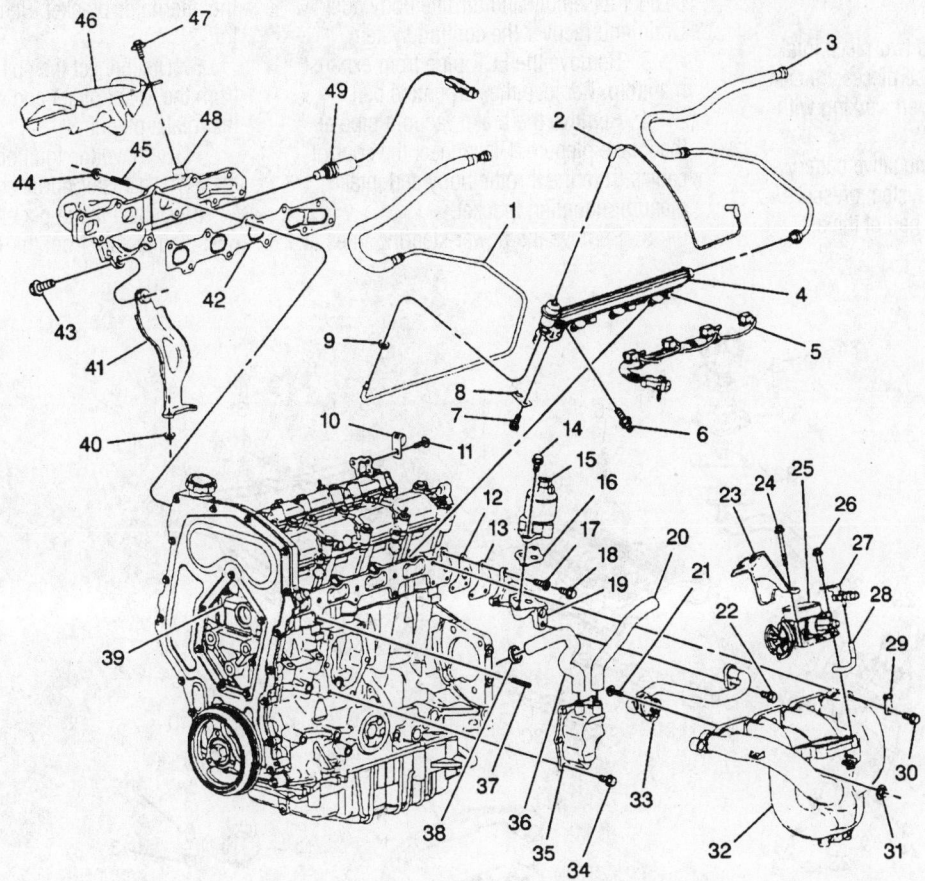

(1) Fuel Return Line
(2) Fuel Pressure Regulator Vacuum Harness
(3) Fuel Line
(4) Fuel Rail (with Injectors)
(5) Fuel Injector Harness
(6) Bolt, Fuel Rail to Cylinder Head
(7) Bolt, Fuel Line Retainer
(8) Fuel Line Retainer
(9) Fuel Line Seal
(10) Fuel Line Retainer
(11) Bolt, Fuel Line Retainer
(12) Gasket, EGR Adapter
(13) EGR Passage Cover (Export Only)
(14) Bolt, EGR Valve
(15) EGR Valve
(16) Gasket, EGR Valve
(17) Bolt, EGR Valve Adapter to Cylinder Head
(18) Bolt, EGR Passage Cover (Export Only)
(19) EGR Adapter
(20) Oil/Air Separator Fresh Air Tube
(21) Bolt, EGR Tube to Intake Manifold
(22) Bolt, EGR Tube to EGR Adapter
(23) Throttle Cable Bracket
(24) Bolt, Throttle Body to Intake Manifold
(25) Throttle Body
(26) Bolt, Throttle Body to Intake Manifold

(27) Manifold Absolute Pressure Sensor
(28) MAP Sensor Vacuum Line
(29) Fuel Rail Bracket
(30) Bolt, Intake Manifold to Cylinder Head
(31) Nut, Intake Manifold to Cylinder Head
(32) Intake Manifold
(33) EGR Pipe
(34) Bolt, Oil/Air Separator to Engine Block
(35) Oil/Air Separator
(36) Oil/Air Separator Foul Air Tube
(37) Stud, Intake Manifold to Cylinder Head
(38) Clamp, Oil/Air Separator Tube
(39) Cylinder Head
(40) Nut, Exhaust Manifold Brace to Oil Pan Stud
(41) Exhaust Manifold Brace
(42) Gasket, Exhaust Manifold
(43) Bolt, Exhaust Manifold Brace to Exhaust Manifold
(44) Nut, Exhaust Manifold to Cylinder Head
(45) Stud, Exhaust Manifold to Cylinder Head
(46) Exhaust Manifold Heat Shield
(47) Bolt, Exhaust Manifold Heat Shield to Exhaust Manifold
(48) Exhaust Manifold
(49) Oxygen Sensor

79222206

Exploded view of the intake and exhaust manifolds—2.4L engine

3.1L Engine

These vehicles use a two-piece intake manifold. Note that these pieces are cast aluminum. Use care when working with light alloy components.

1. Disconnect the negative battery cable. Relieve the fuel system pressure.

2. Remove the top half of the air cleaner assembly and throttle body duct. Drain and recover the cooling system.

3. Remove the EGR pipe from exhaust manifold. Remove the serpentine belt.

4. Remove the brake vacuum pipe at the intake plenum. Disconnect the control cables from the throttle body and intake plenum mounting bracket.

5. Remove the power steering lines at the alternator bracket. Remove the alternator.

6. Disconnect the spark plug wires from the spark plugs and wire retainers on the intake plenum.

7. Remove the Ignition assembly and the EVAP canister purge solenoid together.

8. Detach the upper engine wiring harness connectors from the Throttle Position

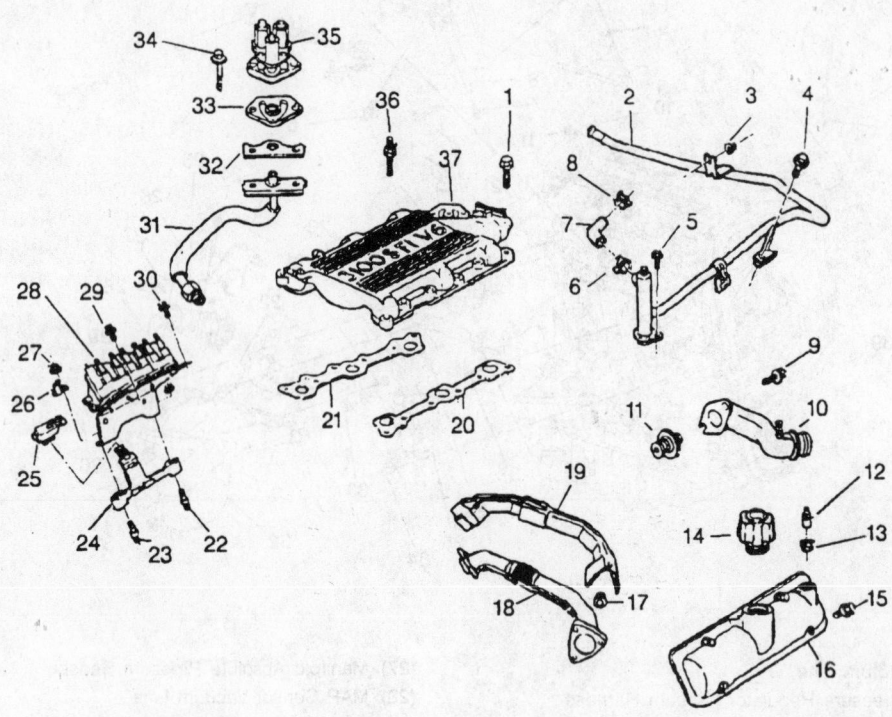

(1) Upper Intake Manifold Bolt
(2) Thermostat Bypass Pipe
(3) Thermostat Bypass Pipe Nut
(4) Thermostat Bypass Pipe Bolt
(5) Thermostat Bypass Pipe Screw
(6) Thermostat Bypass Hose Clamp
(7) Thermostat Bypass Hose
(8) Thermostat Bypass Hose Clamp
(9) Coolant Outlet Bolt
(10) Coolant Outlet Assembly
(11) Thermostat
(12) PCV Valve
(13) PCV Valve Grommet
(14) Oil Fill Cap
(15) Rocker Arm Cover Bolt
(16) Rocker Arm Cover
(17) Exhaust Crossover Nut
(18) Exhaust Crossover Pipe

(19) Exhaust Crossover Upper Heat Shield
(20) Upper Intake Manifold Gasket
(21) Upper Intake Manifold Gasket
(22) EVAP Canister Purge Valve Bracket Stud
(23) EVAP Canister Purge Valve Bracket Stud
(24) EVAP Canister Purge Valve Bracket
(25) EVAP Canister Purge Valve
(26) Spark Plug Wire Support
(27) Electronic Ignition System Nut
(28) Electronic Ignition System
(29) Electronic Ignition System Bolt
(30) Electronic Ignition System Bolt
(31) EGR Valve Pipe Assembly
(32) EGR Valve Pipe Gasket
(33) EGR Valve Gasket
(34) EGR Valve Assembly Bolt
(35) EGR Valve Assembly
(36) Upper Intake Manifold Stud
(37) Upper Intake Manifold

Exploded view of the upper and lower intake manifolds—3.1L engine

7922Z207

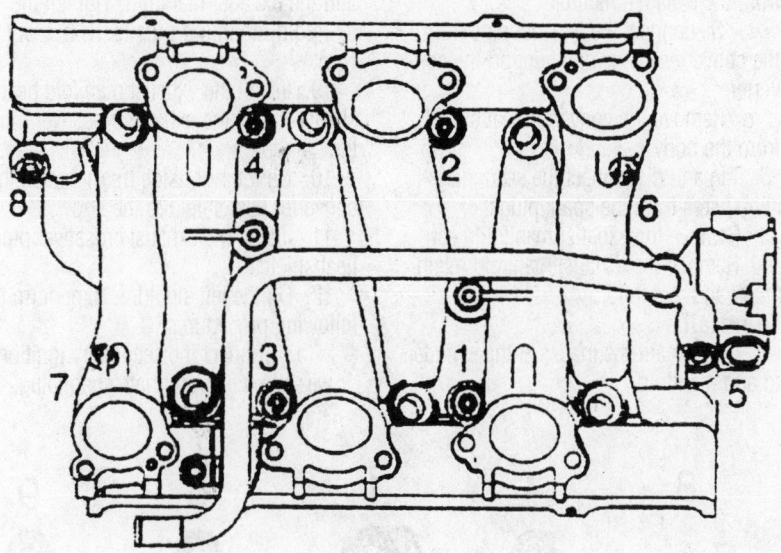

Lower intake manifold tightening sequence—3.1L engine

Sensor (TPS), Idle Air Control (IAC), fuel Injectors, coolant temperature sensor, MAP sensor and Camshaft Position (CMP) sensor.

9. Disconnect the vacuum lines from the vacuum modulator, fuel pressure regulator and PCV valve.

10. Remove the MAP sensor from upper intake manifold. Remove the upper intake plenum mounting bolts and the plenum.

11. Disconnect the fuel lines from the fuel rail and fuel line bracket. Install engine support fixture special tool J-28467-A or an equivalent.

12. Remove the right side engine mount. Remove the power steering mounting bolts and support the pump aside without disconnecting the power steering lines.

13. Disconnect the coolant inlet pipe from coolant outlet housing. Remove the coolant bypass hose from the water pump and the cylinder head.

14. Disconnect the upper radiator hose at thermostat housing.

15. Remove the thermostat housing. Remove both rocker arm covers.

16. Remove the lower intake manifold bolts. Be sure the washers on the four center bolts are installed in their original locations.

➡ **When removing the valvetrain components they should be kept in order for installation the original locations.**

17. Remove the rocker arm nuts, rocker arms and pushrods. Remove the intake manifold from the engine.

To install:

18. Clean gasket material from all mating surfaces. Remove all excess RTV sealant from front and rear ridges of cylinder block.

19. Place a 3mm bead of RTV, on each ridge, where the front and rear of the intake manifold contact the block.

20. Using a new gasket, install the intake manifold to the engine.

21. Install the pushrods, rocker arms and mounting nuts. Be sure the pushrods are properly seated in the valve lifters and rocker arms.

22. Install rocker arm nuts and tighten the rocker arm nuts to 18 ft. lbs. (24 Nm).

23. Install lower the intake manifold attaching bolts. Apply sealant PN 12345739 or equivalent to the threads of bolts, and tighten bolts to 115 inch lbs. (13 Nm).

24. Complete the installation by reversing the removal procedures.

25. Install the upper intake manifold and tighten the mounting bolts to 18 ft. lbs. (25 Nm).

26. Fill the cooling system. An engine oil and filter change is recommended.

27. Connect the negative battery cable. Start the vehicle and verify no leaks.

Exhaust Manifold

REMOVAL & INSTALLATION

2.4L Engine

1. Disconnect the negative battery cable. Disconnect the oxygen sensor.

2. Raise and safely support the vehicle.

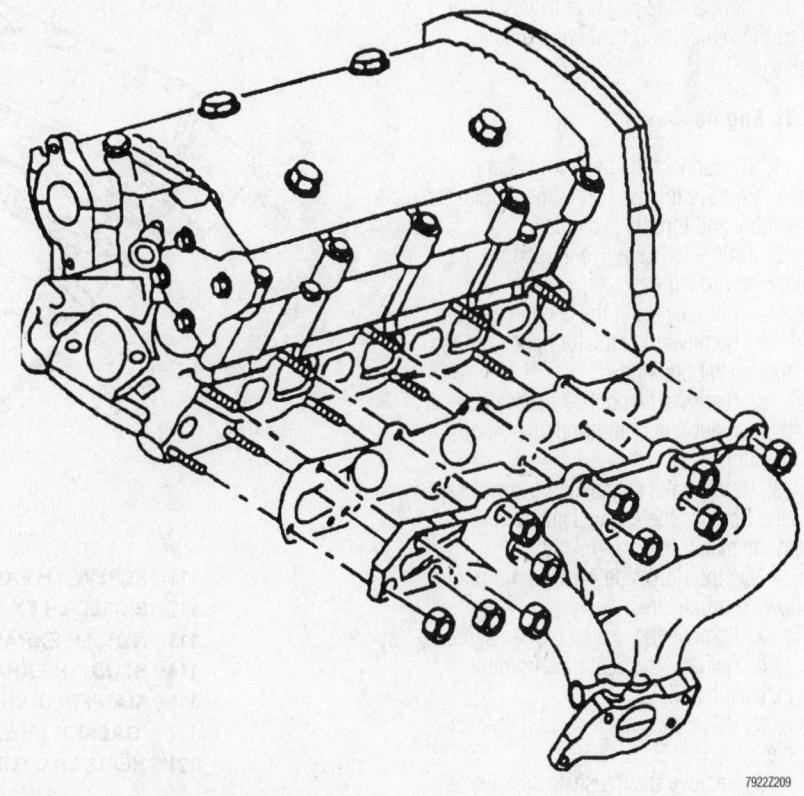

Exploded view of the exhaust manifold—2.4L engine

3. Remove the bolt that attaches the exhaust manifold brace to the manifold.

4. Remove the manifold to exhaust pipe spring loaded bolts using a 13mm box wrench. The nuts should be loosened alternately and evenly to prevent the pipe from binding the nuts.

5. Pull down and back on the exhaust pipe to disengage it from the exhaust manifold bolts.

➡**Do not bend the exhaust flex coupler more than 3 degrees in any direction. Movement of more than 3 degrees will damage the flex coupler.**

6. Lower the vehicle. Remove the exhaust manifold mounting nuts and manifold.

To install:

7. Clean all the gasket surfaces completely. Install the exhaust manifold and new gaskets.

8. Install the exhaust manifold-to-cylinder head retaining nuts and tighten in sequence to 110 inch lbs. (13 Nm).

9. Raise and safely support the vehicle. Install the heat shields. Install exhaust manifold brace to the manifold bolt.

10. Install the manifold-to-exhaust pipe nuts and tighten evenly to 26 ft. lbs. (35 Nm).

11. Lower the vehicle. Connect the oxygen sensor.

12. Connect the negative battery cable. Start the vehicle and verify no exhaust leaks.

3.1L Engine

1. Disconnect the negative battery cable. Remove the top half of the air cleaner assembly and throttle cable duct.

2. On the left side (front), perform the following procedures:

a. Partially drain the cooling system.

b. Remove the radiator hose from the thermostat housing.

c. Remove the coolant bypass hose at the coolant pump and from the exhaust manifold.

3. Remove the exhaust crossover heat shield. Remove the exhaust crossover pipe from the manifold.

4. On the right side (rear), perform the following procedures:

a. Remove the heated oxygen sensor.

b. Remove the EGR pipe from the exhaust manifold.

c. Raise and safely support the vehicle.

d. Remove the transaxle oil fill tube and lever indicator assembly.

e. Disconnect the front exhaust pipe from the exhaust manifold.

f. Disconnect the exhaust pipe from the converter flange and support the converter.

g. Remove the converter heat shield from the body.

5. Tag and disconnect the secondary ignition wires from the spark plugs.

6. Remove the exhaust manifold heat shield. Remove the exhaust manifold retaining nuts and manifold.

To install:

7. Clean mating surfaces at the cylinder head and manifold.

8. Install the exhaust manifold gasket and the exhaust manifold. Tighten the manifold mounting nuts to 12 ft. lbs. (16 Nm).

9. Install the exhaust manifold heat shield. Install the exhaust crossover pipe to the manifold.

10. On the right side (rear), install the converter heat shield to the body.

11. Install the exhaust crossover pipe heat shield.

12. On the left side (front), perform the following procedures:

a. Connect the secondary ignition wires to the appropriate spark plug.

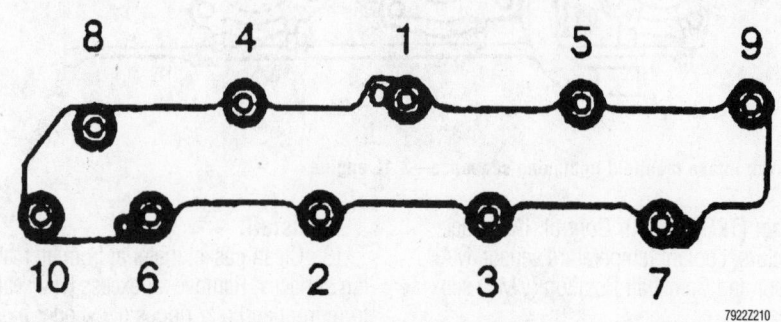

Exhaust manifold tightening sequence—2.4L engine

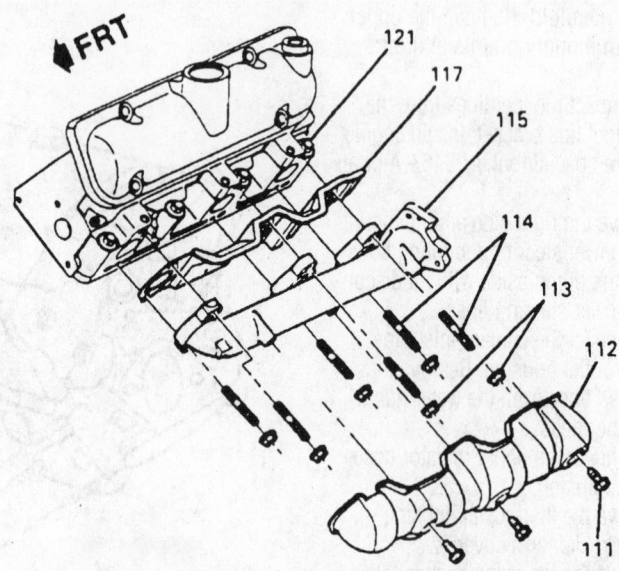

111	SCREW, LH EXHAUST MANIFOLD HEAT SHIELD
112	SHIELD LH EXHAUST MANIFOLD
113	NUT, LH EXHAUST MANIFOLD
114	STUD, LH EXHAUST MANIFOLD
115	MANIFOLD, LH EXHAUST
117	GASKET, LH EXHAUST MANIFOLD
121	HEAD, LH CYLINDER

Exploded view of the left-hand exhaust manifold—3.1L engine

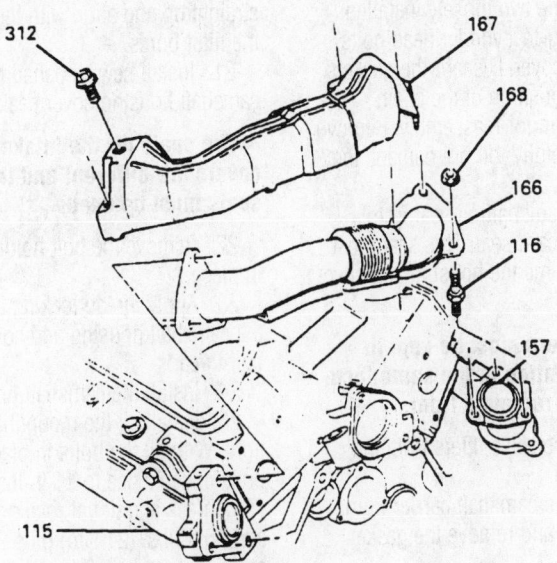

115 MANIFOLD, LEFT HAND EXHAUST
116 STUD, EXHAUST CROSSOVER
157 MANIFOLD, RIGHT HAND EXHAUST
166 CROSSOVER PIPE, EXHAUST
167 SHIELD, EXHAUST CROSSOVER UPPER HEAT
168 NUT, EXHAUST CROSSOVER
312 BOLT/SCREW, EXHAUST CROSSOVER
UPPER HEAT SHIELD

79222212

Exploded view of the exhaust crossover pipe—3.1L engine

b. Connect the coolant bypass pipe to the coolant pump and exhaust manifold.

c. Install the radiator hose to the coolant outlet housing.

13. On the right side (rear), perform the following procedures:

a. Connect the exhaust pipe to the exhaust manifold.

b. Install the transaxle oil level indicator and fill tube assembly.

c. Safely lower the vehicle.

d. Connect the heated oxygen sensor.

e. Connect the EGR pipe to exhaust manifold.

14. Install the top half of the air cleaner and the throttle body duct. Connect the negative battery cable.

Camshaft and Valve Lifters

REMOVAL & INSTALLATION

2.4L Engine

➡**Anytime the camshaft housing-to-cylinder head bolts are loosened or removed, the camshaft housing-to-cylinder head gasket must be replaced.**

1. Disconnect the negative battery cable.

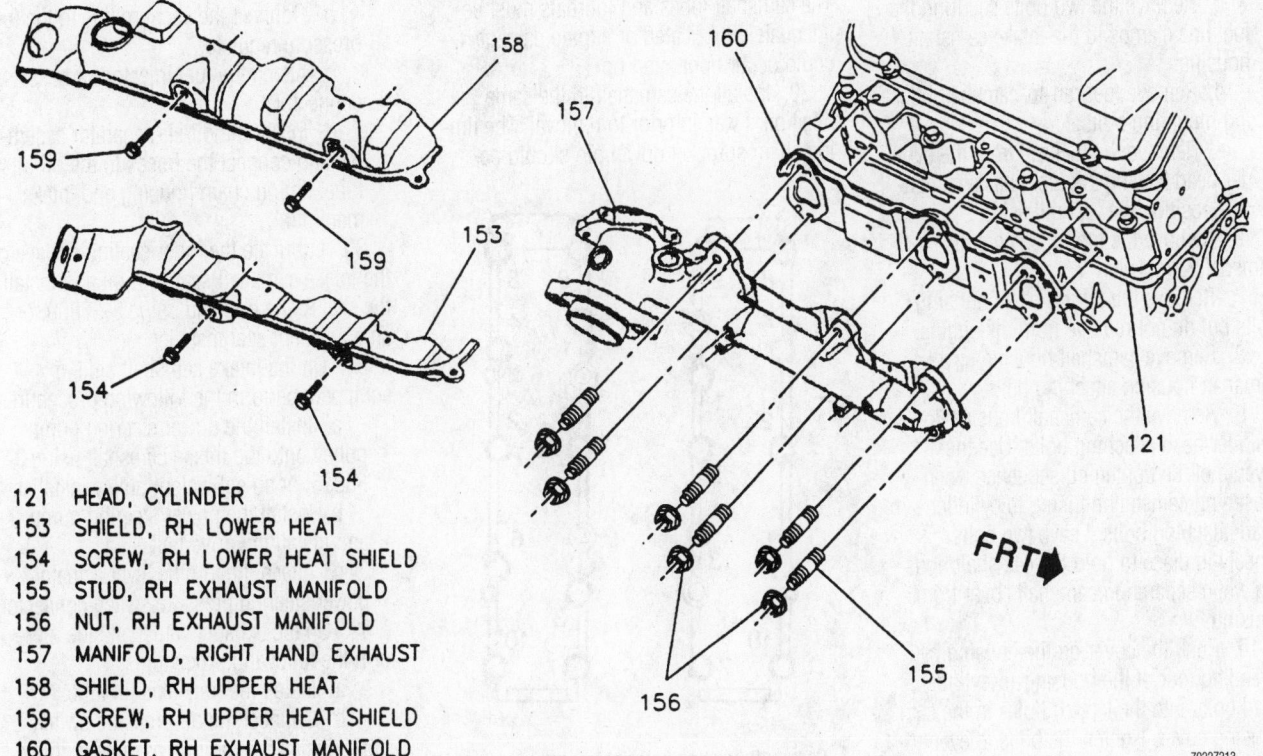

121 HEAD, CYLINDER
153 SHIELD, RH LOWER HEAT
154 SCREW, RH LOWER HEAT SHIELD
155 STUD, RH EXHAUST MANIFOLD
156 NUT, RH EXHAUST MANIFOLD
157 MANIFOLD, RIGHT HAND EXHAUST
158 SHIELD, RH UPPER HEAT
159 SCREW, RH UPPER HEAT SHIELD
160 GASKET, RH EXHAUST MANIFOLD

79222213

Exploded view of the right-hand exhaust manifold—3.1L engine

2. Detach the 11-pin connector from the ignition assembly cover.

3. Remove the four ignition assembly cover-to-camshaft housing bolts and remove assembly by pulling straight up. Use a spark plug boot wire remover to remove connector assemblies.

4. On the intake camshaft side, if equipped, perform the following procedures:

a. Remove the throttle lever actuator power steering pressure switch connector.

b. Loosen the power steering pump adjusting stud bolt. Loosen the power steering pump pivot bolts and remove drive belt.

c. Remove the two power steering pump bracket-to-transaxle bolts.

d. Remove the front power steering pump bracket-to-engine block bolt.

e. Remove the power steering pump assembly from the engine and position it aside. DO NOT disconnect the power steering lines from the pump.

5. On the intake camshaft side, if equipped, perform the following procedures:

a. Remove oil/air separator bolts and hoses. Leave the hoses attached to the separator but disconnect from the oil fill tube, timing chain housing and intake manifold.

b. Disconnect the vacuum line from fuel pressure regulator and disconnect the fuel injector wiring harness.

c. Remove the two bolts securing the fuel line clamps to the intake camshaft housing.

d. Remove fuel rail-to-camshaft housing mounting bolts.

e. Remove the fuel rail from the cylinder head leaving the fuel lines attached and position the fuel rail aside.

6. Remove the timing chain and camshaft sprockets.

7. Remove the timing chain housing bolts but do not remove from the engine.

8. Remove camshaft housing cover-to-camshaft housing attaching bolts.

9. Remove the camshaft housing-to-cylinder head attaching bolts. Use the reverse of the tightening sequence when loosening camshaft housing to cylinder head attaching bolts. Leave two bolts loosely in place to hold the camshaft housing while separating camshaft cover from housing.

10. Push the cover off the housing by threading four of the housing-to-cylinder head bolts into the tapped holes in the cam housing cover. Tighten the bolts in evenly so the cover does not bind on the dowel pins.

11. Remove the two loosely installed camshaft housing-to-cylinder head bolts and remove the cover. Discard the gaskets.

12. Note the position of the chain sprocket dowel pin for reassembly. Remove the camshaft carefully; do not damage the camshaft oil seal.

13. Remove camshaft oil seal from camshaft and discard seal. This seal must be replaced any time the housing and cover are separated.

➡ **The valve lifters must be kept in order for installation in the same locations they were removed from.**

14. Remove the valve lifters from the camshaft housing.

15. Remove the camshaft carrier from the cylinder head and remove the gasket.

To install:

16. Clean all the gasket surfaces completely.

17. Install a new gasket on the cylinder head and install the camshaft housing over the dowel pins. Install one bolt loosely to hold the housing in place.

➡ **If the camshaft was replaced the valve lifters must also be replaced.**

18. Install the lifters into their original bores.

19. Lubricate camshaft lobes, journals and lifters with camshaft and lifter prelube. The camshaft lobes and journals must be adequately lubricated or engine damage could occur upon start up.

20. Install the camshaft in the same position it was in prior to removal. The timing chain sprocket dowel pin should be

straight up and align with the center line of the lifter bores.

21. Install new camshaft housing-to-camshaft housing cover seals into the cover.

➡ **The seals for the intake and exhaust covers are different and the correct seals must be used.**

22. Remove the bolt holding the housing in place.

23. Apply thread locking compound to the camshaft housing and cover attaching bolt threads.

24. Install the camshaft housing cover.

25. Install all the mounting bolts finger-tight. With all the bolts in place tighten the bolts in sequence to 11 ft. lbs. (15 Nm) plus 75 degrees additional rotation (Long bolts) and 16 ft. lbs. (21 Nm) plus 25 degrees additional rotation (Short bolts).

26. Install the timing chain housing mounting bolts.

27. Install the timing chain and sprockets.

28. Install new fuel injector O-ring seals lubricated with oil and install the fuel rail and tighten the mounting bolts to 19 ft. lbs. (26 Nm).

29. On the intake camshaft side, if equipped, perform the following procedures:

a. Install the fuel line clamp mounting bolts on top of the intake camshaft housing.

b. Connect the vacuum line to the fuel pressure regulator.

c. Attach the fuel injector harness connector.

d. Install the oil/air separator assembly and connect the hoses to the oil fill tube, timing chain housing and intake manifold.

30. Lubricate the inner sealing surface of the intake camshaft seal with oil and install the seal to the housing using J-36009, or an equivalent seal installer.

31. On the intake camshaft side, if equipped, perform the following procedures:

a. Install the power steering pump pulley onto the intake camshaft using J-36015, or an equivalent pulley installer.

b. Install the power steering pump and install the drive belt.

c. Attach the throttle lever actuator power steering pressure switch connector.

32. On the exhaust camshaft side, perform the following procedures:

a. Install the transaxle fill tube.

b. Connect the oil pressure switch.

33. Install the ignition assembly on the camshaft housing and tighten the mounting bolts to 13 ft. lbs. (18 Nm).

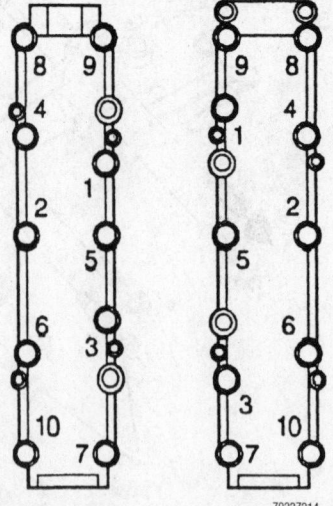

79222Z214

Camshaft housing bolt tightening sequence—2.4L engine

34. Attach the 11-pin connector to ignition assembly.

35. Connect the negative battery cable.

36. Start the vehicle and verify proper operation and no leaks.

3.1L Engine

1. Relieve the fuel system pressure. Disconnect the negative battery cable. Remove the engine assembly.

➡**When removing valvetrain components they must be marked for installation in the same location they are removed from. When the camshaft is being replaced the valve lifters should also be replaced.**

2. Remove the rocker arm covers and the intake manifold.

3. Remove the rocker arm retaining nuts, rocker arm balls, rocker arms and pushrods.

4. Remove the guide bolts from the lifter guide and remove the guide.

5. Remove the valve lifter(s) from the lifter bores.

6. Remove the crankshaft balancer and front cover. Remove the timing chain and sprockets.

7. Remove the oil pump driven gear bolt and gear. Remove the bolts and camshaft thrust plate.

8. Carefully remove the camshaft. Avoid marring the camshaft bearing surfaces.

To install:

9. Coat the camshaft with lubricant 1052365 or equivalent, and install the camshaft.

10. Install the camshaft thrust plate and tighten bolts to 89 inch lbs. (10 Nm).

11. Install the oil pump driven gear and tighten bolt to 27 ft. lbs. (36 Nm).

12. Install the timing chain and sprocket.

13. Install the camshaft thrust button and front cover. Install the crankshaft balancer.

14. Lubricate the bearing surfaces with Molykote® or equivalent.

➡**Installation of a new camshaft or a wear pattern on the old valve lifter will require the replacement of the camshaft and lifters together. If camshaft replacement is not necessary, be sure to install the used valve lifters in their original position upon reinstallation.**

15. Install the lifters in their original locations.

16. Install the lifter guide and lifter guide bolts and tighten the guide bolts to 89 inch lbs. (10 Nm).

17. Install the pushrods, rocker arms, rocker balls and rocker arm nuts.

18. Tighten the rocker arm nuts to 89 inch lbs. (10 Nm) plus an additional 30 degrees.

19. Install the intake manifold and rocker arm covers.

20. Install the engine assembly. Connect the negative battery cable.

21. Adjust the valves, as required. Start the engine and verify no oil leaks.

Valve Lash

ADJUSTMENT

2.4L Engine

The 2.4L engine are equipped with hydraulic valve lifters that do not require periodic valve lash adjustment. Adjustment to zero lash is maintained automatically by hydraulic pressure in the lifters. Also, the rocker arm retaining nuts are tightened to a specific torque value (refer to the rocker arm procedure) to provide proper rocker arm placement.

3.1L Engine

➡**These engines come from the factory with non-adjustable valve lash, BUT if the valves and the valve seats are reconditioned adjustable rocker arm studs must be installed. The adjustment procedure is ONLY for adjustable rocker arm studs.**

The engines are originally equipped with hydraulic valve lifters. Hydraulic valve lifters are not adjustable. The replacement rocker arm stud is adjustable and the adjustment can be performed as listed below. If valve system noise is present, check the tighten on the rocker arm nuts. The correct torque is 14–20 ft. lbs. (19–27 Nm). If noise is still present, check the condition of the camshaft, lifters, rocker arms, pushrods and valves.

1. Install the rocker arms, balls and nuts. Tighten the rocker arm nuts until all of the lash (free-play) is eliminated.

2. Adjust the valves when the lifter is on the base circle of a camshaft lobe as follows:

a. Place the engine in the No. 1 firing position. This can be determined by placing a finger on the No. 1 rocker arm as the engine alignment mark on the front face of the of the torsional dampener pulley aligns with the arrow on the front cover. If the valves don't move freely, the engine is in the No. 1 firing position. If the valves move freely, the engine is in the No. 4 firing position and the engine should be rotated one full rotation to reach the No. 1 position.

b. With the engine in the No. 1 firing order, adjust the following valves:

- Exhaust—1, 2, 3
- Intake—1, 5, 6

c. Loosen the adjusting nut several turns backwards, then tighten until all lash (free-play) is removed. After all of the valve lash is removed turn the nut an additional 1½ turns, this will center the lifter plunger.

d. Turn the engine one complete revolution until the timing tab and the alignment mark are again aligned, this will place the engine in the No. 4 firing position.

e. With the engine in the No. 4 firing order, adjust the following valves:

- Exhaust—4, 5, 6
- Intake—2, 3, 4

Oil Pan

REMOVAL & INSTALLATION

2.4L Engine

1. Disconnect the negative battery cable. Raise and safely support the vehicle.

2. Properly drain the engine oil and cooling system. Remove the flywheel inspection cover.

3. Remove the right front wheel and tire assembly. Remove the right inner fender splash shield.

4. Remove the serpentine belt. Disconnect the engine mount strut from the engine mount strut bracket.

5. Remove the A/C compressor from the mounting bracket and support it aside. Do NOT disconnect the refrigerant lines.

6. Unfasten the engine mount strut bracket bolts, then remove the bracket. Remove the radiator outlet pipe bolts.

7. Disconnect the radiator and A/C outlet pipes from the suspension supports. Remove the exhaust manifold brace.

8. Remove the oil pan-to-flywheel cover bolt and nut and remove the flywheel cover stud and spacer.

9. Disconnect the radiator outlet pipe from the lower radiator hose and oil pan.

10. If equipped, detach the oil level sensor wire. Unfasten the oil pan bolts, then remove the oil pan.

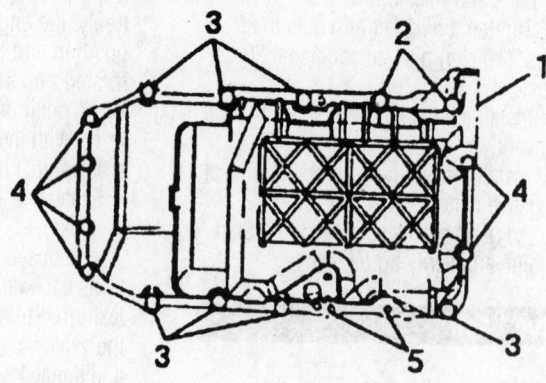

◀FRT

1 Oil pan
2 Oil pan bolt (M8 x 1.25 x 80)
  24 Nm (18 lb. ft.)
3 Oil pan bolt (M8 x 1.25 x 22)
  24 Nm (18 lb. ft.)
4 Oil pan bolt (M6 x 1.00 x 25)
  12 Nm (106 lb. in.)
5 Stud end oil pan bolt
  26 Nm (19 lb. ft.)

7922Z215

Oil pan mounting bolt locations—2.4L engine

To install:

11. Install a new gasket, the oil pan and bolts. Tighten the chain housing and carrier seal bolts to 106 inch lbs. (12 Nm). Tighten the oil pan-to-block bolts to 17 ft. lbs. (23 Nm).

12. Install the flywheel cover spacer, stud nut and bolt. Tighten the nut to 41 ft. lbs. (56 Nm), the stud to 115 inch lbs. (13 Nm) and the bolt to 41 ft. lbs. (56 Nm).

13. Install the oil pan-to-transaxle nut and tighten to 41 ft. lbs. (56 Nm).

14. Complete the installation by reversing the removal procedures.

15. Install the engine mount strut bracket and tighten bolts to 49 ft. lbs. (66 Nm).

16. Fill the crankcase with oil. A filter change is recommended.

17. Fill the cooling system. Connect the negative battery cable, then start the engine and check for leaks.

3.1L Engine

1. Disconnect the negative battery cable. Remove the serpentine belt.

2. If equipped, remove the upper A/C compressor bolts. Raise and safely support the vehicle.

3. Properly drain the engine oil. Remove the right front wheel and tire assembly.

4. Remove the engine mount strut from the suspension support.

5. Remove the cotter pin and castle nut from the lower ball joint, the separate the ball joint from the steering knuckle.

6. Remove the right side sway bar link. Detach the ABS sensor from the right sub-frame.

7. Remove the right side subframe bolts and the right side subframe and control arm as an assembly.

8. If equipped, remove the lower A/C compressor bolts and position the compressor aside. Do NOT disconnect the refrigerant lines or allow the compressor to hang unsupported.

9. Remove the engine mount strut bracket from the engine. Remove the oil filter, the starter and the flywheel cover.

10. Remove the oil pan flange bolts, the oil pan side bolts and the oil pan.

To install:

11. Clean the gasket mating surfaces. Install a new gasket on the oil pan. Apply silicone sealer to the portion of the pan that contacts the rear of the block.

12. Position the oil pan, then install the mounting bolts finger-tight.

13. With all the bolts in place, tighten the oil pan flange bolts to 18 ft. lbs. (25

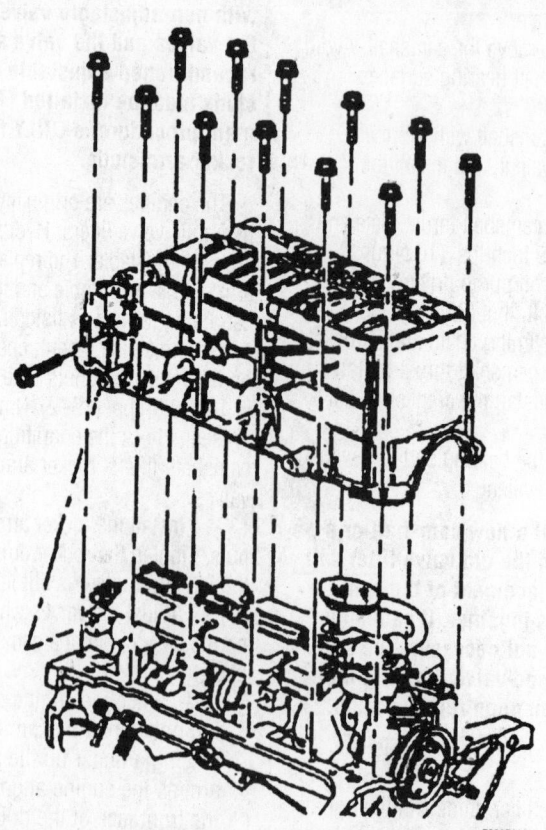

7922Z216

Exploded view of the oil pan mounting—3.1L engine

Nm) and the oil pan side bolts to 34 ft. lbs. (50 Nm).

14. Complete the installation by reversing the removal procedures.

15. Tighten the following item to:

a. Engine-to-transaxle brace, bolts to 68 ft. lbs. (93 Nm)

b. Engine mount strut bracket-to-engine bolts to 85 ft. lbs. (115 Nm)

c. Right side subframe/control arm assembly bolts to 89 ft. lbs. (120 Nm)

d. Right side sway bar line bolt to 22 ft. lbs. (30 Nm)

e. Ball joint-to-steering knuckle nut to 48 ft. lbs. (60 Nm). Install a new cotter pin.

f. Engine mount strut bracket bolt-to-frame bolt to 89 ft. lbs. (120 Nm)

16. Fill the crankcase to the correct level. A filter change is recommended. Connect the negative battery cable, then start the engine and check for leaks.

➡**Whenever the vehicle subframe is removed or lowered, the wheel alignment should be checked.**

17. Check the front end alignment and adjust as required.

Oil Pump

REMOVAL & INSTALLATION

2.4L Engine

➡**Please note that the transaxle must be removed from the vehicle to service the oil pump.**

1. Disconnect the negative battery cable. Install engine support fixture J-28467-A or equivalent.

2. Properly drain the engine oil, then remove the oil pan. Remove the transaxle and the flywheel.

3. Remove the balance shaft chain cover and chain guide. Remove the oil pump bolts and the oil pump cover.

4. Pull the housing to disconnect the pump gear from the balance shaft. Remove the housing assembly from the balance shaft assembly.

To install:

5. Clean all of the parts in suitable cleaning solvent. Remove all varnish sludge and dirt.

6. Lubricate the gears with clean engine oil. Assemble the geroter gear into the housing.

➡**Fill oil pump cavities with petroleum jelly prior to installation. This seals the pump and acts like a "prime" so the pump will draw oil as soon as the engine begins to turn. This will ensure that there is oil pressure immediately on start-up and will prevent engine damage.**

7. Install the pump housing to the balance shaft assembly, the pump cover to the oil pump housing.

8. Install the oil pump-to-block bolts and tighten to 40 ft. lbs. (54 Nm).

9. Install the balance shaft chain guide and chain. Adjust the chain tension inserting a 0.40 in. (1mm) brass feeler, between the chain guide and the chain.

➡**A brass feeler gauge must be used to ensure that correct measurements are obtained. If a steel gauge is used, it will not bend to conform to the guide and will allow for incorrect measurements.**

10. Press the guide against the chain using about 3 lbs. of force. Tighten the chain tensioner fastener to 115 inch lbs. (13 Nm).

11. Install the balance shaft chain cover and tighten the nut and bolt to 115 inch lbs. (13 Nm).

12. Install the flywheel and the transaxle. Install the oil pan, then carefully lower the vehicle.

13. Fill the crankcase with clean engine oil. A filter change is recommended.

14. Remove the engine support fixture. Connect the negative battery cable, then start the engine and verify oil pressure and no leaks.

3.1L Engine

1. Disconnect the negative battery cable.

2. Raise and safely support the vehicle.

3. Drain the engine oil into a suitable container. Remove the oil pan.

4. Remove the crankshaft oil deflector bolts and deflector.

5. Remove the oil pump bolts, the oil pump and pump driveshaft.

To install:

6. Install the oil pump and pump driveshaft. Tighten the oil pump mounting bolts to 30 ft. lbs. (41 Nm).

7. Install the crankshaft oil deflector and tighten the nuts to 18 ft. lbs. (25 Nm).

8. Install the oil pan, then carefully lower the vehicle.

9. Fill the crankcase to the correct level with oil. An filter change is recommended.

10. Connect the negative battery cable, then, start the engine, check the oil pressure and check for leaks.

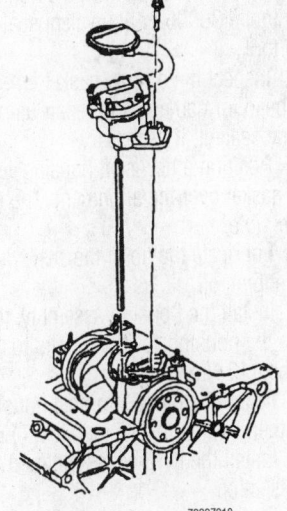

79222Z218

Exploded view of the oil pump mounting— 3.1L engine

Rear Main Seal

REMOVAL & INSTALLATION

2.4L Engine

1. Disconnect the negative battery cable.
2. Remove the transaxle assembly.

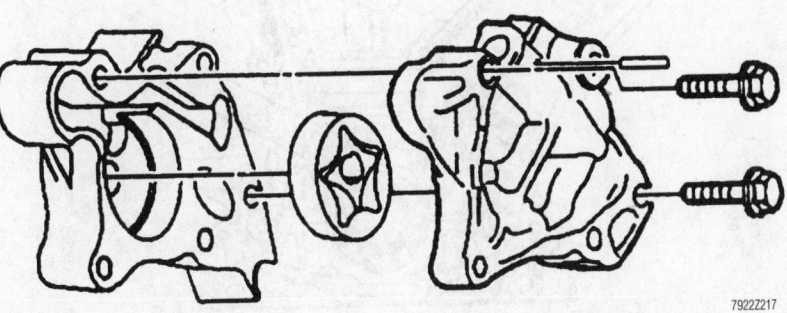

79222Z217

Exploded view of the oil pump components—2.4L engine

3. If equipped with a manual transaxle, remove the pressure plate and clutch disc.

4. Unfasten the flywheel-to-crankshaft bolts, then remove the flywheel.

5. Disconnect the oil pan-to-seal housing bolts

6. Unfasten the seal housing-to-block bolts, then remove the seal housing and gasket.

7. To support the seal housing for seal removal, place two blocks of equal thickness on a flat surface, position the seal housing and blocks so the transaxle side of the seal housing is supported across the dowel pin and center bolt holes on both sides of the seal opening.

➡The seal housing could be damaged if not properly supported during seal removal.

8. Drive the seal evenly out the transaxle side of the seal housing using a small prytool in the relief grooves on the crankshaft side of the seal housing. Discard the seal.

❊❊ WARNING

Be careful not to damage the seal housing sealing surface. If damaged, it may result in an oil leak.

To install:

9. Press a new seal into the housing using tool J 36005 or equivalent seal installation tool.

10. Inspect the oil pan gasket inner silicone bead for damage and repair using a silicone sealant, if necessary.

11. Position a new seal housing-to-block gasket over the alignment. The gasket is reversible.

12. Lubricate the lip of the seal with clean engine oil.

13. Install the housing assembly, then tighten the housing-to-block bolts to 106 inch lbs. (12 Nm).

14. Install the oil pan-to-seal housing bolts, then tighten to 106 inch lbs. (12 Nm).

15. Install the flywheel as outlined later in this section.

16. For vehicles equipped with a manual transaxle, install the clutch, pressure plate and clutch cover assembly.

17. Install the transaxle assembly.

18. Connect the negative battery cable, then start the engine and check for leaks.

3.1L Engine

1. Support the engine with J-28467 engine support or equivalent.

2. Remove the transaxle.

3. Remove the flywheel.

4. Insert a suitable prytool in through the dust lip and pry out the seal by moving the tool around the seal until it is removed.

❊❊ WARNING

Use care not to damage the crankshaft seal surface with a prytool.

To install:

5. Before installing, lubricate the seal bore to seal surface with engine oil.

6. Install the new seal using tool J-34686.

7. Slide the new seal over the mandrel until the dust lip bottoms squarely against the tool collar.

8. Align the dowel pin of the tool with the dowel pin hole in the crankshaft and attach the tool to the crankshaft. Tighten the attaching screws to 2–5 ft. lbs. (2.7–6.8 Nm).

9. Tighten the T-handle of the tool to push the seal into the bore. Continue until the tool collar is flush against the block.

10. Loosen the T-handle completely. Remove the attaching screws and the tool.

➡Check to see that the seal is squarely seated in the bore.

11. Install the flywheel and transaxle.

12. Start the engine and check for leaks.

Timing Chain, Sprockets, Front Cover and Seal

REMOVAL & INSTALLATION

2.4L Engine

➡It is recommended that the entire procedure be reviewed before attempting to service the timing chain.

❊❊ CAUTION

Never open, service or drain the radiator or cooling system when hot; serious burns can occur from the steam and hot coolant.

1. Disconnect the negative battery cable. Drain the cooling system into a suitable container. Remove the coolant surge tank.

2. Remove the serpentine drive belt using a 13mm wrench that is at least 24 in. (61cm) long.

3. Disconnect and remove the alternator from the mounting bracket.

4. Install a suitable engine support, J-28467-A or the equivalent.

5. Remove the upper cover fasteners. Disconnect the upper cover vent hose.

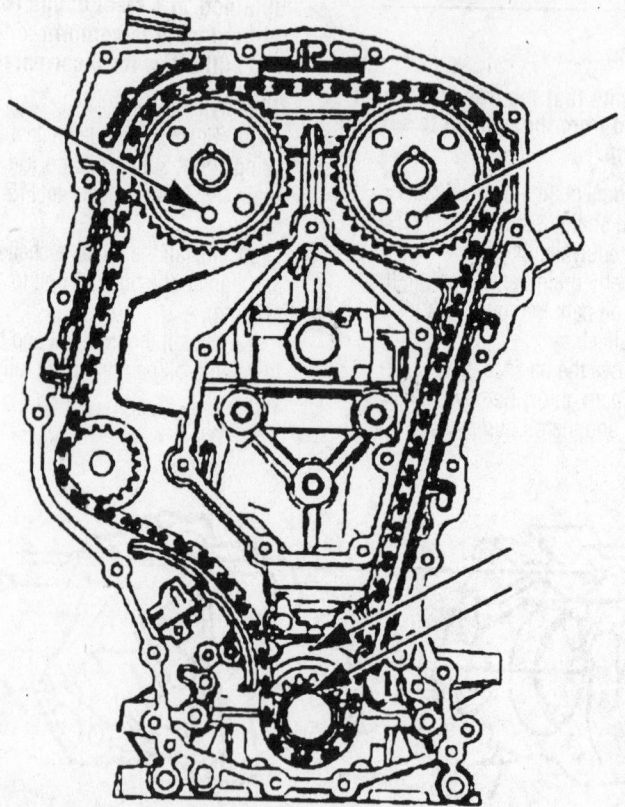

Timing chain and sprocket alignment positions—2.4L DOHC engine

7922Z219

6. Remove the right engine mount and bracket. Raise and safely support the vehicle.

7. Remove the right front tire and wheel assembly. Remove the right lower splash shield from the right wheel house.

8. Remove the crankshaft balancer assembly using the following procedure:

a. Remove the torsional damper mounting bolt and washer while holding the crankshaft in place using J-38122, or the equivalent.

b. Remove the balancer using a suitable puller, J-24420-B or the equivalent.

9. Remove the lower cover fasteners. Lower the vehicle.

10. Remove the front cover and gaskets. Remove the crankshaft oil slinger.

11. Using a suitable seal driver, tap seal out of the front cover.

12. Rotate the crankshaft clockwise, as viewed from front of engine (normal rotation) until the camshaft sprocket's timing dowel pin holes align with the holes in the timing chain housing. The mark on the crankshaft sprocket should align with the mark on the cylinder block. The crankshaft sprocket keyway should point upwards and align with the center line of the cylinder bores. This is the normal timed position.

13. Remove the timing chain guides. Raise and safely support the vehicle.

14. Gently pry off timing chain tensioner spring retainer and remove spring.

➡**Two styles of tensioner are used. Early production engines will have a spring post and late production ones will not. Both styles are identical in operation and are interchangeable.**

15. Remove the timing chain tensioner shoe retainer.

16. Be sure all the slack in the timing chain is above the tensioner assembly; remove the chain tensioner shoe. The timing chain must be disengaged from the wear grooves in the tensioner shoe in order to remove the shoe. Slide a prybar under the timing chain while pulling shoe outward.

17. If difficulty is encountered removing chain tensioner shoe, proceed as follows:

a. Lower the vehicle.

b. Hold the intake camshaft sprocket with a holding tool and remove the sprocket bolt and washer.

c. Remove the washer from the bolt and rethread the bolt back into the camshaft by hand, the bolt provides a surface to push against.

d. Remove intake camshaft sprocket using a 3-jaw puller in the three relief

holes in the sprocket. Do not attempt to pry the sprocket off the camshaft or damage to the sprocket or chain housing could occur.

18. Remove the tensioner assembly attaching bolts and the tensioner.

✳✳ CAUTION

The tensioner piston is spring loaded and could fly out causing personal injury.

19. Remove the chain housing to block stud, which is actually the timing chain tensioner shoe pivot.

20. Remove the timing chain.

To install:

21. Tighten intake camshaft sprocket attaching bolt and washer, while holding the sprocket with tool J36013, if removed.

22. Install the special tool through holes in camshaft sprockets into holes in timing chain housing. This positions the camshafts for correct timing.

23. If the camshafts are out of position and must be rotated more than ⅛ turn in order to install the alignment dowel pins:

a. The crankshaft must be rotated 90 degrees clockwise off TDC in order to give the valves adequate clearance to open.

b. Once the camshafts are in position and the dowels installed, rotate the crankshaft counterclockwise back to TDC. Do not rotate the crankshaft clockwise to TDC or valve and piston damage could occur.

24. Install the timing chain over the exhaust camshaft sprocket, around the idler sprocket and around the crankshaft sprocket.

25. Remove the alignment dowel pin from the intake camshaft. Using a dowel pin remover tool, rotate the intake camshaft sprocket counterclockwise enough to slide the timing chain over the intake camshaft sprocket. Release the camshaft sprocket wrench. The length of chain between the two camshaft sprockets will tighten. If properly timed, the intake camshaft alignment dowel pin should slide in easily. If the dowel pin does not fully index, the camshafts are not timed correctly and the procedure must be repeated.

26. Leave the alignment dowel pins installed.

27. With slack removed from chain between intake camshaft sprocket and crankshaft sprocket, the timing marks on the crankshaft and the cylinder block should be aligned. If marks are not aligned, move the

chain one tooth forward or rearward, remove slack and recheck marks.

28. Tighten the chain housing to block stud. The stud is installed under the timing chain. Tighten to 19 ft. lbs. (26 Nm).

29. Reload timing chain tensioner assembly to its 0 position as follows:

a. Assemble restraint cylinder, spring and nylon plug into plunger. Index slot in restraint cylinder with peg in plunger. While rotating the restraint cylinder clockwise, push the restraint cylinder into the plunger until it bottoms. Keep rotating the restraint cylinder clockwise but allow the spring to push it out of the plunger. The pin in the plunger will lock the restraint in the loaded position.

b. Install tool J36589 or equivalent, onto plunger assembly.

c. Install plunger assembly into tensioner body with the long end toward the crankshaft when installed.

30. Install the tensioner assembly to the chain housing. Recheck plunger assembly installation. It is correctly installed when the long end is toward the crankshaft.

31. Install and tighten timing chain tensioner bolts and tighten to 10 ft. lbs. (14 Nm).

32. Install the tensioner shoe and tensioner shoe retainer. Remove special tool J36589 and squeeze plunger assembly into the tensioner body to unload the plunger assembly.

33. Lower vehicle and remove the alignment dowel pins. Rotate crankshaft clockwise two full rotations. Align crankshaft timing mark with mark on cylinder block and reinstall alignment dowel pins. Alignment dowel pins will slide in easily if engine is timed correctly.

✳✳ WARNING

If the engine is not correctly timed, severe engine damage could occur.

34. Install the timing chain guides and crankshaft oil slinger.

35. Install a new seal using a suitable seal driver and lubricate the inner lip of the seal with engine oil.

36. Install the front cover and gaskets on the engine and tighten the mounting nuts and bolts to 106 inch lbs. (12 Nm).

37. Install the torsional damper as follows:

a. Coat the seal contact area on the crankshaft damper with clean engine oil.

b. Line up the damper on the crankshaft so the notch in the damper lines up with the crankshaft key.

c. Tap the balancer into place using a rubber mallet.

d. Install the damper mounting bolt and washer and tighten the bolt to 129 ft. lbs. (175 Nm) plus 90 degrees rotation.

38. Install the right front lower splash shield. Install the tire and wheel assembly and tighten to specification.

39. Safely lower the vehicle. Install the right engine mount bracket.

40. Install the right engine mount. Connect the upper cover vent hose.

41. Remove the engine support. Install the alternator and attach the electrical connectors.

42. Install the serpentine belt. Install coolant surge tank.

43. Refill the cooling system. Connect the negative battery cable and check for leaks.

3.1L Engine

❄❄ CAUTION

Never open, service or drain the radiator or cooling system when hot; serious burns can occur from the steam and hot coolant.

1. Disconnect the negative battery cable. Drain the cooling system into a suitable container.

2. Remove the right engine mount bracket. Remove the serpentine belt.

3. Remove the crankshaft balancer as follows:

a. Raise and safely support the vehicle.

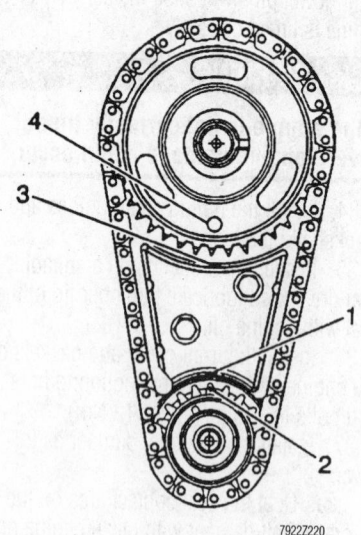

Timing chain and sprocket timing mark alignment—3.1L engine

b. Remove the right front tire and wheel assembly.

c. Remove the right inner fender well splash shield.

d. Remove the flywheel cover and install a flywheel holding tool.

e. Remove the balancer mounting bolt and washer.

f. Using a suitable puller, J-24420-B or the equivalent remove the balancer from the crankshaft.

4. Remove the serpentine belt tensioner mounting bolt and tensioner. Remove the oil pan.

5. Remove coolant bypass pipe from the water pump and the intake manifold.

6. Disconnect the lower radiator hose to from the front cover outlet. Remove the front cover bolts and cover.

7. Using a suitable seal driver, tap seal out of the front cover.

8. Rotate the crankshaft until the timing marks on the camshaft and crankshaft sprockets are in alignment at their closest approach.

9. Remove the camshaft sprocket bolt, sprocket and timing chain.

10. Remove the crankshaft sprocket, using a gear puller J-5825-A or equivalent. Remove the timing chain damper bolts and damper.

To install:

11. Install the timing chain damper and tighten the mounting bolts to 15 ft. lbs. (21 Nm).

12. Install the crankshaft sprocket onto the crankshaft making sure the notch in the sprocket fits over the crankshaft key. Fully seat the sprocket on the crankshaft using J-38612, or an equivalent gear installer.

13. Be sure the timing mark on the crankshaft sprocket is pointing straight up.

14. Install the timing chain over the camshaft sprocket and hold the sprocket in such a way, that the timing mark is pointing down, and the timing chain is hanging down off the sprocket.

15. Loop the timing chain under the crankshaft sprocket and install the camshaft sprocket on the camshaft. The sprocket will only fit on the camshaft if the dowel on the camshaft lines up with the hole in the sprocket.

16. Verify that the marks are aligned (the camshaft sprocket will be at the 6 o'clock position and the crankshaft sprocket will be in the 12 o'clock position).

17. Tighten the camshaft sprocket mounting bolt to 81 ft. lbs. (110 Nm).

18. Lubricate the timing chain components with engine oil. Clean all gasket surfaces completely.

19. Apply a thin bead of sealer around the gasket sealing area of the front cover. Install a new front cover seal on the front cover.

20. Install a new seal using a suitable seal driver and lubricate the inner lip of the seal with engine oil.

21. Install the front cover on the engine and install the mounting bolts. Tighten the small bolts to 15 ft. lbs. (21 Nm) and the large bolts to 35 ft. lbs. (47 Nm).

22. Connect the radiator hose to coolant outlet.

23. Install coolant bypass pipe to the water pump and the intake manifold. Install the oil pan.

24. Install crankshaft balancer as follows:

a. Coat the seal contact surface of the crankshaft balancer with clean engine oil.

b. Line up the notch in the balancer with the crankshaft key and slide the balancer on until the key is in the balancer.

c. Using J-29113 or an equivalent puller, seat the balancer on the crankshaft.

d. Install the balancer mounting bolt and washer and tighten to 76 ft. lbs. (103 Nm).

e. Install the flywheel cover.

f. Install the right inner fender well splash shield.

g. Install the tire and wheel assembly and tighten to specification.

25. Install the serpentine belt tensioner and tighten the mounting bolt to 40 ft. lbs. (54 Nm). Install the serpentine belt.

26. Install the right engine mount bracket and tighten the bracket-to-mount bolts to 96 ft. lbs. (130 Nm).

27. Refill the cooling system with the correct amount and type of coolant. Check the engine oil level. An oil and filter change is recommended.

28. Connect the negative battery cable. Start the engine and verify that there are no leaks.

FUEL SYSTEM

Fuel System Service Precautions

Safety is the most important factor when performing not only fuel system maintenance but any type of maintenance. Failure to conduct maintenance and repairs in a safe manner may result in serious personal injury or death. Maintenance and testing of the vehicle's fuel system components can be accomplished safely and effectively by adhering to the following rules and guidelines.

- To avoid the possibility of fire and personal injury, always disconnect the negative battery cable unless the repair or test procedure requires that battery voltage be applied.

- Always relieve the fuel system pressure prior to disconnecting any fuel system component (injector, fuel rail, pressure regulator, etc.), fitting or fuel line connection. Exercise extreme caution whenever relieving fuel system pressure to avoid exposing skin, face and eyes to fuel spray. Please be advised that fuel under pressure may penetrate the skin or any part of the body that it contacts.

- Always place a shop towel or cloth around the fitting or connection prior to loosening to absorb any excess fuel due to spillage. Ensure that all fuel spillage (should it occur) is quickly removed from engine surfaces. Ensure that all fuel soaked cloths or towels are deposited into a suitable waste container.

- Always keep a dry chemical (Class B) fire extinguisher near the work area.

- Do not allow fuel spray or fuel vapors to come into contact with a spark or open flame.

- Always use a back-up wrench when loosening and tightening fuel line connection fittings. This will prevent unnecessary stress and torsion to fuel line piping. Always follow the proper torque specifications.

- Always replace worn fuel fitting O-rings with new. Do not substitute fuel hose or equivalent, where fuel pipe is installed.

Fuel System Pressure

RELIEVING

☀☀ CAUTION

The fuel injection system remains under pressure, even after the engine has been turned OFF. The fuel system pressure must be relieved before disconnecting any fuel lines. Failure to do so may result in fire and/or personal injury.

2.4L Engine

1. Loosen the fuel filler cap in order to relieve the pressure in the tank (do not tighten at this time).
2. Raise and safely support the vehicle.
3. Detach the fuel pump electrical connector.
4. Start and run the vehicle until it stalls, the engage the starter for an additional three seconds to ensure the relief of any remaining pressure.

5. Disconnect the negative battery cable.
6. Once the tests or repairs are completed, reattach the fuel pump electrical connector.
7. Connect the negative battery cable.
8. Lower the vehicle.
9. Tighten the fuel filler cap.
10. Prime the fuel system by cycling the ignition switch **ON** for 2 seconds, **OFF** for 10 seconds, then **ON** again. Repeat, if necessary to build system pressure.

3.1L Engine

1. Disconnect the negative battery cable in order to avoid possible fuel discharge if an accidental attempt is made to start the engine.
2. Loosen the fuel tank filler cap in order to relieve fuel tank pressure.
3. Connect tool J-34730-A or equivalent fuel pressure gauge to the fuel pressure test port connection. Wrap a towel around the fuel pressure connection when installing the fuel pressure gauge in order to avoid fuel spillage.
4. Install the bleed hose into an approved container and open the valve in order to bleed the fuel system pressure. The fuel pipe connections are now safe for servicing.
5. Drain any fuel remaining in the fuel pressure gauge into an approved container.

Fuel Filter

REMOVAL & INSTALLATION

All Engines

☀☀ CAUTION

The fuel injection system remains under pressure, even after the engine has been turned OFF. The fuel system pressure must be relieved before disconnecting any fuel lines. Failure to do so may result in fire and/or personal injury.

1. Relieve the fuel system pressure.
2. Disconnect the negative battery cable.
3. Raise and safely support the vehicle.
4. Using a back-up wrench on the fuel filter disconnect the fuel line from the filter.
5. Disconnect the quick-connect fitting from the fuel filter by compressing the tabs while pulling outward on the line.
6. Remove the fuel filter from the mounting bracket.

   To install:
7. Install the fuel filter in the mounting bracket.
8. Using a back-up wrench on the fuel filter, tighten the fuel line fitting to 20 ft. lbs. (27 Nm).

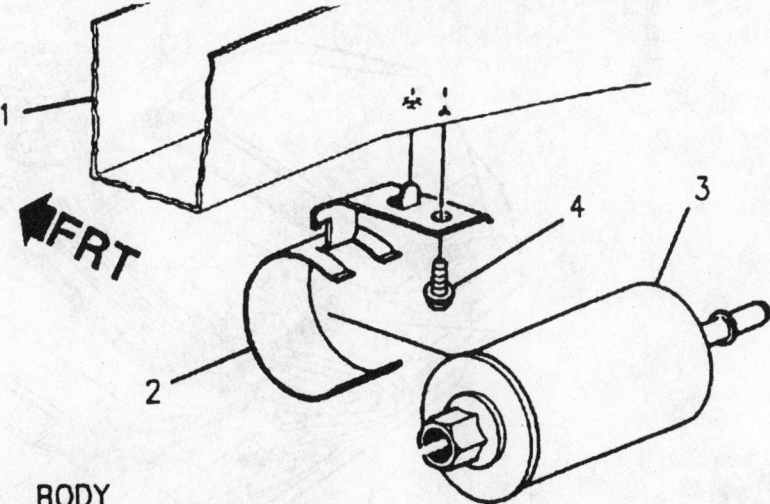

1 BODY
2 FUEL FILTER BRACKET
3 FUEL FILTER
4 SCREW – FULLY DRIVEN, SEATED AND NOT STRIPPED

79222221

Exploded view of the fuel filter mounting

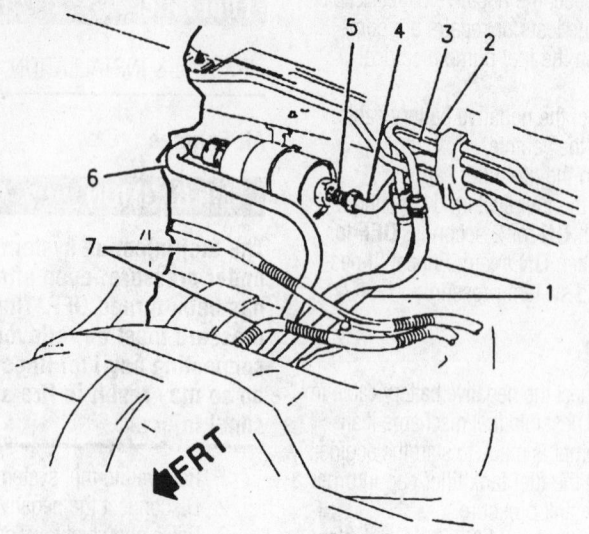

1 HOSE, PART OF FUEL SENDER
2 FUEL VAPOR PIPE
3 FUEL RETURN PIPE
4 FUEL FEED PIPE
5 FUEL FEED PIPE NUT
   27 N•m (20 LBS. FT.)
6 HOSE, PART OF FUEL SENDER
7 ABS AND FUEL SENDER HARNESS

79222222

Fuel filter mounting location and component identification

9. Connect the quick-connect fitting to the fuel filter.
10. Lower the vehicle.
11. Connect the negative battery cable.
12. Pressurize the fuel system and verify no leaks.

Fuel Pump

REMOVAL & INSTALLATION

All Engines

✱✱ CAUTION

The fuel injection system remains under pressure, even after the engine has been turned OFF. The fuel system pressure must be relieved before disconnecting any fuel lines. Failure to do so may result in fire and/or personal injury.

1. Relieve the fuel system pressure. Disconnect the negative battery cable.
2. Drain fuel tank. Remove the fuel tank from the vehicle.
3. While holding the modular fuel sender assembly down, remove the snapring from the designated slots located on the retainer.

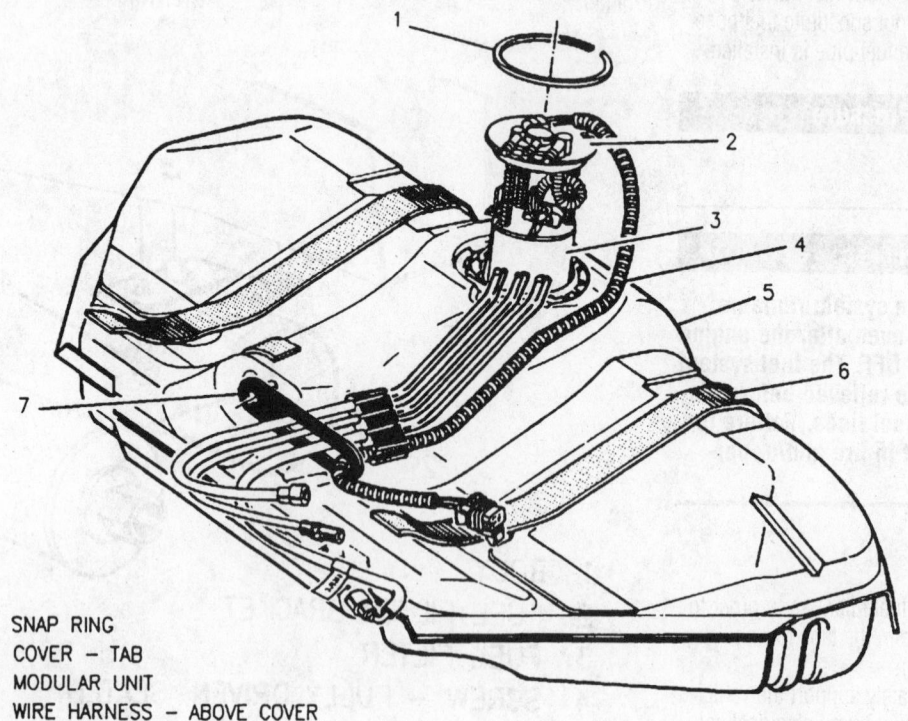

1 SNAP RING
2 COVER – TAB
3 MODULAR UNIT
4 WIRE HARNESS – ABOVE COVER
5 FUEL TANK
6 TANK ISOLATION STRIPS (3)
7 RUBBER ISOLATOR

79222223

Exploded view of the fuel sender assembly mounting

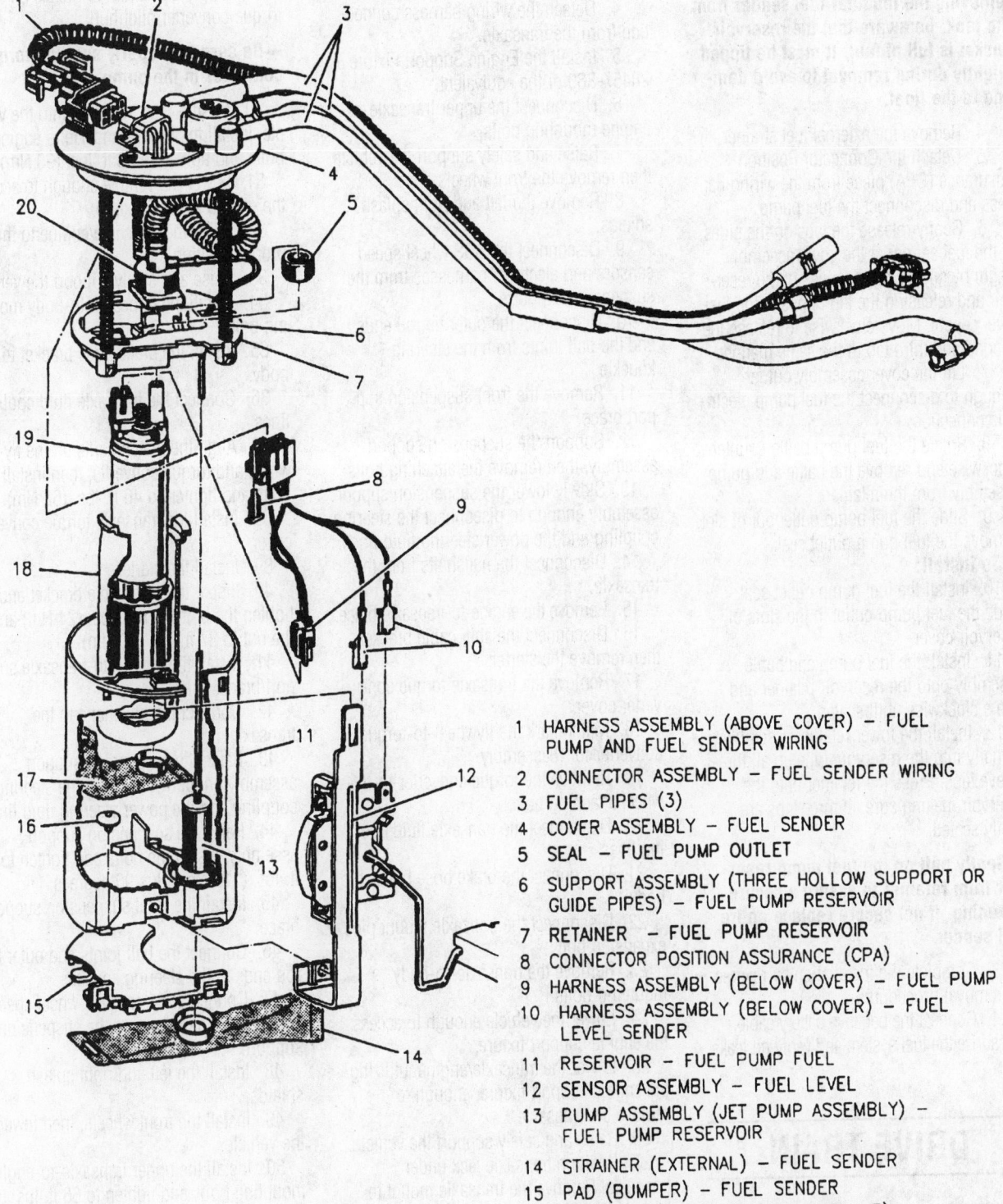

1 HARNESS ASSEMBLY (ABOVE COVER) – FUEL
 PUMP AND FUEL SENDER WIRING

2 CONNECTOR ASSEMBLY – FUEL SENDER WIRING

3 FUEL PIPES (3)

4 COVER ASSEMBLY – FUEL SENDER

5 SEAL – FUEL PUMP OUTLET

6 SUPPORT ASSEMBLY (THREE HOLLOW SUPPORT OR
 GUIDE PIPES) – FUEL PUMP RESERVOIR

7 RETAINER – FUEL PUMP RESERVOIR

8 CONNECTOR POSITION ASSURANCE (CPA)

9 HARNESS ASSEMBLY (BELOW COVER) – FUEL PUMP

10 HARNESS ASSEMBLY (BELOW COVER) – FUEL
 LEVEL SENDER

11 RESERVOIR – FUEL PUMP FUEL

12 SENSOR ASSEMBLY – FUEL LEVEL

13 PUMP ASSEMBLY (JET PUMP ASSEMBLY) –
 FUEL PUMP RESERVOIR

14 STRAINER (EXTERNAL) – FUEL SENDER

15 PAD (BUMPER) – FUEL SENDER

16 VALVE (SECONDARY UMBRELLA VALVE) –
 FUEL PUMP RESERVOIR INLET CHECK

17 STRAINER – FUEL PUMP FUEL

18 BAFFLE (ISOLATOR CUP) – FUEL PUMP

19 PUMP ASSEMBLY (ROLLERVANE) – FUEL

20 OUTLET – FUEL PUMP

79222224

Exploded view of the fuel pump assembly

➡ **The modular fuel sender assembly may spring up from its position. When removing the modular fuel sender from the tank, be aware that the reservoir bucket is full of fuel. It must be tipped slightly during removal to avoid damage to the float.**

4. Remove the external fuel strainer.

5. Detach the Connector Position Assurance (CPA) piece from the wiring harness and disconnect the fuel pump.

6. Gently release the tabs on the sides of the fuel sender at the cover assembly. Begin by squeezing the sides of the reservoir and releasing the tab opposite the fuel level sensor. Move clockwise to release the second and third tab in the same manner.

7. Lift the cover assembly out far enough to disconnect the fuel pump electrical connection.

8. Rotate the fuel pump baffle counterclockwise and remove the baffle and pump assembly from the retainer.

9. Slide the fuel pump outlet out of slot. Remove the fuel pump outlet seal.

To install:

10. Install the fuel pump outlet seal. Slide the fuel pump outlet in the slots of reservoir cover.

11. Install the fuel pump and baffle assembly onto the reservoir retainer and rotate clockwise until seated.

12. Install the lower retainer assembly partially into the reservoir. Line up all three sleeve tabs. Press the retainer onto the reservoir making sure all three tabs are firmly seated.

➡ **Gently pull on the fuel pump reservoir from retainer to assure a secure fastening. If not secure replace entire fuel sender.**

13. Complete the installation by reverse the removal procedures.

14. Connect the negative battery cable. Pressurize the fuel system and verify no leaks.

DRIVE TRAIN

Transaxle

REMOVAL & INSTALLATION

1. Disconnect the negative battery cable.
2. Remove the air cleaner assembly.

3. Disconnect the shift linkage from the transaxle.

4. Detach the wiring harness connection from the transaxle.

5. Install the Engine Support Fixture J 28467-360 or the equivalent.

6. Disconnect the upper transaxle-to-engine mounting bolts.

7. Raise and safely support the vehicle, then remove the front wheels.

8. Remove the left and right splash shields.

9. Disconnect the ABS wheel speed sensors and electrical harnesses from the suspension supports.

10. Disconnect the outer tie rod ends and the ball joints from the steering knuckle.

11. Remove the front suspension support brace.

12. Support the suspension support assembly, then remove the attaching bolts.

13. Slowly lower the suspension support assembly enough to disconnect the steering coupling and the power steering fluid lines.

14. Disconnect the halfshafts from the transaxle.

15. Remove the engine-to-transaxle brace.

16. Disconnect the shift cable bracket, then remove the starter.

17. Remove the transaxle torque converter cover.

18. Matchmark the flywheel-to-torque converter for reassembly.

19. Remove the torque converter-to-flywheel attaching bolts.

20. Disconnect the transaxle fluid cooling lines.

21. Disconnect the brake hose bracket to body.

22. Disconnect the transaxle mount pipe expansion bolt.

23. Remove the transaxle-to-body mounting bolts.

24. Lower the vehicle enough to access the engine support fixture.

25. Lower the transaxle/engine, utilizing the engine support fixture, enough to remove the transaxle.

26. Raise and safely support the vehicle.

27. Position transaxle jack under transaxle. Remove the transaxle mount to body bolts.

28. Remove the remaining engine to transaxle bolts. Remove the transaxle.

➡ **Transaxle cooler and lines should be flushed with J-35944 or equivalent whenever the transaxle has been removed for overhaul or replacement.**

To install:

29. Apply a thin film of grease on the torque converter pilot hub.

➡ **Be sure to properly seat the torque converter in the pump.**

30. Position the transaxle into the vehicle. Install the lower transaxle to engine bolts and tighten to 66 ft. lbs. (90 Nm).

31. Lower the vehicle enough to access the engine support fixture.

32. Raise the transaxle/engine to the proper position.

33. Raise and safely support the vehicle.

34. Install the transaxle-to-body mounting bolts.

35. Install the brake hose bracket to the body.

36. Connect the transaxle fluid cooling lines.

37. Align the matchmarks on the flywheel and torque converter, then install the bolts and tighten to 46 ft. lbs. (62 Nm).

38. Install the transaxle torque converter cover.

39. Install the starter.

40. Install the shift cable bracket and tighten the bolt to 18 ft. lbs. (24 Nm) and the nut to 37 ft. lbs. (50 Nm).

41. Install the engine-to-transaxle support brace.

42. Connect the halfshafts to the transaxle.

43. Raise the suspension support assembly enough to connect the steering coupling and the power steering fluid lines.

44. Raise the suspension support assembly, install the bolts and tighten to 71 ft. lbs. (110 Nm) plus 90 degrees.

45. Install the front suspension support brace.

46. Connect the ball joints and outer tie rod ends to the steering knuckle.

47. Connect the front ABS wheel speed sensors and harnesses to the suspension support.

48. Install the left and right splash shields.

49. Install the front wheels, then lower the vehicle.

50. Install the upper transaxle-to-engine mounting bolts and tighten to 66 ft. lbs. (90 Nm).

51. Remove the engine support fixture.

52. Attach the wiring harness to the transaxle.

53. Connect the shift linkage to the transaxle.

54. Install the air cleaner assembly, then connect the negative battery cable.

55. Fill the transaxle with Dexron® III or equivalent.

56. Apply the brakes, start the engine and shift the transaxle from reverse to drive, then recheck the fluid level.

57. Install the transaxle mount pipe expansion bolt.

Halfshaft

REMOVAL & INSTALLATION

1. Raise and safely support vehicle. Remove the wheel and tire assembly.

2. Disconnect the tie rod from the steering knuckle.

3. Insert a drift pin through the opening in the caliper and into the ventilation openings in the brake rotor to prevent the rotor from turning.

4. Remove the drive shaft hub nut and washer.

5. Remove the lower ball joint cotter pin and nut and loosen the joint using tool J-38892 or equivalent. If removing the right axle, turn the wheel to the left. If removing the left axle, turn the wheel to the right.

6. Disconnect the stabilizer link.

7. Separate the joint, with a prybar between the suspension support.

8. Disengage the axle from the hub and bearing using J-28733-A or equivalent puller.

9. Separate the hub and bearing assembly from the drive axle and move the strut and knuckle assembly rearward.

10. Disconnect the inner joint from the transaxle using tool J-28468 or J-33008 attached to J-29794 and J-2619–01 or their equivalents.

To install:

11. Insert the drive axle into the transaxle, by placing a suitable tool into the groove on the joint housing and tapping until seated.

✵✵ CAUTION

Be careful not to damage the axle seal or dislodge the transaxle seal garter spring when installing the axle.

12. Verify that the drive axle is seated into the transaxle by grasping on the housing and pulling outward.

13. Connect the tie rod to the steering knuckle and tighten the nut to 89 inch lbs. (10 Nm) plus 210 degrees.

14. Install the drive axle into the hub and bearing assembly.

15. Install the lower ball joint to the knuckle. Tighten the ball joint to steering knuckle nut to 89 inch lbs. (10 Nm) plus 180 degrees, then install the cotter pin.

16. Install the washer and new drive shaft nut. Again insert a drift pin into the caliper and rotor to prevent the rotor from

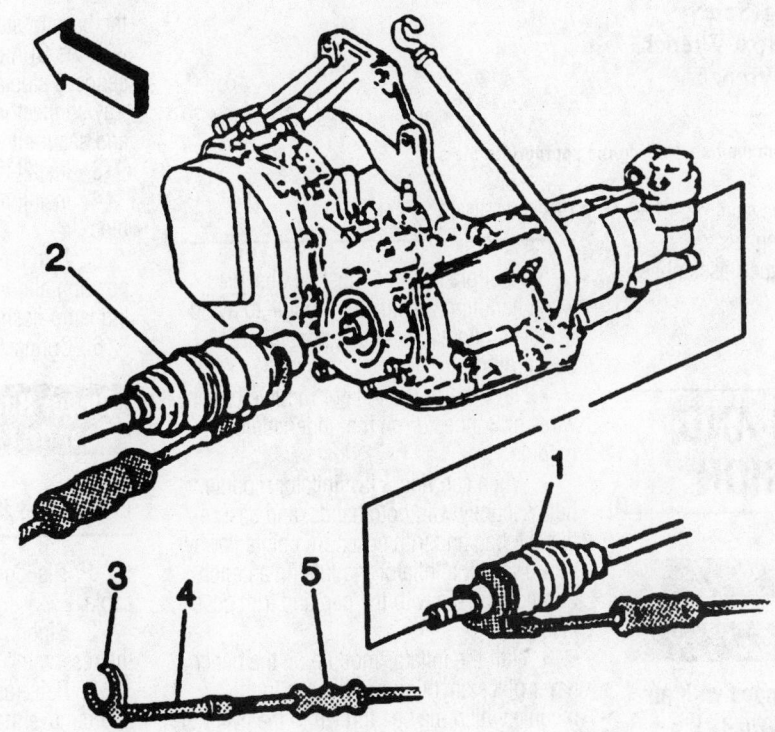

(1) Drive Axle, Right Side
(2) Drive Axle, Left Side
(3) J 28468 or J 33008
(4) J 29794
(5) J 2619-01

Removing the left and right halfshafts

79222225

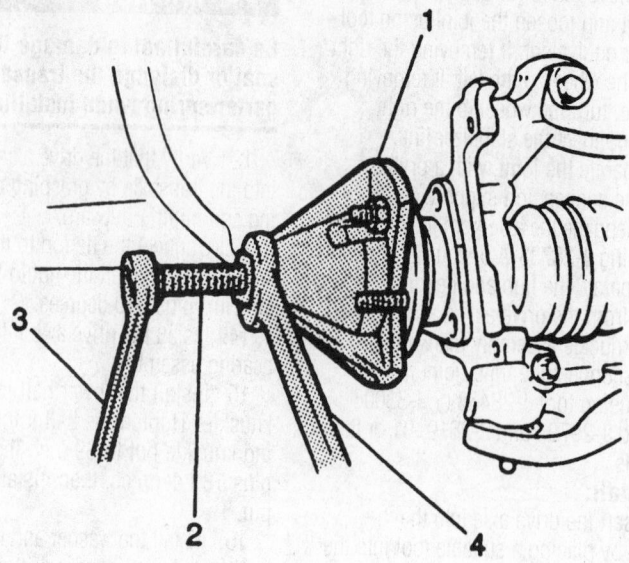

(1) J 28733A
(2) Forcing Screw
(3) Turn Box Wrench
(4) Hold Wrench

79222226

Removing the halfshaft from the hub utilizing the appropriate tools

turning and tighten the drive shaft nut to 30 ft. lbs. (40 Nm) plus 235 degrees.

17. Install the tire and wheel assembly. Lower the vehicle.

STEERING AND SUSPENSION

Air Bag

✳✳ CAUTION

Some vehicles are equipped with an air bag system, also known as the Supplemental Inflatable Restraint (SIR) or Supplemental Restraint System (SRS). The system must be disabled before performing service on or around system components, steering column, instrument panel components, wiring and sensors. Failure to follow safety and disabling procedures could result in accidental air bag deployment, possible personal injury and unnecessary system repairs.

PRECAUTIONS

Several precautions must be observed when handling the inflator module to avoid accidental deployment and possible personal injury.

• Never carry the inflator module by the wires or connector on the underside of the module.

• When carrying a live inflator module, hold securely with both hands, and ensure that the bag and trim cover are pointed away.

• Place the inflator module on a bench or other surface with the bag and trim cover facing up.

• With the inflator module on the bench, never place anything on or close to the module which may be thrown in the event of an accidental deployment.

DISARMING

✳✳ CAUTION

The Supplemental Restraint System (SRS) must be disarmed before performing service procedures around the air bag or SRS wiring. Failure to do so may cause accidental deploy-

ment of the air bag, resulting in unnecessary SRS repairs and/or personal injury.

1. Disconnect the negative battery cable.
2. Turn the steering wheel so the vehicle's wheels are pointing straight ahead.
3. Turn the ignition switch to the **LOCK** position and remove the key.
4. Remove the **AIR BAG** fuse from the fuse block.
5. Remove the left sound insulator.
6. Remove the Connector Position Assurance (CPA) clip from the yellow 2-way connector at the base of the steering column, and detach the connector. If equipped with a passenger's side air bag, remove the CPA and detach the yellow 2-way connector from the passenger air bag lead.

REARMING

1. Turn the ignition switch to the **LOCK** position and remove the key.
2. Attach the yellow 2-way connector at the base of steering column and secure it with the CPA clip. If equipped with a passenger's side air bag, attach the yellow 2-way connector at the passenger air bag lead and secure it with the CPA clip.
3. Install the left sound insulator.
4. Install the **AIR BAG** fuse in the fuse block.
5. Turn the ignition switch to the **RUN** position and verify that the **AIR BAG** warning lamp flashes 7 times, then turns OFF.
6. Connect the negative battery cable.

Power Rack and Pinion Steering Gear

REMOVAL & INSTALLATION

1. Disconnect the negative battery cable.
2. Siphon the power steering fluid from the reservoir.
3. Raise and safely support the vehicle, and remove the left front wheel.
4. Remove the lower pinch bolt on the intermediate shaft assembly and push the shaft up toward the steering column.
5. Remove the tie rods from the steering knuckle.
6. Remove the steering gear mounting bolts.
7. Remove the transaxle-to-crossmember mounting bolt.
8. Remove the rear crossmember-to-body mounting bolts in order to provide

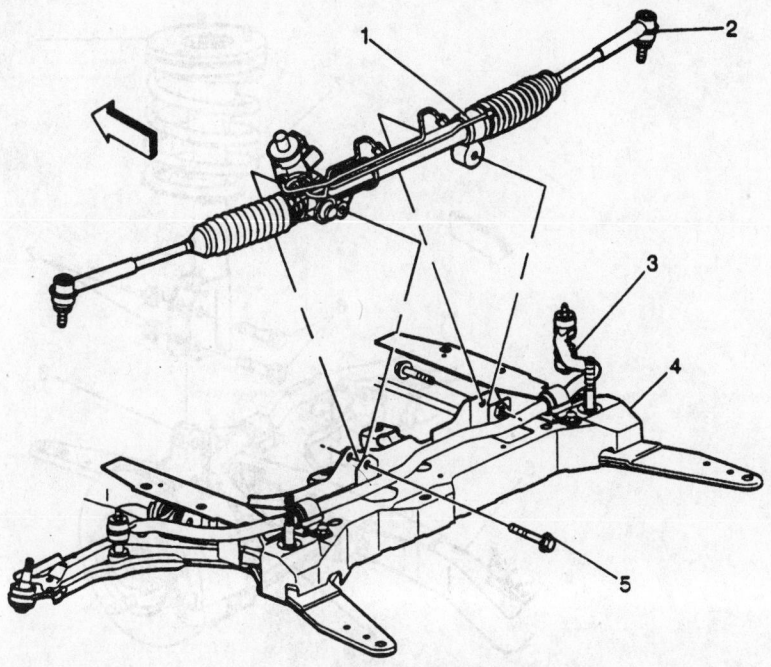

(1) Gear Assembly, Power Steering
(2) Tie Rod
(3) Shaft, Front Stabilizer

(4) Frame Assembly, Drivetrain and Front Crossmember
(5) Bolt, Power Steering Gear Assembly

79222227

Exploded view of the power rack and pinion steering gear mounting

clearance to remove the power steering pipes/hoses.

9. Loosen the front crossmember bolts.

10. Disconnect the power steering pipes from the steering gear.

11. Remove the steering gear through the left wheel opening.

To install:

12. Install the steering gear through the left wheel opening.

13. Connect the power steering pipes to the steering gear.

14. Tighten the front crossmember bolts to 71 ft. lbs. (110 Nm) plus 90 degrees.

15. Install and tighten the rear crossmember-to-body bolts to 71 ft. lbs. (110 Nm) plus 90 degrees.

16. Install the transmission mount-to-crossmember bolt and tighten to 49 ft. lbs. (66 Nm).

17. Install the steering gear mounting bolts and tighten to 89 ft. lbs. (120 Nm).

18. Connect the tie rod ends to the steering knuckle and tighten to the tie rod nut to 15 ft. lbs. (20 Nm) plus 180 degrees.

19. Install the lower pinch bolt on the intermediate shaft and tighten to 16 ft. lbs. (22 Nm).

20. Install the left front wheel, then lower the vehicle.

21. Connect the negative battery cable.

22. Refill and bleed the power steering system. Check the front end alignment and adjust as necessary.

Strut

REMOVAL & INSTALLATION

Front

1. Raise and safely support the vehicle. Remove the front tire and wheel assembly.

2. Remove the cotter pin and castle nut from the outer tie rod end and separate the ball stud from the strut arm using J-24319-01, or an equivalent tie rod puller.

3. Remove the bolt from the brake line bracket and separate the bracket from the strut.

4. Scribe reference marks on the front strut and steering knuckle for installation purposes.

5. Remove the strut lower mounting bracket nut and through-bolts.

6. Lower the vehicle until the upper strut plate mounting bolts are accessible.

7. Support the lower control arm.

8. Hold the strut and remove the strut plate mounting nuts and bolt. Remove the strut from the vehicle.

❊❊ WARNING

If the strut cartridge is being replaced, a strut compressor tool J-34013-A or equivalent must be used to remove the coil spring from the strut assembly. Failure to use a strut compressing tool can result in personal injury and/or part damage.

To install:

9. Install the strut assembly into the vehicle and install the upper strut plate mounting nuts.

10. Connect the lower strut bracket to the steering knuckle and install the through-bolts.

11. Raise the vehicle.

12. Install the through-bolt nuts and tighten to 133 ft. lbs. (180 Nm) with the reference marks in alignment.

13. Connect the tie rod end to the strut arm and tighten the castle nut to 15 ft. lbs. (20 Nm) plus 90 degrees.

14. Connect the brake line bracket to the strut and tighten the bolt to 10 ft. lbs. (14 Nm).

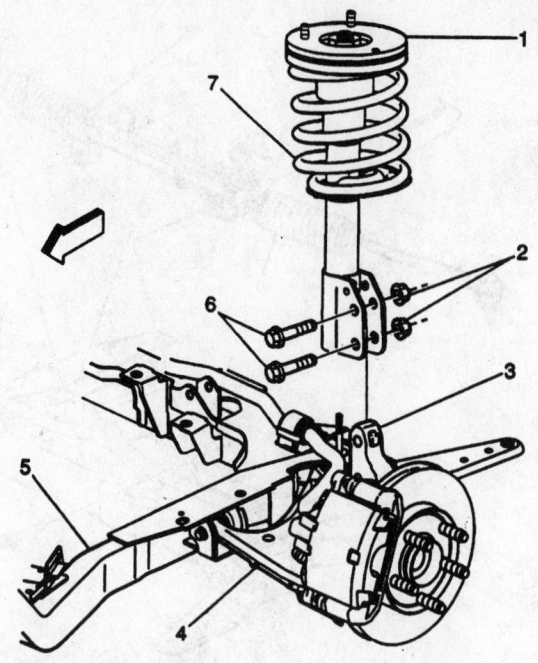

(1) Bearing, Strut Mount
(2) Nuts, Strut Mount
(3) Knuckle
(4) Arm, Control

(5) Crossmember
(6) Bolts, Strut Mount
(7) Spring, Strut Coil

79222228

Exploded view and component identification of the front strut

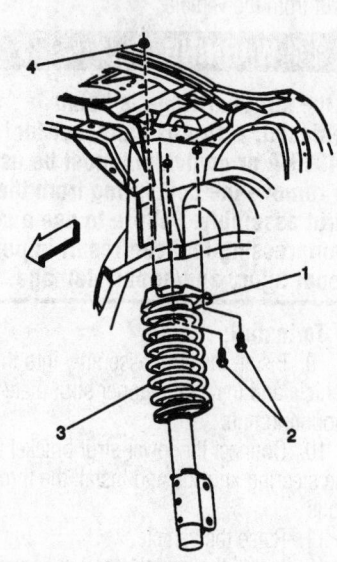

(1) Mount, Strut Upper
(2) Bolt, Strut Mount
(3) Spring, Coil
(4) Nut

79222229

Exploded view of the upper strut mounting

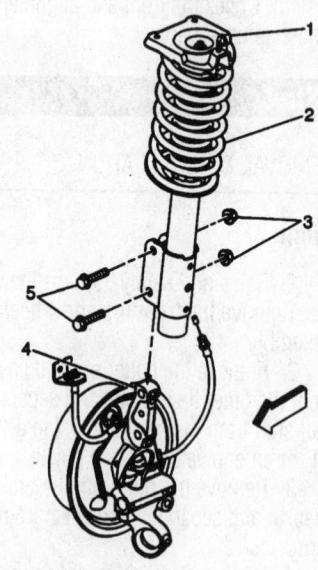

(1) Mount, Strut Upper
(2) Spring, Coil
(3) Nuts, Strut Bolt
(4) Knuckle
(5) Bolt, Strut Mount

79222230

Exploded view of the lower strut mounting

15. Install the tire and wheel assembly. Lower the vehicle.

16. Tighten the upper strut plate mounting nuts and bolt to 18 ft. lbs. (25 Nm).

17. Check the front end alignment and adjust as necessary.

Rear

1. Raise and safely support the vehicle and remove the rear wheels.

2. Matchmark the strut-to-knuckle position.

3. Remove the upper strut mounting nuts from in the trunk area.

4. Remove the upper strut mounting bolts from inside the fender well.

5. Remove the strut-to-knuckle mounting bolts, then remove the strut from the vehicle.

To install:

6. Install the strut to the vehicle, and loosely install the strut-to-knuckle mounting bolts.

7. Install the strut-to-body bolts inside the fender well.

➡**Align the matchmarks to ensure proper alignment.**

8. Install the strut mount-to-body nuts in the trunk area.

9. Tighten the strut-to-knuckle mounting bolts to 89 ft. lbs. (120 Nm).

10. Install the rear wheels, then lower the vehicle.

11. Check the alignment.

Coil Spring

REMOVAL & INSTALLATION

Front and Rear

1. Remove the strut assembly from the vehicle, as outlined earlier in this section.

2. Mount the strut compressor J-34013 in holding fixture J-3289-20.

3. Mount the strut assembly into the compressor. Note that the strut compressor has strut mounting holes drilled for specific vehicle lines.

4. Compress the strut approximately ½ its height after initial contact with the top cap.

❋❋ WARNING

Never bottom the spring or dampener rod!

5. Remove the nut from the strut dampener shaft and place alignment/guiding rod

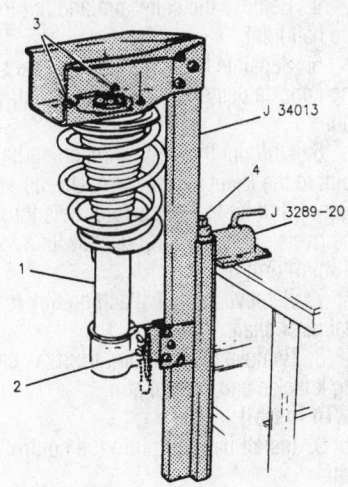

1 STRUT ASSEMBLY
2 INSTALL LOCKING PINS THROUGH STRUT ASSEMBLY
3 TIGHTEN NUTS UNTIL FLUSH WITH STRUT COMPRESSOR
4 COMPRESSOR FORCING SCREW

79222Z231

View of the strut assembly mounted in a compressor

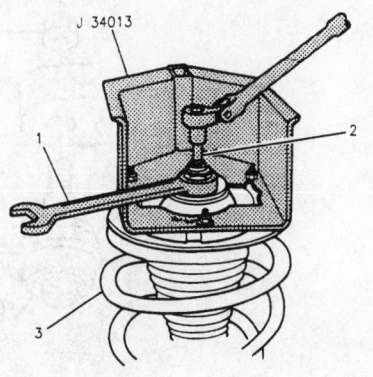

1 WRENCH
2 SOCKET
3 STRUT ASSEMBLY

79222Z232

Use a socket and a wrench to remove the dampener shaft nut spring cap while compressing the spring

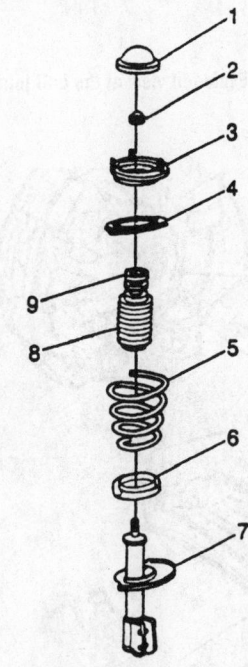

(1) Cap
(2) Nut, Strut Mount
(3) Bearing, Strut Mount
(4) Insulator, Spring (Upper)
(5) Spring, Strut Coil
(6) Insulator, Spring (Lower)
(7) Strut
(8) Bumper, Jounce
(9) Shield, Strut Dust

79222Z233

Exploded view of the front strut

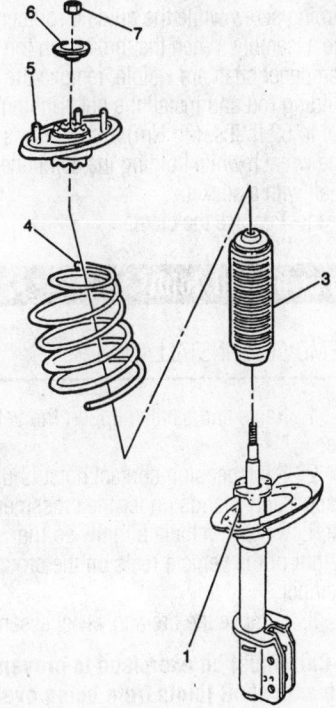

(1) Strut
(2) Dust Shield, Strut
(3) Spring, Strut Coil
(4) Mount, Strut Upper
(5) Washer
(6) Nut

79222Z234

Exploded view of the rear strut

J-34013-27 on top of the dampener shaft. Use the rod to guide the dampener shaft straight down through the spring cap while compressing the spring. Remove the components.

➡**Be careful to avoid chipping or cracking the spring coating when handling the front suspension coil spring assembly.**

To install:

6. Install the bearing cap into the strut compressor if previously removed.

7. Mount the strut assembly in strut compressor, using bottom locking pin only. Extend the dampener shaft and install clamp J-34013–20 on the dampener shaft.

8. Install the spring over the dampener and swing the assembly up so the upper locking pin can be installed.

9. Install all shields, bumpers and insulators on the spring seat. Install the spring seat on top of the spring. Be sure the flat on the upper spring seat is facing in the proper direction. The spring seat flat should be facing the same direction as the centerline of the strut assembly spindle.

10. Install the guiding rod and turn the forcing screw while the guiding rod centers the assembly. When the threads on the dampener shaft are visible, remove the guiding rod and install the nut. Tighten the nut to 52 ft. lbs. (70 Nm). Use a crow's foot line wrench while holding the dampener shaft with a socket.

11. Remove the clamp.

Lower Ball Joint

REMOVAL & INSTALLATION

1. Raise and safely support the vehicle.

2. If suspension contact hoist is used, place safety stands under the crossmember. Lower the vehicle slightly so the weight of the vehicle rests on the crossmember.

3. Remove the tire and wheel assembly.

➡ **Care must be exercised to prevent the axle shaft joints from being over-extended. When either end of the shaft is disconnected, over-extension of the joint could result in separation of internal components and possible joint failure. Failure to observe this can result**

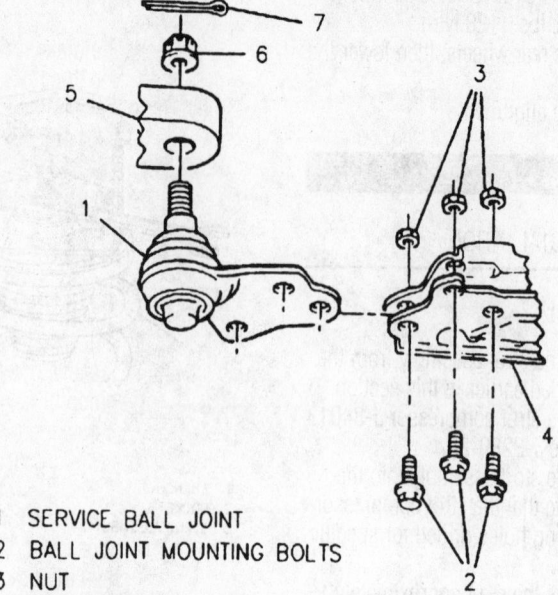

```
1   SERVICE BALL JOINT
2   BALL JOINT MOUNTING BOLTS
3   NUT
4   LOWER CONTROL ARM
5   STEERING KNUCKLE
6   NUT – 55 N·m (41 LBS. FT.) MINIMUM TORQUE
         65 N·m (48 LBS. FT.) MAXIMUM TORQUE
         TO INSTALL PIN
7   PIN
```
79222Z236

Exploded view of the ball joint mounting

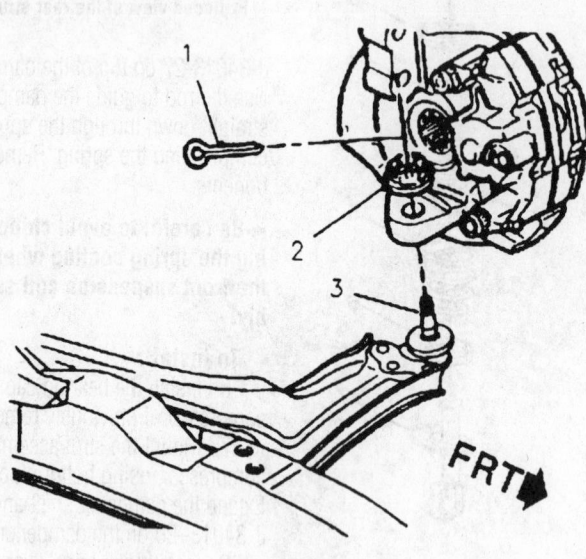

```
1   PIN
2   NUT – 55 N·m (41 LBS. FT.) MINIMUM TORQUE
         65 N·m (48 LBS. FT.) MAXIMUM TORQUE TO
         INSTALL PIN
3   LOWER BALL JOINT
```
79222Z235

Exploded view of the ball joint-to-knuckle mounting

in interior joint or boot damage and possible joint failure.

4. Remove the cotter pin and nut from the ball joint.

5. Separate the ball joint from the steering knuckle using tool J-38892 or equivalent.

6. Drill out the 3 rivets retaining ball joint to the lower control arm. Use an 1/8 in. (3mm) drill bit to make a pilot hole through the rivets. Finish drilling rivets with a 1/2 in. (13mm) drill bit.

7. Remove the nut attaching link to the stabilizer shaft.

8. Remove the ball joint from the steering knuckle and control arm.

To install:

9. Install ball joint into the control arm.

10. Install 3 new bolts and nuts (supplied with new ball joint) and tighten.

11. Position ball joint stud through the steering knuckle and tighten nut to 50 ft. lbs. (65 Nm) and install cotter pin.

12. Connect stabilizer link to stabilizer shaft and tighten nut to 13 ft. lbs. (17 Nm).

13. Install tire and wheel assembly.

14. Lower the vehicle.

15. Check front wheel alignment and adjust, if necessary.

Wheel Bearings

ADJUSTMENT

These vehicles are equipped with sealed hub and bearing assemblies. The hub and bearing assemblies are non-serviceable. If the assembly is damaged, the complete unit must be replaced.

REMOVAL & INSTALLATION

Front

1. Raise and safely support the vehicle. Remove the front tire and wheel assembly.

2. Insert a drift punch through the caliper and into the rotor cooling fins. This keeps the assembly from turning while the axle nut is being loosened.

3. Remove the axle nut and washer. Remove the punch.

4. Remove the caliper mounting bolts and remove the caliper from the steering knuckle. Support the caliper aside. DO NOT allow the caliper to hang unsupported from the brake hose.

5. Remove the brake rotor.

6. Remove the three bolts securing the hub bearing to the steering knuckle. Remove the backing plate.

7. Using J-28733-A, or an equivalent front hub remover, separate the halfshaft from the hub bearing assembly.

8. Remove the hub bearing assembly.

To install:

9. Install the hub bearing assembly over the end of the axle making sure the splines engage smoothly.

10. Install the backing plate and the three hub bearing mounting bolts. Tighten the bolts to 70 ft. lbs. (95 Nm).

11. Install the brake rotor.

12. Install the caliper onto the steering knuckle and tighten the caliper mounting bolts to 38 ft. lbs. (51 Nm).

13. Install a drift punch into the rotor cooling fins. This keeps the assembly from turning while the axle nut is being tightened. Tighten the axle nut to 192 ft. lbs. (260 Nm).

14. Install the front tire and wheel assembly and tighten the wheel nuts to 100 ft. lbs. (140 Nm). Lower the vehicle.

Rear

1. Raise and safely support the vehicle. Remove the wheel and tire assembly. Remove the brake drum.

2. Remove the four hub bearing assembly mounting nuts from behind the backing plate while holding the bolts from the front. The mounting bolts can be removed from the backing plate once the nuts have been removed. The top rear bolt will not clear the brake shoes and must be removed with the bearing assembly.

3. Disconnect the ABS speed sensor wire from the hub bearing assembly. Remove the hub bearing assembly.

To install:

7. Position the hub bearing assembly on the backing plate and insert the four mounting bolts. Connect the ABS wheel speed sensor, if equipped.

8. Install the hub bearing mounting nuts from behind the backing plate and

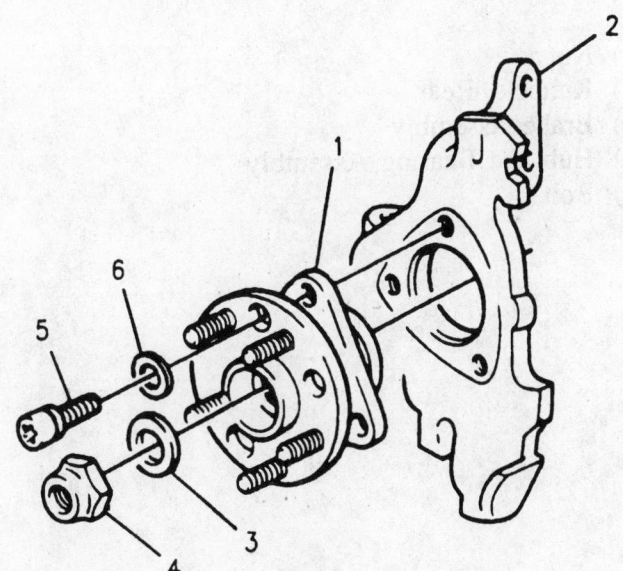

1   HUB AND BEARING ASSEMBLY
2   STEERING KNUCKLE
3   WASHER
4   DRIVE AXLE NUT — 260 N·m (192 LBS. FT.)
5   HUB AND BEARING RETAINING BOLT
6   WASHER

79222237

Exploded view of the front hub and bearing assembly

tighten the nuts to 38 ft. lbs. (52 Nm) while holding the bolts.

9. Install the brake drum.

10. Install the tire and wheel assembly and tighten the wheel bolts to 100 ft. lbs. (140 Nm). Lower the vehicle.

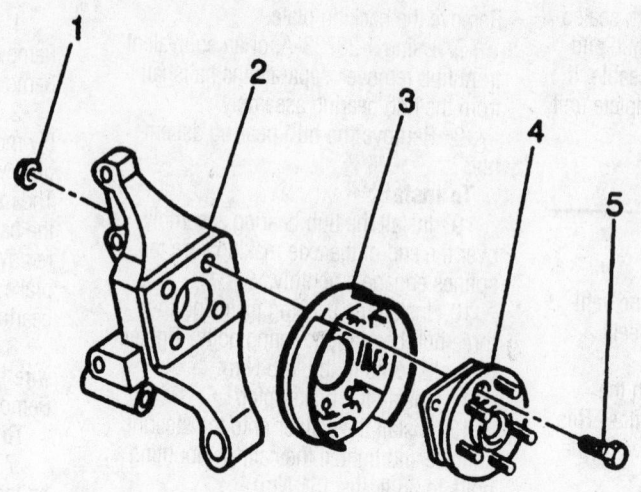

(1) Nut
(2) Knuckle, Rear
(3) Brake Assembly
(4) Hub and Bearing Assembly
(5) Bolt

79222238

Hub and bearing components

GENERAL MOTORS N BODY

36

Buick-Skylark • **Oldsmobile**-Achieva • **Pontiac**-Grand Am

DRIVE TRAIN 36-22
ENGINE REPAIR 36-2
FUEL SYSTEM 36-20
STEERING AND
SUSPENSION 36-26

A
Air Bag 36-26
  Disarming 36-26
  Precautions 36-26
  Rearming 36-26

C
Camshaft and Valve Lifter 36-12
  Removal & Installation 36-12
Clutch 36-24
  Removal & Installation 36-24
Coil Spring 36-29
  Removal & Installation 36-29
Cylinder Head 36-5
  Removal & Installation 36-5

E
Engine Assembly 36-2
  Removal & Installation 36-2
Exhaust Manifold 36-11
  Removal & Installation 36-11

F
Fuel Filter 36-21
  Removal & Installation 36-21
Fuel Pump 36-21
  Removal & Installation 36-21

Fuel System Pressure 36-20
  Relieving 36-20
Fuel System Service
  Precautions 36-20

H
Halfshaft 36-25
  Removal & Installation 36-25
Hydraulic Clutch System 36-25
  Bleeding 36-25

I
Ignition Timing 36-2
  Adjustment 36-2
Intake Manifold 36-8
  Removal & Installation 36-8

L
Lower Ball Joint 36-30
  Removal & Installation 36-30

O
Oil Pan 36-14
  Removal & Installation 36-14
Oil Pump 36-15
  Removal & Installation 36-15

P
Power Rack and Pinion Steering
  Gear 36-26
  Removal & Installation 36-26

R
Rear Main Seal 36-16
  Removal & Installation 36-16
Rocker Arms 36-8
  Removal & Installation 36-8

S
Shock Absorber 36-28
  Removal & Installation 36-28
Strut 36-27
  Removal & Installation 36-27

T
Timing Chain, Sprockets and Front
  Cover 36-17
  Removal & Installation 36-17
Transaxle 36-22
  Removal & Installation 36-22

V
Valve Lash 36-14
  Adjustment 36-14

W
Water Pump 36-3
  Removal & Installation 36-3
Wheel Bearings 36-31
  Adjustment 36-31
  Removal & Installation 36-31

ENGINE REPAIR

➡ **Disconnecting the negative battery cable on some vehicles may interfere with the functions of the on board computer systems and may require the computer to undergo a relearning process, once the negative battery cable is reconnected.**

Ignition Timing

ADJUSTMENT

The ignition timing is not adjustable, and is set according to engine demand electronically. The Powertrain Control Module controls the ignition timing for all driving conditions. The PCM monitors input signals from the following components engine coolant sensor, intake air temperature, mass air flow sensor, park/neutral switch, throttle position sensor and the vehicle speed sensor.

Engine Assembly

REMOVAL & INSTALLATION

❈❈ CAUTION

The fuel injection system remains under pressure, even after the engine has been turned OFF. The fuel system pressure must be relieved before disconnecting any fuel lines. Failure to do so may result in fire and/or personal injury.

2.3L and 2.4L Engines

The engine and transaxle are removed as a unit from under the vehicle.

1. Relieve the fuel system pressure. Disconnect the negative battery cable.

❈❈ CAUTION

Never open, service or drain the radiator or cooling system when hot; serious burns can occur from the steam and hot coolant.

2. Drain the cooling system into a suitable container. If equipped with A/C, recover the refrigerant.
3. Remove the left sound insulator and disconnect the clutch pushrod from pedal assembly (manual transaxle only).
4. Disconnect the heater hose and upper radiator hose from the thermostat housing. Remove the air cleaner-to-throttle body duct, the upper radiator support and cooling fan assembly.
5. If equipped with A/C, disconnect the refrigerant line and discard the O-rings.
6. Disconnect the vacuum hoses from the front of the engine. Label and detach the electrical connectors from the necessary components.
7. Disconnect the power brake booster vacuum lines. Disconnect the throttle cables; remove the cable bracket and position aside.
8. Remove the power steering pump rear bracket and pump; position the pump aside with the lines attached.
9. Disconnect and cap the fuel lines from the fuel rail.
10. Disconnect the shift cable from the transaxle and the TV cable on automatic transaxles.
11. Remove the exhaust manifold and heat shield.
12. Remove the lower radiator hose. Remove the coolant recovery tank and position it aside leaving the hoses attached.
13. Install an engine support fixture, J-28467-A or equivalent. Remove the right engine mount.

❈❈ CAUTION

The EPA warns that prolonged contact with used engine oil may cause a number of skin disorders, including cancer! You should make every effort to minimize your exposure to used engine oil. Protective gloves should be worn when changing the oil. Wash your hands and any other exposed skin areas as soon as possible after exposure to used engine oil. Soap and water, or waterless hand cleaner should be used.

14. Raise and safely support the vehicle. Drain the engine crankcase into a suitable container.
15. Remove the front tire and wheel assemblies. Remove the right lower splash shield.
16. Disconnect the electrical connections from the ABS wheel speed sensors. Disconnect the ground connections from the transaxle.
17. Remove the cotter pins and castle nuts from the lower ball joints and separate the ball joints from the steering knuckles.
18. Support the suspension crossmembers, then remove the mounting bolts and crossmembers.
19. Disconnect the heater outlet hose from the radiator outlet pipe.
20. Disconnect the halfshafts from the transaxle. If equipped with a manual transaxle, disconnect the intermediate shaft.
21. Position a suitable engine table under the power train and lower the vehicle until the powertrain is resting on the table.
22. Remove the engine strut and transaxle mount bolts. Remove the engine support fixture.
23. Raise the vehicle until it is clear of the engine. Separate the transaxle from the engine.
   To install:
24. Install the transaxle onto the engine and tighten the bolts to 55 ft. lbs. (75 Nm). Position the powertrain assembly and the engine table under the vehicle.
25. Lower the vehicle and install the engine support fixture. Install the engine strut and transaxle mounts.
26. Raise the vehicle and remove the engine table. Connect the halfshafts and intermediate shaft to the transaxle.
27. Complete the installation by reversing the removal procedures.
28. Tighten the ball joint-to-steering knuckle nuts to 41 ft. lbs. (55 Nm). Install new cotter pins.

❈❈ WARNING

Operating the engine without the proper amount and type of engine oil will result in severe engine damage.

29. Refill the cooling system, the crankcase and fill the remainder of the underhood fluids.
30. Connect the negative battery cable. Recharge the A/C system. Start the vehicle and verify no leaks.

➡ **Whenever the vehicle sub-frame is removed or lowered, the wheel alignment should be checked.**

3.1L Engine

Please note that the engine and transaxle are removed as an assembly from under the vehicle.

1. Relieve the fuel system pressure. Disconnect the negative battery cable.

❈❈ CAUTION

Never open, service or drain the radiator or cooling system when hot; serious burns can occur from the steam and hot coolant.

2. Remove the top half of the air cleaner assembly and the throttle body inlet duct. Drain the cooling system.

3. Remove the upper and lower radiator hoses. Disconnect the coolant inlet line from the coolant surge tank.

4. Disconnect the vacuum hoses from the EVAP canister purge valve, vacuum modulator and power brake booster.

5. Disconnect the heater outlet hose from the water pump. Remove the serpentine drive belt.

6. Disconnect the control cables from the throttle body lever and intake manifold bracket and position the cables aside.

7. Label and detach the electrical connectors from the necessary components.

8. Remove the alternator. Disconnect and cap the power steering lines from the power steering pump.

9. Disconnect and cap the fuel lines from the fuel rail. Remove the cooling fan assembly.

10. Disconnect the shift control cable from the transaxle shift lever and cable bracket. Disconnect the transaxle vent tube from the transaxle.

11. Disconnect the vacuum hose from the vacuum reservoir. Install an engine support fixture, J-28467-A, or the equivalent.

12. Loosen, but do not remove the top two A/C compressor mounting bolts. Raise and safely support the vehicle.

❋❋ CAUTION

The EPA warns that prolonged contact with used engine oil may cause a number of skin disorders, including cancer! You should make every effort to minimize your exposure to used engine oil. Protective gloves should be worn when changing the oil. Wash your hands and any other exposed skin areas as soon as possible after exposure to used engine oil. Soap and water, or waterless hand cleaner should be used.

13. Drain the engine crankcase into a suitable container.

14. Remove the front tire and wheel assemblies. Remove the right and left inner fender splash shields.

15. Remove the engine mount strut. Disconnect the ABS sensor wires from the wheel sensors and suspension member supports, if equipped.

16. Remove the cotter pins and castle nuts from the lower ball joints. Using a suit-

able tool, separate the lower ball joints from the steering knuckles.

17. Remove the lower suspension support assemblies with the lower control arms attached. Disconnect the halfshafts from the transaxle and support aside.

18. Remove the oil filter and oil filter adapter. Remove the flywheel cover. Remove the starter. Disconnect the heater hoses from the heater core.

19. Remove the A/C compressor lower mounting bolts and remove the compressor from the mounting bracket and position aside. DO NOT disconnect the refrigerant lines from the compressor or allow the lines top support the weight of the compressor.

20. Remove the vacuum reservoir tank. Disconnect the exhaust pipe from the exhaust manifold and position the pipe aside.

21. Remove the engine mount strut bracket from the engine. Disconnect the transaxle cooler lines from the radiator.

22. Remove the transaxle oil fill tube. Lower the vehicle until the power train assembly is resting on a suitable engine table.

23. Remove the transaxle mount-to-body bolts, and the intermediate bracket from the right engine mount.

24. Remove the engine support fixture. Raise the vehicle leaving the powertrain assembly on the engine table.

25. Separate the engine and transaxle assemblies.

To install:

26. Connect the transaxle to the engine and tighten the mounting bolts to 55 ft. lbs. (75 Nm).

27. Position the powertrain assembly under the vehicle and lower the vehicle into position. Loosely install the serpentine belt.

28. Complete the installation by reversing the removal procedures.

29. Tighten the following items to:
- Engine strut bracket-to-engine bolts: 44 ft. lbs. (60 Nm)
- Exhaust pipe-to-exhaust manifold bolts: 18 ft. lbs. (25 Nm)
- Ball joints-to-steering knuckle nuts: 41 ft. lbs. (55 Nm)

❋❋ WARNING

Operating the engine without the proper amount and type of engine oil will result in severe engine damage.

30. Connect the negative battery cable. Check and fill all the engine fluids as nec-

essary. Start the vehicle and bleed the power steering system.

Water Pump

REMOVAL & INSTALLATION

2.3L and 2.4L Engines

1. Disconnect the negative battery cable.

❋❋ CAUTION

Never open, service or drain the radiator or cooling system when hot; serious burns can occur from the steam and hot coolant.

2. Drain the cooling system.

3. Detach the oxygen sensor (O_2) electrical connector.

4. Remove the upper and lower exhaust manifold heat shields.

5. Raise and safely support the vehicle.

6. Remove the exhaust manifold brace-to-manifold bolt.

7. Loosen the exhaust pipe-to-manifold spring loaded nuts.

8. Remove the radiator outlet pipe-to-water pump cover bolts.

9. Raise and safely support the vehicle.

10. Disconnect the exhaust pipe from the manifold.

11. Disconnect the radiator outlet pipe from the oil pan and transaxle.

12. If equipped with a manual transaxle, remove the exhaust manifold brace. If equipped with an automatic transaxle, leave the lower radiator hose attached and pull down on the radiator pipe to disconnect it from the water pump.

13. Lower the vehicle.

14. Remove the exhaust manifold.

15. Remove the water pump cover-to-engine bolts.

16. Remove the three water pump-to-timing chain housing nuts.

17. Remove the water pump and cover assembly.

18. Remove the five water pump cover-to-pump bolts and remove the cover.

To install:

19. Install the water pump-to-engine components and tighten the bolts finger-tight.

20. Lubricate the splines of the water pump drive with chassis grease.

21. To complete the installation, reverse the removal procedures.

22. With all the bolts in place tighten in the following sequence:

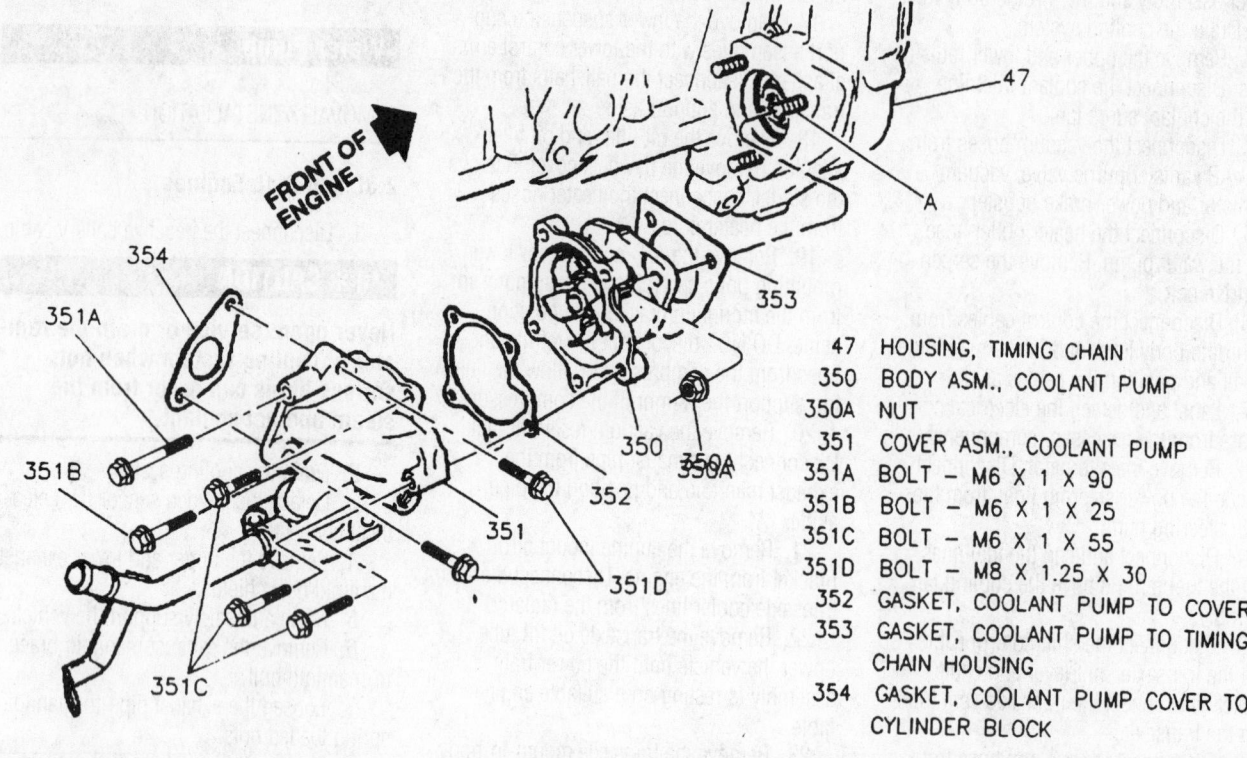

FRONT OF ENGINE

47	HOUSING, TIMING CHAIN
350	BODY ASM., COOLANT PUMP
350A	NUT
351	COVER ASM., COOLANT PUMP
351A	BOLT — M6 X 1 X 90
351B	BOLT — M6 X 1 X 25
351C	BOLT — M6 X 1 X 55
351D	BOLT — M8 X 1.25 X 30
352	GASKET, COOLANT PUMP TO COVER
353	GASKET, COOLANT PUMP TO TIMING CHAIN HOUSING
354	GASKET, COOLANT PUMP COVER TO CYLINDER BLOCK

79222302

Exploded view of the water pump mounting—2.3L and 2.4L engines

- Water pump-to-timing chain housing: 19 ft. lbs. (26 Nm)
- Water pump cover-to-water pump: 106–124 inch lbs. (12–14 Nm)
- Water pump cover-to-engine block (bottom first): 19 ft. lbs. (26 Nm)
- Radiator outlet pipe-to-pump cover: 10 ft. lbs. (14 Nm).

23. Refill the cooling system. Connect the negative battery cable. Start the vehicle and verify no leaks.

3.1L Engine

> **⁂ CAUTION**
>
> **Never open, service or drain the radiator or cooling system when hot; serious burns can occur from the steam and hot coolant.**

1. Disconnect the negative battery cable. Drain the cooling system into a suitable container.

2. Loosen, but do not remove, the water pump pulley bolts. Remove the serpentine belt.

3. Remove the water pump pulley bolts and pulley. Remove the five water pump mounting bolts and pump.

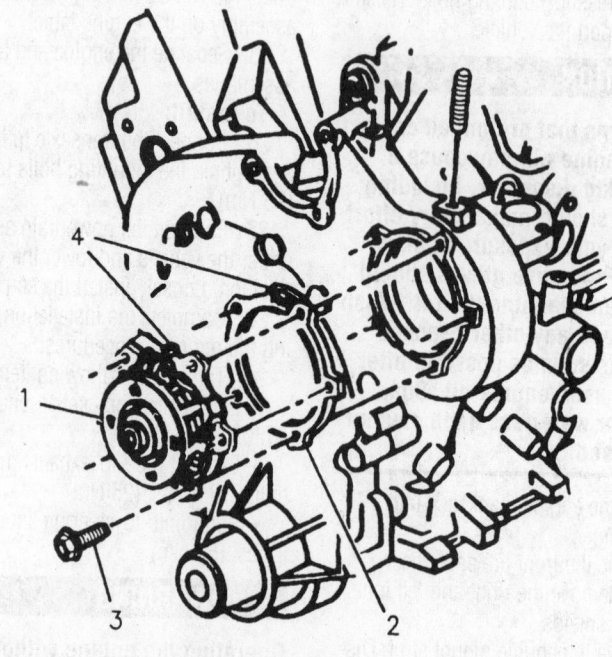

1	COOLANT PUMP
2	GASKET
3	BOLT — 10 N·m (89 LBS. IN.)
4	LOCATOR (MUST BE VERTICAL)

79222303

Exploded view of the water pump mounting—3.1L engine

To install:

4. Clean all the gasket surfaces completely. Apply a thin bead of sealer around the outside edge of the water pump along the gasket sealing area and install the gasket onto the water pump.

5. Install the water pump and tighten the bolts to 89 inch lbs. (10 Nm).

6. Install the water pump pulley and tighten the pulley bolts finger-tight. Install the serpentine belt.

7. Tighten the water pump pulley bolts to 18 ft. lbs. (25 Nm). Connect the negative battery cable.

8. Refill and bleed the cooling system. Test run the engine and check for leaks.

Cylinder Head

REMOVAL & INSTALLATION

※ CAUTION

The fuel injection system remains under pressure, even after the engine has been turned OFF. The fuel system pressure must be relieved before disconnecting any fuel lines. Failure to do so may result in fire and/or personal injury.

2.3L Engine

※ CAUTION

Never open, service or drain the radiator or cooling system when hot; serious burns can occur from the steam and hot coolant.

1. Disconnect the negative battery cable. Drain the cooling system.

2. Disconnect heater inlet and throttle body heater hoses from water outlet. Remove the exhaust manifold.

3. Remove the intake and exhaust camshaft housings. Remove the oil fill tube bolt and unseat the fill tube assembly from the engine.

4. Detach the injector harness electrical connector. Remove the oil fill tube.

5. Disconnect the throttle body air intake duct. Disconnect the power brake booster vacuum line.

6. Remove the throttle cable bracket. Remove the throttle body and position aside with all the lines and cables attached.

7. Disconnect the MAP sensor vacuum hose. Remove the intake manifold brace.

8. Remove the fuel rail bolts and the fuel rail from the intake manifold with the fuel lines attached.

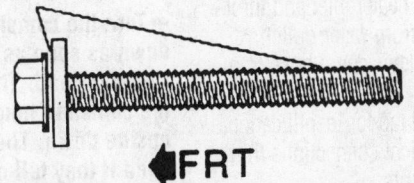

SPARINGLY APPLY CLEAN ENGINE OIL HERE

◀FRT

A TIGHTEN THE BOLTS TO THE FOLLOWING N•m (LB. FT.) SPECIFICATION IN SEQUENCE:
   BOLTS 1 THROUGH 8: 40 N•m (30 LB. FT.)
   BOLTS 9 AND 10: 35 N•m (26 LB. FT.)
B THEN TURN ALL 10 BOLTS AN ADDITIONAL 90 DEGREES IN SEQUENCE

79222304

Cylinder head bolt tightening sequence—2.3L engine

9. Disconnect the electrical connections from the following components:
- MAP sensor
- MAT sensor
- EVAP purge solenoid

10. Disconnect the upper radiator hose from the thermostat housing. Disconnect the coolant sensor.

11. Remove the cylinder head bolts in reverse order of the installation sequence. Remove the cylinder head and gasket.

To install:

12. Clean all gasket surfaces completely. Install a new cylinder head gasket and the cylinder head.

13. Coat the cylinder head bolt threads with clean engine oil and allow the oil to drain off before installing.

14. Install the cylinder head bolts and tighten, in sequence, as follows:
   a. No's 1 through 8: 30 ft. lbs. (40 Nm)
   b. No's 9 and 10: 26 ft. lbs. (35 Nm)
   c. All 10 bolts: an additional 90 degree turn

15. Connect the heater inlet and throttle body heater hoses to the water outlet. Connect the upper radiator hose to the water outlet.

16. Connect the two coolant sensor connections. Install the fuel rail on the intake manifold. Install the intake manifold brace.

17. Connect the electrical connections to the following components:
- MAP sensor
- MAT sensor
- EVAP purge solenoid

18. Connect the MAP sensor vacuum hose to the intake manifold. Install the throttle body.

19. Install the throttle cable bracket. Connect the throttle body air intake duct.

20. Lubricate the O-ring on the oil fill tube with engine oil and install oil fill tube into the block and bolt.

21. Install the exhaust and intake camshaft housings. Attach the injector harness electrical connector.

22. Install the exhaust manifold. Fill all fluids to their proper levels.

23. Connect the negative battery cable. Start the vehicle and verify no leaks.

2.4L Engine

※ CAUTION

Never open, service or drain the radiator or cooling system when hot; serious burns can occur from the steam and hot coolant.

1. Disconnect the negative battery cable. Drain the cooling system.

2. Disconnect the heater inlet and throttle body heater hoses from water outlet. Remove the power brake vacuum hose from the throttle body.

3. Disconnect and tag for identification, if necessary, the electrical connections from the following components:
- MAP sensor
- MAT sensor
- EVAP purge solenoid
- Camshaft sensor

4. Remove the stud-ended bolt from the alternator. Remove the intake manifold brace. Remove the intake manifold.

5. Install the stud-ended alternator bolt back into the engine. Install engine support fixtures J-28467-400 and J-28467-A or equivalent.

6. Remove the exhaust manifold. Remove the ignition coil and module assembly.

7. Disconnect the camshaft position sensor. Remove the power steering pump.

8. Disconnect the vacuum line from the fuel pressure regulator, fuel injector and fuel injector wiring harness.

9. Disconnect the fuel line clamp from the bracket on top of the intake camshaft housing. Remove the fuel line rail and position it aside (leaving the fuel lines attached).

10. Disconnect the timing chain housing at the intake camshaft housing, but do not remove from the vehicle.

11. Remove the electrical connection from the oil switch.

12. If equipped with an automatic transaxle, remove the transaxle fluid level indicator tube assembly from the exhaust camshaft cover and position it aside.

➡ Any time the camshaft housing to cylinder head bolts are loosened or removed, the camshaft housing to cylinder head gasket must be replaced.

13. Remove the intake camshaft housing bolts, the housing and gasket.

➡ Turn the camshaft housing upside down as soon as it is removed from the cylinder head. The lifters will fall out of the camshaft housing if it is not turned upside down. The lifters can be damaged if they fall out and hit a hard surface.

14. Remove the exhaust camshaft housing from the cylinder head (remove the bolts in reverse of the tightening sequence).

15. Remove the radiator inlet (upper) hose from the coolant outlet. Disconnect the coolant temperature sensor electrical connections.

16. Remove the cylinder head bolts (in reverse of the tightening sequence), the cylinder head and gasket.

➡ Do not use abrasive pads to clean the cylinder head or block surfaces. An abrasive pad may damage the cylinder head and block. Use a razor or scraper to clean the gasket surfaces.

To install:

17. Clean all gasket surfaces completely. Be sure the threaded holes in the engine block for the cylinder head bolts are clean.

18. Install a new cylinder head gasket on the cylinder block and carefully position the cylinder head in place.

19. New replacement head bolts are recommended. Coat the cylinder head bolt threads with clean engine oil and allow the oil to drain off before installing.

20. Install the cylinder head bolts and tighten, in sequence, as follows:

   a. No's 1 through 8: 40 ft. lbs. (65 Nm)

   b. No's 9 and 10: 30 ft. lbs. (40 Nm)

   c. All 10 bolts: an additional 90 degrees (¼ turn).

21. Install the intake and exhaust camshaft housings. Tighten the long bolts

to 11 ft. lbs. (15 Nm) plus an additional 90 degrees (¼ turn). Tighten the short bolts to 11 ft. lbs. (15 Nm) plus an additional 30 degrees (⅙ turn).

22. Install the timing chain and timing chain housing. Connect the upper radiator hose to the water outlet.

23. Connect the two coolant sensor connections. Install the fuel rail on the intake manifold and the intake manifold brace.

24. Connect the electrical connections to the following components:
- MAP sensor
- MAT sensor
- EVAP purge solenoid
- Camshaft position sensor

25. Connect the MAP sensor vacuum hose to the intake manifold. Install the power steering pump.

26. Install the throttle body and the throttle cable bracket. Connect the throttle body air intake duct.

27. Lubricate the O-ring on the oil fill tube with engine oil. Install oil fill tube into the block and the mounting.

28. Install the ignition coil and module assembly. Attach the injector harness electrical connector.

29. Remove the engine support fixture. Install the exhaust manifold.

30. Fill all fluids to their proper levels. An oil and filter change is recommended.

31. Connect the negative battery cable. Start the vehicle and verify no leaks.

3.1L Engine

LEFT (FRONT) CYLINDER HEAD

✳✳ CAUTION

Never open, service or drain the radiator or cooling system when hot; serious burns can occur from the steam and hot coolant.

1. Relieve the fuel system pressure. Disconnect the negative battery cable. Drain the cooling system.

2. Remove the top half of the air cleaner assembly and the throttle body air inlet duct.

3. Remove the exhaust crossover pipe heat shield and crossover pipe.

4. Disconnect the spark plug wires from spark plugs and wire looms and route the wires aside. Remove the rocker arm covers.

5. Remove upper intake plenum and lower intake manifold. Remove the left side exhaust manifold and oil level indicator tube.

Cylinder head bolt tightening sequence—2.4L engine

7922Z305

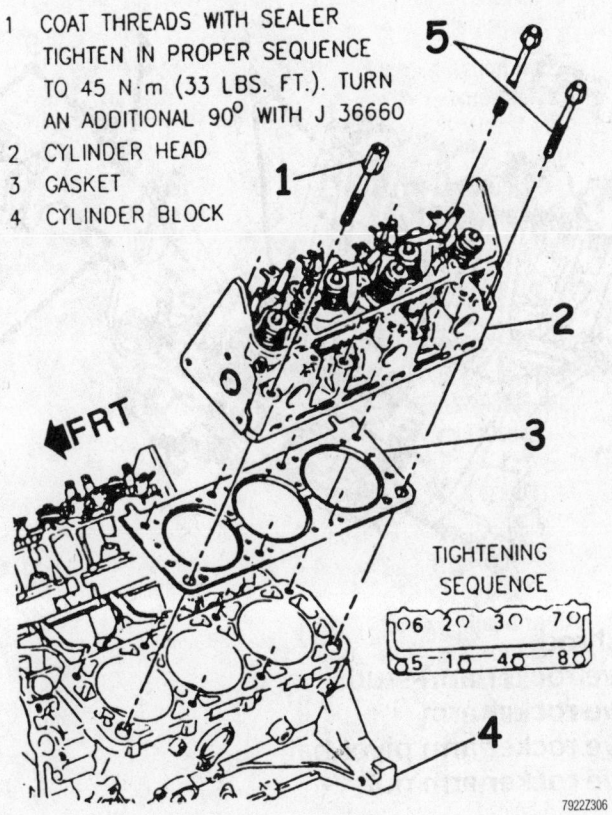

1 COAT THREADS WITH SEALER TIGHTEN IN PROPER SEQUENCE TO 45 N·m (33 LBS. FT.). TURN AN ADDITIONAL 90° WITH J 36660
2 CYLINDER HEAD
3 GASKET
4 CYLINDER BLOCK

FRT

TIGHTENING SEQUENCE

7922Z306

Cylinder head bolt tightening sequence—3.1L engine

➡**When removing the valvetrain components they must be kept in order for installation in the same locations they were removed from.**

6. Remove rocker arms nut, rocker arms, balls and pushrods.

7. Remove the cylinder head bolts evenly and the cylinder head.

To install:

8. Clean all the gasket surfaces completely. Clean the threads on the cylinder head bolts and block threads.

9. Place the cylinder head gasket in position over the dowel pins on the cylinder block so the words **THIS SIDE UP** are showing.

10. Coat the bolt threads with sealer and install finger-tight.

11. On 1995–97 models, tighten the cylinder head bolts in sequence to 33 ft. lbs. (45 Nm). With all the bolts tightened make a second pass tightening all the bolts an additional 90 degrees (¼ turn).

12. On 1998–99 models, tighten the cylinder head bolts in sequence to 37 ft. lbs. (50 Nm). With all the bolts tightened make a second pass tightening all the bolts an additional 90 degrees (¼ turn).

13. Install the pushrods, rocker arms, balls and rocker arm nuts. Tighten the rocker arm nuts to 18 ft. lbs. (25 Nm).

14. Install the lower intake manifold, upper intake plenum and the rocker arm covers.

15. Install the oil level indicator tube. Connect the spark plug wires to spark plugs and wire looms.

16. Install the left side exhaust manifold. Install the exhaust crossover pipe and crossover pipe heat shield.

17. Refill the cooling system. Drain the engine oil and refill with the proper type of engine oil. A filter change is recommended.

18. Install the top half of the air cleaner assembly and the throttle body air inlet duct.

19. Connect negative battery cable. Start the engine and verify no leaks.

RIGHT (REAR) CYLINDER HEAD

✱✱ CAUTION

Never open, service or drain the radiator or cooling system when hot; serious burns can occur from the steam and hot coolant.

1. Relieve the fuel system pressure. Disconnect the negative battery cable. Drain the cooling system.

2. Remove the top half of the air cleaner assembly and remove the throttle body air inlet duct.

3. Remove the exhaust crossover pipe heat shield and crossover pipe. Raise and safely support the vehicle.

4. Disconnect the Oxygen (O_2) sensor. Disconnect the exhaust pipe from the exhaust manifold.

5. Remove the right side exhaust manifold. Lower the vehicle.

6. Disconnect the spark plug wires from spark plugs and wire looms and route the wires aside.

7. Remove the rocker arm covers, upper intake plenum and lower intake manifold.

➡**When removing the valvetrain components they must be kept in order for installation in the same locations they were removed from.**

8. Remove rocker arms nut, rocker arms, balls and pushrods.

9. Remove the cylinder head bolts evenly and the cylinder head.

To install:

10. Clean all the gasket surfaces completely. Clean the threads on the cylinder head bolts and block threads.

11. Place the cylinder head gasket in position over the dowel pins on the cylinder block so the words **THIS SIDE UP** are showing.

12. Coat the bolt threads with sealer and install finger-tight.

13. On 1995–97 models, tighten the cylinder head bolts in sequence to 33 ft. lbs. (45 Nm). With all the bolts tightened make a second pass tightening all the bolts an additional 90 degrees (¼ turn).

14. On 1998–99 models, tighten the cylinder head bolts in sequence to 37 ft. lbs. (50 Nm). With all the bolts tightened make a second pass tightening all the bolts an additional 90 degrees (¼ turn).

15. Install the pushrods, rocker arms, balls and rocker arm nuts. Tighten the rocker arm nuts to 18 ft. lbs. (25 Nm).

16. Install the lower intake manifold and upper intake plenum. Install the rocker arm covers.

17. Connect the spark plug wires to spark plugs and wire looms. Raise vehicle and safely support.

18. Install the exhaust manifold. Connect the exhaust pipe to the exhaust manifold.

19. Lower the vehicle. Connect the Oxygen (O_2) sensor.

20. Install the exhaust crossover pipe and heat shield. Refill the cooling system.

21. Drain the engine oil and refill with the proper type of engine oil. A filter change is recommended.

22. Install the top half of the air cleaner assembly and the throttle body air inlet duct.

23. Connect negative battery cable. Start the engine and verify no leaks.

Rocker Arms

REMOVAL & INSTALLATION

2.3L and 2.4L Engines

The 2.3L and 2.4L engines are not equipped with rocker arms. The camshafts directly actuate the valves.

3.1L Engine

1. Disconnect the negative battery cable.

2. For the left side (front), perform the following procedures:

✲✲ CAUTION

Never open, service or drain the radiator or cooling system when hot; serious burns can occur from the steam and hot coolant.

    a. Drain the cooling system to a level below the coolant pipe on the front of the engine.

    b. Remove the coolant bypass hose clamp at the coolant tube.

    c. Remove the two bolts and nut securing the coolant tube to the cylinder head and position the tube aside.

    d. Disconnect the PCV valve from the rocker arm cover.

3. For the right side (rear), perform the following procedures:

    a. Disconnect the spark plug wires from the spark plugs and upper intake plenum wire retainer and position aside.

    b. Disconnect the power brake booster vacuum pipe from the intake plenum.

    c. Remove the serpentine belt.

    d. Remove the alternator.

    e. Disconnect and remove the ignition assembly and EVAP canister purge solenoid as an assembly.

4. Remove the four rocker arm cover bolts and remove the rocker arm cover.

5. Remove the rocker arm nuts, balls, rocker arms and pushrods.

To install:

6. Clean all the gasket surfaces completely.

7. Coat all the valvetrain components with engine oil prior to installation.

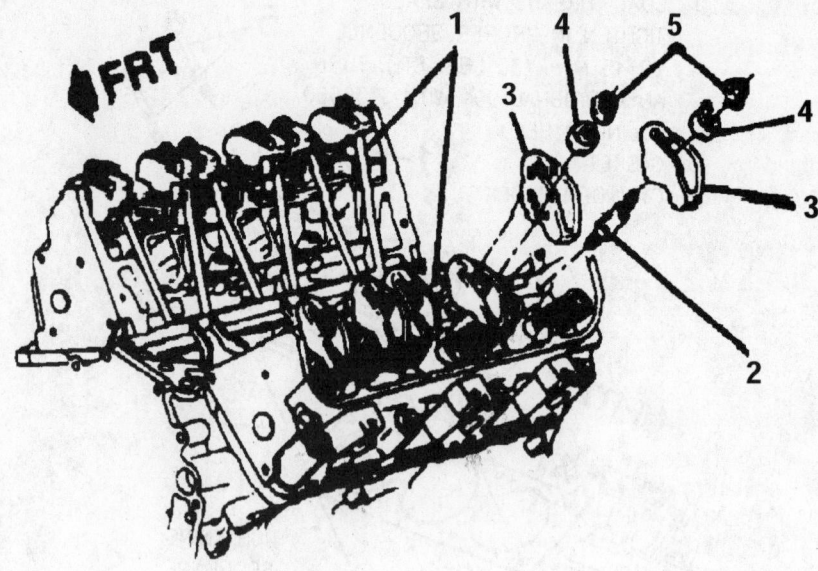

1 Pushrod
2 Valve rocker arm stud
3 Valve rocker arm
4 Valve rocker arm pivot ball
5 Valve rocker arm nut

79222307

Rocker arm components—3.1L engine

8. Install the pushrods and install the rocker arms on the studs. Install the rocker arm balls and mounting nuts. Be sure the pushrods are properly seated in the lifter and rocker arm. Tighten the mounting nuts to 18 ft. lbs. (24 Nm) for 1995 vehicles or to 89 inch lbs. (10 Nm) plus an additional 30 degrees for 1996–99 vehicles.

9. Install the rocker arm cover using a new gasket and tighten the rocker cover bolts to 90 inch lbs. (10 Nm).

10. For the left side (front), perform the following procedures:

    a. Connect the PCV valve to the rocker arm cover.

    b. Position the coolant tube and connect the thermostat bypass hose.

    c. Install the coolant tube mounting nut and bolts. Tighten the screw at the water pump to 106 inch lbs. (12 Nm), the bolt at the corner of the cylinder head to 18 ft. lbs. (25 Nm) and the nut to 18 ft. lbs. (25 Nm).

11. For the right side (rear), perform the following procedures:

    a. Install the alternator.

    b. Install the serpentine belt.

    c. Connect the power brake booster vacuum pipe to the plenum.

    d. Install the EVAP solenoid and ignition assembly.

    e. Connect the spark plug wires to the wire retainers on the plenum and the spark plugs.

12. Refill the cooling system.

13. Connect the negative battery cable.

14. Start the vehicle and verify no leaks.

Intake Manifold

REMOVAL & INSTALLATION

Starting with the 1996 Model Year, these vehicles were filled at the factory with a new type of antifreeze/coolant called GM Goodwrench DEX-COOL®. When adding coolant to vehicles, it is important that you use GM Goodwrench DEX-COOL (orange-colored, silicate-free) coolant. **Propylene glycol is not recommended for use in GM vehicles.** A 50/50 mixture of DEX-COOL and clean water will provide all the recommended protection. **DO NOT use DEX-COOL in pre-1996 vehicles. DO NOT mix DEX-COOL with any other type of antifreeze.**

The fuel injection system remains under pressure, even after the engine has been turned OFF. The fuel system pressure must be relieved before disconnecting any fuel lines. Failure to do so may result in fire and/or personal injury.

2.3L and 2.4L Engines

1. Relieve the fuel system pressure. Disconnect the negative battery cable.

Never open, service or drain the radiator or cooling system when hot; serious burns can occur from the steam and hot coolant.

2. Drain the cooling system to a level below the intake manifold. Disconnect the vacuum hose from the MAP sensor.

3. Detach the following electrical connectors from the MAP sensor, Intake Air Temperature (IAT) sensor, EVAP canister purge solenoid and fuel injectors.

4. Disconnect the vacuum hoses from the intake manifold, fuel pressure regulator and EVAP canister purge solenoid to the canister.

5. Remove the air cleaner duct. Remove the vent tube-to-air cleaner duct. Remove the control cable mounting bracket from the intake manifold.

6. Disconnect the power brake vacuum line from the intake manifold and remove the vacuum line mounting bracket at the power steering pump.

7. Disconnect the coolant lines from the throttle body. Remove the crankcase air/oil separator mounting bolts from the intake manifold.

8. Disconnect the hoses from the separator and the separator. Remove the oil fill tube.

9. Remove the intake manifold support brace, the intake manifold nuts/bolts and discard the old gasket.

To install:

10. Clean all the gasket surfaces completely. Install the intake manifold gasket and intake manifold.

11. Tighten the intake manifold fasteners in sequence, to 18 ft. lbs. (25 Nm).

12. Install the intake manifold brace and tighten the mounting bolts to 19 ft. lbs. (26 Nm). The brace-to-block bolts must be tightened first, then the brace-to-manifold bolt.

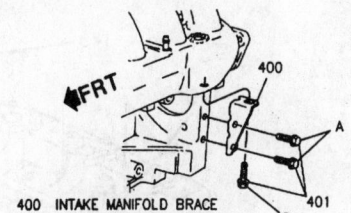

400 INTAKE MANIFOLD BRACE
401 BOLTS, 26 N·m (19 LB. FT.)
A FINGER START ALL BOLTS
B PUSH BRACE AGAINST MANIFOLD WITH FINGERS
C TIGHTEN BOLTS "A" TO SPECIFICATION
D TIGHTEN BOLTS "B" TO SPECIFICATION

79222309

Exploded view of the intake manifold brace mounting—2.3L and 2.4L engines

13. Complete the installation by reversing the removal procedures.

14. Lubricate a new oil fill tube O-ring seal with engine oil and install tube between No. 1 and 2 intake tubes. Rotate as necessary to gain clearance for oil/air separator nipple on fill tube.

15. Refill the cooling system. Connect the negative battery cable.

3.1L Engine

These vehicles use a two-piece intake manifold. Note that these pieces are cast aluminum. Use care when working with light alloy components.

1. Disconnect the negative battery cable. Relieve the fuel system pressure.

Never open, service or drain the radiator or cooling system when hot; serious burns can occur from the steam and hot coolant.

2. Remove the top half of the air cleaner assembly and throttle body duct. Drain and recover the cooling system.

3. Remove the EGR pipe from exhaust manifold. Remove the serpentine belt.

4. Remove the brake vacuum pipe at the intake plenum. Disconnect the control cables from the throttle body and intake plenum mounting bracket.

5. Remove the power steering lines at the alternator bracket. Remove the alternator.

6. Disconnect the spark plug wires from the spark plugs and wire retainers on the intake plenum.

7. Remove the Ignition assembly and the EVAP canister purge solenoid together.

8. Detach the upper engine wiring harness connectors from the Throttle Position Sensor (TPS), Idle Air Control (IAC), fuel Injectors, coolant temperature sensor, MAP sensor and Camshaft Position (CMP) sensor.

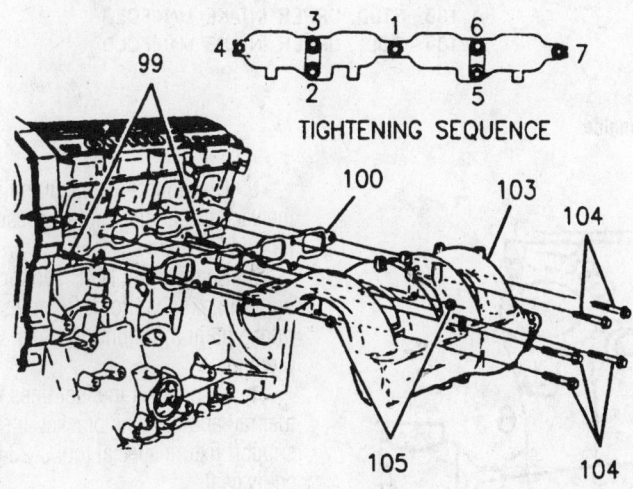

TIGHTENING SEQUENCE

99 STUD – 11 N·m (96 LBS. IN.)
100 INTAKE MANIFOLD GASKET
103 INTAKE MANIFOLD
104 BOLT – 26 N·m (19 LBS. FT.)
105 NUT – 26 N·m (19 LBS. FT.)

7922Z308

Exploded view and bolt tightening sequence for the intake manifold—2.3L and 2.4L engines

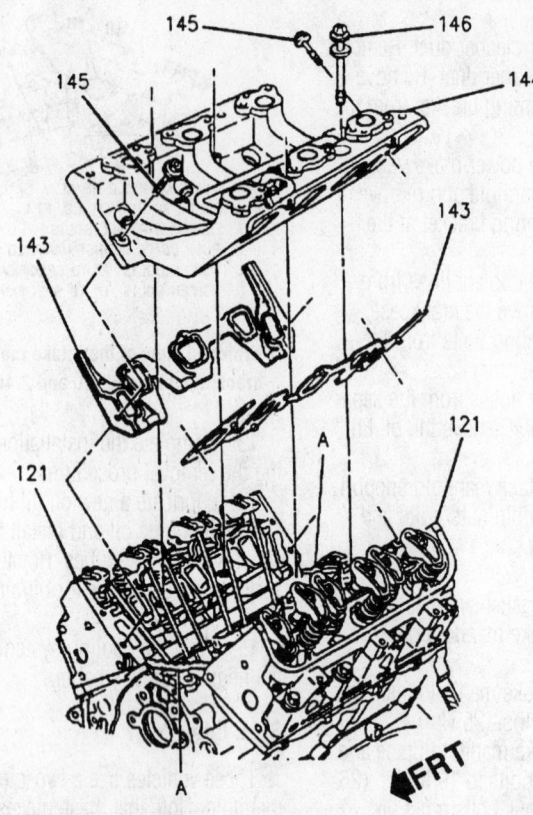

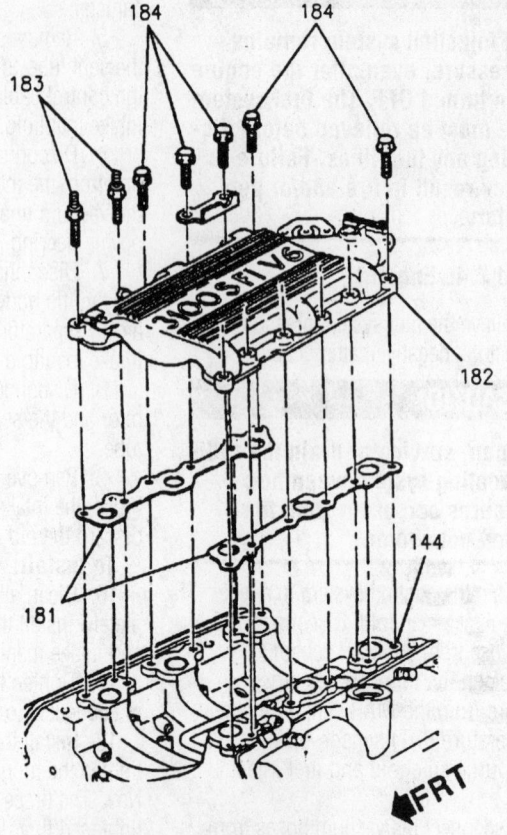

A   APPLY SEALANT
121 HEAD ASSEMBLY, CYLINDER
143 GASKET, LOWER INTAKE MANIFOLD
144 BOLT, LOWER INTAKE
145 BOLT, LOWER INTAKE MANIFOLD
146 BOLT, LOWER INTAKE MANIFOLD

144 MANIFOLD, LOWER INTAKE
181 GASKET, UPPER INTAKE MANIFOLD
182 MANIFOLD, UPPER INTAKE
183 STUD, UPPER INTAKE MANIFOLD
184 BOLT, UPPER INTAKE MANIFOLD

7922Z310

Exploded view of the upper and lower intake manifolds—3.1L engine

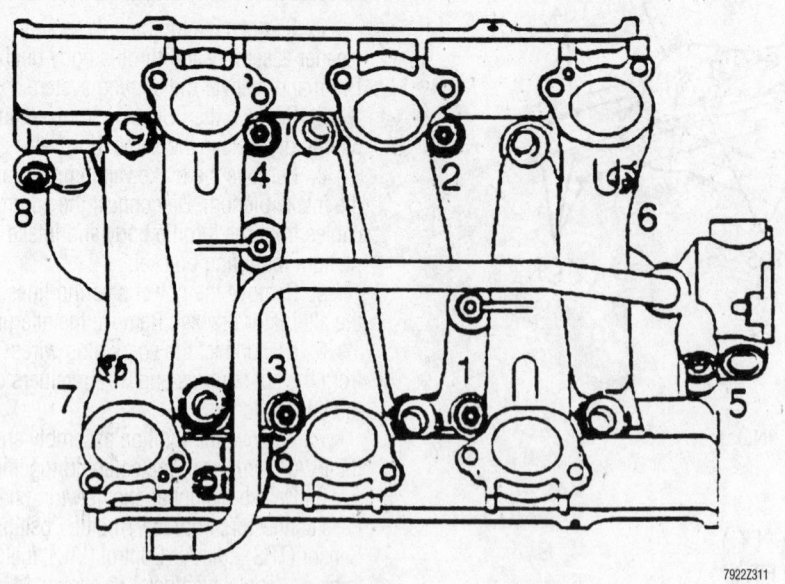

7922Z311

Lower intake manifold tightening sequence—3.1L engine

9. Disconnect the vacuum lines from the vacuum modulator, fuel pressure regulator and PCV valve.

10. Remove the MAP sensor from upper intake manifold. Remove the upper intake plenum mounting bolts and the plenum.

11. Disconnect the fuel lines from the fuel rail and fuel line bracket. Install engine support fixture special tool J-28467-A or an equivalent.

12. Remove the right side engine mount. Remove the power steering mounting bolts and support the pump aside without disconnecting the power steering lines.

13. Disconnect the coolant inlet pipe from coolant outlet housing. Remove the coolant bypass hose from the water pump and the cylinder head.

14. Disconnect the upper radiator hose at thermostat housing.

15. Remove the thermostat housing. Remove both rocker arm covers.

16. Remove the lower intake manifold bolts. Be sure the washers on the four center bolts are installed in their original locations.

➡**When removing the valvetrain components they should be kept in order for installation the original locations.**

17. Remove the rocker arm nuts, rocker arms and pushrods. Remove the intake manifold from the engine.

To install:

18. Clean gasket material from all mating surfaces. Remove all excess RTV sealant from front and rear ridges of cylinder block.

19. Place a 3mm bead of RTV, on each ridge, where the front and rear of the intake manifold contact the block.

20. Using a new gasket, install the intake manifold to the engine.

21. Install the pushrods, rocker arms and mounting nuts. Be sure the pushrods are properly seated in the valve lifters and rocker arms.

22. Install rocker arm nuts and tighten the rocker arm nuts to 18 ft. lbs. (24 Nm).

23. Install lower the intake manifold attaching bolts. Apply sealant PN 12345739 or equivalent to the threads of bolts, and tighten bolts to 115 inch lbs. (13 Nm).

24. Complete the installation by reversing the removal procedures.

25. Install the upper intake manifold and tighten the mounting bolts to 18 ft. lbs. (25 Nm).

26. Fill the cooling system. An engine oil and filter change is recommended.

27. Connect the negative battery cable. Start the vehicle and verify no leaks.

Exhaust Manifold

REMOVAL & INSTALLATION

2.3L and 2.4L Engines

➡**There are two different exhaust manifolds used on the Quad 4 engine. While the manifolds vary greatly in appearance, the removal and installation procedures are the same. The cast iron manifold is a single outlet design and the sheet metal manifold is a tubular construction with a dual outlet flange.**

1. Disconnect the negative battery cable. Disconnect the oxygen sensor. Raise and safely support the vehicle.

2. Remove the bolt that attaches the exhaust manifold brace to the manifold.

3. Remove the manifold to exhaust pipe spring loaded bolts using a 13mm box

wrench. The nuts should be loosened alternately and evenly to prevent the pipe from binding the nuts.

4. Pull down and back on the exhaust pipe to disengage it from the exhaust manifold bolts.

➡**On the 2.4L engine, do not bend the exhaust flex coupler more than 3 degrees in any direction. Movement of more than 3 degrees will damage the flex coupler.**

5. On the 2.3L engine, remove the upper and lower exhaust manifold heat shields.

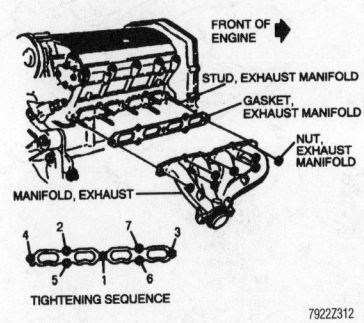

Exploded view of the exhaust manifold assembly mounting—2.3L engine

6. Lower the vehicle. Remove the exhaust manifold mounting nuts and manifold.

To install:

7. Clean all the gasket surfaces completely. Install the exhaust manifold and new gaskets.

8. Install the exhaust manifold-to-cylinder head retaining nuts and tighten in sequence to 31 ft. lbs. (42 Nm) on the 2.3L engine or to 110 inch lbs. (13 Nm) on the 2.4L engine.

9. Raise and safely support the vehicle. Install the heat shields. Install exhaust manifold brace to the manifold bolt.

10. Install the manifold-to-exhaust pipe nuts and tighten evenly to 19 ft. lbs. (26 Nm) on the 2.3L engine or to 26 ft. lbs. (35 Nm) on the 2.4L engine.

11. Lower the vehicle. Connect the oxygen sensor.

12. Connect the negative battery cable. Start the vehicle and verify no exhaust leaks.

3.1L Engine

1. Disconnect the negative battery cable. Remove the top half of the air cleaner assembly and throttle cable duct.

2. On the left side (front), perform the following procedures:

a. Partially drain the cooling system.

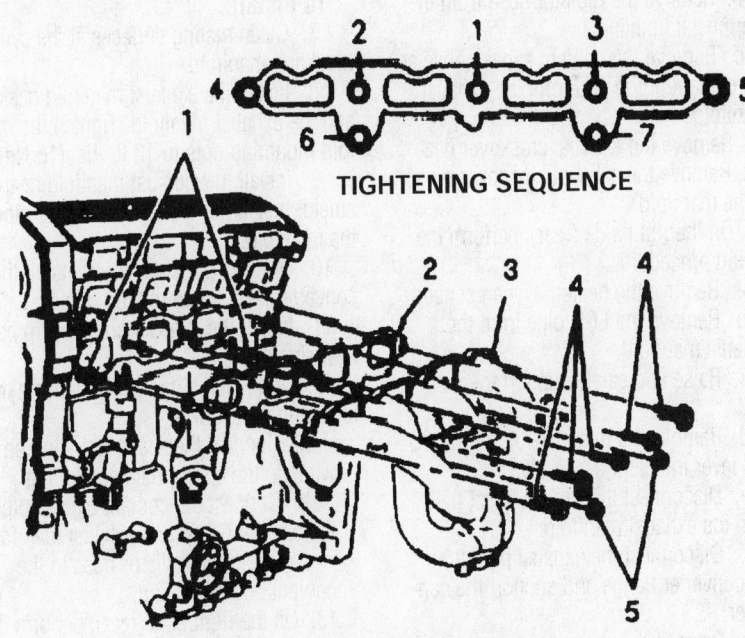

1. Stud - 100 in. lbs. (12Nm)
2. Intake manifold gasket
3. Intake manifold
4. Bolt - 17 ft. lbs. (24 Nm)
5. Nut - 17 ft. lbs. (24 Nm)

Exploded view of the exhaust manifold assembly mounting—2.4L engine

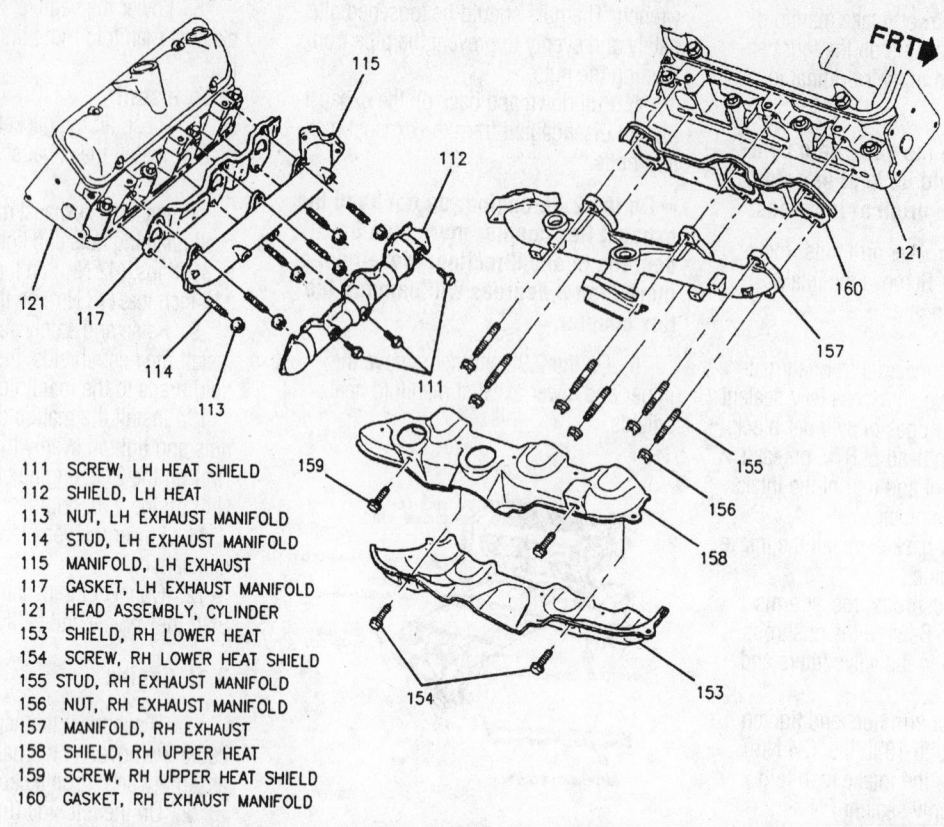

111 SCREW, LH HEAT SHIELD
112 SHIELD, LH HEAT
113 NUT, LH EXHAUST MANIFOLD
114 STUD, LH EXHAUST MANIFOLD
115 MANIFOLD, LH EXHAUST
117 GASKET, LH EXHAUST MANIFOLD
121 HEAD ASSEMBLY, CYLINDER
153 SHIELD, RH LOWER HEAT
154 SCREW, RH LOWER HEAT SHIELD
155 STUD, RH EXHAUST MANIFOLD
156 NUT, RH EXHAUST MANIFOLD
157 MANIFOLD, RH EXHAUST
158 SHIELD, RH UPPER HEAT
159 SCREW, RH UPPER HEAT SHIELD
160 GASKET, RH EXHAUST MANIFOLD

Exploded view of the exhaust manifold mounting—3.1L engine

b. Remove the radiator hose from the thermostat housing.

c. Remove the coolant bypass hose at the coolant pump and from the exhaust manifold.

3. Remove the exhaust crossover heat shield. Remove the exhaust crossover pipe from the manifold.

4. On the right side (rear), perform the following procedures:

a. Remove the heated oxygen sensor.

b. Remove the EGR pipe from the exhaust manifold.

c. Raise and safely support the vehicle.

d. Remove the transaxle oil fill tube and lever indicator assembly.

e. Disconnect the front exhaust pipe from the exhaust manifold.

f. Disconnect the exhaust pipe from the converter flange and support the converter.

g. Remove the converter heat shield from the body.

5. Tag and disconnect the secondary ignition wires from the spark plugs.

6. Remove the exhaust manifold heat shield. Remove the exhaust manifold retaining nuts and manifold.

To install:

7. Clean mating surfaces at the cylinder head and manifold.

8. Install the exhaust manifold gasket and the exhaust manifold. Tighten the manifold mounting nuts to 12 ft. lbs. (16 Nm).

9. Install the exhaust manifold heat shield. Install the exhaust crossover pipe to the manifold.

10. On the right side (rear), install the converter heat shield to the body.

11. Install the exhaust crossover pipe heat shield.

12. On the left side (front), perform the following procedures:

a. Connect the secondary ignition wires to the appropriate spark plug.

b. Connect the coolant bypass pipe to the coolant pump and exhaust manifold.

c. Install the radiator hose to the coolant outlet housing.

13. On the right side (rear), perform the following procedures:

a. Connect the exhaust pipe to the exhaust manifold.

b. Install the transaxle oil level indicator and fill tube assembly.

c. Safely lower the vehicle.

d. Connect the heated oxygen sensor.

e. Connect the EGR pipe to exhaust manifold.

14. Install the top half of the air cleaner and the throttle body duct. Connect the negative battery cable.

Camshaft and Valve Lifter

REMOVAL & INSTALLATION

2.3L and 2.4L Engines

➡**Anytime the camshaft housing-to-cylinder head bolts are loosened or removed, the camshaft housing-to-cylinder head gasket must be replaced.**

1. Disconnect the negative battery cable.

2. Detach the 11-pin connector from the ignition assembly cover.

3. Remove the four ignition assembly cover-to-camshaft housing bolts and remove assembly by pulling straight up. Use a spark plug boot wire remover to remove connector assemblies.

4. On the intake camshaft side, if equipped, perform the following procedures:

a. Remove the throttle lever actuator power steering pressure switch connector.

b. Loosen the power steering pump adjusting stud bolt. Loosen the power steering pump pivot bolts and remove drive belt.

c. Remove the two power steering pump bracket-to-transaxle bolts.

d. Remove the front power steering pump bracket-to-engine block bolt.

e. Remove the power steering pump assembly from the engine and position it aside. DO NOT disconnect the power steering lines from the pump.

5. On the intake camshaft side, if equipped, perform the following procedures:

a. Remove oil/air separator bolts and hoses. Leave the hoses attached to the separator but disconnect from the oil fill tube, timing chain housing and intake manifold.

b. Disconnect the vacuum line from fuel pressure regulator and disconnect the fuel injector wiring harness.

c. Remove the two bolts securing the fuel line clamps to the intake camshaft housing.

d. Remove fuel rail-to-camshaft housing mounting bolts.

e. Remove the fuel rail from the cylinder head leaving the fuel lines attached and position the fuel rail aside.

6. Remove the timing chain and camshaft sprockets.

7. Remove the timing chain housing bolts but do not remove from the engine.

8. Remove camshaft housing cover-to-camshaft housing attaching bolts.

9. Remove the camshaft housing-to-cylinder head attaching bolts. Use the reverse of the tightening sequence when loosening camshaft housing to cylinder head attaching bolts. Leave two bolts loosely in place to hold the camshaft housing while separating camshaft cover from housing.

10. Push the cover off the housing by threading four of the housing-to-cylinder head bolts into the tapped holes in the cam housing cover. Tighten the bolts in evenly so the cover does not bind on the dowel pins.

11. Remove the two loosely installed camshaft housing-to-cylinder head bolts and remove the cover. Discard the gaskets.

12. Note the position of the chain sprocket dowel pin for reassembly. Remove the camshaft carefully; do not damage the camshaft oil seal.

13. Remove camshaft oil seal from camshaft and discard seal. This seal must be replaced any time the housing and cover are separated.

➡ **The valve lifters must be kept in order for installation in the same locations they were removed from.**

14. Remove the valve lifters from the camshaft housing.

15. Remove the camshaft carrier from the cylinder head and remove the gasket.

To install:

16. Clean all the gasket surfaces completely.

17. Install a new gasket on the cylinder head and install the camshaft housing over the dowel pins. Install one bolt loosely to hold the housing in place.

➡ **If the camshaft was replaced the valve lifters must also be replaced.**

18. Install the lifters into their original bores.

19. Lubricate camshaft lobes, journals and lifters with camshaft and lifter prelube. The camshaft lobes and journals must be adequately lubricated or engine damage could occur upon start up.

20. Install the camshaft in the same position it was in prior to removal. The timing chain sprocket dowel pin should be straight up and align with the center line of the lifter bores.

21. Install new camshaft housing-to-camshaft housing cover seals into the cover.

➡ **The seals for the intake and exhaust covers are different and the correct seals must be used.**

22. Remove the bolt holding the housing in place.

23. Apply thread locking compound to the camshaft housing and cover attaching bolt threads.

24. Install the camshaft housing cover.

25. Install all the mounting bolts finger-tight. With all the bolts in place tighten the bolts in sequence to 11 ft. lbs. (15 Nm) plus 75 degrees additional rotation (long bolts) and 16 ft. lbs. (21 Nm) plus 25 degrees additional rotation (short bolts).

26. Install the timing chain housing mounting bolts.

27. Install the timing chain and sprockets.

28. Install new fuel injector O-ring seals lubricated with oil and install the fuel rail and tighten the mounting bolts to 19 ft. lbs. (26 Nm).

29. On the intake camshaft side, if equipped, perform the following procedures:

a. Install the fuel line clamp mounting bolts on top of the intake camshaft housing.

b. Connect the vacuum line to the fuel pressure regulator.

c. Attach the fuel injector harness connector.

d. Install the oil/air separator assembly and connect the hoses to the oil fill tube, timing chain housing and intake manifold.

30. Lubricate the inner sealing surface of the intake camshaft seal with oil and install the seal to the housing using J-36009, or an equivalent seal installer.

31. On the intake camshaft side, if equipped, perform the following procedures:

a. Install the power steering pump pulley onto the intake camshaft using J-36015, or an equivalent pulley installer.

b. Install the power steering pump and install the drive belt.

c. Attach the throttle lever actuator power steering pressure switch connector.

32. On the exhaust camshaft side, perform the following procedures:

a. Install the transaxle fill tube.

b. Connect the oil pressure switch.

33. Install the ignition assembly on the camshaft housing and tighten the mounting bolts to 13 ft. lbs. (18 Nm).

34. Attach the 11-pin connector to ignition assembly.

35. Connect the negative battery cable.

36. Start the vehicle and verify proper operation and no leaks.

3.1L Engine

1. Relieve the fuel system pressure. Disconnect the negative battery cable. Remove the engine assembly.

➡ **When removing valvetrain components they must be marked for installation in the same location they are**

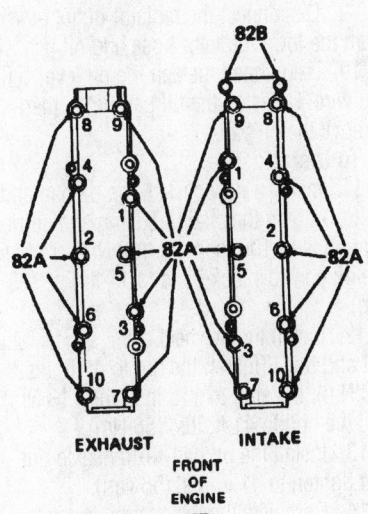

Camshaft housing bolt tightening sequence—2.3L and 2.4L engines

removed from. When the camshaft is being replaced the valve lifters should also be replaced.

2. Remove the rocker arm covers and the intake manifold.

3. Remove the rocker arm retaining nuts, rocker arm balls, rocker arms and pushrods.

4. Remove the guide bolts from the lifter guide and remove the guide.

5. Remove the valve lifter(s) from the lifter bores.

6. Remove the crankshaft balancer and front cover. Remove the timing chain and sprockets.

7. Remove the oil pump driven gear bolt and gear. Remove the bolts and camshaft thrust plate.

8. Carefully remove the camshaft. Avoid marring the camshaft bearing surfaces.

To install:

9. Coat the camshaft with lubricant 1052365 or equivalent, and install the camshaft.

10. Install the camshaft thrust plate and tighten bolts to 89 inch lbs. (10 Nm).

11. Install the oil pump driven gear and tighten bolt to 27 ft. lbs. (36 Nm).

12. Install the timing chain and sprocket.

13. Install the camshaft thrust button and front cover. Install the crankshaft balancer.

14. Lubricate the bearing surfaces with Molykote® or equivalent.

➡**Installation of a new camshaft or a wear pattern on the old valve lifter will require the replacement of the camshaft and lifters together. If camshaft replacement is not necessary, be sure to install the used valve lifters in their original position upon reinstallation.**

15. Install the lifters in their original locations.

16. Install the lifter guide and lifter guide bolts and tighten the guide bolts to 89 inch lbs. (10 Nm).

17. Install the pushrods, rocker arms, rocker balls and rocker arm nuts.

18. Tighten the rocker arm nuts to 18 ft. lbs. (25 Nm) on 1995 engines or to 89 inch lbs. (10 Nm) plus an additional 30 degrees on 1996-99 engines.

19. Install the intake manifold and rocker arm covers.

20. Install the engine assembly. Connect the negative battery cable.

21. Adjust the valves, as required. Start the engine and verify no oil leaks.

Valve Lash

ADJUSTMENT

2.3L and 2.4L Engines

The 2.3L and 2.4L of the engines are equipped with hydraulic valve lifters that do not require periodic valve lash adjustment. Adjustment to zero lash is maintained automatically by hydraulic pressure in the lifters.

3.1L Engine

➡**These engines come from the factory with non-adjustable valve lash, BUT if the valves and the valve seats are reconditioned adjustable rocker arm studs must be installed. The adjustment procedure is ONLY for adjustable rocker arm studs.**

The engines are originally equipped with hydraulic valve lifters. Hydraulic valve lifters are not adjustable. The replacement rocker arm stud is adjustable and the adjustment can be performed as listed below. If valve system noise is present, check the tighten on the rocker arm nuts. The correct torque is 14–20 ft. lbs. (19–27 Nm). If noise is still present, check the condition of the camshaft, lifters, rocker arms, pushrods and valves.

1. Install the rocker arms, balls and nuts. Tighten the rocker arm nuts until all of the lash (free-play) is eliminated.

2. Adjust the valves when the lifter is on the base circle of a camshaft lobe as follows:

a. Place the engine in the No. 1 firing position. This can be determined by placing a finger on the No. 1 rocker arm as the engine alignment mark on the front face of the of the torsional dampener pulley aligns with the arrow on the front cover. If the valves don't move freely, the engine is in the No. 1 firing position. If the valves move freely, the engine is in the No. 4 firing position and the engine should be rotated one full rotation to reach the No. 1 position.

b. With the engine in the No. 1 firing order, adjust the following valves:
- Exhaust—1, 2, 3
- Intake—1, 5, 6

c. Loosen the adjusting nut several turns backwards, then tighten until all lash (free-play) is removed. After all of the valve lash is removed turn the nut an additional 1½ turns, this will center the lifter plunger.

d. Turn the engine one complete revolution until the timing tab and the alignment mark are again aligned, this will place the engine in the No. 4 firing position.

e. With the engine in the No. 4 firing order, adjust the following valves:
- Exhaust—4, 5, 6
- Intake—2, 3, 4

Oil Pan

REMOVAL & INSTALLATION

2.3L and 2.4L Engines

1. Disconnect the negative battery cable. Raise and safely support the vehicle.

2. Properly drain the engine oil and cooling system. Remove the flywheel inspection cover.

3. Remove the right front wheel and tire assembly. Remove the right inner fender splash shield.

4. Remove the serpentine belt. Disconnect the engine mount strut from the engine mount strut bracket.

5. Remove the A/C compressor from the mounting bracket and support it aside. Do NOT disconnect the refrigerant lines.

6. Unfasten the engine mount strut bracket bolts, then remove the bracket. Remove the radiator outlet pipe bolts.

7. Disconnect the radiator and A/C outlet pipes from the suspension supports. Remove the exhaust manifold brace.

8. Remove the oil pan-to-flywheel cover bolt and nut and remove the flywheel cover stud and spacer.

9. Disconnect the radiator outlet pipe from the lower radiator hose and oil pan.

10. If equipped, detach the oil level sensor wire. Unfasten the oil pan bolts, then remove the oil pan.

To install:

11. Install a new gasket, the oil pan and bolts. Tighten the chain housing and carrier seal bolts to 106 inch lbs. (12 Nm). Tighten the oil pan-to-block bolts to 17 ft. lbs. (23 Nm).

12. Install the flywheel cover spacer, stud nut and bolt. Tighten the nut to 41 ft. lbs. (56 Nm), the stud to 115 inch lbs. (13 Nm) and the bolt to 41 ft. lbs. (56 Nm).

13. Install the oil pan-to-transaxle nut and tighten to 41 ft. lbs. (56 Nm).

14. Complete the installation by reversing the removal procedures.

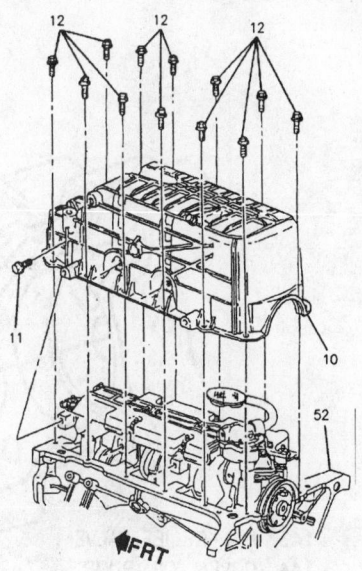

FRT

38   OIL PAN
39A   BOLT, OIL PAN (M8 X 1.25 X 80)
      24 N•m (18 LB. FT.)
39B   BOLT, OIL PAN (M8 X 1.25 X 22)
      24 N•m (18 LB. FT.)
39C   BOLT, OIL PAN (M6 X 1.00 X 25)
      12 N•m (106 LB. IN.)
39D   BOLT, STUD END OIL PAN
      26 N•m (19 LB. FT.)

7922Z316

Oil pan mounting bolt locations—2.3L and 2.4L engines

10   PAN, OIL
11   BOLT, OIL PAN SIDE
12   BOLT, OIL PAN RETAINING
52   BLOCK, ENGINE

7922Z317

Exploded view of the oil pan mounting—3.1L engine

15. Install the engine mount strut bracket and tighten bolts to 49 ft. lbs. (66 Nm).

16. Fill the crankcase with oil. A filter change is recommended.

17. Fill the cooling system. Connect the negative battery cable, then start the engine and check for leaks.

3.1L Engine

1. Disconnect the negative battery cable. Remove the serpentine belt.

2. If equipped, remove the upper A/C compressor bolts. Raise and safely support the vehicle.

3. Properly drain the engine oil. Remove the right front wheel and tire assembly.

4. Remove the engine mount strut from the suspension support.

5. Remove the cotter pin and castle nut from the lower ball joint, the separate the ball joint from the steering knuckle.

6. Remove the right side sway bar link. Detach the ABS sensor from the right subframe.

7. Remove the right side subframe bolts and the right side subframe and control arm as an assembly.

8. If equipped, remove the lower A/C compressor bolts and position the compressor aside. Do NOT disconnect the refrigerant lines or allow the compressor to hang unsupported.

9. Remove the engine mount strut bracket from the engine. Remove the oil filter, the starter and the flywheel cover.

10. Remove the oil pan flange bolts, the oil pan side bolts and the oil pan.

To install:

11. Clean the gasket mating surfaces. Install a new gasket on the oil pan. Apply silicone sealer to the portion of the pan that contacts the rear of the block.

12. Position the oil pan, then install the mounting bolts finger-tight.

13. With all the bolts in place, tighten the oil pan flange bolts to 18 ft. lbs. (25 Nm) and the oil pan side bolts to 34 ft. lbs. (50 Nm).

14. Complete the installation by reversing the removal procedures.

15. Tighten the following item to:

   a. Engine-to-transaxle brace, bolts to 68 ft. lbs. (93 Nm)

   b. Engine mount strut bracket-to-engine bolts to 85 ft. lbs. (115 Nm)

   c. Right side subframe/control arm assembly bolts to 89 ft. lbs. (120 Nm)

   d. Right side sway bar line bolt to 22 ft. lbs. (30 Nm)

   e. Ball joint-to-steering knuckle nut to 48 ft. lbs. (60 Nm). Install a new cotter pin.

   f. Engine mount strut bracket bolt-to-frame bolt to 89 ft. lbs. (120 Nm)

16. Fill the crankcase to the correct level. A filter change is recommended. Connect the negative battery cable, then start the engine and check for leaks.

➡**Whenever the vehicle subframe is removed or lowered, the wheel alignment should be checked.**

17. Check the front end alignment and adjust as required.

Oil Pump

REMOVAL & INSTALLATION

2.3L and 2.4L Engines

➡**Please note that the transaxle must be removed from the vehicle to service the oil pump.**

1. Disconnect the negative battery cable. Install engine support fixture J-28467-A or equivalent.

2. Properly drain the engine oil, then remove the oil pan. Remove the transaxle and the flywheel.

3. Remove the balance shaft chain cover and chain guide. Remove the oil pump bolts and the oil pump cover.

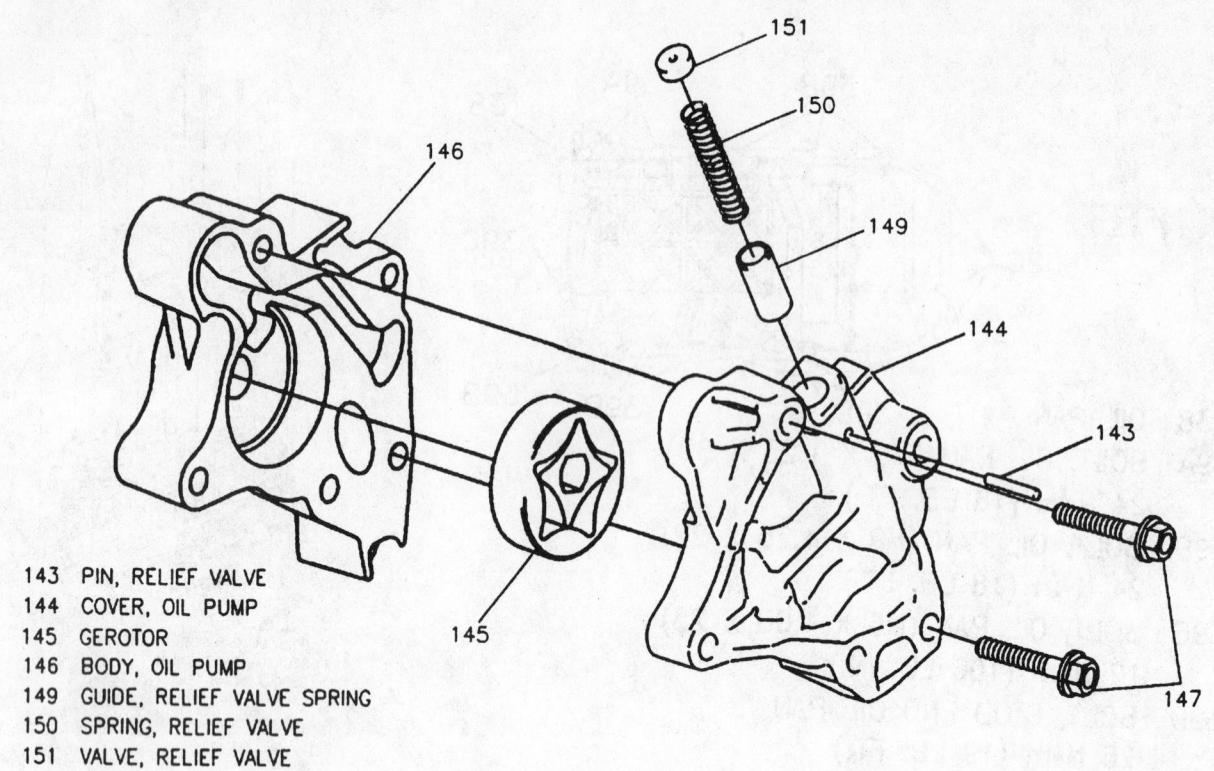

143 PIN, RELIEF VALVE
144 COVER, OIL PUMP
145 GEROTOR
146 BODY, OIL PUMP
149 GUIDE, RELIEF VALVE SPRING
150 SPRING, RELIEF VALVE
151 VALVE, RELIEF VALVE

7922Z318

Exploded view of the oil pump components—2.3L and 2.4L engines

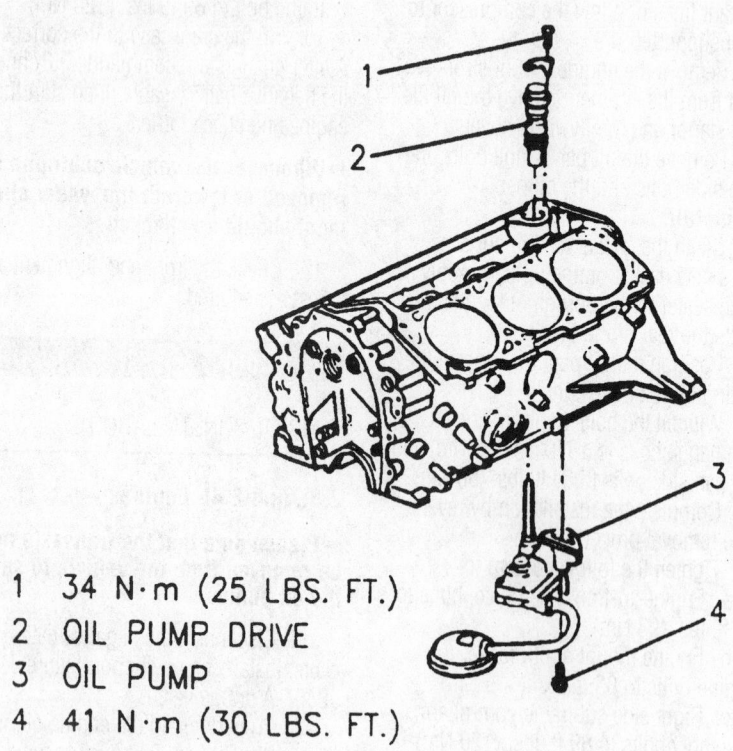

1   34 N·m (25 LBS. FT.)
2   OIL PUMP DRIVE
3   OIL PUMP
4   41 N·m (30 LBS. FT.)

7922Z319

Exploded view of the oil pump mounting—3.1L engine

4. Pull the housing to disconnect the pump gear from the balance shaft. Remove the housing assembly from the balance shaft assembly.

To install:

5. Clean all of the parts in suitable cleaning solvent. Remove all varnish sludge and dirt.

6. Lubricate the gears with clean engine oil. Assemble the geroter gear into the housing.

➡**Fill oil pump cavities with petroleum jelly prior to installation. This seals the pump and acts like a "prime" so the pump will draw oil as soon as the engine begins to turn. This will ensure that there is oil pressure immediately on start-up and will prevent engine damage.**

7. Install the pump housing to the balance shaft assembly, the pump cover to the oil pump housing.

8. Install the oil pump-to-block bolts and tighten to 40 ft. lbs. (54 Nm).

9. Install the balance shaft chain guide and chain. Adjust the chain tension inserting a 0.40 in. (1mm) brass feeler, between the chain guide and the chain.

→A brass feeler gauge must be used to ensure that correct measurements are obtained. If a steel gauge is used, it will not bend to conform to the guide and will allow for incorrect measurements.

10. Press the guide against the chain using about 3 lbs. of force. Tighten the chain tensioner fastener to 115 inch lbs. (13 Nm).

11. Install the balance shaft chain cover and tighten the nut and bolt to 115 inch lbs. (13 Nm).

12. Install the flywheel and the transaxle. Install the oil pan, then carefully lower the vehicle.

13. Fill the crankcase with clean engine oil. A filter change is recommended.

14. Remove the engine support fixture. Connect the negative battery cable, then start the engine and verify oil pressure and no leaks.

3.1L Engine

1. Disconnect the negative battery cable.

2. Raise and safely support the vehicle.

3. Drain the engine oil into a suitable container. Remove the oil pan.

4. Remove the crankshaft oil deflector bolts and deflector.

5. Remove the oil pump bolts, the oil pump and pump driveshaft.

To install:

6. Install the oil pump and pump driveshaft. Tighten the oil pump mounting bolts to 30 ft. lbs. (41 Nm).

7. Install the crankshaft oil deflector and tighten the nuts to 18 ft. lbs. (25 Nm).

8. Install the oil pan, then carefully lower the vehicle.

9. Fill the crankcase to the correct level with oil. An filter change is recommended.

10. Connect the negative battery cable;, then, start the engine, check the oil pressure and check for leaks.

Rear Main Seal

REMOVAL & INSTALLATION

2.3L and 2.4L Engines

1. Disconnect the negative battery cable.

2. Remove the transaxle assembly.

3. If equipped with a manual transaxle, remove the pressure plate and clutch disc.

4. Unfasten the flywheel-to-crankshaft bolts, then remove the flywheel.

5. Disconnect the oil pan-to-seal housing bolts

6. Unfasten the seal housing-to-block bolts, then remove the seal housing and gasket.

7. To support the seal housing for seal removal, place two blocks of equal thickness on a flat surface, position the seal housing and blocks so the transaxle side of the seal housing is supported across the dowel pin and center bolt holes on both sides of the seal opening.

→The seal housing could be damaged if not properly supported during seal removal.

8. Drive the seal evenly out the transaxle side of the seal housing using a small prytool in the relief grooves on the crankshaft side of the seal housing. Discard the seal.

✳✳ WARNING

Be careful not to damage the seal housing sealing surface. If damaged, it may result in an oil leak.

To install:

9. Press a new seal into the housing using tool J 36005 or equivalent seal installation tool.

10. Inspect the oil pan gasket inner silicone bead for damage and repair using a silicone sealant, if necessary.

11. Position a new seal housing-to-block gasket over the alignment. The gasket is reversible.

12. Lubricate the lip of the seal with clean engine oil.

13. Install the housing assembly, then tighten the housing-to-block bolts to 106 inch lbs. (12 Nm).

14. Install the oil pan-to-seal housing bolts, then tighten to 106 inch lbs. (12 Nm).

15. Install the flywheel as outlined later in this section.

16. For vehicles equipped with a manual transaxle, install the clutch, pressure plate and clutch cover assembly.

17. Install the transaxle assembly.

18. Connect the negative battery cable, then start the engine and check for leaks.

3.1L Engine

1. Support the engine with J-28467 engine support or equivalent.

2. Remove the transaxle.

3. Remove the flywheel.

4. Insert a suitable prytool in through the dust lip and pry out the seal by moving the tool around the seal until it is removed.

✳✳ WARNING

Use care not to damage the crankshaft seal surface with a prytool.

To install:

5. Before installing, lubricate the seal bore to seal surface with engine oil.

6. Install the new seal using tool J-34686.

7. Slide the new seal over the mandrel until the dust lip bottoms squarely against the tool collar.

8. Align the dowel pin of the tool with the dowel pin hole in the crankshaft and attach the tool to the crankshaft. Tighten the attaching screws to 2–5 ft. lbs. (2.7–6.8 Nm).

9. Tighten the T-handle of the tool to push the seal into the bore. Continue until the tool collar is flush against the block.

10. Loosen the T-handle completely. Remove the attaching screws and the tool.

→Check to see that the seal is squarely seated in the bore.

11. Install the flywheel and transaxle.

12. Start the engine and check for leaks.

Timing Chain, Sprockets, Front Cover and Seal

REMOVAL & INSTALLATION

2.3L and 2.4L Engines

→It is recommended that the entire procedure be reviewed before attempting to service the timing chain.

1. Disconnect the negative battery cable. Drain the cooling system into a suitable container. Remove the coolant surge tank.

✳✳ CAUTION

Never open, service or drain the radiator or cooling system when hot; serious burns can occur from the steam and hot coolant.

2. Remove the serpentine drive belt using a 13mm wrench that is at least 24 in. (61cm) long.

3. Disconnect and remove the alternator from the mounting bracket.

4. Install a suitable engine support, J-28467-A or the equivalent.

5. Remove the upper cover fasteners. Disconnect the upper cover vent hose.

6. Remove the right engine mount and bracket. Raise and safely support the vehicle.

7. Remove the right front tire and wheel assembly. Remove the right lower splash shield from the right wheel house.

8. Remove the crankshaft balancer assembly using the following procedure:

   a. Remove the torsional damper mounting bolt and washer while holding the crankshaft in place using J-38122, or the equivalent.

   b. Remove the balancer using a suitable puller, J-24420-B or the equivalent.

9. Remove the lower cover fasteners. Lower the vehicle.

10. Remove the front cover and gaskets. Remove the crankshaft oil slinger.

11. Using a suitable seal driver, tap the oil seal out of the front cover.

12. Rotate the crankshaft clockwise, as viewed from front of engine (normal rotation) until the camshaft sprocket's timing dowel pin holes align with the holes in the timing chain housing. The mark on the crankshaft sprocket should align with the mark on the cylinder block. The crankshaft sprocket keyway should point upwards and align with the center line of the cylinder bores. This is the normal timed position.

13. Remove the timing chain guides. Raise and safely support the vehicle.

14. Gently pry off timing chain tensioner spring retainer and remove spring.

➡**Two styles of tensioner are used. Early production engines will have a spring post and late production ones will not. Both styles are identical in operation and are interchangeable.**

15. Remove the timing chain tensioner shoe retainer.

16. Be sure all the slack in the timing chain is above the tensioner assembly; remove the chain tensioner shoe. The timing chain must be disengaged from the wear grooves in the tensioner shoe in order to remove the shoe. Slide a prybar under the timing chain while pulling shoe outward.

17. If difficulty is encountered removing chain tensioner shoe, proceed as follows:

   a. Lower the vehicle.

b. Hold the intake camshaft sprocket with a holding tool and remove the sprocket bolt and washer.

c. Remove the washer from the bolt and thread the bolt back into the camshaft by hand, the bolt provides a surface to push against.

d. Remove intake camshaft sprocket using a 3-jaw puller in the three relief holes in the sprocket. Do not attempt to pry the sprocket off the camshaft or damage to the sprocket or chain housing could occur.

18. Remove the tensioner assembly attaching bolts and the tensioner.

✳✳ CAUTION

The tensioner piston is spring loaded and could fly out causing personal injury.

19. Remove the chain housing to block stud, which is actually the timing chain tensioner shoe pivot.

20. Remove the timing chain.

To install:

21. Tighten intake camshaft sprocket attaching bolt and washer, while holding the sprocket with tool J36013, if removed.

22. Install the special tool through holes in camshaft sprockets into holes in timing chain housing. This positions the camshafts for correct timing.

23. If the camshafts are out of posi-

tion and must be rotated more than ⅛ turn in order to install the alignment dowel pins:

   a. The crankshaft must be rotated 90 degrees clockwise off TDC in order to give the valves adequate clearance to open.

   b. Once the camshafts are in position and the dowels installed, rotate the crankshaft counterclockwise back to TDC. Do not rotate the crankshaft clockwise to TDC or valve and piston damage could occur.

24. Install the timing chain over the exhaust camshaft sprocket, around the idler sprocket and around the crankshaft sprocket.

25. Remove the alignment dowel pin from the intake camshaft. Using a dowel pin remover tool, rotate the intake camshaft sprocket counterclockwise enough to slide the timing chain over the intake camshaft sprocket. Release the camshaft sprocket wrench. The length of chain between the two camshaft sprockets will tighten. If properly timed, the intake camshaft alignment dowel pin should slide in easily. If the dowel pin does not fully index, the camshafts are not timed correctly and the procedure must be repeated.

26. Leave the alignment dowel pins installed.

27. With slack removed from chain between intake camshaft sprocket and

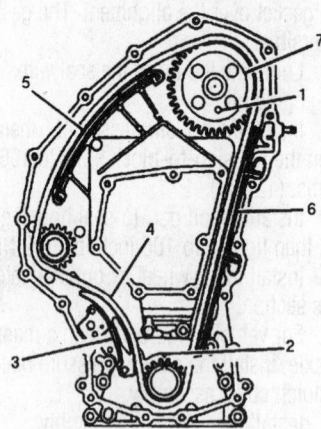

1. Camshaft timing alignment pin locations
2. Crankshaft gear timing marks
3. Shoe and tensioner assembly
4. Timing chain
5. RH timing chain guide
6. LH timing chain guide
7. Camshaft sprocket

79222320

Timing chain and sprockets in the "Timed Position"—2.3L SOHC engine

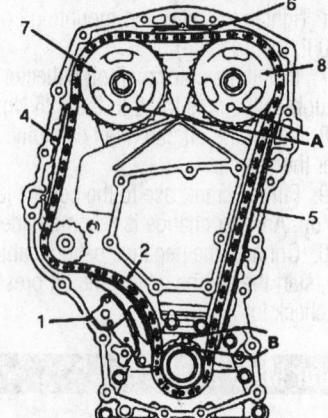

A. Camshaft timing alignment pin locations
B. Crankshaft gear timing marks
1. Shoe asm. timing chain tensioner
2. Timing chain
3. Timing chain tensioner
4. R.H. timing chain guide
5. L.H. timing chain guide
6. Upper timing chain guide
7. Exhaust camshaft sprocket
8. Intake camshaft sprocket

79222321

Timing chain and sprocket alignment positions—2.3L and 2.4L DOHC engines

crankshaft sprocket, the timing marks on the crankshaft and the cylinder block should be aligned. If marks are not aligned, move the chain one tooth forward or rearward, remove slack and recheck marks.

28. Tighten the chain housing to block stud. The stud is installed under the timing chain. Tighten to 19 ft. lbs. (26 Nm).

29. Reload timing chain tensioner assembly to its 0 position as follows:

a. Assemble restraint cylinder, spring and nylon plug into plunger. Index slot in restraint cylinder with peg in plunger. While rotating the restraint cylinder clockwise, push the restraint cylinder into the plunger until it bottoms. Keep rotating the restraint cylinder clockwise but allow the spring to push it out of the plunger. The pin in the plunger will lock the restraint in the loaded position.

b. Install tool J36589 or equivalent, onto plunger assembly.

c. Install plunger assembly into tensioner body with the long end toward the crankshaft when installed.

30. Install the tensioner assembly to the chain housing. Recheck plunger assembly installation. It is correctly installed when the long end is toward the crankshaft.

31. Install and tighten timing chain tensioner bolts and tighten to 10 ft. lbs. (14 Nm).

32. Install the tensioner shoe and tensioner shoe retainer. Remove special tool J36589 and squeeze plunger assembly into the tensioner body to unload the plunger assembly.

33. Lower vehicle and remove the alignment dowel pins. Rotate crankshaft clockwise two full rotations. Align crankshaft timing mark with mark on cylinder block and reinstall alignment dowel pins. Alignment dowel pins will slide in easily if engine is timed correctly.

✳✳ WARNING

If the engine is not correctly timed, severe engine damage could occur.

34. Install the timing chain guides and crankshaft oil slinger.

35. Using a suitable seal installer, tap the new seal into the front cover and lubricate the lip of the seal.

36. Install the front cover and gaskets on the engine and tighten the mounting nuts and bolts to 106 inch lbs. (12 Nm).

37. Install the torsional damper as follows:

a. Coat the seal contact area on the crankshaft damper with clean engine oil.

b. Line up the damper on the crankshaft so the notch in the damper lines up with the crankshaft key.

c. Tap the balancer into place using a rubber mallet.

d. Install the damper mounting bolt and washer and tighten the bolt to 129 ft. lbs. (175 Nm) plus 90 degrees rotation.

38. Install the right front lower splash shield. Install the tire and wheel assembly and tighten to 100 ft. lbs. (140 Nm).

39. Safely lower the vehicle. Install the right engine mount bracket.

40. Install the right engine mount. Connect the upper cover vent hose.

41. Remove the engine support. Install the alternator and attach the electrical connectors.

42. Install the serpentine belt. Install the coolant surge tank.

43. Refill the cooling system. Connect the negative battery cable and check for leaks.

3.1L Engine

✳✳ CAUTION

Never open, service or drain the radiator or cooling system when hot; serious burns can occur from the steam and hot coolant.

1. Disconnect the negative battery cable. Drain the cooling system into a suitable container.

2. Remove the right engine mount bracket. Remove the serpentine belt.

3. Remove the crankshaft balancer as follows:

a. Raise and safely support the vehicle.

b. Remove the right front tire and wheel assembly.

c. Remove the right inner fender well splash shield.

d. Remove the flywheel cover and install a flywheel holding tool.

e. Remove the balancer mounting bolt and washer.

f. Using a suitable puller, J-24420-B or the equivalent remove the balancer from the crankshaft.

4. Remove the serpentine belt tensioner mounting bolt and tensioner. Remove the oil pan.

5. Remove coolant bypass pipe from the water pump and the intake manifold.

6. Disconnect the lower radiator hose to from the front cover outlet. Remove the front cover bolts and cover.

7. Using a suitable seal driver, tap the oil seal out of the front cover.

8. Rotate the crankshaft until the timing marks on the camshaft and crankshaft sprockets are in alignment at their closest approach.

9. Remove the camshaft sprocket bolt, sprocket and timing chain.

10. Remove the crankshaft sprocket, using a gear puller J-5825-A or equivalent. Remove the timing chain damper bolts and damper.

To install:

11. Install the timing chain damper and tighten the mounting bolts to 15 ft. lbs. (21 Nm).

12. Install the crankshaft sprocket onto the crankshaft making sure the notch in the sprocket fits over the crankshaft key. Fully seat the sprocket on the crankshaft using J-38612, or an equivalent gear installer.

13. Be sure the timing mark on the crankshaft sprocket is pointing straight up.

14. Install the timing chain over the camshaft sprocket and hold the sprocket in

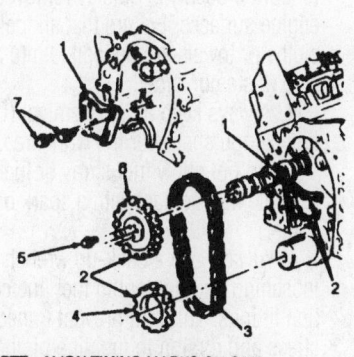

NOTE - ALIGN TIMING MARKS ON CAM & CRANK SPROCKETS USING ALIGNMENT MARKS ON DAMPER STAMPING OR CAST ALIGNMENT MARKS ON CYL & CASE

1. Damper
2. Alignment marks
3. Timing chain
4. Crank sprocket
5. 28 Nm (21 lb. ft.)
6. Camshaft sprocket
7. 21 Nm (15 lb. ft.)

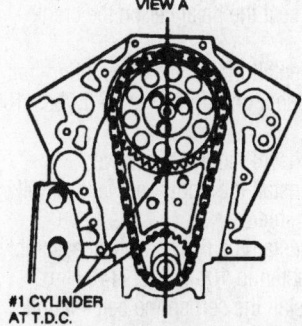

VIEW A

#1 CYLINDER AT T.D.C.

NOTE - CAMSHAFT SPROCKET MARK AT 6 O'CLOCK
CRANKSHAFT SPROCKET MARK AT 12 O'CLOCK

79227322

Exploded view of the timing chain and sprocket timing mark alignment—3.1L engine

such a way, that the timing mark is pointing down, and the timing chain is hanging down off the sprocket.

15. Loop the timing chain under the crankshaft sprocket and install the camshaft sprocket on the camshaft. The sprocket will only fit on the camshaft if the dowel on the camshaft lines up with the hole in the sprocket.

16. Verify that the marks are aligned (the camshaft sprocket will be at the 6 o'clock position and the crankshaft sprocket will be in the 12 o'clock position).

17. On 1995 vehicles, tighten the camshaft sprocket mounting bolt to 74 ft. lbs. (100 Nm). On 1996–99 vehicles, tighten the bolt to 81 ft. lbs. (110 Nm).

18. Lubricate the timing chain components with engine oil. Clean all gasket surfaces completely.

19. Apply a thin bead of sealer around the gasket sealing area of the front cover. Install a new front cover seal on the front cover.

20. Install the front cover on the engine and install the mounting bolts. On 1995 vehicles, tighten the small bolts to 18 ft. lbs. (24 Nm) and the large bolts to 41 ft. lbs. (55 Nm). On 1996–99 vehicles, tighten the small bolts to 15 ft. lbs. (21 Nm) and the large bolts to 35 ft. lbs. (47 Nm).

21. Connect the radiator hose to the coolant outlet.

22. Install coolant bypass pipe to the water pump and the intake manifold. Install the oil pan.

23. Install crankshaft balancer as follows:

a. Coat the seal contact surface of the crankshaft balancer with clean engine oil.

b. Line up the notch in the balancer with the crankshaft key and slide the balancer on until the key is in the balancer.

c. Using J-29113 or an equivalent puller, seat the balancer on the crankshaft.

d. Install the balancer mounting bolt and washer and tighten to 76 ft. lbs. (103 Nm).

e. Install the flywheel cover.

f. Install the right inner fender well splash shield.

g. Install the tire and wheel assembly and tighten to 100 ft. lbs. (140 Nm).

24. Install the serpentine belt tensioner and tighten the mounting bolt to 40 ft. lbs. (54 Nm). Install the serpentine belt.

25. Install the right engine mount bracket and tighten the bracket-to-mount bolts to 96 ft. lbs. (130 Nm).

26. Refill the cooling system with the correct amount and type of coolant. Check the engine oil level. An oil and filter change is recommended.

27. Connect the negative battery cable. Start the engine and verify that there are no leaks.

FUEL SYSTEM

Fuel System Service Precautions

Safety is the most important factor when performing not only fuel system maintenance but any type of maintenance. Failure to conduct maintenance and repairs in a safe manner may result in serious personal injury or death. Maintenance and testing of the vehicle's fuel system components can be accomplished safely and effectively by adhering to the following rules and guidelines.

• To avoid the possibility of fire and personal injury, always disconnect the negative battery cable unless the repair or test procedure requires that battery voltage be applied.

• Always relieve the fuel system pressure prior to disconnecting any fuel system component (injector, fuel rail, pressure regulator, etc.), fitting or fuel line connection. Exercise extreme caution whenever relieving fuel system pressure to avoid exposing skin, face and eyes to fuel spray. Please be advised that fuel under pressure may penetrate the skin or any part of the body that it contacts.

• Always place a shop towel or cloth around the fitting or connection prior to loosening to absorb any excess fuel due to spillage. Ensure that all fuel spillage (should it occur) is quickly removed from engine surfaces. Ensure that all fuel soaked cloths or towels are deposited into a suitable waste container.

• Always keep a dry chemical (Class B) fire extinguisher near the work area.

• Do not allow fuel spray or fuel vapors to come into contact with a spark or open flame.

• Always use a back-up wrench when loosening and tightening fuel line connection fittings. This will prevent unnecessary stress and torsion to fuel line piping. Always follow the proper torque specifications.

• Always replace worn fuel fitting O-rings with new. Do not substitute fuel hose or equivalent, where fuel pipe is installed.

Fuel System Pressure

RELIEVING

> ### ✳✳ CAUTION
>
> **The fuel injection system remains under pressure, even after the engine has been turned OFF. The fuel system pressure must be relieved before disconnecting any fuel lines. Failure to do so may result in fire and/or personal injury.**

2.3L and 2.4L Engines

1. Loosen the fuel filler cap in order to relieve the pressure in the tank (do not tighten at this time).

2. Raise and safely support the vehicle.

3. Detach the fuel pump electrical connector.

4. Start and run the vehicle until it stalls, the engage the starter for an additional three seconds to ensure the relief of any remaining pressure.

5. Disconnect the negative battery cable.

6. Once the tests or repairs are completed, reattach the fuel pump electrical connector.

7. Connect the negative battery cable.

8. Lower the vehicle.

9. Tighten the fuel filler cap.

10. Prime the fuel system by cycling the ignition switch **ON** for 2 seconds, **OFF** for 10 seconds, then **ON** again. Repeat, if necessary to build system pressure.

3.1L Engine

1. Disconnect the negative battery cable in order to avoid possible fuel discharge if an accidental attempt is made to start the engine.

2. Loosen the fuel tank filler cap in order to relieve fuel tank pressure.

3. Connect tool J-34730-A or equivalent fuel pressure gauge to the fuel pressure test port connection. Wrap a towel around the fuel pressure connection when installing the fuel pressure gauge in order to avoid fuel spillage.

4. Install the bleed hose into an approved container and open the valve in order to bleed the fuel system pressure. The fuel pipe connections are now safe for servicing.

5. Drain any fuel remaining in the fuel pressure gauge into an approved container.

Fuel Filter

REMOVAL & INSTALLATION

All Engines

> ※※ **CAUTION**
>
> **The fuel injection system remains under pressure, even after the engine has been turned OFF. The fuel system pressure must be relieved before disconnecting any fuel lines. Failure to do so may result in fire and/or personal injury.**

1. Relieve the fuel system pressure.
2. Disconnect the negative battery cable.
3. Raise and safely support the vehicle.
4. Using a back-up wrench on the fuel filter disconnect the fuel line from the filter.

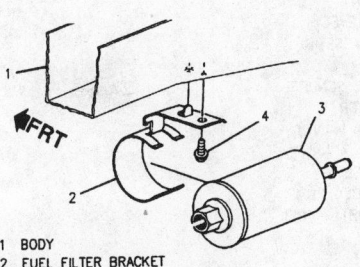

1 BODY
2 FUEL FILTER BRACKET
3 FUEL FILTER
4 SCREW – FULLY DRIVEN, SEATED AND NOT STRIPPED

79222323

Exploded view of the fuel filter mounting

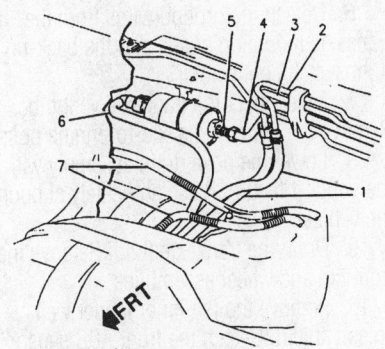

1 HOSE, PART OF FUEL SENDER
2 FUEL VAPOR PIPE
3 FUEL RETURN PIPE
4 FUEL FEED PIPE
5 FUEL FEED PIPE NUT
  27 N·m (20 LBS. FT.)
6 HOSE, PART OF FUEL SENDER
7 ABS AND FUEL SENDER HARNESS

79222324

Fuel filter mounting location and component identification

5. Disconnect the quick-connect fitting from the fuel filter by compressing the tabs while pulling outward on the line.
6. Remove the fuel filter from the mounting bracket.

To install:

7. Install the fuel filter in the mounting bracket.
8. Using a back-up wrench on the fuel filter, tighten the fuel line fitting to 20 ft. lbs. (27 Nm).
9. Connect the quick-connect fitting to the fuel filter.
10. Lower the vehicle.
11. Connect the negative battery cable.
12. Pressurize the fuel system and verify no leaks.

Fuel Pump

REMOVAL & INSTALLATION

All Engines

> ※※ **CAUTION**
>
> **The fuel injection system remains under pressure, even after the engine has been turned OFF. The fuel system pressure must be relieved before disconnecting any fuel lines. Failure to do so may result in fire and/or personal injury.**

1. Relieve the fuel system pressure. Disconnect the negative battery cable.
2. Drain fuel tank. Remove the fuel tank from the vehicle.
3. While holding the modular fuel sender assembly down, remove the snaphing from the designated slots located on the retainer.

➡The modular fuel sender assembly may spring up from its position. When removing the modular fuel sender from the tank, be aware that the reservoir bucket is full of fuel. It must be tipped slightly during removal to avoid damage to the float.

4. Remove the external fuel strainer.
5. Detach the Connector Position Assurance (CPA) piece from the wiring harness and disconnect the fuel pump.
6. Gently release the tabs on the sides of the fuel sender at the cover assembly. Begin by squeezing the sides of the reservoir and releasing the tab opposite the fuel level sensor. Move clockwise to release the second and third tab in the same manner.
7. Lift the cover assembly out far enough to disconnect the fuel pump electrical connection.
8. Rotate the fuel pump baffle counterclockwise and remove the baffle and pump assembly from the retainer.
9. Slide the fuel pump outlet out of slot. Remove the fuel pump outlet seal.

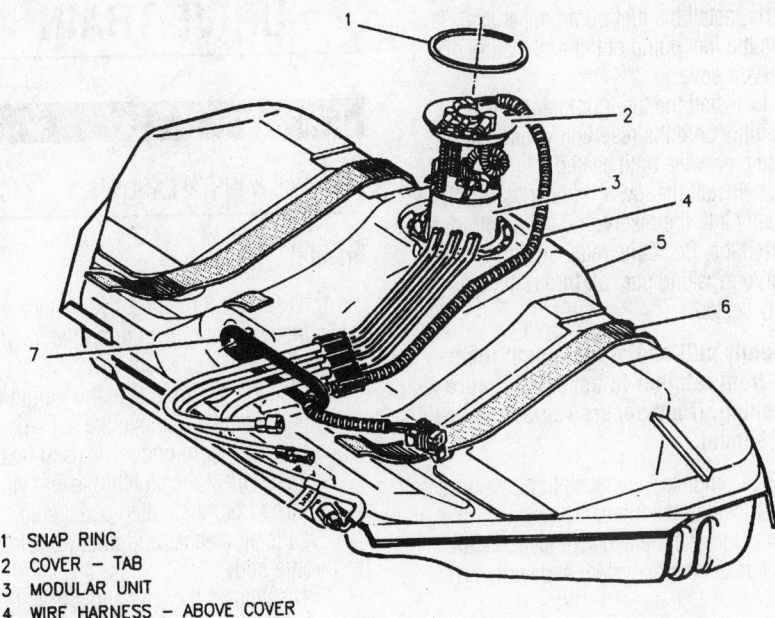

1 SNAP RING
2 COVER – TAB
3 MODULAR UNIT
4 WIRE HARNESS – ABOVE COVER
5 FUEL TANK
6 TANK ISOLATION STRIPS (3)
7 RUBBER ISOLATOR

79222325

Exploded view of the fuel sender assembly mounting to the tank

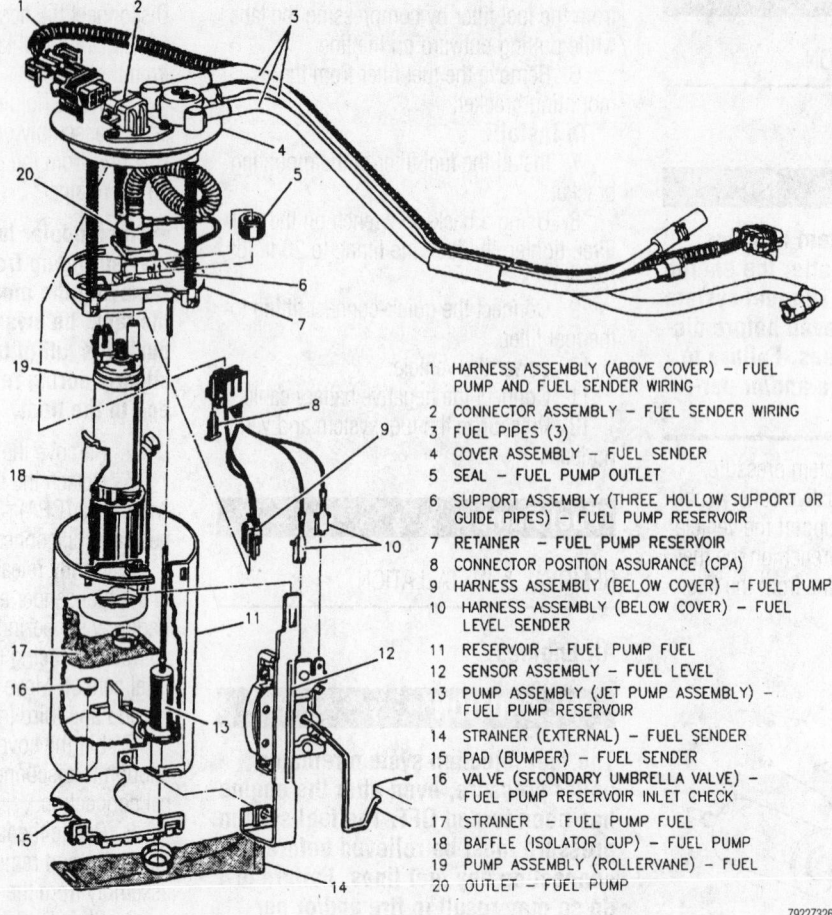

1. HARNESS ASSEMBLY (ABOVE COVER) – FUEL PUMP AND FUEL SENDER WIRING
2. CONNECTOR ASSEMBLY – FUEL SENDER WIRING
3. FUEL PIPES (3)
4. COVER ASSEMBLY – FUEL SENDER
5. SEAL – FUEL PUMP OUTLET
6. SUPPORT ASSEMBLY (THREE HOLLOW SUPPORT OR GUIDE PIPES) – FUEL PUMP RESERVOIR
7. RETAINER – FUEL PUMP RESERVOIR
8. CONNECTOR POSITION ASSURANCE (CPA)
9. HARNESS ASSEMBLY (BELOW COVER) – FUEL PUMP
10. HARNESS ASSEMBLY (BELOW COVER) – FUEL LEVEL SENDER
11. RESERVOIR – FUEL PUMP FUEL
12. SENSOR ASSEMBLY – FUEL LEVEL
13. PUMP ASSEMBLY (JET PUMP ASSEMBLY) – FUEL PUMP RESERVOIR
14. STRAINER (EXTERNAL) – FUEL SENDER
15. PAD (BUMPER) – FUEL SENDER
16. VALVE (SECONDARY UMBRELLA VALVE) – FUEL PUMP RESERVOIR INLET CHECK
17. STRAINER – FUEL PUMP FUEL
18. BAFFLE (ISOLATOR CUP) – FUEL PUMP
19. PUMP ASSEMBLY (ROLLERVANE) – FUEL
20. OUTLET – FUEL PUMP

79222326

Exploded view of the fuel pump assembly

To install:

10. Install the fuel pump outlet seal. Slide the fuel pump outlet in the slots of reservoir cover.

11. Install the fuel pump and baffle assembly onto the reservoir retainer and rotate clockwise until seated.

12. Install the lower retainer assembly partially into the reservoir. Line up all three sleeve tabs. Press the retainer onto the reservoir making sure all three tabs are firmly seated.

➡Gently pull on the fuel pump reservoir from retainer to assure a secure fastening. If not secure replace entire fuel sender.

13. Complete the installation by reverse the removal procedures.

14. Connect the negative battery cable. Pressurize the fuel system and verify no leaks.

DRIVE TRAIN

Transaxle

REMOVAL & INSTALLATION

Manual

1. Disconnect the negative battery cable. Install an engine support fixture, J-28467-A, or the equivalent.

2. Use the fixture to take the weight off the engine mounts. Remove the left side sound insulator from under the dash board.

3. Disconnect the clutch master cylinder pushrod from the clutch pedal stud. Remove the air cleaner and duct work from the throttle body.

4. Disconnect the shift cable from the manual shift lever on the transaxle. Discon-

nect the wiring harness from the transaxle mount bracket.

5. Remove the upper transaxle mount-to-transaxle bolts. Disconnect the clutch master cylinder from the slave cylinder.

6. Detach the ground wires from the transaxle mounting studs, and the back-up light switch connector.

7. Disconnect the transaxle vent tube. Remove the upper transaxle-to-engine bolts.

8. Lower the powertrain assembly with the support fixture. Raise and safely support the vehicle.

9. Drain the transaxle fluid. Remove the front tire and wheel assemblies.

10. Remove the left inner fender well splash shield. Detach the front ABS sensor connectors and unroute the left side sensor wiring.

11. Remove the flywheel housing cover bolts. Disconnect the vehicle speed sensor from the transaxle.

12. Remove the cotter pins and castle nuts and using a suitable ball joint separator, disconnect the ball joints from the steering knuckles.

13. Remove the left side link kit. Remove the left side U-bolt from the sway bar.

14. Remove the left side suspension support bolts and remove the suspension support.

15. Disconnect the halfshafts from the transaxle. Remove the front lower transaxle mount.

16. Support the transaxle with a suitable jack. Remove the remainder of the engine-to-transaxle bolts.

17. Carefully slide the transaxle away from the engine. Once there is sufficient clearance, lower the transaxle from the vehicle.

To install:

18. Raise the transaxle into position with the jack and install the mounting bolts loosely. DO NOT tighten the bolts until at least four of the bolts are in place. Tighten the lower bolts to 55 ft. lbs. (75 Nm).

19. Install the front transaxle mount and tighten the bolts to 55 ft. lbs. (75 Nm).

20. Complete the installation by reversing the removal procedures.

21. Install the upper transaxle-to-engine bolts and tighten to 55 ft. lbs. (75 Nm).

22. Install the rear transaxle mount and tighten the bolts to 55 ft. lbs. (75 Nm) on Isuzu transaxle or 96 ft. lbs. (130 Nm) on NVG transaxle.

23. Connect the negative battery cable. Refill the transaxle with Synchromesh Transaxle Fluid.

Automatic

3T40 AND 4T60-E TRANSAXLES

1. Disconnect the negative battery cable. Remove the air intake duct.

2. For 3T40 transaxle, remove the cable control cover.

3. For 4T60-E transaxle, disconnect the shift linkage from the transaxle.

4. Disconnect the vacuum modulator line from the modulator.

5. For 3T40 transaxle, disconnect the throttle cable from the throttle body.

6. Detach the electrical connectors from the TCC, Park/Neutral position switch and Shift solenoid.

7. For 3T40 transaxle, remove the power steering pump and set it aside leaving the hoses attached.

✳✳ CAUTION

When servicing requires that the "T" latch type wiring connector be

detached from the switch, use care to ensure proper reassembly of both the connector and the "T" latch. Failure to do so may result in intermittent loss of switch functions.

8. For 3T40 transaxle, remove the oil fill tube.

9. Install engine support fixture J-28467-A or equivalent. Remove the top (2) transaxle to engine bolts.

10. For 4T60-E transaxle, disconnect the rubber hose from the transaxle vent pipe.

11. Remove the remaining upper transaxle to engine bolts. Raise and safely support the vehicle.

12. Remove both front tire and wheel assemblies. Remove right and left engine splash shields.

13. Detach the ABS wheel speed sensor connectors and remove the harness from the left side suspension support.

14. Separate both ball joints from the control arms. Remove the left side stabilizer shaft link pin bolt.

15. Remove the left side stabilizer shaft frame bushing clamp nuts. Remove the left side suspension support assembly.

16. Remove both halfshaft. Remove the engine to transaxle brace.

17. Remove the starter motor. Remove the transaxle converter cover.

18. Remove the heater core hose pipe brace to transaxle nut and bolt. Remove the torque converter to flywheel bolts.

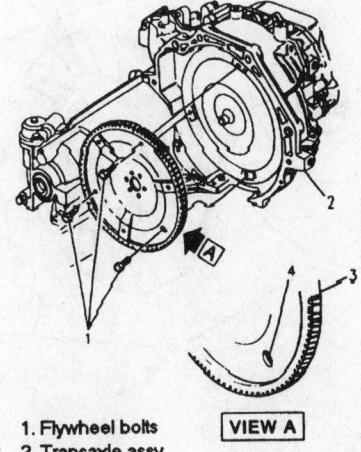

1. Flywheel bolts
2. Transaxle assy
3. Flywheel
4. Net slot

VIEW A

79227327

Flywheel net slot view—vehicles with the 4T60-E transaxle

➡Using a scribe mark the flywheel to torque converter relationship to assure proper reassembly.

19. For 4T60-E transaxle, remove the oil level indicator and fill tube.

20. Disconnect the transaxle cooler lines and plug openings to prevent excess oil leakage.

21. For 4T60-E transaxle, disconnect the vehicle speed sensor and remove the vacuum reservoir tank.

22. Position transaxle jack under transaxle. Remove the transaxle mount to body bolts.

23. Remove the remaining engine to transaxle bolts. Remove the transaxle.

➡Transaxle cooler and lines should be flushed with J-35944 or equivalent whenever the transaxle has been removed for overhaul or replacement.

To install:

24. Apply a thin film of grease on the torque converter pilot hub.

➡Be sure to properly seat the torque converter in the pump.

25. Position the transaxle into the vehicle. Install the lower transaxle to engine bolts.

26. Install the transaxle mount to body bolts and tighten to 66 ft. lbs. (90 Nm). Install the transaxle cooler lines.

27. Install the torque converter to flywheel bolts and tighten the bolts to 46 ft. (62 Nm).

➡For 4T60-E transaxle, note that the flywheel has one oval shaped bolt opening. This is for the so-called "net slot bolt." Hand start and tighten the net slot bolt first, then tighten the remaining bolts.

28. Install the oil level indicator and fill tube.

29. For 4T60-E transaxle, install the vacuum reserve tank.

30. Install the starter. Install the transaxle converter cover.

31. For 4T60-E transaxle, attach the electrical connector to the vehicle speed sensor.

32. Install the halfshafts. Install the left side suspension support assembly.

33. Install the left stabilizer shaft frame bushing nuts. Install the stabilizer shaft link pin bolt.

34. Install the engine to transaxle brace. Connect the ball joints to the control arms.

35. Attach the ABS wheel speed sensor harness and connectors. Install the right and left side splash shields.

36. Install the heater core pipe brace to transaxle nut and bolt. Install the tire/wheel assemblies. Lower the vehicle.

37. Install the upper transaxle to engine bolts and tighten to 66 ft. lbs. (90 Nm). Connect the shift linkage to the transaxle.

38. Attach the electrical connector to the torque converter clutch.

39. Attach the electrical connector to the park/neutral and back-up lamp switch.

40. For 4T60-E transaxle, install the wiring harness and nut securing it to the transaxle.

41. For 3T40 transaxle, install the throttle cable to the throttle body.

42. Remove the engine support fixture.

43. For 4T60-E transaxle, connect the rubber hose to the vent pipe.

44. Connect the vacuum line to the modulator. Install the air intake duct.

45. Connect the negative battery cable. Fill the transaxle with clean transaxle fluid.

46. Verify proper shift linkage adjustment. Start the engine and check for leaks.

47. Road test the vehicle verify proper operation and check the transaxle fluid level.

➡**Whenever the vehicle sub-frame is removed or lowered, the wheel alignment should be checked.**

Clutch

REMOVAL & INSTALLATION

1. Disconnect the negative battery cable.
2. If necessary, disconnect the hydraulic line from the clutch actuator (slave cylinder).
3. Remove the transaxle from the vehicle.
4. If any clutch components are to be reused, use the following procedure:

    a. Matchmark (a dots of paint or marks made with a center punch) the clutch pressure plate to the flywheel. This is to retain the balance of the original parts. If all parts are to be replaced with new, this step is not necessary.

    b. If the pressure plate is to be reused, loosen the pressure plate mounting bolts by turning each bolt one full turn until all the spring pressure is removed. This helps avoid warping the pressure plate.

    c. Remove the bolts the remainder of the way and remove the clutch disc and pressure plate.

To install:

5. Apply a small amount of high-temperature grease to the pilot bearing as well

as the tip of the transaxle input shaft and the clutch splines.

6. Install J-29074, or an equivalent clutch alignment tool into the flywheel and install the clutch disc onto the tool. Be sure the clutch is installed in the correct direction. The light side of the disc should be visible. Many replacement discs will also be marked "Flywheel Side" as an air in getting the disc correctly positioned.

7. Install the pressure plate and loosely install the mounting bolts. New replacement bolts are recommended.

8. Make a first pass tightening the bolts in sequence to 12 ft. lbs. (16 Nm). With all the bolts tight make a second pass tightening each bolt to 15 ft. lbs. (20 Nm). Make a third pass tightening the bolts 30 degrees additional if equipped with a 3.1L engine and 45 degrees additional if equipped with a 2.3L or 2.4L engine.

9. Remove the clutch alignment tool.
10. Lubricate the inside diameter of the actuator (slave cylinder/throwout bearing) with clutch bearing lubricant.
11. Install the transaxle assembly into the vehicle.
12. If necessary, connect the clutch master cylinder hydraulic line to the actuator (slave cylinder).

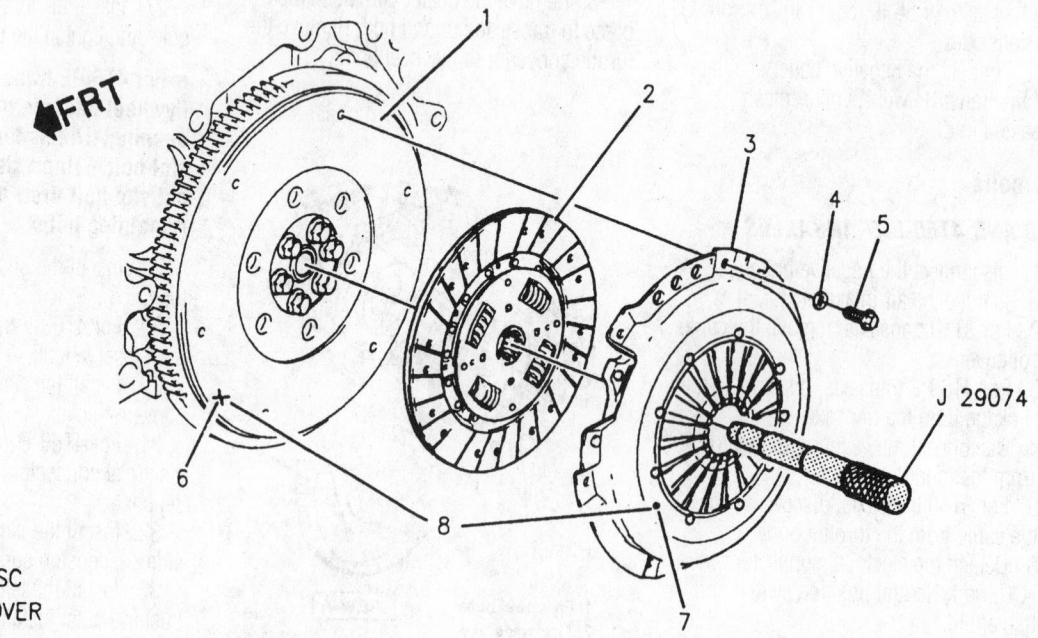

1. FLYWHEEL
2. CLUTCH DISC
3. CLUTCH COVER
4. WASHER
5. BOLT
6. FLYWHEEL "HEAVY SIDE" IDENTIFICATION
7. CLUTCH COVER "LIGHT SIDE" IDENTIFICATION
8. ALIGN IDENTIFICATION MARKS ON ASSEMBLY

Exploded view of the clutch components—showing the clutch disk alignment tool

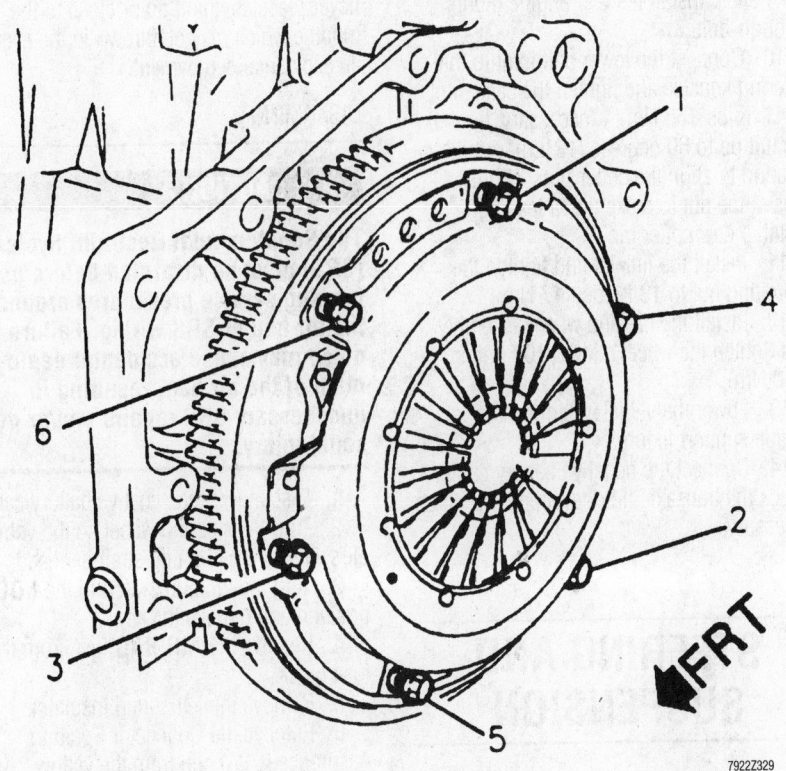

Clutch cover tightening sequence

13. Bleed the clutch hydraulic system.
14. Connect the negative battery cable. Road test vehicle to verify correct operation and easy shifting.

Hydraulic Clutch System

BLEEDING

With Bleeder Screw

1. Be sure the reservoir is full of DOT 3 fluid and is kept topped off throughout this procedure.
2. Loosen the bleed screw, located on the actuator cylinder body next to the inlet connection.
3. When a steady stream of fluid comes out the bleeder, tighten it to 17 inch lbs. (2 Nm).
4. Refill the fluid reservoir.
5. To check the system, start the engine and wait 10 seconds.
6. Depress the clutch pedal and shift into **R**. If there is any gear clash, air may still be present.

Without Bleeder Screw

1. Remove the actuator cylinder from the transaxle.
2. Loosen the master cylinder attaching nuts to the ends of the studs.
3. Remove the reservoir cap and diaphragm.
4. Depress the actuator cylinder pushrod about ¾ in. into its bore and hold the position.
5. Install the reservoir diaphragm and cap while holding the actuator pushrod.
6. Release the pushrod when the diaphragm and cap are properly installed.
7. With the actuator lower than the master cylinder, hold the actuator vertically with the pushrod end facing the ground.
8. Press the actuator pushrod into its bore with ½ in. strokes. Check the reservoir for bubbles. Continue until no bubbles enter the reservoir.
9. Install the master cylinder and actuator. Refill the fluid reservoir.
10. To check the system, start the engine and wait 10 seconds.

11. Depress the clutch pedal and shift into reverse. If there is any gear clash, air may still be present.

Halfshaft

REMOVAL & INSTALLATION

Left and Right Halfshafts

1. Disconnect the negative battery cable. Raise and safely support the vehicle. Remove the wheel assembly.
2. Insert a drift or a punch through the caliper and into the rotor cooling fins to keep the axle from turning.
3. Remove the hub nut and washer and remove the drift.
4. Remove the cotter pin and castle nut from the lower ball joint. Using a suitable ball joint separator, disconnect the ball joint from the steering knuckle.
5. Disconnect the ABS sensor wire. Remove the sway bar link kit.
6. Unseat the halfshaft from the hub bearing assembly using a suitable puller, J-28733-A, or the equivalent.
7. Pivot the strut and knuckle assembly off the halfshaft and position it aside.
8. Place a pan under the transaxle.
9. Disconnect the inner joint from the transaxle using axle puller, J-28468 or J-33008 in conjunction with J-29794 and J-2619-01 or their equivalents.

To install:
10. Insert the halfshaft into the transaxle and using a non-ferrous drift positioned in

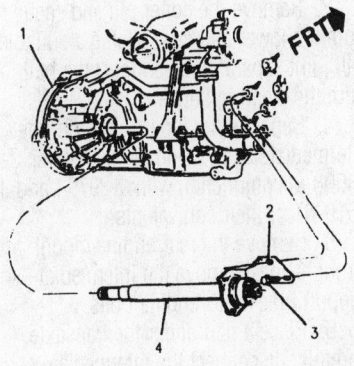

1	TRANSAXLE
2	INTERMEDIATE SHAFT SUPPORT BRACKET
3	BOLT
4	INTERMEDIATE SHAFT

Intermediate shaft components

the groove in the inboard joint, tap the joint in until it is seated in the transaxle.

11. Verify the joint is seated properly by grasping the inboard joint and pulling on it firmly. DO NOT pull on the axle shaft or damage to the inner joint will result.

12. Position the strut and knuckle assembly and guide the axle into the hub assembly.

13. Connect the lower ball joint to the steering knuckle and tighten the castle nut to 41 ft. lbs. (55 Nm). If necessary tighten the nut up to 60 degrees (⅙ turn) additional rotation to align the cotter pins. NEVER loosen the nut to make the holes align. Install a new cotter pin.

14. Install the washer and hub nut.

15. Insert a drift punch through the caliper and into the rotor cooling fins. Tighten the hub nut to 185 ft. lbs. (260 Nm).

16. Remove the drift. Install the link kit and tighten the mounting nut to 13 ft. lbs. (17 Nm).

17. Install the tire and wheel assembly and tighten the wheel nuts to 100 ft. lbs. (140 Nm).

18. Lower the vehicle. Connect the negative battery cable. Check the transaxle fluid level and top off as necessary.

Intermediate Shaft

1. Install an engine support fixture J-28467-A or equivalent. Raise and safely support the vehicle.

2. Remove the right side wheel assembly. Remove the sway bar link kit.

3. Remove the cotter pin and castle nut from the lower ball joint. Using a suitable ball joint separator, disconnect the ball joint from the steering knuckle.

4. Separate the inner joint from the intermediate shaft using an axle puller, J-33008 in conjunction with J-29794 and J-2619-01 or their equivalents.

5. Remove the rear engine mount through-bolt. Remove the intermediate shaft support bracket-to-engine bolts.

6. Place a pan under the transaxle. Carefully disconnect the intermediate shaft from the transaxle and remove the assembly from the vehicle.

To install:

7. Insert the intermediate shaft assembly into the transaxle and position the shaft bracket so the mounting bolts can be installed finger-tight.

8. With all the bolts installed tighten them to 49 ft. lbs. (66 Nm). Coat the splines of the intermediate shaft with chassis grease.

9. Connect the halfshaft to the intermediate shaft. Install the rear engine mount through-bolt.

10. Connect the lower ball joint to the steering knuckle and tighten the castle nut to 41 ft. lbs. (55 Nm). If necessary, tighten the nut up to 60 degrees (⅙ turn) additional rotation to align the cotter pins. NEVER loosen the nut to make the holes align. Install a new cotter pin.

11. Install the link kit and tighten the mounting nut to 13 ft. lbs. (17 Nm).

12. Install the tire and wheel assembly and tighten the wheel nuts to 100 ft. lbs. (140 Nm).

13. Lower the vehicle. Remove the engine support fixture.

14. Connect the negative battery cable. Check the transaxle fluid level and top off as necessary.

STEERING AND SUSPENSION

Air Bag

✳✳ CAUTION

Some vehicles are equipped with an air bag system, also known as the Supplemental Inflatable Restraint (SIR) or Supplemental Restraint System (SRS). The system must be disabled before performing service on or around system components, steering column, instrument panel components, wiring and sensors. Failure to follow safety and disabling procedures could result in accidental air bag deployment, possible personal injury and unnecessary system repairs.

PRECAUTIONS

Several precautions must be observed when handling the inflator module to avoid accidental deployment and possible personal injury.

• Never carry the inflator module by the wires or connector on the underside of the module.

• When carrying a live inflator module, hold securely with both hands, and ensure that the bag and trim cover are pointed away.

• Place the inflator module on a bench or other surface with the bag and trim cover facing up.

• With the inflator module on the bench, never place anything on or close to the module which may be thrown in the event of an accidental deployment.

DISARMING

✳✳ CAUTION

The Supplemental Restraint System (SRS) must be disarmed before performing service procedures around the air bag or SRS wiring. Failure to do so may cause accidental deployment of the air bag, resulting in unnecessary SRS repairs and/or personal injury.

1. Disconnect the negative battery cable.

2. Turn the steering wheel so the vehicle's wheels are pointing straight ahead.

3. Turn the ignition switch to the **LOCK** position and remove the key.

4. Remove the **AIR BAG** fuse from the fuse block.

5. Remove the left sound insulator.

6. Remove the Connector Position Assurance (CPA) clip from the yellow 2-way connector at the base of the steering column, and detach the connector. If equipped with a passenger's side air bag, remove the CPA and detach the yellow 2-way connector from the passenger air bag lead.

REARMING

1. Turn the ignition switch to the **LOCK** position and remove the key.

2. Attach the yellow 2-way connector at the base of steering column and secure it with the CPA clip. If equipped with a passenger's side air bag, attach the yellow 2-way connector at the passenger air bag lead and secure it with the CPA clip.

3. Install the left sound insulator.

4. Install the **AIR BAG** fuse in the fuse block.

5. Turn the ignition switch to the **RUN** position and verify that the **AIR BAG** warning lamp flashes 7 times, then turns OFF.

6. Connect the negative battery cable.

Power Rack and Pinion Steering Gear

REMOVAL & INSTALLATION

1. Remove the left sound insulator from under the dashboard. Remove the upper pinch bolt from the steering shaft coupling assembly.

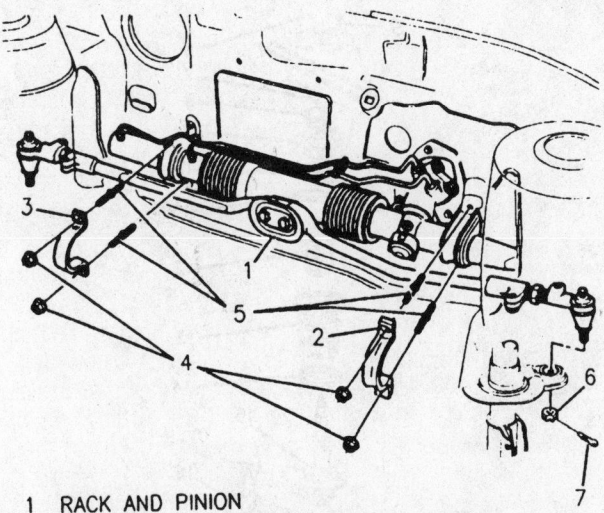

1 RACK AND PINION
2 L.H. CLAMP – HORIZONTAL SLOT AT TOP
3 R.H. CLAMP – HORIZONTAL SLOT AT TOP
4 NUT – 30 N·m (22 LBS.FT.) – HAND START ALL NUTS. TIGHTEN LEFT HAND SIDE CLAMP NUTS FIRST, THEN TIGHTEN RIGHT SIDE NUTS.
5 STUD – 18 N·m (13 LBS. FT.) – AFTER SECOND REUSE OF STUD, THREAD LOCKING KIT NO. 1052624 MUST BE USED.
6 NUT – 60 N·m (44 LBS. FT.)
7 COTTER PIN

7922Z331

Rack and pinion mounting components

2. Remove the power steering line retainer, if equipped. Remove the power brake booster mounting nuts and disconnect the brake pedal pushrod from the pedal stud.

3. Pull the booster away from the firewall and position away from the rack and pinion mounting clamp. DO NOT remove the master cylinder or allow the brake lines to kink.

4. Raise and safely support the vehicle. Remove the left tire and wheel assembly.

5. Remove the cotter pins and castle nuts from the outer tie rod ends and separate the tie rod ends from the strut arms using J-24319–01, or an equivalent tie rod puller.

6. Remove the four clamp mounting bolts and the two rack and pinion mounting clamps. Disconnect and cap the power steering lines from the rack and pinion unit.

7. Pull the rack away from the firewall and remove the lower steering shaft pinch bolt and separate the steering shaft from the rack and pinion stub shaft.

8. Remove the dash seal from the rack and pinion housing. Remove the rack and pinion unit out through the hood.

9. If when removing the rack and pinion mounting clamp nuts the studs came out, remove the nuts from the studs and install the studs back into the firewall. Tighten the studs to 15 ft. lbs. (20 Nm).

To install:

10. Install the rack and pinion assembly through the engine compartment and position it on the firewall. Install the dash seal on the firewall.

11. Connect the steering shaft coupling to the stub shaft and tighten the pinch bolt to 30 ft. lbs. (41 Nm).

12. Connect the power steering lines to the rack and pinion assembly and tighten the fittings to 20 ft. lbs. (27 Nm).

13. Install the rack and pinion clamps and loosely install the mounting nuts. Tighten the left side clamps first, then the right to 22 ft. lbs. (30 Nm).

14. Install the power booster on the firewall. Raise and safely support the vehicle.

15. Connect the tie rod ends to the strut arms and tighten the castle nuts. Install the left front tire and wheel assembly.

16. Connect the booster pushrod to the brake pedal and tighten the booster mount-

ing nuts to 22 ft. lbs. (30 Nm). Install the power steering line retainer.

17. Install the steering column upper pinch bolt and tighten to 30 ft. lbs. (41 Nm). Install the left side sound insulator.

18. Refill and bleed the power steering system. Check the front end alignment and adjust as necessary.

Strut

These vehicles are designed with front struts and rear shocks.

REMOVAL & INSTALLATION

Front

1. Raise and safely support the vehicle. Remove the front tire and wheel assembly.

2. Remove the cotter pin and castle nut from the outer tie rod end and separate the ball stud from the strut arm using J-24319-01, or an equivalent tie rod puller.

3. Remove the bolt from the brake line bracket and separate the bracket from the strut.

4. Scribe reference marks on the front strut and steering knuckle for installation purposes.

5. Remove the strut lower mounting bracket nut and through-bolts.

6. Lower the vehicle until the upper strut plate mounting bolts are accessible.

7. Support the lower control arm.

8. Hold the strut and remove the strut plate mounting nuts and bolt. Remove the strut from the vehicle.

✳✳ WARNING

If the strut cartridge is being replaced, a strut compressor tool J-34013-A or equivalent must be used to remove the coil spring from the strut assembly. Failure to use a strut compressing tool can result in personal injury and/or part damage.

To install:

9. Install the strut assembly into the vehicle and install the upper strut plate mounting nuts.

10. Connect the lower strut bracket to the steering knuckle and install the through-bolts.

11. Raise the vehicle.

12. Install the through-bolt nuts and tighten to 133 ft. lbs. (180 Nm) with the reference marks in alignment.

13. Connect the tie rod end to the strut arm and tighten the castle nut to 44 ft. lbs. (60 Nm). If necessary tighten the nut up to 60

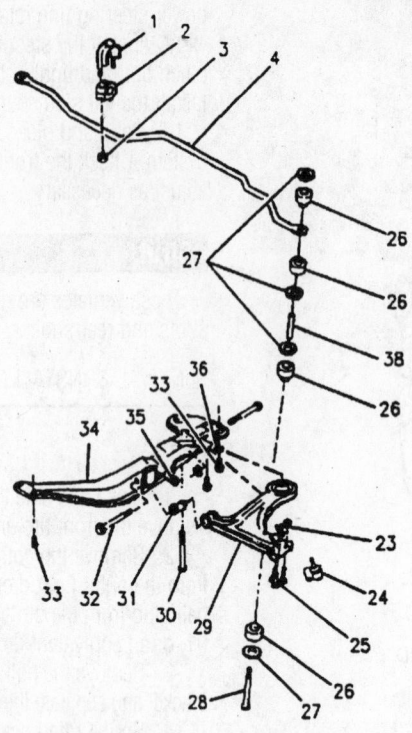

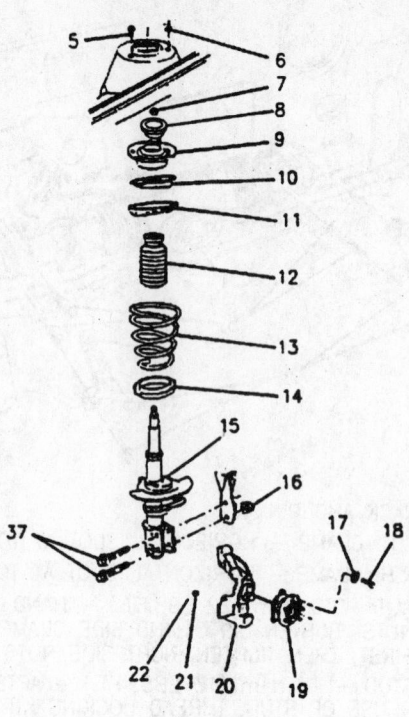

1	CLAMP, STABILIZER SHAFT	20	STEERING KNUCKLE
2	INSULATOR, STABILIZER SHAFT	21	NUT, BALL JOINT
3	NUT	22	COTTER PIN
4	STABILIZER SHAFT	23	NUT
5	BOLT	24	BALL JOINT
6	NUT	25	BOLT
7	NUT, STRUT DAMPENER SHAFT	26	INSULATOR, STABILIZER LINK
8	RATE WASHER	27	WASHER, STABILIZER LINK
9	STRUT MOUNT	28	BOLT, STABILIZER LINK
10	UPPER SPRING SEAT	29	CONTROL ARM
11	UPPER SPRING INSULATOR	30	BOLT
12	DUST TUBE ASSEMBLY	31	BUSHING, CONTROL ARM
13	SPRING	32	BOLT
14	LOWER SPRING INSULATOR	33	BOLT
15	STRUT	34	SUSPENSION SUPPORT
16	NUT	35	NUT
17	WASHER	36	WASHER
18	BOLT	37	BOLT
19	HUB AND BEARING ASSEMBLY	38	SPACER, STABILIZER LINK

79222332

Exploded view of the front suspension

degrees (⅙ turn) additional to align the cotter pin holes. NEVER loosen the nut to make the holes align. Install a new cotter pin.

14. Connect the brake line bracket to the strut and tighten the bolt to 10 ft. lbs. (14 Nm).

15. Install the tire and wheel assembly. Lower the vehicle.

16. Tighten the upper strut plate mounting nuts and bolt to 18 ft. lbs. (25 Nm).

17. Check the front end alignment and adjust as necessary.

Shock Absorber

REMOVAL & INSTALLATION

Rear

❊❊ WARNING

When doing this procedure, if both shocks are to be replaced, only remove one shock at a time or dam-

age can occur to the axle assembly and/or personal injury.

1. Raise and safely support the vehicle with the wheel about 6 in. off the ground. Support the rear axle assembly with a safety stand or hydraulic floor jack.

2. From inside the trunk, remove the cap from the upper shock mounting plate. Remove the two shock plate mounting nuts and remove the reinforcement from the shock tower.

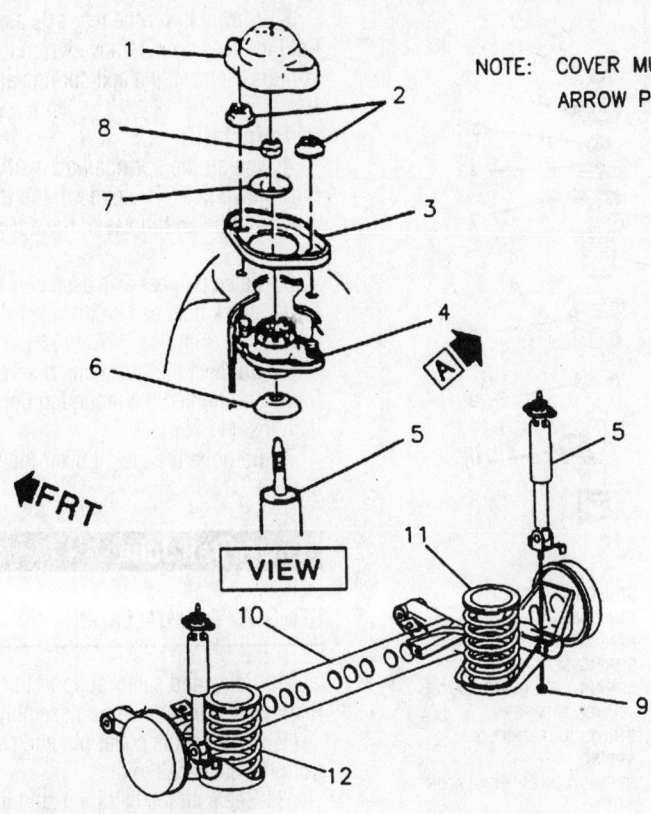

NOTE: COVER MUST BE INSTALLED SO THAT
ARROW POINTS TO LEFT SIDE OF VEHICLE

1 COVER (MUST BE INSTALLED
   WITH ARROW POINTING TO
   LEFT OF VEHICLE)
2 NUT
3 REINFORCEMENT
4 UPPER SHOCK ABSORBER
   MOUNT
5 SHOCK ABSORBER
6 WASHER
7 WASHER
8 NUT
9 NUT
10 REAR AXLE; SPRING-ON-CENTER TYPE
11 INSULATOR, COIL SPRING
12 COIL SPRING

Exploded view of the rear suspension

3. From under the suspension, remove the lower shock mounting nut. Compress the shock and remove it from the vehicle.

4. If the shock is being replaced, remove the upper shock insulator, washer and mount.

To install:

5. Install the mount, washer and insulator on the shock and tighten the mounting nut to 21 ft. lbs. (29 Nm).

6. Install the shock assembly into the vehicle. Be sure the alignment tab on the lower mount is pointing toward the rear of the vehicle.

7. Install the mounting nut and tighten to 35 ft. lbs. (47 Nm).

8. Install the reinforcement and upper shock plate mounting nuts and tighten to 18 ft. lbs. (24 Nm). Install the cap over the mounting plate.

9. Remove the safety stand or hydraulic floor jack from under the axle. Lower the vehicle.

Coil Spring

REMOVAL & INSTALLATION

Front

1. Remove the strut assembly from the vehicle, as outlined earlier in this section.

2. Mount the strut compressor J-34013 in holding fixture J-3289-20.

3. Mount the strut assembly into the compressor. Note that the strut compressor has strut mounting holes drilled for specific vehicle lines.

4. Compress the strut approximately ½ its height after initial contact with the top cap.

❋❋ WARNING

Never bottom the spring or dampener rod!

5. Remove the nut from the strut dampener shaft and place alignment/guiding rod J-34013-27 on top of the dampener shaft. Use the rod to guide the dampener shaft straight down through the spring cap while compressing the spring. Remove the components.

➡**Be careful to avoid chipping or cracking the spring coating when handling the front suspension coil spring assembly.**

To install:

6. Install the bearing cap into the strut compressor if previously removed.

7. Mount the strut assembly in strut compressor, using bottom locking pin only. Extend the dampener shaft and install clamp J-34013-20 on the dampener shaft.

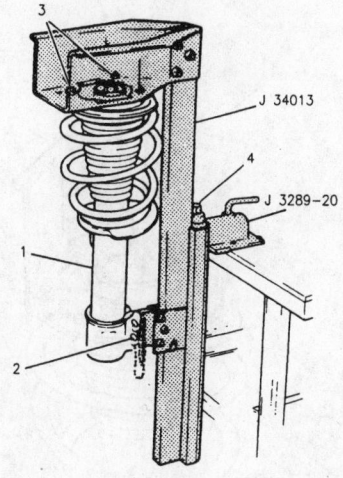

1 STRUT ASSEMBLY
2 INSTALL LOCKING PINS THROUGH
   STRUT ASSEMBLY
3 TIGHTEN NUTS UNTIL FLUSH WITH
   STRUT COMPRESSOR
4 COMPRESSOR FORCING SCREW

View of the front strut assembly mounted in a compressor

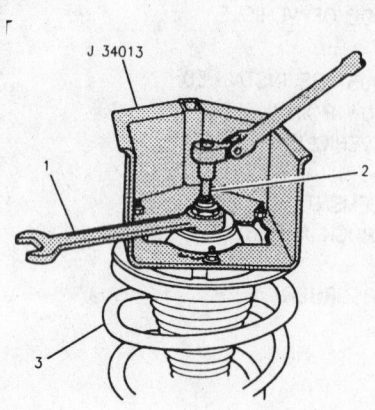

J 34013

1 WRENCH
2 SOCKET
3 STRUT ASSEMBLY

7922Z335

Use a socket and a wrench to remove the dampener shaft nut spring cap while compressing the spring

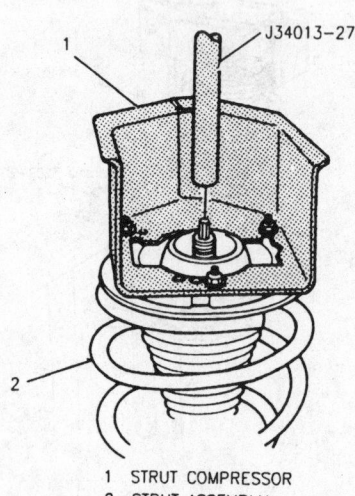

1 STRUT COMPRESSOR
2 STRUT ASSEMBLY

7922Z336

Install the rod to guide the dampener shaft straight down through the spring cap while compressing the spring

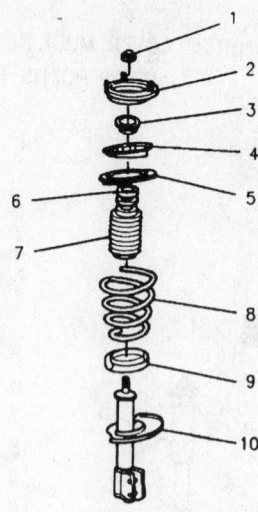

1 STRUT MOUNT NUT
2 STRUT MOUNT
3 RATE WASHER
4 SPRING SEAT
5 SPRING UPPER INSULATOR
6 JOUNCE BUMPER
7 STRUT DUST SHIELD
8 SPRING
9 SPRING LOWER INSULATOR
10 STRUT

7922Z337

Exploded view of the front strut assembly

8. Install the spring over the dampener and swing the assembly up so the upper locking pin can be installed.

9. Install all shields, bumpers and insulators on the spring seat. Install the spring seat on top of the spring. Be sure the flat on the upper spring seat is facing in the proper direction. The spring seat flat should be facing the same direction as the centerline of the strut assembly spindle.

10. Install the guiding rod and turn the forcing screw while the guiding rod centers the assembly. When the threads on the dampener shaft are visible, remove the guiding rod and install the nut. Tighten the nut to 52 ft. lbs. (70 Nm). Use a crow's foot line wrench while holding the dampener shaft with a socket.

11. Remove the clamp.

Rear

1. Raise and safely support the vehicle. Support the rear axle with a suitable screw type jack.

2. Remove the lower shock absorber mounting nuts. Remove the right and left brake line mounting bolts from the floor of the vehicle.

3. Slowly lower the rear axle assembly until the tension is removed from the coil springs. Remove the coil springs and insulators.

To install:

4. Install the springs and insulator into the vehicle. The ends of the lowest coil must be within 9/16 of the spring stop.

5. Raise the rear axle assembly until the lower shock nuts can be installed. Tighten the nuts to 35 ft. lbs. (47 Nm).

6. Connect the brake line brackets to the frame and tighten the mounting bolts to 97 inch lbs. (11 Nm).

7. Remove the jack. Lower the vehicle.

Lower Ball Joint

REMOVAL & INSTALLATION

1. Raise and safely support the vehicle. Remove the tire and wheel assembly.

2. Remove the cotter pin and castle nut from the outer ball joint.

3. Separate the ball joint stud from the steering knuckle using J-29330, or an equivalent ball joint separator.

4. Drill a 1/8 in. pilot hole in the center of each of the three ball joint mounting rivets.

5. Using a 1/2 in. drill bit, drill the heads off the rivets.

6. With a hammer and punch, knock the rivets out of the control arm. Remove the sway bar link kit.

7. Pull the control arm down so the ball stud clears the steering knuckle. Slide the ball joint out of the control arm.

To install:

8. Position the ball joint into the lower control arm and install the bolts. The nuts must be on top of the control arm. Tighten the mounting bolts to the specification provided with the ball joint service kit.

9. Connect the ball joint to the steering knuckle and tighten the castle nut to 41 ft. lbs. (55 Nm). If necessary to align the cotter pin holes tighten the nut up to 60 degrees (1/6 turn) additional rotation. NEVER loosen the nut to make the holes align. Install a new cotter pin.

10. Install the link kit and tighten the nut to 22 ft. lbs. (30 Nm).

11. Install the tire and wheel assembly and tighten the wheel nuts to 100 ft. lbs. (140 Nm).

12. Lower the vehicle.

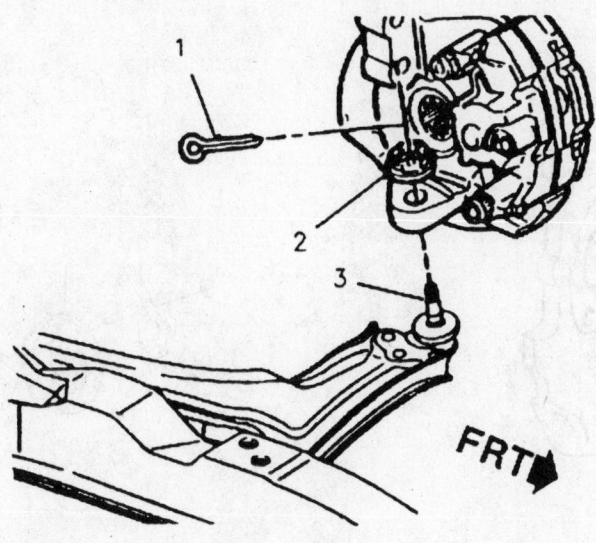

1   PIN
2   NUT – 55 N·m (41 LBS. FT.) MINIMUM TORQUE
    65 N·m (48 LBS. FT.) MAXIMUM TORQUE TO
    INSTALL PIN
3   LOWER BALL JOINT

7922Z339

Exploded view of the ball joint-to-knuckle mounting

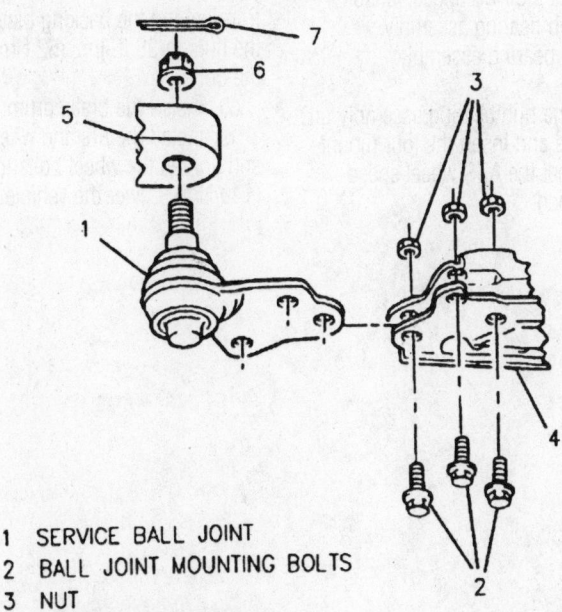

1   SERVICE BALL JOINT
2   BALL JOINT MOUNTING BOLTS
3   NUT
4   LOWER CONTROL ARM
5   STEERING KNUCKLE
6   NUT – 55 N·m (41 LBS. FT.) MINIMUM TORQUE
    65 N·m (48 LBS. FT.) MAXIMUM TORQUE
    TO INSTALL PIN
7   PIN

7922Z338

Exploded view of the replacement ball joint mounting

Wheel Bearings

ADJUSTMENT

These vehicles are equipped with sealed hub and bearing assemblies. The hub and bearing assemblies are non-serviceable. If the assembly is damaged, the complete unit must be replaced.

REMOVAL & INSTALLATION

Front

1. Raise and safely support the vehicle. Remove the front tire and wheel assembly.
2. Insert a drift punch through the caliper and into the rotor cooling fins. This keeps the assembly from turning while the axle nut is being loosened.
3. Remove the axle nut and washer. Remove the punch.
4. Remove the caliper mounting bolts and remove the caliper from the steering knuckle. Support the caliper aside. DO NOT allow the caliper to hang unsupported from the brake hose.
5. Remove the brake rotor.
6. Remove the three bolts securing the hub bearing to the steering knuckle. Remove the backing plate.
7. Using J-28733-A, or an equivalent front hub remover, separate the halfshaft from the hub bearing assembly.
8. Remove the hub bearing assembly.

   To install:
9. Install the hub bearing assembly over the end of the axle making sure the splines engage smoothly.
10. Install the backing plate and the three hub bearing mounting bolts. Tighten the bolts to 70 ft. lbs. (95 Nm).
11. Install the brake rotor.
12. Install the caliper onto the steering knuckle and tighten the caliper mounting bolts to 38 ft. lbs. (51 Nm).
13. Install a drift punch into the rotor cooling fins. This keeps the assembly from turning while the axle nut is being tightened. Tighten the axle nut to 192 ft. lbs. (260 Nm).
14. Install the front tire and wheel assembly and tighten the wheel nuts to 100 ft. lbs. (140 Nm). Lower the vehicle.

Rear

1. Raise and safely support the vehicle. Remove the wheel and tire assembly.
2. Remove the brake drum.

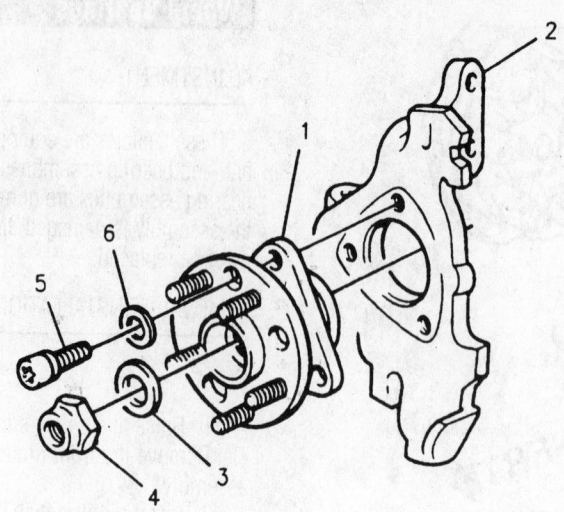

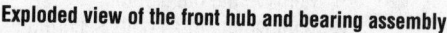

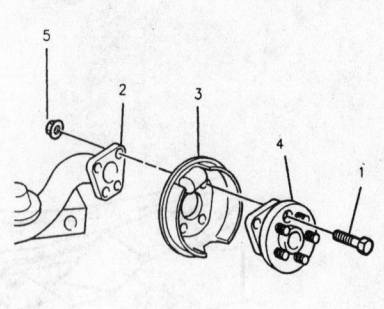

1   HUB AND BEARING ASSEMBLY
2   STEERING KNUCKLE
3   WASHER
4   DRIVE AXLE NUT — 260 N·m (192 LBS. FT.)
5   HUB AND BEARING RETAINING BOLT
6   WASHER

7922Z340

Exploded view of the front hub and bearing assembly

1   BOLT
2   REAR AXLE ASSEMBLY
3   BACKING PLATE
4   HUB AND BEARING ASSEMBLY
5   LOCKNUT

7922Z341

Hub and bearing components

3. Remove the four hub bearing assembly mounting nuts from behind the backing plate while holding the bolts from the front. The mounting bolts can be removed from the backing plate once the nuts have been removed. The top rear bolt will not clear the brake shoes and must be removed with the bearing assembly.

4. Disconnect the ABS speed sensor wire from the hub bearing assembly. Remove the hub bearing assembly.

To install:

5. Position the hub bearing assembly on the backing plate and insert the four mounting bolts. Connect the ABS wheel speed sensor, if equipped.

6. Install the hub bearing mounting nuts from behind the backing plate and tighten the nuts to 38 ft. lbs. (52 Nm) while holding the bolts.

7. Install the brake drum.

8. Install the tire and wheel assembly and tighten the wheel bolts to 100 ft. lbs. (140 Nm). Lower the vehicle.

GENERAL MOTORS V BODY 37

Cadillac-Catera

DRIVE TRAIN 37-9
ENGINE REPAIR 37-2
FUEL SYSTEM 37-8
STEERING AND
  SUSPENSION 37-10

A
Air Bag..............................37-10
  Disarming.......................37-10
  Precautions 37-10
  Rearming.......................37-10

C
Camshaft 37-6
  Removal & Installation ...37-6
Coil Spring 37-12
  Removal & Installation ...37-12
Cylinder Head 37-3
  Removal & Installation ...37-3

E
Engine Assembly 37-2
  Removal & Installation ...37-2
Exhaust Manifold.............37-5
  Removal & Installation ...37-5

F
Fuel Filter 37-8
  Removal & Installation ...37-8
Fuel Pump 37-8
  Removal & Installation ...37-8
Fuel System Pressure 37-8
  Relieving 37-8
Fuel System Service
  Precautions..................37-8

H
Halfshaft.........................37-9
  Removal & Installation ...37-9

I
Ignition Timing 37-2
  Adjustment 37-2
Intake Manifold Assembly.............37-4
  Removal & Installation ...37-4

L
Lower Ball Joint...............37-13
  Removal & Installation ...37-13

O
Oil Pan............................37-7
  Removal & Installation ...37-7

Oil Pump 37-7

R
Rear Main Seal 37-7
  Removal & Installation ...37-7
Recirculating Ball Power Steering
  Gear............................37-10
  Removal & Installation ...37-10

S
Shock Absorber 37-12
  Removal & Installation ...37-12
Strut................................37-11
  Removal & Installation ...37-11

T
Transaxle Assembly 37-9
  Removal & Installation ...37-9

V
Valve Lash 37-7

W
Water Pump 37-3
  Removal & Installation ...37-3
Wheel Bearings...............37-13
  Adjustment 37-13
  Removal & Installation ...37-14

ENGINE REPAIR

➡**Disconnecting the negative battery cable on some vehicles may interfere with the functions of the on board computer systems and may require the computer to undergo a relearning process, once the negative battery cable is reconnected.**

Ignition Timing

ADJUSTMENT

➡**The 3.0L DOHC engine used in the Catera utilizes a Distributorless Ignition System (DIS). The ignition timing is determined by inputs from the knock, oxygen and coolant temperature sensors. No ignition timing adjustment is possible.**

Engine Assembly

REMOVAL & INSTALLATION

1. Disconnect the negative battery cable.

❈❈ CAUTION

Never open, service or drain the radiator or cooling system when hot; serious burns can occur from the steam and hot coolant. Always drain coolant into a sealable container. Coolant should be reused unless it is contaminated or is several years old.

2. Drain the engine coolant into a suitable container.
3. Remove the left and right air inlet grilles.
4. Matchmark the relationship of the hinges to the hood, then remove the hood.
5. Remove the battery.
6. Properly discharge and recover the A/C refrigerant using an approved recycling station.
7. Detach the black wiring harness connector.
8. Tag and remove the ground wire from the negative battery cable.
9. Remove the four power supply wires from the positive battery cable.
10. Siphon the power steering fluid from the reservoir, then disconnect the hoses from the pump.
11. Remove the threaded fitting for the power brake booster vacuum pipe from the intake plenum.

12. Tag the hoses and connectors, then remove the switch-over valve from the intake plenum.
13. Remove the ECM.
14. Remove the relays, then remove the wiring harness from the electrical center box.
15. Detach the blue and white wiring harness connectors.
16. Disconnect the accelerator and cruise control cables from the throttle body.

❈❈ CAUTION

Observe all applicable safety precautions when working around fuel. Whenever servicing the fuel system, always work in a well ventilated area. Do not allow fuel spray or vapors to come in contact with a spark or open flame. Keep a dry chemical fire extinguisher near the work area. Always keep fuel in a container specifically designed for fuel storage; also, always properly seal fuel containers to avoid the possibility of fire or explosion.

17. Properly relieve the fuel system pressure.
18. Disconnect the fuel lines from the fuel rail. Be sure to use a back-up wrench to prevent damage to the fuel rail.
19. Disconnect the following hoses:
    • Coolant hose from the throttle body
    • Vacuum hose from the purge valve on the engine ventilation chamber
    • Vacuum hose from the heater control valve
    • Coolant reservoir hose from the coolant inlet pipe
    • Coolant hoses from the heater core
    • Heater hose quick-disconnects
20. Remove the resonance chamber and the radiator.
21. Remove the A/C hose bracket and position it out of the way.
22. Disconnect the refrigerant line from the A/C compressor. Be sure to cap all open lines to prevent moisture from entering the system.
23. Attach an engine support fixture to support the engine. The two silicone engine mounts are not strong enough to support the engine after the transmission has been removed.
24. Raise and safely support the vehicle.
25. Remove the splash shield.
26. Detach the A/C compressor electrical connector.
27. Remove the transmission from the vehicle.

28. Lower the vehicle and attach a hoist to the engine assembly. Attach the hoist to the three engine lifting shackles. Raise the engine slightly and remove the engine support fixture.
29. Remove the engine mount nuts and lift the engine out of the vehicle. Raise the engine slowly after being certain all wiring, cables and hoses have been removed.

To install:

30. Position the engine in the vehicle. Install the engine mount nuts and tighten them to 41 ft. lbs. (55 Nm).
31. Install the engine support fixture and remove the chain hoist.
32. Raise and safely support the vehicle.
33. Install the transmission assembly.
34. Connect the wiring to the A/C compressor.
35. Install the splash shield. Tighten the bolts until they are fully seated. Be careful not to strip them.
36. Lower the vehicle and remove the engine support fixture.
37. Connect the A/C compressor and install the hose bracket. Tighten the bracket bolts to 71 inch lbs. (8 Nm).
38. Install the radiator and resonance chamber.
39. Connect all coolant, vacuum and fuel lines. Tighten the fuel supply and return lines to 11 ft. lbs. (15 Nm).
40. Connect the accelerator and cruise control cables to the throttle body.
41. Attach the blue and white wiring harness connectors.
42. Install the wiring harness to the electrical center box, then install the relays.
43. Install the ECM.
44. Install the switch-over valve on the intake plenum. Tighten the bolts to 71 inch lbs. (8 Nm).
45. Connect the hoses to the power steering pump. Tighten the discharge hose to 21 ft. lbs. (28 Nm).
46. Install the power brake booster threaded fitting in the intake plenum just until seated. Be certain not to strip the threads by over-tightening.
47. Connect the four power supply wires to the positive battery cable and the ground wire to the negative battery cable.
48. Attach the black wiring harness connector.
49. Install the battery and the hood.
50. Install the air inlet grills and the wiper arms.
51. Reprogram all applicable accessories.
52. Fill the bleed the cooling system. Use only GM DEX-COOL® or equivalent

coolant. When refilling the cooling system, add two crushed engine coolant supplement sealant pellets PN 3634621 or equivalent into the coolant reservoir.

53. Fill the power steering reservoir with Dexron III® automatic transmission fluid and bleed the system.

54. Recharge the A/C system and check for leaks.

Water Pump

REMOVAL & INSTALLATION

1. Disconnect the negative battery cable.

❊❊ CAUTION

Never open, service or drain the radiator or cooling system when hot; serious burns can occur from the steam and hot coolant. Always drain coolant into a sealable container. Coolant should be reused unless it is contaminated or is several years old.

2. Drain the coolant into a suitable container.

3. Remove the resonance chamber.

4. Remove the front timing belt cover.

5. Remove the water pump.

To install:

6. Clean the water pump mounting surface on the engine block.

7. Apply silicone grease to the water pump O-ring and install the water pump on the engine block. Tighten the mounting bolts to 18 ft. lbs. (21 Nm).

8. Install the front timing belt cover.

9. Install the resonance chamber.

10. Connect the negative battery cable.

11. Refill the cooling system through the reservoir tank. The cooling system will bleed automatically during warm-up. Use only GM DEX-COOL® or equivalent coolant. When refilling the cooling system, add two crushed engine coolant supplement sealant pellets PN 3634621 or equivalent into the coolant reservoir.

12. Check the coolant level in the reservoir after the engine has cooled and add coolant as needed.

Cylinder Head

REMOVAL & INSTALLATION

1. Disconnect the negative battery cable.

❊❊ CAUTION

Never open, service or drain the radiator or cooling system when hot; serious burns can occur from the steam and hot coolant. Always drain coolant into a sealable container. Coolant should be reused unless it is contaminated or is several years old.

2. Drain the engine coolant and relieve the fuel system pressure.

3. Remove the intake plenum.

4. Remove the resonance chamber.

5. Remove the intake manifold and spacer.

6. Remove the coolant bridge.

7. Remove the left valve cover and front timing belt cover.

8. Position the crankshaft 60° Before Top Dead Center (BTDC) to avoid contact between the valves and the pistons when the cylinder head is installed.

9. Remove the timing belt.

10. Remove the timing belt tensioner bracket.

11. Install the camshaft gear holding tools Nos. J 42069–1 and J42069–2 on the

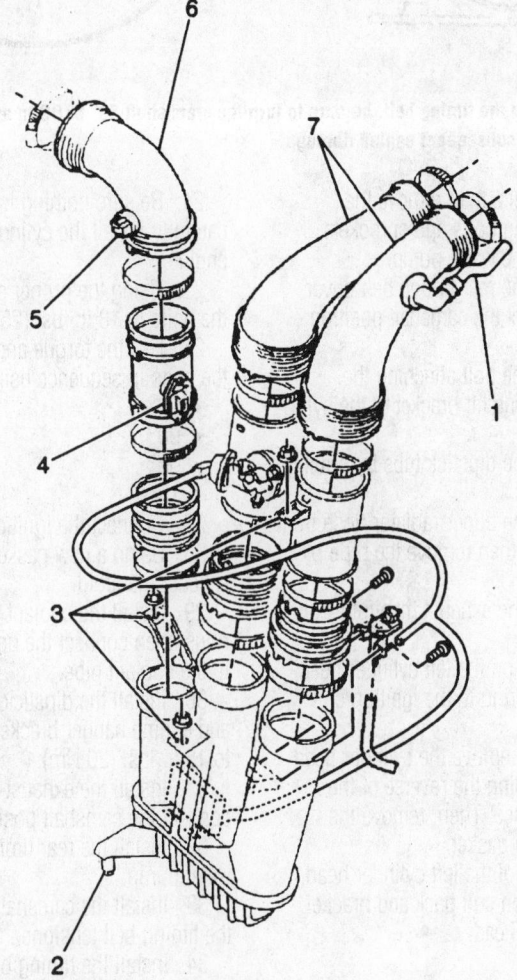

(1) Idle Air Control (IAC) Inlet Hose
(2) Resonance Chamber Guide Pin
(3) Resonance Chamber Nut
(4) Mass Air Flow (MAF) Sensor
(5) Intake Air Temperature (IAT) Sensor
(6) Resonance Chamber Air Intake Hose
(7) Intake Plenum Air Inlet Hose

79227406

Resonance chamber and related components

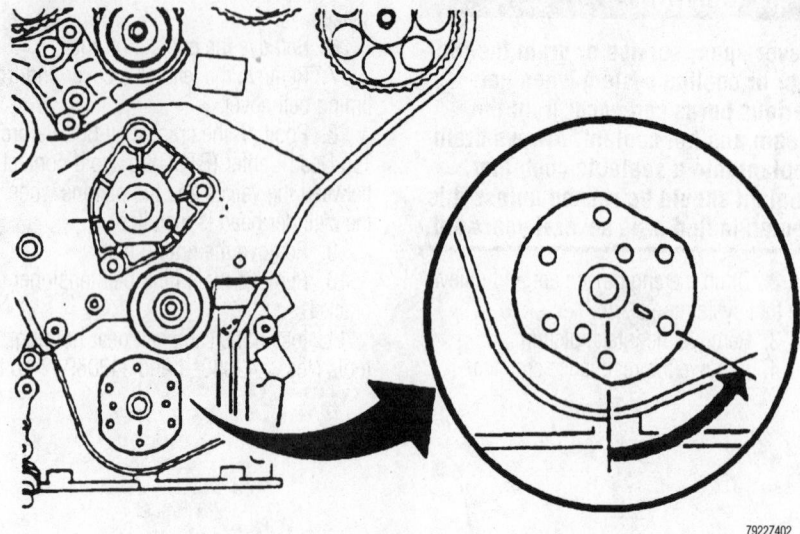

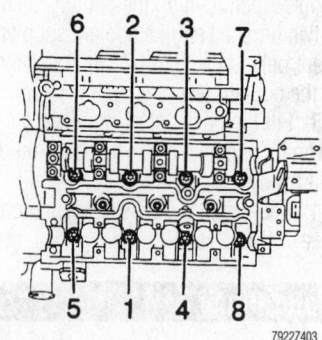

Tighten the cylinder head bolts according to the sequence shown

Before removing the timing belt, be sure to turn the crankshaft 60° BTDC to avoid valve-to-piston contact and subsequent engine damage

camshaft sprockets, then remove the sprocket mounting bolts and sprockets.

12. Remove the water pump.

13. Remove the rear timing belt cover.

14. Disconnect the camshaft position sensor.

15. Remove the bolt attaching the coolant pipe/engine lift bracket to the cylinder head.

16. Remove the dipstick tube by pulling it up firmly.

17. Remove the upper radiator hose from the coolant pipe, then remove the pipe by twisting it.

18. Separate the exhaust manifold from the cylinder head.

19. If working on the left cylinder head, disconnect the wiring to the ignition coil pack.

20. Gradually remove the cylinder head bolts in stages using the reverse of the tightening sequence. Then, remove the cylinder head and gasket.

21. If working of the left cylinder head, remove the ignition coil pack and bracket from the cylinder head.

To install:

22. Clean the surface of the engine block and the bottom of the cylinder head.

23. Place a new gasket on the engine with "OBEN/TOP" facing up and towards the front for the left side or towards the rear for the right side of the engine.

➡**Be sure to use new cylinder head bolts when installing the cylinder head. The old bolts have been stretched by the method used to install them and are not reusable.**

24. Be sure nothing is in the way and carefully install the cylinder head on the engine.

25. Using the proper sequence, tighten the bolts to 18 ft. lbs. (25 Nm).

26. Use the torque angle meter, tighten the bolts in sequence using four stages:

- 90°
- 90°
- 90°
- 15°

27. Connect the ignition coil pack.

28. Using a new gasket, install the exhaust manifold.

29. Install the coolant pipe with new O-rings, then connect the upper radiator hose to the coolant pipe.

30. Install the dipstick tube, coolant pipe and engine hanger bracket. Tighten the bolt to 15 ft. lbs. (20 Nm).

31. Install the exhaust camshaft, then connect the camshaft position sensor.

32. Install the rear timing belt cover and water pump.

33. Install the camshaft sprockets and the timing belt tensioner.

34. Install the timing belt and cover.

35. Using new O-rings and gaskets, install the valve cover. Tighten the bolts to 71 inch lbs. (8 Nm).

36. Install the coolant bridge, intake manifold spacer and the intake manifold.

37. Install the resonance chamber and intake plenum.

38. Connect the negative battery cable.

39. Drain and refill the engine oil, then replace the filter.

40. Refill the cooling system.

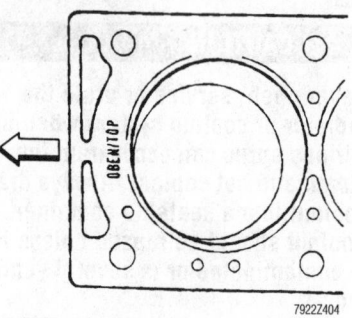

Proper head gasket installation position- left cylinder head

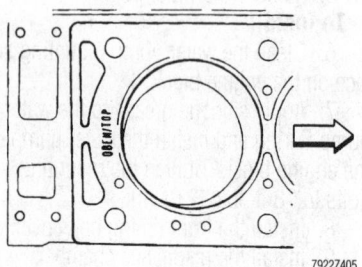

Proper head gasket installation position- right cylinder head

41. Inspect the engine compartment to be sure everything has been assembled.

42. Start the engine and check for leaks.

Intake Manifold Assembly

REMOVAL & INSTALLATION

Intake Plenum

1. Disconnect the negative battery cable.

2. Remove the air intake ducts and the Idle Air Control (IAC) hose from the throttle body assembly.

3. Disconnect the accelerator and cruise control cables from the throttle body, then remove the cables with the bracket.

4. Disconnect the Throttle Position Sensor (TPS) wiring.

5. Remove the threaded brake booster vacuum hose connection from the intake plenum.

6. Remove the wiper arms, them remove the air inlet grilles

7. Remove the wiring harness (channel) from the intake plenum.

8. Disconnect the wiring and vacuum hose from the switch-over valve solenoid.

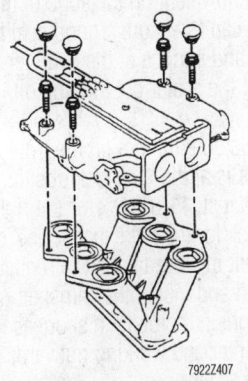

Intake plenum mounting bolt locations

9. Disconnect the vacuum hose from the fuel pressure regulator.

10. Remove the throttle body.

11. Remove the caps and bolts attaching the intake plenum to the intake manifold, then remove the plenum. and O-rings.

12. Place masking tape over the openings on the intake manifold to prevent foreign objects from falling into the engine.

To install:

13. Remove the masking tape and position new O-rings on the intake manifold.

14. Be sure the guide pins are aligned and install the intake plenum. Tighten the bolts to 71 inch lbs. (8 Nm). Install the caps on the bolt heads.

15. Install the throttle body. Tighten the bolts to 71 inch lbs. (8 Nm).

16. Connect the vacuum hose to the fuel pressure regulator and the switch-over valve solenoid.

17. Attach the electrical connector to the switch-over valve solenoid.

18. Position the wiring harness on the plenum and install the two bolts. Tighten the bolts to 71 inch lbs. (8 Nm).

19. Connect the vacuum hoses and crankcase vent hose to the intake plenum.

20. Install the air inlet grilles and the wiper arms.

21. Connect the brake booster hose. Tighten the fitting just until seated.

22. Connect the TPS wiring.

23. Install the accelerator and cruise control cables with bracket. Tighten the bracket bolts to 71 inch lbs. (8 Nm).

24. Install the air inlet ducts and the IAC hose.

25. Connect the negative battery cable.

26. Start the engine and check for proper performance.

Intake Manifold

➥**The fuel injector connectors are numbered. Be sure to attach the correct connector to the proper fuel injector when installing the intake manifold.**

1. Disconnect the negative battery cable.

2. Remove the intake plenum.

3. Properly relieve the fuel system pressure.

4. Disconnect the fuel lines from the fuel delivery rail assembly.

5. Detach the connectors from the fuel injectors.

6. Remove the intake manifold mounting bolts and the intake manifold.

7. Place masking tape over the openings on the intake spacer to prevent foreign objects from falling into the engine.

To install:

8. Remove the masking tape and position new gaskets (if equipped) on the intake spacer.

9. Install the intake manifold. tighten the bolts to 15 ft. lbs. (20 Nm).

10. Connect the fuel injectors and the fuel lines. Tighten the fuel lines to 11 ft. lbs. (15 Nm).

11. Install the intake plenum.

12. Connect the negative battery cable.

Exhaust Manifold

REMOVAL & INSTALLATION

Left Side

1. Remove the engine assembly.

2. Remove the heat shields from the left exhaust manifold.

3. Remove the AIR injection pipe from the manifold to allow access to the exhaust manifold mounting bolts.

4. Remove the dipstick tube.

5. Remove the exhaust manifold mounting nuts and the manifold with gasket.

6. Clean the cylinder head and exhaust manifold mating surfaces.

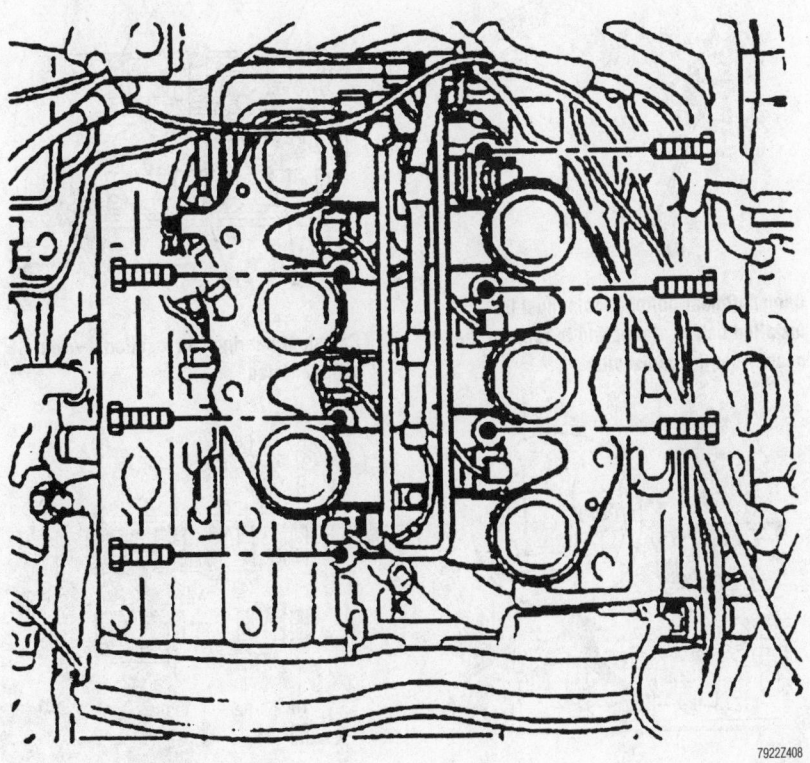

Intake manifold mounting bolt locations

To install:

7. Install the exhaust manifold. Tighten the nuts to 15 ft. lbs. (20 Nm).

8. Install the dipstick tube. Tighten the bolt to 15 ft. lbs. (20 Nm).

9. Coat the AIR injection pipe mounting bolts with high temperature anti-seize compound, then install the pipe to the manifold. Tighten the bolts to 15 ft. lbs. (20 Nm).

10. Coat the heat shield mounting bolts with high temperature anti-seize compound and install the heat shield on the manifold. Tighten the bolts to 71 inch lbs. (8 Nm).

11. Install the engine assembly in the vehicle.

Right Side

1. Remove the transmission assembly.

2. Disconnect the radiator hose, then remove the coolant intake pipe from the right side of the engine.

3. Raise and safely support the vehicle.

4. Remove the exhaust manifold lower heat shield and the catalytic converter hanger bolt.

5. Remove the two lower rear exhaust manifold mounting nuts.

6. Lower the vehicle to the floor.

7. Remove the two lower front exhaust manifold mounting nuts.

8. Remove the bolts attaching the upper heat shield to the manifold and leave the shield in place.

9. Remove the AIR injection pipe from the manifold to allow access to the exhaust manifold mounting bolts.

10. Remove the upper nuts from the exhaust manifold and remove the manifold.

11. Clean the cylinder head and exhaust manifold mating surfaces.

To install:

12. Install the exhaust manifold with the three upper nuts. Tighten the nuts to 15 ft. lbs. (20 Nm).

13. Coat the AIR injection pipe mounting bolts with high temperature anti-seize compound, then install the pipe to the manifold. Tighten the bolts to 15 ft. lbs. (20 Nm).

14. Coat the upper heat shield mounting bolts with high temperature anti-seize compound and install the upper heat shield on the manifold. Tighten the bolts to 71 inch lbs. (8 Nm).

15. Install the two lower front exhaust manifold mounting nuts. Tighten the nuts to 15 ft. lbs. (20 Nm).

16. Raise and safely support the vehicle.

17. Install the two lower rear exhaust manifold mounting bolts. Tighten the nuts to 15 ft. lbs. (20 Nm).

18. Install the catalytic converter hanger

bolt. Tighten the bolt to 15 ft. lbs. (20 Nm).

19. Install the lower exhaust manifold lower heat shield. Tighten the bolts to 71 inch lbs. (8 Nm).

20. Lower the vehicle to the floor and install the coolant intake pipe.

21. Install the transmission assembly.

Camshaft

REMOVAL & INSTALLATION

1. Disconnect the negative battery cable.

2. Remove the intake plenum.

3. Remove the resonance chamber.

4. Remove the valve cover.

5. Remove the front timing belt cover.

6. Position the crankshaft 60° Before Top Dead Center (BTDC) to avoid contact between the valves and the pistons when the camshaft is installed.

7. Remove the timing belt.

8. Install a tool, such as J 42069–1, to hold the camshaft gear. Be sure the camshaft is not under load from the lifters and remove the camshaft gear.

9. Gradually loosen the camshaft bearing caps bolt sequentially, starting in the

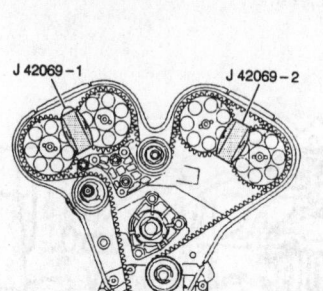

Camshaft gear holding tools must be installed before attempting to remove the gears from the camshafts

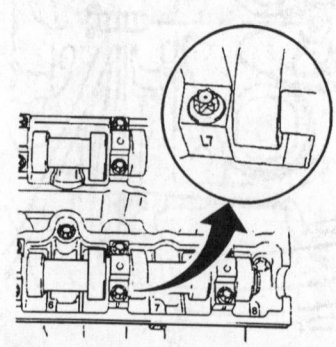

The camshaft bearing caps have identification marks stamped on the side

center and working outward in a spiral. Note the identification marks on the caps.

10. Remove the camshaft from the cylinder head.

To install:

11. Lubricate the camshaft lobes and lifters with PN 1234501 or equivalent. Lubricate the camshaft bearing journals with assembly fluid PN 1052367 or equivalent.

✳✳ WARNING

The bearing caps must be installed in their original positions.

12. Place a small amount of Locktite® 572 or equivalent on the edge of the front bearing cap to ensure a good seal between the cap and surface of the cylinder head. Don't let the sealer get into the oil journal of the cap.

13. Be sure the pin on the front of the camshaft is at the 1 o'clock position for the right exhaust; 11 o'clock for the right intake; 12 o'clock for the left exhaust or 7 o'clock for the left intake camshaft to reduce the lifter load and install the camshaft on the head. Tighten the bolts in sequence starting in the center and working outwards to 71 inch lbs. (8 Nm).

14. Coat the lip of a new camshaft seal

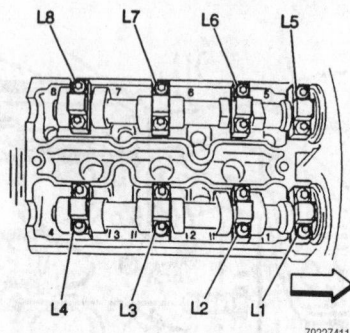

Camshaft bearing cap locations—right cylinder head

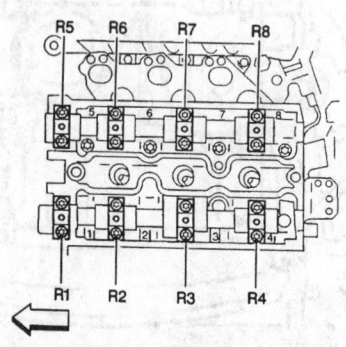

Camshaft bearing cap locations—left cylinder head

with chassis grease and install the seal using J 35268–A or equivalent seal driver.

15. Install the camshaft gear. Tighten the bolt to 37 ft. lbs. (50 Nm) + 60° + 15° using a torque angle meter.

16. Rotate the crankshaft clockwise back to TDC.

17. Install the timing belt.

18. Install the timing belt cover.

19. Install the valve cover.

20. Install the intake plenum.

21. Connect the negative battery cable.

Valve Lash

➡**Valve lash is automatically adjusted by the valve lifters. No adjustment is possible.**

Oil Pan

REMOVAL & INSTALLATION

1. Disconnect the negative battery cable.

2. Raise and safely support the vehicle.

3. Remove the splash shield.

✳✳ CAUTION

The EPA warns that prolonged contact with used engine oil may cause a number of skin disorders, including cancer! You should make every effort to minimize your exposure to used engine oil. Protective gloves should be worn when changing the oil. Wash your hands and any other exposed skin areas as soon as possible after exposure to used engine oil. Soap and water, or waterless hand cleaner should be used.

4. Drain the engine oil.

5. Disconnect the oil level sensor wiring and detach the connector from the C-clip.

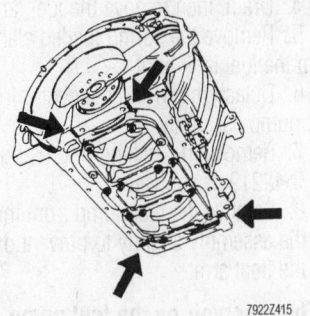

Be sure to apply sealant to the areas shown prior to installing the oil pan housing

6. Remove the oil pan bolts and the oil pan.

7. If necessary, remove the oil level sensor from the oil pan.

To install:

8. Clean the oil pan and housing sealing surfaces.

9. If removed, install the oil level sensor on the oil pan. Tighten the bolts just until seated.

10. Install a new O-ring on the oil level sensor connector.

11. Place the new gasket on the pan, then position the pan near the housing and attach the oil level sensor connector.

12. Apply sealant PN 12345997 or equivalent to the rear main bearing cap area and to the area where the oil pump and engine block meet.

13. Install the oil pan and tighten the bolts to 71 inch lbs. (8 Nm).

14. Install the splash shield.

15. Lower the vehicle to the floor.

✳✳ WARNING

Operating the engine without the proper amount and type of engine oil will result in severe engine damage.

16. Refill the engine with the correct amount of oil.

17. Connect the negative battery cable and reprogram the necessary accessories.

Oil Pump

1. Disconnect the negative battery cable.

✳✳ CAUTION

Never open, service or drain the radiator or cooling system when hot; serious burns can occur from the steam and hot coolant. Always drain coolant into a sealable container. Coolant should be reused unless it is contaminated or is several years old.

2. Drain the engine coolant.

3. Remove the resonance chamber.

4. Remove the timing belt cover and the timing belt.

5. Remove the rear timing belt cover.

6. Remove the crankshaft drive gear.

7. Remove the lower generator mounting bolt.

8. Remove the oil pan and housing.

9. Remove the oil pump mounting bolts and the oil pump.

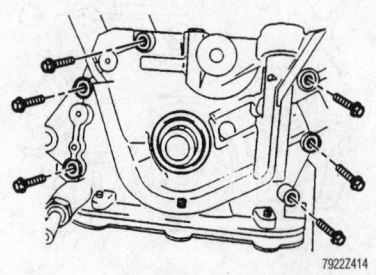

Oil pump mounting bolt locations

To install:

10. Clean the back of the oil pump and the engine block mounting surface. Be sure to remove all old gasket material.

11. Using a new gasket, install the oil pump. Tighten the bolts to 53 inch lbs. (6 Nm).

12. Apply sealant PN 12345997 or equivalent to the area where the oil pump and engine block meet and also to the rear main bearing cap area as shown.

13. Install the oil pan housing. Tighten the bolts to 11 ft. lbs. (15 Nm).

14. Install the lower generator mounting bolt. Tighten the bolt to 30 ft. lbs. (40 Nm).

15. Using a new bolt, install the crankshaft drive gear. Tighten the bolt to 184 ft. lbs. (250 Nm) + 45° + 15° using a torque angle meter. Use tool No. J 42065 to hold the crankshaft drive gear from turning.

16. Install the rear timing belt cover, timing belt and front timing belt cover.

17. Install the resonance chamber.

18. Connect the negative battery cable.

19. Refill the cooling system with the correct type and amount of coolant. Use DEX-COOL® or equivalent coolant only.

Rear Main Seal

REMOVAL & INSTALLATION

1. Disconnect the negative battery cable.

2. Remove the transmission and flex-plate.

3. Punch a small hole in the oil seal housing and install a sheet metal screw into the hole. Use pliers to remove the oil seal.

To install:

4. Coat the lip of the new seal with chassis grease.

5. Install the seal using Seal Installer No. J 42067 or an equivalent threaded seal installer to push the seal into position.

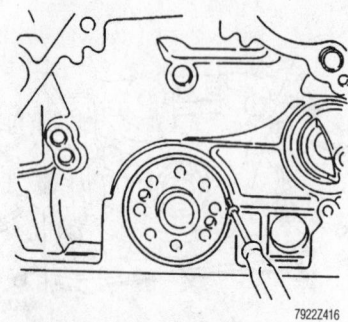

Thread a sheet metal screw into the metal part of the seal and remove the seal with pliers

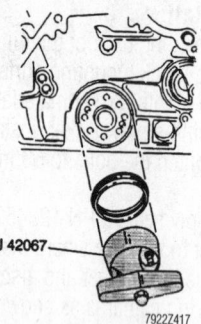

Press the rear main seal into position using a threaded seal installer like J 42067

6. Install the flexplate using new bolts. Tighten the bolts to 48 ft. lbs. (65 Nm) + 30° + 15° using a torque angle meter.
7. Install the transmission assembly.
8. Connect the negative battery cable.

FUEL SYSTEM

Fuel System Service Precautions

Safety is the most important factor when performing not only fuel system maintenance but any type of maintenance. Failure to conduct maintenance and repairs in a safe manner may result in serious personal injury or death. Maintenance and testing of the vehicle's fuel system components can be accomplished safely and effectively by adhering to the following rules and guidelines.

• To avoid the possibility of fire and personal injury, always disconnect the negative battery cable unless the repair or test procedure requires that battery voltage be applied.

• Always relieve the fuel system pressure prior to disconnecting any fuel system component (injector, fuel rail, pressure regulator, etc.), fitting or fuel line connection.

Exercise extreme caution whenever relieving fuel system pressure to avoid exposing skin, face and eyes to fuel spray. Please be advised that fuel under pressure may penetrate the skin or any part of the body that it contacts.

• Always place a shop towel or cloth around the fitting or connection prior to loosening to absorb any excess fuel due to spillage. Ensure that all fuel spillage (should it occur) is quickly removed from engine surfaces. Ensure that all fuel soaked cloths or towels are deposited into a suitable waste container.

• Always keep a dry chemical (Class B) fire extinguisher near the work area.

• Do not allow fuel spray or fuel vapors to come into contact with a spark or open flame.

• Always use a back-up wrench when loosening and tightening fuel line connection fittings. This will prevent unnecessary stress and torsion to fuel line piping. Always follow the proper torque specifications.

• Always replace worn fuel fitting O-rings with new. Do not substitute fuel hose or equivalent, where fuel pipe is installed.

Fuel System Pressure

RELIEVING

1. Disconnect the negative battery cable.
2. Loosen the fuel filler cap to relieve the tank pressure.
3. Wrap a shop towel around the fuel pressure valve fitting (located on the side or end of the fuel rail assembly) to catch any fuel spray and connect J-34730–1 or an equivalent fuel gauge.
4. Place the bleed hose into a suitable container, then open the valve to bleed the fuel system pressure.
5. Close the valve and disconnect the fuel gauge. Drain any remaining fuel from the gauge into the bleed container.

Fuel Filter

REMOVAL & INSTALLATION

➡**Keep a container and shop rag nearby in order to catch any spilled or leaking fuel.**

1. Remove the fuel tank cap to relieve the pressure in the tank.
2. Properly relieve the fuel system pressure.

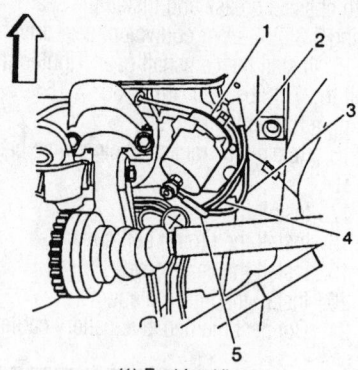

(1) Fuel feed line
(2) Fuel tank
(3) Fuel filter
(4) EVAP Vapor line
(5) Fuel Return line
(6) EVAP Vapor line
(7) Right rear axle shaft

Fuel line and filter identification

3. Hold the filter firmly and remove both fuel line attaching bolts.
4. Remove the filter from the bracket.
To install:
5. Clean the bolts and fittings on bolt fuel lines.
6. Install the lines to the filter. Be sure the filter is in the proper direction.
7. Install the fuel filter in the bracket.
8. Tighten the fuel tank cap.
9. Crank the engine for a few seconds, if the engine starts, turn it **OFF**.
10. Check for fuel leaks and repair if necessary.

Fuel Pump

REMOVAL & INSTALLATION

1. Disconnect the negative battery cable.
2. Properly relieve the fuel system pressure.
3. Raise and safely support the vehicle.
4. Drain, then remove the fuel tank.
5. Remove the spring loaded clamp from the fuel tank boot.
6. Detach the wiring connector from the fuel pump assembly.
7. Remove the locking ring using tool No. J 42219 or equivalent.
8. Remove the fuel pump from the tank. Tilt the assembly slightly to prevent damaging the float arm.

➡**The reservoir on the fuel pump assembly still contains fuel. Drain the fuel into a container after removing the pump.**

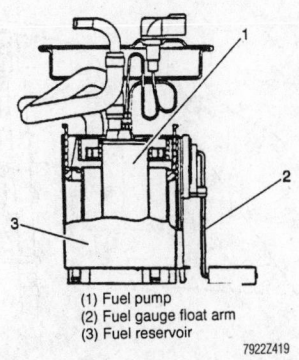

(1) Fuel pump
(2) Fuel gauge float arm
(3) Fuel reservoir

79222419

Fuel pump module components

9. Remove the seal from the fuel pump module. The seal must pass over the reservoir and over the float arm.

To install:

10. Lubricate the inside diameter of a new seal with engine oil and install it on the fuel pump module.

11. Install the pump module in the tank. Be sure to align the hole on the module with the mark on the fuel tank. Tighten the locking ring to 37 ft. lbs. (50 Nm).

12. Attach the wiring harness connector.

13. Install the spring loaded clamp on the fuel tank boot.

14. Install the fuel tank in the vehicle.

15. Connect the negative battery cable.

DRIVE TRAIN

Transaxle Assembly

REMOVAL & INSTALLATION

1. Disconnect the negative battery cable.

2. Raise and safely support the vehicle.

3. Disconnect the shift lever rod from the transmission.

4. Remove the three bolts attaching the driveshaft to the flange on the rear of the transmission. Use a flat bladed to pry the driveshaft away from the flange. Position the driveshaft out of the way.

5. Remove the two hole plugs from the oil pan and bell housing.

6. Matchmark the torque converter to the flexplate, then remove the flexplate-to-torque converter bolts.

7. Disconnect the fluid cooler hoses from the center pipes.

8. Disconnect the oxygen sensor wiring and remove the catalytic converters.

9. Remove the four transmission housing-to-oil pan bolts.

10. Position a transmission jack under the transmission.

11. Remove the two nuts securing the transmission mount to the center of the crossmember.

12. Raise the transmission slightly and remove the crossmember.

13. Lower the transmission in order to gain access to the remaining transmission-to-engine mounting bolts and electrical connectors.

14. Detach the electrical connectors and remove the vent hose from the transmission.

15. Support the engine with a block of wood and a jack.

16. Remove the remaining transmission-to-engine mounting bolts and remove the transmission from the vehicle.

To install:

❋❋ WARNING

The transmission fluid cooler and lines must be flushed out prior to installing a new or reconditioned transmission.

17. Clean the bolt holes in the engine block with a 12 x 1.75mm tap.

18. Apply thread locking compound to the transmission-to-engine mounting bolts.

19. Position the transmission on the rear of the engine block and install the mounting bolts. Tighten the bolts to 44 ft. lbs. (60 Nm).

20. Remove the jack from the engine.

21. Attach the electrical connectors and vent hose to the transmission.

22. Raise the transmission into position and install the crossmember. Tighten the crossmember-to-body bolts to 33 ft. lbs. (45 Nm). Lower the transmission onto the crossmember. Tighten the transmission mount-to-crossmember nuts to 15 ft. lbs. (20 Nm). Remove the transmission jack.

23. Install the catalytic converters and connect the oxygen sensor wiring.

24. Connect the fluid cooler hoses to the center pipes.

25. Align the matchmarks made earlier on the flexplate and torque converter. Install new flexplate-to-torque converter bolts. Be sure to remove the sealing compound from the firs three threads. Tighten the bolts to 22 ft. lbs. (30 Nm).

26. Install the two access hose plugs.

27. Install the driveshaft to the drive flange. Tighten the bolts to 70 ft. lbs. (95 Nm).

28. Connect the shift lever rod to the transmission. Adjust the rod as follows:

a. Place the shift control lever in the **PARK**

b. Loosen the adjusting bolt on the rod

c. Hold the selector lever on the transmission toward the rear stop to eliminate any free-play

d. Tighten the adjusting bolt to 71 inch lbs. (8 Nm)

29. Fill the transmission pan with DEXRON® III to the lower edge of the fill plug. Tighten the plug to 33 ft. lbs. (45 Nm).

30. Lower the vehicle to the floor.

31. Connect the negative battery cable.

32. Start the vehicle and check for leaks. Be sure the vehicle will only start in the **PARK** and **NEUTRAL** positions.

Halfshaft

REMOVAL & INSTALLATION

1. Place the gear selector in **NEUTRAL**.

2. Raise and safely support the vehicle.

3. Remove the rear wheel.

4. Install a tool such as Hub Holding tool No. J 42066 on the outer hub.

5. Remove the bolts attaching the halfshaft outer flange to the hub assembly, then pry the halfshaft away from the hub.

6. Separate the inner side of the halfshaft from the differential using tool No. J 42071 or equivalent. Once the initial connection is broken, the halfshaft can be removed by hand.

To install:

7. Lubricate the splined portion of the halfshaft with differential oil.

8. Install the halfshaft into the differential by tapping on the outer end of the shaft with a rubber/plastic hammer until the halfshaft is seated in the differential.

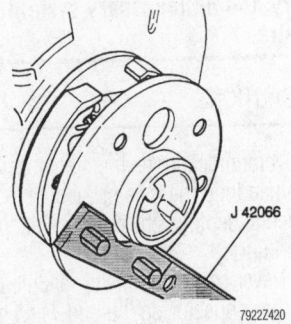

79222420

Prevent the outer hub from turning by installing a tool such as Hub Holding tool J 42066

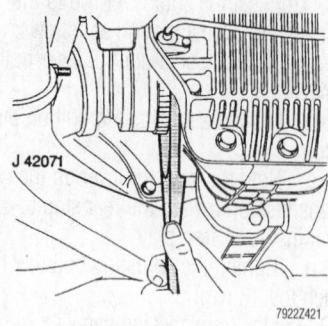

Pry the halfshaft out of the differential using tool J 42071 or equivalent

9. Install the outer end of the halfshaft on the hub assembly. Tighten the bolts to 37 ft. lbs. (50 Nm) + 67° using a torque angle meter while holding the hub stationary.

10. Install the rear wheel.
11. Lower the vehicle to the floor.

STEERING AND SUSPENSION

Air Bag

✳✳ CAUTION

These vehicles are equipped with an air bag system, also known as the Supplemental Inflatable Restraint (SIR) or Supplemental Restraint System (SRS). The system must be disabled before performing service on or around system components, steering column, instrument panel components, wiring and sensors. Failure to follow safety and disabling procedures could result in accidental air bag deployment, possible personal injury and unnecessary system repairs.

PRECAUTIONS

Several precautions must be observed when handling the inflator module to avoid accidental deployment and possible personal injury.

• Never carry the inflator module by the wires or connector on the underside of the module.

• When carrying a live inflator module, hold securely with both hands, and ensure that the bag and trim cover are pointed away.

• Place the inflator module on a bench or other surface with the bag and trim cover facing up.

• With the inflator module on the bench, never place anything on or close to the module which may be thrown in the event of an accidental deployment.

DISARMING

1. Turn the steering wheel to align the wheels in the straight-ahead position.
2. Turn the ignition switch to the **LOCK** position and remove the key.
3. Wait one minute until the capacitors in the Sensing and Diagnostic Module (SDM) discharge.
4. Disconnect the negative battery cable.

REARMING

1. Turn the ignition switch to the **LOCK** position and remove the key.
2. Reconnect the negative battery cable.
3. Wait at least one minute for the capacitors to recharge.
4. While staying away from the inflator modules, turn the key to the run position. The air bag warning lamp should turn on for 3–4 seconds, then turn off.

Recirculating Ball Power Steering Gear

REMOVAL & INSTALLATION

1. Center the front wheels, then turn the ignition key to the **LOCK** position.
2. Disconnect the negative battery cable.
3. Remove the windshield wiper assembly.

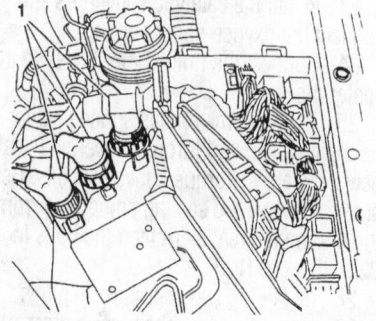

Body harness electrical connector locations (1)

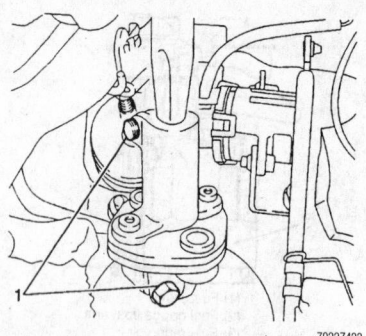

Matchmark the coupler to the steering gear before removing the coupler bolts (1)

4. Detach the three body harness electrical connectors.
5. Remove the ECM from electrical box.

✳✳ CAUTION

Never open, service or drain the radiator or cooling system when hot; serious burns can occur from the steam and hot coolant. Always drain coolant into a sealable container. Coolant should be reused unless it is contaminated or is several years old.

6. Drain the engine coolant.
7. Remove the radiator hose.

➡**Only properly MVAC-trained, EPA-certified technicians should service the A/C system or its components.**

8. Properly discharge and recover the A/C refrigerant using an approved recycling station.
9. Remove the A/C evaporator line extension bolt.
10. Siphon the power steering fluid from the reservoir, then remove the mounting bolt and position the reservoir out of the way.
11. Disconnect the brake booster vacuum hose from the plenum.
12. Siphon the brake fluid from the reservoir, then remove the brake lines from the master cylinder.
13. Disconnect the brake fluid level sensor from the master cylinder cap.
14. Remove the driver's side sound insulator panel.
15. Matchmark the steering coupler and input shaft of the steering gear.
16. Remove the coupler bolts, then spread the clamp and pull the steering shaft upward and away from the steering gear.
17. Remove the brake booster pushrod from the brake pedal.
18. Remove the knee bolster from the lower instrument panel.
19. Remove the fuse and relay panel

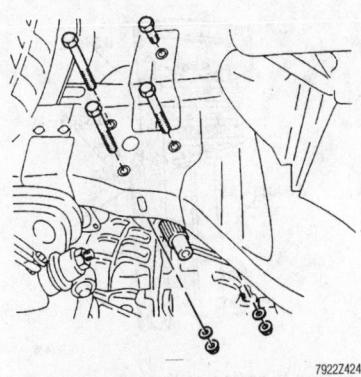

Steering gear mounting bolt locations

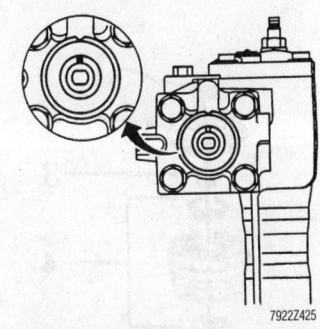

Align the mark on the stub shaft with the "V" mark on the steering gear mounting bolt locations

mounting screws and position the panels out of the way of the brake booster mounting nuts.

20. Remove the nuts, then remove the brake booster with the master cylinder attached.

21. Disconnect the A/C evaporator line quick-connect fitting.

22. Remove the power steering hoses from the steering gear and remove the Electronic Brake Traction Control Module/Brake Pressure Modulator Valve (EBTCM/BPMV) assembly.

23. Remove the upper heat shield mounting bolt.

24. Raise and safely support the vehicle.

25. Matchmark the Pitman arm position on the steering gear cross-shaft and remove the Pitman arm using a suitable puller.

26. Remove the lower heat shield mounting nuts.

27. Remove the three lower mounting bolts from the steering gear.

28. Lower the vehicle to the floor and remove the heat shield.

29. Disconnect the power steering fluid flow control actuator.

30. Remove the upper mounting bolt and the steering gear.

To install:

31. Find the center of travel of the steering gear by turning the stubshaft and counting the number of turns from lock-to-lock. Divide the total number in half and turn the stub shaft from either lock position by this amount. Finally align the mark on the stub shaft to the "V" mark on the steering gear.

32. Position the steering gear on the frame and install the shims and upper mounting bolt. Do not tighten the bolt at this time.

33. Connect the power steering fluid flow control actuator.

34. Install the heat shield with the upper mounting bolt. Tighten the bolt to 71 ft. lbs. (8Nm).

35. Raise and safely support the vehicle.

36. Install the steering gear lower mounting bolts. Tighten all mounting bolts to 30 ft. lbs. (40 Nm).

37. Install the lower heat shield mounting nuts. Tighten the nuts to 11 ft. lbs. (15 Nm).

38. Be sure the rubber steering gear shaft protector is in place, then install the Pitman arm. Tighten the nut to 118 ft. lbs. (160 Nm).

39. Lower the vehicle to the floor.

40. Install the EBTCM/BPMV assembly.

41. Connect the fluid lines to the steering gear. Tighten the fittings to 21 ft. lbs. (28 Nm).

42. Install new O-rings on the A/C evaporator line. Lubricate the O-rings with mineral oil of 525 viscosity and attach the quick-connect.

43. Install the brake booster assembly. Tighten the nuts to 15 ft. lbs. (20 Nm).

44. Install the fuse and relay panel, then install the knee bolster.

45. Connect the brake booster pushrod to the brake pedal.

46. Align the matchmarks and connect the steering shaft to the steering gear. Tighten the bolts to 16 ft. lbs. (22 Nm).

47. Install the sound insulator.

48. Connect the brake lines to the master cylinder. Tighten the fittings to 12 ft. lbs. (16 Nm).

49. Connect the brake fluid level sensor.

50. Install the power steering fluid reservoir. Tighten the clamp bolt to 62 inch lbs. (7 Nm).

51. Install new O-rings on the A/C evaporator-to-cowl connection. Lubricate the O-rings with mineral oil of 525 viscosity and connect the fitting. Tighten the bolt to 15 ft. lbs. (20 Nm).

52. Install the upper radiator hose and install the ECM.

53. Install the windshield wiper assembly.

54. Connect the negative battery cable.

55. Refill the cooling system.

56. Refill the power steering system and bleed the system using the following substeps:

    a. Turn the ignition switch **OFF**

    b. Raise the front wheels off the ground

    c. Turn the steering wheel to the full left position

    d. Fill the reservoir to the "FULL COLD" level

    e. Have an assistant monitor the fluid level and condition while turning the steering wheel from lock-to-lock at least 20 times or until no air bubbles are present. Refill the reservoir as needed.

    f. Start the engine and allow it to run for two minutes.

    g. Turn the front wheels to the center position and lower the vehicle to the floor.

    h. Turn the steering wheel from lock-to-lock and verify smooth operation.

57. Recharge and leak test the A/C system.

58. Bleed the brake system if necessary.

Strut

REMOVAL & INSTALLATION

Front

1. Raise and safely support the vehicle.

2. Remove the front wheel.

3. Detach the ABS speed sensor and break wear indicator wires from the strut.

4. Remove the brake hose from the strut.

5. Remove the brake caliper and suspend it out of the way with wire.

6. Remove the stabilizer bar link from the strut.

7. Mark the location of the strut on the knuckle and remove the lower strut mounting bolts.

8. Lower the vehicle to gain access to the upper strut mounting.

9. Remove the upper support plate protective cap and nut, then remove the strut from the vehicle.

To install:

10. Position the strut assembly in the strut tower.

11. Install the upper support plate and nut. Tighten the nut to 41 ft. lbs. (55 Nm).

12. Install the protective cap and raise the vehicle.

13. Using new bolts, install the strut on the knuckle. Be sure to align the matchmarks and install the bolts from the front.

Tighten the bolts to 66 ft. lbs. (90 Nm) + 52° with a torque angle meter.

14. Connect the stabilizer bar link to the strut. Tighten the nut to 48 ft. lbs. (65 Nm).

15. Clean out the brake caliper mounting bolt holes with a M12 x 1.5mm tap. Apply Locktite® 272 or equivalent thread locker to the bolts and install the caliper on the knuckle. Tighten the bolts to 70 ft. lbs. (95 Nm) + 37° with a torque angle meter.

16. Install the brake hose, ABS sensor and brake wear indicator wires to the strut assembly.

17. Install the front wheel.

18. Lower the vehicle to the floor.

19. Check and adjust the front wheel alignment if necessary.

Shock Absorber

REMOVAL & INSTALLATION

Rear

1. Position the rear seat backs forward to allow access to the upper shock absorber mounting.

2. Remove the protective cap, nut, washer and grommet.

3. Raise and safely support the vehicle.

4. Detach the Automatic Level Control (ALC) air line from the shock absorber.

5. Remove the lower shock absorber mounting bolt and the shock absorber.

To install:

6. If installing a new shock absorber, remove the cap from the air line connection.

7. Lower the vehicle and install the shock absorber upper mounting. Tighten the nut to 15 ft. lbs. (20 Nm). Install the protective cap.

8. Place the rear seats into the proper position.

9. Raise and safely support the vehicle.

10. Install the shock absorber to the control arm. Tighten the bolt to 81 ft. lbs. (110 Nm).

11. Connect the ALC air line to the shock absorber.

12. Lower the vehicle to the floor.

Coil Spring

REMOVAL & INSTALLATION

Front

1. Remove the strut assembly.

2. Compress the coil spring with a coil spring compressor.

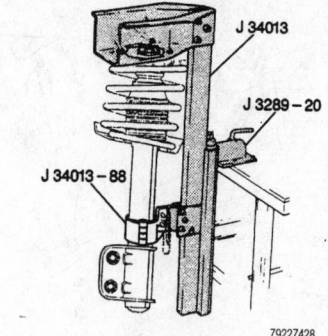

Be sure to compress the coil spring before removing the upper bearing support nut

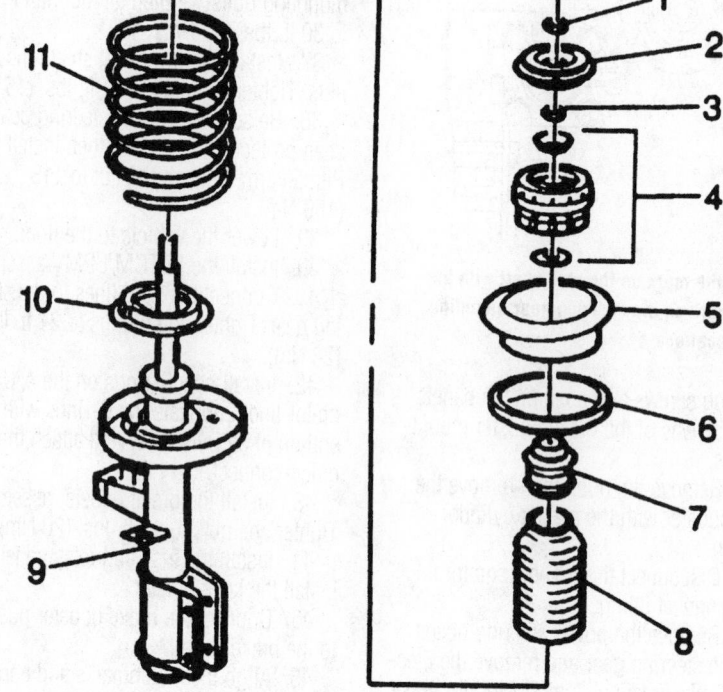

(1) Upper Support Plate Nut
(2) Upper Support Plate
(3) Upper Bearing Support Nut
(4) Bearing and Bearing Plate Assembly
(5) Upper Spring Support Plate
(6) Upper Insulator
(7) Strut Bumper
(8) Strut Cover
(9) Strut
(10) Lower Insulator
(11) Spring

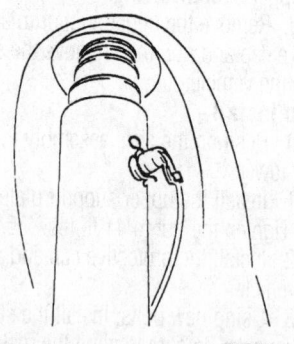

Air line connection to the shock absorber for Automatic Level Control (ALC)

Exploded view of the strut assembly

3. Remove the upper bearing support nut, bearing and plate assembly.

4. Decompress and remove the spring.

To install:

5. Compress the spring and position it on the strut. Be sure the spring is properly positioned between the upper and lower insulators.

6. Install the bearing and plate assembly. Tighten the nut to 52 ft. lbs. (70 Nm).

7. Slowly decompress the spring.

8. Install the strut assembly.

Rear

1. Raise and safely support the vehicle.

2. Remove the brake lines from the lower control arm without opening the brake system.

3. Disconnect the stabilizer link from the lower control arm.

4. Support the exhaust system with a jack, then remove the rubber hangers and lower the exhaust system out of the way.

5. Disconnect the wiring from the rear ABS wheel speed sensors.

6. Support the lower control arm with a jack. Raise the jack slightly to remove the pressure exerted by the spring on the shock absorber, then remove the lower shock absorber mounting bolt and remove the jack.

7. Support the differential with a jack and remove the rear axle cradle mount-to-body bolts.

8. Lower the rear axle cradle and remove the springs with the seats attached.

9. Remove the seat from the spring.

To install:

10. Install the seat on the spring.

11. Position the spring on the lower control arm. Be sure the spring is located properly on the seat.

12. Raise the differential and install the rear axle cradle mounting bolts. Tighten the bolts to 48 ft. lbs. (65 Nm).

13. Remove the jack and place it under the lower control arm. Raise the control arm enough to install the lower shock absorber mounting bolt. Install the bolt and tighten it to 81 ft. lbs. (110 Nm).

14. Connect the ABS wheel speed sensor wiring.

15. Raise the exhaust system into position and install the rubber hangers.

16. Connect the stabilizer link to the lower control arm. Tighten the bolts to 15 ft. lbs. (20 Nm).

17. Install the brake lines on the lower control arms with the retainers.

18. Lower the vehicle to the floor and check the wheel alignment.

REMOVAL & INSTALLATION

1. Raise and safely support the vehicle.

2. Remove the front wheel.

3. Remove the brake pad wear indicator rubber sleeve from the strut assembly.

4. Remove the ABS wheel speed sensor.

5. Detach the brake hose from the strut.

6. Remove the brake caliper from the knuckle and support it with a piece of wire.

7. Disconnect the tie rod end from the knuckle.

8. Disconnect the stabilizer link from the stabilizer bar.

9. Matchmark the position of the strut on the knuckle and remove the bolts attaching the strut to the knuckle.

10. Remove the pinch bolt securing the ball joint stud to the knuckle.

11. Loosen the lower control arm horizontal and vertical bolts. Remove the horizontal bolt and rotate the control arm from the support bracket.

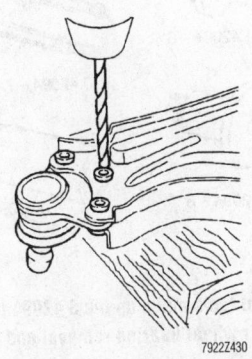

79222430

Drill out the rivets to remove the ball joint from the lower control arm

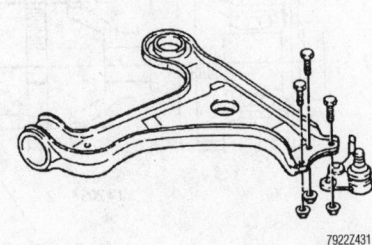

79222431

The replacement ball joint will be bolted to the lower control arm with the bolts supplied in the kit

12. Remove the vertical control arm mounting bolt and the control arm.

13. Drill out the rivets attaching the ball joint to the lower control arm and remove the ball joint.

To install:

14. Install the new ball joint on the lower control arm. Be sure to install the bolts from the top of the control arm. Tighten the bolts to 26 ft. lbs. (35 Nm).

➡**New bolts must be used to reinstall the lower control arm on the body.**

15. Install the lower control arm vertical bolt. Do not tighten the bolt at this time.

16. Position the front of the control arm in the support bracket. Install the bolt from the rear of the vehicle. Hold the control arm horizontal and tighten both bolts to 103 ft. lbs. (140 Nm).

17. Install the ball joint stud to the knuckle using a new bolt. Tighten the bolt to 74 ft. lbs. (100 Nm)

18. Using new bolts, install the strut on the knuckle. Be sure to align the matchmarks and install the bolts from the front. Tighten the bolts to 66 ft. lbs. (90 Nm) + 52° with a torque angle meter.

19. Connect the stabilizer link to the stabilizer bar. Tighten the nut to 48 ft. lbs. (65 Nm) while holding the link with a 17mm wrench.

20. Connect the tie rod end to the knuckle with a new self-locking nut. Tighten the nut to 44 ft. lbs. (60 Nm).

21. Clean out the brake caliper mounting bolt holes with a M12 x 1.5mm tap. Apply Locktite® 272 or equivalent thread locker to the bolts and install the caliper on the knuckle. Tighten the bolts to 70 ft. lbs. (95 Nm) + 37° with a torque angle meter.

22. Install the brake hose, ABS sensor and brake wear indicator wires to the strut assembly.

23. Install the wheel speed sensor. Tighten the bolts to 71 inch lbs. (8 Nm).

24. Install the front wheel.

25. Lower the vehicle to the floor.

26. Check and adjust the front wheel alignment if necessary.

ADJUSTMENT

The wheel bearings used by the Catera are not adjustable.

REMOVAL & INSTALLATION

Front

1. Raise and safely support the vehicle.
2. Remove the front wheel.
3. Remove the brake pad wear indicator wire from the strut assembly.
4. Detach the brake hose from the strut.
5. Remove the brake caliper from the knuckle and support it with a piece of wire.
6. Remove the set screw, then the brake rotor.
7. Remove the wheel hub dust cap.
8. Remove the spindle nut and the wheel bearing assembly.

To install:

9. Install the wheel bearing assembly onto the spindle. Use a socket to press the outer bearing ring into the hub if necessary. Tighten the nut to 236 ft. lbs. (320 Nm).
10. Install the brake rotor. Tighten the set screw to 35 inch lbs. (4 Nm).
11. Clean out the brake caliper mounting bolt holes with a M12 x 1.5mm tap. Apply Locktite® 272 or equivalent thread locker to the bolts and install the caliper on the knuckle. Tighten the bolts to 70 ft. lbs. (95 Nm) + 37° with a torque angle meter.
12. Install the brake hose and the ABS sensor wire to the strut assembly.
13. Install the front wheel.
14. Lower the vehicle to the floor.

Rear

→**Several special tools are necessary to remove the rear wheel bearing from the knuckle. The tools required are as follows:**

- J 36660 Torque Angle Meter
- J 42066 Holding tool
- J 42094–1 Holding Fixture

- J 42094–2 Spacer
- J 42094–3 Threaded Driver
- J 42094–4 Threaded Arbor
- J 42094–5 Ball Head
- J 42094–6 Bearing Remover
- J 42094–7 Threaded Spacer Pin
- J 42094–8 Bearing Installer
- J 42094–9 Hub Installer
- J 42094–10 Thrust Bearing
- J 42072 Triple Hex Head (10mm) deep socket

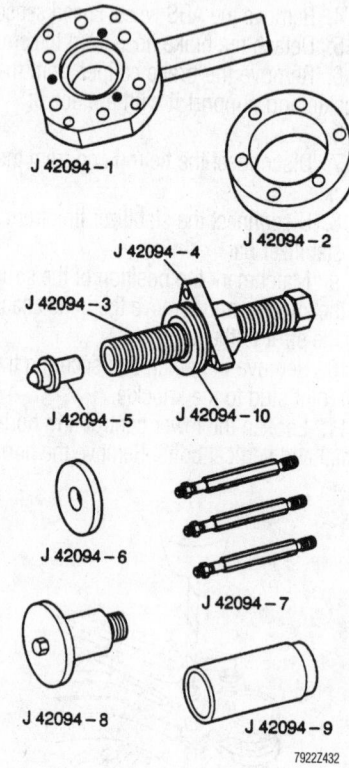

Several tools make up the J 42094 tool kit for rear wheel bearing removal and installation

The tool numbers mentioned are Kent-Moore tools used by GM. Equivalent tools may be available from other sources.

1. Raise and safely support the vehicle.
2. Remove the rear wheel.
3. Remove the halfshaft from the wheel hub and suspend it out of the way with wire.
4. Remove the brake hose from the clip on the lower control arm, them remove the caliper and suspend it out of the way with wire.
5. Remove the set screw, then the brake rotor.
6. Using socket, J 42072, back out three of the four brake backing plate bolts about nine turns to allow for installation of the wheel hub puller pins. The pins will screw into the same holes from the other side.
7. Attach spacer, J 42094–2 and holding fixture, J 42094–1 to the flange on the inside of the hub. Halfshaft bolts may be used if necessary.
8. Remove the wheel hub nut.
9. Attach the threaded arbor, J 42094–4 to the holding fixture with three supplied bolts using thrust bearing, J 42094–10 as a spacer. Thread driver, J 42094–3, into the

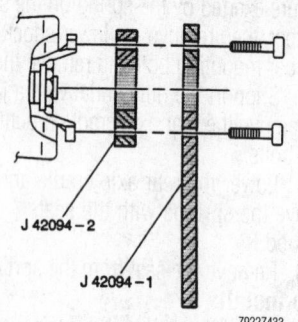

Attach the Spacer and Holding Fixture to the flange before removing the hub retaining nut

Cut-away view of the hub/wheel bearing assembly and steering knuckle

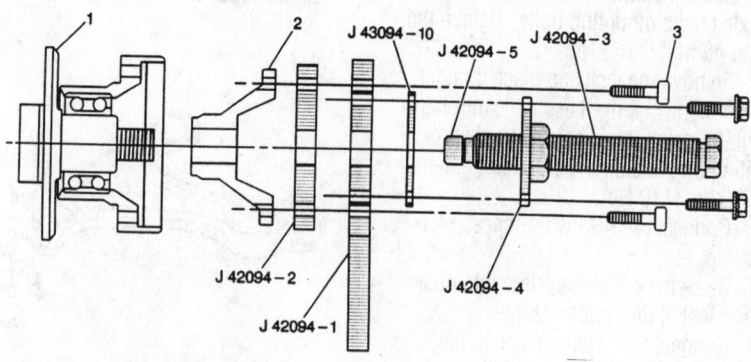

Set up the special tools as shown to remove the flange

threaded arbor, J 42094–4. Place the ball head, J 42094–5 onto J 42094–3 once sufficiently installed.

10. Turn the threaded driver to remove the flange.

11. Screw three spacer pins, J 42094–7, into the backing plate.

12. Attach holding fixture J 42094–1 with the stem pointed upward to the spacer pins with nuts and press out the wheel hub with the threaded driver.

13. Remove the wheel bearing retainer with snap ring pliers.

14. Attach bearing remover J 42094–6 to the end of the threaded driver and remove the wheel bearing by turning the driver clockwise.

To install:

15. Assembly wheel bearing installer J 42094–8 through the bearing.

16. Thread J 42094–8 into the threaded driver J 42094–3.

17. Turn J 42094–3 until the bearing is fully seated.

18. Install the wheel bearing retainer.

19. Insert rear wheel hub installer, J 42094–9 on the shaft of the threaded driver, J 42094–3 and attach holding fixture, J 42094–1, to the threaded spacer pins, J 42094–7.

20. Thread the rear wheel hub into the threaded driver, J 42094–3. Be sure the threaded driver is centered in the bearing to prevent binding of the hub as it is drawn into the bearing.

21. Hold the threaded driver, J 42094–3, and turn the arbor, J 42094–4 clockwise to pull the hub into the bearing.

22. Remove the tools from the vehicle.

23. Connect the threaded arbor, J 42094–4, threaded driver, J 42094–3, holding fixture, J 42094–1, spacer, J 42094–2, and thrust bearing, J 42094–10, to the wheel flange with the halfshaft mounting bolts.

24. Align the splines and position the flange on the hub.

25. Press the flange onto the hub by turning the threaded arbor, J 42094–4, clockwise while counter-holding with the threaded driver, J 42094–3.

26. Remove the threaded arbor, J 42094–4, threaded driver, J 42094–3, and the thrust bearing from the wheel flange. Leave the holding fixture and spacer, J 42094–2, attached.

27. Tighten the hub nut to 221 ft. lbs. (300 Nm). Remove the holding fixture and spacer, then pry the retaining washer over one side of the nut to lock it in place.

28. Install the brake backing plate

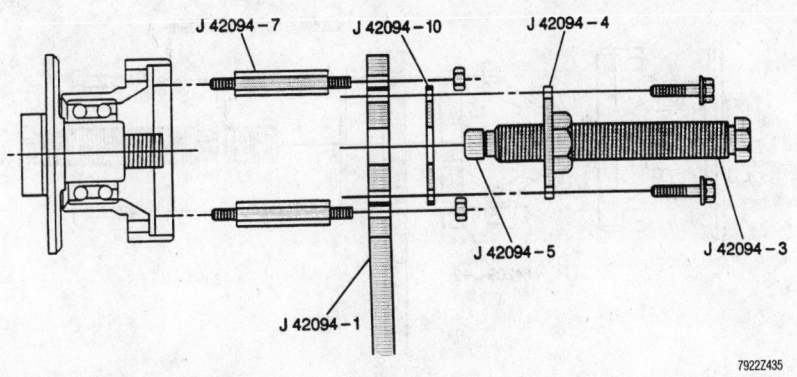

Set up the special tools as shown to remove the hub

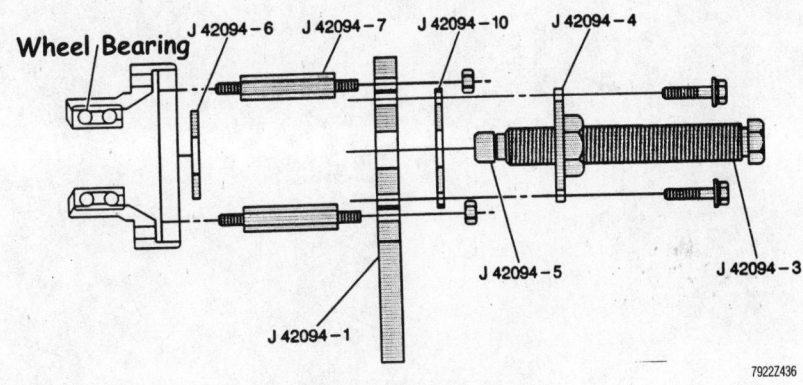

Set up the special tools as shown to remove the bearing assembly

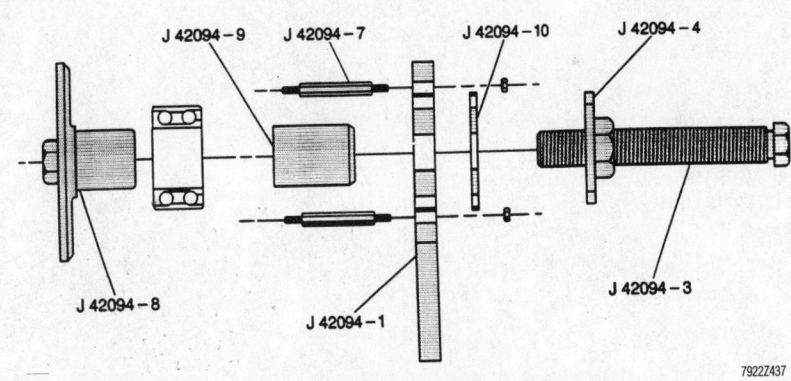

Set up the special tools as shown to install the bearing assembly

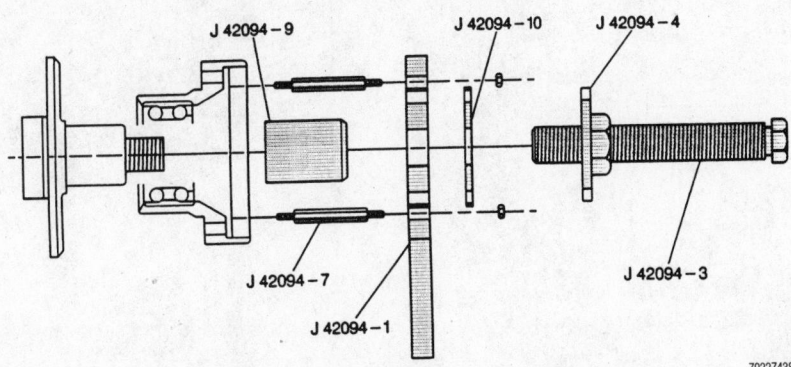

Set up the special tools as shown to pull the hub into the bearing assembly

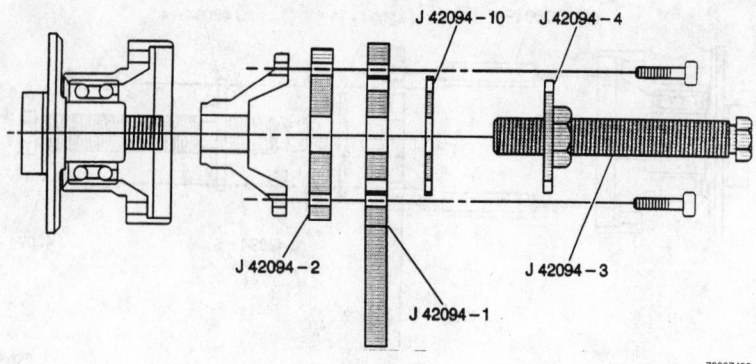

J 42094 – 10   J 42094 – 4

J 42094 – 2

J 42094 – 1

J 42094 – 3

7922Z439

Set up the special tools as shown to install the flange on the hub

mounting bolts. Tighten them to 37 ft. lbs. (50 Nm) + 40° with a torque angle meter.

29. Install the brake rotor. Tighten the set screw to 35 inch lbs. (4 Nm).

30. Install the caliper. Tighten the bolts to 59 ft. lbs. (80 Nm).

31. Install the brake line on the control arm with the retainer.

32. Install the halfshaft. Tighten the bolts to 37 ft. lbs. (50 Nm) + 70° with a torque angle meter.

33. Install the rear wheel.

34. Lower the vehicle to the floor.

GENERAL MOTORS W BODY

38

Buick-Century • Regal • **Chevrolet**-Lumina • Monte Carlo
Oldsmobile-Cutlass Supreme • Intrigue • **Pontiac**-Grand Prix

DRIVE TRAIN **38-25**
ENGINE REPAIR **38-2**
FUEL SYSTEM **38-23**
STEERING AND
SUSPENSION **38-27**

A

Air Bag..................................38-27
  Arming 38-28
  Disarming.............................38-27
  Precautions 38-27

C

Camshaft 38-15
  Removal & Installation 38-15
Coil Spring 38-32
  Removal & Installation.............38-32
Cylinder Head 38-5
  Removal & Installation 38-5

D

Distributor.............................38-2

E

Engine Assembly 38-2
  Removal & Installation 38-2
Exhaust Manifold......................38-12
  Removal & Installation 38-12

F

Fuel Filter 38-24
  Removal & Installation 38-24

Fuel Pump 38-24
  Removal & Installation 38-24
Fuel System Pressure 38-24
  Relieving 38-24
Fuel System Service
  Precautions............................38-23

H

Halfshaft...............................38-27
  Removal & Installation 38-27

I

Ignition Timing 38-2
  Adjustment 38-2
Intake Manifold......................38-9
  Removal & Installation 38-9

L

Lower Ball Joint.........................38-33
  Removal & Installation 38-33

O

Oil Pan...................................38-17
  Removal & Installation 38-17
Oil Pump 38-18
  Removal & Installation 38-18

P

Power Rack and Pinion
  Steering Gear...........................38-28
  Removal & Installation 38-28

R

Rear Main Seal 38-20
  Removal & Installation 38-20
Rocker Arms 38-8
  Removal & Installation 38-8

S

Strut....................................38-30
  Removal & Installation 38-30

T

Timing Chain, Sprockets and
  Front Cover 38-20
  Removal & Installation 38-20
Transaxle Assembly 38-25
  Removal & Installation 38-25

V

Valve Lash 38-17
  Adjustment...........................38-17

W

Water Pump 38-4
  Removal & Installation 38-4
Wheel Bearings.........................38-33
  Adjustment...........................38-33
  Removal & Installation 38-33

ENGINE REPAIR

➡**Disconnecting the negative battery cable on some vehicles may interfere with the functions of the on board computer systems and may require the computer to undergo a relearning process, once the negative battery cable is reconnected.**

Distributor

The 3.1L, 3.4L and 3.8L engines all utilize a Distributorless Ignition System (DIS). This system uses three twin tower coils which fire two spark plugs simultaneously. One spark plug is located in a cylinder on the compression stroke and the other in a cylinder on the exhaust stroke. This is known as the waste spark method of distribution.

Ignition Timing

ADJUSTMENT

The ignition timing is not adjustable, and is set according to engine demand electronically. The Powertrain Control Module (PCM) controls the ignition timing for all driving conditions.

Engine Assembly

REMOVAL & INSTALLATION

✳✳ CAUTION

Fuel injection systems on these vehicles remain under pressure, even after the engine has been turned OFF. The fuel system pressure must be relieved before disconnecting any fuel lines. Failure to do so may result in fire and/or personal injury.

3.1L Engine

1. Disconnect the negative battery cable. Remove the hood panel with an assistant.
2. Drain the engine coolant.
3. Remove the air cleaner and duct assembly.
4. Disconnect the transaxle fluid filler tube assembly.
5. Remove the engine mount strut brackets.
6. Raise and safely support the vehicle. Drain the engine oil.

7. Remove the front exhaust pipe and exhaust manifold heat shield.
8. Remove the lower the rear transaxle bolts.
9. Disconnect the electrical connector from the vehicle speed sensor.
10. Remove the engine mount frame side nuts.
11. Remove the flywheel inspection cover and the starter assembly.
12. Remove the torque converter bolts.
13. Remove the transaxle mount assembly. Safely lower the vehicle.
14. Remove the coolant reservoir recovery assembly.
15. Remove the serpentine belt.
16. Remove the accelerator control cable bracket and move the cables assemblies aside.
17. Remove the power brake booster vacuum hose from the upper intake manifold assembly.
18. Remove the plastic cover from the shock tower.
19. Remove the alternator front and rear braces.
20. Remove the alternator assembly from the vehicle.
21. Disconnect the fuel feed and return pipe assemblies.
22. Remove the tie straps around the ignition wiring harness assembly, engine mount strut and air conditioning compressor bracket.
23. Remove the power steering pump pulley assembly.
24. Remove the power steering pump bolts/screws and move the power steering pump aside.
25. Remove the heater outlet and inlet hose assemblies.
26. Remove the upper radiator hose assembly from the engine assembly.
27. Remove the lower radiator hose assembly from the water pump assembly.
28. Remove the engine mount strut and air conditioning compressor bracket bolts and move the compressor assembly aside.
29. Remove the automatic transaxle vacuum modulator pipe assembly.
30. Disconnect the electrical connectors from the knock sensor, oxygen sensor, coolant temperature sensor, camshaft sensor, crankshaft and wheel speed sensors.
31. Disconnect the electrical connectors from the ignition coil assembly.
32. Disconnect the electrical connectors from the idle air control valve and throttle position sensor.
33. Remove the vacuum hoses from the upper intake manifold assembly.
34. Remove the attaching transaxle bolts.

35. Install a safety bolt between the alternator bracket and the engine lift bracket assemblies.
36. Suitably support the transaxle assembly with floor stands.
37. Attach a suitable engine lifting device.
38. Remove the engine assembly from the vehicle.

To install:

39. Install the engine into the vehicle with a suitable engine lifting device.
40. Remove the engine lifting device. Remove the floor stands from the transaxle.
41. Remove the safety bolt from the alternator bracket and front engine lift bracket assemblies.
42. Install the transaxle bolts.
43. Install the remaining components in the reverse order of removal. Tighten the following:
   • Torque converter bolts to 46 ft. lbs. (63 Nm)
   • Engine mount frame side nuts to drive train bolts to 32 ft. lbs. (43 Nm)
44. Refill the engine with clean oil.
45. Install the engine mount strut brackets.
46. Install the transaxle fluid filler tube assembly.
47. Install the air cleaner and duct assembly. Install the hood panel.
48. Refill the engine coolant and add two engine coolant sealant pellets GM 3634621 or equivalent.
49. Bleed the cooling system using the recommended procedure.
50. Start the engine and verify correct idle and performance.

3.4L Engines

1. Remove the air cleaner assembly.
2. Remove the hood assembly with the help of an assistant.
3. Drain the cooling system. Remove the coolant recovery reservoir.
4. Remove the heater hoses from the engine. Remove the engine torque strut bracket and strut.
5. Remove the cooling fans.
6. Remove the radiator hoses from the engine. Detach the transaxle cooler lines.
7. Remove the radiator and attaching hoses.
8. Remove the control cables from the bracket and throttle body.
9. Remove the exhaust crossover.
10. Remove the ground straps from the bell housing.
11. Remove the fuel injector cover.
12. Remove the power steering lines at the pump and front cover.

13. Remove the serpentine drive belt. Remove the upper A/C compressor bolts.

14. Safely raise and support the vehicle. Remove the right front tire and wheel assembly.

15. Remove the right front splash shield.

16. Remove the lower A/C compressor bolts and reposition the compressor.

17. Remove the flywheel inspection cover. Remove the engine oil filter.

18. Remove the starter assembly.

19. Remove the front exhaust pipe and converter.

20. Remove the motor mount nuts from the subframe.

21. Disconnect the electrical connections from the rear of engine.

22. Remove the right ball joint nut and separate the ball joint from the control arm.

23. Remove the outer tie rod.

24. Remove the drive axle assembly.

25. Remove the transaxle shield.

26. Suitably support the transaxle with an appropriate jacking fixture.

27. Remove the motor mount bracket to the transaxle bolts and nuts.

28. Remove the transaxle support and safely lower the vehicle.

29. Remove the plastic cover from the shock tower.

30. Detach the electrical quick connectors from the PCM.

31. Disconnect all necessary vacuum lines.

32. Support the transaxle.

33. Install a suitable engine lifting device.

34. Remove the engine assembly from vehicle after disconnecting the alternator connections.

To install:

35. Install the engine assembly into the vehicle with a suitable engine lifting device.

36. Connect the electrical connections to the alternator.

37. Remove the engine lifting device. Remove the support from the transaxle.

38. Install the bell housing bolts.

39. Connect all necessary vacuum lines.

40. Attach the quick connects near the PCM.

41. Install the plastic cover to the shock tower.

42. Safely raise and support the vehicle. Safely support the transaxle with a jacking fixture.

43. Install the motor mount bracket to transaxle attaching nuts and tighten nuts to 43 ft. lbs. (58 Nm).

44. Install the remaining components in the reverse order of removal. Tighten the following:

- Ball joint-to-control arm nuts to 63 ft. lbs. (85 Nm)
- Torque converter bolts to 46 ft. lbs. (63 Nm)

45. Refill the engine with clean oil.

46. Refill the cooling system.

47. Refill the power steering fluid.

48. Check and refill the transaxle fluid. Install the hood assembly.

49. Install the air cleaner assembly.

50. Connect the negative battery cable. Start the engine and check all fluids for proper level and leaks.

51. Bleed the cooling system and power steering system using the recommended procedure.

3.8L Engines

1. Relieve the fuel system pressure. Disconnect the negative battery cable.

2. Remove the hood from the vehicle. Remove the air cleaner assembly.

3. Disconnect and cap the fuel lines from the fuel rail and mounting brackets.

4. Drain the engine coolant and remove the recovery bottle.

5. Remove the inner fender electrical cover and the fuel injector sight cover.

6. Disconnect the throttle and cruise control (if equipped) cables from the throttle body and mounting bracket.

7. Remove the rear heat shield from the crossover pipe.

8. Remove the throttle cable mounting bracket and disconnect any vacuum lines from the bracket.

9. Disconnect the exhaust crossover from the manifolds.

10. Remove the torque strut mounting bolt and disconnect the strut from the engine bracket.

11. Remove the right side engine cooling fan.

12. Disconnect the vacuum line from the transaxle module.

13. Remove the serpentine belt.

14. Remove the power steering pump and alternator assemblies.

15. Tag and disconnect all electrical connections from the engine.

16. Disconnect the upper and lower radiator hoses as well as the heater hoses from the engine.

17. Remove the transaxle-to-engine bolts and disconnect the ground wires.

18. Raise and support the vehicle safely. Remove the right front wheel and inner splash shield.

19. Remove the flywheel cover, scribe a mark on the torque converter and flywheel

and remove the flywheel-to-torque converter bolts.

20. Disconnect the wiring harness clamps from the frame near the radiator.

21. Remove the A/C compressor from the bracket, lay aside and secure to the frame.

22. Remove the starter.

23. Support the transaxle with a jack and remove the transaxle-to-engine bolt, through the wheel well, using a long extension.

24. Attach a lifting device and remove the engine mount-to-frame nuts.

25. Drain the engine oil and remove the oil filter.

26. Disconnect the oil cooler pipes from the hose connections.

27. Disconnect the exhaust pipe from the manifold.

28. Lower the vehicle and remove the engine assembly from the vehicle.

To install:

29. With an assistant, install a lifting device onto the engine and position into the vehicle.

30. Support the transaxle, install the transaxle-to-engine bolts and ground wiring harness and tighten to 46 ft. lbs. (62 Nm).

31. Install the heater and upper and lower radiator hoses to the engine.

32. Install all electrical connections to the engine.

33. Install the alternator, power steering pump and serpentine belt.

34. Install the vacuum line to the transaxle module.

35. Install the engine torque strut and bolt and tighten to 41 ft. lbs. (56 Nm).

36. Install the exhaust crossover pipe.

37. Install the throttle cable mounting bracket and vacuum lines.

38. Install the heat shield to the crossover pipe and the throttle cables to the throttle body and mounting bracket.

39. Install the inner fender electrical cover and the coolant recovery bottle.

40. Install the fuel hoses to the fuel rail and mounting brackets.

41. Raise and support the vehicle safely.

42. Connect the front exhaust pipe to the manifold.

43. Install the oil filter and oil cooler pipes.

44. Install the engine mount nuts to the frame and tighten to 32 ft. lbs. (43 Nm).

45. Install the transaxle to engine bolt through the wheel well and tighten to 46 ft. lbs. (62 Nm).

46. Install the starter motor assembly and connect the electrical connectors.

47. Install the air conditioner compressor to the bracket.

48. Install the wiring harness clamps to the frame near the radiator.

49. Align the scribe marks, install the torque converter to flywheel bolts and tighten to 46 ft. lbs. (62 Nm).

50. Install the flywheel cover and the inner fender splash shield.

51. Install the right front wheel assembly and lower the vehicle.

52. Refill the cooling system and bleed the power steering system.

53. Install the right side cooling fan.

54. Install the fuel injector sight shield and the air cleaner assembly.

55. Connect the negative battery cable and install the hood.

56. Check and add fluids as required. Test drive vehicle and recheck for leaks and correct levels.

Water Pump

REMOVAL & INSTALLATION

3.1L and 3.4L Engines

1. Disconnect the negative battery cable. Drain the cooling system.

2. Remove the coolant reservoir and lay aside.

3. Remove the serpentine belt guard, bolts and nuts.

4. Loosen the water pump pulley bolts. Remove the serpentine belt.

5. Remove the water pulley bolts and pulley.

6. Remove the water pump attaching bolts, then remove the water pump and gasket.

To install:

7. Clean all pump mating surfaces.

8. Inspect the pump. There should be a locator tab to identify the top of the pump. This locator must be in the vertical position when the pump is installed. Install the water

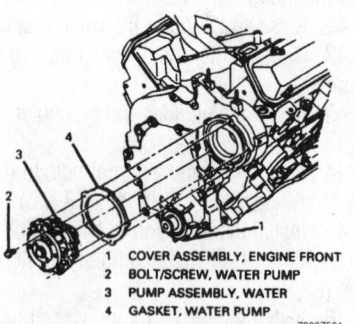

Exploded view of the water pump mounting—3.1L (VIN M) engine

1	COVER ASSEMBLY, ENGINE FRONT
2	BOLT/SCREW, WATER PUMP
3	PUMP ASSEMBLY, WATER
4	GASKET, WATER PUMP

79222501

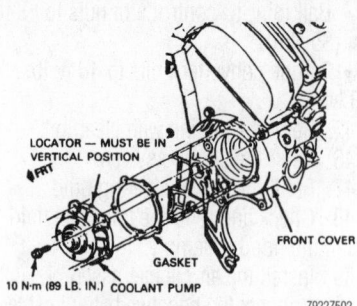

Water pump, gasket and mounting bolts—3.4L (VIN X) engine

LOCATOR — MUST BE IN VERTICAL POSITION
FRT
FRONT COVER
GASKET
10 N·m (89 LB. IN.) COOLANT PUMP
79222502

pump and gasket. Tighten the attaching bolts to 89 inch lbs. (10 Nm).

9. Install the remaining components in the reverse order of removal. Tighten the pulley bolts to 18 ft. lbs. (25 Nm).

10. Connect the negative battery cable. Install the air cleaner assembly.

11. Refill the cooling system with the proper coolant. Bleed the cooling system and check for leaks with the engine running at idle.

3.8L Engines

VIN L AND VIN K MODELS

1. Disconnect the negative battery cable. Drain the cooling system.

2. Remove the coolant recovery reservoir.

3. Remove the serpentine belt. If additional access is needed, remove the inner fender electrical cover.

4. For VIN L engines, remove the alternator and position aside, then remove the serpentine belt tensioner pulley.

5. For VIN K engines, remove the power steering pump pulley using pulley remover J 25034-B or the equivalent.

6. Remove the water pump pulley.

7. Remove the water pump attaching bolts. Note that there are different length bolts. Use care to keep them organized for proper assembly. Remove the water pump from the vehicle.

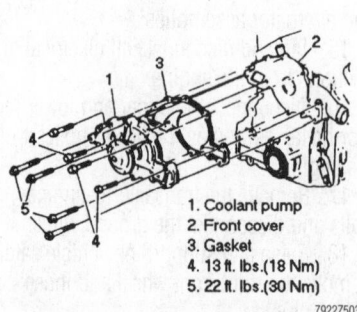

Exploded view of the water pump assembly mounting—3.8L (VIN L and K) engines

1. Coolant pump
2. Front cover
3. Gasket
4. 13 ft. lbs.(18 Nm)
5. 22 ft. lbs.(30 Nm)
79222503

To install:

8. Clean the water pump mating surfaces. Install the water pump using a new gasket.

9. Clean the pump bolt threads well. Install the attaching water pump bolts using care to install the proper bolts in the proper locations. Tighten long bolts to 22 ft. lbs. (30 Nm) and short bolts to 13 ft. lbs. (18 Nm) plus an additional 80 degrees using a torque angle meter.

10. Install the water pump pulley and tighten the bolts to 115 inch lbs. (13 Nm).

11. For VIN L engines, install the serpentine belt tensioner pulley, then install the alternator.

12. For VIN K engines, install the power steering pump pulley using J 25033-B or the equivalent to press the pulley onto the shaft.

13. Install the serpentine belt.

14. Install the coolant recovery reservoir.

15. Refill the cooling system with the correct amount and type of coolant.

16. Connect the negative battery cable. Start the engine and bleed the cooling system using the recommended procedure.

VIN 1 MODELS

1. Disconnect the negative battery cable.

2. Drain the cooling system into an approved container.

3. Remove the accessory drive and supercharger belts.

4. Remove the alternator and brace.

5. Disconnect the hoses and pipes from the water pump.

6. Remove the pulley bracket assembly.

7. Raise and safely support the vehicle.

8. Remove the power steering pump and lines.

9. Lower the vehicle.

10. Support the engine using engine support fixture J 28467-A or equivalent, and remove the front engine mount.

11. Remove the water pump pulley.

12. Remove the water pump mounting bolts and remove the water pump from the vehicle.

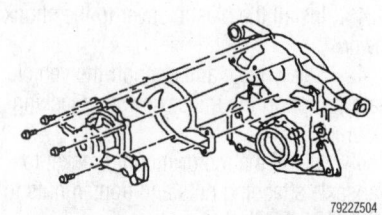

Exploded view of the water pump assembly mounting—3.8L (VIN 1) engines

79222504

To install:

13. Clean all sealing surfaces.

14. Apply a thin bead of sealant to the gasket mating surface of the water pump and install a new gasket to the pump.

15. Install the water pump on the engine. Tighten the pump-to-front cover bolts to 11 ft. lbs. (15 Nm) plus an additional 80° turn with a torque angle meter.

16. Install the water pump pulley and tighten the bolts to 9.5 ft. lbs. (13 Nm).

17. Install the front engine mount.

18. Raise and safely support the vehicle.

19. Install the power steering pump and lines following the proper procedure.

20. Lower the vehicle.

21. Install the pulley bracket assembly.

22. Connect the hoses and pipes from the water pump.

23. Install the alternator and brace.

24. Install the supercharger and accessory drive and belts.

25. Connect the negative battery cable.

26. Fill and bleed the cooling system following the proper procedure.

27. Run the engine and check for leaks.

28. Recheck the coolant level when the engine has cooled.

Cylinder Head

REMOVAL & INSTALLATION

✳✳ CAUTION

Fuel injection systems on these vehicles remain under pressure, even after the engine has been turned OFF. The fuel system pressure must be relieved before disconnecting any fuel lines. Failure to do so may result in fire and/or personal injury.

3.1L Engine

LEFT (FRONT) CYLINDER HEAD

1. Disconnect the negative battery cable. Relieve the fuel system pressure.

2. Drain the cooling system into a suitable container.

3. Remove the rocker arm covers.

4. Remove upper intake plenum and lower intake manifold.

5. Remove the exhaust crossover pipe.

6. Disconnect the spark plug wires from spark plugs and wire looms and route the wires out of the way.

➡ When removing the valvetrain components use care to identify any components that will be reused. Valvetrain components must be kept in order for installation in the same locations from which they were removed.

7. Remove rocker arms nut, rocker arms, balls and pushrods.

8. Remove oil level indicator tube.

9. Remove any A/C compressor bolts accessible from the top.

10. Raise and safely support the vehicle. Remove the lower A/C compressor mounting bolts.

11. Disconnect the A/C compressor electrical connections and reposition the A/C compressor.

12. Remove the A/C compressor lower bracket bolts.

13. Lower the vehicle.

14. Remove the A/C compressor upper bracket bolts.

15. Remove the compressor brackets.

16. Remove the cylinder head bolts evenly.

17. Remove the cylinder head.

To install:

18. Clean all the gasket surfaces completely. Clean the threads on the cylinder head bolts and block threads. New replacement head bolts are recommended.

19. Place the cylinder head gasket in position over the dowel pins on the cylinder block so the words **THIS SIDE UP** or other gasket identification are showing.

20. Coat the bolt threads with sealer and install finger-tight.

21. Tighten the cylinder head bolts in sequence to 33 ft. lbs. (45 Nm). With all the bolts tightened make a second pass tightening all the bolts an additional 90 degrees (¼ turn).

22. Install the remaining components in the reverse order of removal. Tighten the following:

• A/C compressor bracket bolts to 35 ft. lbs. (47 Nm)

• A/C compressor mounting bolts to 18 ft. lbs. (25 Nm)

• Rocker arm nuts to 20 ft. lbs. (27 Nm)

23. Refill the cooling system.

24. Connect negative battery cable.

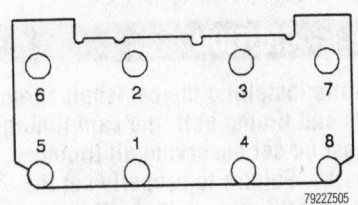

Tighten the cylinder head bolts in the order shown to prevent damaging the head—3.1L (VIN M) engine

25. An oil and filter change are recommended since coolant can enter the oil system when the head is being removed.

26. Start vehicle and verify no leaks.

RIGHT (REAR) CYLINDER HEAD

1. Disconnect the negative battery cable. Relieve the fuel system pressure.

2. Drain the cooling system into a suitable container.

3. Remove the rocker arm covers.

4. Remove upper intake plenum and lower intake manifold.

5. Disconnect the electrical connector from the ignition assembly.

6. Remove the alternator.

7. Remove the exhaust crossover pipe. Detach the O_2 sensor connector.

8. Raise and safely support the vehicle. Disconnect the exhaust pipe from the exhaust manifold.

9. Lower the vehicle.

10. Remove the exhaust manifold.

11. Disconnect the spark plug wires from spark plugs and wire looms and route the wires out of the way.

➡ When removing the valvetrain components use care to identify any components that will be reused. Valvetrain components must be kept in order for installation in the same locations from which they were removed.

12. Remove rocker arms nut, rocker arms, balls and pushrods.

13. Remove the cylinder head bolts evenly. Remove the cylinder head.

To install:

14. Clean all the gasket surfaces completely. Clean the threads on the cylinder head bolts and block threads. New replacement head bolts are recommended.

15. Place the cylinder head gasket in position over the dowel pins on the cylinder block so the words

16. **THIS SIDE UP** or other gasket identification is showing.

17. Coat the bolt threads with sealer and install finger-tight.

18. Tighten the cylinder head bolts in sequence to 33 ft. lbs. (45 Nm). With all the bolts tightened make a second pass tightening all the bolts an additional 90 degrees.

19. Install the pushrods, rocker arms, balls and rocker arm nuts. Tighten the rocker arm nuts to 20 ft. lbs. (27 Nm).

20. Install the remaining components in the reverse order of removal.

21. Refill the cooling system.

22. An oil and filter change are recommended since coolant can enter the oil system when the head is being removed.

23. Connect negative battery cable.
24. Start vehicle and verify no leaks.

3.4L Engine

LEFT (FRONT) CYLINDER HEAD

1. Relieve the fuel system pressure. Remove the air cleaner and duct assembly.
2. Disconnect the negative battery cable. Drain the cooling system.
3. Remove the upper and lower intake manifold components.
4. Remove the left side cam carrier cover using the following procedure.

   a. Disconnect oil/air breather hose from cam carrier cover.

   b. Disconnect the spark plug wires from the front spark plugs.

   c. Disconnect the rear spark plug wires from the clips on the front camshaft cover.

   d. Remove the four camshaft cover mounting bolts.

   e. Remove the camshaft cover and discard the old gaskets and O-rings.

5. Remove the left side cam carrier using the following procedure.

   a. Remove the camshaft timing belt.

   b. Remove the exhaust crossover pipe and reposition aside.

   c. Remove the engine torque strut.

   d. Remove the front engine lift bracket.

   e. Install six, 6-inch sections of fuel line hose under camshaft and between lifters. This will hold lifters in the carrier. For this procedure use ³⁄₁₆ in. (8mm) vacuum/fuel line hose for exhaust valves and ⁵⁄₃₂ in. (5.5mm) vacuum/fuel line hose for the intake valves.

   f. Remove the camshaft carrier bolts and remove the carrier. Discard the gasket.

6. Remove the front exhaust manifold.
7. Remove the oil level indicator tube mounting bolt and remove the tube assembly.
8. Detach the coolant temperature sensor electrical connector.
9. Remove the cylinder head mounting bolts and remove cylinder head. Discard gasket.

To install:

10. Clean all parts well. Be sure the mating surfaces on the head and block are clean. Clean the engine block threaded holes and be sure all oil is removed from the bolt holes. Inspect the cylinder head bolt threads and be sure they are clean. New replacement head bolts are recommended. If the cylinder head is being replaced with a new head, transfer the manifold studs and temperature sensor.

11. Install a new cylinder head gasket on the engine block with the metal tabs (factory type gasket) between the cylinders facing UP.
12. Install the cylinder head on the engine block with the dowel pins lined up. Install the mounting bolts finger-tight.
13. With all the bolts in place tighten the mounting bolts in sequence to 37 ft. lbs. (50 Nm) for 1994 vehicles and 44 ft. lbs. (60 Nm) for 1995–97 vehicles. Make a second pass over each bolt in sequence and tighten an additional 90 degrees (¼ turn) rotation using a torque angle meter.
14. Connect the electrical connector to the coolant temperature sender.
15. Install the oil level tube assembly and tighten the mounting bolt to 89 inch lbs. (10 Nm).
16. Install the front exhaust manifold and tighten the attaching nuts to 18 ft. lbs. (25 Nm).
17. Clean any oil from the cam carrier to cylinder head bolt holes (bolt holes closest to the exhaust manifold).
18. Install the special camshaft holding tool J38613-A or equivalent and tighten the mounting bolt to 22 ft. lbs. (30 Nm).

➡ **The use of petroleum jelly (never chassis grease) in the lifter bores along with the use of the lifter hold down hoses will help keep the lifters in place.**

19. Install a new camshaft carrier gasket on the cylinder head and install the camshaft carrier on the cylinder head.
20. Install the camshaft carrier mounting bolts and tighten to 20 ft. lbs. (27 Nm).
21. Remove the lifter hold down hoses.
22. Install the upper and lower intake manifold assembly. If the thermostat housing was removed, install with a new gasket.
23. Install the front engine lift bracket.
24. Install the engine torque strut and tighten the bolt to 39 ft. lbs. (53 Nm).
25. Install the exhaust manifold crossover pipe.
26. Install the camshaft sprockets, if removed. Install the timing belt using great care to align all timing marks.

✳✳ CAUTION

While installing the camshaft sprockets and timing belt, the cam timing must be set observing all timing marks. Failure to properly set the camshaft timing will result in serious engine damage.

27. Remove the special tools J-38613 camshaft holding clamps, if used.

28. Install the front engine coolant pipe. Connect the heater hose pipe to the intake plenum.
29. Connect the upper radiator hose.
30. Install the new O-rings and gasket on the camshaft carrier cover.

➡ **Before tightening the cover mounting bolts, fully seat the bolt insulators in the camshaft cover.**

31. Install the cover on the camshaft carrier and tighten the mounting bolts to 97 inch lbs. (11 Nm).
32. Connect the rear spark plug wires to the camshaft cover.
33. Connect the front spark plug wires to the spark plugs.
34. Connect the oil/air breather hose to the camshaft cover.
35. Drain the engine oil since coolant can contaminate the oil when a cylinder head is removed. Refill with the proper quantity and quality engine oil. A filter change is recommended,
36. Connect the negative battery cable. Install the air cleaner and air duct assembly.
37. Refill the cooling system.
38. Start the vehicle and verify no oil or coolant leaks. Bleed the cooling system as required.

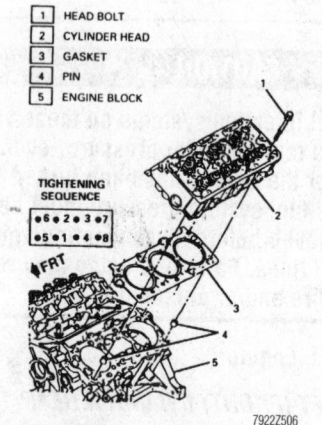

1	HEAD BOLT
2	CYLINDER HEAD
3	GASKET
4	PIN
5	ENGINE BLOCK

79222506

Exploded view of the cylinder head mounting and bolt tightening sequence—3.4L (VIN X) engines

RIGHT (REAR) CYLINDER HEAD

1. Relieve the fuel system pressure. Remove the air cleaner and duct assembly.
2. Disconnect the negative battery cable. Drain the cooling system.
3. Remove the upper intake manifold. Remove the right timing belt cover.
4. Remove the right rear spark plug wires. Disconnect oil/air breather hose from cam carrier cover.

5. Remove the four camshaft cover mounting bolts.

6. Remove the camshaft cover and discard the old gaskets and O-rings.

7. Remove the camshaft timing belt.

8. Install six 6 in. sections of fuel line hose under camshaft and between lifters. This will hold lifters in the carrier. For this procedure use ³⁄₁₆ in. (8mm) vacuum/fuel line hose for exhaust valves and ⁵⁄₃₂ in. (5.5mm) vacuum/fuel line hose for the intake valves.

9. Remove the lower intake manifold.

10. Remove the right camshaft carrier mounting bolts.

11. Remove the right camshaft carrier and gasket from the engine.

12. Remove the exhaust crossover pipe from the engine.

13. Raise and safely support the vehicle. Disconnect the front exhaust pipe at the manifold.

14. Lower the vehicle.

15. Disconnect the electrical connector from the oxygen (O₂) sensor.

16. Remove the rear timing belt tensioner bracket.

17. Remove the cylinder head mounting bolts and remove cylinder head and gasket.

To install:

18. Clean all parts well. Be sure the mating surfaces on the head and block are clean. Clean the engine block threaded holes and be sure all oil is removed from the bolt holes. Inspect the cylinder head bolt threads and be sure they are clean. New replacement head bolts are recommended. If the cylinder head is being replaced with a new head, transfer the manifold studs and temperature sensor.

19. Install a new cylinder head gasket on the engine block with the metal tabs (factory type gasket) between the cylinders facing UP.

20. Install the cylinder head on the engine block, with attached exhaust manifold, with the dowel pins lined up. Install the mounting bolts finger-tight. With all the bolts in place tighten the mounting bolts in sequence to 37 ft. lbs. (50 Nm) for 1994 vehicles or to 44 ft. lbs. (60 Nm) for 1995–97 vehicles. Make a second pass over each bolt in sequence and tighten an additional 90 degrees (¼ turn) rotation using special tool J 36660 or equivalent angle measuring equipment.

21. Install the rear timing belt tensioner bracket.

22. Connect the electrical connector to the oxygen (O₂) sensor.

23. Raise and safely support the vehicle.

24. Connect the front exhaust pipe to the exhaust manifold.

25. Lower the vehicle.

26. Install the exhaust crossover pipe.

➡**Remove oil from the cam hold down tool hole in the camshaft carrier before installing and tightening the bolt.**

27. Install the special camshaft hold down tool J 38613 A or equivalent and tighten to 22 ft. lbs. (30 Nm).

➡**The use of petroleum jelly (never chassis grease) in the lifter bores along with the use of the lifter hold down hoses, will help keep the lifters in place.**

28. Install a new camshaft carrier gasket on the cylinder head and install the camshaft carrier on the cylinder head.

29. Install the mounting bolts and tighten to 20 ft. lbs. (27 Nm).

30. Remove the lifter hold down hoses.

31. Install the camshaft sprockets and timing belt.

✳✳ CAUTION

After installing the camshaft sprockets and timing belt the Cam Timing must be set. Failure to properly set the camshaft timing will result in serious engine damage.

32. Remove the camshaft hold down tool J-38613 or equivalent.

33. Install the lower intake manifold.

34. Install the new O-rings and gasket on the cam carrier cover.

➡**Before tightening the cover mounting bolts, fully seat the bolt insulators in the camshaft cover.**

35. Install the cover on the camshaft carrier and tighten the mounting bolts to 97 inch lbs. (11 Nm).

36. Connect the oil/air breather hose to the camshaft cover.

37. Connect the rear spark plug wires to the camshaft cover and spark plugs.

38. Install the right timing belt cover. Install the upper intake manifold.

39. Connect the negative battery cable. Install the air cleaner and air duct assembly.

40. Drain the engine oil since coolant can contaminate the oil when a cylinder head is removed. Refill with the proper quantity and quality engine oil. A filter change is recommended,

41. Refill the cooling system.

42. Start the vehicle and verify no leaks.

3.8L Engines

LEFT (FRONT) CYLINDER HEAD

1. Remove the air cleaner assembly. Disconnect the negative battery cable.

2. Following proper procedures, relieve the fuel system pressure.

3. Drain the cooling system and remove the intake manifold.

4. Remove the valve covers and remove the rocker arm assemblies, pedestals, valve guide plates and pushrods, keeping everything in order for reinstallation.

5. Tag and disconnect the ignition wires and remove the spark plugs.

6. Remove the engine torque strut from the attaching bracket.

7. Remove the vacuum line from the transaxle module.

8. Remove the left exhaust manifold.

9. Remove the spark plugs.

10. Remove the alternator front mount bracket and ignition module with the mount bracket. Lay the ignition module aside.

11. Remove the cylinder head bolts and discard. Remove the cylinder head from the vehicle.

12. Clean all gasket mating surfaces and the cylinder head bolt holes in the block.

To install:

13. Clean all parts well. If the cylinder head is being replaced with a new one, transfer the torque strut bracket and exhaust support bracket with the transaxle vacuum line. If the cylinder head is to be serviced, these parts must be removed, after servicing, installed.

14. Inspect the head gasket and mating surfaces for leaks, corrosion and blow-by.

15. **The factory specifies new replacement cylinder head bolts** when removing and installing a cylinder head. Clean the remains of sealer from all threaded openings. The cylinder block bolt holes should be cleaned using a ⁷⁄₁₆ 14 tap.

16. Place the cylinder head gasket on the engine block dowels with the note **THIS SIDE UP** facing the cylinder head and the arrow facing the front of the engine. Position the cylinder head on the engine block.

✳✳ CAUTION

To prevent damage to the gasket, when installing the cylinder head, do not slide the cylinder head on the gasket.

17. Install the new replacement cylinder head bolts and tighten as follows:

   a. Tighten the cylinder head bolts, in sequence, to 37 ft. lbs. (50 Nm).

b. Rotate each bolt 120 degrees (⅓ turn) in sequence using special tool J 36660 or equivalent.

c. Rotate the center 4 bolts an additional 30 degrees (¹⁄₁₂ turn) in sequence using special tool J 36660 or equivalent.

18. Install the pushrods, rocker arm assemblies, valve guide retainers and the valve covers.

19. Install the exhaust manifold and tighten manifold studs to 32 ft. lbs. (48 Nm).

20. Install the intake manifold using the recommended procedure.

21. Install the alternator front mount bracket and ignition module with bracket.

22. Install the spark plugs and wires.

23. Install the engine torque strut and tighten bolt to 32 ft. lbs. (53 Nm).

24. Refill the cooling system.

25. It is recommended that the engine oil be drained and crankcase refilled with clean engine oil since coolant and debris will get into the oil pan when removing a cylinder head. A filter change is also recommended.

26. Connect the negative battery cable.

27. Install the air cleaner assembly.

28. Start the engine check for proper operation. Verify no leaks.

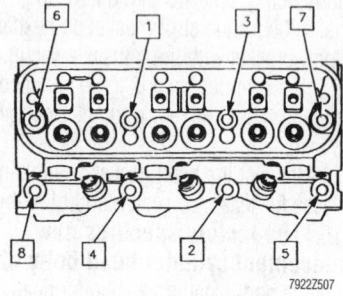

Cylinder head bolt tightening sequence— 3.8L engines

RIGHT (REAR) CYLINDER HEAD

1. Remove the air cleaner assembly. Disconnect the negative battery cable.

2. Relieve the fuel system pressure.

3. Drain the cooling system and disconnect the exhaust crossover pipe.

4. Remove the intake manifold.

5. Raise and safely support the vehicle.

6. Remove the right side exhaust pipe from the manifold.

7. Safely lower the vehicle.

8. Remove the right rear valve cover. Remove the serpentine drive belt.

9. Remove the belt tensioner pulley. Remove the heater hose from the engine.

10. Remove the power steering pump mounting bracket and lay the pump aside.

11. Tag and disconnect the ignition wires and remove the spark plugs.

12. Remove the exhaust manifold from the cylinder head only.

13. Disconnect the electrical connection from the oxygen sensor.

14. Remove the rocker arm assemblies, pedestals, valve guide retainers and pushrods, keeping everything in order for reinstallation.

15. Remove the cylinder head bolts in reverse order of installation and remove the cylinder head. Discard the cylinder head bolts.

To install:

16. Clean all parts well. Inspect the head gasket and mating surfaces for leaks, corrosion and blow-by.

17. **The factory specifies new head bolts** be used when removing and installing a cylinder head. Clean the remains of sealer from all threaded openings. The cylinder block bolt holes should be cleaned using a ⁷⁄₁₆ 14 tap.

18. Place the cylinder head gasket on the engine block dowels with the note **THIS SIDE UP** facing the cylinder head and the arrow facing the front of the engine. Position the cylinder head on the engine block.

✳✳ CAUTION

To prevent damage to the gasket, when installing the cylinder head, do not slide the cylinder head on the gasket.

19. Install the new cylinder head bolts and tighten as follows:

a. Tighten the cylinder head bolts, in sequence, to 37 ft. lbs. (50 Nm).

b. Rotate each bolt 120 degrees (⅓ turn) in sequence using special tool J 36660 or equivalent.

c. Rotate the center 4 bolts an additional 30 degrees (¹⁄₁₂ turn) in sequence using special tool J 36660 or equivalent.

20. Connect the electrical connector to the oxygen sensor.

21. Install the exhaust manifold and intake manifold.

22. Install the pushrods, valve guide retainers and rocker arm assemblies.

23. Install the right rear valve cover. Install the spark plugs and wires in the proper order.

24. Install the power steering pump bracket and tighten the bolts to 35 ft. lbs. (47 Nm).

25. Install the belt tensioner pulley and serpentine belt.

26. Install the exhaust crossover pipe. Raise and safely support the vehicle.

27. Install the front exhaust pipe to the manifold.

28. Refill the cooling system.

29. It is recommended that the engine oil be drained and crankcase refilled with clean engine oil since coolant and debris will get into the oil pan when removing a cylinder head. A filter change is also recommended.

30. Connect the negative battery cable. Install the air cleaner.

31. Start the engine check for proper operation. Verify no leaks.

Rocker Arms

REMOVAL & INSTALLATION

3.1L Engine

1. Remove the rocker arm cover and gasket.

2. Remove the rocker arm bolts and the rocker arms.

➡ **The exhaust pushrods are approximately ¼ in. (6.35mm) longer than the intake pushrods. Place all valvetrain components in a rack so they can be reinstalled in the same location from which they were removed.**

3. Remove the pushrod assemblies.

➡ **Use care when removing pushrods so they do not fall down into the lifter valley.**

To install:

4. Clean all the gasket surfaces completely.

5. Coat the bearing surface of the rocker arms, rocker arm bolts and pushrods with prelube No. 1052365 or equivalent prior to installation.

➡ **Be sure to install all valvetrain components in their original locations.**

✳✳ WARNING

If the cylinder head has been milled as part of an overhaul, shims may be necessary under the rocker arm pedestals.

6. Install the rocker arms. Tighten the bolts to 89 inch lbs. (10 Nm) + 30° with a torque angle meter.

7. Apply a new gasket to the valve rocker arm cover. Apply sealant GM 12345739 or equivalent at the cylinder head to lower manifold joint. Install the attaching

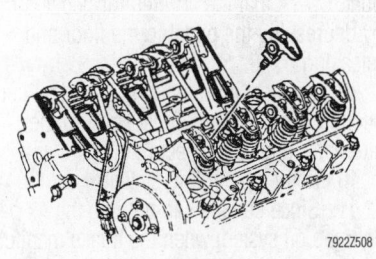

Rocker arm mounting—3.1L engine

rocker arm cover bolts to 89 inch lbs. (10 Nm).

8. Install the remaining components in the reverse order of removal.

3.4L Engine

The overhead camshaft design of the 3.4L engine uses only hydraulic lifters between the camshaft and the valves. No rocker arm assemblies are used.

3.8L Engine

❋❋ WARNING

The rocker arm bolts have been permanently stretched during installation. New original type bolts must be used each time the rocker arm assemblies have been removed for any reason.

LEFT SIDE (FRONT) ROCKER ARMS

1. Disconnect the negative battery cable.
2. Remove the engine lift bracket from the exhaust manifold studs.
3. Remove the fuel injector sight shield.
4. Remove the serpentine belt.
5. Remove the alternator-to-brace bolt (lower bolt).
6. Remove the nut and remove the alternator brace from the intake manifold.
7. Disconnect the spark plug wires from the spark plugs. Disconnect the spark plug wire cover from the rocker arm cover and position aside.
8. Remove the rocker arm cover bolt and remove the cover.

➡️**The rocker arms, pushrods, pedestals and bolts must be kept in order for installation in the same locations they were removed from.**

9. Remove the rocker arm bolt(s), pedestal(s), rocker arm(s) and pushrod(s).
To install:
10. Clean all gasket surfaces completely.

11. Apply thread lock compound to the rocker arm bolt(s).
12. Seat the pushrod in the lifter and install the rocker arm, pedestal and mounting bolt. Tighten the mounting bolt to 11 ft. lbs. (15 Nm) + 90° using a torque angle meter.
13. Install the rocker arm cover with a new gasket and tighten the mounting bolts to 89 inch lbs. (10 Nm).
14. Installation of the remaining components is the reverse of the removal procedure.
15. Connect the negative battery cable. Start the vehicle and verify no oil leaks.

RIGHT SIDE (REAR) ROCKER ARMS

1. Disconnect the negative battery cable. Remove the coolant recovery bottle.
2. Remove the serpentine belt.
3. Remove the power steering pump mounting bolts and pull the power steering pump forward.
4. Remove the fuel injector sight shield. Remove the canister purge valve from the bracket.
5. Remove the power steering pump support braces.
6. Remove the engine lift bracket from the exhaust manifold studs.
7. Disconnect the spark plug wires from the spark plugs. Disconnect the spark plug wire cover from the rocker arm cover and position aside.
8. Remove the rocker arm cover bolt and remove the cover.

➡️**The rocker arms, pushrods, pedestals and bolts must be kept in order for installation in the same locations they were removed from.**

9. Remove the rocker arm bolt(s), pedestal(s), rocker arm(s) and pushrod(s).
To install:
10. Clean all gasket surfaces completely.
11. Apply thread lock compound to the rocker arm bolt(s).

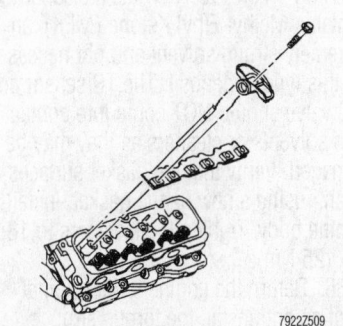

Exploded view of the rocker arm, pushrod and pushrod guide plate assembly—3.8L engine

12. Seat the pushrod in the lifter and install the rocker arm, pedestal and mounting bolt. Tighten the mounting bolt to 11 ft. lbs. (15 Nm) + 90° using a torque angle meter.
13. Install the remaining components in the reverse order of removal. Tighten the rocker arm cover bolts to 89 inch lbs. (10 Nm) and the engine lift bracket bolt to 41 ft. lbs. (55 Nm).
14. Connect the negative battery cable.
15. Start the vehicle and verify no oil leaks.

Intake Manifold

REMOVAL & INSTALLATION

❋❋ CAUTION

Fuel injection systems on these vehicles remain under pressure, even after the engine has been turned OFF. The fuel system pressure must be relieved before disconnecting any fuel lines. Failure to do so may result in fire and/or personal injury.

3.1L Engine

1. Disconnect the negative battery cable.
2. Drain the engine coolant. Remove the coolant recovery bottle.
3. Remove the air cleaner and duct assembly.
4. Remove the serpentine belt.
5. Disconnect the throttle and cruise control cables from the throttle body. Remove retaining brackets and set cable assemblies aside.
6. Disconnect the automatic transaxle vacuum modulator pipe. Disconnect the power brake booster vacuum hose from the manifold assembly.
7. Identify and tag for identification any remaining vacuum lines and disconnect from the intake manifold.

➡️**On some vehicles it will be necessary to remove the torque strut-to-engine bracket bolts, swing the torque struts aside and rotate the engine forward using tool J 41131 for clearance.**

8. Disconnect the electrical connectors from the ignition coil assembly.
9. Remove the front and rear alternator braces.
10. Disconnect any remaining electrical connectors from the intake manifold. Remove the MAP sensor.

11. Remove the spark plug wires from the spark plugs.

12. Remove the electronic ignition system and the EVAP canister purge valve mounting bracket mounting bracket.

13. Remove the EGR tube assembly from the right exhaust manifold.

14. Remove the thermostat bypass pipe nut from the upper intake manifold.

15. Remove the upper intake manifold studs and bolts, then remove the upper intake manifold and gaskets.

16. If the lower intake manifold needs to be removed, remove the fuel injector rail bolts and remove the fuel injector rail assembly.

17. Remove the heater inlet pipe assembly, upper radiator hose and tie straps retaining the heater outlet pipe and ignition wiring assembly. Disconnect the heater pipe from the heater core to the coolant pump.

18. Remove the power steering pump bolts and pump.

19. Remove the rocker arm covers.

20. Remove the lower intake manifold retaining bolts and remove the lower intake manifold and gasket.

To install:

21. Clean all parts well. Use care in cleaning old gasket material from the machined aluminum surfaces on the plenum and manifold as sharp tools may damage sealing surfaces.

22. Clean the mating surfaces to the intake manifold and engine block. Remove any loose pieces of RTV sealer.

23. Apply sealant GM 12345739 or equivalent at the engine block to manifold mating surface. The bead should be 3.0mm wide and 5.0mm thick.

24. Install the lower intake manifold and retaining bolts. Apply sealant GM 12345382 or equivalent to the threads of the bolts. Tighten the bolts in sequence to 115 inch lbs. (13 Nm).

25. Install the valve rocker covers.

26. Connect the heater pipe from the heater core to the coolant pump. Install new

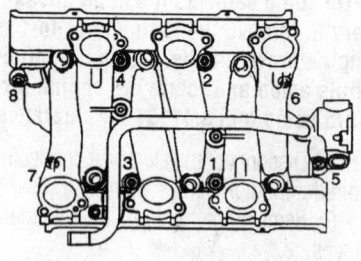

Be sure to tighten the lower intake manifold bolts in the sequence shown—3.1L engine

79227510

tie straps around the heater outlet pipe and ignition harness assembly. Connect the upper radiator hose to the engine and the heater inlet pipe to the manifold assembly.

27. Install the power steering pump and pulley.

28. Remove the injector O-ring seals from both the spray tip ends and the fuel rail end of each injector. Discard the seals. With the spray tip end O-ring removed, the O-ring back-up piece may slip off of the injector. Be sure to retain the O-ring back-up for reuse. Be sure that the O-ring back-up piece is in place on the spray tip end of the injector before installing a new O-ring. Lubricate new injector O-ring seals with clean engine oil and install on the injector assembly.

29. Install the fuel rail assembly to the intake manifold. Tilt the rail assembly to install the injectors. Install the fuel rail attaching bolts and tighten to 89 inch lbs. (10 Nm).

30. Connect the injector electrical connectors.

31. Install new O-rings on the fuel lines and install the fuel feed and return pipes. Tighten the fuel rail nuts to 13 ft. lbs. (17 Nm). Use a back-up wrench on the fittings to prevent them from turning.

32. Using new gaskets, install the intake manifold plenum. Be sure to route the MAP sensor electrical connector to the outside of the plenum gasket. Tighten the bolts to 18 ft. lbs. (25 Nm).

33. Install the serpentine drive belt. Install the coolant recovery tank.

34. Install the MAP sensor, braces to the alternator, ignition coil front bolts and the EGR to plenum bolts. Connect the vacuum lines as noted during removal.

35. If the throttle body was removed from the upper intake manifold, inspect the throttle body before installation. Throttle body bore and valve deposits may be cleaned using carburetor cleaner and a parts cleaning brush. DO NOT use a cleaner that contains Methyl Ethyl Ketone (MEK), an extremely strong solvent and not necessary for this type of deposit. The TP sensor and IAC valve should NOT come into contact with solvents or cleaners as they may be damaged. Verify that the gasket surfaces are clean,, using a new flange gasket, install the throttle body. Tighten the fasteners to 18 ft. lbs. (25 Nm).

36. Return the engine to its original location and install the torque strut.

37. Connect the throttle and cruise control cables.

38. Connect the IAC valve and TP sensor electrical connectors. Connect the air inlet

duct. Check that the accelerator pedal is free by depressing the pedal to the floor and releasing.

39. Connect all remaining electrical connections and vacuum lines. Be sure the alternator braces are secure.

40. Refill the cooling system.

41. Since coolant can get into the engine's oil system when the intake manifold is removed, change the engine oil and filter.

42. Connect the negative battery cable.

43. Turn the key to the **ON** position several times to pressurize the fuel system and check for fuel leaks.

44. After the engine is running, bleed the cooling system and check for proper idle quality.

3.4L Engine

1. Relieve the fuel system pressure. Remove the air cleaner and duct work.

2. Disconnect the negative battery cable. Drain the cooling system.

3. Disconnect the control cables from the throttle body lever and the intake plenum mounting bracket.

4. Remove the fuel rail cover bolts and remove the fuel rail cover.

5. Disconnect and cap the fuel lines from the fuel rail.

6. For 1995 vehicles, disconnect the heater hose from the intake manifold. Disconnect the PCV valve and vacuum hose from the plenum.

7. For 1996–97 vehicles, remove the emission control purge solenoid and bracket.

8. For 95 vehicles, disconnect the AIR solenoid, EGR valve and Throttle Position Sensor electrical connectors.

9. For 1996–97 vehicles, tag and disconnect the spark plug wires.

10. For 1995 vehicles, remove the EGR bolts and position the EGR valve away from the plenum.

11. For 1995 vehicles, remove the fuel line bracket from the plenum. Loosen the throttle body heater hose clamp at the plenum.

12. For 1995 vehicles, disconnect the canister purge solenoid and MAP sensor electrical connectors.

13. For 1996–97 vehicles, remove the control module mounting bolts, then remove the module.

14. Tag and disconnect the vacuum hoses from the plenum.

15. If necessary, remove the wiring loom bracket for the rear spark plug wires.

16. Remove the plenum support bracket nuts, then remove the plenum mounting bolts and remove the intake plenum.

17. Disconnect the vacuum line at the pressure regulator.

18. Remove the fuel rail assembly mounting bolts.

19. Disconnect the fuel injector electrical connectors.

20. Remove the fuel rail assembly.

21. Disconnect the radiator hose from thermostat housing.

22. Remove the heater pipe nut from the throttle body.

23. Remove the intake manifold mounting bolts and remove the intake manifold.

To install:

24. Clean all the gasket surfaces completely.

25. Install the intake manifold gasket and intake manifold.

26. For 1995 vehicles, insert new rubber isolators into the manifold flange. Tighten the mounting bolts to 18 ft. lbs. (25 Nm). Start with the center bolts and work outwards in a circular pattern.

27. For 1996–97 vehicles, install the M8 x 50mm bolts with the washer to the vertical holes in the intake manifold. Tighten these bolts to 62 inch lbs. (7 Nm). Install the lower intake manifold mounting bolts. Insert the rubber insulator fully into the manifold flange before tightening any fasteners. Tighten the bolts as outlined in the following sub-steps:

  a. Draw the lower manifold into place by tightening the bolts gradually, starting with the middle bolts and working in a circular pattern. DO NOT tighten one side more than the other.

  b. Tighten these bolts to 116 inch lbs. (13 Nm).

  c. Remove the 2 bolts in the vertical holes of the intake manifold.

28. Install the remaining components in the reverse order of removal. Tighten the following:

• Fuel rail bolts to 89 inch lbs. (10 Nm)

• Intake plenum and plenum support bracket nuts, starting in the center and working outwards in a circular pattern, to the mounting bolts to 18 ft. lbs. (25 Nm)

29. Fill and bleed the cooling system using the recommended procedure.

30. Connect the negative battery cable.

31. Install the air cleaner assembly.

32. Pressurize the fuel system and verify no leaks.

3.8L (VIN L) Engine

1. Relieve the fuel system pressure. Disconnect the negative battery cable.

2. Remove the air cleaner assembly and fuel injector sight shield.

3. Disconnect the throttle cables from the throttle body lever and intake manifold bracket.

✳✳ CAUTION

Never open, service or drain the radiator or cooling system when hot; serious burns can occur from the steam and hot coolant. Always drain coolant into a sealable container. Coolant should be reused unless it is contaminated or is several years old.

4. Remove the coolant recovery reservoir.

5. Remove the electrical cover from the right inner fender.

6. Remove the exhaust heat shield.

7. Disconnect the EGR adapter from the upper intake manifold pipe.

8. Disconnect and cap the fuel lines from the fuel rail and remove the lines from the manifold bracket.

9. Remove the serpentine belt.

10. Remove the alternator and rear alternator brace from the intake manifold.

11. Remove the throttle body cable bracket from the intake manifold.

12. Disconnect the electrical connectors from the throttle body.

13. Disconnect the fuel injector electrical connectors.

14. Disconnect the vacuum hoses from the canister purge solenoid, transaxle module and intake manifold connections.

15. Disconnect the vacuum hose quick connect at the upper intake manifold.

16. Remove the power steering pump mounting bolts and pull the pump forward.

17. Remove the serpentine belt tensioner pulley from the mounting bracket.

18. Remove the power steering pump support bracket.

19. Tag and disconnect the spark plug wires from the coil assembly and position them out of the way.

20. Drain the cooling system.

21. Disconnect the coolant bypass from the intake manifold.

22. Disconnect the heater pipes from the intake manifold and front cover.

23. Remove the solenoid valve mounting bracket and power steering pump brace from the intake manifold.

24. Disconnect the upper radiator hose from the thermostat housing.

25. Remove the thermostat housing and thermostat from the intake manifold.

26. Disconnect the electrical connector from the coolant sensor and remove the sensor from the intake manifold.

27. Remove the intake manifold mounting bolts and remove the intake manifold.

To install:

28. Clean all the gasket surfaces completely. Be sure all threaded opening are clean of dirt and old sealer. A thread-cutting tap can be used to clean threaded bolt holes.

29. Install the intake manifold with new gaskets.

30. Apply thread lock compound to the intake bolts and install the mounting bolts and tighten to 88 inch lbs. (10 Nm). Make a second pass to verify all bolts are tightened to 88 inch lbs. (10 Nm). Do not overtighten or the light alloy manifold will be damaged.

31. Install the remaining components in the reverse order of removal:

• Thermostat housing bolts to 20 ft. lbs. (27 Nm)

• Power steering pump support bracket bolts to 37 ft. lbs. (50 Nm)

• Serpentine belt pulley bolt to 33 ft. lbs. (45 Nm)

32. Refill the cooling system. An oil and filter change is recommended.

33. Install the air cleaner assembly and the fuel injector sight shield.

34. Connect the negative battery cable.

35. Pressurize the fuel system by turn the ignition switch to the **ON** position to start the fuel pump and verify no fuel leaks.

36. Start the vehicle and verify no oil, vacuum or coolant leaks.

3.8L (VIN K) Engine

1. Disconnect the negative battery cable.

2. Remove the plastic engine cover clipped to the fuel rail.

3. Remove the air cleaner and duct assembly. Tag and remove the spark plug wires.

✳✳ CAUTION

Observe all applicable safety precautions when working around fuel. Whenever servicing the fuel system, always work in a well ventilated area. Do not allow fuel spray or vapors to come in contact with a spark or open flame. Keep a dry chemical fire extinguisher near the work area. Always keep fuel in a container specifically designed for fuel storage; also, always properly seal fuel containers to avoid the possibility of fire or explosion.

4. Properly relieve the fuel system pressure. Detach the fuel injection wiring harness.

5. Remove the canister purge electrical connector.

6. Remove the fuel rail. Remove the EGR adapter to the upper intake pipe.

7. Remove the throttle body assembly.

8. Remove the upper manifold mounting bolts and remove the manifold from the vehicle.

9. Remove the inner fender electrical cover.

10. Remove the exhaust shield.

11. Remove the alternator and the brace. Remove the control cable bracket.

12. Disconnect the control cables, vacuum lines and electrical connectors from the throttle body and intake manifold.

13. Remove the vacuum lines from the canister purge and the transaxle module at the intake connection.

14. Remove the power steering pump and move forward.

15. Remove the belt tensioner mounting bracket.

16. Remove the steering pump support bracket.

17. Remove the coolant bypass hose. Remove the heater pipes from the intake and the front cover.

18. Remove the alternator support brace.

19. Remove the upper radiator hose from the thermostat housing.

20. Detach the connector from the temperature sensor.

21. Remove the lower manifold mounting bolts and remove the lower manifold.

Remove the old sealing gaskets and clean the surfaces correctly.

To install:

22. Install the new lower intake gaskets and apply sealer to the corners of the end seals. Be sure to coat the threads of the intake bolts with P/N 12345336 or the equivalent. Tighten the bolts to 89 inch lbs. (10 Nm) using the proper sequence.

23. Install the remaining components in the reverse order of removal. Tighten the

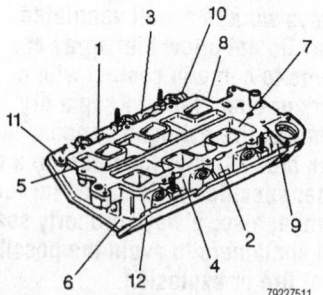

Always tighten the lower intake manifold bolts in the sequence shown—3.8L (VIN 1 and K) engines

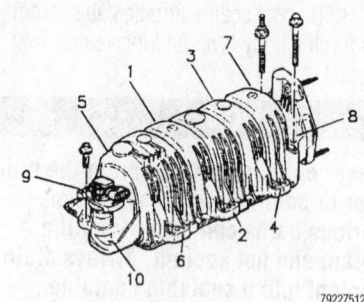

Upper intake manifold bolt tightening sequence—3.8L (VIN K) engine

upper manifold bolts, in sequence, to 11 ft. lbs. (15 Nm).

24. Fill and bleed the cooling system using the recommended procedure.

25. Install the air cleaner and duct assembly. Connect the negative battery cable.

26. Cycle the ignition key several times to ensure there are no fuel leaks before starting the vehicle.

27. Verify the proper idle quality and be sure there is no oil or coolant leaks.

3.8L (VIN 1) Engine

1. Disconnect the negative battery cable.

✳✳ CAUTION

Observe all applicable safety precautions when working around fuel. Whenever servicing the fuel system, always work in a well ventilated area. Do not allow fuel spray or vapors to come in contact with a spark or open flame. Keep a dry chemical fire extinguisher near the work area. Always keep fuel in a container specifically designed for fuel storage; also, always properly seal fuel containers to avoid the possibility of fire or explosion.

2. Relieve the fuel system pressure using the recommended procedure.

3. Raise and safely support vehicle.

4. Remove the front splash shield.

✳✳ CAUTION

Never open, service or drain the radiator or cooling system when hot; serious burns can occur from the steam and hot coolant. Also, when draining engine coolant, keep in mind that cats and dogs are attracted to ethylene glycol antifreeze and could drink any that

is left in an uncovered container or in puddles on the ground. This will prove fatal in sufficient quantities. Always drain coolant into a sealable container. Coolant should be reused unless it is contaminated or is several years old.

5. Drain the radiator coolant into a suitable container. Close the radiator drain.

6. Reinstall the front splash shield.

7. Lower the vehicle.

8. Remove the supercharger assembly.

9. Remove the thermostat housing.

10. Disconnect the EGR tube at the intake manifold.

11. Disconnect the coolant temperature sensor.

12. Remove the intake manifold.

To install:

13. Thoroughly clean all sealing surfaces.

14. Clean all old sealant from the intake manifold bolts and bolt holes.

15. Install new gaskets and manifold seals. Apply sealant to the ends of the manifold seals.

16. Install the intake manifold.

17. Install the intake manifold bolts and tighten, in sequence, to 11 ft. lbs. (15 Nm).

18. Install the electrical connector to the temperature sensor.

19. Install the EGR tube to the intake manifold.

20. Install the thermostat housing.

21. Install the supercharger assembly.

22. Connect the negative battery cable.

23. Fill and bleed the cooling system using the recommended procedure.

24. Run the engine and check for leaks and proper engine operation.

Exhaust Manifold

REMOVAL & INSTALLATION

3.1L Engine

LEFT SIDE (FRONT) MANIFOLD

1. Disconnect the negative battery cable. Remove the air cleaner and duct assembly.

2. Remove the coolant recovery bottle.

➡ For some 1995–97 vehicles, it will be necessary to rotate the engine forward by removing the torque strut and installing tool J 41131 or equivalent, then rotate the engine assembly forward.

To prevent shearing of the rubber bushing, loosen the bolts on the engine strut before swinging the struts.

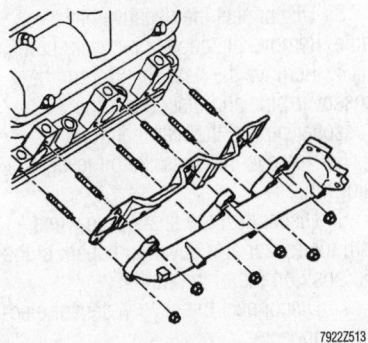

Exploded view of the left exhaust manifold mounting—3.1L engine

3. Relieve the accessory drive belt tension and remove the belt.

4. Disconnect the upper and lower radiator hoses from the engine assembly.

5. Remove the tie straps securing the heater outlet hose and the ignition wiring harness in position.

6. Remove the transaxle modulator vacuum pipe assembly.

7. Disconnect the heater pipe running from the water pump to the heater core.

8. Remove the heat shield and crossover pipe at the manifold.

9. Remove the engine mount strut.

10. Remove the air conditioning compressor bracket.

11. Remove the exhaust manifold mounting bolts and remove the manifold.

To install:

12. Clean the gasket mounting surfaces. Install the exhaust manifold to the engine, loosely install the mounting nuts.

13. Install the exhaust crossover pipe. Tighten the exhaust manifold nuts to 12 ft. lbs. (16 Nm) and the crossover pipe bolts to 18 ft. lbs. (25 Nm).

14. Install the exhaust manifold shield and tighten to 89 inch lbs. (10 Nm).

15. Install the engine mount strut and A/C compressor bracket.

16. Connect the heater pipe at the water pump and the heater core.

17. Connect the transaxle modulator vacuum pipe assembly.

18. Using new tie straps, secure the heater outlet pipe and ignition wiring harness in position.

19. Connect the upper and lower radiator hoses to the engine.

20. Install the serpentine belt.

21. Install the coolant recovery bottle and refill the cooling system. Properly bleed the cooling system.

22. Install the air cleaner and duct assembly.

23. Connect the negative battery cable. Start the engine and check for exhaust leaks.

RIGHT SIDE (REAR) MANIFOLD

1. Disconnect the negative battery cable. Remove the air cleaner and duct assembly.

2. Remove the coolant recovery bottle.

3. Disconnect the upper radiator hose from the engine.

4. Remove the tie straps securing the heater outlet hose and the ignition wiring harness in position.

5. Remove the transaxle modulator vacuum pipe assembly.

6. Disconnect the heater pipe running from the water pump to the heater core.

7. Remove the heat shield and crossover pipe at the manifold.

8. Disconnect the oxygen sensor electrical connector(s).

➡ **For some 1995–97 vehicles, it will be necessary to rotate the engine forward by removing the torque strut and installing tool J 41131 or equivalent, then rotate the engine assembly forward.**

To prevent shearing of the rubber bushing, loosen the bolts on the engine strut before swinging the struts.

9. Raise and properly support the vehicle.

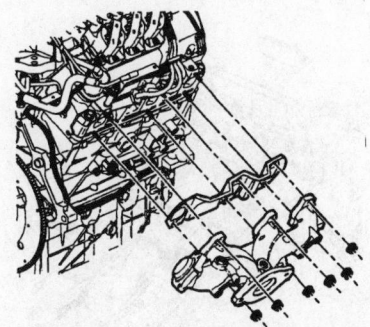

Exploded view of the right exhaust manifold mounting—3.1L engine

10. While supporting the rear or the engine carrier assembly, remove the rear mounting bolts and lower the rear portion of the carrier assembly.

11. Disconnect the front exhaust pipe from the manifold.

12. Disconnect the EGR tube from the manifold assembly.

13. Remove the oxygen sensor.

14. Remove the automatic transaxle dipstick and filler tube from the transaxle.

15. Remove the exhaust manifold heat shields and mounting bolts. Remove the manifold assembly from the vehicle.

To install:

16. Clean the gasket mounting surfaces.

17. Install the exhaust manifold to the engine and tighten the mounting nuts to 12 ft. lbs. (16 Nm).

18. Install the automatic transaxle dipstick and filler tube from the transaxle.

19. Install the oxygen sensor and connect the EGR tube the manifold.

20. Raise the rear engine carrier assembly into position. Install the mounting bolts and tighten to 107 ft. lbs. (145 Nm).

21. Lower the vehicle.

22. Install the exhaust crossover pipe and tighten the crossover pipe bolts to 18 ft. lbs. (25 Nm)

23. Connect the heater pipe at the water pump and the heater core.

24. Connect the transaxle modulator vacuum pipe assembly.

25. Using new tie straps, secure the heater outlet pipe and ignition wiring harness in position.

26. Connect the upper radiator hose to the engine.

27. Install the coolant recovery bottle and refill the cooling system. Properly bleed the cooling system.

28. Install the air cleaner and duct assembly.

29. Connect the negative battery cable.

30. Start the engine and check for exhaust leaks.

3.4L Engine

LEFT SIDE (FRONT) MANIFOLD

1. Remove the air cleaner assembly. Disconnect the negative battery cable.

2. Remove the exhaust crossover pipe. remove the cooling fans.

3. If equipped with a manual transaxle, disconnect the AIR hose from the AIR pipe on the exhaust manifold.

4. Remove the exhaust manifold nuts, then remove the exhaust manifold, gasket and heat shield.

To install:

5. Clean all gasket mating surfaces completely. Install a new exhaust manifold gasket.

6. Install the exhaust manifold and heat shield and tighten the mounting nuts to 115 inch lbs. (13 Nm).

7. Install the remaining components in the reverse order of removal. Tighten the exhaust crossover bolts to 18 ft. lbs. (25 Nm).

8. Connect the negative battery cable. Start the engine and check for exhaust leaks.

RIGHT SIDE (REAR) MANIFOLD

1. Remove the air cleaner and duct assembly.

2. Disconnect the negative battery cable. Remove the exhaust crossover pipe.

3. Remove the EGR tube from the exhaust manifold.

4. Raise and safely support the vehicle. Remove the front exhaust pipe and converter assembly.

5. Remove the oxygen sensor.

6. Remove the exhaust pipe front heat shield.

7. Remove the rear alternator brace. Remove the transmission dipstick tube.

8. Remove the intermediate shaft from the steering gear.

9. Remove the exhaust manifold attaching nuts.

10. Install a jacking fixture to support the rear cradle.

11. Remove the rear cradle bolts.

12. Safely lower the engine cradle.

13. Remove the steering gear heat shield.

14. Remove the exhaust manifold and heat shield and attaching gasket.

To install:

15. Clean all gasket surfaces completely.

16. Install the exhaust manifold gasket, manifold and heat shields.

17. Install exhaust mounting manifold nuts and tighten to 115 inch lbs. (13 Nm).

18. Install the remaining components in the reverse order of removal. Tighten the following:

• Rear cradle bolts to 125 ft. lbs. (170 Nm)

• Intermediate shaft-to-steering gear pinch bolt to 35 ft. lbs. (47 Nm)

• Exhaust crossover pipe and tighten the nuts to 18 ft. lbs. (25 Nm)

19. Connect the negative battery cable. Install the air cleaner and duct assembly.

20. Start the engine and check for exhaust leaks.

➡ **Whenever the vehicle sub-frame is removed or lowered, the wheel alignment should be checked.**

3.8L Engine

LEFT SIDE (FRONT) MANIFOLD

1. Remove the air cleaner and air duct assembly.

2. Disconnect the negative battery cable. Remove the torque struts.

3. Remove the right engine cooling fan.

4. Remove the heat shield from the crossover pipe and disconnect the crossover pipe at the rear manifold.

5. Drain the cooling system.

6. Disconnect the upper radiator hose from the thermostat housing.

7. Remove the fuel rail sight cover.

8. Disconnect the EGR adapter-to-intake manifold pipe.

9. Remove the EGR valve.

10. Remove the front engine lift hook.

11. Remove the torque strut bracket from the left cylinder head.

12. Remove the oil level indicator tube.

13. Disconnect the spark plug wires from the rocker arm cover and spark plugs and position out of the way.

14. Remove the A/C compressor support brace from the exhaust manifold stud.

15. Remove the manifold mounting bolts and remove the exhaust manifold.

To install:

16. Clean all the gasket surfaces completely.

17. Install the exhaust manifold and loosely install the mounting bolts. Starting with the center bolts and working outward tighten the bolts to 38 ft. lbs. (52 Nm).

18. Install the spark plugs and tighten to 11 ft. lbs. (15 Nm).

19. Install the remaining components in the reverse order of removal. Tighten the following:

• Engine lift hook nut to 22 ft. lbs. (30 Nm)

• Crossover pipe-to-rear exhaust manifold bolts to 22 ft. lbs. (30 Nm)

• Tighten the strut mounting nuts to 41 ft. lbs. (56 Nm)

20. Connect the negative battery cable. Install the air cleaner and duct assembly.

21. Start the vehicle and verify no leaks.

RIGHT SIDE (REAR) MANIFOLD

1. Remove the air cleaner and air duct assembly.

2. Remove the fuel rail sight cover.

3. Disconnect the negative battery cable. Remove the coolant recovery bottle.

4. Remove the heat shield from the crossover pipe and disconnect the crossover pipe at the rear manifold.

5. Remove the transaxle oil level indicator tube.

6. Disconnect the spark plug wires from the rocker arm cover and spark plugs and position out of the way.

7. Disconnect the oxygen sensor electrical connector.

8. Remove the torque struts.

9. Remove the rear engine lift hook. Remove the spark plugs.

10. Raise and safely support the vehicle. Remove the front exhaust pipe and converter assembly.

11. Remove the rear engine mount-to-frame nuts.

12. Lower the vehicle.

13. Using a floor jack raise the right rear corner of the engine to provide clearance.

14. Remove the manifold mounting bolts and remove the exhaust manifold.

To install:

15. Clean all the gasket surfaces completely.

16. Install the exhaust manifold and loosely install the mounting bolts. Starting with the center bolts and working outward tighten the bolts to 38 ft. lbs. (52 Nm).

17. Connect the crossover pipe to the rear exhaust manifold and tighten the mounting bolts to 22 ft. lbs. (30 Nm).

18. Install the remaining components in the reverse order of removal. Tighten the following:

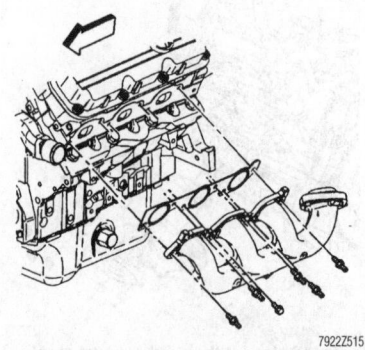

Exploded view of the left exhaust manifold mounting—3.8L engine

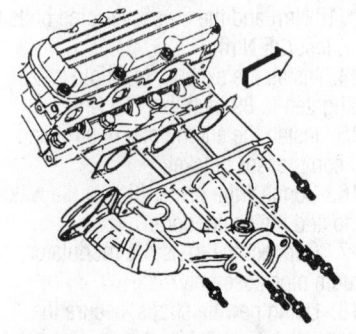

Exploded view of the right exhaust manifold mounting—3.8L engine

- Rear engine mount nuts to 50 ft. lbs. (68 Nm)
- Engine lift hook nut to 22 ft. lbs. (30 Nm)
- Spark plugs to 11 ft. lbs. (15 Nm)
- Torque the strut mounting nuts to 41 ft. lbs. (56 Nm)

19. Connect the negative battery cable. Install the air cleaner and duct assembly.

20. Start the vehicle and verify no leaks.

Camshaft

REMOVAL & INSTALLATION

❋❋ CAUTION

Fuel injection systems on these vehicles remain under pressure, even after the engine has been turned OFF. The fuel system pressure must be relieved before disconnecting any fuel lines. Failure to do so may result in fire and/or personal injury.

3.1L Engine

1. Relieve the fuel system pressure. Disconnect the negative battery cable.

2. Remove the engine assembly from the vehicle using the recommended procedure, and mount on a suitable engine stand.

➡️**When removing valvetrain components they must be marked for installation in the same location they are removed from. When the camshaft is being replaced the valve lifters should also be replaced.**

3. Remove the intake manifold, valve cover, rocker arms, pushrods and valve lifters.

4. Remove the using a puller to draw the crankshaft balancer from the crankshaft nose.

5. Remove the timing chain front cover.

6. It is good practice to set the engine to Top Dead Center No. 1 cylinder (firing position) before disassembling the timing chain and sprockets. This should align all the timing marks and serve as a point of reference for later work. Remove the timing chain and sprockets.

❋❋ CAUTION

If the camshaft was replaced the valve liters must also be replaced. The old lifters have already developed a wear pattern from the old camshaft and if installed on the new camshaft they will cause premature wear of the camshaft lobes.

7. Remove the oil pump driven gear so the camshaft can be pulled out in Step 10.

8. Remove the two bolts and remove the camshaft thrust plate.

9. Carefully remove the camshaft. Avoid marring the camshaft bearing surfaces.

To install:

10. Coat the camshaft with lubricant GM part number 1052365 or equivalent camshaft break-in lubricant or quality engine oil supplement, and install the camshaft.

11. Install the camshaft thrust plate and tighten the mounting bolts to 89 inch lbs. (10 Nm).

12. Install the oil pump driven gear and tighten the mounting bolt to 27 ft. lbs. (36 Nm).

13. Install the timing chain and sprocket.

14. Verify that all timing marks are aligned. It is good practice to turn the crankshaft several complete rotations to be sure the timing marks still align when the crankshaft is returned to Top Dead Center No. 1 cylinder compression stroke (firing position). This is most important. If the valvetrain is not correctly timed, the engine will be damaged on start-up. When satisfied that all marks are correctly aligned, install the camshaft thrust button and front cover.

15. Install the crankshaft balancer and tighten the bolt to 75 ft. lbs. (103 Nm).

16. Install the intake manifold, valve cover, rocker arms, pushrods and valve lifters.

17. Install the engine assembly into the vehicle using the recommended procedure.

18. Connect the negative battery cable.

19. Verify that all fluid levels (coolant, engine oil, etc.) are full and correct. An oil and filter change is recommended.

➡️**The only time valve adjustment in needed is if there was a valve job performed or the rocker studs have been replaced with an adjustable rocker arm stud. The rocker arm stud installed from the factory should be shouldered and not need any adjustment.**

20. Adjust the valves, as required.

21. Start the engine and verify no oil leaks.

3.4L Engine

The 3.4L VIN X engine has aluminum cam carriers which house the camshafts with the actual aluminum of the carrier serving as the camshaft bearing surfaces. Use care when working with these light alloy parts. They can be damaged or broken if carelessly handled.

DOHC engines have numerous valvetrain timing marks which must be aligned or serious engine damage will result. It may be helpful to set the engine to TDC, compression stroke of No. 1 cylinder (firing position) before beginning work. This gives a known point-of-reference. In this way as the work progresses, timing marks can be observed and noted which should save time at assembly. In addition, note the routing of the timing belt so that it can eventually be properly installed.

LEFT SIDE CAMSHAFT

1. Remove the air cleaner and duct assembly.

2. Disconnect the negative battery cable. Drain the cooling system.

3. Remove the left side camshaft cover. Remove the timing belt assembly.

4. Install six 6 in. sections of vacuum/fuel line hose under camshaft and between lifters. This will hold lifters in the carrier. For this procedure use 5/16 in. (8mm) vacuum/fuel line hose for exhaust valves and 7/32 in. (5.5mm) vacuum/fuel line hose for the intake valves.

5. Remove the exhaust crossover pipe. Remove the upper radiator hose.

6. Disconnect the heater pipe hose from the intake plenum.

7. Remove the front exhaust manifold.

8. Disconnect the engine front coolant pipe and position out of the way.

9. Remove the camshaft carrier mounting bolts.

10. Remove the camshaft carrier and gasket from the engine.

11. Place the camshaft carrier on a suitable work surface.

12. Remove the hoses securing the lifters and remove the lifters from the camshaft carrier.

13. Clean the oil out of the camshaft hold down tool mounting hole and install J-38613, or an equivalent camshaft hold down tool. Tighten the tool mounting bolt to 22 ft. lbs. (30 Nm).

14. Remove the camshaft sprocket bolts and washers, while holding the camshaft from turning using J-38614, or an equivalent camshaft sprocket holding tool.

15. Using a suitable sprocket puller, J-38616, or the equivalent, remove the camshaft sprockets.

16. Remove the six thrust plate cover screws and remove the cover and gasket.

17. Remove the two camshaft thrust plate bolts and remove the thrust plate.

18. Remove tool J-38613.

✳✳ CAUTION

The camshaft bearing journals are all the same diameter and care must be used when removing the camshaft to prevent damage to the camshaft and/or camshaft carrier.

19. Remove the camshaft by carefully sliding it out the back of the camshaft carrier.

20. Remove the camshaft oil seal using a suitable prying tool. Do not allow the prying tool to scratch the camshaft carrier.

To install:

21. Coat the lips of the camshaft seals with clean engine oil and using J-38619, or an equivalent seal installer, seat the seals in their recesses.

22. Coat the camshaft journals and lobes with engine lube, GM 1052637 or the equivalent.

23. Carefully install the camshaft into the camshaft carrier. Be sure the camshaft does not distort the camshaft seal during installation.

24. Install the thrust plate and tighten the two mounting bolts to 89 inch lbs. (10 Nm).

25. Install the thrust plate cover and gasket and tighten the six mounting bolts to 89 inch lbs. (10 Nm).

26. Install the camshaft holding tool and tighten the mounting bolt to 22 ft. lbs. (30 Nm).

27. Install a new camshaft carrier gasket on the cylinder head and install the camshaft carrier on the cylinder head.

28. Install the mounting bolts and tighten to 18 ft. lbs. (25 Nm).

29. Install the lifters under the camshafts. Install the camshaft sprockets and timing belt.

✳✳ CAUTION

After installing the camshaft sprockets and timing belt the Cam Timing must be set. Failure to properly set the camshaft timing will result in serious engine damage.

30. Remove the J-38613.

31. Install the front engine coolant pipe. Install the front exhaust manifold.

32. Connect the heater hose pipe to the intake plenum.

33. Connect the upper radiator hose.

34. Install the exhaust crossover pipe. Install the camshaft cover.

➡**Before tightening the cover mounting bolts, fully seat the bolt insulators in the camshaft cover.**

35. Connect the negative battery cable. Install the air cleaner and air duct assembly.

36. Refill the cooling system.

37. Start the vehicle and verify no leaks and proper engine performance.

RIGHT SIDE CAMSHAFT

1. Remove the air cleaner and duct assembly.

2. Disconnect the negative battery cable. Drain the cooling system.

3. Remove the right side camshaft cover.

4. Remove the timing belt.

5. Install six 6 in. sections of vacuum/fuel line hose under camshaft and between lifters. This will hold lifters in the carrier. For this procedure use 5/16 in. (8mm) fuel line hose for exhaust valves and 7/32 in. (5.5mm) vacuum/fuel line hose for the intake valves.

6. Remove the front and rear engine lift brackets.

7. Remove the camshaft carrier mounting bolts, then remove the carrier and gasket from the engine.

8. Place the camshaft carrier on a suitable work surface.

9. Remove the hoses securing the lifters and remove the lifters from the camshaft carrier.

10. Clean the oil out of the camshaft hold down tool mounting hole and install J-38613, or an equivalent camshaft hold down tool. Tighten the tool mounting bolt to 22 ft. lbs. (30 Nm).

11. Remove the camshaft sprocket bolts and washers, while holding the camshaft from turning using J-38614, or an equivalent camshaft sprocket holding tool.

12. Using a suitable sprocket puller, J-38616, or the equivalent, remove the camshaft sprockets.

13. Remove the six thrust plate cover screws and remove the cover and gasket.

14. Remove the two camshaft thrust plate bolts and remove the thrust plate.

15. Remove the J-38613.

✳✳ CAUTION

The camshaft bearing journals are all the same diameter and care must be used when removing the camshaft to prevent damage to the camshaft and/or camshaft carrier.

16. Remove the camshaft by carefully sliding it out the back of the camshaft carrier.

17. Remove the camshaft oil seal using a suitable prying tool. Do not allow the prying tool to scratch the camshaft carrier.

To install:

18. Coat the lips of the camshaft seals with clean engine oil and using J-38619, or an equivalent seal installer, seat the seals in their recesses.

19. Coat the camshaft journals and lobes with engine lube, GM 1052637 or the equivalent.

20. Carefully install the camshaft into the camshaft carrier. Be sure the camshaft does not distort the camshaft seal during installation.

21. Install the thrust plate and tighten the two mounting bolts to 89 inch lbs. (10 Nm).

22. Install the thrust plate cover and gasket and tighten the six mounting bolts to 89 inch lbs. (10 Nm).

23. Install the camshaft holding tool and tighten the mounting bolt to 22 ft. lbs. (30 Nm).

24. Install a new camshaft carrier gasket on the cylinder head and install the camshaft carrier on the cylinder head.

25. Install the mounting bolts and tighten to 18 ft. lbs. (25 Nm).

26. Install the lifters under the camshafts.

27. Install the camshaft sprockets and timing belt.

✳✳ CAUTION

After installing the camshaft sprockets and timing belt the Cam Timing must be set. Failure to properly set the camshaft timing will result in serious engine damage.

28. Remove tool J-38613.

29. Install the front and rear engine lift brackets.

30. Install the camshaft cover.

➡**Before tightening the cover mounting bolts, fully seat the bolt insulators in the camshaft cover.**

31. Connect the negative battery cable. Install the air cleaner and air duct assembly.

32. Fill and bleed the cooling system using the recommended procedure.

33. Start the vehicle and verify no leaks and proper engine performance.

3.8L Engine

1. Disconnect the negative battery cable.

2. Relieve the fuel system pressure using the recommended procedure.

3. Remove the engine assembly from the vehicle and mount on a suitable engine stand.

4. If equipped, remove the supercharger.

5. Remove the intake manifold.

6. Remove the rocker arm covers.

7. Remove the rocker arm assemblies, pushrods and lifters. Identify all parts as they are removed, so they can be reinstalled in their original locations.

8. Remove the crankshaft balancer center bolt, using a suitable puller, remove the balancer from the crankshaft.

9. Remove the timing chain front cover.

10. Set the engine to Top Dead Center (TDC) No. 1 cylinder (firing position) to align the timing marks, before disassembling the timing chain and sprockets.

➡**Align the timing marks of the camshaft and crankshaft sprockets to avoid burring the camshaft journals by the crankshaft.**

11. Remove the camshaft sprocket and timing chain.

12. Remove the camshaft thrust plate bolts and remove the thrust plate.

13. Carefully remove the camshaft from the engine block.

14. Inspect the camshaft lobes and journals for wear and/or damage; replace as necessary.

➡**If the camshaft was replaced the lifters must also be replaced. The old lifters have developed a wear pattern and will cause the new camshaft to wear prematurely.**

To install:

15. Coat the camshaft lobes and bearings with lubricant GM part number 1052365 or equivalent camshaft break-in prelube prior to installation.

16. Carefully install the camshaft into the engine.

17. Install the camshaft thrust plate and tighten the mounting bolts to 11 ft. lbs. (15 Nm).

18. Install the camshaft sprocket and timing chain. Be sure the timing marks are

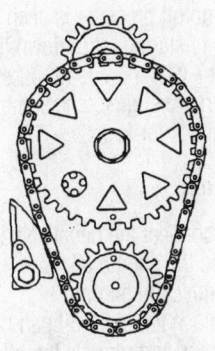

79222517

The timing marks should face each other if the chain and gears are installed properly

aligned. Tighten the camshaft sprocket retaining bolt to 74 ft. lbs. (100 Nm) plus an additional 90 degree (¼) turn.

19. Install the timing chain front cover.

20. Install the crankshaft balancer and tighten the mounting bolt to 111 ft. lbs. (150 Nm) plus an additional 76 degree turn.

21. Coat the valve lifters with camshaft prelube and install the lifters in the lifter bores.

22. Install the lifter guides and lifter guide retainer. Tighten the retainer mounting bolts to 22 ft. lbs. (30 Nm).

23. Install the pushrods and rocker arms and tighten the rocker arm bolts to 11 ft. lbs. (15 Nm) plus an additional 90 degree turn.

24. Install the rocker arm covers.

25. Install the intake manifold.

26. If equipped, install the supercharger.

27. Install the engine in the vehicle.

28. Connect the negative battery cable.

29. Verify that all fluid levels are full and correct.

30. Start the engine and check for leaks. Check engine operation.

Valve Lash

ADJUSTMENT

The valve clearance cannot be adjusted on these engines. The hydraulic lifters function to maintain a zero clearance when the valves are opening and closing. Any clearance is instantaneously taken up by the hydraulic action. "Valve lifter noise" complaints may require cleaning of sludge coated or varnished valve lifters, or replacement.

Oil Pan

REMOVAL & INSTALLATION

3.1L Engine

1. Disconnect the negative battery cable. Remove the hood.

2. Remove the torque struts and A/C compressor/torque strut mounting bracket.

3. Remove the cooling fan assemblies.

4. Install an engine support fixture, J-28467-A, or the equivalent.

5. Raise and safely support the vehicle.

6. Disconnect the front pipe from the exhaust manifold.

7. Remove the pinch bolt from the intermediate steering shaft and disconnect the steering shaft from the rack and pinion unit.

8. Drain the engine oil.

9. Disconnect and remove the oil level indicator.

10. Remove the engine splash shield.

11. Support the frame assembly with suitable floor stands.

12. Remove the transaxle-to-frame nuts and the engine mount-to-frame nuts.

13. Remove the frame mounting bolts and lower the fame slightly.

14. Remove the front engine mount and engine bracket.

15. Remove the flywheel cover.

16. Remove the starter.

17. Remove the transaxle mount from the oil pan.

18. Remove the oil pan mounting bolts and remove the oil pan.

To install:

19. Clean all gasket surfaces completely. Install the oil pan in position with new gaskets.

20. Tighten the oil pan rail bolts to 18 ft. lbs. (25 Nm) and the oil pan side bolts to 37 ft. lbs. (50 Nm).

21. Install the remaining components in the reverse order of removal. Tighten the frame mounting bolts to 103 ft. lbs. (140 Nm) and the intermediate shaft-to-rack and pinion unit pinch bolt to 35 ft. lbs. (47 Nm).

22. Refill the crankcase with oil and the power steering if disconnecting the pressure lines for clearance.

23. Connect the negative battery cable. Start the vehicle and verify no oil leaks.

➡**Whenever the vehicle sub-frame is removed or lowered, the wheel alignment should be checked.**

3.4L Engine

1. Remove the air cleaner assembly. Disconnect the negative battery cable. Drain the cooling system.

2. Remove the coolant recovery tank.

3. Install an engine support fixture J-28467-A, J-28467-90, J-36462, or their equivalents.

4. Raise and safely support the vehicle. Remove the front tire and wheel assemblies.

5. Drain the engine oil.

6. Remove the steering rack and pinion gear assembly retaining bolts and hang the steering gear from the body.

7. Remove the cotter pins and castle nuts from the lower ball joints and separate the lower ball joints from the lower control arms.

8. Remove the power steering cooler line clamps at the frame.

9. Remove the engine mount nuts at the frame.

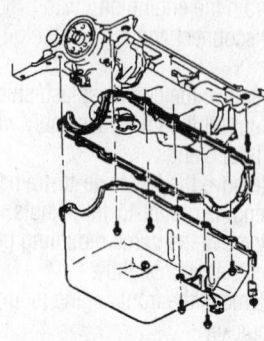

79222518

Exploded view of the oil pan mounting— 3.4L engine

10. Support the frame with a suitable jack and remove the frame mounting bolts and remove the frame assembly.

11. Remove the oil filter.

12. Remove the engine oil cooler as follows:

a. Disconnect the oil pressure sender and crankshaft sensor electrical connectors.

b. Disconnect the outlet hose and position aside.

c. Disconnect the inlet hose and position aside.

d. Remove the mounting bolts.

e. Remove the oil cooler.

13. Remove the starter.

14. Remove the flywheel cover.

15. Remove the oil pan retaining nuts and bolts.

16. Remove the oil pan and gasket.

To install:

17. Clean all the gasket surfaces completely. Install a new oil pan gasket and apply sealer to the gasket near the rear main bearing cap.

18. Install the oil pan and fasteners and tighten as follows:

- Oil pan nuts: 97 inch lbs. (11 Nm).
- Rear mounting bolts: 20 ft. lbs. (27 Nm).
- Remainder of bolts: 97 inch lbs. (11 Nm).

19. Install the remaining components in the reverse order of removal. Tighten the following:

- Engine oil cooler bolts to 24 ft. lbs. (33 Nm)
- Frame mounting bolts to 103 ft. lbs. (140 Nm)
- Ball joint-to-control arm castle nuts to 63 ft. lbs. (85 Nm)
- Steering gear mounting bolts 59 ft. lbs. (80 Nm)

20. Install the coolant recovery tank. Refill the crankcase with oil.

21. Fill and bleed the cooling system. Connect the negative battery cable.

22. Install the air cleaner assembly.

23. Start the vehicle and verify no oil leaks.

➡ **Whenever the vehicle sub-frame is removed or lowered, the wheel alignment should be checked.**

24. Check the front end alignment and correct if necessary.

3.8L Engine

1. Disconnect the negative battery cable. Remove the torque struts.

2. Raise and safely support the vehicle.

3. Disconnect the front exhaust pipe from the exhaust manifold.

4. Remove the right front tire and wheel assembly.

5. Remove the right inner fender well splash shield.

6. Drain the engine oil.

7. Disconnect and cap the engine oil cooler pipes.

8. Safely support the weight of the engine since the engine mounts will be disconnected. Use a suitable lifting/holding device.

9. Remove the engine mount-to-frame nuts from both front engine mounts.

10. Remove the flywheel inspection cover.

11. Raise the powertrain assembly at the transaxle to provide clearance using a suitable transmission jack.

➡ **The oil level sensor should be removed before removing the oil pan or damage to the oil level sensor may occur.**

12. Remove the oil level sensor.

13. Remove the oil pan mounting bolts.

14. Lower the oil pan and disconnect the oil pump screen assembly and lower it into the oil pan.

15. Remove the oil pan and pump screen together.

To install:

16. Clean all parts well, especially the gasket sealing flanges on the engine block and the oil pan.

17. Install a new oil pan gasket onto the oil pan.

18. Place the oil pump screen in the oil pan and position the pan under the engine.

19. Mount the screen on the oil pump and tighten the mounting bolt to 10 ft. lbs. (14 Nm).

20. Raise the oil pan into position and tighten the mounting bolts to 10 ft. lbs. (14 Nm).

21. Install the oil level sensor.

22. Install the remaining components in the reverse order of removal. Tighten the following:

- Engine mount nuts to 50 ft. lbs. (68 Nm)
- Tighten the strut mounting bolts to 41 ft. lbs. (56 Nm)

❄❄ WARNING

Operating the engine without the proper amount and type of engine oil will result in severe engine damage.

23. Refill the crankcase with clean engine oil.

24. Connect the negative battery cable. Start the vehicle and verify no oil leaks.

Oil Pump

REMOVAL & INSTALLATION

3.1L Engine

1. Disconnect the negative battery cable. Raise and safely support the vehicle.

2. Drain the engine oil, then remove the oil pan.

3. Remove the crankshaft oil deflector bolts, then remove the deflector.

4. Remove the oil pump retaining bolts and remove the oil pump and pump driveshaft.

To install:

5. Install the oil pump and pump driveshaft. Tighten the oil pump mounting bolts to 30 ft. lbs. (41 Nm).

6. Install the crankshaft oil deflector and mounting bolts. Tighten the mounting bolts to 18 ft. lbs. (25 Nm).

7. Install the oil pan.

8. Lower the vehicle.

9. Fill the crankcase to the correct level with oil.

10. It is good practice to install a reliable mechanical oil pressure gauge so that actual pump pressure can be read after startup. If no oil pressure is seen within a short time of startup, shut down the engine to determine why there is no oil pressure.

11. Start the engine, check the oil pressure and check for leaks.

3.4L Engine

1. Disconnect the negative battery cable. Drain the engine oil.

2. Remove the oil pan.

3. Remove the eight oil pan baffle mounting nuts and remove the oil pan baffle.

4. Remove the oil pump retaining bolt.

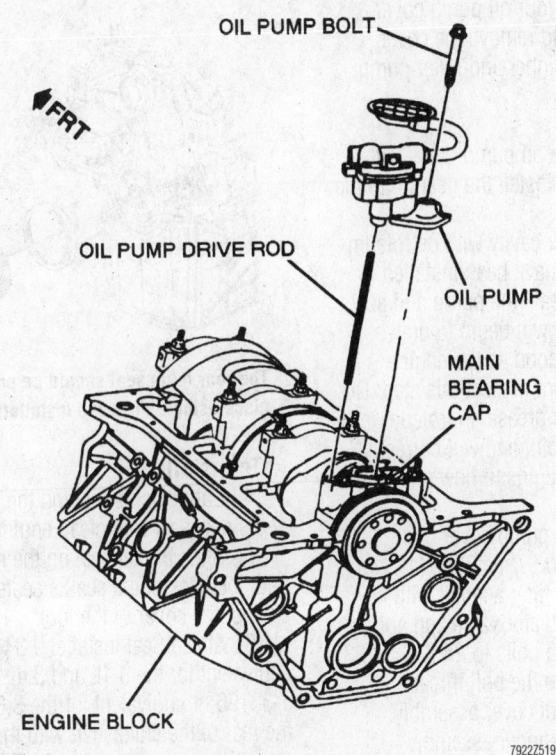

OIL PUMP BOLT

OIL PUMP DRIVE ROD

OIL PUMP

MAIN BEARING CAP

ENGINE BLOCK

Oil pump and driveshaft components—3.1L engine

5. Remove the oil pump with pick-up and drive shaft extension.

To install:

6. Clean all parts well. Clean the inside of the oil pan, the gasket flanges on pan and gasket rail on the block.

7. Install the oil pump with the drive-shaft extension and pick-up assembly onto the rear main bearing cap making sure the driveshaft extension engages in the oil pump drive.

8. Install the oil pump retaining bolt and tighten to 40 ft. lbs. 54 (Nm).

9. Install the oil pan baffle and tighten the mounting nuts to 18 ft. lbs. (25 Nm).

10. Install the oil pan following the recommended procedure.

11. Refill the crankcase with oil.

12. Connect the negative battery cable.

13. Start the vehicle and verify proper oil pressure, 15 lbs. minimum at 1100 rpm at normal operating temperature.

3.8L Engines

VIN L MODELS

1. Disconnect the negative battery cable. Raise and safely support the vehicle.

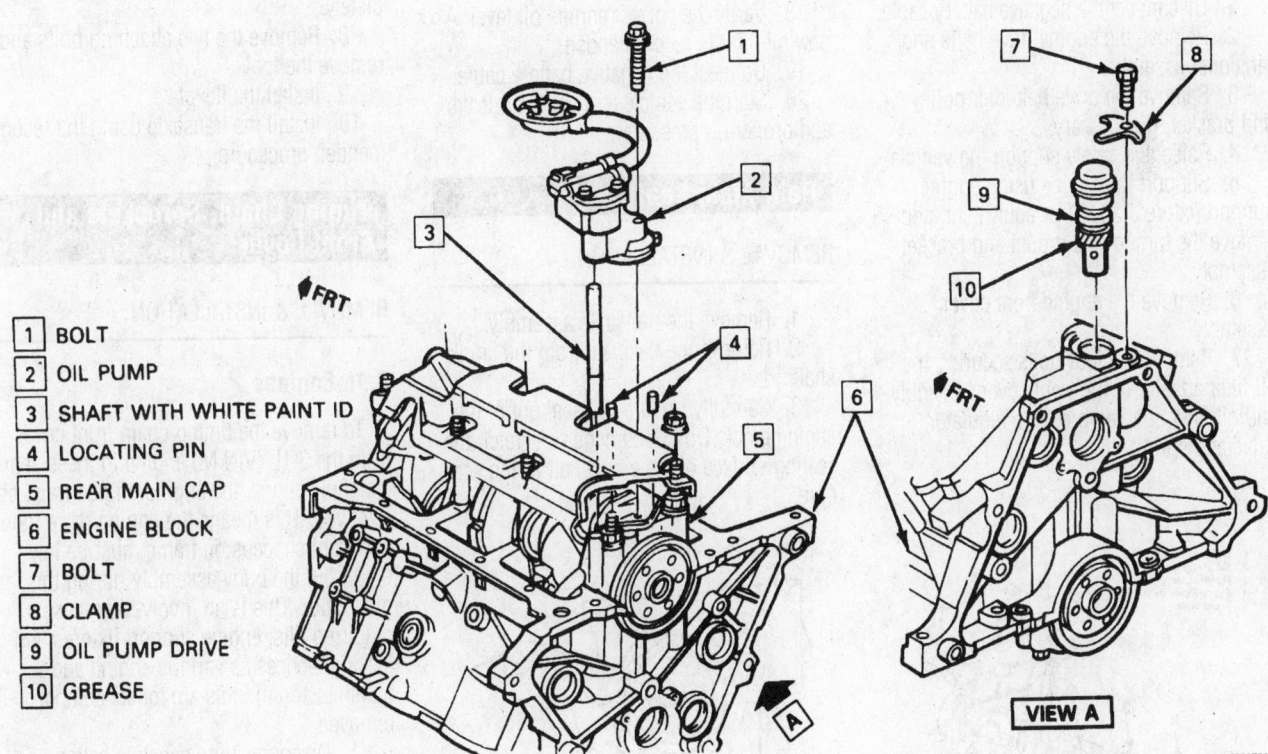

1	BOLT
2	OIL PUMP
3	SHAFT WITH WHITE PAINT ID
4	LOCATING PIN
5	REAR MAIN CAP
6	ENGINE BLOCK
7	BOLT
8	CLAMP
9	OIL PUMP DRIVE
10	GREASE

VIEW A

Oil pump and drivegear assembly—3.4L engine

2. Drain the engine oil.

3. Remove the front cover assembly.

4. Remove the four bolts securing the oil filter adapter to the front cover assembly and oil filter adapter, pressure regulator valve and spring.

5. Remove the four oil pump cover attaching screws and remove the cover.

6. Remove the inner and outer pump gears.

To install:

7. Lubricate the gears with petroleum jelly and install the gears into the housing.

8. Pack the gear cavity with petroleum jelly after the gears have been installed in the housing.

9. Install the oil pump cover and screws and tighten to 97 inch lbs. (11 Nm).

10. Install the oil filter adapter with new gasket, pressure regulator valve and spring. Tighten the mounting bolts to 24 ft. lbs. (33 Nm).

11. Install the front cover assembly.

12. Refill the crankcase with clean engine oil.

13. Connect the negative battery cable. Start the vehicle and verify no leaks and proper oil pressure.

VIN K AND 1 MODELS

1. Disconnect the negative battery cable.

2. Remove the engine drive belts and tensioner assembly.

3. Remove the drive belt idler pulley and bracket, if necessary.

4. Raise and safely support the vehicle.

5. Support the engine using engine support fixture J 28467 or equivalent, and remove the torque axis mount and bracket assembly.

6. Remove the engine front cover assembly.

7. Remove the four bolts securing the oil filter adapter to the front cover assembly and oil filter adapter, pressure regulator valve and spring.

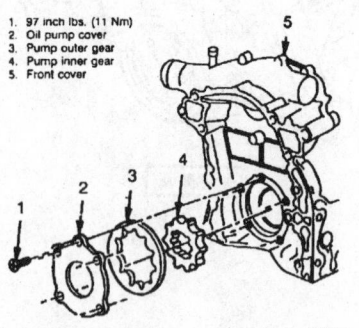

1. 97 inch lbs. (11 Nm)
2. Oil pump cover
3. Pump outer gear
4. Pump inner gear
5. Front cover

Oil pump assembly—3.8L engine

8. Remove the four oil pump cover attaching screws and remove the cover.

9. Remove the inner and outer pump gears and inspect.

To install:

10. Lubricate the oil pump gears with petroleum jelly and install the gears into the housing.

11. Pack the gear cavity with petroleum jelly after the gears have been installed in the housing. This seals the pump and acts like a "prime" so oil will begin to drawn from the oil pan as soon as the engine begins to turn. Do not neglect this step. DO NOT use any type of grease. Petroleum jelly has a low melting point and will correctly dissipate when oil begins to flow and it is no longer needed.

12. Install the oil pump cover and screws and tighten to 97 inch lbs. (11 Nm).

13. Install the oil filter adapter with new gasket, pressure regulator valve and spring. Tighten the mounting bolts to 24 ft. lbs. (33 Nm). Apply sealant to the bolt threads.

14. Install the front cover assembly.

15. Install the tensioner assembly.

16. Install the drive belt idler pulley and bracket, if removed.

17. Install the torque axis mount assembly and remove the engine support fixture.

18. Verify the correct engine oil level. A new oil filter is recommended.

19. Connect the negative battery cable.

20. Start the vehicle and verify no leaks and proper oil pressure.

Rear Main Seal

REMOVAL & INSTALLATION

1. Remove the transaxle assembly.

2. Remove the flexplate from the crankshaft.

3. Carefully pry the old seal out of the engine block. Do not damage or scratch the sealing surface of the crankshaft or the seal bore.

Carefully pry the seal from the bore without scratching the crankshaft seal surface

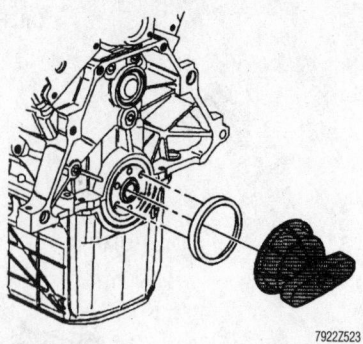

The rear main seal should be pressed into place using a threaded installation tool

To install:

4. Lubricate the lip and the outer edge of the new seal with clean engine oil.

5. Slide the oil seal on the mandrel until the back of the seal is seated squarely against the collar of the tool.

6. Attach Seal Installer J 34686 or equivalent for the 3.1L and 3.4L engines or J-38196 or equivalent for the 3.8L engine to the rear of the crankshaft with the two mounting bolts, then turn the T-handle until the oil seal is fully seated into the rear of the engine.

7. Loosen the T-handle of the tool completely.

8. Remove the two attaching bolts and remove the tool

9. Install the flexplate.

10. Install the transaxle using the recommended procedure.

Timing Chain, Sprockets and Front Cover

REMOVAL & INSTALLATION

3.1L Engines

To remove the timing chain front cover from the 3.1L (VIN M) engine in these vehicle applications, the engine oil pan must be removed. This means the engine/drive train and front suspension frame must be lowered from the body assembly during this procedure. This is an involved process requiring lifts, engine support fixtures and jacking devices as various engine and transmission mounts are loosened and/or removed.

1. Disconnect the negative battery cable. Drain the engine coolant.

2. Remove the serpentine drive belt.

3. Remove the hood assembly. An assistant may be required to avoid damage to vehicle's paintwork.

➡ **Do not allow the hood to fold back onto the windshield. Glass and paint damage may result from improper handling of the hood.**

4. Remove the engine mount strut and air conditioning compressor bracket and engine mount strut bracket.

5. Remove the electric engine coolant fan assemblies.

6. Install a suitable lifting fixture that will safely support the engine with the vehicle both on the floor or raised on a lift.

7. Raise and safely support vehicle. Disconnect the front exhaust manifold pipe.

8. Remove the intermediate steering shaft bolt/screw and move the cover aside.

9. Drain the engine oil. Remove the engine splash shield.

10. Remove the engine crankshaft balancer. Use the following procedure:

➡ **The inertia weight section of the crankshaft balancer is assembled to the hub with a rubber sleeve. The removal and installation procedures, with the proper tools MUST be followed or movement of the inertia weight section on the hub will destroy the tuning of the torsional damper.**

a. Remove the right engine splash shield.

b. Remove the crankshaft balancer center bolt and washer.

c. Draw the balancer off the crankshaft using the appropriate puller.

11. Support the drive train and front suspension frame assembly with floor stands.

12. Remove the transaxle mount frame side bracket nuts from the drive train and suspension frame assembly. Use the following procedure:

a. With the vehicle safely supported, remove the left front tire and wheel assembly.

b. Remove the left front bumper fascia splash shield assembly.

c. Suitably support the transaxle assembly with floor stands.

d. Remove the transaxle mount side bracket nuts.

13. Remove the engine mount frame side nuts from the drive train and front suspension frame assembly.

14. Remove the rear drive train and front suspension frame bolts and screws.

15. Lower the drive train and front suspension frame assembly.

16. Remove the engine mount assembly. Remove the flywheel inspection cover.

17. Remove the starter motor assembly. Note any shims between the starter and the engine block so they can be reused at assembly.

18. Remove the transaxle mount assembly from the engine oil pan assembly.

19. Disconnect the oil level sensor. Remove the engine oil pan assembly. Do not miss the bolt going through the side of the pan and threads into the front main bearing cap.

20. Once the pan has been removed, raise the drive train and front suspension frame assembly back into place and install the bolts as required to temporarily secure the frame assembly. Lower the vehicle.

21. Remove the coolant recovery reservoir assembly.

22. Remove the serpentine drive belt shield.

23. Remove the power steering pump pulley, using special tools. Disconnect the ignition control wiring harness, then remove the pump mounting bolts and position the pump aside.

24. Remove the thermostat bypass pipe clip nut from the upper intake manifold. Remove the pipe from the water pump.

25. Remove the coolant pump pulley assembly. Remove the lower radiator outlet hose from the water pump.

26. Remove the serpentine drive belt tensioner.

27. If the front cover is only going to be removed and not replaced with a new cover, the 24X Crankshaft Position Sensor can be left in place on the front cover. Just disconnect the wiring harness. If the cover is going to be replaced due to damage, remove the sensor from the front cover.

➡ **Note that this engine uses two crankshaft position sensors; one, called the 3X Crankshaft Position Sensor, is in the side of the block to read off the crankshaft. The other sensor is called the 24X Crankshaft Position Sensor and it is mounted to the bottom of the front cover. Only the 24X Crankshaft Position Sensor needs to be removed if the front cover is to be replaced.**

28. Remove the bolts from the front cover, noting their locations, and separate the cover from the engine block. If the cover is being replaced, transfer the water pump to the new cover, making sure the locator on the top of the pump is installed at the top (12 o'clock) position.

29. Rotate the crankshaft until the timing marks on the camshaft and crankshaft sprockets are in alignment at their closest approach.

30. Remove the camshaft sprocket mounting bolt and remove the camshaft sprocket and timing chain.

31. Remove the crankshaft sprocket, using a gear puller J-5825-A or equivalent.

32. Remove the two mounting bolts from the timing chain damper and remove the damper.

To install:

33. Clean all parts and gasket mating surfaces well. Note that the oil pan bolts may have small O-rings under the bolt head flange which should be inspected carefully. A damaged seal will cause an oil leak.

34. Install the timing chain damper and tighten the mounting bolts to 15 ft. lbs. (21 Nm).

35. Install the crankshaft sprocket onto the crankshaft making sure the notch in the sprocket fits over the crankshaft key. Fully seat the sprocket on the crankshaft using J-38612, or equivalent.

36. Be sure the timing mark on the crankshaft sprocket is pointing straight up.

37. Install the timing chain over the camshaft sprocket and hold the sprocket in such a way, that the timing mark is pointing down, and the timing chain is hanging down off the sprocket.

38. Loop the timing chain under the crankshaft sprocket and install the camshaft sprocket on the camshaft. The sprocket will only fit on the camshaft if, the dowel on the camshaft lines up with the hole in the sprocket.

39. Verify that the marks are aligned (the camshaft sprocket will be at the 6 o'clock position and the crankshaft sprocket will be in the 12 o'clock position).

40. On 1995 vehicles, tighten the camshaft sprocket mounting bolt to 74 ft. lbs. (100 Nm). On 1996–99 vehicles tighten the bolt to 81 ft. lbs. (110 Nm).

41. Lubricate the timing chain components with engine oil.

42. The crankshaft front seal should be carefully removed and a replacement installed with the seal lip facing the engine. With the front cover removed, the timing chain and sprockets may be inspected. If the timing chain and sprockets are to be replaced, turn the engine crankshaft until the timing marks on the crankshaft sprocket and camshaft sprocket are aligned. There should also be timing marks on the engine block and the timing chain damper (guide). Remove the camshaft sprocket center bolt and lift off the sprocket and chain. Use a puller to remove the crankshaft sprocket. Replace the chain and sprockets as a set, using care to align all timing marks as before.

43. If the cover is being replaced with a service part or if the coolant pump is being changed out to a service replacement pump,

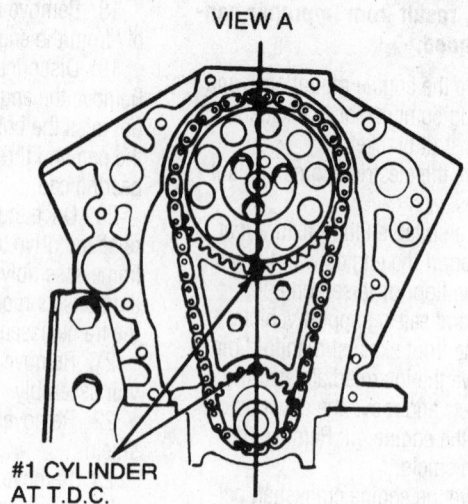

VIEW A

#1 CYLINDER
AT T.D.C.

NOTE - ALIGN TIMING MARKS ON CAM
& CRANK SPROCKETS USING ALIGNMENT
MARKS ON DAMPER STAMPING OR CAST
ALIGNMENT MARKS ON CYL & CASE

NOTE - CAMSHAFT SPROCKET
MARK AT 6 O'CLOCK
CRANKSHAFT SPROCKET
MARK AT 12 O'CLOCK

1. Damper
2. Alignment marks
3. Timing chain
4. Crank sprocket
5. 28 Nm (21 lb. ft.)
6. Camshaft sprocket
7. 21 Nm (15 lb. ft.)

Timing chain and sprocket timing mark alignment—3.1L engines

install the coolant pump, using a new gasket, to the front cover taking care that the locator on the top of the pump is installed at the top (12 o'clock) position.

44. Coat both sides of the lower tabs of a new front cover gasket with sealer. Install the front cover assembly. Tighten the large bolts to 35 ft. lbs. (47 Nm) and the small bolts to 15 ft. lbs. (21 Nm).

45. If removed, install the 24X Crankshaft Position Sensor to the front cover and connect the wiring harness.

46. Install the serpentine drive belt tensioner. Tighten the attaching bolt to 37 ft. lbs. (50 Nm).

47. Connect the lower radiator outlet hose assembly to the coolant pump assembly. Install the pulley on the coolant pump.

48. Connect the thermostat bypass pipe to the coolant pump assembly with the pipe clip nut.

49. Position the power steering pump and install the bolts. Tighten the bolts to 25 ft. lbs. (34 Nm). Using the proper tools, install the pulley onto the power steering pump. The face of the pulley hub MUST be flush with the pump drive shaft. DO NOT use an arbor press to install the pulley.

50. Install the serpentine drive belt shield.

51. Install the coolant reservoir. Lube the reservoir hose with clean water and route it through the hole in the ECM heat shield and up to the radiator overflow fitting until the hose end butts against the filler neck.

52. Raise and safely support vehicle. Support the drive train and front suspension frame assembly with suitable floor stands.

53. Remove the rear drive train and front suspension frame bolts. Lower the drive train and front suspension frame assembly. This should give clearance for oil pan installation.

54. Apply a small amount of sealer on either side of the rear main bearing cap, where the seal surface on the cap meets the cylinder block (2 locations). Using a new gasket, install the oil pan and retaining bolts. Tighten the oil pan retaining bolts to 18 ft. lbs. (25 Nm). Install the oil pan side bolts to 37 ft. lbs. (50 Nm).

55. Install the transaxle mount assembly to the engine oil pan assembly.

56. Install the starter motor. Tighten the bolts to 32 ft. lbs. (43 Nm). Install the flywheel inspection cover.

➡ Before attaching the electrical leads, tighten the inner nuts on the solenoid terminals. If the nuts are not tight, the solenoid cap may be damaged during

installation of the leads. Tighten the inner nut of BAT terminal to 84 inch lbs. (9.5 Nm).

57. Install the engine mount.

58. Raise the drive train and front suspension frame assembly and install the retainer bolts. Install the engine mount frame side nuts. Tighten the nuts to 32 ft. lbs. (43 Nm). Remove the floor stands from the drive train and front suspension frame assembly.

59. Lubricate the front cover crankshaft seal with clean engine oil. Apply a small amount of sealant to the keyway of the balancer assembly. Install the harmonic balancer using the proper tool to pull the balancer onto the crankshaft. Install the center bolt and tighten to 76 ft. lbs. (103 Nm).

➡ The inertia weight section of the crankshaft balancer is assembled to the hub with a rubber sleeve. The removal and installation procedures, with the proper tools MUST be followed or movement of the inertia weight section on the hub will destroy the tuning of the torsional damper.

60. Install the engine splash shield.

61. Connect the intermediate steering shaft bolt. Tighten to 35 ft. lbs. (48 Nm). Install the cover shield.

62. Connect the front exhaust manifold pipe. Lower vehicle.

63. Remove the engine support fixtures.

64. Install the electric cooling fan assemblies.

65. Install the engine mount strut and air conditioning compressor bracket and engine mount strut bracket.

66. Install the hood panel assembly.

➡**Use care when working around the hood. Glass and paint damage may result from improper handling of the hood.**

67. Install the right engine splash shield and the front wheel assemblies.

68. Install the serpentine drive belt.

69. Refill the crankcase with engine oil and refill the cooling system. Connect the negative battery cable.

70. Locate the bleeder screw on the thermostat neck. Be sure it is closed but DO NOT overtighten. The air bleeder screw is made of soft brass. Start engine and allow to warm up. Open the bleeder screw as necessary to remove air from the system. Watch for oil or coolant leaks.

➡**The low coolant indicator lamp may come on after this procedure. After operating the vehicle so that the engine heats up and cools down three times, if the low coolant indicator lamp does NOT go out, or fails to come on at ignition check and coolant is at proper level (engine cold, coolant is level to the base of the radiator neck), electrical troubleshooting of the system is required. If at any time the TEMP warning indicator comes on, immediate action is required.**

71. Road test the vehicle.

➡**Whenever the vehicle sub-frame is removed or lowered, the wheel alignment should be checked.**

3.8L Engine

1. Disconnect the negative battery cable. Drain the cooling system.

2. Disconnect the coolant hoses from the timing chain front cover.

3. Support the engine using support fixture J 28467-A or equivalent, then remove the engine mount.

4. Remove the drive belt(s) and belt tensioner.

5. Raise and safely support the vehicle. Remove the right front wheel. Remove the right inner fender access panel.

6. Detach the connectors at the camshaft position sensor, crankshaft position sensor and the oil pressure sensor.

7. Keep the flywheel from turning using J 37096 or equivalent,. Remove the crankshaft balancer retaining bolts, then use puller tool J 38197 or equivalent, to remove the balancer from the crankshaft.

8. Remove the crankshaft position sensor shield and the crankshaft position sensor.

9. Remove the oil pan-to-front cover bolts.

10. Remove the front cover attaching bolts, then remove the cover.

11. Align the timing marks on the camshaft and crankshaft sprockets so they are as close together as possible.

12. Remove the timing chain damper.

13. Remove the crankshaft sprocket retaining bolts, then remove the sprocket and timing chain.

14. Remove the crankshaft sprocket.

➡**Do NOT rotate the camshaft or crankshaft while the timing chain and sprockets are removed.**

15. Thoroughly clean all gasket mating surfaces.

To install:

16. Assembly the timing chain and sprockets with the timing marks aligned. Install the timing chain and sprockets to the camshaft and crankshaft.

17. Install the camshaft sprocket bolt and tighten to 74 ft. lbs. (100 Nm) plus an additional 90 degree turn. Recheck the camshaft and crankshaft sprocket timing marks to be sure they are still aligned.

18. Install the timing chain damper and tighten to 14 ft. lbs. (19 Nm).

19. Remove the screws and the oil pump cover from the back of the timing chain front cover. Pack the space around the oil pump gears completely full of petroleum jelly. There must be no air space left inside the pump. Reinstall the pump cover and tighten the screws to 97 inch lbs. (11 Nm).

20. Using new gaskets, install the timing chain front cover to the block. Tighten the

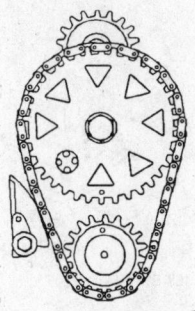

79222525

Install the timing chain and sprockets with the timing marks aligned and close together—3.8L engine

bolts to 22 ft. lbs. (30 Nm) on 1995 vehicles or to 11 ft. lbs. (15 Nm) plus an additional 40 degree turn on 1996–99 vehicles. Tighten the oil pan-to-front cover bolts to 125 inch lbs. (14 Nm).

21. Install the crankshaft position sensor and tighten the bolts to 14–28 ft. lbs. (20–40 Nm). Install the crankshaft position sensor shield.

22. Keep the flywheel from turning using holder tool J 37096 or equivalent. Install the crankshaft balancer and tighten the bolt to 111 ft. lbs. (150 Nm) plus an additional 76 degree turn.

23. Attach the connectors to the camshaft position sensor, crankshaft position sensor and oil pressure sensor.

24. Install the right inner fender access panel and the right front wheel.

25. Lower the vehicle.

26. Install the tensioner assembly.

27. Install the drive belt(s).

28. Install the engine mount and remove the engine support fixture.

29. Connect the coolant hoses.

30. Connect the negative battery cable. Fill and bleed the cooling system.

31. Start the vehicle and check for leaks and proper engine operation.

FUEL SYSTEM

Fuel System Service Precautions

Safety is the most important factor when performing not only fuel system maintenance but any type of maintenance. Failure to conduct maintenance and repairs in a safe manner may result in serious personal injury or death. Maintenance and testing of the vehicle's fuel system components can be accomplished safely and effectively by adhering to the following rules and guidelines.

• To avoid the possibility of fire and personal injury, always disconnect the negative battery cable unless the repair or test procedure requires that battery voltage be applied.

• Always relieve the fuel system pressure prior to disconnecting any fuel system component (injector, fuel rail, pressure regulator, etc.), fitting or fuel line connection. Exercise extreme caution whenever relieving fuel system pressure to avoid exposing skin, face and eyes to fuel spray. Please be advised that fuel under pressure may penetrate the skin or any part of the body that it contacts.

• Always place a shop towel or cloth around the fitting or connection prior to loosening to absorb any excess fuel due to spillage. Ensure that all fuel spillage (should it occur) is quickly removed from engine surfaces. Ensure that all fuel soaked cloths or towels are deposited into a suitable waste container.

• Always keep a dry chemical (Class B) fire extinguisher near the work area.

• Do not allow fuel spray or fuel vapors to come into contact with a spark or open flame.

• Always use a back-up wrench when loosening and tightening fuel line connection fittings. This will prevent unnecessary stress and torsion to fuel line piping. Always follow the proper torque specifications.

• Always replace worn fuel fitting O-rings with new. Do not substitute fuel hose or equivalent, where fuel pipe is installed.

Fuel System Pressure

RELIEVING

✳✳ CAUTION

Fuel injection systems on these vehicles remain under pressure, even after the engine has been turned OFF. The fuel system pressure must be relieved before disconnecting any fuel lines. Failure to do so may result in fire and/or personal injury. Please note that even after relieving fuel system pressure, a small amount of fuel may be released when servicing the fuel pipes or connections. In order to reduce the chance of personal injury, cover the fuel pipe fittings with a shop towel before disconnecting, to catch any fuel that may have leaked out.

1. Disconnect the negative battery cable to prevent possible discharge of fuel if an accidental attempt is made to start the engine.

2. Loosen the fuel filler cap to relieve tank pressure.

3. This procedure calls for GM special tool J 34730-1A or equivalent. It is a fuel pressure test gauge with a line equipped with a fitting to connect to the to the fuel pressure test connection and another hose to discharge into an approved gasoline container. Wrap a shop towel around the pressure test fitting connection while connecting gauge to avoid spillage.

4. Install the bleed hose into an approved container and open the valve to bleed fuel system pressure. The fuel connections are now safe for servicing.

5. Drain any fuel remaining in the gauge into an approved container.

6. Reconnect the negative battery cable unless addition service work is being performed.

Fuel Filter

REMOVAL & INSTALLATION

✳✳ CAUTION

Fuel injection systems on these vehicles remain under pressure, even after the engine has been turned OFF. The fuel system pressure must be relieved before disconnecting any fuel lines. Failure to do so may result in fire and/or personal injury.

1. Disconnect the negative battery cable. Relieve the fuel system pressure.

2. Raise and safely support the vehicle.

3. Note that one side of the fuel filter uses a quick-connect fitting. Special handling is required. To remove, grasp both sides of the fitting. Twist the female connector ¼ turn in each direction to loosen any dirt within the fitting.

✳✳ CAUTION

Safety glasses must be worn when using compressed air, as flying dirt particles may cause eye injury.

4. Using compressed air, blow dirt out of the fitting.

5. If equipped with a plastic fitting, squeeze the plastic retainer release tabs and pull the connection apart.

6. If equipped with a metal fitting, choose the correct special tool J 37088 A ,

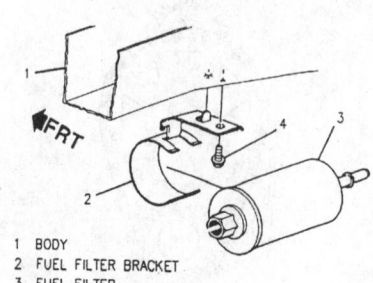

1 BODY
2 FUEL FILTER BRACKET
3 FUEL FILTER
4 SCREW – FULLY DRIVEN, SEATED
   AND NOT STRIPPED

79222526

Common fuel filter mounting

J 39504 or equivalent tool set for the correct size fitting. Insert tool into the female connector, then push/pull inward to release the locking tabs. Carefully pull the connection apart.

7. Remove the fuel feed pipe nut to the outlet side of the fuel filter.

8. Remove the fuel filter from the holder.

To install:

9. Install the fuel filter into the holder.

10. Use a back-up wrench on the inlet fitting while tightening the fuel filter outlet nut to 22 ft. lbs. (30 Nm).

11. Apply a few drops of clean engine oil to the male pipe end.

12. Push both sides of the fitting together to cause the retaining tabs/fingers to snap into place.

13. Once installed, pull on both sides of the fitting to be sure the connection is secure.

14. Safely lower the vehicle.

15. Tighten the fuel filler cap.

16. Turn the ignition switch to the **ON** position to pressurize the fuel system to check for leaks.

Fuel Pump

REMOVAL & INSTALLATION

✳✳ CAUTION

Fuel injection systems on these vehicles remain under pressure, even after the engine has been turned OFF. The fuel system pressure must be relieved before disconnecting any fuel lines. Failure to do so may result in fire and/or personal injury.

1. Relieve the fuel system pressure. Disconnect the negative battery cable.

✳✳ CAUTION

Observe all applicable safety precautions when working around fuel. Do not allow fuel spray or fuel vapors to come in contact with a spark or open flame. Keep a dry chemical (Class B) fire extinguisher near the work area. Never drain or store fuel in an open container due to the possibility of fire or explosion.

2. On the Lumina, Monte Carlo, 1995–96 Buick, Oldsmobile and Pontiac vehicles, drain and remove the fuel tank using the recommended procedure. On all others, remove the spare tire, then remove the fuel pump access panel.

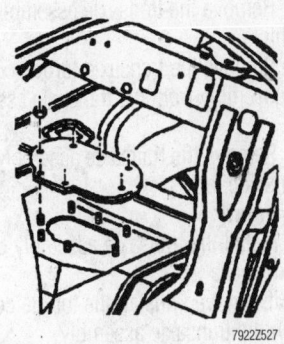

Fuel pump access panel—1997–99 Buick, Olds and Pontiac models

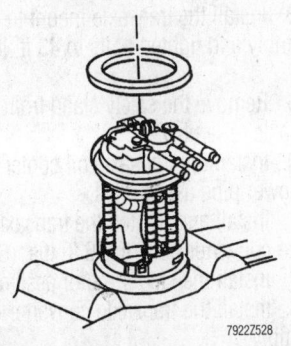

Remove the fuel pump from the tank after removing the locking ring

3. Disconnect the quick connects from the fuel sender assembly and remove the fuel lines.

4. Using J-35731, or an equivalent spanner wrench, remove the fuel sender lock ring.

5. Lift the sender assembly carefully out of the fuel tank.

6. Remove the sender O-ring from the top of the tank and discard.

☀☀ WARNING

DO NOT run the fuel pump unless it is submerged in fuel. Running the pump dry will cause serious damage to the fuel pump and may cause the pump to explode due to the oxygen in the air.

7. Note position of fuel pump strainer on fuel pump and while supporting the pump assembly in one hand twist the fuel strainer off the pump and discard the strainer.

8. Disconnect the fuel pump electrical connector.

9. Remove the clamp from the fuel line at the top of the pump.

10. Hold the fuel sender upside down on a work bench and pull the fuel pump out of the lower mounting bracket. Once the pump is clear of the lower mounting bracket tilt the pump outward and disconnect the pump from the sender assembly.

To install:

11. Install the rubber bumper and insulator on the fuel pump.

12. Hold the fuel sender upside down and install the fuel pump between fuel pulse dampener and mounting bracket.

13. Connect the fuel pump electrical connector.

14. Install the clamp on the fuel line.

15. Install a new fuel pump strainer on outer edge of ferrule until fully seated. The strainer must be facing in the same direction as prior to removal.

16. Install a new O-ring on top of the fuel tank and install the sender assembly.

17. Install the lock ring and using J-35731, or an equivalent spanner wrench.

18. On the Lumina, Monte Carlo, 1995–96 Buick, Oldsmobile and Pontiac vehicles, install and refill the fuel tank. On all others, install the fuel pump access panel. Tighten the nuts to 8 ft. lbs. (11 Nm) and install the spare tire and cover.

19. Connect the negative battery cable. Turn the ignition switch to the **ON** position to pressurize the fuel system and check for leaks.

DRIVE TRAIN

For information on CV-joint boot replacement, please refer to the unit repair section.

Transaxle Assembly

REMOVAL & INSTALLATION

➡These transaxles were used in a variety of General Motors vehicles. Due to model year, vehicle model and installed options, the removal and installation procedures may vary slightly. The procedures given here should suffice for most all vehicles using these transaxles.

3T40 3-Speed Automatic Transaxles

1. Remove the air cleaner assembly.

2. Disconnect the negative battery cable.

3. Remove the coolant recovery reservoir.

4. Remove the shift control and TV cables at the transaxle.

5. Remove the throttle cable bracket and brake booster hose if equipped.

6. Remove both torque struts at the engine.

7. Remove the left torque strut bracket.

8. Remove the transaxle oil cooler lines at the transaxle and plug lines.

9. Install engine support special tools J 28467-a, J 28467 90 and J 36462 or equivalent.

10. Safely raise the vehicle.

11. Remove the tire and wheel assembly.

12. Remove the caliper bracket assemblies and rotors.

13. Remove both lower engine splash shields.

14. Remove the axle assemblies.

15. Remove the tie rods and ball joints.

16. Remove the rack and pinion heat shield and electrical connector.

17. Remove the bolts holding the main engine harness to the transaxle case.

18. Wire the rack and pinion assembly to the exhaust and remove the rack and pinion bolts from the frame.

19. Remove the bolts holding the power steering lines to the frame.

20. Remove the engine and transaxle mounts from the frame.

21. Support the frame with jackstands at each end.

22. Remove the frame bolts.

23. With an assistant, remove the frame and jackstands.

24. Remove the flywheel cover.

25. Remove the torque converter bolts.

26. Remove the starter and hang with mechanics wire.

27. Remove the ground cable at the transaxle.

28. Remove the transaxle fill tube bolt.

29. Remove the transaxle mount bracket.

30. Remove the transaxle to engine brace.

31. Safely lower the vehicle.

32. Remove the electrical connector to the transaxle.

33. Remove the transaxle fill tube.

34. Using special tool J 28467-a or equivalent and lower left side of engine approximately four inches.

35. Safely raise the vehicle.

36. Remove the fuel line bracket from the transaxle.

37. Remove the transaxle to engine bolts.

38. Remove the transaxle from the vehicle.

To install:

39. Apply a small amount of MP grease on the torque converter pilot hub.

40. Position the transaxle in the vehicle.

41. Install the transaxle to engine bolts and tighten to 55 ft. lbs. (75 Nm). Remove the transaxle jack.

42. Install the fuel line bracket to the transaxle.

43. Safely lower the vehicle.

44. Using special tool J 28467-a or equivalent, raise the engine to its proper position.

45. Install the transaxle oil fill tube.

46. Connect the electrical connector to the transaxle.

47. Safely raise the vehicle.

48. Install and tighten the transaxle engine brace to 35 ft. lbs. (47 Nm).

49. Install the transaxle mount bracket.

50. Install the transaxle oil fill tube bolt.

51. With an assistant, position and support the frame under the vehicle.

52. Install new frame to body bolts and tighten bolts to 125 ft. lbs. (170 Nm).

53. Remove the frame supports.

54. Safely lower the vehicle and position engine and transaxle mount to the frame.

55. Safely raise the vehicle.

56. Install and tighten the torque converter to flywheel bolts to 46 ft. lbs. (63 Nm).

57. Install the flywheel cover.

58. Install the starter assembly.

59. Install the ground cable to the transaxle.

60. Install the ball joints and tie rod rods.

61. Install the axle assemblies.

62. Install the rotors and caliper bracket assemblies.

63. Install the rack and pinion with attaching lines to the frame.

64. Install the rack and pinion electrical connector and heat shields.

65. Install both lower engine splash shields.

66. Install the main engine harness to the transaxle case.

67. Install the tire and wheel assemblies.

68. Safely lower the vehicle.

69. Remove the engine support tools.

70. Install the throttle cable bracket and brake booster line, if equipped.

71. Install the torque strut bracket.

72. Install the transaxle oil cooler lines to the transaxle.

73. Install the torque struts.

74. Install the shift control and TV cables to the transaxle.

75. Install the coolant recovery reservoir.

76. Connect the negative battery cable.

77. Install the air cleaner assembly.

78. Adjust the shift linkage and TV cables.

79. Start engine and check engine and transaxle oil level. Add oil as necessary.

4T60-E And 4T65 4-Speed Automatic Transaxles

1. Remove the hood assembly, scribe a mark at the hinge area for installation reference.

2. Remove the transaxle fluid level indicator (dipstick) assembly.

3. Remove the engine mount strut brackets.

4. Remove the electric engine cooling fan assemblies.

5. Install engine support tools J 28467-A, J 28467-90 and J 36462 or the equivalent.

6. Remove the air intake duct assembly.

7. Disconnect the electrical connections from the transaxle assembly.

8. Remove the upper transaxle to engine bolts.

9. Safely raise and support the vehicle.

10. Remove the tire and wheel assembly. Mark the position of the wheel to the wheel studs, prior to removal, for installation reference.

11. Remove the left front bumper fascia splash shield assembly.

12. Remove the front exhaust manifold pipe.

13. Remove the steering gear heat shield assembly.

14. Remove the steering gear bolts.

15. Remove the front lower control arm assemblies from the front suspension strut assemblies.

16. Remove the power steering fluid cooling pipe assembly.

17. Safely support the drive train and front suspension frame assembly with safety stands.

18. Remove the transaxle mount side bracket nuts.

19. Remove the engine mount frame side nuts.

20. Remove the transaxle brace.

21. Remove the drive train and front suspension frame assembly.

22. Disconnect the electrical connector from the vehicle speed sensor.

23. Remove the front wheel driveshaft assemblies from the transaxle.

24. Remove the transaxle converter cover assembly.

25. Remove the starter assembly.

26. Remove the transaxle torque converter bolts.

27. Remove the transaxle oil cooler upper and lower pipe assemblies.

28. Install a transaxle jack.

29. Remove the transaxle mount assembly from the engine assembly.

30. Remove the transaxle fluid filler tube assembly.

31. Remove the transaxle assembly from the vehicle.

32. Remove the transaxle torque converter assembly from the transaxle assembly.

33. Remove the transaxle assembly from the transaxle jack.

To install:

34. Install the transaxle assembly on the transaxle jack.

35. If removed, install the torque converter in the transaxle assembly.

36. Install the transaxle assembly to the engine. Tighten the lower rear transaxle to engine bolts to 55 ft. lbs. (75 Nm).

37. Install the dipstick tube assembly.

38. Install the transaxle mount to engine assembly and tighten bolts to 43 ft. lbs. (58 Nm).

39. Remove the safety stand from the transaxle assembly.

40. Install the transaxle oil cooler upper and lower pipe assemblies.

41. Install and tighten the transaxle torque converter bolts to 46 ft. lbs. (63 Nm).

42. Install the starter motor assembly.

43. Install the transaxle converter cover assembly.

44. Install and tighten the transaxle converter cover bolts to 89 inch lbs. (10 Nm).

45. Install the front wheel drive shaft assemblies to the transaxle.

46. Install and tighten the drive train and front suspension frame bolts to 107 ft. lbs. (145 Nm).

47. Install the transaxle brace and bracket. Tighten the attaching bolts to 39 ft. lbs. (53 Nm).

48. Remove the safety stands from the drive train and front suspension frame assembly.

49. Install the power steering fluid cooling pipe assembly.

50. Install the lower control arm assemblies to the front suspension strut assemblies.

51. Install the steering gear bolts and heat shield assembly.

52. Install the front exhaust manifold pipe to the engine.

53. Install the left front bumper fascia splash shield assembly.

54. Install the tire and wheel assembly, aligning the marks made on the removal.

55. Safely lower the vehicle.

56. Install and tighten the upper transaxle to engine bolts to 55 ft. lbs. (75 Nm).

57. Install the transaxle range selector assembly to the transaxle assembly.

58. Attach the electrical connectors to the transaxle assembly.

59. Install the air intake duct assembly.

60. Remove the engine support fixture tools.

61. Install the engine cooling fan assemblies.

62. Install the engine mount strut brackets.

63. Install the transaxle fluid level indicator assembly.

64. Install the hood panel assembly.

65. Start the engine and check for leaks.

66. Check the fluid level in the transaxle and adjust the shift linkage.

67. Road test to verify proper transaxle shifting and smooth operation.

Halfshaft

REMOVAL & INSTALLATION

➡**If vehicle is equipped with ABS brakes, use care to avoid damage to the ABS toothed ring. Damage to the ring may cause the self-diagnostic feature of the ABS system to set a system fault code.**

1. Disconnect the negative battery cable.

2. Raise and safely support the vehicle. Remove the wheel and tire assembly.

3. Remove the front wheel drive axle nut.

4. Remove the brake calipers, bracket assemblies and hang caliper with mechanic's wire from the strut.

5. Remove the brake rotors.

6. Remove the four hub/bearing retaining bolts and hub.

7. Remove the ABS sensor mounting bolt and position the sensor out of the way, if equipped with ABS brakes.

8. Place a drain pan under the transaxle.

➡**Use care when removing the halfshaft. Tripot joints can be damaged if the drive axle is over-extended. It is important to handle the halfshaft in a manner to prevent over-extending.**

9. Remove the halfshaft from the vehicle using the appropriate procedure for each side and transmission model:

   a. To remove the right side halfshaft, use special tool J 33008, J 29794 and J 2619-01 or equivalent. Separate the halfshaft from the transaxle.

   b. To remove the left side halfshaft on a 3T40 transaxle, use special tools J 33008, J 29794 and J 2619-01 or equivalent. Separate the axle from the transaxle.

   c. To remove the left side halfshaft, using the frame for leverage, separate the halfshaft from the transaxle with a suitable prying tool in the groove provided on the inner joint.

10. Remove the halfshaft/bearing assembly through the knuckle.

To install:

11. Install the halfshaft/bearing assembly through the knuckle and into the transaxle. Remove special tool J 37292-a or equivalent and discard.

12. Properly position the ABS sensor and install the mounting bolt, if removed.

13. Loosely secure the bearing to the knuckle bolts.

14. Seat the halfshaft into the transaxle, using a suitable prying tool in the groove provided on the inner joint. Carefully pry against the frame or the lower control arm to seat the halfshaft.

15. Verify that the snapring is seated by tapping on the inner groove with a suitable prying tool. Grasp the inner housing of the axle shaft and pull outboard. Do not pull on the axle shaft. If the snapring is properly seated, the axle will remain in place.

16. Install the hub and bearing assembly to the axle with a new nut and washer. Tighten the bolts to 60 ft. lbs. (80 Nm).

17. Install the brake rotor.

18. Install the brake caliper and attaching bracket. and tighten caliper slide bolts to 80 ft. lbs. (108 Nm).

19. Install the wheel and tire assembly.

20. Safely lower the vehicle.

➡**Do not reuse the front wheel drive axle nut. Always use a new front wheel drive axle nut of similar design when installing the nut. Do not use a Nylock or free spinning style nut.**

21. Install the front wheel drive axle nut and tighten to 150 ft. lbs. (205 Nm).

22. Connect the negative battery cable.

STEERING AND SUSPENSION

Air Bag

✳✳ CAUTION

These vehicles are equipped with an air bag system, also known as the Supplemental Inflatable Restraint (SIR) or Supplemental Restraint System (SRS). The system must be disabled before performing service on or around system components, steering column, instrument panel components, wiring and sensors. Failure to follow safety and disabling procedures could result in accidental air bag deployment, possible personal injury and unnecessary system repairs.

PRECAUTIONS

Several precautions must be observed when handling the inflator module to avoid accidental deployment and possible personal injury.

• Never carry the inflator module by the wires or connector on the underside of the module.

• When carrying a live inflator module, hold securely with both hands, and ensure that the bag and trim cover are pointed away.

• Place the inflator module on a bench or other surface with the bag and trim cover facing up.

• With the inflator module on the bench, never place anything on or close to the module which may be thrown in the event of an accidental deployment.

DISARMING

✳✳ CAUTION

The Supplemental Restraint System (SRS) must be disarmed before performing service around the air bag or SRS wiring. Failure to do so may cause accidental deployment of the air bag, resulting in unnecessary SRS repairs and/or personal injury.

1. Turn the steering wheel so the front wheels are in the straight ahead position.

2. Turn the ignition switch to the **LOCK** position.

3. Disconnect the negative battery cable.

4. Remove the Air Bag fuse from the fuse panel.

➡**The position of the fuse on the panel varies according to model and year. Consult the vehicle owner's manual for fuse location.**

5. Remove the left-hand sound insulator (trim panel under the instrument panel).

6. Remove the Connector Position Assurance (CPA) clip and detach the yellow 2-way connector at the base of the steering column.

7. Remove the CPA and detach the passenger side yellow 2-way connector, if equipped. Positions for the connector vary from behind the glove box to removing the right side sound insulator and finding the yellow connector.

ARMING

1. Turn the steering wheel so the front wheels are in the straight ahead position.
2. Turn the ignition switch to the **LOCK** position.
3. Disconnect the negative battery cable.
4. Attach the yellow 2-way connector at the base of the steering column and install the CPA.
5. Install the left-hand sound insulator.
6. Attach the yellow 2-way connector on the right side and install the CPA, if equipped. Install the sound insulator and/or glove box.
7. Install the air bag fuse.
8. Connect the negative battery cable.
9. Turn the ignition switch to the **RUN** position. Verify that the INFLATABLE RESTRAINT indicator lamp flashes 7–9 times, then remains OFF. If the lamp does not function as specified, there is a malfunction in the air bag system.

Power Rack and Pinion Steering Gear

REMOVAL & INSTALLATION

Except 3.4L Engine

1. Disconnect the negative battery cable.
2. Raise and safely support the vehicle.
3. Remove the front wheels.
4. Remove the electrical connection from the steering gear pressure switch.
5. Remove the intermediate shaft pinch bolt at the steering gear and disconnect the intermediate shaft from the rack an pinion unit.

❋❋ CAUTION

Failure to disconnect the intermediate shaft from the rack and pinion stub shaft may result in damage to the steering gear. This damage may cause a loss of steering control and may cause personal injury.

➡**Set the steering shaft so the block tooth on the upper steering shaft is at the 12 o'clock position. The wheels should be straight ahead. Set the ignition key lock to the LOCK position. Failure to follow these procedures could result in damage to the SIR coil assembly.**

6. Remove the cotter pins and castle nuts from the outer tire rod ends and separate the tie rods from the steering knuckles using a tie rod puller J 35917 or the equivalent.
7. Support the rear of the subframe with a suitable adjustable jack.

➡**DO NOT lower the frame too far. Engine components near the firewall may be damaged.**

8. Remove the rear frame bolts and lower the rear of the frame up to 5 in. (128mm).
9. Remove the heat shield, pipe retaining clip and the fluid pipes from the rack assembly. Use flare nut wrenches to remove the fluid pipes.
10. Remove the rack mounting bolts and nuts.
11. Remove the rack assembly out through the left wheel opening.
   To install:
12. Install the rack assembly through the left wheel opening.
13. Install the mounting bolts and nuts and tighten to 59 ft. lbs. (80 Nm).
14. Connect the power steering fluid lines with new O-rings to the rack and pinion assembly. Tighten the fittings to 20 ft. lbs. (27 Nm).
15. Install the pipe retaining clips and heat shield.
16. Raise the frame and install the rear bolts and tighten to 103 ft. lbs. (140 Nm).
17. Connect the tie rod ends to the steering knuckles and tighten the castle nuts to 40 ft. lbs. (54 Nm). If necessary to align the cotter pin holes, tighten the nuts slightly until the cotter pins can be installed. NEVER loosen the nuts to align the holes.
18. Connect the intermediate shaft-to-stub shaft and tighten the lower pinch bolt to 35–40 ft. lbs. (47–54 Nm).
19. Install the front wheels and tighten to specifications.
20. Lower the vehicle.
21. Connect the negative battery cable.
22. Refill and bleed the power steering system.

➡**Whenever the vehicle sub-frame is removed or lowered, the wheel alignment should be checked.**

23. Check the wheel alignment and adjust if required.

3.4L Engine

1. Disconnect the negative battery cable. Remove the air cleaner and duct assembly.
2. Install the engine support fixtures J 28467 A, J 28467 90 and J 36462 or equivalent.
3. Raise and safely support the vehicle.
4. Remove the left front wheel.
5. Loosen the right side engine splash shield.
6. Remove both left and right tie rod nuts and tie rods from the steering knuckles.
7. Remove the intermediate shaft pinch bolt at the steering gear and disconnect the intermediate shaft from the rack an pinion unit.

❋❋ CAUTION

Failure to disconnect the intermediate shaft from the rack and pinion stub shaft may result in damage to the steering gear. This damage may cause a loss of steering control and may cause personal injury.

➡**Set the steering shaft so that the block tooth on the upper steering shaft is at the 12 o'clock position. The wheels should be straight ahead. Set the ignition key lock to the LOCK position. Failure to follow these procedures could result in damage to the SIR coil assembly.**

8. Remove the electrical connection from the steering gear pressure switch.
9. Remove the exhaust pipe and catalytic converter assembly.
10. Support frame at the center rear using safety stands.

➡**DO NOT lower the frame too far. Engine components near the firewall may be damaged.**

11. Remove the rear frame bolts and lower the rear of the frame up to 3 in. (76.mm).
12. Remove the heat shield, power steering line retaining clip and the fluid lines from the rack assembly. Use flare nut wrenches to remove the fluid pipes.
13. Remove the rack and pinion mounting bolts and nuts.
14. Remove the rack and pinion assembly out through the wheel opening.
15. Replace the stub shaft seals.
   To install:
16. Install the rack and pinion assembly through the wheel opening.

17. Install the mounting bolts and nuts and tighten to 59 ft. lbs. (80 Nm).

18. Connect the power steering lines with new O-rings to the rack and pinion assembly. Tighten the fittings to 20 ft. lbs. (27 Nm).

19. Install the power steering line retaining clips and heat shield.

20. Raise the frame and align the steering gear stub shaft to the intermediate steering shaft. Install the rear bolts and tighten to 103 ft. lbs. (140 Nm).

21. Remove the safety stands at the frame center rear location.

22. Install the exhaust pipe and catalytic converter assembly.

23. Connect the electrical connection from the steering gear pressure switch.

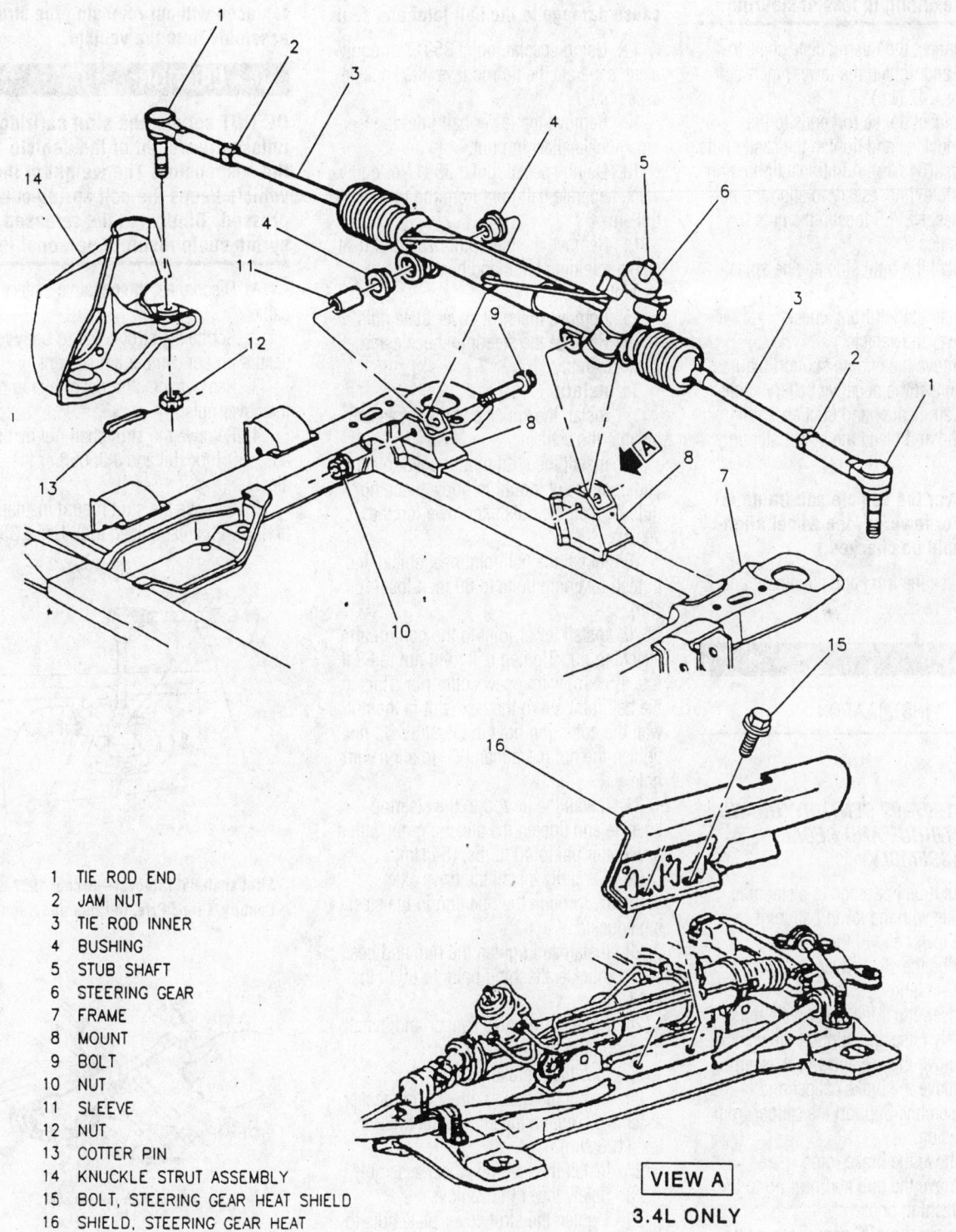

1 TIE ROD END
2 JAM NUT
3 TIE ROD INNER
4 BUSHING
5 STUB SHAFT
6 STEERING GEAR
7 FRAME
8 MOUNT
9 BOLT
10 NUT
11 SLEEVE
12 NUT
13 COTTER PIN
14 KNUCKLE STRUT ASSEMBLY
15 BOLT, STEERING GEAR HEAT SHIELD
16 SHIELD, STEERING GEAR HEAT

VIEW A
3.4L ONLY

79222529

Common rack and pinion steering gear mounting—1997–99 models shown

⁂ CAUTION

When installing the intermediate shaft be sure the shaft is seated prior to the pinch bolt installation. If the pinch bolt is inserted into the coupling before the intermediate shaft installation, the two mating surfaces may disengage resulting in loss of steering.

24. Connect the intermediate shaft-to-stub shaft and tighten the lower pinch bolt to 35 ft. lbs. (47 Nm).
25. Connect the tie rod ends to the steering knuckles and tighten the castle nuts to 40 ft. lbs. (54 Nm). Additional tightening is permissible if necessary to align the cotter pin holes. NEVER loosen the nuts to align the holes.
26. Install the right side engine splash shield.
27. Install the left front wheel.
28. Lower the vehicle.
29. Remove the engine support fixtures.
30. Connect the negative battery cable. Install the air cleaner and duct assembly.
31. Refill and bleed the power steering system.

➡ **Whenever the vehicle sub-frame is removed or lowered, the wheel alignment should be checked.**

32. Check the front end alignment and adjust as necessary.

Strut

REMOVAL & INSTALLATION

Front

EXCEPT 1997–99 CENTURY, GRAND PRIX, INTRIGUE AND REGAL—STRUT ASSEMBLY

The strut tube is welded to a stamped steel knuckle with the lower ball joint riveted to the lower end of the knuckle.
1. Scribe the strut cover plate for proper reinstallation.
2. Loosen the three cover plate nuts.
3. Safely raise and support the vehicle.
4. Remove the tire and wheel assembly.
5. Remove the brake caliper and bracket assembly. Support the caliper from the suspension.
6. Remove the brake rotor.
7. Remove the hub and bearing to the knuckle attaching bolts.
8. Remove the ABS sensor mounting bolt and position the ABS sensor aside to prevent damage, if equipped with ABS.

9. Separate the axle from the transaxle and remove the axle.
10. Remove the tie rod to steering knuckle attaching nut.

➡ **Use only the recommended tools (a ball joint press or puller, not a "pickle fork") for separating the ball joints. Failure to use recommended tools may cause damage to the ball joint and seal.**

11. Using special tool J 35917 or equivalent, separate the tie rod from the knuckle assembly.
12. Remove the lower ball joint to steering knuckle attaching nut.
13. Using special tool J 35917 or equivalent, separate ball joint from the lower control arm.
14. Remove the ball joint heat joint heat shield retaining bolts and heat shield assembly.
15. Remove the strut cover plate nuts.
16. Remove the steering knuckle and strut assembly.

To install:

17. Install the knuckle and strut assembly into the vehicle.
18. Install the strut mount cover plate and upper strut mount to body attaching nuts Tighten the nuts after lowering the vehicle.
19. Install the ball joint heat shield and tighten retaining bolts to 89 inch lbs. (10 Nm).
20. Install lower joint to the control arm attaching nut. Tighten ball joint nut to 63 ft. lbs. (85 Nm) with a new cotter pin. Tighten the ball joint nut to the next slot in the nut with the cotter pin hole in the stud. Do not tighten the nut more than 60° to align with hole.
21. Install the tie rod to the steering knuckle and tighten the attaching nut with a new cotter pin to 40 ft. lbs. (54 Nm).
22. Carefully install the drive axle assembly through the opening in the steering knuckle.
23. Install and tighten the hub and bearing to knuckle attaching bolts to 60 ft. lbs. (80 Nm).
24. Position the ABS sensor and install the mounting bolts, if removed.
25. Install the brake rotor.
26. Install the brake caliper and bracket assembly. Tighten the slide bolts to 80 ft. lbs. (108 Nm).
27. Install the tire and wheel assembly.
28. Safely lower the vehicle.
29. Tighten the strut cover plate nuts to 18 ft. lbs. (24 Nm) after aligning the scribe marks.
30. Road test the vehicle.

EXCEPT 1997–99 CENTURY, GRAND PRIX, INTRIGUE AND REGAL—STRUT CARTRIDGE

➡ The Lumina, Monte Carlo, 1995–96 Century, Regal, Grand Prix, 1995–97 Cutlass Supreme models utilize a replaceable cartridge within the strut assembly. This cartridge can be replaced without removing the strut assembly from the vehicle.

⁂ CAUTION

DO NOT service the strut cartridge unless the weight of the vehicle is on the suspension. The weight of the vehicle keeps the coil spring compressed. Otherwise the released coil spring could result in personal injury.

1. Disconnect the negative battery cable.
2. Scribe the strut cover to body to assure proper camber adjustment.
3. Remove the strut cover by removing the cover nuts.
4. Remove the strut shaft nut by using a no. 50 Torx® bit and J 35668 or the equivalent.
5. Remove the strut mount insulator by prying with a flat bladed tool. Use J 35668

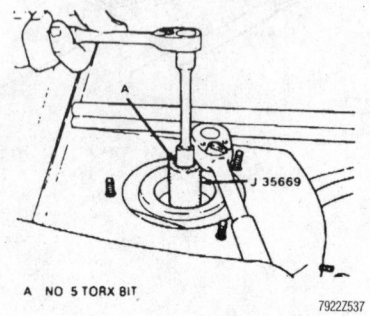

A NO 5 TORX BIT

79222537

Strut shaft nut removal—except 1997–99 Century, Grand Prix, Intrigue and Regal

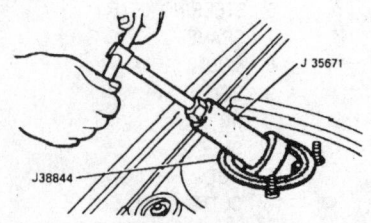

Strut closure nut removal

79222532

Strut closure nut removal—except 1997–99 Century, Grand Prix, Intrigue and Regal

or the equivalent to apply pressure on the strut as necessary to relieve the side load (compression) on the bushing.

6. Remove the bumper by attaching J 35668 or the equivalent to the strut and pulling out the bumper.

7. Install J 38844 or the equivalent in the correct position and compress the strut down into the cartridge.

8. Remove the strut closure nut by unscrewing the closure nut using J 35671 or the equivalent.

9. Remove the cartridge and remove the oil in the strut housing using a suction pump to remove all the oil.

To install:

10. Install the self contained replacement cartridge into the strut housing.

11. Install the strut cartridge closure nut using J 35671 or the equivalent and tighten the nut to 82 ft. lbs. (110 Nm).

12. Install J 35668 or the equivalent and compress the shaft down into the cartridge.

13. Remove J 38844 or the equivalent strut alignment tool.

14. Install the strut bumper and raise the strut and remove J 35668 or the equivalent.

15. Install the strut mount insulator as follows:

a. Use a soap solution to lubricate the bushing for ease of installation.

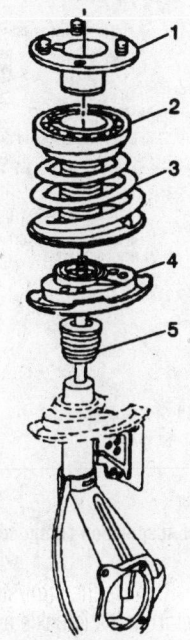

(1) Front Suspension Strut Mount Retainer
(2) Front Spring Upper Insulator
(3) Front Spring
(4) Front Spring Seat
(5) Front Suspension Strut Bumper

79222531

Knuckle and strut assembly with replaceable strut cartridge

b. If necessary, install J 35668 or the equivalent after the bushing is partially installed and position the strut as required to assist in the bushing installation.

16. Install the strut shaft nut using the No. 50 Torx® bit and J 35669 or the equivalent and tighten the nut to 59 ft. lbs. (80 Nm).

17. Install the strut cover mount and aligning the scribe marks, then tighten the bolts to 24 ft. lbs. (33 Nm).

18. A four wheel alignment is recommended after any steering/suspension repairs are performed.

1997–99 CENTURY, GRAND PRIX, INTRIGUE AND REGAL

1. Remove the three strut-to-body nuts.
2. Raise and safely support the vehicle allowing the control arms to hang free.
3. Remove the front wheel.
4. Matchmark the position of the strut on the steering knuckle for reference when installing the strut.

✳✳ WARNING

To prevent damage to the ball joint or halfshaft, support the knuckle assembly when removing the strut.

5. Remove the bolts securing the strut assembly to the knuckle and remove the strut.

To install:

6. Position the strut in the body and install the three retaining nuts. Tighten the nuts to 30 ft. lbs. (41 Nm).

7. Attach the lower end of the strut to the knuckle assembly. Install the bolts from the front and tighten the nuts to 90 ft. lbs. (123 Nm).

8. Install the front wheel.

9. Lower the vehicle to the floor.

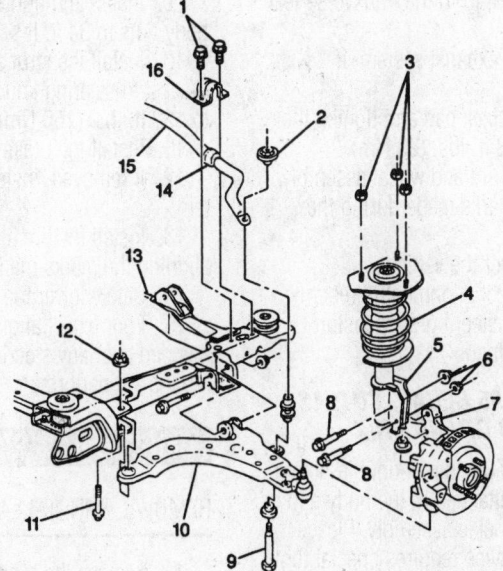

(1) Front Stabilizer Shaft Insulator Clamp Bolt/screw
(2) Front Stabilizer Shaft Link Nut
(3) Front Suspension Strut Mount Nut
(4) Front Suspension Spring
(5) Front Suspension Strut
(6) Strut To Knuckle Nut
(7) Front Steering Knuckle
(8) Strut To Knuckle Bolt/screw
(9) Front Stabilizer Shaft Link]
(10) Front Lower Control Arm
(11) Front Lower Control Arm Bolt/screw
(12) Front Lower Cotrol Arm Nut
(13) Frame
(14) Front Stabilizer Shaft Insulator
(15) Front Stabilizer Shaft
(16) Front Stabilizer Shaft Clamp

79222530

Exploded view of the front suspension without replaceable strut cartridge

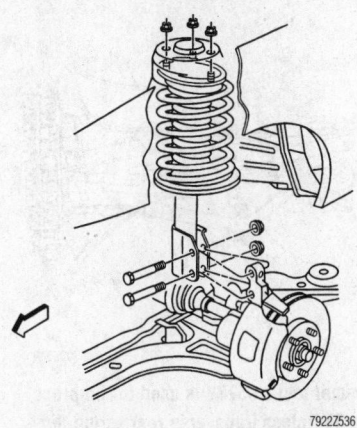

79222536

Front strut mounting—1997–99 Century, Grand Prix, Intrigue and Regal

Rear

1995–96 REGAL, CUTLASS SUPREME AND GRAND PRIX

1. Raise and safely support the vehicle.

2. Remove the tire and wheel assembly. Mark the position of the wheel to the wheel studs, prior to removal, for installation reference.

3. Scribe a reference line from the strut to the knuckle.

4. Remove the jack pad.

5. Remove the exhaust system, if equipped with dual exhausts.

6. Install special tool J 35778 or equivalent. This Y-shaped tool spans the width of the fiberglass transverse leaf spring. Fully compress the special tool to take tension off the outboard suspension components. **Do not** remove the spring or retention plates. Just relieve the tension so the strut can be removed.

➡**Do not use corrosive cleaning agents, silicone lubricants, engine degreasers, solvents, etc. on or near the fiberglass rear transverse spring. These materials could cause extensive damage to the spring.**

7. Remove the brake hose bracket at the strut assembly.

8. Remove the strut to body bolts.

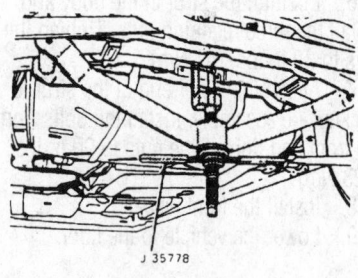

J 35778

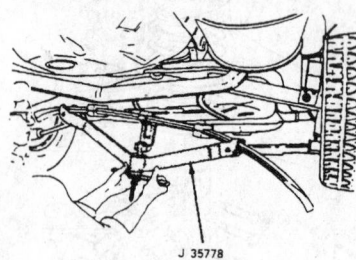

J 35778

79222534

Special tool J 35778 is used to compress the fiberglass transverse rear spring during rear suspension service—1995–96 Regal, Cutlass Supreme and Grand Prix

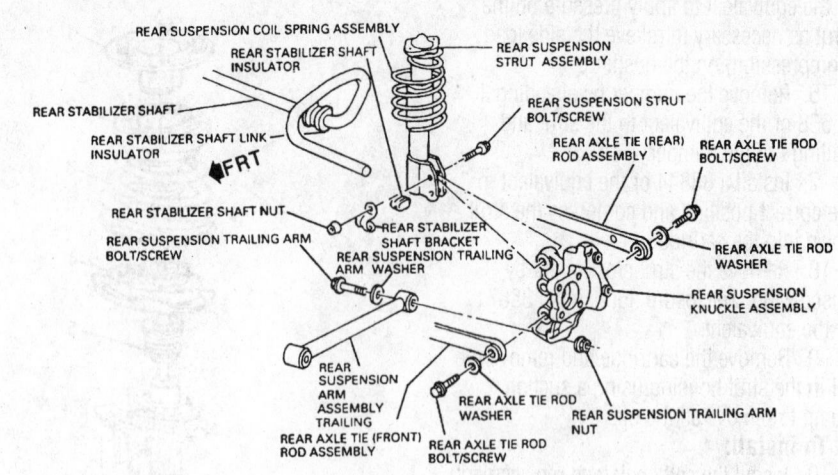

79222533

Rear suspension components—except 1995–96 Regal, Cutlass Supreme and Grand Prix

9. Remove the strut/stabilizer shaft bracket from the knuckle assembly.

To install:

10. Install the strut assembly to the body bolts. Tighten the bolts to 34 ft. lbs. (46 Nm).

11. Install the strut/stabilizer shaft bracket to the knuckle. Align the scribe marks to ensure proper alignment. Tighten the attaching bolts to 122 ft. lbs. (165 Nm).

12. Install the brake hose bracket.

13. Remove the special tool J 35778 or equivalent, if used, from the transverse leaf spring.

14. Install the exhaust system, if removed.

15. Install the jack pad and tighten the attaching bolts 18 ft. lbs. (25 Nm).

16. Install the tire and wheel assembly, aligning up the marks made during the removal.

17. Safely lower the vehicle.

18. A four wheel alignment is recommended after any steering/suspension repairs are performed.

EXCEPT 1995–96 REGAL, CUTLASS SUPREME AND GRAND PRIX

These vehicles use direct double-acting struts which are attached to the body and to the rear wheel knuckle assembly. Please note that strut service requires special tools to compress and hold the components during removal and installation. This procedure calls out the GM-recommended special tool numbers. Use care if using substitutes.

1. Disconnect the negative battery cable. Raise and safely support the vehicle.

2. Remove the rear wheel assembly. Mark the position of the wheel to the wheel studs, prior to removal, for installation reference.

3. Make a scribe mark to the strut-to-knuckle for proper installation.

4. Remove the brake hose bracket at the strut.

5. If equipped with dual exhaust, remove the exhaust system.

6. Remove the strut mount to body nuts.

7. Remove the strut stabilizer shaft bracket from the knuckle.

8. Remove the strut from the vehicle.

To install:

9. Install and tighten the strut mount to body nuts to 34 ft. lbs. (46 Nm).

10. Install the strut stabilizer shaft bracket to steering knuckle and tighten nuts to 122 ft. lbs. (165 Nm).

11. Install the brake hose bracket.

12. If removed, install the exhaust system.

13. Install the tire and wheel assembly, aligning the marks made on the removal.

14. Safely lower the vehicle.

15. Four wheel alignment is recommended after any steering/suspension repairs are performed.

Coil Spring

REMOVAL & INSTALLATION

1. Remove the strut assembly from the vehicle.

✳✳ CAUTION

Do not over compress the coil spring. Only compress the spring until it comes away from the seat.

2. Compress the coil spring using a suitable spring compressor.

3. Remove the strut rod nut.

4. Remove the upper spring seat and coil spring.

To install:

5. Carefully compress the coil spring and place it on the strut assembly in the same orientation as the original spring.

6. Install the upper spring seat. On front struts with the replaceable cartridge, tighten the nut to 81 ft. lbs. (110 Nm). On all other models, tighten the nut to 63 ft. lbs. (85 Nm). Tighten the nut on all rear strut assemblies to 55 ft. lbs. (75 Nm).

7. Install the strut in the vehicle.

Lower Ball Joint

REMOVAL & INSTALLATION

Except 1997–99 Century, Grand Prix, Intrigue and Regal

1. Raise and safely support the vehicle. Remove the tire and wheel assembly.

2. Remove the ball joint heat shield retaining bolts and the ball joint heat shield.

3. Remove the lower ball joint cotter pin and nut.

4. Loosen, but do not remove the stabilizer shaft bushing assembly bolts.

5. Remove the ball joint front the lower control arm using special tool J 35917 or equivalent ball joint press tool.

6. Drill out the four rivets retaining the ball joint to the steering knuckle. Use an ⅛ in. drill bit to make a pilot hole through the rivets. Finish drilling the rivets using a ½ in. drill bit.

➡**Do not damage the drive axle boots when drilling out the ball joint rivets.**

7. Remove the ball joint from the steering knuckle assembly.

To install:

8. Install the ball joint in the steering knuckle assembly.

9. Install four new ball joints bolts and nuts.

10. Install the ball joint to the lower control arm and tighten the nut to 63 ft. lbs. (85 Nm). Tighten the nut to align slot in the ball joint nut with the cotter pin opening in the stud. Do not tighten the ball joint nut more than 60 degrees (one flat, or ⅙ turn) to align the hole and never loosen the nut anytime during installation. When ball joint nut installation is satisfactory, install a new cotter pin.

11. Install the ball joint heat shield.

12. Install the tire and wheel assembly.

13. Safely lower, then road test the vehicle.

➡**A four wheel alignment is recommended after any steering/suspension repairs are performed.**

1997–99 Century, Grand Prix, Intrigue and Regal

1. Raise and safely support the vehicle.

2. Remove the front wheel.

3. Disconnect the tie rod end from the knuckle.

4. Remove the stabilizer bar link from the lower control arm.

5. Remove the halfshaft outer retaining nut and remove the halfshaft.

6. Remove the ABS speed sensor jumper harness.

7. Remove the control arm mounting bolts.

8. Disconnect the ball joint stud from the knuckle.

9. Drill out or grind of the heads of the rivets and remove the ball joint from the control arm.

To install:

10. Install the new ball joint of the control arm. Be sure to install the bolts from the top of the control arm. Tighten the nuts to 50 ft. lbs. (68 Nm).

11. Install the control arm on the frame but do not tighten the fasteners at this time.

12. Install the ball joint stud to the knuckle. Tighten the nut to 40 ft. lbs. (55 Nm). Tighten the nut slightly if necessary to align the cotter pin holes and install the cotter pin.

13. Connect the stabilizer bar link to the control arm. Tighten the nut to 17 ft. lbs. (23 Nm).

14. Install the ABS speed sensor jumper harness.

15. Install the halfshaft.

16. Place a jack under the lower control arm, raise the jack to support the weight of the vehicle and tighten the control arm nuts to 83 ft. lbs. (113 Nm). Remove the jack.

17. Install the front wheel and lower the vehicle to the floor.

Wheel Bearings

ADJUSTMENT

The wheel bearings are not adjustable. If a wheel bearing is out of specification, it must be replaced. Using a dial indicator, check for looseness. If it exceeds 0.005 in. (0.1270mm) on drum or disc brakes the bearing wear is excessive and the hub and bearing should be replaced.

REMOVAL & INSTALLATION

Front

➡**Do not remove the drive axle nut at this time. Failure to follow the proper sequence of removal steps may cause permanent bearing damage.**

1. Loosen the hub nut one full turn. Do not remove.

2. Raise and safely support the vehicle. Remove the tire and wheel assembly.

3. Remove the caliper mounting bracket-to-steering knuckle mounting bolts and remove the caliper and bracket assembly and support the assembly out of the way. DO NOT disconnect the brake hose from the caliper or allow the hose to support the weight of the caliper.

4. Remove the brake rotor.

5. Remove the hub nut and washer.

6. Using special tool J 28733-a or an equivalent hub puller, push the halfshaft out of the hub assembly about an inch.

7. Remove the hub assembly-to-knuckle attaching bolts.

8. If equipped, remove the ABS sensor mounting bolt and position the ABS sensor out of the way.

9. Remove the hub and bearing assembly from the vehicle.

To install:

10. Clean all parts well. Install the hub and bearing assembly onto the axle shaft splines. Be sure the splines engage smoothly. DO NOT force the hub over the axle splines or the splines will be damaged.

11. Seat the hub and bearing assembly against the steering knuckle and install the mounting bolts. Tighten the bolts alternately and evenly to 52 ft. lbs. (70 Nm).

12. Be sure the ABS sensor is clean. The small magnet in its end will attract rust and metal chips that can degrade its performance. Install the ABS sensor and tighten the mounting bolt.

13. Install the brake rotor.

14. Install the caliper and bracket assembly on the steering knuckle and tighten the caliper bracket mounting bolts to 79 ft. lbs. (107 Nm).

15. Install the tire and wheel assembly. Lower the vehicle.

16. Install a new drive axle nut and washer. On 1995 vehicles, tighten the nut to 184 ft. lbs. (250 Nm). On 1996–99 vehicles, tighten the drive axle nut to 151 ft. lbs. (205 Nm).

Rear

The rear suspension uses a non-service-able hub/bearing assembly with integral wheel speed sensor that is bolted to the rear suspension knuckle. This hub/bearing assembly is a sealed, maintenance free unit. If the hub/bearing is damaged, the complete assembly must be replaced.

1. Raise and safely support the vehicle.
2. Remove the rear wheel.
3. Remove the brake hose bracket and the brake drum or caliper.
4. If equipped, detach the ABS electrical connector.

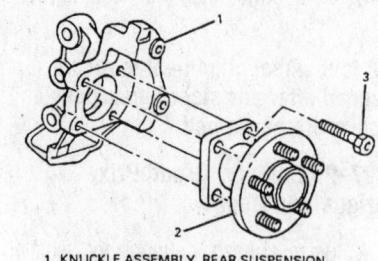

1 KNUCKLE ASSEMBLY, REAR SUSPENSION
2 HUB AND BEARING ASSEMBLY
3 BOLT/SCREW, WHEEL

7922Z535

The rear hub/bearing assembly is bolted to the knuckle

5. Remove the hub and bearing mounting bolts and remove the bearing assembly.

To install:

6. Install the hub and bearing assembly, tighten the bolt to 55 ft. lbs. (75 Nm).
7. If equipped, attach the ABS connector.
8. Install the brake caliper or drum and the brake hose bracket.
9. Install the rear wheel.
10. Safely lower the vehicle to the floor.
11. A four wheel alignment is recommended after any steering/suspension repairs have been performed.

GENERAL MOTORS Y BODY 39

Chevrolet-Corvette

DRIVE TRAIN 39-30
ENGINE REPAIR 39-2
FUEL SYSTEM 39-28
STEERING AND
SUSPENSION 39-36

A

Air Bag.......................... 39-36
Disarming.................... 39-36
Precautions 39-36
Automatic Transmission
Assembly.......................... 39-30
Removal & Installation 39-30

C

Camshaft 39-17
Removal & Installation 39-17
Clutch 39-34
Adjustment 39-34
Removal & Installation 39-34
Cylinder Head 39-8
Removal & Installation 39-8

D

Distributor 39-2
Removal & Installation 39-2

E

Engine Assembly 39-2
Removal & Installation 39-2
Exhaust Manifold 39-15
Removal & Installation 39-15

F

Fuel Filter 39-28
Removal & Installation 39-28

Fuel Pump 39-29
Removal & Installation 39-29
Fuel System Pressure 39-28
Relieving 39-28
Fuel System Service
Precautions..................... 39-28

H

Halfshaft........................ 39-35
Removal & Installation 39-35
Hydraulic Clutch System 39-35
Bleeding 39-35

I

Ignition Timing 39-2
Adjustment.................... 39-2
Intake Manifold 39-12
Removal & Installation 39-12

L

Lower Ball Joint 39-39
Removal & Installation 39-39
Lower Control Arm 39-39
Removal & Installation 39-39

M

Manual Transmission
Assembly...................... 39-32
Removal & Installation 39-32

O

Oil Pan......................... 39-22
Removal & Installation 39-22
Oil Pump 39-23
Removal & Installation 39-23

P

Power Rack and Pinion Steering
Gear........................... 39-37
Removal & Installation 39-37

R

Rear Main Seal 39-25
Removal & Installation 39-25
Rocker Arms 39-11
Removal & Installation 39-11

S

Shock Absorber 39-37
Removal & Installation 39-37
Steering Knuckle............... 39-38
Removal & Installation 39-38

T

Timing Chain, Sprockets, Front
Cover and Seal 39-25
Removal & Installation 39-25

U

Upper Ball Joint 39-38
Removal & Installation 39-38

V

Valve Lash..................... 39-21
Adjustment.................... 39-21

W

Water Pump.................... 39-6
Removal & Installation 39-6
Wheel Bearings................ 39-39
Adjustment.................... 39-39
Removal & Installation 39-39

ENGINE REPAIR

➡️Disconnecting the negative battery cable on some vehicles may interfere with the functions of the on board computer systems and may require the computer to undergo a relearning process, once the negative battery cable is reconnected.

Distributor

All 1997–99 5.7L (VINS J and G) engines use a direct ignition system. This system eliminates the need for a distributor and has several advantages. Some of the advantages are, no moving parts, no load on the engine, more coil cool down time between firing and increased coil saturation time which produces a stronger spark.

The direct ignition system consists of
• Eight ignition coils and modules (one for each cylinder)
• Eight ignition control circuits
• Camshaft position sensor
• 1X Camshaft reluctor wheel
• 24X Crankshaft reluctor wheel
• Crankshaft position sensor
• Powertrain Control Module (PCM)

The 24X crankshaft position sensor is the most critical part of the system, if it is damaged, the engine will not start. Ignition timing is controlled by the PCM. There are no timing marks on the crankshaft balancer or timing chain cover.

REMOVAL & INSTALLATION

➡️This procedure is for 5.7L (VIN P and 5) engines only. The 5.7 (VIN G and J) engines do not utilize a distributor.

The 5.7L (VIN P and 5) engines utilize a distributor which is driven off the front of the camshaft and is mounted flush to the front of the engine, behind the water pump.

1. Be sure the ignition is in the **OFF** or **LOCK** position. Disconnect the negative battery cable.
2. Disconnect the Intake Air Temperature (IAT) sensor.
3. Remove the intake air duct.
4. Remove the accessory drive belt.

❊❊ CAUTION

Never open, service or drain the radiator or cooling system when hot; serious burns can occur from the steam and hot coolant. Also, when draining engine coolant, keep in mind that cats and dogs are attracted

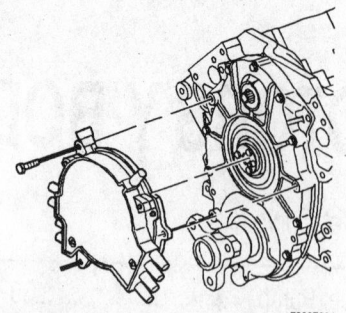

79222601

The distributor is located under the water pump assembly—5.7L (VIN P and 5) engines

to ethylene glycol antifreeze and could drink any that is left in an uncovered container or in puddles on the ground. This will prove fatal in sufficient quantities. Always drain coolant into a sealable container. Coolant should be reused unless it is contaminated or is several years old.

5. Drain the engine coolant.
6. Disconnect the Engine Coolant Temperature (ECT) sensor.
7. Remove the water pump and crankshaft balancer.
8. Remove the drive belt tensioner.

➡️The spark plug wire boots should be twisted ½ turn in each direction while removing. Do not pull on the wires to remove them from the spark plugs. Pull on the boots or use a tool specifically designed for this purpose.

9. Remove the spark plug wiring harnesses from the distributor.
10. Unplug the 4-terminal electrical connector from the distributor.
11. Disconnect the distributor vacuum harness from the distributor.
12. Loosen and remove the distributor attaching bolts.
13. Pull the distributor assembly forward until the driveshaft disengages from the end of the camshaft.
14. Mark the top surface of the driveshaft to assure ease of alignment during installation.

To install:

➡️Replace the O-rings on the coupling shaft or ignition system performance may suffer. Lubricate the O-rings and the end of the camshaft.

❊❊ WARNING

Don't try to fully seat the distributor using the distributor retainers. If the

distributor will not seat by hand, it's not properly aligned with the camshaft. Rotate the crankshaft until the engine is at the No. 1 cylinder TDC (camshaft sprocket pin at the 9 o'clock position). Rotate the distributor coupling until the camshaft sprocket pin slot aligns with the distributor base timing mark. Install the distributor using hand pressure to fully seat the distributor.

15. Install the distributor assembly into position with the driveshaft in the end of the camshaft. Rotate the distributor coupling until the camshaft sprocket pin slot aligns with the camshaft sprocket pin. Slide the distributor onto the end of the camshaft until fully seated on the engine front cover.
16. Install the distributor mounting bolts. Tighten to 97–106 inch lbs. (11–12 Nm).
17. Engage all electrical connections and vacuum hoses on the distributor that were unplugged during removal.
18. Install the drive belt tensioner.
19. Install the crankshaft balancer and water pump.
20. Connect the ECT sensor.
21. Install the drive belt.
22. Install the air duct.
23. Connect the IAT sensor.
24. Refill the cooling system.
25. Connect the negative battery cable.

Ignition Timing

ADJUSTMENT

On these vehicles, base timing is preset when the engine is manufactured. All timing changes are, then controlled directly by the PCM based on information from the ignition and knock sensor systems. No adjustments are necessary or possible.

Engine Assembly

REMOVAL & INSTALLATION

5.7L (VIN P and 5) Engines

1. Disconnect the negative battery cable and properly relieve fuel system pressure.
2. Drain the coolant into a suitable container.
3. Remove the air intake duct.
4. Unplug the electrical harness and vacuum connections from the top of the engine.
5. Disconnect the upper radiator, radiator hose and heater hoses from the pump,

then remove the throttle body coolant hose from the radiator tee and from the right side of the throttle body.

6. Remove the power steering pump and support aside.

7. Remove the alternator and support aside.

8. Remove the left wheel well center panel. Remove the serpentine drive belt.

9. Remove the air conditioning compressor from the bracket and position aside.

10. Unplug the electrical connector and remove the cover from the wiper motor.

11. Disconnect the AIR diverter valve hose.

12. Disconnect and plug the fuel lines at the fuel rail.

13. Remove the hoses from the power steering fluid reservoir. Plug the openings to prevent system contamination or excessive fluid loss.

14. Disconnect the accelerator cable from the throttle body.

15. Raise and support the vehicle safely.

16. Remove the starter motor.

17. Remove the left and right catalytic converters, then remove the exhaust pipe and muffler assembly.

18. Remove the transmission.

19. If equipped with a manual transmission, remove the clutch cover and plate, then remove the flywheel.

20. Remove the ground leads from the rear of the engine, then disengage the electrical connectors from the oil level, knock, oil temperature and coolant temperature sensors.

21. Remove the nuts from the engine mount studs, then lower the vehicle.

22. Install a suitable lifting device and carefully remove the engine from the vehicle.

To install:

23. Lower the engine into position in the vehicle.

24. Remove the lifting device from the engine, then raise and support the vehicle safely.

25. Install the nuts on the engine mount studs.

26. Engage the sensor electrical connectors, then connect the ground leads to the rear of the engine.

27. If equipped with a manual transmission, install the flywheel, then install the clutch cover and plate.

28. Install the transmission.

29. Install the right and left catalytic converters, then install the exhaust pipe and muffler assembly.

30. Install the starter and lower the vehicle.

31. Connect the accelerator cable to the throttle body, then unplug the openings and install the hose to the power steering reservoir.

32. Unplug and connect the fuel lines to the fuel rail.

33. Connect the AIR diverter valve hose.

34. Install the electrical connector and cover to the wiper motor.

35. Install the air conditioning compressor.

36. Install the serpentine drive belt, then install the left wheel well center panel.

37. Install the alternator and the power steering pump.

38. Connect the coolant hose to the right side of the throttle body and connect the throttle body hose to the radiator tee. Connect the heater and radiator hoses.

39. Engage the electrical harness and all vacuum connections to the top of the engine.

40. Install the air intake duct and properly fill the engine cooling system.

41. Check all fluid levels, connect the negative battery cable and tighten the fuel filler cap.

42. Reset the CHANGE OIL indicator:

 a. Turn the ignition **ON** but do not start the engine.

 b. Depress the ENG MET button on the trip monitor, then within 5 seconds, depress the button a 2nd time. Within another 5 seconds, depress and hold the GAUGES button.

 c. While holding the GAUGES button and watch the CHANGE OIL light, it should begin to flash. Continue to hold the gauges button until the flashing stops and the light goes out indicating that the indicator is reset.

 d. If the indicator does not reset, turn the ignition **OFF** and restart the procedure.

43. Check and adjust the ASR control cables, as necessary.

44. Start the engine and check for leaks, then bleed the power steering system.

5.7L (VIN J) Engine

1. Disconnect the battery negative cable and properly relieve fuel system pressure.

2. Raise and support the vehicle safely.

3. Drain engine coolant into a suitable container and drain the engine oil.

4. Remove the complete exhaust system assembly, then remove the driveshaft.

5. Position a suitable transmission support stand under transmission and remove the transmission support beam.

6. Remove transmission from the vehicle.

7. Remove the clutch actuator cylinder, left side converter shield and clutch housing cover, then remove the clutch cover and disc.

8. Install a suitable engine lift hook to rear of engine.

9. Remove the AIR tube center section from the AIR hose and oil pan.

10. Unplug the oxygen sensor electrical connector.

11. Remove the power steering lower hose from the oil cooler.

12. Remove the negative battery cable from the cylinder case.

13. Remove the nuts attaching the engine mounts to the driveline and suspension frame, then lower the vehicle.

14. Remove the air cleaner assembly and air duct.

15. Disconnect the engine oil cooler lines from the oil filter housing.

16. Raise the rear of the engine.

17. Disconnect the fuel lines from the fuel rail.

18. Remove the evaporator housing panel and the resistor.

19. Remove the bolts attaching the right bulkhead connector.

20. Remove the engine right side wiring harness.

21. Remove the instrument panel right lower sound insulator pan.

22. Disengage the bulkhead wiring harness connectors from under the dash.

23. Remove the air bleed hose from the plenum.

24. Remove the radiator upper and lower hoses, then disconnect the power steering pump vacuum line(s).

25. Properly discharge and recover the air conditioning system.

26. Remove the air conditioning suction and discharge line flange from the compressor, then remove the air conditioning compressor-to-accumulator line from the accumulator.

27. Remove the air conditioning accumulator and position aside.

28. Remove the air conditioning accumulator bracket from the vehicle.

29. Disconnect and plug the power steering pressure line at the power steering gear.

30. Disconnect the throttle body linkage shield, then remove the throttle body cable to plenum retainer.

31. Disconnect the accelerator and cruise control cable or the control cable from the throttle body.

32. Install a suitable engine lift hook to front of the engine.

33. Remove the ECM from the ECM bracket, then disconnect ECM harness connector.

34. Remove the left front fender attaching bolts, shims and seal. Remove the left fender.

35. Disconnect the positive cable from the battery, remove the battery hold-down clamp, and remove the battery from the vehicle.

36. Disengage the engine left side bulkhead block electrical connector.

37. Disconnect the engine wiring harness fusible links at the junction block.

38. Disengage the engine harness connectors from the following:
- Secondary injector modules
- Positive battery cable at junction block
- Differential pressure switch vacuum and electrical connectors
- Air conditioning cutout relay
- Air conditioning high blower relay
- Transmission shift solenoid relay
- Fuel pump fuse
- Forward light link connector
- Positive battery lead
- Air conditioning blower resistor
- Air conditioning pressure sensors
- Air conditioning cooling fan switch
- Windshield washer pump
- Low coolant sensor
- Blower motor
- ESC knock sensor
- ESC knock sensor relay

39. Disconnect hoses from the vacuum pump, then tag and disconnect the front and rear vacuum connections.

40. Reposition engine harness aside and remove the braided ground strap from the left side frame rail.

41. Reposition the positive battery cable aside and remove the left side plenum panel screen.

42. Disconnect the brake booster vacuum hose.

43. Remove the windshield wiper motor from the vehicle.

44. Remove the MAP sensor and the MAP sensor bracket from the plenum.

45. Disconnect the AIR hose from the left exhaust manifold.

46. Using an engine lifting device, carefully remove the engine from the vehicle.

47. Transfer the following parts to the new engine, as necessary:
- Oil level indicator tube
- The exhaust manifolds
- The converter heat shields
- The wire pack heat shields
- Engine mounts

To install:

48. Install the engine mounts to the drive train and to the suspension frame, finger-tighten only at this time.

49. Position the engine into the vehicle using the lifting device.

50. Install the engine mount/bracket bolts, then remove the lifting device and lifting brackets.

51. Connect the AIR hose to the left exhaust manifold.

52. Install the MAP sensor and bracket to the plenum.

53. Install the wiper motor.

54. Install the left side plenum panel screen.

55. Route the left side wiring harness into position, then install the braided ground strap to the frame rail.

56. Install the left side bulkhead block connector.

57. Engage the engine harness fusible links and relays.

58. Install the battery and hold-down clamp.

59. Connect the battery positive cable to the battery, then install the left front fender.

60. Install the ECM to the ECM bracket, then engage the wiring harness electrical connector.

61. Remove the front engine lift hook, then connect power brake booster vacuum hose to the plenum.

62. Connect the cruise control and throttle cables or the control cable to the throttle body. Adjust the cables or the ASR cable adjuster, as applicable.

63. Install the cable shield, then install cable retainers to the plenum.

64. Connect the power steering pressure line to the power steering gear.

65. Connect the engine oil cooler lines to the engine.

66. Install the accumulator bracket, then install the accumulator.

67. Connect the air conditioning lines.

68. Attach the vacuum line(s) to the power steering pump.

69. Connect the radiator upper and lower hoses.

70. Connect the air bleed hose to the plenum.

71. Install the bulkhead wire connector to the bulkhead.

72. Engage the evaporator housing panel resistor connector.

73. Install the hose onto the vacuum pump, then install the front and rear vacuum connections.

74. Engage the engine harness connectors to the following:
- Air conditioning blower resistor
- Air conditioning pressure sensor
- Air conditioning cooling fan
- Windshield washer pump
- Low coolant sensor
- Blower motor
- ESC knock sensor
- ESC knock sensor relay
- Differential pressure switch

75. Connect the fuel lines to the fuel rail.

76. Install the engine right side wiring harness under the dash.

77. Install the instrument panel right sound insulator panel.

78. Raise and support the vehicle safely, then tighten the engine/bracket bolts and nuts to 40 ft. lbs. (54 Nm).

79. Install the power steering hose to power steering oil cooler.

80. Install the oxygen sensor wire connectors.

81. Connect the AIR tube center section to the AIR hose and oil pan.

82. Connect the negative battery cable to the engine and suitably support the engine, then remove the rear engine lift hook.

83. Install the clutch cover and disc, then install the clutch housing to the cylinder block and install the housing cover.

84. Install the left side converter shield to the housing, then position the actuator cylinder and install the retaining nuts.

85. Install the transmission and support beam.

86. Install the driveshaft.

87. Install the complete exhaust system assembly.

88. Lower the vehicle and add the proper type and amount of engine oil.

89. Tighten the fuel filler cap and connect the negative battery cable.

90. Properly fill the engine cooling system and check for leaks.

91. Recharge the air conditioning system.

92. If equipped, reset the CHANGE OIL indicator:

a. Turn the ignition **ON** but do not start the engine.

b. Depress the ENG MET button on the trip monitor, then within 5 seconds, depress the button a 2nd time. Within another 5 seconds, depress and hold the GAUGES button.

c. While holding the GAUGES button and watch the CHANGE OIL light, it should begin to flash. Continue to hold the gauges button until the flashing stops and the light goes out indicating that the indicator is reset.

d. If the indicator does not reset, turn the ignition **OFF** and restart the procedure.

5.7L (VIN G) Engine

The following tools will be required in addition to the basic hand tools:

- Transverse leaf spring compressor and adapters
- Ball joint separator
- Engine support table
- Driveshaft support strap
- Fuel pressure gauge

1. Disconnect the negative battery cable.

2. Properly discharge and recover the air conditioning system.

3. Raise and safely support the vehicle and drain the coolant, then lower the vehicle.

4. Disconnect the wiring from the Intake Air Temperature (IAT) and Mass Air Flow (MAF) sensors.

5. Remove the fuel pressure regulator purge line from the intake air duct.

6. Remove the air cleaner assembly.

7. Remove the accessory drive belts.

8. Remove the fuel rail covers.

✳✳ CAUTION

Observe all applicable safety precautions when working around fuel. Whenever servicing the fuel system, always work in a well ventilated area. Do not allow fuel spray or vapors to come in contact with a spark or open flame. Keep a dry chemical fire extinguisher near the work area. Always keep fuel in a container specifically designed for fuel storage; also, always properly seal fuel containers to avoid the possibility of fire or explosion.

9. Relieve the fuel system pressure.

10. Disconnect and cap the fuel lines from the fuel rail.

11. Remove the coolant hoses from the water pump.

12. Disconnect the wiring harness from the following:
- Fuel injectors
- Ignition coil
- EVAP solenoid
- Electric throttle motor
- Throttle position sensor
- ECT sensor
- A/C compressor

13. Disconnect the wiring from the generator

14. Remove the generator rear bracket, then remove the generator.

15. Disconnect the vacuum hose from the brake booster.

16. Disconnect the intermediate shaft from the steering rack and position it on the left frame rail.

17. Disconnect the AIR hose from the left exhaust manifold.

➡**Do not remove the bell housing-to-driveshaft support bolts until specified in the procedure.**

18. Raise and safely support the vehicle.

19. Remove the front wheels.

20. Remove the intermediate exhaust pipe.

21. Remove the under covers.

22. Remove the ground wires from the right side of the engine block.

23. Remove the starter.

24. Disconnect the oil level sensor, crankshaft position sensor and the right front oxygen sensor.

25. Remove the refrigerant lines from the rear of the compressor.

26. Disconnect the engine oil temperature sensor and all ground wires from the left side of the engine.

27. Disconnect the left front oxygen sensor.

28. Remove the front stabilizer bar from the cradle.

29. Remove both cooling fans.

30. Compress the front transverse spring using tool J 33432-A or equivalent spring compressor.

31. Disconnect the tie rod ends from the steering knuckles.

32. If equipped, detach the ABS, Real Time Damping (RTD) and Electronic Variable Orifice (EVO) connectors at the cradle.

33. Remove the lower shock absorber mounting bolts.

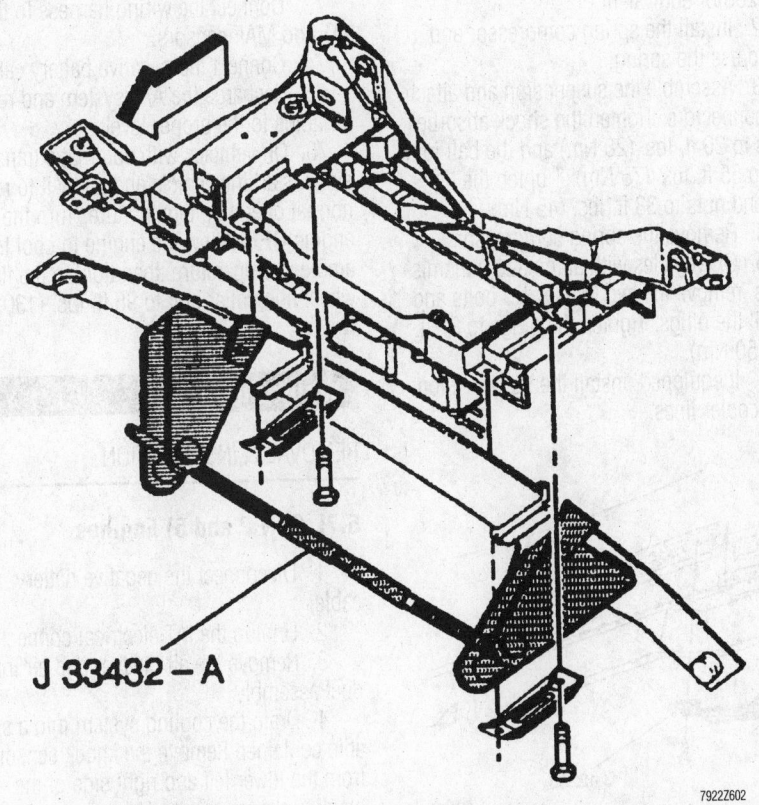

J 33432 - A

79222602

Use a compressor such as J 33432-A to compress the front transverse spring—5.7L (VIN G) engine

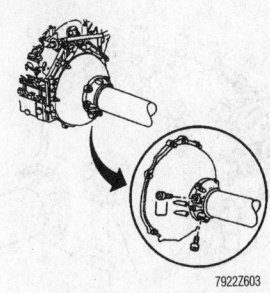

79222603

On A/T vehicles, remove the plugs from the driveshaft support assembly and install two M10 X 1.5 bolts into the plug holes—5.7L (VIN G) engine

34. Separate the lower ball joints from the knuckles.

35. Remove the spring compressor from the spring.

36. If equipped, disconnect the transmission fluid cooler lines from the radiator and at the junction near the bell housing.

37. On vehicles with automatic transmissions, remove the two plugs from the driveshaft support assembly and install two M10 X 1.5 bolts into the plug holes. The bolts must be at least 55mm in length or longer. Tighten the bolts to 26 ft. lbs. (35 Nm).

38. On vehicles with automatic transmission, remove the bell housing inspection cover, then turn the flywheel hub collar to access the bolt and loosen it.

✳✳ CAUTION

Never use J 42203 or equivalent to support the weight of the engine assembly.

39. Install J 42203 or equivalent to the under cover flange and support the engine assembly with a lift such as J 39580.

40. Remove the front cradle nuts BY HAND and lower the engine assembly slightly.

41. Disconnect the wiring harness from the following and remove the harness from the retainers on the bell housing:
- Engine oil pressure gauge
- Camshaft position sensor
- MAP sensor
- Knock sensor
- Ground wire on the rear of the left cylinder
- Any remaining connectors

42. Remove the front driveshaft support assembly bolts.

43. Move the engine assembly forward to clear the driveshaft.

44. Carefully raise the vehicle off of the engine and cradle assembly.

✳✳ WARNING

Do not disconnect the power steering lines when removing the steering components from the engine. Remove them from the engine and fasten them to the cradle.

45. Remove the engine from the cradle.
To install:

46. Install the engine on the cradle. Tighten the nuts to 40 ft. lbs. (54 Nm).

47. Install the power steering pump and reservoir on the engine. Tighten the mounting bolts to 18 ft. lbs. (25 Nm).

48. Position the assembly under the vehicle, then partially lower the vehicle on the engine and attach the connectors and wiring harness.

✳✳ WARNING

Do not force the driveshaft into the engine assembly.

49. Move the assembly rearward onto the driveshaft and install the bolts. Tighten the bolts to 37 ft. lbs. (50 Nm).

50. Lower the vehicle onto the cradle and install the mounting nuts BY HAND. Tighten them to 81 ft. lbs. (110 Nm).

51. Raise the vehicle and remove bracket J 42203 or equivalent.

52. Install the spring compressor and compress the spring.

53. Assemble the suspension and attach the connectors. Tighten the shock absorber bolts to 20 ft. lbs. (28 Nm) and the ball joint nut to 55 ft. lbs. (75 Nm). Tighten the tie rod end nuts to 33 ft. lbs. (45 Nm).

54. Remove the spring compressor.

55. On vehicles with automatic transmissions, remove the two M10 X 1.5 bolts and install the plugs. Tighten the plugs to 37 ft. lbs. (50 Nm).

56. If equipped, install the transmission fluid cooler lines.

57. Install the cooling fans.

58. Install the stabilizer bar. Tighten the bolt to 40 ft. lbs. (54 Nm).

59. Attach the connectors and grounds on the left side of the engine.

60. Connect the lines to the rear of the A/C compressor. Tighten the bolt to 26 ft. lbs. (35 Nm).

61. Attach the connectors and grounds on the right side of the engine.

62. Install the starter and tighten the bolts to 37 ft. lbs. (50 Nm).

63. Install the under covers and the front wheels. Tighten the wheel nuts to 100 ft. lbs. (140 Nm).

64. Connect the AIR hose to the exhaust manifold.

65. Install the intermediate shaft to the steering gear.

66. Connect the vacuum hose to the brake booster.

67. Install the generator and brackets.

68. Reposition the harness and attach all electrical connectors on the engine.

69. Connect the coolant hoses to the water pump.

70. Connect the fuel lines.

71. Install the fuel rail covers and drive belts.

72. Install the intake air duct and regulator purge line.

73. Connect the wiring harness to the IAT and MAF sensors.

74. Connect the negative battery cable.

75. Recharge the A/C system and refill all fluids to the proper level.

76. On vehicles with automatic transmissions, start the vehicle and allow it to reach normal operating temperature. Turn the engine **OFF**. Allow the engine to cool to ambient temperature, then tighten the flywheel hub collar bolt to 96 ft. lbs. (130 Nm).

Water Pump

REMOVAL & INSTALLATION

5.7L (VIN P and 5) Engines

1. Disconnect the negative battery cable.

2. Unplug the IAT electrical connection.

3. Remove the air cleaner and air intake duct assembly.

4. Drain the cooling system into a suitable container. Remove the knock sensors from the lower left and right side of the block to assure proper draining.

5. Remove the upper and lower radiator hoses and the heater hose from the water

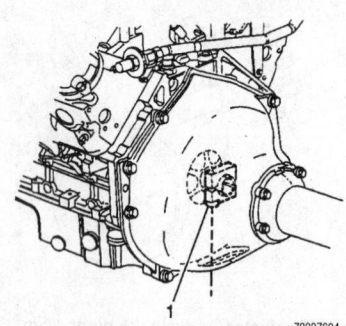

79222604

Loosen the bolt on the flywheel hub collar after turning it for access—5.7L (VIN G) engine

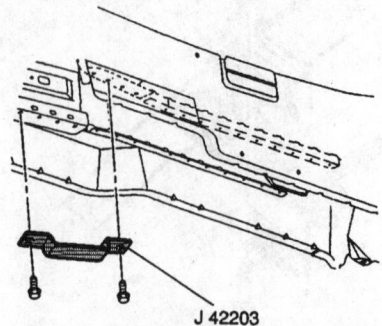

J 42203

79222605

Install bracket J 42203 or equivalent to the under cover flange—5.7L (VIN G) engine

pump. Remove the throttle body hose from the tee fitting.

6. Unplug the coolant sensor electrical connection and remove the sensor wiring harness from the retainer on the front of the coolant pump.

7. Use a box wrench or socket on the tensioner pulley bolt to rotate the tensioner and relieve belt tension, then remove the serpentine drive belt from the alternator pulley. This should create sufficient room to work, if additional room is necessary, the belt can be completely removed.

8. Remove the 6 bolts securing the water pump flanges to the engine block, then remove the water pump from the vehicle.

9. Remove and discard the old gaskets from the mating surfaces.

10. If replacing the pump, remove the coolant sensor from the old pump.

To install:

11. If replacing the pump, install the coolant sensor on the new pump and tighten to 17 ft. lbs. (23 Nm).

12. Thoroughly clean all gasket mating surfaces and apply a light coat of grease to the seals and splines before assembling the coupling to the water pump. The white band on the coupling should be positioned towards the engine.

13. Install the new gaskets with the tabs up, the coolant pump with the drive coupling and the mounting bolts. Tighten the bolts to 30 ft. lbs. (41 Nm).

14. Install the serpentine drive belt.

15. Install the coolant sensor wiring harness to the retainer on the front of the pump, then engage the sensor electrical connection.

16. Connect the heater hose and the upper and lower radiator hoses to the water pump.

17. Install the throttle body hose at the tee.

18. If removed, install the knock sensors.

19. Open the bleed valves on the thermostat housing and the throttle body. Fill the cooling system through the radiator surge tank until a solid stream of coolant comes out of the bleeds.

20. Close all bleeds and continue to fill the surge tank until the coolant is level at the base of the surge tank neck.

21. Install the radiator pressure cap and check the coolant recovery reservoir for the proper level of coolant, add as necessary.

22. Install the air cleaner and intake duct assembly.

23. Engage the IAT electrical connection and clean any excess coolant from the engine compartment.

24. Connect the negative battery cable, start the engine and check for leaks.

25. If the low coolant indicator lamp is lit, the engine must be cycled from cold to normal operating temperature and back to cold 3 times. If the lamp does not go out after this and coolant is at the proper level, the indicator system must be repaired.

5.7L (VIN J) Engine

1. Disconnect the negative battery cable.

2. Drain engine coolant into a suitable container.

3. Remove the air cleaner and intake duct assembly.

4. Remove the screws attaching the throttle body extension to the throttle body, then remove the throttle body extension and gasket.

5. Remove clamps and hoses from the coolant outlets, radiator inlet and inlet pipe.

6. Remove the inlet pipe assembly and hose from the vehicle.

7. Remove the serpentine drive belt, then remove the tensioner retaining bolt and remove the tensioner from the pump. It is not necessary to remove the water pump pulley.

8. Remove the engine hose clamp, then the hose from the water pump.

9. Remove the alternator lower bracket mounting bolts, then remove the bracket from the vehicle.

10. Remove the water pump attaching bolts (noting the position and size of each bolt) and remove the bolt attaching the air conditioning compressor to the water pump. Remove the water pump from the vehicle.

To install:

11. Thoroughly clean the pump and front cover sealing surfaces.

12. Install the water pump, new gasket and bolts, finger-tight only.

13. Install and finger-tighten the bolt attaching air conditioning compressor to the pump.

14. Tighten air conditioning compressor bolt and water pump attaching bolts to 20 ft. lbs. (26 Nm).

15. Install engine hose and clamp.

16. Apply Loctite® 565 to the bolt threads, then install the alternator mounting bolts. Tighten the bolts to 39 ft. lbs. (52 Nm) and the bracket bolts to 20 ft. lbs. (26 Nm).

17. Install the belt tensioner and tighten the retaining bolt to 45 ft. lbs. (60 Nm).

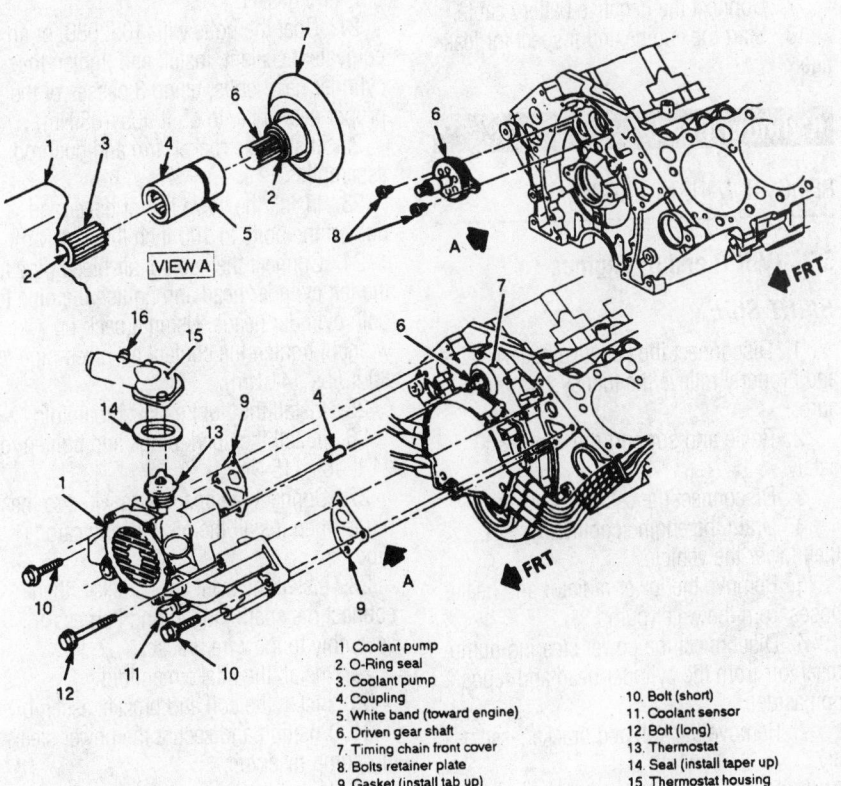

1 Coolant pump	
2 O-Ring seal	
3 Coolant pump	
4 Coupling	
5 White band (toward engine)	10. Bolt (short)
6 Driven gear shaft	11. Coolant sensor
7 Timing chain front cover	12. Bolt (long)
8 Bolts retainer plate	13. Thermostat
9 Gasket (install tab up)	14. Seal (install taper up)
	15. Thermostat housing

79222606

Exploded view of the water pump mounting—5.7L (VIN P and 5) engines

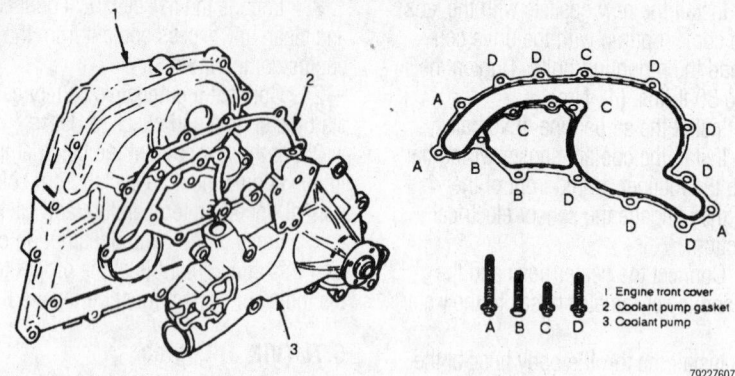

Water pump assembly mounting and bolt locations—5.7L (VIN J) engine

18. Install the serpentine drive belt.
19. Install the hose and inlet pipe assembly.
20. Install throttle body extension and gasket. Tighten bolts to 53 inch lbs. (6 Nm).
21. Install air cleaner and intake duct assembly, then connect the negative battery cable.
22. Refill the cooling system with the proper type and quantity of antifreeze and inspect the system for leaks.

5.7L (VIN G) Engine

1. Disconnect the negative battery cable.
2. Disconnect the wiring harness from the Intake Air Temperature (IAT) and Mass Air Flow (MAF) sensors.
3. Remove the fuel pressure regulator purge line from the intake air duct.
4. Remove the air cleaner assembly.
5. Remove the accessory drive belts.
6. Disconnect the radiator and heater hoses from the water pump.
7. Remove the water pump pulley.
8. Remove the six mounting bolts and the water pump assembly.

To install:

9. Clean the sealing surfaces on the water pump and engine block.

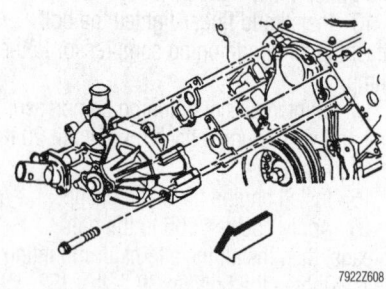

Water pump assembly mounting—5.7L (VIN G) engine

10. Be sure the tabs on the new gaskets are pointing up and install the water pump on the engine. Tighten the mounting bolts to 30 ft. lbs. (41 Nm).
11. Install the water pump pulley. Tighten the bolts first to 89 inch lbs. (10 Nm), then to 18 ft. lbs. (25 Nm).
12. Connect the radiator and heater hoses.
13. Install the drive belts.
14. Install the intake air duct and fuel regulator purge line.
15. Attach the IAT and MAF sensor connectors.
16. Refill the cooling system with the proper type of coolant.
17. Connect the negative battery cable.
18. Start the engine and inspect for leakage.

Cylinder Head

REMOVAL & INSTALLATION

5.7L (VIN P and 5) Engines

RIGHT SIDE

1. Disconnect the negative battery cable and properly relieve the fuel system pressure.
2. Raise and support the vehicle safely.
3. Disconnect the catalytic converter.
4. Drain the engine cooling system, then lower the vehicle.
5. Remove the lower radiator and heater hoses from the water pump.
6. Disconnect the power steering pump reservoir from the cylinder head and reposition aside.
7. Remove the coil and bracket assembly.
8. Remove the intake manifold.
9. Remove the spark plug wires from the clips, then remove the front wire bracket.
10. Remove the oil level indicator tube.
11. Disconnect the spark plug wires from the plugs, then remove the spark plugs from the cylinder head.
12. Remove the right exhaust manifold.
13. Using a back-up wrench on the pipe fitting, disconnect the coolant air bleed pipe from the left cylinder head.
14. Remove the right valve rocker cover, then remove the rocker arm and pushrod assemblies.
15. Remove the cylinder head bolts.
16. Remove the cylinder head along with the coolant air bleed pipe, then remove the head gasket.
17. If necessary, remove the coolant air bleed pipe from the cylinder head.

To install:

18. Thoroughly clean the cylinder head and cylinder case mating surfaces. Be sure both surfaces are free of any foreign matter, nicks or scratches. The threads in both the bolts holes and on the bolts must be clean and free of old sealer.
19. If removed, install the coolant air bleed pipe to the cylinder head, finger-tight.
20. Position the new gasket in place on the cylinder case with the yellow tab facing up. Install the cylinder head over the dowel pins and gasket.
21. Coat the bolts with 1052080, or an equivalent sealant. Install and tighten the cylinder head bolts, using 3 passes of the proper sequence, to 65 ft. lbs. (88 Nm).
22. Install the rocker arm and pushrod assemblies.
23. Install the valve rocker cover and tighten the bolts to 100 inch lbs. (11 Nm).
24. Connect the coolant air bleed pipe to the left cylinder head and tighten the pipe to both cylinder heads. Using a back-up wrench, tighten the coolant air bleed pipe to 30 ft. lbs. (41 Nm).
25. Install the right exhaust manifold.
26. Install the spark plugs and tighten to 11 ft. lbs. (15 Nm).
27. Connect the spark plug wires to the plugs, then install the oil level indicator tube.
28. Install the front wire bracket, then connect the spark plug wiring harness assembly to the wire bracket.
29. Install the intake manifold.
30. Install the coil and bracket assembly.
31. Position and secure the power steering pump reservoir.
32. Install the lower radiator and heater hoses to the water pump.

33. Raise and support the vehicle safely.

34. Connect the catalytic converter, then lower the vehicle.

35. Properly fill the cooling system.

36. Tighten the fuel filler cap and connect the negative battery cable.

LEFT SIDE

1. Disconnect the negative battery cable and properly relieve the fuel system pressure.

2. Raise and support the vehicle safely.

3. Disconnect the catalytic converter.

4. Drain the engine cooling system, then lower the vehicle.

5. Remove the upper radiator hose.

6. Remove the serpentine drive belt.

7. Remove the intake manifold.

8. Remove the left wheel well lower center panel.

9. Disconnect the air conditioning compressor from the bracket and position aside. Use care not to kink or damage the refrigerant lines. Remove the compressor and alternator brace.

10. Remove the spark plug wire bracket, disconnect the wires from the spark plugs and remove the spark plugs from the cylinder head.

11. Remove the left exhaust manifold.

12. Remove the remaining alternator brace, then remove the alternator.

13. Disconnect the AIR diverter valve hose.

14. Remove the left valve rocker cover.

15. Remove the drive belt idler pulley, then remove the drive belt tensioner.

16. Disconnect the power steering lines from the pump, then remove the pump. Plug the openings to prevent system contamination or excessive fluid loss.

17. Remove the spark plug and coil wires from the distributor.

18. Remove the accessory mounting bracket.

19. Remove the rocker arm and pushrod assemblies.

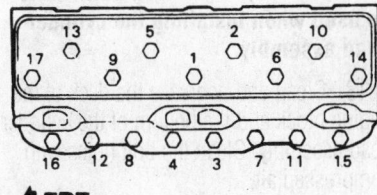

Cylinder head bolt torque sequence—5.7L (VIN P and 5) engines

20. Disconnect the coolant air bleed pipe from the cylinder head.

21. Remove the cylinder head bolts, then remove the cylinder head and gasket.

To install:

22. Thoroughly clean the cylinder head and cylinder case mating surfaces. Be sure both surfaces are free of any foreign matter, nicks or scratches. The threads in both the bolts holes and on the bolts must be clean and free of old sealer.

23. Position the new gasket in place on the cylinder case with the yellow tab facing up. Install the cylinder head over the dowel pins and gasket.

24. Coat the bolts with 1052080, or an equivalent sealant. Install and tighten the cylinder head bolts, using 3 passes of the proper sequence, to 65 ft. lbs. (88 Nm).

25. Connect the coolant air bleed pipe to the cylinder head and tighten to 30 ft. lbs. (41 Nm).

26. Install the rocker arm and pushrod assemblies.

27. Install the accessory mounting bracket and bolts. Tighten the bolts to 31 ft. lbs. (42 Nm).

28. Connect the spark plug and coil wires to the distributor.

29. Install the power steering pump, then remove the plugs from the openings and connect the lines.

30. Install the drive belt tensioner, then install the idler pulley. Tighten the tensioner and pulley bolts to 24 ft. lbs. (33 Nm).

31. Install the left valve rocker cover and bolts. Tighten the bolts to 100 inch lbs. (11 Nm).

32. Connect the AIR diverter valve hose and install the alternator lower brace.

33. Install the left exhaust manifold.

34. Install the spark plugs and tighten to 11 ft. lbs. (15 Nm). Connect the spark plug wires to the plugs and insert the wires into the brackets.

35. Install the air conditioning compressor and alternator brace, then install the compressor.

36. Install the left wheel well lower center panel.

37. Install the intake manifold.

38. Install the serpentine drive belt and the upper radiator hose.

39. Raise and safely support the vehicle, then connect the catalytic converter and lower the vehicle.

40. Properly fill the engine cooling system.

41. Tighten the fuel filler cap and connect the negative battery cable.

42. Bleed the power steering system.

5.7L (VIN J) Engine

RIGHT SIDE

1. Disconnect the negative battery cable and properly relieve fuel system pressure.

2. Drain engine coolant into a suitable container.

3. Remove the intake plenum assembly.

4. Remove the right injector housing.

5. Remove the right bank valve lifters.

6. Remove the alternator assembly.

7. Disconnect the right exhaust manifold from the cylinder head. It is not necessary to completely remove the exhaust manifold from the vehicle for cylinder head removal.

8. If raised, lower the vehicle for underhood access.

9. Remove the vacuum hose from secondary port throttle valve actuator.

10. Remove the access plug from the right cylinder head.

11. Remove the top bolt attaching the right secondary timing chain fixed guide.

12. Remove cylinder head bolts, then remove the cylinder head and gasket from the vehicle.

To install:

13. Thoroughly clean the cylinder head and cylinder case mating surfaces. Be sure both surfaces are free of any foreign matter, nicks or scratches. The threads in both the bolts holes and on the bolts must be clean and free of old sealer.

➡**Cylinder head gaskets are not interchangeable between cylinder banks.**

14. Install the cylinder head locating dowels into block, if loosened or removed, then position the new gasket in place on the cylinder case.

15. Install the cylinder head over the dowels. Coat bolt threads and washers with clean engine oil and insert.

16. Tighten the cylinder head bolts in sequence as follows:
- 1st pass-45 ft. lbs. (60 Nm)
- 2nd pass-74 ft. lbs. (100 Nm)
- 3rd pass-118 ft. lbs. (160 Nm)

17. Apply Loctite® 262 to the fixed guide top bolt threads, install the bolt and tighten to 19 ft. lbs. (26 Nm).

18. Install the access plug into the cylinder head and tighten to 15 ft. lbs. (20 Nm).

19. Connect the vacuum hose to the actuator.

20. Raise and support vehicle, drain the engine oil.

21. Install the exhaust manifold.

22. If still supported, lower the vehicle for underhood access.

23. Install the alternator.
24. Install valve lifters.
25. Install the right injector housing assembly.
26. Install the plenum assembly.
27. Fill the engine crankcase with the proper type and amount of engine oil.
28. Tighten the fuel filler cap and properly refill the cooling system.
29. Connect the negative battery cable.
30. If equipped, reset the CHANGE OIL indicator:

a. Turn the ignition **ON** but do not start the engine.

b. Depress the ENG MET button on the trip monitor, then within 5 seconds, depress the button a 2nd time. Within another 5 seconds, depress and hold the GAUGES button.

c. While holding the GAUGES button and watch the CHANGE OIL light, it should begin to flash. Continue to hold the gauges button until the flashing stops and the light goes out indicating that the indicator is reset.

d. If the indicator does not reset, turn the ignition **OFF** and restart the procedure.

LEFT SIDE

1. Disconnect the negative battery cable and properly relieve fuel system pressure.
2. Drain engine coolant into a suitable container.
3. Remove the intake plenum assembly.
4. Remove the left injector housing.
5. Remove the vacuum hose from the secondary port throttle valve actuator.
6. Remove the power brake booster assembly.
7. Remove the left bank valve lifters.
8. Remove the AIR control valve hoses, then disengage the electrical connector.
9. Remove the camshaft position sensor.
10. Disconnect the left exhaust manifold from the cylinder head. It is not necessary to completely remove the exhaust manifold from the vehicle for cylinder head removal.
11. Remove the access plug from the left cylinder head.
12. Remove the bolt attaching the left secondary timing chain guide.
13. Remove the cylinder head bolts. Remove the cylinder head and gasket from the vehicle.

To install:

14. Thoroughly clean the cylinder head and cylinder case mating surfaces. Be sure both surfaces are free of any foreign matter, nicks or scratches. The threads in both the bolts holes and on the bolts must be clean and free of old sealer.

➡ **Cylinder head gaskets are not interchangeable between cylinder banks.**

15. Install the cylinder head locating dowels into block, if loosened or removed, then position the new gasket in place on the cylinder case.
16. Install the cylinder head over the dowels. Coat bolt threads and washers with clean engine oil and insert.
17. Tighten the cylinder head bolts in sequence as follows:

- 1st pass-45 ft. lbs. (60 Nm)
- 2nd pass-74 ft. lbs. (100 Nm)
- 3rd pass-118 ft. lbs. (160 Nm)

18. Apply Loctite® 262 to the fixed guide bolt threads, install the bolt and tighten to 19 ft. lbs. (26 Nm).
19. Install the access plug into the cylinder head and tighten to 15 ft. lbs. (20 Nm).
20. Connect the vacuum hose to the actuator.
21. Raise and support vehicle, drain the engine oil and lower the vehicle.
22. Install the exhaust manifold.
23. Install the camshaft position sensor.
24. Connect the AIR control valve hoses and electrical connector.
25. Install the valve lifters.
26. Install the left injector housing assembly.
27. Install the plenum assembly.
28. Fill the engine crankcase with the proper type and amount of engine oil.
29. Tighten the fuel filler cap and properly refill the cooling system.
30. Connect the negative battery cable.
31. If equipped, reset the CHANGE OIL indicator:

a. Turn the ignition **ON** but do not start the engine.

b. Depress the ENG MET button on the trip monitor, then within 5 seconds, depress the button a 2nd time. Within another 5 seconds, depress and hold the GAUGES button.

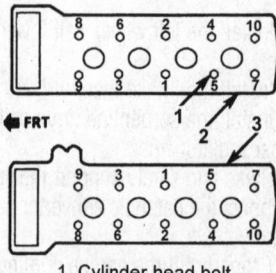

1 Cylinder head bolt
2 Cylinder head RH
3 Cylinder head LH

79227610

Cylinder head torque sequence—5.7L (VIN J) engine

c. While holding the GAUGES button and watch the CHANGE OIL light, it should begin to flash. Continue to hold the gauges button until the flashing stops and the light goes out indicating that the indicator is reset.

d. If the indicator does not reset, turn the ignition **OFF** and restart the procedure.

5.7L (VIN G) Engine

1. Disconnect the negative battery cable.

❊❊❊ CAUTION

Never open, service or drain the radiator or cooling system when hot; serious burns can occur from the steam and hot coolant. Coolant should be reused unless it is contaminated or is several years old.

2. Drain the engine coolant.
3. Relieve the fuel system pressure.
4. Remove the drive belts.
5. Remove the cylinder head covers.
6. Remove the rocker arms, pedestals and pushrods. Be sure to keep them in order so they can be installed in their original positions.
7. Remove the exhaust manifolds.

➡ **The intake manifold, throttle body, fuel rail and injectors can be removed as an assembly.**

8. Remove the intake manifold.
9. Remove the vapor vent pipe.
10. Remove the spark plugs.
11. For the left cylinder head remove the power steering pump by first removing the pulley. Then, remove the lower accessory mounting bracket and the ground wire from the back of the cylinder head.
12. Remove the cylinder head bolts and the head assembly. Be sure to place the cylinder head on wooden blocks to prevent damage.

To install:

➡ **New M11 cylinder head bolts must be used when installing the cylinder head assembly.**

13. Clean and degrease the deck of the engine block and the bottom of the cylinder head assembly. Clean the bolt holes with compressed air.

➡ **The tab on the edge of the head gasket will be closer to the front of the engine when properly installed.**

14. Position the new gaskets on the engine block.

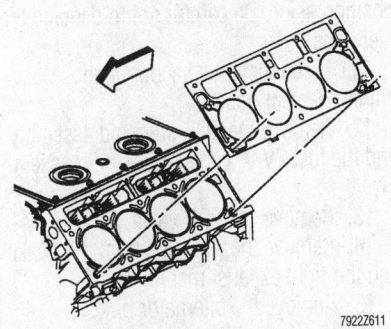

79222611

Be sure the tab on the edge of the gasket is closer to the front of the engine when installed—5.7L (VIN G) engine

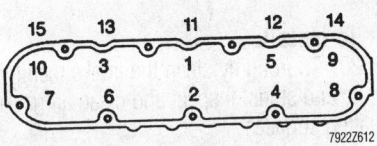

79222612

Always tighten the cylinder head bolts according to the sequence shown to prevent damaging the head and causing leaks—5.7L (VIN G) engine

15. Carefully place the cylinder head on the engine and install the bolts finger-tight.

16. Tighten the M11 bolts in sequence to 22 ft. lbs. (30 Nm).

17. Tighten the M11 bolts in sequence again 76° using a torque angle wrench.

18. Tighten the M11 bolts (1–8) in sequence 76°, and bolts 9–10 in sequence 34° using a torque angle wrench.

19. Tighten the M8 bolts (11–15) to 22 ft. lbs. (30 Nm) starting at the center bolt and alternately working outward until all of the bolts have been tightened.

20. For the left cylinder head, install the ground wire, lower accessory mounting bracket and the power steering pump.

21. Install the spark plugs. If using new plugs, tighten them to 15 ft. lbs. (20 Nm). Tighten used plugs to 11 ft. lbs. (15 Nm).

22. Install the vapor vent pipe.

23. Install the intake manifold assembly.

24. Install the exhaust manifolds.

25. Use new gaskets and install the cylinder head covers. Tighten the bolts to 106 inch lbs. (12 Nm).

26. Install the drive belts.

27. Connect the negative battery cable.

28. Refill the cooling system.

29. Start the engine and inspect for leaks.

Rocker Arms

REMOVAL & INSTALLATION

5.7L (VIN P and 5) Engines

1. Disconnect the negative battery cable.

2. Remove the right valve rocker cover as follows:

   a. Remove the fuel rail cover and the fuel rail bolts.

   b. Disconnect the fuel pressure regulator vacuum hose.

   c. Remove the fuel injector and rail assembly from the manifold and reposition.

   d. Remove the fuel rail cover studs and position the wiring harness aside.

   e. Remove the AIR pipe and check valve from the intake and exhaust manifolds.

   f. Disconnect the crankcase vent hose.

   g. Remove the valve rocker cover bolts, cover and gasket. Replace the gasket as necessary.

3. Remove the left rocker arm cover as follows:

   a. If not done already, remove the fuel rail cover.

   b. Remove the alternator brace bolts, then remove the brace.

   c. Remove the remaining alternator bolts and position the alternator aside.

   d. Disconnect the AIR diverter valve hose from the check valve.

   e. Position the wiring harness aside.

   f. Remove the valve rocker cover bolts, cover and gasket. Replace the gasket as necessary.

4. Remove the rocker arm nuts, rocker arm balls, rocker arms and pushrods. If the valvetrain components are to be reused, mark or arrange the assemblies in a rack to assure installation in their original locations.

To install:

5. Coat the bearing surfaces of the rocker arms and rocker arm balls with 1052365 or equivalent pre-lube, prior to installation.

6. Install the pushrods making certain they seat in the lifter sockets.

7. Install the rocker arms, rocker arm balls and rocker arm nuts in their original positions.

8. Tighten the rocker arm nuts until all lash is eliminated.

9. Adjust the valve lash.

10. Thoroughly clean the gasket mating surfaces and install the valve rocker arm covers in the reverse order of removal. Tighten the valve rocker cover bolts to 100 inch lbs. (11 Nm). For the right valve cover, be sure to tighten the AIR pipe-to-exhaust manifold fitting to 25 ft. lbs. (34 Nm).

11. Connect the battery negative cable, start the engine and inspect for leaks.

5.7L (VIN J) Engine

This engine utilizes an overhead cam design that eliminates the use of any rocker arm assembly. In many applications, this design improves and smoothes engine operation.

5.7L (VIN G) Engine

LEFT SIDE

➡**Always keep the rocker arms and pushrods in order so they can be installed in their original positions.**

1. Disconnect the negative battery cable.

2. Remove the fuel rail cover.

3. Relieve the fuel system pressure.

4. Disconnect the fuel lines from the fuel rail assembly.

5. Remove the generator rear bracket.

6. Disconnect the generator and coolant temperature sensor wiring.

7. Disconnect the vacuum hose from the brake booster.

8. Remove the PCV hose from the rocker arm cover.

9. Remove the Secondary Air Injection (SAI) hose from the check valve.

10. Remove the spark plug wires from the coils.

11. Detach the main ignition coil connector.

12. Disconnect the lines from the EVAP purge solenoid and remove the solenoid.

13. Remove the mounting bolts and the rocker arm cover.

14. Remove the rocker arm bolt and the rocker arm. If removing more than one rocker arm, be sure to keep them in order. They must be installed in their original positions.

To install:

15. If removed, install the rocker arm pivot support.

16. Lubricate the rocker arms and pushrods with clean engine oil.

17. Lubricate the flange and washer surface of the bolt rocker arm mounting bolts.

18. Install the pushrod. Be sure it is seated in the lifter socket.

19. Install the rocker arms but do not tighten the bolts at this time.

20. Rotate the crankshaft so the No. 1 piston is at Top Dead Center (TDC) on compression.

➡**The engine firing order is 1, 8, 7, 2, 6, 5, 4, 3. The cylinders are numbered 1, 3, 5 and 7 on the left bank and 2, 4, 6 and 8 on the right bank.**

21. With the engine in this position, tighten the exhaust valve rocker arm bolts on cylinders 1, 2, 7 and 8. Then, tighten the intake valve rocker arm bolts on cylinders 1, 3, 4 and 5. Tighten the bolts to 22 ft. lbs. 30 Nm).

22. Rotate the crankshaft one revolution (360°).

23. With the engine in this position, tighten the exhaust valve rocker arm bolts on cylinders 3, 4, 5 and 6. Then, tighten the intake valve rocker arm bolts on cylinders 2, 6, 7 and 8. Tighten the bolts to 22 ft. lbs. 30 Nm).

24. Install the rocker arm using a new gasket. Tighten the bolts to 106 inch lbs. (12 Nm).

25. Install the EVAP purge solenoid.

26. Attach the ignition coil connector.

27. Connect the spark plug wires.

28. Connect the SAI hose to the check valve.

29. Install the PCV hose.

30. Connect the vacuum hose to the brake booster.

31. Connect the generator and coolant temperature sensor.

32. Install the generator rear bracket.

33. Connect the fuel lines.

34. Install the fuel rail cover.

35. Connect the negative battery cable.

RIGHT SIDE

➡**If the oil fill tube is removed from the rocker arm cover, a new tube must be installed during assembly.**

1. Disconnect the negative battery cable.

2. Remove the rocker arm cover bolts.

3. Remove the rocker arm cover.

➡**Always keep the rocker arms and pushrods in order so they can be installed in their original positions.**

4. Remove the rocker arm bolt and the rocker arm. If removing more than one rocker arm, be sure to keep them in order. They must be installed in their original positions.

To install:

5. If removed, install the rocker arm pivot support.

6. Lubricate the rocker arms and pushrods with clean engine oil.

7. Lubricate the flange and washer surface of the bolt rocker arm mounting bolts.

8. Install the pushrod. Be sure it is seated in the lifter socket.

9. Install the rocker arms but do not tighten the bolts at this time.

10. Rotate the crankshaft so the No. 1 piston is at Top Dead Center (TDC) on compression.

➡**The engine firing order is 1, 8, 7, 2, 6, 5, 4, 3. The cylinders are numbered 2, 4, 6 and 8 on the right bank.**

11. With the engine in this position, tighten the exhaust valve rocker arm bolts on cylinders 2 and 8. Then, tighten the intake valve rocker arm bolts on cylinder No. 4. Tighten the bolts to 22 ft. lbs. 30 Nm).

12. Rotate the crankshaft one revolution (360°).

13. With the engine in this position, tighten the exhaust valve rocker arm bolts on cylinders 4 and 6. Then, tighten the intake valve rocker arm bolts on cylinders 2, 6 and 8. Tighten the bolts to 22 ft. lbs. 30 Nm).

14. Install the rocker arm using a new gasket. Tighten the bolts to 106 inch lbs. (12 Nm).

Intake Manifold

REMOVAL & INSTALLATION

5.7L (VIN P and 5) Engines

1. Disconnect the negative battery cable.

2. Drain engine coolant into a suitable container.

3. Remove the throttle body air duct.

4. Remove the fuel rail covers.

5. Disengage the wiring harness connectors from the fuel injectors. Disengage and reposition the left and right wiring harnesses.

6. Remove the accelerator cable bracket, then disconnect the cables from the throttle body.

7. Disconnect the AIR diverter valve hoses.

8. Remove the electrical ground strap from the intake manifold.

9. Remove the fuel rail bolts and disconnect the vacuum hose from the fuel pressure regulator.

10. Carefully remove the fuel rail and injector assembly from the manifold and

position aside. Be careful not to damage the fuel lines.

11. Disconnect the vacuum and crankcase vent hoses.

12. Remove the EGR solenoid assembly and the fuel EVAP canister solenoid assembly.

13. Remove the EGR valve.

14. Remove the AIR pipe from the intake and the right exhaust manifold.

15. Remove the alternator brace.

16. Disconnect the coolant hoses from the throttle body.

17. Remove the throttle body bolts, the throttle body and gasket.

18. Remove the intake manifold bolts and studs.

19. Remove the intake manifold and gaskets.

To install:

20. Thoroughly clean the intake manifold bolts and studs. Inspect and clean all gasket mating surfaces.

21. Apply a ³⁄₁₆ in. (5mm) bead of RTV sealer to the front and rear of the cylinder block. Extend the bead ½ in. (13mm) up each cylinder head to seal and retain the gaskets.

22. Position the new gaskets and install the intake manifold.

23. Install the manifold bolts and studs, then tighten using 2 passes of the proper sequence. First, tighten the bolts/studs to 71 inch lbs. (8 Nm), then tighten them to 35 ft. lbs. (48 Nm).

24. Install the throttle body, gasket and retaining bolts. Tighten the throttle body bolts to 19 ft. lbs. (26 Nm).

25. Connect the coolant hoses to the throttle body.

26. Install the alternator brace.

27. Install the accelerator cables and bracket. Tighten the bracket bolts to 90 inch lbs. (10 Nm).

28. Install the AIR pipe. Tighten the exhaust manifold fitting and the bracket-to-

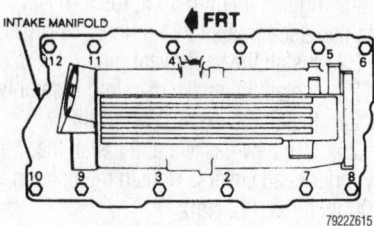

Intake manifold bolt torque sequence—
5.7L (VIN P and 5) engines

cylinder head bolt to 25 ft. lbs. (34 Nm) and tighten the flange-to-intake manifold bolts to 19 ft. lbs. (26 Nm).

29. Install the EGR valve, then EGR solenoid and bracket. Tighten valve bolts to 16 ft. lbs. (22 Nm) and the solenoid bracket nut to 25 ft. lbs. (34 Nm).

30. Install the fuel EVAP canister purge solenoid and bracket, then tighten the nut to 15 ft. lbs. (20 Nm).

31. Connect the vacuum and crankcase vent hoses.

32. Install the fuel injector and fuel rail assembly to the intake manifold, connect the fuel pressure regulator vacuum hose and install the fuel rail bolts. Tighten the bolts to 15 ft. lbs. (20 Nm).

33. Connect the electrical ground strap to the intake manifold.

34. Connect the AIR diverter valve hoses.

35. Position the left and right wiring harnesses, then engage the fuel injector electrical connectors.

36. Install the throttle body air duct.

37. Install the fuel rail covers.

38. Properly fill the engine cooling system.

39. Connect the negative battery cable, then adjust the ASR accelerator and cruise control cables, as necessary.

5.7L (VIN J) Engine

UPPER INTAKE MANIFOLD (PLENUM)

The 5.7L (VIN J) engine does not utilize a single intake manifold assembly like the other 5.7L Corvette engine. Instead it uses an intake plenum mated to a right and left fuel injector housing.

1. Disconnect the negative battery cable and properly relieve fuel system pressure.

2. Drain the cooling system into a suitable container.

3. Remove the air intake duct.

4. Remove the throttle cable cover and attaching hardware.

5. Remove the throttle and cruise control cables or the ASR control cable from the throttle body. Remove the cable hold-down clamp(s) and set the cables aside.

6. Remove the fresh air hose from the left and right side of the of the throttle body extension.

7. Disengage the electrical connectors from the IAC, TPS and the IAT or MAT sensors.

8. Disconnect the coolant air bleed hose from the plenum.

9. Remove the power brake booster hose, then remove the vacuum hose located between the fuel pressure regulator and the plenum.

10. Tag and remove the left and right vacuum hoses at the mid-plenum.

11. Tag and remove the MAP sensor vacuum hose.

12. Disconnect the fuel lines from the fuel rail assembly and discard the O-rings.

13. Remove the plenum assembly attaching bolts.

14. Remove the EVAP purge solenoid/PCV dual hose fitting from the plenum.

15. Remove the EVAP purge canister hose from the plenum.

16. Remove the upper EGR pipe bolts, then remove the pipe.

17. Lift the plenum and disengage the ignition module electrical connections, then remove the plenum assembly and discard the gaskets.

18. Cover the intake ports to prevent dirt or other contaminants from entering.

To install:

19. If the plenum is being replaced, transfer the MAP sensor and bracket, the throttle body, the throttle body extension and the ignition module to the new plenum.

20. Remove the tape or other cover from the intake ports and position the plenum assembly on the injector housings with MAP sensor over the fuel pressure regulator. Engage the electrical connectors to the ignition module and MAP sensor, then install the MAP sensor vacuum hose.

21. Be sure the remaining vacuum hoses and electrical connectors are accessible, then position the new plenum gaskets between the plenum and injector housing, Install the plenum attaching bolts and tighten the bolts in the proper sequence to 20 ft. lbs. (26 Nm).

22. Install the vacuum hoses to mid-plenum.

23. Install new O-rings, then reconnect the fuel lines to the fuel rail assembly. Tighten the fuel line fittings to 20 ft. lbs. (26 Nm).

24. Install the vacuum hose between the pressure regulator and the plenum.

25. Connect the power brake booster vacuum hose to the plenum.

26. Install the fresh air hose onto the left and right side of the throttle body extension.

27. Install the EVAP purge solenoid/PCV hose fitting to the plenum, then install the EVAP canister connection to the rear right side.

28. Engage the wiring harness connectors to the TPS, IAC and IAT or MAT sensors.

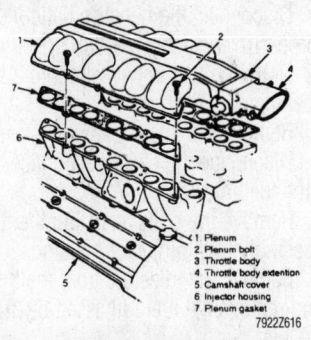

Exploded view of the upper intake manifold (plenum) mounting—5.7L (VIN J) engine

1 Plenum
2 Plenum bolt
3 Throttle body
4 Throttle body extention
5 Camshaft cover
6 Injector housing
7 Plenum gasket

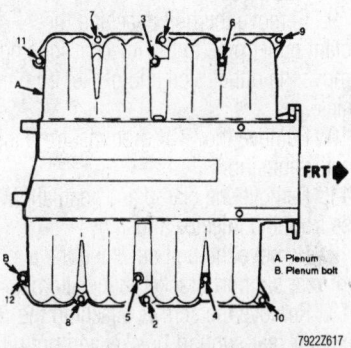

Upper intake manifold (plenum) bolt torque sequence—5.7L (VIN J) engine

A. Plenum
B. Plenum bolt

29. Install the screws retaining the cable hold-down clamps to the plenum and tighten to 18 inch lbs. (2 Nm).

30. Install the coolant air bleed hose to the plenum.

31. Connect the throttle and cruise control cables or the ASR control cable to the throttle. Be sure the cables do not hold the throttle open and adjust, as necessary.

32. Install the cable shield, screw and nuts to the throttle body.

33. Connect the upper EGR pipe. Tighten the screw and nuts to 27 inch lbs. (3 Nm), then tighten the EGR pipe bolts to 89 inch lbs. (10 Nm).

34. Install the air intake duct, tighten the fuel filler cap and connect the negative battery terminal.

35. Properly refill the engine cooling system, then start the engine and check for leaks.

RIGHT LOWER INTAKE MANIFOLD (INJECTOR HOUSING)

The 5.7L (VIN J) engine does not utilize a single intake manifold assembly like the other 5.7L Corvette engines. Instead it uses an intake plenum mated to a right and left fuel injector housing.

1. Disconnect the negative battery cable and properly relieve fuel system pressure.

2. Drain the cooling system into a suitable container.

3. Remove the intake plenum assembly.

4. Disengage the electrical connectors from the fuel injectors.

5. Remove the bolts attaching the fuel rail assembly to the injector housing.

6. Remove the injectors from the housing and remove the fuel rail assembly from the vehicle.

7. Disconnect the hose from the right coolant outlet pipe.

8. Remove the oil pressure sensor from the oil filter housing.

9. Remove the bolt attaching the coolant outlet pipe to the injector housing. Remove the outlet pipe and gasket from the vehicle.

10. Remove the PCV grommet from the injector housing.

11. Remove the clamp and ventilation hose from the injector housing.

12. Remove the bolt attaching the alternator rear support bracket to the alternator.

13. Remove the screws attaching the alternator rear support bracket and right side ventilation pipe to the injector housing.

14. Remove the ventilation pipe and bracket from the vehicle.

15. Remove the injector housing attaching bolts, then remove the injector housing and gasket from the vehicle.

To install:

16. Thoroughly clean all gasket mating surfaces and position the a new housing gasket.

17. Install the injector housing, rear alternator bracket, right ventilation pipe and the housing bolts. Be sure the spark plug wiring harness retainer is secured by the injector housing rear bolt and tighten the fasteners to 19 ft. lbs. (26 Nm).

18. Install the ventilation hose.

19. Install PCV grommet into the injector housing.

20. Install a new gasket, the coolant outlet pipe and the retaining screws. Tighten screws to 89 inch lbs. (10 Nm).

21. Install the oil pressure sensor. Apply Loctite® to sensor threads.

22. Install the hose and clamp onto the right coolant outlet pipe.

23. Install new injector lower O-rings and install the fuel rail assembly to the injector housing. Tighten the retaining bolts to 19 ft. lbs. (26 Nm).

24. Engage the injector electrical connectors.

25. Install intake plenum assembly.

26. Connect the negative battery cable and refill the engine cooling system. Start the engine and check for leaks.

LEFT LOWER INTAKE MANIFOLD (INJECTOR HOUSING)

The 5.7L (VIN J) engine does not utilize a single intake manifold assembly like the other 5.7L Corvette engines. Instead it uses an intake plenum mated to a right and left fuel injector housing.

1. Disconnect the negative battery cable and properly relieve fuel system pressure.

2. Drain the cooling system into a suitable container.

3. Remove the intake plenum assembly.

4. Disengage the electrical connectors from the fuel injectors.

5. Remove the screws attaching the fuel rail assembly to the injector housing.

6. Remove the injectors from the housing and remove the fuel rail assembly from the vehicle.

7. Disconnect the hose from the left coolant outlet pipe.

8. Remove the bolts attaching the coolant outlet pipe to the injector housing. Remove the outlet pipe and gasket from the vehicle.

9. Remove the PCV grommet from the injector housing.

10. Remove the clamp and ventilation hose from the injector housing.

11. Disengage the electrical connectors from the coolant temperature sensor and the cooling fan switch.

12. Remove the injector housing attaching bolts, then remove the injector housing and gasket from the vehicle.

To install:

13. Thoroughly clean all gasket mating surfaces and position the a new housing gasket.

14. Install the injector housing and secure using the housing retaining bolts. Be sure the spark plug wiring harness retainer is secured by the injector housing rear bolt and tighten the fasteners to 19 ft. lbs. (26 Nm).

15. Install the ventilation hose and clamp.

16. Install PCV grommet into the injector housing.

17. Engage the electrical connectors to the coolant temperature sensor and the cooling fan switch.

18. Install a new gasket, the coolant outlet pipe and the retaining screws. Tighten screws to 89 inch lbs. (10 Nm).

19. Install the hose and clamp onto the left coolant outlet pipe.

20. Install new injector lower O-rings and install the fuel rail assembly to the injector housing. Tighten the retaining bolts to 19 ft. lbs. (26 Nm).

21. Engage the injector electrical connectors.

22. Install intake plenum assembly.

23. Connect the negative battery cable and refill the engine cooling system. Start the engine and check for leaks.

5.7L (VIN G) Engine

➡**The intake manifold, throttle body, fuel injectors and rail may be removed from the engine as an assembly.**

1. Disconnect the negative battery cable.

2. Drain the cooling system.

3. Detach the Intake Air Temperature (IAT) sensor connector.

4. Detach the Mass Air Flow (MAF) sensor connector.

5. Disconnect the fuel regulator purge line from the intake air duct and remove the air cleaner assembly.

✳✳ CAUTION

Observe all applicable safety precautions when working around fuel. Whenever servicing the fuel system, always work in a well ventilated area. Do not allow fuel spray or vapors to come in contact with a spark or open flame. Keep a dry chemical fire extinguisher near the work area. Always keep fuel in a container specifically designed for fuel storage; also, always properly seal fuel containers to avoid the possibility of fire or explosion.

6. Remove the fuel rail covers and relieve the fuel system pressure.

7. Disconnect the fuel lines from the fuel rail.

8. Remove the vacuum and crankcase vent hoses.

9. Remove the coolant hose from the throttle body.

10. Disconnect the wiring from the fuel injectors.

11. Detach any remaining electrical connections on the intake manifold.

12. Remove the vapor vent pipe clamp and hose from the throttle body.

13. Remove the Manifold Absolute Pressure (MAP) sensor from the grommet on the intake manifold.

14. Remove the intake manifold mounting bolts and fuel rail stop bracket.

15. Remove the intake manifold from the engine.

16. Remove and discard the gaskets.

To install:

17. Install new gaskets on the cylinder heads and engine.

18. Position the intake manifold on the engine.

19. Apply a 0.20 in. (5mm) band of threadlocker PN 12345383 or equivalent to the threads of the intake manifold bolts.

❋❋ CAUTION

The fuel rail stop bracket must be installed on the intake manifold in its original position. Failure to reinstall this part may result in fuel spray causing personal injury during a crash.

20. Install the fuel stop bracket and the intake manifold bolts. Tighten the bolts first to 44 inch lbs. (5 Nm) in the sequence shown. Then, make a second pass and tighten the bolts to 89 inch lbs. (8 Nm) using the same sequence.

21. Lubricate the grommet with engine oil and install the MAP sensor.

22. Install the vapor vent pipe and clamp.

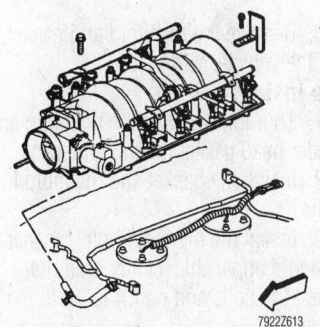

79222613

Be sure to reinstall the fuel stop bracket when installing the intake manifold—5.7L (VIN G) engine

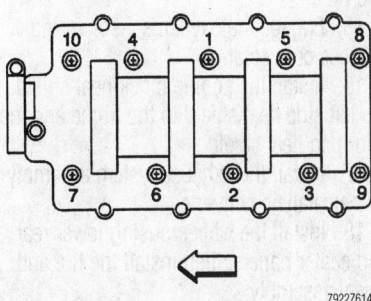

79222614

To avoid damage to the intake manifold, tighten the bolts in the sequence shown—5.7L (VIN G) engine

23. Attach the fuel injector and remaining electrical connectors.

24. Connect the coolant hose to the throttle body.

25. Install the crankcase vent and vacuum hoses.

26. Connect the fuel lines.

27. Install the fuel rail covers.

28. Install the air cleaner assembly and fuel regulator purge line.

29. Connect the wiring harness to the MAF and IAT sensors.

30. Connect the negative battery cable.

❋❋ WARNING

Use only DEX-COOL® or equivalent coolant when refilling the cooling system. If silicate coolant is used, premature corrosion may develop in the engine, heater core and radiator.

31. Refill and bleed the cooling system.

32. Start the engine and check for leaks.

Exhaust Manifold

REMOVAL & INSTALLATION

5.7L (VINS P and 5) Engine

RIGHT SIDE

1. Disconnect the negative battery cable.

2. Raise and support the vehicle safely.

3. For convertibles, if necessary for access, remove the underbody crossbrace.

4. Remove the nuts and disconnect the catalytic converter from the exhaust manifold. It may be necessary to loosen and remove the entire exhaust assembly.

5. Lower the vehicle.

6. Remove the fuel rail covers and disengage the fuel injector electrical connectors.

7. Remove the vacuum hose from the fuel pressure regulator.

8. Remove the fuel rail bolts, then remove the fuel injector/rail assembly from the intake manifold and position aside. Be careful not to damage the fuel lines.

9. Disconnect the spark plug wires from the plugs, then disconnect the wire clips from the supports. If necessary for clearance, or to prevent the possibility of breakage during manifold removal, remove the spark plugs.

10. Remove the front spark plug bracket and bolt.

11. Remove the oil level indicator and guide tube.

12. Remove the AIR pipe, gasket and

check valve as an assembly from the intake manifold, exhaust manifolds and the cylinder head.

13. Remove the exhaust manifold studs and bolts.

14. Remove the heat shields, exhaust manifold and gasket.

To install:

15. Thoroughly clean the manifold and cylinder head gasket mating surfaces.

16. Install the exhaust manifold gasket, manifold and heat shields.

17. Install the exhaust manifold studs and bolts. Tighten the fasteners to 26 ft. lbs. (35 Nm).

18. Install the AIR pipe, gasket and check valve assembly with the retaining bolts. Tighten the AIR pipe-to-exhaust manifold fitting and the bracket bolt to 25 ft. lbs. (34 Nm) and tighten the pipe flange bolts to 19 ft. lbs. (26 Nm).

19. Apply 1052080 or equivalent, sealer to the oil level indicator guide tube ½ in. (13mm) below the bead. Install the level indicator and guide tube into the block.

20. Install the front spark plug bracket and bolt. Tighten to 108 inch lbs. (12 Nm).

21. If removed, install the spark plugs and tighten to 11 ft. lbs. (15 Nm).

22. Install the spark plug wires and clips.

23. Install the fuel injectors and fuel rail assembly to the intake manifold. Tighten the fuel rail bolts to 15 ft. lbs. (20 Nm).

24. Connect the fuel pressure regulator vacuum hose.

25. Engage the wiring harness connectors to the fuel injectors.

26. Install the fuel rail covers.

27. Raise and support the vehicle safely.

28. If removed, install the exhaust assembly.

29. Connect the catalytic converter and nuts to the exhaust manifold.

30. Tighten catalytic converter nuts to 15 ft. lbs. (21 Nm).

31. If removed on a convertible, install the underbody crossbrace.

32. Lower the vehicle.

33. Connect the negative battery cable.

LEFT SIDE

1. Disconnect the negative battery cable.

2. Raise and support the vehicle safely.

3. For convertibles, if necessary for access, remove the underbody crossbrace.

4. Remove the nuts and disconnect the catalytic converter from the exhaust manifold. It may be necessary to loosen and remove the entire exhaust assembly.

5. Lower the vehicle.

6. Remove the air intake duct and the serpentine drive belt.

7. Remove the ASR adjuster assembly from the wheel well center panel and reposition out of the way.

8. Remove the mounting bolts and reposition the alternator and the A/C compressor.

9. Remove the AIR pipe, check valve and hose as an assembly from the exhaust manifold.

10. Remove the spark plug wires from the plugs and the clips from the supports, then position the wires aside.

11. Remove the spark plug wire supports. If necessary for clearance, or to prevent the possibility of breakage during manifold removal, remove the spark plugs.

12. Remove the exhaust manifold studs and bolts.

13. Remove the heat shields, exhaust manifold and gasket.

To install:

14. Thoroughly clean the manifold and cylinder head gasket mating surfaces.

15. Install the exhaust manifold gasket, manifold and heat shields.

16. Install the exhaust manifold studs and bolts. Tighten the fasteners to 26 ft. lbs. (35 Nm).

17. Install the spark plug wire supports and tighten to 108 inch lbs. (12 Nm).

18. If removed, install the spark plugs and tighten to 11 ft. lbs. (15 Nm).

19. Connect the spark plug wires and clips.

20. Install the AIR pipe, check valve and hose assembly. Tighten the AIR pipe-to-exhaust manifold fitting and the bracket bolt to 25 ft. lbs. (34 Nm).

21. Reposition and the install the air conditioning compressor and alternator.

22. Install the serpentine drive belt and the air intake duct.

23. Install the ASR adjuster assembly, then check and adjust the control cable, as necessary.

24. Raise and support the vehicle safely.

25. If removed, install the exhaust assembly.

26. Connect the catalytic converter and nuts to the exhaust manifold.

27. Tighten catalytic converter nuts to 15 ft. lbs. (21 Nm).

28. If removed on a convertible, install the underbody crossbrace.

29. Lower the vehicle.

30. Connect the negative battery cable.

5.7L (VIN J) Engine

RIGHT SIDE

1. Disconnect the negative battery cable, then raise and support the vehicle safely.

2. Remove the right tire and wheel assembly, then remove the wheel house lower rear and center panels.

3. Disconnect the exhaust system assembly from the catalytic converter.

4. If equipped, remove the engine block heat shield.

5. Disconnect the catalytic converter from the manifold.

6. Disengage the oxygen sensor wiring harness connector.

7. Remove the rear exhaust manifold bolts, spacers and nut.

8. Disconnect the lower EGR pipe from the manifold.

9. Lower the vehicle.

10. Disconnect the AIR check valve and hose from the manifold.

11. Remove the retaining bolt, then remove the oil level indicator and guide tube from the vehicle.

12. Remove the remaining exhaust manifold attaching bolts and spacers.

13. Remove the exhaust manifold and gasket from the vehicle. If the manifold is being replaced, transfer the oxygen sensor and heat shields to the new manifold, as necessary.

To install:

14. Thoroughly clean the manifold and cylinder head gasket mating surfaces.

15. Install the gasket and manifold to the engine using the front and center manifold bolts and spacers.

16. Install the oil level indicator and guide tube, then tighten the manifold bolts to 18 ft. lbs. (24 Nm).

17. Install the AIR check valve and hose.

18. Raise and support the vehicle safely.

19. Install the rear manifold bolts, spacers and nut. Tighten the bolts and nut to 18 ft. lbs. (24 Nm).

20. Install the wiring harness connector to the oxygen sensor.

21. Connect the catalytic converter and bolts to the manifold. Tighten the bolts to 17 ft. lbs. (23 Nm).

22. If equipped, install the engine block heat shield.

23. Connect the exhaust system assembly.

24. Install the manifold outer heat shields.

25. Install the lower EGR pipe to the manifold.

26. Lower the vehicle sufficiently for access.

27. Install the wheelhousing lower rear and center panels.

28. Install the tire and wheel assembly, then lower the vehicle completely.

29. Connect the negative battery cable.

LEFT SIDE

1. Disconnect the negative battery cable, then raise and support the vehicle safely.

2. Remove the right tire and wheel assembly, then remove the wheel house lower rear and center panels.

3. Disconnect the exhaust assembly from the catalytic converter.

4. Remove the left floor pan heat shield, the left heat shield from the frame and the engine block heat shield.

5. Disengage the converter oxygen sensor electrical connector.

6. Disconnect the AIR check valves, hoses and pipes from the manifold.

7. Remove the manifold outer heat shield and remove the catalytic converter from the exhaust manifold.

8. Remove the exhaust manifold bolts, spacers and nut.

9. If applicable, remove the center stud nut.

10. Remove the manifold and gasket from the vehicle.

To install:

11. Thoroughly clean the manifold and cylinder head gasket mating surfaces.

12. Install the gasket and manifold to the engine.

13. Install the manifold bolts, spacer, nut, and if applicable, center stud nut. Tighten the bolts and nut(s) to 18 ft. lbs. (24 Nm).

14. Install the catalytic converter and bolts to the manifold. Tighten to 17 ft. lbs. (23 Nm) and install the manifold outer heat shield. Install the AIR check valve, hoses and pipe.

15. Engage the oxygen sensor wiring harness connector.

16. Install the engine block heat shield, the left side heat shield to the frame and the floor pan heat shield.

17. Install the exhaust system assembly to the catalytic converter.

18. Install the wheelhousing lower rear and center panels, then install the tire and wheel assembly.

19. Lower the vehicle and connect the negative battery cable.

5.7L (VIN G) Engine

LEFT SIDE

1. Disconnect the negative battery cable.
2. Raise and safely support the vehicle.
3. Disconnect the intermediate pipe from the exhaust manifold flange.
4. Disconnect the wiring from the oxygen sensor.
5. Lower the vehicle to the floor.
6. Remove the left fuel rail cover.
7. Relieve the fuel system pressure.
8. Disconnect the fuel line from the fuel rail.
9. Remove the rear generator bracket.
10. Remove the drive belt from the generator, then remove the generator.
11. Remove the Secondary Air Injection (SAI) hose from the pipe on the exhaust manifold.
12. Remove the SAI pipe from the exhaust manifold.
13. Remove the spark plug wires from the plugs and remove the spark plugs.
14. Remove the No. 5 ignition coil bolts and position the coil aside.
15. Remove the exhaust manifold mounting bolts and the manifold. Discard the gasket.

To install:

16. Clean the bolt holes and the mounting surface of the cylinder head.
17. Apply threadlocker PN 12345493 or equivalent to the threads of the bolts. Do not apply it to the first three threads.
18. Install the exhaust manifold using a new gasket.
19. Beginning with the center two bolts, tighten the bolts to 11 ft. lbs. (15 Nm) alternating from side-to-side until all the bolts are tight. Then, tighten the bolts again in the same sequence to 18 ft. lbs. (25 Nm).
20. Bend over the exposed portion of the gasket at the rear of the cylinder head.
21. Install the AIR pipe using a new gasket. Tighten the bolts to 15 ft. lbs. (20 Nm).
22. If removed, install the oxygen sensor in the manifold, tighten it to 30 ft. lbs. (42 Nm).
23. Connect the air hose to the SAI pipe.
24. Install the No. 5 ignition coil.
25. Install the spark plugs and wires.
26. Install the generator, rear bracket and drive belt.
27. Connect the fuel lines.
28. Install the fuel rail cover.
29. Install the intermediate pipe to the manifold using a new gasket. Tighten the nuts to 15 ft. lbs. (20 Nm)
30. Connect the negative battery cable.

Right Side

1. Disconnect the negative battery cable.
2. Raise the vehicle and disconnect the intermediate pipe from the exhaust manifold.
3. Disconnect the wiring from the oxygen sensor.
4. Lower the vehicle and remove the AIR pipe with check valve from the manifold.
5. Disconnect the spark plug wires from the spark plugs.
6. Remove the exhaust manifold mounting bolts and the manifold.

To install:

7. Clean the bolt holes and the mounting surface of the cylinder head.
8. Apply threadlocker PN 12345493 or equivalent to the threads of the bolts. Do not apply it to the first three threads.
9. Install the exhaust manifold using a new gasket.
10. Beginning with the center two bolts, tighten the bolts to 11 ft. lbs. (15 Nm) alternating from side-to-side until all the bolts are tight. Then, tighten the bolts again in the same sequence to 18 ft. lbs. (25 Nm).
11. Bend over the exposed portion of the gasket at the rear of the cylinder head.
12. Install the AIR pipe using a new gasket. Tighten the bolts to 15 ft. lbs. (20 Nm).
13. If removed, install the oxygen sensor in the manifold, tighten it to 30 ft. lbs. (42 Nm).
14. Connect the spark plug wires to the plugs.
15. Install the intermediate pipe to the manifold using a new gasket. Tighten the nuts to 15 ft. lbs. (20 Nm)
16. Connect the negative battery cable.

Camshaft

REMOVAL & INSTALLATION

5.7L (VINS P and 5) Engine

1. Disconnect battery negative cable and remove the air cleaner assembly.
2. Remove the timing chain front cover.
3. Remove the intake manifold.
4. Remove the retaining bolt and lift the oil pump driveshaft assembly from the rear of the lifter valley.
5. Remove the rocker arm and pushrod assemblies.
6. Remove the camshaft sprocket from the engine.
7. Remove the valve lifters.
8. Remove the high fill reservoir hose from the radiator.
9. Remove the relay bracket from the left side of the radiator support.
10. Remove the AIR pump intake duct and bolts, then reposition the AIR pump.
11. Remove the retaining nuts and screws, then remove the upper radiator support.
12. Remove the radiator.
13. Raise and support the vehicle safely.
14. Unplug the cooling fan electrical connector.
15. Remove the lower fan shroud bolts and lower the vehicle.
16. Remove the fan shroud and fan assembly.
17. Disconnect the A/C condenser line bracket at the front crossmember.
18. Raise the front of the engine with a suitable lifting device.

➡When raising and supporting the engine, NEVER place a jack under the oil pan, crankshaft pulley or any sheetmetal. There is a minimal clearance between the oil pan and the pump screen. Jacking against the pan could cause sufficient deformation to damage the oil pick-up unit.

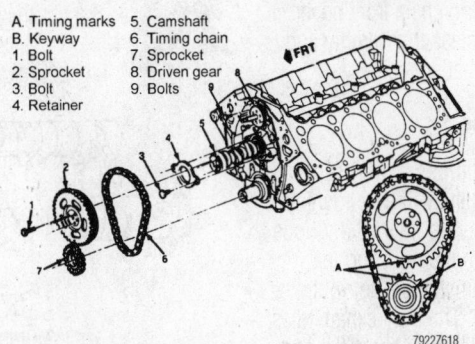

A. Timing marks
B. Keyway
1. Bolt
2. Sprocket
3. Bolt
4. Retainer
5. Camshaft
6. Timing chain
7. Sprocket
8. Driven gear
9. Bolts

79222618

Exploded view of the camshaft mounting and the valve timing marks—5.7L (VINS P and 5) engines

19. Remove the camshaft retainer bolts and retainer.

20. Install three ⁵⁄₁₆ –18 x 4 in. bolts into the camshaft bolt holes.

21. Using the bolts, carefully rotate the camshaft and pull from the bearings. All camshaft journals are the same diameter so care must be used to avoid damaging the bearings. Remove the camshaft from the vehicle.

To install:

22. Inspect the camshaft and bearings, replace as necessary.

23. If installing a new camshaft, coat the lobes with Molykote® or equivalent pre-lube and be sure to replace all lifters to assure camshaft durability.

24. Lubricate all camshaft journals with clean engine oil and carefully insert the camshaft into the engine block.

25. Install the camshaft retainer and tighten the bolts to 108 inch lbs. (12 Nm).

26. Lower the front of the engine and connect the A/C condenser line bracket to the front crossmember.

27. Install the fan and shroud assembly.

28. Raise and support the vehicle safely, then install the lower fan shroud bolts.

29. Engage the cooling fan electrical connections and lower the vehicle.

30. Install the radiator, followed by the upper radiator support, nuts and screws.

31. Install the AIR pump, bolts and intake duct.

32. Install the relay bracket to the left side of the radiator support.

33. Connect the high fill reservoir hose to the radiator.

34. Install the valve lifters.

35. Install the camshaft sprocket.

36. Install the valve rocker arm and pushrod assemblies.

37. Install the oil pump driveshaft assembly and bolt. Tighten the bolt to 13 ft. lbs. (18 Nm).

38. Install the intake manifold.

39. Install the timing chain front cover.

40. Install the air cleaner assembly and connect the negative battery cable.

5.7L (VIN J) Engine

The VIN J engine utilizes 4 overhead camshafts. Certain shafts will have identifying bands between the first journal and lobe to distinguish between the right and left, intake and exhaust camshafts. The right intake has 1 flat band. The right exhaust has 1 raised band. The left intake has 1 flat and 1 raised band. The left exhaust has 2 raised bands.

1. Disconnect battery negative cable and drain the engine coolant into a suitable container.

2. To gain access to the right camshafts, remove the oil filter housing and right camshaft cover as follows:

a. Remove the air intake duct.

b. Remove the hoses and clamps from the coolant outlets, radiator inlet and inlet pipe.

c. Remove the hoses and inlet pipe assembly from the vehicle.

d. Remove the water pump pulley.

e. Release the belt tensioner and remove the serpentine belt.

f. Remove the retaining bolt and the belt tensioner from the engine.

g. Remove the oil filter.

h. Disengage the electrical connectors from the oil pressure sensor, oil temperature sensor and the low oil pressure switch.

i. Remove the oil pressure sensor from the oil filter housing.

j. Remove the alternator bracket from the oil filter housing.

k. Disconnect and plug the oil cooler lines from the filter housing.

l. Remove the oil filter housing mounting bolts and remove the assembly.

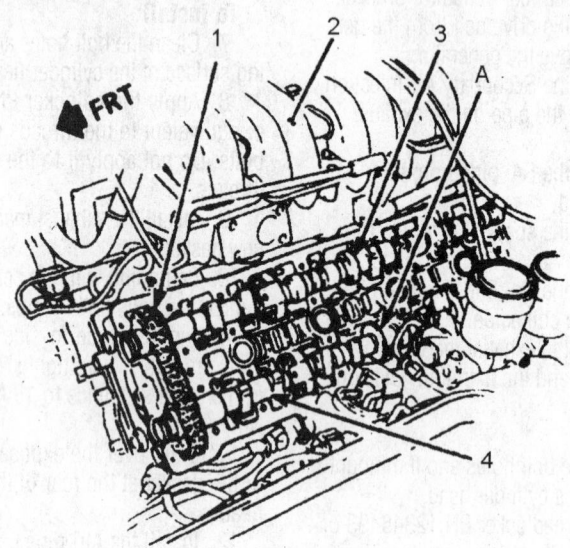

A. Camshaft sensor reluctor disc
1. Camshaft secondary timing chain
2. Plenum
3. Intake camshaft LH
4. Exhaust camshaft LH

Left cylinder head camshaft assembly—5.7L (VIN J) engine

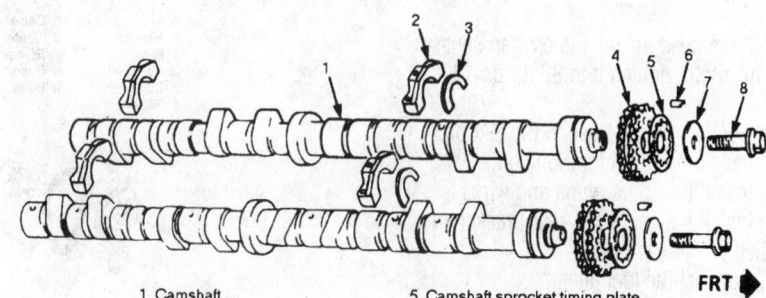

1. Camshaft	5. Camshaft sprocket timing plate
2. Camshaft retainer	6. Camshaft sprocket pin
3. Camshaft thrust washer	7. Camshaft sprocket washer
4. Camshaft sprocket	8. Camshaft sprocket bolt

Exploded view of the camshafts and related components—5.7L (VIN J) engine

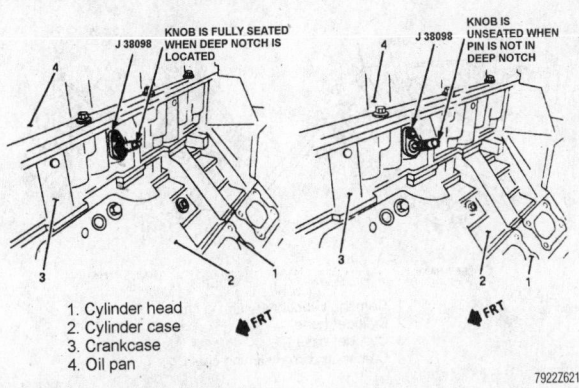

1. Cylinder head
2. Cylinder case
3. Crankcase
4. Oil pan

79222621

Crankshaft timing slot locator tool installed—5.7L (VIN J) engine

➡**If equipped with a 1 piece front cover/oil filter housing gasket, cut the old gasket along the front cover.**

m. Remove spark plug wires from plugs in the right cylinder head.

n. Disengage the electrical connector from the blower motor resistor block.

o. Remove the screws attaching the evaporator housing quarter panel, then remove the panel.

p. Remove the bolts attaching the coolant outlet pipe bracket to the alternator bracket and the coolant outlet to the injector housing, then position aside.

q. Remove the upper EGR pipe bolts and pipe.

r. Remove the bolt attaching the fresh air pipe bracket to the injector housing.

s. Remove the camshaft cover attaching bolts and the camshaft cover.

3. To gain access to the left camshafts, remove the air conditioning compressor and left valve cover as follows:

a. Properly discharge the air conditioning system.

b. Remove the throttle body assembly and the serpentine drive belt.

c. Remove the engine oil temperature sensor.

d. Remove the alternator assembly.

e. Remove the refrigerant hose from the A/C compressor, then immediately cap or plug the openings to prevent system contamination and damage.

f. Remove the compressor mounting bolts and electrical connection.

g. Remove the compressor from the vehicle.

h. Remove the power steering pump from the engine.

i. Remove the spark plug wires from the plugs in the left cylinder head.

j. Remove the ventilation breather pipe from the camshaft cover.

k. Remove the throttle and cruise control cable or control cable hold-down clamps from the plenum.

l. If not done already, remove the throttle body extension and coolant outlet pipe.

m. Remove the vacuum hose from the power brake booster, if necessary, remove the booster assembly.

n. Remove the left camshaft cover attaching bolts and remove the cover.

4. Raise and support the vehicle safely.

5. Disengage the electrical connector from the crankshaft ignition timing sensor.

6. Remove the ignition timing sensor from the cylinder case.

7. Install the crankshaft timing slot locator tool J-38098 or equivalent, into the ignition timing sensor opening. Be sure the tool head is fully seated with the indicating pin inserted into the deep notch of the crankshaft timing disc.

8. Lower vehicle.

9. Remove the bolts attaching the secondary timing chain tensioner housing to the cylinder head, then remove the housing, O-ring and tensioner from the cylinder case.

10. Remove the bolts and washers attaching the camshaft to the sprockets.

➡**Install a wrench on the rear camshaft hex when removing the sprocket bolts, to prevent the camshafts from exerting force on the crankshaft timing slot locator tool.**

11. Remove the camshaft timing plates and pins.

12. Remove the camshaft retainers and thrust washers.

13. Remove the camshafts and sprockets from the vehicle. Install timing chain retainers J-38099 or equivalent, to retain secondary chain loops.

14. Remove lifters from bores and inspect. Be sure any lifters, to be reused, are retained in proper order so each one can be returned to its original bore.

| 1 | CYLINDER HEAD |
| 2 | CAMSHAFT SECONDARY TIMING CHAIN |

79222627

Install the timing chain retainers to hold the chains in position—5.7L (VIN J) engine

To install:

15. Inspect the camshaft bearing journals for wear or damage.

16. Inspect the camshaft bearing surfaces in the cylinder head and camshaft cover for wear or damage.

➡**The camshaft cover and cylinder head must be replaced as a set if excessive wear or damage to the bearing surfaces is found.**

17. Install the each camshaft and lifter assembly, 1 at a time:

a. Lubricate lifters and bores with clean engine oil, then install lifters into bores. If a camshaft is replaced, new lifters must also be used.

b. Install the camshaft sprocket onto the secondary timing chain, while removing the timing chain retainer.

c. Slide the camshaft into the sprocket, noting the position of the alignment hole for timing pin tool installation. Position the camshaft in the neutral position, no valves opened.

d. Lubricate camshaft journals, lobes, thrust washers and retainers with clean engine oil.

e. Install the camshaft thrust washers, retainers and bolts. Tighten bolts to 89 inch lbs. (10 Nm).

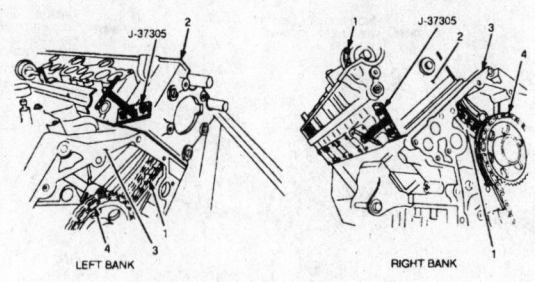

1. Camshaft secondary timing chain
2. Cylinder head
3. Cylinder case
4. Camshaft primary timing chain

79222624

Secondary timing chain pre-tensioner tool—5.7L (VIN J) engine

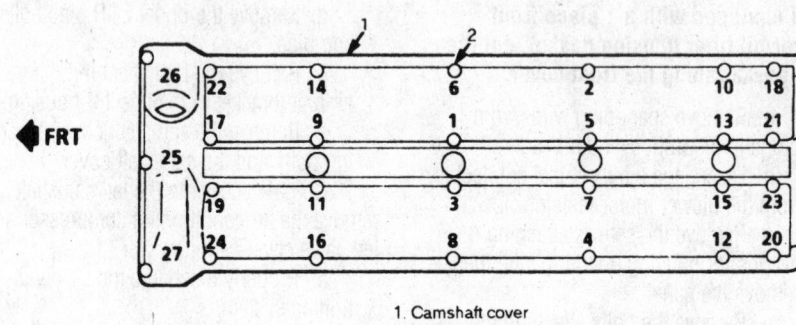

1. Camshaft cover
2. Camshaft cover bolt

79222625

Be sure to tighten the camshaft (valve) cover bolts in the correct sequence—5.7L (VIN J) engine

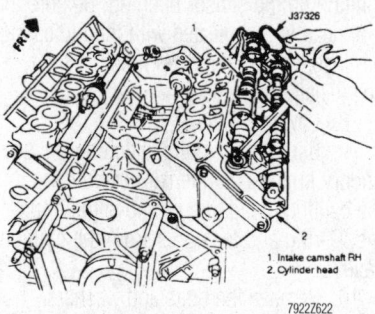

1. Intake camshaft RH
2. Cylinder head

79222622

Installing the camshaft timing pins—5.7L (VIN J) engine

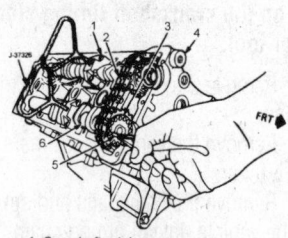

1. Camshaft retainer
2. Camshaft secondary timing chain
3. Camshaft sprocket timing plate
4. Cylinder head
5. Camshaft sprocket pin
6. Camshaft

79222623

Installing the camshaft sprocket pin—5.7L (VIN J) engine

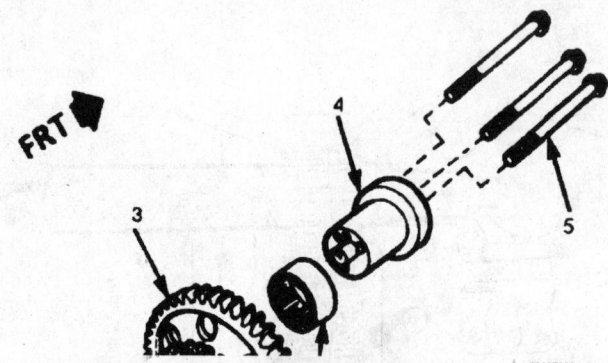

79222626

Exploded view of the timing chain idler sprocket—5.7L (VIN J) engine

f. Repeat Steps a-e for the remaining camshafts.

18. Install timing pins J-37326 into camshaft retainers and the indexing holes in the camshafts. Camshafts can be rotated using the cast hex at the camshaft rear.

19. Install the camshaft secondary chain pre-tensioner, J-37305 or equivalent. Hand-tighten to remove slack from the timing chain, but do not overtighten.

20. Install timing plates, pins and washers. If no holes line up on the timing plate, reverse the plate.

21. Apply Loctite® 262 or equivalent, on the NEW camshaft sprocket bolts, then

install and finger-tighten the bolts. New camshaft bolts should be used each time the camshaft is removed. Tighten the bolts to 18 ft. lbs. (25 Nm) and turn 80–85 degrees using a torque angle meter. A back-up wrench should be used on the rear camshaft hex to prevent damaging the timing pins.

22. Remove timing pins J-37326.

23. Remove the secondary timing chain pre-tensioner tool and install the new secondary timing chain tensioner, housing, new O-ring and bolts. Lubricate tensioner with engine oil. Be sure the oil hole in the tensioner piston be installed in a vertical position and that the fork on the end of the

tensioner is properly engaged onto the chain guide. After installing, use a blunt punch to release the plunger. Tighten chain tensioner bolts to 89 inch lbs. (10 Nm).

24. Raise and support the vehicle safely.

25. Remove crankshaft timing slot locator J-38098 from the cylinder case.

26. Install the crankshaft position sensor into the cylinder case and tighten the retainer(s) to 71 inch lbs. (8 Nm).

27. Engage the timing sensor electrical connector and lower the vehicle.

28. Apply Permabond® A136 or equivalent, to the camshaft covers and Loctite® 565 or equivalent, to the end plugs. Install the end plugs and new spark plug bore O-rings prior to cover installation.

29. Install the camshaft covers in the reverse order of removal. The camshaft cover retainers must be tightened in the proper sequence in order to assure proper camshaft operation. Tighten the M8 bolts to 15 ft. lbs. (20 Nm), repeat 3 times. Tighten the M6 screws to 89 inch lbs. (10 Nm). Also, be sure to install a new coolant outlet cover gasket and tighten the cover screws to 89 inch lbs. (10 Nm).

30. For the right bank camshafts, install oil filter housing assembly.

31. For the left bank camshafts, install the air conditioning compressor assembly.

32. Reconnect the battery negative cable and properly fill the engine cooling system.

5.7L (VIN G) Engine

1. Disconnect the negative battery cable.

2. Drain the cooling system.

3. Relieve the fuel system pressure, then disconnect the lines from the fuel rail assembly.

4. Remove the cylinder head covers.

5. Remove the rocker arms and pushrods. Keep them in order so they can be installed in their original positions.

6. Remove the cylinder heads and the valve lifters.

7. Remove the radiator assembly.

8. Remove the engine front cover.

9. Remove the oil pump.

10. Remove the camshaft sprocket and timing chain.

11. Remove the camshaft sensor bolt and sensor from the engine.

12. Remove the camshaft retainer plate.

❊❊ WARNING

All camshaft bearing are the same diameter, extreme care must be taken during camshaft removal or installation to avoid damaging them.

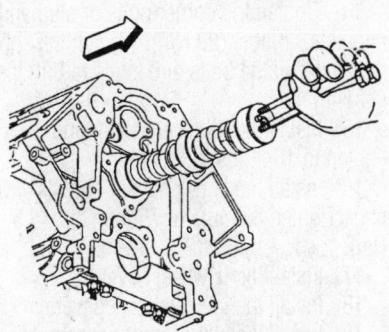

79222628

Install three M8–1.25 x 100mm long bolts into the camshaft to use as a handle

13. Install three M8–1.25 x 100mm long bolts in the front of the camshaft to use as a handle.

14. Carefully withdraw the camshaft from the engine, be sure not to damage the camshaft bearings.

To install:

15. Lubricate the camshaft with clean engine oil.

16. Install three M8–1.25 x 100mm long bolts in the front of the camshaft to use as a handle.

17. Carefully install the camshaft into the engine and remove the three bolts.

18. Install the camshaft retainer using a new gasket. Tighten the bolts to 18 ft. lbs. (25 Nm).

19. Install the camshaft sensor. Use a new O-ring if the old one is damaged. Tighten the bolt to 18 ft. lbs. (25 Nm).

20. Install the timing chain. Be sure the valve timing marks are facing each other.

21. Install the oil pump.

22. Install the front cover.

23. Install the radiator assembly.

24. Install the cylinder heads, valve lifters, pushrods and rocker arms.

25. Install all remaining components.

26. Connect the negative battery cable.

27. Refill the cooling system with DEX-COOL® or equivalent coolant.

28. Start the engine and check for leaks.

Valve Lash

ADJUSTMENT

5.7L (VIN P and 5) Engines

The valve lash on the VIN P and 5 engine is adjusted whenever the rocker arm assemblies have been removed.

➡The 5.7L (VIN P) engine utilize hydraulic lifters which normally require

very little maintenance or adjustment. These components are simple in design and are best maintained through regular, scheduled engine oil changes. If the engine is running well and no audible clicking sounds are heard from the valvetrain, there is no need to remove or disassemble the valve lifters.

1. Disconnect the negative battery cable.

2. Remove the valve rocker covers.

3. Tighten the nuts slowly until all lash is eliminated.

4. Adjust the valves when the lifter is on the base circle of the camshaft lobe. Slowly turn or crank the engine until the mark on the vibration damper is in the 12 o'clock position (aligned with the timing cover mark, if equipped) and the engine is in the No. 1 firing position.

➡The No. 1 firing position may be determined by watching the No. 1 cylinder valves as the mark on the damper approaches the 12 o'clock position. If both the intake and exhaust valves are closed as the mark comes up to the timing tab, the engine is in the No. 1 firing position. If either valve opens as the timing mark approaches the top of its travel, the engine is in No. 6 firing position and should be turned over 1 full revolution in order to reach the No. 1 firing position.

5. With the engine in the No. 1 firing position, adjust the following valves:
- Exhaust—1, 3, 4, 8
- Intake—1, 2, 5, 7

6. Back out the rocker arm adjusting nut until lash is felt at the pushrod, then turn the adjusting nut inward until all lash is removed. This can be determined by rotating pushrod while turning the adjusting nut. When play has been removed, the pushrod will not turn. Then, tighten the adjusting nut 1 full additional turn.

7. Slowly turn or crank the engine 1 revolution until the vibration damper mark is at 12 o'clock again and the No. 1 cylinder valves open. This is the No. 6 firing position.

8. With the engine in this position, adjust the following valves:
- Exhaust—2, 5, 6, 7
- Intake—3, 4, 6, 8

9. Install the valve rocker arm covers.

10. Connect the battery negative cable.

5.7L (VIN J) Engine

This engine is equipped with hydraulic lifters which are installed in bores directly below the camshaft lobes. The lifters maintain 0 lash between the camshaft lobes and the valve stem. The lifter and installation position is non-adjustable, therefore upon failure, the lifter assembly must be replaced.

5.7L (VIN G) Engine

The 5.7L (VIN G) engine uses rocker arms, pushrods and hydraulic lifters which are similar to the other pushrod motors in this section (VIN P and 5). The lifters maintain correct valve lash so that no periodic adjustment is necessary or possible. But, unlike the other pushrod motors which use adjustment nuts to retain the rocker arms (and therefore must be adjusted during installation), the rocker arms on the 5.7L (VIN G) are bolted into position so that no initial adjustment is possible either. If the valves are noisy, suspect low oil pressure or worn valvetrain components.

Oil Pan

REMOVAL & INSTALLATION

5.7L (VIN P) Engine

1. Disconnect the negative battery cable.
2. Raise and support the vehicle safely, then drain the engine oil.
3. Disengage the oil level sensor electrical connector and remove the sensor assembly from the side of the oil pan.
4. Remove the oil filter, then remove the oil filter adapter bolts and the adapter assembly.
5. Remove the starter motor assembly.
6. Remove the left catalytic converter.
7. Remove the flywheel cover.
8. Remove the knock sensor retaining nuts and shields.
9. Remove the oil pan bolts, nuts and studs. Be sure to note the location of stud bolts.
10. Remove the oil pan, reinforcements and gasket.

To install:

11. Thoroughly clean all gasket mating surfaces and apply a small amount of 1052914 or equivalent sealer, to the front cover and cylinder block junction and the rear seal retainer and cylinder block junction. Extend the bead of sealer approximately 1 in. (25mm) in either direction of these junctions.
12. Install the gasket onto the oil pan and reinforcements.
13. Install the gasket, pan and reinforcement assembly to the cylinder block with the bolts, studs and nuts.

14. Tighten the corner bolts or stud and nuts to 15 ft. lbs. (20 Nm). Tighten the remainder of the bolts and studs to 8 ft. lbs. (11 Nm).
15. Install the oil level sensor and tighten to 16 ft. lbs. (22 Nm).
16. Install the knock sensor shields and nuts. Tighten the nuts to 75 inch lbs. (8.5 Nm).
17. Install the flywheel cover.
18. Install the left catalytic converter.
19. Install the starter motor assembly.
20. Engage the wiring harness to the oil level sensor terminal.
21. Install the oil filter adapter and tighten the retainers to 17 ft. lbs. (23 Nm), then install the oil filter.
22. Lower the vehicle and properly fill the crankcase with clean engine oil.
23. Connect the negative battery cable.
24. Reset the CHANGE OIL indicator:
    a. Turn the ignition **ON** but do not start the engine.
    b. Depress the ENG MET button on the trip monitor, then within 5 seconds, depress the button a 2nd time. Within another 5 seconds, depress and hold the GAUGES button.
    c. While holding the GAUGES button and watch the CHANGE OIL light, it should begin to flash. Continue to hold the gauges button until the flashing stops and the light goes out indicating that the indicator is reset.
    d. If the indicator does not reset, turn the ignition **OFF** and restart the procedure.

5.7L (VIN J) Engine

1. Disconnect negative battery cable and remove the oil lever indicator from the guide tube.
2. Raise and support the vehicle safely, then drain the engine oil.
3. Remove the clutch housing cover attaching bolts, then remove the cover from the vehicle.
4. If equipped, remove the left and right wiring harness heat shields from the oil pan.
5. Disconnect the low oil sensor connection and remove the sensor from the pan.
6. Remove the bolts attaching the AIR pipe bracket to the oil pan, then remove the left and right converter heat shields.
7. Remove the nuts attaching the engine mounts at the front crossmember rear brace on the left and right sides. Remove the bolts attaching the front crossmember to the rear braces.

8. Remove the bolts attaching the left front crossmember rear brace to the left front side member, then remove the brace from the vehicle.
9. Remove the bolts attaching the right front crossmember rear brace to the right front side member, then remove the brace from the vehicle.
10. Remove the bolts attaching the oil pan and crankcase. Remove the oil pan and gasket from the vehicle.

To install:

11. Apply Loctite® 242 to the oil pan screw threads.
12. Install the oil pan and new gasket to the engine crankcase. Tighten the oil pan front screws to 106 inch lbs. (12 Nm). Tighten the oil pan bolts to 23 ft. lbs. (31 Nm).
13. Install the front crossmember rear braces and bolts retaining the braces to the front crossmember bolts. Finger-tighten the bolts.
14. Install the bolts retaining the left front crossmember rear brace to the left front side member, finger-tight.
15. Install the bolts retaining the right front crossmember rear brace to the left front side member, finger-tight.
16. Tighten the left and right front crossmember rear brace to front crossmember bolts to 59 ft. lbs. (80 Nm), then tighten the left and right front crossmember rear brace to front side member bolts to 46 ft. lbs. (62 Nm).
17. Install the nuts retaining the engine mounts to the front crossmember and tighten to 40 ft. lbs. (54 Nm).
18. Install the converter heat shields and screws.
19. Install the bolts retaining the AIR pipe bracket to the oil pan and tighten to 89 inch lbs. (10 Nm).
20. Install the oil level sensor in the pan and tighten to 18 ft. lbs. (25 Nm), then engage the wiring harness to the sensor.
21. Install the left and right wiring harness heat shields, if equipped, and tighten the bolts to 89 inch lbs. (10 Nm).
22. Install the clutch housing cover and tighten the bolts to 80 inch lbs. (9 Nm).
23. Lower the vehicle and insert the oil level indicator into the guide tube.
24. Properly fill the crankcase with clean engine oil.
25. Connect the negative battery cable.
26. If equipped, reset the CHANGE OIL indicator:
    a. Turn the ignition **ON** but do not start the engine.
    b. Depress the ENG MET button on the trip monitor, then within 5 seconds,

depress the button a 2nd time. Within another 5 seconds, depress and hold the GAUGES button.

c. While holding the GAUGES button and watch the CHANGE OIL light, it should begin to flash. Continue to hold the gauges button until the flashing stops and the light goes out indicating that the indicator is reset.

d. If the indicator does not reset, turn the ignition **OFF** and restart the procedure.

5.7L (VIN G) Engine

1. Disconnect the negative battery cable.

2. Remove the generator.

3. Support the engine with a suitable support fixture such as Engine Support Fixture J 41803.

4. Raise and safely support the vehicle.

5. Remove the front wheels.

6. Matchmark the adjustment, then remove the tie rod ends from the steering knuckles.

7. Remove the stabilizer bar brackets from the cradle.

8. Remove the power steering fluid cooler tube from the cradle and position it upward.

9. Remove the steering rack assembly mounting bolts and position the rack upward.

10. Install Spring Compressor J 33432 or equivalent and compress the spring.

11. Remove the lower shock absorber mounting bolts.

12. Remove the lower ball joint nut, then disconnect the ball joint from the knuckle.

13. Remove the spring compressor.

14. Detach all electrical connectors at the front crossmember.

15. Remove the motor mount-to-front crossmember nuts.

16. Remove the front crossmember.

☀☀ CAUTION

The EPA warns that prolonged contact with used engine oil may cause a number of skin disorders, including cancer! You should make every effort to minimize your exposure to used engine oil. Protective gloves should be worn when changing the oil. Wash your hands and any other exposed skin areas as soon as possible after exposure to used engine oil. Soap and water, or waterless hand cleaner should be used.

17. Drain the engine oil.
18. Remove the oil filter.

19. If equipped with automatic transmission, remove the fluid cooler line bracket near the oil pan.

20. Remove the flywheel housing to oil pan bolts.

21. Remove the flywheel housing cover.

22. Remove the engine oil level sensor.

23. Disconnect the engine oil temperature sensor.

24. Remove the left and right closeout covers.

25. Remove the oil pan mounting bolts and the oil pan.

To install:

26. Apply a 0.20 in. (5mm) bead of sealant PN 12378190 or equivalent directly on the tabs of the front and rear cover gaskets that extend onto the oil pan mounting surface.

27. Position a new gasket on the oil pan. It is not necessary to rivet the new gasket on the oil pan.

28. Install the oil pan on the engine but do not tighten the bolts.

29. Install the flywheel housing-to-oil pan bolts. Tighten the bolts to 37 ft. lbs. (50 Nm).

30. Tighten the oil pan-to-block and oil pan-to-front cover bolts to 18 ft. lbs. (25 Nm). Tighten the oil pan-to-rear cover bolts to 106 inch lbs. (12 Nm).

31. If removed, install the oil level sensor, tighten the sensor to 26 ft. lbs. (35 Nm).

32. Install the left and right closeout covers. Tighten the bolts to 106 inch lbs. (12 Nm).

33. Connect the oil temperature sensor.

34. Install the transmission fluid cooler line bracket.

35. Install a new oil filter.

36. Install the front crossmember. Tighten the nuts to 81 ft. lbs. (110 Nm).

37. Install the motor mount on the crossmember. Tighten the nuts to 40 ft. lbs. (54 Nm).

38. Attach the electrical connectors.

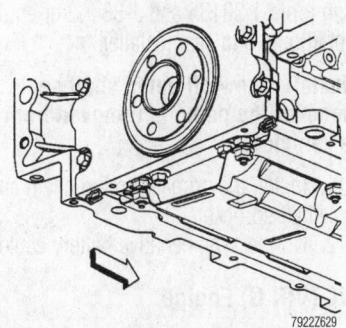

Apply sealant to the areas where the front and rear covers attach to the engine block—5.7L (VIN G) engine

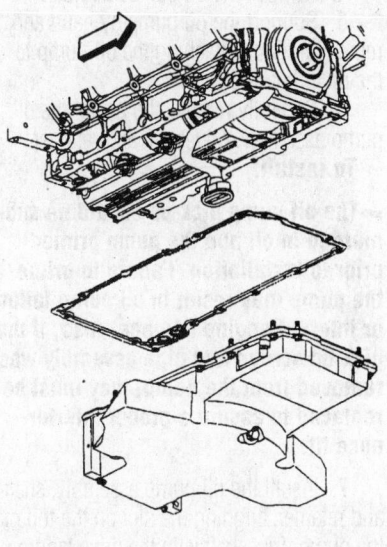

Exploded view of the oil pan mounting— 5.7L (VIN G) engine

39. Install spring compressor J 33432 or equivalent and compress the spring.

40. Install the lower shock absorber mounting bolts. Tighten them to 20 ft. lbs. (28 Nm).

41. Install the lower ball joints to the steering knuckles.

42. Install the power steering gear and fluid cooling tube.

43. Install the stabilizer bar brackets to the cradle. Tighten the bolts to 40 ft. lbs. (54 Nm).

44. Connect the tie rod ends to the knuckles. Tighten the nuts to 33 ft. lbs. (45 Nm).

45. Install the front wheels. Tighten the nuts to 100 ft. lbs. (140 Nm).

46. Lower the vehicle to the floor and remove the engine support fixture.

47. Install the generator.

48. Connect the negative battery cable.

49. Refill the engine with the proper type of engine oil.

50. Start the engine and check for leaks.

51. Check the front wheel alignment and adjust as needed.

Oil Pump

REMOVAL & INSTALLATION

5.7L (VIN P and 5) Engines

1. Disconnect the negative battery cable.

2. Raise and support the vehicle safely.

3. Drain the engine oil and remove the oil pan.

4. Remove the oil pan baffle nuts.

5. Support the oil pump by hand and remove the bolt attaching the oil pump to the main bearing cap.

6. Carefully remove the baffle, the oil pump assembly, driveshaft and retainer.

To install:

➡The oil pump pick-up should be submerged in oil and the pump primed prior to installation. Failure to prime the pump may result in oil pump failure or internal engine damage. Also, if the pick-up screen and pipe assembly was removed from the pump, they must be replaced to assure a proper interference fit.

7. Install the oil pump assembly, shaft and retainer, aligning the slot on the top of the pump driveshaft with the drive tang on the lower end of the distributor driveshaft.

8. Install the oil pan baffle, then install the bolt to the main bearing cap, followed by the baffle nuts. Tighten the retaining bolt to 65 ft. lbs. (88 Nm) and the baffle nuts to 25 ft. lbs. (34 Nm).

9. Install the oil pan and lower the vehicle.

10. Properly fill the engine crankcase with clean engine oil.

11. Connect the negative battery cable.

12. If equipped, reset the CHANGE OIL indicator:

   a. Turn the ignition **ON** but do not start the engine.

   b. Depress the ENG MET button on the trip monitor, then within 5 seconds, depress the button a 2nd time. Within another 5 seconds, depress and hold the GAUGES button.

   c. While holding the GAUGES button and watch the CHANGE OIL light, it should begin to flash. Continue to hold the gauges button until the flashing stops and the light goes out indicating that the indicator is reset.

   d. If the indicator does not reset, turn the ignition **OFF** and restart the procedure.

5.7L (VIN J) Engine

1. Disconnect battery negative cable.

2. Remove the primary timing chain and crankshaft sprocket.

3. Remove bolts attaching the oil pump to the cylinder case, then remove the oil pump from the vehicle.

4. Remove O-rings from crankshaft, if applicable, the oil pump.

5. Remove the oil pick-up seal.

To install:

6. Install new O-rings onto the crankshaft and oil pump, as applicable.

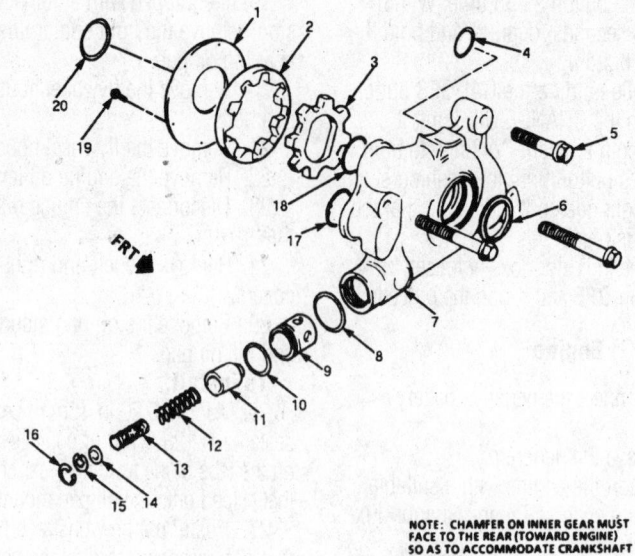

1: Oil pump plate
2. Outer gear
3. Inner gear
4. Oil pump body o-ring
5. Oil pump bolt
6. Oil pump crankshaft seal
7. Oil pump body
8. Oil pressure regulation valve o-ring
9. Oil pressure regulation valve housing
10. Oil pressure regulation valve o-ring
11. Oil pressure regulation valve
12. Oil pressure regulation valve outer spring
13. Oil pressure regulation valve inner spring
14. Oil pressure regulation valve stop
15. Oil pressure regulation valve retainer
16. Oil pressure regulation valve retainer
17. Oil filter feed return o-ring
18. Oil filter feed return o-ring
19. Oil pump plate screw
20. Oil pump crankshaft o-ring

NOTE: CHAMFER ON INNER GEAR MUST FACE TO THE REAR (TOWARD ENGINE) SO AS TO ACCOMMODATE CRANKSHAFT O-RING 20.

79222Z631

Exploded view of the oil pump assembly—5.7L (VIN J) engine

7. If applicable, install the oil pick-up assembly seal.

8. Apply Loctite® 262 to the oil pump bolts and install them along with the oil pump, finger-tight.

➡Be sure the 2 flats of the pump drive gear are aligned with the 2 flats on the crankshaft. Do not force pump onto crankshaft.

9. Using oil pump aligning tool J-38135 or equivalent pump aligner/seal installer, align oil pump on the crankshaft. Tighten the oil pump bolts to 19 ft. lbs. (26 Nm).

10. Install a new oil pump shaft seal using tools J-38135 and J-38463 or equivalent aligner and seal installer.

➡Install a new oil pump shaft seal whenever the pump is removed from the vehicle.

11. Install the primary timing chain and crankshaft sprocket.

12. Connect the negative battery cable.

5.7L (VIN G) Engine

1. Disconnect the negative battery cable.

2. Drain the engine oil and coolant.

3. Remove the engine front cover.

4. Remove the oil pan.

5. Remove the oil pump pick-up and O-ring.

6. Remove the oil pump mounting bolts.

7. Remove the oil pump.

To install:

8. Inspect the oil passages in the pump and on the mounting surface. Be sure they are clean and free of debris.

9. Align the splined surface of the crankshaft sprocket and the oil pump drive gear. Install the oil pump. Tighten the bolts to 18 ft. lbs. (25 Nm).

10. Install a new O-ring on the pick-up tube and install it into the pump by hand. Don't use the bolt to force the tube into the pump. After the tube is seated in the oil pump, tighten the bolt to 106 inch lbs. (12 Nm). Tighten the nut to 18 ft. lbs. (25 Nm).

11. Install the oil pan.

12. Install the front cover.

13. Refill the cooling system and the engine oil with the correct type and amount of fluids.

14. Connect the negative battery cable.

15. Start the engine and check for leaks.

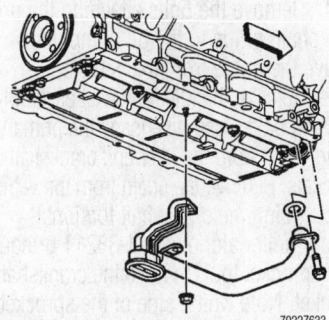

Be sure to seat the tube into the pump before installing the bolt—5.7L (VIN G) engine

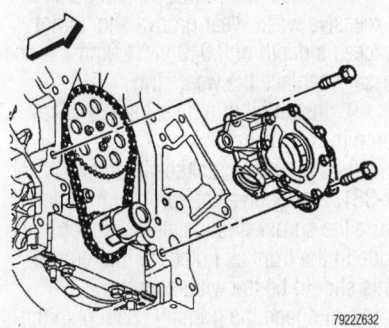

Oil pump assembly mounting—5.7L (VIN G) engine

Rear Main Seal

REMOVAL & INSTALLATION

✳✳ CAUTION

The EPA warns that prolonged contact with used engine oil may cause a number of skin disorders, including cancer! You should make every effort to minimize your exposure to used engine oil. Protective gloves should be worn when changing the oil. Wash your hands and any other exposed skin areas as soon as possible after exposure to used engine oil. Soap and water, or waterless hand cleaner should be used.

➡**The rear main seal is a one piece unit. It can be removed or installed without removing the oil pan or crankshaft.**

1. Raise and safely support the vehicle.
2. On 1995–96 models, remove the transmission. On 1997–99 models, remove the driveshaft support and bell housing.
3. If equipped with a manual transmission, remove the clutch and pressure plate.

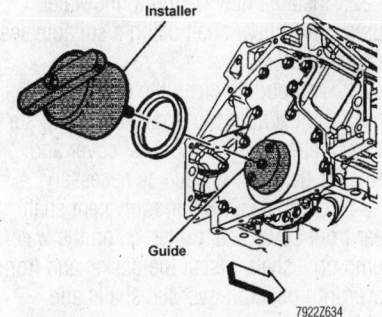

Use a suitable seal installer to press the seal on the crankshaft—5.7L (VIN G) engine shown, other engines are similar

4. Remove the flywheel assembly.
5. Using a suitable prytool, carefully pry the old seal out.
6. Inspect the crankshaft for nicks or burrs, correct as required.
   To install:
7. Clean the area and coat the seal and crankshaft with engine oil. Install the seal using a suitable seal installer.
8. Install the flywheel.
9. If equipped with a manual transmission, install the clutch and pressure plate.
10. Install the transmission or bell housing and driveshaft support.
11. Check the fluid levels, start the engine and check for leaks.

Timing Chain, Sprockets, Front Cover and Seal

REMOVAL & INSTALLATION

5.7L (VIN P and 5) Engines

1. Disconnect the negative battery cable.
2. Drain the engine oil and coolant into suitable containers.
3. Remove the throttle body air intake duct.

4. Remove the serpentine drive belt.
5. Remove the water pump assembly.
6. Remove the crankshaft balancer and hub.
   a. If not done already, raise and support the vehicle safely, then remove the motor mount nuts.
   b. Remove the power steering fluid cooler, then raise the engine sufficiently for tool access to the crankshaft balancer.

➡**When raising and supporting the engine, NEVER place a jack under the oil pan, crankshaft pulley or any sheet-metal. There is a minimal clearance between the oil pan and the pump screen. Jacking against the pan could cause sufficient deformation to damage the oil pick-up unit.**

   c. Remove the balancer bolts, then remove the balancer from the hub.
   d. Disconnect the power steering line from the steering gear.
   e. Matchmark the crankshaft hub to the engine front cover, then remove the hub bolt and washer.
   f. Remove the crankshaft hub using J-39046, or an equivalent hub removal/installation tool. To preserve the relationship between the hub and crankshaft, DO NOT crank the engine over once the hub has been removed. If the hub is not matchmarked and installed in the original position, an engine imbalance could result.
7. Remove the distributor assembly.
8. Remove the oil pan assembly.
9. Remove the engine front cover bolts.
10. Remove the engine front cover and gasket.
11. Rotate the crankshaft until the timing marks on the timing chain sprockets are aligned nearest each other. The camshaft sprocket mark should be at the 6 o'clock position while the mark on the crankshaft sprocket should be at the 12 o'clock position.

A. Timing marks
B. Keyway
1. Bolt
2. Sprocket
3. Bolt
4. Retainer
5. Camshaft
6. Timing chain
7. Sprocket
8. Driven gear
9. Bolts

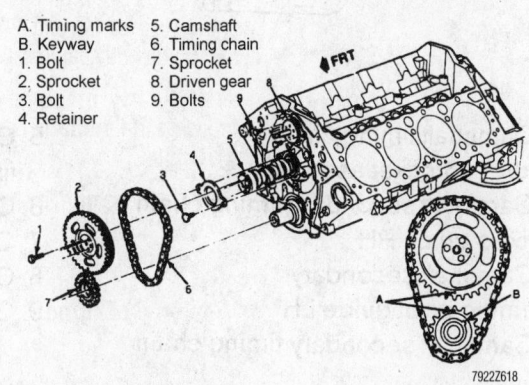

Exploded view of the timing chain and gear assembly—5.7L (VIN P and 5) engines

12. Remove the camshaft sprocket bolts.

13. Remove the camshaft sprocket and timing chain.

➡ **To prevent piston or valve damage, do not turn the crankshaft after the timing chain has been removed.**

14. Remove the water pump bearing retainer bolts, then remove the driveshaft assembly using J-39243 or equivalent driven gear assembly remover.

15. Remove the crankshaft sprocket using J-5825-A or equivalent crankshaft sprocket remover.

16. If necessary, remove the crankshaft key.

To install:

17. If removed, install the crankshaft key.

18. Install the crankshaft sprocket using a suitable installation tool.

19. Install the water pump driveshaft assembly using a suitable tool. Install the retainer bolts and tighten to 108 inch lbs. (12 Nm).

20. Align the timing marks and install the camshaft sprocket and timing chain. The gears of the camshaft sprocket and water pump driveshaft must mesh or damage to the thrust plate retainer could occur.

21. Install the camshaft sprocket bolts and tighten to 21 ft. lbs. (28 Nm).

22. Install a new O-ring to the water pump driven gear shaft using a suitable seal installation tool.

23. Thoroughly clean the engine front cover and cylinder block gasket mating surfaces. Inspect the engine front cover and seals for damage, replace as necessary.

24. Using J-39087 or equivalent shaft gear front cover seal protector, on the water pump driveshaft, install the gasket and front cover into position over the shafts and guide pins.

25. Install the engine front cover bolts and tighten to 100 inch lbs. (11 Nm).

26. Install the oil pan and gasket.

27. Install the distributor assembly.

28. Install and connect all remaining components.

29. Connect the negative battery cable.

30. Refill the cooling system, start the engine and check for leaks.

5.7L (VIN J) Engine

PRIMARY TIMING CHAIN AND CRANKSHAFT SPROCKET

1. Disconnect battery negative cable.

2. Remove the timing chain front cover assembly.

3. Remove the left and right intake camshafts.

4. Remove the bolts attaching the primary chain guide to the oil pump, then remove the guide from the vehicle.

5. Remove the idler sprocket assembly attaching bolts, then disengage the primary timing chain from the idler and crankshaft sprockets. Remove the chain from the vehicle.

6. Using the crankshaft torsional damper puller along with J-38211 or equivalent sprocket tool, remove the crankshaft sprocket. Note which side of the sprocket faces forward for installation purposes.

7. Remove the key and oil pump seal seat from the crankshaft.

To install:

8. Inspect the primary chain guide for excessive wear. Wear groove should not exceed a depth of 0.040 in. (1.0mm). If necessary, replace the wear strip.

9. Install oil pump seal seat and key onto the crankshaft.

10. Install the crankshaft sprocket using J-38132 or equivalent sprocket installer. Be sure the sprocket is installed with the same side to the front as noted during removal, this should be the wide shoulder.

11. Engage the primary chain onto the idler and crankshaft sprocket.

12. Apply Loctite® 262 or equivalent, to the idler sprocket assembly bolts and tighten to 19 ft. lbs. (26 Nm).

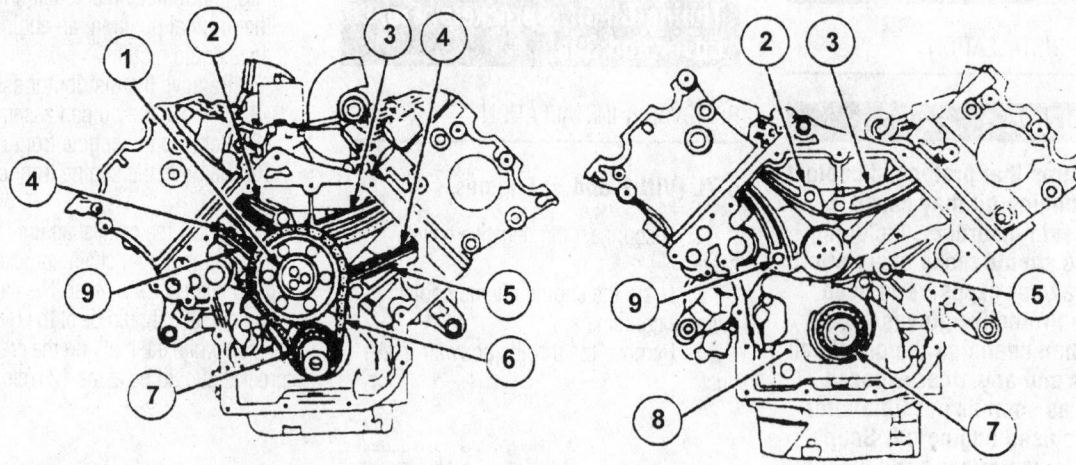

1. Camshaft timing chain idler sprocket assembly
2. Camshaft secondary timing chain fixed guide RH
3. Camshaft secondary timing pivot guide LH
4. Camshaft secondary timing chain
5. Camshaft secondary timing chain fixed guide LH
6. Camshaft primary timing chain
7. Crankshaft sprocket
8. Oil pump
9. Camshaft timing chain pivot guide RH

7922Z635

Primary and secondary timing chain assembly—5.7L (VIN J) engine

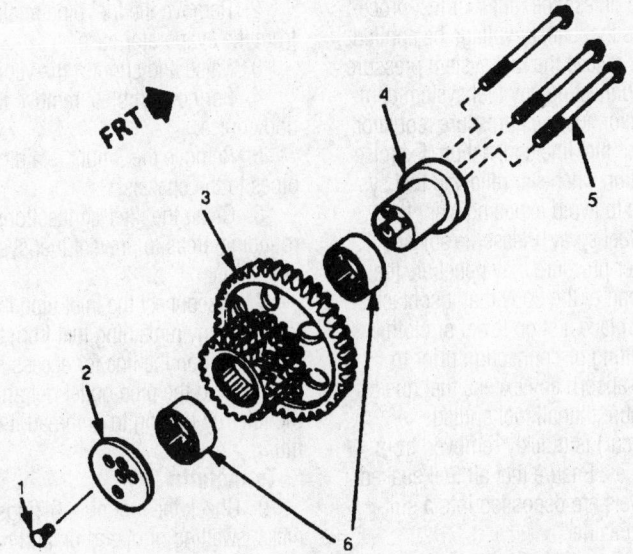

1. Camshaft idler sprocket assembly screw
2. Camshaft timing chain idler sprocket washer
3. Camshaft timing chain idler sprocket
4. Camshaft timing chain idler sprocket shaft
5. Camshaft idler sprocket bolt
6. Camshaft timing chain idler sprocket bearing

79222626

Timing chain idler sprocket assembly—5.7L (VIN J) engine

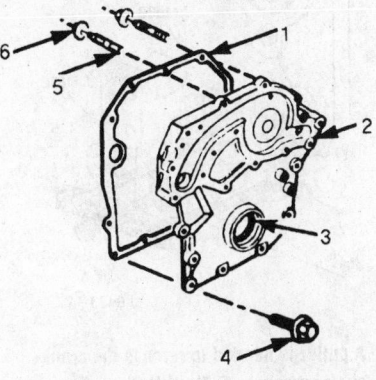

1. Engine front cover gasket
2. Engine front cover
3. Engine front cover seal
4. Engine front cover bolt
5. Engine front cover stud
6. Engine front cover stud nut

79222637

Exploded view of the front engine cover mounting—5.7L (VIN J) engine

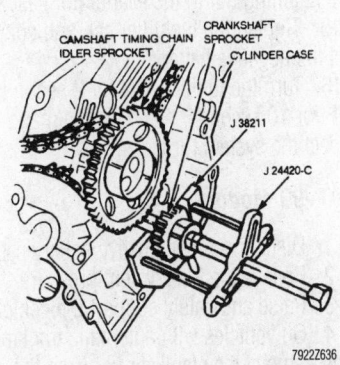

79222636

A special puller is needed to remove the crankshaft sprocket—5.7L (VIN J) engine

13. Apply Loctite® 262 or equivalent, to the primary chain guide bolts. Install the guide and bolts. Push the guide so the slack is removed from the chain and tighten the bolts to 89 inch lbs. (10 Nm).

➡**When installing guide, do not use any leverage tools, finger pressure is sufficient.**

14. Install the left and right intake camshafts.
15. Install the timing chain front cover.
16. Connect the negative battery cable.

SECONDARY TIMING CHAINS AND IDLER SPROCKET ASSEMBLY

1. Disconnect battery negative cable.
2. Remove the camshafts.
3. Remove the primary timing chain and crankshaft sprocket.
4. Disengage the left and right secondary chains from the idler sprocket.
5. Remove the idler sprocket assembly.
6. Remove the left and right secondary chains from the vehicle.

To install:

7. Inspect the chains and sprockets for abnormal wear or damage. If abnormal wear or damage is present on either the secondary timing chain, cam sprockets or idler sprockets, the entire assembly must be replaced.
8. Inspect the idler sprocket shaft bearings for wear or damage. If necessary, replace idler sprocket shaft bearings as follows:

   a. Remove the idler sprocket screw, washer and shaft.

   b. Using tool No. J-37328 or equivalent, remove bearings from idler sprocket.

   c. When installing bearings, ensure the manufacture's name and part numbers are visible from either end of the sprocket assembly.

d. Using a press, carefully push in the bearings until they are flush with idler sprocket. Apply minimum pressure to obtain a fit 0.0–0.05 in. (0.0–1.3mm) below the surface.

9. Install the shorter (inner) secondary chain through the right head and install J-38099 or equivalent timing chain retaining tool.
10. Locate the right chain onto the rear idler sprocket.
11. Install the longer (outer) secondary chain through the left head and install No. J-38099 or equivalent timing chain retaining tool.
12. Locate the left chain onto the middle idler sprocket.
13. Install the primary timing chain.
14. Install the camshafts.
15. Connect the negative battery cable.

5.7L (VIN G) Engine

1. Remove the oil pump.
2. Rotate the crankshaft until the timing marks are aligned. The marks should face each other.
3. Remove the camshaft sprocket bolts.
4. Remove the camshaft sprocket and timing chain assembly.
5. Use a puller such as J 8433–1 and adapters to remove the crankshaft sprocket if necessary.

To install:

6. If the crankshaft sprocket has been removed, install it using tool No. J 41665–1 or equivalent.

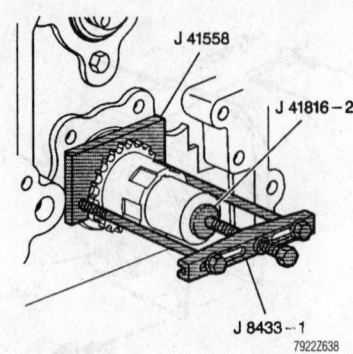

A puller is needed to remove the crankshaft sprocket—5.7L (VIN G) engine

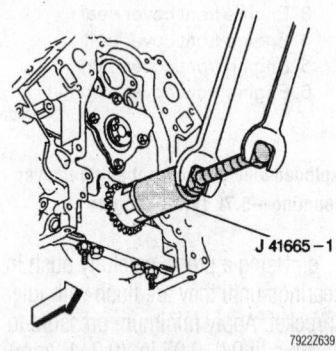

Be sure to install the crankshaft sprocket properly—5.7L (VIN G) engine

7. Be sure the timing mark on the crankshaft sprocket is at the 12 O'clock position. If not rotate the crankshaft to the correct position.

8. Position the chain on the camshaft sprocket. Install the chain and sprocket assembly so the timing marks are aligned. Tighten the camshaft sprocket bolts to 18 ft. lbs. (25 Nm).

9. Install the oil pump, front cover and remaining components.

FUEL SYSTEM

Fuel System Service Precautions

Safety is the most important factor when performing not only fuel system maintenance but any type of maintenance. Failure to conduct maintenance and repairs in a safe manner may result in serious personal injury or death. Maintenance and testing of the vehicle's fuel system components can be accomplished safely and effectively by adhering to the following rules and guidelines.

• To avoid the possibility of fire and personal injury, always disconnect the negative battery cable unless the repair or test procedure requires that battery voltage be applied.

• Always relieve the fuel system pressure prior to disconnecting any fuel system component (injector, fuel rail, pressure regulator, etc.), fitting or fuel line connection. Exercise extreme caution whenever relieving fuel system pressure to avoid exposing skin, face and eyes to fuel spray. Please be advised that fuel under pressure may penetrate the skin or any part of the body that it contacts.

• Always place a shop towel or cloth around the fitting or connection prior to loosening to absorb any excess fuel due to spillage. Ensure that all fuel spillage (should it occur) is quickly removed from engine surfaces. Ensure that all fuel soaked cloths or towels are deposited into a suitable waste container.

• Always keep a dry chemical (Class B) fire extinguisher near the work area.

• Do not allow fuel spray or fuel vapors to come into contact with a spark or open flame.

• Always use a back-up wrench when loosening and tightening fuel line connection fittings. This will prevent unnecessary stress and torsion to fuel line piping. Always follow the proper torque specifications.

• Always replace worn fuel fitting O-rings with new. Do not substitute fuel hose or equivalent, where fuel pipe is installed.

Fuel System Pressure

RELIEVING

1. Disconnect the negative battery cable.

2. Loosen the fuel filler cap to relieve the tank pressure.

3. Wrap a shop towel around the fuel pressure valve fitting (located on the side or end of the fuel rail assembly) to catch any fuel spray and connect J-34730–1 or an equivalent fuel gauge.

4. Place the bleed hose into a suitable container, then open the valve to bleed the fuel system pressure.

5. Close the valve and disconnect the fuel gauge. Drain any remaining fuel from the gauge into the bleed container.

Fuel Filter

REMOVAL & INSTALLATION

1995–96 Models

1. Disconnect the battery negative cable and properly relieve the fuel system pressure.

2. Remove the fuel pipe retaining nut from the evaporator case.

3. Raise and support the vehicle safely.

4. For convertibles, remove the underbody brace.

5. Remove the 3 nuts retaining the fuel pipes to the chassis.

6. Clean the filter connections and surrounding areas to prevent fuel system contamination.

7. Disconnect the inlet pipe from the filter, drain any remaining fuel from the line and reposition the line for access to the filter.

8. Hold the pipe outlet nut and remove the filter by turning to unthread it from the fitting.

To install:

9. Check the fuel pipe O-rings for cuts, nicks, swelling or distortion and replace, if damaged.

10. Install the fuel filter onto the fuel outlet pipe nut.

11. Connect the fuel inlet pipe to the filter and tighten the fitting to 20 ft. lbs. (27 Nm).

12. Install the 3 chassis-to-fuel pipe retaining nuts.

13. For convertibles, install the underbody brace, tighten the retaining nuts to 20 ft. lbs. (27 Nm) and the retaining bolts to 47 ft. lbs. (63 Nm).

14. Lower the vehicle and install the fuel pipe retaining nut to the evaporator case.

15. Tighten the fuel filler cap and connect the negative battery cable.

16. Turn the ignition **ON** for 2 seconds, **OFF** for 10 seconds, then **ON** again and inspect the system for leaks.

1997–99 Models

1. Disconnect the negative battery cable.

2. Relieve the fuel system pressure.

3. Raise and safely support the vehicle.

4. On vehicles with automatic transmissions, remove the stabilizer bar from the rear cradle and detach the intermediate pipe from the muffler.

5. Clean the filter connections, then depress the locking tabs and detach the quick-connect fittings from the filter.

6. Remove the fuel filter and bracket.

To install:

7. Remove the protective caps from the new filter.

8. Install the new plastic connector retainers on the ends of the filter.

9. Install the filter into the bracket. Apply a few drops of oil to the male end of the fitting and attach the fuel lines.

10. Install the filter and bracket on the mounting stud. Tighten the nut to 40 inch lbs. (4.5 Nm).

11. On vehicles with automatic transmissions, connect the exhaust pipe to the muffler and install the stabilizer bar.

12. Lower the vehicle to the floor.

13. Connect the negative battery cable.

14. Turn the ignition switch **ON** for two seconds, then **OFF** for 10 seconds. Turn the ignition switch back **ON** and inspect for leaks.

Fuel Pump

REMOVAL & INSTALLATION

1995–96 Models

➡️**Vehicles equipped with the 5.7L (VIN J) engine use a fuel sender assembly which is equipped with 2 fuel pumps. The strainers and pumps are not serviced separately from the sender, and if 1 component is damaged, the sender assembly must be replaced as a unit.**

1. Disconnect the negative battery cable.

2. Properly relieve the fuel system pressure and drain the fuel tank.

3. Remove the 4 filler door bezel attaching screws, then remove the filler door bezel.

4. Lift the fuel tank filler neck housing and disconnect the drain hose from the nipple. Remove filler neck housing.

5. Clean the area around all fuel fittings to prevent system contamination, then disconnect and plug the fuel pipes and fuel vapor pipe.

6. Unplug the sending unit electrical connector, remove the attaching bolts and remove the sending unit assembly from the vehicle.

7. If equipped with the VIN J engine, replace fuel sender assembly.

8. If equipped with the VIN P or 5 engine, service the sender assembly, as necessary:

a. Note the position of the fuel strainer on the pump.

b. Support the pump with one hand and grasp the strainer with the other. Turn the strainer in one direction, pull the strainer off the pump and discard it.

c. Unplug the fuel pump electrical connection.

d. Place the fuel sender assembly upside down on a flat bench.

e. Pull the fuel pump downward to remove it from the mounting bracket, then tilt the pump outward and remove it from the pulse dampener.

f. Note the position of the dampener on the inlet tube, then remove the damp-ener from the tube. Shake the dampener and listen for fuel, if fuel is heard inside the dampener, it must be replaced.

To install:

9. If equipped with the VIN P or 5 engine, assemble the sender for installation, as necessary:

a. Install the fuel pulse dampener in the same position as noted during disassembly.

b. Assemble the rear bumper and insulator onto the fuel pump.

c. Position the fuel sender assembly upside down on a flat bench and install the fuel pump between the fuel pulse dampener and mounting bracket.

d. Engage the pump electrical connector.

e. Install the new fuel strainer into the same position as noted during disassembly. Push on the outer edge of ferrule until fully seated.

10. Position a new gasket on the fuel tank with the notch facing forward in the right corner of the fuel tank.

11. Carefully fold the strainer to allow it to fit through the opening in the tank. Be sure the strainer unfolds in the tank and lower the fuel sender assembly into position.

12. Install the fuel sender assembly attaching screws and tighten alternately and evenly to 45 inch lbs. (5 Nm).

13. Engage the fuel sender assembly electrical connector.

14. Connect all sender assembly fuel and vapor hoses.

15. Connect the fuel drain hose to the nipple on the rubber filler neck housing, then position the housing around the fuel tank filler neck.

16. Install the filler door bezel with the attaching screws.

17. Add fuel, tighten the filler cap and connect the negative battery cable.

18. Turn the ignition **ON** for 2 seconds, **OFF** for 10 seconds, then **ON** again and inspect the system for leaks.

1997–99 Models

1997–99 models have two fuel tanks (right and left). The fuel pump is part of the fuel sender assembly in the left fuel tank. The right fuel tank contains a fuel sender assembly with a siphon jet pump which supplies fuel to the left tank through the fuel sender feed pipe.

1. Disconnect the negative battery cable.

2. Properly relieve the fuel system pressure and drain the fuel tank.

3. Raise and safely support the vehicle.

4. Remove the rear wheel.

5. Clean the area around the fuel sender assembly.

6. Mark each fuel line to help identify them during installation.

7. Detach the fuel lines and electrical connector.

8. Support the fuel tank and remove the tank shield.

9. Remove the fuel sender mounting bolts, lift out the sender slightly and remove the float arm retaining clip and float arm.

10. Remove the fuel sender assembly and gasket.

To install:

RIGHT SIDE

➡️**Always install a new fuel strainer before installing the fuel sender assembly. A strainer that has been exposed to fuel will not unfold completely and may interfere with the full travel of the float arm.**

1. Install new fuel strainers on the siphon jet pump. The small strainer goes on the float side. Secure the strainers with a small tie wrap through the tab opening.

2. Install the new gasket on the fuel sender assembly using the new bolts to hold it in place.

❄❄ WARNING

Do not damage the float arm during installation.

3. Fold the long strainer in half on itself, then squeeze both strainers upward toward each other. Insert the fuel sender in the tank. It may be necessary to turn the sender assembly to ease installation. Look into the opening to ensure that the long strainer is about one in. from the opening, if not, rotate the sender assembly back-and-forth until the strainer is visible.

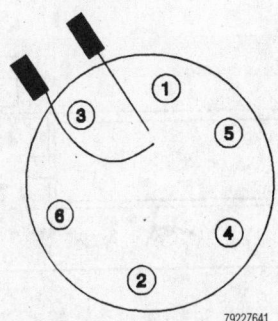

79222641

Tighten the fuel pump mounting bolts in the sequence shown—1997–99 5.7L (VIN G) engine

✲✲ WARNING

The mounting bolts break easily. Do not overtighten them.

4. Install the float arm and clip.
5. Using a torque wrench, tighten the mounting bolts to 27 inch lbs. (3 Nm).
6. Connect the fuel lines and attach the electrical connector.
7. Install the fuel tank shield. Tighten the bolts to in sequence19 ft. lbs. (25 Nm).
8. Install the wheel.
9. Lower the vehicle to the floor.
10. Refill the fuel tank.
11. Connect the negative battery cable.
12. Turn the ignition **ON** for 2 seconds, **OFF** for 10 seconds, then **ON** again and inspect the system for leaks.

LEFT SIDE

➡**Always install a new fuel strainer before installing the fuel pump assembly. A strainer that has been exposed to fuel will not unfold completely and may interfere with the full travel of the float arm.**

1. Install a new fuel strainer on the pump.
2. Install the new gasket on the fuel pump using the new bolts to hold it in place.

✲✲ WARNING

Do not damage the float arm during installation.

3. Fold the strainer three times so the strainer is about the same size as the opening in the tank. Insert the fuel pump in the tank. It may be necessary to turn the fuel pump to ease installation. Look into the opening to ensure that the long strainer is about one in. from the opening, if not, rotate the pump assembly back-and-forth until the strainer is visible.

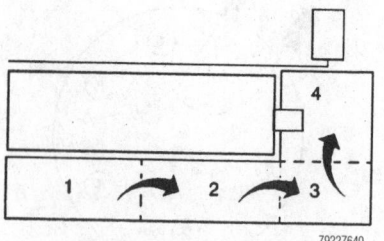

Fold the strainer on itself three times so it will fit into the opening in the fuel tank—1997–99 5.7L (VIN G) engine

---

4. Install the float arm and retaining clip.

✲✲ WARNING

The mounting bolt break easily. Do not overtighten them.

5. Using a torque wrench, tighten the mounting bolts in sequence to 27 inch lbs. (3 Nm).
6. Connect the fuel lines and attach the electrical connector.
7. Install the fuel tank shield. Tighten the bolts to 19 ft. lbs. (25 Nm).
8. Install the wheel.
9. Lower the vehicle to the floor.
10. Refill the fuel tank.
11. Connect the negative battery cable.
12. Turn the ignition **ON** for 2 seconds, **OFF** for 10 seconds, then **ON** again and inspect the system for leaks.

DRIVE TRAIN

Automatic Transmission Assembly

REMOVAL & INSTALLATION

1995–96 Models

The engine must be supported before removing the transmission assembly in order to prevent the vapor blow pipe located across the rear of the engine from contacting the dash panel.

1. Disconnect the negative battery cable and remove the transmission fluid level indicator.
2. Disconnect the TV cable at the throttle lever or the adjuster assembly.
3. Raise and support the vehicle safely.
4. If equipped, remove the upper and lower underbody braces.
5. Remove the complete exhaust system.
6. Support the transmission with a suitable jack.
7. Remove the driveline support beam.
8. Matchmark and remove the driveshaft.
9. Disengage the speedometer electrical connector, then disconnect the shift control cable and the remaining electrical leads from the transmission.
10. Remove the torque converter cover and mark the relationship of the converter to

---

the flywheel, then remove the converter-to-flywheel bolts.
11. Disconnect the oil cooler pipes at the transmission. Plug the openings to prevent system contamination or excessive fluid loss.
12. Disconnect the TV cable at the transmission.
13. Remove the transmission-to-engine mounting bolts and fasten the torque converter to the transmission using a converter restraining tool or a length of wire.
14. Carefully move the transmission rearward, downward and out from under the vehicle. If interference is encountered with cables, cooler lines, etc., remove the component(s) before finally lowering the transmission.

To install:

15. Flush the transmission oil cooler lines using J-35944 or an equivalent transmission cooler and line flushing tool.
16. Install a converter restraint tool to hold the torque converter in place.
17. Support the transmission with a suitable jack, then raise the transmission into position and remove the torque converter holding tool.
18. Install and tighten the transmission to engine bolts to 35 ft. lbs. (47 Nm).
19. Connect the TV cable to the transmission.
20. Remove the plugs, then connect the oil cooler pipes to the transmission.
21. Align the marks made during removal and start the torque converter to flywheel bolts by hand. Tighten the bolts to 46 ft. lbs. (62 Nm).
22. Install converter cover and tighten screws to 89 inch lbs. (10 Nm).
23. Engage the electrical connectors to the transmission.
24. Connect the shift control cable.
25. Engage the speedometer electrical connector.
26. Align the marks made earlier and install the driveshaft, then the driveline support beam.
27. Install the exhaust system, if equipped, the underbody braces.
28. Lower the vehicle and install the oil level indicator.
29. Connect the TV cable to the throttle lever or to the adjuster assembly.
30. Connect the negative battery cable.
31. Check and add the proper type and amount of transmission fluid.
32. Because the driveline support beam was removed, check clearance between the air intake duct and the throttle body. If the air duct becomes dislodged from the throttle body, a driveability problem could occur.

1997–99 Models

1. Disconnect the negative battery cable.
2. Shift the transmission into **Neutral**.
3. Raise and safely support the vehicle.
4. Remove the rear wheels.
5. Remove the intermediate exhaust pipe.
6. Tie the left muffler out of the way and remove the right muffler.
7. Remove the driveshaft tunnel undercover.
8. Remove the bell housing inspection plug using a flat bladed tool.

9. Matchmark the torque converter to the flexplate, them remove the attaching bolts.
10. Remove the two plug bolts from the front of the driveshaft support assembly. Install two M10 x 1.5 x 55mm or longer bolts into the bolt holes. Tighten the bolts to 26 ft. lbs. (35 Nm). These bolts must remain installed until instructed to remove them in order to maintain the position of the input shaft bearing.
11. Rotate the flywheel to gain access to the driveshaft clamp bolt, then loosen it.
12. Shift the transmission into **PARK**, then remove the shifter cable bracket at the

transmission. Detach the cable from the transmission linkage.
13. Remove the Electronic Brake Traction Control Module (EBTCM) and position it out of the way.
14. Remove the rear transverse spring.
15. Support the control arm with a jack. Remove the lower shock absorber mounting from the control arm and the tie rod end and ball joint from the knuckle. Remove the jack. Do this for both control arms.
16. Support the transmission securely with a transmission jack.
17. Remove the transmission-to-differential lower nut.
18. Remove the rear suspension crossmember. Be sure to remove the nuts by hand.
19. Remove the transaxle mount and bracket.
20. Remove the halfshafts from the differential and secure them out of the way.
21. Remove the wiring harness from the driveshaft support. Lower the assembly slightly to gain access to the connectors and detach them.
22. Remove the wiring harness from the differential housing.

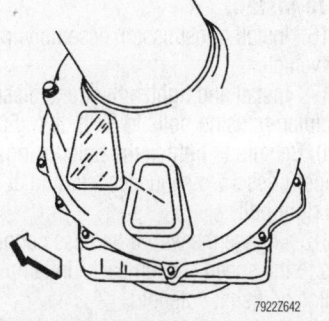

Remove the inspection plug, then remove the flexplate-to-torque converter bolts—1997–99 A/T models

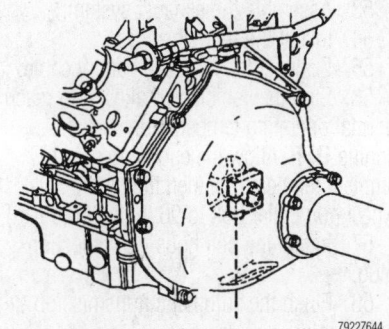

Rotate the flywheel to gain access to the clamp bolt, then loosen it

❉❉ WARNING

Do not lower the top rear portion of the differential past the bottom of the storage compartment or the PCV pipe will hit the dash panel possibly causing damage.

23. Lower the transmission assembly enough to remove the wiring harness from the top of the assembly.
24. Disconnect the transmission fluid cooler lines at the junction near the bell housing.
25. Support the engine with a block of wood and a jack.
26. Remove the driveshaft support-to-flexplate housing bolts. It may be necessary to bend the wiring harness bracket toward the tunnel wall for greater clearance when removing the driveshaft assembly.
27. Pry the front driveshaft support out of the flexplate housing while an assistant moves the assembly rearward.
28. Carefully lower the assembly away from the vehicle while simultaneously adjusting the angle.
29. Attach a chain hoist to the assembly, then remove it from the jack and place it on a workbench.
30. Remove the driveshaft-to-transmission mounting bolts. Carefully pry the driveshaft assembly away from the transmission.

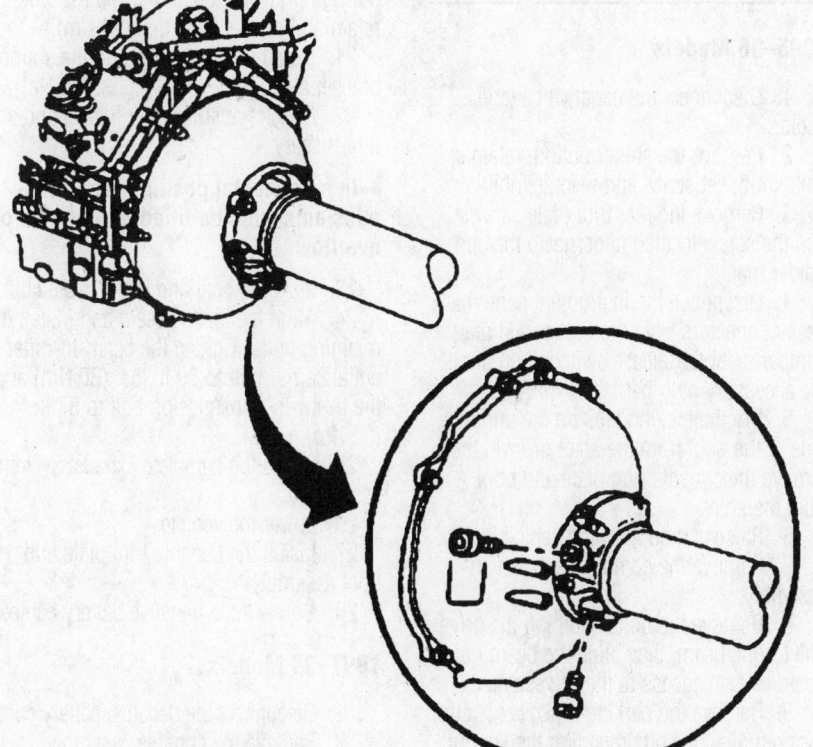

Remove the plugs and install two M10 x 1.5 x 55mm or longer bolts into the bolt holes to secure the bearing—1997–99 A/T models

31. Remove the transmission-to-differential mounting bolts and separate the differential from the transmission. Be careful not to damage the output shaft seal on the differential plate.

To install:

32. If removed, install the differential plate on the transmission. Be sure to position the square-lip seal flush with the transmission case.

33. Install the differential on the transmission. Tighten the bolts and nuts to 37 ft. lbs. (50 Nm).

34. Install the driveshaft support assembly on the transmission. Tighten the fasteners to 37 ft. lbs. (50 Nm).

35. Using a chain hoist, place the assembly on a transmission jack.

36. Carefully raise the assembly into the vehicle while placing the wiring harness loosely into the harness retaining slots.

37. Align the assembly for installation into the engine. The driveshaft will slide into the rear of the engine as long as the angles are the same. Install the driveshaft assembly in the engine. Reposition the wiring harness bracket to align with the appropriate hole in the driveshaft assembly. Tighten the bolts to 37 ft. lbs. (50 Nm).

38. Install the wiring harness on the top of the driveshaft assembly.

39. Connect the fluid cooling lines. Tighten the fittings to 20 ft. lbs. (27 Nm).

40. Install the wiring harness to the left side of the transmission. Reattach the connectors.

41. Raise the transmission to installed height and remove the jack from the engine.

42. Install the suspension crossmember. Tighten the nuts to 81 ft. lbs. (110 Nm).

43. Install the transmission mount to the crossmember. Tighten the nuts to 37 ft. lbs. (50 Nm).

44. Install the transmission-to-differential lower nut. Tighten the nut to 37 ft. lbs. (50 Nm).

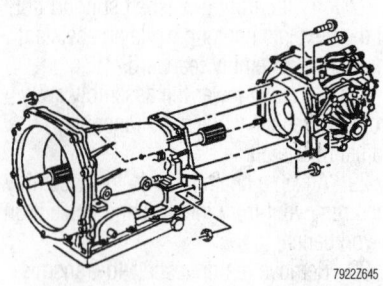

Transmission-to-differential mounting bolt locations

45. Assembly the suspension.

46. Install the rear spring.

47. Install the EBTCM in the proper position. Tighten the bracket mounting bolts to 37 ft. lbs. (50 Nm).

48. Install the shifter cable and bracket.

49. Align the matchmarks and install the flexplate-to-torque converter bolts. Tighten the bolts to 47 ft. lbs. (63 Nm). Install the inspection plug.

50. Hand-tighten the driveshaft hub clamp bolt.

51. Remove the two M10 x 1.5 x 55mm long bolts and install the plugs. Tighten the plugs to 37 ft. lbs. (50 Nm).

52. Install the tunnel undercover.

53. Assemble the exhaust system.

54. Install the rear wheels.

55. Connect the negative battery cable.

56. Start the vehicle and allow it to reach normal operating temperature. Turn the engine **OFF**. Allow the engine to cool to ambient temperature, then tighten the flywheel hub collar bolt to 96 ft. lbs. (130 Nm).

57. Install the bell housing inspection plug.

58. Flush the automatic transmission fluid cooler.

Manual Transmission Assembly

REMOVAL & INSTALLATION

1995–96 Models

1. Disconnect the negative battery cable.

2. Remove the shifter button, retainer, shift knob, set screw and reverse inhibitor.

3. Remove the rear trim plate screws and the screw located underneath the cup holder mat.

4. Disengage the instrument panel harness connectors from the lighter and rear compartment lid release switch, then unclip the accessory plug harness.

5. Pry the locking tabs on the underside of the boot from the shaft groove, then remove the console trim plate and boot from the shaft.

6. Raise and support the vehicle safely.

7. Remove the complete exhaust assembly.

8. Remove the bolts retaining the driveline torque beam, then slide the beam outboard to gain access to the driveshaft.

9. Remove the parking brake cable clip, then remove the bolts retaining the support bracket.

10. To maintain drive train balance, matchmark the relationship between the

driveshaft and the differential carrier yoke, then remove the bolts attaching the driveshaft to the yoke.

11. Slide the driveline torque beam rearward until it make contact with the rear exhaust hanger.

12. Support the transmission using an adjustable transmission jack.

13. Disengage the electrical connectors from the speed sensor, back-up lamp switch and the computer aided shift solenoid.

14. Remove the transmission to clutch housing attaching bolts.

15. Carefully lower the transmission and remove the transmission assembly from the vehicle.

To install:

16. Install transmission assembly into the vehicle.

17. Install and tighten the transmission to clutch housing bolts to 37 ft. lbs. (50 Nm). Be sure to tighten the bolts using the proper crisscross sequence, starting at the top right bolt.

18. Engage the wiring harness connectors to the speed sensor, back-up lamp switch and shift solenoid.

19. Slide the driveline torque beam forward and onto the transmission extension housing.

20. Install the driveshaft, aligning the matchmarks made on the shaft and yoke during removal. Tighten the shaft-to-yoke retaining bolts to 18 ft. lbs. (24 Nm).

21. Install the bolts retaining the support bracket and tighten to 18 ft. lbs. (25 Nm).

22. Check transmission oil level and add if necessary.

➡**In a horizontal position, the transmission should be filled to the point of overflow.**

23. Install the parking brake cable clip.

24. Align the torque beam and install the retaining bolts. Tighten the beam-to-differential carrier bolt to 60 ft. lbs. (80 Nm) and the beam-to-transmission bolt to 37 ft. lbs. (50 Nm).

25. Install the complete exhaust system assembly.

26. Lower the vehicle.

27. Install the console trim plate and boot assembly.

28. Connect the negative battery cable.

1997–99 Models

1. Disconnect the negative battery cable.

2. Remove the console assembly.

3. Pry out the shift control knob button.

4. Pry out the shift control knob retainer.

5. Unscrew and remove the knob.

6. Remove the boot by grabbing both sides and pulling it in toward the lever.

7. Remove the instrument panel accessory trim plate.

8. Remove the shift control closeout boot, then remove the shift control assembly.

9. Remove the left lower instrument panel trim.

10. Remove the clutch master cylinder pushrod retainer and the pushrod from the pedal.

11. Raise and safely support the vehicle.

12. Detach the clutch hydraulic line from the rear of the engine, then disconnect the hose.

13. Remove the rear wheels.

14. Remove the intermediate pipe and secure the mufflers out of the way.

15. Remove the driveshaft tunnel undercover.

16. Remove the Electronic Brake Traction Control Module (EBTCM) and position it out of the way.

17. Remove the rear spring.

18. Disconnect the lower shock absorber, tie rod end and lower ball joint.

19. Remove the wiring harness and brake lines from the suspension crossmember.

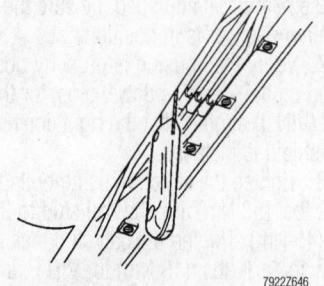

Insert a flat bladed tool between the shifter bracket and brake liner retainer before lowering the transmission assembly—5.7L (VIN G) engine

20. Place a transmission jack under the transmission.

21. Remove the transmission-to-differential lower nut.

22. Remove the transmission mount from the crossmember.

23. Place a jack under the suspension crossmember, remove the mounting nuts and the crossmember.

24. Remove the mount from the differential.

25. Remove the halfshafts from the differential and position them out of the way.

26. Remove the wiring harness from the driveshaft support assembly. Lower the assembly to gain access to the connectors.

27. Remove the harness clip from the top of the differential.

28. Detach the electrical connectors from the transmission.

29. Insert a putty knife or other flat bladed tool between the edge of the shifter bracket and the brake line retainer on the side of the driveshaft tunnel.

✳✳ WARNING

Do not lower the top rear portion of the differential past the bottom of the storage compartment or the PCV pipe will hit the dash panel possibly causing damage.

30. Lower the transmission assembly enough to remove the wiring harness from the top of the assembly.

31. Support the engine with a block of wood and a jack.

32. Remove the driveshaft support-to-bell housing bolts. It may be necessary to bend the wiring harness bracket toward the tunnel wall for greater clearance when removing the driveshaft assembly.

33. Pry the front driveshaft support out of the bell housing while an assistant moves the assembly rearward.

34. Carefully lower the assembly away from the vehicle while simultaneously adjusting the angle.

35. Attach a chain hoist to the assembly, then remove it from the jack and place it on a workbench.

36. Remove the driveshaft-to-transmission mounting bolts. Carefully pry the driveshaft assembly away from the transmission while guiding the shift rod through the opening in the driveshaft support.

37. Remove the roll pin to detach the shift rod from the transmission.

38. Remove the transmission-to-differential mounting bolts and separate the differential from the transmission.

To install:

39. Install the differential on the transmission. Tighten the bolts to 37 ft. lbs. (50 Nm).

40. Install the driveshaft support to the transmission. Tighten the bolts to 37 ft. lbs. (50 Nm).

41. To help connect the shift rod to the shift control assembly, place a rubber band on the shift rod just behind the clamp, then tape the rod to the driveshaft support with masking tape.

42. Place the assembly on the transmission jack with the chain hoist.

43. Begin to raise the assembly into the vehicle while loosely installing the wiring harness along the driveshaft assembly retaining slots.

44. Have an assistant guide the front of the driveshaft to the bell housing.

45. Insert a putty knife or other flat bladed tool between the edge of the shifter bracket and the brake line retainer on the side of the driveshaft tunnel.

46. Be sure the driveshaft assembly is at the same angle as the engine before trying to install it. Carefully push the driveshaft splines into the clutch disc. Turn the shaft with a screwdriver if needed to align the splines. Tighten the bolts to 37 ft. lbs. (50 Nm).

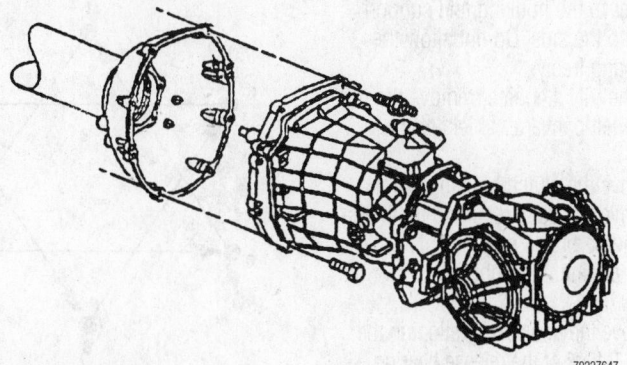

Driveshaft support-to-transmission mounting—5.7L (VIN G) engine

Place a rubber band on the shift rod, then tape the rod to the driveshaft support assembly—5.7L (VIN G) engine

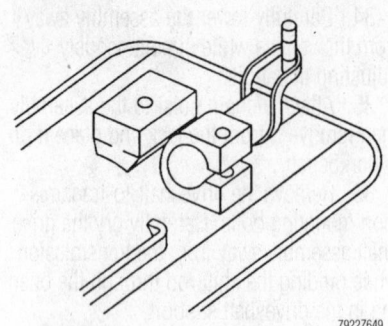

Pull the shift rod up to break the tape and hook the rubber band on the rear stud— 5.7L (VIN G) engine

47. Reposition the wiring harness bracket to align with the appropriate hole in the driveshaft assembly.

48. Install the wiring harness to the retainer on the top of the transmission and attach the connectors.

49. Raise the transmission to the installed position. Remove the putty knife and the jack from under the engine.

50. Install the halfshafts in the differential.

51. If removed, install the transmission mount on differential. Tighten the bolts to 37 ft. lbs. (50 Nm).

52. Install the suspension crossmember. Tighten the nuts to 81 ft. lbs. (110 Nm) and remove the jacks. Tighten the transmission mount nuts and the transmission-to-differential lower nut to 37 ft. lbs. (50 Nm).

53. Install the wiring harness and brake line retainers to the suspension crossmember.

54. Assemble the suspension and install the spring.

55. Install the EBTCM in the correct position. Tighten the mounting bracket bolts to 37 ft. lbs. (50 Nm).

56. Ensure that the wiring harness is pulled into the retaining slots along the driveshaft support assembly.

57. Connect the clutch hydraulic line and secure it in the retainer.

58. Install the driveshaft tunnel undercover.

59. Assemble the exhaust system.

60. Install the rear wheels and lower the vehicle.

61. Connect the master cylinder pushrod to the pedal and install the instrument panel trim panel.

62. Pull up the shift rod to break the tape and hook the rubber band on the rear stud on the top of the driveshaft tunnel.

63. Install the shift control assembly. Break the rubber band to remove it.

64. Install the closeout boot. Tighten the nuts to 106 inch lbs. (12 Nm).

65. Install the instrument panel accessory trim plate.

66. Install the boot, knob and retainer and button on the shift lever.

67. Install the console.

68. Connect the negative battery cable.

69. Bleed the clutch system.

Clutch

ADJUSTMENT

1995–96 Models

The hydraulic clutch system on these models requires no periodic adjustment.

1997–99 Models

1. Raise and safely support the vehicle.
2. Remove the flywheel inspection cover.
3. Have an assistant depress the clutch pedal until the tension is released from the stepped adjusting ring.
4. Using two screwdrivers, rotate the stepped adjusting ring counterclockwise until fully adjusted out.
5. Continue to hold them in this position and have the assistant release the clutch pedal.
6. Remove the screwdrivers.
7. Install the inspection cover and lower the vehicle.

REMOVAL & INSTALLATION

1995–96 Models

1. Disconnect the negative battery cable, then raise and support the vehicle safely.
2. Remove the complete exhaust system.
3. Remove the transmission assembly.
4. Except for the VIN J engine, disconnect the ground wire attached to the left clutch housing stud.
5. Remove the nuts attaching the clutch slave cylinder to the housing and support the cylinder to the side. Do not allow the cylinder to hang freely.
6. For the VIN J engine, remove the nut retaining the left converter shield to the housing.
7. Remove the clutch housing cover.
8. Remove the bolts retaining the housing to the engine block, if applicable on the 5.7L (VIN J) engine, the right side converter heat shield.
9. Remove the housing by aligning the fork onto the 2 flats of the release bearing and push the fork away from the bearing with a twisting motion. Remove the clutch housing, for 5.7L (VIN P and 5) engine with magnesium housings, the aluminum spacers.

➡**Excessive clutch wear may require removal of the ball stud locking screw and loosening of the ball stud to disengage the fork and housing.**

10. Mark the alignment of the clutch cover and flywheel for installation purposes.

11. Loosen the clutch cover bolts evenly, 1 turn at a time until spring pressure is released. Failure to properly release spring pressure may result in damage to the clutch cover assembly and the flywheel.

12. Remove the clutch plate and disc assembly.

To install:

13. Inspect flywheel, clutch plate and disc for heat stress, cracks or worn parts and replace as necessary.

14. Install the clutch assembly using a suitable universal clutch disc alignment tool.

15. Be sure the marks made earlier are in alignment, then install the cover assembly-to-flywheel bolts. Tighten the bolts in the proper sequence, 1 turn at a time, until spring pressure is properly attained and the bolts are tightened to 30 ft. lbs. (41 Nm).

16. Position the clutch housing to the engine block and engage the fork onto the release bearing. If equipped, be sure the aluminum spacer is in position.

17. Verify the housing is properly positioned on the 2 engine dowel pins, for the 5.7L (VIN J) engine, that the right converter heat shield is installed.

18. Tighten the clutch housing bolts to 37 ft. lbs. (50 Nm) and the ball stud to 33 ft. lbs. (45 Nm). Tighten the ball stud locking screw to 11 ft. lbs. (15 Nm) for VIN P and 5 engines or to 16 ft. lbs. (22 Nm) for VIN J engine, as applicable.

19. If equipped, install the ground harness connection to the housing.

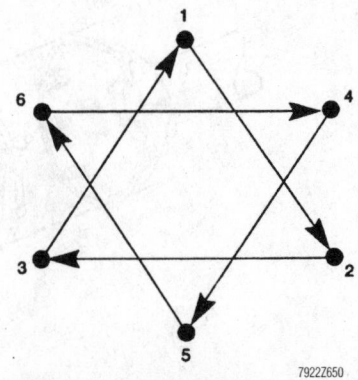

Clutch cover assembly torque sequence

20. Install the housing cover and tighten the bolts to 80 inch lbs. (9 Nm).

21. For the VIN J engine, install the left heat shield and tighten the retaining nut to 12 inch lbs. (1.4 Nm).

22. Install the clutch slave cylinder and tighten the retaining nuts to 19 ft. lbs. (25 Nm).

23. Install the transmission assembly.

24. Install the exhaust system and lower the vehicle.

25. Connect the battery negative cable and check clutch for proper operation.

1997–99 Models

1. Disconnect the negative battery cable.

2. Raise and safely support the vehicle.

3. Remove the exhaust system.

4. Remove the driveshaft support and transmission as an assembly.

5. Remove the inspection cover.

6. Remove the pressure plate bolts. It will be necessary to rotate the flywheel to access all the bolts.

7. Remove the pressure plate and disc.

To install:

8. Adjust the pressure plate using the following sub-steps:

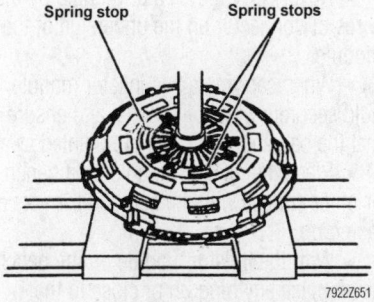

There are three stepped adjusting ring tension spring stops on the pressure plate assembly

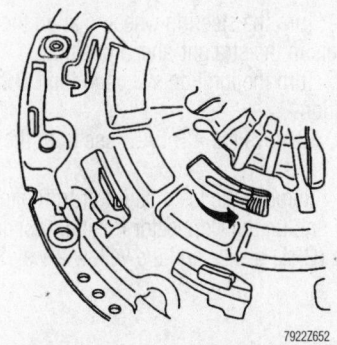

Rotate the stepped adjusting ring counter-clockwise to compress the springs

a. Place the pressure plate assembly on a press.

b. Compress the diaphragm fingers until the tension is released from the stepped adjusting ring.

c. Using two screwdrivers, rotate the stepped adjusting ring counterclockwise until fully adjusted out.

d. Continue to hold them in this position and releasing the press.

e. Remove the screwdrivers.

9. Position the clutch disc and pressure plate on the flywheel.

10. Install the clutch disc alignment tool through the disc to keep it in place.

11. Install the pressure plate bolts finger-tight. Turn the flywheel to access all the bolt holes.

12. Tighten the bolts evenly using three steps to 37 ft. lbs. (50 Nm).

13. Remove the alignment tool and install the inspection cover.

14. Install the driveshaft support and transmission assembly.

15. Install the exhaust system.

16. Connect the negative battery cable.

17. Bleed the clutch system.

Hydraulic Clutch System

BLEEDING

1995–96 Models

1. Disconnect the negative battery cable and remove the ECM from the mounting bracket to access the master cylinder for filling. Fill the master cylinder reservoir with the proper grade and type of fresh brake fluid or hydraulic clutch fluid.

2. Prior to bleeding the actuator, most of the air can be removed as follows:

a. Remove the master cylinder cap and moisture barrier.

b. Install the master cylinder cover.

c. Lightly stroke the clutch pedal to release trapped air through the master cylinder.

d. Remove the master cylinder cap and install the moisture barrier.

e. Install the master cylinder cap.

3. Raise and support the vehicle safely.

4. Remove the actuator cylinder attaching stud nuts.

5. Remove the pushrod and actuator cylinder from the clutch housing and the hydraulic line from the retaining clip.

6. Lower cylinder slightly for access and disconnect the hydraulic hose fitting from the actuator cylinder.

7. Remove the bleed screw dust cap.

8. Position a drain pan or attach a clear plastic hose.

9. Support the slave cylinder in a horizontal position, with the bleeder screw vertical.

10. Fully depress the clutch pedal and open the bleeder screw.

11. Close the bleed screw and release the clutch pedal.

12. Repeat Steps 11 and 12 until all the air is expelled from the system. Check the fluid reservoir and replenish, as required during the procedure. Be sure the reservoir is kept sufficiently full to prevent air from being drawn into the system.

13. Tighten the bleeder screw and install the dust cover.

14. Install the hydraulic line into the retaining clip, position the actuator cylinder and tighten the stud nuts to 19 ft. lbs. (25 Nm).

15. Lower the vehicle.

16. Install the ECM and connect the negative battery cable.

1997–99 Models

1. Fill the clutch master cylinder with clutch hydraulic fluid PN12345347 or equivalent.

2. Raise and safely support the vehicle with an assistant in it.

3. Remove the intermediate exhaust pipe.

4. Remove the driveshaft tunnel undercover.

5. Have the assistant depress and hold the clutch pedal down.

6. Loosen the bleeder screw on the actuator cylinder to release the air, then tighten the screw. Do not allow the clutch pedal to be released until the bleeder screw is closed or air will be drawn into the system.

7. Repeat steps 5 and 6 until the air has been purged from the system. Check the master cylinder fluid level often and refill as needed during the procedure.

8. Install the driveshaft tunnel undercover.

9. Install the intermediate exhaust pipe.

10. Lower the vehicle to the floor.

Halfshaft

REMOVAL & INSTALLATION

1995–96 Models

1. Raise and safely support the vehicle.

2. Remove the rear transverse spring from the knuckle.

3. Remove the tie rod end from the knuckle.

4. Remove the spindle rod mounting bracket from the differential.

5. Matchmark the U-joint to the yokes, then remove the halfshaft universal joint retainers at both ends of the shaft.

6. Push the wheel assembly outward while supporting the halfshaft, then remove the shaft.

To install:

7. Hold the wheel assembly outward and place the shaft in position. Be sure to align the marks made earlier. Tighten the retaining bolts to 26 ft. lbs. (35 Nm).

8. Install the spindle rod bracket on the differential carrier. Tighten the bolts to 59 ft. lbs. (80 Nm).

9. Connect the tie rod end to the knuckle. Tighten the nut to 33 ft. lbs. (45 Nm).

10. Install the spring to the knuckle. Tighten the nut until the slot in the nut aligns with the hole in the bolt, then insert the cotter pin.

11. Lower the vehicle to the floor.

1997–99 Models

1. Shift the transmission into PARK or NEUTRAL for manual transmissions.

2. Apply the parking brake and raise the vehicle.

3. Remove the rear wheel.

4. Insert a large drift through the brake rotor cooling fins to prevent the hub assembly from turning, then remove the axle nut.

5. Release the parking brake.

6. Remove the rear leaf spring.

7. Disconnect the outer tie rod end from the knuckle and position it to the rear.

8. Disconnect the ABS wheel speed sensor.

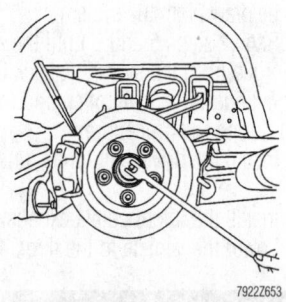

Insert a large drift through the cooling fins to keep the hub assembly from turning while removing the retaining nut—1997–99 models

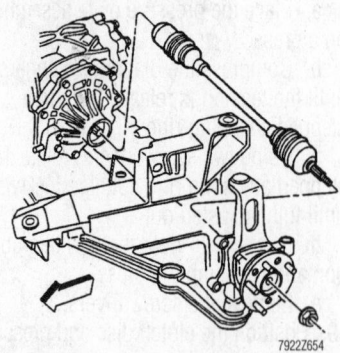

Exploded view of the halfshaft mounting—1997–99 models

9. Disconnect the parking brake cable from the lever and bracket.

➡Be sure to support the halfshaft until it is removed. Don't let it hang by the Constant Velocity (CV) joint.

10. Attach a puller such as J 42129 on the wheel studs and start to push the axle shaft into the hub assembly. This will provide clearance for the ball joint to be separated from the knuckle assembly. Disconnect the ball joint from the knuckle, then continue to disengage the axle shaft from the hub assembly.

11. Support the knuckle and upper control arm while positioning it toward the front.

12. Disengage the halfshaft assembly from the differential by inserting a suitable tool between the CV joint and differential and prying them apart.

To install:

13. Install the halfshaft on the differential output shaft. Use light force to be sure it is fully seated.

14. Begin to install the halfshaft through the hub assembly but do not install completely. This will provide clearance for installing the ball joint.

15. Install the ball joint to the knuckle and insert the halfshaft through the hub assembly completely.

16. Install the parking brake cable and connect the ABS wheel speed sensor.

17. Connect the tie rod end to the knuckle.

18. Install the rear spring.

19. Install the halfshaft retaining nut. Tighten the nut to 118 ft. lbs. (160 Nm).

20. Install the rear wheel and lower the vehicle to the floor.

STEERING AND SUSPENSION

Air Bag

✳✳ CAUTION

All vehicles are equipped with an air bag system, also known as the Supplemental Inflatable Restraint (SIR) or Supplemental Restraint System (SRS). The system must be disabled before performing service on or around system components, steering column, instrument panel components, wiring and sensors. Failure to follow safety and disabling procedures could result in accidental air bag deployment, possible personal injury and unnecessary system repairs.

PRECAUTIONS

Several precautions must be observed when handling the inflator module to avoid accidental deployment and possible personal injury.

• Never carry the inflator module by the wires or connector on the underside of the module.

• When carrying a live inflator module, hold securely with both hands, and ensure that the bag and trim cover are pointed away.

• Place the inflator module on a bench or other surface with the bag and trim cover facing up.

• With the inflator module on the bench, never place anything on or close to the module which may be thrown in the event of an accidental deployment.

DISARMING

1. Turn the steering wheel to align the wheels in the straight-ahead position.

2. Turn the ignition switch to the **LOCK** position.

3. Remove the AIR BAG fuse from the fuse block.

4. Remove the left side lower trim panel, then unplug the Connector Position Assurance (CPA) device and the yellow 2-way SIR

harness wire connector at the base of the steering column.

After the necessary repairs have been made, re-enable the air bag system as follows:

5. Turn the ignition switch to the **LOCK** position.

6. Engage the yellow 2-way connector and the CPA device at the base of the steering column.

7. Install the left side lower trim panel, then install the SIR fuse to the fuse block.

8. Turn the ignition switch to the RUN position.

9. Verify the SIR indicator light flashes 7–9 times, then turns OFF. If not, inspect system for malfunction.

Power Rack and Pinion Steering Gear

REMOVAL & INSTALLATION

1995–96 Models

1. Disconnect the negative battery cable and position a drain pan under the vehicle to catch fluid.

2. For VIN P engine, remove the air intake duct, then remove the serpentine drive belt and the drive belt idler pulley.

3. Remove the power steering gear inlet hose assembly from the steering gear.

4. Remove the power steering gear outlet hose assembly from the steering gear.

➡**If equipped with a power steering fluid cooling pipe, disconnect fluid cooling pipe outlet hose from the fluid cooling pipe.**

5. Remove the steering gear coupling shield.

6. Disconnect the intermediate shaft from the power steering gear and lower steering shaft, then position aside.

7. Raise and support the vehicle safely.

8. Remove the front tire and wheel assemblies.

9. Remove both outer tie rods from the knuckles using a suitable puller.

10. If equipped, remove the power steering cooler assembly.

11. Remove the stabilizer shaft.

12. Remove the steering gear to frame attaching clamp nuts, then remove the bolts and clamp from the vehicle.

13. Remove the power steering gear attaching nut and bolt.

14. Remove the power steering gear from the vehicle.

15. If necessary, remove the outer tie rods, rack and pinion boots, and the inner tie rods from the power steering gear.

To install:

16. If removed, install the inner tie rods, boots and outer tie rods.

17. Install the power steering gear, nuts and bolts. Tighten the attaching nut to 30 ft. lbs. (40 Nm). Tighten the steering gear clamp nuts to 18 ft. lbs. (25 Nm).

18. Install the stabilizer shaft, if applicable, the power steering cooler assembly.

19. Install both outer tie rods to the steering knuckle.

20. Install tire and wheel assemblies, then lower the vehicle.

21. Install the intermediate shaft and the steering gear coupling shield.

22. Install the power steering gear outlet and inlet hose assemblies to the power steering gear. Tighten fittings to 21 ft. lbs. (28 Nm).

23. For VIN P engine, install the drive belt idler pulley and the serpentine drive belt, then install the air intake duct.

24. Remove the drain and fill the power steering reservoir.

25. Connect the negative battery cable, bleed the system and check for proper operation.

1997–99 Models

1. Disconnect the negative battery cable.

2. Drain the power steering fluid into a suitable container.

3. Remove the intermediate shaft cover.

4. Disconnect the inlet hose from the steering gear.

5. Remove the steering fluid cooler tube from the steering gear.

6. Remove the lower steering coupling from the steering gear input shaft.

7. Raise and safely support the vehicle.

8. Remove the front wheels and disconnect the tie rod ends from the knuckles.

9. Detach the Magnasteer electrical connector.

10. Remove the stabilizer bar.

11. Remove the steering gear mounting bolt and the steering gear through the left wheel opening.

To install:

12. Install the steering gear to the crossmember. Tighten the nuts to 74 ft. lbs. (100 Nm).

13. Install the stabilizer bar.

14. Connect the tie rods to the knuckles. Tighten the nuts to 33 ft. lbs. (45 Nm).

15. Connect the lower steering coupling to the steering gear input shaft.

16. Attach the Magnasteer electrical connector.

17. Install the fluid cooler on the steering gear. Tighten the fitting to 20 ft. lbs. (27 Nm).

18. Connect the inlet hose to the steering gear. Hold the hose against the steering gear while tightening the fitting to 20 ft. lbs. (27 Nm).

19. Install the intermediate shaft cover.

20. Install the front wheels.

21. Lower the vehicle to the floor.

22. Connect the negative battery cable.

23. Refill and bleed the power steering system.

24. Check the toe angle of the front wheels and adjust them if necessary.

Shock Absorber

REMOVAL & INSTALLATION

Front

WITHOUT SELECTIVE RIDE CONTROL

1. Raise and support the vehicle safely.
2. Remove the tire and wheel assemblies.

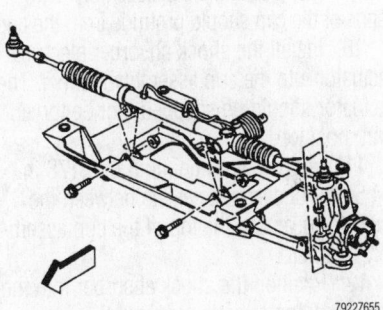

Exploded view of the power steering gear mounting—1997–99 models

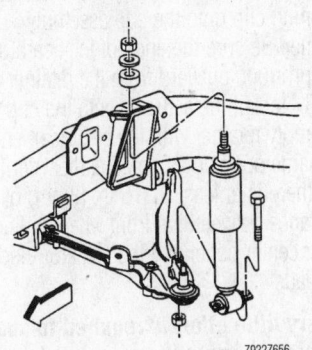

Exploded view of the front shock absorber mounting—1997–99 models

3. Disconnect the shock absorber from the lower control arm and the shock tower. If necessary, remove the front wheelhousing lower center panel to access the upper mount nut.

4. Remove the insulator and retainers from the shock absorber and the shock absorber from the vehicle.

5. Installation is the reverse of the removal procedure. Tighten the upper and lower mount nuts to 19 ft. lbs. (26 Nm).

WITH SELECTIVE RIDE CONTROL

1. Disconnect the negative battery cable.

2. Raise and safely support vehicle, then remove the tire and wheel assemblies.

3. Safely support the lower control arm with a jackstand.

4. Remove the actuator retaining clip, then remove the actuator from the cup retainer. Note the position of the actuator electrical leads for installation purposes.

5. Remove the shock absorber upper mounting nuts.

6. Remove the cup retainer, then the upper insulator retainer and insulator.

7. Remove the shock absorber lower mounting bolts, nuts, then compress the shock absorber and remove it from the vehicle. If necessary, remove the lower insulator from the shock.

To install:

8. If removed, install the lower insulator to the shock absorber, compress the shock and install into the vehicle.

9. Install the shock absorber lower mounting nuts and bolts, then tighten the bolts to 19 ft. lbs. (26 Nm).

10. Install the upper insulator and retainer, then install the cup assembly retainer.

11. Install the upper mounting nut and tighten the 31 ft. lbs. (42 Nm). The selector gear should be at least 0.178 in. (4.5mm) above the top of the cup assembly retainer.

12. Install and properly seat the actuator retaining clip onto the cup assembly retainer. Be sure the ends of the actuator clip protrude outward from the retainer.

13. Install the actuator onto the cup assembly retainer with the electrical leads in the same position as noted earlier. Verify that there is at least 0.315 in. (8mm) of clearance between the front wheelhousing lower center panel and the actuator electrical leads.

➡**Very little effort is required to snap the actuator onto the retainer, do not force it into position.**

14. Remove the jackstand, then install the tire and wheel assembly.

15. Lower the vehicle and connect the negative battery cable.

Rear

WITHOUT SELECTIVE RIDE CONTROL

1. Raise and support the vehicle safely. Support the knuckle with a jackstand.

2. Remove the shock absorber lower mounting nut and washer.

3. Remove the shock absorber upper bracket mounting bolt.

4. Disconnect the shock absorber from the lower mounting stud.

5. If necessary, remove the shock absorber upper bracket retaining nut and remove the bracket assembly.

6. Installation is the reverse of the removal procedure.

7. Tighten the upper bracket retaining nut, if removed, to 19 ft. lbs. (26 Nm). With the suspension at proper trim height, tighten the upper bracket mounting bolts to 22 ft. lbs. (30 Nm) and the lower mounting nut to 61 ft. lbs. (83 Nm).

WITH SELECTIVE RIDE CONTROL

1. Disconnect the negative battery cable.

2. Raise and support the vehicle safely.

3. Support the rear knuckle with a jackstand.

4. Disconnect the shock absorber lower mounting nut and washer.

5. Remove the shock absorber upper bracket mounting bolt.

6. Disconnect the shock absorber from the mounting stud and support. Do not allow the shock to hang from the actuator harness.

7. Remove the actuator retaining clip and remove the actuator from the shock.

8. Remove the shock absorber from vehicle.

To install:

9. Install and properly seat the actuator retaining clip onto the cup assembly. The ends of the clip should protrude from the cup.

10. Install the shock absorber electrical actuator into the cup assembly retainer. The actuator should be snapped, not be forced into position.

11. Verify that a minimum of 0.178 in. (4.5mm) of clearance exists between the selector gear and the top of the cup assembly retainer.

12. Position the shock absorber into the frame and onto the lower mounting stud.

13. Install the shock absorber upper bracket mounting bolt.

14. With the suspension held at the proper trim height. Tighten the upper

bracket mounting bolts to 22 ft. lbs. (30 Nm) and the lower mounting nut to 61 ft. lbs. (83 Nm).

15. Lower the vehicle to the floor.

16. Connect the negative battery cable.

REMOVAL & INSTALLATION

1995–96 Models

1. Raise and support the vehicle safely.

2. Safely support the lower control arm with a jackstand.

3. Remove the tire and wheel assemblies.

4. Using J-33436 or equivalent ball joint removal tool, separate the ball joint from the knuckle.

5. Remove the upper ball joint from the control arm as follows:

   a. Center punch the rivet.

   b. Drill a pilot hole, then drill the rivet head.

   c. Punch out the rivet.

To install:

6. Install a new ball joint into the upper control arm and position so the cotter pin can be installed from the rear to the front of the vehicle.

7. Install and tighten the mounting nuts to 13 ft. lbs. (18 Nm).

8. Position the ball stud into the steering knuckle, then install the upper ball joint stud washer and nut. Tighten the upper control arm ball stud nut to 33 ft. lbs. (45 Nm). Tighten the nut additionally as necessary to insert the cotter pin but do not exceed 63 ft. lbs. (85 Nm).

9. Install a new cotter pin from the rear to the front of the vehicle.

10. Remove the jackstand and lubricate the ball joint.

11. Install the tire and wheel assembly, then lower the vehicle.

1997–99 Models

The upper ball joint is a part of the steering knuckle assembly and must be replaced as a complete unit. Refer to the steering knuckle procedure.

Steering Knuckle

REMOVAL & INSTALLATION

1. Raise and safely support the vehicle.
2. Remove the front wheel.
3. Remove the brake caliper and rotor.

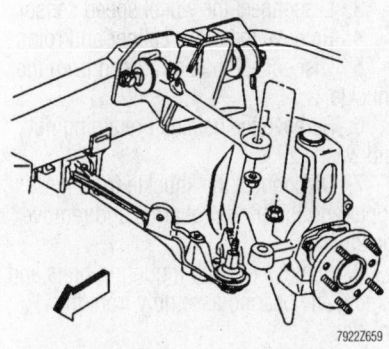

Exploded view of the steering knuckle removal—1997–99 models

4. Detach the wheel speed sensor electrical connector.

5. Support the lower control arm using a jack.

6. Disconnect the tie rod end from the steering knuckle using a ball joint separator.

7. Separate the upper ball joint stud from the upper control arm with a ball joint separator.

8. Separate the lower ball joint stud from the steering knuckle.

9. Remove the knuckle from the vehicle.

To install:

10. Install the knuckle on the lower control arm ball joint stud. Tighten the nut to 55 ft. lbs. (75 Nm).

11. Install the upper ball joint stud in the upper control arm. Tighten the nut to 44 ft. lbs. (60 Nm).

12. Remove the jack from the lower control arm.

13. Connect the tie rod end to the steering knuckle. Tighten the nut to 33 ft. lbs. (45 Nm).

14. Attach the wheel speed sensor electrical connector.

15. Install the brake rotor and caliper.

16. Install the front wheel and lower the vehicle to the floor.

17. Check and adjust the wheel alignment if necessary.

Lower Ball Joint

REMOVAL & INSTALLATION

1995–96 Models

1. Raise and support the vehicle safely.

2. Safely support the lower control arm with a jackstand.

3. Remove the tire and wheel assembly.

4. Using J-33436 or equivalent ball joint removal tool, separate the ball joint from the knuckle.

5. Press the upper ball joint from the control arm using tool J-9519-E or an equivalent removal tool.

To install:

6. Position the ball stud so the cotter pin may be installed from the rear to the front of the vehicle and press into the control arm using J-9519-E or equivalent.

7. Position the ball joint into the steering knuckle, then install the washer and nut. Tighten the lower control arm ball stud nut to 50 ft. lbs. (68 Nm). Tighten the ball stud nut additionally, as necessary to insert a cotter pin, but do not exceed 88 ft. lbs. (120 Nm) to align the cotter pin holes.

8. Install a new cotter pin from the rear to the front of the vehicle.

9. Remove the jackstand and lubricate the ball joint.

10. Install the tire and wheel assembly, then lower the vehicle.

1997–99 Models

The lower ball joint is an integral part of the lower control arm. If the ball joint is bad, a new lower control arm must be installed.

Lower Control Arm

REMOVAL & INSTALLATION

1. Raise and safely support the vehicle.

2. Remove the front wheel.

3. Install a transverse leaf spring compressor on the front spring and compress the spring.

4. Support the control arm with a jack and remove the lower shock absorber mounting bolts from the control arm to be replaced.

5. Remove the stabilizer bar link from the control arm to be replaced.

6. Separate the lower control arm ball joint stud from the knuckle.

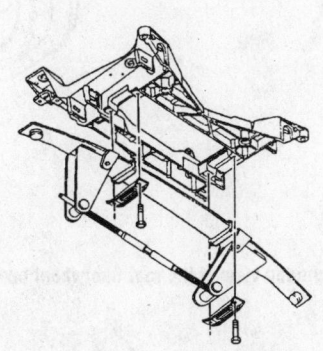

Spring compressor mounted on the transverse spring—1997–99 models

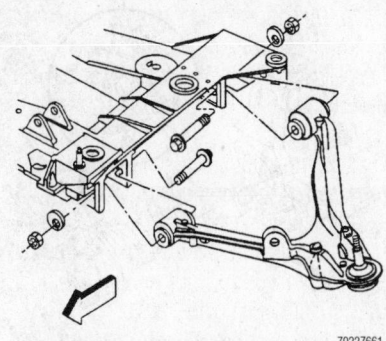

Lower control arm mounting bolts—1997–99 models

7. Mark the position of the cam bolts attaching the lower control arm to the frame for reference when installing the control arm.

8. Remove the cam bolts and the lower control arm.

To install:

9. Install the control arm to the frame. Align the marks made during removal but do not tighten the cam bolts completely at this time.

10. Install the ball joint stud to the steering knuckle. Tighten the nut to 44 ft. lbs. (60 Nm).

11. Support the control arm with a jack and install the lower shock absorber mounting bolts. Tighten the nuts to 21 ft. lbs. (28 Nm).

12. Install the stabilizer link to the control arm. Tighten the nut to 55 ft. lbs. (75 Nm).

13. Remove the spring compressor from the transverse spring.

14. Install the front wheel and remove the jack from under the control arm.

15. Perform a front wheel alignment. Tighten the lower control arm nuts to 125 ft. lbs. (170 Nm).

Wheel Bearings

ADJUSTMENT

No periodic wheel bearing adjustment is necessary. The wheel bearings are a sealed unit which must be replaced if loosed or noisy.

REMOVAL & INSTALLATION

Front

1. Disconnect the negative battery cable, then raise and support the vehicle safely.

2. Remove the tire and wheel assembly.

3. Remove the caliper and support it aside, then remove the rotor.

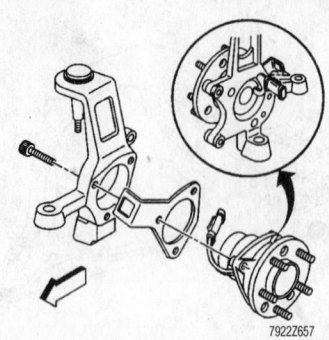

Exploded view of the front hub/wheel bearing and knuckle assembly—1997–99 models

4. Disengage the ABS speed sensor electrical connector.

5. Remove the ABS speed sensor cable bracket.

6. Remove the wheel hub/bearing/ speed sensor assembly.

To install:

7. Install the hub/bearing/speed sensor assembly onto the vehicle. Be sure the speed sensor cable connection is facing rearward.

8. Tighten the assembly mounting nuts to 46 ft. lbs. (62 Nm).

9. Engage the ABS electrical connector and install the cable bracket.

10. Install the brake rotor and caliper.

11. Install the tire and wheel assembly, then lower the vehicle.

12. Connect the negative battery cable. The bearings do not require adjustment.

Rear

1995–96 MODELS

1. Disconnect the negative battery cable, then raise and support the vehicle safely.

2. Remove the tire and wheel assembly.

3. Remove the wheel speed sensor.

4. Remove the brake caliper and parking brake assembly, then remove the rotor.

5. Remove the wheel hub mounting bolts.

6. Remove the cotter pin, wheel nut retainer, spindle nut and washer.

7. Remove the wheel hub and bearing, caliper mounting plate and wheel spindle washer from the vehicle.

To install:

8. Inspect the wheel hub and bearing seal, replace if necessary. Also inspect the wheel spindle washer and replace, if damaged or excessively worn.

9. Install the wheel hub and bearing, caliper mounting plate and the wheel spindle washer. The washer flat should firmly seat against the shoulder of the wheel spindle. The lip of the washer should face the wheel spindle splines prior to hub and bearing installation.

10. Install the wheel hub mounting bolts and tighten to 66 ft. lbs. (90 Nm).

11. Install the washer and spindle nut, then tighten the nut to 164 ft. lbs. (223 Nm). The vehicle should not rest on the tires or move until the spindle nut is tightened.

12. Install the wheel retainer and a new cotter pin.

13. Install the brake rotor, then install the caliper and parking brake assembly.

14. Install the wheel speed sensor, then install the wheel and tire assembly.

15. Lower the vehicle and connect the negative battery cable.

1997–99 MODELS

1. Raise and safely support the vehicle.

2. Remove the rear wheel.

3. Disconnect the wheel speed sensor.

4. Remove the brake caliper and rotor.

5. Disconnect the tie rod end from the knuckle.

6. Remove the halfshaft retaining nut and washer.

7. Disconnect the knuckle from the upper and lower control arms and remove the knuckle.

8. Remove the hub mounting bolts and remove the bearing assembly from the knuckle.

To install:

9. Install the bearing assembly on the knuckle. Tighten the mounting bolts to 70 ft. lbs. (95 Nm).

10. Install the knuckle assembly to the upper and lower control arms.

11. Install the halfshaft retaining nut.

12. Connect the tie rod end to the knuckle.

13. Install the brake rotor and caliper.

14. Connect the wheel speed sensor and install the rear wheel.

15. Lower the vehicle to the floor.

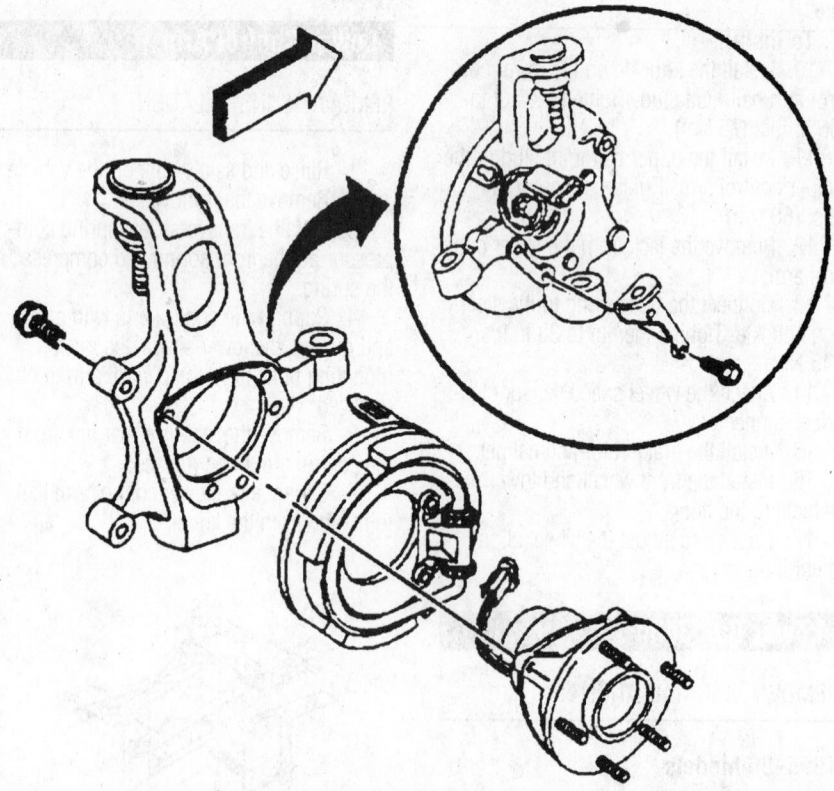

Exploded view of the rear hub/wheel bearing and knuckle assembly—1997–99 models

GEO/CHEVROLET

40

Metro • Prizm

DRIGE TRAIN 40-19
ENGINE REPAIR 40-2
FUEL SYSTEM 40-16
STEERING AND
  SUSPENSION 40-25

A

Air Bag.................................40-25
  Disarming..........................40-25
  Precautions 40-25

C

Camshaft 40-11
  Removal & Installation 40-11
Clutch 40-22
  Adjustments 40-22
  Removal & Installation 40-22
Coil Spring 40-32
  Removal & Installation 40-32
Cylinder Head 40-6
  Removal & Installation 40-6

D

Distributor..........................40-2
  Removal 40-2
  Installation........................40-2

E

Engine Assembly 40-3
  Removal & Installation 40-3
Exhaust Manifold.................40-10
  Removal & Installation 40-10

F

Front Crankshaft Seal 40-11
  Removal & Installation 40-11

Fuel Filter 40-17
  Removal & Installation 40-17
Fuel Pump 40-18
  Removal & Installation 40-18
Fuel System Pressure 40-17
  Relieving 40-17
Fuel System Service
  Precautions.......................40-16

H

Halfshaft............................40-23
  Removal & Installation 40-23
Hydraulic Clutch System 40-23
  Bleeding 40-23

I

Ignition Timing 40-2
  Adjustments 40-2
Intake Manifold 40-9
  Removal & Installation 40-9

L

Lower Ball Joints 40-32
  Removal & Installation 40-32

O

Oil Pan...............................40-15
  Removal & Installation 40-15
Oil Pump 40-15
  Removal & Installation 40-15

R

Rack and Pinion Steering Gear 40-25
  Removal & Installation 40-25
Rear Main Seal 40-16
  Removal & Installation 40-16

Rocker Arm/Shafts 40-9
  Removal & Installation 40-9

S

Strut..................................40-27
  Removal & Installation 40-27
Shock Absorber 40-30
  Removal & Installation 40-30

T

Transaxle Assembly 40-19
  Removal & Installation 40-19
Transfer Case Assembly.............40-40
  Removal & Installation 40-40
Turbocharger.......................40-4
  Removal & Installation 40-4

U

Upper Ball Joint......................40-18
  Removal & Installation 40-18

V

Valve Lash 40-14
  Adjustment.......................40-14

W

Water Pump 40-5
  Removal & Installation 40-5
Wheel Bearings.....................40-33
  Adjustment.......................40-33
  Removal & Installation 40-33

ENGINE REPAIR

➥**Disconnecting the negative battery cable on some vehicles may interfere with the functions of the on board computer systems and may require the computer to undergo a relearning process, once the negative battery cable is reconnected.**

Distributor

REMOVAL

1. Disconnect the negative battery cable.
2. Tag and disconnect the spark plug wires, wiring harness and vacuum line at the distributor, if equipped.
3. Remove the distributor cap.
4. Mark the position of the distributor rotor in relation to the distributor body. Mark the position of the distributor body in relation to the cylinder head.
5. Mark the distributor position on the housing and engine. Remove the hold-down bolt(s) and the distributor from the cylinder head. Do not rotate the engine after the distributor has been removed.
6. Remove the distributor from the engine and remove the O-ring from the distributor shaft.
7. Inspect all components of the assembly, including the cap and rotor, for cracks, terminal corrosion or wear and replace, if necessary.

INSTALLATION

Timing Not Disturbed

1. Align the reference marks made during removal. Position carefully and be sure the drive gear engage properly within the slot. Install the hold-down bolts but do not tighten.
2. Reconnect the wiring and the plug wires. Connect vacuum hoses, if equipped.
3. Install the distributor cap. Connect the battery negative cable. Check and/or adjust the ignition timing.
4. Tighten the distributor hold-down bolt(s).

Timing Disturbed

1. Remove the No. 1 spark plug.
2. Place a thumb over the spark plug hole. Have someone rotate the engine by hand, using a wrench on the crankshaft pulley, until compression is felt.

3. Align the timing mark on the crankshaft pulley with the **0** degrees mark on the timing scale attached to the front of the engine. This places the engine at TDC on the compression stroke.
4. Turn the distributor shaft until the rotor points to the No. 1 spark plug tower on the cap.
5. Install the distributor to the cylinder head, aligning the distributor housing-to-cylinder head marks made during the removal procedure.
6. Connect the spark plug wires, electrical wires and vacuum advance hose, if equipped.
7. Install the distributor cap.
8. Reconnect the negative battery cable.
9. Check and/or adjust ignition timing.
10. Tighten the distributor hold-down bolt.

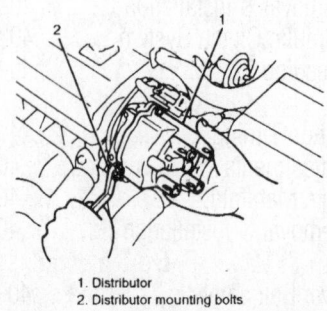

1. Distributor
2. Distributor mounting bolts

79222Z701

Be sure to tighten the distributor mounting bolts to prevent the timing from changing—Prizm

Ignition Timing

ADJUSTMENT

➥**Metro and 1995 Prizm models have adjustable timing, whereas 1996–99 Prizm models have non-adjustable timing set at 10 degrees BTDC.**

Metro and 1995 Prizm Models

1. Run the engine until it reaches normal operating temperature. Stop the engine, but keep the ignition switch in the **ON** position for approximately 5 seconds. Start the engine again.
2. Run the engine at 2000 RPM for approximately 5 minutes. After 5 minutes, allow it to run at idle speed.
3. Be sure all accessories are turned OFF.
4. Fully engage the parking brake.
5. Be sure the shift lever is placed in the **NEUTRAL** or **PARK** depending on transaxle type.

6. Connect a tachometer to the negative terminal of the ignition coil for 1.0L and 1.3L engines, or to the **IG** terminal in the Data Link Connector (DLC), located next to the left strut tower for 1.6L and 1.8L engines. Connect a timing light to the No. 1 spark plug wire. Check the engine idle speed and adjust if needed.
7. Refer to the underhood Vehicle Emission Control Information label for ignition timing specifications.
8. Remove the cap from the diagnostic check connector, located next to the ignition coil or left side strut tower, and insert a fused jumper wire between appropriate terminals.
9. For 1.0L and 1.3L engines, proceed as follows;
  a. On four terminal connectors, hold the connector with the locking tab at the top and jump the lower two terminals (**C** and **D**).
  b. On six terminal connectors, hold the connector with the locking tab at the top and jump the two terminals at the lower left (**D** and **E**).
10. For 1.6L and 1.8L engines, proceed as follows;
  a. Using a fused jumper wire, connect terminals **E1** and **TE1** together at the DLC.
11. Aim the timing light at the timing marks.
12. Loosen the distributor hold-down bolt and rotate the distributor until the correct timing marks are aligned.

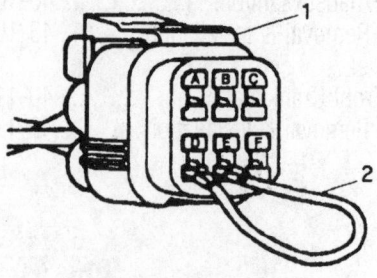

A. Blank (cavity 1)
B. Diagnostic request terminal (cavity 2)
C. Diagnostic output terminal (cavity 3)
D. Ground terminal(cavity 4)
E. Test switch terminal(cavity 5)
F. Duty check terminal(cavity 6)
1. Duty check DLC
2. Jumper

79222Z702

Jumping terminals in the six terminal Data Link Connector (DLC)—Metro and 1995 Prizm models

13. Tighten the distributor hold-down bolt to 11–15 ft. lbs. (15–20 Nm) and recheck the timing. Be sure the timing advances according to engine speed.

14. Remove the diagnostic check jumper. Remove the timing light and tachometer.

1996–99 Prizm Models

➡The ignition timing is on 1996–97 1.6L and 1.8L engines is NOT adjustable. If the ignition timing is out of specification a possible engine electrical failure may have occurred. The following procedure may be used to check the timing:

1. Warm the engine to normal operating temperature. Turn all electrical accessories OFF.

2. Connect a tachometer or engine scan tool and check the engine idle speed and be sure it is 650–750 rpm.

3. Connect a timing light. Remove the cap on the Data Link Connector (DLC). Using a fused jumper wire, connect terminals **E1** and **TE1**.

4. Start the engine and check timing. With the jumper wire connected, the timing should be at 10 degrees BTDC.

5. Remove the jumper wire from the DLC and reinstall the cap.

6. Shut the engine **OFF** and disconnect all test equipment.

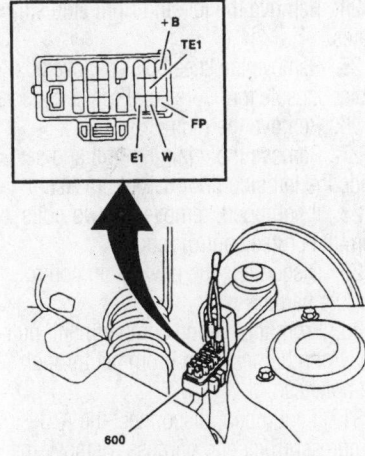

+B	SYSTEM VOLTAGE
E1	GROUND TERMINAL
FP	FUEL PUMP TERMINAL
TE1	DIAGNOSTIC REQUEST TERMINAL
W	MALFUNCTION INDICATOR LAMP (MIL) TERMINAL
600	DATA LINK CONNECTOR (DLC) (UNDER HOOD)

79222Z703

DLC pin locations—1996–99 Prizm models

Engine Assembly

REMOVAL & INSTALLATION

✳✳ CAUTION

The fuel injection system remains under pressure, even after the engine has been turned OFF. The fuel system pressure must be relieved before disconnecting any fuel lines. Failure to do so may result in fire and/or personal injury. Also, some models covered by this manual may be equipped with a Supplemental Restraint System (SRS), which uses an air bag. Whenever working near any of the SRS components, such as the impact sensors, the air bag module, steering column and instrument panel, properly disable the SRS.

Metro

1. Properly relieve the fuel system pressure. Disconnect the negative battery cable.

✳✳ CAUTION

Observe all applicable safety precautions when working around fuel. Whenever servicing the fuel system, always work in a well ventilated area. Do not allow fuel spray or vapors to come in contact with a spark or open flame. Keep a dry chemical fire extinguisher near the work area. Always keep fuel in a container specifically designed for fuel storage; also, always properly seal fuel containers to avoid the possibility of fire or explosion.

2. Scribe a hood hinge-to-hood outline, then, with the aid of an assistant remove the hood.

✳✳ CAUTION

Never open, service or drain the radiator or cooling system when hot; serious burns can occur from the steam and hot coolant.

3. Drain the cooling system.

4. Remove the air cleaner. Remove the radiator and cooling fan.

5. Label and unplug the engine electrical connections and vacuum lines.

6. Disconnect and plug the fuel return and feed lines.

7. Disconnect the heater inlet and outlet hoses.

8. Disconnect the following cables:
   a. Accelerator cable from the throttle body.
   b. Clutch cable from manual transaxle vehicles.
   c. The gear select and the oil pressure control cable from automatic transaxle vehicles.
   d. The speedometer cable from the transaxle.

9. Using Engine Support Assembly tool J-28467-A and J-28467–89 or equivalents, support the engine.

10. Raise and safely support the vehicle.

11. Disconnect the exhaust pipe from the manifold.

12. Disconnect the gear shift control shaft and extension from manual transaxle vehicles.

✳✳ CAUTION

The EPA warns that prolonged contact with used engine oil may cause a number of skin disorders, including cancer! You should make every effort to minimize your exposure to used engine oil. Protective gloves should be worn when changing the oil. Wash your hands and any other exposed skin areas as soon as possible after exposure to used engine oil. Soap and water, or waterless hand cleaner should be used.

13. Drain the engine and transaxle oil.

14. Remove the left and right halfshafts. It is not necessary to remove the halfshaft from the knuckles.

15. Remove the A/C compressor and belt. Position the compressor aside, leaving the hoses connected.

16. Disconnect and plug the power steering hoses from the pump, if equipped.

17. Remove the engine rear torque rod bracket from automatic transaxle vehicles.

18. Lower the vehicle.

19. Attach suitable lifting equipment to the engine.

20. Remove the right side engine mount from the bracket.

21. On automatic transaxles, remove the transaxle rear mounting nut.

22. On manual transaxles, remove the transaxle rear mount from the body.

23. Remove the transaxle left side mounting bracket.

24. Lift the engine and transaxle assembly out of the vehicle.

To install:

25. Install the engine and transaxle assembly into the vehicle and leave the hoist connected.

26. On automatic transaxles, install the transaxle rear mounting nut.

27. On manual transaxles, install the transaxle rear mount to the body.

28. Install the transaxle left side mounting bracket. Tighten the bolts to 41 ft. lbs. (55 Nm).

29. Install the transaxle right side engine mount and bracket. Tighten the bolts to 41 ft. lbs. (55 Nm).

30. Remove the lifting device and support the engine assembly with tool J-28467-A and J-28467–89 or equivalents.

31. Raise and safely support the vehicle.

32. Install the rear engine torque rod bracket to the transaxle, if equipped.

33. Connect the left and right halfshafts.

34. Connect the gear shift control shaft and the extension to the transaxle, if equipped with a manual transaxle.

35. Connect the exhaust pipe to the manifold. Tighten the to 37 ft. lbs. (50 Nm).

36. Install the A/C compressor and belt. Tighten the compressor-to-mounting bracket bolts to 21 ft. lbs. (28 Nm).

37. Reconnect the power steering hoses to the power steering pump, if equipped.

88. Lower the vehicle.

39. Install the remaining components.

40. Install the hood. Tighten the hood mounting bolts to 20 ft. lbs. (27 Nm).

41. Adjust the clutch pedal free-play, gear select cable and accelerator cable play.

42. Refill the engine, transaxle and power steering pump, if equipped with the proper type and quantity of oil. Refill and bleed the cooling system.

❋❋ WARNING

Operating the engine without the proper amount and type of engine oil will result in severe engine damage.

43. Reconnect the negative battery cable. Start the engine and check for leaks. Make any necessary adjustments.

Prizm

❋❋ CAUTION

Observe all applicable safety precautions when working around fuel. Whenever servicing the fuel system, always work in a well ventilated area. Do not allow fuel spray or vapors to come in contact with a spark or open flame. Keep a dry chemical fire extinguisher near the work area. Always keep fuel in a container specifically designed for fuel storage; also, always properly seal fuel containers to avoid the possibility of fire or explosion.

1. Properly relieve the fuel system pressure.

2. Disconnect the battery cables, negative cable first. Remove the battery.

3. Scribe a hood hinge-to-hood outline. Disconnect the windshield washer fluid lines from the hood. Support the hood and remove the hood hinge bolts and remove the hood.

4. Raise and safely support the vehicle.

❋❋ CAUTION

The EPA warns that prolonged contact with used engine oil may cause a number of skin disorders, including cancer! You should make every effort to minimize your exposure to used engine oil. Protective gloves should be worn when changing the oil. Wash your hands and any other exposed skin areas as soon as possible after exposure to used engine oil. Soap and water, or waterless hand cleaner should be used.

5. Drain the engine oil and cooling system.

❋❋ CAUTION

Never open, service or drain the radiator or cooling system when hot; serious burns can occur from the steam and hot coolant.

6. Remove the left and right splash shields.

7. Drain the transaxle fluid into a suitable container.

8. Lower the vehicle.

9. Disconnect the accelerator cable from the throttle lever. If equipped with an automatic transaxle, disconnect the throttle cable from the bracket.

10. Remove the radiator and cooling fan.

11. Remove the air cleaner.

12. Remove coolant reservoir support bracket.

13. Remove the windshield washer reservoir.

14. If equipped, remove the cruise control actuator and bracket as follows:

a. Unclip the actuator cover and remove.

b. Disconnect the accelerator cable from the actuator.

c. Remove the three bolts and actuator.

d. Remove the bolts and actuator bracket.

15. Label and disconnect the manifold absolute pressure sensor hose, brake booster hose and A/C solenoid vacuum valve hose.

16. Disconnect the A/C SV valve, MAP sensor, data link connector and A/C pressure switch wiring harnesses. Also disconnect the ground wires from the intake manifold and fenders.

17. Remove the bolts, then disconnect the four wiring connectors and remove the fuse and relay box.

18. Disconnect the hose from the evaporative emission canister. Loosen the canister bracket bolt and remove the canister.

19. Loosen the hose clamps and disconnect the heater hoses from the thermostat housing.

20. Disconnect the fuel feed line from the fuel rail. Discard the gaskets.

21. Disconnect the fuel return hose from fuel pressure regulator.

22. If equipped with a manual transaxle, remove the clutch slave cylinder, and disconnect the shift select and shift control cables from the transaxle.

23. If equipped with an automatic transaxle, disconnect the shift select cable from the transaxle.

24. Remove the left and right side sill plates.

25. Remove the knee bolster, glove box, center console trim bezel, and center console.

26. Remove the radio.

27. Remove the retaining bolt and set aside the left side floor carpet bracket.

28. If equipped, remove the two bolts from the cruise control module.

29. Disconnect the powertrain control module harnesses.

30. From the engine compartment, pull the wiring harness and grommet through the bulkhead.

31. If equipped, disconnect the A/C compressor belt and harness. Remove the compressor and suspend.

32. If equipped, remove the power steering and suspend without disconnecting the hoses.

33. Raise and safely support the vehicle.

34. Disconnect the oxygen sensor harness, then remove the front exhaust pipe.

35. Remove the left and right halfshafts. It is not necessary to remove the halfshafts from the steering knuckles.

36. Attach a suitable engine hoist to the engine/transaxle assembly.

37. Remove the front and rear transaxle mounts.

38. Lower the vehicle.

39. Remove the left transaxle mount.

40. Carefully lift the engine/transaxle assembly from the vehicle.

41. Remove the starter.

To install:

42. Install the starter.

43. Lower the engine/transaxle assembly into the vehicle.

44. Install the left transaxle mounts. Tighten the bolts to 41 ft. lbs. (56 Nm), the through-bolt to 64 ft. lbs. (87 Nm) and the reinforcement bolts to 15 ft. lbs. (21 Nm).

45. Raise and safely support the vehicle.

46. Install the front and rear transaxle mounts. Tighten the front mount bolts to 47 ft. lbs. (64 Nm) and the through-bolt to 64 ft. lbs. (87 Nm). Tighten the rear transaxle nuts to 42 ft. lbs. (52 Nm) and the through-bolt to 64 ft. lbs. (87 Nm).

47. Remove the engine hoist.

48. Raise and safely support the vehicle.

49. Install the left and right halfshafts.

50. Install the front exhaust pipe and connect the oxygen sensor.

51. If equipped, install the power steering pump and drive belt.

52. Lower the vehicle.

53. If equipped, install the A/C compressor and drive belt.

54. Insert the wiring harness through the bulkhead into the passenger compartment and install the grommet to the bulkhead.

55. Connect the wiring to the PCM.

56. If equipped, position the cruise control module and tighten the bolts to 44 inch lbs. (5 Nm).

57. Position the left side floor carpet bracket and tighten the bolt to 44 inch lbs. (5 Nm).

58. Install the radio, knee bolster, glove box, center console and trim bezel.

59. Install the left and right sill plates.

60. Connect the shift cable(s) to the transaxle.

61. If equipped with manual transaxle, install the clutch slave cylinder.

62. Connect the fuel return hose to the fuel pressure regulator.

63. Connect the fuel feed pipe to the fuel rail, using new gaskets. Tighten the bolt to 22 ft. lbs. (29 Nm).

64. Connect the heater hoses to the thermostat housing.

65. Install the EVAP canister to the bracket and tighten the bolt to 89 inch lbs. (10 Nm). Connect the hose to the canister.

66. Install the fuse and relay box. Connect the wiring connectors and tighten the bolts to 89 inch lbs. (10 Nm).

67. Connect the engine ground wires to the intake manifold and fenders. Connect the A/C pressure switch, DLC, MAP sensor and A/C SV valve wiring harnesses.

68. Connect the hoses to the A/C SV valve (if equipped), brake booster and MAP sensor.

69. If equipped, install the cruise control actuator and bracket as follows:

   a. Install the actuator bracket and tighten the bolts to 18 ft. lbs. (25 Nm).

   b. Install the actuator and tighten the bolts to 89 inch lbs. (10 Nm).

   c. Connect the accelerator cable to the actuator.

   d. Clip the actuator cover in place.

70. Install the windshield washer reservoir and secure with bolt. Connect the wiring harness and hose.

71. Install the coolant reservoir and tighten the bolts to 11 ft. lbs. (15 Nm).

72. Install the remaining components.

73. Install the battery and cables. Connect the positive cable first.

74. Install the hood.

✳✳ WARNING

Operating the engine without the proper amount and type of engine oil will result in severe engine damage.

75. Fill the engine with the proper type and quantity of oil.

76. Fill and bleed the cooling system.

77. Run the engine and check for leaks and proper operation.

Water Pump

REMOVAL & INSTALLATION

Metro

1. Disconnect the negative battery cable.

✳✳ CAUTION

Never open, service or drain the radiator or cooling system when hot; serious burns can occur from the steam and hot coolant.

2. Drain the cooling system.

3. Remove the air cleaner.

4. Remove the suction pipe bracket for the A/C compressor, if equipped.

5. Loosen but do not remove the four water pump pulley bolts.

6. Raise and safely support the vehicle.

7. Remove the lower splash shield.

8. Remove the A/C compressor drive belt, if equipped.

9. Remove the alternator drive belt.

10. Remove the crankshaft and water pump pulleys.

11. Remove the timing belt.

12. Remove the oil level dipstick and tube.

13. Remove the alternator adjusting bracket from the water pump.

14. Remove the water pump rubber seals.

15. Remove the water pump mounting bolts and nuts and remove the water pump from the engine.

To install:

16. Clean the gasket mating surfaces thoroughly.

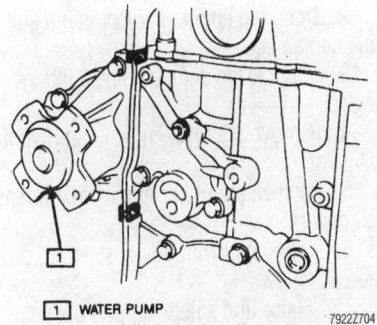

| 1 | WATER PUMP   7922Z704

To ensure a tight seal, be sure gasket surfaces are properly prepared—Metro

17. Check the water pump by hand for smooth operation. If the pump does not operate smoothly or is noisy, replace it.

18. Install the pump using a new gasket. Tighten the bolts to 115 inch lbs. (13 Nm).

19. Install new rubber seals.

20. Install the upper alternator adjusting bracket and tighten the bolt to 17 ft. lbs. (23 Nm).

21. Install the oil level dipstick and tube.

22. Install the timing belt.

23. Install the water pump and crankshaft pulleys. Leave the water pump pulley bolts hand-tight.

24. Install the alternator drive belt.

25. Install the lower alternator cover plate and tighten the bolts to 89 inch lbs. (10 Nm).

26. Install the A/C compressor drive belt, if equipped.

27. Install the lower splash shield and lower the vehicle.

28. Tighten the water pump pulley mounting bolts to 18 ft. lbs. (24 Nm).

29. Adjust the water pump drive belt tension and tighten the alternator adjustment bolt to 17 ft. lbs. (23 Nm).

30. Install the suction pipe bracket for the A/C compressor, if equipped.

31. Install the air cleaner.

32. Refill the cooling system.

33. Connect the negative battery cable.

34. Start the engine and check for leaks.

Prizm

1. Disconnect the negative battery cable.

✳✳ CAUTION

Never open, service or drain the radiator or cooling system when hot; serious burns can occur from the steam and hot coolant.

2. Drain the engine coolant into a suitable container.

3. Support the engine using a lifting device.

4. Remove the right engine mount and insulator.

5. Remove the upper and middle timing belt covers.

6. If equipped with power steering, proceed as follows:

   a. Raise and safely support the vehicle.

   b. Remove the plastic cover from the front transaxle mount.

   c. Remove the two mount-to-chassis bolts.

   d. Remove the nut and through-bolt, then remove the front transaxle mount from the vehicle.

   e. Lower the vehicle.

   f. Remove the coolant reservoir.

   g. Disconnect the upper hose from the radiator.

   h. Unclip the wiring connectors from the shroud.

   i. Remove the cooling fan assembly.

7. Remove the two nuts, one bolt and the engine wiring harness retainer.

8. If equipped, unclip the crankshaft position sensor electrical connector from the dipstick tube.

9. Remove the dipstick tube and dipstick. Immediately plug the hole in the block.

10. Disconnect the cooling fan switch electrical connector.

11. Remove the two nuts securing the engine coolant inlet pipe to the cylinder block.

12. Loosen the hose clamps and remove the coolant inlet pipe from the water pump.

13. Loosen the hose clamp and remove the coolant inlet hose from the water pump.

14. Remove the water pump bolts and remove the assembly. Discard the O-ring.

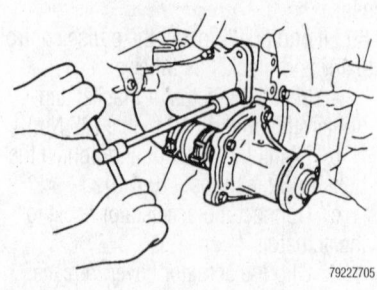

79222705

Remove the coolant inlet pipe from the block, then the pump—Prizm

To install:

15. Install a new O-ring. Install the water pump, and tighten the retaining bolts evenly to 10 ft. lbs. (14 Nm).

16. Connect the coolant hose to the water pump and secure the clamp.

17. Install the engine coolant inlet pipe. Secure to the hose with clamps and nuts. Tighten the nuts to 11 ft. lbs. (15 Nm).

18. Connect the cooling fan switch electrical connector.

19. Install the dipstick and tube. Tighten the retaining bolt to 84 inch lbs. If equipped, clip the crankshaft position sensor electrical connector to the dipstick tube.

20. Install the engine wiring harness retainer and tighten the nuts and bolt to 89 inch lbs. (10 Nm).

21. If equipped with power steering, proceed as follows:

   a. Install the cooling fan assembly and tighten the bolts to 52 inch lbs. (6 Nm).

   b. Attach the wiring connectors to the shroud.

   c. Connect the upper hose to the radiator.

   d. Install the coolant reservoir.

   e. Raise and safely support the vehicle.

   f. Install the front transaxle mount and tighten the through-bolt nut to 64 ft. lbs. (87 Nm).

   g. Install the mount-to-chassis bolts and tighten to 47 ft. lbs. (64 Nm).

   h. Install the plastic cover to the front transaxle mount.

   i. Lower the vehicle.

22. Install the upper and middle timing belt covers.

23. Support the engine using a lifting device. Install the right engine mount and insulator.

24. Remove the engine lifting equipment.

25. Connect the negative battery cable.

26. Fill the cooling system. Start the engine and check for leaks.

Cylinder Head

REMOVAL & INSTALLATION

✳✳ CAUTION

The fuel injection system remains under pressure, even after the engine has been turned OFF. The fuel system pressure must be relieved before disconnecting any fuel lines. Failure to do so may result in fire and/or personal injury.

Metro

1. Properly relieve the fuel system pressure.

✳✳ CAUTION

Observe all applicable safety precautions when working around fuel. Whenever servicing the fuel system, always work in a well ventilated area. Do not allow fuel spray or vapors to come in contact with a spark or open flame. Keep a dry chemical fire extinguisher near the work area. Always keep fuel in a container specifically designed for fuel storage; also, always properly seal fuel containers to avoid the possibility of fire or explosion.

2. Disconnect the negative battery cable.

✳✳ CAUTION

Never open, service or drain the radiator or cooling system when hot; serious burns can occur from the steam and hot coolant.

3. Drain the cooling system.

4. Remove the intake and exhaust manifolds.

5. Remove the timing belt.

6. Disconnect and label the spark plug wires. Be sure to pull firmly on the wire boot, not the plug wire.

7. Remove the distributor and case from the cylinder head.

8. Remove the cylinder head cover.

9. Loosen and remove the cylinder head bolts by reversing the tightening sequence.

10. Remove the cylinder head from the engine. Discard the cylinder head gasket and rubber seal.

11. Clean all gasket mating surfaces.

12. Check the cylinder head flatness using a straightedge and feeler gauge. Check the surface across the combustion chambers, on each side of the combustion chambers, and diagonally. If distortion exceeds 0.002 in. (0.05mm) at any location, the cylinder head should be resurfaced.

To install:

13. Install the cylinder head gasket with the TOP indicator facing upward and toward the crankshaft pulley.

14. Install the cylinder head. Lubricate the cylinder head bolts with clean engine oil and install them finger-tight.

15. Tighten the cylinder head bolts in three even stages, following the tightening sequence, to 54 ft. lbs. (73 Nm).

16. Install the rubber seal between the water pump and the cylinder head.

17. Install the cylinder head cover and secure with nuts and new seal washers. Tighten nuts to 44 inch lbs. (5 Nm).

18. Install the timing belt.

19. Install the distributor.

20. Connect the spark plug wires.

21. Install the intake manifold, using a new gasket.

22. Install the exhaust manifold using a new gasket.

23. Refill the cooling system.

24. Connect the negative battery cable.

25. Start the engine and allow it to reach normal operating temperature. Check for leaks and adjust the ignition timing, if necessary.

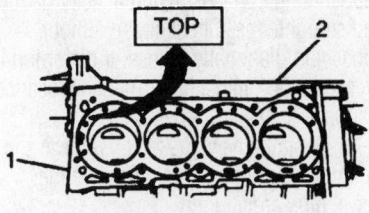

1. Cylinder head gasket
2. Cylinder block

Cylinder head gasket positioning—1.3L engine

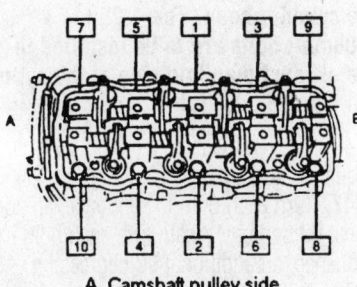

A. Camshaft pulley side
B. Distributor side

Cylinder head bolt torque sequence—1.3L engine

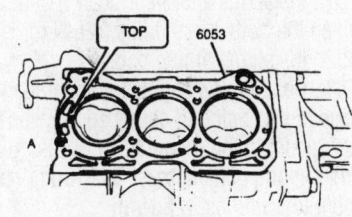

A CRANKSHAFT PULLEY SIDE
6053 CYLINDER HEAD GASKET

Cylinder head gasket positioning—1.0L engine

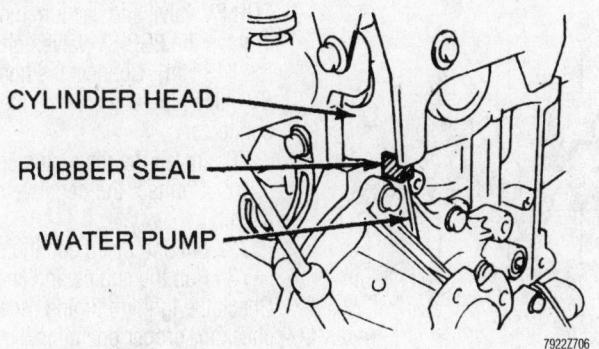

Install the rubber seal as shown—Metro

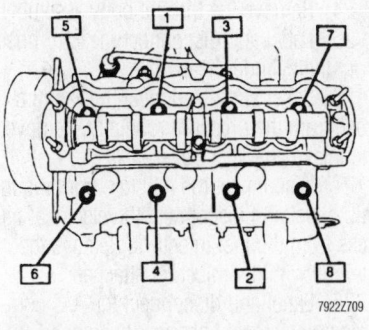

Cylinder head bolt torque sequence—1.0L engine

Prizm

1. Properly relieve the fuel system pressure.

2. Disconnect the negative battery cable.

3. Drain the engine coolant into a suitable container.

4. Raise and safely support the vehicle.

5. Remove the right side engine splash shield.

6. Lower the vehicle.

7. Remove the accelerator cable from the bracket and throttle body.

8. Disconnect the vacuum hose from the evaporative emissions canister.

9. Unfasten the intake air temperature sensor harness from the air cleaner cap. Loosen the air cleaner hose clamp and disconnect four clips to remove the cover.

10. Remove the alternator.

11. Disconnect the spark plug wires and remove the distributor case.

12. Disconnect the oxygen sensor harness, then detach the bolts and nuts and remove the front exhaust pipe.

13. Remove the exhaust manifold.

14. Remove the thermostat housing.

15. Unfasten the ground strap connector.

16. Disconnect the manifold absolute pressure sensor harness and A/C pressure switch harness, if equipped.

17. Remove the engine wiring harness from the right side fender.

18. Label and disconnect all of the hoses from the intake chamber.

19. If equipped, label and unfasten the A/C solenoid vacuum valve electrical connector.

20. Disconnect the hoses and harnesses from the EGR Switching Valve (SV) and vacuum modulator, then remove.

21. Remove the engine wire clamp and the intake manifold brace.

22. Disconnect the fuel return line from the pressure regulator. Remove the air pipe from the intake manifold.

23. Remove the air pipe and EGR valve.

24. Remove the throttle body assembly.

25. Label and disconnect the PCV hoses from the cylinder head cover.

26. Label and disconnect the vacuum hose from the pressure regulator. Remove the intake chamber cover.

27. Disconnect the fuel line from the fuel rail. Label and disconnect the electrical harness from the fuel injectors. Remove the fuel rail with the injectors attached.

28. Label and disconnect the A/C compressor electrical harness, if equipped. Label and disconnect the oil pressure switch, crankshaft position sensor and fan thermostat electrical harness. Remove the engine wiring harness cover. Disconnect the engine ground strap and remove the engine wiring harness retainer.

29. Remove the intake manifold.

30. If equipped with cruise control, remove the actuator cover, actuator mounting bolts and actuator bracket.

31. Support the engine using a suitable jack.

32. Remove the two nuts from the right-hand engine mount studs.

33. If equipped, remove the 2-piece A/C line bracket from the engine mount.

34. Remove the engine mount reinforcement bracket and mount, if equipped.

❊❊ WARNING

It may be necessary to lower the engine to gain clearance for engine mount removal. If equipped, be careful not to bend the A/C line too much or a leak may develop.

35. Remove the timing belt.

36. Remove the alternator mounting bracket.

37. Remove the oil level dipstick and tube.

38. Remove the coolant inlet pipe.

39. Remove the camshafts.

40. Uniformly loosen and remove the 10 cylinder head bolts in several passes, in reverse order of tightening.

❊❊ WARNING

Cylinder head warpage or cracking could result from removing the bolts in the incorrect order.

41. Lift off the cylinder head assembly from the dowels.

To install:

42. Clean all gasket mating surfaces, using care not to damage the aluminum components.

43. Check the engine block and cylinder head mating surfaces, using a feeler gauge and straightedge. Check along all four edges and diagonally across. If distortion exceeds 0.002 in. (0.05mm) at any location, resurface the cylinder head.

44. Install a new gasket, then lower the cylinder head to the block. Be sure the dowel pins are aligned.

45. Coat the cylinder head bolt threads and the underside of each bolt head with clean engine oil and install finger-tight.

➡**The cylinder head bolts are in lengths of 3.54 in. (90mm) and 4.25 in. (108mm). The 3.54 in. (90mm) bolts are to be installed in the intake side of the cylinder head. The 4.25 in. (108mm) bolts are to be installed in the exhaust manifold side of the cylinder head.**

46. Tighten the cylinder head bolts, in sequence, to 22 ft. lbs. (29 Nm).

47. Turn each bolt, in sequence, an additional 90 degrees. Again turn each bolt, in sequence, an additional 90 degrees.

48. Install the camshafts.

49. Install the coolant inlet pipe. Tighten the nuts to 11 ft. lbs. (15 Nm).

50. Install the engine oil level dipstick tube and tighten the bolt to 84 inch lbs. (9.5 Nm).

51. Install the alternator bracket and tighten the bolts to 19 ft. lbs. (26 Nm).

52. Install the timing belt.

53. Position the engine mount, and locate the A/C pipe, if equipped. Tighten the bolts closet to the mount to 47 ft. lbs. (64 Nm), and the bolts farthest away from the mount to 18 ft. lbs. (25 Nm).

54. Install the rubber insulator to the mount and tighten the bolt to 18 ft. lbs. (25 Nm).

55. Install the reinforcement bracket and tighten the nut and bolt to 18 ft. lbs. (25 Nm).

56. Attach the A/C pipe bracket, if equipped, to the engine mount and tighten the bolt to 89 inch lbs. (10 Nm).

57. Install the nuts to the engine mount studs and tighten to 38 ft. lbs. (52 Nm).

58. If equipped, install the cruise control actuator bracket, actuator and cover.

59. Install the intake manifold and ground strap.

60. Install the engine wiring harness with the retainer and tighten the bolts and nuts to 89 inch lbs. (10 Nm). Secure the ground strap with the bolt, and tighten to 89 inch lbs. (10 Nm). Connect the fan thermostat, oil pressure switch, if equipped, A/C compressor harnesses. Install the engine wire cover and tighten the bolts to 89 inch lbs. (10 Nm).

61. Install the fuel rail and injectors. Tighten the fuel rail bolts to 11 ft. lbs. (15 Nm) and connect the fuel injector electrical connectors.

62. Connect the fuel line to the fuel rail, using new gaskets. Tighten the union bolt to 22 ft. lbs. (29 Nm).

63. Install the intake chamber cover using a new gasket. Tighten the nuts and bolts to 14 ft. lbs. (19 Nm). Connect the vacuum hose to the pressure regulator.

64. Connect the PCV hoses to the cylinder head cover.

65. Install the throttle body assembly, using a new gasket. Tighten the bolts and nuts to 16 ft. lbs. (22 Nm). Connect the throttle position sensor harness. Install the accelerator cable bracket and tighten the bolts to 98 inch lbs. (11 Nm).

66. Install the EGR valve, using a new gasket. Tighten the nuts to 115 inch lbs. (13 Nm).

67. Connect the fuel return hose to the pressure regulator.

68. Install the intake manifold brace. Tighten the 12mm bolts to 14 ft. lbs. (19 Nm) and the 14mm bolts to 29 ft. lbs. (39 Nm).

69. Install the engine wire clamp and the EGR SV valve and vacuum modulator. Tighten the EGR SV valve bolt to 115 inch lbs. (13 Nm). Connect the hose to the EGR SV valve pipe and connect the electrical connector.

70. Install the remaining components.

71. Connect the negative battery cable.

72. Fill and bleed cooling system.

73. Run the engine and check for leaks. Check the ignition timing. Road test and check for proper operation.

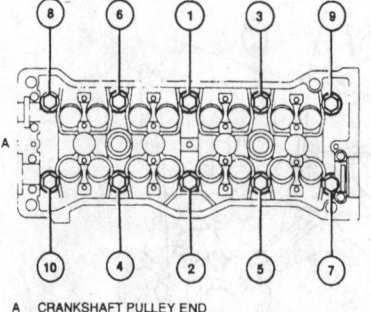

A CRANKSHAFT PULLEY END

79222711

Cylinder head bolt torque sequence—Prizm

Rocker Arms/Shafts

REMOVAL & INSTALLATION

All engines except the 1.3L do not use rocker arms/shafts, the camshaft directly actuates the valves.

1.3L Engine

1. Disconnect the negative battery cable.
2. If equipped, remove the A/C compressor drive belt, then the remove the compressor and mounting bracket. Place off to the side.
3. Disconnect the spark plug wires at the plugs.
4. Remove the air cleaner and cylinder head cover.
5. Remove the rocker shaft retaining bolts. Lift out the exhaust and intake rocker arm shafts, with the springs and rocker arms attached.
6. If necessary, remove the rocker arms, washers and springs from the rocker shafts. Note the order in which the components are removed; they must be re-assembled in the same order and position.

➡The intake and exhaust rocker arm shafts are NOT the same. The two can be distinguished by looking at the ends of the shafts, which are different. Install the intake rocker arm shaft with the stepped end toward the distributor side.

7. Inspect the rockers arms, shafts and lash adjusters for wear and/or damage; replace as necessary.

To install:

8. If disassembled earlier, assemble the springs, washers and rocker arms on to the rocker shafts.
9. Lubricate the rocker arms and shafts with clean engine oil. Position the intake and exhaust rocker arm shafts in their original positions. Secure with the retaining bolts. Tighten the shaft bolts, in the correct sequence, to 97 inch lbs. (11 Nm).
10. Apply a small amount of silicone sealer to the corners of the cylinder head cover gasket mating surface. Install a new head cover gasket and position the cylinder head cover to the head.
11. Install new seal washers and the nuts. Tighten the head cover nuts to 44 inch lbs. (5 Nm).
12. Install the remaining components.
13. Reconnect the negative battery cable. Run the engine and check for leaks and proper engine operation.

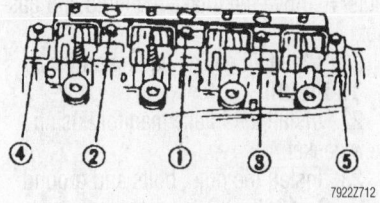

Rocker arm shaft retaining bolt tightening sequence—1.3L engines

Intake Manifold

REMOVAL & INSTALLATION

❋❋ CAUTION

Some models covered by this manual may be equipped with a Supplemental Restraint System (SRS), which uses an air bag. Whenever working near any of the SRS components, such as the impact sensors, the air bag module, steering column and instrument panel, properly disable the SRS.

Metro

❋❋ CAUTION

Observe all applicable safety precautions when working around fuel. Whenever servicing the fuel system, always work in a well ventilated area. Do not allow fuel spray or vapors to come in contact with a spark or open flame. Keep a dry chemical fire extinguisher near the work area. Always keep fuel in a container specifically designed for fuel storage; also, always properly seal fuel containers to avoid the possibility of fire or explosion.

1. Relieve the fuel system pressure.
2. Disconnect the negative battery cable.

❋❋ CAUTION

Never open, service or drain the radiator or cooling system when hot; serious burns can occur from the steam and hot coolant.

3. Drain the cooling system.
4. Remove the air cleaner.
5. Disconnect the electrical harnesses to the intake manifold.
6. Disconnect the fuel return and feed hoses from the throttle body.

7. Disconnect the water hoses from the throttle body and the intake manifold.
8. Disconnect all vacuum hoses to the intake manifold.
9. Disconnect the PCV hose from the valve cover.
10. Disconnect the accelerator cable from the throttle body.
11. Remove the intake manifold fasteners and remove with the throttle body attached.
12. Remove and discard the intake manifold gasket. Clean all gasket mating surfaces.

To install:

13. Install the intake manifold to the cylinder head, using a new gasket. Install the fasteners and tighten evenly, to 17 ft. lbs. (23 Nm), starting in the center working towards the ends.
14. Connect the PCV hose to the valve cover.
15. Connect the vacuum and coolant hoses.
16. Install the fuel feed and return hoses.
17. Reconnect the electrical harnesses to the proper locations.
18. Install the accelerator cable and adjust as follows:
    a. Be sure the cable locknut and adjusting nut are loose.
    b. Adjust the cable end-play to 0.4–0.6 in. (10–15mm), then tighten the lock and adjusting nuts.
    c. If equipped with automatic transaxle, loosen the cable locknut and adjusting nut at the cable retaining bracket on the transaxle. Adjust the cable end-play to 0.4–0.6 in. (10–15mm), then tighten the locknut and adjusting nut.
19. Install the air cleaner.
20. Refill the cooling system and reconnect the negative battery cable.
21. Start the engine. Check for leaks.

Prizm

❋❋ CAUTION

Observe all applicable safety precautions when working around fuel. Whenever servicing the fuel system, always work in a well ventilated area. Do not allow fuel spray or vapors to come in contact with a spark or open flame. Keep a dry chemical fire extinguisher near the work area. Always keep fuel in a container specifically designed for fuel storage; also, always properly seal fuel containers to avoid the possibility of fire or explosion.

1. Relieve the fuel system pressure.
2. Disconnect the negative battery cable.
3. Label and disconnect the following hoses:
   a. Vacuum sensor hose from the fuel filter.
   b. Brake booster vacuum hose.
   c. If equipped, A/C vacuum hoses.
4. If equipped, disconnect the A/C actuator electrical connector.
5. If equipped with an EGR Switching Valve (SV), proceed as follows:
   a. Disconnect the EGR SV electrical connector and EGR temperature sensor.
   b. Disconnect the vacuum hose from the EGR valve.
   c. Remove bolt and EGR SV valve.
6. Remove the engine wire clamp.
7. Remove the intake manifold brace.
8. Disconnect the fuel return hose from the pressure regulator.
9. Remove the mounting bolt and nut and the air pipe from the intake manifold.
10. Remove the throttle body as follows:
    a. Disconnect the throttle position sensor harness.
    b. Loosen the locknut and adjusting nut, then remove the accelerator cable from the throttle lever.
    c. If equipped with an automatic transaxle, loosen the locknut and adjusting nut and remove the throttle cable from the bracket.
    d. Remove the two bolts and the accelerator bracket.
    e. Remove the two bolts, two nuts and the throttle body with the gasket.
11. Remove the fuel inlet hose clamp bolt, intake chamber brace and gasket.
12. Disconnect the PCV hoses from the valve cover.
13. Disconnect the vacuum hose from the fuel pressure regulator.
14. Remove the bolts, nuts and the intake chamber cover with gasket.
15. Remove the union bolt and gaskets and disconnect the fuel inlet pipe from the fuel rail.
16. Disconnect the fuel injector harnesses.
17. Remove the bolts and remove the fuel rail with the injectors attached.

⁂ WARNING

Be careful not to drop the fuel injectors when removing the fuel rail. Damage could result.

18. Remove the insulators and spacers from the cylinder head.
19. Remove the nuts and bolt and the engine wiring harness retainer.
20. Remove the bolts, ground strap and nuts. Remove the intake manifold and gasket.
21. Clean all gasket mating surfaces.

To install:

22. Install the intake manifold using a new gasket.
23. Install the nuts, bolts and ground strap. On 1995 vehicles, tighten the nuts and bolts to 14 ft. lbs. (19 Nm). On 1996–99 vehicles, tighten the nuts to 9 ft. lbs. (13 Nm) and bolts to 14 ft. lbs. (19 Nm).
24. Install the engine wiring harness retainer, and tighten the nuts and bolts to 89 inch lbs. (10 Nm).
25. Install new insulators and spacers to the cylinder head.
26. Install the fuel rail with the injectors. Loosely install the retaining bolts.

➡Check the fuel injectors for smooth rotation. If the injectors do not rotate smoothly, replace the O-rings and recheck. Be sure the electrical connector is pointed upward.

27. Tighten the fuel rail retaining bolts to 11 ft. lbs. (15 Nm).
28. Connect the injector harnesses.
29. Connect the fuel inlet pipe to the fuel rail with the union bolt. Use new gaskets. Tighten the union bolt to 22 ft. lbs. (29 Nm).
30. Install the intake chamber cover using a new gasket. Tighten the nuts and bolts to 14 ft. lbs. (19 Nm).
31. Connect the vacuum hose to the pressure regulator. Connect the PCV hoses to the valve cover.
32. Install the intake chamber brace with a new gasket. Tighten the bolt to 21 ft. lbs. (28 Nm).
33. Install the fuel inlet hose clamp bolt and tighten to 89 inch lbs. (10 Nm).
34. Install the throttle body using a new gasket. Tighten the nuts and bolts to 17 ft. lbs. (23 Nm).
35. Install the accelerator cable bracket and tighten the bolts to 11 ft. lbs. (15 Nm).
36. Connect the throttle cable, if equipped, and accelerator cable. Tighten the locknut and adjusting nut to 11 ft. lbs. (15 Nm).
37. If equipped, install the EGR valve, vacuum modulator and vacuum hose assembly, using new gaskets. Tighten the EGR valve-to-intake manifold bolts and nuts to 115 inch lbs. (13 Nm) and tighten the EGR valve-to-EGR pipe bolts and nuts to 43 ft. lbs. (59 Nm). Connect the two vacuum hoses.
38. Connect the fuel return hose to the pressure regulator.
39. Install the intake manifold brace and engine wire clamps.
40. If equipped, install the EGR SV valve and tighten the bolt to 115 inch lbs. (13 Nm). Connect the EGR SV valve and EGR TP sensors. Connect the vacuum hose to the EGR valve.
41. Connect the remaining components.
42. Connect the negative battery cable.
43. Run the engine and check for leaks.

Exhaust Manifold

REMOVAL & INSTALLATION

⁂ CAUTION

Do not service any part of the exhaust system while it is still hot. Allow the system to cool down before performing any service.

Metro

1. Disconnect the negative battery cable.
2. Raise and safely support the vehicle.
3. Disconnect the front pipe/catalytic converter from the exhaust manifold.
4. Lower the vehicle.
5. Disconnect the oxygen sensor harness.
6. Remove the exhaust manifold heat shield.
7. Remove the manifold retaining bolts and engine hanger. Remove the manifold and gasket from the cylinder head.
8. Clean all gasket mating surfaces.

To install:

9. Install a new gasket, followed by the exhaust manifold, and engine hanger. Install the exhaust manifold bolts and nuts, tighten evenly, to 17 ft. lbs. (23 Nm), working from the center towards the ends.
10. Install the manifold heat shield and tighten the retaining bolts to 11 ft. lbs. (15 Nm).
11. Raise and safely support the vehicle.
12. Connect the exhaust pipe to the manifold with a new seal. Tighten the bolts to 30–43 ft. lbs. (40–60 Nm).
13. Lower the vehicle.
14. Reconnect the oxygen sensor harness.
15. Reconnect the negative battery cable. Start the engine and check for leaks.

Prizm

1. Disconnect the negative battery cable.
2. Disconnect the oxygen sensor, if equipped, sub-oxygen sensor connectors.
3. Disconnect the exhaust pipe from the exhaust manifold or, if equipped, from the Warm Up Three Way Catalytic Converter (WU TWC).

4. Remove the five bolts and the upper heat shield.

5. Remove the bolts and the exhaust manifold brace.

6. Remove the nuts, then the exhaust manifold and gasket. If equipped, remove the manifold with the WU TWC attached.

7. Remove the three bolts and the lower heat insulator from the exhaust manifold.

8. If equipped, remove the four nuts and two WU TWC heat insulators.

9. If equipped, remove the two bolts, two nuts and WU TWC from the exhaust manifold.

To install:

10. Clean all gasket mating surfaces.

11. If equipped, install the bolts, nuts and WU TWC to the exhaust manifold. Tighten the bolts and nuts to 80 inch lbs. (9 Nm).

12. If equipped, install the nuts and WU TWC heat insulators. Tighten the nuts to 80 inch lbs. (9 Nm).

13. Install the three bolts and the lower heat insulator to the manifold and tighten to 80 inch lbs. (9 Nm).

14. Install a new gasket and the exhaust manifold to the cylinder head. Tighten the nuts evenly, to 25 ft. lbs. (34 Nm), working from the center towards the ends.

15. Install the exhaust manifold brace and tighten the bolts to 29 ft. lbs. (39 Nm).

16. Install the upper heat insulator and tighten the bolts to 80 inch lbs. (9 Nm).

17. Using a new gasket, connect the exhaust pipe to the exhaust manifold or WU TWC, as required. Install the nuts and tighten to 32 ft. lbs. (43 Nm).

18. Connect the oxygen sensor harness, if equipped, sub-oxygen sensor connector.

19. Connect the negative battery cable. Run the engine and check for exhaust leaks.

Front Crankshaft Seal

REMOVAL & INSTALLATION

❊❊ CAUTION

Some models covered by this manual may be equipped with a Supplemental Restraint System (SRS), which uses an air bag. Whenever working near any of the SRS components, such as the impact sensors, the air bag module, steering column and instrument panel, properly disable the SRS.

Metro

1. Disconnect the negative battery cable.
2. Raise and safely support the vehicle.

❊❊ CAUTION

The EPA warns that prolonged contact with used engine oil may cause a number of skin disorders, including cancer! You should make every effort to minimize your exposure to used engine oil. Protective gloves should be worn when changing the oil. Wash your hands and any other exposed skin areas as soon as possible after exposure to used engine oil. Soap and water, or waterless hand cleaner should be used.

3. Drain the engine oil.

4. Remove the timing belt and crankshaft sprocket.

5. Remove the oil pan and strainer.

6. Remove the oil pump bolts and pump assembly. Identify the bolts so they can be reinstalled in their proper locations.

7. Remove the crankshaft oil seal from the oil pump using a suitable removal tool.

To install:

8. Install a new crankshaft oil seal into the oil pump body using a suitable seal installer tool. Lubricate the lip of the new seal with clean engine oil.

9. Using new gaskets, install the oil pump on the engine. Use Oil Seal Protector tool J-34853 to aid installation of the oil seal over the crankshaft without damaging the seal. The lip of the seal must **NOT** be turned out.

➡**Take care not to damage the front seal on the crankshaft snout.**

10. Apply Loctite® pipe sealant or equivalent to the pump bolts and install. Tighten to 97 inch lbs. (11 Nm).

➡**After tightening, the oil pump gasket may bulge out. Trim the edge with a knife, making sure the edge is smooth with the cylinder block.**

11. Install the rubber seal between the oil and coolant pumps.

12. Install the crankshaft sprocket. With the crankshaft locked, tighten the bolt to 81 ft. lbs. (110 Nm).

13. Install the oil strainer and pan.

14. Install the timing belt.

15. Lower the vehicle and fill the engine with clean engine oil.

❊❊ WARNING

Operating the engine without the proper amount and type of engine oil will result in severe engine damage.

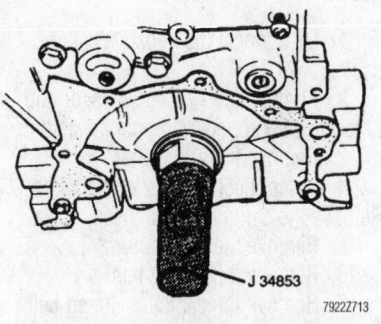

J 34853

79222713

To prevent damage to the lip of the seal use an oil seal protector as shown—Metro

16. Connect the negative battery cable, start the engine and check for leaks.

Prizm

1. Disconnect the negative battery cable.
2. Remove the timing belt and crankshaft sprocket.
3. Cut off the oil seal lip.

❊❊ WARNING

When removing the oil seal, be careful not to damage the crankshaft or the oil pump housing.

4. Carefully pry out the oil seal from the pump housing.

To install:

5. Apply clean engine oil to the new oil seal and lubricate the lip of the new oil seal with multi-purpose grease.

6. Install the new oil seal to the oil pump using a suitable seal driver.

7. Install the timing belt and crankshaft sprocket.

8. Connect the negative battery cable. Check the engine oil level.

9. Run the engine and check for leaks.

Camshaft

REMOVAL & INSTALLATION

❊❊ CAUTION

Some models covered by this manual may be equipped with a Supplemental Restraint System (SRS), which uses an air bag. Whenever working near any of the SRS components, such as the impact sensors, the air bag module, steering column and instrument panel, properly disable the SRS.

Metro

1. Disconnect the negative battery cable.

2. Remove the A/C compressor and bracket, if equipped, and position aside.

3. Remove the air cleaner.

4. Disconnect the spark plug wires. Remove the cylinder head cover.

5. Remove the distributor.

6. Remove the timing belt.

7. Remove the camshaft timing belt sprocket. Lock the camshaft with a 0.39 in. (10mm) rod inserted into the hole in the camshaft, before loosening the sprocket retaining bolt.

✳✳ WARNING

The mating surface of the cylinder head and cover must not be damaged during this procedure. Place a clean shop cloth between the rod and mating surfaces and use care not to bump the rod when loosening.

8. On 1.3L engines, remove the rocker arms and shafts from the cylinder head. Keep all parts in order so they can be reinstalled in their original locations. The intake and exhaust rocker arm shafts are different. Be sure to identify them as they are removed.

9. Turn the crankshaft until the crankshaft sprocket timing mark is 60 degrees to the left of the arrow mark on the oil pump case.

10. Remove the camshaft caps from the cylinder head.

11. Remove the camshaft from the cylinder head.

12. On 1.3L engines remove the camshaft from the cylinder head by removing it from the flywheel end.

13. Inspect the camshaft journals and lobes for wear and/or damage and replace, as necessary.

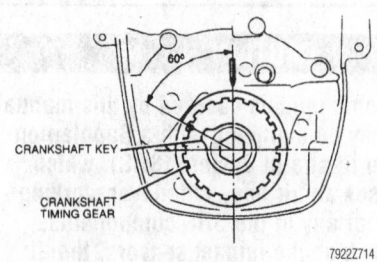

Position the crankshaft as shown before removing the camshaft—Metro with 1.0L engine

14. Using a micrometer, measure the camshaft lobe height. If the camshaft lobe height is below the minimum limit, replace the camshaft. The minimum camshaft specifications are as follows:

- **Metro XFi** —1.5562 in. (39.528mm)
- **Metro Standard and LSi** —1.5872 in. (40.315mm)

15. On 1.3L engines, inspect the camshaft for excessive wear or scoring and replace, if necessary. Using a micrometer, measure the cam lobe height; if it is less than 1.49575 in. (38.036mm), replace the camshaft.

To install:

16. Fill the oil passage in the cylinder head with clean engine oil. Pour engine oil through the camshaft journal oil holes and check that engine oil comes out from the oil holes in the HVL adjuster bores.

17. Install the camshaft to the cylinder head. After applying engine oil to the camshaft journal and all around the cam, position the camshaft so the camshaft timing sprocket pin hole in camshaft is at the lower position.

18. On 1.0L engines, install the camshaft bearing caps to the camshaft and the cylinder head as follows:

a. Apply clean engine oil to the sliding surface of each bearing cap against the camshaft journal.

b. Apply silicone sealant to the mating surface of the No. 1 and No. 3 bearing cap which will mate with the cylinder head.

c. There are marks provided on each camshaft bearing cap indicating the position and direction for installation. Install the bearing cap as indicated by the marks.

d. Camshaft bearing cap No. 1 is installed first. It retains the camshaft in the proper position and thrust direction. Apply clean engine oil to the retaining bolts and install, but do not tighten fully.

e. Install the remaining camshaft bearing caps and bolts.

f. Tighten the bolts evenly, repeating the tightening sequence 3–4 times. Tightened to 8 ft. lbs. (11 Nm).

➡**On 1.3L engines, exhaust and intake rocker shafts are different. To distinguish between the two, the end dimensions are different. Install the intake rocker arm shaft with the stepped end facing the distributor side.**

19. On 1.3L engines, install the intake rocker arm shaft and arms. Tighten the rocker arm shaft retaining bolts to 97 inch lbs. (11 Nm).

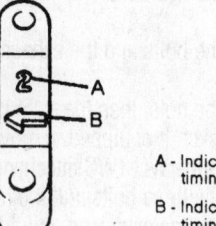

A - Indicates position from timing belt

B - Indicates direction to timing belt

Directional markings for the camshaft bearing caps—Metro with 1.0L engine

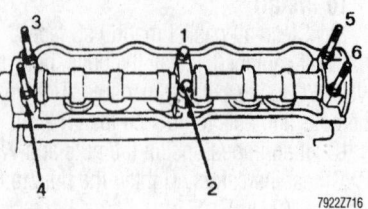

Torque sequence for the camshaft bearing cap bolts—Metro with 1.0L engine

20. On 1.3L engines, install the exhaust rocker arm shaft and arms. Tighten the rocker arm shaft retaining bolts to 97 inch lbs. (11 Nm).

21. Install the camshaft oil seal. After applying engine oil to the seal lip, press-fit the camshaft oil seal until the seal surface until flush with the housing.

22. Install the camshaft timing belt sprocket after installing the dowel pin. Tighten the retaining bolt to 44 ft. lbs. (60 Nm).

23. Using a new valve cover gasket, apply a small amount of silicone sealant to the corners of the cylinder gasket, install the valve cover the cover nuts. Tighten the cylinder head cover nuts to 44 inch lbs. (5 Nm).

24. Install the timing belt.

25. Install the distributor to the cylinder head. Connect the spark plug wires.

26. Install the A/C compressor and bracket, if equipped.

27. Install the air cleaner and connect the negative battery cable.

28. Start the engine and check. Adjust the ignition timing.

➡**When the engine is started, if air is trapped in the HVL adjuster, the valve may make a tapping sound when the engine is operated after the HVL adjuster is installed. In such a case, run the engine at 2000 rpm until the air is purged and the tapping sound ceases.**

Prizm

1. Disconnect the negative battery cable.

2. Disconnect the spark plug wires.
3. Remove the valve cover.
4. Remove the timing belt.
5. Set the exhaust camshaft so the knock pin is slightly above the cylinder head (10 O'clock position). This angle allows the No. 1 and No. 3 cylinder cam lobes of the intake camshaft to push the lash adjusters evenly.
6. Remove the two bolts and the front bearing cap of the intake camshaft.
7. Secure the intake camshaft end gear to the sub gear with a service bolt. The service bolt should match the following specifications:
   - Thread diameter: 6.0mm
   - Thread pitch: 1.0mm
   - Bolt length: 16mm
8. Uniformly loosen each intake camshaft bearing cap bolt in several passes in the proper sequence.

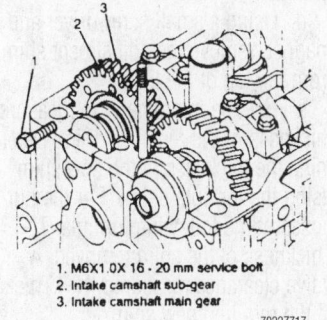

1. M6X1.0X 16 - 20 mm service bolt
2. Intake camshaft sub-gear
3. Intake camshaft main gear

7922Z717

Fastening the sub-gear to the main gear—Prizm

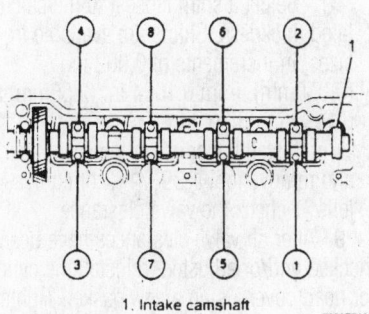

1. Intake camshaft

7922Z718

Intake camshaft bearing cap removal sequence—Prizm

9. Remove the bearing caps and intake camshaft.

➡ If the camshaft cannot be removed straight and level, install and retighten the No. 3 bearing cap. Alternately loosen the bolts on the bearing cap a little at a time while pulling upwards on the camshaft gear. DO NOT attempt to pry or force the cam loose.

10. Turn the exhaust camshaft approximately 105 degrees, so the guide pin is just past the 5 O'clock position. This angle allows the No. 1 and the No. 3 cylinder cam lobes of the exhaust camshaft to push the lash adjusters evenly.
11. Loosen the exhaust camshaft bearing cap bolts uniformly in several passes in sequence.
12. Remove the bearing caps and exhaust camshaft.

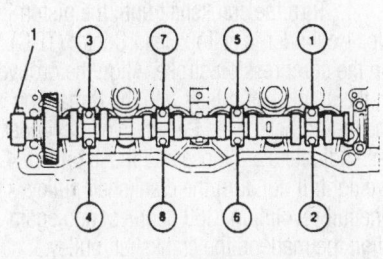

1. Exhaust camshaft

7922Z719

Exhaust camshaft bearing cap removal sequence—Prizm

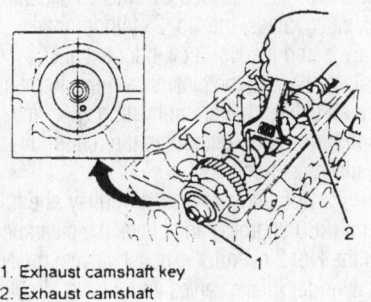

1. Exhaust camshaft key
2. Exhaust camshaft

7922Z720

Positioning the exhaust camshaft for removal—Prizm

➡ If the camshaft cannot be removed straight and level, install and retighten the No. 3 bearing cap. Alternately loosen the bolts on the bearing cap a little at a time while pulling upwards on the camshaft gear. DO NOT attempt to pry or force the cam loose.

To install:

13. Apply multi-purpose grease to the thrust portion of the camshaft.
14. Place the exhaust camshaft on the cylinder head so the cam lobes press evenly on the lash adjusters for cylinders Nos. 1 and 3. This will place the guide pin on the camshaft slightly counter clockwise at about 5 O'clock.
15. Apply a light coat of clean engine oil to the camshaft bearing cap bolts. Install the five bearing caps in position according to the number cast into the cap. The arrow should point towards the pulley end of the motor.
16. Tighten the bearing cap bolts uniformly and in several passes in the proper sequence to 9 ft. lbs. (13 Nm)

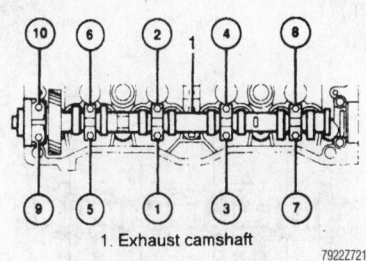

1. Exhaust camshaft

7922Z721

Exhaust camshaft bearing cap tightening sequence—Prizm

17. Apply multi-purpose grease to a new exhaust camshaft oil seal.
18. Install the exhaust camshaft seal using a seal driver. Be very careful not to install the seal on a slant.
19. Set the intake camshaft so the guide pin is slightly above the cylinder head.
20. Apply multi-purpose grease to the thrust portion of the intake camshaft.
21. Hold the intake camshaft next to the exhaust camshaft and engage the gears by matching the alignment marks.

➡ **DO NOT use the TDC timing marks for the timing belt.**

22. Keeping the gears engaged, roll the intake camshaft down and into the bearing journals. This angle allows the No. 1 and the No. 3 cylinder cam lobes of the camshaft to push the lash adjusters evenly.

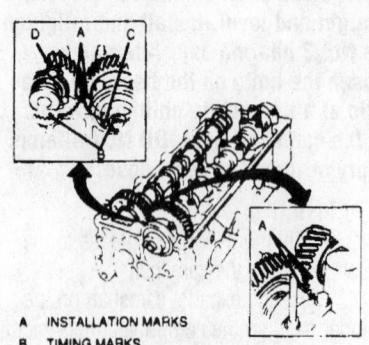

A   INSTALLATION MARKS
B   TIMING MARKS
C   EXHAUST CAMSHAFT KEY
D   INTAKE CAMSHAFT SERVICE BOLT

7922Z722

Engaging the intake camshaft with the exhaust camshaft—Prizm

23. Apply a light coat of clean engine oil to the camshaft bearing cap bolts and install the bearing caps. Observe the numbers on each cap and make certain the arrows point to the pulley end of the motor.

24. Uniformly tighten each of the bearing cap bolts in several passes in the proper sequence. Tighten each bolt to 9 ft. lbs. (13 Nm)

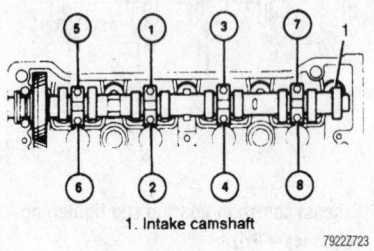

1. Intake camshaft

7922Z723

Intake camshaft bearing cap tightening sequence—Prizm

➡ **If the No. 1 bearing cap does not fit properly, push the camshaft gear backwards by prying apart the cylinder head and camshaft gear with a suitable tool.**

25. Turn the exhaust camshaft clockwise, and set it with the guide pin facing upward. Check that the timing marks of the camshaft gears are aligned. The camshaft assembly installation marks should now be in the 12 O'clock position.

26. Secure the exhaust camshaft and install the timing belt pulley. Tighten the bolt to 43 ft. lbs. (59 Nm).

27. Install the timing belt.

28. Install the spark plug wires and connect the negative battery cable.

29. Start the engine, check for leaks. Check the ignition timing.

Valve Lash

ADJUSTMENT

Metro

Hydraulic Valve Lash (HVL) adjusters, located between the camshaft and valve stems, are used to adjust the valve clearance to zero lash automatically at all times. Adjustment is not required.

Prizm

➡ **The following tools and part are needed for this procedure; Valve clearance adjustment tool set J-39871 or equivalent, a feeler gauge, a 0–1 inch micrometer and a selection of valve adjustment shims. Check and adjust the valve clearance with the engine cold.**

1. Disconnect the negative battery cable.

2. Disconnect the ignition wires, then remove the spark plugs.

3. Remove the retaining nuts and remove the cylinder head cover.

4. Turn the crankshaft until the piston in No. 1 cylinder is at Top Dead Center (TDC) on the compression stroke. Align the groove in the crankshaft pulley with the **0** mark on the timing belt cover. Be sure No. 1 cylinder lash adjusters are loose and those on No. 4 are tight. If not, turn the crankshaft pulley one full revolution (360 degrees) and again align the mark on the crankshaft pulley.

5. The intake valve clearance should be 0.006–0.010 in. (0.15–0.25mm) and the exhaust valve clearance should be 0.010–0.014 in. (0.25–0.35mm).

6. Using a feeler gauge, measure the clearance between the camshaft and valve lifter shim at the No. 1 cylinder intake and exhaust valves, the No. 2 cylinder intake valves and the No. 3 cylinder exhaust valves. Record the clearance measurements for all valves that are not within specification, in order to determine the required replacement shims.

7. Rotate the crankshaft pulley one full turn (360 degrees) and check the clearance at the No. 2 cylinder exhaust valves, the No. 3 cylinder intake valves and the No. 4 cylinder intake and exhaust valves. Record the

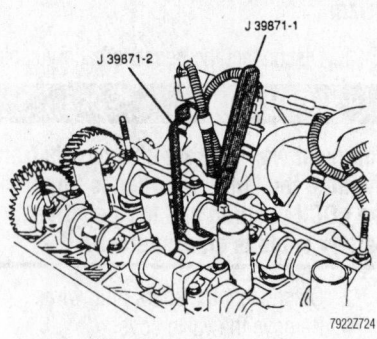

7922Z724

Using the valve clearance adjustment tool set to hold the lifter—Prizm

clearance measurements for all valves that are not within specification, in order to determine the required replacement shims.

8. For valves requiring adjustment, proceed as follows:

a. Be sure the base of the camshaft lobe is directly over the valve (camshaft lobe pointing away from the valve).

b. Insert tool J-3987–1 or equivalent, between the camshaft and lifter adjustment shim, to compress the valve spring and push the lifter down.

c. Insert tool J-39871–2 or equivalent, between the camshaft and the lifter, to hold the lifter away from the camshaft. Position the bottom edge of the tool on the lifter.

d. Using a small screwdriver and a magnet, remove the adjustment shim from the top of the lifter.

e. Use the micrometer to measure the thickness of the removed shim. Determine the thickness of the new shim using the formula below. For the purposes of the following formula, T = Thickness of the shim removed; A = Valve clearance measured; N = Thickness of the required new shim.

• For the intake camshaft valves: $N = T + (A—0.008$ in. $(0.20mm))$

• For the exhaust camshaft valves: $N = T + (A—0.010$ in. $(0.25mm))$

f. Select a shim closest to the calculated thickness. Shims are available in 16 sizes, in increments of 0.002 in. (0.050mm), from 0.1004 in. (2.55mm) to 0.1299 in. (3.30mm).

g. Install the shim on the valve lifter and remove tool J-39871–2 or equivalent. Recheck the valve clearance.

9. After all valve clearances have been checked and/or adjusted, reinstall the cylinder head cover, using a new gasket. Tighten the nuts to 53 inch lbs. (6 Nm).

10. Install the spark plugs and connect the ignition wires.

11. Connect the negative battery cable.

12. Run the engine and check operation.

Oil Pan

REMOVAL & INSTALLATION

1. Disconnect the negative battery cable.

2. Raise and safely support the vehicle.

3. Drain the engine oil.

4. Remove the flywheel dust cover.

5. On Prizm, remove the splash shields.

6. Separate the support hanger from the catalytic converter. Remove the front exhaust pipe between the exhaust manifold and the resonator/tail pipe assembly.

7. On Prizm, perform the following;

   a. Remove the bolts and lower the front end of the center support.

   b. Remove the lower engine reinforcement brace-to-engine bolts and remove the brace.

8. Remove the crankshaft position sensor.

9. On Prizm, remove the three brace-to-transaxle bolts. Remove the 14 bolts from the powertrain reinforcement brace. Remove the six Torx® bolts and powertrain reinforcement brace from the vehicle.

➡**When removing the powertrain reinforcement brace, do not pry on the oil pump body or the rear main seal retainer. Be careful not to damage the contact surfaces of the powertrain reinforcement brace, cylinder block, oil pump or the rear main seal retainer.**

10. Remove the oil pan bolts, nuts and oil pan from the vehicle.

11. Remove the oil pump strainer.

12. Clean all gasket material from the mating surfaces.

   To install:

13. Install the oil pump strainer with a new seal. Secure with bolt and tighten to 97 inch lbs. (11 Nm).

14. Apply a continuous bead of RTV sealant to the engine oil pan and install. Tighten the bolts and nuts to 97 inch lbs. (11 Nm) starting in the center and working outward.

15. Install the exhaust pipe with a new seal to the manifold and secure with bolts. Do not fully tighten the bolts.

16. On Prizm, apply GM 1050026 gasket paste or equivalent, to the powertrain reinforcement brace-to-cylinder block mating surface.

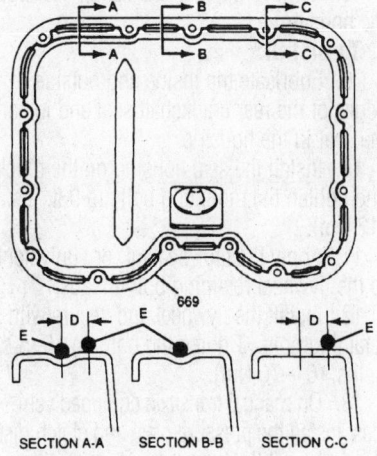

SECTION A-A **SECTION B-B** **SECTION C-C**

D   6.0 mm (0.24")
E   SILICONE SEALER
669   OIL PAN

79222Z725

To ensure a leak-free seal, apply sealer as shown—Prizm 1.8L engine

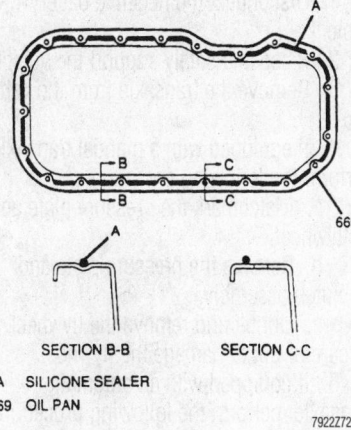

SECTION B-B   **SECTION C-C**

A   SILICONE SEALER
669   OIL PAN

79222Z726

To ensure a leak-free seal, apply sealer as shown—Prizm 1.6L engine

17. Install the powertrain reinforcement brace. Tighten the Torx® bolts to 12 ft. lbs. (16 Nm). Tighten the remaining bolts to 69 inch lbs. (7.8 Nm). Install the three brace-to-transaxle bolts and tighten to 17 ft. lbs. (23 Nm).

18. On Prizm, perform the following;

   a. Install the lower engine reinforcement brace to the vehicle and tighten the bolts to 47 ft. lbs. (64 Nm).

   b. Install the center support and tighten the bolts to 45 ft. lbs. (61 Nm).

19. Install the front pipe to the resonator/muffler/tail pipe assembly using a new gasket and secure with nuts.

20. Install the support hanger to the catalytic converter. If equipped, tighten the hanger bolts to 11 ft. lbs. (15 Nm).

21. Tighten the front exhaust pipe-to-exhaust manifold bolts to 37 ft. lbs. (50 Nm). Tighten the front exhaust pipe-to-resonator/muffler/tail pipe assembly nuts to 26 ft. lbs. (35 Nm).

22. Install the flywheel dust cover.

23. Lower the vehicle and fill the engine with oil.

24. Reconnect the negative battery cable, start the engine and check for leaks.

Oil Pump

REMOVAL & INSTALLATION

1. Disconnect the negative battery cable.

2. Raise and safely support the vehicle.

3. Drain the engine oil.

4. Remove the water pump belt and pulley, alternator belt, body and bracket.

5. Remove the air conditioning compressor and mounting bracket, if equipped.

6. On Prizm, if equipped, unclip the crankshaft position sensor electrical connector from the dipstick tube. Remove the bolt and the dipstick tube from the oil pump. Remove the dipstick tube O-ring from the oil pump.

7. Remove the timing belt and crankshaft pulley.

8. Disconnect the engine oil level gauge.

9. Remove the crankshaft timing belt sprocket.

10. Remove the oil pan and strainer assembly.

11. Remove the oil pump bolts and the oil pump assembly. Identify the bolts so they can be reinstalled in their proper locations. If necessary, gently tap on the back side of the oil pump with a plastic or rubber faced hammer to free it from the cylinder block.

12. Remove the crankshaft oil seal from the oil pump.

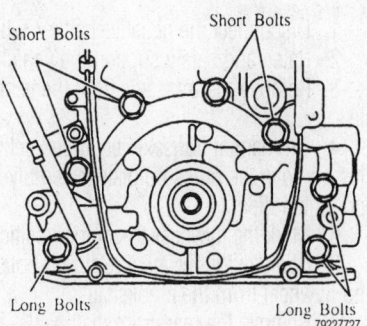

Short Bolts   Short Bolts

Long Bolts   Long Bolts

79222Z727

Oil pump mounting bolt identification—Metro

13. Remove the pump rotor plate pins from the oil pump. Clean the gasket mating surfaces thoroughly.

To install:

14. Install the pump rotor plate pins into the pump body.

15. Install a new crankshaft oil seal using an oil/grease seal installer tool.

16. Lubricate the oil pump with fresh engine oil.

17. Using new gaskets, install the oil pump. Install the crankshaft seal.

18. Apply Loctite® pipe sealant or equivalent to the oil pump bolts and install. On Metro, tighten the bolts to 97 inch lbs. (11 Nm) or on Prizm, tighten the bolts to 16 ft. lbs. (21 Nm).

➡️**After tightening, the oil pump gasket may bulge. Trim the edge with a knife, making sure the edge is smooth with the cylinder block.**

19. Install the rubber seal between the oil and coolant pump.

20. Install the crankshaft belt gear and guide. Tighten the pulley bolt to 81 ft. lbs. (110 Nm).

21. On Prizm, install a new dipstick tube O-ring to the oil pump. Install the dipstick tube and secure. Tighten the bolt to 84 inch lbs. (9.5 Nm). If equipped, clip the crankshaft position sensor electrical connector to the dipstick tube.

22. Install the engine oil level gauge.

23. Install the oil strainer and oil pan.

24. Install the timing belt.

25. Install the remaining components.

26. Lower the vehicle and fill the engine with clean engine oil.

27. Connect the negative battery cable. Start the engine and check for leaks.

Rear Main Seal

REMOVAL & INSTALLATION

Metro

1. Disconnect the negative battery cable.
2. Raise and safely support the vehicle.
3. Remove the transaxle from the vehicle.
4. On manual transaxle equipped vehicles, remove the pressure plate assembly and clutch disc.
5. Mark the flywheel-to-engine position.
6. Remove the flywheel retaining bolts and flywheel from the crankshaft.
7. Remove the rear crankshaft seal housing attaching bolts and remove the seal housing.

8. Remove the rear crankshaft seal from the housing.

To install:

9. Lubricate the inside and outside edges of the rear crankshaft seal and install the seal in the housing.

10. Install the seal housing on the block and tighten the mounting bolts to 9 ft. lbs. (12 Nm).

11. Apply Loctite® sealant, or equivalent, to the flywheel retaining bolt threads.

12. Install the flywheel and secure with retaining bolts. Tighten the bolts to 45–52 ft. lbs. (61–70 Nm).

13. On manual transaxle equipped vehicles, install the pressure plate and clutch disc.

14. Install the transaxle into the vehicle.

15. Lower the vehicle.

16. Reconnect the negative battery cable.

17. Check the engine oil level.

18. Run the engine and check for leaks.

Prizm

1. Disconnect the negative battery cable.
2. Raise and safely support the vehicle.
3. Remove the transaxle from the vehicle.
4. If equipped with a manual transaxle, perform the following procedures:

   a. Matchmark the pressure plate and flywheel.

   b. Remove the pressure plate and clutch assembly.

   c. Unbolt and remove the flywheel. Be careful not to damage the flywheel.

5. If equipped with an automatic transaxle, perform the following procedures:

   a. Matchmark the torque converter flexplate and crankshaft.

   b. Unbolt and remove the torque converter flexplate.

6. Remove the bolts holding the rear end plate to the engine and the remove the rear end plate.

7. Loosen the rear oil seal retainer bolts in a crisscross pattern. Loosen and remove the two bolts securing the oil pan to the rear oil seal retainer,, then remove the rear oil seal retainer.

8. Using a small prybar, pry the rear oil seal retainer from the mating surfaces. Don't pry around the crankshaft, or it may be damaged. Be sure not to damage any of the oil seal retainer surfaces. If necessary, wrap a cloth around the edge of the prybar.

9. Using a hammer and small punch or a seal driver, drive the oil seal from the rear bearing retainer.

To install:

10. Clean the gasket mounting surfaces.

The contact surfaces are completely free of oil and foreign matter.

11. Clean the oil seal mounting surface.

12. Using multi-purpose grease, lubricate the lip of the new seal.

13. Use a properly sized seal installation driver to tap the new oil seal straight into the retainer. When properly installed, the oil seal should be flush with the surface of the seal retainer.

14. Position a new gasket on the retainer and coat it lightly with gasket sealer. Fit the seal retainer into place on the engine; be careful when installing the oil seal over the crankshaft.

15. Install the retaining bolts and tighten them to 7 ft. lbs. (10 Nm) in a crisscross pattern.

16. Install the rear end plate. Tighten the bolts to 7 ft. lbs. (10 Nm).

17. Install either the flexplate (automatic transaxle) or the flywheel (manual transaxle), carefully observing the matchmarks made earlier. The front flexplate spacer should be installed so that its chamfered internal edge is facing the crankshaft. The rear flexplate spacer should be installed with its curved side facing out. Apply a small amount of thread locking compound to the first 2–3 threads of each flexplate/flywheel bolt before installation.

18. Tighten the flexplate (automatic transaxle) or the flywheel (manual transaxle) bolts to the correct torque specification following a two-step crisscross pattern.

• Vehicles with manual transaxles: 58 ft. lbs. (78 Nm)

• Vehicles with automatic transaxles: 47 ft. lbs. (64 Nm).

19. Install the clutch disc and pressure plate, if equipped with manual transaxle.

20. Install the transaxle.

21. Lower the vehicle.

22. Reconnect the negative battery cable.

23. Check the engine oil level and add oil as necessary.

24. Start the engine and check for leaks.

FUEL SYSTEM

Fuel System Service Precautions

Safety is the most important factor when performing not only fuel system maintenance but any type of maintenance. Failure to conduct maintenance and repairs in a safe manner may result in serious personal injury or death. Maintenance and testing of the vehicle's fuel system components can be accomplished safely and effectively by adhering to the following rules and guidelines.

• To avoid the possibility of fire and personal injury, always disconnect the negative battery cable unless the repair or test procedure requires that battery voltage be applied.

• Always relieve the fuel system pressure prior to disconnecting any fuel system component (injector, fuel rail, pressure regulator, etc.), fitting or fuel line connection. Exercise extreme caution whenever relieving fuel system pressure to avoid exposing skin, face and eyes to fuel spray. Please be advised that fuel under pressure may penetrate the skin or any part of the body that it contacts.

• Always place a shop towel or cloth around the fitting or connection prior to loosening to absorb any excess fuel due to spillage. Ensure that all fuel spillage (should it occur) is quickly removed from engine surfaces. Ensure that all fuel soaked cloths or towels are deposited into a suitable waste container.

• Always keep a dry chemical (Class B) fire extinguisher near the work area.

• Do not allow fuel spray or fuel vapors to come into contact with a spark or open flame.

• Always use a back-up wrench when loosening and tightening fuel line connection fittings. This will prevent unnecessary stress and torsion to fuel line piping. Always follow the proper torque specifications.

• Always replace worn fuel fitting O-rings with new. Do not substitute fuel hose or equivalent, where fuel pipe is installed.

Fuel System Pressure

RELIEVING

✳✳ CAUTION

The fuel injection system remains under pressure, even after the engine has been turned OFF. The fuel system pressure must be relieved before disconnecting any fuel lines. Failure to do so may result in fire and/or personal injury.

1. Remove the fuel filler cap.
2. On Metro, perform the following;
   a. Remove the control relay box cover from the relay box.

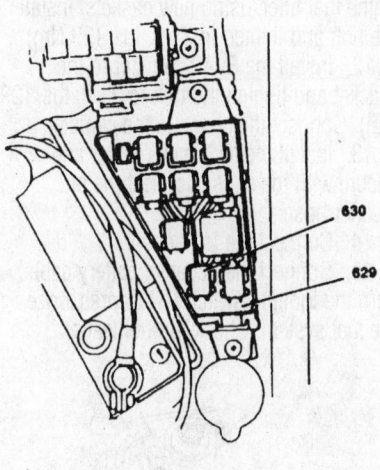

629 **RELAY BOX**
630 **FUEL PUMP RELAY**

79222728

View of the relay box showing the location of the fuel pump relay—Metro

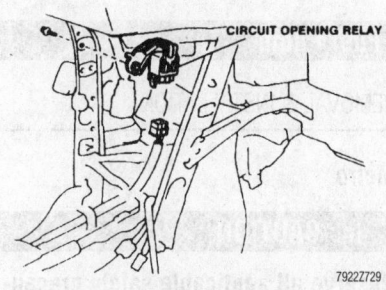

CIRCUIT OPENING RELAY

79222729

Leave the circuit opening relay unpluged during fuel system servicing—Prizm

b. Disconnect the fuel pump relay from the relay box connector.
3. On Prizm, perform the following;
   a. Remove the center trim bezel from the center console by gently prying around the edges.
   b. Remove the radio from the center console.
   c. Disconnect the circuit opening relay, located in the center console.
4. Attempt to start the engine, if it starts, let it run until the engine stalls due to lack of fuel.

5. Engage the starter for a few seconds to assure relief of remaining fuel pressure.
6. Disconnect the negative battery cable.
7. Continue with the required service procedure(s).

Fuel Filter

REMOVAL & INSTALLATION

Metro

✳✳ CAUTION

Observe all applicable safety precautions when working around fuel. Whenever servicing the fuel system, always work in a well ventilated area. Do not allow fuel spray or vapors to come in contact with a spark or open flame. Keep a dry chemical fire extinguisher near the work area. Always keep fuel in a container specifically designed for fuel storage; also, always properly seal fuel containers to avoid the possibility of fire or explosion.

1. Properly relieve the fuel system pressure.
2. Disconnect the negative battery cable.
3. Raise and safely support the vehicle.
4. Remove the bolt securing the parking brake cable bracket from the underbody.
5. Disconnect the fuel feed hose from the fuel filter.
6. Remove the two fuel filter mounting bracket bolts, and remove the fuel filter from the frame.
7. Disconnect the outlet hose from the fuel filter.
8. Remove the fuel filter from the bracket by removing the mounting bolt.
   To install:
9. Install the fuel filter on the bracket. Install the mounting bolt and tighten to 11 ft. lbs. (15 Nm).

➡Be sure the matchmarks between the fuel filter and the mounting bracket are aligned before tightening the bolt.

10. Connect the fuel feed outlet hose to the fuel filter. Secure with a new clamp.
11. Install the fuel filter on the frame and install the two bracket bolts. Tighten the bracket bolts to 11 ft. lbs. (15 Nm).
12. Connect the fuel inlet hose to the fuel filter. Secure with a new clamp.
13. Position the brake cable bracket and install the bolt. Tighten the cable bracket bolt to 11 ft. lbs. (15 Nm).

14. Lower the vehicle.

15. Reconnect the negative battery cable.

16. Turn the ignition switch to **ON** , then back to **LOCK** to pressurize the fuel system. Check for leaks and correct as necessary.

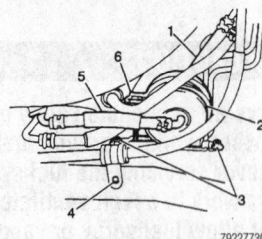

1. Fuel Feed Outlet Hose
2. Fuel Filter
3. Fuel Filter Mounting Bolts
4. Parking Brake Cable Bracket Bolt
5. Fuel Feed Inlet Hose
6. Fuel Filter Mounting Bracket to Fuel Filter Bolt

79222Z730

Component view of the fuel filter mounting—Metro

Prizm

✳✳ CAUTION

Observe all applicable safety precautions when working around fuel. Whenever servicing the fuel system, always work in a well ventilated area. Do not allow fuel spray or vapors to come in contact with a spark or open flame. Keep a dry chemical fire extinguisher near the work area. Always keep fuel in a container specifically designed for fuel storage; also, always properly seal fuel containers to avoid the possibility of fire or explosion.

1. Properly relieve the fuel system pressure.

2. Disconnect the negative battery cable.

3. Disconnect the Intake Air Temperature (IAT) sensor.

4. Loosen the air cleaner clamp and release the four air cleaner housing cover clips. Remove the hose and air cleaner cover.

5. Disconnect the hoses from the Evaporative Emissions (EVAP) canister. Remove the bolt and canister from the bracket.

6. Remove the bolt and disconnect the fuel outlet hose from the top of the fuel filter. Discard the gaskets.

7. Remove the two bolts and remove the fuel filter from the bracket.

8. Remove the nut and fuel inlet pipe from the bottom of the fuel filter.

To install:

9. Install the fuel filter to the bracket and tighten the bolts to 43 inch lbs. (5 Nm).

10. Connect the fuel inlet pipe to the bottom of the fuel filter and tighten the nut to 22 ft. lbs. (30 Nm).

11. Install the fuel outlet hose to the top of the fuel filter, using new gaskets. Install the bolt and tighten to 21 ft. lbs. (29 Nm).

12. Install the EVAP canister to the bracket and tighten the bolt to 21 ft. lbs. (29 Nm). Connect the hoses to the canister.

13. Install the air cleaner cover and secure with the clips. Connect the air cleaner hose.

14. Connect the IAT sensor.

15. Connect the negative battery cable. Turn the ignition switch **ON** to pressurize the fuel system. Check for fuel leaks.

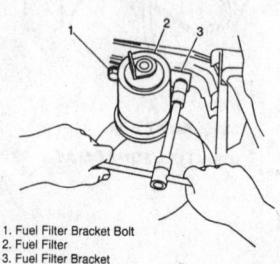

1. Fuel Filter Bracket Bolt
2. Fuel Filter
3. Fuel Filter Bracket

79222Z731

To remove the filter from a Prizm, detach the fuel lines and remove the two bracket bolts, then slide the filter out of the bracket—be sure to check for fuel leaks after replacing the filter

Fuel Pump

REMOVAL & INSTALLATION

Metro

✳✳ CAUTION

Observe all applicable safety precautions when working around fuel. Whenever servicing the fuel system, always work in a well ventilated area. Do not allow fuel spray or vapors to come in contact with a spark or open flame. Keep a dry chemical fire extinguisher near the work area. Always keep fuel in a container specifically designed for fuel storage; also, always properly seal fuel containers to avoid the possibility of fire or explosion.

1. Properly relieve the fuel system pressure.

2. Disconnect the negative battery cable.

3. Remove the filler cap. Drain the fuel tank by pumping the fuel out through the filler neck into a suitable container.

4. Reinstall the fuel filler cap.

5. Raise and safely support the vehicle.

6. Remove the fuel tank from the vehicle. Clean all dirt from the fuel pump area, to prevent fuel system contamination.

7. Remove the fuel feed and return clamps and hoses from the pump assembly.

8. Remove the screws from the pump assembly and remove with the gasket from the fuel tank.

9. Remove the screws from the pump motor assembly. Remove the fuel pump motor connectors and detach the pump motor from the pump assembly.

To install:

10. Install the two pump motor connectors. Install the pump motor on the pump assembly and tighten the screw.

11. Install the assembly with the gasket on the fuel tank. Install the screws and tighten.

12. Install the fuel feed and return hoses and clamps on the fuel pump assembly.

13. Install the fuel tank in the vehicle.

14. Lower the vehicle and reconnect the negative battery cable.

15. Refill the fuel tank. Turn the ignition key **ON** and allow the fuel system to pressurize. Check for leaks.

16. Start the engine and check for leaks.

Prizm

✳✳ CAUTION

Observe all applicable safety precautions when working around fuel. Whenever servicing the fuel system, always work in a well ventilated area. Do not allow fuel spray or vapors to come in contact with a spark or open flame. Keep a dry chemical fire extinguisher near the work area. Always keep fuel in a container specifically designed for fuel storage; also, always properly seal fuel containers to avoid the possibility of fire or explosion.

1. Properly relieve the fuel system pressure.

2. Use a pump to drain the fuel through the filler neck.

3. Remove the rear seat cushion to gain access to the service panel.

4. Remove the four screws, then the service access panel.

5. Disconnect the fuel pump and sending wiring.

6. Remove the nut and fuel feed hose from the sender.

7. Remove the clamp and fuel return hose from the sender.

8. Remove the bolts and remove the sender assembly from the tank.

9. Remove the pump from the bracket by pulling off the lower side of the pump from the bracket.

10. Disconnect the fuel pump electrical connector.

11. Remove the rubber cushion from the fuel pump.

12. Remove the clamp and disconnect the strainer from the pump.

To install:

13. Install the clamp and connect the strainer to the pump.

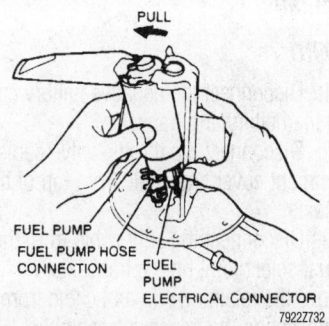

PULL

FUEL PUMP
FUEL PUMP HOSE
CONNECTION   FUEL
PUMP
ELECTRICAL CONNECTOR

79222732

Remove the fuel pump and the sender assembly from the fuel tank slowly to keep from splashing fuel—Prizm

14. Install the rubber cushion to the pump.

15. Connect the fuel pump electrical connector.

16. Install the pump to the bracket.

17. Install the fuel sender assembly to the tank, using a new gasket. Secure with the bolts, tightened to 35 inch lbs. (4 Nm).

18. Install the clamp and fuel return hose to the sender.

19. Install the nut and fuel feed hose to the sender. Tighten the nut to 22 ft. lbs. (30 Nm).

20. Connect the fuel pump electrical connector.

21. Refill the fuel tank.

22. Connect the negative battery cable. Turn the ignition **ON** to pressurize the fuel system. Check for fuel leaks.

23. Install the service access panel and rear seat cushion.

DRIVE TRAIN

Transaxle Assembly

REMOVAL & INSTALLATION

Manual

METRO

1. Disconnect the negative battery cable and ground strap at the transaxle.

2. Loosen the clutch cable adjusting nuts, then remove the cable from the bracket.

3. Disconnect the wiring harness attached to the transaxle.

4. Remove the speedometer cable boot, case clip and cable from the transaxle.

5. Remove the transaxle retaining bolts.

6. Remove the starter.

7. Remove the vacuum hose from the pressure sensor.

8. Install a suitable engine support.

9. Raise and safely support the vehicle. Drain the transaxle oil.

10. Remove the gear shift control shaft bolt and nut, then detach the control shaft from the gear shift shaft.

11. Remove the extension rod nut and rod with washers.

12. Remove the exhaust pipe front and rear flange bolts.

13. Remove the clutch housing lower plate.

14. Remove the wheels.

15. Remove both the tie rod ends.

16. Remove both halfshafts at the transaxle.

17. Support the transaxle with a suitable jack and remove the transaxle retaining bolts and nuts.

18. Remove the two rear engine mounting bolts.

19. Remove the bolts and nuts from the transaxle left-hand bracket and remove.

20. Lower the transaxle with the engine attached. Carefully pull the transaxle straight out toward the left side to disconnect the input shaft from the clutch cover. Lower and remove the transaxle assembly.

To install:

21. While the transaxle is being raised into the correct position, install the right side halfshaft.

22. Install the transaxle to the engine block using the retainer nuts and bolts. Install the left side bracket with the bolts and nuts. Tighten them to 37 ft. lbs. (50 Nm).

23. Install the rear engine mounting nuts and tighten to 37 ft. lbs. (50 Nm).

24. Lower the transaxle supporting jack. Tighten the transaxle to engine bolt and nut to 37 ft. lbs. (50 Nm).

25. Install the left halfshaft to the transaxle.

26. Install both tie rod ends.

27. Install the front wheels.

28. Install the clutch housing lower plate.

29. Install the exhaust pipe front and rear flange nuts.

30. Install the extension rod nut and washers. Tighten the nut to 24 ft. lbs. (33 Nm).

31. Install the control shaft to gear shift and the gear shift control shaft bolt and nut. Tighten the gear shaft bolt and nut to 13–15 ft. lbs. (18–20 Nm).

32. Refill the transaxle.

33. Lower the vehicle.

34. Remove the engine support fixture.

35. Install the vacuum hose to the pressure sensor.

36. Install the starter motor.

37. Install the transaxle retaining bolts. Tighten the retaining bolts to 37 ft. lbs. (50 Nm).

38. Install the speedometer cable, case clip and speedometer cable boot.

39. Install the clutch cable into the bracket, Adjust the cable free-play.

40. Connect the negative battery cable and ground strap to the transaxle.

PRIZM

1. Install Engine Support Fixture J-28467 or equivalent.

2. Disconnect the battery cables, negative cable first. Remove the battery and tray.

3. Remove the air cleaner.

4. Disconnect the reverse light and ground strap at the transaxle.

5. Remove the actuator mounting bolts and actuator line bracket.

6. Remove the shift cable retainers and end clips.

7. Remove the shift cables and place out of the way.

8. Remove the left transaxle mount cover and bracket. Remove the through-bolt from the mount.

9. Remove the two upper transaxle-to-engine bolts.

10. Remove the upper starter bolt. Remove the speedometer cable or disconnect the vehicle speed sensor (VSS), as applicable.

11. Raise and safely support the vehicle.

12. Remove the splash shields. Drain the transaxle oil.

13. Disconnect the starter wires. Remove the bottom starter bolt and starter.

14. Remove both halfshafts.

15. Remove the bolts holding the center crossmember to the radiator support.

16. For 1995 models, remove the following:

   a. Remove the front, center and rear mounting bolts.

   b. Remove the bolts holding the center crossmember to the main crossmember.

   c. Remove the exhaust hanger bracket nuts and the exhaust hanger.

17. For 1996–99 models, remove the following:

   a. Remove the bolts holding the center crossmember to the radiator support.

   b. Remove the front mount shield.

   c. Remove the front mounting bolts and rear mounting nuts.

   d. Remove the exhaust pipe support bolts.

18. Support the main crossmember. Remove the bolts holding the main crossmember to the body.

19. Remove the bolts holding the lower control arm brackets to the body.

❊❊ CAUTION

The crossmembers are loose and free to fall. Be sure they are properly supported.

20. Slowly lower the main crossmember while holding onto the center crossmember.

21. Remove the through-bolt and mount, at the front transaxle mount

22. Remove the front mounting bracket from the transaxle.

23. Remove the center mount from the transaxle.

24. Remove the inspection cover bolt.

25. Remove the two lower transaxle bracket-to-transaxle bolts.

26. Lower the vehicle.

27. Remove the transaxle mount.

28. Raise and safely support the vehicle.

29. Support the transaxle with a suitable jack.

30. Remove the lower front and rear bolts holding the transaxle to the engine.

31. Remove the transaxle assembly from the engine and carefully lower it from the vehicle.

   To install:

32. Raise the transaxle into position, making sure the input shaft aligns with the clutch splines.

33. Install the lower front and rear bolts holding the transaxle to the engine. Do not tighten them fully.

34. Remove the jack from under the transaxle.

35. Install the starter and wiring.

36. Lower the vehicle.

37. Install the left transaxle mounting bolts.

38. Raise the transaxle and install the through-bolt. Tighten the through-bolt to 64 ft. lbs. (87 Nm). Tighten the transaxle mounting bolts to 45 ft. lbs. (61 Nm).

39. Install the mount cover and bracket. Tighten the cover bolts to 45 ft. lbs. (61 Nm).

40. Install the upper transaxle to engine bolts and tighten them to 34 ft. lbs. (46 Nm).

41. Connect the speedometer cable or VSS connector.

42. Position the shift cables into the brackets and connect the cable retainers and end clips.

43. Install the two actuator mounting bolts, then install the actuator line bracket and bolt. Tighten the line and mounting bolts to 15 ft. lbs. (20 Nm).

44. Connect the ground strap and reverse lights harnesses.

45. Install the air cleaner.

46. Raise and safely support the vehicle.

47. Tighten the lower transaxle mount bolts to 45 ft. lbs. (61 Nm).

48. Install the center mount and bolts. Tighten them to 45 ft. lbs. (61 Nm). Install the inspection cover bolt and tighten it to 45 ft. lbs. (61 Nm).

49. Install the front mount bracket on the transaxle.

50. Install the front mount and through-bolt loosely.

➡**When installing the front mount, weight on the mount must go toward the transaxle.**

51. Position the central crossmember over the center and rear transaxle mount studs. Secure with nuts.

52. Loosely install the bolts holding the center crossmember to the radiator support.

53. Loosely install the front mount bolts.

54. Raise the main crossmember into position over the rear mount studs and align the bolts. Install the rear mount nuts loosely. Install the main crossmember to underbody bolts loosely.

55. Install the lower control arm bracket bolts loosely.

56. Loosely install the bolts holding center crossmember to the main crossmember.

57. Install the exhaust hanger bracket and hardware.

58. Tighten the components below to the correct torque specification:

   a. Main crossmember to underbody bolts: 152 ft. lbs. (206 Nm).

   b. Lower control arm bolts: 94 ft. lbs. (127 Nm) for 1995 models or 161 ft. lbs. (218 Nm) for 1996–99 models.

   c. Center crossmember to radiator support bolts: 45 ft. lbs. (61 Nm).

   d. Front, center and rear mount bolts: 45 ft. lbs. (61 Nm).

   e. Exhaust bracket bolts: 115 inch lbs. (13 Nm).

   f. Front mount through-bolt: 64 ft. lbs. (87 Nm).

59. Install the remaining components.

60. Install the halfshafts.

61. Install the battery tray and battery.

62. Connect the battery cables to the battery, negative cable last.

Automatic

METRO

1. Disconnect the negative battery cable from the battery and transaxle.

2. Disconnect the throttle valve cable adjustment cover and cable from top of the transaxle.

3. Disconnect the shift cable from the manual select joint on the transaxle. Remove the retaining clip and cable from the bracket on the transaxle assembly. Loosen the adjustment nuts.

4. Remove the accelerator cable from the bracket on top of the transaxle. Remove the engine harness bracket from the rear of the transaxle.

5. Disconnect the harnesses to the vehicle speed sensor, shift solenoid and transaxle range switch.

6. Remove the speedometer cable from the gear case.

7. Disconnect and plug the inlet and outlet fluid cooler lines.

8. Remove the starter motor. Drain the transaxle fluid.

9. Remove the rear transaxle/engine mount. Remove the upper transaxle-to-engine mount bolts from the transaxle.

10. Support the engine from the top using engine support tool J-28467-A and support fixture adapters J-28467–89 or equivalents.

11. Raise and support the vehicle safely. Remove the right and left splash shields.

12. Separate both ball joints from the steering knuckles.

13. Remove the right and left halfshaft assemblies. Remove the rear engine torque rod/engine mount from the transaxle case.

➡**Make alignment marks on the torque converter and drive plate for assembly reference.**

14. Remove the flywheel cover from the transaxle case. Lock the flywheel in place and remove the flywheel-to-torque converter bolts.

15. Remove the muffler mounting from the rear engine exhaust hanger.

16. Lower the engine slightly.

17. Support the transaxle and remove the transaxle-to-engine bolts. Slide the transaxle off the engine and lower it from the vehicle.

To install:

18. Use grease to lubricate the cup around the center of the torque converter.

19. Measure the distance between the torque converter and the edge of the transaxle housing; it should be at least 0.85 in. (21.4mm). If the distance is less than specified, the torque converter is improperly installed. Remove and reinstall it.

20. Position the transaxle to the engine. Install the lower transaxle-to-engine retaining bolt and tighten to 40 ft. lbs. (55 Nm).

21. Install the left transaxle mount and secure with the bracket bolts and nuts. Tighten the bolts and nuts to 40 ft. lbs. (55 Nm).

22. Remove the transaxle support jack.

23. Raise the engine assembly slightly.

24. Install the rear engine mount bracket to the transaxle. Install the bracket bolts and nuts and tighten to 40 ft. lbs. (55 Nm). Install the rear engine mount-to-bulkhead bolt and tighten to 40 ft. lbs. (55 Nm).

25. Install the muffler mounting to the rear exhaust hanger.

26. Install the flywheel-to-torque converter bolts. Lock the flywheel in place and tighten the flywheel-to-torque converter bolts to 14 ft. lbs. (19 Nm).

27. Install the flywheel inspection cover to the transaxle and tighten the bolts to 89 inch lbs. (10 Nm).

28. Install the rear engine torque rod assembly and tighten the bolts to 40 ft. lbs. (55 Nm).

29. Install the right and left side halfshaft assemblies.

30. Install the ball joints to the steering knuckles.

31. Install the right and left splash shields.

32. Remove the engine support fixture from the vehicle.

33. Install the upper transaxle-to-engine mounting bolts and tighten to 40 ft. lbs. (55 Nm).

34. Install the through-bolt and nut to the rear engine mount and tighten to 40 ft. lbs. (55 Nm).

35. Install the upper rear transaxle mount-to-bulkhead bolt and tighten to 40 ft. lbs. (55 Nm).

36. Install the starter motor.

37. Connect the inlet and outlet fluid cooler lines using new hose clamps.

38. Install the speedometer cable into the gear case.

39. Connect the harness to the VSS, shift solenoid and transaxle range switch.

40. Install the remaining components.

41. Connect the negative battery cable to the transaxle and battery.

42. Refill and check the transaxle fluid level.

PRIZM

1. Disconnect both battery cables, negative cable first, then remove the battery.

2. Disconnect the intake air temperature sensor, park neutral position switch, vehicle speed sensor and solenoid wiring harness, if equipped with a 4-speed transaxle.

3. Remove the air cleaner housing cover and filter, then remove the housing from the vehicle.

4. Disconnect the ground wire from the transaxle.

5. Remove the throttle valve cable from the throttle linkage and bracket.

6. Disconnect the vehicle speed sensor harness.

7. Remove the nut and disconnect the shift select cable from the manual lever. Remove the E-clip and disconnect the cable from the cable bracket.

8. Remove the bolt and the TV cable guide bracket from the transaxle case.

9. Remove the upper transaxle-to-engine bolts. Remove the starter motor.

10. Install engine/transaxle support fixture J-28467-A or equivalent.

11. Remove the bolts and left mounting bracket.

12. Raise and safely support the vehicle.

13. Drain the transaxle fluid.

14. On 3-speed transaxles, remove the differential filler plug at the rear of the transaxle and drain the differential fluid.

15. Remove the right and left splash shields.

16. Disconnect the fluid cooler hoses from the lines at the transaxle.

17. Remove the front wheels.

18. Remove the right and left half-shafts.

19. Remove the bolts from the exhaust pipe support. Remove the nut and exhaust pipe support from the center crossmember.

20. Disconnect the oxygen sensor harness. Remove the front exhaust pipe from the vehicle.

21. Remove the plastic cover from the front transaxle mounting cavity in the center crossmember.

22. Remove the bolts/nuts from the front and rear transaxle mounts.

23. Remove the bolts and nuts, then remove the front suspension and center crossmembers.

24. Support the transaxle with a suitable jack.

25. On 3-speed transaxles, remove the bolts and the lower engine reinforcement brace.

26. Remove the flywheel access cover from the rear of the engine.

27. Remove the six flywheel-to-torque converter bolts.

28. Remove the drain pan from underneath the transaxle fluid pan.

29. Remove the lower transaxle-to-engine bolts from the transaxle.

30. Remove the transaxle by carefully moving the transaxle away from the engine toward the left side of the engine compartment and slowly lowering the jack.

To install:

31. Before installing the transaxle assembly, perform the following:

a. Apply grease around the pilot shaft at the center of the torque converter.

b. Measure the distance between the outside edge of the torque converter housing and the torque converter lug. The distance should be more than 0.906 in. (23mm). If it is less than 0.906 in. (23mm), the torque converter is improperly installed. Remove and properly seat on the input shaft.

32. Install the transaxle to the engine.

33. Install the lower transaxle-to-engine bolts and tighten to 47 ft. lbs. (64 Nm).

34. Install the flywheel-to-torque converter bolts. Tighten the bolts to 14 ft. lbs. (19 Nm).

35. Install the flywheel access cover.

36. On 3-speed transaxles, install the lower engine reinforcement brace and tighten the bolts to 47 ft. lbs. (64 Nm).

37. Remove the jack from under the transaxle.

38. Install the front suspension and center crossmembers. Tighten the front suspension crossmember bolts to 152 ft. lbs. (206 Nm) and the center crossmember bolts to 45 ft. lbs. (61 Nm).

39. Install the front and rear transaxle mount nuts. Tighten the rear nuts to 42 ft. lbs. (57 Nm), and the front to 47 ft. lbs. (64 Nm).

40. Install the plastic access cover to the center crossmember.

41. Install the front exhaust pipe and connect the oxygen sensor harness.

42. Install the halfshafts and the front wheels.

43. Install the bolts to the left transaxle bracket and tighten to 35 ft. lbs. (48 Nm).

44. Connect the fluid cooler hoses to the lines at the transaxle; secure with hose clamps.

45. Install the splash shields.

46. Install the starter.

✳✳ WARNING

The differential portion of the 3-speed transaxle is separated from the rest of the transaxle and must be drained and refilled separately. The differential cannot be drained or refilled through the transaxle drain plug or filler tube. If the differential portion is drained and not refilled as outlined in this procedure, damage to the differential due to lack of lubrication will result.

47. On 3-speed transaxle, refill the differential with 1 ½ qts. (1.4 L) of Dexron® III automatic transmission fluid, into the differential filler plug hole. The fluid level should be even with the bottom of the differential filler plug hole. Install the filler plug and tighten to 29 ft. lbs. (39 Nm).

48. Lower the vehicle.

49. Install the left transaxle bracket reinforcement and tighten the bolts to 15 ft. lbs. (21 Nm).

50. Remove the engine/transaxle support fixture.

51. Install the upper transaxle-to-engine bolts and tighten to 47 ft. lbs. (64 Nm).

52. Install the TV guide cable bracket onto the transaxle and tighten the bolt to 71 inch lbs. (8 Nm).

53. Install the shift select cable into the bracket at the transaxle and secure with clip.

54. Install the shift select cable to the manual lever and secure with the nut and lockwasher. Adjust the shift select cable.

55. Connect the wiring connector to the VSS.

56. Install the TV cable to the throttle linkage. Adjust the cable as follows:

 a. Be sure the throttle valve is fully closed.

 b. Measure the distance between the end of the outer cable boot and the end of the TV cable stopper, it should be 0–0.04 in. (0–1mm).

 c. If the distance is greater than specified, loosen the TV cable locknut and

tighten the adjust nut until within specification.

 d. If the distance is less than specified, loosen the TV cable adjust nut and tighten the locknut within specification.

 e. After adjustment is completed, tighten the locknut and adjust nuts to 71 inch lbs. (8 Nm).

57. Connect the ground strap to the transaxle and tighten the bolt to 115 inch lbs. (13 Nm).

58. Connect the PNP switch, IAT sensor and solenoid harness if equipped with a 4-speed transaxle.

59. Install the air cleaner and filter. Tighten the bolts to 106 inch lbs. (12 Nm). Tighten the air cleaner intake tube clamp.

60. Install the battery and connect the cables, negative cable last.

61. Refill the transaxle with the proper amount of Dexron® III automatic transmission fluid.

62. Operate the vehicle and check for leaks.

Clutch

ADJUSTMENT

These vehicles are equipped with a hydrulic clutch actuating system that is self-adjusting.

REMOVAL & INSTALLATION

1. Disconnect the negative battery cable.

2. Raise and safely support the vehicle.

3. Remove the transaxle assembly.

4. Install a clutch pilot tool into the pilot bearing to support the clutch.

5. Matchmark the clutch cover and flywheel.

6. On Prizm, unfasten the release fork bearing clips. Withdraw the release bearing hub and release bearing. Remove the release fork and support.

7. Loosen the clutch cover-to-flywheel bolts evenly, using a crisscross pattern, until the spring pressure is released.

8. Remove the clutch disc and clutch cover.

To install:

9. Clean the flywheel mating surfaces of all oil, grease and metal deposits.

10. Check the diaphragm spring and pressure plate for wear or damage. If the spring or plate is excessively worn, replace the clutch cover assembly.

11. Check the pilot bearing for smooth operation. If the bearing does not spin freely, replace it.

12. Inspect the disc, pressure plate and flywheel for damage and wear using a caliper to measure depth and width

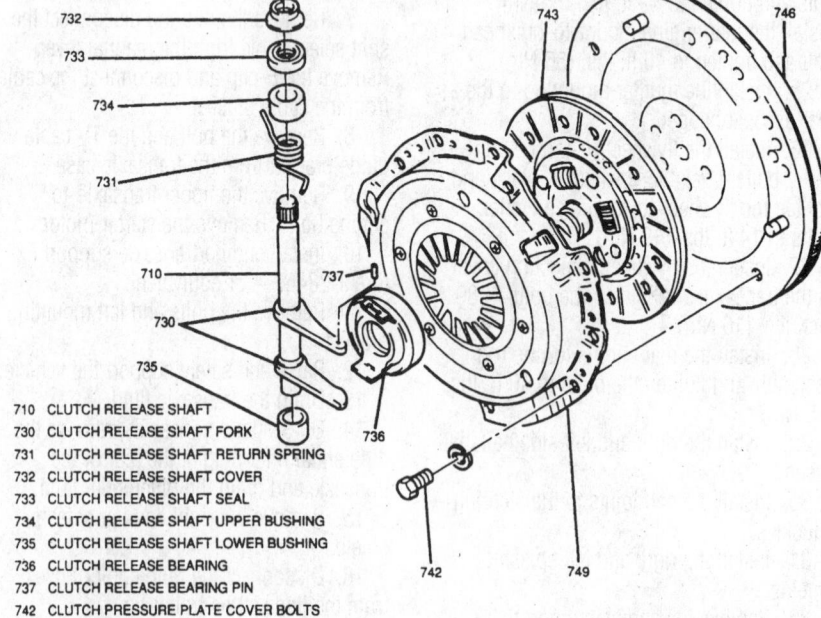

710	CLUTCH RELEASE SHAFT
730	CLUTCH RELEASE SHAFT FORK
731	CLUTCH RELEASE SHAFT RETURN SPRING
732	CLUTCH RELEASE SHAFT COVER
733	CLUTCH RELEASE SHAFT SEAL
734	CLUTCH RELEASE SHAFT UPPER BUSHING
735	CLUTCH RELEASE SHAFT LOWER BUSHING
736	CLUTCH RELEASE BEARING
737	CLUTCH RELEASE BEARING PIN
742	CLUTCH PRESSURE PLATE COVER BOLTS
743	CLUTCH DISC
746	FLYWHEEL
749	CLUTCH PRESSURE PLATE COVER

Exploded view of the clutch assembly—Metro

79222Z733

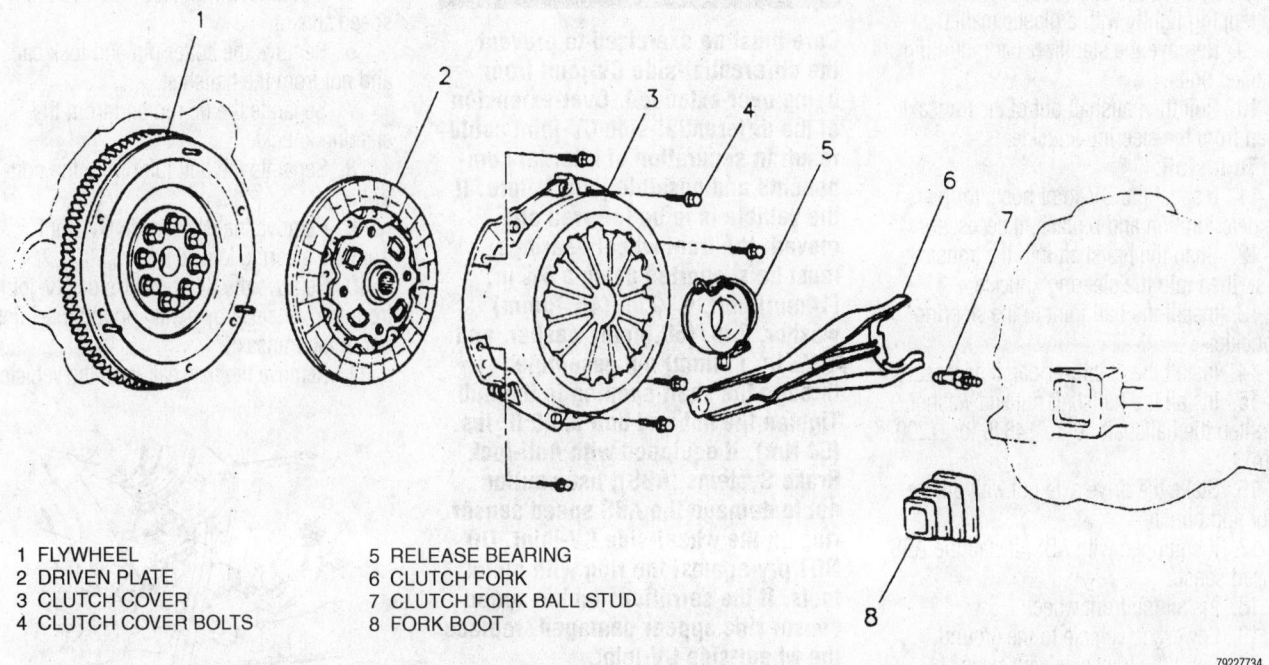

1 FLYWHEEL
2 DRIVEN PLATE
3 CLUTCH COVER
4 CLUTCH COVER BOLTS
5 RELEASE BEARING
6 CLUTCH FORK
7 CLUTCH FORK BALL STUD
8 FORK BOOT

79222Z734

Exploded view of the clutch assembly—Prizm

and a dial indicator to measure run-out.

    a. The minimum clutch disc rivet head depth is 0.012 in. (0.3mm).

    b. The maximum clutch disc runout is 0.031 in. (0.8mm).

    c. The maximum pressure plate spring depth is 0.024 in. (0.6mm).

    d. The maximum pressure plate spring width is 0.197 in. (5.0mm).

    e. The maximum flywheel runout is 0.004 in. (0.1mm).

13. When reassembling, apply a thin coating of multi-purpose grease to the release bearing hub and release fork contact points. Also, pack the groove inside the clutch hub with multi-purpose grease and lubricate the pivot points of the release fork.

14. Position the clutch disc and clutch cover with the matchmarks aligned and support with pilot tool.

15. Install the clutch cover bolts and tighten evenly to 17 ft. lbs. (23 Nm) for Metro or 14 ft. lbs. (19 Nm) for Prizm, in a crisscross pattern. Remove the pilot tool.

16. On Prizm, install the release bearing, fork, and the boot.

17. Lightly lubricate the splines, pilot bearing surface of the input shaft and release bearing with grease.

18. Install the transaxle.

19. Adjust the clutch cable.

20. Connect the negative battery cable. Check clutch operation.

Hydraulic Clutch System

BLEEDING

➡**If any maintenance on the clutch system was performed or the system is suspected of containing air, bleed the system.**

1. Fill the clutch reservoir with brake fluid. Check the reservoir level frequently and add fluid as needed.

2. Connect one end of a vinyl tube to the bleeder plug on the slave cylinder and submerge the other end into a clear container half-filled with clean brake fluid.

3. Slowly pump the clutch pedal several times.

4. Repeat Steps 2 and 3 until all of the air bubbles are removed from the system.

5. Tighten the bleeder plug when no more air bubbles emerge from the tube.

6. Refill the master cylinder to the proper level.

7. Check the system for leaks.

Halfshaft

REMOVAL & INSTALLATION

Metro

1. Disconnect the negative battery cable.

2. Unstake and remove the halfshaft nut with the vehicle's weight on the ground.

3. Raise and safely support the vehicle.

4. Remove the front wheel.

5. If equipped with ABS, remove the wheel speed sensor.

6. Remove the ball joint from the steering knuckle.

7. Drain the transaxle fluid.

➡**On 4-door sedan vehicles, the right side halfshaft is stabilized by a center support bearing between the transaxle and the inner tripod joint. It is not necessary to remove the right inner drive axle and center support bearing assembly. The right side halfshaft assembly can be separated from the center support bearing by lightly tapping with a plastic mallet.**

8. Using a suitable prytool, pry on the inboard joints of the halfshaft to detach the

snapring lock in the differential. On 4-door sedans, separate the right side halfshaft assembly from the center support bearing by tapping lightly with a plastic mallet.

9. Remove the stabilizer bar mounting bracket bolts.

10. Pull the halfshaft out of the transaxle, then from the steering knuckle.

To install:

11. Inspect the CV-joint boots for tears or deterioration and replace, if necessary.

12. Snap the halfshaft into the transaxle first, then into the steering knuckle.

13. Install the ball joint to the steering knuckle.

14. Install the stabilizer bar, if necessary.

15. Install the halfshaft nut and washer. Tighten the halfshaft nut to 148 ft. lbs. (200 Nm).

16. Stake the drive axle nut with a hammer and punch.

17. If equipped with ABS, install the ABS speed sensor.

18. Install the front wheel.

19. Lower the vehicle to the ground.

20. Refill the transaxle with clean transaxle fluid.

21. Connect the negative battery cable.

22. Road test the vehicle.

Prizm

✳✳ WARNING

Care must be exercised to prevent the differential-side CV-joint from being over-extended. Over-extension of the differential-side CV-joint could result in separation of internal components and possible joint failure. If the vehicle is to be lowered and moved, the front wheel bearings must be supported using a 9/16 in. (14mm) bolt, 1 3/4 in. (44.45mm) washer, 2 in. (50.8mm) washer, and a 9/16 in. (14mm) nut assembled through the shaft opening in the hub. Tighten the nut and bolt to 40 ft. lbs. (54 Nm). If equipped with Anti-lock Brake Systems (ABS), use caution not to damage the ABS speed sensor ring on the wheel-side CV-joint. DO NOT pry against the ring with metal tools. If the serrations on the speed sensor ring appear damaged, replace the wheel-side CV-joint.

1. Disconnect the negative battery cable.
2. Raise and safely support the vehicle.
3. Remove the front wheel.

4. Remove the six bolts and splash shield from the vehicle.

5. If equipped with ABS, remove the speed sensor.

6. Remove the cotter pin and lock cap and nut from the halfshaft.

7. Separate the tie rod end from the steering knuckle.

8. Separate the ball joint from the control arm.

9. Remove the wheel-side CV-joint from the steering knuckle.

10. Remove the differential-side CV-joint from the transaxle by gently prying the joint out of the transaxle.

11. Remove the halfshaft from the vehicle.

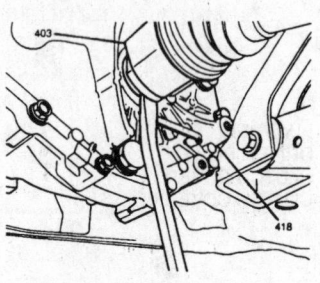

403 DIFFERENTIAL-SIDE JOINT HOUSING
418 TRANSAXLE ASSEMBLY

79222736

Prying out the inner halfshaft from the transaxle—Prizm

To install:

12. Inspect the CV-joint boots for tears or deterioration. Replace as necessary.

13. Inspect the front wheel bearing inner and outer oil seals for damage or deterioration. Replace as necessary.

14. Inspect the halfshaft fluid seal at the transaxle for leakage or damage. Replace as necessary.

15. Install the differential-side CV-joint into the transaxle.

16. Install the wheel-side CV-joint into the steering knuckle.

17. Install the ball joint to the control arm.

18. Install the tie rod end.

19. Install the nut onto the halfshaft and tighten to 167 ft. lbs. (228 Nm).

20. Install the lock cap onto the halfshaft and secure with a new cotter pin.

21. Install the front wheel.

22. If equipped, install the ABS speed sensor.

23. Install the splash shield and secure with the six bolts. Tighten the splash shield bolts to 71 inch lbs. (8 Nm). Lower the vehicle.

24. Connect the negative battery cable.

25. Check the wheel alignment.

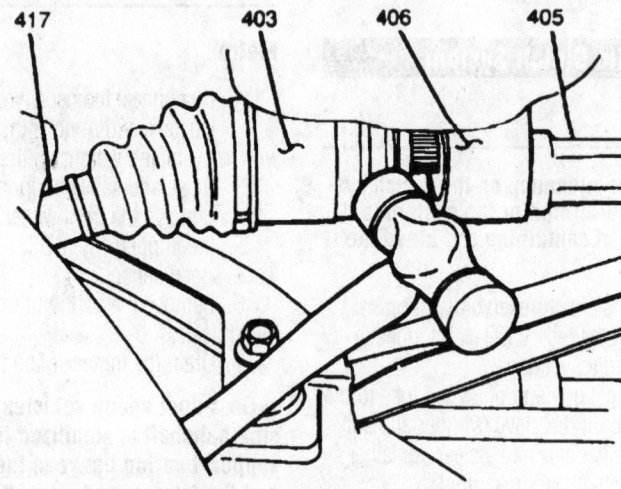

403 DIFFERENTIAL-SIDE JOINT HOUSING
405 RIGHT INNER DRIVE AXLE SHAFT
406 RIGHT INNER DRIVE AXLE SUPPORT ARBOR
417 RIGHT DRIVE AXLE ASSEMBLY

79222735

Tapping out the right inner halfshaft from the transaxle—4-door models

STEERING AND SUSPENSION

Air Bag

❄❄ CAUTION

All vehicles are equipped with an Air Bag system, also known as the Supplemental Inflatable Restraint (SIR) or Supplemental Restraint System (SRS). The system must be disabled before performing service on or around system components, steering column, instrument panel components, wiring and sensors. Failure to follow safety and disabling procedures could result in accidental Air Bag deployment, possible personal injury and unnecessary system repairs.

PRECAUTIONS

Several precautions must be observed when handling the inflator module to avoid accidental deployment and possible personal injury.

• Never carry the inflator module by the wires or connector on the underside of the module.

• When carrying a live inflator module, hold securely with both hands, and ensure that the bag and trim cover are pointed away.

• Place the inflator module on a bench or other surface with the bag and trim cover facing up.

• With the inflator module on the bench, never place anything on or close to the module which may be thrown in the event of an accidental deployment.

DISARMING

Metro

❄❄ WARNING

When performing service on or around the Air Bag system components or wiring, disable the Air Bag system. Failure to follow the procedures could result in possible deployment, personal injury or unneeded system repairs.

1. Disconnect the negative battery cable.

2. Turn the steering wheel so the wheels are pointing straight ahead.

3. Turn the ignition switch to the **LOCK** position and remove the key.

4. Remove the **AIR BAG-IG** fuse from the Air Bag fuse box located near the junction/fuse box.

5. Remove the left side steering wheel side cap and disconnect the yellow connector for the driver side Air Bag (inflator) module.

6. Pull out the glove box while pushing in on the stoppers from the left and the right sides. Disconnect the yellow connector for the passenger Air Bag (inflator) module.

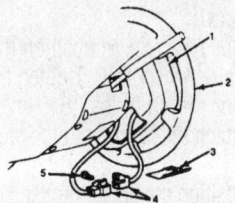

1 Inflator module housing
2 Steering wheel
3 Rear plastic access cover
4 SIR harness connector
5 Connector position assurance (CPA)

7922Z737

Location of the SIR harness connector—Metro

To arm:

7. Connect the negative battery cable.

8. Turn the ignition switch to the **LOCK** position and remove the key.

9. Connect the yellow connector for the passenger side Air Bag (inflator) module and the yellow connector for the driver Air Bag (inflator) module. Be sure to lock each connector with the lock lever.

10. Install the glove box assembly.

11. Install the left side steering wheel side cover.

12. Install the **AIR BAG-IG** fuse to the Air Bag fuse box.

13. Turn the ignition **ON** and verify that the **AIR BAG** warning lamp flashes seven times, then turns off. If the system does not operate as described, diagnosis and repairs to the Air Bag system are necessary.

Prizm

❄❄ CAUTION

The Air Bag system must be disarmed before performing service on components or wiring. Failure to do so may cause accidental deployment, resulting in unnecessary repairs and/or personal injury.

➡**The center sensor assembly can maintain sufficient voltage to cause deployment for up to 2 minutes after the ignition switch is turned to the LOCK position or the battery is disconnected.**

1. Disconnect the negative battery cable.

2. Turn the steering wheel so the front wheels are in the straight-ahead position.

3. Turn the ignition switch to **LOCK**.

4. Remove the IGN fuse and CIG and RADIO fuse from junction block 1.

5. Remove the Connector Position Assurance (CPA) and disconnect the yellow 2-way connector at the base of the steering column.

To arm:

6. Connect the negative battery cable.

7. Turn the ignition switch to the **LOCK** position and remove the key.

8. Attach the yellow connector for the driver Air Bag (inflator) module. Be sure to lock the connector with the lock lever.

9. Install the IGN fuse and CIG and RADIO fuse from junction block 1.

10. Turn the ignition **ON** and verify that the **AIR BAG** warning lamp flashes seven times, then turns off. If the system does not operate as described, diagnosis and repairs to the Air Bag system are necessary.

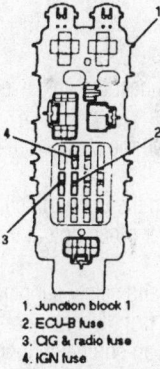

1. Junction block 1
2. ECU-B fuse
3. CIG & radio fuse
4. IGN fuse

7922Z738

Connector view and terminal identification of junction block 1—Prizm

Rack and Pinion Steering Gear

REMOVAL & INSTALLATION

Manual

METRO

1. Disconnect the negative battery cable.

2. Slide the driver's seat back as far as possible and pull off the front part of the floor mat on the driver's side and remove the steering shaft joint cover.

3. Loosen the steering shaft upper joint bolt, but do not remove. Remove the steering shaft lower joint bolt and disconnect the lower joint from the pinion.

4. Raise and safely support the vehicle. Remove the front wheels.

5. Remove the tie rod ends from the steering knuckles.

6. Remove the rack and pinion mounting bolts, brackets and the rack from the vehicle.

To install:

7. Install the rack, brackets and mounting bolts into the vehicle at the bulkhead. Be sure to align the pinion shaft through the floor opening. Tighten the bolts to 18 ft. lbs. (25 Nm).

8. Install the right and left tie rod ends to the steering knuckles.

9. Install the front wheels. Lower the vehicle.

10. Connect the steering shaft to the steering gear. Install the lower steering shaft-to-steering gear clinch bolt and tighten to 18 ft. lbs. (25 Nm). Tighten the steering shaft upper joint bolt to 18 ft. lbs. (25 Nm).

11. Install the steering joint cover.

12. Place the driver's side floor mat back into position.

13. Check and adjust the front wheel alignment.

PRIZM

1. Disconnect the negative battery cable.

2. Remove the cover from the intermediate shaft.

3. Loosen the upper pinch bolt. Remove the lower pinch bolt at the pinion shaft.

4. Raise and safely support the vehicle. Remove both front wheels.

5. Install an engine support and tension the engine without raising it.

6. Remove the bolts holding the center crossmember to the radiator support.

7. Remove the covers from the front and center mount bolts.

8. Remove the front mount bolts, and the center mount bolts.

9. Support the crossmember and remove the rear mount bolts.

10. Remove the bolts holding the center crossmember to the main crossmember.

11. Use a floor jack and piece of wood to support the main crossmember.

12. Remove the bolts holding the main crossmember to the body.

❄❄ CAUTION

When the fasteners are removed, be sure the crossmembers are properly supported.

---

13. Remove the bolts holding the lower control arm brackets to the body.

14. Slowly lower the main crossmember while holding onto the center crossmember.

15. Remove both tie rod joints from the steering knuckles.

16. Remove the nuts and bolts attaching the steering rack to the body.

17. Remove the rack through the right side wheel well.

To install:

18. Place the rack in position through the right side wheel well. Tighten the bracket bolts to 45 ft. lbs. (61 Nm).

19. Attach the tie rods to the steering knuckles.

20. Position the center crossmember over the center and rear transaxle mount studs; start the two nuts on the center mount.

21. Loosely install the bolts holding the center crossmember to the radiator support.

22. Loosely install the front mounting bolts.

23. Raise the main crossmember into position over the rear mount studs and align the underbody bolts. Install the two rear mount nuts loosely.

24. Install the main crossmember to underbody bolts loosely.

25. Install the lower control arm bracket bolts loosely.

26. Loosely install the bolts holding the center crossmember to the main crossmember.

27. Tighten the fasteners as follows:
- Main crossmember-to-underbody bolts: 152 ft. lbs. (206 Nm)
- Center crossmember-to-radiator support bolts: 45 ft. lbs. (61 Nm)
- Lower A-frame-to-center bolts: 161 ft. lbs. (218 Nm)
- Lower A-frame-to-outer bolts: 109 ft. lbs. (147 Nm)
- Front, center and rear mount bolts: 45 ft. lbs. (61 Nm)

28. Install the covers on the front and center mount bolts.

29. Install the front wheels. Lower the vehicle.

30. Connect the yoke to the pinion and tighten both the upper and lower bolts to 26 ft. lbs.

31. Install the yoke cover.

32. Check and adjust the front end alignment as necessary. Check steering operation.

Power

METRO

1. Disconnect the negative battery cable.

2. Slide the driver's seat back. Pull off the front part of the floor mat on the driver's side and remove the steering shaft joint cover.

3. Loosen the steering shaft upper joint bolt, but do not remove. Remove the steering shaft lower joint bolt and disconnect the lower joint from the pinion.

4. Raise and safely support the vehicle. Remove the front wheels.

5. Remove the tie rod ends.

6. Disconnect the exhaust pipe at the manifold.

7. If equipped with manual transaxle, separate the shift linkage and extension rod from the transaxle.

8. If equipped with automatic transaxle, remove the engine rear torque rod with bracket from the transaxle.

9. Disconnect the power steering lines from the rack and pinion assembly. Plug the openings to prevent system contamination.

10. Remove the rack and pinion mounting bolts, brackets and the rack from the vehicle.

To install:

11. Install the rack, brackets and mounting bolts into the vehicle at the bulkhead. Be sure to align the pinion shaft through the floor opening. Tighten bolts to 18 ft. lbs. (25 Nm).

12. Unplug and reconnect the power steering lines to the rack and pinion assembly. Tighten the inlet and outlet fluid lines to 25 ft. lbs. (35 Nm). Tighten the remaining lines to 18 ft. lbs. (25 Nm).

13. If equipped, install the shift linkage and extension rod to the manual transaxle.

14. If equipped, install the engine rear torque rod with torque rod bracket to the automatic transaxle.

15. Connect the front exhaust pipe to the manifold using a new seal, and secure with bolts. Tighten the bolts to 37 ft. lbs. (50 Nm).

16. Install the tie rod ends to the steering knuckles.

17. Install the front wheels. Lower the vehicle.

18. Connect the steering shaft to the steering gear. Install the lower steering shaft-to-steering gear clinch bolt and tighten to 18 ft. lbs. (25 Nm). Tighten the steering shaft upper joint bolt to 18 ft. lbs. (25 Nm).

19. Install the steering joint cover.

20. Place the driver's side floor mat back into the original position.

21. Bleed the system.

22. Check and adjust the front wheel alignment.

PRIZM

1. Disconnect the negative battery cable.
2. Place a drain pan under the steering rack and pinion unit.
3. Remove the cover from the intermediate shaft.
4. Loosen the upper pinch bolt. Remove the lower pinch bolt at the pinion shaft.
5. Raise and safely support the vehicle.
6. Remove both front wheels.
7. Install an engine support and tension the engine without raising it.
8. Remove the bolts holding the center crossmember to the radiator support.
9. Remove the covers from the front and center mount bolts.
10. Remove the front mount bolts, then the center mount bolts, then the rear mount bolts.
11. Remove the bolts holding the center crossmember to the main crossmember.
12. Use a floor jack and piece of wood to support the main crossmember.
13. Remove the bolts holding the main crossmember to the body.
14. Remove the bolts holding the lower control arm brackets to the body.

❊❊ CAUTION

Be sure the crossmembers are properly supported.

15. Slowly lower the main crossmember while holding onto the center crossmember.
16. Remove the tie rods.
17. Remove both ball joints.
18. Disconnect the fluid pressure and return lines from the rack.
19. Remove the nuts and bolts attaching the steering rack to the body.
20. Remove the rack and pinion unit through the right side wheel well.

To install:

21. Place the rack and pinion unit in position through the right wheel well and tighten the bracket bolts to 44 ft. lbs. (59 Nm).
22. Connect the fluid lines to the rack and tighten to 33 ft. lbs. (45 Nm). Make certain the fittings are correctly threaded before tightening.
23. Attach the tie rod ends.
24. Position the center crossmember and rear transaxle mount over the studs. Start the nuts on the center mount.
25. Loosely install the bolts holding the center crossmember to the radiator support.
26. Loosely install the front mount bolts.
27. Raise the main crossmember into

position over the rear mount studs and align the underbody bolts. Install the rear mount nuts loosely.

28. Install the main crossmember to underbody bolts loosely.
29. Install the lower control arm bracket bolts loosely.
30. Tighten the components in the order listed to the following torque specification:
  • Main crossmember to underbody bolts: 152 ft. lbs. (205 Nm).
  • Center crossmember-to-radiator support bolts: 45 ft. lbs. (61 Nm)
  • Lower A-frame-to-center bolts: 161 ft. lbs. (218 Nm)
  • Lower A-frame-to-outer bolts: 109 ft. lbs. (147 Nm)
  • Front, center and rear mount bolts: 45 ft. lbs. (61 Nm)
31. Install the covers on the front and center mount bolts.
32. Install the front wheels. Lower the vehicle to the ground.
33. Connect the yoke to the pinion and tighten both the upper and lower bolts to 26 ft. lbs.
34. Install the yoke cover.
35. Add power steering fluid to the reservoir and bleed the system.
36. Check and adjust the front end alignment, as necessary.

Strut

REMOVAL & INSTALLATION

Front

1. Disconnect the negative battery cable.
2. Remove the upper strut support nuts from the engine compartment.
3. Raise and safely support the vehicle.
4. Remove the front wheel.
5. If equipped with ABS, remove the speed sensor.
6. Remove the brake hose clip, then the hose from the strut.
7. Support the lower control arm with a floor jack.
8. Remove the strut-to-steering knuckle bolts, then remove the strut assembly from the vehicle.

To install:

9. Install the strut assembly onto the vehicle. Install the upper support nuts loosely.
10. Install the strut-to-steering knuckle bolts and tighten to 59 ft. lbs. (80 Nm) for Metro models and 203 ft. Lbs. (275 Nm) for Prizm models.

11. Tighten the upper strut support nuts to 24 ft. lbs. (34 Nm) for Metro models and 29 ft. Lbs. (39 Nm) for Prizm models.
12. Install the brake hose and brake hose clip. Be sure the brake hose is not twisted when installing.
13. If equipped with ABS, attach the wheel speed sensor.
14. Install the front wheels. Lower the vehicle.
15. Connect the negative battery cable. Check front wheel alignment.

Rear

PRIZM

1. Raise and safely support the vehicle.
2. Remove the wheel.
3. Disconnect the brake line from the wheel cylinder (at the backing plate). Plug the lines to prevent leakage.
4. Remove the clip from the brake hose and remove the hose from the strut bracket.
5. Disconnect the sway bar link from the strut.
6. Remove the strut mounting bolts from the rear knuckle.
7. Remove the seat back side cushion (Sedan) or the rear sill side panel (Hatchback) to gain access to the upper strut mount.
8. Remove the nuts holding the upper strut mount to the body. Do not loosen the center strut piston nut.
9. Remove the strut assembly from the vehicle.

To install:

10. Place the strut in position on the vehicle and install the three upper retaining nuts, tightening them to 29 ft. lbs. (39 Nm).
11. Reinstall the seat back side cushion or the rear sill side panel.
12. Install the strut into the rear knuckle assembly; install the bolts and nuts and tighten them to 105 ft. lbs. (142 Nm).
13. Reconnect the sway bar link to the strut. Tighten the bolts to 32 ft. lbs. (44 Nm).
14. Connect the brake hose into the bracket and install the clip.
15. Connect the metal line to the wheel cylinder. Make certain the fittings are properly threaded before tightening them.
16. Bleed the brake system.
17. Install the wheel. Lower the vehicle.
18. Check the wheel alignment.

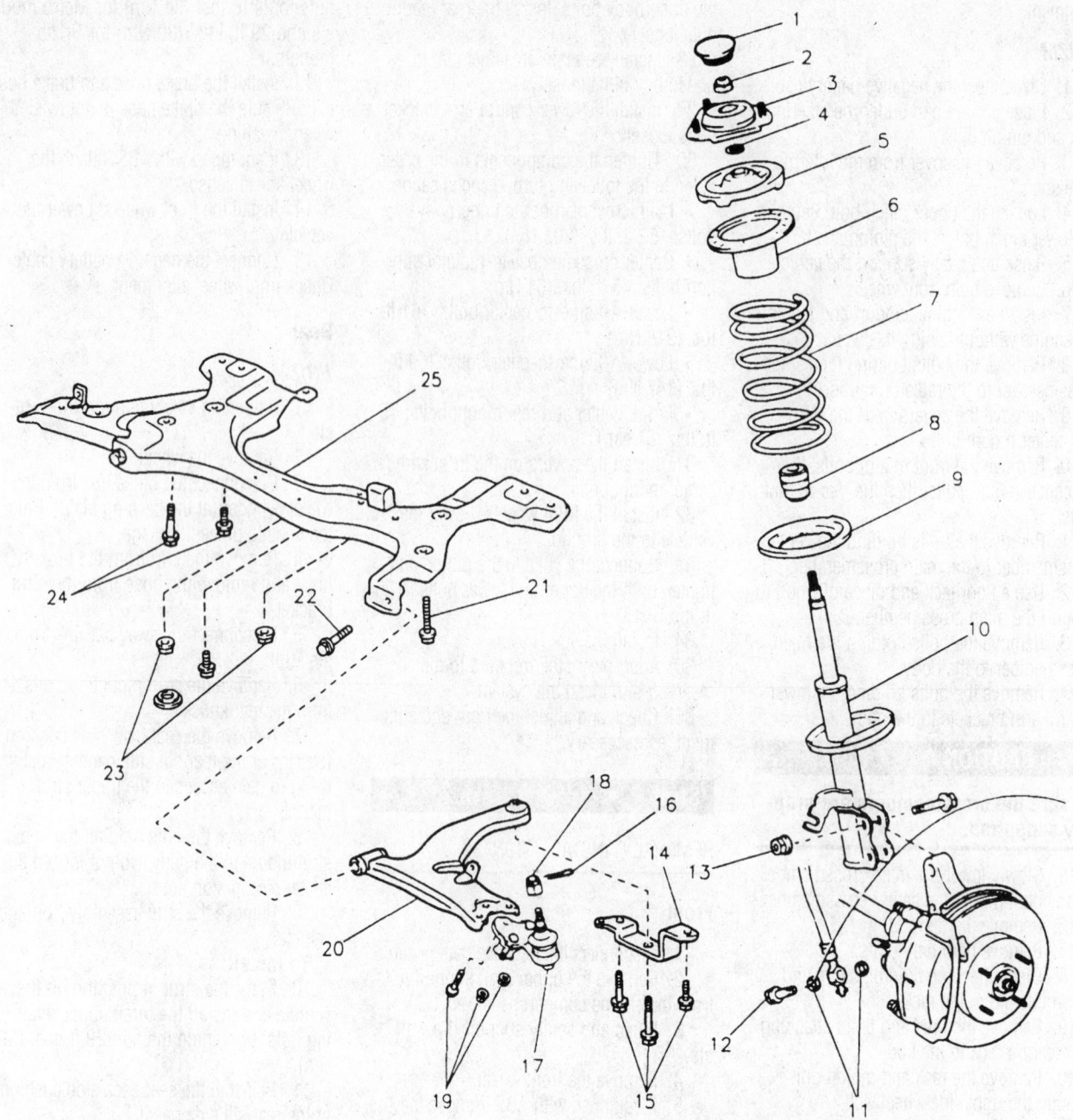

1   DUST CAP
2   STRUT ROD PISTON NUT
3   STRUT SUPPORT
4   DUST SEAL
5   SPRING SEAT
6   UPPER INSULATOR
7   COIL SPRING
8   SPRING BUMPER
9   LOWER INSULATOR
10 STRUT
11 BRAKE LINE GASKETS
12 BRAKE LINE-TO-CALIPER BOLT
13 STRUT MOUNTING NUT AND BOLT

14 LOWER CONTROL ARM RETAINING BRACKET
15 LOWER CONTROL ARM RETAINING BRACKET BOLTS
16 COTTER PIN
17 BALL JOINT
18 BALL JOINT CASTLE NUT
19 BALL JOINT MOUNTING NUT AND BOLT
20 CONTROL ARM
21 CROSSMEMBER MOUNTING BOLTS
22 CROSSMEMBER-TO-CONTROL ARM BOLT
23 CROSSMEMBER MOUNTING NUTS
24 CROSSMEMBER MOUNTING BOLTS
25 SUSPENSION CROSSMEMBER

79222739

Exploded view of the front suspension assembly—Prizm

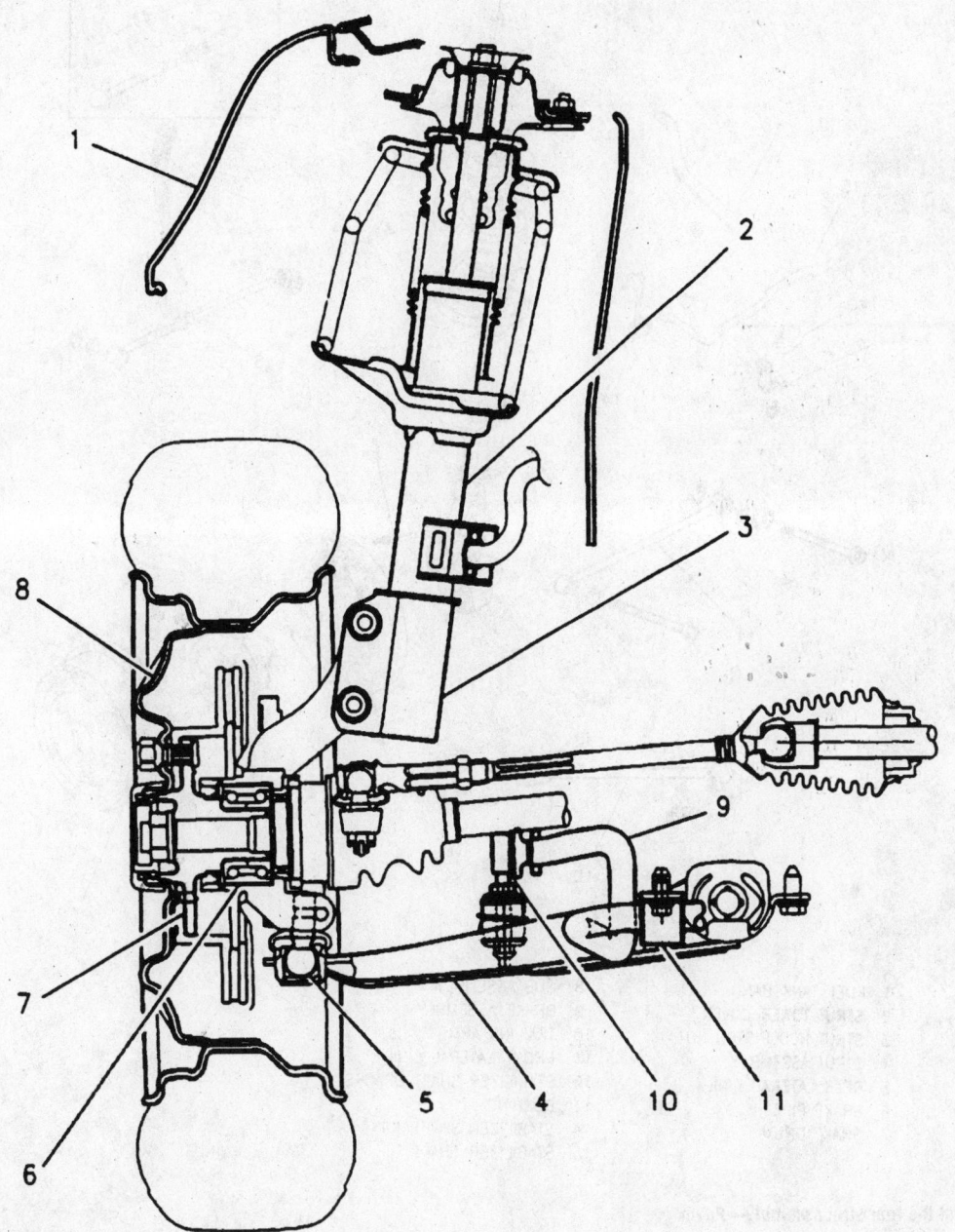

1	VEHICLE BODY	7	HUB
2	STRUT ASSEMBLY	8	WHEEL
3	STEERING KNUCKLE	9	SWAY BAR
4	CONTROL ARM	10	SWAY BAR LINK
5	BALL STUD	11	SWAY BAR BRACKET
6	WHEEL BEARING		

79222744

Cut-away view of the front suspension—Metro

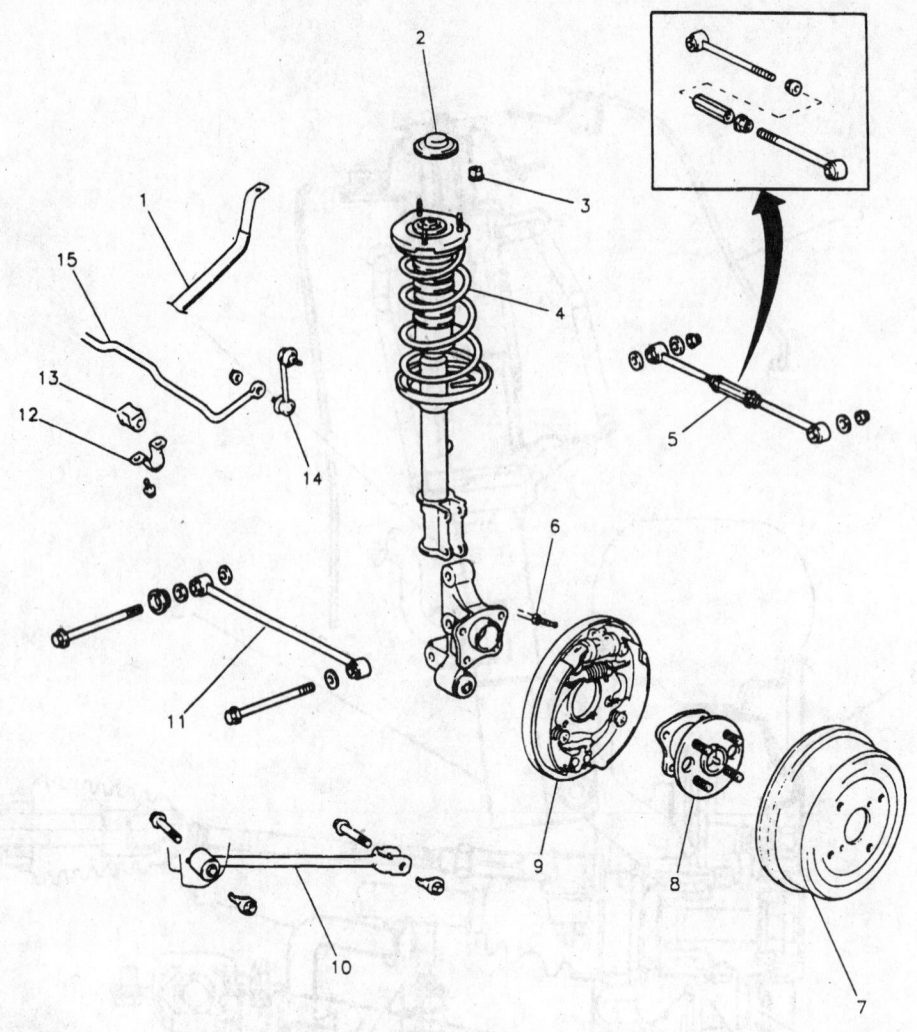

1 FUEL TANK BAND	8 HUB ASSEMBLY
2 STRUT TOWER COVER	9 BRAKE ASSEMBLY
3 STRUT ROD PISTON NUT	10 TRAILING ARM
4 STRUT ASSEMBLY	11 FRONT LATERAL LINK
5 REAR LATERAL LINK	12 STABILIZER SHAFT BRACKET
6 BRAKE PIPE	13 BUSHING
7 BRAKE DRUM	14 STABILIZER SHAFT JOINT
	15 STABILIZER SHAFT

7922Z740

Exploded view of the rear strut assembly—Prizm

Shock Absorber

REMOVAL & INSTALLATION

Rear

METRO

1. Open the vehicle hatchback or deck lid.

2. Raise and safely support the vehicle just high enough so the top and the bottom of the shock can be accessed. Remove the wheels.

3. Place a floor jack under the suspension arm for support.

4. Remove the lower shock-to-knuckle bolt.

➡**Do not open the knuckle slit wider than necessary. Do not lower the jack more than necessary during the strut removal or the coil spring may become unseated.**

5. Remove the trim cover from the upper shock tower and remove the upper support nuts.

6. Remove the shock from the knuckle. Compress the shock and remove. If the shock is hard to remove, open the slit of the knuckle by inserting a wedge.

To install:

7. Install the shock into the vehicle. Position the bottom of the alignment projection inside the knuckle opening.

8. Install the upper support nuts and tighten to 24 ft. lbs. (30 Nm). Install the trim cover to the upper shock tower.

9. Install the shock-to-knuckle bolt and tighten to 44 ft. lbs. (60 Nm). Remove the floor jack.

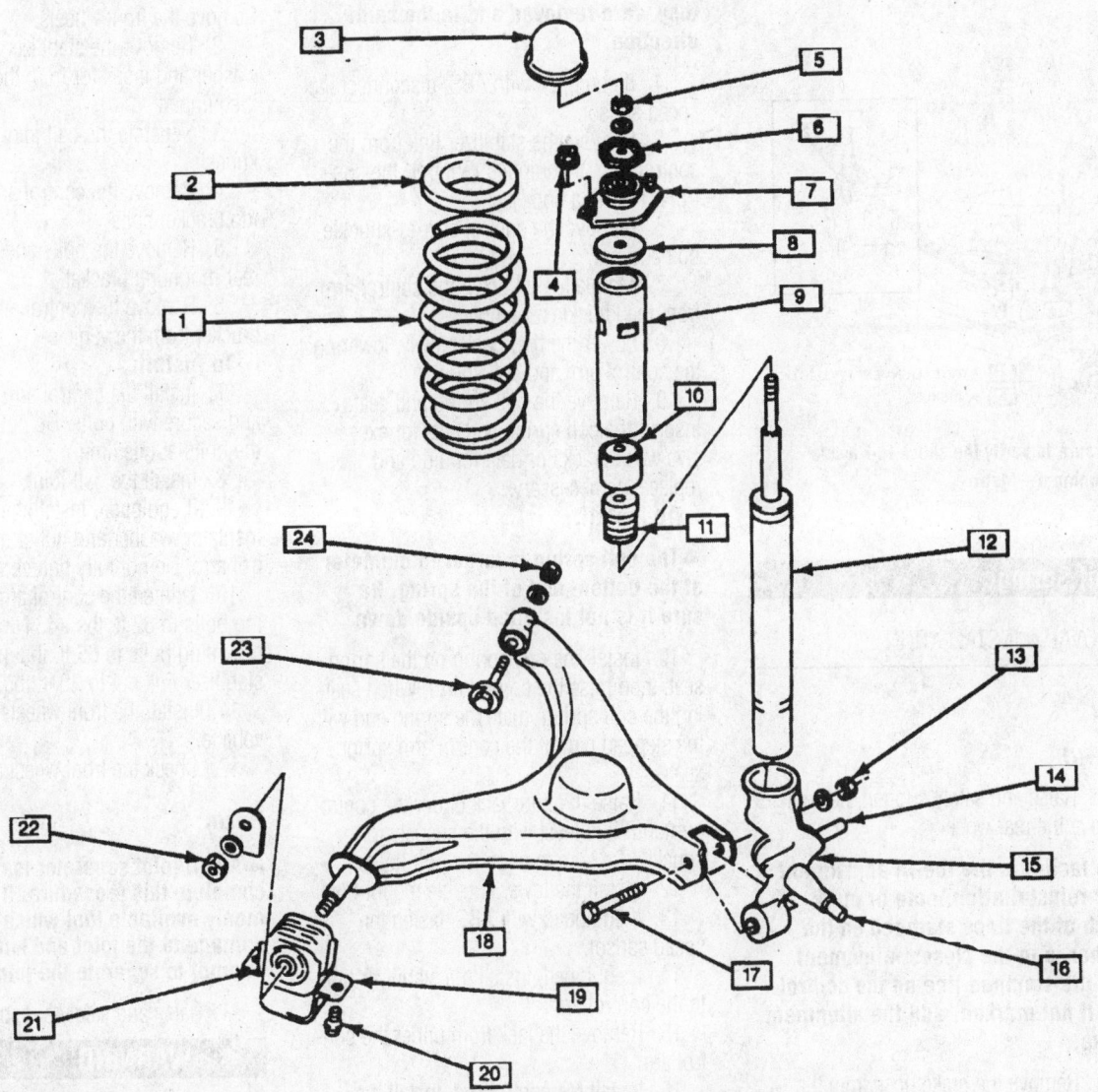

1	COIL SPRING
2	COIL SPRING UPPER SEAT
3	STRUT UPPER CAP
4	STRUT UPPER SUPPORT NUT
5	STRUT TOP NUT
6	STRUT UPPER SEAT
7	STRUT UPPER SUPPORT
8	STRUT LOWER SEAT
9	STRUT DUST COVER
10	BUMPER STOPPER CAP
11	BUMPER STOPPER
12	STRUT (SHOCK ABSORBER)
13	SUSPENSION ARM-TO-KNUCKLE NUT

14	SUSPENSION KNUCKLE CONTROL ROD STUD (CONTROL ROD NOT SHOWN)
15	SUSPENSION KNUCKLE
16	WHEEL SPINDLE
17	SUSPENSION ARM-TO-KNUCKLE BOLT
18	SUSPENSION ARM
19	SUSPENSION ARM FRONT BRACKET
20	BRACKET BOLT
21	SUSPENSION ARM FRONT BUSHING
22	FRONT BUSHING NUT
23	SUSPENSION ARM REAR MOUNTING BOLT
24	SUSPENSION ARM REAR MOUNTING NUT

79222Z741

Exploded view of the rear suspension—Metro

10. Install the rear wheels. Lower the vehicle.

11. Check the wheel alignment.

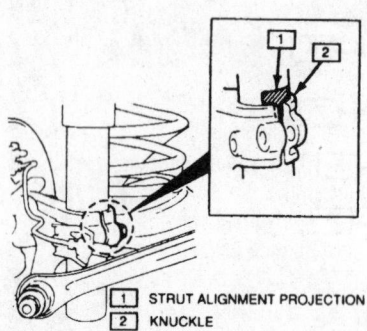

☐ 1 STRUT ALIGNMENT PROJECTION
☐ 2 KNUCKLE

7922Z742

Be sure to verify the shock-to-knuckle alignment—Metro

Coil Spring

REMOVAL & INSTALLATION

Rear

METRO

1. Raise and safely support the vehicle. Remove the rear wheel.

➡To facilitate the toe-in adjustment after reinstallation, note or mark which of the lines stamped on the washer is in the closest alignment with the stamped line on the control rod. If not marked, add the alignment marks.

2. Remove the brake hose from the bracket on the control rod.

3. Remove the control rod inside bolt (body center side) and outside (wheel side) stud bolts.

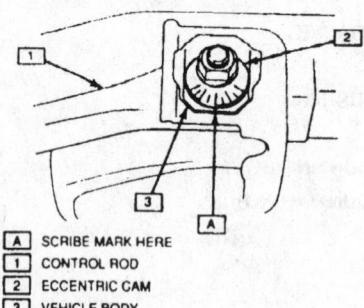

☐ A SCRIBE MARK HERE
☐ 1 CONTROL ROD
☐ 2 ECCENTRIC CAM
☐ 3 VEHICLE BODY

7922Z743

Be sure to matchmark the eccentric cam—Metro

➡The control rods are not interchangeable. Be sure the control rods are installed on the same side from which they were removed and in the same direction.

4. If equipped with ABS, disconnect the speed sensor.

5. Remove the stabilizer link from the control arm, if equipped. Support the control arm with a floor jack.

6. Remove the control arm-to-knuckle bolt and nut.

7. Manually pull down the control arm from the knuckle assembly.

8. Lower the floor jack slowly, lowering the control arm and coil spring.

9. Remove the coil spring and seat. Inspect the coil spring and seat for excessive wear, cracks or deterioration and replace, as necessary.

To install:

➡The coil spring is larger in diameter at the bottom end of the spring. Be sure it is not installed upside down.

10. Install the coil spring on the spring seat, then raise the control arm. When seating the coil spring, mate the spring end with the stepped part of the control arm spring seat.

11. Raise the floor jack under the control arm until it is level with the suspension knuckle. Be sure to align the bolt holes.

12. Install the lower knuckle mount bolt.

13. If equipped with ABS, fasten the speed sensor.

14. If equipped, install the stabilizer link to the control arm.

15. Remove the jack from under the control arm.

16. Install the control rod. Install the inside and outside control rod bolts, Align the marks on the eccentric cam and body that were made before removal. Tighten the control rod inside and outside nuts to 59 ft. lbs. (80 Nm).

17. Install the brake hose to the control rod bracket and secure with the E-clip.

18. Remove the floor jack.

19. Install the rear wheels. Lower the vehicle.

Lower Ball Joints

REMOVAL & INSTALLATION

Metro

The lower ball joint is an integral part of the lower control arm assembly and is not serviceable as an individual component. If

the lower ball joint is defective, the entire lower control arm must be replaced.

1. Raise and safely support the vehicle. Remove the front wheels.

2. Remove the stabilizer bar link nut, washer and insulator from the control arm, if equipped.

3. Separate the ball joint from the knuckle.

4. Remove the control arm front mounting bracket bolts.

5. Remove the bolts and control arm rear mounting bracket.

6. Remove the control arm and the front bracket from the vehicle.

To install:

7. Install the control arm on the vehicle and secure with bolts. Do not fully tighten the bolts at this time.

8. Install the ball joint.

9. If equipped, install the stabilizer link insulator, washer and nut to the lower control arm. Do not fully tighten the nut.

10. Tighten the control arm rear mounting bolts to 32 ft. lbs. (43 Nm); front mounting bolts to 66 ft. lbs. (90 Nm) and stabilizer link nut to 20 ft. lbs. (28 Nm).

11. Install the front wheels. Lower the vehicle.

12. Check the front wheel alignment.

Prizm

➡A ball joint separator is required to complete this procedure. It is a commonly available tool which prevents damage to the joint and knuckle. Do not attempt to separate the joint without it.

1. Raise and safely support the vehicle.

✳✳ WARNING

Do not allow the halfshaft joints to over-extend or the CV-joints may become disconnected.

2. Remove the wheel.

3. Remove the cotter pin from the ball joint nut.

4. Loosen the castle nut but do not remove it. Unscrew it just to the top of the threads and install ball joint separator tool 34754 or equivalent.

5. Use the separator to loosen the ball joint from the steering knuckle.

6. Remove the nuts and bolt holding the ball joint to the control arm.

7. Remove the ball joint from the control arm and steering knuckle.

To install:

8. Install the ball joint to the control arm and tighten the bolts and nut to 105 ft. lbs. (142 Nm).

9. Carefully install the ball joint on the steering knuckle. Use a new castle nut and tighten it to 94 ft. lbs. (128 Nm).

10. Install the cotter pin. If the holes are not in alignment, tighten the nut enough so the cotter pin can be installed. NEVER loosen the castle nut to install a cotter pin.

11. Install the wheel.

12. Lower the vehicle.

13. Check the front wheel alignment.

Wheel Bearings

ADJUSTMENT

Front

The front wheel bearings on all models cannot be adjusted. To check for a loose wheel bearing, proceed as follows:

1. Raise and safely support the vehicle.

2. Remove the front wheel.

3. Compress the caliper piston to free the caliper assembly.

4. Mount a suitable dial indicator to the knuckle or strut, with the indicator foot contacting the center of the hub next to the nut.

5. Push and pull the brake rotor by hand. If rotor movement exceeds 0.016 in. (0.04mm) for Metro or 0.020 in. (0.05mm) for Prizm, replace the wheel bearings.

6. Install the wheel and lower the vehicle.

7. Apply the brakes several times before moving the vehicle, to seat the caliper piston.

Rear

The rear wheel bearings cannot be adjusted. To check for a loose wheel bearing, proceed as follows:

1. Raise and safely support the vehicle.

2. Remove the rear wheel.

3. Mount a suitable dial indicator to the knuckle or strut, with the indicator foot contacting the center of the hub next to the hub nut.

4. Push and pull the brake drum/hub assembly by hand. If hub movement exceeds 0.012 in. (0.03mm) on Metro or 0.020 in. (0.05mm) on Prizm, replace the wheel bearings.

5. Install the wheel and lower the vehicle.

REMOVAL & INSTALLATION

Front

METRO

1. Raise and safely support the vehicle.

2. Remove the front wheel.

3. Remove the brake caliper and carrier and rotor. Do not let the caliper hang from the brake hose.

4. If equipped, disconnect the ABS wheel speed sensor.

5. Unstake and remove the hub nut and washer. Discard the nut.

6. Remove the hub from the steering knuckle using tools J-2619–01 and J-34866 or equivalent.

7. Separate the tie rod end from the steering knuckle.

8. Separate the ball joint from the steering knuckle.

9. Remove the strut-to-steering knuckle bolts and remove the steering knuckle from the vehicle. Support the halfshaft to prevent damage to the CV-joint.

10. Remove the inner and outer oil seals using removal tools J-26941 and J-2619–01 or equivalent.

11. Remove the inner bearing race and bearing from the steering knuckle.

12. Remove the inner wheel bearing oil seal using special tools J-26941 and J-2619–01, or equivalent.

13. Remove the snapring from the outer side of the steering knuckle.

14. Remove the outer wheel bearing from the steering knuckle.

15. Remove the outer bearing race from the hub using split plate bearing removal tool J-22912–01, hub race puller pilot tool J-38424 or equivalents, and a press.

16. Support the steering knuckle using support tool J-41392 or equivalent, and remove the wheel bearing assembly from the steering knuckle using wheel bearing removal tool J-41391 and driver handle J-7079–2 or equivalents.

17. Inspect the wheel bearings for excessive wear, corrosion or contamination by dirt or water; replace as necessary.

To install:

18. Support the steering knuckle using support tool J-41392 or equivalent.

19. Install the wheel bearing assembly into the steering knuckle using the wheel bearing race/outer seal installer tool J-39722 and driver handle J-7079–2 or equivalents.

20. Install the outer bearing into the steering knuckle.

21. Install the snapring into the outer side of the steering knuckle.

22. Install the inner wheel bearing and inner bearing race into the steering knuckle.

23. Install the outer bearing race onto the wheel hub using wheel hub race installer tool J-36050 or equivalent, and a press.

24. Apply a coating of wheel bearing grease to the lip of the inner oil seal.

25. If equipped with ABS, be sure to align the hole in the seal to the hole in the steering knuckle for the wheel speed sensor.

26. Install the inner oil seal to the inner side of the steering knuckle using installer tool J-41392 and driver handle J-7079–2 or equivalents.

27. Install the outer oil seal to the outer side of the steering knuckle using installer tool J-39722 and driver handle J-7079–2 or equivalents.

28. Apply a light coat of wheel bearing grease to the outside of the hub shaft.

29. Install the wheel hub and outer bearing race to the steering knuckle using a press.

30. Install the steering knuckle on the vehicle, inserting the CV-joint into the hub, and secure with the strut-to-steering knuckle bolts. Tighten the strut-to-steering knuckle bolts to 59 ft. lbs. (80 Nm).

31. Install the ball joint ball. Connect the tie rod end to the steering knuckle with the castle nut. Install a new cotter pin.

32. Install a new hub nut and washer. Tighten the hub nut to 129 ft. lbs. (175 Nm). Using a hammer and dull punch, stake the nut.

33. Install the brake rotor, brake caliper carrier and caliper.

34. If equipped, install the ABS wheel speed sensor. Connect the sensor electrical connector.

35. Install the front wheel and lug nuts.

36. Lower the vehicle. Check the wheel alignment.

PRIZM

1. Raise and safely support the vehicle. Remove the wheel.

2. Disconnect the ABS speed sensor, if equipped.

3. Remove the halfshaft nut.

4. Remove the brake caliper and support it out of the way. Remove the disc brake rotor.

5. Separate the tie rod end from the knuckle.

6. Separate the ball joint from the knuckle.

7. Remove the two nuts and bolts securing the steering knuckle to the strut.

8. Remove the steering knuckle and hub assembly.

9. Clamp the knuckle in a vise with protected jaws.

10. Remove the dust deflector.

11. Use a slide hammer and extractor tool J-26941 or equivalent, to remove the inner oil seal.

12. Remove the snapring.

13. Using hub puller tools J-25287 and J-35378 or equivalents, remove the hub from the knuckle.

14. Remove the brake splash shield.

15. Using a split plate bearing remover, puller pilot and a shop press, remove the inside and outside inner bearing races from the hub.

16. Remove the outer oil seal.

17. Position the steering knuckle on support tool J-35379 or equivalent. Remove the bearing from the knuckle using a plastic hammer and remover tool J-35399 or equivalent, and driver handle J-8092 or equivalent.

To install:

18. Clean and inspect the hub and knuckle for cracks, wear or other damage; replace as necessary.

19. Press a new bearing into the steering knuckle using installation tool J-37777 or equivalent, and driver handle J-8092 or equivalent.

20. Install a new snapring into the steering knuckle.

21. Install a new inner oil seal onto the steering knuckle using seal installer J-35737-2 or equivalent.

22. Insert the side lip of a new outer oil seal into seal installer J-35737-01 or equivalent, and drive the outer oil seal into the steering knuckle.

23. Apply multi-purpose grease to the outer oil seal lip.

24. Install the brake splash shield and tighten the bolts to 73 inch lbs. (8.3 Nm).

25. Place new inside and outside inner races on the hub bearing. Press the hub into the knuckle using installer J-35399 or equivalent, and driver handle J-8092 or equivalent.

26. Apply GM lubricant 1050109 or equivalent, to the contact surfaces of the inner oil seal lip and drive axle.

27. Drive a new dust deflector into the steering knuckle using a hammer and dull chisel.

28. Place the steering knuckle and hub assembly onto the halfshaft and install the knuckle-to-strut bolts and nuts.

29. Attach the ball joint to the control arm.

30. Attach the tie rod end to the knuckle.

31. Install the disc brake rotor and the brake caliper.

32. Install the ABS sensor, if equipped.

33. Install the nut on the end of the halfshaft.

34. Install the wheel and lower the vehicle.

35. Check the wheel alignment.

Rear

METRO WITHOUT ABS

1. Raise and safely support the vehicle.

2. Remove the rear wheel.

3. Pry off the dust cap.

4. Unstake and remove the spindle nut and washer. Discard the spindle nut; it must not be reused.

5. Gently tap on the brake drum/hub assembly with a hammer to break loose and remove the drum/hub assembly from the spindle.

6. Place the brake drum/hub assembly on wooden blocks on a workbench.

7. Drive the inner bearing out from the drum/hub assembly using a hammer and punch. The bearing spacer can be moved from side to side to enable placement of the punch. Work around the perimeter of the inner bearing by moving this spacer.

8. Remove the bearing spacer.

9. Drive the outer bearing out from the inboard side of the drum/hub assembly using the same method.

10. Inspect the wheel bearings for wear, corrosion or contamination and replace, as necessary.

To install:

➡**The outer bearings is smaller in diameter than the inner one.**

11. Install the outer bearing in the drum/hub assembly using installation tools J-34842 and J-7079-2 or equivalents.

12. Install the bearing spacer in the brake drum hub with the inner lip toward the outer bearing.

➡**Be sure the bearing spacer is properly installed. The assembly will not fit on the spindle if the spacer lip is toward the inner bearing.**

13. Install the inner bearing in the drum/hub assembly using installation tools J-34842 and J-7079-2 or equivalents.

14. Install the brake drum/hub assembly on the spindle.

15. Install the washer and a new spindle nut.

16. Tighten the spindle nut to 74 ft. lbs. (100 Nm).

17. Stake the spindle nut.

18. Install the dust cap.

19. Install the wheel. Lower the vehicle.

METRO WITH ABS

1. Raise and safely support the vehicle.

2. Remove the rear wheel.

3. Remove the two screws and remove the brake drum.

4. Pry off the dust cap.

5. Unstake and remove the spindle nut and washer.

6. Remove the hub assembly from the spindle.

To install:

7. Install the hub assembly on the spindle.

8. Install the washer and spindle nut.

9. Tighten the spindle nut to 120 ft. lbs. (170 Nm).

10. Stake the spindle nut.

11. Install the dust cap.

12. Install the wheel. Tighten the lug nuts, in a cross sequence, to 44 ft. lbs. (60 Nm).

13. Lower the vehicle.

PRIZM

1. Raise and safely support the vehicle.

2. Remove the wheel.

3. Remove the brake drum or rotor.

4. Check bearing end-play using a dial indicator. If end-play exceeds 0.002 in. (0.05mm), the hub must be replaced.

5. Disconnect and remove the ABS wheel speed sensor, if equipped.

6. Remove the four bolts securing the hub to the knuckle and remove the hub.

7. Remove the O-ring from the backing plate.

To install:

8. Install a new O-ring onto the backing plate.

9. Install the hub to the knuckle with the mounting bolts and tighten to 59 ft. lbs. (80 Nm).

10. Install the wheel speed sensor, if equipped.

11. Install the brake drum or rotor.

12. Install the wheel and lower the vehicle.

13. Check the rear wheel alignment.

SATURN

SC1 • SC2 • SL • SL1 • SL2 • SW1 • SW2

41

DRIVE TRAIN **41-18**
ENGINE REPAIR **41-2**
FUEL SYSTEM **41-14**
STEERING AND
SUSPENSION **41-22**

A
Air Bag 41-22
  Disarming 41-22
  Precautions 41-22

C
Camshaft and Lifters 41-9
  Removal & Installation 41-9
Clutch 41-20
  Adjustment 41-20
  Removal & Installation 41-20
Coil Spring 41-24
  Removal & Installation 41-24
Cylinder Head 41-4
  Removal & Installation 41-4

E
Engine Assembly 41-2
  Removal & Installation 41-2
Exhaust Manifold 41-8
  Removal & Installation 41-8

F
Front Crankshaft Seal 41-8

Fuel Filter 41-15
  Removal & Installation 41-15
Fuel Pump 41-16
  Removal & Installation 41-16
Fuel System Pressure 41-15
  Relieving 41-15
Fuel System Service
  Precautions 41-14

H
Halfshafts 41-21
  Removal & Installation 41-21
Hydraulic Clutch System 41-21
  Bleeding 41-21

I
Ignition Timing 41-2
  Adjustment 41-2
Intake Manifold 41-7
  Removal & Installation 41-7

O
Oil Pan 41-10
  Removal & Installation 41-10
Oil Pump 41-10
  Removal & Installation 41-10

R
Rack and Pinion Steering Gear 41-22
  Removal & Installation 41-22

Rear Main Seal 41-11
  Removal & Installation 41-11
Rocker Arms/Shafts 41-6
  Removal & Installation 41-6

S
Strut 41-23
  Removal & Installation 41-23

T
Timing Chain, Sprockets, Front
  Cover and Seal 41-11
  Removal & Installation 41-11
Transaxle Assembly 41-18
  Removal & Installation 41-18

V
Valve Lash 41-10
  Adjustment 41-10

W
Water Pump 41-4
  Removal & Installation 41-4
Wheel Bearings 41-24
  Adjustment 41-24
  Removal & Installation 41-24

ENGINE REPAIR

➡**Disconnecting the negative battery cable on some vehicles may interfere with the functioning of the on-board computer system, and may require the computer to undergo a relearning process once the negative battery cable is reconnected.**

Ignition Timing

ADJUSTMENT

The 1.9L Engines utilizes a Distributorless Ignition System (DIS). This system uses three twin tower coils which fire two spark plugs simultaneously. One spark plug is located in a cylinder on the compression stroke and the other in a cylinder on the exhaust stroke. This is known as the waste spark method of distribution. The Powertrain Control Module controls the igntion timing, no adjustment is necessary or possible.

Engine Assembly

REMOVAL & INSTALLATION

It is virtually impossible to list each individual wire and hose which must be disconnected, simply because so many different model and engine combinations have been manufactured. Careful observation and common sense are the best possible approaches to any repair procedure.

Removal and installation of the engine can be made easier if you follow these basic points:

• If you have to drain any of the fluids, use a suitable container.

• Always tag any wires or hoses, if possible, the components they came from before disconnecting them.

• Because there are so many bolts and fasteners involved, store and label the retainers from components separately in muffin pans, jars or coffee cans. This will prevent confusion during installation.

• After unbolting the transaxle, always be sure it is properly supported.

• When unbolting the engine mounts, always be sure the engine is properly supported. When removing the engine, be sure that any lifting devices are properly attached to the engine. It is recommended that if your engine is supplied with lifting hooks, your lifting apparatus be attached to them.

• Lift the engine from its compartment slowly, checking that no hoses, wires or other components are still connected.

• After the engine is clear of the compartment, place it on an engine stand or workbench.

• After the engine has been removed, you can perform a partial or full teardown of the engine using the procedures outlined in this manual.

➡**The manufacturer recommends that the engine and transaxle be removed as a complete unit. Disconnect the cradle and lower the entire assembly instead of lifting the assembly out of the vehicle. Both the SOHC and DOHC engines are removed or installed in the same manner. It is however possible to remove the engine or transaxle on some models with automatic transaxles.**

1. Properly disable the SIR system. Disconnect the negative, then the positive battery cable.

2. Drain the engine coolant from the radiator and engine block into a suitable clean container.

✳✳ CAUTION

Never open, service or drain the radiator or cooling system when hot; serious burns can occur from the steam and hot coolant.

3. Properly relieve the fuel system pressure.

4. Remove the complete air cleaner/intake duct assembly, as applicable.

5. Disconnect and label the all electrical plugs and vacuum lines, as applicable for engine removal.

6. Disengage all connectors from the automatic or manual transaxle. If access is difficult, wait until the vehicle is safely raised and supported, then unplug the connections from underneath the vehicle.

7. Disconnect the accelerator cable assembly.

8. Separate the fuel supply and return lines at their connectors. Plug the lines to prevent fuel contamination or loss. The lines may be tied to the master cylinder lines to help prevent fuel spillage and to keep them out of the way.

9. Disconnect the upper radiator hose and the cylinder head outlet and the de-aeration hose at the engine.

10. Remove the serpentine drive belt and if equipped, remove the air conditioning compressor from its brackets and with the hoses attached. Support the compressor from the front crossbar.

➡**It is not necessary to discharge the A/C compressor during engine removal, but be careful not to kink, damage or rupture the refrigerant lines.**

11. If equipped, disconnect the automatic transaxle cooler lines at the transaxle by pinching the plastic connector tabs and carefully pulling back on the lines. Plug the openings to prevent fluid loss or contamination.

12. Disconnect the automatic transaxle shifter cable or the manual shifter cables from the transaxle.

13. If equipped with a manual transaxle, remove the two hydraulic slave cylinder retaining nuts from the clutch housing studs, then slide the cylinder and bracket assembly from the studs. Rotate the clutch actuator ¼ turn counterclockwise while pushing toward the housing to disengage the bayonet connector and remove it from the clutch housing. Support the clutch hydraulic system to the battery tray; being sure not to kink or pinch the hydraulic lines.

14. Using a length of an appropriate wire, tie the radiator, condenser and fan to the front crossbar. Route the wire around the two fan shroud supports and the crossbar.

15. Raise and support the vehicle safely.

16. Remove the front wheels and remove the fasteners connecting the side and front fender shields to the cradle.

17. Remove the brake caliper bracket attaching bolts (2 on each side) and hang the caliper assemblies from the shock tower springs using wire. Do not hang the assembly by the brake hose or damage to the brake hydraulic system may occur. The springs and shocks will remain with the body when the powertrain is lowered.

18. Disconnect the struts from the knuckles on each side of the vehicle (2 bolts per side). This will allow the knuckle and hub assembly to remain with the powertrain cradle when lowered. The stabilizer bar will remain attached to the cradle and the lower control arms.

19. Using a suitable hose clamp tool, disconnect the lower radiator and heater return hoses from the engine. Disconnect the heater inlet hose at the front of the dash or the engine.

20. Disconnect the steering shaft and pressure switch connectors at the gear, as applicable.

21. Disconnect the front exhaust pipe at the manifold, catalytic converter and powertrain stiffening bracket.

22. Remove the flywheel cover and torque converter bolts, if equipped.

23. Remove the alternator and starter shields.

24. Unclip the brake lines from the rear side of the cradle.

25. Carefully remove the electrical harness from the engine and transaxle, then lay the electrical harness on top of the underhood junction block and battery cover.

26. Place a 1 in. x 1 in. x 2 in. long block of wood between the torque strut and cradle to ease removal and installation of the torque engine mount. Remove the three right side upper engine torque axis to front cover nuts and the two mount to midrail bracket nuts, allowing the powertrain to rest on the block of wood.

➡**Placing a block of wood under the torque axis mount prior to removing the upper mount will allow the engine to rest on the wood, thus preventing the engine from shifting. If the engine is not to be removed from the cradle, this will allow you to install the engine and the upper mount without jacking or raising the engine.**

27. Place a powertrain support dolly under the cradle. Use two 4 in. x 4 in. x 36 in. pieces of wood to support the cradle on the dolly.

28. Remove the two right side front engine mount, torque strut brackets to cradle nuts.

29. Remove the four cradle attaching bolts and carefully lower the complete powertrain assembly from the vehicle. Verify that all necessary components are disconnected and free before complete removal.

30. Attach the two washers located between the cradle and body, to the cradle. They must be repositioned and installed during cradle installation.

31. Disconnect the spark plug wires at the ignition module.

32. If applicable, remove the power steering pump and bracket. Support the assembly, in an upright position, from the cradle or the steering gear.

33. Install a suitable engine lifting device to the service support brackets.

34. Remove the front mount assembly and disconnect the motion restrictor bracket, if applicable.

35. Place a ½ in. x 1 in. x 3 in. block of wood under the axle shaft and remove the starter support bracket bolt, intake manifold support brace (on DOHC engines), and three axle shaft bracket support bolts. Allow the bracket to rotate rearward. Lift the engine slightly for clearance, as necessary.

36. Place a 4 in. x 4 in. x 6 in. long block of wood under the transaxle housing for support.

37. Remove the engine strut bracket and torque strut from the front of the engine as an assembly. Lift the engine slightly as necessary for removal.

38. Remove the four transaxle attaching bolts/studs and separate the assembly. Manual transaxles will require the engine to be moved about 4 inches (100mm) forward in the cradle to disengage the input shaft.

39. Carefully lift the engine off the cradle.

To install:

40. If removed, install the transaxle on the engine.

41. Tighten the front engine mount-to-engine bolts to 41 ft. lbs. (55 Nm).

42. Tighten the engine mount torque strut-to-cradle bracket fasteners to 52 ft. lbs. (70 Nm). Hand-tighten the cradle fasteners, but do not torque until the upper midrail mount is installed.

43. Tighten the axle shaft bracket fasteners to 41 ft. lbs. (55 Nm) and the starter bracket to 80 inch lbs. (9 Nm).

44. Carefully lift the powertrain and cradle into position. Be sure the radiator grommets are correctly aligned and that the two washers are reinstalled between the cradle and body at each rear cradle attachment position. If necessary, use two ³⁄₁₆ in. x 18 in. long guide pins in the forward cradle holes (located next to the attaching holes) to help align the cradle. Tighten the cradle to body fasteners to 151 ft. lbs. (205 Nm).

45. Tighten the steering shaft U-joint bolt to 35 ft. lbs. (47 Nm).

46. Tighten the starter solenoid connector to 44 inch lbs. (5 Nm).

47. Tighten the alternator and starter battery connectors to 89 inch lbs. (10 Nm).

48. Install the following components, if previously removed:

a. Tighten the oil pressure sensor to 26 ft. lbs. (35 Nm).

b. Tighten the knock sensor to 133 inch lbs. (15 Nm).

c. Tighten the crankshaft position sensor mounting fastener(s) to 80 inch lbs. (9 Nm).

d. Tighten canister purge valve mounting screw to 22 ft. lbs. (30 Nm).

49. Connect the wiring harness PCM ground and tighten to 89 inch lbs. (10 Nm).

50. Connect the wiring harness to the transaxle case/engine block and tighten to 18 ft. lbs. (25 Nm).

51. Tighten the engine stiffening bracket bolts to 35 ft. lbs. (47 Nm).

52. Tighten the exhaust pipe-to-intake manifold fasteners to 23 ft. lbs. (31 Nm), the pipe-to-stiffener bracket fasteners to 35

ft. lbs. (47 Nm), the pipe-to-support bracket fasteners to 23 ft. lbs. (31 Nm) and the pipe-to-catalytic converter fasteners to 18 ft. lbs. (25 Nm).

53. Tighten the cylinder block drain plug to 27 ft. lbs. (36 Nm), then close the radiator drain.

54. Tighten the steering knuckle-to-strut attachment bolts to 148 ft. lbs. (200 Nm).

55. Tighten the brake caliper assembly bolts to 81 ft. lbs. (110 Nm).

56. If equipped with a manual transaxle, tighten the hydraulic clutch slave cylinder fasteners to 19 ft. lbs. (25 Nm).

57. Tighten the A/C compressor-to-front bracket bolts to 40 ft. lbs. (54 Nm) and the compressor-to-rear bracket bolts to 22 ft. lbs. (30 Nm).

58. Tighten the two engine mounts-to-midrail bracket nuts to 37 ft. lbs. (50 Nm). Tighten the torque axis mount-to-front cover nuts uniformly to 37 ft. lbs. (50 Nm).

59. Tighten the strut bracket-to-cradle nuts to 37 ft. lbs. (50 Nm).

60. If equipped with automatic transaxle, tighten the torque converter-to-flexplate bolts to 52 ft. lbs. (70 Nm). Tighten the dust cover fasteners to 89 inch lbs. (10 Nm).

61. Tighten the wheel lugs, in a crisscross pattern to 103 ft. lbs. (140 Nm).

62. Lower the vehicle to the ground.

63. Apply a drop of clean engine oil to the male ends of the fuel line connectors and attach the line quick-connect fittings. Be sure the lines are not kinked or damaged.

➡**Check the upper cooling module grommets for binding or misalignment. The module retaining pins must be centered in the grommets supported by the brackets. If the grommets are pinched, loosen the brackets and reposition them. It is extremely important that the cooling module be able to move freely.**

64. Thoroughly inspect the transaxle and engine compartment area to be sure that all wires, hoses and lines have been connected. Also inspect to be sure that all components and fasteners have been properly installed.

65. Connect the positive, then the negative battery cables.

66. Fill the engine cooling system, then check all engine and transaxle fluids, add or fill as necessary.

67. Prime the fuel system by cycling the ignition **ON** for 5 seconds and **OFF** for 10 seconds a few times without cranking the engine. Start the engine and check for leaks.

68. Perform a short road test and check the engine again for leaks. Be sure the cooling system is filled to the surge tank FULL COLD line.

Water Pump

REMOVAL & INSTALLATION

1. Allow a sufficient amount of time for the engine to cool down.
2. Disconnect the negative battery cable. Drain the engine coolant from the radiator and block drains into a suitable clean container.

※※ CAUTION

Never open, service or drain the radiator or cooling system when hot; serious burns can occur from the steam and hot coolant.

3. Remove the serpentine drive belt.
4. Raise and support the vehicle safely. Remove the right front tire and inner wheel well splash shield.
5. If access to the water pump is desired from underhood, remove the air conditioning compressor bolts and position the compressor aside with the refrigerant lines intact.
6. Spray the water pump hub with penetrating oil to loosen any rust or corrosion that might bind the pulley and damage it during removal.
7. Remove the water pump pulley bolts and allow the pulley to hang freely on the hub. A 1 in. (25.4mm) block of wood or a hammer handle may be wedged between the pump and crankshaft pulleys to hold the assembly while loosening the retaining bolts.
8. Move the pulley outward or remove as necessary for access and remove the 6 water pump flange bolts. Carefully pull the pump and pulley assembly away from the engine and remove the assembly from the vehicle. If necessary, a gasket scraper may be inserted under the flange, but be careful not to damage the machined aluminum block sealing surface.

To install:
9. Thoroughly clean the gasket mating surfaces of all old gasket material. Apply a small amount of gasket sealant at the outer edges of the bolt holes to hold the gasket in place, then install the gasket onto the water pump assembly.
10. Install the pump assembly with the small bump located next to one of the attaching bolts in the 11 o'clock position. Install and tighten the bolts in a crisscross sequence as shown to 22 ft. lbs. (30 Nm).

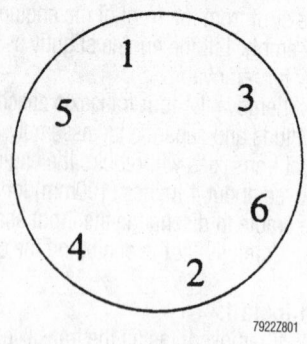

Water pump bolt torque sequence

11. Install or reposition the pump pulley, as applicable and tighten the bolts to 19 ft. lbs. (25 Nm). If the pump hub exposed through the pulley is rusty, clean it with a wire brush and apply a thin coat of primer to prevent the pulley from rusting onto the hub.
12. Install the serpentine drive belt, the right splash shield and right tire assembly.
13. If repositioned, install the air conditioning compressor.
14. Close the radiator drain plug and install the cylinder block drain plug. Tighten the block plug to 26 ft. lbs. (35 Nm).
15. Connect the negative battery cable and properly fill the engine cooling system.
16. Operate the engine and check for coolant leaks.

Cylinder Head

REMOVAL & INSTALLATION

※※ WARNING

Only remove the cylinder head when the engine is cold. Warpage may result if the cylinder head is removed while the engine is hot.

1. Disconnect the negative battery cable and remove the coolant bottle cap. Drain the engine coolant from the radiator and block drains into a suitable clean container.

※※ CAUTION

When draining the coolant, keep in mind that cats and dogs are attracted by ethylene glycol antifreeze, and are quite likely to drink any that is left in an uncovered container or in puddles on the ground. This will prove fatal in sufficient quantity. Always drain the coolant into a sealable container. Coolant may be reused unless it is contaminated or several years old.

2. Remove the air cleaner/inlet duct assembly. Disconnect the PCV valve and fresh air hose from the camshaft cover.
3. Disconnect the accelerator cable from the throttle body and the bracket from the intake manifold.
4. Properly relieve the fuel system pressure.
5. Label and disengage the vacuum hoses and electrical connectors from the cylinder head assembly and components, as applicable. Long nose pliers are necessary to disconnect the coolant temperature connectors. When disconnected, position the electrical harness over the underhood junction block and battery cover.
6. Disconnect the upper radiator hose at the cylinder head outlet, the de-aeration hose at the intake manifold, or the heater hose at the intake manifold or front of dash.
7. Remove the bolt which retains the fuel lines to the intake manifold assembly. Disconnect the fuel feed and return lines from the fuel supply rail.
8. Unclip the lower splash shield for access. Place a 1 in. **x** 1 in. **x** 2 in. long block of wood between the torque strut and cradle to ease removal and installation of the torque engine mount. Remove the three right side upper engine torque axis to front cover nuts and the two mount to midrail bracket nuts, allowing the powertrain to rest on the block of wood.

➥**Placing a block of wood under the torque axis mount prior to removing the upper mount will allow the engine to rest on the wood, thus preventing the engine from shifting. This will allow you to install the mount during assembly without jacking or raising the engine.**

9. Remove the serpentine drive belt and belt tensioner. It is not necessary to remove the water pump pulley, however, it will be necessary to remove the idler pulley to access the engine front cover.
10. For SOHC engines, disconnect the de-aeration line at the cylinder head water outlet and from the support bracket.
11. Remove the fasteners and the camshaft cover, then inspect the cover's silicone insulators for cracks or deterioration and replace, as necessary. Be sure to cover the valvetrain area to prevent dirt and debris from entering the engine.
12. Remove the power steering pump bracket attaching bolts (3 for the SOHC engine or 5 for the DOHC engine) and position the assembly next to the right side front of the dash panel away from the intake manifold and cylinder head. It is not necessary to remove the water pump pulley.

13. If equipped, remove the three air conditioning compressor front bracket bolts attached to the cylinder head and block, then remove the rear bracket bolts from the compressor or engine. Do not discharge the system or disconnect the refrigerant lines. Support the compressor out of the way, from the vehicle front support bar.

14. Raise and support the front of the vehicle safely.

15. Drain the engine oil into a suitable drain pan.

16. Remove the right side tire and splash shield.

17. For DOHC engines, remove the intake manifold support brace bolt attached to the intake manifold next to the alternator.

18. Remove the crankshaft damper/pulley assembly. Use a strap wrench or a block of wood wedged between the pulley spoke and the rear lower side of the front cover to hold the assembly while removing the bolt. Then, use a 3-jaw puller on the jaw slots cast into the pulley to remove the assembly.

19. Disconnect the exhaust pipe from the manifold, then remove and discard the gasket.

20. Install a crankshaft sprocket retainer tool, such as SA9104E or equivalent, with the flat side toward the sprocket. Properly remove the engine front/timing chain cover.

✳✳ WARNING

Be sure to properly install the crankshaft sprocket retainer tool. Failure to hold the crankshaft timing sprocket in place will cause timing chain damage.

21. Rotate the crankshaft clockwise to the position 90 degrees from Top Dead Center (TDC) so the timing mark and keyway align with the main bearing cap split line. This will be sure pistons will not contact the valves upon assembly.

22. Remove the timing chain, tensioner,

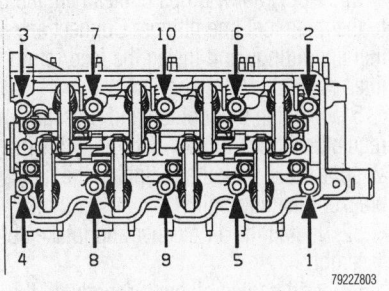

Gradually remove the cylinder head bolts in the sequence shown to prevent warping the head

guides, camshaft sprocket(s) and chain. Use a ⅞ in. (21mm) wrench to hold the camshaft when removing the sprocket bolts.

23. Use a 6 point socket to remove the 10 cylinder head bolts in several passes of the proper sequence. Failure to follow the proper sequence or removal of the head when hot could result in head warpage or cracking. Also, the use of a 12 point socket on the cylinder head bolts may round the bolt heads.

24. Lift the cylinder head from the dowels, if necessary use a small prybar for leverage between the cylinder head and block bosses. Be careful not to damage the sealing surfaces if prying is necessary to remove the head from the block.

25. If necessary, remove the intake manifold or the exhaust manifold by loosening the mounting nuts in the proper sequence. If any cylinder head studs back out, the threads should be cleaned, the studs carefully installed, then tightened to 106 inch lbs. (12 Nm).

To install:

➡**Before installing the cylinder head, it should be cleaned and inspected for excessive wear or damage.**

26. If removed, install the intake manifold and/or the exhaust manifold and new

gasket(s). Tighten to specification in the proper sequence.

27. Clean the gasket mating surfaces. Be careful not to damage the aluminum components. Be sure the block bolt holes are clean of any residual sealer, oil or foreign matter.

28. Using a dial gauge at four points around each cylinder, check that the cylinder liners are flush or do not deviate more than 0.0005 in. (0.013mm).

29. Be sure the crankshaft is still 90 degrees past TDC and that the camshaft(s) are properly positioned with the dowel pin(s) at the 12 o'clock position to prevent valve damage. Install the cylinder head gasket and carefully guide the head into place over the dowels.

30. If the head bolts and/or the block were replaced, install the bolts and tighten in sequence to 48 ft. lbs. (65 Nm) to insure proper clamp load, then remove the bolts.

31. Coat the cylinder head bolts with clean engine oil and thread the bolts by hand until finger-tight. Tighten the bolts in sequence to 22 ft. lbs. (30 Nm).

32. Tighten the cylinder head bolts again, in sequence to 33 ft. lbs. (45 Nm) for SOHC engines or to 37 ft. lbs. (50 Nm) for DOHC engines. Install a torque angle gauge tool and calibrate the tool to zero. In sequence, tighten each cylinder head bolt an additional 90 degrees.

33. Install the timing chain, sprockets, guides and tensioner. Then, install the front cover assembly. Refer to the appropriate procedures in this section.

34. Position a new gasket, then connect the exhaust pipe to the manifold. Install and tighten the fasteners to 23 ft. lbs. (31 Nm).

35. If not already done, remove the crankshaft sprocket retainer tool. Apply a thin film of RTV sealant to the damper/pulley assembly flange and washer only. Install the crankshaft damper/pulley assembly and tighten the bolt to 158 ft. lbs. (214 Nm)

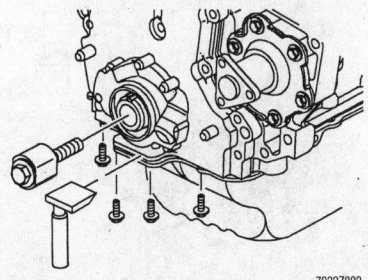

Crankshaft gear retaining and oil pan removal tool shown—be sure to install the crankshaft tool with the flat side toward the gear

Rotate the crankshaft clockwise to 90 degrees past TDC (the crankshaft sprocket timing mark will be at three o'clock) to prevent valve damage during assembly

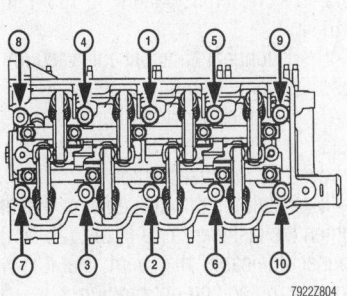

Tighten the cylinder head bolts in the correct sequence to promote good cylinder sealing and to prevent leaks

while holding the pulley with a strap wrench or block of wood.

36. For DOHC vehicles, install the intake manifold support brace bolts next to the alternator, then tighten the block bolt to 33 ft. lbs. (45 Nm) and tighten the manifold bolt to 22 ft. lbs. (30 Nm).

37. Apply a small drop of RTV across the cylinder head and front cover T-joints. Inspect the old camshaft cover gasket and replace if damaged. Install the gasket and the camshaft cover. Tighten the fasteners uniformly to 22 ft. lbs. (30 Nm) for SOHC vehicles or in proper sequence to 89 inch lbs. (10 Nm) for DOHC vehicles.

38. Install the drive belt tensioner and tighten the bolt to 22 ft. lbs. (30 Nm). For 1992–98 vehicles, install the idler pulley and tighten the fasteners to 20 ft. lbs. (27 Nm).

39. If not done during removal, drain the engine oil and change the filter, then install the drain plug and tighten to 26 ft. lbs. (35 Nm).

40. If removed, verify the gaps on all spark plugs and install. Tighten to 20 ft. lbs. (27 Nm).

41. Install the power steering pump assembly to the bracket, then tighten the bolts to 22 ft. lbs. (30 Nm).

42. If equipped, install the air conditioning compressor and bolts. Tighten the rear bracket bolts to 19 ft. lbs. (25 Nm), then tighten the front bracket bolts to 35 ft. lbs. (47 Nm).

43. Install the accessory drive belt making sure the belt is properly aligned on the pulley.

44. Install the two mount to midrail bracket nuts and tighten to 37 ft. lbs. (50 Nm). Install the three upper mount to engine front cover nuts and tighten them uniformly to 37 ft. lbs. (50 Nm). Remove the support block of wood after the assembly is installed.

45. Install the splash shield, then install the tire and wheel assembly. Tighten the lug nuts, in a crisscross pattern, to 103 ft. lbs. (140 Nm).

46. Reconnect all applicable vacuum hoses disconnected during removal.

47. Position the wiring harness and install all applicable wire connectors removed during disassembly:

48. Install the throttle cable bracket and tighten the fastener to 19 ft. lbs. (25 Nm). Connect the cable, then verify that it is properly routed and not binding.

49. Connect the upper radiator hose at the cylinder head outlet, the de-aeration hose at the intake manifold, or the heater hose to the intake manifold or front of dash.

50. Apply a few drops of clean engine oil to the male fuel line fittings. Connect any fuel line fittings and install the feed/return lines.

51. Tighten the fuel rail and pressure regulator fittings to 133 inch lbs. (15 Nm), as applicable. Install and tighten the fuel bracket retaining bolt.

52. Install the air cleaner and intake duct assembly.

53. Add engine oil and properly fill the engine cooling system.

54. Thoroughly inspect the engine compartment area, especially the cylinder head, to be sure that all wires, hoses and lines have been connected. Also inspect to be sure that all components and fasteners have been properly installed.

55. Connect the negative battery cable.

56. Prime the fuel system by cycling the ignition a few times without cranking the engine, then start the engine and check for leaks.

57. Operate the engine at idle for 3–5 minutes and listen or unusual noises. If the lifters are noisy or the cylinders are misfiring, warm the engine to normal operating temperature running at less than 2000 rpm. Once the engine is warm and the thermostat has opened, cycle the engine between idle and 3000 rpm for 10 minutes or drive the vehicle at least 5 miles to purge air from the lifters. If air cannot be purged, the faulty lifters must be replaced.

58. Verify proper coolant level. Add coolant, if necessary, after the engine has cooled.

Rocker Arms/Shafts

Only the SOHC engine has a cylinder head that uses rocker arms in its valvetrain.

REMOVAL & INSTALLATION

1. Disconnect the negative battery cable.

2. Remove the rocker arm cover, then inspect the cover silicone isolators for cracks or deterioration and replace, as necessary.

3. Uniformly remove the rocker arm assembly bolts, then carefully remove the two shaft and rocker arm assemblies. The shafts may be unsnapped from the lifter guideplates leaving the lifters and plates in the cylinder head. If this cannot be accomplished, remove the guideplates and lifters, but be sure to reposition the lifters in the same bores and guideplates during installation.

4. If necessary, remove the rocker arms from the shafts.

To install:

5. If removed, oil the shafts and install the rocker arms into position on the shafts.

6. Snap one end of each lifter guide plate retaining spring onto the rocker arm shaft between the No. 1-no. 2 and the No. 3-no. 4 cylinder rocker arms.

7. Install the rocker arm shaft assemblies. To prevent valve or piston damage, be sure the rocker arm tangs are squarely seated on the lifter plungers and the retaining springs are positioned in the guide plate slots.

➡If difficulty is encountered aligning the rocker arms on the valves and lifters, use a flat piece of wood, cardboard or an extension bar of suitable length on top of the shafts and rocker arms to hold both assemblies in position.

8. Tighten the 5 rocker arm bolts on each shaft in a uniform sequence to 18 ft. lbs. (25 Nm). Verify the proper position and seating of all rocker components.

9. Apply a small drop of RTV to each cylinder head and front cover T-joint. Inspect the rocker arm cover gasket and

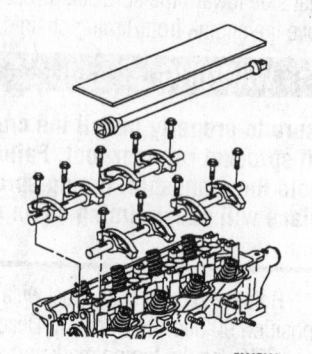

Exploded view of the rocker arms, shafts and lifter guide plates

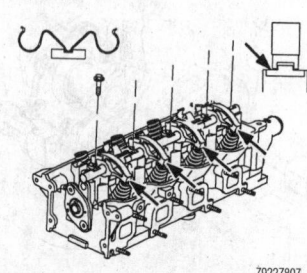

The rocker arm tangs or slots must be properly positioned on the plungers and the springs must be positioned on the guide plate slots to prevent valve or piston damage

replace if necessary. Install the gasket and rocker arm cover, then tighten the fasteners uniformly to 22 ft. lbs. (30 Nm). Be sure all cover hoses and components are reconnected after installation.

10. Connect the negative battery cable, start the engine and check for leaks.

Intake Manifold

REMOVAL & INSTALLATION

SOHC Engine

1. Disconnect the negative battery cable and drain the engine coolant from the radiator and engine block, into a suitable container.

2. Remove the air cleaner, disconnecting the fresh air tube at the valve cover. Remove the PCV tube and hose.

3. Properly relieve the fuel system pressure at the test port.

4. Remove the retaining screw from the fuel line retaining bracket.

5. Disconnect the fuel line(s) at the quick-connect fitting(s) by compressing the two tangs of the retainer and pulling.

6. Plug the line(s) to prevent fuel contamination or loss.

7. Disconnect the throttle cable from the throttle body, then remove the throttle cable bracket attaching nuts and position the assembly aside.

8. Label and disconnect the wiring from the intake manifold, throttle body and valve cover components, as follows:

    a. The fuel injectors.
    b. Idle Air Control (IAC) valve.
    c. Throttle Position Sensor (TPS).
    d. Exhaust Gas Recirculation (EGR) solenoid.
    e. Manifold Absolute Pressure (MAP) sensor.

9. Remove the wiring tube and lay the harness away from the manifold onto the Under Hood Junction Block (UHJB).

10. Label and disconnect all vacuum hoses from the throttle body unit. Disconnect the vacuum line from the brake booster.

11. Disconnect the heater hose from the intake manifold and the de-aeration line fitting at the cylinder head coolant outlet. Remove the two clamps and lay the line onto the coolant bottle.

12. Remove the intake manifold support bracket bolt located next to the starter. If necessary, the bolt can be removed from below the vehicle.

13. Remove the serpentine drive belt, then remove the power steering pump and

support the pump next to the right side dash panel sufficiently away from the intake manifold and cylinder head.

14. Remove the manifold retaining nuts, then remove the manifold and throttle body assembly. It may be much easier to access the lower manifold nuts from under the vehicle. Remove and discard the old gasket from the mating surfaces.

To install:

15. Thoroughly clean all gasket mating surfaces. Be careful not to damage or score the aluminum surface. If replaced, use Loctite® 290 or equivalent to seal the new PCV valve inlet tube into the manifold.

16. Position the new gasket, then install the manifold and retaining nuts. Tighten the nuts in sequence to 22 ft. lbs. (30 Nm).

17. Install the power steering pump and tighten the fasteners to 27 ft. lbs. (38 Nm).

18. Install the serpentine drive belt.

19. Connect the coolant hose, the de-aeration line and clamps, then install the manifold support bracket bolt. Tighten the bolt to 22 ft. lbs. (30 Nm).

20. Lubricate the male ends of the fuel lines with a few drops of clean engine oil, then connect the fuel supply and return lines.

21. Secure the fuel line(s) in the retaining bracket and tighten the mounting screw to 36 inch lbs. (4 Nm).

22. Reposition the wiring harness and connect the wiring and vacuum hoses to their original locations. The harness leads to the TPS and EGR solenoid must be routed between the intake manifold runners.

23. Inspect the air cleaner/throttle body unit gasket and replace, if necessary. Install the air cleaner and fresh air tubes, then install the PCV valve hose.

24. Connect the negative battery cable, close the radiator drain and install the engine block drain plug. Tighten the block drain plug to 26 ft. lbs. (35 Nm). Fill the engine cooling system.

Upper Side				
8	4	1		5
7	3	2	6	9
Lower Side				

79222808

Intake manifold torque sequence—SOHC engine

25. Prime the fuel system by cycling the ignition switch **ON** for 5 seconds, then **OFF** for 10 seconds and repeating two times. Start the engine and check for leaks.

DOHC Engine

1. Disconnect the negative battery cable and drain the engine coolant from the radiator and engine block, into a suitable container.

2. Remove the air inlet tube, disconnecting the fresh air tube at the camshaft cover. Lift the resonator upward to disengage it from the engine service support bracket. Remove the PCV valve and hose.

3. Properly relieve the fuel system pressure at the test port.

4. Remove the retaining screw from the fuel line retaining bracket.

5. Disconnect the fuel line(s) at the quick-connect fitting(s) by compressing the two tangs of the retainer and pulling.

6. Plug the line(s) to prevent fuel contamination or loss.

7. Disconnect the throttle cable from the throttle body, then remove the throttle cable bracket attaching nuts and position the assembly aside.

8. Label and disconnect the wiring from the intake manifold and surrounding components, as follows:

    a. The fuel injectors.
    b. Idle Air Control (IAC) valve.
    c. Throttle Position Sensor (TPS).
    d. Manifold Absolute Pressure (MAP) sensor.

9. Disconnect the heater and de-aeration hoses from the intake manifold outlets. The heater hose may be removed at the firewall, if necessary. Disconnect the EGR solenoid vacuum hose.

10. Position the wiring harness over the brake master cylinder, then remove the intake manifold support bracket bolt attached to the manifold next to the brake master cylinder.

11. Remove the serpentine drive belt. Remove the power steering pump assembly with the support bracket, then remove the upper pump bracket attachment bolts and position the pump away from the manifold and cylinder head, near the right dash panel.

12. Remove the three upper intake manifold attachment nuts, then raise and support the vehicle safely.

13. Remove the lower power steering unit support bracket. Remove the intake manifold support bracket bolt located next to the alternator, then the lower bracket bolt and remove from the vehicle.

14. Disconnect the canister purge solenoid and brake booster vacuum hoses.

15. Remove the intake manifold attaching stud, remove the supports and lower the vehicle.

16. Remove the remaining fasteners and the intake manifold assembly, then remove and discard the old gasket.

To install:

17. Thoroughly clean the gasket mating surfaces. Be careful not to score or damage the aluminum sealing surfaces. If installing a new coolant de-aeration tube elbow into the manifold use Loctite® 290 or equivalent to achieve a proper seal.

18. Position the new gasket, then install the intake manifold and retaining nuts. Tighten the nuts in sequence to 22 ft. lbs. (30 Nm).

19. Install the power steering pump and brackets. Tighten the fasteners to 28 ft. lbs. (38 Nm).

20. Install the serpentine drive belt making sure the belt is properly aligned on the pulleys.

21. Connect the heater hose and de-aeration line to the manifold.

22. Position the manifold support brackets and install the bolts. Tighten the right block bolt to 41 ft. lbs. (55 Nm), then tighten the left block bolt and the support bracket to intake manifold bolts to 22 ft. lbs. (30 Nm).

23. Lubricate the male ends of the fuel lines with a few drops of clean engine oil, then connect the fuel supply and return lines.

24. Secure the fuel line(s) in the retaining bracket and tighten the mounting screw to 36 inch lbs. (4 Nm).

25. Connect the throttle cable to the throttle body and install the support bracket. Tighten the bracket retaining bolts to 19 ft. lbs. (25 Nm). Verify that the cable

locking tangs are fully engaged when assembled.

26. Position the wiring harness and connect all electrical connectors and vacuum hoses in their original locations.

27. Install the PCV hose, the air inlet tube and resonator.

28. Connect the negative battery cable, close the radiator drain and install the engine block drain plug. Tighten the block drain plug to 26 ft. lbs. (35 Nm). Fill the engine cooling system.

29. Prime the fuel system by cycling the ignition switch **ON** for 5 seconds, then **OFF** for 10 seconds and repeating two times. Start the engine and check for leaks.

Exhaust Manifold

REMOVAL & INSTALLATION

1. Disconnect the negative battery cable, then raise and support the vehicle safely.

2. If equipped, remove the two front exhaust pipe to engine support bracket mounting fasteners.

3. Remove the pipe-to-manifold nuts and lower the pipe.

4. Remove the supports and lower the vehicle.

➡**When performing the next step, do NOT disconnect the refrigerant lines.**

5. If equipped, remove the air conditioning compressor and bracket from the engine, which will first require the removal of the serpentine belt, then position them aside.

6. Disconnect the oxygen sensor connector. If necessary, use a 19mm, 6-point crow's foot wrench to remove the oxygen sensor from the manifold.

7. Remove the manifold retaining nuts and remove the manifold from the cylinder head. Remove and discard the old gaskets from the mating surfaces.

To install:

8. Thoroughly clean the gasket mating surfaces, being careful not to score or damage the aluminum surface.

9. Install the new gasket with the smooth side facing the manifold, then install the manifold and attaching nuts. Tighten the nuts in sequence to 16 ft. lbs. (22 Nm) for the SOHC engine or to 23 ft. lbs. (31 Nm) for the DOHC engine.

10. If replacing the oxygen sensor, coat the threads with nickel based anti-seize compound and tighten to 33 ft. lbs. (45

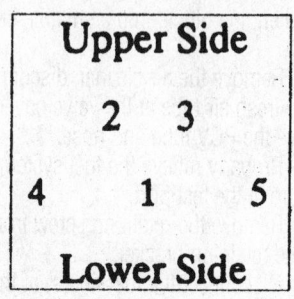

Fasten the exhaust manifold bolts in the sequence shown—SOHC engine

Tighten the exhaust manifold bolts according the sequence shown—DOHC engine

Nm). Connect the oxygen sensor electrical connector.

11. Install the air conditioning compressor and brackets. Tighten all fasteners except the front bracket-to-compressor fasteners to 19 ft. lbs. (25 Nm). Tighten the front bracket-to-compressor fasteners to 40 ft. lbs. (54 Nm).

12. Raise and support the vehicle safely, then install a new gasket onto the studs between the pipe and manifold.

13. Connect the pipe and manifold, then tighten the fasteners in a crosswise pattern to 23 ft. lbs. (31 Nm).

14. If necessary, position the exhaust pipe to engine support bracket in place and install the two mounting fasteners. Tighten the mounting fasteners to 23 ft. lbs. (31 Nm).

15. Lower the vehicle.

16. Connect the negative battery cable, start the engine and check for exhaust leaks.

Front Crankshaft Seal

The front crankshaft seal is located in the timing chain front cover. Refer to the timing chain procedure for information about removing the front cover and replacing the seal.

Tighten the intake manifold bolts in the correct sequence to ensure a good seal to the cylinder head—DOHC engine

Camshaft and Lifters

REMOVAL & INSTALLATION

SOHC Engine

1. Disconnect the negative, then the positive battery cable. Remove the battery cover and battery from the vehicle.

2. Remove the timing chain front cover.

3. Remove the timing chain and camshaft sprocket.

4. Remove the rocker arm/shaft assemblies.

5. Remove the lifters, and label or position them for assembly in their original locations.

6. Drive the camshaft plug inward, then remove it from the cylinder head with a magnet.

7. Carefully pull the camshaft from the rear of the cylinder head through the oversized camshaft plug hole. Turn the camshaft back and forth slowly while withdrawing to help prevent journal or bearing damage.

To install:

8. Clean and inspect all parts prior to installation. Lubricate the camshaft and carefully insert it through the hole at the rear of the cylinder head.

9. Coat a new rear cylinder head plug with Loctite® 242 or equivalent and install it using a standard bushing driver.

10. Install the valve lifters into their original bores, or if the camshaft has been replaced, install new lifters.

11. Install the rocker arm/shaft assemblies.

12. Install the timing chain and camshaft sprocket.

13. Install the timing chain front cover.

14. Install the battery and tighten the battery hold-down nut and screw to 80 inch lbs. (9 Nm). Connect the positive battery cable only, at this time.

15. Connect the positive, then the negative battery cable, start the engine and check for leaks.

DOHC Engine

➡ **Be very careful when working around the camshaft sprockets and timing chain cover during this procedure. If a bolt or washer is accidentally dropped between the front cover and engine assembly, the cover will have to be removed for retrieval.**

1. Disconnect the negative battery cable and remove the serpentine drive belt.

2. Disconnect the spark plug wires from the plugs, remove the EGR valve solenoid attachment screw and remove the PCV fresh air hose.

3. Remove the camshaft cover, then inspect the cover's silicone insulators for cracks or deterioration, and replace, as necessary.

4. Turn the crankshaft clockwise until the mark on the crankshaft pulley is in alignment with the pointer on the front cover and the No. 1 cylinder is at Top Dead Center (TDC) of the compression stroke. Both camshaft dowel pins will be at the 12 o'clock position and the timing pin holes will be aligned when the No. 1 cylinder is at TDC. If necessary, the right wheel and splash shield can be removed to help observe the timing marks.

5. Carefully remove each camshaft sprocket's retaining bolt. Use a ⅞ in. (21mm) open end wrench to hold the camshaft from turning while removing the bolts.

6. Position a front angled support fixture in front of the camshaft sprockets.

7. Attach the camshaft sprocket adapters to the end of each camshaft using the pilot bolts, but do not tighten the bolts. The front angled support should come between the sprocket adapters and camshaft sprockets.

8. Remove the upper timing chain guide and both front camshaft bearing caps.

9. Secure the support fixture using ⅞ in. bolts/blocks and align the two holes in each camshaft sprocket, adapter and the front support fixture. Install the four nuts, but do not tighten. The steel blocks should be installed against the rearward side of the camshaft sprocket. Tighten the sprocket pilot bolts to 19 ft. lbs. (25 Nm) while holding the camshafts from turning with an open end wrench.

10. Move each camshaft sprocket off the end of the camshaft by rocking the sprocket

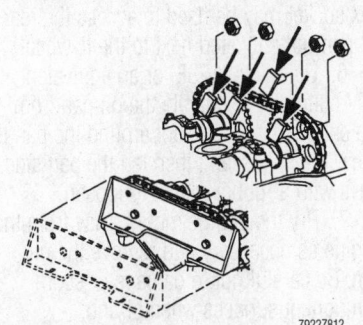

79222812

Install the camshaft support fixture to securely hold the camshafts in position— DOHC engine

forward or by carefully prying between the end of the camshaft and the sprocket. Then, tighten the four nuts and bolts with blocks from the side of the support fixture to 19 ft. lbs. (25 Nm).

11. Install the two bolts retaining the support fixture to the engine front cover and tighten the bolts to 89 inch lbs. (10 Nm). Then, remove each camshaft sprocket pilot bolt while holding the camshafts with a wrench.

12. Carefully pry between the sprocket and the end of the camshaft to move the camshaft rearward. Pry only enough to remove its end from inside the sprocket pilot otherwise camshaft or lifter damage may occur.

13. Uniformly loosen and remove the remaining camshaft bearing cap bolts. To prevent bolt/cap damage, do not use power tools and make several passes. Then, remove each camshaft. Position the bearing caps for installation in their original locations.

14. Pull the lifters out to remove them, always place them in the order in which they were removed and with the camshaft side facing down. Oil will drain out of the lifter if it is placed valve side down.

To install:

15. Install the lifters in their original locations.

16. Clean and inspect all parts prior to installation. Oil the camshaft and install with the **IN** camshaft on the intake side and **EX** camshaft on the exhaust side.

➡ **The dowel pin in each camshaft must be located at the 12 o'clock position during installation to prevent valve and piston damage.**

17. Install all bearing caps, except for the forward pair, in their original positions, making sure the arrows on the caps are pointing forward toward the camshaft sprockets. Lightly oil each of the cap bolts, then install and uniformly tighten the bolts to 124 inch lbs. (14 Nm).

18. Install one camshaft sprocket pilot bolt in each camshaft and tighten to 124 inch lbs. (14 Nm) in order to pull the camshaft fully forward and align the sprocket support for installation of the sprocket onto the camshaft.

19. Remove the four sprocket support bolt/blocks and nuts.

20. Verify that the camshafts are fully positioned forward and install the two forward bearing caps and the upper chain guide. The caps are marked **E1** or **I1** for exhaust or intake and must be positioned

with their arrows pointing towards the sprockets. Tighten the cap bolts to 124 inch lbs. (14 Nm).

21. Be sure the camshaft dowel pin aligns with the slot in each camshaft sprocket. If necessary, rotate the camshaft slightly (1–2 degrees) and move each sprocket from the adapter onto the end of the camshaft. Fully seat each sprocket on the end of each camshaft.

22. Remove the two sprocket pilot bolts and adapters while using a wrench on the camshaft flats to assure the camshaft cannot move.

23. Remove the support angled fixture.

24. Install the camshaft sprocket retaining bolts and washers. Hold the camshafts and tighten the bolts to 76 ft. lbs. (103 Nm).

25. Verify all visible timing marks and holes are in alignment. Turn the crankshaft clockwise until the mark on the crankshaft pulley aligns with the mark on the front cover. Check timing by inserting ³⁄₁₆ in. drill bits through the camshaft sprocket alignment holes, into the cylinder head. If the alignment pins cannot be inserted, turn the crankshaft 360 degrees clockwise and repeat. If the pins cannot be inserted within 1–2 degrees of either TDC position, the camshafts are not properly timed. Do not

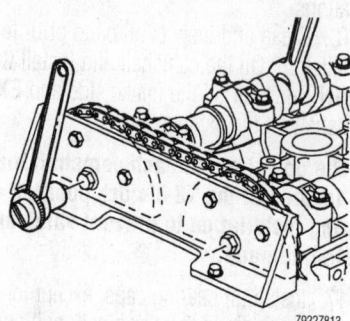

Hold the camshaft with a wrench while install the sprocket bolts—DOHC engine shown

Intake Side		
6	8	9
12		
3	1 2	4
11		
5	7	10
Exhaust Side		

Tighten the camshaft cover bolts in the sequence shown—DOHC engine

start the engine until the camshafts are timed.

26. Apply a small drop of RTV across the cylinder head and front cover T-joints. Inspect the old camshaft cover gasket and replace if damaged. Install the gasket and the camshaft cover. Tighten the fasteners in proper sequence to 89 inch lbs. (10 Nm).

27. Install the right splash shield and wheel, if removed to observe the timing marks.

28. Install the PCV and fresh air hoses, the EGR valve solenoid attaching screw and the spark plug wires.

29. Install the serpentine drive belt and connect the negative battery cable.

30. Start the engine and check for leaks.

Valve Lash

ADJUSTMENT

Both the SOHC and DOHC engines utilize hydraulic lash adjusters which do not require any adjustment. There is no provision for adjusting the valve lash.

Oil Pan

REMOVAL & INSTALLATION

1. Raise the vehicle and support it safely. Remove the plug and drain the engine oil from the pan and crankcase.

2. Disconnect the fasteners from the exhaust manifold flange and the pipe rear flange, then remove the front exhaust pipe and gaskets from the vehicle.

3. Remove the engine stiffening bracket and the flywheel cover.

4. Remove the right wheel and splash shield, then loosen the four front motor mount bolts. Back the bolts out about ½ in. (12mm).

5. Remove all the oil pan bolts. For vehicles with a manual transaxle, an 8mm flex socket may be used to access the rear oil pan bolts located next to the flywheel.

6. Using SA9123E, or an equivalent RTV cutter tool, separate the oil pan from the engine. Drive the tool around the pan to shear the RTV seam, then tap the pan sideways with a rubber mallet to loosen.

7. Pry the engine mount away from the engine as necessary and remove the oil pan. Be careful not to damage or score component surfaces when prying.

To install:

8. Carefully clean the gasket mating surfaces with a scraper and solvent.

9. Apply a 0.16 in. (4mm) bead of RTV sealer to the pan flange. Be sure the RTV is

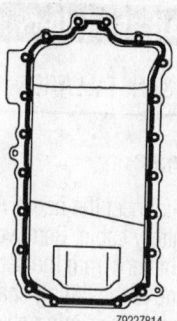

Apply a 0.16 in (4mm) bead of RTV to the oil pan flange to the inner side of the bolt holes

applied to the inner side of the flange from the bolt holes as shown.

10. Install the oil pan within three minutes of RTV application and tighten the bolts to 80 inch lbs. (9 Nm).

11. Tighten the front mount bolts to 37 ft. lbs. (50 Nm).

12. Install the right splash shield and wheel.

13. Install the engine stiffening bracket and the flywheel cover.

14. Install the exhaust pipe. Tighten the pipe to manifold nuts in a crosswise pattern to 23 ft. lbs. (31 Nm) and the pipe to converter bolts to 33 ft. lbs. (45 Nm).

15. Carefully lower the vehicle, then fill the engine crankcase with clean engine oil immediately in order to prevent an attempt to start the engine without oil.

16. Start the engine and check for leaks.

Oil Pump

REMOVAL & INSTALLATION

1. Disconnect the negative battery cable and drain the engine oil.

2. Remove the timing chain front cover which contains the oil pump assembly.

3. Remove the oil pump cover Torx® bolts using a suitable impact driver. Because the pump cover screws are coated with a sealant to prevent oil leakage, they must be replaced when removed.

4. Remove the drive rotor and driven rotor.

5. If necessary, remove the relief valve using tool SA9103E or equivalent, to pull the valve from the bore. Because the puller jaws will damage the relief valve sealing seat, the valve cannot be used again when removed.

To install:

6. If removed, install a new relief valve into the cover bore. Coat the valve with

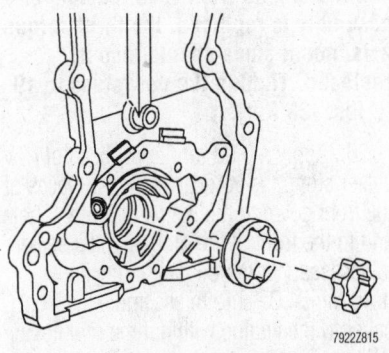

79222815

Exploded view of the oil pump assembly

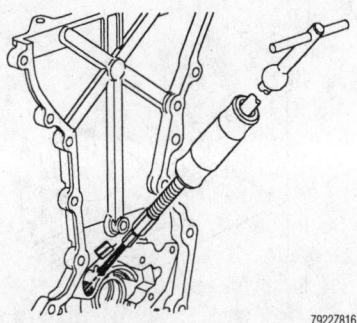

79222816

Only use SA9103E or an equivalent puller to remove the relief valve, if replacement is necessary

clean engine oil and tap it into the bore using a hammer and SA9103E or an equivalent installer tool.

➡**Whenever the oil pump is installed, the assembly must be packed with petroleum jelly in order to prime the pump.**

7. Install the drive and driven rotors into the pump with the chamfer toward the front oil seal.

8. Make sure the front cover bolt holes are clean, then install the pump body cover and secure using new bolts that are coated with sealant to prevent oil leakage. Tighten the bolts to 97 inch lbs. (11 Nm).

9. Install the timing chain cover and oil pump assembly to the front of the engine.

10. Properly fill the engine crankcase, then start the engine and check for leaks.

Rear Main Seal

REMOVAL & INSTALLATION

Both engines use a one-piece round seal mounted in a seal carrier.

1. Disconnect the negative battery cable.
2. Remove the transaxle assembly from the vehicle.

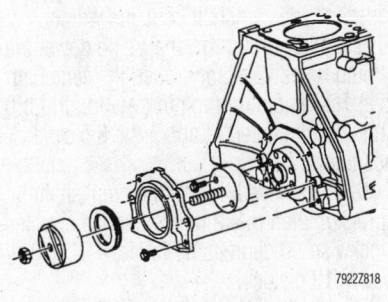

79222818

Exploded view of the rear main seal installation

3. As applicable, remove the clutch and flywheel assembly or the flexplate.

4. Use the prying tangs provided in the carrier to remove the seal with a small suitable prybar and hammer. Be careful not to damage the crankshaft oil seal lip contact surface.

To install:

5. Clean the carrier and crankshaft with solvent and a rag to prevent seal lip damage during installation. Check for scores or damage to the sealing surfaces.

6. Apply a light coat of clean engine oil to the seal lip and the carrier inner diameter. Install using SA9121E or an equivalent seal installer. The tool is designed to prevent seal lip from rolling during installation and will seat the seal 0.04 in. (1mm) lower than the factory seal. Never tap on the seal or seal installer with a hammer.

7. Install the flywheel or flexplate assembly.

8. Install the transaxle assembly into the vehicle.

9. Connect the negative battery cable, start the engine and check for leaks.

Timing Chain, Sprockets, Front Cover and Seal

REMOVAL & INSTALLATION

SOHC Engine

1. Disconnect the negative battery cable, then raise and support the vehicle sufficiently to work both under the vehicle and under the hood.

2. Remove the plug from the oil pan and drain the engine oil into a suitable container. Remove the right wheel and splash shield.

3. Place a 1 in. **x** 1 in. **x** 2 in. long block of wood between the torque strut and cradle to ease removal and installation of the torque engine mount. Remove the three

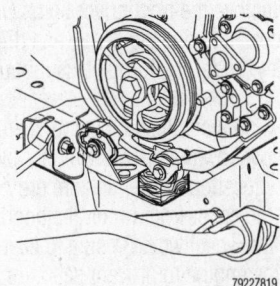

79222819

Place a 1 in. x 1 in. x 2 in. long piece of wood between the torque strut and cradle before removal of the torque engine mount—SOHC engine

right side upper engine torque axis to front cover nuts and the two mount to midrail bracket nuts, allowing the powertrain to rest on the block of wood.

➡**Placing a block of wood under the torque axis mount prior to removing the upper mount will allow the engine to rest on the wood, thus preventing the engine from shifting. This will allow you to install the mount during assembly without jacking or raising the engine.**

4. Remove the serpentine drive belt, belt tensioner and idler pulley.

5. Remove the power steering pump attaching bolts and support the assembly aside with the lines attached. If equipped, and if necessary for access to the front cover bolts, separate the A/C compressor from the bracket and support the assembly aside. Again, keep the compressor lines attached and do not discharge the system.

6. Remove the camshaft cover. Cover the valvetrain assemblies to protect them from foreign debris or dirt.

7. Using a strap wrench or a piece of wood wedged between the damper spoke and the lower side of the engine front cover, hold the damper and remove the bolt. With a suitable 3-jaw puller and the slots cast into the damper, pull the crankshaft damper/pulley assembly from the crankshaft.

8. Install the special oil seal replacement tool SA9104E or equivalent, to be sure the front crankshaft timing sprocket is held firmly in place and prevent guide damage. Install with the flat side towards the crankshaft sprocket.

9. Remove the front four oil pan bolts, then using a suitable RTV cutting tool, cut the front seal away from the front cover.

10. Spray the two dowel pin holes with penetrating oil to facilitate front cover removal from the dowel pins.

11. Remove the front cover bolts. One bolt is located above the serpentine drive belt pulley, under the torque axis mount flange.

12. Using a small suitable tool, carefully pry the cover away from the cylinder block at the pry location tabs which are provided. Remove the cover from under the hood or through the wheel well. Be sure to cover the front of the engine to prevent debris or dirt from entering the oil galley openings and oil pan. If necessary, pry the front cover oil seal from the cover for replacement.

➡**During timing chain and sprocket removal, position the crankshaft 90 degrees past Top Dead Center (TDC), to be sure the pistons will not contact the valves upon assembly.**

13. Carefully rotate the crankshaft clockwise so the timing mark on the crankshaft sprocket and keyway align with the main bearing cap split line (90 degrees past TDC).

14. Remove bolts, then remove the timing guides and tensioner.

15. Remove the camshaft sprocket bolt, using a ⅞ in. (21mm) wrench to hold the camshaft. Then, remove the timing chain and camshaft sprocket. Remove the crankshaft sprocket, if necessary.

To install:

16. Inspect the chain for wear and damage. Check the inside diameter of the chain, it should be no more than 16.77 in. (426mm). Inspect the chain guides for wear or cracks and the timing sprockets for teeth or key wear. Replace components as necessary.

17. Verify that the crankshaft keyway is positioned 90 degrees clockwise past TDC (keyway at three o'clock). The keyway should align with the split between the bearing cap and engine block.

18. Bring the camshaft up to No. 1 TDC by loosely installing the sprocket and rotat-

ing the sprocket until the timing pin can be inserted. The camshaft contains wrench flats to assist in turning the shaft. The dowel pin should be at 12 o'clock when the camshaft is at TDC and a timing pin (³⁄₁₆ in. drill bit) should, then insert at about the 8 o'clock position.

19. If removed, install the crankshaft sprocket, then rotate the crankshaft counterclockwise 90 degrees up to No. 1 TDC (keyway at 12 o'clock).

20. Position the chain under the crankshaft sprocket and over the camshaft sprocket. If necessary remove the camshaft sprocket, then slide the camshaft sprocket into position with the chain already engaged. The timing chain should be positioned so that one silver link plate aligns with the reference mark on the camshaft sprocket and the other aligns with the downward tooth (at the 6 o'clock position) on the crankshaft sprocket. The letters FRT on the camshaft sprocket must face forward, away from the cylinder head and excess chain slack should be located on the tensioner side of the block.

21. Temporarily install a timing pin to verify proper alignment of the camshaft and sprocket, then install and tighten the sprocket bolt to 75 ft. lbs. (102 Nm). Again, use a wrench on the camshaft flats to hold the shaft in position while tightening the bolt. Do not allow the camshaft retaining bolt to torque against the timing pin or cylinder head damage will result.

22. Install the chain guides with the words FRONT facing out. Install the fixed guide first and verify the chain is snug against the guide, then install the pivot guide. Tighten the bolts to 19 ft. lbs. (26 Nm) and verify that the pivot guide moves freely.

23. Retract the tensioner plunger and pin the ratchet lever using a ⅛ in. No. 31 drill bit inserted in the alignment hole at the bottom front of the component. Install the tensioner and tighten the bolts to 14 ft. lbs. (19 Nm), then remove the drill bit.

24. Make one final check to verify all components are properly timed, then remove all timing pins.

25. Be sure the oil galleys are clear. Carefully clean the gasket mating surfaces with a scraper or wire brush and carburetor solvent, brake clean or alcohol. Use a ³⁄₁₆ in. drill bit and tap handle to clean the front cover holes. If removed, seat a new front cover oil seal using the installation tool with a suitable press, or wait until the cover is installed, then use a threaded installation tool and the crankshaft's threads to pull the seal into position.

➡**If the engine front cover casting or assembly is replaced, the three torque axis mount studs should also be replaced. Tighten the new studs to 19 ft. lbs. (25 Nm).**

26. Apply a 0.08 in. (2mm) bead of RTV sealer along the vertical sealing surfaces of the front cover to the inside of the bolt holes and to the front of the oil pan. Extra sealer is necessary at the oil pan and cylinder head joints. Be sure to assemble the front cover to the engine within three minutes of RTV application.

27. If removed, install the crankshaft

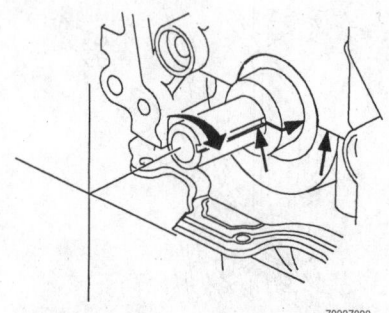

At 90 degrees past TDC, the crankshaft sprocket keyway will align with the main bearing cap split line

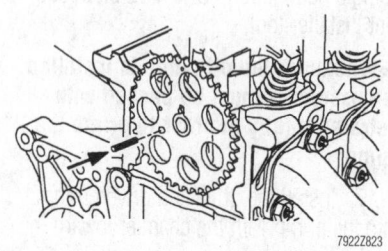

Insert a timing pin to ensure that the camshaft is at No. 1 TDC—SOHC engine

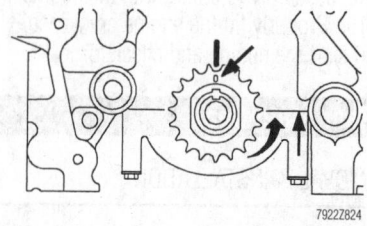

When the camshaft is at TDC, rotate the crankshaft counterclockwise 90 degrees to achieve TDC

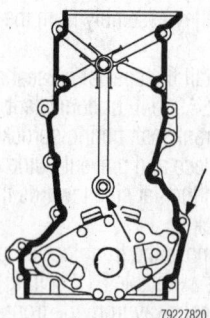

Apply a 0.08 in. (2mm) bead of RTV sealer along the vertical sealing surfaces of the front cover—SOHC engine

sprocket retaining tool to align the oil pump and crankshaft during cover installation. Position the front cover to the engine and install the bolts. Tighten the perimeter bolts starting at the center and working outwards on both sides to 19 ft. lbs. (25 Nm).

28. Install and tighten the front cover center or inner bolts to 89 inch lbs. (10 Nm) except for the upper inside bolt which should be tightened to 22 ft. lbs. (30 Nm). Install the four oil pan front bolts and tighten to 80 inch lbs. (9 Nm).

29. After front cover installation, spray 6–12 squirts of oil through the front oil seal drain back hole to verify that it is not plugged.

30. Apply a thin film of RTV between the damper/pulley assembly flange and washer only; the washer and bolt head flange are designed to prevent oil leakage.

31. Remove the crankshaft retaining tool and position the crankshaft damper/pulley assembly, then secure using the wood or strap wrench (as accomplished during removal) and tighten the bolt to 159 ft. lbs. (215 Nm).

32. Apply a small drop of RTV across the cylinder head and front cover T-joints. Inspect the old camshaft cover gasket and replace if damaged. Install the gasket and the camshaft cover. Tighten the fasteners uniformly to 22 ft. lbs. (30 Nm).

33. Position and install the A/C compressor assembly and/or the power steering pump assembly, as applicable.

34. Install the idler pulley if removed, then install the belt tensioner and the serpentine drive belt.

35. Install the two engine mounts to midrail bracket nuts and tighten to 37 ft. lbs. (50 Nm). Next install the three mount to front cover nuts, tighten them uniformly to 37 ft. lbs. (50 Nm) in order to prevent front cover damage. Then, remove the block of wood from under the torque strut.

36. Install the splash shield and the wheel assembly, then carefully lower the vehicle.

37. Immediately fill the engine crankcase with clean engine oil and connect the negative battery cable.

38. Start the engine and check for leaks.

DOHC Engine

1. Disconnect the negative battery cable, then raise and support the vehicle sufficiently to work both under the vehicle and under the hood.

2. Remove the plug from the oil pan and drain the engine oil into a suitable container. Remove the right wheel and splash shield.

3. Place a 1 in. **x** 1 in. **x** 2 in. long block of wood between the torque strut and cradle to ease removal and installation of the torque engine mount. Remove the three right side upper engine torque axis to front cover nuts and the two mount to midrail bracket nuts, allowing the powertrain to rest on the block of wood.

➡**Placing a block of wood under the torque axis mount prior to removing the upper mount will allow the engine to rest on the wood, thus preventing the engine from shifting. This will allow you to install the mount during assembly without jacking or raising the engine.**

4. Remove the serpentine drive belt, belt tensioner and idler pulley.

5. Remove the power steering pump attaching bolts and support the assembly aside with the lines attached. If equipped, and if necessary for access to the front cover bolts, separate the A/C compressor from the bracket and support the assembly aside. Again, keep the compressor lines attached and do not discharge the system.

6. Remove the camshaft cover. Cover the valvetrain assemblies to protect them from foreign debris or dirt.

7. Using a strap wrench or a piece of wood wedged between the damper spoke and the lower side of the engine front cover, hold the damper and remove the bolt. With a suitable 3-jaw puller and the slots cast into the damper, pull the crankshaft damper/pulley assembly from the crankshaft.

8. Install the special oil seal replacement tool SA9104E or equivalent, to be sure the front crankshaft timing sprocket is held firmly in place and prevent guide damage. Install with the flat side towards the crankshaft sprocket.

9. Remove the front four oil pan bolts, then using a suitable RTV cutting tool, cut the front seal away from the front cover.

10. Spray the two dowel pin holes with penetrating oil to facilitate front cover removal from the dowel pins.

11. Remove the front cover bolts. One bolt is located above the serpentine drive belt pulley, under the torque axis mount flange.

12. Using a small suitable tool, carefully pry the cover away from the cylinder block at the pry location tabs which are provided. Remove the cover from under the hood or through the wheel well. Be sure to cover the front of the engine to prevent debris or dirt from entering the oil galley openings and oil pan. If necessary, pry the front cover oil seal from the cover for replacement.

➡**During timing chain and sprocket removal, position the crankshaft 90 degrees past Top Dead Center (TDC) to be sure the pistons will not contact the valves upon assembly.**

13. Carefully rotate the crankshaft clockwise so the timing mark on the crankshaft sprocket and keyway align with the main bearing cap split line.

14. Remove the bolts, then remove the timing guides and tensioner.

15. Remove the camshaft sprocket bolts, using a ⅞ in. (21mm) wrench to hold the camshaft. Then, remove the timing chain and camshaft sprocket. Remove the crankshaft sprocket, if necessary.

To install:

16. Inspect the chain for wear and damage. Check the inside diameter of the chain, it should be no more than 23.15 in. (588mm). Inspect the chain guides for wear or cracks and the timing sprockets for teeth or key wear. Replace components as necessary.

17. Verify that the crankshaft is positioned 90 degrees clockwise past TDC. The crankshaft keyway should be at three o'clock aligned with the main bearing cap split line to prevent piston and valve damage.

18. Install the camshaft sprockets, retaining bolts and washers. Be sure the letters FRT on the sprockets face forward, away from the cylinder block. Use the wrench flats provided on the camshafts to hold the shaft and tighten the bolts to 75 ft. lbs. (102 Nm).

19. Bring the camshafts up to No. 1 TDC by rotating the camshafts and sprocket until the dowel pins are at 12 o'clock. Install a ⅛ in. drill bit into the hole in the sprocket about 9 o'clock.

20. If removed, install the crankshaft sprocket, then rotate the crankshaft counterclockwise 90 degree up to No. 1 TDC (keyway and sprocket timing mark at 12 o'clock, in alignment with the block timing mark).

21. Position the timing chain under the crankshaft sprocket and over the camshaft sprockets so two silver link plates align with the reference marks on the camshaft sprockets and another two plates align with the downward tooth (at 6 o'clock position) on the crankshaft sprocket. Excess chain slack should be located on the tensioner side of the cylinder block.

22. Verify that the crankshaft reference mark aligns with the cylinder block mark at 12 o'clock and that the timing pins are installed in the holes at about the 9 o'clock position. Remove the timing pins from the camshaft sprockets.

23. Install the timing chain fixed guide to the right of the block face toward the water pump. Tighten the bolts to 21 ft. lbs. (28 Nm) and verify the chain is snug against the guide.

24. Install the pivoting chain guide and check for clearance between the block and head. Tighten the bolt to 19 ft. lbs. (26 Nm) and verify the guide pivots freely.

25. Install the two forward camshaft bearing caps and the upper timing chain guide, then tighten the retaining bolts to 124 inch lbs. (14 Nm).

26. Retract the tensioner plunger and pin the ratchet lever using a ⅛ in. (3.18mm) No. 31 drill bit inserted in the alignment hole at the lower front of the component. Install the tensioner and tighten the bolts to 14 ft. lbs. (19 Nm), then remove the drill bit.

27. Make one final check to verify all components are properly timed, then remove all timing pins.

28. Be sure the oil galleys are clear. Carefully clean the gasket mating surfaces with a scraper or wire brush and carburetor solvent, brake clean or alcohol. Use a ³⁄₁₆ in. drill bit and tap handle to clean the front cover holes. If removed, seat a new front cover oil seal using the installation tool with a suitable press, or wait until the cover is installed, then use a threaded installation tool and the crankshaft's threads to pull the seal into position.

➡**If the engine front cover casting or assembly is replaced, the three torque axis mount studs should also be replaced. Tighten the new studs to 19 ft. lbs. (25 Nm).**

29. Apply a 0.08 in. (2mm) bead of RTV sealer along the vertical sealing surfaces of the front cover to the inside of the bolt holes and to the front of the oil pan. Extra sealer is necessary at the oil pan and cylinder head joints. Aply a thin bead around the two

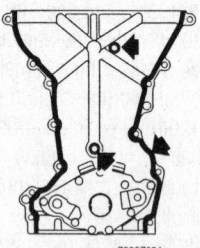

79227821

Apply a 0.08 in. (2mm) bead of RTV sealer along the vertical sealing surfaces of the front cover to the inside of the bolt holes and to the front of the oil pan— DOHC engine

inner cover bolt holes. Be sure to assemble the front cover to the engine within three minutes of RTV application.

30. If removed, install the crankshaft sprocket retaining tool to align the oil pump and crankshaft during cover installation. Position the front cover to the engine and install the bolts. Tighten the perimeter bolts starting at the center and working outwards on both sides to 22 ft. lbs. (30 Nm).

31. Install and tighten the front cover center or inner bolts to 89 inch lbs. (10 Nm) except for the upper inside bolt which should be tightened to 22 ft. lbs. (30 Nm). Install the four oil pan front bolts and tighten to 80 inch lbs. (9 Nm).

32. After front cover installation, spray 6–12 squirts of oil through the front oil seal drain back hole to verify that it is not plugged.

33. Apply a thin film of RTV between the damper/pulley assembly flange and washer only; the washer and bolt head flange are designed to prevent oil leakage.

34. Remove the crankshaft retaining tool and position the crankshaft damper/pulley assembly, then secure using the wood or strap wrench (as accomplished during removal) and tighten the bolt to 159 ft. lbs. (215 Nm).

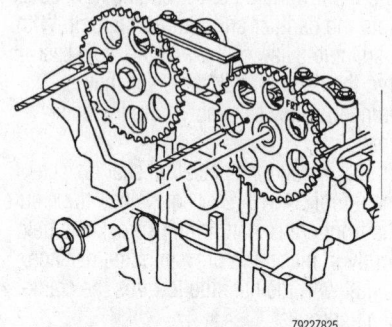

79227825

Use drill bits as timing pins to verify that the camshafts are at TDC—DOHC engine

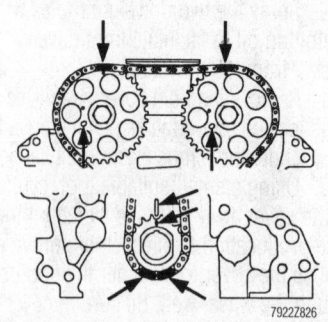

79227826

Be sure that the silver link plates and reference marks are all in alignment as shown—DOHC engine

35. Apply a small drop of RTV across the cylinder head and front cover T-joints. Inspect the old camshaft cover gasket and replace if damaged. Install the gasket and the camshaft cover. Tighten the fasteners uniformly to 22 ft. lbs. (30 Nm) for SOHC vehicles or in proper sequence to 89 inch lbs. (10 Nm) for DOHC vehicles.

36. Position and install the A/C compressor assembly and/or the power steering pump assembly, as applicable.

37. Install the idler pulley if removed, then install the belt tensioner and the serpentine drive belt.

38. Install the two engine mounts to midrail bracket nuts and tighten to 37 ft. lbs. (50 Nm). Next install the three mount to front cover nuts, tighten them uniformly to 37 ft. lbs. (50 Nm) in order to prevent front cover damage. Then, remove the block of wood from under the torque strut.

39. Install the splash shield and the wheel assembly, then carefully lower the vehicle.

40. Immediately fill the engine crankcase with clean engine oil and connect the negative battery cable.

41. Start the engine and check for leaks.

FUEL SYSTEM

Fuel System Service Precautions

Safety is the most important factor when performing not only fuel system maintenance but any type of maintenance. Failure to conduct maintenance and repairs in a safe manner may result in serious personal injury or death. Maintenance and testing of the vehicle's fuel system components can be accomplished safely and effectively by adhering to the following rules and guidelines.

• To avoid the possibility of fire and personal injury, always disconnect the negative battery cable unless the repair or test procedure requires that battery voltage be applied.

• Always relieve the fuel system pressure prior to disconnecting any fuel system component (injector, fuel rail, pressure regulator, etc.), fitting or fuel line connection. Exercise extreme caution whenever relieving fuel system pressure to avoid exposing skin, face and eyes to fuel spray. Please be advised that fuel under pressure may penetrate the skin or any part of the body that it contacts.

- Always place a shop towel or cloth around the fitting or connection prior to loosening to absorb any excess fuel due to spillage. Ensure that all fuel spillage (should it occur) is quickly removed from engine surfaces. Ensure that all fuel soaked cloths or towels are deposited into a suitable waste container.

- Always keep a dry chemical (Class B) fire extinguisher near the work area.

- Do not allow fuel spray or fuel vapors to come into contact with a spark or open flame.

- Always use a back-up wrench when loosening and tightening fuel line connection fittings. This will prevent unnecessary stress and torsion to fuel line piping. Always follow the proper torque specifications.

- Always replace worn fuel fitting O-rings with new. Do not substitute fuel hose or equivalent, where fuel pipe is installed.

Fuel System Pressure

RELIEVING

1. Unless battery voltage is necessary for testing, disconnect the negative battery cable. This will prevent the fuel pump from running and causing a fuel spill through the disconnected components if the ignition key is accidentally turned **ON**.

2. Remove the air cleaner assembly, for access.

3. Wrap a shop rag around the fuel test port fitting, located at the lower rear of the engine, then remove the cap and connect the fuel pressure gauge tool SA9127E or equivalent.

4. Install the bleed hose from the pressure gauge into an approved container and open the valve to bleed the system pressure.

5. After the pressure is bled, remove the gauge from the test port and recap it.

6. Install the air cleaner assembly.

7. After servicing the vehicle, connect the negative battery cable and prime the fuel system as follows:

    a. Turn the ignition **ON** for 5 seconds, then **OFF** for 10 seconds.

    b. Repeat the **ON/OFF** cycle two more times.

    c. Crank the engine until it starts.

    d. If the engine does not readily start, repeat Steps a-c.

    e. Run the engine and check for leaks.

Fuel Filter

On 1995–97 models, the fuel filter is attached to the vehicle frame in the lower left portion of the engine compartment. On 1998–99 models, the fuel filter and fuel pressure regulator are one integral component of the new returnless fuel injection system, and is located underneath the vehicle at the forward edge of the left side of the fuel tank.

REMOVAL & INSTALLATION

1995–97 Models

1. Disconnect the negative battery cable from the battery.

2. Remove the fresh air intake hose from the camshaft or rocker cover and remove the air inlet tube.

3. Properly relieve the fuel system pressure as follows:

    a. Remove the air cleaner or air intake duct, as applicable.

    b. Wrap a shop rag around the fuel test port fitting at the lower rear of the engine, remove the cap and connect pressure gauge tool SA9127E or equivalent.

    c. Install the bleed hose into an approved container and open the valve to bleed the system pressure. After the system pressure is bled, remove the gauge from the pressure test port and recap it.

4. Clean the female end of the quick-connect fitting by spraying it with penetrating oil, then detach the large underhood fuel line connection located near the intake manifold support brace on the left side of the vehicle using the tool supplied with the replacement filter, SA9157E or equivalent.

5. Raise and support the vehicle safely.

6. Disengage the quick-connect fitting at the fuel filter inlet (rear of the filter) by pinching the two plastic tangs together and pulling on the supply line.

7. Loosen the fuel filter band clamp nut, but do not completely remove.

8. Carefully push or pull the filter out of the assembly and discard the filter in an appropriate container.

To install:

9. If the band clamp was removed, clip the fuel return and vapor lines in place and install two new band clamp nuts. Be sure all lines are in place and will not interfere with or be damaged by filter installation, then tighten the bracket nuts to 27 inch lbs. (3 Nm).

10. Clean the female end of the filter inlet quick-connect fitting by holding the line facing downward and spraying penetrating oil up into the fitting. Be careful not to bend or kink the line.

11. If not already installed, insert a new snap lock retainer into the female end of the filter inlet quick-connect fitting.

12. Route the filter's nylon outlet line through the band clamp and insert the filter far enough into the band clamp to connect the outlet line to the engine's fuel line attachment. Lubricate the male end of the connector with clean engine oil, snap the connector together and pull on the line to verify proper fitting.

13. Position the filter in the band clamp assembly with the filter's forward edge located ¼ inch (6.35mm) from the front of the band clamp.

14. Lubricate the male end of the fuel supply line with clean engine oil. Snap the line into the fuel filter and pull back to verify that the fitting is secure. Tighten the band clamp nut to 89 inch lbs. (10 Nm).

15. Lower the vehicle and install the air cleaner or intake duct, as applicable. Connect the air inlet tube and fresh air hose.

16. Connect the negative battery cable, then prime the fuel system as follows:

    a. Turn the ignition **ON** for 5 seconds, then **OFF** for 10 seconds.

    b. Repeat the **ON/OFF** cycle two more times.

    c. Crank the engine until it starts.

    d. If it does not start, repeat Steps a-c.

    e. Run the engine and check for leaks.

1998–99 Models

1. Disconnect the negative battery cable from the battery.

2. Properly relieve the fuel system pressure as follows:

    a. Remove the air cleaner or air intake duct, as applicable.

    b. Wrap a shop rag around the fuel test port fitting on the fuel supply line at the fuel rail. Remove the protective cap and connect pressure gauge tool SA9127E or equivalent.

    c. Install the bleed hose into an approved container and open the valve to bleed the system pressure. After the system pressure is bled, remove the gauge from the pressure test port and recap it.

3. Raise and support the vehicle safely.

4. Loosen the fuel filter/pressure regulator bracket retaining screws.

※ WARNING

Exercise extreme care when opening the retaining clip. The fuel lines must be retained in this clip; if damaged, the fuel tank assembly must be replaced, since the fitting is not serviced separately.

5. Disengage the quick-connect fitting on the left side of the fuel tank by pinching the two plastic tangs together and pulling on the supply line.

6. Disengage the EVAP purge line at the 90° quick-connect fitting.

7. Slide the outlet of the fuel filter/regulator out of the support on the fuel tank bracket and disengage the fuel feed line at the 90° quick-connect fitting.

8. Pivot the fuel filter/regulator down while moving the leg of the mounting bracket out from under the parking brake lines.

➡**It is not necessary to separate the fuel filter/regulator from the mounting bracket, since both items are serviced as an assembly.**

9. Disengage the fuel feed and return line quick-connect fittings on the fuel filter/regulator. Discard the filter in an appropriate container.

To install:

10. Install new fuel line retainers (3) into the female portion of the quick-connect fuel line fittings.

11. To ease installation, lubricate the male ends of the fuel filter/regulator with clean engine oil.

12. Install the fuel feed and return lines onto the fuel filter/regulator and snap them closed. Pull back to verify that the fittings are secure.

13. Install the fuel feed, return, and EVAP purge lines into the fuel tank's retaining clip.

14. Slide the leg of the fuel filter/regula-

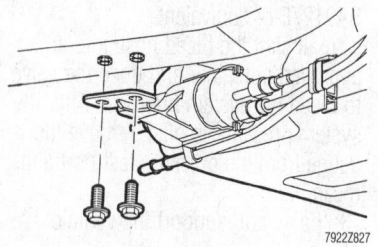

The fuel filter/regulator bracket is held to the frame with two bolts—1998–99 models

tor mounting bracket under the parking brake lines and pivot upward.

※ WARNING

Be sure to route the chassis fuel feed and purge lines above the parking brake cable. The parking brake cable must be firmly secured to the underbody to support the fuel and purge lines.

15. Install the 90° fuel feed line quick-connect fitting onto the fuel filter/regulator outlet and snap it closed. Pull back to verify that the fitting is secure.

16. Install the 90° EVAP purge line quick-connect fitting to the purge line and snap it closed. Pull back to verify that the fitting is secure.

17. Install the fuel feed outlet pipe of the fuel filter/regulator into the retaining clip on the fuel tank bracket.

18. Install the fuel filter/regulator bracket mounting screws. Tighten the bracket mounting screws to 71 inch lbs. (8 Nm).

19. Lower the vehicle and install the air cleaner or intake duct, as applicable. Connect the air inlet tube and fresh air hose.

20. Connect the negative battery cable, then prime the fuel system as follows:

a. Turn the ignition **ON** for 5 seconds, then **OFF** for 10 seconds.

b. Repeat the **ON/OFF** cycle two more times.

c. Crank the engine until it starts.

d. If it does not start, repeat Steps a–c.

e. Run the engine and check for leaks.

Fuel Pump

REMOVAL & INSTALLATION

1995–97 Models

To prevent excessive fuel spillage, whenever the tank is removed from the vehicle, it should be no more than ¾ full. Removal of the fuel pump module assembly requires removal of the fuel tank.

※ CAUTION

The following procedure will produce a small fuel spill and fumes. Be sure there is proper ventilation and be sure to take the appropriate fire safety precautions.

1. Disconnect the negative battery cable, then properly relieve the fuel system pressure. Refer to the procedure in this section.

2. Remove the fuel filler cap, then raise and support the vehicle safely, with the rear of the vehicle approximately 28 in. (711mm) higher than the front (to keep any fuel in the tank forward and away from the fill hose).

3. Clean the area surrounding the filler neck to avoid fuel system contamination, then position a container with a minimum 12 in. (300mm) diameter opening under the filler neck to catch any escaping fuel. Loosen the filler neck clamp at the rear of the fuel tank, wrap a shop rag around the neck tube and carefully remove the tube from the tank.

4. Remove the filler neck, then siphon the fuel from the tank into an approved gasoline container.

5. Remove the filler neck bracket fastener at the left side of the rear frame rail, then loosen the fuel vent hose retaining clamp at the tank.

➡**It is easier to remove hoses from the tank than to pull them from the steel vent and fill tubes.**

6. Disconnect the fuel pressure and return line quick-connect fittings by pinching the two plastic tangs together, then grasp both ends of one fuel line connection and twist ¼ turn in each direction while pulling them apart. Disconnect the fuel vent hose by holding the line and by pushing on the rubber connector with a small open end wrench.

➡**Do not allow the fuel tank retaining straps to become bent during tank removal, otherwise strap damage and/or breakage may occur. Always utilize an assistant when removing the fuel tank to prevent damage.**

7. With the aid of an assistant, remove the two support strap fasteners at the rear of the tank, then lower the tank and support panel approximately 8 in. (203mm). Reach upward and unplug the electrical connector from the top of the tank, then remove the tank and support panel from the vehicle.

8. Clean the area surrounding the fuel pump module and spray the cam lockring tangs with a suitable penetrating oil to loosen the fitting.

9. Using SA9156E, or an equivalent fuel module lockring removal tool, and a ½ in. breaker bar of approximately 18 in. (457mm) length, remove the pump unit locking ring from the tank. Attempting to use a 12 in. or shorter breaker bar may cause lockring damage.

10. Lift and tilt the unit out at a 45 degree angle, being careful not to bend the sending unit float arm. Remove and discard the unit-to-tank O-ring.

11. The sending unit is the only portion of the module that may be serviced. The filter may be cleaned with mineral spirits, but must be replaced as an assembly with the module if damaged. If necessary, remove the sending unit from the module as follows:

a. Unplug the two electrical connections using needlenose pliers or by pressing down the locking tab and pulling the connectors from the terminal.

b. Using a small suitable tool, push in on the sender assembly attaching tang, then lift upward and remove the sender.

To install:

12. If removed, install the sender on the fuel pump module by positioning the tang in the locator slot and snapping the unit into place. Attach the two sending unit electrical connectors to their terminals.

13. Install a new O-ring to the opening in the top of the fuel tank, then carefully insert the pump module into the tank at a 45 degree angle to prevent sending unit and float damage. The filter and flow arm must be directed toward the front of the tank.

14. Align the pump locator tabs with the fuel tank slots, then install the cam lockring using the ring service tool.

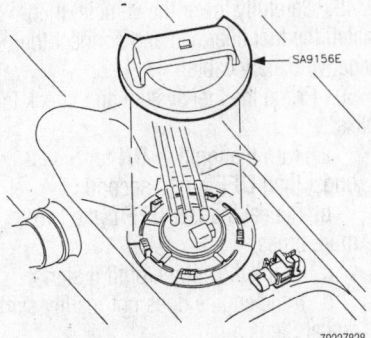

SA9156E

7922Z828

Use the special tool to remove the fuel module lockring—1995–97 models shown

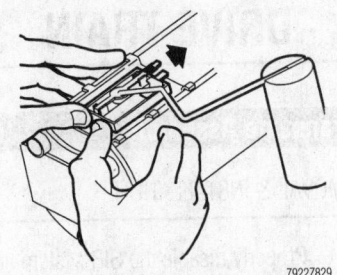

7922Z829

Press the sender toward the top of the module to release the bottom attachment tang and remove the sender—1995–97 models shown

15. With the aid of an assistant, position the tank and support panel so the wires can be connected to the module, install the module electrical connector, then secure the tank in place using the retaining straps. Tighten the strap bolts to 35 ft. lbs. (47 Nm).

16. Loosely install the filler tube, vent lines and clamps. Align the filler neck and tank so the fender will not be deflected or pushed outward by the hose. When everything is properly positioned, install the fill neck tube and bracket fastener. Vent lines should be installed into the rubber boot until the tube white marks align with the side of the boot.

17. Tighten the clamps to 18 inch lbs. (2 Nm) and tighten the fill neck bracket fastener to 53 inch lbs. (6 Nm). If the clamps must be replaced, original equipment or equivalent parts must be used, because the original parts are designed to prevent hose damage.

18. Lubricate the male ends of the fuel supply and return quick-connect fittings with a few drops of clean engine oil. Push the connectors together until the retaining tabs snap into place, then pull on opposite ends of each connection to verify that the connection is secure.

19. Carefully lower the vehicle, then install the fuel filler cap and connect the negative battery cable.

20. Prime the fuel system and check for leaks:

a. Turn the ignition **ON** for 5 seconds, then **OFF** for 10 seconds.

b. Repeat the **ON/OFF** cycle two more times.

c. Crank the engine until it starts.

d. If the engine does not readily start, repeat Steps a–e.

e. Run the engine and check for leaks.

1998–99 Models

To prevent excessive fuel spillage, whenever the tank is removed from the vehicle it should be no more than ½ full. Removal of the fuel pump module assembly requires the removal of the fuel tank.

✳✳ CAUTION

The following procedure will produce a small fuel spill and fumes. Be sure there is proper ventilation and be sure to take the appropriate fire safety precautions.

1. Disconnect the negative battery cable, then properly relieve the fuel system pressure. Refer to the procedure in this section.

2. Remove the fuel filler cap and rubber closeout grommet (inside fuel filler door). Remove the mounting fastener (using a T-30 Torx® bit) at the upper end of the fuel tank filler pipe.

3. Raise and support the vehicle safely, with the rear of the vehicle at a comfortable working height.

4. Remove the wheel house inner fender liner.

5. Disengage the wiring harness connector from the EVAP canister vent solenoid.

6. Remove the fuel tank filler pipe lower bracket mounting fastener located at the underbody left side rail.

7. Disconnect the EVAP canister vent tube at the ⅝ in. quick-connect fitting.

8. Loosen the fuel tank filler pipe hose clamp located closest to the tank.

9. Disconnect the filler pipe hose from the fuel tank, making sure that it is as straight and level as possible so as to prevent it from falling into the fuel tank. Remove the filler pipe from the vehicle. Be careful when disconnecting the filler hose as a residual amount of fuel may be left sitting in the filler pipe due to the inlet check valve on the fuel tank.

10. If there is more than three gallons of fuel in the tank, insert a siphon hose into the filler neck and drain into an approved gasoline container.

✳✳ CAUTION

Whenever fuel line fittings are loosened or disconnected, always wrap a shop towel around the fitting and have a suitable container available to catch any fuel spill.

11. Disengage the fuel feed line from the outlet tube of the filter/pressure regulator. If necessary, refer to Section 1 for service information on the fuel filter/pressure regulator.

12. Disengage the fuel vapor/canister purge line at the fitting adjacent to the filter/pressure regulator.

13. Remove the fuel filter/pressure regulator bracket from under the brake lines.

14. With the aid of an assistant, remove the two fuel tank retaining strap mounting bolts. Lower the tank just enough to disengage the two electrical connectors from the fuel pump and pressure sensor.

15. Carefully remove the fuel tank from the vehicle.

16. If fuel pump module replacement/service is necessary:

17. Clean the area surrounding the fuel pump module and spray the cam lockring tangs with a suitable penetrating oil to loosen the fitting.

18. Using SA9156E, or an equivalent fuel module lockring removal tool, and a ½ in. breaker bar of approximately 18 in. (457mm) length, remove the pump unit locking ring from the tank. Attempting to use a 12 in. or shorter breaker bar may cause lockring damage.

19. Lift and tilt the unit out at a 45 degree angle, being careful not to bend the sending unit float arm. Remove and discard the unit to tank O-ring.

20. The sending unit is the only portion of the module that may be serviced. The filter may be cleaned with mineral spirits, but must be replaced as an assembly with the module if damaged. If necessary, remove the sending unit from the module as follows:

a. Unplug the electrical connector from the fuel level sending unit by disengaging the locking tab and pulling the connector from the terminal.

b. Holding the fuel pump module in both hands, and applying firm upward pressure at the bottom of the fuel sending unit using both thumbs, push the sender toward the top of the module to disengage the bottom attachment clip.

To install:

21. Install the sending unit to the pump module by positioning the attachment clip in the locator slot and snapping the component into place. Verify that the sending unit float arm has the correct relationship to the fuel pump module by holding the component up on a flat, horizontal surface and measuring. Using a ⁵⁄₃₂ in. diameter drill bit, it must easily pass between the float and the horizontal surface with no more than ¹⁄₁₆ in. clearance. If bending the float arm is required to meet the clearance specification, it must be performed at the 90° bend near the level sender while supporting the short section of the arm.

22. Connect the wiring harness from the fuel pump module to the sending unit. Press firmly to engage.

23. Install the fuel pump assembly as follows:

24. Before installing the pump module, inspect the fuel tank and clean the seal groove of any debris or foreign matter.

➡**The correct fuel pump module O-ring seal is green in color. It is incorrect to use the older black seal.**

25. Install a new O-ring seal to the opening in the top of the fuel tank, then carefully insert the pump module into the tank at a 45 degree angle to prevent sending unit and float damage.

26. Align the pump locator tabs with the fuel tank slots by rotating the module 90° counterclockwise. The lines from the pump unit should face the 10 o'clock position.

27. Install the cam lockring using the ring service tool.

28. Connect the fuel pump vapor line to the tank vent pipe.

29. Connect the feed and return lines to the fuel filter/pressure regulator.

30. Place the feed, return and EVAP canister purge lines into the fuel tank retaining clip and snap closed.

31. Install the loose retainer clip around the fuel feed, return and purge lines and snap shut.

32. With the aid of an assistant, position the fuel tank so the wires can be connected to the pump module and pressure sensor.

33. Properly install the fuel tank under the vehicle by making certain that the small white locator button on the left side of the tank is positioned tightly against the left side rail.

34. Install the retaining straps and shield. Tighten the strap bolts to 35 ft. lbs. (47 Nm).

35. Place the fuel filter/pressure regulator in position and install the mounting screws in the bracket. Tighten the mounting screws to 71 inch lbs. (8 Nm).

36. Install a new retainer into the female portion of the 90° quick-connect fitting of the underbody fuel line.

37. Lubricate the outlet end of the filter/pressure regulator with clean engine oil and connect to the underbody fuel feed line.

38. Be sure that the fuel filler check valve is in correct position at the end of the filler pipe. Using plain water, lightly wipe the outside of the fuel tank inlet connector.

39. Position the fuel tank filler pipe into the wheel opening with the top of the pipe protruding out of the fender/filler door opening.

40. Insert the filler pipe into the fuel tank opening and loosely install the lower bracket attachment screw. Be sure that the EVAP vent solenoid pipe is installed in the correct position on the filler pipe bracket.

41. Lower the vehicle and install the fuel tank filler pipe upper bracket mounting fastener. Tighten the mounting fastener to 31–35 inch lbs. (3.5–4.5 Nm).

42. Install the rubber closeout grommet to the filler pipe and install the fuel cap.

43. Raise the vehicle to a comfortable working height.

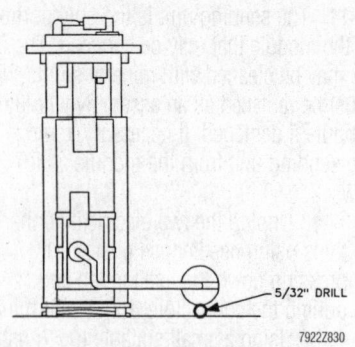

Use a ⁵⁄₃₂ in. diameter drill bit to measure float arm clearance—1998–99 models

44. The fuel pipe connecting hose should be installed to within ¼ in. (6mm) of the stops on the fuel tank inlet connector and the hose clamp should be positioned within ³⁄₁₆ in. (4mm) of the end of the connecting hose. Tighten the hose clamp to 35 inch lbs. (4 Nm).

45. Connect the EVAP canister vent pipe to the canister vent hose.

46. Tighten the lower mounting screw on the filler pipe to the underbody. Tighten the mounting screw to 71 inch lbs. (8 Nm).

47. Engage the wiring harness connector to the EVAP vent solenoid.

48. Install the wheel house inner fender liner.

49. Carefully lower the vehicle, then install the fuel filler cap and connect the negative battery cable.

50. Prime the fuel system and check for leaks:

a. Turn the ignition **ON** for 5 seconds, then **OFF** for 10 seconds.

b. Repeat the **ON/OFF** cycle two more times.

c. Crank the engine until it starts.

d. If the engine does not readily start, repeat Steps a-c.

51. Run the engine and check for leaks.

DRIVE TRAIN

Transaxle Assembly

REMOVAL & INSTALLATION

1. Properly disable the SIR system and disconnect the negative battery cable.

2. Remove the two air inlet duct fasteners, disengage the air temperature sensor connector and remove the air inlet duct. For the DOHC engine, loosen the flex tube to air

box clamp, remove the three air box fasteners and remove the air box.

3. Remove the transaxle strut-to-cradle bracket through-bolt located on the radiator side of the transaxle.

4. Disengage the wiring harness connectors from the vehicle and turbine speed sensor, transaxle temperature sensor, selector switch and actuator connector from the transaxle.

5. On automatic transaxles, remove the two ground terminals from the top two converter housing studs.

6. Remove the ground wire from the neutral (selector) switch and unclip the oxygen sensor wire retainer.

7. Remove the top two transaxle housing studs.

8. Remove the four Electronic Ignition (EI) coil to converter housing bolts, then wire the coil to the cylinder head coolant outlet. Discard the old coil retaining bolts and replace with new bolts upon installation.

9. For manual transaxles, remove the shifter cables from the shift arms and clutch housing, taking care not to damage the cable boot.

❊❊ WARNING

On vehicles with manual transaxles, do not use power tools when loosening and removing the slave cylinder. The use of a power tool to remove the slave cylinder could result in breaking off the hydraulic line.

10. Wire the radiator to the upper radiator support in order to hold the assembly in place when the cradle is removed.

11. Install SA9105E or an equivalent engine support bar assembly.

12. Raise and support the vehicle safely.

13. On manual transaxles, remove the drain plug from the transaxle housing and drain the transaxle fluid. The drain plug is on the lower cowl side of the housing and is inserted from the engine side of the vehicle.

14. Remove the front wheels and engine splash shields from the vehicle. For coupes, remove the left and right lower facia braces.

15. Remove the front engine strut cradle bracket-to-cradle nuts from below the cradle.

16. Remove the transaxle mount-to-cradle nut from under the cradle.

17. Remove and discard the cotter pin

from the lower ball joints. Loosen the ball joint nut until the top of the nut is even with the top of the threads.

18. Use tool SA9132S to separate the lower control arm ball joint from the steering knuckle, then remove the nut. Do not use a wedge tool or seal damage may occur.

➡ **The outer CV-joint for vehicles equipped with ABS contains a speed sensor ring. Use of an incorrect tool to separate the control arm from the knuckle may result in damage and loss of the ABS system.**

19. Remove the front exhaust pipe nuts at the manifold, then disconnect the pipe from the support bracket.

20. Remove the front pipe-to-catalytic converter bolts and lower the pipe from the vehicle.

21. Remove the engine-to-transaxle stiffening bracket bolts and remove the bracket.

22. Remove the steering rack-to-cradle bolts and wire the gear for support when the cradle is removed.

23. Remove the brake line from the retainer at the rear of the cradle.

24. On manual transaxles, remove the clutch housing dust cover.

25. On automatic transaxles, remove the torque converter dust cover, then remove the converter-to-flywheel bolts.

26. Position two 4 inch x 4 inch x 36 inch pieces of wood onto a powertrain support dolly and position the dolly under the vehicle.

27. Remove the four cradle-to-body bolts and carefully lower the cradle from the vehicle with the support dolly. Tape or wire the two large washers from the rear cradle-to-body attachments in position to prevent loss.

28. On automatic transaxles, squeeze the plastic tabs at the transaxle cooler line connectors and pull the lines out of the connectors. The plastic retainer should remain on the lines. Connect 1 end of a ⅜ in. rubber hose over each cooler line to prevent fluid contamination or loss.

29. If necessary for the transaxle to clear the body, lower the vehicle enough to adjust the engine support assembly and lower the transaxle side of the assembly until the valve body cover clears the frame.

30. Raise and support the vehicle safely, then support the transaxle securely with a suitable jack.

31. Use a prybar to separate the left side

axle from the transaxle. Remove the axle sufficiently to install SA91112T or an equivalent seal protector around the axle and into the seal to prevent the seal from being cut by the shaft spline.

32. On automatic transaxles, remove the two bottom converter housing-to-engine bolts and lower the transaxle sufficiently to reach the shifter cable.

33. On manual transaxles, remove the two bottom clutch housing-to-engine bolts and install a guide bolt into the bottom rear clutch housing bolt hole from the side of the engine block.

34. Separate the transaxle only sufficiently enough to install an axle seal protector on the remaining engaged axle.

35. On automatic transaxles, disconnect the transaxle shifter cable, then squeeze the retaining tabs to release the cable from the converter housing.

36. Carefully lower the transaxle from the vehicle. Use SA9165T or an equivalent transaxle cooler cleaning tool to clean the cooler and lines.

To install:

37. Place the transaxle assembly securely onto the jack and position under the vehicle. Install axle seal protectors into seals on both sides.

38. On automatic transaxles, raise the transaxle sufficiently, then connect the shifter cable to the gear selector lever and to the converter housing.

39. Raise the transaxle into the vehicle and verify that the intermediate shaft splines line up with the differential side gear splines, then install the two lower transaxle housing-to-engine bolts and tighten the bolts to 96 ft. lbs. (130 Nm). The bolts should not be used to draw the transaxle to the engine.

40. Be sure the axle seal protectors are installed into the transaxle. Carefully install the axles to the transaxle, after the splines clear the seal, but before the axle snaps into place remove the seal protector. Push the axle all of the way into the transaxle and install the snapring. Remove the transaxle jack.

41. Clean and lubricate the ball joint threads, then raise the cradle up on the support dolly and place the ball joints into the knuckles. Verify the correct positioning of the lower control arm bar studs to the knuckles, the cooling module support bushings, the engine strut bracket and the transaxle mount.

42. Insert ⁹⁄₁₆ in. round steel rods into the cradle-to-body alignment holes near the front cradle to body fastener holes. Guide the cradle into position making sure all mount studs are properly guided into their holes.

43. Be sure the washers are in place, then install the two rear cradle to body bolts. Verify proper cradle positioning and install the two front cradle bolts, then tighten the four cradle bolts to 151 ft. lbs. (205 Nm).

44. Remove the support dolly and lower the vehicle sufficiently for underhood access. Remove the engine support bar assembly.

45. Install the transaxle strut-to-cradle bracket through-bolt and nut, then tighten the fasteners to 52 ft. lbs. (70 Nm). If removed, install the strut cradle bracket-to-cradle bolt and also tighten to 52 ft. lbs. (70 Nm).

46. On manual transaxles, connect the shift control cables to the shift arms and the clutch housing, then install the cable retainers.

47. Remove the radiator assembly support wire.

48. Use a 6 x 1.0mm tap to clean the sealant from the ignition module mounting holes in the transaxle. Install the ignition module, then secure using the new bolts with sealant. Use extreme caution to assure proper bolt installation. Tighten the bolts to 71 inch lbs. (8 Nm) and verify that the bolt heads are properly seated on the ignition module.

49. Install the two top transaxle housing-to-engine studs and tighten to 66 ft. lbs. (90 Nm). Connect the two ground terminals to the studs and tighten to 19 ft. lbs. (25 Nm).

50. Engage the actuator circuit connector and tighten to 22 inch lbs. (2.5 Nm). Engage the wiring harness connectors to the vehicle and turbine speed sensors, the transaxle oil temperature sensor and the selector switch connectors.

51. Connect the ground wire to the neutral (selector) switch and clip the oxygen sensor wire to the converter housing.

52. On automatic transaxles, unplug the transaxle cooler lines and press them into the transaxle connectors until they bottom out.

53. On automatic transaxles, adjust the shifter cable, then for DOHC vehicles, install the air box and tighten the fasteners to 89 inch lbs. (10 Nm). Connect the flex tube to the air box, align the arrows and tighten the clamp to 18 inch lbs. (2 Nm).

54. Install the air inlet duct and fasteners, then engage the air temperature sensor electrical connector.

55. Raise and support the vehicle safely.

56. Install the transaxle mount-to-cradle nut and the transaxle strut cradle bracket-to-cradle nut, then tighten the nuts to 52 ft. lbs. (70 Nm).

57. Install the two engine strut cradle bracket-to-cradle fasteners from under the cradle and tighten to 52 ft. lbs. (72 Nm).

58. If applicable, install the nuts to the front transaxle-to-cradle studs and tighten to 35 ft. lbs. (48 Nm).

59. Remove the steering gear support wire and position the gear to the cradle. Install the gear bolts and nuts, then tighten the fasteners to 40 ft. lbs. (54 Nm).

60. Connect the brake line to the cradle retainer.

61. On automatic transaxles, install the torque converter-to-flexplate bolts and tighten the bolts to 52 ft. lbs. (70 Nm). Install the converter housing dust cover and tighten the bolts to 89 inch lbs. (10 Nm).

62. Install the powertrain stiffening bracket and tighten the bracket bolts to 35 ft. lbs. (47 Nm).

63. Position the exhaust manifold front pipe into the vehicle, then install the gasket and manifold retaining nuts. Tighten the nuts in a crosswise pattern to 23 ft. lbs. (31 Nm). Install the front pipe to the catalytic converter and tighten the bolts to 33 ft. lbs. (45 Nm). Finally, install the front pipe to the transaxle support bracket and tighten the fasteners to 23 ft. lbs. (31 Nm).

64. On manual transaxles, install the transaxle drain plug and tighten to 40 ft. lbs. (45 Nm).

➡If the torque converter flange threads (automatic only) are damaged use the Saturn 21010753 converter fastener kit in place of the self tapping screws in order to provide proper clamp load and prevent exhaust leaks.

65. Install the nuts onto the ball joint studs and tighten to 55 ft. lbs. (75 Nm). Continue to tighten the nuts as necessary and install a new cotter pin.

66. Install the center and both wheel splash shields.

67. For coupes, install the right left lower facia braces and J-nuts. Tighten the fasteners to 89 inch lbs. (10 Nm).

68. Install the tire and wheel assemblies, then lower the vehicle.

69. Connect the negative battery cable and fill the transaxle with the proper type and amount of fluid.

70. Properly enable the SIR system.

71. Warm the engine and check the transaxle fluid. Check and adjust vehicle alignment, as necessary.

Clutch

ADJUSTMENT

The hydraulic clutch system is self-adjusting; therefore, no manual clutch pedal adjustments are necessary or possible.

REMOVAL & INSTALLATION

1. Properly disable the SIR system and disconnect the negative battery cable.

2. Remove the transaxle from the vehicle.

3. Unsnap the release fork from the ball stud, then remove the fork and bearing from the vehicle. Slide the bearing from the fork. The bearing should be checked for excessive play and for minimal bearing drag. It should be replaced if no/little drag or excessive play is found.

➡**The release bearing is packed with grease and should not be washed with solvent.**

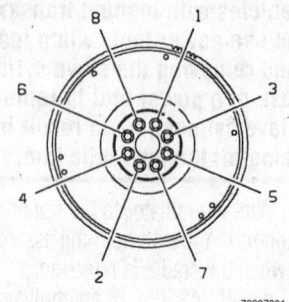

If removed, install the flywheel and tighten the mounting bolts in a crisscross pattern to specifications

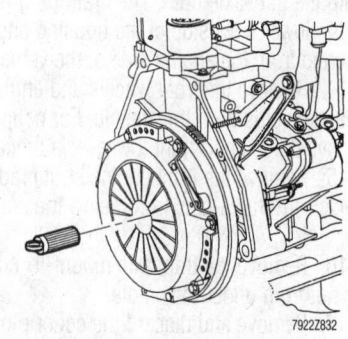

Install a proper clutch alignment tool before tightening the pressure plate retaining bolts

4. Using a feeler gauge, measure the distance between the pressure plate and flywheel surfaces in order to determine clutch face thickness. Replace the clutch disc if it is not within specification, 0.205–0.287 in. (5.2–7.3mm).

5. Remove the pressure plate-to-flywheel bolts in a progressive crisscross pattern to prevent warping the cover, then remove the pressure plate and clutch disc.

6. Inspect the pressure plate, as follows:

a. Check for excessive wear, chatter marks, cracks or overheating (indicated by a blue discoloration). Black random spots on the friction surface of the pressure plate is normal.

b. Check the plate for warpage using a straightedge and a feeler gauge; the maximum allowable warpage is 0.006 in. (0.15mm).

c. Replace the plate, if necessary.

7. Inspect the clutch disc, as follows:

a. Check the disc face for oil or burnt spots.

b. Check the disc for loose damper springs, hub or rivets.

c. Replace the disc, if necessary.

8. Check the flywheel, as follows:

a. Check the ring gear for wear or damage.

b. Check the friction surface for excessive wear, chatter marks, cracks or overheating (indicated by a blue discoloration). Black random spots on the friction surface of the pressure plate is normal.

c. Check flywheel thickness; the minimum allowable is 1.102 in. (28mm).

d. Measure flywheel run-out using a dial indicator, positioned for at least two flywheel revolutions. Push the crankshaft forward to take up thrust bearing clearance. Maximum flywheel run-out is 0.006 in. (0.15mm).

e. Check the flywheel for warpage using a straightedge and a feeler gauge; the maximum allowable warpage is 0.006 in. (0.15mm).

f. Replace the flywheel, if necessary.

9. If necessary, remove the flywheel retaining bolts and remove the flywheel from the crankshaft.

To install:

10. If removed, install the flywheel and tighten the bolts in a crisscross sequence to 59 ft. lbs. (80 Nm).

11. Install the clutch disc and pressure

plate with the yellow dot on the pressure plate aligned as close as possible to the mark on the flywheel. The clutch disc is labeled FLYWHEEL SIDE in order to help correctly position the disc. Start the pressure plate bolts.

12. Install clutch alignment tool SA9145T or equivalent in the clutch disc, and push in until it bottoms out in the crankshaft.

13. Tighten the pressure plate bolts using multiple passes of a crisscross sequence to 18 ft. lbs. (25 Nm) and remove the alignment tool.

14. Lubricate the fork pivot point with high temperature grease and install the release bearing to the fork. Do not lubricate the release bearing or bearing quill.

15. Snap the release bearing and fork onto the ball stud.

16. Lubricate the splines of the input shaft lightly with a high temperature grease.

17. Install the transaxle assembly.

18. Connect the negative battery cable and properly enable the SIR system.

Hydraulic Clutch System

BLEEDING

The clutch hydraulic assembly has been filled with fluid and bled of air at the factory. Do not attempt to bleed the hydraulic system. While the unit does not require periodic checking, it must be serviced, when necessary, as a complete assembly. The system is full when the reservoir is half full.

Only DOT 3 brake fluid should be added to the system. If the fluid level drops, inspect the system, including the slave cylinder, for leakage. A slight wetting of the slave cylinder surface is normal. Fill the clutch master cylinder reservoir with brake fluid. Be careful not to spill brake fluid on the painted surface of the vehicle.

Halfshafts

REMOVAL & INSTALLATION

1. Remove the wheel cover or the center cap for access to the halfshaft nut. Have an assistant depress the brake pedal and loosen the front halfshaft nut.

2. Raise the vehicle and support it safely, then remove the corresponding wheel and splash shield.

3. If removing the left side halfshaft or

right side intermediate shaft, loosen the plug and drain the transaxle fluid into a clean container.

4. Remove the halfshaft nut and washer.

5. Remove and discard the cotter pin from the lower control arm ball joints. Back off the ball joint nut until the top of the nut is even with the top of the threads.

6. Use tool SA9132S or equivalent to loosen the lower control arm ball joint in the steering knuckle, then remove the nut. Do not use a wedge tool or seal damage may occur. The ball joint will be completely removed from the knuckle after the tie rod ball joint is separated.

✳✳ WARNING

The outer CV-joint for vehicles equipped with ABS contains a speed sensor ring. Use of an incorrect tool to separate the control arm from the knuckle may result in damage and loss of the ABS system.

7. Remove the tie rod cotter pin and loosen the castle nut, then separate the tie rod end from the knuckle using Tie Rod Separator SA91100C or equivalent. Once loosened sufficiently, remove the castle nut and separate the components completely. Do not use a wedge-type tool or the seal may be damaged.

8. You may wish to place a cloth over the sway bar to protect the surface, then use a prybar to leverage the knuckle and separate the lower control arm ball joint. Position a cloth at the prybar contact point with the cradle, then push down on the bar and separate the ball joint from the knuckle. Be sure the knuckle does not contact and damage the ball stud seal.

9. While pulling the knuckle/strut assembly away from the halfshaft, pull the end of the halfshaft from the wheel hub. If difficulty is encountered, tap on the end of the halfshaft using a block of wood and a hammer. Support the halfshaft assembly using a length of mechanic's wire or with a jackstand.

10. If removing the right halfshaft, disconnect the halfshaft from the intermediate shaft by tapping the inner joint with a hammer and a block of wood. Remove the halfshaft from the vehicle.

11. If removing the left halfshaft, disconnect the halfshaft by inserting a large prybar into the space between the inner joint and transaxle. Pry the halfshaft from the transaxle, being careful not to contact and damage the transaxle oil seal. Remove the halfshaft from the vehicle.

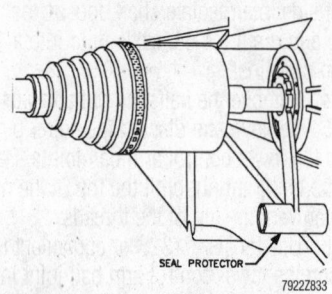

SEAL PROTECTOR

79222833

Failure to use a seal protector may allow the halfshaft splines to damage the transaxle seal, causing a need for replacement

To install:

12. If installing the left side halfshaft, install SA91112T or equivalent transaxle seal protector. Install the halfshaft into the transaxle; after the splines have safely passed the transaxle oil seal, remove the seal protector and fully seat the halfshaft.

13. If installing the right side halfshaft, insert the shaft onto the intermediate shaft and push firmly to engage the circlip. Install the shaft assembly into the transaxle.

14. Insert the outer end of the halfshaft into the wheel hub. Be careful not to damage the CV-joint boot.

15. Thoroughly clean and lubricate the ball joint stud threads of the lower control arm and tie rod end.

16. Install the lower control arm ball stud to the steering knuckle, then install the nut, but do not tighten at this time.

17. Attach the tie rod end to the steering knuckle, then install the nut. Tighten the nut to 33 ft. lbs. (45 Nm) and install a new cotter pin. If necessary, tighten the nut additionally, but do not back it off to insert the cotter pin.

18. Tighten the lower control arm ball stud nut to 55 ft. lbs. (75 Nm); tighten additionally if necessary to align the holes, then install a new cotter pin.

19. Install the washer and a new halfshaft nut, then tighten the nut to 148 ft. lbs. (200 Nm).

20. If equipped with ABS, inspect the wheel speed sensor signal for proper operation.

21. Install the inner splash shield and wheel.

22. Lower the vehicle and properly fill the transaxle.

23. Check and adjust the front end alignment as necessary.

➡ **Refer to the Unit Repair Section for information on how to replace the CV-joint boot(s).**

STEERING AND SUSPENSION

Air Bag

✳✳ CAUTION

All vehicles are equipped with an air bag system, also known as the Supplemental Inflatable Restraint (SIR) or Supplemental Restraint System (SRS). The system must be disabled before performing service on or around system components, steering column, instrument panel components, wiring and sensors. Failure to follow safety and disabling procedures could result in accidental air bag deployment, possible personal injury and unnecessary system repairs.

PRECAUTIONS

Several precautions must be observed when handling the inflator module to avoid accidental deployment and possible personal injury.

• Never carry the inflator module by the wires or connector on the underside of the module.

• When carrying a live inflator module, hold securely with both hands, and ensure that the bag and trim cover are pointed away.

• Place the inflator module on a bench or other surface with the bag and trim cover facing up.

• With the inflator module on the bench, never place anything on or close to the module which may be thrown in the event of an accidental deployment.

DISARMING

1. Align the steering wheel so the vehicle wheels are pointing in the straight-ahead position.

2. Turn the ignition switch to the **LOCK** position.

3. Remove the SIR or AIR BAG fuse from the fuse block.

4. Remove the Connector Position Assurance (CPA) device, then disengage the yellow 2-way SIR wiring harness connector at the base of the steering column.

5. After the repairs, enable the system as follows:

a. Turn the ignition switch to the **LOCK** position.

b. Engage the yellow 2-way connector at the base of the steering column, then install the CPA device.

c. Reinstall the SIR or AIR BAG fuse.

d. Turn the ignition switch to the **RUN** position.

e. Verify the SIR indicator light flashes 7–9 times, if not, inspect system for malfunction.

Rack and Pinion Steering Gear

REMOVAL & INSTALLATION

1. Disconnect the negative battery cable, then raise and support the vehicle safely.

2. Remove both front tires and the left inner splash shield.

3. Remove and discard the tie rod cotter pins, then remove the castle nuts. Disconnect the tie rod ends using SA91100C or an equivalent separator tool. Do not use a wedge-type tool or seal damage may occur.

4. Loosen the intermediate shaft cover from the steering gear and move up enough

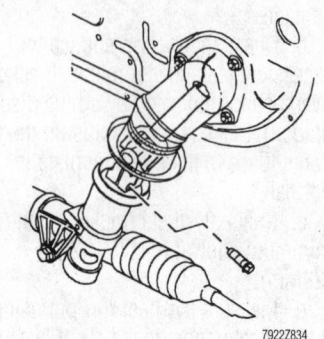

79222834

The pinch bolt is located under the intermediate shaft cover

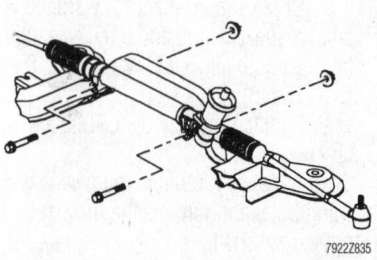

79222835

Remove the steering gear-to-cradle fasteners, then remove the gear through the left fenderwell

to access the pinch bolt. Remove the pinch bolt.

5. On vehicles with power steering, place a suitable container under the steering assembly. Disconnect the pressure and return lines at the steering gear and allow the system to drain.

6. Remove the steering gear-to-cradle fasteners, then remove the gear through the left fenderwell.

To install:

7. Install the steering gear and tighten the steering gear fasteners to 37 ft. lbs. (50 Nm). Be sure to use new nuts because the torque retention of the old nuts may be insufficient.

8. Position the intermediate steering shaft to the gear and tighten the pinch bolt to 35 ft. lbs. (47 Nm).

9. On vehicles with power steering, connect the pressure and return hoses, then tighten the fittings to 20 ft. lbs. (28 Nm).

10. Thoroughly clean and lubricate the threads of the tie rod ends, then install the ends into the steering knuckles. Install the castle nuts and tighten to 33 ft. lbs. (45 Nm), then install new cotter pins. If necessary, tighten the nut additionally to install the pin, do not back off.

11. Install the left inner splash shield and install the front wheels.

12. Lower the vehicle and connect the negative battery cable.

13. Check alignment and adjust vehicle toe, as necessary.

14. On vehicles with power steering, bleed the power steering system.

Strut

REMOVAL & INSTALLATION

Front

> ❊❊ **CAUTION**
>
> **The MacPherson strut is under extreme spring pressure. Do not remove the strut shaft center nut at the top without using an approved spring compressor. Personal injury may result if this caution is not followed.**

1. If equipped with an Anti-lock Brake System (ABS), disconnect the negative battery cable, then raise and support the vehicle safely. Be sure the vehicle is at a height where underhood access is still possible.

2. Remove the front wheel.

3. If equipped, unplug and disconnect the ABS wire from the strut wiring bracket. Note the wiring position for assembly purposes, then place the ABS wiring out of the way to prevent damage. If the strut is being replaced, drill the rivet head retaining the ABS wiring bracket to the strut and remove the bracket.

4. Loosen the two steering knuckle-to-strut housing bolts, but do not remove them at this time. For reassembly purposes, matchmark the strut position to the steering knuckle.

5. Remove and discard the three upper strut-to-body nuts.

6. Place a rag over the CV-joint seal to protect it from damage, then remove the two steering knuckle-to-strut housing bolts.

7. Remove the strut assembly from the vehicle.

To install:

8. Position the strut in the vehicle and install three new upper mount nuts. New nuts must be used because the torque retention of the old nut may be insufficient. Tighten the nuts to 21 ft. lbs. (29 Nm).

9. Install the knuckle bolts, also using new nuts. Push the bottom of the strut inward while tightening the fasteners to 126 ft. lbs. (170 Nm).

10. If equipped with ABS and the strut was replaced, install the ABS wiring bracket to the strut using a new rivet. Connect the ABS wiring to the bracket and install the wiring to the speed sensor connector. Be sure the wiring is positioned as noted during removal.

11. Install the wheel assembly, remove the supports and lower the vehicle.

12. Connect the negative battery cable, then check and adjust the alignment as necessary.

Rear

> ❊❊ **CAUTION**
>
> **The MacPherson strut is under extreme spring pressure. Do not remove the strut shaft support nut at the top center of the assembly without using an approved spring compressor. Personal injury may result if this caution is not followed.**

1. On coupes, remove the rear seat cushion bottom, left or right rocker panel interior moldings and the left or right rear sail interior panels.

2. On sedans, remove the left or right C-pillar interior molding.

3. Fold down the rear seat backs and

remove the rear seat side bolsters from the vehicle, if equipped.

4. On coupes, remove the rear deck package shelf retainers that attach the shelf to the side of the cargo area.

5. If equipped, remove the speaker grill fasteners and grills from the shelf, then remove the seat belt bezel and separate the seat belts from the shelf. Remove the rear package shelf cover.

6. If equipped with an Anti-lock Brake System (ABS), disconnect the negative battery cable.

7. Raise the rear of the vehicle and support it safely, then remove the appropriate rear wheel.

8. If equipped, disconnect the ABS wiring from the strut wiring bracket. If the strut is being replaced, drill out the rivet head, if required, retaining the ABS wiring bracket to the strut and remove the bracket. In either case, note the position of the wiring and move the wiring to prevent damage. If necessary, unplug the wheel speed sensor connector.

9. For reassembly purposes, matchmark the strut position to the knuckle. Then, loosen the two strut-to-knuckle bolts, but do not remove at this time.

10. Position a floor jack under the rear

Remove the three nuts to remove the top of the strut from the body

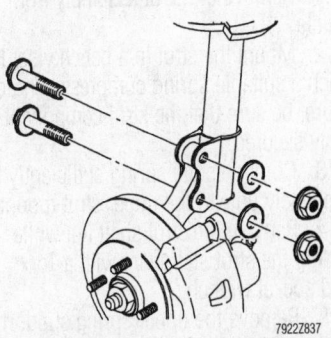

Matchmark, then remove the two bolts that secure the strut to the knuckle assembly

knuckle, then raise the jack only enough to support the knuckle. If a 2nd floor jack is not available, position a jackstand under the knuckle to support the strut.

11. Remove the three strut-to-body mounting nuts.

12. Slowly raise the vehicle using another jack, lowering the strut from the body. If another jack is not available, lower the jack holding the strut, but do so carefully.

13. Unfasten the strut-to-knuckle bolts and remove the strut assembly from the vehicle.

To install:

14. Install three new strut-to-upper mount nuts and tighten the nuts to 21 ft. lbs. (29 Nm).

➡**New nuts must be used, because the torque retention of the old fasteners may not be sufficient.**

15. Install the knuckle bolts with new nuts, then push the bottom of the strut inward, aligning the marks made earlier. Tighten the fasteners to 126 ft. lbs. (170 Nm).

16. If the strut was replaced, install the ABS wiring bracket to the strut using a new rivet if necessary. Connect the ABS wiring to the bracket, if unplugged, connect the wiring harness to the speed sensor. Be sure the wiring is positioned as noted during removal to prevent damage.

17. Install the wheel assembly, remove the supports and lower the vehicle.

18. Install the interior components.

19. Connect the negative battery cable, then check and adjust the rear alignment as necessary.

Coil Spring

REMOVAL & INSTALLATION

Front

1. Remove the strut assembly from the vehicle.

2. Mount the strut in a bench vise, then attach a suitable spring compressor/holding fixture; be sure that the strut component is firmly secured.

3. Compress the spring sufficiently to completely unload the upper strut mount.

4. Remove the strut shaft nut while holding the strut stationary with a Torx® head socket wrench.

5. Remove the upper spring support and inspect the rubber for cracks or deterioration.

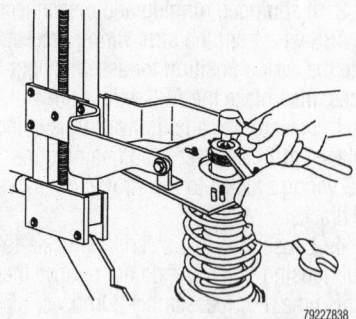

Remove the strut shaft nut while the coil spring is compressed to unload the upper strut mount

6. Remove the spring from the strut and inspect the spring for damage.

7. Remove the dust shield assembly and inspect for cracks or deterioration.

8. Remove the strut from the vise or applicable holding fixture and retract the strut shaft, checking for smooth, even resistance.

9. If replacing the coil spring, carefully release the spring compressor.

To install:

10. Secure the strut in the bench vise, or applicable holding fixture.

11. Extend the strut shaft to the limit of its travel.

12. Install the dust shield assembly onto the strut, then install the spring with the compressor tool installed.

13. Install the spring isolator and the strut mount to the top of the assembly.

14. Guide the strut shaft through the upper strut mount assembly. Compress the coil until the washer and shaft nut can be installed to the end of the shaft, but do not overcompress and damage the spring.

15. Tighten the shaft to the nut using a Torx® head socket wrench and a torque wrench, while holding the nut steady with an open end wrench. Tighten the fastener to 37 ft. lbs. (50 Nm).

16. Release the spring compressor tool and remove the strut from the fixture.

17. Install the strut assembly in the vehicle.

Rear

1. Remove the strut assembly from the vehicle.

2. Mount the strut in a bench vise, then attach a suitable spring compressor/holding fixture; be sure that the strut component is firmly secured.

3. Compress the spring sufficiently to completely unload the upper strut mount.

4. Remove the strut shaft nut while holding the strut stationary with a Torx® head socket wrench.

5. Remove the upper spring support and inspect the rubber for cracks or deterioration.

6. Remove the spring from the strut and inspect the spring for damage.

7. Remove the dust shield assembly and inspect for cracks or deterioration.

8. Remove the strut from the vise or applicable holding fixture and retract the strut shaft, checking for smooth, even resistance.

9. If replacing the coil spring, carefully release the spring compressor.

To assemble:

10. Secure the strut in the bench vise, or applicable holding fixture.

11. Extend the strut shaft to the limit of its travel.

12. Install the dust shield assembly onto the strut, then install the spring with the compressor tool installed.

13. Install the spring isolator and the strut mount to the top of the assembly.

14. Guide the strut shaft through the upper strut mount assembly. Compress the coil until the washer and shaft nut can be installed to the end of the shaft, but do not overcompress and damage the spring.

15. Tighten the shaft to the nut using a Torx® head socket wrench and a torque wrench, while holding the nut steady with an open end wrench. Tighten the fastener to 37 ft. lbs. (50 Nm).

16. Release the spring compressor tool and remove the strut from the fixture.

17. Install the strut using the procedure in this manual.

Wheel Bearings

ADJUSTMENT

All Saturn wheel bearing assemblies are an integral component of the wheel hub assembly. They are sealed at the factory and do not require any adjustment or maintainence.

REMOVAL & INSTALLATION

Front

1. If equipped with ABS, disconnect the negative battery cable.

2. Loosen the front halfshaft nut, while an assistant depresses the brake pedal, then raise and support the vehicle safely.

3. Remove the wheel assembly.

4. Remove the brake caliper mounting

bracket bolts and suspend the assembly from the strut spring with wire.

5. Loosen the strut-to-knuckle bolts, but do not remove at this time.

6. Remove the rotor, axle nut and washer.

7. Remove and discard the cotter pin from the lower control arm ball joint. Back the ball joint nut until the top of the nut is even with the top of the threads.

8. Use tool SA9132S to separate the lower control arm ball joint from the steering knuckle, then remove the nut. Do not use a wedge tool or seal damage may occur.

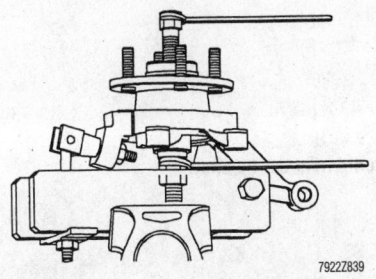

Tighten the hub driver screw to remove the hub while the assembly is held firmly in a vise

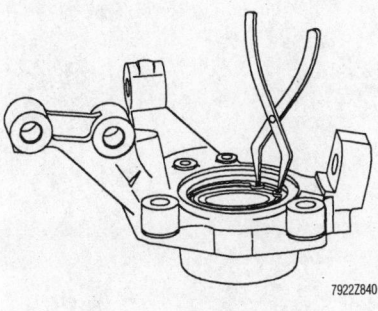

Remove the bearing retainer snapring before pressing out the bearing

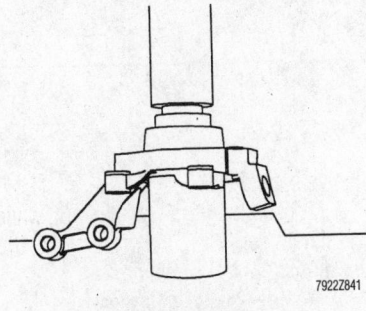

Carefully press the bearing out of the knuckle after removing the snapring

➡ **The outer CV-joint for vehicles equipped with ABS contains a speed sensor ring. Use of an incorrect tool to separate the control arm from the knuckle may result in damage and loss of the ABS system.**

9. Remove the tie rod cotter pin and castle nut, then separate the tie rod end from the knuckle using a tie rod separator SA91100C or equivalent. Do not use a wedge-type tool.

10. If equipped, disengage the ABS wheel speed sensor electrical connector.

11. Suspend the halfshaft from the body with wire, then remove the two knuckle-to-strut fasteners and remove the knuckle/hub assembly from the vehicle. If difficulty is encountered, position a block of wood on the end of the halfshaft and tap on the wood with a hammer to free the hub assembly.

12. If necessary, disassemble the knuckle hub assembly as follows:

   a. If equipped, remove the ABS wheel speed sensor from the knuckle.

➡ **Any time the hub or bearing is separated from the steering knuckle, a new bearing must be used upon assembly.**

   b. Install wheel bearing removing tools SA9159S or equivalent, to the knuckle and secure the assembly in a vice.

   c. Hold the hub driver with a wrench and tighten the hub driver screw to remove the hub. If the inner bearing race is pulled out with the hub, remove the race with a bearing race remover.

   d. Remove the assembly from the vice and separate the wheel hub removal tool from the knuckle.

   e. Remove the bearing retainer snapring.

   f. Position the knuckle in a shop press on a knuckle support tube and press the bearing from the knuckle with a suitable small driver.

To install:

13. If necessary, assemble the knuckle hub assembly as follows:

   a. Use a suitable large driver and press in the new bearing until seats.

   b. Use the small driver and the knuckle support tube to press in the hub assembly. The small driver must be used to support the bearing inner race with its small (pilot) side facing towards the press and away from the bearing.

   c. Install the bearing retainer snapring.

   d. If equipped, install the ABS wheel speed sensor into the knuckle and tighten the fastener to 6 ft. lbs. (8 Nm).

➡ **Service knuckles may not have holes for brake dust shield mounting. The dust shield is no longer required and does not have to be reinstalled. Also, should the shield become damaged it may be removed, there is no need to repair or replace it. But, should a shield be removed and discarded, the shield should also be removed from the opposite side to maintain balance/symmetry.**

14. Thoroughly clean and lubricate the ball joint stud threads of the lower control arm and tie rod end. Install the knuckle/hub assembly onto the axle shaft. Then, install the washer with a new nut, but do not tighten the nut at this time.

15. Install the lower control arm ball stud through the knuckle bore and install the nut, but do not tighten at this time.

16. Install the steering knuckle-to-strut fasteners, but do not tighten at this time.

17. Install the tie rod end and nut, then tighten the nut to 33 ft. lbs. (45 Nm) and install a new cotter pin. If necessary, tighten the nut additionally, do not back off to insert the cotter pin.

18. Push inward on the bottom of the strut and tighten the knuckle fasteners to 148 ft. lbs. (200 Nm).

19. Tighten the lower control arm ball stud nut to 55 ft. lbs. (75 Nm), tighten additionally if necessary and install a new cotter pin.

20. Install the rotor onto the hub, then install the caliper mount bracket onto the knuckle. Tighten the mount bracket assembly bolts to 81 ft. lbs. (110 Nm).

21. If equipped, engage the ABS electrical connector to the wheel speed sensor.

22. While an assistant depresses the brake pedal, tighten the halfshaft nut to 148 ft. lbs. (200 Nm).

23. Install the wheel assembly and lower the vehicle.

24. Connect the negative battery cable, check and adjust the alignment, as necessary.

Rear

Unlike the front wheel bearings, which may be removed from the hub for replacement, the rear wheel hub and bearing assembly is not serviceable. If damaged or worn, the hub and bearing assembly must be replaced as a unit.

1. If equipped with ABS, disconnect the negative battery cable.

2. Raise and support the vehicle safely, then remove the rear wheel.

3. If equipped, unplug the ABS speed sensor connector.

4. On disc brake equipped models, remove the caliper assembly-to-knuckle mounting bolts and support it with a wire from the strut, then remove the rotor.

5. On drum brake equipped models, remove the brake drum.

6. Remove the four hub/bearing-to-knuckle bolts, then remove the assembly from the vehicle.

To install:

7. Install the brake backing plate, hub/bearing assembly and retaining bolts. Tighten the bolts to 63 ft. lbs. (85 Nm).

8. Install the brake drum or rotor and caliper assembly. If applicable, tighten the caliper retaining bolts to 63 ft. lbs. (85 Nm).

9. If equipped, engage the ABS speed sensor connector.

10. Install the wheel assembly and lower the vehicle. If applicable, connect the negative battery cable.

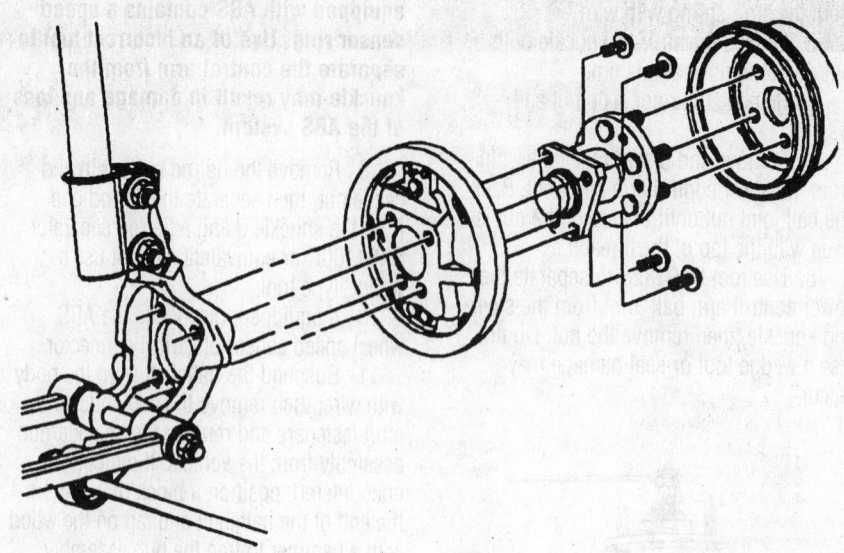

Exploded view of the rear hub/bearing assembly—drum brake set-up shown

Total Car Care, continued

Sentra/Pulsar/NX 1982-96
PART NO. 8263/52700

Stanza/200SX/240SX 1982-92
PART NO. 8262/52750

240SX/Altima 1993-98
PART NO. 52752

Datsun/Nissan Z and ZX 1970-88
PART NO. 8846/52800

RENAULT
Coupes/Sedans/Wagons 1975-85
PART NO. 58300

SATURN
Coupes/Sedans/Wagons 1991-98
PART NO. 8419/62300

SUBARU
ff-1/1300/1400/1600/1800/Brat 1970-84
PART NO. 8790/64300

Coupes/Sedans/Wagons 1985-96
PART NO. 8259/64302

SUZUKI
Samurai/Sidekick/Tracker 1986-98
PART NO. 66500

TOYOTA
Camry 1983-96
PART NO. 8265/68200

Celica/Supra 1971-85
PART NO. 68250

Celica 1986-93
PART NO. 8413/68252

Corolla 1970-87
PART NO. 8586/68300

Corolla 1988-97
PART NO. 8414/68302

Cressida/Corona/Crown/MkII 1970-82
PART NO. 68350

Cressida/Van 1983-90
PART NO. 68352

Toyota Trucks 1970-88
PART NO. 8578/68600

Pick-Ups/Land Cruiser/4Runner
1989-96
PART NO. 8163/68602

Previa 1991-98
PART NO. 68640

Tercel 1984-94
PART NO. 8595/68700

VOLKSWAGEN
Air-Cooled 1949-69
PART NO. 70200

Air-Cooled 1970-81
PART NO. 70202

Front Wheel Drive 1974-89
PART NO. 8663/70400

Golf/Jetta/Cabriolet 1990-93
PART NO. 8429/70402

VOLVO
Coupes/Sedans/Wagons 1970-89
PART NO. 8786/72300

Coupes/Sedans/Wagons 1990-98
PART NO. 8428/72302

Total Service Series

Auto Detailing
PART NO. 8394

Auto Body Repair
PART NO. 7898

Automatic Transmission Repair
1980-84
PART NO. 7890

Automatic Transmissions/
Transaxles Diagnosis and Repair
PART NO. 8944

Brake System Diagnosis and Repair
PART NO. 8945

Chevrolet Engine Overhaul Manual
PART NO. 8794

Easy Car Care
PART NO. 8042

Engine Code Manual
PART NO. 8851

Ford Engine Overhaul Manual
PART NO. 8793

Fuel Injection and Feedback
Carburetors 1977-85
PART NO. 7488

Fuel Injection Diagnosis
and Repair
PART NO. 8946

Motorcycle Repair
PART NO. 9099

Small Engine Repair
(Up to 20 Hp)
PART NO. 8325

Collector's Hard-Cover Manuals

Auto Repair Manual 1993-97
PART NO. 7919

Auto Repair Manual 1988-92
PART NO. 7906

Auto Repair Manual 1980-87
PART NO. 7670

Auto Repair Manual 1972-79
PART NO. 6914

Auto Repair Manual 1964-71
PART NO. 5974

Auto Repair Manual 1954-63
PART NO. 5652

Auto Repair Manual 1940-53
PART NO. 5631

Import Car Repair Manual 1993-97
PART NO. 7920

Import Car Repair Manual 1988-92
PART NO.7907

Import Car Repair Manual 1980-87
PART NO. 7672

Truck and Van Repair Manual 1993-97
PART NO. 7921

Truck and Van Repair Manual 1991-95
PART NO. 7911

Truck and Van Repair Manual 1986-90
PART NO. 7902

Truck and Van Repair Manual 1979-86
PART NO. 7655

Truck and Van Repair Manual 1971-78
PART NO. 7012

Truck Repair Manual 1961-71
PART NO. 6198

Motorcycle and ATV Repair Manual
1945-85
PART NO. 7635

System-Specific Manuals

Guide to Air Conditioning Repair and
Service 1982-85
PART NO. 7580

Guide to Automatic Transmission
Repair 1984-89
PART NO. 8054

Guide to Automatic Transmission
Repair 1984-89
Domestic cars and trucks
PART NO. 8053

Guide to Automatic Transmission
Repair 1980-84
Domestic cars and trucks
PART NO. 7891

Guide to Automatic Transmission
Repair 1974-80
Import cars and trucks
PART NO. 7645

Guide to Brakes, Steering, and
Suspension 1980-87
PART NO. 7819

Guide to Fuel Injection and Electronic
Engine Controls 1984-88
Guide to Electronic Engine Controls
1978-85
PART NO. 7535

Guide to Engine Repair and Rebuilding
PART NO. 7643

Guide to Vacuum Diagrams 1980-86
Domestic cars and trucks
PART NO. 7821

Multi-Vehicle Spanish Repair Manuals

Auto Repair Manual 1992-96
PART NO. 8947

Import Repair Manual 1992-96
PART NO. 8948

Truck and Van Repair Manual
1992-96
PART NO. 8949

Auto Repair Manual 1987-91
PART NO. 8138

Auto Repair Manual
1980-87
PART NO. 7795

Auto Repair Manual
1976-83
PART NO. 7476